CRC Handbook
of
Chemistry and Physics

A Ready-Reference Book of Chemical and Physical Data

HANDBOOK OF CHEMISTRY AND PHYSICS
1997-1998
78th
EDITION
CRC PRESS, INC.

Editor-in-Chief

David R. Lide, Ph.D.
Former Director, Standard Reference Data
National Institute of Standards and Technology

Associate Editor

H. P. R. Frederikse, Ph.D.
(Retired)
Ceramics Division
National Institute of Standards and Technology

CRC Press
Boca Raton   New York

PREFACE
78th Edition

The enhancements in the 78th Edition of the *Handbook of Chemistry and Physics* include:

- A new section on "Definitions of Scientific Terms", which contains concise definitions for about 800 important terms in chemistry, physics, and related disciplines. The section stresses physical quantities, units of measure, classes of compounds and materials, and the basic concepts, laws, and theories of physical science. Cross references to other parts of the *Handbook* are given when appropriate.
- A completely revised and updated table of vapor pressures, covering about 1800 organic and inorganic substances. The data are presented in SI units in a uniform format. The table is designed to cover a wide range from very low to ambient pressures, and references are given to data sources and to more specialized *Handbook of Chemistry and Physics* tables of vapor pressure.
- A new table of quantitative data on the solubility of organic compounds in water at specified temperatures. All data have been taken from evaluated sources and converted to a standard format. Henry's Law constants, which play an important role in environmental modeling, are given in the same table.
- A table of selected values of octanol-water partition coefficients, widely used in predicting biological effects of chemical substances.
- A comprehensive table of flash points and other data related to flammability.
- Several new tables dealing with properties of solid materials, including phase diagrams, electron work functions, and ionic radii and polarizability in crystals.
- An updated table of "Ionization Energies of Gas-Phase Molecules".
- Other new or revised tables on index of refraction, dielectric constant, speed of sound in various media, and chemical nomenclature rules.

Several older tables that appeared in the last edition have been dropped in order to make room for the new material. These include:

- Biochemical Symbols and Abbreviations
- Gibbs Energy of Formation of Metal Oxides
- Properties of Carbohydrates
- Ion Exchange Resins
- Emissivity of Various Materials
- Ultraviolet Spectra of Common Liquids
- Radiative Transition Probabilities for X-ray Lines
- Viscosity of Liquid Metals
- Recommended Daily Dietary Allowances

The Editor will appreciate comments on the removal of these tables (as well as others removed from recent editions of the *Handbook*); if there is sufficient interest, an effort will be made to restore them in abbreviated form. Suggestions on new topics of current interest (and on older tables whose usefulness is questionable) are also invited. Address all comments to Editor, Handbook of Chemistry and Physics, CRC Press LLC, 2000 Corporate Blvd. N.W., Boca Raton, FL 33431. Comments may also be sent by electronic mail to dlide@earthlink.net.

The *Handbook of Chemistry and Physics* is dependent on the efforts of many contributors throughout the world. The list of current contributors follows this Preface. Many valuable suggestions have come from the Editorial Advisory Board. I am particularly indebted to Robert Goldberg, as well as to Associate Editor Hans Frederikse, who has greatly strengthened the coverage of solid state physics and materials science in the *Handbook*. Finally, the staff at CRC Press who are responsible for the production of the book are instrumental to its success. I wish particularly to thank Paul Gottehrer for his meticulous attention to detail and his dedication to assuring the high quality of the *Handbook* over many years, as well as James Brody, with whom I have had exceedingly pleasant and effective interactions since I became Editor.

David R. Lide
February 23, 1997

Note on the Ordering of Chemical Compounds: The decision on the order in which to list chemical compounds in a table is always difficult. An alphabetical list by name has the disadvantage that several different synonyms are often in common use, with the result that a reader may conclude incorrectly that a compound is not present if he looks it up under the wrong name. Listing by line formula is satisfactory for simple inorganic compounds, but is cumbersome for organics. A listing by molecular formula is attractive because clear rules can be given for locating a compound, but the user may have to go to some effort to determine the molecular formula. In this book the choice is made on pragmatic grounds. The long tables, "Physical Constants of Organic Compounds" and "Physical Constants of Inorganic Compounds", are ordered by systematic name, but indexes to synonyms, formulas, and CAS Registry Numbers are provided. If the table is very short and includes only common substances, the listing is usually alphabetical by common name or formula. The remaining tables are ordered by molecular formula using a modification of the Hill convention. In this convention the molecular formula is written with C first, H second, and then all other elements in alphabetical order of their chemical symbols. For tables with organic compounds only, the sequence of entries is determined by the alphabetical order of elements in the molecular formula and the number of atoms of each element, in ascending order, e.g., C_3H_7Cl, C_3H_7N, C_3H_7NO, $C_3H_7NO_2$, etc. (For organic compounds, a quick way to determine the molecular formula is to use the "Physical Constants of Organic Compounds" table, which starts on page **3**-1, and its synonym index on page **3**-586.) In tables containing non-carbon compounds, those are listed first, followed by a separate listing of compounds that do contain carbon. This is in contrast to the strict Hill convention as followed by Chemical Abstracts Service, where the molecular formulas beginning with A and B precede the formulas for carbon-containing compounds, while those beginning with D... Z follow. For tabular displays, as opposed to an index, it appears more convenient to the user if the non-carbon compounds are listed as a block, rather than being split by the longer list of carbon compounds.

This Edition is Dedicated to my Grandson, David Austell Whitcomb.

CURRENT CONTRIBUTORS

Douglas Ambrose
Department of Chemistry
University College London
London WC1H OAJ
England

Lev I. Berger
California Institute of Electronics
and Materials Science
2115 Flame Tree Way
Hemet, California 92545

George W. Burns
National Institute of Standards and
Technology
Gaithersburg, Maryland 20899

R.W. Clay
Department of Physics and
Mathematical Physics
The University of Adelaide
Adelaide, South Australia 5005
Australia

A. K. Covington
Department of Chemistry
University of Newcastle
Newcastle upon Tyne NE1 7RU
England

Hans Dolezalek
1812 Drury Lane
Alexandria, Virginia 22307

K.F. Eckerman
Oak Ridge National Laboratory
Oak Ridge, Tennessee 37831

Robert B. Fox
6115 Wiscassett Rd.
Bethesda, Maryland 20816

H. P. R. Frederikse
9625 Dewmar Lane
Kensington, Maryland 20895

J.R. Fuhr
Atomic Physics Division
National Institute of Standards and
Technology
Gaithersburg, Maryland 20899

A.C. Gregory
Department of Physics and
Mathematical Physics
The University of Adelaide
Adelaide, South Australia 5005
Australia

Karl A. Gschneidner
Ames Laboratory
Energy and Mineral Resources
Research Institute
Iowa State University
Ames, Iowa 50011

C. R. Hammond
17 Greystone Rd.
West Hartford, Connecticut 06107

Robert F. Hampson
Chemical Kinetics Division
National Institute of Standards and
Technology
Gaithersburg, Maryland 20899

Norman E. Holden
National Nuclear Data Center
Brookhaven National Laboratory
Upton, New York 11973

Henry V. Kehiaian
ITODYS
1 rue Guy de la Brosse
75005 Paris
France

J. A. Kerr
Atmospheric Chemistry Group
EAWAG
CH-8600 Dubendorf-Zurich
Switzerland

M.V. Korobov
Department of Chemistry
Moscow State University
119899 Moscow
Russia

William F. Krupke
Lawrence Livermore Laboratory
Livermore, California 94550

Kozo Kuchitsu
Department of Chemistry
Josai University
1-1 Keyakidai
Sakado 350-02
Japan

John Latham
Department of Physics
University of Manchester
Manchester M60 1QD
England

Sharon G. Lias
National Institute of Standards and
Technology
Gaithersburg, Maryland 20899

Arthur Maki
15012 24 Ave., S. E.
Mill Creek, Washington 98012

B. W. Mangum
Temperature and Pressure Division
National Insitute of Standards and
Technology
Gaithersburg, Maryland 20899

Thomas M. Miller
Phillips Laboratory/GPID
29 Randolph Rd.
Hanscom AFB, Massachusetts
01731-3010

Dale L. Perry
Lawrence Berkeley Laboratory
1 Cyclotron Road
Berkeley, California 94720

Sidney L. Phillips
Camatx/Basic Chemistry
171 El Toyonal
Orinda, California 94563

C. N. R. Rao
Solid State and Structural Chemistry
Unit
Indian Institute of Science
Bangalore 560-012
India

CURRENT CONTRIBUTORS (continued)

A. K. Raychaudhuri
Solid State and Structural Chemistry
    Unit
Indian Institute of Science
Bangalore 560-012
India

Hannes Tammet
Air Electricity Laboratory
18 Ulikooli St.
Tartu 202400
Estonia

B. N. Taylor
Physics Laboratory
National Institute of Standards and
    Technology
Gaithersburg, Maryland 20899

Thomas G. Trippe
Particle Data Group
Lawrence Berkeley Laboratory
1 Cyclotron Road
Berkeley, California 94720

Martin A. Uman
Department of Electrical
    Engineering
University of Florida
Gainesville, Florida 32611

Petr Vanýsek
Department of Chemistry
Northern Illinois University
DeKalb, Illinois 60115

Wolfgang L. Wiese
Atomic Physics Division
National Institute of Standards and
    Technology
Gaithersburg, Maryland 20899

Christian Wohlfarth
Institut für Physikalische Chemie
Martin Luther University
D-06217 Merseburg
Germany

TABLE OF CONTENTS

SECTION 10: ATOMIC, MOLECULAR, AND OPTICAL PHYSICS

SECTION 11: NUCLEAR AND PARTICLE PHYSICS

SECTION 12: PROPERTIES OF SOLIDS

SECTION 13: POLYMER PROPERTIES

SECTION 14: GEOPHYSICS, ASTRONOMY, AND ACOUSTICS

SECTION 15: PRACTICAL LABORATORY DATA

SECTION 16: HEALTH AND SAFETY INFORMATION

APPENDIX A: MATHEMATICAL TABLES

INDEX

Section 1
Basic Constants, Units, and Conversion Factors

FUNDAMENTAL PHYSICAL CONSTANTS

The 1986 CODATA Recommended Values

By E. Richard Cohen and Barry N. Taylor as published in the *Journal of Research of the National Bureau of Standards*, 92, 85, 1987. Discussions of the background, data selection and evaluation procedures are presented in CODATA Bulletin Number 63, November 1986, "The 1986 Adjustment of the Fundamental Physical Constants", a Report of the CODATA Task Group on Fundamental Physical Constants (36 pages) published by Pergamon Press.

The 1986 recommended values of the fundamental physical constants are given in five tables. Table 1 is an abbreviated list containing the quantities which should be of greatest interest to most users. Table 2 is a more complete compilation. Table 3 is a list of related "maintained units and standard values." Table 4 contains a number of scientifically, technologically, and metrologically useful energy conversion factors. Table 5 is an extended covariance matrix containing the variances, covariances, and correlation coefficients of the unknowns and a number of different constants (included for convenience) from which the like quantities of other constants may be calculated. (B. N. Taylor, W. H. Parker, and D. N. Langenberg, *Rev. Mod. Phys.*, 41, 375, 1969. Such a matrix is necessary because the variables in a least-square adjustment are correlated.

Table 1
SUMMARY OF THE 1986 RECOMMENDED VALUES
OF THE FUNDAMENTAL PHYSICAL CONSTANTS

Quantity	Symbol	Value	Units	Relative Uncertainty (ppm)
speed of light in vacuum	c	299 792 458	$m\,s^{-1}$	(exact)
permeability of vacuum	μ_0	$4\pi \times 10^{-7}$	$N\,A^{-2}$	
		$=12.566\,370\,614\ldots$	$10^{-7}\,N\,A^{-2}$	(exact)
permittivity of vacuum	ϵ_0	$1/\mu_0 c^2$		
		$=8.854\,187\,817\ldots$	$10^{-12}\,F\,m^{-1}$	(exact)
Newtonian constant of gravitation	G	6.672 59(85)	$10^{-11}\,m^3\,kg^{-1}\,s^{-2}$	128
Planck constant	h	6.626 0755(40)	$10^{-34}\,J\,s$	0.60
$h/2\pi$	$\hbar$	1.054 572 66(63)	$10^{-34}\,J\,s$	0.60
elementary charge	e	1.602 177 33(49)	$10^{-19}\,C$	0.30
magnetic flux quantum, $h/2e$	Φ_0	2.067 834 61(61)	$10^{-15}\,Wb$	0.30
electron mass	m_e	9.109 3897(54)	$10^{-31}\,kg$	0.59
proton mass	m_p	1.672 6231(10)	$10^{-27}\,kg$	0.59
proton-electron mass ratio	m_p/m_e	1836.152 701(37)		0.020
fine-structure constant, $\mu_0 c e^2/2h$	α	7.297 353 08(33)	10^{-3}	0.045
inverse fine-structure constant	α^{-1}	137.035 9895(61)		0.045
Rydberg constant, $m_e c \alpha^2/2h$	R_∞	10 973 731.534(13)	m^{-1}	0.0012
Avogadro constant	N_A, L	6.022 1367(36)	$10^{23}\,mol^{-1}$	0.59
Faraday constant, $N_A e$	F	96 485.309(29)	$C\,mol^{-1}$	0.30
molar gas constant	R	8.314 510(70)	$J\,mol^{-1}\,K^{-1}$	8.4
Boltzmann constant, R/N_A	k	1.380 658(12)	$10^{-23}\,J\,K^{-1}$	8.5
Stefan–Boltzmann constant, $(\pi^2/60)k^4/\hbar^3 c^2$	σ	5.670 51(19)	$10^{-8}\,W\,m^{-2}\,K^{-4}$	34
	Non-SI units used with SI			
electron volt, $(e/C)\,J = \{e\}\,J$	eV	1.602 177 33(49)	$10^{-19}\,J$	0.30
(unified) atomic mass unit, $1\ u = m_u = \frac{1}{12}m(^{12}C)$	u	1.660 5402(10)	$10^{-27}\,kg$	0.59

NOTE: An abbreviated list of the fundamental constants of physics and chemistry based on a least-squares adjustment with 17 degrees of freedom. The digits in parentheses are the one-standard-deviation uncertainty in the last digits of the given value. Since the uncertainties of many entries are correlated, the full covariance matrix must be used in evaluating the uncertainties of quantities computed from them.

Table 2
THE 1986 RECOMMENDED VALUES OF THE
FUNDAMENTAL PHYSICAL CONSTANTS

Quantity	Symbol	Value	Units	Relative Uncertainty (ppm)
	GENERAL CONSTANTS			
	Universal Constants			
speed of light in vacuum	c	299 792 458	$m\,s^{-1}$	(exact)
permeability of vacuum	μ_0	$4\pi \times 10^{-7}$	$N\,A^{-2}$	
		$=12.566\,370\,614\ldots$	$10^{-7}\,N\,A^{-2}$	(exact)
permittivity of vacuum	ϵ_0	$1/\mu_0 c^2$		
		$=8.854\,187\,817\ldots$	$10^{-12}\,F\,m^{-1}$	(exact)
Newtonian constant of gravitation	G	6.672 59(85)	$10^{-11}\,m^3\,kg^{-1}\,s^{-2}$	128
Planck constant	h	0.020 0755(40)	$10^{-34}\,J\,s$	0.00
in electron volts, $h/\{e\}$		4.135 6692(12)	$10^{-15}\,eV\,s$	0.30
$h/2\pi$	$\hbar$	1.054 572 66(63)	$10^{-34}\,J\,s$	0.60
in electron volts, $\hbar/\{e\}$		6.582 1220(20)	$10^{-16}\,eV\,s$	0.30

Quantity	Symbol	Value	Units	Relative Uncertainty (ppm)
Planck mass, $(\hbar c/G)^{\frac{1}{2}}$	m_P	2.17671(14)	10^{-8} kg	64
Planck length, $\hbar/m_Pc = (\hbar G/c^3)^{\frac{1}{2}}$	l_P	1.61605(10)	10^{-35} m	64
Planck time, $l_P/c = (\hbar G/c^5)^{\frac{1}{2}}$	t_P	5.39056(34)	10^{-44} s	64

Electromagnetic Constants

Quantity	Symbol	Value	Units	Relative Uncertainty (ppm)
elementary charge	e	1.60217733(49)	10^{-19} C	0.30
	e/h	2.41798836(72)	10^{14} A J^{-1}	0.30
magnetic flux quantum, $h/2e$	Φ_o	2.06783461(61)	10^{-15} Wb	0.30
Josephson frequency-voltage ratio	$2e/h$	4.8359767(14)	10^{14} Hz V^{-1}	0.30
quantized Hall conductance	e^2/h	3.87404614(17)	10^{-5} S	0.045
quantized Hall resistance, $h/e^2 = \mu_o c/2\alpha$	R_H	25812.8056(12)	Ω	0.045
Bohr magneton, $e\hbar/2m_e$	μ_B	9.2740154(31)	10^{-24} J T^{-1}	0.34
in electron volts, $\mu_B/\{e\}$		5.78838263(52)	10^{-5} eV T^{-1}	0.089
in hertz, μ_B/h		1.39962418(42)	10^{10} Hz T^{-1}	0.30
in wavenumbers, μ_B/hc		46.686437(14)	m^{-1} T^{-1}	0.30
in kelvins, μ_B/k		0.6717099(57)	K T^{-1}	8.5
nuclear magneton, $e\hbar/2m_p$	μ_N	5.0507866(17)	10^{-27} J T^{-1}	0.34
in electron volts, $\mu_N/\{e\}$		3.15245166(28)	10^{-8} eV T^{-1}	0.089
in hertz, μ_N/h		7.6225914(23)	MHz T^{-1}	0.30
in wavenumbers, μ_N/hc		2.54262281(77)	10^{-2} m^{-1} T^{-1}	0.30
in kelvins, μ_N/k		3.658246(31)	10^{-4} K T^{-1}	8.5

ATOMIC CONSTANTS

Quantity	Symbol	Value	Units	Relative Uncertainty (ppm)
fine-structure constant, $\mu_o ce^2/2h$	α	7.29735308(33)	10^{-3}	0.045
inverse fine-structure constant	α^{-1}	137.0359895(61)		0.045
Rydberg constant, $m_e c\alpha^2/2h$	R_∞	10973731.534(13)	m^{-1}	0.0012
in hertz, $R_\infty c$		3.2898419499(39)	10^{15} Hz	0.0012
in joules, $R_\infty hc$		2.1798741(13)	10^{-18} J	0.60
in eV, $R_\infty hc/\{e\}$		13.6056981(40)	eV	0.30
Bohr radius, $\alpha/4\pi R_\infty$	a_o	0.529177249(24)	10^{-10} m	0.045
Hartree energy, $e^2/4\pi\epsilon_o a_o = 2R_\infty hc$	E_h	4.3597482(26)	10^{-18} J	0.60
in eV, $E_h/\{e\}$		27.2113961(81)	eV	0.30
quantum of circulation	$h/2m_e$	3.63694807(33)	10^{-4} m^2 s^{-1}	0.089
	h/m_e	7.27389614(65)	10^{-4} m^2 s^{-1}	0.089

Electron

Quantity	Symbol	Value	Units	Relative Uncertainty (ppm)
electron mass	m_e	9.1093897(54)	10^{-31} kg	0.59
		5.48579903(13)	10^{-4} u	0.023
in electron volts, $m_e c^2/\{e\}$		0.51099906(15)	MeV	0.30
electron-muon mass ratio	m_e/m_μ	4.83633218(71)	10^{-3}	0.15
electron-proton mass ratio	m_e/m_p	5.44617013(11)	10^{-4}	0.020
electron-deuteron mass ratio	m_e/m_d	2.72443707(6)	10^{-4}	0.020
electron-α-particle mass ratio	m_e/m_α	1.37093354(3)	10^{-4}	0.021
electron specific charge	$-e/m_e$	$-1.75881962(53)$	10^{11} C kg^{-1}	0.30
electron molar mass	$M(e), M_e$	5.48579903(13)	10^{-7} kg/mol	0.023
Compton wavelength, $h/m_e c$	λ_C	2.42631058(22)	10^{-12} m	0.089
$\lambda_C/2\pi = \alpha a_o = \alpha^2/4\pi R_\infty$	λ_C	3.86159323(35)	10^{-13} m	0.089
classical electron radius, $\alpha^2 a_o$	r_e	2.81794092(38)	10^{-15} m	0.13
Thomson cross section, $(8\pi/3)r_e^2$	σ_e	0.66524616(18)	10^{-28} m^2	0.27
electron magnetic moment	μ_e	928.47701(31)	10^{-26} J T^{-1}	0.34
in Bohr magnetons	μ_e/μ_B	1.001159652193(10)		1×10^{-5}
in nuclear magnetons	μ_e/μ_N	1838.282000(37)		0.020
electron magnetic moment anomaly, $\mu_e/\mu_B - 1$	a_e	1.159652193(10)	10^{-3}	0.0086
electron g-factor, $2(1 + a_e)$	g_e	2.002319304386(20)		1×10^{-5}
electron-muon magnetic moment ratio	μ_e/μ_μ	206.766967(30)		0.15
electron-proton magnetic moment ratio	μ_e/μ_p	658.2106881(66)		0.010

Muon

Quantity	Symbol	Value	Units	Relative Uncertainty (ppm)
muon mass	m_μ	1.8835327(11)	10^{-28} kg	0.61
		0.113428913(17)	u	0.15
in electron volts, $m_\mu c^2/\{e\}$		105.658389(34)	MeV	0.32
muon-electron mass ratio	m_μ/m_e	206.768262(30)		0.15
muon molar mass	$M(\mu), M_\mu$	1.13428913(17)	10^{-4} kg/mol	0.15
muon magnetic moment	μ_μ	4.4904514(15)	10^{-26} J T^{-1}	0.33
in Bohr magnetons,	μ_μ/μ_B	4.84197097(71)	10^{-3}	0.15
in nuclear magnetons,	μ_μ/μ_N	8.8905981(13)		0.15

Quantity	Symbol	Value	Units	Relative Uncertainty (ppm)
muon magnetic moment anomaly, $[\mu_\mu/(e\hbar/2m_\mu)] - 1$	a_μ	1.1659230(84)	10^{-3}	7.2
muon g-factor, $2(1 + a_\mu)$	g_μ	2.002331846(17)		0.0084
muon–proton magnetic moment ratio	μ_μ/μ_p	3.18334547(47)		0.15

Proton

Quantity	Symbol	Value	Units	Relative Uncertainty (ppm)
proton mass	m_p	1.6726231(10)	10^{-27} kg	0.59
		1.007276470(12)	u	0.012
in electron volts, $m_p c^2/\{e\}$		938.27231(28)	MeV	0.30
proton–electron mass ratio	m_p/m_e	1836.152701(37)		0.020
proton–muon mass ratio	m_p/m_μ	8.8802444(13)		0.15
proton specific charge	e/m_p	9.5788309(29)	10^7 C kg^{-1}	0.30
proton molar mass	$M(p), M_p$	1.007276470(12)	10^{-3} kg/mol	0.012
proton Compton wavelength, $h/m_p c$	$\lambda_{C,p}$	1.32141002(12)	10^{-15} m	0.089
$\lambda_{C,p}/2\pi$	$\lambda_{C,p}$	2.10308937(19)	10^{-16} m	0.089
proton magnetic moment	μ_p	1.41060761(47)	10^{-26} J T^{-1}	0.34
in Bohr magnetons	μ_p/μ_B	1.521032202(15)	10^{-3}	0.010
in nuclear magnetons	μ_p/μ_N	2.792847386(63)		0.023
diamagnetic shielding correction for protons in pure water, spherical sample, 25 °C, $1 - \mu_p'/\mu_p$	σ_{H_2O}	25.689(15)	10^{-6}	—
shielded proton moment (H_2O, sph., 25 °C)	μ_p'	1.41057138(47)	10^{-26} J T^{-1}	0.34
in Bohr magnetons	μ_p'/μ_B	1.520993129(17)	10^{-3}	0.011
in nuclear magnetons	μ_p'/μ_N	2.792775642(64)		0.023
proton gyromagnetic ratio	γ_p	26752.2128(81)	10^4 s^{-1} T^{-1}	0.30
	$\gamma_p/2\pi$	42.577469(13)	MHz T^{-1}	0.30
uncorrected (H_2O, sph., 25 °C)	γ_p'	26751.5255(81)	10^4 s^{-1} T^{-1}	0.30
	$\gamma_p'/2\pi$	42.576375(13)	MHz T^{-1}	0.30

Neutron

Quantity	Symbol	Value	Units	Relative Uncertainty (ppm)
neutron mass	m_n	1.6749286(10)	10^{-27} kg	0.59
		1.008664904(14)	u	0.014
in electron volts, $m_n c^2/\{e\}$		939.56563(28)	Mev	0.30
neutron–electron mass ratio	m_n/m_e	1838.683662(40)		0.022
neutron–proton mass ratio	m_n/m_p	1.001378404(9)		0.009
neutron molar mass	$M(n), M_n$	1.008664904(14)	10^{-3} kg/mol	0.014
neutron Compton wavelength, $h/m_n c$	$\lambda_{C,n}$	1.31959110(12)	10^{-15} m	0.089
$\lambda_{C,n}/2\pi$	$\lambda_{C,n}$	2.10019445(19)	10^{-16} m	0.089
neutron magnetic moment *	μ_n	0.96623707(40)	10^{-26} J T^{-1}	0.41
in Bohr magnetons	μ_n/μ_B	1.04187563(25)	10^{-3}	0.24
in nuclear magnetons	μ_n/μ_N	1.91304275(45)		0.24
neutron–electron magnetic moment ratio	μ_n/μ_e	1.04066882(25)	10^{-3}	0.24
neutron–proton magnetic moment ratio	μ_n/μ_p	0.68497934(16)		0.24

Deuteron

Quantity	Symbol	Value	Units	Relative Uncertainty (ppm)
deuteron mass	m_d	3.3435860(20)	10^{-27} kg	0.59
		2.013553214(24)	u	0.012
in electron volts, $m_d c^2/\{e\}$		1875.61339(57)	MeV	0.30
deuteron–electron mass ratio	m_d/m_e	3670.483014(75)		0.020
deuteron–proton mass ratio	m_d/m_p	1.999007496(6)		0.003
deuteron molar mass	$M(d), M_d$	2.013553214(24)	10^{-3} kg/mol	0.012
deuteron magnetic moment *	μ_d	0.43307375(15)	10^{-26} J T^{-1}	0.34
in Bohr magnetons,	μ_d/μ_B	0.4669754479(91)	10^{-3}	0.019
in nuclear magnetons,	μ_d/μ_N	0.857438230(24)		0.028
deuteron–electron magnetic moment ratio	μ_d/μ_e	0.4664345460(91)	10^{-3}	0.019
deuteron–proton magnetic moment ratio	μ_d/μ_p	0.3070122035(51)		0.017

PHYSICO-CHEMICAL CONSTANTS

Quantity	Symbol	Value	Units	Relative Uncertainty (ppm)
Avogadro constant	N_A, L	6.0221367(36)	10^{23} mol^{-1}	0.59
atomic mass constant, $\frac{1}{12}m(^{12}C)$	m_u	1.6605402(10)	10^{-27} kg	0.59
in electron volts, $m_u c^2/\{e\}$		931.49432(28)	MeV	0.30
Faraday constant	F	96485.309(29)	C mol^{-1}	0.30
molar Planck constant	$N_A h$	3.99031323(36)	10^{-10} J s mol^{-1}	0.089
	$N_A hc$	0.11962658(11)	J m mol^{-1}	0.089
molar gas constant	R	8.314510(70)	J mol^{-1} K^{-1}	8.4

Quantity	Symbol	Value	Units	Relative Uncertainty (ppm)
Boltzmann constant, R/N_A	k	1.380658(12)	10^{-23} J K^{-1}	8.5
in electron volts, $k/\{e\}$		8.617385(73)	10^{-5} eV K^{-1}	8.4
in hertz, k/h		2.083674(18)	10^{10} Hz K^{-1}	8.4
in wavenumbers, k/hc		69.50387(59)	m^{-1} K^{-1}	8.4
molar volume (ideal gas), RT/p				
$T = 273.15$ K, $p = 101325$ Pa	V_m	22.41410(19)	L/mol	8.4
Loschmidt constant, N_A/V_m	n_0	2.686763(23)	10^{25} m^{-3}	8.5
$T = 273.15$ K, $p = 100$ kPa	V_m	22.71108(19)	L/mol	8.4
Sackur-Tetrode constant (absolute entropy constant), **				
$\frac{5}{2} + \ln\{(2\pi m_u kT_1/h^2)^{\frac{3}{2}} kT_1/p_0\}$				
$T_1 = 1$ K, $p_0 = 100$ kPa	S_0/R	−1.151693(21)		18
$p_0 = 101325$ Pa		−1.164856(21)		18
Stefan–Boltzmann constant, $(\pi^2/60)k^4/\hbar^3 c^2$	σ	5.67051(19)	10^{-8} W m^{-2} K^{-4}	34
first radiation constant, $2\pi hc^2$	c_1	3.7417749(22)	10^{-16} W m^2	0.60
second radiation constant, hc/k	c_2	0.01438769(12)	m K	8.4
Wien displacement law constant, $b = \lambda_{max} T = c_2/4.96511423\ldots$	b	2.897756(24)	10^{-3} m K	8.4

NOTE: This list of the fundamental constants of physics and chemistry is based on a least-squares adjustment with 17 degrees of freedom. The digits in parentheses are the one-standard-deviation uncertainty in the last digits of the given value. Since the uncertainties of many of these entries are correlated, the full covariance matrix must be used in evaluating the uncertainties of quantities computed from them.

* The scalar magnitude of the neutron moment is listed here. The neutron magnetic dipole is directed oppositely to that of the proton, and corresponds to the dipole associated with a spinning negative charge distribution. The vector sum, $\mu_d = \mu_p + \mu_n$, is approximately satisfied.

** The entropy of an ideal monatomic gas of relative atomic weight A_r is given by $S = S_0 + \frac{3}{2}R$ in $A_r - R$ in $(p/p_0) + \frac{5}{2}R$ in(T/K).

Table 3
MAINTAINED UNITS AND STANDARD VALUES

Quantity	Symbol	Value	Units	Relative Uncertainty (ppm)
electron volt, $(e/C) J = \{e\} J$	eV	1.60217733(49)	10^{-19} J	0.30
(unified) atomic mass unit, $1 u = m_u = \frac{1}{12}m(^{12}C)$	u	1.6605402(10)	10^{-27} kg	0.59
standard atmosphere	atm	101325	Pa	(exact)
standard acceleration of gravity	g_n	9.80665	m s^{-2}	(exact)
'As–Maintained' Electrical Units				
BIPM maintained ohm, Ω_{69-BI} $\Omega_{BI85} \equiv \Omega_{69-BI}$(1 Jan 1985)	Ω_{BI85}	$1 - 1.563(50)\times10^{-6}$ = 0.999998437(50)	Ω Ω	0.050
Drift rate of Ω_{69-BI}	$\dfrac{d\Omega_{69-BI}}{dt}$	−0.0566(15)	$\mu\Omega$/a	—
BIPM maintained volt, $V_{76-BI} \equiv 483594$ GHz$(h/2e)$	V_{76-BI}	$1 - 7.59(30)\times10^{-6}$ = 0.99999241(30)	V V	0.30
BIPM maintained ampere, $A_{BIPM} = V_{76-BI}/\Omega_{69-BI}$	A_{BI85}	$1 - 6.03(30)\times10^{-6}$ = 0.99999397(30)	A A	0.30
X-Ray Standards				
Cu x-unit : $\lambda(CuK\alpha_1) \equiv 1537.400$ xu	$xu(CuK\alpha_1)$	1.00207789(70)	10^{-13} m	0.70
Mo x-unit : $\lambda(MoK\alpha_1) \equiv 707.831$ xu	$xu(MoK\alpha_1)$	1.00209938(45)	10^{-13} m	0.45
Å* : $\lambda(WK\alpha_1) \equiv 0.209100$ Å*	Å*	1.00001481(92)	10^{-10} m	0.92

Table 3
MAINTAINED UNITS AND STANDARD VALUES
(continued)

Quantity	Symbol	Value	Units	Relative Uncertainty (ppm)
lattice spacing of Si (in vacuum, 22.5 °C), +	a	0.543 101 96(11)	nm	0.21
$d_{220} = a/\sqrt{8}$	d_{220}	0.192 015 540(40)	nm	0.21
molar volume of Si, $M(\text{Si})/\rho(\text{Si}) = N_A a^3/8$	$V_m(\text{Si})$	12.058 8179(89)	cm³/mol	0.74

NOTE: A summary of "maintained" units and "standard" values and their relationship to SI units, based on a least-squares adjustment with 17 degrees of freedom. The digits in parentheses are the one-standard-deviation uncertainty in the last digits of the given value. Since the uncertainties of many of these entries are correlated, the full covariance matrix must be used in evaluating the uncertainties of quantities computed from them.

+ The lattice spacing of single-crystal Si can vary by parts in 10^7 depending on the preparation process. Measurements at PTB indicate also the possibility of distortions from exact cubic symmetry of the order of 0.2 ppm.

Table 4
ENERGY CONVERSION FACTORS

	J	kg	m⁻¹	Hz
1 J =	1	$1/\{c^2\}$ 1.112 650 06×10⁻¹⁷	$1/\{hc\}$ 5.034 1125(30)×10²⁴	$1/\{h\}$ 1.509 188 97(90)×10³³
1 kg =	$\{c^2\}$ 8.987 551 787×10¹⁶	1	$\{c/h\}$ 4.524 4347(27)×10⁴¹	$\{c^2/h\}$ 1.356 391 40(81)×10⁵⁰
1 m⁻¹ =	$\{hc\}$ 1.986 4475(12)×10⁻²⁵	$\{h/c\}$ 2.210 2209(13)×10⁻⁴²	1	$\{c\}$ 299 792 458
1 Hz =	$\{h\}$ 6.626 0755(40)×10⁻³⁴	$\{h/c^2\}$ 7.372 5032(44)×10⁻⁵¹	$1/\{c\}$ 3.335 640 952×10⁻⁹	1
1 K =	$\{k\}$ 1.380 658(12)×10⁻²³	$\{k/c^2\}$ 1.536 189(13)×10⁻⁴⁰	$\{k/hc\}$ 69.503 87(59)	$\{k/h\}$ 2.083 674(18)×10¹⁰
1 eV =	$\{e\}$ 1.602 177 33(49)×10⁻¹⁹	$\{e/c^2\}$ 1.782 662 70(54)×10⁻³⁶	$\{e/hc\}$ 806 554.10(24)	$\{e/h\}$ 2.417 988 36(72)×10¹⁴
1 u =	$\{m_u c^2\}$ 1.492 419 09(88)×10⁻¹⁰	$\{m_u\}$ 1.660 5402(10)×10⁻²⁷	$\{m_u c/h\}$ 7.513 005 63(67)×10¹⁴	$\{m_u c^2/h\}$ 2.252 342 42(20)×10²³
1 hartree =	$\{2R_\infty hc\}$ 4.359 7482(26)×10⁻¹⁸	$\{2R_\infty h/c\}$ 4.850 8741(29)×10⁻³⁵	$\{2R_\infty\}$ 21 947 463.067(26)	$\{2R_\infty c\}$ 6.579 683 8999(78)×10¹⁵

	K	eV	u	hartree
1 J =	$1/\{k\}$ 7.242 924(61)×10²²	$1/\{e\}$ 6.241 5064(19)×10¹⁸	$1/\{m_u c^2\}$ 6.700 5308(40)×10⁹	$1/\{2R_\infty hc\}$ 2.293 7104(14)×10¹⁷
1 kg =	$\{c^2/k\}$ 6.509 616(55)×10³⁹	$\{c^2/e\}$ 5.609 5862(17)×10³⁵	$1/\{m_u\}$ 6.022 1367(36)×10²⁶	$\{c/2R_\infty h\}$ 2.061 4841(12)×10³⁴
1 m⁻¹ =	$\{hc/k\}$ 0.014 387 69(12)	$\{hc/e\}$ 1.239 842 44(37)×10⁻⁶	$\{h/m_u c\}$ 1.331 025 22(12)×10⁻¹⁵	$1/\{2R_\infty\}$ 4.556 335 2672(54)×10⁻⁸
1 Hz =	$\{h/k\}$ 4.799 216(41)×10⁻¹¹	$\{h/e\}$ 4.135 6692(12)×10⁻¹⁵	$\{h/m_u c^2\}$ 4.439 822 24(40)×10⁻²⁴	$1/\{2R_\infty c\}$ 1.519 829 8508(18)×10⁻¹⁶
1 K =	1	$\{k/e\}$ 8.617 385(73)×10⁻⁵	$\{k/m_u c^2\}$ 9.251 140(78)×10⁻¹⁴	$\{k/2R_\infty hc\}$ 3.166 829(27)×10⁻⁶
1 eV =	$\{e/k\}$ 11 604.45(10)	1	$\{e/m_u c^2\}$ 1.073 543 85(33)×10⁻⁹	$\{e/2R_\infty hc\}$ 0.036 749 309(11)
1 u =	$\{m_u c^2/k\}$ 1.080 9478(91)×10¹³	$\{m_u c^2/e\}$ 931.494 32(28)×10⁶	1	$\{m_u c/2R_\infty h\}$ 3.423 177 25(31)×10⁷
1 hartree =	$\{2R_\infty hc/k\}$ 3.157 733(27)×10⁵	$\{2R_\infty hc/e\}$ 27.211 3961(81)	$\{2R_\infty h/m_u c\}$ 2.921 262 69(26)×10⁻⁸	1

NOTE: To use this table note that all entries on the same line are equal; the unit at the top of a column applies to all of the values beneath it. Example: 1 eV = 806544.10 m⁻¹.

Table 5
EXPANDED COVARIANCE AND CORRELATION COEFFICIENT MATRIX FOR THE 1986 RECOMMENDED SET OF FUNDAMENTAL PHYSICAL CONSTANTS

	α^{-1}	K_V	K_Ω	μ_μ/μ_p	e	h	m_e	N_A	F
α^{-1}	1997	−1062	925	3267	−3059	−4121	−127	127	−2932
K_V	−0.080	87988	90	−1737	89050	177038	174914	−174914	−85864
K_Ω	0.416	0.006	2477	1513	−835	−744	1105	−1105	−1939
μ_μ/μ_p	0.498	−0.040	0.207	21523	−5004	−6742	−208	208	−4796
e	−0.226	0.989	−0.055	−0.112	92109	181159	175042	−175042	−82933
h	−0.154	0.997	−0.025	−0.077	0.997	358197	349956	−349956	−168797
m_e	−0.005	0.997	0.038	−0.002	0.975	0.989	349702	−349702	−174660
N_A	0.005	−0.997	−0.038	0.002	−0.975	−0.989	−1.000	349702	174660
F	−0.217	−0.956	−0.129	−0.108	−0.902	−0.931	−0.975	0.975	91727

The elements of the covariance matrix appear on and above the major diagonal in (parts in $10^9)^2$; correlation coefficients appear in *italics* below the diagonal. The values are given to as many as six digits only as a matter of consistency. The correlation coefficient between m_e and N_A appears as −1.000 in this table because the auxiliary constants were considered to be exact in carrying out the least-squares adjustment. When the uncertainties of m_p/m_e and M_p are properly taken into account, the correlation coefficient is −0.999 and the variances of m_e and N_A are slightly increased.

STANDARD ATOMIC WEIGHTS (1995)

This table of atomic weights is reprinted from the 1995 report of the IUPAC Commission on Atomic Weights and Isotopic Abundances. The Standard Atomic Weights apply to the elements as they exist naturally on Earth, and the uncertainties take into account the isotopic variation found in most laboratory samples. Further comments on the variability are given in the footnotes.

The number in parentheses following the atomic weight value gives the uncertainty in the last digit. An entry in brackets indicates the mass number of the longest-lived isotope of an element that has no stable isotopes and for which a Standard Atomic Weight cannot be defined because of wide variability in isotopic composition (or complete absence) in nature.

REFERENCE

IUPAC Commission on Atomic Weights and Isotopic Abundances, Atomic Weights of the Elements, 1995, *Pure Appl. Chem.*, 68, 2339, 1996.

Name	Symbol	At. no.	Atomic Weight	Footnotes		
Actinium	Ac	89	[227]			
Aluminum	Al	13	26.981538(2)			
Americium	Am	95	[243]			
Antimony	Sb	51	121.760(1)	g		
Argon	Ar	18	39.948(1)	g		r
Arsenic	As	33	74.921560(2)			
Astatine	At	85	[210]			
Barium	Ba	56	137.327(7)			
Berkelium	Bk	97	[247]			
Beryllium	Be	4	9.012182(3)			
Bismuth	Bi	83	208.98038(2)			
Boron	B	5	10.811(7)	g	m	r
Bromine	Br	35	79.904(1)			
Cadmium	Cd	48	112.411(8)	g		
Calcium	Ca	20	40.078(4)	g		
Californium	Cf	98	[251]			
Carbon	C	6	12.0107(8)	g		r
Cerium	Ce	58	140.116(1)	g		
Cesium	Cs	55	132.90545(2)			
Chlorine	Cl	17	35.4527(9)		m	
Chromium	Cr	24	51.9961(6)			
Cobalt	Co	27	58.933200(9)			
Copper	Cu	29	63.546(3)			r
Curium	Cm	96	[247]			
Dysprosium	Dy	66	162.50(3)	g		
Einsteinium	Es	99	[252]			
Erbium	Er	68	167.26(3)	g		
Europium	Eu	63	151.964(1)	g		
Fermium	Fm	100	[257]			
Fluorine	F	9	18.9984032(5)			
Francium	Fr	87	[223]			
Gadolinium	Gd	64	157.25(3)	g		
Gallium	Ga	31	69.723(1)			
Germanium	Ge	32	72.61(2)			
Gold	Au	79	196.96655(2)			
Hafnium	Hf	72	178.49(2)			
Hahnium*	Ha	105	[262]			
Helium	He	2	4.002602(2)	g		r
Holmium	Ho	67	164.93032(2)			
Hydrogen	H	1	1.00794(7)	g	m	r
Indium	In	49	114.818(3)			
Iodine	I	53	126.90447(3)			
Iridium	Ir	77	192.217(3)			
Iron	Fe	26	55.845(2)			

Name	Symbol	At. no.	Atomic Weight	Footnotes		
Krypton	Kr	36	83.80(1)	g	m	
Lanthanum	La	57	138.9055(2)	g		
Lawrencium	Lr	103	[262]			
Lead	Pb	82	207.2(1)	g		r
Lithium	Li	3	6.941(2)	g	m	r
Lutetium	Lu	71	174.967(1)	g		
Magnesium	Mg	12	24.3050(6)			
Manganese	Mn	25	54.938049(9)			
Mendelevium	Md	101	[258]			
Mercury	Hg	80	200.59(2)			
Molybdenum	Mo	42	95.94(1)	g		
Neodymium	Nd	60	144.24(3)	g		
Neon	Ne	10	20.1797(6)	g	m	
Neptunium	Np	93	[237]			
Nickel	Ni	28	58.6934(2)			
Niobium	Nb	41	92.90638(2)			
Nitrogen	N	7	14.00674(7)	g		r
Nobelium	No	102	[259]			
Osmium	Os	76	190.23(3)	g		
Oxygen	O	8	15.9994(3)	g		r
Palladium	Pd	46	106.42(1)	g		
Phosphorus	P	15	30.973761(2)			
Platinum	Pt	78	195.078(2)			
Plutonium	Pu	94	[244]			
Polonium	Po	84	[209]			
Potassium	K	19	39.0983(1)	g		
Praseodymium	Pr	59	140.90765(2)			
Promethium	Pm	61	[145]			
Protactinium	Pa	91	231.03588(2)			
Radium	Ra	88	[226]			
Radon	Rn	86	[222]			
Rhenium	Re	75	186.207(1)			
Rhodium	Rh	45	102.90550(2)			
Rubidium	Rb	37	85.4678(3)	g		
Ruthenium	Ru	44	101.07(2)	g		
Rutherfordium**	Rf	104	[261]			
Samarium	Sm	62	150.36(3)	g		
Scandium	Sc	21	44.955910(8)			
Selenium	Se	34	78.96(3)			
Silicon	Si	14	28.0855(3)			r
Silver	Ag	47	107.8682(2)	g		
Sodium	Na	11	22.989770(2)			
Strontium	Sr	38	87.62(1)	g		r
Sulfur	S	16	32.066(6)	g		r
Tantalum	Ta	73	180.9479(1)			
Technetium	Tc	43	[98]			
Tellurium	Te	52	127.60(3)	g		
Terbium	Tb	65	158.92534(2)			
Thallium	Tl	81	204.3833(2)			
Thorium	Th	90	232.0381(1)	g		
Thulium	Tm	69	168.93421(2)			
Tin	Sn	50	118.710(7)	g		
Titanium	Ti	22	47.867(1)			
Tungsten	W	74	183.84(1)			
Uranium	U	92	238.0289(1)	g	m	
Vanadium	V	23	50.9415(1)			
Xenon	Xe	54	131.29(2)	g	m	
Ytterbium	Yb	70	173.04(3)	g		

Name	Symbol	At. no.	Atomic Weight	Footnotes
Yttrium	Y	39	88.90585(2)	
Zinc	Zn	30	65.39(2)	
Zirconium	Zr	40	91.224(2)	g

*	Current IUPAC name is unnilpentium, symbol Unp.
**	Current IUPAC name is unnilquadium, symbol Unq.
g	geological specimens are known in which the element has an isotopic composition outside the limits for normal material. The difference between the atomic weight of the element in such specimens and that given in the table may exceed the stated uncertainty.
m	modified isotopic compositions may be found in commercially available material because it has been subjected to an undisclosed or inadvertent isotopic fractionation. Substantial deviations in atomic weight of the element from that given the table can occur.
r	range in isotopic composition of normal terrestrial material prevents a more precise atomic weight being given; the tabulated atomic weight value should be applicable to any normal material.

ATOMIC MASSES AND ABUNDANCES

This table lists the mass (in atomic mass units, symbol u) and the natural abundance (in percent) of the stable nuclides and a few important radioactive nuclides. A complete table of all nuclides may be found in Section 10 ("Table of the Isotopes").

Numbers in parentheses give the uncertainty in the last digit(s) of the stated values. The uncertainty in the natural abundance includes both the estimated measurement uncertainty and the reported range of variation in different terrestrial sources of the element (see Reference 3 for more details). A * in the Abundance column indicates a radioactive nuclide not present in nature or which varies so widely that a meaningful natural abundance cannot be defined.

REFERENCES

1. Holden, N. E., "Table of the Isotopes", in Lide, D. R., Ed., *CRC Handbook of Chemistry and Physics*, 74th Ed., CRC Press, Boca Raton FL, 1993.
2. Wapstra, A. H., and Audi, G., *Nucl. Phys.*, A432, 1, 1985.
3. IUPAC Commission on Atomic Weights and Isotopic Abundances, *Pure Appl. Chem.*, 63, 991, 1991.

Z	Isotope	Mass in u	Abundance in %	Z	Isotope	Mass in u	Abundance in %
1	^{1}H	1.007825035(12)	99.985(1)	20	^{42}Ca	41.9586176(13)	0.647(9)
1	^{2}H	2.014101779(24)	0.015(1)	20	^{43}Ca	42.9587662(13)	0.135(6)
1	^{3}H	3.01604927(4)	*	20	^{44}Ca	43.9554806(14)	2.086(12)
2	^{3}He	3.01602931(4)	0.000137(3)	20	^{46}Ca	45.953689(4)	0.004(3)
2	^{4}He	4.00260324(5)	99.999863(3)	20	^{48}Ca	47.952533(4)	0.187(4)
3	^{6}Li	6.0151214(7)	7.5(2)	21	^{45}Sc	44.9559100(14)	100
3	^{7}Li	7.0160030(9)	92.5(2)	22	^{46}Ti	45.9526294(14)	8.0(1)
4	^{9}Be	9.0121822(4)	100	22	^{47}Ti	46.9517640(11)	7.3(1)
5	^{10}B	10.0129369(3)	19.9(2)	22	^{48}Ti	47.9479473(11)	73.8(1)
5	^{11}B	11.0093054(4)	80.1(2)	22	^{49}Ti	48.9478711(11)	5.5(1)
6	^{12}C	12 (by definition)	98.90(3)	22	^{50}Ti	49.9447921(12)	5.4(1)
6	^{13}C	13.003354826(17)	1.10(3)	23	^{50}V	49.9471609(17)	0.250(2)
6	^{14}C	14.003241982(27)	*	23	^{51}V	50.9439617(17)	99.750(2)
7	^{14}N	14.003074002(26)	99.634(9)	24	^{50}Cr	49.9460464(17)	4.345(13)
7	^{15}N	15.00010897(4)	0.366(9)	24	^{52}Cr	51.9405098(17)	83.789(18)
8	^{16}O	15.99491463(5)	99.762(15)	24	^{53}Cr	52.9406513(17)	9.501(17)
8	^{17}O	16.9991312(4)	0.038(3)	24	^{54}Cr	53.9388825(17)	2.365(7)
8	^{18}O	17.9991603(9)	0.200(12)	25	^{55}Mn	54.9380471(16)	100
9	^{19}F	18.99840322(15)	100	26	^{54}Fe	53.9396127(15)	5.8(1)
10	^{20}Ne	19.9924356(22)	90.48(3)	26	^{56}Fe	55.9349393(16)	91.72(30)
10	^{21}Ne	20.9938428(21)	0.27(1)	26	^{57}Fe	56.9353958(16)	2.1(1)
10	^{22}Ne	21.9913831(18)	9.25(3)	26	^{58}Fe	57.9332773(16)	0.28(1)
11	^{23}Na	22.9897677(10)	100	27	^{59}Co	58.9331976(16)	100
12	^{24}Mg	23.9850423(3)	78.99(3)	28	^{58}Ni	57.9353462(16)	68.077(9)
12	^{25}Mg	24.9858374(8)	10.00(1)	28	^{60}Ni	59.9307884(16)	26.223(8)
12	^{26}Mg	25.9825937(8)	11.01(2)	28	^{61}Ni	60.9310579(16)	1.140(1)
13	^{27}Al	26.9815386(8)	100	28	^{62}Ni	61.9283461(16)	3.634(2)
14	^{28}Si	27.9769271(7)	92.23(1)	28	^{64}Ni	63.9279679(17)	0.926(1)
14	^{29}Si	28.9764949(7)	4.67(1)	29	^{63}Cu	62.9295989(16)	69.17(3)
14	^{30}Si	29.9737707(7)	3.10(1)	29	^{65}Cu	64.9277929(20)	30.83(3)
15	^{31}P	30.9737620(6)	100	30	^{64}Zn	63.9291448(19)	48.6(3)
16	^{32}S	31.97207070(25)	95.02(9)	30	^{66}Zn	65.9260347(17)	27.9(2)
16	^{33}S	32.97145854(23)	0.75(4)	30	^{67}Zn	66.9271291(17)	4.1(1)
16	^{34}S	33.96786665(22)	4.21(8)	30	^{68}Zn	67.9248459(18)	18.8(4)
16	^{36}S	35.96708062(27)	0.02(1)	30	^{70}Zn	69.925325(4)	0.6(1)
17	^{35}Cl	34.968852721(69)	75.77(7)	31	^{69}Ga	68.925580(3)	60.108(9)
17	^{37}Cl	36.96590262(11)	24.23(7)	31	^{71}Ga	70.9247005(25)	39.892(9)
18	^{36}Ar	35.96754552(29)	0.337(3)	32	^{70}Ge	69.9242497(16)	21.23(4)
18	^{38}Ar	37.9627325(9)	0.063(1)	32	^{72}Ge	71.9220789(16)	27.66(3)
18	^{40}Ar	39.9623837(14)	99.600(3)	32	^{73}Ge	72.9234626(16)	7.73(1)
19	^{39}K	38.9637074(12)	93.2581(44)	32	^{74}Ge	73.9211774(15)	35.94(2)
19	^{40}K	39.9639992(12)	0.0117(1)	32	^{76}Ge	75.9214016(17)	7.44(2)
19	^{41}K	40.9618254(12)	6.7302(44)	33	^{75}As	74.9215942(17)	100
20	^{40}Ca	39.9625906(13)	96.941(18)	34	^{74}Se	73.9224746(16)	0.89(2)

Z	Isotope	Mass in u	Abundance in %	Z	Isotope	Mass in u	Abundance in %
34	^{76}Se	75.9192120(16)	9.36(11)	49	^{113}In	112.904061(4)	4.3(2)
34	^{77}Se	76.9199125(16)	7.63(6)	49	^{115}In	114.903880(4)	95.7(2)
34	^{78}Se	77.9173076(16)	23.78(9)	50	^{112}Sn	111.904826(5)	0.97(1)
34	^{80}Se	79.9165196(19)	49.61(10)	50	^{114}Sn	113.902784(4)	0.65(1)
34	^{82}Se	81.9166978(23)	8.73(6)	50	^{115}Sn	114.903348(3)	0.34(1)
35	^{79}Br	78.9183361(26)	50.69(7)	50	^{116}Sn	115.901747(3)	14.53(1)
35	^{81}Br	80.916289(6)	49.31(7)	50	^{117}Sn	116.902956(3)	7.68(7)
36	^{78}Kr	77.920396(9)	0.35(2)	50	^{118}Sn	117.901609(3)	24.23(11)
36	^{80}Kr	79.916380(9)	2.25(2)	50	^{119}Sn	118.903310(3)	8.59(4)
36	^{82}Kr	81.913482(6)	11.6(1)	50	^{120}Sn	119.9021991(29)	32.59(10)
36	^{83}Kr	82.914135(4)	11.5(1)	50	^{122}Sn	121.9034404(30)	4.63(3)
36	^{84}Kr	83.911507(4)	57.0(3)	50	^{124}Sn	123.9052743(17)	5.79(5)
36	^{86}Kr	85.910616(5)	17.3(2)	51	^{121}Sb	120.9038212(29)	57.36(8)
37	^{85}Rb	84.911794(3)	72.165(20)	51	^{123}Sb	122.9042160(24)	42.64(8)
37	^{87}Rb	86.909187(3)	27.835(20)	52	^{120}Te	119.904048(21)	0.096(2)
38	^{84}Sr	83.913430(4)	0.56(1)	52	^{122}Te	121.903054(3)	2.603(4)
38	^{86}Sr	85.9092672(28)	9.86(1)	52	^{123}Te	122.9042710(22)	0.908(2)
38	^{87}Sr	86.9088841(28)	7.00(1)	52	^{124}Te	123.902823(2)	4.816(6)
38	^{88}Sr	87.9056188(28)	82.58(1)	52	^{125}Te	124.904433(3)	7.139(6)
39	^{89}Y	88.905849(3)	100	52	^{126}Te	125.903314(3)	18.95(1)
40	^{90}Zr	89.9047026(26)	51.45(3)	52	^{128}Te	127.904463(4)	31.69(1)
40	^{91}Zr	90.9056439(26)	11.22(4)	52	^{130}Te	129.906229(5)	33.80(1)
40	^{92}Zr	91.9050386(26)	17.15(2)	53	^{127}I	126.904473(5)	100
40	^{94}Zr	93.9063148(28)	17.38(4)	54	^{124}Xe	123.9058942(22)	0.10(1)
40	^{96}Zr	95.908275(4)	2.80(2)	54	^{126}Xe	125.904281(8)	0.09(1)
41	^{93}Nb	92.9063772(27)	100	54	^{128}Xe	127.9035312(17)	1.91(3)
42	^{92}Mo	91.906808(4)	14.84(4)	54	^{129}Xe	128.9047801(21)	26.4(6)
42	^{94}Mo	93.9050853(26)	9.25(3)	54	^{130}Xe	129.9035094(17)	4.1(1)
42	^{95}Mo	94.9058411(22)	15.92(5)	54	^{131}Xe	130.905072(5)	21.2(4)
42	^{96}Mo	95.9046785(22)	16.68(5)	54	^{132}Xe	131.904144(5)	26.9(5)
42	^{97}Mo	96.9060205(22)	9.55(3)	54	^{134}Xe	133.905395(8)	10.4(2)
42	^{98}Mo	97.9054073(22)	24.13(7)	54	^{136}Xe	135.907214(8)	8.9(1)
42	^{100}Mo	99.907477(6)	9.63(3)	55	^{133}Cs	132.905429(7)	100
43	^{98}Tc	97.907215(4)	*	56	^{130}Ba	129.906282(8)	0.106(2)
44	^{96}Ru	95.907599(8)	5.52(6)	56	^{132}Ba	131.905042(9)	0.101(2)
44	^{98}Ru	97.905287(7)	1.88(6)	56	^{134}Ba	133.904486(7)	2.417(27)
44	^{99}Ru	98.9059389(23)	12.7(1)	56	^{135}Ba	134.905665(7)	6.592(18)
44	^{100}Ru	99.9042192(24)	12.6(1)	56	^{136}Ba	135.904553(7)	7.854(36)
44	^{101}Ru	100.9055819(24)	17.0(1)	56	^{137}Ba	136.905812(6)	11.23(4)
44	^{102}Ru	101.9043485(25)	31.6(2)	56	^{138}Ba	137.905232(6)	71.70(7)
44	^{104}Ru	103.905424(6)	18.7(2)	57	^{138}La	137.907105(6)	0.0902(2)
45	^{103}Rh	102.905500(4)	100	57	^{139}La	138.906347(5)	99.9098(2)
46	^{102}Pd	101.905634(5)	1.02(1)	58	^{136}Ce	135.907140(50)	0.19(1)
46	^{104}Pd	103.904029(6)	11.14(8)	58	^{138}Ce	137.905985(12)	0.25(1)
46	^{105}Pd	104.905079(6)	22.33(8)	58	^{140}Ce	139.905433(4)	88.48(10)
46	^{106}Pd	105.903478(6)	27.33(3)	58	^{142}Ce	141.909241(4)	11.08(10)
46	^{108}Pd	107.903895(4)	26.46(9)	59	^{141}Pr	140.907647(4)	100
46	^{110}Pd	109.905167(20)	11.72(9)	60	^{142}Nd	141.907719(4)	27.13(12)
47	^{107}Ag	106.905092(6)	51.839(7)	60	^{143}Nd	142.909810(4)	12.18(6)
47	^{109}Ag	108.904757(4)	48.161(7)	60	^{144}Nd	143.910083(4)	23.80(12)
48	^{106}Cd	105.906461(7)	1.25(4)	60	^{145}Nd	144.912570(4)	8.30(6)
48	^{108}Cd	107.904176(6)	0.89(2)	60	^{146}Nd	145.913113(4)	17.19(9)
48	^{110}Cd	109.903005(4)	12.49(12)	60	^{148}Nd	147.916889(4)	5.76(3)
48	^{111}Cd	110.904182(3)	12.80(8)	60	^{150}Nd	149.920887(4)	5.64(3)
48	^{112}Cd	111.902758(3)	24.13(14)	61	^{145}Pm	144.912743(4)	*
48	^{113}Cd	112.904400(3)	12.22(8)	62	^{144}Sm	143.911998(4)	3.1(1)
48	^{114}Cd	113.903357(3)	28.73(28)	62	^{147}Sm	146.914895(4)	15.0(2)
48	^{116}Cd	115.904754(4)	7.49(12)	62	^{148}Sm	147.914819(4)	11.3(1)

Z	Isotope	Mass in u	Abundance in %	Z	Isotope	Mass in u	Abundance in %
62	^{149}Sm	148.917181(4)	13.8(1)	74	^{184}W	183.950928(3)	30.67(15)
62	^{150}Sm	149.917273(4)	7.4(1)	74	^{186}W	185.954357(4)	28.6(2)
62	^{152}Sm	151.919729(4)	26.7(2)	75	^{185}Re	184.952951(3)	37.40(2)
62	^{154}Sm	153.922206(4)	22.7(2)	75	^{187}Re	186.955744(3)	62.60(2)
63	^{151}Eu	150.919847(8)	47.8(15)	76	^{184}Os	183.952488(4)	0.02(1)
63	^{153}Eu	152.921225(4)	52.2(15)	76	^{186}Os	185.953830(4)	1.58(30)
64	^{152}Gd	151.919786(4)	0.20(1)	76	^{187}Os	186.955741(3)	1.6(3)
64	^{154}Gd	153.920861(4)	2.18(3)	76	^{188}Os	187.955860(3)	13.3(7)
64	^{155}Gd	154.922618(4)	14.80(5)	76	^{189}Os	188.958137(4)	16.1(8)
64	^{156}Gd	155.922118(4)	20.47(4)	76	^{190}Os	189.958436(4)	26.4(12)
64	^{157}Gd	156.923956(4)	15.65(3)	76	^{192}Os	191.961467(4)	41.0(8)
64	^{158}Gd	157.924019(4)	24.84(12)	77	^{191}Ir	190.960584(4)	37.3(5)
64	^{160}Gd	159.927049(4)	21.86(4)	77	^{193}Ir	192.962917(4)	62.7(5)
65	^{159}Tb	158.925342(4)	100	78	^{190}Pt	189.959917(7)	0.01(1)
66	^{156}Dy	155.924277(8)	0.06(1)	78	^{192}Pt	191.961019(5)	0.79(6)
66	^{158}Dy	157.924403(5)	0.10(1)	78	^{194}Pt	193.962655(4)	32.9(6)
66	^{160}Dy	159.925193(4)	2.34(6)	78	^{195}Pt	194.964766(4)	33.8(6)
66	^{161}Dy	160.926930(4)	18.9(2)	78	^{196}Pt	195.964926(4)	25.3(6)
66	^{162}Dy	161.926795(4)	25.5(2)	78	^{198}Pt	197.967869(6)	7.2(2)
66	^{163}Dy	162.928728(4)	24.9(2)	79	^{197}Au	196.966543(4)	100
66	^{164}Dy	163.929171(4)	28.2(2)	80	^{196}Hg	195.965807(5)	0.15(1)
67	^{165}Ho	164.930319(4)	100	80	^{198}Hg	197.966743(4)	9.97(8)
68	^{162}Er	161.928775(4)	0.14(1)	80	^{199}Hg	198.968254(4)	16.87(10)
68	^{164}Er	163.929198(4)	1.61(2)	80	^{200}Hg	199.968300(4)	23.10(16)
68	^{166}Er	165.930290(4)	33.6(2)	80	^{201}Hg	200.970277(4)	13.18(8)
68	^{167}Er	166.932046(4)	22.95(15)	80	^{202}Hg	201.970617(4)	29.86(20)
68	^{168}Er	167.932368(4)	26.8(2)	80	^{204}Hg	203.973467(5)	6.87(4)
68	^{170}Er	169.935461(4)	14.9(2)	81	^{203}Tl	202.972320(5)	29.524(14)
69	^{169}Tm	168.934212(4)	100	81	^{205}Tl	204.974401(5)	70.476(14)
70	^{168}Yb	167.933894(5)	0.13(1)	82	^{204}Pb	203.973020(5)	1.4(1)
70	^{170}Yb	169.934759(4)	3.05(6)	82	^{206}Pb	205.974440(4)	24.1(1)
70	^{171}Yb	170.936323(3)	14.3(2)	82	^{207}Pb	206.975872(4)	22.1(1)
70	^{172}Yb	171.936378(3)	21.9(3)	82	^{208}Pb	207.976627(4)	52.4(1)
70	^{173}Yb	172.938208(3)	16.12(21)	83	^{209}Bi	208.980374(5)	100
70	^{174}Yb	173.938859(3)	31.8(4)	84	^{209}Po	208.982404(5)	*
70	^{176}Yb	175.942564(4)	12.7(2)	85	^{210}At	209.987126(12)	*
71	^{175}Lu	174.940770(3)	97.41(2)	86	^{222}Rn	222.017570(3)	*
71	^{176}Lu	175.942679(3)	2.59(2)	87	^{223}Fr	223.019733(4)	*
72	^{174}Hf	173.940044(4)	0.162(3)	88	^{226}Ra	226.025402(3)	*
72	^{176}Hf	175.941406(4)	5.206(5)	89	^{227}Ac	227.027750(3)	*
72	^{177}Hf	176.943217(3)	18.606(4)	90	^{232}Th	232.038054(2)	100
72	^{178}Hf	177.943696(3)	27.297(4)	91	^{231}Pa	231.035880(3)	*
72	^{179}Hf	178.9458122(29)	13.629(6)	92	^{234}U	234.0409468(24)	0.0055(5)
72	^{180}Hf	179.9465457(30)	35.100(7)	92	^{235}U	235.0439242(24)	0.7200(12)
73	^{180}Ta	179.947462(4)	0.012(2)	92	^{238}U	238.0507847(23)	99.2745(60)
73	^{181}Ta	180.947992(3)	99.988(2)	93	^{237}Np	237.0481678(23)	*
74	^{180}W	179.946701(5)	0.13(4)	94	^{239}Pu	239.052157(2)	*
74	^{182}W	181.948202(3)	26.3(2)	94	^{244}Pu	244.064199(5)	*
74	^{183}W	182.950220(3)	14.3(1)				

ELECTRON CONFIGURATION OF NEUTRAL ATOMS IN THE GROUND STATE

Atomic no.	Element	K 1 s	L 2 s	L 2 p	M 3 s	M 3 p	M 3 d	N 4 s	N 4 p	N 4 d	N 4 f	O 5 s	O 5 p	O 5 d	O 5 f	P 6 s	P 6 p	P 6 d	Q 7 s
1	H	1																	
2	He	2																	
3	Li	2	1																
4	Be	2	2																
5	B	2	2	1															
6	C	2	2	2															
7	N	2	2	3															
8	O	2	2	4															
9	F	2	2	5															
10	Ne	2	2	6															
11	Na	2	2	6	1														
12	Mg	2	2	6	2														
13	Al	2	2	6	2	1													
14	Si	2	2	6	2	2													
15	P	2	2	6	2	3													
16	S	2	2	6	2	4													
17	Cl	2	2	6	2	5													
18	Ar	2	2	6	2	6													
19	K	2	2	6	2	6		1											
20	Ca	2	2	6	2	6		2											
21	Sc	2	2	6	2	6	1	2											
22	Ti	2	2	6	2	6	2	2											
23	V	2	2	6	2	6	3	2											
24	Cr	2	2	6	2	6	5*	1											
25	Mn	2	2	6	2	6	5	2											
26	Fe	2	2	6	2	6	6	2											
27	Co	2	2	6	2	6	7	2											
28	Ni	2	2	6	2	6	8	2											
29	Cu	2	2	6	2	6	10*	1											
30	Zn	2	2	6	2	6	10	2											
31	Ga	2	2	6	2	6	10	2	1										
32	Ge	2	2	6	2	6	10	2	2										
33	As	2	2	6	2	6	10	2	3										
34	Se	2	2	6	2	6	10	2	4										
35	Br	2	2	6	2	6	10	2	5										
36	Kr	2	2	6	2	6	10	2	6										
37	Rb	2	2	6	2	6	10	2	6			1							
38	Sr	2	2	6	2	6	10	2	6			2							
39	Y	2	2	6	2	6	10	2	6	1		2							
40	Zr	2	2	6	2	6	10	2	6	2		2							
41	Nb	2	2	6	2	6	10	2	6	4*		1							
42	Mo	2	2	6	2	6	10	2	6	5		1							
43	Tc	2	2	6	2	6	10	2	6	5		2							
44	Ru	2	2	6	2	6	10	2	6	7		1							
45	Rh	2	2	6	2	6	10	2	6	8		1							
46	Pd	2	2	6	2	6	10	2	6	10*									
47	Ag	2	2	6	2	6	10	2	6	10		1							
48	Cd	2	2	6	2	6	10	2	6	10		2							
49	In	2	2	6	2	6	10	2	6	10		2	1						
50	Sn	2	2	6	2	6	10	2	6	10		2	2						
51	Sb	2	2	6	2	6	10	2	6	10		2	3						
52	Te	2	2	6	2	6	10	2	6	10		2	4						
53	I	2	2	6	2	6	10	2	6	10		2	5						
54	Xe	2	2	6	2	6	10	2	6	10		2	6						
55	Cs	2	2	6	2	6	10	2	6	10		2	6			1			
56	Ba	2	2	6	2	6	10	2	6	10		2	6			2			

Atomic no.	Element	K 1 s	L 2 s	L 2 p	M 3 s	M 3 p	M 3 d	N 4 s	N 4 p	N 4 d	N 4 f	O 5 s	O 5 p	O 5 d	O 5 f	P 6 s	P 6 p	P 6 d	Q 7 s
57	La	2	2	6	2	6	10	2	6	10		2	6	1		2			
58	Ce	2	2	6	2	6	10	2	6	10	1*	2	6	1		2			
59	Pr	2	2	6	2	6	10	2	6	10	3	2	6			2			
60	Nd	2	2	6	2	6	10	2	6	10	4	2	6			2			
61	Pm	2	2	6	2	6	10	2	6	10	5	2	6			2			
62	Sm	2	2	6	2	6	10	2	6	10	6	2	6			2			
63	Eu	2	2	6	2	6	10	2	6	10	7	2	6			2			
64	Gd	2	2	6	2	6	10	2	6	10	7	2	6	1		2			
65	Tb	2	2	6	2	6	10	2	6	10	9*	2	6			2			
66	Dy	2	2	6	2	6	10	2	6	10	10	2	6			2			
67	Ho	2	2	6	2	6	10	2	6	10	11	2	6			2			
68	Er	2	2	6	2	6	10	2	6	10	12	2	6			2			
69	Tm	2	2	6	2	6	10	2	6	10	13	2	6			2			
70	Yb	2	2	6	2	6	10	2	6	10	14	2	6			2			
71	Lu	2	2	6	2	6	10	2	6	10	14	2	6	1		2			
72	Hf	2	2	6	2	6	10	2	6	10	14	2	6	2		2			
73	Ta	2	2	6	2	6	10	2	6	10	14	2	6	3		2			
74	W	2	2	6	2	6	10	2	6	10	14	2	6	4		2			
75	Re	2	2	6	2	6	10	2	6	10	14	2	6	5		2			
76	Os	2	2	6	2	6	10	2	6	10	14	2	6	6		2			
77	Ir	2	2	6	2	6	10	2	6	10	14	2	6	7		2			
78	Pt	2	2	6	2	6	10	2	6	10	14	2	6	9		1			
79	Au	2	2	6	2	6	10	2	6	10	14	2	6	10		1			
80	Hg	2	2	6	2	6	10	2	6	10	14	2	6	10		2			
81	Tl	2	2	6	2	6	10	2	6	10	14	2	6	10		2	1		
82	Pb	2	2	6	2	6	10	2	6	10	14	2	6	10		2	2		
83	Bi	2	2	6	2	6	10	2	6	10	14	2	6	10		2	3		
84	Po	2	2	6	2	6	10	2	6	10	14	2	6	10		2	4		
85	At	2	2	6	2	6	10	2	6	10	14	2	6	10		2	5		
86	Rn	2	2	6	2	6	10	2	6	10	14	2	6	10		2	6		
87	Fr	2	2	6	2	6	10	2	6	10	14	2	6	10		2	6		1
88	Ra	2	2	6	2	6	10	2	6	10	14	2	6	10		2	6		2
89	Ac	2	2	6	2	6	10	2	6	10	14	2	6	10		2	6	1	2
90	Th	2	2	6	2	6	10	2	6	10	14	2	6	10		2	6	2	2
91	Pa	2	2	6	2	6	10	2	6	10	14	2	6	10	2*	2	6	1	2
92	U	2	2	6	2	6	10	2	6	10	14	2	6	10	3	2	6	1	2
93	Np	2	2	6	2	6	10	2	6	10	14	2	6	10	4	2	6	1	2
94	Pu	2	2	6	2	6	10	2	6	10	14	2	6	10	6*	2	6		2
95	Am	2	2	6	2	6	10	2	6	10	14	2	6	10	7	2	6		2
96	Cm	2	2	6	2	6	10	2	6	10	14	2	6	10	7*	2	6	1	2
97	Bk	2	2	6	2	6	10	2	6	10	14	2	6	10	9	2	6		2
98	Cf	2	2	6	2	6	10	2	6	10	14	2	6	10	10	2	6		2
99	Es	2	2	6	2	6	10	2	6	10	14	2	6	10	11	2	6		2
100	Fm	2	2	6	2	6	10	2	6	10	14	2	6	10	12	2	6		2
101	Md	2	2	6	2	6	10	2	6	10	14	2	6	10	13	2	6		2
102	No	2	2	6	2	6	10	2	6	10	14	2	6	10	14	2	6		2
103	Lr	2	2	6	2	6	10	2	6	10	14	2	6	10	14	2	6	1	2
104	Rf	2	2	6	2	6	10	2	6	10	14	2	6	10	14	2	6	2	2

* Note irregularity.

REFERENCE

W. L. Wiese and G. A. Martin, in *A Physicist's Desk Reference*, American Institute of Physics, New York, 1989, 94.

INTERNATIONAL TEMPERATURE SCALE OF 1990 (ITS-90)

B. W. Mangum

A new temperature scale, the International Temperature Scale of 1990 (ITS-90), was officially adopted by the Comité International des Poids et Mesures (CIPM), meeting 26—28 September 1989 at the Bureau International des Poids et Mesures (BIPM). The ITS-90 was recommended to the CIPM for its adoption following the completion of the final details of the new scale by the Comité Consultatif de Thermométrie (CCT), meeting 12—14 September 1989 at the BIPM in its 17th Session. The ITS-90 became the official international temperature scale on 1 January 1990. The ITS-90 supersedes the present scales, the International Practical Temperature Scale of 1968 (IPTS-68) and the 1976 Provisional 0.5 to 30 K Temperature Scale (EPT-76).

The ITS-90 extends upward from 0.65 K, and temperatures on this scale are in much better agreement with thermodynamic values that are those on the IPTS-68 and the EPT-76. The new scale has subranges and alternative definitions in certain ranges that greatly facilitate its use. Furthermore, its continuity, precision, and reproducibility throughout its ranges are much improved over that of the present scales. The replacement of the thermocouple with the platinum resistance thermometer at temperatures below 961.78°C resulted in the biggest improvement in reproducibility.

The ITS-90 is divided into four primary ranges:

1. Between 0.65 and 3.2 K, the ITS-90 is defined by the vapor pressure-temperature relation of ^{3}He, and between 1.25 and 2.1768 K (the λ point) and between 2.1768 and 5.0 K by the vapor pressure-temperature relations of ^{4}He. T_{90} is defined by the vapor pressure equations of the form:

$$T_{90}/K = A_0 + \sum_{i=1}^{9} A_i \left[\left(\ln(p/Pa) - B \right)/C \right]^i$$

The values of the coefficients A_i, and of the constants A_o, B, and C of the equations are given below.
2. Between 3.0 and 24.5561 K, the ITS-90 is defined in terms of a ^{3}He or ^{4}He constant volume gas thermometer (CVGT). The thermometer is calibrated at three temperatures — at the triple point of neon (24.5561 K), at the triple point of equilibrium hydrogen (13.8033 K), and at a temperature between 3.0 and 5.0 K, the value of which is determined by using either ^{3}He or ^{4}He vapor pressure thermometry.
3. Between 13.8033 K (–259.3467°C) and 1234.93 K (961.78°C), the ITS-90 is defined in terms of the specified fixed points given below, by resistance ratios of platinum resistance thermometers obtained by calibration at specified sets of the fixed points, and by reference functions and deviation functions of resistance ratios which relate to T_{90} between the fixed points.
4. Above 1234.93 K, the ITS-90 is defined in terms of Planck's radiation law, using the freezing-point temperature of either silver, gold, or copper as the reference temperature.

Full details of the calibration procedures and reference functions for various subranges are given in:

The International Temperature Scale of 1990, *Metrologia*, 27, 3, 1990; errata in *Metrologia*, 27, 107, 1990.

Defining Fixed Points of the ITS-90

Material[a]	Equilibrium state[b]	Temperature	
		T_{90} (K)	t_{90} (°C)
He	VP	3 to 5	–270.15 to –268.15
e-H$_2$	TP	13.8033	–259.3467
e-H$_2$ (or He)	VP (or CVGT)	≈17	≈ –256.15
e-H$_2$ (or He)	VP (or CVGT)	≈20.3	≈ –252.85
Ne[c]	TP	24.5561	–248.5939
O$_2$	TP	54.3584	–218.7916
Ar	TP	83.8058	–189.3442
Hg[c]	TP	234.3156	–38.8344
H$_2$O	TP	273.16	0.01
Ga[c]	MP	302.9146	29.7646
In[c]	FP	429.7485	156.5985
Sn	FP	505.078	231.928
Zn	FP	692.677	419.527
Al[c]	FP	933.473	660.323
Ag	FP	1234.93	961.78
Au	FP	1337.33	1064.18
Cu[c]	FP	1357.77	1084.62

INTERNATIONAL TEMPERATURE SCALE OF 1990 (ITS-90) (continued)

Defining Fixed Points of the ITS-90 (continued)

[a] e-H_2 indicates equilibrium hydrogen, that is, hydrogen with the equilibrium distribution of its ortho and para states. Normal hydrogen at room temperature contains 25% para hydrogen and 75% ortho hydrogen.

[b] VP indicates vapor pressure point; CVGT indicates constant volume gas thermometer point; TP indicates triple point (equilibrium temperature at which the solid, liquid, and vapor phases coexist); FP indicates freezing point, and MP indicates melting point (the equilibrium temperatures at which the solid and liquid phases coexist under a pressure of 101 325 Pa, one standard atmosphere). The isotopic composition is that naturally occurring.

[c] Previously, these were secondary fixed points.

Values of Coefficients in the Vapor Pressure Equations for Helium

Coef.or constant	^{3}He 0.65—3.2 K	^{4}He 1.25—2.1768 K	^{4}He 2.1768—5.0 K
A_0	1.053 447	1.392 408	3.146 631
A_1	0.980 106	0.527 153	1.357 655
A_2	0.676 380	0.166 756	0.413 923
A_3	0.372 692	0.050 988	0.091 159
A_4	0.151 656	0.026 514	0.016 349
A_5	−0.002 263	0.001 975	0.001 826
A_6	0.006 596	−0.017 976	−0.004 325
A_7	0.088 966	0.005 409	−0.004 973
A_8	−0.004 770	0.013 259	0
A_9	−0.054 943	0	0
B	7.3	5.6	10.3
C	4.3	2.9	1.9

CONVERSION OF TEMPERATURES FROM THE 1948 AND 1968 SCALES TO ITS-90

This table gives temperature corrections from older scales to the current International Temperature Scale of 1990 (see the preceding table for details on ITS-90). The first part of the table may be used for converting Celsius temperatures in the range -180 to 4000°C from IPTS-68 or IPTS-48 to ITS-90. Within the accuracy of the corrections, the temperature in the first column may be identified with either t_{68}, t_{48}, or t_{90}. The second part of the table is designed for use at lower temperatures to convert values expressed in kelvins from EPT-76 or IPTS-68 to ITS-90.

The references give analytical equations for expressing these relations. Note that Reference 1 supersedes Reference 2 with respect to corrections in the 630 to 1064°C range.

REFERENCES

1. Burns, G. W. et al., in *Temperature: Its Measurement and Control in Science and Industry,* Vol. 6, Schooley, J. F., Ed., American Institute of Physics, New York, 1993.
2. Goldberg, R. N. and Weir, R. D., *Pure and Appl. Chem.,* 1545, 1992.

$t/°C$	$t_{90}\text{-}t_{68}$	$t_{90}\text{-}t_{48}$	$t/°C$	$t_{90}\text{-}t_{68}$	$t_{90}\text{-}t_{48}$	$t/°C$	$t_{90}\text{-}t_{68}$	$t_{90}\text{-}t_{48}$
-180	0.008	0.020	270	-0.039	0.028	720	0.00	0.45
-170	0.010	0.017	280	-0.039	0.030	730	0.02	0.49
-160	0.012	0.007	290	-0.039	0.032	740	0.03	0.53
-150	0.013	0.000	300	-0.039	0.034	750	0.03	0.56
-140	0.014	0.001	310	-0.039	0.035	760	0.04	0.60
-130	0.014	0.008	320	-0.039	0.036	770	0.05	0.63
-120	0.014	0.017	330	-0.040	0.036	780	0.05	0.66
-110	0.013	0.026	340	-0.040	0.037	790	0.05	0.69
-100	0.013	0.035	350	-0.041	0.036	800	0.05	0.72
-90	0.012	0.041	360	-0.042	0.035	810	0.05	0.75
-80	0.012	0.045	370	-0.043	0.034	820	0.04	0.76
-70	0.011	0.045	380	-0.045	0.032	830	0.04	0.79
-60	0.010	0.042	390	-0.046	0.030	840	0.03	0.81
-50	0.009	0.038	400	-0.048	0.028	850	0.02	0.83
-40	0.008	0.032	410	-0.051	0.024	860	0.01	0.85
-30	0.006	0.024	420	-0.053	0.022	870	0.00	0.87
-20	0.004	0.016	430	-0.056	0.019	880	-0.02	0.87
-10	0.002	0.008	440	-0.059	0.015	890	-0.03	0.89
0	0.000	0.000	450	-0.062	0.012	900	-0.05	0.90
10	-0.002	-0.006	460	-0.065	0.009	910	-0.06	0.92
20	-0.005	-0.012	470	-0.068	0.007	920	-0.08	0.93
30	-0.007	-0.016	480	-0.072	0.004	930	-0.10	0.94
40	-0.010	-0.020	490	-0.075	0.002	940	-0.11	0.96
50	-0.013	-0.023	500	-0.079	0.000	950	-0.13	0.97
60	-0.016	-0.026	510	-0.083	-0.001	960	-0.15	0.97
70	-0.018	-0.026	520	-0.087	-0.002	970	-0.16	0.99
80	-0.021	-0.027	530	-0.090	-0.001	980	-0.18	1.00
90	-0.024	-0.027	540	-0.094	0.000	990	-0.19	1.02
100	-0.026	-0.026	550	-0.098	0.002	1000	-0.20	1.04
110	-0.028	-0.024	560	-0.101	0.007	1010	-0.22	1.05
120	-0.030	-0.023	570	-0.105	0.011	1020	-0.23	1.07
130	-0.032	-0.020	580	-0.108	0.018	1030	-0.23	1.10
140	-0.034	-0.018	590	-0.112	0.025	1040	-0.24	1.12
150	-0.036	-0.016	600	-0.115	0.035	1050	-0.25	1.14
160	-0.037	-0.012	610	-0.118	0.047	1060	-0.25	1.17
170	-0.038	-0.009	620	-0.122	0.060	1070	-0.25	1.19
180	-0.039	-0.005	630	-0.125	0.075	1080	-0.26	1.20
190	-0.039	-0.001	640	-0.11	0.12	1090	-0.26	1.20
200	-0.040	0.003	650	-0.10	0.15	1100	-0.26	1.2
210	-0.040	0.007	660	-0.09	0.19	1200	-0.30	1.4
220	-0.040	0.011	670	-0.07	0.24	1300	-0.35	1.5
230	-0.040	0.014	680	-0.05	0.29	1400	-0.39	1.6
240	-0.040	0.018	690	-0.04	0.32	1500	-0.44	1.8
250	-0.040	0.021	700	-0.02	0.37	1600	-0.49	1.9
260	-0.040	0.024	710	-0.01	0.41	1700	-0.54	2.1

CONVERSION OF TEMPERATURES FROM THE 1948 AND 1968 SCALES TO ITS-90 (continued)

$t/°C$	$t_{90}-t_{68}$	$t_{90}-t_{48}$
1800	-0.60	2.2
1900	-0.66	2.3
2000	-0.72	2.5
2100	-0.79	2.7
2200	-0.85	2.9
2300	-0.93	3.1
2400	-1.00	3.2
2500	-1.07	3.4
2600	-1.15	3.7
2700	-1.24	3.8
2800	-1.32	4.0
2900	-1.41	4.2
3000	-1.50	4.4
3100	-1.59	4.6
3200	-1.69	4.8
3300	-1.78	5.1
3400	-1.89	5.3
3500	-1.99	5.5
3600	-2.10	5.8
3700	-2.21	6.0
3800	-2.32	6.3
3900	-2.43	6.6
4000	-2.55	6.8

T/K	$T_{90}-T_{76}$	$T_{90}-T_{68}$
5	-0.0001	
6	-0.0002	
7	-0.0003	
8	-0.0004	
9	-0.0005	
10	-0.0006	
11	-0.0007	
12	-0.0008	
13	-0.0010	
14	-0.0011	-0.006
15	-0.0013	-0.003
16	-0.0014	-0.004
17	-0.0016	-0.006
18	-0.0018	-0.008
19	-0.0020	-0.009
20	-0.0022	-0.009
21	-0.0025	-0.008
22	-0.0027	-0.007
23	-0.0030	-0.007
24	-0.0032	-0.006
25	-0.0035	-0.005
26	-0.0038	-0.004
27	-0.0041	-0.004

T/K	$T_{90}-T_{76}$	$T_{90}-T_{68}$
28		-0.005
29		-0.006
30		-0.006
31		-0.007
32		-0.008
33		-0.008
34		-0.008
35		-0.007
36		-0.007
37		-0.007
38		-0.006
39		-0.006
40		-0.006
41		-0.006
42		-0.006
43		-0.006
44		-0.006
45		-0.007
46		-0.007
47		-0.007
48		-0.006
49		-0.006
50		-0.006
51		-0.005
52		-0.005
53		-0.004
54		-0.003
55		-0.002
56		-0.001
57		0.000
58		0.001
59		0.002
60		0.003
61		0.003
62		0.004
63		0.004
64		0.005
65		0.005
66		0.006
67		0.006
68		0.007
69		0.007
70		0.007
71		0.007
72		0.007
73		0.007
74		0.007
75		0.008
76		0.008

T/K	$T_{90}-T_{76}$	$T_{90}-T_{68}$
77		0.008
78		0.008
79		0.008
80		0.008
81		0.008
82		0.008
83		0.008
84		0.008
85		0.008
86		0.008
87		0.008
88		0.008
89		0.008
90		0.008
91		0.008
92		0.008
93		0.008
94		0.008
95		0.008
96		0.008
97		0.009
98		0.009
99		0.009
100		0.009
110		0.011
120		0.013
130		0.014
140		0.014
150		0.014
160		0.014
170		0.013
180		0.012
190		0.012
200		0.011
210		0.010
220		0.009
230		0.008
240		0.007
250		0.005
260		0.003
270		0.001
273.16		0.000
300		-0.006
400		-0.031
500		-0.040
600		-0.040
700		-0.055
800		-0.089
900		-0.124

INTERNATIONAL SYSTEM OF UNITS (SI)

The International System of units (SI) was adopted by the 11th General Conference on Weights and Measures (CGPM) in 1960. It is a coherent system of units built from seven *SI base units*, one for each of the seven dimensionally independent base quantities: they are the meter, kilogram, second, ampere, kelvin, mole, and candela, for the dimensions length, mass, time, electric current, thermodynamic temperature, amount of substance, and luminous intensity, respectively, The definitions of the SI base units are given below. The *SI derived units* are expressed as products of powers of the base units, analogous to the corresponding relations between physical quantities but with numerical factors equal to unity.

In the International System there is only one SI unit for each physical quanitity. This is either the appropriate SI base unit itself or the appropriate SI derived unit. However, any of the approved decimal prefixes, called *SI prefixes*, may be used to construct decimal multiples or submultiples of SI units.

It is recommended that only SI units be used in science and technology (with SI prefixes where appropriate). Where there are special reasons for making an exception to this rule, it is recommended always to define the units used in terms of SI units. This section was reprinted with the permission of IUPAC.

Definitions of SI Base Units

Meter — the meter is the length of path travelled by light in vacuum during a time interval of 1/299 792 458 of a second (17th CGPM, 1983).

Kilogram — The kilogram is the unit of mass; it is equal to the mass of the international prototype of the kilogram (3rd CGPM, 1901).

Second — The second is the duration of 9 192 631 770 periods of the radiation corresponding to the transition between the two hyperfine levels of the ground state of the cesium-133 atom (13th CGPM, 1967).

Ampere — The ampere is that constant current which, if maintained in two straight parallel conductors of infinite length, of negligible circular cross-section, and placed 1 meter apart in vacuum, would produce between these conductors a force equal to 2×10^{-7} newton per meter of length (9th CGPM, 1948).

Kelvin — The kelvin, unit of thermodynamic temperature, is the fraction 1/273.16 of the thermodynamic temperature of the triple point of water (13th CGPM, 1967)

Mole — The mole is the amount of substance of a system which contains as many elementary entities as there are atoms in 0.012 kilogram of carbon-12. When the mole is used, the elementary entities must be specified and may be atoms, molecules, ions, electrons, other particles, or specified groups of such particles (14th CGPM, 1971).

Examples of the use of the mole:

1 mol of H_2 contains about 6.022×10^{23} H_2 molecules, or 12.044×10^{23} H atoms

1 mol of HgCl has a mass of 236.04 g

1 mol of Hg_2Cl_2 has a mass of 472.08 g

1 mol of Hg_2^{2+} has a mass of 401.18 g and a charge of 192.97 kC

1 mol of $Fe_{0.91}S$ has a mass of 82.88 g

1 mol of e^- has a mass of 548.60 μg and a charge of –96.49 kC

1 mol of photons whose frequency is 10^{14} Hz has energy of about 39.90 kJ

Candela — The candela is the luminous intensity, in a given direction, of a source that emits monochromatic radiation of frequency 540×10^{12} hertz and that has a radiant intensity in that direction of (1/683) watt per steradian (16th CGPM, 1979).

Names and Symbols for the SI Base Units

Physical quantity	Name of SI unit	Symbol for SI unit
length	meter	m
mass	kilogram	kg
time	second	s
electric current	ampere	A
thermodynamic temperature	kelvin	K
amount of substance	mole	mol
luminous intensity	candela	cd

SI Derived Units with Special Names and Symbols

Physical quantity	Name of SI unit	Symbol for SI unit	Expression in terms of SI base units	
frequency[1]	hertz	Hz	s^{-1}	
force	newton	N	$m\,kg\,s^{-2}$	
pressure, stress	pascal	Pa	$N\,m^{-2}$	$= m^{-1}\,kg\,s^{-2}$
energy, work, heat	joule	J	$N\,m$	$= m^2\,kg\,s^{-2}$
power, radiant flux	watt	W	$J\,s^{-1}$	$= m^2\,kg\,s^{-3}$
electric charge	coulomb	C	$A\,s$	
electric potential, electromotive force	volt	V	$J\,C^{-1}$	$= m^2\,kg\,s^{-3}\,A^{-1}$
electric resistance	ohm	Ω	$V\,A^{-1}$	$= m^2\,kg\,s^{-3}\,A^{-2}$
electric conductance	siemens	S	Ω^{-1}	$= m^{-2}\,kg^{-1}\,s^3\,A^2$
electric capacitance	farad	F	$C\,V^{-1}$	$= m^{-2}\,kg^{-1}\,s^4\,A^2$
magnetic flux density	tesla	T	$V\,s\,m^{-2}$	$= kg\,s^{-2}\,A^{-1}$
magnetic flux	weber	Wb	$V\,s$	$= m^2\,kg\,s^{-2}\,A^{-1}$
inductance	henry	H	$V\,A^{-1}\,s$	$= m^2\,kg\,s^{-2}\,A^{-2}$
Celsius temperature[2]	degree Celsius	°C	K	
luminous flux	lumen	lm	$cd\,sr$	
illuminance	lux	lx	$cd\,sr\,m^{-2}$	
activity[3] (radioactive)	becquerel	Bq	s^{-1}	
absorbed dose[3] (of radiation)	gray	Gy	$J\,kg^{-1}$	$= m^2\,s^{-2}$
dose equivalent[3] (dose equivalent index)	sievert	Sv	$J\,kg^{-1}$	$= m^2\,s^{-2}$
plane angle[4]	radian	rad	1	$= m\,m^{-1}$
solid angle[4]	steradian	sr	1	$= m^2\,m^{-2}$

(1) For radial (circular) frequency and for angular velocity the unit rad s⁻¹, or simply s⁻¹, should be used, and this may not be simplified to Hz. The unit Hz should be used only for frequency in the sense of cycles per second.

(2) The Celsius temperature θ is defined by the equation:

$$\theta/°C = T/K - 273.15$$

The SI unit of Celsius temperature interval is the degree Celsius, °C, which is equal to the kelvin, K. °C should be treated as a single symbol, with no space between the ° sign and the letter C. (The symbol °K, and the symbol °, should no longer be used.)

(3) The units gray and sievert are admitted for reasons of safeguarding human health.

(4) The units radian and steradian are described as 'SI supplementary units'. However, in chemistry, as well as in physics, they are usually treated as dimensionless derived units, and this was recognized by CIPM in 1980. Since they are then of dimension 1, this leaves open the possibility of including them or omitting them in expressions of SI derived units, In practice this means that rad and sr may be used when appropriate and may be omitted if clarity is not lost thereby.

SI Prefixes

To signify decimal multiples and submultiples of SI units the following prefixes may be used.

Factor	Prefix	Symbol	Factor	Prefix	Symbol
10^{24}	yotta	Y	10^{-1}	deci	d
10^{21}	zetta	Z	10^{-2}	centi	c
10^{18}	exa	E	10^{-3}	milli	m
10^{15}	peta	P	10^{-6}	micro	μ
10^{12}	tera	T	10^{-9}	nano	n
10^{9}	giga	G	10^{-12}	pico	p
10^{6}	mega	M	10^{-15}	femto	f
10^{3}	kilo	k	10^{-18}	atto	a
10^{2}	hecto	h	10^{-21}	zepto	z
10^{1}	deka	da	10^{-24}	yocto	y

Prefix symbols should be printed in roman (upright) type with no space between the prefix and the unit symbol.

Example kilometer, km

When a prefix is used with a unit symbol, the combination is taken as a new symbol that can be raised to any power without the use of parentheses.

Examples

$$1 \text{ cm}^3 = (0.01 \text{ m})^3 = 10^{-6} \text{ m}^3$$
$$1 \text{ } \mu\text{s}^{-1} = (10^{-6} \text{ s})^{-1} = 10^6 \text{ s}^{-1}$$
$$1 \text{ V/cm} = 100 \text{ V/m}$$
$$1 \text{ mmol/dm}^3 = \text{mol m}^{-3}$$

A prefix should never be used on its own, and prefixes are not to be combined into compound prefixes.

Example pm, not $\mu\mu$m

The names and symbols of decimal multiples and sub-multiples of the SI base unit of mass, the kg, which already contains a prefix, are constructed by adding the appropriate prefix to the word gram and symbol g.

Examples mg, not μkg; Mg, not kkg

The SI prefixes are not to be used with °C.

Units in Use Together with the SI

These units are not part of the SI, but it is recognized that they will continue to be used in appropriate contexts. SI prefixes may be attached to some of these units, such as milliliter, ml; millibar, mbar; megaelectronvolt, MeV; kilotonne, ktonne.

Physical quantity	Name of unit	Symbol for unit	Value in SI units
time	minute	min	60 s
time	hour	h	3600 s
time	day	d	86 400 s
plane angle	degree	°	$(\pi/180)$ rad
plane angle	minute	′	$(\pi/10\ 800)$ rad
plane angle	second	″	$(\pi/648\ 000)$ rad
length	ångström[1]	Å	10^{-10} m
area	barn	b	10^{-28} m^2
volume	litre	l, L	dm^3 $= 10^{-3}$ m^3
mass	tonne	t	Mg $= 10^3$ kg
pressure	bar[1]	bar	10^5 Pa $= 10^5$ N m^{-2}
energy	electronvolt[2]	eV $(= e \times$ V$)$	$\approx 1.60218 \times 10^{-19}$ J
mass	unified atomic mass unit[2,3]	u $(= m_a(^{12}$C$)/12)$	$\approx 1.66054 \times 10^{-27}$ kg

(1) The ångström and the bar are approved by CIPM for 'temporary use with SI units', until CIPM makes a further recommendation, However, they should not be introduced where they are not used at present.

(2) The values of these units in terms of the corresponding SI units are not exact, since they depend on the values of the physical constants e (for the electronvolt) and N_A (for the unified atomic mass unit), which are determined by experiment.

(3) The unified atomic mass unit is also sometimes called the dalton, with symbol Da, although the name and symbol have not been approved by CGPM.

Atomic Units

For the purposes of quantum mechanical calculations of electronic wavefunctions, it is convenient to regard certain fundamental constants (and combinations of such constants) as though they were units. They are customarily called *atomic units* (abbreviated: au), and they may be regarded as forming a coherent system of units for the calculation of electronic properties in theoretical chemistry, although there is no authority from CGPM for treating them as units. The first five atomic units in the table below have special names and symbols. Only four of these are independent; all others may be derived by multiplication and division in the usual way, and the table includes a number of examples.

The relation of atomic units to the corresponding SI units involves the values of the fundamental physical constants, and is therefore not exact. The numerical values in the table are based on the 1986 CODATA values of the fundamental constants. The numerical results of calculations in theoretical chemistry are frequently quoted in atomic units, or as numerical values in the form (*physical quantity*)/(*atomic unit*), so that the reader may make the conversion using the current best estimates of the physical constants.

Physical quantity	Name of unit	Symbol for unit	Definition and value of unit in SI
mass	electron rest mass	m_e	$m_e \approx 9.1095 \times 10^{-31}$ kg
charge	elementary charge	e	$e \approx 1.6022 \times 10^{-19}$ C
action	Planck constant/2π	$\hbar$	$\hbar = h/2\pi \approx 1.0546 \times 10^{-34}$ J s
length	bohr	a_0	$4\pi\varepsilon_0\hbar^2/m_e e^2 \approx 5.2918 \times 10^{-11}$ m
energy	hartree	E_h	$\hbar^2/m_e a_0^2 \approx 4.3598 \times 10^{-18}$ J
time	au of time	$\hbar/E_h$	$\approx 2.4189 \times 10^{-17}$ s
velocity[1]	au of velocity	$a_0 E_h/\hbar$	$\approx 2.1877 \times 10^{6}$ m s^{-1}
force	au of force	E_h/a_0	$\approx 8.2389 \times 10^{-8}$ N
momentum, linear	au of momentum	$\hbar/a_0$	$\approx 1.9929 \times 10^{-24}$ N s
electric current	au of current	$eE_h/\hbar$	$\approx 6.6236 \times 10^{-3}$ A
electric field	au of electric field	E_h/ea_0	$\approx 5.1422 \times 10^{11}$ V m^{-1}
electric dipole moment	au of electric dipole	ea_0	$\approx 8.4784 \times 10^{-30}$ C m
magnetic flux density	au of magnetic flux density	$\hbar/ea_0^2$	$\approx 2.3505 \times 10^{5}$ T
magnetic dipole moment[2]	au of magnetic dipole	$e\hbar/m_e$	$= 2\mu_B \approx 1.8548 \times 10^{-23}$ J T^{-1}

(1) The numerical value of the speed of light, when expressed in atomic units, is equal to the reciprocal of the fine structure constant α; $c/(\text{au of velocity}) = c\hbar/a_0 E_h = \alpha^{-1}$ 137.04.

(2) The atomic unit of magnetic dipole moment is twice the Bohr magneton, μ_B.

REFERENCES

Quantities, Units and Symbols in Physical Chemistry, Ian Mills, Editor, Blackwell Scientific Publications, Oxford, 1987.

Symbols, Units, Nomenclature and Fundamental Constants in Physics, E. R. Cohen and P. Giacomo, Editors, International Union of Pure and Applied Physics Document 25, 1987; also appears in Physica 146A, 1—8, 1987.

ISO Standards Handbook 2, Units of Measurement, International Organization for Standardization, Geneva, 1982.

CONVERSION FACTORS

The following table gives conversion factors from various units of measure to SI units. It is reproduced from NIST Special Publication 811, *Guide for the Use of the International System of Units* (Superintendent of Documents, U.S. Government Printing Office, 1991), which in turn was derived from IEEE Std 268—1982, *IEEE Standard Metric Practice* (© 1982 by the Institute of Electrical and Electronics Engineers, Inc.).

The SI values are expressed in terms of the base, supplementary, and derived units of SI in order to provide a coherent presentation of the conversion factors and facilitate computations (see the table "International System of Units" in this Section). Powers of ten can be avoided by using SI prefixes and shifting the decimal point if necessary. Conversion from a non-SI unit to a different non-SI unit may be carried out by using this table in two stages, e.g.,

1 cal (thermochemical) = 4.184 J
1 Btu (mean) = 1.05587 E+03 J

Thus, 1 Btu (mean) = (1.05587 E+03/4.184) cal (thermochemical)
$\qquad\qquad$ = 252.359 cal (thermochemical).

Conversion factors are presented for ready adaptation to computer readout and electronic data transmission. The factors are written as a number equal to or greater than one and less than ten with six or less decimal places. This number is followed by the letter E (for exponent), plus or minus symbol, and two digits which indicate the power of 10 by which the number must be multiplied to obtain the correct value. For example:

$$3.523\,907\ \text{E–02 is } 3.523\,907 \times 10^{-2}$$

or

$$0.035\,239\,07$$

Similarly:

$$3.386\,389\ \text{E+03 is } 3.386\,389 \times 10^{3}$$

or

$$3\,386.389$$

An asterisk (*) after the sixth decimal place indicates that the conversion factor is exact and that all subsequent digits are zero. All other conversion factors have been rounded to the figures given in accordance with accepted practice. Where less than six decimal places are shown, more precision is not warranted.

To convert from	to	Multiply by
abampere	ampere (A)	1.000 000*E + 01
abcoulomb	coulomb (C)	1.000 000*E + 01
abfarad	farad (F)	1.000 000*E + 09
abhenry	henry (H)	1.000 000*E − 09
abmho	siemens (S)	1.000 000*E + 09
abohm	ohm (Ω)	1.000 000*E − 09
abvolt	volt (V)	1.000 000*E − 08
acre foot	meter³ (m³)	1.233 5 E + 03
acre	meter² (m²)	4.046 873 E + 03
ampere hour	coulomb (C)	3.600 000*E + 03
ångström	meter (m)	1.000 000*E − 10
are	meter² (m²)	1.000 000*E + 02
astronomical unit	meter (m)	1.495 979 E + 11
atmosphere (standard)	pascal (Pa)	1.013 250*E + 05
atmosphere (technical = 1 kgf/cm²)	pascal (Pa)	9.806 650*E + 04
bar	pascal (Pa)	1.000 000*E + 05
barn	meter² (m²)	1.000 000*E − 28
barrel (for petroleum, 42 gal)	meter³ (m³)	1.589 873 E − 01
board foot	meter³ (m³)	2.359 737 E − 03
British thermal unit (International Table)	joule (J)	1.055 056 E + 03
British thermal unit (mean)	joule (J)	1.055 87 E + 03

To convert from	to	Multiply by
British thermal unit (thermochemical)	joule (J)	1.054 350 E+03
British thermal unit (39°F)	joule (J)	1.059 67 E+03
British thermal unit (59°)	joule (J)	1.054 80 E+03
British thermal unit (60°)	joule (J)	1.054 68 E+03
Btu (International Table) · ft/(h · ft² · °F) (thermal conductivity)	watt per meter kelvin [W/(m · K)]	1.730 735 E+00
Btu (thermochemical) · ft/(h · ft² · °F) (thermal conductivity)	watt per meter kelvin [W/(m · K)]	1.729 577 E+00
Btu (International Table) · in/(h · ft² · °F) (thermal conductivity)	watt per meter kelvin [W/(m · K)]	1.442 279 E−01
Btu (thermochemical) · in/(h · ft² · °F) (thermal conductivity)	watt per meter kelvin [W/(m · K)]	1.441 314 E−01
Btu (International Table) · in/(s · ft² · °F) (thermal conductivity)	watt per meter kelvin [W/(m · K)]	5.192 204 E+02
Btu (thermochemical) · in/(s · ft² · °F) (thermal conductivity)	watt per meter kelvin [W/(m · K)]	5.188 732 E+02
Btu (International Table)/h	watt (W)	2.930 711 E−01
Btu (International Table)/s	watt (W)	1.055 056 E+03
Btu (thermochemical)/h	watt (W)	2.928 751 E−01
Btu (thermochemical)/min	watt (W)	1.757 250 E+01
Btu (thermochemical)/s	watt (W)	1.054 350 E+03
Btu (International Table)/ft²	joule per meter² (J/m²)	1.135 653 E+04
Btu (thermochemical)/ft²	joule per meter² (J/m²)	1.134 893 E+04
Btu (International Table)/(ft² · h)	watt per meter² (W/m²)	3.154 591 E+00
Btu (International Table)/(ft² · s)	watt per meter² (W/m²)	1.135 653 E+04
Btu (thermochemical)/(ft² · h)	watt per meter² (W/m²)	3.152 481 E+00
Btu (thermochemical)/(ft² · min)	watt per meter² (W/m²)	1.891 489 E+02
Btu (thermochemical)/(ft² · s)	watt per meter² (W/m²)	1.134 893 E+04
Btu (thermochemical)/(in² · s)	watt per meter² (W/m²)	1.634 246 E+06
Btu (International Table)/(h · ft² · °F)	watt per meter² kelvin [W/(m² · K)]	5.678 263 E+00
Btu (thermochemical)/(h · ft² · °F)	watt per meter² kelvin [W/(m² · K)]	5.674 466 E+00
Btu (International Table)/(s · ft² · °F)	watt per meter² kelvin [W/(m² · K)]	2.044 175 E+04
Btu (thermochemical)/(s · ft² · °F)	watt per meter² kelvin [W/(m² · K)]	2.042 808 E+04
Btu (International Table)/lb	joule per kilogram (J/kg)	2.326 000*E+03
Btu (thermochemical)/lb	joule per kilogram (J/kg)	2.324 444 E+03
Btu (International Table)/(lb · °F) (specific heat capacity)	joule per kilogram kelvin (J/(kg · K)]	4.186 800*E+03
Btu (thermochemical)/(lb · °F) (specific heat capacity)	joule per kilogram kelvin [J/(kg · K)]	4.184 000*E+03
Btu (International Table)/ft³	joule per meter³ (J/m³)	3.725 895 E+04
Btu (thermochemical)/ft³	joule per meter³ (J/m³)	3.723 402 E+04
bushel	meter³ (m³)	3.523 907 E−02
calorie (International Table)	joule (J)	4.186 800*E+00
calorie (mean)	joule (J)	4.190 02 E+00
calorie (thermochemical)	joule (J)	4.184 000*E+00
calorie (15 °C)	joule (J)	4.185 80 E+00
calorie (20 °C)	joule (J)	4.181 90 E+00
calorie (kilogram, International Table)	joule (J)	4.186 800*E+03
calorie (kilogram, mean)	joule (J)	4.190 02 E+03
calorie (kilogram, thermochemical)	joule (J)	4.184 000*E+03
cal (thermochemical)/cm²	joule per meter² (J/m²)	4.184 000*E+04
cal (International Table)/g	joule per kilogram (J/kg)	4.186 800*E+03
cal (thermochemical)/g	joule per kilogram (J/kg)	4.184 000*E+03
cal (International Table)/(g · °C)	joule per kilogram kelvin [J/(kg · K)]	4.186 800*E+03
cal (thermochemical)/(g · °C)	joule per kilogram kelvin [J/(kg · K)]	4.184 000*E+03

To convert from	to	Multiply by
cal (thermochemical)/min	watt (W)	6.973 333 E−02
cal (thermochemical)/s	watt (W)	4.184 000*E+00
cal (thermochemical)/(cm² · min)	watt per meter² (W/m²)	6.973 333 E−02
cal (thermochemical)/(cm² · s)	watt per meter² (W/m²)	4.184 000 E+04
cal (thermochemical)/(cm · s · °C)	watt per meter kelvin [W/(m · K)]	4.184 000 E+02
cd/in²	candela per meter² (cd/m²)	1.550 003 E+03
carat (metric)	kilogram (kg)	2.000 000*E−04
centimeter of mercury (0 °C)	pascal (Pa)	1.333 22 E+03
centimeter of water (4 °C)	pascal (Pa)	9.806 38 E+01
centipoise	pascal second (Pa . s)	1.000 000*E−03
centistokes	meter² per second (m²/s)	1.000 000*E−06
chain	meter² (m²)	5.067 075 E+01
circular mil	meter² (m²)	5.067 075 E−10
clo	kelvin meter² per watt (K · m²/W)	2.003 712 E−01
cup	milliliter (mL)	2.366 E+02
curie	becquerel (Bq)	3.700 000*E+10
darcy[a]	meter² (m²)	9.869 233 E−13
day	second (s)	8.640 000*E+04
day (sidereal)	second (s)	8.616 409 E+04
degree (angle)	radian (rad)	1.745 329 E−02
degree Celsius	kelvin (K)	$T_K = t_{°C} + 273.15$
degree centigrade	[see note below]	
degree Fahrenheit	degree Celsius (°C)	$t_{°C} = (t_{°F} - 32)/1.8$
degree Fahrenheit	kelvin (K)	$T_K = (t_{°F} + 459.67)/1.8$
degree Rankine	kelvin (K)	$T_K = T_{°R}/1.8$
°F · h · ft²/Btu (International Table)	kelvin meter² per watt (K · m²/W)	1.761 102 E−01
°F · h · ft²/Btu (thermochemical)	kelvin meter² per watt (K · m²/W)	1.762 280 E−01
°F · h · ft²/[Btu (International Table) · in] (thermal resistivity)	kelvin meter² per watt (K · m²/W)	6.933 472 E+00
°F · h · ft²/[Btu (thermochemical) · in] (thermal resistivity)	kelvin meter² per watt (K · m²/W)	6.938 112 E+00
denier	kilogram per meter (kg/m)	1.111 111 E−07
dyne	newton (N)	1.000 000*E−05
dyne · cm	newton meter (N · m)	1.000 000*E−07
dyne/cm²	pascal (Pa)	1.000 000*E−01
electronvolt	joule (J)	1.602 19 E−19
EMU of capacitance	farad (F)	1.000 000*E+09
EMU of current	ampere (A)	1.000 000*E+01
EMU of electric potential	volt (V)	1.000 000*E−08
EMU of inductance	henry (H)	1.000 000*E−09
EMU of resistance	ohm (Ω)	1.000 000*E−09
ESU of capacitance	farad (F)	1.112 650 E−12
ESU of current	ampere (A)	3.335 641 E−10
ESU of electric potential	volt (V)	2.997 925 E+02
ESU of inductance	henry (H)	8.987 552 E+11
ESU of resistance	ohm (Ω)	8.987 552 E+11
erg	joule (J)	1.000 000*E−07
erg/cm² · s	watt per meter² (W/m²)	1.000 000*E−03
erg/s	watt (W)	1.000 000*E−07
faraday (based on carbon-12)	coulomb (C)	9.648 70 E+04
faraday (chemical)	coulomb (C)	9.649 57 E+04
faraday (physical)	coulomb (C)	9.652 19 E+04

To convert from	to	Multiply by
fathom	meter (m)	1.828 8 E + 00
fermi (femtometer)	meter (m)	1.000 000*E − 15
fluid ounce (US)	meter³ (m³)	2.957 353 E − 05
foot	meter (m)	3.048 000*E − 01
foot (US survey)	meter (m)	3.048 006 E − 01
foot of water (39.2 °F)	pascal (Pa)	2.988 98 E + 03
ft²	meter² (m²)	9.290 304*E − 02
ft²/h (thermal diffusivity)	meter² per second (m²/s)	2.580 640*E − 05
ft²/s	meter² per second (m²/s)	9.290304*E–02
ft³ (volume; section modulus)	meter³ (m³)	2.831 685 E − 02
ft³/min	meter³ per second (m³/s)	4.719 474 E − 04
ft³/s	meter³ per second (m³/s)	2.831 685 E − 02
ft⁴ (second moment of area)[b]	meter⁴ (m⁴)	8.630 975 E − 03
ft/h	meter per second (m/s)	8.466 667 E − 05
ft/min	meter per second (m/s)	5.080 000*E − 03
ft/s	meter per second (m/s)	3.048 000*E − 01
ft/s²	meter per second² (m/s²)	3.048 000*E − 01
footcandle	lux (lx)	1.076 391 E + 01
footlambert	candela per meter² (cd/m²)	3.426 259 E + 00
ft · lbf	joule (J)	1.355 818 E + 00
ft · lbf/h	watt (W)	3.766 161 E − 04
ft · lbf/min	watt (W)	2.259 697 E − 02
ft · lbf/s	watt (W)	1.355 818 E + 00
ft · poundal	joule (J)	4.214 011 E − 02
g, standard acceleration of free fall	meter per second² (m/s²)	9.806 650*E + 00
gal	meter per second² (m/s²)	1.000 000*E − 02
gallon (Canadian liquid)	meter³ (m³)	4.546 090*E − 03
gallon (UK liquid)	meter³ (m³)	4.546 090*E − 03
gallon (US liquid)	meter³ (m³)	3.785 412 E − 03
gallon (US liquid) per day	meter³ per second (m³/s)	4.381 264 E − 08
gallon (US liquid) per minute	meter³ per second (m³/s)	6.309 020 E − 05
gallon (US liquid) per (hp · h) (SFC, specific fuel consumption)	meter³ per joule (m³/J)	1.410 089 E − 09
gamma	tesla (T)	1.000 000*E − 09
gauss	tesla (T)	1.000 000*E − 04
gilbert	ampere (A)	7.957 747 E − 01
gill (UK)	meter³ (m³)	1.420 654 E − 04
gill (US)	meter³ (m³)	1.182 941 E − 04
grade	degree (angular)	9.000 000*E − 01
grade	radian (rad)	1.570 796 E − 02
grain	kilogram (kg)	6.479 891*E − 05
grain/gal (US liquid)	kilogram per meter³ (kg/m³)	1.711 806 E − 02
gram	kilogram (kg)	1.000 000*E − 03
g/cm³	kilogram per meter³ (kg/m³)	1.000 000*E + 03
gram-force/cm²	pascal (PA)	9.806 650*E + 01
hectare	meter² (m²)	1.000 000*E + 04
horsepower (550 ft · lbf/s)	watt (W)	7.456 999 E + 02
horsepower (boiler)	watt (W)	9.809 50 E + 03
horsepower (electric)	watt (W)	7.460 000*E + 02
horsepower (metric)	watt (W)	7.354 99 E + 02
horsepower (water)	watt (W)	7.460 43 E + 02
horsepower (UK)	watt (W)	7.457 0 E + 02
hour	second (s)	3.600 000*E + 03
hour (sidereal)	second (s)	3.590 170 E + 03
hundredweight (long)	kilogram (kg)	5.080 235 E + 01
hundredweight (short)	kilogram (kg)	4.535 924 E + 01

To convert from	to	Multiply by
inch..........	meter (m)	2.540 000*E − 02
inch of mercury (32 °F) [c]........	pascal (Pa)	3.386 38 E + 03
inch of mercury (60 °F) [c]........	pascal (Pa)	3.376 85 E + 03
inch of water (39.2 °F)	pascal (Pa)	2.490 82 E + 02
inch of water (60 °F)........	pascal (Pa).............	2.488 4 E + 02
in^2........	meter^2 (m^2)........	6.451 600*E − 04
in^3 (volume; section modulus) [d]	meter^3 (m^3)........	1.638 706 E − 05
in^3/min........	meter^3 per second (m^3/s)........	2.731 177 E − 07
in^4 (second moment of area) [b]	meter^4 (m^4)........	4.162 314 E − 07
in/s	meter per second (m/s)	2.540 000*E − 02
in/s^2........	meter per second^2 (m/s^2)........	2.540 000*E − 02
kayser........	1 per meter (1/m)	1.000 000*E + 02
kelvin........	degree Celsius	$t_{°C} = T_K − 273.15$
kilocalorie (International Table)........	joule (J)	4.186 800*E + 03
kilocalorie (mean)	joule (J)	4.190 02 E + 03
kilocalorie (thermochemical)	joule (J)	4.184 000*E + 03
kilocalorie (thermochemical)/min	watt (W)	6.973 333 E + 01
kilocalorie (thermochemical)/s	watt (W)	4.184 000*E + 03
kilogram-force (kgf)	newton (N)	9.806 650*E + 00
kgf · m	newton meter (N · m)........	9.806 650*E + 00
kgf · s^2/m (mass)	kilogram (kg)	9.806 650*E + 00
kgf/cm^2........	pascal (Pa)	9.806 650*E + 04
kgf/m^2........	pascal (Pa)	9.806 650*E + 00
kgf/mm^2........	pascal (Pa)	9.806 650*E + 06
km/h........	meter per second (m/s)	2.777 778 E − 01
kilopond (1 kp = 1 kgf)	newton (N)	9.806 650*E + 00
kW · h	joule (J)	3.600 000*E + 06
kip (1000 lbf)	newton (N)	4.448 222 E + 03
kip/in^2 (ksi)	pascal (Pa)	6.894 757 E + 06
knot (international)........	meter per second (m/s)	5.144 444 E − 01
lambert	candela per meter^2 (cd/m^2)	$1/\pi$ *E + 04
lambert	candela per meter^2 (cd/m^2)	3.183 099 E + 03
langley........	joule per meter^2 (J/m^2)	4.184 000*E + 04
light year [e]	meter (m)	9.460 73 E + 15
liter [f]	meter^3 (m^3)	1.000 000*E − 03
lumen per ft^2........	lumen per meter^2 (lm/m^2)	1.076 391 E + 01
maxwell........	weber (Wb)........	1.000 000*E − 08
mho	siemens (S)	1.000 000*E + 00
microinch	meter (m)	2.540 000*E − 08
micron........	meter (m)	1.000 000*E − 06
mil........	meter (m)	2.540 000*E − 05
mile (international)........	meter (m)	1.609 344*E + 03
mile (US statute)	meter (m)	1.609 3 E + 03
mile (international nautical)........	meter (m)	1.852 000*E + 03
mile (US nautical)........	meter (m)	1.852 000*E + 03
mi^2 (international)........	meter^2 (m^2)........	2.589 988 E + 06
mi^2 (US statute)	meter^2 (m^2)........	2.589 998 E + 06
mi/h (international)........	meter per second (m/s)	4.470 400*E − 01
mi/h (international)........	kilometer per hour (km/h)	1.609 344*E + 00
mi/min (international)	meter per second (m/s)	2.682 240*E + 01
mi/s (international)	meter per second (m/s)........	1.609 344*E + 03
millibar	pascal (Pa)	1.000 000*E + 02
millimeter of mercury (0 °C) [c]	pascal (Pa)	1.333 22 E + 02
minute (angle)	radian (rad)	2.908 882 E − 04
minute........	second (s)	6.000 000*E + 01
minute (sidereal)	second (s)	5.983 617 E + 01
oersted........	ampere per meter (A/m)	7.957 747 E + 01
ohm centimeter........	ohm meter (Ω · m)	1.000 000*E − 02

To convert from	to	Multiply by
ohm circular-mil per ft	ohm meter ($\Omega \cdot$ m)	1.662 426 E$-$09
ounce (avoirdupois)	kilogram (kg)	2.834 952 E$-$02
ounce (troy or apothecary)	kilogram (kg)	3.110 348 E$-$02
ounce (UK fluid)	meter^3 (m^3)	2.841 307 E$-$05
ounce (US fluid)	meter^3 (m^3)	2.957 353 E$-$05
ounce-force	newton (N)	2.780 139 E$-$01
ozf $\cdot$ in	newton meter (N $\cdot$ m)	7.061 552 E$-$03
oz (avoirdupois)/gal (UK liquid)	kilogram per meter^3 (kg/m^3)	6.236 023 E$+$00
oz (avoirdupois)/gal (US liquid)	kilogram per meter^3 (kg/m^3)	7.489 152 E$+$00
oz (avoirdupois)/in^3	kilogram per meter^3 (kg/m^3)	1.729 994 E$+$03
oz (avoirdupois)/ft^2	kilogram per meter^2 (kg/m^2)	3.051 517 E$-$01
oz (avoirdupois)/yd^2	kilogram per meter^2 (kg/m^2)	3.390 575 E$-$02
parsec	meter (m)	3.085 678 E$+$16
peck (US)	meter^3 (m^3)	8.809 768 E$-$03
pennyweight	kilogram (kg)	1.555 174 E$-$03
perm (0 °C)	kilogram per pascal second meter^2 [kg/Pa $\cdot$ s $\cdot$ m^2)]	5.721 35 E$-$11
perm (23 °C)	kilogram per pascal second meter^2 [kg/(Pa $\cdot$ s $\cdot$ m^2)]	5.745 25 E$-$11
perm $\cdot$ in (0 °C)	kilogram per pascal second meter [kg/(Pa $\cdot$ s $\cdot$ m)]	1.453 22 E$-$12
perm $\cdot$ in (23 °C)	kilogram per pascal second meter [kg/(Pa $\cdot$ s $\cdot$ m)]	1.459 29 E$-$12
phot	lumen per meter^2 (lm/m^2)	1.000 000*E$+$04
pica (printer's)	meter (m)	4.217 518 E$-$03
pint (US dry)	meter^3 (m^3)	5.506 105 E$-$04
pint (US liquid)	meter^3 (m^3)	4.731 765 E$-$04
point (printer's)	meter (m)	3.514 598*E$-$04
poise (absolute viscosity)	pascal second (Pa $\cdot$ s)	1.000 000*E$-$01
pound (avoirdupois)[g]	kilogram (kg)	4.535 924 E$-$01
pound (troy or apothecary)	kilogram (kg)	3.732 417 E$-$01
lb/ft	kilogram per meter (kg/m)	1.488 164 E$+$00
lb $\cdot$ ft^2 (moment of inertia)	kilogram meter^2 (kg $\cdot$ m^2)	4.214 011 E$-$02
lb $\cdot$ in^2 (moment of inertia)	kilogram meter^2 (kg $\cdot$ m^2)	2.926 397 E$-$04
lb/ft $\cdot$ h	pascal second (Pa $\cdot$ s)	4.133 789 E$-$04
lb/ft $\cdot$ s	pascal second (Pa $\cdot$ s)	1.488 164 E$+$00
lb/ft^2	kilogram per meter^2 (kg/m^2)	4.882 428 E$+$00
lb/ft^3	kilogram per meter^3 (kg/m^3)	1.601 846 E$+$01
lb/gal (UK liquid)	kilogram per meter^3 (kg/m^3)	9.977 633 E$+$01
lb/gal (US liquid)	kilogram per meter^3 (kg/m^3)	1.198 264 E$+$02
lb/h	kilogram per second (kg/s)	1.259 979 E$-$04
lb/hp $\cdot$ h (SFC, specific fuel consumption	kilogram per joule (kg/J)	1.689 659 E$-$07
lb/in	kilogram per meter (kg/m)	1.785 797 E$+$01
lb/in^3	kilogram per meter^3 (kg/m^3)	2.767 990 E$+$04
lb/min.	kilogram per second (kg/s)	7.559 873 E$-$03
lb/s	kilogram per second (kg/s)	4.535 924 E$-$01
lb/yd^3	kilogram per meter^3 (kg/m^3)	5.932 764 E$-$01
poundal	newton (N)	1.382 550 E$-$01
poundal/ft^2	pascal (Pa)	1.488 164 E$+$00
poundal $\cdot$ s/ft^2	pascal second (Pa $\cdot$ s)	1.488 164 E$+$00
pound-force (lbf)[h]	newton (N)	4.448 222 E$+$00
lbf $\cdot$ ft	newton meter (N $\cdot$ m)	1.355 818 E$+$00
lbf $\cdot$ ft/in	newton meter per meter (N $\cdot$ m/m)	5.337 866 E$+$01
lbf $\cdot$ in	newton meter (N $\cdot$ m)	1.129 848 E$-$01
lbf $\cdot$ in/in	newton meter per meter (N $\cdot$ m/m)	4.448 222 E$+$00
lbf $\cdot$ s/ft^2	pascal second (Pa $\cdot$ s)	4.788 026 E$+$01

To convert from	to	Multiply by
lbf · s/in².	pascal second (Pa · s)	6.894 757 E+03
lbf/ft	newton per meter (N/m)	1.459 390 E+01
lbf/ft²	pascal (Pa)	4.788 026 E+01
lbf/in.	newton per meter (N/m)	1.751 268 E+02
lbf/in² (psi)	pascal (Pa)	6.894 757 E+03
lbf/lb (thrust/weight [mass] ratio)	newton per kilogram (N/kg)	9.806 650 E+00
quad	joule (J)	1.055 E+18
quart (US dry)	meter³ (m³)	1.101 221 E−03
quart (US liquid)	meter³ (m³)	9.463 529 E−04
rad (absorbed dose)	gray (Gy)	1.000 000*E−02
rem (dose equivalent)	sievert (Sv)	1.000 000*E−02
rhe	1 per pascal second [1/Pa · s)]	1.000 000*E+01
rod	meter (m)	5.029 210 E+00
roentgen	coulomb per kilogram (C/kg)	2.58 E−04
second (angle)	radian (rad)	4.848 137 E−06
second (sidereal)	second (s)	9.972 696 E−01
shake	second (s)	1.000 000*E−08
slug	kilogram (kg)	1.459 390 E+01
slug/ft · s	pascal second (Pa · s)	4.788 026 E+01
slug/ft³	kilogram per meter³ (kg/m³)	5.153 788 E+02
statampere	ampere (A)	3.335 641 E−10
statcoulomb	coulomb (C)	3.335 641 E−10
statfarad	farad (F)	1.112 650 E−12
stathenry	henry (H)	8.987 552 E+11
statmho	siemens (S)	1.112 650 E−12
statohm	ohm (Ω)	8.987 552 E+11
statvolt	volt (V)	2.997 925 E+02
stere	meter³ (m³)	1.000 000*E+00
stilb	candela per meter² (cd/m²)	1.000 000*E+04
stokes (kinematic viscosity)	meter² per second (m²/s)	1.000 000*E−04
tablespoon	milliliter (mL)	1.479 E+01
teaspoon	meter³ (m³) milliliter (mL)	4.929 E+00
tex	kilogram per meter (kg/m)	1.000 000*E−06
therm (EEG) [i]	joule (J)	1.055 060*E+08
therm (US) [i]	joule (J)	1.054 804*E+08
ton (assay)	kilogram (kg)	2.916 667 E−02
ton (long, 2240 lb)	kilogram (kg)	1.016 047 E+03
ton (metric)	kilogram (kg)	1.000 000*E+03
ton (explosive energy of one ton of TNT)	joule (J)	4.184 E+09 [j]
ton of refrigeration (12 000 Btu/h)	watt (W)	3.517 E+03
ton (register)	meter³ (m³)	2.831 685 E+00
ton (short, 2000 lb)	kilogram (kg)	9.071 847 E+02
ton (long)/yd³	kilogram per meter³ (kg/m³)	1.328 939 E+03
ton (short)/yd³	kilogram per meter³ (kg/m³)	1.186 553 E+03
ton (short)/h	kilogram per second (kg/s)	2.519 958 E−01
ton-force (2000 lbf)	newton (N)	8.896 443 E+03
tonne	kilogram (kg)	1.000 000*E+03
torr (mmHg, 0 °C) [c]	pascal (Pa)	1.333 22 E+02
unit pole	weber (Wb)	1.256 637 E−07
W · h	joule (J)	3.600 000*E+03
W · s	joule (J)	1.000 000*E+00
W/cm²	watt per meter² (W/m²)	1.000 000*E+04
W/in²	watt per meter² (W/m²)	1.550 003 E+03

To convert from	to	Multiply by
yard	meter (m)	9.144 000*E − 01
yd^2	meter^2 (m^2)	8.361 274 E − 01
yd^3	meter^3 (m^3)	7.645 549 E − 01
yd^3/min.	meter^3 per second (m^3/s)	1.274 258 E − 02
year (365 days)	second (s)	3.153 600*E + 07
year (sidereal)	second (s)	3.155 815 E + 07
year (tropical)	second (s)	3.155 693 E + 07

Note: The centigrade temperature scale is obsolete. The unit, degree centigrade, is only approximately equal to the degree Celsius.

[a] The darcy is a unit for measuring permeability of porous solids.

[b] This is sometimes called the moment of section or area moment of ineria of a plane section about a specified axis.

[c] Conversion factors for mercury manometer pressure units are calculated using the standard value for the acceleration of gravity and the density of mercury at the stated temperature. Higher levels of precision are not justified because the definitions of the units do not take into account the compressibility of mercury or the density value change caused by the revised practical temperature scale, ITS-90.

[d] The exact conversion factor is 1.638 706 4*E–05.

[e] This conversion factor is based on the astronomical unit of time of one day (86 400 seconds); an interval of 36 525 days is one Julian century. (See the Astronomical Almanac for the Year 1991, page K6, U.S. Government Printing Office, Washington, D.C.)

[f] In 1964 the General Conference on Weights and Measures reestablished the name liter as a special name for the cubic decimeter. Between 1901 and 1964, the liter was slightly larger (1.000 028 dm^3); in the use of high-accuracy volume data of that time interval, this fact must be kept in mind.

[g] The exact conversion factor is 4.535 923 7*E–01.

[h] The exact conversion factor is 4.448 221 615 260 5*E+00.

[i] The therm (EEC) is legally defined in the Council Directive of 20 December 1979, Council of the European Communities. The therm (US) is legally defined in the Federal Register of July 27, 1968. Although the therm (EEC), which is based on the International Table Btu, is frequently used by engineers in the US, the therm (US) is the legal unit used by the US natural gas industry.

[j] Defined (not measured) value.

Note concerning the foot:

The U.S. Metric Law of 1866 gave the relationship, 1 meter equals 39.37 inches. Since 1893 the U.S. yard has been derived from the meter. In 1959 a refinement was made in the definition of the yard to bring the U.S. yard and the yard used in other countries into agreement. The U.S. yard was changed from 3600/3937 m to 0.9144 m exactly. The new length is shorter by exactly two parts in a million. At the same time it was decided that any data in feet derived from and published as a result of geodetic surveys within the United States would remain with the old standard (1 ft = 1200/3937 m) until further decision This foot is named the U.S. survey foot and has the following relationships:

1 rod (pole or perch) = 16 $\frac{1}{2}$ feet
1 chain = 66 feet
1 mile (U.S. statute) = 5280 feet

CONVERSION OF TEMPERATURES

From	To	
°Celsius	°Fahrenheit	$t_F = (t_C \times 1.8) + 32$
	Kelvin	$T_K = t_C + 273.15$
	°Rankine	$T_R = (t_C + 273.15) \times 1.8$
°Fahrenheit	°Celsius	$t_C = \dfrac{t_F - 32}{1.8}$
	Kelvin	$T_k = \dfrac{t_F - 32}{1.8} + 273.15$
	°Rankine	$T_R = t_F + 459.67$
Kelvin	°Celsius	$t_C = T_K - 273.15$
	°Rankine	$T_R = T_K \times 1.8$
°Rankine	°Fahrenheit	$t_F = T_R - 459.67$
	Kelvin	$T_K = \dfrac{T_R}{1.8}$

DESIGNATION OF LARGE NUMBERS

	U.S.A.	Other countries
10^6	million	million
10^9	billion	milliard
10^{12}	trillion	billion
10^{15}	quadrillion	billiard
10^{18}	quintillion	trillion

CONVERSION FACTORS FOR ENERGY UNITS

$E = h\nu = hc\bar{\nu} = kT; \quad E_m = LE$

		wavenumber $\bar{\nu}$	frequency ν	energy E			molar energy E_m		temperature T
		cm⁻¹	MHz	aJ	eV	E_h	kJ/mol	kcal/mol	K
$\bar{\nu}$:	1 cm⁻¹	$\hat{=}$ 1	2.997925×10^4	1.986447×10^{-5}	1.239842×10^{-4}	4.556335×10^{-6}	11.96266×10^{-3}	2.85914×10^{-3}	1.438769
ν:	1 MHz	$\hat{=}$ 3.33564×10^{-5}	1	6.626076×10^{-10}	4.135669×10^{-9}	1.519830×10^{-10}	3.990313×10^{-7}	9.53708×10^{-8}	4.79922×10^{-5}
E:	1 aJ	$\hat{=}$ 50341.1	1.509189×10^9	1	6.241506	0.2293710	602.2137	143.9325	7.24292×10^4
	1 eV	$\hat{=}$ 8065.54	2.417988×10^8	0.1602177	1	3.674931×10^{-2}	96.4853	23.0605	1.16045×10^4
	1 E_h	$\hat{=}$ 219474.63	6.579684×10^9	4.359748	27.2114	1	2625.500	627.510	3.15773×10^5
E_m:	1 kJ/mol	$\hat{=}$ 83.5935	2.506069×10^6	1.660540×10^{-3}	1.036427×10^{-2}	3.808798×10^{-4}	1	0.239006	120.272
	1 kcal/mol	$\hat{=}$ 349.755	1.048539×10^7	6.947700×10^{-3}	4.336411×10^{-2}	1.593601×10^{-3}	4.184	1	503.217
T:	1 K	$\hat{=}$ 0.695039	2.08367×10^4	1.380658×10^{-5}	8.61738×10^{-5}	3.16683×10^{-6}	8.31451×10^{-3}	1.98722×10^{-3}	1

Examples of the use of this table: 1 aJ $\hat{=}$ 50341 cm⁻¹
1 eV $\hat{=}$ 96.4853 kJ mol⁻¹

The symbol $\hat{=}$ should be read as meaning "corresponds to" or "is equivalent to".

CONVERSION FACTORS FOR PRESSURE UNITS

	Pa	kPa	MPa	bar	atmos	Torr	μmHg	psi
Pa	1	0.001	0.000001	0.00001	9.8692×10^{-6}	0.0075006	7.5006	0.0001450377
kPa	1000	1	0.001	0.01	0.0098692	7.5006	7500.6	0.1450377
MPa	1000000	1000	1	10	9.8692	7500.6	7500600	145.0377
bar	100000	100	0.1	1	0.98692	750.06	750060	14.50377
atmos	101325	101.325	0.101325	1.01325	1	760	760000	14.69594
Torr	133.322	0.133322	0.00013322	0.00133322	0.00131579	1	1000	0.0193672
μmHg	0.133322	0.000133322	1.33322×10^{-7}	1.33322×10^{-6}	1.31579×10^{-6}	0.001	1	1.933672×10^{-5}
psi	6894.757	6.894757	0.006894757	0.06894757	0.068046	51.7151	51715.1	1

To convert a pressure value from a unit in the left hand column to a new unit, multiply the value by the factor appearing in the column for the new unit. For example:

1 kPa = 9.8692×10^{-3} atmos
1 Torr = 1.33322×10^{-4} MPa

Notes: μmHg is often referred to as "micron"
Torr is essentially identical to mmHg
psi is pounds per square inch

CONVERSION FACTORS FOR THERMAL CONDUCTIVITY UNITS

MULTIPLY by appropriate factor to OBTAIN →

OBTAIN ↓	Btu_{IT} h^{-1} ft^{-1} F^{-1}	Btu_{IT} in. h^{-1} ft^{-2} F^{-1}	Btu_{th} h^{-1} ft^{-1} F^{-1}	Btu_{th} in. h^{-1} ft^{-2} F^{-1}	cal_{IT} s^{-1} cm^{-1} C^{-1}	cal_{th} s^{-1} cm^{-1} C^{-1}	$kcal_{th}$ h^{-1} m^{-1} C^{-1}	J s^{-1} cm^{-1} K^{-1}	W cm^{-1} K^{-1}	W m^{-1} K^{-1}	mW cm^{-1} K^{-1}
Btu_{IT} h^{-1} ft^{-1} F^{-1}	1	12	1.00067	12.0080	4.13379×10^{-3}	4.13656×10^{-3}	1.48916	1.73073×10^{-2}	1.73073×10^{-2}	1.73073	17.3073
Btu_{IT} in. h^{-1} ft^{-2} F^{-1}	8.33333×10^{-2}	1	8.33891×10^{-2}	1.00067	3.44482×10^{-4}	3.44713×10^{-4}	0.124097	1.44228×10^{-3}	1.44228×10^{-3}	0.144228	1.44228
Btu_{th} h^{-1} ft^{-1} F^{-1}	0.999331	11.9920	1	12	4.13102×10^{-3}	4.13379×10^{-3}	1.48816	1.72958×10^{-2}	1.72958×10^{-2}	1.72958	17.2958
Btu_{th} in. h^{-1} ft^{-2} F^{-1}	8.32776×10^{-2}	0.999331	8.33333×10^{-2}	1	3.44252×10^{-4}	3.44482×10^{-4}	0.124014	1.44131×10^{-3}	1.44131×10^{-3}	0.144131	1.44131
cal_{IT} s^{-1} cm^{-1} C^{-1}	2.41909×10^{2}	2.90291×10^{3}	2.42071×10^{2}	2.90485×10^{3}	1	1.00067	3.60241×10^{2}	4.1868	4.1868	4.1868×10^{2}	4.1868×10^{-3}
cal_{th} s^{-1} cm^{-1} C^{-1}	2.41747×10^{2}	2.90096×10^{3}	2.41909×10^{2}	2.90291×10^{3}	0.999331	1	3.6×10^{2}	4.184	4.184	4.184×10^{2}	4.184×10^{-3}
$kcal_{th}$ h^{-1} m^{-1} C^{-1}	0.671520	8.05824	0.671969	8.06363	2.77592×10^{-3}	2.77778×10^{-3}	1	1.16222×10^{-2}	1.16222×10^{-2}	1.16222	11.6222
J s^{-1} cm^{-1} K^{-1}	57.7789	6.93347×10^{2}	57.8176	6.93811×10^{2}	0.238846	0.239006	86.0421	1	1	1×10^{2}	1×10^{3}
W cm^{-1} K^{-1}	57.7789	6.93347×10^{2}	57.8176	6.93811×10^{2}	0.238846	0.239006	86.0421	1	1	1×10^{2}	1×10^{3}
W m^{-1} K^{-1}	0.577789	6.93347	0.578176	6.93811	2.38846×10^{-3}	2.39006×10^{-3}	0.860421	1×10^{-2}	1×10^{-2}	1	10
mW cm^{-1} K^{-1}	5.77789×10^{-2}	0.693347	5.78176×10^{-2}	0.693811	2.38846×10^{-4}	2.39006×10^{-4}	8.60421×10^{-2}	1×10^{-3}	1×10^{-3}	0.1	1

CONVERSION FACTORS FOR ELECTRICAL RESISTIVITY UNITS

To convert from multiply by appropriate factor to →
Obtain →

Obtain	abΩ cm	μΩ cm	Ω cm	StatΩ cm	Ω m	Ω cir. mil ft⁻¹	Ω in.	Ω ft
abohm centimeter	1	1×10^{-3}	10^{-9}	1.113×10^{-21}	10^{-11}	6.015×10^{-3}	3.937×10^{-10}	3.281×10^{-11}
microohm centimeter	10^3	1	10^{-6}	1.113×10^{-18}	10^{-6}	6.015	3.937×10^{-7}	3.281×10^{-6}
ohm centimeter	10^8	10^6	1	1.113×10^{-12}	1×10^{-2}	6.015×10^6	3.937×10^{-1}	3.281×10^{-2}
statohm centimeter (esu)	8.987×10^{20}	8.987×10^{17}	8.987×10^{11}	1	8.987×10^9	5.406×10^{18}	3.538×10^{11}	2.949×10^{10}
ohm meter	10^{11}	10^8	10^2	1.113×10^{-10}	1	6.015×10^8	3.937×10^1	3.281
ohm circular mil per foot	1.662×10^2	1.662×10^{-1}	1.662×10^{-7}	1.850×10^{-19}	1.662×10^{-9}	1	6.54×10^{-6}	5.45×10^{-9}
ohm inch	2.54×10^9	2.54×10^6	2.54	2.827×10^{-12}	2.54×10^{-2}	1.528×10^7	1	8.3×10^{-2}
ohm foot	3.048×10^{10}	3.048×10^7	3.048×10^{-1}	3.3924×10^{-11}	3.048×10^{-1}	1.833×10^8	12	1

CONVERSION FACTORS FOR CHEMICAL KINETICS

Equivalent second order rate constants

A \ B	cm³ mol⁻¹ s⁻¹	dm³ mol⁻¹ s⁻¹	m³ mol⁻¹ s⁻¹	cm³ molecule⁻¹ s⁻¹	(mm Hg)⁻¹ s⁻¹	atm⁻¹ s⁻¹	ppm⁻¹ min⁻¹	m² kN⁻¹ s⁻¹
1 cm³ mol⁻¹ s⁻¹ =	1	10^{-3}	10^{-6}	1.66×10^{-24}	$1.604 \times 10^{-5} T^{-1}$	$1.219 \times 10^{-2} T^{-1}$	2.453×10^{-9}	$1.203 \times 10^{-4} T^{-1}$
1 dm³ mol⁻¹ s⁻¹ =	10^3	1	10^{-3}	1.66×10^{-21}	$1.604 \times 10^{-2} T^{-1}$	$12.19 T^{-1}$	2.453×10^{-6}	$1.203 \times 10^{-1} T^{-1}$
1 m³ mol⁻¹ s⁻¹ =	10^6	10^3	1	1.66×10^{-18}	$16.04 T^{-1}$	$1.219 \times 10^4 T^{-1}$	2.453×10^{-3}	$120.3 T^{-1}$
1 cm³ molecule⁻¹ s⁻¹ =	6.023×10^{23}	6.023×10^{20}	6.023×10^{17}	1	$9.658 \times 10^{18} T^{-1}$	$7.34 \times 10^{21} T^{-1}$	1.478×10^{15}	$7.244 \times 10^{19} T^{-1}$
1 (mm Hg)⁻¹ s⁻¹ =	$6.236 \times 10^4 T$	$62.36 T$	$6.236 \times 10^{-2} T$	$1.035 \times 10^{-19} T$	1	760	4.56×10^{-2}	7.500
1 atm⁻¹ s⁻¹ =	$82.06 T$	$8.206 \times 10^{-2} T$	$8.206 \times 10^{-5} T$	$1.362 \times 10^{-22} T$	1.316×10^{-3}	1	6×10^{-5}	9.869×10^{-3}
1 ppm⁻¹ min⁻¹ = at 298 K, 1 atm total pressure	4.077×10^8	4.077×10^5	407.7	6.76×10^{-16}	21.93	1.667×10^4	1	164.5
1 m² kN⁻¹ s⁻¹ =	$8314 T$	$8.314 T$	$8.314 \times 10^{-3} T$	$1.38 \times 10^{-20} T$	0.1333	101.325	6.079×10^{-3}	1

To convert a rate constant from one set of units A to a new set B find the conversion factor for the row A under column B and multiply the old value by it, e.g. to convert cm³ molecule⁻¹ s⁻¹ to m³ mol⁻¹ s⁻¹ multiply by 6.023×10^{17}.

Table adapted from High Temperature Reaction Rate Data No. 5, The University, Leeds (1970).

Equivalent third order rate constants

A \ B	cm⁶ mol⁻² s⁻¹	dm⁶ mol⁻² s⁻¹	m⁶ mol⁻² s⁻¹	cm⁶ molecule⁻² s⁻¹	(mm Hg)⁻² s⁻¹	atm⁻² s⁻¹	ppm⁻² min⁻¹	m⁴ kN⁻² s⁻¹
1 cm⁶ mol⁻² s⁻¹ =	1	10^{-6}	10^{-12}	2.76×10^{-48}	$2.57 \times 10^{-10} T^{-2}$	$1.48 \times 10^{-4} T^{-2}$	1.003×10^{-19}	$1.447 \times 10^{-8} T^{-2}$
1 dm⁶ mol⁻² s⁻¹ =	10^6	1	10^{-6}	2.76×10^{-42}	$2.57 \times 10^{-4} T^{-2}$	$148 T^{-2}$	1.003×10^{-13}	$1.447 \times 10^{-2} T^{-2}$
1 m⁶ mol⁻² s⁻¹ =	10^{12}	10^6	1	2.76×10^{-36}	$257 T^{-2}$	$1.48 \times 10^8 T^{-2}$	1.003×10^{-7}	$1.447 \times 10^4 T^{-2}$
1 cm⁶ molecule⁻² s⁻¹ =	3.628×10^{47}	3.628×10^{41}	3.628×10^{35}	1	$9.328 \times 10^{37} T^{-2}$	$5.388 \times 10^{43} T^{-2}$	3.64×10^{28}	$5.248 \times 10^{39} T^{-2}$
1 (mm Hg)⁻² s⁻¹ =	$3.89 \times 10^9 T^2$	$3.89 \times 10^3 T^2$	$3.89 \times 10^{-3} T^2$	$1.07 \times 10^{-38} T^2$	1	5.776×10^5	3.46×10^{-5}	56.25
1 atm⁻² s⁻¹ =	$6.733 \times 10^3 T^2$	$6.733 \times 10^{-3} T^2$	$6.733 \times 10^{-9} T^2$	$1.86 \times 10^{-44} T^2$	1.73×10^{-6}	1	6×10^{-11}	9.74×10^{-5}
1 ppm⁻² min⁻¹ = at 298 K, 1 atm total pressure	9.97×10^{18}	9.97×10^{12}	9.97×10^6	2.75×10^{-29}	2.89×10^4	1.667×10^{10}	1	1.623×10^6
1 m⁴ kN⁻² s⁻¹ =	$6.91 \times 10^7 T^2$	$6.91 T^2$	$69.1 \times 10^{-5} T^2$	$1.904 \times 10^{-40} T^2$	0.0178	1.027×10^4	6.16×10^{-7}	1

From *J. Phys. Chem. Ref. Data*, 9, 470, 1980, by permission of the authors and the copyright owner, the American Institute of Physics.

CONVERSION FACTORS FOR IONIZING RADIATION

CONVERSION BETWEEN SI AND OTHER UNITS

Quantity	Symbol for quantity	Expression in SI units	Expression in symbols for SI units	Special name for SI units	Symbols using special names	Conventional units	Symbol for conventional unit	Value of conventional unit in SI units
Activity	A	1 per second	s^{-1}	becquerel	Bq	curie	Ci	3.7×10^{10} Bq
Absorbed dose	D	joule per kilogram	$J\,kg^{-1}$	gray	Gy	rad	rad	0.01 Gy
Absorbed dose rate	$\dot{D}$	joule per kilogram second	$J\,kg^{-1}\,s^{-1}$		Gy s⁻¹	rad	rad s⁻¹	0.01 Gy s⁻¹
Average energy per ion pair	W	joule	J			electronvolt	eV	1.602×10^{-19} J
Dose equivalent	H	joule per kilogram	$J\,kg^{-1}$	sievert	Sv	rem	rem	0.01 Sv
Dose equivalent rate	$\dot{H}$	joule per kilogram second	$J\,kg^{-1}\,s^{-1}$		Sv s⁻¹	rem per second	rem s⁻¹	0.01 Sv s⁻¹
Electric current	I	ampere	A			ampere	A	1.0 A
Electric potential difference	U, V	watt per ampere	$W\,A^{-1}$	volt	V	volt	V	1.0 A
Exposure	X	coulomb per kilogram	$C\,kg^{-1}$			roentgen	R	2.58×10^{-4} C kg⁻¹
Exposure rate	$\dot{X}$	coulomb per kilogram second	$C\,kg^{-1}\,s^{-1}$			roentgen	R s⁻¹	2.58×10^{-4} C kg⁻¹ s⁻¹
Fluence	ϕ	1 per meter squared	m^{-2}			1 per centimeter squared	cm⁻²	1.0×10^{4} m⁻²
Fluence rate	Φ	1 per meter squared second	$m^{-2}\,s^{-1}$			1 per centimeter squared second	cm⁻² s⁻¹	1.0×10^{4} m⁻² s⁻¹
Kerma	K	joule per kilogram	$J\,kg^{-1}$	gray	Gy	rad	rad	0.01 Gy
Kerma rate	$\dot{K}$	joule per kilogram second	$J\,kg^{-1}\,s^{-1}$		Gy s⁻¹	rad per second	rad s⁻¹	0.01 Gy s⁻¹
Lineal energy	y	joule per meter	$J\,m^{-1}$			kiloelectron volt per micrometer	keV μm⁻¹	1.602×10^{-10} J m⁻¹
Linear energy transfer	L	joule per meter	$J\,m^{-1}$			kiloelectron volt per micrometer	keV μm⁻¹	1.602×10^{-10} J m⁻¹
Mass attenuation coefficient	μ/ρ	meter squared per kilogram	$m^2\,kg^{-1}$			centimeter squared per gram	cm² g⁻¹	0.1 m² kg⁻¹
Mass energy transfer coefficient	μ_{tr}/ρ	meter squared per kilogram	$m^2\,kg^{-1}$			centimeter squared per gram	cm² g⁻¹	0.1 m² kg⁻¹
Mass energy absorption coefficient	μ_{en}/ρ	meter squared per kilogram	$m^2\,kg^{-1}$			centimeter squared per gram	cm² g⁻¹	0.1 m² kg⁻¹
Mass stopping power	S/ρ	joule meter squared per kilogram	$J\,m^2\,kg^{-1}$			MeV centimeter squared per gram	MeV cm² g⁻¹	1.602×10^{-14} J m² kg⁻¹
Power	P	joule per second	$J\,s^{-1}$	watt	W	watt	W	1.0 W
Pressure	p	newton per meter squared	$N\,m^{-2}$	pascal	Pa	torr	torr	(101325/760)Pa
Radiation chemical yield	G	mole per joule	$mol\,J^{-1}$			molecules per 100 electron volts	molecules (100 eV)⁻¹	1.04×10^{-7} mol J⁻¹
Specific energy	z	joule per kilogram	$J\,kg^{-1}$	gray	Gy	rad	rad	0.01 Gy

CONVERSION FACTORS FOR IONIZING RADIATION (continued)

CONVERSION OF RADIOACTIVITY UNITS FROM MBq TO mCi AND μCi

MBq	mCi	MBq	mCi	MBq	μCi	MBq	μCi
7000	189.	500	13.5	30	810	1	27
6000	162.	400	10.8	20	540	0.9	24
5000	135.	300	8.1	10	270	0.8	21.6
4000	108.	200	5.4	9	240	0.7	18.9
3000	81.	100	2.7	8	220	0.6	16.2
2000	54.	90	2.4	7	189	0.5	13.5
1000	27.	80	2.16	6	162	0.4	10.8
900	24.	70	1.89	5	135	0.3	8.1
800	21.6	60	1.62	4	108	0.2	5.4
700	18.9	50	1.35	3	81	0.1	2.7
600	16.2	40	1.08	2	54		

CONVERSION OF RADIOACTIVITY UNITS FROM mCi AND μCi TO MBq

mCi	MBq	mCi	MBq	μCi	MBq	μCi	MBq
200	7400	10	370	1000	37.0	80	2.96
150	5550	9	333	900	33.3	70	2.59
100	3700	8	296	800	29.6	60	2.22
90	3330	7	259	700	25.9	50	1.85
80	2960	6	222	600	22.2	40	1.48
70	2590	5	185	500	18.5	30	1.11
60	2220	4	148	400	14.8	20	0.74
50	1850	3	111	300	11.1	10	0.37
40	1480	2	74.0	200	7.4	5	0.185
30	1110	1	37.0	100	3.7	2	0.074
20	740			90	3.33	1	0.037

CONVERSION OF RADIOACTIVITY UNITS

100 TBq (10^{14} Bq)	=	2.7 kCi (2.7×10^3 Ci)		100 kBq (10^5 Bq)	=	2.7 μCi (2.7×10^{-6} Ci)
10 TBq (10^{13} Bq)	=	270 Ci (2.7×10^2 Ci)		10 kBq (10^4 Bq)	=	270 nCi (2.7×10^{-7} Ci)
1 TBq (10^{12} Bq)	=	27 Ci (2.7×10^1 Ci)		1 kBq (10^3 Bq)	=	27 nCi (2.7×10^{-8} Ci)
100 GBq (10^{11} Bq)	=	2.7 Ci (2.7×10^0 Ci)		100 Bq (10^2 Bq)	=	2.7 nCi (2.7×10^{-9} Ci)
10 GBq (10^{10} Bq)	=	270 mCi (2.7×10^{-1} Ci)		10 Bq (10^1 Bq)	=	270 pCi (2.7×10^{-10} Ci)
1 GBq (10^9 Bq)	=	27 mCi (2.7×10^{-2} Ci)		1 Bq (10^0 Bq)	=	27 pCi (2.7×10^{-11} Ci)
100 MBq (10^8 Bq)	=	2.7 mCi (2.7×10^{-3} Ci)		100 mBq (10^{-1} Bq)	=	2.7 pCi (2.7×10^{-12} Ci)
10 MBq (10^7 Bq)	=	270 μCi (2.7×10^{-4} Ci)		10 mBq (10^{-2} Bq)	=	270 fCi (2.7×10^{-13} Ci)
1 MBq (10^6 Bq)	=	27 μCi (2.7×10^{-5} Ci)		1 mBq (10^{-3} Bq)	=	27 fCi (2.7×10^{-14} Ci)

CONVERSION OF ABSORBED DOSE UNITS

SI Units		Conventional
100 Gy (10^2 Gy)	=	10,000 rad (10^4 rad)
10 Gy (10^1 Gy)	=	1,000 rad (10^3 rad)
1 Gy (10^0 Gy)	=	100 rad (10^2 rad)
100 mGy (10^{-1} Gy)	=	10 rad (10^1 rad)
10 mGy (10^{-2} Gy)	=	1 rad (10^0 rad)
1 mGy (10^{-3} Gy)	=	100 mrad (10^{-1} rad)
100 μGy (10^{-4} Gy)	=	10 mrad (10^{-2} rad)
10 μGy (10^{-5} Gy)	=	1 mrad (10^{-3} rad)
1 μGy (10^{-6} Gy)	=	100 μrad (10^{-4} rad)
100 nGy (10^{-7} Gy)	=	10 μrad (10^{-5} rad)
10 nGy (10^{-8} Gy)	=	1 μrad (10^{-6} rad)
1 nGy (10^{-9} Gy)	=	100 nrad (10^{-7} rad)

CONVERSION OF DOSE EQUIVALENT UNITS

100 Sv (10^2 Sv)	=	10,000 rem (10^4 rem)
10 Sv (10^1 Sv)	=	1,000 rem (10^3 rem)
1 Sv (10^0 Sv)	=	100 rem (10^2 rem)
100 mSv (10^{-1} Sv)	=	10 rem (10^1 rem)
10 mSv (10^{-2} Sv)	=	1 rem (10^0 rem)
1 mSv (10^{-3} Sv)	=	100 mrem (10^{-1} rem)
100 μSv (10^{-4} Sv)	=	10 mrem (10^{-2} rem)
10 μSv (10^{-5} Sv)	=	1 mrem (10^{-3} rem)
1 μSv (10^{-6} Sv)	=	100 μrem (10^{-4} rem)
100 nSv (10^{-7} Sv)	=	10 μrem (10^{-5} rem)
10 nSv (10^{-8} Sv)	=	1 μrem (10^{-6} rem)
1 nSv (10^{-9} Sv)	=	100 nrem (10^{-7} rem)

VALUES OF THE GAS CONSTANT IN DIFFERENT UNIT SYSTEMS

In SI units the value of the gas constant, R, is:

$R = 8.314510$ Pa m^3 K^{-1} mol^{-1}

$\quad = 8314.510$ Pa L K^{-1} mol^{-1}

$\quad = 0.08314510$ bar L K^{-1} mol^{-1}

This table gives the appropriate value of R for use in the ideal gas equation, $PV = nRT$, when the variables are expressed in other units. The following conversion factors for pressure units were used in generating the table:

   1 atm = 101325 Pa

   1 psi = 6894.757 Pa

   1 torr (mmHg) = 133.322 Pa [at 0°C]

   1 in Hg = 3386.38 Pa [at 0°C]

   1 in H$_2$O = 249.082 Pa [at 4°C]

   1 ft H$_2$O = 2988.98 Pa [at 4°C]

The advice of Prabir K. Chandra is appreciated.

Units of V, T, n			Units of P						
V	T	n	kPa	atm	psi	mmHg	in Hg	in H$_2$O	ft H$_2$O
ft^3	K	mol	0.2936241	0.00289785	0.0425866	2.20237	0.0867074	1.17882	0.0982355
		lb·mol	133.1857	1.31444	19.3169	998.978	39.3298	534.706	44.5589
	°R	mol	0.1631245	0.00160991	0.0236592	1.22354	0.0481708	0.654903	0.0545753
		lb·mol	73.99204	0.730245	10.7316	554.987	21.8499	297.059	24.7549
cm^3	K	mol	8314.510	82.0578	1205.92	62364.1	2455.28	33380.6	2781.72
		lb·mol	3771398	37220.8	546995	28287900	1113700	15141200	1261770
	°R	mol	4619.172	45.5877	669.954	34646.7	1364.04	18544.8	1545.40
		lb·mol	2095221	20678.2	303886	15715500	618720	8411770	700982
L	K	mol	8.314510	0.0820578	1.20592	62.3641	2.45528	33.3806	2.78172
		lb·mol	3771.398	37.2208	546.995	28287.9	1113.70	15141.2	1261.77
	°R	mol	4.619172	0.0455877	0.669954	34.6467	1.36404	18.5448	1.54540
		lb·mol	2095.221	20.6782	303.886	15715.5	618.720	8411.77	700.982
m^3	K	mol	0.008314510	0.0000820578	0.00120592	0.0623641	0.00245528	0.0333806	0.00278172
		lb·mol	3.771398	0.0372208	0.546995	28.2879	1.11370	15.1412	1.26177
	°R	mol	0.004619172	0.0000455877	0.000669954	0.0346467	0.00136404	0.0185448	0.00154540
		lb·mol	2.095221	0.0206782	0.303886	15.7155	0.618720	8.41177	0.700982

Section 2
Symbols, Terminology, and Nomenclature

SYMBOLS AND TERMINOLOGY FOR PHYSICAL AND CHEMICAL QUANTITIES

The International Organization for Standardization (ISO), International Union of Pure and Applied Chemistry (IUPAC), and the International Union of Pure and Applied Physics (IUPAP) have jointly developed a set of recommended symbols for physical and chemical quantities. Consistent use of these recommended symbols helps assure unambiguous scientific communication. The list below is reprinted from Reference 1 with permission from IUPAC. Full details may be found in the following references:

1. Ian Mills, Ed., *Quantities, Units, and Symbols in Physical Chemistry*, Blackwell Scientific Publications, Oxford, 1988.
2. E. R. Cohen and P. Giacomo, *Symbols, Units, Nomenclature, and Fundamental Constants in Physics*, Document IUPAP-25, 1987; also published in *Physica*, 146A, 1—68, 1987.
3. *ISO Standards Handbook 2: Units of Measurement*, International Organization of Standardization, Geneva, 1982.

GENERAL RULES

The value of a physical quantity is expressed as the product of a numerical value and a unit, e.g.:

$T = 300$ K
$V = 26.2$ cm^3
$C_p = 45.3$ J mol^{-1} K^{-1}

The symbol for a physical quantity is always given in italic (sloping) type, while symbols for units are given in roman type. Column headings in tables and axis labels on graphs may conveniently be written as the physical quantity symbol divided by the unit symbol, e.g.:

T/K
V/cm^3
C_p/J mol K^{-1}

The values in the table or graph axis are then pure numbers.
Subscripts to symbols for physical quantities should be italic if the subscript refers to another physical quantity or to a number, e.g.:

C_p — heat capacity at constant pressure
B_n — nth virial coefficient

Subscripts which have other meanings should be in roman type:

m_p — mass of the proton
E_k — kinetic energy

The following tables give the recommended symbols for the major classes of physical and chemical quantities. The expression in the Definition column is given as an aid in identifying the quantity but is not necessarily the complete or unique definition. The SI Unit gives one (not necessarily unique) expression for the coherent SI unit for the quantity. Other equivalent unit expressions, including those which involve SI prefixes, may be used.

Name	Symbol	Definition	SI unit
SPACE AND TIME			
cartesian space coordinates	x, y, z		m
spherical polar coordinates	r, θ, ϕ		m, 1, 1
generalized coordinate	q, q_i		(varies)
position vector	r	$r = xi + yj + zk$	m
length	l		m
special symbols:			
height	h		
breadth	b		
thickness	d, δ		
distance	d		

Name	Symbol	Definition	SI unit
radius	r		
diameter	d		
path length	s		
length of arc	s		
area	A, A_s, S		m^2
volume	$V, (v)$		m^3
plane angle	$\alpha, \beta, \gamma, \theta, \phi \ldots$	$\alpha = s/r$	rad, 1
solid angle	ω, Ω	$\omega = A/r^2$	sr, 1
time	t		s
period	T	$T = t/N$	s
frequency	v, f	$v = 1/T$	Hz
circular frequency, angular frequency	ω	$\omega = 2\pi v$	rad s^{-1}, s^{-1}
characteristic time interval, relaxation time, time constant	τ, T	$\tau = \lvert dt/d\ln x \rvert$	s
angular velocity	ω	$\omega = d\phi/dt$	rad s^{-1}, s^{-1}
velocity	$v, \boldsymbol{u}, \boldsymbol{w}, \boldsymbol{c}, \dot{\boldsymbol{r}}$	$v = dr/dt$	m s^{-1}
speed	v, u, w, c	$v = \lvert v \rvert$	m s^{-1}
acceleration	$\boldsymbol{a}, (\boldsymbol{g})$	$\boldsymbol{a} = dv/dt$	m s^{-2}

CLASSICAL MECHANICS

Name	Symbol	Definition	SI unit
mass	m		kg
reduced mass	μ	$\mu = m_1 m_2/(m_1 + m_2)$	kg
density, mass density	ρ	$\rho = m/V$	kg m^{-3}
relative density	d	$d = \rho/\rho^\bullet$	1
surface density	ρ_A, ρ_S	$\rho_A = m/A$	kg m^{-2}
specific volume	v	$v = V/m = 1/\rho$	m^3 kg^{-1}
momentum	$\boldsymbol{p}$	$\boldsymbol{p} = mv$	kg m s^{-1}
angular momentum, action	$\boldsymbol{L}$	$\boldsymbol{L} = \boldsymbol{r} \times \boldsymbol{p}$	J s
moment of inertia	I, J	$I = \sum m_i r_i^2$	kg m^2
force	$\boldsymbol{F}$	$\boldsymbol{F} = d\boldsymbol{p}/dt = m\boldsymbol{a}$	N
torque, moment of a force	$\boldsymbol{T}, (\boldsymbol{M})$	$\boldsymbol{T} = \boldsymbol{r} \times \boldsymbol{F}$	N m
energy	E		J
potential energy	E_p, V, Φ	$E_p = -\int \boldsymbol{F} \cdot d\boldsymbol{s}$	J
kinetic energy	E_k, T, K	$E_k = \frac{1}{2}mv^2$	J
work	W, w	$W = \int \boldsymbol{F} \cdot d\boldsymbol{s}$	J
Hamilton function	H	$H(q, p)$ $= T(q, p) + V(q)$	J
Lagrange function	L	$L(q, \dot{q})$ $= T(q, \dot{q}) - V(q)$	J

Name	Symbol	Definition	SI unit
pressure	p, P	$p = F/A$	Pa, N m^{-2}
surface tension	γ, σ	$\gamma = \mathrm{d}W/\mathrm{d}A$	N m^{-1}, J m^{-2}
weight	$G, (W, P)$	$G = mg$	N
gravitational constant	G	$F = Gm_1 m_2/r^2$	N m^2 kg^{-2}
normal stress	σ	$\sigma = F/A$	Pa
shear stress	τ	$\tau = F/A$	Pa
linear strain, relative elongation	ε, e	$\varepsilon = \Delta l/l$	1
modulus of elasticity, Young's modulus	E	$E = \sigma/\varepsilon$	Pa
shear strain	γ	$\gamma = \Delta x/d$	1
shear modulus	G	$G = \tau/\gamma$	Pa
volume strain, bulk strain	θ	$\theta = \Delta V/V_0$	1
bulk modulus, compression modulus	K	$K = -V_0(\mathrm{d}p/\mathrm{d}V)$	Pa
viscosity, dynamic viscosity	η, μ	$\tau_{x,z} = \eta(\mathrm{d}v_x/\mathrm{d}z)$	Pa s
fluidity	ϕ	$\phi = 1/\eta$	m kg^{-1} s
kinematic viscosity	ν	$\nu = \eta/\rho$	m^2 s^{-1}
friction coefficient	$\mu, (f)$	$F_{\mathrm{frict}} = \mu F_{\mathrm{norm}}$	1
power	P	$P = \mathrm{d}W/\mathrm{d}t$	W
sound energy flux	P, P_a	$P = \mathrm{d}E/\mathrm{d}t$	W
acoustic factors,			
reflection factor	ρ	$\rho = P_r/P_0$	1
acoustic absorption factor	$\alpha_a, (\alpha)$	$\alpha_a = 1 - \rho$	1
transmission factor	τ	$\tau = P_{tr}/P_0$	1
dissipation factor	δ	$\delta = \alpha_a - \tau$	1

ELECTRICITY AND MAGNETISM

Name	Symbol	Definition	SI unit
quantity of electricity, electric charge	Q		C
charge density	ρ	$\rho = Q/V$	C m^{-3}
surface charge density	σ	$\sigma = Q/A$	C m^{-2}
electric potential	V, ϕ	$V = \mathrm{d}W/\mathrm{d}Q$	V, J C^{-1}
electric potential difference	$U, \Delta V, \Delta\phi$	$U = V_2 - V_1$	V
electromotive force	E	$E = \int(F/Q)\cdot\mathrm{d}s$	V
electric field strength	E	$E = F/Q = -\mathrm{grad}\ V$	V m^{-1}
electric flux	Ψ	$\Psi = \int D\cdot\mathrm{d}A$	C
electric displacement	D	$D = \varepsilon E$	C m^{-2}
capacitance	C	$C = Q/U$	F, C V^{-1}
permittivity	ε	$D = \varepsilon E$	F m^{-1}

Name	Symbol	Definition	SI unit
permittivity of vacuum	ε_0	$\varepsilon_0 = \mu_0^{-1} c_0^{-2}$	$F\,m^{-1}$
relative permittivity	ε_r	$\varepsilon_r = \varepsilon/\varepsilon_0$	1
dielectric polarization (dipole moment per volume)	P	$P = D - \varepsilon_0 E$	$C\,m^{-2}$
electric susceptibility	χ_e	$\chi_e = \varepsilon_r - 1$	1
electric dipole moment	p, μ	$p = Qr$	$C\,m$
electric current	I	$I = dQ/dt$	A
electric current density	j, J	$I = \int j \cdot dA$	$A\,m^{-2}$
magnetic flux density, magnetic induction	B	$F = Qv \times B$	T
magnetic flux	Φ	$\Phi = \int B \cdot dA$	Wb
magnetic field strength	H	$B = \mu H$	$A\,m^{-1}$
permeability	μ	$B = \mu H$	$N\,A^{-2}, H\,m^{-1}$
permeability of vacuum	μ_0		$H\,m^{-1}$
relative permeability	μ_r	$\mu_r = \mu/\mu_0$	1
magnetization (magnetic dipole moment per volume)	M	$M = B/\mu_0 - H$	$A\,m^{-1}$
magnetic susceptibility	$\chi, \kappa, (\chi_m)$	$\chi = \mu_r - 1$	1
molar magnetic susceptibility	χ_m	$\chi_m = V_m \chi$	$m^3\,mol^{-1}$
magnetic dipole moment	m, μ	$E_p = -m \cdot B$	$A\,m^2, J\,T^{-1}$
electrical resistance	R	$R = U/I$	Ω
conductance	G	$G = 1/R$	S
loss angle	δ	$\delta = (\pi/2) + \phi_I - \phi_U$	1, rad
reactance	X	$X = (U/I)\sin\delta$	Ω
impedance (complex impedance)	Z	$Z = R + iX$	Ω
admittance (complex admittance)	Y	$Y = 1/Z$	S
susceptance	B	$Y = G + iB$	S
resistivity	ρ	$\rho = E/j$	$\Omega\,m$
conductivity	κ, γ, σ	$\kappa = 1/\rho$	$S\,m^{-1}$
self-inductance	L	$E = -L(dI/dt)$	H
mutual inductance	M, L_{12}	$E_1 = L_{12}(dI_2/dt)$	H
magnetic vector potential	A	$B = \nabla \times A$	$Wb\,m^{-1}$
Poynting vector	S	$S = E \times H$	$W\,m^{-2}$

QUANTUM MECHANICS

Name	Symbol	Definition	SI unit
momentum operator	$\hat{p}$	$\hat{p} = -i\hbar\nabla$	$m^{-1}\,J\,s$
kinetic energy operator	$\hat{T}$	$\hat{T} = -(\hbar^2/2m)\nabla^2$	J

Name	Symbol	Definition	SI unit
hamiltonian operator	$\hat{H}$	$\hat{H} = \hat{T} + V$	J
wavefunction, state function	Ψ, ψ, ϕ	$\hat{H}\psi = E\psi$	$(m^{-3/2})$
probability density	P	$P = \psi^*\psi$	(m^{-3})
charge density of electrons	ρ	$\rho = -eP$	$(C\,m^{-3})$
probability current density	S	$S = -i\hbar(\psi^*\nabla\psi - \psi\nabla\psi^*)/2m_e$	$(m^{-2}\,s^{-1})$
electric current density of electrons	j	$j = -eS$	$(A\,m^{-2})$
matrix element of operator $\hat{A}$	$A_{ij}, \langle i|\hat{A}|j\rangle$	$A_{ij} = \int\psi_i^*\hat{A}\psi_j d\tau$	(varies)
expectation value of operator $\hat{A}$	$\langle A\rangle, \bar{A}$	$\langle A\rangle = \int\psi^*\hat{A}\psi d\tau$	(varies)
hermitian conjugate of $\hat{A}$	$\hat{A}^\dagger$	$(\hat{A}^\dagger)_{ij} = (A_{ji})^*$	(varies)
commutator of $\hat{A}$ and $\hat{B}$	$[\hat{A}, \hat{B}], [\hat{A}, \hat{B}]_-$	$[\hat{A}, \hat{B}] = \hat{A}\hat{B} - \hat{B}\hat{A}$	(varies)
anticommutator	$[\hat{A}, \hat{B}]_+$	$[\hat{A}, \hat{B}]_+ = \hat{A}\hat{B} + \hat{B}\hat{A}$	(varies)
spin wavefunction	$\alpha; \beta$		1
coulomb integral	H_{AA}	$H_{AA} = \int\psi_A^*\hat{H}\psi_A d\tau$	J
resonance integral	H_{AB}	$H_{AB} = \int\psi_A^*\hat{H}\psi_B d\tau$	J
overlap integral	S_{AB}	$S_{AB} = \int\psi_A^*\psi_B d\tau$	1

ATOMS AND MOLECULES

Name	Symbol	Definition	SI unit
nucleon number, mass number	A		1
proton number, atomic number	Z		1
neutron number	N	$N = A - Z$	1
electron rest mass	m_e		kg
mass of atom, atomic mass	m_a, m		kg
atomic mass constant	m_u	$m_u = m_a(^{12}C)/12$	kg
mass excess	Δ	$\Delta = m_a - Am_u$	kg
elementary charge, proton charge	e		C
Planck constant	h		J s
Planck constant/2π	$\hbar$	$\hbar = h/2\pi$	J s
Bohr radius	a_0	$a_0 = 4\pi\varepsilon_0\hbar^2/m_e e^2$	m
Hartree energy	E_h	$E_h = \hbar^2/m_e a_0^2$	J
Rydberg constant	R_∞	$R_\infty = E_h/2hc$	m^{-1}
fine structure constant	α	$\alpha = e^2/4\pi\varepsilon_0\hbar c$	1
ionization energy	E_i		J

Name	Symbol	Definition	SI unit
electron affinity	E_{ea}		J
dissociation energy	E_d, D		J
from the ground state	D_0		J
from the potential minimum	D_e		J
principal quantum number (H atom)	n	$E = -hcR/n^2$	1
angular momentum quantum numbers	see under Spectroscopy		
magnetic dipole moment of a molecule	m, μ	$E_p = -m \cdot B$	$J\,T^{-1}$
magnetizability of a molecule	ξ	$m = \xi B$	$J\,T^{-2}$
Bohr magneton	μ_B	$\mu_B = e\hbar/2m_e$	$J\,T^{-1}$
nuclear magneton	μ_N	$\mu_N = (m_e/m_p)\mu_B$	$J\,T^{-1}$
magnetogyric ratio (gyromagnetic ratio)	γ	$\gamma = \mu/L$	$C\,kg^{-1}$
g factor	g		1
Larmor circular frequency	ω_L	$\omega_L = (e/2m)B$	s^{-1}
Larmor frequency	ν_L	$\nu_L = \omega_L/2\pi$	Hz
longitudinal relaxation time	T_1		s
transverse relaxation time	T_2		s
electric dipole moment of a molecule	p, μ	$E_p = -p \cdot E$	C m
quadrupole moment of a molecule	$Q; \Theta$	$E_p = \frac{1}{2}Q:V'' = \frac{1}{3}\Theta:V''$	$C\,m^2$
quadrupole moment of a nucleus	eQ	$eQ = 2\langle\Theta_{zz}\rangle$	$C\,m^2$
electric field gradient tensor	q	$q_{\alpha\beta} = -\partial^2 V/\partial\alpha\partial\beta$	$V\,m^{-2}$
quadrupole interaction energy tensor	χ	$\chi_{\alpha\beta} = eQq_{\alpha\beta}$	J
electric polarizability of a molecule	α	$p\,(\text{induced}) = \alpha E$	$C\,m^2\,V^{-1}$
activity (of a radioactive substance)	A	$A = -dN_B/dt$	Bq
decay (rate) constant, disintegration (rate) constant	λ	$A = \lambda N_B$	s^{-1}
half life	$t_{\frac{1}{2}}, T_{\frac{1}{2}}$		s
mean life	τ		s
level width	Γ	$\Gamma = \hbar/\tau$	J

Name	Symbol	Definition	SI unit
disintegration energy	Q		J
cross section (of a nuclear reaction)	σ		m^2

SPECTROSCOPY

Name	Symbol	Definition	SI unit
total term	T	$T = E_{tot}/hc$	m^{-1}
transition wavenumber	$\tilde{\nu}, (\nu)$	$\tilde{\nu} = T' - T''$	m^{-1}
transition frequency	ν	$\nu = (E' - E'')/h$	Hz
electronic term	T_e	$T_e = E_e/hc$	m^{-1}
vibrational term	G	$G = E_{vib}/hc$	m^{-1}
rotational term	F	$F = E_{rot}/hc$	m^{-1}
spin orbit coupling constant	A	$T_{s.o.} = A\langle \hat{L} \cdot \hat{S} \rangle$	m^{-1}
principal moments of inertia	$I_A; I_B; I_C$	$I_A \leqslant I_B \leqslant I_C$	$kg\,m^2$
rotational constants,			
in wavenumber	$\tilde{A}; \tilde{B}; \tilde{C}$	$\tilde{A} = h/8\pi^2 c I_A$	m^{-1}
in frequency	$A; B; C$	$A = h/8\pi^2 I_A$	Hz
inertial defect	Δ	$\Delta = I_C - I_A - I_B$	$kg\,m^2$
asymmetry parameter	κ	$\kappa = \dfrac{(2B - A - C)}{(A - C)}$	1
centrifugal distortion constants,			
S reduction	$D_J; D_{JK}; D_K; d_1; d_2$		m^{-1}
A reduction	$\Delta_J; \Delta_{JK}; \Delta_K; \delta_J; \delta_K$		m^{-1}
harmonic vibration wavenumber	$\omega_e; \omega_r$		m^{-1}
vibrational anharmonicity constant	$\omega_e x_e; x_{rs}; g_{tt'}$		m^{-1}
vibrational quantum numbers	$v_r; l_t$		1
Coriolis zeta constant	ζ_{rs}^{α}		1
angular momentum quantum numbers	see additional information below		
degeneracy, statistical weight	g, d, β		1
electric dipole moment of a molecule	p, μ	$E_p = -p \cdot E$	C m
transition dipole moment of a molecule	M, R	$M = \int \psi' p \psi'' d\tau$	C m
molecular geometry, interatomic distances,			
equilibrium distance	r_e		m
zero-point average distance	r_z		m
ground state distance	r_0		m

Name	Symbol	Definition	SI unit
substitution structure distance	r_s		m
vibrational coordinates,			
internal coordinates	R_i, r_i, θ_j, etc.		(varies)
symmetry coordinates	S_i		(varies)
normal coordinates			
mass adjusted	Q_r		kg$^{\frac{1}{2}}$ m
dimensionless	q_r		1
vibrational force constants,			
diatomic	$f, (k)$	$f = \partial^2 V/\partial r^2$	J m^{-2}
polyatomic,			
internal coordinates	f_{ij}	$f_{ij} = \partial^2 V/\partial r_i \partial r_j$	(varies)
symmetry coordinates	F_{ij}	$F_{ij} = \partial^2 V/\partial S_i \partial S_j$	(varies)
dimensionless normal coordinates	$\phi_{rst...}$, $k_{rst...}$		m^{-1}
nuclear magnetic resonance (NMR),			
magnetogyric ratio	γ	$\gamma = \mu/I\hbar$	C kg^{-1}
shielding constant	σ_A	$B_A = (1 - \sigma_A)B$	1
chemical shift, δ scale	δ	$\delta = 10^6(v - v_0)/v_0$	1
(indirect) spin–spin coupling constant	J_{AB}	$\hat{H}/h = J_{AB}\hat{I}_A \cdot \hat{I}_B$	Hz
direct (dipolar) coupling constant	D_{AB}		Hz
longitudinal relaxation time	T_1		s
transverse relaxation time	T_2		s
electron spin resonance, electron paramagnetic resonance (ESR, EPR),			
magnetogyric ratio	γ	$\gamma = \mu/s\hbar$	C kg^{-1}
g factor	g	$hv = g\mu_B B$	1
hyperfine coupling constant,			
in liquids	a, A	$\hat{H}_{hfs}/h = a\hat{S} \cdot \hat{I}$	Hz
in solids	T	$\hat{H}_{hfs}/h = \hat{S} \cdot T \cdot \hat{I}$	Hz

| | Operator | Quantum number symbol | | |
Angular momentum	*symbol*	*Total*	*Z-axis*	*z-axis*
electron orbital	$\hat{L}$	L	M_L	Λ
one electron only	$\hat{l}$	l	m_l	λ

Angular momentum	Operator symbol	Quantum number symbol Total	Z-axis	z-axis
electron spin	$\hat{S}$	S	M_S	Σ
one electron only	$\hat{s}$	s	m_s	σ
electron orbital + spin	$\hat{L} + \hat{S}$			$\Omega = \Lambda + \Sigma$
nuclear orbital (rotational)	$\hat{R}$	R		K_R, k_R
nuclear spin	$\hat{I}$	I	M_I	
internal vibrational				
spherical top	$\hat{l}$	$l(l\zeta)$		K_l
other	$\hat{j}, \hat{\pi}$			$l(l\zeta)$
sum of $R + L(+j)$	$\hat{N}$	N		K, k
sum of $N + S$	$\hat{J}$	J	M_J	K, k
sum of $J + I$	$\hat{F}$	F	M_F	

ELECTROMAGNETIC RADIATION

Name	Symbol	Definition	SI unit
wavelength	λ		m
speed of light			
in vacuum	c_0		$m\ s^{-1}$
in a medium	c	$c = c_0/n$	$m\ s^{-1}$
wavenumber in vacuum	$\tilde{\nu}$	$\tilde{\nu} = \nu/c_0 = 1/n\lambda$	m^{-1}
wavenumber (in a medium)	σ	$\sigma = 1/\lambda$	m^{-1}
frequency	ν	$\nu = c/\lambda$	Hz
circular frequency, pulsatance	ω	$\omega = 2\pi\nu$	s^{-1}, $rad\ s^{-1}$
refractive index	n	$n = c_0/c$	1
Planck constant	h		J s
Planck constant/2π	$\hbar$	$\hbar = h/2\pi$	J s
radiant energy	Q, W		J
radiant energy density	ρ, w	$\rho = Q/V$	$J\ m^{-3}$
spectral radiant energy density			
in terms of frequency	ρ_ν, w_ν	$\rho_\nu = d\rho/d\nu$	$J\ m^{-3}\ Hz^{-1}$
in terms of wavenumber	$\rho_{\tilde{\nu}}, w_{\tilde{\nu}}$	$\rho_{\tilde{\nu}} = d\rho/d\tilde{\nu}$	$J\ m^{-2}$
in terms of wavelenglth	ρ_λ, w_λ	$\rho_\lambda = d\rho/d\lambda$	$J\ m^{-4}$
Einstein transition probabilities			
spontaneous emission	A_{nm}	$dN_n/dt = -A_{nm}N_n$	s^{-1}
stimulated emission	B_{nm}	$dN_n/dt = -\rho_{\tilde{\nu}}(\tilde{\nu}_{nm}) \times B_{nm}N_n$	$s\ kg^{-1}$
stimulated absorption	B_{mn}	$dN_n/dt = \rho_{\tilde{\nu}}(\tilde{\nu}_{nm})B_{mn}N_m$	$s\ kg^{-1}$
radiant power, radiant energy per time	Φ, P	$\Phi = dQ/dt$	W

Name	Symbol	Definition	SI unit
radiant intensity	I	$I = d\Phi/d\Omega$	W sr^{-1}
radiant exitance, (emitted radiant flux)	M	$M = d\Phi/dA_{source}$	W m^{-2}
irradiance, (radiant flux received)	$E, (I)$	$E = d\Phi/dA$	W m^{-2}
emittance	ε	$\varepsilon = M/M_{bb}$	1
Stefan–Boltzmann constant	σ	$M_{bb} = \sigma T^4$	W m^{-2} K^{-4}
first radiation constant	c_1	$c_1 = 2\pi h c_0^2$	W m^2
second radiation constant	c_2	$c_2 = h c_0/k$	K m
transmittance, transmission factor	τ, T	$\tau = \Phi_{tr}/\Phi_0$	1
absorptance, absorption factor	α	$\alpha = \Phi_{abs}/\Phi_0$	1
reflectance, reflection factor	ρ	$\rho = \Phi_{refl}/\Phi_0$	1
(decadic) absorbance	A	$A = -\lg(1 - \alpha_i)$	1
napierian absorbance	B	$B = -\ln(1 - \alpha_i)$	1
absorption coefficient			
(linear) decadic	a, K	$a = A/l$	m^{-1}
(linear) napierian	α	$\alpha = B/l$	m^{-1}
molar (decadic)	ε	$\varepsilon = a/c = A/cl$	m^2 mol^{-1}
molar napierian	κ	$\kappa = \alpha/c = B/cl$	m^2 mol^{-1}
absorption index	k	$k = \alpha/4\pi\tilde{\nu}$	1
complex refractive index	$\hat{n}$	$\hat{n} = n + ik$	1
molar refraction	R, R_m	$R = \dfrac{(n^2 - 1)}{(n^2 + 2)} V_m$	m^3 mol^{-1}
angle of optical rotation	α		1, rad

SOLID STATE

Name	Symbol	Definition	SI unit
lattice vector	$\mathbf{R}, \mathbf{R}_0$		m
fundamental translation vectors for the crystal lattice	$\mathbf{a}_1; \mathbf{a}_2; \mathbf{a}_3,$ $\mathbf{a}; \mathbf{b}; \mathbf{c}$	$\mathbf{R} = n_1\mathbf{a}_1 + n_2\mathbf{a}_2 + n_3\mathbf{a}_3$	m
(circular) reciprocal lattice vector	$\mathbf{G}$	$\mathbf{G} \cdot \mathbf{R} = 2\pi m$	m^{-1}
(circular) fundamental translation vectors for the reciprocal lattice	$\mathbf{b}_1; \mathbf{b}_2; \mathbf{b}_3,$ $\mathbf{a}^*; \mathbf{b}^*; \mathbf{c}^*$	$\mathbf{a}_i \cdot \mathbf{b}_k = 2\pi\delta_{ik}$	m^{-1}
lattice plane spacing	d		m
Bragg angle	θ	$n\lambda = 2d \sin\theta$	1, rad

Name	Symbol	Definition	SI unit
order of reflection	n		1
order parameters			
short range	σ		1
long range	s		1
Burgers vector	b		m
particle position vector	r, R_j		m
equilibrium position vector of an ion	R_0		m
displacement vector of an ion	u	$u = R - R_0$	m
Debye–Waller factor	B, D		1
Debye circular wavenumber	q_D		m^{-1}
Debye circular frequency	ω_D		s^{-1}
Grüneisen parameter	γ, Γ	$\gamma = \alpha V / \kappa C_V$	1
Madelung constant	$\alpha, \mathscr{M}$	$E_{coul} = \dfrac{\alpha N_A z_+ z_- e^2}{4\pi\varepsilon_0 R_0}$	1
density of states	N_E	$N_E = dN(E)/dE$	$J^{-1} m^{-3}$
(spectral) density of vibrational modes	N_ω, g	$N_\omega = dN(\omega)/d\omega$	$s\, m^{-3}$
resistivity tensor	ρ_{ik}	$E = \rho \cdot j$	$\Omega\, m$
conductivity tensor	σ_{ik}	$\sigma = \rho^{-1}$	$S\, m^{-1}$
thermal conductivity tensor	λ_{ik}	$J_q = -\lambda \cdot \mathrm{grad}\, T$	$W\, m^{-1} K^{-1}$
residual resistivity	ρ_R		$\Omega\, m$
relaxation time	τ	$\tau = l/v_F$	s
Lorenz coefficient	L	$L = \lambda/\sigma T$	$V^2 K^{-2}$
Hall coefficient	A_H, R_H	$E = \rho \cdot j + R_H(B \times j)$	$m^3 C^{-1}$
thermoelectric force	E		V
Peltier coefficient	Π		V
Thomson coefficient	$\mu, (\tau)$		$V\, K^{-1}$
work function	Φ	$\Phi = E_\infty - E_F$	J
number density, number concentration	$n, (p)$		m^{-3}
gap energy	E_g		J
donor ionization energy	E_d		J
acceptor ionization energy	E_a		J
Fermi energy	E_F, ε_F		J
circular wave vector, propagation vector	k, q	$k = 2\pi/\lambda$	m^{-1}
Bloch function	$u_k(r)$	$\psi(r) = u_k(r)\exp(i k \cdot r)$	$m^{-3/2}$
charge density of electrons	ρ	$\rho(r) = -e\psi^*(r)\psi(r)$	$C\, m^{-3}$

Name	Symbol	Definition	SI unit
effective mass	m^*		kg
mobility	μ	$\mu = v_{drift}/E$	$m^2\,V^{-1}\,s^{-1}$
mobility ratio	b	$b = \mu_n/\mu_p$	1
diffusion coefficient	D	$dN/dt = -DA(dn/dx)$	$m^2\,s^{-1}$
diffusion length	L	$L = \sqrt{D\tau}$	m
characteristic (Weiss) temperature	θ, θ_W		K
Curie temperature	T_C		K
Néel temperature	T_N		K

STATISTICAL THERMODYNAMICS

Name	Symbol	Definition	SI unit
number of entities	N		1
number density of entities, number concentration	n, C	$n = N/V$	m^{-3}
Avogadro constant	L, N_A		mol^{-1}
Boltzmann constant	k, k_B		$J\,K^{-1}$
gas constant (molar)	R	$R = Lk$	$J\,K^{-1}\,mol^{-1}$
molecular position vector	$r\,(x, y, z)$		m
molecular velocity vector	$c(c_x, c_y, c_z),$ $u(u_x, u_y, u_z)$	$c = dr/dt$	$m\,s^{-1}$
molecular momentum vector	$p(p_x, p_y, p_z)$	$p = mc$	$kg\,m\,s^{-1}$
velocity distribution function (Maxwell)	$f(c_x)$	$f(c_x) = (m/2\pi kT)^{\frac{1}{2}}$ $\times \exp(-mc_x^2/2kT)$	$m^{-1}\,s$
speed distribution function (Maxwell–Boltzmann)	$F(c)$	$F(c) = (m/2\pi kT)^{3/2}$ $\times 4\pi c^2\exp(-mc^2/2kT)$	$m^{-1}\,s$
average speed	$\bar{c}, \bar{u},$ $\langle c\rangle, \langle u\rangle$	$\bar{c} = \int cF(c)dc$	$m\,s^{-1}$
generalized coordinate	q		(m)
generalized momentum	p	$p = \partial L/\partial \dot{q}$	$(kg\,m\,s^{-1})$
volume in phase space	Ω	$\Omega = (1/h)\int p\,dq$	1
probability	P		1
statistical weight, degeneracy	g, d, W, ω, β		1
density of states	$\rho(E)$	$\rho(E) = dN/dE$	J^{-1}
partition function, sum over states, for a single molecule	q, z	$q = \sum_i g_i\exp(-\varepsilon_i/kT)$	1
for a canonical ensemble (system, or assembly)	Q, Z		1

Name	Symbol	Definition	SI unit
microcanonical ensemble	Ω		1
grand (canonical ensemble)	Ξ		1
symmetry number	σ, s		1
reciprocal temperature parameter	β	$\beta = 1/kT$	J^{-1}
characteristic temperature	Θ		K

GENERAL CHEMISTRY

Name	Symbol	Definition	SI unit
number of entities (e.g. molecules, atoms, ions, formula units)	N		1
amount (of substance)	n	$n_B = N_B/L$	mol
Avogadro constant	L, N_A		mol^{-1}
mass of atom, atomic mass	m_a, m		kg
mass of entity (molecule, or formula unit)	m_f, m		kg
atomic mass constant	m_u	$m_u = m_a(^{12}C)/12$	kg
molar mass	M	$M_B = m/n_B$	$kg\,mol^{-1}$
relative molecular mass (relative molar mass, molecular weight)	M_r	$M_{r,B} = m_B/m_u$	1
molar volume	V_m	$V_{m,B} = V/n_B$	$m^3\,mol^{-1}$
mass fraction	w	$w_B = m_B/\Sigma m_i$	1
volume fraction	ϕ	$\phi_B = V_B/\Sigma V_i$	1
mole fraction, amount fraction, number fraction	x, y	$x_B = n_B/\Sigma n_i$	1
(total) pressure	p, P		Pa
partial pressure	p_B	$p_B = y_B p$	Pa
mass concentration (mass density)	γ, ρ	$\gamma_B = m_B/V$	$kg\,m^{-3}$
number concentration, number density of entities	C, n	$C_B = N_B/V$	m^{-3}
amount concentration, concentration	c	$c_B = n_B/V$	$mol\,m^{-3}$
solubility	s	$s_B = c_B$ (saturated solution)	$mol\,m^{-3}$
molality (of a solute)	$m, (b)$	$m_B = n_B/m_A$	$mol\,kg^{-1}$

Name	Symbol	Definition	SI unit
surface concentration	Γ	$\Gamma_B = n_B/A$	mol m^{-2}
stoichiometric number	v		1
extent of reaction, advancement	ξ	$\Delta\xi = \Delta n_B/v_B$	mol
degree of dissociation	α		1

CHEMICAL THERMODYNAMICS

Name	Symbol	Definition	SI unit
heat	q, Q		J
work	w, W		J
internal energy	U	$\Delta U = q + w$	J
enthalpy	H	$H = U + pV$	J
thermodynamic temperature	T		K
Celsius temperature	θ, t	$\theta/°C = T/K - 273.15$	°C
entropy	S	$dS \geqslant dq/T$	J K^{-1}
Helmholtz energy, (Helmholtz function)	A	$A = U - TS$	J
Gibbs energy, (Gibbs function)	G	$G = H - TS$	J
Massieu function	J	$J = -A/T$	J K^{-1}
Planck function	Y	$Y = -G/T$	J K^{-1}
surface tension	γ, σ	$\gamma = (\partial G/\partial A_s)_{T,p}$	$\text{J m}^{-2}, \text{N m}^{-1}$
molar quantity X	X_m	$X_m = X/n$	(varies)
specific quantity X	x	$x = X/m$	(varies)
pressure coefficient	β	$\beta = (\partial p/\partial T)_V$	Pa K^{-1}
relative pressure coefficient	α_p	$\alpha_p = (1/p)(\partial p/\partial T)_V$	K^{-1}
compressibility,			
isothermal	κ_T	$\kappa_T = -(1/V)(\partial V/\partial p)_T$	Pa^{-1}
isentropic	κ_S	$\kappa_S = -(1/V)(\partial V/\partial p)_S$	Pa^{-1}
linear expansion coefficient	α_l	$\alpha_l = (1/l)(\partial l/\partial T)$	K^{-1}
cubic expansion coefficient	α, α_V, γ	$\alpha = (1/V)(\partial V/\partial T)_p$	K^{-1}
heat capacity,			
at constant pressure	C_p	$C_p = (\partial H/\partial T)_p$	J K^{-1}
at constant volume	C_V	$C_V = (\partial U/\partial T)_V$	J K^{-1}
ratio of heat capacities	$\gamma, (\kappa)$	$\gamma = C_p/C_V$	1
Joule–Thomson coefficient	μ, μ_{JT}	$\mu = (\partial T/\partial p)_H$	K Pa^{-1}
second virial coefficient	B	$pV_m = RT(1 + B/V_m + \cdots)$	$\text{m}^3 \text{mol}^{-1}$
compression factor (compressibility factor)	Z	$Z = pV_m/RT$	1

Name	Symbol	Definition	SI unit
partial molar quantity X	X_B, (X'_B)	$X_B = (\partial X / \partial n_B)_{T, p, n_{j \neq B}}$	(varies)
chemical potential (partial molar Gibbs energy)	μ	$\mu_B = (\partial G / \partial n_B)_{T, p, n_{j \neq B}}$	$\mathrm{J\,mol^{-1}}$
absolute activity	λ	$\lambda_B = \exp(\mu_B / RT)$	1
standard chemical potential	$\mu^{\bullet}$, μ°		$\mathrm{J\,mol^{-1}}$
standard partial molar enthalpy	$H_B^{\bullet}$	$H_B^{\bullet} = \mu_B^{\bullet} + T S_B^{\bullet}$	$\mathrm{J\,mol^{-1}}$
standard partial molar entropy	$S_B^{\bullet}$	$S_B^{\bullet} = -(\partial \mu_B^{\bullet} / \partial T)_p$	$\mathrm{J\,mol^{-1}\,K^{-1}}$
standard reaction Gibbs energy (function)	$\Delta_r G^{\bullet}$	$\Delta_r G^{\bullet} = \sum_B \nu_B \mu_B^{\bullet}$	$\mathrm{J\,mol^{-1}}$
affinity of reaction	A, $(\mathscr{A})$	$A = -(\partial G / \partial \xi)_{p, T}$ $= -\sum_B \nu_B \mu_B$	$\mathrm{J\,mol^{-1}}$
standard reaction enthalpy	$\Delta_r H^{\bullet}$	$\Delta_r H^{\bullet} = \sum_B \nu_B H_B^{\bullet}$	$\mathrm{J\,mol^{-1}}$
standard reaction entropy	$\Delta_r S^{\bullet}$	$\Delta_r S^{\bullet} = \sum_B \nu_B S_B^{\bullet}$	$\mathrm{J\,mol^{-1}\,K^{-1}}$
equilibrium constant	$K^{\bullet}$, K	$K^{\bullet} = \exp(-\Delta_r G^{\bullet} / RT)$	1
equilibrium constant, pressure basis	K_p	$K_p = \prod_B p_B^{\nu_B}$	$\mathrm{Pa}^{\Sigma \nu}$
concentration basis	K_c	$K_c = \prod_B c_B^{\nu_B}$	$(\mathrm{mol\,m^{-3}})^{\Sigma \nu}$
molality basis	K_m	$K_m = \prod_B m_B^{\nu_B}$	$(\mathrm{mol\,kg^{-1}})^{\Sigma \nu}$
fugacity	f, $\tilde{p}$	$f_B = \lambda_B \lim_{p \to 0} (p_B / \lambda_B)_T$	Pa
fugacity coefficient	ϕ	$\phi_B = f_B / p_B$	1
activity and activity coefficient referenced to Raoult's law, (relative) activity	a	$a_B = \exp\left[\dfrac{\mu_B - \mu_B^{*}}{RT}\right]$	1
activity coefficient	f	$f_B = a_B / x_B$	1
activities and activity coefficients referenced to Henry's law, (relative) activity, molality basis	a_m	$a_{m, B} = \exp\left[\dfrac{\mu_B - \mu_B^{\bullet}}{RT}\right]$	1
concentration basis	a_c	$a_{c, B} = \exp\left[\dfrac{\mu_B - \mu_B^{\bullet}}{RT}\right]$	1

Name	Symbol	Definition	SI Unit
mole fraction basis	a_x	$a_{x,B} = \exp\left[\dfrac{\mu_B - \mu_B^{\bullet}}{RT}\right]$	1
activity coefficient,			
molality basis	γ_m	$a_{m,B} = \gamma_{m,B} m_B / m^{\bullet}$	1
concentration basis	γ_c	$a_{c,B} = \gamma_{c,B} c_B / c^{\bullet}$	1
mole fraction basis	γ_x	$a_{x,B} = \gamma_{x,B} x_B$	1
ionic strength,			
molality basis	I_m, I	$I_m = \frac{1}{2}\Sigma m_B z_B^2$	$mol\ kg^{-1}$
concentration basis	I_c, I	$I_c = \frac{1}{2}\Sigma c_B z_B^2$	$mol\ m^{-3}$
osmotic coefficient,			
molality basis	ϕ_m	$\phi_m = (\mu_A^* - \mu_A)/(RTM_A\Sigma m_B)$	1
mole fraction basis	ϕ_x	$\phi_x = (\mu_A - \mu_A^*)/(RT\ln x_A)$	1
osmotic pressure	Π	$\Pi = c_B RT$ (ideal dilute solution)	Pa

(i) *Symbols used as subscripts to denote a chemical process or reaction*
These symbols should be printed in roman (upright) type, without a full stop (period).

vaporization, evaporation (liquid→gas)	vap
sublimation (solid→gas)	sub
melting, fusion (solid→liquid)	fus
transition (between two phases)	trs
mixing of fluids	mix
solution (of solute in solvent)	sol
dilution (of a solution)	dil
adsorption	ads
displacement	dpl
immersion	imm
reaction in general	r
atomization	at
combustion reaction	c
formation reaction	f

(ii) *Recommended superscripts*

standard	⊖, o
pure substance	*
infinite dilution	∞
ideal	id
activated complex, transition state	‡
excess quantity	E

Name	Symbol	Definition	SI unit

CHEMICAL KINETICS

Name	Symbol	Definition	SI unit
rate of change of quantity X	$\dot{X}$	$\dot{X} = dX/dt$	(varies)
rate of conversion	$\dot{\xi}$	$\dot{\xi} = d\xi/dt$	$mol\ s^{-1}$
rate of concentration change (due to chemical reaction)	r_B, v_B	$r_B = dc_B/dt$	$mol\ m^{-3}\ s^{-1}$
rate of reaction (based on amount concentration)	v	$v = \dot{\xi}/V$ $= v_B^{-1} dc_B/dt$	$mol\ m^{-3}\ s^{-1}$
partial order of reaction	n_B	$v = k\Pi c_B^{n_B}$	1
overall order of reaction	n	$n = \Sigma n_B$	1
rate constant, rate coefficient	k	$v = k\Pi c_B^{n_B}$	$(mol^{-1}\ m^3)^{n-1} s^{-1}$
Boltzmann constant	k, k_B		$J\ K^{-1}$
half life	$t_{\frac{1}{2}}$	$c(t_{\frac{1}{2}}) = c_0/2$	s
relaxation time	τ	$\tau = 1/(k_1 + k_{-1})$	s
energy of activation, activation energy	E_a, E	$E_a = RT^2\ d \ln k/dT$	$J\ mol^{-1}$
pre-exponential factor	A	$k = A \exp(-E_a/RT)$	$(mol^{-1}\ m^3)^{n-1} s^{-1}$
volume of activation	$\Delta^{\ddagger}V$	$\Delta^{\ddagger}V = -RT \times$ $(\partial \ln k/\partial p)_T$	$m^3\ mol^{-1}$
collision diameter	d	$d_{AB} = r_A + r_B$	m
collision cross-section	σ	$\sigma_{AB} = \pi d_{AB}^2$	m^2
collision frequency	Z_A		s^{-1}
collision number	Z_{AB}, Z_{AA}		$m^{-3} s^{-1}$
collision frequency factor	z_{AB}, z_{AA}	$z_{AB} = Z_{AB}/Lc_A c_B$	$m^3\ mol^{-1} s^{-1}$
standard enthalpy of activation	$\Delta^{\ddagger}H^{\circ}, \Delta H^{\ddagger}$		$J\ mol^{-1}$
standard entropy of activation	$\Delta^{\ddagger}S^{\circ}, \Delta S^{\ddagger}$		$J\ mol^{-1}\ K^{-1}$
standard Gibbs energy of activation	$\Delta^{\ddagger}G^{\circ}, \Delta G^{\ddagger}$		$J\ mol^{-1}$
quantum yield, photochemical yield	ϕ		1

ELECTROCHEMISTRY

Name	Symbol	Definition	SI unit
elementary charge (proton charge)	e		C
Faraday constant	F	$F = eL$	$C\ mol^{-1}$

Name	Symbol	Definition	SI unit
charge number of an ion	z	$z_B = Q_B/e$	1
ionic strength	I_c, I	$I_c = \frac{1}{2}\sum c_i z_i^2$	$\mathrm{mol\ m^{-3}}$
mean ionic activity	$a_{\pm}$	$a_{\pm} = m_{\pm}\gamma_{\pm}/m^{\ominus}$	1
mean ionic molality	$m_{\pm}$	$m_{\pm}^{(\nu_+ + \nu_-)} = m_+^{\nu_+} m_-^{\nu_-}$	$\mathrm{mol\ kg^{-1}}$
mean ionic activity coefficient	$\gamma_{\pm}$	$\gamma_{\pm}^{(\nu_+ + \nu_-)} = \gamma_+^{\nu_+}\gamma_-^{\nu_-}$	1
charge number of electrochemical cell reaction	$n, (z)$		1
electric potential difference (of a galvanic cell)	$\Delta V, E, U$	$\Delta V = V_R - V_L$	V
emf, electromotive force	E	$E = \lim_{I\to 0} \Delta V$	V
standard emf, standard potential of the electrochemical cell reaction	$E^{\ominus}$	$E^{\ominus} = -\Delta_r G^{\ominus}/nF$ $= (RT/nF)\ln K^{\ominus}$	V
standard electrode potential	$E^{\ominus}$		V
emf of the cell, potential of the electrochemical cell reaction	E	$E = E^{\ominus} - (RT/nF)$ $\times \sum \nu_i \ln a_i$	V
pH	pH	$\mathrm{pH} \approx -\lg\left[\dfrac{c(H^+)}{\mathrm{mol\ dm^{-3}}}\right]$	1
inner electric potential	ϕ	$\nabla\phi = -E$	V
outer electric potential	ψ	$\psi = Q/4\pi\varepsilon_0 r$	V
surface electric potential	χ	$\chi = \phi - \psi$	V
Galvani potential difference	$\Delta\phi$	$\Delta_\alpha^\beta\phi = \phi^\beta - \phi^\alpha$	V
volta potential difference	$\Delta\psi$	$\Delta_\alpha^\beta\psi = \psi^\beta - \psi^\alpha$	V
electrochemical potential	$\tilde{\mu}$	$\tilde{\mu}_B^\alpha = (\partial G/\partial n_B^\alpha)$	$\mathrm{J\ mol^{-1}}$
electric current	I	$I = dQ/dt$	A
(electric) current density	j	$j = I/A$	$\mathrm{A\ m^{-2}}$
(surface) charge density	σ	$\sigma = Q/A$	$\mathrm{C\ m^{-2}}$
electrode reaction rate constant	k	$k_{ox} = I_a/(nFA\prod_i c_i^{\,n_i})$	(varies)
mass transfer coefficient, diffusion rate constant	k_d	$k_{d,B} = \lvert\nu_B\rvert I_{l,B}/nFcA$	$\mathrm{m\ s^{-1}}$
thickness of diffusion layer	δ	$\delta_B = D_B/k_{d,B}$	m

Name	Symbol	Definition	SI unit
transfer coefficient (electrochemical)	α	$\alpha_c = \dfrac{-|\nu|RT}{nF}\dfrac{\partial \ln|I_c|}{\partial E}$	1
overpotential	η	$\eta = E_I - E_{I=0} - IR_u$	V
electrokinetic potential (zeta potential)	ζ		V
conductivity	$\kappa, (\sigma)$	$\kappa = j/E$	$S\,m^{-1}$
conductivity cell constant	K_{cell}	$K_{cell} = \kappa R$	m^{-1}
molar conductivity (of an electrolyte)	Λ	$\Lambda_B = \kappa/c_B$	$S\,m^2\,mol^{-1}$
ionic conductivity, molar conductivity of an ion	λ	$\lambda_B = |z_B|Fu_B$	$S\,m^2\,mol^{-1}$
electric mobility	$u, (\mu)$	$u_B = v_B/E$	$m^2\,V^{-1}\,s^{-1}$
transport number	t	$t_B = j_B/\Sigma j_i$	1
reciprocal radius of ionic atmosphere	κ	$\kappa = (2F^2I/\varepsilon RT)^{\frac{1}{2}}$	m^{-1}

COLLOID AND SURFACE CHEMISTRY

Name	Symbol	Definition	SI unit
specific surface area	a, a_s, s	$a = A/m$	$m^2\,kg^{-1}$
surface amount of B, adsorbed amount of B	n_B^s, n_B^a		mol
surface excess of B	n_B^σ		mol
surface excess concentration of B	$\Gamma_B, (\Gamma_B^\sigma)$	$\Gamma_B = n_B^\sigma/A$	$mol\,m^{-2}$
total surface excess concentration	$\Gamma, (\Gamma^\sigma)$	$\Gamma = \sum_i \Gamma_i$	$mol\,m^{-2}$
area per molecule	a, σ	$a_B = A/N_B^\sigma$	m^2
area per molecule in a filled monolayer	a_m, σ_m	$a_{m,B} = A/N_{m,B}$	m^2
surface coverage	θ	$\theta = N_B^\sigma/N_{m,B}$	1
contact angle	θ		1, rad
film thickness	t, h, δ		m
thickness of (surface or interfacial) layer	τ, δ, t		m
surface tension, interfacial tension	γ, σ	$\gamma = (\partial G/\partial A_s)_{T,p}$	$N\,m^{-1}, J\,m^{-2}$
film tension	Σ_f	$\Sigma_f = 2\gamma_f$	$N\,m^{-1}$
reciprocal thickness of the double layer	κ	$\kappa = [2F^2I_c/\varepsilon RT]^{\frac{1}{2}}$	m^{-1}
average molar masses			
number-average	M_n	$M_n = \Sigma n_i M_i/\Sigma n_i$	$kg\,mol^{-1}$
mass-average	M_m	$M_m = \Sigma n_i M_i^2/\Sigma n_i M_i$	$kg\,mol^{-1}$

Name	Symbol	Definition	SI unit
Z-average	M_Z	$M_Z = \Sigma n_i M_i^3 / \Sigma n_i M_i^2$	$kg\ mol^{-1}$
sedimentation coefficient	s	$s = v/a$	s
van der Waals constant	λ		J
retarded van der Waals constant	β, B		J
van der Waals–Hamaker constant	A_H		J
surface pressure	π^s, π	$\pi^s = \gamma^0 - \gamma$	$N\ m^{-1}$

TRANSPORT PROPERTIES

Name	Symbol	Definition	SI unit
flux (of a quantity X)	$J_X, \dot{J}$	$J_X = A^{-1}\,dX/dt$	(varies)
volume flow rate	$q_V, \dot{V}$	$q_v = dV/dt$	$m^3\ s^{-1}$
mass flow rate	$q_m, \dot{m}$	$q_m = dm/dt$	$kg\ s^{-1}$
mass transfer coefficient	k_d		$m\ s^{-1}$
heat flow rate	ϕ	$\phi = dq/dt$	W
heat flux	J_q	$J_q = \phi/A$	$W\ m^{-2}$
thermal conductance	G	$G = \phi/\Delta T$	$W\ K^{-1}$
thermal resistance	R	$R = 1/G$	$K\ W^{-1}$
thermal conductivity	λ, k	$\lambda = J_q/(dT/dl)$	$W\ m^{-1}\ K^{-1}$
coefficient of heat transfer	$h, (k, K, \alpha)$	$h = J_q/\Delta T$	$W\ m^{-2}\ K^{-1}$
thermal diffusivity	a	$a = \lambda/\rho c_p$	$m^2\ s^{-1}$
diffusion coefficient	D	$D = J_n/(dc/dl)$	$m^2\ s^{-1}$

The following symbols are used in the definitions of the dimensionless quantities: mass (m), time (t), volume (V), area (A), density (ρ), speed (v), length (l), viscosity (η), pressure (p), acceleration of free fall (g), cubic expansion coefficient (α), temperature (T), surface tension (γ), speed of sound (c), mean free path (λ), frequency (f), thermal diffusivity (a), coefficient of heat transfer (h), thermal conductivity (k), specific heat capacity at constant pressure (c_p), diffusion coefficient (D), mole fraction (x), mass transfer coefficient (k_d), permeability (μ), electric conductivity (κ), and magnetic flux density (B).

Name	Symbol	Definition	SI unit
Reynolds number	Re	$Re = \rho vl/\eta$	1
Euler number	Eu	$Eu = \Delta p/\rho v^2$	1
Froude number	Fr	$Fr = v/(lg)^{\frac{1}{2}}$	1
Grashof number	Gr	$Gr = l^3 g\alpha\Delta T\rho^2/\eta^2$	1
Weber number	We	$We = \rho v^2 l/\gamma$	1
Mach number	Ma	$Ma = v/c$	1
Knudsen number	Kn	$Kn = \lambda/l$	1
Strouhal number	Sr	$Sr = lf/v$	1
Fourier number	Fo	$Fo = at/l^2$	1
Péclet number	Pe	$Pe = vl/a$	1
Rayleigh number	Ra	$Ra = l^3 g\alpha\Delta T\rho/\eta a$	1
Nusselt number	Nu	$Nu = hl/k$	1
Stanton number	St	$St = h/\rho vc_p$	1

Name	Symbol	Definition	SI units
Fourier number for mass transfer	Fo^*	$Fo^* = Dt/l^2$	1
Péclet number for mass transfer	Pe^*	$Pe^* = vl/D$	1
Grashof number for mass transfer	Gr^*	$Gr^* = l^3 g \left(\dfrac{\partial \rho}{\partial x}\right)_{T,p} \left(\dfrac{\Delta x \rho}{\eta}\right)$	1
Nusselt number for mass transfer	Nu^*	$Nu^* = k_d l/D$	1
Stanton number for mass transfer	St^*	$St^* = k_d/v$	1
Prandtl number	Pr	$Pr = \eta/\rho a$	1
Schmidt number	Sc	$Sc = \eta/\rho D$	1
Lewis number	Le	$Le = a/D$	1
magnetic Reynolds number	Rm, Re_m	$Rm = v\mu\kappa l$	1
Alfvén number	Al	$Al = v(\rho\mu)^{\frac{1}{2}}/B$	1
Hartmann number	Ha	$Ha = Bl(\kappa/\eta)^{\frac{1}{2}}$	1
Cowling number	Co	$Co = B^2/\mu\rho v^2$	1

NOMENCLATURE OF CHEMICAL COMPOUNDS

The International Union of Pure and Applied Chemistry (IUPAC) maintains several commissions that deal with the naming of chemical substances. In general, the approach of IUPAC is to present rules for arriving at names in a systematic manner, rather than recommending a unique name for each compound. Thus there are often several alternative "IUPAC names", depending on which nomenclature system is used, each of which may have advantages in specific applications. However, each of these names will be unambiguous.

Organizations such as the Chemical Abstacts Service and the Beilstein Institute that prepare indexes to the chemical literature must adopt a system for selecting unique names in order to avoid excessive cross referencing. Chemical Abstracts Service uses a system which groups together compounds derived from a single parent compound. Thus most index names are inverted (e.g., Benzene, bromo rather than bromobenzene; Acetic acid, sodium salt rather than sodium acetate). In this *Handbook* the CAS Index Names are used only in the table "Physical Constants of Organic Compounds". Other tables use more familiar names which, with a few possible exceptions, conform to one of the IUPAC naming systems.

Recommended names for the most common substituent groups, ligands, ions, and organic rings are given in the two following tables, "Nomenclature for Inorganic Ions and Ligands" and "Organic Substituent Groups and Ring Systems". For the basics of macromolecular nomenclature, see "Naming Organic Polymers" in Section 13.

Some of the most useful recent guides to chemical nomenclature, prepared by IUPAC and other organizations such as the International Union of Biochemistry and Molecular Biology (IUBMB) and the American Chemical Society are listed below . These books contain citations to the more detailed nomenclature documents in each area.

Inorganic Chemistry

International Union of Pure and Applied Chemistry, *Nomenclature of Inorganic Chemistry, Recommendations 1990,* edited by Leigh, G.J., Blackwell Scientific Publications, Oxford, 1990.

Block, B.P., Powell, W.H., and Fernelius, W.C., *Inorganic Chemical Nomenclature, Principles and Practice*, American Chemical Society, Washington, 1990.

Organic Chemistry

International Union of Pure and Applied Chemistry, *A Guide to IUPAC Nomenclature of Organic Compounds, Recommendations 1993,* edited by Panico, R., Powell, W.H., and Richer, J.-C., Blackwell Scientific Publications, Oxford, 1993.

International Union of Pure and Applied Chemistry, *Glossary of Class Names of Organic Compounds and Reactive Intermediates Based on Structure*, edited by Moss, G.P., Smith, P.A.S., and Tavernier, D., *Pure & Appl. Chem*, 67, 1307, 1995.

Rhodes, P.H., *The Organic Chemist's Desk Reference*, Chapman & Hall, London, 1995.

International Union of Pure and Applied Chemistry, *Basic Terminology of Stereochemistry*, edited by Moss, G.P., *Pure & Applied Chemistry*, 68, 2193, 1966.

Macromolecular Chemistry

International Union of Pure and Applied Chemistry, *Compendium of Macromolecular Nomenclature*, edited by, Metanomski, W.V., Blackwell Scientific Publications, Oxford, 1991.

International Union of Pure and Applied Chemistry, *Glossary of Basic Terms in Polymer Science*, edited by Jenkins, A.D., Kratochvil, P., Stepto, R.F.T., and Suter, U.W., *Pure & Appl. Chem*, 68, in press.

Biochemistry

International Union of Biochemistry and Molecular Biology, *Biochemical Nomenclature and Related Documents*, 2nd Edition, 1992, Portland Press, London, 1993; includes recommendations of the IUPAC-IUBMB Joint Commission on Biochemical Nomenclature.

International Union of Biochemistry and Molecular Biology, *Enzyme Nomenclature, 1992*, Academic Press, Orlando, FL, 1992.

IUPAC-IUBMB Joint Commission on Biochemical Nomenclature, *Nomenclature of Carbohydrates*, *Recommendations 1996*, edited by McNaught, A.D., *Pure & Appl. Chem.*, 68, 1919, 1996.

General

Chemical Abstracts Service, *Naming and Indexing Chemical Substances for Chemical Abstracts, Appendix IV, Chemical Abstracts 1994 Index Guide.*

NOMENCLATURE FOR INORGANIC IONS AND LIGANDS

See the table *Nomenclature of Chemical Compounds* for references. The assistance of Warren H. Powell in preparing this list is gratefully acknowledged.

Group	As cation	As anion	As ligand	As prefix in organic compounds
H	hydrogen	hydride	hydrido	
F	fluorine	fluoride	fluoro	fluoro
Cl	chlorine	chloride	chloro	chloro
Br	bromine	bromide	bromo	bromo
I	iodine	iodide	iodo	iodo
ClO	chlorosyl	hypochlorite	hypochlorito	chlorosyl
ClO_2	chloryl	chlorite	chlorito	chloryl
ClO_3	perchloryl	chlorate	chlorato	perchloryl
ClO_4		perchlorate		
IO	iodosyl	hypoiodite		iodoso
IO_2	iodyl			iodyl; iodoxy
O		oxide	oxo	oxo
O_2		peroxide (O_2^{2-})	peroxo	peroxy
		hyperoxide (O_2^-)		
HO		hydroxide	hydroxo	hydroxy
HO_2		hydrogen peroxide	hydrogen peroxo	hydroperoxy
S		sulfide	thio; sulfido	thio; thioxo
HS		hydrogen sulfide	thiolo	mercapto
S_2		disulfide	disulfido	
SO	sulfinyl; thionyl			sulfinyl
SO_2	sulfonyl; sulfuryl	sulfoxylate		sulfonyl
SO_3		sulfite	sulfito	
HSO_3		hydrogen sulfite	hydrogen sulfito	
S_2O_3		thiosulfate	thiosulfato	
SO_4		sulfate	sulfato	
Se		selenide	seleno	seleno; selenoxo
SeO	seleninyl			seleninyl
SeO_2	selenonyl			selenonyl
SeO_3		selenite	selenito	
SeO_4		selenate	selenato	
Te		telluride	telluro	telluro; telluroxo
CrO_2	chromyl			
UO_2	uranyl			
NpO_2	neptunyl			
PuO_2	plutoryl			
AmO_2	americyl			
N		nitride	nitrido	
N_3		azide	azido	
NH		imide	imido	imino
NH_2		amide	amido	amino
NHOH		hydroxylamide	hydroxylamido	hydroxyamino
N_2H_3		hydrazide	hydrazido	hydrazino; diazanyl
NO	nitrosyl		nitrosyl	nitroso
NO_2	nitryl		nitro	nitro
ONO		nitrite	nitrito	
NS	thionitrosyl			
NO_3		nitrate	nitrato	
N_2O_3		hyponitrite	hyponitrito	
P		phosphide	phosphido	phosphinidyne
PO	phosphoryl			phosphoroso; phosphinylidyne
PO_2	phospho			
PS	thiophosphoryl			phosphinothioylidyne; thiophosphorozo
PH_2O_3		hypophosphite	hypophosphito	
PHO_3		phosphite	phosphito	

Group	As cation	As anion	As ligand	As prefix in organic compounds
PO_4		phosphate	phosphato	
AsO_4		arsenate	arsenato	
VO	vanadyl			
CO	carbonyl		carbonyl	carbonyl
CS	thiocarbonyl			thiocarbonyl
CH_3O		methanolate	methoxo	methoxy
C_2H_5O		ethanolate	ethoxo	ethoxy
CH_3S		methanethiolate	methanethiolato	methylthio
C_2H_5S		ethanethiolate	ethanethiolato	ethylthio
CN	cyanogen	cyanide	cyano	cyano
OCN		cyanate	cyanato	cyanato
SCN		thiocyanate	thiocyanato	thiocyanato
$SeCN$		selenocyanate	selenocyanato	selenocyanato
$TeCN$		tellurocyanate	tellurocyanato	tellurocyanato
CO_3		carbonate	carbonato	
HCO_3		hydrogen carbonate	hydrogen carbonato	carboxycarbonyl
C_2O_4		oxalate	oxalato	

ORGANIC SUBSTITUENT GROUPS AND RING SYSTEMS

The first part of this table lists substituent groups and their line formulas. A substituent group is defined by IUPAC as a group that replaces one or more hydrogen atoms attached to a parent structure. Such groups are sometimes called radicals, but IUPAC now reserves the term radical for a free molecular species with unpaired electrons. IUPAC does not recommend some of these names, which are marked here with asterisks (e.g., amyl*), but they are included in this list because they are often encountered in the older literature. Substituent group names which are formed by systematic rules (e.g., methyl from methane, ethyl from ethane, etc.) are included here only for the first few members of a homologous series.

In the second part of the table a number of common organic ring compounds are shown, with the conventional numbering of the ring positions indicated.

The help of Warren H. Powell in preparing this table is greatly appreciated. Pertinent references may be found in the table *Nomenclature of Chemical Compounds*.

SUBSTITUENT GROUPS

Name	Formula	Name	Formula
acetamido (acetylamino)	CH_3CONH-	cyanamido (cyanoamino)	$NCNH-$
acetoacetyl	CH_3COCH_2CO-	cyanato	$NCO-$
acetonyl	CH_3COCH_2-	cyano	$NC-$
acetyl	CH_3CO-	decanedioyl	$-OC(CH_2)_8CO-$
acryloyl* (1-oxo-2-propenyl)	$CH_2=CHCO-$	decanoyl	$CH_3(CH_2)_8CO-$
alanyl (from alanine)	$CH_3CH(NH_2)CO-$	diazo	$N_2=$
β-alanyl	$H_2N(CH_2)_2CO-$	diazoamino	$-NHN=N-$
allyl (2-propenyl)	$CH_2=CHCH_2-$	disilanyl	H_3SiSiH_2-
allylidene (2-propenylidene)	$CH_2=CHCH=$	disiloxanyloxy	$H_3SiOSiH_2O-$
amidino (aminoiminomethyl)	$H_2NC(=NH)-$	disulfinyl	$-S(O)S(O)-$
amino	H_2N-	dithio	$-SS-$
amyl* (pentyl)	$CH_3(CH_2)_4-$	enanthoyl* (heptanoyl)	$CH_3(CH_2)_5CO-$
anilino (phenylamino)	C_6H_5NH-	epoxy	$-O-$
anisidino	$CH_3OC_6H_4NH-$	ethenyl (vinyl)	$CH_2=CH-$
anthranoyl (2-aminobenzoyl)	$2-H_2NC_6H_4CO-$	ethynyl	$HC≡C-$
arsino	AsH_2-	ethoxy	C_2H_5O-
azelaoyl (from azelaic acid)	$-OC(CH_2)_7CO-$	ethyl	CH_3CH_2-
azido	N_3-	ethylene	$-CH_2CH_2-$
azino	$=N-N=$	ethylidene	$CH_3CH=$
azo	$-N=N-$	ethylthio	C_2H_5S-
azoxy	$-N(O)=N-$	formamido (formylamino)	$HCONH-$
benzal* (benzylidene)	$C_6H_5CH=$	formyl	$HCO-$
benzamido (benzoylamino)	C_6H_5CONH-	furmaroyl (from fumaric acid)	$-OCCH=CHCO-$
benzhydryl (diphenylmethyl)	$(C_6H_5)_2CH-$	furfuryl (2-furanylmethyl)	$OC_4H_3CH_2-$
benzoxy* (benzoyloxy)	C_6H_5COO-	furfurylidene (2-furanylmethylene)	$OC_4H_3CH=$
benzoyl	C_6H_5CO-	glutamoyl (from glutamic acid)	$-OC(CH_2)_2CH(NH_2)CO-$
benzyl	$C_6H_5CH_2-$	glutaryl (from glutaric acid)	$-OC(CH_2)_3CO-$
benzylidene	$C_6H_5CH=$	glycylamino	H_2NCH_2CONH-
benzylidyne	$C_6H_5C=$	glycoloyl; glycolyl (hydroxyacetyl)	$HOCH_2CO-$
biphenylyl	$C_6H_5C_6H_5-$	glycyl (aminoacetyl)	H_2NCH_2CO-
biphenylene	$-C_6H_4-C_6H_4-$	glyoxyloyl; glyoxylyl (oxoacetyl)	$HCOCO-$
butoxy	C_4H_9O-	guanidino	$H_2NC(=NH)NH-$
sec-butoxy (1-methylpropoxy)	$C_2H_5CH(CH_3)O-$	guanyl (aminoiminomethyl)	$H_2NC(=NH)-$
tert-butoxy (1,1-dimethylethoxy)	$(CH_3)_3CO-$	heptadecanoyl	$CH_3(CH_2)_{15}CO-$
butyl	$CH_3(CH_2)_3-$	heptanamido	$CH_3(CH_2)_5CONH-$
sec-butyl (1-methylpropyl)	$CH_3CH_2CH(CH_3)-$	heptanedioyl	$-OC(CH_2)_5CO-$
tert-butyl (1,1-dimethylethyl)	$(CH_3)_3C-$	heptanoyl	$CH_3(CH_2)_5CO-$
butyryl (1-oxobutyl)	$CH_3(CH_2)_2CO-$	hexadecanoyl	$CH_3(CH_2)_{14}CO-$
caproyl* (hexanoyl)	$CH_3(CH_2)_4CO-$	hexamethylene (1,6-hexanediyl)	$-(CH_2)_6-$
capryl* (decanoyl)	$CH_3(CH_2)_8CO-$	hexanedioyl	$-OC(CH_2)_4CO-$
caprytoyl* (octanoyl)	$CH_3(CH_2)_6CO-$	hippuryl (N-benzoylglycyl)	$C_6H_5CONHCH_2CO-$
carbamido (carbamoylamino)	$H_2NCONH-$	hydrazino	H_2NNH-
carbamoyl (aminocarbonyl)	H_2NCO-	hydrazo	$-HNNH-$
carbamyl (aminocarbonyl)	H_2NCO-	hydrocinnamoyl	$C_6H_5(CH_2)_2CO-$
carbazoyl (hydrazinocarbonyl)	$H_2NNHCO-$	hydroperoxy	$HOO-$
carbethoxy (ethoxycarbonyl)	C_2H_5OCO-	hydroxyamino	$HONH-$
carbonyl	$=C=O$	hydroxy	$HO-$
carboxy	$HOOC-$	imino	$HN=$
cetyl* (hexadecyl)	$CH_3(CH_2)_{15}-$	iodoso* (iodosyl)	$OI-$
chloroformyl (chlorcarbonyl)	$ClCO-$	iodyl	O_2I-
cinnamoyl	$C_6H_5CH=CHCO-$	isoamyl* (isopentyl; 3-methylbutyl)	$(CH_3)_2CH(CH_2)_2-$
cinnamyl (3-phenyl-2-propenyl)	$C_6H_5CH=CHCH_2-$	isobutenyl (2-methyl-1-propenyl)	$(CH_3)_2C=CH-$
cinnamylidene	$C_6H_5CH=CHCH=$	isobutoxy (2-methylpropoxy)	$(CH_3)_2CHCH_2O-$
cresyl* (hydroxymethylphenyl)	$HO(CH_3)C_6H_4-$	isobutyl (2-methylpropyl)	$(CH_3)_2CHCH_2-$
crotonoyl	$CH_3CH=CHCO-$	isobutylidene (3-methylpropylidene)	$(CH_3)_2CHCH=$
crotyl (2-butenyl)	$CH_3CH=CHCH_2-$	isobutyryl (2-methyl-1-oxopropyl)	$(CH_3)_2CHCO-$

isocyanato	OCN-
isocyano	CN-
isohexyl (4-methylpentyl)	$(CH_3)_2CH(CH_2)_3-$
isoleucyl (from isoleucine)	$C_2H_5CH(CH_3)CH(NH_2)CO-$
isonitroso* (hydroxyamino)	HON=
isopentyl (3-methylbutyl)	$(CH_3)_2CH(CH_2)_2-$
isopentylidene (3-methylbutylidene)	$(CH_3)_2CHCH_2CH=$
isopropenyl (1-methylethenyl)	$CH_2=C(CH_3)-$
isopropoxy (1-methylethoxy)	$(CH_3)_2CHO-$
isopropyl (1-methylethyl)	$(CH_3)_2CH-$
isopropylidene (1-methylethylidene)	$(CH_3)_2C=$
isothiocyanato (isothiocyano)	SCN-
isovaleryl* (3-methyl-1-oxobutyl)	$(CH_3)_2CHCH_2CO-$
lactoyl (from lactic acid)	$CH_3CH(OH)CO-$
lauroyl (from lauric acid)	$CH_3(CH_2)_{10}CO-$
lauryl (dodecyl)	$CH_3(CH_2)_{11}-$
leucyl (from leucine)	$(CH_3)_2CHCH_2CH(NH_2)CO-$
levulinoyl (from levulinic acid)	$CH_3CO(CH_2)_2CO-$
malonyl (from malonic acid)	$-OCCH_2CO-$
mandeloyl (from mandelic acid)	$C_6H_5CH(OH)CO-$
mercapto	HS-
mesityl	$2,4,6-(CH_3)_3C_6H_2-$
methacryloyl (from methacrylic acid)	$CH_2=C(CH_3)CO-$
methallyl (2-methyl-2-propenyl)	$CH_2=C(CH_3)CH_2-$
methionyl (from methionine)	$CH_3SCH_2CH_2CH(NH_2)CO-$
methoxy	CH_3O-
methyl	H_3C-
methylene	$H_2C=$
methylthio	CH_3S-
myristoyl (from myristic acid)	$CH_3(CH_2)_{12}CO-$
myristyl (tetradecyl)	$CH_3(CH_2)_{13}-$
naphthyl	$(C_{10}H_7)-$
naphthylene	$-(C_{10}H_6)-$
neopentyl (2,2-dimethylpropyl)	$(CH_3)_3CCH_2-$
nitramino (nitroamino)	O_2NNH-
nitro	O_2N-
nitrosamino (nitrosoamino)	ONNH-
nitrosimino (nitrosoimino)	ONN=
nitroso	ON-
nonanoyl (from nonanoic acid)	$CH_3(CH_2)_7CO-$
oleoyl (from oleic acid)	$CH_3(CH_2)_7CH=CH(CH_2)_7CO-$
oxalyl (from oxalic acid)	-OCCO-
oxo	O=
palmitoyl (from palmitic acid)	$CH_3(CH_2)_{14}CO-$
pentamethylene (1,5-pentanediyl)	$-(CH_2)_5-$
pentyl	$CH_3(CH_2)_4-$
tert-pentyl	$CH_3CH_2C(CH_3)_2-$
phenacyl	$C_6H_5COCH_2-$
phenacylidene	$C_6H_5COCH=$
phenethyl (2-phenylethyl)	$C_6H_5CH_2CH_2-$
phenoxy	C_6H_5O-
phenyl	C_6H_5-
phenylene (benzenediyl)	$-C_6H_4-$
phosphino* (phosphanyl)	H_2P-
phosphinyl* (phosphinoyl)	$H_2P(O)-$
phospho	O_2P-
phosphono	$(HO)_2P(O)-$
phthaloyl (from phthalic acid)	$1,2-C_6H_4(CO-)_2$
picryl (2,4,6-trinitrophenyl)	$2,4,6-(NO_2)_3C_6H_2-$
pimeloyl (from pimelic acid)	$-OC(CH_2)_5CO-$
piperidino (1-piperidinyl)	$C_5H_{10}N-$
pivaloyl (from pivalic acid)	$(CH_3)_3CCO-$
prenyl (3-methyl-2-butenyl)	$(CH_3)_2C=CHCH_2-$
propargyl (2-propynyl)	$HC'CCH_2-$
1-propenyl	$-CH=CHCH_3$
2-propenyl (allyl)	$CH_2=CHCH_2-$
propionyl* (propanyl)	CH_3CH_2CO-
propoxy	$CH_3CH_2CH_2O-$
propyl	$CH_3CH_2CH_2-$
propylidene	$CH_3CH_2CH=$
pyrryl (pyrrolyl)	C_3H_4N-
salicyloyl (2-hydroxybenzoyl)	$2-HOC_6H_4CO-$
selenyl* (selanyl; hydroseleno)	HSe-
seryl (from serine)	$HOCH_2CH(NH_2)CO-$
siloxy	H_3SiO-
silyl	H_3Si-
silylene	$H_2Si=$
sorboyl (from sorbic acid)	$CH_3CH=CHCH=CHCO-$
stearoyl (from stearic acid)	$CH_3(CH_2)_{14}CO-$
stearyl (octadecyl)	$CH_3(CH_2)_{17}-$
styryl (2-phenylethenyl)	$C_6H_5CH=CH-$
suberoyl (from suberic acid)	$-OC(CH_2)_6CO-$
succinyl (from succinic acid)	$-OCCH_2CH_2CO-$
sulfamino (sulfoamino)	$HOSO_2NH-$
sulfamoyl (sulfamyl)	H_2NSO_2-
sulfanilyl [(4-aminophenyl)sulfonyl]	$4-H_2NC_6H_4SO_2-$
sulfeno	HOS-
sulfhydryl (mercapto)	HS-
sulfinyl	OS=
sulfo	HO_3S-
sulfonyl (sulfuryl)	$-SO_2-$
terephthaloyl	$1,4-C_6H_4(CO-)_2$
tetramethylene	$-(CH_2)_4-$
thienyl (from thiophene)	$(C_4H_3S)-$
thiocarbonyl (carbothionyl)	=CS
thiocarboxy	HOSC-
thiocyanato (thiocyano)	NCS-
thionyl* (sulfinyl)	-SO-
threonyl (from threonine)	$CH_3CH(OH)CH(NH_2)CO-$
toluidino [(methylphenyl)amino]	$CH_3C_6H_4NH-$
toluoyl (methylbenzoyl)	$CH_3C_6H_4CO-$
tolyl (methylphenyl)	$CH_3C_6H_4-$
α-tolyl (benzyl)	$C_6H_5CH_2-$
tolylene (methylphenylene)	$-(CH_3C_6H_3)-$
tosyl [(4-methylphenyl) sulfonyl)]	$4-CH_3C_6H_4SO_2-$
triazano	$H_2NNHNH-$
trimethylene (1,3-propanediyl)	$-(CH_2)_3-$
trityl (triphenylmethyl)	$(C_6H_5)_3C-$
valeryl* (pentanoyl)	$CH_3(CH_2)_3CO-$
valyl (from valine)	$(CH_3)_2CHCH(NH_2)CO-$
vinyl (ethenyl)	$CH_2=CH-$
vinylidene (ethenylidene)	$CH_2=C=$
xylidino [(dimethylphenyl)amino]	$(CH_3)_2C_6H_3NH-$
xylyl (dimethylphenyl)	$(CH_3)_2C_6H_3-$
xylylene [phenelenebis(methylene)]	$-CH_2C_6H_4CH_2-$

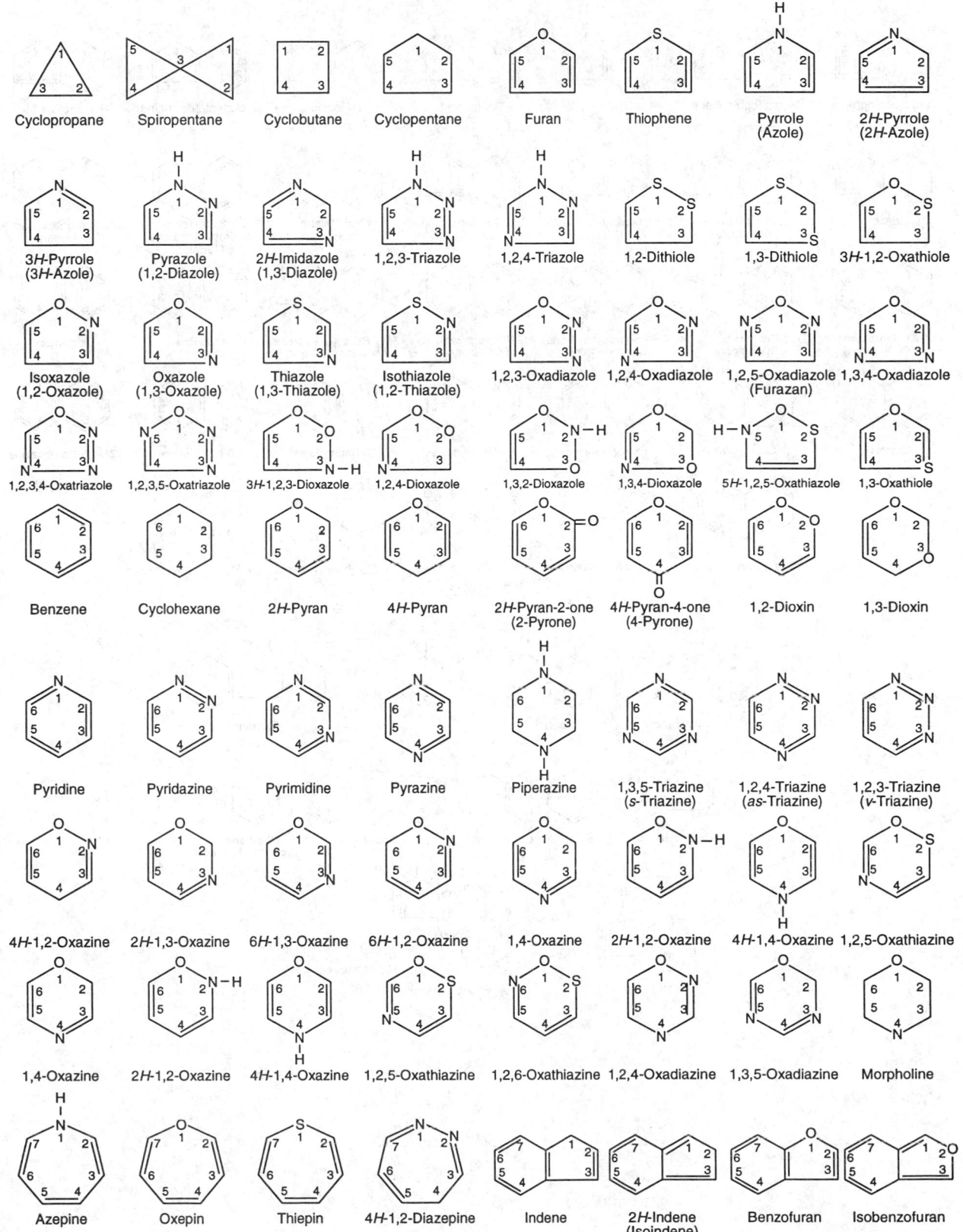

Cyclopropane   Spiropentane   Cyclobutane   Cyclopentane   Furan   Thiophene   Pyrrole (Azole)   2H-Pyrrole (2H-Azole)

3H-Pyrrole (3H-Azole)   Pyrazole (1,2-Diazole)   2H-Imidazole (1,3-Diazole)   1,2,3-Triazole   1,2,4-Triazole   1,2-Dithiole   1,3-Dithiole   3H-1,2-Oxathiole

Isoxazole (1,2-Oxazole)   Oxazole (1,3-Oxazole)   Thiazole (1,3-Thiazole)   Isothiazole (1,2-Thiazole)   1,2,3-Oxadiazole   1,2,4-Oxadiazole   1,2,5-Oxadiazole (Furazan)   1,3,4-Oxadiazole

1,2,3,4-Oxatriazole   1,2,3,5-Oxatriazole   3H-1,2,3-Dioxazole   1,2,4-Dioxazole   1,3,2-Dioxazole   1,3,4-Dioxazole   5H-1,2,5-Oxathiazole   1,3-Oxathiole

Benzene   Cyclohexane   2H-Pyran   4H-Pyran   2H-Pyran-2-one (2-Pyrone)   4H-Pyran-4-one (4-Pyrone)   1,2-Dioxin   1,3-Dioxin

Pyridine   Pyridazine   Pyrimidine   Pyrazine   Piperazine   1,3,5-Triazine (s-Triazine)   1,2,4-Triazine (as-Triazine)   1,2,3-Triazine (v-Triazine)

4H-1,2-Oxazine   2H-1,3-Oxazine   6H-1,3-Oxazine   6H-1,2-Oxazine   1,4-Oxazine   2H-1,2-Oxazine   4H-1,4-Oxazine   1,2,5-Oxathiazine

1,4-Oxazine   2H-1,2-Oxazine   4H-1,4-Oxazine   1,2,5-Oxathiazine   1,2,6-Oxathiazine   1,2,4-Oxadiazine   1,3,5-Oxadiazine   Morpholine

Azepine   Oxepin   Thiepin   4H-1,2-Diazepine   Indene   2H-Indene (Isoindene)   Benzofuran   Isobenzofuran

Benzo[b]thiophene   Benzo[c]thiophene   Indole   3*H*-Indole   1*H*-Indole   Cyclopenta[*b*]pyridine   Pyrano[3,4-b]-pyrrole

Indazole   Benzisoxazole (Indoxazine)   Benzoxazole   2,1-Benzisoxazole   Naphthalene   1,2,3,4-Tetra-hydronaphthalene (Tetralin)   Octahydronaphthalene (Decalin)

2*H*-1-Benzopyran (2*H*-Chromene)   2*H*-1-Benzopyran-2-one (Coumarin)   4*H*-1-Benzopyran-4-one (Chromen-4-one)   1*H*-2-Benzopyran-1-one (Isocoumarin)   3*H*-2-Benzopyran-1-one (Isochromen-3-one)   Quinoline   Isoquinoline

Cinnoline   Quinazoline   1,8-Naphtyridine   1,7-Naphtyridine   1,5-Naphtyridine   1,6-Naphtyridine   2*H*-1,3-Benzoxazine

2*H*-1,4-Benzoxazine   1*H*-2,3-Benzoxazine   4*H*-3,1-Benzoxazine   2*H*-1,2-Benzoxazine   4*H*-1,4-Benzoxazine

Anthracene   Phenanthrene   Phenalene   Fluorene

Carbazole   Xanthene   Acridine   Norpinane (Bicyclo[3.1.1]heptane)

7*H*-Purine   Steroid ring system

R = Nearly always methyl
R´ = Usually methyl
R´´ = Various groups

SCIENTIFIC ABBREVIATIONS AND SYMBOLS

This table lists some symbols, abbreviations, and acronyms commonly encountered in the physical sciences. Entries in italic type are symbols for physical quantities; for more details see the table "Symbols and Terminology for Physical and Chemical Quantities" in this section. Additional information on units may be found in the table "International System of Units" in Section 1. Abbreviations are listed in alphabetical order without regard to case. Entries beginning with Greek letters fall at the end of the table.

Publication practices vary with regard to the use of capital letters or lower case for many abbreviations. An effort has been made to follow the most common practices in this table, but there are many exceptions. Likewise, policies on the use of periods in an abbreviation vary considerably. Periods are generally omitted in this table unless they are necessary for clarity. Periods should never appear in SI units.

a	absorption coefficient, acceleration, activity		C	coulomb
a_0	Bohr radius		°C	degree Celsius
A	ampere		C	capacitance, heat capacity, number concentration
Å	ångstrom		cal	calorie
A	absorbance, area, Helmholtz energy, mass number		calc	calculated
			cc	cubic centimeter
AAS	atomic absorption spectroscopy		CCD	charge-coupled device
abs	absolute		cd	candela
ac	alternating current		CD	circular dichroism
ADP	adenosine diphosphate		cfm	cubic feet per minute
ae	eon (10^9 years)		cgs	centimeter-gram-second system
AES	atomic emission spectroscopy, Auger electron spectroscopy		Ci	curie
			CI	configuration interaction
AF	audio frequency		cir	circular
A_H	Hall coefficient		cm	centimeter
AI	artificial intelligence		c.m.	center of mass
Al	Alfen number		CDNO	complete neglect of differential overlap
AM	amplitude modulation		Co	Cowling number
AMP	adenosine monophosphate		conc	concentrated, concentration
amu	atomic mass unit (recommended symbol is u)		const	constant
antilog	antilogarithm		cos	cosine
AO	atomic orbital		cosh	hyperbolic cosine
APW	augmented plane wave		cot	cotangent
aq	aqueous		coth	hyperbolic cotangent
A_r	atomic weight (relative atomic mass)		cp	candle power
ASCII	American National Standard Code for Information Interchange		cP	centipoise
			CP	chemically pure
atm	standard atmosphere		cpd	contact potential difference
ATP	adenosine triphosphate		cps	cycles per second
ATR	attenuated total internal reflection		CPU	central processing unit
at.wt.	atomic weight		csc	cosecant
AU	astronomical unit		ct	carat
av	average		cu	cubic
avdp	avoirdupois		cw	continuous wave
b	barn		cwt	hundredweight (112 pounds)
B	magnetic flux density, second virial coefficient, susceptance		cyl	cylinder
			d	day, deuteron
bar	bar		D	debye unit
bbl	barrel		D	diffusion coefficient, dissociation energy, electric displacement
bcc	body centered cubic			
BCS	Bardeen-Cooper-Schrieffer		Da	dalton
Bé	Baumé		dB	decibel
BeV	billion electronvolt		dc	direct current
Bhn	Brinell hardness number		deg	degree
Bi	biot		den	density
BOD	biochemical oxygen demand		dev	deviation
bp	boiling point		diam	diameter
Bq	becquerel		dil	dilute
Btu	British thermal unit		dm	decimeter
bu	bushel		DNA	deoxyribonucleic acid
c	amount concentration, specific heat, velocity		doz	dozen
c_0	speed of light in vacuum		dpm	disintegrations per minute

dps	disintegrations per second
dr	dram
DSC	differential scanning calorimetry
DTA	differential thermal analysis
dyn	dyne
e	electron, base of natural logarithms
e	elementary charge, linear strain
E	electric field strength, electromotive force, energy, modulus of elasticity
E_h	Hartree energy
emf	electromotive force
emu	electromagnetic unit system
ENDOR	electron-nuclear double resonance
EPR	electron paramagnetic resonance
eq,eqn	equation
eqQ	quadrupole coupling constant
erg	erg
ESCA	electron spectroscopy for chemical analysis
e.s.d.	estimated standard deviation
ESR	electron spin resonance
esu	electrostatic unit system
ET	ephemeris time
e.u.	entropy unit
Eu	Euler number
eV	electronvolt
exp	exponential function
expt	experimental
ext	external
f	activity coefficient, aperture ratio, focal length, force constant, frequency, fugacity
F	farad
°F	degree Fahrenheit
F	Faraday constant, force, angular momentum
fcc	face centered cubic
FET	field effect transistor
FIR	far infrared
fl	fluid
FM	frequency modulation
Fo	Fourier number
fp	freezing point
fpm	feet per minute
fps	feet per second, foot-pound-second system
Fr	franklin
Fr	Froude number
ft	foot
FT	Fourier transform
ft-lb	foot pound
FTIR	Fourier transform infrared spectroscopy
g	gram
g	acceleration due to gravity, degeneracy, Landé g-factor
G	gauss
G	electrical conductance, Gibbs energy, gravitational constant, sheer modulus
gal	gallon
Gal	gal, galileo
GC	gas chromatography
GeV	gigaelectronvolt
GLC	gas-liquid chromatography
GMT	Greenwich mean time

gpm	gallons per minute
gps	gallons per second
gr	grain
Gr	Grashof number
GTO	gaussian type atomic orbital
GUT	grand unified theory
Gy	gray, gigayear
h	helion, hour
h	Planck constant
H	henry
H	enthalpy, hamiltonian function, magnetic field
H_0	Hubble constant
ha	hectare
Ha	Hartmann number
hav	haversine
hcp	hexagonal closed packed
HF	high frequency
hfs	hyperfine structure
HMO	Hückel molecular orbital
HOMO	highest occupied molecular orbital
hp	horsepower
HPLC	high-performance liquid chromatography
hr	hour
Hz	hertz
i	square root of minus one
I	electric current, ionic strength, moment of inertia, nuclear spin angular momentum, radiant intensity
IAT	international atomic time
IC	integrated circuit
ICR	ion cyclotron resonance
ID	inside diameter
IF	intermediate frequency
Im	imaginary part
IMPATT	impact ionization avalanche transit time
in.	inch
int	internal
I/O	input/output
IPTS	International Practical Temperature Scale
IR	infrared
ITS	International Temperature Scale (1990)
j	angular momentum, electric current density
J	joule
J	angular momentum, electric current density, flux, Massieu function
k	absorption index, Boltzmann constant, rate constant, thermal conductivity, wave vector
K	kelvin
K	absorption coefficient, bulk modulus, equilibrium constant, kinetic energy
kb	kilobar, kilobases (DNA or RNA)
kcal	kilocalorie
keV	kiloelectronvolt
kg	kilogram
kgf	kilogram force
kJ	kilojoule
km	kilometer
Kn	Knudsen number
kPa	kilopascal
kt	karat

kV	kilovolt		MSL	mean sea level
kva	kilovolt ampere		mV	millivolt
kW	kilowatt		mW	milliwatt
kwh	kilowatt hour		MW	megawatt, microwave, molecular weight
l	liter		Mx	maxwell
l	angular momentum, length		n	neutron
L	liter, lambert		n	amount of substance, number density, principal quantum number, refractive index
L	Avogadro constant, inductance, Lagrange function		N	newton
lat.	latitude		N	angular momentum, neutron number
lb	pound		N_A	Avogadro constant
lbf	pound force		NAD	nicotinamide adenine dinucleotide
LCAO	linear combination of atomic orbitals		NADP	nicotinamide adenine dinucleotide phosphate
Le	Lewis function		nbp	normal boiling point
LED	light emitter diode		N_E	density of states
LEED	low-energy electron diffraction		ng	nanogram
liq	liquid		NIR	near infrared
lm	lumen		nm	nanometer
ln	logarithm (natural)		NMR	nuclear magnetic resonance
log	logarithm (common)		NQR	nuclear quadrupole resonance
long.	longitude		ns	nanosecond
LST	local sidereal time		NTP	normal temperature and pressure
LT	local time		Nu	Nusselt number
lx	lux		obs	observed
ly	langley		OD	outside diameter
l.y.	light year		Oe	oersted
m	meter, molal (as in 0.1 m solution)		ORD	optical rotatory dispersion
m	magnetic dipole moment, mass, molality, angular momentum component		oz	ounce
			p	proton
M	molar (as in 0.1 M solution)		p	dielectric polarization, electric dipole moment, momentum, pressure
M	magnetization, molar mass, mutual inductance, torque, angular momentum component		P	poise
Ma	Mach number		P	power, pressure, probability, sound energy flux
max	maximum		Pa	pascal
MESFET	metal-semiconductor field-effect transistor		pc	parsec
MeV	megaelectronvolt		PD	potential difference
MF	molecular formula		pdl	poundal
mg	milligram		Pe	Péclet number
MHD	magnetohydrodynamics		pe	probable error
mi	mile		PES	photoelectron spectroscopy
min	minimum, minute		pf	power factor
MIR	mid-infrared		pg	picogram
MKS	meter-kilogram-second system		pH	negative log of hydrogen ion concentration
MKSA	meter-kilogram-second-ampere system		pI	isoelectric point
ml,mL	milliliter		pK	negative log of ionization constant
mm	millimeter		pm	picometer
mmf	magnetomotive force		ppb	parts per billion
mmHg	millimeter of mercury		ppm	parts per million
MO	molecular orbital		ppt	parts per thousand
mol	mole		Pr	Prandtl number
mol.wt.	molecular weight		ps	picosecond
MOS	metal-oxide semiconductor		PS	photoelectron spectroscopy
MOSFET	metal-oxide semiconductor field-effect transistor		psi	pounds per square inch
mp	melting point		psia	pounds per square inch absolute
MPa	megapascal		psig	pounds per square inch gage
Mpc	megaparsec		pt	pint
M_r	molecular weight (relative molar mass)		PVT	pressure-volume-temperature
mRNA	messenger RNA		q	electric field gradient, flow rate, heat, wave vector (phonons)
ms	millisecond			
MS	mass spectroscopy			

Q	electric charge, heat, partition function, quadrupole moment, radiant energy, vibrational normal coordinate	Sv	sievert
		t	metric tonne, triton
		t	Celsius temperature, thickness, time, transport number
QCD	quantum chromodynamics		
QED	quantum electrodynamics	T	tesla
Q.E.D.	quod erat demonstrandum (which was to be proved)	T	kinetic energy, period, term value, temperature (thermodynamic), torque, transmittance
QSO	quasi-stellar object	tan	tangent
qt	quart	tanh	hyperbolic tangent
quad	quadrillion Btu (= $1.055 \cdot 10^{18}$ J)	TE	transverse electric
q.v.	quod vide (which you should see)	TED	transferred electron device
r	position vector, radius	TEM	transverse electromagnetic
°R	degree Rankine	temp	temperature
R	electrical resistance, gas constant, molar refraction, Rydberg constant	TFD	Thomas-Fermi-Dirac method
		TL	thermoluminescence
RA	right ascension	TLC	thin-layer chromatography
rad	radian	TM	transverse magnetic
RAM	random access memory	Torr	torr
Re	real part	tRNA	transfer RNA
rem	roentgen-equivalent-man	u	unified atomic mass unit
RF	radiofrequency	u	Bloch function, electric mobility, velocity
rms	root-mean-square	U	electric potential difference, internal energy
RNA	ribonucleic acid	UHF	ultrahigh frequency
ROM	read only memory	UPS	ultraviolet photoelectron spectroscopy
rpm	revolutions per minute	UT	universal time
rps	revolutions per second	UV	ultraviolet
RRS	resonance Raman spectroscopy	v	reaction rate, specific volume, velocity, vibrational quantum number
RS	Raman spectroscopy		
Ry	rydberg	V	volt
s	second	V	electric potential, potential energy, volume
s	path length, solubility, spin angular momentum, symmetry number	VB	valence band, valence bond
		VHF	very high frequency
S	siemens	VIS	visible region of the spectrum
S	area, entropy, probability current density, Poynting vector, symmetry coordinate, spin angular momentum	VSLI	very large scale integrated (circuit)
		VUV	vacuum ultraviolet
		v/v	volume for volume
Sc	Schmidt number	w	energy density, mass fraction, velocity, work
SCF	self-consistent field	W	watt
SCR	silicon-controlled rectifier	W	radiant energy, statistical weight, work
sd	standard deviation	Wb	weber
sec	secant, second	We	Weber number
SEM	scanning electron microscope	WKB	Wentzel-Kramers-Brillouin method
SERS	surface-enhanced Raman spectroscopy	wt	weight
Sh	Sherwood number	w/v	weight for volume
SI	International System of Units	w/w	weight for weight
sin	sine	x	mole fraction
sinh	hyperbolic sine	X	X unit
SMOW	Standard Mean Ocean Water	X	reactance
SNU	solar neutrino unit	XPS	x-ray photoelectron spectroscopy
sp gr	specific gravity	XRF	x-ray fluorescence
sq	square	XRS	x-ray spectroscopy
sr	steradian	Y	admittance, Planck function, Young's modulus
Sr	Strouhal number		
St	stoke	yd	yard
St	Stanton number	yr	year
std	standard	z	charge number (of an ion), collision frequency factor
STEM	scanning transmission electron microscope		
STO	Slater type orbital	Z	atomic number, compression factor, collision number, impedance, partition function
STP	standard temperature and pressure		

ZULU	Greenwich mean time		μF	microfarad
α	alpha particle		μg	microgram
α	absorption coefficient, degree of dissociation, electric polarizability, expansion coefficient, fine structure constant		μm	micrometer
			μs	microsecond
β	beta particle		ν	frequency, kinematic velocity, stoichiometric number
γ	photon		$\tilde{\nu}$	wavenumber
γ	activity coefficient, conductivity, magnetogyric ratio, mass concentration, ratio of heat capacities, surface tension		ν_e	neutrino
			π	pion
Γ	Gruneisin parameter, level width, surface concentration		Π	osmotic pressure, Peltier coefficient
δ	chemical shift, Dirac delta function, Kronecker delta, loss angle		ρ	density, reflectance, resistivity
			σ	electrical conductivity, cross section, normal stress, shielding constant (NMR), Stefan-Boltzmann constant, surface tension
Δ	inertia defect, mass defect		τ	transmittance, chemical shift, shear stress, relaxation time
ε	emittance, Levi-Civita symbol, linear strain, molar absorption coefficient, permittivity			
η	overpotential, viscosity		ϕ	electrical potential, fugacity coefficient, osmotic coefficient, quantum yield, wavefunction
κ	compressibility, conductivity, magnetic susceptibility, molar absorption coefficient		Φ	magnetic flux, potential energy, radiant power, work function
λ	absolute activity, radioactive decay constant, thermal conductivity, wavelength		χ	magnetic susceptibility
			χ_e	electric susceptibility
Λ	angular momentum, ionic conductivity		ψ	wavefunction
μ	muon		ω	circular frequency, angular velocity, harmonic vibration wavenumber, statistical weight
μ	chemical potential, electric dipole moment, electric mobility, friction coefficient, Joule-Thompson coefficient, magnetic dipole moment, mobility, permeability		Ω	ohm
			Ω	axial angular momentum, solid angle

GREEK, RUSSIAN, AND HEBREW ALPHABETS

The following table presents the Hebrew, Greek, and Russian alphabets, their letters, the names of the letters, and the English equivalents.

HEBREW[1,3]			GREEK[4]			RUSSIAN	
א	aleph	' [2]	Α α alpha	a		А а	a
ב	beth	b, bh	Β β beta	b		Б б	b
						В в	v
ג	gimel	g, gh	Γ γ gamma	g, n		Г г	g
ד	daleth	d, dh	Δ δ delta	d		Д д	d
ה	he	h	Ε ε epsilon	e		Е е	e
ו	waw	w	Ζ ζ zeta	z		Ж ж	zh
ז	zayin	z	Η η eta	ē		З з	z
						И и Й й	i, ĭ
ח	heth	ḥ	Θ θ theta	th		К к	k
ט	teth	ṭ	Ι ι iota	i		Л л	l
			Κ κ kappa	k		М м	m
י	yodh	y	Λ λ lambda	l		Н н	n
כ ך	kaph	k, kh	Μ μ mu	m		О о	o
ל	lamedh	l	Ν ν nu	n		П п	p
מ ם	mem	m	Ξ ξ xi	x		Р р	r
נ ן	nun	n	Ο ο omicron	o		С с	s
ס	samekh	s	Π π pi	p		Т т	t
ע	ayin	'	Ρ ρ rho	r, rh		У у	u
פ ף	pe	p, ph	Σ σ ς sigma	s		Ф ф	f
						Х х	kh
צ ץ	sadhe	ṣ	Τ τ tau	t		Ц ц	ts
ק	qoph	q	Υ υ upsilon	y, u		Ч ч	ch
						Ш ш	sh
ר	resh	r	Φ φ phi	ph		Щ щ	shch
						Ъ ъ [5]	"
שׂ	sin	ś	Χ χ chi	ch		Ы ы	y
						Ь ь [6]	'
שׁ	shin	sh	Ψ ψ psi	ps		Э э	e
						Ю ю	yu
ת	taw	t, th	Ω ω omega	ō		Я я	ya

[1] Where two forms of a letter are given, the second one is the form used at the end of a word.

[2] Not represented in transliteration when initial.

[3] The Hebrew letters are primarily consonants; a few of them are also used secondarily to represent certain vowels, when provided at all, is by means of a system of dots or strokes adjacent to the consonated characters.

[4] The letter gamma is transliterated "n" only before velars; the letter upsilon is transliterated "u" only as the final element in diphthongs.

[5] This sign indicates that the immediately preceding consonant is not palatized even though immediately followed by a palatized vowel.

[6] This sign indicates that the immediately preceding consonant is palatized even though not immediately followed by a palatized vowel.

DEFINITIONS OF SCIENTIFIC TERMS

Brief definitions of selected terms of importance in chemistry, physics, and related fields of science are given in this section. The selection process emphasizes the following types of terms:
- Physical quantities
- Units of measure
- Classes of chemical compounds and materials
- Important theories, laws, and basic concepts.

Individual chemical compounds are not included.

Definitions have taken wherever possible from the recommendations of international or national bodies, especially the International Union of Pure and Applied Chemistry (IUPAC) and International Organization for Standardization (ISO). For physical quantities and units, the recommended symbol is also given. The source of such definitions is indicated by the reference number in brackets following the definition. In many cases these official definitions have been edited in the interest of stylistic consistency and economy of space. The user is referred to the original source for further details.

* An asterisk following a term indicates that further information can be found by consulting the index of this handbook under the entry for that term.

REFERENCES

1. *ISO Standards Handbook 2, Units of Measurement*, International Organization for Standardization, Geneva, 1992.
2. *Quantities, Units, and Symbols in Physical Chemistry, Second Edition*, International Union of Pure and Applied Chemistry, Blackwell Scientific Publications, Oxford, 1993.
3. *Compendium of Chemical Terminology*, International Union of Pure and Applied Chemistry, Blackwell Scientific Publications, Oxford, 1987.
4. *A Guide to IUPAC Nomenclature of Organic Compounds*, International Union of Pure and Applied Chemistry, Blackwell Scientific Publications, Oxford, 1993.
5. *Glossary of Class Names of Organic Compounds and Reactive Intermediates Based on Structure, Pure and Applied Chemistry*, 67, 1307, 1995.
6. *Compendium of Analytical Nomenclature*, International Union of Pure and Applied Chemistry, Blackwell Scientific Publications, Oxford, 1987.
7. *Nomenclature of Inorganic Chemistry*, International Union of Pure and Applied Chemistry, Blackwell Scientific Publications, Oxford, 1990.
8. *Glossary of Basic Terms in Polymer Science, Pure and Applied Chemistry*, 68, 2287, 1996.
9. *The International Temperature Scale of 1990, Metrologia*, 27, 107, 1990.
10. *Compilation of ASTM Standard Definitions*, American Society of Testing and Materials, Philadelphia, 1990.
11. *ASM Metals Reference Book*, American Society for Metals, Metals Park, OH, 1983.

Ab initio **method** - An approach to quantum-mechanical calculations on molecules which starts with the Schrödinger equation and carries out a complete integration, without introducing empirical factors derived from experimental measurement.

Absorbance (A) - Defined as $-\log(1-\alpha) = \log(1/\tau)$, where α is the absorptance and τ the transmittance of a medium through which a light beam passes. [2]

Absorbed dose (D) - For any ionizing radiation, the mean energy imparted to an element of irradiated matter divided by the mass of that element. [1]

Absorptance (α) - Ratio of the radiant or luminous flux in a given spectral interval absorbed in a medium to that of the incident radiation. Also called absorption factor. [1]

Absorption coefficient (a) - The relative decrease in the intensity of a collimated beam of electromagnetic radiation, as a result of absorption by a medium, during traversal of an infinitesimal layer of the medium, divided by the length traversed. [1]

Absorption coefficient, molar (ε) - Absorption coefficient divided by amount-of-substance concentration of the absorbing material in the sample solution ($\varepsilon = a/c$). The SI unit is m^2/mol. Also called extinction coefficient, but usually in units of $mol^{-1}dm^3cm^{-1}$. [2]

Acceleration - Rate of change of velocity with respect to time.

Acceleration due to gravity (g)* - The standard value (9.80665 m/s^2) of the acceleration experienced by a body in the earth's gravitational field. [1]

Acenes - Polycyclic aromatic hydrocarbons consisting of fused benzene rings in a rectilinear arrangement. [5]

Acid - Historically, a substance that yields an H^+ ion when it dissociates in solution, resulting in a pH<7. In the Brönsted definition, an acid is a substance that donates a proton in any type of reaction. The most general definition, due to G.N. Lewis, classifies any chemical species capable of accepting an electron pair as an acid.

Acid dissociation constant (K_a)* - The equilibrium constant for the dissociation of an acid HA through the reaction $HA + H_2O \rightleftharpoons A^- + H_3O^+$. The quantity $pK_a = -\log K_a$ is often used to express the acid dissociation constant.

Actinides - The elements of atomic number 89 through 103, e.g., Ac, Th, Pa, U, Np, Pu, Am, Cm, Bk, Cf, Es, Fm, Md, No, Lr. [7]

Activation energy* - In general, the energy that must be added to a system in order for a process to occur, even though the process may already be thermodynamically possible. In chemical kinetics, the activation energy is the height of the potential barrier separating the products and reactants. It determines the temperature dependence of the reaction rate.

Activity - For a mixture of substances, the absolute activity λ of substance B is defined as $\lambda_B = \exp(\mu_B/RT)$, where μ_B is the chemical potential of substance B, R the gas constant, and T the thermodynamic temperature. The relative activity a is defined as $a_B = \exp[(\mu_B-\mu_B°)/RT]$, where $\mu_B°$ designates the chemical potential in the standard state. [2]

Activity coefficient (γ)* - Ratio of the activity a_B of component B of a mixture to the concentration of that component. The value of γ depends on the method of stating the composition. For mole fraction x_B, the relation is $a_B = \gamma_B x_B$; for molarity c_B, it is $a_B = \gamma_B c_B/c°$, where $c°$ is the standard state composition (typically chosen as 1 mol/L); for molality

m_B, it is $a_B = \gamma_B m_B/m°$, where $m°$ is the standard state molality (typically 1 mol/kg). [2]

Activity, of radioactive substance (A) - The average number of spontaneous nuclear transitions from a particular energy state occurring in an amount of a radionuclide in a small time interval divided by that interval. [1]

Acyl groups - Groups formed by removing the hydroxy groups from oxoacids that have the general structure $RC(=O)(OH)$ and replacement analogues of such acyl groups. [5]

Adiabatic process - A thermodynamic process in which no heat enters or leaves the system.

Admittance (Y) - Reciprocal of impedance. $Y = G + iB$, where G is conductance and B is susceptance. [1]

Adsorption - A process in which molecules of gas, of dissolved substances in liquids, or of liquids adhere in an extremely thin layer to surfaces of solid bodies with which they are in contact. [10]

Albedo* - The ratio of the light reflected or scattered from a surface to the intensity of incident light. The term is often used in reference to specific types of terrain or to entire planets.

Alcohols - Compounds in which a hydroxy group, -OH, is attached to a saturated carbon atom. [5]

Aldehydes - Compounds $RC(=O)H$, in which a carbonyl group is bonded to one hydrogen atom and to one R group. [5]

Aldoses - Aldehydic parent sugars (polyhydroxyaldehydes $H[CH(OH)]_nC(=O)H$, $n>1$) and their intramolecular hemiacetals. [5]

Aldoximes - Oximes of aldehydes: $RCH=NOH$. [5]

Alfvén number (Al) - A dimensionless quantity used in plasma physics, defined by $Al = v(\rho\mu)^{1/2}/B$, where ρ is density, v is velocity, μ is permeability, and B is magnetic flux density. [2]

Alfven waves - Very low frequency waves which can exist in a plasma in the presence of a uniform magnetic field. Also called magnetohydrodynamic waves.

Alicyclic compounds - Aliphatic compounds having a carbocyclic ring structure which may be saturated or unsaturated, but may not be a benzenoid or other aromatic system. [5]

Aliphatic compounds - Acyclic or cyclic, saturated or unsaturated carbon compounds, excluding aromatic compounds. [5]

Alkali metals - The elements lithium, sodium, potassium, rubidium, cesium, and francium.

Alkaline earth metals - The elements calcium, strontium, barium, and radium. [7]

Alkaloids - Basic nitrogen compounds (mostly heterocyclic) occurring mostly in the plant kingdom (but not excluding those of animal origin). Amino acids, peptides, proteins, nucleotides, nucleic acids, and amino sugars are not normally regarded as alkaloids. [5]

Alkanes - Acyclic branched or unbranched hydrocarbons having the general formula C_nH_{2n+2}, and therefore consisting entirely of hydrogen atoms and saturated carbon atoms. [5]

Alkenes - Acyclic branched or unbranched hydrocarbons having one carbon-carbon double bond and the general formula C_nH_{2n}. Acyclic branched or unbranched hydrocarbons having more than one double bond are alkadienes, alkatrienes, etc. [5]

Alkoxides - Compounds, ROM, derivatives of alcohols, ROH, in which R is saturated at the site of its attachment to oxygen and M is a metal or other cationic species. [5]

Alkyl groups - Univalent groups derived from alkanes by removal of a hydrogen atom from any carbon atom: C_nH_{2n+1}-. The groups derived by removal of a hydrogen atom from a terminal carbon atom of unbranched alkanes form a subclass of normal alkyl (n-alkyl) groups. The groups RCH_2-, R_2CH-, and R_3C- (R not equal to H) are primary, secondary, and tertiary alkyl groups, respectively. [5]

Alkynes - Acyclic branched or unbranched hydrocarbons having a carbon-carbon triple bond and the general formula C_nH_{2n-2}, $RC\equiv CR'$.

Acyclic branched or unbranched hydrocarbons having more than one triple bond are known as alkadiynes, alkatriynes, etc. [5]

Allotropy - The occurrence of an element in two or more crystalline forms.

Allylic groups - The group $CH_2=CHCH_2$- (allyl) and derivatives formed by substitution. The term 'allylic position' or 'allylic site' refers to the saturated carbon atom. A group, such as -OH, attached at an allylic site is sometimes described as "allylic". [5]

Amagat volume unit - A non-SI unit previously used in high pressure science. It is defined as the molar volume of a real gas at one atmosphere pressure and 273.15 K. The approximate value is 22.4 L/mol.

Amides - Derivatives of oxoacids $R(C=O)(OH)$ in which the hydroxy group has been replaced by an amino or substituted amino group. [5]

Amine oxides - Compounds derived from tertiary amines by the attachment of one oxygen atom to the nitrogen atom: $R_3N^+-O^-$. By extension the term includes the analogous derivatives of primary and secondary amines. [5]

Amines - Compounds formally derived from ammonia by replacing one, two, or three hydrogen atoms by hydrocarbyl groups, and having the general structures RNH_2 (primary amines), R_2NH (secondary amines), R_3N (tertiary amines). [5]

Amino acids* - Compounds containing both a carboxylic acid group (-COOH) and an amino group (-NH_2). The most important are the α-amino acids, in which the -NH_2 group in attached to the C atom adjacent to the -COOH group. In the ß-amino acids, there is an intervening carbon atom. [4]

Ampere (A)* - The SI base unit of electric current. [1]

Ampere's law - The defining equation for the magnetic induction B, viz., $dF = IdI \times B$, where dF is the force produced by a current I flowing in an element of the conductor dI pointing in the direction of the current.

Ångström (Å) - A unit of length used in spectroscopy, crystallography, and molecular structure, equal to 10^{-10} m.

Angular momentum (L) - The angular momentum of a particle about a point is the vector product of the radius vector from this point to the particle and the momentum of the particle; i.e., $L = r \times p$. [1]

Angular velocity (ω) - The angle through which a body rotates per unit time.

Anilides - Compounds derived from oxoacids $R(C=O)(OH)$ by replacing the -OH group by the -NHPh group or derivative formed by ring substitution. Also used for salts formed by replacement of a nitrogen-bound hydrogen of aniline by a metal. [5]

Anion - A negatively charged atomic or molecular particle.

Antiferroelectricity* - An effect analogous to antiferromagnetism in which electric dipoles in a crystal are ordered in two sublattices that are polarized in opposite directions, leading to zero net polarization. The effect vanishes above a critical temperature.

Antiferromagnetism* - A type of magnetism in which the magnetic moments of atoms in a solid are ordered into two antiparallel aligned sublattices. Antiferromagnets are characterized by a zero or small positive magnetic susceptibility. The susceptibility increases with temperature up to a critical value, the Néel temperature, above which the material becomes paramagnetic.

Antiparticle - A particle having the same mass as a given elementary particle and a charge equal in magnitude but opposite in sign.

Appearance potential* - The lowest energy which must be imparted to the parent molecule to cause it to produce a particular specified parent ion. This energy, usually stated in eV, may be imparted by electron impact, photon impact, or in other ways. More properly called appearance energy. [3]

Appearance potential spectroscopy (APS) - See Techniques for Materials Characterization, page **12**-1.

Are (a) - A unit of area equal to 100 m^2. [1]

Arenes - Monocyclic and polycyclic aromatic hydrocarbons. See aromatic compounds. [5]

Aromatic compounds - Compounds whose structure includes a cyclic delocalized π-electron system. Historical use of the term implies a ring containing only carbon (e.g., benzene, naphthalene), but it is often generalized to include heterocyclic structures such as pyridine and thiophene. [5]

Arrhenius equation - A key equation in chemical kinetics which expresses the rate constant k as $k = A\exp(-E_a/RT)$, where E_a is the activation energy, R the molar gas constant, and T the temperature. A is called the preexponential factor and, for simple gas phase reactions, may be identified with the collision frequency.

Arsines - AsH_3 and compounds derived from it by substituting one, two or three hydrogen atoms by hydrocarbyl groups. $RAsH_2$, R_2AsH, R_3As (R not equal to H) are called primary, secondary and tertiary arsines, respectively. [5]

Aryl groups - Groups derived from arenes by removal of a hydrogen atom from a ring carbon atom. Groups similarly derived from heteroarenes are sometimes subsumed in this definition. [5]

Astronomical unit (AU)* - The mean distance of the earth from the sun, equal to $1.49597870 \times 10^{11}$ m.

Atomic absorption spectroscopy (AAS) - See Techniques for Materials Characterization, page **12-1**.

Atomic emission spectroscopy (AES) - See Techniques for Materials Characterization, page **12-1**.

Atomic force microscopy (AFM) - See Techniques for Materials Characterization, page **12-1**.

Atomic mass* - The mass of a nuclide, normally expressed in unified atomic mass units (u).

Atomic mass unit (u)* - A unit of mass used in atomic, molecular, and nuclear science, defined as the mass of one atom of ^{12}C divided by 12. Its approximate value is 1.66054×10^{-27} kg. Also called the unified atomic mass unit. [1]

Atomic number (Z) - A characteristic property of an element, equal to the number of protons in the nucleus.

Atomic weight (A_r)* - The ratio of the average mass per atom of an element to 1/12 of the mass of nuclide ^{12}C. An atomic weight can be defined for a sample of any given isotopic composition. The standard atomic weight refers to a sample of normal terrestrial isotopic composition. The term relative atomic mass is synonymous with atomic weight. [2]

Attenuated total reflection (ATR) - See Techniques for Materials Characterization, page **12-1**.

Auger effect - An atomic process in which an electron from a higher energy level fills a vacancy in an inner shell, transferring the released energy to another electron which is ejected.

Aurora - An atmospheric phenomenon in which streamers of light are produced when electrons from the sun are guided into the thermosphere by the earth's magnetic field. It occurs in the polar regions at altitudes of 95—300 km.

Avogadro constant (N_A)* - The number of elementary entities in one mole of a substance.

Azeotrope - A liquid mixture in a state where the variation of vapor pressure with composition at constant temperature (or, alternatively, the variation of normal boiling point with composition) shows either a maximum or a minimum. Thus when an azeotrope boils the vapor has the same composition as the liquid.

Azides - Compounds bearing the group $-N_3$, viz. $-N=N^+=N^-$; usually attached to carbon, e.g. PhN_3, phenyl azide or azidobenzene. Also used for salts of hydrazoic acid, HN_3, e.g. NaN_3, sodium azide. [5]

Azines - Condensation products, $R_2C=NN=CR_2$, of two moles of a carbonyl compound with one mole of hydrazine. [5]

Azo compounds - Derivatives of diazene (diimide), $HN=NH$, wherein both hydrogens are substituted by hydrocarbyl groups, e.g., $PhN=NPh$, azobenzene or diphenyldiazene. [5]

Balmer series - The series of lines in the spectrum of the hydrogen atom which corresponds to transitions between the state with principal quantum number $n = 2$ and successive higher states. The wavelengths are given by $1/\lambda = R_H(1/4 - 1/n^2)$, where $n = 3, 4, ...$ and R_H is the Rydberg constant for hydrogen. The first member of the series ($n = 2 \rightleftharpoons 3$), which is often called the H_α line, falls at a wavelength of 6563 Å.

Bar (bar) - A unit of pressure equal to 10^5 Pa.

Bardeen-Cooper-Schrieffer (BCS) theory - A theory of superconductivity which is based upon the formation of electron pairs as a result of an electron-lattice interaction. The theory relates the superconducting transition temperature to the density of states and the Debye temperature.

Barn (b) - A unit used for expressing cross sections of nuclear processes, equal to 10^{-28} m^2.

Barrel - A unit of volume equal to 158.9873 L.

Baryon - Any elementary particle built up from three quarks. Examples are the proton, neutron, and various short-lived hyperons. Baryons have odd half-integer spins.

Base - Historically, a substance that yields an OH$^-$ ion when it dissociates in solution, resulting in a pH>7. In the Brönsted definition, a base is a substance capable of accepting a proton in any type of reaction. The more general definition, due to G.N. Lewis, classifies any chemical species capable of donating an electron pair as a base.

Becquerel (Bq)* - The SI unit of radioactivity (disintegrations per unit time), equal to s^{-1}. [1]

Beer's law - An approximate expression for the change in intensity of a light beam that passes through an absorbing medium, viz., $\log(I/I_0) = -\varepsilon cl$, where I_0 is the incident intensity, I is the final intensity, ε is the molar (decadic) absorption coefficient, c is the molar concentration of the absorbing substance, and l is the path length. Also called the Beer-Lambert law

Binding energy* - A generic term for the energy required to decompose a system into two or more of its constituent parts. In nuclear physics, the binding energy is the energy difference between a nucleus and the separated nucleons of which it is composed (the energy equivalent of the mass defect). In atomic physics, it is the energy required to remove an electron from an atom.

Biot (Bi) - A name sometimes used for the unit of current in the emu system.

Birefringence - A property of certain crystals in which two refracted rays result from a single incident light ray. One, the ordinary ray, follows the normal laws of refraction, while the other, the extraordinary ray, exhibits a variable refractive index which depends on the direction in the crystal.

Black body radiation* - The radiation emitted by a perfect black body, i.e., a body which absorbs all radiation incident on it and reflects none. The wavelength dependence of the radiated energy density ρ (energy per unit volume per unit wavelength range) is given by the Planck formula

$$\rho = \frac{8\pi hc}{\lambda^5 \left(e^{hc/\lambda kT} - 1\right)}$$

where λ is the wavelength, h is Planck's constant, c is the speed of light, k is the Boltzmann constant, and T is the temperature.

Black hole - A very dense object, formed in a supernova explosion, whose gravitational field is so large that no matter or radiation can escape from the object.

Bloch wave function - A solution of the Schrödinger equation for an electron moving in a spatially periodic potential; used in the band theory of solids.

Bohr magneton (μ_B)* - The atomic unit of magnetic moment, defined as

$eh/4\pi m_e$, where h is Planck's constant, m_e the electron mass, and e the elementary charge. It is the moment associated with a single electron spin.

Bohr, bohr radius (a_0)* - The radius of the lowest orbit in the Bohr model of the hydrogen atom, defined as $\varepsilon_0 h^2/\pi m_e e^2$, where ε_0 is the permittivity of a vacuum, h is Planck's constant, m_e the electron mass, and e the elementary charge. It is customarily taken as the unit of length when using atomic units.

Boiling point - The temperature at which the liquid and gas phases of a substance are in equilibrium at a specified pressure. The normal boiling point is the boiling point at normal atmospheric pressure (101.325 kPa).

Boltzmann constant (k)* - The molar gas constant R divided by Avogadro's constant.

Boltzmann distribution - An expression for the equilibrium distribution of molecules as a function of their energy, in which the number of molecules in a state of energy E is proportional to $\exp(-E/kT)$, where k is the Boltzmann constant and T is the temperature.

Bond strength - See Dissociation energy.

Born-Haber cycle* - A thermodynamic cycle in which a crystalline solid is converted to gaseous ions and then reconverted to the solid. The cycle permits calculation of the lattice energy of the crystal.

Bose-Einstein distribution - A modification of the Boltzmann distribution which applies to a system of particles that are bosons. The number of particles of energy E is proportional to $[e^{(E-\mu)/kT}-1]^{-1}$, where μ is a normalization constant, k is the Boltzmann constant, and T is the temperature.

Boson - A particle that obeys Bose-Einstein Statistics; specifically, any particle with spin equal to zero or an integer. This includes the photon, pion, deuteron, and all nuclei of even mass number.

Boyle's law - The empirical law, exact only for an ideal gas, which states that the volume of a gas is inversely proportional to its pressure at constant temperature.

Bragg angle (θ) - Defined by the equation $n\lambda = 2d\sin\theta$, which relates the angle θ between a crystal plane and the diffracted x-ray beam, the wavelength λ of the x-rays, the crystal plane spacing d, and the diffraction order n (any integer).

Bravais lattices* - The 14 distinct crystal lattices that can exist in three dimensions. They include three in the cubic crystal system, two in the tetragonal, four in the orthorhombic, two in the monoclinic, and one each in the triclinic, hexagonal, and trigonal systems.

Breakdown voltage - The potential difference at which an insulating substance undergoes a physical or chemical change that causes it to become a conductor, thus allowing current to flow through the sample.

Bremsstrahlung - Electromagnetic radiation generated when the velocity of a charged particle is reduced (literally, "braking radiation"). An example is the x-ray continuum resulting from collisions of electrons with the target in an x-ray tube.

Brewster angle - The angle of incidence for which the maximum degree of plane polarization occurs when a beam of unpolarized light is incident on the surface of a medium of refractive index n. At this angle, the angle between the reflected and refracted beams is 90°. The value of the Brewster angle is $\tan^{-1}n$.

Brillouin scattering - The scattering of light by acoustic phonons in a solid or liquid.

Brillouin zone - A region of allowed wave vectors and energy levels in a crystalline solid, which plays a part in the propagation of waves through the lattice.

British thermal unit (Btu) - A non-SI unit of energy, equal to approximately 1055 J. Several values of the Btu, defined in slightly different ways, have been used.

Brownian motion - The random movements of small particles suspended in a fluid, which arise from collisions with the fluid molecules.

Brunauer-Emmett-Teller method (BET) - See Techniques for Materials Characterization, page 12-1.

Buffer* - A solution designed to maintain a constant pH when small amounts of a strong acid or base are added. Buffers usually consist of a fairly weak acid and its salt with a strong base. Suitable concentrations are chosen so that the pH of the solution remains close to the pK_a of the weak acid.

Calorie (cal) - A non-SI unit of energy, originally defined as the heat required to raise the temperature of 1 g of water by 1°C. Several calories of slightly different values have been used. The thermochemical calorie is now defined as 4.184 J.

Candela (cd)* - The SI base unit of luminous intensity. [1]

Capacitance (C) - Ratio of the charge acquired by a body to the change in potential. [1]

Carbamates - Salts or esters of carbamic acid, $H_2NC(=O)OH$, or of N-substituted carbamic acids: $R_2NC(=O)OR'$, (R' = hydrocarbyl or a cation). The esters are often called urethanes or urethans, a usage that is strictly correct only for the ethyl esters. [5]

Carbenes - The electrically neutral species H_2C: and its derivatives, in which the carbon is covalently bonded to two univalent groups of any kind or a divalent group and bears two nonbonding electrons, which may be spin-paired (singlet state) or spin-non-paired (triplet state). [5]

Carbinols - An obsolete term for substituted methanols, in which the name carbinol is synonymous with methanol. [5]

Carbohydrates - Originally, compounds such as aldoses and ketoses, having the stoichiometric formula $C_n(H_2O)_n$ (hence "hydrates of carbon"). The generic term carbohydrate now includes mono-, oligo-, and polysaccharides, as well as their reaction products and derivatives. [5]

Carboranes - A contraction of carbaboranes. Compounds in which a boron atom in a polyboron hydride is replaced by a carbon atom with maintenance of the skeletal structure. [5]

Carboxylic acids - Oxoacids having the structure $RC(=O)OH$. The term is used as a suffix in systematic name formation to denote the $-C(=O)OH$ group including its carbon atom. [5]

Carnot cycle - A sequence of reversible changes in a heat engine using a perfect gas as the working substance, which is used to demonstrate that entropy is a state function. The Carnot cycle also provides a means to calculate the efficiency of a heat engine.

Catalyst - A substance that participates in a particular chemical reaction and thereby increases its rate but without a net change in the amount of that substance in the system. [3]

Catenanes, catena compounds - Hydrocarbons having two or more rings connected in the manner of links of a chain, without a covalent bond. More generally, the class catena compounds embraces functional derivatives and hetero analogues. [5]

Cation - A positively charged atomic or molecular particle.

Centipoise (cP) - A common non-SI unit of viscosity, equal to mPa s.

Centrifugal distortion - An effect in molecular spectroscopy in which rotational levels are lowered in energy, relative to the values of a rigid rotor, as the rotational angular momentum increases. The effect may be understood classically as a stretching of the bonds in the molecule as it rotates faster, thus increasing the moment of inertia.

Ceramic - A nonmetallic material of very high melting point.

Cerenkov radiation - Light emitted when a beam of charged particles travels through a medium at a speed greater than the speed of light in the medium. It is typically blue in color.

Cgs system of units - A system of units based upon the centimeter, gram, and second. The cgs system has been supplanted by the International System (SI).

Chalcogens - The Group VIA elements (oxygen, sulfur, selenium, tellurium, and polonium). Compounds of these elements are called chalcogenides. [7]

Chaotic system - A complex system whose behavior is governed by deterministic laws but whose evolution can vary drastically when small changes are made in the initial conditions.

Charge - See Electric charge.

Charles' law - The empirical law, exact only for an ideal gas, which states that the volume of a gas is directly proportional to its temperature at constant pressure.

Charm - A quantum number introduced in particle physics to account for certain properties of elementary particles and their reactions.

Chelate - A compound characterized by the presence of bonds from two or more bonding sites within the same ligand to a central metal atom. [3]

Chemical potential - For a mixture of substances, the chemical potential of constituent B is defined as the partial derivative of the Gibbs energy G with respect to the amount (number of moles) of B, with temperature, pressure, and amounts of all other constituents held constant. Also called partial molar Gibbs energy. [2]

Chemical shift* - A small change in the energy levels (and hence in the spectra associated with these levels) resulting from the effects of chemical binding in a molecule. The term is used in fields such as NMR, Mössbauer, and photoelectron spectroscopy, where the energy levels are determined primarily by nuclear or atomic effects.

Chiral molecule - A molecule which cannot be superimposed on its mirror image. A common example is an organic molecule containing a carbon atom to which four different atoms or groups are attached. Such molecules exhibit optical activity, i.e., they rotate the plane of a polarized light beam.

Chlorocarbons - Compounds consisting solely of chlorine and carbon. [5]

Chromatography* - A method for separation of the components of a sample in which the components are distributed between two phases, one of which is stationary while the other moves. In gas chromatography the gas moves over a liquid or solid stationary phase. In liquid chromatography the liquid mixture moves through another liquid, a solid, or a gel. The mechanism of separation of components may be adsorption, differential solubility, ion-exchange, permeation, or other mechanisms. [6]

Clapeyron equation - A relation between pressure and temperature of two phases of a pure substance that are in equilibrium, viz., $dp/dT = \Delta_{trs}S/\Delta_{trs}V$, where $\Delta_{trs}S$ is the difference in entropy between the phases and $\Delta_{trs}V$ the corresponding difference in volume.

Clathrates - Inclusion compounds in which the guest molecule is in a cage formed by the host molecule or by a lattice of host molecules. [5]

Clausius (Cl) - A non-SI unit of entropy or heat capacity defined as cal/K = 4.184 J/K. [2]

Clausius-Clapeyron equation - An approximation to the Clapeyron equation applicable to liquid-gas and solid-gas equilibrium, in which one assumes an ideal gas with volume much greater than the condensed phase volume. For the liquid-gas case, it takes the form $d(\ln p)/dT = \Delta_{vap}H/RT^2$ where R is the molar gas constant and $\Delta_{vap}H$ is the molar enthalpy of vaporization. For the solid-gas case, $\Delta_{vap}H$ is replaced by the molar enthalpy of sublimation, $\Delta_{sub}H$.

Clausius-Mosotti equation - A relation between the dielectric constant ε_r at optical frequencies and the polarizability α:

$$\frac{\varepsilon_r - 1}{\varepsilon_r + 2} = \frac{\rho N_A \alpha}{3M\varepsilon_0}$$

where ρ is density, N_A is Avogadro's number, M is molar mass, and ε_0 is the permittivity of a vacuum.

Clebsch-Gordon coefficients - A set of coefficients used to describe the vector coupling of angular momenta in atomic and nuclear physics.

Codon - A set of three bases, chosen from the four primary bases found in the DNA molecule (uracil, cytosine, adenine, and guanine), which specifies the production of a particular amino acid or carries some other genetic instruction. For example, the codon UCA specifies the amino acid serine, CAG specifies glutamine, etc. There are a total of 64 codons.

Coercive force - The magnetizing force at which the magnetic flux density is equal to zero. [10]

Coercivity* - The maximum value of coercive force that can be attained when a magnetic material is symmetrically magnetized to saturation induction. [10]

Coherent anti-Stokes Raman spectroscopy (CARS) - See Techniques for Materials Characterization, page 12-1.

Colloid - Molecules or polymolecular particles dispersed in a medium that have, at least in one direction, a dimension roughly between 1 nm and 1 μm. [3]

Color center - A defect in a crystal that gives rise to optical absorption, thus changing the color of the material. A common type is the F-center, which results when an electron occupies the site of a negative ion.

Compressibility (κ)* - The fractional change of volume as pressure is increased, viz., $\kappa = -(1/V)(dV/dp)$. [1]

Compton wavelength (λ_C)* - In the scattering of electromagnetic radiation by a free particle (e.g., electron, proton), $\lambda_C = h/mc$ is the increase in wavelength, at a 90° scattering angle, corresponding to the transfer of energy from radiation to particle. Here h is Planck's constant, c the speed of light, and m the mass of the particle.

Conductance (G)* - For direct current, the reciprocal of resistance. More generally, the real part of admittance. [1]

Conductivity, electrical (σ)* - The reciprocal of the resistivity. [1]

Conductivity, thermal - See Thermal conductivity.

Congruent transformation - A phase transition (melting, vaporization, etc.) in which the substance preserves its exact chemical composition.

Constitutional repeating unit (CRU) - In polymer science, the smallest constitutional unit, the repetition of which constitutes a regular macromolecule, i.e., a macromolecule with all units connected identically with respect to directional sense. [8]

Copolymer - A polymer derived from more than one species of monomer. [8]

Coriolis effect - The deviation from simple trajectories when a mechanical system is described in a rotating coordinate system. It affects the motion of projectiles on the earth and in molecular spectroscopy leads to an important interaction between the rotational and vibrational motions. The effect may be described by an additional term in the equations of motion, called the Coriolis force.

Cosmic rays* - High energy nuclear particles, electrons, and photons, originating mostly outside the solar system, which continually bombard the earth's atmosphere.

Coulomb (C)* - The SI unit of electric charge, equal to A s. [1]

Coulomb's law - The statement that the force F between two electrical charges q_1 and q_2 separated by a distance r is $F = (4\pi\varepsilon_0)^{-1}q_1q_2/r^2$, where ε_0 is the permittivity of a vacuum.

Covalent bond - A chemical bond between two atoms whose stability results from the sharing of two electrons, one from each atom.

Cowling number (Co) - A dimensionless quantity used in plasma physics, defined by $Co = B^2/\mu\rho v^2$, where ρ is density, v is velocity, μ is permeability, and B is magnetic flux density. [2]

CPT theorem - A theorem in particle physics which states that any local Lagrangian theory that is invariant under proper Lorentz transformations is also invariant under the combined operations of charge conjugation, C, space inversion, P, and time reversal, T, taken in any order.

Critical point* - In general, the point on the phase diagram of a two-phase system at which the two coexisting phases have identical properties and therefore represent a single phase. At the liquid-gas critical point of a pure substance, the distinction between liquid and gas vanishes, and the vapor pressure curve ends. The coordinates of this point are called the critical temperature and critical pressure. Above the critical temperature, it is not possible to liquefy the substance.

Cross section (σ)* - A measure of the probability of collision (or other interaction) between a beam of particles and a target which it encounters. In rough terms it is the effective area the target particles present to the incident ones; however, the precise definition depends on the nature of the interaction. A general definition of σ is the number of encounters per unit time divided by nv, where n is the concentration of incident particles and v their velocity.

Crosslink - In polymer science, a small region in a macromolecule from which at least four chains emanate, and formed by reactions involving sites or groups on existing macromolecules or by interactions between existing macromolecules. [8]

Crown compounds - Macrocyclic polydentate compounds, usually uncharged, in which three or more coordinating ring atoms (usually oxygen or nitrogen) are or may become suitably close for easy formation of chelate complexes with metal ions or other cationic species. [5]

Crust* - The outer layer of the solid earth, above the Mohorovicic discontinuity. Its thickness averages about 35 km on the continents and about 7 km below the ocean floor.

Cryoscopic constant (E_f)* - The constant that expresses the amount by which the freezing point T_f of a solvent is lowered by a non-dissociating solute, through the relation $\Delta T_f = E_f m$, where m is the molality of the solute.

Curie (Ci) - A non-SI unit of radioactivity (disintegrations per unit time), equal to 3.7×10^{10} s^{-1}.

Curie temperature (T_C)* - For a ferromagnetic material, the critical temperature above which the material becomes paramagnetic. Also applied to the temperature at which the spontaneous polarization disappears in a ferroelectric solid. [1]

Cyanohydrins - Alcohols substituted by a cyano group, most commonly, but not limited to, examples having a CN and an OH group attached to the same carbon atom. They are formally derived from aldehydes or ketones by the addition of hydrogen cyanide. [5]

Cycloalkanes - Saturated monocyclic hydrocarbons (with or without side chains). See alicyclic compounds. Unsaturated monocyclic hydrocarbons having one endocyclic double or one triple bond are called cycloalkenes and cycloalkynes, respectively. [5]

Cyclotron resonance - The resonant absorption of energy from a system in which electrons or ions that are orbiting in a uniform magnetic field are subjected to radiofrequency or microwave radiation. The resonance frequency is given by $v = eH/2\pi m^* c$, where e is the elementary charge, H is the magnetic field strength, m^* is the effective mass of the charged particle, and c is the speed of light. The effect occurs in both solids (involving electrons or holes) and in low pressure gasses (involving ions)

Dalton (Da) - A name sometimes used in biochemistry for the unified atomic mass unit (u).

De Broglie wavelength - The wavelength associated with the wave representation of a moving particle, given by h/mv, where h is Planck's constant, m the particle mass, and v the velocity.

De Haas-Van Alphen effect - An effect observed in certain metals and semiconductors at low temperatures and high magnetic fields, characterized by a periodic variation of magnetic susceptibility with field strength.

Debye equation* - The relation between the relative permittivity (dielectric constant) ε_r, polarizability α, and permanent dipole moment μ in a dielectric material whose molecules are free to rotate. It takes the form

$$\frac{\varepsilon_r - 1}{\varepsilon_r + 2} = \frac{\rho N_A}{3M\varepsilon_0}\left(\alpha + \frac{\mu^2}{3kT}\right)$$

where ρ is density, N_A is Avogadro's number, M is molar mass, and ε_0 is the permittivity of a vacuum.

Debye length - In the Debye-Hückel theory of ionic solutions, the effective thickness of the cloud of ions of opposite charge which surrounds each given ion and shields the Coulomb potential produced by that ion.

Debye temperature (θ_D)* - In the Debye model of the heat capacity of a crystalline solid, $\theta_D = h v_D / k$, where h is Planck's constant, k is the Boltzmann constant, and v_D is the maximum vibrational frequency the crystal can support. For $T \ll \theta_D$, the heat capacity is proportional to T^3.

Debye unit (D) - A non-SI unit of electric dipole moment used in molecular physics, equal to 3.335641×10^{-30} C m.

Debye-Waller factor (D) - The factor by which the intensity of a diffraction line is reduced because of lattice vibrations. [1]

Defect - Any departure from the regular structure of a crystal lattice. A Frenkel defect results when an atom or ion moves to an interstitial position and leaves behind a vacancy. A Schottky defect involves either a vacancy where the atom has moved to the surface or a structure where a surface atom has moved to an interstitial position.

Degree of polymerization - The number of monomeric units in a macromolecule or an oligomer molecule. [8]

Dendrite - A tree-like crystalline pattern often observed, for example, in ice crystals and alloys in which the crystal growth branches repeatedly.

Density (ρ)* - In the most common usage, mass density or mass per unit volume. More generally, the amount of some quantity (mass, charge, energy, etc.) divided by a length, area, or volume.

Density of states (N_E, ρ) - The number of one-electron states in an infinitesimal interval of energy, divided by the range of that interval and by volume. [1]

Dew point* - The temperature at which liquid begins to condense as the temperature of a gas mixture is lowered. In meteorology, it is the temperature at which moisture begins to condense on a surface in contact with the air.

Diamagnetism - A type of magnetism characterized by a negative magnetic susceptibility, so that the material, when placed in an external magnetic field, becomes weakly magnetized in the direction opposite to the field. This magnetization is independent of temperature.

Diazo compounds - Compounds having the divalent diazo group, $=N^+=N^-$, attached to a carbon atom, e.g., $CH_2=N_2$ diazomethane. [5]

Dielectric constant (ε)* - Ratio of the electric displacement in a medium to the electric field strength. Also called permittivity. [1]

Dienes - Compounds that contain two fixed double bonds (usually assumed to be between carbon atoms). Dienes in which the two double-bond units are linked by one single bond are termed conjugated. [5]

Differential scanning calorimetry (DSC) - See Techniques for Materials Characterization, page 12-1.

Differential thermal analysis (DTA) - See Techniques for Materials Characterization, page 12-1.

Diffusion* - The migration of atoms, molecules, ions, or other particles as a result of some type of gradient (concentration, temperature, etc.).

Diopter - A unit used in optics, formally equal to m^{-1}. It is used in expressing dioptic power, which is the reciprocal of the focal length of a lens.

Dipole moment, electric (p,μ)* - For a distribution of equal positive and negative charge, the magnitude of the dipole moment vector is the positive charge multiplied by the distance between the centers of positive and negative charge distribution. The direction is given by the line from the center of negative charge to the center of positive charge.

Dipole moment, magnetic (m,μ) - Formally defined in electromagnetic theory as a vector quantity whose vector product with the magnetic flux density equals the torque. The magnetic dipole generated by a current I flowing in a small loop of area A has a magnetic moment of magnitude IA. In atomic and nuclear physics, a magnetic moment is associated with the angular momentum of a particle; e.g., an electron with orbital angular momentum l exhibits a magnetic moment of $-el/2m_e$ where e is the elementary charge and m_e the mass of the electron. [1]

Disaccharides - Compounds in which two monosaccharides are joined by a glycosidic bond. [5]

Dislocation - An extended displacement of a crystal from a regular lattice. An edge dislocation results when one portion of the crystal has partially slipped with respect to the other, resulting in an extra plane of atoms extending through part of the crystal. A screw dislocation transforms successive atomic planes into the surface of a helix.

Dispersion - Splitting of a beam of light (or other electromagnetic radiation) of mixed wavelengths into the constituent wavelengths as a result of the variation of refractive index of the medium with wavelength.

Dissociation constant* - The equilibrium constant for a chemical reaction in which a compound dissociates into its constituent parts.

Dissociation energy (D_e)* - For a diatomic molecule, the difference between the energies of the free atoms at rest and the minimum in the potential energy curve. The term bond dissociation energy (D_0), which can be applied to polyatomic molecules as well, is used for the difference between the energies of the fragments resulting when a bond is broken and the energy of the original molecule in its lowest energy state. The term bond strength implies differences in enthalpy rather than energy.

Domain - A small region of a solid in which the magnetic or electric moments of the individual units (atoms, molecules, or ions) are aligned in the same direction.

Domain wall - The transition region between adjacent ferromagnetic domains, generally a layer with a thickness of a few hundred ångström units. Also called Bloch wall.

Doppler effect - The change in the apparent frequency of a wave (sound, light, or other) when the source of the wave is moving relative to the observer.

Dose equivalent (H) - The product of the absorbed dose of radiation at a point of interest in tissue and various modifying factors which depend on the type of tissue and radiation. [1]

Drift velocity - The velocity of charge carriers (electrons, ions, etc.) moving under the influence of an electric field in a medium which subjects the carriers to some frictional force.

Dyne (dyn) - A non-SI (cgs) unit of force, equal to 10^{-5} N.

Ebullioscopic constant (E_b)* - The constant that expresses the amount by which the boiling point T_b of a solvent is raised by a non-dissociating solute, through the relation $\Delta T_b = E_b\, m$, where m is the molality of the solute.

Eddy currents - Circulating currents set up in conducting bulk materials or sheets by varying magnetic fields.

Effinghausen effect - The appearance of a temperature gradient in a current carrying conductor that is placed in a transverse magnetic field. The direction of the gradient is perpendicular to the current and the field.

Eigenvalue - An allowed value of the constant a in the equation $Au = au$, where A is an operator acting on a function u (which is called an eigenfunction). In quantum mechanics, the outcome of any observation is an eigenvalue of the corresponding operator. Also called characteristic value.

Einstein - A non-SI unit used in photochemistry, equal to one mole of photons.

Einstein temperature (θ_V) - In the Einstein theory of the heat capacity of a crystalline solid, $\theta_V = h\nu/k$, where h is Planck's constant, k is the Boltzmann constant, and ν is the vibrational frequency of the crystal.

Einstein transition probability - A constant in the Einstein relation $A_{ij} + B_{ij}\rho$ for the probability of a transition between two energy levels i and j in a radiation field of energy density ρ. The A_{ij} coefficient describes the probability of spontaneous emission, while B_{ij} and B_{ji} govern the probability of stimulated emission and absorption, respectively ($B_{ij} = B_{ji}$).

Elastic limit - The greatest stress which a material is capable of sustaining without any permanent strain remaining after complete release of the stress. [10]

Elastic modulus - See Young's modulus.

Electric charge (Q) - The quantity of electricity; i.e., the property that controls interactions between bodies through electrical forces.

Electric current (I) - The charge passing through a circuit per unit time. [1]

Electric displacement (D) - A vector quantity whose magnitude equals the electric field strength multiplied by the permittivity of the medium and whose direction is the same as that of the field strength.

Electric field strength (E) - The force exerted by an electric field on a point charge divided by the electric charge. [1]

Electric potential (V) - A scalar quantity whose gradient is equal to the negative of the electric field strength.

Electrical conductance - See Conductance.

Electrical resistance - See Resistance.

Electrical resistivity - See Resistivity.

Electrochemical series* - An arrangement of reactions which produce or consume electrons in an order based on standard electrode potentials. A common arrangement places metals in decreasing order of their tendency to give up electrons.

Electrode potential* - The electromotive force of a cell in which the electrode on the left is the standard hydrogen electrode and that on the right is the electrode in question. [2]

Electrolysis - The decomposition of a substance as a result of passing an electric current between two electrodes immersed in the sample.

Electromotive force (emf) - The energy supplied by a source divided by the charge transported through the source. [1]

Electron* - An elementary particle in the family of leptons, with negative charge and spin of 1/2.

Electron affinity* - The energy difference between the ground state of a gas-phase atom or molecule and the lowest state of the corresponding negative ion.

Electron cyclotron resonance (ECR) - See Techniques for Materials Characterization, page 12-1.

Electron energy loss spectroscopy (EELS) - See Techniques for Materials Characterization, page 12-1.

Electron nuclear double resonance (ENDOR) - See Techniques for Materials Characterization, page 12-1.

Electron paramagnetic resonance (EPR) - See Techniques for Materials Characterization, page 12-1.

Electron probe microanalysis (EPMA) - See Techniques for Materials Characterization, page 12-1.

Electron spectroscopy for chemical analysis (ESCA) - See Techniques for Materials Characterization, page 12-1.

Electron spin (s) - The quantum number, equal to 1/2, that specifies the intrinsic angular momentum of the electron.

Electron stimulated desorption (ESD) - See Techniques for Materials Characterization, page 12-1.

Electron volt (eV)* - A non-SI unit of energy used in atomic and nuclear physics, equal to approximately 1.602177×10^{-19} J. The electron volt is defined as the kinetic energy acquired by an electron upon acceleration through a potential difference of 1 V. [1]

Electronegativity* - A parameter originally introduced by Pauling which describes, on a relative basis, the power of an atom or group of atoms to attract electrons from the same molecular entity. [3]

Electrophoresis - The motion of macromolecules or colloidal particles in an electric field. [3]

Emissivity (ε)* - Ratio of the radiant flux emitted per unit area to that of an ideal black body at the same temperature. Also called emittance. [1]

Emu - The electromagnetic system of units, based upon the cm, g, and s plus the emu of current (sometimes called the abampere).

Enantiomers - A chiral molecule and its non-superposable mirror image. The two forms rotate the plane of polarized light by equal amounts in opposite directions. Also called optical isomers.

Energy (E, U)* - The characteristic of a system that enables it to do work.

Energy gap* - In the theory of solids, the region between two energy bands, in which no bound states can occur.

Enols, alkenols - The term refers specifically to vinylic alcohols, which have the structure $HOCR'=CR_2$. Enols are tautomeric with aldehydes ($R' = H$) or ketones (R' not equal to H). [5]

Enthalpy (H)* - A thermodynamic function, especially useful when dealing with constant-pressure processes, defined by $H = E + PV$, where E is energy, P pressure, and V volume. [1]

Enthalpy of combustion* - The enthalpy change in a combustion reaction. Its negative is the heat released in combustion.

Enthalpy of formation, standard* - The enthalpy change for the reaction in which a substance is formed from its constituent elements, each in its standard reference state (normally refers to 1 mol, sometimes to 1 g, of the substance).

Enthalpy of fusion* - The enthalpy change in the transition from solid to liquid state.

Enthalpy of sublimation - The enthalpy change in the transition from solid to gas state.

Enthalpy of vaporization* - The enthalpy change in the transition from liquid to gas state.

Entropy (S)* - A thermodynamic function defined such that when a small quantity of heat dQ is received by a system at temperature T, the entropy of the system is increased by dQ/T, provided that no irreversible change takes place in the system. [1]

Entropy unit (e.u.) - A non-SI unit of entropy, equal to 4.184 J/K mol.

Ephemeris time - Time measured in tropical years from January 1, 1900.

Epoxy compounds - Compounds in which an oxygen atom is directly attached to two adjacent or non-adjacent carbon atoms of a carbon chain or ring system; thus cyclic ethers. [5]

Equation of continuity - Any of a class of equations that express the fact that some quantity (mass, charge, energy, etc.) cannot be created or destroyed. Such equations typically specify that the rate of increase of the quantity in a given region of space equals the net current of the quantity flowing into the region.

Equation of state* - An equation relating the pressure, volume, and temperature of a substance or system.

Equilibrium constant (K)* - For a chemical reaction $a\text{A} + b\text{B} \rightleftharpoons c\text{C} + d\text{D}$, the equilibrium constant is defined by:

$$K = \frac{a_\text{C}{}^c \cdot a_\text{D}{}^d}{a_\text{A}{}^a \cdot a_\text{B}{}^b}$$

where a_i is the activity of component i. To a certain approximation, the activities can be replaced by concentrations. The equilibrium constant is related to $\Delta_r G°$, the standard Gibbs energy change in the reaction, by $RT \ln K = -\Delta_r G°$.

Equivalent conductance - See Conductivity, electrical

Erg (erg) - A non-SI (cgs) unit of energy, equal to 10^{-7} J.

Esters - Compounds formally derived from an oxoacid RC(=O)(OH) and an alcohol, phenol, heteroarenol, or enol by linking, with formal loss of water from an acidic hydroxy group of the former and a hydroxy group of the latter. [5]

Esu - The electrostatic system of units, based upon the cm, g, and s plus the esu of charge (sometimes called the statcoulomb or franklin).

Ethers - Compounds with formula ROR, where R is not equal to H. [5]

Euler number (Eu) - A dimensionless quantity used in fluid mechanics, defined by $Eu = \Delta p/\rho v^2$, where p is pressure, ρ is density, and v is velocity. [2]

Eutectic - The point on a two-component solid-liquid phase diagram which represents the lowest melting point of any possible mixture. A liquid having the eutectic composition will freeze at a single temperature without change of composition.

Excitance (M) - Radiant energy flux leaving an element of a surface divided by the area of that element. [1]

Exciton - A localized excited state consisting of a bound electron-hole pair in a molecular or ionic crystal. The exciton can propagate through the crystal.

Exosphere - The outermost part of the earth's atmosphere, beginning at about 500 to 1000 km above the surface. It is characterized by densities so low that air molecules can escape into outer space.

Expansion coefficient - See thermal expansion coefficient.

Extended electron energy loss fine structure (EXELFS) - See Techniques for Materials Characterization, page 12-1.

Extended x-ray absorption fine structure (EXAFS) - See Techniques for Materials Characterization, page 12-1.

Extinction coefficient - See Absorption coefficient, molar

F-Center - See Color center

Fahrenheit temperature ($°F$) - The temperature scale based on the assignment of $32°F = 0°C$ and a temperature interval of $°F = (5/9)°C$; i.e., $t/°F = (9/5)t/°C + 32$.

Farad (F)* - The SI unit of electric capacitance, equal to C/V. [1]

Faraday constant (F)* - The electric charge of 1 mol of singly charged positive ions; i.e., $F = N_\text{A}e$, where N_A is Avogadro's constant and e is the elementary charge. [1]

Faraday effect* - The rotation of the plane of plane-polarized light by a medium placed in a magnetic field parallel to the direction of the light beam. The effect can be observed in solids, liquids, and gasses.

Fatty acids - Aliphatic monocarboxylic acids derived from or contained in esterified form in an animal or vegetable fat, oil, or wax. Natural fatty acids commonly have a chain of 4 to 28 carbons (usually unbranched and even-numbered), which may be saturated or unsaturated. By extension, the term is sometimes used to embrace all acyclic aliphatic carboxylic acids. [5]

Fermat's principle - The law that a ray of light traversing one or more media will follow a path which minimizes the time required to pass between two given points.

Fermi (f) - Name sometimes used in nuclear physics for the femtometer.

Fermi level - The highest energy of occupied states in a solid at zero temperature. Sometimes called Fermi energy. The Fermi surface is the surface in momentum space formed by electrons occupying the Fermi level.

Fermi resonance - An effect observed in vibrational spectroscopy when an overtone of one fundamental vibration closely coincides in energy with another fundamental of the same symmetry species. It leads to a splitting of vibrational bands.

Fermi-Dirac distribution - A modification of the Boltzmann distribution which takes into account the Pauli exclusion principle. The number of particles of energy E is proportional to $[e^{(E-\mu)/kT}+1]^{-1}$, where μ is a normalization constant, k the Boltzmann constant, and T the temperature. The distribution is applicable to a system of fermions.

Fermion - A particle that obeys Fermi-Dirac statistics. Specifically, any particle with spin equal to an odd multiple of 1/2. Examples are the electron, proton, neutron, muon, etc.

Ferrimagnetism* - A type of magnetism in which the magnetic moments of atoms in a solid are ordered into two nonequivalent sublattices with unequal magnetic moments, leading to a nonzero magnetic susceptibility.

Ferrite - A ferrimagnetic material of nominal formula MFe_2O_4, where M is a divalent metal; widely used in microwave switches and other solid state devices.

Ferroelectricity* - The retention of electric polarization by certain materials after the external field that produced the polarization has been removed.

Ferromagnetism* - A type of magnetism in which the magnetic moments of atoms in a solid are aligned within domains which can in turn be aligned with each other by a weak magnetic field. Some ferromagnetic materials can retain their magnetization when the external field is removed, as long as the temperature is below a critical value, the Curie temperature. They are characterized by a large positive magnetic susceptibility.

Fick's law - The statement that the flux J of a diffusing substance is proportional to the concentration gradient, i.e., $J = -D(dc/dx)$, where D is called the diffusion coefficient.

Field - A mathematical construct which describes the interaction between particles resulting from gravity, electromagnetism, or other physical phenomena. In classical physics a field is described by equations. Quantum field theory introduces operators to represent the physical observables.

Field emission microscopy (FEM) - See Techniques for Materials Characterization, page 12-1.

Field ion microscopy (FIM) - See Techniques for Materials Characterization, page 12-1.

Fine structure - The splitting in spectral lines that results from interactions of the electron spin with the orbital angular momentum.

Fine structure constant (α)* - Defined as $e^2/2hc\varepsilon_0$, where e is the elementary charge, h Planck's constant, c the speed of light, and ε_0 the permittivity of a vacuum. It is a measure of the strength of the electromagnetic interaction between particles.

First radiation constant (c_1)* - Constant ($=2\pi hc^2$) in the equation for the radiant excitance M_λ of a black body:

$$M_\lambda = \frac{c_1 \lambda^{-5} \Delta\lambda}{e^{c_2/\lambda T} - 1}$$

where λ is the wavelength, T is the temperature, and $c_2 = hc/k$ is the second radiation constant.

Flash point - The lowest temperature at which vapors above a volatile combustible substance will ignite in air when exposed to a flame. [10]

Fluence (F) - Term used in photochemistry to specify the energy per unit area delivered in a given time interval, for example by a laser pulse. [2]

Fluorocarbons - Compounds consisting solely of fluorine and carbon. [5]

Fluxoid - The quantum of magnetic flux in superconductivity theory, equal to $hc/2e$, where h is Planck's constant, c the velocity of light, and e the elementary charge.

Force (F) - The rate of change of momentum with time. [1]

Force constants (f, k)* - In molecular vibrations, the coefficients in the expression of the potential energy in terms of atom displacements from their equilibrium positions. In a diatomic molecule, $f = d^2V/dr^2$, where $V(r)$ is the potential energy and r is the interatomic distance. [2]

Fourier number (Fo) - A dimensionless quantity used in fluid mechanics, defined by $Fo = at/l^2$, where a is thermal diffusivity, t is time, and l is length. [2]

Fourier transform infrared spectroscopy (FTIR) - A technique for obtaining an infrared spectrum by use of an interferometer in which the path length of one of the beams is varied. A Fourier transformation of the resulting interferogram yields the actual spectrum. The technique is also used for NMR and other types of spectroscopy.

Fractals - Geometrical objects that are self-similar under a change of scale; i.e., they appear similar at all levels of magnification. They can be considered to have fractional dimensionality. Examples occur in diverse fields such as geography (rivers and shorelines), biology (trees), and solid state physics (amorphous materials).

Franck-Condon principle - An important principle in molecular spectroscopy which states that the nuclei in a molecule remain essentially stationary while an electronic transition is taking place. The physical interpretation rests on the fact that the electrons move much more rapidly than the nuclei because of their much smaller mass.

Franklin (Fr) - Name sometimes given to the unit of charge in the esu system.

Fraunhofer diffraction - Diffraction of light in situations where the source and observation point are so far removed that the wave surfaces may be considered planar.

Fraunhofer lines - Sharp absorption lines in the spectrum of sunlight, caused by absorption of the solar blackbody radiation by atoms near the sun's surface.

Free radical - See Radicals. The term "free radical" is often used more broadly for molecules that have a paramagnetic ground state (e.g., O_2) and sometimes for any transient or highly reactive molecular species.

Freezing point - See Melting point

Frequency (ν)* - Number of cycles of a periodic phenomenon divided by time. [1]

Fresnel diffraction - Diffraction of light in a situation where the source and observation point are sufficiently close together that the curvature of the wave surfaces must be taken into account.

Froude number (Fr) - A dimensionless quantity used in fluid mechanics, defined by $Fr = v/(lg)^{1/2}$, where v is velocity, l is length, and g is acceleration due to gravity. [2]

Fugacity (f_B) - For a gas mixture, the fugacity of component B is defined as the absolute activity λ_B times the limit, as the pressure p approaches zero at constant temperature, of p_B/λ_B. [2]

Fullerenes - Compounds composed solely of an even number of carbon atoms, which form a cage-like fused-ring polycyclic system with twelve five-membered rings and the rest six-membered rings. The archetypal example is [60]fullerene, where the atoms and bonds delineate a truncated icosahedron. The term has been broadened to include any closed cage structure consisting entirely of three-coordinate carbon atoms. [5]

Fulvalenes - The hydrocarbon fulvalene and its derivatives formed by substitution (and by extension, analogues formed by replacement of one or more carbon atoms of the fulvalene skeleton by a heteroatom). [5]

Fulvenes - The hydrocarbon fulvene and its derivatives formed by substitution (and by extension, analogues formed by replacement of one or more carbon atoms of the fulvene skeleton by a heteroatom). [5]

Fundamental vibrational frequencies* - In molecular spectroscopy, the characteristic vibrational frequencies obtained when the vibrational energy is expressed in normal coordinates. They determine the primary features of the infrared and Raman spectra of the molecule.

γ - Name sometimes used for microgram.

γ-rays* - Electromagnetic radiation (photons) with energy greater than about 0.1 MeV (wavelength less than about 1 pm).

g-Factor of the electron* - The proportionality factor in the equation relating the magnetic moment μ of an electron to its total angular momentum quantum number J, i.e., $\mu = -g\mu_B J$, where μ_B is the Bohr magneton. Also called Landé factor.

Gal - A non-SI unit of acceleration, equal to 0.01 m/s. Also called galileo.

Gallon (US) - A unit of volume equal to 3.785412 L.

Gallon (UK, Imperial) - A unit of volume equal to 4.546090 L.

Gauss (G) - A non-SI unit of magnetic flux density (B) equal to 10^{-4} T.

Gaussian system of units - A hybrid system used in electromagnetic theory, which combines features of both the esu and emu systems.

Gel - A colloidal system with a finite, but usually rather small, yield stress (the sheer stress at which yielding starts abruptly). [3]

Genetic code* - The set of relations between each of the 64 codons of DNA and a specific amino acid (or other genetic instruction).

Gibbs energy (G)* - An important function in chemical thermodynamics, defined by $G = H-TS$, where H is the enthalpy, S the entropy, and T the thermodynamic temperature. Sometimes called Gibbs free energy and, in older literature, simply "free energy". [2]

Gibbs phase rule - The relation $F = C - P + 2$, where C is the number of components in a mixture, P is the number of phases, and F is the degrees of freedom, i.e., the number of intensive variables that can be changed independently without affecting the number of phases.

Glass transition temperature* - The temperature at which an amorphous polymer is transformed, in a reversible way, from a viscous or rubbery condition to a hard and relatively brittle one. [10]

Glow discharge mass spectroscopy (GDMS) - See Techniques for Materials Characterization, page **12**-1.

Gluon - A hypothetical particle postulated to take part in the binding of quarks, in analogy to the role of the photon in electromagnetic interactions.

Glycerides - Esters of glycerol (propane-1,2,3-triol) with fatty acids, widely distributed in nature. They are by long-established custom subdivided into triglycerides, 1,2- or 1,3-diglycerides, and 1- or 2-monoglycerides, according to the number and positions of acyl groups. [5]

Glycols - Dihydric alcohols in which two hydroxy groups are on different carbon atoms, usually but not necessarily adjacent. Also called diols. [5]

Grain (gr) - A non-SI unit of mass, equal to 64.79891 mg.

Grain boundary - The interface between two regions of different crystal orientation.

Grashof number (Gr) - A dimensionless quantity used in fluid mechanics, defined by $Gr = l^3 g\alpha\Delta T\rho^2/\eta^2$, where T is temperature, ρ is density, l is length, η is viscosity, α is cubic expansion coefficient, and g is acceleration of gravity. [2]

Gravitational constant (G)* - The universal constant in the equation for the gravitational force between two particles, $F = Gm_1m_2/r^2$, where r is the distance between the particles and m_1 and m_2 are their masses. [1]

Gray (Gy)* - The SI unit of absorbed dose of radiation, equal to J/kg. [1]

Gregorian calendar - The modification of the Julian calendar introduced in 1582 by Pope Gregory XII which specified that a year divisible by 100 is a leap year only if divisible by 400.

Grignard reagents - Organomagnesium halides, RMgX, having a carbon-magnesium bond (or their equilibrium mixtures in solution with $R_2Mg + MgX_2$). [5]

Gruneisen parameter (γ) - Defined by $\gamma = \alpha_V/\kappa\, c_V\, \rho$, where α_V is the cubic thermal expansion coefficient, κ is the isothermal compressibility, c_V is the specific heat capacity at constant volume, and ρ is the mass density. γ is independent of temperature for most crystalline solids. [1]

Gyromagnetic ratio (γ) - Ratio of the magnetic moment of a particle to its angular momentum. Also called magnetogyric ratio.

Hadron - Any elementary particle that can take part in the strong interaction. Hadrons are subdivided into baryons, with odd half integer spins, and mesons, which have zero or integral spin.

Hall effect* - The development of a transverse potential difference V in a conducting material when subjected to a magnetic field H perpendicular to the direction of the current. The potential difference is given by $V = R_H BJt$, where B is the magnetic induction, J the current density, t the thickness of the specimen in the direction of the potential difference, and R_H is called the Hall coefficient.

Halocarbon - A compound containing no elements other than carbon, hydrogen, and one or more halogens. In common practice, the term is used mainly for compounds of no more than four or five carbon atoms.

Halogens - The elements F, Cl, Br, I, and At. Compounds of these elements are called halogenides or halides. [7]

Hamiltonian (H) - An expression for the total energy of a mechanical system in terms of the momenta and positions of constituent particles. In quantum mechanics, the Hamiltonian operator appears in the eigenvalue equation $H\psi = E\psi$, where E is an energy eigenvalue and ψ the corresponding eigenfunction.

Hardness* - The resistance of a material to deformation, indentation, or scratching. Hardness is measured on various scales, such as Mohs, Brinell, Knoop, Rockwell, and Vickers. [10]

Hartmann number (Ha) - A dimensionless quantity used in plasma physics, defined by $Ha = Bl(\kappa/\eta)^{1/2}$, where B is magnetic flux density, l is length, κ is electric conductivity, and η is viscosity. [2]

Hartree (E_h)* - An energy unit used in atomic and molecular science, equal to approximately $4.3597482 \times 10^{-18}$ J.

Hartree-Fock method - A iterative procedure for solving the Schrödinger equation for an atom or molecule in which the equation is solved for each electron in an initial assumed potential from all the other electrons. The new potential that results is used to repeat the calculation and the procedure continued until convergence is reached. Also called self-consistent field (SCF) method.

Heat capacity* - Defined in general as dQ/dT, where dQ is the amount of heat that must be added to a system to increase its temperature by a small amount dT. The heat capacity at constant pressure is $C_p = (\partial H/\partial T)_p$; that at constant volume is $C_V = (\partial E/\partial T)_V$, where H is enthalpy, E is internal energy, p is pressure, V is volume, and T is temperature. An upper case C normally indicates the molar heat capacity, while a lower case c is used for the specific (per unit mass) heat capacity. [1]

Heat of formation, vaporization, etc. - See corresponding terms under Enthalpy.

Hectare (ha) - A unit of area equal to 10^4 m². [1]

Heisenberg uncertainty principle - The statement that two observable properties of a system that are complementary, in the sense that their quantum-mechanical operators do not commute, cannot be specified simultaneously with absolute precision. An example is the position and momentum of a particle; according to this principle, the uncertainties in position Δq and momentum Δp must satisfy the relation $\Delta p\Delta q \geq h/4\pi$, where h is Planck's constant.

Heitler-London model - An early quantum-mechanical model of the hydrogen atom which introduced the concept of the exchange interaction between electrons as the primary reason for stability of the chemical bond.

Helicon - A low-frequency wave generated when a metal at low temperature is exposed to a uniform magnetic field and a circularly polarized electric field.

Helmholz energy (A) - A thermodynamic function defined by $A = E-TS$, where E is the energy, S the entropy, and T the thermodynamic temperature. [2]

Hemiacetals - Compounds having the general formula $R_2C(OH)OR'$ (R' not equal to H). [5]

Henry (H)* - The SI unit of inductance, equal to Wb/A. [1]

Henry's law * - An expression which applies to an ideal dilute solution in which one or more gasses are dissolved, viz., $p_i = H_i x_i$, where p_i is the partial pressure of component i above the solution, x_i is its mole fraction in the solution, and H_i is the Henry's law constant (a characteristic of the given gas and solvent, as well as the temperature).

Hermitian operator - An operator A that satisfies the relation $\int u_m{}^* A u_n dx = (\int u_n{}^* A u_m\ dx)^*$, where * indicates the complex conjugate. The eigenvalues of Hermitian operators are real, and eigenfunctions belonging to different eigenvalues are orthogonal.

Hertz (Hz) - The SI unit of frequency, equal to s^{-1}. [1]

Heterocyclic compounds - Cyclic compounds having as ring members atoms of at least two different elements, e.g., quinoline, 1,2-thiazole, bicyclo[3.3.1]tetrasiloxane. [5]

Heusler alloys - Alloys of manganese, copper, aluminum, nickel, and sometimes other metals which find important uses as permanent magnets.

Holography - A technique for creating a three-dimensional image of a object by recording the interference pattern between a light beam diffracted from the object and a reference beam. The image can be reconstructed from this pattern by a suitable optical system.

Homopolymer - A polymer derived from one species of (real, implicit, or hypothetical) monomer. [8]

Hooke's law - The statement that the ratio of stress to strain is a constant in a totally elastic medium.

Horse power - A non-SI unit of energy, equal to approximately 746 W.

Hubble constant - The ratio of the recessional velocity of an extragalactic object to the distance of that object. Its value is about $2 \times 10^{-18}\ s^{-1}$.

Huckel theory - A simple approximation for calculating the energy of conjugated molecules in which only the resonance integrals between neighboring bonds are considered. Also called CNDO method (complete neglect of differential overlap).

Hume-Rothery rules - A set of empirical rules for predicting the occurrence of solid solutions in metallic systems. The rules involve size, crystal structure, and electronegativity.

Hund's rules - A series of rules for predicting the sequence of energy states in atoms and molecules. One of the important results is that when two electrons exist in different orbitals, the state with their spins parallel (triplet state) lies at lower energy than the state with antiparallel spins (singlet).

Hydrazines - Hydrazine (diazane), H_2NNH_2, and its hydrocarbyl derivatives. When one or more substituents are acyl groups, the compound is a hydrazide. [5]

Hydrocarbon - A compound containing only carbon and hydrogen. [5]

Hydrolysis - A reaction occurring in water in which a chemical bond is cleaved and a new bond formed with the oxygen atom of water.

Hyperfine structure - Splitting of energy levels and spectral lines into several closely spaced components as a result of interaction of nuclear spin angular momentum with other angular momenta in the atom or molecule.

Hysteresis* - An irreversible response of a system (parameter A) as a function of an external force (parameter F), usually symmetric with respect to the origin of the A vs. F graph after the initial application of the force. A common example is magnetic induction vs. magnetic field strength in a ferromagnet.

Ideal gas law - The equation of state $pV = RT$, which defines an ideal gas, where p is pressure, V molar volume, T temperature, and R the molar gas constant.

Ideal solution - A solution in which solvent-solvent and solvent-solute interactions are identical, so that properties such as volume and enthalpy are exactly additive. Ideal solutions follow Raoult's law, which states that the vapor pressure p_i of component i is $p_i = x_i p_i{}^*$, where x_i is the mole fraction of component i and $p_i{}^*$ the vapor pressure of the pure substance i.

Ignition temperature* - The lowest temperature at which combustion of a material will occur spontaneously under specified conditions. Sometimes called autoignition temperature, kindling point. [10]

Imides - Diacyl derivatives of ammonia or primary amines, especially those cyclic compounds derived from diacids. Also used for salts having the anion RN_2^-. [5]

Impedence (Z) - The complex representation of potential difference divided by the complex representation of current. In terms of reactance X and resistance R, the impedance is given by $Z = R + iX$. [1]

Index of refraction (n)* - For a non-absorbing medium, the ratio of the velocity of electromagnetic radiation *in vacuo* to the phase velocity of radiation of a specified frequency in the medium. [1]

Inductance - The ratio of the electromagnetic force induced in a coil by a current to the rate of change of the current.

Inductive coupled plasma mass spectroscopy (ICPMS) - See Techniques for Materials Characterization, page 12-1.

Inertial defect - In molecular spectroscopy, the quantity $I_c-I_a-I_b$ for a molecule whose equilibrium configuration is planar, where I_a, I_b, and I_c are the effective principal moments of inertia. The inertial defect for a rigid planar molecule would be zero, but vibration-rotation interactions in a real molecule lead to a positive inertial defect.

Insulator - A material in which the highest occupied energy band (valence band) is completely filled with electrons, while the next higher band (conduction band) is empty. Solids with an energy gap of 5 eV or more are generally considered as insulators at room temperature. Their conductivity is less than 10^{-6} S/m and increases with temperature.

Intercalation compounds - Compounds resulting from reversible inclusion, without covalent bonding, of one kind of molecule in a solid matrix of another compound, which has a laminar structure. The host compound, a solid, may be macromolecular, crystalline, or amorphous. [5]

International System of Units (SI)* - The unit system adopted by the General Conference on Weights and Measures in 1960. It consists of seven base units (meter, kilogram, second, ampere, kelvin, mole, candela), plus derived units and prefixes. [1]

International Temperature Scale (ITS-90)* - The official international temperature scale adopted in 1990. It consists of a set of fixed points and equations which enable the thermodynamic temperature to be determined from operational measurements. [9]

Ion - An atomic or molecular particle having a net electric charge. [3]

Ion exchange - A process involving the adsorption of one or several ionic species accompanied by the simultaneous desorption (displacement) of one or more other ionic species. [3]

Ion neutralization spectroscopy (INS) - See Techniques for Materials Characterization, page 12-1.

Ionic strength (I) - A measure of the total concentration of ions in a solution, defined by $I = 1/2\Sigma_i\ z_i^2 m_i$, where z_i is the charge of ionic species and m_i is its molality. For a 1-1 electrolyte at molality m, $I = m$.

Ionization constant* - The equilibrium constant for a reaction in which a substance in solution dissociates into ions.

Ionization potential* - The minimum energy required to remove an electron from an isolated atom or molecule (in its vibrational ground state) in the gaseous phase. More properly called ionization energy. [3]

Irradiance (E) - The radiant energy flux incident on an element of a surface, divided by the area of that element. [1]

Isentropic process - A thermodynamic process in which the entropy of the system does not change.

Ising model - A model describing the coupling between two atoms in a ferromagnetic lattice, in which the interaction energy is proportional to the negative of the product of the spin components along a specified axis.

Isobar - A line connecting points of equal pressure on a graphical representation of a physical system.

Isochore - A line or surface of constant volume on a graphical representation of a physical system.

Isoelectric point* - The pH of a solution or dispersion at which the net charge on the macromolecules or colloidal particles is zero. In electrophoresis there is no motion of the particles in an electric field at the isoelectric point.

Isomers - In chemistry, compounds that have identical molecular formulas but differ in the nature or sequence of bonding of their atoms or in the arrangement of their atoms in space. In physics, nuclei of the same atomic number Z and mass number A but in different energy states. [3]

Isomorphs - Substances of different chemical nature but having the same crystal structure.

Isotactic macromolecule - A tactic macromolecule, essentially comprising only one species of repeating unit which has chiral or prochiral atoms in the main chain in a unique arrangement with respect to its adjacent constitutional units. [8]

Isotherm - A line connecting points of equal temperature on a graphical representation of a physical system.

Isothermal process - A thermodynamic process in which the temperature of the system does not change.

Isotones - Nuclides having the same neutron number N but different atomic number Z. [3]

Isotopes - Two or more nuclides with the same atomic number Z but different mass number A. The term is sometimes used synonymously with nuclide, but it is preferable to reserve the word nuclide for a species of specific Z and A. [3]

Jahn-Teller effect - An interaction of vibrational and electronic motions in a nonlinear molecule which removes the degeneracy of certain electronic energy levels. It can influence the spectrum, crystal structure, and magnetic properties of the substance.

Johnson noise - Electrical noise generated by random thermal motion of electrons in a conductor or semiconductor. Also called thermal noise.

Josephson effect - The tunneling of electron pairs through a thin insulating layer which separates two superconductors. When a potential difference is applied to the superconductors, an alternating current is generated whose frequency is precisely proportional to the potential difference. This effect has important applications in metrology and determination of fundamental physical constants.

Joule (J)* - The SI unit of energy, equal to N m. [1]

Joule-Thomson coefficient (μ) - A parameter which describes the temperature change when a gas expands adiabatically through a nozzle from a high pressure to a low pressure region. It is defined by $\mu = (\partial T/\partial p)_H$, where H is enthalpy.

Julian calendar - The calendar introduced by Julius Caeser in 46 B.C. which divided the year into 365 days with a leap year of 366 days every fourth year.

Julian date (JD) - The number of days elapsed since noon Greenwich Mean Time on January 1, 4713 B.C. Thus January 1, 2000, 0h (midnight) will be JD 2,451,543.5. This dating system was introduced by Joseph Scaliger in 1582.

Kaon - One of the elementary particles in the family of mesons. Kaons have a spin of zero and may be neutral or charged.

Kelvin (K)* - The SI base unit of thermodynamic temperature. [1]

Kepler's laws - The three laws of planetary motion, which established the elliptical shape of planetary orbits and the relation between orbital dimensions and the period of rotation.

Kerr effect* - An electrooptical effect in which birefringence is induced in a liquid or gas when a strong electric field is applied perpendicular to the direction of an incident light beam. The Kerr constant k is given by $n_1 - n_2 = k\lambda E^2$, where λ is the wavelength, E is the electric field

strength, and n_1 and n_2 are the indices of refraction of the ordinary and extraordinary rays, respectively.

Ketenes - Compounds in which a carbonyl group is connected by a double bond to an alkylidene group: $R_2C=C=O$. [5]

Ketones - Compounds in which a carbonyl group is bonded to two carbon atoms: $R_1R_2C=O$ (neither R may be H). [5]

Kilogram (kg)* - The SI base unit of mass. [1]

Kinetic energy (E_k, T) - The energy associated with the motion of a system of particles in a specified reference frame. For a single particle of mass m moving at velocity v, $E_k = 1/2mv^2$.

Kirchhoff's laws - Basic rules for electric circuits, which state (a) the algebraic sum of the currents at a network node is zero and (b) the algebraic sum of the voltage drops around a closed path is zero.

Klein-Gordon equation - A relativistic extension of the Schrödinger equation.

Klein-Nishima formula - An expression for the scattering cross section of a photon by an unbound electron, based upon the Dirac electron theory.

Knight shift - The change in magnetic resonance frequency of a nucleus in a metal relative to the same nucleus in a diamagnetic solid. The effect is due to the polarization of the conduction electrons in the metal.

Knudsen number (Kn) - A dimensionless quantity used in fluid mechanics, defined by $Kn = \lambda/l$, where λ is mean free path and l is length. [2]

Kondo effect - A large increase in electrical resistance observed at low temperatures in certain dilute alloys of a magnetic metal in a nonmagnetic material.

Kramers-Kronig relation - A set of equations relating the real and imaginary parts of the index of refraction of a medium

Lactams - Cyclic amides of amino carboxylic acids, having a 1-azacycloalkan-2-one structure, or analogues having unsaturation or heteroatoms replacing one or more carbon atoms of the ring. [5]

Lactones - Cyclic esters of hydroxy carboxylic acids, containing a 1-oxacycloalkan-2-one structure, or analogues having unsaturation or heteroatoms replacing one or more carbon atoms of the ring. [5]

Lagrangian function (L) - A function used in classical mechanics, defined as the kinetic energy minus the potential energy for a system of particles.

Lamb shift - The small energy difference between the $^2S_{1/2}$ and $^2P_{1/2}$ levels in the hydrogen atom, which results from interactions between the electron and the radiation field.

Laminar flow - Smooth, uniform, non-turbulent flow of a gas or liquid in parallel layers, with little mixing between layers. It is characterized by small values of the Reynolds number.

Landé g-factor - See g-Factor of the electron

Langevin function - The mathematical function $L(x) = (e^x + e^{-x})/(e^x - e^{-x}) - 1/x$, which occurs in the expression for the average dipole moment of a group of rotating polar molecules in an electric field: $\mu_{av} = \mu L(\mu E/kT)$, where μ is the electric dipole moment of a single molecule, E is the electric field strength, k is the Boltzmann constant, and T is the temperature.

Lanthanides - The elements of atomic number 57 through 71, which share common chemical properties: La, Ce, Pr, Nd, Pm, Sm, Eu, Gd, Tb, Dy, Ho, Er, Tm, Yb, Lu. [7]

Larmor frequency (ν_L) - The precession frequency of a magnetic dipole in an applied magnetic field. In particular, a nucleus in a magnetic field of strength B has a Larmor frequency of $\gamma B/2\pi$, where γ is the magnetogyric ratio of the nucleus.

Laser* - A device in which an optical cavity is filled with a medium where a population inversion can be produced by some means. When the resonant frequency of the cavity bears the proper relation to the separation of the inverted energy levels, stimulated emission occurs, producing a highly monochromatic, coherent beam of light.

Laser ionization mass spectroscopy (LIMS) - See Techniques for Materials Characterization, page 12-1.

Lattice constants* - Parameters specifying the dimensions of a unit cell in a crystal lattice, specifically the lengths of the cell edges and the angles between them.

Lattice energy* - The energy per ion pair required to separate completely the ions in a crystal lattice at a temperature of absolute zero.

Laue diagram - A diffraction pattern produced when an x-ray beam passes through a thin slice of a crystal and impinges on a detector behind the crystal.

Lenz's law - The statement that the current induced in a circuit by a change in magnetic flux is so directed as to oppose the change in flux

Leonard-Jones potential - A simple but useful function for approximating the interaction between two neutral atoms or molecules separated by a distance r by writing the potential energy as $U(r) = 4\varepsilon\{(r_0/r)^{12} - (r_0/r)^6\}$, where ε and r_0 are adjustable parameters. In this form the depth of the potential well is ε and the minimum occurs at $2^{1/6}r_0$. The $(1/r)^{12}$ term is often replaced by other powers of $1/r$.

Lepton - One of the class of elementary particles that do not take part in the strong interaction. Included are the electron, muon, and neutrino. All leptons have a spin of 1/2.

Lewis number (*Le*) - A dimensionless quantity used in fluid mechanics, defined by $Le = a/D$, where a is thermal diffusivity and D is diffusion coefficient. [2]

Ligand field theory - A description of the structure of crystals containing a transition metal ion surrounded by nonmetallic ions (ligands). It is based on construction of molecular orbitals involving the d-orbitals of the central metal ion and combinations of atomic orbitals of the ligands.

Light year (l.y.) - A unit of distance used in astronomy, defined as the distance light travels in one year in a vacuum. Its approximate value is 9.46073×10^{15} m.

Lignins - Macromolecular constituents of wood related to lignans, composed of phenolic propylbenzene skeletal units, linked at various sites and apparently randomly. [5]

Ligroin - The petroleum fraction consisting mostly of C_7 and C_8 hydrocarbons and boiling in the range 90-140°C; commonly used as a laboratory solvent.

Lipids - A loosely defined term for substances of biological origin that are soluble in nonpolar solvents. They consist of saponifiable lipids, such as glycerides (fats and oils) and phospholipids, as well as nonsaponifiable lipids, principally steroids. [5]

Lipoproteins - Clathrate complexes consisting of a lipid enwrapped in a protein host without covalent binding, in such a way that the complex has a hydrophilic outer surface consisting of all the protein and the polar ends of any phospholipids. [5]

Liter (L)* - A synonym for cubic decimeter. [1]

Lithosphere* - The outer layer of the solid earth, extending from the base of the mantle to the surface of the crust.

Lorentz contraction - The reduction in length of a moving body in the direction of motion, given by the factor $(1-v^2/c^2)^{1/2}$, where v is the velocity of the body and c the velocity of light. Also known as the FitzGerald-Lorentz contraction.

Lorentz force - The force exerted on a point charge Q moving at velocity v in the presence of external fields E and B. It is given (in SI units) by $F = Q(E + v \times B)$.

Loss angle (δ) - For a dielectric material in an alternating electromagnetic field, δ is the phase difference between the current and the potential difference. The function tan δ is a measure of the ratio of the power dissipated in the dielectric to the power stored.

Low energy electron diffraction (LEED) - See Techniques for Materials Characterization, page 12-1.

Lumen (lm)* - The SI unit of luminous flux, equal to cd sr. [1]

Luminous flux (Φ) - The intensity of light from a source multiplied by the solid angle. The SI unit is lumen. [1]

Lux (lx)* - The SI unit of illuminance, equal to cd sr m^{-2}. [1]

Lyddane-Sachs-Teller relation - A relation between the phonon frequencies and dielectric constants of an ionic crystal which states that $(\omega_T/\omega_L)^2 = \varepsilon(\infty)/\varepsilon(0)$, where ω_T is the angular frequency of transverse optical phonons, ω_L that of longitudinal optical phonons, $\varepsilon(0)$ is the static dielectric constant, and $\varepsilon(\infty)$ the dielectric constant at optical frequencies.

Lyman series - The series of lines in the spectrum of the hydrogen atom which corresponds to transitions between the ground state (principal quantum number $n = 1$) and successive excited states. The wavelengths are given by $1/\lambda = R_H(1-1/n^2)$, where $n = 2,3,4,...$ and R_H is the Rydberg constant for hydrogen. The first member of the series ($n = 1\leftrightarrow2$), which is often called the Lyman-α line, falls at a wavelength of 1216 Å, and the series converges at 912 Å, the ionization limit of hydrogen.

Mach number (*Ma*) - A dimensionless quantity used in fluid mechanics, defined by $Ma = v/c$, where v is velocity and c is the speed of sound. [2]

Macromolecule - A molecule of high relative molecular mass (molecular weight), the structure of which essentially comprises the multiple repetition of units derived, actually or conceptually, from molecules of low relative molecular mass [8]

Madelung constant* - A constant characteristic of a particular crystalline material which gives a measure of the electrostatic energy binding the ions in the crystal.

Magnetic field strength (*H*) - An axial vector quantity, the curl of which is equal to the current density, including the displacement current. [1]

Magnetic induction (*B*) - An axial vector quantity such that the force exerted on an element of current is equal to the vector product of this element and the magnetic induction. [1]

Magnetic moment - See Dipole moment, magnetic.

Magnetic susceptibility (χ_m, κ)* - Defined by $\chi_m = (\mu-\mu_0)/\mu_0$, where μ is the permeability of the medium and μ_0 the permeability of a vacuum. [1]

Magnetization (*M*) - Defined by $M = (B/\mu_0)-H$, where B is magnetic induction, H magnetic field strength, and μ_0 the permeability of a vacuum. [1]

Magnetogyric ratio (γ) - Ratio of the magnetic moment of a particle to its angular momentum. Also called gyromagnetic ratio.

Magneton - See Bohr magneton, Nuclear magneton.

Magnetostriction* - The change in dimensions of a solid sample when it is placed in a magnetic field.

Magnon - A quantum of magnetic energy associated with a spin wave in a ferromagnetic or antiferromagnetic crystal.

Mantle - The layer of the earth between the crust and the liquid outer core, which begins about 2900 km below the earth's surface.

Maser - A device in which a microwave cavity is filled with a medium where a population inversion can be produced by some means. When the resonant frequency of the cavity bears the proper relation to the separation of the inverted energy levels, the device can serve as an amplifier or oscillator at that frequency.

Mass (*m*)* - Quantity of matter. Mass can also be defined as "resistance to acceleration".

Mass defect (*B*) - Defined by $B = Zm(^1H) + Nm_n - m_a$, where Z is the atomic number, $m(^1H)$ is the mass of the hydrogen atom, N is the neutron number, m_n is the rest mass of the neutron, and m_a is the mass of the atom in question. Thus Bc^2 can be equated to the binding energy of the nucleus if the binding energy of atomic electrons is neglected. [1]

Mass excess (Δ) - Defined by $\Delta = m_a - Am_u$, where m_a is the mass of the atom, A the number of nucleons, and m_u the unified atomic mass constant ($m_u = 1$ u). [1]

Mass fraction (w_B) - The ratio of the mass of substance B to the total mass of a mixture. [1]

Mass number (A) - A characteristic property of a specific isotope of an element, equal to the sum of the number of protons and neutrons in the nucleus.

Mass spectrometry - An analytical technique in which ions are separated according to the mass/charge ratio and detected by a suitable detector. The ions may be produced by electron impact on a gas, a chemical reaction, energetic vaporization of a solid, etc. [6]

Massieu function - A thermodynamic function defined by $J = -A/T$, where A is the Helmholz energy and T the thermodynamic temperature. [2]

Matthiessen's rule - The statement that the electrical resistivity ρ of a metal can be written as $\rho = \rho_L + \rho_i$, where ρ_L is due to scattering of conduction electrons by lattice vibrations and ρ_i to scattering by impurities and imperfections. If the impurity concentration is small, ρ_i is temperature independent.

Maxwell (Mx)* - A non-SI unit of magnetic field strength (H) equal to 10^{-8} Wb. [1]

Maxwell's equations - The fundamental equations of electromagnetism. In a form appropriate to SI units, they are:
curl $H = \partial D/\partial t + j$
div $B = 0$
curl $E = -\partial B/\partial t$
div $D = \rho$
where H is the magnetic field strength, B the magnetic induction, E the electric field strength, D the electric displacement, j the current density, ρ the charge density, and t is time.

Maxwell-Boltzmann distribution - An expression for the fraction of molecules $f(v)$ in a gas that have velocity v within a specified interval. It takes the form
$$f(v) = 4\pi (M/2\pi RT)^{3/2} v^2 e^{-Mv^2/2RT}$$
where M is the molar mass, R the molar gas constant, and T the temperature.

Mean free path* - The average distance a gas molecule travels between collisions.

Meissner effect - The complete exclusion of magnetic induction from the interior of a superconductor.

Melting point* - The temperature at which the solid and liquid phases of a substance are in equilibrium at a specified pressure (normally taken to be atmospheric unless stated otherwise).

Mercaptans - A traditional term abandoned by IUPAC, synonymous with thiols. This term is still widely used. [5]

Meson - Any elementary particle that has zero or integral spin. Mesons are responsible for the forces between protons and neutrons in the nucleus.

Mesosphere - The part of the earth's atmosphere extending from the top of the stratosphere (about 50 km above the surface) to 80-90 km. It is characterized by a decrease in temperature with increasing altitude.

Metal - A material in which the highest occupied energy band (conduction band) is only partially filled with electrons. The electrical conductivity of metals generally decreases with temperature.

Metallocenes - Organometallic coordination compounds in which one atom of a transition metal such as iron, ruthenium or osmium is bonded to and only to the face of two cyclopentadienyl ligands which lie in parallel planes. [5]

Meter (m)* - The SI base unit of length. [1]

Methine group - In organic compounds, the -C= group. [5]

Mho - An archaic name for the SI unit siemens (reciprocal ohm).

Micelle - A particle formed by the aggregation of surfactant molecules (typically, 10 to 100 molecules) in solution. For aqueous solutions, the hydrophilic end of the molecule is on the surface of the micelle, while the hydrophobic end (often a hydrocarbon chain) points toward the center. At the critical micelle concentration (cmc) the previously dissolved molecules aggregate into a micelle.

Micron (μ) - An obsolete name for micrometer.

Mie scattering - The scattering of light by spherical dielectric particles whose diameter is comparable to the wavelength of the light.

Milky way - The band of light in the night sky resulting from the stars in the galactic plane. The term is also used to denote the galaxy in which the sun is located.

Miller indices (hkl) - A set of indices used to label planes in a crystal lattice. [2]

Millimeter of mercury (mmHg) - A non-SI unit of pressure, equal to 133.322 Pa. The name is generally considered interchangeable with torr.

Mobility (μ)* - In solid state physics, the drift velocity of electrons or holes in a solid divided by the applied electric field strength. The term is used in a similar sense in other fields.

Molality (m) - A measure of concentration of a solution in which one states the amount of substance (i.e., number of moles) of solute per kilogram of solvent. Thus a 0.1 molal solution (often written as 0.1 m) has $m = 0.1$ mol/kg.

Molar mass - The mass of one mole of a substance. It is normally expressed in units of g/mol, in which case its numerical value is identical with the molecular weight (relative molecular mass). [1]

Molar quantity - It is often convenient to express an extensive quantity (e.g., volume, enthalpy, heat capacity, etc.) as the actual value divided by amount of substance (number of moles). The resulting quantity is called molar volume, molar enthalpy, etc

Molar refraction (R) - A property of a dielectric defined by the equation $R = V_m[(n^2-1)/(n^2+2)]$, where n is the index of refraction of the medium (at optical wavelengths) and V_m the molar volume. It is related to the polarizability α of the molecules that make up the medium by the Lorenz-Lorentz equation, $R = N_A\alpha/3\epsilon_0$, where N_A is Avogadro's constant and ϵ_0 is the permittivity of a vacuum.

Molarity (c) - A measure of concentration of a solution in which one states the amount of substance (i.e., number of moles) of solute per liter of solution. Thus a 0.1 molar solution (often referred to as 0.1 M) has a concentration $c = 0.1$ mol/L.

Mole (mol)* - The SI base unit of amount of substance. [1]

Mole fraction (x_B) - The ratio of the amount of substance (number of moles) of substance B to the total amount of substance in a mixture. [1]

Molecular orbital - See Orbital.

Molecular weight (M_r)* - The ratio of the average mass per molecule or specified entity of a substance to 1/12 of the mass of nuclide ^{12}C. Also called relative molar (or molecular) mass. [1]

Moment of inertia (I) - The moment of inertia of a body about an axis is the sum (or integral) of the products of its elements of mass and the squares of their distances from the axis. [1]

Momentum (p) - The product of mass and velocity. [1]

Monomer - A substance consisting of molecules which can undergo polymerization, thereby contributing constitutional units to the essential structure of a macromolecule. [8]

Monosaccharides - A term which includes aldoses, ketoses, and a wide variety of derivatives. [5]

Mössbauer effect - The recoilless emission of γ-rays from nuclei bound in a crystal under conditions where the recoil energy associated with the γ emission is taken up by the crystal as a whole. This results in a very narrow line width, which can be exploited in various types of precise measurements.

Muon* - An unstable elementary particle of spin 1/2 and mass about 200 times that of the electron.

Naphtha - The petroleum fraction consisting mostly of C_6 to C_8 hydrocarbons and boiling in the range 80-120°C. Solvents derived from this fraction include ligroin and petroleum ether.

Nautical mile - A non-SI unit of length, equal to exactly 1852 m.

Navier-Stokes equations - A set of complex equations for the motion of a viscous fluid subject to external forces.

Néel temperature (T_N)* - The critical temperature above which an antiferromagnetic substance becomes paramagnetic. [1]

Nernst effect - The production of an electric field in a conductor subject to an applied magnetic field and containing a transverse temperature gradient. The electric field is perpendicular to the magnetic field and the temperature gradient.

Network - In polymer science, a highly ramified macromolecule in which essentially each constitutional unit is connected to each other constitutional unit and to the macroscopic phase boundary by many permanent paths through the macromolecule, the number of such paths increasing with the number of intervening bonds. The paths must on the average be coextensive with the macromolecule. [8]

Neutrino - A stable elementary particle in the lepton family. Neutrinos have zero (or at least near-zero) rest mass and spin 1/2.

Neutron* - An elementary particle on spin 1/2 and zero charge. The free neutron has a mean lifetime of 887 seconds. Neutrons and protons, which are collectively called nucleons, are the constituents of the nucleus.

Neutron activation analysis (NAA) - See Techniques for Materials Characterization, page **12-1**.

Neutron number (N) - A characteristic property of a specific isotope of an element, equal to the number of neutrons in the nucleus.

Newton (N)* - The SI unit of force, equal to m kg s^{-2}. [1]

Nitriles - Compounds having the structure RC≡N; thus C-substituted derivatives of hydrocyanic acid, HC≡N. [5]

Nitrosamines - N-Nitroso amines: compounds of the structure R_2NNO. Compounds RNHNO are not ordinarily isolatable, but they, too, are nitrosamines. The name is a contraction of N-nitrosoamine and, as such, does not require the N locant. [5]

Nuclear magnetic resonance (NMR)* - A widely used technique in which the resonant absorption of radiofrequency radiation by magnetic nuclei in a magnetic field is measured. The results give important information on the local environment of each nucleus.

Nuclear magneton (μ_N)* - The unit of nuclear magnetic moment, defined as $eh/4\pi m_p$, where h is Planck's constant, m_p the proton mass, and e the elementary charge.

Nuclear quadrupole resonance (NQR) - See Techniques for Materials Characterization, page **12-1**.

Nuclear reaction analysis (NRA) - See Techniques for Materials Characterization, page **12-1**.

Nuclear spin (I) - The quantum number that specifies the intrinsic angular momentum of a particular nucleus. The magnitude of the angular momentum is given by $[I(I+1)]^{1/2} h/2\pi$, where h is Planck's constant.

Nucleic acids* - Macromolecules, the major organic matter of the nuclei of biological cells, made up of nucleotide units, and hydrolyzable into certain pyrimidine or purine bases (usually adenine, cytosine, guanine, thymine, uracil), D-ribose or 2-deoxy-D-ribose. [5]

Nucleon - A collective term for the proton and neutron.

Nucleosides - Ribosyl or deoxyribosyl derivatives (rarely, other glycosyl derivatives) of certain pyrimidine or purine bases. They are thus glycosylamines or N-glycosides related to nucleotides by the lack of phosphorylation. [5]

Nucleotides - Compounds formally obtained by esterification of the 3′ or 5′ hydroxy group of nucleosides with phosphoric acid. They are the monomers of nucleic acids and are formed from them by hydrolytic cleavage. [5]

Nuclide - A species of atoms in which each atom has identical atomic number Z and identical mass number A. [3]

Nusselt number (Nu) - A dimensionless quantity used in fluid mechanics, defined by $Nu = hl/k$, where h is coefficient of heat transfer, l is length, and k is thermal conductivity. [2]

Nyquist theorem - An expression for the mean square thermal noise voltage across a resistor, given by $4RkT\Delta f$ where R is the resistance, k the Boltzmann constant, T the temperature, and Δf the frequency band within which the voltage is measured.

Octanol-water partition coefficient (P)* - A measure of the way in which a compound will partition itself between the octanol and water phases in the two-phase octanol-water system, and thus an indicator of certain types of biological activity. Specifically, P is the ratio of the concentration (in moles per liter) of the compound in the octanol phase to that in the water phase at infinite dilution. The quantity normally reported is log P.

Oersted (Oe) - A non-SI unit of magnetic field (H), equal to 79.57747 A/m.

Ohm (Ω)* - The SI unit of electric resistance, equal to V/A. [1]

Ohm's law - A relation among electric current I, potential difference V, and resistance R, viz., $I = V/R$. The resistance is constant at constant temperature to high precision for many materials.

Olefins - Acyclic and cyclic hydrocarbons having one or more carbon-carbon double bonds, apart from the formal ones in aromatic compounds. The class olefins subsumes alkenes and cycloalkenes and the corresponding polyenes. [5]

Oligomer - A substance consisting of molecules of intermediate relative molecular mass (molecular weight), the structure of which essentially comprises the multiple repetition of units derived, actually or conceptually, from molecules of low relative molecular mass. In contrast to a polymer, the properties of an oligomer can vary significantly with the removal of one or a few of its units. [8]

Oligopeptides - Peptides containing from three to nine amino groups. [5]

Onsager relations - An important set of equations in the thermodynamics of irreversible processes. They express the symmetry between the transport coefficients describing reciprocal processes in systems with a linear dependence of flux on driving forces.

Optical rotary power - Angle by which the plane of polarization of a light beam is rotated by an optically active medium, divided by path length and by concentration of the active constituent. Depending on whether mass or molar concentration is used, the modifier "specific" or "molar" is attached. [2]

Orbital - A one-electron wavefunction. Atomic orbitals are classified as s-, p-, d-, or f-orbitals according to whether the angular momentum quantum number $l = 0, 1, 2,$ or 3. Molecular orbitals, which are usually constructed as linear combinations of atomic orbitals, describe the distribution of electrons over the entire molecule.

Oscillator strength (f) - A measure of the intensity of a spectroscopic transition, defined by

$$f = \frac{8\pi^2 Me\nu}{3he^2}\left|\mu_{ij}\right|^2$$

where ν is the frequency, μ_{ij} the transition dipole moment, m_e the mass of the electron, e the elementary charge, and h Planck's constant.

Osmosis - The flow of a solvent in a system in which two solutions of different concentration are separated by a semipermeable membrane which cannot pass solute molecules. The solvent will flow from the side of lower concentration to that of higher concentration, thus tending to equalize the concentrations. The pressure that must be applied to the more concentrated side to stop the flow is called the osmotic pressure.

Osmotic coefficient (ϕ) - Defined by $\phi = \ln a_A/(M_A \Sigma m_B)$, where M_A is the molar mass of substance A (normally the solvent), a_A is its activity, and the m_B are molalities of the solutes. [1]

Osmotic pressure (Π) - The excess pressure necessary to maintain osmotic equilibrium between a solution and the pure solvent separated by a membrane permeable only to the solvent. In an ideal dilute solution $\Pi = c_B RT$, where c_B is the amount-of-substance concentration of the solute, R is the molar gas constant, and T the temperature. [1,2]

Ostwald dilution law - A relation for the concentration dependence of the molar conductivity Λ of an electrolyte solution, viz.,

$$\frac{1}{\Lambda} = \frac{1}{\Lambda^\circ} + \frac{\Lambda c}{K\left(\Lambda^\circ\right)^2}$$

where c is the solute concentration, K is the equilibrium constant for dissociation of the solute, and Λ° is the conductivity at $c\Lambda = 0$.

Ounce (oz) - A non-SI unit of mass. The avoirdupois once equals 28.34952 g, while the troy ounce equals 31.10348 g.

Overpotential (η) - In an electrochemical cell, the difference between the potential of an electrode and its zero-current value.

Oximes - Compounds of structure R_2C=NOH derived from condensation of aldehydes or ketones with hydroxylamine. Oximes from aldehydes may be called aldoximes; those from ketones may be called ketoximes. [5]

Oxo compounds - Compounds containing an oxygen atom, =O, doubly bonded to carbon or another element. The term thus embraces aldehydes, carboxylic acids, ketones, sulfonic acids, amides and esters. [5]

Ozonides - The 1,2,4-trioxolanes formed by the reaction of ozone at a carbon-carbon double bond, or the analogous compounds derived from acetylenic compounds. [5]

Pair production - A process in which a photon is converted into a particle and its antiparticle (e.g., an electron and positron) in the electromagnetic field of a nucleus.

Paraffins - Obsolescent term for saturated hydrocarbons, commonly but not necessarily acyclic. Still widely used in the petrochemical industry, where the term designates acyclic saturated hydrocarbons, and stands in contradistinction to naphthenes. [5]

Paramagnetism* - A type of magnetism characterized by a positive magnetic susceptibility, so that the material becomes weakly magnetized in the direction of an external field. The magnetization disappears when the field in removed. In the simplest approximation (Curie's law) the susceptibility is inversely proportional to temperature.

Parity - The property of a quantum-mechanical wave function that describes its behavior under the symmetry operation of coordinate inversion. A parity of +1 (or even) is assigned if the wave function does not change sign when the signs of all the coordinates are changed; the parity is -1 (or odd) if the wave function changes sign under this operation.

Parsec (pc) - A unit of distance defined as the distance at which 1 astronomical unit (AU) subtends an angle of 1 second of arc. It is equal to 206264.806 AU or 3.085678×10^{16} m.

Particle induced x-ray emission (PIXE) - See Techniques for Materials Characterization, page 12-1.

Partition function (q, z) - For a single molecule, $q = \Sigma_i g_i \exp(\varepsilon_i/kT)$, where ε_i is an energy level of degeneracy g_i, k the Boltzmann constant, and T the absolute temperature; the summation extends over all energy states. For a system of N non-interacting molecules which are indistinguishable, as in an ideal gas, the canonical partition function $Q = q^N/N!$.

Pascal (Pa)* - The SI unit of pressure, equal to N/m^2. [1]

Paschen series - The series of lines in the spectrum of the hydrogen atom which corresponds to transitions between the state with principal quantum number $n = 3$ and successive higher states. The wavelengths are given by $1/\lambda = R_H(1/9 - 1/n^2)$, where $n = 4,5,6,\ldots$ and R_H is the Rydberg constant. The first member of the series ($n = 3\leftrightarrow4$), which is often called the P_α line, falls in the infrared at a wavelength of 1.875 μm.

Paschen-Back effect - In atomic spectroscopy, the decoupling of electron spin from orbital angular momentum as the strength of an external magnetic field is increased.

Pauli exclusion principle - The statement that two electrons in an atom cannot have identical quantum numbers; thus if there are two electrons in the same orbital, their spin quantum numbers must be of opposite sign.

Pearson symbol - A code for designating crystallographic information, including the crystal system, the lattice type, and the number of atoms per unit cell.

Péclet number (Pe) - A dimensionless quantity used in fluid mechanics, defined by $Pe = vl/a$, where v is velocity, l is length, and a is thermal diffusivity. [2]

Peltier effect - The absorption or generation of heat (depending on the current direction) which occurs when an electric current is passed through a junction between two materials.

Peptides - Amides derived from two or more amino carboxylic acid molecules (the same or different) by formation of a covalent bond from the carbonyl carbon of one to the nitrogen atom of another with formal loss of water. [5]

Permeability (μ) - Magnetic induction divided by magnetic field strength; i.e. $\mu = B/H$. The relative permeability $\mu_r = \mu/\mu_0$, where μ_0 is the permeability of a vacuum. [1]

Permittivity (ε) - Ratio of the electric displacement in a medium to the electric field strength. Also called dielectric constant. [1]

Peroxides - Compounds of structure ROOR in which R may be any organic group. In inorganic chemistry, salts of the anion O_2^{-2} [5]

Peroxy acids - Acids in which an acidic -OH group has been replaced by an -OOH group; e.g. $CH_3C(=O)OOH$ peroxyacetic acid, $PhS(=O)_2OOH$ benzeneperoxysulfonic acid. [5]

Petroleum ether - The petroleum fraction consisting of C_5 and C_6 hydrocarbons and boiling in the range 35-60°C; commonly used as a laboratory solvent.

pH* - A convenient measure of the acid-base character of a solution, usually defined by pH = -log $[c(H^+)/\text{mol L}^{-1})]$, where $c(H^+)$ is the concentration of hydrogen ions. The more precise definition is in terms af activity rather than concentration. [2]

Phenols - Compounds having one or more hydroxy groups attached to a benzene or other arene ring. [5]

Phonon - A quantum of energy associated with a vibrational mode of a crystal lattice.

Phosphines - PH_3 and compounds derived from it by substituting one, two or three hydrogen atoms by hydrocarbyl groups. RPH_2, R_2PH and R_3P (R not equal to H) are called primary, secondary and tertiary phosphines, respectively. [5]

Phosphonium compounds - Salts (and hydroxides) $[R_4P]^+X^-$ containing tetracoordinate phosphonium ion and the associated anion. [5]

Phosphonium ylides - Compounds having the structure R_3P^+-C^-R_2 $\rightleftharpoons$ R_3P=CR_2. Also known as Wittig reagents. [5]

Phosphoresence - The process by which a molecule is excited by light to a higher electronic state and then undergoes a radiationless transition to a state of different multiplicity from which it decays, after some delay, to the ground state. The emitted light is normally of longer wavelength than the exciting light because vibrational energy has been dissipated.

Photoelectric effect - The complete absorption of a photon by a solid with the emission of an electron.

Photon - An elementary particle of zero mass and spin 1/2. The photon is involved in electromagnetic interactions and is the quantum of electromagnetic radiation.

Photon stimulated desorption (PSD) - See Techniques for Materials Characterization, page **12**-1.

Pinacols - Tetra(hydrocarbyl)ethane-1,2-diols, $R_2C(OH)C(OH)R_2$, of which the tetramethyl example is the simplest one and is itself commonly known as pinacol. [5]

Pion - An elementary particle in the family of mesons. Pions have zero spin and may be neutral or charged. They participate in the strong interaction which holds the nucleus together.

pK* - The negative logarithm (base 10) of an equilibrium constant K. For pK_a, see Acid dissociation constant.

Planck constant (h)* - The elementary quantum of action, which relates energy to frequency through the equation $E = h\nu$.

Planck distribution - See Black body radiation

Planck function - A thermodynamic function defined by $Y = -G/T$, where G is Gibbs energy and T thermodynamic temperature. [2]

Plasma - A highly ionized gas in which the charge of the electrons is balanced by the charge of the positive ions, so that the system as a whole is electrically neutral.

Plasmon - A quantum associated with a plasma oscillation in the electron gas of a solid.

Point group* - A group of symmetry operations (rotations, reflections, etc.) that leave a molecule invariant. Every molecular conformation can be assigned to a specific point group, which plays a major role in determining the spectrum of the molecule.

Poise (P) - A non-SI unit of viscosity, equal to 0.1 Pa s.

Poiseuille's equation - A formula for the rate of flow of a viscous fluid through a tube:

$$\frac{dV}{dt} = \frac{\left(p_1{}^2 - p_2{}^2\right)\pi r^4}{16l\eta p_0}$$

where V is the volume as measured at pressure p_0; p_1 and p_2 are the pressures at each end of the tube; r is the radius and l the length of the tube; and η is the viscosity.

Poisson ratio (μ) - The absolute value of the ratio of the transverse strain to the corresponding axial strain resulting from uniformly distributed axial stress below the proportional limit (i.e., where Hooke's law is valid). [10]

Polariton - A quantum associated with the coupled modes of photons and optical phonons in an ionic crystal.

Polarizability (α)* - The change in dipole moment of a molecule produced by an external electric field; specifically, $\alpha_{ab} = \partial p_a/\partial E_b$, where p_a is the dipole moment component on the a axis and E_b is the component of the electric field strength along the b axis. [2]

Polymer - A substance composed of molecules of high relative molecular mass (molecular weight), the structure of which essentially comprises the multiple repetition of units derived, actually or conceptually, from molecules of low relative molecular mass. A single molecule of a polymer is called a macromolecule. [8]

Polypeptides - Peptides containing 10 or more amino acid residues. See also Peptides. [5]

Polysaccharides - Compounds consisting of a large number of monosaccharides linked glycosidically. This term is commonly used only for those containing more than ten monosaccharide residues. Also called glycans. [5]

Porphyrins - Natural pigments containing a fundamental skeleton of four pyrrole nuclei united through the α-positions by four methine groups to form a macrocyclic structure (porphyrin is designated porphine in Chemical Abstracts indexes). [5]

Positron - The antiparticle of the electron. It has the same mass and spin as an electron, and an equal but opposite charge.

Positronium - The hydrogen-like "atom" formed from a positron nucleus and an electron. Its lifetime is very short because of annihilation of the positron and electron.

Potential - See Electric potential

Potential energy (E_p, V, U) - The portion of the energy of a system that is associated with its position in a force field.

Pound (lb) - A non-SI unit of mass, equal to 0.4535924 kg.

Power (P) - Rate of energy transfer. For electrical circuits, this is equal to the product of current and potential difference, $P = IV$. [1]

Poynting vector (S) - For electromagnetic radiation, the vector product of the electric field strength and the magnetic field strength. [1]

Prandtl number (Pr) - A dimensionless quantity used in fluid mechanics, defined by $Pr = \eta/\rho a$, where η is viscosity, ρ is density, and a is thermal diffusivity. [2]

Pressure* - Force divided by area. [1]

Proteins - Naturally occurring and synthetic polypeptides having molecular weights greater than about 10,000 (the limit is not precise). See also Peptides. [5]

Proton* - A stable elementary particle of unit positive charge and spin 1/2. Protons and neutrons, which are collectively called nucleons, are the constituents of the nucleus.

Pulsar - A neutron star which rotates rapidly and emits electromagnetic radiation in regular pulses at a frequency related to the rotation period.

Purine bases* - Purine and its substitution derivatives, especially naturally occurring examples. [5]

Pyrimidine bases* - Pyrimidine and its substitution derivatives, especially naturally occurring examples. [5]

Q-switching - A technique for obtaining very high power from a laser by keeping the Q factor of the laser cavity low while the population inversion builds up, then suddenly increasing the Q to initiate the stimulated emission.

Quad - A unit of energy defined as 10^{15} Btu, equal to approximately 1.055056×10^{18} J.

Quadrupole moment - A coefficient of the third term (after monopole and dipole) in the power series expansion of the electric potential of an array of charges. A nucleus of spin greater than 1/2 has a non-vanishing nuclear quadrupole moment which can interact with the electric field gradient of the surrounding electrons. Molecular quadrupole moments have an influence on intermolecular forces.

Quality factor (Q) - The ratio of the absolute value of the reactance of an electrical system to the resistance; thus a measure of the energy stored per cycle relative to the energy dissipated.

Quantum yield - In photochemistry, the number of moles transformed in a specific process, either physically (e.g., by emission of photons) or chemically, per mole of photons absorbed by the system. [3]

Quark - An elementary entity which has not been directly observed but is considered a constituent of protons, neutrons, and other hadrons.

Quasar - An extragalactic object emitting electromagnetic radiation at a very high power level and showing a very large red shift, thus indicating that the object is receding at a speed approaching the speed of light.

Quasicrystal - A solid having conventional crystalline properties but whose lattice does not display translational periodicity.

Quaternary ammonium compounds - Derivatives of ammonium compounds, $NH_4{}^+Y^-$, in which all four of the hydrogens bonded to nitrogen have been replaced with hydrocarbyl groups. Compounds having a carbon-nitrogen double bond (i.e. $R_2C=N^+R_2Y^-$) are more accurately called iminium compounds. [5]

Quinones - Compounds having a fully conjugated cyclic dione structure, such as that of benzoquinones, derived from aromatic compounds by conversion of an even number of -CH= groups into -C(=O)- groups with any necessary rearrangement of double bonds. [5]

Racemic mixture - A mixture of equal amounts of a pair of enantiomers (optical isomers); such a mixture is not optically active.

Rad - A non-SI unit of absorbed dose of radiation, equal to 0.01 Gy.

Radiance (L) - The radiant intensity in a given direction from an element of a surface, divided by the area of the orthogonal projection of this element on a plane perpendicular to the given direction. [1]

Radiant intensity (I) - The radiant energy flux leaving an element of a source within an element of solid angle, divided by that element of solid angle. [1]

Radicals - Molecular entities possessing an unpaired electron, such as $\cdot CH_3$, $\cdot SnH_3$, $\cdot Cl$. (In these formulas the dot, symbolizing the unpaired electron, should be placed so as to indicate the atom of highest spin density, if this is possible). [5]

Raman effect - The inelastic scattering of light by a molecule, in which the incident photon either gives up to, or receives energy from, one of the internal vibrational modes of the molecule. The scattered light thus has either a lower frequency (Stokes radiation) or higher frequency (anti-Stokes radiation) than the incident light. These shifts provide a measure of the normal vibrational frequencies of the molecule.

Rankine cycle - A thermodynamic cycle which can be used to calculate the ideal performance of a heat engine that uses a condensable vapor as the working fluid (e.g., a steam engine or a heat pump).

Rankine temperature - A thermodynamic temperature scale based on a temperature interval $°R = (5/9)$ K; i.e., $T/°R = (9/5)T/K = t/°F + 459.67$.

Raoult's law - The expression for the vapor pressure p_i of component i in an ideal solution, viz., $p_i = x_i p_{i0}$, where x_i is the mole fraction of component i and p_{i0} the vapor pressure of the pure substance i.

Rare earth elements - The elements Sc, Y, and the lanthanides (La, Ce, Pr, Nd, Pm, Sm, Eu, Gd, Tb, Dy, Ho, Er, Tm, Yb, Lu). [7]

Rayleigh number (Ra) - A dimensionless quantity used in fluid mechanics, defined by $Ra = l^3 g \alpha \Delta T \rho / \eta a$, where l is length, g is acceleration of gravity, α is cubic expansion coefficient, T is temperature, ρ is density, η is viscosity, and a is thermal diffusivity. [2]

Rayleigh scattering - The scattering of light by particles which are much smaller than the wavelength of the light. It is characterized by a scattered intensity which varies as the inverse fourth power of the wavelength.

Rayleigh wave - A guided elastic wave along the surface of a solid; also called surface acoustic wave.

Reactance (X) - The imaginary part of impedance. For an inductive reactance L and a capacitive reactance C in series, the reactance is $X = L\omega - 1/(C\omega)$, where ω is 2π times the frequency of the current. [1]

Red shift - A displacement of a spectral line toward longer wavelengths. This can occur through the Doppler effect (e.g., in the light from receding galaxies) or, in the general theory of relativity, from the effects of a star's gravitational field.

Reflectance (ρ) - Ratio of the radiant or luminous flux at a given wavelength that is reflected to that of the incident radiation. Also called reflection factor. [1]

Reflection high energy electron diffraction (RHEED) - See Techniques for Materials Characterization, page **12-1**.

Relative humidity* - The ratio of the partial pressure of water vapor in air to the saturation vapor pressure of water at the same temperature, expressed as a percentage. [10]

Relative molar mass - See Molecular weight.

Rem - A non-SI unit of dose equivalent, equal to 0.01 Sv.

Resistance (R) - Electric potential difference divided by current when there is no electromotive force in the conductor. This definition applies to direct current. More generally, resistance is defined as the real part of impedance. [1]

Resistivity (ρ) - Electric field strength divided by current density when there is no electromotive force in the conductor. Resistivity is an intrinsic property of a material. For a conductor of uniform cross section with area A and length L, and whose resistance is R, the resistivity is given by $\rho = RA/L$. [1]

Reynolds number (Re) - A dimensionless quantity used in fluid mechanics, defined by $Re = \rho v l / \eta$, where ρ is density, v is velocity, l is length, and η is viscosity. [2]

Rheology - The study of the flow of liquids and deformation of solids. Rheology addresses such phenomena as creep, stress relaxation, anelasticity, nonlinear stress deformation, and viscosity.

Ribonucleic acids (RNA) - Naturally occurring polyribonucleotides. See also nucleic acids, nucleosides, nucleotides, ribonucleotides. [5]

Ribonucleotides - Nucleotides in which the glycosyl group is a ribosyl group. See also nucleotides. [5]

Roentgen (R) - A unit used for expressing the charge (positive or negative) liberated by x-ray or γ radiation in air, divided by the mass of air. A roentgen is defined as 2.58×10^{-4} C/kg.

Rotational constants - In molecular spectroscopy, the constants appearing in the expression for the rotational energy levels as a function of the angular momentum quantum numbers. These constants are proportional to the reciprocals of the principal moments of inertia, averaged over the vibrational motion.

Rutherford back scattering (RBS) - See Techniques for Materials Characterization, page **12-1**.

Rydberg constant (R_∞)* - The fundamental constant which appears in the equation for the energy levels of hydrogen-like atoms; i.e., $E_n = hcR_\infty Z^2 \mu / n^2$, where h is Planck's constant, c the speed of light, Z the atomic number, μ the reduced mass of nucleus and electron, and n the principal quantum number ($n = 1, 2, \ldots$).

Rydberg series - A regular series of lines in the spectrum of an atom or molecule, with the spacing between successive lines becoming smaller as the frequency increases (wavelength decreases). The series eventually converges to a limit which usually corresponds to the complete removal of an electron from the atom or molecule.

Sackur-Tetrode equation* - An equation for the molar entropy S_m of an ideal monatomic gas: $S_m = R\ln(e^{5/2} V/N_A \Lambda^3)$, where R is the molar gas constant, V is the volume, and N_A is Avogadro's number. The constant Λ is given by $\Lambda = h/(2\pi mkT)^{1/2}$, where h is Planck's constant, m the atomic mass, k the Boltzmann constant, and T the temperature.

Salinity (S)* - A parameter used in oceanography to describe the concentration of dissolved salts in seawater. It is defined in terms of electrical conductivity relative to a standard solution of KCl. When expressed in units of parts per thousand, S may be roughly equated to the concentration of dissolved material in grams per kilogram of seawater.

Salt - An ionic compound formed by the reaction of an acid and a base.

Scanned probe microscopy (SPM) - See Techniques for Materials Characterization, page **12-1**.

Scanning electron microscopy (SEM) - See Techniques for Materials Characterization, page **12-1**.

Scanning laser acoustic microscopy (SLAM) - See Techniques for Materials Characterization, page **12-1**.

Scanning transmission electron microscopy (STEM) - See Techniques for Materials Characterization, page **12-1**.

Scanning tunneling microscopy (STM) - See Techniques for Materials Characterization, page **12-1**.

Schiff bases - Imines bearing a hydrocarbyl group on the nitrogen atom: $R_2C=NR'$ (R' not equal to H). Considered by many to be synonymous with azomethines. [5]

Schmidt number (Sc) - A dimensionless quantity used in fluid mechanics, defined by $Sc = \eta/\rho D$, where η is viscosity, ρ is density, and D is diffusion coefficient. [2]

Schottky barrier - A potential barrier associated with a metal-semiconductor contact. It forms the basis for the rectifying device known as the Schottly diode.

Schrödinger equation - The basic equation of wave mechanics which, for systems not dependent on time, takes the form:

$$-(\hbar/2m)\nabla^2\psi + V\psi = E\psi$$

where ψ is the wavefunction, V is the potential energy expressed as a function of the spatial coordinates, E is an energy eigenvalue, ∇^2 is the Laplacian operator, $\hbar$ is Planck's constant divided by 2π, and m is the mass.

Second (s)* - The SI base unit of time. [1]

Second radiation constant (c_2)* - See First radiation constant.

Secondary ion mass spectroscopy (SIMS) - See Techniques for Materials Characterization, page 12-1.

Seebeck effect - The development of a potential difference in a circuit where two different metals or semiconductors are joined and their junctions maintained at different temperatures. It is the basis of the thermocouple.

Selenides - Compounds having the structure RSeR (R not equal to H). They are thus selenium analogues of ethers. Also used for metal salts of H_2Se. [5]

Semicarbazones - Compounds having the structure $R_2C=NNHC(=O)NH_2$, formally derived by condensation of aldehydes or ketones with semicarbazide [$NH_2NHC(=O)NH_2$]. [5]

Semiconductor - A material in which the highest occupied energy band (valence band) is completely filled with electrons at $T = 0$ K, and the energy gap to the next highest band (conduction band) ranges from 0 to 4 or 5 eV. With increasing temperature electrons are excited into the conduction band, leading to an increase in the electrical conductivity.

Semiquinones - Radical anions having the structure -O-Z-O· where Z is an ortho- or para-arylene group or analogous heteroarylene group; they are formally generated by the addition of an electron to a quinone. [5]

SI units* - The International System of Units adopted in 1960 and recommended for use in all scientific and technical fields. [1]

Siemens (S)* - The SI unit of electric conductance, equal to Ω^{-1}. [1]

Sievert (Sv)* - The SI unit of dose equivalent (of radiation), equal to J/kg. [1]

Silanes - Saturated silicon hydrides, analogues of the alkanes; i.e. compounds of the general formula Si_nH_{2n+2}. Silanes may be subdivided into silane, oligosilanes, and polysilanes. Hydrocarbyl derivatives are often referred to loosely as silanes. [5]

Silicones - Polymeric or oligomeric siloxanes, usually considered unbranched, of general formula [-OSiR$_2$-]$_n$ (R not equal to H). [5]

Siloxanes - Saturated silicon-oxygen hydrides with unbranched or branched chains of alternating silicon and oxygen atoms (each silicon atom is separated from its nearest silicon neighbors by single oxygen atoms). [5]

Skin effect - The concentration of high frequency alternating currents near the surface of a conductor.

Slater orbital - A particular mathematical expression for the radial part of the wave function of a single electron, which is used in quantum-mechanical calculations of the energy and other properties of atoms and molecules.

Small angle neutron scattering (SANS) - See Techniques for Materials Characterization, page 12-1.

Snell's law - The relation between the angle of incidence i and the angle of refraction r of a light beam which passes from a medium of refractive index n_0 to a medium of index n_1, viz., sin i/sin $r = n_1/n_0$.

Solar constant* - The mean radiant energy flux from the sun on a unit surface normal to the direction of the rays at the mean distance of the earth from the sun. The value is approximately 1373 W/m^2.

Solar wind - The stream of high velocity hydrogen and helium ions emitted by the sun which flows through the solar system and beyond.

Soliton - A spatially localized wave in a solid or liquid that can interact strongly with other solitons but will afterwards regain its original form.

Solubility* - A quantity expressing the maximum concentration of some material (the solute) that can exist in another liquid or solid material (the solvent) at thermodynamic equilibrium at specified temperature and pressure. Common measures of solubility include the mass of solute per unit mass of solution (mass fraction), mole fraction of solute, molality, molarity, and others.

Solubility product constant (K_{sp})* - The equilibrium constant for the dissolution of a sparsely soluble salt into its constituent ions.

Space group* - A group of symmetry operations (reflections, rotations, etc.) that leave a crystal invariant. A total of 230 space groups have been identified.

Spark source mass spectroscopy (SSMS) - See Techniques for Materials Characterization, page 12-1.

Specific gravity - Ratio of the mass density of a material to that of water. Since one must specify the temperature of both the sample and the water to have a precisely defined quantity, the use of this term is now discouraged.

Specific heat - Heat capacity divided by mass. See Heat capacity.

Specific quantity - It is often convenient to express an extensive quantity (e.g., volume, enthalpy, heat capacity, etc.) as the actual value divided by mass. The resulting quantity is called specific volume, specific enthalpy, etc.

Specific rotation [α]$^\theta_\lambda$ - For an optically active substance, defined by $[\alpha]^\theta_\lambda = \alpha/\gamma l$, where α is the angle through which plane polarized light is rotated by a solution of mass concentration γ and path length l. Here θ is the Celsius temperature and λ the wavelength of the light at which the measurement is carried out. Also called specific optical rotatory power. [2]

Spin (s, I)* - A measure of the intrinsic angular momentum of a particle, which it possesses independent of its orbital motion. The symbol s is used for the spin quantum number of an electron, while I is generally used for nuclear spin.

Spiro compounds - Compounds having one atom (usually a quaternary carbon) as the only common member of two rings. [5]

Stacking fault - An error in the normal sequence of layer growth in a crystal.

Standard mean ocean water (SMOW) - A standard sample of pure water of accurately known isotopic composition which is maintained by the International Atomic Energy Agency. It is used for precise calibration of density and isotopic composition measurements.

Standard reduction potential ($E°$) - The zero-current potential of a cell in which the specified reduction reaction occurs at the right-hand electrode and the left-hand electrode is the standard hydrogen electrode. Also called Standard electrode potential.

Standard state - A defined state (specified temperature, pressure, concentration, etc.) for tabulating thermodynamic functions and carrying out thermodynamic calculations. The standard state pressure is usually taken as 100,000 Pa (1 bar), but various standard state temperatures are used. [2]

Stanton number (St) - A dimensionless quantity used in fluid mechanics, defined by $St = h/\rho v c_p$, where h is coefficient of heat transfer, ρ is density, v is velocity, and c_p is specific heat capacity at constant pressure. [2]

Stark effect - The splitting of an energy level of an atom or molecule, and hence a splitting of spectral lines arising from that level, as a result of the application of an external electric field.

Statistical weight (g) - The number of distinct states corresponding to the same energy level. Also called degeneracy.

Stefan-Boltzmann constant (σ)* - Constant in the equation for the radiant exitance M (radiant energy flux per unit area) from a black body at thermodynamic temperature T, viz. $M = \sigma T^4$. [1]

Stibines - SbH_3 and compounds derived from it by substituting one, two or three hydrogen atoms by hydrocarbyl groups: R_3Sb. $RSbH_2$, R_2SbH, and R_3Sb (R not equal to H) are called primary, secondary and tertiary stibines, respectively. [5]

Stochastic process - A process which involves random variables and whose outcome can thus be described only in terms of probabilities.

Stoichiometric number (v) - The number appearing before the symbol for each compound in the equation for a chemical reaction. By convention, it is negative for reactants and positive for products. [2]

Stokes (St) - A non-SI unit of kinematic viscosity, equal to 10^{-4} m^2/s.

Stokes' law - The statement, valid under certain conditions, that the viscous force F experienced by a sphere of radius a moving at velocity v in a medium of viscosity η is given by $F = -6\pi\eta av$.

Strain - The deformation of a body that results from an applied stress.

Stratosphere - The part of the earth's atmosphere extending from the top of the troposphere (typically 10 to 15 km above the surface) to about 50 km. It is characterized by an increase in temperature with increasing altitude.

Stress - Force per unit area (pressure) applied to a body. Tensile stress tends to stretch or compress the body in the direction of the applied force. Sheer stress results from a tangential force which tends to twist the body.

Strong interaction - The short range (order of 1 fm) attractive forces between protons, neutrons, and other hadrons which are responsible for the stability of the nucleus.

Strouhal number (Sr) - A dimensionless quantity used in fluid mechanics, defined by $Sr = lf/v$, where l is length, f is frequency, and v is velocity. [2]

Structure factor - In x-ray crystallography, the sum of the scattering factors of all the atoms in a unit cell, weighted by an appropriate phase factor. The intensity of a given reflection is proportional to the square of the structure factor.

Sublimation pressure - The pressure of a gas in equilibrium with a solid at a specified temperature.

Sulfides - Compounds having the structure RSR (R not equal to H). Such compounds were once called thioethers. In an inorganic sense, salts or other derivatives of hydrogen sulfide. [5]

Sulfones - Compounds having the structure, $RS(=O)_2R$ (R not equal to H), e.g. $C_2H_5S(=O)_2CH_3$, ethyl methyl sulfone. [5]

Sulfonic acids - $HS(=O)_2OH$, sulfonic acid, and its S-hydrocarbyl derivatives. [5]

Sulfoxides - Compounds having the structure $R_2S=O$ (R not equal to H), e.g. $Ph_2S=O$, diphenyl sulfoxide. [5]

Superconductor - A material that experiences a nearly total loss of electrical resistivity below a critical temperature T_c. The effect can occur in pure metals, alloys, semiconductors, organic compounds, and certain inorganic solids.

Superfluid - A fluid with near-zero viscosity and extremely high thermal conductivity. Liquid helium exhibits these properties below 2.186 K (the λ point).

Supernova - A star in the process of exploding because of instabilities which follow the exhaustion of its nuclear fuel.

Surface analysis by laser ionization (SALI) - See Techniques for Materials Characterization, page 12-1.

Surface tension (γ, σ)* - The force per unit length in the plane of the interface between a liquid and a gas, which resists an increase in the area of that surface. It can also be equated to the surface Gibbs energy per unit area.

Surfactant - A substance which lowers the surface tension of the medium in which it is dissolved, and/or the interfacial tension with other phases, and accordingly is positively adsorbed at the liquid-vapor or other interfaces. [3]

Susceptance (B) - Imaginary part of admittance. [1]

Svedberg - A non-SI unit of time, used to express sedimentation coefficients, equal to 10^{-13} s.

Syndiotactic macromolecule - A tactic macromolecule, essentially comprising alternating enantiomeric configurational base units which have chiral or prochiral atoms in the main chain in a unique arrangement with respect to their adjacent constitutional units. In this case the repeating unit consists of two configurational base units that are enantiomeric. [8]

Tacticity - The orderliness of the succession of configurational repeating units of a macromolecule or oligomer molecule. In a tactic macromolecule essentially all the configurational repeating units are identical with respect to directional sense. See Configurational repeating unit, Isotactic, Syndiotactic. [8]

Tautomerism - Isomerism of the general form G-X-Y=Z $\rightleftharpoons$ X=Y-Z-G, where the isomers (called tautomers) are readily interconvertible; the atoms connecting the groups X, Y, Z are typically any of C, H, O, or S, and G is a group which becomes an electrofuge (i.e., a group that does not carry away the bonding electron pair when it leaves its position in the molecule) or nucleofuge (a group that does carry away the bonding electrons when leaving) during isomerization. The commonest case, when the electrofuge is H^+, is also known as prototropy. A common example, written so as to illustrate the general pattern given above, is keto-enol tautomerism, such as

$$H-O-C(CH_3)=CH-CO_2Et \text{ (enol)} \rightleftharpoons (CH_3)C(=O)-CH_2-CO_2Et \text{ (keto)}$$

In some cases the interconversion rate between tautomers is slow enough to permit isolation of the separate keto and enol forms. [5]

Tensile strength* - In tensile testing, the ratio of maximum load a body can bear before breaking to original cross-sectional area. Also called ultimate strength. [11]

Terpenes - Hydrocarbons of biological origin having carbon skeletons formally derived from isoprene $[CH_2=C(CH_3)CH=CH_2]$. [5]

Terpenoids - Natural products and related compounds formally derived from isoprene units. They contain oxygen in various functional groups. The skeleton of terpenoids may differ from strict additivity of isoprene units by the loss or shift of a methyl (or other) group. [5]

Tesla (T)* - The SI unit of magnetic flux density (B), equal to V s/m^2. [1]

Thermal conductivity* - Rate of heat flow divided by area and by temperature gradient. [1]

Thermal diffusivity - Thermal conductivity divided by density and by specific heat capacity at constant pressure. [1]

Thermal expansion coefficient (α)* - The linear expansion coefficient is defined by $\alpha_l = (1/l)(dl/dT)$; the volume expansion coefficient by $\alpha_V = (1/V)(dV/dT)$. [1]

Thermionic emission - The emission of electrons from a solid as a result of heat. The effect requires a high enough temperature to impart sufficient kinetic energy to the electrons to exceed the work function of the solid.

Thermodynamic laws - The foundation of the science of thermodynamics:

First law: The internal energy of an isolated system is constant; if energy is supplied to the system in the form of heat dq and work dw, then the change in energy $dU = dq + dw$.

Second law: No process is possible in which the only result is the transfer of heat from a reservoir and its complete conversion to work.

Third law: The entropy of a perfect crystal approaches zero as the thermodynamic temperature approaches zero.

Thermoelectric power - For a bar of a pure material whose ends are at different temperatures, the potential difference divided by the difference in temperature of the ends. See also Seeback effect.

Thermogravimetric analysis (TGA) - See Techniques for Materials Characterization, page **12**-1.

Thermosphere - The layer of the earth's atmosphere extending from the top of the mesosphere (typically 80-90 km above the surface) to about 500 km. It is characterized by a rapid increase in temperature with increasing altitude up to about 200 km, followed by a leveling off in the 300-500 km region.

Thiols - Compounds having the structure RSH (R not equal to H). Also known by the term mercaptans (abandoned by IUPAC); e.g. CH_3CH_2SH, ethanethiol. [5]

Thomson coefficient (μ, τ) - The heat power developed in the Thomson effect (whereby heat is evolved in a conductor when a current is flowing in the presence of a temperature gradient), divided by the current and the temperature difference. [1]

Tonne (t) - An alternative name for megagram (1000 kg). [1]

Torque (T) - For a force F that produces a torsional motion, $T = r \times F$, where r is a vector from some reference point to the point of application of the force.

Torr - A non-SI unit of pressure, equal to 133.322 Pa. The name is generally considered interchangeable with millimeter of mercury.

Townsend coefficient - In a radiation counter, the number of ionizing collisions by an electron per unit path length in the direction of an applied electric field.

Transducer - Any device that converts a signal from acoustical, optical, or some other form of energy into an electrical signal (or vice versa) while preserving the information content of the original signal.

Transistor - A voltage amplifier using controlled electron currents inside a semiconductor.

Transition metals - Elements characterized by a partially filled d subshell. The First Transition Series comprises Sc, Ti, V, Cr, Mn, Fe, Co, Ni, Cu. The Second and Third Transition Series include the lanthanides and actinides, respectively. [7]

Transition probability* - See Einstein transition probability.

Transmittance (τ) - Ratio of the radiant or luminous flux at a given wavelength that is transmitted to that of the incident radiation. Also called transmission factor. [1]

Tribology - The study of frictional forces between solid surfaces.

Triple point* - The point in p,T space where the solid, liquid, and gas phases of a substance are in thermodynamic equilibrium. The corresponding temperature and pressure are called the triple point temperature and triple point pressure.

Troposphere - The lowest part of the earth's atmosphere, extending to 10-15 km above the surface. It is characterized by a decrease in temperature with increasing altitude. The exact height varies with latitude and season.

Tunnel diode - A device involving a p-n junction in which both sides are so heavily doped that the Fermi level on the p-side lies in the valence band and on the n-side in the conduction band. This leads to a current-voltage curve with a maximum, so that the device exhibits a negative resistance in some regions.

Ultraviolet photoelectron spectroscopy (UPS) - See Techniques for Materials Characterization, page **12**-1.

Umklapp process - A process involving the interaction of three or more waves (lattice or electron) in a solid in which the sum of the wave vectors does not equal zero.

Unified atomic mass unit (u)* - A unit of mass used in atomic, molecular, and nuclear science, defined as the mass of one atom of ^{12}C divided by 12. Its approximate value is 1.66054×10^{-27} kg. [1]

Universal time (t_U, UT) - Mean solar time counted from midnight at the Greenwich meridian. Also called Greenwich mean time (GMT). The interval of mean solar time is based on the average, over one year, of the time between successive transits of the sun across the observer's meridian.

Vacancy - A missing atom or ion in a crystal lattice.

Van Allen belts - Two toroidal regions above the earth's atmosphere containing protons and electrons. The outer belt at about 25,000 km above the surface is probably of solar origin. The inner belt at about 3000 km contains more energetic particles from outside the solar system.

Van der Waals' equation* - An equation of state for fluids which takes the form:

$$pV_m = RT\left(\frac{1}{V_m - b} - \frac{a}{V_m{}^2}\right)$$

where p is pressure, V_m is molar volume, T is temperature, R is the molar gas constant, and a and b are characteristic parameters of the substance which describe the effect of attractive and repulsive intermolecular forces, respectively.

Van der Waals' force - The weak attractive force between two molecules which arises from electric dipole interactions. It can lead to the formation of stable but weakly bound dimer molecules or clusters.

Van't Hoff equation - The equation expressing the temperature dependence of the equilibrium constant K of a chemical reaction:

$$\frac{d\ln k}{dT} = \frac{\Delta_r H^\circ}{RT^2}$$

where $\Delta_r H^\circ$ is the standard enthalpy of reaction, R the molar gas constant, and T the temperature. Also called van't Hoff isochore.

Vapor pressure* - The pressure of a gas in equilibrium with a liquid (or, in some usage, a solid) at a specified temperature.

Varistor - A device that utilizes the properties of certain metal oxides with small amounts of impurities, which show abrupt nonlinearities at specific voltages where the material changes from a semiconductor to an insulator.

Velocity (v) - Rate of change of distance with time.

Verdet constants (V)* - Angle of rotation of a plane polarized light beam passing through a medium in a magnetic field, divided by the field strength and by the path length.

Virial equation of state* - An equation relating the pressure p, molar volume V_m, and temperature T of a real gas in the form of an expansion in powers of the molar volume, viz., $pV_m = RT(1 + BV_m^{-1} + CV_m^{-2} + \ldots)$, where R is the molar gas constant. B is called the second virial coefficient, C the third virial coefficient, etc. The virial coefficients are functions of temperature.

Viscosity (η)* - The proportionality factor between sheer rate and sheer stress, defined through the equation $F = \eta\,A(dv/dx)$, where F is the tangential force required to move a planar surface of area A at velocity v relative to a parallel surface separated from the first by a distance x. Sometimes called dynamic or absolute viscosity. The term kinematic viscosity (symbol v) is defined as η divided by the mass density.

Volt (V)* - The SI unit of electric potential, equal to W/A. [1]

Volume fraction (ϕ_j) - Defined as $V_j/\Sigma_i V_i$, where V_j is the volume of the specified component and the V_i are the volumes of all the components of a mixture prior to mixing. [2]

Watt (W)* - The SI unit of power, equal to J/s. [1]

Wave function - A function of the coordinates of all the particles in a quantum mechanical system (and, in general, of time) which fully describes the state of the system. The product of the wave function and its complex conjugate is proportional to the probability of finding a particle at a particular point in space.

Weak interaction - The weak forces (order of 10^{-12} of the strong interaction) between elementary particles which are responsible for beta decay and other nuclear effects.

Weber (Wb)* - The SI unit of magnetic flux, equal to V s. [1]

Weber number (We) - A dimensionless quantity used in fluid mechanics, defined by $We = \rho v^2 l / \gamma$, where ρ is density, v is velocity, l is length, and γ is surface tension. [2]

Weight - That force which, when applied to a body, would give it an acceleration equal to the local acceleration of gravity. [1]

Wiedeman-Franz law - The law stating that the thermal conductivity k and electrical conductivity σ of a pure metal are related by $k = L\sigma T$, where T is the temperature and L (called the Lorenz ratio) has the approximate value 2.45×10^{-8} V^2/K^2.

Wien displacement law - The relation, which can be derived from the Planck formula for black body radiation, that $\lambda_{max} T = 0.0028978$ m K, where λ_{max} is the wavelength of maximum radiance at temperature T.

Wigner-Seitz method - A method of calculating electron energy levels in a solid using a model in which each electron is subject to a spherically symmetric potential.

Wittig reagents - See phosphonium ylides.

Work (W) - Force multiplied by the displacement in the direction of the force. [1]

Work function (Φ)* - The energy difference between an electron at rest at infinity and an electron at the Fermi level in the interior of a substance. It is thus the minimum energy required to remove an electron from the interior of a solid to a point just outside the surface. [1]

X unit (X) - A unit of length used in x-ray crystallography, equal to approximately 1.002×10^{-13} m.

X-ray photoelectron spectroscopy (XPS) - See Techniques for Materials Characterization, page **12**-1.

Yield strength - The stress at which a material exhibits a specified deviation (often chosen as 0.2% for metals) from proportionality of stress and strain. [11]

Young's modulus (E) - In tension or compression of a body below its elastic limit, the ratio of stress to corresponding strain. Since strain is normally expressed on a fractional basis, Young's modulus has dimensions of pressure. Also called elastic modulus. [11]

Zeeman effect - The splitting of an energy level of an atom or molecule, and hence a splitting of spectral lines arising from that level, as a result of the application of an external magnetic field.

Zener diode - A control device utilizing a p-n junction with a well defined reverse-bias avalanche breakdown voltage.

Zeotrope - A liquid mixture that shows no maximum or minimum when vapor pressure is plotted against composition at constant temperature. See Azeotrope.

Zero-point energy - The energy possessed by a quantum mechanical system as a result of the uncertainty principle even when it is in its lowest energy state; e.g., the difference between the lowest energy level of a harmonic oscillator and the minimum in the potential well.

Zeta potential (ζ) - The electric potential at the surface of a colloidal particle relative to the potential in the bulk medium at a long distance. Also called electrokinetic potential.

Zwitterions - Neutral compounds having formal unit electrical charges of opposite sign. Some chemists restrict the term to compounds with the charges on non-adjacent atoms. Sometimes referred to as inner salts, dipolar ions (a misnomer). [5]

Section 3
Physical Constants of Organic Compounds

PHYSICAL CONSTANTS OF ORGANIC COMPOUNDS

The basic physical constants for over 12,000 organic compounds are presented in this table, along with structures and references to other sources of information. An effort has been made to include the compounds most frequently encountered in the laboratory, the workplace, and the environment. The selection was based mainly on the appearance of the compounds in various specialized tables in this *Handbook* and in other widely used reference sources, such as the *Merck Index* and the *DIPPR Database of Pure Compound Properties*. The occurrence of a compound on regulatory lists of hazardous chemicals was also taken into consideration, as was the availability of reliable physical constant data. Clearly, criteria of this type are somewhat subjective, and compounds considered important by some users have undoubtedly been omitted. Suggestions for additional compounds or other improvements are welcomed.

The data in the table have been taken from many sources, including both compilations and the primary literature. Where conflicts were found, the value deemed most reliable was chosen. Some of the useful compilations of physical property data are listed at the end of this introduction.

The table is arranged alphabetically by the primary name, which is generally the Index Name from the 8th or 9th Collective Index of Chemical Abstracts Service (CAS). In a few cases, especially pesticides and pharmaceuticals, the common name is used rather than the more complex systematic name. By convention, CAS Index Names are written in inverted order, e.g., chloromethane is listed as methane, chloro and ethyl acetate as acetic acid, ethyl ester. Furthermore, certain important compounds are listed under Index Names which differ from the names by which they are commonly known (e.g. aniline appears as benzenamine and acetone as 2-propanone). In order to facilitate the location of compounds in the table, three indexes are provided:

- **Synonym Index:** Includes common synonyms, but not the primary name by which the table is arranged.
- **Molecular Formula Index:** Lists compounds by molecular formula in the Hill order (see Preface to this *Handbook*).
- **CAS Registry Number Index:** Lists compounds by Chemical Abstracts Service Registry Number.

Two lines of data appear for each compound. The explanation of the data fields follows.

Top Line:
- **No.:** An identification number used in the indexes and to identify the structure diagrams.
- **Name:** Primary name, generally the CAS Index Name.
- **Mol. Form.:** The molecular formula written in the Hill convention.
- **CAS RN:** The Chemical Abstracts Service Registry Number assigned by CAS as a unique identifier for the compound.
- **Merck No:** Monograph Number in *The Merck Index, Eleventh Edition*. It should be noted that this is not a unique identifier for a single compound, since several derivatives or isomers of a compound may be included in the same Monograph.
- **Beil. Ref:** Citation to the *Beilstein Handbook of Organic Chemistry*. An entry of 5-18-11-01234, for example, indicates that the compound may be found in the 5th Series, Volume 18, Subvolume 11, page 1234.
- **Solubility:** Solubility in common solvents on a relative scale: 1 = insoluble; 2 = slightly soluble; 3 = soluble; 4 = very soluble; 5 = miscible; 6 = decomposes. See List of Abbreviations for the solvent abbreviations.

Bottom line:
- **Synonym:** A synonym in common use. When the primary name is non-systematic, the systematic name appears here.
- **Mol. Wt.:** Molecular weight (relative molar mass) as calculated with the 1991 IUPAC Standard Atomic Weights.
- **mp/°C:** Normal melting point in °C. Although some values are quoted to 0.1°C, uncertainties are typically several degrees Celsius. A value is sometimes followed by "dec", indicating decomposition is observed at the stated temperature (so that it is probably not a true melting point). See the List of Abbreviations for other abbreviations.
- **bp/°C:** Boiling point in °C. When available, the normal boiling point is given first, without a superscript. This is the temperature at which the liquid phase is in equilibrium with the vapor at a pressure of 760 mmHg (101.325 kPa). Boiling point values at reduced pressure are also given in many cases; here the superscript indicates the pressure in mmHg. A "dec" or "exp" following the value indicates decomposition or explosion has been observed at the boiling point. A simple entry of "exp" (sometimes followed by a temperature) indicates explosion may occur on heating, even below the boiling point. An entry of "sub" indicates that no boiling point is available, but measurable vapor (sublimation) pressure has been observed upon heating the solid. A temperature may be given, but no precise meaning can be attached because the pressure is not specified.
- **den/g cm^{-3}:** density (mass per unit volume) in g/cm^{-3}. The superscript indicates the temperature in °C. Values are given only for the liquid and solid phases, and all values are true densities, not specific gravities. The number of decimal places gives a rough estimate of the accuracy of the value.
- **n_D:** Refractive index, at the temperature indicated by the superscript. Unless otherwise indicated, all values refer to a wavelength of 589 nm (sodium D line). Values are given only for liquids and solids.

Structures are given, when available, in the section following the main table, using the **No.** in the first column as the linking identifier.

LIST OF ABBREVIATIONS

Ac	Acetyl		EtOH	Ethanol
ace	Acetone		exp	Explodes
acid	Acids		gl	Glacial
AcOEt	Ethyl acetate		HOAc	Acetic acid
alk	Alkalis		hp	Heptane
anh	Anhydrous		hx	Hexane
aq	Aqueous		lig	Ligroin
bz	Benzene		MeOH	Methanol
chl	Chloroform		os	Organic solvents
con	Concentrated		peth	Petroleum ether
ctc	Carbon tetrachloride		$PhNO_2$	Nitrobenzene
cyhex	Cyclohexane		PrOH	1-Propanol
dec	Decomposes		py	Pyridine
dil	Dilute		sub	Sublimes
diox	Dioxane		sulf	Sulfuric acid
DMF	Dimethyl formamide		temp.	Temperature
DMSO	Dimethyl sulfoxide		tfa	Trifluoroacetic acid
DMSO-d_6	Deuterated dimethyl sulfoxide		tol	Toluene
eth	Ethyl ether		xyl	Xylene

REFERENCES

1. Lide, D. R., and Milne, G. W. A., Editors, *Handbook of Data on Organic Compounds, Third Edition*, CRC Press, Boca Raton, FL, 1993.
2. Luckenbach, R., Editor, *Beilstein Handbook of Organic Chemistry*, Springer-Verlag, Heidelberg. Data can also be accessed through the on-line Beilstein Database.
3. Budavari, S., Editor, *The Merck Index, Eleventh Edition*, Merck & Co., Rahway, NJ, 1989.
4. Daubert, T. E., Danner, R. P., Sibul, H. M., and Stebbins, C. C., *Physical and Thermodynamic Properties of Pure Compounds: Data Compilation*, extant 1994 (core with 4 supplements), Taylor & Francis, Bristol, PA (also available as database).
5. *TRC Thermodynamic Tables*, Thermodynamic Research Center, Texas A & M University, College Station, TX.
6. Riddick, J. A., Bunger, W. B., and Sakano, T. K., *Organic Solvents, Fourth Edition*, John Wiley & Sons, New York, 1986.
7. *Physical Constants of Hydrocarbon and Non-Hydrocarbon Compounds,* ASTM Data Series DS 4B, ASTM, Philadelphia, 1988.
8. Stevenson, R. M., and Malanowski, S., *Handbook of the Thermodynamics of Organic Compounds*, Elsevier, New York, 1987.
9. Buckingham, J., Editor, *Dictionary of Organic Compounds, Fifth Edition* (and Supplements), Chapman and Hall, New York, 1982-1993.
10. Lide, D. R., and Kehiaian, H. V., *Handbook of Thermophysical and Thermochemical Data*, CRC Press, Boca Raton, FL, 1994.

No.	Name Synonym	Mol. Form. Mol. Wt.	CAS RN mp/°C	Merck No. bp/°C	Beil. Ref. den/g cm^{-3}	Solubility n_D
1	Abate Phosphorothioic acid, O,O'-(thiodi-4,1-phenylene) O,O,O',O'-tetramethyl ester	$C_{16}H_{20}O_6P_2S_3$ 466.48	3383-96-8 30	9075	1.32	
2	1-Acenaphthylenamine, 1,2-dihydro- 1-Acenaphthenamine	$C_{12}H_{11}N$ 169.23	40745-44-6 135	sub	2-12-00-00764	bz 4; EtOH 4
3	Acenaphthylene Acenaphthalene	$C_{12}H_8$ 152.20	208-96-8 92.5	280; 150[28]	4-05-00-02138 0.8987[16]	H_2O 1; EtOH 4; eth 4; bz 4
4	Acenaphthylene, 5-bromo-1,2-dihydro- 5-Bromoacenaphthene	$C_{12}H_9Br$ 233.11	2051-98-1 52	335	4-05-00-01839 1.4392[52]	H_2O 1; EtOH 2; chl 2 1.6565[54]
5	Acenaphthylene, 5-chloro-1,2-dihydro- 5-Chloroacenaphthene	$C_{12}H_9Cl$ 188.66	5209-33-6 70.5	319; 163[13]	4-05-00-01837 1.1954[70]	1.6169[100]
6	Acenaphthylene, 1,2-dihydro- Acenaphthene	$C_{12}H_{10}$ 154.21	83-32-9 93.4	23 279	4-05-00-01834 1.222[20]	H_2O 1; EtOH 2; bz 4; chl 2 1.6048[95]
7	Acenaphthylene, 1,2-dihydro-5-iodo- 5-Iodoacenaphthene	$C_{12}H_9I$ 280.11	6861-64-9 64.3		2-05-00-00497 1.500[20]	bz 4; EtOH 4 1.6909[65]
8	1,2-Acenaphthylenedione	$C_{12}H_6O_2$ 182.18	82-86-0 261	sub	4-07-00-02498 1.480[20]	H_2O 1; EtOH 2; bz 2; lig 3
9	Acenaphthylene, 1,2,2a,3,4,5-hexahydro-	$C_{12}H_{14}$ 158.24	480-72-8 12.8	249	4-05-00-01566 1.0290[20]	chl 3
10	Acephate Phosphoramidothioic acid, acetyl-, O,S-dimethyl ester	$C_4H_{10}NO_3PS$ 183.17	30560-19-1 88		1.35[20]	
11	Acetaldehyde Ethanal	C_2H_4O 44.05	75-07-0 -123	32 20.1	4-01-00-03094 0.7834[18]	H_2O 5; EtOH 5; eth 5; bz 5 1.3316[20]
12	Acetaldehyde, chloro- Chloroacetaldehyde	C_2H_3ClO 78.50	107-20-0 -16.3	2108 85.5	4-01-00-03134	eth 3
13	Acetaldehyde, chloro-, trimer Chloroacetaldehyde, trimer	$C_6H_9Cl_3O_3$ 235.49	65438-35-9 87	142[10]	4-19-00-04718	eth 4
14	Acetaldehyde, ethoxy- α-Ethoxyacetaldehyde	$C_4H_8O_2$ 88.11	22056-82-2		3-01-00-03181 0.942[20]	H_2O 4; ace 4; EtOH 4 1.3956[20]
15	Acetaldehyde, hydroxy- Glycolaldehyde	$C_2H_4O_2$ 60.05	141-46-8 97		4-01-00-03955 1.366[100]	chl 3 1.4772[19]
16	Acetaldehyde, methoxy- Methoxyacetaldehyde	$C_3H_6O_2$ 74.08	10312-83-1	92	4-01-00-03955 1.005[25]	H_2O 4; ace 4; eth 4; EtOH 4 1.3950[20]
17	Acetaldehyde, oxime Acetaldoxime	C_2H_5NO 59.07	107-29-9 45	35 115	4-01-00-03121 0.9656[20]	H_2O 3; EtOH 5; eth 5; chl 3 1.4264[20]
18	Acetaldehyde, phenylhydrazone	$C_8H_{10}N_2$ 134.18	935-07-9 99.5	150[40]; 135[21]	3-15-00-00079	EtOH 4
19	Acetaldehyde, tetramer Metaldehyde II (tetramer)	$C_4H_{12}O_4$ 124.14	66056-06-2 47	110	4-19-00-05643	ace 4; bz 4; eth 4; EtOH 4
20	Acetaldehyde, tribromo- Bromal	C_2HBr_3O 280.74	115-17-3	1372 174	4-01-00-03155 2.6649[25]	ace 4; eth 4; EtOH 4 1.5939[20]
21	Acetaldehyde, trichloro- Chloral	C_2HCl_3O 147.39	75-87-6 -57.5	9538 97.8	4-01-00-03142 1.512[20]	H_2O 4; EtOH 3; eth 3 1.4580[20]
22	Acetamide Ethanamide	C_2H_5NO 59.07	60-35-5 81	36 222.0	4-02-00-00399 0.9986[85]	H_2O 4; EtOH 4 1.4278
23	Acetamide, N-acetyl-	$C_4H_7NO_2$ 101.11	625-77-4 79	223.5	4-02-00-00416	H_2O 3; EtOH 3; eth 3; chl 3
24	Acetamide, N-acetyl-N-ethyl-	$C_6H_{11}NO_2$ 129.16	1563-83-3	197	4-04-00-00352 1.0092[20]	EtOH 4 1.4513[20]
25	Acetamide, N-acetyl-N-methyl-	$C_5H_9NO_2$ 115.13	1113-68-4 -25	195; 114.5[61]	4-04-00-00183 1.0663[25]	H_2O 5; eth 1 1.4502[25]
26	Acetamide, N-acetyl-N-phenyl-	$C_{10}H_{11}NO_2$ 177.20	1563-87-7 37.5	200[100]; 142[11]	4-12-00-00385	H_2O 2; EtOH 3; bz 3; tol 3
27	Acetamide, N-(aminocarbonyl)-	$C_3H_6N_2O_2$ 102.09	591-07-1 218	sub 180	3-03-00-00119	H_2O 2; EtOH 3; eth 2
28	Acetamide, N-(2-aminoethyl)-	$C_4H_{10}N_2O$ 102.14	1001-53-2 51	136[27]	4-04-00-01193	H_2O 3; EtOH 3; eth 1; bz 3
29	Acetamide, N-(4-aminophenyl)- p-Aminoacetanilide	$C_8H_{10}N_2O$ 150.18	122-80-5 166.5	422 267	4-13-00-00137	H_2O 3; EtOH 4; eth 4
30	Acetamide, N-[(4-aminophenyl)sulfonyl]- Sulfacetamide	$C_8H_{10}N_2O_3S$ 214.25	144-80-9 183	8870	4-14-00-02662	H_2O 2; EtOH 3; eth 1; ace 4
31	Acetamide, N-[4-(aminosulfonyl)phenyl]- Acetylsulfanilamide	$C_8H_{10}N_2O_3S$ 214.25	121-61-9 219.5	97	4-14-00-02704	H_2O 3; EtOH 3; ace 3
32	Acetamide, N-[5-(aminosulfonyl)-1,3,4-thiadiazol-2-yl]- Acetazolamide	$C_4H_6N_4O_3S_2$ 222.25	59-66-5 260.5	45	4-27-00-08219	H_2O 2
33	Acetamide, N-(aminothioxomethyl)- Acetylthiourea	$C_3H_6N_2OS$ 118.16	591-08-2 165		4-03-00-00354	H_2O 2; EtOH 3; eth 2
34	Acetamide, N-[4-[[(aminothioxomethyl)hydrazono]methyl]phenyl]- Thiacetazone	$C_{10}H_{12}N_4OS$ 236.30	104-06-3 225 dec	9218	4-14-00-00075	H_2O 1; os 1; CS_2 1
35	Acetamide, N-[1,1'-biphenyl]-2-yl-	$C_{14}H_{13}NO$ 211.26	2113-47-5 121	355	4-12-00-03224	H_2O 1; EtOH 4; eth 4
36	Acetamide, N-bromo- N-Bromoacetamide	C_2H_4BrNO 137.96	79-15-2 103.5	1386	4-02-00-00417	eth 4

No.	Name / Synonym	Mol. Form. / Mol. Wt.	CAS RN / mp/°C	Merck No. / bp/°C	Beil. Ref. / den/g cm^{-3}	Solubility / n_D
37	Acetamide, N-(4-bromophenyl)- p-Bromoacetanilide	C_8H_8BrNO 214.06	103-88-8 168	1387	4-12-00-01504 1.717[25]	H_2O 1; EtOH 3; eth 2; bz 2
38	Acetamide, N-butyl-	$C_6H_{13}NO$ 115.18	1119-49-9 	229	4-04-00-00565 0.8960[25]	1.4388[25]
39	Acetamide, N-butyl-N-phenyl-	$C_{12}H_{17}NO$ 191.27	91-49-6 24.5	281	4-12-00-00379 0.9912[20]	chl 2 1.5146[20]
40	Acetamide, N,N'-carbonylbis-	$C_5H_8N_2O_3$ 144.13	638-20-0 154.5	sub	3-03-00-00121	H_2O 2; EtOH 2; eth 3; ace 3
41	Acetamide, 2-chloro- Chloroacetamide	C_2H_4ClNO 93.51	79-07-2 121	2109 225	4-02-00-00490	H_2O 3; EtOH 4; eth 2
42	Acetamide, 2-chloro-N,N-bis(1-methylethyl)-	$C_8H_{16}ClNO$ 177.67	7403-66-9 48.5	163[55]; 86[2.7]	4-04-00-00516	1.4619[25]
43	Acetamide, 2-chloro-N-3-chloroallyl- Chloroacetamide, N-3-chloroallyl	$C_5H_7Cl_2NO$ 168.02	100130-25-4 52.5	112[0.5]	4-04-00-01087	peth 4
44	Acetamide, 2-chloro-N,N-di-2-propenyl- Allidochlor	$C_8H_{12}ClNO$ 173.64	93-71-0 250	116[1]; 92[0.7]	4-04-00-01064	H_2O 2; EtOH 3 1.4932[25]
45	Acetamide, 2-chloro-N-hexyl- α-Chloro-N-hexylacetamide	$C_8H_{16}ClNO$ 177.67	5326-81-8 108.5	100[0.2]	4-04-00-00713	peth 4
46	Acetamide, 2-chloro-N-(3-methoxypropyl)- α-Chloro-N-(3-methoxypropyl)acetamide	$C_6H_{12}ClNO_2$ 165.62	1709-03-1 30	88[0.5]	4-04-00-01645	1.4712[25]
47	Acetamide, 2-chloro-N-(1-methylpropyl)-	$C_8H_{12}ClNO$ 149.62	32322-73-9 45.5	68[0.7]	4-04-00-00621	peth 4
48	Acetamide, N-(4-chlorophenyl)-	C_8H_8ClNO 169.61	539-03-7 179	333	4-12-00-01178 1.385[22]	H_2O 1; EtOH 3; eth 4; ctc 2
49	Acetamide, 2-cyano- Cyanoacetamide	$C_3H_4N_2O$ 84.08	107-91-5 121.5	2695	4-02-00-01891	H_2O 4
50	Acetamide, N,N-diacetyl- Triacetamide	$C_6H_9NO_3$ 143.14	641-06-5 79		4-02-00-00416	eth 4
51	Acetamide, 2,2-dichloro-	$C_2H_3Cl_2NO$ 127.96	683-72-7 99.4	234	4-02-00-00505	H_2O 3; EtOH 3; eth 3; ace 2
52	Acetamide, 2,2-dichloro-N-[2-hydroxy-1-(hydroxymethyl)-2-(4-nitrophenyl)ethyl]-, [R-(R*,R*)]- Chloramphenicol	$C_{11}H_{12}Cl_2N_2O_5$ 323.13	56-75-7 150.5	2068 sub	3-13-00-02268	ace 4; EtOH 4; chl 4
53	Acetamide, N,N-diethyl-	$C_6H_{13}NO$ 115.18	685-91-6 	185.5	4-04-00-00349 0.9130[17]	H_2O 3; EtOH 3; eth 5; ace 5 1.4374[17]
54	Acetamide, 2-(diethylamino)-N-(2,6-dimethylphenyl)- Lidocaine	$C_{14}H_{22}N_2O$ 234.34	137-58-6 68.5	5359 180-2[4]	4-12-00-02538	bz 4; eth 4; EtOH 4; chl 4
55	Acetamide, N,N-dimethyl- N,N-Dimethylethanamide	C_4H_9NO 87.12	127-19-5 -20	3216 165	4-04-00-00180 0.9366[25]	H_2O 5; EtOH 5; eth 5; ace 5 1.4380[20]
56	Acetamide, N-(2,4-dimethylphenyl)-	$C_{10}H_{13}NO$ 163.22	2050-43-3 129.3	170[10]	4-12-00-02548	EtOH 4; chl 4
57	Acetamide, N,N-diphenyl- Diphenylacetamide	$C_{14}H_{13}NO$ 211.26	519-87-9 103	3315 sub	4-12-00-00381	H_2O 2; EtOH 3; eth 2; chl 2
58	Acetamide, N,N-dipropyl-	$C_8H_{17}NO$ 143.23	1116-24-1 	209.5	4-04-00-00476 0.8992[17]	EtOH 4 1.4419[17]
59	Acetamide, N-(2-ethoxyphenyl)-	$C_{10}H_{13}NO_2$ 179.22	581-08-8 79	>240	3-13-00-00779	H_2O 1; EtOH 3; eth 3; chl 3
60	Acetamide, N-(4-ethoxyphenyl)- Phenacetin	$C_{10}H_{13}NO_2$ 179.22	62-44-2 137.5	7155	4-13-00-01092	H_2O 2; EtOH 3; eth 2; ace 3 1.571
61	Acetamide, N-ethyl- N-Ethylacetamide	C_4H_9NO 87.12	625-50-3 	205; 104[18]	4-04-00-00347 0.942[4]	H_2O 5; EtOH 5; chl 3; HOAc 3 1.4338[20]
62	Acetamide, N-ethyl-N-phenyl-	$C_{10}H_{13}NO$ 163.22	529-65-7 55	260	4-12-00-00379 0.9938[60]	H_2O 3; eth 3; ctc 3
63	Acetamide, 2-fluoro- Fluoroacetic acid amide	C_2H_4FNO 77.06	640-19-7 108	4095 sub	4-02-00-00454	H_2O 3; ace 3; chl 2
64	Acetamide, N-9H-fluoren-2-yl- 2-(Acetylamino)fluorene	$C_{15}H_{13}NO$ 223.27	53-96-3 193	4083	4-12-00-03373	H_2O 1; EtOH 3; eth 3; HOAc 3
65	Acetamide, N-(2-hydroxyethyl)-	$C_4H_9NO_2$ 103.12	142-26-7 63.5	166-7[8]	4-04-00-01535 1.1079[25]	H_2O 5; ace 3; bz 2; lig 2 1.4674[20]
66	Acetamide, N-(2-hydroxy-1-naphthalenyl)-	$C_{12}H_{11}NO_2$ 201.22	117-93-1 235 dec	sub	4-13-00-02095	ace 4; bz 4; eth 4; EtOH 4
67	Acetamide, N-(4-hydroxyphenyl)- Acetaminophen	$C_8H_9NO_2$ 151.17	103-90-2 170	40	4-13-00-01091 1.293[21]	H_2O 1; EtOH 4
68	Acetamide, 2-mercapto-N-2-naphthalenyl- Thionalide	$C_{12}H_{11}NOS$ 217.29	93-42-5 111.5	9273	3-12-00-03052	H_2O 1; EtOH 4; os 4
69	Acetamide, N-[2-(5-methoxy-1H-indol-3-yl)ethyl]- Melatonin	$C_{13}H_{16}N_2O_2$ 232.28	73-31-4 117	5695	5-22-12-00042	
70	Acetamide, N-(2-methoxyphenyl)-	$C_9H_{11}NO_2$ 165.19	93-26-5 87.5	304	4-13-00-00833	H_2O 4; EtOH 4; eth 3; ace 3
71	Acetamide, N-(4-methoxyphenyl)- p-Acetanisidine	$C_9H_{11}NO_2$ 165.19	51-66-1 131	43	4-13-00-01092	ace 4; EtOH 4; chl 4
72	Acetamide, N-methyl-	C_3H_7NO 73.09	79-16-3 28	205	4-04-00-00176 0.9371[25]	ace 4; bz 4; eth 4; EtOH 4 1.4301[20]
73	Acetamide, N-[(methylamino)carbonyl]-	$C_4H_8N_2O_2$ 116.12	623-59-6 180.5	dec	4-04-00-00207	H_2O 3; EtOH 2; eth 2; chl 3

No.	Name Synonym	Mol. Form. Mol. Wt.	CAS RN mp/°C	Merck No. bp/°C	Beil. Ref. den/g cm^{-3}	Solubility n_D
74	Acetamide, N-(1-methylethyl)-N-phenyl- N-Isopropylacetanilide	$C_{11}H_{15}NO$ 177.25	5461-51-8 40.5	264	3-12-00-00466	lig 4
75	Acetamide, N-methyl-N-(2-methylphenyl)-	$C_{10}H_{13}NO$ 163.22	573-26-2 55.5	260	3-12-00-01854	EtOH 3; chl 3
76	Acetamide, N-(2-methylphenyl)-	$C_9H_{11}NO$ 149.19	120-66-1 110	296	4-12-00-01755 1.168^{15}	H_2O 2; EtOH 3; eth 3; ace 3
77	Acetamide, N-(3-methylphenyl)-	$C_9H_{11}NO$ 149.19	537-92-8 65.5	303	4-12-00-01823 1.141^{15}	H_2O 2; EtOH 4; eth 4; chl 3
78	Acetamide, N-(4-methylphenyl)-	$C_9H_{11}NO$ 149.19	103-89-9 152	307	4-12-00-01902 1.2120^{15}	eth 4; EtOH 4
79	Acetamide, N-methyl-N-phenyl- N-Methylacetanilide	$C_9H_{11}NO$ 149.19	579-10-2 103	5931 256	4-12-00-00378 1.0036^{105}	H_2O 3; EtOH 3; eth 3; chl 3 1.576
80	Acetamide, 2-nitro-N-phenyl- α-Nitroacetanilide	$C_8H_8N_2O_3$ 180.16	10151-95-8 94	100$^{0.1}$	1-12-00-00193 1.419^{15}	bz 4; EtOH 4; lig 4
81	Acetamide, N-(2-nitrophenyl)-	$C_8H_8N_2O_3$ 180.16	552-32-9 94	100$^{0.1}$	4-12-00-01574 1.419^{15}	H_2O 3; EtOH 3; eth 4; bz 3
82	Acetamide, N-(3-nitrophenyl)-	$C_8H_8N_2O_3$ 180.16	122-28-1 155	100$^{0.0074}$	4-12-00-01595	H_2O 3; EtOH 3; eth 1; chl 3
83	Acetamide, N-(4-nitrophenyl)-	$C_8H_8N_2O_3$ 180.16	104-04-1 216	100$^{0.008}$	4-12-00-01632	H_2O 2; EtOH 3; eth 2; chl 2
84	Acetamide, N-(5-nitro-2-propoxyphenyl)- 5'-Nitro-2'-propoxyacetanilide	$C_{11}H_{14}N_2O_4$ 238.24	553-20-8 102.5	6550	3-13-00-00881	
85	Acetamide, N-(5-nitro-2-thiazolyl)- Aminitrozole	$C_5H_5N_3O_3S$ 187.18	140-40-9 264.5	421	4-27-00-04676	alk 3
86	Acetamide, N-phenyl- Acetanilide	C_8H_9NO 135.17	103-84-4 114.3	42 304	4-12-00-00373 1.2190^{15}	H_2O 2; EtOH 4; eth 3; ace 4
87	Acetamide, N-(phenylmethyl)-	$C_9H_{11}NO$ 149.19	588-46-5 61	157^2	4-12-00-02230	EtOH 4; eth 4
88	Acetamide, N-phenyl-N-propyl- N-Propylacetanilide	$C_{11}H_{15}NO$ 177.25	2437-98-1 49	268	4-12-00-00379	eth 4; EtOH 4; lig 4
89	Acetamide, N-3-pyridinyl-	$C_7H_8N_2O$ 136.15	5867-45-8 133	326.5	5-22-09-00017	H_2O 3; EtOH 3; eth 2; bz 2
90	Acetamide, N-4-quinolinyl-	$C_{11}H_{10}N_2O$ 186.21	32433-28-6 176	sub	5-22-10-00250	EtOH 4
91	Acetamide, N,N'-(sulfonyldi-4,1-phenylene)bis- Acedapsone	$C_{16}H_{16}N_2O_4S$ 332.38	77-46-3 290	16	4-13-00-01361	EtOH 4
92	Acetamide, N-(5,6,7,9-tetrahydro-10-hydroxy-1,2,3-trimethoxy-9-oxobenzo[a]heptalen-7-yl)-, (S)- Colchiceine	$C_{21}H_{23}NO_6$ 385.42	477-27-0 178.5	2469	4-14-00-00943 1.24^{25}	H_2O 2; EtOH 4; eth 1; bz 1
93	Acetamide, N-(5,6,7,9-tetrahydro-1,2,3,10-tetramethoxy-9-oxobenzo[a]heptalen-7-yl)-, (S)- Colchicine	$C_{22}H_{25}NO_6$ 399.44	64-86-8 156	2470	4-14-00-00946	H_2O 4; EtOH 4
94	Acetamide, 2,2,2-tribromo- Tribromoacetamide	$C_2H_2Br_3NO$ 295.76	594-47-8 121.5	sub	3-02-00-00485	H_2O 4; eth 4; EtOH 4
95	Acetamide, 2,2,2-trichloro-	$C_2H_2Cl_3NO$ 162.40	594-65-0 142	240	4-02-00-00520	H_2O 2; EtOH 4; eth 4
96	Acetamide, 2,2,2-trichloro-N,N-diethyl-	$C_6H_{10}Cl_3NO$ 218.51	2430-00-4 27	112^{12}	4-04-00-00351	1.4900^{24}
97	Acetamide, 2,2,2-trichloro-N,N-dimethyl-	$C_4H_6Cl_3NO$ 190.46	7291-33-0 12	231 dec	4-04-00-00182 1.390^{20}	bz 4; chl δ 1.5017^{25}
98	Acetamide, 2,2,2-trichloro-N-phenyl-	$C_8H_6Cl_3NO$ 238.50	2563-97-5 96	169^{19}	4-12-00-00377	EtOH 4
99	Acetamide, 2,2,2-trifluoro-N,N-dimethyl-	$C_4H_6F_3NO$ 141.09	1547-87-1	133	4-04-00-00182 1.232^{21}	1.3611^{25}
100	Acetic acid Ethanoic acid	$C_2H_4O_2$ 60.05	64-19-7 16.6	47 117.9	4-02-00-00094 1.0492^{20}	H_2O 5; EtOH 5; eth 5; ace 5 1.3720^{20}
101	Acetic acid, (acetyloxy)-	$C_4H_6O_4$ 118.09	13831-30-6 67.5	145^{12}	4-03-00-00576	H_2O 4; EtOH 4; eth 4; ace 4
102	Acetic acid, (acetyloxy)-, ethyl ester	$C_6H_{10}O_4$ 146.14	623-86-9	179	4-03-00-00584 1.0880^{20}	eth 4; EtOH 4; HOAc 4 1.4112^{20}
103	Acetic acid, [2-(aminocarbonyl)phenoxy]- Salicylamide O-acetic acid	$C_9H_9NO_4$ 195.17	25395-22-6	8298	4-10-00-00184	
104	Acetic acid, aminooxo- Oxamic acid	$C_2H_3NO_3$ 89.05	471-47-6 210 dec	6870	4-02-00-01857	H_2O 2; EtOH 1; eth 1
105	Acetic acid, aminooxo-, hydrazide Semioxamazide	$C_2H_5N_3O_2$ 103.08	515-96-8 221 dec	8397	4-02-00-01868	H_2O 2; EtOH 1; eth 1; alk 4
106	Acetic acid, ammonium salt Ammonium, acetate	$C_2H_7NO_2$ 77.08	631-61-8 114	514 dec	1.17^{20}	H_2O 4; EtOH 3; ace 2
107	Acetic acid, anhydride Acetic anhydride	$C_4H_6O_3$ 102.09	108-24-7 -73	48 139.5	4-02-00-00386 1.082^{20}	H_2O 4; EtOH 3; eth 5; bz 3 1.3901^{20}
108	Acetic acid, anhydride with nitric acid Acetyl nitrate	$C_2H_3NO_4$ 105.05	591-09-3 92	22^{70}; exp 60	4-02-00-00435 1.24^{15}	
109	Acetic acid, azido-	$C_2H_3N_3O_2$ 101.06	18523-48-3 16	116^{12}; 92^3	4-02-00-00541 1.354^{33}	
110	Acetic acid, barium salt	$C_4H_6BaO_4$ 255.42	543-80-6	975	4-02-00-00094 2.468^{25}	H_2O 3

PHYSICAL CONSTANTS OF ORGANIC COMPOUNDS (continued)

No.	Name / Synonym	Mol. Form. / Mol. Wt.	CAS RN / mp/°C	Merck No. / bp/°C	Beil. Ref. / den/g cm^{-3}	Solubility / n_D
111	Acetic acid, bromo- / Bromoacetic acid	$C_2H_3BrO_2$ / 138.95	79-08-3 / 50	1388 / 208	4-02-00-00526 / 1.9335[50]	H$_2$O 5; EtOH 5; eth 5; ace 3 / 1.4804[50]
112	Acetic acid, bromochloro- / Bromochloroacetic acid	$C_2H_2BrClO_2$ / 173.39	5589-96-8 / 31.5	215; 104[11] /	4-02-00-00532 / 1.9848[31]	
113	Acetic acid, bromochloro-, ethyl ester / Ethyl bromochloroacetate	$C_4H_6BrClO_2$ / 201.45	22524-32-9 /	174 dec /	4-02-00-00532 / 1.5890[22]	eth 4; EtOH 4 / 1.4639[24]
114	Acetic acid, bromo-, 1,1-dimethylethyl ester	$C_6H_{11}BrO_2$ / 195.06	5292-43-3 /	73-4[25] /	3-02-00-00482 /	eth 4; EtOH 4 / 1.4430[20]
115	Acetic acid, bromo-, ethyl ester / Ethyl bromacetate	$C_4H_7BrO_2$ / 167.00	105-36-2 /	168.5 /	4-02-00-00527 / 1.5032[20]	H$_2$O 1; EtOH 5; eth 5; ace 3 / 1.4489[20]
116	Acetic acid, bromofluoro- / Bromofluoroacetic acid	$C_2H_2BrFO_2$ / 156.94	359-25-1 / 50	183 /	4-02-00-00531 /	H$_2$O 4; EtOH 4; chl 4 /
117	Acetic acid, bromo-, methyl ester / Methyl 2-bromoacetate	$C_3H_5BrO_2$ / 152.98	96-32-2 /	132 /	4-02-00-00527 / 1.6350[20]	H$_2$O 1; EtOH 3; eth 3; ace 3 / 1.4520[20]
118	Acetic acid, bromo-, 2-methylpropyl ester	$C_6H_{11}BrO_2$ / 195.06	59956-48-8 /	188; 74.5[10] /	1-02-00-00096 / 1.3269[20]	ace 4; eth 4; EtOH 4 /
119	Acetic acid, bromo-, phenyl ester	$C_8H_7BrO_2$ / 215.05	620-72-4 / 32	140[20] /	3-06-00-00599 /	eth 4; EtOH 4
120	Acetic acid, bromo-, propyl ester / Propyl bromoacetate	$C_5H_9BrO_2$ / 181.03	35223-80-4 /	176 /	4-02-00-00528 / 1.4099[20]	ace 4; eth 4; EtOH 4 / 1.4518[20]
121	Acetic acid, butyl ester / Butyl acetate	$C_6H_{12}O_2$ / 116.16	123-86-4 / -78	1535 / 126.1	4-02-00-00143 / 0.8825[20]	H$_2$O 2; EtOH 5; eth 5; ace 3 / 1.3941[20]
122	Acetic acid, calcium salt	$C_4H_6CaO_4$ / 158.17	62-54-4 / dec	1643 /	4-02-00-00094 /	H$_2$O 3; EtOH 2 / 1.55
123	Acetic acid, chloro- / Chloroacetic acid (β)	$C_2H_3ClO_2$ / 94.50	79-11-8 / 63	2111 / 189.3	4-02-00-00474 / 1.4043[40]	H$_2$O 4; EtOH 3; eth 3; bz 3 / 1.4351[55]
124	Acetic acid, chloro-, anhydride / Chloroacetic anhydride	$C_4H_4Cl_2O_3$ / 170.98	541-88-8 / 46	2112 / 203	4-02-00-00487 / 1.5497[20]	
125	Acetic acid, chloro-, butyl ester	$C_6H_{11}ClO_2$ / 150.60	590-02-3 /	183 /	4-02-00-00482 / 1.0704[20]	eth 4; EtOH 4 / 1.4297[20]
126	Acetic acid, chloro-, 2-chloroethyl ester	$C_4H_6Cl_2O_2$ / 157.00	3848-12-2 /	203; 102[20] /	4-02-00-00481 / 1.3600[25]	eth 4 / 1.4619[25]
127	Acetic acid, chlorodifluoro- / Chlorodifluoroacetic acid	$C_2HClF_2O_2$ / 130.48	76-04-0 / 25	122 /	4-02-00-00497 /	chl 3 / 1.3559[20]
128	Acetic acid, chloro-, 1,1-dimethylethyl ester / tert-Butyl chloroacetate	$C_6H_{11}ClO_2$ / 150.60	107-59-5 /	1563 / 150; 50[10]	3-02-00-00444 /	H$_2$O 6 / 1.4260[20]
129	Acetic acid, chloro-, 1,2-ethanediyl ester	$C_6H_8Cl_2O_4$ / 215.03	6941-69-1 / 45.5	157[8] /	3-02-00-00448 /	H$_2$O 1; eth 4
130	Acetic acid, chloro-, ethyl ester / Ethyl chloroacetate	$C_4H_7ClO_2$ / 122.55	105-39-5 / -21	3741 / 144.3	4-02-00-00481 / 1.1585[20]	H$_2$O 1; EtOH 5; eth 5; ace 5 / 1.4215[20]
131	Acetic acid, chlorofluoro- / Chlorofluoroacetic acid	$C_2H_2ClFO_2$ / 112.49	471-44-3 /	162 /	4-02-00-00493 / 1.532[25]	/ 1.4085[25]
132	Acetic acid, chlorofluoro-, butyl ester / Chlorofluoroacetic acid, butyl ester	$C_6H_{10}ClFO_2$ / 168.60	368-34-3 /	165.5 /	3-02-00-00453 / 1.124[25]	/ 1.4067[25]
133	Acetic acid, chlorofluoro-, ethyl ester	$C_4H_6ClFO_2$ / 140.54	401-56-9 /	129 /	4-02-00-00493 / 1.225[20]	/ 1.3927[20]
134	Acetic acid, chlorofluoro-, methyl ester / Methyl chlorofluoroacetate	$C_3H_4ClFO_2$ / 126.51	433-52-3 /	116 /	4-02-00-00493 / 1.323[25]	/ 1.3903[25]
135	Acetic acid, chlorofluoro-, propyl ester / Chlorofluoroacetic acid, propyl ester	$C_5H_8ClFO_2$ / 154.57	348-70-9 /	147 /	3-02-00-00453 / 1.170[25]	/ 1.3994[25]
136	Acetic acid, chloro-, 2-hydroxyethyl ester	$C_4H_7ClO_3$ / 138.55	35280-53-6 /	240 dec; 86[1.5] /	3-02-00-00447 / 1.3300[20]	H$_2$O 4; EtOH 4 / 1.4609[20]
137	Acetic acid, chloro-, 2-methoxyethyl ester	$C_5H_9ClO_3$ / 152.58	13361-36-9 /	86-6[9] /	3-02-00-00447 / 1.2015[20]	eth 4 / 1.4382[20]
138	Acetic acid, chloro-, methyl ester / Chloroacetic acid, methyl ester	$C_3H_5ClO_2$ / 108.52	96-34-4 / -32.1	5965 / 129.5	4-02-00-00480 / 1.236[20]	ace 4; bz 4; eth 4; EtOH 4 / 1.4218[20]
139	Acetic acid, chloro-, 1-methylethyl ester / Isopropyl chloroacetate	$C_5H_9ClO_2$ / 136.58	105-48-6 /	150.5 /	4-02-00-00482 / 1.0888[20]	eth 4 / 1.4382[20]
140	Acetic acid, chloro-, 4-methylphenyl ester	$C_9H_9ClO_2$ / 184.62	15150-39-7 / 32	254; 162[41] /	4-06-00-02112 / 1.1840[35]	eth 4; EtOH 4
141	Acetic acid, (4-chloro-2-methylphenoxy)- / MCPA	$C_9H_9ClO_3$ / 200.62	94-74-6 / 120	5645 /	4-06-00-01991 /	H$_2$O 2; EtOH 4; eth 4; bz 3
142	Acetic acid, chloro-, 1-methylpropyl ester	$C_6H_{11}ClO_2$ / 150.60	17696-64-9 /	163.5 /	3-02-00-00443 / 1.060[20]	eth 4; EtOH 4 / 1.4251[19]
143	Acetic acid, chloro-, 2-methylpropyl ester	$C_6H_{11}ClO_2$ / 150.60	13361-35-8 /	170 /	4-02-00-00483 / 1.0612[20]	ace 4; eth 4 / 1.4255[20]
144	Acetic acid, chlorooxo-, ethyl ester	$C_4H_5ClO_3$ / 136.53	4755-77-5 /	137 /	4-02-00-01853 / 1.2226[20]	bz 4; eth 4
145	Acetic acid, chlorooxo-, methyl ester	$C_3H_3ClO_3$ / 122.51	5781-53-3 /	119 /	3-02-00-01583 / 1.3316[20]	/ 1.4189[20]
146	Acetic acid, (4-chlorophenoxy)- / (p-Chlorophenoxy)acetic acid	$C_8H_7ClO_3$ / 186.59	122-88-3 / 156.5	/	4-06-00-00845 /	H$_2$O 4; chl 2
147	Acetic acid, chloro-, phenyl ester / Phenyl chloroacetate	$C_8H_7ClO_2$ / 170.60	620-73-5 / 44.5	232.5 /	3-06-00-00598 / 1.2202[44]	H$_2$O 1; EtOH 4; eth 4 / 1.5146[44]
148	Acetic acid, chloro-, phenylmethyl ester	$C_9H_9ClO_2$ / 184.62	140-18-1 /	147.5[9]; 85[0.4] /	4-06-00-02264 / 1.1223[4]	eth 4; EtOH 4 / 1.5426[18]
149	Acetic acid, chloro-, propyl ester / Propyl chloroacetate	$C_5H_9ClO_2$ / 136.58	5396-24-7 /	161 /	4-02-00-00482 / 1.1022[20]	eth 4 / 1.4261[20]

No.	Name Synonym	Mol. Form. Mol. Wt.	CAS RN mp/°C	Merck No. bp/°C	Beil. Ref. den/g cm^{-3}	Solubility n_D
150	Acetic acid, cyano- Cyanoacetic acid	$C_3H_3NO_2$ 85.06	372-09-8 66	2696 108[0.15]; 160 dec	4-02-00-01888	H_2O 3; EtOH 3; eth 3; chl 2
151	Acetic acid, cyano-, butyl ester	$C_7H_{11}NO_2$ 141.17	5459-58-5	231; 115[15]	4-02-00-01891 1.0010[20]	1.4200[20]
152	Acetic acid, cyano-, ethyl ester Ethyl cyanoacetate	$C_5H_7NO_2$ 113.12	105-56-6 -22.5	3744 205	4-02-00-01889 1.0654[20]	eth 4; EtOH 4 1.4175[20]
153	Acetic acid, cyano-, hydrazide Cyacetacide	$C_3H_5N_3O$ 99.09	140-87-4 114.5	2688	4-02-00-01895	H_2O 4; EtOH 4
154	Acetic acid, cyano-, methyl ester Methyl cyanoacetate	$C_4H_5NO_2$ 99.09	105-34-0 -22.5	200.5	4-02-00-01889 1.1225[25]	eth 4; EtOH 4 1.4176[20]
155	Acetic acid, cyclohexyl ester Cyclohexyl acetate	$C_8H_{14}O_2$ 142.20	622-45-7	173; 96[75]	4-06-00-00036 0.968[20]	eth 4; EtOH 4 1.442[20]
156	Acetic acid, cyclohexylidene-, methyl ester Methyl cyclohexylideneacetate	$C_9H_{14}O_2$ 154.21	40203-74-5	99[20]	4-09-00-00124 1.0021[19]	1.4838[19]
157	Acetic acid, decyl ester Decyl acetate	$C_{12}H_{24}O_2$ 200.32	112-17-4 -15	244	4-02-00-00168 0.8671[20]	H_2O 1; EtOH 3; eth 3; bz 3 1.4273[20]
158	Acetic acid, diazo-, ethyl ester Diazoacetic ester	$C_4H_6N_2O_2$ 114.10	623-73-4 -22	2980 140 dec	4-03-00-01495 1.0852[18]	H_2O 2; EtOH 5; eth 5; bz 5 1.4605[20]
159	Acetic acid, dibromo- Dibromoacetic acid	$C_2H_2Br_2O_2$ 217.84	631-64-1 49	195[250]; 130[16]	4-02-00-00533	H_2O 4; EtOH 4; eth 3
160	Acetic acid, dibromo-, ethyl ester Ethyl dibromoacetate	$C_4H_6Br_2O_2$ 245.90	617-33-4	194	4-02-00-00533 1.8991[20]	H_2O 1; EtOH 5; eth 5 1.5017[13]
161	Acetic acid, dibromofluoro-, ethyl ester Ethyl dibromofluoroacetate	$C_4H_5Br_2FO_2$ 263.89	565-53-7	83[33]	4-02-00-00534 1.894[25]	1.4633[25]
162	Acetic acid, dichloro- Dichloroacetic acid	$C_2H_2Cl_2O_2$ 128.94	79-43-6 13.5	3037 194; 102[20]	4-02-00-00498 1.5634[20]	H_2O 5; EtOH 5; eth 5; ace 3 1.4658[20]
163	Acetic acid, dichloro-, butyl ester Butyl dichloroacetate	$C_8H_{10}Cl_2O_2$ 185.05	29003-73-4	193.5	4-02-00-00502 1.1820[20]	eth 4; EtOH 4 1.4420[20]
164	Acetic acid, dichloro-, ethyl ester Ethyl dichloroacetate	$C_4H_6Cl_2O_2$ 157.00	535-15-9	155; 56[10]	4-02-00-00501 1.2827[20]	H_2O 2; EtOH 5; eth 5; ace 3 1.4386[20]
165	Acetic acid, dichloro-, 2-hydroxyethyl ester	$C_4H_6Cl_2O_3$ 173.00	17704-30-2	81-2[0.5]	3-02-00-00460 1.438[20]	EtOH 4 1.4735[20]
166	Acetic acid, dichloro-, methyl ester Methyl dichloroacetate	$C_3H_4Cl_2O_2$ 142.97	116-54-1 -51.9	142.9	4-02-00-00501 1.3774[20]	H_2O 1; EtOH 3; ctc 3 1.4420[20]
167	Acetic acid, dichloro-, 1-methylethyl ester	$C_5H_8Cl_2O_2$ 171.02	25006-60-4	163.5	4-02-00-00502 1.2053[20]	eth 4; EtOH 4 1.4328[20]
168	Acetic acid, [2,3-dichloro-4-(2-methylene-1-oxobutyl)phenoxy]- Ethacrynic acid	$C_{13}H_{12}Cl_2O_4$ 303.14	58-54-8 122.5	3669		
169	Acetic acid, (2,4-dichlorophenoxy)- 2,4-D	$C_8H_6Cl_2O_3$ 221.04	94-75-7 140.5	2802 160[0.4]	4-06-00-00908	H_2O 1; EtOH 3; bz 2
170	Acetic acid, (2,4-dichlorophenoxy)-, 2-butoxyethyl ester 2,4-D 2-Butoxyethyl ester	$C_{14}H_{18}Cl_2O_4$ 321.20	1929-73-3	156-62[1]	4-06-00-00911 1.232[20]	
171	Acetic acid, (2,4-dichlorophenoxy)-, butyl ester 2,4-D Butyl ester	$C_{12}H_{14}Cl_2O_3$ 277.15	94-80-4	2802	3-06-00-00706	
172	Acetic acid, (2,4-dichlorophenoxy)-, methyl ester 2,4-D Methyl ester	$C_9H_8Cl_2O_3$ 235.07	1928-38-7 119	141[18]	4-06-00-00909	
173	Acetic acid, (2,4-dichlorophenoxy)-, 1-methylethyl ester 2,4-D Isopropyl ester	$C_{11}H_{12}Cl_2O_3$ 263.12	94-11-1 5	140[1]	4-06-00-00910 1.26[25]	1.5209[25]
174	Acetic acid, diethoxy-, ethyl ester	$C_8H_{16}O_4$ 176.21	6065-82-3	199	4-03-00-01494 0.985[25]	1.4100[20]
175	Acetic acid, difluoro- Difluoroacetic acid	$C_2H_2F_2O_2$ 96.03	381-73-7 -1	133	4-02-00-00455 1.526[25]	1.3470[20]
176	Acetic acid, difluoro-, ethyl ester Ethyl difluoroacetate	$C_4H_6F_2O_2$ 124.09	454-31-9	100	4-02-00-00455 1.1893[9]	H_2O 1
177	Acetic acid, 1,1-dimethylethyl ester tert-Butyl acetate	$C_6H_{12}O_2$ 116.16	540-88-5	1537 95.1	4-02-00-00151 0.8665[20]	EtOH 3; eth 3; chl 3; HOAc 3 1.3855[20]
178	Acetic acid, dodecyl ester	$C_{14}H_{28}O_2$ 228.38	112-66-3 1.3	265; 180[40]	4-02-00-00170 0.8652[22]	1.4439[20]
179	Acetic acid ethenyl ester Vinyl acetate	$C_4H_6O_2$ 86.09	108-05-4 -93.2	9896 72.5	4-02-00-00176 0.9317[20]	H_2O 1; EtOH 5; eth 3; ace 3 1.3959[20]
180	Acetic acid, ethoxy- 2-Ethoxyacetic acid	$C_4H_8O_3$ 104.11	627-03-2	206.5	4-03-00-00574 1.1021[20]	H_2O 4; EtOH 4; eth 4; chl 3 1.4194[20]
181	Acetic acid, ethoxy-, ethyl ester	$C_6H_{12}O_3$ 132.16	817-95-8	158	4-03-00-00581 0.9702[20]	EtOH 3; eth 3; ace 3 1.4039[20]
182	Acetic acid, ethoxy-, methyl ester Methyl ethoxyacetate	$C_5H_{10}O_3$ 118.13	17640-26-5	147	4-03-00-00578 1.0112[15]	ace 4; eth 4; EtOH 4
183	Acetic acid, ethoxy-, 5-methyl-2-(1-methylethyl)cyclohexyl ester, (1α,2β,5α)-	$C_{14}H_{26}O_3$ 242.36	579-94-2	155[20]	2-06-00-00047 0.9545[20]	eth 4; EtOH 4; chl 4
184	Acetic acid, (ethoxysulfonyl)-, ethyl ester Acetic acid, sulfo, diethyl ester	$C_6H_{12}O_5S$ 196.22	59376-54-4	116[0.5]	3-04-00-00058 1.232[16]	1.436[21]
185	Acetic acid, 2-ethylbutyl ester 2-Ethylbutyl acetate	$C_8H_{16}O_2$ 144.21	10031-87-5 <-100	162.5	4-02-00-00161 0.8790[20]	H_2O 1; EtOH 3; eth 3; ctc 3 1.4109[20]
186	Acetic acid ethyl ester Ethyl acetate	$C_4H_8O_2$ 88.11	141-78-6 -83.6	3713 77.1	4-02-00-00127 0.9003[20]	H_2O 3; EtOH 5; eth 5; ace 4 1.3723[20]

No.	Name / Synonym	Mol. Form. / Mol. Wt.	CAS RN / mp/°C	Merck No. / bp/°C	Beil. Ref. / den/g cm^{-3}	Solubility / n_D
187	Acetic acid, 2-ethylhexyl ester / 2-Ethylhexyl acetate	$C_{10}H_{20}O_2$ / 172.27	103-09-3 / -80	6683 / 199	4-02-00-00166 / 0.8718[20]	H_2O 1; EtOH 3; eth 3 / 1.4204[20]
188	Acetic acid, (ethylthio)-	$C_4H_8O_2S$ / 120.17	627-04-3 / -8.5	164[83]; 109[5]	4-03-00-00603 / 1.1497[20]	H_2O 4; EtOH 4; eth 4
189	Acetic acid, fluoro- / Fluoroethanoic acid	$C_2H_3FO_2$ / 78.04	144-49-0 / 35.2	4096 / 168	4-02-00-00446 / 1.3693[36]	H_2O 3; EtOH 3
190	Acetic acid, fluoro-, 1,1-dimethylethyl ester	$C_6H_{11}FO_2$ / 134.15	406-74-6 /	/ 132	4-02-00-00450 / 0.9904[20]	1.386[20]
191	Acetic acid, fluoro-, ethyl ester	$C_4H_7FO_2$ / 106.10	459-72-3 /	/ 120	4-02-00-00447 / 1.098[25]	H_2O 4 / 1.3755[20]
192	Acetic acid, fluoro-, methyl ester / Methyl fluoroacetate	$C_3H_5FO_2$ / 92.07	453-18-9 / -35	/ 104.5	4-02-00-00447 / 1.1613[15]	
193	Acetic acid, heptyl ester / Heptyl acetate	$C_9H_{18}O_2$ / 158.24	112-06-1 / -50.2	/ 193	4-02-00-00162 / 0.8750[15]	H_2O 1; EtOH 3; eth 3; ctc 3 / 1.4150[20]
194	Acetic acid, hexyl ester	$C_8H_{16}O_2$ / 144.21	142-92-7 / -80.9	/ 171.5	4-02-00-00159 / 0.8779[15]	eth 4; EtOH 4 / 1.4092[20]
195	Acetic acid, hydrazide	$C_2H_6N_2O$ / 74.08	1068-57-1 / 67	/ 137[25]	4-02-00-00435	H_2O 3; EtOH 3; eth 2
196	Acetic acid, hydroxy- / Glycolic acid	$C_2H_4O_3$ / 76.05	79-14-1 / 79.5	4394 / 100	4-03-00-00571	H_2O 3; EtOH 3; eth 3
197	Acetic acid, hydroxy-, ethyl ester	$C_4H_8O_3$ / 104.11	623-50-7 /	/ 160	4-03-00-00580 / 1.0826[23]	eth 4; EtOH 4 / 1.4180[20]
198	Acetic acid, hydroxy-, methyl ester	$C_3H_6O_3$ / 90.08	96-35-5 /	149; 52[17]	4-03-00-00578 / 1.1677[18]	H_2O 3; EtOH 5; eth 5
199	Acetic acid, hydroxy-, propyl ester / Hydroxyacetic acid, propyl ester	$C_5H_{10}O_3$ / 118.13	90357-58-7 /	/ 170.5	4-03-00-00588 / 1.0631[18]	1.4231[18]
200	Acetic acid, iodo- / Iodoacetic acid	$C_2H_3IO_2$ / 185.95	64-69-7 / 82.5	4918 / dec	4-02-00-00534	H_2O 3; EtOH 3; eth 2; chl 2
201	Acetic acid, iodo-, ethyl ester	$C_4H_7IO_2$ / 214.00	623-48-3 /	/ 179	4-02-00-00535 / 1.8173[13]	EtOH 3; eth 3 / 1.5079[13]
202	Acetic acid, mercapto- / Thioglycolic acid	$C_2H_4O_2S$ / 92.12	68-11-1 / -16.5	9265 / 120[20]	4-03-00-00600 / 1.3253[20]	H_2O 5; EtOH 5; eth 5; chl 2 / 1.5080[20]
203	Acetic acid, mercapto-, ethyl ester	$C_4H_8O_2S$ / 120.17	623-51-8 /	/ 157	4-03-00-00616 / 1.0964[15]	EtOH 3; eth 3; ctc 2 / 1.4582[20]
204	Acetic acid, mercapto-, methyl ester	$C_3H_6O_2S$ / 106.15	2365-48-2 /	42-3[10]	4-03-00-00614	eth 4; EtOH 4 / 1.4657[20]
205	Acetic acid, methoxy-	$C_3H_6O_3$ / 90.08	625-45-6 /	/ 203.5	4-03-00-00574 / 1.1768[20]	H_2O 3; EtOH 3; eth 3 / 1.4168[20]
206	Acetic acid, methoxy-, ethyl ester / Ethyl methoxyacetate	$C_5H_{10}O_3$ / 118.13	3938-96-3 /	/ 142	3-03-00-00381 / 1.0118[15]	H_2O 2; EtOH 4; eth 4; ctc 3 / 1.4050[20]
207	Acetic acid, methoxy-, methyl ester	$C_4H_8O_3$ / 104.11	6290-49-9 /	/ 131	4-03-00-00578 / 1.0511[20]	H_2O 2; EtOH 4; eth 4; ace 3 / 1.3962[20]
208	Acetic acid, methyl ester / Methyl acetate	$C_3H_6O_2$ / 74.08	79-20-9 / -98	5932 / 56.8	4-02-00-00122 / 0.9342[20]	H_2O 4; eth 4; EtOH 4 / 1.3614[20]
209	Acetic acid, 1-methylethyl ester / Isopropyl acetate	$C_5H_{10}O_2$ / 102.13	108-21-4 / -73.4	5093 / 88.6	4-02-00-00141 / 0.8718[20]	H_2O 3; EtOH 3; eth 5; ace 3 / 1.3773[20]
210	Acetic acid, [[5-methyl-2-(1-methylethyl)cyclohexyl]oxy]-, [1R-(1α,2β,5α)]- / l-Methoxyacetic acid	$C_{12}H_{22}O_3$ / 214.30	40248-63-3 / 54	171[11]	4-06-00-00157	H_2O 2; EtOH 4; eth 3; ace 4
211	Acetic acid, 2-methylphenyl ester / o-Cresyl acetate	$C_9H_{10}O_2$ / 150.18	533-18-6 /	2588 / 208	4-06-00-01960 / 1.0533[15]	eth 4; EtOH 4 / 1.5002[20]
212	Acetic acid, 3-methylphenyl ester / m-Cresyl acetate	$C_9H_{10}O_2$ / 150.18	122-46-3 / 12	2587 / 212	4-06-00-02047 / 1.043[20]	bz 4; eth 4; EtOH 4 / 1.4978[20]
213	Acetic acid, 4-methylphenyl ester	$C_9H_{10}O_2$ / 150.18	140-39-6 /	/ 212.5	4-06-00-02112 / 1.0512[17]	H_2O 2; EtOH 3; eth 3; ctc 2 / 1.5163[22]
214	Acetic acid, 1-methylpropyl ester / sec-Butyl acetate	$C_6H_{12}O_2$ / 116.16	105-46-4 / -98.9	1536 / 112	4-02-00-00148 / 0.8748[20]	H_2O 1; EtOH 3; eth 3; ctc 2 / 1.3888[20]
215	Acetic acid, 2-methylpropyl ester / Isobutyl acetate	$C_6H_{12}O_2$ / 116.16	110-19-0 / -98.8	5014 / 116.5	4-02-00-00149 / 0.8712[20]	H_2O 2; EtOH 5; eth 5; ace 3 / 1.3902[20]
216	Acetic acid, (methylthio)-	$C_3H_6O_2S$ / 106.15	2444-37-3 /	130-1[27]	4-03-00-00603 / 1.221[20]	1.495[20]
217	Acetic acid, (2-naphthalenyloxy)- / 2-Naphthoxyacetic acid	$C_{12}H_{10}O_3$ / 202.21	120-23-0 / 156	6317	4-06-00-04274	H_2O 3; EtOH 3; eth 3; HOAc 3
218	Acetic acid, nitro-, ethyl ester / Ethyl nitroacetate	$C_4H_7NO_4$ / 133.10	626-35-7 /	106[25]; 83[6]	4-02-00-00537 / 1.1953[20]	H_2O 2; EtOH 5; eth 4; dil alk 3 / 1.4250[20]
219	Acetic acid, 2-nitrophenyl ester	$C_8H_7NO_4$ / 181.15	610-69-5 / 40.5	253 dec; 141[11]	4-06-00-01256	H_2O 3; EtOH 4; eth 4; ace 4
220	Acetic acid, nonyl ester / Nonyl acetate	$C_{11}H_{22}O_2$ / 186.29	143-13-5 / -26	6597 / 210	4-02-00-00167 / 0.8785[15]	1.426[20]
221	Acetic acid, octadecyl ester	$C_{20}H_{40}O_2$ / 312.54	822-23-1 / 34.5	208[9]	4-02-00-00171 / 0.8510[30]	EtOH 4
222	Acetic acid, octyl ester	$C_{10}H_{20}O_2$ / 172.27	112-14-1 / -38.5	/ 210	4-02-00-00165 / 0.8705[20]	H_2O 1; EtOH 3; eth 3; ctc 2 / 1.4150[20]
223	Acetic acid, oxo- / Glyoxylic acid	$C_2H_2O_3$ / 74.04	298-12-4 / 98	4407	4-03-00-01489	H_2O 4; EtOH 2; eth 2; bz 2
224	Acetic acid, oxo(phenylamino)-, ethyl ester	$C_{10}H_{11}NO_3$ / 193.20	1457-85-8 / 65.5	/ 280	3-12-00-00550	ace 4; bz 4; eth 4; EtOH 4

No.	Name / Synonym	Mol. Form. / Mol. Wt.	CAS RN / mp/°C	Merck No. / bp/°C	Beil. Ref. / den/g cm^{-3}	Solubility / n_D
225	Acetic acid, [[(4-oxo-2-phenyl-4H-1-benzopyran-7-yl)oxy]-, ethyl ester Efloxate	$C_{19}H_{16}O_5$ 324.33	119-41-5 123.7	3490	4-18-00-00695	chl 3
226	Acetic acid, 2,2'-oxybis-	$C_4H_6O_5$ 134.09	110-99-6 148	dec	4-03-00-00577	H_2O 4; eth 4; EtOH 4
227	Acetic acid, pentyl ester Pentyl acetate	$C_7H_{14}O_2$ 130.19	628-63-7 -70.8	149.2	4-02-00-00152 0.8756[20]	H_2O 2; EtOH 5; eth 5; ctc 3 1.4023[20]
228	Acetic acid, phenoxy- Phenoxyacetic acid	$C_8H_8O_3$ 152.15	122-59-8 98.5	7223 285 dec	4-06-00-00634	H_2O 3; EtOH 4; eth 4; bz 4
229	Acetic acid, phenoxy-, ethyl ester Ethyl phenoxyacetate	$C_{10}H_{12}O_3$ 180.20	2555-49-9	109-12[3]	4-06-00-00635 1.1082[20]	1.5080[20]
230	Acetic acid, phenoxy-, methyl ester Methyl phenoxyacetate	$C_9H_{10}O_3$ 166.18	2065-23-8	245	4-06-00-00635 1.1493[20]	eth 4; EtOH 4 1.5155[20]
231	Acetic acid, phenyl ester Phenyl acetate	$C_8H_8O_2$ 136.15	122-79-2	7238 196; 75[8]	4-06-00-00611 1.0780[20]	H_2O 2; EtOH 5; eth 5; ctc 3 1.5035[20]
232	Acetic acid, 2-phenylethyl ester	$C_{10}H_{12}O_2$ 164.20	103-45-7 -31.1	232.6	4-06-00-03073 1.0883[20]	eth 4; EtOH 4 1.5171[20]
233	Acetic acid, phenylmethyl ester Benzyl acetate	$C_9H_{10}O_2$ 150.18	140-11-4 -51.3	1137 213	4-06-00-02262 1.0550[20]	H_2O 2; EtOH 5; eth 3; ace 3 1.5232[20]
234	Acetic acid, potassium salt Potassium acetate	$C_2H_3KO_2$ 98.14	127-08-2 292	7580	4-02-00-00094 1.57[25]	H_2O 4; EtOH 3; eth 1; ace 1
235	Acetic acid, 2-propenyl ester Allyl acetate	$C_5H_8O_2$ 100.12	591-87-7	103.5	4-02-00-00180 0.9275[20]	H_2O 2; EtOH 5; eth 5; ace 3 1.4049[20]
236	Acetic acid, propyl ester Propyl acetate	$C_5H_{10}O_2$ 102.13	109-60-4 -93	7853 101.5	4-02-00-00138 0.8878[20]	H_2O 2; EtOH 5; eth 5; ctc 3 1.3842[20]
237	Acetic acid, sodium salt Sodium acetate	$C_2H_3NaO_2$ 82.03	127-09-3 324	8513	4-02-00-00094 1.528[25]	H_2O 4; EtOH 2 1.464
238	Acetic acid, sulfo- Sulfoacetic acid	$C_2H_4O_5S$ 140.12	123-43-3 85	8932 245 dec	4-04-00-00102	H_2O 4; ace 4; EtOH 4
239	Acetic acid, 2,2'-sulfonylbis- Sulfonyldiacetic acid	$C_4H_6O_6S$ 182.15	123-45-5 187	8940	3-03-00-00424	H_2O 4; EtOH 4; eth 3; sulf 3
240	Acetic acid, 2,2'-thiobis- Thiodiglycolic acid	$C_4H_6O_4S$ 150.16	123-93-3 129	9260	4-03-00-00612	H_2O 2; EtOH 4; bz 3
241	Acetic acid, tribromo- Tribromoacetic acid	$C_2HBr_3O_2$ 296.74	75 06 7 132	0521 245 dec	4-02-00-00534	H_2O 3; EtOH 3; eth 3
242	Acetic acid, tribromo-, ethyl ester Ethyl tribromoacetate	$C_4H_5Br_3O_2$ 324.79	599-99-5	225	3-02-00-00485 2.2260[20]	EtOH 3; eth 3 1.5438[13]
243	Acetic acid, trichloro- Trichloroacetic acid	$C_2HCl_3O_2$ 163.39	76-03-9 57.5	9539 196.5	4-02-00-00508 1.6126[64]	H_2O 4; EtOH 3; eth 3; ctc 2 1.4603[61]
244	Acetic acid, trichloro-, anhydride Trichloroacetic anhydride	$C_4Cl_6O_3$ 308.76	4124-31-6	223 dec	4-02-00-00518 1.6908[20]	eth 4; HOAc 4
245	Acetic acid, trichloro-, butyl ester Butyl trichloroacetate	$C_6H_9Cl_3O_2$ 219.49	3657-07-6	204	4-02-00-00515 1.2778[20]	ctc 3 1.4525[25]
246	Acetic acid, trichloro-, 2-chloroethyl ester	$C_4H_4Cl_4O_2$ 225.89	4974-21-4	217	4-02-00-00515 1.5357[20]	eth 4 1.4813[20]
247	Acetic acid, trichloro-, 1,1-dimethylethyl ester	$C_6H_9Cl_3O_2$ 219.49	1860-21-5 25.5	54-5[7]	4-02-00-00516 1.2363[25]	eth 4; EtOH 4 1.4398[25]
248	Acetic acid, trichloro-, ethyl ester Ethyl trichloroacetate	$C_4H_5Cl_3O_2$ 191.44	515-84-4 167.5	9539	4-02-00-00514 1.3836[20]	H_2O 1; EtOH 3; eth 3; bz 3 1.4505[20]
249	Acetic acid, trichloro-, 2-hydroxyethyl ester	$C_4H_5Cl_3O_3$ 207.44	33560-17-7	130-4[12]	3-02-00-00474 1.532[20]	EtOH 4 1.4775[20]
250	Acetic acid, trichloro-, 2-methoxyethyl ester 2-Methoxyethyl trichloroacetate	$C_5H_7Cl_3O_3$ 221.47	35449-34-4 14.5	98-9[17]	3-02-00-00474 1.3826[20]	bz 4; eth 4; EtOH 4 1.4563[20]
251	Acetic acid, trichloro-, 1-methylpropyl ester	$C_6H_9Cl_3O_2$ 219.49	4484-80-4	93-4[24]	3-02-00-00472 1.2636[20]	bz 4; eth 4; EtOH 4 1.4483[20]
252	Acetic acid, trichloro-, 2-methylpropyl ester Isobutyl trichloroacetate	$C_6H_9Cl_3O_2$ 219.49	33560-15-5	188	4-02-00-00515 1.2636[20]	bz 4; eth 4; EtOH 4 1.4483[20]
253	Acetic acid, trichloro-, methyl ester Methyl trichloroacetate	$C_3H_3Cl_3O_2$ 177.41	598-99-2 -17.5	153.8	4-02-00-00513 1.4874[20]	H_2O 1; EtOH 4; eth 4; ctc 3 1.4572[20]
254	Acetic acid, trichloro-, 1-methylethyl ester Isopropyl trichloroacetate	$C_5H_7Cl_3O_2$ 205.47	3974-99-0	175; 66[15]	4-02-00-00515 1.3034[20]	bz 4; eth 4; EtOH 4 1.4428[20]
255	Acetic acid, trichloro-, 3-methylbutyl ester	$C_7H_{11}Cl_3O_2$ 233.52	57392-55-9	217	4-02-00-00516 1.2314[20]	eth 4; EtOH 4 1.4521[20]
256	Acetic acid, (2,4,5-trichlorophenoxy)- 2,4,5-T	$C_8H_5Cl_3O_3$ 255.48	93-76-5 153	8999	4-06-00-00973	H_2O 1; EtOH 3; bz 4
257	Acetic acid, (2,4,5-trichlorophenoxy)-, 2-butoxyethyl ester 2,4,5-T Butoxyethyl ester	$C_{14}H_{17}Cl_3O_4$ 355.64	2545-59-7	163-6[1]	4-06-00-00976 1.280[20]	ctc 3
258	Acetic acid, (2,4,5-trichlorophenoxy)-, butyl ester 2,4,5-T Butyl ester	$C_{12}H_{13}Cl_3O_3$ 311.59	93-79-8 28.5		4-06-00-00975	
259	Acetic acid, (2,4,5-trichlorophenoxy)-, 1-methylethyl ester 2,4,5-T Isopropyl ester	$C_{11}H_{11}Cl_3O_3$ 297.56	93-78-7 45		4-06-00-00974	chl 4
260	Acetic acid, (2,4,5-trichlorophenoxy)-, 1-methylpropyl ester 2,4,5-T sec-Butyl ester	$C_{12}H_{13}Cl_3O_3$ 311.59	61792-07-2 39		4-06-00-00975	
261	Acetic acid, trichloro-, propyl ester	$C_5H_7Cl_3O_2$ 205.47	13313-91-2	187	4-02-00-00515 1.3221[20]	eth 4; EtOH 4 1.4501[20]
262	Acetic acid, trichloro-, trichloromethyl ester	$C_3Cl_6O_2$ 280.75	6135-29-1 34	191.5	2-03-00-00015 1.6733[35]	bz 4; eth 4; chl 4; lig 4

No.	Name Synonym	Mol. Form. Mol. Wt.	CAS RN mp/°C	Merck No. bp/°C	Beil. Ref. den/g cm^{-3}	Solubility n_D
263	Acetic acid, trifluoro- Trifluoroacetic acid	$C_2HF_3O_2$ 114.02	76-05-1 -15.2	9595 73	4-02-00-00458 1.5351[25]	H_2O 3; EtOH 3; eth 3; ace 3 1.5351[25]
264	Acetic acid, trifluoro-, anhydride	$C_4F_6O_3$ 210.03	407-25-0 -65	39.5	4-02-00-00469 1.490[25]	1.269[25]
265	Acetic acid, trifluoro-, butyl ester Butyl trifluoroacetate	$C_6H_9F_3O_2$ 170.13	367-64-6	102	4-02-00-00464 1.0268[22]	chl 3 1.353[22]
266	Acetic acid, trifluoro-, 1,1-dimethylethyl ester	$C_6H_9F_3O_2$ 170.13	400-52-2	83	4-02-00-00465	chl 3 1.3300[25]
267	Acetic acid, trifluoro-, ethyl ester	$C_4H_5F_3O_2$ 142.08	383-63-1	61	4-02-00-00463 1.194[20]	1.308[20]
268	Acetic acid, trifluoro-, 2-methylpentyl ester 2-Methylpentyl trifluoroacetate	$C_8H_{13}F_3O_2$ 198.19	10042-29-2	140	4-02-00-00466 1.0504[20]	chl 2
269	Acetic acid, trifluoro-, propyl ester Propyl trifluoroacetate	$C_5H_7F_3O_2$ 156.10	383-66-4	82.5	4-02-00-00464 1.1285[25]	chl 3 1.3233[22]
270	Acetochlor Acetamide, 2-chloro-N-(ethoxymethyl)-N-(2-ethyl-6-methylphenyl)-	$C_{14}H_{20}ClNO_2$ 269.77	34256-82-1			
271	Acetonitrile Methyl cyanide	C_2H_3N 41.05	75-05-8 -43.8	62 81.6	4-02-00-00419 0.7857[20]	H_2O 5; EtOH 5; eth 5; ace 5 1.3442[30]
272	Acetonitrile, amino- Aminoacetonitrile	$C_2H_4N_2$ 56.07	540-61-4	423 58[15]	4-04-00-02363	EtOH 4
273	Acetonitrile, (butylamino)-	$C_6H_{12}N_2$ 112.17	3010-04-6	85[9]	4-04-00-02382 0.891[25]	1.4337[20]
274	Acetonitrile, chloro- Chloromethyl cyanide	C_2H_2ClN 75.50	107-14-2	126.5	4-02-00-00492 1.1930[20]	eth 4; EtOH 4 1.4202[25]
275	Acetonitrile, (cyclohexylamino)-	$C_8H_{14}N_2$ 138.21	1074-58-4 18	76[1]	4-12-00-00073 0.9657[25]	
276	Acetonitrile, cyclohexylidene- Cyclohexylideneacetonitrile	$C_8H_{11}N$ 121.18	4435-18-1	107-8[22]	4-09-00-00124 0.9483[15]	eth 4; EtOH 4 1.4382[25]
277	Acetonitrile, dibromo- Dibromoacetonitrile	C_2HBr_2N 198.84	3252-43-5	67-9[24]	4-02-00-00533 2.296[25]	1.5393[20]
278	Acetonitrile, dichloro- Dichloroacetonitrile	C_2HCl_2N 109.94	3018-12-0	112.5	4-02-00-00506 1.369[20]	MeOH 3 1.4391[25]
279	Acetonitrile, dichlorofluoro- Dichlorofluoroacetonitrile	C_2Cl_2FN 127.93	353-82-2 -110	33	4-02-00-00508 1.3909[25]	1.3682[20]
280	Acetonitrile, (diethylamino)-	$C_6H_{12}N_2$ 112.17	3010-02-4	170	4-04-00-02378 0.8660[20]	H_2O 3 1.4260[20]
281	Acetonitrile, (dimethylamino)- Glycinonitrile, N,N-dimethyl-	$C_4H_8N_2$ 84.12	926-64-7	137.5	4-04-00-02368 0.8649[20]	H_2O 4; EtOH 4 1.4095[20]
282	Acetonitrile, hydroxy- Glyconitrile	C_2H_3NO 57.05	107-16-4 <-72	183 dec; 119[24]	4-03-00-00598	H_2O 4; EtOH 4; eth 4; bz 1 1.4117[19]
283	Acetonitrile, iodo- Iodoacetonitrile	C_2H_2IN 166.95	624-75-9	185	2-02-00-00206 2.307[25]	1.5744[20]
284	Acetonitrile, methoxy-	C_3H_5NO 71.08	1738-36-9	119	3-03-00-00399 0.9492[20]	H_2O 2; EtOH 3; eth 3; ace 3 1.3831[20]
285	Acetonitrile, (methyleneamino)- Methyleneaminoacetonitrile	$C_3H_4N_2$ 68.08	109-82-0 129	5976		
286	Acetonitrile, (methylthio)- Cyanomethyl methyl sulfide	C_3H_5NS 87.15	35120-10-6	61-3[15]	4-03-00-00630 1.039[25]	1.4826[20]
287	Acetonitrile, phenoxy- Phenoxyacetonitrile	C_8H_7NO 133.15	3598-14-9	239.5	4-06-00-00640 1.0991[20]	eth 4; EtOH 4 1.5246[20]
288	Acetonitrile, trichloro- Trichloroacetonitrile	C_2Cl_3N 144.39	545-06-2 -42	9540 85.7	4-02-00-00524 1.4403[25]	H_2O 1 1.4409[20]
289	Acetonitrile, trifluoro- Trifluoroacetonitrile	C_2F_3N 95.02	353-85-5	-68.8	4-02-00-00472	
290	Acetonitrile, trinitro- Trinitroacetonitrile	$C_2N_4O_6$ 176.05	630-72-8 41.5	exp 220	0-02-00-00229	eth 4
291	Acetyl bromide Acetic acid, bromide	C_2H_3BrO 122.95	506-96-7 -96	76 76	4-02-00-00398 1.6625[16]	eth 5; ace 3; bz 5; chl 5 1.4486[20]
292	Acetyl bromide, bromo- Bromoacetyl bromide	$C_2H_2Br_2O$ 201.85	598-21-0	148.5	4-02-00-00530 2.312[22]	ace 3; ctc 3 1.5449[20]
293	Acetyl bromide, trichloro- Trichloroacetyl bromide	C_2BrCl_3O 226.28	34069-94-8	143	3-02-00-00476 1.898[15]	bz 4; eth 4
294	Acetyl chloride Ethanoyl chloride	C_2H_3ClO 78.50	75-36-5 -112.8	79 50.7	4-02-00-00395 1.1051[20]	eth 5; ace 5; bz 5; ctc 3 1.3886[20]
295	Acetyl chloride, chloro- Chloroacetyl chloride	$C_2H_2Cl_2O$ 112.94	79-04-9 -22	2054 106	4-02-00-00488 1.4202[20]	eth 5; ace 3; ctc 3 1.4530[20]
296	Acetyl chloride, chlorofluoro- Chlorofluoroacetyl chloride	C_2HCl_2FO 130.93	359-32-0	69.5	3-02-00-00453 1.468[25]	1.3992[25]
297	Acetyl chloride, dichloro- Dichloroacetyl chloride	C_2HCl_3O 147.39	79-36-7	3040 108	4-02-00-00504 1.5315[16]	H_2O 6; EtOH 6; eth 5 1.4591[20]
298	Acetyl chloride, ethoxy-	$C_4H_7ClO_2$ 122.55	14077-58-8	125	3-03-00-00396 1.1170[25]	ace 4; eth 4 1.4204[20]
299	Acetyl chloride, methoxy-	$C_3H_5ClO_2$ 108.52	38870-89-2	112.5	3-03-00-00396 1.1871[20]	eth 3; ace 3; ctc 3; chl 4 1.4199[20]
300	Acetyl chloride, trichloro- Trichloroacetyl chloride	C_2Cl_4O 181.83	76-02-8	117.9	4-02-00-00519 1.6202[20]	eth 5 1.4695[20]
301	Acetyl fluoride Acetic acid, fluoride	C_2H_3FO 62.04	557-99-3 -84	20.8	4-02-00-00393 1.032[25]	EtOH 5; eth 5; bz 3; chl 3

No.	Name / Synonym	Mol. Form. / Mol. Wt.	CAS RN / mp/°C	Merck No. / bp/°C	Beil. Ref. / den/g cm⁻³	Solubility / n_D
302	Acetyl iodide	C_2H_3IO / 169.95	507-02-8 / 88	108	4-02-00-00399 / 2.0673^{20}	eth 4 / 1.5491^{20}
303	Acetyl isothiocyanate / Isothiocyanic acid, acetyl	C_3H_3NOS / 101.13	13250-46-9	132.5	3-03-00-00278 / 1.1523^{13}	eth 3; CS_2 3 / 1.5231^{18}
304	Acifluorfen / Benzoic acid, 5-[2-chloro-4-(trifluoromethyl)phenoxy]-2-nitro-	$C_{14}H_7ClF_3NO_5$ / 361.66	50594-66-6 / 150	105		
305	Aconitane-3,8,13,14,15-pentol, 20-ethyl-1,6,16-trimethoxy-4-(methoxymethyl)-, / Aconine	$C_{25}H_{41}NO_9$ / 499.60	509-20-6 / 132	110	5-21-06-00310	H_2O 3; EtOH 3; eth 2; chl 3
306	Aconitane-3,8,13,14,15-pentol, 20-ethyl-1,6,16-trimethoxy-4-(methoxymethyl)-, 8-acetate 14-benzoate, / Aconitine	$C_{34}H_{47}NO_{11}$ / 645.75	302-27-2 / 204	113	5-21-06-00310	bz 4; EtOH 4; chl 4
307	Aconitane-3,8,13,14-tetrol, 20-ethyl-1,6,16-trimethoxy-4-(methoxymethyl)-, 8-acetate 14-benzoate, (1α,3alp / Indaconitine	$C_{34}H_{47}NO_{10}$ / 629.75	4491-19-4 / 202-3 dec	4842	5-21-06-00308	eth 4; EtOH 4; chl 4
308	Aconitane-3,8,13,14-tetrol, 20-ethyl-1,6,16-trimethoxy-4-(methoxymethyl)-, 8-acetate 14-(3,4-dimethoxybenzoate),(1α,3α,6α,14α,16β)- / Pseudoaconitine	$C_{36}H_{51}NO_{12}$ / 689.80	127-29-7 / 214	7924	5-21-06-00309	eth 4; EtOH 4
309	Aconitane-4,8,9-triol, 20-ethyl-1,14,16-trimethoxy-, 4-[2-(acetylamino)benzoate], (1α,14α,16β)- / Lappaconitine	$C_{32}H_{44}N_2O_8$ / 584.71	32854-75-4 / 217.5	5237	5-21-06-00291	H_2O 1; EtOH 2; eth 2; bz 3
310	Aconitane-8,13,14-triol, 20-ethyl-1,6,16-trimethoxy-4-(methoxymethyl)-, 8-acetate 14-(3,4-dimethoxybenzoate), / Bikhaconitine	$C_{36}H_{51}NO_{11}$ / 673.80	6078-26-8 / 164	1234	5-21-06-00303	eth 4; EtOH 4; chl 4
311	Aconitane-8,13,14-triol, 1,6,16-trimethoxy-4-(methoxymethyl)-20-methyl-, 8-acetate 14-benzoate, / Delphinine	$C_{33}H_{45}NO_9$ / 599.72	561-07-9 / 199	2867	5-21-06-00302	H_2O 1; chl 3; ace 3; eth 3
312	4-Acridinamine	$C_{13}H_{10}N_2$ / 194.24	578-07-4 / 108	183.5	5-22-11-00008	ace 4; bz 4; eth 4; EtOH 4
313	9-Acridinamine / Aminacrine	$C_{13}H_{10}N_2$ / 194.24	90-45-9 / 241	418	5-22-11-00008	EtOH 3; ace 3; dil HCl 4
314	Acridine / Dibenzo[b,e]pyridine	$C_{13}H_9N$ / 179.22	260-94-6 / 106(form a); 110(form b)	117 / 345.5	5-20-08-00199 / 1.005^{20}	H_2O 2; EtOH 4; eth 4; bz 4
315	Acridine, 9-chloro- / 9-Chloroacridine	$C_{13}H_8ClN$ / 213.67	1207-69-8 / 121	sub	5-20-08-00210	H_2O 4; EtOH 4
316	3,6-Acridinediamine / Proflavine	$C_{13}H_{11}N_3$ / 209.25	92-62-6 / 285	7780	5-22-11-00322	H_2O 3; EtOH 4; eth 2; bz 2
317	3,9-Acridinediamine, 7-ethoxy- / Ethacridine	$C_{15}H_{15}N_3O$ / 253.30	442-16-0 / 226	3668	5-22-12-00243	
318	Acridine, 9-phenyl- / 9-Phenylacridine	$C_{19}H_{13}N$ / 255.32	602-56-2 / 184	404	5-20-08-00564	H_2O 1; EtOH 2; eth 3; bz 4
319	Adenosine / β-D-Ribofuranoside, adenine-9	$C_{10}H_{13}N_5O_4$ / 267.24	58-61-7 / 235.5	143	4-26-00-03598	H_2O 2; EtOH 1
320	5'-Adenylic acid / Adenosine 5'-monophosphate	$C_{10}H_{14}N_5O_7P$ / 347.22	61-19-8 / 195 dec	148	4-26-00-03615	H_2O 4; EtOH 3
321	Ajmalan-17,21-diol, (17R,21α)- / Ajmaline	$C_{20}H_{26}N_2O_2$ / 326.44	4360-12-7 / 206	184	4-23-00-03212	H_2O 1; EtOH 3; eth 2; bz 2
322	Alachlor / Acetamide, 2-chloro-N-(2,6-diethylphenyl)-N-(methoxymethyl)-	$C_{14}H_{20}ClNO_2$ / 269.77	15972-60-8 / 40	193 / $100^{0.02}$	1.133^{25}	
323	β-Alanine / 3-Aminopropanoic acid	$C_3H_7NO_2$ / 89.09	107-95-9 / 200 dec	196	4-04-00-02526 / 1.437^{19}	H_2O 3; EtOH 2; eth 1; ace 1
324	DL-Alanine / DL-2-Aminopropanoic acid	$C_3H_7NO_2$ / 89.09	302-72-7 / 300 dec	195 / sub 250	4-04-00-02481 / 1.424^{25}	H_2O 3; EtOH 4
325	L-Alanine / (S)-2-Aminopropanoic acid	$C_3H_7NO_2$ / 89.09	56-41-7 / 300 dec	195 / sub 250	4-04-00-02480 / 1.432^{22}	H_2O 3; EtOH 2; eth 1; ace 1
326	Alanine, 3-amino- / 2,3-Diaminopropionic acid	$C_3H_8N_2O_2$ / 104.11	515-94-6 / 110	2962	4-04-00-02581	H_2O 4
327	DL-Alanine, N-benzoyl-	$C_{10}H_{11}NO_3$ / 193.20	1205-02-3 / 165.5 dec		4-09-00-00794	H_2O 3; EtOH 3; eth 2; DMSO 2
328	β-Alanine, N,N-diethyl-, ethyl ester / Ethyl 3-(diethylamino)propionate	$C_9H_{19}NO_2$ / 173.26	5515-83-3 / $83-4^{12}$		4-04-00-02538 / 0.881^{25}	1.4253^{20}
329	β-Alanine, N-(2,4-dihydroxy-3,3-dimethyl-1-oxobutyl)-, (R)- / Pantothenic acid	$C_9H_{17}NO_5$ / 219.24	79-83-4	6964	4-04-00-02569	H_2O 4; bz 4; eth 4
330	β-Alanine, N-(3-ethoxy-3-oxopropyl)-N-methyl-, ethyl ester / Propionic acid, 3,3'-iminodi-N-methyl, diethyl ester	$C_{11}H_{21}NO_4$ / 231.29	6315-60-2 / $136-8^4$		3-04-00-01278 / 1.0172^{20}	eth 4; EtOH 4 / 1.4421^{20}
331	Alanine, 2-methyl- / α-Aminoisobutyric acid	$C_4H_9NO_2$ / 103.12	62-57-7 / 335	457 / sub 280	4-04-00-02616	H_2O 4; EtOH 2; eth 1

No.	Name Synonym	Mol. Form. Mol. Wt.	CAS RN mp/°C	Merck No. bp/°C	Beil. Ref. den/g cm⁻³	Solubility n_D
332	β-Alanine, N-methyl-, ethyl ester	$C_6H_{13}NO_2$ 131.17	2213-08-3	80[21]	3-04-00-01264 1.0064[20]	eth 4; EtOH 4 1.4443[20]
333	L-Alanine, 3-(2-propenylsulfinyl)-, (S)- Alliin	$C_6H_{11}NO_3S$ 177.22	556-27-4 165	251	4-04-00-03147	H₂O 4
334	Aldicarb Propanal, 2-methyl-2-(methylthio)-, O-[(methylamino)carbonyl]oxime	$C_7H_{14}N_2O_2S$ 190.27	116-06-3 99	216	1.195[25]	
335	Aldoxycarb Propanal, 2-methyl-2-(methylsulfonyl)-, O-[(methylamino)carbonyl]oxime	$C_7H_{14}N_2O_4S$ 222.27	1646-88-4			
336	Allethrin 2-Methyl-4-oxo-3-(2-propenyl)-2-cyclopenten-1-yl 2,2-dimethyl-3-(2-methyl-1-propenyl) cyclopropane carboxylate	$C_{19}H_{26}O_3$ 302.41	584-79-2		1.010[20]	
337	D-Allose	$C_6H_{12}O_6$ 180.16	2595-97-3 128	279	4-01-00-04299	H₂O 4
338	D-Altrose	$C_6H_{12}O_6$ 180.16	1990-29-0 103.5	319	4-01-00-04300	H₂O 4
339	Aluminum, trimethyl Trimethyl aluminum	C_3H_9Al 72.09	75-24-1 15.4	130; 20[8]	0.752[20]	
340	Aluminum, tris(2-hydroxypropanoato-O1,O2)- Aluminum lactate	$C_9H_{15}AlO_9$ 294.19	18917-91-4	350		H₂O 3
341	Ametryn N-Ethyl-N'-(1-methylethyl)-6-(methylthio)1,3,5-triazine-2,4-diamine	$C_9H_{17}N_5S$ 227.33	834-12-8 88	402		
342	Amitraz N-Methylbis(2,4-xylyliminomethyl)amine	$C_{19}H_{23}N_3$ 293.41	33089-61-1 86	503	1.128[20]	
343	Androsta-1,4-dien-3-one, 17-hydroxy-, (17β)- Boldenone	$C_{19}H_{26}O_2$ 286.41	846-48-0 165	1327		
344	Androsta-1,4-dien-3-one, 17-hydroxy-17-methyl-, (17β)- Methandrostenolone	$C_{20}H_{28}O_2$ 300.44	72-63-9 166	5862	4-08-00-01119	
345	Androstane	$C_{19}H_{32}$ 260.46	24887-75-0 50	668 60[0.003]	4-05-00-01211	ace 4; eth 4; EtOH 4; peth 4
346	Androstane-17-carboxylic acid, (5β,17β)- Etiocholanic acid	$C_{20}H_{32}O_2$ 304.47	438-08-4 228.5	3824 sub 160	3-09-00-02644	
347	Androstan-17-one, 3-hydroxy-, (3β,5α)- Epiandrosterone	$C_{19}H_{30}O_2$ 290.45	481-29-8 178	3562	4-08-00-00641	
348	Androstan-17-one, 3-hydroxy-, (3α,5α)- Androsterone	$C_{19}H_{30}O_2$ 290.45	53-41-8 185	673	4-08-00-00642	H₂O 2; EtOH 3; eth 3; ace 3
349	Androstan-3-one, 17-hydroxy-, (5α,17β)- Stanolone	$C_{19}H_{30}O_2$ 290.45	521-18-6 181	8753 sub 135	4-08-00-00634	
350	Androstan-3-one, 17-hydroxy-17-methyl-, (5α,17β)- Mestanolone	$C_{20}H_{32}O_2$ 304.47	521-11-9 192.5	5816	4-08-00-00656	AcOEt 2
351	Androstan-3-one, 2-methyl-17-(1-oxopropoxy)-, (2α,5α,17β)- Dromostanolone propionate	$C_{23}H_{36}O_3$ 360.54	521-12-0 128	3436	4-08-00-00654	
352	Androst-4-ene-3,17-dione 4-Androstene-3,17-dione	$C_{19}H_{26}O_2$ 286.41	63-05-8 143(form a); 173(form b)	671	4-07-00-02381	
353	Androst-4-ene-3,11,17-trione Adrenosterone	$C_{19}H_{24}O_3$ 300.40	382-45-6 222	165 sub	4-07-00-02796	H₂O 2; EtOH 3; eth 3; ace 3
354	Androst-4-en-3-one, 4-chloro-17-hydroxy-, (17β)- Clostebol	$C_{19}H_{27}ClO_2$ 322.87	1093-58-9 189	2409	4-08-00-00989	
355	Androst-4-en-3-one, 9-fluoro-11,17-dihydroxy-17-methyl-, (11β,17β)- Fluoxymesterone	$C_{20}H_{29}FO_3$ 336.45	76-43-7 270	4113	4-08-00-02057	
356	Androst-4-en-3-one, 17-hydroxy-, (17β)- Testosterone	$C_{19}H_{28}O_2$ 288.43	58-22-0 155	9109	4-08-00-00974	H₂O 1; EtOH 3; eth 3; ace 3
357	Androst-4-en-3-one, 17-hydroxy-7,17-dimethyl-, (7β,17β)- Calusterone	$C_{28}H_{48}O$ 400.69	17021-26-0 157.5	1730	4-08-00-01029	
358	Androst-4-en-3-one, 17-hydroxy-17-methyl-, (17β)- 17-Methyltestosterone	$C_{20}H_{30}O_2$ 302.46	58-18-4 163.5	6044	4-08-00-01010	eth 4; EtOH 4
359	Androst-4-en-3-one, 17-(1-oxopropoxy)-, (17β)- Testosterone-17-propionate	$C_{22}H_{32}O_3$ 344.49	57-85-2 120	9115	3-08-00-00897	eth 4; py 4; EtOH 4
360	Anilazine 2,4-Dichloro-6-(o-chloroanilino)-s-triazine	$C_9H_5Cl_3N_4$ 275.52	101-05-3 160	685	1.8[20]	
361	Aniline, 3,3'-azobis[N,N-dimethyl- Azobenzene, 3,3'-bis(dimethylamino)	$C_{16}H_{20}N_4$ 268.36	21232-53-1 118	sub	0-16-00-00305	bz 4; EtOH 4; chl 4
362	Aniline, p-isopentyl-	$C_{11}H_{17}N$ 163.26	104177-72-2	257	4-12-00-02858 0.928[15]	1.5305[20]
363	Anisole, 3-isopropyl-2-methyl- o-Cresol, 3-isopropyl, methyl ether	$C_{11}H_{16}O$ 164.25	31202-12-7 -0.5	67[3]	4-06-00-03328 0.9540[25]	1.5148[20]
364	Anisole, 2,3,5-trinitro- 2,3,5-Trinitroanisole	$C_7H_5N_3O_7$ 243.13	7539-25-5 107.4		2-06-00-00253 1.618[15]	ace 4; bz 4; EtOH 4

No.	Name Synonym	Mol. Form. Mol. Wt.	CAS RN mp/°C	Merck No. bp/°C	Beil. Ref. den/g cm^{-3}	Solubility n_D
365	2-Anthracenamine	C$_{14}$H$_{11}$N 193.25	613-13-8 238.8	sub	4-12-00-03435	H$_2$O 1; EtOH 3; os 2; con sulf 1
366	Anthracene	C$_{14}$H$_{10}$ 178.23	120-12-7 215.0	712 339.9	4-05-00-02281 1.28[25]	H$_2$O 1; EtOH 2; eth 2; ace 2
367	Anthracene, 9,10-bis(phenylmethyl)- 9,10-Dibenzylanthracene	C$_{28}$H$_{22}$ 358.48	3613-42-1 246.5		3-05-00-02640 1.17871[16]	H$_2$O 1; eth 2; bz 2; lig 1
368	1-Anthracenecarboxylic acid 1-Anthroic acid	C$_{15}$H$_{10}$O$_2$ 222.24	607-42-1 251.5	sub	4-09-00-02670	H$_2$O 1; EtOH 3; eth 3; bz 2
369	2-Anthracenecarboxylic acid 2-Anthroic acid	C$_{15}$H$_{10}$O$_2$ 222.24	613-08-1 281	sub	4-09-00-02671	HOAc 4
370	9-Anthracenecarboxylic acid 9-Anthroic acid	C$_{15}$H$_{10}$O$_2$ 222.24	723-62-6 217 dec	sub	4-09-00-02671	H$_2$O 1; EtOH 3
371	2-Anthracenecarboxylic acid, 9,10-dihydro-4,5-dihydroxy-9,10-dioxo- Rhein	C$_{15}$H$_8$O$_6$ 284.22	478-43-3 321	8175 sub	4-10-00-04088	H$_2$O 2; EtOH 2; eth 2; ace 2
372	2-Anthracenecarboxylic acid, 9,10-dihydro-9,10-dioxo-	C$_{15}$H$_8$O$_4$ 252.23	117-78-2 291	sub	4-10-00-03200	EtOH 2; eth 1; ace 3; bz 1
373	2-Anthracenecarboxylic acid, 7-β-D-glucopyranosyl-9,10-dihydro-3,5,6,8-tetrahydroxy-1-methyl-9,10-dioxo- Carminic acid	C$_{22}$H$_{20}$O$_{13}$ 492.39	1260-17-9 136 dec	1850	3-10-00-04874	H$_2$O 3; EtOH 3; eth 2; bz 1
374	Anthracene, 1-chloro- 1-Chloroanthracene	C$_{14}$H$_9$Cl 212.68	4985-70-0 83.5		4-05-00-02292 1.1707[100]	H$_2$O 1; EtOH 3; eth 3; bz 3 1.6959[100]
375	Anthracene, 9,10-dibromo- 9,10-Dibromoanthracene	C$_{14}$H$_8$Br$_2$ 336.03	523-27-3 226	3000 sub	4-05-00-02295	H$_2$O 1; EtOH 2; eth 2; bz 2
376	Anthracene, 9,10-dihydro- 9,10-Dihydroanthracene	C$_{14}$H$_{12}$ 180.25	613-31-0 111	305	4-05-00-02182 1.215[20]	H$_2$O 1; EtOH 3; eth 3; bz 3
377	Anthracene, 1,3-dimethyl- 1,3-Dimethylanthracene	C$_{16}$H$_{14}$ 206.29	610-46-8 83	143[2]	4-05-00-02327	eth 4; EtOH 4
378	Anthracene, 2,10-dimethyl- Anthracene, 9,3-dimethyl	C$_{16}$H$_{14}$ 206.29	27532-76-9 84	140-5[2]	4-05-00-02328	H$_2$O 1; EtOH 4; eth 4
379	9,10-Anthracenedione Anthraquinone	C$_{14}$H$_8$O$_2$ 208.22	84-65-1 286	717 377	4-07-00-02556 1.438[20]	H$_2$O 1; EtOH 2; eth 2; bz 2
380	9,10-Anthracenedione, 1-amino- 1-Aminoanthraquinone	C$_{14}$H$_9$NO$_2$ 223.23	82-45-1 253.5	428 sub	4-14-00-00429	ace 4; bz 4; EtOH 4; chl 4
381	9,10-Anthracenedione, 2-amino- 2-Aminoanthraquinone	C$_{14}$H$_9$NO$_2$ 223.23	117-79-3 304.5	sub	4-14-00-00447	H$_2$O 1; EtOH 1; eth 1; ace 3
382	9,10-Anthracenedione, 3-amino-1,2-dihydroxy-	C$_{14}$H$_9$NO$_4$ 255.23	3963-78-8 >300	sub	2-14-00-00185	aq NH$_3$ 4
383	9,10-Anthracenedione, 2-amino-1-hydroxy- Anthraquinone, 2-amino-1-hydroxy	C$_{14}$H$_9$NO$_3$ 239.23	568-99-0 258	sub	4-14-00-00891	bz 4; eth 4; EtOH 4
384	9,10-Anthracenedione, 1-amino-2-methyl- C.I. Solvent Orange 35	C$_{15}$H$_{11}$NO$_2$ 237.26	82-28-0 205.5		4-14-00-00496	H$_2$O 1; EtOH 3; eth 2; bz 3
385	9,10-Anthracenedione, 1-bromo-	C$_{14}$H$_7$BrO$_2$ 287.11	632-83-7 186.5	sub	3-07-00-04076	bz 4; EtOH 4
386	9,10-Anthracenedione, 2-bromo-	C$_{14}$H$_7$BrO$_2$ 287.11	572-83-8 205.8	sub	3-07-00-04076	bz 4
387	9,10-Anthracenedione, 1-chloro-	C$_{14}$H$_7$ClO$_2$ 242.66	82-44-0 163	sub	4-07-00-02559	H$_2$O 1; EtOH 2; eth 5; bz 3
388	9,10-Anthracenedione, 2-chloro-	C$_{14}$H$_7$ClO$_2$ 242.66	131-09-9 211	sub	4-07-00-02559	H$_2$O 1; EtOH 2; eth 1; bz 2
389	9,10-Anthracenedione, 1,5-diamino-	C$_{14}$H$_{10}$N$_2$O$_2$ 238.25	129-44-2 319	sub	4-14-00-00479	H$_2$O 1; EtOH 2; eth 2; ace 2
390	9,10-Anthracenedione, 2,7-dibromo- 2,7-Dibromoanthraquinone	C$_{14}$H$_6$Br$_2$O$_2$ 366.01	605-42-5 249	sub	2-07-00-00718	bz 4; HOAc 4
391	9,10-Anthracenedione, 1,2-dihydroxy- Alizarin	C$_{14}$H$_8$O$_4$ 240.22	72-48-0 289.5	237	4-08-00-03256	H$_2$O 2; EtOH 3; eth 3; ace 3
392	9,10-Anthracenedione, 1,4-dihydroxy- Quinizarin	C$_{14}$H$_8$O$_4$ 240.22	81-64-1 200	8094	4-08-00-03260	H$_2$O 3; EtOH 3; eth 3; bz 3
393	9,10-Anthracenedione, 1,5-dihydroxy- Anthrarufin	C$_{14}$H$_8$O$_4$ 240.22	117-12-4 280	719 sub	4-08-00-03268	H$_2$O 1; EtOH 2; eth 2; ace 2
394	9,10-Anthracenedione, 1,7-dihydroxy- 9,10-Anthraquinone, 1,7-dihydroxy	C$_{14}$H$_8$O$_4$ 240.22	569-08-4 293.3	sub	4-08-00-03270	H$_2$O 1; EtOH 3; eth 3; bz 3
395	9,10-Anthracenedione, 1,8-dihydroxy- Danthron	C$_{14}$H$_8$O$_4$ 240.22	117-10-2 193	2813 sub	4-08-00-03271	H$_2$O 1; EtOH 3; eth 3; ace 3
396	9,10-Anthracenedione, 2,7-dihydroxy-	C$_{14}$H$_8$O$_4$ 240.22	572-93-0 353.8	sub	4-08-00-03273	H$_2$O 1; EtOH 3; eth 2; bz 2
397	9,10-Anthracenedione, 1,8-dihydroxy-3-(hydroxymethyl)- Aloe-emodin	C$_{15}$H$_{10}$O$_5$ 270.24	481-72-1 223.5	303 sub	4-08-00-03578	bz 4; eth 4; EtOH 4
398	9,10-Anthracenedione, 1,8-dihydroxy-3-methyl- Chrysophanic acid	C$_{15}$H$_{10}$O$_4$ 254.24	481-74-3 196	2263 sub	4-08-00-03277 0.92[25]	bz 4; HOAc 4
399	9,10-Anthracenedione, 1,2-dihydroxy-3-nitro- Alizarine Orange	C$_{14}$H$_7$NO$_6$ 285.21	568-93-4 244 dec	240 sub	4-08-00-03259	H$_2$O 2; EtOH 3; bz 3; chl 3
400	9,10-Anthracenedione, 1,2-dihydroxy-4-nitro-	C$_{14}$H$_7$NO$_6$ 285.21	2243-71-2 289 dec	sub	4-08-00-03259	bz 4; EtOH 4; chl 4; HOAc 4

No.	Name Synonym	Mol. Form. Mol. Wt.	CAS RN mp/°C	Merck No. bp/°C	Beil. Ref. den/g cm^{-3}	Solubility n_D
401	9,10-Anthracenedione, 1,8-dihydroxy-2,4,5,7-tetranitro- Chrysamminic acid	$C_{14}H_4N_4O_{12}$ 420.21	517-92-0 exp	2253 dec	1-08-00-00723	eth 4; EtOH 4
402	9,10-Anthracenedione, 1,4-dimethyl-	$C_{16}H_{12}O_2$ 236.27	1519-36-4 140.5	sub	4-07-00-02583	H_2O 1; EtOH 2; bz 3; xyl 3
403	9,10-Anthracenedione, 2,3-dimethyl-	$C_{16}H_{12}O_2$ 236.27	6531-35-7 210.8	sub	3-07-00-04131	EtOH 3; bz 3; chl 2; xyl 3
404	9,10-Anthracenedione, 2,6-dimethyl-	$C_{16}H_{12}O_2$ 236.27	3837-38-5 242	sub	3-07-00-04132	EtOH 4; HOAc 4; tol 4
405	9,10-Anthracenedione, 1,5-dinitro-	$C_{14}H_6N_2O_6$ 298.21	82-35-9 385	sub	4-07-00-02562	H_2O 1; EtOH 2; eth 2; bz 2
406	9,10-Anthracenedione, 1,2,3,5,6,7-hexahydroxy- Rufigallol	$C_{14}H_8O_8$ 304.21	82-12-2	8267 sub	4-08-00-03750	H_2O 1; EtOH 2; eth 2; ace 3
407	9,10-Anthracenedione, 1-hydroxy-	$C_{14}H_8O_3$ 224.22	129-43-1 193.8	sub	4-08-00-02586	H_2O 1; EtOH 3; eth 3; bz 3
408	9,10-Anthracenedione, 2-hydroxy-	$C_{14}H_8O_3$ 224.22	605-32-3 306	sub	4-08-00-02598	H_2O 1; EtOH 3; eth 3; KOH 3
409	9,10-Anthracenedione, 2-methyl- 2-Methylanthraquinone	$C_{15}H_{10}O_2$ 222.24	84-54-8 177	5943 sub	4-07-00-02574	bz 4; EtOH 4; HOAc 4
410	9,10-Anthracenedione, 1-nitro-	$C_{14}H_7NO_4$ 253.21	82-34-8 231.5	270^7	4-07-00-02561	H_2O 1; EtOH 2; eth 2; ace 3
411	9,10-Anthracenedione, 2-nitro-	$C_{14}H_7NO_4$ 253.21	605-27-6 184.5	sub	4-07-00-02561	H_2O 1; EtOH 2; eth 2; ace 3
412	9,10-Anthracenedione, 1,2,5,6-tetrahydroxy-	$C_{14}H_8O_6$ 272.21	632-77-9 340	sub	3-08-00-04288	py 4; EtOH 4; HOAc 4
413	9,10-Anthracenedione, 1,2,5,8-tetrahydroxy- Quinalizarin	$C_{14}H_8O_6$ 272.21	81-61-8 >275	8059	4-08-00-03683	H_2O 2; EtOH 2; eth 2; ace 2
414	9,10-Anthracenedione, 1,3,5,7-tetrahydroxy-	$C_{14}H_8O_6$ 272.21	632-82-6 360	sub	4-08-00-03683	ace 4; bz 4; EtOH 4; HOAc 4
415	9,10-Anthracenedione, 1,2,3-trihydroxy- Anthragallol	$C_{14}H_8O_5$ 256.21	602-64-2 313	713 sub 290	4-08-00-03567	H_2O 2; EtOH 3; eth 3; CS_2 3
416	9,10-Anthracenedione, 1,2,4-trihydroxy- Purpurin	$C_{14}H_8O_5$ 256.21	81-54-9 259	7962 sub	4-08-00-03568	H_2O 2; EtOH 4; eth 3; bz 4
417	9,10-Anthracenedione, 1,2,6-trihydroxy-	$C_{14}H_8O_5$ 256.21	82-29-1 330	459	4-08-00-03568	H_2O 3; EtOH 3; eth 2; bz 3
418	9,10-Anthracenedione, 1,2,7-trihydroxy-	$C_{14}H_8O_5$ 256.21	602-65-3 374	462	4-08-00-03569	bz 4; EtOH 4; HOAc 4
419	9,10-Anthracenedione, 1,3,8-trihydroxy-6-methyl- Emodin	$C_{15}H_{10}O_5$ 270.24	518-82-1 257	3518 sub	4-08-00-03575	eth 4; EtOH 4
420	Anthracene, 9-ethyl- 9-Ethylanthracene	$C_{16}H_{14}$ 206.29	605-83-4 60		4-05-00-02326 1.0413^{99}	H_2O 1; EtOH 3; eth 3 1.6767^{99}
421	Anthracene, 9-ethyl-9,10-dihydro- 9-Ethyl-9,10-dihydroanthracene	$C_{16}H_{16}$ 208.30	605-82-3	321.5	0-05-00-00649 1.048^{18}	bz 4; eth 4; EtOH 4
422	Anthracene, 1,2,3,4,5,6-hexahydro- 1,2,3,4,5,6-Hexahydroanthracene	$C_{14}H_{16}$ 184.28	6109-22-4 68.5	160^{15}	3-05-00-01676	bz 4
423	Anthracene, 1-methyl- 1-Methylanthracene	$C_{15}H_{12}$ 192.26	610-48-0 85.5	199.5	4-05-00-02310 1.0471^{99}	H_2O 1; EtOH 3; eth 3; bz 3 1.6802^{99}
424	Anthracene, 2-methyl- 2-Methylanthracene	$C_{15}H_{12}$ 192.26	613-12-7 209	sub	4-05-00-02311 1.80^0	H_2O 1; EtOH 2; eth 2; ace 1
425	Anthracene, 9-methyl- 9-Methylanthracene	$C_{15}H_{12}$ 192.26	779-02-2 81.5	$196-7^{12}$	4-05-00-02312 1.065^{99}	EtOH 3; eth 3; ace 3; bz 3 1.6959^{99}
426	Anthracene, 9-nitro- 9-Nitroanthracene	$C_{14}H_9NO_2$ 223.23	602-60-8 146	275^{17}	4-05-00-02296	H_2O 1; EtOH 2; ace 4; chl 2
427	Anthracene, 1,2,3,4,5,6,7,8-octahydro-	$C_{14}H_{18}$ 186.30	1079-71-6 78	294	4-05-00-01584 0.9703^{80}	H_2O 1; EtOH 3; bz 4; ctc 2 1.5372^{80}
428	Anthracene, 9-phenyl- 9-Phenylanthracene	$C_{20}H_{14}$ 254.33	602-55-1 156	417	4-05-00-02628	H_2O 1; EtOH 3; eth 3; bz 3
429	1,2,10-Anthracenetriol Anthrarobin	$C_{14}H_{10}O_3$ 226.23	577-33-3 208	718	4-06-00-07601	H_2O 2; EtOH 4; eth 4; ace 4
430	1,8,9-Anthracenetriol Anthralin	$C_{14}H_{10}O_3$ 226.23	1143-38-0 179	714	4-06-00-07602	H_2O 1; EtOH 3; eth 2; ace 3
431	1-Anthracenol	$C_{14}H_{10}O$ 194.23	610-50-4 158	234^{13}	4-06-00-04928	H_2O 1; EtOH 4; eth 4; NaOH 3
432	9-Anthracenol Anthranol	$C_{14}H_{10}O$ 194.23	529-86-2 152	716	4-06-00-04930	
433	9(10H)-Anthracenone Anthrone	$C_{14}H_{10}O$ 194.23	90-44-8 155	721	4-06-00-04930	ace 3; bz 3; con sulf 3; dil alk 3
434	5,9,14,18-Anthrazinetetrone, 6,15-dihydro- Indanthrene	$C_{28}H_{14}N_2O_4$ 442.43	81-77-6 470-500 dec	4846	5-24-09-00504	H_2O 1; EtOH 1; eth 1; ace 1
435	8'-Apo-β,ψ-carotenal, 3-hydroxy-, (3R)- β-Citraurin	$C_{30}H_{40}O_2$ 432.65	650-69-1 147	2326	3-08-00-01599	H_2O 1; EtOH 4; eth 4; ace 4
436	β-D-Arabinopyranose Arabinose (D)(β)	$C_5H_{10}O_5$ 150.13	6748-95-4 155.5		4-01-00-04215 1.625^{25}	H_2O 4
437	DL-Arginine	$C_6H_{14}N_4O_2$ 174.20	7200-25-1 217 dec	805	3-04-00-01359	H_2O 1; EtOH 1; eth 1; bz 1
438	L-Arginine	$C_6H_{14}N_4O_2$ 174.20	74-79-3 244 dec	805	4-04-00-02648	H_2O 3; EtOH 2; eth 1

No.	Name Synonym	Mol. Form. Mol. Wt.	CAS RN mp/°C	Merck No. bp/°C	Beil. Ref. den/g cm^{-3}	Solubility n_D
439	Arsenic acid, triethyl ester Ethyl arsenate	$C_6H_{15}AsO_4$ 226.10	15606-95-8	236.5	4-01-00-01359 1.3021[20]	1.4343[20]
440	Arsenous acid, triethyl ester Ethyl arsenite	$C_6H_{15}AsO_3$ 210.10	3141-12-6	165.5	4-01-00-01358 1.2239[20]	1.4369[13]
441	Arsine, diethyl- Diethylarsine	$C_4H_{11}As$ 134.05	692-42-2	105	3-04-00-01797 1.1338[24]	ace 4; bz 4; eth 4; EtOH 4 1.4709
442	Arsine, dimethyl- Dimethylarsine	C_2H_7As 106.00	593-57-7 -136.1	36	4-04-00-03665 1.208[29]	ace 4; bz 4; eth 4; EtOH 4
443	Arsine, diphenyl- Diphenylarsine	$C_{12}H_{11}As$ 230.14	829-83-4	174[25]	4-16-00-01138 1.30[25]	eth 4; EtOH 4
444	Arsine, ethyl- Ethylarsine	C_2H_7As 106.00	593-59-9	36	3-04-00-01797 1.214[22]	eth 4; EtOH 4
445	Arsine, ethyldifluoro-	$C_2H_5AsF_2$ 141.98	430-40-0 -38.7	94.3	3-04-00-01799 1.708[17]	
446	Arsine, methyl- Methylarsine	CH_5As 91.97	593-52-2 -143	2	3-04-00-01795	ace 4; eth 4; EtOH 4
447	Arsine, oxophenyl- Oxophenylarsine	C_6H_5AsO 168.03	637-03-6 145	6903	4-16-00-01160	H_2O 1; EtOH 2; eth 1; bz 4
448	Arsine, triethyl- Triethylarsine	$C_6H_{15}As$ 162.11	617-75-4	138.5	4-04-00-03666 1.150[20]	ace 4; eth 4; EtOH 4 1.467[20]
449	Arsine, trimethyl- Trimethylarsine	C_3H_9As 120.03	593-88-4 -87.3	52	4-04-00-03665 1.144[15]	bz 4; eth 4; EtOH 4
450	Arsine, triphenyl- Triphenylarsine	$C_{18}H_{15}As$ 306.24	603-32-7 61	360	4-16-00-01139 1.2634[18]	H_2O 1; EtOH 2; eth 4; bz 4 1.6888[21]
451	Arsinic acid, dimethyl- Cacodylic acid	$C_2H_7AsO_2$ 138.00	75-60-5 195	1603	4-04-00-03681	H_2O 4; EtOH 3; eth 1
452	Arsinous acid, dimethyl-, anhydride Cacodyl oxide	$C_4H_{12}As_2O$ 225.98	503-80-0 -25	150	3-04-00-01814 1.4816[15]	eth 4; EtOH 4 1.5225[9]
453	Arsinous chloride, dimethyl- Dimethylchloroarsine	C_2H_6AsCl 140.44	557-89-1 <-45	109; 77[37]	4-04-00-03670 1.505[12]	EtOH 4 1.5203[12]
454	Arsinous chloride, diphenyl- Chlorodiphenylarsine	$C_{12}H_{10}AsCl$ 264.59	712-48-1 44	337	3-16-00-00944 1.4820[16]	ace 4; bz 4; eth 4; EtOH 4 1.6332[56]
455	Arsonic acid, [4-[(aminocarbonyl)amino]phenyl]- Carbarsone	$C_7H_9AsN_2O_4$ 260.08	121-59-5 174	1788	3-16-00-01112	H_2O 2; EtOH 2; eth 1; chl 1
456	Arsonic acid, (4-aminophenyl)- Arsanilic acid	$C_6H_8AsNO_3$ 217.06	98-50-0 232	818	4-16-00-01190 1.9571[10]	H_2O 3, EtOH 3; eth 3; ace 1
457	Arsonic acid, ethyl- Ethanearsonic acid	$C_2H_7AsO_3$ 154.00	507-32-4 99.5	3677 209-11[12]	4-04-00-03682	H_2O 4; EtOH 4
458	Arsonic acid, methyl- Methanearsonic acid	CH_5AsO_3 139.97	124-58-3 160.5	5864	4-04-00-03682	H_2O 3; EtOH 3
459	Arsonic acid, (4-nitrophenyl)- Nitarsone	$C_6H_6AsNO_5$ 247.04	98-72-6 >310 dec	6483	4-16-00-01185	H_2O 2; EtOH 2; DMSO-d_6 2
460	Arsonic acid, phenyl- Benzenearsonic acid	$C_6H_7AsO_3$ 202.04	98-05-5 158- dec	1075	4-16-00-01183	H_2O 4; EtOH 4
461	Arsonic acid, propyl- 1-Propanearsonic acid	$C_3H_9AsO_3$ 168.02	107-34-6 134.5	7810	4-04-00-03682	H_2O 4; EtOH 4; eth 1
462	Arsonous dichloride, phenyl- Dichlorophenylarsine	$C_6H_5AsCl_2$ 222.93	696-28-6 -19	255	4-16-00-01161 1.6516[20]	bz 4; eth 4; EtOH 4 1.6386[15]
463	Arsonous difluoride, methyl- Difluoromethylarsine	CH_3AsF_2 127.95	420-24-6 -29.7	76.5	3-04-00-01796 1.924[18]	
464	Arsonous diiodide, phenyl-	$C_6H_5AsI_2$ 405.84	6380-34-3 15	205[14] 185[10]	3-16-00-00958 1.6264[15]	
465	DL-Ascorbic acid	$C_6H_8O_6$ 176.13	62624-30-0 168.5		5-18-05-00026	H_2O 4; EtOH 3; eth 1; bz 1
466	L-Ascorbic acid Vitamin C	$C_6H_8O_6$ 176.13	50-81-7 190-2 dec	855	5-18-05-00026 1.65[25]	H_2O 4; EtOH 3; eth 1; bz 1
467	L-Ascorbic acid, 6-deoxy- 6-Desoxy-L-ascorbic acid	$C_6H_8O_5$ 160.13	528-81-4 168	2880 sub 160	4-18-00-02301	H_2O 4; ace 4; EtOH 4
468	L-Asparagine α-Aminosuccinamic acid	$C_4H_8N_2O_3$ 132.12	70-47-3	859	4-04-00-03004 1.543[15]	H_2O 3; EtOH 1; eth 1; MeOH 1
469	DL-Asparagine, monohydrate	$C_4H_{10}N_2O_4$ 150.13	69833-18-7 182.5	859 214 dec	2-04-00-00900 1.4540[15]	H_2O 2; EtOH 1; eth 1; bz 1
470	DL-Aspartic acid	$C_4H_7NO_4$ 133.10	617-45-8 275	208	2-04-00-00900 1.6622[13]	H_2O 2; EtOH 1; eth 1; bz 1
471	L-Aspartic acid L-Aminosuccinic acid	$C_4H_7NO_4$ 133.10	56-84-8 270	862	4-04-00-02998 1.6603[13]	H_2O 1; EtOH 1; eth 1; bz 1
472	L-Aspartic acid, N-[2-[(2-amino-2-oxoethyl)amino]-2-carboxyethyl]- Lycomarasmine	$C_9H_{15}N_3O_7$ 277.23	7611-43-0 227-9 dec	5491	3-04-00-01521	
473	Aspidospermidine, 1-acetyl-17-methoxy- Aspidospermine	$C_{22}H_{30}N_2O_2$ 354.49	466-49-9 208	872 220[2]	5-23-12-00210	H_2O 2; EtOH 3; eth 2; bz 3
474	Asulam Methyl [(4-aminophenyl)sulfonyl]carbamate	$C_8H_{10}N_2O_4S$ 230.24	3337-71-1 144			
475	Atisine Anthorine	$C_{22}H_{33}NO_2$ 343.51	466-43-3 58.5	881		eth 4; EtOH 4; chl 4
476	Atrazine 6-Chloro-N-ethyl-N'-(1-methylethyl)-1,3,5-triazine-2,4-diamine	$C_8H_{14}ClN_5$ 215.69	1912-24-9 173	886		

No.	Name / Synonym	Mol. Form. / Mol. Wt.	CAS RN / mp/°C	Merck No. / bp/°C	Beil. Ref. / den/g cm⁻³	Solubility / n_D
477	Avermectin B1a / Abamectin	$C_{48}H_{72}O_{14}$ / 873.09	71751-41-2 / 152	1		
478	7-Azabicyclo[4.1.0]heptane / 2,3-Tetramethyleneaziridine	$C_6H_{11}N$ / 97.16	286-18-0 / 21	/ 150; 50[22]	5-20-04-00304 / 0.9480[27]	bz 4; eth 4; EtOH 4
479	9-Azabicyclo[3.3.1]nonan-3-one, 9-methyl- / Pseudopelletierine	$C_9H_{15}NO$ / 153.22	552-70-5 / 54	7936 / 246	5-21-07-00059 / 1.001[100]	H_2O 4; eth 4; EtOH 4 / 1.4760[100]
480	1-Azabicyclo[2.2.2]octane / Quinuclidine	$C_7H_{13}N$ / 111.19	100-76-5 / 158	8109	5-20-04-00335	H_2O 4; ace 4; eth 4; EtOH 4
481	8-Azabicyclo[3.2.1]octane-2-carboxylic acid, 3-(benzoyloxy)-8-methyl-, [1R-(exo,exo)]- / Benzoylecgonine	$C_{16}H_{19}NO_4$ / 289.33	519-09-5 / 195	1125	5-22-05-00053	bz 4; EtOH 4
482	8-Azabicyclo[3.2.1]octane-2-carboxylic acid, 3-(benzoyloxy)-8-methyl-, methyl ester, [1R-(exo,exo)]- / Cocaine	$C_{17}H_{21}NO_4$ / 303.36	50-36-2 / 98	2450 / 187[0.1]	5-22-05-00054	H_2O 2; EtOH 4; eth 4; ace 3 / 1.5022[98]
483	8-Azabicyclo[3.2.1]octane-2-carboxylic acid, 3-hydroxy-8-methyl-, [1R-(exo,exo)]- / Ecgonine	$C_9H_{15}NO_3$ / 185.22	481-37-8 / 205	3467	5-22-05-00053	H_2O 4; EtOH 4
484	8-Azabicyclo[3.2.1]octane, 8-methyl- / Tropane	$C_8H_{15}N$ / 125.21	529-17-9 /	9689 / 166	5-20-04-00331 / 0.9251[15]	
485	1-Azabicyclo[2.2.2]octan-3-ol / 3-Quinuclidinol	$C_7H_{13}NO$ / 127.19	1619-34-7 / 221	8110 / sub 120	5-21-01-00271	ace 3
486	8-Azabicyclo[3.2.1]octan-3-ol, 8-methyl-, exo- / Pseudotropine	$C_8H_{15}NO$ / 141.21	135-97-7 / 109	7937 / 241	5-21-01-00219	H_2O 4; EtOH 4; eth 2; bz 3
487	8-Azabicyclo[3.2.1]octan-3-ol, 8-methyl-, benzoate (ester), endo-	$C_{15}H_{19}NO_2$ / 245.32	19145-60-9 / 41.5	/ 177.5	5-21-01-00226	H_2O 4; eth 4; EtOH 4
488	8-Azabicyclo[3.2.1]octan-3-ol, 8-methyl-, benzoate (ester), exo- / Tropacocaine	$C_{15}H_{19}NO_2$ / 245.32	537-26-8 / 49	9686 / dec	5-21-01-00226 / 1.0426[100]	bz 4; eth 4; EtOH 4; peth 4 / 1.5080[100]
489	8-Azabicyclo[3.2.1]octan-3-ol, 8-methyl-, endo- / Tropine	$C_8H_{15}NO$ / 141.21	120-29-6 / 64	9693 / 233	5-21-01-00219 / 1.016[100]	H_2O 4; eth 4; EtOH 4 / 1.4811[100]
490	8-Azabicyclo[3.2.1]octan-3-one, 8-methyl-	$C_8H_{13}NO$ / 139.20	532-24-1 / 43	/ 227; 113[25]	5-21-07-00032 / 1.9872[100]	EtOH 3; eth 3; ace 3; bz 3 / 1.4598[100]
491	8-Azabicyclo[3.2.1]oct-2-ene-2-carboxylic acid, 8-methyl-, (1R)- / Ecgonidine	$C_9H_{13}NO_2$ / 167.21	484-93-5 / 226-30 dec	3466	5-22-01-00446	H_2O 4
492	1H-Azepine, hexahydro- / Hexamethylenimine	$C_6H_{13}N$ / 99.18	111-49-9 /	/ 138	5-20-04-00003 / 0.8643[22]	H_2O 3; EtOH 4; eth 4 / 1.4631[20]
493	2H-Azepin-2-one, hexahydro- / Caprolactam	$C_6H_{11}NO$ / 113.16	105-60-2 / 69.3	1762 / 270	5-21-06-00444	H_2O 4; bz 4; EtOH 4; chl 4
494	Azetidine	C_3H_7N / 57.10	503-29-7 /	/ 63	5-20-01-00136 / 0.8436[20]	ace 4; bz 4; eth 4; EtOH 4 / 1.4287[25]
495	2-Azetidinecarboxylic acid	$C_4H_7NO_2$ / 101.11	2517-04-6 / 217 dec	923	5-22-01-00015	
496	2-Azetidinone	C_3H_5NO / 71.08	930-21-2 / 73.5	/ 106[15]	5-21-06-00312	eth 4; EtOH 4; chl 4
497	Azinphos-methyl / O,O-Dimethyl S-[(4-oxo-1,2,3-benzotriazin-3(4H)-yl)methyl]phosphorodithioate	$C_{10}H_{12}N_3O_3PS_2$ / 317.33	86-50-0 / 73	926	/ 1.44[20]	
498	Aziridine / Ethyleneimine	C_2H_5N / 43.07	151-56-4 / -77.9	3760 / 56	5-20-01-00003 / 0.832[25]	H_2O 5; EtOH 3; eth 4; chl 2
499	1-Aziridineethanol	C_4H_9NO / 87.12	1072-52-2 /	/ 168	5-20-01-00014 / 1.088[25]	1.4560[20]
500	Aziridine, 1-methyl-	C_3H_7N / 57.10	1072-44-2 /	/ 27.5	5-20-01-00007 / 0.7572[19]	H_2O 5 / 1.3885[19]
501	Aziridine, 2-methyl- / 1,2-Propylenimine	C_3H_7N / 57.10	75-55-8 /	/ 67	5-20-01-00150 / 0.812[16]	
502	Azocine, octahydro-	$C_7H_{15}N$ / 113.20	1121-92-2 / 29	/ 51-3[15]	5-20-04-00156 / 0.896[25]	1.4720[20]
503	1H-Azonine, octahydro-	$C_8H_{17}N$ / 127.23	5661-71-2 / -19	/ 91[24]	5-20-04-00203 / 0.83[25]	1.4760[20]
504	Azulene / Bicyclo[5.3.0]decapentaene	$C_{10}H_8$ / 128.17	275-51-4 / 99	939 / 270 dec; 125[10]	4-05-00-01636	H_2O 1; EtOH 3; eth 3; ace 3
505	Azulene, 1,4-dimethyl-7-(1-methylethyl)- / Guaiazulene	$C_{15}H_{18}$ / 198.31	489-84-9 / 31.5	4464 / 167-8[12]	4-05-00-01751 / 0.973[20]	EtOH 3; eth 3; AcOEt 3
506	Azulene, 4,8-dimethyl-2-(1-methylethyl)-	$C_{15}H_{18}$ / 198.31	529-08-8 / 32.3	/ 122[1.0]	4-05-00-01754 / 0.9735[19]	eth 4; EtOH 4
507	5-Azulenemethanol, 1,2,3,4,5,6,7,8-octahydro-α,α,3,8-tetramethyl-, / Guaiol	$C_{15}H_{26}O$ / 222.37	489-86-1 / 91	4466 / 288 dec; 165[17]	3-06-00-00412 / 0.9074[100]	H_2O 1; EtOH 3; eth 3 / 1.4716[100]
508	Bacitracin	$C_{66}H_{103}N_{17}O_{16}S$ / 1422.71	1405-87-4 / 223	948		H_2O 4; EtOH 4
509	Balan / Benzenamine, N-butyl-N-ethyl-2,6-dinitro-4-(trifluoromethyl)-	$C_{13}H_{16}F_3N_3O_4$ / 335.28	1861-40-1 / 66	1048 / 121[0.5]; 148[7]		

No.	Name / Synonym	Mol. Form. / Mol. Wt.	CAS RN / mp/°C	Merck No. / bp/°C	Beil. Ref. / den/g cm^{-3}	Solubility / n_D
510	Barban Carbamic acid, (3-chlorophenyl)-, 4-chloro-2-butynyl ester	$C_{11}H_9Cl_2NO_2$ 258.10	101-27-9 75	969		
511	Basic copper carbonate Copper, [μ-[carbonato(2-)-O:O']]dihydroxydi-	$CH_2Cu_2O_5$ 221.12	12069-69-1 200 dec	2634	3.8[20]	
512	Bayleton 2-Butanone, 1-(4-chlorophenoxy)-3,3-dimethyl-1-(1H-1,2,4-triazol-1-yl)-	$C_{14}H_{16}ClN_3O_2$ 293.75	43121-43-3 82	9507	1.22[20]	
513	Benalaxyl Methyl N-(2,6-dimethylphenyl)-N-(phenylacetyl)-DL-alaninate	$C_{20}H_{23}NO_3$ 325.41	71626-11-4 79		1.27[25]	
514	Bendiocarb 1,3-Benzodioxol-4-ol, 2,2-dimethyl-, methylcarbamate	$C_{11}H_{13}NO_4$ 223.23	22781-23-3 130	1044	1.25[20]	
515	Benomyl Carbamic acid, [1-[(butylamino)carbonyl]-1H-benzimidazol-2-yl]-, methyl ester	$C_{14}H_{18}N_4O_3$ 290.32	17804-35-2 dec	1053		
516	Bensulfuron-methyl 2-[[[[(4,6-Dimethoxypyrimidin-2-yl)amino]carbonyl]amino]sulfonyl]methyl]benzoic acid, methyl ester	$C_{16}H_{18}N_4O_7S$ 410.41	83055-99-6 187			
517	Bensulide Phosphorodithioic acid, O,O-bis(1-methylethyl) S-[2-[(phenylsulfonyl)amino]ethyl] ester	$C_{14}H_{24}NO_4PS_3$ 397.52	741-58-2 34.4		1.224 [20]	
518	Bentazon 1H-2,1,3-Benzothiadiazin-4(3H)-one, 3-(1-methylethyl)-, 2,2-dioxide	$C_{10}H_{12}N_2O_3S$ 240.28	25057-89-0 138	1060		
519	Benz[a]anthracene 1,2-Benzanthracene	$C_{18}H_{12}$ 228.29	56-55-3 84	1069	4-05-00-02549	EtOH 4
520	Benz[a]anthracene, 7,12-dimethyl- 9,10-Dimethyl-1,2-benzanthracene	$C_{20}H_{16}$ 256.35	57-97-6 122.5	3224	4-05-00-02587	ace 4; bz 4
521	Benz[a]anthracene, 11-methyl-	$C_{19}H_{14}$ 242.32	6111-78-0 118	360 dec	4-05-00-02568	H_2O 1; EtOH 3; HOAc 3
522	Benz[a]anthracene, 3-methyl- 3-Methylbenz[a]anthracene	$C_{19}H_{14}$ 242.32	2498-75-1 156.5	160[0.1]	4-05-00-02562	H_2O 1; EtOH 3; ace 3; bz 3
523	Benz[a]anthracene, 8-methyl-	$C_{19}H_{14}$ 242.32	2381-31-9 156.5	272[3]; 160[0.1]	4-05-00-02566 1.2310[0]	H_2O 1; EtOH 3; eth 3; bz 3
524	Benzadox Acetic acid, [(benzoylamino)oxy]-	$C_9H_9NO_4$ 195.17	5251-93-4 140			
525	Benzaldehyde Benzenecarboxaldehyde	C_7H_6O 106.12	100-52-7 -26	1065 179.0	4-07-00-00505 1.0415[10]	H_2O 2; EtOH 5; eth 5; ace 4 1.5463[20]
526	Benzaldehyde, 2-bromo-	C_7H_5BrO 185.02	6630-33-7 21.5	230	4-07-00-00578	H_2O 1; EtOH 4; bz 4; ctc 2 1.5925[20]
527	Benzaldehyde, 3-bromo-	C_7H_5BrO 185.02	3132-99-8	234.5	4-07-00-00579	H_2O 1; EtOH 4; eth 4; ctc 2 1.5935[20]
528	Benzaldehyde, 4-bromo-	C_7H_5BrO 185.02	1122-91-4 67	66-8[2]	4-07-00-00579	H_2O 1; EtOH 4; bz 4; chl 2
529	Benzaldehyde, 2-chloro-	C_7H_5ClO 140.57	89-98-5 12.4	211.9	4-07-00-00561 1.2483[20]	H_2O 2; EtOH 3; eth 3; ace 3 1.5662[20]
530	Benzaldehyde, 3-chloro-	C_7H_5ClO 140.57	587-04-2 17.5	213.5	4-07-00-00566 1.2410[20]	H_2O 3; EtOH 3; eth 3; ace 3 1.5650[20]
531	Benzaldehyde, 4-chloro-	C_7H_5ClO 140.57	104-88-1 47.5	213.5	4-07-00-00568 1.196[61]	H_2O 3; EtOH 4; eth 4; ace 3 1.555[61]
532	Benzaldehyde, 3-chloro-4-hydroxy-	$C_7H_5ClO_2$ 156.57	2420-16-8 139	150[14]	4-08-00-00296	eth 4; EtOH 4
533	Benzaldehyde, 5-chloro-2-hydroxy-	$C_7H_5ClO_2$ 156.57	635-93-8 100.3	105[12]	4-08-00-00224	H_2O 1; EtOH 4; eth 3; alk 3
534	Benzaldehyde, 3,5-dibromo-2-hydroxy- 3,5-Dibromosalicylaldehyde	$C_7H_4Br_2O_2$ 279.92	90-59-5 86	3011 sub	3-08-00-00188	bz 4; eth 4; chl 4
535	Benzaldehyde, 2,5-dichloro- 2,5-Dichlorobenzaldehyde	$C_7H_4Cl_2O$ 175.01	6361-23-5 58	232	4-07-00-00576	bz 4; eth 4; EtOH 4; chl 4
536	Benzaldehyde, 3,4-dichloro- 3,4-Dichlorobenzaldehyde	$C_7H_4Cl_2O$ 175.01	6287-38-3 44	247.5	4-07-00-00576	H_2O 1; EtOH 3; eth 3; ctc 2
537	Benzaldehyde, 3,5-dichloro- 3,5-Dichlorobenzaldehyde	$C_7H_4Cl_2O$ 175.01	10203-08-4 65	240	4-07-00-00577	ace 4; bz 4; eth 4; EtOH 4
538	Benzaldehyde, 3,4-diethoxy- 3,4-Diethoxybenzaldehyde	$C_{11}H_{14}O_3$ 194.23	2029-94-9 22	279; 200[50]	4-08-00-01766 1.0100[22]	EtOH 4
539	Benzaldehyde, 4-(diethylamino)-	$C_{11}H_{15}NO$ 177.25	120-21-8 41	172[10]	4-14-00-00066	H_2O 4; EtOH 3; eth 3; bz 3
540	Benzaldehyde, 2,3-dihydroxy-	$C_7H_6O_3$ 138.12	24677-78-9 108	235; 120[16]	3-08-00-01979	ace 4; EtOH 4; HOAc 4
541	Benzaldehyde, 2,4-dihydroxy- β-Resorcylaldehyde	$C_7H_6O_3$ 138.12	95-01-2 135	8159 226[22]	4-08-00-01753	H_2O 3; EtOH 4; eth 4; bz 2
542	Benzaldehyde, 3,4-dihydroxy- Protocatechualdehyde	$C_7H_6O_3$ 138.12	139-85-5 153 dec	7908	4-08-00-01762	H_2O 3; EtOH 4; eth 4
543	Benzaldehyde, 2,4-dimethoxy- 2,4-Dimethoxybenzaldehyde	$C_9H_{10}O_3$ 166.18	613-45-6 72	290; 165[10]	4-08-00-01754	H_2O 1; EtOH 3; eth 3; bz 3
544	Benzaldehyde, 2,5-dimethoxy- 2,5-Dimethoxybenzaldehyde	$C_9H_{10}O_3$ 166.18	93-02-7 52	270; 146[10]	4-08-00-01759	H_2O 2; EtOH 3; eth 3

No.	Name Synonym	Mol. Form. Mol. Wt.	CAS RN mp/°C	Merck No. bp/°C	Beil. Ref. den/g cm^{-3}	Solubility n_D
545	Benzaldehyde, 3,4-dimethoxy- Veratraldehyde	$C_9H_{10}O_3$ 166.18	120-14-9 43	9852 281; 155[10]	4-08-00-01765	H_2O 2; EtOH 4; eth 4; chl 2
546	Benzaldehyde, 3,5-dimethoxy- 3,5-Dimethoxybenzaldehyde	$C_9H_{10}O_3$ 166.18	7311-34-4 46.3	151[16]	4-08-00-01786	H_2O 2; EtOH 3; bz 3; peth 2
547	Benzaldehyde, 2,4-dimethyl- 2,4-Dimethylbenzaldehyde	$C_9H_{10}O$ 134.18	15764-16-6 -9	218	4-07-00-00706	EtOH 3; eth 3; ace 3; bz 3
548	Benzaldehyde, 2,5-dimethyl- Isoxylaldehyde	$C_9H_{10}O$ 134.18	5779-94-2	220	4-07-00-00707 0.9500[20]	EtOH 4; eth 3; ace 3; bz 3
549	Benzaldehyde, 3,5-dimethyl-	$C_9H_{10}O$ 134.18	5779-95-3 9	221	4-07-00-00707	ace 4; bz 4; eth 4; EtOH 4
550	Benzaldehyde, 4-(dimethylamino)- Ehrlich's reagent	$C_9H_{11}NO$ 149.19	100-10-7 74.5	3219 176-7[17]	4-14-00-00051 1.0254[100]	H_2O 2; EtOH 3; eth 3; ace 3
551	Benzaldehyde, 2,4-dinitro- 2,4-Dinitrobenzaldehyde	$C_7H_4N_2O_5$ 196.12	528-75-6 72	3265 200[15]	3-07-00-00923	H_2O 2, EtOH 3; eth 3; bz 3
552	Benzaldehyde, 2-ethoxy-	$C_9H_{10}O_2$ 150.18	613-69-4 21	248	4-08-00-00180	EtOH 5; eth 5; chl 2
553	Benzaldehyde, 3-ethoxy-	$C_9H_{10}O_2$ 150.18	22924-15-8	245.5	0-08-00-00060 1.0768[20]	EtOH 3; eth 3; bz 3 1.5408[20]
554	Benzaldehyde, 4-ethoxy-	$C_9H_{10}O_2$ 150.18	10031-82-0 13.5	249	4-08-00-00254 1.08[21]	EtOH 4; eth 4; bz 4
555	Benzaldehyde, 3-ethoxy-4-hydroxy- Ethylvanillin	$C_9H_{10}O_3$ 166.18	121-32-4 77.5	3815	4-08-00-01765	H_2O 2; EtOH 3; eth 3; bz 3
556	Benzaldehyde, 5-ethoxy-2-hydroxy- 5-Ethoxy-2-hydroxybenzaldehyde	$C_9H_{10}O_3$ 166.18	80832-54-8 51.5	230	3-08-00-02004	eth 4; EtOH 4; chl 4
557	Benzaldehyde, 4-ethoxy-3-methoxy-	$C_{10}H_{12}O_3$ 180.20	120-25-2 64.5	168[13]	4-08-00-01766	H_2O 2; EtOH 3; eth 3; bz 3
558	Benzaldehyde, 2-ethyl-	$C_9H_{10}O$ 134.18	22927-13-5	209	4-07-00-00696 1.0216[17]	1.5425[17]
559	Benzaldehyde, 2-fluoro-	C_7H_5FO 124.11	446-52-6 -44.5	175	4-07-00-00559 1.178[25]	1.5234[20]
560	Benzaldehyde, 3-fluoro-	C_7H_5FO 124.11	456-48-4	173	4-07-00-00559 1.17[25]	1.5206[20]
561	Benzaldehyde, 4-fluoro-	C_7H_5FO 124.11	459-57-4 -10	181.5	4-07-00-00559 1.181[19]	
562	Benzaldehyde, 2-(β-D-glucopyranosyloxy)- Helicin	$C_{13}H_{16}O_7$ 284.27	618-65-5 175	4544	5-17-07-00150	H_2O 4; EtOH 4
563	Benzaldehyde, hydrazone Benzylidene hydrazine	$C_7H_8N_2$ 120.15	5281-18-5 16	140[14]	4-07-00-00530	EtOH 3
564	Benzaldehyde, 2-hydroxy- Salicylaldehyde	$C_7H_6O_2$ 122.12	90-02-8 -7	8295 197	4-08-00-00176 1.1674[20]	H_2O 2; EtOH 5; eth 5; ace 4 1.5740[20]
565	Benzaldehyde, 3-hydroxy-	$C_7H_6O_2$ 122.12	100-83-4 108	240	4-08-00-00240 1.1179[130]	H_2O 2; EtOH 3; eth 3; ace 3
566	Benzaldehyde, 4-hydroxy- p-Hydroxybenzaldehyde	$C_7H_6O_2$ 122.12	123-08-0 117	4741	4-08-00-00251 1.129[130]	H_2O 2; EtOH 4; eth 4; ace 2 1.5705[130]
567	Benzaldehyde, 2-hydroxy-4,6-dimethoxy-	$C_9H_{10}O_4$ 182.18	708-76-9 70	193[25]; 165[10]	4-08-00-02717	H_2O 1; EtOH 4; eth 4; bz 4
568	Benzaldehyde, 3-hydroxy-4,5-dimethoxy-	$C_9H_{10}O_4$ 182.18	29865-90-5 62.5	179[12]	4-08-00-02718	bz 4; EtOH 4; lig 4
569	Benzaldehyde, 4-hydroxy-3,5-dimethoxy- Syringaldehyde	$C_9H_{10}O_4$ 182.18	134-96-3 113	8996 192-3[14]	4-08-00-02718	H_2O 2; EtOH 4; eth 4; bz 4
570	Benzaldehyde, 2-hydroxy-3-methoxy-	$C_8H_8O_3$ 152.15	148-53-8 44.5	265.5	4-08-00-01747	H_2O 2; EtOH 4; eth 4; ctc 4
571	Benzaldehyde, 2-hydroxy-5-methoxy-	$C_8H_8O_3$ 152.15	672-13-9 4	247.5	4-08-00-01759	eth 4; EtOH 4
572	Benzaldehyde, 3-hydroxy-4-methoxy-	$C_8H_8O_3$ 152.15	621-59-0 114	179[15]	4-08-00-01764 1.196[25]	H_2O 2; EtOH 3; eth 3; bz 3
573	Benzaldehyde, 3-hydroxy-5-methoxy- 3-Hydroxy-5-methoxybenzaldehyde	$C_8H_8O_3$ 152.15	57179-35-8 130.5	sub 110	3-08-00-02073	bz 4; EtOH 4; HOAc 4
574	Benzaldehyde, 4-hydroxy-3-methoxy- Vanillin	$C_8H_8O_3$ 152.15	121-33-5 81.5	9839 285	4-08-00-01763 1.056[25]	H_2O 2; EtOH 4; eth 4; ace 4
575	Benzaldehyde, 2-hydroxy-3-methyl- m-Tolualdehyde, 2-hydroxy	$C_8H_8O_2$ 136.15	824-42-0 17	208.5	4-08-00-00430	eth 4; EtOH 4; chl 4
576	Benzaldehyde, 2-hydroxy-4-methyl-	$C_8H_8O_2$ 136.15	698-27-1 60.5	223	4-08-00-00434	eth 4; EtOH 4; chl 4
577	Benzaldehyde, 2-hydroxy-5-methyl-	$C_8H_8O_2$ 136.15	613-84-3 56	217.5	4-08-00-00432 1.0913[59]	eth 4; EtOH 4; chl 4 1.547[59]
578	Benzaldehyde, 2-hydroxy-6-methyl-	$C_8H_8O_2$ 136.15	18362-36-2 32.3	230	4-08-00-00430	bz 4
579	Benzaldehyde, 2-hydroxy-, oxime Salicylaldoxime	$C_7H_7NO_2$ 137.14	94-67-7 57	8296	1-08-00-00520	H_2O 2; EtOH 4; eth 4; bz 4
580	Benzaldehyde, 2-iodo-	C_7H_5IO 232.02	26260-02-6 37	129[14]	4-07-00-00582	H_2O 2; ace 3
581	Benzaldehyde, 3-iodo-	C_7H_5IO 232.02	696-41-3 57	125[13]	4-07-00-00583	EtOH 3; ace 3; bz 3
582	Benzaldehyde, 4-iodo-	C_7H_5IO 232.02	15164-44-0 77.5	265	4-07-00-00584	H_2O 2; EtOH 3; bz 3
583	Benzaldehyde, 2-methoxy-	$C_8H_8O_2$ 136.15	135-02-4 37.5	243.5	4-08-00-00180 1.1326[20]	H_2O 1; EtOH 3; eth 4; ace 4 1.5600[20]
584	Benzaldehyde, 3-methoxy-	$C_8H_8O_2$ 136.15	591-31-1	231	4-08-00-00241 1.1187[20]	H_2O 1; EtOH 3; eth 4; ace 4 1.5530[20]

No.	Name Synonym	Mol. Form. Mol. Wt.	CAS RN mp/°C	Merck No. bp/°C	Beil. Ref. den/g cm⁻³	Solubility n_D
585	Benzaldehyde, 4-methoxy- p-Anisaldehyde	$C_8H_8O_2$ 136.15	123-11-5 0	693 248; 134[12]	4-08-00-00252 1.119[15]	H_2O 1; EtOH 5; eth 5; ace 4 1.5730[20]
586	Benzaldehyde, 4-methoxy-3-methyl-	$C_9H_{10}O_2$ 150.18	32723-67-4	80-5[1]	4-08-00-00432 1.025[25]	1.5670[20]
587	Benzaldehyde, 2-methyl- o-Tolualdehyde	C_8H_8O 120.15	529-20-4	9453 200; 94[10]	4-07-00-00667 1.0328[20]	H_2O 2; EtOH 3; eth 3; ace 4 1.5462[20]
588	Benzaldehyde, 3-methyl-	C_8H_8O 120.15	620-23-5	199	4-07-00-00669 1.0189[21]	H_2O 2; EtOH 5; eth 5; ace 4 1.5413[21]
589	Benzaldehyde, 4-methyl- p-Tolualdehyde	C_8H_8O 120.15	104-87-0	204-5.106[10]	4-07-00-00672 1.0194[17]	H_2O 2; EtOH 5; eth 5; ace 5 1.5454[20]
590	Benzaldehyde, 4-(1-methylethyl)- Cuminaldehyde	$C_{10}H_{12}O$ 148.20	122-03-2	2623 235.5	4-07-00-00723 0.9755[20]	H_2O I; EtOH 3; eth 3; ctc 2 1.5301[20]
591	Benzaldehyde, 2-nitro-	$C_7H_5NO_3$ 151.12	552-89-6 43.5	153[23]	4-07-00-00584 1.2844[20]	H_2O 2; EtOH 4; eth 4; ace 4
592	Benzaldehyde, 3-nitro-	$C_7H_5NO_3$ 151.12	99-61-6 58.5	164[23]	4-07-00-00591 1.2792[20]	H_2O 2; EtOH 3; eth 3; ace 4
593	Benzaldehyde, 4-nitro-	$C_7H_5NO_3$ 151.12	555-16-8 107	sub	4-07-00-00598 1.496[25]	H_2O 2; EtOH 4; eth 2; bz 3
594	Benzaldehyde, oxime, (E)-	C_7H_7NO 121.14	622-31-1 35	119[10]	4-07-00-00527 1.145[20]	H_2O 3; EtOH 4; eth 4
595	Benzaldehyde, oxime, (Z)-	C_7H_7NO 121.14	622-32-2 36.5	200	4-07-00-00527 1.1111[20]	bz 4; eth 4; EtOH 4 1.5908[20]
596	Benzaldehyde, pentafluoro-	C_7HF_5O 196.08	653-37-2 20	167	4-07-00-00561	1.4506[20]
597	Benzaldehyde, 3-phenoxy-	$C_{13}H_{10}O_2$ 198.22	39515-51-0	140[0.1]	4-08-00-00242 1.147[25]	1.5954[20]
598	Benzaldehyde, 2,3,5,6-tetramethyl- Durene aldehyde	$C_{11}H_{14}O$ 162.23	17432-37-0 20	135[11]	3-07-00-01148	1.5560[30]
599	Benzaldehyde, 2,4,6-trimethyl-	$C_{10}H_{12}O$ 148.20	487-68-3 14	238.5	4-07-00-00730 1.0154[25]	H_2O 1; EtOH 3; eth 3; ace 3
600	Benzamide Benzoic acid amide	C_7H_7NO 121.14	55-21-0 129.1	1067 290	4-09-00-00725 1.0792[130]	H_2O 2; EtOH 4; eth 2; bz 2
601	Benzamide, 2-amino-	$C_7H_8N_2O$ 136.15	88-68-6 110.5 dec	300	4-14-00-01010	H_2O 3; EtOH 3; eth 2; bz 2
602	Benzamide, 4-amino-N-[2-(diethylamino)ethyl]-, monohydrochloride Procainamide hydrochloride	$C_{13}H_{22}ClN_3O$ 271.79	614-39-1 167	7762	4-14-00-01154	H_2O 4; EtOH 3; eth 1; bz 1
603	Benzamide, N-[(4-aminophenyl)sulfonyl]- Sulfabenzamide	$C_{13}H_{12}N_2O_3S$ 276.32	127-71-9 181.5	8868	4-14-00-02664	
604	Benzamide, 2-bromo-	C_7H_6BrNO 200.03	4001-73-4 160.5	sub	4-09-00-01013	H_2O 2; EtOH 3; eth 2; tfa 2
605	Benzamide, 3-bromo-	C_7H_6BrNO 200.03	22726-00-7 155.3	sub	4-09-00-01015	H_2O 2; EtOH 3
606	Benzamide, 5-bromo-N,2-dihydroxy- 5-Bromosalicylhydroxamic acid	$C_7H_6BrNO_3$ 232.03	5798-94-7 232 dec	1424	4-10-00-00221	
607	Benzamide, N-butyl-4-chloro-2-hydroxy- Buclosamide	$C_{11}H_{14}ClNO_2$ 227.69	575-74-6 91.5	1450		
608	Benzamide, 5-chloro-N-(3,4-dichlorophenyl)-2-hydroxy- 3',4',5-Trichlorosalicylanilide	$C_{13}H_8Cl_3NO_2$ 316.57	642-84-2 247	9558	4-12-00-01273	
609	Benzamide, 3,5-dibromo-N-(4-bromophenyl)-2-hydroxy- Tribromsalan	$C_{13}H_8Br_3NO_2$ 449.92	87-10-5 227	9529		
610	Benzamide, 3,5-dichloro-N-(3,4-dichlorophenyl)-2-hydroxy- 3,3',4',5-Tetrachlorosalicylanilide	$C_{13}H_7Cl_4NO_2$ 351.01	1154-59-2 161	9127		
611	Benzamide, N,N-diethyl-4-hydroxy-3-methoxy- Ethamivan	$C_{12}H_{17}NO_3$ 223.27	304-84-7 95	3673	4-10-00-01481	chl 3
612	Benzamide, N,N-diethyl-3-methyl- DEET	$C_{12}H_{17}NO$ 191.27	134-62-3	2848 160[19]; 111[1]	4-09-00-01716 0.996[20]	H_2O 4; bz 4; eth 4; EtOH 4 1.5212[20]
613	Benzamide, N,2-dihydroxy- Salicylhydroxamic acid	$C_7H_7NO_3$ 153.14	89-73-6 168	8300 sub	4-10-00-00192	H_2O 2; EtOH 4; eth 4; HOAc 3
614	Benzamide, 3,5-dinitro- Nitromide	$C_7H_5N_3O_5$ 211.13	121-81-3 184	6533	4-09-00-01351	H_2O 4
615	Benzamide, 2-ethoxy- Ethenzamide	$C_9H_{11}NO_2$ 165.19	938-73-8 133	3685	4-10-00-00175	H_2O 2; EtOH 4; eth 4; chl 2
616	Benzamide, 2-hydroxy- Salicylamide	$C_7H_7NO_2$ 137.14	65-45-2 142	8297 181.5[14]	4-10-00-00169 1.175[140]	H_2O 2; EtOH 3; eth 2
617	Benzamide, 2-hydroxy-N-phenyl- Salicylanilide	$C_{13}H_{11}NO_2$ 213.24	87-17-2 136.5	8299	4-12-00-00906	H_2O 3; EtOH 2; eth 2; bz 2
618	Benzamide, 4-methoxy-	$C_8H_9NO_2$ 151.17	3424-93-9 166.5	295	4-10-00-00433	H_2O 4; EtOH 4
619	Benzamide, 2-methyl- o-Toluamide	C_8H_9NO 135.17	527-85-5 147	9454	4-09-00-01701	H_2O 2; EtOH 4; eth 2; bz 2
620	Benzamide, N-methyl-	C_8H_9NO 135.17	613-93-4 82	291; 167[12]	4-09-00-00727	EtOH 3; ace 3
621	Benzamide, N-(2-methylphenyl)-	$C_{14}H_{13}NO$ 211.26	584-70-3 145.5		4-12-00-01757 1.205[15]	H_2O 2; EtOH 3; eth 3

No.	Name / Synonym	Mol. Form. / Mol. Wt.	CAS RN / mp/°C	Merck No. / bp/°C	Beil. Ref. / den/g cm^{-3}	Solubility / n_D
622	Benzamide, N-(3-methylphenyl)- m-Benzotoluidide	C$_{14}$H$_{13}$NO 211.26	582-77-4 125		4-12-00-01825 1.170[15]	EtOH 4
623	Benzamide, N-(4-methylphenyl)-	C$_{14}$H$_{13}$NO 211.26	582-78-5 158		4-12-00-01910 1.202[15]	eth 4; EtOH 4
624	Benzamide, 2-nitro-	C$_7$H$_6$N$_2$O$_3$ 166.14	610-15-1 176.6	317	4-09-00-01049	H$_2$O 3; EtOH 3; eth 3
625	Benzamide, 3-nitro-	C$_7$H$_6$N$_2$O$_3$ 166.14	645-09-0 142.7	312.5	4-09-00-01061	H$_2$O 3; EtOH 3; eth 3
626	Benzamide, N-phenyl- Benzanilide	C$_{13}$H$_{11}$NO 197.24	93-98-1 163	1068 sub 117	4-12-00-00417 1.315[25]	H$_2$O 1; EtOH 2; eth 2; HOAc 2
627	Benzamide, N,N,4-trimethyl-	C$_{10}$H$_{13}$NO 163.22	14062-78-3 41	156[10]	0-09-00-00486	H$_2$O 2; EtOH 2
628	Benz[b]indeno[1,2-d]pyran-9(6H)-one, 6a,7-dihydro-3,4,6a,10-tetrahydroxy- Hematein	C$_{16}$H$_{12}$O$_6$ 300.27	475-25-2 250 dec	4553	4-18-00-03343	H$_2$O 1; EtOH 2; eth 1; bz 1
629	Benz[b]indeno[1,2-d]pyran-3,4,6a,9,10(6H)-pentol, 7,11b-dihydro-, cis-(+)- Hematoxylin	C$_{16}$H$_{14}$O$_6$ 302.28	517-28-2 140	4556	2-17-00-00273	H$_2$O 2; EtOH 3; eth 2; alk 3
630	Benz[c]acridine 12-Azabenz[a]anthracene	C$_{17}$H$_{11}$N 229.28	225-51-4 132		5-20-08-00519	bz 4; eth 4; EtOH 4
631	Benz[e]acephenanthrylene Benzo[b]fluoranthene	C$_{20}$H$_{12}$ 252.32	205-99-2 168		4-05-00-02686	H$_2$O 1; bz 5
632	1H-Benz[e]indene, 2,3-dihydro-	C$_{13}$H$_{12}$ 168.24	4944-94-9	294.5	4-05-00-01865 1.066[20]	H$_2$O 1; HOAc 3 1.6290[20]
633	Benzenamine Aniline	C$_6$H$_7$N 93.13	62-53-3 -6.0	687 184.1	4-12-00-00223 1.0217[20]	H$_2$O 3; EtOH 5; eth 5; ace 5 1.5863[20]
634	Benzenamine, 4-[(4-aminophenyl)(4-imino-2,5-cyclohexadien-1-ylidene)methyl]-2-methyl-, monohydrochloride Magenta I	C$_{20}$H$_{20}$ClN$_3$ 337.85	632-99-5 200 dec	5528		H$_2$O 2; EtOH 2; eth 1
635	Benzenamine, 4,4'-azobis- p-Diaminoazobenzene	C$_{12}$H$_{12}$N$_4$ 212.25	538-41-0 250.5	2954	4-16-00-00515	H$_2$O 2; EtOH 3; bz 4; chl 4
636	Benzenamine, 2,6-bis(1-methylethyl)-	C$_{12}$H$_{19}$N 177.29	24544-04-5 -45	257	3-12-00-02764 0.94[25]	1.5332[20]
637	Benzenamine, 3,5-bis(trifluoromethyl)-	C$_8$H$_5$F$_6$N 229.12	328-74-5	85[15]	3-12-00-02498 1.487[25]	1.4335[20]
638	Benzenamine, 2-bromo- o-Bromoaniline	C$_6$H$_6$BrN 172.02	615-36-1 32	229	4-12-00-01487 1.578[20]	H$_2$O 1; EtOH 3; eth 3 1.6113[20]
639	Benzenamine, 3-bromo- m-Bromoaniline	C$_6$H$_6$BrN 172.02	591-19-5 18.5	251	4-12-00-01491 1.5793[20]	H$_2$O 2; EtOH 3; eth 3 1.6260[20]
640	Benzenamine, 4-bromo- p-Bromoaniline	C$_6$H$_6$BrN 172.02	106-40-1 66.4	1392 dec	4-12-00-01497 1.4970[100]	H$_2$O 1; EtOH 3; eth 3; chl 2
641	Benzenamine, 2-bromo-4,6-dichloro-	C$_6$H$_4$BrCl$_2$N 240.91	697-86-9 83.5	273	2-12-00-00355	bz 4; EtOH 4; chl 4
642	Benzenamine, 4-bromo-N,N-diethyl-	C$_{10}$H$_{14}$BrN 228.13	2052-06-4 38	270	4-12-00-01500	H$_2$O 1; EtOH 4; eth 4
643	Benzenamine, 2-bromo-N,N-dimethyl-	C$_8$H$_{10}$BrN 200.08	698-00-0 	107-8[14]	4-12-00-01487 1.3839[25]	EtOH 4 1.5748[25]
644	Benzenamine, 3-bromo-N,N-dimethyl-	C$_8$H$_{10}$BrN 200.08	16518-62-0 11	259	4-12-00-01492 1.3651[20]	EtOH 4; HOAc 4; con acid 3
645	Benzenamine, 4-bromo-N,N-dimethyl-	C$_8$H$_{10}$BrN 200.08	586-77-6 55	264	4-12-00-01499 1.3220[100]	H$_2$O 1; EtOH 3; eth 4; os 3
646	Benzenamine, 2-bromo-4,6-dinitro-	C$_6$H$_4$BrN$_3$O$_4$ 262.02	1817-73-8 153.5	sub	4-12-00-01734	EtOH 4; ace 4; HOAc 3
647	Benzenamine, 2-bromo-4-methyl-	C$_7$H$_8$BrN 186.05	583-68-6 26	240	4-12-00-01992 1.510[20]	H$_2$O 1; EtOH 3; eth 3 1.5999[20]
648	Benzenamine, 2-bromo-5-methyl- 2-Bromo-5-methylaniline	C$_7$H$_8$BrN 186.05	53078-85-6 46	129[15]	3-12-00-01998 1.470[25]	eth 4; EtOH 4
649	Benzenamine, 3-bromo-4-methyl-	C$_7$H$_8$BrN 186.05	7745-91-7 26	255.5	4-12-00-01991	eth 4
650	Benzenamine, 4-bromo-2-methyl-	C$_7$H$_8$BrN 186.05	583-75-5 59.5	240	4-12-00-01804	H$_2$O 2; EtOH 3; eth 4; chl 2
651	Benzenamine, 4-bromo-3-methyl-	C$_7$H$_8$BrN 186.05	6933-10-4 81	240	4-12-00-01856	H$_2$O 1; EtOH 4; chl 2
652	Benzenamine, 5-bromo-2-methyl-	C$_7$H$_8$BrN 186.05	39478-78-9 33	255 dec; 165[25]	3-12-00-01923	eth 4; EtOH 4
653	Benzenamine, 2-bromo-6-nitro- Aniline, 2-bromo-6-nitro-	C$_6$H$_5$BrN$_2$O$_2$ 217.02	59255-95-7 74.5		2-12-00-00402 1.988[25]	EtOH 4
654	Benzenamine, 4-bromo-2-nitro-	C$_6$H$_5$BrN$_2$O$_2$ 217.02	875-51-4 111.5	sub	3-12-00-01670	EtOH 4
655	Benzenamine, 4-butyl-	C$_{10}$H$_{15}$N 149.24	104-13-2 	261	4-12-00-02807 0.945[20]	ctc 2
656	Benzenamine, N-butyl- N-Butylaniline	C$_{10}$H$_{15}$N 149.24	1126-78-9 -14.4	243.5	4-12-00-00256 0.9323[20]	eth 4; EtOH 4 1.5341[20]
657	Benzenamine, 4,4'-carbonimidoylbis[N,N-dimethyl- 4,4'-Dimethylaminobenzophenonimide	C$_{17}$H$_{21}$N$_3$ 267.37	492-80-8 136		4-14-00-00256	H$_2$O 1; EtOH 3; eth 2

No.	Name / Synonym	Mol. Form. / Mol. Wt.	CAS RN / mp/°C	Merck No. / bp/°C	Beil. Ref. / den/g cm^{-3}	Solubility / n_D
658	Benzenamine, 2-chloro- / o-Chloroaniline	C$_6$H$_6$ClN / 127.57	95-51-2 / -14	2118 / 208.8	4-12-00-01115	H$_2$O 1; EtOH 5; eth 3; ace 3 / 1.5895[20]
659	Benzenamine, 3-chloro- / m-Chloroaniline	C$_6$H$_6$ClN / 127.57	108-42-9 / -10.4	2118 / 230.5	4-12-00-01137 / 1.2161[20]	H$_2$O 1; EtOH 5; eth 5; ace 5 / 1.5941[20]
660	Benzenamine, 4-chloro- / p-Chloroaniline	C$_6$H$_6$ClN / 127.57	106-47-8 / 72.5	2118 / 232	4-12-00-01166 / 1.429[19]	H$_2$O 3; EtOH 3; eth 3; chl 3 / 1.5546[87]
661	Benzenamine, 4-chloro-N,N-diethyl-	C$_{10}$H$_{14}$ClN / 183.68	2873-89-4 / 45.5	/ 252; 113[6]	4-12-00-01170	EtOH 3
662	Benzenamine, 2-chloro-N,N-dimethyl-	C$_8$H$_{10}$ClN / 155.63	698-01-1 /	/ 205	4-12-00-01116 / 1.1067[20]	bz 4; EtOH 4 / 1.5578[20]
663	Benzenamine, 4-chloro-N,N-dimethyl-	C$_8$H$_{10}$ClN / 155.63	698-69-1 / 35.5	/ 231	4-12-00-01169 / 1.0480[100]	EtOH 3
664	Benzenamine, 2-chloro-6-fluoro- / Aniline, 2-chloro-6-fluoro	C$_6$H$_5$ClFN / 145.56	363-51-9 /	/ 91[30]	4-12-00-01238 / 1.316[23]	1.5511[23]
665	Benzenamine, 2-chloro-, hydrochloride / Aniline, 2-chloro, hydrochloride	C$_6$H$_7$Cl$_2$N / 164.03	137-04-2 / 235		3-12-00-01281 / 1.505[18]	H$_2$O 4
666	Benzenamine, 2-chloro-5-methoxy-	C$_7$H$_8$ClNO / 157.60	2401-24-3 / 27	/ 142[18]	4-13-00-00999	1.5848[25]
667	Benzenamine, 4-chloro-2-methoxy- / 4-Chloro-o-anisidine	C$_7$H$_8$ClNO / 157.60	93-50-5 / 52	/ 260	4-13-00-00885	EtOH 3; eth 3; bz 3; chl 3
668	Benzenamine, 2-chloro-4-methyl- / 2-Chloro-p-toluidine	C$_7$H$_8$ClN / 141.60	615-65-6 / 7	/ 220	3-12-00-02152 / 1.151[20]	EtOH 2; bz 2 / 1.5748[22]
669	Benzenamine, 2-chloro-5-methyl- / 6-Chloro-m-toluidine	C$_7$H$_8$ClN / 141.60	95-81-8 / 29.5	/ 229	4-12-00-01849	EtOH 4
670	Benzenamine, 2-chloro-N-methyl-	C$_7$H$_8$ClN / 141.60	932-32-1 /	/ 218	4-12-00-01116 / 1.1735[11]	EtOH 3; ace 3; bz 3 / 1.5780[25]
671	Benzenamine, 3-chloro-2-methyl- / 3-Chloro-o-toluidine	C$_7$H$_8$ClN / 141.60	87-60-5 / 1	/ 245	4-12-00-01800	H$_2$O 3; EtOH 3; eth 1; bz 1 / 1.5880[20]
672	Benzenamine, 3-chloro-4-methyl- / 3-Chloro-p-toluidine	C$_7$H$_8$ClN / 141.60	95-74-9 / 26	/ 243	4-12-00-01985	EtOH 3; ctc 2
673	Benzenamine, 4-chloro-2-methyl-	C$_7$H$_8$ClN / 141.60	95-69-2 / 30.3	/ 244	4-12-00-01796	EtOH 3; ctc 2
674	Benzenamine, 4-chloro-3-methyl- / 4-Chloro-m-toluidine	C$_7$H$_8$ClN / 141.60	7149-75-9 / 83.5	/ 241	4-12-00-01850	ace 4; bz 4; EtOH 4
675	Benzenamine, 4-chloro-N-methyl-	C$_7$H$_8$ClN / 141.60	932-96-7 /	/ 240	4-12-00-01168 / 1.169[11]	EtOH 3; ace 3; bz 3 / 1.5835[20]
676	Benzenamine, 5-chloro-2-methyl- / 5-Chloro-o-toluidine	C$_7$H$_8$ClN / 141.60	95-79-4 / 26	/ 239; 140[38]	4-12-00-01794	EtOH 4
677	Benzenamine, 5-chloro-2-nitro-	C$_6$H$_5$ClN$_2$O$_2$ / 172.57	1635-61-6 / 127.8	/ sub	4-12-00-01673	eth 4; EtOH 4
678	Benzenamine, 3-chloro-N-phenyl-	C$_{12}$H$_{10}$ClN / 203.67	101-17-7 /	/ 338	4-12-00-01141 / 1.2[25]	H$_2$O 1; EtOH 3; eth 3; bz 3 / 1.6513[20]
679	Benzenamine, 4-chloro-N-phenyl-	C$_{12}$H$_{10}$ClN / 203.67	1205-71-6 / 74	/ 335	4-12-00-01170	bz 4; eth 4; EtOH 4; lig 4
680	Benzenamine, N-cyclohexyl-	C$_{12}$H$_{17}$N / 175.27	1821-36-9 / 16	/ 279; 192[73]	4-12-00-00268 / 1.0155[20]	H$_2$O 1; EtOH 3; eth 3; bz 3 / 1.5610[20]
681	Benzenamine, 2,4-dibromo-	C$_6$H$_5$Br$_2$N / 250.92	615-57-6 / 79.5	/ 156[74]	4-12-00-01532 / 2.260[20]	EtOH 3; eth 3; chl 3; HOAc 3
682	Benzenamine, 2,6-dibromo-	C$_6$H$_5$Br$_2$N / 250.92	608-30-0 / 87.5	/ 263	4-12-00-01535	EtOH 4; eth 4; bz 4; chl 4
683	Benzenamine, 3,4-dibromo- / 3,4-Dibromoaniline	C$_6$H$_5$Br$_2$N / 250.92	615-55-4 / 80.5	/ 100	4-12-00-01535	EtOH 4
684	Benzenamine, N,N-dibutyl-	C$_{14}$H$_{23}$N / 205.34	613-29-6 / -32.2	/ 274.8	4-12-00-00256 / 0.9037[20]	H$_2$O 1; EtOH 5; eth 5; ace 4 / 1.5186[20]
685	Benzenamine, 2,3-dichloro- / 2,3-Dichloroaniline	C$_6$H$_5$Cl$_2$N / 162.02	608-27-5 / 24	/ 252	0-12-00-00621	EtOH 3; eth 4; ace 3; bz 2
686	Benzenamine, 2,4-dichloro- / 2,4-Dichloroaniline	C$_6$H$_5$Cl$_2$N / 162.02	554-00-7 / 63.5	/ 245	4-12-00-01241 / 1.567[20]	H$_2$O 2; EtOH 3; eth 3; chl 2
687	Benzenamine, 2,5-dichloro- / 2,5-Dichloroaniline	C$_6$H$_5$Cl$_2$N / 162.02	95-82-9 / 50	/ 251	4-12-00-01250	H$_2$O 2; EtOH 3; eth 3; bz 3
688	Benzenamine, 2,6-dichloro- / 2,6-Dichloroaniline	C$_6$H$_5$Cl$_2$N / 162.02	608-31-1 / 39		2-12-00-00337	H$_2$O 2; EtOH 3; eth 3
689	Benzenamine, 3,4-dichloro- / 3,4-Dichloroaniline	C$_6$H$_5$Cl$_2$N / 162.02	95-76-1 / 72	3041 / 272	4-12-00-01257	EtOH 3; eth 3; bz 2; chl 2
690	Benzenamine, 3,5-dichloro- / 3,5-Dichloroaniline	C$_6$H$_5$Cl$_2$N / 162.02	626-43-7 / 52	/ 261	4-12-00-01274	H$_2$O 1; EtOH 3; eth 3; ctc 3
691	Benzenamine, 2,6-dichloro-4-ethoxy- / Aniline, 2,6-dichloro-4-ethoxy	C$_8$H$_9$Cl$_2$NO / 206.07	51225-20-8 / 46	/ 275	2-13-00-00276	bz 4; eth 4; EtOH 4
692	Benzenamine, 2,6-dichloro-4-nitro- / 2,6-Dichloro-4-nitroaniline	C$_6$H$_4$Cl$_2$N$_2$O$_2$ / 207.02	99-30-9 / 191		4-12-00-01681	EtOH 3; DMSO-d$_6$ 2; acid 3
693	Benzenamine, 3,4-diethoxy- / 3,4-Diethoxyaniline	C$_{10}$H$_{15}$NO$_2$ / 181.23	39052-12-5 / 48	/ 132-5[0.45]	4-13-00-02507	bz 4; eth 4; EtOH 4
694	Benzenamine, 2,6-diethyl-	C$_{10}$H$_{15}$N / 149.24	579-66-8 / 3	/ 243	4-12-00-02841 / 0.906[25]	1.5452[20]
695	Benzenamine, N,N-diethyl- / N,N-Diethylaniline	C$_{10}$H$_{15}$N / 149.24	91-66-7 / -38.8	3102 / 216.3	4-12-00-00252 / 0.9307[20]	H$_2$O 2; EtOH 3; eth 4; ace 3 / 1.5409[20]
696	Benzenamine, N,N-diethyl-2-methyl-	C$_{11}$H$_{17}$N / 163.26	606-46-2 / -60	/ 209	4-12-00-01748 / 0.9286[20]	H$_2$O 2; EtOH 5; eth 5; ctc 3 / 1.5153[20]
697	Benzenamine, N,N-diethyl-4-methyl-	C$_{11}$H$_{17}$N / 163.26	613-48-9 /	/ 229	4-12-00-01875 / 0.9242[16]	H$_2$O 2; EtOH 5; eth 5

No.	Name Synonym	Mol. Form. Mol. Wt.	CAS RN mp/°C	Merck No. bp/°C	Beil. Ref. den/g cm^{-3}	Solubility n_D
698	Benzenamine, N,N-diethyl-4-nitro-	C$_{10}$H$_{14}$N$_2$O$_2$ 194.23	2216-15-1 77.5		4-12-00-01618 1.225[25]	EtOH 3; lig 2
699	Benzenamine, N,N-diethyl-4-nitroso- Aniline, N,N-diethyl-4-nitroso	C$_{10}$H$_{14}$N$_2$O 178.23	120-22-9 87.5		4-12-00-01559 1.24[15]	H$_2$O 2; EtOH 3; eth 3; ace 3
700	Benzenamine, 2,4-difluoro- 2,4-Difluoroaniline	C$_6$H$_5$F$_2$N 129.11	367-25-9 -7.5	3131 170	4-12-00-01112 1.268[25]	1.5063[20]
701	Benzenamine, 2,4-diiodo- 2,4-Diiodoaniline	C$_6$H$_5$I$_2$N 344.92	533-70-0 95.5	3173	3-12-00-01508 2.748[25]	ace 4; bz 4; eth 4; EtOH 4
702	Benzenamine, 2,5-dimethoxy-	C$_8$H$_{11}$NO$_2$ 153.18	102-56-7 82.5	270	4-13-00-02548	H$_2$O 3; EtOH 3; chl 3; lig 3
703	Benzenamine, 2,6-dimethoxy- 2,6-Dimethoxyaniline	C$_8$H$_{11}$NO$_2$ 153.18	2734-70-5 76.5	146[23]	4-13-00-02526	bz 4; eth 4; EtOH 4; lig 4
704	Benzenamine, 2,3-dimethyl- 2,3-Xylidine	C$_8$H$_{11}$N 121.18	87-59-2 <-15	221.5	4-12-00-02497 0.9931[20]	H$_2$O 2; EtOH 4; eth 4; ctc 3 1.5684[20]
705	Benzenamine, 2,4-dimethyl- 2,4-Xylidine	C$_8$H$_{11}$N 121.18	95-68-1 -14.3	214	4-12-00-02545 0.9723[20]	H$_2$O 2; EtOH 3; eth 3; bz 3 1.5569[20]
706	Benzenamine, 2,5-dimethyl- 2,5-Xylidine	C$_8$H$_{11}$N 121.18	95-78-3 15.5	214	4-12-00-02567 0.9790[21]	H$_2$O 2; eth 3; ctc 3 1.5591[21]
707	Benzenamine, 2,6-dimethyl- 2,6-Xylidine	C$_8$H$_{11}$N 121.18	87-62-7 11.2	215	4-12-00-02521 0.9842[20]	eth 4; EtOH 4 1.5610[20]
708	Benzenamine, 3,4-dimethyl- 3,4-Xylidine	C$_8$H$_{11}$N 121.18	95-64-7 51	228	4-12-00-02502 1.076[18]	H$_2$O 2; eth 3; chl 2; lig 4
709	Benzenamine, 3,5-dimethyl- 3,5-Xylidine	C$_8$H$_{11}$N 121.18	108-69-0 9.8	220.5	4-12-00-02561 0.9706[20]	H$_2$O 2; eth 3; ctc 3 1.5581[20]
710	Benzenamine, N,2-dimethyl-	C$_8$H$_{11}$N 121.18	611-21-2 	207.5	4-12-00-01746 0.9709[20]	H$_2$O 1; EtOH 5; eth 5; ace 3 1.5649[20]
711	Benzenamine, N,3-dimethyl-	C$_8$H$_{11}$N 121.18	696-44-6 	206.5	4-12-00-01815	H$_2$O 1; EtOH 5; eth 5; ace 3 1.5557[25]
712	Benzenamine, N,4-dimethyl-	C$_8$H$_{11}$N 121.18	623-08-5 	210	4-12-00-01874 0.9348[55]	H$_2$O 1; EtOH 5; eth 5; ace 3 1.5568[20]
713	Benzenamine, N,N-dimethyl- N,N-Dimethylaniline	C$_8$H$_{11}$N 121.18	121-69-7 2.4	3223 194.1	4-12-00-00243 0.9557[20]	H$_2$O 2; EtOH 3; eth 3; ace 3 1.5582[20]
714	Benzenamine, 2-(1,1-dimethylethyl)-	C$_{10}$H$_{15}$N 149.24	6310-21-0 	234	4-12-00-02823 0.977[15]	EtOH 4; eth 4; bz 4; acid 3 1.5453[20]
715	Benzenamine, 4-(1,1-dimethylethyl)-	C$_{10}$H$_{15}$N 149.24	769-92-6 17	241	4-12-00-02828 0.9525[15]	H$_2$O 2; EtOH 5; eth 5; bz 4 1.5380[20]
716	Benzenamine, N-(1,1-dimethylethyl)- N-tert-Butylaniline	C$_{10}$H$_{15}$N 149.24	937-33-7 	215; 95[19]	4-12-00-00257	EtOH 3; ace 4; bz 4; chl 4 1.5270[20]
717	Benzenamine, N,N-dimethyl-, hydrochloride	C$_8$H$_{12}$ClN 157.64	5882-44-0 90		4-12-00-00243 1.1156[19]	H$_2$O 4; EtOH 4; chl 4
718	Benzenamine, N,N-dimethyl-2-nitro-	C$_8$H$_{10}$N$_2$O$_2$ 166.18	610-17-3 -20	146[20]	4-12-00-01564 1.1794[20]	H$_2$O 3; EtOH 4; eth 3; chl 4 1.6102[20]
719	Benzenamine, N,N-dimethyl-3-nitro-	C$_8$H$_{10}$N$_2$O$_2$ 166.18	619-31-8 60.5	282.5	4-12-00-01591 1.313[17]	H$_2$O 1; EtOH 3; eth 3
720	Benzenamine, N,N-dimethyl-4-nitroso- p-Nitroso-N,N-dimethylaniline	C$_8$H$_{10}$N$_2$O 150.18	138-89-6 92.5	6559	4-12-00-01558 1.145[20]	H$_2$O 2; EtOH 3; eth 3; chl 3
721	Benzenamine, N,N-dimethyl-4-(phenylazo)- 4-(Dimethylamino)azobenzene	C$_{14}$H$_{15}$N$_3$ 225.29	60-11-7 117	3218 dec	3-16-00-00340	H$_2$O 1; EtOH 4; eth 3; chl 2
722	Benzenamine, N,N-dimethyl-4-(2-phenylethenyl)-, (E)-	C$_{16}$H$_{17}$N 223.32	838-95-9 150	 145[0.1]	4-12-00-03400	H$_2$O 1; EtOH 3; chl 3
723	Benzenamine, N,N-dimethyl-4-[[3-(trifluoromethyl)phenyl]azo]- Azobenzene, 4-dimethylamino-3'-trifluoromethyl	C$_{15}$H$_{14}$F$_3$N$_3$ 293.29	328-96-1 80	 76[8]	4-16-00-00487	chl 3
724	Benzenamine, 2,3-dinitro- 2,3-Dinitroaniline	C$_6$H$_5$N$_3$O$_4$ 183.12	602-03-9 128		4-12-00-01689 1.646[50]	H$_2$O 1; EtOH 3; eth 2
725	Benzenamine, 2,4-dinitro- 2,4-Dinitroaniline	C$_6$H$_5$N$_3$O$_4$ 183.12	97-02-9 180	3263	4-12-00-01689 1.615[14]	H$_2$O 1; EtOH 2; ace 2; HCl 2
726	Benzenamine, 2,6-dinitro- 2,6-Dinitroaniline	C$_6$H$_5$N$_3$O$_4$ 183.12	606-22-4 141.5	3264	4-12-00-01729	H$_2$O 1; EtOH 2; eth 3; bz 3
727	Benzenamine, 3,5-dinitro- 3,5-Dinitroaniline	C$_6$H$_5$N$_3$O$_4$ 183.12	618-87-1 163		4-12-00-01729 1.601[50]	H$_2$O 1; EtOH 3; eth 3; ace 2
728	Benzenamine, N,N-diphenyl- Triphenylamine	C$_{18}$H$_{15}$N 245.32	603-34-9 127	365	4-12-00-00276 0.774[0]	H$_2$O 1; EtOH 2; eth 3; bz 3 1.353[16]
729	Benzenamine, 4-(diphenylmethyl)-	C$_{19}$H$_{17}$N 259.35	603-38-3 84.5	248[12]	4-12-00-03470	bz 4; eth 4; lig 4
730	Benzenamine, N,N-dipropyl-	C$_{12}$H$_{19}$N 177.29	2217-07-4 	242	4-12-00-00255 0.9104[20]	H$_2$O 1; EtOH 3; eth 3; ace 3 1.5271[20]
731	Benzenamine, 4,4'-(1,2-ethanediyl)bis-	C$_{14}$H$_{16}$N$_2$ 212.29	621-95-4 137	sub	4-13-00-00401	H$_2$O 1; EtOH 4
732	Benzenamine, 4,4'-(1,2-ethenediyl)bis-, (E)-	C$_{14}$H$_{14}$N$_2$ 210.28	7314-06-9 231	sub	4-13-00-00453	H$_2$O 2; bz 2; MeOH 3; CS$_2$ 2
733	Benzenamine, 2-ethenyl-	C$_8$H$_9$N 119.17	3867-18-3 	112[20]; 97[8]	4-12-00-02914 1.0181[20]	ace 3; bz 3 1.6124[20]
734	Benzenamine, 3-ethenyl-	C$_8$H$_9$N 119.17	15411-43-5 	112-5[12]	4-12-00-02915 1.0198[20]	ace 3; bz 3 1.6069[26]
735	Benzenamine, 4-ethenyl-	C$_8$H$_9$N 119.17	1520-21-4 23.5	116[9]	4-12-00-02915 1.010[20]	ace 3; bz 3 1.6250[22]

No.	Name / Synonym	Mol. Form. / Mol. Wt.	CAS RN / mp/°C	Merck No. / bp/°C	Beil. Ref. / den/g cm⁻³	Solubility / n_D
736	Benzenamine, 2-ethoxy- / o-Phenetidine	$C_8H_{11}NO$ / 137.18	94-70-2 / <-21	7187 / 232.5	4-13-00-00807	H_2O 2; EtOH 3; eth 3; ctc 2 / 1.5560[20]
737	Benzenamine, 4-ethoxy- / p-Phenetidine	$C_8H_{11}NO$ / 137.18	156-43-4 / 2.4	7188 / 254	4-13-00-01017 / 1.0652[16]	H_2O 2; EtOH 3; eth 3; chl 3 / 1.5528[20]
738	Benzenamine, 3-ethoxy-N,N-diethyl-	$C_{12}H_{19}NO$ / 193.29	1864-92-2	286; 97[0.6]	4-13-00-00970	EtOH 3; bz 3; HOAc 3 / 1.5325[25]
739	Benzenamine, 2-ethoxy-5-nitro- / o-Phenetidine, 5-nitro	$C_8H_{10}N_2O_3$ / 182.18	136-79-8 / 96.5	6538 / 205-6[14]	4-13-00-00896	eth 4; EtOH 4
740	Benzenamine, 2-ethyl- / 2-Ethylaniline	$C_8H_{11}N$ / 121.18	578-54-1 / -43	209.5	4-12-00-02411 / 0.983[22]	H_2O 2; EtOH 4; eth 4; chl 2 / 1.5584[22]
741	Benzenamine, 3-ethyl-	$C_8H_{11}N$ / 121.18	587-02-0 / -64	214; 94[6]	4-12-00-02417 / 0.9896[25]	eth 4; EtOH 4
742	Benzenamine, 4-ethyl- / p-Ethylaniline	$C_8H_{11}N$ / 121.18	589-16-2 / -4.8	217.5	4-12-00-02419 / 0.9679[20]	H_2O 2; EtOH 4; eth 4; ctc 2 / 1.5554[20]
743	Benzenamine, N-ethyl- / N-Ethylaniline	$C_8H_{11}N$ / 121.18	103-69-5 / -63.5	3722 / 203.0	4-12-00-00250 / 0.9625[20]	H_2O 1; EtOH 5; eth 5; ace 4 / 1.5559[20]
744	Benzenamine, 2-ethyl-N-(2-ethylphenyl)-	$C_{16}H_{19}N$ / 225.33	64653-59-4 / 29	336[603,] 173[10]	4-12-00-02413	H_2O 1; EtOH 4; eth 4; chl 2 / 1.5550[25]
745	Benzenamine, 2-ethyl-6-methyl-	$C_9H_{13}N$ / 135.21	24549-06-2 / -33	231	4-12-00-02638 / 0.968[25]	1.5525[20]
746	Benzenamine, N-ethyl-2-methyl-	$C_9H_{13}N$ / 135.21	94-68-8 / <-15	216	4-12-00-01747 / 0.948[25]	EtOH 3; eth 3 / 1.5456[20]
747	Benzenamine, N-ethyl-3-methyl-	$C_9H_{13}N$ / 135.21	102-27-2	221	4-12-00-01816	EtOH 3; eth 3 / 1.5451[20]
748	Benzenamine, N-ethyl-4-methyl- / N-Ethyl-4-toluidine	$C_9H_{13}N$ / 135.21	622-57-1	217	4-12-00-01875 / 0.9391[16]	EtOH 3; eth 3
749	Benzenamine, N-ethyl-N-methyl-	$C_9H_{13}N$ / 135.21	613-97-8	204	4-12-00-00252 / 0.92[55]	H_2O 1; EtOH 5; eth 5; ctc 3
750	Benzenamine, N-ethyl-N-nitroso-	$C_8H_{10}N_2O$ / 150.18	612-64-6	119-20[15]	4-16-00-00865 / 1.0874[20]	H_2O 1; HOAc 3
751	Benzenamine, N-ethyl-N-phenyl-	$C_{14}H_{15}N$ / 197.28	606-99-5	295.5; 147[10]	4-12-00-00276 / 1.0377[20]	eth 4; EtOH 4 / 1.6095[20]
752	Benzenamine, 2-fluoro- / o-Fluoroaniline	C_6H_6FN / 111.12	348-54-9 / -34.8	175, 55[12]	4-12-00-01099 / 1.1513[21]	H_2O 1; EtOH 3; eth 3; ctc 2 / 1.5421[20]
753	Benzenamine, 3-fluoro- / m-Fluoroaniline	C_6H_6FN / 111.12	372-19-0	188	4-12-00-01101 / 1.1561[19]	H_2O 2; EtOH 3; eth 3; chl 2 / 1.5436[20]
754	Benzenamine, 4-fluoro- / p-Fluoroaniline	C_6H_6FN / 111.12	371-40-4 / -0.8	4098 / 182; 85[19]	4-12-00-01104 / 1.1725[20]	H_2O 2; EtOH 3; eth 3; ctc 2 / 1.5195[20]
755	Benzenamine, 4-fluoro-2-methyl-	C_7H_8FN / 125.15	452-71-1 / 14 2	94[16]	4-12-00-01786 / 1.1263[18]	eth 3; ace 3; bz 3; ctc 3 / 1.5363[18]
756	Benzenamine, hydrochloride / Aniline, hydrochloride	C_6H_8ClN / 129.59	142-04-1 / 198	687	4-12-00-00223 / 1.2215[4]	H_2O 4; EtOH 4; eth 1; chl 1
757	Benzenamine, N-hydroxy- / Phenylhydroxylamine	C_6H_7NO / 109.13	100-65-2 / 83.5	7266	4-15-00-00004	bz 4; eth 4; EtOH 4; chl 4
758	Benzenamine, N-hydroxy-4-methyl-	C_7H_9NO / 123.15	623-10-9 / 96	117 dec	4-15-00-00019	eth 4; EtOH 4; chl 4
759	Benzenamine, N-hydroxy-N-nitroso-, ammonium salt / Cupferron	$C_6H_9N_3O_2$ / 155.16	135-20-6 / 163.5	2625	4-16-00-00889	DMSO 2
760	Benzenamine, 3-iodo-	C_6H_6IN / 219.02	626-01-7 / 33	145[15]	4-12-00-01543	H_2O 1; EtOH 3; chl 3 / 1.6811[20]
761	Benzenamine, 4-iodo- / p-Iodoaniline	C_6H_6IN / 219.02	540-37-4 / 67.5	4920	4-12-00-01544	H_2O 2; EtOH 3; eth 3; os 3
762	Benzenamine, 2-iodo-4-methyl- / 2-Iodo-4-methylaniline	C_7H_8IN / 233.05	29289-13-2 / 40	dec	4-12-00-01996	ace 4; bz 4; eth 4; EtOH 4
763	Benzenamine, 5-iodo-2-methyl- / o-Toluidine, 5-iodo	C_7H_8IN / 233.05	83863-33-6 / 48.5	273 dec	1-12-00-00391	EtOH 4
764	Benzenamine, N,N'-methanetetraylbis-	$C_{13}H_{10}N_2$ / 194.24	622-16-2 / 169	331; 175[20]	4-12-00-00866	H_2O 2; EtOH 2; eth 2; bz 3
765	Benzenamine, N,N'-methanetetraylbis[4-methyl-	$C_{15}H_{14}N_2$ / 222.29	726-42-1 / 58.5	221[20]	4-12-00-01938 / 1.1500[20]	
766	Benzenamine, 2-methoxy- / o-Anisidine	C_7H_9NO / 123.15	90-04-0 / 6.2	224	4-13-00-00806 / 1.0923[20]	H_2O 2; EtOH 3; eth 3; ace 3 / 1.5715[10]
767	Benzenamine, 3-methoxy- / m-Anisidine	C_7H_9NO / 123.15	536-90-3 / -1	251	4-13-00-00953 / 1.096[20]	H_2O 2; EtOH 3; eth 3; ace 3 / 1.5794[20]
768	Benzenamine, 4-methoxy- / p-Anisidine	C_7H_9NO / 123.15	104-94-9 / 57.2	243	4-13-00-01015 / 1.071[57]	H_2O 3; EtOH 4; eth 4; ace 3 / 1.5559[60]
769	Benzenamine, 2-methoxy-N,N-dimethyl-	$C_9H_{13}NO$ / 151.21	700-75-4	210	4-13-00-00816 / 1.0160[23]	chl 4
770	Benzenamine, 2-methoxy-5-methyl- / 5-Methyl-o-anisidine	$C_8H_{11}NO$ / 137.18	120-71-8 / 53	235	3-13-00-01577	H_2O 2; EtOH 3; eth 3; bz 3
771	Benzenamine, 3-methoxy-4-methyl-	$C_8H_{11}NO$ / 137.18	16452-01-0 / 56.5	251	4-13-00-01666	bz 4; eth 4; EtOH 4; lig 4
772	Benzenamine, 4-methoxy-2-methyl-	$C_8H_{11}NO$ / 137.18	102-50-1 / 29.5	248.5	4-13-00-01698 / 1.065[25]	EtOH 4 / 1.5647[20]
773	Benzenamine, 5-methoxy-2-methyl-	$C_8H_{11}NO$ / 137.18	50868-72-9 / 47	253; 140[20]	4-13-00-01711	eth 4

No.	Name / Synonym	Mol. Form. / Mol. Wt.	CAS RN / mp/°C	Merck No. / bp/°C	Beil. Ref. / den/g cm^{-3}	Solubility / n_D
774	Benzenamine, 2-methoxy-6-methyl-	$C_8H_{11}NO$ 137.18	50868-73-0 31	119[16]	3-13-00-01549	EtOH 4
775	Benzenamine, 2-methoxy-5-nitro- 5-Nitro-o-anisidine	$C_7H_8N_2O_3$ 168.15	99-59-2 118		4-13-00-00896 1.2068[15]	H_2O 3; EtOH 4; eth 3; ace 4
776	Benzenamine, 3-methoxy-4-nitro- m-Anisidine, 4-nitro-	$C_7H_8N_2O_3$ 168.15	16292-88-9 169	sub	4-13-00-01005	ace 4; EtOH 4
777	Benzenamine, 5-methoxy-2-nitro-	$C_7H_8N_2O_3$ 168.15	16133-49-6 131	sub	4-13-00-01003	EtOH 4
778	Benzenamine, 2-methyl- o-Toluidine	C_7H_9N 107.16	95-53-4 -16.3	9462 200.3	4-12-00-01744 0.9984[20]	H_2O 2; EtOH 5; eth 5; ctc 5 1.5725[20]
779	Benzenamine, 3-methyl- m-Toluidine	C_7H_9N 107.16	108-44-1 -31.2	9462 203.3	4-12-00-01813 0.9889[20]	ace 4; bz 4; eth 4; EtOH 4 1.5681[20]
780	Benzenamine, 4-methyl- p-Toluidine	C_7H_9N 107.16	106-49-0 43.7	9462 200.4	4-12-00-01866 0.9619[20]	H_2O 2; EtOH 4; eth 3; ace 3 1.5534[45]
781	Benzenamine, N-methyl- N-Methylaniline	C_7H_9N 107.16	100-61-8 -57	5941 196.2	4-12-00-00241 0.9891[20]	H_2O 1; EtOH 3; eth 3; ctc 3 1.5684[20]
782	Benzenamine, 4-(6-methyl-2-benzothiazolyl)-	$C_{14}H_{12}N_2S$ 240.33	92-36-4 194.8	434	4-27-00-05052	EtOH 2; eth 2; bz 2; HOAc 2
783	Benzenamine, N-(3-methylbutyl)-	$C_{11}H_{17}N$ 163.26	2051-84-5	254.5	4-12-00-00259 0.891[55]	eth 4; EtOH 4 1.5305[20]
784	Benzenamine, 4,4'-methylenebis- p,p'-Diaminodiphenylmethane	$C_{13}H_{14}N_2$ 198.27	101-77-9 92.5	2958 398; 257[18]	4-13-00-00390	H_2O 2; EtOH 4; eth 4; bz 4
785	Benzenamine, 4,4'-methylenebis[2-chloro- 4,4'-Methylenebis[2-chloroaniline]	$C_{13}H_{12}Cl_2N_2$ 267.16	101-14-4	5978	1-13-00-00074	ctc 3
786	Benzenamine, 4,4'-methylenebis[N,N-dimethyl- Michler's Base	$C_{17}H_{22}N_2$ 254.38	101-61-1 91.5	6099 390 dec; 183[3]	4-13-00-00390	H_2O 1; EtOH 2; eth 4; bz 4
787	Benzenamine, 2-(1-methylethenyl)-	$C_9H_{11}N$ 133.19	52562-19-3	95[13]	4-12-00-02925 0.976[25]	1.5722[20]
788	Benzenamine, 2-(1-methylethyl)- 2-Isopropylaniline	$C_9H_{13}N$ 135.21	643-28-7	221; 95[13]	4-12-00-02630 0.976012	H_2O 1; eth 3; bz 3; ctc 3
789	Benzenamine, 4-(1-methylethyl)- Cumidine	$C_9H_{13}N$ 135.21	99-88-7	2622 225	4-12-00-02632 0.953[20]	
790	Benzenamine, N-(1-methylethyl)- N-Isopropylaniline	$C_9H_{13}N$ 135.21	768-52-5	203	4-12-00-00255 0.9526[25]	EtOH 3; eth 3; ace 3; bz 3 1.5380[20]
791	Benzenamine, 2-methyl-, hydrochloride o-Toluidine hydrochloride	$C_7H_{10}ClN$ 143.62	636-21-5 215		4-12-00-01744	H_2O 4; EtOH 4
792	Benzenamine, N-methyl-, hydrochloride	$C_7H_{10}ClN$ 143.62	2739-12-0 122.7		4-12-00-00241 1.0660[131]	H_2O 4; EtOH 3; eth 1; bz 1
793	Benzenamine, 2-methyl-5-(1-methylethyl)-	$C_{10}H_{15}N$ 149.24	2051-53-8 -16	241	3-12-00-02733 0.9942[20]	ctc 3; CS_2 3 1.5387[20]
794	Benzenamine, 4-methyl-N-(1-methylethyl)-	$C_{10}H_{15}N$ 149.24	10436-75-6	220	4-12-00-01876 0.9226[20]	1.5332
795	Benzenamine, 2-methyl-N-(2-methylphenyl)-	$C_{14}H_{15}N$ 197.28	617-00-5 52.5	316; 192[23]	4-12-00-01750	H_2O 2
796	Benzenamine, 3-methyl-N-(3-methylphenyl)-	$C_{14}H_{15}N$ 197.28	626-13-1 -12	319.5	2-12-00-00467	H_2O 2; EtOH 3; eth 3; peth 3
797	Benzenamine, 4-methyl-N-(4-methylphenyl)-	$C_{14}H_{15}N$ 197.28	620-93-9 79.8	330.5	4-12-00-01879	eth 4; peth 4
798	Benzenamine, 2-methyl-4-[(2-methylphenyl)azo]- 2',3-Dimethyl-4-aminoazobenzene	$C_{14}H_{15}N_3$ 225.29	97-56-3 102		4-16-00-00528	eth 4; EtOH 4
799	Benzenamine, 2-methyl-3-nitro-	$C_7H_8N_2O_2$ 152.15	603-83-8 92	305	4-12-00-01811 1.3780[15]	H_2O 2; EtOH 3; eth 3; bz 3
800	Benzenamine, 2-methyl-4-nitro-	$C_7H_8N_2O_2$ 152.15	99-52-5 133.5		4-12-00-01809 1.1586[140]	H_2O 2; EtOH 3; bz 3; HOAc 3
801	Benzenamine, 2-methyl-5-nitro- 5-Nitro-o-toluidine	$C_7H_8N_2O_2$ 152.15	99-55-8 105.5		4-12-00-01807	H_2O 2; EtOH 3; eth 3; ace 3
802	Benzenamine, 2-methyl-6-nitro-	$C_7H_8N_2O_2$ 152.15	570-24-1 96		4-12-00-01807 1.1900[100]	H_2O 2; EtOH 3; eth 3; bz 3
803	Benzenamine, 4-methyl-2-nitro-	$C_7H_8N_2O_2$ 152.15	89-62-3 116.3		4-12-00-02000 1.16[121]	H_2O 2; EtOH 3; chl 3
804	Benzenamine, N-methyl-2-nitro-	$C_7H_8N_2O_2$ 152.15	612-28-2 38	158[18]	4-12-00-01564	H_2O 2; EtOH 3; eth 3; ace 3
805	Benzenamine, N-methyl-4-nitro-	$C_7H_8N_2O_2$ 152.15	100-15-2 152	dec	4-12-00-01616 1.201[155]	H_2O 1; EtOH 3; eth 2; bz 3
806	Benzenamine, N-methyl-N-nitroso-	$C_7H_8N_2O$ 136.15	614-00-6 14.7	225 dec; 121[13]	4-16-00-00864 1.1240[20]	H_2O 1; EtOH 3; eth 3 1.5769[20]
807	Benzenamine, N-methyl-N-phenyl- Methyldiphenylamine	$C_{13}H_{13}N$ 183.25	552-82-9 -7.5	5973 293.5	4-12-00-00275 1.0476[20]	H_2O 1; EtOH 3; ctc 3; MeOH 2 1.6193[20]
808	Benzenamine, 3-methyl-N-phenyl-	$C_{13}H_{13}N$ 183.25	1205-64-7 30	316; 183[17]	4-12-00-01818	bz 4; eth 4; EtOH 4 1.6350[20]
809	Benzenamine, 2-methyl-N-(phenylmethylene)-	$C_{14}H_{13}N$ 195.26	5877-55-4	314	4-12-00-01751 1.041[25]	H_2O 1; ace 3 1.6310[25]
810	Benzenamine, 3-methyl-N-(phenylmethylene)-	$C_{14}H_{13}N$ 195.26	5877-58-7 31	315	3-12-00-01958 1.0390[25]	ace 3; os 3 1.6353[25]

No.	Name / Synonym	Mol. Form. / Mol. Wt.	CAS RN / mp/°C	Merck No. / bp/°C	Beil. Ref. / den/g cm^{-3}	Solubility / n_D
811	Benzenamine, 4-methyl-N-(phenylmethylene)-	$C_{14}H_{13}N$ 195.26	2272-45-9 35	318; 178[11]	2-12-00-00496	ace 4
812	Benzenamine, 2-(1-methylpropyl)-	$C_{10}H_{15}N$ 149.24	55751-54-7	120-2[16]	4-12-00-02814 0.9574[20]	EtOH 3; ace 3; bz 3; ctc 2
813	Benzenamine, 4-(1-methylpropyl)-	$C_{10}H_{15}N$ 149.24	30273-11-1	238; 118[15]	4-12-00-02814 0.949[15]	bz 4; eth 4 1.5360[29]
814	Benzenamine, 4-(2-methylpropyl)-	$C_{10}H_{15}N$ 149.24	30090-17-6	238; 112[11]	4-12-00-02818 0.949[15]	ace 4; bz 4; EtOH 4
815	Benzenamine, 4-methyl-N-propyl- N-Propyl-p-toluidine	$C_{10}H_{15}N$ 149.24	54837-90-0	235	4-12-00-01876 0.9243[20]	1.5367
816	Benzenamine, N-(1-methylpropyl)- N-sec-Butylaniline	$C_{10}H_{15}N$ 149.24	6068-69-5	225	4-12-00-00257	chl 2; os 3 1.5333[20]
817	Benzenamine, N-(2-methylpropyl)-	$C_{10}H_{15}N$ 149.24	588-47-6	232; 113[11]	4-12-00-00257 0.940[18]	H_2O 1; eth 4; bz 4 1.5328[20]
818	Benzenamine, N-methyl-N,2,4,6-tetranitro- Nitramine	$C_7H_5N_5O_8$ 287.15	479-45-8 131.5	6488 exp 180	4-16-00-00895 1.57[10]	H_2O 1; EtOH 2; eth 2; ace 3
819	Benzenamine, 2-(methylthio)- Aniline, 2-methylthio	C_7H_9NS 139.22	2987-53-3	234	4-13-00-00910 1.111[25]	1.6239[20]
820	Benzenamine, 4-(methylthio)-	C_7H_9NS 139.22	104-96-1	272.5	4-13-00-01289 1.1379[20]	EtOH 3; eth 3; ace 3; bz 3 1.6395[20]
821	Benzenamine, nitrate Aniline nitrate	$C_6H_8N_2O_3$ 156.14	542-15-4 190 dec	687	4-12-00-00223 1.356[4]	H_2O 4; eth 4; EtOH 4
822	Benzenamine, 2-nitro- o-Nitroaniline	$C_6H_6N_2O_2$ 138.13	88-74-4 71.2	6504 284	4-12-00-01563 1.442[15]	H_2O 2; EtOH 3; eth 4; ace 4
823	Benzenamine, 3-nitro- m-Nitroaniline	$C_6H_6N_2O_2$ 138.13	99-09-2 114	6503 306 dec	4-12-00-01589 0.991[25]	H_2O 2; EtOH 3; eth 3; ace 3
824	Benzenamine, 4-nitro- p-Nitroaniline	$C_6H_6N_2O_2$ 138.13	100-01-6 147	6505 332	4-12-00-01613 1.424[20]	H_2O 1; EtOH 3; eth 3; ace 3
825	Benzenamine, N-nitro-	$C_6H_6N_2O_2$ 138.13	645-55-6 47.3	exp	4-16-00-00889	H_2O 4; bz 4; eth 4; EtOH 4
826	Benzenamine, 4-nitro-N-phenyl-	$C_{12}H_{10}N_2O_2$ 214.22	836-30-6 135.3	211[30]	4-12-00-01619	H_2O 1; EtOH 4; ace 2; HOAc 4
827	Benzenamine, 5-nitro-2-propoxy- Aniline, 5-nitro-2-propoxy	$C_9H_{12}N_2O_3$ 196.21	553-79-7 49	6551	4-12-00-00897	EtOH 4
828	Benzenamine, 4-nitroso-	$C_6H_6N_2O$ 122.13	659-49-4 173.5	dec	4-12-00-01558	H_2O 3; EtOH 3; bz 2
829	Benzenamine, 4-nitroso-N-phenyl- p-Nitrosodiphenylamine	$C_{12}H_{10}N_2O$ 198.22	156-10-5 143	6560	4-12-00-01560	H_2O 2; EtOH 4; eth 4; bz 4
830	Benzenamine, N-nitroso-N-phenyl- N-Nitroso-N-diphenylamine	$C_{12}H_{10}N_2O$ 198.22	86-30-6 66.5		4-16-00-00865	EtOH 2; bz 3; chl 2
831	Benzenamine, 4-octyl-	$C_{14}H_{23}N$ 205.34	16245-79-7 20	310; 138[5]	4-12-00-02887 0.9128[20]	eth 4
832	Benzenamine, 4,4'-oxybis- Bis(p-aminophenyl) ether	$C_{12}H_{12}N_2O$ 200.24	101-80-4 189 dec		4-13-00-01038	
833	Benzenamine, 2,3,4,5,6-pentamethyl-	$C_{11}H_{17}N$ 163.26	2243-30-3 152.5	277.5	3-12-00-02758	eth 4; EtOH 4
834	Benzenamine, 2-phenoxy-	$C_{12}H_{11}NO$ 185.23	2688-84-8 45.8	308; 172[14]	4-13-00-00809	EtOH 3; eth 3; ace 3; bz 3
835	Benzenamine, 3-phenoxy-	$C_{12}H_{11}NO$ 185.23	3586-12-7 37	315; 180[10]	4-13-00-00955 1.1583[25]	EtOH 3; eth 3; ace 3; bz 3
836	Benzenamine, N-phenyl- Diphenylamine	$C_{12}H_{11}N$ 169.23	122-39-4 52.9 302	3317	4-12-00-00271 1.158[22]	H_2O 1; EtOH 4; eth 3; ace 4
837	Benzenamine, 4-(phenylazo)- p-Aminoazobenzene	$C_{12}H_{11}N_3$ 197.24	60-09-3 127	430 >360	2-16-00-00149	H_2O 2; EtOH 3; eth 3; bz 3
838	Benzenamine, 2-(2-phenylethyl)-	$C_{14}H_{15}N$ 197.28	5697-85-8 33	193[19]	4-12-00-03311 1.0430[20]	
839	Benzenamine, 4-(phenylmethyl)-	$C_{13}H_{13}N$ 183.25	1135-12-2 34.5	300	4-12-00-03281 1.038[25]	eth 4; EtOH 4; lig 4
840	Benzenamine, N-(phenylmethylene)- Benzylideneaniline	$C_{13}H_{11}N$ 181.24	538-51-2 54	1152 310	2-12-00-00113 1.038[55]	H_2O 1; EtOH 3; eth 3; chl 2 1.600[100]
841	Benzenamine, N-2-propenyl-	$C_9H_{11}N$ 133.19	589-09-3	219; 106[12]	4-12-00-00265 0.982[25]	H_2O 2; EtOH 3; eth 5; ace 3 1.563[20]
842	Benzenamine, 2-propyl-	$C_9H_{13}N$ 135.21	1821-39-2	226; 116[15]	4-12-00-02580 0.9602[20]	H_2O 1; EtOH 3; eth 3 1.5427[20]
843	Benzenamine, N-propyl-	$C_9H_{13}N$ 135.21	622-80-0	222; 98[11]	4-12-00-00254 0.9443[20]	eth 4; EtOH 4 1.5428[20]
844	Benzenamine, N-sulfinyl-	C_6H_5NOS 139.18	1122-83-4	200	4-12-00-01075 1.236[25]	1.6270[20]
845	Benzenamine, 4,4'-sulfinylbis- 4,4'-Sulfinyldianiline	$C_{12}H_{12}N_2OS$ 232.31	119-59-5 175 dec		8927 4-13-00-01306	H_2O 3; EtOH 3
846	Benzenamine, 4,4'-sulfonylbis- Dapsone	$C_{12}H_{12}N_2O_2S$ 248.31	80-08-0 175.5	2820	4-13-00-01306	EtOH 3; DMSO-d$_6$ 2
847	Benzenamine, 2,3,4,5-tetramethyl- Aniline, 2,3,4,5-tetramethyl	$C_{10}H_{15}N$ 149.24	2217-45-0 69	259.5	3-12-00-02743	eth 4; EtOH 4; lig 4
848	Benzenamine, 2,3,4,6-tetramethyl-	$C_{10}H_{15}N$ 149.24	488-71-1 23.5	255	4-12-00-02849 0.978[24]	EtOH 4
849	Benzenamine, N,N,2,6-tetramethyl-	$C_{10}H_{15}N$ 149.24	769-06-2 -36	196; 88[20]	4-12-00-02522 0.9147[20]	

No.	Name / Synonym	Mol. Form. / Mol. Wt.	CAS RN / mp/°C	Merck No. / bp/°C	Beil. Ref. / den/g cm^{-3}	Solubility / n_D
850	Benzenamine, 4,4'-thiobis- Bis(4-aminophenyl) sulfide	$C_{12}H_{12}N_2S$ 216.31	139-65-1 108.5		4-13-00-01306	H_2O 2; EtOH 4; eth 4; bz 4
851	Benzenamine, 2,4,6-tribromo- 2,4,6-Tribromoaniline	$C_6H_4Br_3N$ 329.82	147-82-0 122	9522 300	4-12-00-01538 2.35^{20}	H_2O 1; EtOH 2; eth 3; chl 3
852	Benzenamine, 2,3,4-trichloro- 2,3,4-Trichloroaniline	$C_6H_4Cl_3N$ 196.46	634-67-3 73	 292	0-12-00-00626	EtOH 4
853	Benzenamine, 2,4,5-trichloro-	$C_6H_4Cl_3N$ 196.46	636-30-6 96.5	 270	4-12-00-01277	EtOH 3; eth 3; CS_2 4; lig 2
854	Benzenamine, 2,4,6-trichloro-	$C_6H_4Cl_3N$ 196.46	634-93-5 78.5	 262	4-12-00-01281	H_2O 1; EtOH 3; eth 3; chl 3
855	Benzenamine, 2-(trifluoromethyl)- o-(Trifluoromethyl)aniline	$C_7H_6F_3N$ 161.13	88-17-5 35.5	68^{15}	4-12-00-01786 1.282^{25}	1.4810^{20}
856	Benzenamine, 3-(trifluoromethyl)- m-(Trifluoromethyl)aniline	$C_7H_6F_3N$ 161.13	98-16-8 5.5	$187; 74-5^{10}$	4-12-00-01843 1.304^{12}	H_2O 2; EtOH 3; eth 3 1.4787^{20}
857	Benzenamine, 4-(trifluoromethyl)-	$C_7H_6F_3N$ 161.13	455-14-1 38	117.5^{60}	4-12-00-01982 1.283^{27}	1.4815^{25}
858	Benzenamine, 2,4,5-trimethyl-	$C_9H_{13}N$ 135.21	137-17-7 68	234.5	4-12-00-02645 0.957^{25}	EtOH 4
859	Benzenamine, 2,4,6-trimethyl- Mesitylamine	$C_9H_{13}N$ 135.21	88-05-1 -5	232.5	4-12-00-02648 0.9633^{25}	ctc 2 1.5495^{20}
860	Benzenamine, N,N,2-trimethyl- N,N-Dimethyl-o-toluidine	$C_9H_{13}N$ 135.21	609-72-3 -60	194.1	4-12-00-01747 0.9286^{20}	eth 4; EtOH 4 1.5152^{20}
861	Benzenamine, N,N,3-trimethyl-	$C_9H_{13}N$ 135.21	121-72-2	212	4-12-00-01815 0.9410^{20}	EtOH 5; eth 5 1.5492^{20}
862	Benzenamine, N,N,4-trimethyl-	$C_9H_{13}N$ 135.21	99-97-8	211	4-12-00-01874 0.9366^{20}	H_2O 1; EtOH 5; eth 5; ctc 3 1.5366^{20}
863	Benzenamine, 2,4,6-trinitro-	$C_6H_4N_4O_6$ 228.12	489-98-5 193.5	exp	4-12-00-01735 1.762^{10}	H_2O 1; EtOH 2; eth 2; ace 3
864	Benzenamine, 2,4,6-trinitro-N-(2,4,6-trinitrophenyl)- Dipicrylamine	$C_{12}H_5N_7O_{12}$ 439.21	131-73-7 244 dec	3340	4-12-00-01737	H_2O 1; EtOH 1; eth 2; ace 2
865	Benzenaminium, N-ethyl-3-hydroxy-N,N-dimethyl-, chloride Ammonium, dimethyl ethyl 3-hydroxyphenyl,chloride	$C_{10}H_{16}ClNO$ 201.70	116-38-1	3487	4-13-00-00969	
866	Benzenaminium, N,N,N-trimethyl-, iodide Phenyl trimethyl ammonium iodide	$C_9H_{14}IN$ 263.12	98-04-4 224	7289	4-12-00-00249	H_2O 4; EtOH 3; ace 2; chl 1
867	Benzene [6]Annulene	C_6H_6 78.11	71-43-2 5.5	1074 80.0	4-05-00-00583 0.8765^{20}	H_2O 2; EtOH 5; eth 5; ace 5 1.5011^{20}
868	Benzeneacetaldehyde Phenylacetaldehyde	C_8H_8O 120.15	122-78-1 33.5	7236 195	4-07-00-00664 1.0272^{20}	H_2O 2; EtOH 5; eth 5; ace 3 1.5255^{20}
869	Benzeneacetaldehyde, 3,4-dimethoxy-	$C_{10}H_{12}O_3$ 180.20	5703-21-9	$125^{1.5}$	4-08-00-01828 1.148^{20}	1.5426^{20}
870	Benzeneacetaldehyde, 2-methoxy-	$C_9H_{10}O_2$ 150.18	33567-59-8	$245; 117^{17}$	1-08-00-00543 1.0897^{20}	ace 4; EtOH 4 1.5393^{20}
871	Benzeneacetaldehyde, 4-methoxy-	$C_9H_{10}O_2$ 150.18	5703-26-4	255.5	4-08-00-00426 1.096^{20}	1.5359^{20}
872	Benzeneacetaldehyde, α-methyl-	$C_9H_{10}O$ 134.18	93-53-8	203.5	 1.0089^{20}	EtOH 4 1.5176^{20}
873	Benzeneacetaldehyde, 2-methyl-	$C_9H_{10}O$ 134.18	10166-08-2	$221; 92^{10}$	1-07-00-00163 1.0241^{10}	eth 4; EtOH 4; chl 4
874	Benzeneacetaldehyde, 4-methyl-	$C_9H_{10}O$ 134.18	104-09-6 40	221.5	4-07-00-00705 1.0052^{20}	eth 4; EtOH 4; chl 4 1.5255^{20}
875	Benzeneacetaldehyde, 2-nitro-	$C_8H_7NO_3$ 165.15	1969-73-9 22.5	135^5	2-07-00-00229	bz 4; eth 4; EtOH 4
876	Benzeneacetaldehyde, α-oxo-, aldoxime Isonitrosoacetophenone	$C_8H_7NO_2$ 149.15	532-54-7 129	5077	4-07-00-02132	H_2O 2; chl 3
877	Benzeneacetaldehyde, α-phenyl-	$C_{14}H_{12}O$ 196.25	947-91-1	315 dec; 157^7	4-07-00-01400 1.1061^{21}	H_2O 1; EtOH 4; eth 4; bz 4 1.5920^{21}
878	Benzeneacetamide α-Phenylacetamide	C_8H_9NO 135.17	103-81-1 157	7237	4-09-00-01632	H_2O 2; EtOH 3; eth 2; bz 2
879	Benzeneacetamide, N-(aminocarbonyl)- Phenacemide	$C_9H_{10}N_2O_2$ 178.19	63-98-9 215	7154	4-09-00-01636	bz 4; eth 4; EtOH 4
880	Benzeneacetamide, 4-butoxy-N-hydroxy- Bufexamac	$C_{12}H_{17}NO_3$ 223.27	2438-72-4 154	1462		
881	Benzeneacetamide, α-ethyl- α-Phenylbutyramide	$C_{10}H_{13}NO$ 163.22	90-26-6 86	7249 185^{16}	4-09-00-01818	H_2O 3; ace 2; ctc 3
882	Benzeneacetamide, α-(1-methylethyl)-	$C_{11}H_{15}NO$ 177.25	5470-47-3 111.5	180^{14}	3-09-00-02518	eth 4; EtOH 4
883	Benzeneacetic acid Phenylacetic acid	$C_8H_8O_2$ 136.15	103-82-2 76.7	7239 265.5	4-09-00-01614 1.228^6	H_2O 2; EtOH 4; eth 4; ace 3
884	Benzeneacetic acid, α-acetyl-, ethyl ester	$C_{12}H_{14}O_3$ 206.24	5413-05-8	156^{22}	4-10-00-02776 1.0855^{20}	eth 4; EtOH 4 1.5176^{20}
885	Benzeneacetic acid, α-amino- α-Phenylglycine	$C_8H_9NO_2$ 151.17	69-91-0 292 dec	7263 sub 255	4-14-00-01317	alk 3; os 2
886	Benzeneacetic acid, 4-amino- p-Aminophenylacetic acid	$C_8H_9NO_2$ 151.17	1197-55-3 200 dec	475	3-14-00-01182	H_2O 1; EtOH 2; DMSO-d_6 2
887	Benzeneacetic acid, anhydride	$C_{16}H_{14}O_3$ 254.29	1555-80-2 73.3	195^{12}	4-09-00-01629	eth 4; chl 4

No.	Name Synonym	Mol. Form. Mol. Wt.	CAS RN mp/°C	Merck No. bp/°C	Beil. Ref. den/g cm^{-3}	Solubility n_D
888	Benzeneacetic acid, 4-bromo	$C_8H_7BrO_2$ 215.05	1878-68-8 116	sub	3-09-00-02275	H_2O 2; EtOH 4; eth 4; CS_2 4
889	Benzeneacetic acid, 4-bromo-α-hydroxy-, (±)- Acetic acid, (4-bromophenyl)-hydroxy(DL)	$C_8H_7BrO_3$ 231.05	7021-04-7 119	1412	3-10-00-00480	H_2O 4; EtOH 4; eth 4; bz 4
890	Benzeneacetic acid, α-butyl-	$C_{12}H_{16}O_2$ 192.26	24716-09-4 	182[20]; 118[0.1]	4-09-00-01897 1.0225[19]	1.5071[19]
891	Benzeneacetic acid, α-chloro-, ethyl ester, (S)- Acetic acid, chlorophenyl-, ethyl ester, L-(+)-	$C_{10}H_{11}ClO_2$ 198.65	10606-73-2 	162[45]; 133[12]	3-09-00-02266 1.1594[20]	1.5152[20]
892	Benzeneacetic acid, α-chloro-α-phenyl-, ethyl ester	$C_{16}H_{15}ClO_2$ 274.75	52460-86-3 43.5	185[14]	4-09-00-02508	eth 4; EtOH 4
893	Benzeneacetic acid, α-cyano-, ethyl ester	$C_{11}H_{11}NO_2$ 189.21	4553-07-5 	275 dec; 165[20]	3-09-00-04262 1.091[20]	ace 4; bz 4; eth 4; EtOH 4 1.5012[25]
894	Benzeneacetic acid, 4-[2-(diethylamino)-2-oxoethoxy]-3-methoxy-, propyl ester Acetic acid, amide,(4(carboxy(propyl ester)methyl)-2-methoxyphenoxy)-N,N-diethyl	$C_{18}H_{27}NO_5$ 337.42	1421-14-3 	7813 210-2[0.7]		H_2O 1; EtOH 3; chl 3
895	Benzeneacetic acid, 2,5-dihydroxy- Homogentisic acid	$C_8H_8O_4$ 168.15	451-13-8 153	4658	4-10-00-01506	H_2O 4; EtOH 4; eth 4; bz 1
896	Benzeneacetic acid, α,4-dihydroxy-3-methoxy- Vanilmandelic acid	$C_9H_{10}O_5$ 198.18	55-10-7 131-3 dec	9840	3-10-00-02100	H_2O 4; ace 4; eth 4
897	Benzeneacetic acid, α-ethyl-	$C_{10}H_{12}O_2$ 164.20	90-27-7 47.5	271	2-09-00-00356	eth 3; bz 3; ctc 3
898	Benzeneacetic acid, α-ethyl-, 2-(diethylamino)ethyl ester Butethamate	$C_{16}H_{25}NO_2$ 263.38	14007-64-8 	1516 167-9[11]	3-09-00-02463	1.4909[20]
899	Benzeneacetic acid, ethyl ester Ethyl phenylacetate	$C_{10}H_{12}O_2$ 164.20	101-97-3 -29.4	3794 227	4-09-00-01618 1.0333[20]	eth 4; EtOH 4 1.4980[20]
900	Benzeneacetic acid, α-ethyl-, methyl ester	$C_{11}H_{14}O_2$ 178.23	2204-71-5 77.5	228	3-09-00-02462	eth 4; EtOH 4
901	Benzeneacetic acid, 4-fluoro- (p-Fluorophenyl)acetic acid	$C_8H_7FO_2$ 154.14	405-50-5 86	4105 164[2]	4-09-00-01672	
902	Benzeneacetic acid, α-(hydroxymethyl)- 8-methyl-8-azabicyclo[3.2.1]oct-3-yl ester endo-(±)- Atropine	$C_{17}H_{23}NO_3$ 289.37	51-55-8 118.5	891 sub 95	5-21-01-00235	H_2O 4; EtOH 4; eth 1; chl 2
903	Benzeneacetic acid, α-hydroxy-, (S)-	$C_8H_8O_3$ 152.15	17199-29-0 134		4-10-00-00564 1.341[25]	H_2O 4; eth 4; EtOH 4; chl 4
904	Benzeneacetic acid, 2-hydroxy-	$C_8H_8O_3$ 152.15	614-75-5 148	240	4-10-00-00536	H_2O 2; eth 3; chl 2
905	Benzeneacetic acid, 3-hydroxy-	$C_8H_8O_3$ 152.15	621-37-4 132	190[11]	4-10-00-00541	H_2O 4; EtOH 4; eth 4; bz 3
906	Benzeneacetic acid, 4-hydroxy-	$C_8H_8O_3$ 152.15	156-38-7 152	sub	4-10-00-00543	H_2O 2; EtOH 4; eth 4
907	Benzeneacetic acid, α-hydroxy-, ethyl ester, (±)-	$C_{10}H_{12}O_3$ 180.20	4358-88-7 35.5	254	4-10-00-00569 1.1258[20]	EtOH 3; eth 3; ctc 2; CS_2 3
908	Benzeneacetic acid, α-(hydroxymethyl)-, 9-methyl-3-oxa-9-azatricyclo[3.3.1.02,4]non-7-yl ester, [(7S)-(1α,2β,4β,5α,7β)]- Scopolamine	$C_{17}H_{21}NO_4$ 303.36	51-34-3	8361	2-27-00-00064	
909	Benzeneacetic acid, α-hydroxy-, ethyl ester	$C_{10}H_{12}O_3$ 180.20	774-40-3 35	150[21]	2-10-00-00121 1.1270[20]	eth 4; EtOH 4
910	Benzeneacetic acid, α-hydroxy-, methyl ester, (±)-	$C_9H_{10}O_3$ 166.18	4358-87-6 58	250 dec; 144[20]	2-10-00-00120 1.1756[20]	EtOH 4; chl 4
911	Benzeneacetic acid, α-hydroxy-, 8-methyl-8-azabicyclo[3.2.1]oct-3-yl ester, endo-(±)- Homatropine	$C_{16}H_{21}NO_3$ 275.35	87-00-3 99.5	4649	5-21-01-00234	H_2O 2; EtOH 3; eth 3; ace 3
912	Benzeneacetic acid, α-hydroxy-, 3-methylbutyl ester Mandelic acid isoamyl ester	$C_{13}H_{18}O_3$ 222.28	5421-04-5 	5600 172[11]	4-10-00-00571	
913	Benzeneacetic acid, 4-hydroxy-3-methoxy- Homovanillic acid	$C_9H_{10}O_4$ 182.18	306-08-1 143.5	4662	4-10-00-01509	
914	Benzeneacetic acid, α-hydroxy-α-methyl- Atrolactic acid	$C_9H_{10}O_3$ 166.18	515-30-0 116.5	889	2-10-00-00157	H_2O 4; ace 4; bz 4; EtOH 4
915	Benzeneacetic acid, α-(hydroxymethyl)- Tropic acid	$C_9H_{10}O_3$ 166.18	529-64-6 118	9691	3-10-00-00564	eth 4; EtOH 4
916	Benzeneacetic acid, α-(hydroxymethyl)-, 8-azabicyclo[3.2.1]oct-3-yl ester, [3(S)-endo]- Norhyoscyamine	$C_{16}H_{21}NO_3$ 275.35	537-29-1 140.5	8623	5-21-01-00219	EtOH 4; chl 4
917	Benzeneacetic acid, α-(hydroxymethyl)-, 8-methyl-8-azabicyclo[3.2.1]oct-3-yl ester, [3(S)-endo]- Hyoscyamine	$C_{17}H_{23}NO_3$ 289.37	101-31-5 108.5	4795	5-21-01-00235	H_2O 2; EtOH 4; eth 2; bz 2

No.	Name / Synonym	Mol. Form. / Mol. Wt.	CAS RN / mp/°C	Merck No. / bp/°C	Beil. Ref. / den/g cm^{-3}	Solubility / n_D
918	Benzeneacetic acid, α-hydroxy-α-phenyl- Benzilic acid	$C_{14}H_{12}O_3$ 228.25	76-93-7 151	1089 180 dec	4-10-00-01256	H_2O 2; EtOH 4; eth 4; ace 2
919	Benzeneacetic acid, α-hydroxy-α-phenyl-, ethyl ester	$C_{16}H_{16}O_3$ 256.30	52182-15-7 34	201[21]	4-10-00-01258 1.5620[20]	eth 4; EtOH 4
920	Benzeneacetic acid, α-hydroxy-α-phenyl-, methyl ester	$C_{15}H_{14}O_3$ 242.27	76-89-1 75.8	187[13]	4-10-00-01258	eth 4; EtOH 4
921	Benzeneacetic acid, 2-methoxy-	$C_9H_{10}O_3$ 166.18	93-25-4 124	100-1[2]	4-10-00-00536	H_2O 3; EtOH 4; eth 4; ace 4
922	Benzeneacetic acid, 4-methoxy-	$C_9H_{10}O_3$ 166.18	104-01-8 87	138[2]	4-10-00-00544	H_2O 1; EtOH 4; eth 3; bz 3
923	Benzeneacetic acid, 4-methoxy-, ethyl ester	$C_{11}H_{14}O_3$ 194.23	14062-18-1	138-40[70]	4-10-00-00549 1.097[25]	1.5075[20]
924	Benzeneacetic acid, α-methyl-, (±)-	$C_9H_{10}O_2$ 150.18	2328-24-7 <-20	263	4-09-00-01779 1.1[0]	1.5237[20]
925	Benzeneacetic acid, 3-methyl-	$C_9H_{10}O_2$ 150.18	621-36-3 62	120-3[26]	4-09-00-01792	H_2O 3; chl 3
926	Benzeneacetic acid, 4-methyl-	$C_9H_{10}O_2$ 150.18	622-47-9 93	265	4-09-00-01795	bz 4; eth 4; EtOH 4
927	Benzeneacetic acid, α-methylene- Atropic acid	$C_9H_8O_2$ 148.16	492-38-6 106.5	890 267 dec	4-09-00-02050	H_2O 2; EtOH 3; eth 3; bz 3
928	Benzeneacetic acid, α-methylene-, 8-methyl-8-azabicyclo[3.2.1]oct-3-yl ester, *endo*- Apoatropine	$C_{17}H_{21}NO_2$ 271.36	500-55-0 62	770	5-21-01-00228	H_2O 2; EtOH 4; eth 4; ace 4
929	Benzeneacetic acid, methyl ester	$C_9H_{10}O_2$ 150.18	101-41-7	216.5	4-09-00-01617 1.0622[16]	H_2O 1; EtOH 5; eth 5; ace 3 1.5075[20]
930	Benzeneacetic acid, α-(1-methylethyl)-	$C_{11}H_{14}O_2$ 178.23	3508-94-9 63	159[14]	3-09-00-02518	eth 4; EtOH 4
931	Benzeneacetic acid, α-methyl-α-phenyl-	$C_{15}H_{14}O_2$ 226.27	5558-66-7 176	sub	4-09-00-02535	H_2O 2; EtOH 3; eth 4; bz 3
932	Benzeneacetic acid, 2-methylpropyl ester	$C_{12}H_{16}O_2$ 192.26	102-13-6	247	4-09-00-01619 0.999[18]	H_2O 1; EtOH 3; eth 3
933	Benzeneacetic acid, 4-nitro- *p*-Nitrophenylacetic acid	$C_8H_7NO_4$ 181.15	104-03-0 154	6543	4-09-00-01687	H_2O 2; EtOH 3; eth 3; bz 3
934	Benzeneacetic acid, α-oxo-	$C_8H_6O_3$ 150.13	611-73-4 66	163[15]	4-10-00-02737	H_2O 4; EtOH 3; eth 3; ctc 2
935	Benzeneacetic acid, α-oxo-, ethyl ester	$C_{10}H_{10}O_3$ 178.19	1603-79-8	256.5	4-10-00-02738 1.1222[25]	1.5190[25]
936	Benzeneacetic acid, α-phenyl- Diphenylacetic acid	$C_{14}H_{12}O_2$ 212.25	117-34-0 147	3316 194[25]	4-09-00-02492 1.257[15]	H_2O 2; EtOH 4; eth 3; chl 3
937	Benzeneacetic acid, α-phenyl-, anhydride	$C_{28}H_{22}O_3$ 406.48	1760-46-9 98	222[15]	4-09-00-02497	bz 4; chl 4
938	Benzeneacetic acid, α-phenyl-, 2-(diethylamino)ethyl ester, hydrochloride Adiphenine hydrochloride	$C_{20}H_{26}ClNO_2$ 347.88	50-42-0 113.5	151	3-09-00-03297	H_2O 4; EtOH 2; eth 2
939	Benzeneacetic acid, α-phenyl-, ethyl ester	$C_{16}H_{16}O_2$ 240.30	3468-99-3 60	195[25]	4-09-00-02493 1.1860[20]	H_2O 1; EtOH 4; eth 4; CS_2 4
940	Benzeneacetic acid, 2-phenylethyl ester	$C_{16}H_{16}O_2$ 240.30	102-20-5 26.5	177-8[4.5]	3-09-00-02183 1.077[25]	EtOH 4
941	Benzeneacetic acid, α-(phenylmethylene)- α-Phenylcinnamic acid	$C_{15}H_{12}O_2$ 224.26	3368-16-9 172.5	7253 sub	3-09-00-03438	eth 4; EtOH 4
942	Benzeneacetic acid, α-(phenylmethylene)-, ethyl ester	$C_{17}H_{16}O_2$ 252.31	24446-63-7 33.5	214-5[28]	3-09-00-03416 1.0971[18]	1.5972[18]
943	Benzeneacetic acid, α-propyl-, (±)- Valeric acid, 2-phenyl-, (±)-	$C_{11}H_{14}O_2$ 178.23	7782-21-0 58	280	4-09-00-01868	eth 4; EtOH 4
944	Benzeneacetic acid, 2,3,6-trichloro- Chlorfenac	$C_8H_5Cl_3O_2$ 239.48	85-34-7 161	2085	4-09-00-01681	
945	Benzeneacetonitrile Benzyl cyanide	C_8H_7N 117.15	140-29-4 -23.8	1145 233.5	4-09-00-01663 1.0205[15]	1.5211[25]
946	Benzeneacetonitrile, 4-amino-	$C_8H_8N_2$ 132.17	3544-25-0 46	312	4-14-00-01312	H_2O 2; EtOH 3; CS_2 2; os 3
947	Benzeneacetonitrile, α-bromo- α-Bromobenzyl cyanide	C_8H_6BrN 196.05	5798-79-8 29	1400 242 dec; 133[12]	3-09-00-02278 1.539[29]	H_2O 1; EtOH 4; eth 4; ace 4
948	Benzeneacetonitrile, 2-bromo-	C_8H_6BrN 196.05	19472-74-3 1	140[13]	4-09-00-01682	H_2O 1; EtOH 4
949	Benzeneacetonitrile, 2-chloro-	C_8H_6ClN 151.60	2856-63-5 24	251	4-09-00-01674 1.1737[18]	
950	Benzeneacetonitrile, 3-chloro-	C_8H_6ClN 151.60	1529-41-5	134-6[10]	4-09-00-01675 1.283[25]	1.5437[20]
951	Benzeneacetonitrile, α-ethyl-	$C_{10}H_{11}N$ 145.20	769-68-6	241	4-09-00-01822 0.977[14]	H_2O 1; EtOH 3; eth 3; bz 3
952	Benzeneacetonitrile, 2-fluoro-	C_8H_6FN 135.14	326-62-5	232; 102[10]	4-09-00-01671 1.059[25]	1.5009[20]
953	Benzeneacetonitrile, 4-fluoro-	C_8H_6FN 135.14	459-22-3	119-20[18]	4-09-00-01673 1.126[25]	1.5002[20]

No.	Name Synonym	Mol. Form. Mol. Wt.	CAS RN mp/°C	Merck No. bp/°C	Beil. Ref. den/g cm^{-3}	Solubility n_D
954	Benzeneacetonitrile, α-[(6-O-β-D-glucopyranosyl-β-D-glucopyranosyl)oxy]-, (R)- Amygdalin	$C_{20}H_{27}NO_{11}$ 457.43	29883-15-6 224.5	629	5-17-08-00118	H$_2$O 4; EtOH 2; eth 1; chl 1
955	Benzeneacetonitrile, α-(β-D-glucopyranosyloxy)- Mandelonitrile glucoside	$C_{14}H_{17}NO_6$ 295.29	138-53-4 122	5602		H$_2$O 4; EtOH 4
956	Benzeneacetonitrile, α-hydroxy-, (±)-	C_8H_7NO 133.15	613-88-7 22	170	3-10-00-00473 1.1165[20]	eth 4; EtOH 4; chl 4
957	Benzeneacetonitrile, 2-methoxy-	C_9H_9NO 147.18	7035-03-2 69.8	141[15]	4-10-00-00538	bz 4
958	Benzeneacetonitrile, 4-methoxy-	C_9H_9NO 147.18	104-47-2	286.5	4-10-00-00555 1.0845[20]	EtOH 3; eth 3; chl 3 1.5309[20]
959	Benzeneacetonitrile, α-methyl-	C_9H_9N 131.18	1823-91-2	231	4-09-00-01783 0.9854[20]	eth 4; EtOH 4 1.5095[25]
960	Benzeneacetonitrile, 2-methyl-	C_9H_9N 131.18	22364-68-7	244	4-09-00-01791 1.0156[22]	bz 4; eth 4; EtOH 4 1.5252[20]
961	Benzeneacetonitrile, 3-methyl-	C_9H_9N 131.18	2947-60-6	248; 133[15]	4-09-00-01792 1.0022[22]	H$_2$O 1; EtOH 3; eth 3; bz 3 1.5233[20]
962	Benzeneacetonitrile, 4-methyl-	C_9H_9N 131.18	2947-61-7 18	242.5	4-09-00-01796 0.992[25]	H$_2$O 1; EtOH 3; eth 3; bz 3 1.5190[20]
963	Benzeneacetonitrile, α-(1-methylethyl)-	$C_{11}H_{13}N$ 159.23	5558-29-2	247	4-09-00-01879 0.967[15]	bz 4; EtOH 4 1.5038[25]
964	Benzeneacetonitrile, α-(2-methylpropyl)- Valeronitrile, 4-methyl-2-phenyl-	$C_{12}H_{15}N$ 173.26	5558-31-6	264; 137[15]	4-09-00-01903 0.942[16]	bz 4; eth 4; EtOH 4
965	Benzeneacetonitrile, 2-nitro-	$C_8H_6N_2O_2$ 162.15	610-66-2 84	178[12]; 138[1]	3-09-00-02283	ace 4; bz 4; eth 4; EtOH 4
966	Benzeneacetonitrile, 3-nitro-	$C_8H_6N_2O_2$ 162.15	621-50-1 63	180[15]	4-09-00-01687	bz 4; eth 4; EtOH 4; chl 4
967	Benzeneacetonitrile, 4-nitro- p-Nitrobenzyl cyanide	$C_8H_6N_2O_2$ 162.15	555-21-5 117	6512 195-7[12]	4-09-00-01689	H$_2$O 2; EtOH 3; eth 3; bz 3
968	Benzeneacetonitrile, α-oxo-	C_8H_5NO 131.13	613-90-1 32.5	206	4-10-00-02743	H$_2$O 1; EtOH 4; eth 4; chl 2
969	Benzeneacetonitrile, α-phenyl-	$C_{14}H_{11}N$ 193.25	86-29-3 74.3	184[16]	4-09-00-02505	EtOH 3; eth 4; chl 3; lig 2
970	Benzeneacetonitrile, α-propyl-	$C_{11}H_{13}N$ 159.23	5558-78-1	255; 126[13]	4-09-00-01869 0.9425[20]	bz 4; EtOH 4 1.5000[20]
971	Benzeneacetyl chloride Phenylacetyl chloride	C_8H_7ClO 154.60	103-80-0	170[250; 105[24]	4-09-00-01631 1.1682[20]	eth 4 1.5325[20]
972	Benzeneacetyl chloride, α-(acetyloxy)-	$C_{10}H_9ClO_3$ 212.63	1638-63-7 36	152[33; 125[10]	1-10-00-00089	eth 4; bz 4; chl 4; lig 4
973	Benzeneacetyl chloride, α-chloro-	$C_8H_6Cl_2O$ 189.04	2912-02-1	120[23]	2-09-00-00308 1.196[25]	1.5440[20]
974	Benzeneacetyl chloride, α-ethyl-	$C_{10}H_{11}ClO$ 182.65	36854-57-6	110-2[10]	4-09-00-01817	ctc 3 1.5169[20]
975	Benzeneacetyl chloride, α-phenyl-	$C_{14}H_{11}ClO$ 230.69	1871-76-7 56.5	170[16]	3-09-00-03300	lig 3
976	Benzene, azido-	$C_6H_5N_3$ 119.13	622-37-7 -27.5	70[11]	4-05-00-00759 1.0860[20]	H$_2$O 1; EtOH 2; eth 2 1.5589[25]
977	Benzene, 1-azido-4-bromo-	$C_6H_4BrN_3$ 198.02	2101-88-4 20	105[10]; 69[2]	4-05-00-00761 1.5827[25]	H$_2$O 1; EtOH 2; eth 4; bz 4
978	Benzene, 1-azido-4-chloro-	$C_6H_4ClN_3$ 153.57	3296-05-7 20	96[20]	4-05-00-00760 1.2634[25]	H$_2$O 1; eth 3
979	Benzene, 1-azido-2-iodo-	$C_6H_4IN_3$ 245.02	54467-95-7	90-100[0.9]	0-05-00-00278 1.8893[25]	ace 4; EtOH 4 1.6631[25]
980	Benzene, (azidomethyl)-	$C_7H_7N_3$ 133.15	622-79-7	108[23]; 74[11]	4-05-00-00877 1.0655[25]	H$_2$O 1; EtOH 5; eth 5 1.5341[25]
981	Benzene, 1-azido-2-methyl-	$C_7H_7N_3$ 133.15	31656-92-5 <-10	90.5[30]; 76[12]	4-05-00-00876 1.0648[25]	eth 4
982	Benzene, 1-azido-4-methyl-	$C_7H_7N_3$ 133.15	2101-86-2	180 dec; 80[10]	4-05-00-00876 1.0527[23]	eth 4; EtOH 4
983	Benzene, 1,2-bis(bromomethyl)-	$C_8H_8Br_2$ 263.96	91-13-4 95	128-30[4.5]	4-05-00-00929 1.988[25]	H$_2$O 1; EtOH 3; eth 3; ctc 3
984	Benzene, 1,3-bis(bromomethyl)-	$C_8H_8Br_2$ 263.96	626-15-3 77	135-40[20]	4-05-00-00946 1.959[25]	H$_2$O 1; EtOH 3; eth 3; chl 3
985	Benzene, 1,4-bis(bromomethyl)-	$C_8H_8Br_2$ 263.96	623-24-5 144.5	245	4-05-00-00970 2.012[25]	H$_2$O 1; EtOH 4; eth 2; bz 3
986	Benzene, 1,2-bis(chloromethyl)-	$C_8H_8Cl_2$ 175.06	612-12-4 55	239.5	4-05-00-00927 1.393[25]	H$_2$O 1; EtOH 4; eth 4; ctc 3
987	Benzene, 1,3-bis(chloromethyl)-	$C_8H_8Cl_2$ 175.06	626-16-4 34.2	251.5	4-05-00-00944 1.302[20]	H$_2$O 1; EtOH 4; eth 4; chl 2
988	Benzene, 1,4-bis(chloromethyl)-	$C_8H_8Cl_2$ 175.06	623-25-6 100	245 dec; 135[16]	4-05-00-00967 1.417[25]	H$_2$O 1; EtOH 4; eth 4; ace 4
989	Benzene, 1,4-bis(dichloromethyl)-	$C_8H_6Cl_4$ 243.95	7398-82-5 9.5		3-05-00-00857 1.606[25]	H$_2$O 1; EtOH 3; eth 3; bz 3
990	Benzene, 1,4-bis(1,1-dimethylethyl)- Benzene, 1,4-di-tert-butyl	$C_{14}H_{22}$ 190.33	1012-72-2 79.5	238; 109[15]	4-05-00-01163 0.9850[20]	H$_2$O 1; EtOH 3; eth 3

No.	Name / Synonym	Mol. Form. / Mol. Wt.	CAS RN / mp/°C	Merck No. / bp/°C	Beil. Ref. / den/g cm^{-3}	Solubility / n_D
991	Benzene, 1,4-bis(2-hydroxybenzylideneamino)-2-nitro Ethionine	$C_{20}H_{15}N_3O_4$ 361.36	13073-35-3 195	3693		
992	Benzene, 1,2-bis(1-methylethyl)-	$C_{12}H_{18}$ 162.27	577-55-9 -57	204	4-05-00-01125 0.8701[20]	H_2O 1; EtOH 5; eth 5; ace 5 1.4960[20]
993	Benzene, 1,3-bis(1-methylethyl)-	$C_{12}H_{18}$ 162.27	99-62-7 -63.1	203.2	4-05-00-01125 0.8559[20]	H_2O 1; EtOH 5; eth 5; ace 5 1.4883[20]
994	Benzene, 1,4-bis(1-methylethyl)-	$C_{12}H_{18}$ 162.27	100-18-5 -17	210.3	4-05-00-01126 0.8568[20]	H_2O 1; EtOH 5; eth 5; ace 5 1.4898[20]
995	Benzene, 1,4-bis(phenylmethyl)-	$C_{20}H_{18}$ 258.36	793-23-7 87.5	225[18]	4-05-00-02512	EtOH 3; eth 2; bz 3; chl 3
996	Benzene, 1,4-bis(trichloromethyl)-	$C_8H_4Cl_6$ 312.84	68-36-0 109	1313	4-05-00-00968	chl 3
997	Benzene, 1,3-bis(trifluoromethyl)-	$C_8H_4F_6$ 214.11	402-31-3	116	4-05-00-00943 1.3790[25]	1.3916[25]
998	Benzene, bromo- Bromobenzene	C_6H_5Br 157.01	108-86-1 -30.6	1394 156.0	4-05-00-00670 1.4950[20]	H_2O 1; EtOH 4; eth 4; bz 4 1.5597[20]
999	Benzene, 1-bromo-2-(bromomethyl)-	$C_7H_6Br_2$ 249.93	3433-80-5 31	129[19]	4-05-00-00836	eth 4; EtOH 4; HOAc 4
1000	Benzene, 1-bromo-4-(bromomethyl)- p-Bromobenzyl bromide	$C_7H_6Br_2$ 249.93	589-15-1 63	1397	4-05-00-00836	H_2O 2; EtOH 3; eth 4; bz 3
1001	Benzene, (3-bromo-3-butenyl)- 1-Butene, 2-bromo-4-phenyl	$C_{10}H_{11}Br$ 211.10	62692-40-4	118[21]; 90[5]	3-05-00-01209 1.2907[20]	ace 4 1.5450[20]
1002	Benzene, (4-bromo-2-butenyl)- 2-Butene, 1-bromo-4-phenyl-	$C_{10}H_{11}Br$ 211.10	40734-75-6	126-30[10.5]	4-05-00-01377 1.2660[20]	eth 4 1.5678[20]
1003	Benzene, (4-bromobutoxy)-	$C_{10}H_{13}BrO$ 229.12	1200-03-9 41	154[18]	4-06-00-00558	EtOH 2; ctc 2
1004	Benzene, 1-bromo-2-chloro-	C_6H_4BrCl 191.45	694-80-4 -12.3	204	4-05-00-00680 1.6387[25]	H_2O 1; bz 4; ctc 2 1.5809[20]
1005	Benzene, 1-bromo-3-chloro-	C_6H_4BrCl 191.45	108-37-2 -21.5	196	4-05-00-00680 1.6302[20]	H_2O 1; EtOH 4; eth 4 1.5771[20]
1006	Benzene, 1-bromo-4-chloro-	C_6H_4BrCl 191.45	106-39-8 68	196	4-05-00-00681 1.576[71]	H_2O 1; EtOH 2; eth 3; bz 3 1.5531[70]
1007	Benzene, 1-bromo-3-(chloromethyl)-	C_7H_6BrCl 205.48	932-77-4 32.5	119[18]	4-05-00-00832	EtOH 4
1008	Benzene, 1-bromo-4-(chloromethyl)- p-Bromobenzyl chloride	C_7H_6BrCl 205.48	589-17-3 50	1398 236	4-05-00-00832	H_2O 1; EtOH 4; eth 4; peth 3
1009	Benzene, 1-bromo-4-cyclohexyl-	$C_{12}H_{15}Br$ 239.16	25109-28-8	160[25]	3-05-00-01258 1.283[25]	bz 4; chl 4 1.5584[20]
1010	Benzene, 1-bromo-2,3-dichloro-	$C_6H_3BrCl_2$ 225.90	56961-77-4 60	245	0-05-00-00209	bz 4; eth 4; chl 4
1011	Benzene, 1-bromo-2,4-dichloro-	$C_6H_3BrCl_2$ 225.90	1193-72-2 26	235	3-05-00-00564	bz 4; eth 4; chl 4
1012	Benzene, 1-bromo-3,5-dichloro-	$C_6H_3BrCl_2$ 225.90	19752-55-7 83	232	2-05-00-00162	H_2O 1; EtOH 3; eth 3; bz 4
1013	Benzene, 2-bromo-1,3-dichloro-	$C_6H_3BrCl_2$ 225.90	19393-92-1 65	242	0-05-00-00210	bz 4; eth 4; chl 4
1014	Benzene, 2-bromo-1,4-dichloro-	$C_6H_3BrCl_2$ 225.90	1435-50-3 35	226; 119[20]	3-05-00-00564	bz 4; eth 4; EtOH 4; lig 4
1015	Benzene, 4-bromo-1,2-dichloro-	$C_6H_3BrCl_2$ 225.90	18282-59-2 25	237	3-05-00-00564	H_2O 1; EtOH 2; eth 4; bz 4
1016	Benzene, 4-bromo-1,2-dimethoxy-	$C_8H_9BrO_2$ 217.06	2859-78-1	254.5	4-06-00-05621 1.702[25]	1.5743[20]
1017	Benzene, 1-bromo-2,3-dimethyl-	C_8H_9Br 185.06	576-23-8	214	4-05-00-00928 1.365[25]	ace 4; bz 4; eth 4 1.5587[20]
1018	Benzene, 1-bromo-2,4-dimethyl-	C_8H_9Br 185.06	583-70-0 -17	205	4-05-00-00945 1.3419[20]	H_2O 1; EtOH 4; eth 4; ace 4 1.5501[20]
1019	Benzene, 1-bromo-3,5-dimethyl-	C_8H_9Br 185.06	556-96-7	204	4-05-00-00945 1.362[20]	eth 4; ace 3; bz 3 1.5462[22]
1020	Benzene, 2-bromo-1,3-dimethyl-	C_8H_9Br 185.06	576-22-7	203.5; 100[20]	4-05-00-00945	eth 4; ace 3; bz 3 1.5552[20]
1021	Benzene, 2-bromo-1,4-dimethyl-	C_8H_9Br 185.06	553-94-6 9	199; 88-9[13]	4-05-00-00969 1.3582[18]	H_2O 1; EtOH 4; bz 3 1.5514[18]
1022	Benzene, 4-bromo-1,2-dimethyl-	C_8H_9Br 185.06	583-71-1 -0.2	214.5	4-05-00-00928 1.3708[20]	H_2O 1; EtOH 4; eth 4 1.5530[20]
1023	Benzene, 1-bromo-4-(1,1-dimethylethyl)-	$C_{10}H_{13}Br$ 213.12	3972-65-4 19	231.5	4-05-00-01050 1.2286[20]	H_2O 1; eth 3; bz 3; chl 3 1.5436[20]
1024	Benzene, 1-bromo-2,3-dinitro- 1-Bromo-2,3-dinitrobenzene	$C_6H_3BrN_2O_4$ 247.00	19613-76-4 101.5	320	4-05-00-00749	EtOH 4
1025	Benzene, (1-bromoethenyl)-	C_8H_7Br 183.05	98-81-7 -44	86[14]; 71[3]	4-05-00-01349 1.4025[23]	1.5881[20]
1026	Benzene, 1-bromo-2-ethenyl-	C_8H_7Br 183.05	2039-88-5 -52.8	209.2; 98[20]	4-05-00-01349 1.4160[20]	1.5927[20]
1027	Benzene, 1-bromo-3-ethenyl-	C_8H_7Br 183.05	2039-86-3	90-4[20]	3-05-00-01176 1.4059[20]	1.5933[20]
1028	Benzene, 1-bromo-4-ethenyl-	C_8H_7Br 183.05	2039-82-9 7.7	212; 103[20]	4-05-00-01349 1.3984[20]	H_2O 1; chl 4; HOAc 3 1.5947[20]
1029	Benzene, (2-bromoethenyl)-, (E)-	C_8H_7Br 183.05	588-72-7 7	219 dec; 108[20]	4-05-00-01349 1.4269[16]	H_2O 1; EtOH 5; eth 5; chl 3 1.6093[20]

No.	Name / Synonym	Mol. Form. / Mol. Wt.	CAS RN / mp/°C	Merck No. / bp/°C	Beil. Ref. / den/g cm^{-3}	Solubility / n_D
1030	Benzene, 1,1'-(bromoethenylidene)bis-	$C_{14}H_{11}Br$ 259.15	13249-58-6 41.5	175[11.5]	4-05-00-02177	EtOH 2; eth 3; ace 3; CS_2 3
1031	Benzene, (2-bromoethenyl)-, (Z)-	C_8H_7Br 183.05	588-73-8 -7	55-6[2]	4-05-00-01349 1.4322[10]	1.5990[22]
1032	Benzene, 1-bromo-4-ethoxy-	C_8H_9BrO 201.06	588-96-5 4	231	4-06-00-01045 1.4071[25]	H_2O 1; EtOH 4; eth 4; chl 3 1.5517[20]
1033	Benzene, (2-bromoethoxy)-	C_8H_9BrO 201.06	589-10-6 39	240 dec; 128[20]	4-06-00-00556 1.3555[20]	H_2O 1; EtOH 4; eth 4
1034	Benzene, (1-bromoethyl)-, (±)-	C_8H_9Br 185.06	38661-81-3	202-3. 85[13]	4-05-00-00906 1.3605[20]	EtOH 3; eth 3; bz 3; ctc 2 1.5612[20]
1035	Benzene, 1-bromo-2-ethyl-	C_8H_9Br 185.06	1973-22-4 -67.9	199.3	4-05-00-00905 1.3548[20]	ace 4; bz 4; eth 4; EtOH 4 1.5472[20]
1036	Benzene, 1-bromo-3-ethyl-	C_8H_9Br 185.06	2725-82-8	202	4-05-00-00906 1.3493[20]	1.5465[20]
1037	Benzene, 1-bromo-4-ethyl-	C_8H_9Br 185.06	1585-07-5 -43.5	204	4-05-00-00906 1.3423[20]	ace 4; bz 4; eth 4; EtOH 4 1.5445[20]
1038	Benzene, (2-bromoethyl)-	C_8H_9Br 185.06	103-63-9	219	4-05-00-00907 1.3587[20]	H_2O 1; eth 3; bz 3; ctc 2 1.5372[20]
1039	Benzene, (1-bromoethyl)-, (R)-	C_8H_9Br 185.06	1459-14-9	203	4-05-00-00906 1.3108[23]	H_2O 1; EtOH 3; eth 3 1.5612[20]
1040	Benzene, 1-bromo-2-ethynyl- o-Bromophenylacetylene	C_8H_5Br 181.03	766-46-1	92-3[20]	3-05-00-01351 1.4434[25]	1.5962[25]
1041	Benzene, 1-bromo-3-ethynyl- m-Bromophenylacetylene	C_8H_5Br 181.03	766-81-4	85-6[16]	3-05-00-01351 1.4466[25]	1.5896[25]
1042	Benzene, 1-bromo-4-ethynyl-	C_8H_5Br 181.03	766-96-1 64.5	89[16]	4-05-00-01529	chl 3
1043	Benzene, 1-bromo-3-fluoro-	C_6H_4BrF 175.00	1073-06-9	150	4-05-00-00678 1.7081[20]	ctc 3 1.5257[20]
1044	Benzene, 1-bromo-4-fluoro-	C_6H_4BrF 175.00	460-00-4 -17.4	151.5	4-05-00-00678 1.593[15]	H_2O 1; EtOH 3; eth 3; chl 3 1.5310[15]
1045	Benzene, 1-bromo-4-(hexyloxy)-	$C_{12}H_{17}BrO$ 257.17	30752-19-3	156-7[13]	4-06-00-01047 1.2306[20]	1.5262[20]
1046	Benzene, 1-bromo-2-iodo-	C_6H_4BrI 282.91	583-55-1 9.5	257; 120[15]	4-05-00-00698 2.2570[25]	H_2O 1; EtOH 2; ace 3; HOAc 2 1.6618[25]
1047	Benzene, 1-bromo-3-iodo- m-Bromoiodobenzene	C_6H_4BrI 282.91	591-18-4 -9.3	252; 120[18]	4-05-00-00698	H_2O 1; EtOH 2; HOAc 2
1048	Benzene, 1-bromo-4-iodo-	C_6H_4BrI 282.91	589-87-7 92	252	4-05-00-00698	H_2O 1; EtOH 2; eth 3; chl 2
1049	Benzene, 1-bromo-4-isocyanato- p-Bromophenyl isocyanate	C_7H_4BrNO 198.02	2493-02-9	1418 226	1-12-00-00321	eth 4
1050	Benzene, 1-bromo 2 methoxy-	C_7H_7BrO 187.04	578-57-4 2.5	216	4-06-00-01037 1.5018[20]	H_2O 1; EtOH 4; eth 4 1.5727[20]
1051	Benzene, 1-bromo-3-methoxy-	C_7H_7BrO 187.04	2398-37-0	211; 105[16]	4-06-00-01043	H_2O 1; EtOH 3; eth 3; bz 3 1.5635[20]
1052	Benzene, 1-bromo-4-methoxy-	C_7H_7BrO 187.04	104-92-7 13.5	215	4-06-00-01044 1.4564[20]	H_2O 2; EtOH 4; eth 4; ctc 3 1.5642[20]
1053	Benzene, (bromomethyl)- Benzyl bromide	C_7H_7Br 171.04	100-39-0 -3	1142 201	4-05-00-00829 1.4380[25]	H_2O 1; EtOH 5; eth 5; ctc 3 1.5752[20]
1054	Benzene, 1-bromo-2-methyl- o-Bromotoluene	C_7H_7Br 171.04	95-46-5 -27.8	181.7	4-05-00-00825 1.4232[20]	H_2O 1; EtOH 4; eth 4; bz 4 1.5565[20]
1055	Benzene, 1-bromo-3-methyl- m-Bromotoluene	C_7H_7Br 171.04	591-17-3 -39.8	183.7	4-05-00-00827 1.4099[20]	H_2O 1; EtOH 3; eth 5; ace 3 1.5510[20]
1056	Benzene, 1-bromo-4-methyl- p-Bromotoluene	C_7H_7Br 171.04	106-38-7 28.5	1429 184.3	4-05-00-00827 1.3959[35]	H_2O 1; EtOH 3; eth 3; ace 3 1.5477[20]
1057	Benzene, 1-(bromomethyl)-3,5-dimethyl- Mesitylene, α-bromo-	$C_9H_{11}Br$ 199.09	27129-86-8 40	230; 118[22]	4-05-00-01027	bz 4; eth 4; EtOH 4; chl 4
1058	Benzene, 1,1'-(bromomethylene)bis- Bromodiphenylmethane	$C_{13}H_{11}Br$ 247.13	776-74-9 45	184[20]; 152[2]	4-05-00-01850	EtOH 3; bz 4; chl 3
1059	Benzene, 1-bromo-4-(1-methylethyl)-	$C_9H_{11}Br$ 199.09	586-61-8 -22.5	218.7	4-05-00-00994 1.3145[20]	H_2O 1; eth 3; bz 3; ctc 2 1.5569[20]
1060	Benzene, (2-bromo-1-methylethyl)-	$C_9H_{11}Br$ 199.09	1459-00-3 -59.3	188.5	4-05-00-00995 1.3020[20]	
1061	Benzene, 1,1',1''-(bromomethylidyne)tris-	$C_{19}H_{15}Br$ 323.23	596-43-0 153	230[15]	4-05-00-02500 1.5500[20]	
1062	Benzene, 1-(bromomethyl)-2-methyl-	C_8H_9Br 185.06	89-92-9 21	217; 108[16]	4-05-00-00928 1.3811[23]	H_2O 1; EtOH 3; eth 3; ace 3 1.5730[20]
1063	Benzene, 1-(bromomethyl)-3-methyl-	C_8H_9Br 185.06	620-13-3	212.5	4-05-00-00945 1.3711[23]	H_2O 1; EtOH 4; eth 4; os 3 1.5660[20]
1064	Benzene, 1-(bromomethyl)-4-methyl-	C_8H_9Br 185.06	104-81-4 35	220	4-05-00-00969 1.324[25]	H_2O 1; EtOH 3; eth 4; chl 4
1065	Benzene, 2-bromo-1-methyl-4-(1-methylethyl)-	$C_{10}H_{13}Br$ 213.12	2437-76-5	234.3	4-05-00-01063 1.2689[18]	eth 4; EtOH 4; chl 4 1.5360[20]
1066	Benzene, 1-bromo-2-methyl-4-nitro-	$C_7H_6BrNO_2$ 216.03	7149-70-4 78	141[17]	4-05-00-00860	eth 4
1067	Benzene, 1-(bromomethyl)-3-nitro-	$C_7H_6BrNO_2$ 216.03	3958-57-4 59.3	162[13]	4-05-00-00860	H_2O 1; EtOH 3

No.	Name / Synonym	Mol. Form. / Mol. Wt.	CAS RN / mp/°C	Merck No. / bp/°C	Beil. Ref. / den/g cm⁻³	Solubility / n_D
1068	Benzene, 1-bromo-3-methyl-2-nitro- / 3-Bromo-2-nitrotoluene	$C_7H_6BrNO_2$ / 216.03	52414-97-8 / 28.5	153[30]; 129[10]	4-05-00-00860	eth 4; EtOH 4
1069	Benzene, 1-bromo-3-methyl-5-nitro- / Toluene, 3-bromo-5-nitro-	$C_7H_6BrNO_2$ / 216.03	52488-28-5 / 83	269.5	3-05-00-00752	eth 4; EtOH 4
1070	Benzene, 1-bromo-2-methyl-3-nitro- / Toluene, 2-bromo-6-nitro-	$C_7H_6BrNO_2$ / 216.03	55289-35-5 / 42	143[22]	3-05-00-00751	EtOH 4
1071	Benzene, 2-bromo-1-methyl-3-nitro-	$C_7H_6BrNO_2$ / 216.03	41085-43-2 / 41.5	161[30]	4-05-00-00860	EtOH 4
1072	Benzene, 2-bromo-1-methyl-4-nitro-	$C_7H_6BrNO_2$ / 216.03	7745-93-9 / 78	175[38]; 150[20]	4-05-00-00861	eth 4
1073	Benzene, 2-bromo-4-methyl-1-nitro- / 2-Bromo-4-methyl-1-nitrobenzene	$C_7H_6BrNO_2$ / 216.03	40385-54-4 / 37.8	269; 154[20]	4-05-00-00861	eth 4; EtOH 4
1074	Benzene, 4-bromo-1-methyl-2-nitro- / 4-Bromo-2-nitrotoluene	$C_7H_6BrNO_2$ / 216.03	60956-26-5 / 47	256.5	4-05-00-00860	eth 4; EtOH 4
1075	Benzene, 1-bromo-2-nitro-	$C_6H_4BrNO_2$ / 202.01	577-19-5 / 43	258	4-05-00-00728 / 1.6245[80]	H_2O 1; EtOH 4; eth 3; ace 3
1076	Benzene, 1-bromo-3-nitro-	$C_6H_4BrNO_2$ / 202.01	585-79-5 / 56	265	4-05-00-00729 / 1.7036[20]	H_2O 2; EtOH 3; eth 3; bz 3 / 1.5979[20]
1077	Benzene, 1-bromo-4-nitro- / p-Nitrobromobenzene	$C_6H_4BrNO_2$ / 202.01	586-78-7 / 127	256	4-05-00-00729 / 1.948[25]	H_2O 1; EtOH 3; eth 3; bz 3
1078	Benzene, bromopentafluoro- / Bromopentafluorobenzene	C_6BrF_5 / 246.96	344-04-7 / -31	137	4-05-00-00680 / 1.981[25]	1.4490[20]
1079	Benzene, 1-bromo-4-phenoxy- / 4-Bromophenyl phenyl ether	$C_{12}H_9BrO$ / 249.11	101-55-3 / 18.72	124-8[3-4]	4-06-00-01047 / 1.6088[20]	H_2O 1; eth 3; ctc 3 / 1.6084[20]
1080	Benzene, 1-bromo-2-(2-phenylethenyl)-, (E)- / trans-2-Bromostilbene	$C_{14}H_{11}Br$ / 259.15	54737-45-0 / 34	143[0.2]	4-05-00-02164	H_2O I; EtOH 3; peth 3 / 1.6822[25]
1081	Benzene, 1-bromo-2-(2-phenylethenyl)-, (Z)- / Stilbene, 2-bromo (cis)	$C_{14}H_{11}Br$ / 259.15	4877-77-4	121[0.5]	4-05-00-02164	H_2O 1; EtOH 3; peth 3 / 1.6404[25]
1082	Benzene, 1-bromo-4-(1-propenyl)-	C_9H_9Br / 197.07	4489-23-0 / 35	238.5	2-05-00-00372 / 1.332[36]	1.590[36]
1083	Benzene, 1-bromo-4-(2-propenyl)-	C_9H_9Br / 197.07	2294-43-1	224; 96[12]	4-05-00-01363 / 1.324[15]	1.559[15]
1084	Benzene, (3-bromo-1-propenyl)-	C_9H_9Br / 197.07	4392-24-9 / 34	130[10]	3-05-00-01188 / 1.3428[30]	EtOH 4 / 1.613[20]
1085	Benzene, (3-bromopropoxy)-	$C_9H_{11}BrO$ / 215.09	588-63-6 / 10.7	127[18]	4-06-00-00557 / 1.364[16]	eth 4
1086	Benzene, (1-bromopropyl)-, (±)- / Benzene, (1-bromopropyl) (DL)	$C_9H_{11}Br$ / 199.09	63790-14-7	105[17]	2-05-00-00305 / 1.3098[19]	1.5517[19]
1087	Benzene, (2-bromopropyl)-	$C_9H_{11}Br$ / 199.09	2114-39-8	107-9[16]	4-05-00-00981 / 1.2908[16]	chl 4 / 1.5450[20]
1088	Benzene, (3-bromopropyl)-	$C_9H_{11}Br$ / 199.09	637-59-2	219.5; 117[25]	4-05-00-00982 / 1.3106[25]	H_2O 1; eth 4 / 1.5440[25]
1089	Benzene, 1-bromo-2-(trifluoromethyl)-	$C_7H_4BrF_3$ / 225.01	392-83-6	167.5	4-05-00-00831 / 1.652[25]	1.4817[20]
1090	Benzene, 1-bromo-3-(trifluoromethyl)-	$C_7H_4BrF_3$ / 225.01	401-78-5 / 1	151.5	4-05-00-00831 / 1.613[25]	1.4716[20]
1091	Benzene, 1-bromo-4-(trifluoromethyl)-	$C_7H_4BrF_3$ / 225.01	402-43-7	160	4-05-00-00831 / 1.607[25]	1.4705[25]
1092	Benzene, 1-bromo-2,3,5-trimethyl- / 1-Bromo-2,3,5-trimethylbenzene	$C_9H_{11}Br$ / 199.09	31053-99-3 / <-15	238	4-05-00-01014	bz 4 / 1.5516[20]
1093	Benzene, 1-bromo-2,4,5-trimethyl-	$C_9H_{11}Br$ / 199.09	5469-19-2 / 73	234	4-05-00-01014	H_2O 1; EtOH 4
1094	Benzene, 2-bromo-1,3,5-trimethyl-	$C_9H_{11}Br$ / 199.09	576-83-0 / -1	225	4-05-00-01027 / 1.3191[10]	H_2O 1; eth 4; bz 3; ctc 2 / 1.5510[20]
1095	Benzene, 1,1'-(1,3-butadiene-1,4-diyl)bis-, (E,E)-	$C_{18}H_{14}$ / 206.29	538-81-8 / 154.3	352	4-05-00-02319	bz 4; eth 4; EtOH 4; peth 4
1096	Benzene, 1,1'-(1,3-butadiene-1,4-diyl)bis-, (Z,Z)-	$C_{18}H_{14}$ / 206.29	5807-76-1 / 70.5		4-05-00-02318 / 0.9697[100]	H_2O 1; EtOH 2; eth 4; bz 4 / 1.6183[100]
1097	Benzene, 1,3-butadienyl-, (E)-	$C_{10}H_{10}$ / 130.19	16939-57-4 / 4.5	76[11]	4-05-00-01536 / 0.9286[20]	H_2O I; EtOH 3; eth 3; ace 3 / 1.6089[25]
1098	Benzene, 1,3-butadienyl-, (Z)-	$C_{10}H_{10}$ / 130.19	31915-94-3 / -57	86[11]	/ 0.9334[20]	1.6095[20]
1099	Benzene, 1,1'-(1,4-butanediyl)bis- / 1,4-Diphenylbutane	$C_{16}H_{18}$ / 210.32	1083-56-3 / 52.5	317	4-05-00-01937 / 0.9880[20]	H_2O 1; EtOH 3; eth 3; chl 3
1100	Benzenebutanoic acid	$C_{10}H_{12}O_2$ / 164.20	1821-12-1 / 52	290	4-09-00-01811	H_2O 3; EtOH 3; eth 3
1101	Benzenebutanoic acid, 4-[bis(2-chloroethyl)amino]- / Chlorambucil	$C_{14}H_{19}Cl_2NO_2$ / 304.22	305-03-3 / 65	2064	4-14-00-01715	
1102	Benzenebutanoic acid, α,2-diamino-γ-oxo- / Kynurenine	$C_{10}H_{12}N_2O_3$ / 208.22	343-65-7 / 191 dec	5207	4-14-00-02562	
1103	Benzenebutanoic acid, 4-hydroxy-γ-(4-hydroxyphenyl)-γ-methyl- / Diphenolic acid	$C_{17}H_{18}O_4$ / 286.33	126-00-1 / 171.5	3312	4-10-00-01890	H_2O 4; ace 4; EtOH 4

No.	Name Synonym	Mol. Form. Mol. Wt.	CAS RN mp/°C	Merck No. bp/°C	Beil. Ref. den/g cm^{-3}	Solubility n_D
1104	Benzenebutanoic acid, γ-methyl-	$C_{11}H_{14}O_2$ 178.23	16433-43-5 13	210[85]; 165[12]	4-09-00-01866 1.0554[15]	eth 4; EtOH 4 1.5167[20]
1105	Benzenebutanol, 4-methoxy-	$C_{11}H_{16}O_2$ 180.25	52244-70-9 3.5	160[8]; 125[1.5]	4-06-00-06005	1.5267[20]
1106	Benzenebutanol, α-methyl-	$C_{11}H_{16}O$ 164.25	2344-71-0	134-5[16]	3-06-00-01954 0.9643[25]	1.518[19]
1107	Benzenebutanol, β-methyl-, (±)- 1-Butanol, 2-methyl-4-phenyl (DL)	$C_{11}H_{16}O$ 164.25	116783-11-0	135[11]	4-06-00-03380 0.9719[20]	ace 4; bz 4; eth 4; EtOH 4 1.5173[16]
1108	Benzenebutanol, β-phenyl-, (R)- 1-Butanol, 2,4-diphenyl-, (-)-	$C_{16}H_{18}O$ 226.32	17297-04-0 51.5	178[1]; 145[0.1]	chl 3 1.5686[25]	
1109	Benzene, 1-butenyl-, (E)-	$C_{10}H_{12}$ 132.21	1005-64-7 -43.1	198.7	4-05-00-01374 0.9019[20]	H_2O 1; EtOH 3; eth 3; bz 3 1.5420[20]
1110	Benzene, 1-butenyl-, (Z)-	$C_{10}H_{12}$ 132.21	1560-09-4	189	3-05-00-01205 0.9065[25]	H_2O 1; EtOH 3; eth 3; bz 3 1.5390[20]
1111	Benzene, 2-butenyl-	$C_{10}H_{12}$ 132.21	1560-06-1	176	3-05-00-01207 0.8831[20]	1.5101[20]
1112	Benzene, 3-butenyl-	$C_{10}H_{12}$ 132.21	768-56-9 -70	177	4-05-00-01378 0.8831[20]	H_2O 1; eth 3; bz 3 1.5059[20]
1113	Benzene, 1,1'-(1-butenylidene)bis-	$C_{16}H_{16}$ 208.30	1726-14-3 -1.2	296	4-05-00-02209 0.9934[20]	1.5604[20]
1114	Benzene, butoxy-	$C_{10}H_{14}O$ 150.22	1126-79-0 -19.4	210	4-06-00-00558 0.9351[20]	eth 3; ace 3 1.4969[20]
1115	Benzene, (butoxymethyl)-	$C_{11}H_{16}O$ 164.25	588-67-0	223	4-06-00-02231 0.9227[20]	ace 4; eth 4; EtOH 4 1.4833[20]
1116	Benzene, 1-butoxy-3-methyl-	$C_{11}H_{16}O$ 164.25	23079-65-4	229.2	4-06-00-02040 0.9406[70]	eth 4 1.4970[20]
1117	Benzene, 1-butoxy-4-methyl-	$C_{11}H_{16}O$ 164.25	10519-06-9	229.5	3-06-00-01354 0.9205[25]	eth 3 1.4970[20]
1118	Benzene, butyl- Butylbenzene	$C_{10}H_{14}$ 134.22	104-51-8 -87.9	1549 183.3	4-05-00-01033 0.8601[20]	H_2O 1; EtOH 5; eth 5; ace 5 1.4898[20]
1119	Benzene, (1-butylhexadecyl)- Eicosane, 5-phenyl-	$C_{26}H_{46}$ 358.65	2400-04-6 30.2	197[1]	4-05-00-01223 0.8549[20]	1.4796[20]
1120	Benzene, 1,1'-butylidenebis- 1,1-Diphenylbutane	$C_{16}H_{18}$ 210.32	719-79-9 27	287	4-05-00-01944 0.9928[20]	H_2O 1; EtOH 3; eth 3; bz 3 1.5664[20]
1121	Benzene, 1-butyl-2-methyl-	$C_{11}H_{16}$ 148.25	1595-11-5	208	3-05-00-00999 0.8710[20]	ace 4; bz 4; eth 4 1.4960[20]
1122	Benzene, 1-butyl-3-methyl-	$C_{11}H_{16}$ 148.25	1595-04-6	205	3-05-00-00999 0.8590[20]	ace 4; bz 4; eth 4 1.4910[20]
1123	Benzene, 1-butyl-4-methyl-	$C_{11}H_{16}$ 148.25	1595-05-7 -85	207	4-05-00-01094 0.8586[20]	ace 4; bz 4; eth 4 1.4916[20]
1124	Benzene, 3-butynyl-	$C_{10}H_{10}$ 130.19	16520-62-0	190	4-05-00-01535 0.9258[20]	1.5208[20]
1125	Benzenecarbodithioic acid, 2-hydroxy- Dithiosalicylic acid	$C_7H_6OS_2$ 170.26	527-89-9 49	3381	2-10-00-00078	bz 4; eth 4; EtOH 4
1126	Benzenecarboperoxoic acid Perbenzoic acid	$C_7H_6O_3$ 138.12	93-59-4 42	7109 100[14]	4-09-00-00715	ace 4; bz 4; eth 4; EtOH 4
1127	Benzenecarboperoxoic acid, 3-chloro- 3-Chloroperoxybenzoic acid	$C_7H_5ClO_3$ 172.57	937-14-4 92-4 dec		4-09-00-00972	
1128	Benzenecarboperoxoic acid, 1,1-dimethylethyl ester Benzoyl tert-butyl peroxide	$C_{11}H_{14}O_3$ 194.23	614-45-9	75-6[0.2]	4-09-00-00715 1.021[25]	1.4990[20]
1129	Benzenecarbothioamide, N-phenyl-	$C_{13}H_{11}NS$ 213.30	636-04-4 102	dec	4-12-00-00423	H_2O 1; EtOH 4; eth 3; bz 3
1130	Benzenecarbothioic acid	C_7H_6OS 138.19	98-91-9 24	85-7[10]	4-09-00-01364 1.28[20]	ace 4; bz 4; eth 4; EtOH 4 1.6040[20]
1131	Benzene, chloro- Chlorobenzene	C_6H_5Cl 112.56	108-90-7 -45.2	2121 131.7	4-05-00-00640 1.1058[20]	H_2O 1; EtOH 5; eth 5; bz 4 1.5241[20]
1132	Benzene, 1-chloro-2-(chloromethyl)-	$C_7H_6Cl_2$ 161.03	611-19-8 -17	217	4-05-00-00816 1.2699[0]	H_2O 1; EtOH 2; eth 4; bz 4 1.5530[20]
1133	Benzene, 1-chloro-3-(chloromethyl)-	$C_7H_6Cl_2$ 161.03	620-20-2	216; 110[25]	4-05-00-00816 1.2695[15]	EtOH 4 1.5554[20]
1134	Benzene, 1-chloro-4-(chloromethyl)- 4-Chlorobenzyl chloride	$C_7H_6Cl_2$ 161.03	104-83-6 31	223	4-05-00-00816	ctc 2
1135	Benzene, 1-chloro-4-(chloromethyl)-2-nitro-	$C_7H_5Cl_2NO_2$ 206.03	57403-35-7	122[10]	4-05-00-00858 1.32[25]	1.5855[20]
1136	Benzene, 1-chloro-4-[(chloromethyl)thio]-	$C_7H_6Cl_2S$ 193.10	7205-90-5	128-9[12]	4-06-00-01593 1.346[25]	1.6055[20]
1137	Benzene, 1-chloro-2,3-dimethyl-	C_8H_9Cl 140.61	608-23-1	188	4-05-00-00926 1.053[20]	bz 4; chl 4 1.526[20]
1138	Benzene, 1-chloro-2,4-dimethyl-	C_8H_9Cl 140.61	95-66-9 -32.5	184	4-05-00-00943 1.0579[20]	H_2O 1; ace 3; bz 4 1.5230[25]
1139	Benzene, 2-chloro-1,3-dimethyl-	C_8H_9Cl 140.61	6781-98-2 -35	186	4-05-00-00943 1.053[20]	H_2O 1; ace 3; bz 4; chl 3 1.526[20]
1140	Benzene, 2-chloro-1,4-dimethyl-	C_8H_9Cl 140.61	95-72-7 1.6	187	4-05-00-00965 1.0589[15]	H_2O 1; ace 3; bz 4; ctc 3
1141	Benzene, 4-chloro-1,2-dimethyl-	C_8H_9Cl 140.61	615-60-1 -6	194	3-05-00-00816 1.0682[15]	H_2O 1; ace 3; bz 4; ctc 3
1142	Benzene, 1-chloro-4-(1,1-dimethylethyl)-	$C_{10}H_{13}Cl$ 168.67	3972-56-3	213	4-05-00-01048 1.0075[18]	1.5123[20]

No.	Name / Synonym	Mol. Form. / Mol. Wt.	CAS RN / mp/°C	Merck No. / bp/°C	Beil. Ref. / den/g cm^{-3}	Solubility / n_D
1143	Benzene, (2-chloro-1,1-dimethylethyl)- / Neophyl chloride	$C_{10}H_{13}Cl$ / 168.67	515-40-2 /	6375 / 223; 105[18]	4-05-00-01048 / 1.047[20]	ace 4; bz 4; eth 4; EtOH 4 / 1.5247[20]
1144	Benzene, 1-chloro-2,4-dinitro- / 1-Chloro-2,4-dinitrobenzene	$C_6H_3ClN_2O_4$ / 202.55	97-00-7 / 53	2135 / 315	4-05-00-00744 / 1.4982[75]	H_2O 1; EtOH 2; eth 3; bz 3 / 1.5857[60]
1145	Benzene, 2-chloro-1,3-dinitro- / 2-Chloro-1,3-dinitrobenzene	$C_6H_3ClN_2O_4$ / 202.55	606-21-3 / 88	2136 / 315	4-05-00-00744 / 1.6867[16]	H_2O 1; EtOH 3; eth 3; chl 2
1146	Benzene, 1,1'-(1-chloro-1,2-ethenediyl)bis-, (E)-	$C_{14}H_{11}Cl$ / 214.69	948-98-1 / 53.5	/ 322; 128[2]	4-05-00-02161 /	H_2O 1; EtOH 3; eth 3; bz 3
1147	Benzene, 1,1'-(1-chloro-1,2-ethenediyl)bis-, (Z)-	$C_{14}H_{11}Cl$ / 214.69	948-99-2 /	/ 160-2[12]	4-05-00-02162 /	H_2O 1; EtOH 3; eth 3; bz 3 / 1.6281[19]
1148	Benzene, (1-chloroethenyl)-	C_8H_7Cl / 138.60	618-34-8 / -23	/ 199	4-05-00-01345 / 1.1016[18]	eth 4; EtOH 4 / 1.5612[20]
1149	Benzene, 1-chloro-2-ethenyl- / o-Chlorostyrene	C_8H_7Cl / 138.60	2039-87-4 / -63.1	/ 188.7	4-05-00-01345 / 1.1000[20]	EtOH 3; eth 3; ace 3; ctc 3 / 1.5649[20]
1150	Benzene, 1-chloro-3-ethenyl-	C_8H_7Cl / 138.60	2039-85-2 /	/ 62-3[6]	4-05-00-01345 / 1.1168[20]	H_2O 1; EtOH 3; eth 3 / 1.5625[20]
1151	Benzene, 1-chloro-4-ethenyl-	C_8H_7Cl / 138.60	1073-67-2 / 15.9	/ 192	4-05-00-01345 / 1.0868[20]	H_2O 1; EtOH 3; eth 3; ace 5 / 1.5660[20]
1152	Benzene, 1-(1-chloroethenyl)-2,4-dimethyl- / Styrene, α-chloro-2,4-dimethyl	$C_{10}H_{11}Cl$ / 166.65	74346-30-8 /	/ 104-5[19]	2-05-00-00381 / 1.044[13]	1.5446[13]
1153	Benzene, (2-chloroethenyl)-, (E)-	C_8H_7Cl / 138.60	4110-77-4 /	/ 199	4-05-00-01346 / 1.1095[18]	H_2O 1; EtOH 3; eth 3; ace 3 / 1.5648[20]
1154	Benzene, 1,1',1''-(1-chloro-1-ethenyl-2-ylidene)tris[4-methoxy- / Chlorotrianisene	$C_{23}H_{21}ClO_3$ / 380.87	569-57-3 / 115	2173 /	4-06-00-07650 /	
1155	Benzene, (2-chloroethenyl)-, (Z)-	C_8H_7Cl / 138.60	4604-28-8 /	/ 59[3.5]	4-05-00-01346 / 1.1046[25]	1.5762[25]
1156	Benzene, 1-chloro-2-ethoxy-	C_8H_9ClO / 156.61	614-72-2 /	/ 210	4-06-00-00785 / 1.1288[15]	bz 4; eth 4; EtOH 4 / 1.5284[25]
1157	Benzene, 1-chloro-3-ethoxy- / m-Chlorophenyl ethyl ether	C_8H_9ClO / 156.61	2655-83-6 /	/ 207	4-06-00-00811 / 1.1712[20]	H_2O 1; EtOH 4; eth 4; bz 4
1158	Benzene, 1-chloro-4-ethoxy-	C_8H_9ClO / 156.61	622-61-7 / 21	/ 213	4-06-00-00823 / 1.1254[20]	EtOH 3; eth 3; bz 4; ctc 2 / 1.5252[20]
1159	Benzene, (2-chloroethoxy)-	C_8H_9ClO / 156.61	622-86-6 / 28	/ 218.5	4-06-00-00556 /	H_2O 1; EtOH 4; eth 4; ace 4
1160	Benzene, (1-chloroethyl)-, (±)-	C_8H_9Cl / 140.61	38661-82-4 /	/ 81[17]	4-05-00-00898 / 1.0620[20]	EtOH 3; eth 4; bz 3 / 1.5276[20]
1161	Benzene, 1-chloro-2-ethyl-	C_8H_9Cl / 140.61	89-96-3 / -82.7	/ 178.4	4-05-00-00897 / 1.0569[20]	H_2O 1; ace 3; bz 3; ctc 3 / 1.5218[20]
1162	Benzene, 1-chloro-3-ethyl-	C_8H_9Cl / 140.61	620-16-6 / -55	/ 183.8	4-05-00-00897 / 1.0529[20]	ace 4; bz 4; eth 4; EtOH 4 / 1.5195[20]
1163	Benzene, 1-chloro-4-ethyl-	C_8H_9Cl / 140.61	622-98-0 / -62.6	/ 184.4	4-05-00-00897 / 1.0455[20]	H_2O 1; EtOH 5; eth 5; ace 5 / 1.5175[20]
1164	Benzene, (2-chloroethyl)-	C_8H_9Cl / 140.61	622-24-2 /	/ 197.5; 92[20]	4-05-00-00899 / 1.069[25]	H_2O 1; EtOH 3; eth 3; ace 3 / 1.5276[20]
1165	Benzene, [(2-chloroethyl)thio]-	C_8H_9ClS / 172.68	5535-49-9 /	/ 117-8[11]	4-06-00-01469 / 1.1769[25]	eth 4; chl 4 / 1.5828[20]
1166	Benzene, 1-chloro-2-fluoro-	C_6H_4ClF / 130.55	348-51-6 / -43	/ 137.6	4-05-00-00652 / 1.2233[30]	H_2O 1; ace 3; bz 3 / 1.4918[30]
1167	Benzene, 1-chloro-3-fluoro-	C_6H_4ClF / 130.55	625-98-9 /	/ 127.6	4-05-00-00652 / 1.221[25]	1.4911
1168	Benzene, 1-chloro-4-fluoro-	C_6H_4ClF / 130.55	352-33-0 / -26.8	/ 130	4-05-00-00652 / 1.4990[15]	H_2O 1; EtOH 3; eth 3; bz 3 / 1.4990[15]
1169	Benzene, 1-chloro-3-fluoro-2-methyl-	C_7H_6ClF / 144.58	443-83-4 /	/ 154	3-05-00-00691 / 1.191[25]	1.5026[20]
1170	Benzene, 2-chloro-4-fluoro-1-methyl-	C_7H_6ClF / 144.58	452-73-3 /	/ 152.5	3-05-00-00691 / 1.1972[20]	1.4985[25]
1171	Benzene, 1-chloro-2-iodo-	C_6H_4ClI / 238.45	615-41-8 / 0.7	/ 234.5	4-05-00-00695 / 1.9515[25]	H_2O 1; ace 3; ctc 2 / 1.6331[25]
1172	Benzene, 1-chloro-4-iodo-	C_6H_4ClI / 238.45	637-87-6 / 57	/ 227	4-05-00-00696 / 1.886[27]	H_2O 1; EtOH 3; chl 2
1173	Benzene, 1-chloro-2-iodo-4-methyl- / Toluene, 4-chloro-3-iodo-	C_7H_6ClI / 252.48	2401-22-1 / -10	/ 249	3-05-00-00726 / 1.7940[15]	1.614[23]
1174	Benzene, 4-chloro-1-iodo-2-methyl- / Toluene, 5-chloro-2-iodo	C_7H_6ClI / 252.48	23399-70-4 / -21	/ 240	3-05-00-00726 / 1.702[17]	1.616[23]
1175	Benzene, 4-chloro-2-iodo-1-methyl-	C_7H_6ClI / 252.48	33184-48-4 / -25	/ 242.5	3-05-00-00726 / 1.8358[15]	1.620[23]
1176	Benzene, 1-chloro-2-isocyanato-	C_7H_4ClNO / 153.57	3320-83-0 / 30.5	/ 200; 115[43]	4-12-00-01129 /	ctc 2
1177	Benzene, 1-chloro-4-isothiocyanato-	C_7H_4ClNS / 169.63	2131-55-7 / 46	/ 249.5	4-12-00-01214 /	H_2O 1; EtOH 3
1178	Benzene, 1-chloro-2-methoxy-	C_7H_7ClO / 142.58	766-51-8 / -26.8	/ 198.5	4-06-00-00785 / 1.1911[20]	H_2O 1; EtOH 3; eth 3; chl 2 / 1.5480[20]
1179	Benzene, 1-chloro-3-methoxy-	C_7H_7ClO / 142.58	2845-89-8 /	/ 193.5	4-06-00-00811 / 1.1759[12]	H_2O 1; EtOH 3; eth 3 / 1.5365[20]
1180	Benzene, 1-chloro-4-methoxy-	C_7H_7ClO / 142.58	623-12-1 / <-18	/ 197.5	4-06-00-00822 / 1.201[20]	H_2O 1; EtOH 4; eth 4; ctc 3 / 1.5390[20]
1181	Benzene, [(chloromethoxy)methyl]-	C_8H_9ClO / 156.61	3587-60-8 /	/ 103[13]	4-06-00-02252 / 1.1350[20]	1.5192[20]

No.	Name / Synonym	Mol. Form. / Mol. Wt.	CAS RN / mp/°C	Merck No. / bp/°C	Beil. Ref. / den/g cm^{-3}	Solubility / n_D
1182	Benzene, (chloromethyl)- / Benzyl chloride	C_7H_7Cl / 126.59	100-44-7 / -45	1143 / 179	4-05-00-00809 / 1.1004[20]	H_2O 1; EtOH 5; eth 5; ctc 2 / 1.5391[20]
1183	Benzene, 1-chloro-2-methyl- / o-Chlorotoluene	C_7H_7Cl / 126.59	95-49-8 / -35.6	2172 / 159.0	4-05-00-00805 / 1.0825[20]	H_2O 1; EtOH 3; eth 5; ace 5 / 1.5268[20]
1184	Benzene, 1-chloro-3-methyl- / m-Chlorotoluene	C_7H_7Cl / 126.59	108-41-8 / -47.8	2172 / 161.8	4-05-00-00806 / 1.075[20]	H_2O 1; EtOH 3; eth 5; bz 3 / 1.5214[19]
1185	Benzene, 1-chloro-4-methyl- / p-Chlorotoluene	C_7H_7Cl / 126.59	106-43-4 / 7.5	2172 / 162.4	4-05-00-00806 / 1.0697[20]	H_2O 1; EtOH 3; eth 5; ctc 3 / 1.5150[20]
1186	Benzene, 1-(chloromethyl)-2,4-dimethyl-	$C_9H_{11}Cl$ / 154.64	824-55-5 /	/ 215.5; 110[20]	4-05-00-01012 / 1.0580[19]	bz 4; eth 4; EtOH 4
1187	Benzene, 1,1'-(chloromethylene)bis-	$C_{13}H_{11}Cl$ / 202.68	90-99-3 / 16	/ 140[3]	4-05-00-01847 / 1.140[25]	chl 3 / 1.5951[20]
1188	Benzene, 1-(2-chloro-1-methylethenyl)-4-methyl- / Styrene, β-chloro, α-4-dimethyl	$C_{10}H_{11}Cl$ / 166.65	30926-60-4 /	/ 107-8[10]	2-05-00-00381 / 1.058[20]	/ 1.5549[23]
1189	Benzene, (1-chloro-1-methylethyl)-	$C_9H_{11}Cl$ / 154.64	934-53-2 /	/ 98[1]	4-05-00-00992 / 1.192[25]	/ 1.5290[25]
1190	Benzene, 1-chloro-2-(1-methylethyl)-	$C_9H_{11}Cl$ / 154.64	2077-13-6 / -74.4	/ 191.1	4-05-00-00992 / 1.0341[20]	ace 4; bz 4; eth 4; EtOH 4 / 1.5168[20]
1191	Benzene, 1-(chloromethyl)-4-ethyl-	$C_9H_{11}Cl$ / 154.64	1467-05-6 /	/ 95-6[15]	4-05-00-01004 /	bz 4; EtOH 4; chl 4 / 1.5290[20]
1192	Benzene, 1-chloro-4-(1-methylethyl)-	$C_9H_{11}Cl$ / 154.64	2621-46-7 / -12.3	/ 198.3	4-05-00-00992 / 1.0208[20]	H_2O 1; EtOH 5; eth 5; ace 5 / 1.5117[20]
1193	Benzene, (2-chloro-1-methylethyl)-	$C_9H_{11}Cl$ / 154.64	824-47-5 /	/ 85[13]	4-05-00-00992 / 1.047[19]	/ 1.5245[19]
1194	Benzene, 1-(chloromethyl)-2-fluoro-	C_7H_6ClF / 144.58	345-35-7 /	/ 86[40]	4-05-00-00813 / 1.216[25]	
1195	Benzene, 1-(chloromethyl)-4-fluoro-	C_7H_6ClF / 144.58	352-11-4 /	/ 82[26]	4-05-00-00813 / 1.207[25]	/ 1.5150[20]
1196	Benzene, 1,1',1''-(chloromethylidyne)tris-	$C_{19}H_{15}Cl$ / 278.78	76-83-5 / 113.5	/ 310	4-05-00-02497 /	H_2O 1; EtOH 2; eth 4; ace 3 / 1.5130
1197	Benzene, 1-(chloromethyl)-4-methoxy-	C_8H_9ClO / 156.61	824-94-2 / 24.5	/ 262.5	4-06-00-02137 / 1.261[20]	ace 4; bz 4; eth 4 / 1.580[20]
1198	Benzene, 1-(chloromethyl)-2-methyl-	C_8H_9Cl / 140.61	552-45-4 /	/ 198; 90[20]	4-05-00-00926 / 1.063[25]	eth 4; EtOH 4 / 1.5410[25]
1199	Benzene, 1-(chloromethyl)-3-methyl-	C_8H_9Cl / 140.61	620-19-9 /	/ 195.5	4-05-00-00943 / 1.064[20]	H_2O 1; EtOH 3; eth 3 / 1.5345[20]
1200	Benzene, 1-(chloromethyl)-4-methyl-	C_8H_9Cl / 140.61	104-82-5 /	/ 201; 90[20]	4-05-00-00966 / 1.0512[20]	H_2O 1; EtOH 3; eth 5 / 1.5380
1201	Benzene, 2-chloro-1-methyl-4-(1-methylethyl)-	$C_{10}H_{13}Cl$ / 168.67	4395-79-3 /	/ 216.5	4-05-00-01062 / 1.0104[25]	H_2O 1; ace 3; bz 3 / 1.5078[20]
1202	Benzene, 2-chloro-4-methyl-1-(1-methylethyl)-	$C_{10}H_{13}Cl$ / 168.67	4395-80-6 /	/ 215; 113[35]	3-05-00-00956 / 1.018[15]	/ 1.5180[20]
1203	Benzene, 1-(chloromethyl)-2-nitro-	$C_7H_6ClNO_2$ / 171.58	612-23-7 / 51		4-05-00-00854 /	H_2O 1; EtOH 3; eth 3; ace 4 / 1.5557[62]
1204	Benzene, 1-chloro-2-methyl-3-nitro-	$C_7H_6ClNO_2$ / 171.58	83-42-1 / 37.8	/ 238	4-05-00-00854 /	H_2O 1; EtOH 3 / 1.5377[69]
1205	Benzene, 1-chloro-2-methyl-4-nitro-	$C_7H_6ClNO_2$ / 171.58	13290-74-9 / 42.5	/ 249	4-05-00-00855 /	eth 4
1206	Benzene, 1-(chloromethyl)-3-nitro-	$C_7H_6ClNO_2$ / 171.58	619-23-8 / 46	/ 173[34]	4-05-00-00855 /	ace 4; bz 4; eth 4; EtOH 4 / 1.5557[62]
1207	Benzene, 1-chloro-3-methyl-5-nitro- / 3-Chloro-5-nitrotoluene	$C_7H_6ClNO_2$ / 171.58	16582-38-0 / 61		4-05-00-00855 /	EtOH 4 / 1.5404[69]
1208	Benzene, 1-(chloromethyl)-4-nitro- / 4-Nitrobenzyl chloride	$C_7H_6ClNO_2$ / 171.58	100-14-1 / 71		4-05-00-00856 /	H_2O 1; EtOH 3; eth 3; ace 4 / 1.5647[62]
1209	Benzene, 1-chloro-4-methyl-2-nitro-	$C_7H_6ClNO_2$ / 171.58	89-60-1 / 7	/ 261; 118[11]	4-05-00-00855 /	H_2O 1; ctc 3 / 1.5572[20]
1210	Benzene, 2-chloro-1-methyl-4-nitro-	$C_7H_6ClNO_2$ / 171.58	121-86-8 / 66.5	/ 260	4-05-00-00855 /	H_2O 2; EtOH 3; eth 3; chl 2 / 1.5470[69]
1211	Benzene, 2-chloro-4-methyl-1-nitro-	$C_7H_6ClNO_2$ / 171.58	38939-88-7 / 24	/ 146[15]	4-05-00-00856 /	H_2O 1; ctc 3
1212	Benzene, 4-chloro-1-methyl-2-nitro-	$C_7H_6ClNO_2$ / 171.58	89-59-8 / 38	/ 242; 115.5[11]	4-05-00-00853 / 1.2559[80]	H_2O 1; EtOH 3; eth 3; chl 2
1213	Benzene, 4-chloro-2-methyl-1-nitro-	$C_7H_6ClNO_2$ / 171.58	5367-28-2 / 25.5		4-05-00-00854 /	H_2O 1; ctc 3 / 1.5496[65]
1214	Benzene, 1-chloro-2-nitro- / o-Chloronitrobenzene	$C_6H_4ClNO_2$ / 157.56	88-73-3 / 32.5	2151 / 245.5	4-05-00-00721 / 1.368[242]	H_2O 1; EtOH 3; eth 3; ace 4
1215	Benzene, 1-chloro-3-nitro- / m-Chloronitrobenzene	$C_6H_4ClNO_2$ / 157.56	121-73-3 / 44.4	2151 / 235.5	4-05-00-00722 / 1.343[50]	H_2O 1; EtOH 3; eth 3; bz 3 / 1.5374[80]
1216	Benzene, 1-chloro-4-nitro- / p-Chloronitrobenzene	$C_6H_4ClNO_2$ / 157.56	100-00-5 / 83.5	2151 / 242	4-05-00-00723 / 1.2979[90]	H_2O I; EtOH 2; eth 3; chl 3 / 1.5376[100]
1217	Benzene, 1-chloro-2-nitro-4-(trifluoromethyl)- / 2-Nitro-4-(trifluoromethyl)-1-chlorobenzene	$C_7H_3ClF_3NO_2$ / 225.55	121-17-5 / -2.5	/ 222; 95[10]	4-05-00-00857 / 1.511[25]	/ 1.4893[20]
1218	Benzene, 4-chloro-1-nitro-2-(trifluoromethyl)- / 4-Chloro-3-(trifluoromethyl)nitrobenzene	$C_7H_3ClF_3NO_2$ / 225.55	777-37-7 / 22	/ 232	4-05-00-00857 / 1.527[25]	/ 1.5083[26]
1219	Benzene, 4-chloro-1-nitro-2-(trifluoromethyl)- / 4-Chloro-2-(trifluoromethyl)nitrobenzene	$C_7H_3ClF_3NO_2$ / 225.55	118-83-2 /		3-05-00-00748 / 1.526[25]	/ 1.4980[20]
1220	Benzene, chloropentafluoro- / Chloropentafluorobenzene	C_6ClF_5 / 202.51	344-07-0 /	/ 117.9	4-05-00-00654 / 1.568[25]	/ 1.4256[20]

No.	Name Synonym	Mol. Form. Mol. Wt.	CAS RN mp/°C	Merck No. bp/°C	Beil. Ref. den/g cm^{-3}	Solubility n_D
1221	Benzene, 1-chloro-4-phenoxy- 4-Chlorophenyl phenyl ether	$C_{12}H_9ClO$ 204.66	7005-72-3	284.5	4-06-00-00826 1.2026[15]	1.599
1222	Benzene, 1-chloro-2-(2-phenylethenyl)-, (E)-	$C_{14}H_{11}Cl$ 214.69	1657-52-9 39.5	209[30]; 139[2]	4-05-00-02161	H_2O 1; EtOH 3; chl 3
1223	Benzene, 1-chloro-3-(2-phenylethenyl)-, (E)-	$C_{14}H_{11}Cl$ 214.69	14064-43-8 73.5	177[0.2]	4-05-00-02161	H_2O 1; EtOH 3; chl 3
1224	Benzene, 1-chloro-4-(phenylmethyl)-	$C_{13}H_{11}Cl$ 202.68	831-81-2 7.5	299; 147-8[8]	4-05-00-01847 1.1247[20]	ace 4
1225	Benzene, 1-chloro-4-(phenylsulfonyl)- Sulphenone	$C_{12}H_9ClO_2S$ 252.72	80-00-2 94	8969	4-06-00-01587	H_2O 1; EtOH 2; eth 3; ace 4
1226	Benzene, (1-chloro-1-propenyl)-	C_9H_9Cl 152.62	35673-03-1	90.5[0.9]	2-05-00-00372 1.085[20]	ace 4; bz 4 1.5635[15]
1227	Benzene, (2-chloro-1-propenyl)-	C_9H_9Cl 152.62	13099-50-8	120[20]; 61[1]	3 05-00-01186 1.0738[19]	ace 4; bz 4; chl 4 1.5565[19]
1228	Benzene, (3-chloro-2-propenyl)-	C_9H_9Cl 152.62	6268-37-7	211	4-05-00-01363 1.073[13]	H_2O 1; eth 3; ace 3; bz 3 1.545[14]
1229	Benzene, 1-chloro-3-(1-propenyl)-, (E)-	C_9H_9Cl 152.62	23204-80-0 8.5	106.7[13]	1.0926[20]	H_2O 1; EtOH 5; eth 5; ace 4 1.5851[20]
1230	Benzene, (3-chloro-1-propenyl)-, (E)-	C_9H_9Cl 152.62	21087-29-6 8.5	106.7[13]	4-05-00-01360 1.0926[20]	ace 4; bz 4; eth 4; EtOH 4 1.5851[20]
1231	Benzene, 1-(3-chloro-2-propenyl)-4-methoxy- Benzene, 1(3-chloroallyl)-4-methoxy	$C_{10}H_{11}ClO$ 182.65	54644-23-4	120[11]	3-06-00-02416 1.155[8]	1.553[8]
1232	Benzene, 1-chloro-4-(2-propenyloxy)-	C_9H_9ClO 168.62	13997-70-1	106.7[12]	4-06-00-00825 1.131[15]	bz 4; eth 4; EtOH 4 1.5348[25]
1233	Benzene, (3-chloropropoxy)-	$C_9H_{11}ClO$ 170.64	3384-04-1 12	250	4-06-00-00557 1.1167[20]	ace 4; eth 4 1.5235[25]
1234	Benzene, (2-chloropropyl)-, (R)- R-(-)-2-Chloro-1-phenylpropane	$C_9H_{11}Cl$ 154.64	55449-46-2	94[17]	4-05-00-00980 1.038[19]	H_2O 1; ace 3; bz 3; chl 3 1.5198[22]
1235	Benzene, (3-chloropropyl)-	$C_9H_{11}Cl$ 154.64	104-52-9	219.5	4-05-00-00980 1.056[21]	ctc 2 1.5160[25]
1236	Benzene, [(3-chloropropyl)thio]-	$C_9H_{11}ClS$ 186.71	4911-65-3	116-7[4]	3-06-00-00981 1.1536[20]	ace 4; py 4 1.5752[20]
1237	Benzene, 1-chloro-2-(trichloromethyl)- 1-Chloro-2-(trichloromethyl)benzene	$C_7H_4Cl_4$ 229.92	2136-89-2 29.4	264.3	4-05-00-00823 1.5187[20]	H_2O 1; eth 3; ace 3; ctc 2 1.5836[20]
1238	Benzene, 1-chloro-3-(trichloromethyl)- 1-Chloro-3-(trichloromethyl)benzene	$C_7H_4Cl_4$ 229.92	2136-81-4 0.6	255	4-05-00-00823 1.495[14]	H_2O 1; eth 3; ace 3; ctc 2 1.4461[20]
1239	Benzene, 1-chloro-4-(trichloromethyl)- 1-Chloro-4-(trichloromethyl)benzene	$C_7H_4Cl_4$ 229.92	5216-25-1	245	4-05-00-00823 1.4463[20]	ace 4; eth 4
1240	Benzene, 1-chloro-2-(trifluoromethyl)-	$C_7H_4ClF_3$ 180.56	88-16-4 -6	152.2	4-05-00-00814 1.2540[30]	chl 3 1.4513[25]
1241	Benzene, 1-chloro-3-(trifluoromethyl)- Benzotrifluoride, 3-chloro	$C_7H_4ClF_3$ 180.56	98-15-7 -56	137.5	4-05-00-00814 1.3311[25]	1.4438[25]
1242	Benzene, 1-chloro-4-(trifluoromethyl)-	$C_7H_4ClF_3$ 180.56	98-56-6 -33	138.5	4-05-00-00815 1.3340[25]	1.4431[30]
1243	Benzene, 2-chloro-1,3,5-trimethyl-	$C_9H_{11}Cl$ 154.64	1667-04-5 <-20	205	4-05-00-01026 1.0337[30]	eth 4; EtOH 4 1.5212[30]
1244	Benzene, 2-chloro-1,3,5-trinitro- Picryl chloride	$C_6H_2ClN_3O_6$ 247.55	88-88-0 83	7390	4-05-00-00757 1.797[20]	H_2O 1; EtOH 3; eth 2; ace 4
1245	Benzene, cyclobutyl-	$C_{10}H_{12}$ 132.21	4392-30-7	190[155]; 101[41]	4-05-00-01387 0.9378[20]	1.5277[20]
1246	Benzene, 1-cyclohexen-1-yl-	$C_{12}H_{14}$ 158.24	771-98-2 -11	252	4-05-00-01557 0.9939[20]	MeOH 4 1.5718[20]
1247	Benzene, 2-cyclohexen-1-yl-	$C_{12}H_{14}$ 158.24	15232-96-9	115-7[16]	4-05-00-01558 0.9800[20]	1.5530[20]
1248	Benzene, cyclohexyl- Cyclohexylbenzene	$C_{12}H_{16}$ 160.26	827-52-1 7.3	240.1	4-05-00-01424 0.9427[20]	H_2O 1; EtOH 4; eth 3; ctc 2 1.5329[20]
1249	Benzene, (1-cyclohexylethyl)-	$C_{14}H_{20}$ 188.31	4413-16-5	265	4-05-00-01454 0.9773[17]	1.549[17]
1250	Benzene, 1-cyclohexyl-2-methoxy-	$C_{13}H_{18}O$ 190.29	2206-48-6	268	4-06-00-03899 1.007[18]	1.5365[18]
1251	Benzene, 1-cyclohexyl-2-nitro- Cyclohexane, 2-nitrophenyl	$C_{12}H_{15}NO_2$ 205.26	7137-56-6 45	174[16]; 113[0.5]	3-05-00-01259 1.111[23]	1.5472[25]
1252	Benzene, 1-cyclohexyl-4-nitro-	$C_{12}H_{15}NO_2$ 205.26	5458-48-0 58.5	210[25]	4-05-00-01426	bz 4; eth 4
1253	Benzene, (cyclohexyloxy)-	$C_{12}H_{16}O$ 176.26	2206-38-4	261	4-06-00-00565 1.0077[20]	H_2O 1; ace 3; bz 3 1.520[22]
1254	Benzene, cyclopentyl-	$C_{11}H_{14}$ 146.23	700-88-9	219	4-05-00-01409 0.9462[20]	eth 4 1.5280[20]
1255	Benzene, (3-cyclopentylpropyl)-	$C_{14}H_{20}$ 188.31	2883-12-7	272	4-05-00-01455 0.9233[19]	1.5130[19]
1256	Benzene, cyclopropyl- Cyclopropylbenzene	C_9H_{10} 118.18	873-49-4 -31	173.6; 80[37]	4-05-00-01370 0.9317[20]	H_2O 1; eth 3; ace 3; chl 3 1.5285[20]
1257	Benzenedecanoic acid, ι-octyl-	$C_{24}H_{40}O_2$ 360.58	1938-17-6 40.5	199-205[0.09]	2-09-00-00376 0.9310[25]	1.4894[20]
1258	Benzene, decyl- Decylbenzene	$C_{16}H_{26}$ 218.38	104-72-3 -14.4	298	4-05-00-01193 0.8555[20]	ace 4; bz 4; eth 4; EtOH 4 1.4832[20]
1259	1,4-Benzenediacetic acid, dimethyl ester	$C_{12}H_{14}O_4$ 222.24	36076-25-2 51.5	189[15]	2-09-00-00624	EtOH 3; eth 3

No.	Name Synonym	Mol. Form. Mol. Wt.	CAS RN mp/°C	Merck No. bp/°C	Beil. Ref. den/g cm^{-3}	Solubility n_D
1260	1,2-Benzenediamine o-Phenylenediamine	$C_6H_8N_2$ 108.14	95-54-5 102.5	7255 257	4-13-00-00038	H_2O 3; EtOH 4; eth 3; bz 3
1261	1,3-Benzenediamine m-Phenylenediamine	$C_6H_8N_2$ 108.14	108-45-2 63.5	7254 285	4-13-00-00079 1.0096[58]	H_2O 4; EtOH 3; eth 3; bz 3 1.6339[58]
1262	1,4-Benzenediamine p-Phenylenediamine	$C_6H_8N_2$ 108.14	106-50-3 146	7256 267	4-13-00-00104	H_2O 2; EtOH 3; eth 3; bz 3
1263	1,4-Benzenediamine, N-(4-aminophenyl)- 4,4'-Diaminodiphenylamine	$C_{12}H_{13}N_3$ 199.26	537-65-5 158	2957 dec	4-13-00-00203	eth 4; EtOH 4
1264	1,2-Benzenediamine, 4-chloro-	$C_6H_7ClN_2$ 142.59	95-83-0 76		4-13-00-00068	H_2O 2; EtOH 4; eth 4; bz 3
1265	1,3-Benzenediamine, N,N-diethyl-	$C_{10}H_{16}N_2$ 164.25	26513-20-2 152	277	2-13-00-00026	EtOH 4
1266	1,4-Benzenediamine, N,N-diethyl- N,N-Diethyl-p-phenylenediamine	$C_{10}H_{16}N_2$ 164.25	93-05-0	261	4-13-00-00109	bz 4
1267	1,2-Benzenediamine, N,N-dimethyl-	$C_8H_{12}N_2$ 136.20	2836-03-5	218; 90[22]	4-13-00-00042 0.995[22]	H_2O 2; EtOH 4; eth 4; ace 4
1268	1,3-Benzenediamine, N,N-dimethyl-	$C_8H_{12}N_2$ 136.20	2836-04-6 <-20	270; 138[10]	4-13-00-00080 0.995[25]	H_2O 2; EtOH 4; eth 4
1269	1,2-Benzenediamine, N,N'-dimethyl- o-Phenylenediamine, N,N'-dimethyl	$C_8H_{12}N_2$ 136.20	3213-79-4 34.5	247	4-13-00-00042	1.5914[25]
1270	1,4-Benzenediamine, N,N-dimethyl- Dimethyl-p-phenylenediamine	$C_8H_{12}N_2$ 136.20	99-98-9 53	3242 263	4-13-00-00106 1.036[20]	H_2O 3; EtOH 4; eth 4; bz 4
1271	1,4-Benzenediamine, N,N'-diphenyl- N,N'-Diphenyl-p-phenylenediamine	$C_{18}H_{16}N_2$ 260.34	74-31-7 144	3331	4-13-00-00116	EtOH 2; eth 2; bz 2; chl 2
1272	1,2-Benzenediamine, 4-ethoxy-	$C_8H_{12}N_2O$ 152.20	1197-37-1 71.5	295	3-13-00-01362	H_2O 4; EtOH 4; eth 3; chl 3
1273	1,2-Benzenediamine, 4-methoxy-	$C_7H_{10}N_2O$ 138.17	102-51-2 51	200[21]; 168[11]	4-13-00-01448	eth 4
1274	1,3-Benzenediamine, 4-methoxy- 4-Methoxy-m-phenylenediamine	$C_7H_{10}N_2O$ 138.17	615-05-4 67.5	5920	4-13-00-01425	EtOH 3; eth 3; DMSO-d$_6$ 2
1275	1,2-Benzenediamine, 3-methyl-	$C_7H_{10}N_2$ 122.17	2687-25-4 63.5	255	3-13-00-00277	ace 4; bz 4; EtOH 4
1276	1,2-Benzenediamine, 4-methyl- Toluene-3,4-diamine	$C_7H_{10}N_2$ 122.17	496-72-0 89.5	265	4-13-00-00260	H_2O 4; lig 3
1277	1,3-Benzenediamine, 2-methyl- 2,6-Diaminotoluene	$C_7H_{10}N_2$ 122.17	823-40-5 106		3-13-00-00291	H_2O 3; EtOH 3; bz 3
1278	1,3-Benzenediamine, 4-methyl- Toluene-2,4-diamine	$C_7H_{10}N_2$ 122.17	95-80-7 99	292	4-13-00-00235	H_2O 4; EtOH 4; eth 4; bz 4
1279	1,4-Benzenediamine, 2-methyl-	$C_7H_{10}N_2$ 122.17	95-70-5 64	273.5	4-13-00-00246	H_2O 3; EtOH 3; eth 3; bz 2
1280	1,4-Benzenediamine, N-methyl-	$C_7H_{10}N_2$ 122.17	623-09-6 36	258	4-13-00-00106	H_2O 4; EtOH 4; eth 4; ace 4
1281	1,2-Benzenediamine, 4-nitro- 4-Nitro-o-phenylenediamine	$C_6H_7N_3O_2$ 153.14	99-56-9 199.5	6544	4-13-00-00075	acid 3
1282	1,2-Benzenediamine, N-phenyl-	$C_{12}H_{12}N_2$ 184.24	534-85-0 79.5	313	4-13-00-00043	H_2O 2; ace 3; bz 3; chl 3
1283	1,4-Benzenediamine, N-phenyl- p-Aminodiphenylamine	$C_{12}H_{12}N_2$ 184.24	101-54-2 66	354	4-13-00-00113	H_2O 2; EtOH 4; eth 3; chl 2
1284	1,3-Benzenediamine, 4-(phenylazo)-, monohydrochloride Chrysoidine hydrochloride	$C_{12}H_{13}ClN_4$ 248.71	532-82-1 118.5	2262	2-16-00-00203	ace 4
1285	1,3-Benzenediamine, N,N,N',N'-tetraethyl- N,N,N',N'-Tetraethyl-m-phenylenediamine	$C_{14}H_{24}N_2$ 220.36	64287-26-9	148[9]	2-13-00-00026 0.9522[12]	eth 4; EtOH 4 1.5537[12]
1286	1,2-Benzenediamine, N,N,N',N'-tetramethyl-	$C_{10}H_{16}N_2$ 164.25	704-01-8 8.9	215.5	4-13-00-00042 0.9560[20]	
1287	1,4-Benzenediamine, N,N,N',N'-tetramethyl- Tetramethyl-p-phenylenediamine	$C_{10}H_{16}N_2$ 164.25	100-22-1 51	9159 260	4-13-00-00107	H_2O 2; EtOH 4; eth 4; bz 4
1288	Benzene, 1,2-dibromo- o-Dibromobenzene	$C_6H_4Br_2$ 235.91	583-53-9 7.1	225	4-05-00-00682 1.9843[20]	H_2O 1; EtOH 3; eth 5; ace 5 1.6155[20]
1289	Benzene, 1,3-dibromo- m-Dibromobenzene	$C_6H_4Br_2$ 235.91	108-36-1 -7	218	4-05-00-00682 1.9523[20]	H_2O 1; EtOH 3; eth 5 1.6083[17]
1290	Benzene, 1,4-dibromo- p-Dibromobenzene	$C_6H_4Br_2$ 235.91	106-37-6 87.3	3001 218.5	4-05-00-00683 2.261[17]	H_2O 1; EtOH 3; eth 4; ace 4 1.5742
1291	Benzene, 1,3-dibromo-5-(bromomethyl)- 3,5-Dibromobenzyl bromide	$C_7H_5Br_3$ 328.83	56908-88-4 96	173[19]	3-05-00-00719	EtOH 4
1292	Benzene, 1,2-dibromo-3-chloro- 1,2-Dibromo-3-chlorobenzene	$C_6H_3Br_2Cl$ 270.35	104514-49-0 73.5	264	1-05-00-00117	bz 4; eth 4; chl 4
1293	Benzene, 1,2-dibromo-4-chloro- 1,2-Dibromo-4-chlorobenzene	$C_6H_3Br_2Cl$ 270.35	60956-24-3 35.5	256	0-05-00-00212	bz 4; eth 4; chl 4
1294	Benzene, 1,3-dibromo-2-chloro- 1,3-Dibromo-2-chlorobenzene	$C_6H_3Br_2Cl$ 270.35	19230-27-4 72	265	3-05-00-00569	bz 4; eth 4; chl 4
1295	Benzene, 1,4-dibromo-2-chloro- 1,4-Dibromo-2-chlorobenzene	$C_6H_3Br_2Cl$ 270.35	3460-24-0 40.5	259	0-05-00-00212	bz 4; eth 4; chl 4
1296	Benzene, 2,4-dibromo-1-chloro-	$C_6H_3Br_2Cl$ 270.35	29604-75-9 27	258	0-05-00-00212	bz 4; eth 4; chl 4
1297	Benzene, 1,2-dibromo-3,4-dimethyl- 3,4-Dibromo-o-xylene	$C_8H_8Br_2$ 263.96	24932-49-8 7	277	2-05-00-00285 1.782[15]	EtOH 4
1298	Benzene, 1,5-dibromo-2,4-dimethyl-	$C_8H_8Br_2$ 263.96	615-87-2 70	255.5	3-05-00-00839	EtOH 4

No.	Name / Synonym	Mol. Form. / Mol. Wt.	CAS RN / mp/°C	Merck No. / bp/°C	Beil. Ref. / den/g cm^{-3}	Solubility / n_D
1299	Benzene, (1,2-dibromoethyl)-	$C_8H_8Br_2$ 263.96	93-52-7 75	133[19]	4-05-00-00909	EtOH 3; eth 3; bz 3; chl 3
1300	Benzene, 2,4-dibromo-1-fluoro-	$C_6H_3Br_2F$ 253.90	1435-53-6 69	103-5[32]	3-05-00-00569 2.047[20]	1.5840[20]
1301	Benzene, (dibromomethyl)-	$C_7H_6Br_2$ 249.93	618-31-5	156[23]	4-05-00-00836 1.8365[28]	H_2O 1; EtOH 5; eth 5 1.6147[20]
1302	Benzene, 1,2-dibromo-3-methyl- 2,3-Dibromotoluene	$C_7H_6Br_2$ 249.93	61563-25-5 30.5	126-8[19]	4-05-00-00835 1.823[15]	1.5984[25]
1303	Benzene, 1,2-dibromo-4-methyl- 3,4-Dibromotoluene	$C_7H_6Br_2$ 249.93	60956-23-2 -10	240; 123[19]	4-05-00-00836 1.8197[25]	1.5979[25]
1304	Benzene, 1,3-dibromo-2-methyl- 2,6-Dibromotoluene	$C_7H_6Br_2$ 249.93	69321-60-4 5.5	246	4-05-00-00836 1.812[22]	
1305	Benzene, 1,4-dibromo-2-methyl- 2,5-Dibromotoluene	$C_7H_6Br_2$ 249.93	615-59-8 5.6	236	4-05-00-00835 1.8127[17]	H_2O 1 1.5982[18]
1306	Benzene, 2,4-dibromo-1-methyl-	$C_7H_6Br_2$ 249.93	31543-75-6 -9.7	103-4[11]	4-05-00-00835 1.8176[25]	1.5964[25]
1307	Benzene, 1,2-dibromo-4-nitro- 3,4-Dibromonitrobenzene	$C_6H_3Br_2NO_2$ 280.90	5411-50-7 58.5	296	4-05-00-00732 2.354[8]	bz 4; EtOH 4; HOAc 4 1.9835[111]
1308	Benzene, 1,3-dibromo-2-nitro- 2,6-Dibromonitrobenzene	$C_6H_3Br_2NO_2$ 280.90	13402-32-9 84	sub	3-05-00-00621 2.211[8]	ace 4; bz 4; EtOH 4
1309	Benzene, 1,3-dibromo-5-nitro-	$C_6H_3Br_2NO_2$ 280.90	6311-60-0 106		3-05-00-00621 2.36[8]	bz 4; eth 4; EtOH 4
1310	Benzene, 1,4-dibromo-2-nitro-	$C_6H_3Br_2NO_2$ 280.90	3460-18-2 85.5		4-05-00-00732 2.368[8]	H_2O 1; ace 3; bz 3; chl 2
1311	Benzene, 2,4-dibromo-1-nitro-	$C_6H_3Br_2NO_2$ 280.90	51686-78-3 62		3-05-00-00620 2.356[8]	ace 4; bz 4
1312	Benzene, 1,2-dibromo-3,4,5,6-tetrafluoro-	$C_6Br_2F_4$ 307.87	827-08-7 12	198	2.260[20]	1.5151[25]
1313	Benzene, 2,4-dibromo-1,3,5-trimethyl- Mesitylene, 2,4-dibromo-	$C_9H_{10}Br_2$ 277.99	6942-99-0 65.5	285	3-05-00-00921	ctc 2
1314	Benzene, 1,4-dibutoxy-	$C_{14}H_{22}O_2$ 222.33	104-36-9 45.5	158[15]	4-06-00-05721	ctc 3
1315	1,3-Benzenedicarbonitrile 1,3-Dicyanobenzene	$C_8H_4N_2$ 128.13	626-17-5 162	sub	4-09-00-03297 0.992[40]	H_2O 2; EtOH 4; eth 3; bz 3
1316	1,4-Benzenedicarbonitrile	$C_8H_4N_2$ 128.13	623-26-7 224	sub	4-09-00-03328	H_2O 1; EtOH 2; eth 2; bz 3
1317	1,3-Benzenedicarbonitrile, 2,4,5,6-tetrachloro- Chlorothalonil	$C_8Cl_4N_2$ 265.91	1897-45-6 250	2167 350	1.7[25]	H_2O 1; ace 2; cyhex 2
1318	1,2-Benzenedicarbonyl dichloride Phthaloyl chloride	$C_8H_4Cl_2O_2$ 203.02	88-95-9 15.5	7349 281.1	4-09-00-03261 1.4089[20]	1.5684[20]
1319	1,3-Benzenedicarbonyl dichloride	$C_8H_4Cl_2O_2$ 203.02	99-63-8 43.5	276	4-09-00-03295 1.3880[17]	H_2O 2; EtOH 2; eth 3 1.570[47]
1320	1,4-Benzenedicarbonyl dichloride	$C_8H_4Cl_2O_2$ 203.02	100-20-9 83.5	258; 125[9]	4-09-00-03318	eth 3
1321	1,2-Benzenedicarbonyl difluoride	$C_8H_4F_2O_2$ 170.12	445-69-2 42.5	227	4-09-00-03260 1.3066[50]	H_2O 1; peth 3
1322	1,3-Benzenedicarboxaldehyde	$C_8H_6O_2$ 134.13	626-19-7 89.5	246; 136[13]	4-07-00-02139	H_2O 2; EtOH 4; eth 2; ace 3
1323	1,4-Benzenedicarboxaldehyde	$C_8H_6O_2$ 134.13	623-27-8 117	246	4-07-00-02140	H_2O 2; EtOH 4; eth 3; chl 3
1324	1,2-Benzenedicarboxamide Phthalamide	$C_8H_8N_2O_2$ 164.16	88-96-0 222	7343 dec	4-09-00-03265	H_2O 2; EtOH 2; eth 1
1325	1,2-Benzenedicarboxamide, N,N,N',N'-tetraethyl- N,N,N',N'-Tetraethylphthalamide	$C_{16}H_{24}N_2O_2$ 276.38	83-81-8 36	9137 204[16]	4-09-00-03266	
1326	1,3-Benzenedicarboxamide, N,N,N',N'-tetraethyl- Isophthalamide, N,N,N',N'-tetraethyl	$C_{16}H_{24}N_2O_2$ 276.38	13698-87-8 86	242[12]	4-09-00-03296	ace 4; bz 4; eth 4; EtOH 4
1327	1,2-Benzenedicarboxylic acid Phthalic acid	$C_8H_6O_4$ 166.13	88-99-3 230 dec	7345 dec	4-09-00-03167 2.18[191]	H_2O 2; EtOH 3; eth 2; chl 1
1328	1,3-Benzenedicarboxylic acid Isophthalic acid	$C_8H_6O_4$ 166.13	121-91-5 347	5083 sub	4-09-00-03292	H_2O 2; EtOH 3; eth 1; bz 1
1329	1,4-Benzenedicarboxylic acid Terephthalic acid	$C_8H_6O_4$ 166.13	100-21-0	9093 sub 300	4-09-00-03301	H_2O 1; EtOH 1; eth 1; ctc 2
1330	1,3-Benzenedicarboxylic acid, 2-amino-	$C_8H_7NO_4$ 181.15	39622-79-2 340	sub 267	4-14-00-01899	eth 4; EtOH 4
1331	1,3-Benzenedicarboxylic acid, 5-amino-	$C_8H_7NO_4$ 181.15	99-31-0 360	sub	1-14-00-00636	H_2O 1; EtOH 2
1332	1,2-Benzenedicarboxylic acid, 3-benzoyl-	$C_{15}H_{10}O_5$ 270.24	602-82-4 140.5	dec	3-10-00-04007	bz 4; EtOH 4
1333	1,2-Benzenedicarboxylic acid, bis(2-ethoxyethyl) ester	$C_{16}H_{22}O_6$ 310.35	605-54-9 34	345	2-09-00-00597 1.1229[21]	
1334	1,2-Benzenedicarboxylic acid, bis(2-ethylhexyl) ester Bis(2-ethylhexyl) phthalate	$C_{24}H_{38}O_4$ 390.56	117-81-7 -55	1262 384	4-09-00-03181 0.981[25]	ctc 2 1.4853[20]
1335	1,2-Benzenedicarboxylic acid, bis(3-methylbutyl) ester Diisoamyl phthalate	$C_{18}H_{26}O_4$ 306.40	605-50-5	5006 334 dec	4-09-00-03179 1.0209[16]	EtOH 4 1.4871[20]

No.	Name Synonym	Mol. Form. Mol. Wt.	CAS RN mp/°C	Merck No. bp/°C	Beil. Ref. den/g cm^{-3}	Solubility n_D
1336	1,2-Benzenedicarboxylic acid, bis(2-methylpropyl) ester	C$_{16}$H$_{22}$O$_4$ 278.35	84-69-5 296.5; 159[4]		4-09-00-03177 1.0490[15]	ctc 3
1337	1,2-Benzenedicarboxylic acid, bis(phenylmethyl) ester	C$_{22}$H$_{18}$O$_4$ 346.38	523-31-9 44.0	277[15]	4-09-00-03219	H$_2$O 1; EtOH 4; eth 4; lig 1
1338	1,4-Benzenedicarboxylic acid, bis(2-methylpropyl) ester	C$_{16}$H$_{22}$O$_4$ 278.35	18699-48-4 55	180[6]	4-09-00-03305	eth 3
1339	1,2-Benzenedicarboxylic acid, dibutyl ester Dibutyl phthalate	C$_{16}$H$_{22}$O$_4$ 278.35	84-74-2 -35	1586 340	4-09-00-03175 1.0465[20]	H$_2$O 1; EtOH 5; eth 5; bz 5 1.4911[20]
1340	1,2-Benzenedicarboxylic acid, 3,5-dichloro- Phthalic acid, 3,5-dichloro-	C$_8$H$_4$Cl$_2$O$_4$ 235.02	25641-98-9 164 dec	sub	4-09-00-03270	ace 4; eth 4; EtOH 4
1341	1,2-Benzenedicarboxylic acid, dicyclohexyl ester	C$_{20}$H$_{26}$O$_4$ 330.42	84-61-7 66		4-09-00-03189 1.383[20]	H$_2$O 1; EtOH 3; eth 3; chl 2 1.431[20]
1342	1,2-Benzenedicarboxylic acid, diethyl ester Diethyl phthalate	C$_{12}$H$_{14}$O$_4$ 222.24	84-66-2 -40.5	7345 295	4-09-00-03172 1.232[14]	H$_2$O I; EtOH 5; eth 5; ace 3 1.5000[21]
1343	1,2-Benzenedicarboxylic acid, dimethyl ester Dimethyl phthalate	C$_{10}$H$_{10}$O$_4$ 194.19	131-11-3 5.5	3243 283.7	4-09-00-03170 1.1905[20]	H$_2$O 1; EtOH 5; eth 5; bz 3 1.5138[20]
1344	1,2-Benzenedicarboxylic acid, dioctyl ester Dioctyl phthalate	C$_{24}$H$_{38}$O$_4$ 390.56	117-84-0 25		4-09-00-03180	
1345	1,2-Benzenedicarboxylic acid, diphenyl ester Phenyl phthalate	C$_{20}$H$_{14}$O$_4$ 318.33	84-62-8 73	7278 253[14]	4-09-00-03217	H$_2$O 1; EtOH 2; eth 2; ctc 2
1346	1,3-Benzenedicarboxylic acid, diethyl ester Diethyl isophthalate	C$_{12}$H$_{14}$O$_4$ 222.24	636-53-3 11.5		4-09-00-03294 1.1239[17]	H$_2$O 1 1.508[18]
1347	1,3-Benzenedicarboxylic acid, dimethyl ester Dimethyl isophthalate	C$_{10}$H$_{10}$O$_4$ 194.19	1459-93-4 67.5	282	4-09-00-03293 1.194[20]	H$_2$O 2 1.5168[20]
1348	1,3-Benzenedicarboxylic acid, 4,6-dimethyl-	C$_{10}$H$_{10}$O$_4$ 194.19	2790-09-2 266	sub	3-09-00-04298	EtOH 4
1349	1,4-Benzenedicarboxylic acid, 2,5-dichloro- 2,5-Dichloroterephthalic acid	C$_8$H$_4$Cl$_2$O$_4$ 235.02	13799-90-1 306	sub	4-09-00-03335	H$_2$O 2; EtOH 3; eth 4
1350	1,4-Benzenedicarboxylic acid, diethyl ester Diethyl terephthalate	C$_{12}$H$_{14}$O$_4$ 222.24	636-09-9 44	302	4-09-00-03304 1.0989[45]	H$_2$O 1; EtOH 4; eth 3
1351	1,4-Benzenedicarboxylic acid, dimethyl ester Dimethyl terephthalate	C$_{10}$H$_{10}$O$_4$ 194.19	120-61-6 141	288	4-09-00-03303 1.075[141]	H$_2$O 2; EtOH 2; eth 3; chl 3
1352	1,3-Benzenedicarboxylic acid, 4-hydroxy- 4-Hydroxyisophthalic acid	C$_8$H$_6$O$_5$ 182.13	636-46-4 310	4758	4-10-00-02091	H$_2$O 1; EtOH 4; eth 4; chl 1
1353	1,2-Benzenedicarboxylic acid, 3,4,5,6-tetrachloro-, monoethyl ester Phthalic acid, tetrachloro-, monoethyl ester	C$_{10}$H$_6$Cl$_4$O$_4$ 331.97	602-21-1 94.5	250 dec	3-09-00-04206	eth 4; EtOH 4
1354	Benzene, 1,2-dichloro- o-Dichlorobenzene	C$_6$H$_4$Cl$_2$ 147.00	95-50-1 -16.7	3044 180	4-05-00-00654 1.3059[20]	H$_2$O 1; EtOH 3; eth 3; ace 5 1.5515[20]
1355	Benzene, 1,3-dichloro- m-Dichlorobenzene	C$_6$H$_4$Cl$_2$ 147.00	541-73-1 -24.8	3043 173	4-05-00-00657 1.2884[20]	H$_2$O 1; EtOH 3; eth 3; ace 5 1.5459[20]
1356	Benzene, 1,4-dichloro- p-Dichlorobenzene	C$_6$H$_4$Cl$_2$ 147.00	106-46-7 52.7	3045 174	4-05-00-00658 1.2475[55]	H$_2$O 1; EtOH 5; eth 3; ace 5 1.5285[20]
1357	Benzene, 1,2-dichloro-4-(chloromethyl)-	C$_7$H$_5$Cl$_3$ 195.47	102-47-6 37.5	241	4-05-00-00820	H$_2$O 1; EtOH 3; ctc 3; os 3
1358	Benzene, 1,3-dichloro-5-(chloromethyl)-	C$_7$H$_5$Cl$_3$ 195.47	3290-06-0 36	60[0.35]	3-05-00-00699	EtOH 4
1359	Benzene, 1,2-dichloro-4-(dichloromethyl)- Benzylidene chloride, 3,4-dichloro	C$_7$H$_4$Cl$_4$ 229.92	56961-84-3	257	3-05-00-00702 1.515[22]	bz 4; eth 4; EtOH 4
1360	Benzene, 1,2-dichloro-3,4-dimethyl- o-Xylene, 3,4-dichloro	C$_8$H$_8$Cl$_2$ 175.06	68266-67-1 9	231	4-05-00-00926 1.2240[20]	
1361	Benzene, 1,3-dichloro-2,5-dimethyl-	C$_8$H$_8$Cl$_2$ 175.06	38204-89-6 15	222	3-05-00-00855 1.2156[20]	
1362	Benzene, 1,4-dichloro-2,5-dimethyl-	C$_8$H$_8$Cl$_2$ 175.06	1124-05-6 71	222	3-05-00-00855	chl 3
1363	Benzene, 1,5-dichloro-2,4-dimethyl-	C$_8$H$_8$Cl$_2$ 175.06	2084-45-9 68.5	223	3-05-00-00835	bz 4; eth 4; chl 4
1364	Benzene, 2,3-dichloro-1,4-dimethyl-	C$_8$H$_8$Cl$_2$ 175.06	34840-79-4 -2	230	3-05-00-00855 1.2310[20]	
1365	Benzene, 1,1'-(1,2-dichloro-1,2-ethenediyl)bis-, (E)-	C$_{14}$H$_{10}$Cl$_2$ 249.14	951-86-0 143.5	183[18]	1-05-00-00304	H$_2$O 1; EtOH 4; eth 4
1366	Benzene, 1,1'-(1,2-dichloro-1,2-ethenediyl)bis-, (Z)-	C$_{14}$H$_{10}$Cl$_2$ 249.14	5216-32-0 67.5	179[18]	1-05-00-00304	eth 4; EtOH 4
1367	Benzene, 1,2-dichloro-3-ethenyl- 2,3-Dichlorostyrene	C$_8$H$_6$Cl$_2$ 173.04	2123-28-6	92-4[4.5]	3-05-00-01174 1.2826[20]	1.5848[20]
1368	Benzene, 1,2-dichloro-4-ethenyl-	C$_8$H$_6$Cl$_2$ 173.04	2039-83-0	95[5]	4-05-00-01347 1.256[20]	1.5857[20]
1369	Benzene, 1,3-dichloro-2-ethenyl- 2,6-Dichlorostyrene	C$_8$H$_6$Cl$_2$ 173.04	28469-92-3 89	64-5[3]	3-05-00-01174 1.2631[20]	1.5752[20]
1370	Benzene, 1,3-dichloro-5-ethenyl- 3,5-Dichlorostyrene	C$_8$H$_6$Cl$_2$ 173.04	2155-42-2	59[1]	3-05-00-01175 1.225[25]	1.5745[25]
1371	Benzene, 1,4-dichloro-2-ethenyl- 2,5-Dichlorostyrene	C$_8$H$_6$Cl$_2$ 173.04	1123-84-8	72-3[2]	4-05-00-01347 1.4045[20]	1.5798[20]
1372	Benzene, (2,2-dichloroethenyl)-	C$_8$H$_6$Cl$_2$ 173.04	698-88-4	225	4-05-00-01348 1.2531[20]	ace 3; chl 3 1.5852[20]

No.	Name / Synonym	Mol. Form. / Mol. Wt.	CAS RN / mp/°C	Merck No. / bp/°C	Beil. Ref. / den/g cm^{-3}	Solubility / n_D
1373	Benzene, 2,4-dichloro-1-ethenyl- / 2,4-Dichlorostyrene	$C_8H_6Cl_2$ / 173.04	2123-27-5 /	/ 81[6]	4-05-00-01347 / 1.243[25]	/ 1.5828[20]
1374	Benzene, 1,1'-(dichloroethenylidene)bis[4-chloro- / 2,2-Dichloro-1,1-bis(4-chlorophenyl)ethylene	$C_{14}H_8Cl_4$ / 318.03	72-55-9 / 89		4-05-00-02177 /	
1375	Benzene, 2,4-dichloro-1-ethoxy-	$C_8H_8Cl_2O$ / 191.06	5392-86-9 / 32	/ 237	4-06-00-00885 /	H_2O 1; EtOH 3; eth 3; bz 3
1376	Benzene, (1,2-dichloroethyl)-	$C_8H_8Cl_2$ / 175.06	1074-11-9 /	/ 233.5	4-05-00-00901 / 1.240[15]	H_2O 1; ace 3; bz 3 / 1.5554[15]
1377	Benzene, 1,4-dichloro-2-ethyl- / 2,5-Dichloroethylbenzene	$C_8H_8Cl_2$ / 175.06	54484-63-8 /	/ 213.5	4-05-00-00901 / 1.239[25]	eth 2; bz 3; chl 2; peth 2
1378	Benzene, 1,1'-(2,2-dichloroethylidene)bis[4-chloro- / 1,1-Dichloro-2,2-bis(p-chlorophenyl)ethane	$C_{14}H_{10}Cl_4$ / 320.04	72-54-8 / 109.5	3049 /	4-05-00-01884 /	chl 2
1379	Benzene, (dichlorofluoromethyl)-	$C_7H_5Cl_2F$ / 179.02	498-67-9 / -26.8	/ 179	4-05-00-00818 / 1.3138[11]	EtOH 4 / 1.5180[11]
1380	Benzene, 1,4-dichloro-2-iodo-	$C_6H_3Cl_2I$ / 272.90	29682-41-5 / 22	/ 253	3-05-00-00580 /	bz 4; eth 4; EtOH 4; chl 4
1381	Benzene, 2,4-dichloro-1-methoxy-	$C_7H_6Cl_2O$ / 177.03	553-82-2 / 28.5	/ 232; 125[10]	4-06-00-00885 /	chl 2
1382	Benzene, (dichloromethyl)- / Benzal chloride	$C_7H_6Cl_2$ / 161.03	98-87-3 / -17	1064 / 205	4-05-00-00817 / 1.26[25]	eth 4; EtOH 4 / 1.5502[20]
1383	Benzene, 1,2-dichloro-3-methyl- / 2,3-Dichlorotoluene	$C_7H_6Cl_2$ / 161.03	32768-54-0 / 6	/ 207.5	4-05-00-00815 / 1.2458[20]	bz 4 / 1.5511[20]
1384	Benzene, 1,2-dichloro-4-methyl- / 3,4-Dichlorotoluene	$C_7H_6Cl_2$ / 161.03	95-75-0 / -15.2	/ 208.9	4-05-00-00815 / 1.2564[20]	H_2O 1; EtOH 5; eth 5; ace 5 / 1.5471[20]
1385	Benzene, 1,3-dichloro-2-methyl- / 2,6-Dichlorotoluene	$C_7H_6Cl_2$ / 161.03	118-69-4 / 25.8	/ 198	4-05-00-00815 / 1.2686[20]	H_2O 1; chl 3 / 1.5507[20]
1386	Benzene, 1,3-dichloro-5-methyl- / 3,5-Dichlorotoluene	$C_7H_6Cl_2$ / 161.03	25186-47-4 / 26	/ 201.5	4-05-00-00816 /	H_2O 1
1387	Benzene, 1,4-dichloro-2-methyl- / 2,5-Dichlorotoluene	$C_7H_6Cl_2$ / 161.03	19398-61-9 / 5	/ 200	3-05-00-00694 / 1.2535[20]	H_2O 1; bz 3 / 1.5449[20]
1388	Benzene, 2,4-dichloro-1-methyl- / 2,4-Dichlorotoluene	$C_7H_6Cl_2$ / 161.03	95-73-8 / -13.5	/ 201	4-05-00-00815 / 1.2476[20]	H_2O 1; ctc 3 / 1.5511[20]
1389	Benzene, 1,1'-(dichloromethylene)bis-	$C_{13}H_{10}Cl_2$ / 237.13	2051-90-3 /	/ 305 dec; 190[21]	4-05-00-01848 / 1.235[18]	eth 3; bz 3; ctc 3
1390	Benzene, 1-(dichloromethyl)-4-methyl-	$C_8H_8Cl_2$ / 175.06	23063-36-7 / 48.5	/ 105[18]	3-05-00-00855 /	EtOH 2; bz 2; chl 2
1391	Benzene, 1,2-dichloro-3-nitro-	$C_6H_3Cl_2NO_2$ / 192.00	3209-22-1 / 61.5	/ 257.5	4-05-00-00725 / 1.721[14]	H_2O 1; EtOH 3; eth 3; ace 3
1392	Benzene, 1,2-dichloro-4-nitro-	$C_6H_3Cl_2NO_2$ / 192.00	99-54-7 / 43	/ 255.5	4-05-00-00726 / 1.4558[75]	H_2O 1; EtOH 3; eth 3; ctc 2
1393	Benzene, 1,3-dichloro-2-nitro- / 2,6-Dichloronitrobenzene	$C_6H_3Cl_2NO_2$ / 192.00	601-88-7 / 72.5	/ 130[8]	4-05-00-00726 / 1.603[17]	eth 4; EtOH 4
1394	Benzene, 1,3-dichloro-5-nitro-	$C_6H_3Cl_2NO_2$ / 192.00	618-62-2 / 65.4		4-05-00-00727 / 1.4000[100]	H_2O 1; EtOH 3; eth 3 / 1.4000[100]
1395	Benzene, 1,4-dichloro-2-nitro-	$C_6H_3Cl_2NO_2$ / 192.00	89-61-2 / 56	/ 267	4-05-00-00726 / 1.439[75]	H_2O 1; EtOH 3; eth 3; bz 3 / 1.4390[75]
1396	Benzene, 2,4-dichloro-1-nitro-	$C_6H_3Cl_2NO_2$ / 192.00	611-06-3 / 34	/ 258.5	4-05-00-00726 / 1.4790[80]	H_2O 1; EtOH 3; eth 3; chl 2 / 1.5512[70]
1397	Benzene, (3,3-dichloro-2-propenyl)-	$C_9H_8Cl_2$ / 187.07	38862-78-1 / 59	/ 242; 142[30]	4-05-00-01363 / 1.2028[20]	eth 3; bz 3; chl 3; peth 3
1398	Benzene, 1,2-dichloro-3,4,5,6-tetrafluoro / 1,2-Dichlorotetrafluorobenzene	$C_6Cl_2F_4$ / 218.97	1198-59-0 /			
1399	Benzene, 1,3-dichloro-2,4,5,6-tetrafluoro- / 1,3-Dichlorotetrafluorobenzene	$C_6Cl_2F_4$ / 218.97	1198-61-4 /			
1400	Benzene, 1,4-dichloro-2,3,5,6-tetrafluoro- / 1,4-Dichlorotetrafluorobenzene	$C_6Cl_2F_4$ / 218.97	1198-62-5 /			
1401	Benzene, 1,2-dichloro-4-(trichloromethyl)-	$C_7H_3Cl_5$ / 264.36	13014-24-9 / 25.8	/ 283.1	4-05-00-00824 / 1.5913[20]	/ 1.5886[20]
1402	Benzene, 2,4-dichloro-1,3,5-trimethyl- / Mesitylene, 2,4-dichloro-	$C_9H_{10}Cl_2$ / 189.08	57386-83-1 / 59	/ 243.5	3-05-00-00919 /	EtOH 3; eth 2; bz 2
1403	Benzene, 1,2-diethenyl- / o-Divinylbenzene	$C_{10}H_{10}$ / 130.19	91-14-5 /	/ 82[14]	4-05-00-01540 / 0.9325[22]	ace 3; bz 3 / 1.5767[20]
1404	Benzene, 1,3-diethenyl- / m-Divinylbenzene	$C_{10}H_{10}$ / 130.19	108-57-6 / -52.3	/ 121[76]; 52[3]	3-05-00-01367 / 0.9294[20]	ace 3; bz 3 / 1.5760[20]
1405	Benzene, 1,4-diethenyl- / p-Divinylbenzene	$C_{10}H_{10}$ / 130.19	105-06-6 / 31	/ 95[18]; 34[0.2]	4-05-00-01541 / 0.913[40]	ace 3; bz 3 / 1.5835[25]
1406	Benzene, 1,2-diethoxy-	$C_{10}H_{14}O_2$ / 166.22	2050-46-6 / 44	/ 219	4-06-00-05565 / 1.0075[20]	EtOH 3; eth 4; ctc 3 / 1.5083[25]
1407	Benzene, 1,3-diethoxy-	$C_{10}H_{14}O_2$ / 166.22	2049-73-2 / 12.4	/ 235	4-06-00-05664 / 1.0170[15]	H_2O 1; EtOH 3; eth 3
1408	Benzene, 1,4-diethoxy-	$C_{10}H_{14}O_2$ / 166.22	122-95-2 / 72	/ 246	4-06-00-05720 /	EtOH 4; eth 3; bz 3; ctc 3
1409	Benzene, (diethoxymethoxy)-	$C_{11}H_{16}O_3$ / 196.25	14444-77-0 /	/ 103-4[10]	4-06-00-00610 / 1.014[25]	/ 1.4822[20]
1410	Benzene, 1,4-diethoxy-2-methyl- / 2,5-Diethoxytoluene	$C_{11}H_{16}O_2$ / 180.25	41901-72-8 / 24.5	/ 248	3-06-00-04499 / 1.0134[15]	bz 4; eth 4; EtOH 4

No.	Name / Synonym	Mol. Form. / Mol. Wt.	CAS RN / mp/°C	Merck No. / bp/°C	Beil. Ref. / den/g cm⁻³	Solubility / n_D
1411	Benzene, 1-(diethoxymethyl)-4-(1-methylethyl)-	$C_{14}H_{22}O_2$ 222.33	35364-90-0	265; 142[18.5]	2-07-00-00247 0.944[19]	1.484[19]
1412	Benzene, (3,3-diethoxy-1-propynyl)-	$C_{13}H_{16}O_2$ 204.27	6142-95-6	99-100[2]	4-07-00-01173 0.991[25]	1.5170[20]
1413	Benzene, 1,2-diethyl- o-Diethylbenzene	$C_{10}H_{14}$ 134.22	135-01-3 -31.2	184	4-05-00-01065 0.8800[20]	H_2O 1; EtOH 5; eth 5; ace 5 1.5035[20]
1414	Benzene, 1,3-diethyl- m-Diethylbenzene	$C_{10}H_{14}$ 134.22	141-93-5 -83.9	181.1	4-05-00-01066 0.8602[20]	H_2O 1; EtOH 5; eth 5; ace 5 1.4955[20]
1415	Benzene, 1,4-diethyl- p-Diethylbenzene	$C_{10}H_{14}$ 134.22	105-05-5 -42.83	183.7	4-05-00-01067 0.8620[20]	H_2O 1; EtOH 5; eth 5; ace 5 1.4967[20]
1416	Benzene, 1,2-diethyl-3-methyl-	$C_{11}H_{16}$ 148.25	13632-93-4	206.6	4-05-00-01105 0.8910[20]	ace 4; bz 4; eth 4; EtOH 4 1.5105[20]
1417	Benzene, 1,2-diethyl-4-methyl- Toluene, 3,4-diethyl-	$C_{11}H_{16}$ 148.25	13732-80-4	203.6	4-05-00-01106 0.8762[20]	ace 4; bz 4; eth 4; EtOH 4 1.5039[20]
1418	Benzene, 1,3-diethyl-2-methyl- Toluene, 2,6-diethyl-	$C_{11}H_{16}$ 148.25	13632-95-6	208.8	4-05-00-01105 0.8907[20]	ace 4; bz 4; eth 4; EtOH 4 1.5106[20]
1419	Benzene, 1,3-diethyl-5-methyl-	$C_{11}H_{16}$ 148.25	2050-24-0 -74.1	205	4-05-00-01106 0.8748[20]	H_2O 1; EtOH 5; eth 5; ace 5 1.5027[20]
1420	Benzene, 1,4-diethyl-2-methyl-	$C_{11}H_{16}$ 148.25	13632-94-5	207.1	4-05-00-01106 0.8758[20]	ace 4; bz 4; eth 4; EtOH 4 1.5034[20]
1421	Benzene, 2,4-diethyl-1-methyl-	$C_{11}H_{16}$ 148.25	1758-85-6	205	4-05-00-01105 0.8748[20]	ace 4; bz 4; eth 4; EtOH 4 1.5027[20]
1422	Benzene, 1,2-difluoro- o-Difluorobenzene	$C_6H_4F_2$ 114.09	367-11-3 -34	94	4-05-00-00637 1.1599[18]	H_2O 1; ace 3; bz 3; chl 3 1.4451[18]
1423	Benzene, 1,3-difluoro- m-Difluorobenzene	$C_6H_4F_2$ 114.09	372-18-9 -69.0	82.6	4-05-00-00637 1.1572[20]	H_2O 1; ace 3; bz 3 1.4374[20]
1424	Benzene, 1,4-difluoro- p-Difluorobenzene	$C_6H_4F_2$ 114.09	540-36-3 -13	3132 89	4-05-00-00637 1.1701[20]	H_2O 1; ace 3; bz 3; ctc 2 1.4422[20]
1425	Benzene, (difluoromethyl)-	$C_7H_6F_2$ 128.12	455-31-2	140	4-05-00-00801 1.1370[19]	H_2O 1; EtOH 3 1.4577[20]
1426	Benzene, 2,4-difluoro-1-nitro-	$C_6H_3F_2NO_2$ 159.09	446-35-5 9.8	207	4-05-00-00720 1.457[14]	chl 2 1.5149[14]
1427	Benzene, 1,2-diiodo-	$C_6H_4I_2$ 329.91	615-42-9 27	287; 100[3]	4-05-00-00700 2.54[20]	H_2O 2; EtOH 2 1.7179[20]
1428	Benzene, 1,3-diiodo-	$C_6H_4I_2$ 329.91	626-00-6 40.4	285	4-05-00-00700 2.47[25]	eth 4; EtOH 4; chl 4
1429	Benzene, 1,4-diiodo-	$C_6H_4I_2$ 329.91	624-38-4 131.5	285	4-05-00-00700	H_2O 1; EtOH 3; eth 4; chl 2
1430	Benzene, 1,3-diisocyanato-2-methyl- Toluene-2,6-diisocyanate	$C_9H_6N_2O_2$ 174.16	91-08-7 18.3			H_2O 6; ace 3; bz 3
1431	Benzene, 2,4-diisocyanato-1-methyl- Toluene 2,4-diisocyanate	$C_9H_6N_2O_2$ 174.16	584-84-9 20.5	9456 251	4-13-00-00243 1.2244[20]	ace 4; bz 4; eth 4
1432	Benzene, 1,4-diisothiocyanato- Bitoscanate	$C_8H_4N_2S_2$ 192.27	4044-65-9 132	1318	4-13-00-00174	
1433	1,3-Benzenedimethanamine m-Xylenediamine	$C_8H_{12}N_2$ 136.20	1477-55-0	247	4-13-00-00293 1.052[20]	H_2O 4; eth 4; EtOH 4
1434	1,3-Benzenedimethanol	$C_8H_{10}O_2$ 138.17	626-18-6 57	156[13]	4-06-00-05966 1.1610[18]	H_2O 4; eth 4; EtOH 4
1435	1,4-Benzenedimethanol	$C_8H_{10}O_2$ 138.17	589-29-7 117.5	138-43[1]	4-06-00-05971	H_2O 4; ace 4; eth 4; EtOH 4
1436	Benzene, 1,2-dimethoxy- Veratrole	$C_8H_{10}O_2$ 138.17	91-16-7 22.5	9857 206	4-06-00-05564 1.080[25]	H_2O 2; EtOH 3; eth 3; ctc 3 1.5827[21]
1437	Benzene, 1,3-dimethoxy-	$C_8H_{10}O_2$ 138.17	151-10-0 -52	217.5	4-06-00-05663 1.0521[25]	H_2O 2; EtOH 3; eth 3; bz 3 1.5231[20]
1438	Benzene, 1,4-dimethoxy-	$C_8H_{10}O_2$ 138.17	150-78-7 59	212.6	4-06-00-05718 1.0375[55]	H_2O 2; EtOH 3; eth 4; bz 4
1439	Benzene, 1,2-dimethoxy-3-methyl-	$C_9H_{12}O_2$ 152.19	4463-33-6	204	4-06-00-05860 1.0335[20]	eth 4; EtOH 4 1.5121[25]
1440	Benzene, 1,2-dimethoxy-4-methyl-	$C_9H_{12}O_2$ 152.19	494-99-5 24	220	4-06-00-05879 1.0509[25]	H_2O 1; ctc 2; os 4 1.5257[25]
1441	Benzene, 1,3-dimethoxy-5-methyl-	$C_9H_{12}O_2$ 152.19	4179-19-5	244	3-06-00-04533 1.0478[15]	bz 4; eth 4; EtOH 4 1.5234[20]
1442	Benzene, 1,4-dimethoxy-2-(1-methylethyl)-	$C_{11}H_{16}O_2$ 180.25	4132-71-2	114-6[15]	4-06-00-05990 1.0129[17]	bz 4; eth 4 1.5103[17]
1443	Benzene, 1-(dimethoxymethyl)-3-nitro-	$C_9H_{11}NO_4$ 197.19	3395-79-7	162-4[19]	4-07-00-00592 1.209[15]	bz 4
1444	Benzene, 1-(dimethoxymethyl)-4-nitro-	$C_9H_{11}NO_4$ 197.19	881-67-4 24	294	4-07-00-00600	bz 4
1445	Benzene, 1,2-dimethoxy-4-nitro-	$C_8H_9NO_4$ 183.16	709-09-1 98	230[15]	4-06-00-05627 1.1888[133]	H_2O 1; EtOH 4; eth 4; chl 3
1446	Benzene, 1,3-dimethoxy-5-nitro-	$C_8H_9NO_4$ 183.16	16147-07-2 89		3-06-00-04347 1.1693[133]	bz 4; EtOH 4
1447	Benzene, 1,4-dimethoxy-2-nitro-	$C_8H_9NO_4$ 183.16	89-39-4 72.5		4-06-00-05786 1.1666[132]	H_2O 1; EtOH 3; bz 3; chl 3
1448	Benzene, 2,4-dimethoxy-1-nitro- 1,3-Dimethoxy-4-nitrobenzene	$C_8H_9NO_4$ 183.16	4920-84-7 76.5		4-06-00-05691 1.1876[132]	H_2O 1; EtOH 3; os 3; lig 1
1449	Benzene, 1,2-dimethoxy-4-(1-propenyl)-	$C_{11}H_{14}O_2$ 178.23	93-16-3 18	270.5	2-06-00-00918 1.0521[20]	1.5616[20]

No.	Name Synonym	Mol. Form. Mol. Wt.	CAS RN mp/°C	Merck No. bp/°C	Beil. Ref. den/g cm[-3]	Solubility n_D
1450	Benzene, 1,2-dimethoxy-4-(2-propenyl)-	$C_{11}H_{14}O_2$ 178.23	93-15-2 -4	254.7	4-06-00-06337 1.0396[20]	H$_2$O 1; EtOH 3; eth 3 1.5340[20]
1451	Benzene, 1,2-dimethoxy-4-(1-propenyl)-, (Z)-	$C_{11}H_{14}O_2$ 178.23	6380-24-1 70	270.5	4-06-00-06325 1.0521[20]	H$_2$O 1; ace 3; bz 3 1.5616[20]
1452	Benzene, 1,2-dimethyl- o-Xylene	C_8H_{10} 106.17	95-47-6 -25.2	9988 144.5	4-05-00-00917 0.8802[10]	H$_2$O 1; EtOH 5; eth 5; ace 5 1.5055[20]
1453	Benzene, 1,3-dimethyl- m-Xylene	C_8H_{10} 106.17	108-38-3 -47.8	9988 139.1	4-05-00-00932 0.8642[20]	H$_2$O 1; EtOH 5; eth 5; ace 5 1.4972[10]
1454	Benzene, 1,4-dimethyl- p-Xylene	C_8H_{10} 106.17	106-42-3 13.2	9988 138.3	4-05-00-00951 0.8611[20]	H$_2$O 1; EtOH 5; eth 5; ace 5 1.4958[20]
1455	Benzene, (1,1-dimethylbutyl)- Pentane, 2-methyl-2-phenyl	$C_{12}H_{18}$ 162.27	1985-57-5 	205.5	4-05-00-01118 0.8796[10]	1.4955[16]
1456	Benzene, 1,1'-(1,2-dimethyl-1,2-ethenediyl)bis-, (E)-	$C_{16}H_{16}$ 208.30	782-06-9 107	156[16]	4-05-00-02207 0.987[20]	H$_2$O 1; EtOH 2; eth 3, MeOH 2
1457	Benzene, 1,1'-(1,2-dimethyl-1,2-ethenediyl)bis-, (Z)-	$C_{16}H_{16}$ 208.30	782-05-8 67.5		4-05-00-02207 1.004[20]	H$_2$O 1; EtOH 3; HOAc 3; MeOH 3 1.5612[78]
1458	Benzene, (1,1-dimethylethoxy)-	$C_{10}H_{14}O$ 150.22	6669-13-2 -24	185.5	4-06-00-00559 0.9214[20]	
1459	Benzene, (1,1-dimethylethyl)- tert-Butylbenzene	$C_{10}H_{14}$ 134.22	98-06-6 -57.8	1551 169.1	4-05-00-01045 0.8665[20]	H$_2$O 1; EtOH 4; eth 4; ace 5 1.4927[20]
1460	Benzene, 1-(1,1-dimethylethyl)-3,5-dimethyl-	$C_{12}H_{18}$ 162.27	98-19-1 -18	207	4-05-00-01130 0.8668[20]	ctc 3
1461	Benzene, 1-(1,1-dimethylethyl)-3,5-dimethyl-2,4,6-trinitro- 2,4,6-Trinitro-3,5-dimethyl-tert-butylbenzene	$C_{12}H_{15}N_3O_6$ 297.27	81-15-2 110		4-05-00-01132	H$_2$O 1; EtOH 2; eth 3; chl 3
1462	Benzene, 1-(1,1-dimethylethyl)-4-methoxy-	$C_{11}H_{16}O$ 164.25	5396-38-3 	238	4-06-00-03297 0.9436[20]	1.5039[20]
1463	Benzene, 1-(1,1-dimethylethyl)-2-methoxy-4-methyl-3,5-dinitro-	$C_{12}H_{16}N_2O_5$ 268.27	83-66-9 85	185[16]	4-06-00-03402	H$_2$O 1; EtOH 2; eth 3; chl 3
1464	Benzene, 1-(1,1-dimethylethyl)-2-methyl-	$C_{11}H_{16}$ 148.25	1074-92-6 -50.3	200.4	4-05-00-01096 0.8897[20]	ace 4; bz 4; eth 4; EtOH 4 1.5076[20]
1465	Benzene, 1-(1,1-dimethylethyl)-3-methyl-	$C_{11}H_{16}$ 148.25	1075-38-3 -41.4	189.3	4-05-00-01096 0.8657[20]	ace 4; bz 4; eth 4; EtOH 4 1.4944[20]
1466	Benzene, 1-(1,1-dimethylethyl)-4-methyl- p-tert-Butyltoluene	$C_{11}H_{16}$ 148.25	98-51-1 -52	190	4-05-00-01097 0.8612[20]	H$_2$O 1; EtOH 2; eth 4; ace 3 1.4918[20]
1467	Benzene, 1-(1,1-dimethylethyl)-3-nitro-	$C_{10}H_{13}NO_2$ 179.22	23132-52-7 2	254; 136[16]	4-05-00-01052 1.0643[25]	
1468	Benzene, 1-(1,5-dimethyl-4-hexenyl)-4-methyl- ar-Curcumene	$C_{15}H_{22}$ 202.34	644-30-4 	140[19]	4-05-00-01465 0.8805[20]	H$_2$O 1; bz 3 1.4989[20]
1469	Benzene, 1,2-dimethyl-3-(1-methylethyl)-	$C_{11}H_{16}$ 148.25	22539-65-7 	202.6	4-05-00-01103 0.888[20]	ace 4; bz 4; eth 4; EtOH 4 1.508[20]
1470	Benzene, 1,2-dimethyl-4-(1-methylethyl)-	$C_{11}H_{16}$ 148.25	4132-77-8 	200	4-05-00-01103 0.8710[20]	1.4951[20]
1471	Benzene, 1,3-dimethyl-2-(1-methylethyl)-	$C_{11}H_{16}$ 148.25	14411-75-7 	199	4-05-00-01103 0.890[20]	ace 4; bz 4; eth 4; EtOH 4 1.509[20]
1472	Benzene, 2,4-dimethyl-1-(1-methylethyl)-	$C_{11}H_{16}$ 148.25	4706-89-2 -82	197	4-05-00-01104 0.8734[20]	1.4998[25]
1473	Benzene, 2,4-dimethyl-1-(1-methylpropyl)-	$C_{12}H_{18}$ 162.27	1483-60-9 	84[8]	4-05-00-01129 0.8664[20]	1.4939[25]
1474	Benzene, 1,2-dimethyl-3-nitro-	$C_8H_9NO_2$ 151.17	83-41-0 15	240	4-05-00-00930 1.1402[20]	H$_2$O 1; EtOH 3; ctc 3 1.5441[20]
1475	Benzene, 1,2-dimethyl-4-nitro- 4-Nitro-o-xylene	$C_8H_9NO_2$ 151.17	99-51-4 30.5	251; 143[21]	4-05-00-00930 1.112[15]	H$_2$O 1; EtOH 5 1.5202[20]
1476	Benzene, 1,3-dimethyl-2-nitro-	$C_8H_9NO_2$ 151.17	81-20-9 15	226	4-05-00-00948 1.112[15]	H$_2$O 1; EtOH 4; ctc 3 1.5202[20]
1477	Benzene, 1,3-dimethyl-5-nitro-	$C_8H_9NO_2$ 151.17	99-12-7 75	274	4-05-00-00948	H$_2$O 1; EtOH 4; eth 4
1478	Benzene, 1,4-dimethyl-2-nitro-	$C_8H_9NO_2$ 151.17	89-58-7 -25	240.5	4-05-00-00971 1.132[15]	H$_2$O 1; EtOH 3 1.5413[20]
1479	Benzene, 2,4-dimethyl-1-nitro-	$C_8H_9NO_2$ 151.17	89-87-2 9	247; 122[18]	4-05-00-00948 1.135[15]	H$_2$O 1; eth 3; ace 3; bz 3 1.5473[25]
1480	Benzene, (1,1-dimethylpropyl)-	$C_{11}H_{16}$ 148.25	2049-95-8 	192.4	4-05-00-01090 0.8748[20]	1.4958[20]
1481	Benzene, (1,2-dimethylpropyl)-	$C_{11}H_{16}$ 148.25	4481-30-5 	187	4-05-00-01092 0.8672[16]	1.4972[16]
1482	Benzene, 1,3-dimethyl-2-propyl-	$C_{11}H_{16}$ 148.25	17059-45-9 	207.6	4-05-00-01102 0.8856[20]	ace 4; bz 4; eth 4; EtOH 4 1.5063[20]
1483	Benzene, 1,3-dimethyl-5-propyl-	$C_{11}H_{16}$ 148.25	3982-64-7 -59.1	202.2	4-05-00-01102 0.8607[20]	ace 4; bz 4; eth 4; EtOH 4 1.4952[20]
1484	Benzene, (2,2-dimethylpropyl)-	$C_{11}H_{16}$ 148.25	1007-26-7 	186	4-05-00-01092 0.8581[18]	1.4884[18]
1485	Benzene, 1,2-dimethyl-3-propyl- o-Xylene, 3-propyl	$C_{11}H_{16}$ 148.25	17059-44-8 	210.7	4-05-00-01101 0.8864[20]	ace 4; bz 4; eth 4; EtOH 4 1.5075[20]
1486	Benzene, 1,4-dimethyl-2-propyl-	$C_{11}H_{16}$ 148.25	3042-50-0 	204.3	4-05-00-01102 0.8717[20]	ace 4; bz 4; eth 4; EtOH 4 1.4999[20]
1487	Benzene, 2,4-dimethyl-1-propyl- m-Xylene, 4-propyl-	$C_{11}H_{16}$ 148.25	61827-85-8 	206.6	4-05-00-01102 0.8723[20]	ace 4; bz 4; eth 4; EtOH 4 1.4998[20]

No.	Name Synonym	Mol. Form. Mol. Wt.	CAS RN mp/°C	Merck No. bp/°C	Beil. Ref. den/g cm^{-3}	Solubility n_D
1488	Benzene, 1-(1,1-dimethylpropyl)-4-methoxy-	$C_{12}H_{18}O$ 178.27	2050-03-5	238	4-06-00-03383 0.9436[20]	1.5039[20]
1489	Benzene, 1,4-dimethyl-2,3,5-trinitro- p-Xylene, 2,3,5-trinitro-	$C_8H_7N_3O_6$ 241.16	602-27-7 139.5	exp 410	3-05-00-00864 1.59[19]	EtOH 4
1490	Benzene, 2,4-dimethyl-1,3,5-trinitro-	$C_8H_7N_3O_6$ 241.16	632-92-8 184		4-05-00-00950 1.604[19]	H_2O 1; EtOH 2; eth 2; ace 3
1491	Benzene, 1,2-dinitro- o-Dinitrobenzene	$C_6H_4N_2O_4$ 168.11	528-29-0 118.5	318; 194[30]	4-05-00-00738 1.3119[120]	H_2O 1; EtOH 3; bz 3; chl 3 1.565[17]
1492	Benzene, 1,3-dinitro- m-Dinitrobenzene	$C_6H_4N_2O_4$ 168.11	99-65-0 90	291; 167[14]	4-05-00-00739 1.5751[18]	H_2O 2; EtOH 4; eth 3; ace 4
1493	Benzene, 1,4-dinitro- p-Dinitrobenzene	$C_6H_4N_2O_4$ 168.11	100-25-4 174	297; 183[34]	4-05-00-00741 1.625[18]	H_2O 1; EtOH 2; ace 3; bz 3
1494	Benzene, 1,1'-(1,2-dinitro-1,2-ethenediyl)bis-, (Z)-	$C_{14}H_{10}N_2O_4$ 270.24	1796-05-0 108.5	dec	4-05-00-02171	H_2O 1; EtOH 3; eth 3; ace 4
1495	Benzene, 2,4-dinitro-1-phenoxy-	$C_{12}H_8N_2O_5$ 260.21	2486-07-9 71	230[27]	4-06-00-01375	eth 4; EtOH 4
1496	Benzene, 2,4-dinitro-1-(2-phenylethenyl)-, (E)- Stilbene, 2,4-dinitro (trans)	$C_{14}H_{10}N_2O_4$ 270.24	56456-42-9 144	exp 412	4-05-00-02170	EtOH 1; bz 2; chl 3; CS_2 3
1497	1,2-Benzenediol Pyrocatechol	$C_6H_6O_2$ 110.11	120-80-9 105	8009 245	4-06-00-05557 1.1493[22]	H_2O 4; bz 4; eth 4; EtOH 4 1.604
1498	1,3-Benzenediol Resorcinol	$C_6H_6O_2$ 110.11	108-46-3 111	8158 178[16]	4-06-00-05658 1.2717[25]	H_2O 3; EtOH 3; eth 3; bz 2
1499	1,4-Benzenediol p-Hydroquinone	$C_6H_6O_2$ 110.11	123-31-9 172.3	4738 287	4-06-00-05712 1.328[15]	H_2O 3; EtOH 4; eth 3; ace 4
1500	1,2-Benzenediol, 4-(2-amino-1-hydroxyethyl)-, (R)- Norepinephrine	$C_8H_{11}NO_3$ 169.18	51-41-2 217 dec	6612		H_2O 2; EtOH 2; eth 2; alk 4
1501	1,4-Benzenediol, 2,5-bis(1,1-dimethylpropyl)- 2,5-Di-tert-pentylhydroquinone	$C_{16}H_{26}O_2$ 250.38	79-74-3 180	3300	3-06-00-04748	
1502	1,3-Benzenediol, 4-bromo-	$C_6H_5BrO_2$ 189.01	6626-15-9 103	140[12]	4-06-00-05687	H_2O 4; EtOH 2; eth 4; bz 2
1503	1,4-Benzenediol, 2-bromo-	$C_6H_5BrO_2$ 189.01	583-69-7 111.5	sub	4-06-00-05780	H_2O 4; EtOH 4; eth 4; bz 4
1504	1,2-Benzenediol, 3-chloro-	$C_6H_5ClO_2$ 144.56	4018-65-9 48.5	110[11]	4-06-00-05613	lig 4
1505	1,2-Benzenediol, 4-chloro-	$C_6H_5ClO_2$ 144.56	2138-22-9 90.5	139[10.5]	4-06-00-05614	H_2O 4; ace 4; eth 4; EtOH 4
1506	1,3-Benzenediol, 5-chloro-	$C_6H_5ClO_2$ 144.56	52780-23-1 118	sub	4-06-00-05684	H_2O 3; eth 3; ace 3; os 3
1507	1,4-Benzenediol, 2-chloro-	$C_6H_5ClO_2$ 144.56	615-67-8 108	263	4-06-00-05767	H_2O 4; EtOH 3; eth 3; bz 4
1508	1,2-Benzenediol, diacetate	$C_{10}H_{10}O_4$ 194.19	635-67-6 64.5	142-3[9]	4-06-00-05582	H_2O 1; EtOH 4; eth 4; chl 4
1509	1,4-Benzenediol, diacetate	$C_{10}H_{10}O_4$ 194.19	1205-91-0 123.5		4-06-00-05741 0.8731[25]	H_2O 3; EtOH 4; eth 4; chl 4
1510	1,3-Benzenediol, 4,6-dichloro-	$C_6H_4Cl_2O_2$ 179.00	137-19-9 113	254	4-06-00-05685	H_2O 4; EtOH 4; eth 4; ace 4
1511	1,2-Benzenediol, 4,5-dimethyl-	$C_8H_{10}O_2$ 138.17	2785-74-2 87.5	160[26]	4-06-00-05950	H_2O 4; EtOH 4; eth 4; bz 2
1512	1,3-Benzenediol, 2,5-dimethyl-	$C_8H_{10}O_2$ 138.17	488-87-9 163	278.5	4-06-00-05970	H_2O 3; EtOH 3; eth 3
1513	1,3-Benzenediol, 4,5-dimethyl-	$C_8H_{10}O_2$ 138.17	527-55-9 136	284	4-06-00-05948 1.1994[20]	H_2O 3; EtOH 4; eth 3; bz 2
1514	1,3-Benzenediol, 4,6-dimethyl-	$C_8H_{10}O_2$ 138.17	615-89-4 126.5	276	4-06-00-05963	H_2O 3; EtOH 3; eth 3
1515	1,2-Benzenediol, 4,4'-(2,3-dimethyl-1,4- butanediyl)bis- Nordihydroguaiaretic acid	$C_{18}H_{22}O_4$ 302.37	500-38-9 185.5	6610	4-06-00-07771	H_2O 2; EtOH 3; eth 3; ace 3
1516	1,2-Benzenediol, 4-(1,1-dimethylethyl)-	$C_{10}H_{14}O_2$ 166.22	98-29-3 54.3	285; 160[22]	4-06-00-06014	tfa 3
1517	1,3-Benzenediol, 2,4-dinitro- 2,4-Dinitroresorcinol	$C_6H_4N_2O_6$ 200.11	519-44-8 147.5	3278	4-06-00-05696	H_2O 2; EtOH 2
1518	1,3-Benzenediol, 4-ethyl-	$C_8H_{10}O_2$ 138.17	2896-60-8 98.5	160[24; 131[15]	4-06-00-05924	H_2O 2; EtOH 2; eth 2
1519	1,3-Benzenediol, 4-hexyl- 4-Hexylresorcinol	$C_{12}H_{18}O_2$ 194.27	136-77-6 68	4633 334	4-06-00-06048	ace 4; eth 4; EtOH 4; chl 4
1520	1,4-Benzenediol, 2-(hydroxymethyl)- Gentisyl alcohol	$C_7H_8O_3$ 140.14	495-08-9 100	4292 sub 75	4-06-00-07380	H_2O 4; EtOH 4; chl 4
1521	1,2-Benzenediol, 4-[1-hydroxy-2- (methylamino)ethyl]-, (R)- Epinephrine	$C_9H_{13}NO_3$ 183.21	51-43-4 211.5	3569	4-13-00-02927	H_2O 2; EtOH 1; HOAc 3; min acid 3
1522	1,2-Benzenediol, 4-[1-hydroxy-2-[(1- methylethyl)amino]ethyl]- Ethanol, 1(3,4-dihydroxyphenyl)- 2(isopropylamino)	$C_{11}H_{17}NO_3$ 211.26	7683-59-2 170.5	5105	3-13-00-02387	
1523	1,2-Benzenediol, 4-iodo- 4-Iodocatechol	$C_6H_5IO_2$ 236.01	76149-14-9 92	sub	4-06-00-05625	ace 4; bz 4; eth 4; EtOH 4
1524	1,3-Benzenediol, 5-iodo- Resorcinol, 5-iodo	$C_6H_5IO_2$ 236.01	64339-43-1 92.5	sub	2-06-00-00821	EtOH 4

No.	Name / Synonym	Mol. Form. / Mol. Wt.	CAS RN / mp/°C	Merck No. / bp/°C	Beil. Ref. / den/g cm^{-3}	Solubility / n_D
1525	1,2-Benzenediol, 3-methoxy-	$C_7H_8O_3$ 140.14	934-00-9 42.8	 163[48]; 129[10]	4-06-00-07329	chl 3
1526	1,3-Benzenediol, 5-methoxy-	$C_7H_8O_3$ 140.14	2174-64-3 80.3	 213[16]	4-06-00-07362	H_2O 2; EtOH 4; eth 4; bz 2
1527	1,2-Benzenediol, 3-methyl-	$C_7H_8O_2$ 124.14	488-17-5 68	 241	4-06-00-05860	H_2O 3; EtOH 3; bz 3; chl 3
1528	1,2-Benzenediol, 4-methyl-	$C_7H_8O_2$ 124.14	452-86-8 65	 251	4-06-00-05878 1.1287[74]	H_2O 3; EtOH 3; eth 3; ace 3 1.5425[74]
1529	1,3-Benzenediol, 2-methyl-	$C_7H_8O_2$ 124.14	608-25-3 120	 264	4-06-00-05877	H_2O 4; bz 4; eth 4; EtOH 4
1530	1,3-Benzenediol, 4-methyl-	$C_7H_8O_2$ 124.14	496-73-1 106	 268.5	4-06-00-05864	H_2O 3; EtOH 3; eth 3; bz 2
1531	1,3-Benzenediol, 5-methyl- Orcinol	$C_7H_8O_2$ 124.14	504-15-4 107.5	6819 289.5	4-06-00-05892 1.290[4]	H_2O 3; EtOH 3; eth 3; bz 3
1532	1,4-Benzenediol, 2-methyl-	$C_7H_8O_2$ 124.14	95-71-6 128	 163[11]	4-06-00-05866	H_2O 4; EtOH 4; eth 4; ace 3
1533	1,2-Benzenediol, 4-[2-(methylamino)ethyl]- Deoxyepinephrine	$C_9H_{13}NO_2$ 167.21	501-15-5 188.5	2885	4-13-00-02606	
1534	1,3-Benzenediol, 4-(3-methylbutyl)-	$C_{11}H_{16}O_2$ 180.25	15116-17-3 67	 177[7]	3-06-00-04700	bz 4; eth 4; EtOH 4
1535	1,2-Benzenediol, 4-(1-methylethyl)-	$C_9H_{12}O_2$ 152.19	2138-43-4 78	 271	3-06-00-04632	peth 3
1536	1,2-Benzenediol, 3-methyl-6-(1-methylethyl)-	$C_{10}H_{14}O_2$ 166.22	490-06-2 48	 267	4-06-00-06019	ace 4; eth 4
1537	1,3-Benzenediol, 2-[3-methyl-6-(1-methylethenyl)-2-cyclohexen-1-yl]-5-pentyl-, (1R-trans)- Cannabidiol	$C_{21}H_{30}O_2$ 314.47	13956-29-1 67	1750 187-90[2]	3-06-00-05362 1.040[40]	H_2O 1; EtOH 3; eth 3; bz 3 1.5404[20]
1538	1,4-Benzenediol, 2-methyl-5-(1-methylethyl)-	$C_{10}H_{14}O_2$ 166.22	2217-60-9 148	 290	4-06-00-06019	eth 4; EtOH 4
1539	1,3-Benzenediol, 4-(2-methylpropyl)- Resorcinol, 4-isobutyl-	$C_{10}H_{14}O_2$ 166.22	18979-62-9 62.5	 166[6]	3-06-00-04667	eth 4; EtOH 4
1540	1,2-Benzenediol, monoacetate	$C_8H_8O_3$ 152.15	2848-25-1 57.5	 189[102]; 148[25]	4-06-00-05581	H_2O 4; ace 4; EtOH 4; chl 4
1541	1,3-Benzenediol, 4-nitro-	$C_6H_5NO_4$ 155.11	3163-07-3 122	 178[11]	4-06-00-05691	EtOH 3; eth 3; bz 3; ctc 2
1542	1,3-Benzenediol, 4-[(4-nitrophenyl)azo]- Magneson	$C_{12}H_9N_3O_4$ 259.22	74-39-5 200	5578	4-16-00-00266	H_2O 1; EtOH 2; bz 2; HOAc 2
1543	1,2-Benzenediol, 3-pentadecyl- 3-Pentadecylcatechol	$C_{21}H_{36}O_2$ 320.52	492-89-7 59.5	7061	4-06-00-06109	bz 4; eth 4; EtOH 4
1544	1,3-Benzenediol, 4-pentyl-	$C_{11}H_{16}O_2$ 180.25	533-24-4 72.5	 168[6]	3-06-00-04693	bz 4; eth 4; EtOH 4
1545	1,3-Benzenediol, 5-(2-phenylethenyl)-, (E)- Pinosylvin	$C_{14}H_{12}O_2$ 212.25	22139-77-1 156	7419	3-06-00-05577	ace 4; bz 4; chl 4; HOAc 4
1546	1,4-Benzenediol, 2-(phenylmethyl)-	$C_{13}H_{12}O_2$ 200.24	1706-73-6 105.8	 230[13]	3-06-00-05404	eth 4; EtOH 4
1547	1,2-Benzenediol, 3-(2-propenyl)-	$C_9H_{10}O_2$ 150.18	1125-74-2	 141-5[16]	4-06-00-06334 1.1241[20]	1.5656[20]
1548	1,2-Benzenediol, 4-propyl-	$C_9H_{12}O_2$ 152.19	2525-02-2 60	 177[30]; 111[0.2]	3-06-00-04613 1.100[18]	1.4440[18]
1549	1,3-Benzenediol, 4-propyl-	$C_9H_{12}O_2$ 152.19	18979-60-7 82.5	 172[14]	4-06-00-05975	H_2O 3; EtOH 2; eth 3
1550	1,3-Benzenediol, 5-propyl-	$C_9H_{12}O_2$ 152.19	500-49-2 82.8	 169[8]; 148[3]	3-06-00-04622	ace 4; bz 4; eth 4; EtOH 4
1551	1,4-Benzenediol, 2,3,5,6-tetrabromo-	$C_6H_2Br_4O_2$ 425.70	2641-89-6 244		3-06-00-04440 3.023[21]	eth 4; EtOH 4
1552	1,4-Benzenediol, 2,3,5,6-tetramethyl- Durohydroquinone	$C_{10}H_{14}O_2$ 166.22	527-18-4 233	3451	4-06-00-06028	EtOH 3; eth 2
1553	1,3-Benzenediol, 2,4,6-trimethyl-	$C_9H_{12}O_2$ 152.19	608-98-0 150.8	 274.5	4-06-00-06000	bz 4; eth 4; EtOH 4
1554	1,3-Benzenediol, 2,4,6-trinitro- Styphnic acid	$C_6H_3N_3O_8$ 245.11	82-71-3 175.5	8828 sub	4-06-00-05699	eth 4; EtOH 4
1555	Benzene, 1,2-dipropoxy-	$C_{12}H_{18}O_2$ 194.27	6280-98-4	 235.5	4-06-00-05566 0.9554[33]	chl 2 1.4950[27]
1556	Benzene, 1,3-dipropoxy- Resorcinol dipropyl ether	$C_{12}H_{18}O_2$ 194.27	56106-37-7	 253	3-06-00-04308 1.033[20]	1.5138[83]
1557	Benzene, 1,4-dipropyl-	$C_{12}H_{18}$ 162.27	4815-57-0	 109[23]	4-05-00-01124 0.8563[19]	1.4917[19]
1558	1,3-Benzenedisulfonamide, 4-amino-6-chloro- Chloraminophenamide	$C_6H_8ClN_3O_4S_2$ 285.73	121-30-2 254.5	2067	4-14-00-02810	
1559	1,3-Benzenedisulfonamide, 4,5-dichloro- 1,3-Benzenedisulfonic acid, diamide,4,5-dichloro-	$C_6H_6ClN_2O_4$ 205.58	120-97-8 228.7	3067	4-11-00-00555	
1560	1,3-Benzenedisulfonic acid, 4-hydroxy- Phenoldisulfonic acid	$C_6H_6O_7S_2$ 254.24	96-77-5 >100 dec	7207	2-11-00-00139	H_2O 4; EtOH 4
1561	1,2-Benzenedithiol	$C_6H_6S_2$ 142.25	17534-15-5 28.5	 238.5	4-06-00-05651	EtOH 4; eth 4; bz 4; AcOEt 3

No.	Name / Synonym	Mol. Form. / Mol. Wt.	CAS RN / mp/°C	Merck No. / bp/°C	Beil. Ref. / den/g cm^{-3}	Solubility / n_D
1562	1,3-Benzenedithiol	$C_6H_6S_2$ 142.25	626-04-0 27	245	4-06-00-05705	bz 4; eth 4; EtOH 4
1563	1,3-Benzenedithiol, 4-chloro-	$C_6H_5ClS_2$ 176.69	58593-78-5	145-6[13]	4-06-00-05708 1.393[25]	1.6704[20]
1564	1,2-Benzenedithiol, 4-methyl- Toluene-3,4-dithiol	$C_7H_8S_2$ 156.27	496-74-2 29	9457	4-06-00-05890	chl 3
1565	Benzene, docosyl- Docosane, 1-phenyl-	$C_{28}H_{50}$ 386.71	5634-22-0 51	448	3-05-00-01142 0.8544[20]	
1566	Benzene, dodecyl- 1-Phenyldodecane	$C_{18}H_{30}$ 246.44	123-01-3 3	328	4-05-00-01200 0.8551[20]	H_2O 1 1.4824[20]
1567	Benzene, 1,1'-dodecylidenebis[4-methyl- Dodecane, 1,1-di-(4-tolyl)	$C_{26}H_{38}$ 350.59	55268-62-7	197[1.0]	3-05-00-01925 0.9117[20]	1.5001[20]
1568	Benzene, eicosyl-	$C_{26}H_{46}$ 358.65	2398-68-7 44	429	4-05-00-01222 0.8545[20]	1.4805[20]
1569	Benzeneethanamine Phenylethylamine	$C_8H_{11}N$ 121.18	64-04-0	7186 197.5	4-12-00-02453 0.9580[24]	H_2O 3; eth 4; eth 4; ctc 3 1.5290[25]
1570	Benzeneethanamine, 4-bromo- Phenethylamine, 4-bromo	$C_8H_{10}BrN$ 200.08	73918-56-6	63-72[0.2]	4-12-00-02492 1.29[25]	1.5750[20]
1571	Benzeneethanamine, 3,4-dimethoxy-	$C_{10}H_{15}NO_2$ 181.23	120-20-7	163-5[14]	4-13-00-02604	ctc 3 1.5464[20]
1572	Benzeneethanamine, α,α-dimethyl- Phentermine	$C_{10}H_{15}N$ 149.24	122-09-8	7232 205; 100[21]	4-12-00-02820	
1573	Benzeneethanamine, N,α-dimethyl-, (S)- Methamphetamine	$C_{10}H_{15}N$ 149.24	537-46-2	5859	4-12-00-02589	
1574	Benzeneethanamine, N,β-dimethyl- Phenylpropylmethylamine	$C_{10}H_{15}N$ 149.24	93-88-9	7280 207.5	3-12-00-02689 0.915[25]	bz 4; eth 4; EtOH 4
1575	Benzeneethanamine, N,α-dimethyl-N-(phenylmethyl)-, (+)- Benzphetamine	$C_{17}H_{21}N$ 239.36	156-08-1	1130 127[0.02]	4-12-00-02594	eth 4; EtOH 4; MeOH 4; chl 4 1.5515[19]
1576	Benzeneethanamine, N-ethyl-α-methyl- N-Ethylamphetamine	$C_{11}H_{17}N$ 163.26	457-87-4	3720 105-6[14]	3-12-00-02668	1.4986[25]
1577	Benzeneethanamine, 2-methoxy- Phenethylamine, 2-methoxy	$C_9H_{13}NO$ 151.21	2045-79-6	236.5	3-13-00-01625 1.0400[25]	1.5422[20]
1578	Benzeneethanamine, α-methyl-, (±)- Amphetamine	$C_9H_{13}N$ 135.21	300-62-9	616 203	4-12-00-02587 0.9306[25]	H_2O 2, EtOH 3; eth 2; chl 3 1.518[26]
1579	Benzeneethanamine, β-methyl-	$C_9H_{13}N$ 135.21	582-22-9	210	4-12-00-02634 0.9433[4]	bz 4; eth 4; EtOH 4 1.5255[20]
1580	Benzeneethanamine, 4-methyl-	$C_9H_{13}N$ 135.21	3261-62-9	214	4-12-00-02641 0.93[25]	1.5257[20]
1581	Benzeneethanamine, N-methyl-	$C_9H_{13}N$ 135.21	589-08-2	206	4-12-00-02453 0.93[25]	1.5162[20]
1582	Benzeneethanamine, α-methyl-, (S)-, sulfate (2:1) Dextroamphetamine sulfate	$C_{18}H_{28}N_2O_4S$ 368.50	51-63-8 >300	2932	3-12-00-02666 1.15[25]	H_2O 4
1583	Benzeneethanamine, α-methyl-N-(2,2,2-trichloroethylidene)- Amphecloral	$C_{11}H_{12}Cl_3N$ 264.58	5581-35-1	613 96.[0.5]		1.530
1584	Benzeneethanamine, α-phenyl-	$C_{14}H_{15}N$ 197.28	25611-78-3	311; 175[15]	4-12-00-03313 1.031[15]	eth 4; EtOH 4
1585	Benzeneethanamine, N-(2-phenylethyl)-	$C_{16}H_{19}N$ 225.33	6308-98-1 29	360; 190[15]	4-12-00-02459	eth 4; EtOH 4 1.5550[25]
1586	Benzeneethanamine, 3,4,5-trimethoxy- Mescaline	$C_{11}H_{17}NO_3$ 211.26	54-04-6 35.5	5808 180[12]	4-13-00-02919	H_2O 3; EtOH 3; eth 1; bz 3
1587	Benzene, 1,1'-(1,2-ethanediyl)bis- Dibenzyl	$C_{14}H_{14}$ 182.27	103-29-7 52.5	1219 284	4-05-00-01868 0.9780[25]	H_2O 1; EtOH 3; eth 3; CS_2 3 1.5476[60]
1588	Benzene, 1,1'-[1,2-ethanediylbis(oxy)]bis-	$C_{14}H_{14}O_2$ 214.26	104-66-5 98	182[12]	4-06-00-00573	H_2O 1; EtOH 2; eth 3; chl 3
1589	Benzene, 1,1'-(1,2-ethanediyl)bis[3-methyl-	$C_{16}H_{18}$ 210.32	4662-96-8	298	4-05-00-01943 0.9703[22]	1.5566[22]
1590	Benzene, 1,1'-(1,2-ethanediyl)bis[4-methyl- 1,2-Di-p-tolylethane	$C_{16}H_{18}$ 210.32	538-39-6 85	3386 178[18]	4-05-00-01943	H_2O 1; EtOH 2; bz 3; peth 3
1591	Benzene, 1,1',1'',1'''-(1,2-ethanediylidene)tetrakis-	$C_{26}H_{22}$ 334.46	632-50-8 214.5	360	4-05-00-02746	EtOH 2; bz 3; HOAc 3
1592	Benzeneethanol Phenethyl alcohol	$C_8H_{10}O$ 122.17	60-12-8 -27	7185 218.2	4-06-00-03067 1.0202[20]	H_2O 2; EtOH 5; eth 5 1.5325[20]
1593	Benzeneethanol, β-amino-	$C_8H_{11}NO$ 137.18	7568-92-5 58.3	261; 160[17]	4-13-00-01835	H_2O 4; EtOH 4
1594	Benzeneethanol, 2-amino-	$C_8H_{11}NO$ 137.18	5339-85-5	147-8[3.5]	3-13-00-01679 1.045[25]	H_2O 4 1.5849[20]
1595	Benzeneethanol, 4-bromo-	C_8H_9BrO 201.06	4654-39-1	138[9]	4-06-00-03082 1.436[25]	1.5735[20]
1596	Benzeneethanol, 3-chloro-	C_8H_9ClO 156.61	5182-44-5	133-7[13]	4-06-00-03079 1.181[25]	1.5491[20]
1597	Benzeneethanol, 4-chloro-	C_8H_9ClO 156.61	1875-88-3	259	4-06-00-03079 1.1804[20]	1.5487[20]
1598	Benzeneethanol, 4-chloro-α,α-dimethyl-	$C_{10}H_{13}ClO$ 184.67	5468-97-3 34	126[12]	4-06-00-03291	1.5310[20]
1599	Benzeneethanol, 2,5-dimethoxy-	$C_{10}H_{14}O_3$ 182.22	7417-19-8	176[18]	4-06-00-07393 1.055[25]	1.5395[20]

No.	Name / Synonym	Mol. Form. / Mol. Wt.	CAS RN / mp/°C	Merck No. / bp/°C	Beil. Ref. / den/g cm⁻³	Solubility / n_D
1600	Benzeneethanol, α,α-dimethyl-	$C_{10}H_{14}O$ 150.22	100-86-7 24	215	4-06-00-03290 0.9787[16]	1.5173[16]
1601	Benzeneethanol, α,α-diphenyl-	$C_{20}H_{18}O$ 274.36	4428-13-1 89.5	222[11]	4-06-00-05057	H_2O 1; EtOH 4; eth 2; chl 2
1602	Benzeneethanol, β-ethyl-	$C_{10}H_{14}O$ 150.22	2035-94-1	125[25]; 81[2]	2-06-00-00488 0.9891[6]	1.43[16]
1603	Benzeneethanol, 2-ethyl- Phenethyl alcohol, 2-ethyl	$C_{10}H_{14}O$ 150.22	22545-12-6	249.7; 127[20]	4-06-00-03351 0.9972[20]	1.5304[20]
1604	Benzeneethanol, 4-ethyl- Phenethyl alcohol, 4-ethyl	$C_{10}H_{14}O$ 150.22	22545-13-7 7.9	250	4-06-00-03355 0.9907[20]	1.5229[20]
1605	Benzeneethanol, α-ethyl-α-methyl-, (±)- 2-Butanol, 3-chloro (erythro, DL)	C_4H_9ClO 108.57	116783-12-1	235; 56.1[30]	3-06-00-01958 1.0610[25]	eth 4; EtOH 4; chl 4 1.4397[25]
1606	Benzeneethanol, 3-hydroxy-	$C_8H_{10}O_2$ 138.17	13398-94-2	168-73[4]	4-06-00-05938 1.082[25]	1.5643[20]
1607	Benzeneethanol, α-methyl-	$C_9H_{12}O$ 136.19	698-87-3	125[25]	4-06-00-03192 0.991[20]	1.5190[20]
1608	Benzeneethanol, β-methyl-	$C_9H_{12}O$ 136.19	1123-85-9	105-6[11]	2-06-00-00477 0.975[25]	H_2O 1; EtOH 3 1.5582[2]
1609	Benzeneethanol, 2-methyl-	$C_9H_{12}O$ 136.19	19819-98-8 2	243.5	4-06-00-03234 1.016[25]	1.5355[20]
1610	Benzeneethanol, 4-methyl-	$C_9H_{12}O$ 136.19	699-02-5	244.5; 94[6]	4-06-00-03244 1.0028[20]	1.5267[20]
1611	Benzeneethanol, β-(1-methylethyl)- 1-Butanol, 3-methyl-2-phenyl	$C_{11}H_{16}O$ 164.25	90499-41-5	130[15]	4-06-00-03390 0.9694[25]	ace 4; bz 4; eth 4; EtOH 4 1.5137[20]
1612	Benzeneethanol, 2-nitro-	$C_8H_9NO_3$ 167.16	15121-84-3 2	267	3-06-00-01714 1.19[25]	1.5637[20]
1613	Benzeneethanol, α-phenyl-	$C_{14}H_{14}O$ 198.26	614-29-9 67	177[15]	2-06-00-00637 1.0360[70]	
1614	Benzeneethanol, α-phenyl-, (S)-	$C_{14}H_{14}O$ 198.26	5773-56-8 67.5	168[10]	4-06-00-04701 1.0358[70]	H_2O 1; EtOH 4; eth 4
1615	Benzeneethanol, β-phenyl- 2,2-Diphenylethanol	$C_{14}H_{14}O$ 198.26	1883-32-5 64.5	193[20]; 144[1]	4-06-00-04724	EtOH 2; eth 3; ace 3; chl 3
1616	Benzeneethanol, α,α,β-trimethyl- 2-Butanol, 2-methyl-3-phenyl	$C_{11}H_{16}O$ 164.25	3280-08-8 46	197; 106[12]	4-06-00-03389 0.9794[20]	ace 4; bz 4; eth 4; EtOH 4 1.5193[20]
1617	Benzeneethanol, α,β,β-trimethyl-	$C_{11}H_{16}O$ 164.25	2977-31-3	197	4-06-00-03387	ace 4; bz 4; eth 4; EtOH 4 1.5161[13]
1618	Benzene, 1,1',1'',1'''-(1-ethanyl-2-ylidyne)tetrakis- 1,1,1,2-Tetraphenylethane	$C_{26}H_{22}$ 334.46	2294-94-2 144.8	278[21]	4-05-00-02747	bz 4
1619	Benzene, 1,1'-(1,2-ethenediyl)bis-, (E)- trans-Stilbene	$C_{14}H_{12}$ 180.25	103-30-0 123	8774 307; 166[12]	4-05-00-02156 0.9707[20]	H_2O 1; EtOH 2; eth 4; bz 4 1.6264[17]
1620	Benzene, 1,1'-(1,2-ethenediyl)bis-, (Z)- cis-Stilbene	$C_{14}H_{12}$ 180.25	645-49-8 -5	8774 141[12]	4-05-00-02155 1.0143[20]	H_2O 1; EtOH 3; eth 3; ace 3 1.6130[20]
1621	Benzene, 1,1'-(1,2-ethenediyl)bis[4-methoxy-	$C_{16}H_{16}O_2$ 240.30	4705-34-4 214.5	sub	4-06-00-06823	H_2O 1; ace 3; bz 3; HOAc 3
1622	Benzene, 1,1'-(1,2-ethenediyl)bis[2-methyl- 2,2'-Dimethylstilbene	$C_{16}H_{16}$ 208.30	10311-74-7 83	176-80[10]	4-05-00-02208	eth 4; EtOH 4; MeOH 4
1623	Benzene, 1,1',1'',1'''-(1,2-ethenediylidene)tetrakis-	$C_{28}H_{20}$ 332.44	632-51-9 225	420	4-05-00-02780 1.155[0]	H_2O 1; EtOH 2; eth 2; bz 4
1624	Benzene, ethenyl- Styrene	C_8H_8 104.15	100-42-5 -31	8830 145	4-05-00-01334 0.9060[20]	H_2O 1; EtOH 3; eth 3; ace 3 1.5468[20].
1625	Benzene, (1-ethenyl-3-butenyl)- 1,5-Hexadiene, 3-phenyl	$C_{12}H_{14}$ 158.24	1076-66-0	78[8]	4-05-00-01555 0.8911[25]	1.5141[25]
1626	Benzene, 1-ethenyl-2,4-dimethyl-	$C_{10}H_{12}$ 132.21	2234-20-0	79-80[12]	4-05-00-01386 0.905[20]	1.539[20]
1627	Benzene, 1-ethenyl-3,5-dimethyl-	$C_{10}H_{12}$ 132.21	5379-20-4	57-8[4]	4-05-00-01386 0.894[25]	1.5382[20]
1628	Benzene, 2-ethenyl-1,4-dimethyl-	$C_{10}H_{12}$ 132.21	2039-89-6	69[10]	4-05-00-01385 0.9072[17]	1.5236[17]
1629	Benzene, 4-ethenyl-1,2-dimethyl-	$C_{10}H_{12}$ 132.21	27831-13-6	94-6[20]	4-05-00-01385 0.906[25]	1.5463[20]
1630	Benzene, 1-ethenyl-2-ethyl-	$C_{10}H_{12}$ 132.21	7564-63-8 -75.5	187.3; 68[12]	3-05-00-01216 0.9017[25]	1.5380[20]
1631	Benzene, 1-ethenyl-3-ethyl-	$C_{10}H_{12}$ 132.21	7525-62-4 -101	190.0	3-05-00-01217 0.8945[20]	1.5351[20]
1632	Benzene, 1-ethenyl-4-ethyl-	$C_{10}H_{12}$ 132.21	3454-07-7 -49.7	192.3; 86[20]	4-05-00-01384 0.8884[25]	1.5376[20]
1633	Benzene, 1-ethenyl-2-fluoro-	C_8H_7F 122.14	394-46-7	46[32]	4-05-00-01342 1.0282[20]	bz 4; eth 4; EtOH 4 1.5200[20]
1634	Benzene, 1-ethenyl-3-fluoro-	C_8H_7F 122.14	350-51-6	30-1[4]	4-05-00-01343 1.0177[20]	H_2O 1; EtOH 3; eth 3; bz 3 1.5170[20]
1635	Benzene, 1-ethenyl-4-fluoro-	C_8H_7F 122.14	405-99-2 -34.5	67.4[50]; 30[4]	4-05-00-01343 1.0220[20]	H_2O 1; EtOH 3; eth 3; bz 3 1.5150[20]
1636	Benzene, 1,1'-ethenylidenebis- 1,1-Diphenylethene	$C_{14}H_{12}$ 180.25	530-48-3 8.2	3323 277	4-05-00-02173 1.0232[20]	H_2O 1; eth 3; chl 3 1.6085[20]
1637	Benzene, 1-ethenyl-2-methoxy-	$C_9H_{10}O$ 134.18	612-15-7 29	197; 83[12]	4-06-00-03771 1.0049[17]	ace 4; bz 4; eth 4; EtOH 4 1.5388[20]
1638	Benzene, 1-ethenyl-3-methoxy-	$C_9H_{10}O$ 134.18	626-20-0	90-3[15]	4-06-00-03773 0.999[16]	H_2O 1; EtOH 3; eth 3; bz 3 1.5586[23]

No.	Name / Synonym	Mol. Form. / Mol. Wt.	CAS RN / mp/°C	Merck No. / bp/°C	Beil. Ref. / den/g cm^{-3}	Solubility / n_D
1639	Benzene, 1-ethenyl-4-methoxy-	C$_9$H$_{10}$O 134.18	637-69-4 	 205; 91[13]	4-06-00-03775 1.0001[13]	H$_2$O 1; EtOH 3; eth 3; bz 3 1.5642[13]
1640	Benzene, 1-ethenyl-2-methyl- o-Methylstyrene	C$_9$H$_{10}$ 118.18	611-15-4 -68.5	 169.8	4-06-00-01367 0.9077[25]	H$_2$O 1; bz 3; chl 3 1.5437[20]
1641	Benzene, 1-ethenyl-3-methyl- m-Methylstyrene	C$_9$H$_{10}$ 118.18	100-80-1 -86.3	 164	4-05-00-01367 0.9076[25]	H$_2$O 1; EtOH 3; eth 3; bz 3 1.5411[20]
1642	Benzene, 1-ethenyl-4-methyl- p-Methylstyrene	C$_9$H$_{10}$ 118.18	622-97-9 -34.1	 172.8	4-05-00-01369 0.9173[25]	H$_2$O 1; bz 3 1.5420
1643	Benzene, 1-ethenyl-4-(1-methylethyl)-	C$_{11}$H$_{14}$ 146.23	2055-40-5 -44.7	 204.1	4-05-00-01408 0.8850[20]	ace 4; bz 4; eth 4; EtOH 4 1.5289[20]
1644	Benzene, 1-ethenyl-3-nitro-	C$_8$H$_7$NO$_2$ 149.15	586-39-0 -10	 120-1[11]	4-05-00-01351 1.1552[32]	H$_2$O 1; EtOH 3; eth 3; bz 3 1.5836[20]
1645	Benzene, 1-ethenyl-4-nitro-	C$_8$H$_7$NO$_2$ 149.15	100-13-0 29	 dec	4-05-00-01351	EtOH 4; eth 4; chl 3; HOAc 3
1646	Benzene, (ethenyloxy)-	C$_8$H$_8$O 120.15	766-94-9 	 155.5	4-06-00-00561 0.9770[20]	H$_2$O 1; eth 4 1.5224[20]
1647	Benzene, 2-ethenyl-1,3,5-trimethyl-	C$_{11}$H$_{14}$ 146.23	769-25-5 -37	 209	4-05-00-01408 0.9057[20]	ctc 3 1.5296[20]
1648	Benzene, 1,1',1''-(1-ethenyl-2-ylidene)tris-	C$_{20}$H$_{16}$ 256.35	58-72-0 72.5	 220-1[14]	4-05-00-02575 1.0373[78]	H$_2$O 1; EtOH 3; eth 4; chl 3 1.6292[78]
1649	Benzene, ethoxy- Phenetole	C$_8$H$_{10}$O 122.17	103-73-1 -29.5	7189 169.8	4-06-00-00554 0.9651[20]	H$_2$O 1; EtOH 3; eth 3; ctc 3 1.5076[20]
1650	Benzene, 2-ethoxy-1,3-dinitro- Phenetole, 2,6-dinitro-	C$_8$H$_8$N$_2$O$_5$ 212.16	13027-43-5 60.5	 137[3]	3-06-00-00868	eth 4
1651	Benzene, (2-ethoxyethenyl)-	C$_{10}$H$_{12}$O 148.20	17655-74-2 0	 224.5	 0.979[20]	 1.5496[20]
1652	Benzene, (2-ethoxyethoxy)- Ethane, 1-ethoxy-2-phenoxy-	C$_{10}$H$_{14}$O$_2$ 166.22	19594-02-6 	 230	4-06-00-00572 1.018[11]	H$_2$O 1
1653	Benzene, 1-ethoxy-4-ethyl-	C$_{10}$H$_{14}$O 150.22	1585-06-4 	 210	3-06-00-01665 0.9385[17]	ace 4; bz 4; EtOH 4
1654	Benzene, 1-ethoxy-2-fluoro-	C$_8$H$_9$FO 140.16	451-80-9 -16.7	 171.4	4-06-00-00771 1.0874[17]	H$_2$O 1; ace 3; bz 3; ctc 2 1.4932[17]
1655	Benzene, 1-ethoxy-3-fluoro-	C$_8$H$_9$FO 140.16	458-03-7 -27.5	 172; 65.2[15]	3-06-00-00669 1.0716[16]	H$_2$O 1; bz 3 1.4847[17]
1656	Benzene, 1-ethoxy-4-fluoro-	C$_8$H$_9$FO 140.16	459-26-7 -8.5	 173; 54[7]	4-06-00-00774 1.0715[18]	H$_2$O 1; bz 3; ctc 2; chl 3 1.4826[18]
1657	Benzene, 1-ethoxy-4-iodo-	C$_8$H$_9$IO 248.06	699-08-1 29	 250	4-06-00-01077	bz 4; eth 4; EtOH 4; chl 4
1658	Benzene, (ethoxymethyl)- Benzyl ethyl ether	C$_9$H$_{12}$O 136.19	539-30-0 	1147 186	4-06-00-02229 0.9478[20]	H$_2$O 1; EtOH 5; eth 5 1.4955[20]
1659	Benzene, 1-ethoxy-2-methyl-	C$_9$H$_{12}$O 136.19	614-71-1 	 184	4-06-00-01944 0.9592[13]	eth 4; EtOH 4 1.508[13]
1660	Benzene, 1-ethoxy-3-methyl-	C$_9$H$_{12}$O 136.19	621-32-9 	 192	4-06-00-02039 0.949[20]	H$_2$O 1; EtOH 3; eth 3 1.513[20]
1661	Benzene, 1-ethoxy-4-methyl-	C$_9$H$_{12}$O 136.19	622-60-6 	 188.5	4-06-00-02099 0.9509[18]	H$_2$O 1; EtOH 3; eth 3; ctc 2 1.5058[18]
1662	Benzene, 1-ethoxy-2-nitro-	C$_8$H$_9$NO$_3$ 167.16	610-67-3 2.1	 267	4-06-00-01250 1.1903[15]	eth 4; EtOH 4 1.5425[20]
1663	Benzene, 1-ethoxy-3-nitro-	C$_8$H$_9$NO$_3$ 167.16	621-52-3 36	 264; 169[70]	4-06-00-01271	H$_2$O 1; EtOH 3; eth 3
1664	Benzene, 1-ethoxy-4-nitro-	C$_8$H$_9$NO$_3$ 167.16	100-29-8 60	 283	4-06-00-01283 1.1176[100]	H$_2$O 2; EtOH 2; eth 4; ace 5
1665	Benzene, 1-ethoxy-2-(1-propenyl)- Phenetole, 2-propenyl	C$_{11}$H$_{14}$O 162.23	67191-37-1 	 231; 120[17]	0-06-00-00565 0.9731[24]	 1.544[24]
1666	Benzene, (3-ethoxy-1-propenyl)-	C$_{11}$H$_{14}$O 162.23	1476-07-9 	 127-8[22]	3-06-00-02404 0.970[15]	 1.547[15]
1667	Benzene, 1-ethoxy-2-propyl- Phenetole, 2-propyl	C$_{11}$H$_{16}$O 164.25	101144-90-5 	 213; 100[16]	0-06-00-00499 0.9240[20]	 1.494[20]
1668	Benzene, ethyl- Ethylbenzene	C$_8$H$_{10}$ 106.17	100-41-4 -94.9	3723 136.1	4-05-00-00885 0.8670[20]	H$_2$O 1; EtOH 5; eth 5; chl 2 1.4959[20]
1669	Benzene, (1-ethylbutyl)-	C$_{12}$H$_{18}$ 162.27	4468-42-2 -55.4	 209	4-05-00-01117 0.8239[25]	 1.4859[20]
1670	Benzene, 1-ethyl-2,3-dimethyl- 3-Ethyl-o-xylene	C$_{10}$H$_{14}$ 134.22	933-98-2 -49.5	 194	4-05-00-01069 0.8881[25]	 1.5117[20]
1671	Benzene, 1-ethyl-2,4-dimethyl-	C$_{10}$H$_{14}$ 134.22	874-41-9 -62.9	 188.4	4-05-00-01070 0.8763[20]	ace 4; bz 4; eth 4; EtOH 4 1.5038[20]
1672	Benzene, 1-ethyl-3,5-dimethyl-	C$_{10}$H$_{14}$ 134.22	934-74-7 -84.3	 183.6	4-05-00-01071 0.8608[25]	H$_2$O 1; EtOH 5; eth 5; ace 5 1.4981[20]
1673	Benzene, 2-ethyl-1,3-dimethyl-	C$_{10}$H$_{14}$ 134.22	2870-04-4 -16.2	 190	4-05-00-01069 0.8864[25]	 1.5107[20]
1674	Benzene, 2-ethyl-1,4-dimethyl-	C$_{10}$H$_{14}$ 134.22	1758-88-9 -53.7	 186.9	4-05-00-01070 0.8732[25]	H$_2$O 1; EtOH 5; eth 5; ace 5 1.5043[20]
1675	Benzene, 4-ethyl-1,2-dimethyl-	C$_{10}$H$_{14}$ 134.22	934-80-5 -66.9	 189.5	4-05-00-01069 0.8706[25]	H$_2$O 1; EtOH 5; eth 5; ace 5 1.5031[20]
1676	Benzene, 1,1'-(1-ethyl-1,2-ethanediyl)bis-	C$_{16}$H$_{18}$ 210.32	5223-59-6 	 290; 152[11]	4-05-00-01939 0.9777[20]	H$_2$O 1; EtOH 3; eth 3; bz 3 1.5554[20]
1677	Benzene, 1,1'-ethylidenebis- 1,1-Diphenylethane	C$_{14}$H$_{14}$ 182.27	612-00-0 -17.9	 272.6	4-05-00-01880 0.9997[20]	H$_2$O 1; EtOH 5; eth 5; bz 3 1.5756[20]
1678	Benzene, 1,1'-ethylidenebis[4-chloro- 1,1-Bis(4-chlorophenyl)ethane	C$_{14}$H$_{12}$Cl$_2$ 251.15	3547-04-4 56	 320	4-05-00-01881	

No.	Name Synonym	Mol. Form. Mol. Wt.	CAS RN mp/°C	Merck No. bp/°C	Beil. Ref. den/g cm^{-3}	Solubility n_D
1679	Benzene, 1,1',1''-ethylidynetris-	C$_{20}$H$_{18}$ 258.36	5271-39-6 95	207[18]	4-05-00-02510	H$_2$O 1; EtOH 2; eth 3; bz 3
1680	Benzene, 1-ethyl-2-iodo- o-Ethyliodobenzene	C$_8$H$_9$I 232.06	18282-40-1 226		4-05-00-00910 1.6189[10]	ace 4; bz 4 1.5941[22]
1681	Benzene, 1-ethyl-4-iodo-	C$_8$H$_9$I 232.06	25309-64-2 -17	209	3-05-00-00801 1.6095[10]	ace 4; bz 4 1.5909[22]
1682	Benzene, 1-ethyl-2-methoxy-	C$_9$H$_{12}$O 136.19	14804-32-1	187; 80[14]	3-06-00-01656 0.9636[19]	bz 4; eth 4 1.5142[20]
1683	Benzene, 1-ethyl-3-methoxy-	C$_9$H$_{12}$O 136.19	10568-38-4	197; 74[10]	3-06-00-01662 0.9575[18]	H$_2$O 1; eth 4; bz 3; ctc 3 1.5102
1684	Benzene, 1-ethyl-4-methoxy-	C$_9$H$_{12}$O 136.19	1515-95-3	198	4-06-00-03021 0.9624[15]	bz 4; eth 4 1.5120[20]
1685	Benzene, 1-ethyl-2-methyl- o-Ethyltoluene	C$_9$H$_{12}$ 120.19	611-14-3 -80.8	165.2	4-05-00-00999 0.8807[20]	H$_2$O 1; EtOH 5; eth 5; ace 5 1.5046[20]
1686	Benzene, 1-ethyl-3-methyl- m-Ethyltoluene	C$_9$H$_{12}$ 120.19	620-14-4 -95.5	161.3	4-05-00-01001 0.8645[20]	H$_2$O 1; EtOH 4; eth 4; ace 5 1.4966[20]
1687	Benzene, 1-ethyl-4-methyl- p-Ethyltoluene	C$_9$H$_{12}$ 120.19	622-96-8 -62.3	162	4-05-00-01003 0.8614[20]	H$_2$O 1; EtOH 4; eth 4; ace 5 1.4959[20]
1688	Benzene, 1-ethyl-2-(1-methylethyl)- Benzene, 1-ethyl-2-isopropyl	C$_{11}$H$_{16}$ 148.25	18970-44-0	193	4-05-00-01100 0.888[20]	ace 4; bz 4; eth 4; EtOH 4 1.508[20]
1689	Benzene, 1-ethyl-3-(1-methylethyl)- Benzene, 1-ethyl-3-isopropyl	C$_{11}$H$_{16}$ 148.25	4920-99-4 <-20	192	3-05-00-01005 0.859[20]	H$_2$O 1; EtOH 3; eth 3; bz 4 1.4921[20]
1690	Benzene, 1-ethyl-4-(1-methylethyl)- Benzene, 1-ethyl-4-isopropyl	C$_{11}$H$_{16}$ 148.25	4218-48-8 <-20	196.6	4-05-00-01100 0.8585[20]	H$_2$O 1; eth 3; bz 4 1.4923[20]
1691	Benzene, (1-ethyl-1-methylpropyl)- Pentane, 3-methyl-3-phenyl	C$_{12}$H$_{18}$ 162.27	1985-97-3 205		4-05-00-01120 0.8773[15]	1.4972[16]
1692	Benzene, 1-ethyl-2-nitro-	C$_8$H$_9$NO$_2$ 151.17	612-22-6 -12.3	232.5	4-05-00-00911 1.1207[20]	H$_2$O 1; EtOH 4; eth 4; ace 3 1.5356[20]
1693	Benzene, 1-ethyl-3-nitro-	C$_8$H$_9$NO$_2$ 151.17	7369-50-8 -37.9	242.5	4-05-00-00911 1.1345[25]	ace 4; eth 4; EtOH 4
1694	Benzene, 1-ethyl-4-nitro-	C$_8$H$_9$NO$_2$ 151.17	100-12-9 -12.3	245.5	4-05-00-00912 1.1192[20]	H$_2$O 1; EtOH 4; eth 4; ace 3 1.5455[20]
1695	Benzene, (1-ethyloctadecyl)- 3-Phenyleicosane	C$_{26}$H$_{46}$ 358.65	2400-02-4 29.3	202[1]	4-05-00-01222 0.8546[20]	1.4796[20]
1696	Benzene, 1-ethyl-4-(phenylmethyl)-	C$_{15}$H$_{16}$ 196.29	620-85-9 -24	297	4-05-00-01926 0.9777[20]	EtOH 3; eth 3; chl 3 1.5616[20]
1697	Benzene, 1,1'-(1-ethyl-1,3-propanediyl)bis- Pentane, 1,3-diphenyl-	C$_{17}$H$_{20}$ 224.35	838-45-9	304; 113[0.8]	4-05-00-01963 0.9734[21]	1.553[21]
1698	Benzene, 1,1'-(2-ethyl-1,3-propanediyl)bis- Propane, 1,1-dibenzyl	C$_{17}$H$_{20}$ 224.35	1520-45-2	304.5	4-05-00-01964 0.9734[21]	1.553[21]
1699	Benzene, (1-ethyl-1-propenyl)- 3-Phenyl-2-pentene	C$_{11}$H$_{14}$ 146.23	4701-36-4	198; 92[18]	4-05-00-01404 0.9173[14]	1.5266[15]
1700	Benzene, (1-ethyl-2-propenyl)-	C$_{11}$H$_{14}$ 146.23	19947-22-9	191.5; 71[12]	3-05-00-01240 0.8458[23]	1.5030[21]
1701	Benzene, (1-ethylpropyl)-	C$_{11}$H$_{16}$ 148.25	1196-58-3	187.5	4-05-00-01090 0.8649[20]	1.4880[20]
1702	Benzene, 1-ethyl-2-propyl-	C$_{11}$H$_{16}$ 148.25	16021-20-8	203	4-05-00-01099 0.8744[20]	1.4992[20]
1703	Benzene, 1-ethyl-4-propyl- 1-Ethyl-4-propylbenzene	C$_{11}$H$_{16}$ 148.25	20024-90-2	205	4-05-00-01100 0.8594[20]	ace 4; bz 4; eth 4; EtOH 4 1.4921[21]
1704	Benzene, (ethylsulfonyl)-	C$_8$H$_{10}$O$_2$S 170.23	599-70-2 42	160[12]	4-06-00-01469 1.1410[20]	bz 4; eth 4; EtOH 4; chl 4
1705	Benzene, (ethylthio)-	C$_8$H$_{10}$S 138.23	622-38-8	205	4-06-00-01468 1.0211[20]	EtOH 3 1.5670[20]
1706	Benzene, 1-(ethylthio)-3-methyl- Ethyl m-tolyl sulfide	C$_9$H$_{12}$S 152.26	34786-24-8	219	3-06-00-01332 0.9987[20]	1.5590[20]
1707	Benzene, 1-(ethylthio)-4-methyl-	C$_9$H$_{12}$S 152.26	622-63-9	220	4-06-00-02156 0.9996[20]	1.555[20]
1708	Benzene, 1-ethyl-2,3,4-trimethyl- Benzene, 4-ethyl-1,2,3-trimethyl	C$_{11}$H$_{16}$ 148.25	61827-86-9	220.4	4-05-00-01107 0.9019[20]	ace 4; bz 4; eth 4; EtOH 4 1.5180[20]
1709	Benzene, 1-ethyl-2,3,5-trimethyl- 1-Ethyl-2,3,5-trimethylbenzene	C$_{11}$H$_{16}$ 148.25	18262-85-6	213	4-05-00-01107 0.8897[20]	ace 4; bz 4; eth 4; EtOH 4 1.5118[20]
1710	Benzene, 1-ethyl-2,4,5-trimethyl- 1-Ethyl-2,4,5-trimethylbenzene	C$_{11}$H$_{16}$ 148.25	17851-27-3 -13.5	213	4-05-00-01108 0.883[20]	ace 4; bz 4; eth 4; EtOH 4 1.5075[20]
1711	Benzene, 2-ethyl-1,3,4-trimethyl- Benzene, 3-ethyl-1,2,4-trimethyl	C$_{11}$H$_{16}$ 148.25	61827-87-0	216.6	4-05-00-01107 0.895[20]	ace 4; bz 4; eth 4; EtOH 4 1.5133[20]
1712	Benzene, 2-ethyl-1,3,5-trimethyl-	C$_{11}$H$_{16}$ 148.25	3982-67-0 -15.5	212.4	4-05-00-01107 0.883[20]	ace 4; bz 4; eth 4; EtOH 4 1.5074[20]
1713	Benzene, 5-ethyl-1,2,3-trimethyl- 5-Ethyl-1,2,3-trimethylbenzene	C$_{11}$H$_{16}$ 148.25	31366-00-4	215.8	4-05-00-01107 0.8863[20]	ace 4; eth 4; EtOH 4 1.5101[20]
1714	Benzene, 1,1'-(1,2-ethynediyl)bis- Diphenylacetylene	C$_{14}$H$_{10}$ 178.23	501-65-5 62.5	9428 300	4-05-00-02276 0.9657[100]	H$_2$O 1; EtOH 2; eth 4; ctc 2
1715	Benzene, ethynyl- Ethynylbenzene	C$_8$H$_6$ 102.14	536-74-3 -44.8	3817 143	4-05-00-01525 0.9300[20]	H$_2$O 1; EtOH 5; eth 5; ace 3 1.5470[20]
1716	Benzene, (ethynyloxy)-	C$_8$H$_6$O 118.14	4279-76-9 -36	61-2[25]	4-06-00-00565 1.0614[20]	eth 4; EtOH 4 1.5125[20]
1717	Benzene, fluoro- Fluorobenzene	C$_6$H$_5$F 96.10	462-06-6 -42.2	4099 84.7	4-05-00-00632 1.0225[20]	bz 4; eth 4; EtOH 4; lig 4 1.4684[30]
1718	Benzene, 1-fluoro-2,4-dinitro- 2,4-Dinitrophenyl fluoride	C$_6$H$_3$FN$_2$O$_4$ 186.10	70-34-8 25.8	4101 296	4-05-00-00742 1.4718[54]	EtOH 3; chl 2 1.5690[20]

No.	Name / Synonym	Mol. Form. / Mol. Wt.	CAS RN / mp/°C	Merck No. / bp/°C	Beil. Ref. / den/g cm^{-3}	Solubility / n_D
1719	Benzene, 1-fluoro-2-iodo-	C_6H_4FI 222.00	348-52-7 -41.5	188.6	4-05-00-00693	ace 3; bz 3; chl 3 1.5910[20]
1720	Benzene, 1-fluoro-4-iodo-	C_6H_4FI 222.00	352-34-1 -27	183	4-05-00-00694 1.9523[15]	H_2O 1; EtOH 3; eth 3; ace 3 1.5270[22]
1721	Benzene, 1-fluoro-3-isothiocyanato-	C_7H_4FNS 153.18	404-72-8	227	3-12-00-01276 1.27[25]	1.6186[20]
1722	Benzene, 1-fluoro-2-methoxy-	C_7H_7FO 126.13	321-28-8 -39	154.5	4-06-00-00771 1.5489[17]	H_2O 1; eth 3; ctc 3 1.4969[17]
1723	Benzene, 1-fluoro-3-methoxy-	C_7H_7FO 126.13	456-49-5 -35	159; 51[14]	4-06-00-00772 1.104[25]	1.4876[20]
1724	Benzene, 1-fluoro-4-methoxy-	C_7H_7FO 126.13	459-60-9 -45	157	4-06-00-00773 1.1781[18]	eth 3 1.4886[18]
1725	Benzene, (fluoromethyl)-	C_7H_7F 110.13	350-50-5 -35	140; 40[14]	4-05-00-00800 1.0228[25]	ctc 3 1.4892[25]
1726	Benzene, 1-fluoro-2-methyl- o-Fluorotoluene	C_7H_7F 110.13	95-52-3 -62	4108 115	4-05-00-00799 1.0041[13]	H_2O 1; EtOH 4; eth 4 1.4704[20]
1727	Benzene, 1-fluoro-3-methyl- m-Fluorotoluene	C_7H_7F 110.13	352-70-5 -87	4108 115	4-05-00-00799 0.9974[20]	H_2O 1; EtOH 4; eth 4 1.4691[20]
1728	Benzene, 1-fluoro-4-methyl- p-Fluorotoluene	C_7H_7F 110.13	352-32-9 -56	4108 116.6	4-05-00-00800 0.9975[20]	H_2O 1; EtOH 4; eth 4 1.4699[20]
1729	Benzene, 1-fluoro-4-methyl-2-nitro- Toluene, 4-fluoro-3-nitro	$C_7H_6FNO_2$ 155.13	446-11-7 26.5	241	3-05-00-00743 1.2619[28]	1.5237[28]
1730	Benzene, 2-fluoro-4-methyl-1-nitro- 3-Fluoro-4-nitrotoluene	$C_7H_6FNO_2$ 155.13	446-34-4 53.2	97[3]	4-05-00-00852 1.4380[25]	
1731	Benzene, 4-fluoro-1-methyl-2-nitro-	$C_7H_6FNO_2$ 155.13	446-10-6 27	213; 138[83]	4-05-00-00851 1.2686[20]	1.5218[20]
1732	Benzene, 4-fluoro-2-methyl-1-nitro-	$C_7H_6FNO_2$ 155.13	446-33-3 27.5	217; 97-8[10]	4-05-00-00851 1.272[28]	1.5271[20]
1733	Benzene, 1-fluoro-2-nitro- o-Fluoronitrobenzene	$C_6H_4FNO_2$ 141.10	1493-27-2 -6	215 dec	4-05-00-00718 1.3285[18]	eth 4; EtOH 4 1.5489[17]
1734	Benzene, 1-fluoro-3-nitro- m-Fluoronitrobenzene	$C_6H_4FNO_2$ 141.10	402-67-5 41	199; 86[19]	4-05-00-00719 1.3254[19]	H_2O 1; EtOH 3; eth 3; bz 2 1.5262[15]
1735	Benzene, 1-fluoro-4-nitro- p-Fluoronitrobenzene	$C_6H_4FNO_2$ 141.10	350-46-9 21	205	4-05-00-00719 1.3300[20]	H_2O 1; EtOH 3; eth 3; ctc 2 1.5316[20]
1736	Benzene, 1-fluoro-2-(trichloromethyl)-	$C_7H_4Cl_3F$ 213.47	488-98-2	75[5]	3-05-00-00701 1.453[25]	1.5432[20]
1737	Benzene, 1-fluoro-2-(trifluoromethyl)-	$C_7H_4F_4$ 164.10	392-85-8	114.5	4-05-00-00804 1.293[25]	1.4040[25]
1738	Benzene, 1-fluoro-3-(trifluoromethyl)-	$C_7H_4F_4$ 164.10	401-80-9 -81.5	101.5	4-05-00-00804 1.3021[17]	
1739	Benzene, 1-fluoro-4-(trifluoromethyl)-	$C_7H_4F_4$ 164.10	402-44-8 -41.7	103.5	4-05-00-00804 1.293[25]	1.4025[20]
1740	Benzene, 2-fluoro-1,3,5-trimethyl-	$C_9H_{11}F$ 138.18	392-69-8 -36.7	171.5	4-05-00-01025 0.9745[25]	1.4809[25]
1741	Benzene, heneicosyl- Heneicosylbenzene	$C_{27}H_{48}$ 372.68	40775-09-5 48	439	4-05-00-01226 0.8545[20]	1.4804[20]
1742	Benzene, heptacosyl- Heptacosylbenzene	$C_{33}H_{60}$ 456.84	61828-25-9 64	397[100]; 262[1]	4-05-00-01243 0.8543[20]	1.4797[20]
1743	Benzene, heptadecyl- Heptadecane, 1-phenyl-	$C_{23}H_{40}$ 316.57	14752-75-1 32	397	3-05-00-01123 0.8546[20]	1.4810[20]
1744	Benzene, 1,1'-(1-heptenylidene)bis- 1,1-Diphenyl-1-heptene	$C_{19}H_{22}$ 250.38	1530-20-7		4-05-00-02252 0.9673[18]	EtOH 2 1.5648[18]
1745	Benzene, heptyl-	$C_{13}H_{20}$ 176.30	1078-71-3 -48	240; 109[10]	4-05-00-01143 0.8567[20]	H_2O 1; bz 3; chl 3 1.4865[20]
1746	Benzene, 1,1'-heptylidenebis- 1,1-Diphenylheptane	$C_{19}H_{24}$ 252.40	1530-05-8 14	333.5; 137[1]	4-05-00-01992 0.9497[20]	
1747	Benzene, (heptyloxy)-	$C_{13}H_{20}O$ 192.30	32395-96-3 -33.5	267	4-06-00-00560 0.9170[15]	ace 4; eth 4; EtOH 4 1.4912[20]
1748	Benzenehexacarboxylic acid Mellitic acid	$C_{12}H_6O_{12}$ 342.17	517-60-2 286-8 dec	5706	4-09-00-03825	H_2O 4; EtOH 3; sulf 3
1749	Benzene, hexachloro- Hexachlorobenzene	C_6Cl_6 284.78	118-74-1 231.8	4600 325	4-05-00-00670 2.044[23]	H_2O 1; EtOH 2; eth 3; bz 4 1.5691[23]
1750	Benzene, hexacosyl- Hexacosane, 1-phenyl-	$C_{32}H_{58}$ 442.81	13024-80-1 62	388[100]; 250[1]	4-05-00-01242 0.8543[20]	1.4798[20]
1751	Benzene, hexadecyl-	$C_{22}H_{38}$ 302.54	1459-09-2 27	385	4-05-00-01216 0.8547[20]	H_2O 1; EtOH 2; eth 4; bz 4 1.4813[20]
1752	Benzene, (hexadecyloxy)-	$C_{22}H_{38}O$ 318.54	35021-70-6 41.8	200[1]	3-06-00-00555 0.8434[82]	1.4556[82]
1753	Benzene, 1,3-hexadienyl- 1-Phenyl-1,3-hexadiene	$C_{12}H_{14}$ 158.24	41635-77-2	128[16]	3-05-00-01378 0.9253[12]	1.6025[12]
1754	Benzene, 1,5-hexadienyl- 1,5-Hexadiene, 1-phenyl-	$C_{12}H_{14}$ 158.24	1009-81-0	102-3[8]	4-05-00-01554 0.9005[25]	1.5421[25]
1755	Benzene, 3,5-hexadienyl- 6-Phenyl-1,3-hexadiene	$C_{12}H_{14}$ 158.24	39669-95-9	100.5[11]	3-05-00-01378 0.9304[13]	1.5446[13]
1756	Benzene, hexaethyl- Hexaethylbenzene	$C_{18}H_{30}$ 246.44	604-88-6 129	298	4-05-00-01208 0.8305[130]	H_2O 1; EtOH 3; eth 4; bz 4 1.4736[130]
1757	Benzene, hexafluoro- Perfluorobenzene	C_6F_6 186.06	392-56-3 5.3	80.2	4-05-00-00640 1.6184[20]	1.3777[20]

No.	Name Synonym	Mol. Form. Mol. Wt.	CAS RN mp/°C	Merck No. bp/°C	Beil. Ref. den/g cm^{-3}	Solubility n_D
1758	Benzene, hexamethyl- Mellitene	$C_{12}H_{18}$ 162.27	87-85-4 166.5	263.4	4-05-00-01137 1.0630[25]	H_2O 1; EtOH 3; eth 3; ace 3
1759	Benzenehexanoic acid	$C_{12}H_{16}O_2$ 192.26	5581-75-9 23	206[30]	4-09-00-01895	1.5164[21]
1760	Benzene, 2-hexenyl- 2-Hexene, 1-phenyl	$C_{12}H_{16}$ 160.26	67590-77-6	108[16]	0-05-00-00501 0.8898[16]	1.5058[16]
1761	Benzene, 3-hexenyl-	$C_{12}H_{16}$ 160.26	35008-86-7	222	4-05-00-01418 0.8790[20]	1.5039[20]
1762	Benzene, 5-hexenyl-	$C_{12}H_{16}$ 160.26	1588-44-9	94-5[10]	3-05-00-01251 0.8839[20]	1.5033[20]
1763	Benzene, hexyl-	$C_{12}H_{18}$ 162.27	1077-16-3 -61	226.1	4-05-00-01115 0.8575[20]	H_2O 1; eth 5; bz 3; peth 3 1.4864[20]
1764	Benzene, (1-hexylheptyl)- 7-Phenyltridecane	$C_{19}H_{32}$ 260.46	2400-01-3	183-4[20]	4-05-00-01210 0.8723[20]	1.4931[18]
1765	Benzene, (hexyloxy)-	$C_{12}H_{18}O$ 178.27	1132-66-7 -19	240	4-06-00-00560 0.9174[20]	H_2O 1; eth 3 1.4921[20]
1766	Benzene, iodo- Iodobenzene	C_6H_5I 204.01	591-50-4 -31.3	4922 188.4	4-05-00-00688 1.8308[20]	H_2O 1; EtOH 3; eth 5; ace 5 1.6200[20]
1767	Benzene, 1-iodo-2,3-dimethoxy- 3-Iodoveratrole	$C_8H_9IO_2$ 264.06	25245-33-4 45.5	144-5[14]	3-06-00-04262 1.7799[20]	1.6127[20]
1768	Benzene, 4-iodo-1,2-dimethoxy-	$C_8H_9IO_2$ 264.06	5460-32-2 35	174[30]	4-06-00-05625	bz 4; EtOH 4
1769	Benzene, 1-iodo-2,3-dimethyl-	C_8H_9I 232.06	31599-60-7	229	3-05-00-00820 1.6395[20]	ace 4 1.6074[20]
1770	Benzene, 1-iodo-2,4-dimethyl-	C_8H_9I 232.06	4214-28-2	231 dec; 111[14]	4-05-00-00947 1.6282[16]	H_2O 1; ace 3; bz 3 1.6008[16]
1771	Benzene, 1-iodo-3,5-dimethyl- m-Xylene, 5-iodo-	C_8H_9I 232.06	22445-41-6	230.5	3-05-00-00840 1.6085[18]	ace 4 1.5967[18]
1772	Benzene, 2-iodo-1,3-dimethyl-	C_8H_9I 232.06	608-28-6 11.2	229.5	4-05-00-00947 1.6158[20]	H_2O 1; ace 3; bz 3 1.6035[20]
1773	Benzene, 2-iodo-1,4-dimethyl-	C_8H_9I 232.06	1122-42-5	227 dec	4-05-00-00970 1.6168[17]	H_2O 1; ace 3; bz 3 1.5992[17]
1774	Benzene, 4-iodo-1,2-dimethyl- 4-Iodo-o-xylene	C_8H_9I 232.06	31599-61-8	231.5	3-05-00-00821 1.6334[18]	ace 4 1.6049[18]
1775	Benzene, 1-iodo-2-methoxy- o-Iodoanisole	C_7H_7IO 234.04	529-28-2	4921 241; 91[2]	4-06-00-01070 1.8[20]	EtOH 4; eth 4; ace 4; bz 4
1776	Benzene, 1-iodo-4-methoxy-	C_7H_7IO 234.04	696-62-8 53	238; 138[25]	4-06-00-01075	EtOH 3; eth 3; chl 3
1777	Benzene, (iodomethyl)-	C_7H_7I 218.04	620-05-3 24.5	93[10]	4-05-00-00842 1.7335[25]	bz 4; eth 4; EtOH 4 1.6334[25]
1778	Benzene, 1-iodo-2-methyl-	C_7H_7I 218.04	615-37-2	211.5	4-05-00-00838 1.713[20]	H_2O 1; EtOH 5; eth 5 1.6079[20]
1779	Benzene, 1-iodo-3-methyl-	C_7H_7I 218.04	625-95-6 -27.2	213	4-05-00-00839 1.705[20]	H_2O 1; EtOH 5; eth 5 1.6053[20]
1780	Benzene, 1-iodo-4-methyl-	C_7H_7I 218.04	624-31-7 36.5	211	4-05-00-00840 1.678[20]	H_2O 1; EtOH 3; eth 3; chl 2
1781	Benzene, 2-iodo-1-methyl-4-(1-methylethyl)- 2-Iodo-p-cymene	$C_{10}H_{13}I$ 260.12	56739-95-8	139[23]	3-05-00-00960 1.4205[17]	1.5800[17]
1782	Benzene, 1-iodo-4-methyl-1-(1-methylethyl)- p-Cymene, 3-iodo-	$C_{10}H_{13}I$ 260.12	4395-81-7	123[13]; 80[5]	3-05-00-00960 1.4113[17]	1.5690[17]
1783	Benzene, 1-iodo-2-nitro-	$C_6H_4INO_2$ 249.01	609-73-4 54	290; 162[18]	4-05-00-00732 1.9186[75]	H_2O 1; EtOH 3; eth 3
1784	Benzene, 1-iodo-3-nitro-	$C_6H_4INO_2$ 249.01	645-00-1 38.5	280	4-05-00-00733 1.9477[50]	H_2O 1; EtOH 3; eth 3
1785	Benzene, 1-iodo-4-nitro-	$C_6H_4INO_2$ 249.01	636-98-6 174.7	288	4-05-00-00734 1.8090[155]	H_2O 1; EtOH 3; HOAc 3
1786	Benzene, isocyanato- Phenyl isocyanate	C_7H_5NO 119.12	103-71-9 163; 55[13]	7267	4-12-00-00864 1.0956[20]	eth 4; chl 2 1.5368[20]
1787	Benzene, 1-isocyanato-2-methyl-	C_8H_7NO 133.15	614-68-6 185		4-12-00-01768	H_2O 1; eth 3 1.5282[20]
1788	Benzene, 1-isocyanato-2-nitro-	$C_7H_4N_2O_3$ 164.12	3320-86-3 41	137[18]	3-12-00-01535	bz 4; eth 4; chl 4
1789	Benzene, 1-isocyanato-3-nitro-	$C_7H_4N_2O_3$ 164.12	3320-87-4 51	130[11]	4-12-00-01604	bz 4; eth 4; chl 4
1790	Benzene, 1-isocyanato-4-nitro-	$C_7H_4N_2O_3$ 164.12	100-28-7 57	162[20]; 137[11]	4-12-00-01653	bz 4; eth 4; chl 4
1791	Benzene, (isocyanomethyl)-	C_8H_7N 117.15	10340-91-7	199 dec; 93[55]	4-12-00-02219 0.972[15]	1.5193[20]
1792	Benzene, isothiocyanato- Phenyl isothiocyanate	C_7H_5NS 135.19	103-72-0 -21	7268 221	4-12-00-00867 1.1303[20]	H_2O 1; EtOH 3; eth 3; ctc 3 1.6492[23]
1793	Benzene, 1-isothiocyanato-2-methoxy-	C_8H_7NOS 165.22	3288-04-8	131-2[11]	4-13-00-00855 1.1878[20]	1.6458[20]
1794	Benzene, (isothiocyanatomethyl)-	C_8H_7NS 149.22	622-78-6	243	4-12-00-02276 1.1246[16]	H_2O 1; EtOH 5; eth 3 1.6049[15]
1795	Benzenemethanamine Benzylamine	C_7H_9N 107.16	100-46-9 185; 90[12]	1139	4-12-00-02155 0.9813[20]	H_2O 5; EtOH 5; eth 5; ace 4 1.5401[20]
1796	Benzenemethanamine, 2-amino-	$C_7H_{10}N_2$ 122.17	4403-69-4 61	269	4-13-00-00265	EtOH 4

No.	Name / Synonym	Mol. Form. / Mol. Wt.	CAS RN / mp/°C	Merck No. / bp/°C	Beil. Ref. / den/g cm⁻³	Solubility / n_D
1797	Benzenemethanamine, N-[bis(4-methoxyphenyl)methylene]- / Benzylimidobis(p-methoxyphenyl)methane	$C_{22}H_{21}NO_2$ / 331.41	524-96-9 / 90	1153	3-12-00-02254	eth 4; chl 4
1798	Benzenemethanamine, N,N-bis(phenylmethyl)-	$C_{21}H_{21}N$ / 287.40	620-40-6 / 91.5	385	4-12-00-02183 / 0.9912[95]	H_2O 2; EtOH 2; eth 3; ctc 3
1799	Benzenemethanamine, 4-bromo- / Benzyl amine, 4-bromo	C_7H_8BrN / 186.05	3959-07-7 / 20	250	3-12-00-02350	eth 4
1800	Benzenemethanamine, N-3-butynyl-N-methyl-	$C_{12}H_{15}N$ / 173.26	15240-91-2	127[16]	4-12-00-02171 / 0.9372[20]	1.5202[20]
1801	Benzenemethanamine, N-(2-chloroethyl)-N-(phenylmethyl)-, hydrochloride / N-(2-Chloroethyl)dibenzylamine hydrochloride	$C_{16}H_{19}Cl_2N$ / 296.24	55-43-6	2138	4-12-00-02180	
1802	Benzenemethanamine, 3-chloro-N-methyl-	$C_8H_{10}ClN$ / 155.63	39191-07-6	88[4]	3-12-00-02342	chl 3 / 1.5350[25]
1803	Benzenemethanamine, 2,4-dichloro-	$C_7H_7Cl_2N$ / 176.04	95-00-1	124-6[13]	4-12-00-02376	chl 3 / 1.5762[25]
1804	Benzenemethanamine, N,N-diethyl-α-phenyl- / N,N-Diethylbenzhydrylamine	$C_{17}H_{21}N$ / 239.36	519-72-2 / 58.5	3103 170[17]	4-12-00-03283	
1805	Benzenemethanamine, 3,4-dimethoxy-	$C_9H_{13}NO_2$ / 167.21	5763-61-1	156[12]; 120[3]	4-13-00-02582 / 1.143[25]	chl 3
1806	Benzenemethanamine, α,α-dimethyl-	$C_9H_{13}N$ / 135.21	585-32-0	196.5	4-12-00-02633 / 0.9423[20]	1.5181[25]
1807	Benzenemethanamine, N,N-dimethyl- / Dimethylbenzylamine	$C_9H_{13}N$ / 135.21	103-83-3	181	4-12-00-02161 / 0.915[0]	H_2O 2; EtOH 5; eth 5 / 1.5011[20]
1808	Benzenemethanamine, 2-ethoxy- / 2-Ethoxybenzylamine	$C_9H_{13}NO$ / 151.21	37806-29-4	69-74[0.2]	4-13-00-01676 / 1.015[25]	1.5326[20]
1809	Benzenemethanamine, α-ethyl-	$C_9H_{13}N$ / 135.21	2941-20-0	206; 100[16]	4-12-00-02582 / 0.9346[25]	1.5173[23]
1810	Benzenemethanamine, N-ethyl-	$C_9H_{13}N$ / 135.21	14321-27-8	194	4-12-00-02163 / 0.9342[17]	H_2O 2; EtOH 3; eth 3; bz 3 / 1.5117[20]
1811	Benzenemethanamine, N-ethyl-N-phenyl- / Ethylbenzylaniline	$C_{15}H_{17}N$ / 211.31	92-59-1 / 35	3728 288; 185[22]	4-12-00-02176 / 1.001[55]	H_2O 1; EtOH 3; eth 3; chl 3 / 1.5943[23]
1812	Benzenemethanamine, 2-methoxy-	$C_8H_{11}NO$ / 137.18	6850-57-3	229	4-13-00-01675 / 1.051[25]	1.5475[20]
1813	Benzenemethanamine, 4-methoxy-	$C_8H_{11}NO$ / 137.18	2393-23-9	236.5	4-13-00-01719 / 1.050[15]	H_2O 2; EtOH 2; eth 2 / 1.5462[20]
1814	Benzenemethanamine, α-methyl-, (±)-	$C_8H_{11}N$ / 121.18	618-36-0 / 32	187	4-12-00-02424 / 0.9395[15]	H_2O 3; EtOH 5; eth 5; chl 3 / 1.5238[25]
1815	Benzenemethanamine, 2-methyl-	$C_8H_{11}N$ / 121.18	89-93-0 / -30	206; 81[15]	4-12-00-02518 / 0.9766[19]	1.5436[19]
1816	Benzenemethanamine, 3-methyl-	$C_8H_{11}N$ / 121.18	100-81-2	203.5	4-12-00-02565 / 0.966[25]	1.5360[20]
1817	Benzenemethanamine, 4-methyl-	$C_8H_{11}N$ / 121.18	104-84-7 / 12.5	195	4-12-00-02574 / 0.952[20]	1.5340[20]
1818	Benzenemethanamine, N-methyl-	$C_8H_{11}N$ / 121.18	103-67-3	180.5	4-12-00-02161 / 0.9442[18]	H_2O 4
1819	Benzenemethanamine, 4-(1-methylethyl)-	$C_{10}H_{15}N$ / 149.24	4395-73-7	227	4-12-00-02836	eth 4; EtOH 4 / 1.5182[17]
1820	Benzenemethanamine, N-(1-methylethyl)-	$C_{10}H_{15}N$ / 149.24	102-97-6	200; 93[10]	4-12-00-02165 / 0.892[25]	1.5025[20]
1821	Benzenemethanamine, N-(2-methylphenyl)-	$C_{14}H_{15}N$ / 197.28	5405-13-0 / 60	302.5	4-12-00-02177 / 1.0142[65]	ace 4; EtOH 4; chl 4 / 1.5861[65]
1822	Benzenemethanamine, α-methyl-N-(1-phenylethyl)-	$C_{16}H_{19}N$ / 225.33	10024-74-5	296.5	4-12-00-02428 / 1.018[15]	1.573
1823	Benzenemethanamine, N-methyl-N-2-propynyl- / Pargyline	$C_{11}H_{13}N$ / 159.23	555-57-7	6988 96-7[11]	2-12-00-00548 / 0.944[25]	1.5213[20]
1824	Benzenemethanamine, N-(3-methylphenyl)- / m-Toluidine, N-benzyl	$C_{14}H_{15}N$ / 197.28	5405-17-4	312	4-12-00-02178 / 1.0083[65]	ace 4; bz 4; eth 4 / 1.5845[65]
1825	Benzenemethanamine, N-(4-methylphenyl)- / p-Toluidine, N-benzyl	$C_{14}H_{15}N$ / 197.28	5405-15-2 / 19.5	319; 181[10]	4-12-00-02178 / 1.0064[65]	bz 4; eth 4; EtOH 4; chl 4 / 1.5832[65]
1826	Benzenemethanamine, α-phenyl- / Benzhydrylamine	$C_{13}H_{13}N$ / 183.25	91-00-9 / 34	1085 304; 176[23]	4-12-00-03282 / 1.0633[20]	H_2O 2; ace 3 / 1.5963
1827	Benzenemethanamine, N-phenyl- / Benzylaniline	$C_{13}H_{13}N$ / 183.25	103-32-2 / 37.5	1140 306.5	4-12-00-02172 / 1.0298[65]	eth 4; EtOH 4 / 1.6118[25]
1828	Benzenemethanamine, N-(phenylmethyl)- / Dibenzylamine	$C_{14}H_{15}N$ / 197.28	103-49-1 / -26	2993 300 dec; 270[250]	4-12-00-02179 / 1.0256[22]	H_2O 1; EtOH 4; eth 4; ctc 3 / 1.5781[20]
1829	Benzenemethanamine, N-phenyl-N-(phenylmethyl)-	$C_{20}H_{19}N$ / 273.38	91-73-6 / 69	226[10]	3-12-00-02225 / 1.0444[80]	H_2O 1; EtOH 2; eth 3; bz 3 / 1.6065[80]
1830	Benzenemethanamine, α-propyl-	$C_{10}H_{15}N$ / 149.24	2941-19-7	221; 108[16]	3-12-00-02716 / 0.9366[20]	EtOH 4
1831	Benzene, 1,1',1'',1'''-methanetetrayltetrakis- / Tetraphenylmethane	$C_{25}H_{20}$ / 320.43	630-76-2 / 282	431	4-05-00-02741	H_2O 1; EtOH 1; eth 1; bz 3
1832	Benzenemethanethiol / Thiobenzyl alcohol	C_7H_8S / 124.21	100-53-8 / -30	9249 194.5	4-06-00-02632 / 1.058[20]	H_2O 1; EtOH 4; eth 4; ctc 2 / 1.5151[20]

No.	Name Synonym	Mol. Form. Mol. Wt.	CAS RN mp/°C	Merck No. bp/°C	Beil. Ref. den/g cm^{-3}	Solubility n_D
1833	Benzenemethanethiol, 4-chloro-	C$_7$H$_7$ClS 158.65	6258-66-8 19.5	113[17]	4-06-00-02770 1.202[25]	1.5893[20]
1834	Benzenemethanethiol, α-methyl-, (S)- Ethanethiol, 1-phenyl (1)	C$_8$H$_{10}$S 138.23	33877-11-1	199; 83[10]	4-06-00-03062 1.022[20]	bz 4; eth 4; EtOH 4 1.5593[20]
1835	Benzenemethanimine, α-phenyl-	C$_{13}$H$_{11}$N 181.24	1013-88-3	282	4-07-00-01365 1.0847[19]	eth 4 1.6191[19]
1836	Benzenemethanol Benzyl alcohol	C$_7$H$_8$O 108.14	100-51-6 -15.2	1138 205.3	4-06-00-02222 1.0419[24]	H$_2$O 3; EtOH 3; eth 3; ace 3 1.5396[20]
1837	Benzenemethanol, 2-amino-	C$_7$H$_9$NO 123.15	5344-90-1 83.5	273	3-13-00-01615	H$_2$O 3; EtOH 3; eth 3; bz 4
1838	Benzenemethanol, α-(1-aminoethyl)-, hydrochloride, (R*,S*)-(±)- Phenylpropanolamine hydrochloride	C$_9$H$_{14}$ClNO 187.67	154-41-6 194	7279	2-13-00-00371	H$_2$O 4; EtOH 3; eth 1; bz 1
1839	Benzenemethanol, α-(aminomethyl)- Phenylethanolamine	C$_8$H$_{11}$NO 137.18	7568-93-6 56.5	7258 160[17]	4-13-00-01801	H$_2$O 4; EtOH 3
1840	Benzenemethanol, α-(1-aminopropyl)- α-(α-Aminopropyl)benzyl alcohol	C$_{10}$H$_{15}$NO 165.24	5897-76-7 79.5	483	3-13-00-01791	
1841	Benzenemethanol, 2-[bis(4- hydroxyphenyl)methyl]- Phenolphthalol	C$_{20}$H$_{18}$O$_3$ 306.36	81-92-5 201.5	7211	4-06-00-07623	
1842	Benzenemethanol, α,α-bis(1-methylethyl)-	C$_{13}$H$_{20}$O 192.30	4397-05-1	229; 158[60]	4-06-00-03462 0.9755[20]	eth 4 1.5239[20]
1843	Benzenemethanol, 5-bromo-2-hydroxy- Bromosaligenin	C$_7$H$_7$BrO$_2$ 203.04	2316-64-5 113	1426	3-06-00-04541	bz 4; eth 4; EtOH 4; chl 4
1844	Benzenemethanol, α-(bromomethyl)-	C$_8$H$_9$BrO 201.06	2425-28-7	109-10[2]	4-06-00-03053 1.4994[20]	H$_2$O 1 1.5800[17]
1845	Benzenemethanol, 2-bromo-α-methyl-	C$_8$H$_9$BrO 201.06	5411-56-3	109-10[2]	4-06-00-03053 1.4994[20]	1.5800[17]
1846	Benzenemethanol, α-butyl- 1-Phenyl-1-pentanol	C$_{11}$H$_{16}$O 164.25	583-03-9	3921 141[25]; 102[3]	2-06-00-00503 0.9655[20]	ace 4; eth 4; EtOH 4 1.4086[25]
1847	Benzenemethanol, α-butyl-α-methyl- 2-Hexanol, 2-phenyl-	C$_{12}$H$_{18}$O 178.27	4396-98-9	129-30[4]	3-06-00-01995 0.9522[20]	1.5091[20]
1848	Benzenemethanol, 2-chloro-	C$_7$H$_7$ClO 142.58	17849-38-6 73	230	4-06-00-02589	H$_2$O 2; EtOH 4; eth 4; lig 4
1849	Benzenemethanol, 4-chloro-	C$_7$H$_7$ClO 142.58	873-76-7 75	235	4-06-00-02593	bz 4; eth 4; EtOH 4
1850	Benzenemethanol, 4-chloro-α-(4- chlorophenyl)-α-methyl- 1,1-Bis(4-chlorophenyl)ethanol	C$_{14}$H$_{12}$Cl$_2$O 267.15	80-06-8 70	2086	4-06-00-04718	H$_2$O 1; EtOH 1; eth 3; bz 3
1851	Benzenemethanol, 4-chloro-α-(4- chlorophenyl)-α-(trichloromethyl)- 1,1-Bis(p-chlorophenyl)-2,2,2-trichloroethanol	C$_{14}$H$_9$Cl$_5$O 370.49	115-32-2 77.5	3075	4-06-00-04722	H$_2$O 1; os 1
1852	Benzenemethanol, α-(chloromethyl)-	C$_8$H$_9$ClO 156.61	1674-30-2	128[17]	4-06-00-03045 1.1926[20]	EtOH 3; eth 4 1.5523[20]
1853	Benzenemethanol, 2,4-dichloro- 2,4-Dichlorobenzyl alcohol	C$_7$H$_6$Cl$_2$O 177.03	1777-82-8 59.5	3048 150[25]	4-06-00-02597	chl 3
1854	Benzenemethanol, 2,4-dichloro-α-methyl-	C$_8$H$_8$Cl$_2$O 191.06	1475-13-4	125-6[7]	4-06-00-03048 1.293[25]	1.5605[20]
1855	Benzenemethanol, α,α-diethyl-	C$_{11}$H$_{16}$O 164.25	1565-71-5 <-17	223.5	4-06-00-03381 0.9831[20]	EtOH 4 1.5165[20]
1856	Benzenemethanol, 3,4-dimethoxy-	C$_9$H$_{12}$O$_3$ 168.19	93-03-8	298; 172[12]	4-06-00-07381 1.178[17]	H$_2$O 3; EtOH 3 1.555[17]
1857	Benzenemethanol, α,α-dimethyl- α-Cumyl alcohol	C$_9$H$_{12}$O 136.19	617-94-7 36	202	4-06-00-03219 0.9735[20]	H$_2$O 1; EtOH 3; eth 3; bz 3 1.5325[20]
1858	Benzenemethanol, α,3-dimethyl- 1-(3-Methylphenyl)ethyl alcohol	C$_9$H$_{12}$O 136.19	7287-81-2	112[12]	3-06-00-01823 0.9974[15]	eth 4; EtOH 4 1.5240[20]
1859	Benzenemethanol, α,4-dimethyl- 1-(4-Methylphenyl)ethanol	C$_9$H$_{12}$O 136.19	536-50-5	3226 219	4-06-00-03242 0.9668[25]	H$_2$O 1; EtOH 4; eth 4 1.5246[20]
1860	Benzenemethanol, 2,4-dimethyl-	C$_9$H$_{12}$O 136.19	16308-92-2 22	232; 150[44]	4-06-00-03250 1.0310[20]	1.5339[20]
1861	Benzenemethanol, 3,5-dimethyl-	C$_9$H$_{12}$O 136.19	27129-87-9	219.5	0-06-00-00521 0.927[25]	1.5312[20]
1862	Benzenemethanol, 4-(dimethylamino)-	C$_9$H$_{13}$NO 151.21	1703-46-4 69	123[1]	4-13-00-01771 1.059[14]	1.5727[14]
1863	Benzenemethanol, α-[1- (dimethylamino)ethyl]-, [R-(R*,S*)]- (1R,2S)-N-Methylephedrine	C$_{11}$H$_{17}$NO 179.26	552-79-4 87.5	5987	H$_2$O 1; EtOH 3; eth 3; MeOH 3
1864	Benzenemethanol, α-(1,1-dimethylethyl)-	C$_{11}$H$_{16}$O 164.25	3835-64-1 45	111[15]	1-06-00-00270	eth 4; EtOH 4
1865	Benzenemethanol, 4-(1,1-dimethylethyl)-	C$_{11}$H$_{16}$O 164.25	877-65-6	140[20]	4-06-00-03402 0.928[25]	1.5179[20]
1866	Benzenemethanol, α,α-diphenyl- Triphenylmethanol	C$_{19}$H$_{16}$O 260.34	76-84-6 164.2	9653 380	4-06-00-05014 1.199[0]	H$_2$O 1; EtOH 4; eth 4; ace 3
1867	Benzenemethanol, α,α-dipropyl- 4-Heptanol, 4-phenyl-	C$_{13}$H$_{20}$O 192.30	4436-96-8	134[20]	2-06-00-00513 0.9470[15]	1.516[10]
1868	Benzenemethanol, α-ethenyl- 1-Phenylallyl alcohol	C$_9$H$_{10}$O 134.18	4393-06-0	2-06-00-00529 1.0249[21]	H$_2$O 2; EtOH 3; eth 3; bz 3 1.5406[20]
1869	Benzenemethanol, α-ethenyl-α-methyl- 3-Buten-2-ol, 2-phenyl-	C$_{10}$H$_{12}$O 148.20	6051-52-1	92-5[7]	4-06-00-03839 0.9965[21]	1.5277[21]

No.	Name / Synonym	Mol. Form. / Mol. Wt.	CAS RN / mp/°C	Merck No. / bp/°C	Beil. Ref. / den/g cm^{-3}	Solubility / n_D
1870	Benzenemethanol, α-ethyl- α-Ethylbenzyl alcohol	$C_9H_{12}O$ 136.19	93-54-9	3727 219	4-06-00-03183 0.9915[25]	bz 4; eth 4; EtOH 4; MeOH 4 1.5169[23]
1871	Benzenemethanol, α-[1-(ethylamino)ethyl]-, (R*,S*)- N-Ethylnorephedrine	$C_{11}H_{17}NO$ 179.26	37025-57-3 51.5	143[18]	4-13-00-01887	bz 4
1872	Benzenemethanol, α-ethyl-3-hydroxy- 1-(m-Hydroxyphenyl)-1-propanol	$C_9H_{12}O_2$ 152.19	55789-02-1 107	177[13]	3-06-00-04623	eth 4; EtOH 4
1873	Benzenemethanol, α-ethyl-4-methoxy-	$C_{10}H_{14}O_2$ 166.22	5349-60-0	143[20]	3-06-00-04624	ctc 3 1.5277[20]
1874	Benzenemethanol, α-ethyl-α-methyl-, (R)-	$C_{10}H_{14}O$ 150.22	1006-06-0 -13	211; 113[23]	0.984[25]	H$_2$O 1; EtOH 3; eth 3 1.5185[20]
1875	Benzenemethanol, α-ethyl-α-propyl- 3-Hexanol, 3-phenyl	$C_{12}H_{18}O$ 178.27	20731-93-5	134[27]	3-06-00-02000 0.964[20]	1.5100[20]
1876	Benzenemethanol, α-ethynyl- 1-Phenylpropargyl alcohol	C_9H_8O 132.16	4187-87-5 22	114[12]	4-06-00-04066 1.0655[20]	1.5508[20]
1877	Benzenemethanol, α-ethynyl-, carbamate Carfimate	$C_{10}H_9NO_2$ 175.19	3567-38-2 86.5	1845	4-06-00-04067	
1878	Benzenemethanol, α-ethynyl-α-methyl-	$C_{10}H_{10}O$ 146.19	127-66-2 52.3	217.5; 102[12]	4-06-00-04073 1.0314[20]	
1879	Benzenemethanol, 3-fluoro-	C_7H_7FO 126.13	456-47-3	201; 104[22]	4-06-00-02589 1.164[25]	1.5095[20]
1880	Benzenemethanol, 4-fluoro-	C_7H_7FO 126.13	459-56-3 23	210	4-06-00-02589	1.5080[20]
1881	Benzenemethanol, α-hexyl- 1-Phenyl-1-heptanol	$C_{13}H_{20}O$ 192.30	614-54-0	275	4-06-00-03454 0.946[25]	1.5024[20]
1882	Benzenemethanol, 2-hydroxy- Salicyl alcohol	$C_7H_8O_2$ 124.14	90-01-7 87	8294 sub	4-06-00-05896 1.1613[25]	H$_2$O 3; EtOH 3; eth 3; bz 3
1883	Benzenemethanol, 3-hydroxy-	$C_7H_8O_2$ 124.14	620-24-6 73	300 dec	4-06-00-05907 1.161[25]	H$_2$O 4; EtOH 4; eth 4; chl 2
1884	Benzenemethanol, 4-hydroxy- p-Hydroxybenzyl alcohol	$C_7H_8O_2$ 124.14	623-05-2 124.5	252	4-06-00-05909	H$_2$O 4; EtOH 4; eth 3; bz 4
1885	Benzenemethanol, 2-hydroxy-3-methoxy-	$C_8H_{10}O_3$ 154.17	4383-05-5 61.5	162[12]	4-06-00-07378	chl 2
1886	Benzenemethanol, 4-hydroxy-3-methoxy-	$C_8H_{10}O_3$ 154.17	498-00-0 115	dec	4-06-00-07381	H$_2$O 3; EtOH 3; eth 3; bz 3
1887	Benzenemethanol, 4-hydroxy-α-[(methylamino)methyl]- Synephrine	$C_9H_{13}NO_2$ 167.21	94-07-5 184.5	8994	2-13-00-00491	
1888	Benzenemethanol, 2-iodo-	C_7H_7IO 234.04	5159-41-1 92	147-9[32]	4-06-00-02605	1.6349[20]
1889	Benzenemethanol, 2-methoxy-	$C_8H_{10}O_2$ 138.17	612-16-8	249	4-06-00-05896 1.0386[25]	H$_2$O 1; EtOH 3; eth 5 1.5455[20]
1890	Benzenemethanol, 3-methoxy-	$C_8H_{10}O_2$ 138.17	6971-51-3	252	4-06-00-05907 1.112[25]	1.5440[20]
1891	Benzenemethanol, 4-methoxy- Anise alcohol	$C_8H_{10}O_2$ 138.17	105-13-5 25	695 259.1	4-06-00-05909 1.109[26]	H$_2$O 3; EtOH 4; eth 4; ctc 3 1.5420[25]
1892	Benzenemethanol, 4-methoxy-, acetate	$C_{10}H_{12}O_3$ 180.20	104-21-2 84	270; 150[23]	4-06-00-05914 1.105[25]	ctc 3
1893	Benzenemethanol, 2-methoxy-α-methyl-	$C_9H_{12}O_2$ 152.19	13513-82-1 -85.1	125	3-06-00-04563 0.9647[20]	ace 4; bz 4; eth 4; EtOH 4 1.4024[20]
1894	Benzenemethanol, 3-methoxy-α-methyl-	$C_9H_{12}O_2$ 152.19	23308-82-9	133[15]	4-06-00-05929 1.0781[19]	eth 4; EtOH 4 1.5325[20]
1895	Benzenemethanol, 4-methoxy-α-methyl-	$C_9H_{12}O_2$ 152.19	3319-15-1	310 dec; 140[17]	4-06-00-05930 1.0794[20]	ctc 3 1.5310[25]
1896	Benzenemethanol, α-methyl- α-Methylbenzyl alcohol	$C_8H_{10}O$ 122.17	98-85-1 20	205	2-06-00-00444 1.013[25]	H$_2$O 1; EtOH 4; eth 4 1.5265[20]
1897	Benzenemethanol, 2-methyl- o-Methylbenzyl alcohol	$C_8H_{10}O$ 122.17	89-95-2 38	224; 118[20]	4-06-00-03109 1.023[40]	eth 4; EtOH 4; chl 4
1898	Benzenemethanol, 3-methyl- m-Methylbenzyl alcohol	$C_8H_{10}O$ 122.17	587-03-1 <-20	215.5	4-06-00-03162 0.9157[17]	H$_2$O 2; EtOH 4; eth 4; chl 3
1899	Benzenemethanol, 4-methyl- p-Methylbenzyl alcohol	$C_8H_{10}O$ 122.17	589-18-4 61.5	217	4-06-00-03171 0.978[22]	eth 4; EtOH 4
1900	Benzenemethanol, α-[1-(methylamino)ethyl]-, (R*,S*)-(±)- (±)-Ephedrine	$C_{10}H_{15}NO$ 165.24	90-81-3 76.5	3561 135[12]	4-13-00-01881 1.1220[20]	H$_2$O 3; EtOH 3; eth 3; bz 3
1901	Benzenemethanol, α-(1-methylethyl)- 1-Phenyl-2-methylpropyl alcohol	$C_{10}H_{14}O$ 150.22	611-69-8	223	4-06-00-03289 0.9869[14]	H$_2$O 1; EtOH 3; ace 3 1.5193[14]
1902	Benzenemethanol, 4-(1-methylethyl)- Cumic alcohol	$C_{10}H_{14}O$ 150.22	536-60-7 28	2621 249	4-06-00-03348 0.9818[20]	H$_2$O 1; EtOH 5; eth 5; bz 4 1.5210[20]
1903	Benzenemethanol, α-[1-[methyl(3-phenyl-2-propenyl)amino]ethyl]- Cinnamedrine	$C_{19}H_{23}NO$ 281.40	90-86-8 75	2299	3-13-00-01734	
1904	Benzenemethanol, α-methyl-α-propyl-	$C_{11}H_{16}O$ 164.25	4383-18-0	216	2-06-00-00505 0.9723[22]	EtOH 4
1905	Benzenemethanol, α-(2-methylpropyl)-	$C_{11}H_{16}O$ 164.25	1565-86-2	236; 112[9]	4-06-00-03380 0.9537[19]	ace 4; bz 4; eth 4; EtOH 4 1.5080[18]
1906	Benzenemethanol, 2-nitro-	$C_7H_7NO_3$ 153.14	612-25-9 74	270; 168[20]	4-06-00-02608	H$_2$O 2; EtOH 3; eth 3

No.	Name Synonym	Mol. Form. Mol. Wt.	CAS RN mp/°C	Merck No. bp/°C	Beil. Ref. den/g cm^{-3}	Solubility n_D
1907	Benzenemethanol, 3-nitro-	$C_7H_7NO_3$ 153.14	619-25-0 30.5	175-80[3]	4-06-00-02609 1.296[19]	H_2O 3; EtOH 3; eth 3; chl 2
1908	Benzenemethanol, 4-nitro- p-Nitrobenzyl alcohol	$C_7H_7NO_3$ 153.14	619-73-8 96.5	255 dec; 185[12]	4-06-00-02611	H_2O 2; EtOH 3; eth 3; ace 2
1909	Benzenemethanol, α-pentyl-	$C_{12}H_{18}O$ 178.27	4471-05-0	170[50]; 128[5]	4-06-00-03417 0.9497[25]	EtOH 3; eth 3 1.5105[20]
1910	Benzenemethanol, α-phenyl- Benzohydrol	$C_{13}H_{12}O$ 184.24	91-01-0 69	1100 298; 180[20]	4-06-00-04648	H_2O 2; EtOH 4; eth 4; ctc 4
1911	Benzenemethanol, α-2-propenyl-	$C_{10}H_{12}O$ 148.20	936-58-3	228.5	4-06-00-03838 1.004[18]	1.5289[21]
1912	Benzenemethanol, α-propyl-, (R)-	$C_{10}H_{14}O$ 150.22	22144-60-1 16	232	3-06-00-01845 0.9740[20]	eth 4; EtOH 4 1.5139[20]
1913	Benzenemethanol, α-[(2-pyridinylamino)methyl]- Phenyramidol	$C_{13}H_{14}N_2O$ 214.27	553-69-5 83.5	7292	5-22-08-00301	
1914	Benzenemethanol, 3,4,5-trimethoxy- 3,4,5-Trimethoxybenzyl alcohol	$C_{10}H_{14}O_4$ 198.22	3840-31-1	228[25]	4-06-00-07695 1.232[25]	1.5439[20]
1915	Benzene, methoxy- Anisole	C_7H_8O 108.14	100-66-3 -37.5	699 153.7	4-06-00-00548 0.9940[20]	H_2O 1; EtOH 3; eth 3; ace 4 1.5174[20]
1916	Benzene, 1-methoxy-2,3-dimethyl-	$C_9H_{12}O$ 136.19	2944-49-2 29	199	4-06-00-03097 0.9596[40]	H_2O 1; EtOH 4; eth 4; ace 3 1.5120[40]
1917	Benzene, 1-methoxy-2,4-dimethyl-	$C_9H_{12}O$ 136.19	6738-23-4	192	4-06-00-03127 0.9740[16]	H_2O 1; EtOH 3; eth 3; bz 3 1.5190[16]
1918	Benzene, 1-methoxy-3,5-dimethyl-	$C_9H_{12}O$ 136.19	874-63-5	194; 89[15]	4-06-00-03142 0.9627[15]	H_2O 1; EtOH 3; eth 3; bz 3 1.5110[20]
1919	Benzene, 2-methoxy-1,3-dimethyl-	$C_9H_{12}O$ 136.19	1004-66-6	182.5	4-06-00-03114 0.9619[14]	H_2O 1; EtOH 3; eth 3; bz 3 1.5053[14]
1920	Benzene, 2-methoxy-1,4-dimethyl-	$C_9H_{12}O$ 136.19	1706-11-2	194	4-06-00-03165 0.9693[13]	H_2O 1; EtOH 3; eth 3; bz 3 1.5182[15]
1921	Benzene, 4-methoxy-1,2-dimethyl-	$C_9H_{12}O$ 136.19	4685-47-6	202	4-06-00-03100 0.9744[14]	H_2O 1; EtOH 3; eth 3; bz 3 1.5198[14]
1922	Benzene, 1-methoxy-2,3-dinitro- Anisole, 2,3-dinitro	$C_7H_6N_2O_5$ 198.14	16315-07-4 119		4-06-00-01369 1.23[137]	EtOH 4; lig 4
1923	Benzene, 1-methoxy-2,4-dinitro-	$C_7H_6N_2O_5$ 198.14	119-27-7 94.5	206[12]	4-06-00-01372 1.3364[131]	H_2O 2; EtOH 3; eth 3; ace 3 1.546[15]
1924	Benzene, 1-methoxy-3,5-dinitro- 3,5-Dinitroanisole	$C_7H_6N_2O_5$ 198.14	5327-44-6 105.3		4-06-00-01385 1.558[12]	ace 4; bz 4; MeOH 4
1925	Benzene, 2-methoxy-1,3-dinitro-	$C_7H_6N_2O_5$ 198.14	3535-67-9 118		3-06-00-00868 1.3000[128]	EtOH 4
1926	Benzene, 2-methoxy-1,4-dinitro- Anisole, 2,5-dinitro-	$C_7H_6N_2O_5$ 198.14	3962-77-4 97	136-8[2]	4-06-00-01383 1.476[18]	H_2O 2; EtOH 3; ace 3; bz 3
1927	Benzene, 4-methoxy-1,2-dinitro-	$C_7H_6N_2O_5$ 198.14	4280-28-8 71	168[3]	4-06-00-01384 1.3332[110]	EtOH 4
1928	Benzene, (2-methoxyethenyl)-	$C_9H_{10}O$ 134.18	4747-15-3	211.5	2-06-00-00521 0.9894[23]	1.5620[24]
1929	Benzene, 1-methoxy-2-(3-methoxyphenoxy)-	$C_{14}H_{14}O_3$ 230.26	1655-71-6 54.8	327.5	4-06-00-05670	bz 4; eth 4; EtOH 4
1930	Benzene, (methoxymethyl)- Benzyl methyl ether	$C_8H_{10}O$ 122.17	538-86-3 -52.6	1154 170	4-06-00-02229 0.9634[20]	H_2O 1; EtOH 4; eth 4; bz 3 1.5008[20]
1931	Benzene, 1-methoxy-2-methyl-	$C_8H_{10}O$ 122.17	578-58-5 -34.1	171	4-06-00-01943 0.985[25]	H_2O 1; EtOH 3; eth 3; ace 3 1.5161[20]
1932	Benzene, 1-methoxy-3-methyl-	$C_8H_{10}O$ 122.17	100-84-5 -47	175.5	4-06-00-02039 0.969[25]	H_2O 1; EtOH 3; eth 3; ace 3 1.5130[20]
1933	Benzene, 1-methoxy-4-methyl-	$C_8H_{10}O$ 122.17	104-93-8 -32	175.5	4-06-00-02098 0.969[25]	H_2O 1; EtOH 3; eth 3; chl 3 1.5112[20]
1934	Benzene, 4-methoxy-1-methyl-2-nitro-	$C_8H_9NO_3$ 167.16	17484-36-5 17	266.5	3-06-00-01384 1.207[25]	1.5525[20]
1935	Benzene, 1-methoxy-2-nitro- 2-Nitroanisole	$C_7H_7NO_3$ 153.14	91-23-6 10.5	277; 144[4]	4-06-00-01249 1.2540[20]	H_2O 1; EtOH 5; eth 5; ctc 3 1.5161[20]
1936	Benzene, 1-methoxy-3-nitro-	$C_7H_7NO_3$ 153.14	555-03-3 38.5	258	4-06-00-01270 1.373[18]	H_2O 1; EtOH 3; eth 4
1937	Benzene, 1-methoxy-4-nitro- Anisole, 4-nitro	$C_7H_7NO_3$ 153.14	100-17-4 54	274	4-06-00-01282 1.2192[60]	H_2O 1; EtOH 4; eth 4; ctc 3 1.5070[60]
1938	Benzene, 1-methoxy-2-phenoxy-	$C_{13}H_{12}O_2$ 200.24	1695-04-1 79	288; 91[7]	4-06-00-05571	bz 4; eth 4; EtOH 4
1939	Benzene, 1-methoxy-4-(2-phenylethenyl)-, (E)-	$C_{15}H_{14}O$ 210.28	1694-19-5 136.5	142.5[15]	4-06-00-04855	H_2O 1; EtOH 4; eth 4; ace 4
1940	Benzene, 1-methoxy-4-(1-phenylethyl)-	$C_{15}H_{16}O$ 212.29	2605-18-7	180-2[19]	4-06-00-04711 1.0473[20]	bz 4; eth 4; HOAc 4 1.5725[20]
1941	Benzene, 1-methoxy-2-(1-propenyl)-	$C_{10}H_{12}O$ 148.20	10577-44-3	224; 104[13]	4-06-00-03793 0.9962[15]	1.560[15]
1942	Benzene, 1-methoxy-4-(1-propenyl)- Anethole	$C_{10}H_{12}O$ 148.20	104-46-1 21.3	675 234; 115[12]	2-06-00-00523 0.9882[20]	ace 4; bz 4; eth 4; EtOH 4 1.5615[20]
1943	Benzene, 1-methoxy-4-(2-propenyl)- Estragole	$C_{10}H_{12}O$ 148.20	140-67-0	3657 215.5	4-06-00-03817 0.965[25]	EtOH 4; chl 4 1.5195[20]
1944	Benzene, 1-methoxy-4-(1-propenyl)-, (E)-	$C_{10}H_{12}O$ 148.20	4180-23-8 21.35	234; 115[12]	4-06-00-03796 0.9882[20]	H_2O 2; EtOH 5; eth 5; ace 3 1.5615[20]
1945	Benzene, 1-methoxy-4-propyl-	$C_{10}H_{14}O$ 150.22	104-45-0	211.5	4-06-00-03181 0.9472[20]	H_2O 2; EtOH 3; eth 4; ace 3 1.5045[20]

No.	Name / Synonym	Mol. Form. / Mol. Wt.	CAS RN / mp/°C	Merck No. / bp/°C	Beil. Ref. / den/g cm^{-3}	Solubility / n_D
1946	Benzene, 2-methoxy-1,3,5-trinitro- / Methyl picrate	$C_7H_5N_3O_7$ / 243.13	606-35-9 / 69		4-06-00-01456 / 1.4947[80]	H_2O 1; EtOH 4; eth 3; bz 4 / 1.4947[80]
1947	Benzene, methyl- / Toluene	C_7H_8 / 92.14	108-88-3 / -94.9	9455 / 110.6	4-05-00-00766 / 0.8669[20]	H_2O 1; EtOH 5; eth 5; ace 3 / 1.4961[20]
1948	Benzene, 1-methyl-2,4-bis(1-methylethyl)- / Toluene, 2,4-diisopropyl-	$C_{13}H_{20}$ / 176.30	1460-98-6 /	/ 82[7]	4-05-00-01152 / 0.8636[20]	ctc 3; CS_2 3 / 1.4912[20]
1949	Benzene, 1-methyl-3,5-bis(1-methylethyl)-	$C_{13}H_{20}$ / 176.30	3055-14-9 / -60.8	/ 216.5	4-05-00-01152 / 0.8668[20]	/ 1.4950[20]
1950	Benzene, (1-methyl-1-butenyl)- / 2-Phenyl-2-pentene	$C_{11}H_{14}$ / 146.23	53172-84-2 /	/ 202	4-05-00-01402 / 0.8950[26]	/ 1.5196[26]
1951	Benzene, [(3-methylbutoxy)methyl]-	$C_{12}H_{18}O$ / 178.27	122-73-6 /	/ 237; 118[19]	3-06-00-01457 / 0.9098[20]	eth 4; EtOH 4 / 1.4792[20]
1952	Benzene, (1-methylbutyl)-	$C_{11}H_{16}$ / 148.25	2719-52-0 /	/ 199	4-05-00-01088 / 0.8594[21]	H_2O 1; EtOH 3; eth 3 / 1.4875[25]
1953	Benzene, (2-methylbutyl)-	$C_{11}H_{16}$ / 148.25	3968-85-2 /	/ 195.4	2-05-00-00332 / 0.8584[20]	/ 1.4873[20]
1954	Benzene, (3-methylbutyl)-	$C_{11}H_{16}$ / 148.25	2049-94-7 /	/ 195	4-05-00-01089 / 0.856[20]	H_2O 1; EtOH 3; eth 3; bz 4 / 1.4867[10]
1955	Benzene, (1-methylcyclopropyl)-	$C_{10}H_{12}$ / 132.21	2214-14-4 /	/ 91[50]	4-05-00-01388 /	ace 3; bz 3; chl 3 / 1.5160[20]
1956	Benzene, 1-methyl-2,4-dinitro- / 2,4-Dinitrotoluene	$C_7H_6N_2O_4$ / 182.14	121-14-2 / 71	/ 300 dec	4-05-00-00865 / 1.3208[71]	H_2O 1; EtOH 3; eth 3; ace 4 / 1.442
1957	Benzene, 1-methyl-3,5-dinitro-	$C_7H_6N_2O_4$ / 182.14	618-85-9 / 93	/ sub	4-05-00-00867 / 1.2772[111]	H_2O 2; EtOH 3; eth 3; bz 3
1958	Benzene, 2-methyl-1,3-dinitro- / 2,6-Dinitrotoluene	$C_7H_6N_2O_4$ / 182.14	606-20-2 / 66		4-05-00-00866 / 1.2833[111]	EtOH 3; chl 3 / 1.479
1959	Benzene, 2-methyl-1,4-dinitro-	$C_7H_6N_2O_4$ / 182.14	619-15-8 / 52.5		4-05-00-00866 / 1.282[111]	EtOH 3; bz 3; CS_2 4
1960	Benzene, 4-methyl-1,2-dinitro- / 3,4-Dinitrotoluene	$C_7H_6N_2O_4$ / 182.14	610-39-9 / 58.3		4-05-00-00866 / 1.2594[111]	H_2O 1; EtOH 3; chl 2; CS_2 3
1961	Benzene, 1,1'-methylenebis- / Diphenylmethane	$C_{13}H_{12}$ / 168.24	101-81-5 / 25.2	3329 / 265.0	4-05-00-01841 / 1.001[26]	H_2O 1; EtOH 3; eth 3; chl 3 / 1.5753[20]
1962	Benzene, 1,1'-methylenebis[2-bromo- / Diphenylmethane, 2,2'-dibromo-	$C_{13}H_{10}Br_2$ / 326.03	61592-89-0 /	/ 234-5[40]	3-05-00-01795 / 1.6197[20]	/ 1.6300[20]
1963	Benzene, 1,1'-methylenebis[4-chloro-	$C_{13}H_{10}Cl_2$ / 237.13	101-76-8 / 55.5	/ 186-90[18]	4-05-00-01848 / 1.365[17]	EtOH 3
1964	Benzene, 1,1'-methylenebis[4-methyl-	$C_{15}H_{16}$ / 196.29	4957-14-6 / 28	/ 286	4-05-00-01929 / 0.9800[20]	eth 4; EtOH 4
1965	Benzene, 1,1'-[methylenebis(oxy)]bis[4-chloro- / Bis(p-chlorophenoxy)methane	$C_{13}H_{10}Cl_2O_2$ / 269.13	555-89-5 / 70.5	1258 / 189-94[6]	4-06-00-00833 /	ace 4; bz 4
1966	Benzene, 1,1'-(1-methylene-1,2-ethanediyl)bis-	$C_{15}H_{14}$ / 194.28	948-97-0 / 48	/ 289	4-05-00-02191 / 1.1014[20]	/ 1.5903[20]
1967	Benzene, (1-methylene-2-propenyl)-	$C_{10}H_{10}$ / 130.19	2288-18-8 /	/ 60-1[17]	4-05-00-01539 / 0.9266[20]	H_2O 1; eth 3; bz 3; chl 3 / 1.5489[20]
1968	Benzene, (1-methylenepropyl)-	$C_{10}H_{12}$ / 132.21	2039-93-2 /	/ 182	4-05-00-01380 / 0.887[25]	/ 1.5288[20]
1969	Benzene, 1,1'-(1-methyl-1,2-ethanediyl)bis-	$C_{15}H_{16}$ / 196.29	5814-85-7 / 52	/ 280.5	4-05-00-01915 / 0.9807[20]	/ 1.5700[20]
1970	Benzene, 1,1'-[(1-methyl-1,2-ethanediyl)bis(oxy)]bis- / Propane, 1,2-diphenoxy-	$C_{15}H_{16}O_2$ / 228.29	69813-63-4 / 32	/ 175-8[12]	2-06-00-00151 / 1.0748[33]	ace 4; bz 4; eth 4; EtOH 4 / 1.5542[33]
1971	Benzene, 1,1'-(1-methyl-1,2-ethenediyl)bis-	$C_{15}H_{14}$ / 194.28	779-51-1 / 82	/ 285.5	4-05-00-02189 / 0.9857[17]	EtOH 3; eth 3; bz 3; lig 2 / 1.5635[17]
1972	Benzene, (1-methylethenyl)- / α-Methylstyrene	C_9H_{10} / 118.18	98-83-9 / -23.2	/ 165.4	4-05-00-01364 / 0.9106[20]	H_2O 1; EtOH 3; eth 3; ace 5 / 1.5386[20]
1973	Benzene, 1-(1-methylethenyl)-4-(1-methylethyl)-	$C_{12}H_{16}$ / 160.26	2388-14-9 / -30.6	/ 220.8	4-05-00-01422 / 0.8936[20]	ace 4; bz 4; eth 4; EtOH 4 / 1.5238[20]
1974	Benzene, (1-methylethoxy)-	$C_9H_{12}O$ / 136.19	2741-16-4 / -33	/ 176.8	4-06-00-00557 / 0.9408[25]	H_2O 3; EtOH 3; ace 3; bz 3 / 1.4975[20]
1975	Benzene, (1-methylethyl)- / Cumene	C_9H_{12} / 120.19	98-82-8 / -96.0	2619 / 152.4	4-05-00-00985 / 0.8618[20]	H_2O 1; EtOH 5; eth 5; ace 5 / 1.4915[20]
1976	Benzene, 1,1'-(1-methylethylidene)bis-	$C_{15}H_{16}$ / 196.29	778-22-3 / 29	/ 282.5	4-05-00-01925 / 0.9980[20]	
1977	Benzene, 1-(1-methylethyl)-2-nitro-	$C_9H_{11}NO_2$ / 165.19	6526-72-3 /	/ 103[9]	4-05-00-00997 / 1.101[12]	H_2O 1; ace 3; bz 3; HOAc 3 / 1.5259[20]
1978	Benzene, 1-(1-methylethyl)-4-nitro-	$C_9H_{11}NO_2$ / 165.19	1817-47-6 /	/ 122[9]	4-05-00-00997 / 1.0830[20]	H_2O 1; ace 3; bz 3; lig 3 / 1.5367[20]
1979	Benzene, 1-(1-methylethyl)-4-(phenylmethyl)-	$C_{16}H_{18}$ / 210.32	886-58-8 / 13.4	/ 310	4-05-00-01951 / 1.007[18]	
1980	Benzene, [(1-methylethyl)thio]-	$C_9H_{12}S$ / 152.26	3019-20-3 /	/ 208	4-06-00-01471 / 0.9852[20]	/ 1.5464[20]
1981	Benzene, (1-methylheptyl)- / 2-Phenyloctane	$C_{14}H_{22}$ / 190.33	777-22-0 /	/ 123-5[20]	4-05-00-01157 / 0.8611[20]	/ 1.4837[20]
1982	Benzene, 1,1',1''-methylidynetris- / Triphenylmethane	$C_{19}H_{16}$ / 244.34	519-73-3 / 94	9655 / 359; 200[10]	4-05-00-02495 / 1.014[99]	H_2O 1; EtOH 2; eth 4; bz 3 / 1.5839[99]
1983	Benzene, 1,1',1''-[methylidynetris(oxy)]tris-	$C_{19}H_{16}O_3$ / 292.33	16737-44-3 / 76.5	/ 269[50] dec	4-06-00-00611 /	eth 4; EtOH 4
1984	Benzene, 1-methyl-2-(1-methylethyl)-	$C_{10}H_{12}$ / 132.21	7399-49-7 /	/ 175	3-05-00-01214 / 0.9180[15]	/ 1.5112[30]

No.	Name / Synonym	Mol. Form. / Mol. Wt.	CAS RN / mp/°C	Merck No. / bp/°C	Beil. Ref. / den/g cm⁻³	Solubility / n_D
1985	Benzene, 1-methyl-3-(1-methylethenyl)-	$C_{10}H_{12}$ / 132.21	1124-20-5 /	185 /	4-05-00-01383 / 0.9010[25]	/ 1.5335[20]
1986	Benzene, 1-methyl-4-(1-methylethenyl)-	$C_{10}H_{12}$ / 132.21	1195-32-0 / -20	/ 185.3	4-05-00-01383 / 0.8936[23]	/ 1.5283[23]
1987	Benzene, 1-methyl-3-(1-methylethoxy)-	$C_{10}H_{14}O$ / 150.22	19177-04-9 /	/ 195	4-06-00-02040 / 0.931[20]	/ 1.4959[20]
1988	Benzene, 1-methyl-2-(1-methylethyl)- / o-Cymene	$C_{10}H_{14}$ / 134.22	527-84-4 / -71.5	2770 / 178.1	4-05-00-01057 / 0.8766[20]	H₂O 1; EtOH 5; eth 5; ace 5 / 1.5006[20]
1989	Benzene, 1-methyl-3-(1-methylethyl)- / m-Cymene	$C_{10}H_{14}$ / 134.22	535-77-3 / -63.7	2770 / 175.1	4-05-00-01058 / 0.8610[20]	H₂O 1; EtOH 5; eth 5; ace 5 / 1.4930[20]
1990	Benzene, 1-methyl-4-(1-methylethyl)- / p-Cymene	$C_{10}H_{14}$ / 134.22	99-87-6 / -68.9	2770 / 177.1	4-05-00-01060 / 0.8573[20]	H₂O 1; EtOH 5; eth 5; ace 5 / 1.4909[20]
1991	Benzene, 1-methyl-4-(1-methylethyl)-2-nitro-	$C_{10}H_{13}NO_2$ / 179.22	943-15-7 /	126[10] /	4-05-00-01064 / 1.0744[20]	eth 4; EtOH 4 / 1.5301[20]
1992	Benzene, [2-methyl-1-(1-methylethyl)propyl]-	$C_{13}H_{20}$ / 176.30	21777-84-4 / -24.5	/ 220.7	4-05-00-01146 / 0.8821[10]	/ 1.512
1993	Benzene, 1-methyl-2-[[(4-methylphenyl)thio]- / 2,4'-Ditolyl sulfide	$C_{14}H_{14}S$ / 214.33	4279-70-3 /	173[11] /	4-06-00-02172 / 1.0774[15]	eth 4; EtOH 4
1994	Benzene, 1-methyl-2-(1-methylpropyl)-	$C_{11}H_{16}$ / 148.25	1595-06-8 /	/ 196	4-05-00-01094 / 0.873[20]	ace 4; bz 4; eth 4; EtOH 4 / 1.497[20]
1995	Benzene, 1-methyl-2-(2-methylpropyl)-	$C_{11}H_{16}$ / 148.25	36301-29-8 / -73.3	/ 196	4-05-00-01095 / 0.8649[20]	ace 4; bz 4; eth 4; EtOH 4 / 1.4935[20]
1996	Benzene, 1-methyl-3-(1-methylpropyl)-	$C_{11}H_{16}$ / 148.25	1772-10-7 /	/ 194	4-05-00-01095 / 0.858[20]	ace 4; bz 4; eth 4; EtOH 4 / 1.490[20]
1997	Benzene, 1-methyl-3-(2-methylpropyl)-	$C_{11}H_{16}$ / 148.25	5160-99-6 /	/ 194	3-05-00-01001 / 0.8536[20]	ace 4; bz 4; eth 4; EtOH 4 / 1.4888[20]
1998	Benzene, 1-methyl-4-(1-methylpropyl)-	$C_{11}H_{16}$ / 148.25	1595-16-0 /	/ 197	4-05-00-01095 / 0.8640[19]	bz 4; eth 4; chl 4 / 1.493[20]
1999	Benzene, 1-methyl-4-(2-methylpropyl)-	$C_{11}H_{16}$ / 148.25	5161-04-6 /	/ 196	4-05-00-01096 / 0.8517[20]	ace 4; bz 4; eth 4; EtOH 4 / 1.4874[20]
2000	Benzene, 1-methyl-2-nitro- / o-Nitrotoluene	$C_7H_7NO_2$ / 137.14	88-72-2 / -10	6572 / 222	4-05-00-00845 / 1.1611[19]	H₂O 1; EtOH 5; eth 5; ctc 3 / 1.5450[20]
2001	Benzene, 1-methyl-3-nitro- / m-Nitrotoluene	$C_7H_7NO_2$ / 137.14	99-08-1 / 15.5	6572 / 232	4-05-00-00847 / 1.1581[20]	H₂O 1; EtOH 3; eth 5; bz 3 / 1.5466[20]
2002	Benzene, 1-methyl-4-nitro- / p-Nitrotoluene	$C_7H_7NO_2$ / 137.14	99-99-0 / 51.6	6572 / 238.3	4-05-00-00848 / 1.1038[75]	H₂O 1; EtOH 3; eth 4; ace 4
2003	Benzene, (1-methyl-1-nitroethyl)-	$C_9H_{11}NO_2$ / 165.19	3457-58-7 /	224 dec; 126[15] /	4-05-00-00998 / 1.1025[20]	ace 4; bz 4 / 1.5209[20]
2004	Benzene, 1-methyl-4-(nitromethyl)-	$C_8H_9NO_2$ / 151.17	29559-27-1 / 11.5	150-1[35] /	4-05-00-00972 / 1.1234[20]	H₂O 1; bz 3 / 1.5278[20]
2005	Benzene, 1-methyl-2-(2-nitrophenoxy)- / 2-Nitrophenyl 2-tolyl ether	$C_{13}H_{11}NO_3$ / 229.24	54106-40-0 / 39.5	194-6[14] /	4-06-00-01946 / 1.195[20]	
2006	Benzene, 1-methyl-4-(4-nitrophenoxy)-	$C_{13}H_{11}NO_3$ / 229.24	3402-74-2 / 69	224[25] /	4-06-00-02102	bz 4; HOAc 4
2007	Benzene, (1-methylnonadecyl)- / Eicosane, 2-phenyl-	$C_{26}H_{46}$ / 358.65	2398-66-5 / 29.0	204.5[1] /	4-05-00-01222 / 0.8547[20]	/ 1.4795[20]
2008	Benzene, (1-methyl-1-pentenyl)- / 2-Hexene, 2-phenyl	$C_{12}H_{16}$ / 160.26	20247-89-6 /	66-8[1] /	4-05-00-01419 / 0.910[20]	/ 1.5200[20]
2009	Benzene, (1-methylpentyl)-	$C_{12}H_{18}$ / 162.27	6031-02-3 /	/ 208	4-05-00-01116 / 0.869[15]	/ 1.492[15]
2010	Benzene, (3-methylpentyl)- / (3-Methylpentyl)benzene	$C_{12}H_{18}$ / 162.27	54410-69-4 /	/ 220	0-05-00-00444 / 0.8644[14]	/ 1.4896
2011	Benzene, (4-methylpentyl)-	$C_{12}H_{18}$ / 162.27	4215-86-5 /	/ 215	4-05-00-01116 / 0.8568[16]	bz 4; eth 4; EtOH 4
2012	Benzene, 1-methyl-2-phenoxy-	$C_{13}H_{12}O$ / 184.24	3991-61-5 / 21.7	/ 263.5	4-06-00-01946 / 1.0480[25]	
2013	Benzene, 1-methyl-3-phenoxy-	$C_{13}H_{12}O$ / 184.24	3586-14-9 /	/ 272	4-06-00-02041 / 1.051[25]	/ 1.5727[20]
2014	Benzene, 1-methyl-2-(phenylmethyl)-	$C_{14}H_{14}$ / 182.27	713-36-0 / 6.6	/ 280.5	4-05-00-01892 / 1.0020[20]	/ 1.5763[20]
2015	Benzene, 1-methyl-3-(phenylmethyl)-	$C_{14}H_{14}$ / 182.27	620-47-3 / -28	/ 279	4-05-00-01893 / 0.9913[20]	H₂O 1; EtOH 3; eth 3; bz 3 / 1.5712[20]
2016	Benzene, 1-methyl-4-(phenylmethyl)-	$C_{14}H_{14}$ / 182.27	620-83-7 / -30	/ 286	4-05-00-01894 / 0.9976[20]	eth 4; bz 4; EtOH 4; chl 4 / 1.5712[20]
2017	Benzene, 1-methyl-2-(phenylthio)-	$C_{13}H_{12}S$ / 200.30	13963-35-4 /	308; 164[12] /	4-06-00-02018 / 1.0893[20]	H₂O 1; ace 3; bz 3
2018	Benzene, 1-methyl-3-(phenylthio)-	$C_{13}H_{12}S$ / 200.30	13865-48-0 / -6.5	309.5 /	4-06-00-02080 / 1.0937[15]	H₂O 1; ace 3; bz 3
2019	Benzene, 1-methyl-4-(phenylthio)-	$C_{13}H_{12}S$ / 200.30	3699-01-2 / 15.7	/ 317	4-06-00-02169 / 1.0986[25]	H₂O 1; ace 3; bz 3 / 1.6225[25]
2020	Benzene, 1,1'-(1-methyl-1,3-propanediyl)bis-, (± / Butane, 1,3-diphenyl (DL)	$C_{16}H_{18}$ / 210.32	116783-21-2 /	/ 293	4-05-00-01938 / 0.9722[20]	/ 1.5525[20]
2021	Benzene, 1,1'-(2-methyl-1,3-propanediyl)bis-	$C_{16}H_{18}$ / 210.32	1520-46-3 / -33.7	/ 303	4-05-00-01939 / 0.9669[20]	
2022	Benzene, 1,1'-(3-methyl-1-propene-1,3-diyl)bis- / 1,3-Diphenyl-1-butene	$C_{16}H_{16}$ / 208.30	7614-93-9 / 47.5	/ 311	4-05-00-02205 / 0.9996[20]	/ 1.590[15]
2023	Benzene, 1-methyl-2-(2-propenyl)-	$C_{10}H_{12}$ / 132.21	1587-04-8 /	/ 182.5	3-05-00-01212 / 0.9005[20]	/ 1.5187[20]

No.	Name / Synonym	Mol. Form. / Mol. Wt.	CAS RN / mp/°C	Merck No. / bp/°C	Beil. Ref. / den/g cm^{-3}	Solubility / n_D
2024	Benzene, (2-methyl-1-propenyl)-	$C_{10}H_{12}$ 132.21	768-49-0	99[43.5]	4-05-00-01380 0.9029[20]	1.5388[20]
2025	Benzene, (1-methyl-1-propenyl)-, (E)-	$C_{10}H_{12}$ 132.21	768-00-3 -23.5	194.7	4-05-00-01378 0.9138[25]	H_2O 1; bz 3; chl 3 1.5425[20]
2026	Benzene, 1-methyl-2-(2-propenyloxy)-	$C_{10}H_{12}O$ 148.20	936-72-1	206.5	4-06-00-01946 0.9698[15]	H_2O 1; ctc 2 1.5188[15]
2027	Benzene, 1-methyl-3-(2-propenyloxy)-	$C_{10}H_{12}O$ 148.20	1758-10-7	212.5	4-06-00-02041 0.9564[20]	bz 4 1.5179[20]
2028	Benzene, 1-methyl-4-(2-propenyloxy)-	$C_{10}H_{12}O$ 148.20	23431-48-3	214.5	4-06-00-02101 0.9719[15]	H_2O 1; ace 3; ctc 2 1.5157[24]
2029	Benzene, (1-methyl-1-propenyl)-, (Z)-	$C_{10}H_{12}$ 132.21	767-99-7	194.7; 177[500]	3-05-00-01209 0.9191[25]	H_2O 1; bz 3; chl 3 1.5402[25]
2030	Benzene, (1-methylpropoxy)-	$C_{10}H_{14}O$ 150.22	10574-17-1	194.5	4-06-00-00559 0.9415[20]	H_2O 1; eth 3; ctc 2 1.4926[25]
2031	Benzene, (2-methylpropoxy)-	$C_{10}H_{14}O$ 150.22	1126-75-6	197.5	4-06-00-00559 0.9232[24]	1.4932[14]
2032	Benzene, 1-methyl-4-propoxy- Toluene, 4-propoxy	$C_{10}H_{14}O$ 150.22	5349-18-8 0	210.4	3-06-00-01354 0.9496[0]	
2033	Benzene, [(2-methylpropoxy)methyl]- Benzyl isobutyl ether	$C_{11}H_{16}O$ 164.25	940-49-8	212	3-06-00-01457 0.9233[20]	H_2O 1; eth 4; chl 4 1.4826[20]
2034	Benzene, 1-(2-methylpropoxy)-2-nitro- 1-Isobutoxy-2-nitrobenzene	$C_{10}H_{13}NO_3$ 195.22	56245-02-4	277.5	0-06-00-00218 1.1361[20]	eth 4
2035	Benzene, 1-methylpropyl sec-Butylbenzene	$C_{10}H_{14}$ 134.22	135-98-8 -82.7	1550 173.5	0.8580[25]	H_2O 1; EtOH 5; eth 5; ace 5 1.4895[20]
2036	Benzene, 1-methyl-2-propyl-	$C_{10}H_{14}$ 134.22	1074-17-5 -60.3	185	4-05-00-01056 0.8697[25]	1.4996[20]
2037	Benzene, 1-methyl-3-propyl-	$C_{10}H_{14}$ 134.22	1074-43-7 -82.5	182	4-05-00-01056 0.8569[25]	1.4935[20]
2038	Benzene, 1-methyl-4-propyl-	$C_{10}H_{14}$ 134.22	1074-55-1 -63.6	183.4	4-05-00-01057 0.8544[25]	H_2O 1; EtOH 3; eth 3 1.4922[20]
2039	Benzene, (2-methylpropyl)- Isobutylbenzene	$C_{10}H_{14}$ 134.22	538-93-2 -51.4	5018 172.7	4-05-00-01042 0.8532[20]	H_2O 1; EtOH 5; eth 5; ace 5 1.4866[20]
2040	Benzene, (methylsulfinyl)-	C_7H_8OS 140.21	1193-82-4 32.0	263.5; 140[13]	4-06-00-01467	1.5885[20]
2041	Benzene, (methylthio)-	C_7H_8S 124.21	100-68-5	193	4 06 00 01466 1.0579[20]	H_2O 1; EtOH 3; ace 4; os 3 1.5868[20]
2042	Benzene, [(methylthio)methyl]-	$C_8H_{10}S$ 138.23	766-92-7 -30	210; 120[48]	4-06-00-02633 1.0274[20]	1.5620[20]
2043	Benzene, 1-(methylthio)-2-nitro-	$C_7H_7NO_2S$ 169.20	3058-47-7 64.5		4-06-00-01661 1.2628[78]	H_2O 3; EtOH 3; bz 3; chl 3 1.6246[78]
2044	Benzene, 1-(methylthio)-4-nitro-	$C_7H_7NO_2S$ 169.20	701-57-5 72	135-40[2]	4-06-00-01687 1.2391[80]	H_2O 1; ace 3; bz 3 1.6401[20]
2045	Benzene, 1-methyl-4-(trifluoromethyl)-	$C_8H_7F_3$ 160.14	6140-17-6	129	1.4320[26]	eth 3; ace 3; bz 3; chl 3 1.4320[26]
2046	Benzene, 1-methyl-2,3,4-trinitro- 2,3,4-Trinitrotoluene	$C_7H_5N_3O_6$ 227.13	602-29-9 112	301	4-05-00-00872 1.62[25]	H_2O 1; EtOH 2; eth 3; ace 3
2047	Benzene, 1-methyl-2,4,5-trinitro- 2,4,5-Trinitrotoluene	$C_7H_5N_3O_6$ 227.13	610-25-3 104	exp 300	4-05-00-00872	H_2O 1; EtOH 2; eth 3; ace 3
2048	Benzene, 2-methyl-1,3,5-trinitro- 2,4,6-Trinitrotoluene	$C_7H_5N_3O_6$ 227.13	118-96-7 80.1	9643 exp 240	4-05-00-00873 1.654[25]	H_2O 1; EtOH 3; eth 3; ace 4
2049	Benzene, nitro- Nitrobenzene	$C_6H_5NO_2$ 123.11	98-95-3 5.7	6509 210.8	4-05-00-00708 1.2037[20]	H_2O 2; EtOH 4; eth 4; ace 4 1.5562[20]
2050	Benzene, (2-nitroethenyl)-, (E)-	$C_8H_7NO_2$ 149.15	5153-67-3 60	255	4-05-00-01352	H_2O 1; EtOH 3; eth 4; ace 3
2051	Benzene, (2-nitroethyl)-	$C_8H_9NO_2$ 151.17	6125-24-2 -23	250; 137[16]	4-05-00-00912 1.126[24]	1.5407[19]
2052	Benzene, (nitromethyl)-	$C_7H_7NO_2$ 137.14	622-42-4	226; 135[25]	4-05-00-00850 1.1596[20]	ace 4; eth 4 1.5323[20]
2053	Benzene, 1-nitro-2-phenoxy-	$C_{12}H_9NO_3$ 215.21	2216-12-8 <-20	235[60]; 184[8]	4-06-00-01252 1.2539[22]	bz 4; eth 4; EtOH 4; chl 4 1.575[20]
2054	Benzene, 1-nitro-4-phenoxy-	$C_{12}H_9NO_3$ 215.21	620-88-2 61	320; 225[30]	4-06-00-01287	H_2O 1; EtOH 2; eth 3; bz 3
2055	Benzene, 1-nitro-2-(2-phenylethenyl)-, (E)-	$C_{14}H_{11}NO_2$ 225.25	4264-29-3 73	209[11]	4-05-00-02166	H_2O 1; EtOH 3; eth 4
2056	Benzene, 1-nitro-2-(2-phenylethenyl)-, (Z)-	$C_{14}H_{11}NO_2$ 225.25	52208-62-5 64.8	187[11]	4-05-00-02166	EtOH 3; eth 4; AcOEt 4; HOAc 4
2057	Benzene, 1-nitro-2-(phenylthio)-	$C_{12}H_9NO_2S$ 231.28	4171-83-9 82	210[15]	4-06-00-01663	eth 4; EtOH 4
2058	Benzene, 1-nitro-4-(phenylthio)-	$C_{12}H_9NO_2S$ 231.28	952-97-6 56	288[100]; 240[25]	4-06-00-01694	eth 4; EtOH 4
2059	Benzene, nitroso- Nitrosobenzene	C_6H_5NO 107.11	586-96-9 68.5	58[18]	4-05-00-00702	H_2O 1; EtOH 3; eth 3; bz 3
2060	Benzene, 1-nitro-2-(trifluoromethyl)-	$C_7H_4F_3NO_2$ 191.11	384-22-5 32.5	217; 105[20]	4-05-00-00852	H_2O 1; EtOH 4; bz 4; ctc 2
2061	Benzene, 1-nitro-3-(trifluoromethyl)- 3-(Trifluoromethyl)-1-nitrobenzene	$C_7H_4F_3NO_2$ 191.11	98-46-4 -2.4	202.8; 103[40]	4-05-00-00853 1.4357[15]	H_2O 1; EtOH 3; eth 3; ctc 2 1.4719[20]
2062	Benzene, nonadecyl-	$C_{25}H_{44}$ 344.62	29136-19-4 40	419	3-05-00-01126 0.8545[20]	1.4807[20]

No.	Name / Synonym	Mol. Form. / Mol. Wt.	CAS RN / mp/°C	Merck No. / bp/°C	Beil. Ref. / den/g cm^{-3}	Solubility / n_D
2063	Benzenenonanoic acid, θ-nonyl-	$C_{24}H_{40}O_2$ 360.58	1938-22-3 37	200-4$^{0.09}$	2-09-00-00376 0.9312^{25}	1.4891^{20}
2064	Benzene, nonyl-	$C_{15}H_{24}$ 204.36	1081-77-2 -24	280.5	3-05-00-01075 0.8584^{20}	1.4816^{20}
2065	Benzene, octacosyl- Octacosylbenzene	$C_{34}H_{62}$ 470.87	61828-26-0 66	405$^{100.}$ 268^1	4-05-00-01243 0.8543^{20}	1.4796^{20}
2066	Benzene, octadecyl- Octadecylbenzene	$C_{24}H_{42}$ 330.60	4445-07-2 29	408	4-05-00-01219 0.8546^{20}	1.4809^{20}
2067	Benzene, octyl-	$C_{14}H_{22}$ 190.33	2189-60-8 -36	264.5	4-05-00-01157 0.8562^{20}	H_2O 1; eth 5; bz 5 1.4845^{20}
2068	Benzene, (1-octyldodecyl)- Eicosane, 9-phenyl-	$C_{26}H_{46}$ 358.65	2398-65-4 17.9	196$^{1.0}$	4-05-00-01223 0.8534^{20}	1.4790^{20}
2069	Benzene, (octyloxy)-	$C_{14}H_{22}O$ 206.33	1818-07-1 8	285	4-06-00-00500 0.9131^{15}	H_2O 1; EtOH 3; eth 3 1.4875^{20}
2070	Benzene, (3-octylundecyl)- 9-(2-Phenylethyl)heptadecane	$C_{25}H_{44}$ 344.62	5637-96-7 -26.7	189$^{1.0}$	4-05-00-01221 0.8560^{20}	1.4806^{20}
2071	Benzene, 1,1'-oxybis- Diphenyl ether	$C_{12}H_{10}O$ 170.21	101-84-8 26.8	7259 258.0	4-06-00-00568 1.0661^{30}	H_2O 1; EtOH 3; eth 3; bz 3 1.5787^{25}
2072	Benzene, 1,1'-oxybis[4-bromo-	$C_{12}H_8Br_2O$ 328.00	2050-47-7 60.5	339	4-06-00-01048 1.8^{25}	H_2O 1; EtOH 3; eth 4; bz 3
2073	Benzene, 1,1'-oxybis[4-chloro-	$C_{12}H_8Cl_2O$ 239.10	2444-89-5 30	313	4-06-00-00826 1.1231^{20}	H_2O 1 1.611^{20}
2074	Benzene, 1,1'-oxybis[2-ethyl-	$C_{16}H_{18}O$ 226.32	56911-77-4	318.5	1.0141^{18}	H_2O 1; eth 3; chl 3 1.5488^{18}
2075	Benzene, 1,1'-oxybis[2-methoxy-	$C_{14}H_{14}O_3$ 230.26	1655-70-5 79.5	330.5	3-06-00-04223	H_2O 1; EtOH 4; eth 4; lig 1
2076	Benzene, 1,1'-oxybis[2-methyl-	$C_{14}H_{14}O$ 198.26	4731-34-4	271	4-05-00-01947 1.047^{24}	bz 4; eth 4; EtOH 4
2077	Benzene, 1,1'-oxybis[3-methyl-	$C_{14}H_{14}O$ 198.26	19814-71-2	284; 136^{14}	4-06-00-02043 1.0323^{21}	H_2O 1; EtOH 3; eth 3; bz 3
2078	Benzene, 1,1'-oxybis[4-methyl-	$C_{14}H_{14}O$ 198.26	1579-40-4 51	285	4-06-00-02103	bz 4; eth 4; EtOH 4
2079	Benzene, 1,1'-[oxybis(methylene)]bis- Benzyl ether	$C_{14}H_{14}O$ 198.26	103-50-4 3.6	1146 298	4-06-00-02240 1.0428^{20}	H_2O 1; EtOH 5; eth 5; ctc 3 1.5168^{20}
2080	Benzene, 1,1'-(oxydi-2,1-ethanediyl)bis-	$C_{16}H_{18}O$ 226.32	2396-53-6	318.5; 86$^{0.5}$	4-06-00-03071 1.0141^{18}	eth 4; chl 4 1.5488^{18}
2081	Benzene, 1,1'-(oxydiethylidene)bis-, (R*,R*)- (±)- Bis-(1-phenylethyl)ether (DL)	$C_{16}H_{18}O$ 226.32	53776-69-5	280.2	4-06-00-03034 1.0058^{15}	H_2O 1; eth 3; chl 3 1.5454^{21}
2082	Benzene, pentabromo- Pentabromobenzene	C_6HBr_5 472.59	608-90-2 160.5	sub	4-05-00-00687	bz 4; chl 4
2083	Benzene, pentabromomethyl-	$C_7H_3Br_5$ 486.62	87-83-2 288		3-05-00-00720 2.97^{17}	H_2O 1; EtOH 2; bz 3; HOAc 2
2084	Benzene, pentachloro- Pentachlorobenzene	C_6HCl_5 250.34	608-93-5 86	277	4-05-00-00669 1.8342^{16}	H_2O 1; EtOH 1; eth 2; bz 2
2085	Benzene, pentachloro(dichloromethyl)-	C_7HCl_7 333.25	2136-95-0 119.5	334	4-05-00-00824	EtOH 2
2086	Benzene, pentachloromethyl-	$C_7H_3Cl_5$ 264.36	877-11-2 224.8	301	4-05-00-00823	EtOH 2; eth 2; bz 3; tol 3
2087	Benzene, pentachloronitro- Quintozene	$C_6Cl_5NO_2$ 295.34	82-68-8 144	8108 328 dec	4-05-00-00728 1.718^{25}	H_2O 1; EtOH 2; bz 3; chl 3
2088	Benzene, pentacosyl- Pentacosylbenzene	$C_{31}H_{56}$ 428.79	61828-06-6 59	380$^{100.}$ 244^1	4-05-00-01241 0.8544^{20}	1.4799^{20}
2089	Benzene, pentadecyl-	$C_{21}H_{36}$ 288.52	2131-18-2 22	373	3-05-00-01119 0.8548^{20}	1.4815^{20}
2090	Benzene, pentaethyl- Pentaethylbenzene	$C_{16}H_{26}$ 218.38	605-01-6 <-20	277	4-05-00-01197 0.8971^{19}	1.5127^{20}
2091	Benzene, pentafluoro- Pentafluorobenzene	C_6HF_5 168.07	363-72-4 -47.3	85.7	4-05-00-00639 1.514^{25}	1.3905^{20}
2092	Benzene, pentafluoroiodo-	C_6F_5I 293.96	827-15-6 -29	166	4-05-00-00695 2.212^{20}	1.4950^{25}
2093	Benzene, pentafluoromethoxy-	$C_7H_3F_5O$ 198.09	389-40-2 -37	138.5	4-06-00-00782 1.493^{20}	1.4087^{20}
2094	Benzene, pentafluoromethyl- 2,3,4,5,6-Pentafluorotoluene	$C_7H_3F_5$ 182.09	771-56-2 -29.8	117.5	1.440^{20}	1.4016^{25}
2095	Benzene, pentafluoro(trifluoromethyl)- Perfluorotoluene	C_7F_8 236.06	434-64-0 -65.6	104.5	4-05-00-00804	1.3670^{20}
2096	Benzene, pentaiodo- Pentaiodobenzene	C_6HI_5 707.60	608-96-8 172	sub	4-05-00-00702	bz 4; chl 4
2097	Benzene, pentamethyl- Pentamethylbenzene	$C_{11}H_{16}$ 148.25	700-12-9 54.5	232	4-05-00-01109 0.917^{20}	H_2O 1; EtOH 4; bz 4; chl 3 1.527^{20}
2098	Benzene, 1,1'-(1,5-pentanediyl)bis- 1,5-Diphenylpentane	$C_{17}H_{20}$ 224.35	1718-50-9 -8	327	4-05-00-01963 0.9812^{19}	H_2O 1 1.559^{19}
2099	Benzenepentanoic acid 5-Phenylvaleric acid	$C_{11}H_{14}O_2$ 178.23	2270-20-4 57.5	190^{30}	4-09-00-01864	H_2O 2; EtOH 4; os 3
2100	Benzenepentanoic acid, δ-oxo-, methyl ester	$C_{12}H_{14}O_3$ 206.24	1501-04-8 -2	186^{19}; 147^8	4-10-00-02783	H_2O 4
2101	Benzenepentanol	$C_{11}H_{16}O$ 164.25	10521-91-2	155^{20}	4-06-00-03373 0.9725^{20}	eth 4; EtOH 4 1.5156^{20}

No.	Name / Synonym	Mol. Form. / Mol. Wt.	CAS RN / mp/°C	Merck No. / bp/°C	Beil. Ref. / den/g cm⁻³	Solubility / n_D
2102	Benzenepentanol, α-ethyl- / 7-Phenylheptan-3-ol	$C_{13}H_{20}O$ / 192.30	60012-63-7	/ 277.5; 150[20]	3-06-00-02027 / 0.968[20]	/ 1.5115[20]
2103	Benzenepentanol, α-methyl- / 6-Phenyl-2-hexanol	$C_{12}H_{18}O$ / 178.27	38487-94-4	/ 148[18]	3-06-00-01995 / 0.9567[20]	/ 1.5079[20]
2104	Benzene, 1-pentenyl-	$C_{11}H_{14}$ / 146.23	826-18-6	/ 217	4-05-00-01401 / 0.8782[22]	/ 1.5158[20]
2105	Benzene, 3-pentenyl-	$C_{11}H_{14}$ / 146.23	1745-16-0	/ 201	4-05-00-01401 / 0.8884[16]	/ 1.5089[16]
2106	Benzene, 4-pentenyl- / 1-Pentene, 5-phenyl-	$C_{11}H_{14}$ / 146.23	1075-74-7	/ 201; 77[10]	3-05-00-01238 / 0.8889[20]	/ 1.5065[20]
2107	Benzene, pentyl- / Amylbenzene	$C_{11}H_{16}$ / 148.25	538-68-1 / -75	635 / 205.4	4-05-00-01085 / 0.8585[20]	H_2O 1; EtOH 5; eth 5; ace 5 / 1.4878[20]
2108	Benzene, 1,1'-pentylidenebis- / Pentane, 1,1-diphenyl-	$C_{17}H_{20}$ / 224.35	1726-12-1 / -12	308	4-05-00-01966 / 0.9659[20]	H_2O 1 / 1.5511[20]
2109	Benzene, (pentyloxy)- / Pentyl phenyl ether	$C_{11}H_{16}O$ / 164.25	2050-04-6	/ 200; 111[17]	3-06-00-00552 / 0.9270[20]	ace 4; EtOH 4 / 1.4947[20]
2110	Benzene, 1,1'-(phenylmethylene)bis[4-methyl-	$C_{21}H_{20}$ / 272.39	603-39-4 / 56	/ 218[12]	4-05-00-02520	H_2O 1; EtOH 3; eth 4; bz 4
2111	Benzene, 1-(phenylmethyl)-4-propyl-	$C_{16}H_{18}$ / 210.32	62155-41-3	/ 152-5[10]	4-05-00-01950 / 0.9739[18]	/ 1.3552[20]
2112	Benzene, [(phenylmethyl)sulfonyl]-	$C_{13}H_{12}O_2S$ / 232.30	3112-88-7 / 146		4-06-00-02647 / 1.1261[153]	H_2O 1; EtOH 2; eth 2; bz 2
2113	Benzene, [(phenylmethyl)thio]-	$C_{13}H_{12}S$ / 200.30	831-91-4 / 43.5	/ 197[27]	4-06-00-02644	H_2O 1; EtOH 3; eth 3; con sulf 3
2114	Benzenepropanal / Hydrocinnamic aldehyde	$C_9H_{10}O$ / 134.18	104-53-0 / 47	224; 117[28]	4-07-00-00692 / 1.0190[20]	H_2O 1; EtOH 4; eth 5
2115	Benzenepropanal, 2-methyl- / Hydrocinnamaldehyde, o-methyl-	$C_{10}H_{12}O$ / 148.20	19564-40-0	/ 120[13]	2-07-00-00246 / 0.998[19]	/ 1.522[19]
2116	Benzenepropanal, 4-methyl-	$C_{10}H_{12}O$ / 148.20	5406-12-2	/ 223	2-07-00-00247 / 0.999[14]	/ 1.525[14]
2117	Benzenepropanal, α-methyl-4-(1-methylethyl)- / 3-p-Cumenyl-2-methylpropionaldehyde	$C_{13}H_{18}O$ / 190.29	103-95-7	/ 135[99]; 115[9]	4-07-00-00788 / 0.951[15]	bz 4; eth 4; EtOH 4 / 1.5068[20]
2118	Benzenepropanal, α-methyl-β-oxo-	$C_{10}H_{10}O_2$ / 162.19	16837-43-7 / 119.3	155[125]	3-07-00-03497	bz 4; EtOH 4; MeOH 4
2119	Benzenepropanamine, α-methyl- / 1-Methyl-3-phenylpropylamine	$C_{10}H_{15}N$ / 149.24	22374-89-6 / 143	/ 223; 101[14]	4-12-00-02811 / 0.9289[15]	EtOH 4 / 1.5152[20]
2120	Benzenepropanaminium, γ-(aminocarbonyl)-N-methyl-N,N-bis(1-methylethyl)-γ-phenyl-, iod / Ammonium, (3-carbamoyl-3,3-diphenylpropyl) diisopropyl methyl,iodide	$C_{23}H_{33}IN_2O$ / 480.43	71-81-8	5090	4-14-00-01853	
2121	Benzene, 1,1'-(1,3-propanediyl)bis- / 1,3-Diphenylpropane	$C_{15}H_{16}$ / 196.29	1081-75-0 / 6	/ 300.3	4-05-00-01913 / 1.007[20]	/ 1.5760[20]
2122	Benzene, 1,1'-[1,3-propanediylbis(oxy)]bis-	$C_{15}H_{16}O_2$ / 228.29	726-44-3 / 62	/ 339	4-06-00-00585	H_2O 1; EtOH 3; eth 3
2123	Benzenepropanenitrile / Hydrocinnamonitrile	C_9H_9N / 131.18	645-59-0 / -1	261; 141[25]	4-09-00-01764 / 1.0016[20]	EtOH 3; eth 3; chl 2 / 1.5266[28]
2124	Benzenepropanenitrile, β-oxo- / Benzoylacetonitrile	C_9H_7NO / 145.16	614-16-4 / 80.5	160[10]	4-10-00-02757	H_2O 2; EtOH 3; eth 3; bz 3
2125	Benzenepropanethiol	$C_9H_{12}S$ / 152.26	24734-68-7	/ 109[10]	4-06-00-03206 / 1.01[25]	/ 1.5494[20]
2126	Benzenepropanoic acid / Hydrocinnamic acid	$C_9H_{10}O_2$ / 150.18	501-52-0 / 107	4707 / 279.8	4-09-00-01752 / 1.0712[49]	H_2O 3; EtOH 3; eth 3; bz 4
2127	Benzenepropanoic acid, α-acetyl-α-(phenylmethyl)-, ethyl ester	$C_{20}H_{22}O_3$ / 310.39	42597-26-2 / 57	/ 233	3-10-00-03350	EtOH 2; eth 2
2128	Benzenepropanoic acid, β-(aminomethyl)- / 4-Amino-3-phenylbutyric acid	$C_{10}H_{13}NO_2$ / 179.22	1078-21-3 / 250-5 dec	476	4-14-00-01723	
2129	Benzenepropanoic acid, α,α-dimethyl-, methyl ester / Methyl α,α-dimethylhydrocinnamate	$C_{12}H_{16}O_2$ / 192.26	14248-22-7	/ 121-2[23]	4-09-00-01879 / 1.0043[25]	/ 1.4945[25]
2130	Benzenepropanoic acid, β,β-dimethyl-	$C_{11}H_{14}O_2$ / 178.23	1010-48-6 / 60	/ 167[10]	4-09-00-01874	H_2O 3; ace 3; peth 3 / 1.5182[20]
2131	Benzenepropanoic acid, ethyl ester	$C_{11}H_{14}O_2$ / 178.23	2021-28-5	/ 247.2	4-09-00-01753 / 1.0147[20]	eth 4; EtOH 4 / 1.4954[20]
2132	Benzenepropanoic acid, α-ethyl-β-oxo-, ethyl ester	$C_{13}H_{16}O_3$ / 220.27	24346-56-3	/ 152[7]	3-10-00-03065 / 1.0706[15]	eth 4 / 1.509[15]
2133	Benzenepropanoic acid, α-hydroxy-, (±)- / (±)-3-Phenyllactic acid	$C_9H_{10}O_3$ / 166.18	828-01-3 / 98	149[15]	2-10-00-00154	H_2O 4; ace 4; eth 4; EtOH 4
2134	Benzenepropanoic acid, 4-hydroxy- / p-Hydroxyhydrocinnamic acid	$C_9H_{10}O_3$ / 166.18	501-97-3 / 130.8	208-10[14]	4-10-00-00631	H_2O 3; EtOH 3; eth 3; bz 3
2135	Benzenepropanoic acid, 4-hydroxy-3,5-diiodo-α-phenyl- / Iodoalphionic acid	$C_{15}H_{12}I_2O_3$ / 494.07	577-91-3 / 184	4919	4-10-00-01291	H_2O 1; EtOH 3; eth 3; bz 2
2136	Benzenepropanoic acid, 4-(4-hydroxy-3-iodophenoxy)-3,5-diiodo- / 3,3',5-Triiodothyropropionic acid	$C_{15}H_{11}I_3O_4$ / 635.96	51-26-3	9346	4-10-00-00641	
2137	Benzenepropanoic acid, α-methyl- / α-Methyldihydrocinnamic acid	$C_{10}H_{12}O_2$ / 164.20	1009-67-2 / 36.5	272; 152[12]	2-09-00-00357	eth 4; EtOH 4

No.	Name / Synonym	Mol. Form. / Mol. Wt.	CAS RN / mp/°C	Merck No. / bp/°C	Beil. Ref. / den/g cm^{-3}	Solubility / n_D
2138	Benzenepropanoic acid, β-methyl-, (±)-	$C_{10}H_{12}O_2$ 164.20	772-17-8 46.5	140-5[3]	4-09-00-01814 1.0701[20]	H_2O 2; peth 3 1.5155[20]
2139	Benzenepropanoic acid, methyl ester Methyl dihydrocinnamate	$C_{10}H_{12}O_2$ 164.20	103-25-3	238.5; 91[4]	4-09-00-01753 1.0455[25]	H_2O 1; EtOH 3; eth 3; bz 3
2140	Benzenepropanoic acid, 1-methylethyl ester	$C_{12}H_{16}O_2$ 192.26	22767-95-9	126[11]	2-09-00-00339 0.9860[25]	eth 4; EtOH 4
2141	Benzenepropanoic acid, β-oxo-, ethyl ester Ethyl benzoylacetate	$C_{11}H_{12}O_3$ 192.21	94-02-0 <0	3726 267 dec; 167[20]	4-10-00-02756 1.1202[15]	H_2O 2; EtOH 3; eth 3 1.5317[15]
2142	Benzenepropanoic acid, β-oxo-, methyl ester	$C_{10}H_{10}O_3$ 178.19	614-27-7	265 dec; 151[12]	4-10-00-02755 1.158[29]	ace 4; eth 4; EtOH 4 1.537[20]
2143	Benzenepropanoic acid, α-phenyl-, (±)- Propionic acid, 2,3-diphenyl (DL)	$C_{15}H_{14}O_2$ 226.27	94942-89-9 88(form a); 95(form b)	337.5	4-09-00-02528 1.1495[06]	bz 4; eth 4; EtOH 4
2144	Benzenepropanoic acid, α-(phenylmethyl)-	$C_{16}H_{16}O_2$ 240.30	618-68-8 90	235[18]	4-09-00-02546	bz 4; eth 4; EtOH 4
2145	Benzenepropanoic acid, phenylmethyl ester	$C_{16}H_{16}O_2$ 240.30	22767-96-0	325; 198[20]	4-09-00-01755 1.090[15]	eth 4
2146	Benzenepropanoic acid, β-propyl-	$C_{12}H_{16}O_2$ 192.26	5703-52-6	152[4]	4-09-00-01900 1.025[25]	1.5078
2147	Benzenepropanoic-β-14C acid, α-bromo- Hydrocinnamic acid, α-bromo-	$C_9H_9BrO_2$ 229.07	16503-53-0 52	150[13]	4-09-00-01771 1.4790[25]	ace 4; bz 4; EtOH 4
2148	Benzenepropanol Hydrocinnamyl alcohol	$C_9H_{12}O$ 136.19	122-97-4 <-18	235	4-06-00-03198 0.995[25]	H_2O 3; EtOH 5; eth 5; ctc 3 1.5357[25]
2149	Benzenepropanol, carbamate Phenprobamate	$C_{10}H_{13}NO_2$ 179.22	673-31-4 102	7229		H_2O 1; EtOH 3; chl 3
2150	Benzenepropanol, α,α-dimethyl- Benzyl-tert-butanol	$C_{11}H_{16}O$ 164.25	103-05-9 24.5	121[13]	4-06-00-03380 0.9626[21]	H_2O 1; EtOH 4; eth 4; ace 4 1.5077[21]
2151	Benzenepropanol, β,β-dimethyl-	$C_{11}H_{16}O$ 164.25	13351-61-6 34.5	125[14.5]	2-06-00-00507	eth 4; EtOH 4
2152	Benzenepropanol, α-ethyl-, (S)- 3-Pentanol, 1-phenyl (d)	$C_{11}H_{16}O$ 164.25	71747-37-0 38	143[19]	3-06-00-01953 0.9687[20]	EtOH 4
2153	Benzenepropanol, 4-hydroxy-3-methoxy-	$C_{10}H_{14}O_3$ 182.22	2305-13-7 65	197[15]	4-06-00-07405	eth 4; EtOH 4 1.5545[25]
2154	Benzenepropanol, 4-methoxy-	$C_{10}H_{14}O_2$ 166.22	5406-18-8 26	166[18]	4-06-00-05982	1.5315[20]
2155	Benzenepropanol, α-methyl-	$C_{10}H_{14}O$ 150.22	2344-70-9	239; 123[15]	4-06-00-03275 0.9899[16]	1.517[16]
2156	Benzenepropanol, γ-methyl-	$C_{10}H_{14}O$ 150.22	2722-36-3	139[33]; 110[6]	4-06-00-03286 0.9832[20]	1.4965[20]
2157	Benzenepropanol, γ-methylene- 3-Buten-1-ol, 3-phenyl-	$C_{10}H_{12}O$ 148.20	3174-83-2	123[10]	4-06-00-03840 1.0272[20]	1.5577[20]
2158	Benzenepropanol, β-phenyl-	$C_{15}H_{16}O$ 212.29	3536-29-6 51	310; 190[15]	4-06-00-04754 1.0585[24]	1.5742[19]
2159	Benzenepropanol, γ-propyl- 1-Hexanol, 3-phenyl	$C_{12}H_{18}O$ 178.27	67700-22-5	127[5]	3-06-00-02001 0.955[25]	1.5101[25]
2160	Benzenepropanoyl chloride	C_9H_9ClO 168.62	645-45-4	225 dec; 105[10]	4-09-00-01762 1.135[21]	eth 3; CS_2 3
2161	Benzene, 1,1'-(1-propene-1,3-diyl)bis-	$C_{15}H_{14}$ 194.28	5209-18-7 16	302; 178[15]	4-05-00-02187 1.0120[20]	
2162	Benzene, 1-propenyl-	C_9H_{10} 118.18	637-50-3 -27.1	175.5	3-05-00-01185 0.9019[25]	ace 4; bz 4; eth 4; EtOH 4 1.5508[20]
2163	Benzene, 1-propenyl-, (E)-	C_9H_{10} 118.18	873-66-5 -29.3	178.3	4-05-00-01359 0.9023[25]	H_2O 1; EtOH 5; eth 5; ace 5 1.5506[20]
2164	Benzene, 1-propenyl-, (Z)-	C_9H_{10} 118.18	766-90-5 -61.6	167.5	4-05-00-01359 0.9088[20]	H_2O 1; EtOH 5; eth 5; ace 5 1.5420[20]
2165	Benzene, 2-propenyl-	C_9H_{10} 118.18	300-57-2 -40	156	4-05-00-01362 0.8920[20]	H_2O 1; EtOH 3; eth 3; bz 3 1.5131[20]
2166	Benzene, 1,1'-(1-propenylidene)bis-	$C_{15}H_{14}$ 194.28	778-66-5 52	280-1 149[11]	4-05-00-02192 1.0250[20]	H_2O 3; eth 3; bz 3 1.5880[20]
2167	Benzene, (2-propenyloxy)-	$C_9H_{10}O$ 134.18	1746-13-0	191.7	4-06-00-00562 0.9811[20]	H_2O 1; EtOH 3; eth 5; ctc 2 1.5223[20]
2168	Benzene, propoxy-	$C_9H_{12}O$ 136.19	622-85-5 -27	189.9	4-06-00-00556 0.9474[20]	EtOH 3; eth 3 1.5014[20]
2169	Benzene, propyl- Propylbenzene	C_9H_{12} 120.19	103-65-1 -99.5	7856 159.2	4-05-00-00977 0.8620[20]	H_2O 1; EtOH 5; eth 5; ace 5 1.4920[20]
2170	Benzene, (1-propylheptadecyl)- Eicosane, 4-phenyl-	$C_{26}H_{46}$ 358.65	2400-03-5 31.4	199[1]	4-05-00-01223 0.8546[20]	1.4794[20]
2171	Benzene, 1,1'-propylidenebis- 1,1-Diphenylpropane	$C_{15}H_{16}$ 196.29	1530-03-6 13.3	281.6	4-05-00-01920 0.9951[14]	1.5681[14]
2172	Benzene, (propylthio)-	$C_9H_{12}S$ 152.26	874-79-3 -45	220	4-06-00-01470 0.9995[20]	1.5571[20]
2173	Benzene, 1-propynyl-	C_9H_8 116.16	673-32-5	183	4-05-00-01530 0.942[15]	1.563[15]
2174	Benzeneseleninic acid Phenylseleninic acid	$C_6H_6O_2Se$ 189.07	6996-92-5 124.5		4-11-00-00708 1.93[20]	H_2O 2; bz 1; alk 4
2175	Benzene, 1,1'-selenobis-	$C_{12}H_{10}Se$ 233.17	1132-39-4 2.5	301.5	4-06-00-01779 1.351[20]	H_2O 1; EtOH 5; eth 5; bz 3 1.5500[20]

No.	Name / Synonym	Mol. Form. / Mol. Wt.	CAS RN / mp/°C	Merck No. / bp/°C	Beil. Ref. / den/g cm⁻³	Solubility / n_D
2176	Benzene, 1,1'-selenobis[4-methyl-	$C_{14}H_{14}Se$ 261.23	22077-55-0 69.5	196^{16}	4-06-00-02218	H_2O 1; EtOH 3; eth 2; chl 2
2177	Benzeneselenol	C_6H_6Se 157.07	645-96-5	$183.6; 84^{25}$	4-06-00-01777 1.4865^{15}	H_2O 1; EtOH 3; eth 4; ctc 4
2178	Benzenesulfenyl chloride, 2,4-dinitro- 2,4-Dinitrobenzenesulfenyl chloride	$C_6H_3ClN_2O_4S$ 234.62	528-76-7 99	3267	4-06-00-01772	bz 4; chl 4; HOAc 4; peth 2
2179	Benzenesulfenyl chloride, 4-nitro-	$C_6H_4ClNO_2S$ 189.62	937-32-6 52	$125^{0.1}$	4-06-00-01717	bz 4
2180	Benzenesulfinic acid	$C_6H_6O_2S$ 142.18	618-41-7 84	dec	4-11-00-00003	H_2O 2; EtOH 3; eth 3; bz 3
2181	Benzenesulfinic acid, 2-methyl- o-Toluenesulfinic acid	$C_7H_8O_2S$ 156.21	13165-77-0 80	dec	4-11-00-00008	H_2O 4; ace 4; eth 4; EtOH 4
2182	Benzenesulfinic acid, 4-methyl- p-Toluenesulfinic acid	$C_7H_8O_2S$ 156.21	536-57-2 86.5	9458	4-11-00-00009	H_2O 3; EtOH 4; eth 4; bz 2
2183	Benzene, 1,1'-sulfinylbis-	$C_{12}H_{10}OS$ 202.28	945-51-7 71.2	340^{16}	4-06-00-01489	EtOH 4; eth 4; bz 4; chl 2
2184	Benzene, 1,1'-[sulfinylbis(methylene)]bis-	$C_{14}H_{14}OS$ 230.33	621-08-9 134	210 dec	4-06-00-02651	H_2O 1; EtOH 4; eth 3
2185	Benzenesulfinyl chloride	C_6H_5ClOS 160.62	4972-29-6 38	$71-2^{1.5}$	4-11-00-00005 1.3469^{25}	eth 3; chl 3 1.3470^{25}
2186	Benzenesulfinyl chloride, 4-methyl-	C_7H_7ClOS 174.65	10439-23-3 57	$113^{3.5}$	4-11-00-00010	chl 4
2187	Benzenesulfonamide, 4-amino- Sulfanilamide	$C_6H_8N_2O_2S$ 172.21	63-74-1 165.5	8898	4-14-00-02658 1.08^{25}	H_2O 3; EtOH 3; eth 3; ace 3
2188	Benzenesulfonamide, 4-amino-N-[4-(aminosulfonyl)phenyl- N4-Sulfanilylsulfanilamide	$C_{12}H_{13}N_3O_4S_2$ 327.39	547-52-4 137	8905	4-14-00-02760	H_2O 2; EtOH 3; eth 3; ace 3
2189	Benzenesulfonamide, 4-amino-N-(aminothioxomethyl)- Sulfathiourea	$C_7H_9N_3O_2S_2$ 231.30	515-49-1 182	8921	4-14-00-02671	H_2O 1; EtOH 2
2190	Benzenesulfonamide, 4-amino-N-(aminocarbonyl)- Sulfanilylurea	$C_7H_9N_3O_3S$ 215.23	547-44-4 147 dec	8906	4-14-00-02667	
2191	Benzenesulfonamide, 4-amino-N-(aminoiminomethyl)- Sulfaguanidine	$C_7H_{10}N_4O_2S$ 214.25	57-67-0 191 5	8879	4-14-00-02668	
2192	Benzenesulfonamide, 4-amino-N-[(butylamino)carbonyl]- Carbutamide	$C_{11}H_{17}N_3O_3S$ 271.34	339-43-5 144.5	1839	4-14-00-02667	
2193	Benzenesulfonamide, 4-amino-N-(6-chloro-3-pyridazinyl)- Sulfachlorpyridazine	$C_{10}H_9ClN_4O_2S$ 284.73	80-32-0 187	8871	4-25-00-02069	
2194	Benzenesulfonamide, 4-amino-N-(2,6-dimethoxy-4-pyrimidinyl)- Sulfadimethoxine	$C_{12}H_{14}N_4O_4S$ 310.33	122-11-2 203.5	8876	4-25-00-03483	
2195	Benzenesulfonamide, 4-amino-N-(3,4-dimethyl-5-isoxazolyl)- Sulfisoxazole	$C_{11}H_{13}N_3O_3S$ 267.31	127-69-5 191	8930	4-27-00-04747	
2196	Benzenesulfonamide, 4-amino-N-(4,6-dimethyl-2-pyrimidinyl)- Sulfamethazine	$C_{12}H_{14}N_4O_2S$ 278.33	57-68-1 198.5	8886	4-25-00-02215	H_2O 3; acid 3; alk 3
2197	Benzenesulfonamide, 4-amino-N-(6-methoxy-3-pyridazinyl)- Sulfamethoxypyridazine	$C_{11}H_{12}N_4O_3S$ 280.31	80-35-3 182.5	8890	4-25-00-03332	
2198	Benzenesulfonamide, 4-amino-N-(5-methyl-3-isoxolyl)- Sulfamethoxazole	$C_{10}H_{11}N_3O_3S$ 253.28	723-46-6	8889	4-27-00-04685	
2199	Benzenesulfonamide, 4-amino-N-(4-methyl-2-pyrimidinyl)- Sulfamerazine	$C_{11}H_{12}N_4O_2S$ 264.31	127-79-7 236	8884	4-25-00-02159	H_2O 2; EtOH 2; eth 1; ace 2
2200	Benzenesulfonamide, 4-amino-N-(4-methyl-2-thiazolyl)- Sulfamethylthiazole	$C_{10}H_{11}N_3O_2S_2$ 269.35	515-59-3 237	8891	4-27-00-04715	EtOH 4
2201	Benzenesulfonamide, 4-amino-N-(1-phenyl-1H-pyrazol-5-yl)- Sulfaphenazole	$C_{15}H_{14}N_4O_2S$ 314.37	526-08-9	8910	4-25-00-02029	
2202	Benzenesulfonamide, 4-amino-N-pyrazinyl- Sulfapyrazine	$C_{10}H_{10}N_4O_2S$ 250.28	116-44-9 251	8912	4-25-00-02143	H_2O 1; EtOH 1; eth 1; ace 2
2203	Benzenesulfonamide, 4-amino-N-2-pyridinyl- Sulfapyridine	$C_{11}H_{11}N_3O_2S$ 249.29	144-83-2 192	8913	5-22-08-00424	H_2O 1; EtOH 3; bz 1; ctc 1
2204	Benzenesulfonamide, 4-amino-N-2-pyrimidinyl- Sulfadiazine	$C_{10}H_{10}N_4O_2S$ 250.28	68-35-9 255-6 dec	8874	4-25-00-02097	H_2O 2; EtOH 2; ace 2
2205	Benzenesulfonamide, 4-amino-N-2-quinoxalinyl- Sulfaquinoxaline	$C_{14}H_{12}N_4O_2S$ 300.34	59-40-5 247.5	8914	4-25-00-02606	H_2O 2; EtOH 2; ace 2; aq alk 3
2206	Benzenesulfonamide, 4-amino-N-2-thiazolyl- Sulfathiazole	$C_9H_9N_3O_2S_2$ 255.32	72-14-0 175(form a); 202(form b)	8920	4-27-00-04623	H_2O 2; EtOH 2; DMSO-d_6 2

No.	Name / Synonym	Mol. Form. / Mol. Wt.	CAS RN / mp/°C	Merck No. / bp/°C	Beil. Ref. / den/g cm^{-3}	Solubility / n_D
2207	Benzenesulfonamide, N-[(butylamino)carbonyl]-4-methyl- Tolbutamide	$C_{12}H_{18}N_2O_3S$ 270.35	64-77-7 128.5	9432	4-11-00-00396 1.245[25]	H$_2$O 2; EtOH 3; eth 3; chl 3
2208	Benzenesulfonamide, 4-chloro-N-[(propylamino)carbonyl]- Chloropropamide	$C_{10}H_{13}ClN_2O_3S$ 276.74	94-20-2	2187	4-11-00-00119	
2209	Benzenesulfonamide, 4-[(2,4-diaminophenyl)azo]- Prontosil	$C_{12}H_{14}ClN_5O_2S$ 327.79	103-12-8 249.5	8895	4-16-00-00563	H$_2$O 2; EtOH 3; ace 3; oils 3
2210	Benzenesulfonamide, N,N-dichloro-4-methyl- Dichloramine-T	$C_7H_7Cl_2NO_2S$ 240.11	473-34-7 83	3034	3-11-00-00301	H$_2$O 1; EtOH 3; eth 3; bz 3
2211	Benzenesulfonamide, N,4-dimethyl-	$C_8H_{11}NO_2S$ 185.25	640-61-9 78.5		4-11-00-00377 1.340[25]	eth 4; EtOH 4
2212	Benzenesulfonamide, N,4-dimethyl-N-nitroso- p-Tolylsulfonylmethylnitrosamide	$C_8H_{10}N_2O_3S$ 214.25	80-11-5 60	9469	4-11-00-00478	H$_2$O 1; EtOH 4; eth 4
2213	Benzenesulfonamide, 4-[(phenylmethyl)amino]- N4-Benzylsulfanilamide	$C_{13}H_{14}N_2O_2S$ 262.33	104-22-3 171	1161	3-14-00-02026	
2214	Benzenesulfonic acid Besylic acid	$C_6H_6O_3S$ 158.18	98-11-3 65	1078	4-11-00-00027	H$_2$O 4; EtOH 4; eth 1; bz 2
2215	Benzenesulfonic acid, 2-amino- Orthanilic acid	$C_6H_7NO_3S$ 173.19	88-21-1 >320 dec	6835	4-14-00-02638	H$_2$O 2; EtOH 1; eth 1
2216	Benzenesulfonic acid, 3-amino- Metanilic acid	$C_6H_7NO_3S$ 173.19	121-47-1 dec	5832	4-14-00-02647	H$_2$O 2; EtOH 1; eth 1
2217	Benzenesulfonic acid, 4-amino- Sulfanilic acid	$C_6H_7NO_3S$ 173.19	121-57-3 288	8901	4-14-00-02655 1.485[25]	H$_2$O 2; EtOH 1; eth 1
2218	Benzenesulfonic acid, 2-amino-5-[(4-amino-3-sulfophenyl)(4-imino-3-sulfo-2,5-cyclohexadien-1-ylidene)methyl]-3 Fuchsin, acid	$C_{20}H_{17}N_3Na_2O_9S_3$ 585.55	3244-88-0	103		H$_2$O 2; EtOH 2
2219	Benzenesulfonic acid, sec-butyl ester Benzenesulfonic acid, chloride	$C_{10}H_{14}O_3S$ 214.29	103275-78-1 14.5	251 dec; 120[10]	4-11-00-00032 1.3830[15]	H$_2$O 1; EtOH 4; eth 3; ctc 3
2220	Benzenesulfonic acid, 4-chloro- p-Chlorobenzenesulfonic acid	$C_6H_5ClO_3S$ 192.62	98-66-8 67	2122 147-8[25]	4-11-00-00107	H$_2$O 3; EtOH 3; eth 1; bz 1
2221	Benzenesulfonic acid, 4-chloro-, 4-chlorophenyl ester Ovex	$C_{12}H_8Cl_2O_3S$ 303.17	80-33-1 86.5	6856	4-11-00-00109	H$_2$O 1; EtOH 2; ace 3
2222	Benzenesulfonic acid, 4-[[4-(dimethylamino)phenyl]azo]-, sodium salt Methyl orange	$C_{14}H_{14}N_3NaO_3S$ 327.34	547-58-0 dec	6019		H$_2$O 2; EtOH 2; eth 1; py 2
2223	Benzenesulfonic acid, ethyl ester Ethyl benzenesulfonate	$C_8H_{10}O_3S$ 186.23	515-46-8 156[15]	1078	4-11-00-00030 1.2167[20]	H$_2$O 2; EtOH 3; eth 4; chl 4 1.5081[20]
2224	Benzenesulfonic acid, 4-hydrazino- p-Sulfophenylhydrazine	$C_6H_8N_2O_3S$ 188.21	98-71-5 286	4695	3-15-00-00865	H$_2$O 2; EtOH 2
2225	Benzenesulfonic acid, 2-hydroxy- o-Phenolsulfonic acid	$C_6H_6O_4S$ 174.18	609-46-1 145 dec		4-11-00-00574	H$_2$O 4; EtOH 4
2226	Benzenesulfonic acid, 4-hydroxy- p-Phenolsulfonic acid	$C_6H_6O_4S$ 174.18	98-67-9	7212	4-11-00-00582	H$_2$O 4; EtOH 4
2227	Benzenesulfonic acid, methyl ester Methyl benzenesulfonate	$C_7H_8O_3S$ 172.20	80-18-2 154[20]		4-11-00-00029 1.2730[17]	H$_2$O 2; EtOH 4; eth 4; chl 4 1.5151[20]
2228	Benzenesulfonic acid, 2-methyl-	$C_7H_8O_3S$ 172.20	88-20-0 67.5	128.8[25]	4-11-00-00228	H$_2$O 4; EtOH 3; eth 1
2229	Benzenesulfonic acid, 4-methyl- p-Toluenesulfonic acid	$C_7H_8O_3S$ 172.20	104-15-4 104.5	9459 140[20]	4-11-00-00241	H$_2$O 4; EtOH 3; eth 3
2230	Benzenesulfonic acid, 4-methyl-, butyl ester	$C_{11}H_{16}O_3S$ 228.31	778-28-9 164-6[6]		4-11-00-00250 1.1319[20]	H$_2$O 1; eth 3; ctc 2 1.5050[20]
2231	Benzenesulfonic acid, 4-methyl-, ethyl ester Ethyl p-toluenesulfonate	$C_9H_{12}O_3S$ 200.26	80-40-0 34.5	3814 173[15]	4-11-00-00248 1.166[48]	H$_2$O 1; EtOH 3; eth 3; ctc 2
2232	Benzenesulfonic acid, 4-methyl-, methyl ester	$C_8H_{10}O_3S$ 186.23	80-48-8 28.5	292; 186[22]	4-11-00-00247 1.2087[40]	H$_2$O 1; EtOH 4; eth 3; bz 4
2233	Benzenesulfonic acid, 4-methyl-, propyl ester	$C_{10}H_{14}O_3S$ 214.29	599-91-7 <-20	199[9]	4-11-00-00249 1.144[20]	1.4998[20]
2234	Benzenesulfonic acid, 4-(phenylamino)- N-Phenylsulfanilic acid	$C_{12}H_{11}NO_3S$ 249.29	101-57-5 206	7284	3-14-00-02025	H$_2$O 4; EtOH 4
2235	Benzenesulfonic acid, propyl ester Propyl benzenesulfonate	$C_9H_{12}O_3S$ 200.26	80-42-2 162-3[15]		4-11-00-00031 1.1804[17]	H$_2$O 2; EtOH 3; eth 4; chl 4 1.5035[25]
2236	Benzenesulfonothioic acid, 4-chloro-, S-phenyl ester Benzenesulfonic acid, p-chlorothio-, S-phenyl ester	$C_{12}H_9ClO_2S_2$ 284.79	1142-97-8 -86.8	118.5	4-11-00-00223 0.8341[20]	1.4087[14]
2237	Benzene, 1,1'-sulfonylbis- Diphenyl sulfone	$C_{12}H_{10}O_2S$ 218.28	127-63-9 128.5	3336 379	4-06-00-01490 1.252[20]	H$_2$O 1; EtOH 3; eth 3; bz 3
2238	Benzene, 1,1'-sulfonylbis[4-bromo-	$C_{12}H_8Br_2O_2S$ 376.07	2050-48-8 172		4-06-00-01652 1.88[25]	H$_2$O 1; EtOH 2; HOAc 3
2239	Benzene, 1,1'-sulfonylbis[4-chloro-	$C_{12}H_8Cl_2O_2S$ 287.17	80-07-9 147.9	250[10]	4-06-00-01587	H$_2$O 2; EtOH 3; chl 3
2240	Benzene, 1,1'-sulfonylbis[4-methyl-	$C_{14}H_{14}O_2S$ 246.33	599-66-6 159	406	4-06-00-02174	H$_2$O 3; EtOH 3; eth 2; bz 3
2241	Benzene, 1,1'-[sulfonylbis(methylene)]bis-	$C_{14}H_{14}O_2S$ 246.33	620-32-6 152	290 dec	4-06-00-02651	H$_2$O 1; EtOH 2; ace 4; bz 3

No.	Name / Synonym	Mol. Form. / Mol. Wt.	CAS RN / mp/°C	Merck No. / bp/°C	Beil. Ref. / den/g cm^{-3}	Solubility / n_D
2242	Benzenesulfonyl chloride / Phenylsulfonyl chloride	$C_6H_5ClO_2S$ / 176.62	98-09-9 / 14.5	1080 / 251 dec	4-11-00-00049 / 1.3470[15]	H$_2$O 1; EtOH 4; eth 3; ctc 3
2243	Benzenesulfonyl chloride, 4-(acetylamino)- / Acetylsulfanilyl chloride	$C_8H_8ClNO_3S$ / 233.68	121-60-8 / 149	99 /	4-14-00-02703	EtOH 4; eth 4; bz 3; chl 3
2244	Benzenesulfonyl chloride, 4-bromo- / p-Brosyl chloride	$C_6H_4BrClO_2S$ / 255.52	98-58-8 / 76	1395 / 153[15]	4-11-00-00162	H$_2$O 1; eth 4; chl 3
2245	Benzenesulfonyl chloride, 4-chloro-	$C_6H_4Cl_2O_2S$ / 211.07	98-60-2 / 51	/ 141[15]	4-11-00-00114	eth 4; bz 4
2246	Benzenesulfonyl chloride, 4-fluoro-	$C_6H_4ClFO_2S$ / 194.61	349-88-2 / 30	/ 106[9]	4-11-00-00102	bz 4; eth 4; chl 4
2247	Benzenesulfonyl chloride, 4-iodo- / Pipsyl chloride	$C_6H_4ClIO_2S$ / 302.52	98-61-3 / 85	7458 /	4-11-00-00169	
2248	Benzenesulfonyl chloride, 2-methoxy-	$C_7H_7ClO_3S$ / 206.65	10130-87-7 / 56	/ 126[0.3]	4-11-00-00575	eth 3; peth 3
2249	Benzenesulfonyl chloride, 2-methyl- / o-Toluenesulfonyl chloride	$C_7H_7ClO_2S$ / 190.65	133-59-5 / 10.2	/ 154[36]	4-11-00-00229 / 1.3383[20]	H$_2$O 1; EtOH 3; eth 3; bz 3 / 1.5565[20]
2250	Benzenesulfonyl chloride, 4-methyl- / p-Toluenesulfonyl chloride	$C_7H_7ClO_2S$ / 190.65	98-59-9 / 71	9460 / 145-6[15]	4-11-00-00375	H$_2$O 1; EtOH 3; eth 3; bz 4
2251	Benzenesulfonyl fluoride / Phenylsulfonyl fluoride	$C_6H_5FO_2S$ / 160.17	368-43-4 /	/ 203.5	4-11-00-00049 / 1.3286[20]	EtOH 3; eth 3 / 1.4932[18]
2252	Benzenesulfonyl fluoride, 4-amino- / p-Sulfanilyl fluoride	$C_6H_6FNO_2S$ / 175.18	98-62-4 / 68.5	8904 /	4-14-00-02658	
2253	Benzene, 1,2,3,5-tetrabromo- / 1,2,3,5-Tetrabromobenzene	$C_6H_2Br_4$ / 393.70	634-89-9 / 99.5	/ 329	4-05-00-00686	bz 4; eth 4; EtOH 4
2254	Benzene, 1,2,4,5-tetrabromo- / 2,3,5,6-Tetrabromobenzene	$C_6H_2Br_4$ / 393.70	636-28-2 / 182	/ 3.1[20]	4-05-00-00687 / 3.072[20]	eth 4
2255	Benzene, 1,2,3,4-tetrabromo-5,6-dimethyl-	$C_8H_6Br_4$ / 421.75	2810-69-7 / 262	/ 374.5	3-05-00-00820	bz 4
2256	Benzene, 1,2,3,5-tetrabromo-4-methoxy- / 2,3,4,6-Tetrabromoanisole	$C_7H_4Br_4O$ / 423.72	95970-07-3 / 113.5	/ 340	3-06-00-00766	EtOH 4
2257	1,2,4,5-Benzenetetracarboxylic acid / Pyromellitic acid	$C_{10}H_6O_8$ / 254.15	89-05-4 / 276	8015 /	4-09-00-03804	H$_2$O 2; EtOH 3
2258	1,2,4,5-Benzenetetracarboxylic acid, tetramethyl ester	$C_{14}H_{14}O_8$ / 310.26	635-10-9 / 144	/ sub	4-09-00-03804	EtOH 4
2259	Benzene, 1,2,3,4-tetrachloro- / 1,2,3,4-Tetrachlorobenzene	$C_6H_2Cl_4$ / 215.89	634-66-2 / 47.5	/ 254	4-05-00-00667	H$_2$O 1; EtOH 2; eth 4; CS$_2$ 4
2260	Benzene, 1,2,4,5-tetrachloro- / s-Tetrachlorobenzene	$C_6H_2Cl_4$ / 215.89	95-94-3 / 139.5	/ 244.5	4-05-00-00668 / 1.858[22]	H$_2$O 1; EtOH 2; eth 3; bz 3
2261	Benzene, 1,2,3,5-tetrachloro-4,6-dimethyl-	$C_8H_6Cl_4$ / 243.95	877-09-8 / 223	/	4-05-00-00944 / 1.703[25]	H$_2$O 1; EtOH 1; eth 3; bz 3
2262	Benzene, 1,2,3,5-tetrachloro-4-methyl- / 2,3,4,6-Tetrachlorotoluene	$C_7H_4Cl_4$ / 229.92	875-40-1 / 94	/ 276.5	3-05-00-00702	bz 4; EtOH 4
2263	Benzene, 1,2,4,5-tetrachloro-3-methyl- / 2,3,5,6-Tetrachlorotoluene	$C_7H_4Cl_4$ / 229.92	1006-31-1 / 93.5	/ sub	3-05-00-00702	eth 4; EtOH 4
2264	Benzene, 1,2,4,5-tetrachloro-3-nitro-	$C_6HCl_4NO_2$ / 260.89	117-18-0 / 99.5	/ 304	4-05-00-00728 / 1.744[25]	H$_2$O 1; EtOH 3; bz 3; chl 3
2265	Benzene, tetracosyl- / Tetracosylbenzene	$C_{30}H_{54}$ / 414.76	61828-05-5 / 57	/ 373[100.] 238[1]	4-05-00-01239 / 0.8544[20]	1.4800[20]
2266	Benzene, tetradecyl-	$C_{20}H_{34}$ / 274.49	1459-10-5 / 16	359 /	4-05-00-01212 / 0.8549[20]	1.4818[20]
2267	Benzene, 1,1'-tetradecylidenebis-	$C_{26}H_{38}$ / 350.59	55268-63-8 / 17.9	/ 207[1]	4-05-00-02017 / 0.9187[20]	1.5202[20]
2268	Benzene, 1,2,3,4-tetraethyl- / 1,2,3,4-Tetraethylbenzene	$C_{14}H_{22}$ / 190.33	642-32-0 / 11.8	/ 252; 121.7[14]	4-05-00-01168 / 0.8875[20]	eth 4; EtOH 4 / 1.5125[20]
2269	Benzene, 1,2,4,5-tetraethyl- / 1,2,4,5-Tetraethylbenzene	$C_{14}H_{22}$ / 190.33	635-81-4 / 10	/ 250	4-05-00-01168 / 0.8788[20]	eth 4; EtOH 4 / 1.5054[20]
2270	Benzene, 1,2,3,4-tetrafluoro- / 1,2,3,4-Tetrafluorobenzene	$C_6H_2F_4$ / 150.08	551-62-2 /	/ 94.3	4-05-00-00639	1.4054[20]
2271	Benzene, 1,2,3,5-tetrafluoro- / 1,2,3,5-Tetrafluorobenzene	$C_6H_2F_4$ / 150.08	2367-82-0 / -48	/ 84.4	4-05-00-00639 / 1.319[25]	1.4035[20]
2272	Benzene, 1,2,4,5-tetrafluoro- / 1,2,4,5-Tetrafluorobenzene	$C_6H_2F_4$ / 150.08	327-54-8 / 4.5	/ 90.2	4-05-00-00639 / 1.4255[20]	1.4075[20]
2273	Benzene, 1,2,3,4-tetraiodo- / 1,2,3,4-Tetraiodobenzene	$C_6H_2I_4$ / 581.70	634-68-4 / 136	/ sub	0-05-00-00229	eth 4; EtOH 4
2274	Benzene, 1,2,3,5-tetraiodo- / 1,2,3,5-Tetraiodobenzene	$C_6H_2I_4$ / 581.70	634-92-4 / 148	/ sub	0-05-00-00229	HOAc 4
2275	Benzene, 1,2,4,5-tetraiodo- / 1,2,4,5-Tetraiodobenzene	$C_6H_2I_4$ / 581.70	636-31-7 / 254	/ sub	0-05-00-00229	HOAc 4
2276	Benzene, 1,2,4,5-tetrakis(1-methylethyl)- / 1,2,4,5-Tetraisopropylbenzene	$C_{18}H_{30}$ / 246.44	635-11-0 / 118.4	/ 259; 133[17]	4-05-00-01207 / 0.758[150]	
2277	Benzene, 1,2,3,4-tetramethoxy-5-(2-propenyl)-	$C_{13}H_{18}O_4$ / 238.28	15361-99-6 / 25	/ 145[12]	3-06-00-06691 / 1.087[25]	1.5146[25]
2278	Benzene, 1,2,3,4-tetramethyl-	$C_{10}H_{14}$ / 134.22	488-23-3 / -6.2	/ 205	4-05-00-01072 / 0.9052[20]	H$_2$O 1; EtOH 5; eth 5; ace 5 / 1.5203[20]
2279	Benzene, 1,2,3,5-tetramethyl- / Isodurene	$C_{10}H_{14}$ / 134.22	527-53-7 / -23.7	5050 / 198	4-05-00-01073 / 0.8903[20]	H$_2$O 1; EtOH 5; eth 5; ace 5 / 1.5130[20]
2280	Benzene, 1,2,4,5-tetramethyl- / Durene	$C_{10}H_{14}$ / 134.22	95-93-2 / 79.3	3450 / 196.8	4-05-00-01076 / 0.8380[81]	H$_2$O 1; EtOH 5; eth 5; ace 5 / 1.4790[81]

No.	Name / Synonym	Mol. Form. / Mol. Wt.	CAS RN / mp/°C	Merck No. / bp/°C	Beil. Ref. / den/g cm^{-3}	Solubility / n_D
2281	Benzene, 1,1'-(1,1,4,4-tetramethyl-1,4-butanediyl)bis- Bineophyl	$C_{20}H_{26}$ 266.43	17648-05-4 61.8	146-50[0.5]	4-05-00-01997	chl 3
2282	Benzene, 1,2,4,5-tetramethyl-3,6-dinitro-	$C_{10}H_{12}N_2O_4$ 224.22	5465-13-4 211.5	sub	4-05-00-01080	H_2O 1; EtOH 2; eth 4; bz 3
2283	Benzene, 1,1'-thiobis- Phenyl sulfide	$C_{12}H_{10}S$ 186.28	139-66-2 -25.9	7285 296	4-06-00-01488 1.1136[20]	H_2O 1; EtOH 3; eth 5; bz 5 1.6334[20]
2284	Benzene, 1,1'-thiobis[4-bromo-	$C_{12}H_8Br_2S$ 344.07	3393-78-0 115	268.5[40]	4-06-00-01651 1.84[25]	H_2O 1; EtOH 2; eth 3; ctc 3
2285	Benzene, 1,1'-thiobis[2-methyl-	$C_{14}H_{14}S$ 214.33	4537-05-7 64	285	4-06-00-02019	eth 4; EtOH 4; chl 4
2286	Benzene, 1,1'-thiobis[4-methyl-	$C_{14}H_{14}S$ 214.33	620-94-0 57.3	>300; 175[16]	4-06-00-02173	H_2O 1; EtOH 3; eth 4; ace 3
2287	Benzene, 1,1'-[thiobis(methylene)]bis- Benzyl sulfide	$C_{14}H_{14}S$ 214.33	538-74-9 49.5	1162 dec	4-06-00-02649 1.0583[50]	H_2O 1; EtOH 3; eth 3; CS_2 3
2288	Benzene, 1,1'-thiobis[2,4,6-trinitro- Picryl sulfide	$C_{12}H_4N_6O_{12}S$ 456.26	2217-06-3		4-06-00-01775	
2289	Benzenethiol Phenyl mercaptan	C_6H_6S 110.18	108-98-5 -14.9	9285 169.1	4-06-00-01463 1.0775[20]	H_2O 1; EtOH 3; eth 3; bz 3 1.5893[20]
2290	Benzenethiol, 2-amino-	C_6H_7NS 125.19	137-07-5 26	234	4-13-00-00910	EtOH 3; eth 3 1.4606[20]
2291	Benzenethiol, 4-amino-	C_6H_7NS 125.19	1193-02-8 46	143[17]	4-13-00-01289	H_2O 3; EtOH 3
2292	Benzenethiol, 4-bromo-	C_6H_5BrS 189.08	106-53-6 73	230.5	4-06-00-01650 1.5260[83]	H_2O 2; EtOH 2; eth 4; ctc 4
2293	Benzenethiol, 2-chloro-	C_6H_5ClS 144.62	6320-03-2	205.5	4-06-00-01570 1.2752[10]	H_2O 2; EtOH 2
2294	Benzenethiol, 3-chloro-	C_6H_5ClS 144.62	2037-31-2	206	4-06-00-01576 1.2637[13]	H_2O 1; EtOH 3; eth 3; chl 3
2295	Benzenethiol, 4-chloro-	C_6H_5ClS 144.62	106-54-7 61	206	4-06-00-01581 1.1911[20]	H_2O 1; EtOH 4; eth 4; bz 4 1.5480[20]
2296	Benzenethiol, 4-ethoxy-	$C_8H_{10}OS$ 154.23	699-09-2 1.6	238	4-06-00-05792 1.1070[17]	ace 4; bz 4; eth 4; EtOH 4
2297	Benzenethiol, 3-ethyl- 3-Ethylthiophenol	$C_8H_{10}S$ 138.23	62154-77-2	211	3-06-00-01663 1.038[20]	1.572[30]
2298	Benzenethiol, 4-ethyl-	$C_8H_{10}S$ 138.23	4946-13-8	211	4-06-00-03027 1.038[20]	1.572[20]
2299	Benzenethiol, 3-methoxy-	C_7H_8OS 140.21	15570-12-4	224.5; 114[20]	4-06-00-05701	chl 3 1.5874[20]
2300	Benzenethiol, 4-methoxy-	C_7H_8OS 140.21	696-63-9	228	4-06-00-05790 1.1313[25]	EtOH 3; eth 3; bz 3; chl 2 1.5801[25]
2301	Benzenethiol, 2-methyl-	C_7H_8S 124.21	137-06-4 15	195	4-06-00-02014 1.041[20]	H_2O 1; EtOH 3; eth 4 1.570[20]
2302	Benzenethiol, 3-methyl-	C_7H_8S 124.21	108-40-7 -20	195	4-06-00-02079 1.044[20]	H_2O 1; EtOH 3; eth 5 1.572[20]
2303	Benzenethiol, 4-methyl-	C_7H_8S 124.21	106-45-6 43	195	4-06-00-02153 1.0220[51]	H_2O 1; EtOH 3; eth 4; chl 3
2304	Benzenethiol, pentafluoro-	C_6HF_5S 200.13	771-62-0 -24	143	1.501[25]	1.4645[20]
2305	1,2,3-Benzenetriamine 1,2,3-Triaminobenzene	$C_6H_9N_3$ 123.16	608-32-2 103	336	3-13-00-00551	H_2O 4; eth 4; EtOH 4
2306	1,2,4-Benzenetriamine	$C_6H_9N_3$ 123.16	615-71-4 97.3	340	4-13-00-00506	H_2O 4; EtOH 4; chl 4
2307	Benzene, 1,2,3-tribromo- 1,2,3-Tribromobenzene	$C_6H_3Br_3$ 314.80	608-21-9 87.5		3-05-00-00569 2.658[25]	eth 4
2308	Benzene, 1,2,4-tribromo- 1,2,4-Tribromobenzene	$C_6H_3Br_3$ 314.80	615-54-3 44.5	275	4-05-00-00685	H_2O 1; EtOH 3; eth 4; ace 4
2309	Benzene, 1,3,5-tribromo- 1,3,5-Tribromobenzene	$C_6H_3Br_3$ 314.80	626-39-1 122.8	271	4-05-00-00685	H_2O 1; EtOH 2; eth 3; bz 3
2310	Benzene, 1,2,3-tribromo-5-methoxy- Anisole, 3,4,5-tribromo	$C_7H_5Br_3O$ 344.83	73557-60-5 93.3	303	2-06-00-00195	ace 4; bz 4; EtOH 4
2311	Benzene, 1,2,4-tribromo-5-methoxy- 2,4,5-Tribromoanisole	$C_7H_5Br_3O$ 344.83	95970-10-8 105	307	4-06-00-01067	ace 4; bz 4; EtOH 4
2312	Benzene, 1,2,5-tribromo-3-methoxy- Anisole, 2,3,5-tribromo	$C_7H_5Br_3O$ 344.83	73931-44-9 82	308.5	2-06-00-00192	ace 4; bz 4; EtOH 4
2313	Benzene, 1,3,5-tribromo-2-methoxy-	$C_7H_5Br_3O$ 344.83	607-99-8 88	298	4-06-00-01067 2.491[25]	H_2O 2; EtOH 2; ace 4; bz 4
2314	Benzene, 1,3,5-tribromo-2-methyl- 2,4,6-Tribromotoluene	$C_7H_5Br_3$ 328.83	6320-40-7 70	290	3-05-00-00719 2.479[17]	
2315	Benzene, 1,2,3-tribromo-5-nitro- 3,4,5-Tribromonitrobenzene	$C_6H_2Br_3NO_2$ 359.80	3460-20-6 112	sub	4-05-00-00732 2.645[25]	
2316	Benzene, 1,3,5-tribromo-2-nitro-	$C_6H_2Br_3NO_2$ 359.80	3463-40-9 125	177[11]	4-05-00-00732	eth 4; chl 4
2317	1,3,5-Benzenetricarbonyl trichloride	$C_9H_3Cl_3O_3$ 265.48	4422-95-1 36.3	180[16]	4-09-00-03748	chl 3
2318	1,2,3-Benzenetricarboxylic acid Hemimellitic acid	$C_9H_6O_6$ 210.14	569-51-7 200		4-09-00-03745 1.546[20]	eth 4; EtOH 4
2319	1,2,4-Benzenetricarboxylic acid Trimellitic acid	$C_9H_6O_6$ 210.14	528-44-9 219	9616	4-09-00-03746	H_2O 4; eth 4; EtOH 4
2320	1,3,5-Benzenetricarboxylic acid, 2-chloro- 2-Chloro-1,3,5-benzenetricarboxylic acid	$C_9H_5ClO_6$ 244.59	56961-24-1 285	sub	4-09-00-03749	H_2O 4; eth 4; EtOH 4

No.	Name / Synonym	Mol. Form. / Mol. Wt.	CAS RN / mp/°C	Merck No. / bp/°C	Beil. Ref. / den/g cm^{-3}	Solubility / n_D
2321	Benzene, 1,2,3-trichloro- / 1,2,3-Trichlorobenzene	$C_6H_3Cl_3$ / 181.45	87-61-6 / 53.5	9542 / 218.5	4-05-00-00664 / 1.4533[25]	H_2O 1; EtOH 2; eth 4; bz 4
2322	Benzene, 1,2,4-trichloro- / 1,2,4-Trichlorobenzene	$C_6H_3Cl_3$ / 181.45	120-82-1 / 17	9543 / 213.5	4-05-00-00664 / 1.459[25]	H_2O 1; EtOH 2; eth 4; ctc 2 / 1.5717[20]
2323	Benzene, 1,3,5-trichloro- / 1,3,5-Trichlorobenzene	$C_6H_3Cl_3$ / 181.45	108-70-3 / 63.5	9544 / 208	4-05-00-00666	H_2O 1; EtOH 2; eth 4; bz 4
2324	Benzene, 1,2,4-trichloro-5-(chloromethyl)-	$C_7H_4Cl_4$ / 229.92	3955-26-8 /	/ 273	4-05-00-00822 / 1.547[20]	ace 4; eth 4; EtOH 4
2325	Benzene, 1,2,4-trichloro-5-[(4-chlorophenyl)sulfonyl]- / Tetradifon	$C_{12}H_8Cl_4O_2S$ / 356.06	116-29-0 / 146	9132	4-06-00-01636 / 1.151[20]	
2326	Benzene, 1,2,3-trichloro-4-(dichloromethyl)- / Benzylidene chloride, 2,3,4-trichloro	$C_7H_3Cl_5$ / 264.36	56961-82-1 / 84	/ 278	1-05-00-00153	bz 4
2327	Benzene, 1,2,4-trichloro-5-(dichloromethyl)- / Benzylidene chloride, 2,4,5-trichloro	$C_7H_3Cl_5$ / 264.36	33429-70-8 /	/ 280.5	4-05-00-00824 / 1.5956[20]	bz 4 / 1.5992[20]
2328	Benzene, 1,1'-(2,2,2-trichloroethylidene)bis[4-chloro- / DDT	$C_{14}H_9Cl_5$ / 354.49	50-29-3 / 108.5	2832 / 260; 186[0.05]	4-05-00-01885	H_2O 1; EtOH 2; eth 4; ace 4
2329	Benzene, 1,1'-(2,2,2-trichloroethylidene)bis[4-methoxy- / Methoxychlor	$C_{16}H_{15}Cl_3O_2$ / 345.65	72-43-5 / 87	5913	4-06-00-06691 / 1.41[25]	H_2O 1; EtOH 3; eth 4; bz 4
2330	Benzene, 1,2,4-trichloro-3-methoxy-	$C_7H_5Cl_3O$ / 211.47	50375-10-5 / 45	/ 227	3-06-00-00717	ace 4; EtOH 4
2331	Benzene, 1,2,4-trichloro-5-methoxy-	$C_7H_5Cl_3O$ / 211.47	6130-75-2 / 77.5	/ 254	4-06-00-00963	EtOH 4; ace 4; os 3
2332	Benzene, 1,3,5-trichloro-2-methoxy- / 2,4,6-Trichloroanisole	$C_7H_5Cl_3O$ / 211.47	87-40-1 / 61.5	9541 / 241	4-06-00-01005 / 1.640[25]	EtOH 3; ace 4; bz 3; chl 3
2333	Benzene, (trichloromethyl)- / (Trichloromethyl)benzene	$C_7H_5Cl_3$ / 195.47	98-07-7 / -5	1120 / 221	4-05-00-00820 / 1.3723[20]	H_2O 1; EtOH 3; eth 3; bz 3 / 1.5580[20]
2334	Benzene, 1,2,3-trichloro-4-methyl- / 2,3,4-Trichlorotoluene	$C_7H_5Cl_3$ / 195.47	7359-72-0 / 43.5	/ 244	4-05-00-00819	H_2O 1; EtOH 3; eth 3; ace 3
2335	Benzene, 1,2,3-trichloro-5-methyl- / Toluene, 3,4,5-trichloro-	$C_7H_5Cl_3$ / 195.47	21472-86-6 / 45.5	/ 246	0-05-00-00299	H_2O 1; EtOH 3; ace 3; os 3
2336	Benzene, 1,2,4-trichloro-5-methyl- / 2,4,5-Trichlorotoluene	$C_7H_5Cl_3$ / 195.47	6639-30-1 / 82.4	/ 231	4-05-00-00819	H_2O 1; EtOH 3; ace 3; os 3
2337	Benzene, 1,2,5-trichloro-3-methyl- / 2,3,5-Trichlorotoluene	$C_7H_5Cl_3$ / 195.47	56961-86-5 / 45.5	/ 230	0-05-00-00299	H_2O 1; EtOH 3; ace 3; os 3
2338	Benzene, 1,2,3-trichloro-nitro-	$C_6H_2Cl_3NO_2$ / 226.45	20098-48-0 / 72.5		2-05-00-00187 / 1.807[25]	EtOH 4
2339	Benzene, 1,2,4-trichloro-5-nitro-	$C_6H_2Cl_3NO_2$ / 226.45	89-69-0 / 57.5	/ 288	4-05-00-00728 / 1.790[23]	H_2O 1; EtOH 2; eth 3; bz 3
2340	Benzene, 1,3,5-trichloro-2,4,6-trifluoro- / 1,3,5-Trichloro-2,4,6-trifluorobenzene	$C_6Cl_3F_3$ / 235.42	319-88-0 /	/ 198.4		
2341	Benzene, tricosyl- / Tricosylbenzene	$C_{29}H_{52}$ / 400.73	61828-04-4 / 54	/ 365[100] 231[1]	4-05-00-01237 / 0.8544[20]	1.4801[20]
2342	Benzene, tridecyl- / 1-Phenyltridecane	$C_{19}H_{32}$ / 260.46	123-02-4 / 10	9575 / 346	3-05-00-01108 / 0.8550[20]	1.4821[20]
2343	Benzene, 1,3,5-triethoxy- / 1,3,5-Triethoxybenzene	$C_{12}H_{18}O_3$ / 210.27	2437-88-9 / 43.5	/ 170[24]	4-06-00-07362	eth 4; EtOH 4
2344	Benzene, 1,2,4-triethyl- / 1,2,4-Triethylbenzene	$C_{12}H_{18}$ / 162.27	877-44-1 /	/ 218; 99[15]	4-05-00-01133 / 0.8738[20]	H_2O 1; EtOH 3; eth 3 / 1.5024[20]
2345	Benzene, 1,3,5-triethyl- / 1,3,5-Triethylbenzene	$C_{12}H_{18}$ / 162.27	102-25-0 / -66.5	/ 215.9	4-05-00-01133 / 0.8631[20]	H_2O 1; EtOH 4; eth 4 / 1.4969[20]
2346	Benzene, 1,2,4-trifluoro-	$C_6H_3F_3$ / 132.09	367-23-7 /	/ 90	4-05-00-00638 / 1.264[25]	1.4171[20]
2347	Benzene, 1,3,5-trifluoro- / 1,3,5-Trifluorobenzene	$C_6H_3F_3$ / 132.09	372-38-3 / -5.5	/ 75.5	4-05-00-00638 / 1.277[25]	1.4140[20]
2348	Benzene, (trifluoromethyl)- / (Trifluoromethyl)benzene	$C_7H_5F_3$ / 146.11	98-08-8 / -29.1	1121 / 102.1	4-05-00-00802 / 1.1884[20]	EtOH 5; eth 5; ace 5; bz 5 / 1.4146[20]
2349	Benzene, 1,2,3-triiodo- / 1,2,3-Triiodobenzene	$C_6H_3I_3$ / 455.80	608-29-7 / 116	/ sub	1-05-00-00122	eth 4; EtOH 4; chl 4
2350	Benzene, 1,2,4-triiodo- / 1,2,4-Triiodobenzene	$C_6H_3I_3$ / 455.80	615-68-9 / 91.5	/ sub	3-05-00-00585	EtOH 4; chl 4
2351	Benzene, 1,3,5-triiodo- / 1,3,5-Triiodobenzene	$C_6H_3I_3$ / 455.80	626-44-8 / 184.2	/ sub	4-05-00-00702	H_2O 1; EtOH 2; eth 2; bz 2
2352	Benzene, 1,3,5-triiodo-2-methyl- / 2,4,6-Triiodotoluene	$C_7H_5I_3$ / 469.83	36994-79-3 / 118.5	/ 300 dec	1-05-00-00158	EtOH 4
2353	Benzene, 1,2,3-triiodo-5-nitro-	$C_6H_2I_3NO_2$ / 500.80	53663-23-3 / 167		3-05-00-00626 / 3.256[25]	bz 4; eth 4; EtOH 4; chl 4
2354	Benzene, 1,2,3-trimethoxy-	$C_9H_{12}O_3$ / 168.19	634-36-6 / 48.5	/ 235	4-06-00-07329 / 1.1009[45]	H_2O 1; EtOH 3; eth 3; bz 3
2355	Benzene, 1,3,5-trimethoxy-	$C_9H_{12}O_3$ / 168.19	621-23-8 / 54.5	/ 255.5	4-06-00-07362	H_2O 1; EtOH 3; eth 3; bz 3
2356	Benzene, 1,2,4-trimethoxy-5-(1-propenyl)-	$C_{12}H_{16}O_3$ / 208.26	494-40-6 / 67	/ 296	4-06-00-07476 / 1.165[20]	eth 4; EtOH 4; peth 4 / 1.5683[20]
2357	Benzene, 1,2,3-trimethyl- / 1,2,3-Trimethylbenzene	C_9H_{12} / 120.19	526-73-8 / -25.4	/ 176.1	4-05-00-01007 / 0.8944[20]	H_2O 1; EtOH 5; eth 5; ace 5 / 1.5139[20]
2358	Benzene, 1,2,4-trimethyl- / Pseudocumene	C_9H_{12} / 120.19	95-63-6 / -43.8	7929 / 169.3	4-05-00-01010 / 0.8758[20]	H_2O 1; EtOH 5; eth 5; ace 5 / 1.5048[20]

No.	Name Synonym	Mol. Form. Mol. Wt.	CAS RN mp/°C	Merck No. bp/°C	Beil. Ref. den/g cm^{-3}	Solubility n_D
2359	Benzene, 1,3,5-trimethyl- Mesitylene	C$_9$H$_{12}$ 120.19	108-67-8 -44.7	5810 164.7	4-05-00-01016 0.8652[20]	H$_2$O 1; EtOH 5; eth 5; ace 5 1.4994[20]
2360	Benzene, 1,3,5-trimethyl-2,4-dinitro-	C$_9$H$_{10}$N$_2$O$_4$ 210.19	608-50-4 85.3	exp 418	4-05-00-01029	H$_2$O 1; EtOH 3
2361	Benzene, 1,2,4-trimethyl-5-(1-methylethyl)-	C$_{12}$H$_{18}$ 162.27	10222-95-4 21	221	3-05-00-01038 0.8795[21]	1.5065[21]
2362	Benzene, 1,2,4-trimethyl-5-nitro-	C$_9$H$_{11}$NO$_2$ 165.19	610-91-3 71	265	4-05-00-01015	EtOH 3; chl 3; peth 3
2363	Benzene, 1,3,5-trimethyl-2-nitro-	C$_9$H$_{11}$NO$_2$ 165.19	603-71-4 44	255	4-05-00-01028 1.51[25]	EtOH 4
2364	Benzene, 1,3,5-trimethyl-2,4,6-trinitro-	C$_9$H$_9$N$_3$O$_6$ 255.19	602-96-0 238.2	exp 415	4-05-00-01029	H$_2$O 1; EtOH 2; eth 2; ace 3
2365	Benzene, 1,3,5-trinitro- sym-Trinitrobenzene	C$_6$H$_3$N$_3$O$_6$ 213.11	99-35-4 121.5	9639 315	4-05-00-00755 1.4775[152]	H$_2$O 2; EtOH 2; eth 2; ace 4
2366	1,2,3-Benzenetriol Pyrogallol	C$_6$H$_6$O$_3$ 126.11	87-66-1 133	8010 309	4-06-00-07327 1.453[4]	H$_2$O 4; EtOH 4; eth 4; ace 3 1.561[134]
2367	1,2,4-Benzenetriol Hydroxyhydroquinone	C$_6$H$_6$O$_3$ 126.11	533-73-3 140.5	1081	4-06-00-07338	H$_2$O 4; EtOH 4; eth 4; bz 1
2368	1,3,5-Benzenetriol Phloroglucinol	C$_6$H$_6$O$_3$ 126.11	108-73-6 218.5	7301 sub	4-06-00-07361 1.46[25]	H$_2$O 4; EtOH 4; eth 4; ace 3
2369	1,2,3-Benzenetriol, 5-methyl-	C$_7$H$_8$O$_3$ 140.14	609-25-6 129	sub	4-06-00-07376	bz 4
2370	1,2,4-Benzenetriol, triacetate	C$_{12}$H$_{12}$O$_6$ 252.22	613-03-6 99	300	4-06-00-07342	EtOH 3; chl 3; MeOH 3
2371	1,3,5-Benzenetriol, 2,4,6-trichloro- Trichloro-1,3,5-trihydroxybenzene	C$_6$H$_3$Cl$_3$O$_3$ 229.45	56961-23-0 136	sub	0-06-00-01104	EtOH 4
2372	Benzene, 1,2,4-tris(1-methylethyl)- 1,2,4-Triisopropylbenzene	C$_{15}$H$_{24}$ 204.36	948-32-3	244	4-05-00-01178 0.8574[25]	1.4896[25]
2373	Benzene, 1,3,5-tris(1-methylethyl)-	C$_{15}$H$_{24}$ 204.36	717-74-8 -7.4	238	4-05-00-01178 0.8545[20]	ace 3; bz 3; chl 3 1.4882[20]
2374	Benzene, undecyl-	C$_{17}$H$_{28}$ 232.41	6742-54-7 -5	316	3-05-00-01100 0.8553[20]	1.4828[20]
2375	9,10[1',2']-Benzenoanthracene, 9,10-dihydro- Triptycene	C$_{20}$H$_{14}$ 254.33	477-75-8	9663	4-05-00-02639	
2376	1H-Benzimidazole N,N'-Methenyl-o-phenylenediamine	C$_7$H$_6$N$_2$ 118.14	51-17-2 170.5	1091 >360	5-23-06-00196	H$_2$O 2; EtOH 4; eth 2; bz 1
2377	1H-Benzimidazole, 4,5-dichloro-2- (trifluoromethyl)- Chlorflurazole	C$_8$H$_3$Cl$_2$F$_3$N$_2$ 255.03	3615-21-2 213.5			
2378	1H-Benzimidazole, 1,5-dimethyl- 1,5-Dimethylbenzimidazole	C$_9$H$_{10}$N$_2$ 146.19	10394-35-1 95	300	5-23-06-00380	bz 4; eth 4; EtOH 4
2379	1H-Benzimidazole, 2,5-dimethyl-	C$_9$H$_{10}$N$_2$ 146.19	1792-41-2 203	350	5-23-06-00448	eth 4; EtOH 4
2380	1H-Benzimidazole, 5,6-dimethyl- Dimedazole	C$_9$H$_{10}$N$_2$ 146.19	582-60-5 205.5	3225 sub	5-23-06-00454	H$_2$O 3; EtOH 3; eth 3; chl 3
2381	1H-Benzimidazole, 1-ethyl-2-methyl- 1-Ethyl-2-methylbenzimidazole	C$_{10}$H$_{12}$N$_2$ 160.22	5805-76-5 51	296	5-23-06-00326 1.073[25]	
2382	1H-Benzimidazole, 1-methyl-	C$_8$H$_8$N$_2$ 132.17	1632-83-3 66	286	5-23-06-00215 1.1254[20]	peth 3 1.6013[7]
2383	1H-Benzimidazole, 1-phenyl-	C$_{13}$H$_{10}$N$_2$ 194.24	2622-60-8 97	221[25]	5-23-06-00222	H$_2$O 2; EtOH 3; MeOH 3
2384	1H-Benzimidazole, 2-phenyl- Phenzidole	C$_{13}$H$_{10}$N$_2$ 194.24	716-79-0 293	7245	5-23-08-00437	H$_2$O 2; EtOH 3; bz 2; chl 2
2385	1H-Benzimidazole, 2-(phenylmethyl)- Bendazol	C$_{14}$H$_{12}$N$_2$ 208.26	621-72-7 187	1043	5-23-08-00498	bz 4; EtOH 4; gl HOAc 4
2386	2H-Benzimidazole-2-thione, 1,3-dihydro- 2-Benzimidazolethiol	C$_7$H$_6$N$_2$S 150.20	583-39-1 298	1092	5-24-02-00413	EtOH 4
2387	1,2-Benzisothiazole 1-Thia-2-azaindene	C$_7$H$_5$NS 135.19	272-16-2 38	220	2-27-00-00016	ctc 3
2388	1,2-Benzisothiazol-3(2H)-one, 1,1-dioxide Saccharin	C$_7$H$_5$NO$_3$S 183.19	81-07-2 228 dec	8282 sub	4-27-00-02649 0.828[25]	H$_2$O 2; EtOH 3; eth 2; ace 3
2389	1,2-Benzisoxazole Indoxazene	C$_7$H$_5$NO 119.12	271-95-4	100[26]	4-27-00-01067 1.1727[21]	eth 4 1.5570[20]
2390	2,1-Benzisoxazole Anthranil	C$_7$H$_5$NO 119.12	271-58-9 <-18	215	4-27-00-01068 1.1827[20]	ace 4; EtOH 4 1.5845[20]
2391	Benz[j]aceanthrylene, 1,2-dihydro- Cholanthrene	C$_{20}$H$_{14}$ 254.33	479-23-2 174	2200	4-05-00-02638	H$_2$O 1; EtOH 3; bz 3; HOAc 3
2392	Benz[j]aceanthrylene, 1,2-dihydro-3-methyl- 3-Methylcholanthrene	C$_{21}$H$_{16}$ 268.36	56-49-5 180	5967 280[80]	4-05-00-02648 1.28[20]	H$_2$O 1
2393	11H-Benzo[a]fluorene	C$_{17}$H$_{12}$ 216.28	238-84-6 189.5	405	4-05-00-02473	EtOH 2; eth 3; bz 3; chl 3
2394	Benzo[a]phenazine 7,12-Diazabenz[a]anthracene	C$_{16}$H$_{10}$N$_2$ 230.27	225-61-6 142.5	>360	5-23-09-00494	H$_2$O 1; EtOH 3; eth 3; ace 3
2395	Benzo[a]pyrene 2,3-Benzopyrene	C$_{20}$H$_{12}$ 252.32	50-32-8 176.5	1113	4-05-00-02687	chl 4
2396	2H-Benzo[a]quinolizin-2-one, 1,3,4,6,7,11b- hexahydro-9,10-dimethoxy-3-(2-methylpropyl)- 1,2-Benzoquinolizine-2-one, 9,10-dimethoxy- 1,2,3,4,6,7,11b-heptahydro-3-isobutyl	C$_{19}$H$_{27}$NO$_3$ 317.43	58-46-8 128	9118	5-21-13-00178	chl 3

No.	Name / Synonym	Mol. Form. / Mol. Wt.	CAS RN / mp/°C	Merck No. / bp/°C	Beil. Ref. / den/g cm^{-3}	Solubility / n_D
2397	2H,8H-Benzo[1,2-b:5,4-b']dipyran-2-one, 8,8-dimethyl- / Xanthyletin	$C_{14}H_{12}O_3$ / 228.25	553-19-5 / 131.5	9978 / 140-5$^{0.1}$	5-19-04-00470	EtOH 3; peth 3
2398	2H,8H-Benzo[1,2-b:5,4-b']dipyran-2-one, 5-methoxy-8,8-dimethyl- / Xanthoxyletin	$C_{15}H_{14}O_4$ / 258.27	84-99-1 / 133	9975	5-19-06-00039	H_2O 1; EtOH 3; eth 2; ace 3
2399	Benzo[b]thiophene / Thianaphthene	C_8H_6S / 134.20	95-15-8 / 32	9232 / 221	5-17-02-00006 / 1.1484^{32}	H_2O 1; EtOH 4; eth 3; ace 3 / 1.6374^{37}
2400	Benzo[b]thiophene-2-carboxylic acid / Thionaphthene-2-carboxylic acid	$C_9H_6O_2S$ / 178.21	6314-28-9 / 240.5	9274	5-18-06-00427	eth 4
2401	Benzo[b]thiophene-2,3-dione	$C_8H_4O_2S$ / 164.18	493-57-2 / 121	247	5-17-11-00251	H_2O 1; EtOH 3; bz 3; HOAc 3
2402	Benzo[b]thiophene, 5-methyl-	C_9H_8S / 148.23	14315-14-1 / 20.5	111-5^{12}	5-17-02-00031 / 1.111^{22}	1.615^{22}
2403	Benzo[b]thiophene-4-ol	C_8H_6OS / 150.20	3610-02-4 / 79.8	sub	5-17-04-00199	EtOH 4
2404	Benzo[b]thiophen-3(2H)-one, 2-(3-oxobenzo[b]thien-2(3H)-ylidene)-	$C_{16}H_8O_2S_2$ / 296.37	522-75-8 / 359	sub	5-19-05-00281	H_2O 1; EtOH 1; bz 3; chl 2
2405	7H-Benzo[c]carbazole / 7-Aza-7H-benzo[c]fluorene	$C_{16}H_{11}N$ / 217.27	205-25-4 / 134	448	5-20-08-00455	bz 4; EtOH 4; HOAc 4
2406	Benzo[c]phenanthrene / Tetrahelicene	$C_{18}H_{12}$ / 228.29	195-19-7 / 68	1129	4-05-00-02552	H_2O 1; EtOH 2; lig 2
2407	5H-Benzocyclohepten-5-one, 6,7,8,9-tetrahydro-	$C_{11}H_{12}O$ / 160.22	826-73-3 / 124^7; 108^1		4-07-00-01029 / 1.0780^{20}	EtOH 3 / 1.5698^{20}
2408	5H-Benzocyclohepten-5-one, 2,3,4,6-tetrahydroxy- / Purpurogallin	$C_{11}H_8O_5$ / 220.18	569-77-7 / 274-5 dec	7963	4-08-00-03456	
2409	3H-1,4-Benzodiazepin-2-amine, 7-chloro-N-methyl-5-phenyl-, 4-oxide / Chlordiazepoxide	$C_{16}H_{14}ClN_3O$ / 299.76	58-25-3 / 236.2	2082		
2410	2H-1,4-Benzodiazepin-2-one, 7-chloro-1,3-dihydro-1-methyl-5-phenyl- / Diazepam	$C_{16}H_{13}ClN_2O$ / 284.74	439-14-5 / 132	2977	5-24-04-00300	
2411	2H-1,4-Benzodiazepin-2-one, 7-chloro-1,3-dihydro-5-phenyl- / Nordazepam	$C_{15}H_{11}ClN_2O$ / 270.72	1088-11-5 / 216.5	6608	5-24-04-00291	
2412	1,3,2-Benzodioxaborole	$C_6H_5BO_2$ / 119.92	274-07-7 / 12	88^{156}; 50^{50}	5-19-01-00439 / 1.2700^{20}	1.5070^{20}
2413	1,3,2-Benzodioxaphosphole, 2-chloro-	$C_6H_4ClO_2P$ / 174.52	1641-40-3 / 30	80^{20}	4-06-00-05598 / 1.4650^{20}	1.5712^{20}
2414	1,4-Benzodioxin, 2,3-dihydro-	$C_8H_8O_2$ / 136.15	493-09-4	103^6	5-19-01-00439 / 1.142^{25}	1.5485^{20}
2415	1,3-Benzodioxole	$C_7H_6O_2$ / 122.12	274-09-9	172.5; 77^{27}	5-19-01-00434 / 1.064^{25}	1.5398^{20}
2416	1,3-Benzodioxole-5-acetic acid, methyl ester / Phenylacetic acid, 3,4-methylenedioxy, methyl ester	$C_{10}H_{10}O_4$ / 194.19	326-59-0	153-5^{10}	2-19-00-00295 / 1.246^{20}	1.534
2417	1,3-Benzodioxole-5-carboxaldehyde / Piperonal	$C_8H_6O_3$ / 150.13	120-57-0 / 37	7445 / 263	5-19-04-00225	H_2O 2; EtOH 4; eth 5; ace 3
2418	1,3-Benzodioxole-5-carboxylic acid / Piperonylic acid	$C_8H_6O_4$ / 166.13	94-53-1 / 229	7447	5-19-07-00300	
2419	1,3-Benzodioxole-5-carboxylic acid, ethyl ester	$C_{10}H_{10}O_4$ / 194.19	6951-08-2 / 18.5	285.5; 135^6	5-19-07-00300	eth 4; EtOH 4; peth 4
2420	1,3-Benzodioxole-5-carboxylic acid, methyl ester	$C_9H_8O_4$ / 180.16	326-56-7 / 53	273 dec	5-19-07-00300	eth 4; EtOH 4
2421	1,3-Benzodioxole, 5-(chloromethyl)-	$C_8H_7ClO_2$ / 170.60	20850-43-5 / 20.5	134-5^{14}	5-19-01-00443 / 1.312^{25}	1.5660^{20}
2422	1,3-Benzodioxole, 4,7-dimethoxy-5-(1-propenyl)-, (E)-	$C_{12}H_{14}O_4$ / 222.24	17672-88-7 / 56	303.5	4-19-00-01033	ace 4; bz 4; eth 4; EtOH 4
2423	1,3-Benzodioxole, 4,7-dimethoxy-5-(2-propenyl)- / Apiole	$C_{12}H_{14}O_4$ / 222.24	523-80-8 / 29.5	767 / 294. 179^{35}	5-19-03-00307 / 1.015^{20}	ace 4; bz 4; EtOH 4; lig 4 / 1.5360^{20}
2424	1,3-Benzodioxole-5-ethanamine	$C_9H_{11}NO_2$ / 165.19	1484-85-1	166^{20}	5-19-08-00407 / 1.225^{20}	1.5620^{20}
2425	1,3-Benzodioxole, 5-ethenyl-	$C_9H_8O_2$ / 148.16	7315-32-4	224	5-19-01-00544 / 1.1488^{18}	eth 4; EtOH 4; chl 4 / 1.5802
2426	1,3-Benzodioxole-5-methanamine	$C_8H_9NO_2$ / 151.17	2620-50-0	138-9^{13}	5-19-08-00395 / 1.214^{25}	1.5635^{20}
2427	1,3-Benzodioxole-5-methanol	$C_8H_8O_3$ / 152.15	495-76-1 / 58	157^{16}	5-19-02-00543	H_2O 2; EtOH 3; eth 3; bz 3
2428	1,3-Benzodioxole, 4-methoxy-6-(2-propenyl)- / Myristicin	$C_{11}H_{12}O_3$ / 192.21	607-91-0 / <-20	6247 / 276.5	5-19-02-00631 / 1.1416^{20}	H_2O 1; EtOH 2; eth 3; bz 3 / 1.5403^{20}
2429	1,3-Benzodioxole, 5-(2-propenyl)- / Safrole	$C_{10}H_{10}O_2$ / 162.19	94-59-7 / 11.2	8287 / 234.5	5-19-01-00553 / 1.1000^{20}	H_2O 1; EtOH 4; eth 5; chl 5 / 1.5381^{20}
2430	1,3-Benzodioxole, 5-(1-propenyl)-, (E)-	$C_{10}H_{10}O_2$ / 162.19	4043-71-4 / 6.8	253	5-19-01-00552 / 1.1224^{20}	H_2O 1; EtOH 5; eth 5; ace 4 / 1.5782^{20}

No.	Name Synonym	Mol. Form. Mol. Wt.	CAS RN mp/°C	Merck No. bp/°C	Beil. Ref. den/g cm^{-3}	Solubility n_D
2431	1,3-Benzodioxole, 5-propyl- Dihydrosafrole	$C_{10}H_{12}O_2$ 164.20	94-58-6		5-19-01-00473	ctc 3
2432	[1,3]Benzodioxolo[5,6-c]-1,3-dioxolo[4,5-i]phenanthridinium, 13-methyl- Sanguinarine	$C_{20}H_{15}NO_5$ 349.34	2447-54-3 266	8320		ace 4; bz 4; eth 4; EtOH 4
2433	[1,3]Benzodioxolo[5,6-c]phenanthridinium, 1,2-dimethoxy-12-methyl- Chelerythrine	$C_{21}H_{19}NO_5$ 365.39	34316-15-9	2041 207		chl 4
2434	1,4-Benzodithiin	$C_8H_6S_2$ 166.27	255-50-5	67-70[0.1]	5-19-01-00536 1.2799[20]	1.6754[25]
2435	Benzo[e]-1,3-dioxolo[4,5-l][2]benzazecin-12(5H)-one, 4,6,7,13-tetrahydro-9,10-dimethoxy-5-methyl- Cryptopine	$C_{21}H_{23}NO_5$ 369.42	482-74-6 223	2611	4-27-00-06652 1.315[20]	H_2O 1; EtOH 2; eth 2; bz 2
2436	Benzo[e]pyrene 1,2-Benzpyrene	$C_{20}H_{12}$ 252.32	192-97-2 177.5	1114 311	4-05-00-02689	
2437	Benzo[f]quinoline β-Naphthoquinoline	$C_{13}H_9N$ 179.22	85-02-9 94	1115 352; 202-5[8]	5-20-08-00220	H_2O 2; EtOH 4; eth 4; ace 3
2438	Benzofuran Coumarone	C_8H_6O 118.14	271-89-6 <-18	1098 174	5-17-02-00003 1.0913[25]	H_2O 1; EtOH 3; eth 3 1.5615[17]
2439	2-Benzofurancarboxylic acid Coumarilic acid	$C_9H_6O_3$ 162.14	496-41-3 192.5	2562 312.5	5-18-06-00419	EtOH 4
2440	2-Benzofurancarboxylic acid, ethyl ester	$C_{11}H_{10}O_3$ 190.20	3199-61-9 30.5	276; 161[15]	5-18-06-00420 1.1656[28]	1.564[27]
2441	Benzofuran, 2-(chloromethyl)-2,3-dihydro- 2-(Chloromethyl)-2,3-dihydrobenzofuran	C_9H_9ClO 168.62	53491-32-0 41.5	118-9[11]	4-17-00-00421 1.2183[7]	ace 4; eth 4 1.5620[7]
2442	Benzofuran, 2,3-dihydro- Coumaran	C_8H_8O 120.15	496-16-2 -21.5	2560 188.5	5-17-01-00581 1.058[25]	eth 4; EtOH 4; chl 4 1.5497[20]
2443	Benzofuran, 2,3-dihydro-2-methyl-	$C_9H_{10}O$ 134.18	1746-11-8 	197.5	5-17-01-00597 1.061[25]	
2444	Benzofuran, 2,3-dihydro-5-methyl- 2,3-Dihydro-5-methylbenzofuran	$C_9H_{10}O$ 134.18	76429-68-0 	210.5	4-17-00-00423 1.0463[19]	1.54[20]
2445	Benzofuran, 2,5-dimethyl- 2,5-Dimethylbenzofuran	$C_{10}H_{10}O$ 146.19	29040-46-8 	220	5-17-02-00045 1.031[20]	eth 4 1.5534[20]
2446	Benzofuran, 2,6-dimethyl- 2,6-Dimethylbenzofuran	$C_{10}H_{10}O$ 146.19	24410-51-3 	217.5	5-17-02-00046 1.051[12]	eth 4 1.554[15]
2447	Benzofuran, 2,7-dimethyl- 2,7-Dimethylfuran	$C_{10}H_{10}O$ 146.19	59020-74-5 	216	4-17-00-00512 1.036[20]	eth 4 1.5546[20]
2448	Benzofuran, 3,5-dimethyl- 3,5-Dimethylbenzofuran	$C_{10}H_{10}O$ 146.19	10410-35-2 	220.5	5-17-02-00047 1.036[20]	eth 4 1.550[20]
2449	Benzofuran, 3,6-dimethyl- 3,6-DimethylfuranBenzofuran	$C_{10}H_{10}O$ 146.19	24410-50-2 	222	4-17-00-00512 1.0456[20]	eth 4 1.5505[20]
2450	Benzofuran, 4,6-dimethyl-	$C_{10}H_{10}O$ 146.19	116668-34-9 	219	0-17-00-00063 1.037[20]	eth 4 1.5485[21]
2451	Benzofuran, 4,7-dimethyl- 4,7-Dimethylbenzofuran	$C_{10}H_{10}O$ 146.19	28715-26-6 	216	0-17-00-00063 1.041[17]	eth 4 1.549[17]
2452	Benzofuran, 5,6-dimethyl- 5,6-Dimethylbenzofuran	$C_{10}H_{10}O$ 146.19	24410-52-4 	221	5-17-02-00048 1.060[15]	eth 4 1.5516[15]
2453	Benzofuran, 5,7-dimethyl- 5,7-Dimethylbenzofuran	$C_{10}H_{10}O$ 146.19	64965-91-9 	222	0-17-00-00063 1.0262[18]	eth 4 1.5358[20]
2454	Benzofuran, 6,7-dimethyl- 6,7-Dimethylfuran	$C_{10}H_{10}O$ 146.19	35355-36-3 	218	0-17-00-00063 1.038[20]	eth 4 1.5478[20]
2455	2-Benzofuranmethanol 2-Hydroxymethylbenzofuran	$C_9H_8O_2$ 148.16	55038-01-2 	149-50[14]	5-17-04-00224 1.177[13]	1.5550[11]
2456	Benzofuran, 2-methyl-	C_9H_8O 132.16	4265-25-2 	197.5	5-17-02-00022 1.0540[20]	eth 4; EtOH 4 1.5495[22]
2457	Benzofuran, 3-methyl- 3-Methylbenzo[b]furan	C_9H_8O 132.16	21535-97-7 	197; 86[20]	5-17-02-00025 1.0540[25]	H_2O 1; EtOH 3; eth 3 1.5536[16]
2458	Benzofuran, 5-methyl- 5-Methylbenzo[b]furan	C_9H_8O 132.16	18441-43-5 	198	5-17-02-00030 1.0603[19]	H_2O 1; EtOH 3; eth 3; sulf 3 1.5570[19]
2459	Benzofuran, 7-methyl- 7-Methylbenzo[b]furan	C_9H_8O 132.16	17059-52-8 109	190.5	5-17-02-00032 1.0490[19]	H_2O 1; EtOH 3; eth 3; sulf 3 1.5525[19]
2460	Benzofuran, 4-methyl-7-(1-methylethyl)- Benzofuran, 7-isopropyl-4-methyl	$C_{12}H_{14}O$ 174.24	95835-77-1 	241.5	0-17-00-00065 1.0145[18]	1.5363[16]
2461	2(3H)-Benzofuranone	$C_8H_6O_2$ 134.13	553-86-6 50	249	5-17-10-00006 1.2236[14]	
2462	3(2H)-Benzofuranone	$C_8H_6O_2$ 134.13	7169-34-8 102.5	152[15]	5-17-04-00185	bz 4
2463	2(3H)-Benzofuranone, 3-[2-(diethylamino)ethyl]-3-phenyl- Amolanone	$C_{20}H_{23}NO_2$ 309.41	76-65-3 43.4	603 192-4[2.0]	4-18-00-07995	1.5614[25]
2464	5-Benzofuranpropanol, 2-(1,3-benzodioxol-5-yl)-7-methoxy- Egonol	$C_{19}H_{18}O_5$ 326.35	530-22-3 118	228-30[15]	5-19-09-00583	chl 4
2465	Benzofuran, 4,5,6,7-tetrahydro-3,6-dimethyl-	$C_{10}H_{14}O$ 150.22	494-90-6 86	80[18]	5-17-01-00518 0.972[15]	
2466	6H-Benzofuro[3,2-c][1]benzopyran-6-one, 3,9-dihydroxy- Coumestrol	$C_{15}H_8O_5$ 268.23	479-13-0 385 dec	2565	5-19-06-00405	H_2O 1; EtOH 2; eth 1; ace 2

No.	Name / Synonym	Mol. Form. / Mol. Wt.	CAS RN / mp/°C	Merck No. / bp/°C	Beil. Ref. / den/g cm^{-3}	Solubility / n_D
2467	Benzo[g]-1,3-benzodioxolo[5,6-a]quinolizinium, 5,6-dihydro-9,10-dimethoxy- Berberine	$C_{20}H_{19}NO_5$ 353.37	2086-83-1 145	1169		eth 4; EtOH 4
2468	5H-Benzo[g]-1,3-benzodioxolo[6,5,4-de]quinoline, 6,7,7a,8-tetrahydro-10,11-dimethoxy-7-methyl-, (S)- Dicentrine	$C_{20}H_{21}NO_4$ 339.39	517-66-8	3028	4-27-00-06486	chl 3
2469	5H-Benzo[g]-1,3-benzodioxolo[6,5,4-de]quinoline, 6,7,7a,8-tetrahydro-11-methoxy-7-methyl-, (R)- Laureline	$C_{19}H_{19}NO_3$ 309.36	81-38-9 114	5251	1-27-00-00461	H_2O 1; EtOH 3; eth 3; dil acid 3
2470	5H-Benzo[g]-1,3-benzodioxolo[6,5,4-de]quinolin-12-ol, 6,7,7a,8-tetrahydro-11-methoxy-7-methyl-, (S)- Bulbocapnine	$C_{19}H_{19}NO_4$ 325.36	298-45-3 199.5	1471	2-27-00-00554	H_2O 1; EtOH 3; chl 4; os 3
2471	Benzo[ghi]perylene 1,12-Benzperylene	$C_{22}H_{12}$ 276.34	191-24-2		4-05-00-02766	
2472	Benzo[g]pteridine-2,4(1H,3H)-dione, 7,8-dimethyl- Lumichrome	$C_{12}H_{10}N_4O_2$ 242.24	1086-80-2	5468	4-26-00-02538	
2473	Benzo[g]quinoline 6,7-Benzoquinoline	$C_{13}H_9N$ 179.22	260-36-6 114	202[14]	5-20-08-00195	eth 4; EtOH 4
2474	Benzo[h]quinoline	$C_{13}H_9N$ 179.22	230-27-3 52	339; 233[47]	5-20-08-00215 1.2340[20]	H_2O 2; EtOH 3; eth 3; ace 3
2475	Benzoic acid Benzenecarboxylic acid	$C_7H_6O_2$ 122.12	65-85-0 122.4	1101 249.2	4-09-00-00273 1.265[15]	H_2O 2; EtOH 4; eth 4; ace 3 1.504[132]
2476	Benzoic acid, 2-acetyl-	$C_9H_8O_3$ 164.16	577-56-0 114.5	110-2[2]	4-10-00-02766	H_2O 4; eth 4; EtOH 4
2477	Benzoic acid, 3-acetyl-	$C_9H_8O_3$ 164.16	586-42-5 172	110-2[2]	4-10-00-02769	H_2O 3; EtOH 5
2478	Benzoic acid, 4-acetyl-	$C_9H_8O_3$ 164.16	586-89-0 208	sub	4-10-00-02769	H_2O 4
2479	Benzoic acid, 4-acetyl-, methyl ester	$C_{10}H_{10}O_3$ 178.19	3609-53-8 95	142[4]	3-10-00-03030	H_2O 4
2480	Benzoic acid, 2-(acetyloxy)- Aspirin	$C_9H_8O_4$ 180.16	50-78-2 135	873	4-10-00-00138	H_2O 3; EtOH 4; eth 3; bz 2
2481	Benzoic acid, 2-(acetyloxy)-5-bromo- 5-Bromoacetylsalicylic acid	$C_9H_7BrO_4$ 259.06	1503-53-3 60	1425	2-10-00-00064	H_2O 1; EtOH 4; eth 3
2482	Benzoic acid, 2-(acetyloxy)-, methyl ester Methyl acetylsalicylate	$C_{10}H_{10}O_4$ 194.19	580-02-9 51.5	5934 134-6[9]	4-10-00-00147	eth 4; EtOH 4; chl 4
2483	Benzoic acid, 2-(acetyloxy)-, phenyl ester Phenyl acetylsalicylate	$C_{15}H_{12}O_4$ 256.26	134-55-4 96	7241	4-10-00-00156	
2484	Benzoic acid, 2-amino- o-Anthranilic acid	$C_7H_7NO_2$ 137.14	118-92-3 146.5	433 sub	4-14-00-01004 1.412[20]	H_2O 3; EtOH 3; eth 3; bz 2
2485	Benzoic acid, 3-amino- Aniline 3 carboxylic acid	$C_7H_7NO_2$ 137.14	99-05-8 173	432	4-14-00-01092 1.51[25]	H_2O 2; EtOH 2; eth 3; ace 4
2486	Benzoic acid, 4-amino- Aniline-4-carboxylic acid	$C_7H_7NO_2$ 137.14	150-13-0 188.5	434	4-14-00-01126 1.374[20]	H_2O 3; EtOH 3; eth 3; ace 2
2487	Benzoic acid, 2-amino-5-bromo- 5-Bromoanthranilic acid	$C_7H_6BrNO_2$ 216.03	5794-88-7 219.5	1393	4-14-00-01081	DMSO 3
2488	Benzoic acid, 4-amino-, butyl ester Butamben	$C_{11}H_{15}NO_2$ 193.25	94-25-7 58	1504 173-4[8]	4-14-00-01130	H_2O 1; EtOH 3; eth 3; bz 3
2489	Benzoic acid, 4-(aminocarbonyl)-	$C_8H_7NO_3$ 165.15	6051-43-0 300	sub	4-09-00-03318	H_2O 1; EtOH 1; eth 1; bz 1
2490	Benzoic acid, 5-amino-2-chloro- 5-Amino-2-chlorobenzoic acid	$C_7H_6ClNO_2$ 171.58	89-54-3 188		4-14-00-01116 1.519[15]	EtOH 4
2491	Benzoic acid, 2-amino-3,6-dichloro-	$C_7H_5Cl_2NO_2$ 206.03	3032-32-4 155	sub	4-14-00-01078	ace 4; bz 4; eth 4; EtOH 4
2492	Benzoic acid, 3-amino-2,5-dichloro- Chloramben	$C_7H_5Cl_2NO_2$ 206.03	133-90-4 200	2063		DMSO 2
2493	Benzoic acid, 4-amino-, 2-(diethylamino)ethyl ester Procaine	$C_{13}H_{20}N_2O_2$ 236.31	59-46-1 61	7763	4-14-00-01138	H_2O 2; EtOH 3; eth 3; bz 3
2494	Benzoic acid, 4-[[(2-amino-1,4-dihydro-4-oxo-6-pteridinyl)methyl]formylamino]- Rhizopterin	$C_{15}H_{12}N_6O_4$ 340.30	119-20-0 >300	8181	4-26-00-03949	H_2O 1; EtOH 1; eth 1; aq alk 3
2495	Benzoic acid, 2-amino-, ethyl ester	$C_9H_{11}NO_2$ 165.19	87-25-2 13	268	4-14-00-01008 1.1174[20]	eth 4; EtOH 4 1.5646[20]
2496	Benzoic acid, 3-amino-, ethyl ester	$C_9H_{11}NO_2$ 165.19	582-33-2	294; 160[5]	4-14-00-01093 1.1248[22]	H_2O 2; EtOH 4; eth 4; ctc 3 1.5600[22]
2497	Benzoic acid, 4-amino-, ethyl ester Ethyl aminobenzoate	$C_9H_{11}NO_2$ 165.19	94-09-7 92	3719 310	4-14-00-01129	H_2O 1; EtOH 4; eth 4; chl 3
2498	Benzoic acid, 4-amino-2-hydroxy- p-Aminosalicylic acid	$C_7H_7NO_3$ 153.14	65-49-6 150-1 dec	491	4-14-00-01967	H_2O 3; EtOH 3; eth 3; ace 3
2499	Benzoic acid, 5-amino-2-hydroxy- Mesalamine	$C_7H_7NO_3$ 153.14	89-57-6 283	5807	4-14-00-02058	H_2O 2; EtOH 1
2500	Benzoic acid, 4-amino-2-hydroxy-, hydrazide p-Aminosalicylic acid hydrazide	$C_7H_9N_3O_2$ 167.17	6946-29-8 195	492	4-14-00-01995	EtOH 4
2501	Benzoic acid, 3-amino-4-hydroxy-, methyl ester Orthocaine	$C_8H_9NO_3$ 167.16	536-25-4 143	6836	3-14-00-01477	H_2O 1; EtOH 4; eth 3; bz 2

No.	Name / Synonym	Mol. Form. / Mol. Wt.	CAS RN / mp/°C	Merck No. / bp/°C	Beil. Ref. / den/g cm^{-3}	Solubility / n_D
2502	Benzoic acid, 4-amino-2-hydroxy-, phenyl ester Phenyl p-aminosalicylate	$C_{13}H_{11}NO_3$ 229.24	133-11-9 153	7243	4-14-00-01979	
2503	Benzoic acid, 2-amino-, methyl ester Methyl anthranilate	$C_8H_9NO_2$ 151.17	134-20-3 24.5	5942 256	4-14-00-01008 1.1682[10]	H_2O 2; EtOH 4; eth 4 1.5810
2504	Benzoic acid, 3-amino-, methyl ester	$C_8H_9NO_2$ 151.17	4518-10-9 39	152-3[11]	4-14-00-01093 1.232[20]	EtOH 4; eth 4; bz 4; chl 4
2505	Benzoic acid, 4-amino-, 2-methylpropyl ester Isobutyl p-aminobenzoate	$C_{11}H_{15}NO_2$ 193.25	94-14-4 64.5	5017	4-14-00-01130	
2506	Benzoic acid, 2-amino-3-nitro- 3-Nitroanthranilic acid	$C_7H_6N_2O_4$ 182.14	606-18-8 208.5		4-14-00-01087 1.558[15]	eth 4; EtOH 4
2507	Benzoic acid, 4-amino-, propyl ester Risocaine	$C_{10}H_{13}NO_2$ 179.22	94-12-2 75	8226	4-14-00-01130	bz 4; eth 4; EtOH 4; chl 4
2508	Benzoic acid, 4-(aminosulfonyl)- Carzenide	$C_7H_7NO_4S$ 201.20	138-41-0 290-2 dec	1885	4-11-00-00690	H_2O 1; EtOH 4; eth 2; bz 1
2509	Benzoic acid, ammonium salt	$C_7H_9NO_2$ 139.15	1863-63-4 194.5	515	4-09-00-00273	H_2O 3
2510	Benzoic acid, anhydride Benzoic anhydride	$C_{14}H_{10}O_3$ 226.23	93-97-0 42.5	1102 360	4-09-00-00550 1.989[15]	H_2O 1; EtOH 3; eth 3; chl 2 1.5767[15]
2511	Benzoic acid, 3-benzoyl-	$C_{14}H_{10}O_3$ 226.23	579-18-0 163.3	sub	4-10-00-02982	eth 4; EtOH 4
2512	Benzoic acid, 4-benzoyl-	$C_{14}H_{10}O_3$ 226.23	611-95-0 199	sub	3-10-00-03305	H_2O 2; EtOH 3; eth 3; bz 2
2513	Benzoic acid, 3-(benzoylamino)-	$C_{14}H_{11}NO_3$ 241.25	587-54-2 252.5		4-14-00-01105 1.510[4]	EtOH 4
2514	Benzoic acid, 4-(benzoylamino)-2-hydroxy- Benzoyl-PAS	$C_{14}H_{11}NO_4$ 257.25	13898-58-3 260.5	1127	4-14-00-02029	
2515	Benzoic acid, 2-benzoyl-, ethyl ester	$C_{16}H_{14}O_3$ 254.29	604-61-5 59.5		3-10-00-03291 1.221[64]	H_2O 1; EtOH 4; eth 4; chl 2 1.560[64]
2516	Benzoic acid, 2-benzoyl-, methyl ester	$C_{15}H_{12}O_3$ 240.26	606-28-0 52	351	4-10-00-02977 1.1903[19]	H_2O 1; EtOH 4; eth 4; sulf 3 1.591[20]
2517	Benzoic acid, 2-[bis(4-hydroxyphenyl)methyl]- Phenolphthalin	$C_{20}H_{16}O_4$ 320.34	81-90-3 230.5	7210	4-10-00-01932	eth 4; EtOH 4
2518	Benzoic acid, 2-bromo-	$C_7H_5BrO_2$ 201.02	88-65-3 150	sub	4-09-00-01011 1.929[25]	H_2O 2; EtOH 3; eth 3; ace 3
2519	Benzoic acid, 3-bromo-	$C_7H_5BrO_2$ 201.02	585-76-2 155	>280	4-09-00-01013 1.845[20]	H_2O 1; EtOH 3; eth 3
2520	Benzoic acid, 4-bromo- p-Bromobenzoic acid	$C_7H_5BrO_2$ 201.02	586-76-5 254.5	1396	4-09-00-01017 1.894[20]	H_2O 2; EtOH 3; eth 3
2521	Benzoic acid, 2-bromo-, ethyl ester	$C_9H_9BrO_2$ 229.07	6091-64-1 	254.5	4-09-00-01012 1.4438[15]	ace 4; bz 4; eth 4; EtOH 4 1.5455[15]
2522	Benzoic acid, 3-bromo-, ethyl ester	$C_9H_9BrO_2$ 229.07	24398-88-7 	261	4-09-00-01014 1.4308[19]	ace 4; bz 4; eth 4; EtOH 4 1.5430[19]
2523	Benzoic acid, 4-bromo-, ethyl ester	$C_9H_9BrO_2$ 229.07	5798-75-4 	263; 125[15]	3-09-00-01405 1.4332[17]	H_2O 2; EtOH 3; eth 3; ace 3 1.5438[17]
2524	Benzoic acid, 5-bromo-2-hydroxy-	$C_7H_5BrO_3$ 217.02	89-55-4 169.8	sub 100	4-10-00-00217	H_2O 2; EtOH 4; eth 4; ace 2
2525	Benzoic acid, 3-bromo-, methyl ester	$C_8H_7BrO_2$ 215.05	618-89-3 32	125[15]	4-09-00-01014	H_2O 2; EtOH 3; eth 3
2526	Benzoic acid, 4-bromo-, methyl ester	$C_8H_7BrO_2$ 215.05	619-42-1 81		4-09-00-01017 1.689[25]	EtOH 3; eth 3; ace 3; bz 4
2527	Benzoic acid, 2-bromo-4-nitro- 2-Bromo-4-nitrobenzoic acid	$C_7H_4BrNO_4$ 246.02	16426-64-5 166.5	sub 150	3-09-00-01771	eth 4; EtOH 4
2528	Benzoic acid, 2-bromo-5-nitro- 2-Bromo-5-nitrobenzoic acid	$C_7H_4BrNO_4$ 246.02	943-14-6 180.5	sub	3-09-00-01771	eth 4; EtOH 4
2529	Benzoic acid, 4-bromo-3-nitro- 4-Bromo-3-nitrobenzoic acid	$C_7H_4BrNO_4$ 246.02	6319-40-0 203.5	sub	4-09-00-01236	EtOH 4
2530	Benzoic acid, 5-bromo-2-nitro- 2-Nitro-5-bromobenzoic acid	$C_7H_4BrNO_4$ 246.02	6950-43-2 140		1-09-00-00165 1.920[18]	bz 4; EtOH 4
2531	Benzoic acid, 4-(butylamino)-, 2-(dimethylamino)ethyl ester, monohydrochloride Tetracaine hydrochloride	$C_{15}H_{25}ClN_2O_2$ 300.83	136-47-0 147	9123	4-14-00-01172	
2532	Benzoic acid, butyl ester Butyl benzoate	$C_{11}H_{14}O_2$ 178.23	136-60-7 -22.4	1552 250.3	4-09-00-00290 1.000[20]	H_2O 1; EtOH 5; eth 5; ace 3 1.4940[25]
2533	Benzoic acid, 4,4'-carbonylbis-	$C_{15}H_{10}O_5$ 270.24	964-68-1 365	sub	4-10-00-03497	HOAc 4
2534	Benzoic acid, 2-chloro- o-Chlorobenzoic acid	$C_7H_5ClO_2$ 156.57	118-91-2 140.2	2125 sub	4-09-00-00956 1.544[20]	H_2O 3; EtOH 4; eth 4; ace 4
2535	Benzoic acid, 3-chloro- m-Chlorobenzoic acid	$C_7H_5ClO_2$ 156.57	535-80-8 158	2124 sub	4-09-00-00969 1.496[25]	H_2O 4; EtOH 3; eth 4; bz 2
2536	Benzoic acid, 4-chloro- p-Chlorobenzoic acid	$C_7H_5ClO_2$ 156.57	74-11-3 243	2126	4-09-00-00973	H_2O 1; EtOH 4; eth 2; ace 2
2537	Benzoic acid, 2-chloro-, ethyl ester	$C_9H_9ClO_2$ 184.62	7335-25-3 	243	4-09-00-00957 1.1942[15]	eth 4; EtOH 4 1.5242[15]
2538	Benzoic acid, 3-chloro-, ethyl ester	$C_9H_9ClO_2$ 184.62	1128-76-3 	243	4-09-00-00970 1.1859[15]	H_2O 1; EtOH 3; eth 3 1.5223[20]
2539	Benzoic acid, 4-chloro-, ethyl ester	$C_9H_9ClO_2$ 184.62	7335-27-5 	237.5	4-09-00-00975 1.1873[14]	EtOH 4
2540	Benzoic acid, 3-chloro-4-hydroxy-	$C_7H_5ClO_3$ 172.57	3964-58-7 171	sub	4-10-00-00463	H_2O 2; EtOH 4; eth 4; ace 4

No.	Name Synonym	Mol. Form. Mol. Wt.	CAS RN mp/°C	Merck No. bp/°C	Beil. Ref. den/g cm⁻³	Solubility n_D
2541	Benzoic acid, 3-chloro-2-hydroxy-, ethyl ester Salicylic acid, 3-chloro, ethyl ester	$C_9H_9ClO_3$ 200.62	56961-32-1 21	269.5	0-10-00-00101	EtOH 4
2542	Benzoic acid, 5-chloro-2-hydroxy-, methyl ester	$C_8H_7ClO_3$ 186.59	4068-78-4 50	249 dec; 120^{12}	4-10-00-00209	EtOH 4
2543	Benzoic acid, 4-chloro-, methyl ester	$C_8H_7ClO_2$ 170.60	1126-46-1 43.5		4-09-00-00974 1.382^{20}	EtOH 4
2544	Benzoic acid, 2-chloro-3-nitro- 2-Chloro-3-nitrobenzoic acid	$C_7H_4ClNO_4$ 201.57	3970-35-2 183.5		4-09-00-01225 1.662^{18}	H_2O 3; EtOH 3
2545	Benzoic acid, 2-chloro-5-nitro-	$C_7H_4ClNO_4$ 201.57	2516-96-3 166.5		4-09-00-01228 1.608^{18}	H_2O 2; EtOH 3; eth 3; ace 2
2546	Benzoic acid, 3-chloro-2-nitro- 2-Nitro-3-chlorobenzoic acid	$C_7H_4ClNO_4$ 201.57	4771-47-5 238		4-09-00-01225 1.566^{18}	eth 4; EtOH 4
2547	Benzoic acid, 4-chloro-3-nitro-	$C_7H_4ClNO_4$ 201.57	96-99-1 182.8		4-09-00-01226 1.645^{18}	H_2O 1; EtOH 2; ace 2
2548	Benzoic acid, 4-chloro-3-nitro-, methyl ester Methyl 4-chloro-3-nitrobenzoate	$C_8H_6ClNO_4$ 215.59	14719-83-6 83		4-09-00-01226 1.522^{18}	EtOH 4
2549	Benzoic acid, 5-chloro-2-nitro-, methyl ester Methyl 5-chloro-2-nitrobenzoate	$C_8H_6ClNO_4$ 215.59	51282-49-6 48.5		0-09-00-00401 1.453^{18}	MeOH 4
2550	Benzoic acid, 3-cyano-	$C_8H_5NO_2$ 147.13	1877-72-1 219	sub	4-09-00-03296	H_2O 2; EtOH 3; eth 3
2551	Benzoic acid, cyclohexyl ester	$C_{13}H_{16}O_2$ 204.27	2412-73-9 <-10	285	4-09-00-00296 1.0429^{20}	H_2O 1; EtOH 3; eth 3 1.5200^{20}
2552	Benzoic acid, 2,3-diamino- 2,3-Diaminobenzoic acid	$C_7H_8N_2O_2$ 152.15	603-81-6 201	dec	3-14-00-01172	EtOH 4; HOAc 4
2553	Benzoic acid, 2,4-diamino- 4-Aminoanthranilic acid	$C_7H_8N_2O_2$ 152.15	611-03-0 140	200 dec	4-14-00-01299	EtOH 4; HOAc 4
2554	Benzoic acid, 3,5-diamino- 3,5-Diaminobenzoic acid	$C_7H_8N_2O_2$ 152.15	535-87-5 228	2955	4-14-00-01304	H_2O 2; EtOH 3; eth 4; tfa 2
2555	Benzoic acid, 2,4-dibromo- 2,4-Dibromobenzoic acid	$C_7H_4Br_2O_2$ 279.92	611-00-7 174	sub	4-09-00-01027	eth 4; EtOH 4
2556	Benzoic acid, 2,5-dibromo- 2,5-Dibromobenzoic acid	$C_7H_4Br_2O_2$ 279.92	610-71-9 157	sub	4-09-00-01027	eth 4; EtOH 4; chl 4
2557	Benzoic acid, 2,6-dibromo- 2,6-Dibromobenzoic acid	$C_7H_4Br_2O_2$ 279.92	601-84-3 150.5	209^{16}	4-09-00-01028	H_2O 3; EtOH 4; eth 4; ace 3
2558	Benzoic acid, 3,5-dibromo-2-hydroxy- 3,5-Dibromosalicylic acid	$C_7H_4Br_2O_3$ 295.91	3147-55-5 228	3012	4-10-00-00222	ace 3
2559	Benzoic acid, 3,5-dibromo-2-hydroxy-, methyl ester Methyl 3,5-dibromosalicylate	$C_8H_6Br_2O_3$ 309.94	21702-79-4 149	181^{12}	4-10-00-00223	eth 4
2560	Benzoic acid, 2,6-dibromo-3,4,5-trihydroxy- Dibromogallic acid	$C_7H_4Br_2O_5$ 327.91	602-92-6 150	3006	2-10-00-00347	H_2O 4; eth 4; EtOH 4
2561	Benzoic acid, 2,4-dichloro- 2,4-Dichlorobenzoic acid	$C_7H_4Cl_2O_2$ 191.01	50-84-0 164.2	sub	4-09-00-00998	H_2O 3; EtOH 3; eth 3; ace 2
2562	Benzoic acid, 2,5-dichloro- 2,5-Dichlorobenzoic acid	$C_7H_4Cl_2O_2$ 191.01	50-79-3 154.4	301	4-09-00-01005	H_2O 3; EtOH 3; eth 3
2563	Benzoic acid, 2,6-dichloro- 2,6-Dichlorobenzoic acid	$C_7H_4Cl_2O_2$ 191.01	50-30-6 144	sub	4-09-00-01005	H_2O 3; EtOH 3; eth 3; bz 3
2564	Benzoic acid, 3,5-dichloro- 3,5-Dichlorobenzoic acid	$C_7H_4Cl_2O_2$ 191.01	51-36-5 188	sub	4-09-00-01008	H_2O 2; EtOH 3; eth 3; lig2
2565	Benzoic acid, 4-[(dichloroamino)sulfonyl]- Halazone	$C_7H_5Cl_2NO_4S$ 270.09	80-13-7 195 dec	4503	2-11-00-00220	H_2O 2; chl 2; HOAc 4; peth 1
2566	Benzoic acid, 3,5-dichloro-2-hydroxy-	$C_7H_4Cl_2O_3$ 207.01	320-72-9 220.5	sub	4-10-00-00213	H_2O 2; EtOH 4; eth 3
2567	Benzoic acid, 3,6-dichloro-2-methoxy- Dicamba	$C_8H_6Cl_2O_3$ 221.04	1918-00-9 115	3026	1.57^{25}	
2568	Benzoic acid, 2,3-dihydroxy-	$C_7H_6O_4$ 154.12	303-38-8 204		4-10-00-01414 1.542^{20}	H_2O 3; EtOH 3; eth 3; ace 2
2569	Benzoic acid, 2,4-dihydroxy- β-Resorcylic acid	$C_7H_6O_4$ 154.12	89-86-1 225-7 dec	8160	4-10-00-01420	H_2O 3; EtOH 3; eth 3; bz 3
2570	Benzoic acid, 2,5-dihydroxy- Gentisic acid	$C_7H_6O_4$ 154.12	490-79-9 199.5	4290	4-10-00-01441	H_2O 4; EtOH 4; eth 4; ace 3
2571	Benzoic acid, 3,4-dihydroxy- Protocatechuic acid	$C_7H_6O_4$ 154.12	99-50-3 200-2 dec	7909	4-10-00-01459 1.524^4	H_2O 2; EtOH 3; eth 3; bz 1
2572	Benzoic acid, 2-(3,6-dihydroxy-9H-xanthen-9-yl)- Fluorescin	$C_{20}H_{14}O_5$ 334.33	518-44-5 126	4087	2-18-00-00307	H_2O 1; EtOH 3; eth 3; ace 3
2573	Benzoic acid, 2,4-dihydroxy-6-methyl- o-Orsellinic acid	$C_8H_8O_4$ 168.15	480-64-8 176 dec	6834	4-10-00-01526	EtOH 3; eth 3
2574	Benzoic acid, 2,4-dihydroxy-6-methyl-, ethyl ester	$C_{10}H_{12}O_4$ 196.20	2524-37-0 132	sub	4-10-00-01526	eth 4; EtOH 4
2575	Benzoic acid, 3,4-dihydroxy-5-[(3,4,5-trihydroxybenzoyl)oxy]- Digallic acid	$C_{14}H_{10}O_9$ 322.23	536-08-3 268-70 dec	3135	3-10-00-02086	ace 4; EtOH 4
2576	Benzoic acid, 3,4-dimethoxy- Veratric acid	$C_9H_{10}O_4$ 182.18	93-07-2 181	9854 sub	4-10-00-01460	H_2O 1; EtOH 4; eth 4; chl 2
2577	Benzoic acid, 3,5-dimethoxy- 3,5-Dimethoxybenzoic acid	$C_9H_{10}O_4$ 182.18	1132-21-4 185.5	sub	4-10-00-01501	eth 4; EtOH 4

No.	Name Synonym	Mol. Form. Mol. Wt.	CAS RN mp/°C	Merck No. bp/°C	Beil. Ref. den/g cm⁻³	Solubility n_D
2578	Benzoic acid, 3,4-dimethoxy-, ethyl ester	$C_{11}H_{14}O_4$ 210.23	3943-77-9 43.5	295.5	4-10-00-01469	eth 4; EtOH 4
2579	Benzoic acid, 3,4-dimethoxy-, methyl ester	$C_{10}H_{12}O_4$ 196.20	2150-38-1 60.8	283	4-10-00-01467	bz 4; eth 4; EtOH 4
2580	Benzoic acid, 2,4-dimethyl- 4-Carboxy-1,3-dimethylbenzene	$C_9H_{10}O_2$ 150.18	611-01-8 90	268	4-09-00-01801	H_2O 2; EtOH 3; ace 3; bz 3
2581	Benzoic acid, 2,5-dimethyl-	$C_9H_{10}O_2$ 150.18	610-72-0 132	sub	4-09-00-01802 1.069²¹	H_2O 1; EtOH 3; eth 3; ace 3
2582	Benzoic acid, 2,6-dimethyl- 2,6-Dimethylbenzoic acid	$C_9H_{10}O_2$ 150.18	632-46-2 116	274.5; 155[17]	4-09-00-01798	H_2O 2; EtOH 3; eth 3; lig 2
2583	Benzoic acid, 3,5-dimethyl- Mesitylenic acid	$C_9H_{10}O_2$ 150.18	499-06-9 171.1	sub	4-09-00-01806	H_2O 2; EtOH 4; eth 4
2584	Benzoic acid, 2-(dimethylamino)-	$C_9H_{11}NO_2$ 165.19	610-16-2 72	sub	4-14-00-01017	H_2O 4; eth 4; EtOH 4
2585	Benzoic acid, 4-(dimethylamino)- p-(Dimethylamino)benzoic acid	$C_9H_{11}NO_2$ 165.19	619-84-1 242.5	3221	4-14-00-01164	EtOH 3; eth 2
2586	Benzoic acid, 6-[[6-[2-(dimethylamino)ethyl]-4-methoxy-1,3-benzodioxol-5-yl]acetyl]-2,3-dimethoxy- Narceine	$C_{23}H_{27}NO_8$ 445.47	131-28-2 138	6341	5-19-09-00051	
2587	Benzoic acid, 2-[[4-(dimethylamino)phenyl]azo]- Methyl red	$C_{15}H_{15}N_3O_2$ 269.30	493-52-7 183	6037	4-16-00-00504	H_2O 2; EtOH 3; ace 4; bz 4
2588	Benzoic acid, 4-(1,1-dimethylethyl)- p-tert-Butylbenzoic acid	$C_{11}H_{14}O_2$ 178.23	98-73-7 164.5		4-09-00-01884	H_2O 1; EtOH 4; bz 4; chl 3
2589	Benzoic acid, 2,4-dimethyl-, methyl ester Methyl 2,4-dimethylbenzoate	$C_{10}H_{12}O_2$ 164.20	23617-71-2 -2	232.5	2-09-00-00350	1.5052²⁰
2590	Benzoic acid, 3,4-dinitro- 3,4-Dinitrobenzoic acid	$C_7H_4N_2O_6$ 212.12	528-45-0 166	3268	4-09-00-01242	H_2O 2; EtOH 4; eth 4
2591	Benzoic acid, 3,5-dinitro- 3,5-Dinitrobenzoic acid	$C_7H_4N_2O_6$ 212.12	99-34-3 205	3269	4-09-00-01242	EtOH 4; HOAc 4
2592	Benzoic acid, 3,5-dinitro-, butyl ester Butyl 3,5-dinitrobenzoate	$C_{11}H_{12}N_2O_6$ 268.23	10478-02-1 62.5		4-09-00-01246	EtOH 3; DMSO-d₆ 2 1.488
2593	Benzoic acid, 3,5-dinitro-, ethyl ester Ethyl 3,5-dinitrobenzoate	$C_9H_8N_2O_6$ 240.17	618-71-3 94		4-09-00-01243 1.295¹¹¹	EtOH 3 1.560
2594	Benzoic acid, 3,3'-[(1,6-dioxo-1,6-hexanediyl)diimino]bis[2,4,6-triiodo- Iodipamide	$C_{20}H_{14}I_6N_2O_6$ 1139.77	606-17-7 307 dec	4916	4-14-00-01122	H_2O 1; EtOH 2; eth 2; ace 2
2595	Benzoic acid, 4-[(dipropylamino)sulfonyl]- Probenecid	$C_{13}H_{19}NO_4S$ 285.36	57-66-9 195	7760	4-11-00-00691	
2596	Benzoic acid, 2-ethoxy-	$C_9H_{10}O_3$ 166.18	134-11-2 20.7	211-2³⁹	3-10-00-00098	H_2O 2; EtOH 2; ctc 2
2597	Benzoic acid, 3-ethoxy-	$C_9H_{10}O_3$ 166.18	621-51-2 137	sub	3-10-00-00245	H_2O 3; EtOH 3; eth 3; bz 3
2598	Benzoic acid, 2-ethoxy-, ethyl ester Ethyl 2-ethoxybenzoate	$C_{11}H_{14}O_3$ 194.23	6290-24-0	251	3-10-00-00117 1.1005²⁰	eth 4; EtOH 4
2599	Benzoic acid, 4-ethoxy-, ethyl ester Ethyl 4-ethoxybenzoate	$C_{11}H_{14}O_3$ 194.23	23676-09-7	275	4-10-00-00368 1.076¹²	H_2O 1; EtOH 4; eth 4; ctc 2
2600	Benzoic acid, 2-ethyl-	$C_9H_{10}O_2$ 150.18	612-19-1 68	259	4-09-00-01789 1.0431¹⁰⁰	eth 4; EtOH 4 1.5099¹⁰⁰
2601	Benzoic acid, 3-ethyl-	$C_9H_{10}O_2$ 150.18	619-20-5 47		4-09-00-01791 1.042¹⁰⁰	eth 4; EtOH 4 1.5345¹⁰⁰
2602	Benzoic acid, ethyl ester Ethyl benzoate	$C_9H_{10}O_2$ 150.18	93-89-0 -34	3725 212	4-09-00-00285 1.0511¹⁵	H_2O 1; EtOH 3; eth 5; ace 3 1.5007²⁰
2603	Benzoic acid, 2-fluoro-	$C_7H_5FO_2$ 140.11	445-29-4 126.5		4-09-00-00950 1.460²⁵	H_2O 2; EtOH 4; eth 4; bz 1
2604	Benzoic acid, 3-fluoro-	$C_7H_5FO_2$ 140.11	455-38-9 124		4-09-00-00952 1.474²⁵	H_2O 2; eth 3
2605	Benzoic acid, 4-fluoro- p-Fluorobenzoic acid	$C_7H_5FO_2$ 140.11	456-22-4 185	4100	4-09-00-00953 1.479²⁵	H_2O 2; EtOH 3; eth 3; ace 2
2606	Benzoic acid, 4-fluoro-, ethyl ester	$C_9H_9FO_2$ 168.17	451-46-7 26	210	4-09-00-00953 1.146²⁵	eth 4; EtOH 4 1.4864²⁰
2607	Benzoic acid, 2-formyl-	$C_8H_6O_3$ 150.13	119-67-5 98		4-10-00-02748 1.404²⁵	H_2O 3; EtOH 4; eth 4
2608	Benzoic acid, 6-formyl-2,3-dimethoxy- Opianic acid	$C_{10}H_{10}O_5$ 210.19	519-05-1 150	6806	4-10-00-03863	EtOH 3; eth 3
2609	Benzoic acid, 3-formyl-4-hydroxy-	$C_8H_6O_4$ 166.13	584-87-2 244	sub	4-10-00-03622	eth 4; EtOH 4
2610	Benzoic acid, hexyl ester	$C_{13}H_{18}O_2$ 206.28	6789-88-4	272; 139⁸	4-09-00-00293 0.9793²⁰	H_2O 1; EtOH 3; ace 3
2611	Benzoic acid, hydrazide	$C_7H_8N_2O$ 136.15	613-94-5 115	267 dec	4-09-00-00922	H_2O 3; EtOH 3; eth 2; ace 2
2612	Benzoic acid, 2-hydroxy- Salicylic acid	$C_7H_6O_3$ 138.12	69-72-7 158	8301 211²⁰	4-10-00-00125 1.443²⁰	H_2O 2; EtOH 4; eth 4; ace 4 1.565
2613	Benzoic acid, 4-hydroxy- p-Salicylic acid	$C_7H_6O_3$ 138.12	99-96-7 214.5	4742	4-10-00-00345 1.46²⁵	H_2O 2; EtOH 4; eth 3; ace 3
2614	Benzoic acid, 2-hydroxy-, butyl ester	$C_{11}H_{14}O_3$ 194.23	2052-14-4 -5.9	271	4-10-00-00153 1.0728²⁰	ctc 2 1.5115²⁰
2615	Benzoic acid, 4-hydroxy-, butyl ester Butylparaben	$C_{11}H_{14}O_3$ 194.23	94-26-8 68.5	1583	4-10-00-00375	H_2O 2; EtOH 2; ctc 2

No.	Name Synonym	Mol. Form. Mol. Wt.	CAS RN mp/°C	Merck No. bp/°C	Beil. Ref. den/g cm^{-3}	Solubility n_D
2616	Benzoic acid, 2-hydroxy-, 2-carboxyphenyl ester Salsalate	$C_{14}H_{10}O_5$ 258.23	552-94-3 147	8307	4-10-00-00165	ace 2
2617	Benzoic acid, 2-hydroxy-3,5-diiodo- 3,5-Diiodosalicylic acid	$C_7H_4I_2O_3$ 389.92	133-91-5 235.5	3175	4-10-00-00226	H_2O 2; EtOH 4; eth 4; bz 1
2618	Benzoic acid, 4-hydroxy-3,5-diiodo-	$C_7H_4I_2O_3$ 389.92	618-76-8 237	260 dec	4-10-00-00479	H_2O 1; EtOH 4; eth 4; bz 2
2619	Benzoic acid, 2-hydroxy-, ethyl ester Ethyl salicylate	$C_9H_{10}O_3$ 166.18	118-61-6 45	3804 150-1[10]	4-10-00-00149 1.1326[20]	H_2O 1; EtOH 5; eth 4; ctc 3 1.5296[20]
2620	Benzoic acid, 3-hydroxy-, ethyl ester	$C_9H_{10}O_3$ 166.18	7781-98-8 74		4-10-00-00321 1.0680[131]	H_2O 2; EtOH 3; eth 3; chl 2
2621	Benzoic acid, 4-hydroxy-, ethyl ester Ethylparaben	$C_9H_{10}O_3$ 166.18	120-47-8 117	3792 297.5	4-10-00-00367	H_2O 2; EtOH 4; eth 4; chl 2
2622	Benzoic acid, 2-hydroxy-, 2-hydroxyethyl ester Glycol salicylate	$C_9H_{10}O_4$ 182.18	87-28-5 37	4395 173[15]	3-10-00-00138 1.2526[15]	H_2O 2; EtOH 4; eth 4; bz 4
2623	Benzoic acid, 4-hydroxy-3-iodo- 4-Hydroxybenzoic acid, 3-iodo	$C_7H_5IO_3$ 264.02	37470-46-5 174.3	sub	4-10-00-00478	eth 4; EtOH 4
2624	Benzoic acid, 5-hydroxy-2-iodo- 3-Hydroxybenzoic acid, 6-iodo	$C_7H_5IO_3$ 264.02	57772-57-3 198	sub 160	3-10-00-00261	eth 4; EtOH 4
2625	Benzoic acid, 4-hydroxy-3-methoxy- Vanillic acid	$C_8H_8O_4$ 168.15	121-34-6 211.5	9838 sub	4-10-01459	H_2O 2; EtOH 4; eth 3; DMSO 3
2626	Benzoic acid, 4-hydroxy-3-methoxy-, ethyl ester	$C_{10}H_{12}O_4$ 196.20	617-05-0 44	292	3-10-00-01413	H_2O 1; EtOH 4; eth 4; chl 3
2627	Benzoic acid, 4-hydroxy-3-methoxy-, methyl ester	$C_9H_{10}O_4$ 182.18	3943-74-6 64	286	3-10-00-01410	EtOH 3; chl 2; peth 3
2628	Benzoic acid, 4-hydroxy-3-methoxy-, 2-methylpropyl ester	$C_{12}H_{16}O_4$ 224.26	7152-88-7 56.5	125[2]	3-10-00-01415	chl 2
2629	Benzoic acid, 2-hydroxy-3-methyl- o-Cresotic acid	$C_8H_8O_3$ 152.15	83-40-9 165.5	2585	4-10-00-00601	H_2O 2; EtOH 3; eth 3; bz 3
2630	Benzoic acid, 2-hydroxy-4-methyl- m-Cresotic acid	$C_8H_8O_3$ 152.15	50-85-1 177	2584	4-10-00-00617	H_2O 2; EtOH 3; eth 4; bz 3
2631	Benzoic acid, 2-hydroxy-5-methyl- p-Cresotic acid	$C_8H_8O_3$ 152.15	89-56-5 151	2586	4-10-00-00610	H_2O 2; EtOH 3; eth 3; bz 3
2632	Benzoic acid, 3-hydroxy-5-methyl-	$C_8H_8O_3$ 152.15	585-81-9 210	sub	4-10-00-00609	H_2O 3; EtOH 4; eth 4; chl 2
2633	Benzoic acid, 2-hydroxy-, 3-methylbutyl ester Isopentyl salicylate	$C_{12}H_{16}O_3$ 208.26	87-20-7 278; 151[15]	5007	4-10-00-00153 1.0535[20]	H_2O 1; EtOH 4; eth 3; ctc 2 1.5080[20]
2634	Benzoic acid, 2-hydroxy-, methyl ester Methyl salicylate	$C_8H_8O_3$ 152.15	119-36-8 -8	6038 222.9	4-10-00-00143 1.181[25]	eth 4; EtOH 4; chl 4 1.535[25]
2635	Benzoic acid, 2-hydroxy-, 1-methylethyl ester Isopropyl salicylate	$C_{10}H_{12}O_3$ 180.20	607-85-2 	238	4-10-00-00152 1.0729[20]	H_2O 1; EtOH 5; eth 5 1.5065[20]
2636	Benzoic acid, 3-hydroxy-, methyl ester	$C_8H_8O_3$ 152.15	19438-10-9 73	281; 178[17]	4-10-00-00319 1.1528[100]	EtOH 3; bz 3; chl 2; peth 3
2637	Benzoic acid, 4-hydroxy-, methyl ester Methylparaben	$C_8H_8O_3$ 152.15	99-76-3 131	6021 275 dec	4-10-00-00360	H_2O 2; EtOH 4; eth 4; ace 4
2638	Benzoic acid, 2-hydroxy-3-methyl-, methyl ester	$C_9H_{10}O_3$ 166.18	23287-26-5 29	235	4-10-00-00602 1.1683[25]	1.5354[16]
2639	Benzoic acid, 2-hydroxy-4-methyl-, methyl ester	$C_9H_{10}O_3$ 166.18	4670-56-8 27.5	243	2-10-00-00139 1.1483[15]	1.5378[15]
2640	Benzoic acid, 2-hydroxy-5-methyl-, methyl ester	$C_9H_{10}O_3$ 166.18	22717-57-3 -1	244.5	4-10-00-00611 1.1673[25]	1.5351[15]
2641	Benzoic acid, 2-hydroxy-3-methyl-6-(1-methylethyl)-	$C_{11}H_{14}O_3$ 194.23	4389-53-1 136	sub	4-10-00-00738	eth 4; EtOH 4
2642	Benzoic acid, 2-hydroxy-6-methyl-3-(1-methylethyl)- o-Thymotic acid	$C_{11}H_{14}O_3$ 194.23	548-51-6 127	9342 sub	3-10-00-00629	bz 4; eth 4; EtOH 4
2643	Benzoic acid, 2-hydroxy-, 2-methylpropyl ester Isobutyl salicylate	$C_{11}H_{14}O_3$ 194.23	87-19-4 5.9	261	3-10-00-00121 1.0639[20]	H_2O 1; EtOH 3; eth 3; ctc 3 1.5087[20]
2644	Benzoic acid, 2-hydroxy-, 1-naphthalenyl ester 1-Naphthyl salicylate	$C_{17}H_{12}O_3$ 264.28	550-97-0 83	6334	2-10-00-00052	eth 4
2645	Benzoic acid, 2-hydroxy-, 2-naphthalenyl ester 2-Naphthyl salicylate	$C_{17}H_{12}O_3$ 264.28	613-78-5 95.5	6335	4-10-00-00158 1.11[116]	H_2O 1; EtOH 2; eth 3; bz 3
2646	Benzoic acid, 2-hydroxy-3-nitro- 3-Nitrosalicylic acid	$C_7H_5NO_5$ 183.12	85-38-1 148	6553	4-10-00-00228	H_2O 2; EtOH 4; eth 4; ace 3
2647	Benzoic acid, 2-hydroxy-5-nitro- 5-Nitrosalicylic acid	$C_7H_5NO_5$ 183.12	96-97-9 229.5	6554	4-10-00-00255 1.650[20]	H_2O 2; EtOH 4; eth 4; ace 4
2648	Benzoic acid, 2-hydroxy-5-[(4-nitrophenyl)azo]- Alizarine Yellow R	$C_{13}H_9N_3O_5$ 287.23	2243-76-7 253-4 dec	241	4-16-00-00372	H_2O 4; EtOH 4
2649	Benzoic acid, 2-hydroxy-, pentyl ester	$C_{12}H_{16}O_3$ 208.26	2050-08-0 	270	4-10-00-00153 1.064[15]	H_2O 2; EtOH 5; eth 5 1.506[20]
2650	Benzoic acid, 2-hydroxy-, phenyl ester Phenyl salicylate	$C_{13}H_{10}O_3$ 214.22	118-55-8 130.5	7282 173[12]	4-10-00-00154 1.2614[30]	H_2O 1; EtOH 4; eth 3; ace 4

No.	Name / Synonym	Mol. Form. / Mol. Wt.	CAS RN / mp/°C	Merck No. / bp/°C	Beil. Ref. / den/g cm⁻³	Solubility / n_D
2651	Benzoic acid, 2-hydroxy-, phenylmethyl ester / Benzyl salicylate	$C_{14}H_{12}O_3$ / 228.25	118-58-1 / 1160	320	4-10-00-00157 / 1.1799[20]	H_2O 3; EtOH 3; eth 3; ctc 3 / 1.5805[20]
2652	Benzoic acid, 2-hydroxy-, propyl ester	$C_{10}H_{12}O_3$ / 180.20	607-90-9 / 97	239	4-10-00-00152 / 1.0979[20]	ctc 3; CS_2 3 / 1.5161[20]
2653	Benzoic acid, 4-hydroxy-, propyl ester / Propylparaben	$C_{10}H_{12}O_3$ / 180.20	94-13-3 / 97	7879	4-10-00-00374 / 1.0630[102]	H_2O 1; EtOH 3; eth 3; chl 2 / 1.5050[102]
2654	Benzoic acid, 2-hydroxy-5-[[4-[(2-pyridinylamino)sulfonyl]phenyl]azo]- / Sulfasalazine	$C_{18}H_{14}N_4O_5S$ / 398.40	599-79-1 / 220 dec	8917	5-22-08-00433	
2655	Benzoic acid, 2-hydroxy-5-sulfo- / Sulfosalicylic acid	$C_7H_6O_6S$ / 218.19	97-05-2 / 120	8944	4-11-00-00702	H_2O 5; EtOH 5; eth 4
2656	Benzoic acid, 2,2'-iminobis- / Diphenylamine-2,2'-dicarboxylic acid	$C_{14}H_{11}NO_4$ / 257.25	579-92-0 / 296-7 dec	3318	4-14-00-01058	
2657	Benzoic acid, 2-iodo- / o-Iodobenzoic acid	$C_7H_5IO_2$ / 248.02	88-67-5 / 163	4923 / exp 233	4-09-00-01030 / 2.25[25]	H_2O 2; EtOH 4; eth 4; ace 2
2658	Benzoic acid, 3-iodo-	$C_7H_5IO_2$ / 248.02	618-51-9 / 188.3	sub	4-09-00-01033	H_2O 2; EtOH 4; eth 2
2659	Benzoic acid, 4-iodo-	$C_7H_5IO_2$ / 248.02	619-58-9 / 270	sub	4-09-00-01035 / 2.184[20]	H_2O 2; EtOH 2; eth 3
2660	Benzoic acid, 2-iodo-, methyl ester	$C_8H_7IO_2$ / 262.05	610-97-9	280; 146[16]	4-09-00-01031	EtOH 3 / 1.6052[20]
2661	Benzoic acid, 3-iodo-, methyl ester	$C_8H_7IO_2$ / 262.05	618-91-7 / 54.5	277; 150[18]	4-09-00-01034	H_2O 1; EtOH 3; eth 4; ace 4
2662	Benzoic acid, 4-iodo-, methyl ester	$C_8H_7IO_2$ / 262.05	619-44-3 / 114.8	sub	4-09-00-01036 / 2.0200[10]	EtOH 3; eth 3
2663	Benzoic acid, 2-mercapto- / o-Thiosalicylic acid	$C_7H_6O_2S$ / 154.19	147-93-3 / 168.5	9291 / sub	4-10-00-00272	H_2O 3; EtOH 3; eth 3; lig 2
2664	Benzoic acid, 2-methoxy-	$C_8H_8O_3$ / 152.15	579-75-9 / 101	200	4-10-00-00130	H_2O 2; EtOH 4; eth 4; bz 3
2665	Benzoic acid, 3-methoxy-	$C_8H_8O_3$ / 152.15	586-38-9 / 107	170[10]	4-10-00-00316	H_2O 2; EtOH 3; eth 3; bz 3
2666	Benzoic acid, 4-methoxy- / p-Anisic acid	$C_8H_8O_3$ / 152.15	100-09-4 / 185	696 / 276.5	4-10-00-00346	H_2O 1; EtOH 4; eth 4; chl 3
2667	Benzoic acid, 2-(4-methoxybenzoyl)- / o-(p-Anisoyl)benzoic acid	$C_{15}H_{12}O_4$ / 256.26	1151-15-1 / 146	702	3-10-00-04429	eth 4; EtOH 4; tol 4
2668	Benzoic acid, 4-methoxy-, butyl ester	$C_{12}H_{16}O_3$ / 208.26	6946-35-6	183[40]	3-10-00-00307 / 1.054[16]	/ 1.5141[16]
2669	Benzoic acid, 2-methoxy-, ethyl ester	$C_{10}H_{12}O_3$ / 180.20	7335-26-4	261	4-10-00-00150 / 1.1124[20]	eth 4; EtOH 4 / 1.5224[20]
2670	Benzoic acid, 3-methoxy-, ethyl ester	$C_{10}H_{12}O_3$ / 180.20	10259-22-0	260.5	4-10-00-00321 / 1.0993[20]	eth 4; EtOH 4 / 1.5161[20]
2671	Benzoic acid, 4-methoxy-, ethyl ester	$C_{10}H_{12}O_3$ / 180.20	94-30-4 / 7.5	269.5	4-10-00-00368 / 1.1038[20]	H_2O 1; EtOH 3; eth 3 / 1.5254[20]
2672	Benzoic acid, 3-methoxy-2-(methylamino)-, methyl ester / Damascenine	$C_{10}H_{13}NO_3$ / 195.22	483-64-7 / 28	2809 / 271; 147[10]	4-14-00-02073	bz 4; eth 4; EtOH 4; lig 4
2673	Benzoic acid, 2-methoxy-, methyl ester	$C_9H_{10}O_3$ / 166.18	606-45-1	246.5	4-10-00-00144 / 1.1571[19]	H_2O 1; EtOH 3 / 1.534[19]
2674	Benzoic acid, 3-methoxy-, methyl ester	$C_9H_{10}O_3$ / 166.18	5368-81-0	248	4-10-00-00319 / 1.1310[20]	H_2O 1; EtOH 3 / 1.5224[20]
2675	Benzoic acid, 4-methoxy-, methyl ester	$C_9H_{10}O_3$ / 166.18	121-98-2 / 49	244	4-10-00-00360	H_2O 1; EtOH 3; eth 3; chl 3
2676	Benzoic acid, 2-methyl- / o-Toluic acid	$C_8H_8O_2$ / 136.15	118-90-1 / 103.7	9461 / 259	4-09-00-01697 / 1.062[115]	H_2O 1; EtOH 4; eth 4; chl 3 / 1.512[115]
2677	Benzoic acid, 3-methyl- / m-Toluic acid	$C_8H_8O_2$ / 136.15	99-04-7 / 108.7	9461	4-09-00-01712 / 1.054[112]	H_2O 2; EtOH 4; eth 4; chl 2 / 1.509
2678	Benzoic acid, 4-methyl- / p-Toluic acid	$C_8H_8O_2$ / 136.15	99-94-5 / 179.6	9461	4-09-00-01724	H_2O 1; EtOH 4; eth 4; tfa 2
2679	Benzoic acid, 2-(methylamino)-	$C_8H_9NO_2$ / 151.17	119-68-6 / 180.5	80[0.01]	4-14-00-01015	H_2O 2; EtOH 4; eth 4; bz 4
2680	Benzoic acid, 2-(methylamino)-, ethyl ester / Anthranilic acid, N-methyl-, ethyl ester	$C_{10}H_{13}NO_2$ / 179.22	35472-56-1 / 39	266; 172[45]	2-14-00-00213	eth 4
2681	Benzoic acid, 2-(methylamino)-, methyl ester	$C_9H_{11}NO_2$ / 165.19	85-91-6 / 19	255	4-14-00-01016 / 1.120[15]	H_2O 1; EtOH 3; eth 3 / 1.5839[15]
2682	Benzoic acid, 2-methyl-, anhydride	$C_{16}H_{14}O_3$ / 254.29	607-86-3 / 38.5	>325; 220[11]	4-09-00-01700	eth 4
2683	Benzoic acid, 3-methyl-, anhydride	$C_{16}H_{14}O_3$ / 254.29	21436-44-2 / 71	230[17]	4-09-00-01715	ace 4; bz 4; eth 4; EtOH 4
2684	Benzoic acid, 2-(4-methylbenzoyl)- / 2-(p-Toluyl)benzoic acid	$C_{15}H_{12}O_3$ / 240.26	85-55-2 / 146	9465	4-10-00-02991	H_2O 2; EtOH 4; eth 4; ace 4
2685	Benzoic acid, 3-methyl-, butyl ester / Butyl m-toluate	$C_{12}H_{16}O_2$ / 192.26	6640-77-3	262.3	3-09-00-00397 / 1.0040[20]	H_2O 1; EtOH 3; eth 5 / 1.4950[20]
2686	Benzoic acid, 3,3'-methylenebis[6-hydroxy- / 5,5'-Methylenedisalicylic acid	$C_{15}H_{12}O_6$ / 288.26	122-25-8 / 243.5	5984	4-10-00-02337	ace 4; eth 4; EtOH 4
2687	Benzoic acid, methyl ester / Methyl benzoate	$C_8H_8O_2$ / 136.15	93-58-3 / -15	5947 / 199	4-09-00-00283 / 1.0933[15]	H_2O 1; EtOH 3; eth 5; ctc 3 / 1.5164[20]
2688	Benzoic acid, 2-(1-methylethyl)-	$C_{10}H_{12}O_2$ / 164.20	2438-04-2 / 64	160-1[25]	4-09-00-01839	H_2O 2; EtOH 4; eth 4; bz 3
2689	Benzoic acid, 4-(1-methylethyl)- / Cumic acid	$C_{10}H_{12}O_2$ / 164.20	536-66-3 / 117.5	2620 / sub	4-09-00-01843 / 1.162[4]	H_2O 2; EtOH 4; eth 4; peth 3

No.	Name / Synonym	Mol. Form. / Mol. Wt.	CAS RN / mp/°C	Merck No. / bp/°C	Beil. Ref. / den/g cm^{-3}	Solubility / n_D
2690	Benzoic acid, 2-methyl-, ethyl ester	$C_{10}H_{12}O_2$ 164.20	87-24-1 <-10	227; 113[18]	4-09-00-01699 1.0325[21]	H_2O 1; EtOH 5; eth 5 1.507[22]
2091	Benzoic acid, 3-methyl-, ethyl ester	$C_{10}H_{12}O_2$ 164.20	120-33-2	234	4-09-00-01713 1.0265[21]	eth 4; EtOH 4 1.5052[22]
2692	Benzoic acid, 4-methyl-, ethyl ester	$C_{10}H_{12}O_2$ 164.20	94-08-6	232	4-09-00-01726 1.0269[18]	H_2O 1; EtOH 5; eth 5 1.5089[18]
2693	Benzoic acid, 1-methylethyl ester	$C_{10}H_{12}O_2$ 164.20	939-48-0	216	4-09-00-00289 1.0163[15]	H_2O 1; EtOH 3; eth 3; ace 3 1.4890[20]
2694	Benzoic acid, 4-(1-methylethyl)-, methyl ester	$C_{11}H_{14}O_2$ 178.23	20185-55-1	126[14]	4-09-00-01843 1.018[19]	1.515[19]
2695	Benzoic acid, 4-(1-methylethyl)-, 2-methylpropyl ester / Benzoic acid, 4-isopropyl, isobutyl ester	$C_{14}H_{20}O_2$ 220.31	6315-03-3	155[14]	2-09-00-00359 0.966[19]	1.497[19]
2696	Benzoic acid, 2-methyl-, methyl ester	$C_9H_{10}O_2$ 150.18	89-71-4 <-50	215	4-09-00-01699 1.068[20]	H_2O 1; EtOH 5; eth 5
2697	Benzoic acid, 3-methyl-, methyl ester	$C_9H_{10}O_2$ 150.18	99-36-5	221	4-09-00-01713 1.061[20]	H_2O 1; EtOH 3; ctc 2
2698	Benzoic acid, 4-methyl-, methyl ester	$C_9H_{10}O_2$ 150.18	99-75-2 33.2	220	4-09-00-01726	H_2O 1; EtOH 4; eth 4
2699	Benzoic acid, 2-methylphenyl ester	$C_{14}H_{12}O_2$ 212.25	617-02-7	309; 155[8.5]	4-09-00-00305 1.114[19]	eth 4; EtOH 4
2700	Benzoic acid, 4-methylphenyl ester	$C_{14}H_{12}O_2$ 212.25	614-34-6 71.5	316	4-09-00-00306	eth 4; EtOH 4
2701	Benzoic acid, 2-methylpropyl ester	$C_{11}H_{14}O_2$ 178.23	120-50-3	242	4-09-00-00291 0.9990[20]	H_2O 1; EtOH 5; eth 5; ace 3
2702	Benzoic acid, 2-[(1-naphthalenylamino)carbonyl]- / Naptalam	$C_{18}H_{13}NO_3$ 291.31	132-66-1 185	6338	3-12-00-02876 1.4[20]	H_2O 1; EtOH 2; ace 2; bz 2
2703	Benzoic acid, 2-nitro- / o-Nitrobenzoic acid	$C_7H_5NO_4$ 167.12	552-16-9 147.5		4-09-00-01046 1.575[20]	H_2O 3; EtOH 4; eth 3; ace 4
2704	Benzoic acid, 3-nitro- / m-Nitrobenzoic acid	$C_7H_5NO_4$ 167.12	121-92-6 141		4-09-00-01055 1.494[20]	H_2O 2; EtOH 4; eth 4; ace 4
2705	Benzoic acid, 4-nitro- / p-Nitrobenzoic acid	$C_7H_5NO_4$ 167.12	62-23-7 242	sub	4-09-00-01072 1.610[20]	ace 4; eth 4; EtOH 4; chl 4
2706	Benzoic acid, 4-nitro-, butyl ester	$C_{11}H_{13}NO_4$ 223.23	120-48-9 35.3	160[8]	3-09-00-01544	bz 4; eth 4
2707	Benzoic acid, 2-nitro-, ethyl ester	$C_9H_9NO_4$ 195.17	610-34-4 30	275; 178[23]	4-09-00-01047	eth 4; EtOH 4
2708	Benzoic acid, 3-nitro-, ethyl ester	$C_9H_9NO_4$ 195.17	618-98-4 47	297	4-09-00-01056	H_2O 1; EtOH 4; eth 4
2709	Benzoic acid, 4-nitro-, ethyl ester	$C_9H_9NO_4$ 195.17	99-77-4 57	186.3; 153[8]	4-09-00-01074	H_2O 1; EtOH 3; eth 3
2710	Benzoic acid, 2-nitro-, methyl ester	$C_8H_7NO_4$ 181.15	606-27-9 -13	275	4-09-00-01047 1.2855[20]	H_2O 1; EtOH 3; eth 3; bz 3
2711	Benzoic acid, 3-nitro-, methyl ester	$C_8H_7NO_4$ 181.15	618-95-1 78	279[60]	4-09-00-01056	H_2O 1; EtOH 2; eth 2; MeOH 2
2712	Benzoic acid, pentachloro- / Perchlorobenzoic acid	$C_7HCl_5O_2$ 294.35	1012-84-6 209.3	sub	4-09-00-01011	EtOH 4; tol 4; HOAc 4
2713	Benzoic acid, pentamethyl- / Permethylbenzoic acid	$C_{12}H_{16}O_2$ 192.26	2243-32-5 210.5	sub	4-09-00-01920	H_2O 1; EtOH 4
2714	Benzoic acid, 2-phenoxy-	$C_{13}H_{10}O_3$ 214.22	2243-42-7 113	355	4-10-00-00132 1.1553[50]	H_2O 1; EtOH 4; eth 4; chl 3
2715	Benzoic acid, 2-(phenylamino)- / N-Phenylanthranilic acid	$C_{13}H_{11}NO_2$ 213.24	91-40-7 183.5	7244	4-14-00-01019	H_2O 1; EtOH 4; eth 2; bz 2
2716	Benzoic acid, phenyl ester / Phenyl benzoate	$C_{13}H_{10}O_2$ 198.22	93-99-2 71	7246 314	4-09-00-00303 1.235[20]	H_2O 1; EtOH 3; eth 3; chl 3
2717	Benzoic acid, 2-phenylhydrazide	$C_{13}H_{12}N_2O$ 212.25	532-96-7 168	314	4-15-00-00169	H_2O 2; EtOH 3; eth 2; bz 3
2718	Benzoic acid, 2-(phenylmethyl)-	$C_{14}H_{12}O_2$ 212.25	612-35-1	sub	4-09-00-02516	H_2O 2; EtOH 3; eth 3; bz 3
2719	Benzoic acid, 4-(phenylmethyl)-	$C_{14}H_{12}O_2$ 212.25	620-86-0 158.8	sub	4-09-00-02518	bz 4; eth 4; EtOH 4; chl 4
2720	Benzoic acid, phenylmethyl ester / Benzyl benzoate	$C_{14}H_{12}O_2$ 212.25	120-51-4 21	1141 323.5	4-09-00-00307 1.1121[25]	H_2O 1; EtOH 3; eth 3; ace 3 1.5680[20]
2721	Benzoic acid, 4-[[(phenylmethyl)sulfonyl]amino]- / Carinamide	$C_{14}H_{13}NO_4S$ 291.33	536-95-8 229.5	1163	4-14-00-01264	EtOH 4
2722	Benzoic acid, 2-propenyl ester	$C_{10}H_{10}O_2$ 162.19	583-04-0		4-09-00-00296 1.0569[15]	H_2O 1; EtOH 3; eth 3; ace 3 1.5178[20]
2723	Benzoic acid, o-propyl- / Benzoic acid, 2-propyl	$C_{10}H_{12}O_2$ 164.20	2438-03-1 58	273; 164[20]	4-09-00-01836 1.0020[15]	eth 4; EtOH 4
2724	Benzoic acid, propyl ester	$C_{10}H_{12}O_2$ 164.20	2315-68-6 -51.6	211	4-09-00-00288 1.0230[20]	H_2O 1; EtOH 5; eth 5 1.5000[20]
2725	Benzoic acid, sodium salt	$C_7H_5NaO_2$ 144.11	532-32-1 >300	8527	4-09-00-00273	H_2O 3
2726	Benzoic acid, 2-[[[4-[(2-thiazolylamino)sulfonyl]phenyl]amino]carbonyl]- / Phthalylsulphathiazole	$C_{17}H_{13}N_3O_5S_2$ 403.44	85-73-4 273	7351	4-27-00-04639	H_2O 1; EtOH 2; eth 1; chl 1

No.	Name Synonym	Mol. Form. Mol. Wt.	CAS RN mp/°C	Merck No. bp/°C	Beil. Ref. den/g cm^{-3}	Solubility n_D
2727	Benzoic acid, 2,3,6-trichloro- 2,3,6-Trichlorobenzoic acid	$C_7H_3Cl_3O_2$ 225.46	50-31-7 124.5		4-09-00-01009	H_2O 2; eth 3
2728	Benzoic acid, 2,4,5-trichloro- 2,4,5-Trichlorobenzoic acid	$C_7H_3Cl_3O_2$ 225.46	50-82-8 166.5	sub	4-09-00-01010	H_2O 1; EtOH 3; eth 3
2729	Benzoic acid, 2-[[3-(trifluoromethyl)phenyl]amino]- Flufenamic acid	$C_{14}H_{10}F_3NO_2$ 281.23	530-78-9 133.5	4060	3-14-00-00905	DMSO 3
2730	Benzoic acid, 2,3,4-trihydroxy-	$C_7H_6O_5$ 170.12	610-02-6 221	sub	4-10-00-01971	H_2O 2; EtOH 3; eth 3; ace 3
2731	Benzoic acid, 3,4,5-trihydroxy- Gallic acid	$C_7H_6O_5$ 170.12	149-91-7 253 dec	4251	3-10-00-02070 1.694[6]	H_2O 2; EtOH 4; eth 2; ace 3
2732	Benzoic acid, 3,4,5-trihydroxy-, propyl ester Propyl gallate	$C_{10}H_{12}O_5$ 212.20	121-79-9 130	7872	4-10-00-02003	H_2O 2
2733	Benzoic acid, 2,4,5-trimethoxy-	$C_{10}H_{12}O_5$ 212.20	490-64-2 145	300	4-10-00-01980	H_2O 4; bz 4; EtOH 4; peth 4
2734	Benzoic acid, 3,4,5-trimethoxy-	$C_{10}H_{12}O_5$ 212.20	118-41-2 172.3	225-7[10]	4-10-00-01996	H_2O 2; EtOH 4; eth 4; chl 4
2735	Benzoic acid, 2,4,6-trinitro- 2,4,6-Trinitrobenzoic acid	$C_7H_3N_3O_8$ 257.12	129-66-8 228 dec	9640	4-09-00-01362	H_2O 2; EtOH 3; eth 3; ace 3
2736	1H,5H-Benzo[ij]quinolizine-1,6(7H)-dione, 2,3-dihydro- Julolidine, 1,6-dioxo	$C_{12}H_{11}NO_2$ 201.22	39052-57-8 145.5	190-210[0.3]	4-21-00-05525	EtOH 4
2737	1H,5H-Benzo[ij]quinolizine, 2,3,6,7-tetrahydro- Julolidine	$C_{12}H_{15}N$ 173.26	479-59-4 40	280 dec; 155[17]	5-20-07-00189 1.003[20]	1.568[25]
2738	Benzo[j]fluoranthene Dibenzo[a,jk]fluorene	$C_{20}H_{12}$ 252.32	205-82-3 166		4-05-00-02687	H_2O 1; EtOH 2; HOAc 2
2739	Benzo[k]fluoranthene 2,3,1',8'-Binaphthylene	$C_{20}H_{12}$ 252.32	207-08-9 217	480	4-05-00-02686	H_2O 1; EtOH 3; bz 3; HOAc 3
2740	Benzonitrile Phenyl cyanide	C_7H_5N 103.12	100-47-0 -12.7	1107 191.1	4-09-00-00892 1.0093[15]	H_2O 2; EtOH 5; eth 5; ace 4 1.5289[20]
2741	Benzonitrile, 2-amino-	$C_7H_6N_2$ 118.14	1885-29-6 51	263	4-14-00-01013	H_2O 2; EtOH 4; eth 4; ace 4
2742	Benzonitrile, 3-amino-	$C_7H_6N_2$ 118.14	2237-30-1 54.3	289	4-14-00-01095	H_2O 2; EtOH 4; eth 4; ace 4
2743	Benzonitrile, 5-amino-2-methyl-	$C_8H_8N_2$ 132.17	50670-64-9 88	100-10[22]	3-14-00-01201	EtOH 4
2744	Benzonitrile, 2-bromo-	C_7H_4BrN 182.02	2042-37-7 55.5	252	4-09-00-01013	H_2O 3; EtOH 4; chl 2
2745	Benzonitrile, 3-bromo-	C_7H_4BrN 182.02	6952-59-6 39.5	225	4-09-00-01016	EtOH 4; eth 4; chl 2
2746	Benzonitrile, 4-bromo-	C_7H_4BrN 182.02	623-00-7 114	236	4-09-00-01025	H_2O 3; EtOH 3; eth 3; chl 3
2747	Benzonitrile, 2-chloro-	C_7H_4ClN 137.57	873-32-5 46.3	232	4-09-00-00965	H_2O 3; EtOH 3; eth 3; chl 3
2748	Benzonitrile, 3-chloro-	C_7H_4ClN 137.57	766-84-7 41	100[15]	4-09-00-00973	H_2O 1; EtOH 3; eth 3
2749	Benzonitrile, 4-chloro-	C_7H_4ClN 137.57	623-03-0 95	223; 95[5]	4-09-00-00992 1.1133[17]	H_2O 2; EtOH 3; eth 3; bz 3
2750	Benzonitrile, 2-(chloromethyl)-	C_8H_6ClN 151.60	612-13-5 61	252	4-09-00-01707	EtOH 4
2751	Benzonitrile, 3,5-dibromo-4-hydroxy- Bromoxynil	$C_7H_3Br_2NO$ 276.91	1689-84-5 190	1431	4-10-00-00475	
2752	Benzonitrile, 2,6-dichloro- Dichlobenil	$C_7H_3Cl_2N$ 172.01	1194-65-6 144.5	3029 270	4-09-00-01006	
2753	Benzonitrile, 2,6-dimethoxy- 2,6-Dimethoxybenzonitrile	$C_9H_9NO_2$ 163.18	16932-49-3 119	310	2-10-00-00260	H_2O 2; EtOH 4; eth 2; ace 3
2754	Benzonitrile, 2-ethoxy- o-Ethoxybenzonitrile	C_9H_9NO 147.18	6609-57-0 5	260.5	0-10-00-00097 1.0650[15]	eth 4; EtOH 4; lig 4
2755	Benzonitrile, 4-ethoxy-	C_9H_9NO 147.18	25117-74-2 61.5	258	3-10-00-00344	eth 4; EtOH 4; lig 4
2756	Benzonitrile, 4-fluoro-	C_7H_4FN 121.11	1194-02-1 34.8	188.8	4-09-00-00954 1.1070[55]	chl 2; peth 3 1.4925[55]
2757	Benzonitrile, 3-formyl-	C_8H_5NO 131.13	24964-64-5 76.5	210	4-10-00-02752	H_2O 4; EtOH 4; eth 4; chl 4
2758	Benzonitrile, 4-formyl-	C_8H_5NO 131.13	105-07-7 100.5	133[12]	4-10-00-02754	H_2O 3; EtOH 4; eth 4; chl 4
2759	Benzonitrile, 2-hydroxy-	C_7H_5NO 119.12	611-20-1 98	149[14]	4-10-00-00191 1.1052[100]	H_2O 2; EtOH 4; eth 4; bz 4 1.5372[100]
2760	Benzonitrile, 2-methoxy-	C_8H_7NO 133.15	6609-56-9 24.5	255.5	4-10-00-00191 1.1063[20]	EtOH 3; eth 4
2761	Benzonitrile, 3-methoxy-	C_8H_7NO 133.15	1527-89-5	111-2[13]	4-10-00-00329 1.089[25]	1.5402[20]
2762	Benzonitrile, 4-methoxy-	C_8H_7NO 133.15	874-90-8 61.5	256.5	4-10-00-00441	H_2O 1; EtOH 4; eth 4; bz 3
2763	Benzonitrile, 2-methyl- o-Tolunitrile	C_8H_7N 117.15	529-19-1 -13.5	9463 205	4-09-00-01703 0.9955[20]	H_2O 1; EtOH 5; eth 5; ctc 2 1.5279[20]
2764	Benzonitrile, 3-methyl- m-Tolunitrile	C_8H_7N 117.15	620-22-4 -23	213	4-09-00-01717 1.0316[20]	H_2O 1; EtOH 5; eth 5; ctc 2 1.5252[20]
2765	Benzonitrile, 4-methyl- p-Tolunitrile	C_8H_7N 117.15	104-85-8 29.5	9464 217.0	4-09-00-01738 0.9762[30]	H_2O 1; EtOH 4; eth 4; ctc 2

No.	Name Synonym	Mol. Form. Mol. Wt.	CAS RN mp/°C	Merck No. bp/°C	Beil. Ref. den/g cm^{-3}	Solubility n_D
2766	Benzonitrile, 2-methyl-5-nitro-	$C_8H_6N_2O_2$ 162.15	939-83-3 106	174^{18}	3-09-00-02314	ace 4; bz 4; eth 4; EtOH 4
2767	Benzonitrile, pentafluoro-	C_7F_5N 193.08	773-82-0 2.4	162	4-09-00-00956 1.563^{20}	1.4402^{25}
2768	Benzonitrile, 4-(2-phenylethenyl)-, (Z)- 4-Stilbenecarbonitrile (cis)	$C_{15}H_{11}N$ 205.26	14064-68-7 44.5	$130^{0.1}$	4-09-00-02615	H_2O 2; EtOH 3; chl 3
2769	Benzonitrile, 2-(trifluoromethyl)-	$C_8H_4F_3N$ 171.12	447-60-9 18	205	4-09-00-01705	ctc 3
2770	Benzonitrile, 3-(trifluoromethyl)-	$C_8H_4F_3N$ 171.12	368-77-4 14.5	189	3-09-00-02327 1.2813^{20}	1.4508^{20}
2771	Benzophenone, 3-(dimethylamino)- 3-(Dimethylamino)benzophenone	$C_{15}H_{15}NO$ 225.29	31766-07-1 47	216^{15}	4-14-00-00247	EtOH 3
2772	2H-1-Benzopyran 1,2-Chromene	C_9H_8O 132.16	254-04-6 1112	$132^{102.}$ 91^{13}	5-17-02-00019 1.0993^{16}	H_2O 1 1.5869^{24}
2773	1H-2-Benzopyran-3-carboxylic acid, 6-[2-(4,9-dihydro-8-hydroxy-5,7-dimethoxy-4,9-dioxonaphtho[2,3-b]furan-2-yl Collinomycin	$C_{27}H_{20}O_{12}$ 536.45	27267-69-2 281	2478	5-19-08-00331	ace 4; diox 4; chl 4
2774	3H-2-Benzopyran-7-carboxylic acid, 4,6-dihydro-8-hydroxy-3,4,5-trimethyl-6-oxo-, (3R-trans)- Citrinin	$C_{13}H_{14}O_5$ 250.25	518-75-2 178-9 dec	2329	5-18-09-00061	H_2O 1; EtOH 2; eth 2; ace 3
2775	2H-1-Benzopyran-3-carboxylic acid, 2-oxo- Coumarin-3-carboxylic acid	$C_9H_6O_3$ 162.14	531-81-7 190 dec	2564	5-18-08-00323	EtOH 4
2776	1H-2-Benzopyran, 3,4-dihydro- Isochroman	$C_9H_{10}O$ 134.18	493-05-0 	90^{12}	5-17-01-00595 1.067^{25}	1.5444^{20}
2777	2H-1-Benzopyran, 3,4-dihydro-	$C_9H_{10}O$ 134.18	493-08-3 	215; 98^{18}	5-17-01-00593 1.0610^{20}	H_2O 3; os 5 1.5444^{20}
2778	2H-1-Benzopyran, 3,4-dihydro-2-methyl- 2-Methylchroman	$C_{10}H_{12}O$ 148.20	13030-26-7 	224.5	5-17-01-00609 1.0340^{20}	bz 4; eth 4; EtOH 4 1.532^{13}
2779	2H-1-Benzopyran, 3,4-dihydro-3-methyl- Chroman, 3-methyl	$C_{10}H_{12}O$ 148.20	70401-56-8 	102-4^{15}	4-17-00-00430 1.002^{22}	1.5335^{20}
2780	2H-1-Benzopyran, 3,4-dihydro-6-methyl- Chroman, 6-methyl-	$C_{10}H_{12}O$ 148.20	3722-74-5 	234	4-17-00-00431 1.0374^{14}	1.5392^{25}
2781	2H-1-Benzopyran, 3,4-dihydro-7-methyl- Chroman, 7-methyl-	$C_{10}H_{12}O$ 148.20	3722-73-4 	141-3^{60}	4-17-00-00431 1.028^{26}	1.5380^{20}
2782	2H-1-Benzopyran, 3,4-dihydro-2-phenyl- Flavan	$C_{15}H_{14}O$ 210.28	494-12-2 46.3	332; 190^{16}	5-17-02-00283	bz 4; eth 4; EtOH 4
2783	[1]Benzopyrano[3,4-b]furo[2,3-h][1]benzopyran-6(6aH)-one, 1,2,12,12a-tetrahydro-8,9-dimethoxy-2-(1-methylethen Rotenone	$C_{23}H_{22}O_6$ 394.42	83-79-4 176	8248 210-$20^{0.5}$	5-19-10-00581	H_2O 1; EtOH 3; eth 2; ace 3
2784	[2]Benzopyrano[6,5,4-def][2]benzopyran-1,3,6,8-tetrone	$C_{14}H_4O_6$ 268.18	81-30-1 450	 sub 320	5-19-05-00423	H_2O 1; Na_2CO_3 3; HOAc 3
2785	2H-1-Benzopyran-6-ol, 3,4-dihydro-2,8-dimethyl-2-(4,8,12-trimethyltridecyl)-, [2R-[2R*(4R*,8R*)]]- δ-Tocopherol	$C_{27}H_{46}O_2$ 402.66	119-13-1 	9419 $150^{0.001}$		H_2O 1; EtOH 4; eth 4; ace 4
2786	2H-1-Benzopyran-7-ol, 3,4-dihydro-3-(4-hydroxyphenyl)-, (S)- Equol	$C_{14}H_{14}O_3$ 230.26	531-95-3 189.5	3583	5-17-05-00413	
2787	2H-1-Benzopyran-6-ol, 3,4-dihydro-2,5,7,8-tetramethyl-2-(4,8,12-trimethyltridecyl)-, [2R-[2R*(4R*,8R*)]]- Vitamin E	$C_{29}H_{50}O_2$ 430.71	59-02-9 3.0	9931 $210^{0.1}$	5-17-04-00168 0.950^{25}	H_2O 1; EtOH 3; eth 3; ace 3 1.5045^{25}
2788	2H-1-Benzopyran-6-ol, 3,4-dihydro-2,5,7,8-tetramethyl-2-(4,8,12-trimethyltridecyl)-, acetate, [2R-[2R*(4R*,8R* Vitamin E acetate	$C_{31}H_{52}O_3$ 472.75	58-95-7 -27.5	9932 $184^{0.01}$	5-17-04-00168 0.9533^{21}	H_2O 1; EtOH 2; eth 3; ace 3 1.497^{20}
2789	2H-1-Benzopyran-6-ol, 3,4-dihydro-2,5,8-trimethyl-2-(4,8,12-trimethyltridecyl)- β-Tocopherol	$C_{28}H_{48}O_2$ 416.69	148-03-8 	9417 200-$10^{0.1}$		ace 4; eth 4; EtOH 4; chl 4
2790	2H-1-Benzopyran-6-ol, 3,4-dihydro-2,7,8-trimethyl-2-(4,8,12-trimethyltridecyl)- γ-Tocopherol	$C_{28}H_{48}O_2$ 416.69	7616-22-0 -3	9418 200-$10^{0.1}$		H_2O 1; EtOH 5; eth 5; ace 5
2791	1H-2-Benzopyran-1-one Isocoumarin	$C_9H_6O_2$ 146.15	491-31-6 47	286	5-17-10-00149	H_2O 1; EtOH 4; eth 4; bz 4
2792	2H-1-Benzopyran-2-one Coumarin	$C_9H_6O_2$ 146.15	91-64-5 71	2563 301.7	5-17-10-00143 0.935^{20}	H_2O 3; EtOH 3; eth 4; chl 4
2793	4H-1-Benzopyran-4-one	$C_9H_6O_2$ 146.15	491-38-3 59	sub	5-17-10-00139 1.2900^{20}	H_2O 2; EtOH 3; eth 3; bz 3
2794	4H-1-Benzopyran-4-one, 7-[[6-O-(6-deoxy-α-L-mannopyranosyl)-β-D-glucopyranosyl]oxy]-2,3-dih Hesperidin	$C_{28}H_{34}O_{15}$ 610.57	520-26-3 262	4591	5-18-05-00218	py 4; EtOH 4; HOAc 4
2795	4H-1-Benzopyran-4-one, 7-[[2-O-(6-deoxy-α-L-mannopyranosyl)-β-D-glucopyranosyl]oxy]-2,3-dihydro-5-hy Naringin	$C_{27}H_{32}O_{14}$ 580.54	10236-47-2 	6345	5-18-04-00528	H_2O 2; EtOH 2; eth 1; bz 1

No.	Name / Synonym	Mol. Form. / Mol. Wt.	CAS RN / mp/°C	Merck No. / bp/°C	Beil. Ref. / den/g cm^{-3}	Solubility / n_D
2796	4H-1-Benzopyran-4-one, 3-[(6-deoxy-α-*L*-mannopyranosyl)oxy]-2-(3,4-dihydroxyphenyl)-5,7-dihydroxy- Quercitrin	$C_{21}H_{20}O_{11}$ 448.38	522-12-3 170	8047	5-18-05-00514	H_2O 1; EtOH 3; eth 1; HOAc 3
2797	1H-2-Benzopyran-1-one, 3,4-dihydro-	$C_9H_8O_2$ 148.16	4702-34-5	176[20]; 112[0.4]	5-17-10-00018 1.197[25]	1.5629[25]
2798	2H-1-Benzopyran-2-one, 3,4-dihydro-	$C_9H_8O_2$ 148.16	119-84-6 25	272	5-17-10-00013 1.169[18]	H_2O 1; EtOH 2; eth 2; ctc 2 1.5563[20]
2799	4H-1-Benzopyran-4-one, 2,3-dihydro- 4-Chromanone	$C_9H_8O_2$ 148.16	491-37-2 36.5	160[50]; 127[13]	5-17-10-00014 1.1291[100]	EtOH 3; eth 4; ace 4; bz 4 1.5750
2800	4H-1-Benzopyran-4-one, 2,3-dihydro-5,7-dihydroxy-2-(3-hydroxy-4-methoxyphenyl)-, (*S*)- Hesperetin	$C_{16}H_{14}O_6$ 302.28	520-33-2 227.5	4590 sub 205	4-18-00-03215	eth 4; EtOH 4
2801	4H-1-Benzopyran-4-one, 2,3-dihydro-5,7-dihydroxy-3-(3-hydroxy-4-methoxyphenyl)- Flavanone, 4'-methoxy-3',5,7-trihydroxy	$C_{16}H_{14}O_6$ 302.28	99365-26-1 227.5	sub 205	2-18-00-00204	H_2O 1; EtOH 4; eth 3; bz 2
2802	4H-1-Benzopyran-4-one, 2,3-dihydro-5,7-dihydroxy-2-(4-hydroxyphenyl)-, (*S*)- Naringenin	$C_{15}H_{12}O_5$ 272.26	480-41-1 251	6344	5-18-04-00524	bz 4; eth 4; EtOH 4
2803	4H-1-Benzopyran-4-one, 2,3-dihydro-6-methyl-	$C_{10}H_{10}O_2$ 162.19	39513-75-2 35	141-3[13.5]	5-17-10-00036 1.1245[57]	1.555[57]
2804	2H-1-Benzopyran-2-one, 6,7-dihydroxy- Esculetin	$C_9H_6O_4$ 178.14	305-01-1 276	3645 sub	5-18-03-00202	H_2O 2; EtOH 3; eth 2; ace 3
2805	2H-1-Benzopyran-2-one, 7,8-dihydroxy- Daphnetin	$C_9H_6O_4$ 178.14	486-35-1 262	2816 sub	5-18-03-00211	H_2O 3; EtOH 3; eth 2; bz 2
2806	4H-1-Benzopyran-4-one, 5,7-dihydroxy-2-(4-hydroxyphenyl)- Apigenin	$C_{15}H_{10}O_5$ 270.24	520-36-5 347.5	763	5-18-04-00574	H_2O 1; EtOH 3; py 3; dil alk 4
2807	4H-1-Benzopyran-4-one, 5,7-dihydroxy-3-(4-hydroxyphenyl)- Genistein	$C_{15}H_{10}O_5$ 270.24	446-72-0 301-2 dec	4281	5-18-04-00594	
2808	2H-1-Benzopyran-2-one, 7,8-dihydroxy-6-methoxy- Fraxetin	$C_{10}H_8O_5$ 208.17	574-84-5 231	4178	5-18-04-00332	EtOH 4
2809	4H-1-Benzopyran-4-one, 5,7-dihydroxy-2-(4-methoxyphenyl)- Acacetin	$C_{16}H_{12}O_5$ 284.27	480-44-4 263	9	5-18-04-00575	EtOH 4
2810	4H-1-Benzopyran-4-one, 5,7-dihydroxy-6-methoxy-2-phenyl- Oroxylin A	$C_{16}H_{12}O_5$ 284.27	480-11-5 231.5	6830	5-18-04-00569	ace 4; eth 4; EtOH 4
2811	4H-1-Benzopyran-4-one, 5,7-dihydroxy-2-phenyl- Chrysin	$C_{15}H_{10}O_4$ 254.24	480-40-0 285.5	2261	5-18-04-00076	H_2O 1; EtOH 3; eth 2; ace 3
2812	4H-1-Benzopyran-4-one, 2-(3,4-dihydroxyphenyl)-2,3-dihydro-5,7-dihydroxy-, (*S*)- Eriodictyol	$C_{15}H_{12}O_6$ 288.26	552-58-9 267 dec	3616	5-18-05-00214	EtOH 4; HOAc 4
2813	4H-1-Benzopyran-4-one, 2-(3,4-dihydroxyphenyl)-3,7-dihydroxy- Fisetin	$C_{15}H_{10}O_6$ 286.24	528-48-3 330	4026	5-18-05-00291	H_2O 1; EtOH 3; eth 2; ace 3
2814	4H-1-Benzopyran-4-one, 2-(3,4-dihydroxyphenyl)-5,7-dihydroxy- Luteolin	$C_{15}H_{10}O_6$ 286.24	491-70-3 329-30 dec	5483	5-18-05-00296	H_2O 2; EtOH 3; eth 3; alk 3
2815	4H-1-Benzopyran-4-one, 2-(3,4-dihydroxyphenyl)-3,5-dihydroxy-7-methoxy- Rhamnetin	$C_{16}H_{12}O_7$ 316.27	90-19-7 295	8170	5-18-05-00495	H_2O 2; EtOH 3; ace 3; dil alk 4
2816	4H-1-Benzopyran-4-one, 2-(2,4-dihydroxyphenyl)-3,5,7-trihydroxy- Morin	$C_{15}H_{10}O_7$ 302.24	480-16-0 303.5	6179	5-18-05-00492	H_2O 2; EtOH 4; eth 2; bz 3
2817	4H-1-Benzopyran-4-one, 2-(3,4-dihydroxyphenyl)-3,5,7-trihydroxy- Quercetin	$C_{15}H_{10}O_7$ 302.24	117-39-5 316.5	8044 sub	5-18-05-00494	H_2O 2; EtOH 3; eth 2; ace 3
2818	2H-1-Benzopyran-2-one, 5,7-dimethoxy- Limettin	$C_{11}H_{10}O_4$ 206.20	487-06-9 149	5370 200 dec	5-18-03-00199	H_2O 2; EtOH 4; eth 1; ace 4
2819	4H-1-Benzopyran-4-one, 8-[(dimethylamino)methyl]-7-methoxy-3-methyl-2-phenyl- Dimefline	$C_{20}H_{21}NO_3$ 323.39	1165-48-6 109.5	3190	5-18-12-00083	chl 3
2820	2H-1-Benzopyran-2-one, 3-ethyl- Coumarin, 3-ethyl-	$C_{11}H_{10}O_2$ 174.20	66898-39-3 72.3	299	5-17-10-00185	eth 4; EtOH 4
2821	2H-1-Benzopyran-2-one, 7-(β-*D*-glucopyranosyloxy)- Skimmin	$C_{15}H_{16}O_8$ 324.29	93-39-0 220	8506	5-18-01-00409	H_2O 3; EtOH 3; eth 1; chl 1
2822	2H-1-Benzopyran-2-one, 6-(β-*D*-glucopyranosyloxy)-7-hydroxy- Esculin	$C_{15}H_{16}O_9$ 340.29	531-75-9 205 dec	3646 230 dec	5-18-03-00208	H_2O 2; EtOH 2; eth 2; chl 3

No.	Name Synonym	Mol. Form. Mol. Wt.	CAS RN mp/°C	Merck No. bp/°C	Beil. Ref. den/g cm^{-3}	Solubility n_D
2823	2H-1-Benzopyran-2-one, 8-(β-D-glucopyranosyloxy)-7-hydroxy-6-methoxy- Fraxin	C$_{16}$H$_{18}$O$_{10}$ 370.31	524-30-1 205	4179	5-18-04-00334	
2824	4H-1-Benzopyran-4-one, 3-[4-(β-D-glucopyranosyloxy)phenyl]-5,7-dihydroxy- Sophoricoside	C$_{21}$H$_{20}$O$_{10}$ 432.38	152-95-4 274	8675	5-18-04-00597	
2825	2H-1-Benzopyran-2-one, 6-hydroxy-	C$_9$H$_6$O$_3$ 162.14	6093-68-1 250		5-18-01-00385 1.25^{25}	EtOH 4
2826	2H-1-Benzopyran-2-one, 7-hydroxy- Umbelliferone	C$_9$H$_6$O$_3$ 162.14	93-35-6 230.5	9758 sub	5-18-01-00386	EtOH 4; HOAc 4; chl 4
2827	4H-1-Benzopyran-4-one, 7-hydroxy-3-(4-hydroxyphenyl)- Daidzein	C$_{15}$H$_{10}$O$_4$ 254.24	486-66-8 323 dec	2805 sub	5-18-04-00089	EtOH 3; eth 3
2828	4H-1-Benzopyran-4-one, 5-hydroxy-3-(4-hydroxyphenyl)-7-methoxy- Prunetin	C$_{16}$H$_{12}$O$_5$ 284.27	552-59-0 239.5	7923	5-18-04-00595	
2829	2H-1-Benzopyran-2-one, 7-hydroxy-6-methoxy- Scopoletin	C$_{10}$H$_8$O$_4$ 192.17	92-61-5 204	8363	5-18-03-00203	H$_2$O 2; EtOH 2; eth 2; bz 1
2830	4H-1-Benzopyran-4-one, 7-hydroxy-3-(4-methoxyphenyl)- Formononetin	C$_{16}$H$_{12}$O$_4$ 268.27	485-72-3 256.5	4157	5-18-04-00089	
2831	2H-1-Benzopyran-2-one, 7-hydroxy-4-methyl- Hymecromone	C$_{10}$H$_8$O$_3$ 176.17	90-33-5 195.5	4792	5-18-01-00439	H$_2$O 2; EtOH 3; eth 2; chl 2
2832	2H-1-Benzopyran-2-one, 4-hydroxy-3-(3-oxo-1-phenylbutyl)- Warfarin	C$_{19}$H$_{16}$O$_4$ 308.33	81-81-2 161	9950	5-18-04-00162	H$_2$O 1; EtOH 3; ace 3; diox 3
2833	2H-1-Benzopyran-2-one, 3-methyl- 3-Methylcoumarin	C$_{10}$H$_8$O$_2$ 160.17	2445-82-1 91	292.5	5-17-10-00165	H$_2$O 2; EtOH 3; chl 3; KOH 1
2834	2H-1-Benzopyran-2-one, 5-methyl- 5-Methylcoumarin	C$_{10}$H$_8$O$_2$ 160.17	42286-84-0 65.8	173^{12}	5-17-10-00167	H$_2$O 2; EtOH 4; eth 4; bz 4
2835	2H-1-Benzopyran-2-one, 6-methyl-	C$_{10}$H$_8$O$_2$ 160.17	92-48-8 76.5	304; 174^{14}	5-17-10-00168	EtOH 4; eth 4; bz 4; chl 2
2836	2H-1-Benzopyran-2-one, 7-methyl- 7-Methylcoumarin	C$_{10}$H$_8$O$_2$ 160.17	2445-83-2 128	171.5^{11}	5-17-10-00169	H$_2$O 2; EtOH 4; eth 3; HOAc 4
2837	2H-1-Benzopyran-2-one, 8-methyl- 8-Methylcoumarin	C$_{10}$H$_8$O$_2$ 160.17	1807-36-9 110.5	178^{20}	5-17-10-00170	EtOH 4, eth 4; bz 4; ctc 4
2838	4H-1-Benzopyran-4-one, 3-methyl- Tricromyl	C$_{10}$H$_8$O$_2$ 160.17	85-90-5	9574	5-17-10-00164	chl 3
2839	2H-1-Benzopyran-2-one, 3,3'-methylenebis[4-hydroxy- Dicumarol	C$_{19}$H$_{12}$O$_6$ 336.30	66-76-2 290	3080	5-19-06-00682	
2840	4H-1-Benzopyran-4-one, 2-phenyl- Flavone	C$_{15}$H$_{10}$O$_2$ 222.24	525-82-6 100	4030	5-17-10-00552	H$_2$O 1; EtOH 3; eth 3; ace 3
2841	4H-1-Benzopyran-4-one, 3-phenyl- Isoflavone	C$_{15}$H$_{10}$O$_2$ 222.24	574-12-9 148	5057	5-17-10-00562	
2842	4H-1-Benzopyran-4-one, 3,5,7-trihydroxy-2-(2-hydroxyphenyl)- Datiscetin	C$_{15}$H$_{10}$O$_6$ 286.24	480-15-9 277.5	2822	4-18-00-03281	ace 4; eth 4; EtOH 4
2843	4H-1-Benzopyran-4-one, 3,5,7-trihydroxy-2-(4-hydroxyphenyl)- Kaempferol	C$_{15}$H$_{10}$O$_6$ 286.24	520-18-3 277	5156	5-18-05-00251	H$_2$O 2; EtOH 4; eth 4; ace 4
2844	4H-1-Benzopyran-4-one, 5,6,7-trihydroxy-2-phenyl- Baicalein	C$_{15}$H$_{10}$O$_5$ 270.24	491-67-8 264-5 dec	954	5-18-04-00569	H$_2$O 2; EtOH 3; eth 3; ace 3
2845	2H-1-Benzopyran-2-one, 6,7,8-trimethoxy-	C$_{12}$H$_{12}$O$_5$ 236.22	6035-49-0 104.2	90-100$^{0.2}$	5-18-04-00333	EtOH 3; eth 3
2846	2H-1-Benzopyran-3,5,7-triol, 2-(3,4-dihydroxyphenyl)-3,4-dihydro-, (2S-cis)-	C$_{15}$H$_{14}$O$_6$ 290.27	35323-91-2 242		5-17-08-00447 1.344^4	ace 4; EtOH 4
2847	1-Benzopyrylium, 3,5,7-trihydroxy-2-(4-hydroxy-3,5-dimethoxyphenyl)-, chloride Malvidin chloride	C$_{17}$H$_{15}$ClO$_7$ 366.75	643-84-5 >300	5596	5-17-08-00514	H$_2$O 2; EtOH 3; MeOH 3
2848	1-Benzopyrylium, 3,5,7-trihydroxy-2-(4-hydroxyphenyl)-, chloride Pelargonidin chloride	C$_{15}$H$_{11}$ClO$_5$ 306.70	134-04-3 >350	7014	1-18-00-00410	H$_2$O 3; EtOH 4; chl 2; con sulf 3
2849	1-Benzopyrylium, 3,5,7-trihydroxy-2-(3,4,5-trihydroxyphenyl)-, chloride Delphinidin	C$_{15}$H$_{11}$ClO$_7$ 338.70	528-53-0 >350	2866	5-17-08-00514	H$_2$O 4; EtOH 4; MeOH 4; AcOEt 3
2850	Benzo[rst]pentaphene 1,2:7,8-Dibenzopyrene	C$_{24}$H$_{14}$ 302.38	189-55-9 281.5		4-05-00-02803	
2851	2H-1,2,4-Benzothiadiazine-7-sulfonamide, 3-bicyclo[2.2.1]hept-5-en-2-yl-6-chloro-3,4-dihydro-, 1,1-dioxide Cyclothiazide	C$_{14}$H$_{16}$ClN$_3$O$_4$S$_2$ 389.88	2259-96-3 234	2760		
2852	2H-1,2,4-Benzothiadiazine-7-sulfonamide, 6-chloro-, 1,1-dioxide Chlorothiazide	C$_7$H$_6$ClN$_3$O$_4$S$_2$ 295.73	58-94-6 350 dec	2169	4-27-00-08036	

No.	Name / Synonym	Mol. Form. / Mol. Wt.	CAS RN / mp/°C	Merck No. / bp/°C	Beil. Ref. / den/g cm^{-3}	Solubility / n_D
2853	2H-1,2,4-Benzothiadiazine-7-sulfonamide, 6-chloro-3-(cyclopentylmethyl)-3,4-dihydro-, 1,1-dioxide Cyclopenthiazide	$C_{13}H_{18}ClN_3O_4S_2$ 379.89	742-20-1 238	2750		
2854	2H-1,2,4-Benzothiadiazine-7-sulfonamide, 6-chloro-3-(dichloromethyl)-3,4-dihydro-, 1,1-dioxide Trichloromethiazide	$C_8H_8Cl_3N_3O_4S_2$ 380.66	133-67-5 270 dec	9537		H_2O 2; EtOH 3
2855	2H-1,2,4-Benzothiadiazine-7-sulfonamide, 6-chloro-3,4-dihydro-, 1,1-dioxide Hydrochlorothiazide	$C_7H_8ClN_3O_4S_2$ 297.74	58-93-5 274	4704		
2856	2H-1,2,4-Benzothiadiazine-7-sulfonamide, 6-chloro-3,4-dihydro-3-(2-methylpropyl)-, 1,1-dioxide Buthiazide	$C_{11}H_{16}ClN_3O_4S_2$ 353.85	2043-38-1 221.5	1519		
2857	2H-1,2,4-Benzothiadiazine-7-sulfonamide, 6-chloro-3,4-dihydro-2-methyl-3-[[(2,2,2-trifluoroethyl)thio]methyl]- Polythiazide	$C_{11}H_{13}ClF_3N_3O_4S_3$ 439.89	346-18-9 214	7561		
2858	2H-1,2,4-Benzothiadiazine-7-sulfonamide, 3,4-dihydro-6-(trifluoromethyl)-, 1,1-dioxide Hydroflumethiazide	$C_8H_8F_3N_3O_4S_2$ 331.30	135-09-1 270.5	4716	4-27-00-08035	
2859	2-Benzothiazolamine 2-Aminobenzothiazole	$C_7H_6N_2S$ 150.20	136-95-8 132	435	4-27-00-04824	H_2O 2; EtOH 3; eth 3; chl 3
2860	6-Benzothiazolamine 6-Aminobenzothiazole	$C_7H_6N_2S$ 150.20	533-30-2 87	436	4-27-00-04884	H_2O 1; EtOH 3; eth 1
2861	Benzothiazole Benzosulfonazole	C_7H_5NS 135.19	95-16-9 2	1118 231	4-27-00-01069 1.2460[20]	H_2O 2; EtOH 4; eth 4; ace 3 1.6379[20]
2862	Benzothiazole, 2-chloro- 2-Chlorobenzothiazole	C_7H_4ClNS 169.63	615-20-3 24	248	4-27-00-01072 1.3715[10]	ace 4; eth 4; EtOH 4 1.6338[10]
2863	Benzothiazole, 2-methyl- 2-Methylbenzothiazole	C_8H_7NS 149.22	120-75-2 14	238	4-27-00-01080 1.1763[19]	H_2O 1; EtOH 3; chl 3 1.6092[19]
2864	Benzothiazole, 2-phenyl- 2-Phenylbenzothiazole	$C_{13}H_9NS$ 211.29	883-93-2 115	371	4-27-00-01385	H_2O 1; EtOH 3; eth 3; CS_2 3
2865	2-Benzothiazolesulfonamide, 6-ethoxy- Ethoxzolamide	$C_9H_{10}N_2O_3S_2$ 258.32	452-35-7 189	3711	4-27-00-04404	
2866	2(3H)-Benzothiazolethione 2-Mercaptobenzothiazole	$C_7H_5NS_2$ 167.26	149-30-4 181	5759	4-27-00-02709 1.42[20]	H_2O 1; EtOH 3; eth 2; bz 2
2867	2(3H)-Benzothiazolethione, 3-methyl-	$C_8H_7NS_2$ 181.28	2254-94-6 90	335	4-27-00-02714	H_2O 1; EtOH 2; eth 2; bz 4
2868	2(3H)-Benzothiazolone	C_7H_5NOS 151.19	934-34-9 139	360	4-27-00-02693	H_2O 1; EtOH 4; eth 4
2869	4H-1-Benzothiopyran-4-one, 2,3-dihydro-	C_9H_8OS 164.23	3528-17-4 29	154[12]	5-17-10-00016 1.2487[14]	1.6395[20]
2870	1,2,4-Benzotriazine 1,2,4-Triazanaphthalene	$C_7H_5N_3$ 131.14	254-87-5 75.8	237.5	4-26-00-00166	H_2O 4; eth 4; EtOH 4
2871	1H-Benzotriazole 1,2,3-Triaza-1H-indene	$C_6H_5N_3$ 119.13	95-14-7 100	1119 204[15]	4-26-00-00093	H_2O 2; EtOH 3; bz 3; chl 3
2872	1H-Benzotriazole, 1-methyl-	$C_7H_7N_3$ 133.15	13351-73-0 64.5	270.5	4-26-00-00095	bz 4; EtOH 4; HOAc 4
2873	1H-Benzotriazole, 5-nitro- 5-Nitrobenzotriazole	$C_6H_4N_4O_2$ 164.12	2338-12-7 217		4-26-00-00129	
2874	1,2,3-Benzoxadiazole, 5,7-dinitro- 2,3-Benzoxadiazole, 5,7-dinitro	$C_6H_2N_4O_5$ 210.11	87-31-0 158			EtOH 4
2875	3H-2,1-Benzoxathiol-3-one, 1,1-dioxide	$C_7H_4O_4S$ 184.17	81-08-3 129.5	184[18]	5-19-04-00215	bz 4; chl 4
2876	2-Benzoxazolamine, 5-chloro- Zoxazolamine	$C_7H_5ClN_2O$ 168.58	61-80-3 184.5	10098	4-27-00-04819	EtOH 4
2877	Benzoxazole 1-Oxa-3-azaindene	C_7H_5NO 119.12	273-53-0 31	182.5	4-27-00-01069 1.1754[20]	H_2O 1; EtOH 3; sulf 3 1.5594[20]
2878	Benzoxazole, 2-chloro- 2-Chlorobenzoxazole	C_7H_4ClNO 153.57	615-18-9 7	201.5	2-27-00-00017 1.3453[18]	1.5678[20]
2879	Benzoxazole, 2,5-dimethyl- 2,5-Dimethylbenzoxazole	C_9H_9NO 147.18	5676-58-4 218.5		4-27-00-01101 1.0880[18]	ctc 3 1.5412[20]
2880	Benzoxazole, 2-methyl- 2-Methylbenzoxazole	C_8H_7NO 133.15	95-21-6 9.5	200.5	4-27-00-01078 1.1211[20]	H_2O 1; EtOH 4; eth 5 1.5497[20]
2881	2-Benzoxazolol 2-Hydroxybenzoxazole	$C_7H_5NO_2$ 135.12	69564-68-7 141	230[30]	2-27-00-00233	eth 4; EtOH 4
2882	2(3H)-Benzoxazolone	$C_7H_5NO_2$ 135.12	59-49-4 138	335; 230[30]	4-27-00-02677	H_2O 2; EtOH 3; eth 3; tfa 3
2883	2(3H)-Benzoxazolone, 5-chloro- Chlorzoxazone	$C_7H_4ClNO_2$ 169.57	95-25-0 191.5	2197	4-27-00-02683	EtOH 4; MeOH 4
2884	Benzoyl azide Benzazide	$C_7H_5N_3O$ 147.14	582-61-6 32	exp	4-09-00-00945 1.1680[35]	eth 4; EtOH 4
2885	Benzoyl bromide Benzoic acid, bromide	C_7H_5BrO 185.02	618-32-6 -24	218.5	4-09-00-00724 1.570[15]	eth 5 1.5868[25]
2886	Benzoyl chloride Benzenecarbonyl chloride	C_7H_5ClO 140.57	98-88-4 -1.0	1124 197.2; 71[9]	4-09-00-00721 1.2120[20]	eth 5; bz 3; ctc 3; CS_2 3 1.5537[20]
2887	Benzoyl chloride, 2-bromo-	C_7H_4BrClO 219.46	7154-66-7 11	243	4-09-00-01013	ctc 2 1.5963[20]
2888	Benzoyl chloride, 4-bromo-	C_7H_4BrClO 219.46	586-75-4 42	246; 181[125]	4-09-00-01023	EtOH 4; eth 4; bz 4; lig 4

No.	Name / Synonym	Mol. Form. / Mol. Wt.	CAS RN / mp/°C	Merck No. / bp/°C	Beil. Ref. / den/g cm⁻³	Solubility / n_D
2889	Benzoyl chloride, 4-butyl-	$C_{11}H_{13}ClO$ 196.68	28788-62-7	155-6[22]	3-09-00-02520 1.051[25]	1.5351[20]
2890	Benzoyl chloride, 2-chloro-	$C_7H_4Cl_2O$ 175.01	609-65-4 -4	238	4-09-00-00962	ctc 3 1.5726[16]
2891	Benzoyl chloride, 3-chloro- m-Chlorobenzoyl chloride	$C_7H_4Cl_2O$ 175.01	618-46-2	225	4-09-00-00972	1.5677[20]
2892	Benzoyl chloride, 4-chloro-	$C_7H_4Cl_2O$ 175.01	122-01-0 16	222	4-09-00-00984 1.3770[20]	chl 2 1.5756[20]
2893	Benzoyl chloride, 2,4-dichloro- 2,4-Dichlorobenzoyl chloride	$C_7H_3Cl_3O$ 209.46	89-75-8 16.5	150[34], 111[7.5]	4-09-00-01002	ctc 3 1.5895[20]
2894	Benzoyl chloride, 3,4-dichloro- 3,4-Dichlorobenzoyl chloride	$C_7H_3Cl_3O$ 209.46	3024-72-4 25	242	3-09-00-01379	ctc 2
2895	Benzoyl chloride, 4-(1,1-dimethylethyl)-	$C_{11}H_{13}ClO$ 196.68	1710-98-1	135[20]	4-09-00-01887 1.007[25]	1.5364[20]
2896	Benzoyl chloride, 3,5-dinitro- 3,5-Dinitrobenzoyl chloride	$C_7H_3ClN_2O_5$ 230.56	99-33-2 74	3270 196[12]	4-09-00-01350	eth 3; chl 3
2897	Benzoyl chloride, 2-fluoro-	C_7H_4ClFO 158.56	393-52-2 4	90-2[15]	4-09-00-00951 1.328[25]	1.5365[20]
2898	Benzoyl chloride, 3-fluoro-	C_7H_4ClFO 158.56	1711-07-5 -30	189	4-09-00-00953 1.304[25]	1.5285[20]
2899	Benzoyl chloride, 4-fluoro-	C_7H_4ClFO 158.56	403-43-0 9	82[20]	4-09-00-00954 1.342[25]	1.5296[20]
2900	Benzoyl chloride, 2-hydroxy-	$C_7H_5ClO_2$ 156.57	1441-87-8 19	92[15]	3-10-00-00150 1.3112[20]	eth 4 1.5812[20]
2901	Benzoyl chloride, 4-methoxy- p-Anisoyl chloride	$C_8H_7ClO_2$ 170.60	100-07-2 24.5	703 262.5	4-10-00-00430 1.261[20]	eth 3; ace 3; bz 4; ctc 2 1.580[20]
2902	Benzoyl chloride, 2-methyl-	C_8H_7ClO 154.60	933-88-0	213.5	3-09-00-02304	eth 4; EtOH 4 1.5549[20]
2903	Benzoyl chloride, 3-methyl-	C_8H_7ClO 154.60	1711-06-4 -23	219.5	4-09-00-01716 1.0265[21]	eth 4; EtOH 4 1.505[22]
2904	Benzoyl chloride, 4-methyl-	C_8H_7ClO 154.60	874-60-2 -3	226	4-09-00-01733 1.1686[20]	ctc 3 1.5547[20]
2905	Benzoyl chloride, 3-nitro-	$C_7H_4ClNO_3$ 185.57	121-90-4 36	276.5	4-09-00-01061	eth 4
2906	Benzoyl chloride, 4-nitro- p-Nitrobenzoyl chloride	$C_7H_4ClNO_3$ 185.57	122-04-3 75	6511 203[105], 151[15]	4-09-00-01191	eth 3
2907	Benzoyl chloride, 4-pentyl-	$C_{12}H_{15}ClO$ 210.70	49763-65-7	121[8]	4-09-00-01910 1.036[25]	1.5300[20]
2908	Benzoyl chloride, 4-(pentyloxy)-	$C_{12}H_{15}ClO_2$ 226.70	36823-84-4	196-200[30]	4-10-00-00431 1.087[25]	1.5434[20]
2909	Benzoyl fluoride Benzoic acid, fluoride	C_7H_5FO 124.11	455-32-3 -28	154.5	4-09-00-00721 1.1400[20]	EtOH 4; eth 4; ctc 3
2910	Benzoyl iodide	C_7H_5IO 232.02	618-38-2 3	128[20]	4-09-00-00725 1.746[18]	eth 4; EtOH 4
2911	Berbaman-12-ol, 6,6',7-trimethoxy-2,2'-dimethyl- Berbamine	$C_{37}H_{40}N_2O_6$ 608.73	478-61-5 198.5	1168	2-27-00-00891	H_2O 2; EtOH 3; eth 3; chl 3
2912	Beryllium, bis(2,4-pentanedionato-O,O')-, (T-4)- Beryllium acetylacetonate	$C_{10}H_{14}BeO_4$ 207.23	10210-64-7 108	1180 270		
2913	Bicyclo[2.2.1]hepta-2,5-diene	C_7H_8 92.14	121-46-0 -19.1	89.5	4-05-00-00879 0.9064[20]	H_2O 1; EtOH 3; eth 3; ace 3 1.4702[20]
2914	Bicyclo[2.2.1]heptane	C_7H_{12} 96.17	279-23-2 87.5	105.3	4-05-00-00258	ace 4; bz 4; eth 4; EtOH 4
2915	Bicyclo[4.1.0]heptane Norcarane	C_7H_{12} 96.17	286-08-8	6605 116.5	4-05-00-00257 0.853[25]	1.4564[20]
2916	Bicyclo[2.2.1]heptane-2-carboxaldehyde 2-Norbornanecarboxaldehyde	$C_8H_{10}O$ 122.17	19396-83-9	70-2[22]	4-07-00-00157 1.0227[19]	eth 4 1.4760[25]
2917	Bicyclo[2.2.1]heptane-2-carboxylic acid, 4,7,7-trimethyl-3-oxo- d-Camphocarboxylic acid	$C_{11}H_{16}O_3$ 196.25	18530-30-8 127.5	1737	3-10-00-02925	bz 4; eth 4; EtOH 4
2918	Bicyclo[2.2.1]heptane, 2-chloro-	$C_7H_{11}Cl$ 130.62	29342-53-8 -5	162; 52[11]	4-05-00-00259 1.0603[20]	chl 4 1.4849[20]
2919	Bicyclo[2.2.1]heptane, 2-chloro-1,7,7-trimethyl- Bornane, 2-chloro-	$C_{10}H_{17}Cl$ 172.70	6120-13-4 132.8	207	4-05-00-00319	eth 4; EtOH 4
2920	Bicyclo[2.2.1]heptane, 2-chloro-1,7,7-trimethyl-, endo- Bornyl chloride	$C_{10}H_{17}Cl$ 172.70	464-41-5 132	1341 207.5	4-05-00-00319	bz 4; eth 4; EtOH 4; peth 4
2921	Bicyclo[2.2.1]heptane, 2,2-dimethyl-3-methylene-, (±)-	$C_{10}H_{16}$ 136.24	565-00-4 51.5	158.5	4-05-00-00462 0.879[20]	H_2O 1; EtOH 4; eth 4 1.4551[54]
2922	Bicyclo[2.2.1]heptane, 2,2-dimethyl-5-methylene-	$C_{10}H_{16}$ 136.24	497-32-5	152	4-05-00-00465 0.8591[20]	1.4645[25]
2923	Bicyclo[2.2.1]heptane, 7,7-dimethyl-2-methylene-, (±)- α-Fenchene (DL)	$C_{10}H_{16}$ 136.24	2623-54-3	155	3-05-00-00389 0.8660[20]	ace 4; eth 4; EtOH 4 1.4705[20]

No.	Name / Synonym	Mol. Form. / Mol. Wt.	CAS RN / mp/°C	Merck No. / bp/°C	Beil. Ref. / den/g cm⁻³	Solubility / n_D
2924	Bicyclo[3.1.1]heptane, 6,6-dimethyl-2-methylene-, (1*R*)-	$C_{10}H_{16}$	19902-08-0	7415	4-05-00-00456	bz 4; eth 4; EtOH 4; chl 4
		136.24	-50	164.6	0.8654^{20}	1.4789^{20}
2925	Bicyclo[3.1.1]heptane, 6,6-dimethyl-2-methylene- β-Pinene	$C_{10}H_{16}$	127-91-3	7415		H_2O 1; bz 3; EtOH 3; eth 3
		136.24	-61.5	166	0.860^{25}	1.4768^{25}
2926	Bicyclo[2.2.1]heptane-2,3-dione, 1,7,7-trimethyl-, (1*R*)-	$C_{10}H_{14}O_2$	10334-26-6		4-07-00-02039	H_2O 3; EtOH 4; eth 4; bz 3
		166.22	199	sub		
2927	Bicyclo[2.2.1]heptane-1-methanamine, 7,7-dimethyl- α-Camphylamine	$C_{10}H_{19}N$	54131-44-1		1-12-00-00129	
		153.27		195; 95^{12}	0.8688^{20}	1.4728^{18}
2928	Bicyclo[3.1.1]heptane-2-methanamine, 6,6-dimethyl-, [1*S*-(1α,2β,5α)]-	$C_{10}H_{19}N$	73522-42-6		4-06-00-00384	
		153.27		$94-9^{27}$	0.915^{25}	1.4877^{20}
2929	Bicyclo[2.2.1]heptane-1-methanesulfonic acid, 7,7-dimethyl-2-oxo-, (1*S*)- *D*-Camphorsulfonic acid	$C_{10}H_{16}O_4S$	3144-16-9	1740	4-11-00-00642	H_2O 4; eth 1; HOAc 2
		232.30	195 dec			
2930	Bicyclo[3.1.1]heptane-2-methanol, 6,6-dimethyl-, [1*S*-(1α,2β,5α)]-	$C_{10}H_{18}O$	51152-12-6		4-06-00-00278	
		154.25		$70-2^1$	0.986^{25}	1.4890^{20}
2931	Bicyclo[2.2.1]heptane, 1,3,3-trimethyl-, (+)- Fenchane(*d*)	$C_{10}H_{18}$	116435-30-4		3-05-00-00256	H_2O 1; EtOH 3; eth 3; HOAc 2
		138.25	<-15	151	0.8345^{20}	1.44714^{20}
2932	Bicyclo[2.2.1]heptane, 1,7,7-trimethyl-	$C_{10}H_{18}$	464-15-3		4-05-00-00319	H_2O 1; EtOH 3; eth 3; AcOEt 3
		138.25		161		
2933	Bicyclo[2.2.1]heptane, 2,2,3-trimethyl-	$C_{10}H_{18}$	473-19-8		3-05-00-00263	ace 4; EtOH 4
		138.25	62.5	166	0.8276^{67}	
2934	Bicyclo[3.1.1]heptane, 2,6,6-trimethyl-, (1α,2α,5α)-(±)- Pinane (*DL*)	$C_{10}H_{18}$	112456-46-9		1-05-00-00048	bz 4; eth 4
		138.25		164.5	0.8551^{20}	1.4609^{20}
2935	Bicyclo[3.1.1]heptane, 2,6,6-trimethyl-, [1*R*-(1α,2β,5α)]-	$C_{10}H_{18}$	4795-86-2		4-05-00-00318	bz 4; eth 4
		138.25	-53	169	0.8560^{20}	1.4629^{20}
2936	Bicyclo[4.1.0]heptane, 3,7,7-trimethyl-, [1*S*-(1α,3β,6α)]-	$C_{10}H_{18}$	2778-68-9		4-05-00-00316	
		138.25		169	0.8410^{20}	1.4569
2937	Bicyclo[4.1.0]heptane, 3,7,7-trimethyl-, [1*S*-(1α,3α,6α)]-	$C_{10}H_{18}$	6069-97-2		4-05-00-00316	
		138.25		169	0.841^{20}	1.456
2938	Bicyclo[2.2.1]heptan-2-ol, 1,7-dimethyl-, (*exo,syn*)- β-Santenol	$C_9H_{16}O$	509-12-6		4-06-00-00242	eth 4
		140.23	101.5	192; 60^{15}		
2939	Bicyclo[2.2.1]heptan-2-ol, 1-phenyl-, acetate	$C_{15}H_{18}O_2$	71173-15-4		3-06-00-02755	ctc 2
		230.31	44	159^{13}; $128^{1.0}$		
2940	Bicyclo[2.2.1]heptan-2-ol, 1,3,3-trimethyl-, *endo*-(±)-	$C_{10}H_{18}O$	36386-49-9		3-06-00-00288	eth 4; EtOH 4
		154.25	39	199.5	0.9420^{40}	
2941	Bicyclo[2.2.1]heptan-2-ol, 1,3,3-trimethyl-, (1*S-endo*)- *l*-α-Fenchyl alcohol	$C_{10}H_{18}O$	512-13-0		4-06-00-00278	
		154.25	48	94^{20}	0.9034^{84}	
2942	Bicyclo[2.2.1]heptan-2-ol, 1,7,7-trimethyl-, *endo*-(±)-	$C_{10}H_{18}O$	6627-72-1		4-06-00-00281	H_2O 1; EtOH 4; eth 4; bz 4
		154.25	210.5	sub	1.011^{20}	
2943	Bicyclo[2.2.1]heptan-2-ol, 1,7,7-trimethyl-, *exo*-(±)-	$C_{10}H_{18}O$	24393-70-2		4-06-00-00282	H_2O 1; EtOH 4; eth 4; bz 2
		154.25	212	sub	1.10^{20}	
2944	Bicyclo[3.1.1]heptan-2-ol, 4,6,6-trimethyl-, [1*R*-(1α,2α,4α,5α)]-	$C_{10}H_{18}O$	515-88-8		3-06-00-00284	
		154.25	58	218	0.940^{20}	1.4702^{20}
2945	Bicyclo[2.2.1]heptan-2-ol, 1,7,7-trimethyl-, acetate, *endo*-(±)-	$C_{12}H_{20}O_2$	36386-52-4		4-06-00-00283	ctc 3
		196.29	<-17	223.5	1.4630^{20}	1.4630^{20}
2946	Bicyclo[2.2.1]heptan-2-ol, 1,7,7-trimethyl-, acetate, *exo*-(±)-	$C_{12}H_{20}O_2$	17283-45-3		4-06-00-00283	ace 4; EtOH 4
		196.29		116^{21}	0.9841^{20}	1.4640^{20}
2947	Bicyclo[2.2.1]heptan-2-ol, 4,7,7-trimethyl-, acetate, (1*R-endo*)- Epiborneol, acetate	$C_{12}H_{20}O_2$	22621-74-5		2-06-00-00092	
		196.29	<-15	101^{11}	0.9872^{14}	1.4651^{14}
2948	Bicyclo[2.2.1]heptan-2-ol, 1,7,7-trimethyl-, acetate, *endo*- Borneol, acetate	$C_{12}H_{22}O_2$	76-49-3	1339	3-06-00-00303	
		198.31	29	221		
2949	Bicyclo[2.2.1]heptan-2-ol, 1,7,7-trimethyl-, formate, (1*R-endo*)- Bornyl formate (*d*)	$C_{11}H_{18}O_2$	74219-20-8		3-06-00-00301	
		182.26		90^{10}	1.009^{22}	1.4708^{15}
2950	Bicyclo[2.2.1]heptan-2-one, 3-(acetyl-17*O*)-1,7,7-trimethyl- Camphor, 3-acetyl	$C_{12}H_{18}O_2$	15068-90-3		2-07-00-00563	
		194.27		$127-8^{15}$	1.0314^{19}	1.4949^{17}

No.	Name Synonym	Mol. Form. Mol. Wt.	CAS RN mp/°C	Merck No. bp/°C	Beil. Ref. den/g cm^{-3}	Solubility n_D
2951	Bicyclo[2.2.1]heptan-2-one, 1-(bromomethyl)-7,7-dimethyl-, (1S)- d-Camphor, 10-bromo-	C$_{10}$H$_{15}$BrO 231.13	64161-50-8 78	265 dec	4-07-00-00218	ace 4; bz 4; eth 4; EtOH 4
2952	Bicyclo[2.2.1]heptan-2-one, 3-bromo-1,7,7-trimethyl-, (1R-endo)-	C$_{10}$H$_{15}$BrO 231.13	10293-06-8 76	274 dec	1.449^{25}	bz 4; eth 4; EtOH 4
2953	Bicyclo[2.2.1]heptan-2-one, 3,3-dibromo-1,7,7-trimethyl-, (1R)- α,α'-Dibromo-d-camphor	C$_{10}$H$_{14}$Br$_2$O 310.03	514-12-5 61	3002	4-07-00-00220 1.854^{21}	H$_2$O 1; EtOH 4; eth 4; bz 4
2954	Bicyclo[2.2.1]heptan-2-one, 3,3-dimethyl-, (1S)-	C$_9$H$_{14}$O 138.21	6069-71-2 41.9	193; 76^{12}	4-07-00-00177 0.807^{20}	eth 4
2955	Bicyclo[3.1.1]heptan-2-one, 6,6-dimethyl-, (1R)-	C$_9$H$_{14}$O 138.21	38651-65-9 -1	209	4-07-00-00176 0.9807^{20}	eth 4; EtOH 4 1.4787^{20}
2956	Bicyclo[3.1.1]heptan-3-one, 6,6-dimethyl-2-methylene-, (1S)-	C$_{10}$H$_{14}$O 150.22	19890-00-7 -1.8	67-9^4	0.9875^{15}	1.4950^{20}
2957	Bicyclo[2.2.1]heptan-2-one, 3-hydroxy-1,7,7-trimethyl- 3-Hydroxycamphor	C$_{10}$H$_{16}$O$_2$ 168.24	10373-81-6 205.5	4747	4-08-00-00081	eth 4; EtOH 4; chl 4
2958	Bicyclo[2.2.1]heptan-2-one, 3-methylene- 3-Methylene-2-norbornanone	C$_8$H$_{10}$O 122.17	5597-27-3	69-71^{11}	4-07-00-00301 0.997^{25}	1.4891^{20}
2959	Bicyclo[2.2.1]heptan-2-one, 1,4,7,7-tetramethyl-	C$_{11}$H$_{18}$O 166.26	10309-50-9 168	213	2-07-00-00113	eth 4; EtOH 4
2960	Bicyclo[2.2.1]heptan-2-one, 1,3,3-trimethyl-, (±)-	C$_{10}$H$_{16}$O 152.24	18492-37-0 -18	193.5	3-07-00-00393 0.9492^{15}	H$_2$O 1; EtOH 4; eth 3; ace 3 1.4702^{20}
2961	Bicyclo[2.2.1]heptan-2-one, 1,7,7-trimethyl-, (1R)- (+)-Camphor	C$_{10}$H$_{16}$O 152.24	464-49-3 178.8	1738 207.4	4-07-00-00213 0.990^{25}	H$_2$O 1; EtOH 4; eth 4; ace 3 1.5462
2962	Bicyclo[2.2.1]heptan-2-one, 1,7,7-trimethyl-, oxime, (1S)- (-)-Camphor oxime	C$_{10}$H$_{17}$NO 167.25	36065-15-3 118	252 dec	4-07-00-00215 1.01^{116}	eth 4; EtOH 4
2963	Bicyclo[2.2.1]heptan-2-one, 4,7,7-trimethyl-, (1S)-	C$_{10}$H$_{16}$O 152.24	10292-98-5 186.5	213	3-07-00-00421	H$_2$O 2; EtOH 4; eth 4; peth 4
2964	Bicyclo[3.1.1]heptan-2-one, 4,6,6-trimethyl-, [1R-(1α,4α,5α)]-	C$_{10}$H$_{16}$O 152.24	515-90-2	222	4-07-00-00210 0.961^{20}	1.4752^{20}
2965	Bicyclo[3.1.1]heptan-3-one, 2,6,6-trimethyl-, [1S (1α,2β,5α)]	C$_{10}$H$_{16}$O 152.24	14575-93-0	212	4-07-00-00209 0.964^{25}	1.4735^{25}
2966	Bicyclo[2.2.1]hept-5-ene-2-carbonitrile	C$_8$H$_9$N 119.17	95-11-4 13	82-6^{10}	0.999^{25}	1.4885^{20}
2967	Bicyclo[2.2.1]hept-5-ene-2-carboxaldehyde	C$_8$H$_{10}$O 122.17	5453-80-5	67-70^{12}	2-07-00-00125 1.018^{25}	1.4893^{20}
2968	Bicyclo[2.2.1]hept-2-ene, 2,3-dimethyl- 2-Norbornene, 2,3-dimethyl-	C$_9$H$_{14}$ 122.21	529-16-8	140.5	4-05-00-00427 0.8698^{17}	eth 3; ace 3; bz 3 1.4688^{17}
2969	Bicyclo[3.1.1]hept-2-ene-2-ethanol, 6,6-dimethyl-	C$_{11}$H$_{18}$O 166.26	128-50-7	235; 110^{10}	4-06-00-00398 0.973^{25}	chl 3 1.4930^{20}
2970	Bicyclo[2.2.1]hept-2-ene, 1,7,7-trimethyl-	C$_{10}$H$_{16}$ 136.24	464-17-5 113	146	2-05-00-00105	bz 4; eth 4; EtOH 4
2971	Bicyclo[3.1.1]hept-2-ene, 2,6,6-trimethyl-, (1S)- (-)-α-Pinene	C$_{10}$H$_{16}$ 136.24	7785-26-4 -64	7414 155	4-05-00-00452 0.859^{20}	1.4660^{20}
2972	Bicyclo[3.1.1]hept-2-ene, 2,6,6-trimethyl-, (±)- (±)-α-Pinene	C$_{10}$H$_{16}$ 136.24	2437-95-8 -55	7414 156.2	2-05-00-00097 0.8582^{20}	H$_2$O 1; EtOH 5; eth 5; chl 5 1.4658^{20}
2973	Bicyclo[3.1.1]hept-2-ene, 2,6,6-trimethyl 2-Pinene	C$_{10}$H$_{16}$ 136.24	80-56-8 -64	155.9	0.8539^{25}	1.4632^{25}
2974	Bicyclo[4.1.0]hept-3-ene, 3,7,7-trimethyl-, (1S)-	C$_{10}$H$_{16}$ 136.24	498-15-7	171; 123^{200}	4-05-00-00450 0.8549^{30}	ace 4; bz 4; eth 4 1.469^3
2975	Bicyclo[3.1.1]hept-3-en-2-ol, 4,6,6-trimethyl-, (1α,2β,5α)-	C$_{10}$H$_{16}$O 152.24	1845-30-3 15.5	90^{10}	0.9684^{25}	1.4912^{25}
2976	Bicyclo[3.1.1]hept-3-en-2-ol, 4,6,6-trimethyl-, (1α,2α,5α)-	C$_{10}$H$_{16}$O 152.24	1820-09-3 24	92^{10}	4-06-00-00383 0.9657^{25}	1.4908^{25}
2977	Bicyclo[3.1.1]hept-2-en-6-one, 2,7,7-trimethyl- Chrysanthenone	C$_{10}$H$_{14}$O 150.22	473-06-3	2256 88-9^{12}	4-07-00-00327	EtOH 4 1.4720^{22}
2978	Bicyclo[3.1.1]hept-3-en-2-one, 4,6,6-trimethyl-, (1R)- d-Verbenone	C$_{10}$H$_{14}$O 150.22	18309-32-5 9.8	9862 227.5	4-07-00-00327 0.9978^{20}	H$_2$O 3; EtOH 3; ace 3; bz 3 1.4993^{18}
2979	[Bi-1,4-cyclohexadien-1-yl]-3,3',6,6'-tetrone, 2,2'-dihydroxy-4,4'-dimethyl- Phenicin	C$_{14}$H$_{10}$O$_6$ 274.23	128-68-7 230.5	7194	4-08-00-03660	H$_2$O 2; EtOH 4; chl 4; HOAc 4
2980	Bicyclo[3,1,0]hexane, 1-methyl- 1-Methylbicyclo[3,1,0]hexane	C$_7$H$_{12}$ 96.17	4625-24-5	93.1		

No.	Name Synonym	Mol. Form. Mol. Wt.	CAS RN mp/°C	Merck No. bp/°C	Beil. Ref. den/g cm^{-3}	Solubility n_D
2981	Bicyclo[3.1.0]hexane, 4-methylene-1-(1-methylethyl)-, (±)- 4(10)-Thujene, (±)-	$C_{10}H_{16}$ 136.24	15826-80-9	164; 69[30]	3-05-00-00365 0.8437[20]	H_2O 1; EtOH 4; eth 3; ace 3 1.4676[20]
2982	Bicyclo[3.1.0]hexane, 4-methyl-1-(1-methylethyl)-	$C_{10}H_{18}$ 138.25	471-12-5	157	4-05-00-00317 0.8139[20]	1.4376[20]
2983	Bicyclo[3.1.0]hexan-3-ol, 4-methylene-1-(1-methylethyl)- 4(10)-Thujene-3-ol	$C_{10}H_{16}O$ 152.24	59905-55-4 208		4-06-00-00382 0.9461[25]	1.4871[25]
2984	Bicyclo[3.1.0]hexan-3-ol, 4-methylene-1-(1-methylethyl)-, [1S-(1α,3β,5α)]-	$C_{10}H_{16}O$ 152.24	471-16-9 208		4-06-00-00382 0.9488[19]	eth 3 1.4871[25]
2985	Bicyclo[3.1.0]hexan-3-ol, 4-methyl-1-(1-methylethyl)-	$C_{10}H_{18}O$ 154.25	513-23-5 209		3-06-00-00278 0.9210[20]	1.4621[25]
2986	Bicyclo[3.1.0]hex-2-ene, 2-methyl-5-(1-methylethyl)-	$C_{10}H_{16}$ 136.24	2867-05-2 151		4-05-00-00451 0.8301[20]	1.4515[20]
2987	Bicyclo[3.1.0]hex-3-en-2-one, 4-methyl-1-(1-methylethyl)-	$C_{10}H_{14}O$ 150.22	24545-81-1 220		0-07-00-00159 0.9572[15]	1.4832[5]
2988	1,1'-Bicyclohexyl Cyclohexylcyclohexane	$C_{12}H_{22}$ 166.31	92-51-3 4		238	H_2O 2; EtOH 3; eth 3
2989	[1,1'-Bicyclohexyl]-2-one	$C_{12}H_{20}O$ 180.29	90-42-6 -32	264	4-07-00-00248 0.9696[25]	1.4877[25]
2990	Bicyclo[3.3.1]nonane	C_9H_{16} 124.23	280-65-9 145.5	169.5	4-05-00-00293	EtOH 4
2991	Bicyclo[4.2.0]octane	C_8H_{14} 110.20	278-30-8 136		4-05-00-00276 0.8573[20]	1.4613[20]
2992	Bicyclo[2.2.2]octane, 2-methyl- 2-Methylbicyclo[2.2.2]octane	C_9H_{16} 124.23	766-53-0 33.5	159	4-05-00-00295 0.8664[40]	1.4608[40]
2993	Bicyclo[4.2.0]oct-2-ene	C_8H_{12} 108.18	13367-29-8 138		4-05-00-00413 0.8948[20]	1.4810[20]
2994	[1,1'-Bicyclopentyl]-2-ol Bicyclopentyl, 2-hydroxy	$C_{10}H_{18}O$ 154.25	4884-25-7 20	235.5	3-06-00-00261 0.9785[15]	1.4884[17]
2995	[1,1'-Bicyclopentyl]-2-one	$C_{10}H_{16}O$ 152.24	4884-24-6 -13	232.5	4-07-00-00198 0.9745[21]	1.4763
2996	Bicyclo[7.2.0]undec-4-ene, 4,11,11-trimethyl-8-methylene-, [1R-(1R*,4E,9S*)]- Caryophyllene	$C_{15}H_{24}$ 204.36	87-44-5 122[13.5]	1884	4-05-00-01182 0.9075[20]	bz 4 1.4986[20]
2997	Bifenox Benzoic acid, 5-(2,4-dichlorophenoxy)-2-nitro-, methyl ester	$C_{14}H_9Cl_2NO_5$ 342.13	42576-02-3 85	1228		
2998	Bifenthrin 2-Methylbiphenyl-3-ylmethyl (Z)-(1RS,3RS)-3-(2-chloro-3,3,3-trifluoroprop-1-enyl)- 2,2-dimethylcyclopropanecarboxylate	$C_{23}H_{22}ClF_3O_2$ 422.87	82657-04-3 69	1229	1.2[125]	
2999	[2,2'-Bi-1H-indene]-1,1',3,3'(2H,2'H)-tetrone, 2,2'-dihydroxy- Hydrindantin	$C_{18}H_{10}O_6$ 322.27	5103-42-4 250 dec	4698	4-08-00-03701	
3000	21H-Biline-8,12-dipropanoic acid, 2,17-diethenyl-1,10,19,22,23,24-hexahydro-3,7,13,18-tetramethyl-1,19-dioxo- Bilirubin	$C_{33}H_{36}N_4O_6$ 584.67	635-65-4	1235	4-26-00-03268	H_2O 1; EtOH 2; eth 2; bz 3
3001	21H-Biline-8,12-dipropanoic acid, 3,18-diethenyl-1,19,22,24-tetrahydro-2,7,13,17-tetramethyl-1,19-dioxo- Biliverdine	$C_{33}H_{34}N_4O_6$ 582.66	114-25-0 >300	1236	4-26-00-03272	H_2O 1; EtOH 3; eth 2; bz 3
3002	[2,2'-Bimorphinan]-3,3',6,6'-tetrol, 7,7',8,8'-tetradehydro-4,5:4',5'-diepoxy-17,17'-dimethyl- Pseudomorphine	$C_{34}H_{36}N_2O_6$ 568.67	125-24-6 282.5	7934	4-27-00-08724	H_2O 1; EtOH 1; eth 1; chl 1
3003	Binapacryl 2-Butenoic acid, 3-methyl-, 2-(1-methylpropyl)-4,6-dinitrophenyl ester	$C_{15}H_{18}N_2O_6$ 322.32	485-31-4 70	1237	1.27[20]	
3004	1,1'-Binaphthalene 1,1'-Binaphthyl	$C_{20}H_{14}$ 254.33	604-53-5 160	>360; 240[12]	4-05-00-02634 1.3000[20]	H_2O 1; EtOH 2; eth 3; ace 3
3005	2,2'-Binaphthalene	$C_{20}H_{14}$ 254.33	612-78-2 188.3	452	4-05-00-02636	H_2O 1; EtOH 2; eth 3; bz 3
3006	[1,1'-Binaphthalene]-4,4'-diol	$C_{20}H_{14}O_2$ 286.33	1446-34-0 300	sub	4-06-00-07021	eth 4; EtOH 4
3007	2,2'-Bioxirane, (R*,S*)- Erythritol anhydride	$C_4H_6O_2$ 86.09	564-00-1 -19	3621 138	5-19-01-00184 1.113[18]	H_2O 5; EtOH 4 1.4330[20]
3008	1,1'-Biphenyl Diphenyl	$C_{12}H_{10}$ 154.21	92-52-4 69	3314 256.1	4-05-00-01807 1.04[20]	H_2O 1; EtOH 3; eth 3; bz 4 1.588[77]
3009	[1,1'-Biphenyl]-4-acetic acid Felbinac	$C_{14}H_{12}O_2$ 212.25	5728-52-9 160.5	3893	4-09-00-02520	
3010	[1,1'-Biphenyl]-2-amine	$C_{12}H_{11}N$ 169.23	90-41-5 51	299	4-12-00-03223	H_2O 1; EtOH 3; eth 3; bz 3
3011	[1,1'-Biphenyl]-4-amine p-Biphenylamine	$C_{12}H_{11}N$ 169.23	92-67-1 53.5	1248 302	4-12-00-03241	H_2O 2; EtOH 3; eth 3; ace 3
3012	[1,1'-Biphenyl]-4-amine, 4'-chloro- 4-Amino-4'-chlorodiphenyl	$C_{12}H_{10}ClN$ 203.67	135-68-2 134	445	4-12-00-03269	ace 4; bz 4; eth 4

No.	Name Synonym	Mol. Form. Mol. Wt.	CAS RN	Merck No. bp/°C	Beil. Ref. den/g cm⁻³	Solubility n_D
3013	[1,1'-Biphenyl]-2-amine, 6-methyl- Biphenyl, 2-amino-6-methyl	$C_{13}H_{13}N$ 183.25	76472-83-8 43.5	355; 144[2]	3-12-00-03234 1.9060[0]	bz 4; eth 4; EtOH 4
3014	[1,1'-Biphenyl]-4-amine, 3-methyl-	$C_{13}H_{13}N$ 183.25	63019-98-7 43	190[15]	3-12-00-03238	bz 4; eth 4; EtOH 4
3015	[1,1'-Biphenyl]-4-amine, 4'-methyl- Biphenyl, 4-amino-4'-methyl	$C_{13}H_{13}N$ 183.25	1204-78-0 99	190[18]	2-12-00-00771	ace 4; bz 4; eth 4; EtOH 4
3016	1,1'-Biphenyl, 2-bromo-	$C_{12}H_9Br$ 233.11	2052-07-5 1.5	297	4-05-00-01818 1.2175[26]	eth 4; EtOH 4 1.6248[25]
3017	1,1'-Biphenyl, 3-bromo-	$C_{12}H_9Br$ 233.11	2113-57-7	300; 171[17]	4-05-00-01818	H_2O 1 1.6411[20]
3018	1,1'-Biphenyl, 4-bromo-	$C_{12}H_9Br$ 233.11	92-66-0 91.5	310	4-05-00-01819 0.9327[25]	H_2O 1; EtOH 3; eth 3; bz 3
3019	[1,1'-Biphenyl]-4-butanoic acid	$C_{16}H_{16}O_2$ 240.30	6057-60-9 119.5	190[0.3]	4-09-00-02558	eth 4; EtOH 4
3020	[1,1'-Biphenyl]-2-carbonitrile	$C_{13}H_9N$ 179.22	24973-49-7 41	175[13]	3-09-00-03269	eth 4; EtOH 4
3021	[1,1'-Biphenyl]-2-carboxylic acid	$C_{13}H_{10}O_2$ 198.22	947-84-2 114.3	343.5	4-09-00-02472	H_2O 1; EtOH 4; bz 4; HOAc 4
3022	[1,1'-Biphenyl]-4-carboxylic acid	$C_{13}H_{10}O_2$ 198.22	92-92-2 228	sub	4-09-00-02479	H_2O 1; EtOH 3; eth 3; bz 3
3023	1,1'-Biphenyl, 2-chloro-	$C_{12}H_9Cl$ 188.66	2051-60-7 34	274	4-05-00-01816 1.1499[32]	eth 4; EtOH 4; lig 4
3024	1,1'-Biphenyl, 3-chloro-	$C_{12}H_9Cl$ 188.66	2051-61-8 16	284.5	4-05-00-01816 1.1579[25]	ace 4; eth 4; EtOH 4 1.6181[25]
3025	1,1'-Biphenyl, 4-chloro-	$C_{12}H_9Cl$ 188.66	2051-62-9 78.8	292.9; 146[10]	4-05-00-01816	H_2O 1; EtOH 3; eth 3; lig 3
3026	Biphenyl, 2-chloro-2'-methyl- 2-Chloro-2'-methylbiphenyl	$C_{13}H_{11}Cl$ 202.68	19493-31-3 17	276; 174[50]	3-05-00-01799	1.588[25]
3027	1,1'-Biphenyl, 2,2',3,3',4,4',5,5',6,6'- decafluoro-	$C_{12}F_{10}$ 334.12	434-90-2 67.5	206	4-05-00-01816 1.785[20]	
3028	[1,1'-Biphenyl]-2,2'-diamine	$C_{12}H_{12}N_2$ 184.24	1454-80-4 81	162[4]	4-13-00-00356 1.3090[20]	H_2O 3; ace 3; bz 3
3029	[1,1'-Biphenyl]-2,4'-diamine 2,4-Biphenyldiamine	$C_{12}H_{12}N_2$ 184.24	492-17-1 54.5	1249 363	4-13-00-00360	H_2O 1; EtOH 3; eth 3
3030	[1,1'-Biphenyl]-3,3'-diamine	$C_{12}H_{12}N_2$ 184.24	2050-89-7 93.5	205[0.001]	4-13-00-00363	H_2O 2; eth 4; bz 3
3031	[1,1'-Biphenyl]-4,4'-diamine Benzidine	$C_{12}H_{12}N_2$ 184.24	92-87-5 120	1086 401	4-13-00-00364	H_2O 2; EtOH 3; eth 2; DMSO 2
3032	[1,1'-Biphenyl]-4,4'-diamine, 2,2'-dichloro- o-Dichlorobenzidine	$C_{12}H_{10}Cl_2N_2$ 253.13	84-68-4 165	3046	4-13-00-00384	eth 4; EtOH 4
3033	[1,1'-Biphenyl]-4,4'-diamine, 3,3'-dimethoxy- o-Dianisidine	$C_{14}H_{16}N_2O_2$ 244.29	119-90-4 137	2070	4-13-00-02834	H_2O 1; EtOH 3; eth 3; ace 3
3034	[1,1'-Biphenyl]-4,4'-diamine, 3,3'-dimethyl- o-Tolidine	$C_{14}H_{16}N_2$ 212.29	119-93-7 131.5	9437	4-13-00-00419	H_2O 2; EtOH 4; eth 4; chl 2
3035	[1,1'-Biphenyl]-4,4'-diamine, N,N'-diphenyl- N,N'-Diphenylbenzidine	$C_{24}H_{20}N_2$ 336.44	531-91-9 247	3319	4-13-00-00368	H_2O 1; EtOH 2; eth 2; bz 2
3036	[1,1'-Biphenyl]-4,4'-diamine, tetramethyl-	$C_{16}H_{20}N_2$ 240.35	34314-06-2 198	>360	3-13-00-00429	bz 4; chl 4
3037	1,1'-Biphenyl, 4,4'-dibromo-	$C_{12}H_8Br_2$ 312.00	92-86-4 164	357.5	4-05-00-01820	H_2O 1; EtOH 2; bz 3
3038	[1,1'-Biphenyl]-2,2'-dicarboxylic acid o,o'-Diphenic acid	$C_{14}H_{10}O_4$ 242.23	482-05-3 233.5	3309 sub	4-09-00-03552	H_2O 1; EtOH 3; eth 3; os 3
3039	[1,1'-Biphenyl]-2,2'-dicarboxylic acid, dimethyl ester	$C_{16}H_{14}O_4$ 270.28	5807-64-7 74	204-6[14]	4-09-00-03553	H_2O 1; EtOH 4; eth 3; bz 3
3040	1,1'-Biphenyl, 3,3'-dichloro-	$C_{12}H_8Cl_2$ 223.10	2050-67-1 29	320	3-05-00-01739	bz 4; eth 4; EtOH 4
3041	1,1'-Biphenyl, 4,4'-dichloro-	$C_{12}H_8Cl_2$ 223.10	2050-68-2 149.3	317	4-05-00-01817 1.4420[0]	H_2O 1; EtOH 2; bz 3; chl 2
3042	1,1'-Biphenyl, 2,2'-difluoro-	$C_{12}H_8F_2$ 190.19	388-82-9 118.5		4-05-00-01815 1.393[20]	H_2O 1; EtOH 3; xyl 3
3043	1,1'-Biphenyl, 3,3'-difluoro-	$C_{12}H_8F_2$ 190.19	396-64-5 8	130[14]	4-05-00-01815 1.192[25]	ctc 2 1.5678[20]
3044	1,1'-Biphenyl, 4,4'-difluoro- 4,4'-Difluorodiphenyl	$C_{12}H_8F_2$ 190.19	398-23-2 94.5	3133 254.5	4-05-00-01816	H_2O 1; EtOH 4; eth 3; ace 3
3045	1,1'-Biphenyl, 2,2'-dimethoxy- 2,2'-Dimethoxybiphenyl	$C_{14}H_{14}O_2$ 214.26	4877-93-4 155	307.5	4-06-00-06645 1.268[25]	H_2O 1; EtOH 4; eth 2; bz 4
3046	1,1'-Biphenyl, 3,3'-dimethoxy- 3,3'-Dimethoxybiphenyl	$C_{14}H_{14}O_2$ 214.26	6161-50-8 36	328	4-06-00-06650	H_2O 1; EtOH 4; eth 4; ace 3
3047	1,1'-Biphenyl, 4,4'-dimethoxy-	$C_{14}H_{14}O_2$ 214.26	2132-80-1 175	sub	4-06-00-06651	H_2O 1; EtOH 4; eth 2; bz 4
3048	1,1'-Biphenyl, 2,2'-dimethoxy-5,5'-dimethyl- Biphenyl, 2,2'-dimethoxy-5,5'-dimethyl	$C_{16}H_{18}O_2$ 242.32	7168-55-0 71	188[12]	3-06-00-05447	ace 4; bz 4; eth 4; EtOH 4
3049	1,1'-Biphenyl, 2,5-dimethoxy-2',5-dimethyl- Biphenyl, 2,5-dimethoxy-2',5-dimethyl	$C_{16}H_{18}O_2$ 242.32	72935-12-7 66	168[4]	4-06-00-06704	bz 4; peth 4
3050	1,1'-Biphenyl, 2,2'-dimethyl-	$C_{14}H_{14}$ 182.27	605-39-0 19.5	256	4-05-00-01897 0.9906[20]	H_2O 1; EtOH 4; eth 4; ace 3 1.5752[20]
3051	1,1'-Biphenyl, 2,3-dimethyl-	$C_{14}H_{14}$ 182.27	611-43-8	270	4-05-00-01902 0.9924[20]	H_2O 1; EtOH 4; eth 4; ace 3 1.5810[20]

No.	Name Synonym	Mol. Form. Mol. Wt.	CAS RN mp/°C	Merck No. bp/°C	Beil. Ref. den/g cm^{-3}	Solubility n_D
3052	1,1'-Biphenyl, 2,3-dimethyl- 2,3-Dimethylbiphenyl	$C_{14}H_{14}$ 182.27	3864-18-4 42	141[14]; 98[2]	4-05-00-01897	H_2O 1; eth 3; peth 3 1.5845[23]
3053	1,1'-Biphenyl, 2,4-dimethyl-	$C_{14}H_{14}$ 182.27	611-61-0	277.2	4-05-00-01903 0.9924[20]	H_2O 1; EtOH 4; eth 4; ace 3 1.5826[20]
3054	1,1'-Biphenyl, 2,4-dimethyl- 2,4-Dimethylbiphenyl	$C_{14}H_{14}$ 182.27	4433-10-7	273; 144[20]	4-05-00-01897 0.9947[20]	1.5844[20]
3055	1,1'-Biphenyl, 2,5-dimethyl-	$C_{14}H_{14}$ 182.27	7372-85-2	140[14.5]	4-05-00-01897 0.9931[20]	1.5819[20]
3056	1,1'-Biphenyl, 2,6-dimethyl-	$C_{14}H_{14}$ 182.27	3976-34-9 -5	262.5	4-05-00-01897 0.9907[20]	1.5745[20]
3057	1,1'-Biphenyl, 3,3'-dimethyl-	$C_{14}H_{14}$ 182.27	612-75-9 9	280	4-05-00-01903 0.9995[20]	H_2O 1; EtOH 4; eth 4; ace 3 1.5946[20]
3058	1,1'-Biphenyl, 3,4-dimethyl-	$C_{14}H_{14}$ 182.27	7383-90-6 14.5	289; 153[15]	4-05-00-01905 0.9978[20]	H_2O 1; bz 3 1.5968[20]
3059	1,1'-Biphenyl, 3,4-dimethyl- 3,4-Dimethylbiphenyl	$C_{14}H_{14}$ 182.27	4433-11-8 29.5	282	4-05-00-01903 1.0087[20]	H_2O 1; bz 3; chl 3 1.6036[20]
3060	1,1'-Biphenyl, 3,5-dimethyl- 3,5-Dimethylbiphenyl	$C_{14}H_{14}$ 182.27	17057-88-4 22.5	274.5	4-05-00-01903 0.9990[20]	1.5952[20]
3061	1,1'-Biphenyl, 4,4'-dimethyl-	$C_{14}H_{14}$ 182.27	613-33-2 125	295	4-05-00-01906 0.917[121]	H_2O 1; EtOH 2; eth 3; ace 3
3062	1,1'-Biphenyl, 2,2'-dinitro-	$C_{12}H_8N_2O_4$ 244.21	2436-96-6 126	305	4-05-00-01826 1.45[25]	H_2O 1; EtOH 4; eth 3; ace 2
3063	1,1'-Biphenyl, 2,4-dinitro-	$C_{12}H_8N_2O_4$ 244.21	606-81-5 93.5		4-05-00-01827 1.474[25]	H_2O 1; EtOH 4; eth 3; bz 3
3064	[1,1'-Biphenyl]-2,2'-diol	$C_{12}H_{10}O_2$ 186.21	1806-29-7 109	320	4-06-00-06645 1.3420[20]	H_2O 3; EtOH 3; eth 3; ace 3
3065	[1,1'-Biphenyl]-2,4-diol	$C_{12}H_{10}O_2$ 186.21	611-62-1 162.5	342	3-06-00-05387	eth 4
3066	[1,1'-Biphenyl]-3,3'-diol	$C_{12}H_{10}O_2$ 186.21	612-76-0 124.8	247[18]	3-06-00-05388	H_2O 2; EtOH 3; eth 3; bz 3
3067	[1,1'-Biphenyl]-2,2'-diol, 3,3'-dimethyl-	$C_{14}H_{14}O_2$ 214.26	32750-14-4 113	sub	3-06-00-05445	bz 4; eth 4; EtOH 4
3068	[1,1'-Biphenyl]-2,2'-diol, 5,5'-dimethyl-	$C_{14}H_{14}O_2$ 214.26	15519-73-0 153.5	sub	4-06-00-06705	H_2O 2; EtOH 3; eth 4; ace 3
3069	[1,1'-Biphenyl]-4,4'-diol, 3,3',5,5'-tetramethyl-	$C_{16}H_{18}O_2$ 242.32	2417-04-1 221.8	sub	4-06-00-06744	EtOH 2; bz 2; gl HOAc 2; tol 2
3070	[1,1'-Biphenyl]-4,4'-disulfonic acid 4,4'-Biphenyldisulfonic acid	$C_{12}H_{10}O_6S_2$ 314.34	5314-37-4 72.5	>200	3-11-00-00472	H_2O 4
3071	1,1'-Biphenyl, 3-ethoxy- 3-Ethoxybiphenyl	$C_{14}H_{14}O$ 198.26	54852-73-2 35	305	3-06-00-03313	EtOH 3; eth 3; ace 3; bz 3
3072	1,1'-Biphenyl, 2-ethyl-	$C_{14}H_{14}$ 182.27	1812-51-7 -6	266; 128[11]	4-05-00-01896 0.9966[20]	1.5758[25]
3073	1,1'-Biphenyl, 3-ethyl-	$C_{14}H_{14}$ 182.27	5668-93-9 -27.5	283.5	4-05-00-01896 1.048[25]	1.5859[25]
3074	1,1'-Biphenyl, 2-fluoro-	$C_{12}H_9F$ 172.20	321-60-8 73.5	248	4-05-00-01815 1.2452[25]	EtOH 3; eth 3; chl 3; lig 2
3075	1,1'-Biphenyl, 4-fluoro-	$C_{12}H_9F$ 172.20	324-74-3 74.2	253	4-05-00-01815 1.247[25]	EtOH 2; eth 3; gl HOAc 3
3076	1,1'-Biphenyl, 2,2',4,4',6,6'-hexamethyl-	$C_{18}H_{22}$ 238.37	4482-03-5 103.5	297	4-05-00-01989 1.023[50]	H_2O 1; EtOH 2; eth 3; bz 3
3077	1,1'-Biphenyl, 2-iodo-	$C_{12}H_9I$ 280.11	2113-51-1	190[36]; 158[6]	4-05-00-01820 1.5991[25]	H_2O 1; EtOH 3; eth 3; bz 3 1.6620[20]
3078	1,1'-Biphenyl, 3-iodo-	$C_{12}H_9I$ 280.11	20442-79-9 26.5	188[16]	4-05-00-01821 1.5967[25]	
3079	1,1'-Biphenyl, 4-iodo-	$C_{12}H_9I$ 280.11	1591-31-7 113.5	320; 183[11]	4-05-00-01821	H_2O 1; EtOH 3; eth 3; bz 3
3080	1,1'-Biphenyl, 2-methoxy-	$C_{13}H_{12}O$ 184.24	86-26-0 29	274	4-06-00-04580 1.0233[99]	H_2O 1; EtOH 3; ctc 2; peth 3 1.5641[99]
3081	1,1'-Biphenyl, 4-methoxy-	$C_{13}H_{12}O$ 184.24	613-37-6 90	157[10]	4-06-00-04600 1.0278[100]	H_2O 1; EtOH 3; eth 3 1.5744[100]
3082	1,1'-Biphenyl, 2-methyl-	$C_{13}H_{12}$ 168.24	643-58-3 -0.2	255.5	4-05-00-01855 1.010[22]	H_2O 1; EtOH 3; eth 3 1.5914[20]
3083	1,1'-Biphenyl, 3-methyl-	$C_{13}H_{12}$ 168.24	643-93-6 4.5	272.7	4-05-00-01858 1.0182[17]	H_2O 1; EtOH 3; eth 3; ctc 3 1.5972[20]
3084	1,1'-Biphenyl, 4-methyl-	$C_{13}H_{12}$ 168.24	644-08-6 49.5	267.5	4-05-00-01860 1.015[27]	H_2O 1; EtOH 3; eth 3; ctc 2
3085	1,1'-Biphenyl, 2-nitro- o-Nitrobiphenyl	$C_{12}H_9NO_2$ 199.21	86-00-0 37.2	6513 320	4-05-00-01823 1.44[25]	H_2O 1; EtOH 3; eth 3; chl 3
3086	1,1'-Biphenyl, 3-nitro-	$C_{12}H_9NO_2$ 199.21	2113-58-8 62	227[35]; 143[9]	4-05-00-01823	H_2O 1; EtOH 3; eth 3; HOAc 3
3087	1,1'-Biphenyl, 4-nitro- p-Nitrobiphenyl	$C_{12}H_9NO_2$ 199.21	92-93-3 114	6514 340	4-05-00-01823	H_2O 1; EtOH 2; eth 3; bz 3
3088	[1,1'-Biphenyl]-2-ol o-Phenylphenol	$C_{12}H_{10}O$ 170.21	90-43-7 59	7276 286	4-06-00-04579 1.213[25]	H_2O 1; EtOH 3; eth 4; ace 3
3089	[1,1'-Biphenyl]-3-ol m-Phenylphenol	$C_{12}H_{10}O$ 170.21	580-51-8 78	>300	4-06-00-04597	H_2O 2; EtOH 4; eth 4; bz 4
3090	[1,1'-Biphenyl]-4-ol p-Phenylphenol	$C_{12}H_{10}O$ 170.21	92-69-3 166	7277 305	4-06-00-04600	H_2O 2; EtOH 4; eth 4; chl 4

No.	Name Synonym	Mol. Form. Mol. Wt.	CAS RN mp/°C	Merck No. bp/°C	Beil. Ref. den/g cm^{-3}	Solubility n_D
3091	[1,1'-Biphenyl]-4-ol, acetate	C$_{14}$H$_{12}$O$_2$ 212.25	148-86-7 89	196[13]	3-06-00-03326	H$_2$O 2; EtOH 2; eth 3; bz 3
3092	[1,1'-Biphenyl]-2-ol, 3-chloro- 2-Phenyl-6-chlorophenol	C$_{12}$H$_9$ClO 204.66	85-97-2 6	7251 317 dec	4-06-00-04586 1.24[25]	H$_2$O 1; EtOH 3; eth 3; ace 3 1.6237[30]
3093	1,1'-Biphenyl, 2-(phenylmethyl)-	C$_{19}$H$_{16}$ 244.34	606-97-3 55	283[110]	3-05-00-02323	bz 4; eth 4; EtOH 4
3094	1,1'-Biphenyl, 4-(phenylmethyl)-	C$_{19}$H$_{16}$ 244.34	613-42-3 85	285-6[110]	4-05-00-02502 1.171[0]	H$_2$O 1; EtOH 3; eth 4; bz 4
3095	1,1'-Biphenyl, 2,2',4,4'-tetramethyl-	C$_{16}$H$_{18}$ 210.32	3976-36-1 41	287	3-05-00-01891	EtOH 3; eth 2; bz 2; chl 3
3096	1,1'-Biphenyl, 2,2',5,5'-tetramethyl-	C$_{16}$H$_{18}$ 210.32	3075-84-1 51.5	285	3-05-00-01892	eth 3; bz 3
3097	[1,1'-Biphenyl]-3,3',5,5'-tetrol Diresorcinol	C$_{12}$H$_{10}$O$_4$ 218.21	531-02-2 310	3361	2-06-00-01129	H$_2$O 4; eth 4; EtOH 4
3098	2,3'-Bipiperidine 2,3-Bipiperidyl	C$_{10}$H$_{20}$N$_2$ 168.28	2467-09-6 68.5	269.5	4-23-00-00512	H$_2$O 4; EtOH 4
3099	2,2'-Bipyridine α,α'-Dipyridyl	C$_{10}$H$_8$N$_2$ 156.19	366-18-7 72	3355 273.5	5-23-08-00016	H$_2$O 2; EtOH 4; eth 4; bz 4
3100	2,3'-Bipyridine 2,3'-Bipyridyl	C$_{10}$H$_8$N$_2$ 156.19	581-50-0	295.5	5-23-08-00026 1.140[20]	H$_2$O 1; EtOH 4; eth 4; bz 4 1.6223[20]
3101	2,4'-Bipyridine	C$_{10}$H$_8$N$_2$ 156.19	581-47-5 61.5	281	5-23-08-00027	H$_2$O 2; EtOH 4; eth 4; chl 4
3102	3,3'-Bipyridine 3,3'-Dipyridyl	C$_{10}$H$_8$N$_2$ 156.19	581-46-4 68	291.5	5-23-08-00027 1.1614[20]	H$_2$O 4; EtOH 4; eth 2
3103	3,4'-Bipyridine 3,4'-Bipyridyl	C$_{10}$H$_8$N$_2$ 156.19	4394-11-0 62	297; 144[15]	5-23-08-00028	EtOH 4; peth 4
3104	4,4'-Bipyridine γ,γ'-Dipyridyl	C$_{10}$H$_8$N$_2$ 156.19	553-26-4 111	3356 305	5-23-08-00028	H$_2$O 4; EtOH 4; eth 3; bz 4
3105	2,3'-Bipyridine, 1,2,3,6-tetrahydro-, (S)- Anatabine	C$_{10}$H$_{12}$N$_2$ 160.22	581-49-7	661 145-6[10]	5-23-06-00507 1.0919[19]	H$_2$O 5; EtOH 3; eth 3; bz 3 1.5676[20]
3106	[5,5'-Bipyrimidine]- 2,2',4,4',6,6'(1H,1'H,3H,3'H,5H,5'H)-hexone- 5,5'-dihydroxy Alloxantin	C$_8$H$_6$N$_4$O$_8$ 286.16	76-24-4 253-5 dec	282	4-26-00-02782	H$_2$O 2; EtOH 2; eth 2
3107	2,3'-Biquinoline 2,3'-Biquinolyl	C$_{18}$H$_{12}$N$_2$ 256.31	612-81-7 175.8	>400	5-23-10-00009	bz 4; eth 4; EtOH 4; chl 4
3108	Bis[1,3]benzodioxolo[4,5-c:5',6'-g]azecin- 13(5H)-one, 4,6,7,14-tetrahydro-5,14- dimethyl-, (R)- Corycavamine	C$_{21}$H$_{21}$NO$_5$ 367.40	521-85-7 149	2540	2-27-00-00621	EtOH 4; chl 4
3109	Bis[1,3]benzodioxolo[4,5-c:5',6'-g]azecin- 13(5H)-one, 4,6,7,14-tetrahydro-5-methyl- Protopine	C$_{20}$H$_{19}$NO$_5$ 353.37	130-86-9 208	7912	4-27-00-06881	H$_2$O 1; EtOH 2; eth 2; bz 2
3110	3H-Bis[1]benzopyrano[3,4-b:6',5'-e]pyran- 7(7aH)-one, 13,13a-dihydro-7a-hydroxy-9,10- dimethoxy-3,3-dimethyl Tephrosin	C$_{23}$H$_{22}$O$_7$ 410.42	76-80-2 198	9082	4-19-00-05271	ace 4; eth 4; chl 4
3111	Bismuthine, triphenyl- Triphenylbismuth	C$_{18}$H$_{15}$Bi 440.30	603-33-8 77.6	242[14]	4-16-00-01226 1.715[75]	EtOH 2; eth 3; ace 3; bz 3 1.7040[75]
3112	2,2'-Bithiophene	C$_8$H$_6$S$_2$ 166.27	492-97-7 33	260	5-19-01-00537	H$_2$O 1; EtOH 4; eth 3; ctc 3
3113	Boranamine, N,N,1,1-tetramethyl-	C$_4$H$_{12}$BN 84.96	1113-30-0 -92	65	4-04-00-04367	eth 4; ace 4; os 5
3114	Boranediamine, 1-fluoro-N,N,N',N'- tetramethyl-	C$_4$H$_{12}$BFN$_2$ 117.96	383-90-4 -44.3	106	4-04-00-00303	ace 4; eth 4
3115	Borane, difluoro(4-methylphenyl)-	C$_7$H$_7$BF$_2$ 139.94	768-39-8	128	4-16-00-01674 1.055[25]	bz 4; eth 4 1.4535[25]
3116	Borane, difluorophenyl-	C$_6$H$_5$BF$_2$ 125.91	368-98-9 -36.2	98	4-16-00-01661 1.087[25]	bz 4; eth 4 1.4441[25]
3117	Borane, dimethoxy- Dimethoxyborane	C$_2$H$_7$BO$_2$ 73.89	4542-61-4 -130.6	25.9	4-01-00-01269	H$_2$O 6
3118	Borane, triethyl Triethylborane	C$_6$H$_{15}$B 98.00	97-94-9 -93	95	0.70[23]	EtOH 3; eth 3 1.3971
3119	Borane, trimethyl- Trimethylborane	C$_3$H$_9$B 55.92	593-90-8 -161.5	-20.2		
3120	Borane, trimethyl-, monoammoniate Aminetrimethylboron	C$_3$H$_{12}$BN 72.95	1830-95-1 73.5			
3121	Borane, tripropyl-	C$_9$H$_{21}$B 140.08	1116-61-6 -56	159	4-04-00-04360 0.7204[25]	1.4135[22]
3122	Borane, tris(3-methylbutyl)- Borane, triisopentyl-	C$_{15}$H$_{33}$B 224.24	3062-81-5	119[14]	2-04-00-01023 0.7600[25]	ace 4; eth 4; EtOH 4 1.4321
3123	Borane, tris(2-methylpropyl)-	C$_{12}$H$_{27}$B 182.16	1116-39-8	188; 86[20]	4-04-00-04361 0.7380[25]	bz 4; eth 4; EtOH 4 1.4188[23]
3124	Boric acid, tributyl ester	C$_{12}$H$_{27}$BO$_3$ 230.16	688-74-4	234	4-01-00-01544 0.8567[20]	EtOH 3; eth 4; bz 3; MeOH 4 1.4106[18]
3125	Boric acid, triethyl ester	C$_6$H$_{15}$BO$_3$ 145.99	150-46-9 -84.8	120	4-01-00-01365 0.8546[20]	EtOH 5; eth 5 1.3749[20]
3126	Boric acid, trimethyl ester Trimethyl borate	C$_3$H$_9$BO$_3$ 103.91	121-43-7 -29.3	9626 67.5	4-01-00-01269 0.915[25]	eth 4; MeOH 4 1.3568[20]

No.	Name / Synonym	Mol. Form. / Mol. Wt.	CAS RN / mp/°C	Merck No. / bp/°C	Beil. Ref. / den/g cm^{-3}	Solubility / n_D
3127	Boric acid, tripropyl ester	$C_9H_{21}BO_3$ 188.07	688-71-1 179.5		4-01-00-01436 0.8576[20]	EtOH 4; eth 5; PrOH 3 1.3948[20]
3128	Boric acid, tris(1-methylethyl) ester Isopropyl borate	$C_9H_{21}BO_3$ 188.07	5419-55-6	140; 75[76]	4-01-00-01488 0.8251[20]	EtOH 4; eth 4; bz 4; PrOH 4 1.3777[20]
3129	Borinic acid, dimethylthio-, methyl ester	C_3H_9BS 87.98	19163-05-4 -84	71	4-04-00-04368	ace 4; eth 4
3130	Borinic acid, diphenyl-	$C_{12}H_{11}BO$ 182.03	2622-89-1 57.5	150-5[20]	4-16-00-01639 1.0740[20]	EtOH 3; bz 3; peth 3
3131	3-Bornanecarboxylic acid Camphane-3-carboxylic acid	$C_{11}H_{18}O_2$ 182.26	91965-23-0 90.5	153[13]	3-09-00-00244	EtOH 4; HOAc 4
3132	Boron, (N,N-dimethylmethanamine)trihydro-, (T-4)- Trimethylamineborane	$C_3H_{12}BN$ 72.95	75-22-9 94	172	4-04-00-00134 0.792[25]	eth 4; EtOH 4
3133	Boronic acid, ethyl-	$C_2H_7BO_2$ 73.89	4433-63-0 40	dec	4-04-00-04379	H_2O 4; eth 4; EtOH 4
3134	Boronic acid, (2-methylphenyl)-	$C_7H_9BO_2$ 135.96	16419-60-6 166.5	dec	3-16-00-01277	bz 4; eth 4; EtOH 4
3135	Boronic acid, phenyl- Benzeneboronic acid	$C_6H_7BO_2$ 121.93	98-80-6 219	1076	4-16-00-01654	H_2O 2; EtOH 3; eth 3; bz 3
3136	Boronic acid, propyl-	$C_3H_9BO_2$ 87.91	17745-45-8 107	dec	3-04-00-01964	H_2O 4; eth 4; EtOH 4
3137	Boron, trifluoro[1,1'-oxybis[ethane]]-, (T-4)- Boron trifluoride etherate	$C_4H_{10}BF_3O$ 141.93	109-63-7 -60.4	1352 125.5	4-01-00-01314 1.3572[20]	eth 4; EtOH 4 1.4447[20]
3138	Boron, trifluoro[oxybis[methane]]-, (T-4)- Boron trifluoride compd. with methyl ether	$C_2H_6BF_3O$ 113.88	353-42-4 -14	127 dec	4-01-00-01245 1.2410[20]	1.302[20]
3139	Bromacil 2,4(1H,3H)-Pyrimidinedione, 5-bromo-6-methyl-3-(1-methylpropyl)-	$C_9H_{13}BrN_2O_2$ 261.12	314-40-9 158	1370	1.55[25]	
3140	Bromopropylate Benzeneacetic acid, 4-bromo-α-(4-bromophenyl)-α-hydroxy-, 1-methylethyl ester	$C_{17}H_{16}Br_2O_3$ 428.12	18181-80-1 77	1422	1.59[20]	
3141	Bufa-20,22-dienolide, 16-(acetyloxy)-3,14-dihydroxy-, (3β,5β,16β)- Bufotalin	$C_{26}H_{36}O_6$ 444.57	471-95-4 223 dec	1466	5-18-04-00443	H_2O 1; EtOH 3; chl 3
3142	Bufa-4,20,22-trienolide, 6-(acetyloxy)-3-(β-D-glucopyranosyloxy)-8,14-dihydroxy-, (3β,6β)- Scilliroside	$C_{32}H_{44}O_{12}$ 620.69	507-60-8 169 dec	8356	4-18-00-03178	H_2O 2; EtOH 4; eth 1; ace 2
3143	sec-Bumeton 1,3,5-Triazine-2,4-diamine, N-ethyl-6-methoxy-N'-(1-methylpropyl)-	$C_{10}H_{19}N_5O$ 225.29	26259-45-0 87			
3144	Butachlor Acetamide, N-(butoxymethyl)-2-chloro-N-(2,6-diethylphenyl)-	$C_{17}H_{26}ClNO_2$ 311.85	23184-66-9 <-5	1498 156.5	1.070[25]	
3145	1,2-Butadiene Methylallene	C_4H_6 54.09	590-19-2 -136.2	10.9	4-01-00-00975 0.676[0]	H_2O 1; EtOH 5; eth 5; bz 4 1.4205[1]
3146	1,3-Butadiene Divinyl	C_4H_6 54.09	106-99-0 -108.9	1500 -4.4	4-01-00-00976 *0.6149[25]	H_2O 1; EtOH 3; eth 3; ace 4 1.4292[-25]
3147	1,2-Butadiene, 4-bromo- 4-Bromo-1,2-butadiene	C_4H_5Br 132.99	18668-68-3 110		3-01-00-00929 1.4255[20]	ace 4 1.5248[20]
3148	1,3-Butadiene, 1-bromo- 1-Bromo-1,3-butadiene	C_4H_5Br 132.99	21890-35-7 93		4-01-00-00989 1.416[0]	eth 4; EtOH 4; peth 4
3149	1,3-Butadiene, 2-bromo-	C_4H_5Br 132.99	1822-86-2 42-3[165]		4-01-00-00989 1.397[20]	eth 4; EtOH 4 1.4988[20]
3150	1,2-Butadiene, 4-chloro-	C_4H_5Cl 88.54	25790-55-0 88		4-01-00-00975 0.9891[20]	ace 4; bz 4; eth 4 1.4775[20]
3151	1,3-Butadiene, 1-chloro-	C_4H_5Cl 88.54	627-22-5 68		3-01-00-00949 0.9606[20]	eth 4; EtOH 4; chl 4 1.4712[20]
3152	1,3-Butadiene, 2-chloro- 2-Chloro-1,3-butadiene	C_4H_5Cl 88.54	126-99-8 -130	59.4	4-01-00-00984 0.956[20]	H_2O 2; eth 5; ace 5; bz 5 1.4583[20]
3153	1,3-Butadiene, 1-chloro-2-methyl- Chloromethylbutadiene	C_5H_7Cl 102.56	35383-51-8 107		3-01-00-00974 0.9710[20]	ace 4; EtOH 4 1.4702[20]
3154	1,3-Butadiene, 1-chloro-3-methyl- 4-Chloroisoprene	C_5H_7Cl 102.56	51034-46-9 100.7		4-01-00-01005 0.9543[20]	ace 4; eth 4; EtOH 4; chl 4 1.4719[20]
3155	1,3-Butadiene, 2-chloro-3-methyl-	C_5H_7Cl 102.56	1809-02-5 93		4-01-00-01004 0.9593[20]	H_2O 1; EtOH 3; eth 4; ace 3 1.4686[20]
3156	1,3-Butadiene, 1,1-dichloro- 1,1-Dichloro-1,3-butadiene	$C_4H_4Cl_2$ 122.98	6061-06-9 42-3[90]		4-01-00-00985 1.1831[20]	H_2O 1; eth 4; bz 4; chl 4 1.5022[20]
3157	1,3-Butadiene, 1,2-dichloro- 1,2-Dichloro-1,3-butadiene	$C_4H_4Cl_2$ 122.98	3574-40-1 62[105]; 35[40]		4-01-00-00985 1.1991[20]	ctc 4 1.4960[20]
3158	1,3-Butadiene, 2,3-dichloro-	$C_4H_4Cl_2$ 122.98	1653-19-6 98		4-01-00-00986 1.1829[20]	chl 4 1.4890[20]
3159	1,3-Butadiene, 2,3-dimethyl- Diisopropenyl	C_6H_{10} 82.15	513-81-5 -76	3228 68.8	4-01-00-01023 0.7222[25]	ctc 3 1.4394[20]
3160	1,3-Butadiene, 1-ethoxy-	$C_6H_{10}O$ 98.14	5614-32-4 110.5		4-01-00-02222 0.8154[20]	ace 4; bz 4; eth 4; EtOH 4 1.4529[20]
3161	1,3-Butadiene, 2-ethoxy-	$C_6H_{10}O$ 98.14	4747-05-1 94.5		4-01-00-02223 0.8177[20]	ace 4; bz 4; eth 4; EtOH 4 1.4400[20]
3162	1,3-Butadiene, 2-fluoro- Fluoroprene	C_4H_5F 72.08	381-61-3 12		4-01-00-00982 0.843[4]	1.400[4]
3163	1,3-Butadiene, 1,1,2,3,4,4-hexachloro- Hexachloro-1,3-butadiene	C_4Cl_6 260.76	87-68-3 -21	215	4-01-00-00988 1.556[25]	H_2O 1; EtOH 3; eth 3 1.5542[20]

No.	Name Synonym	Mol. Form. Mol. Wt.	CAS RN mp/°C	Merck No. bp/°C	Beil. Ref. den/g cm^{-3}	Solubility n_D
3164	1,3-Butadiene, 1,1,2,3,4,4-hexafluoro- 1,3-Butadiene, hexafluoro	C_4F_6 162.03	685-63-2 -132	6	4-01-00-00983 1.553[20]	1.378[20]
3165	1,3-Butadiene, 2-iodo-	C_4H_5I 179.99	19221-28-4	112	3-01-00-00956 1.7278[20]	1.5616
3166	1,2-Butadiene, 4-iodo- 4-Iodo-1,2-butadiene	C_4H_5I 179.99	67885-08-9	130	3-01-00-00929 1.7129[20]	1.5709[20]
3167	1,2-Butadiene, 4-methoxy-	C_5H_8O 84.12	36678-06-5	88	4-01-00-02221 0.8286[20]	EtOH 4 1.435[20]
3168	1,3-Butadiene, 1-methoxy-	C_5H_8O 84.12	3036-66-6	91.5	3-01-00-01975 0.8296[20]	H_2O 3; EtOH 3 1.4594[20]
3169	1,3-Butadiene, 2-methoxy-	C_5H_8O 84.12	3588-30-5	75	4-01-00-02223 0.8272[20]	ace 4; bz 4; eth 4; EtOH 4 1.4442[20]
3170	1,2-Butadiene, 3-methyl-	C_5H_8 68.12	598-25-4 -113.6	40.83	4-01-00-01006 0.6806[25]	ace 4; bz 4; eth 4; EtOH 4 1.4203[20]
3171	1,3-Butadiene, 2-methyl- Isoprene	C_5H_8 68.12	78-79-5 -145.9	5087 34.0	4-01-00-01001 0.679[20]	H_2O 1; EtOH 5; eth 5; ace 5 1.4219[20]
3172	1,3-Butadiene, 1,1,2,4,4-pentafluoro-3-(trifluoromethyl)- 1,3-Butadiene, pentafluoro-2-(trifluoromethyl)-	C_5F_8 212.04	384-04-3	39	1.527[0]	ace 4; bz 4; eth 4 1.3000[0]
3173	1,3-Butadiene, 1,2,3,4-tetrachloro-	$C_4H_2Cl_4$ 191.87	1637-31-6 -4	188	4-01-00-00987 1.515[15]	H_2O 1; EtOH 3; eth 3; ace 3 1.5455[20]
3174	1,3-Butadiene, 1,2,3-trichloro- 1,2,3-Trichloro-1,3-butadiene	$C_4H_3Cl_3$ 157.43	1573-58-6 33-4[7]		3-01-00-00954 1.4060[20]	eth 4; chl 4 1.5262[20]
3175	2,3-Butadien-1-ol	C_4H_6O 70.09	18913-31-0	128.5	4-01-00-02221 0.9164[20]	ace 4; bz 4; eth 4; EtOH 4 1.4759[20]
3176	1,3-Butadien-1-ol, acetate	$C_6H_8O_2$ 112.13	1515-76-0	60-1[40]	3-02-00-00295 0.945[25]	1.4690[20]
3177	1,3-Butadiyne	C_4H_2 50.06	460-12-8 -36.4	10.3	4-01-00-01116 0.7364[0]	H_2O 4; EtOH 4; eth 4; ace 4 1.4189[5]
3178	Butanal Butyraldehyde	C_4H_8O 72.11	123-72-8 -99	1591 74.8	4-01-00-03229 0.8016[20]	H_2O 3; EtOH 5; eth 5; ace 4 1.3843[20]
3179	Butanal, 2-bromo-	C_4H_7BrO 151.00	24764-97-4	33[17]	4-01-00-03241 1.469[20]	eth 3; ace 3; bz 3; chl 3 1.4683[20]
3180	Butanal, 2-chloro-	C_4H_7ClO 106.55	28832-55-5	107	4-01-00-03238 1.1072[17]	1.4262[17]
3181	Butanal, 4-chloro-	C_4H_7ClO 106.55	6139-84-0	50-1[13]	4-01-00-03240 1.106[8]	ace 4; eth 4; EtOH 4 1.4466[8]
3182	Butanal, 2,3-dichloro-	$C_4H_6Cl_2O$ 141.00	55775-41-2	58-60[20]	4-01-00-03240 1.2666[21]	ace 4; eth 4; EtOH 4; chl 4 1.4618[21]
3183	Butanal, 2-ethyl- Diethylacetaldehyde	$C_6H_{12}O$ 100.16	97-96-1	117-9[160]	4-01-00-03310 0.8110[20]	H_2O 2; EtOH 5; eth 5; ctc 2 1.4025[20]
3184	Butanal, 3-hydroxy- Aldol	$C_4H_8O_2$ 88.11	107-89-1	217 83[20]	4-01-00-03984 1.103[20]	H_2O 5; EtOH 5; eth 3; ace 4 1.4238[20]
3185	Butanal, 2-methyl-, (±)-	$C_5H_{10}O$ 86.13	57456-98-1	92.5	4-01-00-03285 0.8029[20]	ace 4; eth 4; EtOH 4 1.3869[20]
3186	Butanal, 3-methyl- Isovaleraldehyde	$C_5H_{10}O$ 86.13	590-86-3 -51	5118 92.5	4-01-00-03291 0.7977[20]	H_2O 2; EtOH 3; eth 3 1.3902[20]
3187	Butanal, 3-methyl-, oxime Butyraldehyde, 3-methyl, oxime	$C_5H_{11}NO$ 101.15	626-90-4 48.5	161.3	4-01-00-03293 0.8934[20]	ace 4; eth 4; EtOH 4 1.4367[20]
3188	Butanal, oxime Butyraldehyde, oxime	C_4H_9NO 87.12	110-69-0 -29.5	154	4-01-00-03234 0.923[20]	H_2O 4; EtOH 5; eth 5; ace 4
3189	Butanal, phenylhydrazone Butyraldehyde, phenylhydrazone	$C_{10}H_{14}N_2$ 162.23	940-54-5 94	192[80.] 152[14]	3-15-00-00080	ctc 2
3190	Butanal, 2-(phenylmethylene)-	$C_{11}H_{12}O$ 160.22	28467-92-7	157-8[5]	4-07-00-01024 1.0201[22]	1.578[20]
3191	Butanal, 2,2,3-trichloro- 2,2,3-Trichlorobutyraldehyde	$C_4H_5Cl_3O$ 175.44	76-36-8	9545 164	4-01-00-03241 1.3956[20]	H_2O 4; eth 4; EtOH 4 1.4755[20]
3192	Butanal, 2,3,4-trihydroxy-, [R-(R*,R*)]- D-Erythrose	$C_4H_8O_4$ 120.11	583-50-6	3637	4-01-00-04172	H_2O 3; EtOH 4
3193	Butanal, 2,3,4-trihydroxy-, [S-(R*,S*)]- Threose (D-)	$C_4H_8O_4$ 120.11	95-43-2 129	9317	4-01-00-04173	
3194	Butanamide Butyramide	C_4H_9NO 87.12	541-35-5 114.8	1592 216	4-02-00-00804 0.8850[120]	H_2O 2; EtOH 3; eth 2; bz 1 1.4087[130]
3195	Butanamide, N-(aminocarbonyl)-2-bromo-2-ethyl- Carbromal	$C_7H_{13}BrN_2O_2$ 237.10	77-65-6 118	1837	4-03-00-00117 1.544[25]	H_2O 2; ace 3; bz 3; chl 2
3196	Butanamide, N-(aminocarbonyl)-2-bromo-3-methyl- Bromisovalum	$C_6H_{11}BrN_2O_2$ 223.07	496-67-3 154	1385 sub	3-03-00-00123 1.56[15]	ace 4; bz 4; eth 4; EtOH 4
3197	Butanamide, N-[5-(aminosulfonyl)-1,3,4-thiadiazol-2-yl]- Butazolamide	$C_6H_{10}N_4O_3S_2$ 250.30	16790-49-1	1511	4-27-00-08223	
3198	Butanamide, 2-bromo-2-ethyl- Diethylbromoacetamide	$C_6H_{12}BrNO$ 194.07	511-70-6 67	3105	3-02-00-00755	H_2O 2; EtOH 4; eth 4; bz 4
3199	Butanamide, 2-bromo-2-ethyl-3-methyl- Ibrotamide	$C_7H_{14}BrNO$ 208.10	466-14-8 50.5	4809 sub	4-02-00-00979	ace 4; bz 4; eth 4; EtOH 4
3200	Butanamide, N,N-diethyl-	$C_8H_{17}NO$ 143.23	1114-76-7	206	4-04-00-00354	H_2O 4; EtOH 4 1.4403[25]
3201	Butanamide, N,N-diethyl-3-methyl- Isovaleryl diethylamide	$C_9H_{19}NO$ 157.26	533-32-4	5122 211	4-04-00-00356 0.8764[20]	eth 4; EtOH 4 1.4422[20]

No.	Name Synonym	Mol. Form. Mol. Wt.	CAS RN mp/°C	Merck No. bp/°C	Beil. Ref. den/g cm^{-3}	Solubility n_D
3202	Butanamide, 2,4-dihydroxy-N-(3-hydroxypropyl)-3,3-dimethyl-, (R)- Dexpanthenol	$C_9H_{19}NO_4$ 205.25	81-13-0 dec	2924	4-04-00-01652 1.20^{20}	H$_2$O 4; EtOH 4; eth 2; MeOH 4 1.497^{20}
3203	Butanamide, N,N-dimethyl-	$C_6H_{13}NO$ 115.18	760-79-2 -40	186; 125[100]	4-04-00-00185 0.9064^{25}	ace 4; bz 4; eth 4; EtOH 4 1.4391^{25}
3204	Butanamide, N,N'-(3,3'-dimethyl[1,1'-biphenyl]-4,4'-diyl)bis[3-oxo- N,N'-Bis(acetoacetyl)-3,3'-dimethylbenzidine	$C_{22}H_{24}N_2O_4$ 380.44	91-96-3 212		3-13-00-00490	DMSO 2
3205	Butanamide, N-(4-hydroxyphenyl)- 4'-Hydroxybutyranilide	$C_{10}H_{13}NO_2$ 179.22	101-91-7 139.5	4745	4-13-00-01109	H$_2$O 4; EtOH 4
3206	Butanamide, 3-methyl- Isovaleramide	$C_5H_{11}NO$ 101.15	541-46-8 137	5119 226	4-02-00-00902	H$_2$O 3; EtOH 3; eth 3; peth 4
3207	Butanamide, 3-oxo-N-phenyl- Acetoacetanilide	$C_{10}H_{11}NO_2$ 177.20	102-01-2 86	50	4-12-00-00955	H$_2$O 2; EtOH 3; eth 3; bz 3
3208	Butanamide, N-phenyl-	$C_{10}H_{13}NO$ 163.22	1129-50-6 97	189[15]	4-12-00-00387 1.134^{25}	H$_2$O 1; EtOH 4; eth 4; chl 2
3209	1-Butanamine Butylamine	$C_4H_{11}N$ 73.14	109-73-9 -49.1	1543 77.0	4-04-00-00540 0.7414^{20}	H$_2$O 5; EtOH 3; eth 3 1.4031^{20}
3210	2-Butanamine, (±)- sec-Butylamine, (±)-	$C_4H_{11}N$ 73.14	33966-50-6 <-72	1544 63.5	4-04-00-00618 0.7246^{20}	H$_2$O 3; EtOH 5; eth 5; ace 4 1.3932^{20}
3211	1-Butanamine, N-butyl- Dibutylamine	$C_8H_{19}N$ 129.25	111-92-2 -62	3019 159.6	4-04-00-00550 0.7670^{20}	H$_2$O 3; EtOH 4; eth 4; ace 3 1.4177^{20}
3212	1-Butanamine, N-butyl-N-nitroso- Dibutylnitrosamine	$C_8H_{18}N_2O$ 158.24	924-16-3 105[8]		4-04-00-03389	
3213	1-Butanamine, N,N-dibutyl- Tributylamine	$C_{12}H_{27}N$ 185.35	102-82-9 -70	9530 216.5	4-04-00-00554 0.7770^{20}	H$_2$O 2; EtOH 4; eth 4; ace 3 1.4299^{20}
3214	1-Butanamine, 4,4-diethoxy-	$C_8H_{19}NO_2$ 161.24	6346-09-4 196		4-04-00-01928 0.933^{25}	1.4275^{20}
3215	1-Butanamine, N,N-dimethyl-	$C_6H_{15}N$ 101.19	927-62-8 95		4-04-00-00546 0.7206^{20}	H$_2$O 5; EtOH 5; eth 5; ace 5 1.3970^{20}
3216	2-Butanamine, 2,3-dimethyl-	$C_6H_{15}N$ 101.19	4358-75-2 -70.5	104.5	4-04-00-00733 0.7683^0	1.4096^{17}
3217	2-Butanamine, 3,3-dimethyl-	$C_6H_{15}N$ 101.19	3850-30-4 -20	102	4-04-00-00730 0.7668^{20}	H$_2$O 4 1.4105^{25}
3218	1-Butanamine, N-ethyl- Butylethylamine	$C_6H_{15}N$ 101.19	13360-63-9 107.5		4-04-00-00547 0.7398^{20}	EtOH 5; eth 5; ace 5; bz 5 1.4040^{20}
3219	2-Butanamine, N-ethyl-, (±)- sec-Butyl ethyl amine (DL)	$C_6H_{15}N$ 101.19	116724-10-8 -104.3	98	2-04-00-00636 0.7358^{20}	ace 4; bz 4; eth 4; EtOH 4
3220	1-Butanamine, 3-methyl- Isoamylamine	$C_5H_{13}N$ 87.16	107-85-7 96	4994	4-04-00-00696 0.7505^{20}	H$_2$O 5; EtOH 5; eth 5; ace 3 1.4083^{20}
3221	1-Butanamine, N-methyl- Butylmethylamine	$C_5H_{13}N$ 87.16	110-68-9 91		4-04-00-00546 0.7637^{15}	
3222	2-Butanamine, 2-methyl-	$C_5H_{13}N$ 87.16	594-39-8 -105	77	4-04-00-00694 0.731^{25}	H$_2$O 4; ace 4; eth 4; EtOH 4 1.3954^{25}
3223	2-Butanamine, 3-methyl-	$C_5H_{13}N$ 87.16	598-74-3 -50	85.5	2-04-00-00644 0.7574^{19}	H$_2$O 4; EtOH 3 1.4096^{18}
3224	1-Butanamine, 2-methyl-N,N-bis(2-methylbutyl)-	$C_{15}H_{33}N$ 227.43	620-43-9 232		0-04-00-00179 0.9^{13}	ace 4; bz 4; eth 4; EtOH 4 1.4330^{20}
3225	1-Butanamine, 3-methyl-N,N-bis(3-methylbutyl)-	$C_{15}H_{33}N$ 227.43	645-41-0 235		4-04-00-00700 0.7848^{20}	H$_2$O 1; EtOH 4; eth 5; bz 5 1.4331^{20}
3226	1-Butanamine, 3-methyl-N-(3-methylbutyl)- Diisoamylamine	$C_{10}H_{23}N$ 157.30	544-00-3 -44	3178 188	4-04-00-00699 0.7672^{21}	H$_2$O 1; EtOH 3; eth 5 1.4235^{20}
3227	2-Butanamine, N-(1-methylpropyl)-	$C_8H_{19}N$ 129.25	626-23-3 135		4-04-00-00620 0.7534^{20}	H$_2$O 4; EtOH 3; os 3 1.4162^{20}
3228	1-Butanamine, 1,1,2,2,3,3,4,4,4-nonafluoro-N,N-bis(nonafluorobutyl)-	$C_{12}F_{27}N$ 671.10	311-89-7 178		4-02-00-00819 1.884^{25}	ace 3 1.291^{25}
3229	Butane	C_4H_{10} 58.12	106-97-8 -138.2	1507 -0.5	4-01-00-00236 $^*0.573^{25}$	H$_2$O 3; EtOH 4; eth 4; chl 4 1.3326^{20}
3230	Butane, 2,2-bis(ethylsulfonyl)- Sulfonethylmethane	$C_8H_{18}O_4S_2$ 242.36	76-20-0 76	8937 dec	3-01-00-02790 1.199^{85}	chl 3
3231	Butane, 1-bromo- Butyl bromide	C_4H_9Br 137.02	109-65-9 -112.4	1553 101.6	4-01-00-00258 1.2758^{20}	H$_2$O 1; EtOH 5; eth 5; ace 5 1.4401^{20}
3232	Butane, 2-bromo-, (±)- (±)-sec-Butyl bromide	C_4H_9Br 137.02	5787-31-5 -111.9	1554 91.2	4-01-00-00261 1.2585^{20}	ace 4; eth 4; chl 4 1.4366^{20}
3233	Butane, 1-bromo-4-chloro-	C_4H_8BrCl 171.46	6940-78-9 175; 63[10]		4-01-00-00264 1.488^{20}	H$_2$O 1; EtOH 3; eth 3; ctc 2 1.4885^{20}
3234	Butane, 2-bromo-1-chloro- 2-Bromo-1-chlorobutane	C_4H_8BrCl 171.46	79504-01-1 146.5		4-01-00-00264 1.468^{20}	bz 4; eth 4; EtOH 4; chl 4 1.4880^{20}
3235	Butane, 1-bromo-3,3-dimethyl-	$C_6H_{13}Br$ 165.07	1647-23-0 138		3-01-00-00409 1.1556^{20}	eth 4; EtOH 4; chl 4 1.4440^{20}
3236	Butane, 2-bromo-2,3-dimethyl- 2-Bromo-2,3-dimethylbutane	$C_6H_{13}Br$ 165.07	594-52-5 24.5	133; 87[180]	4-01-00-00374 1.1772^{10}	eth 4; chl 4 1.4517
3237	Butane, 1-bromo-4-fluoro- 1-Bromo-4-fluorobutane	C_4H_8BrF 155.01	462-72-6 135		4-01-00-00263	eth 4; EtOH 4 1.4370^{25}
3238	Butane, 1-bromo-2-methyl-, (S)- d-Amyl bromide	$C_5H_{11}Br$ 151.05	534-00-9 121.6	636	4-01-00-00327 1.2234^{20}	H$_2$O 1; EtOH 3; eth 3; chl 4 1.4451^{20}
3239	Butane, 1-bromo-3-methyl- Isoamyl bromide	$C_5H_{11}Br$ 151.05	107-82-4 -112	4996 120.4	4-01-00-00328 1.2071^{20}	H$_2$O 1; EtOH 3; eth 3; ctc 2 1.4420^{20}

No.	Name Synonym	Mol. Form. Mol. Wt.	CAS RN mp/°C	Merck No. bp/°C	Beil. Ref. den/g cm^{-3}	Solubility n_D
3240	Butane, 2-bromo-2-methyl- *tert*-Amyl bromide	$C_5H_{11}Br$ 151.05	507-36-8	638 108	4-01-00-00327 1.197[18]	1.4421
3241	Butane, 1-chloro- Butyl chloride	C_4H_9Cl 92.57	109-69-3 -123.1	1560 78.6	4-01-00-00246 0.8862[20]	H_2O 1; EtOH 5; eth 5; ctc 2 1.4021[20]
3242	Butane, 2-chloro-, (±)- *sec*-Butyl chloride (*DL*)	C_4H_9Cl 92.57	53178-20-4 -131.3	1561 68.2	4-01-00-00248 0.8732[20]	bz 4; eth 4; chl 4 1.3971[20]
3243	Butane, 3-chloro-1,1-diethoxy- Butyraldehyde, 3-chloro-, diethyl acetal	$C_8H_{17}ClO_2$ 180.67	51786-70-0	70-1[12]	4-01-00-03239 0.9709[20]	bz 4; eth 4; EtOH 4 1.4195[25]
3244	Butane, 3-chloro-1,1-dimethoxy- Butyraldehyde, 3-chloro-, dimethyl acetal	$C_6H_{13}ClO_2$ 152.62	50710-40-2	40-4[7]	4-01-00-03239 1.0024[20]	1.4160[25]
3245	Butane, 1-chloro-2,3-dimethyl- 1-Chloro-2,3-dimethylbutane	$C_6H_{13}Cl$ 120.62	600-06-6	124	4-01-00-00373 0.887[22]	eth 4; EtOH 4; chl 4
3246	Butane, 1-chloro-3,3-dimethyl-	$C_6H_{13}Cl$ 120.62	2855-08-5	115	4-01-00-00369 0.8670[20]	eth 4; EtOH 4; chl 4 1.4161[20]
3247	Butane, 2-chloro-2,3-dimethyl-	$C_6H_{13}Cl$ 120.62	594-57-0 -10.4	112	4-01-00-00373 0.8780[20]	ace 4; EtOH 4 1.4191[20]
3248	Butane, 3-chloro-2,2-dimethyl-	$C_6H_{13}Cl$ 120.62	5750-00-5 0.9	111	4-01-00-00369 0.8767[20]	eth 4 1.4182[20]
3249	Butane, 1-(2-chloroethoxy)-	$C_6H_{13}ClO$ 136.62	10503-96-5	154 dec; 50[11]	4-01-00-01519 0.9335[20]	eth 4 1.4155[20]
3250	Butane, 1-chloro-4-fluoro- 1-Chloro-4-fluorobutane	C_4H_8ClF 110.56	462-73-7	114.7	4-01-00-00249 1.0627[25]	eth 4; EtOH 4 1.4020[25]
3251	Butane, 1-chloro-2-methyl-, (±)- Butane, 1-chloro-2-methyl (*DL*)	$C_5H_{11}Cl$ 106.60	114180-21-1	99.9	4-01-00-00324 0.8810[15]	EtOH 3; eth 3 1.4102[25]
3252	Butane, 1-chloro-3-methyl- Isoamyl chloride	$C_5H_{11}Cl$ 106.60	107-84-6 -104.4	4998 98.9	4-01-00-00325 0.8750[20]	H_2O 2; EtOH 5; eth 5; chl 4 1.4084[20]
3253	Butane, 2-chloro-2-methyl-	$C_5H_{11}Cl$ 106.60	594-36-5 -73.5	85.6	4-01-00-00324 0.8653[20]	H_2O 2; EtOH 3; eth 3; ctc 3 1.4055[20]
3254	Butane, 2-chloro-2,3,3-trimethyl-	$C_7H_{15}Cl$ 134.65	918-07-0 136	sub	4-01-00-00411	eth 4
3255	Butane, decafluoro- Perfluorobutane	C_4F_{10} 238.03	355-25-9 -128.2	-1.9	4-01-00-00245 1.6484[25]	bz 3; chl 3
3256	Butanedial	$C_4H_6O_2$ 86.09	638-37-9	170 dec; 58[9]	4-01-00-03642 1.064[20]	H_2O 4; ace 4; eth 4; EtOH 4 1.4262[18]
3257	Butanediamide Succinamide	$C_4H_8N_2O_2$ 116.12	110-14-5 268 dec	8837 sub 125	4-02-00-01922	
3258	1,4-Butanediamine Putrescine	$C_4H_{12}N_2$ 88.15	110-60-1 27.5	7964 158.5	4-04-00-01283 0.877[25]	H_2O 3 1.4969[20]
3259	1,4-Butanediamine, *N,N*'-bis(3-aminopropyl)- Spermine	$C_{10}H_{26}N_4$ 202.34	71-44-3 29	8699 150[5]	4 04 00 01301	
3260	1,4-Butanediamine, dihydrochloride Butane, 1,4-diamino, dihydrochloride	$C_4H_{14}Cl_2N_2$ 161.07	333-93-7 280 dec	sub	4-04-00-01283	H_2O 4; EtOH 4; eth 1; bz 1
3261	2,3-Butanediamine, (*R**,*R**)-(±)-	$C_4H_{12}N_2$ 88.15	20699-48-3 -22	57-8[60]	4-04-00-01302 0.8499[25]	1.4408[25]
3262	1,4-Butanediamine, *N,N,N*',*N*'-tetraethyl- *N,N,N*',*N*'-Tetraethyl-1,4-butanediamine	$C_{12}H_{28}N_2$ 200.37	69704-44-5	85-6[5]	4-04-00-01286	H_2O 4; eth 4; EtOH 4 1.4383[25]
3263	1,4-Butanediamine, *N,N,N*',*N*'-tetramethyl- *N,N,N*',*N*'-Tetramethyl-1,4-diaminobutane	$C_8H_{20}N_2$ 144.26	111-51-3	9157 168	4-04-00-01284 0.7942[15]	H_2O 5; EtOH 3; eth 3 1.4621[25]
3264	Butane, 1,1-dibromo- 1,1-Dibromobutane	$C_4H_8Br_2$ 215.92	62168-25-6	91[101]; 72[43]	4-01-00-00266 1.80[25]	1.4980[25]
3265	Butane, 1,2-dibromo- α-Butylene dibromide	$C_4H_8Br_2$ 215.92	533-98-2 -65.4	1565 166.3	4-01-00-00266 1.7915[20]	H_2O 1; eth 3; chl 3 1.4025[20]
3266	Butane, 1,3-dibromo- 1,3-Dibromobutane	$C_4H_8Br_2$ 215.92	107-80-2	174	4-01-00-00266 1.800[20]	H_2O 1; eth 3; ctc 2; chl 3 1.507[20]
3267	Butane, 1,4-dibromo- 1,4-Dibromobutane	$C_4H_8Br_2$ 215.92	110-52-1 -16.5	197	4-01-00-00267 1.7890[20]	H_2O 1; ctc 2; chl 3 1.5190[20]
3268	Butane, 2,3-dibromo- 2,3-Dibromobutane	$C_4H_8Br_2$ 215.92	5408-86-6 -24	161	3-01-00-00296 1.7893[22]	H_2O 1; eth 3 1.5133[22]
3269	Butane, 1,2-dibromo-2,3-dimethyl- 1,2-Dibromo-2,3-dimethylbutane	$C_6H_{12}Br_2$ 243.97	29916-45-8	80[17]	3-01-00-00414 1.6033[20]	1.5105[20]
3270	Butane, 1,4-dibromo-2,3-dimethyl- 1,4-Dibromo-2,3-dimethylbutane	$C_6H_{12}Br_2$ 243.97	54462-70-3	106-8[22]	4-01-00-00375 1.620[20]	1.5128[20]
3271	Butane, 2,3-dibromo-2,3-dimethyl- 1,2-Dibromo-1,1,2,2-tetramethylethane	$C_6H_{12}Br_2$ 243.97	594-81-0 171 dec	78[10]	4-01-00-00375 1.8110[21]	
3272	Butane, 1,2-dibromo-2-methyl- 1,2-Dibromo-2-methylbutane	$C_5H_{10}Br_2$ 229.94	10428-64-5	47-8[9]	4-01-00-00330 1.6711[20]	1.5088[20]
3273	Butane, 1,4-dibromo-2-methyl-, (*R*)- Butane, 1,4-dibromo-2-methyl (*d*)	$C_5H_{10}Br_2$ 229.94	69498-28-8	78-9[12]	4-01-00-00330 1.695[17]	1.5128[20]
3274	Butane, 2,3-dibromo-2-methyl- 1,2-Dibromo-1,1,2-trimethylethane	$C_5H_{10}Br_2$ 229.94	594-51-4 7	61-4[17]	4-01-00-00330 1.6717[20]	1.5729[25]
3275	Butane, 1,4-dibromo-1,1,2,2,3,3,4,4-octafluoro- 1,4-Dibromooctafluorobutane	$C_4Br_2F_8$ 359.84	335-48-8	97		
3276	Butane, 1,1-dichloro- Butylidene chloride	$C_4H_8Cl_2$ 127.01	541-33-3 113.8	1571	4-01-00-00250 1.0863[20]	H_2O 1; chl 3 1.4355[20]
3277	Butane, 1,2-dichloro- 1,2-Dichlorobutane	$C_4H_8Cl_2$ 127.01	616-21-7 124.1		4-01-00-00250 1.1116[25]	H_2O 1; eth 3; ctc 2; chl 3 1.4450[20]
3278	Butane, 1,3-dichloro- 1,3-Dichlorobutane	$C_4H_8Cl_2$ 127.01	1190-22-3 134		4-01-00-00250 1.1158[20]	H_2O 1; eth 3; ctc 2; chl 3 1.4445[20]

No.	Name / Synonym	Mol. Form. / Mol. Wt.	CAS RN / mp/°C	Merck No. / bp/°C	Beil. Ref. / den/g cm^{-3}	Solubility / n_D
3279	Butane, 1,4-dichloro- / 1,4-Dichlorobutane	$C_4H_8Cl_2$ / 127.01	110-56-5 / -37.3	/ 161	4-01-00-00250 / 1.1408[20]	chl 4 / 1.4542[20]
3280	Butane, 2,2-dichloro- / 2,2-Dichlorobutane	$C_4H_8Cl_2$ / 127.01	4279-22-5 / -74	/ 104	4-01-00-00251 / 1.1048[25]	H_2O 1; chl 3 / 1.4295
3281	Butane, 2,3-dichloro-, (R*,R*)-(±)-	$C_4H_8Cl_2$ / 127.01	2211-67-8 / -80	/ 118; 53.2[80]	4-01-00-00251 / 1.063[25]	/ 1.4409[25]
3282	Butane, 1,1-dichloro-3-methyl- / 1,1-Dichloro-3-methylbutane	$C_5H_{10}Cl_2$ / 141.04	625-66-1 /	/ 130	4-01-00-00326 / 1.0473[20]	eth 4; EtOH 4 / 1.4344[20]
3283	Butane, 1,2-dichloro-2-methyl- / 1,2-Dichloro-2-methylbutane	$C_5H_{10}Cl_2$ / 141.04	23010-04-0 /	/ 134	3-01-00-00360 / 1.0785[20]	eth 4; EtOH 4; chl 4 / 1.4432[21]
3284	Butane, 1,3-dichloro-3-methyl-	$C_5H_{10}Cl_2$ / 141.04	624-96-4 /	/ 146.7	4-01-00-00326 / 1.0654[20]	eth 4; chl 4 / 1.4455[20]
3285	Butane, 1,4-dichloro-2-methyl- / 1,4-Dichloro-2-methylbutane	$C_5H_{10}Cl_2$ / 141.04	623-34-7 /	/ 168.5	3-01-00-00360 / 1.1003[25]	chl 4 / 1.4562[21]
3286	Butane, 2,3-dichloro-2-methyl- / Amylene dichloride	$C_5H_{10}Cl_2$ / 141.04	507-45-9 /	645 / 129	4-01-00-00325 / 1.0696[15]	eth 4; EtOH 4 / 1.4450[18]
3287	Butane, 2,3-dichloro-1,1,1,2,3,4,4,4-octafluoro-	$C_4Cl_2F_8$ / 270.94	355-20-4 / -68	/ 63	4-01-00-00252 / 1.6801[20]	/ 1.3100[20]
3288	Butane, 1,4-diiodo-	$C_4H_8I_2$ / 309.92	628-21-7 / 5.8	/ 125[15] dec	4-01-00-00276 / 2.3659[15]	H_2O 1; ctc 2; os 3 / 1.6239[15]
3289	Butane, 2,2-dimethyl- / Neohexane	C_6H_{14} / 86.18	75-83-2 / -99	/ 49.7	4-01-00-00367 / 0.6444[25]	H_2O 1; EtOH 3; eth 3; ace 4 / 1.3688[20]
3290	Butane, 2,3-dimethyl- / 2,3-Dimethylbutane	C_6H_{14} / 86.18	79-29-8 / -128.8	/ 57.9	4-01-00-00371 / 0.6616[20]	H_2O 1; EtOH 3; eth 3; ace 4 / 1.3750[20]
3291	Butanedinitrile / Succinonitrile	$C_4H_4N_2$ / 80.09	110-61-2 / 54.5	8843 / 266	4-02-00-01923 / 0.9867[60]	H_2O 4; EtOH 3; eth 2; ace 3 / 1.4173[60]
3292	Butanedinitrile, tetramethyl- / Tetramethylsuccinonitrile	$C_8H_{12}N_2$ / 136.20	3333-52-6 / 170.5		4-02-00-02054 / 1.070[25]	EtOH 3
3293	Butane, 1,4-dinitro- / 1,4-Dinitrobutane	$C_4H_8N_2O_4$ / 148.12	4286-49-1 / 33.5	/ 176[13]	4-01-00-00280	H_2O 1; EtOH 2; eth 3; bz 3
3294	Butanedioic acid / Succinic acid	$C_4H_6O_4$ / 118.09	110-15-6 / 188	8840 / 235 dec	4-02-00-01908 / 1.572[25]	H_2O 2; EtOH 3; eth 3; ace 3 / 1.450
3295	Butanedioic acid, acetyl-, diethyl ester	$C_{10}H_{16}O_5$ / 216.23	1115-30-6 /	/ 255; 133[17]	4-03-00-01826 / 1.081[20]	H_2O 1; EtOH 3; eth 3; bz 3 / 1.4346[20]
3296	Butanedioic acid, bis(1,1-dimethylethyl) ester / Di-tert-butyl succinate	$C_{12}H_{22}O_4$ / 230.30	926-26-1 / 36.5	3024 / 109-10[9]	4-02-00-01916	
3297	Butanedioic acid, bis(1-methylpropyl) ester / Di-sec-butyl succinate	$C_{12}H_{22}O_4$ / 230.30	626-31-3 /	/ 256	3-02-00-01665 / 0.9735[20]	bz 4; eth 4; EtOH 4 / 1.4238[25]
3298	Butanedioic acid, bis(phenylmethyl) ester	$C_{18}H_{18}O_4$ / 298.34	103-43-5 / 49.5	/ 245[15]	4-06-00-02271 / 1.256[25]	H_2O 1; EtOH 3; eth 3; bz 3 / 1.596
3299	Butanedioic acid, bromo-, (±)-	$C_4H_5BrO_4$ / 196.99	584-98-5 / 161		4-02-00-01929 / 2.073[25]	H_2O 3; EtOH 3; HOAc 2
3300	Butanedioic acid, bromo- / Bromosuccinic acid	$C_4H_5BrO_4$ / 196.99	923-06-8 / 161	1427	1-02-00-00268 / 2.073[25]	EtOH 4
3301	Butanedioic acid, dibutyl ester / Dibutyl succinate	$C_{12}H_{22}O_4$ / 230.30	141-03-7 / -29.2	3023 / 274.5	4-02-00-01916 / 0.9752[20]	H_2O 1; EtOH 3; eth 3; bz 3 / 1.4299[20]
3302	Butanedioic acid, 2,3-dichloro-, (±)- / Succinic acid, 2,3-dichloro-, (±)-	$C_4H_4Cl_2O_4$ / 186.98	1114-09-6 / 175 dec		4-02-00-01928 / 1.844[15]	ace 4; eth 4; chl 4
3303	Butanedioic acid, diethyl ester / Diethyl succinate	$C_8H_{14}O_4$ / 174.20	123-25-1 / -21	8840 / 217.7	4-02-00-01914 / 1.0402[20]	H_2O 1; EtOH 5; eth 5; ace 3 / 1.4201[20]
3304	Butanedioic acid, 2,3-dihydroxy-, (R*,S*)- / meso-Tartaric acid	$C_4H_6O_6$ / 150.09	147-73-9 / 147	9040	4-03-00-01218 / 1.666[20]	H_2O 4; EtOH 4
3305	Butanedioic acid, 2,3-dihydroxy-, bis(1-methylethyl) ester, (R*,R*)-(±)- / Tartaric acid, diisopropyl ester (DL)	$C_{10}H_{18}O_6$ / 234.25	58167-01-4 / 34	/ 275; 154[12]	2-03-00-00337 / 1.1166[20]	ace 4; eth 4; EtOH 4
3306	Butanedioic acid, 2,3-dihydroxy- [R-(R*,R*)]-, bis(phenylmethyl) ester	$C_{18}H_{18}O_6$ / 330.34	622-00-4 / 50	/ 250-70[4]	3-06-00-01537 / 1.2036[72]	py 4; EtOH 4
3307	Butanedioic acid, 2,3-dihydroxy-, diethyl ester, (R*,R*)-(±)- / Diethyl DL-tartrate	$C_8H_{14}O_6$ / 206.20	57968-71-5 / 18.7	/ 281; 158[14]	4-03-00-01232 / 1.2046[20]	H_2O 2; EtOH 5; eth 5; ace 3 / 1.4438[20]
3308	Butanedioic acid, 2,3-dihydroxy-, dimethyl ester, (R*,R*)-(±)-	$C_6H_{10}O_6$ / 178.14	608-69-5 / 90	/ 280	4-03-00-01232 / 1.2604[90]	H_2O 4; ace 4; eth 4; EtOH 4
3309	Butanedioic acid, 2,3-dihydroxy- [R-(R*,R*)]- / L-Tartaric acid	$C_4H_6O_6$ / 150.09	87-69-4 / 169	9039	4-03-00-01219	
3310	Butanedioic acid, 2,3-dihydroxy- [R-(R*,R*)]-, dipropyl ester / Tartaric acid, dipropyl ester, (+)-	$C_{10}H_{18}O_6$ / 234.25	2217-14-3 /	/ 303	2-03-00-00337 / 1.1390[20]	H_2O 4; ace 4; eth 4; EtOH 4
3311	Butanedioic acid, 2,3-dihydroxy-, (R*,R*)-(±)- / DL-Tartaric acid	$C_4H_6O_6$ / 150.09	133-37-9 / 206	9038	4-03-00-01229 / 1.788[25]	H_2O 3; EtOH 3; eth 2; bz 1
3312	Butanedioic acid, 2,3-dihydroxy-, [S-(R*,R*)]- / D-Tartaric acid	$C_4H_6O_6$ / 150.09	147-71-7 / 172.5	9037	4-03-00-01229 / 1.7598[20]	DMSO 2 / 1.4955[20]
3313	Butanedioic acid, 2,3-dihydroxy- [R-(R*,R*)]-, dibutyl ester	$C_{12}H_{22}O_6$ / 262.30	87-92-3 / 22	/ 320	4-03-00-01232 / 1.0909[20]	H_2O 4; ace 4; EtOH 4 / 1.4451[20]
3314	Butanedioic acid, 2,3-dihydroxy- [R-(R*,R*)]-, disodium salt / Tartaric acid, disodium salt	$C_4H_4Na_2O_6$ / 194.05	868-18-8 /	8640		H_2O 3

No.	Name / Synonym	Mol. Form. / Mol. Wt.	CAS RN / mp/°C	Merck No. / bp/°C	Beil. Ref. / den/g cm^{-3}	Solubility / n_D
3315	Butanedioic acid, 2,3-dihydroxy- [R-(R*,R*)]-, monoethyl ester / Tartaric acid, ethyl ester	$C_6H_{10}O_6$ / 178.14	608-89-9 / 90	3812	3-03-00-01028	H_2O 4; EtOH 4
3316	Butanedioic acid, [(dimethoxyphosphinothioyl)thio]-, diethyl ester / Malathion	$C_{10}H_{19}O_6PS_2$ / 330.36	121-75-5 / 2.8	5582	4-03-00-01136 / 156$^{0.7}$ dec / 1.2076^{20}	H_2O 2; EtOH 3; eth 3; bz 3 / 1.4960^{20}
3317	Butanedioic acid, 2,3-dimethyl-	$C_6H_{10}O_4$ / 146.14	13545-04-5 / 132 dec	87-90^2	4-02-00-01998	ace 4; eth 4
3318	Butanedioic acid, dimethyl ester	$C_6H_{10}O_4$ / 146.14	106-65-0 / 19	196.4	4-02-00-01913 / 1.1198^{20}	H_2O 2; EtOH 3; eth 4; ace 3 / 1.4197^{20}
3319	Butanedioic acid, dipentyl ester	$C_{14}H_{26}O_4$ / 258.36	645-69-2 / -10.8	171.5^{16}	3-02-00-01665 / 0.9616^{20}	
3320	Butanedioic acid, diphenyl ester / Diphenyl succinate	$C_{16}H_{14}O_4$ / 270.28	621-14-7 / 121	330; 222.5^{15}	3-06-00-00605	H_2O 1; EtOH 3; eth 3; ace 3
3321	Butanedioic acid, dipropyl ester	$C_{10}H_{18}O_4$ / 202.25	925-15-5 / -5.9	250.8	4-02-00-01916 / 1.0020^{20}	ace 4; bz 4; eth 4 / 1.4250^{20}
3322	Butanedioic acid, ethyl-, (R)-	$C_6H_{10}O_4$ / 146.14	4074-24-2 / 96	181.5	4-02-00-01995 / 1.0017^{20}	H_2O 4; eth 4; EtOH 4
3323	Butanedioic acid, ethyl methyl ester	$C_7H_{12}O_4$ / 160.17	627-73-6 / <-20	208.2	0-02-00-00609 / 1.076^{20}	eth 4; EtOH 4
3324	Butanedioic acid, formyl-, diethyl ester	$C_9H_{14}O_5$ / 202.21	5472-38-8 /	130-4^{15}	4-03-00-01819	eth 4; EtOH 4 / 1.4486^{25}
3325	Butanedioic acid, methyl-, (±)-	$C_5H_8O_4$ / 132.12	636-60-2 / 115	dec	4-02-00-01948 / 1.4200^0	H_2O 4; EtOH 4; eth 3; chl 2 / 1.4303
3326	Butanedioic acid, methyl-, diethyl ester	$C_9H_{16}O_4$ / 188.22	4676-51-1 /	217.5	3-02-00-01696 / 1.012^{25}	1.4199^{20}
3327	Butanedioic acid, methyl-, dimethyl ester	$C_7H_{12}O_4$ / 160.17	1604-11-1 /	196	3-02-00-01696 / 1.076^{25}	1.4200^{20}
3328	Butanedioic acid, methylene- / Itaconic acid	$C_5H_6O_4$ / 130.10	97-65-4 / 175	5130 / dec	4-02-00-02228 / 1.632^{25}	H_2O 3; EtOH 3; eth 2; ace 3
3329	Butanedioic acid, methylene-, diethyl ester	$C_9H_{14}O_4$ / 186.21	2409-52-1 / 58.5	228	4-02-00-02230 / 1.0467^{20}	EtOH 5; eth 3; ace 4; bz 3 / 1.4377^{20}
3330	Butanedioic acid, methylene-, dimethyl ester	$C_7H_{10}O_4$ / 158.15	617-52-7 / 38	208	4-02-00-02229 / 1.1241^{18}	EtOH 3; eth 3; ace 4; MeOH 3 / 1.4457^{20}
3331	Butanedioic acid, (1-methylethylidene)-	$C_7H_{10}O_4$ / 158.15	584-27-0 / 160.5	174 dec	4-02-00-02251	eth 4; EtOH 4
3332	Butanedioic acid, (1-methylpropyl)-, diethyl ester / Succinic acid, sec butyl, diethyl ester	$C_{12}H_{22}O_4$ / 230.30	69248-35-7 /	75-6$^{0.2}$	3-02-00-01779 / 0.9745^{25}	1.4293^{25}
3333	Butanedioic acid, monobutyl ester	$C_8H_{14}O_4$ / 174.20	5150-93-6 / 8.6	136.5^3	3-02-00-01665 / 1.0732^{20}	1.4360^{20}
3334	Butanedioic acid, mono(2,2-dimethylhydrazide) / Daminozide	$C_6H_{12}N_2O_3$ / 160.17	1596-84-5 / 154.5	2810		
3335	Butanedioic acid, monoethyl ester / Succinic acid, ethyl ester	$C_8H_{10}O_4$ / 146.14	1070-34-4 / 8	172^{42}; 119^3	4-02-00-01914 / 1.1466^{20}	H_2O 4; eth 4; EtOH 4 / 1.4327^{20}
3336	Butanedioic acid, monomethyl ester	$C_5H_8O_4$ / 132.12	3878-55-5 / 58	151^{20}; 122^4	4-02-00-01913	H_2O 3
3337	Butanedioic acid, monopentyl ester / Succinic acid, pentyl ester	$C_9H_{16}O_4$ / 188.22	97479-78-2 / 17.2	147^3	2-02-00-00551 / 1.0460^{20}	1.4378^{20}
3338	Butanedioic acid, monopropyl ester / Succinic acid, monopropyl ester	$C_7H_{12}O_4$ / 160.17	6946-88-9 / 15	126^3	3-02-00-01664 / 1.1071^{20}	1.4343^{20}
3339	Butanedioic acid, oxo- / Oxalacetic acid	$C_4H_4O_5$ / 132.07	328-42-7 / 161 dec	6863	4-03-00-01808	
3340	Butanedioic acid, oxo-, diethyl ester / Diethyl oxalacetate	$C_8H_{12}O_5$ / 188.18	108-56-5 /	3791 / 131-2^{24}	4-03-00-01809 / 1.131^{20}	H_2O 1; EtOH 5; eth 5; ace 4 / 1.4561^{17}
3341	Butanedioic acid, phenyl-, (±)-	$C_{10}H_{10}O_4$ / 194.19	10424-29-0 / 168	dec	4-09-00-03352	H_2O 2; EtOH 4; eth 4; ace 4
3342	Butanedioic acid, tetrahydroxy- / Dihydroxytartaric acid	$C_4H_6O_8$ / 182.09	76-30-2 / 114.5	3172	3-03-00-01419	
3343	Butanedioic acid, tetramethyl-	$C_8H_{14}O_4$ / 174.20	630-51-3 / 200	sub	4-02-00-02054 / 1.30^{25}	bz 4; EtOH 4; chl 4
3344	1,2-Butanediol, (±)-	$C_4H_{10}O_2$ / 90.12	26171-83-5 /	190.5	4-01-00-02507 / 1.0024^{20}	H_2O 3; EtOH 3; ace 3 / 1.4378^{20}
3345	1,3-Butanediol / 1,3-Butylene glycol	$C_4H_{10}O_2$ / 90.12	107-88-0 / <-50	1566 / 207.5	0-01-00-00477 / 1.0053^{20}	1.4401^{20}
3346	1,4-Butanediol / Tetramethylene glycol	$C_4H_{10}O_2$ / 90.12	110-63-4 / 20.1	235	4-01-00-02515 / 1.0171^{20}	H_2O 5; EtOH 3; eth 2; DMSO 2 / 1.4460^{20}
3347	2,3-Butanediol, (R*,R*)-(±)-	$C_4H_{10}O_2$ / 90.12	6982-25-8 / 7.6	182.5	4-01-00-02525 / 1.0033^{20}	H_2O 5; EtOH 5; eth 3; ace 3 / 1.4310^{25}
3348	2,3-Butanediol, 1,4-diacetate	$C_8H_{14}O_4$ / 174.20	628-67-1 / 12	229	4-02-00-00224 / 1.0479^{15}	1.4251^{15}
3349	2,3-Butanediol, 1,4-dichloro- / 1,4-Dichloro-2,3-butanediol	$C_4H_8Cl_2O_2$ / 159.01	2419-73-0 / 126.5	150^{30}	4-01-00-02531	EtOH 4
3350	2,3-Butanediol, 2,3-dimethyl- / Pinacol	$C_6H_{14}O_2$ / 118.18	76-09-5 / 43	7408 / 174.4	4-01-00-02575	H_2O 2; EtOH 4; eth 4; CS_2 2
3351	2,3-Butanediol, 2,3-dimethyl-, hexahydrate	$C_6H_{26}O_8$ / 226.27	6091-58-3 / 45.4	172	2-01-00-00553 / 0.967^{15}	eth 4; EtOH 4

No.	Name Synonym	Mol. Form. Mol. Wt.	CAS RN mp/°C	Merck No. bp/°C	Beil. Ref. den/g cm^{-3}	Solubility n_D
3352	1,2-Butanediol, 3-methyl- 3-Methyl-1,2-butanediol	$C_5H_{12}O_2$ 104.15	50468-22-9 	 206	4-01-00-02549 0.9987[0]	eth 4; EtOH 4
3353	1,3-Butanediol, 2-methyl- 2-Methyl-1,3-butanediol	$C_5H_{12}O_2$ 104.15	684-84-4 	 200; 115[20]	4-01-00-02546 0.9912[20]	H_2O 3; EtOH 3; eth 3
3354	1,3-Butanediol, 3-methyl-	$C_5H_{12}O_2$ 104.15	2568-33-4 	 202.5	4-01-00-02549 0.9448[20]	H_2O 3; EtOH 3 1.4452[20]
3355	2,3-Butanediol, 2-methyl- 2,3-Dihydroxy-2-methylbutane	$C_5H_{12}O_2$ 104.15	5396-58-7 	 175	4-01-00-02547 0.9920[25]	H_2O 4; eth 4; EtOH 4 1.4375[20]
3356	1,2-Butanediol, 2-phenyl- 2-Phenyl-1,2-butanediol	$C_{10}H_{14}O_2$ 166.22	90925-48-7 56	 165[23]	3-06-00-04667 	H_2O 4; eth 4; EtOH 4
3357	1,3-Butanediol, 1-phenyl- 1-Phenylbutane-1,3-diol	$C_{10}H_{14}O_2$ 166.22	65469-88-7 73.5	 175[21]; 130[2]	4-06-00-06006 1.0720[25]	bz 4; EtOH 4; lig 4
3358	1,3-Butanediol, 3-phenyl-	$C_{10}H_{14}O_2$ 166.22	7133-68-8 	 150[2]; 109[0.5]	4-06-00-06009 1.0865[20]	 1.5341[20]
3359	2,3-Butanedione Diacetyl	$C_4H_6O_2$ 86.09	431-03-8 -2.4	2946 88	4-01-00-03644 0.9808[18]	H_2O 4; EtOH 5; eth 5; ace 4 1.3951[20]
3360	2,3-Butanedione, dioxime Dimethylglyoxime	$C_4H_8N_2O_2$ 116.12	95-45-4 245.5	3235 sub 234	3-01-00-03105 	H_2O 1; EtOH 4; eth 4; bz 2
3361	2,3-Butanedione, monooxime	$C_4H_7NO_2$ 101.11	57-71-6 76.8	 185.5	4-01-00-03646 	H_2O 2; EtOH 4; eth 4; chl 4
3362	1,3-Butanedione, 1-phenyl-	$C_{10}H_{10}O_2$ 162.19	93-91-4 56	 261.5	4-07-00-02151 1.0599[74]	H_2O 1; eth 3; chl 2 1.5678[78]
3363	1,3-Butanedione, 1-(2-pyridinyl)-	$C_9H_9NO_2$ 163.18	40614-52-6 50	 140[15]	4-21-00-04750 	bz 4; eth 4; EtOH 4; chl 4
3364	1,3-Butanedione, 1-(3-pyridinyl)-	$C_9H_9NO_2$ 163.18	3594-37-4 85	 171[15]	5-21-10-00180 	H_2O 2; EtOH 3; ace 3
3365	1,3-Butanedione, 1-(4-pyridinyl)-	$C_9H_9NO_2$ 163.18	75055-73-1 62	 145[18]	5-21-10-00180 	bz 4; eth 4; EtOH 4
3366	Butanedioyl dichloride Succinyl chloride	$C_4H_4Cl_2O_2$ 154.98	543-20-4 20	8844 193.3	4-02-00-01921 1.3748[20]	eth 3; ace 3; bz 3 1.4683[20]
3367	Butanedioyl dichloride, methylene-	$C_5H_4Cl_2O_2$ 166.99	1931-60-8 	 89[17]	3-02-00-01934 	ace 3 1.4919[20]
3368	1,4-Butanedithiol Tetramethylenedithiol	$C_4H_{10}S_2$ 122.26	1191-08-8 -53.9	 195.5	4-01-00-02523 1.0021[0]	H_2O 1; EtOH 4; ctc 2 1.5290[20]
3369	Butane, 1-(ethenyloxy)- Butyl vinyl ether	$C_6H_{12}O$ 100.16	111-34-2 -92	 94	4-01-00-02052 0.7888[20]	H_2O 1; EtOH 4; eth 5; ace 4 1.4026[20]
3370	Butane, 2-(ethenyloxy)-	$C_6H_{12}O$ 100.16	1888-85-3 	 81	4-01-00-02054 0.7715[20]	 1.3970[20]
3371	Butane, 1-(ethenyloxy)-3-methyl- Isopentyl vinyl ether	$C_7H_{14}O$ 114.19	39782-38-2 	 112.5	4-01-00-02055 0.7826[20]	eth 4; EtOH 4 1.4072[20]
3372	Butane, 1-ethoxy- Butyl ethyl ether	$C_6H_{14}O$ 102.18	628-81-9 -124	 92.3	4-01-00-01518 0.7495[20]	H_2O 1; EtOH 5; eth 5; ace 4 1.3818[20]
3373	Butane, 2-ethoxy- sec-Butyl ethyl ether	$C_6H_{14}O$ 102.18	2679-87-0 	 81	4-01-00-01572 0.7503[20]	H_2O 1; EtOH 4; eth 4 1.3802[20]
3374	Butane, 1-ethoxy-3-methyl- Ethyl isopentyl ether	$C_7H_{16}O$ 116.20	628-04-6 	 112.5	4-01-00-01681 0.7688[21]	eth 4; EtOH 4
3375	Butane, 2-ethoxy-2-methyl- Ethyl tert-amyl ether	$C_7H_{16}O$ 116.20	919-94-8 	 102	3-01-00-01626 0.751[18]	eth 4; EtOH 4
3376	Butane, 1-(ethylthio)- Butyl ethyl sulfide	$C_6H_{14}S$ 118.24	638-46-0 -95.1	 144.3	4-01-00-01558 0.8376[20]	EtOH 4; chl 3 1.4492[10]
3377	Butane, 1-(ethynyloxy)-	$C_6H_{10}O$ 98.14	3329-56-4 	 104	4-01-00-02213 0.8200[20]	eth 4; EtOH 4 1.4067
3378	Butane, 1-fluoro- Butyl fluoride	C_4H_9F 76.11	2366-52-1 -134	 32.5	4-01-00-00244 0.7789[20]	EtOH 4 1.3396[20]
3379	Butane, 2-fluoro- sec-Butyl fluoride	C_4H_9F 76.11	359-01-3 -121.4	 25.1	4-01-00-00244 0.7559[25]	
3380	Butane, 1,1,2,2,3,4,4-heptachloro-	$C_4H_3Cl_7$ 299.24	34973-41-6 	 137.5[13.5]	4-01-00-00257 1.739[20]	ace 4; ctc 4 1.5407[20]
3381	Butane, 1-iodo- Butyl iodide	C_4H_9I 184.02	542-69-8 -103	1572 130.6	4-01-00-00271 1.6154[20]	H_2O 1; EtOH 5; eth 5; chl 4 1.5001[20]
3382	Butane, 2-iodo-, (±)- Butane, 2-iodo (DL)	C_4H_9I 184.02	52152-71-3 -104.2	1573 120	4-01-00-00272 1.5920[20]	H_2O 1; EtOH 5; eth 5; chl 4 1.4991[20]
3383	Butane, 1-iodo-2-methyl-, (S)- Butane, 1-iodo-2-methyl (d)	$C_5H_{11}I$ 198.05	29394-58-9 	 148; 132.6[500]	3-01-00-00366 1.5253[20]	eth 4; EtOH 4 1.4977[20]
3384	Butane, 1-iodo-3-methyl- Isoamyl iodide	$C_5H_{11}I$ 198.05	541-28-6 	5002 147	4-01-00-00331 1.5118[20]	H_2O 2; EtOH 5; eth 5; ctc 2 1.4939[20]
3385	Butane, 1-iodo-2-methyl-	$C_5H_{11}I$ 198.05	594-38-7 	 124.5	4-01-00-00331 1.4937[20]	H_2O 1; EtOH 5; eth 5 1.4981[20]
3386	Butane, 1-isocyanato-3-methyl-	$C_6H_{11}NO$ 113.16	1611-65-0 	 137	0-04-00-00186 0.806[20]	H_2O 1; EtOH 5; eth 5 1.406[20]
3387	Butane, 1-isocyano- Butyl isocyanide	C_5H_9N 83.13	2769-64-4 	 120	4-04-00-00562 0.78[20]	eth 4; EtOH 4
3388	Butane, 1-isothiocyanato-	C_5H_9NS 115.20	592-82-5 	 168	4-04-00-00596 0.9546[20]	eth 4; EtOH 4 1.501[20]
3389	Butane, 2-isothiocyanato-, (±)- sec-Butyl isothiocyanate (DL)	C_5H_9NS 115.20	116724-11-9 	 159.5	4-04-00-00624 0.944[12]	eth 4; EtOH 4
3390	Butane, 1-isothiocyanato-3-methyl-	$C_6H_{11}NS$ 129.23	628-03-5 	 183	4-04-00-00707 0.9419[17]	eth 4; EtOH 4
3391	Butane, 1-methoxy- Butyl methyl ether	$C_5H_{12}O$ 88.15	628-28-4 -115.5	 70.1	4-01-00-01518 0.7443[20]	H_2O 1; EtOH 5; eth 5; ace 3 1.3736[20]

No.	Name / Synonym	Mol. Form. / Mol. Wt.	CAS RN / mp/°C	Merck No. / bp/°C	Beil. Ref. / den/g cm^{-3}	Solubility / n_D
3392	Butane, 2-methoxy-, (S)- / sec-Butyl methyl ether (d)	$C_5H_{12}O$ / 88.15	66610-39-7	60.7	4-01-00-01571 / 0.7400[25]	
3393	Butane, 2-methoxy-, (±)- / sec-Butyl methyl ether (DL)	$C_5H_{12}O$ / 88.15	116783-23-4	60	4-01-00-01571 / 0.7415[20]	ace 4; eth 4; EtOH 4 / 1.3680[25]
3394	Butane, 1-methoxy-2-methyl- / Methyl (2-methylbutyl) ether	$C_6H_{14}O$ / 102.18	62016-48-2	90	4-01-00-01667 / 0.754[18]	1.3849[20]
3395	Butane, 1-methoxy-3-methyl-	$C_6H_{14}O$ / 102.18	626-91-5	91	4-01-00-01681 / 0.7517[20]	eth 4; EtOH 4 / 1.3830[20]
3396	Butane, 2-methoxy-2-methyl-	$C_6H_{14}O$ / 102.18	994-05-8	86.3	4-01-00-01671 / 0.7703[20]	eth 4; EtOH 4 / 1.3855[20]
3397	Butane, 2-methyl- / Isopentane	C_5H_{12} / 72.15	78-78-4 / -159.9	27.8	4-01-00-00320 / 0.6201[20]	H_2O 1; EtOH 5; eth 5 / 1.3537[20]
3398	Butane, 1,1'-[methylenebis(oxy)]bis-	$C_9H_{20}O_2$ / 160.26	2568-90-3 / -58.1	179.2	4-01-00-03029 / 0.8339[20]	1.4072[17]
3399	Butane, 1-(1-methylethoxy)- / Butyl isopropyl ether	$C_7H_{16}O$ / 116.20	1860-27-1	109	4-01-00-01519 / 0.7594[15]	H_2O 1; EtOH 3; eth 3; ace 3 / 1.3870[15]
3400	Butane, 3-methyl-1-nitro- / 3-Methyl-1-nitrobutane	$C_5H_{11}NO_2$ / 117.15	627-67-8 / 97	163.5; 72[27]	4-01-00-00332 / 0.9458[25]	
3401	Butane, 1-(2-methylpropoxy)- / Butyl isobutyl ether	$C_8H_{18}O$ / 130.23	17071-47-5	151	4-01-00-01594 / 0.7980[22]	ace 4; eth 4; EtOH 4 / 1.4077[21]
3402	Butane, 1-(methylthio)- / Butyl methyl sulfide	$C_5H_{12}S$ / 104.22	628-29-5 / -97.8	123.5	4-01-00-01557 / 0.8426[20]	EtOH 4; MeOH 4 / 1.4477[20]
3403	Butanenitrile / Propyl cyanide	C_4H_7N / 69.11	109-74-0 / -111.9	1597 / 117.6	4-02-00-00806 / 0.7936[20]	H_2O 2; EtOH 5; eth 5; bz 3 / 1.3842[20]
3404	Butanenitrile, 4-bromo-	C_4H_6BrN / 148.00	5332-06-9	206	4-02-00-00836 / 1.4967[20]	EtOH 3; eth 3; chl 3 / 1.4818[20]
3405	Butanenitrile, 4-chloro-	C_4H_6ClN / 103.55	628-20-6	192	4-02-00-00827 / 1.0934[15]	H_2O 1; EtOH 3; eth 3; ctc 2 / 1.4413[20]
3406	Butanenitrile, 4,4-diethoxy-	$C_8H_{15}NO_2$ / 157.21	18381-45-8	104-6[10]	3-03-00-01210 / 0.937[25]	H_2O 2; EtOH 5; eth 5 / 1.4186[20]
3407	Butanenitrile, 2-ethyl-	$C_6H_{11}N$ / 97.16	617-80-1	145.5	4-02-00-00953	eth 4; EtOH 4 / 1.3891[24]
3408	Butanenitrile, 2-hydroxy-3-methyl-2-(1-methylethyl)- / Diisopropyl ketone, cyanohydrin	$C_8H_{15}NO$ / 141.21	4390-75-4 / 59	111[18]	4-03-00-00884	ace 4; bz 4; eth 4; EtOH 4
3409	Butanenitrile, 2-methyl- / 2-Methylbutanenitrile	C_5H_9N / 83.13	18936-17-9	125	4-02-00-00892 / 0.7913[15]	eth 4; EtOH 4 / 1.3933[20]
3410	Butanenitrile, 3-methyl- / Isobutyl cyanide	C_5H_9N / 83.13	625-28-5 / -101	127.5	4-02-00-00902 / 0.7914[20]	H_2O 2; EtOH 5; eth 5; ace 4 / 1.3927[20]
3411	Butanenitrile, 4-phenoxy-	$C_{10}H_{11}NO$ / 101.20	2243-43-8 / 45.5	288	4-06-00-00646	eth 3
3412	Butane, 1-nitro- / 1-Nitrobutane	$C_4H_9NO_2$ / 103.12	627-05-4	153	4-01-00-00277 / 0.970[25]	H_2O 2; EtOH 5; eth 5; alk 3 / 1.4303[20]
3413	Butane, 2-nitro-, (±)-	$C_4H_9NO_2$ / 103.12	116781-85-2 / -132	140	4-01-00-00278 / 0.9854[17]	chl 3 / 1.4044[20]
3414	Butane, 1,1,1,2,3,4,4,4-octafluoro-2,3-bis(trifluoromethyl)- / Perfluoro-2,3-dimethylbutane	C_8F_{14} / 338.04	354-96-1 / -15	59.8		
3415	Butane, 1,1'-oxybis- / Dibutyl ether	$C_8H_{18}O$ / 130.23	142-96-1 / -95.2	1568 / 140.2	4-01-00-01520 / 0.7684[20]	H_2O 1; EtOH 5; eth 5; ace 4 / 1.3992[20]
3416	Butane, 2,2'-oxybis-, (±)- / sec-Butyl ether, (±)-	$C_8H_{18}O$ / 130.23	17226-28-7	120.5	3-01-00-01533 / 0.756[25]	ace 4; eth 4; EtOH 4 / 1.393[25]
3417	Butane, 1,1'-[oxybis(2,1-ethanediyloxy)]bis-	$C_{12}H_{26}O_3$ / 218.34	112-73-2 / -60	256	4-01-00-02395 / 0.885[25]	1.4235[20]
3418	Butane, 1,1'-oxybis[3-methyl- / Diisoamyl ether	$C_{10}H_{22}O$ / 158.28	544-01-4 / 5000	172.5	4-01-00-01682 / 0.7777[20]	ace 4; EtOH 4; chl 4 / 1.4085[20]
3419	Butane, 1,1,2,3,4-pentachloro- / Butane, 1,1,2,3,4-pentachloro (liquid)	$C_4H_5Cl_5$ / 230.35	77753-24-3	95.5[11]	4-01-00-00256 / 1.561[18]	EtOH 4; ctc 4 / 1.5141[18]
3420	Butane, 1,2,2,3,4-pentachloro- / 1,2,2,3,4-Pentachlorobutane	$C_4H_5Cl_5$ / 230.35	2431-52-9	85[10]	4-01-00-00256 / 1.5543[20]	ace 4; chl 4 / 1.5157[20]
3421	Butane, 1-propoxy- / Butyl propyl ether	$C_7H_{16}O$ / 116.20	3073-92-5	118.1	4-01-00-01519 / 0.7772[0]	H_2O 1; EtOH 4; eth 4
3422	Butane, 1,1'-sulfinylbis-	$C_8H_{18}OS$ / 162.30	2168-93-6 / 32.6	dec	4-01-00-01561 / 0.8317[23]	H_2O 1; EtOH 3; eth 3 / 1.4669[20]
3423	Butane, 1,1'-sulfonylbis-	$C_8H_{18}O_2S$ / 178.30	598-04-9 / 45	291	4-01-00-01561 / 0.9885[47]	H_2O 1; EtOH 3; eth 3
3424	Butane, 1,1,4,4-tetrabromo-	$C_4H_6Br_4$ / 373.71	116779-77-2	138-45[10]	3-01-00-00298 / 2.529[20]	bz 4; eth 4; chl 4 / 1.6077[20]
3425	Butane, 1,2,2,3-tetrabromo-	$C_4H_6Br_4$ / 373.71	116779-78-3 / -2	97.5[7]	3-01-00-00298 / 2.510[20]	
3426	Butane, 1,2,3,4-tetrabromo-, DL- / Butane, 1,2,3,4-tetrabromo (DL)	$C_4H_6Br_4$ / 373.71	2657-65-0 / 118	260; 180[60]	4-01-00-00271	ace 4; eth 4; EtOH 4; lig 4
3427	Butane, 1,1,1,2-tetrachloro- / 1,1,1,2-Tetrachlorobutane	$C_4H_6Cl_4$ / 195.90	39966-95-5	135	4-01-00-00253 / 1.3907[20]	1.4920[25]
3428	Butane, 1,2,2,3-tetrachloro- / 1,2,2,3-Tetrachlorobutane	$C_4H_6Cl_4$ / 195.90	79630-70-9 / -48	182	3-01-00-00286 / 1.4276[18]	ace 4; chl 4 / 1.491[20]
3429	Butane, 1,2,3,3-tetrachloro- / 1,2,3,3-Tetrachlorobutane	$C_4H_6Cl_4$ / 195.90	13138-51-7	90[32]; 56[10]	4-01-00-00254 / 1.4204[20]	ace 4; eth 4; EtOH 4 / 1.4958[20]
3430	Butane, 2,2,3,3-tetramethyl- / 2,2,3,3-Tetramethylbutane	C_8H_{18} / 114.23	594-82-1 / 100.7	106.4	4-01-00-00447 / 0.8242[20]	H_2O 1; eth 3; chl 3 / 1.4695[20]

No.	Name / Synonym	Mol. Form. / Mol. Wt.	CAS RN / mp/°C	Merck No. / bp/°C	Beil. Ref. / den/g cm^{-3}	Solubility / n_D
3431	1,2,3,4-Butanetetrol, (R*,S*)- / Erythritol	$C_4H_{10}O_4$ / 122.12	149-32-6 / 121.5	3620 / 330.5	4-01-00-02807 / 1.451[20]	H_2O 3; eth 1; bz 1
3432	1,2,3,4-Butanetetrol, tetranitrate, (R*,S*)- / Erythrityl tetranitrate	$C_4H_6N_4O_{12}$ / 302.11	7297-25-8 / 61	3622 /	4-01-00-02809 /	EtOH 4
3433	Butane, 1,1'-thiobis- / Dibutyl sulfide	$C_8H_{18}S$ / 146.30	544-40-1 / -79.7	1590 / 185	4-01-00-01559 / 0.8386[20]	eth 4; EtOH 4; chl 4 / 1.4530[20]
3434	Butane, 2,2'-thiobis-	$C_8H_{18}S$ / 146.30	626-26-6 /	/ 165	4-01-00-01586 / 0.8348[20]	H_2O 1; EtOH 4; eth 4 / 1.4506[20]
3435	Butane, 1,1'-thiobis[2-methyl- / Bis(2-methylbutyl) sulfide	$C_{10}H_{22}S$ / 174.35	96034-00-3 /	/ 165; 95[13]	0-01-00-00387 / 0.8348[20]	eth 4; EtOH 4 / 1.4506[20]
3436	Butane, 1,1'-thiobis[3-methyl- / Isoamyl sulfide	$C_{10}H_{22}S$ / 174.35	544-02-5 / -74.6	/ 211	4-01-00-01689 / 0.8323[20]	H_2O 1; EtOH 5; eth 4 / 1.4520[20]
3437	1-Butanethiol / Butyl mercaptan	$C_4H_{10}S$ / 90.19	109-79-5 / -115.7	1575 / 98.5	4-01-00-01555 / 0.8416[20]	H_2O 2; EtOH 4; eth 4; chl 2 / 1.4440[20]
3438	2-Butanethiol, (±)- / 2-Butanethiol (DL)	$C_4H_{10}S$ / 90.19	91840-99-2 / -165	1576 / 85	4-01-00-01584 / 0.8295[20]	EtOH 3; eth 3; bz 3; ctc 2 / 1.4366[20]
3439	1-Butanethiol, 2-methyl-, (S)- / (+)-2-Methyl-1-butanethiol	$C_5H_{12}S$ / 104.22	20089-07-0 /	/ 118.2	4-01-00-01668 / 0.8420[20]	/ 1.4440[20]
3440	1-Butanethiol, 3-methyl- / Isopentyl mercaptan	$C_5H_{12}S$ / 104.22	541-31-1 /	/ 120	4-01-00-01688 / 0.8350[20]	H_2O 1; EtOH 5; eth 5; ctc 3 / 1.4418[20]
3441	2-Butanethiol, 2-methyl- / 2-Methyl-2-butanethiol	$C_5H_{12}S$ / 104.22	1679-09-0 /	/ 99.1	4-01-00-01674 / 0.8120[20]	/ 1.4385[20]
3442	Butane, 1,1,2-tribromo- / 1,1,2-Tribromocyclobutane	$C_4H_7Br_3$ / 294.81	3675-68-1 /	/ 216.2	2-01-00-00084 / 2.1835[20]	eth 4; EtOH 4; chl 4 / 1.5626[17]
3443	Butane, 1,2,2-tribromo- / 1,2,2-Tribromobutane	$C_4H_7Br_3$ / 294.81	3675-69-2 /	/ 213.8	2-01-00-00085 / 2.1691[20]	eth 4; EtOH 4; chl 4 / 1.5624[10]
3444	Butane, 1,2,3-tribromo- / 1,2,3-Tribromobutane	$C_4H_7Br_3$ / 294.81	632-05-3 / -19	/ 220	4-01-00-00271 / 2.1907[20]	eth 4; EtOH 4; chl 4 / 1.5680[20]
3445	Butane, 1,2,4-tribromo- / 1,2,4-Tribromobutane	$C_4H_7Br_3$ / 294.81	38300-67-3 / -18	/ 215	4-01-00-00271 / 2.170[20]	eth 4; EtOH 4; chl 4 / 1.5608[20]
3446	Butane, 1,3,3-tribromo- / 1,3,3-Tribromobutane	$C_4H_7Br_3$ / 294.81	62127-46-2 /	/ 202.5	3-01-00-00298 / 2.1445[20]	eth 4; EtOH 4; chl 4 / 1.5564[20]
3447	Butane, 2,2,3-tribromo- / 2,2,3-Tribromobutane	$C_4H_7Br_3$ / 294.81	62127-47-3 / 1.8	/ 206	3-01-00-00298 / 2.1723[20]	H_2O 1; EtOH 3; eth 3; ctc 2 / 1.5602[20]
3448	Butane, 1,1,3-trichloro- / 1,1,3-Trichlorobutane	$C_4H_7Cl_3$ / 161.46	13279-87-3 /	/ 152	4-01-00-00252 / 1.317[15]	H_2O 1; EtOH 3; eth 3; chl 4 / 1.4600[15]
3449	Butane, 1,2,3-trichloro- / 1,2,3-Trichlorobutane	$C_4H_7Cl_3$ / 161.46	18338-40-4 /	/ 168; 63[20]	3-01-00-00285 / 1.3164[20]	H_2O 1; EtOH 3; eth 3; chl 4 / 1.4790[20]
3450	Butane, 1,2,4-trichloro- / 1,2,4-Trichlorobutane	$C_4H_7Cl_3$ / 161.46	1790-22-3 /	/ 61-3[10]	4-01-00-00253 / 1.3175[20]	H_2O 1; bz 3; ctc 3; chl 4 / 1.4820[20]
3451	Butane, 2,2,3-trichloro- / 2,2,3-Trichlorobutane	$C_4H_7Cl_3$ / 161.46	10403-60-8 /	/ 144	3-01-00-00285 / 1.2699[20]	H_2O 1; chl 4 / 1.4645[20]
3452	Butane, 2,2,3-trichloro-1,1,1,3,4,4,4-heptafluoro-	$C_4Cl_3F_7$ / 287.39	335-44-4 / 4	/ 98	4-01-00-00253 / 1.7484[20]	/ 1.3530[20]
3453	Butane, 1,2,3-trichloro-2-methyl- / Butane, 2-methyl-1,2,3-trichloro	$C_5H_9Cl_3$ / 175.48	62521-69-1 /	/ 184	3-01-00-00361 / 1.2527[20]	chl 4
3454	Butane, 2,2,3-trichloro-3-methyl- / Butane, 2-methyl-2,3,3-trichloro	$C_5H_9Cl_3$ / 175.48	98070-91-8 /	/ 182.5	3-01-00-00361 / 1.215[15]	chl 4; HOAc 4 / 1.473[21]
3455	Butane, 1,1,1-trimethoxy-	$C_7H_{16}O_3$ / 148.20	10138-89-3 /	/ 157	4-01-00-03985 / 0.935[25]	
3456	Butane, 1,3,3-trimethoxy-	$C_7H_{16}O_3$ / 148.20	6607-66-5 /	/ 61-3[20]	4-01-00-03995 / 0.94[25]	/ 1.4096[20]
3457	Butane, 2,2,3-trimethyl- / 2,2,3-Trimethylbutane	C_7H_{16} / 100.20	464-06-2 / -25	/ 80.8	4-01-00-00410 / 0.6901[20]	H_2O 1; EtOH 3; eth 3; ace 4 / 1.3864[20]
3458	1,2,3-Butanetriol / 1-Methylglycerol	$C_4H_{10}O_3$ / 106.12	4435-50-1 /	/ 170[20]	4-01-00-02774 /	H_2O 4; EtOH 4 / 1.4462[20]
3459	1,2,4-Butanetriol	$C_4H_{10}O_3$ / 106.12	3068-00-6 /	/ 172-4[12]	2-01-00-00596 / 1.018[20]	H_2O 4; EtOH 4 / 1.4688[20]
3460	Butanoic acid / Butyric acid	$C_4H_8O_2$ / 88.11	107-92-6 / -5.7	1593 / 163.7	4-02-00-00779 / 0.9577[20]	H_2O 5; EtOH 5; eth 5; ctc 2 / 1.3980[20]
3461	Butanoic acid, 2-acetyl-3-methyl-, ethyl ester	$C_9H_{16}O_3$ / 172.22	1522-46-9 /	/ 201; 97-8[20]	4-03-00-01608 / 0.9648[18]	H_2O 1; EtOH 5; eth 5 / 1.4256[18]
3462	Butanoic acid, 2-acetyl-3-oxo-, ethyl ester	$C_8H_{12}O_4$ / 172.18	603-69-0 /	/ 210	4-03-00-01781 / 1.1045[20]	bz 4; eth 4; EtOH 4 / 1.4690[20]
3463	Butanoic acid, 2-amino-, (±)-	$C_4H_9NO_2$ / 103.12	2835-81-6 / 305 dec	/ sub	3-04-00-01296 / 1.2300[20]	H_2O 4; EtOH 2; eth 1; bz 1
3464	Butanoic acid, 4-amino- / γ-Aminobutyric acid	$C_4H_9NO_2$ / 103.12	56-12-2 / 203 dec	441 /	4-04-00-02600 /	H_2O 4; EtOH 2; eth 1; ace 2
3465	Butanoic acid, 3-amino-4-hydroxy- / γ-Hydroxy-β-aminobutyric acid	$C_4H_9NO_3$ / 119.12	589-44-6 / 216	455 /	4-04-00-03216 /	
3466	Butanoic acid, 4-amino-2-hydroxy-4-oxo-	$C_4H_7NO_4$ / 133.10	66398-52-5 / 147.5	/	3-03-00-00918 / 1.577[18]	H_2O 3; EtOH 2; eth 1; MeOH 2
3467	Butanoic acid, anhydride / Butyric anhydride	$C_8H_{14}O_3$ / 158.20	106-31-0 / -75	1594 / 200	4-02-00-00802 / 0.9668[20]	eth 3; ctc 2 / 1.4070[20]
3468	Butanoic acid, 2-bromo-, (±)- / DL-α-Bromobutyric acid	$C_4H_7BrO_2$ / 167.00	2385-70-8 / -4	1401 / 217 dec; 181[250]	4-02-00-00833 / 1.5641[20]	H_2O 3; EtOH 3; eth 3
3469	Butanoic acid, 2-bromo-, ethyl ester	$C_6H_{11}BrO_2$ / 195.06	533-68-6 /	/ 177; 43[5]	4-02-00-00834 / 1.3273[20]	H_2O 1; EtOH 5; eth 5; chl 3 / 1.4475[20]

No.	Name / Synonym	Mol. Form. / Mol. Wt.	CAS RN / mp/°C	Merck No. / bp/°C	Beil. Ref. / den/g cm⁻³	Solubility / n_D
3470	Butanoic acid, 4-bromo-, ethyl ester	$C_6H_{11}BrO_2$ 195.06	2969-81-5	82^{10}	4-02-00-00836 1.363^{25}	1.4559^{20}
3471	Butanoic acid, 2-bromo-3-methyl-, (±)-	$C_5H_9BrO_2$ 181.03	10323-40-7 44	230 dec; 138^{25}	3-02-00-00706 1.459^{20}	H_2O 2; EtOH 4; eth 3; ace 3
3472	Butanoic acid, 3-bromo-3-methyl- β-Bromoisovaleric acid	$C_5H_9BrO_2$ 181.03	5798-88-9 74	1410	4-02-00-00905	bz 4; eth 4; EtOH 4
3473	Butanoic acid, 2-bromo-, methyl ester	$C_5H_9BrO_2$ 181.03	3196-15-4	168	4-02-00-00834 1.4528^{20}	EtOH 4 1.4029^{25}
3474	Butanoic acid, 4-bromo-, methyl ester	$C_5H_9BrO_2$ 181.03	4897-84-1	186.5	4-02-00-00835 1.42^{25}	EtOH 4 1.4567^{25}
3475	Butanoic acid, 2-bromo-3-methyl-, ethyl ester	$C_7H_{13}BrO_2$ 209.08	609-12-1	186	4-02-00-00906 1.2760^{20}	eth 4; EtOH 4 1.4496^{20}
3476	Butanoic acid, 2-bromo-3-methyl-, methyl ester	$C_6H_{11}BrO_2$ 195.06	26330-51-8	177	3-02-00-00706 1.352^{13}	eth 4; EtOH 4 1.4530^{20}
3477	Butanoic acid, 2-bromo-3-oxo-, ethyl ester	$C_6H_9BrO_3$ 209.04	609-13-2	210 dec; 107^{15}	4-03-00-01551 1.4294^{16}	eth 4; EtOH 4 1.463^{14}
3478	Butanoic acid, 4-bromo-3-oxo-, ethyl ester	$C_6H_9BrO_3$ 209.04	13176-46-0	115^{14}	4-03-00-01551 1.4840^{20}	eth 4; EtOH 4 1.5281^{20}
3479	Butanoic acid, butyl ester Butyl butanoate	$C_8H_{16}O_2$ 144.21	109-21-7 -91.5	1556 166	4-02-00-00789 0.8700^{20}	H_2O 1; EtOH 5; eth 5; ctc 3 1.4075^{20}
3480	Butanoic acid, 2-chloro-	$C_4H_7ClO_2$ 122.55	4170-24-5	189^{627}; 101^{15}	4-02-00-00821 1.1796^{20}	H_2O 2; EtOH 4; eth 4 1.441^{20}
3481	Butanoic acid, 3-chloro-, (±)-	$C_4H_7ClO_2$ 122.55	625-68-3 16	116^{22}	4-02-00-00823 1.1898^{20}	EtOH 3; eth 4; ctc 2 1.4221^{20}
3482	Butanoic acid, 4-chloro-	$C_4H_7ClO_2$ 122.55	627-00-9 16	196^{22}; $68^{0.2}$	4-02-00-00825 1.2236^{20}	EtOH 4 1.4642^{20}
3483	Butanoic acid, 2-chloro-, ethyl ester	$C_6H_{11}ClO_2$ 150.60	7425-45-8	163.5	4-02-00-00822 1.056^{18}	eth 4; EtOH 4 1.4180^{25}
3484	Butanoic acid, 3-chloro-, ethyl ester	$C_6H_{11}ClO_2$ 150.60	7425-48-1		4-02-00-00824 1.0517^{20}	EtOH 4 1.4246^{20}
3485	Butanoic acid, 4-chloro-, ethyl ester	$C_6H_{11}ClO_2$ 150.60	3153-36-4	184	4-02-00-00825 1.0756^{20}	ace 4; eth 4; EtOH 4 1.4311^{20}
3486	Butanoic acid, 2-chloro-2-methyl- Butyric acid, 2-chloro-2-methyl-	$C_5H_9ClO_2$ 136.58	73758-54-0	200 dec	3-02-00-00688 1.1204^{20}	eth 4; EtOH 4 1.4445^{20}
3487	Butanoic acid, 2-chloro-3-methyl-	$C_5H_9ClO_2$ 136.58	921-08-4 21	210; 126^{32}	4-02-00-00904 1.135^{13}	H_2O 1; EtOH 3; eth 3 1.4450^{11}
3488	Butanoic acid, 2-chloro-, methyl ester	$C_5H_9ClO_2$ 136.58	26464-32-4	146	3-02-00-00622 1.0979^{14}	eth 4; EtOH 4 1.4247^{20}
3489	Butanoic acid, 4-chloro-, methyl ester	$C_5H_9ClO_2$ 136.58	3153-37-5	175; 55^4	4-02-00-00825 1.1201^{20}	H_2O 1; EtOH 4; eth 4; ace 3 1.4321^{20}
3490	Butanoic acid, 2-chloro-3-methyl-, ethyl ester Ethyl 2-chloro-3-methylbutanoate	$C_7H_{13}ClO_2$ 164.63	91913-99-4	179	0-02-00-00316 1.021^{13}	eth 4; EtOH 4
3491	Butanoic acid, 2-chloro-2-methyl-, ethyl ester Butyric acid, 2-chloro-2-methyl, ethyl ester	$C_7H_{13}ClO_2$ 164.63	58190-94-6	176	3-02-00-00688 1.069^{14}	eth 4; EtOH 4 1.4388^{11}
3492	Butanoic acid, 2-chloro-2-methyl-3-oxo-, ethyl ester Butyric acid, 2-chloro-2-methyl-3-oxo, ethyl ester	$C_7H_{11}ClO_3$ 178.62	37935-39-0	$116-7^{75}$	2-03-00-00433 1.157^{18}	EtOH 4 1.4382^{18}
3493	Butanoic acid, 4-(4-chloro-2-methylphenoxy)- 4-(4-Chloro-2-methylphenoxy)butyric acid	$C_{11}H_{13}ClO_3$ 228.68	94-81-5 100		4-06-00-01996	
3494	Butanoic acid, 4-chloro-3-oxo-, ethyl ester	$C_6H_9ClO_3$ 164.59	638-07-3 -8	220 dec; 115^{14}	4-03-00-01550 1.218^{25}	1.4520^{20}
3495	Butanoic acid, 2-cyano-2-ethyl-, ethyl ester Cyanoacetic acid, diethyl, ethyl ester	$C_9H_{15}NO_2$ 169.22	1619-56-3	214.5	3-02-00-01761	eth 4; EtOH 4 1.4200^{27}
3496	Butanoic acid, cyclohexyl ester	$C_{10}H_{18}O_2$ 170.25	1551-44-6	213	4-06-00-00037 0.9572^0	H_2O 1; EtOH 3; ctc 2; os 3
3497	Butanoic acid, 2,3-dibromo-	$C_4H_6Br_2O_2$ 245.90	600-30-6 87	$100-10^{20}$	4-02-00-00837	H_2O 2; EtOH 4; eth 4; bz 4
3498	Butanoic acid, 2,3-dibromo-, ethyl ester	$C_6H_{10}Br_2O_2$ 273.95	609-11-0 58.5	113^{30}	3-02-00-00634 1.6800^{20}	H_2O 2; EtOH 3; eth 3; ctc 2
3499	Butanoic acid, 2,4-dibromo-, ethyl ester	$C_6H_{10}Br_2O_2$ 273.95	36847-51-5	$149-50^{52}$	4-02-00-00838 1.6987^{20}	H_2O 1; EtOH 3; eth 3 1.4960^{20}
3500	Butanoic acid, 2,2-dichloro- Butyric acid, 2,2-dichloro-	$C_4H_6Cl_2O_2$ 157.00	13023-00-2	$107-10^{14}$	2-02-00-00254 1.389^{20}	H_2O 2; EtOH 4; eth 4
3501	Butanoic acid, 3,4-dichloro-, butyl ester Butyric acid, 3,4-dichloro, butyl ester	$C_8H_{14}Cl_2O_2$ 213.10	116723-94-5	$127-8^{15}$	4-02-00-00829 1.143^{25}	1.454^{24}
3502	Butanoic acid, 3,4-dichloro-, methyl ester	$C_5H_8Cl_2O_2$ 171.02	819-93-2	90^{15}	4-02-00-00829 1.278^{23}	1.459^{23}
3503	Butanoic acid, 2,2-dichloro-3-oxo-, ethyl ester	$C_6H_8Cl_2O_3$ 199.03	6134-66-3	206; 91^{12}	2-03-00-00427 1.291^{16}	EtOH 4 1.4492^{17}
3504	Butanoic acid, 2,2-diethyl-3-oxo-, ethyl ester	$C_{10}H_{18}O_3$ 186.25	1619-57-4	216; 64^3	4-03-00-01625 0.9717^{18}	H_2O 1; EtOH 5; eth 5; ctc 3 1.4326^{17}
3505	Butanoic acid, 2,3-dihydroxypropyl ester	$C_7H_{14}O_4$ 162.19	557-25-5	280; 117^{10}	4-02-00-00798 1.129^{18}	H_2O 4; EtOH 4 1.4531^{20}
3506	Butanoic acid, 2,2-dimethyl-	$C_6H_{12}O_2$ 116.16	595-37-9 -14	186	4-02-00-00954 0.9276^{20}	H_2O 2; EtOH 3; eth 3 1.4145^{20}

No.	Name Synonym	Mol. Form. Mol. Wt.	CAS RN mp/°C	Merck No. bp/°C	Beil. Ref. den/g cm^{-3}	Solubility n_D
3507	Butanoic acid, 2,3-dimethyl- Butyric acid, 2,3-dimethyl	$C_6H_{12}O_2$ 116.16	14287-61-7 -1.5	191.7	4-02-00-00958 0.9275[20]	eth 4; EtOH 4 1.4146[20]
3508	Butanoic acid, 3,3-dimethyl- tert-Butylacetic acid	$C_6H_{12}O_2$ 116.16	1070-83-3 6.5	1538 190	4-02-00-00955 0.9124[20]	EtOH 3; eth 3 1.4096[20]
3509	Butanoic acid, 1,1-dimethylethyl ester	$C_8H_{16}O_2$ 144.21	2308-38-5	146	4-02-00-00790	ace 4; eth 4; EtOH 4 1.4007[17]
3510	Butanoic acid, 3,3-dimethyl-2-oxo-	$C_6H_{10}O_3$ 130.14	815-17-8 90.5	189; 80[15]	4-03-00-01595	H_2O 2; eth 3; bz 3; chl 3
3511	Butanoic acid, 2,2-dimethyl-3-oxo-, ethyl ester	$C_8H_{14}O_3$ 158.20	597-04-6	182	4-03-00-01594 0.9759[20]	eth 4; EtOH 4 1.4180[20]
3512	Butanoic acid, 1,2-ethanediyl ester	$C_{10}H_{18}O_4$ 202.25	105-72-6 -80	240	4-02-00-00796 1.0005[20]	eth 4; EtOH 4 1.4262[20]
3513	Butanoic acid, ethenyl ester Vinyl butyrate	$C_6H_{10}O_2$ 114.14	123-20-6	116.7; 64[130]	4-02-00-00792 0.9006[20]	
3514	Butanoic acid, 2-ethyl- Diethylacetic acid	$C_6H_{12}O_2$ 116.16	88-09-5 -31.8	3099 194	4-02-00-00950 0.9239[20]	H_2O 2; EtOH 5; eth 5; ctc 2 1.4132[20]
3515	Butanoic acid, ethyl ester Ethyl butanoate	$C_6H_{12}O_2$ 116.16	105-54-4 -98	3733 121.5	4-02-00-00787 0.8844[15]	H_2O 2; EtOH 3; eth 3; ctc 2 1.4000[20]
3516	Butanoic acid, 2-ethyl-2-methyl-	$C_7H_{14}O_2$ 130.19	19889-37-3 <-20	207	4-02-00-00981	EtOH 4 1.4250[20]
3517	Butanoic acid, 2-ethyl-, methyl ester	$C_7H_{14}O_2$ 130.19	816-11-5	137	4-02-00-00950 0.8886[12]	1.4067[12]
3518	Butanoic acid, 2-ethyl-3-oxo-, ethyl ester	$C_8H_{14}O_3$ 158.20	607-97-6	198.0; 80[10]	4-03-00-01592 0.9847[16]	EtOH 5; eth 5 1.4214[25]
3519	Butanoic acid, 2-ethyl-3-oxo-, methyl ester	$C_7H_{12}O_3$ 144.17	51756-08-2	182	4-03-00-01592 0.995[14]	ace 4; eth 4; EtOH 4
3520	Butanoic acid, 2-furanylmethyl ester	$C_9H_{12}O_3$ 168.19	623-21-2	212.5	4-17-00-01247 1.0530[20]	H_2O 2; EtOH 3; eth 5
3521	Butanoic acid, heptafluoro-	$C_4HF_7O_2$ 214.04	375-22-4 -17.5	121	4-02-00-00810 1.651[20]	H_2O 3; eth 3; tol 3; peth 1 1.295[25]
3522	Butanoic acid, heptafluoro-, anhydride	$C_8F_{14}O_3$ 410.06	336-59-4 -43	106.5	4-02-00-00817 1.665[20]	1.285[20]
3523	Butanoic acid, heptafluoro-, ethenyl ester	$C_6H_3F_7O_2$ 240.08	356-28-5	79	4-02-00-00815 1.418[20]	1.309
3524	Butanoic acid, heptafluoro-, ethyl ester	$C_6H_5F_7O_2$ 242.09	356-27-4	95	4-02-00-00813 1.394[20]	H_2O 2; eth 3; ace 3 1.3011[20]
3525	Butanoic acid, heptafluoro-, methyl ester	$C_5H_3F_7O_2$ 228.07	356-24-1 -86	80	4-02-00-00812 1.483[20]	H_2O 2; eth 3; ace 3 1.295[20]
3526	Butanoic acid, heptafluoro-, 1-methylethyl ester Isopropyl heptafluorobutyrate	$C_7H_7F_7O_2$ 256.12	425-23-0	106	4-02-00-00813 1.324[20]	1.310[20]
3527	Butanoic acid, heptyl ester	$C_{11}H_{22}O_2$ 186.29	5870-93-9 -57.5	225.8	4-02-00-00791 0.8637[20]	EtOH 4 1.4231[20]
3528	Butanoic acid, hexyl ester	$C_{10}H_{20}O_2$ 172.27	2639-63-6 -78	208	4-02-00-00791 0.8652[20]	H_2O 1; EtOH 3; chl 2; os 3 1.4160[15]
3529	Butanoic acid, hydrazide Butyrhydrazide	$C_4H_{10}N_2O$ 102.14	3538-65-6 45.5	138[20]; 120[10]	4-02-00-00807	H_2O 4; eth 4; EtOH 4
3530	Butanoic acid, 2-hydroxy-, (±)-	$C_4H_8O_3$ 104.11	600-15-7 44.2	260 dec; 140[14]	3-03-00-00561 1.125[20]	H_2O 3; EtOH 3; eth 3
3531	Butanoic acid, 3-hydroxy-, (±)-	$C_4H_8O_3$ 104.11	625-71-8 49	130[12]; 94[0.1]	4-03-00-00761	H_2O 4; EtOH 4; eth 4; bz 1 1.4424[20]
3532	Butanoic acid, 2-hydroxy-, ethyl ester, (±)- Butyric acid, 2-hydroxy, ethyl ester (DL)	$C_6H_{12}O_3$ 132.16	68057-83-0	167	4-03-00-00756 1.0069[20]	EtOH 4 1.4179[20]
3533	Butanoic acid, 3-hydroxy-, ethyl ester, (±)-	$C_6H_{12}O_3$ 132.16	35608-64-1	185; 76[15]	4-03-00-00762 1.017[20]	H_2O 3; EtOH 3; ctc 2 1.4182[20]
3534	Butanoic acid, 3-hydroxy-, ethyl ester	$C_6H_{12}O_3$ 132.16	5405-41-4	185; 76[15]	4-03-00-00762 1.017[20]	H_2O 4; EtOH 4 1.4182[20]
3535	Butanoic acid, 3-hydroxy-3-methyl-	$C_5H_{10}O_3$ 118.13	625-08-1 <-32	162[12]	4-03-00-00827 0.9384[20]	H_2O 4; eth 4; EtOH 4 1.5081[20]
3536	Butanoic acid, 2-hydroxy-, 2-methylpropyl ester Butyric acid, 2-hydroxy, isobutyl ester	$C_8H_{16}O_3$ 160.21	116723-95-6 196		0-03-00-00302 0.944[15]	1.4182
3537	Butanoic acid, 3-mercapto-	$C_4H_8O_2S$ 120.17	26473-49-4	110[10]; 88[2.5]	4-03-00-00769 1.1371[20]	1.4782[20]
3538	Butanoic acid, 2-methyl-, (±)- (±)-2-Methylbutyric acid	$C_5H_{10}O_2$ 102.13	600-07-7 <-80	177	4-02-00-00889 0.934[20]	H_2O 2; EtOH 5; eth 5; chl 3 1.4051[20]
3539	Butanoic acid, 3-methyl- Isovaleric acid	$C_5H_{10}O_2$ 102.13	503-74-2 -29.3	5120 176.5	4-02-00-00895 0.931[20]	H_2O 3; EtOH 5; eth 5; chl 5 1.4033[20]
3540	Butanoic acid, 2-methyl-, 10-(acetyloxy)-9,10-dihydro-8,8-dimethyl-2-oxo-2H,8H-benzo[1,2-b:3,4-b']dipyran-9-yl Visnadine	$C_{21}H_{24}O_7$ 388.42	477-32-7 85.5	9915	5-19-06-00272	H_2O 1; EtOH 3; eth 3
3541	Butanoic acid, 3-(methylamino)-, ethyl ester Butyric acid, 3-methylamino-, ethyl ester	$C_7H_{15}NO_2$ 145.20	68384-70-3	72[12.5]	3-04-00-01313 0.9282[20]	1.4250[20]
3542	Butanoic acid, 3-methyl-, anhydride	$C_{10}H_{18}O_3$ 186.25	1468-39-9	215	4-02-00-00901 0.9327[20]	eth 4 1.4043[20]
3543	Butanoic acid, 3-methyl-, butyl ester Butyl isovalerate	$C_9H_{18}O_2$ 158.24	109-19-3		3-02-00-00698	1.4058[25]

No.	Name Synonym	Mol. Form. Mol. Wt.	CAS RN mp/°C	Merck No. bp/°C	Beil. Ref. den/g cm⁻³	Solubility n_D
3544	Butanoic acid, 3-methylbutyl ester Isopentyl butanoate	$C_9H_{18}O_2$ 158.24	106-27-4	4997 179	4-02-00-00791 0.865[19]	H_2O 1; EtOH 4; eth 4 1.4110[20]
3545	Butanoic acid, 2-methyl-, butyl ester Butyric acid, (2-methylbutyl) ester	$C_9H_{18}O_2$ 158.24	15706-73-7 	179	0-02-00-00304 0.8620[20]	1.4135[20]
3546	Butanoic acid, 3-methyl-, 1,1-dimethylpropyl ester tert-Amyl isovalerate	$C_{10}H_{20}O_2$ 172.27	542-37-0	647 173.5	0-02-00-00312 0.8729[0]	EtOH 4
3547	Butanoic acid, methyl ester Methyl butanoate	$C_5H_{10}O_2$ 102.13	623-42-7 -85.8	5957 102.8	4-02-00-00786 0.8984[20]	H_2O 2; EtOH 5; eth 5; ctc 2 1.3878[20]
3548	Butanoic acid, 3-methyl-, ethyl ester Ethyl isovalerate	$C_7H_{14}O_2$ 130.19	108-64-5 -99.3	3772 135.0	4-02-00-00898 0.8656[20]	EtOH 4; eth 4 1.3962[20]
3549	Butanoic acid, 1-methylethyl ester Isopropyl butyrate	$C_7H_{14}O_2$ 130.19	638-11-9 	130.5	4-02-00-00789 0.8588[20]	H_2O 1; EtOH 3 1.3936[20]
3550	Butanoic acid, 2-methyl-, ethyl ester, (S)-	$C_7H_{14}O_2$ 130.19	10307-61-6 	133; 35[16]	4-02-00-00890 0.8689[25]	H_2O 1; EtOH 3; bz 3 1.3964[20]
3551	Butanoic acid, 3-methyl-, 3-methylbutyl ester Isoamyl isovalerate	$C_{10}H_{20}O_2$ 172.27	659-70-1 	5003 190.4	4-02-00-00899 0.8583[19]	1.4130[19]
3552	Butanoic acid, 3-methyl-, methyl ester Methyl isovalerate	$C_6H_{12}O_2$ 116.16	556-24-1 	6006 116.5	4-02-00-00897 0.8808[20]	H_2O 1; EtOH 4; eth 4; ace 4 1.3927[20]
3553	Butanoic acid, 3-methyl-, 1-methylethyl ester Butyric, 3-methyl, isopropyl ester	$C_8H_{16}O_2$ 144.21	32665-23-9 	142; 70[55]	3-02-00-00698 0.8538[17]	ace 4; eth 4; EtOH 4 1.3960[20]
3554	Butanoic acid, 3-methyl-, 5-methyl-2-(1-methylethyl)cyclohexyl ester	$C_{15}H_{28}O_2$ 240.39	16409-46-4 	129[9]	2-06-00-00043 0.9089[15]	H_2O 1; EtOH 3; ace 3 1.4486[20]
3555	Butanoic acid, 3-methyl-, 2-methylpropyl ester Isobutyl isovalerate	$C_9H_{18}O_2$ 158.24	589-59-3 	5029 168.5	4-02-00-00898 0.853[20]	H_2O 1; EtOH 5; eth 5; ace 4 1.4057[20]
3556	Butanoic acid, 3-methyl-2-oxo-	$C_5H_8O_3$ 116.12	759-05-7 31.5	170.5	4-03-00-01577 0.9968[20]	H_2O 3; EtOH 3; eth 3 1.3850[16]
3557	Butanoic acid, 2-methyl-3-oxo-, ethyl ester	$C_7H_{12}O_3$ 144.17	609-14-3 	187	4-03-00-01573 0.9941[20]	H_2O 2; EtOH 3; eth 3; ace 4 1.4185[20]
3558	Butanoic acid, 2-methyl-3-oxo-, methyl ester	$C_6H_{10}O_3$ 130.14	17094-21-2 	177.4	3-03-00-01225 1.0217[25]	eth 4; EtOH 4 1.416[24]
3559	Butanoic acid, 1-methylpentyl ester, (S)- Butyric acid (1-methylpentyl) ester (d)	$C_{10}H_{20}O_2$ 172.27	116723-96-7 	85[20]	1-02-00-00120 0.8744[21]	eth 4; EtOH 4
3560	Butanoic acid, 3-methyl-, phenylmethyl ester	$C_{12}H_{16}O_2$ 192.26	103-38-8 	245; 136[25]	4-06-00-02266 0.9983[15]	1.4884[20]
3561	Butanoic acid, 3-methyl-, 1,2,3-propanetriyl ester	$C_{18}H_{32}O_6$ 344.45	620-63-3 	332.5	4-02-00-00900 0.9984[20]	eth 4; EtOH 4 1.4354[20]
3562	Butanoic acid, 3-methyl-, propyl ester Propyl isovalerate	$C_8H_{16}O_2$ 144.21	557-00-6 	155.9	4-02-00-00898 0.8617[20]	eth 4; EtOH 4 1.4031[20]
3563	Butanoic acid, 2-methylpropyl ester Isobutyl butanoate	$C_8H_{16}O_2$ 144.21	539-90-2 	5020 156.9	4-02-00-00790 0.8364[18]	H_2O 2; EtOH 5; eth 5 1.4032[20]
3564	Butanoic acid, 1-methylpropyl ester, (S)- Butyric acid, sec-butyl ester (d)	$C_8H_{16}O_2$ 144.21	116836-55-6 -91.5	152; 54[18]	3-02-00-00600 0.8737[13]	EtOH 3; bz 3; CS_2 3; py 3 1.4011[20]
3565	Butanoic acid, 3-methyl-, 1,7,7-trimethylbicyclo[2.2.1]hept-2-yl ester, (1R-endo)- d-Bornyl isovalerate	$C_{15}H_{26}O_2$ 238.37	53022-14-3 	1342 257.5	2-06-00-08488 0.955[25]	eth 4; EtOH 4
3566	Butanoic acid, octyl ester	$C_{12}H_{24}O_2$ 200.32	110-39-4 -55.6	244.1	4-02-00-00791 0.8692[20]	EtOH 4 1.4267[15]
3567	Butanoic acid, 2-oxo-	$C_4H_6O_3$ 102.09	600-18-0 33	80-2[16]	4-03-00-01524 1.200[17]	H_2O 4; EtOH 4; eth 2 1.3972[20]
3568	Butanoic acid, 3-oxo- Acetoacetic acid	$C_4H_6O_3$ 102.09	541-50-4 36.5	51 100 dec	4-03-00-01527 	H_2O 4; eth 4; EtOH 4
3569	Butanoic acid, 3-oxo-, butyl ester	$C_8H_{14}O_3$ 158.20	591-60-6 -35.6	127[50]; 85[8]	4-03-00-01536 0.9671[25]	H_2O 2; EtOH 5; bz 5; lig 5 1.4137[20]
3570	Butanoic acid, 3-oxo-, 2-chloroethyl ester	$C_6H_9ClO_3$ 164.59	54527-68-3 	198; 107[14]	4-03-00-01534 1.190[20]	bz 4; eth 4; EtOH 4 1.4430[20]
3571	Butanoic acid, 3-oxo-, ethyl ester Ethyl acetoacetate	$C_6H_{10}O_3$ 130.14	141-97-9 -45	3714 180.8	4-03-00-01528 1.0368[10]	H_2O 4; EtOH 5; eth 5; bz 3 1.4171[20]
3572	Butanoic acid, 4-oxo-, ethyl ester	$C_6H_{10}O_3$ 130.14	10138-10-0 	162; 85[12]	4-03-00-01555 1.0490[22]	chl 2 1.4287[20]
3573	Butanoic acid, 3-oxo-, methyl ester Methyl acetoacetate	$C_5H_8O_3$ 116.12	105-45-3 27.5	5933 171.7	4-03-00-01527 1.0762[20]	H_2O 4; EtOH 5; eth 5; ctc 3 1.4184[20]
3574	Butanoic acid, 3-oxo-, 1-methylethyl ester Isopropyl acetoacetate	$C_7H_{12}O_3$ 144.17	542-08-5 -27.3	5094 186	4-03-00-01535 0.9835[20]	eth 4; EtOH 4; lig 4 1.4173[20]
3575	Butanoic acid, 4-oxo-4-(phenylamino)- Succinanilic acid	$C_{10}H_{11}NO_3$ 193.20	102-14-7 148.5	8839	4-12-00-00468 	H_2O 2; EtOH 3; eth 4
3576	Butanoic acid, 3-oxo-2-(phenylmethylene)-, ethyl ester	$C_{13}H_{14}O_3$ 218.25	620-80-4 60.5	296; 180[17]	3-10-00-03158 	H_2O 1; EtOH 2; eth 2; bz 2
3577	Butanoic acid, 3-oxo-, 2-propenyl ester	$C_7H_{10}O_3$ 142.15	1118-84-9 	196; 66.5[14]	4-03-00-01538 1.0366[20]	H_2O 3; EtOH 5; bz 5; lig 3 1.4398[20]
3578	Butanoic acid, 4-oxo-4-[[4-[(2-thiazolylamino)sulfonyl]phenyl]amino]- Succinylsulphathiazole	$C_{13}H_{13}N_3O_5S_2$ 355.40	116-43-8 193.5	8850	4-27-00-04637 	H_2O 1; EtOH 2; eth 1; ace 2
3579	Butanoic acid, pentyl ester Amyl butyrate	$C_9H_{18}O_2$ 158.24	540-18-1 -73.2	639 186.4	4-02-00-00790 0.8713[15]	H_2O 1; EtOH 4; eth 4 1.4123[20]
3580	Butanoic acid, 2-phenoxy-	$C_{10}H_{12}O_3$ 180.20	13794-14-4 98	258	3-06-00-00616 	H_2O 2; EtOH 3; eth 3; ace 3

No.	Name / Synonym	Mol. Form. / Mol. Wt.	CAS RN / mp/°C	Merck No. / bp/°C	Beil. Ref. / den/g cm^{-3}	Solubility / n_D
3581	Butanoic acid, 4-phenoxy-, ethyl ester Ethyl 4-phenoxybutyrate	$C_{12}H_{16}O_3$ 208.26	2364-59-2	170-3[25]	2-06-00-00159 1.042[35]	eth 4; EtOH 4 1.491[35]
3582	Butanoic acid, 2-phenoxyethyl ester	$C_{12}H_{16}O_3$ 208.26	23511-70-8	251; 88[2]	3-06-00-00571 1.0388[21]	ace 4; eth 4; EtOH 4
3583	Butanoic acid, 2-phenoxy-, phenyl ester 2-Phenoxybutyric acid, phenyl ester	$C_{16}H_{16}O_3$ 256.30	116836-19-2 48.5	202-3[25]	4-06-00-00645 1.135[15]	
3584	Butanoic acid, phenyl ester Phenyl butyrate	$C_{10}H_{12}O_2$ 164.20	4346-18-3	225	4-06-00-00615 1.0382[15]	H_2O 1; EtOH 3; eth 3
3585	Butanoic acid, phenylmethyl ester	$C_{11}H_{14}O_2$ 178.23	103-37-7	239	4-06-00-02266 1.0111[20]	H_2O 1; EtOH 4; eth 4; ctc 3 1.4920[20]
3586	Butanoic acid, 1,2,3-propanetriyl ester Tributyrin	$C_{15}H_{26}O_6$ 302.37	60-01-5 -75	9532 307.5	4-02-00-00799 1.0350[20]	H_2O 1; EtOH 3; eth 4; ace 3 1.4359[20]
3587	Butanoic acid, 2-propenyl ester	$C_7H_{12}O_2$ 128.17	2051-78-7	142; 44.5[15]	4-02-00-00793 0.9017[20]	H_2O 1; EtOH 5; eth 5; ctc 2 1.4158[20]
3588	Butanoic acid, propyl ester Propyl butanoate	$C_7H_{14}O_2$ 130.19	105-66-8 -95.2	7858 143.0	4-02-00-00788 0.8730[20]	H_2O 2; EtOH 5; eth 5 1.4001[20]
3589	Butanoic acid, 3-thioxo-, ethyl ester Butyric acid, 3-thioxo, ethyl ester	$C_6H_{10}O_2S$ 146.21	7740-33-2	75[15]	4-03-00-01553 1.0554[31]	eth 4; EtOH 4 1.4712[26]
3590	Butanoic acid, 2,2,3-trichloro- 2,2,3-Trichlorobutyric acid	$C_4H_5Cl_3O_2$ 191.44	5344-55-8 60	237	3-02-00-00629	eth 4
3591	Butanoic acid, 2,2,3-trichloro-, ethyl ester Butyric acid, 2,2,3-trichloro, ethyl ester	$C_6H_9Cl_3O_2$ 219.49	116723-97-8	212	0-02-00-00281 1.3114[20]	eth 4; EtOH 4
3592	Butanoic acid, 4,4,4-trifluoro-3-oxo-, ethyl ester	$C_6H_7F_3O_3$ 184.11	372-31-6	133	4-03-00-01548 1.2586[15]	EtOH 3; eth 3 1.3783[15]
3593	1-Butanol Butyl alcohol	$C_4H_{10}O$ 74.12	71-36-3 -89.8	1540 117.7	4-01-00-01506 0.8098[20]	H_2O 3; EtOH 5; eth 5; ace 4 1.3993[20]
3594	2-Butanol sec-Butyl alcohol	$C_4H_{10}O$ 74.12	78-92-2 -114.7	1541 99.5	2-01-00-00400 0.8063[20]	H_2O 4; EtOH 5; eth 5; ace 4 1.3978[20]
3595	1-Butanol, 2-amino-, (±)-	$C_4H_{11}NO$ 89.14	13054-87-0 -2	178	4-04-00-01705 0.9162[20]	H_2O 5; EtOH 5; eth 5; chl 2 1.4489[25]
3596	1-Butanol, 3-amino-	$C_4H_{11}NO$ 89.14	2867-59-6	82-5[19]	4-04-00-01710	H_2O 4; EtOH 4 1.4534[25]
3597	1-Butanol, 4-amino-	$C_4H_{11}NO$ 89.14	13325-10-5	205; 125[34]	4-04-00-01711 0.967[12]	H_2O 3; EtOH 3; eth 1 1.4625[20]
3598	2-Butanol, 3-amino-	$C_4H_{11}NO$ 89.14	42551-55-3 19	160; 79[20]	3-04-00-00779 0.9399[20]	H_2O 4; eth 4; EtOH 4 1.4445[25]
3599	1-Butanol, 2-amino-3-methyl-, (±)-	$C_5H_{13}NO$ 103.16	16369-05-4	75-7[8]	4-04-00-01777 0.93[25]	1.4543[20]
3600	1-Butanol, 3-bromo-	C_4H_9BrO 153.02	6089-12-9	69-71[20]	4-01-00-01553 1.7107[18]	1.5010[21]
3601	2-Butanol, 3-bromo-, (R*,S*)-(±)- 2-Butanol, 3-bromo-, erythro-(±)-	C_4H_9BrO 153.02	19246-39-0	154; 53[13]	4-01-00-01580 1.4550[20]	eth 4; EtOH 4 1.4780[20]
3602	1-Butanol, 2-chloro-	C_4H_9ClO 108.57	26106-95-6	74-6[25]	4-01-00-01549 1.062[25]	eth 4; EtOH 4 1.4438[20]
3603	1-Butanol, 3-chloro-	C_5H_9ClO 120.58	2203-35-2	175; 66[15]	3-01-00-01517 1.0883[20]	EtOH 3; eth 3; ace 3; ctc 2 1.4518[20]
3604	1-Butanol, 4-chloro-	C_4H_9ClO 108.57	928-51-8	84-5[18]	4-01-00-01550 1.0883[20]	eth 4; EtOH 4 1.4518[20]
3605	2-Butanol, 1-chloro- α-Butylene chlorohydrin	C_4H_9ClO 108.57	1873-25-2	141	4-01-00-01578 1.068[25]	EtOH 3; eth 3 1.4400[20]
3606	2-Butanol, 1-chloro- β-Butylene chlorohydrin	C_4H_9ClO 108.57	563-84-8	139	3-01-00-01538 1.0669[20]	EtOH 4; eth 4 1.4432[20]
3607	2-Butanol, 4-chloro-	C_4H_9ClO 108.57	2203-34-1	67[20]	4-01-00-01578	eth 4; EtOH 4 1.4408[20]
3608	2-Butanol, 1-chloro-2,3-dimethyl- 1-Chloro-2,3-dimethyl-2-butanol	$C_6H_{13}ClO$ 136.62	66235-62-9	163	4-01-00-01730 1.049[20]	1.4459[20]
3609	2-Butanol, 1-chloro-3,3-dimethyl- 1-Chloro-3,3-dimethyl-2-butanol	$C_6H_{13}ClO$ 136.62	36402-31-0 13	157.5	4-01-00-01728 1.0063[25]	1.4432[25]
3610	2-Butanol, 1-chloro-2-methyl-	$C_5H_{11}ClO$ 122.59	74283-48-0	151	4-01-00-01673 1.0161[20]	EtOH 4 1.4469[20]
3611	2-Butanol, 3-chloro-2-methyl-	$C_5H_{11}ClO$ 122.59	21326-62-5	141.5	3-01-00-01627 1.0295[20]	eth 4; EtOH 4 1.4436[20]
3612	1-Butanol, 2,3-dibromo- 2,3-Dibromo-1-butanol	$C_4H_8Br_2O$ 231.91	4021-75-4 32	99.5[10]	4-01-00-01554 1.9475[20]	1.5442[20]
3613	1-Butanol, 3,4-dibromo- 3,4-Dibromobutanol	$C_4H_8Br_2O$ 231.91	87018-30-2	114[11]	3-01-00-01518 1.98[15]	1.548[15]
3614	2-Butanol, 1,4-dibromo- 2-Hydroxy-1,4-dibromobutane	$C_4H_8Br_2O$ 231.91	19398-47-1	114-5[15]	4-01-00-01581 1.953[20]	1.544[20]
3615	1-Butanol, 2,2-dimethyl- 2,2-Dimethyl-1-butanol	$C_6H_{14}O$ 102.18	1185-33-7 <-15	136.5	4-01-00-01726 0.8283[20]	H_2O 2; EtOH 3; eth 3 1.4208[20]
3616	1-Butanol, 2,3-dimethyl-, (±)-	$C_6H_{14}O$ 102.18	20281-85-0	144.5	4-01-00-01729 0.8297[20]	EtOH 3; eth 3; ace 3 1.4195[20]
3617	1-Butanol, 3,3-dimethyl-	$C_6H_{14}O$ 102.18	624-95-3 -60	143	4-01-00-01729 0.844[15]	H_2O 2; EtOH 3; eth 3; ace 3 1.4323[15]
3618	2-Butanol, 2,3-dimethyl-	$C_6H_{14}O$ 102.18	594-60-5 -14	118.4	4-01-00-01729 0.8236[20]	H_2O 3; EtOH 5; eth 5 1.4176[20]
3619	2-Butanol, 3,3-dimethyl-, (±)- 3,3-Dimethylbutan-2-ol (DL-)	$C_6H_{14}O$ 102.18	20281-91-8 5.6	120.4	4-01-00-01727 0.8122[25]	H_2O 2; EtOH 3; eth 5 1.4148[20]
3620	1-Butanol, 2-ethyl- 2-Ethyl-1-butanol	$C_6H_{14}O$ 102.18	97-95-0 <-15	147	4-01-00-01725 0.8326[20]	H_2O 2; EtOH 3; eth 3; chl 3 1.4220[20]

No.	Name / Synonym	Mol. Form. / Mol. Wt.	CAS RN	mp/°C	Merck No.	bp/°C	Beil. Ref.	den/g cm⁻³	Solubility	n_D
3621	1-Butanol, 4-fluoro- / 4-Fluoro-1-butanol	C$_4$H$_9$FO / 92.11	372-93-0			58[15]	4-01-00-01546		ace 4; eth 4; EtOH 4	1.3942[15]
3622	1-Butanol, 2,2,3,3,4,4,4-heptafluoro-	C$_4$H$_3$F$_7$O / 200.06	375-01-9			95	4-01-00-01547	1.600[20]	EtOH 3; ace 3	1.294[20]
3623	2-Butanol, 1-[(2-hydroxyethyl)amino]- / 2-Hydroxyethyl 2-hydroxybutyl amine	C$_6$H$_{15}$NO$_2$ / 133.19	6967-43-7			137[9]	4-04-00-01725	1.0310[20]	H$_2$O 4; ace 4; EtOH 4	1.4690[30]
3624	2-Butanol, 3,3'-iminobis- / Bis-(3-hydroxy-2-butyl)amine	C$_8$H$_{19}$NO$_2$ / 161.24	6959-06-4			112-5[3]	3-04-00-00781	0.9775[20]	H$_2$O 4; ace 4; EtOH 4	1.4162[20]
3625	1-Butanol, 3-methoxy- / 3-Methoxy-1-butanol	C$_5$H$_{12}$O$_2$ / 104.15	2517-43-3			157	4-01-00-02509	0.923[23]	EtOH 4; eth 3; ace 4; chl 2	1.4148[25]
3626	1-Butanol, 2-methyl- / 2-Methyl-1-butanol	C$_5$H$_{12}$O / 88.15	137-32-6		5952	128	4-01-00-01666	0.8150[25]	H$_2$O 2; EtOH 5; eth 5; ace 4	1.4092[20]
3627	1-Butanol, 3-methyl- / Isopentyl alcohol	C$_5$H$_{12}$O / 88.15	123-51-3	-117.2	5081	131.1	4-01-00-01677	0.8104[20]	ace 4; eth 4; EtOH 4	1.4053[20]
3628	2-Butanol, 2-methyl- / tert-Amyl alcohol	C$_5$H$_{12}$O / 88.15	75-85-4	-8.8	7096	102.4	4-01-00-01668	0.8096[20]	H$_2$O 3; EtOH 5; eth 5; ace 4	1.4052[20]
3629	2-Butanol, 3-methyl-, (±)- / (±)-3-Methyl-2-butanol	C$_5$H$_{12}$O / 88.15	70116-68-6		5953	112.9	4-01-00-01675	0.8180[20]	H$_2$O 2; EtOH 5; eth 5; ace 4	1.4089[20]
3630	1-Butanol, 2-methyl-, acetate	C$_7$H$_{14}$O$_2$ / 130.19	624-41-9			140	4-02-00-00156	0.8740[20]	ace 4; eth 4; EtOH 4	1.4040[20]
3631	1-Butanol, 3-methyl-, acetate / Isopentyl acetate	C$_7$H$_{14}$O$_2$ / 130.19	123-92-2	-78.5	4993	142.5	4-02-00-00157	0.876[15]	H$_2$O 2; EtOH 5; eth 5; ace 3	1.4000[20]
3632	1-Butanol, 3-methyl-, benzoate / Isoamyl benzoate	C$_{12}$H$_{16}$O$_2$ / 192.26	94-46-2		4995	261	4-09-00-00292	0.993[15]	EtOH 4	
3633	1-Butanol, 3-methyl-, carbamate	C$_6$H$_{13}$NO$_2$ / 131.17	543-86-2	59		220	4-03-00-00058	0.9438[71]	eth 4; EtOH 4	1.4175[20]
3634	2-Butanol, 2-methyl-, carbamate / tert-Amyl carbamate	C$_6$H$_{13}$NO$_2$ / 131.17	590-60-3	86	641		1-03-00-00014		ace 4; bz 4	
3635	1-Butanol, 3-methyl-, carbonate (2:1)	C$_{11}$H$_{22}$O$_3$ / 202.29	2050-95-5			234; 122[16]	3-03-00-00011	0.9067[20]	H$_2$O 1	1.4174[20]
3636	1-Butanol, 3-methyl-, formate / Isopentyl formate	C$_6$H$_{12}$O$_2$ / 116.16	110-45-2	-93.5	5001	123.5	4-02-00-00030	0.877[20]	H$_2$O 2; EtOH 3; eth 5; ctc 2	1.3967[20]
3637	1-Butanol, 3-methyl-, nitrate / Isoamyl nitrate	C$_5$H$_{11}$NO$_3$ / 133.15	543-87-3		5004	148	4-01-00-01683	0.996[22]		1.4122[21]
3638	1-Butanol, 3-methyl-2-nitro-	C$_5$H$_{11}$NO$_3$ / 133.15	77392-54-2			111[10]	3-01-00-01645	1.0889[25]		1.4430[20]
3639	1-Butanol, 3-methyl-, propanoate / Isopentyl propanoate	C$_8$H$_{16}$O$_2$ / 144.21	105-68-0			160.2	4-02-00-00709	0.8697[20]	eth 4; EtOH 4	1.4069[20]
3640	1-Butanol, 2-nitro- / 2-Nitro-1-butanol	C$_4$H$_9$NO$_3$ / 119.12	609-31-4	-47		105[10]	4-01-00-01555	1.1332[25]	H$_2$O 3; EtOH 5; eth 5; ace 3	1.4390[20]
3641	2-Butanol, 1-nitro- / 2-Hydroxy-1-nitrobutane	C$_4$H$_9$NO$_3$ / 119.12	3156-74-9			204; 75[2]	4-01-00-01582	1.1353[20]	H$_2$O 3; EtOH 4; eth 4; ace 3	1.4435[20]
3642	2-Butanol, 3-nitro- / 3-Nitro-2-butanol	C$_4$H$_9$NO$_3$ / 119.12	6270-16-2			112[38]; 55[0.5]	4-01-00-01583	1.1166[7]		1.4414[20]
3643	2-Butanol, 1,1,1-trichloro- / Ethyl(trichloromethyl)carbinol	C$_4$H$_7$Cl$_3$O / 177.46	6111-61-1			171; 83[22]	4-01-00-01579	1.3630[25]	ace 4; bz 4; eth 4; EtOH 4	1.4800[20]
3644	2-Butanol, 2,3,3-trimethyl-	C$_7$H$_{16}$O / 116.20	594-83-2	17		131	4-01-00-01755	0.8380[25]	ace 4; eth 4; EtOH 4	1.4233[22]
3645	2-Butanone / Methyl ethyl ketone	C$_4$H$_8$O / 72.11	78-93-3	-86.6	5991	79.5	4-01-00-03243	0.8054[20]	H$_2$O 4; EtOH 5; eth 5; ace 5	1.3788[20]
3646	2-Butanone, 1-chloro- / Chloromethyl ethyl ketone	C$_4$H$_7$ClO / 106.55	616-27-3			137.5	4-01-00-03255	1.0850[20]	MeOH 4	1.4372[20]
3647	2-Butanone, 3-chloro-	C$_4$H$_7$ClO / 106.55	4091-39-8			115	4-01-00-03256	1.0554[25]		1.4219[20]
3648	2-Butanone, 4-chloro- / 1-Chloro-3-butanone	C$_4$H$_7$ClO / 106.55	6322-49-2			120 dec	4-01-00-03256	1.0680[23]	eth 4; EtOH 4	1.4284[23]
3649	2-Butanone, 3-chloro-3-methyl- / 3-Chloro-3-methyl-2-butanone	C$_5$H$_9$ClO / 120.58	5950-19-6			117.2	3-01-00-02818	1.0083[20]	eth 4; EtOH 4	1.4204[20]
3650	1-Butanone, 4-chloro-1-phenyl-	C$_{10}$H$_{11}$ClO / 182.65	939-52-6	19.5		130-3[4]	4-07-00-00711	1.137[25]		1.5459[20]
3651	1-Butanone, 4-[4-(4-chlorophenyl)-4-hydroxy-1-piperidinyl]-1-(4-fluorophenyl)- / Haloperidol	C$_{21}$H$_{23}$ClFNO$_2$ / 375.87	52-86-8	151.5	4511		5-21-02-00377			
3652	1-Butanone, 1-cyclohexyl-	C$_{10}$H$_{18}$O / 154.25	1462-27-7			94[13]	4-07-00-00080	0.903[20]		1.4537[20]
3653	2-Butanone, 3,4-dibromo- / 1,2-Dibromoethyl methyl ketone	C$_4$H$_6$Br$_2$O / 229.90	25109-57-3			80-1[10]	3-01-00-02788	1.9662[15]		1.5314[15]
3654	2-Butanone, 1,3-dichloro- / 1,3-Dichloro-2-butanone	C$_4$H$_6$Cl$_2$O / 141.00	16714-77-5			166.5	3-01-00-02786	1.3116[20]	ace 4; bz 4; eth 4; EtOH 4	1.4686[20]
3655	2-Butanone, 3,4-dichloro-	C$_4$H$_6$Cl$_2$O / 141.00	58625-77-7			65-70[16]	3-01-00-02787	1.293[18]		1.4628[18]
3656	2-Butanone, 4-(diethylamino)-	C$_8$H$_{17}$NO / 143.23	3299-38-5			84[30]	4-04-00-01930	0.8630[20]	ace 4; bz 4; eth 4; EtOH 4	1.4333[24]
3657	2-Butanone, 3,3-dimethyl- / Pinacolone	C$_6$H$_{12}$O / 100.16	75-97-8	-52.5	7409	106.1	4-01-00-03310	0.7229[25]	H$_2$O 2; EtOH 3; eth 3; ace 3	1.3952[20]
3658	2-Butanone, 3,3-diphenyl- / 2,2-Diphenyl-3-butanone	C$_{16}$H$_{16}$O / 224.30	2575-20-4	41		310.5	4-07-00-01451	1.069[20]	H$_2$O 1; EtOH 3; eth 3; chl 4	1.5748[20]
3659	2-Butanone, 4-(2-furanyl)-	C$_8$H$_{10}$O$_2$ / 138.17	699-17-2			203	4-17-00-04559	1.036[19]		1.4696[17]
3660	2-Butanone, 1-hydroxy-	C$_4$H$_8$O$_2$ / 88.11	5077-67-8			160; 78[60]	4-01-00-03989	1.0272[20]	H$_2$O 4; EtOH 4; eth 4	1.4189[20]

No.	Name / Synonym	Mol. Form. / Mol. Wt.	CAS RN / mp/°C	Merck No. / bp/°C	Beil. Ref. / den/g cm^{-3}	Solubility / n_D
3661	2-Butanone, 3-hydroxy- / Acetoin	$C_4H_8O_2$ / 88.11	513-86-0 / 15	55 / 143	2-01-00-00870 / 1.0044[20]	H_2O 4; ace 4; EtOH 4 / 1.4171[20]
3662	2-Butanone, 4-hydroxy-	$C_4H_8O_2$ / 88.11	590-90-9	/ 109[30]; 90[11]	4-01-00-03994 / 1.1089[24]	H_2O 5; EtOH 5; eth 5; ace 4 / 1.4585[14]
3663	2-Butanone, 4-(4-hydroxy-3-methoxyphenyl)- / Zingerone	$C_{11}H_{14}O_3$ / 194.23	122-48-5 / 40.5	10072 / 187-8[14]	4-08-00-01866	eth 4
3664	2-Butanone, 4-hydroxy-3-methyl-	$C_5H_{10}O_2$ / 102.13	3393-64-4	/ 90-5[15]	4-01-00-04009 / 0.993[25]	/ 1.4340[20]
3665	1-Butanone, 1-(2-hydroxyphenyl)-	$C_{10}H_{12}O_2$ / 164.20	2887-61-8 / 10	124-6[14]	4-08-00-00497 / 1.0683[24]	/ 1.5379[25]
3666	1-Butanone, 2-hydroxy-1-phenyl- / Butyrophenone, 2-hydroxy-	$C_{10}H_{12}O_2$ / 164.20	16183-46-3	132-3[12]	4-08-00-00502 / 1.077[19]	/ 1.529[19]
3667	1-Butanone, 4-hydroxy-1-phenyl-	$C_{10}H_{12}O_2$ / 164.20	39755-03-8 / 30	187-8[9]	3-08-00-00449	EtOH 4
3668	2-Butanone, 3-methyl- / Methyl isopropyl ketone	$C_5H_{10}O$ / 86.13	563-80-4 / -92	94.3	4-01-00-03287 / 0.8051[20]	H_2O 2; EtOH 5; eth 5; ace 4 / 1.3880[20]
3669	1-Butanone, 3-methyl-1-(2-methylphenyl)- / Isobutyl 2-tolyl ketone	$C_{12}H_{16}O$ / 176.26	58138-81-1	248; 85-6[3]	4-07-00-00765 / 0.9578[20]	/ 1.5104[20]
3670	1-Butanone, 3-methyl-1-(4-methylphenyl)- / Isobutyl 4-tolyl ketone	$C_{12}H_{16}O$ / 176.26	61971-91-3	259; 137[18]	4-07-00-00765 / 0.9557[20]	/ 1.5081[20]
3671	1-Butanone, 3-methyl-1-phenyl-	$C_{11}H_{14}O$ / 162.23	582-62-7	236.5	4-07-00-00739 / 0.9701[16]	H_2O 1; EtOH 5; eth 5; ace 4 / 1.5139[15]
3672	1-Butanone, 1-(4-methylphenyl)- / Butyrophenone, 4-methyl	$C_{11}H_{14}O$ / 162.23	4160-52-5 / 12	251.5	4-07-00-00743 / 0.9745[20]	eth 4; EtOH 4 / 1.5232[20]
3673	2-Butanone, (1-methylpropylidene)hydrazone	$C_8H_{16}N_2$ / 140.23	5921-54-0	171.5	4-01-00-03252 / 0.8404[20]	/ 1.4511[20]
3674	1-Butanone, 1-(1-naphthalenyl)- / 1-Butyronaphthone	$C_{14}H_{14}O$ / 198.26	2876-62-2	317	4-07-00-01305 / 1.0861[25]	bz 4; eth 4; EtOH 4 / 1.596[27]
3675	2-Butanone, oxime	C_4H_9NO / 87.12	96-29-7 / -29.5	152.5	4-01-00-03250 / 0.9232[20]	H_2O 3; EtOH 5; eth 5; chl 3 / 1.4410[20]
3676	1-Butanone, 1-phenyl-	$C_{10}H_{12}O$ / 148.20	495-40-9 / 12	228.5	4-07-00-00709 / 0.988[20]	H_2O 1; EtOH 5; eth 5; ace 4 / 1.5203[20]
3677	2-Butanone, 1-phenyl-	$C_{10}H_{12}O$ / 148.20	1007-32-5	230; 111[16]	4-07-00-00712 / 1.002[20]	H_2O 1; EtOH 3; eth 5; ace 4
3678	2-Butanone, 4-phenyl-	$C_{10}H_{12}O$ / 148.20	2550-26-7 / -13	233.5	4-07-00-00713 / 0.9849[22]	H_2O 1; EtOH 3; eth 3; ace 4 / 1.511[22]
3679	2-Butanone, 1,3,4-trihydroxy-, (S)- / L-Erythrulose	$C_4H_8O_4$ / 120.11	533-50-6 / dec	3641 / dec	4-01-00-04176	H_2O 4; EtOH 4
3680	Butanoyl bromide / Butyryl bromide	C_4H_7BrO / 151.00	5856-82-6	129	4-02-00-00804 / 1.4162[17]	/ 1.596[17]
3681	Butanoyl chloride / Butyryl chloride	C_4H_7ClO / 106.55	141-75-3 / -89	1598 / 102	4-02-00-00803 / 1.0277[20]	eth 5 / 1.4121[20]
3682	Butanoyl chloride, 2-bromo-	C_4H_6BrClO / 185.45	22118-12-3	151	4-02-00-00834 / 1.5320[20]	eth 4 / 1.5320[20]
3683	Butanoyl chloride, 2-chloro-	$C_4H_6Cl_2O$ / 141.00	7623-11-2	130.5	3-02-00-00623 / 1.2360[17]	eth 4 / 1.4475[20]
3684	Butanoyl chloride, 3-chloro- / 3-Chlorobutyryl chloride	$C_4H_6Cl_2O$ / 141.00	1951-11-7	40-1[12]	4-02-00-00824 / 1.2163[20]	/ 1.4509[20]
3685	Butanoyl chloride, 4-chloro-	$C_4H_6Cl_2O$ / 141.00	4635-59-0	173.5	4-02-00-00826 / 1.2581[20]	eth 3 / 1.4616[20]
3686	Butanoyl chloride, 2-chloro-3-methyl- / Butyryl chloride, 2-chloro-3-methyl	$C_5H_8Cl_2O$ / 155.02	35383-59-6	148.5; 56[20]	4-02-00-00904 / 1.135[13]	eth 4
3687	Butanoyl chloride, 4-chloro-3-oxo- / Butyryl chloride, 4-chloro-3-oxo	$C_4H_4Cl_2O_2$ / 154.98	41295-64-1	117-9[17]	3-03-00-01207 / 1.4397[20]	bz 4 / 1.4860[20]
3688	Butanoyl chloride, 2,2-dimethyl-	$C_6H_{11}ClO$ / 134.61	5856-77-9	132	4-02-00-00955 / 0.9810[20]	eth 4 / 1.4245[20]
3689	Butanoyl chloride, 2,3-dimethyl- / Butyryl chloride, 2,3-dimethyl	$C_6H_{11}ClO$ / 134.61	51760-90-8	136; 38[18]	3-02-00-00761 / 0.9795[20]	eth 4
3690	Butanoyl chloride, 3,3-dimethyl-	$C_6H_{11}ClO$ / 134.61	7065-46-5	130; 68[100]	4-02-00-00956 / 0.9696[20]	eth 4 / 1.4210[20]
3691	Butanoyl chloride, 2-ethyl-	$C_6H_{11}ClO$ / 134.61	2736-40-5	140	4-02-00-00952 / 0.9825[20]	eth 4 / 1.4234[20]
3692	Butanoyl chloride, heptafluoro-	C_4ClF_7O / 232.48	375-16-6	38.5	4-02-00-00818 / 1.55[20]	/ 1.288[20]
3693	Butanoyl chloride, 3-methyl- / Isovaleryl chloride	C_5H_9ClO / 120.58	108-12-3	5121 / 114	4-02-00-00901 / 0.9844[20]	eth 3 / 1.4149[20]
3694	Butanoyl chloride, 2-methyl-, (±)- / Butyryl chloride, 2-methyl (DL)	C_5H_9ClO / 120.58	57526-28-0	116	4-02-00-00891 / 0.9917[20]	/ 1.4170[20]
3695	2-Butenal, (E)- / trans-Crotonaldehyde	C_4H_6O / 70.09	123-73-9 / -76	2599 / 102.2	4-01-00-03447 / 0.8516[20]	H_2O 3; EtOH 4; eth 4; ace 4 / 1.4366[20]
3696	2-Butenal, 2-bromo-	C_4H_5BrO / 148.99	24247-53-8	63[15]	3-01-00-02981 / 1.5797[13]	/ 1.5184[20]
3697	2-Butenal, 2-chloro- / α-Chlorocrotonaldehyde	C_4H_5ClO / 104.54	53175-28-3	147.5	3-01-00-02981 / 1.1404[23]	H_2O 2; EtOH 3; eth 3; ctc 3 / 1.4780[25]
3698	2-Butenal, 2-methyl-	C_5H_8O / 84.12	1115-11-3	117; 64[110]	3-01-00-02990 / 0.8710[20]	H_2O 3; EtOH 5; eth 5 / 1.4475[20]
3699	2-Butenal, 2-methyl-, (E)-	C_5H_8O / 84.12	497-03-0	117; 64[119]	4-01-00-03464 / 0.8710[20]	H_2O 4; EtOH 4 / 1.4475[20]
3700	2-Butenal, 3-methyl- / Senecialdehyde	C_5H_8O / 84.12	107-86-8	8399 / 134	4-01-00-03464 / 0.8722[20]	H_2O 3; EtOH 3; eth 3 / 1.4528[20]

No.	Name Synonym	Mol. Form. Mol. Wt.	CAS RN mp/°C	Merck No. bp/°C	Beil. Ref. den/g cm^{-3}	Solubility n_D
3701	2-Butenal, 3-phenyl-	$C_{10}H_{10}O$ 146.19	1196-67-4 95	124[12]; 109[4]	4-07-00-01011	1.5876[23]
3702	2-Butenamide, (E)-	C_4H_7NO 85.11	625-37-6 161.5	sub 140	2-02-00-00392 0.9461[120]	H_2O 2; EtOH 3; eth 2; bz 3 1.4420[165]
3703	3-Butenamide, 3-methyl-4-phenyl- β-Benzalbutyramide	$C_{11}H_{13}NO$ 175.23	7236-47-7 133	1063		
3704	1-Butene 1-Butylene	C_4H_8 56.11	106-98-9 -185.3	1513 -6.2	4-01-00-00765 *0.588[25]	H_2O 1; EtOH 4; eth 4; bz 3 1.3962[20]
3705	2-Butene, (E)- trans-2-Butene	C_4H_8 56.11	624-64-6 -105.5	1514 0.8	4-01-00-00781 *0.599[25]	bz 3 1.3848[-25]
3706	2-Butene, (Z)- cis-2-Butene	C_4H_8 56.11	590-18-1 -138.9	1514 3.7	4-01-00-00778 *0.616[25]	H_2O 1; EtOH 4; eth 4; bz 3 1.3931[-25]
3707	1-Butene, 1-bromo-, (E)- trans-1-Butenyl bromide	C_4H_7Br 135.00	32620-08-9 -100.3	94.7	1.3209[15]	H_2O 1; eth 3; ace 3; bz 3 1.4527[20]
3708	1-Butene, 1-bromo-, (Z)-	C_4H_7Br 135.00	31849-78-2	86.1	1.3265[15]	H_2O 1; eth 3; ace 3; bz 3 1.4536[20]
3709	1-Butene, 2-bromo- 2-Bromo-1-butene	C_4H_7Br 135.00	23074-36-4 -133.4	88	4-01-00-00775 1.3209[15]	H_2O 1; eth 3; ace 3; bz 3 1.4527[20]
3710	1-Butene, 3-bromo- α-Methallyl bromide	C_4H_7Br 135.00	22037-73-6	7[10]	4-01-00-00775 1.2998[25]	1.4618[20]
3711	1-Butene, 4-bromo-	C_4H_7Br 135.00	5162-44-7	98.5	4-01-00-00775 1.3230[20]	bz 4; eth 4; EtOH 4 1.4622[20]
3712	2-Butene, 1-bromo-	C_4H_7Br 135.00	4784-77-4	104.5	4-01-00-00789 1.3371[25]	H_2O 1; EtOH 3; eth 3; bz 4 1.4822[20]
3713	2-Butene, 2-bromo-, (E)-	C_4H_7Br 135.00	3017-71-8 -114.6	85.6	4-01-00-00790 1.3323[15]	H_2O 1; EtOH 3; eth 3; bz 4 1.4602[16]
3714	2-Butene, 2-bromo-, (Z)-	C_4H_7Br 135.00	3017-68-3 -111.5	93.9	4-01-00-00790 1.3416[15]	H_2O 1; EtOH 3; eth 3; bz 4 1.4631[19]
3715	1-Butene, 4-bromo-3-chloro-3,4,4-trifluoro- 4-Bromo-3-chloro-3,4,4-trifluoro-1-butene	$C_4H_3BrClF_3$ 223.42	374-25-4	99.5	4-01-00-00776 1.678[25]	1.4092[25]
3716	1-Butene, 2-bromo-3-methyl- 2-Bromo-3-methyl-1-butene	C_5H_9Br 149.03	31844-96-9	105	3-01-00-00800 1.2328[20]	bz 4; eth 4; chl 4 1.4504[20]
3717	2-Butene, 1-bromo-3-methyl-	C_5H_9Br 149.03	870-63-3	131 dec; 50[40]	4-01-00-00824 1.2817[20]	ace 4; bz 4; eth 4; EtOH 4 1.4930[15]
3718	2-Butene, 2-bromo-3-methyl-	C_5H_9Br 149.03	3017-70-7	119.5	4-01-00-00824 1.2773[20]	ace 4; eth 4; chl 4 1.4738[20]
3719	2-Butene, 2-bromo-3-phenyl- 2-Bromo-3-phenyl-2-butene	$C_{10}H_{11}Br$ 211.10	90841-14-8	120-30[11]	4-05-00-01379 1.3348[20]	bz 4; eth 4; chl 4 1.5811[20]
3720	1-Butene, 1-chloro-, (E)- trans-1-Chloro-1-butene	C_4H_7Cl 90.55	7611-87-2	68	3-01-00-00723 0.9205[15]	H_2O 1; EtOH 5; eth 4; ace 3 1.4223[15]
3721	1-Butene, 1-chloro-, (Z)- cis-1-Chloro-1-butene	C_4H_7Cl 90.55	7611-86-1	63.5	3-01-00-00723 0.9153[15]	ace 4; eth 4; EtOH 4; chl 4 1.4194[15]
3722	1-Butene, 2-chloro- 2-Chloro-1-butene	C_4H_7Cl 90.55	2211-70-3	58.5	4-01-00-00769 0.9107[15]	ace 4; bz 4; eth 4; EtOH 4 1.4165[21]
3723	1-Butene, 3-chloro- 3-Chloro-1-butene	C_4H_7Cl 90.55	563-52-0	2131 64.5	4-01-00-00769 0.8978[20]	eth 4; ace 4; chl 3 1.4149[20]
3724	1-Butene, 4-chloro-	C_4H_7Cl 90.55	927-73-1	75	4-01-00-00771 0.9211[20]	ace 4; eth 4; chl 4 1.4233[20]
3725	2-Butene, 1-chloro-, (E)-	C_4H_7Cl 90.55	4894-61-5	85	4-01-00-00784 0.9295[20]	H_2O 1; ace 3; chl 3 1.4350[20]
3726	2-Butene, 1-chloro-, (Z)-	C_4H_7Cl 90.55	4628-21-1	84.1	4-01-00-00783 0.9426[20]	H_2O 1; EtOH 3; ace 3; chl 3 1.4390[20]
3727	2-Butene, 2-chloro-, (E)-	C_4H_7Cl 90.55	2211-68-9 -105.8	62.8	4-01-00-00785 0.9138[20]	H_2O 1; EtOH 5; ace 3; chl 3 1.4190[20]
3728	2-Butene, 2-chloro-, (Z)-	C_4H_7Cl 90.55	2211-69-0 -117.3	70.6	4-01-00-00785 0.9239[20]	H_2O 1; EtOH 5; ace 3; chl 3 1.4240[20]
3729	1-Butene, 3-chloro-2-(chloromethyl)- 3-Chloro-2-(chloromethyl)-1-butene	$C_5H_8Cl_2$ 139.02	69295-21-2	155	3-01-00-00787 1.1233[20]	ace 4; chl 4 1.4724[20]
3730	2-Butene, 2-chloro-1,1-dimethoxy- Crotonaldehyde-2-chloro, dimethacetal	$C_6H_{11}ClO_2$ 150.60	108365-83-9	58[13]	3-01-00-02981 1.074[18]	1.4466[18]
3731	2-Butene, 1-chloro-2,3-dimethyl- 1-Chloro-2,3-dimethyl-2-butene	$C_6H_{11}Cl$ 118.61	37866-06-1	112	4-01-00-00856 0.9355[20]	ace 4; eth 4; chl 4 1.4605[20]
3732	1-Butene, 1-chloro-2-methyl- 1-Chloro-2-methyl-1-butene	C_5H_9Cl 104.58	23378-11-2	96.5	3-01-00-00787 0.9170[20]	ace 4; eth 4 1.4141[20]
3733	1-Butene, 1-chloro-3-methyl- 1-Chloro-3-methyl-1-butene	C_5H_9Cl 104.58	23010-00-6	87	3-01-00-00799 1.4229[20]	ace 4; eth 4; chl 4
3734	1-Butene, 3-chloro-2-methyl- 3-Chloro-2-methyl-1-butene	C_5H_9Cl 104.58	5166-35-8	94	4-01-00-00819 0.9088[20]	ace 4; eth 4; chl 4 1.4304[20]
3735	2-Butene, 1-chloro-2-methyl- 1-Chloro-2-methyl-2-butene	C_5H_9Cl 104.58	13417-43-1	110	4-01-00-00822 0.9327[20]	ace 4; eth 4; EtOH 4 1.4481[20]
3736	2-Butene, 1-chloro-3-methyl-	C_5H_9Cl 104.58	503-60-6	109	4-01-00-00823 0.9273[20]	ace 4; eth 4; EtOH 4; chl 4 1.4485[20]
3737	2-Butene, 2-chloro-3-methyl- 2-Chloro-3-methyl-2-butene	C_5H_9Cl 104.58	17773-65-8	96	3-01-00-00794 0.9324[20]	ace 4; eth 4; EtOH 4; chl 4 1.4320[20]
3738	1-Butene, 1,1-dibromo- 1,1-Dibromo-1-butene	$C_4H_6Br_2$ 213.90	73383-24-1	53-5[22]	3-01-00-00727 1.8315[20]	1.5168[20]
3739	1-Butene, 1,2-dibromo- 1,2-dibromo-1-butene	$C_4H_6Br_2$ 213.90	55030-56-3 -49.5	150	0-01-00-00204 1.887[25]	eth 4; EtOH 4
3740	1-Butene, 3,4-dibromo- 1,2-Dibromo-3-butene	$C_4H_6Br_2$ 213.90	10463-48-6	52[10]	4-01-00-00776 1.865[24]	1.541[21]

No.	Name / Synonym	Mol. Form. / Mol. Wt.	CAS RN / mp/°C	Merck No. / bp/°C	Beil. Ref. / den/g cm^{-3}	Solubility / n_D
3741	2-Butene, 1,3-dibromo- / 2,4-Dibromo-2-butene	$C_4H_6Br_2$ / 213.90	64930-16-1 /	/ 168 dec	3-01-00-00746 / 1.8768[20]	/ 1.5485[20]
3742	2-Butene, 1,4-dibromo-, (E)-	$C_4H_6Br_2$ / 213.90	821-06-7 / 53.4	/ 203; 74[14]	4-01-00-00791 /	H_2O 2; EtOH 4; ace 3; chl 2
3743	1-Butene, 1,3-dichloro- / 1,3-Dichloro-1-butene	$C_4H_6Cl_2$ / 125.00	52497-07-1 /	/ 125	4-01-00-00772 / 1.1341[24]	ace 4; eth 4; EtOH 4; chl 4 / 1.4647[20]
3744	1-Butene, 2,3-dichloro- / 2,3-Dichloro-1-butene	$C_4H_6Cl_2$ / 125.00	7013-11-8 /	/ 112	4-01-00-00772 / 1.1340[20]	ace 4; eth 4; chl 4 / 1.4580[20]
3745	1-Butene, 3,4-dichloro- / 1,2-Dichloro-3-butene	$C_4H_6Cl_2$ / 125.00	760-23-6 / -61	/ 116	4-01-00-00772 / 1.1170[20]	H_2O 1; EtOH 3; eth 3; bz 4 / 1.4641[20]
3746	2-Butene, 1,1-dichloro- / 1,1-Dichloro-2-butene	$C_4H_6Cl_2$ / 125.00	56800-09-0 /	/ 126	3-01-00-00741 / 1.1310[20]	/ 1.466[18]
3747	2-Butene, 1,2-dichloro- / 1,2-Dichloro-2-butene	$C_4H_6Cl_2$ / 125.00	13602-13-6 /	/ 130.5	3-01-00-00741 / 1.1601[20]	ace 4; eth 4; chl 4 / 1.4734[20]
3748	2-Butene, 1,3-dichloro-, (E)-	$C_4H_6Cl_2$ / 125.00	7415-31-8 /	/ 131; 53[50]	4-01-00-00786 / 1.1585[20]	ace 4; bz 4; eth 4; EtOH 4 / 1.4719[20]
3749	2-Butene, 1,3-dichloro-, (Z)-	$C_4H_6Cl_2$ / 125.00	10075-38-4 /	/ 131; 34[20]	4-01-00-00786 / 1.1605[20]	ace 4; bz 4; eth 4; EtOH 4 / 1.4735[20]
3750	2-Butene, 1,4-dichloro-, (E)- / 1,4-Dichloro-trans-2-butene	$C_4H_6Cl_2$ / 125.00	110-57-6 / 2.0	/ 155.4	4-01-00-00787 / 1.183[25]	ace 4; bz 4; eth 4; EtOH 4 / 1.4871[25]
3751	2-Butene, 1,4-dichloro-, (Z)-	$C_4H_6Cl_2$ / 125.00	1476-11-5 / -48	/ 152.5	4-01-00-00787 / 1.188[25]	ace 4; bz 4; eth 4; EtOH 4 / 1.4887[25]
3752	2-Butene, 2,3-dichloro-, (E)-	$C_4H_6Cl_2$ / 125.00	1587-29-7 /	/ 102	4-01-00-00787 / 1.1416[20]	H_2O 1; EtOH 3; eth 3; ace 3 / 1.4582[20]
3753	2-Butene, 2,3-dichloro-, (Z)-	$C_4H_6Cl_2$ / 125.00	1587-26-4 /	/ 125.5	4-01-00-00787 / 1.1618[20]	H_2O 1; EtOH 3; eth 3; ace 3 / 1.4590[20]
3754	2-Butene, 2,3-dichloro-1,1,1,4,4,4-hexafluoro- / 2,3-Dichlorohexafluoro-2-butene	$C_4Cl_2F_6$ / 232.94	303-04-8 / -67.3	/ 68.5	4-01-00-00787 / 1.6233[20]	/ 1.3459[20]
3755	1-Butene, 3,3-dichloro-2-methyl- / 3,3-Dichloro-2-methyl-1-butene	$C_5H_8Cl_2$ / 139.02	42101-38-2 /	/	3-01-00-00787 / 1.1276[18]	ace 4; eth 4; chl 4 / 1.4737[18]
3756	2-Butene, 1,3-dichloro-2-methyl- / 1,3-Dichloro-2-methyl-2-butene	$C_5H_8Cl_2$ / 139.02	25148-87-2 /	/ 150	3-01-00-00795 / 1.1293[20]	ace 4; chl 4
3757	2-Butene, 1,4-dichloro-2-methyl- / 1,4-Dichloro-2-methyl-2-butene	$C_5H_8Cl_2$ / 139.02	29843-58-1 /	/ 93[50]; 56[10]	3-01-00-00795 / 1.1526[20]	ace 4; chl 4 / 1.4932[20]
3758	2-Butene, 1,1-diethoxy- / Crotonaldehyde, diethyl acetal	$C_8H_{16}O_2$ / 144.21	10602-34-3 /	/ 147.5	3-01-00-02980 / 0.8473[18]	H_2O 5; EtOH 5; ace 5 / 1.4097[20]
3759	1-Butene, 2,3-dimethyl- / 2,3-Dimethyl-1-butene	C_6H_{12} / 84.16	563-78-0 / -157.3	/ 55.6	4-01-00-00852 / 0.6803[20]	H_2O 1; EtOH 3; eth 3; ace 3 / 1.3995[20]
3760	1-Butene, 3,3-dimethyl-	C_6H_{12} / 84.16	558-37-2 / -115.2	/ 41.2	4-01-00-00850 / 0.6529[20]	H_2O 1; EtOH 3; eth 3; ctc 3 / 1.3763[20]
3761	2-Butene, 2,3-dimethyl- / 2,3-Dimethyl-2-butene	C_6H_{12} / 84.16	563-79-1 / -74.6	/ 73.3	4-01-00-00853 / 0.7080[20]	H_2O 1; EtOH 3; eth 3; ace 3 / 1.4122[20]
3762	2-Butenedinitrile, (E)-	$C_4H_2N_2$ / 78.07	764-42-1 / 96.8	/ 186	4-02-00-02219 / 0.9416[111]	H_2O 3; EtOH 3; eth 3; ace 3 / 1.4349[111]
3763	2-Butenedinitrile, 2-chloro-, (Z)- / Fumaronitrile, chloro-	C_4HClN_2 / 112.52	71200-79-8 / 60.5	/ 172; 68[25]	3-02-00-01909 / 1.2499[20]	/ 1.4957[20]
3764	2-Butenedioic acid (E)- / Fumaric acid	$C_4H_4O_4$ / 116.07	110-17-8 / 287 dec	4200 / sub 165	3-02-00-01890 / 1.635[20]	H_2O 2; EtOH 3; eth 2; ace 2
3765	2-Butenedioic acid (Z)- / Maleic acid	$C_4H_4O_4$ / 116.07	110-16-7 / 130.5	5585 /	4-02-00-02199 / 1.590[20]	H_2O 4; EtOH 4; eth 3; ace 4
3766	2-Butenedioic acid (E)-, bis(1-methylethyl) ester / Diisopropyl fumarate	$C_{10}H_{16}O_4$ / 200.23	7283-70-7 / 2.1	/ 280	4-02-00-02209 /	ace 4; eth 4; EtOH 4
3767	2-Butenedioic acid (E)-, bis(2-methylpropyl) ester / Diisobutyl fumarate	$C_{12}H_{20}O_4$ / 228.29	7283-69-4 /	/ 170[160]; 122[5]	4-02-00-02211 / 0.9760[20]	H_2O 1; EtOH 3; eth 3; ace 3 / 1.4432[20]
3768	2-Butenedioic acid (E)-, bis(phenylmethyl) ester / Benzyl fumarate	$C_{18}H_{16}O_4$ / 296.32	538-64-7 / 59	1149 / 210-11[5]	4-06-00-02275 /	eth 4; EtOH 4; chl 4
3769	2-Butenedioic acid, 2-bromo-, (E)-	$C_4H_3BrO_4$ / 194.97	584-99-6 / 141	/ dec	4-02-00-02224 /	H_2O 4; EtOH 4; eth 4
3770	2-Butenedioic acid, 2-bromo-, (Z)- / 2-Bromofumaric acid	$C_4H_3BrO_4$ / 194.97	644-80-4 / 184.3	/ 200 dec	4-02-00-02224 /	H_2O 4; EtOH 4
3771	2-Butenedioic acid, 2-chloro-, diethyl ester, (E)- / Diethyl chloromaleate	$C_8H_{11}ClO_4$ / 206.63	626-10-8 /	/ 235 dec; 125[19]	3-02-00-01928 / 1.1741[20]	ace 4; eth 4; EtOH 4
3772	2-Butenedioic acid, 2-chloro-, diethyl ester, (Z)-	$C_8H_{11}ClO_4$ / 206.63	10302-94-0 /	/ 250 dec; 127[10]	3-02-00-01909 / 1.1880[20]	eth 4; EtOH 4 / 1.4571[20]
3773	2-Butenedioic acid, 2-chloro-, dimethyl ester, (E)- / Dimethyl chloromaleate	$C_6H_7ClO_4$ / 178.57	19393-45-4 /	/ 224; 100[17]	3-02-00-01928 / 1.2899[25]	eth 4; EtOH 4 / 1.4720[18]
3774	2-Butenedioic acid (E)-, dibutyl ester	$C_{12}H_{20}O_4$ / 228.29	105-75-9 /	/ 150[4]	4-02-00-02210 / 0.9869[20]	H_2O 1; ace 3; chl 3 / 1.4469[20]
3775	2-Butenedioic acid (E)-, diethyl ester	$C_8H_{12}O_4$ / 172.18	623-91-6 / 1.5	/ 214	4-02-00-02207 / 1.0452[20]	H_2O 1; ace 3; chl 3 / 1.4412[20]
3776	2-Butenedioic acid (Z)-, diethyl ester / Diethyl maleate	$C_8H_{12}O_4$ / 172.18	141-05-9 / -8.8	3113 / 223	4-02-00-02207 / 1.0662[20]	H_2O 1; EtOH 3; eth 3; chl 2 / 1.4416[20]

No.	Name / Synonym	Mol. Form. / Mol. Wt.	CAS RN / mp/°C	Merck No. / bp/°C	Beil. Ref. / den/g cm^{-3}	Solubility / n_D
3777	2-Butenedioic acid, 2,3-dihydroxy-, (Z)- / Dihydroxymaleic acid	$C_4H_4O_6$ / 148.07	526-84-1 / 155 dec	3170	4-03-00-01975	H_2O 2; EtOH 3; eth 2; MeOH 2
3778	2-Butenedioic acid (E)-, dimethyl ester	$C_6H_8O_4$ / 144.13	624-49-7 / 103.5	193	4-02-00-02205 / 1.37[20]	H_2O 1; ace 3; chl 3 / 1.4062[111]
3779	2-Butenedioic acid (Z)-, dimethyl ester	$C_6H_8O_4$ / 144.13	624-48-6 / -19	202	4-02-00-02204 / 1.1606[20]	H_2O 1; eth 3; ctc 3; lig 1 / 1.4416[20]
3780	2-Butenedioic acid (Z)-, diphenyl ester	$C_{16}H_{12}O_4$ / 268.27	7242-17-3 / 73	226[15]	4-06-00-00628	ace 4; bz 4; eth 4; EtOH 4
3781	2-Butenedioic acid (E)-, di-2-propenyl ester / Diallyl fumarate	$C_{10}H_{12}O_4$ / 196.20	2807-54-7 /	140[3]	3-02-00-01906 / 1.060[25]	ace 4; bz 4; eth 4; EtOH 4 / 1.4670[25]
3782	2-Butenedioic acid (Z)-, di-2-propenyl ester / Diallyl maleate	$C_{10}H_{12}O_4$ / 196.20	999-21-3 /	108-10[3]	4-02-00-02214 / 1.0773[20]	chl 3 / 1.4699[20]
3783	2-Butenedioic acid (E)-, dipropyl ester / Dipropyl fumarate	$C_{10}H_{16}O_4$ / 200.23	14595-35-8 /	110[5]	4-02-00-02209 / 1.0129[20]	EtOH 3; eth 3 / 1.4435[20]
3784	2-Butenedioic acid (Z)-, dipropyl ester	$C_{10}H_{16}O_4$ / 200.23	2432-63-5 /	126[12]	4-02-00-02209 / 1.0245[20]	H_2O 1; EtOH 3; eth 3; ace 3 / 1.4434[20]
3785	2-Butenedioic acid, 2-methyl-, (E)- / Mesaconic acid	$C_5H_6O_4$ / 130.10	498-24-8 / 204.5	5806 / sub	4-02-00-02231 / 1.466[20]	H_2O 2; EtOH 4; eth 3; bz 2
3786	2-Butenedioic acid, 2-methyl-, (Z)- / Citraconic acid	$C_5H_6O_4$ / 130.10	498-23-7 / 93.5	2323	4-02-00-02230 / 1.617[25]	H_2O 4; eth 2; bz 1; chl 2
3787	2-Butenedioic acid, 2-methyl-, diethyl ester, (E)- / Diethyl mesaconate	$C_9H_{14}O_4$ / 186.21	2418-31-7 /	229	4-02-00-02232 / 1.0434[20]	ace 4; bz 4; eth 4; EtOH 4 / 1.4488[20]
3788	2-Butenedioic acid, 2-methyl-, diethyl ester, (Z)-	$C_9H_{14}O_4$ / 186.21	691-83-8 /	228; 120[20]	4-02-00-02232 / 1.0491[20]	eth 4; EtOH 4; HOAc 4 / 1.4467[20]
3789	2-Butenedioic acid, 2-methyl-, dimethyl ester, (E)-	$C_7H_{10}O_4$ / 158.15	617-53-8 /	203.5	4-02-00-02232 / 1.0914[20]	ace 4; eth 4; EtOH 4 / 1.4512[20]
3790	2-Butenedioic acid, 2-methyl-, dimethyl ester, (Z)-	$C_7H_{10}O_4$ / 158.15	617-54-9 /	210.5	4-02-00-02232 / 1.1153[20]	ace 4; eth 4; EtOH 4 / 1.4473[20]
3791	2-Butenedioic acid (E)-, monoethyl ester	$C_6H_8O_4$ / 144.13	2459-05-4 / 70	147[16]	4-02-00-02206 / 1.1109[87]	EtOH 3; ace 3; chl 2
3792	2-Butene-1,4-diol, (E)- / trans-2-Butene-1,4-diol	$C_4H_8O_2$ / 88.11	821-11-4 / 25	131[13]	4-01-00-02660 / 1.0700[20]	H_2O 4; EtOH 4 / 1.4755[20]
3793	2-Butene-1,4-diol, (Z)- / cis-2-Butene-1,4-diol	$C_4H_8O_2$ / 88.11	6117-80-2 / 4	235	4-01-00-02660 / 1.0698[20]	H_2O 3; EtOH 4 / 1.4782[20]
3794	3-Butene-1,2-diol	$C_4H_8O_2$ / 88.11	497-06-3 /	196.5	4-01-00-02658 / 1.0470[20]	H_2O 3; EtOH 3 / 1.4628[21]
3795	2-Butenedioyl dichloride, (E)- / Fumaric acid dichloride	$C_4H_2Cl_2O_2$ / 152.96	627-63-4 /	159	4-02-00-02217 / 1.408[20]	/ 1.5004[18]
3796	2-Butenedioyl dichloride, 2-chloro-, (Z)-	$C_4HCl_3O_2$ / 187.41	17096-37-6 /	185 dec	3 02-00-01909 / 1.564[20]	eth 4; HOAc 4 / 1.5206[20]
3797	1-Butene, 2-ethyl-3-methyl- / 2-Ethyl-3-methylbut-1-ene	C_7H_{14} / 98.19	7357-93-9 /	89	4-01-00-00871 / 0.7150[20]	H_2O 1; eth 3; ace 3; bz 3 / 1.410[20]
3798	2-Butene, 1,1,2,3,4,4-hexachloro- / 2-Butene, 1,1,2,3,4,4-hexachloro (liquid)	$C_4H_2Cl_6$ / 262.78	13045-99-3 / -19	97-8[10]	4-01-00-00789 / 1.650[15]	H_2O 1; chl 3 / 1.5331
3799	1-Butene, 2-methyl- / 2-Methyl-1-butene	C_5H_{10} / 70.13	563-46-2 / -137.5	31.2	4-01-00-00818 / 0.6504[20]	H_2O 1; EtOH 3; eth 3; bz 3 / 1.3378[20]
3800	1-Butene, 3-methyl- / 3-Methyl-1-butene	C_5H_{10} / 70.13	563-45-1 / -168.5	20.1	4-01-00-00825 / 0.6213[25]	H_2O 1; EtOH 5; eth 5; bz 3 / 1.3643[20]
3801	2-Butene, 2-methyl- / 2-Methyl-2-butene	C_5H_{10} / 70.13	513-35-9 / -133.7	644 / 38.5	4-01-00-00820 / 0.6623[20]	H_2O 1; EtOH 3; eth 3; bz 3 / 1.3874[20]
3802	2-Butenenitrile, (E)-	C_4H_5N / 67.09	627-26-9 / -51.5	120	4-02-00-01507 / 0.8239[20]	eth 3; ace 3 / 1.4225[20]
3803	3-Butenenitrile / Allyl cyanide	C_4H_5N / 67.09	109-75-1 / -87	288 / 119	4-02-00-01491 / 0.8341[20]	H_2O 2; EtOH 5; eth 5 / 1.4060[20]
3804	2-Butene, 1,1,1,2,3,4,4,4-octafluoro- / Perfluoro-2-butene	C_4F_8 / 200.03	360-89-4 / -129	1.5	4-01-00-00783 / 1.529[25]	
3805	2-Butene, 1,1,1,4,4-pentachloro- / 1,1,1,4,4-Pentachloro-2-butene	$C_4H_3Cl_5$ / 228.33	77753-21-0 /	78-80[11]	3-01-00-00744 / 1.609[21]	EtOH 4; chl 4 / 1.5538[21]
3806	1-Butene, 1,3,4,4-tetrachloro- / 1,3,4,4-Tetrachloro-1-butene	$C_4H_4Cl_4$ / 193.89	2984-40-9 /	88[20]	4-01-00-00774 / 1.476[20]	chl 4 / 1.4773[20]
3807	1-Butene, 2,3,3,4-tetrachloro- / 2,3,3,4-Tetrachloro-1-butene	$C_4H_4Cl_4$ / 193.89	64346-45-8 /	41-2[7]	3-01-00-00726 / 1.4602[20]	ace 4; chl 4 / 1.5135[20]
3808	1-Butene, 1,3,4,4-tetrachloro-1,2,3,4-tetrafluoro- / 1-Butene, 1,3,4,4-tetrachloro-tetrafluoro	$C_4Cl_4F_4$ / 265.85	357-20-0 /	140.8	4-01-00-00774 / 1.6902[25]	/ 1.429[20]
3809	1-Butene, 2,3,4-trichloro- / 2,3,4-Trichloro-1-butene	$C_4H_5Cl_3$ / 159.44	2431-50-7 /	60[20]	3-01-00-00725 / 1.3430[20]	ace 4; chl 4 / 1.4944[20]
3810	2-Butene, 1,2,4-trichloro- / 1,2,4-Trichloro-2-butene	$C_4H_5Cl_3$ / 159.44	2431-54-1 /	67-9[10]	3-01-00-00744 / 1.3843[20]	bz 4; eth 4; EtOH 4; chl 4 / 1.5175[20]
3811	2-Butene, 1,1,3-triethoxy- / 2-Butenal, 3-ethoxy, diethyl acetal	$C_{10}H_{20}O_3$ / 188.27	69190-65-4 /	192.5	4-01-00-04082 / 0.908[21]	EtOH 4 / 1.430[21]
3812	1-Butene, 2,3,3-trimethyl-	C_7H_{14} / 98.19	594-56-9 / -109.9	77.9	4-01-00-00873 / 0.7050[20]	H_2O 1; eth 3; bz 3; chl 3 / 1.4025[20]
3813	3-Butenoic acid	$C_4H_8O_2$ / 86.09	625-38-7 / -35	169	4-02-00-01491 / 1.0091[20]	H_2O 3; EtOH 5; eth 5 / 1.4239[20]

No.	Name / Synonym	Mol. Form. / Mol. Wt.	CAS RN / mp/°C	Merck No. / bp/°C	Beil. Ref. / den/g cm^{-3}	Solubility / n_D
3814	2-Butenoic acid, (E)- / trans-Crotonic acid	$C_4H_8O_2$ / 86.09	107-93-7 / 72	2600 / 184.7	4-02-00-01498 / 0.9604[77]	H$_2$O 4; EtOH 4; eth 3; ace 3 / 1.4249[77]
3815	2-Butenoic acid, (Z)- / cis-Crotonic acid	$C_4H_8O_2$ / 86.09	503-64-0 / 15	5048 / 169	4-02-00-01497 / 1.0267[20]	H$_2$O 4; EtOH 3 / 1.4483[14]
3816	2-Butenoic acid, 3-amino-, ethyl ester, (E)-	$C_6H_{11}NO_2$ / 129.16	41867-20-3 / 34	/ 212 dec	3-03-00-01199 / 1.0219[19]	H$_2$O 1; EtOH 3; eth 3; bz 3 / 1.4988[22]
3817	2-Butenoic acid, 4-amino-4-oxo-, (Z)- / Maleamic acid	$C_4H_5NO_3$ / 115.09	557-24-4 / 172.5	5583	4-02-00-02218	H$_2$O 4; EtOH 4
3818	2-Butenoic acid, anhydride / Crotonic acid anhydride	$C_8H_{10}O_3$ / 154.17	623-68-7	/ 247; 129[19]	4-02-00-01505 / 1.0397[20]	eth 4 / 1.4745[20]
3819	2-Butenoic acid, 4-bromo-, ethyl ester, (E)-	$C_6H_9BrO_2$ / 193.04	37746-78-4	/ 97-8[15]	4-02-00-01517 / 1.402[16]	EtOH 4 / 1.4925[20]
3820	2-Butenoic acid, 3-chloro-, (E)-	$C_4H_5ClO_2$ / 120.54	6214-28-4 / 61	/ 195	4-02-00-01510 / 1.1969[69]	H$_2$O 2; EtOH 4; chl 3; CS$_2$ 4
3821	2-Butenoic acid, 3-chloro-, (Z)-	$C_4H_5ClO_2$ / 120.54	6213-90-7 / 61		4-02-00-01510 / 1.1905[66]	H$_2$O 2; EtOH 4; peth 4 / 1.4704[66]
3822	2-Butenoic acid, 4-chloro- / γ-Chlorocrotonic acid	$C_4H_5ClO_2$ / 120.54	16197-90-3 / 83	/ 117[13]	4-02-00-01511	eth 4; HOAc 4
3823	2-Butenoic acid, 2-chloro-, ethyl ester / Ethyl β-chlorocrotonate	$C_6H_9ClO_2$ / 148.59	56216-12-7	/ 176	4-02-00-01510 / 1.1133[20]	1.453[20]
3824	2-Butenoic acid, 2-chloro-, ethyl ester, (E)- / 2-Butenoic acid, 2-chloro, ethyl ester (trans)	$C_6H_9ClO_2$ / 148.59	77825-53-7	/ 177; 75[30]	4-02-00-01510 / 1.1135[20]	eth 4; EtOH 4 / 1.4538[20]
3825	2-Butenoic acid, 2-chloro-, ethyl ester, (Z)- / 2-Butenoic acid, 2-chloro, ethyl ester (cis)	$C_6H_9ClO_2$ / 148.59	77825-54-8	/ 75[30]	4-02-00-01510 / 1.1021[18]	eth 4; EtOH 4
3826	2-Butenoic acid, 3-chloro-, ethyl ester / 3-Chlorocrotonic acid ethyl ester	$C_6H_9ClO_2$ / 148.59	38624-62-3	/ 180; 75[14]	4-02-00-01511 / 1.1062[20]	1.459[20]
3827	2-Butenoic acid, 3-chloro-, ethyl ester, (E)-	$C_6H_9ClO_2$ / 148.59	6127-92-0	/ 184	4-02-00-01511 / 1.1062[20]	eth 4; EtOH 4 / 1.4592[20]
3828	2-Butenoic acid, 3-chloro-, ethyl ester, (Z)-	$C_6H_9ClO_2$ / 148.59	6127-93-1	/ 161.4	4-02-00-01511 / 1.0860[20]	eth 4; EtOH 4 / 1.4542[19]
3829	2-Butenoic acid, 4-chloro-, ethyl ester / Crotonic acid, 4-chloro-, ethyl ester	$C_6H_9ClO_2$ / 148.59	15333-22-9	/ 93[10]	4-02-00-01511 / 1.090[18]	1.462[18]
3830	2-Butenoic acid, 2-chloro-, methyl ester, (E)-	$C_5H_7ClO_2$ / 134.56	22038-57-9	/ 161.5; 63[17]	4-02-00-01510 / 1.160[20]	eth 4 / 1.4569[23]
3831	2-Butenoic acid, 3-chloro-, methyl ester, (E)-	$C_5H_7ClO_2$ / 134.56	6372-01-6	/ 64-7[14]	4-02-00-01510 / 1.157[20]	H$_2$O 2; EtOH 4; eth 4 / 1.4630[20]
3832	2-Butenoic acid, 3-chloro-, methyl ester, (Z)-	$C_5H_7ClO_2$ / 134.56	6214-25-1	/ 142.4; 70[24]	4-02-00-01500 / 1.138[20]	H$_2$O 2; eth 4; MeOH 4 / 1.4573[19]
3833	2-Butenoic acid, 4-chloro-, methyl ester / Methyl γ-chlorocrotonate	$C_5H_7ClO_2$ / 134.56	15320-72-6	/ 68[11]	4-02-00-01511 / 1.169[18]	1.467[18]
3834	2-Butenoic acid, 3-cyclohexyl- / Cicrotoic acid	$C_{10}H_{16}O_2$ / 168.24	25229-42-9 / 85.5	2272		
3835	3-Butenoic acid, 2,2-dimethyl- / 2,2-Dimethyl-3-butenoic acid	$C_6H_{10}O_2$ / 114.14	10276-09-2 / -6	/ 185	4-02-00-01574 / 0.9567[15]	1.4305[15]
3836	2-Butenoic acid, 2,3-dimethyl-, ethyl ester / Ethyl 2,3-dimethyl-2-butenoate	$C_8H_{14}O_2$ / 142.20	13979-28-7	/ 155	4-02-00-01576 / 0.9072[19]	1.430[19]
3837	2-Butenoic acid, 2-ethyl-, (E)- / 2-Ethylcrotonic acid	$C_6H_{10}O_2$ / 114.14	1187-13-9 / 45.5	/ 209	3-02-00-01329 / 0.9578[50]	eth 4; EtOH 4 / 1.4475[50]
3838	2-Butenoic acid, ethyl ester	$C_6H_{10}O_2$ / 114.14	10544-63-5	/ 136.5	2-02-00-00394 / 0.9175[20]	eth 4; EtOH 4 / 1.4243[20]
3839	2-Butenoic acid, ethyl ester, (E)- / Ethyl trans-crotonate	$C_6H_{10}O_2$ / 114.14	623-70-1 / 2600	/ 138	4-02-00-01500 / 0.9175[20]	H$_2$O 1; EtOH 3; eth 3 / 1.4243[20]
3840	2-Butenoic acid, ethyl ester, (Z)- / cis-Ethyl crotonate	$C_6H_{10}O_2$ / 114.14	6776-19-8 / 2600	/ 136	4-02-00-01497 / 0.9182[20]	ace 4; eth 4; EtOH 4 / 1.4242[20]
3841	3-Butenoic acid, ethyl ester / Ethyl 3-butenoate	$C_6H_{10}O_2$ / 114.14	1617-18-1	/ 119	4-02-00-01491 / 0.9122[20]	EtOH 3 / 1.4105[20]
3842	3-Butenoic acid, 2-hydroxy- / Vinylglycolic acid	$C_4H_6O_3$ / 102.09	600-17-9 / 43	/ 124[9]	3-03-00-00685 / 1.2020[21]	H$_2$O 4; eth 4; EtOH 4; chl 4
3843	3-Butenoic acid, 2-hydroxy-, ethyl ester / Ethyl 2-hydroxy-3-butenoate	$C_6H_{10}O_3$ / 130.14	91890-87-8	/ 173 dec; 68[15]	3-03-00-00685 / 1.0470[15]	H$_2$O 4; eth 4; EtOH 4 / 1.436[13]
3844	2-Butenoic acid, 2-methyl-, (E)- / Tiglic acid	$C_5H_8O_2$ / 100.12	80-59-1 / 64.5	9365 / 198.5	4-02-00-01552 / 0.9641[76]	eth 4; EtOH 4 / 1.4330[76]
3845	2-Butenoic acid, 2-methyl-, (Z)- / Angelic acid	$C_5H_8O_2$ / 100.12	565-63-9 / 45.5	678 / 185	4-02-00-01551 / 0.9834[49]	H$_2$O 2; EtOH 3; eth 4 / 1.4434[47]
3846	2-Butenoic acid, 3-methyl-	$C_5H_8O_2$ / 100.12	541-47-9 / 69.5	/ 197	4-02-00-01555 / 1.0062[24]	
3847	2-Butenoic acid, methyl ester, (E)- / Methyl trans-crotonate	$C_5H_8O_2$ / 100.12	623-43-8 / -42	2600 / 121	4-02-00-01500 / 0.9444[20]	H$_2$O 1; EtOH 4; eth 4 / 1.4242[20]
3848	2-Butenoic acid, methyl ester, (Z)- / Methyl cis-crotonate	$C_5H_8O_2$ / 100.12	4358-59-2 / 5048	/ 118	4-02-00-01500	1.4175[20]
3849	2-Butenoic acid, 2-methyl-, ethyl ester, (E)-	$C_7H_{12}O_2$ / 128.17	5837-78-5	/ 156	4-02-00-01553 / 0.9200[20]	1.4340[20]
3850	2-Butenoic acid, 3-methyl-, ethyl ester	$C_7H_{12}O_2$ / 128.17	638-10-8	/ 153.5	4-02-00-01556 / 0.9199[21]	1.4345[20]
3851	2-Butenoic acid, 2-methyl-, methyl ester, (E)-	$C_6H_{10}O_2$ / 114.14	6622-76-0	/ 139	4-02-00-01553 / 0.9498[20]	1.4370[20]
3852	2-Butenoic acid, 3-methyl-, methyl ester	$C_6H_{10}O_2$ / 114.14	924-50-5 / 114	/ 136.5	3-02-00-01313 / 0.9337[20]	1.432[20]

No.	Name / Synonym	Mol. Form. / Mol. Wt.	CAS RN / mp/°C	Merck No. / bp/°C	Beil. Ref. / den/g cm^{-3}	Solubility / n_D
3853	2-Butenoic acid, 4-oxo-4-(phenylamino)-, (Z)- Maleanilic acid	$C_{10}H_9NO_3$ 191.19	555-59-9 192 dec	5584	3-12-00-00571 1.418[30]	
3854	2-Butenoic acid, 3-phenyl-, (Z)-	$C_{10}H_{10}O_2$ 162.19	704-79-0 131.5	170[14]	4-09-00-02057	bz 3; CS_2 3; peth 2
3855	3-Butenoic acid, 4-phenyl- Styrylacetic acid	$C_{10}H_{10}O_2$ 162.19	2243-53-0 87	302	4-09-00-02054	H_2O 1; EtOH 4; eth 4; CS_2 4
3856	2-Butenoic acid, 2-propenyl ester Allyl crotonate	$C_7H_{10}O_2$ 126.16	20474-93-5	88-9[70]	4-02-00-01503 0.9440[20]	1.4465[20]
3857	2-Butenoic acid, 2-propenyl ester, (E)-	$C_7H_{10}O_2$ 126.16	5453-44-1	88-9[70]	4-02-00-01503 0.9440[20]	1.4465[20]
3858	3-Buten-2-ol, (±)-	C_4H_8O 72.11	6118-14-5 <-100	97.3	4-01-00-02102 0.8318[20]	1.4137[20]
3859	2-Buten-1-ol, (E)- trans-Crotyl alcohol	C_4H_8O 72.11	504-61-0 <-30	121.2	4-01-00-02107 0.8521[20]	H_2O 4; EtOH 5; eth 5; chl 3 1.4288[20]
3860	2-Buten-1-ol, (Z)- cis-Crotyl alcohol	C_4H_8O 72.11	4088-60-2 	123	4-01-00-02107 0.8662[20]	1.4342[25]
3861	3-Buten-1-ol	C_4H_8O 72.11	627-27-0 	113.5	4-01-00-02105 0.8424[20]	H_2O 3; EtOH 5; eth 5; ace 3 1.4224[20]
3862	2-Buten-1-ol, acetate	$C_6H_{10}O_2$ 114.14	628-08-0 	132	3-02-00-00283 0.9192[20]	bz 4; eth 4; EtOH 4 1.4181[20]
3863	2-Buten-1-ol, 2-chloro-	C_4H_7ClO 106.55	116723-93-4 	159	3-01-00-01900 1.1180[20]	H_2O 4; EtOH 4 1.4682[20]
3864	2-Buten-1-ol, 3-chloro- γ-Chlorocrotyl alcohol	C_4H_7ClO 106.55	40605-42-3 -42.9	161.5	4-01-00-02108 1.095[23]	1.4652[20]
3865	2-Buten-1-ol, 4-chloro- 4-Chloro-2-buten-1-ol	C_4H_7ClO 106.55	7523-44-6 	64-5[2]	4-01-00-02110 	eth 4; EtOH 4 1.4845[20]
3866	3-Buten-2-ol, 1-chloro- 1-Chloro-3-buten-2-ol	C_4H_7ClO 106.55	671-56-7 	145.5	3-01-00-01893 1.111[20]	chl 4 1.4643[20]
3867	3-Buten-1-ol, 2-chloro-	C_4H_7ClO 106.55	75455-41-3 	66-7[30]	4-01-00-02106 1.1044[20]	eth 4; EtOH 4 1.4665[20]
3868	3-Buten-2-ol, 3-chloro- 3-Chloro-3-buten-2-ol	C_4H_7ClO 106.55	6498-47-1 	53-7[19]	3-01-00-01893 1.1138[23]	eth 4; EtOH 4
3869	2-Buten-1-ol, 3-methyl-	$C_5H_{10}O$ 86.13	556-82-1 	140	4-01-00-02129 0.848[25]	1.4412[20]
3870	3-Buten-2-ol, 2-methyl-	$C_5H_{10}O$ 86.13	115-18-4 -28	97	4-01-00-02132 0.82[20]	
3871	3-Buten-2-ol, 3-methyl- Methyl isopropenyl carbinol	$C_5H_{10}O$ 86.13	10473-14-0 	114	4-01-00-02127 0.8531[17]	1.4288[17]
3872	2-Buten-1-ol, 3-phenyl- 2-Butene-1-ol, 3-phenyl	$C_{10}H_{12}O$ 148.20	1504-54-7 	145[15]	4-06-00-03839 1.035[20]	ctc 3 1.5678[20]
3873	3-Buten-2-ol, 4-(2,6,6-trimethyl-1-cyclohexen-1-yl)- β-Ionol	$C_{13}H_{22}O$ 194.32	22029-76-1 	131[15]; 80[0.7]	3-06-00-00401 0.9243[20]	EtOH 3; eth 3; ace 3 1.4969[20]
3874	3-Buten-2-ol, 4-(2,6,6-trimethyl-2-cyclohexen-1-yl)- α-Ionol	$C_{13}H_{22}O$ 194.32	25312-34-9 	127[15]	3-06-00-00402 0.9474[20]	1.4735[20]
3875	3-Buten-2-one Methyl vinyl ketone	C_4H_6O 70.09	78-94-4 81.4	6052	4-01-00-03444 0.864[20]	H_2O 3; EtOH 3; eth 4; ace 4 1.4081[20]
3876	2-Buten-1-one, 1,3-diphenyl- Dypnone	$C_{16}H_{14}O$ 222.29	495-45-4 342.5	3460	4-07-00-01682 1.1080[15]	eth 4; EtOH 4 1.6343[20]
3877	1-Buten-1-one, 2-ethyl- Diethylketene	$C_6H_{10}O$ 98.14	24264-08-2 88.5		4-01-00-03477 0.831[20]	1.4135[11]
3878	3-Buten-2-one, 4-(2-furanyl)-	$C_8H_8O_2$ 136.15	623-15-4 39.5	229 dec; 113[10]	5-17-09-00578 1.0496[57]	H_2O 1; EtOH 4; eth 4; chl 4 1.5788[45]
3879	3-Buten-2-one, 3-methyl- Isopropenyl methyl ketone	C_5H_8O 84.12	814-78-8 -54	98	4-01-00-03462 0.8527[20]	EtOH 4 1.4220[20]
3880	2-Buten-1-one, 1-phenyl-	$C_{10}H_{10}O$ 146.19	495-41-0 20.5	111-2[9]	3-07-00-01410 1.025[15]	1.5626[18]
3881	3-Buten-2-one, 4-phenyl- Benzylideneacetone	$C_{10}H_{10}O$ 146.19	122-57-6 41.5	1151 261	2-07-00-00287 1.0368[15]	bz 4; eth 4; EtOH 4; chl 4
3882	3-Buten-2-one, 4-phenyl-, (E)-	$C_{10}H_{10}O$ 146.19	1896-62-4 42	262	4-07-00-01003 1.0097[45]	H_2O 1; EtOH 4; eth 3; ace 3 1.5836[45]
3883	3-Buten-2-one, 4-(2,5,6,6-tetramethyl-1-cyclohexen-1-yl)- β-Irone	$C_{14}H_{22}O$ 206.33	79-70-9 	4978 85-90[0.1]	4-07-00-00377 0.9434[21]	H_2O 2; EtOH 4; eth 4; bz 4 1.5017[20]
3884	3-Buten-2-one, 4-(2,5,6,6-tetramethyl-2-cyclohexen-1-yl)- α-Irone	$C_{14}H_{22}O$ 206.33	79-69-6 	4977 87-92[0.4]	4-07-00-00363 0.9362[20]	1.5002[20]
3885	3-Buten-2-one, 4-(2,6,6-trimethyl-1-cyclohexen-1-yl)-, (E)-	$C_{13}H_{20}O$ 192.30	79-77-6 	140[18]; 73[0.1]	4-07-00-00361 0.9462[20]	H_2O 2; EtOH 5; eth 5; chl 3 1.5198[20]
3886	3-Buten-2-one, 4-(2,6,6-trimethyl-2-cyclohexen-1-yl)-, (E)-(±)-	$C_{13}H_{20}O$ 192.30	30685-95-1 	146[28]; 75[0.15]	4-07-00-00363 0.9298[21]	ace 4; eth 4; EtOH 4 1.5041[20]
3887	2-Butenoyl chloride	C_4H_5ClO 104.54	10487-71-5 	124.5	2-02-00-00394 1.0905[20]	ace 4 1.460[18]
3888	2-Butenoyl chloride, 3-methyl-	C_5H_7ClO 118.56	3350-78-5 	146	4-02-00-01558 1.065[25]	1.4770[20]

No.	Name / Synonym	Mol. Form. / Mol. Wt.	CAS RN / mp/°C	Merck No. / bp/°C	Beil. Ref. / den/g cm⁻³	Solubility / n_D
3889	1-Buten-3-yne / Vinylacetylene	C_4H_4 / 52.08	689-97-4 /	/ 5.1	4-01-00-01083 / 0.7094^0	H_2O 1; bz 3 / 1.4161^1
3890	1-Buten-3-yne, 4-chloro-	C_4H_3Cl / 86.52	40589-38-6 /	/ 56	4-01-00-01085 / 1.0022^{20}	chl 4 / 1.4656^{20}
3891	1-Buten-3-yne, 1-methoxy-	C_5H_6O / 82.10	2798-73-4 /	/ 123 dec	4-01-00-02301 / 0.906^{20}	H_2O 1; chl 3; os 4 / 1.4818^{20}
3892	1-Buten-3-yne, 2-methyl- / Isopropenylacetylene	C_5H_8 / 66.10	78-80-8 / -113	/ 32	4-01-00-01089 / 0.6801^{11}	chl 3 / 1.4140^{20}
3893	Butralin / Benzenamine, 4-(1,1-dimethylethyl)-N-(1-methylpropyl)-2,6-dinitro-	$C_{14}H_{21}N_3O_4$ / 295.34	33629-47-9 / 60	1532 / $135^{0.5}$		
3894	N-sec-Butyl-2,6-dinitrobenzene / Benzenamine, N-(1-methylpropyl)-2,6-dinitro	$C_{10}H_{13}N_3O_4$ / 239.23	55702-49-3 /			
3895	2-Butynal / Tetrolaldehyde	C_4H_4O / 68.08	1119-19-3 / -26	/ 106.5	4-01-00-03540 / 0.9264^{17}	eth 3; ace 3; chl 3 / 1.446^{19}
3896	3-Butyn-2-amine, 2-methyl-	C_5H_9N / 83.13	2978-58-7 / 18	/ 79.5	4-04-00-01145 / 0.79^{25}	/ 1.4235^{20}
3897	1-Butyne / Ethylacetylene	C_4H_6 / 54.09	107-00-6 / -125.7	/ 8.0	4-01-00-00969 / 0.6783^0	H_2O 1; EtOH 3; eth 3 / 1.3962^{20}
3898	2-Butyne / Dimethylacetylene	C_4H_6 / 54.09	503-17-3 / -32.3	/ 26.9	4-01-00-00971 / 0.691^{20}	H_2O 1; EtOH 3; eth 3; ctc 3 / 1.3921^{20}
3899	1-Butyne, 3-chloro-	C_4H_5Cl / 88.54	21020-24-6 /	/ 68.5	4-01-00-00970 / 1.4218^{25}	/ 1.4218^{25}
3900	2-Butyne, 1-chloro-	C_4H_5Cl / 88.54	3355-17-7 /	/ 103	4-01-00-00973 / 1.0152^{20}	ace 4; eth 4; EtOH 4 / 1.4581^{20}
3901	1-Butyne, 3-chloro-3-methyl-	C_5H_7Cl / 102.56	1111-97-3 / -61	/ 76	4-01-00-01000 / 0.9061^{20}	
3902	2-Butynediamide / Cellocidin	$C_4H_4N_2O_2$ / 112.09	543-21-5 / 217 dec	1958 /	4-02-00-02295 /	H_2O 2; EtOH 2; eth 2; chl 2
3903	2-Butyne-1,4-diamine	$C_4H_8N_2$ / 84.12	53878-96-9 / 44	/ 96^{16}	4-04-00-01392 /	H_2O 4; EtOH 4; MeOH 4
3904	2-Butyne, 1,4-dibromo- / 1,4-Dibromo-2-butyne	$C_4H_4Br_2$ / 211.88	2219-66-1 /	/ 92^{15}	4-01-00-00974 / 2.014^{18}	eth 3; ace 3; chl 4 / 1.588^{18}
3905	2-Butyne, 1,4-dichloro-	$C_4H_4Cl_2$ / 122.98	821-10-3 /	/ 165.5	4-01-00-00973 / 1.258^{20}	eth 3; ace 3; ctc 2; chl 4 / 1.5058^{20}
3906	2-Butyne, 1,4-diiodo-	$C_4H_4I_2$ / 305.88	116529-73-8 / 54.5	/ $70^{0.1}$	4-01-00-00975 /	H_2O 2; EtOH 3; eth 3; ace 3
3907	1-Butyne, 3,3-dimethyl- / tert-Butylacetylene	C_6H_{10} / 82.15	917-92-0 / -78.2	/ 37.7	4-01-00-01022 / 0.6623^{25}	/ 1.3736^{20}
3908	2-Butynedinitrile	C_4N_2 / 76.06	1071-98-3 / 20.5	/ 76.5	4-02-00-02295 / 0.9708^{25}	/ 1.4647^{25}
3909	2-Butynedioic acid, diethyl ester	$C_8H_{10}O_4$ / 170.17	762-21-0 / 1.5	/ 184^{200}	4-02-00-02294 / 1.0075^{20}	EtOH 3; eth 3; ctc 3 / 1.4425^{20}
3910	2-Butynedioic acid, dimethyl ester	$C_6H_6O_4$ / 142.11	762-42-5 /	/ 197 dec; 98^{20}	4-02-00-02291 / 1.1564^{20}	EtOH 3; eth 3; ctc 3 / 1.4434^{20}
3911	2-Butyne-1,4-diol / Bis(hydroxymethyl)acetylene	$C_4H_6O_2$ / 86.09	110-65-6 / 50	/ 238	4-01-00-02687 /	H_2O 4; EtOH 4; eth 2; ace 4 / 1.4804^{20}
3912	3-Butyne-1,2-diol / 1-Butyne-3,4-diol	$C_4H_6O_2$ / 86.09	616-28-4 / 40	/ $64^{0.2}$	4-01-00-02689 /	H_2O 4; EtOH 4
3913	2-Butyne-1,4-diol, diacetate / 1,4-Diacetoxy-2-butyne	$C_8H_{10}O_4$ / 170.17	1573-17-7 /	/ $122-3^{10}$	4-02-00-00244 /	ctc 3 / 1.4611^{20}
3914	2-Butyne, 1,1,1,4,4,4-hexafluoro-	C_4F_6 / 162.03	692-50-2 / -117.4	/ -24.6	4-01-00-00972 /	EtOH 3; eth 3; ace 3; ctc 3
3915	2-Butyne, 1-methoxy-	C_5H_8O / 84.12	2768-41-4 /	/ 99.5	3-01-00-01973 / 0.8496^{20}	bz 4; eth 4; EtOH 4 / 1.4262^{20}
3916	1-Butyne, 3-methyl-	C_5H_8 / 68.12	598-23-2 / -89.7	/ 26.3	4-01-00-00999 / 0.6660^{20}	H_2O 1; EtOH 5; eth 5 / 1.3723^{20}
3917	2-Butynoic acid	$C_4H_4O_2$ / 84.07	590-93-2 / 78	/ 203	4-02-00-01690 / 0.9641^{20}	H_2O 4; eth 4; EtOH 4; chl 4
3918	2-Butynoic acid, ethyl ester	$C_6H_8O_2$ / 112.13	4341-76-8 /	/ 163	4-02-00-01691 / 0.9641^{20}	/ 1.4372^{20}
3919	2-Butyn-1-ol	C_4H_6O / 70.09	764-01-2 / -2.2	/ 148	4-01-00-02220 / 0.9370^{20}	eth 4; EtOH 4 / 1.4530^{20}
3920	3-Butyn-1-ol	C_4H_6O / 70.09	927-74-2 / -63.6	/ 129	4-01-00-02219 / 0.9257^{20}	H_2O 4; EtOH 4 / 1.4409^{20}
3921	3-Butyn-2-ol	C_4H_6O / 70.09	2028-63-9 / -3	/ 106.5	4-01-00-02218 / 0.8618^{20}	H_2O 4; eth 4; EtOH 4 / 1.4207^{20}
3922	3-Butyn-2-ol, 2-methyl- / 1,1-Dimethylpropargyl alcohol	C_5H_8O / 84.12	115-19-5 / 3	5956 / 104	4-01-00-02229 / 0.8618^{20}	/ 1.4207^{20}
3923	3-Butyn-2-one / Ethynyl methyl ketone	C_4H_4O / 68.08	1423-60-5 /	/ 84	4-01-00-03539 / 0.8860^{15}	/ 1.4070^{20}
3924	3-Butyn-2-one, 4-phenyl-	$C_{10}H_8O$ / 144.17	1817-57-8 /	/ $75-6^{0.8}$	4-07-00-01175 / 0.99^{25}	/ 1.5762^{20}
3925	Butyrac 118 / Butanoic acid, 4-(2,4-dichlorophenoxy)-	$C_{10}H_{10}Cl_2O_3$ / 249.09	94-82-6 / 118	2828 /		
3926	Butyric acid, 3,4-dichloro- / 3,4-Dichlorobutyric acid	$C_4H_6Cl_2O_2$ / 157.00	29653-38-1 / 49	/ 123^8	4-02-00-00828 /	eth 4; chl 4; HOAc 4
3927	Butyric acid, 4-fluoro- / 4-Fluorobutyric acid	$C_4H_7FO_2$ / 106.10	462-23-7 /	/ $76-8^5$	4-02-00-00809 /	eth 4; EtOH 4 / 1.3993^{25}

No.	Name Synonym	Mol. Form. Mol. Wt.	CAS RN mp/°C	Merck No. bp/°C	Beil. Ref. den/g cm^{-3}	Solubility n_D
3928	Butyric acid, 2-iodo-, ethyl ester	$C_6H_{11}IO_2$ 242.06	7425-47-0	100[21]; 46[1]	4-02-00-00839 1.570[17]	1.4923[20]
3929	Cadmium, dimethyl- Dimethyl cadmium	C_2H_6Cd 142.48	506-82-1 -4.5	3229 105.5; exp 150	1.9846[18]	peth 3 1.5488
3930	Calcium cyanide Calcyan	C_2CaN_2 92.11	592-01-8	1664		
3931	Camphoric acid, trans-(±)- Isocamphoric acid, (DL)	$C_{10}H_{16}O_4$ 200.23	560-08-7 195		3-09-00-03879 1.249[25]	eth 4
3932	Carbamic acid, (aminocarbonyl)-, ethyl ester	$C_4H_8N_2O_3$ 132.12	626-36-8 196.5	dec	4-03-00-00127	H_2O 1; EtOH 2; eth 1; bz 2
3933	Carbamic acid, butyl-, 2-[[(aminocarbonyl)oxy]methyl]-2-methylpentyl ester 2-Methyl-2-propyltrimethylene butylcarbamate carbamate	$C_{13}H_{26}N_2O_4$ 274.36	4268-36-4	9737		
3934	Carbamic acid, butyl-, butyl ester	$C_9H_{19}NO_2$ 173.26	13105-52-7	88[3]	4-04-00-00577 0.9221[20]	eth 4; EtOH 4 1.4359[20]
3935	Carbamic acid, butyl ester Butyl carbamate	$C_5H_{11}NO_2$ 117.15	592-35-8 53	204 dec; 108[14]	4-03-00-00054	EtOH 4; chl 2
3936	Carbamic acid, butyl-, ethyl ester	$C_7H_{15}NO_2$ 145.20	591-62-8 -22	202; 100[15]	4-04-00-00577 0.9434[26]	1.4278[26]
3937	Carbamic acid, butyl-, methyl ester	$C_6H_{13}NO_2$ 131.17	2594-21-0 -18	92[15]	0-04-00-00158 0.9689[23]	1.4289[23]
3938	Carbamic acid, N-butyl-N-nitro,butyl ester	$C_9H_{18}N_2O_4$ 218.25	110795-27-2	98[3]	1.046[20]	EtOH 5; eth 5 1.448[20]
3939	Carbamic acid, (3-chlorophenyl)-, 1-methylethyl ester Chloropropham	$C_{10}H_{12}ClNO_2$ 213.66	101-21-3 41	2188 149[2]	4-12-00-01149 1.18[30]	1.5388[20]
3940	Carbamic acid, dichloro-, ethyl ester	$C_3H_5Cl_2NO_2$ 157.98	13698-16-3	55-6[15]	3-03-00-00050 1.349[25]	1.4595[20]
3941	Carbamic acid, diethyl- Carbamic acid, N,N-diethyl	$C_5H_{11}NO_2$ 117.15	24579-70-2 -15	171 dec	3-04-00-00222 0.9276[20]	H_2O 4; eth 4; EtOH 4 1.4206[20]
3942	Carbamic acid, (1,1-dimethylethyl)-, ethyl ester	$C_7H_{15}NO_2$ 145.20	1611-50-3 21.5	72[16]	4-04-00-00665 0.943[15]	
3943	Carbamic acid, (1,1-dimethylethyl)-, methyl ester	$C_6H_{13}NO_2$ 131.17	27701-01-5 27	63.3[17]	3-04-00-00325 0.966[15]	
3944	Carbamic acid, (1,1-dimethylethyl)nitro-, ethyl ester	$C_7H_{14}N_2O_4$ 190.20	55696-02-1	56[2] dec	4-04-00-03409 1.049[20]	eth 4; EtOH 4 1.4331[20]
3945	Carbamic acid, dimethyl-, 3-methyl-1-phenyl-1H-pyrazol-5-yl ester Pyrolan	$C_{13}H_{15}N_3O_2$ 245.28	87-47-8 50	8013 160-2[0.2]	5-23-10-00530	ctc 3; CS_2 3
3946	Carbamic acid, diphenyl-, ethyl ester	$C_{15}H_{15}NO_2$ 241.29	603-52-1 72	360	4-12-00-00850	H_2O 4; eth 4; bz 4
3947	Carbamic acid, ethyl-, butyl ester Carbamic acid, N-ethyl, butyl ester	$C_7H_{15}NO_2$ 145.20	16246-07-4	66[3]	0.9396[20]	eth 4; EtOH 4 1.4301[20]
3948	Carbamic acid, ethyl ester Urethan	$C_3H_7NO_2$ 89.09	51-79-6 49	9789 185	4-03-00-00040 0.9862[21]	H_2O 4; EtOH 4; eth 4; bz 4 1.4144[51]
3949	Carbamic acid, ethyl-, ethyl ester	$C_5H_{11}NO_2$ 117.15	623-78-9	176	4-04-00-00365 0.9813[20]	H_2O 4; eth 4; EtOH 4 1.4215[20]
3950	Carbamic acid, ethylnitro-, ethyl ester Carbamic acid, N-ethyl-N-nitro, ethyl ester	$C_5H_{10}N_2O_4$ 162.15	6274-16-4	107[31]	4-04-00-03404 1.161[20]	eth 4; EtOH 4 1.4432[20]
3951	Carbamic acid, ethylnitro-, methyl ester Carbamic acid, N-ethyl-N-nitro, butyl ester	$C_7H_{14}N_2O_4$ 190.20	6162-79-4	79[3]	4-04-00-03403 1.089[20]	H_2O 4; eth 4; EtOH 4 1.4455[20]
3952	Carbamic acid, (3-methylbutyl)-, ethyl ester Urethane, isopentyl	$C_8H_{17}NO_2$ 159.23	1611-52-5 218		2-04-00-00648 0.9322[20]	1.4326[20]
3953	Carbamic acid, methyl ester Methyl carbamate	$C_2H_5NO_2$ 75.07	598-55-0 54	5958 177	4-03-00-00037 1.136[56]	H_2O 4; EtOH 4; eth 3 1.4125[56]
3954	Carbamic acid, methyl-, ethyl ester	$C_4H_9NO_2$ 103.12	105-40-8	170	4-04-00-00200 1.0115[20]	H_2O 4; EtOH 4 1.4183[20]
3955	Carbamic acid, 1-methylethyl ester Isopropyl carbamate	$C_4H_9NO_2$ 103.12	1746-77-6 93	183	4-03-00-00053 0.9951[66]	
3956	Carbamic acid, (1-methylethyl)-, ethyl ester	$C_6H_{13}NO_2$ 131.17	2594-20-9	79[15]	4-04-00-00520 0.9531[20]	eth 4; EtOH 4 1.4229[20]
3957	Carbamic acid, (1-methylethyl)nitro-, ethyl ester Carbamic acid, N-isopropyl-N-nitro, ethyl ester	$C_6H_{12}N_2O_4$ 176.17	62261-05-6	72[7]	4-04-00-03407 1.110[20]	eth 4; EtOH 4 1.4381[20]
3958	Carbamic acid, methylphenyl-, methyl ester	$C_9H_{11}NO_2$ 165.19	28685-60-1 44	235	2-12-00-00235 1.2960[19]	eth 4; lig 4
3959	Carbamic acid, 2-methylpropyl ester Isobutyl carbamate	$C_5H_{11}NO_2$ 117.15	543-28-2 67	5021 207	4-03-00-00056	eth 4; EtOH 4 1.4098[76]
3960	Carbamic acid, (1-methylpropyl)-, ethyl ester Urethane, sec-butyl	$C_7H_{15}NO_2$ 145.20	10212-74-5 -14	194	4-04-00-00622 0.9404[26]	1.4267[26]
3961	Carbamic acid, (2-methylpropyl)-, ethyl ester Isobutyl urethane	$C_7H_{15}NO_2$ 145.20	539-89-9 <-65	5037 110[30]	4-04-00-00647 0.9432[20]	eth 4; EtOH 4 1.4288[20]

No.	Name / Synonym	Mol. Form. / Mol. Wt.	CAS RN / mp/°C	Merck No. / bp/°C	Beil. Ref. / den/g cm^{-3}	Solubility / n_D
3962	Carbamic acid, (1-methylpropyl)-, methyl ester	$C_6H_{13}NO_2$	39076-02-3		0-04-00-00162	
	Carbamic acid, N-sec-butyl, methyl ester	131.17		83[16]	0.9651[25]	1.4263[25]
3963	Carbamic acid, nitro-, ethyl ester	$C_3H_6N_2O_4$	626-37-9		4-03-00-00247	H_2O 4; EtOH 4; eth 1; lig 2
		134.09	64	140 dec	1.0074[20]	
3964	Carbamic acid, nitropropyl-, ethyl ester	$C_6H_{12}N_2O_4$	13855-77-1		4-04-00-03406	eth 4; EtOH 4
	Carbamic acid, N-propyl-N-nitro, ethyl ester	176.17		66[3]	1.121[20]	1.4431[20]
3965	Carbamic acid, phenyl-, ethyl ester	$C_9H_{11}NO_2$	101-99-5	7291	4-12-00-00619	H_2O 1; EtOH 4; eth 4; bz 3
	Phenylurethane	165.19	53	237 dec	1.1064[30]	1.5376[30]
3966	Carbamic acid, phenyl-, 1-methylethyl ester	$C_{10}H_{13}NO_2$	122-42-9	7828	4-12-00-00620	bz 4; EtOH 4
	Propham	179.22	90		1.09[20]	1.4989[91]
3967	Carbamic acid, (phenylmethyl)-, ethyl ester	$C_{10}H_{13}NO_2$	2621-78-5		4-12-00-02250	bz 4; eth 4; EtOH 4; chl 4
		179.22	49	230		
3968	Carbamic acid, phenyl-, 2-methylpropyl ester	$C_{11}H_{15}NO_2$	2291-80-7		3-12-00-00014	bz 4; eth 4; EtOH 4
		193.25	86	216		
3969	Carbamic acid, propyl ester	$C_4H_9NO_2$	627-12-3		4-03-00-00052	ace 4; eth 4; EtOH 4
		103.12	60	196		
3970	Carbamic acid, propyl-, ethyl ester	$C_6H_{13}NO_2$	623-85-8		4-04-00-00480	EtOH 4
		131.17		192	0.9921[15]	
3971	Carbamic chloride	CH_2ClNO	463-72-9	1785	3-03-00-00065	
	Carbamyl chloride	79.49		62 dec		
3972	Carbamic chloride, diethyl-	$C_5H_{10}ClNO$	88-10-8		4-04-00-00379	
	Carbamoyl chloride, N,N-diethyl	135.59		186		
3973	Carbamic chloride, dimethyl-	C_3H_6ClNO	79-44-7		4-04-00-00224	
	Dimethylcarbamoyl chloride	107.54	-33	167	1.168[25]	1.4540[20]
3974	Carbamic chloride, methylphenyl-	C_8H_8ClNO	4285-42-1		4-12-00-00839	eth 4; EtOH 4
		169.61	88.5	280		
3975	Carbamodithioic acid, dimethyl-, 2,4-dinitrophenyl ester	$C_9H_9N_3O_4S_2$	89-37-2		4-06-00-01760	H_2O 1; EtOH 3; ace 3; bz 3
		287.32	152.5		1.54[20]	
3976	Carbamothioic acid, butylethyl-, S-propyl ester	$C_{10}H_{21}NOS$	1114-71-2	7007		ace 4; bz 4; MeOH 4
	Pebulate	203.35		142[20]	0.9458[20]	1.4752[20]
3977	Carbamothioic acid, diphenyl-, S-[2-(diethylamino)ethyl] ester	$C_{19}H_{24}N_2OS$	3735-90-8	7178		eth 4; chl 4; MeOH 4; peth 4
	Phencarbamide	328.48	48.5	120-6[0.01]		
3978	Carbamothioic acid, O-ethyl ester	C_3H_7NOS	625-57-0		4-03-00-00294	eth 4; EtOH 4; chl 4
		105.16	41	dec	1.069[20]	1.520[20]
3979	Carbamothioic acid, S-ethyl ester	C_3H_7NOS	637-98-9		4-03-00-00294	eth 4; EtOH 4
		105.16	109	dec		
3980	Carbamothioic chloride, dimethyl-	C_3H_6ClNS	16420-13-6		3-04-00-00147	eth 4; chl 3; peth 3
		123.61	42.5	98[10]		
3981	9H-Carbazole	$C_{12}H_9N$	86-74-8	1792	5-20-08-00009	H_2O 1; EtOH 2; eth 2; ace 3
	Dibenzo[b,d]pyrrole	167.21	246.2	354.7		
3982	9H-Carbazole-9-acetic acid	$C_{14}H_{11}NO_2$	524-80-1	1793	5-20-08-00033	eth 4; EtOH 4; chl 4; HOAc 4
	(N-Carbazolyl)acetic acid	225.25	215			
3983	9H-Carbazole, 9-acetyl-	$C_{14}H_{11}NO$	574-39-0		5-20-08-00030	
		209.25	69	190[6]	1.158[100]	1.640[100]
3984	9H-Carbazole, 9-butyl-	$C_{16}H_{17}N$	1484-08-8		5-20-08-00018	eth 4
		223.32	58	218-9[19]		
3985	9H-Carbazole, 9-ethyl-	$C_{14}H_{13}N$	86-28-2		5-20-08-00016	H_2O 1; EtOH 4; eth 4
		195.26	68	190[10]	1.059[80]	1.6394[80]
3986	9H-Carbazole, 3-methyl-	$C_{13}H_{11}N$	4630-20-0		5-20-08-00076	bz 4; eth 4
		181.24	208.5	365		
3987	9H-Carbazole, 9-methyl-	$C_{13}H_{11}N$	1484-12-4		5-20-08-00015	eth 4
		181.24	89.34	195[12]		
3988	9H-Carbazole, 9-(phenylmethyl)-	$C_{19}H_{15}N$	19402-87-0		5-20-08-00023	bz 4
		257.33	119	267[24]		
3989	1H-Carbazole, 2,3,4,9-tetrahydro-	$C_{12}H_{13}N$	942-01-8		5-20-07-00468	H_2O 1; EtOH 3; eth 4; bz 4
		171.24	120	327.5		
3990	Carbendazim	$C_9H_9N_3O_2$	10605-21-7	1794		
	Carbamic acid, 1H-benzimidazol-2-yl-, methyl ester	191.19	300 dec		1.45	
3991	Carbofuran	$C_{12}H_{15}NO_3$	1563-66-2	1810		
	7-Benzofuranol, 2,3-dihydro-2,2-dimethyl-, methylcarbamate	221.26	151		1.18	
3992	Carbon dioxide	CO_2	124-38-9	1816	0-03-00-00004	
	Carbonic anhydride	44.01	-56.57 (5.11 atm)	sub -78.5 (1 atm)	*0.720[25]	
3993	Carbon disulfide	CS_2	75-15-0	1818	4-03-00-00395	H_2O 3; EtOH 5; eth 5; chl 3
	Carbon bisulfide	76.14	-111.5	46	1.2632[20]	1.6319[20]
3994	Carbonic acid, bis(1,1-dimethylethyl) ester	$C_9H_{18}O_3$	34619-03-9		4-03-00-00009	EtOH 4
		174.24	40	158		
3995	Carbonic acid, bis(1-methylethyl) ester	$C_7H_{14}O_3$	6482-34-4		3-03-00-00009	EtOH 4
		146.19		147	0.9162[20]	1.3932[20]
3996	Carbonic acid, bis(2-methylphenyl) ester	$C_{15}H_{14}O_3$	617-09-4		3-06-00-01256	HOAc 4
		242.27	60	144[0.5]		
3997	Carbonic acid, bis(2-methylpropyl) ester	$C_9H_{18}O_3$	539-92-4		4-03-00-00009	H_2O 1; EtOH 5; eth 5
		174.24		190	0.9138[20]	1.4072[20]
3998	Carbonic acid, dibutyl ester	$C_9H_{18}O_3$	542-52-9	1558	4-03-00-00008	H_2O 1; EtOH 3; eth 3
	Dibutyl carbonate	174.24		207	0.9251[20]	1.4117[20]

No.	Name / Synonym	Mol. Form. / Mol. Wt.	CAS RN / mp/°C	Merck No. / bp/°C	Beil. Ref. / den/g cm⁻³	Solubility / n_D
3999	Carbonic acid, diethyl ester / Diethyl carbonate	$C_5H_{10}O_3$ / 118.13	105-58-8 / -43	3738 / 126	4-03-00-00005 / 0.9752[20]	H_2O 1; EtOH 3; eth 3; chl 3 / 1.3845[20]
4000	Carbonic acid, dimethyl ester / Methyl carbonate	$C_3H_6O_3$ / 90.08	616-38-6 / 0.5	5960 / 90.5	4-03-00-00003 / 1.065[17]	H_2O 1; EtOH 3; eth 3; ctc 2 / 1.3687[20]
4001	Carbonic acid, diphenyl ester / Phenyl carbonate	$C_{13}H_{10}O_3$ / 214.22	102-09-0 / 83	7250 / 306	4-06-00-00629 / 1.1215[87]	H_2O 1; EtOH 3; eth 3; ctc 3
4002	Carbonic acid, dipropyl ester / Dipropyl carbonate	$C_7H_{14}O_3$ / 146.19	623-96-1 /	/ 168	4-03-00-00006 / 0.9435[20]	H_2O 2; EtOH 5; eth 5 / 1.4008[20]
4003	Carbonic acid, ethyl methyl ester / Ethyl methyl carbonate	$C_4H_8O_3$ / 104.11	623-53-0 / -14	/ 107.5	4-03-00-00004 / 1.012[20]	eth 4; EtOH 4 / 1.3778[20]
4004	Carbonic acid monosodium salt	$CHNaO_3$ / 84.01	144-55-8 /	8528 /		H_2O 3; EtOH 1
4005	Carbonic chloride fluoride / Carbonyl chloride fluoride	CClFO / 82.46	353-49-1 /	/	4-03-00-00031 /	
4006	Carbonic dibromide / Carbonyl bromide	CBr_2O / 187.82	593-95-3 /	/ 64.5	/ 2.52[15]	
4007	Carbonic dichloride / Phosgene	CCl_2O / 98.92	75-44-5 / -127.9	7310 / 8	4-03-00-00031 / *1.3719[25]	bz 3; ctc 3; chl 3; tol 3
4008	Carbonic difluoride / Carbonyl fluoride	CF_2O / 66.01	353-50-4 / -111.2	1826 / -84.5	4-03-00-00021 / 1.139[25]	
4009	Carbonic dihydrazide / Carbohydrazide	CH_6N_4O / 90.09	497-18-7 / 154	1811 /	4-03-00-00240 / 1.616[20]	H_2O 4; EtOH 4
4010	Carbonic dihydrazide, 2,2'-diphenyl- / sym-Diphenylcarbazide	$C_{13}H_{14}N_4O$ / 242.28	140-22-7 / 170 dec	3321 /	4-15-00-00182 /	H_2O 2; EtOH 3; eth 2; ace 3
4011	Carbon monoxide / Carbon oxide	CO / 28.01	630-08-0 / -205	1820 / -191.5	0-01-00-00720 / 0.7909[-19]	H_2O 2; bz 3; HOAc 3
4012	Carbonochloridic acid, butyl ester / Butyl chloroformate	$C_5H_9ClO_2$ / 136.58	592-34-7 /	/ 142	4-03-00-00025 / 1.074[25]	eth 5; ace 3; ctc 2 / 1.4114[20]
4013	Carbonochloridic acid, 2-chloroethyl ester / 2-Chloroethyl chloroformate	$C_3H_4Cl_2O_2$ / 142.97	627-11-2 /	/ 155	4-03-00-00024 / 1.3847[20]	H_2O 1; EtOH 3; eth 3; ace 3 / 1.4483[20]
4014	Carbonochloridic acid, chloromethyl ester / Chloromethyl chloroformate	$C_2H_2Cl_2O_2$ / 128.94	22128-62-7 /	/ 107	4-03-00-00030 / 1.465[15]	ace 4; eth 4 / 1.4286[22]
4015	Carbonochloridic acid, 3-chloropropyl ester	$C_4H_6Cl_2O_2$ / 167.00	628-11-5 /	/ 177	3-03-00-00025 / 1.2926[25]	H_2O 1 / 1.4456[20]
4016	Carbonochloridic acid, dichloromethyl ester	$C_2HCl_3O_2$ / 163.39	22128-63-8 /	/ 110.5	4-03-00-00031 / 1.5600[15]	ace 4
4017	Carbonochloridic acid, 2-ethoxyethyl ester	$C_5H_9ClO_3$ / 152.58	628-64-8 /	/ 67.2[14]	3-03-00-00029 / 1.1341[25]	EtOH 4 / 1.4169[25]
4018	Carbonochloridic acid, ethyl ester / Ethyl chloroformate	$C_3H_5ClO_2$ / 108.52	541-41-3 / -80.6	3742 / 95	4-03-00-00023 / 1.1352[20]	bz 4; eth 4; chl 4 / 1.3974[20]
4019	Carbonochloridic acid, 2-methoxyethyl	$C_4H_7ClO_3$ / 138.55	020-12-6 /	/ 58.7[13]	3-03-00-00029 / 1.190[25]	eth 4 / 1.4163[20]
4020	Carbonochloridic acid, methyl ester / Methyl chlorocarbonate	$C_2H_3ClO_2$ / 94.50	79-22-1 /	5966 / 70.5	4-03-00-00023 / 1.2231[20]	EtOH 5; eth 5; bz 3; ctc 3 / 1.3868[20]
4021	Carbonochloridic acid, 3-methylbutyl ester	$C_6H_{11}ClO_2$ / 150.60	628-50-2 /	/ 154.3	4-03-00-00026 / 1.0288[17]	eth 4; EtOH 4 / 1.4176[20]
4022	Carbonochloridic acid, 1-methylethenyl ester / Isopropenyl chloroformate	$C_4H_5ClO_2$ / 120.54	57933-83-2 /	/ 102	3-03-00-00028 / 1.101[20]	eth 4
4023	Carbonochloridic acid, 1-methylethyl ester / Isopropyl chloroformate	$C_4H_7ClO_2$ / 122.55	108-23-6 /	/ 105	4-03-00-00024 /	eth 4 / 1.4013[20]
4024	Carbonochloridic acid, 2-methylpropyl ester / Isobutyl chlorocarbonate	$C_5H_9ClO_2$ / 136.58	543-27-1 /	5023 / 128.8	4-03-00-00026 / 1.0426[18]	EtOH 3; eth 5; bz 3; chl 3 / 1.4071[18]
4025	Carbonochloridic acid, pentyl ester	$C_6H_{11}ClO_2$ / 150.60	638-41-5 /	/ 60-2[15]	4-03-00-00026 /	eth 3 / 1.4181[18]
4026	Carbonochloridic acid, phenyl ester / Phenyl chloroformate	$C_7H_5ClO_2$ / 156.57	1885-14-9 /	/ 71[9]	4-06-00-00629 /	
4027	Carbonochloridic acid, phenylmethyl ester / Carbobenzoxy chloride	$C_8H_7ClO_2$ / 170.60	501-53-1 /	1807 / 103[20]	4-06-00-02278 / 1.195[25]	eth 3; ace 3; bz 3 / 1.5190[20]
4028	Carbonochloridic acid, propyl ester / Propyl chlorocarbonate	$C_4H_7ClO_2$ / 122.55	109-61-5 /	7860 / 115.2	4-03-00-00024 / 1.0901[20]	EtOH 5; eth 5 / 1.4035[20]
4029	Carbonochloridic acid, trichloromethyl ester / Diphosgene	$C_2Cl_4O_2$ / 197.83	503-38-8 / -57	3339 / 128	4-03-00-00033 / 1.6525[14]	eth 4; EtOH 4 / 1.4566[22]
4030	Carbonocyanidic acid, ethyl ester	$C_4H_5NO_2$ / 99.09	623-49-4 /	/ 115.5	4-02-00-01862 / 1.003[25]	H_2O 1; EtOH 3; eth 3; ctc 3 / 1.3820[20]
4031	Carbonodithioic acid, S,S-diethyl ester	$C_5H_{10}OS_2$ / 150.27	623-80-3 /	/ 197	3-03-00-00339 / 1.085[20]	eth 4; EtOH 4 / 1.5237[18]
4032	Carbonothioic dichloride / Thiophosgene	CCl_2S / 114.98	463-71-8 /	/ 73	4-03-00-00281 / 1.508[15]	H_2O 1; EtOH 3; eth 3 / 1.5442[20]
4033	Carbonothioic dihydrazide / 1,3-Diamino-2-thiourea	CH_6N_4S / 106.15	2231-57-4 / 170 dec		4-03-00-00388 /	H_2O 4
4034	Carbonotrithioic acid / Trithiocarbonic acid	CH_2S_3 / 110.22	594-08-1 / -30	9671 / 57 dec	4-03-00-00428 / 1.47[17]	EtOH 4; tol 4; chl 4
4035	Carbon oxide sulfide / Carbonyl sulfide	COS / 60.08	463-58-1 / -138.8	/ -50	4-03-00-00271 / 1.028[17]	H_2O 2; EtOH 3; KOH 4 / 1.24[-87]
4036	Carbon selenide / Carbon diselenide	CSe_2 / 169.93	506-80-9 / -45.5	1817 / 125.5	4-03-00-00436 / 2.6823[20]	1.8454[20]
4037	Carbophenothion / Phosphorodithioic acid, S-[[(4-chlorophenyl)thio]methyl] O,O-diethyl ester	$C_{11}H_{16}ClO_2PS_3$ / 342.87	786-19-6 /	1827 / 82[0.01]	/ 1.271[20]	

No.	Name / Synonym	Mol. Form. / Mol. Wt.	CAS RN / mp/°C	Merck No. / bp/°C	Beil. Ref. / den/g cm^{-3}	Solubility / n_D
4038	Carbosulfan 2,3-Dihydro-2,2-dimethyl-7-benzofuranyl (dibutylamino)thio methylcarbamate	$C_{20}H_{32}N_2O_3S$ 380.55	55285-14-8	126	1.056^{20}	
4039	Carboxin 1,4-Oxathii$N$-3-carboxamide, 5,6-dihydro-2-methyl-N-phenyl-	$C_{12}H_{13}NO_2S$ 235.31	5234-68-4 94	1832		
4040	Card-20(22)-enolide, 3-[(6-deoxy-α-L-mannopyranosyl)oxy]-1,5,11,14,19-pentahydroxy-, (1β,3β Ouabain	$C_{29}H_{44}O_{12}$ 584.66	630-60-4 200	6854	5-18-05-00625	H_2O 2; EtOH 4
4041	Card-20(22)-enolide, 3-[(O-2,6-dideoxy-β-D-ribo-hexopyranosyl)-(1-4)-O-2,6-dideoxy-β-D-ribo-hexopyranosyl)-(1-4)-2,6-dideoxy-β-D, 12,14-dihydroxy Digoxin	$C_{41}H_{64}O_{14}$ 780.95	20830-75-5 248-50 dec	3150	5-18-04-00381	EtOH 4
4042	Card-20(22)-enolide, 3-[(O-2,6-dideoxy-β-D-ribo-hexopyranosyl)-(1-4)-O-2,6-dideoxy-β-D-ribo-hexopyranosyl)-(1-4)-2,6-dideoxy-β-D, 14,16-dihydroxy Gitoxin	$C_{41}H_{64}O_{14}$ 780.95	4562-36-1 285 dec	4328	5-18-04-00388	
4043	Card-20(22)-enolide, 3-[(O-2,6-dideoxy-β-D-ribo-hexopyranosyl)-(1-4)-O-2,6-dideoxy-β-D-ribo-hexopyranosyl)-(1-4)-2,6-dideoxy-β-D, 14-dihydroxy Digitoxin	$C_{41}H_{64}O_{13}$ 764.95	71-63-6 255.5	3146	5-18-03-00354	H_2O 2; EtOH 4; eth 3; chl 3
4044	Card-20(22)-enolide, 3,14-dihydroxy-, (3β,5β)- Digitoxigenin	$C_{23}H_{34}O_4$ 374.52	143-62-4 253	3145	5-18-03-00348	EtOH 3; MeOH 4
4045	Card-20(22)-enolide, 3-[(6-O-β-D-glucopyranosyl-β-D-glucopyranosyl)oxy]-14-hydroxy-, Uzarin	$C_{35}H_{54}O_{14}$ 698.81	20231-81-6 269	9809	4-18-00-01497	
4046	Card-20(22)-enolide, 3,11,14-trihydroxy-, (3β,5β,11α)- Sarmentogenin	$C_{23}H_{34}O_5$ 390.52	76-28-8 280	8334	5-18-04-00378	H_2O 1; EtOH 3; eth 1; ace 2
4047	Card-20(22)-enolide, 3,12,14-trihydroxy-, (3β,5β,12β)- Digoxigenin	$C_{23}H_{34}O_5$ 390.52	1672-46-4 222	3149	5-18-04-00380	EtOH 4; chl 2; MeOH 4
4048	Card-20(22)-enolide, 3,14,16-trihydroxy-, (3β,5β,16β)- Gitoxigenin	$C_{23}H_{34}O_5$ 390.52	545-26-6 234	4327	5-18-04-00385	H_2O 1; eth 2; chl 3
4049	Card-20(22)-enolide, 3,5,14-trihydroxy-19-oxo-, (3β,5β)- Strophanthidin	$C_{23}H_{32}O_6$ 404.50	66-28-4 173 dec	8818	5-18-05-00126	H_2O 1; EtOH 3; eth 1; ace 3
4050	4-Carene, (1S,3R,6R)-(-)- Δ4-Carene (l)	$C_{10}H_{16}$ 136.24	5208-49-1 169; 64^{20}		4-05-00-00451 0.8441^{30}	ace 4; bz 4; eth 4 1.4740^{30}
4051	β,β-Carotene β-Carotene, all-$trans$-	$C_{40}H_{56}$ 536.88	7235-40-7 183	1860	4-05-00-02617 1.00^{20}	H_2O 1; EtOH 2; eth 3; ace 3
4052	β,ψ-Carotene γ-Carotene	$C_{40}H_{56}$ 536.88	472-93-5 178	1861	4-05-00-02616	H_2O 1; EtOH 1; eth 2; bz 3
4053	ψ,ψ-Carotene $trans$-Lycopene	$C_{40}H_{56}$ 536.88	502-65-8 175	5492	4-01-00-01165	EtOH 2; eth 3; bz 4; chl 4
4054	β,ϵ-Carotene, (6'R)- α-Carotene	$C_{40}H_{56}$ 536.88	7488-99-5 187.5	1859	4-05-00-02620 1.00^{20}	bz 4; eth 4; chl 4
4055	β,β-Carotene, 3'-(acetyloxy)-6',7'-didehydro-5,6-epoxy-5,5',6,6',7,8-hexahydro-3,5'-dihydroxy-8-oxo-, Fucoxanthin	$C_{40}H_{60}O_6$ 636.91	3351-86-8 168	4195	5-18-04-00673	eth 4; EtOH 4
4056	β,β-Carotene-3,3'-diol, 5,6:5',6'-diepoxy-5,5',6,6'-tetrahydro- Violaxanthin	$C_{40}H_{56}O_4$ 600.88	126-29-4 208	9902	5-19-03-00464	EtOH 3; eth 3; CS_2 3; peth 1
4057	β,β-Carotene-3,3'-diol, (3R,3'R)- Zeaxanthin	$C_{40}H_{56}O_2$ 568.88	144-68-3 215.5	10019	4-06-00-07017	H_2O 1; EtOH 2; eth 3; ace 3
4058	β,ϵ-Carotene-3,3'-diol, (3R,3'R,6'R)- Xanthophyll	$C_{40}H_{56}O_2$ 568.88	127-40-2 196	9972	3-06-00-05871	bz 4; eth 4; EtOH 4; peth 4
4059	β,ψ-Caroten-16'-oic acid, 3',4'-didehydro- Torularhodin	$C_{40}H_{52}O_2$ 564.85	514-92-1 211	9476	4-09-00-02767	py 4; chl 4; CS_2 4
4060	ψ,ψ-Caroten-16-ol Lycoxanthin	$C_{40}H_{56}O$ 552.88	19891-74-8 168	5499	4-01-00-02368	H_2O 1; EtOH 2; bz 3; CS_2 3
4061	β,β-Caroten-3-ol, (3R)- Cryptoxanthin	$C_{40}H_{56}O$ 552.88	472-70-8 160	2612	4-06-00-05111	bz 4; chl 4
4062	β,ψ-Caroten-3-ol, (3R)- Rubixanthin	$C_{40}H_{56}O$ 552.88	3763-55-1 160	8265	4-06-00-05110	EtOH 2; bz 3; chl 3; peth 2
4063	β,κ-Caroten-6'-one, 3,3'-dihydroxy-, (3R,3'S,5'R)- Capsanthin	$C_{40}H_{56}O_3$ 584.88	465-42-9 176	1768	4-08-00-02657	
4064	Carpaine 2,15-Dioxa-12,25-diazatricyclo[22.2.2.2^{11,14}]triacontane, carpaine deriv.	$C_{28}H_{50}N_2O_4$ 478.72	3463-92-1 121	1866	2-27-00-00209	ace 4; bz 4; eth 4; EtOH 4

No.	Name Synonym	Mol. Form. Mol. Wt.	CAS RN mp/°C	Merck No. bp/°C	Beil. Ref. den/g cm^{-3}	Solubility n_D
4065	CDEC Carbamodithioic acid, diethyl-, 2-chloro-2-propenyl ester	$C_8H_{14}ClNS_2$ 223.79	95-06-7	8882 129[1]	1.088	
4066	Cedrene 1H-3a,7-Methanoazulene, cedrene deriv.	$C_{15}H_{24}$ 204.36	11028-42-5	262.5	3-05-00-01095 0.9342[20]	H$_2$O 1; bz 3; lig 3 1.5034[20]
4067	Cevane-3,4,12,14,16,17,20-heptol, 3-acetate, (3β,4α,16β)- Sabadine	$C_{29}H_{47}NO_8$ 537.69	124-80-1 258	8281	4-21-00-02883	ace 4; EtOH 4
4068	Cevane-3,4,7,14,15,16,20-heptol, 4,9-epoxy-, (3β,4α,7α,15α,16β)- Germine	$C_{27}H_{43}NO_8$ 509.64	508-65-6 220	4307	5-21-13-00706	bz 3; MeOH 3; alk 3; acid 3
4069	Cevane-3,4,12,14,16,17,20-heptol, 4,9-epoxy-, 3-(2-methyl-2-butenoate), [3β(Z),4α,16β]- Cevadine	$C_{32}H_{49}NO_9$ 591.74	62-59-9 213-4 dec	2025	4-21-00-06820	
4070	Cevane-3,4,6,7,14,15,16,20-octol, 4,9-epoxy-, (3β,4α,6α,7α,15α,16β) Protoverine	$C_{27}H_{43}NO_9$ 525.64	76-45-9 221	7916	4-21-00-06841	H$_2$O 1; EtOH 3; bz 3; aq acid 3
4071	Chelidonine Stylophorine	$C_{20}H_{19}NO_5$ 353.37	476-32-4 135.5	2043 220[0.002]	2-27-00-00615	H$_2$O 1; EtOH 3; eth 3; chl 3
4072	Chinomethionat Carbonic acid, dithio-, cyclic S,S-(6-methyl-2,3-quinoxalinediyl) ester	$C_{10}H_8N_2OS_2$ 234.30	2439-01-2 170	6933		
4073	DL-chiro-Inositol	$C_6H_{12}O_6$ 180.16	18685-70-6 253		4-06-00-07920 1.752[15]	H$_2$O 4; EtOH 2; eth 1; bz 1
4074	D-chiro-Inositol, 2-deoxy- D-Quercitol	$C_6H_{12}O_5$ 164.16	488-73-3 236	8046	4-06-00-07883 1.5845[13]	H$_2$O 4
4075	Chlorbenside Benzene, 1-chloro-4-[[(4-chlorophenyl)methyl]thio]-	$C_{13}H_{10}Cl_2S$ 269.19	103-17-3 75	2074	1.4210[20]	
4076	Chlorbromuron Urea, N'-(4-bromo-3-chlorophenyl)-N-methoxy-N-methyl-	$C_9H_{10}BrClN_2O_2$ 293.55	13360-45-7 96		1.69[20]	
4077	Chlordane 4,7-Methano-1H-indene, 1,2,4,5,6,7,8,8-octachloro-2,3,3a,4,7,7a-hexahydro-	$C_{10}H_6Cl_8$ 409.78	57-74-9 106	2079 175[1]	1.60[25]	
4078	Chlorfenvinphos Phosphoric acid, 2-chloro-1-(2,4-dichlorophenyl)ethenyl diethyl ester	$C_{12}H_{14}Cl_3O_4P$ 359.57	470-90-6 170[0.05]	2087		
4079	Chlorflurecol 9H-Fluorene-9-carboxylic acid, 2-chloro-9-hydroxy-	$C_{14}H_9ClO_3$ 260.68	2464-37-1		1.496[20]	
4080	Chloridazon 3(2H)-Pyridazinone, 5-amino-4-chloro-2-phenyl-	$C_{10}H_8ClN_3O$ 221.65	1698-60-8 205			
4081	Chlorimuron 2-(4-Chloro-6-methoxypyrimidin-2-ylcarbamoylsulfamoyl) benzoic acid, ethyl ester	$C_{15}H_{15}ClN_4O_6S$ 414.83	90982-32-4 186	2092		
4082	Chlorobenzilate Benzeneacetic acid, 4-chloro-α-(4-chlorophenyl)-α-hydroxy-, ethyl ester	$C_{16}H_{14}Cl_2O_3$ 325.19	510-15-6 37	157[0.07]	1.2816[20]	
4083	5-Chloro-3-methyl-4-nitro-1H-pyrazole 1H-Pyrazole, 3-chloro-5-methyl-4-nitro-	$C_4H_4ClN_3O_2$ 161.55	6814-58-0 111			
4084	Chloroneb Benzene, 1,4-dichloro-2,5-dimethoxy-	$C_8H_8Cl_2O_2$ 207.06	2675-77-6 134	268		
4085	Chlorosulfuric acid, ethyl ester	$C_2H_5ClO_3S$ 144.58	625-01-4 152.5; 93[100]		4-01-00-01326 1.3502[25]	eth 4; chl 4; lig 4 1.416[20]
4086	Chlorosulfuric acid, methyl ester	CH_3ClO_3S 130.55	812-01-1 -70	134	4-01-00-01252 1.4805[25]	ace 4; bz 4; eth 4 1.4138[18]
4087	Chlorosulfurous acid, ethyl ester	$C_2H_5ClO_2S$ 128.58	6378-11-6 52.5[44]; 32[16]		3-01-00-01316 1.2766[25]	eth 4 1.4550[25]
4088	Chloroxuron Urea, N'-[4-(4-chlorophenoxy)phenyl]-N,N-dimethyl-	$C_{15}H_{15}ClN_2O_2$ 290.75	1982-47-4 151			
4089	Chlorpyrifos Phosphorothioic acid, O,O-diethyl O-(3,5,6-trichloro-2-pyridinyl) ester	$C_9H_{11}Cl_3NO_3PS$ 350.59	2921-88-2 42	2190		
4090	Chlorpyrifos-methyl Phosphorothioic acid, O,O-dimethyl O-(3,5,6-trichloro-2-pyridinyl) ester	$C_7H_7Cl_3NO_3PS$ 322.54	5598-13-0 43			
4091	Chlorsulfuron 2-Chloro-N-[[(4-methoxy-6-methyl-1,3,5-triaziN-2-yl)amino]carbonyl]benzenesulfonamide	$C_{12}H_{12}ClN_5O_4S$ 357.78	64902-72-3 176	2192		
4092	Chlorthiophos Phosphorothioic acid, O-[2,5-dichloro-4-(methylthio)phenyl] O,O-diethyl ester	$C_{11}H_{15}Cl_2O_3PS_2$ 361.25	21923-23-9 150[0.001]			

No.	Name / Synonym	Mol. Form. / Mol. Wt.	CAS RN / mp/°C	Merck No. / bp/°C	Beil. Ref. / den/g cm^{-3}	Solubility / n_D
4093	Cholane 1H-Cyclopenta[a]phenanthrene, hexadecahydro-10,13-dimethyl-17-(1-methylbutyl)-, [8R-[8α,9β,10α,1	C$_{24}$H$_{42}$ 330.60	548-98-1 90	2198 190$^{0.001}$	4-05-00-01220	
4094	Cholan-24-oic acid Cholanic acid	C$_{24}$H$_{40}$O$_2$ 360.58	25312-65-6 163.5	2199	4-09-00-01992	EtOH 3; chl 3; HOAc 3
4095	Cholan-24-oic acid, 3,12-dihydroxy-, (3α,5β,12α)- Deoxycholic acid	C$_{24}$H$_{40}$O$_4$ 392.58	83-44-3 177	2881	4-10-00-01608	
4096	Cholan-24-oic acid, 3,6-dihydroxy-, (3α,5β,6α)- Hyodeoxycholic acid	C$_{24}$H$_{40}$O$_4$ 392.58	83-49-8 198.5	4794	4-10-00-01600	H$_2$O 2; EtOH 3; eth 2; ace 2
4097	Cholan-24-oic acid, 3,7-dihydroxy-, (3α,5β,7α)- Chenodiol	C$_{24}$H$_{40}$O$_4$ 392.58	474-25-9 119	2044	4-10-00-01604	H$_2$O 1; EtOH 4; eth 3; ace 4
4098	Cholan-24-oic acid, 3,7-dihydroxy-, (3α,5β,7β)- Ursodiol	C$_{24}$H$_{40}$O$_4$ 392.58	128-13-2 203	9801	4-10-00-01604	EtOH 4; eth 2
4099	Cholan-24-oic acid, 3-hydroxy-, (3α,5β)- Lithocholic acid	C$_{24}$H$_{40}$O$_3$ 376.58	434-13-9 186	5423	4-10-00-00785	H$_2$O 1; EtOH 3; eth 2; chl 3
4100	Cholan-24-oic acid, 3,7,12-trihydroxy-, (3α,5β,7α,12α)- Cholic acid	C$_{24}$H$_{40}$O$_5$ 408.58	81-25-4 198	2206	4-10-00-02072	H$_2$O 2; EtOH 3; eth 4; ace 3
4101	Cholan-24-oic acid, 3,7,12-trioxo-, (5β)- Dehydrocholic acid	C$_{24}$H$_{34}$O$_5$ 402.53	81-23-2 237	2858	4-10-00-03478	H$_2$O 1; EtOH 2; eth 1; ace 3
4102	Cholesta-3,5-diene	C$_{27}$H$_{44}$ 368.65	747-90-0 80	260^{13}	4-05-00-01620 0.925^{100}	H$_2$O 1; EtOH 3; eth 5; bz 5
4103	Cholesta-5,7-dien-3-ol, (3β)- 7-Dehydrocholesterol	C$_{27}$H$_{44}$O 384.65	434-16-2 150.5	2857	4-06-00-04153	H$_2$O 1; EtOH 2; eth 3; ace 3
4104	Cholesta-8,24-dien-3-ol, (3β,5α)-	C$_{27}$H$_{44}$O 384.65	128-33-6 110	160$^{0.001}$	3-06-00-02828	ace 3; chl 3; MeOH 3
4105	Cholestane, (5α)- 28,29,30-Trinorlanostane	C$_{27}$H$_{48}$ 372.68	481-21-0 80	2202 250^1	4-05-00-01227 0.9090^{88}	H$_2$O 1; EtOH 2; eth 4; bz 4 1.4887^{88}
4106	Cholestane, (5β)- Coprostane	C$_{27}$H$_{48}$ 372.68	481-20-9 72	2518	4-05-00-01226 0.9119^{87}	eth 4; chl 4 1.4884^{88}
4107	Cholestanol Dihydrocholesterol	C$_{27}$H$_{48}$O 388.68	80-97-7 141.5	2203	4-06-00-03577	eth 4; chl 4
4108	Cholestan-3-ol, (3α,5α)- Epicholestanol	C$_{27}$H$_{48}$O 388.68	516-95-0 185.5	3564	4-06-00-03577	chl 3
4109	Cholestan-3-ol, (3β,5β)- Coprosterol	C$_{27}$H$_{48}$O 388.68	360-68-9 102	2519	4-06-00-03576	H$_2$O 1; EtOH 4; eth 4; bz 4
4110	Cholest-5-en-3-ol (3β)- Cholesterol	C$_{27}$H$_{46}$O 386.66	57-88-5 148.5	2204 360 dec; 233$^{0.5}$	4-06-00-04000 1.067^{20}	H$_2$O 1; EtOH 2; eth 4; ace 2
4111	Cholest-4-en-3-ol, (3β)- Allocholesterol	C$_{27}$H$_{46}$O 386.66	517-10-2 132	253	3-06-00-02604	H$_2$O 1; EtOH 3; eth 4; ace 4
4112	Cholest-5-en-3-ol, (3α)- Epicholesterol	C$_{27}$H$_{46}$O 386.66	474-77-1 141.5	3565	4-06-00-04001	EtOH 2
4113	6-Chrysenamine 6-Aminochrysene	C$_{18}$H$_{13}$N 243.31	2642-98-0 210.5	2258	4-12-00-03492	
4114	Chrysene 1,2-Benzophenanthrene	C$_{18}$H$_{12}$ 228.29	218-01-9 258.2	2259 448	4-05-00-02554 1.274^{20}	H$_2$O 1; EtOH 2; eth 2; ace 2
4115	Chrysene, 5,6-dimethyl- 5,6-Dimethylchrysene	C$_{20}$H$_{16}$ 256.35	3697-27-6 129.3	200$^{0.5}$	4-05-00-02590	H$_2$O 1; EtOH 3; CS$_2$ 3; HOAc 3
4116	5,6-Chrysenedione	C$_{18}$H$_{10}$O$_2$ 258.28	2051-10-7 239.5	sub •	4-07-00-02646	H$_2$O 1; EtOH 3; eth 2; bz 3
4117	Chrysene, 1-methyl- 1-Methylchrysene	C$_{19}$H$_{14}$ 242.32	3351-28-8 256.5	sub 140	4-05-00-02572	H$_2$O 1; EtOH 3
4118	Chrysene, octadecahydro-	C$_{18}$H$_{30}$ 246.44	2090-14-4 115	353	3-05-00-01107	EtOH 4
4119	Cinchonan-6',9-diol, (8α,9R)- Cupreine	C$_{19}$H$_{22}$N$_2$O$_2$ 310.40	524-63-0 202	2626	4-23-00-03260	EtOH 4
4120	Cinchonan-9-ol, (9S)- Cinchonine	C$_{19}$H$_{22}$N$_2$O 294.40	118-10-5 265	2289	5-23-12-00406	
4121	Cinchonan-9-ol, 10,11-dihydro-, (8α,9R)- Hydrocinchonidine	C$_{19}$H$_{24}$N$_2$O 296.41	485-64-3 229	4705		EtOH 4
4122	Cinchonan-9-ol, 10,11-dihydro-, (9S)- Hydrocinchonine	C$_{19}$H$_{24}$N$_2$O 296.41	485-65-4 268.5	4706		H$_2$O 3; EtOH 2; eth 1
4123	Cinchonan-9-ol, 10,11-dihydro-6'-methoxy-, (8α,9R)- Hydroquinine	C$_{20}$H$_{26}$N$_2$O$_2$ 326.44	522-66-7 172.5	4737	5-23-13-00340	ace 4; eth 4; EtOH 4; chl 4
4124	Cinchonan-9-ol, 10,11-dihydro-6'-methoxy-, (9S)- Hydroquinidine	C$_{20}$H$_{26}$N$_2$O$_2$ 326.44	1435-55-8 168.5	4736	5-23-13-00340	EtOH 3; eth 3; ace 3; chl 3
4125	Cinchonan-9-ol, 6'-methoxy-, (9R)- Epiquinidine	C$_{20}$H$_{24}$N$_2$O$_2$ 324.42	572-59-8 113	3570	0-23-00-00505	EtOH 4; eth 3
4126	Cinchonan-9-ol, 6'-methoxy-, (9S)- Quinidine	C$_{20}$H$_{24}$N$_2$O$_2$ 324.42	56-54-2 174	8072	5-23-13-00395	H$_2$O 2; EtOH 3; eth 2; bz 3

No.	Name / Synonym	Mol. Form. / Mol. Wt.	CAS RN / mp/°C	Merck No. / bp/°C	Beil. Ref. / den/g cm^{-3}	Solubility / n_D
4127	Cinchonan-9-ol, 6'-methoxy-, (8α,9R)- Quinine	$C_{20}H_{24}N_2O_2$ 324.42	130-95-0 57	8075	5-23-13-00395	H_2O 2; EtOH 4; eth 3; ace 2 1.625[15]
4128	Cinchonan-9-ol, 6'-methoxy-, carbonate (2:1) (esteR), (8α,9R)- Quinine carbonate	$C_{41}H_{46}N_4O_5$ 674.84	146-06-5 189	8077	4-23-00-03282	
4129	Cinchonan-9-ol, 6'-methoxy-, (8α,9R)-, monoformate (salt) Quinine, formate	$C_{21}H_{26}N_2O_4$ 370.45	130-90-5 149.5	8081	4-23-00-03274	H_2O 4; EtOH 4; chl 4
4130	Cinchonan-9-ol, 6'-methoxy-, monohydrochloride, (8α,9R)- Quinine hydrochloride	$C_{20}H_{25}ClN_2O_2$ 360.88	130-89-2 159	8085		H_2O 4; EtOH 4; chl 4
4131	Cinchonan-9-ol, 6'-methoxy-, (8α,9R)-, mono(2-hydroxybenzoate) (salt) Quinine salicylate	$C_{27}H_{32}N_2O_6$ 480.56	750-90-3 195	8088	4-23-00-03277	EtOH 4; chl 4
4132	Cinchonan-9-ol, 6'-methoxy-, (8α,9R)-, sulfate (2:1) (salt) Quinine sulfate	$C_{40}H_{50}N_4O_8S$ 746.93	804-63-7 235.2	8089		EtOH 4
4133	Cinchonan-9-ol, (8α,9R)- Cinchonidine	$C_{19}H_{22}N_2O$ 294.40	485-71-2 210.5	2288 sub	5-23-12-00406	H_2O 1; EtOH 3; eth 2; bz 1
4134	Cinchonan-9-one, 6'-methoxy-, (8α)- Quininone	$C_{20}H_{22}N_2O_2$ 322.41	84-31-1 108	8093	4-25-00-00204	bz 4; eth 4; EtOH 4
4135	Cinnamic acid, β-propyl- 2-Hexenoic acid, 3-phenyl-	$C_{12}H_{14}O_2$ 190.24	4362-01-0 94	183-4[14]	3-09-00-02811	bz 4
4136	Cinnoline 1,2-Benzodiazine	$C_8H_6N_2$ 130.15	253-66-7 38	2309 114[0.3]	5-23-07-00124	eth 4; EtOH 4
4137	Ciodrin 2-Butenoic acid, 3-[(dimethoxyphosphinyl)oxy]-, 1-phenylethyl ester, (E)-	$C_{14}H_{19}O_6P$ 314.28	7700-17-6 135[0.03]	2603	1.19[25]	
4138	Clethodim	$C_{17}H_{26}ClNO_3S$ 359.92	99129-21-2		1.14[20]	
4139	Clofentezine 3,6-Bis(2-chlorophenyl)-1,2,4,5-tetrazine	$C_{14}H_8Cl_2N_4$ 303.15	74115-24-5 182	2373		
4140	Clomazone 2-(2-Chlorobenzyl)-4,4-dimethyl-1,2-oxazolidin-3-one	$C_{12}H_{14}ClNO_2$ 239.70	81777-89-1	3203	1.192[20]	
4141	Cloprop Propanoic acid, 2-(3-chlorophenoxy)-	$C_9H_9ClO_3$ 200.62	101-10-0			
4142	Clopyralid 3,6-Dichloro-2-pyridinecarboxylic acid	$C_6H_3Cl_2NO_2$ 192.00	1702-17-6 151	2398		
4143	Con-5-enin-3-amine, N,N-dimethyl-, (3β)- Conessine	$C_{24}H_{40}N_2$ 356.60	546-06-5 125.5	2492 165-7[0.1]		H_2O 2; chl 3; HOAc 3
4144	Conhydrine, N-methyl- N-Methylconhydrine	$C_9H_{19}NO$ 157.26	94826-08-1 94-5[11]		2-21-00-00011 0.9400[20]	1.4708[19]
4145	Coronene	$C_{24}H_{12}$ 300.36	191-07-1 437.3	525	4-05-00-02830 1.371[25]	H_2O 1; bz 2; con sulf 1
4146	Corynan-16-carboxylic acid, 16,17,18,19-tetrahydro-17-methoxy-, methyl ester, (16E)- Corynantheine	$C_{22}H_{26}N_2O_3$ 366.46	18904-54-6 165.5	2546	4-25-00-01255	EtOH 4
4147	Coumaphos Coumarin, 3-chloro-7-hydroxy-4-methyl, O-ester with O,O-diethylphosphorothioate	$C_{14}H_{16}ClO_5PS$ 362.77	56-72-4 93	2559	1.474	
4148	o-Cresol, 4-(3-methylpentyl)- Phenol, 2-methyl-4-(3-methyl pentyl)	$C_{13}H_{20}O$ 192.30	882-25-7 35	145-6[11]		ace 4 1.5200[25]
4149	Crotonaldehyde, 2-bromo-, diethyl acetal 2-Butenal, 2-bromo, diethyl acetal	$C_8H_{15}BrO_2$ 223.11	98551-14-5 86[15]		1-01-00-00380 1.2255[21]	1.4565[21]
4150	Crotonic acid, 2-chloro- 2-Chlorocrotonic acid	$C_4H_5ClO_2$ 120.54	600-13-5 99.8	212; 111[14]	4-02-00-01509	eth 4; EtOH 4
4151	Crotonic acid, 4-chloro-, (E)- 2-Butenoic acid, 4-chloro (trans)	$C_4H_5ClO_2$ 120.54	26340-58-9 83	117-8[13]	4-02-00-01511	eth 4
4152	Crotonic acid, 4,4-dichloro- 4,4-Dichlorocrotonic acid	$C_4H_4Cl_2O_2$ 154.98	99980-00-4 100.5	130[18]	4-02-00-01512 1.3331[99]	eth 4; EtOH 4; chl 4 1.4597[99]
4153	Crufomate Phosphoramidic acid, methyl-, 2-chloro-4-(1,1-dimethylethyl)phenyl methyl ester	$C_{12}H_{19}ClNO_3P$ 291.71	299-86-5 60	2607 118[0.01]		
4154	Curan-17-ol, (16α)- Geissoschizoline	$C_{19}H_{26}N_2O$ 298.43	18397-07-4 135 dec	4274	5-23-12-00211	H_2O 1; EtOH 4; eth 4; chl 4
4155	Cyanamide Cyanogenamide	CH_2N_2 42.04	420-04-2 44	2691 140[19]	4-03-00-00145 1.282[20]	H_2O 4; EtOH 4; eth 3; ace 3 1.4418[48]
4156	Cyanamide, diethyl-	$C_5H_{10}N_2$ 98.15	617-83-4 -80.6	188	4-04-00-00381 0.854[20]	H_2O 1; EtOH 3; eth 3 1.4126[25]
4157	Cyanamide, dimethyl-	$C_3H_6N_2$ 70.09	1467-79-4	163.5	4-04-00-00226	ace 4; eth 4; EtOH 4 1.4089[19]
4158	Cyanamide, diphenyl- Diphenylcyanamide	$C_{13}H_{10}N_2$ 194.24	27779-01-7 73.5	237[60]	4-12-00-00857	EtOH 4; lig 4
4159	Cyanamide, di-2-propenyl- Diallylcyanamide	$C_7H_{10}N_2$ 122.17	538-08-9 142[90]; 95[9]	2952	4-04-00-01078	EtOH 3; eth 2; ctc 2
4160	Cyanamide, methylphenyl-	$C_8H_8N_2$ 132.17	18773-77-8 32	137[12]	4-12-00-00840 1.0400[55]	

No.	Name / Synonym	Mol. Form. / Mol. Wt.	CAS RN / mp/°C	Merck No. / bp/°C	Beil. Ref. / den/g cm^{-3}	Solubility / n_D
4161	Cyanazine / Propanenitrile, 2-[[4-chloro-6-(ethylamino)-1,3,5-triazin-2-yl]amino]-2-methyl-	$C_9H_{13}ClN_6$ / 240.70	21725-46-2 / 168	2692		
4162	Cyanic acid / Hydrogen cyanate (HOCN)	CHNO / 43.03	420-05-3 / -86	2693 / 23	4-03-00-00080 / 1.140[20]	H$_2$O 4; bz 4; eth 4; chl 4
4163	Cyanic acid, ethyl ester / Ethyl cyanate	C_3H_5NO / 71.08	627-48-5	/ 162 dec; 30[12]	0.89[20]	EtOH 4 / 1.3788[25]
4164	Cyanic acid, sodium salt / Sodium cyanate	CNNaO / 65.01	917-61-3 / 550	8552	1.893[20]	H$_2$O 3; EtOH 2; eth 1
4165	Cyanogen bromide / Bromine cyanide	CBrN / 105.92	506-68-3 / 52	2700 / 61.5	4-03-00-00092 / 2.015[20]	H$_2$O 3; EtOH 3; eth 3
4166	Cyanogen chloride / Chlorine cyanide	CClN / 61.47	506-77-4 / -6.5	2701 / 13	4-03-00-00090 / 1.186[20]	H$_2$O 3; EtOH 3; eth 4
4167	Cyanogen fluoride / Fluorine cyanide	CFN / 45.02	1495-50-7 / -82	/ -46		
4168	Cycloate / Carbamothioic acid, cyclohexylethyl-, S-ethyl ester	$C_{11}H_{21}NOS$ / 215.36	1134-23-2 / 11.5	/ 145[10]	1.0156[30]	
4169	Cyclobutanamine / Aminocyclobutane	C_4H_9N / 71.12	2516-34-9 /	/ 82	4-12-00-00003 / 0.8328[20]	1.4363[19]
4170	Cyclobutane / Tetramethylene	C_4H_8 / 56.11	287-23-0 / -90.6	2720 / 12.6	4-05-00-00006 / 0.6890[25]	H$_2$O 1; EtOH 4; eth 5; ace 4 / 1.365290
4171	Cyclobutaneacetic acid, 3-(ethoxycarbonyl)-2,2-dimethyl-, ethyl ester / Pinic acid diethyl ester	$C_{13}H_{22}O_4$ / 242.32	28664-03-1 /	/ 142-6[10]	2-09-00-00529 / 1.0104[20]	1.4496[70]
4172	Cyclobutane, 1,2-bis(methylene)-	C_6H_8 / 80.13	14296-80-1 /	/ 64	4-05-00-00389 / 0.7698[20]	1.4232[20]
4173	Cyclobutanecarbonitrile / Cyanocyclobutane	C_5H_7N / 81.12	4426-11-3 /	/ 149.6		
4174	Cyclobutanecarboxylic acid / Cyclobutylcarboxylic acid	$C_5H_8O_2$ / 100.12	3721-95-7 / -2	/ 190; 74-5[2]	4-09-00-00006 / 1.0599[20]	H$_2$O 2; EtOH 5; eth 5 / 1.4400[20]
4175	Cyclobutanecarboxylic acid, ethyl ester / Ethyl cyclobutanecarboxylate	$C_7H_{12}O_2$ / 128.17	14924-53-9 /	/ 159	3-09-00-00007 / 0.928[25]	ctc 2 / 1.4261[20]
4176	Cyclobutanecarboxylic acid, methyl ester / Methyl cyclobutanecarboxylate	$C_6H_{10}O_2$ / 114.14	765-85-5 /	/ 135.5		
4177	Cyclobutane, 1-chloro-1,2,2,3,3,4,4-heptafluoro- / Chloroheptafluorocyclobutane	C_4ClF_7 / 216.49	377-41-3			
4178	Cyclobutane, 1,1-dibromo- / 1,1-Dibromocyclobutane	$C_4H_6Br_2$ / 213.90	33742-81-3 /	/ 160	0-05-00-00017 / 1.930[20]	bz 4; EtOH 4; lig 4 / 1.5362[20]
4179	Cyclobutane, 1,2-dibromo- / 1,2-Dibromocyclobutane	$C_4H_6Br_2$ / 213.90	89033-70-5 / 2.5	/ 172.5	0-05-00-00017 / 1.972[0]	
4180	1,1-Cyclobutanedicarboxylic acid, diethyl ester / Diethyl 1,1-cyclobutanedicarboxylate	$C_{10}H_{16}O_4$ / 200.23	3779-29-1 /	/ 224	4-09-00-02788 / 1.0456[20]	EtOH 4; ctc 2 / 1.4330[26]
4181	1,2-Cyclobutanedicarboxylic acid, dimethyl ester, trans- / Dimethyl trans-1,2-cyclobutanedicarboxylate	$C_8H_{12}O_4$ / 172.18	7371-67-7 /	/ 225; 118[24]	3-09-00-03799 / 1.1276[20]	1.4450[20]
4182	1,3-Cyclobutanedicarboxylic acid, cis-	$C_6H_8O_4$ / 144.13	2398-16-5 / 143.5	/ 252	3-09-00-03801	H$_2$O 4; EtOH 4
4183	1,3-Cyclobutanedicarboxylic acid, 2,2-dimethyl-, cis- / cis-Norpinic acid	$C_8H_{12}O_4$ / 172.18	3211-48-1 / 175	/ sub 100	4-09-00-02813	eth 4; chl 4
4184	1,3-Cyclobutanedicarboxylic acid, 2,2-dimethyl-, dimethyl ester / Norpinic acid, dimethyl ester	$C_{10}H_{16}O_4$ / 200.23	91057-79-3 /	/ 229; 113[11]	3-09-00-03829 / 1.070[14]	1.4459[17]
4185	1,3-Cyclobutanedicarboxylic acid, trans-	$C_6H_8O_4$ / 144.13	7439-33-0 / 171	/ sub	3-09-00-03802	H$_2$O 4; EtOH 4
4186	Cyclobutane, 1,2-diethenyl-, cis-	C_8H_{12} / 108.18	16177-46-1 /	/ 38[38]	4-05-00-00412 / 0.8010[20]	1.4563[20]
4187	Cyclobutane, 1,2-diethenyl-, trans- / trans-1,2-Divinylcyclobutane	C_8H_{12} / 108.18	6553-48-6 /	/ 112.5	4-05-00-00412 / 0.7817[20]	1.4451[20]
4188	Cyclobutane, 1,2-dimethyl-, trans- / trans-1,2-Dimethylcyclobutane	C_6H_{12} / 84.16	15679-02-4 / -122.5	/ 58	4-05-00-00088 / 0.703[20]	1.395[20]
4189	Cyclobutane, 1,3-dimethyl-, cis- / cis-1,3-Dimethylcyclobutane	C_6H_{12} / 84.16	2398-09-6 /	/ 60.5	4-05-00-00089 / 0.7060[25]	1.3933[20]
4190	Cyclobutane, 1,3-dimethyl-, trans- / trans-1,3-Dimethylcyclobutane	C_6H_{12} / 84.16	2398-10-9 /	/ 57.5	4-05-00-00089 / 0.6970[25]	1.3896[20]
4191	Cyclobutane, ethyl- / Ethylcyclobutane	C_6H_{12} / 84.16	4806-61-5 / -142.9	/ 70.8	4-05-00-00087 / 0.7284[20]	H$_2$O 1; EtOH 5; eth 5; ace 3 / 1.4020[20]
4192	Cyclobutane, methyl- / Methylcyclobutane	C_5H_{10} / 70.13	598-61-8 / -161.5	/ 36.3	4-05-00-00021 / 0.6884[20]	H$_2$O 1; EtOH 5; eth 5; ace 3 / 1.3866[20]
4193	Cyclobutane, methylene- / exo-Methylenecyclobutane	C_5H_8 / 68.12	1120-56-5 / -134.7	/ 42.2	4-05-00-00216 / 0.7401[20]	1.4210[20]
4194	Cyclobutane, octafluoro- / Perfluorocyclobutane	C_4F_8 / 200.03	115-25-3 / -40.1	6666 / -5.9	4-05-00-00008 / *1.500[25]	eth 3
4195	Cyclobutane, 1,2,3,4-tetrabromo- / 1,2,3,4-Tetrabromocyclobutane	$C_4H_4Br_4$ / 371.69	101257-79-8 /	/ 110[1.5]	4-05-00-00012 / 2.5672[20]	1.6303[20]
4196	Cyclobutanol / Hydroxycyclobutane	C_4H_8O / 72.11	2919-23-5 /	/ 124	4-06-00-00003 / 0.9218[15]	1.4371[20]

No.	Name / Synonym	Mol. Form. / Mol. Wt.	CAS RN / mp/°C	Merck No. / bp/°C	Beil. Ref. / den/g cm^{-3}	Solubility / n_D
4197	Cyclobutanone	C_4H_6O / 70.09	1191-95-3 / -50.9	99	4-07-00-00003 / 0.9547[0]	H_2O 3; EtOH 4; eth 3; bz 3 / 1.4215[20]
4198	Cyclobutene	C_4H_6 / 54.09	822-35-5 /	2	4-05-00-00207 / 0.733[0]	ace 4; bz 3; peth 3
4199	Cyclobutene, 1,2-dichloro-3,3,4,4-tetrafluoro-	$C_4Cl_2F_4$ / 194.94	377-93-5 / -43.4	67	4-05-00-00209 / 1.534[25]	1.3699[25]
4200	Cyclobutene, hexafluoro-	C_4F_6 / 162.03	697-11-0 / -60	5.5	4-05-00-00208 / 1.602[-20]	1.298[-20]
4201	Cyclobutene, 1-methyl- / 1-Methylcyclobutene	C_5H_8 / 68.12	1489-60-7 /	37	4-05-00-00216 / 0.7244[20]	1.4088[18]
4202	Cyclodecane	$C_{10}H_{20}$ / 140.27	293-96-9 / 10	202	4-05-00-00144 / 0.8538[25]	1.4716[20]
4203	1,2-Cyclodecanedione / Sebacil	$C_{10}H_{18}O_2$ / 168.24	96-01-5 / 40.5	8370 / 104-5[10]	4-07-00-02009	
4204	Cyclodecanol	$C_{10}H_{20}O$ / 156.27	1502-05-2 / 40.5	125[12]	4-06-00-00138 / 0.9606[20]	EtOH 3 / 1.4926[20]
4205	Cyclodecanone	$C_{10}H_{18}O$ / 154.25	1502-06-3 / 28	106-7[13]	4-07-00-00076 / 0.9654[20]	bz 4; eth 4; chl 4 / 1.4806[20]
4206	Cyclodecanone, 2-hydroxy- / Sebacoin	$C_{10}H_{18}O_2$ / 170.25	96-00-4 / 38.5	8371 / 134-8[14]	4-08-00-00037	
4207	Cyclodecene, (E)-	$C_{10}H_{18}$ / 138.25	2198-20-1 /	68-70[10]	4-05-00-00296 / 0.8672[20]	H_2O 1; bz 3; chl 3 / 1.4822[20]
4208	Cyclodecene, (Z)-	$C_{10}H_{18}$ / 138.25	935-31-9 /	195	4-05-00-00295 / 0.8770[20]	1.4854[20]
4209	Cyclodecyne	$C_{10}H_{16}$ / 136.24	3022-41-1 /	204	4-05-00-00430 / 0.8975[20]	1.4903[20]
4210	Cyclododecane	$C_{12}H_{24}$ / 168.32	294-62-2 / 60.4	247	4-05-00-00169 / 0.82[80]	
4211	Cyclododecanone / Cyclodoecanone	$C_{12}H_{22}O$ / 182.31	830-13-7 / 59	126-8[12]	4-07-00-00100 / 0.9059[66]	1.4571[60]
4212	1,5,9-Cyclododecatriene / CDT	$C_{12}H_{18}$ / 162.27	4904-61-4 / -17	240	4-05-00-01114 / 0.84[100]	
4213	1,5,9-Cyclododecatriene, (Z,E,E)- / 1,5,9-Cyclododecatriene, (cis, trans, trans)	$C_{12}H_{18}$ / 162.27	53859-78-2 / -18	100-1[11]	4-05-00-01115 / 0.8910[20]	bz 4; chl 4 / 1.5058[20]
4214	1,5,9-Cyclododecatriene, (Z,Z,Z)-	$C_{12}H_{18}$ / 162.27	4736-48-5 / -1.5	108[17]		bz 4; chl 4 / 1.5100[25]
4215	Cyclododecene, (E)-	$C_{12}H_{22}$ / 166.31	1486-75-5 /	112-5[17]	4-05-00-00332	bz 4; chl 4 / 1.4850[20]
4216	Cyclododecene, (Z)-	$C_{12}H_{22}$ / 166.31	1129-89-1 /	133[35]; 71[2]	4-05-00-00332	bz 4; chl 4 / 1.4840[20]
4217	9-Cycloheptadecen-1-one / 9-Cycloheptadecene-1-one	$C_{17}H_{30}O$ / 250.42	74244-64-7 / 32.5	343; 159[2]	3-07-00-00524 / 0.9170[33]	H_2O 2; EtOH 3; bz 3 / 1.4830[33]
4218	9-Cycloheptadecen-1-one, (Z)- / Civetone	$C_{17}H_{30}O$ / 250.42	542-46-1 / 32.5	2337 / 343; 159[2]	3-07-00-00524	
4219	1,3-Cycloheptadiene	C_7H_{10} / 94.16	4054-38-0 / -110.4	120.5	4-05-00-00390 / 0.868[25]	1.4978[20]
4220	2,4-Cycloheptadien-1-one, 2,6,6-trimethyl- / Eucarvone	$C_{10}H_{14}O$ / 150.22	503-93-5 /	105[22]; 88[16]	4-07-00-00311 / 0.9490[20]	eth 3; ace 3 / 1.5087[20]
4221	Cycloheptane	C_7H_{14} / 98.19	291-64-5 / -8.0	118.4	4-05-00-00092 / 0.8098[20]	H_2O 1; EtOH 4; eth 4; bz 3 / 1.4436[20]
4222	Cycloheptane, bromo- / Cycloheptyl bromide	$C_7H_{13}Br$ / 177.08	2404-35-5 /	101.5[40]; 75[12]	4-05-00-00093 / 1.2887[22]	H_2O 1; eth 4; chl 4 / 1.4996[20]
4223	Cycloheptanecarboxylic acid / Cycloheptylcarboxylic acid	$C_8H_{14}O_2$ / 142.20	1460-16-8 /	259; 130[8]	4-09-00-00038 / 1.0423[20]	ace 3 / 1.4753[20]
4224	Cycloheptanecarboxylic acid, ethyl ester / Ethyl cycloheptanecarboxylate	$C_{10}H_{18}O_2$ / 170.25	32777-26-7 /	74[3]	4-09-00-00038 / 0.9515[20]	1.4482[20]
4225	1,2-Cycloheptanedione	$C_7H_{10}O_2$ / 126.16	3008-39-7 / -40	107-9[17]	4-07-00-01989 / 1.0583[22]	EtOH 3 / 1.4689[22]
4226	Cycloheptane, methyl- / Methylcycloheptane	C_8H_{16} / 112.22	4126-78-7 /	134	4-05-00-00114 / 0.800[20]	bz 4; eth 4; EtOH 4; peth 4 / 1.439[18]
4227	Cycloheptane, 1,1,2-trimethyl- / 1,1,2-Trimethylcycloheptane	$C_{10}H_{20}$ / 140.27	35099-89-9 /	104-5[100]	3-05-00-00127 / 0.8243[20]	1.4527[20]
4228	Cycloheptane, 1,1,4-trimethyl- / 1,1,4-Trimethylcycloheptane	$C_{10}H_{20}$ / 140.27	2158-55-6 /	164	4-05-00-00146 / 0.8011[20]	1.4420[20]
4229	Cycloheptanol	$C_7H_{14}O$ / 114.19	502-41-0 / 2	185	4-06-00-00094 / 0.9554[20]	H_2O 2; EtOH 4; eth 4 / 1.40705[20]
4230	Cycloheptanol, 1-methyl- / 1-Methyl-1-cycloheptanol	$C_8H_{16}O$ / 128.21	3761-94-2 /	183.5	4-06-00-00114 / 0.9392[21]	1.4677[22]
4231	Cycloheptanol, 2-methyl- / 2-Methylcycloheptanol	$C_8H_{16}O$ / 128.21	59777-92-3 /	191	3-06-00-00083 / 0.9492[15]	1.4762[15]
4232	Cycloheptanol, 4-methyl-	$C_8H_{16}O$ / 128.21	90200-61-6 /	105-6[39]	3-06-00-00084 / 0.9134[31]	1.4574
4233	Cycloheptanol, 2,6,6-trimethyl- / 3,3,7-Trimethylcycloheptanol	$C_{10}H_{20}O$ / 156.27	33515-82-1 /	216	1-06-00-00018 / 0.9096[24]	1.4369
4234	Cycloheptanone / Suberone	$C_7H_{12}O$ / 112.17	502-42-1 /	2728 / 178.5	4-07-00-00039 / 0.9508[20]	H_2O 1; EtOH 4; eth 3 / 1.4608[20]
4235	Cycloheptanone, 2-methyl- / 2-Methylcycloheptanone	$C_8H_{14}O$ / 126.20	932-56-9 /	185	4-07-00-00051 / 0.9395[18]	1.461[18]
4236	Cycloheptanone, 2-phenyl- / 2-Phenylcycloheptanone	$C_{13}H_{16}O$ / 188.27	14996-78-2 / 22.2	141[5]; 125[2]	4-07-00-01061	1.5395[20]

No.	Name Synonym	Mol. Form. Mol. Wt.	CAS RN	mp/°C bp/°C	Merck No.	Beil. Ref. den/g cm^{-3}	Solubility n_D
4237	Cycloheptanone, 2,2,6-trimethyl-	$C_{10}H_{18}O$ 154.25	1686-41-5	 191		3-07-00-00138 0.878[14]	1.4421[16]
4238	Cycloheptanone, 2,6,6-trimethyl- Tetrahydroeucarvone	$C_{10}H_{18}O$ 154.25	4436-59-3	 208.5		4-07-00-00079 0.9095[18]	1.4568[18]
4239	Cycloheptasiloxane, tetradecamethyl- Tetradecamethylcycloheptasiloxane	$C_{14}H_{42}O_7Si_7$ 519.08	107-50-6	-26 154[20]		4-04-00-04130 0.9703[20]	1.4040[20]
4240	1,3,5-Cycloheptatriene Tropilidene	C_7H_8 92.14	544-25-2	-79.5 117; 60.5[122]		4-05-00-00765 0.8875[19]	H_2O 1; EtOH 3; eth 3; bz 4 1.5343[20]
4241	1,3,6-Cycloheptatrien-1-carboxylic acid, 5,5-dimethyl- Thujic acid	$C_{10}H_{12}O_2$ 164.20	499-89-8		9325	4-09-00-01810	
4242	2,4,6-Cycloheptatrien-1-one	C_7H_6O 106.12	539-80-0	-7 113[15]; 84[6]		4-07-00-00501 1.095[22]	bz 4; chl 4 1.6172[22]
4243	2,4,6-Cycloheptatrien-1-one, 2-hydroxy-	$C_7H_6O_2$ 122.12	533-75-5	50.8 sub 40		4-08-00-00159	H_2O 3; eth 3; ace 3; os 3
4244	Cycloheptene	C_7H_{12} 96.17	628-92-2	-56 115		4-05-00-00244 0.8228[20]	H_2O 1; EtOH 3; eth 3; bz 3 1.4552[20]
4245	Cycloheptene, 1-methyl- 1-Methylcycloheptene	C_8H_{14} 110.20	1453-25-4	 136		4-05-00-00265 0.8243[22]	1.4581[20]
4246	Cycloheptene, 5-methyl- 5-Methylcycloheptene	C_8H_{14} 110.20	2505-06-8	 69-70[38]		3-05-00-00211 0.7606[31]	1.4202[31]
4247	Cycloheptene, 1,4,4-trimethyl- 1,4,4-Trimethyl-1-cycloheptene	$C_{10}H_{18}$ 138.25	4755-36-6	 165		3-05-00-00232 0.8185[20]	1.4561[20]
4248	1,3-Cyclohexadiene	C_6H_8 80.13	592-57-4	-89 80.5		4-05-00-00382 0.8405[20]	H_2O 1; EtOH 3; eth 4; bz 3 1.4755[20]
4249	1,4-Cyclohexadiene 1,4-Dihydrobenzene	C_6H_8 80.13	628-41-1	-49.2 85.5		4-05-00-00385 0.8471[20]	H_2O 1; EtOH 5; eth 3; bz 3 1.4725[20]
4250	1,3-Cyclohexadiene-1-carboxaldehyde, 2,6,6-trimethyl- Safranal	$C_{10}H_{14}O$ 150.22	116-26-7	 70[1]	8286	4-07-00-00318 0.9734[19]	EtOH 4; peth 4 1.5281[19]
4251	1,4-Cyclohexadiene-1,2-dicarbonitrile, 4,5-dichloro-3,6-dioxo- 2,3-Dichloro-5,6-dicyanobenzoquinone	$C_8Cl_2N_2O_2$ 227.01	84-58-2	 214.5	3052	0-10-00-00902	bz 4; HOAc 4; diox 4
4252	1,3-Cyclohexadiene, 1,3-dimethyl- 1,3-Dimethyl-1,3-cyclohexadiene	C_8H_{12} 108.18	4573-05-1	 135.5		3-05-00-00325 0.8373[20]	1.4776[20]
4253	1,3-Cyclohexadiene, 1,4-dimethyl- 1,4-Dimethyl-1,3-cyclohexadiene	C_8H_{12} 108.18	26120-52-5	 136.5		4-05-00-00409 0.830[20]	1.4792[19]
4254	1,3-Cyclohexadiene, 1,5-dimethyl- 1,5-Dimethyl-1,3-cyclohexadiene	C_8H_{12} 108.18	1453-17-4	 130		1-05-00-00063 0.821[20]	1.471[20]
4255	1,3-Cyclohexadiene, 2,5-dimethyl- 2,5-Dimethyl-1,3-cyclohexadiene	C_8H_{12} 108.18	2050-33-1	 127		3-05-00-00325 0.8245[18]	1.4631[18]
4256	1,3-Cyclohexadiene, 2,6-dimethyl- 1,3-Cyclohexadiene, 3,5-dimethyl	C_8H_{12} 108.18	2050-32-0	 131		0-05-00-00119 0.8225[20]	1.4675[20]
4257	2,5-Cyclohexadiene-1,4-dione p-Benzoquinone	$C_6H_4O_2$ 108.10	106-51-4	115.7 sub	8103	4-07-00-02065 1.318[20]	H_2O 2; EtOH 3; eth 3; chl 3
4258	2,5-Cyclohexadiene-1,4-dione, 2,5-dichloro-3,6-dihydroxy- Chloranilic acid	$C_6H_2Cl_2O_4$ 208.98	87-88-7	283.5	2072	4-08-00-02707	H_2O 3
4259	2,5-Cyclohexadiene-1,4-dione, 2,5-dihydroxy-3,6-dinitro- Nitranilic acid	$C_6H_2N_2O_8$ 230.09	479-22-1	100 dec exp 170	6489	3-08-00-03351	H_2O 4; EtOH 4; eth 1
4260	2,5-Cyclohexadiene-1,4-dione, 2,5-dihydroxy-3-methoxy-6-methyl- Spinulosin	$C_8H_8O_5$ 184.15	85-23-4	202.5 sub 120	8706	3-08-00-03983	H_2O 2; alk 3
4261	2,5-Cyclohexadiene-1,4-dione, 2,5-dihydroxy-3-undecyl- Embelin	$C_{17}H_{26}O_4$ 294.39	550-24-3	142.5	3513	4-08-00-02769	bz 4; eth 4; EtOH 4
4262	2,5-Cyclohexadiene-1,4-dione, 2,6-dimethoxy- 2,6-Dimethoxy-p-quinone	$C_8H_8O_4$ 168.15	530-55-2	256 sub	3214	4-08-00-02710	H_2O 2; EtOH 2; eth 2; tfa 3
4263	2,5-Cyclohexadiene-1,4-dione, 2,3-dimethyl-	$C_8H_8O_2$ 136.15	526-86-3	55 sub		4-07-00-02093	H_2O 2; EtOH 3; eth 3; chl 3
4264	2,5-Cyclohexadiene-1,4-dione, 2,6-dimethyl-	$C_8H_8O_2$ 136.15	527-61-7	72.5 sub		4-07-00-02095 1.0479[28]	chl 3
4265	2,5-Cyclohexadiene-1,4-dione, 3-hydroxy-2-methoxy-5-methyl- Fumigatin	$C_8H_8O_4$ 168.15	484-89-9	116	4201	3-08-00-03374	ace 4; bz 4; eth 4; EtOH 4
4266	2,5-Cyclohexadiene-1,4-dione, 2-methyl-	$C_7H_6O_2$ 122.12	553-97-9	69 sub		4-07-00-02088 1.08[75]	H_2O 2; EtOH 3; eth 3
4267	2,5-Cyclohexadiene-1,4-dione, 2-methyl-5-(1-methylethyl)-	$C_{10}H_{12}O_2$ 164.20	490-91-5	45.5 232		4-07-00-02100	chl 3
4268	2,5-Cyclohexadiene-1,4-dione, 2,3,5,6-tetrachloro- Chloranil	$C_6Cl_4O_2$ 245.88	118-75-2	290 sub	2071	4-07-00-02083	H_2O 1; EtOH 2; eth 3; chl 2
4269	2,5-Cyclohexadiene-1,4-dione, 2,3,5,6-tetrahydroxy- Tetroquinone	$C_6H_4O_8$ 172.09	319-89-1		9177	4-08-00-03604	H_2O 2; EtOH 4; eth 2; ctc 2
4270	2,5-Cyclohexadiene-1,4-dione, 2,3,5,6-tetramethyl- Duroquinone	$C_{10}H_{12}O_2$ 164.20	527-17-3	111.5	3452	4-07-00-02102	H_2O 1; EtOH 3; eth 3; ace 3
4271	1,3-Cyclohexadiene, 2-methyl- 4,5-Dihydrotoluene	C_7H_{10} 94.16	1489-57-2	 107.5		3-06-00-00211 0.8260[18]	1.4662[18]

No.	Name / Synonym	Mol. Form. / Mol. Wt.	CAS RN / mp/°C	Merck No. / bp/°C	Beil. Ref. / den/g cm⁻³	Solubility / n_D
4272	1,3-Cyclohexadiene, 5-methyl-, (±)- 1,3-Cyclohexadiene, 5-methyl (DL)	C_7H_{10} 94.16	116781-86-3		3-05-00-00318 0.8354[20]	bz 4; eth 4; EtOH 4; lig 4 1.4763[20]
				101.5		
4273	1,3-Cyclohexadiene, 1-methyl-4-(1-methylethyl)-	$C_{10}H_{16}$ 136.24	99-86-5		4-05-00-00435 0.8375[19]	H_2O 1; EtOH 5; eth 5 1.477[19]
				175		
4274	1,3-Cyclohexadiene, 2-methyl-5-(1-methylethyl)-, (±)-	$C_{10}H_{16}$ 136.24	13811-01-3		3-05-00-00342 0.8410[22]	H_2O 1; EtOH 1; eth 3 1.4772[19]
				175.5		
4275	1,4-Cyclohexadiene, 1-methyl-4-(1-methylethyl)-	$C_{10}H_{16}$ 136.24	99-85-4		4-05-00-00436 0.849[20]	1.4765[14]
				183		
4276	1,3-Cyclohexadiene, 1,2,3,4,5,5,6,6-octafluoro-	C_8F_8 224.05	377-70-8		4-05-00-00384 1.601[20]	1.3149[20]
				63.5		
4277	1,4-Cyclohexadiene, 1,2,3,3,4,5,6,6-octafluoro-	C_8F_8 224.05	775-51-9		4-05-00-00386 1.6010[25]	1.318[18]
				57.5		
4278	2,5-Cyclohexadien-1-one, 4-[bis(4-hydroxyphenyl)methylene]- Aurin	$C_{19}H_{14}O_3$ 290.32	603-45-2 308-10 dec	899	4-08-00-02646	H_2O 1; EtOH 3; eth 2; bz 1
4279	2,5-Cyclohexadien-1-one, 2,6-dibromo-4-(chloroimino)- 2,6-Dibromoquinone-4-chlorimide	$C_6H_2Br_2CINO$ 299.35	537-45-1 83	3010	4-07-00-02086	EtOH 4
4280	2,5-Cyclohexadien-1-one, 2,6-dichloro-4-(chloroimino)- Gibbs' reagent	$C_8H_2Cl_3NO$ 210.45	101-38-2 66	4315	3-07-00-03377	
4281	2,4-Cyclohexadien-1-one, 3,5-dihydroxy-2,6,6-tris(3-methyl-2-butenyl)-4-(3-methyl-1-oxobutyl)- Lupulon	$C_{26}H_{38}O_4$ 414.59	468-28-0 93	5482	3-07-00-04753	H_2O 1; EtOH 3; peth 3; hx 3
4282	2,4-Cyclohexadien-1-one, 3,5,6-trihydroxy-4,6-bis(3-methyl-2-butenyl)-2-(3-methyl-1-oxobutyl)-, (R)- Humulon	$C_{21}H_{30}O_5$ 362.47	26472-41-3 66.5	4673	4-08-00-03410	H_2O 2; EtOH 3; eth 3; ace 3
4283	Cyclohexanamine Cyclohexylamine	$C_6H_{13}N$ 99.18	108-91-8 -17.7	2735 134	4-12-00-00008 0.8191[20]	H_2O 3; EtOH 4; eth 5; ace 5 1.4372[20]
4284	Cyclohexanamine, N-cyclohexyl- Dicyclohexylamine	$C_{12}H_{23}N$ 181.32	101-83-7 -0.1	3085 256 dec; 114[9]	4-12-00-00022 0.9123[20]	H_2O 2; EtOH 3; eth 3; bz 3 1.4842[20]
4285	Cyclohexanamine, N,N-diethyl-	$C_{10}H_{21}N$ 155.28	91-65-6	 193; 85[20]	4-12-00-00019 0.872[0]	EtOH 3; ctc 2
4286	Cyclohexanamine, N,N-dimethyl- Cyclohexyldimethylamine	$C_8H_{17}N$ 127.23	98-94-2	 162	4-12-00-00018	
4287	Cyclohexanamine, 2-ethyl- 2-Ethylcyclohexylamine	$C_8H_{17}N$ 127.23	6850-36-8	 170.5	3-12-00-00093 0.8744[20]	1.4682[20]
4288	Cyclohexanamine, N-ethyl-	$C_8H_{17}N$ 127.23	5459-93-8	 164	4-12-00-00018 0.868[0]	H_2O 2; EtOH 5; eth 5; ctc 2
4289	Cyclohexanamine, N-(2-ethylhexyl)-	$C_{14}H_{29}N$ 211.39	5432-61-1	 270; 140[14]	3-12-00-00018 0.8473[20]	ctc 2
4290	Cyclohexanamine, 1-ethynyl-	$C_8H_{13}N$ 123.20	30389-18-5	 65-6[20]	4-12-00-00214 0.913[25]	1.4817[20]
4291	Cyclohexanamine, N,N'-methanetetraylbis- Dicyclohexylcarbodiimide	$C_{13}H_{22}N_2$ 206.33	538-75-0 34.5	3086 123[6]; 99[0.5]	4-12-00-00072	
4292	Cyclohexanamine, 2-methyl-, cis-	$C_7H_{15}N$ 113.20	2164-19-4	 153.5	4-12-00-00118 0.8778[20]	1.4688[20]
4293	Cyclohexanamine, 2-methyl-, trans- trans-2-Methylcyclohexylamine	$C_7H_{15}N$ 113.20	931-10-2	 150	4-12-00-00118 0.8685[20]	1.4650[20]
4294	Cyclohexanamine, 3-methyl-, cis-	$C_7H_{15}N$ 113.20	1193-16-4	 153	4-12-00-00121 0.8552[20]	1.4538[20]
4295	Cyclohexanamine, 3-methyl-, trans-	$C_7H_{15}N$ 113.20	1193-17-5	 151.5	4-12-00-00121 0.8572[20]	1.4547[20]
4296	Cyclohexanamine, 4-methyl-, cis-	$C_7H_{15}N$ 113.20	2523-56-0	 153.5	4-12-00-00122 0.8567[20]	1.4559[20]
4297	Cyclohexanamine, 4-methyl-, trans-	$C_7H_{15}N$ 113.20	2523-55-9	 152	4-12-00-00122 0.8543[20]	1.4550[20]
4298	Cyclohexanamine, N-methyl-	$C_7H_{15}N$ 113.20	100-60-7	 147	4-12-00-00017 0.8660[23]	H_2O 2; EtOH 4; eth 5; chl 3 1.4560[20]
4299	Cyclohexanamine, 4,4'-methylenebis-	$C_{13}H_{26}N_2$ 210.36	1761-71-3 15	320	0.92[75]	
4300	Cyclohexanamine, 4,4'-methylenebis-, [cis(cis)]-	$C_{13}H_{26}N_2$ 210.36	6693-31-8 61	 141[2]	4-13-00-00031	H_2O 1 1.5014[27]
4301	Cyclohexanamine, 4,4'-methylenebis-, [trans(cis)]-	$C_{13}H_{26}N_2$ 210.36	6693-30-7 36.5	 127-8[1.2]	4-13-00-00031 0.9608[25]	H_2O 1 1.5046[25]
4302	Cyclohexanamine, 4,4'-methylenebis-, [trans(trans)]-	$C_{13}H_{26}N_2$ 210.36	6693-29-4 64.7	 131[0.8]	4-13-00-00031	H_2O 1 1.5032[25]
4303	Cyclohexanamine, N-(1-methylethyl)-	$C_9H_{19}N$ 141.26	1195-42-2	 60-5[12]	4-12-00-00019 0.859[25]	1.4480[20]

No.	Name / Synonym	Mol. Form. / Mol. Wt.	CAS RN / mp/°C	Merck No. / bp/°C	Beil. Ref. / den/g cm^{-3}	Solubility / n_D
4304	Cyclohexanamine, N-2-propenyl-	$C_9H_{17}N$ 139.24	6628-00-8	65-6[12]	4-12-00-00021 0.962[25]	1.4664[20]
4305	Cyclohexane Hexahydrobenzene	C_6H_{12} 84.16	110-82-7 6.6	2729 80.7	4-05-00-00027 0.7785[20]	H_2O 1; EtOH 5; eth 5; ace 5 1.4266[20]
4306	Cyclohexaneacetic acid Cyclohexylacetic acid	$C_8H_{14}O_2$ 142.20	5292-21-7 33	245	4-09-00-00038 1.0423[18]	H_2O 2; eth 3; ace 3; os 3 1.4775[20]
4307	Δ1,α-Cyclohexaneacetic acid, 3,3-dimethyl- Allocyclogeranic acid	$C_{10}H_{16}O_2$ 168.24	6790-47-2	145[14]	4-09-00-00161 0.9921[17]	1.4765[17]
4308	Cyclohexaneacetic acid, ethyl ester	$C_{10}H_{18}O_2$ 170.25	5452-75-5	211	4-09-00-00038 0.9537[14]	1.451[14]
4309	Cyclohexaneacetic acid, α-ethyl-1-hydroxy- Cyclobutyrol	$C_{10}H_{18}O_3$ 186.25	512-16-3 81.5	2721 164[24]	4-10-00-00049 1.0010[18]	ace 4; eth 4; EtOH 4; chl 4 1.4680[18]
4310	Cyclohexaneacetic acid, methyl ester Methyl cyclohexaneacetate	$C_9H_{16}O_2$ 156.22	14352-61-5	201	4-09-00-00038 0.9896[24]	1.459[14]
4311	Cyclohexaneacetic acid, 2-methyl- Cyclohexanacetic acid, 2-methyl	$C_9H_{16}O_2$ 156.22	6617-04-5	145-7[13]	2-09-00-00013 1.012[25]	1.4656[25]
4312	Cyclohexaneacetic acid, 3-methyl- Cyclohexanacetic acid, 3-methyl	$C_9H_{16}O_2$ 156.22	67451-76-7	148[18]	2-09-00-00013 0.9847[20]	1.495[20]
4313	Cyclohexaneacetonitrile	$C_8H_{13}N$ 123.20	4435-14-7	105[22]	4-09-00-00040 0.9473[21]	eth 4; EtOH 4 1.4787[21]
4314	Cyclohexane, 1,2-bis(methylene)-	C_8H_{12} 108.18	2819-48-9	125; 60-1[90]	4-05-00-00409 0.8229[25]	H_2O 1; EtOH 3; eth 3; ace 4 1.4718[25]
4315	Cyclohexane, bromo- Cyclohexyl bromide	$C_6H_{11}Br$ 163.06	108-85-0 -56.5	2736 166.2	4-05-00-00067 1.3359[20]	H_2O 1; EtOH 5; eth 5; ace 5 1.4957[20]
4316	Cyclohexane, (2-bromoethyl)-	$C_8H_{15}Br$ 191.11	1647-26-3 -57	212	4-05-00-00116 1.2357[20]	1.4899[20]
4317	Cyclohexane, 1-bromo-2-methoxy-	$C_7H_{13}BrO$ 193.08	24618-31-3	72-4[8]	3-06-00-00043 1.3257[20]	1.4871[20]
4318	Cyclohexane, (bromomethyl)-	$C_7H_{13}Br$ 177.08	2550-36-9	76-7[26]	4-05-00-00100 1.2763[25]	bz 4; eth 4; chl 4 1.4907[30]
4319	Cyclohexane, 1-bromo-1-methyl-	$C_7H_{13}Br$ 177.08	931-77-1	158	4-05-00-00100 1.2510[20]	H_2O 1; EtOH 3; chl 3 1.4866[20]
4320	Cyclohexane, 1-bromo-3-methyl- 3-Methylcyclohexyl bromide	$C_7H_{13}Br$ 177.08	13905-48-1	181; 60[11]	3-05-00-00076 1.275[25]	H_2O 1; eth 4; bz 3 1.4979[20]
4321	Cyclohexane, 1-bromo-4-methyl-1-(1-methylethyl)- p-Menthane, 4-bromo	$C_{10}H_{19}Br$ 219.16	116836-10-3 98-9[11]		0-05-00-00051 1.165[20]	1.4872[20]
4322	Cyclohexane, (2-bromo-2-propenyl)- Propylene, 2-bromo-3-cyclohexyl	$C_9H_{15}Br$ 203.12	53608-85-8	90[15]	3-05-00-00224 1.215[17]	eth 4 1.495[17]
4323	Cyclohexane, 1,1'-(1,4-butanediyl)bis- 1,4-Dicyclohexylbutane	$C_{16}H_{30}$ 222.41	6165-44-2 12	305	4-05-00-00357 0.8771[21]	EtOH 2; eth 2; bz 2; chl 2 1.475[21]
4324	Cyclohexanebutanol	$C_{10}H_{20}O$ 156.27	4441-57-0	103-4[4]	4-06-00-00139 0.902[25]	1.4660[20]
4325	Cyclohexane, butyl- Butylcyclohexane	$C_{10}H_{20}$ 140.27	1678-93-9 -74.7	180.9	4-05-00-00146 0.7902[20]	H_2O 1 1.4408[20]
4326	Cyclohexane, 1,1'-butylidenebis- Butane, 1,1-dicyclohexyl-	$C_{16}H_{30}$ 222.41	54890-00-5 -10.5	293	4-05-00-00359 0.8841[16]	1.485[16]
4327	Cyclohexane, 2-butyl-1,1,3-trimethyl- 2-Butyl-1,1,3-trimethylcyclohexane	$C_{13}H_{26}$ 182.35	54676-39-0	94-5[10]	4-05-00-00176 0.8292[19]	1.4563[19]
4328	Cyclohexanecarbonitrile	$C_7H_{11}N$ 109.17	766-05-2			
4329	Cyclohexanecarbonitrile, 1-hydroxy-	$C_7H_{11}NO$ 125.17	931-97-5 35	132[20]	4-10-00-00021 1.0172[20]	H_2O 4; eth 4 1.4693[20]
4330	Cyclohexanecarbonyl chloride	$C_7H_{11}ClO$ 146.62	2719-27-9	180	4-09-00-00026 1.0962[15]	1.4711[29]
4331	Cyclohexanecarboxaldehyde	$C_7H_{12}O$ 112.17	2043-61-0	159.3	4-07-00-00045 0.9035[20]	H_2O 3; eth 3 1.4496[20]
4332	Cyclohexanecarboxylic acid Hexahydrobenzoic acid	$C_7H_{12}O_2$ 128.17	98-89-5 31.5	2730 232.5	4-09-00-00016 1.0334[22]	H_2O 2; EtOH 4; bz 4; ctc 2 1.4530[20]
4333	Cyclohexanecarboxylic acid, 4-(aminomethyl)-, trans- Tranexamic acid	$C_8H_{15}NO_2$ 157.21	1197-18-8 >300	9487	3-14-00-00868	H_2O 4
4334	Cyclohexanecarboxylic acid, ethyl ester	$C_9H_{16}O_2$ 156.22	3289-28-9	196	4-09-00-00018 0.9362[20]	ace 4; eth 4; EtOH 4; chl 4 1.4501[15]
4335	Cyclohexanecarboxylic acid, methyl ester	$C_8H_{14}O_2$ 142.20	4630-82-4	183	4-09-00-00017 0.9954[15]	H_2O 3; eth 3; ace 3 1.4433[20]
4336	Cyclohexanecarboxylic acid, 3-methyl-4-oxo- Cyclohexanone-4-carboxylic acid, 2-methyl	$C_8H_{12}O_3$ 156.18	101567-36-6 93.5	195[20]	4-10-00-02617	bz 4; EtOH 4; chl 4
4337	Cyclohexanecarboxylic acid, propyl ester Propyl cyclohexanecarboxylate	$C_{10}H_{18}O_2$ 170.25	6739-34-0	215.5	4-09-00-00018 0.9530[15]	ace 4; eth 4; EtOH 4; chl 4 1.4486[15]
4338	Cyclohexanecarboxylic acid, 1,3,4,5-tetrahydroxy-, [1R-(1α,3α,4α,5β)]- Quinic acid	$C_7H_{12}O_6$ 192.17	77-95-2 162.5	8071	4-10-00-02257 1.64[25]	H_2O 4; EtOH 4; HOAc 4
4339	Cyclohexane, chloro- Cyclohexyl chloride	$C_6H_{11}Cl$ 118.61	542-18-7 -44	2738 142	4-05-00-00048 1.000[20]	H_2O 1; EtOH 5; eth 5; ace 5 1.4626[20]
4340	Cyclohexane, cyclopentyl-	$C_{11}H_{20}$ 152.28	1606-08-2	215.1	4-05-00-00328 0.8758[20]	1.4725[20]
4341	Cyclohexane, (3-cyclopentylpropyl)- Propane, 1-cyclohexyl-3-cyclopentyl-	$C_{14}H_{26}$ 194.36	2883-07-0	270	4-05-00-00347 0.8751[20]	1.4765[20]

No.	Name / Synonym	Mol. Form. / Mol. Wt.	CAS RN / mp/°C	Merck No. / bp/°C	Beil. Ref. / den/g cm⁻³	Solubility / n_D
4342	Cyclohexane, 1,1,2,2,3,3,4,4,5,6-decafluoro-5,6-bis(trifluoromethyl)- Cyclohexane, 1,2-dimethyl, perfluoro	C_8F_{16} 400.06	306-98-9 -56	101.5	3-05-00-00098 1.829^{25}	1.283^{25}
4343	Cyclohexane, 1,1,2,2,3,4,4,5,5,6-decafluoro-3,6-bis(trifluoromethyl)-	C_8F_{16} 400.06	374-77-6	100.5	4-05-00-00124 1.8503^{20}	1.2897^{20}
4344	Cyclohexane, decyl-	$C_{16}H_{32}$ 224.43	1795-16-0 -1.7	299	4-05-00-00187 0.8186^{20}	1.4534^{20}
4345	Cyclohexane, 1,1-dibromo- 1,1-Dibromocyclohexane	$C_6H_{10}Br_2$ 241.95	10489-97-1	87^{14}	4-05-00-00070 1.7349^{20}	1.5392^{25}
4346	Cyclohexane, 1,2-dibromo-, cis- cis-1,2-Dibromocyclohexane	$C_6H_{10}Br_2$ 241.95	19246-38-9 9.7	115^{14}	4-05-00-00070 1.798^{25}	ace 4; bz 4; eth 4; lig 4 1.5514^{25}
4347	Cyclohexane, 1,2-dibromo-, trans-(±)-	$C_6H_{10}Br_2$ 241.95	5183-77-7 -4	145^{100}; 105^{20}	4-05-00-00071 1.7759^{20}	ace 4; bz 4; eth 4; EtOH 4 1.5445^{19}
4348	Cyclohexane, 1,3-dibromo-, trans- trans-1,3-Dibromocyclohexane	$C_6H_{10}Br_2$ 241.95	29624-17-7 1	116^{16}	4-05-00-00072	bz 4; EtOH 4 1.5480^{20}
4349	Cyclohexane, 1,4-dibromo-, cis- cis-1,4-Dibromocyclohexane	$C_6H_{10}Br_2$ 241.95	16661-99-7	$137-8^{25}$	4-05-00-00072 1.7834^{20}	eth 4 1.5531^{20}
4350	1,2-Cyclohexanedicarboxylic acid, diethyl ester, trans-(±)- Cyclohexane-1,2-dicarboxylic acid, diethyl ester (trans, DL)	$C_{12}H_{20}O_4$ 228.29	96836-97-4	133^{11}	4-09-00-02803 1.040^{20}	eth 4 1.4522^{13}
4351	1,3-Cyclohexanedicarboxylic acid, diethyl ester, cis- Cyclohexane-1,3-dicarboxylic acid, diethyl ester (cis)	$C_{12}H_{20}O_4$ 228.29	62059-56-7	288; 151^{15}	3-09-00-03817 1.045^{20}	1.452^{20}
4352	1,3-Cyclohexanedicarboxylic acid, dimethyl ester, cis- Dimethyl cis-1,3-cyclohexanedicarboxylate	$C_{10}H_{16}O_4$ 200.23	6998-82-9	263	4-09-00-02806 1.0997^{20}	1.4568^{20}
4353	1,3-Cyclohexanedicarboxylic acid, dimethyl ester, trans- Cyclohexane-1,3-dicarboxylic acid, dimethyl ester (trans)	$C_{10}H_{16}O_4$ 200.23	10021-92-8	139^{20}	4-09-00-02806 1.1095^{20}	1.4577^{25}
4354	1,4-Cyclohexanedicarboxylic acid, trans-	$C_8H_{12}O_4$ 172.18	619-82-9 312.5	sub 300	4-09-00-02807	H_2O 2; EtOH 4; eth 2; ace 3
4355	1,4-Cyclohexanedicarboxylic acid, diethyl ester, cis- Cyclohexane-1,4-dicarboxylic acid, diethyl ester (cis)	$C_{12}H_{20}O_4$ 228.29	116724-15-3	151^{13}	2-09-00-00524 1.015^{20}	eth 4
4356	1,4-Cyclohexanedicarboxylic acid, diethyl ester, trans- Cyclohexane-1,4-dicarboxylic acid, diethyl ester (trans)	$C_{12}H_{20}O_4$ 228.29	19145-96-1 43.5		2-09-00-00524 1.0110^{20}	eth 4 1.4337^{64}
4357	1,4-Cyclohexanedicarboxylic acid, dimethyl ester, cis- cis-Dimethyl 1,4-cyclohexanedicarboxylate	$C_{10}H_{16}O_4$ 200.23	3399-21-1 14	132.5^{10}	4-09-00-02808 1.1112^{20}	
4358	Cyclohexane, 1,1-dichloro- 1,1-Dichlorocyclohexane	$C_6H_{10}Cl_2$ 153.05	2108-92-1 -47	171	4-05-00-00050 1.1559^{20}	1.4803^{20}
4359	Cyclohexane, 1,2-dichloro-, cis- cis-1,2-Dichlorocyclohexane	$C_6H_{10}Cl_2$ 153.05	10498-35-8 -1.5	206.9	4-05-00-00050 1.2021^{20}	bz 4 1.4967^{20}
4360	Cyclohexane, 1,2-dichloro-, trans-(±)-	$C_6H_{10}Cl_2$ 153.05	5183-79-9 -6.3	189	4-05-00-00051 1.1839^{20}	1.4902^{20}
4361	Cyclohexane, 1,4-dichloro-, cis- cis-1,4-Dichlorocyclohexane	$C_6H_{10}Cl_2$ 153.05	16749-11-4 18	80.3^{25}	4-05-00-00051 1.1900^{20}	1.4942^{20}
4362	1,3-Cyclohexanedimethanamine Cyclohexane, 1,3-bis(aminomethyl)	$C_8H_{18}N_2$ 142.24	2579-20-6 <-70	220	0.945^{20}	H_2O 4; eth 4; EtOH 4
4363	1,3-Cyclohexanedimethanol Cyclohexane, 1,3-bis-(hydroxymethyl)	$C_8H_{16}O_2$ 144.21	3971-28-6	105.5	4-06-00-05236 1.036^{25}	1.4912^{20}
4364	Cyclohexane, 1,1-dimethyl- 1,1-Dimethylcyclohexane	C_8H_{16} 112.22	590-66-9 -33.3	119.6	4-05-00-00117 0.7809^{20}	H_2O 1; EtOH 3; eth 3; ace 3 1.4290^{20}
4365	Cyclohexane, 1,2-dimethyl-, cis- cis-1,2-Dimethylcyclohexane	C_8H_{16} 112.22	2207-01-4 -49.9	129.8	4-05-00-00118 0.7963^{20}	H_2O 1; EtOH 3; eth 5; ace 5 1.4360^{20}
4366	Cyclohexane, 1,2-dimethyl-, trans- trans-1,2-Dimethylcyclohexane	C_8H_{16} 112.22	6876-23-9 -90	123.5	4-05-00-00118 0.7760^{20}	H_2O 1; EtOH 3; eth 3; ace 5 1.4270^{20}
4367	Cyclohexane, 1,3-dimethyl-, cis- cis-1,3-Dimethylcyclohexane	C_8H_{16} 112.22	638-04-0 -75.6	120.1	4-05-00-00121 0.7660^{20}	H_2O 1; EtOH 5; eth 5; ace 5 1.4229^{20}
4368	Cyclohexane, 1,4-dimethyl-, cis- cis-1,4-Dimethylcyclohexane	C_8H_{16} 112.22	624-29-3 -87.4	124.4	4-05-00-00122 0.7829^{20}	H_2O 1; EtOH 5; eth 5; ace 5 1.4230^{20}
4369	Cyclohexane, (1,1-dimethylethyl)-	$C_{10}H_{20}$ 140.27	3178-22-1 -41.2	171.5	4-05-00-00147 0.8127^{20}	H_2O 1 1.4469^{20}
4370	Cyclohexane, 1,3-dimethyl-, trans- trans-1,3-Dimethylcyclohexane	C_8H_{16} 112.22	2207-03-6 -90.1	124.5	0.79^{15}	1.4284^{25}
4371	Cyclohexane, 1,4-dimethyl-, trans- trans-1,4-Dimethylcyclohexane	C_8H_{16} 112.22	2207-04-7 -36.9	119.4	0.77^{15}	1.4185^{25}
4372	1,2-Cyclohexanediol, cis-	$C_6H_{12}O_2$ 116.16	1792-81-0 100	120^{15}	4-06-00-05193 1.0297^{101}	EtOH 3; ace 3; bz 3; chl 2
4373	1,2-Cyclohexanediol, trans-(±)-	$C_6H_{12}O_2$ 116.16	54383-22-1 105	117^{13}	4-06-00-05194 1.147^{24}	H_2O 4; bz 4; eth 4; EtOH 4
4374	1,4-Cyclohexanediol, trans- trans-1,4-Dihydroxycyclohexane	$C_6H_{12}O_2$ 116.16	6995-79-5 143		4-06-00-05209 1.18^{20}	H_2O 3; EtOH 3; eth 1; ace 2

No.	Name Synonym	Mol. Form. Mol. Wt.	CAS RN mp/°C	Merck No. bp/°C	Beil. Ref. den/g cm^{-3}	Solubility n_D
4375	1,2-Cyclohexanedione 1,2-Dioxocyclohexane	$C_6H_8O_2$ 112.13	765-87-7 40	194	4-07-00-01982 1.1187[21]	H_2O 3; EtOH 3; eth 3; bz 3 1.4995[20]
4376	1,3-Cyclohexanedione	$C_6H_8O_2$ 112.13	504-02-9 105.5	3161	4-07-00-01985 1.0861[91]	H_2O 3; EtOH 3; eth 2; ace 3 1.4576[102]
4377	1,4-Cyclohexanedione Tetrahydroquinone	$C_6H_8O_2$ 112.13	637-88-7 78	132[20]	4-07-00-01986 1.0861[91]	H_2O 3; EtOH 3; eth 3; ace 3
4378	1,3-Cyclohexanedione, 5,5-dimethyl- 5,5-Dimethyldihydroresorcinol	$C_8H_{12}O_2$ 140.18	126-81-8 149-51 dec	3231	4-07-00-01999	H_2O 2; eth 2; ace 3; ctc 3
4379	1,2-Cyclohexanedione, dioxime Nioxime	$C_6H_{10}N_2O_2$ 142.16	492-99-9 192	6477	4-07-00-01982	H_2O 3; ace 3; chl 3; tfa 2
4380	Cyclohexane, docosyl- Docosane, 1-cyclohexyl-	$C_{28}H_{56}$ 392.75	61828-07-7 54.4	449	3-05-00-00167 0.8334[20]	1.4632[20]
4381	Cyclohexane, dodecafluoro- Perfluorocyclohexane	C_6F_{12} 300.05	355-68-0 62.5 (trip. pt.)	52.8 (subl. pt.)	4-05-00-00048	
4382	Cyclohexane, dodecyl-	$C_{18}H_{36}$ 252.48	1795-17-1 12.5	331	4-05-00-00191 0.8223[20]	1.4559[20]
4383	Cyclohexane, dotriacontyl- Dotriacontylcyclohexane	$C_{38}H_{76}$ 533.02	61828-17-9 74.8	527	4-05-00-00204 0.8388[20]	1.4668[20]
4384	Cyclohexane, eicosyl- 1-Cyclohexyleicosane	$C_{26}H_{52}$ 364.70	4443-55-4 48.5	430	4-05-00-00198 0.8318[20]	1.4622[20]
4385	Cyclohexaneethanamine, N,α-dimethyl- Propylhexedrine	$C_{10}H_{21}N$ 155.28	101-40-6	7873 205; 82[10]	3-12-00-00108 0.8501[20]	EtOH 4 1.4600[20]
4386	Cyclohexane, 1,1'-(1,2-ethanediyl)bis- 1,2-Dicyclohexylethane	$C_{14}H_{26}$ 194.36	3321-50-4 11.5	272.5	4-05-00-00344 0.8728[24]	1.4745[24]
4387	Cyclohexaneethanol	$C_8H_{16}O$ 128.21	4442-79-9	208	4-06-00-00119 0.9229[20]	EtOH 3; eth 3; bz 3 1.4641[20]
4388	Cyclohexane, ethenyl-	C_8H_{14} 110.20	695-12-5	128	4-05-00-00268 0.8166[19]	1.455[19]
4389	Cyclohexane, 1-ethenyl-1-methyl-2,4-bis(1-methylethyl)-, (1α,2β,4β)- β-Elemene	$C_{15}H_{24}$ 204.36	33880-83-0	117-24[15.5]	4-05-00-01180 0.8749[20]	1.4935[20]
4390	Cyclohexane, ethyl- Ethylcyclohexane	C_8H_{16} 112.22	1678-91-7 -111.3	131.9	4-05-00-00115 0.7880[20]	H_2O 1; EtOH 3; eth 3; ace 3 1.4330[20]
4391	Cyclohexane, ethylidene- Ethylidenecyclohexane	C_8H_{14} 110.20	1003-64-1	136	4-05-00-00267 0.822[25]	1.4618[20]
4392	Cyclohexane, 1,1'-ethylidenebis- 1,1-Dicyclohexylethane	$C_{14}H_{26}$ 194.36	2319-61-1 -20.9	256; 112[7]	4-05-00-00345 0.9070[25]	1.4887[25]
4393	Cyclohexane, 1-ethylidene-2-methyl- Cyclohexane, 2-methyl-1-ethylidene	C_9H_{16} 124.23	40514-70-3	158	0-05-00-00078 0.823[25]	1.47
4394	Cyclohexane, 1-ethyl-1-methyl-	C_9H_{18} 126.24	4926-90-3	152.2	4-05-00-00135 0.8025[25]	1.4419[20]
4395	Cyclohexane, 1-ethyl-2-methyl-, cis-	C_9H_{18} 126.24	4923-77-7	156	4-05-00-00136 0.8059[25]	1.4456[20]
4396	Cyclohexane, 1-ethyl-2-methyl-, trans- Cyclohexane, 1-methyl-2-ethyl (trans)	C_9H_{18} 126.24	4923-78-8	151.7	4-05-00-00136 0.790[25]	1.4381[20]
4397	Cyclohexane, 1-ethyl-3-methyl-, cis- cis-1-Ethyl-3-methylcyclohexane	C_9H_{18} 126.24	19489-10-2	156	4-05-00-00136 0.8094[20]	1.4432[20]
4398	Cyclohexane, 1-ethyl-3-methyl-, trans-	C_9H_{18} 126.24	4926-76-5	150	4-05-00-00136 0.7972[20]	1.4382[20]
4399	Cyclohexane, 1-ethyl-4-methyl-, cis- cis-1-Ethyl-4-methylcyclohexane	C_9H_{18} 126.24	4926-78-7	152.6	4-05-00-00136 0.7969[20]	1.4374[20]
4400	Cyclohexane, 1-ethyl-4-methyl-, trans- trans-1-Methyl-4-ethylcyclohexane	C_9H_{18} 126.24	6236-88-0 -80.8	149	4-05-00-00136 0.7798[20]	1.4304[20]
4401	Cyclohexane, 1-ethyl-1-methyl-2,4-bis(1-methylethyl)-, [1R-(1α,2β,4β)]-	$C_{15}H_{30}$ 210.40	515-12-8	115-9[10]	4-05-00-00185 0.8509[20]	bz 4; eth 4; peth 4 1.4640[20]
4402	Cyclohexane, 1,1'-(2-ethyl-1,3-propanediyl)bis- Propane, 1,3-dicyclohexyl-2-ethyl	$C_{17}H_{32}$ 236.44	54833-34-0	306	4-05-00-00363 0.8845[20]	1.483[21]
4403	Cyclohexane, fluoro- Cyclohexyl fluoride	$C_6H_{11}F$ 102.15	372-46-3 13	101	4-05-00-00044 0.9279[20]	H_2O 1; py 3 1.4146[20]
4404	Cyclohexane, heptadecyl- Heptadecane, 1-cyclohexyl-	$C_{23}H_{46}$ 322.62	19781-73-8 38	397	3-05-00-00161 0.8290[20]	1.4603[20]
4405	Cyclohexane, heptyl-	$C_{13}H_{26}$ 182.35	5617-41-4 -30	244	4-05-00-00174 0.8109[20]	1.4484[20]
4406	Cyclohexane, 1,2,3,4,5,6-hexachloro-, (1α,2β,3α,4β,5α,6β)- β-Lindane	$C_6H_6Cl_6$ 290.83	319-85-7	60[0.50]	4-05-00-00061 1.89[19]	H_2O 1; EtOH 2; bz 2; chl 2
4407	Cyclohexane, 1,2,3,4,5,6-hexachloro-, (1α,2α,3α,4β,5α,6β)- δ-Lindane	$C_6H_6Cl_6$ 290.83	319-86-8 141.5	60[0.36]	4-05-00-00056	
4408	Cyclohexane, 1,2,3,4,5,6-hexachloro-, (1α,2α,3β,4α,5β,6β)-(±)- Cyclohexane, 1,2,3,4,5,6-hexachloro, (α, DL)	$C_6H_6Cl_6$ 290.83	60291-32-9 159.5	288	4-05-00-00060 1.87[20]	H_2O 1; EtOH 4; bz 3; chl 3
4409	Cyclohexane, 1,2,3,4,5,6-hexachloro-, (1α,2α,3β,4α,5α,6β)- Lindane	$C_6H_6Cl_6$ 290.83	58-89-9 112.5	5379 323.4	4-05-00-00058	ace 4; bz 4
4410	Cyclohexane, hexacosyl- Hexacosylcyclohexane	$C_{32}H_{64}$ 448.86	61828-11-3 64	484	4-05-00-00203 0.8359[20]	1.4649[20]

No.	Name / Synonym	Mol. Form. / Mol. Wt.	CAS RN / mp/°C	Merck No. / bp/°C	Beil. Ref. / den/g cm^{-3}	Solubility / n_D
4411	Cyclohexane, hexadecyl- / Hexadecane, 1-cyclohexyl-	$C_{22}H_{44}$ / 308.59	6812-38-0 / 33.6	385	4-05-00-00196 / 0.8279[20]	1.4596[20]
4412	Cyclohexanehexanoic acid / Caproic acid, 6-cyclohexyl	$C_{12}H_{22}O_2$ / 198.31	4354-56-7 / 34	180[11]	4-09-00-00095 / 0.9626[20]	eth 3; chl 2 / 1.4750[20]
4413	Cyclohexane, hexatriacontyl- / Hexatriacontylcyclohexane	$C_{42}H_{84}$ / 589.13	61828-21-5 / 80	551	4-05-00-00205 / 0.8402[20]	1.4678[20]
4414	Cyclohexane, hexyl-	$C_{12}H_{24}$ / 168.32	4292-75-5 / -43	224	4-05-00-00170 / 0.8076[20]	1.4462[20]
4415	Cyclohexane, iodo- / Cyclohexyl iodide	$C_6H_{11}I$ / 210.06	626-62-0 /	180 dec; 81[20]	4-05-00-00078 / 1.6244[20]	H_2O 1; EtOH 3; eth 3; ace 3 / 1.5477[20]
4416	Cyclohexane, isocyanato- / Cyclohexyl isocyanate	$C_7H_{11}NO$ / 125.17	3173-53-3 /	172	4-12-00-00071 / 0.98[25]	1.4551[20]
4417	Cyclohexane, isothiocyanato-	$C_7H_{11}NS$ / 141.24	1122-82-3 /	220	4-12-00-00073 / 0.996[25]	H_2O 1; EtOH 3; eth 3; ctc 2 / 1.5375[20]
4418	Cyclohexanemethanamine	$C_7H_{15}N$ / 113.20	3218-02-8 /	160	4-12-00-00123 / 0.87[25]	1.4630[20]
4419	Cyclohexanemethanol / Cyclohexylcarbinol	$C_7H_{14}O$ / 114.19	100-49-2 / -43	2737 / 183	4-06-00-00106 / 0.9297[20]	eth 4; EtOH 4 / 1.4644[20]
4420	Cyclohexanemethanol, α-methyl-	$C_8H_{16}O$ / 128.21	1193-81-3 /	189	1-06-00-00012 / 0.928[25]	EtOH 4; eth 4; ctc 2 / 1.4656[20]
4421	Cyclohexanemethanol, α-methyl-, (S)-	$C_8H_{16}O$ / 128.21	3113-98-2 /	105[35]; 82[12]	4-06-00-00117 / 0.9254[20]	1.4635[25]
4422	Cyclohexanemethanol, 2-methyl-, cis- / Cyclohexane, 1-hydroxymethyl-2-methyl (cis)	$C_8H_{16}O$ / 128.21	3937-45-9 /	188.5	4-06-00-00122 / 0.9342[20]	1.4689[20]
4423	Cyclohexanemethanol, 2-methyl-, trans- / Cyclohexane, 1-hydroxymethyl-2-methyl (trans)	$C_8H_{16}O$ / 128.21	3937-46-0 /	192	4-06-00-00122 / 0.9224[20]	1.4665[20]
4424	Cyclohexanemethanol, 4-methyl-	$C_8H_{16}O$ / 128.21	34885-03-5 /	75[2.5]	2-06-00-00030 / 0.9074[20]	1.4617[20]
4425	Cyclohexane, methoxy-	$C_7H_{14}O$ / 114.19	931-56-6 / -74.4	133	4-06-00-00026 / 0.8756[20]	eth 4; EtOH 4 / 1.4355[20]
4426	Cyclohexane, methyl- / Methylcyclohexane	C_7H_{14} / 98.19	108-87-2 / -126.6	100.9	4-05-00-00094 / 0.7694[20]	H_2O 1; EtOH 3; eth 3; ace 5 / 1.4231[20]
4427	Cyclohexane, (3-methylbutyl)-	$C_{11}H_{22}$ / 154.30	54105-76-9 /	196.5	3-05-00-00143 / 0.8023[20]	bz 4; lig 4 / 1.4420[20]
4428	Cyclohexane, methylene- / Methylenecyclohexane	C_7H_{12} / 96.17	1192-37-6 / -106.7	102.5	4-05-00-00250 / 0.8074[20]	H_2O 1; eth 3; bz 3; chl 3 / 1.4523[20]
4429	Cyclohexane, 1,1'-methylenebis- / Dicyclohexylmethane	$C_{13}H_{24}$ / 180.33	3178-23-2 /	110-5[18]	4-05-00-00340 / 0.8750[20]	1.4752[20]
4430	Cyclohexane, 1,1'-(1-methyl-1,2-ethanediyl)bis- / 1,2-Dicyclohexylpropane	$C_{15}H_{28}$ / 208.39	41851-34-7 / -22	284.5	4-05-00-00350 / 0.8724[21]	1.479[21]
4431	Cyclohexane, (1-methylethyl)- / Isopropylcyclohexane	C_9H_{18} / 126.24	696-29-7 / -89.4	154.8	4-05-00-00134 / 0.8023[20]	H_2O 1; EtOH 4; eth 4; ace 5 / 1.4410[20]
4432	Cyclohexane, 1,1'-(1-methylethylidene)bis- / Propane, 2,2-dicyclohexyl	$C_{15}H_{28}$ / 208.39	54934-90-6 / 15.6	286	4-05-00-00350 / 0.9001[23]	1.490[23]
4433	Cyclohexane, 1,1',1''-methylidynetris- / Methane, tricyclohexyl-	$C_{19}H_{34}$ / 262.48	1610-24-8 / 48	325.5	4-05-00-00504 / 0.9273[50]	H_2O 1; EtOH 2; eth 3; bz 3 / 1.4986[40]
4434	Cyclohexane, 1-methyl-4-methylene-	C_8H_{14} / 110.20	2808-80-2 /	122	4-05-00-00271 / 0.7910[19]	1.4465[18]
4435	Cyclohexane, 1-methyl-3-(1-methylethenyl)- / m-Menth-8-ene	$C_{10}H_{18}$ / 138.25	16605-36-0 /	170	3-05-00-00235 / 0.8178[20]	1.4546
4436	Cyclohexane, 1-methyl-2-(1-methylethyl)-	$C_{10}H_{20}$ / 140.27	16580-23-7 /	171	4-05-00-00150 / 0.8134[22]	1.447[21]
4437	Cyclohexane, 1-methyl-3-(1-methylethyl)-	$C_{10}H_{20}$ / 140.27	16580-24-8 /	166.5	3-05-00-00132 / 0.7963[20]	1.44[24]
4438	Cyclohexane, 1-methyl-4-(1-methylethyl)-	$C_{10}H_{20}$ / 140.27	99-82-1 / -87.6	170.7	4-05-00-00151 / 0.797[25]	1.4373[20]
4439	Cyclohexane, 1-methyl-4-(1-methylethyl)-, cis-	$C_{10}H_{20}$ / 140.27	6069-98-3 / -89.9	172	4-05-00-00151 / 0.8039[20]	H_2O 1; EtOH 4; eth 4; bz 3 / 1.4431[20]
4440	Cyclohexane, 1-methyl-4-(1-methylethyl)-, trans-	$C_{10}H_{20}$ / 140.27	1678-82-6 / -86.3	170.6	4-05-00-00151 / 0.7928[20]	bz 4; eth 4; EtOH 4; lig 4 / 1.4366[20]
4441	Cyclohexane, 1-methyl-3-(1-methylethylidene)-	$C_{10}H_{18}$ / 138.25	13828-34-7 /	174	1-05-00-00043 / 0.8214[20]	1.4670[20]
4442	Cyclohexane, 1-methyl-4-(1-methylethylidene)-	$C_{10}H_{18}$ / 138.25	1124-27-2 /	173	4-05-00-00302 / 0.819[21]	1.4568[21]
4443	Cyclohexane, 1-methyl-1-nitro- / 1-Methyl-1-nitrocyclohexane	$C_7H_{13}NO_2$ / 143.19	59368-15-9 / -71	118-20[50]	4-05-00-00103 / 1.0384[20]	1.4598[20]
4444	Cyclohexane, 1-methyl-2-nitro- / 2-Methylnitrocyclohexane	$C_7H_{13}NO_2$ / 143.19	74221-86-6 /	100[20]	4-05-00-00103 / 1.046[25]	1.4608[25]
4445	Cyclohexane, 1-methyl-4-nitro- / 1-Methyl-4-nitrocyclohexane	$C_7H_{13}NO_2$ / 143.19	89895-45-4 /	102[20]	4-05-00-00103 / 1.037[25]	1.4567[23]
4446	Cyclohexane, 1-methyl-2-pentyl- / 1-Methyl-2-pentylcyclohexane	$C_{12}H_{24}$ / 168.32	54411-01-7 /	217.5	4-05-00-00171 / 0.815[20]	1.4487[20]
4447	Cyclohexane, 1,1'-(2-methyl-1,3-propanediyl)bis- / 1,3-Dicyclohexyl-2-methylpropane	$C_{16}H_{30}$ / 222.41	2883-08-1 / 0.6	295.2	4-05-00-00358 / 0.8715[20]	1.4756[20]

No.	Name Synonym	Mol. Form. Mol. Wt.	CAS RN mp/°C	Merck No. bp/°C	Beil. Ref. den/g cm^{-3}	Solubility n_D
4448	Cyclohexane, (1-methylpropyl)- sec-Butylcyclohexane	$C_{10}H_{20}$ 140.27	7058-01-7 	 179.3	4-05-00-00146 0.8131[20]	H_2O 1; ace 3 1.4467[20]
4449	Cyclohexane, 1-methyl-1-propyl- 1-Methyl-1-propylcyclohexane	$C_{10}H_{20}$ 140.27	4258-93-9 	 172	4-05-00-00148 0.8101[20]	 1.4440[20]
4450	Cyclohexane, 1-methyl-2-propyl- 1-Methyl-2-propylcyclohexane	$C_{10}H_{20}$ 140.27	4291-79-6 	 176; 56[13]	4-05-00-00148 0.8130[19]	 1.4468[19]
4451	Cyclohexane, 1-methyl-3-propyl- 1-Methyl-3-propylcyclohexane	$C_{10}H_{20}$ 140.27	4291-80-9 	 164.5	4-05-00-00149 0.7895[21]	 1.4377[20]
4452	Cyclohexane, 1-methyl-4-propyl- 1-Methyl-4-propylcyclohexane	$C_{10}H_{20}$ 140.27	4291-81-0 	 175.5	4-05-00-00149 0.797[20]	 1.4393[20]
4453	Cyclohexane, (2-methylpropyl)-	$C_{10}H_{20}$ 140.27	1678-98-4 -95	 171.3	4-05-00-00147 0.7952[20]	H_2O 1; EtOH 4; eth 4; ace 3 1.4386[20]
4454	Cyclohexane, nitro- Nitrocyclohexane	$C_6H_{11}NO_2$ 129.16	1122-60-7 -34	205; 95[22]	4-05-00-00081 1.0610[20]	H_2O 1; EtOH 3; lig 3 1.4612[19]
4455	Cyclohexane, (nitromethyl)- (Nitromethyl)cyclohexane	$C_7H_{13}NO_2$ 143.19	2625-30-1 	123[40]; 98[10]	4-05-00-00103 1.0482[20]	 1.4705[20]
4456	Cyclohexane, nonadecyl-	$C_{25}H_{50}$ 350.67	22349-03-7 45	 420	4-05-00-00197 0.8310[20]	 1.4616[20]
4457	Cyclohexane, nonyl-	$C_{15}H_{30}$ 210.40	2883-02-5 -10	 282	4-05-00-00183 0.8163[20]	 1.4519[20]
4458	Cyclohexane, octacosyl- Octacosylcyclohexane	$C_{34}H_{68}$ 476.91	61828-13-5 68	 499	4-05-00-00203 0.8370[20]	 1.4656[20]
4459	Cyclohexane, octadecyl- Octadecane, 1-cyclohexyl-	$C_{24}H_{48}$ 336.65	4445-06-1 41.6	409; 175[1]	4-05-00-00197 0.8300[20]	 1.4610[20]
4460	Cyclohexane, octyl-	$C_{14}H_{28}$ 196.38	1795-15-9 -20	 264	4-05-00-00178 0.8138[20]	 1.4503[20]
4461	Cyclohexane, 1,1'-oxybis-	$C_{12}H_{22}O$ 182.31	4645-15-2 -36	 242.5	3-06-00-00019 0.9227[20]	 1.4741[20]
4462	Cyclohexane, pentacosyl- Pentacosylcyclohexane	$C_{31}H_{62}$ 434.83	61828-10-2 62	 476	4-05-00-00202 0.8353[20]	 1.4645[20]
4463	Cyclohexane, pentadecyl-	$C_{21}H_{42}$ 294.56	6006-95-7 29	 373	4-05-00-00195 0.8267[20]	 1.4588[20]
4464	Cyclohexane, 1,1'-(1,5-pentanediyl)bis- Pentane, 1,5-dicyclohexyl-	$C_{17}H_{32}$ 236.44	54833-31-7 -13.6	 325	4-05-00-00362 0.8718[21]	 1.479[21]
4465	Cyclohexane, pentyl- Pentylcyclohexane	$C_{11}H_{22}$ 154.30	4292-92-6 -57.5	 203.7	4-05-00-00164 0.8037[20]	ace 4; bz 4; eth 4; EtOH 4 1.4437[20]
4466	Cyclohexane, 1,1'-(1,3-propanediyl)bis- 1,3-Dicyclohexylpropane	$C_{15}H_{28}$ 208.39	3178-24-3 -17	 291.5	4-05-00-00349 0.8728[24]	 1.4736[24]
4467	Cyclohexanepropanenitrile, 2-oxo-	$C_9H_{13}NO$ 151.21	4594-78-9 	138-42[10]	4-10-00-02623 1.0181[20]	ctc 2 1.4755[20]
4468	Cyclohexanepropanoic acid	$C_9H_{16}O_2$ 156.22	701-97-3 16	 276.5	4-09-00-00056 0.912[25]	H_2O 3; eth 3; ctc 2 1.4638[20]
4469	Cyclohexanepropanol, α-methyl- 3-Cyclohexyl-1-methylpropanol	$C_{10}H_{20}O$ 156.27	10528-67-3 	112[16]	3-06-00-00123 0.903[21]	 1.464[21]
4470	Cyclohexane, 2-propenyl-	C_9H_{16} 124.23	2114-42-3 	 132	4-05-00-00283 0.8135[20]	EtOH 4; eth 3; ace 3; bz 3 1.4500[20]
4471	Cyclohexane, propyl- Propylcyclohexane	C_9H_{18} 126.24	1678-92-8 -94.9	 156.7	4-05-00-00134 0.7936[20]	H_2O 1; EtOH 5; eth 3; ace 5 1.4370[20]
4472	Cyclohexane, 1,1'-propylidenebis- Propane, 1,1-dicyclohexyl-	$C_{15}H_{28}$ 208.39	54934-91-7 -23.5	 283.6	4-05-00-00350 0.8886[23]	 1.485[23]
4473	Cyclohexane, 2-propynyl-	C_9H_{14} 122.21	17715-00-3 	 157.5	4-05-00-00419 0.8449[20]	eth 3 1.4605[20]
4474	Cyclohexane, tetracosyl- Tetracosylcyclohexane	$C_{30}H_{60}$ 420.81	61828-09-9 59.5	 467	4-05-00-00202 0.8347[20]	 1.4641[20]
4475	Cyclohexane, tetradecyl-	$C_{20}H_{40}$ 280.54	1795-18-2 24	 360	4-05-00-00194 0.8254[20]	 1.4579[20]
4476	Cyclohexane, 1,1'-tetradecylidenebis- 1,1-Dicyclohexyltetradecane	$C_{26}H_{50}$ 362.68	55334-08-2 37.6	 406	4-05-00-00374 0.8735[20]	 1.4799[20]
4477	Cyclohexane, 1,1,3,4-tetramethyl- 1,1,3,4-Tetramethylcyclohexane	$C_{10}H_{20}$ 140.27	24612-75-7 	 160	4-05-00-00159 0.7976[20]	 1.4380[20]
4478	Cyclohexane, 1,1,3,5-tetramethyl-, cis- cis-1,1,3,5-Tetramethylcyclohexane	$C_{10}H_{20}$ 140.27	50876-32-9 	 152.5	4-05-00-00159 0.7813[20]	 1.4319[20]
4479	Cyclohexane, tetratriacontyl- Tetratriacontylcyclohexane	$C_{40}H_{80}$ 561.08	61828-19-1 78	 540	4-05-00-00205 0.8395[20]	 1.4673[20]
4480	Cyclohexanethiol Cyclohexyl mercaptan	$C_6H_{12}S$ 116.23	1569-69-3 	 158.9	4-06-00-00072 0.9782[20]	ace 4; bz 4; eth 4; EtOH 4 1.4921[20]
4481	Cyclohexanethione Thiocyclohexanone	$C_6H_{10}S$ 114.21	2720-41-4 	82[17]; 74[11]	3-07-00-00039 0.8135[20]	EtOH 3; eth 3; ace 3 1.5375[20]
4482	Cyclohexane, triacontyl- Triacontylcyclohexane	$C_{36}H_{72}$ 504.97	61828-15-7 71.6	 514	4-05-00-00204 0.8379[20]	 1.4662[20]
4483	Cyclohexane, tricosyl- Tricosylcyclohexane	$C_{29}H_{58}$ 406.78	61828-08-8 57	 459	4-05-00-00202 0.8341[20]	 1.4637[20]
4484	Cyclohexane, tridecyl- Tridecane, 1-cyclohexyl-	$C_{19}H_{38}$ 266.51	6006-33-3 18.5	 346	4-05-00-00192 0.8239[20]	 1.4570[20]
4485	Cyclohexane, 1,1,2-trimethyl- 1,1,2-Trimethylcyclohexane	C_9H_{18} 126.24	7094-26-0 -29	 145.2	4-05-00-00137 0.7963[25]	 1.4382[20]
4486	Cyclohexane, 1,1,3-trimethyl- Cyclogeraniolane	C_9H_{18} 126.24	3073-66-3 -65.7	 136.6	4-05-00-00137 0.7749[25]	 1.4295[20]
4487	Cyclohexane, 1,1,4-trimethyl- 1,1,4-Trimethylcyclohexane	C_9H_{18} 126.24	7094-27-1 	 135	4-05-00-00137 0.7685[25]	 1.4251[20]

No.	Name Synonym	Mol. Form. Mol. Wt.	CAS RN mp/°C	Merck No. bp/°C	Beil. Ref. den/g cm^{-3}	Solubility n_D
4488	Cyclohexane, 1,2,3-trimethyl- 1,2,3-Trimethylcyclohexane	C_9H_{18} 126.24	1678-97-3	142	4-05-00-00137 0.7898[20]	1.4346[20]
4489	Cyclohexane, 1,2,3-trimethyl-, (1α,2β,3α)-	C_9H_{18} 126.24	1678-81-5 -66.9	144	4-05-00-00137 0.777[25]	1.430[20]
4490	Cyclohexane, 1,2,3-trimethyl-, (1α,2α,3α)-	C_9H_{18} 126.24	1839-88-9 -85	151.7	4-05-00-00137 0.7989[25]	1.4405[20]
4491	Cyclohexane, 1,2,3-trimethyl-, (1α,2α,3β)-	C_9H_{18} 126.24	7667-55-2 -85.7	151.2	4-05-00-00137 0.7995[25]	1.4399[20]
4492	Cyclohexane, 1,2,4-trimethyl-, (1α,2α,4α)-	C_9H_{18} 126.24	1678-80-4 -77.5	146.6	4-05-00-00138 0.783[25]	1.4340[20]
4493	Cyclohexane, 1,2,4-trimethyl-, (1α,2β,4α)-	C_9H_{18} 126.24	7667-59-6 -86	142.9	4-05-00-00138 0.7683[25]	1.4266[20]
4494	Cyclohexane, 1,2,4-trimethyl-, (1α,2α,4β)-	C_9H_{18} 126.24	7667-58-5 -91.8	146.7	4-05-00-00138 0.7870[25]	1.4345[20]
4495	Cyclohexane, 1,2,4-trimethyl-, (1α,2β,4β)-	C_9H_{18} 126.24	7667-60-9 -83.5	142.9	4-05-00-00138 0.7870[25]	1.4341[20]
4496	Cyclohexane, 1,3,5-trimethyl-, (1α,3α,5α)-	C_9H_{18} 126.24	1795-27-3 -49.7	138.5	4-05-00-00138 0.7708[20]	H_2O 1; eth 3; bz 3; ctc 3 1.4269[20]
4497	Cyclohexane, 1,3,5-trimethyl-, (1α,3α,5β)- trans-1,3,5-Trimethylcyclohexane	C_9H_{18} 126.24	1795-26-2 -107.4	140.5	4-05-00-00138 0.7794[20]	bz 4; eth 4; lig 4 1.4307[20]
4498	Cyclohexane, undecafluoro- Undecafluorocyclohexane	C_6HF_{11} 282.06	308-24-7	62.0		
4499	Cyclohexane, undecafluoro(trifluoromethyl)- Perfluoro(methylcyclohexane)	C_7F_{14} 350.05	355-02-2 -44.7	76.3	4-05-00-00097 1.7878[25]	ace 3; bz 3; ctc 3; tol 3 1.285[17]
4500	Cyclohexanol Cyclohexyl alcohol	$C_6H_{12}O$ 100.16	108-93-0 25.4	2731 160.8	4-06-00-00020 0.9624[20]	H_2O 3; EtOH 3; eth 3; ace 3 1.4641[20]
4501	Cyclohexanol, 2-amino-, trans-(±)- Cyclohexanol, 2-amino (trans, DL)	$C_6H_{13}NO$ 115.18	33092-82-9 68	219; 105[10]	4-13-00-00540	bz 4; chl 4; HOAc 4
4502	Cyclohexanol, 3-(aminomethyl)-3,5,5-trimethyl- 1-Hydroxy-3-aminomethyl-3,5,5-trimethylcyclohexane	$C_{10}H_{21}NO$ 171.28	15647-11-7 45.5	265	0.969[25]	1.4904[20]
4503	Cyclohexanol, 2-bromo- 2-Bromo-1-cyclohexanol	$C_6H_{11}BrO$ 179.06	24796-87-0	85[10]; 60[1]	4-06-00-00063 1.4604[20]	1.5169[20]
4504	Cyclohexanol, 2-butyl-, trans- trans-2-Butylcyolohexanol	$C_{10}H_{20}O$ 156.27	35242-05-8	111-2[16]	3-06-00-00122 0.9020[20]	H_2O 1; eth 3; ace 3; bz 3 1.4641[20]
4505	Cyclohexanol, 2-chloro-, trans- trans-2-Chlorocyclohexanol	$C_6H_{11}ClO$ 134.61	6628-80-4 29	93[20]	4-06-00-00064 1.146[16]	bz 4; eth 4; EtOH 4; chl 4 1.4899[20]
4506	Cyclohexanol, 2-chloro-, cis-(±)-	$C_6H_{11}ClO$ 134.61	116783-28-9 36.5	93-4[26]	4-06-00-00064 1.1261[25]	bz 4; EtOH 4; chl 4 1.4894[25]
4507	Cyclohexanol, 4-chloro-, trans- trans-4-Chlorocyclohexanol	$C_6H_{11}ClO$ 134.61	29538-77-0 82.5	106[14]	4-06-00-00068 1.1435[13]	EtOH 3; eth 3; bz 3; chl 3 1.4930[17]
4508	Cyclohexanol, 1,2-dimethyl-, cis- Cyclohexanol, 1,2-dimethyl-, stereoisomer	$C_8H_{16}O$ 128.21	19879-11-9 23.2	82.8[25]	4-06-00-00121 0.9250[20]	1.4625[20]
4509	Cyclohexanol, 1,2-dimethyl-, trans- Cyclohexanol, 1,2-dimethyl-, stereoisomer	$C_8H_{16}O$ 128.21	19879-12-0 13.2	74[25]	4-06-00-00121 0.9187[20]	1.4590[20]
4510	Cyclohexanol, 1,3-dimethyl-, cis- cis-1,3-Dimethylcyclohexanol	$C_8H_{16}O$ 128.21	15466-94-1 27.5	84[25]	3-06-00-00092 0.9022[20]	1.4575[20]
4511	Cyclohexanol, 1,3-dimethyl-, trans- trans-1,3-Dimethylcyclohexanol	$C_8H_{16}O$ 128.21	15466-93-0 14.5	77.6[25]	3-06-00-00093 0.8894[30]	1.4507[20]
4512	Cyclohexanol, 1,4-dimethyl-, cis-	$C_8H_{16}O$ 128.21	16980-61-3 24	83.5[25]	3-06-00-00098 0.9011[30]	1.4564[20]
4513	Cyclohexanol, 2,2-dimethyl- 2,2-Dimethylcyclohexanol	$C_8H_{16}O$ 128.21	1193-46-0 8	177	0.9225[20]	1.4648[20]
4514	Cyclohexanol, 2,4-dimethyl- 2,4-Dimethylcyclohexanol	$C_8H_{16}O$ 128.21	69542-91-2	177.5	4-06-00-00124 0.900[20]	1.4500[20]
4515	Cyclohexanol, 2,6-dimethyl- 2,6-Dimethyl-1-cyclohexanol	$C_8H_{16}O$ 128.21	5337-72-4 32.5	176	2-06-00-00028 0.944[25]	1.4600[20]
4516	Cyclohexanol, 3,3-dimethyl- 3,3-Dimethylcyclohexanol	$C_8H_{16}O$ 128.21	767-12-4 11.5	185; 99.5[35]	3-06-00-00089 0.9128[14]	1.4606[15]
4517	Cyclohexanol, 3,4-dimethyl- 3,4-Dimethylcyclohexyl alcohol	$C_8H_{16}O$ 128.21	5715-23-1	189	3-06-00-00091 0.9073[16]	1.458[16]
4518	Cyclohexanol, 3,5-dimethyl- 3,5-Dimethylcyclohexanol	$C_8H_{16}O$ 128.21	5441-52-1 11.6	187	2-06-00-00029 0.898[20]	1.4550[20]
4519	Cyclohexanol, 4,4-dimethyl- 4,4-Dimethylcyclohexanol	$C_8H_{16}O$ 128.21	932-01-4 16	186	4-06-00-00120 0.9250[20]	1.4613[20]
4520	Cyclohexanol, 3-(dimethylamino)-	$C_8H_{17}NO$ 143.23	6890-03-5 73	231	3-13-00-00719 0.9737[25]	EtOH 4 1.4852[20]
4521	Cyclohexanol, 2-(1,1-dimethylethyl)-	$C_{10}H_{20}O$ 156.27	13491-79-7 45	139[95]	3-06-00-00125 0.902[25]	
4522	Cyclohexanol, 1-ethyl- 1-Ethylcyclohexanol	$C_8H_{16}O$ 128.21	1940-18-7 34.5	166	4-06-00-00115 0.9227[25]	H_2O 2; bz 3; ctc 2; peth 3 1.4633[20]
4523	Cyclohexanol, 2-ethyl-, trans-(±)- Cyclohexanol, 2-ethyl (trans, DL)	$C_8H_{16}O$ 128.21	89886-25-9	79[12]	4-06-00-00117 0.9193[21]	ace 4; bz 4; eth 4; peth 4 1.4640[21]
4524	Cyclohexanol, 2-ethyl-, cis-(±)-	$C_8H_{16}O$ 128.21	116697-35-9	181	3-06-00-00085 0.9274[21]	ace 4; bz 4; eth 4; peth 4 1.4655[21]
4525	Cyclohexanol, 4-ethyl- 4-Ethylcyclohexanol	$C_8H_{16}O$ 128.21	4534-74-1	84[10]	2-06-00-00026 0.889[25]	1.4625[20]
4526	Cyclohexanol, 1-ethyl-2-methyl- 1-Ethyl-2-methylcyclohexanol	$C_9H_{18}O$ 142.24	102370-18-3	182; 82[73]	4-06-00-00133 0.9235[20]	1.458[20]

No.	Name Synonym	Mol. Form. Mol. Wt.	CAS RN mp/°C	Merck No. bp/°C	Beil. Ref. den/g cm^{-3}	Solubility n_D
4527	Cyclohexanol, 1-ethyl-3-methyl- 1-Ethyl-3-methylcyclohexanol	$C_9H_{18}O$ 142.24	62067-44-1	88[20]	2-06-00-00035 0.9013[20]	1.459
4528	Cyclohexanol, 1,1'-(1,2-ethynediyl)bis-	$C_{14}H_{22}O_2$ 222.33	78-54-6 112.5	182[13]	4-06-00-06076	H_2O 1; EtOH 4; eth 4; ace 4
4529	Cyclohexanol, 1-ethynyl-	$C_8H_{12}O$ 124.18	78-27-3 31.5	174	4-06-00-00348 0.9873[20]	H_2O 1; EtOH 3; bz 3; chl 2 1.4822[20]
4530	Cyclohexanol, 1-methyl- 1-Methylcyclohexanol	$C_7H_{14}O$ 114.19	590-67-0 25	155; 70[25]	4-06-00-00095 0.9194[20]	H_2O 1; EtOH 3; bz 3; chl 3 1.4595[20]
4531	Cyclohexanol, 2-methyl-, cis-(±)- (±)-cis-2-Methylcyclohexanol	$C_7H_{14}O$ 114.19	615-38-3 7	165	4-06-00-00100 0.9360[20]	EtOH 4 1.4640[20]
4532	Cyclohexanol, 2-methyl-, trans-(±)- (±)-trans-2-Methylcyclohexanol	$C_7H_{14}O$ 114.19	615-39-4 -4	167.5	4-06-00-00100 0.9247[20]	eth 4; EtOH 4 1.4616[20]
4533	Cyclohexanol, 3-methyl-, cis-(±)- cis-3-Methylcyclohexanol	$C_7H_{14}O$ 114.19	5454-79-5 -5.5	168; 94[12]	4-06-00-00100 0.9155[20]	eth 4; EtOH 4 1.4752[20]
4534	Cyclohexanol, 3-methyl-, trans-(±)- trans-3-Methylcyclohexanol	$C_7H_{14}O$ 114.19	7443-55-2 -0.5	167; 84[13]	4-06-00-00100 0.9214[30]	eth 4; EtOH 4 1.4580[20]
4535	Cyclohexanol, 4-methyl-, cis- cis-4-Methylcyclohexanol	$C_7H_{14}O$ 114.19	7731-28-4 -9.2	173	4-06-00-00105 0.9170[20]	eth 4; EtOH 4 1.4614[20]
4536	Cyclohexanol, 4-methyl-, trans- trans-4-Methylcyclohexanol	$C_7H_{14}O$ 114.19	7731-29-5	174	4-06-00-00105 0.9118[21]	H_2O 2; EtOH 5; eth 3 1.4561[20]
4537	Cyclohexanol, 2-methylene-	$C_7H_{12}O$ 112.17	4065-80-9	82-4[13]	4-06-00-00208 0.955[20]	1.4843[20]
4538	Cyclohexanol, 1-(1-methylethyl)-	$C_9H_{18}O$ 142.24	3552-01-0	76-9[10]	4-06-00-00131 0.928[19]	1.4683[19]
4539	Cyclohexanol, 2-(1-methylethyl)-, cis- cis-2-Isopropylcyclohexanol	$C_9H_{18}O$ 142.24	10488-25-2 52.5	77[13]	4-06-00-00131 0.9223[25]	1.4665[25]
4540	Cyclohexanol, 4-(1-methylethyl)-, cis- cis-4-Isopropylcyclohexanol	$C_9H_{18}O$ 142.24	22900-08-9	72[1.9]	3-06-00-00111 0.9195[20]	1.4671[20]
4541	Cyclohexanol, 4-(1-methylethyl)-, trans- trans-4-Isopropylcyclohexanol	$C_9H_{18}O$ 142.24	15890-36-5	79.5[1.8]	3-06-00-00112 0.9140[20]	1.4658[20]
4542	Cyclohexanol, 1-methyl-4-(1-methylethenyl)- β-Terpineol	$C_{10}H_{18}O$ 154.25	138-87-4 32.5	210; 90[10]	4-06-00-00254 0.917[20]	1.4747[20]
4543	Cyclohexanol, 2-methyl-5-(1-methylethenyl)-, [1S-(1α,2β,5α)]-	$C_{10}H_{18}O$ 154.25	22567-21-1	225; 107[14]	3-06-00-00256 0.9274[20]	1.4780[20]
4544	Cyclohexanol, 5-methyl-2-(1-methylethenyl)-, [1R-(1α,2β,5α)]-	$C_{10}H_{18}O$ 154.25	89-79-2 78	93[14]	3-06-00-00257 0.911[20]	H_2O 2; EtOH 3; eth 3 1.4723[20]
4545	Cyclohexanol, 5-methyl-2-(1-methylethenyl)-, acetate, [1R-(1α,2β,5α)]-	$C_{12}H_{20}O_2$ 196.29	57576-09-7 85	112-4[8]	4-06-00-00255 0.925[25]	1.4566[20]
4546	Cyclohexanol, 1-methyl-4-(1-methylethyl)-	$C_{10}H_{20}O$ 156.27	21129-27-1	208.5	4-06-00-00147 0.90[20]	1.4619[20]
4547	Cyclohexanol, 2-methyl-5-(1-methylethyl)-, [1R-(1α,2α,5β)]-	$C_{10}H_{20}O$ 156.27	5563-78-0	217.5; 102[18]	4-06-00-00148 0.9012[20]	eth 4; EtOH 4 1.4632[20]
4548	Cyclohexanol, 5-methyl-2-(1-methylethyl)-, (1α,2β,5α)-(±)- (±)-Menthol	$C_{10}H_{20}O$ 156.27	15356-70-4 38	5723 216	4-06-00-00151 0.903[15]	H_2O 1; EtOH 4; eth 4; ace 4 1.4615[20]
4549	Cyclohexanol, 5-methyl-2-(1-methylethyl)-, acetate, [1R-(1α,2β,5α)]-	$C_{12}H_{22}O_2$ 198.31	2623-23-6	109[10]	4-06-00-00153 0.9185[20]	1.4469[20]
4550	Cyclohexanol, 2-methyl-1-propyl- 2-Methyl-1-propylcyclohexanol	$C_{10}H_{20}O$ 156.27	24580-48-1	97-8[34]	3-06-00-00127 0.919[20]	1.48[20]
4551	Cyclohexanol, 3-methyl-1-propyl- 3-Methyl-1-propylcyclohexanol	$C_{10}H_{20}O$ 156.27	24580-52-7	199; 96-8[20]	1-06-00-00018 0.8903[21]	1.4566[24]
4552	Cyclohexanol, 1-phenyl- 1-Phenylcyclohexan-1-ol	$C_{12}H_{16}O$ 176.26	1589-60-2 63.5	157[28]; 112[5]	4-06-00-03908 1.035[16]	bz 4; eth 4; EtOH 4; lig 4 1.5415[16]
4553	Cyclohexanol, 2-phenyl-, cis-(±)- Cyclohexanol, 2-phenyl (cis, DL)	$C_{12}H_{16}O$ 176.26	40960-73-4 41.5	140-1[16]	4-06-00-03909 1.035[16]	1.5415[16]
4554	Cyclohexanol, 2-phenyl-, trans-(±)- Cyclohexanol, 2-phenyl (trans, DL)	$C_{12}H_{16}O$ 176.26	40960-69-8 56.5	153[16]	4-06-00-03909	EtOH 4; chl 4
4555	Cyclohexanol, 1-(2-propenyl)-	$C_9H_{16}O$ 140.23	1123-34-8	190	4-06-00-00235 0.9341[22]	1.4756[22]
4556	Cyclohexanol, 2-(2-propenyl)-, trans- Cyclohexanol, 2-allyl-, trans-	$C_9H_{16}O$ 140.23	24844-28-8	94-6[15]	4-06-00-00235 0.947[20]	HOAc 4 1.4778[20]
4557	Cyclohexanol, 1-propyl- 1-Hydroxy-1-propylcyclohexane	$C_9H_{18}O$ 142.24	5445-24-9	180 dec; 85[20]	4-06-00-00129 0.934[12]	1.468[12]
4558	Cyclohexanol, 2-propyl-, cis- cis-2-Propylcyclohexanol	$C_9H_{18}O$ 142.24	5857-86-3	84[10]	4-06-00-00130 0.9247[11]	1.4688[11]
4559	Cyclohexanol, 2-propyl-, trans- trans-2-Propylcyclohexanol	$C_9H_{18}O$ 142.24	5846-43-5	90[14]	4-06-00-00130 0.9162[11]	1.4668[11]
4560	Cyclohexanol, 4-propyl-, trans- trans-4-Propylcyclohexanol	$C_9H_{18}O$ 142.24	77866-58-1	208; 103[14]	3-06-00-00106 0.9072[17]	1.4613[17]
4561	Cyclohexanol, 4-propyl-, cis-	$C_9H_{18}O$ 142.24	98790-22-8	99[17]	3-06-00-00105 0.9102[16]	1.4652[16]
4562	Cyclohexanol, 1,2,2-trimethyl- 1,2,2-Trimethylcyclohexanol	$C_9H_{18}O$ 142.24	31720-69-1	81[20]; 76[16]	3-06-00-00114 0.9230[20]	bz 4; eth 4; EtOH 4 1.4680[20]

No.	Name / Synonym	Mol. Form. / Mol. Wt.	CAS RN / mp/°C	Merck No. / bp/°C	Beil. Ref. / den/g cm⁻³	Solubility / n_D
4563	Cyclohexanol, 1,2,6-trimethyl- / 1,2,6-Trimethylcyclohexanol	$C_9H_{18}O$ / 142.24	96244-13-2 /	/ 78[22]	1-06-00-00017 / 0.9126[15]	ace 4; bz 4; eth 4; EtOH 4 / 1.4598[15]
4564	Cyclohexanol, 1,3,5-trimethyl- / 1,3,5-Trimethylcyclohexanol	$C_9H_{18}O$ / 142.24	90760-75-1 /	/ 181	3-06-00-00117 / 0.8876[17]	eth 4; EtOH 4; chl 4 / 1.454[16]
4565	Cyclohexanol, 2,2,5-trimethyl- / 2,2,5-Trimethylcyclohexanol	$C_9H_{18}O$ / 142.24	73210-25-0 /	/ 188	0-06-00-00022 / 0.8955[22]	EtOH 4 / 1.4569[20]
4566	Cyclohexanol, 2,2,6-trimethyl- / 1,1,3-Trimethyl-2-cyclohexanol	$C_9H_{18}O$ / 142.24	10130-91-3 / 51	187; 87[28]	4-06-00-00135 / 0.9128[20]	eth 4; EtOH 4; chl 4 / 1.4600[20]
4567	Cyclohexanol, 2,3,3-trimethyl-	$C_9H_{18}O$ / 142.24	116724-17-5 / 28	/ 197	1-06-00-00016 / 0.9002[25]	ace 4; EtOH 4 / 1.4572[25]
4568	Cyclohexanol, 2,3,6-trimethyl- / 2,3,6-Trimethylcyclohexanol	$C_9H_{18}O$ / 142.24	58210-03-0 /	/ 195	0-06-00-00022 / 0.9117[17]	EtOH 4; chl 4
4569	Cyclohexanol, 3,3,5-trimethyl-, cis- / cis-3,3,5-Trimethylcyclohexanol	$C_9H_{18}O$ / 142.24	933-48-2 / 37.3	202; 92[12]	4-06-00-00135 / 0.900[16]	H_2O 1; EtOH 3; eth 3; chl 3 / 1.4550[16]
4570	Cyclohexanol, 3,3,5-trimethyl-, trans- / trans-3,3,5-Trimethylcyclohexanol	$C_9H_{18}O$ / 142.24	767-54-4 / 55.8	/ 189.2	4-06-00-00135 / 0.8631[60]	H_2O 1; EtOH 3; eth 3; chl 3
4571	Cyclohexanone / Pimelic ketone	$C_6H_{10}O$ / 98.14	108-94-1 / -31	2732 / 155.4	4-07-00-00015 / 0.9478[20]	H_2O 3; EtOH 3; eth 3; ace 3 / 1.4507[20]
4572	Cyclohexanone, 2-acetyl-	$C_8H_{12}O_2$ / 140.18	874-23-7 /	/ 111-2[18]	4-07-00-01997 / 1.0782[25]	ctc 3 / 1.5138[20]
4573	Cyclohexanone, 2,6-bis(phenylmethylene)-	$C_{20}H_{18}O$ / 274.36	897-78-9 / 117.5	185-95[20]	3-07-00-02661 /	EtOH 2; bz 3; HOAc 3
4574	Cyclohexanone, 2-bromo-	C_6H_9BrO / 177.04	822-85-5 /	/ 89-90[14]	4-07-00-00034 / 1.340[25]	/ 1.5085[25]
4575	Cyclohexanone, 2-butyl- / 2-Butylcyclohexanone	$C_{10}H_{18}O$ / 154.25	1126-18-7 /	/ 70[2]	4-07-00-00079 / 0.905[20]	H_2O 1; ctc 2; os 4 / 1.4545[20]
4576	Cyclohexanone, 2-butylidene- / 2-Butylidenecyclohexanone	$C_{10}H_{16}O$ / 152.24	7153-14-2 /	/ 98-100[10]	4-07-00-00181 / 0.935[20]	ace 4; bz 4; eth 4; EtOH 4 / 1.4800[20]
4577	Cyclohexanone, 2-chloro- / 2-Chlorocyclohexanone	C_6H_9ClO / 132.59	822-87-7 / 23	/ 82[15]	4-07-00-00032 / 1.160[20]	eth 3; bz 3; ctc 2; diox 3 / 1.4825[20]
4578	Cyclohexanone, 4-chloro- / 4-Chlorocyclohexanone	C_6H_9ClO / 132.59	21299-26-3 /	/ 95[17]	4-07-00-00032 /	eth 4 / 1.4867[20]
4579	Cyclohexanone, 2-(2-chlorophenyl)-2-(methylamino)-, (±)- / Ketamine	$C_{13}H_{16}ClNO$ / 237.73	6740-88-1 / 92.5	5174		
4580	Cyclohexanone, 2,2-dimethyl- / 2,2-Dimethyl-1-cyclohexanone	$C_8H_{14}O$ / 126.20	1193-47-1 / -20.5	/ 172	4-07-00-00055 / 0.9145[20]	/ 1.4486[20]
4581	Cyclohexanone, 2,3-dimethyl-, cis- / cis-2,3-Dimethylcyclohexanone	$C_8H_{14}O$ / 126.20	1551-88-8 /	178.5; 99[57]	3-07-00-00088 / 0.9159[20]	/ 1.4505[20]
4582	Cyclohexanone, 2,3-dimethyl-, trans- / trans-2,3-Dimethylcyclohexanone	$C_8H_{14}O$ / 126.20	1551-89-9 /	/ 96[57]	4-07-00-00057 / 0.908[22]	/ 1.4511[26]
4583	Cyclohexanone, 2,4-dimethyl-, (2S-trans)- / Cyclohexanone, 2,4-dimethyl (trans, d)	$C_8H_{14}O$ / 126.20	93921-42-7 /	179; 69[17]	3-07-00-00093 / 0.9004[16]	ace 4; bz 4; eth 4 / 1.4488[22]
4584	Cyclohexanone, 2,4-dimethyl-, cis-	$C_8H_{14}O$ / 126.20	116783-29-0 /	/ 177	3-07-00-00093 / 0.910[15]	/ 1.4493[15]
4585	Cyclohexanone, 2,5-dimethyl-, trans-(±)- / trans-(±)-2,5-Dimethylcyclohexanone	$C_8H_{14}O$ / 126.20	66395-23-1 /	/ 172	4-07-00-00059 / 0.9025[20]	eth 4; EtOH 4 / 1.4446[20]
4586	Cyclohexanone, 2,6-dimethyl- / 2,6-Dimethylcyclohexanone	$C_8H_{14}O$ / 126.20	2816-57-1 /	/ 175	2-07-00-00025 / 0.925[25]	/ 1.4460[20]
4587	Cyclohexanone, 3,3-dimethyl- / 3,3-Dimethylcyclohexanone	$C_8H_{14}O$ / 126.20	2979-19-3 /	180; 72-3[25]	4-07-00-00056 / 0.909[15]	/ 1.4482[17]
4588	Cyclohexanone, 3,4-dimethyl- / 3,4-Dimethylcyclohexanone	$C_8H_{14}O$ / 126.20	5465-09-8 /	/ 187	3-07-00-00089 / 0.906[20]	/ 1.4507[20]
4589	Cyclohexanone, 3,5-dimethyl- / 3,5-Dimethylcyclohexanone	$C_8H_{14}O$ / 126.20	2320-30-1 /	/ 182.5	4-07-00-00058 / 0.896[20]	/ 1.4407[20]
4590	Cyclohexanone, 4,4-dimethyl- / 4,4-Dimethylcyclohexanone	$C_8H_{14}O$ / 126.20	4255-62-3 / 39	/ 73[14]	4-07-00-00056 / 0.932[20]	/ 1.4537[24]
4591	Cyclohexanone, 2-[(dimethylamino)methyl]- / 2-[(Dimethylamino)methyl]cyclohexanone	$C_9H_{17}NO$ / 155.24	15409-60-6 /	92[10.5]	4-14-00-00008 / 0.9504[20]	H_2O 2; EtOH 3; eth 3 / 1.4672[20]
4592	Cyclohexanone, 4-(1,1-dimethylpropyl)-	$C_{11}H_{20}O$ / 168.28	16587-71-6 /	124-5[16]	4-07-00-00096 / 0.920[25]	/ 1.4677[20]
4593	Cyclohexanone, 4-ethyl- / 4-Ethylcyclohexanone	$C_8H_{14}O$ / 126.20	5441-51-0 /	/ 193	3-07-00-00083 / 0.895[25]	/ 1.4515[20]
4594	Cyclohexanone, 2-ethylidene- / 2-Ethylidenecyclohexanone	$C_8H_{12}O$ / 124.18	1122-25-4 /	92[20]	4-07-00-00146 / 0.962[20]	H_2O 1; os 4 / 1.4882[20]
4595	Cyclohexanone, 2-ethyl-2-methyl- / 2-Ethyl-2-methylcyclohexanone	$C_9H_{16}O$ / 140.23	17206-52-9 /	/ 195	2-07-00-00031 / 0.9037[17]	/ 1.4515[17]
4596	Cyclohexanone, 2-methyl-, (±)- / (±)-2-Methylcyclohexanone	$C_7H_{12}O$ / 112.17	24965-84-2 / -13.9	/ 165	4-07-00-00041 / 0.9250[20]	H_2O 1; EtOH 3; eth 3 / 1.4483[25]
4597	Cyclohexanone, 3-methyl-, (±)-	$C_7H_{12}O$ / 112.17	625-96-7 / -73.5	169; 65[15]	4-07-00-00043 / 0.9136[20]	H_2O 1; EtOH 3; eth 3 / 1.4456[20]
4598	Cyclohexanone, 4-methyl- / 4-Methylcyclohexanone	$C_7H_{12}O$ / 112.17	589-92-4 / -40.6	/ 170	4-07-00-00044 / 0.9138[20]	H_2O 1; EtOH 3; eth 3; ctc 2 / 1.4451[20]
4599	Cyclohexanone, 2-(1-methylethyl)- / 2-Isopropylcyclohexanone	$C_9H_{16}O$ / 140.23	1004-77-9 / 72.5	83[17]	4-07-00-00066 / 0.922[16]	H_2O 1; EtOH 3; eth 3; ace 3 / 1.4564[15]
4600	Cyclohexanone, 4-(1-methylethyl)- / 4-Isopropylcyclohexanone	$C_9H_{16}O$ / 140.23	5432-85-9 /	139-40[100]	4-07-00-00066 / 0.9158[25]	/ 1.4552[25]
4601	Cyclohexanone, 2-methyl-5-(1-methylethenyl)-, (2S-trans)-	$C_{10}H_{16}O$ / 152.24	619-02-3 / -11	/ 221.5	0.9253[20]	/ 1.4717[20]

No.	Name / Synonym	Mol. Form. / Mol. Wt.	CAS RN / mp/°C	Merck No. / bp/°C	Beil. Ref. / den/g cm⁻³	Solubility / n_D
4602	Cyclohexanone, 5-methyl-2-(1-methylethenyl)-	$C_{10}H_{16}O$ / 152.24	529-00-0	/ 100[18]	4-07-00-00190 / 0.9198[20]	/ 1.4675[20]
4603	Cyclohexanone, 5-methyl-2-(1-methylethenyl)-, *trans*-	$C_{10}H_{16}O$ / 152.24	29606-79-9	/ 100[18]	4-07-00-00190 / 0.9198[20]	/ 1.4675[20]
4604	Cyclohexanone, 2-methyl-5-(1-methylethyl)-, (2S-*trans*)-	$C_{10}H_{18}O$ / 154.25	13163-73-0	/ 220; 96[15]	4-07-00-00085 / 0.904[20]	ace 4; EtOH 3; chl 4 / 1.4553[20]
4605	Cyclohexanone, 5-methyl-2-(1-methylethyl)-, (2R-*cis*)- Menthone(*d*)	$C_{10}H_{18}O$ / 154.25	1196-31-2	/ 204; 85[14]	4-07-00-00087 / 0.8947[20]	H_2O 2; EtOH 3; eth 3; ace 3 / 1.4503[20]
4606	Cyclohexanone, 5-methyl-2-(1-methylethyl)-, (2S-*trans*)- (-)-Menthone	$C_{10}H_{18}O$ / 154.25	14073-97-3 / -6	5724 / 207	/ 0.8954[20]	H_2O 2; EtOH 5; eth 5; ace 3 / 1.4505[20]
4607	Cyclohexanone, 5-methyl-2-(1-methylethylidene)-	$C_{10}H_{16}O$ / 152.24	15932-80-6	/ 95-6[11]	4-07-00-00188 / 0.9367[20]	/ 1.4869[20]
4608	Cyclohexanone, 5-methyl-2-(1-methylethylidene)-, (*R*)- Pulegone	$C_{10}H_{16}O$ / 152.24	89-82-7	7955 / 224	4-07-00-00188 / 0.9346[45]	H_2O 1; EtOH 5; eth 5; ctc 3 / 1.4894[20]
4609	Cyclohexanone, 4-(2-methylpropyl)-	$C_{10}H_{18}O$ / 154.25	42061-72-3	/ 218; 98[12]	4-07-00-00082 / 0.8914[24]	/ 1.4502[24]
4610	Cyclohexanone, oxime (Hydroxyimino)cyclohexane	$C_6H_{11}NO$ / 113.16	100-64-1 / 90	/ 206	4-07-00-00021	H_2O 3; EtOH 3; eth 3; chl 2
4611	Cyclohexanone, 2-(1-piperidinylmethyl)- Pimeclone	$C_{12}H_{21}NO$ / 195.30	534-84-9	7399 / 118-20[14]	5-20-02-00330	
4612	Cyclohexanone, 2-propyl- 2-Propylcyclohexanone	$C_9H_{16}O$ / 140.23	94-65-5 /	/ 197	4-07-00-00064 / 0.927[20]	H_2O 1; EtOH 3; eth 4; ace 3 / 1.4538[20]
4613	Cyclohexanone, 4-propyl-	$C_9H_{16}O$ / 140.23	40649-36-3	/ 213; 97.5[15]	3-07-00-00116 / 0.9072[19]	/ 1.4530[19]
4614	Cyclohexanone, 2,2,3-trimethyl- 2,2,3-Trimethylcyclohexanone	$C_9H_{16}O$ / 140.23	39257-08-4	/ 66-7[10]	4-07-00-00069 / 0.9234[15]	/ 1.4569[15]
4615	Cyclohexanone, 2,2,4-trimethyl- 2,2,4-Trimethylcyclohexanone	$C_9H_{16}O$ / 140.23	35413-38-8	/ 86-7[12]	3-07-00-00125 / 0.8914[20]	/ 1.4520[20]
4616	Cyclohexanone, 2,2,5-trimethyl- 2,2,5-Trimethylcyclohexanone	$C_9H_{16}O$ / 140.23	933-36-8	/ 183	3-07-00-00126 / 0.8871[24]	/ 1.4432[24]
4617	Cyclohexanone, 2,2,6-trimethyl-	$C_9H_{16}O$ / 140.23	2408-37-9 / -31.8	/ 178.5	4-07-00-00069 / 0.9043[18]	/ 1.4470[20]
4618	Cyclohexanone, 2,3,6-trimethyl- 2,3,6-Trimethylcyclohexanone	$C_9H_{16}O$ / 140.23	42185-47-7	/ 190	3-07-00-00126 / 0.9129[18]	/ 1.4464[21]
4619	Cyclohexanone, 2,4,4-trimethyl- 2,4,4-Trimethylcyclohexanone	$C_9H_{16}O$ / 140.23	2230-70-8	/ 191	4-07-00-00070 / 0.902[20]	/ 1.4493[20]
4620	Cyclohexanone, 2,4,5-trimethyl- 2,4,5-Trimethylcyclohexanone	$C_9H_{16}O$ / 140.23	51299-55-9	/ 193	4-07-00-00072 / 0.897[20]	/ 1.4479[20]
4621	Cyclohexanone, 2,4,6-trimethyl- 2,4,6-Trimethylcyclohexanone	$C_9H_{16}O$ / 140.23	90645-54-8	/ 185	3-07-00-00127 / 0.8992[20]	/ 1.4458[20]
4622	Cyclohexanone, 3,3,5-trimethyl- Dihydroisophorone	$C_9H_{16}O$ / 140.23	873-94-9	/ 189	4-07-00-00070 / 0.8919[19]	/ 1.4454[15]
4623	Cyclohexanone, 3,4,4-trimethyl- 3,4,4-Trimethylcyclohexanone	$C_9H_{16}O$ / 140.23	40441-35-8	/ 80-1[13]	3-07-00-00122 / 0.911[25]	/ 1.4552[20]
4624	Cyclohexasiloxane, dodecamethyl- Dodecamethylcyclohexasiloxane	$C_{12}H_{36}O_6Si_6$ / 444.93	540-97-6 / -3	3400 / 245	4-04-00-04129 / 0.9672[25]	H_2O 1 / 1.4015[20]
4625	Cyclohexasiloxane, 2,4,6,8,10,12-hexamethyl- Hexamethylcyclohexasiloxane	$C_6H_{24}O_6Si_6$ / 360.77	6166-87-6 / -79	/	4-04-00-04100 / 1.006[20]	H_2O 1 / 1.3944[20]
4626	Cyclohexene Tetrahydrobenzene	C_6H_{10} / 82.15	110-83-8 / -103.5	2733 / 82.9	4-05-00-00218 / 0.8110[20]	H_2O 1; EtOH 5; eth 5; ace 5 / 1.4465[20]
4627	Cyclohexene, 1-bromo-	C_6H_9Br / 161.04	2044-08-8	/ 165	4-05-00-00236 / 1.3901[20]	eth 3; ace 4; bz 3; chl 4 / 1.5134[20]
4628	Cyclohexene, 3-bromo-	C_6H_9Br / 161.04	1521-51-3	/ 80-2[40]	4-05-00-00237 / 1.3890[20]	H_2O 1; eth 3; bz 3; chl 3 / 1.5320[20]
4629	3-Cyclohexene-1-carbonitrile	C_7H_9N / 107.16	100-45-8	/ 39[1.5]	4-09-00-00116	ctc 3 / 1.4716[20]
4630	1-Cyclohexenecarbonitrile 1-Cyanocyclohexene	C_7H_9N / 107.16	1855-63-6	/ 81[12]		
4631	1-Cyclohexene-1-carboxaldehyde	$C_7H_{10}O$ / 110.16	1192-88-7	/ 72[15]	4-07-00-00132 / 0.9694[20]	EtOH 3; eth 3 / 1.5005[20]
4632	3-Cyclohexene-1-carboxaldehyde	$C_7H_{10}O$ / 110.16	100-50-5 / 2	/ 105	4-07-00-00134 / 0.9692[20]	ace 3; ctc 2; MeOH 3
4633	1-Cyclohexene-1-carboxaldehyde, 4-(1-methylethenyl)- Perillaldehyde	$C_{10}H_{14}O$ / 150.22	2111-75-3	7119	2-07-00-00130	
4634	1-Cyclohexene-1-carboxaldehyde, 4-(1-methylethenyl)-, (*R*)-	$C_{10}H_{14}O$ / 150.22	5503-12-8	/ 238; 99[7]	4-07-00-00316 / 0.953[20]	/ 1.5058[20]
4635	1-Cyclohexene-1-carboxaldehyde, 4-(1-methylethyl)- Phellandral	$C_{10}H_{16}O$ / 152.24	21391-98-0	/ 223	4-07-00-00187 / 0.93[20]	/ 1.4911[20]
4636	1-Cyclohexene-1-carboxaldehyde, 2,6,6-trimethyl- β-Cyclocitral	$C_{10}H_{16}O$ / 152.24	432-25-7	/ 95-100[15]	4-07-00-00192 / 0.959[15]	/ 1.4971[15]

No.	Name Synonym	Mol. Form. Mol. Wt.	CAS RN mp/°C	Merck No. bp/°C	Beil. Ref. den/g cm^{-3}	Solubility n_D
4637	1-Cyclohexene-1-carboxylic acid Cyclohexene-1-carboxylic acid	$C_7H_{10}O_2$ 126.16	636-82-8 38	241	4-09-00-00111 1.109[20]	H_2O 2; EtOH 3; ace 3 1.4902[20]
4638	3-Cyclohexene-1-carboxylic acid	$C_7H_{10}O_2$ 126.16	4771-80-6 17	234.5	4-09-00-00114 1.0820[20]	H_2O 4; EtOH 3; ace 3 1.4814[20]
4639	3-Cyclohexene-1-carboxylic acid, ethyl ester	$C_9H_{14}O_2$ 154.21	15111-56-5	194.5	4-09-00-00114 0.9688[20]	1.4578[20]
4640	3-Cyclohexene-1-carboxylic acid, methyl ester	$C_8H_{12}O_2$ 140.18	6493-77-2	80[20]	4-09-00-00114 0.995[25]	1.4610[20]
4641	2-Cyclohexene-1-carboxylic acid, 2,6-dimethyl-4-oxo-, ethyl ester	$C_{11}H_{16}O_3$ 196.25	6102-15-4	157-8[18]	4-10-00-02674 1.0493[20]	ace 3 1.4773[20]
4642	Cyclohexene, 1-chloro- 1-Chlorocyclohexene	C_6H_9Cl 116.59	930-66-5	142.5	4-05-00-00230 1.0361[19]	eth 3; ace 3; ctc 3; chl 3 1.4797[20]
4643	Cyclohexene, decafluoro- Perfluorocyclohexene	C_6F_{10} 262.05	355-75-9	52.0	4-05-00-00229 1.6650[25]	1.293[20]
4644	1-Cyclohexene-1,2-dicarboxylic acid, diethyl ester Cyclohexene-1,2-dicarboxylic acid, diethyl ester	$C_{12}H_{18}O_4$ 226.27	92687-41-7	160[14]	3-09-00-03940 1.0803[19]	1.4747[19]
4645	1-Cyclohexene-1,3-dicarboxylic acid, diethyl ester Diethyl 1-cyclohexene-1,3-dicarboxylate	$C_{12}H_{18}O_4$ 226.27	38511-07-8	150[12]	3-09-00-03950 1.0772[20]	1.4722[20]
4646	1-Cyclohexene-1,4-dicarboxylic acid, dimethyl ester Cyclohexene-1,4-dicarboxylic acid, dimethyl ester	$C_{10}H_{14}O_4$ 198.22	22646-79-3 39	153[20]	4-09-00-02897	ace 4; eth 4; EtOH 4
4647	Cyclohexene, 1,2-dichloro-3,3,4,4,5,5,6,6-octafluoro-	$C_6Cl_2F_8$ 294.96	336-19-6 -70	113	4-05-00-00231 1.719[20]	1.375[20]
4648	Cyclohexene, 1,2-dimethyl-	C_8H_{14} 110.20	1674-10-8 -84.1	138	4-05-00-00268 0.8220[25]	1.4620[20]
4649	Cyclohexene, 1,3-dimethyl- 1,3-Dimethyl-1-cyclohexene	C_8H_{14} 110.20	2808-76-6	127	4-05-00-00270 0.799[25]	1.449[20]
4650	Cyclohexene, 1,4-dimethyl- 1,4-Dimethyl-1-cyclohexene	C_8H_{14} 110.20	2808-79-9 -59.4	128.5	4-05-00-00270 0.8005[20]	1.4457[20]
4651	Cyclohexene, 1,5-dimethyl- Cyclohexene, 2,4-dimethyl-	C_8H_{14} 110.20	2808-77-7	126	3-05-00-00216 0.8009[25]	1.450[20]
4652	Cyclohexene, 1,6-dimethyl-	C_8H_{14} 110.20	1759-64-4	131	4-05-00-00269 0.815[20]	1.454[20]
4653	Cyclohexene, 3,3-dimethyl- 3,3-Dimethylcyclohexene	C_8H_{14} 110.20	695-28-3	119	4-05-00-00268 0.804[20]	1.445[20]
4654	Cyclohexene, 3,5-dimethyl- 3,5-Dimethylcyclohexene	C_8H_{14} 110.20	823-17-6	124.5	3-05-00-00216 0.8005[18]	1.443[18]
4655	Cyclohexene, 3,6-dimethyl- 3,6-Dimethylcyclohexene	C_8H_{14} 110.20	19550-40-4	125	4-05-00-00271 0.802[25]	1.4433[25]
4656	Cyclohexene, 4,4-dimethyl-	C_8H_{14} 110.20	14072-86-7 -74.4	117	4-05-00-00268 0.7996[20]	1.4420[20]
4657	Cyclohexene, 1-ethenyl-	C_8H_{12} 108.18	2622-21-1	145	4-05-00-00405 0.8623[15]	H_2O 1; eth 3; bz 3; MeOH 4 1.4915[20]
4658	Cyclohexene, 4-ethenyl-	C_8H_{12} 108.18	100-40-3 -108.9	128	4-05-00-00406 0.8299[20]	H_2O 1; eth 3; bz 3; peth 3 1.4639[20]
4659	Cyclohexene, 6-ethenyl-6-methyl-1-(1-methylethyl)-3-(1-methylethylidene)-, (S)-	$C_{15}H_{24}$ 204.36	5951-67-7	120-30[8]	3-05-00-01083 0.8782[20]	ace 4; bz 4 1.5130[26]
4660	Cyclohexene, 1-ethyl-	C_8H_{14} 110.20	1453-24-3 -109.9	137	4-05-00-00266 0.8176[25]	1.4567[20]
4661	Cyclohexene, 3-ethyl- 3-Ethyl-1-cyclohexene	C_8H_{14} 110.20	2808-71-1	134	4-05-00-00266 0.814[20]	1.451[20]
4662	Cyclohexene, 4-ethyl- 4-Ethyl-1-cyclohexene	C_8H_{14} 110.20	3742-42-5	133	4-05-00-00267 0.810[20]	1.449[20]
4663	Cyclohexene, 1-ethyl-4-methyl- 1-Ethyl-4-methylcyclohexene	C_9H_{16} 124.23	62088-36-2	149	4-05-00-00286 0.8169[16]	1.453[16]
4664	1-Cyclohexene-1-methanol, 4-(1-methylethenyl)-	$C_{10}H_{16}O$ 152.24	536-59-4	244; 12.5[12]	2-06-00-00102 0.9690[20]	1.5005[20]
4665	1-Cyclohexene-1-methanol, 2,4,4-trimethyl- β-Cyclolavandulol	$C_{10}H_{18}O$ 154.25	103985-40-6	105-6[15]	4-06-00-00259 0.9153[20]	1.4814[20]
4666	1-Cyclohexene-1-methanol, 2,6,6-trimethyl- Cyclogeranyl alcohol	$C_{10}H_{18}O$ 154.25	472-20-8 43.5		3-06-00-00259 0.945[19]	1.487[19]
4667	2-Cyclohexene-1-methanol, 2,6,6-trimethyl- α-Cyclogeraniol	$C_{10}H_{18}O$ 154.25	6627-74-3	215	4-06-00-00257 0.9382[20]	1.4843[20]
4668	3-Cyclohexene-1-methanol, α,α,4-trimethyl-, (±)-	$C_{10}H_{18}O$ 154.25	2438-12-2 40.5	220	4-06-00-00252 0.9337[20]	ace 4; bz 4; eth 4; EtOH 4 1.4831[20]
4669	3-Cyclohexene-1-methanol, α,α,4-trimethyl-, acetate	$C_{12}H_{20}O_2$ 196.29	80-26-2	140[40]; 105[11]	4-06-00-00253 0.9659[21]	H_2O 1; EtOH 3; eth 3; bz 3 1.4689[21]
4670	Cyclohexene, 1-methyl-	C_7H_{12} 96.17	591-49-1 -120.4	110.3	4-05-00-00245 0.8102[20]	H_2O 1; eth 3; bz 3; ctc 3 1.4503[20]

No.	Name / Synonym	Mol. Form. / Mol. Wt.	CAS RN / mp/°C	Merck No. / bp/°C	Beil. Ref. / den/g cm^{-3}	Solubility / n_D
4671	Cyclohexene, 3-methyl-, (±)- / (±)-3-Methylcyclohexene	C_7H_{12} / 96.17	56688-75-6 / -115.5	104	4-05-00-00247 / 0.7990[20]	bz 4; eth 4; chl 4; peth 4 / 1.4414[20]
4672	Cyclohexene, 4-methyl-	C_7H_{12} / 96.17	591-47-9 / -115.5	102.7	4-05-00-00248 / 0.7991[20]	H_2O 1; EtOH 3; eth 3 / 1.4414[20]
4673	Cyclohexene, 3-methylene-6-(1-methylethyl)- / β-Phellandrene	$C_{10}H_{16}$ / 136.24	555-10-2 /	7152 / 171.5	2-05-00-00087 / 0.8520[20]	H_2O 1; EtOH 1; eth 3 / 1.4788[20]
4674	Cyclohexene, 3-methylene-6-(1-methylethyl)-, (+)-	$C_{10}H_{16}$ / 136.24	6153-16-8 /	171.5	4-05-00-00436 / 0.8520[20]	eth 4 / 1.4788[20]
4675	Cyclohexene, 4-methylene-1-(1-methylethyl)-	$C_{10}H_{16}$ / 136.24	99-84-3 /	173.5	4-05-00-00437 / 0.838[22]	1.4754[22]
4676	Cyclohexene, 1-methyl-3-(1-methylethenyl)-	$C_{10}H_{16}$ / 136.24	38738-60-2 /	175	4-05-00-00434 / 0.8479[18]	1.4760[18]
4677	Cyclohexene, 1-methyl-4-(1-methylethenyl)- / Limonene	$C_{10}H_{16}$ / 136.24	138-86-3 / -95	5371 / 178	4-05-00-00440 / 0.8402[21]	H_2O 1; EtOH 5; eth 5; ctc 3 / 1.4727[19]
4678	Cyclohexene, 1-methyl-5-(1-methylethenyl)-	$C_{10}H_{16}$ / 136.24	13898-73-2 /	176; 95[20]	4-05-00-00435 / 0.8481[20]	eth 4; EtOH 4 / 1.4804[20]
4679	Cyclohexene, 5-methyl-1-(1-methylethenyl)-, (±)- / Cyclohexene, 1-isopropenyl-5-methyl (DL)	$C_{10}H_{16}$ / 136.24	105065-24-5 /	184	3-05-00-00336 / 0.8594[20]	1.4975
4680	Cyclohexene, 1-methyl-4-(1-methylethyl)-	$C_{10}H_{18}$ / 138.25	5502-88-5 /	174.5	4-05-00-00300 / 0.8457[15]	1.4735[20]
4681	Cyclohexene, 1-methyl-4-(1-methylethyl)-, (R)-	$C_{10}H_{18}$ / 138.25	1195-31-9 /	173	4-05-00-00299 / 0.8246[18]	bz 4; EtOH 4; peth 4 / 1.4563[18]
4682	Cyclohexene, 3-methyl-6-(1-methylethyl)-, (3R-trans)-	$C_{10}H_{18}$ / 138.25	5113-93-9 /	167; 55-6[12]	4-05-00-00300 / 0.824[20]	1.461[20]
4683	Cyclohexene, 4-methyl-1-(1-methylethyl)-, (R)-	$C_{10}H_{18}$ / 138.25	619-52-3 /	167.5	4-05-00-00301 / 0.8078[20]	H_2O 1; EtOH 3; eth 3; bz 3 / 1.4503[15]
4684	Cyclohexene, 1-methyl-4-(1-methylethylidene)- / Terpinolene	$C_{10}H_{16}$ / 136.24	586-62-9 /	186	4-05-00-00437 / 0.8632[15]	H_2O 1; EtOH 5; eth 5; bz 3 / 1.4883[20]
4685	Cyclohexene, 1-methyl-4-(5-methyl-1-methylene-4-hexenyl)-, (S)-	$C_{15}H_{24}$ / 204.36	495-61-4 /	129-30[10.5]	4-05-00-01174 / 0.8673[20]	1.4880[20]
4686	5-Cyclohexene-1,2,3,4-tetrone, 5,6-dihydroxy- / Rhodizonic acid	$C_6H_2O_6$ / 170.08	118-76-3 / 248 dec	sub	4-08-00-03609	EtOH 3
4687	Cyclohexene, 1,2,3-trimethyl- / 1,2,3-Trimethylcyclohexene	C_9H_{16} / 124.23	72312-48-2 /	150	1-05-00-00040 / 0.8347[11]	1.463[11]
4688	Cyclohexene, 1,3,5-trimethyl- / 1,3,5-Trimethylcyclohexene	C_9H_{16} / 124.23	3643-64-9 /	141	3-05-00-00226 / 0.8025[11]	1.449[13]
4689	Cyclohexene, 1,4,4-trimethyl- / 1,4,4-Trimethylcyclohexene	C_9H_{16} / 124.23	3419-71-4 /	140	1-05-00-00039 / 0.8032[19]	1.444[23]
4690	Cyclohexene, 1,4,5-trimethyl- / 1,4,5-Trimethylcyclohexene	C_9H_{16} / 124.23	20030-32-4 /	145	2-05-00-00050 / 0.805[20]	1.4482[20]
4691	Cyclohexene, 1,5,5-trimethyl- / α-Cyclogeraniolene	C_9H_{16} / 124.23	503-46-8 /	140	4-05-00-00286 / 0.7981[23]	1.4461[21]
4692	Cyclohexene, 1,6,6-trimethyl- / 1,6,6-Trimethylcyclohexene	C_9H_{16} / 124.23	69745-49-9 /	147	3-05-00-00225 / 0.8217[20]	1.456[20]
4693	Cyclohexene, 1,5,6-trimethyl-	C_9H_{16} / 124.23	116724-18-6 /	140	3-05-00-00225 / 0.829[25]	1.4572[25]
4694	2-Cyclohexen-1-ol	$C_6H_{10}O$ / 98.14	822-67-3 /	164	4-06-00-00196 / 0.9923[15]	EtOH 3; ace 3 / 1.4790[25]
4695	3-Cyclohexen-1-ol	$C_6H_{10}O$ / 98.14	822-66-2 /	164; 68[16]	4-06-00-00200 / 0.9845[23]	H_2O 2; eth 3; ace 3 / 1.4851[22]
4696	2-Cyclohexen-1-ol, 2-methyl-5-(1-methylethenyl)-	$C_{10}H_{16}O$ / 152.24	99-48-9 /	228	2-06-00-00102 / 0.9484[25]	1.4942[25]
4697	2-Cyclohexen-1-ol, 3-methyl-6-(1-methylethyl)-	$C_{10}H_{18}O$ / 154.25	491-04-3 /	97-8[15.5]	2-06-00-00065 / 0.9119[25]	1.4729[25]
4698	2-Cyclohexen-1-ol, 6-methyl-3-(1-methylethyl)- / Carvenol	$C_{10}H_{18}O$ / 154.25	586-27-6 /	220	4-06-00-00253 / 0.925[22]	1.479
4699	3-Cyclohexen-1-ol, 4-methyl-1-(1-methylethyl)-	$C_{10}H_{18}O$ / 154.25	562-74-3 /	209	4-06-00-00250 / 0.926[20]	1.4785[19]
4700	2-Cyclohexen-1-ol, 1,3,5-trimethyl- / Cyclohexen-3-ol, 1,3,5-trimethyl	$C_9H_{16}O$ / 140.23	33843-55-9 / 46	87-90[17]	1-06-00-00036 / 0.9132[20]	1.4735[19]
4701	2-Cyclohexen-1-ol, 2,4,4-trimethyl- / 2,4,4-Trimethyl-2-cyclohexen-1-ol	$C_9H_{16}O$ / 140.23	73741-61-4 /	193	1-06-00-00036 / 0.9310[25]	bz 4; EtOH 4
4702	2-Cyclohexen-1-ol, 3,5,5-trimethyl- / Isophorol	$C_9H_{16}O$ / 140.23	470-99-5 /	69[5]	3-06-00-00230 / 0.9128[20]	1.4717[20]
4703	2-Cyclohexen-1-one	C_6H_8O / 96.13	930-68-7 / -53	170	4-07-00-00124 / 0.9620[25]	EtOH 4; ace 3 / 1.4883[20]
4704	3-Cyclohexen-1-one	C_6H_8O / 96.13	4096-34-8 /	170	4-07-00-00126 / 0.9620[25]	ace 4; EtOH 4 / 1.4842[18]
4705	2-Cyclohexen-1-one, 2,3-dimethyl-	$C_8H_{12}O$ / 124.18	1122-20-9 /	93-6[20]	4-07-00-00148 / 0.9695[20]	EtOH 3; eth 3 / 1.4995[20]
4706	2-Cyclohexen-1-one, 2,5-dimethyl- / 2,5-Dimethyl-2-cyclohexenone	$C_8H_{12}O$ / 124.18	14845-35-3 /	189.5	1-07-00-00051 / 0.938[22]	eth 4; EtOH 4 / 1.4753[22]
4707	2-Cyclohexen-1-one, 3,5-dimethyl-	$C_8H_{12}O$ / 124.18	1123-09-7 /	208.5	4-07-00-00149 / 0.9400[20]	EtOH 3; eth 3 / 1.4812[20]

No.	Name / Synonym	Mol. Form. / Mol. Wt.	CAS RN / mp/°C	Merck No. / bp/°C	Beil. Ref. / den/g cm⁻³	Solubility / n_D
4708	2-Cyclohexen-1-one, 3,6-dimethyl-	$C_8H_{12}O$ 124.18	15329-10-9	[75][19]	4-07-00-00150 [1.007][18]	H_2O 1; EtOH 3; eth 3 [1.4805][18]
4709	2-Cyclohexen-1-one, 2-hydroxy-3-methyl-6-(1-methylethyl)- Diosphenol	$C_{10}H_{16}O_2$ 168.24	490-03-9 83	3292 [109][10]		
4710	2-Cyclohexen-1-one, 2-methyl-	$C_7H_{10}O$ 110.16	1121-18-2	178.5	4-07-00-00131 [0.966][20]	bz 3 [1.4833][20]
4711	2-Cyclohexen-1-one, 3-methyl-	$C_7H_{10}O$ 110.16	1193-18-6 -21	201	4-07-00-00130 [0.9693][20]	H_2O 5; bz 3 [1.49475][20]
4712	2-Cyclohexen-1-one, 5-methyl- 5-Methyl-2-cyclohexenone	$C_7H_{10}O$ 110.16	7214-50-8	185	4-07-00-00133 [0.947][25]	ace 4; bz 4; EtOH 4 [1.4739][25]
4713	3-Cyclohexen-1-one, 3-methyl-	$C_7H_{10}O$ 110.16	31883-98-4 -21	201	4-07-00-00131 [0.9693][20]	bz 4 [1.4947][20]
4714	3-Cyclohexen-1-one, 4-methyl-	$C_7H_{10}O$ 110.16	5259-65-4	[171; 74][17]	4-07-00-00131 [0.9551][20]	EtOH 3; ace 3; bz 3 [1.4652][20]
4715	2-Cyclohexen-1-one, 2-methyl-5-(1-methylethenyl)-, (S)- (+)-Carvone	$C_{10}H_{14}O$ 150.22	2244-16-8 <15	1883 231	4-07-00-00315 [0.965][20]	H_2O 2; EtOH 4; eth 3; chl 3 [1.4989][20]
4716	2-Cyclohexen-1-one, 3-methyl-6-(1-methylethyl)-, (±)- (±)-Piperitone	$C_{10}H_{16}O$ 152.24	6091-52-7 -19	7443 232.5	4-07-00-00186 [0.9331][20]	ace 4; EtOH 4 [1.4845][20]
4717	2-Cyclohexen-1-one, 6-methyl-3-(1-methylethyl)-, (±)- Carvenone (DL)	$C_{10}H_{16}O$ 152.24	23733-68-8	235.5	4-07-00-00187 [0.9263][20]	ace 4 [1.4826][20]
4718	2-Cyclohexen-1-one, 3-methyl-6-(1-methylethylidene)- Piperitenone	$C_{10}H_{14}O$ 150.22	491-09-8	[120-2][14]	4-07-00-00314 [0.9774][20]	EtOH 4; eth 4 [1.5294][20]
4719	2-Cyclohexen-1-one, 3,4,4-trimethyl- Cyclohexen-3-one, 1,6,6-trimethyl	$C_9H_{14}O$ 138.21	17299-41-1	[217; 98][13]	4-07-00-00164 [0.944][25]	[1.4889][25]
4720	2-Cyclohexen-1-one, 3,4,6-trimethyl- Cyclohexen-3-one, 1,4,6-trimethyl	$C_9H_{14}O$ 138.21	7474-10-4	213	4-07-00-00170 [0.943][17]	[1.484][17]
4721	2-Cyclohexen-1-one, 3,5,5-trimethyl- Isophorone	$C_9H_{14}O$ 138.21	78-59-1 -8.1	215.2	4-07-00-00165 [0.9255][20]	[1.4766][18]
4722	2-Cyclohexen-1-one, 3,6,6-trimethyl- 3,6,6-Trimethyl-2-cyclohexenone	$C_9H_{14}O$ 138.21	23438-77-9	[208; 86-8][15]	3-07-00-00286 [0.927][24]	[1.4798][20]
4723	Cyclohexylamine, N,4-dimethyl- Cyclohexane, 1-methylamino-4-methyl	$C_8H_{17}N$ 127.23	90226-23-6 -15	196	4-05-00-00133 [0.8188][22]	[1.4175][22]
4724	Cyclononane	C_9H_{18} 126.24	293-55-0 11	178.4	4-05-00-00133 [0.8463][25]	[1.4666][20]
4725	Cyclononanone	$C_9H_{16}O$ 140.23	3350-30-9 34	[148][24]; [94][12]	4-07-00-00062 [0.9560][20]	EtOH 3 [1.4729][20]
4726	Cyclononasiloxane, octadecamethyl- Octadecamethylcyclononasiloxane	$C_{18}H_{54}O_9Si_9$ 667.39	556-71-8	[188][20]	3-04-00-01886	bz 4; lig 4 [1.4070][20]
4727	Cyclononene, (E)-	C_9H_{16} 124.23	3958-38-1	[94-6][30]	4-05-00-00280 [0.8615][20]	[1.4799][20]
4728	Cyclononene, (Z)-	C_9H_{16} 124.23	933-21-1	168	4-05-00-00280 [0.8671][20]	bz 4 [1.4805][20]
4729	Cyclononyne	C_9H_{14} 122.21	6573-52-0	178	4-05-00-00419 [0.8972][20]	[1.4890][20]
4730	2,4(1H)-Cyclo-3,4-secoakuammilanium, 3,17-dihydroxy-16-(methoxycarbonyl)-4-methyl-, (3β,16R)- Echitamine	$C_{22}H_{30}N_2O_5$ 402.49	6871-44-9 206	3474		H_2O 3; EtOH 3; eth 3; chl 3
4731	1,3-Cyclooctadecadiene	$C_{18}H_{32}$ 248.45	6568-58-7	[115][3]	4-05-00-00503 [0.8814][25]	[1.4899][20]
4732	1,4-Cyclooctadiene	C_8H_{12} 108.18	1073-07-0 -53	145	4-05-00-00403 [0.8754][20]	
4733	1,5-Cyclooctadiene	C_8H_{12} 108.18	111-78-4 -56.4	150.8	3-05-00-00321 [0.8818][25]	bz 4 [1.4905][25]
4734	1,5-Cyclooctadiene, (Z,Z)-	C_8H_{12} 108.18	1552-12-1 -70	151; [51-2][25]	4-05-00-00403 [0.8818][25]	bz 4 [1.4905][25]
4735	Cyclooctanamine Aminocyclooctane	$C_8H_{17}N$ 127.23	5452-37-9 -48	190	4-12-00-00128 [0.928][25]	[1.4804][20]
4736	Cyclooctane	C_8H_{16} 112.22	292-64-8 14.8	149	4-05-00-00111 [0.8349][20]	H_2O 1; bz 3; lig 3 [1.4586][20]
4737	Cyclooctane, butyl- Butylcyclooctane	$C_{12}H_{24}$ 168.32	16538-93-5	[90-1][10]	4-05-00-00170 [0.8260][25]	[1.4585][25]
4738	Cyclooctanecarboxaldehyde Formylcyclooctane	$C_9H_{16}O$ 140.23	6688-11-5	[96][15]	3-07-00-00112 [0.94][25]	[1.4748][20]
4739	Cyclooctanol	$C_8H_{16}O$ 128.21	696-71-9 25.1	[99][16]	4-06-00-00113 [0.9740][20]	EtOH 3 [1.4871][20]
4740	Cyclooctanone	$C_8H_{14}O$ 126.20	502-49-8 29	196	4-07-00-00049 [0.9581][20]	H_2O 1; EtOH 3; ace 3; bz 3 [1.4694][20]
4741	Cyclooctasiloxane, hexadecamethyl- Hexadecamethylcyclooctasiloxane	$C_{16}H_{48}O_8Si_8$ 593.24	556-68-3 31.5	290	3-04-00-01886 [1.177][25]	bz 4; lig 4 [1.4060][20]
4742	1,3,5,7-Cyclooctatetraene [8]Annulene	C_8H_8 104.15	629-20-9 -4.7	140.5	4-05-00-01331 [0.9206][20]	EtOH 3; eth 3; ace 3; bz 3 [1.5381][20]
4743	1,3,5,7-Cyclooctatetraene, 1-bromo-	C_8H_7Br 183.05	7567-22-8	[52-3][18]	4-05-00-01334 [1.4206][25]	ace 3 [1.5870][25]
4744	1,3,5,7-Cyclooctatetraene, 1-butyl-	$C_{12}H_{16}$ 160.26	13402-37-4	[98][20]	4-05-00-01417 [0.8876][25]	[1.5083][25]

No.	Name Synonym	Mol. Form. Mol. Wt.	CAS RN mp/°C	Merck No. bp/°C	Beil. Ref. den/g cm^{-3}	Solubility n_D
4745	1,3,5,7-Cyclooctatetraene, 1-chloro- Chlorocyclooctatetraene	C$_8$H$_7$Cl 138.60	29554-49-2 	 50-1$^{5.5}$	4-05-00-01334 1.1199^{25}	H$_2$O 1; ace 3; bz 3; chl 3 1.5542^{25}
4746	1,3,5,7-Cyclooctatetraene, 1-ethyl-	C$_{10}$H$_{12}$ 132.21	13402-35-2 	 81^{37}	4-05-00-01373 0.8996^{25}	 1.5187^{25}
4747	1,3,5,7-Cyclooctatetraene, 1-methyl-	C$_9$H$_{10}$ 118.18	2570-12-9 	 84.5^{67}	4-05-00-01358 0.8978^{25}	H$_2$O 1; EtOH 3; eth 3; bz 3 1.5249^{25}
4748	1,3,5,7-Cyclooctatetraene, propyl- Cyclooctatetraene, propyl	C$_{11}$H$_{14}$ 146.23	13402-36-3 	 73^9	4-05-00-01400 0.8870^{25}	 1.5131^{25}
4749	1,3,5-Cyclooctatriene	C$_8$H$_{10}$ 106.17	1871-52-9 -83	 145.5	4-05-00-00882 0.8971^{25}	 1.5035^{25}
4750	1,3,6-Cyclooctatriene	C$_8$H$_{10}$ 106.17	3725-30-2 -62	 68^{60}	4-05-00-00883 0.8940^{25}	
4751	Cyclooctene, (E)-	C$_8$H$_{14}$ 110.20	931-89-5 -59	 143	4-05-00-00263 0.8483^{20}	EtOH 3; ctc 2; chl 3 1.4741^{25}
4752	Cyclooctene, (Z)-	C$_8$H$_{14}$ 110.20	931-87-3 -12	 138	3-05-00-00209 0.8472^{20}	EtOH 3; eth 3; ctc 3 1.4698^{20}
4753	2-Cycloocten-1-ol	C$_8$H$_{14}$O 126.20	3212-75-7 	 74^2	4-06-00-00217 0.9756^{25}	 1.4959^{25}
4754	3-Cycloocten-1-one	C$_8$H$_{12}$O 124.18	4734-90-1 	 200.5	4-07-00-00142 0.990^{20}	ace 4; eth 4; chl 4 1.4953^{25}
4755	Cyclooctyne	C$_8$H$_{12}$ 108.18	1781-78-8 	 158	4-05-00-00401 0.868^{12}	 1.4850^{20}
4756	15H-Cyclopenta[a]phenanthrene, 16,17- dihydro- 1,2-Cyclopentenophenanthrene	C$_{17}$H$_{14}$ 218.30	482-66-6 135.5	2749	4-05-00-02423	H$_2$O 1; EtOH 3; peth 3
4757	Cyclopenta[c]furo[3',2':4,5]furo[2,3- h][1]benzopyran-1,11-dione, 2,3,6a,8,9,9a- hexahydro-4-methoxy-, (6aR- Aflatoxin B2	C$_{17}$H$_{14}$O$_6$ 314.29	7220-81-7 287.5	168	5-19-10-00552	
4758	Cyclopenta[c]pyran-1(4aH)-one, 5,6,7,7a- tetrahydro-4,7-dimethyl- Nepetalactone	C$_{10}$H$_{14}$O$_2$ 166.22	490-10-8 	6383 71-2$^{0.05}$	4-17-00-04622 1.0663^{25}	 1.4859^{25}
4759	Cyclopentadecane	C$_{15}$H$_{30}$ 210.40	295-48-7 61.3	4-05-00-00181 0.8364^{61}		 1.4592^{61}
4760	Cyclopentadecanol Exaltol	C$_{15}$H$_{30}$O 226.40	4727-17-7 80.5	 177^{11}; 145$^{0.3}$	4-06-00-00185 0.930^{20}	 1.4555^{98}
4761	Cyclopentadecanone	C$_{15}$H$_{28}$O 224.39	502-72-7 63	 120$^{0.3}$	4-07-00-00113 0.8895^{25}	H$_2$O 2; EtOH 3; ace 3 1.4637^{60}
4762	Cyclopentadecanone, 2-methyl- 2-Methylcyclopentadecanone	C$_{16}$H$_{30}$O 238.41	52914-66-6 	 171-3^{12}	3-07-00-00208 0.9213^{16}	 1.4812^{16}
4763	Cyclopentadecanone, 3-methyl- Muscone	C$_{16}$H$_{30}$O 238.41	541-91-3 	6222 329; 130$^{0.5}$	4-07-00-00118 0.9221^{17}	ace 4; eth 4; EtOH 4 1.4802^{17}
4764	1,3-Cyclopentadiene Pyropentylene	C$_5$H$_6$ 66.10	542-92-7 -85	2744 41	4-05-00-00377 0.8021^{20}	H$_2$O 1; EtOH 5; eth 5; ace 3 1.4440^{20}
4765	1,3-Cyclopentadiene, 5-ethylidene-	C$_7$H$_8$ 92.14	3839-50-7 	 32^{17}	4-05-00-00878 0.865^{18}	 1.526^{20}
4766	1,3-Cyclopentadiene, 1,2,3,4,5,5-hexachloro- Perchlorocyclopentadiene	C$_5$Cl$_6$ 272.77	77-47-4 -9	 239; 48$^{0.3}$	4-05-00-00381 1.7019^{25}	 1.5658^{20}
4767	1,3-Cyclopentadiene, 5-methylene- Fulvene	C$_6$H$_6$ 78.11	497-20-1 	 7-8^{56}	4-05-00-00764 0.8241^{20}	H$_2$O 1; bz 3; chl 3 1.4920^{20}
4768	1,3-Cyclopentadiene, 5-(1-methylethylidene)-	C$_8$H$_{10}$ 106.17	2175-91-9 	 49-50^{11}	4-05-00-00974 0.881^{20}	 1.5474^{20}
4769	1,3-Cyclopentadiene, 5-(2-propenylidene)-, (E)- Fulvene, 6-vinyl(trans)	C$_8$H$_8$ 104.15	116862-65-8 -35	 45^{12} 20$^{0.15}$	 0.898^{20}	H$_2$O 1; bz 3; chl 3
4770	1,3-Cyclopentadiene, 1,2,3,4-tetrachloro-5,5- dimethoxy-	C$_7$H$_6$Cl$_4$O$_2$ 263.93	2207-27-4 109	4-07-00-00286 1.501^{25}		 1.5282^{20}
4771	Cyclopentanamine Cyclopentylamine	C$_5$H$_{11}$N 85.15	1003-03-8 -82.7	 108	4-12-00-00004 0.8689^{20}	ace 3; bz 3; chl 3 1.4728^{25}
4772	Cyclopentanamine, 1-methyl-	C$_6$H$_{13}$N 99.18	40571-45-7 	 121	4-12-00-00113 0.8025^{25}	H$_2$O 4 1.4408^{25}
4773	Cyclopentanamine, 2-methyl-	C$_6$H$_{13}$N 99.18	41223-14-7 	 123	0-12-00-00007 0.801^{20}	H$_2$O 4
4774	Cyclopentane Pentamethylene	C$_5$H$_{10}$ 70.13	287-92-3 -93.8	2746 49.3	4-05-00-00014 0.7457^{20}	H$_2$O I; EtOH 5; eth 5; ace 5 1.4065^{20}
4775	Cyclopentaneacetic acid 2-Cyclopentylacetic acid	C$_7$H$_{12}$O$_2$ 128.17	1123-00-8 13.5	 228	4-09-00-00033 1.0216^{18}	 1.4523^{18}
4776	Cyclopentaneacetic acid, 3-methyl- 3-Methylcyclopentaneacetic acid	C$_8$H$_{14}$O$_2$ 142.20	63370-69-4 	 237	3-09-00-00060 0.9818^{20}	 1.4504^{20}
4777	Cyclopentane, bromo- Cyclopentyl bromide	C$_5$H$_9$Br 149.03	137-43-9 	 137.5	4-05-00-00019 1.3873^{20}	ctc 2 1.4886^{20}
4778	Cyclopentane, butyl- Butylcyclopentane	C$_9$H$_{18}$ 126.24	2040-95-1 -108	 156.6	4-05-00-00139 0.7846^{20}	ace 4; bz 4; eth 4; EtOH 4 1.4316^{20}
4779	Cyclopentanecarbonitrile Cyanocyclopentane	C$_6$H$_9$N 95.14	4254-02-8 			
4780	Cyclopentanecarboxaldehyde	C$_6$H$_{10}$O 98.14	872-53-7 	 133.5	4-07-00-00037 0.9371^{20}	H$_2$O 4; eth 4; EtOH 4 1.4432^{20}
4781	Cyclopentanecarboxylic acid Cyclopentanoic acid	C$_6$H$_{10}$O$_2$ 114.14	3400-45-1 -7	 212; 104^{11}	4-09-00-00011 1.0527^{20}	H$_2$O 2; ctc 2; MeOH 3 1.4532^{20}

No.	Name / Synonym	Mol. Form. / Mol. Wt.	CAS RN / mp/°C	Merck No. / bp/°C	Beil. Ref. / den/g cm^{-3}	Solubility / n_D
4782	Cyclopentanecarboxylic acid, 1-amino- / Cycloleucine	$C_6H_{11}NO_2$ / 129.16	52-52-8 / 330 dec	2740	4-14-00-00974	
4783	Cyclopentanecarboxylic acid, 1-methyl- / 1-Methylcyclopentanecarboxylic acid	$C_7H_{12}O_2$ / 128.17	5217-05-0	219	4-09-00-00035 / 1.0218^{20}	eth 4; EtOH 4 / 1.4529^{20}
4784	Cyclopentanecarboxylic acid, 2-methyl-5-(1-methylethyl)- / Puleganic acid	$C_{10}H_{18}O_2$ / 170.25	528-23-4 / -18	152^{25}; 139^{12}	4-09-00-00080 / 0.9642^{20}	1.4524^{24}
4785	Cyclopentanecarboxylic acid, 2-oxo-, ethyl ester	$C_8H_{12}O_3$ / 156.18	611-10-9	221; 110^{16}	4-10-00-02602 / 1.0781^{21}	eth 3; bz 3 / 1.4519^{20}
4786	Cyclopentanecarboxylic acid, 1,2,2,3-tetramethyl- / Campholic acid	$C_{10}H_{18}O_2$ / 170.25	464-88-0 / 106	254; 146^{12}	2-09-00-00019	H_2O 2; EtOH 3; eth 3
4787	Cyclopentanecarboxylic acid, 3-isopropyl-1-methyl-	$C_{12}H_{22}O_2$ / 198.31	104068-42-0 / 18.5	257	4-09-00-00079 / 0.9698^{19}	1.4563^{20}
4788	Cyclopentanecarboxylic acid, 2-methyl-5-(1-methylethyl)-, ethyl ester / Cyclopentanecarboxylic acid, 2-isopropyl-5-methyl, ethyl ester	$C_{12}H_{22}O_2$ / 198.31	116530-99-5	145^4	1-09-00-00017 / 0.9178^{12}	1.4405^{12}
4789	Cyclopentane, chloro- / Cyclopentyl chloride	C_5H_9Cl / 104.58	930-28-9	114	4-05-00-00018 / 1.0051^{20}	H_2O 1; eth 3; ace 3; bz 3 / 1.4510^{20}
4790	Cyclopentane, decyl- / Decane, 1-cyclopentyl-	$C_{15}H_{30}$ / 210.40	1795-21-7 / -22	279	4-05-00-00186 / 0.8110^{20}	ace 4; bz 4; eth 4; EtOH 4 / 1.4486^{20}
4791	1,1-Cyclopentanedicarboxylic acid, diethyl ester	$C_{11}H_{18}O_4$ / 214.26	4167-77-5	210	4-09-00-02793 / 1.0186^{20}	1.4370^{25}
4792	1,2-Cyclopentanedicarboxylic acid, dimethyl ester, trans-(±)- / Cyclopentane-1,2-dicarboxylic acid, dimethyl ester (DL, trans)	$C_9H_{14}O_4$ / 186.21	80656-12-8	$118-9^{17}$	4-09-00-02794 / 1.1130^{19}	1.4498^{17}
4793	1,3-Cyclopentanedicarboxylic acid, cis-	$C_7H_{10}O_4$ / 158.15	876-05-1 / 121	300 dec	4-09-00-02795	H_2O 3; EtOH 3; eth 3; ace 3
4794	1,3-Cyclopentanedicarboxylic acid, 1,2,2-trimethyl-, (1R-cis)-	$C_{10}H_{16}O_4$ / 200.23	124-83-4 / 189.1		4-09-00-02851 / 1.186^{20}	H_2O 2; EtOH 4; eth 4; ace 3
4795	1,3-Cyclopentanedicarboxylic acid, 1,2,2-trimethyl-, dimethyl ester, (1R-cis)- / Camphoric acid, dimethyl ester, (+)-	$C_{12}H_{20}O_4$ / 228.29	15797-21-4 / <-16	264	4-09-00-02851 / 1.015^{20}	eth 4; EtOH 4 / 1.4627^{19}
4796	Cyclopentane, 1,1-diethyl- / 1,1-Diethylcyclopentane	C_9H_{18} / 126.24	2721-38-2 / -95.6	150.5	4-05-00-00141 / 0.7988^{25}	1.4388^{20}
4797	Cyclopentane, 1,2-diethyl-, cis- / cis-1,2-Diethylcyclopentane	C_9H_{18} / 126.24	932-39-8 / -118	153.6	4-05-00-00141 / 0.7920^{25}	1.4355^{20}
4798	Cyclopentane, 1,2-diethyl-, trans- / trans-1,2-Diethylcyclopentane	C_9H_{18} / 126.24	932-40-1 / -95	147.5	4-05-00-00141 / 0.7792^{25}	bz 4; eth 4; peth 4 / 1.4295^{20}
4799	Cyclopentane, 1,3-diethyl- / 1,3-Diethylcyclopentane	C_9H_{18} / 126.24	19398-75-5	150	0-05-00-00045 / 0.783^{25}	1.430^{20}
4800	Cyclopentane, 1,1-dimethyl- / 1,1-Dimethylcyclopentane	C_7H_{14} / 98.19	1638-26-2 / -69.8	87.5	4-05-00-00105 / 0.7499^{25}	1.4136^{20}
4801	Cyclopentane, 1,2-dimethyl-, cis- / cis-1,2-Dimethylcyclopentane	C_7H_{14} / 98.19	1192-18-3 / -54	99.5	4-05-00-00106 / 0.7680^{25}	1.4222^{20}
4802	Cyclopentane, 1,2-dimethyl-, trans- / trans-1,2-Dimethylcyclopentane	C_7H_{14} / 98.19	822-50-4 / -117.6	91.9	4-05-00-00107 / 0.7468^{25}	1.4120^{20}
4803	Cyclopentane, 1,3-dimethyl-, cis- / cis-1,3-Dimethylcyclopentane	C_7H_{14} / 98.19	2532-58-3 / -133.7	90.8	4-05-00-00107 / 0.7402^{25}	1.4089^{20}
4804	Cyclopentane, 1,3-dimethyl-, trans- / trans-1,3-Dimethylcyclopentane	C_7H_{14} / 98.19	1759-58-6 / -134	91.7	4-05-00-00108 / 0.7443^{25}	1.4107^{20}
4805	Cyclopentane, (1,1-dimethylethyl)-	C_9H_{18} / 126.24	3875-52-3 / -95.8	144.9	3-05-00-00124 / 0.7870^{25}	1.4338^{20}
4806	1,3-Cyclopentanediol, cis-	$C_5H_{10}O_2$ / 102.13	16326-97-9 / 31	105^5; 90^1	3-06-00-04057 / 1.099^{15}	1.4792^{20}
4807	1,3-Cyclopentanediol, trans- / Cyclopentane-1,3-diol (trans)	$C_5H_{10}O_2$ / 102.13	16326-98-0 / 40	$80-5^{0.1}$	3-06-00-04057 / 1.094^{25}	1.4830^{20}
4808	Cyclopentane, docosyl- / Docosylcyclopentane	$C_{27}H_{54}$ / 378.73	62016-55-1 / 45	433	4-05-00-00201 / 0.8295^{20}	bz 4; eth 4 / 1.4608^{20}
4809	Cyclopentane, dodecyl-	$C_{17}H_{34}$ / 238.46	5634-30-0 / -5	312	3-05-00-00156 / 0.8158^{20}	ace 4; bz 4; eth 4; EtOH 4 / 1.4518^{20}
4810	Cyclopentaneethanamine, N,α-dimethyl- / Cyclopentamine	$C_9H_{19}N$ / 141.26	102-45-4 / 171	2745	4-12-00-00137	1.4500^{20}
4811	Cyclopentaneethanol	$C_7H_{14}O$ / 114.19	766-00-7	183; 96^{24}	4-06-00-00110 / 0.9180^{20}	eth 4 / 1.4577^{20}
4812	Cyclopentane, ethyl- / Ethylcyclopentane	C_7H_{14} / 98.19	1640-89-7 / -138.4	103.5	4-05-00-00104 / 0.7665^{20}	H_2O 1; EtOH 5; eth 5; ace 5 / 1.4198^{20}
4813	Cyclopentane, ethylidene- / Ethylidenecyclopentane	C_7H_{12} / 96.17	2146-37-4 / -129.5	115	4-05-00-00253 / 0.8020^{20}	1.4481^{20}
4814	Cyclopentane, 1-ethyl-1-methyl- / 1-Ethyl-1-methylcyclopentane	C_8H_{16} / 112.22	16747-50-5 / -143.8	121.6	4-05-00-00126 / 0.7767^{25}	ace 4; bz 4; eth 4; EtOH 4 / 1.4272^{20}
4815	Cyclopentane, 1-ethyl-2-methyl-, cis- / cis-1-Methyl-2-ethylcyclopentane	C_8H_{16} / 112.22	930-89-2 / -106	128	4-05-00-00126 / 0.7852^{20}	1.4293^{20}

No.	Name / Synonym	Mol. Form. / Mol. Wt.	CAS RN / mp/°C	Merck No. / bp/°C	Beil. Ref. / den/g cm^{-3}	Solubility / n_D
4816	Cyclopentane, 1-ethyl-2-methyl-, *trans*-	C_8H_{16} 112.22	930-90-5 -105.9	121.2	4-05-00-00126 0.7649[25]	1.4219[20]
4817	Cyclopentane, 1-ethyl-3-methyl-, *cis*- 1-Methyl-*cis*-3-ethylcyclopentane	C_8H_{16} 112.22	2613-66-3	121	4-05-00-00127 0.7724[20]	ace 4; bz 4; eth 4; EtOH 4 1.4203[20]
4818	Cyclopentane, 1-ethyl-3-methyl-, *trans*- 1-Methyl-*trans*-3-ethylcyclopentane	C_8H_{16} 112.22	2613-65-2 -108	121	4-05-00-00127 0.7619[20]	1.4186[20]
4819	Cyclopentane, heneicosyl-	$C_{26}H_{52}$ 364.70	6703-82-8 42	420	4-05-00-00200 0.8286[20]	1.4602[20]
4820	Cyclopentane, heptyl- Heptane, 1-cyclopentyl-	$C_{12}H_{24}$ 168.32	5617-42-5 -53	224	4-05-00-00173 0.8010[20]	ace 4; bz 4; eth 4; EtOH 4 1.4421[20]
4821	Cyclopentane, hexacosyl- Hexacosylcyclopentane	$C_{31}H_{62}$ 434.83	62016-58-4 56	468	4-05-00-00203 0.8326[20]	1.4628[20]
4822	Cyclopentane, hexadecyl- Hexadecane, 1-cyclopentyl-	$C_{21}H_{42}$ 294.56	6812-39-1 21	368	4-05-00-00195 0.8228[20]	ace 4; bz 4; eth 4; EtOH 4 1.4564[20]
4823	Cyclopentane, hexyl-	$C_{11}H_{22}$ 154.30	4457-00-5 -73	203	4-05-00-00168 0.7965[20]	ace 4; bz 4; eth 4; EtOH 4 1.4392[20]
4824	Cyclopentane, iodo- Cyclopentyl iodide	C_5H_9I 196.03	1556-18-9	166.5	4-05-00-00020 1.7096[20]	H_2O 1; eth 3; bz 3; ctc 2 1.5447[20]
4825	Cyclopentanemethanol	$C_6H_{12}O$ 100.16	3637-61-4	163	4-06-00-00092 0.9332[20]	1.4579[20]
4826	Cyclopentane, methyl- Methylcyclopentane	C_8H_{12} 84.16	96-37-7 -142.5	71.8	4-05-00-00084 0.7486[20]	H_2O 1; EtOH 5; eth 5; ace 5 1.4097[20]
4827	Cyclopentane, methylene- Methylenecyclopentane	C_6H_{10} 82.15	1528-30-9	75.5	4-05-00-00241 0.7787[20]	bz 3; chl 3 1.4355[20]
4828	Cyclopentane, (1-methylethyl)- Isopropylcyclopentane	C_8H_{16} 112.22	3875-51-2 -111.4	126.5	4-05-00-00125 0.7765[20]	H_2O 1; EtOH 5; eth 3; ace 5 1.4258[20]
4829	Cyclopentane, 1-methyl-2-(1-methylethyl)- Cyclopentane, 1-isopropyl-2-methyl-	C_9H_{18} 126.24	89223-57-4	143	4-05-00-00141 0.7832[15]	1.4279[20]
4830	Cyclopentane, 1-methyl-1-nitro- 1-Methyl-1-nitrocyclopentane	$C_6H_{11}NO_2$ 129.16	30168-50-4	181; 92[40]	4-05-00-00087 1.0395[20]	1.4504[20]
4831	Cyclopentane, 1-methyl-2-nitro- 1-Methyl-2-nitrocyclopentane	$C_6H_{11}NO_2$ 129.16	102153-88-8	185 dec; 98[40]	1-05-00-00011 1.0381[22]	1.4488[22]
4832	Cyclopentane, (1-methylpropyl)-	C_9H_{18} 126.24	4850-32-2	154.4	3-05-00-00124 0.7905[25]	1.4357[20]
4833	Cyclopentane, 1-methyl-1-propyl- 1-Methyl-1-propylcyclopentane	C_9H_{18} 126.24	16631-63-3	146	4-05-00-00140 0.795[25]	1.437[20]
4834	Cyclopentane, 1-methyl-2-propyl-, *cis*- *cis*-1-Methyl-2-propylcyclopentane	C_9H_{18} 126.24	932-43-4 -104	152.6	4-05-00-00140 0.7881[25]	1.4343[20]
4835	Cyclopentane, 1-methyl-2-propyl-, *trans*- *trans*-1-Methyl-2-propylcyclopentane	C_9H_{18} 126.24	932-44-5 -123	146.4	4-05-00-00140 0.7735[25]	1.4274[20]
4836	Cyclopentane, 1-methyl-3-propyl-, *cis*- *cis*-1-Methyl-3-propylcyclopentane	C_9H_{18} 126.24	2443-04-1	148	3-05-00-00125 0.776[25]	1.426[20]
4837	Cyclopentane, 1-methyl-3-propyl-, *trans*- *trans*-1-Methyl-3-propylcyclopentane	C_9H_{18} 126.24	2443-03-0	148	3-05-00-00125 0.776[25]	1.426[20]
4838	Cyclopentane, (2-methylpropyl)- Isobutylcyclopentane	C_9H_{18} 126.24	3788-32-7 -115.2	148	4-05-00-00140 0.7769[25]	1.4298[20]
4839	Cyclopentane, nitro- Nitrocyclopentane	$C_5H_9NO_2$ 115.13	2562-38-1	90-1[40]	4-05-00-00020 1.0776[23]	H_2O I; bz 3; chl 3 1.4538[20]
4840	Cyclopentane, (nitromethyl)- (Nitromethyl)cyclopentane	$C_6H_{11}NO_2$ 129.16	2625-31-2	110[35]	1-05-00-00011 1.0713[20]	1.4587[20]
4841	Cyclopentane, nonyl-	$C_{14}H_{28}$ 196.38	2882-98-6 -29	262	4-05-00-00181 0.8081[20]	ace 4; bz 4; eth 4; EtOH 4 1.4467[20]
4842	Cyclopentane, octadecyl- Octadecane, 1-cyclopentyl-	$C_{23}H_{46}$ 322.62	62016-53-9 30	391	4-05-00-00196 0.8254[20]	ace 4; bz 4; eth 4; EtOH 4 1.4581[20]
4843	Cyclopentane, octyl- Octane, 1-cyclopentyl-	$C_{13}H_{28}$ 182.35	1795-20-6 -44	243	4-05-00-00176 0.8048[20]	1.4446[20]
4844	Cyclopentane, pentadecyl- Pentadecane, 1-cyclopentyl-	$C_{20}H_{40}$ 280.54	4669-01-6 17	355	4-05-00-00195 0.8213[20]	ace 4; bz 4; eth 4; EtOH 4 1.4554[20]
4845	Cyclopentane, pentatriacontyl- Pentatriacontylcyclopentane	$C_{40}H_{80}$ 561.08	61827-98-3 74	531	4-05-00-00205 0.8374[20]	1.4660[20]
4846	Cyclopentane, pentyl-	$C_{10}H_{20}$ 140.27	3741-00-2 -83	180	4-05-00-00160 0.7912[20]	ace 4; bz 4; eth 4; EtOH 4 1.4356[20]
4847	Cyclopentanepropanoic acid	$C_8H_{14}O_2$ 142.20	140-77-2	130-2[12]	4-09-00-00046 0.996[25]	1.4570[20]
4848	Cyclopentane, 2-propenyl-	C_8H_{14} 110.20	3524-75-2 -110.7	125	4-05-00-00272 0.793[25]	chl 3 1.4412[20]
4849	Cyclopentane, propyl- Propylcyclopentane	C_8H_{16} 112.22	2040-96-2 -117.3	131	4-05-00-00125 0.7763[20]	H_2O 1; EtOH 5; eth 5; ace 5 1.4266[20]
4850	Cyclopentane, tetradecyl- Tetradecane, 1-cyclopentyl-	$C_{19}H_{38}$ 266.51	1795-22-8 9	341	3-05-00-00158 0.8196[20]	1.4543[20]
4851	Cyclopentane, 1,1,3,3-tetramethyl- 1,1,3,3-Tetramethylcyclopentane	C_9H_{18} 126.24	50876-33-0 -88.4	118	4-05-00-00143 0.7469[25]	1.4125[20]
4852	Cyclopentanethiol Cyclopentyl mercaptan	$C_5H_{10}S$ 102.20	1679-07-8	132.1	4-06-00-00015 0.9550[20]	
4853	Cyclopentane, triacontyl- Triacontylcyclopentane	$C_{35}H_{70}$ 490.94	61827-93-8 65	498	4-05-00-00204 0.8350[20]	1.4644[20]
4854	Cyclopentane, tridecyl- Tridecane, 1-cyclopentyl-	$C_{18}H_{36}$ 252.48	6006-34-4 5	327	4-05-00-00192 0.8178[20]	ace 4; bz 4; eth 4; EtOH 4 1.4531[20]
4855	Cyclopentane, 1,1,2-trimethyl- Camphocean	C_8H_{16} 112.22	4259-00-1 -21.6	114; 53[100]	4-05-00-00127 0.7660[20]	1.4199[20]

No.	Name / Synonym	Mol. Form. / Mol. Wt.	CAS RN / mp/°C	Merck No. / bp/°C	Beil. Ref. / den/g cm^{-3} / n_D	Solubility
4856	Cyclopentane, 1,1,3-trimethyl- / 1,1,3-Trimethylcyclopentane	C$_8$H$_{16}$ / 112.22	4516-69-2 / -142.4	/ 104.9	3-05-00-00111 / 0.7439[25] / 1.4112[20]	
4857	Cyclopentane, 1,2,3-trimethyl-, (1α,2β,3α)- / Cyclopentane, 1,2,3-trimethyl-, trans-1,2,cis-1,3-	C$_8$H$_{16}$ / 112.22	19374-46-0 / -112.7	/ 110.4	4-05-00-00128 / 0.7492[25] / 1.4138[20]	
4858	Cyclopentane, 1,2,3-trimethyl-, (1α,2α,3α)- / Cyclopentane, 1,2,3-trimethyl-, cis-1,2,cis-1,3-	C$_8$H$_{16}$ / 112.22	2613-69-6 / -116.4	/ 123	4-05-00-00128 / 0.7751[25] / 1.4262[20]	
4859	Cyclopentane, 1,2,3-trimethyl-, (1α,2α,3β)-	C$_8$H$_{16}$ / 112.22	15890-40-1 / -112	/ 117.5	4-05-00-00128 / 0.7661[25] / 1.4218[20]	
4860	Cyclopentane, 1,2,4-trimethyl-, (1α,2β,4α)-	C$_8$H$_{16}$ / 112.22	16883-48-0 / -130.8	/ 109.3	4-05-00-00129 / 0.7430[25] / 1.4106[20]	
4861	Cyclopentane, 1,2,4-trimethyl-, (1α,2α,4α)- / Cyclopentane, 1,2,4-trimethyl-, cis-1,2,cis-1,4-	C$_8$H$_{16}$ / 112.22	2613-72-1 / -132.3	/ 116.8	4-05-00-00129 / 0.758[25] / 1.4186[20]	
4862	Cyclopentane, 1,2,4-trimethyl-, (1α,2α,4β)-	C$_8$H$_{16}$ / 112.22	4850-28-6 / -132.6	/ 116.7	4-05-00-00129 / 0.7592[25] / 1.4186[20]	
4863	Cyclopentane, undecyl-	C$_{16}$H$_{32}$ / 224.43	6785-23-5 / -10	/ 296	4-05-00-00189 / 0.8135[20] / 1.4503[20]	ace 4; bz 4; eth 4; EtOH 4
4864	Cyclopentanol / Cyclopentyl alcohol	C$_5$H$_{10}$O / 86.13	96-41-3 / -19	2747 / 140.4	4-06-00-00005 / 0.9488[20] / 1.4530[20]	H$_2$O 2; EtOH 3; eth 3; ace 3
4865	Cyclopentanol, acetate / Cyclopentyl acetate	C$_7$H$_{12}$O$_2$ / 128.17	933-05-1 /	/ 51-2[12]	4-06-00-00007 / 0.9522[16]	
4866	Cyclopentanol, 1-ethynyl- / 1-Ethynyl-1-cyclopentanol	C$_7$H$_{10}$O / 110.16	17356-19-3 / 27	/ 157.5	4-06-00-00339 / 0.962[25] / 1.4751[20]	
4867	Cyclopentanol, 1-methyl-	C$_6$H$_{12}$O / 100.16	1462-03-9 / 36	/ 136; 53[30]	4-06-00-00086 / 0.9044[23] / 1.4429[23]	
4868	Cyclopentanol, 2-methyl-, cis- / cis-2-Methylcyclopentanol	C$_6$H$_{12}$O / 100.16	25144-05-2 /	/ 148.5	3-06-00-00055 / 0.9379[16] / 1.4504[16]	
4869	Cyclopentanol, 2-methyl-, trans- / trans-2-Methylcyclopentanol	C$_6$H$_{12}$O / 100.16	25144-04-1 /	/ 150.5	4-06-00-00091 / 0.9248[16] / 1.4499[16]	
4870	Cyclopentanol, 3-methyl- / 3-Methylcyclopentanol	C$_6$H$_{12}$O / 100.16	18729-48-1 /	/ 148.5	2-06-00-00015 / 0.9158[16] / 1.4487[16]	eth 4; EtOH 4
4871	Cyclopentanol, 1-phenyl-	C$_{11}$H$_{14}$O / 162.23	10487-96-4 /	/ 132-3[18]	4-06-00-03884 / 1.0530[20] / 1.5479[20]	
4872	Cyclopentanol, 1-propyl-	C$_8$H$_{16}$O / 128.21	1604-02-0 / -37.5	/ 173.5	3-06-00-00101 / 0.9040[25] / 1.4502[25]	
4873	Cyclopentanol, 2-propyl-, cis- / cis-2-Propylcyclopentanol	C$_8$H$_{16}$O / 128.21	25172-42-3 /	/ 79-80[12]	3-06-00-00101 / 0.9163[9] / 1.4600[9]	
4874	Cyclopentanol, 2-propyl-, trans- / trans-2-Propylcyclopentanol	C$_8$H$_{16}$O / 128.21	25172-43-4 /	/ 78-9[10]	3-06-00-00101 / 0.9016[9] / 1.4565[9]	
4875	Cyclopentanol, 1,2,2-trimethyl- / 1,2,2-Trimethylcyclopentanol	C$_8$H$_{16}$O / 128.21	1121-95-5 /	/ 156; 60[15]	1-06-00-00014 / 0.9101[20] / 1.4513[20]	
4876	Cyclopentanol, 1,2,4-trimethyl- / 1,2,4-Trimethylcyclopentanol	C$_8$H$_{16}$O / 128.21	34103-97-4 /	/ 158	1-06-00-00014 / 0.8850[21] / 1.4424[21]	
4877	Cyclopentanol, 1,2,5-trimethyl- / 1,2,5-Trimethylcyclopentanol	C$_8$H$_{16}$O / 128.21	33840-38-9 /	/ 56-60[8]	1-06-00-00014 / 0.9121[15] / 1.4554[16]	
4878	Cyclopentanone / Adipic ketone	C$_5$H$_8$O / 84.12	120-92-3 / -51.3	2748 / 130.5	4-07-00-00005 / 0.9487[20] / 1.4366[20]	H$_2$O 1; EtOH 3; eth 5; ace 3
4879	Cyclopentanone, 2-acetyl- / 2-Acetylcyclopentanone	C$_7$H$_{10}$O$_2$ / 126.16	1670-46-8 /	/ 72-5[20]	4-07-00-01993 / 1.0431[25] / 1.4906[20]	
4880	Cyclopentanone, 2-chloro-	C$_5$H$_7$ClO / 118.56	694-28-0 /	/ 72-4[12]	4-07-00-00012 / 1.185[25] / 1.4750[20]	
4881	Cyclopentanone, 2-cyclopentylidene-	C$_{10}$H$_{14}$O / 150.22	825-25-2 /	/ 135[25]	4-07-00-00320 / 1.0179[18] / 1.5215[18]	
4882	Cyclopentanone, 2,5-dimethyl- / 2,5-Dimethylcyclopentanone	C$_7$H$_{12}$O / 112.17	4041-09-2 /	/ 146.5	4-07-00-00048 / 0.882[25] / 1.4310[20]	
4883	Cyclopentanone, 2-ethyl-5-methyl- / 2-Ethyl-5-methylcyclopentanone	C$_8$H$_{14}$O / 126.20	51686-60-3 /	/ 164.5	3-06-00-02552 / 0.900[12] / 1.4360[22]	
4884	Cyclopentanone, 4-isopropyl-2-methyl- / Cyclopentanone, 3-isopropyl-5-methyl	C$_9$H$_{16}$O / 140.23	90645-61-7 /	/ 191.5	2-07-00-00034 / 0.8862[20] / 1.4413[19]	
4885	Cyclopentanone, 2-methyl-	C$_6$H$_{10}$O / 98.14	1120-72-5 / -75	/ 139.5	4-07-00-00036 / 0.9139[20] / 1.4364[20]	H$_2$O 3; EtOH 4; eth 4; ace 4
4886	Cyclopentanone, 3-methyl-, (±)- / (±)-3-Methylcyclopentanone	C$_6$H$_{10}$O / 98.14	6195-92-2 / -58.4	/ 144	4-07-00-00037 / 0.913[22] / 1.4329[20]	H$_2$O 3; EtOH 4; eth 4; ace 4
4887	Cyclopentanone, 2-(1-methylethyl)- / 2-Isopropylcyclopentanone	C$_8$H$_{14}$O / 126.20	14845-55-7 /	/ 175	4-07-00-00060 / 0.9105[20] / 1.4425[15]	
4888	Cyclopentanone, 3-methyl-2-(3-methylbutyl)- / Cyclopentanone, 2-isopentyl-3-methyl	C$_{11}$H$_{20}$O / 168.28	52033-97-3 /	/ 98-9[8]	3-07-00-00181 / 0.8938[20] / 1.4537[20]	
4889	Cyclopentanone, 2-methyl-2-(1-methylethyl)- / Cyclopentanone, 2-isopropyl-2-methyl-	C$_9$H$_{16}$O / 140.23	32116-65-7 /	/ 97.5[45]	3-07-00-00130 / 0.9067[16] / 1.4495[16]	
4890	Cyclopentanone, 2-methyl-5-(1-methylethyl)-	C$_9$H$_{16}$O / 140.23	6784-18-5 /	/ 182	3-07-00-00130 / 0.889[20] / 1.4402[20]	
4891	Cyclopentanone, 4-methyl-2-(1-methylethyl)- / Cyclopentanone, 2-isopropyl-4-methyl	C$_9$H$_{16}$O / 140.23	69770-98-5 /	/ 183	1-07-00-00026 / 0.8850[20] / 1.4392[20]	
4892	Cyclopentanone, oxime	C$_5$H$_9$NO / 99.13	1192-28-5 / 57.8	/ 196	4-07-00-00008 /	H$_2$O 4; bz 4
4893	Cyclopentanone, 2-(phenylmethyl)- / 2-Benzylcyclopentanone	C$_{12}$H$_{14}$O / 174.24	2867-63-2 /	/ 152[18]; 84[0.3]	4-07-00-01043 / 1.038[20] / 1.534[13]	
4894	Cyclopentanone, 2-propyl- / 2-Propylcyclopentanone	C$_8$H$_{14}$O / 126.20	1193-70-0 / -68.3	/ 183	4-07-00-00060 / 0.9017[20] / 1.4429[20]	

No.	Name / Synonym	Mol. Form. / Mol. Wt.	CAS RN / mp/°C	Merck No. / bp/°C	Beil. Ref. / den/g cm^{-3}	Solubility / n_D
4895	Cyclopentanone, 3-propyl- 3-Propylcyclopentanone	$C_8H_{14}O$ 126.20	82322-93-8	 190.5	3-07-00-00099 0.9041[20]	 1.4456[12]
4896	Cyclopentanone, 2,2,4-trimethyl- 2,2,4-Trimethylcyclopentanone	$C_8H_{14}O$ 126.20	28056-54-4 -40.6	 158	4-07-00-00061 0.877[25]	 1.4300[20]
4897	Cyclopentanone, 2,2,5-trimethyl- 2,2,5-Trimethylcyclopentanone	$C_8H_{14}O$ 126.20	4573-09-5	 152	3-07-00-00107 0.8781[20]	 1.4306[20]
4898	Cyclopentanone, 2,3,5-trimethyl- 2,3,5-Trimethylcyclopentanol	$C_8H_{14}O$ 126.20	90112-26-8	 158	1-07-00-00021 0.8778[19]	 1.4316[19]
4899	Cyclopentanone, 2,4,4-trimethyl- 2,4,4-Trimethylcyclopentanone	$C_8H_{14}O$ 126.20	4694-12-6 -25.6	 160.5	4-07-00-00061 0.8785[18]	 1.433[18]
4900	1H-Cyclopentapyrimidine-2,4(3H,5H)-dione, 3-cyclohexyl-6,7-dihydro- Lenacil	$C_{13}H_{18}N_2O_2$ 234.30	2164-08-1 290	5318	5-24-07-00375 1.32[25]	py 4
4901	Cyclopentasiloxane, decamethyl- Decamethylcyclopentasiloxane	$C_{10}H_{30}O_5Si_5$ 370.77	541-02-6 -38	2841 210	4-04-00-04128 0.9593[20]	H_2O 1 1.3982[20]
4902	Cyclopentasiloxane, 2,4,6,8,10-pentamethyl-	$C_5H_{20}O_5Si_5$ 300.64	6166-86-5 -108	 169	4-04-00-04099 0.9985[20]	 1.3912[20]
4903	Cyclopentene	C_5H_8 68.12	142-29-0 -135.1	 44.2	4-05-00-00209 0.7720[20]	H_2O 1; EtOH 3; eth 3; bz 3 1.4225[20]
4904	3-Cyclopentene-1-acetic acid	$C_7H_{10}O_2$ 126.16	767-03-3 19	 93-4[2.5]	 1.047[25]	 1.4676[20]
4905	1-Cyclopentene-1-acetonitrile	C_7H_9N 107.16	22734-04-9	 124[100]	3-09-00-00151 0.951[25]	 1.4670[20]
4906	Cyclopentene, 1-butyl- 1-Butyl-1-cyclopentene	C_9H_{16} 124.23	2423-01-0 -88.4	 157	4-05-00-00289 0.8035[25]	 1.4486[20]
4907	1-Cyclopentenecarbonitrile 1-Cyanocyclopentene	C_6H_7N 93.13	3047-38-9			
4908	1-Cyclopentene-1-carboxaldehyde	C_6H_8O 96.13	6140-65-4 -32	 146	4-07-00-00127 0.970[21]	 1.4872[17]
4909	1-Cyclopentene-1-carboxylic acid	$C_6H_8O_2$ 112.13	1560-11-8 123.5	 210	4-09-00-00109	H_2O 3; EtOH 3; eth 3; peth 3
4910	1-Cyclopentene-1-carboxylic acid, 2,3,3-trimethyl-	$C_9H_{14}O_2$ 154.21	5587-63-3 135	 255.5	4-09-00-00150	eth 4; EtOH 4
4911	1-Cyclopentene-1-carboxylic acid, 2,3,3-trimethyl-, methyl ester	$C_{10}H_{16}O_2$ 168.24	1460-88-4	 203.5	0-09-00-00058 0.9656[20]	 1.4571[20]
4912	2-Cyclopentene-1-carboxylic acid, 1,2,3-trimethyl- Laurolenic acid	$C_9H_{14}O_2$ 154.21	6894-69-5 6.5	 192[100]	2-09-00-00040 1.0318[25]	 1.4766
4913	3-Cyclopentene-1-carboxylic acid, 2,2,3-trimethyl- α-Campholytic acid	$C_9H_{14}O_2$ 154.21	6709-22-4 40.5	 241.5; 162[45]	4-09-00-00150 1.0145[18]	 1.4712[17]
4914	Cyclopentene, 3-chloro-	C_5H_7Cl 102.56	96-40-2	 25-31[30]	4-05-00-00212 1.0577[15]	eth 4; EtOH 4; chl 4 1.4708[26]
4915	1-Cyclopentene-1,2-dicarboxylic acid, diethyl ester Diethyl 1-cyclopentene-1,2-dicarboxylate	$C_{11}H_{16}O_4$ 212.25	70202-92-5	 140[16]	3-09-00-03938 1.0805[20]	 1.4652[20]
4916	1-Cyclopentene-1,3-dicarboxylic acid, diethyl ester Cyclopentene-1,3-dicarboxylic acid, diethyl ester	$C_{11}H_{16}O_4$ 212.25	30689-41-9	 168[21]	3-09-00-03939 1.1121[24]	 1.4564
4917	Cyclopentene, 1,2-dichloro-3,3,4,4,5,5-hexafluoro-	$C_5Cl_2F_6$ 244.95	706-79-6 -105.8	 90.7	4-05-00-00213 1.6546[20]	 1.3676[20]
4918	Cyclopentene, 1,2-dimethyl-	C_7H_{12} 96.17	765-47-9 -90.4	 105.8	4-05-00-00254 0.7928[25]	 1.4448[20]
4919	Cyclopentene, 1,3-dimethyl- 1,3-Dimethylcyclopentene	C_7H_{12} 96.17	62184-82-1	 92	4-05-00-00255 0.766[20]	 1.428[20]
4920	Cyclopentene, 1,4-dimethyl- 1,4-Dimethylcyclopentene	C_7H_{12} 96.17	19550-48-2	 93.2	4-05-00-00255 0.779[20]	 1.4283[20]
4921	Cyclopentene, 1,5-dimethyl-	C_7H_{12} 96.17	16491-15-9 -118	 99	4-05-00-00254 0.780[20]	 1.4331[20]
4922	Cyclopentene, 4,4-dimethyl-	C_7H_{12} 96.17	19037-72-0	 88	4-05-00-00254 0.766[25]	 1.423[20]
4923	Cyclopentene, 1,5-dimethyl-4-(1-methylethyl)- Cyclopentene, 2,3-dimethyl-4-isopropyl	$C_{10}H_{18}$ 138.25	6912-05-6	 163	1-05-00-00046 0.8085[22]	 1.4503[22]
4924	Cyclopentene, 1-ethyl-	C_7H_{12} 96.17	2146-38-5 -118.5	 106.3	4-05-00-00252 0.7936[25]	 1.4412[20]
4925	Cyclopentene, 3-ethyl-	C_7H_{12} 96.17	694-35-9	 97.8	4-05-00-00252 0.7784[25]	 1.4315[20]
4926	Cyclopentene, 4-ethyl- 4-Ethylcyclopentene	C_7H_{12} 96.17	3742-38-9	 98.3	4-05-00-00253 0.778[25]	 1.431[20]
4927	Cyclopentene, 4-isopropyl-1-methyl- 4-Isopropyl-1-methylcyclopentene	C_9H_{16} 124.23	90769-70-3	 142.5	1-05-00-00041 0.7945[21]	 1.4403[21]
4928	Cyclopentene, 1-methyl-	C_6H_{10} 82.15	693-89-0 -126.5	 75.5	4-05-00-00239 0.7748[25]	 1.4322[20]
4929	Cyclopentene, 3-methyl- 3-Methylcyclopentene	C_6H_{10} 82.15	1120-62-3	 64.9	4-05-00-00240 0.7572[25]	 1.4216[20]
4930	Cyclopentene, 4-methyl-	C_6H_{10} 82.15	1759-81-5 -160.8	 65.7	4-05-00-00240 0.7634[25]	 1.4209[20]

No.	Name / Synonym	Mol. Form. / Mol. Wt.	CAS RN / mp/°C	Merck No. / bp/°C	Beil. Ref. / den/g cm^{-3}	Solubility / n_D
4931	Cyclopentene, 3-methyl-1-(1-methylethyl)- / Pulegene	C_9H_{16} / 124.23	51115-02-7	/ 138.5	1-05-00-00040 / 0.791[22]	/ 1.4380[23]
4932	Cyclopentene, octachloro- / Perchlorocyclopentene	C_5Cl_8 / 343.68	706-78-5 / 40	/ 283	4-05-00-00214 / 1.8200[50]	H_2O 1; EtOH 4 / 1.5660[50]
4933	1-Cyclopentene-1-pentanoic acid, δ-hydroxy-3,5-bis(1-methylpropyl)-β-oxo-	$C_{18}H_{30}O_4$ / 310.43	53109-18-5 / 183		3-10-00-04194 / 1.269[20]	eth 4; EtOH 4
4934	Cyclopentene, 1-pentyl-	$C_{10}H_{18}$ / 138.25	4291-98-9 / -83	/ 179	4-05-00-00306 / 0.8085[25]	
4935	Cyclopentene, 1-propyl- / 1-Propyl-1-cyclopentene	C_8H_{14} / 110.20	3074-61-1 / -100.3	/ 131.2	4-05-00-00271 / 0.7978[25]	/ 1.4452[20]
4936	2-Cyclopentene-1-tridecanoic acid, (S)- / Chaulmoogric acid	$C_{18}H_{32}O_2$ / 280.45	29106-32-9 / 68.5	2037 / 247-8[20]	4-09-00-00216	eth 4; chl 4
4937	Cyclopentene, 1,2,3-trimethyl-	C_8H_{14} / 110.20	473-91-6	/ 121.6	3-05-00-00219 / 0.8039[15]	/ 1.4464[16]
4938	Cyclopentene, 1,5,5-trimethyl- / 1,5,5-Trimethylcyclopentene	C_8H_{14} / 110.20	62184-83-2	/ 109	3-05-00-00219 / 0.7824[20]	/ 1.4324[20]
4939	4-Cyclopentene-1,2,3-trione, 4,5-dihydroxy- / Croconic acid	$C_5H_2O_5$ / 142.07	488-86-8 / 150	/ dec	4-08-00-03342	H_2O 4; EtOH 3
4940	2-Cyclopentene-1-undecanoic acid, (R)- / Hydnocarpic acid	$C_{16}H_{28}O_2$ / 252.40	459-67-6 / 60.5	4679	4-09-00-00215	EtOH 4; chl 4; peth 4
4941	2-Cyclopenten-1-one / Cyclopenten-3-one	C_5H_6O / 82.10	930-30-3	/ 136; 40[12]	4-07-00-00122 / 0.989[15]	eth 4; EtOH 4 / 1.4629[15]
4942	2-Cyclopenten-1-one, 2,3-dihydroxy- / Reductic acid	$C_5H_6O_3$ / 114.10	80-72-8 / 212	8134	4-08-00-01714	H_2O 3; EtOH 3; eth 2; ace 2
4943	2-Cyclopenten-1-one, 2,4-dimethyl-	$C_7H_{10}O$ / 110.16	23048-13-7	/ 165	4-07-00-00138 / 0.9423[20]	/ 1.4655[20]
4944	2-Cyclopenten-1-one, 2-methyl-	C_6H_8O / 96.13	1120-73-6	/ 157	4-07-00-00127 / 0.9808[16]	/ 1.4762[15]
4945	2-Cyclopenten-1-one, 3-methyl- / 3-Methyl-2-cyclopentenone	C_6H_8O / 96.13	2758-18-1	/ 157.5	4-07-00-00126 / 0.9712[20]	/ 1.4714[20]
4946	2-Cyclopenten-1-one, 5-methyl- / Cyclopenten-3-one, 4-methyl	C_6H_8O / 96.13	14963-40-7	/ 140	3-07-00-00229 / 0.942[18]	/ 1.4460[18]
4947	2-Cyclopenten-1-one, 3-(1-methylethyl)-	$C_8H_{12}O$ / 124.18	1619-28-9	/ 214	2-07-00-00061 / 0.9378[20]	/ 1.4788[20]
4948	2-Cyclopenten-1-one, 5-methyl-2-(1-methylethyl)-	$C_9H_{14}O$ / 138.21	5587-79-1	/ 189	3-07-00-00291 / 0.9143[20]	/ 1.4660[20]
4949	2-Cyclopenten-1-one, 3-methyl-2-(2-pentenyl)-, (Z)- / Jasmone	$C_{11}H_{16}O$ / 164.25	488-10-8	5141 / 258; 134[12]	4-07-00-00337 / 0.9437[22]	H_2O 2; EtOH 3; eth 3; ctc 3 / 1.4979[22]
4950	2-Cyclopenten-1-one, 3-methyl-2-pentyl-	$C_{11}H_{18}O$ / 166.26	1128-08-1	/ 143[22]; 116[12]	4-07-00-00230 / 0.9165[18]	/ 1.4767[20]
4951	2-Cyclopenten-1-one, 3-phenyl-	$C_{11}H_{10}O$ / 158.20	3810-26-2 / -23	/ 234.2	4-07-00-01180 / 0.9711[20]	EtOH 3; eth 2; ace 3; chl 3 / 1.5440[20]
4952	Cycloprate / Cyclopropanecarboxylic acid, hexadecyl ester	$C_{20}H_{38}O_2$ / 310.52	54460-46-7	/ 154[0.05]		
4953	Cyclopropanamine / Cyclopropylamine	C_3H_7N / 57.10	765-30-0 / -35.4	/ 50.5	4-12-00-00003 / 0.8240[20]	H_2O 5; EtOH 3; eth 3; chl 3 / 1.4210[20]
4954	Cyclopropa[d]naphthalene, 1,1a,4,4a,5,6,7,8-octahydro-2,4a,8,8-tetramethyl-, [1aS-(1aα,4aβ,8aR*)]- / Thujopsene	$C_{15}H_{24}$ / 204.36	470-40-6	9327	4-05-00-01189	
4955	Cyclopropane / Trimethylene	C_3H_6 / 42.08	75-19-4 / -127.4	2755 / -32.8	4-05-00-00003 / *0.617[25]	H_2O 3; EtOH 4; eth 4; bz 3 / 1.3799[-42]
4956	Cyclopropanecarbonitrile / Cyclopropyl cyanide	C_4H_5N / 67.09	5500-21-0	/ 135.1	4-09-00-00005 / 0.8946[20]	eth 3; ctc 2; hx 3 / 1.4229[20]
4957	Cyclopropanecarbonyl chloride, 2-phenyl-, trans- / trans-2-Phenylcyclopropanecarbonyl chloride	$C_{10}H_9ClO$ / 180.63	939-87-7	/ 108-10[2]	3-09-00-02774 / 1.163[25]	/ 1.5560[20]
4958	Cyclopropanecarboxylic acid	$C_4H_6O_2$ / 86.09	1759-53-1 / 18.5	/ 183	4-09-00-00003 / 1.0885[20]	H_2O 3; EtOH 3; eth 3; ctc 2 / 1.4390[20]
4959	Cyclopropanecarboxylic acid, 2,2-dimethyl-3-(2-methyl-1-propenyl)-, (1R-cis)-	$C_{10}H_{16}O_2$ / 168.24	26771-11-9 / 41	/ 135[12]; 95[0.1]	4-09-00-00168	eth 4; EtOH 4; chl 4
4960	Cyclopropanecarboxylic acid, 2,2-dimethyl-3-(2-methyl-1-propenyl)-, 2-methyl-4-oxo-3-(2,4-pentadienyl)-2-cyclopenten-1-yl ester, [1R-[1α[S*(Z)],3β]]- / Pyrethrin 1	$C_{21}H_{28}O_3$ / 328.45	121-21-1	7978 / 170[0.1] dec	3-09-00-00215 / 1.5192[18]	H_2O 1; EtOH 3; eth 3; ctc 3 / 1.5192[18]
4961	Cyclopropanecarboxylic acid, ethyl ester / Ethyl cyclopropylcarboxylate	$C_6H_{10}O_2$ / 114.14	4606-07-9	/ 134	4-09-00-00003 / 0.9608[15]	/ 1.4190[20]
4962	Cyclopropanecarboxylic acid, 3-(3-methoxy-2-methyl-3-oxo-1-propenyl)-2,2-dimethyl-, 2-methyl-4-oxo-3-(2,4-pentadienyl)-2-cyclopenten-1-yl ester [1R-[1α[S*(Z)],3β(E)]]- / Pyrethrin 2	$C_{22}H_{28}O_5$ / 372.46	121-29-9	7978 / 200[0.1] dec	3-09-00-03988 / 1.5258[20]	H_2O 1; EtOH 3; eth 3; ctc 3
4963	Cyclopropanecarboxylic acid, methyl ester / Methyl cyclopropanecarboxylate	$C_5H_8O_2$ / 100.12	2868-37-3 / 114.9		4-09-00-00003 / 0.9848[20]	ace 3; chl 3 / 1.4144[19]

No.	Name / Synonym	Mol. Form. / Mol. Wt.	CAS RN / mp/°C	Merck No. / bp/°C	Beil. Ref. / den/g cm⁻³	Solubility / n_D
4964	Cyclopropanecarboxylic acid, 2-methyl- / 2-Methylcyclopropanecarboxylic acid	$C_5H_8O_2$ / 100.12	29555-02-0	191; 96.5[14]	2-09-00-00006 / 1.0267[20]	1.4441[20]
4965	Cyclopropanecarboxylic acid, 2,2-dimethyl-3-(2-methyl-1-propenyl)-, methyl ester, trans-(±)- / Chrysanthemum monocarboxylic acid, methyl ester (DL, trans)	$C_{11}H_{18}O_2$ / 182.26	15543-70-1	108[20]; 95[13]	4-09-00-00169 / 0.9274[25]	1.4614[25]
4966	Cyclopropane, (chloromethyl)-	C_4H_7Cl / 90.55	5911-08-0 / -90.9	88	4-05-00-00013 / 0.98[25]	1.4350[20]
4967	1,1-Cyclopropanedicarboxylic acid, diethyl ester	$C_9H_{14}O_4$ / 186.21	1559-02-0	216; 100[12]	4-09-00-02786 / 1.0535[25]	EtOH 4; eth 4 / 1.4345[18]
4968	1,2-Cyclopropanedicarboxylic acid, diethyl ester, cis- / Cyclopropane-1,2-dicarboxylic acid, diethyl ester (cis)	$C_9H_{14}O_4$ / 186.21	710-43-0		2-09-00-00513 / 1.062[12]	eth 4; EtOH 4 / 1.4450[20]
4969	1,2-Cyclopropanedicarboxylic acid, dimethyl ester / Cyclopropane-1,2-dicarboxylic acid, dimethyl ester	$C_7H_{10}O_4$ / 158.15	702-28-3	219; 110[3]	1.1584[16]	eth 4; EtOH 4 / 1.4472[14]
4970	1,2-Cyclopropanedicarboxylic acid, trans-(±)- / Cyclopropane-1,2-dicarboxylic acid (trans, DL)	$C_5H_6O_4$ / 130.10	58616-95-8 / 175	210[30]	4-09-00-02786 / 1.4600[20]	H_2O 4; eth 4; EtOH 4
4971	Cyclopropane, 1,1-dimethyl- / gem-Dimethylcyclopropane	C_5H_{10} / 70.13	1630-94-0 / -109	20.6	4-05-00-00024 / 0.6604[20]	H_2O 1; EtOH 3; eth 4; sulf 4 / 1.3668[20]
4972	Cyclopropane, 1,2-dimethyl-, cis- / cis-1,2-Dimethylcyclopropane	C_5H_{10} / 70.13	930-18-7 / -140.9	37.0	4-05-00-00025 / 0.6889[25]	H_2O 1; EtOH 3; eth 4; ctc 2 / 1.3829[20]
4973	Cyclopropane, 1,2-dimethyl-, trans- / trans-1,2-Dimethylcyclopropane	C_5H_{10} / 70.13	2402-06-4 / -149.6	28.2	4-05-00-00026 / 0.6648[25]	eth 4; EtOH 4 / 1.3713[20]
4974	Cyclopropane, ethyl- / Ethylcyclopropane	C_5H_{10} / 70.13	1191-96-4 / -149.2	35.9	4-05-00-00023 / 0.6790[25]	1.3786[20]
4975	Cyclopropane, 1-ethyl-1-methyl- / Cyclopropane, 1-methyl-1-ethyl	C_6H_{12} / 84.16	53778-43-1 / -130.2	56.8	4-05-00-00091 / 0.6968[25]	1.3887[20]
4976	Cyclopropane, 1-ethyl-2-methyl-, cis- / cis-1-Methyl-2-ethylcyclopropane	C_6H_{12} / 84.16	19781-68-1	58.7	3-05-00-00063 / 0.6935[20]	1.3846[20]
4977	Cyclopropane, 1-ethyl-2-methyl-, trans- / trans-1-Methyl-2-ethylcyclopropane	C_6H_{12} / 84.16	19781-69-2	59	3-05-00-00063 / 0.7146[20]	ctc 3 / 1.3953[20]
4978	Cyclopropanemethanol, 2,2-dimethyl-3-(2-methyl-1-propenyl)-	$C_{10}H_{18}O$ / 154.25	5617-92-5	66-9[0.07]	4-06-00-00264 / 0.888[25]	1.4757[20]
4979	Cyclopropanemethanol, α-methyl-	$C_5H_{10}O$ / 86.13	765-42-4 / -32.1	123.5	4-06-00-00020 / 0.8805[20]	1.4316[20]
4980	Cyclopropane, methoxy- / Cyclopropyl methyl ether	C_4H_8O / 72.11	540-47-6 / -119	2756 / 44.7	3-06-00-00003 / 0.8100[20]	H_2O 4; bz 4; eth 4; EtOH 4 / 1.3802[20]
4981	Cyclopropane, methyl- / Methylcyclopropane	C_4H_8 / 56.11	594-11-6 / -177.3	0.7	4-05-00-00013 / 0.6912[-20]	eth 4; EtOH 4
4982	Cyclopropane, (1-methylethenyl)-	C_6H_{10} / 82.15	4663-22-3	70	4-05-00-00243 / 0.7500[20]	1.4252[20]
4983	Cyclopropane, (1-methylethyl)-	C_6H_{12} / 84.16	3638-35-5 / -112.9	58.3	4-05-00-00090 / 0.6936[25]	1.3865[20]
4984	Cyclopropane, propyl- / Propane, 1-cyclopropyl-	C_6H_{12} / 84.16	2415-72-7	69.1	4-05-00-00090 / 0.7062[25]	1.3930[20]
4985	Cyclopropane, 1,1,2-trimethyl- / 1,1,2-Trimethylcyclopropane	C_6H_{12} / 84.16	4127-45-1 / -138.2	54	4-05-00-00091 / 0.6897[25]	1.3864[20]
4986	Cyclopropane, 1,2,3-trimethyl-, (1α,2α,3α)-	C_6H_{12} / 84.16	4806-58-0	66	3-05-00-00063 / 0.7130[25]	1.3970[20]
4987	Cyclopropane, 1,2,3-trimethyl-, (1α,2α,3β)-	C_6H_{12} / 84.16	4806-59-1	59.7	3-05-00-00063 / 0.6929[25]	1.3873[20]
4988	Cyclopropanone	C_3H_4O / 56.06	5009-27-8 / stable only at low temp.			
4989	1H-Cycloprop[e]azulen-4-ol, decahydro-1,1,4,7-tetramethyl-, (1aR,4R,4aS,7R,7aS,7bS)- / Ledol	$C_{15}H_{26}O$ / 222.37	577-27-5 / 105	5313 / 292	0.9078[100]	ace 4; eth 4; EtOH 4 / 1.4667[110]
4990	Cyclopropene	C_3H_4 / 40.06	2781-85-3	-36 dec		
4991	Cyclotetrasiloxane, ethenylheptamethyl-	$C_9H_{24}O_4Si_4$ / 308.63	3763-39-1	84[20]	0.9505[20]	ctc 3; CS_2 3 / 1.4034[20]
4992	Cyclotetrasiloxane, heptamethyl- / Heptamethylcyclotetrasiloxane	$C_7H_{22}O_4Si_4$ / 282.59	15721-05-8 / -27	165	4-04-00-04125 / 0.9583[20]	H_2O I / 1.3965[20]
4993	Cyclotetrasiloxane, octamethyl- / Octamethylcyclotetrasiloxane	$C_8H_{24}O_4Si_4$ / 296.62	556-67-2 / 17.5	6668 / 175.8	4-04-00-04125 / 0.9561[20]	H_2O I; ctc 3 / 1.3968[20]
4994	Cyclotetrasiloxane, octaphenyl-	$C_{48}H_{40}O_4Si_4$ / 793.19	546-56-5 / 200.5	330[1]	4-16-00-01530	H_2O I; EtOH 2; bz 3; chl 3
4995	Cyclotetrasiloxane, 2,4,6,8-tetrabutyl-2,4,6,8-tetramethyl- / Siloxane, butylmethyl(cyclic tetramer)	$C_{20}H_{48}O_4Si_4$ / 464.94	14685-29-1	291	4-04-00-04173 / 0.9230[20]	1.4300[20]
4996	Cyclotetrasiloxane, 2,4,6,8-tetraethenyl-2,4,6,8-tetramethyl-	$C_{12}H_{24}O_4Si_4$ / 344.66	2554-06-5 / -43.5	224; 111[12]	4-04-00-04184 / 0.9875[20]	ctc 3; CS_2 3

No.	Name Synonym	Mol. Form. Mol. Wt.	CAS RN mp/°C	Merck No. bp/°C	Beil. Ref. den/g cm^{-3}	Solubility n_D
4997	Cyclotetrasiloxane, 2,4,6,8-tetraethyl-2,4,6,8-tetramethyl- Cyclotetrasiloxane, 2,4,6,8-tetramethyl-2,4,6,8-tetraethyl	$C_{12}H_{32}O_4Si_4$ 352.73	7623-01-0 -43.5	245; 111[12]	4-04-00-04154 0.9600[20]	ctc 3; CS_2 3
4998	Cyclotetrasiloxane, 2,4,6,8-tetraethyl-2,4,6,8-tetraphenyl- Cyclotetrasiloxane, 1,3,5,7-tetraethyl-1,3,5,7-tetraphenyl	$C_{32}H_{40}O_4Si_4$ 601.01	18758-34-4 106	212[0.1]	3-16-00-01211 1.1000[20]	1.5430[25]
4999	Cyclotetrasiloxane, 2,4,6,8-tetramethyl-	$C_4H_{16}O_4Si_4$ 240.51	2370-88-9 -65	134.5	4-04-00-04099 0.9912[20]	H_2O I 1.3870[20]
5000	Cyclotetrasiloxane, 2,4,6,8-tetramethyl-2,4,6,8-tetraphenyl-	$C_{28}H_{32}O_4Si_4$ 544.90	77-63-4 99	237[115]	3-16-00-01211 1.1183[20]	H_2O 1; ace 5; hp 5 1.5461[20]
5001	Cyclotrisilazane, 2,2,4,4,6,6-hexamethyl- 1,1,3,3,5,5-Hexamethylcyclotrisilazane	$C_6H_{21}N_3Si_3$ 219.51	1009-93-4 -10	188	4-04-00-04114 0.9196[20]	1.448[20]
5002	Cyclotrisiloxane, hexamethyl- Dimethylsiloxane cyclic trimer	$C_6H_{18}O_3Si_3$ 222.46	541-05-9 64.5	134	4-04-00-04123 1.1200[20]	H_2O 1
5003	Cyclotrisiloxane, 2,4,6-triethyl-2,4,6-trimethyl- 2,4,6-Triethyl-2,4,6-trimethylcyclotrisiloxane	$C_9H_{24}O_3Si_3$ 264.54	15901-49-2 -3	199	4-04-00-04154	ctc 3; CS_2 3
5004	Cyclotrisiloxane, 2,4,6-triethyl-2,4,6-triphenyl- 2,4,6-Triethyl-2,4,6-triphenylcyclotrisiloxane	$C_{24}H_{30}O_3Si_3$ 450.76	546-33-8 177.5	166[0.025]	3-16-00-01211 1.0952[25]	H_2O 1 1.5402[25]
5005	Cyclotrisiloxane, 2,4,6-trimethyl-2,4,6-triphenyl-	$C_{21}H_{24}O_3Si_3$ 408.68	546-45-2 100	190[1.5]	3-16-00-01211 1.1062[20]	1.5397[20]
5006	1,4,8-Cycloundecatriene, 2,6,6,9-tetramethyl-, (E,E,E)- Humulene	$C_{15}H_{24}$ 204.36	6753-98-6 123[10]	4672	4-05-00-01171 0.8905[20]	1.5038[20]
5007	7,20-Cycloveatchane-1,12,15-triol, 21-ethyl-4-methyl-16-methylene-, (1α,12α,15β)- Luciculine	$C_{22}H_{35}NO_3$ 361.52	5008-52-6 149	6286 165[0.02]	4-21-00-02584	EtOH 4
5008	Cyfluthrin Cyano(4-fluoro-3-phenoxyphenyl)methyl 3-(2,2-dichloroethenyl)-2,2-dimethylcyclopropanecarboxylate	$C_{22}H_{18}Cl_2FNO_3$ 434.29	68359-37-5 60	2764		
5009	Cygon Phosphorodithioic acid, O,O-dimethyl S-[2-(methylamino)-2-oxoethyl] ester	$C_5H_{12}NO_3PS_2$ 229.26	60-51-5 52	3209 117[0.1]	1.277[65]	
5010	Cyhalothrin	$C_{23}H_{19}ClF_3NO_3$ 449.86	91465-08-6 49.2			
5011	Cyhexatin Stannane, tricyclohexylhydroxy-	$C_{18}H_{34}OSn$ 385.18	13121-70-5 196	2767		
5012	Cypermethrin Cyclopropanecarboxylic acid, 3-(2,2-dichloroethenyl)-2,2-dimethyl-, cyano(3-phenoxy-phenyl)methyl ester	$C_{22}H_{19}Cl_2NO_3$ 416.30	52315-07-8 70	2775	1.25[20]	
5013	Cyprazine 1,3,5-Triazine-2,4-diamine, 6-chloro-N-cyclopropyl-N'-(1-methylethyl)-	$C_9H_{14}ClN_5$ 227.70	22936-86-3 167			
5014	L-Cysteine Propanoic acid, 2-amino-3-mercapto-, (R)-	$C_3H_7NO_2S$ 121.16	52-90-4 240 dec	2787	4-04-00-03144	H_2O 4; ace 4; EtOH 4
5015	L-Cysteine, N-acetyl- Acetylcysteine	$C_5H_9NO_3S$ 163.20	616-91-1 109.5	82	4-04-00-03160	
5016	L-Cysteine, S-(2-amino-2-carboxyethyl)-, (R)- L-Lanthionine	$C_6H_{12}N_2O_4S$ 208.24	922-55-4 293-5 dec	5234	3-04-00-01593	
5017	L-Cysteine, S-(carboxymethyl)- Carbocysteine	$C_5H_9NO_4S$ 179.20	638-23-3 206	1809	4-04-00-03150	
5018	L-Cysteine, S,S'-methylenebis- Djenkolic acid	$C_7H_{14}N_2O_4S_2$ 254.33	498-59-9 300-50 dec	3393	3-04-00-01591	
5019	L-Cysteine, methyl ester, hydrochloride Methylcysteine hydrochloride	$C_4H_{10}ClNO_2S$ 171.65	18598-63-5 140.5	5668	3-04-00-01601	
5020	L-Cystine 3,3'-Dithiobis(2-aminopropanoic acid)	$C_6H_{12}N_2O_4S_2$ 240.30	56-89-3 260-1 dec	2788	4-04-00-03155 1.677[25]	H_2O 2; EtOH 1; eth 1; bz 1
5021	Cytidine 4-Amino-1-β-D-ribofuranosyl-2(1H)-pyrimidinone	$C_9H_{13}N_3O_5$ 243.22	65-46-3 230-1 dec	2792	4-25-00-03667	H_2O 4; EtOH 2
5022	3'-Cytidylic acid Cytidine 3'-monophosphate	$C_9H_{14}N_3O_8P$ 323.20	84-52-6 233 dec	2794	4-25-00-03669	H_2O 3; EtOH 3
5023	1,3-Decadiene 1-Hexyl-1,3-butadiene	$C_{10}H_{18}$ 138.25	2051-25-4 169		4-01-00-01056 0.752[30]	bz 4
5024	1,9-Decadiene α,ω-Decadiene	$C_{10}H_{18}$ 138.25	1647-16-1 167		4-01-00-01056 0.75[25]	1.4325[20]
5025	2,4-Decadienoic acid, (E,E)- 2,4-Decadienoic acid (trans-2, trans-4)	$C_{10}H_{16}O_2$ 168.24	30361-33-2 49.5	121[0.3]	4-02-00-01730	1.5058[31]
5026	2,4-Decadienoic acid, methyl ester, (E,E)-	$C_{11}H_{18}O_2$ 182.26	7328-33-8 70[0.2]		4-02-00-01731 0.982[22]	1.4918[22]
5027	2,4-Decadienoic acid, methyl ester, (E,Z)-	$C_{11}H_{18}O_2$ 182.26	4493-42-9 71[0.15]		4-02-00-01731 0.9128[22]	1.4874[22]

No.	Name / Synonym	Mol. Form. / Mol. Wt.	CAS RN / mp/°C	Merck No. / bp/°C	Beil. Ref. / den/g cm⁻³	Solubility / n_D
5028	2,4-Decadienoic acid, methyl ester, (Z,E)- / 2,4-Decadienoic acid, methyl ester (cis-2, trans-4)	$C_{11}H_{18}O_2$ / 182.26	108965-84-0	76-80[0.5]	4-02-00-01731 / 0.9131[23]	1.4876[23]
5029	2,4-Decadienoic acid, methyl ester, (Z,Z)- / 2,4-Decadienoic acid, methyl ester (cis-2, cis-4)	$C_{11}H_{18}O_2$ / 182.26	108965-86-2	78[0.6]	4-02-00-01731 / 0.9095[23]	1.4830[23]
5030	Decanal / Capraldehyde	$C_{10}H_{20}O$ / 156.27	112-31-2 / -5	208.5	4-01-00-03366 / 0.830[15]	H_2O 1; EtOH i; eth 3; ace 3 / 1.4287[20]
5031	Decanamide / Capramide	$C_{10}H_{21}NO$ / 171.28	2319-29-1 / 108		4-02-00-01050 / 0.822[130]	ace 4; eth 4; EtOH 4 / 1.4621[110]
5032	1-Decanamine	$C_{10}H_{23}N$ / 157.30	2016-57-1 / 17	220.5	4-04-00-00783 / 0.7936[20]	H_2O 2; EtOH 5; eth 5; ace 5 / 1.4369[20]
5033	Decane / n-Decane	$C_{10}H_{22}$ / 142.28	124-18-5 / -29.7	174.1	4-01-00-00464 / 0.7300[20]	H_2O 1; EtOH 5; eth 3; ctc 2 / 1.4102[20]
5034	Decane, 1-bromo-	$C_{10}H_{21}Br$ / 221.18	112-29-8 / -29.2	240.6	4-01-00-00470 / 1.0702[20]	H_2O 1; eth 4; ctc 3; chl 4 / 1.4557[20]
5035	Decane, 1-bromo-10-fluoro- / 1-Bromo-10-fluorodecane	$C_{10}H_{20}BrF$ / 239.17	334-61-2	131-2[11]	4-01-00-00471 / 1.152[20]	eth 4; EtOH 4 / 1.4512[25]
5036	Decane, 1-chloro-	$C_{10}H_{21}Cl$ / 176.73	1002-69-3 / -31.3	223.4	4-01-00-00469 / 0.8705[20]	H_2O 1; eth 4; ctc 3; chl 4 / 1.4379[20]
5037	Decane, 1-chloro-10-fluoro- / 1-Chloro-10-fluorodecane	$C_{10}H_{20}ClF$ / 194.72	334-62-3	115[9]	4-01-00-00469 / 0.957[20]	eth 4; EtOH 4 / 1.4333[25]
5038	Decane, 1,10-dibromo- / Decamethylene dibromide	$C_{10}H_{20}Br_2$ / 300.08	4101-68-2 / 28	161[9]; 128[4]	4-01-00-00471 / 1.335[30]	H_2O 1; EtOH 2; eth 3 / 1.4927[25]
5039	Decane, 5,6-dibromo- / 5,6-Dibromodecane	$C_{10}H_{20}Br_2$ / 300.08	77928-86-0	119[9]	4-01-00-00472 / 1.3484[20]	1.4912[20]
5040	Decane, 1,10-dichloro- / Decamethylene dichloride	$C_{10}H_{20}Cl_2$ / 211.17	2162-98-3 / 15.6	167-8[28]	4-01-00-00469 / 0.9945[25]	1.4586[25]
5041	Decanedinitrile	$C_{10}H_{16}N_2$ / 164.25	1871-96-1	204[16]	4-02-00-02089 / 0.9313[20]	H_2O 1; chl 3 / 1.4474[20]
5042	Decanedioic acid / Sebacic acid	$C_{10}H_{18}O_4$ / 202.25	111-20-6 / 130.8	8369 / 295[100]; 232[10]	4-02-00-02078 / 1.2705[20]	H_2O 2; EtOH 3; eth 3; bz 1 / 1.422[133]
5043	Decanedioic acid, bis(2-ethylhexyl) ester / Bis(2-ethylhexyl) sebacate	$C_{26}H_{50}O_4$ / 426.68	122-62-3 / -48	1263 / 256[5]	4-02-00-02083 / 0.912[25]	ace 4; bz 4; EtOH 4 / 1.451[25]
5044	Decanedioic acid, dibutyl ester / Dibutyl sebacate	$C_{18}H_{34}O_4$ / 314.47	109-43-3 / -10	344.5	4-02-00-02081 / 0.9405[15]	H_2O 1; eth 3; ctc 3 / 1.4433[15]
5045	Decanedioic acid, diethyl ester	$C_{14}H_{26}O_4$ / 258.36	110-40-7 / 5	305; 188[19]	4-02-00-02080 / 0.9646[20]	H_2O 2; EtOH 3; ace 3; bz 1 / 1.4306[20]
5046	Decanedioic acid, dimethyl ester	$C_{12}H_{22}O_4$ / 230.30	106-79-6 / 38	175[20]; 144[5]	4-02-00-02080 / 0.9882[28]	H_2O 1; EtOH 3; eth 3; ace 3 / 1.4355[28]
5047	1,10-Decanediol / Decamethylene glycol	$C_{10}H_{22}O_2$ / 174.28	112-47-0 / 74	2842 / 192[20]	4-01-00-02613	H_2O 2; EtOH 4; eth 2; DMSO 3
5048	2,9-Decanediol	$C_{10}H_{22}O_2$ / 174.28	14021-92-2 / 33	114[5]	3-01-00-02232	1.4505[21]
5049	Decanedioyl dichloride	$C_{10}H_{16}Cl_2O_2$ / 239.14	111-19-3 / -2.5	220[75]; 165[11]	4-02-00-02088 / 1.1212[20]	1.4684[18]
5050	Decane, docosafluoro- / Perfluorodecane	$C_{10}F_{22}$ / 538.07	307-45-9	144.2		
5051	Decane, 1-(ethenyloxy)- / Decyl vinyl ether	$C_{12}H_{24}O$ / 184.32	765-05-9 / -41	101[10]	4-01-00-02057 / 0.812[20]	1.4346[20]
5052	Decane, 1-fluoro- / Decyl fluoride	$C_{10}H_{21}F$ / 160.28	334-56-5 / -35	186.2	4-01-00-00468 / 0.8194[20]	eth 4 / 1.4085
5053	Decane, 1-iodo-	$C_{10}H_{21}I$ / 268.18	2050-77-3 / -16.3	132[15]	4-01-00-00472 / 1.2546[20]	H_2O 1; EtOH 3; eth 3; ctc 3 / 1.4858[20]
5054	Decane, 2-methyl- / 2-Methyldecane	$C_{11}H_{24}$ / 156.31	6975-98-0 / -48.9	189.3	4-01-00-00491 / 0.7368[20]	1.4154[20]
5055	Decane, 3-methyl- / 3-Methyldecane	$C_{11}H_{24}$ / 156.31	13151-34-3 / -92.9	188.1	4-01-00-00491 / 0.7422[20]	1.4177[20]
5056	Decane, 4-methyl- / 4-Methyldecane	$C_{11}H_{24}$ / 156.31	2847-72-5 / -77.5	187		1.4352[20]
5057	Decanenitrile / Caprinitrile	$C_{10}H_{19}N$ / 153.27	1975-78-6 / -17.9	243. 106[10]	4-02-00-01051 / 0.8199[20]	ace 4; eth 4; EtOH 4; chl 4 / 1.4296[20]
5058	1-Decanethiol / Decyl mercaptan	$C_{10}H_{22}S$ / 174.35	143-10-2 / -26	240.6	4-01-00-01821 / 0.8443[20]	H_2O 1; EtOH 3; eth 3 / 1.4509[20]
5059	Decanoic acid / Capric acid	$C_{10}H_{20}O_2$ / 172.27	334-48-5 / 31.9	1759 / 268.7	4-02-00-01041 / 0.8858[40]	ace 4; bz 4; eth 4; EtOH 4 / 1.4288[40]
5060	Decanoic acid, anhydride	$C_{20}H_{38}O_3$ / 326.52	2082-76-0 / 24.7		4-02-00-01049 / 0.8865[25]	eth 4; EtOH 4 / 1.400[25]
5061	Decanoic acid, 2-bromo-	$C_{10}H_{19}BrO_2$ / 251.16	2623-95-2 / 4	140-1[2]	4-02-00-01054 / 1.1912[24]	eth 4 / 1.4595[24]
5062	Decanoic acid, decyl ester	$C_{20}H_{40}O_2$ / 312.54	1654-86-0 / 9.7	219[15]	4-02-00-01045 / 0.8586[20]	eth 4 / 1.4423[20]
5063	Decanoic acid, ethyl ester / Ethyl caprate	$C_{12}H_{24}O_2$ / 200.32	110-38-3 / -20	3734 / 241.5	4-02-00-01044 / 0.8650[20]	eth 4; EtOH 4; chl 4 / 1.4256[20]
5064	Decanoic acid, 10-fluoro- / Capric acid, 10-fluoro	$C_{10}H_{19}FO_2$ / 190.26	334-59-8 / 49	135-8[10]	4-02-00-01051	eth 4; EtOH 4; lig 4
5065	Decanoic acid, 2-hexyl- / 2-Hexyldecanoic acid	$C_{16}H_{32}O_2$ / 256.43	25354-97-6 / 4630	140-50[0.02]	4-02-00-01189	1.4432[24]

No.	Name Synonym	Mol. Form. Mol. Wt.	CAS RN mp/°C	Merck No. bp/°C	Beil. Ref. den/g cm^{-3}	Solubility n_D
5066	Decanoic acid, methyl ester	$C_{11}H_{22}O_2$ 186.29	110-42-9 -18	224	4-02-00-01044 0.8730[20]	H_2O 1; EtOH 4; eth 4; ctc 2 1.4259[20]
5067	Decanoic acid, 1-methylethyl ester	$C_{13}H_{26}O_2$ 214.35	2311-59-3	121[10]	4-02-00-01045 0.8543[20]	1.4221[25]
5068	Decanoic acid, 2-octyl-	$C_{18}H_{36}O_2$ 284.48	619-39-6 38.5	215[13]	4-02-00-01254 0.8447[70]	eth 4; EtOH 4
5069	Decanoic acid, propyl ester	$C_{13}H_{26}O_2$ 214.35	30673-60-0	128.5[10]	4-02-00-01045 0.8623[20]	1.4280[20]
5070	1-Decanol Capric alcohol	$C_{10}H_{22}O$ 158.28	112-30-1 6.9	2847 231.1	4-01-00-01815 0.8297[20]	H_2O 1; EtOH 5; eth 5; ace 5 1.4372[20]
5071	2-Decanol, (±)- 2-Decanol (DL)	$C_{10}H_{22}O$ 158.28	74742-10-2 -2.4	211	4-01-00-01823 0.8250[20]	EtOH 3; eth 5; ace 5; bz 3 1.4326[25]
5072	4-Decanol 1-Propylheptyl alcohol	$C_{10}H_{22}O$ 158.28	2051-31-2 -11	210.5	4-01-00-01824 0.8261[20]	H_2O 1; EtOH 3; ctc 3 1.4320[20]
5073	1-Decanol, 10-chloro-	$C_{10}H_{21}ClO$ 192.73	51309-10-5 12.5	185-9[15]	4-01-00-01821 0.9630[25]	eth 4; EtOH 4 1.4578[20]
5074	1-Decanol, 10-fluoro- 10-Fluoro-1-decanol	$C_{10}H_{21}FO$ 176.27	334-64-5 22	136-7[15]	4-01-00-01821 0.919[20]	eth 4; EtOH 4 1.4322[25]
5075	2-Decanone Methyl octyl ketone	$C_{10}H_{20}O$ 156.27	693-54-9 14	210; 96[12]	4-01-00-03367 0.8248[20]	H_2O 1; EtOH 3; eth 3; ctc 2 1.4255[20]
5076	3-Decanone Ethyl heptyl ketone	$C_{10}H_{20}O$ 156.27	928-80-3 2.5	203	4-01-00-03368 0.8251[20]	EtOH 3; eth 3; ctc 3 1.4252[20]
5077	4-Decanone Hexyl propyl ketone	$C_{10}H_{20}O$ 156.27	624-16-8 -9	206.5	4-01-00-03368 0.824[20]	H_2O 1; EtOH 5; eth 5 1.4240[21]
5078	Decanoyl chloride Caprinoyl chloride	$C_{10}H_{19}ClO$ 190.71	112-13-0 -34.5	95	4-02-00-01050 0.919[25]	eth 3; ctc 3 1.4410[20]
5079	Decasiloxane, docosamethyl- Decasiloxane, dicosamethyl	$C_{22}H_{66}O_9Si_{10}$ 755.62	556-70-7	183.4	3-04-00-01881 0.925[20]	bz 4; lig 4 1.3988[20]
5080	2-Decenal	$C_{10}H_{18}O$ 154.25	3913-71-1	230	4-01-00-03511 0.845[17]	1.4533[17]
5081	3-Decenal	$C_{10}H_{18}O$ 154.25	58474-80-9	93-4[14]	4-01-00-03512 0.850[15]	1.4462[15]
5082	1-Decene	$C_{10}H_{20}$ 140.27	872-05-9 -66.3	170.5	3-01-00-00858 0.7408[20]	H_2O 1; EtOH 5; eth 5 1.4215[20]
5083	4-Decene	$C_{10}H_{20}$ 140.27	19689-18-0	170.6	4-01-00-00902 0.7404[20]	1.4243[20]
5084	5-Decene, (E)-	$C_{10}H_{20}$ 140.27	7433-56-9 -73	171	4-01-00-00902 0.7401[20]	H_2O 1; EtOH 5; eth 5; ctc 2 1.4243[20]
5085	5-Decene, (Z)-	$C_{10}H_{20}$ 140.27	7433-78-5 -112	171; 73[20]	3-01-00-00859 0.7445[20]	H_2O 1; EtOH 5; eth 5; ctc 2 1.4258[20]
5086	1-Decene, 2-bromo- 2-Bromo-1-decene	$C_{10}H_{19}Br$ 219.16	3017-67-2	115-6[22]	3-01-00-00859 1.0844[20]	1.4629[20]
5087	2-Decene, 1-bromo- 1-Bromo-2-decene	$C_{10}H_{19}Br$ 219.16	14304-30-4	121[17]	4-01-00-00902 1.074[18]	lig 4 1.4716[18]
5088	2-Decenoic acid Δ 2-Decenoic acid	$C_{10}H_{18}O_2$ 170.25	3913-85-7 12	165[15]	4-02-00-01606 0.9280[18]	1.4616[20]
5089	3-Decenoic acid	$C_{10}H_{18}O_2$ 170.25	15469-77-9 18	154-63[11]	4-02-00-01606 0.914[15]	1.4510[18]
5090	4-Decenoic acid Deconic acid telomer	$C_{10}H_{18}O_2$ 170.25	26303-90-2	149[13]	4-02-00-01607 0.9197[20]	bz 4; eth 4 1.4497[20]
5091	9-Decenoic acid Caproleic acid	$C_{10}H_{18}O_2$ 170.25	14436-32-9	158[21]; 142[4]	4-02-00-01605 0.9238[15]	eth 4; EtOH 4 1.4507[15]
5092	9-Decen-1-ol Decylenic alcohol	$C_{10}H_{20}O$ 156.27	13019-22-2	236	4-01-00-02184 0.876[25]	1.4480[20]
5093	3-Decen-2-one Heptylidene acetone	$C_{10}H_{18}O$ 154.25	10519-33-2	102-3[15.3]	4-01-00-03512 0.8473[20]	1.4480[20]
5094	1-Decen-3-yne	$C_{10}H_{16}$ 136.24	33622-26-3	76[20]	4-01-00-01105 0.7873[20]	1.4620[20]
5095	1-Decen-4-yne	$C_{10}H_{16}$ 136.24	24948-66-1	73-4[22]	3-01-00-01049 0.7880[20]	1.445[20]
5096	2-Decen-4-yne	$C_{10}H_{16}$ 136.24	116668-40-7	55[5]	3-01-00-01049 0.7850[25]	1.4609[25]
5097	1-Decyne Octylacetylene	$C_{10}H_{18}$ 138.25	764-93-2 -44	174	4-01-00-01054 0.7655[20]	H_2O 1; EtOH 3; eth 3; os 3 1.4265[20]
5098	3-Decyne	$C_{10}H_{18}$ 138.25	2384-85-2	177	4-01-00-01055 0.7619[25]	1.4315[20]
5099	4-Decyne	$C_{10}H_{18}$ 138.25	2384-86-3	74.5[19]	3-01-00-01017 0.772[17]	1.436[17]
5100	5-Decyne Dibutylacetylene	$C_{10}H_{18}$ 138.25	1942-46-7 -73	177; 78.8[25]	4-01-00-01055 0.7690[20]	H_2O 1; EtOH 3; eth 3 1.4331[20]
5101	4-Decyne, 3,3-dimethyl- 3,3-Dimethyl-4-decyne	$C_{12}H_{22}$ 166.31	70732-45-5	86[20]	3-01-00-01026 0.7731[20]	1.4399[20]
5102	Deltamethrin Cyano(3-phenoxyphenyl)methyl-3-(2,2-dibromoethenyl)-2,2-dimethylcyclopropanecarboxylate	$C_{22}H_{19}Br_2NO_3$ 505.21	52918-63-5 99	2869		
5103	Demeton S methyl Phosphorothioic acid, S-[2-(ethylthio)ethyl] O,O-dimethyl ester	$C_6H_{15}O_3PS_2$ 230.29	919-86-8	89[0.15]; 118[1]	1.207[20]	

No.	Name Synonym	Mol. Form. Mol. Wt.	CAS RN mp/°C	Merck No. bp/°C	Beil. Ref. den/g cm^{-3}	Solubility n_D
5104	Desmedipham Carbamic acid, [3-[[(phenylamino)carbonyl]oxy]phenyl]-, ethyl ester	C$_{16}$H$_{16}$N$_2$O$_4$ 300.31	13684-56-5 120			
5105	Dialifor Phosphorodithioic acid, S-[2-chloro-1-(1,3-dihydro-1,3-dioxo-2H-isoindol-2-yl)ethyl]	C$_{14}$H$_{17}$ClNO$_4$P S$_2$ 393.85	10311-84-9 68	2949		
5106	Diallate Carbamothioic acid, bis(1-methylethyl)-, S-(2,3-dichloro-2-propenyl) ester	C$_{10}$H$_{17}$Cl$_2$NOS 270.22	2303-16-4 150^9	2950		
5107	8,8'-Diapo-ψ,ψ-carotenedioic acid Crocetin (trans)	C$_{20}$H$_{24}$O$_4$ 328.41	27876-94-4 286	2592	3-02-00-02018	H$_2$O 2; EtOH 1; eth 1; bz 1
5108	6,6'-Diapo-ψ,ψ-carotenedioic acid, monomethyl ester, 9-cis- Bixin	C$_{25}$H$_{30}$O$_4$ 394.51	6983-79-5 198	1320	3-02-00-02021	H$_2$O 1; EtOH 3; eth 2; ace 3
5109	Diarsine, tetraethyl- Tetraethyldiarsenic	C$_8$H$_{20}$As$_2$ 266.09	612-08-8 186		3-04-00-01832 1.1388^{24}	eth 4; EtOH 4 1.4709
5110	Diarsine, tetramethyl- Cacodyl	C$_4$H$_{12}$As$_2$ 209.98	471-35-2 -6	1602 165	4-04-00-03683 1.447^{15}	eth 4; EtOH 4
5111	1,4-Diazabicyclo[2.2.2]octane Triethylenediamine	C$_6$H$_{12}$N$_2$ 112.17	280-57-9 159	9584	5-23-03-00487	chl 3
5112	Diazene, bis(2-ethoxyphenyl)- Azobenzene, 2,2'-diethoxy-	C$_{16}$H$_{18}$N$_2$O$_2$ 270.33	613-43-4 131	240 dec	4-16-00-00138	H$_2$O 1; EtOH 3; HCl 3
5113	Diazene, bis(4-ethoxyphenyl)-	C$_{16}$H$_{18}$N$_2$O$_2$ 270.33	588-52-3 162	dec	4-16-00-00174	H$_2$O 1; EtOH 2; eth 3; bz 3
5114	Diazene, bis(2-methylphenyl)-	C$_{14}$H$_{14}$N$_2$ 210.28	584-90-7 55.5		2-16-00-00019 1.0215^{85}	H$_2$O 1; EtOH 3; eth 4; bz 4 1.6180^{65}
5115	Diazene, bis(3-methylphenyl)-, (E)-	C$_{14}$H$_{14}$N$_2$ 210.28	51437-67-3 54		4-16-00-00070 1.0123^{66}	EtOH 4; eth 4; bz 4 1.6152^{66}
5116	Diazene, bis(2-methylphenyl)-, 1-oxide, (Z)- Azoxybenzene, 2,2'-dimethyl (trans)	C$_{14}$H$_{14}$N$_2$O 226.28	51284-68-5 60		4-16-00-00064 1.0215^{85}	bz 4; eth 4; EtOH 4 1.6180^{65}
5117	Diazene, bis(2-methylphenyl)-, 1-oxide, (E)- Azoxybenzene, 2,2'-dimethyl(trans)	C$_{14}$H$_{14}$N$_2$O 226.28	116723-89-8		2-16-00-00318 1.0215^{85}	H$_2$O 1; EtOH 3; eth 3; bz 3 1.6180^{65}
5118	Diazene, bis(3-methylphenyl)-, 1-oxide, (Z)- Azoxybenzene, 3,3'-dimethyl (trans)	C$_{14}$H$_{14}$N$_2$O 226.28	71297-97-7 39		4-16-00-00071 1.0123^{66}	bz 4; eth 4; EtOH 4; lig 4 1.6152^{66}
5119	Diazene, bis(3-methylphenyl)-, 1-oxide, (E)- Azoxybenzene, 3,3'-dimethyl(trans)	C$_{14}$H$_{14}$N$_2$O 226.28	116723-90-1		2-16-00-00319 1.0123^{66}	H$_2$O 1; EtOH 3; eth 3; bz 3 1.6152^{66}
5120	Diazene, (4-bromophenyl)phenyl-, 1-oxide Azoxybenzene, 4-bromo-	C$_{12}$H$_9$BrN$_2$O 277.12	16109-68-5 93.5		3-16-00-00584 1.4138^{100}	H$_2$O 1; EtOH 3; bz 3; MeOH 3
5121	Diazenecarbothioic acid, phenyl-, 2-phenylhydrazide Dithizone	C$_{13}$H$_{12}$N$_4$S 256.33	60-10-6 165-9 dec	3383	4-16-00-00018	H$_2$O 1; EtOH 2; eth 2; chl 3
5122	Diazenecarboxylic acid, phenyl-; 2-phenylhydrazide Diphenylcarbazone	C$_{13}$H$_{12}$N$_4$O 240.26	538-62-5 157 dec	3322	4-16-00-00017	H$_2$O 1; EtOH 4; bz 4; chl 4
5123	Diazene, dibutyl Azobutane	C$_8$H$_{18}$N$_2$ 142.24	2159-75-3 60^{18}			
5124	Diazenedicarboxamide Azodicarbonamide	C$_2$H$_4$N$_4$O$_2$ 116.08	123-77-3 212 dec	932	3-03-00-00234	
5125	Diazene, dimethyl- Azomethane	C$_2$H$_6$N$_2$ 58.08	503-28-6 -78	1.5	3-04-00-01747 0.743^0	ace 4; eth 4; EtOH 4 1.4199^{19}
5126	Diazene, dimethyl-, (E)-	C$_2$H$_6$N$_2$ 58.08	4143-41-3 -78	1.5	4-04-00-03366 0.743^0	ctc 3; hp 3
5127	Diazene, di-1-naphthalenyl- 1,1'-Azobisnaphthalene	C$_{20}$H$_{14}$N$_2$ 282.34	487-10-5 190	sub	2-16-00-00026	H$_2$O 1; EtOH 2; ace 3; bz 4
5128	Diazene, di-2-naphthalenyl- 2,2'-Azonaphthalene	C$_{20}$H$_{14}$N$_2$ 282.34	582-08-1 208	sub	2-16-00-00026	H$_2$O 1; EtOH 2; eth 2; bz 3
5129	Diazene, diphenyl-, (E)- trans-Azobenzene	C$_{12}$H$_{10}$N$_2$ 182.22	17082-12-1 68.5	930 293	4-16-00-00007 1.203^{20}	H$_2$O 2; EtOH 3; eth 3; bz 3 1.6266^{78}
5130	Diazene, diphenyl-, (Z)- cis-Azobenzene	C$_{12}$H$_{10}$N$_2$ 182.22	1080-16-6 71	930	4-16-00-00007	H$_2$O 2; EtOH 3; eth 3; bz 3
5131	Diazene, diphenyl-, 1-oxide, (Z)- cis-Azoxybenzene	C$_{12}$H$_{10}$N$_2$O 198.22	20972-43-4 87	937	4-16-00-00010 1.166^{20}	1.633^{20}
5132	Diazene, dipropyl Azopropane	C$_6$H$_{14}$N$_2$ 114.19	821-67-0 114			
5133	Diazene, (4-ethoxyphenyl)phenyl-	C$_{14}$H$_{14}$N$_2$O 226.28	7466-38-8 85	339.5	2-16-00-00040 1.0400^{100}	EtOH 4; eth 4; ace 3; bz 4 1.6419^{100}
5134	Diazene, ethylphenyl-	C$_8$H$_{10}$N$_2$ 134.18	935-08-0 177		4-16-00-00007 0.9628^{22}	bz 4; eth 4; EtOH 4 1.5579
5135	Diazene, (2-methoxyphenyl)phenyl-	C$_{13}$H$_{12}$N$_2$O 212.25	6319-21-7 41	195-7^{14}	3-16-00-00081 1.0728^{100}	H$_2$O 1; EtOH 3; ace 3; os 4
5136	Diazene, (3-methoxyphenyl)phenyl-	C$_{13}$H$_{12}$N$_2$O 212.25	34238-81-8 33.3	193^{15}	4-16-00-00155 1.1023^{53}	H$_2$O 1; EtOH 3; ace 3; os 4
5137	Diazene, (4-methoxyphenyl)phenyl-	C$_{13}$H$_{12}$N$_2$O 212.25	2396-60-3 56	340	2-16-00-00040 1.12^{75}	H$_2$O 1; EtOH 3; eth 3; ace 3
5138	Diazene, (3-methylphenyl)phenyl- Azobenzene, 3-methyl	C$_{13}$H$_{12}$N$_2$ 196.25	17478-66-9 18.5	175^{19}	3-16-00-00047 1.065^{20}	
5139	Diazene, (4-methylphenyl)phenyl- p-Methylazobenzene	C$_{13}$H$_{12}$N$_2$ 196.25	949-87-1 71.5	312	3-16-00-00049	bz 4; eth 4; chl 4

No.	Name Synonym	Mol. Form. Mol. Wt.	CAS RN mp/°C	Merck No. bp/°C	Beil. Ref. den/g cm^{-3}	Solubility n_D
5140	1H-1,4-Diazepine, 1-[(4-chlorophenyl)phenylmethyl]hexahydro-4-methyl- Homochlorocyclizine	$C_{19}H_{23}ClN_2$ 314.86	848-53-3	4653	5-23-03-00248	
5141	1H-1,4-Diazepine, hexahydro-1-methyl-	$C_6H_{14}N_2$ 114.19	4318-37-0	154	5-23-03-00241 0.9111[20]	1.4769[20]
5142	Dibenz[a,h]anthracene 1,2:5,6-Dibenzanthracene	$C_{22}H_{14}$ 278.35	53-70-3 269.5	2989	4-05-00-02722	H_2O 1; EtOH 2; ace 3; bz 3
5143	Dibenz[a,j]acridine 7-Azadibenz[a,j]anthracene	$C_{21}H_{13}N$ 279.34	224-42-0 216		5-20-08-00656	
5144	5H-Dibenz[b,f]azepine-5-carboxamide Carbamazepine	$C_{15}H_{12}N_2O$ 236.27	298-46-4 190.2	1783	5-20-08-00247	
5145	Dibenz[c,e]oxepin-5,7-dione	$C_{14}H_8O_3$ 224.22	6050-13-1 217	sub	5-17-11-00525	H_2O 1; eth 2
5146	10H-Dibenzo[a,d]cyclohepten-10-one, 5,11-dihydro-	$C_{15}H_{12}O$ 208.26	6374-70-5 30	203-4[7]	4-07-00-01675 1.1635[20]	1.6324[20]
5147	5H-Dibenzo[a,d]cyclohepten-5-one, 10,11-dihydro-	$C_{15}H_{12}O$ 208.26	1210-35-1 30	203-4[7]	4-07-00-01674 1.1635[20]	1.6324[20]
5148	13H-Dibenzo[a,g]fluorene	$C_{21}H_{14}$ 266.34	207-83-0 174.5	195-200[0.5]	4-05-00-02693	H_2O 1; EtOH 2; eth 2; bz 3
5149	6H-Dibenzo[a,g]quinolizine, 5,8,13,13a-tetrahydro-2,3,9,10-tetramethoxy-13-methyl-, (13S-trans)- Corydaline (d)	$C_{22}H_{27}NO_4$ 369.46	518-69-4 136	2543	5-21-06-00173	bz 4; eth 4; EtOH 4; chl 4
5150	6H-Dibenzo[a,g]quinolizin-2-ol, 5,8,13,13a-tetrahydro-3,9,10-trimethoxy-13-methyl-, (13S-trans)- Isocorybulbine	$C_{21}H_{25}NO_4$ 355.43	22672-74-8 187.5	5045	5-21-06-00173 1.045[20]	H_2O 1; EtOH 3; chl 3; acid 3
5151	6H-Dibenzo[a,g]quinolizin-3-ol, 5,8,13,13a-tetrahydro-2,9,10-trimethoxy-13-methyl-, (13S-trans)- Corybulbine	$C_{21}H_{25}NO_4$ 355.43	518-77-4 237.5	2539	5-21-06-00173	H_2O 1; EtOH 2; ace 3
5152	13H-Dibenzo[a,h]fluoren-13-one 1,2:6,7-Dibenzo-9-fluorenone	$C_{21}H_{12}O$ 280.33	4599-94-4 214	sub 200	3-07-00-02897	
5153	11H-Dibenzo[b,e][1,4]dioxepin-11-one, 2,4,7,9-tetrachloro-3-hydroxy-8-methoxy-1,6-dimethyl- Diploicin	$C_{16}H_{10}Cl_4O_5$ 424.06	527-93-5 232	3345	5-19-06-00364	
5154	6H-Dibenzo[b,d]pyran-1-ol, 6,6,9-trimethyl-3-pentyl- Cannabinol	$C_{21}H_{26}O_2$ 310.44	521-35-7 77	1751 185[0.05]	5-17-04-00567	H_2O 1; EtOH 3; eth 3; ace 3
5155	4H-Dibenzo[de,g]quinoline-10,11-diol, 5,6,6a,7-tetrahydro-6-methyl-, (R)- Apomorphine	$C_{17}H_{17}NO_2$ 267.33	58-00-4 195 dec	776	5-21-05-00333	H_2O 2; EtOH 3; eth 3; ace 3
5156	4H-Dibenzo[de,g]quinoline-2,9-diol, 5,6,6a,7-tetrahydro-1,10-dimethoxy-6-methyl-, (S)- Boldine	$C_{19}H_{21}NO_4$ 327.38	476-70-0 163	1328	5-21-06-00118	EtOH 4; chl 4
5157	4H-Dibenzo[de,g]quinoline, 5,6,6a,7-tetrahydro-1,2,9,10-tetramethoxy-6-methyl-, (S)- d-Glaucine	$C_{21}H_{25}NO_4$ 355.43	475-81-0 120	4332	5-21-06-00122	ace 4; EtOH 4; chl 4
5158	4H-Dibenzo[de,g]quinolin-1-ol, 5,6,6a,7-tetrahydro-2,11-dimethoxy-6-methyl-, (S)- Isothebaine	$C_{19}H_{21}NO_3$ 311.38	568-21-8 203.5	5116	5-21-05-00658	H_2O 1; EtOH 5; eth 2; chl 5
5159	4H-Dibenzo[de,g]quinolin-10-ol, 5,6,6a,7-tetrahydro-1,2-dimethoxy-, (R)- Tsuduranine	$C_{18}H_{19}NO_3$ 297.35	517-97-5 204	9710	5-21-05-00655	ace 4; eth 4; EtOH 4
5160	4H-Dibenzo[de,g]quinolin-11-ol, 5,6,6a,7-tetrahydro-10-methoxy-6-methyl-, (R)- Apocodeine	$C_{18}H_{19}NO_2$ 281.35	641-36-1 123.5	771	5-21-05-00334	EtOH 2; eth 3; ace 3; bz 3
5161	4H-Dibenzo[de,g]quinolin-11-ol, 5,6,6a,7-tetrahydro-1,2,10-trimethoxy-6-methyl-, (S)- Isocorydine	$C_{20}H_{23}NO_4$ 341.41	475-67-2 185	5046	5-21-06-00132	chl 4
5162	4H-Dibenzo[de,g]quinolin-1-ol, 5,6,6a,7-tetrahydro-2,10,11-trimethoxy-6-methyl-, (S)- Corydine	$C_{20}H_{23}NO_4$ 341.41	476-69-7 149	2545	5-21-06-00134	eth 4; EtOH 4; chl 4
5163	Dibenzofuran 2,2'-Biphenylene oxide	$C_{12}H_8O$ 168.19	132-64-9 86.5	287	5-17-02-00234 1.0886[99]	H_2O 2; EtOH 3; eth 4; ace 3 1.6079[99]
5164	Dibenzofuran, 2-bromo- 2-Bromodibenzofuran	$C_{12}H_7BrO$ 247.09	86-76-0 110	220[40]	5-17-02-00238	EtOH 3; HOAc 2
5165	Dibenzofuran, 3-bromo- 3-Bromodibenzofuran	$C_{12}H_7BrO$ 247.09	26608-06-0 121	220[40]	4-17-00-00588	H_2O 1; EtOH 3; eth 3
5166	Dibenzofuran, 3-nitro- 3-Nitrodibenzofuran	$C_{12}H_7NO_3$ 213.19	5410-97-9 181.5	180-5[3]	5-17-02-00239	H_2O 1; EtOH 2; eth 2; HOAc 3
5167	Dibenzothiophene	$C_{12}H_8S$ 184.26	132-65-0 99.5	332.5	5-17-02-00239	H_2O 3; EtOH 4; bz 4; chl 3
5168	Diborane(6), mu-(methylthio)- Diborane, methylthio	CH_8B_2S 73.76	91572-15-5 -101.5	53	4-01-00-01287	ace 4; bz 4

No.	Name / Synonym	Mol. Form. / Mol. Wt.	CAS RN / mp/°C	Merck No. / bp/°C	Beil. Ref. / den/g cm^{-3}	Solubility / n_D
5169	Dicarbonic acid, diethyl ester / Pyrocarbonic acid diethyl ester	$C_6H_{10}O_5$ / 162.14	1609-47-8 /	8008 / 93-4[18]	4-03-00-00018 / 1.120[20]	ace 4; EtOH 4; lig 4 / 1.3960[20]
5170	Dichlorvos / Phosphoric acid, 2,2-dichloroethenyl dimethyl ester	$C_4H_7Cl_2O_4P$ / 220.98	62-73-7 /	3069 / 140[20]; 84[1]	/ 1.415[25]	
5171	Diclofop-methyl / Propanoic acid, 2-[4-(2,4-dichlorophenoxy)phenoxy]-, methyl ester	$C_{16}H_{14}Cl_2O_4$ / 341.19	51338-27-3 / 40	3072 / 176[0.1]		
5172	Dicrotophos / Phosphoric acid, 3-(dimethylamino)-1-methyl-3-oxo-1-propenyl dimethyl ester, (E)-	$C_8H_{16}NO_5P$ / 237.19	141-66-2 /	3077 / 400; 130[0.1]	/ 1.216[15]	
5173	Diethatyl-ethyl / Glycine, N-(chloroacetyl)-N-(2,6-diethylphenyl)-, ethyl ester	$C_{16}H_{22}ClNO_3$ / 311.81	38727-55-8 /			
5174	Difenoconazole / cis,trans-3-Chloro-4-[4-methyl-2-(1H-1,2,4-triazol-1-ylmethyl)-1,3-dioxolan-2-yl]phenyl 4-chlorophenyl ether	$C_{19}H_{17}Cl_2N_3O_3$ / 406.27	119446-68-3 / 76	/ 220[0.03]		
5175	Difenzoquat methyl sulfate / 1H-Pyrazolium, 1,2-dimethyl-3,5-diphenyl-, methyl sulfate	$C_{17}H_{17}N_2 \cdot CH_3O_4S$ / 360.43	43222-48-6 / 157			
5176	Diflubenzuron / Benzamide, N-[[(4-chlorophenyl)amino]carbonyl]-2,6-difluoro-	$C_{14}H_9ClF_2N_2O_2$ / 310.69	35367-38-5 / 239	3128		
5177	1,4:5,8-Dimethanonaphthalene, 1,2,3,4,10,10-hexachloro-1,4,4a,5,8,8a-hexahydro-, / Aldrin	$C_{12}H_8Cl_8$ / 364.91	309-00-2 / 104	219	3-05-00-01385 /	H_2O 1; EtOH 3; eth 3; ace 3
5178	2,7:3,6-Dimethanonaphth[2,3-b]oxirene, 3,4,5,6,9,9-hexachloro-1a,2,2a,3,6,6a,7,7a-octahydro-,(1aα,2β,2aα,3β,6β,6aα,7β,7aα)- / Dieldrin	$C_{12}H_8Cl_6O$ / 380.91	60-57-1 / 175.5	3093 /	5-17-02-00087 / 1.75[25]	H_2O 1; EtOH 2; ace 3; bz 3
5179	Dimethipin / 2,3-Dihydro-5,6-dimethyl-1,4-dithiin, 1,1,4,4-tetraoxide	$C_6H_{10}O_4S_2$ / 210.28	55290-64-7 / 165			
5180	Dimethyl tetrachloroterephthalate / 1,4-Benzenedicarboxylic acid, 2,3,5,6-tetrachloro-, dimethyl ester	$C_{10}H_6Cl_4O_4$ / 331.97	1861-32-1 / 155	2830		
5181	Dinitramine / 1,3-Benzenediamine, N3,N3-diethyl-2,4-dinitro-6-(trifluoromethyl)-	$C_{11}H_{13}F_3N_4O_4$ / 322.24	29091-05-2 / 98			
5182	Dinocap / 2-Butenoic acid, 2-(1-methylheptyl)-4,6-dinitrophenyl ester	$C_{18}H_{24}N_2O_6$ / 364.40	6119-92-2 /	/ 136[0.01]		
5183	Dinoseb / Phenol, 2-(1-methylpropyl)-4,6-dinitro-	$C_{10}H_{12}N_2O_5$ / 240.22	88-85-7 / 40	3282 /	1.265[45]	
5184	2,3-Dioxabicyclo[2.2.2]oct-5-ene, 1-methyl-4-(1-methylethyl)- / Ascaridole	$C_{10}H_{16}O_2$ / 168.24	512-85-6 / 3.3	852 / 115[15]; 39[0.2]	5-19-01-00319 / 1.0103[20]	H_2O 1; EtOH 3; ace 3; bz 3 / 1.4769[20]
5185	1,7-Dioxadispiro[4.0.4.2]dodeca-3,9-diene-2,8-dione, trans- / Anemonin	$C_{10}H_8O_4$ / 192.17	508-44-1 / 158	674 /	5-19-05-00101 /	chl 4
5186	1,3-Dioxane / 1,3-Dioxacyclohexane	$C_4H_8O_2$ / 88.11	505-22-6 / -45	/ 106.1	5-19-01-00011 / 1.0342[20]	H_2O 5; EtOH 5; eth 5; ace 5 / 1.4165[20]
5187	1,4-Dioxane / 1,4-Dioxacyclohexane	$C_4H_8O_2$ / 88.11	123-91-1 / 11.8	3294 / 101.5	5-19-01-00016 / 1.0337[20]	H_2O 5; EtOH 5; eth 5; ace 5 / 1.4224[20]
5188	1,4-Dioxane, 2,3-dichloro-	$C_4H_6Cl_2O_2$ / 157.00	95-59-0 / 30	/ 80-2[10]	4-19-00-00030 / 1.468[20]	H_2O 1; eth 4; ace 4; bz 4 / 1.4928[20]
5189	1,3-Dioxane, 2,4-dimethyl-	$C_6H_{12}O_2$ / 116.16	766-20-1 /	/ 116.5	5-19-01-00072 / 0.9392[20]	H_2O 2; acid 4 / 1.4136[20]
5190	1,3-Dioxane-4,6-dione, 2,2-dimethyl- / Meldrum's acid	$C_6H_8O_4$ / 144.13	2033-24-1 / 94	5696 /	5-19-05-00008	
5191	1,3-Dioxane-2-ethanol, 4-hydroxy-α,6-dimethyl- / Paraldol	$C_8H_{11}O_4$ / 171.17	19404-07-0 / 90	/ 90[15]	4-01-00-03987 / 1.116[20]	H_2O 4; eth 4; EtOH 4 / 1.4610[20]
5192	1,4-Dioxane, 2,2,3,3,5,5,6-heptachloro- / 1,4-Dioxane, heptachloro	$C_4HCl_7O_2$ / 329.22	6629-96-5 / 55	/ 123-8[8]	4-19-00-00032 /	H_2O 1; eth 4; ace 4; bz 4
5193	1,3-Dioxane, 4-methyl-	$C_5H_{10}O_2$ / 102.13	1120-97-4 / -44.5	/ 114	5-19-01-00054 / 0.9758[20]	H_2O 2; os 4 / 1.4159[20]
5194	1,3-Dioxane, 4-methyl-4-phenyl- / m-Dioxane, 4-methyl-4-phenyl-	$C_{11}H_{14}O_2$ / 178.23	1200-73-3 / 37.5	/ 256	5-19-01-00481 / 1.0864[20]	H_2O 1; chl 2; os 3 / 1.5240[20]
5195	1,3-Dioxane, 2-phenyl-	$C_{10}H_{12}O_2$ / 164.20	772-01-0 / 41	/ 253	5-19-01-00459 / 1.6053[60]	EtOH 4; eth 4
5196	1,3-Dioxane, 4-phenyl-	$C_{10}H_{12}O_2$ / 164.20	772-00-9 /	/ 247	5-19-01-00464 / 1.1038[20]	H_2O 1; os 3 / 1.5306[18]
5197	1,3-Dioxan-5-ol, 2-methyl-	$C_5H_{10}O_3$ / 118.13	3774-03-6 /	/ 176; 79[18]	4-19-00-00624 / 1.0705[17]	H_2O 4 / 1.4375[17]
5198	1,3,2-Dioxathiane, 4-methyl-, 2-oxide / 1,3-Butanediol, sulfite	$C_4H_8O_3S$ / 136.17	4426-51-1 / 5	/ 185	4-01-00-02510 / 1.2352[20]	EtOH 4; eth 4; ace 4; bz 4 / 1.4661[20]

No.	Name / Synonym	Mol. Form. / Mol. Wt.	CAS RN / mp/°C	Merck No. / bp/°C	Beil. Ref. / den/g cm^{-3}	Solubility / n_D
5199	1,3,2-Dioxathiolane, 2,2-dioxide	$C_2H_4O_4S$ 124.12	1072-53-3 99	sub	3-01-00-02110	ace 4; bz 4; eth 4; EtOH 4
5200	1,3,2-Dioxathiolane, 2-oxide	$C_2H_4O_3S$ 108.12	3741-38-6 -11	173	4-01-00-02409 1.4402[20]	H_2O 4; EtOH 4; eth 4; ace 4 1.4463[20]
5201	Dioxathion Phosphorodithioic acid, S,S'-1,4-dioxane-2,3-diyl O,O,O',O'-tetraethyl ester	$C_{12}H_{26}O_6P_2S_4$ 456.55	78-34-2 -20	3296	1.257[26]	
5202	1,4-Dioxin	$C_4H_4O_2$ 84.07	290-67-5	76	5-19-01-00275 1.115[20]	eth 3; ace 3; bz 3 1.4350[20]
5203	1,4-Dioxin, 2,3-dihydro-	$C_4H_6O_2$ 86.09	543-75-9	94.1	5-19-01-00181 1.0836[20]	ctc 3 1.4372[20]
5204	1,3-Dioxolane 1,3-Dioxacyclopentane	$C_3H_6O_2$ 74.08	646-06-0 -95	78	5-19-01-00006 1.060[20]	H_2O 5; EtOH 3; eth 3; ace 3 1.3974[20]
5205	1,3-Dioxolane, 2-(3-bromophenyl)-	$C_9H_9BrO_2$ 229.07	17789-14-9	132.5	5-19-01-00448 1.514[25]	1.5627[20]
5206	1,3-Dioxolane-4-carboxaldehyde, 2,2-dimethyl-	$C_6H_{10}O_3$ 130.14	5736-03-8	74[50]	5-19-04-00041	H_2O 4 1.4189[25]
5207	1,3-Dioxolane-4-methanol, 2,2-dimethyl- Isopropylidene glycerol	$C_6H_{12}O_3$ 132.16	100-79-8	5101 82[10]	5-19-02-00362 1.064[20]	1.4383[20]
5208	1,3-Dioxolane-4-methanol, 2-methyl-	$C_5H_{10}O_3$ 118.13	3773-93-1	187	5-19-02-00350 1.1243[17]	EtOH 4 1.4413[17]
5209	1,3-Dioxolane-4-methanol, 2-phenyl-	$C_{10}H_{12}O_3$ 180.20	1708-39-0	143-4[2]	5-19-02-00562 1.1916[17]	1.5389[17]
5210	1,3-Dioxolane, 2-methyl-	$C_4H_8O_2$ 88.11	497-26-7	81.5	5-19-01-00042 0.9811[20]	H_2O 4; EtOH 5; eth 5 1.4035[17]
5211	1,3-Dioxolan-2-one Ethylene glycol carbonate	$C_3H_4O_3$ 88.06	96-49-1 36.4	248	5-19-04-00006 1.321[39]	H_2O 5; EtOH 5; eth 5; bz 5 1.4148[50]
5212	1,3-Dioxolan-4-one, 2,5-bis(trichloromethyl)- Chloralide	$C_5H_2Cl_6O_3$ 322.79	554-21-2 116	272.5	4-19-00-01571	eth 4; HOAc 4
5213	1,3-Dioxolan-2-one, 4-methyl-	$C_4H_6O_3$ 102.09	108-32-7 -48.8	242	5-19-04-00021 1.2047[20]	H_2O 4; EtOH 4; eth 4; ace 4 1.4189[20]
5214	1,3-Dioxolo[4,5-g]isoquinoline, 5,6,7,8-tetrahydro-4-methoxy-6-methyl- Hydrocotarnine	$C_{12}H_{15}NO_3$ 221.26	550-10-7 56	4715	4-27-00-06395	H_2O 1; EtOH 3; eth 3; ace 3
5215	1,3-Dioxolo[4,5-g]isoquinoline, 5,6,7,8-tetrahydro-6-methyl- Hydrohydrastinine	$C_{11}H_{13}NO_2$ 191.23	494-55-3 66	4732 303	4-27-00-06306	ace 4; bz 4; eth 4; EtOH 4
5216	1,3-Dioxolo[4,5-g]isoquinolin-5-ol, 5,6,7,8-tetrahydro-4-methoxy-6-methyl- Cotarnine	$C_{12}H_{15}NO_4$ 237.26	82-54-2 132-3 dec	2551	4-27-00-06448	H_2O 2; EtOH 3; eth 3; bz 3
5217	1,3-Dioxolo[4,5-g]isoquinolin-5-ol, 5,6,7,8-tetrahydro-6-methyl- Hydrastinine	$C_{11}H_{13}NO_3$ 207.23	6592-85-4 116.5	4689	4-27-00-06396	H_2O 3; EtOH 4; eth 4; chl 4
5218	1,3-Dioxolo[4,5-h]quinolin-8(9H)-one, 6-methoxy-9-methyl- Casimiroin	$C_{12}H_{11}NO_4$ 233.22	477-89-4	1894	4-27-00-06618	chl 2
5219	Diphenamid Benzeneacetamide, N,N-dimethyl-α-phenyl-	$C_{16}H_{17}NO$ 239.32	957-51-7 135	3305	1.17[23.3]	
5220	Diphosphoramide, octamethyl- Schradan	$C_8H_{24}N_4O_3P_2$ 286.25	152-16-9 17	8351 188-22[0.3]	4-04-00-00288 1.1343[25]	H_2O 4; EtOH 4; chl 4 1.462[25]
5221	Diphosphoric acid, tetraethyl ester Tetraethyl pyrophosphate	$C_8H_{20}O_7P_2$ 290.19	107-49-3 170 dec	9138 155[3]	4-01-00-01342 1.1847[20]	H_2O 5; EtOH 5; eth 5; ace 5 1.4180[20]
5222	Dipropetryn 1,3,5-Triazine-2,4-diamine, 6-(ethylthio)-N,N'-bis(1-methylethyl)-	$C_{11}H_{21}N_5S$ 255.39	4147-51-7 105	3349		
5223	5H,10H-Dipyrrolo[1,2-a:1',2'-d]pyrazine-5,10-dione Pyrocoll	$C_{10}H_6N_2O_2$ 186.17	484-73-1 268	sub	5-24-08-00263	eth 4; EtOH 4; chl 4
5224	Diquat dibromide Dipyrido[1,2-a:2',1'-c]pyrazinediium, 6,7-dihydro-, dibromide	$C_{12}H_{12}N_2 \cdot 2Br$ 344.05	85-00-7 337	3359	1.24[20]	
5225	Diselenide, diphenyl Diphenyl diselenide	$C_{12}H_{10}Se_2$ 312.13	1666-13-3 63.5	202[11]	4-06-00-01781 1.557[80]	EtOH 3; eth 3; xyl 3; MeOH 3 1.743[20]
5226	Disilane, 1,2-dichloro-1,1,2,2-tetramethyl-	$C_4H_{12}Cl_2Si_2$ 187.22	4342-61-4	49-50[18]	4-04-00-04280 1.010[20]	1.4548[20]
5227	Disilane, 1,2-difluoro-1,1,2,2-tetramethyl- 1,2-Difluorotetramethyldisilane	$C_4H_{12}F_2Si_2$ 154.31	661-68-7	94	4-04-00-04279 0.9120[20]	1.3837[20]
5228	Disilane, hexamethyl- Permethyldisilane	$C_6H_{18}Si_2$ 146.38	1450-14-2 13.5	113.5	4-04-00-04277 0.7247[22]	H_2O 1; eth 3; ace 3; bz 3 1.4229[20]
5229	Disiloxane, 1,3-bis(bromomethyl)-1,1,3,3-tetramethyl- 1,3-Bis(bromomethyl)tetramethyldisiloxane	$C_6H_{16}Br_2OSi_2$ 320.17	2351-13-5	102-4[15]	4-04-00-04027 1.3918[25]	1.4719[25]
5230	Disiloxane, 1,3-bis(chloromethyl)-1,1,3,3-tetramethyl-	$C_6H_{16}Cl_2OSi_2$ 231.27	2362-10-9 -90	204; 92[21]	4-04-00-04026 1.0381[20]	1.4398[20]
5231	Disiloxane, 1,3-bis(dichloromethyl)-1,1,3,3-tetramethyl- 1,3-Bis-(dichloromethyl)tetramethyl disiloxane	$C_6H_{14}Cl_4OSi_2$ 300.16	2943-70-6	149-50[50]	4-04-00-04031 1.2213[20]	1.4660[20]
5232	Disiloxane, 1,3-dichloro-1,1,3,3-tetramethyl-	$C_4H_{12}Cl_2OSi_2$ 203.22	2401-73-2 -37.5	138	4-04-00-04119 1.038[20]	

No.	Name Synonym	Mol. Form. Mol. Wt.	CAS RN mp/°C	Merck No. bp/°C	Beil. Ref. den/g cm^{-3}	Solubility n_D
5233	Disiloxane, 1,3-diethenyl-1,1,3,3-tetramethyl-	$C_8H_{18}OSi_2$ 186.40	2627-95-4 -99.7	39	4-04-00-04080 0.811[20]	1.4123[20]
5234	Disiloxane, 1,3-dimethoxy-1,1,3,3-tetramethyl- Bis(methoxydimethylsilyl) oxide	$C_6H_{18}O_3Si_2$ 194.38	18187-24-1	139	4-04-00-04117 0.9048[20]	1.3835[20]
5235	Disiloxane, hexaethyl- 1,1,1,3,3,3-Hexaethyldisiloxane	$C_{12}H_{30}OSi_2$ 246.54	994-49-0	233; 129[30]	4-04-00-04055 0.8589[0]	1.4340[20]
5236	Disiloxane, hexamethyl- Hexamethyldisiloxane	$C_6H_{18}OSi_2$ 162.38	107-46-0 -66	99	4-04-00-04018 0.7638[20]	H_2O 1 1.3774[20]
5237	Disiloxane, 1,1,3,3-tetramethyl-	$C_4H_{14}OSi_2$ 134.33	3277-26-7	72	4-04-00-03991 0.7572[20]	1.3700[20]
5238	Disiloxane, 1,1,3,3-tetramethyl-1,3-diphenyl-	$C_{16}H_{22}OSi_2$ 286.52	56-33-7	155-8[13]	4-16-00-01475 0.9763[20]	ctc 3 1.5176[20]
5239	Disiloxane, 1,1,1-trimethyl-3,3,3-triphenyl- 1,1,1-Trimethyl-3,3,3-triphenyldisiloxane	$C_{21}H_{24}OSi_2$ 348.59	799-53-1 51	349	4-16-00-01486 1.0320[25]	ctc 3; CS_2 3
5240	Dispiro[5.1.5.1]tetradecane Dispiro[5,1,5,1]tetradecane	$C_{14}H_{24}$ 192.34	184-97-4 10	131[10]	4-05-00-00491 0.872[26]	1.4735[27]
5241	Distannoxane, hexabutyl- Bis(tributyltin) oxide	$C_{24}H_{54}OSn_2$ 596.11	56-35-9 45	180[2]		
5242	Disulfide, bis(1,1-dimethylethyl)	$C_8H_{18}S_2$ 178.36	110-06-5 -5	88[21]	4-01-00-01638 0.9226[20]	1.4899[20]
5243	Disulfide, bis(3-methylbutyl)	$C_{10}H_{22}S_2$ 206.42	2051-04-9	247	4-01-00-01689 0.9192[20]	1.4864[20]
5244	Disulfide, bis(1-methylethyl)	$C_6H_{14}S_2$ 150.31	4253-89-8 -69	177	4-01-00-01503 0.9435[20]	1.4916[20]
5245	Disulfide, bis(3-methylphenyl)	$C_{14}H_{14}S_2$ 246.40	20333-41-9 -21	177[3]	4-06-00-02085	EtOH 3; eth; ace 3
5246	Disulfide, bis(4-methylphenyl)	$C_{14}H_{14}S_2$ 246.40	103-19-5 47.5	210-15[20]	4-06-00-02206 1.114[51]	H_2O 1; EtOH 3; eth 4; ace 3
5247	Disulfide, bis(3-nitrophenyl) Nitrophenide	$C_{12}H_8N_2O_4S_2$ 308.34	537-91-7 84	6539	4-06-00-01686	EtOH 2; eth 3; chl 2
5248	Disulfide, bis(phenylmethyl) Dibenzyl disulfide	$C_{14}H_{14}S_2$ 246.40	150-60-7 71.5	2995	4-06-00-02760	H_2O 2; EtOH 3; eth 3; bz 3
5249	Disulfide, bis(trifluoromethyl)	$C_2F_6S_2$ 202.14	372-64-5 34.6		4-03-00-00278	EtOH 4; peth 4
5250	Disulfide, diacetyl Acetyl disulfide	$C_4H_6O_2S_2$ 150.22	592-22-3 20	105[18]	4-02-00-00564	eth 4; EtOH 4
5251	Disulfide, dibenzoyl	$C_{14}H_{10}O_2S_2$ 274.36	644-32-6 134.5	dec	3-09-00-01977	H_2O 1; EtOH 2; eth 2; CS_2 3
5252	Disulfide, dibutyl	$C_8H_{18}S_2$ 178.36	629-45-8	117[20]; 85[3]	4-01-00-01562 0.9371[20]	H_2O 1; EtOH 5; eth 5 1.4923[20]
5253	Disulfide, diethyl Diethyl disulfide	$C_4H_{10}S_2$ 122.26	110-81-6 -101.5	154.1	4-01-00-01397 0.9931[20]	H_2O 2; EtOH 5; eth 5 1.5073[20]
5254	Disulfide, dimethyl Dimethyl disulfide	$C_2H_6S_2$ 94.20	624-92-0 -85	109.8	4-01-00-01281 1.0625[20]	H_2O 1; EtOH 5; eth 5 1.5289[20]
5255	Disulfide, di-1-naphthalenyl	$C_{20}H_{14}S_2$ 318.46	39178-11-5 91		4-06-00-04245 1.144[20]	H_2O 1; EtOH 2; eth 4; chl 2
5256	Disulfide, di-2-naphthalenyl	$C_{20}H_{14}S_2$ 318.46	5586-15-2 139.5		4-06-00-04320 1.144[145]	H_2O 1; EtOH 4; eth 4; lig 1 1.4555[20]
5257	Disulfide, dipentyl	$C_{10}H_{22}S_2$ 206.42	112-51-6	119[7]	4-01-00-01654 0.9221[20]	H_2O 1; eth 3 1.4889[20]
5258	Disulfide, diphenyl	$C_{12}H_{10}S_2$ 218.34	882-33-7 62	310	4-06-00-01560 1.353[20]	H_2O 1; EtOH 3; eth 3; bz 3
5259	Disulfide, dipropyl	$C_6H_{14}S_2$ 150.31	629-19-6 -85.6	193.5	4-01-00-01454 0.9599[20]	1.4981[20]
5260	Disulfoton Phosphorodithioic acid, O,O-diethyl S-[2- (ethylthio)ethyl] ester	$C_8H_{19}O_2PS_3$ 274.41	298-04-4 -25	3371 108[0.01]; 128[1]	1.144[20]	
5261	1,2-Dithiane	$C_4H_8S_2$ 120.24	505-20-4 32.5	80[14]; 60[5]	5-19-01-00010	eth 3; bz 3; chl 3 1.5981[25]
5262	1,3-Dithiane	$C_4H_8S_2$ 120.24	505-23-7 54	89[14]	5-19-01-00013	bz 4; eth 4; chl 4 1.5981[25]
5263	1,4-Dithiane	$C_4H_8S_2$ 120.24	505-29-3 112.3	199.5	5-19-01-00036	H_2O 2; EtOH 3; eth 3; ctc 3
5264	4H-1,3,5-Dithiazine, dihydro-5-methyl- 1,3,5-Dithiazine, 4,5-dihydro-5-methyl	$C_4H_9NS_2$ 135.25	6302-94-9 65	185	4-27-00-06285	eth 4; EtOH 4; HOAc 4
5265	4H-1,3,5-Dithiazine, dihydro-2,4,6-trimethyl-, (2α,4α,6α)-	$C_6H_{13}NS_2$ 163.31	638-17-5 46	dec	2-27-00-00525 1.0613[50]	eth 4; EtOH 4
5266	1,3-Dithiolane 1,3-Dithiacyclopentane	$C_3H_6S_2$ 106.21	4829-04-3 -50	175	5-19-01-00008 1.259[17]	EtOH 3; eth 3; xyl 3 1.5975[15]
5267	Dithiopyr S,S'-Dimethyl-2-difluoromethyl-4-isobutyl-6- trifluoromethylpyridine-3,5-dicarbothioate	$C_{15}H_{16}F_5NO_2S_2$ 401.42	97886-45-8 65			
5268	Docosane n-Docosane	$C_{22}H_{46}$ 310.61	629-97-0 44.4	368.6	4-01-00-00572 0.7944[20]	H_2O 1; EtOH 3; eth 4; chl 3 1.4455[20]
5269	Docosane, 5-butyl- 5-n-Butyldocosane	$C_{26}H_{54}$ 366.71	55282-16-1 208	244[10]	4-01-00-00584 0.8058[20]	1.4503[20]
5270	Docosane, 7-butyl- 7-Butyldocosane	$C_{26}H_{54}$ 366.71	55282-15-0 3.2	241[10]	4-01-00-00584 0.8046[20]	1.4499[20]

No.	Name Synonym	Mol. Form. Mol. Wt.	CAS RN mp/°C	Merck No. bp/°C	Beil. Ref. den/g cm^{-3}	Solubility n_D
5271	Docosane, 9-butyl- 9-Butyldocosane	$C_{26}H_{54}$ 366.71	55282-14-9 1.3	242.5[10]	4-01-00-00584 0.8044[20]	1.4498[20]
5272	Docosane, 11-butyl- 11-Butyldocosane	$C_{26}H_{54}$ 366.71	13475-78-8	242.5[10]	4-01-00-00584 0.8041[20]	1.4499[20]
5273	Docosane, decyl- Docosane, 11-decyl	$C_{32}H_{66}$ 450.88	28261-97-4 1.0	286[10]	4-01-00-00597 0.8131[20]	1.4546[20]
5274	Docosane, 7-hexyl- 7-Hexyldocosane	$C_{28}H_{58}$ 394.77	55373-86-9 19.3	258[10]	4-01-00-00590 0.8080[20]	1.4517[20]
5275	Docosane, 9-octyl- 9-Octyldocosane	$C_{30}H_{62}$ 422.82	55319-83-0 8.6	272.5[10]	4-01-00-00594 0.8114[20]	1.4537[20]
5276	Docosanoic acid Behenic acid	$C_{22}H_{44}O_2$ 340.59	112-85-6 81	1033 306[60]	4-02-00-01290 0.8223[90]	H_2O 2; EtOH 2; eth 2 1.4270[100]
5277	Docosanoic acid, ethyl ester	$C_{24}H_{48}O_2$ 368.64	5908-87-2 50	240[10]	4-02-00-01292	eth 4; EtOH 4
5278	Docosanoic acid, methyl ester	$C_{23}H_{46}O_2$ 354.62	929-77-1 54		4-02-00-01291	eth 4; EtOH 4 1.4339[60]
5279	1-Docosanol	$C_{22}H_{46}O$ 326.61	661-19-8 72.5	180[0.22]	4-01-00-01906	H_2O 2; EtOH 4; eth 2; chl 3
5280	1-Docosene	$C_{22}H_{44}$ 308.59	1599-67-3 38	367	3-01-00-00883 0.794[25]	
5281	13-Docosenoic acid	$C_{22}H_{42}O_2$ 338.57	1072-39-5 34.7	242[5]	4-02-00-01676 0.8532[20]	1.4444[70]
5282	13-Docosenoic acid, (E)- Brassidic acid	$C_{22}H_{42}O_2$ 338.57	506-33-2 60	1359 282[30], 256[10]	4-02-00-01677 0.8585[57]	1.4347[100]
5283	13-Docosenoic acid, (Z)- Erucic acid	$C_{22}H_{42}O_2$ 338.57	112-86-7 33.5	3619 265[15]	4-02-00-01676 0.860[55]	H_2O 1; EtOH 3; eth 4; ctc 3 1.4758[20]
5284	13-Docosen-1-ol, (Z)- Erucyl alcohol	$C_{22}H_{44}O$ 324.59	629-98-1 35	240[10]	4-01-00-02209 0.8416[33]	bz 4; EtOH 4; peth 4
5285	1,11-Dodecadiene	$C_{12}H_{22}$ 166.31	5876-87-9	208.5	4-01-00-01067 0.7702[20]	1.4400[20]
5286	3,9-Dodecadiyne	$C_{12}H_{18}$ 162.27	61827-89-2			
5287	5,7-Dodecadiyne Dibutylbutadiyne	$C_{12}H_{18}$ 162.27	1120-29-2	103[8]	4-01-00-01130	
5288	Dodecanal Lauraldehyde	$C_{12}H_{24}O$ 184.32	112-54-9 44.5	185[100], 100[2.5]	4-01-00-03380 0.8352[15]	H_2O 1; EtOH 2; eth 3 1.435[22]
5289	Dodecanamide	$C_{12}H_{25}NO$ 199.34	1120-16-7 110	199[12]	4-02-00-01103	H_2O 1; EtOH 3; eth 2; ace 3 1.4287[110]
5290	Dodecanamide, N,N-diethyl-	$C_{16}H_{33}NO$ 255.44	3352-87-2	166-7[2]	4-04-00-00357 0.847[25]	chl 3 1.4545[20]
5291	1-Dodecanamine	$C_{12}H_{27}N$ 185.35	124-22-1 28.3	259	4-04-00-00794 0.8015[20]	H_2O 2; EtOH 5; eth 5; bz 5 1.4421[20]
5292	Dodecane	$C_{12}H_{26}$ 170.34	112-40-3 -9.6	216.3	4-01-00-00498 0.7487[20]	H_2O 1; EtOH 4; eth 4; ace 4 1.4216[20]
5293	Dodecane, 1-bromo- Lauryl bromide	$C_{12}H_{25}Br$ 249.23	143-15-7 -9.5	5260 276	4-01-00-00502 1.0399[20]	H_2O 1; EtOH 3; eth 3; ace 5 1.4583[20]
5294	Dodecane, 1-bromo-12-fluoro- 1-Bromo-12-fluorododecane	$C_{12}H_{24}BrF$ 267.22	353-29-7	85.8[0.15]	4-01-00-00502	ace 4; eth EtOH 4 1.4524[25]
5295	Dodecane, 1-chloro-	$C_{12}H_{25}Cl$ 204.78	112-52-7 -9.3	260	4-01-00-00501 0.8687[20]	H_2O 1; EtOH 4; ace 5; bz 3 1.4433[20]
5296	Dodecane, 1,12-dibromo-	$C_{12}H_{24}Br_2$ 328.13	3344-70-5 41	215[15]	4-01-00-00503	H_2O 1; EtOH 4; eth 3; chl 4
5297	Dodecane, 1,1-dimethoxy- Lauraldehyde, dimethyl acetal	$C_{14}H_{30}O_2$ 230.39	14620-52-1	132-4[5]	4-01-00-03381	eth 4; EtOH 4 1.4310[25]
5298	Dodecane, 1,12-dimethoxy- 1,12-Dodecanediol, dimethyl ether	$C_{14}H_{30}O_2$ 230.39	73120-52-2 11.5	155-6[15]	4-01-00-02627 0.8563[22]	1.436[22]
5299	Dodecanedioic acid	$C_{12}H_{22}O_4$ 230.30	693-23-2 128	222[25]	4-02-00-02126 1.15[25]	tfa 3
5300	Dodecanedioic acid, dimethyl ester	$C_{14}H_{26}O_4$ 258.36	1731-79-9 31.9	170[10]	4-02-00-02126	H_2O 1; AcOEt 3
5301	1,12-Dodecanediol 1,12-Dihydroxydodecane	$C_{12}H_{26}O_2$ 202.34	5675-51-4 81.3	189[12]	4-01-00-02627	tfa 3
5302	Dodecane, 1-iodo-	$C_{12}H_{25}I$ 296.23	4292-19-7 0.3	298.2	4-01-00-00503 1.1999[20]	H_2O 1; EtOH 3; eth 5; ace 5 1.4840[20]
5303	Dodecanenitrile Lauronitrile	$C_{12}H_{23}N$ 181.32	2437-25-4 4	277; 198[100]	4-02-00-01104 0.8240[20]	H_2O 1; EtOH 5; eth 5; ace 5 1.4361[20]
5304	1-Dodecanethiol	$C_{12}H_{26}S$ 202.40	112-55-0	142-5[15]	4-01-00-01851 0.8435[20]	H_2O 1; EtOH 3; eth 3; chl 3 1.4589[20]
5305	Dodecanoic acid Lauric acid	$C_{12}H_{24}O_2$ 200.32	143-07-7 43.2	5254 91.4	4-02-00-01082 0.8679[50]	H_2O 1; EtOH 4; eth 4; ace 3 1.4183[82]
5306	Dodecanoic acid, anhydride	$C_{24}H_{46}O_3$ 382.63	645-66-9 41.8		4-02-00-01101 0.8533[70]	EtOH 4 1.4292[70]
5307	Dodecanoic acid, 2-bromo-	$C_{12}H_{23}BrO_2$ 279.22	111-56-8 32	157-9[2]	4-02-00-01106 1.1474[74]	bz 4; eth 4; EtOH 4; lig 4 1.4585[24]
5308	Dodecanoic acid, 2-bromoethyl ester	$C_{14}H_{27}BrO_2$ 307.27	6309-50-8	176[12]; 145[3]	4-02-00-01092 1.088[25]	1.4547[25]
5309	Dodecanoic acid, 2-bromo-, methyl ester	$C_{13}H_{25}BrO_2$ 293.24	617-60-7	169-71[12]	4-02-00-01107 1.113[15]	1.4572[25]

No.	Name Synonym	Mol. Form. Mol. Wt.	CAS RN mp/°C	Merck No. bp/°C	Beil. Ref. den/g cm⁻³	Solubility n_D
5310	Dodecanoic acid, 2,3-dihydroxypropyl ester, (±)-	$C_{15}H_{30}O_4$	40738-26-9		4-02-00-01096	EtOH 2; eth 4; ace 4; bz 3
	Glycerol, 1-laurate (DL)	274.40	63	186[2]	0.9248[97]	1.4350[86]
5311	Dodecanoic acid, 1,2-ethanediyl ester	$C_{28}H_{50}O_4$	624-04-4	4393	4-02-00-01094	eth 4; EtOH 4
	Ethylene glycol dilaurate	426.68	56.6	188[20]		
5312	Dodecanoic acid, ethyl ester	$C_{14}H_{28}O_2$	106-33-2	3774	4-02-00-01092	H_2O 1; EtOH 4; eth 5; ctc 2
	Ethyl laurate	228.38	-10	271; 154[15]	0.8618[20]	1.4311[20]
5313	Dodecanoic acid, 2-(2-hydroxyethoxy)ethyl ester	$C_{16}H_{32}O_4$	141-20-8	3110		EtOH 5; eth 5; ace 5; bz 3
	Diethylene glycol monolaurate	288.43	17.5	>270	0.96[25]	
5314	Dodecanoic acid, 2-methyl-	$C_{13}H_{26}O_2$	2874-74-0		4-02-00-01121	
		214.35	22	153[1]	0.890[18]	
5315	Dodecanoic acid, methyl ester	$C_{13}H_{26}O_2$	111-82-0		4-02-00-01090	H_2O 1; EtOH 5; eth 5; ace 5
	Methyl laurate	214.35	5.2	267	0.8702[20]	1.4319[20]
5316	Dodecanoic acid, 1-methylethyl ester	$C_{15}H_{30}O_2$	10233-13-3		4-02-00-01092	eth 4; EtOH 4
		242.40		196[60]; 117[2]	0.8536[20]	1.4280[25]
5317	Dodecanoic acid, phenyl ester	$C_{18}H_{28}O_2$	4228-00-6		4-06-00-00618	ace 4; eth 4; EtOH 4
	Phenyl laurate	276.42	24.5	210[15]	0.9354[30]	
5318	Dodecanoic acid, phenylmethyl ester	$C_{19}H_{30}O_2$	140-25-0		4-06-00-02267	bz 4; eth 4; EtOH 4; peth 4
		290.45	8.5	209-11[12]	0.9429[25]	1.4812[24]
5319	Dodecanoic acid, 1,2,3-propanetriyl ester	$C_{39}H_{74}O_6$	538-24-9		4-02-00-01098	H_2O 1; EtOH 3; eth 3; ace 4
		639.01			0.8986[55]	1.4404[60]
5320	Dodecanoic acid, propyl ester	$C_{15}H_{30}O_2$	3681-78-5		4-02-00-01092	
	Propyl laurate	242.40		205[60]; 124[2]	0.8600[20]	1.4335[20]
5321	1-Dodecanol	$C_{12}H_{26}O$	112-53-8	3402	4-01-00-01844	H_2O 1; EtOH 3; eth 3; bz 2
	Lauryl alcohol	186.34	24	259	0.8309[24]	
5322	2-Dodecanol	$C_{12}H_{26}O$	10203-28-8		3-01-00-01793	
		186.34	19	252	0.8286[20]	1.4400[20]
5323	3-Dodecanol	$C_{12}H_{26}O$	10203-30-2		4-01-00-01854	
		186.34	25	130[15]	0.8223[32]	
5324	6-Dodecanol	$C_{12}H_{26}O$	6836-38-0		3-01-00-01794	eth 4; EtOH 4
		186.34	30	225; 119[9]	0.8201[40]	
5325	2-Dodecanone	$C_{12}H_{24}O$	6175-49-1		4-01-00-03382	H_2O 1; EtOH 3; eth 3; ace 3
	Decyl methyl ketone	184.32	21	246.5	0.8198[20]	1.4330[20]
5326	6-Dodecanone	$C_{12}H_{24}O$	6064-27-3		4-01-00-03383	
	Amyl hexyl ketone	184.32	10	112[9]		1.4302[20]
5327	1-Dodecanone, 1-phenyl-	$C_{18}H_{28}O$	1674-38-0		4-07-00-00847	H_2O 1; ace 3; ctc 2
		260.42	47	201[9]; 181[5]	0.8794[18]	1.4700[18]
5328	Dodecanoyl chloride	$C_{12}H_{23}ClO$	112-16-3		4-02-00-01103	eth 4
		218.77	-17	145[18]	0.9169[25]	1.4458[20]
5329	1,3,6,10-Dodecatetraene, 3,7,11-trimethyl-, (E,E)-	$C_{15}H_{24}$	502-61-4	3883	3-01-00-01067	H_2O 1; eth 3; ace 3; peth 5
	α-Farnesene	204.36		129-32[12]	0.8410[20]	1.4836[20]
5330	2,6,10-Dodecatrienal, 3,7,11-trimethyl-	$C_{15}H_{24}O$	19317-11-4		4-01-00-03603	
		220.35		172-4[14]	0.893[18]	1.4995
5331	1,6,10-Dodecatriene, 7,11-dimethyl-3-methylene-, (E)-	$C_{15}H_{24}$	18794-84-8	3884	4-01-00-01133	ace 4; eth 4; chl 4
	β-Farnesene	204.36		121-2[9]	0.8363[20]	1.4899[20]
5332	1,6,10-Dodecatrien-3-ol, 3,7,11-trimethyl-, [S-(Z)]-	$C_{15}H_{26}O$	142-50-7	6388	4-01-00-02336	EtOH 4; eth 3; ace 3; os 3
		222.37		276; 70[0.1]	0.8778[20]	1.4898[20]
5333	2,6,10-Dodecatrien-1-ol, 3,7,11-trimethyl-, (E,E)-	$C_{15}H_{26}O$	106-28-5		4-01-00-02335	H_2O 1; EtOH 4; eth 3; ace 3
		222.37		160[10]	0.8846[20]	1.4877[20]
5334	2,6,10-Dodecatrien-1-ol, 3,7,11-trimethyl-, (Z,E)-	$C_{15}H_{26}O$	3790-71-4		4-01-00-02335	ace 4; eth 4; EtOH 4
		222.37		156[12]; 120[0.3]	0.8908[20]	1.4877[20]
5335	1-Dodecene	$C_{12}H_{24}$	112-41-4		4-01-00-00914	H_2O 1; EtOH 3; eth 3; ace 3
		168.32	-35.2	213.8	0.7584[20]	1.4300[20]
5336	2-Dodecenedioic acid, (E)-	$C_{12}H_{20}O_4$	6402-36-4	9493	4-02-00-02279	eth 4; EtOH 4; chl 4
	Traumatic acid	228.29	165.5			
5337	2-Dodecenoic acid	$C_{12}H_{22}O_2$	4412-16-2		4-02-00-01619	
		198.31	17.1	155[3]; 127[0.15]	0.9265[20]	1.4629[25]
5338	4-Dodecenoic acid	$C_{12}H_{22}O_2$	505-92-0		4-02-00-01619	bz 4; eth 4; chl 4
	Linderic acid	198.31	1.3	171[13]	0.9081[15]	1.4529[20]
5339	5-Dodecenoic acid	$C_{12}H_{22}O_2$	2761-84-4		4-02-00-01619	
		198.31	1.3	170-2[13]	0.9081[20]	
5340	11-Dodecenoic acid	$C_{12}H_{22}O_2$	65423-25-8		4-02-00-01618	
		198.31	20	171[13]; 144[3]	0.9014[20]	1.4510[20]
5341	11-Dodecenoic acid, methyl ester	$C_{13}H_{24}O_2$	29972-79-0		4-02-00-01618	
	Methyl 11-dodecenoate	212.33		138-9[13]	0.8789[22]	1.4414[20]
5342	1-Dodecen-3-yne	$C_{12}H_{20}$	74744-36-8		4-01-00-01112	
		164.29		78[4]	0.7858[25]	1.4510[25]
5343	1-Dodecyne	$C_{12}H_{22}$	765-03-7		4-01-00-01066	
	Decylacetylene	166.31	-19	215	0.7788[20]	1.4340[20]
5344	2-Dodecyne	$C_{12}H_{22}$	629-49-2		0-01-00-00261	
		166.31	-9	105[15]	0.7917[15]	1.4828[20]
5345	3-Dodecyne	$C_{12}H_{22}$	6790-27-8		3-01-00-01025	ace 4; eth 4
		166.31		95[12]	0.7871[20]	1.4442[20]

No.	Name Synonym	Mol. Form. Mol. Wt.	CAS RN mp/°C	Merck No. bp/°C	Beil. Ref. den/g cm^{-3}	Solubility n_D
5346	6-Dodecyne	$C_{12}H_{22}$ 166.31	6975-99-1	210; 100[14]	4-01-00-01067 0.7871[20]	ace 4; eth 4; EtOH 4 1.4442[20]
5347	Dodine Guanidine, dodecyl-, monoacetate	$C_{13}H_{29}N_3 \cdot C_2H_4O_2$ 287.45	2439-10-3 136	3406		
5348	Dotriacontane Bicetyl	$C_{32}H_{66}$ 450.88	544-85-4 69.7	467	4-01-00-00595 0.8124[20]	H_2O 1; EtOH 2; eth 3; bz 4 1.4550[20]
5349	Eburnamenine-14-carboxylic acid, 14,15- dihydro-14-hydroxy-, methyl ester, (3α,14β,16α)- Vincamine	$C_{21}H_{26}N_2O_3$ 354.45	1617-90-9 231.5	9888	4-25-00-01252	
5350	Eicosane	$C_{20}H_{42}$ 282.55	112-95-8 36.8	343	4-01-00-00563 0.7886[20]	H_2O 1; eth 3; ace 4; bz 3 1.4425[20]
5351	Eicosane, 1-cyclopentyl- Cyclopentane, eicosyl	$C_{25}H_{50}$ 350.67	22331-38-0 38	413	4-05-00-00198 0.8276[20]	ace 4; bz 4; eth 4; EtOH 4 1.4595[20]
5352	Eicosanedioic acid 1,18-Octadecanedicarboxylic acid	$C_{20}H_{38}O_4$ 342.52	2424-92-2 125.5	233-4[2]	4-02-00-02185	eth 3
5353	Eicosanedioic acid, diethyl ester	$C_{24}H_{46}O_4$ 398.63	42235-39-2 54.5	240[12]; 230[2]	3-02-00-01881	eth 4; EtOH 4
5354	Eicosane, 9-octyl- Eicosane, 9-ocytl	$C_{28}H_{58}$ 394.77	13475-77-9 0.5	257[10]; 199[0.5]	4-01-00-00590 0.8075[20]	1.4515[20]
5355	Eicosanoic acid Arachidic acid	$C_{20}H_{40}O_2$ 312.54	506-30-9 75.4	791 328 dec; 204[1]	4-02-00-01275 0.8240[100]	H_2O 1; EtOH 2; eth 4; bz 3 1.425[100]
5356	Eicosanoic acid, ethyl ester	$C_{22}H_{44}O_2$ 340.59	18281-05-5 50	295[100]; 186[2]	3-02-00-01067	H_2O 1; EtOH 3; eth 3; bz 3
5357	Eicosanoic acid, methyl ester	$C_{21}H_{41}O_2$ 325.56	1120-28-1 54.5	215[10]	4-02-00-01276	bz 4; eth 4; EtOH 4; chl 4 1.4317[60]
5358	1-Eicosanol Arachic alcohol	$C_{20}H_{42}O$ 298.55	629-96-9 66.1	309; 222[3]	4-01-00-01900 0.8405[20]	H_2O 1; EtOH 2; ace 4; bz 3 1.4350[20]
5359	2-Eicosanol 1-Methyl-1-nonadecanol	$C_{20}H_{42}O$ 298.55	4340-76-5 63.5	357	4-01-00-01902 0.8378[20]	ace 4; bz 4 1.4912[80]
5360	5,8,11,14-Eicosatetraenoic acid 5,8,11,14-Eicosotetraenoic acid	$C_{20}H_{32}O_2$ 304.47	7771-44-0 -49.5	169[0.15]	4-02-00-01802 0.9219[20]	H_2O 1; EtOH 3; eth 3; ace 3 1.4824[20]
5361	5,8,11,14-Eicosatetraenoic acid, (all-Z)- Arachidonic acid	$C_{20}H_{32}O_2$ 304.47	506-32-1 -49.5	792 163[1]	4-02-00-01802 0.9082[20]	ace 4; eth 4; EtOH 4; peth 4 1.4824[20]
5362	1-Eicosene	$C_{20}H_{40}$ 280.54	3452-07-1 28.5	341; 151[2]	4-01-00-00934 0.7882[30]	H_2O 1; bz 3; peth 3 1.4440[30]
5363	11-Eicosenoic acid Δ 11-Eicosenoic acid	$C_{20}H_{38}O_2$ 310.52	2462-94-4 24	267[15]	4-02-00-01673 0.0026[25]	EtOH 4; MeOH 4
5364	9-Eicosenoic acid	$C_{20}H_{38}O_2$ 310.52	506-31-0 23	220[6]; 170[0.1]	4-02-00-01672 0.8882[25]	1.4597[25]
5365	9-Eicosenoic acid, (Z)-	$C_{20}H_{38}O_2$ 310.52	29204-02-2 24.5	220[6]	4-02-00-01672 0.8882[25]	
5366	1-Eicosyne	$C_{20}H_{38}$ 278.52	765-27-5 36	340	4-01-00-01077 0.8073[20]	bz 4; peth 4 1.4501[20]
5367	Emetan-6'-ol, 7',10,11-trimethoxy- Cephaeline	$C_{28}H_{38}N_2O_4$ 466.62	483-17-0 115.5	1970	5-23-13-00611	ace 4; EtOH 4; MeOH 4; chl 4
5368	Emetan, 6',7',10,11-tetramethoxy- Emetine	$C_{29}H_{40}N_2O_4$ 480.65	483-18-1 74	3517	4-23-00-03422	H_2O 1; EtOH 3; eth 3; ace 3
5369	Endosulfan 5,9-Methano-2,4,3-benzodioxathiepin, 6,7,8,9,10,10-hexachloro-1,5,5a,6,9,9a- hexahydro-, 3-oxide	$C_9H_6Cl_6O_3S$ 406.93	115-29-7 106	3529 106[0.7]	1.745[20]	
5370	Endothall, disodium salt 7-Oxabicyclo[2.2.1]heptane-2,3-dicarboxylic acid, disodium salt	$C_8H_{10}Na_2O_5$ 232.14	145-73-3 144	3530	1.431[20]	
5371	4,7-Epoxyisobenzofuran-1,3-dione, hexahydro-3a,7a-dimethyl-, (3aα,4β,7β,7aα)- Cantharidin	$C_{10}H_{12}O_4$ 196.20	56-25-7 218	1755 sub 84	5-19-05-00051	H_2O 1; EtOH 2; eth 2; ace 2
5372	EPTC Carbamothioic acid, dipropyl-, S-ethyl ester	$C_9H_{19}NOS$ 189.32	759-94-4 127[20]	3580	0.9546[30]	
5373	Ergoline-8-carboxamide, 9,10-didehydro-N-(2- hydroxy-1-methylethyl)-6-methyl-, [8α(S)]- Ergometrinine	$C_{19}H_{23}N_3O_2$ 325.41	479-00-5 195-7 dec	3599	4-25-00-00950	chl 4
5374	Ergoline-8-carboxamide, 9,10-didehydro-6- methyl-, (8β)- Lysergamide	$C_{16}H_{17}N_3O$ 267.33	478-94-4 137.5	5505	4-25-00-00937	EtOH 2; ace 2; os 2
5375	Ergoline-8-carboxylic acid, 9,10-didehydro-6- methyl-, (8α)- Isolysergic acid	$C_{16}H_{16}N_2O_2$ 268.32	478-95-5 218 dec	5065	4-25-00-00935	H_2O 2; EtOH 2; py 3
5376	Ergoline-8-carboxylic acid, 9,10-didehydro-6- methyl-, (8β)- Lysergic acid	$C_{16}H_{16}N_2O_2$ 268.32	82-58-6 240 dec	5506	4-25-00-00934	H_2O 2; EtOH 3; eth 2; bz 2
5377	Ergostane, (5α)-	$C_{28}H_{50}$ 386.71	511-20-6 85	3602	4-05-00-01234	ace 4; eth 4; chl 4
5378	Ergostane, (5β)- Coproergostane	$C_{28}H_{50}$ 386.71	511-21-7 64	2516	3-05-00-01143	eth 4; chl 4

No.	Name Synonym	Mol. Form. Mol. Wt.	CAS RN mp/°C	Merck No. bp/°C	Beil. Ref. den/g cm^{-3}	Solubility n_D
5379	Ergostan-3-ol, (3β,5α)- Ergostanol	C$_{28}$H$_{50}$O 402.70	6538-02-9 144.5	3603	4-06-00-03602	H$_2$O 1; eth 3; chl 3
5380	Ergosta-5,7,9(11),22-tetraen-3-ol, (3β,22E)- Dehydroergosterol	C$_{28}$H$_{42}$O 394.64	516-85-8 146	2861 230$^{0.5}$	4-06-00-04838	ace 4; bz 4; eth 4; EtOH 4
5381	Ergosta-5,7,22-trien-3-ol, (3β,10α,22E)- Pyrocalciferol	C$_{28}$H$_{44}$O 396.66	128-27-8 94	8007	3-06-00-03098	H$_2$O 1; EtOH 3; chl 3; MeOH 3
5382	Ergosta-5,7,22-trien-3-ol, (3β,9β,10α,22E)- Lumisterol	C$_{28}$H$_{44}$O 396.66	474-69-1 118	5471	4-06-00-04408	H$_2$O 1; EtOH 3; eth 4; ace 4
5383	Ergosta-5,7,22-trien-3-ol, (3β,22E)- Ergosterol	C$_{28}$H$_{44}$O 396.66	57-87-4 170	3607 250$^{0.01}$	4-06-00-04407	H$_2$O 1; EtOH 2; eth 2; bz 3
5384	Ergost-7-en-3-ol, (3β,5α)- γ-Ergostenol	C$_{28}$H$_{48}$O 400.69	516-78-9 146	3606	4-06-00-04029	eth 3
5385	Ergost-8(14)-en-3-ol, (3β,5α)- α-Ergostenol	C$_{28}$H$_{48}$O 400.69	632-32-6 131	3604	4-06-00-04031	EtOH 2; eth 3; bz 3; chl 3
5386	Ergost-5-en-3-ol, (3β,24R)- Campesterol	C$_{28}$H$_{48}$O 400.69	474-62-4 157.5	1735	3-06-00-02680	
5387	Ergotaman-3',6',18-trione, 12'-hydroxy-2',5'-bis(1-methylethyl)-, Ergocornine	C$_{31}$H$_{39}$N$_5$O$_5$ 561.68	564-36-3 182-4 dec	3591	4-25-00-00963	H$_2$O 1; EtOH 3; ace 3; bz 3
5388	Ergotaman-3',6',18-trione, 12'-hydroxy-2',5'-bis(1-methylethyl)-, Ergocorninine	C$_{31}$H$_{39}$N$_5$O$_5$ 561.68	564-37-4 228 dec	3592	4-25-00-00963	ace 4; bz 4; EtOH 4; chl 4
5389	Ergotaman-3',6',18-trione, 12'-hydroxy-2'-(1-methylethyl)-5'-(2-methylpropyl)- Ergocryptine	C$_{32}$H$_{41}$N$_5$O$_5$ 575.71	511-09-1 212-4 dec	3595	4-25-00-00964	H$_2$O 1; EtOH 3; chl 3
5390	Ergotaman-3',6',18-trione, 12'-hydroxy-2'-(1-methylethyl)-5'-(2-methylpropyl)-, (5'α,8α)- Ergocryptinine	C$_{33}$H$_{41}$N$_5$O$_5$ 587.72	511-10-4 245 dec	3596	4-25-00-00964	ace 4; chl 4
5391	Ergotaman-3',6',18-trione, 12'-hydroxy-2'-(1-methylethyl)-5'-(phenylmethyl)- Ergocristine	C$_{35}$H$_{39}$N$_5$O$_5$ 609.73	511-08-0 175 dec	3593	4-25-00-00966	H$_2$O 1; EtOH 3; ace 3; chl 3
5392	Ergotaman-3',6',18-trione, 12'-hydroxy-2'-(1-methylethyl)-5'-(phenylmethyl)- Ergocristinine	C$_{35}$H$_{39}$N$_5$O$_5$ 609.73	511-07-9 237-8 dec	3594	4-25-00-00967	H$_2$O 1; EtOH 2; ace 2; chl 2
5393	Ergotaman-3',6',18-trione, 12'-hydroxy-2'-methyl-5'-(2-methylpropyl)-, Ergosine	C$_{30}$H$_{37}$N$_5$O$_5$ 547.65	561-94-4 228 dec	3601	4-25-00-00961	ace 3; chl 3; MeOH 2
5394	Ergotaman-3',6',18-trione, 12'-hydroxy-2'-methyl-5'-(phenylmethyl)-, (5'a)- Ergotamine	C$_{33}$H$_{55}$N$_5$O$_5$ 601.83	113-15-5 213-4 dec	3609	4-25-00-00964	bz 4; eth 4; chl 4
5395	Ergotaman-3',6',18-trione, 12'-hydroxy-2'-methyl-5'-(phenylmethyl)-, Ergotaminine	C$_{33}$H$_{35}$N$_5$O$_5$ 581.67	639-81-6 252 dec	3610	4-25-00-00966	H$_2$O 1; EtOH 2; ace 2; bz 2
5396	Erythromycin Propiocine	C$_{37}$H$_{67}$NO$_{13}$ 733.94	114-07-8 191	3626	5-18-10-00398	ace 4; eth 4; EtOH 4; chl 4
5397	Estra-1,3,5,7,9-pentaen-17-one, 3-hydroxy- Equilenin	C$_{18}$H$_{18}$O$_2$ 266.34	517-09-9 258.5	3581 sub 170	4-08-00-01420	EtOH 2; ace 2; chl 2
5398	Estra-1,3,5(10),7-tetraen-17-one, 3-hydroxy- Equilin	C$_{18}$H$_{20}$O$_2$ 268.36	474-86-2 239	3582 sub 170	4-08-00-01366	H$_2$O 2; EtOH 3; ace 3; diox 3
5399	Estra-1,3,5(10)-triene-3,17-diol (17β)- Estradiol	C$_{18}$H$_{24}$O$_2$ 272.39	50-28-2 178.5	3653	4-06-00-06611	ace 4; EtOH 4; Diox 4
5400	Estra-1,3,5(10)-triene-3,17-diol, (17α)- α-Estradiol	C$_{18}$H$_{24}$O$_2$ 272.39	57-91-0 221.5	3654	4-06-00-06611	H$_2$O 1; EtOH 3; eth 2; ace 3
5401	Estra-1,3,5(10)-triene-3,17-diol, (8α,17β)- Isoestradiol	C$_{18}$H$_{24}$O$_2$ 272.39	517-04-4 181	5051	4-06-00-06611	EtOH 3; diox 3
5402	Estra-5,7,9-triene-3,17-diol, (3β,17β)- Hexahydroequilenin	C$_{18}$H$_{24}$O$_2$ 272.39	517-07-7 168.5	4607 sub 150	4-06-00-06618	
5403	Estra-1,3,5(10)-triene-3,17-diol (17β)-, 3-benzoate Estradiol benzoate	C$_{25}$H$_{28}$O$_3$ 376.50	50-50-0 196	3655	4-09-00-00406	
5404	Estra-1,3,5(10)-triene-3,16,17-triol, (16α,17β)- Estriol	C$_{18}$H$_{24}$O$_3$ 288.39	50-27-1 288 dec	3659	4-06-00-07550 1.27^{25}	EtOH 3; eth 2; bz 2; tfa 2
5405	Estra-1,3,5(10)-triene-3,16,17-triol, (16β,17β)- 16-Epiestriol	C$_{18}$H$_{24}$O$_3$ 288.39	547-81-9 290	3567	4-06-00-07549	
5406	Estra-1,3,5(10)-trien-17-one, 3-hydroxy- Estrone	C$_{18}$H$_{22}$O$_2$ 270.37	53-16-7 260.2	3660	3-08-00-01171 1.236^{25}	H$_2$O 1; EtOH 2; eth 2; ace 3
5407	Estra-1,3,5(10)-trien-17-one, 3-hydroxy-, (8α)- 8-Isoestrone	C$_{18}$H$_{22}$O$_2$ 270.37	517-06-6 247	5052	3-08-00-01170	eth 4; Diox 4
5408	Estr-4-en-3-one, 17-(1-oxo-3-phenylpropoxy)-, (17β)- Nandrolone phenpropionate	C$_{27}$H$_{34}$O$_3$ 406.57	62-90-8	6283	4-09-00-01758	
5409	Ethalfluralin Benzenamine, N-ethyl-N-(2-methyl-2-propenyl)-2,6-dinitro-4-(trifluoromethyl)-	C$_{13}$H$_{14}$F$_3$N$_3$O$_4$ 333.27	55283-68-6 57	3671 256 dec		
5410	Ethanamine Ethylamine	C$_2$H$_7$N 45.08	75-04-7 -80.5	3718 16.5	4-04-00-00307 *0.677^{25}	H$_2$O 5; EtOH 5; eth 5 1.3663^{20}
5411	Ethanamine, 2-[(3-butyl-1-isoquinolinyl)oxy]-N,N-dimethyl- Dimethisoquin	C$_{17}$H$_{24}$N$_2$O 272.39	86-80-6 146	3207 155-7^3	5-21-03-00414	H$_2$O 3; EtOH 3 1.5486^{20}

No.	Name / Synonym	Mol. Form. / Mol. Wt.	CAS RN / mp/°C	Merck No. / bp/°C	Beil. Ref. / den/g cm⁻³	Solubility / n_D
5412	Ethanamine, 2-chloro-N,N-bis(2-chloroethyl)- 2,2',2''-Trichlorotriethylamine	$C_6H_{12}Cl_3N$ 204.53	555-77-1 -4	9560 143-4[15]	4-04-00-00447	bz 4; eth 4; EtOH 4
5413	Ethanamine, 2-chloro-N-(2-chloroethyl)-N-methyl- Mechlorethamine	$C_5H_{11}Cl_2N$ 156.05	51-75-2 -60	5655 87[18]; 64[5]		H_2O 2; ctc 5; DMF 5
5414	Ethanamine, 2,2-diethoxy-	$C_6H_{15}NO_2$ 133.19	645-36-3 -78	163	4-04-00-01918 0.9159[25]	H_2O 4; eth 4; EtOH 4; chl 4 1.4123[25]
5415	Ethanamine, 2,2-diethoxy-N,N-diethyl-	$C_{10}H_{23}NO_2$ 189.30	3616-57-7	194.5	4-04-00-01919 0.863[25]	H_2O 3; EtOH 3; eth 3; ctc 2 1.4189[20]
5416	Ethanamine, 2,2-diethoxy-N,N-dimethyl-	$C_8H_{19}NO_2$ 161.24	3616-56-6	170.5	3-04-00-00870 0.885[7]	H_2O 4; EtOH 3; eth 3; ace 3 1.4129[20]
5417	Ethanamine, N,N-diethyl- Triethylamine	$C_6H_{15}N$ 101.19	121-44-8 -114.7	9582 89	4-04-00-00322 0.7275[20]	H_2O 3; EtOH 3; eth 3; ace 4 1.4010[20]
5418	Ethanamine, N,N-diethyl-, hydrochloride	$C_6H_{16}ClN$ 137.65	554-68-7 260 dec	sub 245	4-04-00-00322 1.0689[21]	H_2O 4; EtOH 4; chl 4
5419	Ethanamine, 2,2-dimethoxy-	$C_4H_{11}NO_2$ 105.14	22483-09-6	135-9[95]	4-04-00-01918 0.965[25]	1.4170[20]
5420	Ethanamine, 2,2-dimethoxy-N-methyl-	$C_5H_{13}NO_2$ 119.16	122-07-6	140	2-04-00-00759 0.928[25]	1.4115[20]
5421	Ethanamine, N,N-dimethyl-	$C_4H_{11}N$ 73.14	598-56-1 -140	36.5	4-04-00-00312 0.675[20]	1.3705[25]
5422	Ethanamine, 2-(diphenylmethoxy)-N,N-dimethyl- Diphenhydramine	$C_{17}H_{21}NO$ 255.36	58-73-1	3308	4-06-00-04658	
5423	Ethanamine, 2-ethoxy-	$C_4H_{11}NO$ 89.14	110-76-9	108	4-04-00-01411 0.8512[20]	H_2O 5; EtOH 5; eth 5; ace 3 1.4101[20]
5424	Ethanamine, 2-ethoxy-N,N-dimethyl-	$C_6H_{15}NO$ 117.19	26311-17-1	121	4-04-00-01425 0.806[20]	1.406[20]
5425	Ethanamine, 2-ethoxy-N-(2-ethoxyethyl)- Bis-(2-ethoxyethyl) amine	$C_8H_{19}NO_2$ 161.24	124-21-0 -50	194	4-04-00-01516 0.883[25]	1.4213[20]
5426	Ethanamine, 2-ethoxy-N-methyl- Methyl(2-ethoxyethyl)amine	$C_5H_{13}NO$ 103.16	38256-94-9	115	3-04-00-00647 0.8363[20]	ace 4; bz 4; eth 4; EtOH 4 1.4147[20]
5427	Ethanamine, N-ethyl- Diethylamine	$C_4H_{11}N$ 73.14	109-89-7 -49.8	3100 55.5	4-04-00-00313 0.7056[20]	H_2O 4; EtOH 5; eth 3; ctc 3 1.3864[20]
5428	Ethanamine, N-ethyl-, hydrochloride	$C_4H_{12}ClN$ 109.60	660-68-4 228.5		4-04-00-00313 1.0477[22]	H_2O 4; EtOH 4
5429	Ethanamine, N-ethyl-N-hydroxy-	$C_4H_{11}NO$ 89.14	3710-84-7 10	133	4-04-00-03304 0.8669[20]	1.4195[20]
5430	Ethanamine, N-ethyl-N-methyl-	$C_5H_{13}N$ 87.16	616-39-7 -196	66	4-04-00-00321 0.703[25]	H_2O 4; EtOH 4; eth 4 1.3879[25]
5431	Ethanamine, N-ethyl-N-nitro-	$C_4H_{10}N_2O_2$ 118.14	7119-92-8	206.5	4-04-00-03403 1.057[15]	eth 4; EtOH 4
5432	Ethanamine, N-ethyl-N-nitroso- N-Nitrosodiethylamine	$C_4H_{10}N_2O$ 102.14	55-18-5	6557 176.9	4-04-00-03386 0.9422[20]	H_2O 3; EtOH 3; eth 3; chl 2 1.4386[20]
5433	Ethanamine, hydrochloride	C_2H_8ClN 81.54	557-66-4 109.5	3718	4-04-00-00307 1.2160[20]	H_2O 4; EtOH 4
5434	Ethanamine, N-hydroxy- N-Ethylhydroxylamine	C_2H_7NO 61.08	624-81-7 59-60 dec		4-04-00-03304 0.9079[20]	H_2O 4; EtOH 4 1.4152[66]
5435	Ethanamine, N-methyl-, hydrochloride Ethyl methyl amine, hydrochloride	$C_3H_{10}ClN$ 95.57	624-60-2 128		2-04-00-00589 1.0874[20]	H_2O 4; EtOH 4; eth 1; chl 3
5436	Ethanamine, 1,1,2,2,2-pentafluoro-N,N-bis(pentafluoroethyl)-	$C_6F_{15}N$ 371.05	359-70-6	70.3	4-02-00-00471 1.736[20]	1.262[25]
5437	Ethanamine, N-(phenylmethylene)-	$C_9H_{11}N$ 133.19	6852-54-6	196	4-07-00-00518 0.937[20]	H_2O 1; EtOH 3; eth 3 1.5378[15]
5438	Ethanaminium, 2-(acetyloxy)-N,N,N-trimethyl-, bromide Acetyl choline bromide	$C_7H_{16}BrNO_2$ 226.11	66-23-9 146	80	4-04-00-01446	H_2O 4
5439	Ethanaminium, 2-(acetyloxy)-N,N,N-trimethyl-, chloride Acetyl choline chloride	$C_7H_{16}ClNO_2$ 181.66	60-31-1 150	81	4-04-00-01446	H_2O 3; EtOH 3; eth 1
5440	Ethanaminium, 2-[(aminocarbonyl)oxy]-N,N,N-trimethyl-, chloride Carbamoylcholine chloride	$C_6H_{15}ClN_2O_2$ 182.65	51-83-2 210 dec	1780	4-04-00-01455	H_2O 4; EtOH 2; eth 1; chl 1
5441	Ethanaminium, 2,2',2''-[1,2,3-benzenetriyltris(oxy)]tris[N,N,N-triethyl-, tr Gallamine triethiodide	$C_{30}H_{60}I_3N_3O_3$ 891.54	65-29-2 147.5	4249	4-06-00-07333	H_2O 4; EtOH 4; eth 2; ace 2
5442	Ethanaminium, 2-chloro-N,N,N-trimethyl-, chloride	$C_5H_{13}Cl_2N$ 158.07	999-81-5 239 dec	2103	4-04-00-00445	
5443	Ethanaminium, 2-[(cyclohexylhydroxyphenylacetyl)oxy]-N,N-diethyl-N-methyl-, bromide Oxyphenonium bromide	$C_{21}H_{34}BrNO_3$ 428.41	50-10-2 191.5	6928	3-10-00-00911	H_2O 4; EtOH 2
5444	Ethanaminium, N-[4-[[4-(diethylamino)phenyl]phenylmethylene]-2,5-cyclohexadien-1-ylidene]-N-ethyl-, sulfate (1 Brilliant green	$C_{27}H_{34}N_2O_4S$ 482.64	633-03-4	1367	4-13-00-02281	H_2O 4; EtOH 4
5445	Ethanaminium, N,N-diethyl-N-methyl-2-[(9H-xanthen-9-ylcarbonyl)oxy]-, bromide Methantheline bromide	$C_{21}H_{28}BrNO_3$ 420.35	53-46-3 174.5	5869	5-18-06-00590	H_2O 3; EtOH 3; eth 1; chl 3

No.	Name / Synonym	Mol. Form. / Mol. Wt.	CAS RN / mp/°C	Merck No. / bp/°C	Beil. Ref. / den/g cm^{-3}	Solubility / n_D
5446	Ethanaminium, 2-hydroxy-N,N,N-trimethyl-, chloride Choline chloride	$C_5H_{14}ClNO$ 139.63	67-48-1 305-5 dec	2208	4-04-00-01443	H_2O 4; EtOH 4
5447	Ethanaminium, N,N,N-triethyl-, bromide Tetraethylammonium bromide	$C_8H_{20}BrN$ 210.16	71-91-0 285-7 dec	9133	4-04-00-00331 1.3970[20]	H_2O 4; EtOH 4; chl 4; MeOH 4
5448	Ethane	C_2H_6 30.07	74-84-0 -182.8	3676 -88.6	4-01-00-00108 0.5446[-89]	bz 4
5449	Ethane, 1,2-bis(2-chloroethoxy)-	$C_6H_{12}Cl_2O_2$ 187.07	112-26-5 	232	4-01-00-02379 1.195[20]	ctc 3 1.4592[25]
5450	Ethane, 1,1-bis(ethylthio)-	$C_6H_{14}S_2$ 150.31	14252-42-7 	184	4-01-00-03162 0.9706[20]	eth 3; ace 3; chl 3 1.5025[20]
5451	Ethane, 1,2-bis(ethylthio)-	$C_6H_{14}S_2$ 150.31	5395-75-5 180	217; 108[27]	4-01-00-02452 0.9815[20]	EtOH 3; eth 3 1.5118[20]
5452	Ethane, 1,2-bis(methylthio)-	$C_4H_{10}S_2$ 122.26	6628-18-8 	183; 79[11]	4-01-00-02451 1.0371[20]	H_2O 4; ace 4; eth 4; EtOH 4 1.5292[20]
5453	Ethanebis(thioic) acid, S,S-diethyl ester	$C_8H_{10}O_2S_2$ 178.28	615-85-0 27	235	1-02-00-00244 1.0565[21]	eth 4
5454	Ethane, bromo- Ethyl bromide	C_2H_5Br 108.97	74-96-4 -118.6	3730 38.5	4-01-00-00150 1.4604[20]	H_2O 2; EtOH 5; eth 5; chl 5 1.4239[20]
5455	Ethane, 1-bromo-1-chloro- 1-Bromo-1-chloroethane	C_2H_4BrCl 143.41	593-96-4 	83	4-01-00-00155 1.667[10]	1.4660[2-]
5456	Ethane, 1-bromo-2-chloro-	C_2H_4BrCl 143.41	107-04-0 -16.7	107	4-01-00-00155 1.7392[20]	H_2O 2; EtOH 3; eth 3; chl 3 1.4908[20]
5457	Ethane, 1-bromo-2-chloro-1,1,2-trifluoro- 1-Bromo-2-chloro-1,1,2-trifluoroethane	$C_2HBrClF_3$ 197.38	354-06-3 	52.5	4-01-00-00156 1.8636[25]	1.3738[20]
5458	Ethane, 2-bromo-2-chloro-1,1,1-trifluoro- Halothane	$C_2HBrClF_3$ 197.38	151-67-7 	4517 50.2; 20[243]	1.871[20]	H_2O 2; peth 3 1.3697[0]
5459	Ethane, 2-bromo-1,1-diethoxy-	$C_6H_{13}BrO_2$ 197.07	2032-35-1 	66-7[18]	4-01-00-03151 1.310[25]	EtOH 3; eth 3 1.4387[20]
5460	Ethane, 2-bromo-1,1-dimethoxy-	$C_4H_9BrO_2$ 169.02	7252-83-7 	149	3-01-00-02672 1.430[20]	eth 3; ace 3; chl 3 1.4450[20]
5461	Ethane, 1-bromo-2-ethoxy- 2-Bromoethyl ethyl ether	C_4H_9BrO 153.02	592-55-2 	127.5	4-01-00-01386 1.3852[0]	H_2O 2; EtOH 5; eth 5 1.4447[20]
5462	Ethane, 1-bromo-2-fluoro-	C_2H_4BrF 126.96	762-49-2 	71.5	4-01-00-00154 1.7044[25]	eth 4; EtOH 4 1.4236[20]
5463	Ethane, 1-bromo-2-iodo- 1-Bromo-2-iodoethane	C_2H_4BrI 234.86	590-16-9 28	163	4-01-00-00168 2.516[29]	eth 4
5464	Ethane, (bromomethoxy)-	C_3H_7BrO 138.99	53588-92-4 	110	3-01-00-02594 1.4402[20]	eth 4 1.4515[20]
5465	Ethane, 1-bromo-2-methoxy-	C_3H_7BrO 138.99	6482-24-2 	110 dec	4-01-00-01386 1.4623[20]	1.44753[20]
5466	Ethane, 2-bromo-1,1,1-trifluoro-	$C_2H_2BrF_3$ 162.94	421-06-7 -93.9	26	4-01-00-00154 1.7881[20]	1.3331[20]
5467	Ethane, chloro- Ethyl chloride	C_2H_5Cl 64.51	75-00-3 -138.7	3740 12.3	4-01-00-00124 *0.8902[25]	H_2O 2; EtOH 4; eth 5; chl 2 1.3676[20]
5468	Ethane, 1-chloro-1-(2-chloroethoxy)- 1-Chloroethyl 2-chloroethyl ether	$C_4H_8Cl_2O$ 143.01	1462-34-6 	56[17]	3-01-00-02655 1.1867[20]	eth 4; EtOH 4; chl 4 1.4473[20]
5469	Ethane, 2-chloro-1,1-diethoxy-	$C_6H_{13}ClO_2$ 152.62	621-62-5 	157.4	4-01-00-03134 1.0180[20]	H_2O 2; EtOH 5; eth 5; ctc 2 1.4170[20]
5470	Ethane, 1-chloro-1,1-difluoro 1-Chloro-1,1-difluoroethane	$C_2H_3ClF_2$ 100.50	75-68-3 -130.8	-9.7	1.107[25]	H_2O 1; bz 3
5471	Ethane, 2-chloro-1,1-difluoro 1-Chloro-2,2-difluoroethane	$C_2H_3ClF_2$ 100.50	338-65-8 	35.1		
5472	Ethane, 2-chloro-1,1-dimethoxy-	$C_4H_9ClO_2$ 124.57	97-97-2 	127.5	4-01-00-03134 1.068[20]	EtOH 2; eth 2; bz 2; ctc 2 1.4150[20]
5473	Ethane, 1-chloro-1-ethoxy-	C_4H_9ClO 108.57	7081-78-9 	93.5	4-01-00-03119 0.9655[20]	1.4053[20]
5474	Ethane, 1-chloro-2-ethoxy-	C_4H_9ClO 108.57	628-34-2 	107.5	4-01-00-01375 0.9895[20]	H_2O 2; eth 5; chl 3 1.4113[20]
5475	Ethane, 1-chloro-1-fluoro- 1-Chloro-1-fluoroethane	C_2H_4ClF 82.50	1615-75-4 	16.2	4-01-00-00127	
5476	Ethane, 1-chloro-2-fluoro-	C_2H_4ClF 82.50	762-50-5 	59	4-01-00-00127 1.1747[20]	eth 4; EtOH 4 1.3775[20]
5477	Ethane, 1-chloro-2-iodo-	C_2H_4ClI 190.41	624-70-4 -15.6	140	4-01-00-00167 2.1644[25]	
5478	Ethane, (chloromethoxy)- Chloromethyl ethyl ether	C_3H_7ClO 94.54	3188-13-4 	83	4-01-00-03047 1.0188[15]	1.4040[20]
5479	Ethane, 1-chloro-1-methoxy-	C_3H_7ClO 94.54	1538-87-0 	73	4-01-00-03119 0.9902[20]	eth 4 1.4004[20]
5480	Ethane, 1-chloro-2-methoxy-	C_3H_7ClO 94.54	627-42-9 	92.5	4-01-00-01375 1.0345[20]	H_2O 4; eth 4 1.4111[20]
5481	Ethane, 1-chloro-2-(methylthio)-	C_3H_7ClS 110.61	542-81-4 	60[30]; 44[20]	4-01-00-01406 1.1225[20]	EtOH 3; eth 3; ace 3 1.4902[20]
5482	Ethane, 1-chloro-1-nitro- 1-Chloro-1-nitroethane	$C_2H_4ClNO_2$ 109.51	598-92-5 	124.5	4-01-00-00172 1.2837[20]	H_2O 1; EtOH 3; ctc 3; alk 3 1.4224[20]
5483	Ethane, 1-chloro-2-nitro- 2-Chloro-1-nitroethane	$C_2H_4ClNO_2$ 109.51	625-47-8 	173.5	4-01-00-00173 1.3550[20]	1.407[7]
5484	Ethane, chloropentafluoro- Chloropentafluoroethane	C_2ClF_5 154.47	76-15-3 -99.4	-37.9	4-01-00-00129 1.5678[-42]	H_2O 1; EtOH 3; eth 3 1.2678[-42]

No.	Name Synonym	Mol. Form. Mol. Wt.	CAS RN mp/°C	Merck No. bp/°C	Beil. Ref. den/g cm^{-3}	Solubility n_D
5485	Ethane, 1-chloro-1,1,2,2-tetrafluoro- 1-Chloro-1,1,2,2-tetrafluoroethane	C_2HClF_4 136.48	354-25-6 -117	-10	4-01-00-00129	
5486	Ethane, 1-chloro-1,2,2,2-tetrafluoro- 1-Chloro-1,2,2,2-tetrafluoroethane	C_2HClF_4 136.48	2837-89-0	-12		
5487	Ethane, 1-chloro-1,1,2-trifluoro- 1-Chloro-1,1,2-trifluoroethane	$C_2H_2ClF_3$ 118.49	421-04-5	12		
5488	Ethane, 1-chloro-1,2,2-trifluoro- 1-Chloro-1,2,2-trifluoroethane	$C_2H_2ClF_3$ 118.49	431-07-2	17		
5489	Ethane, 2-chloro-1,1,1-trifluoro- 2-Chloro-1,1,1-trifluoroethane	$C_2H_2ClF_3$ 118.49	75-88-7 -105.5	6.1	4-01-00-00128 1.389[0]	1.3090[0]
5490	Ethane, 1-chloro-1,1,2-trifluoro-2-iodo- Ethane, 1-chloro-2-iodo-1,1,2-trifluoro	C_2HClF_3I 244.38	354-26-7	77[630]	4-01-00-00167 2.181[20]	1.4320[20]
5491	Ethanedial Glyoxal	$C_2H_2O_2$ 58.04	107-22-2 15	4405 50.4	4-01-00-03625 1.14[20]	H_2O 4; EtOH 3; eth 3 1.3826[20]
5492	Ethanedial, dioxime	$C_2H_4N_2O_2$ 88.07	557-30-2 178 dec	sub	3-01-00-03084	H_2O 4; EtOH 4; eth 4
5493	Ethanediamide Oxamide	$C_2H_4N_2O_2$ 88.07	471-46-5 350 dec	6871	4-02-00-01860 1.667[20]	H_2O 2; EtOH 2; eth 1
5494	Ethanediamide, N,N '-diethyl-	$C_6H_{12}N_2O_2$ 144.17	615-84-9 175		4-04-00-00359 1.169[4]	EtOH 4
5495	Ethanediamide, N,N '-diphenyl-	$C_{14}H_{12}N_2O_2$ 240.26	620-81-5 254	>360	4-12-00-00463	bz 4
5496	1,2-Ethanediamine Ethylenediamine	$C_2H_8N_2$ 60.10	107-15-3 11.1	3752 117	4-04-00-01166 0.8979[20]	H_2O 4; EtOH 5; eth 1; bz 1 1.4565[20]
5497	1,2-Ethanediamine, N-(2-aminoethyl)- Diethylenetriamine	$C_4H_{13}N_3$ 103.17	111-40-0 -39	207	4-04-00-01238 0.9569[20]	H_2O 5; EtOH 5; eth 1; lig 3 1.4810[25]
5498	1,2-Ethanediamine, N-(2-aminoethyl)-N '-[2-[(2-aminoethyl)amino]ethyl]- Tetraethylenepentamine	$C_8H_{23}N_5$ 189.30	112-57-2 341.5		4-04-00-01244	H_2O 3 1.5042[20]
5499	1,2-Ethanediamine, N,N '-bis(2-aminoethyl)- Trientine	$C_6H_{18}N_4$ 146.24	112-24-3 12	9579 266.5	4-04-00-01242	H_2O 3; EtOH 3; acid 3 1.4971[20]
5500	1,2-Ethanediamine, N,N '-bis(phenylmethyl)- Benzathine	$C_{16}H_{20}N_2$ 240.35	140-28-3 26	1072 195[4]	4-12-00-02321 1.024[20]	bz 4; eth 4; EtOH 4 1.5635[20]
5501	1,2-Ethanediamine, N,N '-diethyl-	$C_6H_{16}N_2$ 116.21	111-74-0 146		4-04-00-01174 0.8280[20]	H_2O 4; eth 4; EtOH 4; tol 4 1.4340[20]
5502	1,2-Ethanediamine, N,N-diethyl- N,N-Diethylethylenediamine	$C_6H_{16}N_2$ 116.21	100-36-7 144		4-04-00-01175 0.8280[20]	H_2O 5; EtOH 3; eth 3; ctc 3 1.4340[20]
5503	1,2-Ethanediamine, N,N '-dimethyl-	$C_4H_{12}N_2$ 88.15	110-70-3 120		4-04-00-01171 0.828[15]	EtOH 3; eth 3; dil HCl 3
5504	1,2-Ethanediamine, N,N-dimethyl-N '-(phenylmethyl)-	$C_{11}H_{18}N_2$ 178.28	103-55-9 122-4[11]		4-12-00-02320 0.922[25]	1.5089[20]
5505	1,2-Ethanediamine, N,N-dimethyl-N '-(phenylmethyl)-N '-2-pyridinyl- Tripelennamine	$C_{16}H_{21}N_3$ 255.36	91-81-6 167-72[0.1]	9651	5-22-08-00378	
5506	1,2-Ethanediamine, N,N-dimethyl-N '-phenyl-N '-(phenylmethyl)- Phenbenzamine	$C_{17}H_{22}N_2$ 254.38	961-71-7 179-80[7]	7176	4-12-00-02324 1.016[25]	1.5794[25]
5507	1,2-Ethanediamine, N,N-dimethyl-N '-2-pyridinyl-N '-(2-thienylmethyl)- Methapyrilene	$C_{14}H_{19}N_3S$ 261.39	91-80-5 173-5[3]	5871	5-22-08-00401	1.5915[20]
5508	1,2-Ethanediamine, N,N '-diphenyl- 1,2-Dianilinoethane	$C_{14}H_{16}N_2$ 212.29	150-61-8 74	2969 229[12]; 178[2]	4-12-00-00986	H_2O 1; EtOH 3; eth 3; tfa 2
5509	1,2-Ethanediamine, N-ethyl-	$C_4H_{12}N_2$ 88.15	110-72-5 129		4-04-00-01174 0.837[25]	1.4385[20]
5510	1,2-Ethanediamine, N '-ethyl-N,N-dimethyl- Ethylenediamine, N,N-dimethyl-N '-ethyl	$C_6H_{15}N_2$ 115.20	123-83-1 134.5		4-04-00-01174 0.738[25]	1.4222[20]
5511	1,2-Ethanediamine, N-[(4-methoxyphenyl)methyl]-N ',N '-dimethyl-N-2-pyridinyl- Pyrilamine	$C_{17}H_{23}N_3O$ 285.39	91-84-9 201[5]	7996	5-22-08-00381	
5512	1,2-Ethanediamine, N-methyl-	$C_3H_{10}N_2$ 74.13	109-81-9 115		4-04-00-01170 0.841[25]	1.4395[20]
5513	1,2-Ethanediamine, monohydrate	$C_2H_{10}N_2O$ 78.11	6780-13-8 10	118	4-04-00-01166 0.964[20]	H_2O 4 1.4500[20]
5514	1,2-Ethanediamine, N-1-naphthalenyl- N-(1-Naphthyl)ethylenediamine	$C_{12}H_{14}N_2$ 186.26	551-09-7 204[9]	6329	4-12-00-03101 1.114[25]	1.6648[25]
5515	1,2-Ethanediamine, N,N,N ',N '-tetraethyl-	$C_{10}H_{24}N_2$ 172.31	150-77-6 192		4-04-00-01177 0.808[25]	1.4343[20]
5516	1,2-Ethanediamine, N,N,N ',N '-tetramethyl- N,N,N ',N '-Tetramethyldiaminoethane	$C_6H_{16}N_2$ 116.21	110-18-9 -55	121	4-04-00-01172 0.77[25]	1.4179[20]
5517	Ethane, 1,1-dibromo- Ethylidene dibromide	$C_2H_4Br_2$ 187.86	557-91-5 -63	109	4-01-00-00157 2.0554[20]	H_2O 1; EtOH 3; eth 4; ace 3 1.5128[20]
5518	Ethane, 1,2-dibromo- Ethylene dibromide	$C_2H_4Br_2$ 187.86	106-93-4 9.9	3753 131.6	4-01-00-00158 2.1791[20]	ace 4; bz 4; eth 4; EtOH 4 1.5387[20]
5519	Ethane, 1,2-dibromo-1-chloro-1,2,2-trifluoro- 1,2-Dibromo-1-chloro-1,2,2-trifluoroethane	$C_2Br_2ClF_3$ 276.28	354-51-8 50	93		
5520	Ethane, 1,2-dibromo-1,1-dichloro-	$C_2H_2Br_2Cl_2$ 256.75	75-81-0 -26	195	4-01-00-00161 2.135[20]	ace 4; bz 4; eth 4; EtOH 4 1.5662[20]
5521	Ethane, 1,2-dibromo-1,2-dichloro-	$C_2H_2Br_2Cl_2$ 256.75	683-68-1 -26	195	4-01-00-00161 2.135[20]	H_2O 1; EtOH 3; eth 3; ace 3 1.5662[20]

No.	Name / Synonym	Mol. Form. / Mol. Wt.	CAS RN / mp/°C	Merck No. / bp/°C	Beil. Ref. / den/g cm⁻³	Solubility / n_D
5522	Ethane, 1,2-dibromo-1,1-difluoro- / Genetron 132b-B2	$C_2H_2Br_2F_2$ / 223.84	75-82-1 / -61.3	92.5	4-01-00-00160 / 2.2238[20]	1.4456[20]
5523	Ethane, 1,2-dibromo-1-ethoxy-	$C_4H_8Br_2O$ / 231.91	2983-26-8 /	80[20]	4-01-00-03152 / 1.7320[20]	EtOH 4; chl 4 / 1.5044[20]
5524	Ethane, 1,2-dibromo-1,1,2,2-tetrachloro-	$C_2Br_2Cl_4$ / 325.64	630-25-1 / 202.5		3-01-00-00190 / 2.713[25]	
5525	Ethane, 1,2-dibromo-1,1,2,2-tetrafluoro- / 1,2-Dibromotetrafluoroethane	$C_2Br_2F_4$ / 259.82	124-73-2 / -110.4	47.3	4-01-00-00160 / 2.149[25]	1.361[25]
5526	Ethane, 1,2-dibromo-1,1,2-trifluoro- / Halon 2302	$C_2HBr_2F_3$ / 241.83	354-04-1 /	76	4-01-00-00160 / 2.274[27]	1.4191[24]
5527	Ethane, 1,2-dibutoxy- / Ethylene glycol dibutyl ether	$C_{10}H_{22}O_2$ / 174.28	112-48-1 / -69.1	203.3	0.8319[25]	
5528	Ethane, 1,1-dichloro- / Ethylidene dichloride	$C_2H_4Cl_2$ / 98.96	75-34-3 / -96.9	3766 / 57.4	4-01-00-00130 / 1.1757[20]	H_2O 2; EtOH 4; eth 4; ace 3 / 1.4164[20]
5529	Ethane, 1,2-dichloro- / Ethylene dichloride	$C_2H_4Cl_2$ / 98.96	107-06-2 / -35.5	3754 / 83.5	4-01-00-00131 / 1.2351[20]	H_2O 2; EtOH 4; eth 5; ace 3 / 1.4448[20]
5530	Ethane, 1,1-dichloro-2,2-diethoxy-	$C_6H_{12}Cl_2O_2$ / 187.07	619-33-0 /	183.5	4-01-00-03140 / 1.1383[14]	EtOH 3; eth 3; ctc 2
5531	Ethane, 1,2-dichloro-1,1-difluoro-	$C_2H_2Cl_2F_2$ / 134.94	1649-08-7 / -101.2	46.8	4-01-00-00135 / 1.4163[20]	1.36193[20]
5532	Ethane, 1,2-dichloro-1,2-difluoro-	$C_2H_2Cl_2F_2$ / 134.94	431-06-1 / -101.2	46.8	4-01-00-00135 / 1.4163[20]	1.3619[20]
5533	Ethane, 1,2-dichloro-1-ethoxy-	$C_4H_8Cl_2O$ / 143.01	623-46-1 /	145	4-01-00-03136 / 1.1370[20]	chl 2 / 1.4435[20]
5534	Ethane, 1,1-dichloro-1-fluoro- / 1,1-Dichloro-1-fluoroethane	$C_2H_3Cl_2F$ / 116.95	1717-00-6 / -103.5	32	4-01-00-00134 / 1.250[10]	1.3600[10]
5535	Ethane, 1,2-dichloro-1-fluoro-	$C_2H_3Cl_2F$ / 116.95	430-57-9 / -60	73.7	4-01-00-00134 / 1.3814[20]	1.4132[20]
5536	Ethane, 1,1-dichloro-1-nitro- / Ethide	$C_2H_3Cl_2NO_2$ / 143.96	594-72-9 /	123.5	4-01-00-00173	ctc 3
5537	Ethane, 1,1-dichloro-1,2,2,2-tetrafluoro- / 1,1-Dichlorotetrafluoroethane	$C_2Cl_2F_4$ / 170.92	374-07-2 / -56.6	4	4-01-00-00136 / *1.455[25]	bz 4; eth 4; EtOH 4 / 1.3092[0]
5538	Ethane, 1,2-dichloro-1,1,2,2-tetrafluoro- / Refrigerant 114	$C_2Cl_2F_4$ / 170.92	76-14-2 / -94	2608 / 3.8	4-01-00-00137 / *1.455[25]	eth 4; EtOH 4 / 1.3092[0]
5539	Ethane, 1,2-dichloro-1,1,2-trifluoro-	$C_2HCl_2F_3$ / 152.93	354-23-4 / -78	28	4-01-00-00136 / 1.50[25]	
5540	Ethane, 2,2-dichloro-1,1,1-trifluoro- / 2,2-Dichloro-1,1,1-trifluoroethane	$C_2HCl_2F_3$ / 152.93	306-83-2 / -107	27.82	4-01-00-00135 / 1.4638[25]	
5541	Ethane, 1,1-diethoxy- / Acetal	$C_6H_{14}O_2$ / 118.18	105-57-7 / -100	31 / 102.2	4-01-00-03103 / 0.8254[20]	H_2O 3; EtOH 5; eth 5; ace 4 / 1.3834[20]
5542	Ethane, 1,2-diethoxy- / Ethylene glycol diethyl ether	$C_6H_{14}O_2$ / 118.18	629-14-1 /	119.4	4-01-00-02379 / 0.8484[20]	ace 4; bz 4; eth 4; EtOH 4 / 1.3860[20]
5543	Ethane, 1,1-diethoxy-2-iodo-	$C_6H_{13}IO_2$ / 244.07	51806-20-3 /	132[90]; 100[10]	4-01-00-03156 / 1.489[22]	1.4734[22]
5544	Ethane, 1,1-difluoro-	$C_2H_4F_2$ / 66.05	75-37-6 / -117	-24.9	4-01-00-00120 / *0.896[20]	1.3011[-72]
5545	Ethane, 1,2-difluoro- / Freon 152	$C_2H_4F_2$ / 66.05	624-72-6 /	30.7	4-01-00-00121	bz 4; eth 4; chl 4
5546	Ethane, 1,1-diiodo- / 1,1-Diiodoethane	$C_2H_4I_2$ / 281.86	594-02-5 /	179.5	4-01-00-00169 / 2.84[25]	ace 4; eth 4; EtOH 4; chl 4 / 1.673[20]
5547	Ethane, 1,2-diiodo-	$C_2H_4I_2$ / 281.86	624-73-7 / 83	200	4-01-00-00169 / 3.325[20]	H_2O 2; EtOH 3; eth 3; ace 3 / 1.871[20]
5548	Ethane, 1,1-dimethoxy- / Dimethyl acetal	$C_4H_{10}O_2$ / 90.12	534-15-6 / -113.2	3215 / 64.5	4-01-00-03103 / 0.8501[20]	H_2O 3; EtOH 3; eth 3; ace 4 / 1.3668[20]
5549	Ethane, 1,2-dimethoxy- / Ethylene glycol dimethyl ether	$C_4H_{10}O_2$ / 90.12	110-71-4 / -58	3213 / 85	4-01-00-02376 / 0.8691[20]	H_2O 3; EtOH 3; eth 3; ace 3 / 1.3796[20]
5550	Ethanedinitrile / Cyanogen	C_2N_2 / 52.04	460-19-5 / -27.9	2698 / -21.1	4-02-00-01863 / 0.9537[-21]	H_2O 3; EtOH 3; eth 3
5551	Ethane, 1,1-dinitro- / 1,1-Dinitroethane	$C_2H_4N_2O_4$ / 120.06	600-40-8 /	185.5	4-01-00-00174 / 1.349[24]	H_2O 2; EtOH 3; eth 3
5552	Ethane, 1,2-dinitro- / 1,2-Dinitroethane	$C_2H_4N_2O_4$ / 120.06	7570-26-5 / 39.5	94-6[5]	4-01-00-00175 / 1.4597[20]	eth 4; EtOH 4 / 1.4468[20]
5553	Ethanedioic acid / Oxalic acid	$C_2H_2O_4$ / 90.04	144-62-7 / 189.5 dec	6865 / sub 157	4-02-00-01819 / 1.900[17]	H_2O 3; EtOH 4; eth 2; bz 1
5554	Ethanedioic acid, bis(2-chloroethyl) ester	$C_6H_8Cl_2O_4$ / 215.03	7208-92-6 / 45	150[4]	4-02-00-01849	bz 4; EtOH 4
5555	Ethanedioic acid, bis(3-methylbutyl) ester	$C_{12}H_{22}O_4$ / 230.30	2051-00-5 / -9	267.5	4-02-00-01851 / 0.968[11]	eth 4; EtOH 4
5556	Ethanedioic acid, bis(1-methylethyl) ester	$C_8H_{14}O_4$ / 174.20	615-81-6 /	191	4-02-00-01850 / 1.001[20]	eth 4; EtOH 4 / 1.4100[20]
5557	Ethanedioic acid, bis(2-methylpropyl) ester / Diisobutyl oxalate	$C_{10}H_{18}O_4$ / 202.25	2050-61-5 /	229	4-02-00-01850 / 0.9737[20]	ace 4; eth 4; EtOH 4 / 1.4180[20]
5558	Ethanedioic acid, dibutyl ester	$C_{10}H_{18}O_4$ / 202.25	2050-60-4 / -30.5	241; 96[2]	4-02-00-01850 / 0.9873[20]	H_2O 1; EtOH 3; eth 3 / 1.4234[20]
5559	Ethanedioic acid, dicyclohexyl ester / Dicyclohexyl oxalate	$C_{14}H_{22}O_4$ / 254.33	620-82-6 / 43.5	190[73]	3-06-00-00026	H_2O 1; EtOH 4; eth 4; MeOH 3
5560	Ethanedioic acid, diethyl ester / Diethyl oxalate	$C_6H_{10}O_4$ / 146.14	95-92-1 / -40.6	3115 / 185.7	4-02-00-01848 / 1.0785[20]	H_2O 2; EtOH 5; eth 5; ace 5 / 1.4101[20]

No.	Name / Synonym	Mol. Form. / Mol. Wt.	CAS RN / mp/°C	bp/°C	Merck No.	Beil. Ref. / den/g cm^{-3}	Solubility / n_D
5561	Ethanedioic acid, dihydrate / Oxalic acid dihydrate	$C_2H_6O_6$ / 126.07	6153-56-6 / 101.5			4-02-00-01819 / 1.653[18]	H_2O 3; EtOH 3; eth 2
5562	Ethanedioic acid, dihydrazide	$C_2H_6N_4O_2$ / 118.10	996-98-5 / 243 dec			4-02-00-01868 / 1.458[22]	H_2O 3; EtOH 2; eth 2; bz 2
5563	Ethanedioic acid, dimethyl ester / Dimethyl oxalate	$C_4H_6O_4$ / 118.09	553-90-2 / 54.3	6020 / 163.5		4-02-00-01847 / 1.1716[60]	H_2O 2; EtOH 3; eth 3; ace 3 / 1.3798[2]
5564	Ethanedioic acid, di-2-propenyl ester / Diallyl oxalate	$C_8H_{10}O_4$ / 170.17	615-99-6 /	217		4-02-00-01851 / 1.1582[20]	H_2O 1; EtOH 3; ace 3; bz 3 / 1.4481[20]
5565	Ethanedioic acid, dipropyl ester	$C_8H_{14}O_4$ / 174.20	615-98-5 / -44.3	211		4-02-00-01849 / 1.0188[20]	H_2O 2; EtOH 5; eth 3 / 1.4158[20]
5566	Ethanedioic acid, disodium salt / Oxalic acid, disodium salt	$C_2Na_2O_4$ / 134.00	62-76-0 / 260 dec		8601	4-02-00-01819 / 2.34[25]	H_2O 2; EtOH 1; eth 1
5567	Ethanedioic acid, ethyl methyl ester	$C_5H_8O_4$ / 132.12	615-52-1 /	173.7		0-02-00-00535 / 1.1555[0]	eth 4; EtOH 4
5568	Ethanedioic acid, lead (II) salt / Lead oxalate	C_2O_4Pb / 295.22	814-93-7 / 300 dec		5294	4-02-00-01819 / 5.28	H_2O I; EtOH 1; HNO_3 3
5569	1,2-Ethanediol / Ethylene glycol	$C_2H_6O_2$ / 62.07	107-21-1 / -13	197.3	3755	4-01-00-02369 / 1.1088[20]	H_2O 5; EtOH 5; eth 3; ace 5 / 1.4318[20]
5570	1,1-Ethanediol, diacetate / Ethylidene diacetate	$C_6H_{10}O_4$ / 146.14	542-10-9 / 18.9	169	3767	4-02-00-00282 / 1.070[25]	eth 4; EtOH 4 / 1.3985[25]
5571	1,2-Ethanediol, diacetate / Ethylene glycol diacetate	$C_6H_{10}O_4$ / 146.14	111-55-7 / -31	190	3756	4-02-00-00217 / 1.1043[20]	H_2O 4; EtOH 5; eth 5; ace 5 / 1.4159[20]
5572	1,2-Ethanediol, dibenzoate	$C_{16}H_{14}O_4$ / 270.28	94-49-5 / 73.5	360 dec		4-09-00-00356	H_2O 1; eth 3; chl 3
5573	1,2-Ethanediol, diformate	$C_4H_6O_4$ / 118.09	629-15-2 /	174		4-02-00-00037 / 1.193[0]	H_2O 3; EtOH 3; eth 3 / 1.3580
5574	1,2-Ethanediol, dinitrate / Ethylene glycol dinitrate	$C_2H_4N_2O_6$ / 152.06	628-96-6 / -22.3	198.5		4-01-00-02413 / 1.4918[20]	eth 4; EtOH 4
5575	1,2-Ethanediol, 1,2-diphenyl-, (R*,R*)-(±)-	$C_{14}H_{14}O_2$ / 214.26	655-48-1 / 122.5	>300		4-06-00-06682	H_2O 1; EtOH 4; eth 4; ace 3
5576	1,2-Ethanediol, dipropanoate	$C_8H_{14}O_4$ / 174.20	123-80-8 /	211		4-02-00-00715 / 1.020[15]	H_2O 2; EtOH 5; eth 5; ace 4
5577	1,2-Ethanediol, monoacetate / Ethylene glycol, monoacetate	$C_4H_8O_3$ / 104.11	542-59-6 /	188	3757	4-02-00-00214 / 1.108[15]	H_2O 5; EtOH 5; eth 5
5578	1,2-Ethanediol, monobenzoate	$C_9H_{10}O_3$ / 166.18	94-33-7 / 45	150[10]		4-09-00-00354 / 1.1101[30]	EtOH 4
5579	1,2-Ethanediol, 1-phenyl- / Styrene glycol	$C_8H_{10}O_2$ / 138.17	93-56-1 / 67.5	273	8831	4-06-00-05939	H_2O 4; eth 4; bz 4; EtOH 4
5580	1,2-Ethanediol, 1,1,2,2-tetraphenyl- / Benzopinacol	$C_{26}H_{22}O_2$ / 366.46	464-72-2 / 182		1110	4-06-00-07053	H_2O 1; EtOH 2; eth 3; ace 3
5581	1,1-Ethanediol, 2,2,2-tribromo- / Acetaldehyde, tribromo, hydrate	$C_2H_3Br_3O_2$ / 298.76	507-42-6 / 53.5	dec	1373	4-01-00-03155 / 2.5661[40]	eth 4; EtOH 4
5582	1,1-Ethanediol, 2,2,2-trichloro- / Chloral hydrate	$C_2H_3Cl_3O_2$ / 165.40	302-17-0 / 57	96 dec	2081	4-01-00-03143 / 1.9081[20]	H_2O 4; bz 4; eth 4; EtOH 4
5583	Ethanedione, di-2-furanyl-, dioxime / α-Furildioxime	$C_{10}H_8N_2O_4$ / 220.18	522-27-0 / 167		4218	4-19-00-02008	EtOH 2; eth 2; bz 2; lig 2
5584	Ethanedione, diphenyl- / Benzil	$C_{14}H_{10}O_2$ / 210.23	134-81-6 / 94.8	347	1087	4-07-00-02502 / 1.084[102]	H_2O 1; EtOH 4; eth 4; ace 3
5585	Ethanedioyl dichloride / Oxalyl chloride	$C_2Cl_2O_2$ / 126.93	79-37-8 / -16	63.5	6867	4-02-00-01853 / 1.4785[20]	eth 3 / 1.4316[20]
5586	Ethane, 1,1-diproxy- / 1,2-Dipropoxyethane	$C_8H_{18}O_2$ / 146.23	18854-56-3				
5587	1,2-Ethanedisulfonic acid / Ethylenedisulfonic acid	$C_2H_6O_6S_2$ / 190.20	110-04-3 / 173		3678	4-04-00-00078	Diox 4
5588	Ethanedithioamide / Rubeanic acid	$C_2H_4N_2S_2$ / 120.20	79-40-3 / 170 dec		8256	4-02-00-01871	H_2O 2; EtOH 2; con sulf 3
5589	Ethane(dithioic) acid	$C_2H_4S_2$ / 92.19	594-03-6 /	66[85]; 37[15]		4-02-00-00572 / 1.24[20]	H_2O 3; EtOH 4; eth 4; ace 4
5590	1,2-Ethanedithiol / Ethylenedimercaptan	$C_2H_6S_2$ / 94.20	540-63-6 / -41.2	146	3679	4-01-00-02450 / 1.234[20]	H_2O 1; EtOH 3; eth 3; ace 3 / 1.5590[20]
5591	Ethane, 1-ethoxy-2-methoxy- / 1-Ethoxy-2-methoxyethane	$C_5H_{12}O_2$ / 104.15	5137-45-1 /	102.1		3-01-00-02078 / 0.8529[20]	1.3868[20]
5592	Ethane, 1,1'-[ethylidenebis(oxy)]bis[2-chloro-	$C_6H_{12}Cl_2O_2$ / 187.07	14689-97-5 /	195		4-01-00-03104 / 1.1737[20]	eth 4; EtOH 4 / 1.4527[20]
5593	Ethane, fluoro- / Ethyl fluoride	C_2H_5F / 48.06	353-36-6 / -143.2	-37.6		4-01-00-00120 / 0.7182[20]	H_2O 3; EtOH 4; eth 4 / 1.2656[20]
5594	Ethane, hexabromo- / Hexabromoethane	C_2Br_6 / 503.45	594-73-0 /	200 dec		3-01-00-00193 / 3.823[20]	EtOH 2; eth 2; CS_2 2 / 1.863
5595	Ethane, hexachloro- / Hexachloroethane	C_2Cl_6 / 236.74	67-72-1 / 187 (triple point)		4601	4-01-00-00148 / 2.091[20]	H_2O 1; EtOH 4; eth 4; bz 3
5596	Ethane, hexafluoro- / Perfluoroethane	C_2F_6 / 138.01	76-16-4 / -100.7	-78.1		4-01-00-00123 / 1.590[-78]	H_2O 1; EtOH 2; eth 2
5597	Ethane, iodo- / Ethyl iodide	C_2H_5I / 155.97	75-03-6 / -111.1	72.5	3769	4-01-00-00163 / 1.9358[20]	H_2O 1; EtOH 5; eth 3; chl 3 / 1.5133[20]
5598	Ethane, isocyanato- / Ethyl isocyanate	C_3H_5NO / 71.08	109-90-0 /	60		4-04-00-00402 / 0.9031[20]	H_2O 1; EtOH 5; eth 5 / 1.3808[20]
5599	Ethane, isocyano-	C_3H_5N / 55.08	624-79-3 / <-66	79		4-04-00-00342 / 0.7402[20]	H_2O 3; EtOH 5; eth 5; ace 3 / 1.3622[20]
5600	Ethane, isothiocyanato- / Ethyl isothiocyanate	C_3H_5NS / 87.15	542-85-8 / -5.9	131.5	3771	4-04-00-00403 / 0.9990[20]	H_2O 1; EtOH 5; eth 5 / 1.5130[20]

No.	Name / Synonym	Mol. Form. / Mol. Wt.	CAS RN / mp/°C	Merck No. / bp/°C	Beil. Ref. / den/g cm⁻³	Solubility / n_D
5601	Ethane, 1,1',1",1"'-[methanetetrayltetrakis(oxy)]tetrakis-	$C_9H_{20}O_4$ 192.26	78-09-1	159.5	4-03-00-00006 0.9186[20]	EtOH 5; eth 5; ctc 3 1.3905[25]
5602	Ethane, methoxy- Ethyl methyl ether	C_3H_8O 60.10	540-67-0 -113	3783 7.4	4-01-00-01314 0.7251[0]	H_2O 3; EtOH 5; eth 5; ace 3 1.3420[4]
5603	Ethane, 1,1'-[methylenebis(oxy)]bis- Diethoxymethane	$C_5H_{12}O_2$ 104.15	462-95-3 -66.5	88	4-01-00-03027 0.8319[20]	H_2O 3; EtOH 5; eth 5; ace 4 1.3748[18]
5604	Ethane, 1,1',1"-[methylidynetris(oxy)]tris- Ethyl orthoformate	$C_7H_{16}O_3$ 148.20	122-51-0	143; 60[20]	4-02-00-00025 0.8909[20]	EtOH 3; eth 3 1.3922[20]
5605	Ethane, 1,1',1"-[methylidynetris(thio)]tris-	$C_7H_{16}S_3$ 196.40	6267-24-9	235 dec; 127[12]	4-02-00-00093 1.053[20]	eth 4; EtOH 4 1.5410[15]
5606	Ethane, (methylthio)- Ethyl methyl sulfide	C_3H_8S 76.16	624-89-5 -105.9	66.7	4-01-00-01392 0.8422[20]	H_2O 1; EtOH 5; eth 3; chl 3 1.4404[20]
5607	Ethane, nitro- Nitroethane	$C_2H_5NO_2$ 75.07	79-24-3 -89.5	6518 114.0	4-01-00-00170 1.0448[25]	H_2O 2; EtOH 5; eth 5; ace 3 1.3917[20]
5608	Ethane, 1,1'-oxybis- Diethyl ether	$C_4H_{10}O$ 74.12	60-29-7 -116.3	3762 34.5	4-01-00-01314 0.7138[20]	H_2O 2; EtOH 5; eth 5; ace 4 1.3526[20]
5609	Ethane, 1,1'-oxybis[2-bromo-	$C_4H_8Br_2O$ 231.91	5414-19-7	115[32]	4-01-00-01386 1.8222[27]	1.5131[27]
5610	Ethane, 1,1'-oxybis[1-chloro-	$C_4H_8Cl_2O$ 143.01	6986-48-7	116.5	4-01-00-03120 1.1060[25]	eth 4; EtOH 4; chl 4 1.4186[25]
5611	Ethane, 1,1'-oxybis[2-chloro- Bis(2-chloroethyl) ether	$C_4H_8Cl_2O$ 143.01	111-44-4 -51.9	3055 178.5	4-01-00-01375 1.22[20]	H_2O 1; EtOH 3; eth 3; ace 3 1.451[20]
5612	Ethane, 1,1'-oxybis[2-ethoxy- Diethylene glycol diethyl ether	$C_8H_{18}O_3$ 162.23	112-36-7 -45	3108 188	4-01-00-02394 0.9063[20]	H_2O 4; EtOH 4; eth 3; os 4 1.4115[20]
5613	Ethane, 1,1'-oxybis[2-methoxy- Diethylene glycol dimethyl ether	$C_6H_{14}O_3$ 134.18	111-96-6 -68	3148 162	4-01-00-02393 0.9434[20]	H_2O 5; EtOH 5; eth 5 1.4097[20]
5614	Ethane, 1,1'-[oxybis(methyleneoxy)]bis- Bis(ethoxymethyl) ether	$C_6H_{14}O_3$ 134.18	5648-29-3	140.6		
5615	Ethane, 1,1'-oxybis[1,2,2,2-tetrachloro- Bis(1,2,2,2-tetrachloroethyl) ether	$C_4H_2Cl_8O$ 349.68	41284-12-2 40.2	188; 130[11]	2-01-00-00681	bz 4; peth 4
5616	Ethane, pentabromo- 1,1,1,2,2-Pentabromoethane	C_2HBr_5 424.55	75-95-6 56.5	210[300]	3-01-00-00193 3.312[20]	H_2O 1; EtOH 3; eth 4; chl 3
5617	Ethane, pentachloro- Pentachloroethane	C_2HCl_5 202.29	76-01-7 -29	7058 159.8	4-01-00-00147 1.6796[20]	H_2O 1; EtOH 5; eth 5 1.5025[20]
5618	Ethane, pentachlorofluoro-	C_2Cl_5F 220.28	354-56-3 101.3	135	4-01-00-00148 1.74[25]	H_2O 1; EtOH 3; eth 3
5619	Ethaneperoxoic acid Peroxyacetic acid	$C_2H_4O_3$ 76.05	79-21-0 -0.2	7107 110	4-02-00-00390 1.226[15]	H_2O 4; EtOH 3; eth 4; sulf 4 1.3974[20]
5620	Ethane, 1,1'-selenobis-	$C_4H_{10}Se$ 137.08	627-53-2 55	108	4-01-00-01411 1.2300[20]	1.4768[20]
5621	Ethane, 1,1'-sulfinylbis-	$C_4H_{10}OS$ 106.19	70-29-1 14	104[25]; 90[15]	4-01-00-01395 1.0092[22]	H_2O 4; eth 4; EtOH 4
5622	Ethanesulfonic acid Ethylsulfonic acid	$C_2H_6O_3S$ 110.13	594-45-6 -17	123[1]	4-04-00-00033 1.3341[25]	H_2O 4; EtOH 4 1.4335[20]
5623	Ethanesulfonic acid, 2-amino- Taurine	$C_2H_7NO_3S$ 125.15	107-35-7 328	9043	4-04-00-03289	H_2O 4
5624	Ethanesulfonic acid, 2-(methylamino)- N-Methyltaurine	$C_3H_9NO_3S$ 139.18	107-68-6 241.5	6043	3-04-00-01699	H_2O 4; EtOH 1; eth 1
5625	Ethanesulfonic acid, 2-[[(3α,5β,7α,12α)-3,7,12-trihydroxy-24-oxocholan-24-yl]amino]- Taurocholic acid	$C_{26}H_{45}NO_7S$ 515.71	81-24-3 125 dec	9044	4-10-00-02078	H_2O 4; EtOH 4; eth 2; AcOEt 2
5626	Ethane, 1,1'-sulfonylbis-	$C_4H_{10}O_2S$ 122.19	597-35-3 73.5	248	4-01-00-01396 1.357[20]	H_2O 3; eth 3; bz 4; peth 1
5627	Ethanesulfonyl chloride	$C_2H_5ClO_2S$ 128.58	594-44-5	174	4-04-00-00034 1.357[22]	eth 4; CS_2 3 1.4531[20]
5628	Ethanesulfonyl chloride, 2-bromo- 2-Bromoethanesulfonyl chloride	$C_2H_4BrClO_2S$ 207.48	54429-56-0	102[13]	4-04-00-00037 1.921[20]	1.5242[20]
5629	Ethanesulfonyl chloride, 2-chloro-	$C_2H_4Cl_2O_2S$ 163.02	1622-32-8	201.5	4-04-00-00036 1.555[20]	1.4920[20]
5630	Ethane, 1,1'-tellurobis-	$C_4H_{10}Te$ 185.72	627-54-3	137.5	4-01-00-01412 1.599[15]	EtOH 4 1.5182[15]
5631	Ethane, 1,1,1,2-tetrabromo- 1,2,2,2-Tetrabromoethane	$C_2H_2Br_4$ 345.65	630-16-0 0	112[18] dec	4-01-00-00162 2.8747[20]	ace 4; bz 4; eth 4; EtOH 4 1.6277[20]
5632	Ethane, 1,1,2,2-tetrabromo- Acetylene tetrabromide	$C_2H_2Br_4$ 345.65	79-27-6	9121 243.5; 151[54]	4-01-00-00162 2.9655[20]	H_2O 1; EtOH 5; eth 5; ace 3 1.6353[20]
5633	1,1,2,2-Ethanetetracarboxylic acid, tetraethyl ester	$C_{14}H_{22}O_8$ 318.32	632-56-4 77	305 dec	4-02-00-02415 1.064[80]	EtOH 3 1.4105[80]
5634	Ethane, 1,1,1,2-tetrachloro- 1,1,1,2-Tetrachloroethane	$C_2H_2Cl_4$ 167.85	630-20-6 -70.2	130.5	4-01-00-00143 1.5406[20]	H_2O 2; EtOH 5; eth 5; ace 3 1.4821[20]
5635	Ethane, 1,1,2,2-tetrachloro- Acetylene tetrachloride	$C_2H_2Cl_4$ 167.85	79-34-5 -43.8	9125 146.5	4-01-00-00144 1.5953[20]	H_2O 2; EtOH 5; eth 5; ace 3 1.4940[20]
5636	Ethane, 1,1,1,2-tetrachloro-2,2-difluoro- 1,1-Difluoroperchloroethane	$C_2Cl_4F_2$ 203.83	76-11-9 40.6	91.5	4-01-00-00146 1.649[25]	H_2O 1; EtOH 3; eth 3; chl 3
5637	Ethane, 1,1,2,2-tetrachloro-1,2-difluoro- Tetrachloro-1,2-difluoroethane	$C_2Cl_4F_2$ 203.83	76-12-0 26	93	4-01-00-00146 1.6447[25]	H_2O 1; EtOH 3; eth 3; chl 3 1.4130[25]
5638	Ethane, 1,1,2,2-tetrachloro-1-fluoro-	C_2HCl_4F 185.84	354-14-3 -82.6	116	4-01-00-00145 1.5497[17]	1.4390[20]

No.	Name / Synonym	Mol. Form. / Mol. Wt.	CAS RN / mp/°C	Merck No. / bp/°C	Beil. Ref. / den/g cm^{-3}	Solubility / n_D
5639	Ethane, 1,1,1,2-tetrachloro-2-fluoro- / 1,1,1,2-Tetrachloro-2-fluoroethane	C_2HCl_4F / 185.84	354-11-0 / -95.3	116.5		
5640	Ethane, 1,1,1,2-tetrafluoro- / 1,1,1,2-Tetrafluoroethane	$C_2H_2F_4$ / 102.03	811-97-2 / -103.3	-26.08	4-01-00-00123 / 1.2072[25]	H_2O 1; eth 3
5641	Ethane, 1,1,2,2-tetrafluoro- / 1,1,2,2-Tetrafluoroethane	$C_2H_2F_4$ / 102.03	359-35-3 / -89	-19.9		
5642	Ethane, 1,1,2,2-tetrafluoro-1,2-dinitro- / Ethane, 1,2-dinitro-1,1,2,2-tetrafluoro	$C_2F_4N_2O_4$ / 192.03	356-16-1 / -41.5	58.5	3-01-00-00203 / 1.6024[25]	H_2O 1; ace 3 / 1.3265[25]
5643	Ethanethioamide / Thioacetamide	C_2H_5NS / 75.13	62-55-5 / 115.5	9246	4-01-00-00055	H_2O 4; EtOH 4; eth 2; bz 2
5644	Ethane, 1,1'-thiobis- / Diethyl sulfide	$C_4H_{10}S$ / 90.19	352-93-2 / -103.9	3809 / 92.1	4-01-00-01394 / 0.8362[20]	H_2O 2; EtOH 3; eth 3; ctc 2 / 1.4430[20]
5645	Ethane, 1,1'-thiobis[2-chloro- / Mustard gas	$C_4H_8Cl_2S$ / 159.08	505-60-2 / 13.5	6225 / 216	4-01-00-01407 / 1.2741[20]	1.5313[20]
5646	Ethanethioic acid / Thioacetic acid	C_2H_4OS / 76.12	507-09-5 / <-17	9247 / 93; 26[35]	4-02-00-00542 / 1.064[20]	H_2O 3; EtOH 4; eth 5; ace 4 / 1.4648[20]
5647	Ethanethioic acid, S-ethyl ester	C_4H_8OS / 104.17	625-60-5	116.4	4-02-00-00543 / 0.9792[20]	H_2O 1; EtOH 4; eth 4 / 1.4583[21]
5648	Ethanethiol / Ethyl mercaptan	C_2H_6S / 62.14	75-08-1 / -147.8	3680 / 35.1	4-01-00-01390 / 0.8315[25]	H_2O 2; EtOH 3; eth 3; ace 3 / 1.4310[20]
5649	Ethanethiol, 2-amino- / Cysteamine	C_2H_7NS / 77.15	60-23-1 / 99.5	2785 / dec	4-04-00-01570	H_2O 4; EtOH 4
5650	Ethanethiol, 2-(butylamino)- / 2-(Butylamino)ethanethiol	$C_6H_{15}NS$ / 133.26	5842-00-2	112-5[10]	4-04-00-01602 / 0.901[25]	1.4711[20]
5651	Ethanethiol, 2-chloro- / 2-Chloroethanethiol	C_2H_5ClS / 96.58	4325-97-7	113	4-01-00-01406 / 1.1826[20]	eth 4; EtOH 4; Diox 4 / 1.4929[20]
5652	Ethanethiol, 2,2'-thiobis-	$C_4H_{10}S_3$ / 154.32	3570-55-6	135-6[10]	4-01-00-02455 / 1.183[25]	1.5982[20]
5653	Ethane, 1,1,2-tribromo- / 1,1,2-Tribromoethane	$C_2H_3Br_3$ / 266.76	78-74-0 / -29.3	188.93	4-01-00-00161 / 2.6210[20]	H_2O 1; EtOH 3; eth 3; bz 3 / 1.5933[20]
5654	1,1,2-Ethanetricarboxylic acid, triethyl ester	$C_{11}H_{18}O_6$ / 246.26	7459-46-3	99[05]	4-02-00-02364 / 1.074[25]	1.4290[20]
5655	Ethane, 1,1,1-trichloro- / Methylchloroform	$C_2H_3Cl_3$ / 133.40	71-55-6 / -30.4	9549 / 74.0	4-01-00-00138 / 1.3390[20]	H_2O 2; EtOH 3; eth 5; chl 3 / 1.4379[20]
5656	Ethane, 1,1,2-trichloro- / 1,1,2-Trichloroethane	$C_2H_3Cl_3$ / 133.40	79-00-5 / -36.6	9550 / 113.8	4-01-00-00139 / 1.4397[20]	H_2O 1; EtOH 3; eth 5; chl 3 / 1.4714[20]
5657	Ethane, 1,1,1-trichloro-2,2-diethoxy- / Chloral, diethyl acetal	$C_6H_{11}Cl_3O_2$ / 221.51	599-97-3	197	4-01-00-03144 / 1.266[20]	H_2O 2; EtOH 5; eth 5; glycerol 5 / 1.4586[25]
5658	Ethane, 1,1,1-trichloro-2,2-difluoro / 1,1,1-Trichloro-2,2-difluoroethane	$C_2HCl_3F_2$ / 109.30	354-12-1	73		
5659	Ethane, 1,2,2-trichloro-1,1-difluoro-	$C_2HCl_3F_2$ / 169.38	354-21-2	71.9	4-01-00-00141 / 1.5447[20]	1.3889[20]
5660	Ethane, 1,2,2-trichloro-1,2-difluoro- / 1,2,2-Trichloro-1,2-difluoroethane	$C_2HCl_3F_2$ / 169.38	354-15-4	72.5		
5661	Ethane, 1,1,2-trichloro-2-fluoro-	$C_2H_2Cl_3F$ / 151.39	359-28-4	102	4-01-00-00141 / 1.5393[20]	H_2O 1 / 1.4390[20]
5662	Ethane, 1,1,1-trichloro-2,2,2-trifluoro- / 1,1,1-Trichlorotrifluoroethane	$C_2Cl_3F_3$ / 187.38	354-58-5 / 14.2	46.1	4-01-00-00142 / 1.5790[20]	H_2O 1; EtOH 3; eth 3; chl 3 / 1.3610[35]
5663	Ethane, 1,1,2-trichloro-1,2,2-trifluoro- / 1,1,2-Trichlorotrifluoroethane	$C_2Cl_3F_3$ / 187.38	76-13-1 / -35	47.7	4-01-00-00142 / 1.5635[25]	H_2O 1; EtOH 3; eth 5; bz 5 / 1.3557[25]
5664	Ethane, 1,1,1-triethoxy-	$C_8H_{18}O_3$ / 162.23	78-39-7	145	4-02-00-00137 / 0.8847[25]	H_2O 1; EtOH 5; eth 5; ctc 5 / 1.3980[20]
5665	Ethane, 1,1,1-trifluoro- / Methylfluoroform	$C_2H_3F_3$ / 84.04	420-46-2 / -111.3	-47.5	4-01-00-00122	eth 3; chl 3
5666	Ethane, 1,1,2-trifluoro- / 1,1,2-Trifluoroethane	$C_2H_3F_3$ / 84.04	430-66-0 / -84	5	4-01-00-00123	
5667	Ethane, 1,1,1-trifluoro-2-iodo-	$C_2H_2F_3I$ / 209.94	353-83-3	54.5	4-01-00-00166 / 2.13[25]	1.4009[20]
5668	Ethane, 1,1,1-trifluoro-2-nitro-	$C_2H_2F_3NO_2$ / 129.04	819-07-8	96	4-01-00-00172 / 1.3914[20]	eth 4 / 1.3394[20]
5669	Ethane, 1,1,1-trimethoxy-	$C_5H_{12}O_3$ / 120.15	1445-45-0	108	4-02-00-00127 / 0.9438[25]	eth 4; EtOH 4 / 1.3859[25]
5670	Ethanimidamide, N,N'-bis(4-ethoxyphenyl)-, monohydrochloride / Holocaine, hydrochloride	$C_{18}H_{23}ClN_2O_2$ / 334.85	620-99-5 / 191	7153	3-13-00-01069	H_2O 4; EtOH 4; chl 4
5671	Ethanimidamide, monohydrochloride / Acetamidine, hydrochloride	$C_2H_7ClN_2$ / 94.54	124-42-5 / 177.5	37	4-02-00-00428	H_2O 4; EtOH 4
5672	Ethanol / Ethyl alcohol	C_2H_6O / 46.07	64-17-5 / -114.1	3716 / 78.2	4-01-00-01289 / 0.7893[20]	H_2O 5; EtOH 5; eth 5; ace 5 / 1.3611[20]
5673	Ethanol, 2-[2-(acetyloxy)ethoxy]-	$C_6H_{12}O_4$ / 148.16	2093-20-1	114-5[7]	2-02-00-00155 / 1.1208[20]	1.4320[20]
5674	Ethanol, 1-amino- / Acetaldehyde ammonia	C_2H_7NO / 61.08	75-39-8 / 97	33 / 110 dec		H_2O 3; eth 2
5675	Ethanol, 2-amino- / Ethanolamine	C_2H_7NO / 61.08	141-43-5 / 10.5	3681 / 171	4-04-00-01406 / 1.0180[20]	H_2O 5; EtOH 5; eth 2; bz 2 / 1.4541[20]
5676	Ethanol, 2-[(2-aminoethyl)amino]- / N-(Aminoethyl)ethanolamine	$C_4H_{12}N_2O$ / 104.15	111-41-1	239; 105[10]	4-04-00-01558 / 1.0286[20]	H_2O 5; EtOH 5; ace 3; bz 2 / 1.4863[20]
5677	Ethanol, 2-[(2-amino-2-methylpropyl)amino]-	$C_6H_{16}N_2O$ / 132.21	68750-16-3 / 5	236		chl 3 / 1.4698[25]

No.	Name / Synonym	Mol. Form. / Mol. Wt.	CAS RN / mp/°C	Merck No. / bp/°C	Beil. Ref. / den/g cm^{-3}	Solubility / n_D
5678	Ethanol, 2-[2-[2-(3-aminopropoxy)ethoxy]ethoxy]-	$C_9H_{21}NO_4$ 207.27	49542-66-7 -50	184[10]	1.0663[20]	H_2O 4; EtOH 4 1.4668[20]
5679	Ethanol, 2,2'-[(3-aminopropyl)imino]bis- N-(3-Aminopropyl)diethanolamine	$C_7H_{18}N_2O_2$ 162.23	4985-85-7	160[1]	3-04-00-00720	
5680	Ethanol, 1-amino-2,2,2-trichloro- Chloral ammonia	$C_2H_4Cl_3NO$ 164.42	507-47-1 73	2057 100 dec	4-01-00-03147	bz 4; eth 4; EtOH 4
5681	Ethanol, 2-azido-	$C_2H_5N_3O$ 87.08	1517-05-1	75[40]	4-01-00-01389 1.146[24]	H_2O 4
5682	Ethanol, 2-[bis(3-methylbutyl)amino]- Diisopentyl 2-hydroxyethylamine	$C_{12}H_{27}NO$ 201.35	3574-43-4	248	3-04-00-00685 0.8492[20]	eth 4; EtOH 4; chl 4 1.4435[20]
5683	Ethanol, 2-[bis(1-methylethyl)amino]- N,N-Diisopropyl-2-aminoethanol	$C_8H_{19}NO$ 145.25	96-80-0	190	4-04-00-01504 0.826[25]	1.4417[20]
5684	Ethanol, 2-[bis(2-methylpropyl)amino]- Ethanolamine, N,N diisobutyl	$C_{10}H_{23}NO$ 173.30	4535-66-4	214	4-04-00-01509 0.8407[20]	eth 4; EtOH 4; lig 4; chl 4 1.4355[20]
5685	Ethanol, 2-bromo- Ethylene bromohydrin	C_2H_5BrO 124.97	540-51-2	3749 150; 51[4]	4-01-00-01385 1.7629[20]	H_2O 5; EtOH 5; eth 5; os 4 1.4915[20]
5686	Ethanol, 2-bromo-, acetate	$C_4H_7BrO_2$ 167.00	927-68-4 -13.8	162.5	4-02-00-00136 1.514[20]	H_2O 4; EtOH 5; eth 5; chl 4 1.457[23]
5687	Ethanol, 2-butoxy- Ethylene glycol monobutyl ether	$C_6H_{14}O_2$ 118.18	111-76-2 -74.8	1559 168.4	4-01-00-02380 0.9015[20]	H_2O 5; EtOH 5; eth 5; ctc 2 1.4198[20]
5688	Ethanol, 2-(2-butoxyethoxy)- Diethylene glycol monobutyl ether	$C_8H_{18}O_3$ 162.23	112-34-5 -68	1557 231	4-01-00-02394 0.9553[20]	H_2O 5; EtOH 4; eth 4; ace 4 1.4306[20]
5689	Ethanol, 2-(2-butoxyethoxy)-, acetate	$C_{10}H_{20}O_4$ 204.27	124-17-4 -32	245	3-02-00-00308 0.985[20]	ace 4; eth 4; EtOH 4 1.4262[20]
5690	Ethanol, 2-[2-(2-butoxyethoxy)ethoxy]-	$C_{10}H_{22}O_4$ 206.28	143-22-6	278	4-01-00-02402 0.9890[20]	EtOH 4; MeOH 4 1.4389[20]
5691	Ethanol, 2-(butylamino)-	$C_6H_{15}NO$ 117.19	111-75-1	199; 91-2[11]	4-04-00-00682 0.8907[20]	H_2O 4; EtOH 4; eth 4 1.4437[20]
5692	Ethanol, 2,2'-(butylimino)bis-	$C_8H_{19}NO_2$ 161.24	102-79-4	275; 80[35]	4-04-00-01520 0.9592[20]	chl 3 1.4625[20]
5693	Ethanol, 2-chloro- Ethylene chlorohydrin	C_2H_5ClO 80.51	107-07-3 -67.5	3750 128.6	4-01-00-01372 1.2019[20]	H_2O 5; EtOH 5; eth 2; chl 3 1.4419[20]
5694	Ethanol, 1-chloro-, acetate	$C_4H_7ClO_2$ 122.55	5912-58-3	123	4-02-00-00282 1.110[20]	eth 4 1.409[20]
5695	Ethanol, 2-chloro-, acetate β-Chloroethyl acetate	$C_4H_7ClO_2$ 122.55	542-58-5	2137 145	4-02-00-00136 1.178[20]	H_2O 1; EtOH 5; eth 5; ctc 3 1.4234[20]
5696	Ethanol, 2-chloro-, benzoate	$C_9H_9ClO_2$ 184.62	939-55-9 118	255	4-09-00-00288 1.1789[20]	eth 4; EtOH 4
5697	Ethanol, 2-chloro-, carbonate (2:1) Bis(2-chloroethyl) carbonate	$C_5H_8Cl_2O_3$ 187.02	623-97-2 8	241	4-03-00-00006 1.3506[20]	H_2O 1 1.461[20]
5698	Ethanol, 2-(2-chloroethoxy)-	$C_4H_9ClO_2$ 124.57	628-89-7	79-81[5]	3-01-00-02078 1.18[25]	H_2O 4; EtOH 5; eth 5 1.4529[20]
5699	Ethanol, 2-chloro-, methanesulfonate	$C_3H_7ClO_3S$ 158.61	3570-58-9	125-6[9]	4-04-00-00012 1.39[25]	chl 3 1.4562[20]
5700	Ethanol, 2-[2-[4-[(4-chlorophenyl)phenylmethyl]-1-piperazinyl]ethoxy]- Hydroxyzine	$C_{21}H_{27}ClN_2O_2$ 374.91	68-88-2	4786	5-23-01-00462	
5701	Ethanol, 2-chloro-, phosphate (3:1)	$C_6H_{12}Cl_3O_4P$ 285.49	115-96-8	330; 194[10]	4-01-00-01379 1.39[25]	ctc 3 1.4721[20]
5702	Ethanol, 2-chloro-, phosphite (3:1)	$C_6H_{12}Cl_3O_3P$ 269.49	140-08-9	112-5[2]	4-01-00-01378 1.328[25]	1.4868[20]
5703	Ethanol, 2-(dibutylamino)-	$C_{10}H_{23}NO$ 173.30	102-81-8	114[18]	4-04-00-01506	
5704	Ethanol, 2,2-dichloro- 2,2-Dichloroethanol	$C_2H_4Cl_2O$ 114.96	598-38-9	146	4-01-00-01383 1.4040[25]	H_2O 2; EtOH 3; eth 3; ctc 2 1.4626[20]
5705	Ethanol, 2,2-diethoxy-	$C_6H_{14}O_3$ 134.18	621-63-6	167	4-01-00-03958 0.9[25]	eth 4; EtOH 4 1.4073[20]
5706	Ethanol, 2-(diethylamino)- 2-(Diethylamino)ethanol	$C_6H_{15}NO$ 117.19	100-37-8	3101 163	4-04-00-01471 0.8921[20]	H_2O 5; EtOH 3; eth 3; ace 3 1.4412[20]
5707	Ethanol, 2-[2-(diethylamino)ethoxy]-	$C_8H_{19}NO_2$ 161.24	140-82-9	221.5; 92[7]	4-04-00-01474 0.9421[25]	1.4480[20]
5708	Ethanol, 2,2-difluoro-	$C_2H_4F_2O$ 82.05	359-13-7 -28.2	95.5	4-01-00-01369 1.3084[17]	ace 4; eth 4; EtOH 4 1.3345[11]
5709	Ethanol, 2-(dimethylamino)- Deanol	$C_4H_{11}NO$ 89.14	108-01-0 -59	2834 134	4-04-00-01424 0.8866[20]	H_2O 5; EtOH 5; eth 5; chl 3 1.4300[20]
5710	Ethanol, 2-[(1,1-dimethylethyl)amino]-	$C_6H_{15}NO$ 117.19	4620-70-6 44	176.5; 72[14]	4-04-00-01509 0.8818[20]	
5711	Ethanol, 2,2',2'',2'''-[(4,8-di-1-piperidinylpyrimido[5,4-d]pyrimidine-Dipyridamole	$C_{24}H_{40}N_8O_4$ 504.63	58-32-2 163	3354	4-26-00-03840	
5712	Ethanol, 2-(dipropylamino)-	$C_8H_{19}NO$ 145.25	3238-75-3	196	4-04-00-01502 0.8576[20]	eth 4; EtOH 4 1.4402[20]
5713	Ethanol, 2,2'-[1,2-ethanediylbis(oxy)]bis- Triethylene glycol	$C_6H_{14}O_4$ 150.17	112-27-6 -7	9585 285	4-01-00-02400 1.1274[15]	H_2O 5; EtOH 5; eth 2; bz 5 1.4531[20]
5714	Ethanol, 2,2'-[1,2-ethanediylbis(oxy)]bis-, diacetate Triethylene glycol, diacetate	$C_{10}H_{18}O_6$ 234.25	111-21-7 -50	286	4-02-00-00215 1.1153[20]	H_2O 4; eth 4; EtOH 4
5715	Ethanol, 2,2'-(1,2-ethanediyldiimino)bis-	$C_6H_{16}N_2O_2$ 148.21	4439-20-7 97.5	135-7[1]	4-04-00-01559	H_2O 3

No.	Name Synonym	Mol. Form. Mol. Wt.	CAS RN mp/°C	Merck No. bp/°C	Beil. Ref. den/g cm⁻³	Solubility n_D
5716	Ethanol, 2-(ethenyloxy)-	$C_4H_8O_2$ 88.11	764-48-7 	 141.6	4-01-00-02387 0.9821[20]	H_2O 3; EtOH 3; eth 3; bz 3 1.4564[17]
5717	Ethanol, 2-ethoxy- Ethylene glycol monoethyl ether	$C_4H_{10}O_2$ 90.12	110-80-5 -70	3707 135	4-01-00-02377 0.9297[20]	H_2O 4; ace 4; eth 4; EtOH 4 1.4080[20]
5718	Ethanol, 2-ethoxy-, acetate Ethylene glycol monoethyl ether acetate	$C_6H_{12}O_3$ 132.16	111-15-9 -61.7	3708 156.4	4-02-00-00214 0.9740[20]	H_2O 4; ace 4; eth 4; EtOH 4 1.4054[20]
5719	Ethanol, 2-ethoxy-, carbonate (2:1)	$C_9H_{18}O_5$ 206.24	2049-74-3 	 245.5	3-03-00-00017 1.0439[20]	ace 4; eth 4; EtOH 4 1.4227[20]
5720	Ethanol, 2-(2-ethoxyethoxy)- Diethylene glycol monoethyl ether	$C_6H_{14}O_3$ 134.18	111-90-0 	1806 196	4-01-00-02393 0.9885[20]	H_2O 5; EtOH 5; eth 4; ace 5 1.4300[20]
5721	Ethanol, 2-(2-ethoxyethoxy)-, acetate Diethylene glycol monoethyl ether acetate	$C_8H_{16}O_4$ 176.21	112-15-2 -25	3-02-00-00308 218.5	1.0096[20]	H_2O 4; ace 4; eth 4; EtOH 4 1.4213[20]
5722	Ethanol, 2-(ethylamino)-	$C_4H_{11}NO$ 89.14	110-73-6 	 169.5	4-04-00-01465 0.914[20]	H_2O 4; EtOH 4; eth 4; chl 3 1.444[20]
5723	Ethanol, 2,2'-(ethylimino)bis-	$C_6H_{15}NO_2$ 133.19	139-87-7 -50	3-04-00-00693 247	1.0135[20]	H_2O 4; EtOH 4; eth 2 1.4663[20]
5724	Ethanol, 2-[ethyl(3-methylphenyl)amino]-	$C_{11}H_{17}NO$ 179.26	91-88-3 	4-12-00-01819 118[1.5]	 	ctc 3 1.5540[20]
5725	Ethanol, 2-(ethylsulfonyl)- Ethylsulfonylethyl alcohol	$C_4H_{10}O_3S$ 138.19	513-12-2 	3810 	4-01-00-02430 	chl 2
5726	Ethanol, 2-(ethylthio)- 2-(Ethylthio)ethanol	$C_4H_{10}OS$ 106.19	110-77-0 -100	3813 184	4-01-00-02430 1.0166[20]	H_2O 2; EtOH 3; ace 4 1.4867[20]
5727	Ethanol, 2-fluoro- Ethylene fluorohydrin	C_2H_5FO 64.06	371-62-0 -26.4	 103.5	4-01-00-01366 1.1040[20]	H_2O 5; EtOH 5; eth 5; ace 4 1.3647[18]
5728	Ethanol, 2-(hexyloxy)-	$C_8H_{18}O_2$ 146.23	112-25-4 -45.1	 208	4-01-00-02383 0.8878[20]	H_2O 2; EtOH 4; eth 4 1.4291[20]
5729	Ethanol, 2-hydrazino- 2-Hydrazinoethanol	$C_2H_8N_2O$ 76.10	109-84-2 -70	4696 219; 120[17.5]	4-04-00-03348 1.119[25]	H_2O 4; EtOH 4; MeOH 4
5730	Ethanol, 2,2'-iminobis- Diethanolamine	$C_4H_{11}NO_2$ 105.14	111-42-2 28	3097 268.8	4-04-00-01514 1.0966[20]	H_2O 4; EtOH 4; eth 2; bz 2 1.4776[20]
5731	Ethanol, 2-iodo-	C_2H_5IO 171.97	624-76-0 	 176 dec	4-01-00-01387 2.1967[20]	H_2O 4; eth 4; EtOH 4 1.5713[20]
5732	Ethanol, 2-mercapto- 2-Mercaptoethanol	C_2H_6OS 78.14	60-24-2 	5760 158; 55[13]	4-01-00-02428 1.1143[20]	H_2O 3; EtOH 3; eth 3; bz 3 1.4996[20]
5733	Ethanol, 2-methoxy- Ethylene glycol monomethyl ether	$C_3H_8O_2$ 76.10	109-86-4 -85.1	5961 124.1	4-01-00-02375 0.9647[20]	H_2O 4; EtOH 4; eth 5; ace 3 1.4024[20]
5734	Ethanol, methoxy-, acetate Acetic acid, 2-methoxyethyl ester	$C_5H_{10}O_3$ 118.13	32718-56-2 	 145; 40[12]	2-02-00-00154 1.007[419]	H_2O 3; EtOH 3; eth 3; ctc 2 1.4002[20]
5735	Ethanol, 2-methoxy-, acetate 2-Methoxyethyl acetate	$C_5H_{10}O_3$ 118.13	110-49-6 -70	5962 143	4-02-00-00214 1.007[419]	H_2O 4; EtOH 4 1.4002[20]
5736	Ethanol, 2-methoxy-, carbonate (2:1) Bis(2-methoxyethyl) carbonate	$C_7H_{14}O_5$ 178.19	020-04-6 	 231	3-03-00-00017 1.0988[20]	EtOH 3; eth 3; ace 3 1.4204[20]
5737	Ethanol, 2-(2-methoxyethoxy)- 2-(2-Methoxyethoxy)ethanol	$C_5H_{12}O_3$ 120.15	111-77-3 	5959 193	4-01-00-02392 1.035[20]	H_2O 5; EtOH 4; eth 4; ace 5 1.4264[20]
5738	Ethanol, 2-(methylamino)- N-Methyl-2-ethanolamine	C_3H_9NO 75.11	109-83-1 	5939 158	4-04-00-01422 0.937[20]	H_2O 5; EtOH 5; eth 5 1.4385[20]
5739	Ethanol, 2-[(3-methylbutyl)amino]- N-Isopentylethanolamine	$C_7H_{17}NO$ 131.22	34240-76-1 	 210; 82[5.5]	2-04-00-00728 0.8822[20]	H_2O 4; eth 4; EtOH 4 1.4447[20]
5740	Ethanol, 2-(1-methylethoxy)- 2-Isopropoxyethanol	$C_5H_{12}O_2$ 104.15	109-59-1 	 145	4-01-00-02380 0.9030[20]	H_2O 5; EtOH 5; eth 5; ace 3 1.4095[20]
5741	Ethanol, 2-[(1-methylethyl)amino]-	$C_5H_{13}NO$ 103.16	109-56-8 128.5	 173	3-04-00-00681 0.8970[20]	H_2O 5; EtOH 5; eth 5 1.4395[20]
5742	Ethanol, 2,2'-(methylimino)bis-	$C_5H_{13}NO_2$ 119.16	105-59-9 -21	 247	4-04-00-01517 1.038[25]	H_2O 4 1.4685[20]
5743	Ethanol, 2-(methylphenylamino)-	$C_9H_{13}NO$ 151.21	93-90-3 	 150[14]	4-12-00-00280 0.9993[15]	H_2O 3; EtOH 4; eth 4; ace 4
5744	Ethanol, 2-[(2-methylphenyl)amino]-	$C_9H_{13}NO$ 151.21	136-80-1 	 285.5	4-12-00-01750 1.0794[20]	eth 4; EtOH 4 1.5675[20]
5745	Ethanol, 2-[(4-methylphenyl)amino]-	$C_9H_{13}NO$ 151.21	2933-74-6 42.5	 287	4-12-00-01880 	bz 4; eth 4; EtOH 4; chl 4
5746	Ethanol, 2-(2-methylpropoxy)-	$C_6H_{14}O_2$ 118.18	4439-24-1 	 160	4-01-00-02382 0.8900[20]	 1.4143[20]
5747	Ethanol, 2-[(2-methylpropyl)amino]- Ethanolamine, N-isobutyl	$C_6H_{15}NO$ 117.19	17091-40-6 	 199; 90[16]	3-04-00-00683 0.8818[20]	H_2O 4; EtOH 4; eth 4 1.4402[20]
5748	Ethanol, 2-[(2-methylpropyl)amino]-, 4-aminobenzoate (ester) Butethamine	$C_{13}H_{20}N_2O_2$ 236.31	2090-89-3 	1517 		
5749	Ethanol, 2-(methylthio)-	C_3H_8OS 92.16	5271-38-5 	 68-70[20]	4-01-00-02429 1.063[20]	H_2O 4; eth 4; EtOH 4 1.4861[30]
5750	Ethanol, 2-(2-naphthalenylamino)- Ethanolamine, N-(β-naphthyl)	$C_{12}H_{13}NO$ 187.24	36190-77-9 52	 197-8[3]	4-12-00-03131 	eth 4
5751	Ethanol, 2,2',2''-nitrilotris- Triethanolamine	$C_6H_{15}NO_3$ 149.19	102-71-6 20.5	9581 335.4	4-04-00-01524 1.1242[20]	H_2O 5; EtOH 5; eth 2; bz 2 1.4852[20]
5752	Ethanol, 2-nitro- 2-Nitroethanol	$C_2H_5NO_3$ 91.07	625-48-9 -80	 194; 102[10]	4-01-00-01388 1.270[15]	H_2O 5; EtOH 5; eth 5; bz 1 1.4438[19]
5753	Ethanol, 2,2'-oxybis- Diethylene glycol	$C_4H_{10}O_3$ 106.12	111-46-6 -10.4	3109 245.8	4-01-00-02390 1.1197[15]	H_2O 3; EtOH 3; eth 3; chl 3 1.4472[20]
5754	Ethanol, 2,2'-oxybis-, diacetate Diethylene glycol, diacetate	$C_8H_{14}O_5$ 190.20	628-68-2 18	 200	4-02-00-00216 1.1068[15]	EtOH 4 1.4348[20]
5755	Ethanol, 2,2'-oxybis-, dibenzoate	$C_{18}H_{18}O_5$ 314.34	120-55-8 33.5	 280[24]; 250[1]	4-09-00-00356 1.1690[15]	H_2O 4; EtOH 4

No.	Name / Synonym	Mol. Form. / Mol. Wt.	CAS RN / mp/°C	Merck No. / bp/°C	Beil. Ref. / den/g cm^{-3}	Solubility / n_D
5756	Ethanol, 2,2'-[oxybis(2,1-ethanediyloxy)]bis- Tetraethylene glycol	$C_8H_{18}O_5$ 194.23	112-60-7 -6.2	328	4-01-00-02403 1.1285[15]	H$_2$O 4; EtOH 3; eth 3; ctc 3 1.4577[20]
5757	Ethanol, 2-(pentyloxy)-	$C_7H_{16}O_2$ 132.20	6196-58-3 	182	4-01-00-02383 0.8918[15]	1.4213[25]
5758	Ethanol, 2-phenoxy- 2-Phenoxyethanol	$C_8H_{10}O_2$ 138.17	122-99-6 14	7226 245	4-06-00-00571 1.102[22]	H$_2$O 1; EtOH 3; eth 3; chl 3 1.534[20]
5759	Ethanol, 2-(phenylamino)-	$C_8H_{11}NO$ 137.18	122-98-5 	279.5; 150[10]	4-12-00-00277 1.0945[20]	H$_2$O 2; EtOH 4; eth 4; chl 4 1.5760[20]
5760	Ethanol, 2,2'-(phenylimino)bis-	$C_{10}H_{15}NO_2$ 181.23	120-07-0 57	200[10]	4-12-00-00285 1.201[60]	ace 4; bz 4; eth 4; EtOH 4
5761	Ethanol, 2-(phenylmethoxy)-	$C_9H_{12}O_2$ 152.19	622-08-2 <-75	256	4-06-00-02241 1.0640[20]	H$_2$O 4; eth 4; EtOH 4 1.5233[20]
5762	Ethanol, 2-[(phenylmethyl)amino]-	$C_9H_{13}NO$ 151.21	104-63-2 	153-6[12]	4-12-00-02184 1.065[25]	1.5430[20]
5763	Ethanol, 2-(2-propenyloxy)-	$C_5H_{10}O_2$ 102.13	111-45-5 	158.5	4-01-00-02388 0.9580[20]	H$_2$O 5; EtOH 4; bz 3; ctc 3 1.4358[20]
5764	Ethanol, 2-propoxy- 2-Propoxyethanol	$C_5H_{12}O_2$ 104.15	2807-30-9 	149.8	4-01-00-02379 0.9112[20]	H$_2$O 3; EtOH 4; eth 4 1.4133[20]
5765	Ethanol, 2-(propylamino)-	$C_5H_{13}NO$ 103.16	16369-21-4 	183	3-04-00-00681 0.9005[20]	1.4428[20]
5766	Ethanol, 2,2'-thiobis- 2,2'-Thiodiethanol	$C_4H_{10}O_2S$ 122.19	111-48-8 -10.2	9259 282	4-01-00-02437 1.1793[25]	H$_2$O 5; EtOH 5; eth 3; bz 2 1.5211[20]
5767	Ethanol, 2,2,2-tribromo- Tribromoethanol	$C_2H_3Br_3O$ 282.76	75-80-9 81	9525 92-3[10]	3-01-00-01362	bz 4; eth 4; EtOH 4
5768	Ethanol, 2,2,2-trichloro- 2,2,2-Trichloroethanol	$C_2H_3Cl_3O$ 149.40	115-20-8 19	9551 152; 52[11]	4-01-00-01383 	H$_2$O 2; EtOH 5; eth 5; ctc 2 1.4861[20]
5769	Ethanol, 2,2,2-trichloro-1-ethoxy- Chloral alcoholate	$C_4H_7Cl_3O_2$ 193.46	515-83-3 56.5	2056 115.5	4-01-00-03144 1.143[40]	H$_2$O 3; EtOH 3; eth 3
5770	Ethanol, 2,2,2-trifluoro-	$C_2H_3F_3O$ 100.04	75-89-8 -43.5	74	4-01-00-01370 1.3842[20]	EtOH 4; eth 3; ace 3; bz 3 1.2907[22]
5771	Ethanone, 2-(acetyloxy)-1-phenyl-	$C_{10}H_{10}O_3$ 178.19	2243-35-8 49	270	4-08-00-00368 1.1169[65]	H$_2$O 1; EtOH 4; eth 4; bz 2 1.5036[65]
5772	Ethanone, 1-(3-aminophenyl)-	C_8H_9NO 135.17	99-03-6 98.5	289.5	4-14-00-00096	H$_2$O 2; EtOH 3
5773	Ethanone, 1-(4-aminophenyl)-	C_8H_9NO 135.17	99-92-3 106	294; 195[15]	4-14-00-00100	eth 4; EtOH 4
5774	Ethanone, 2-amino-1-phenyl- Phenacylamine	C_8H_9NO 135.17	613-89-8 20	7159 251	4-14-00-00114	H$_2$O 1; eth 3; ctc 2 1.6160[20]
5775	Ethanone, 1-(2-benzofuranyl)-	$C_{10}H_8O_2$ 160.17	1646-26-0 76	126[11]	5-17-10-00172	H$_2$O 3
5776	Ethanone, 1-[1,1'-biphenyl]-4-yl-	$C_{14}H_{12}O$ 196.25	92-91-1 121	326	4-07-00-01407 1.2510[0]	H$_2$O 1; EtOH 4; ace 4; chl 2
5777	Ethanone, 2-bromo-1-(4-bromophenyl)- p-Bromophenacyl bromide	$C_8H_6Br_2O$ 277.94	99-73-0 111	1415	4-07-00-00652	H$_2$O 1; EtOH 4; eth 3; chl 3
5778	Ethanone, 2-bromo-1-(4-chlorophenyl)- p-Chlorophenacyl bromide	C_8H_6BrClO 233.49	536-38-9 96.5	2153	4-07-00-00651	
5779	Ethanone, 2-bromo-1-(4-methylphenyl)-	C_9H_9BrO 213.07	619-41-0 51	157[14]	4-07-00-00705	eth 4; EtOH 4
5780	Ethanone, 1-(3-bromophenyl)-	C_8H_7BrO 199.05	2142-63-4 7.5	133[19]	4-07-00-00646	H$_2$O 1; ace 3; bz 3 1.5755[20]
5781	Ethanone, 1-(4-bromophenyl)- p-Bromoacetophenone	C_8H_7BrO 199.05	99-90-1 50.5	1390 257; 130[11]	4-07-00-00647 1.647[25]	H$_2$O 1; EtOH 3; eth 3; bz 3 1.647
5782	Ethanone, 2-bromo-1-phenyl- ω-Bromacetophenone	C_8H_7BrO 199.05	70-11-1 50.5	1391 135[18]	4-07-00-00649 1.647[20]	H$_2$O 1; EtOH 4; eth 4; bz 4
5783	Ethanone, 1-(5-bromo-2-thienyl)-	C_6H_5BrOS 205.08	5370-25-2 94.5	103[4]	5-17-09-00392	EtOH 2; ctc 3
5784	Ethanone, 2-chloro-1-(4-chlorophenyl)-	$C_8H_6Cl_2O$ 189.04	937-20-2 101.5	270	4-07-00-00644	EtOH 3; bz 3; MeOH 3
5785	Ethanone, 2-chloro-1,2-diphenyl-	$C_{14}H_{11}ClO$ 230.69	447-31-4 68.5	dec	3-07-00-02106	EtOH 3; chl 2; alk 1
5786	Ethanone, 2-chloro-1-(4-hydroxy-3-methoxyphenyl)-	$C_9H_9ClO_3$ 200.62	6344-28-1 102	208[22]	3-08-00-02114	H$_2$O 2; EtOH 3
5787	Ethanone, 2-chloro-1-(4-methylphenyl)-	C_9H_9ClO 168.62	4209-24-9 57.5	261.5	4-07-00-00704	eth 4; EtOH 4
5788	Ethanone, 1-(3-chlorophenyl)-	C_8H_7ClO 154.60	99-02-5 	244; 129[30]	4-07-00-00638 1.2130[40]	EtOH 3; eth 3; ace 3 1.5494[20]
5789	Ethanone, 1-(4-chlorophenyl)- p-Chloroacetophenone	C_8H_7ClO 154.60	99-91-2 20	2114 232	4-07-00-00639 1.1922[20]	H$_2$O 1; EtOH 5; eth 3; chl 3 1.5550[20]
5790	Ethanone, 2-chloro-1-phenyl- ω-Chloroacetophenone	C_8H_7ClO 154.60	532-27-4 56.5	2115 247	4-07-00-00641 1.324[15]	H$_2$O 1; EtOH 4; eth 4; ace 3
5791	Ethanone, 1-(5-chloro-2-thienyl)-	C_8H_5ClOS 160.62	6310-09-4 52	117[17]; 88[4.5]	5-17-09-00391	H$_2$O 4; eth 4; EtOH 4
5792	Ethanone, 1-cyclobutyl-	$C_8H_{10}O$ 98.14	3019-25-8 	138	4-07-00-00037 0.9020[20]	H$_2$O 2 1.4322[19]
5793	Ethanone, 1-(1-cyclohexen-1-yl)-	$C_8H_{12}O$ 124.18	932-66-1 73	201.5	4-07-00-00143 0.9655[20]	EtOH 3; eth 3 1.4881[20]
5794	Ethanone, 1-cyclohexyl-	$C_8H_{14}O$ 126.20	823-76-7 	180.5	4-07-00-00053 0.9176[20]	H$_2$O 1; eth 3 1.4565[16]
5795	Ethanone, 1-cyclopentyl-	$C_7H_{12}O$ 112.17	6004-60-0 	158.5	4-07-00-00047 0.916[20]	eth 3 1.4409[20]

No.	Name Synonym	Mol. Form. Mol. Wt.	CAS RN mp/°C	Merck No. bp/°C	Beil. Ref. den/g cm^{-3}	Solubility n_D
5796	Ethanone, 1-cyclopropyl- Cyclopropyl methyl ketone	C_5H_8O 84.12	765-43-5 -68.3	111.3	4-07-00-00014 0.8984[20]	H_2O 4; eth 4; EtOH 4 1.4251[20]
5797	Ethanone, 1-(3,5-dibromophenyl)-	$C_8H_6Br_2O$ 277.94	14401-73-1 68	198[15]	3-07-00-00984	bz 4; eth 4; EtOH 4
5798	Ethanone, 2,2-dibromo-1-phenyl-	$C_8H_6Br_2O$ 277.94	13665-04-8 36.5	159[13]	4-07-00-00653	EtOH 3; eth 3; chl 3
5799	Ethanone, 1-(2,4-dichlorophenyl)-	$C_8H_6Cl_2O$ 189.04	2234-16-4 33.5		4-07-00-00643	H_2O 1 1.5640[20]
5800	Ethanone, 1-(2,5-dichlorophenyl)-	$C_8H_6Cl_2O$ 189.04	2476-37-1 12	118[12]	4-07-00-00643 1.321[30]	1.5595[30]
5801	Ethanone, 1-(3,4-dichlorophenyl)-	$C_8H_6Cl_2O$ 189.04	2642-63-9 76	135[12]	4-07-00-00643	H_2O 1; ctc 3; lig 3
5802	Ethanone, 2,2-dichloro-1-phenyl-	$C_8H_6Cl_2O$ 189.04	2648-61-5 20.5	249	4-07-00-00644 1.340[16]	EtOH 3; bz 3; ctc 3 1.5686[20]
5803	Ethanone, 1-(2,4-dihydroxyphenyl)- Resacetophenone	$C_8H_8O_3$ 152.15	89-84-9 146	8143	4-08-00-01792 1.18[141]	H_2O 1; EtOH 3; eth 2; bz 2
5804	Ethanone, 1-(3,4-dihydroxyphenyl)-2-(methylamino)- Adrenalone	$C_9H_{11}NO_3$ 181.19	99-45-6 235-6 dec	161	4-14-00-00832	H_2O 2; EtOH 2; eth 2
5805	Ethanone, 1-(3,4-dimethoxyphenyl)-	$C_{10}H_{12}O_3$ 180.20	1131-62-0 51	287	4-08-00-01815	H_2O 4; bz 4; EtOH 4; chl 4
5806	Ethanone, 1-[3-(dimethylamino)phenyl]-	$C_{10}H_{13}NO$ 163.22	18992-80-8 43	148[13]	4-14-00-00096	ace 4; eth 4
5807	Ethanone, 1-[10-[3-(dimethylamino)propyl]-10H-phenothiazin-2-yl]- Acepromazine	$C_{19}H_{22}N_2OS$ 326.46	61-00-7 27 230[0.5]		4-27-00-02911	
5808	Ethanone, 1-[4-(1,1-dimethylethyl)phenyl]-	$C_{12}H_{16}O$ 176.26	943-27-1 136-8[20]		4-07-00-00769 0.9705[25]	1.518[15]
5809	Ethanone, 1-(2,4-dimethylphenyl)-	$C_{10}H_{12}O$ 148.20	89-74-7 228		4-07-00-00729 1.0121[15]	eth 4; EtOH 4 1.5340[20]
5810	Ethanone, 1-(2,5-dimethylphenyl)-	$C_{10}H_{12}O$ 148.20	2142-73-6 -18.1	232.5	4-07-00-00728 0.9963[19]	H_2O 1; EtOH 4; eth 4; bz 4 1.5291[20]
5811	Ethanone, 1-(3,4-dimethylphenyl)-	$C_{10}H_{12}O$ 148.20	3637-01-2 -3	246.5	4-07-00-00728 1.0090[14]	H_2O 1; EtOH 4; eth 4; bz 4 1.5413[15]
5812	Ethanone, 1,2-diphenyl-	$C_{14}H_{12}O$ 196.25	451-40-1 60	320	4-07-00-01393 1.201[0]	H_2O 2; EtOH 3; eth 3; ctc 3
5813	Ethanone, 2-ethoxy-1,2-diphenyl-	$C_{16}H_{16}O_2$ 240.30	574-09-4 62	194-5[20]	3-08-00-01279 1.1016[17]	bz 4; eth 4; EtOH 4; lig 4 1.5727[17]
5814	Ethanone, 2-ethoxy-1-phenyl- 2-Ethoxyacetophenone	$C_{10}H_{12}O_2$ 164.20	14869-39-7 43	243.5	4-08-00-00365 1.0036[78]	eth 4; EtOH 4; lig 4
5815	Ethanone, 1-(4-fluorophenyl)-	C_8H_7FO 138.14	403-42-9 -45	196	4-07-00-00635 1.1382[25]	H_2O 1; bz 3; chl 3 1.5081[25]
5816	Ethanone, 2-fluoro-1-phenyl-	C_8H_7FO 138.14	450-95-3 29	90-1[12]	4-07-00-00636 1.152[20]	1.5200[20]
5817	Ethanone, 1-(2-furanyl)-	$C_6H_6O_2$ 110.11	1192-62-7 33	175	5-17-09-00381 1.098[20]	H_2O 1; EtOH 3; eth 3 1.5017[20]
5818	Ethanone, 1-[4-(β-D-glucopyranosyloxy)phenyl]- Picein	$C_{14}H_{18}O_7$ 298.29	530-14-3 195.5	7367	5-17-07-00151	H_2O 2; EtOH 3; eth 3; chl 1
5819	Ethanone, 1-(1H-indol-3-yl)-	$C_{10}H_9NO$ 159.19	703-80-0 192.3	144[10]	5-21-08-00297	EtOH 4
5820	Ethanone, 1-(1H-pyrrol-2-yl)-	C_6H_7NO 109.13	1072-83-9 90	220	5-21-07-00204	H_2O 3; EtOH 3; eth 3
5821	Ethanone, 1-(1-hydroxycyclohexyl)-	$C_8H_{14}O_2$ 142.20	1123-27-9 125.5; 91[11]		4-08-00-00023 1.0248[25]	eth 4; EtOH 4 1.4670[25]
5822	Ethanone, 2-hydroxy-1,2-diphenyl-, (±)- (±)-Benzoin	$C_{14}H_{12}O_2$ 212.25	579-44-2 137	1103 344; 194[12]	4-08-00-01279 1.310[20]	EtOH 4; chl 4
5823	Ethanone, 1-(2-hydroxy-4-methoxyphenyl)-	$C_9H_{10}O_3$ 166.18	552-41-0 52.5	158[20]	4-08-00-01793 1.3102[81]	bz 4; eth 4; EtOH 4; chl 4 1.5452[81]
5824	Ethanone, 1-(4-hydroxy-3-methoxyphenyl)- Apocynin	$C_9H_{10}O_3$ 166.18	498-02-2 115	772 297; 234[15]	4-08-00-01814	H_2O 2; EtOH 3; eth 4; ace 3
5825	Ethanone, 1-[6-hydroxy-2-(1-methylethenyl)-5-benzofuranyl]- Euparin	$C_{13}H_{12}O_3$ 216.24	532-48-9 121.5	3857	5-18-02-00029	eth 3; bz 3; chl 3; NaOH 2
5826	Ethanone, 1-(2-hydroxy-4-methylphenyl)-	$C_9H_{10}O_2$ 150.18	6921-64-8 21	245	4-08-00-00480 1.1012[10]	1.5527[13]
5827	Ethanone, 1-(2-hydroxy-5-methylphenyl)-	$C_9H_{10}O_2$ 150.18	1450-72-2 50	210; 120[20]	4-08-00-00472 1.0797[53]	bz 4; eth 4; EtOH 4; chl 4
5828	Ethanone, 1-(4-hydroxy-2-methylphenyl)-	$C_9H_{10}O_2$ 150.18	875-59-2 128	313	4-08-00-00469 1.0592[135]	eth 4; EtOH 4 1.5369[135]
5829	Ethanone, 1-(1-hydroxy-2-naphthalenyl)-	$C_{12}H_{10}O_2$ 186.21	711-79-5 101	325 dec	4-08-00-01175	bz 4; HOAc 4
5830	Ethanone, 1-(2-hydroxyphenyl)-	$C_8H_8O_2$ 136.15	118-93-4 5	218	4-08-00-00320 1.1307[20]	eth 4; EtOH 4; HOAc 4 1.5584[20]
5831	Ethanone, 1-(3-hydroxyphenyl)-	$C_8H_8O_2$ 136.15	121-71-1 96	296; 153[5]	4-08-00-00334 1.0992[109]	H_2O 2; EtOH 4; eth 4; bz 4 1.5348[109]
5832	Ethanone, 1-(4-hydroxyphenyl)-	$C_8H_8O_2$ 136.15	99-93-4 109.5	147-8[3]	4-08-00-00339 1.1090[109]	H_2O 2; EtOH 4; eth 4 1.5577[109]
5833	Ethanone, 2-hydroxy-1-phenyl-	$C_8H_8O_2$ 136.15	582-24-1 90	125[12]; 56[1]	4-08-00-00365 1.0963[99]	H_2O 3; EtOH 3; eth 3; chl 3

No.	Name Synonym	Mol. Form. Mol. Wt.	CAS RN mp/°C	Merck No. bp/°C	Beil. Ref. den/g cm^{-3}	Solubility n_D
5834	Ethanone, 2-[6-(2-hydroxy-2-phenylethyl)-1-methyl-2-piperidinyl]-1-phenyl-, [2R-[2α,6α(S*)]]- Lobeline	C$_{22}$H$_{27}$NO$_2$ 337.46	90-69-7 130.5	5432	5-21-12-00627	H$_2$O 2; EtOH 3; eth 3; ace 4
5835	Ethanone, 1-(3-hydroxy-2-thienyl)-	C$_6$H$_6$O$_2$S 142.18	5556-07-0 51.5	47[0.2]	5-18-01-00121 1.5000[20]	EtOH 3 1.5795[20]
5836	Ethanone, 1-(3-iodophenyl)-	C$_8$H$_7$IO 246.05	14452-30-3	129[8]; 117[4]	4-07-00-00654	bz 3 1.622[20]
5837	Ethanone, 1-(4-iodophenyl)-	C$_8$H$_7$IO 246.05	13329-40-3 86	153[18]	4-07-00-00654	EtOH 3; eth 2; bz 3; HOAc 3
5838	Ethanone, 2-iodo-1-phenyl-	C$_8$H$_7$IO 246.05	4636-16-2 34.4	154[15]	4-07-00-00654	ace 3
5839	Ethanone, 2-methoxy-1,2-diphenyl-	C$_{15}$H$_{14}$O$_2$ 226.27	3524-62-7 49.5	188-9[15]	4-08-00-01280 1.1278[14]	bz 4; eth 4; EtOH 4
5840	Ethanone, 1-(3-methoxyphenyl)-	C$_9$H$_{10}$O$_2$ 150.18	586-37-8 95.5	240	4-08-00-00334 1.0343[19]	H$_2$O 3; EtOH 3; ace 3; ctc 3 1.5410[20]
5841	Ethanone, 1-(4-methoxyphenyl)-	C$_9$H$_{10}$O$_2$ 150.18	100-06-1 38.5	258	4-08-00-00340 1.0818[41]	H$_2$O 2; EtOH 3; eth 3; ace 3 1.547[41]
5842	Ethanone, 2-methoxy-1-phenyl-	C$_9$H$_{10}$O$_2$ 150.18	4079-52-1 8	245; 125[19]	4-08-00-00365 1.0897[20]	H$_2$O 2; EtOH 3; ace 3 1.5393[20]
5843	Ethanone, 1-(1-methylcyclohexyl)-	C$_9$H$_{16}$O 140.23	2890-62-2	186.2	4-07-00-00067 0.9504[20]	1.4543[20]
5844	Ethanone, 1-(3-methylcyclohexyl)-	C$_9$H$_{16}$O 140.23	7193-78-4	197	3-07-00-00120 0.912[19]	1.4517[19]
5845	Ethanone, 1-(4-methylcyclohexyl)-	C$_9$H$_{16}$O 140.23	1879-06-7	196	3-07-00-00121 0.9055[18]	1.4509[18]
5846	Ethanone, 1-(1-methylcyclopentyl)-	C$_8$H$_{14}$O 126.20	13388-93-7	48.4[10]	3-07-00-00102 0.9104[20]	1.4430[20]
5847	Ethanone, 1-(2-methylcyclopentyl)-	C$_8$H$_{14}$O 126.20	1601-00-9	170.5; 59.6[15]	3-07-00-00103 0.9222[4]	1.4434[16]
5848	Ethanone, 1-[4-(1-methylethyl)phenyl]-	C$_{11}$H$_{14}$O 162.23	645-13-6	254	4-07-00-00747 0.9753[15]	1.5235[20]
5849	Ethanone, 1-(1-methyl-1H-pyrrol-2-yl)-	C$_7$H$_9$NO 123.15	932-16-1	201[252]; 93[22]	5-21-07-00205 1.0445[15]	EtOH 3; bz 3; chl 3 1.5403[15]
5850	Ethanone, 1-[2-methyl-5-(1-methylethyl)phenyl]-	C$_{12}$H$_{16}$O 176.26	1202-08-0 <-20	249.5	4-07-00-00772 0.956[20]	1.5181[20]
5851	Ethanone, 1-(2-methylphenyl)-	C$_9$H$_{10}$O 134.18	577-16-2	214	4-07-00-00697 1.026[20]	1.5276[20]
5852	Ethanone, 1-(3-methylphenyl)-	C$_9$H$_{10}$O 134.18	585-74-0	220	4-07-00-00699 1.0165[0]	EtOH 3; eth 3; ace 3; ctc 2 1.533[15]
5853	Ethanone, 1-(4-methylphenyl)-	C$_9$H$_{10}$O 134.18	122-00-9 28	226; 93.5[7]	4-07-00-00701 1.0051[20]	bz 4; eth 4; EtOH 4; chl 4 1.5335[20]
5854	Ethanone, 2,2'-(1-methyl-2,6-piperidinediyl)bis[1-phenyl-, cis- Lobelanine	C$_{22}$H$_{25}$NO$_2$ 335.45	579-21-5 99	5430	5-21-11-00432	ace 4; bz 4; EtOH 4; chl 4
5855	Ethanone, 1-(6-methyl-3-pyridinyl)-	C$_8$H$_9$NO 135.17	36357-38-7 17.6	144[50]	5-21-07-00425 1.0168[25]	H$_2$O 4 1.5302[25]
5856	Ethanone, 1-(5-methyl-2-thienyl)-	C$_7$H$_8$OS 140.21	13679-74-8 27.5	232.5	5-17-09-00425 1.1185[25]	eth 3; ace 3; bz 3; chl 2 1.5604
5857	Ethanone, 1-(1-naphthalenyl)-	C$_{12}$H$_{10}$O 170.21	941-98-0 34	297	4-07-00-01292 1.1171[21]	H$_2$O 1; EtOH 3; eth 3; ace 3 1.6280[22]
5858	Ethanone, 1-(2-naphthalenyl)-	C$_{12}$H$_{10}$O 170.21	93-08-3 56	302	4-07-00-01294	EtOH 2; ctc 2
5859	Ethanone, 1-(1-naphthalenyl)-2-phenyl- Benzyl 1-naphthyl ketone	C$_{18}$H$_{14}$O 246.31	605-85-6 66.5	194-6[0.05]	4-07-00-01790	eth 4
5860	Ethanone, 1-(3-nitrophenyl)-	C$_8$H$_7$NO$_3$ 165.15	121-89-1 81	202; 167[18]	4-07-00-00656	H$_2$O 4; EtOH 2; eth 4; chl 2
5861	Ethanone, 1-(4-nitrophenyl)-	C$_8$H$_7$NO$_3$ 165.15	100-19-6 81.8	165[5]	4-07-00-00657	eth 4; EtOH 4
5862	Ethanone, 2-nitro-1-phenyl-	C$_8$H$_7$NO$_3$ 165.15	614-21-1 106	158[16]; 142[10]	4-07-00-00659	eth 4; EtOH 4 1.5468[30]
5863	Ethanone, 1-(2-nitrophenyl)- Acetophene, 2-nitro	C$_8$H$_7$N$_3$O 161.16	577-59-3 28.5	178[32]; 158[16]	4-07-00-00655 1.2370[25]	H$_2$O 1; EtOH 4; eth 4; chl 4 1.5468[20]
5864	Ethanone, 1-phenyl- Acetophenone	C$_8$H$_8$O 120.15	98-86-2 20	65 202	4-07-00-00619 1.0281[20]	H$_2$O 1; EtOH 3; eth 3; ace 3 1.5372[20]
5865	Ethanone, 1,1'-(1,3-phenylene)bis-	C$_{10}$H$_{10}$O$_2$ 162.19	6781-42-6 32	152[15]	4-07-00-02156	H$_2$O 2; EtOH 3; bz 3; chl 3
5866	Ethanone, 1-phenyl-, oxime	C$_8$H$_9$NO 135.17	613-91-2 60	245	3-07-00-00954 1.0515[78]	H$_2$O 2; EtOH 4; eth 4; ace 4
5867	Ethanone, 1-(2-pyridinyl)-	C$_7$H$_7$NO 121.14	1122-62-9	192	5-21-07-00385 1.077[25]	EtOH 3; eth 3; ctc 2; HOAc 3 1.5203[20]
5868	Ethanone, 1-(3-pyridinyl)- Methyl 3-pyridyl ketone	C$_7$H$_7$NO 121.14	350-03-8 13.5	6034 220	5-21-07-00394	H$_2$O 3; EtOH 3; eth 3; acid 3 1.5341[20]
5869	Ethanone, 1-(4-pyridinyl)-	C$_7$H$_7$NO 121.14	1122-54-9 16	212	5-21-07-00400 1.097[25]	EtOH 2; eth 2; acid 2 1.5282[25]

No.	Name Synonym	Mol. Form. Mol. Wt.	CAS RN mp/°C	Merck No. bp/°C	Beil. Ref. den/g cm^{-3}	Solubility n_D
5870	Ethanone, 1-(2-thienyl)-	C_6H_6OS 126.18	88-15-3 10.5	213.5	5-17-09-00387 1.1679[20]	H_2O 2; EtOH 5; eth 5; ctc 3 1.5667[20]
5871	Ethanone, 2,2,2-trichloro-1-phenyl-	$C_8H_5Cl_3O$ 223.49	2902-69-4	256.5	4-07-00-00645 1.425[16]	eth 4; EtOH 4
5872	Ethanone, 2,2,2-trifluoro-1-phenyl-	$C_8H_5F_3O$ 174.12	434-45-7 -40	153	4-07-00-00637 1.279[20]	1.4583[20]
5873	Ethanone, 1-(2,3,4-trihydroxyphenyl)- Gallacetophenone	$C_8H_8O_4$ 168.15	528-21-2 173	4248	4-08-00-02721	H_2O 3; EtOH 4; eth 3; ace 4
5874	Ethanone, 1-(2,3,4-trimethoxyphenyl)-	$C_{11}H_{14}O_4$ 210.23	13909-73-4 15.8	296	4-08-00-02721	1.5384[20]
5875	Ethanone, 1-(2,4,5-trimethylphenyl)-	$C_{11}H_{14}O$ 162.23	2040-07-5 10.5	246.5	4-07-00-00751 1.0039[15]	H_2O 1; EtOH 4; eth 4; bz 4 1.541[15]
5876	Ethanone, 1-(2,4,6-trimethylphenyl)-	$C_{11}H_{14}O$ 162.23	1667-01-2	241; 120[12]	4-07-00-00749 0.9754[20]	H_2O 1; EtOH 3; eth 3; ace 3 1.5175[20]
5877	Ethanone, 1-(3,4,5-trimethylphenyl)-	$C_{11}H_{14}O$ 162.23	2047-21-4 4.7	101.5[3]	4-07-00-00749 1.0037[25]	1.5420[25]
5878	Ethenaminium, N,N,N-trimethyl-, hydroxide Neurine	$C_5H_{13}NO$ 103.16	463-88-7	6393	3-04-00-00442	H_2O 4; eth 4; EtOH 4
5879	Ethene Ethylene	C_2H_4 28.05	74-85-1 -169	3748 -103.7	4-01-00-00677 0.5678[-104]	H_2O 1; EtOH 2; eth 3; ace 2 1.363[100]
5880	Ethene, bromo- Vinyl bromide	C_2H_3Br 106.95	593-60-2 -137.8	15.8	4-01-00-00718 1.4933[20]	H_2O 1; EtOH 3; eth 3; ace 3 1.4380[20]
5881	Ethene, 1-bromo-2-chloro- Ethylene, 1-bromo-2-chloro-	C_2H_2BrCl 141.39	3018-09-5 -86.7	84.6	3-01-00-00671 1.7972[15]	1.4982
5882	Ethene, 2-bromo-1,1-dichloro-	C_2HBrCl_2 175.84	5870-61-1 -88.5	107.5	4-01-00-00720 1.9053[15]	
5883	Ethene, 1-bromo-1,2-difluoro- 1-Bromo-1,2-difluoroethylene	C_2HBrF_2 142.93	358-99-6 19		0-01-00-00189 1.8434[25]	1.3846[0]
5884	Ethene, chloro- Vinyl chloride	C_2H_3Cl 62.50	75-01-4 -153.7	9898 -13.3	4-01-00-00700 0.9106[20]	H_2O 2; EtOH 3; eth 4 1.3700[20]
5885	Ethene, 2-chloro-1,1-difluoro- 1,1-Difluoro-2-chloroethylene	C_2HClF_2 98.48	359-10-4 -138.5	-18.5	4-01-00-00703	
5886	Ethene, 1-chloro-2-ethoxy- 2-Chlorovinyl ethyl ether	C_4H_7ClO 106.55	928-56-3	120	4-01-00-02061 1.0386[20]	1.4385[20]
5887	Ethene, (2-chloroethoxy)- 2-Chloroethyl vinyl ether	C_4H_7ClO 106.55	110-75-8 -70	2139 108	4-01-00-02051 1.0495[20]	EtOH 4; eth 4; chl 2 1.4378[20]
5888	Ethene, 1-chloro-2-iodo-	C_2H_2ClI 188.40	20244-71-7 -38.2	119	3-01-00-00674 2.2298[25]	
5889	Ethene, chlorotrifluoro- Chlorotrifluoroethylene	C_2ClF_3 116.47	79-38-9 -158	-27.8	4-01-00-00704 1.54[-60]	bz 3; chl 3 1.38[0]
5890	Ethene, 1,1-dibromo-	$C_2H_2Br_2$ 185.85	593-92-0	92	4-01-00-00720 2.1779[21]	EtOH 3; eth 3; ace 3; bz 3
5891	Ethene, 1,2-dibromo-, (E)- trans-1,2-Dibromoethylene	$C_2H_2Br_2$ 185.85	590-12-5 -6.5	108	4-01-00-00721 2.2308[20]	H_2O 1; EtOH 4; eth 4; ace 3 1.5505[18]
5892	Ethene, 1,2-dibromo-, (Z)- cis-1,2-Dibromoethylene	$C_2H_2Br_2$ 185.85	590-11-4 -53	112.5	4-01-00-00720 2.2464[20]	H_2O 1; EtOH 4; eth 4; ace 3 1.5428[20]
5893	Ethene, 1,1-dibromo-2-ethoxy- Ethylene, 1,1-dibromo-2-ethoxy	$C_4H_6Br_2O$ 229.90	77295-79-5	172; 74[15]	2-01-00-00473 1.7697[18]	eth 4
5894	Ethene, 1,1-dichloro- Vinylidene chloride	$C_2H_2Cl_2$ 96.94	75-35-4 -122.5	9900 31.6	4-01-00-00706 1.213[20]	H_2O 1; EtOH 3; eth 4; ace 3 1.4249[20]
5895	Ethene, 1,2-dichloro-, (E)- trans-1,2-Dichloroethylene	$C_2H_2Cl_2$ 96.94	156-60-5 -49.8	86 48.7	4-01-00-00709 1.2565[20]	H_2O 2; EtOH 5; eth 5; ace 5 1.4454[20]
5896	Ethene, 1,2-dichloro-, (Z)- cis-1,2-Dichloroethylene	$C_2H_2Cl_2$ 96.94	156-59-2 -80	86 60.1	4-01-00-00707 1.2837[20]	H_2O 2; EtOH 5; eth 5; ace 5 1.4490[20]
5897	Ethene, 1,1-dichloro-2,2-difluoro-	$C_2Cl_2F_2$ 132.92	79-35-6 -116	19	4-01-00-00711 1.555[-20]	1.383[-20]
5898	Ethene, 1,2-dichloro-1,2-difluoro-	$C_2Cl_2F_2$ 132.92	598-88-9 -130.5	21.1	4-01-00-00711 1.4950[0]	1.3777[0]
5899	Ethene, 1,2-dichloro-1-ethoxy-	$C_4H_6Cl_2O$ 141.00	42345-82-4	128.2	3-01-00-02950 1.1972[25]	1.4558[17]
5900	Ethene, 1,1-dichloro-2-fluoro- 1,1-Dichloro-2-fluoroethylene	C_2HCl_2F 114.93	359-02-4 -108.8	37.5	4-01-00-00711 1.3732[16]	1.4031[16]
5901	Ethene, 1,1-diethoxy-	$C_6H_{12}O_2$ 116.16	2678-54-8	68[100]	4-01-00-03420 0.7932[20]	1.3643[21]
5902	Ethene, 1,1-difluoro- Vinylidene fluoride	$C_2H_2F_2$ 64.03	75-38-7 -144	-85.7	4-01-00-00696	eth 4; EtOH 4
5903	Ethene, 1,2-difluoro-, (Z)- cis-1,2-Difluoroethylene	$C_2H_2F_2$ 64.03	1630-77-9			
5904	Ethene, 1,2-diiodo-, (Z)-	$C_2H_2I_2$ 279.85	590-26-1 -14	72.5[16]	4-01-00-00724 3.0625[20]	eth 3; chl 3
5905	Ethene, ethoxy- Ethyl vinyl ether	C_4H_8O 72.11	109-92-2 -115.8	35.5	4-01-00-02049 0.7589[20]	H_2O 2; EtOH 3; eth 5; ctc 2 1.3767[20]
5906	Ethene, fluoro- Vinyl fluoride	C_2H_3F 46.04	75-02-5 -160.5	-72	4-01-00-00694	H_2O 1; EtOH 3; ace 3
5907	Ethene, iodo- Iodoethylene	C_2H_3I 153.95	593-66-8	56	4-01-00-00722 2.037[20]	eth 4; EtOH 4 1.5385[20]
5908	Ethene, methoxy- Methyl vinyl ether	C_3H_6O 58.08	107-25-5 -122	5.5	4-01-00-02049 0.7725[0]	H_2O 2; EtOH 4; eth 4; ace 4 1.3730[0]
5909	Ethene, (methylsulfonyl)-	$C_3H_6O_2S$ 106.15	3680-02-2	122-4[24]	4-01-00-02065 1.2117[20]	eth 3; ace 3 1.4636[20]

No.	Name / Synonym	Mol. Form. / Mol. Wt.	CAS RN / mp/°C	Merck No. / bp/°C	Beil. Ref. / den/g cm^{-3}	Solubility / n_D
5910	Ethene, (methylthio)-	C$_3$H$_6$S 74.15	1822-74-8	69.5	4-01-00-02065 0.9026^{20}	eth 3; ace 3; chl 3 1.4837^{20}
5911	Ethene, nitro-	C$_2$H$_3$NO$_2$ 73.05	3638-64-0 -55.5	98.5	4-01-00-00725 1.2212^{14}	EtOH 4; eth 4; ace 4; bz 4 1.4282^{20}
5912	Ethene, 1,1'-oxybis- Divinyl ether	C$_4$H$_6$O 70.09	109-93-3 -101	9899 28.3	4-01-00-02058 0.773^{20}	H$_2$O 1; EtOH 5; eth 5; ace 5 1.3989^{20}
5913	Ethene, 1,1'-sulfonylbis-	C$_4$H$_6$O$_2$S 118.16	77-77-0 -26	234.5	4-01-00-02068 1.177^{25}	1.4765^{20}
5914	Ethene, tetrabromo- Tetrabromoethylene	C$_2$Br$_4$ 343.64	79-28-7 56.5	226	4-01-00-00722	H$_2$O 1; EtOH 3; eth 3; ace 3
5915	Ethenetetracarbonitrile Tetracyanoethylene	C$_6$N$_4$ 128.09	670-54-2 199	9129 223	4-02-00-02450 1.348^{25}	eth 2; ace 3; bz 2; ctc 2 1.560^{25}
5916	Ethenetetracarboxylic acid, tetraethyl ester	C$_{14}$H$_{20}$O$_8$ 316.31	6174-95-4 58	328 dec; 203^{13}	4-02-00-02450	H$_2$O 1; EtOH 4; eth 4
5917	Ethene, tetrachloro- Tetrachloroethylene	C$_2$Cl$_4$ 165.83	127-18-4 -22.3	9126 121.3	4-01-00-00715 1.6227^{20}	H$_2$O 1; EtOH 5; eth 5; bz 5 1.5053^{20}
5918	Ethene, tetrafluoro- Tetrafluoroethylene	C$_2$F$_4$ 100.02	116-14-3 -142.5	-75.9	4-01-00-00698 1.519^{-76}	H$_2$O 1
5919	Ethene, tetraiodo- Tetraiodoethylene	C$_2$I$_4$ 531.64	513-92-8 187	9151 sub	4-01-00-00724 2.983^{20}	bz 4; chl 4
5920	Ethene, 1,1'-thiobis-	C$_4$H$_6$S 86.16	627-51-0 20	84	4-01-00-02068 0.9174^{15}	H$_2$O 2; EtOH 5; eth 5; ace 3
5921	Ethene, tribromo-	C$_2$HBr$_3$ 264.74	598-16-3	164	4-01-00-00722 2.708^{20}	H$_2$O 2; EtOH 4; eth 3; ace 3 1.6045^{16}
5922	Ethene, trichloro- Trichloroethylene	C$_2$HCl$_3$ 131.39	79-01-6 -84.7	9552 87.2	4-01-00-00712 1.4642^{20}	H$_2$O 2; EtOH 5; eth 5; ace 3 1.4773^{20}
5923	Ethene, trichlorofluoro-	C$_2$Cl$_3$F 149.38	359-29-5 -108.9	71	4-01-00-00715 1.5460^{20}	chl 3 1.4379^{20}
5924	Ethene, trifluoro- Trifluoroethylene	C$_2$HF$_3$ 82.03	359-11-5	-51	4-01-00-00697 1.26^{-70}	H$_2$O 1; EtOH 2; eth 3
5925	Ethene, trifluoroiodo-	C$_2$F$_3$I 207.92	359-37-5	30	4-01-00-00723 2.284^{25}	1.4143^0
5926	1,4-Etheno-3H,7H-benzo[1,2-c:3,4-c']dipyran-3,7-dione, 9-(3-furanyl)-1,4,4a,5,6,6a,9,10,10a,10b-decahydro-4-hydroxy-4a,10a-dimethyl-, [1R-(1α,4β,4aα,6aβ,9β,10aα,10b)]- Columbin	C$_{20}$H$_{22}$O$_6$ 358.39	546-97-4	2489	5-19-10-00546	
5927	Ethenone Ketene	C$_2$H$_2$O 42.04	463-51-4 -151	5177 -49.8	4-01-00-03418	eth 2; ace 2
5928	Ethenone, diphenyl- Diphenylketene	C$_{14}$H$_{10}$O 194.23	525-06-4	3327 267.5	4-07-00-01653 1.1107^{13}	1.615^{14}
5929	Ethephon Phosphonic acid, (2-chloroethyl)-	C$_2$H$_6$ClO$_3$P 144.49	16672-87-0 74	3686	1.2	
5930	Ethiolate Carbamothioic acid, diethyl-, S-ethyl ester	C$_7$H$_{15}$NOS 161.27	2941-55-1	52$^{0.5}$; 142^{87}	0.9791^{30}	
5931	Ethion Phosphorodithioic acid, S,S'-methylene O,O,O',O'-tetraethyl ester	C$_9$H$_{22}$O$_4$P$_2$S$_4$ 384.48	563-12-2 -13	3691 165$^{0.3}$	1.22^{20}	
5932	Ethirimol 4(1H)-Pyrimidinone, 5-butyl-2-(ethylamino)-6-methyl-	C$_{11}$H$_{19}$N$_3$O 209.29	23947-60-6 160	3695	1.21^{25}	
5933	Ethofumesate 2-Ethoxy-2,3-dihydro-3,3-dimethyl-5-benzofuranyl methanesulfonate	C$_{13}$H$_{18}$O$_5$S 286.35	26225-79-6 71	3697	1.14	
5934	Ethoprop Phosphorodithioic acid, O-ethyl S,S-dipropyl ester	C$_8$H$_{19}$O$_2$PS$_2$ 242.34	13194-48-4	3702 88$^{0.2}$	1.094^{20}	
5935	Ethylene, 1-bromo-1,2-dichloro-, trans- Ethylene, 1-bromo-1,2-dichloro (dis)	C$_2$HBrCl$_2$ 175.84	6795-75-1 -83.5	113.8	3-01-00-00671 1.9133^{15}	1.5218^{15}
5936	Ethyne Acetylene	C$_2$H$_2$ 26.04	74-86-2 -80.7 (trip. pt.)	84 -84.7 (subl. point)	4-01-00-00939 *0.377^{25}	H$_2$O 2; EtOH 2; ace 3; bz 3
5937	Ethyne, bromo- Bromoacetylene	C$_2$HBr 104.93	593-61-3	4.7	3-01-00-00919	eth 4
5938	Ethyne, chloro- Chloroacetylene	C$_2$HCl 60.48	593-63-5 -126	-30	4-01-00-00957	EtOH 2
5939	Ethyne, dibromo-	C$_2$Br$_2$ 183.83	624-61-3 -24	exp 76	3-01-00-00919	ace 4; bz 4; eth 4; EtOH 4
5940	Ethyne, dichloro- Dichloroacetylene	C$_2$Cl$_2$ 94.93	7572-29-4 -66	33	4-01-00-00957 1.261^{20}	EtOH 3; eth 3; ace 3 1.42790^{20}
5941	Ethyne, diiodo- Diiodoacetylene	C$_2$I$_2$ 277.83	624-74-8 81.5	exp	4-01-00-00958	ace 4; bz 4; eth 4; EtOH 4
5942	Ethyne, ethoxy-	C$_4$H$_6$O 70.09	927-80-0	50	4-01-00-02211 0.8000^{20}	1.3796^{20}
5943	Ethyne, fluoro- Fluoroacetylene	C$_2$HF 44.03	2713-09-9		4-01-00-00957	
5944	Ethyne, methoxy-	C$_3$H$_4$O 56.06	6443-91-0		4-01-00-02211 0.8001^{20}	eth 4; EtOH 4 1.3812^{20}

No.	Name / Synonym	Mol. Form. / Mol. Wt.	CAS RN / mp/°C	Merck No. / bp/°C	Beil. Ref. / den/g cm^{-3}	Solubility / n_D
5945	Etrimfos Phosphorothioic acid, O-(6-ethoxy-2-ethyl-4-pyrimidinyl) O,O-dimethyl ester	$C_{10}H_{17}N_2O_4PS$ 292.30	38260-54-7 -3.35	3847	1.195^{20}	
5946	Famphur Phosphorothioic acid, O-[4-[(dimethylamino)sulfonyl]phenyl] O,O-dimethyl ester	$C_{10}H_{16}NO_5PS_2$ 325.35	52-85-7 53	3882		
5947	Fenamiphos Phosphoramidic acid, (1-methylethyl)-, ethyl 3-methyl-4-(methylthio)phenyl ester	$C_{13}H_{22}NO_3PS$ 303.36	22224-92-6 49	3901	1.15^{20}	
5948	Fenarimol 5-Pyrimidinemethanol, α-(2-chlorophenyl)-α-(4-chlorophenyl)-	$C_{17}H_{12}Cl_2N_2O$ 331.20	60168-88-9 118	3903		
5949	Fenbuconazole 4-(4-Chlorophenyl)-2-phenyl-2-(1H-1,2,4-triazol-1-ylmethyl)butyronitrile	$C_{20}H_{19}ClN_4$ 350.85	114369-43-6 125			
5950	Fenbutatin oxide Distannoxane, hexakis(2-methyl-2-phenylpropyl)-	$C_{60}H_{78}OSn_2$ 1052.70	13356-08-6 138	3907		
5951	Fenitrothion Phosphorothioic acid, O,O-dimethyl O-(3-methyl-4-nitrophenyl) ester	$C_9H_{12}NO_5PS$ 277.24	122-14-5	3922 118$^{0.05}$; 164^1	1.3227^{25}	
5952	Fenoxycarb Ethyl 2-(4-phenoxyphenoxy)ethylcarbamate	$C_{17}H_{19}NO_4$ 301.34	79127-80-3 53			
5953	Fenpropathrin Cyano(3-phenoxyphenyl)methyl 2,2,3,3-tetramethylcyclopropanecarboxylate	$C_{22}H_{23}NO_3$ 349.43	64257-84-7 47		1.15^{25}	
5954	Fensulfothion Phosphorothioic acid, O,O-diethyl O-[4-(methylsulfinyl)phenyl] ester	$C_{11}H_{17}O_4PS_2$ 308.36	115-90-2	3943 140$^{0.01}$	1.202^{20}	
5955	Fenthion Phosphorothioic acid, O,O-dimethyl O-[3-methyl-4-(methylthio)phenyl] ester	$C_{10}H_{15}O_3PS_2$ 278.33	55-38-9 7.5	3945 87$^{0.01}$	1.246^{20}	
5956	Fenvalerate Benzeneacetic acid, 4-chloro-α-(1-methylethyl)-, cyano(3-phenoxyphenyl)methyl ester	$C_{25}H_{22}ClNO_3$ 419.91	51630-58-1 dec	3952	1.15^{25}	
5957	Ferbam Iron, tris(dimethylcarbamodithioato-S,S')-, (OC-6-11)-	$C_9H_{18}FeN_3S_6$ 416.51	14484-64-1 180 dec	3954		
5958	Ferrate(2-), chloro[7,12-diethenyl-3,8,13,17-tetramethyl-21H,23H-porphine-2,18-dipropanoato(4-)-N21,N22,N23, Hemin	$C_{34}H_{32}ClFeN_4O_4$ 651.95	16009-13-5 >300	4563		
5959	Ferrate(2-), [7,12-diethenyl-3,8,13,17-tetramethyl-21H,23H-porphine-2,18-dipropanoato(4-)-N21,N22,N23,N24 Hematin	$C_{34}H_{33}FeN_4O_5$ 633.51	15480-90-4 >200	4554	4-26-00-03047	H_2O 1; EtOH 3; eth 1; alk 3
5960	Ferrocene Bis(cyclopentadienyl)iron	$C_{10}H_{10}Fe$ 186.04	102-54-5 172.5	3985 249		H_2O 1
5961	Ferrocene, 1,1'-bis(1,1,3,3-tetramethyl-3-phenyldisiloxanyl)- Disiloxane, 1-(cyclopentadienyl)-1,1,3,3-tetramethyl-3-phenyl-, Fe deriv.	$C_{30}H_{42}FeO_2Si_4$ 602.85	12321-04-9 18.7	 200-5$^{0.03}$	1.1063^{25}	1.5473^{25}
5962	Fluazipop-butyl (±)-2-[4-[[5-(Trifluoromethyl)-2-pyridinyl]oxy]phenoxy]propanoic acid, butyl ester	$C_{19}H_{20}F_3NO_4$ 383.37	69806-50-4 13	4049 170$^{0.5}$	1.21^{20}	
5963	Fluazipop-P-butyl (R)-2-[4-(5-Trifluoromethyl-2-pyridyloxy)phenoxy]propionic acid, butyl ester	$C_{19}H_{20}F_3NO_4$ 383.37	79241-46-6 5			
5964	Fluchloralin Benzenamine, N-(2-chloroethyl)-2,6-dinitro-N-propyl-4-(trifluoromethyl)-	$C_{12}H_{13}ClF_3N_3O_4$ 355.70	33245-39-5 42	4052		
5965	Flucythrinate Cythrin	$C_{26}H_{23}F_2NO_4$ 451.47	70124-77-5 108$^{0.35}$	4055	1.189^{22}	
5966	Fluoranthene 1,2-(1,8-Naphthylene)benzene	$C_{16}H_{10}$ 202.26	206-44-0 107.8	384	4-05-00-02463 1.252^0	H_2O 1; EtOH 3; eth 3; bz 3
5967	9H-Fluorene 2,2'-Methylenebiphenyl	$C_{13}H_{10}$ 166.22	86-73-7 114.8	4081 295	4-05-00-02142 1.203^0	H_2O 1; EtOH 2; eth 3; ace 3
5968	9H-Fluorene, 2-bromo- 2-Bromofluorene	$C_{13}H_9Br$ 245.12	1133-80-8 113.5	 185^{135}	4-05-00-02147	H_2O 1; EtOH 3; chl 4; HOAc 3
5969	9H-Fluorene-2-carboxylic acid, 9-oxo-	$C_{14}H_8O_3$ 224.22	784-50-9 341	sub 340	4-10-00-03040	H_2O 1; EtOH 3; HOAc 3
5970	9H-Fluorene, 9-[(2-chlorophenyl)methylene]- Fluorene, 9-(2-chlorobenzylidene)	$C_{20}H_{13}Cl$ 288.78	1643-49-8 69.5	180$^{0.7}$	4-05-00-02631	EtOH 4; HOAc 4
5971	9H-Fluorene-2,7-diamine 2,7-Diaminofluorene	$C_{13}H_{12}N_2$ 196.25	525-64-4 166	4082	4-13-00-00449	H_2O 1; EtOH 3; chl 3
5972	9H-Fluorene, 2,7-dichloro- 2,7-Dichlorofluorene	$C_{13}H_8Cl_2$ 235.11	7012-16-0 128	sub	4-05-00-02147	bz 4; chl 4

PHYSICAL CONSTANTS OF ORGANIC COMPOUNDS (continued)

No.	Name / Synonym	Mol. Form. / Mol. Wt.	CAS RN / mp/°C	Merck No. / bp/°C	Beil. Ref. / den/g cm⁻³	Solubility / n_D
5973	1H-Fluorene, dodecahydro-	$C_{13}H_{22}$ / 178.32	5744-03-6 /	/ 260	4-05-00-00489 / 0.92^{25}	/ 1.5012^{20}
5974	9H-Fluorene, 9-methyl- / 9-Methylfluorene	$C_{14}H_{12}$ / 180.25	2523-37-7 / 46.5	/ $154-6^{15}$	4-05-00-02185 / 1.0263^{66}	H_2O 1; EtOH 3; eth 3; ace 3 / 1.610^{66}
5975	9H-Fluoren-9-one	$C_{13}H_8O$ / 180.21	486-25-9 / 84	/ 341.5	4-07-00-01629 / 1.1300^{99}	H_2O 1; EtOH 3; eth 4; ace 3 / 1.6309^{99}
5976	9H-Fluoren-9-one, 2-chloro-	$C_{13}H_7ClO$ / 214.65	3096-47-7 / 125.5	/ sub	4-07-00-01634 /	EtOH 4
5977	9H-Fluoren-9-one, 2-nitro-	$C_{13}H_7NO_3$ / 225.20	3096-52-4 / 224.3	/ sub	4-07-00-01636 /	EtOH 2; ace 3; sulf 3; HOAc 3
5978	9H-Fluoren-9-one, 1,2,3,4-tetrahydro- / Phentydrone	$C_{13}H_{12}O$ / 184.24	634-19-5 / 81.5	7235 / $139-40^{0.05}$	4-07-00-01302 /	
5979	9H-Fluoren-9-one, 2,4,7-trinitro- / 2,4,7-Trinitrofluorenone	$C_{13}H_5N_3O_7$ / 315.20	129-79-3 / 176	9641 /	4-07-00-01638 /	H_2O 2; ace 4; bz 4; chl 4
5980	Fluorodifen / Benzene, 2-nitro-1-(4-nitrophenoxy)-4-(trifluoromethyl)-	$C_{13}H_7F_3N_2O_5$ / 328.20	15457-05-3 / 94	/	/	
5981	Fluridone / 4(1H)-Pyridinone, 1-methyl-3-phenyl-5-[3-(trifluoromethyl)phenyl]-	$C_{19}H_{14}F_3NO$ / 329.32	59756-60-4 / 155	/	/	
5982	Fluroxypyr / [(4-Amino-3,5-dichloro-6-fluoro-2-pyridyl)oxy]acetic acid	$C_7H_5Cl_2FN_2O_3$ / 255.03	69377-81-7 / 232	4128 /	/	
5983	Fluvalinate / 2-[[2-Chloro-4-(trifluoromethyl)phenyl]amino]-3-methylbutanoic acid, (±)-cyano(3-phenoxyphenyl)methyl ester	$C_{26}H_{22}ClF_3N_2O_3$ / 502.92	102851-06-9 /	/ >450	/ 1.29^{25}	
5984	Folpet / 1H-Isoindole-1,3(2H)-dione, 2-[(trichloromethyl)thio]-	$C_9H_4Cl_3NO_2S$ / 296.56	133-07-3 / 177	4142 /	/	
5985	Fomesafen / 5-[2-Chloro-4-(trifluoromethyl)phenoxy]-N-(methylsulfonyl)-2-nitrobenzamide	$C_{15}H_{10}ClF_3N_2O_6S$ / 438.77	72178-02-0 / 220	/	/ 1.28^{20}	
5986	Fonofos / Phosphonodithioic acid, ethyl-, O-ethyl S-phenyl ester	$C_{10}H_{15}OPS_2$ / 246.33	944-22-9 /	4147 / $130^{0.1}$	/ 1.16^{25}	
5987	Formaldehyde / Methanal	CH_2O / 30.03	50-00-0 / -92	4148 / -19.1	4-01-00-03017 / 0.815^{-20}	H_2O 3; EtOH 3; eth 5; ace 5
5988	Formaldehyde, oxime	CH_3NO / 45.04	75-17-2 / 2.5	/ 109^{15}	4-01-00-03055 / 1.133^{25}	H_2O 3; EtOH 4; eth 4
5989	Formamide / Methanamide	CH_3NO / 45.04	75-12-7 / 2.55	4151 / 220	4-02-00-00045 / 1.1334^{20}	H_2O 5; EtOH 5; eth 2; ace 3 / 1.4472^{20}
5990	Formamide, N-cyclohexyl-	$C_7H_{13}NO$ / 127.19	766-93-8 / 39.5	/ $260; 113^{0.7}$	4-12-00-00038 / 1.0123^{17}	
5991	Formamide, N,N-diethyl-	$C_5H_{11}NO$ / 101.15	617-84-5 /	/ 177.5	4-04-00-00346 / 0.9080^{19}	H_2O 5; EtOH 4; eth 4; ace 5 / 1.4321^{25}
5992	Formamide, N,N-dimethyl- / N,N-Dimethylformamide	C_3H_7NO / 73.09	68-12-2 / -60.4	3232 / 153	4-04-00-00171 / 0.944^{25}	H_2O 5; EtOH 5; eth 5; ace 5 / 1.4305^{20}
5993	Formamide, N,N-diphenyl-	$C_{13}H_{11}NO$ / 197.24	607-00-1 / 73.5	/ $337.5; 189^{13}$	4-12-00-00370 /	H_2O 1; EtOH 3; eth 3; bz 3
5994	Formamide, N-ethyl- / N-Ethylformamide	C_3H_7NO / 73.09	627-45-2 /	/ 198	4-04-00-00346 / 0.9552^{20}	H_2O 5; EtOH 5; eth 5 / 1.4320^{20}
5995	Formamide, N-methyl- / N-Methylformamide	C_2H_5NO / 59.07	123-39-7 / -3.8	/ 199.5	4-04-00-00170 / 1.011^{19}	H_2O 4; ace 4; EtOH 4 / 1.4319^{20}
5996	Formamide, N-(2-methylphenyl)-	C_8H_9NO / 135.17	94-69-9 / 62	/ 288	3-12-00-01852 / 1.086^{55}	H_2O 3; EtOH 4
5997	Formamide, N-(3-methylphenyl)-	C_8H_9NO / 135.17	3085-53-8 / <-18	/ 290 dec; 176^{17}	3-12-00-01962 /	H_2O 3; NaOH 3
5998	Formamide, N-methyl-N-phenyl-	C_8H_9NO / 135.17	93-61-8 / 14.5	/ 243	4-12-00-00370 / 1.0948^{20}	H_2O 2; EtOH 3; ace 3; ctc 2 / 1.5589^{20}
5999	Formamide, N-phenyl- / Formanilide	C_7H_7NO / 121.14	103-70-8 / 46	4152 / 271	4-12-00-00368 / 1.1186^{50}	H_2O 3; EtOH 4; eth 3; bz 3
6000	Formamide, N-(2,2,2-trichloro-1-hydroxyethyl)- / Chloral formamide	$C_3H_4Cl_3NO_2$ / 192.43	515-82-2 / 120	2060 /	3-02-00-00059 /	ace 4; eth 4; EtOH 4
6001	Formetanate hydrochloride / Methanimidamide, N,N-dimethyl-N'-[3-[[(methylamino)carbonyl]oxy]phenyl]-, monohydrochloride	$C_{11}H_{16}ClN_3O_2$ / 257.72	23422-53-9 /	/	/	
6002	Formic acid / Methanoic acid	CH_2O_2 / 46.03	64-18-6 / 8.3	4153 / 101	4-02-00-00003 / 1.220^{20}	H_2O 5; EtOH 5; eth 5; ace 4 / 1.3714^{20}
6003	Formic acid, ammonium salt / Ammonium formate	CH_5NO_2 / 63.06	540-69-2 /	548 /	4-02-00-00003 /	
6004	Formic acid, butyl ester / Butyl formate	$C_5H_{10}O_2$ / 102.13	592-84-7 / -91.5	/ 106.1	4-02-00-00028 / 0.8885^{20}	H_2O 2; EtOH 5; eth 5; ace 3 / 1.3912^{20}
6005	Formic acid, cyclohexyl ester	$C_7H_{12}O_2$ / 128.17	4351-54-6 /	/ 163	4-06-00-00035 / 1.0057^0	H_2O 1; EtOH 3; eth 4; HCOOH 3 / 1.4430^{20}

No.	Name / Synonym	Mol. Form. / Mol. Wt.	CAS RN / mp/°C	Merck No. / bp/°C	Beil. Ref. / den/g cm⁻³ / n_D	Solubility
6006	Formic acid, ethenyl ester / Vinyl formate	$C_3H_4O_2$ / 72.06	692-45-5		4-02-00-00033	
6007	Formic acid, ethyl ester / Ethyl formate	$C_3H_6O_2$ / 74.08	109-94-4 / -79.6	3763 / 54.4	4-02-00-00023 / 0.9168[20] / 1.3598[10]	H_2O 3; EtOH 5; eth 5; ace 4
6008	Formic acid, heptyl ester / Heptyl formate	$C_8H_{16}O_2$ / 144.21	112-23-2	178.1	4-02-00-00031 / 0.8784[20] / 1.4140[20]	H_2O 1; EtOH 5; eth 5
6009	Formic acid, hexyl ester / Hexyl formate	$C_7H_{14}O_2$ / 130.19	629-33-4 / -62.6	155.5	4-02-00-00031 / 0.8813[20] / 1.4071[20]	H_2O 1; EtOH 5; eth 5
6010	Formic acid, methyl ester / Methyl formate	$C_2H_4O_2$ / 60.05	107-31-3 / -99	5994 / 31.7	4-02-00-00020 / 0.9742[20] / 1.3433[20]	H_2O 4; EtOH 5; eth 3; chl 3
6011	Formic acid, 1-methylethyl ester / Isopropyl formate	$C_4H_8O_2$ / 88.11	625-55-8	68.2	4-02-00-00027 / 0.8728[20] / 1.3678[20]	H_2O 2; EtOH 5; eth 5; ace 4
6012	Formic acid, 1-methylpropyl ester	$C_5H_{10}O_2$ / 102.13	589-40-2	97	4-02-00-00029 / 0.8846[20] / 1.3865[20]	H_2O 2; EtOH 5; eth 5; ace 3
6013	Formic acid, 2-methylpropyl ester / Isobutyl formate	$C_5H_{10}O_2$ / 102.13	542-55-2 / -95.8	5026 / 98.2	4-02-00-00029 / 0.8776[20] / 1.3857[20]	H_2O 2; EtOH 5; eth 5; ace 4
6014	Formic acid, octyl ester	$C_9H_{18}O_2$ / 158.24	112-32-3 / -39.1	198.8	4-02-00-00031 / 0.8744[20] / 1.4208[15]	H_2O 1; EtOH 3; eth 5; ctc 2
6015	Formic acid, pentyl ester / Pentyl formate	$C_6H_{12}O_2$ / 116.16	638-49-3 / -73.5	130.4	4-02-00-00030 / 0.8853[20] / 1.3992[20]	H_2O 2; EtOH 5; eth 5
6016	Formic acid, phenylmethyl ester / Benzyl formate	$C_8H_8O_2$ / 136.15	104-57-4	1148 / 203; 84[10]	4-06-00-02262 / 1.081[20] / 1.5154[20]	H_2O 1; EtOH 5; eth 5; ace 3
6017	Formic acid, 2-propenyl ester / Allyl formate	$C_4H_6O_2$ / 86.09	1838-59-1	83.6	3-02-00-00046 / 0.9460[20]	H_2O 2; EtOH 3; eth 5
6018	Formic acid, propyl ester / Propyl formate	$C_4H_8O_2$ / 88.11	110-74-7 / -92.9	7871 / 80.9	4-02-00-00026 / 0.9058[20] / 1.3779[20]	H_2O 2; EtOH 5; eth 5; ctc 2
6019	Formyl fluoride / Fluoroformaldehyde	CHFO / 48.02	1493-02-3 / -142.2	4162 / -26.5	4-02-00-00042 / 1.1950[-30]	
6020	Fosetyl-Al / Ethyl hydrogen phosphonate, aluminum salt	$C_6H_{18}AlO_9P_3$ / 354.11	39148-24-8 / >300	4167		
6021	Fosthiazate / (RS)-S-sec-Butyl O-ethyl 2-oxo-1,3-thiazolidin-3-ylphosphonothioate	$C_9H_{18}NO_3PS_2$ / 283.35	98886-44-3			
6022	D-Fructopyranose, 3 O α-D-glucopyranosyl-D-Turanose	$C_{12}H_{22}O_{11}$ / 342.30	547-25-1 / 168	9728	5-17-07-00213	H_2O 4; EtOH 3; MeOH 3
6023	β-D-Fructose / β-Levulose	$C_6H_{12}O_6$ / 180.16	53188-23-1 / 103 dec	4185	4-01-00-04401 / 1.60[20]	H_2O 4; EtOH 3; ace 4; MeOH 3
6024	Fulminic acid / Carbyloxime	CHNO / 43.03	506-85-4 / unstable in pure form			eth 3
6025	Furan / Oxacyclopentadiene	C_4H_4O / 68.08	110-00-9 / -85.6	4206 / 31.5	5-17-01-00291 / 0.9514[20] / 1.4214[20]	H_2O 1; EtOH 4; eth 4; ace 3
6026	2-Furanacetic acid	$C_6H_6O_3$ / 126.11	2745-26-8 / 68.5	102[0.4]	5-18-06-00206	H_2O 3; bz 3; MeOH 3; peth 3
6027	2-Furanacetic acid, butyl ester / 2-Furylacetic acid, butyl ester	$C_{10}H_{14}O_3$ / 182.22	4915-23-5	110-11[13]	4-18-00-04062 / 1.0232[25] / 1.4558[25]	
6028	2-Furanacetic acid, ethyl ester	$C_8H_{10}O_3$ / 154.17	4915-21-3	88[15]	5-18-06-00206 / 1.0763[25] / 1.4571[25]	
6029	2-Furanacetic acid, methyl ester	$C_7H_8O_3$ / 140.14	4915-22-4	87-8[21]	5-18-06-00206 / 1.1250[20] / 1.4638[25]	
6030	3-Furanacetic acid, tetrahydro-2,2-dimethyl-5-oxo- / Terpenylic acid	$C_8H_{12}O_4$ / 172.18	26754-48-3 / 90	9100	5-18-08-00032	H_2O 4
6031	2-Furanacetonitrile / 2-Furylacetonitrile	C_6H_5NO / 107.11	2745-25-7	75-80[20]	5-18-06-00207 / 1.0854[25] / 1.4693[20]	eth 4; EtOH 4
6032	Furan, 2,5-bis(1,1-dimethylethyl)-	$C_{12}H_{20}O$ / 180.29	4789-40-6	210	5-17-01-00439 / 0.837[20] / 1.4369[20]	H_2O 1; EtOH 3; eth 3
6033	Furan, 2-bromo-	C_4H_3BrO / 146.97	584-12-3	103	5-17-01-00295 / 1.6500[20] / 1.4980[20]	H_2O 2; EtOH 3; eth 3; ace 3
6034	Furan, 3-bromo- / 3-Bromofuran	C_4H_3BrO / 146.97	22037-28-1	103	5-17-01-00295 / 1.6606[20] / 1.4958[20]	ace 4; bz 4; eth 4; EtOH 4
6035	Furan, 2-(bromomethyl)-	C_5H_5BrO / 161.00	4437-18-7	32-4[2]	4-17-00-00268 / 1.557[20] / 1.5380[20]	
6036	Furan, 2-(bromomethyl)tetrahydro-	C_5H_9BrO / 165.03	1192-30-9	170; 70[22]	5-17-01-00080 / 1.3653[20] / 1.4850[20]	EtOH 3; eth 3
6037	Furan, 2-(butoxymethyl)-	$C_9H_{14}O_2$ / 154.21	56920-82-2	190	2-17-00-00114 / 0.9516[20] / 1.4522[20]	eth 4; EtOH 4
6038	Furan, 2,2'-sec-butylidenedi- / Butane, 2,2-di(2-furyl)	$C_{12}H_{14}O_2$ / 190.24	100121-73-1	64-6[1]	4-19-00-00287 / 1.0330[20] / 1.4970[20]	eth 4; EtOH 4
6039	2-Furancarbonitrile	C_5H_3NO / 93.08	617-90-3	147	5-18-06-00122 / 1.0822[20] / 1.4798[20]	EtOH 3; eth 3
6040	2-Furancarbonyl chloride	$C_5H_3ClO_2$ / 130.53	527-69-5 / -2	173	5-18-06-00110 / 1.324[25] / 1.5310[20]	H_2O 1; eth 3; ctc 2; chl 3
6041	2-Furancarboxaldehyde / Furfural	$C_5H_4O_2$ / 96.09	98-01-1 / -36.5	4214 / 161.7	5-17-09-00292 / 1.1594[20] / 1.5261[20]	H_2O 3; EtOH 4; eth 5; ace 4
6042	3-Furancarboxaldehyde	$C_5H_4O_2$ / 96.09	498-60-2	145; 70-2[43]	5-17-09-00372 / 1.1092[20] / 1.4945[20]	
6043	2-Furancarboxaldehyde, 5-bromo-	$C_5H_3BrO_2$ / 174.98	1899-24-7 / 83.5	201; 112[16]	5-17-09-00320	eth 4; EtOH 4

No.	Name / Synonym	Mol. Form. / Mol. Wt.	CAS RN / mp/°C	Merck No. / bp/°C	Beil. Ref. / den/g cm^{-3}	Solubility / n_D
6044	2-Furancarboxaldehyde, 5-(hydroxymethyl)- 5-(Hydroxymethyl)-2-furaldehyde	$C_6H_6O_3$ 126.11	67-47-0 31.5	4764 114-6[1]	5-18-01-00130 1.2062[25]	H_2O 3; EtOH 3; eth 2; bz 3 1.5627[18]
6045	2-Furancarboxaldehyde, 5-methyl-	$C_6H_6O_2$ 110.11	620-02-0	187; 89[26]	5-17-09-00404 1.1072[18]	H_2O 3; EtOH 4; eth 5; ctc 2 1.5264[20]
6046	2-Furancarboxaldehyde, 5-nitro-	$C_5H_3NO_4$ 141.08	698-63-5 35.5	128-32[10]	5-17-09-00323	H_2O 2; peth 3
6047	2-Furancarboxaldehyde, oxime, (E)-	$C_5H_5NO_2$ 111.10	620-03-1	205 dec; 98[9]	5-17-09-00306 1.1550[80]	H_2O 3; EtOH 4; eth 4; bz 4
6048	2-Furancarboxaldehyde, oxime, (Z)-	$C_5H_5NO_2$ 111.10	1450-58-4 91	100[12]	5-17-09-00306 1.3800[20]	H_2O 2; EtOH 4; eth 4; bz 4
6049	2-Furancarboxaldehyde, tetrahydro-	$C_5H_8O_2$ 100.12	7681-84-7	142; 46[29]	5-17-09-00027 1.0727[20]	H_2O 4; eth 4 1.4366[20]
6050	2-Furancarboxylic acid 2-Furoic acid	$C_5H_4O_3$ 112.08	88-14-2 133.5	4219 231	5-18-06-00102	H_2O 3; EtOH 3; eth 4; ace 2
6051	3-Furancarboxylic acid	$C_5H_4O_3$ 112.08	488-93-7 122.5	sub 105	5-18-06-00196	H_2O 2; EtOH 3; eth 4; AcOEt 3
6052	2-Furancarboxylic acid, butyl ester	$C_9H_{12}O_3$ 168.19	583-33-5	233	4-18-00-03918 1.0555[20]	H_2O 1; EtOH 3; eth 3; bz 3 1.4740
6053	2-Furancarboxylic acid, 5-chloro-, ethyl ester Ethyl 5-chloro-2-furancarboxylate	$C_7H_7ClO_3$ 174.58	4301-39-7 2	217; 108[23]	5-18-06-00131 1.2418[25]	
6054	3-Furancarboxylic acid, 4,5-dihydro-5-oxo-	$C_5H_4O_4$ 128.08	585-68-2 164	dec	4-18-00-05333	H_2O 4; EtOH 3; MeOH 3
6055	3-Furancarboxylic acid, 2,5-dimethyl-, ethyl ester	$C_9H_{12}O_3$ 168.19	29113-63-1	208.5; 96[19]	5-18-06-00233 1.0478[23]	1.4686[20]
6056	2-Furancarboxylic acid, ethyl ester	$C_7H_8O_3$ 140.14	614-99-3 34.5	196.8	5-18-06-00104 1.1174[21]	H_2O 1; EtOH 5; eth 5; ace 5 1.4797[21]
6057	2-Furancarboxylic acid, 2-furanylmethyl ester	$C_{10}H_8O_4$ 192.17	615-11-2 27.5	122[2]	4-18-00-03936 1.2347[25]	ace 4; bz 4; eth 4; EtOH 4 1.5280[20]
6058	2-Furancarboxylic acid, hexyl ester α-Furoic acid, hexyl ester	$C_{11}H_{16}O_3$ 196.25	39251-86-0	105-7[1]	2-18-00-00267 1.0170[20]	H_2O 1; EtOH 3; eth 3; bz 5
6059	2-Furancarboxylic acid, methyl ester	$C_6H_6O_3$ 126.11	611-13-2	181.3	5-18-06-00103 1.1786[21]	H_2O 1; EtOH 3; eth 3; bz 3 1.4860[20]
6060	2-Furancarboxylic acid, 5-methyl-	$C_6H_6O_3$ 126.11	1917-15-3 109.5	105[1]	5-18-06-00217	H_2O 2; EtOH 3; eth 3; bz 2
6061	2-Furancarboxylic acid, 1-methylethyl ester	$C_8H_{10}O_3$ 154.17	6270-34-4	198.5	0-18-00-00275 1.0655[24]	H_2O 1; EtOH 3; eth 3; ace 3 1.4682[24]
6062	2-Furancarboxylic acid, 2-methylpropyl ester	$C_9H_{12}O_3$ 168.19	20279-53-2	222	2-18-00-00266 1.0388[20]	ace 4; bz 4; eth 4; EtOH 4 1.4676[20]
6063	3-Furancarboxylic acid, methyl ester	$C_6H_6O_3$ 126.11	13129-23-2	160	5-18-06-00196 1.1733[15]	H_2O 1; EtOH 3; ace 3 1.4676[20]
6064	3-Furancarboxylic acid, 2-methyl-, ethyl ester	$C_8H_{10}O_3$ 154.17	28921-35-9	85-7[20]	5-18-06-00214 1.0102[25]	eth 4 1.4620[25]
6065	3-Furancarboxylic acid, 5-methyl- 3-Furoic acid, 5-methyl-	$C_6H_6O_3$ 126.11	21984-93-0 119	sub	5-18-06-00216	H_2O 4; eth 4; EtOH 4
6066	2-Furancarboxylic acid, 5-nitro-	$C_5H_3NO_5$ 157.08	645-12-5 186	sub	5-18-06-00139	H_2O 3; EtOH 3; eth 3; ace 2
6067	2-Furancarboxylic acid, octyl ester α-Furoic acid, octyl ester	$C_{13}H_{20}O_3$ 224.30	39251-88-2	126-7[1]	2-18-00-00267 0.9885[20]	H_2O 1; EtOH 3
6068	2-Furancarboxylic acid, pentyl ester 2-Furoic acid, pentyl ester	$C_{10}H_{14}O_3$ 182.22	4996-48-9	95-7[1]	2-18-00-00266 1.0335[20]	H_2O 1; EtOH 5
6069	2-Furancarboxylic acid, phenylmethyl ester	$C_{12}H_{10}O_3$ 202.21	5380-40-5	179-81[18]	4-18-00-03921 1.1623[22]	ace 4; eth 4 1.5550[20]
6070	2-Furancarboxylic acid, 2-propenyl ester	$C_8H_8O_3$ 152.15	4208-49-5	207.5	5-18-06-00105 1.115[25]	eth 3; ace 3; ctc 2 1.4945[20]
6071	2-Furancarboxylic acid, propyl ester	$C_8H_{10}O_3$ 154.17	615-10-1	210.9; 112[20]	5-18-06-00104 1.0745[20]	H_2O 1; EtOH 3; eth 3; ace 3 1.4787[20]
6072	2-Furancarboxylic acid, tetrahydro-	$C_5H_8O_3$ 116.12	16874-33-2 21	145[25]	5-18-06-00010 1.1912[20]	H_2O 4 1.4612[20]
6073	Furan, 2-chloro- 2-Chlorofuran	C_4H_3ClO 102.52	3187-94-8	78	5-17-01-00294 1.1923[20]	ace 4; eth 4; EtOH 4 1.4569[20]
6074	Furan, 3-chloro- 3-Chlorofuran	C_4H_3ClO 102.52	50689-17-3	80	5-17-01-00295 1.2094[20]	ace 4; eth 4 1.4601[20]
6075	Furan, 2-(chloromethyl)-	C_5H_5ClO 116.55	617-88-9	49[26]	5-17-01-00323 1.1783[20]	bz 4; eth 4; EtOH 4 1.4941[20]
6076	Furan, 2-(cyclohexyloxy)- Cyclohexyl 2-furyl ether	$C_{10}H_{14}O_2$ 166.22	99172-57-3	118-9[28]	4-17-00-01219 1.0200[28]	ace 4; eth 4; EtOH 4 1.4861[28]
6077	Furan, 2,3-dibromo- 2,3-Dibromofuran	$C_4H_2Br_2O$ 225.87	30544-34-4	167	5-17-01-00295 1.9938[25]	1.5458[25]
6078	Furan, 2,5-dibromo- 2,5-Dibromofuran	$C_4H_2Br_2O$ 225.87	32460-00-7 9.5	164.5	5-17-01-00295 2.27[20]	1.5455[20]
6079	2,3-Furandicarboxylic acid Isocitrate lactone	$C_6H_4O_5$ 156.09	4282-24-0 226	sub	5-18-07-00043	H_2O 4; EtOH 4; eth 2; AcOEt 4
6080	2,4-Furandicarboxylic acid	$C_6H_4O_5$ 156.09	4282-28-4 266	sub	5-18-07-00045	H_2O 2; EtOH 4; ace 4; chl 2
6081	2,5-Furandicarboxylic acid Dehydromucic acid	$C_6H_4O_5$ 156.09	3238-40-2 342	sub	5-18-07-00047 1.7400[20]	H_2O 2; EtOH 2

No.	Name Synonym	Mol. Form. Mol. Wt.	CAS RN mp/°C	Merck No. bp/°C	Beil. Ref. den/g cm^{-3}	Solubility n_D
6082	3,4-Furandicarboxylic acid, diethyl ester 2,3-Bis(ethoxycarbonyl)furan	$C_{10}H_{12}O_5$ 212.20	30614-77-8	155[13]	5-18-07-00053 1.14[25]	1.4717[20]
6083	2,5-Furandicarboxylic acid, dimethyl ester Dimethyl 2,5-furandicarboxylate	$C_8H_8O_5$ 184.15	4282-32-0 112	154[15]	5-18-07-00048	eth 4; EtOH 4; chl 4
6084	Furan, 2-(diethoxymethyl)-	$C_9H_{14}O_3$ 170.21	13529-27-6	191.5	5-17-09-00299 0.9976[20]	EtOH 4 1.4451[20]
6085	Furan, 2,2-diethyltetrahydro- 2,2-Diethyltetrahydrofuran	$C_8H_{16}O$ 128.21	1193-35-7	146	4-17-00-00107 0.8703[20]	ace 4; bz 4; eth 4; EtOH 4 1.4317[20]
6086	Furan, 2,3-dihydro- 2,3-Dihydrofuran	C_4H_6O 70.09	1191-99-7	54.5	5-17-01-00172 0.927[25]	1.4239[20]
6087	Furan, 2,5-dihydro-2,5-dimethoxy- 2,5-Dimethoxy-2,5-dihydrofuran	$C_6H_{10}O_3$ 130.14	332-77-4	161	5-17-05-00073 1.073[25]	1.4339[20]
6088	Furan, 2,4-dimethyl- 3,5-Dimethylfuran	C_6H_8O 96.13	3710-43-8	94	5-17-01-00343 0.8993[20]	1.4371[20]
6089	Furan, 2,5-dimethyl- 2,5-Dimethylfuran	C_6H_8O 96.13	625-86-5 -62.8	93.5	5-17-01-00344 0.8883[20]	H_2O 1; EtOH 3; eth 3; ace 3 1.4363[20]
6090	Furan, 2-(1,1-dimethylethyl)- 2-tert-Butylfuran	$C_8H_{12}O$ 124.18	7040-43-9	119.5	5-17-01-00371 0.869[20]	ace 4; eth 4; EtOH 4 1.4373[20]
6091	2,5-Furandione Maleic anhydride	$C_4H_2O_3$ 98.06	108-31-6 52.8	5586 202	5-17-11-00055 1.314[60]	H_2O 3; eth 3; ace 3; chl 3
6092	2,5-Furandione, 3-chloro-	C_4HClO_3 132.50	96-02-6 33	196	5-17-11-00063 1.5375[25]	1.4980[20]
6093	2,5-Furandione, dihydro- Succinic anhydride	$C_4H_4O_3$ 100.07	108-30-5 119	8841 261	5-17-11-00006 1.2[20]	H_2O 1; EtOH 3; eth 2; chl 3
6094	2,5-Furandione, dihydro-3-methyl-	$C_5H_6O_3$ 114.10	4100-80-5 34	239	5-17-11-00011 1.22[25]	
6095	2,5-Furandione, dihydro-3-methylene-	$C_5H_4O_3$ 112.08	2170-03-8 69	139[30], 114[18]	5-17-11-00066	eth 2; chl 4
6096	2,5-Furandione, dihydro-3-phenyl-, (±)- Succinic anhydride, phenyl (DL)	$C_{10}H_8O_3$ 176.17	112489-85-7 54	204[22]	5-17-11-00275	ace 4; bz 4; eth 4; EtOH 4
6097	2,5-Furandione, 3,4-dimethyl-	$C_6H_6O_3$ 126.11	766-39-2 96	223	5-17-11-00069 1.107[100]	H_2O 2; EtOH 4; eth 4; bz 4
6098	2,5-Furandione, 3-methyl-	$C_5H_4O_3$ 112.08	616-02-4 7.5	213.5	5-17-11-00065 1.2489[16]	ace 4; eth 4; EtOH 4 1.4710[21]
6099	2,5-Furandione, 3,3,4,4-tetrafluorodihydro-	$C_4F_4O_3$ 172.04	699-30-9	54.5	5-17-11-00007 1.6209[20]	1.3240[20]
6100	Furan, 2,5-diphenyl-	$C_{16}H_{12}O$ 220.27	955-83-9 91	344	5-17-02-00398	H_2O 1; EtOH 4; eth 4; ace 3
6101	Furan, 2,2'-[dithiobis(methylene)]bis-	$C_{10}H_{10}O_2S_2$ 226.32	4437-20-1 10	167[13], 112[0.5]	4-17-00-01258	EtOH 4
6102	Furan, 2-ethenyl-	C_6H_6O 94.11	1407-18-9 -94	99.5	5-17-01-00477 0.9445[19]	1.4992[19]
6103	Furan, 2-ethoxy- Ethyl 2-furyl ether	$C_6H_8O_2$ 112.13	5809-07-4	125.5	4-17-00-01210 0.9849[23]	1.4500[23]
6104	Furan, 2-(ethoxymethyl)-	$C_7H_{10}O_2$ 126.16	6270-56-0	149	5-17-03-00340 0.9844[20]	H_2O 1; EtOH 3; eth 3 1.4523[20]
6105	Furan, 2-ethyl- 2-Ethylfuran	C_6H_8O 96.13	3208-16-0	92	5-17-01-00338 0.911[15]	EtOH 3; eth 3; ace 3; bz 3 1.4466[23]
6106	Furan, 2-ethyl-5-methyl- 2-Methyl-5-ethylfuran	$C_7H_{10}O$ 110.16	1703-52-2	118.5	5-17-01-00357 0.8883[20]	bz 4; eth 4; EtOH 4 1.4473[20]
6107	Furan, 2-ethyltetrahydro- 2-Ethyltetrahydrofuran	$C_6H_{12}O$ 100.16	1003-30-1	109	5-17-01-00095 0.8570[19]	ace 4; bz 4; eth 4; EtOH 4 1.4147[19]
6108	Furan, 2-ethynyl-	C_6H_4O 92.10	18649-64-4	105.5	4-17-00-00396 0.9919[20]	1.5055[20]
6109	Furan, 2,2,3,3,4,4,5-heptafluorotetrahydro-5-(nonafluorobutyl)- Perfluoro-2-butyltetrahydrofuran	$C_8F_{16}O$ 416.06	335-36-4	102.6		
6110	Furan, 2-iodo- 2-Furyl iodide	C_4H_3IO 193.97	54829-48-0	43-5[15]	5-17-01-00295 2.024[20]	eth 4 1.5661[20]
6111	Furan, 3-iodo- 3-Iodofuran	C_4H_3IO 193.97	29172-20-1	133; 37-8[22]	5-17-01-00296 2.045[20]	H_2O 1; eth 3 1.5610[20]
6112	Furan, 2-isopropoxy- 2-Furyl isopropyl ether	$C_7H_{10}O_2$ 126.16	98272-34-5	135.5	4-17-00-01219 0.9689[20]	1.4419[20]
6113	2-Furanmethanamine Furfurylamine	C_5H_7NO 97.12	617-89-0	145.5	5-18-09-00541 1.0995[20]	H_2O 5; EtOH 5; eth 3; chl 3 1.4908[20]
6114	2-Furanmethanamine, N-(2-furanylmethyl)- Difurfurylamine	$C_{10}H_{11}NO_2$ 177.20	18240-50-1	135-42[15]	4-18-00-07092 1.1045[20]	eth 4 1.5168[20]
6115	2-Furanmethanamine, N-methyl-	C_6H_9NO 111.14	4753-75-7	149	5-18-09-00541 0.989[25]	1.4729[20]
6116	2-Furanmethanamine, tetrahydro- Tetrahydrofurfurylamine	$C_5H_{11}NO$ 101.15	4795-29-3	153	5-18-09-00488 0.9752[20]	H_2O 4; eth 4; EtOH 4 1.4551[20]
6117	2-Furanmethanethiol	C_5H_6OS 114.17	98-02-2	157	5-17-03-00351 1.1319[20]	H_2O 1; chl 2 1.5329[20]
6118	2-Furanmethanol Furfuryl alcohol	$C_5H_6O_2$ 98.10	98-00-0 -31	4215 171	5-17-03-00338 1.1296[20]	H_2O 5; EtOH 4; eth 4; chl 3 1.4869[20]
6119	2-Furanmethanol, acetate	$C_7H_8O_3$ 140.14	623-17-6	179	5-17-03-00344 1.1175[20]	H_2O 1; EtOH 3; eth 3 1.4327[20]
6120	2-Furanmethanol, 5-(aminomethyl)tetrahydro-	$C_6H_{13}NO_2$ 131.17	589-14-0 <-70	103[0.3]	1.1021[25]	1.4870[25]

No.	Name / Synonym	Mol. Form. / Mol. Wt.	CAS RN / mp/°C	Merck No. / bp/°C	Beil. Ref. / den/g cm^{-3}	Solubility / n_D
6121	2-Furanmethanol, α-methyl-	$C_6H_8O_2$ 112.13	4208-64-4	162.5	5-17-03-00374 1.0739[25]	1.4827[15]
6122	2-Furanmethanol, 5-methyl-	$C_6H_8O_2$ 112.13	3857-25-8	195 dec; 81[23]	5-17-03-00387 1.0769[20]	eth 4; EtOH 4 1.4853[20]
6123	2-Furanmethanol, propanoate	$C_8H_{10}O_3$ 154.17	623-19-8	195	2-17-00-00115 1.1085[20]	H_2O 2; EtOH 3; eth 5; ace 3
6124	2-Furanmethanol, tetrahydro- Tetrahydrofurfuryl alcohol	$C_5H_{10}O_2$ 102.13	97-99-4 <-80	9146 178	5-17-03-00115 1.0524[20]	ace 4; eth 4 1.4520[20]
6125	2-Furanmethanol, tetrahydro-, acetate	$C_7H_{12}O_3$ 144.17	637-64-9	193; 89[18]	5-17-03-00119 1.0624[20]	H_2O 4; eth 4; EtOH 4; chl 4 1.4350[25]
6126	2-Furanmethanol, tetrahydro-, propanoate	$C_8H_{14}O_3$ 158.20	637-65-0	205.5	5-17-03-00119 1.044[20]	eth 4; EtOH 4; chl 4
6127	Furan, 2-methoxy- 2-Methoxyfuran	$C_5H_6O_2$ 98.10	25414-22-6	110.5	5-17-03-00283 1.0646[25]	1.4468[25]
6128	Furan, 2-(methoxymethyl)-	$C_6H_8O_2$ 112.13	13679-46-4	132	5-17-03-00340 1.0163[20]	H_2O 1; EtOH 3; eth 4 1.4570[20]
6129	Furan, 2-(methoxymethyl)-5-nitro- 2-(Methoxymethyl)-5-nitrofuran	$C_6H_7NO_4$ 157.13	586-84-5 104-5[3]	5917	4-17-00-01254 1.281[20]	EtOH 4 1.5325[20]
6130	Furan, 2-methyl- 2-Methylfuran	C_5H_6O 82.10	534-22-5 -87.5	65	5-17-01-00322 0.9132[20]	H_2O 2; EtOH 3; eth 3; ctc 2 1.4342[20]
6131	Furan, 3-methyl- 3-Methylfuran	C_5H_6O 82.10	930-27-8	66	5-17-01-00330 0.923[18]	H_2O 1; EtOH 3; eth 3 1.4330[19]
6132	Furan, 2,2'-methylenebis-	$C_9H_8O_2$ 148.16	1197-40-6	94[22]	5-19-01-00545 1.102[20]	H_2O 1; EtOH 4; eth 4; ace 4 1.5049[20]
6133	Furan, 3-(4-methyl-3-pentenyl)-	$C_{10}H_{14}O$ 150.22	539-52-6	185.5	5-17-01-00515 0.9017[20]	1.4705[21]
6134	Furan, 2-nitro-	$C_4H_3NO_3$ 113.07 30	609-39-2	134[123]; 84[13]	5-17-01-00296	H_2O 3; EtOH 3; eth 3
6135	Furan, 2-(octyloxy)- 2-Furyl octyl ether	$C_{12}H_{20}O_2$ 196.29	100314-90-7	129-30[18]	4-17-00-01219 0.9214[28]	eth 4 1.4520[20]
6136	2-Furanol, 3,4-bis(1,3-benzodioxol-5-ylmethyl)tetrahydro- Cubebin	$C_{20}H_{20}O_6$ 356.38 131.5	18423-69-3	2616	5-19-12-00048	eth 4; EtOH 4; chl 4
6137	3-Furanol, tetrahydro-	$C_4H_8O_2$ 88.11	453-20-3	181	5-17-03-00064 1.09[25]	1.4500[20]
6138	2-Furanol, tetrahydro-, benzoate	$C_{11}H_{12}O_3$ 192.21	3333-44-6	302; 139[2]	5-17-03-00060 1.137[20]	eth 4; EtOH 4
6139	2(3H)-Furanone, 3-acetyldihydro- α-Acetylbutyrolactone	$C_6H_8O_3$ 128.13	517-23-7 77	107-8[5]	5-17-11-00016 1.1846[20]	H_2O 4 1.4585[20]
6140	2(3H)-Furanone, 5-butyldihydro-	$C_8H_{14}O_2$ 142.20	104-50-7	132-3[20]	5-17-09-00073 0.9796[19]	EtOH 3; ctc 2 1.4451[19]
6141	2(3H)-Furanone, 5-(chloromethyl)dihydro- Valeric acid, 5-chloro-4-hydroxy-, γ-lactone	$C_5H_7ClO_2$ 134.56	39928-72-8	132-5[12]	5-17-09-00025 1.625[18]	EtOH 2; eth 2; bz 2
6142	2(3H)-Furanone, dihydro- γ-Butyrolactone	$C_4H_6O_2$ 86.09	96-48-0 -43.3	1596 204	5-17-09-00007 1.1284[16]	ace 4; bz 4; eth 4; EtOH 4 1.4341[20]
6143	2(3H)-Furanone, dihydro-3-hydroxy-4,4-dimethyl-, (R)- Pantolactone	$C_6H_{10}O_3$ 130.14 92	599-04-2	6963	5-18-01-00022	
6144	2(3H)-Furanone, dihydro-3-(hydroxyphenylmethyl)-4-[(1-methyl-1H-imidazol-5-yl)methyl]-, Isopilosine	$C_{16}H_{18}N_2O_3$ 286.33 187	491-88-3	5085	1-27-00-00612	EtOH 4
6145	2(3H)-Furanone, dihydro-3-methyl-	$C_5H_8O_2$ 100.12	1679-47-6	78-81[10]	5-17-09-00028 1.063[25]	1.4325[20]
6146	2(3H)-Furanone, dihydro-5-methyl-, (±)-	$C_5H_8O_2$ 100.12	57129-69-8 -31	206	5-17-09-00024 1.0465[25]	H_2O 5; EtOH 3; ace 3; ctc 2 1.4328[20]
6147	2(3H)-Furanone, dihydro-3-methylene- α-Methylene butyrolactone	$C_5H_6O_2$ 98.10	547-65-9	5981 85-6[10]	5-17-09-00127 1.1206[20]	H_2O 3; EtOH 4; eth 3; ace 3 1.4650[20]
6148	2(3H)-Furanone, dihydro-5-methylene-	$C_5H_6O_2$ 98.10	10008-73-8	85-6[10]	5-17-09-00124 1.1206[20]	ace 4; bz 4; eth 4; EtOH 4 1.4650[20]
6149	2(3H)-Furanone, dihydro-5-pentyl- 4-Hydroxynonanoic acid lactone	$C_9H_{16}O_2$ 156.22	104-61-0		5-17-09-00084	
6150	2(3H)-Furanone, 5-ethyldihydro-	$C_6H_{10}O_2$ 114.14	695-06-7 -18	215.5	5-17-09-00040 1.0261[20]	H_2O 4; EtOH 4 1.4495[20]
6151	2(3H)-Furanone, 3-ethyldihydro-4-[(1-methyl-1H-imidazol-5-yl)methyl]-, (3S-cis)- Pilocarpine	$C_{11}H_{16}N_2O_2$ 208.26 34	92-13-7	7395 260[5]	2-27-00-00694	H_2O 3; EtOH 3; eth 2; bz 2
6152	2(3H)-Furanone, 5-heptyldihydro- 4-Hydroxyundecanoic acid lactone	$C_{11}H_{20}O_2$ 184.28	104-67-6	286	5-17-09-00098 0.9494[20]	EtOH 4 1.4512[20]
6153	2(3H)-Furanone, 5-methyl-	$C_5H_6O_2$ 98.10	591-12-8 18	56[12]	5-17-09-00120 1.084[20]	H_2O 3; EtOH 3; eth 3; ctc 2 1.4476[20]
6154	2(5H)-Furanone, 5-methyl-	$C_5H_6O_2$ 98.10	591-11-7 <-17	209; 98[15]	5-17-09-00121 1.0810[20]	H_2O 5; EtOH 3; eth 3 1.4454[20]
6155	2(5H)-Furanone, 5-methylene- Protoanemonin	$C_5H_4O_2$ 96.09	108-28-1	7907 73[11]	5-17-09-00371	H_2O 2; chl 3
6156	Furan, 2,2'-[oxybis(methylene)]bis-	$C_{10}H_{10}O_3$ 178.19	4437-22-3	101[2]	5-17-03-00347 1.1405[20]	H_2O 1 1.5088[20]
6157	Furan, 2-phenoxy- 2-Furyl phenyl ether	$C_{10}H_8O_2$ 160.17	60698-31-9	105-6[18]	4-17-00-01219 1.1010[23]	eth 3 1.5418[23]

No.	Name Synonym	Mol. Form. Mol. Wt.	CAS RN	mp/°C bp/°C	Merck No.	Beil. Ref. den/g cm^{-3}	Solubility n_D
6158	Furan, 2-(phenoxymethyl)- Furfuryl phenyl ether	$C_{11}H_{10}O_2$ 174.20	4437-23-4	139-40[20]		5-17-03-00340 1.123[17]	1.5535[17]
6159	Furan, 2-phenyl- 2-Phenylfuran	$C_{10}H_8O$ 144.17	17113-33-6	107-8[18]		5-17-02-00139 1.083[20]	ace 4; bz 4 1.5920[20]
6160	Furan, 2-(phenylthio)- 2-Furyl phenyl sulfide	$C_{10}H_8OS$ 176.24	16003-14-8	119-20[8]		5-17-03-00285 1.1341[26]	eth 4; EtOH 4 1.5976[20]
6161	2-Furanpropanoic acid	$C_7H_8O_3$ 140.14	935-13-7	58 229		5-18-06-00224	H_2O 4; eth 4; chl 4
6162	2-Furanpropanoic acid, methyl ester	$C_8H_{10}O_3$ 154.17	37493-31-5	89[15]		5-18-06-00224 1.088[20]	1.4662[20]
6163	2-Furanpropanoic acid, β-oxo-, ethyl ester	$C_9H_{10}O_4$ 182.18	615-09-8	142-3[10]		5-18-08-00143 1.165[17]	H_2O 1; EtOH 3; eth 3
6164	2-Furanpropanoic acid, tetrahydro- Propionic acid, 3-(2-tetrahydrofuryl)	$C_7H_{12}O_3$ 144.17	935-12-6	263		5-18-06-00033 1.1135[20]	ace 4; EtOH 4 1.4578[25]
6165	2-Furanpropanoic acid, tetrahydro-, ethyl ester	$C_9H_{16}O_3$ 172.22	4525-36-4	222; 73[2]		5-18-06-00034 1.023[7]	ace 4; EtOH 4 1.440[20]
6166	2-Furanpropanol, tetrahydro-α-[2-(tetrahydro-2-furanyl)ethyl]- 1,5-Bis(2-tetrahydrofuryl)-3-pentanol	$C_{13}H_{24}O_3$ 228.33	6265-26-5	116[0.05]		4-19-00-00704	chl 2 1.4541[25]
6167	Furan, 2-propyl- 2-Propylfuran	$C_7H_{10}O$ 110.16	4229-91-8	115		5-17-01-00354 0.8876[20]	EtOH 3; eth 3; ace 3 1.4549[20]
6168	Furan, tetrahydro- Tetrahydrofuran	C_4H_8O 72.11	109-99-9	-108.3 65	9144	5-17-01-00027 0.8892[20]	H_2O 3; EtOH 4; eth 4; ace 4 1.4050[20]
6169	Furan, tetrahydro-2,5-dimethoxy-	$C_6H_{12}O_3$ 132.16	696-59-3	145.7		5-17-05-00004 1.02[25]	1.4180[20]
6170	Furan, tetrahydro-2-(methoxymethyl)-	$C_6H_{12}O_2$ 116.16	19354-27-9	140		5-17-03-00116 0.952[25]	1.4270[20]
6171	Furan, tetrahydro-2-methyl- 2-Methyltetrahydrofuran	$C_5H_{10}O$ 86.13	96-47-9	78		5-17-01-00078 0.8552[20]	H_2O 3, EtOH 4; eth 4; acc 4 1.4059[21]
6172	Furan, tetrahydro-3-methyl- 3-Methyltetrahydrofuran	$C_5H_{10}O$ 86.13	13423-15-9	86.5		5-17-01-00081 0.8642[20]	ace 4; bz 4; eth 4; EtOH 4 1.4122[20]
6173	Furan, tetrahydro-2-propyl-	$C_7H_{14}O$ 114.19	3208-22-8	136		5-17-01-00116 0.8547[20]	1.4242[20]
6174	Furan, tetrahydro-2,2,4,4-tetramethyl- 2,2,4,4-Tetramethyltetrahydrofuran	$C_8H_{16}O$ 128.21	3358-28-9	121		4-17-00-00109	H_2O 1; EtOH 5; eth 5 1.4074[20]
6175	Furan, tetraphenyl- 2,3,4,5-Tetraphenylfuran	$C_{28}H_{20}O$ 372.47	1056-77-5	175 220		4-17-00-00810	H_2O 1; EtOH 3; eth 3; ace 3
6176	Furazan, dimethyl-	$C_4H_6N_2O$ 98.10	4975-21-7	-7 156		4-27-00-07096 1.0528[14]	eth 4; EtOH 4 1.4327[20]
6177	3aH-Furo[2,3-b]indol-3a-ol, 8a-(5-ethenyl-1-azabicyclo[2.2.2]oct-2-yl)-2,3,8,8a-tetrahydro- Conquinamine	$C_{19}H_{24}N_2O_2$ 312.41	464-86-8	123	2505	2-27-00-00667	H_2O 2; EtOH 3; eth 3; chl 2
6178	3aH-Furo[2,3-b]indol-3a-ol, 8a-(5-ethenyl-1-azabicyclo[2.2.2]oct-2-yl)-2,3,8,8a-tetrahydro- Quinamine	$C_{19}H_{24}N_2O_2$ 312.41	464-85-7	185.5	8060	2-27-00-00667	H_2O 1; EtOH 4; eth 3; ace 3
6179	Furo[2,3-b]quinoline, 4,8-dimethoxy- Fagarine	$C_{13}H_{11}NO_3$ 229.24	524-15-2	142	3879	4-27-00-02211	H_2O 2; EtOH 3; eth 3; bz 3
6180	Furo[2,3-b]quinoline, 4-methoxy- Dictamnine	$C_{12}H_9NO_2$ 199.21	484-29-7	133.5	3079	4-27-00-02030	H_2O 2; EtOH 4; eth 3; chl 3
6181	Furo[2,3-b]quinoline, 4,7,8-trimethoxy- Skimmianine	$C_{14}H_{13}NO_4$ 259.26	83-95-4	177	8505	4-27-00-02296	H_2O 1; EtOH 3; eth 2; chl 3
6182	Furo[2,3-b]quinolin-4(2H)-one, 3,9-dihydro-8-methoxy-9-methyl-2-(1-methylethyl)-, (R)- Lunacrine	$C_{16}H_{19}NO_3$ 273.33	82-40-6		5473	4-27-00-03566	chl 3
6183	4H-Furo[3,2-c]pyran-2(6H)-one, 4-hydroxy- Patulin	$C_7H_6O_4$ 154.12	149-29-1	111	7002	5-18-03-00005	H_2O 3; EtOH 3; eth 3; ace 3
6184	1H,12H-Furo[3',2':4,5]furo[2,3-h]pyrano[3,4-c][1]benzopyran-1,12-dione, 3,4,7,4,7a,10a-tetrahydro-5-meth Aflatoxin G1	$C_{17}H_{12}O_7$ 328.28	1165-39-5	245		5-19-12-00081	
6185	5H-Furo[3,2-g][1]benzopyran-5-one, 4,9-dimethoxy-7-methyl- Khellin	$C_{14}H_{12}O_5$ 260.25	82-02-0	154-5 dec 180-200[0.05]	5189	5-19-06-00320	H_2O 1; EtOH 3; eth 2; ace 3
6186	5H-Furo[3,2-g][1]benzopyran-5-one, 4-methoxy-7-methyl- Visnagin	$C_{13}H_{10}O_4$ 230.22	82-57-5	144.5	9916	5-19-06-00030	H_2O 2; EtOH 2; chl 4
6187	7H-Furo[3,2-g][1]benzopyran-7-one, 9-methoxy- Methoxsalen	$C_{12}H_8O_4$ 216.19	298-81-7	148	5911	5-19-06-00015	H_2O 2; EtOH 4; eth 2; ace 2
6188	7H-Furo[3,2-g][1]benzopyran-7-one, 3-methoxy-2-(1-methylethyl)- Peucedanin	$C_{15}H_{14}O_4$ 258.27	133-26-6	85 276-81[17]	7144	5-19-06-00042	H_2O 2; EtOH 3; eth 3; bz 2
6189	7H-Furo[3,2-g][1]benzopyran-7-one, 9-[(3-methyl-2-butenyl)oxy]- Imperatorin	$C_{16}H_{14}O_4$ 270.28	482-44-0	102	4839	5-19-06-00016	H_2O 2; EtOH 3; eth 3; bz 3
6190	3-Furoic acid, 5-bromo-, ethyl ester β-Furoic acid, 5-bromo, ethyl ester	$C_7H_7BrO_3$ 219.03	32460-20-1	17 235; 135[34]		5-18-06-00198 1.528[20]	eth 4; EtOH 4
6191	Furo[3',4':6,7]naphtho[2,3-d]-1,3-dioxol-6(5aH)-one, 5,8,8a-tetrahydro-9-hydroxy-5-(3,4,5-trimethoxyphenyl)- Picropodophyllin	$C_{22}H_{22}O_8$ 414.41	477-47-4	228	7385	5-19-10-00665	ace 4; bz 4; eth 4; EtOH 4

No.	Name / Synonym	Mol. Form. / Mol. Wt.	CAS RN / mp/°C	Merck No. / bp/°C	Beil. Ref. / den/g cm^{-3}	Solubility / n_D
6192	Furo[3',4':6,7]naphtho[2,3-d]-1,3-dioxol-6(5aH)-one, 5,8,8a,9-tetrahydro-9-hydroxy-5-(3,4,5-trimetho / Podophyllotoxin	$C_{22}H_{22}O_8$ / 414.41	518-28-5 / 183	7520	5-19-10-00666	H_2O 2; EtOH 4; eth 1; ace 3
6193	Galactaric acid / Mucic acid	$C_6H_{10}O_8$ / 210.14	526-99-8 / 255 dec	4237	4-03-00-01292	
6194	Galactitol / Dulcose	$C_6H_{14}O_6$ / 182.17	608-66-2 / 189.5	4238 / 275-80[1]	4-01-00-02844 / 1.47[20]	H_2O 3; EtOH 2; eth 1; bz 1
6195	β-D-Galactopyranoside, (2α,3β,5α,15β,25R)-2,15-dihydroxyspirostan-3-yl / Digitonin	$C_{56}H_{92}O_{29}$ / 1229.33	11024-24-1 / 237.5	3144	5-19-03-00603	
6196	β-D-Galactopyranoside, (3β,5α,22β,25S)-spirosolan-3-yl O-β- / Tomatine	$C_{50}H_{83}NO_{21}$ / 1034.20	17406-45-0 / 270	9471	4-27-00-01954	EtOH 4; diox 4
6197	D-Galactose	$C_6H_{12}O_6$ / 180.16	59-23-4 / 170	4241	4-01-00-04336	H_2O 4; EtOH 2; eth 1; bz 1
6198	Galactose, 6-deoxy-3-O-methyl- / Digitalose	$C_7H_{14}O_5$ / 178.19	4481-08-7 / 119	3142	4-01-00-04270	H_2O 4
6199	D-Galacturonic acid	$C_6H_{10}O_7$ / 194.14	685-73-4 / 166	4242	1-03-00-00306	H_2O 3; EtOH 3; eth 1
6200	Galanthamine / Lycoremine	$C_{17}H_{21}NO_3$ / 287.36	357-70-0 / 126.5	4245	4-27-00-02184	ace 4; EtOH 4; chl 4
6201	Galanthan-1,2-diol, 3,12-didehydro-9,10-[methylenebis(oxy)]-, (1α,2β)- / Lycorine	$C_{16}H_{17}NO_4$ / 287.32	476-28-8 / 280 sub	5498	2-27-00-00547	H_2O 1; EtOH 2; eth 2; chl 2
6202	Gardona / Phosphoric acid, 2-chloro-1-(2,4,5-trichlorophenyl)ethenyl dimethyl ester	$C_{10}H_9Cl_4O_4P$ / 365.96	961-11-5			
6203	Gelsemine / Spiro[3,5,8-ethanylylidene-1H-pyrano[3,4-c]pyridine-10,3'-[3H]indol]-2'(1'H)-one, 5-ethenyl	$C_{20}H_{22}N_2O_2$ / 322.41	509-15-9 / 178	4277	4-27-00-07526	ace 4; bz 4; eth 4; EtOH 4
6204	Gibb-3-ene-1,10-dicarboxylic acid, 2,4a,7-trihydroxy-1-methyl-8-methylene-, 1,4a-lactone, / Gibberellic acid	$C_{19}H_{22}O_6$ / 346.38	77-06-5 / 234	4313	5-18-09-00269	ace 4; EtOH 4; MeOH 4
6205	D-Glucaric acid / D-Tetrahydroxyadipic acid	$C_6H_{10}O_8$ / 210.14	87-73-0 / 125.5	4344	4-03-00-01291	H_2O 4; EtOH 4; eth 2; chl 2
6206	D-Glucitol / Sorbitol	$C_6H_{14}O_6$ / 182.17	50-70-4 / 111	8680 / 295[3.5]	4-01-00-02839 / 1.489[20]	H_2O 4; ace 4 / 1.3330[20]
6207	D-Glucitol, 1-amino-1-deoxy- / Glucamine	$C_6H_{15}NO_5$ / 181.19	488-43-7 / 127	4343	4-04-00-01913	H_2O 4; EtOH 4
6208	D-Glucitol, 1-deoxy-1-(methylamino)- / N-Methylglucamine	$C_7H_{17}NO_5$ / 195.22	6284-40-8 / 128.5	5995	4-04-00-01914	H_2O 3
6209	D-Gluconic acid / Gluconic acid	$C_6H_{12}O_7$ / 196.16	526-95-4 / 131	4350	4-03-00-01255	H_2O 3; EtOH 2; eth 1; bz 1
6210	D-Gluconic acid, 4-O-β-D-galactopyranosyl- / Lactobionic acid	$C_{12}H_{22}O_{12}$ / 358.30	96-82-2	5219	5-17-07-00436	H_2O 4; EtOH 2; eth 1; HOAc 2
6211	Gluconic acid, β-lactone / Oxetane, gluconic acid deriv.	$C_6H_{10}O_6$ / 178.14	3087-62-5 / 153		2-18-00-00190 / 1.610[-5]	H_2O 2; EtOH 2; eth 1; ace 1
6212	D-Gluconic acid, δ-lactone / δ-D-Gluconolactone	$C_6H_{10}O_6$ / 178.14	90-80-2	4351	5-18-05-00011	
6213	D-Gluconic acid, monosodium salt / Gluconic acid, sodium salt	$C_6H_{11}NaO_7$ / 218.14	527-07-1	8569	4-03-00-01255	H_2O 3
6214	β-D-Glucopyranose, 4-O-β-D-galactopyranosyl-	$C_{12}H_{22}O_{11}$ / 342.30	5965-66-2 / 254		5-17-07-00196 / 1.59[20]	H_2O 4; EtOH 2; eth 1; chl 1
6215	α-L-Glucopyranose, 2-(methylamino)-2-deoxy- / N-Methyl-α-L-glucosamine	$C_7H_{15}NO_5$ / 193.20	42852-95-9	5996		
6216	α-D-Glucopyranose, pentaacetate	$C_{16}H_{22}O_{11}$ / 390.34	604-68-2 / 113.3 sub		5-17-07-00318	H_2O 2; EtOH 3; chl 3
6217	β-D-Glucopyranose, pentaacetate	$C_{16}H_{22}O_{11}$ / 390.34	604-69-3 / 134 sub		5-17-07-00319 / 1.274[20]	H_2O 1; EtOH 2; eth 2; bz 3
6218	α-D-Glucopyranoside, β-D-fructofuranosyl / Sucrose	$C_{12}H_{22}O_{11}$ / 342.30	57-50-1 / 185.5	8855	5-17-08-00399 / 1.5805[17]	H_2O 3; EtOH 2; eth 1; py 3 / 1.5376
6219	α-D-Glucopyranoside, β-D-fructofuranosyl O-α-D-galactopyranosyl-(1 / Raffinose	$C_{18}H_{32}O_{16}$ / 504.44	512-69-6 / 80	8120	4-17-00-03801 / 1.465[25]	H_2O 4; EtOH 2; eth 1; MeOH 4
6220	α-D-Glucopyranoside, α-D-glucopyranosyl / Trehalose	$C_{12}H_{22}O_{11}$ / 342.30	99-20-7 / 203	9496	1.58[24]	H_2O 4; EtOH 3; eth 1; bz 1
6221	α-D-Glucopyranoside, O-α-D-glucopyranosyl-(13)-β-D-fructofuranosyl / Melezitose	$C_{18}H_{32}O_{16}$ / 504.44	597-12-6 / 153	5698	5-17-08-00414 / 1.5565[25]	H_2O 4
6222	β-D-Glucopyranoside, 2-(hydroxymethyl)phenyl / Salicin	$C_{13}H_{18}O_7$ / 286.28	138-52-3 / 207	8293 / 240 dec	5-17-07-00113 / 1.434[20]	H_2O 4; EtOH 4; HOAc 4
6223	β-D-Glucopyranoside, 2-(hydroxymethyl)phenyl, 6-benzoate / Populin	$C_{20}H_{22}O_8$ / 390.39	99-17-2 / 180	7571	5-17-07-00343	

No.	Name / Synonym	Mol. Form. / Mol. Wt.	CAS RN / mp/°C	Merck No. / bp/°C	Beil. Ref. / den/g cm^{-3}	Solubility / n_D
6224	β-D-Glucopyranoside, 4-hydroxyphenyl / Arbutin	$C_{12}H_{16}O_7$ / 272.25	497-76-7 / 199.5	799	5-17-07-00110	H_2O 4; EtOH 3; eth 2; bz 1
6225	β-D-Glucopyranoside, 4-(3-hydroxy-1-propenyl)-2,6-dimethoxyphenyl / Syringin	$C_{17}H_{24}O_9$ / 372.37	118-34-3 / 192	8997	4-17-00-03002	EtOH 4
6226	β-D-Glucopyranoside, 4-(3-hydroxy-1-propenyl)-2-methoxyphenyl / Coniferin	$C_{16}H_{22}O_8$ / 342.35	531-29-3 / 186	2498	4-17-00-02999	H_2O 3; EtOH 2; eth 1; py 3
6227	α-D-Glucopyranoside, methyl / α-Methylglucoside	$C_7H_{14}O_6$ / 194.18	97-30-3 / 168	5997 $200^{0.2}$	5-17-07-00010 / 1.46^{30}	H_2O 4
6228	α-D-Glucopyranoside, methyl 2,3,4,6-tetra-O-methyl- / α-D-Glucose, pentamethyl ether	$C_{11}H_{22}O_6$ / 250.29	605-81-2 / $180^{0.4}$		5-17-07-00029 / 1.0944^{20}	H_2O 4; ace 4; eth 4; EtOH 4 / 1.4466^{20}
6229	α-D-Glucopyranoside, 1,3,4,6-tetra-O-acetyl-β-D-fructofuranosyl, tetraacetate / Octaacetyl sucrose	$C_{28}H_{38}O_{19}$ / 678.60	126-14-7 / 86.5	8856 250^1	5-17-08-00410 / 1.27^{16}	H_2O 2; EtOH 3; eth 3; ace 3 / 1.4660
6230	α-D-Glucopyranosiduronic acid, (3β,20β)-20-carboxy-11-oxo-30-norolean-12-en-3-yl 2-O- / Glycyrrhizic acid	$C_{42}H_{62}O_{16}$ / 822.94	1405-86-3 / 220 dec	4401	4-18-00-05156	H_2O 4
6231	β-D-Glucopyranosiduronic acid, 2,2,2-trichloroethyl / Urochloralic acid	$C_8H_{11}Cl_3O_7$ / 325.53	97-25-6 / 142	9798	4-18-00-05113	H_2O 4; EtOH 4
6232	β-D-Glucose	$C_6H_{12}O_6$ / 180.16	28905-12-6 / 150		4-01-00-04306 / 1.5620^{18}	H_2O 4; EtOH 2; eth 1; py 3
6233	D-Glucose, 2-amino-2-deoxy- / D-Glucosamine	$C_6H_{13}NO_5$ / 179.17	3416-24-8	4352	4-04-00-02017	H_2O 4
6234	D-Glucose, 6-O-α-L-arabinopyranosyl- / Vicianose	$C_{11}H_{20}O_{10}$ / 312.27	14116-69-9 / 210 dec	9879	5-17-06-00112	H_2O 4
6235	D-Glucose, 6-deoxy- / Quinovose	$C_6H_{12}O_5$ / 164.16	7658-08-4 / 139.5	8106	4-01-00-04260	H_2O 4; EtOH 4
6236	D-Glucose, 6-O-(6-deoxy-α-L-mannopyranosyl)- / Rutinose	$C_{12}H_{22}O_{10}$ / 326.30	90-74-4 / 189-92 dec	8277	5-17-06-00269	H_2O 4; EtOH 4
6237	D-Glucose, 6-O-β-D-xylopyranosyl- / Primeverose	$C_{11}H_{20}O_{10}$ / 312.27	26531-85-1 / 210	7752	4-17-00-02447	H_2O 4; MeOH 4
6238	D-Glucuronic acid	$C_6H_{10}O_7$ / 194.14	6556-12-3 / 165	4360	4-03-00-01996	H_2O 4, EtOH 4
6239	D-Glucuronic acid, γ-lactone / D-Glucuronolactone	$C_6H_8O_6$ / 176.13	32449-92-6 / 177.5	4362	5-18-05-00033 / 1.76^{20}	H_2O 3; EtOH 2; bz 1; MeOH 2
6240	DL-Glutamic acid	$C_5H_9NO_4$ / 147.13	617-65-2 / 199 dec		4-04-00-03028 / 1.4601^{20}	H_2O 2; EtOH 1; eth 2, CS_2 1
6241	L-Glutamic acid / (S)-2-Aminopentanedioic acid	$C_5H_9NO_4$ / 147.13	56-86-0 / 224 dec	4363 sub 175	4-04-00-03028 / 1.538^{20}	
6242	L-Glutamic acid, N-(4-aminobenzoyl)- / N-(p-Aminobenzoyl)-L-(+)-glutamic acid	$C_{12}H_{14}N_2O_5$ / 266.25	4271-30-1 / 173	437	4-14-00-01153	
6243	L-Glutamic acid, N-[4-[[(2-amino-1,4-dihydro-4-oxo-6-pteridinyl)methyl]amino]benzoyl]- / Folic acid	$C_{19}H_{19}N_7O_6$ / 441.40	59-30-3 / 250 dec	4140	4-26-00-03944	py 4; EtOH 4; HOAc 4
6244	L-Glutamic acid, N-[4-[[(2-amino-5-formyl-1,4,5,6,7,8-hexahydro-4-oxo-6-pteridinyl)methyl]amino]benzoyl]- / Folinic acid	$C_{20}H_{23}N_7O_7$ / 473.45	58-05-9 / 245 dec	4141	4-26-00-03881	H_2O 2
6245	L-Glutamic acid, 5-ethyl ester / γ-Ethyl L-glutamate	$C_7H_{13}NO_4$ / 175.18	1119-33-1 / 191	4364	4-04-00-03033	H_2O 2
6246	DL-Glutamic acid, 3-hydroxy-	$C_5H_9NO_5$ / 163.13	5985-23-9 / 209	dec	4-04-00-03253	H_2O 4; EtOH 1; eth 1; bz 1
6247	L-Glutamic acid, 5-[2-[4-(hydroxymethyl)phenyl]hydrazide] / Agaritine	$C_{12}H_{17}N_3O_4$ / 267.28	2757-90-6 / 205-9 dec	175		H_2O 4
6248	L-Glutamic acid, monosodium salt / Glutamic acid, monosodium salt	$C_5H_8NNaO_4$ / 169.11	142-47-2	6165		H_2O 3
6249	L-Glutamine / 2-Aminoglutaramic acid	$C_5H_{10}N_2O_3$ / 146.15	56-85-9 / 185-6 dec	4365	4-04-00-03038	H_2O 3; EtOH 1; eth 1; bz 1
6250	Glycine / Aminoacetic acid	$C_2H_5NO_2$ / 75.07	56-40-6 / 262 dec	4386	4-04-00-02349 / 1.161^{20}	H_2O 4; EtOH 1; eth 1; ace 2
6251	Glycine, N-acetyl- / Aceturic acid	$C_4H_7NO_3$ / 117.10	543-24-8 / 206	74	4-04-00-02399	H_2O 4; ace 4; EtOH 4
6252	Glycine, N-(4-aminobenzoyl)- / p-Aminohippuric acid	$C_9H_{10}N_2O_3$ / 194.19	61-78-9 / 198.5	454	4-14-00-01152	ace 4; bz 4; EtOH 4
6253	Glycine, N-(aminoiminomethyl)- / Glycocyamine	$C_3H_7N_3O_2$ / 117.11	352-97-6 / >300	4391	4-04-00-02414	H_2O 2; EtOH 2; eth 2
6254	Glycine, N-(aminoiminomethyl)-N-methyl- / Creatine	$C_4H_9N_3O_2$ / 131.13	57-00-1 / 303 dec	2570	4-04-00-02426 / 1.33^{25}	H_2O 3; EtOH 2; eth 1
6255	Glycine, N-benzoyl- / Hippuric acid	$C_9H_9NO_3$ / 179.18	495-69-2 / 191.5	4636	4-09-00-00778 / 1.371^{20}	H_2O 3; EtOH 3; eth 2; bz 2
6256	Glycine, N,N-bis(carboxymethyl)- / Nitrilotriacetic acid	$C_6H_9NO_6$ / 191.14	139-13-9 / 242 dec	6499	4-04-00-02441	H_2O 2; EtOH 3; DMSO-d_6 2

No.	Name Synonym	Mol. Form. Mol. Wt.	CAS RN mp/°C	Merck No. bp/°C	Beil. Ref. den/g cm^{-3}	Solubility n_D
6257	Glycine, N,N-bis[2-[bis(carboxymethyl)amino]ethyl]- Pentetic acid	$C_{14}H_{23}N_3O_{10}$ 393.35	67-43-6	7083	4-04-00-02454	
6258	Glycine, N,N-bis(ethoxycarbonyl)-, ethyl ester Aminoacetic acid, N,N-(dicarboethoxy), ethyl ester	$C_{10}H_{17}NO_6$ 247.25	127665-32-1 36.5	152[10]	0-04-00-00365	bz 4; eth 4; EtOH 4
6259	Glycine, N,N-bis(2-hydroxyethyl)- Bicine	$C_6H_{13}NO_4$ 163.17	150-25-4 193-5 dec	1221	4-04-00-02390	H_2O 4; EtOH 1
6260	Glycine, N-(carboxymethyl)- Iminodiacetic acid	$C_4H_7NO_4$ 133.10	142-73-4 247.5	4832	4-04-00-02428	H_2O 2; EtOH 1; eth 1
6261	Glycine, N,N-dimethyl- N,N-Dimethylglycine	$C_4H_9NO_2$ 103.12	1118-68-9 185.5	3233	4-04-00-02365	H_2O 4; EtOH 3; eth 3; ace 3
6262	Glycine, N,N-1,2-ethanediylbis[N-(carboxymethyl)- Edetic acid	$C_{10}H_{16}N_2O_8$ 292.25	60-00-4 245 dec	3484	4-04-00-02449	
6263	Glycine, ethyl ester	$C_4H_9NO_2$ 103.12	459-73-4	149; 58[18]	4-04-00-02355 1.0275[10]	H_2O 5; EtOH 5; eth 5; ace 5 1.4242[10]
6264	Glycine, N-(N-L-γ-glutamyl-L-cysteinyl)- Glutatione	$C_{10}H_{17}N_3O_6S$ 307.33	70-18-8 195	4369	4-04-00-03165	H_2O 4; EtOH 1; eth 1; DMF 3
6265	Glycine, N-glycyl- 2-(Aminoacetamido)acetic acid	$C_4H_8N_2O_3$ 132.12	556-50-3 215 dec	4399	4-04-00-02459	H_2O 3
6266	Glycine, hydrazide Aminoacetic acid hydrazide	$C_2H_7N_3O$ 89.10	14379-80-7 80.5	150 dec	3-04-00-01121	EtOH 2; eth 1; chl 3
6267	Glycine, N-(4-hydroxyphenyl)- N-(4-Hydroxyphenyl)glycine	$C_8H_9NO_3$ 167.16	122-87-2 245-7 dec	4771	4-13-00-01210	H_2O 2; EtOH 2; eth 1; ace 2
6268	Glycine, N-methyl- Sarcosine	$C_3H_7NO_2$ 89.09	107-97-1 212-3 dec	8331	4-04-00-02363	
6269	Glycine, N-methyl-N-(1-oxododecyl)-, sodium salt Gardol	$C_{15}H_{28}NNaO_3$ 293.38	137-16-6	4267		H_2O 2
6270	Glycine, N-phenyl- Phenylaminoacetic acid	$C_8H_9NO_2$ 151.17	103-01-5 127.5	7262	4-12-00-00872	H_2O 4; EtOH 4
6271	Glycine, N-phenyl-, ethyl ester	$C_{10}H_{13}NO_2$ 179.22	2216-92-4 58	273.5	4-12-00-00873	eth 4; EtOH 4
6272	Glycine, N-(phenylmethyl)-, ethyl ester	$C_{11}H_{15}NO_2$ 193.25	6436-90-4	175-9[50]	4-12-00-02277	EtOH 4; eth 4; bz 4 1.5041[20] H_2O 2; EtOH 4; eth 2
6273	Glycine, N-[(3α,5β,7α,12α)-3,7,12-trihydroxy-24-oxocholan-24-yl]- Glycocholic acid	$C_{26}H_{43}NO_6$ 465.63	475-31-0 166.5	4390	4-10-00-02077	
6274	Glycogen	$(C_6H_{10}O_5)_x$ 162.14	9005-79-2	4392		H_2O 4; EtOH 1; eth 1
6275	Glyodin 1H-Imidazole, 2-heptadecyl-4,5-dihydro-, monoacetate	$C_{20}H_{40}N_2 \cdot C_2H_4O_2$ 368.60	556-22-9	4404	1.035[20]	
6276	Glyphosate Glycine, N-(phosphonomethyl)-	$C_3H_8NO_5P$ 169.07	1071-83-6 230 dec	4408		
6277	Glyphosine Glycine, N,N-bis(phosphonomethyl)-	$C_4H_{11}NO_8P_2$ 263.08	2439-99-8	4409		
6278	Guanidine Aminomethanamidine	CH_5N_3 59.07	113-00-8 50	4475	4-03-00-00148	H_2O 4; EtOH 4
6279	Guanidine, N,N'-bis(2-methylphenyl)-	$C_{15}H_{17}N_3$ 239.32	97-39-2 179		4-12-00-01764 1.10[20]	H_2O 2; EtOH 2; eth 4; chl 3
6280	Guanidine, cyano- Dicyanodiamide	$C_2H_4N_4$ 84.08	461-58-5 211	3082	4-03-00-00160 1.404[14]	H_2O 3; EtOH 1; eth 1; ace 3
6281	Guanidine, N,N'-dimethyl-N''-(phenylmethyl)- Bethanidine	$C_{10}H_{15}N_3$ 177.25	55-73-2 196	1208		
6282	Guanidine, N,N'-diphenyl- 1,3-Diphenylguanidine	$C_{13}H_{13}N_3$ 211.27	102-06-7 150	3325 170 dec	4-12-00-00769 1.13[20]	H_2O 2; EtOH 3; eth 4; ctc 3
6283	Guanidine, (3-methyl-2-butenyl)- Galegine	$C_6H_{13}N_3$ 127.19	543-83-9 62.5 dec	4246	3-04-00-00465	H_2O 4; EtOH 4
6284	Guanidine, N-methyl-N'-nitro-N-nitroso- N-Methyl-N'-nitro-N-nitrosoguanidine	$C_2H_5N_5O_3$ 147.09	70-25-7	6017	4-04-00-03386	DMSO 3
6285	Guanidine, monohydrochloride	CH_6ClN_3 95.53	50-01-1 182.3		4-03-00-00148 1.354[20]	H_2O 4; EtOH 4
6286	Guanidine, mononitrate	$CH_6N_4O_3$ 122.08	506-93-4 217 dec		4-03-00-00148	H_2O 4; EtOH 4
6287	Guanidine, nitro- Nitroguanidine	$CH_4N_4O_2$ 104.07	556-88-7 239 dec	6529	4-03-00-00249	H_2O 2; EtOH 2; eth 1; alk 4
6288	Guanidine, N,N',N''-triphenyl-	$C_{19}H_{17}N_3$ 287.36	101-01-9 146.5 dec		4-12-00-00866 1.163[20]	H_2O 2; EtOH 3
6289	Guanosine 2-Amino-1,9-dihydro-9-β-D-ribofuranosyl-6H-purin-6-one	$C_{10}H_{13}N_5O_5$ 283.24	118-00-3 239 dec	4480	4-26-00-03901	H_2O 2; EtOH 1; eth 1; HOAc 4
6290	D-Gulose Gulose, D-	$C_6H_{12}O_6$ 180.16	4205-23-6 dec	4490	4-01-00-04333	H_2O 4
6291	Haloxyfop-(2-ethoxyethyl) (RS)-2-[4-(3-Chloro-5-trifluoromethyl-2-pyridyloxy)phenoxy]propionic acid, ethyl ester	$C_{17}H_{15}ClF_3NO_4$ 389.76	87237-48-7 60		1.34	

No.	Name / Synonym	Mol. Form. / Mol. Wt.	CAS RN / mp/°C	Merck No. / bp/°C	Beil. Ref. / den/g cm^{-3}	Solubility / n_D
6292	7H-Benz[de]anthracen-7-one / Benzanthrone	$C_{17}H_{10}O$ / 230.27	82-05-3 / 170	1070		bz 2
6293	Helenine	$C_{15}H_{20}O_2$ / 232.32	1407-14-3 / 78	275; 192[10]	4-17-00-05030	bz 4; eth 4; EtOH 4
6294	Heneicosane	$C_{21}H_{44}$ / 296.58	629-94-7 / 40.5	356.5	4-01-00-00569 / 0.7919[20]	H_2O 1; EtOH 2; peth 3 / 1.4441[20]
6295	Heneicosane, 11-decyl-	$C_{31}H_{64}$ / 436.85	55320-06-4 / 10.0	282.0[10]	4-01-00-00595 / 0.8116[20]	1.4540[20]
6296	Heneicosane, 11-(2,2-dimethylpropyl)-	$C_{26}H_{54}$ / 366.71	55282-10-5 / -21	238.0[10]	4-01-00-00585 / 0.8031[20]	1.4491[20]
6297	Heneicosane, 11-phenyl- / Benzene, (1-decylundecyl)-	$C_{27}H_{48}$ / 372.68	6703-80-6 / 20.8	205[1.0]	4-05-00-01226 / 0.8531[20]	1.4788[20]
6298	10-Heneicosene, 11-phenyl-	$C_{27}H_{46}$ / 370.66	6703-78-2 / 48.2	203[1.0] 190[0.5]	4-05-00-01506 / 0.8638[20]	1.4922[20]
6299	Hentriacontane / Untriacontane	$C_{31}H_{64}$ / 436.85	630-04-6 / 67.9	458	4-01-00-00594 / 0.781[68]	EtOH 2; eth 2; bz 2; chl 2 / 1.4278[90]
6300	16-Hentriacontanone	$C_{31}H_{62}O$ / 450.83	502-73-8 / 83.8	250[0.1]	4-01-00-03413 / 0.7947[91]	H_2O 1; EtOH 2; eth 3; ace 2 / 1.4297[94]
6301	Heptachlor epoxide / 2,5-Methano-2H-indeno[1,2-b]oxirene, 2,3,4,5,6,7,7-heptachloro-1a,1b,5,5a,6,6a-hexahydro-, (1aα,1bβ,2α,5α,5aβ,6β,6aα)-	$C_{10}H_5Cl_7O$ / 389.32	1024-57-3 / 160			
6302	Heptacosane	$C_{27}H_{56}$ / 380.74	593-49-7 / 59.5	442	4-01-00-00586 / 0.7796[60]	H_2O 1; EtOH 1; eth 2 / 1.4345[65]
6303	14-Heptacosanone	$C_{27}H_{54}O$ / 394.73	542-50-7 / 77.5	238[0.1]	4-01-00-03411 / 0.7986[81]	
6304	Heptadecanal / Margaric aldehyde	$C_{17}H_{34}O$ / 254.46	629-90-3 / 36	204[26]	4-01-00-03395	bz 4; eth 4
6305	1-Heptadecanamine	$C_{17}H_{37}N$ / 255.49	4200-95-7 / 49	336	4-04-00-00824 / 0.8510[20]	H_2O 1; EtOH 3; eth 3 / 1.4510[20]
6306	Heptadecane	$C_{17}H_{36}$ / 240.47	629-78-7 / 22	302.0	4-01-00-00548 / 0.7780[20]	H_2O 1; EtOH 2; eth 3; ctc 2 / 1.4369[20]
6307	Heptadecane, 1-bromo- / 1-Bromoheptadecane	$C_{17}H_{35}Br$ / 319.37	3508-00-7 / 32	349	4-01-00-00549 / 0.9916[20]	H_2O 1; chl 4 / 1.4625[20]
6308	Heptadecane, 1,17-dibromo-	$C_{17}H_{34}Br_2$ / 398.26	81726-82-1 / 38	208[3]	3-01-00-00564	chl 4
6309	Heptadecane, 9-hexyl- / 9-Hexylheptadecane	$C_{23}H_{48}$ / 324.63	55124-79-3 / -19.4	213[10]; 157[0.5]	4-01-00-00578 / 0.7976[20]	1.4465[20]
6310	Heptadecane, 2-methyl- / 2-Methylheptadecane	$C_{18}H_{38}$ / 254.50	1560-89-0 / 5.7	311	4-01-00-00556 / 0.7838[15]	1.4394[16]
6311	Heptadecane, 9-methyl- / 9-Methylheptadecane	$C_{18}H_{38}$ / 254.50	18869-72-2	172-4[10]	3-01-00-00568 / 0.7810[20]	1.4388[20]
6312	Heptadecanenitrile	$C_{17}H_{33}N$ / 251.46	5399-02-0 / 34	349	4-02-00-01195 / 0.8315[20]	H_2O 1; EtOH 2; eth 4; chl 2 / 1.4467[20]
6313	Heptadecane, 9-octyl-	$C_{25}H_{52}$ / 352.69	7225-64-1 / -13.8	231.5[10]	4-01-00-00583 / 0.8020[20]	1.4487[20]
6314	Heptadecanoic acid / Margaric acid	$C_{17}H_{34}O_2$ / 270.46	506-12-7 / 61.3	5634 227[100]	4-02-00-01193 / 0.8532[60]	EtOH 2; eth 3; ace 3; bz 3 / 1.4342[60]
6315	Heptadecanoic acid, ethyl ester	$C_{19}H_{38}O_2$ / 298.51	14010-23-2 / 28	185[5]	4-02-00-01194 / 0.8517[30]	ace 4; bz 4; eth 4; EtOH 4
6316	Heptadecanoic acid, methyl ester	$C_{18}H_{36}O_2$ / 284.48	1731-92-6 / 30	185[9]; 152[0.05]	4-02-00-01194	H_2O 1; EtOH 3; eth 4; ace 3
6317	1-Heptadecanol / Heptadecyl alcohol	$C_{17}H_{36}O$ / 256.47	1454-85-9 / 53.8	333	4-01-00-01884 / 0.8475[20]	H_2O 1; EtOH 3; eth 3
6318	2-Heptadecanol	$C_{17}H_{36}O$ / 256.47	16813-18-6 / 54	155[1]	4-01-00-01885	eth 4; EtOH 4 / 1.4407[37]
6319	9-Heptadecanol	$C_{17}H_{36}O$ / 256.47	624-08-8 / 61	174[9]	4-01-00-01886 / 0.8532[60]	H_2O 1; EtOH 3; eth 3; ace 3 / 1.4262[80]
6320	2-Heptadecanone / Pentadecyl methyl ketone	$C_{17}H_{34}O$ / 254.46	2922-51-2 / 48	320	4-01-00-03395 / 0.8049[48]	H_2O 1; EtOH 2; eth 4; ace 3
6321	9-Heptadecanone	$C_{17}H_{34}O$ / 254.46	540-08-9 / 53	251.5; 142[1.5]	4-01-00-03396 / 0.8140[48]	EtOH 2; MeOH 3
6322	1-Heptadecene / Hexahydroaplotaxene	$C_{17}H_{34}$ / 238.46	6765-39-5 / 11.5	300	4-01-00-00929 / 0.7852[20]	H_2O 1; eth 4; bz 3; lig 5 / 1.4432[20]
6323	8-Heptadecene, 9-octyl-	$C_{25}H_{50}$ / 350.67	24306-18-1	227[10]	4-01-00-00936 / 0.8086[20]	1.4554[20]
6324	9-Heptadecenoic acid	$C_{17}H_{32}O_2$ / 268.44	10136-52-4 / 12.5	175[0.5]	4-02-00-01632 / 0.8942[20]	1.4598[20]
6325	2,4-Heptadienal, (E,E)-	$C_7H_{10}O$ / 110.16	4313-03-5	84.5	4-01-00-03549 / 0.881[25]	1.5315[20]
6326	1,2-Heptadiene / Butylallene	C_7H_{12} / 96.17	2384-90-9	105.5	4-01-00-01027 / 0.7306[18]	1.432[18]
6327	1,4-Heptadiene	C_7H_{12} / 96.17	5675-22-9	93	3-01-00-00999 / 0.7270[20]	bz 4; eth 4; peth 4 / 1.4370[20]
6328	1,5-Heptadiene / 1-Methyldiallyl	C_7H_{12} / 96.17	1541-23-7	94	3-01-00-00999 / 0.7186[20]	ace 4; bz 4; eth 4; EtOH 4 / 1.4200[20]

No.	Name / Synonym	Mol. Form. / Mol. Wt.	CAS RN / mp/°C	bp/°C	Merck No.	Beil. Ref. / den/g cm⁻³	Solubility / n_D
6329	2,4-Heptadiene	C_7H_{12} 96.17	628-72-8	108		4-01-00-01029 0.7384[20]	H_2O 1; EtOH 3; eth 3; ace 3 1.4578[20]
6330	1,3-Heptadiene, 2,6-dimethyl-	C_9H_{16} 124.23	2436-84-2	141		4-01-00-01051 0.7648[10]	1.4606[22]
6331	1,5-Heptadiene, 2,6-dimethyl-	C_9H_{16} 124.23	6709-39-3 -70	143		4-01-00-01051 0.7648[25]	
6332	2,4-Heptadiene, 2,4-dimethyl-	C_9H_{16} 124.23	74421-05-9	138		4-01-00-01051 0.7750[4]	1.4587[4]
6333	2,4-Heptadiene, 3,5-dimethyl-	C_9H_{16} 124.23	101935-28-8	142		4-01-00-01052 0.7728[20]	1.4487[20]
6334	1,6-Heptadiene-3,5-dione, 1,7-bis(4-hydroxy-3-methoxyphenyl)-, (E,E)- Curcumin	$C_{21}H_{20}O_6$ 368.39	458-37-7 183		2681	4-08-00-03697	EtOH 4; HOAc 4
6335	1,5-Heptadiene, 3-methyl- 3-Methyl-1,5-heptadiene	C_8H_{14} 110.20	4894-62-6 -57	111.		4-01-00-01040 0.7250[25]	
6336	1,5-Heptadien-4-ol	$C_7H_{12}O$ 112.17	5638-26-6	156; 68[24]		4-01-00-02248 0.8598[20]	ace 4 1.4510[25]
6337	1,6-Heptadien-4-ol 4-Hydroxy-1,6-heptadiene	$C_7H_{12}O$ 112.17	2883-45-6	151		4-01-00-02249 0.864[25]	1.4505[20]
6338	1,6-Heptadien-4-ol, 4-methyl- 1,6-Heptadiene-4-ol, 4-methyl	$C_8H_{14}O$ 126.20	25201-40-5	158.4		4-01-00-02260 0.8611[20]	ace 4; EtOH 4 1.4500[23]
6339	3,5-Heptadien-2-one	$C_7H_{10}O$ 110.16	3916-64-1	88[28]		4-01-00-03549 0.8946[19]	eth 4 1.5177[19]
6340	2,5-Heptadien-4-one, 2,6-dimethyl- Phorone	$C_9H_{14}O$ 138.21	504-20-1 28	197.5	7307	4-01-00-03564 0.8850[20]	H_2O 2; EtOH 3; eth 3; ace 3 1.4998[20]
6341	1,6-Heptadien-3-yne 1,6-Heptadiene-3-yne	C_7H_8 92.14	5150-80-1	110		3-01-00-01061 0.787[25]	bz 3; peth 3 1.4694[25]
6342	1,5-Heptadiyne	C_7H_8 92.14	764-56-7	26[30]		2-01-00-00247 0.8100[21]	bz 4; peth 4 1.4521[21]
6343	1,6-Heptadiyne	C_7H_8 92.14	2396-63-6 -85	112		4-01-00-01121 0.8164[17]	H_2O 1; bz 3; HOAc 3 1.451[17]
6344	Heptanal Heptaldehyde	$C_7H_{14}O$ 114.19	111-71-7 -43.3	152.8	4578	4-01-00-03314 0.8132[25]	H_2O 2; EtOH 5; eth 5; ctc 2 1.4113[20]
6345	Heptanal, 5-methyl-	$C_8H_{16}O$ 128.21	75579-88-3	72[25]		4-01-00-03345 0.8164[25]	1.4144[25]
6346	Heptanal, oxime Enanthaldoxime	$C_7H_{15}NO$ 129.20	629-31-2 57.5	195		4-01-00-03316 0.8583[55]	H_2O 2; EtOH 3; eth 3 1.4210[20]
6347	Heptanal, 2-(phenylmethylene)-	$C_{14}H_{18}O$ 202.30	122-40-7 80	174-5[20]		2-07-00-00310 0.9711[20]	H_2O 1; ace 3; ctc 3 1.5381[20]
6348	Heptanamide	$C_7H_{15}NO$ 129.20	628-62-6 96	254		4-02-00-00963 0.8521[110]	H_2O 3; EtOH 3; eth 3; chl 2 1.4217[110]
6349	1-Heptanamine Heptylamine	$C_7H_{17}N$ 115.22	111-68-2 -18	156		4-04-00-00734 0.7754[20]	H_2O 2; EtOH 5; eth 5; chl 2 1.4251[20]
6350	2-Heptanamine Tuaminoheptane	$C_7H_{17}N$ 115.22	123-82-0	142	9712	3-04-00-00374 0.7665[19]	H_2O 2; EtOH 3; eth 3; chl 2 1.4199[19]
6351	4-Heptanamine	$C_7H_{17}N$ 115.22	16751-59-0	139.5		4-04-00-00745 0.767[20]	1.4172[20]
6352	1-Heptanamine, N-heptyl- Diheptylamine	$C_{14}H_{31}N$ 213.41	2470-68-0 31.5	271; 135[9]		4-04-00-00736 0.7956[21]	H_2O 2; EtOH 3; eth 4
6353	2-Heptanamine, 6-methyl-, (±)-	$C_8H_{19}N$ 129.25	5984-58-7	155	6678	4-04-00-00764 0.767[25]	1.4209[20]
6354	2-Heptanamine, 6-methyl-N-(3-methylbutyl)- Octamylamine	$C_{13}H_{29}N$ 199.38	502-59-0	100-1[7]	6671	3-04-00-00387	
6355	Heptane	C_7H_{16} 100.20	142-82-5 -90.6	98.5	4580	4-01-00-00376 0.6837[20]	H_2O 1; EtOH 4; eth 5; ace 5 1.3878[20]
6356	Heptane, 1-bromo- Heptyl bromide	$C_7H_{15}Br$ 179.10	629-04-9 -58	179		4-01-00-00391 1.1400[20]	H_2O 1; EtOH 4; eth 4; ctc 2 1.4502[20]
6357	Heptane, 2-bromo- 2-Heptyl bromide	$C_7H_{15}Br$ 179.10	1974-04-5 47	166		3-01-00-00431 1.1277[20]	H_2O 1; bz 4; ctc 3; chl 3 1.4503[20]
6358	Heptane, 4-bromo- 4-Heptyl bromide	$C_7H_{15}Br$ 179.10	998-93-6	84.6[72]; 60[18]		4-01-00-00392 1.1351[20]	H_2O 1; bz 3; ctc 3; chl 3 1.4495[20]
6359	Heptane, 1-bromo-7-fluoro-	$C_7H_{14}BrF$ 197.09	334-42-9	85[11]		4-01-00-00392	eth 4; EtOH 4 1.4463[20]
6360	Heptane, 1-chloro- Heptyl chloride	$C_7H_{15}Cl$ 134.65	629-06-1 -69.5	159		4-01-00-00389 0.8758[20]	H_2O 1; EtOH 5; eth 5; ctc 2 1.4256[20]
6361	Heptane, 2-chloro- 2-Chloroheptane	$C_7H_{15}Cl$ 134.65	1001-89-4	46[19]		4-01-00-00390 0.8672[20]	H_2O 1; eth 4; bz 3; chl 3 1.4221[20]
6362	Heptane, 3-chloro- 3-Chloroheptane	$C_7H_{15}Cl$ 134.65	999-52-0	144; 48.3[20]		4-01-00-00390 0.8960[20]	bz 4; eth 4 1.4228[20]
6363	Heptane, 4-chloro- 4-Chloroheptane	$C_7H_{15}Cl$ 134.65	998-95-8	144		4-01-00-00390 0.8710[20]	bz 4; eth 4 1.4237[20]
6364	Heptane, 3-chloro-3-ethyl-	$C_9H_{19}Cl$ 162.70	28320-89-0	46[3]		3-01-00-00510 0.8856[20]	EtOH 4; chl 4 1.4400[20]
6365	Heptane, 1-chloro-7-fluoro-	$C_7H_{14}ClF$ 152.64	334-43-0	70[10]		4-01-00-00390 0.993[20]	bz 4; eth 4; EtOH 4; chl 4 1.4222[25]
6366	Heptane, 2-chloro-2-methyl- 2-Chloro-2-methylheptane	$C_8H_{17}Cl$ 148.68	4325-49-9	50[15]		4-01-00-00428 0.8568[25]	bz 4; eth 4; EtOH 4; chl 4 1.4240[25]
6367	Heptane, 2-chloro-6-methyl- 2-Chloro-6-methylheptane	$C_8H_{17}Cl$ 148.68	2350-19-8	74[35]		3-01-00-00472 0.8769[20]	bz 4; eth 4; EtOH 4; chl 4 1.4260[15]
6368	Heptane, 3-(chloromethyl)-	$C_8H_{17}Cl$ 148.68	123-04-6	172		4-01-00-00430 0.8769[20]	H_2O 1; EtOH 3; eth 3; ace 3 1.4319[20]

No.	Name / Synonym	Mol. Form. / Mol. Wt.	CAS RN / mp/°C	Merck No. / bp/°C	Beil. Ref. / den/g cm^{-3}	Solubility / n_D
6369	Heptane, 3-chloro-3-methyl- / 3-Methyl-3-chloroheptane	$C_8H_{17}Cl$ / 148.68	5272-02-6	64[27]	4-01-00-00430 / 0.8764[20]	bz 4; eth 4; EtOH 4; chl 4 / 1.4317[20]
6370	Heptane, 4-chloro-4-methyl- / 2-Chloro-2-propylpentane	$C_8H_{17}Cl$ / 148.68	61764-94-1	50[12]	3-01-00-00477 / 0.8690[20]	bz 4; eth 4; EtOH 4; chl 4 / 1.4310[15]
6371	1,7-Heptanediamine	$C_7H_{18}N_2$ / 130.23	646-19-5 / 28.9	224	4-04-00-01354	EtOH 3; eth 3; ace 3
6372	Heptane, 1,2-dibromo- / 1,2-Dibromoheptane	$C_7H_{14}Br_2$ / 258.00	42474-21-5	228	3-01-00-00432 / 1.5086[20]	1.4986[20]
6373	Heptane, 1,5-dibromo- / 1,5-Dibromoheptane	$C_7H_{14}Br_2$ / 258.00	1622-10-2	113-5[11]	4-01-00-00393 / 1.536[15]	1.5041[15]
6374	Heptane, 1,7-dibromo- / Heptamethylene dibromide	$C_7H_{14}Br_2$ / 258.00	4549-31-9 / 41.7	263	4-01-00-00393 / 1.5306[20]	H_2O 1; eth 3; ace 3; bz 3 / 1.5034[20]
6375	Heptane, 2,3-dibromo- / 2,3-Dibromoheptane	$C_7H_{14}Br_2$ / 258.00	21266-88-6	101[17]	3-01-00-00432 / 1.5139[20]	1.4992[20]
6376	Heptane, 3,4-dibromo- / 3,4-Dibromoheptane	$C_7H_{14}Br_2$ / 258.00	21266-90-0	107[24]	3-01-00-00433 / 1.5182[20]	1.5010[20]
6377	Heptane, 1,1-dichloro- / 1,1-Dichloroheptane	$C_7H_{14}Cl_2$ / 169.09	821-25-0	191	4-01-00-00390 / 1.011[20]	H_2O 1; eth 3; bz 3; ctc 3 / 1.4440[20]
6378	Heptane, 1,2-dichloro- / 1,2-Dichloroheptane	$C_7H_{14}Cl_2$ / 169.09	10575-87-8	68-72[7]	3-01-00-00430 / 1.064[20]	bz 4; eth 4; chl 4 / 1.4490[20]
6379	Heptane, 1,7-dichloro- / 1,7-Dichloroheptane	$C_7H_{14}Cl_2$ / 169.09	821-76-1	124-5[35]	4-01-00-00390 / 1.0408[25]	1.4565[25]
6380	Heptane, 2,2-dichloro- / 2,2-Dichloroheptane	$C_7H_{14}Cl_2$ / 169.09	65786-09-6	77[25]	3-01-00-00430 / 1.012[20]	bz 4; eth 4; chl 4 / 1.4440[20]
6381	Heptane, 4,4-dichloro-	$C_7H_{14}Cl_2$ / 169.09	89796-76-9	86[27]	2-01-00-00117 / 1.008[17]	eth 4; chl 4 / 1.448[17]
6382	Heptane, 2,6-dichloro-2,6-dimethyl-	$C_9H_{18}Cl_2$ / 197.15	35951-38-1 / 43	93[16]	3-01-00-00513	bz 4; chl 4
6383	Heptane, 2,2-dimethyl- / 2,2-Dimethylheptane	C_9H_{20} / 128.26	1071-26-7 / -113	132.7	4-01-00-00457 / 0.7105[20]	H_2O 1; eth 3; ace 4; bz 5 / 1.4016[20]
6384	Heptane, 2,3-dimethyl- / 2,3-Dimethylheptane	C_9H_{20} / 128.26	3074-71-3 / -116	140.5	4-01-00-00457 / 0.7260[20]	H_2O 1; EtOH 5; eth 5; ace 5 / 1.4088[20]
6385	Heptane, 2,4-dimethyl- / 2,4-Dimethylheptane	C_9H_{20} / 128.26	2213-23-2	132.9	4-01-00-00457 / 0.7115[25]	H_2O 1; EtOH 5; eth 5; ace 5 / 1.4034[20]
6386	Heptane, 2,5-dimethyl- / 2,5-Dimethylheptane	C_9H_{20} / 128.26	2216-30-0	136	4-01-00-00457 / 0.7198[20]	ace 4; bz 4; eth 4; EtOH 4 / 1.4033[20]
6387	Heptane, 2,6-dimethyl- / 2,6-Dimethylheptane	C_9H_{20} / 128.26	1072-05-5 / -102.9	135.2	4-01-00-00458 / 0.7089[20]	chl 2 / 1.4011[20]
6388	Heptane, 3,3-dimethyl- / 3,3-Dimethylheptane	C_9H_{20} / 128.26	4032-86-4	137.3	4-01-00-00458 / 0.7254[20]	H_2O 1; EtOH 5; eth 3; ace 4 / 1.4087[20]
6389	Heptane, 3,4-dimethyl- / 3,4-Dimethylheptane	C_9H_{20} / 128.26	922-28-1	140.6	3-01-00-00514 / 0.7314[20]	H_2O 1; eth 3; ace 4; bz 5 / 1.4108[20]
6390	Heptane, 3,5-dimethyl- / 3,5-Dimethylheptane	C_9H_{20} / 128.26	926-82-9	136	4-01-00-00458 / 0.7225[20]	H_2O 1; eth 3; ace 4; bz 5 / 1.4083[20]
6391	Heptane, 4,4-dimethyl- / 4,4-Dimethylheptane	C_9H_{20} / 128.26	1068-19-5	135.2	4-01-00-00458 / 0.7221[20]	H_2O 1; eth 3; ace 4; bz 5 / 1.4076[20]
6392	Heptanedinitrile	$C_7H_{10}N_2$ / 122.17	646-20-8 / -31.4	155[14]	4-02-00-02006 / 0.9491[8]	H_2O 1; EtOH 5; eth 5; chl 5 / 1.4472[20]
6393	Heptanedioic acid / Pimelic acid	$C_7H_{12}O_4$ / 160.17	111-16-0 / 106	7401 / 272[100]; 212[10]	4-02-00-02003 / 1.329[15]	H_2O 3; EtOH 3; eth 3; bz 1
6394	Heptanedioic acid, diethyl ester / Diethyl pimelate	$C_{11}H_{20}O_4$ / 216.28	2050-20-6 / -24	254; 140[15]	4-02-00-02004 / 0.9945[20]	eth 4; EtOH 4 / 1.4305[20]
6395	Heptanedioic acid, dimethyl ester	$C_9H_{16}O_4$ / 188.22	1732-08-7 / -21	120[10]; 80[1]	4-02-00-02004 / 1.0625[20]	H_2O 2; EtOH 3; eth 3; bz 3 / 1.4309[20]
6396	Heptanedioic acid, monoethyl ester	$C_9H_{16}O_4$ / 188.22	33018-91-6 / 10	182[18]; 160[4]	3-02-00-01742	1.4415[20]
6397	1,4-Heptanediol	$C_7H_{16}O_2$ / 132.20	40646-07-9	243; 127[4]	4-01-00-02579 / 0.9542[20]	1.4510[25]
6398	1,5-Heptanediol	$C_7H_{16}O_2$ / 132.20	60096-09-5	135-6[11]	3-01-00-02213 / 0.9705[20]	eth 4; EtOH 4; chl 4 / 1.4571[22]
6399	1,7-Heptanediol	$C_7H_{16}O_2$ / 132.20	629-30-1 / 22.5	262	4-01-00-02580 / 0.9569[25]	eth 4; EtOH 4 / 1.4520[25]
6400	2,4-Heptanediol	$C_7H_{16}O_2$ / 132.20	20748-86-1	107-8[8]	4-01-00-02580 / 0.926[20]	1.4386[25]
6401	2,4-Heptanediol, 3-methyl- / 3-Methylheptan-2,4-diol	$C_8H_{18}O_2$ / 146.23	6964-04-1	115[3]	0-01-00-00491 / 0.928[20]	H_2O 1; EtOH 3; ctc 2 / 1.4459[20]
6402	2,3-Heptanedione / Acetyl valeryl	$C_7H_{12}O_2$ / 128.17	96-04-8	46[13]	4-01-00-03697 / 0.9191[8]	1.4150[18]
6403	2,6-Heptanedione	$C_7H_{12}O_2$ / 128.17	13505-34-5 / 33.5	222; 97[11]	4-01-00-03699 / 0.9399[37]	bz 4; eth 4; EtOH 4 / 1.4277[37]
6404	3,5-Heptanedione, 2,2,6,6-tetramethyl- / Dipivaloylmethane	$C_{11}H_{20}O_2$ / 184.28	1118-71-4	72-3[6]	4-01-00-03729 / 0.883[25]	ctc 2 / 1.4589[20]
6405	Heptane, 3-[(ethenyloxy)methyl]-	$C_{10}H_{20}O$ / 156.27	103-44-6 / -100	76[32]	4-01-00-02056 / 0.8108[20]	1.4247[25]
6406	Heptane, 1-ethoxy-	$C_9H_{20}O$ / 144.26	1969-43-3 / -68.3	166	4-01-00-01733 / 0.790[16]	
6407	Heptane, 3-ethyl- / 3-Ethylheptane	C_9H_{20} / 128.26	15869-80-4 / -114.9	143.0	4-01-00-00456 / 0.7225[25]	1.4093[20]
6408	Heptane, 4-ethyl- / 4-Ethylheptane	C_9H_{20} / 128.26	2216-32-2	141.2	4-01-00-00457 / 0.7241[25]	H_2O 1; EtOH 5; eth 3; ace 5 / 1.4096[20]

No.	Name Synonym	Mol. Form. Mol. Wt.	CAS RN mp/°C	bp/°C	Merck No.	Beil. Ref. den/g cm^{-3}	Solubility n_D
6409	Heptane, 3-ethyl-2-methyl- Heptane, 2-methyl-3-ethyl	C$_{10}$H$_{22}$ 142.28	14676-29-0	163		4-01-00-00480 0.7398^{25}	1.4174^{20}
6410	Heptane, 3-ethyl-3-methyl-	C$_{10}$H$_{22}$ 142.28	17302-01-1	163.8		4-01-00-00481 0.7463^{25}	1.4208^{20}
6411	Heptane, 3-ethyl-4-methyl- 3-Ethyl-4-methylheptane	C$_{10}$H$_{22}$ 142.28	52896-91-0	165		4-01-00-00481 0.7468^{25}	1.4207^{20}
6412	Heptane, 3-ethyl-5-methyl- 3-Ethyl-5-methylheptane	C$_{10}$H$_{22}$ 142.28	52896-90-9	158.2		4-01-00-00481 0.7368^{25}	1.4164^{20}
6413	Heptane, 4-ethyl-2-methyl- 4-Ethyl-2-methylheptane	C$_{10}$H$_{22}$ 142.28	52896-88-5	158		4-01-00-00481 0.7322^{25}	1.4137^{20}
6414	Heptane, 4-ethyl-3-methyl- 4-Ethyl-3-methylheptane	C$_{10}$H$_{22}$ 142.28	52896-89-6	165		4-01-00-00481 0.7466^{25}	1.4206^{20}
6415	Heptane, 4-ethyl-4-methyl- 4-Ethyl-4-methylheptane	C$_{10}$H$_{22}$ 142.28	17302-04-4	160.8		4-01-00-00481 0.7472^{25}	1.4210^{25}
6416	Heptane, 5-ethyl-2-methyl-	C$_{10}$H$_{22}$ 142.28	13475-78-0	159.7		4-01-00-00481 0.7318^{25}	1.4134^{20}
6417	Heptane, 1-fluoro-	C$_7$H$_{15}$F 118.19	661-11-0 -73	117.9		4-01-00-00387 0.8062^{20}	H$_2$O 1; eth 3; ace 3; bz 3 1.3854^{20}
6418	Heptane, hexadecafluoro- Perfluoroheptane	C$_7$F$_{16}$ 388.05	335-57-9 -78	82.5		4-01-00-00388 1.7333^{20}	ace 4; eth 4; EtOH 4; chl 4 1.2618^{20}
6419	Heptane, 1-iodo-	C$_7$H$_{15}$I 226.10	4282-40-0 -48.2	204		4-01-00-00393 1.3791^{20}	H$_2$O 1; EtOH 3; eth 3; ace 3 1.4904^{20}
6420	Heptane, 2-iodo- 2-Iodoheptane	C$_7$H$_{15}$I 226.10	18589-29-2	98^{50}		4-01-00-00393 1.304^{20}	H$_2$O 1; ace 3; bz 3; ctc 3 1.4826
6421	Heptane, 1-methoxy-	C$_8$H$_{18}$O 130.23	629-32-3	151		3-01-00-01682 0.7862^{15}	ace 4; eth 4; EtOH 4 1.4073^{20}
6422	Heptane, 2-methyl- 2-Methylheptane	C$_8$H$_{18}$ 114.23	592-27-8 -108.9	117.6		4-01-00-00428 0.6980^{20}	H$_2$O 1; EtOH 5; eth 3; ace 5 1.3949^{20}
6423	Heptane, 3-methyl-, (S)-	C$_8$H$_{18}$ 114.23	6131-25-5 -120	116.5		4-01-00-00429 0.7075^{16}	H$_2$O 1; EtOH 3; eth 3; ace 5 1.4002^{18}
6424	Heptane, 4-methyl- 4-Methylheptane	C$_8$H$_{18}$ 114.23	589-53-7 -121	117.7		4-01-00-00431 0.7046^{20}	H$_2$O 1; EtOH 5; eth 3; ace 5 1.3979^{20}
6425	Heptane, 3-methylene-	C$_8$H$_{16}$ 112.22	1632-16-2	120		4-01-00-00884 0.7270^{20}	H$_2$O 1; eth 4; bz 4; peth 4 1.4157^{20}
6426	Heptane, 4-methylene- 2-Propyl-1-pentene	C$_8$H$_{16}$ 112.22	15918-08-8	117.7		4-01-00-00885 0.7198^{25}	1.4136^{20}
6427	Heptane, 4-(1-methylethyl)- 4-Isopropylheptane	C$_{10}$H$_{22}$ 142.28	52896-87-4	158.9		4-01-00-00481 0.7354^{25}	1.4153^{20}
6428	Heptanenitrile	C$_7$H$_{13}$N 111.19	629-08-3	183; 70-2^{10}		4-02-00-00963 0.8106^{20}	H$_2$O 1; eth 3; ace 3; bz 3 1.4104^{30}
6429	Heptane, 1-nitro- 1-Nitroheptane	C$_7$H$_{15}$NO$_2$ 145.20	693-39-0	194		4-01-00-00395 0.9476^{17}	eth 4; EtOH 4
6430	Heptane, 1,1'-oxybis-	C$_{14}$H$_{30}$O 214.39	629-64-1	258.5		4-01-00-01733 0.8008^{20}	eth 4; EtOH 4 1.4275^{20}
6431	Heptane, 3,3'-[oxybis(methylene)]bis- Ether, dihexyl,2,2'-diethyl	C$_{16}$H$_{34}$O 242.45	10143-60-9	269.4		4-01-00-01785	ctc 2 1.4325^{20}
6432	Heptane, 1,1,1,2,2,3,3,4,4,5,5,6,6,7,7-pentadecafluoro- 1H-Pentadecafluoroheptane	C$_7$HF$_{15}$ 370.06	375-83-7	96.0		4-01-00-00388 1.725^{25}	1.2690^{25}
6433	Heptane, 2,2,4,6,6-pentamethyl- 2,2,4,6,6-Pentamethylheptane	C$_{12}$H$_{26}$ 170.34	13475-82-6 -67	177.8		4-01-00-00510 0.7463^{20}	1.4440^{20}
6434	Heptane, 4-propyl- 4-Propylheptane	C$_{10}$H$_{22}$ 142.28	3178-29-8	157.5		4-01-00-00480 0.7321^{25}	1.4135^{20}
6435	Heptane, 1,1'-thiobis-	C$_{14}$H$_{30}$S 230.46	629-65-2 70	298		4-01-00-01739 0.8416^{20}	H$_2$O 1; eth 3 1.4606^{20}
6436	1-Heptanethiol Heptyl mercaptan	C$_7$H$_{16}$S 132.27	1639-09-4 -43	177		4-01-00-01738 0.8427^{20}	H$_2$O 1; EtOH 5; eth 5; chl 3 1.4521^{20}
6437	Heptane, 2,2,3-trimethyl- 2,2,3-Trimethylheptane	C$_{10}$H$_{22}$ 142.28	52896-92-1	157.8		4-01-00-00481 0.7385^{25}	1.4168^{20}
6438	Heptane, 2,2,4-trimethyl- 2,2,4-Trimethylheptane	C$_{10}$H$_{22}$ 142.28	14720-74-2	148.3		4-01-00-00481 0.7237^{25}	H$_2$O 1; bz 3; ctc 3; chl 3 1.4092^{20}
6439	Heptane, 2,2,5-trimethyl- 2,2,5-Trimethylheptane	C$_{10}$H$_{22}$ 142.28	20291-95-6	150.8		4-01-00-00482 0.7243^{25}	1.4101^{20}
6440	Heptane, 2,2,6-trimethyl- 2,2,6-Trimethylheptane	C$_{10}$H$_{22}$ 142.28	1190-83-6 -105	148.9		4-01-00-00482 0.7200^{25}	1.4078^{20}
6441	Heptane, 2,3,3-trimethyl- 2,3,3-Trimethylheptane	C$_{10}$H$_{22}$ 142.28	52896-93-2	160.2		4-01-00-00482 0.7450^{25}	1.4202^{20}
6442	Heptane, 2,3,4-trimethyl- 2,3,4-Trimethylheptane	C$_{10}$H$_{22}$ 142.28	52896-95-4	161		4-01-00-00482 0.7447^{25}	1.4195^{20}
6443	Heptane, 2,3,5-trimethyl- 2,3,5-Trimethylheptane	C$_{10}$H$_{22}$ 142.28	20278-85-7	160.7		4-01-00-00482 0.7413^{25}	1.4169^{20}
6444	Heptane, 2,3,6-trimethyl- 2,3,6-Trimethylheptane	C$_{10}$H$_{22}$ 142.28	4032-93-3	156.0		4-01-00-00482 0.7305^{25}	1.4131^{20}
6445	Heptane, 2,4,4-trimethyl- 2,4,4-Trimethylheptane	C$_{10}$H$_{22}$ 142.28	4032-92-2	151.0		4-01-00-00482 0.7308^{25}	1.4142^{20}
6446	Heptane, 2,4,5-trimethyl- 2,4,5-Trimethylheptane	C$_{10}$H$_{22}$ 142.28	20278-84-6	156.5		4-01-00-00483 0.7373^{25}	1.4160^{20}
6447	Heptane, 2,4,6-trimethyl- 2,4,6-Trimethylheptane	C$_{10}$H$_{22}$ 142.28	2613-61-8	147.6		4-01-00-00483 0.7190^{25}	1.4071^{20}
6448	Heptane, 2,5,5-trimethyl- 2,5,5-Trimethylheptane	C$_{10}$H$_{22}$ 142.28	1189-99-7	152.8		4-01-00-00483 0.7362^{25}	1.4149^{20}

No.	Name / Synonym	Mol. Form. / Mol. Wt.	CAS RN / mp/°C	Merck No. / bp/°C	Beil. Ref. / den/g cm⁻³	Solubility / n_D
6449	Heptane, 3,3,4-trimethyl- / 3,3,4-Trimethylheptane	$C_{10}H_{22}$ / 142.28	20278-87-9	161 9	4-01-00-00483 / 0.7527[25]	1.4236[20]
6450	Heptane, 3,3,5-trimethyl- / 3,3,5-Trimethylheptane	$C_{10}H_{22}$ / 142.28	7154-80-5	155.7	4-01-00-00483 / 0.7248[20]	H_2O 1; bz 3; ctc 3; chl 3 / 1.4170[20]
6451	Heptane, 3,4,4-trimethyl- / 3,4,4-Trimethylheptane	$C_{10}H_{22}$ / 142.28	20278-88-0	161.1	4-01-00-00484 / 0.7535[25]	1.4235[20]
6452	Heptane, 3,4,5-trimethyl- / 3,4,5-Trimethylheptane	$C_{10}H_{22}$ / 142.28	20278-89-1	162.5	4-01-00-00484 / 0.7519[25]	1.4229[20]
6453	1,4,7-Heptanetriol	$C_7H_{16}O_3$ / 148.20	3920-53-4 / -35	231[25]; 146[1]	4-01-00-02787 / 1.075[18]	H_2O 3; EtOH 3; ace 3 / 1.4725[20]
6454	2,4,6-Heptanetrione	$C_7H_{10}O_3$ / 142.15	626-53-9 / 49	121[10]	4-01-00-03783 / 1.0599[40]	H_2O 4; eth 4; EtOH 4 / 1.4930[20]
6455	Heptanoic acid / Enanthylic acid	$C_7H_{14}O_2$ / 130.19	111-14-8 / -7.5	4581 / 222.2	4-02-00-00958 / 0.9181[20]	H_2O 2; EtOH 3; eth 3; ace 3 / 1.4170[20]
6456	Heptanoic acid, anhydride	$C_{14}H_{26}O_3$ / 242.36	626-27-7 / -12.4	269.5	4-02-00-00962 / 0.9321[20]	H_2O 1; EtOH 3; eth 3 / 1.4335[15]
6457	Heptanoic acid, 2-bromo- / 2-Bromoheptanoic acid	$C_7H_{13}BrO_2$ / 209.08	2624-01-3	250 dec; 147[12]	4-02-00-00967 / 1.319[15]	ace 4; eth 4 / 1.471[18]
6458	Heptanoic acid, 7-bromo-	$C_7H_{13}BrO_2$ / 209.08	30515-28-7 / 31	280; 165[12]	4-02-00-00968	ace 4; bz 4; eth 4; EtOH 4
6459	Heptanoic acid, butyl ester / Butyl enanthate	$C_{11}H_{22}O_2$ / 186.29	5454-28-4 / -67.5	226.2	3-02-00-00768 / 0.8638[20]	ace 4; bz 4; eth 4; EtOH 4 / 1.4204[20]
6460	Heptanoic acid, 2,2-dimethyl-6-oxo- / Geronic acid	$C_9H_{16}O_3$ / 172.22	461-11-0	278; 169[12]	3-03-00-01264 / 1.0211[20]	eth 4; EtOH 4 / 1.4488[20]
6461	Heptanoic acid, 4,4-dimethyl-6-oxo- / Isogeronic acid	$C_9H_{16}O_3$ / 172.22	471-04-5 / 50	178[20]; 162[10]	3-03-00-01266	H_2O 4; bz 4; eth 4; EtOH 4
6462	Heptanoic acid, ethyl ester / Ethyl oenanthate	$C_9H_{18}O_2$ / 158.24	106-30-9 / -66.1	3790 / 187	4-02-00-00960 / 0.8817[20]	H_2O 2; EtOH 3; eth 3; ctc 2 / 1.4100[20]
6463	Heptanoic acid, 7-fluoro-	$C_7H_{13}FO_2$ / 148.18	334-28-1	133[10]	4-02-00-00964 / 1.039[20]	1.4207[25]
6464	Heptanoic acid, heptyl ester / Heptyl heptanoate	$C_{14}H_{28}O_2$ / 228.38	624-09-9 / -33	277	4-02-00-00961 / 0.8649[20]	H_2O 1; EtOH 3; eth 3 / 1.4320[20]
6465	Heptanoic acid, hexyl ester / Hexyl heptanoate	$C_{13}H_{26}O_2$ / 214.35	1119-06-8 / -48	261	3-02-00-00768 / 0.8611[20]	ace 4; bz 4; eth 4; EtOH 4 / 1.429[15]
6466	Heptanoic acid, methyl ester / Methyl heptanoate	$C_8H_{16}O_2$ / 144.21	106-73-0 / -56	4581 / 174	4-02-00-00960 / 0.8815[20]	H_2O 2; EtOH 3; eth 3; ace 2 / 1.4152[20]
6467	Heptanoic acid, 2-methylpropyl ester	$C_{11}H_{22}O_2$ / 186.29	7779-80-8	208	1-02-00-00145 / 0.8593[20]	ace 4; bz 4; eth 4; EtOH 4
6468	Heptanoic acid, octyl ester / Octyl enanthate	$C_{15}H_{30}O_2$ / 242.40	5132-75-2 / -22.5	290	3-02-00-00768 / 0.8596[20]	ace 4; bz 4; eth 4; EtOH 4 / 1.4349[15]
6469	Heptanoic acid, 6-oxo-	$C_7H_{12}O_3$ / 144.17	3128-07-2 / 40.2	251[280]; 135[1]	4-03-00-01598	H_2O 4; ace 4; eth 4; EtOH 4 / 1.4306[25]
6470	Heptanoic acid, pentyl ester / Amyl heptanoate	$C_{12}H_{24}O_2$ / 200.32	7493-82-5 / -50	245.4	4-02-00-00960 / 0.8623[20]	ace 4; bz 4; eth 4; EtOH 4 / 1.4263[15]
6471	Heptanoic acid, propyl ester / Propyl heptanoate	$C_{10}H_{20}O_2$ / 172.27	7778-87-2 / -63.5	208	3-02-00-00767 / 0.8641[15]	ace 4; bz 4; eth 4; EtOH 4 / 1.4183[15]
6472	1-Heptanol / Heptyl alcohol	$C_7H_{16}O$ / 116.20	111-70-6 / -34	4582 / 176.4	4-01-00-01731 / 0.8219[20]	H_2O 2; EtOH 5; eth 5; ctc 2 / 1.4249[20]
6473	2-Heptanol, (±)-	$C_7H_{16}O$ / 116.20	52390-72-4	4583 / 159	4-01-00-01740 / 0.8167[20]	H_2O 2; EtOH 3; eth 3; ctc 2 / 1.4210[20]
6474	3-Heptanol, (S)-	$C_7H_{16}O$ / 116.20	26549-25-7 / -70	157; 66[18]	2-01-00-00444 / 0.8227[20]	H_2O 2; EtOH 3; eth 3; ctc 2 / 1.4201[20]
6475	4-Heptanol / Dipropylcarbinol	$C_7H_{16}O$ / 116.20	589-55-9 / -41.2	156	4-01-00-01743 / 0.8183[20]	H_2O 1; EtOH 3; eth 3 / 1.4205[20]
6476	4-Heptanol, acetate	$C_9H_{18}O_2$ / 158.24	5921-84-6	171	4-02-00-00163 / 0.8741[0]	eth 4 / 1.4105[19]
6477	1-Heptanol, 7-chloro- / Heptamethylene chlorohydrin	$C_7H_{15}ClO$ / 150.65	55944-70-2 / 11	150[20]	4-01-00-01738 / 0.9998[15]	EtOH 4; peth 4 / 1.4537[25]
6478	2-Heptanol, 1-chloro- / 1-Chloro-2-heptanol	$C_7H_{15}ClO$ / 150.65	53660-21-2	93[13]	4-01-00-01741 / 0.9885[20]	EtOH 3; eth 3; ace 3 / 1.4499[20]
6479	2-Heptanol, 2,6-dimethyl- / 2,6-Dimethyl-2-heptanol	$C_9H_{20}O$ / 144.26	13254-34-7	173	2-01-00-00457 / 0.8186[20]	1.4242[20]
6480	3-Heptanol, 2,3-dimethyl- / Butylisopropylmethylcarbinol	$C_9H_{20}O$ / 144.26	19549-71-4	173; 75[16]	3-01-00-01752 / 0.8395[20]	1.4355[20]
6481	3-Heptanol, 2,6-dimethyl- / 2,6-Dimethyl-3-heptanol	$C_9H_{20}O$ / 144.26	19549-73-6	175	3-01-00-01753 / 0.8212[20]	1.4246[20]
6482	3-Heptanol, 3,5-dimethyl-	$C_9H_{20}O$ / 144.26	19549-74-7	95-7[50]	4-01-00-01811 / 0.8177[25]	1.4251[20]
6483	3-Heptanol, 3,6-dimethyl- / 3,6-Dimethylheptan-3-ol	$C_9H_{20}O$ / 144.26	1573-28-0	173	4-01-00-01810 / 0.8285[16]	eth 4; EtOH 4; lig 4 / 1.4326[16]
6484	4-Heptanol, 2,4-dimethyl- / 2,4-Dimethyl-4-heptanol	$C_9H_{20}O$ / 144.26	19549-77-0	171; 70[26]	4-01-00-01810 / 0.8215[20]	1.4292[20]
6485	4-Heptanol, 2,6-dimethyl- / Diisobutylcarbinol	$C_9H_{20}O$ / 144.26	108-82-7	174.5	4-01-00-01810 / 0.8114[20]	H_2O 1; EtOH 3; eth 3; ctc 2 / 1.4242[20]
6486	4-Heptanol, 3,5-dimethyl- / 3,5-Dimethyl-4-heptanol	$C_9H_{20}O$ / 144.26	19549-79-2	186	4-01-00-01811 / 0.836[18]	1.4283[20]
6487	4-Heptanol, 4-ethyl-	$C_9H_{20}O$ / 144.26	597-90-0	182	4-01-00-01809 / 0.8350[20]	eth 4; EtOH 4 / 1.4332[20]

No.	Name / Synonym	Mol. Form. / Mol. Wt.	CAS RN / mp/°C	Merck No. / bp/°C	Beil. Ref. / den/g cm⁻³	Solubility / n_D
6488	3-Heptanol, 3-ethyl-2-methyl-	$C_{10}H_{22}O$ 158.28	66719-37-7	193	1-01-00-00214 0.8455^{20}	1.4378
6489	3-Heptanol, 3-ethyl-6-methyl-	$C_{10}H_{22}O$ 158.28	66719-40-2	$83-6^{15}$	0-01-00-00427 0.852^{25}	1.4409
6490	3-Heptanol, 5-ethyl-4-methyl-	$C_{10}H_{22}O$ 158.28	66731-94-0	$86-8^{14}$	3-01-00-01770 0.86^{17}	1.4565^{17}
6491	1-Heptanol, 7-fluoro-	$C_7H_{15}FO$ 134.19	408-16-2	$98-9^{12}$	4-01-00-01736 0.956^{20}	eth 4; EtOH 4 1.4197^{25}
6492	1-Heptanol, 3-methyl- 3-Methyl-1-heptanol	$C_8H_{18}O$ 130.23	1070-32-2	101^{20}	4-01-00-01782 0.824^{24}	1.4295^{25}
6493	1-Heptanol, 4-methyl- 4-Methyl-1-heptanol	$C_8H_{18}O$ 130.23	817-91-4	188	4-01-00-01789 0.8065^{25}	EtOH 4 1.4253^{25}
6494	1-Heptanol, 6-methyl- 6-Methyl-1-heptanol	$C_8H_{18}O$ 130.23	1653-40-3 -106	188; 95.8^{20}	4-01-00-01782 0.8176^{25}	H_2O 1; EtOH 3; eth 3 1.4251^{25}
6495	2-Heptanol, 2-methyl- 2-Methyl-2-heptanol	$C_8H_{18}O$ 130.23	625-25-2 -50.4	156	4-01-00-01780 0.8142^{20}	H_2O 1; EtOH 3; eth 3 1.4250^{20}
6496	2-Heptanol, 3-methyl- 3-Methyl-2-heptanol	$C_8H_{18}O$ 130.23	31367-46-1	166.1	0.8177^{25}	H_2O 1; EtOH 3; eth 3; ctc 3 1.4199^{25}
6497	2-Heptanol, 5-methyl- 5-Methyl-2-heptanol	$C_8H_{18}O$ 130.23	54630-50-1 -61	170	4-01-00-01783 0.8174^{21}	
6498	2-Heptanol, 6-methyl-	$C_8H_{18}O$ 130.23	4730-22-7 -105	174	3-01-00-01728 0.8218^{20}	1.4238^{10}
6499	3-Heptanol, 2-methyl-, (±)-	$C_8H_{18}O$ 130.23	100296-26-2 -85	167.5	4-01-00-01781 0.8235^{20}	H_2O 2; EtOH 3; eth 3; ctc 3 1.4265^{20}
6500	3-Heptanol, 3-methyl- 2-Ethyl-2-hexanol	$C_8H_{18}O$ 130.23	5582-82-1 -83	163	4-01-00-01783 0.8282^{20}	H_2O 1; EtOH 3; eth 3; ctc 3 1.4279^{20}
6501	3-Heptanol, 4-methyl- 4-Methyl-3-heptanol	$C_8H_{18}O$ 130.23	14979-39-6 -123	170	4-01-00-01789 0.827^{25}	1.4300^{20}
6502	3-Heptanol, 5-methyl-	$C_8H_{18}O$ 130.23	18720-65-5 -91.2	172	4-01-00-01783 0.8425^{25}	1.433^{24}
6503	4-Heptanol, 2-methyl- 6-Methyl-4-heptanol	$C_8H_{18}O$ 130.23	21570-35-4 -81	164	4-01-00-01781 0.8207^{20}	eth 4; EtOH 4 1.4203
6504	4-Heptanol, 3-methyl-, (R*,S*)-(±)- 4-Methyl-3-heptanol(DL)	$C_8H_{18}O$ 130.23	92737-91-2	162	3-01-00-01731 0.8335^{25}	H_2O 2; EtOH 3; eth 3; ctc 3 1.4211^{25}
6505	4-Heptanol, 4-methyl- 4-Hydroxy-4-methylheptane	$C_8H_{18}O$ 130.23	598-01-6 -82	161	4-01-00-01789 0.8248^{20}	H_2O 1; EtOH 3; eth 3; ctc 3 1.4258^{20}
6506	2-Heptanol, 6-methyl-, acetate Acetic acid, (6-methyl-2-heptyl) ester	$C_{10}H_{20}O_2$ 172.27	67952-57-2	187	0-02-00-00135 0.8474^{20}	EtOH 4 1.413^{20}
6507	3-Heptanol, 6-methyl-, acetate Acetic acid, (6-methyl-3-heptyl) ester	$C_{10}H_{20}O_2$ 172.27	32764-34-4	184.5	0-02-00-00135 0.8554^{20}	EtOH 4 1.4160^{20}
6508	4-Heptanol, 4-propyl-	$C_{10}H_{22}O$ 158.28	2198-72-3	191	4-01-00-01831 0.8337^{21}	bz 4; eth 4; EtOH 4 1.4355^{21}
6509	4-Heptanol, 2,4,6-trimethyl-	$C_{10}H_{22}O$ 158.28	60836-07-9	181; 79^{12}	4-01-00-01833 0.823^{21}	1.4334^{18}
6510	2-Heptanone Methyl pentyl ketone	$C_7H_{14}O$ 114.19	110-43-0 -35	4584 151.0	4-01-00-03318 0.8111^{20}	H_2O 4; EtOH 3; eth 3 1.4088^{20}
6511	3-Heptanone Ethyl butyl ketone	$C_7H_{14}O$ 114.19	106-35-4 -39	147	4-01-00-03321 0.8183^{20}	H_2O 1; EtOH 5; eth 5; ctc 2 1.4057^{20}
6512	4-Heptanone Dipropyl ketone	$C_7H_{14}O$ 114.19	123-19-3 -33	3351 144	4-01-00-03323 0.8174^{20}	H_2O 1; EtOH 5; eth 5; ctc 3 1.4069^{20}
6513	2-Heptanone, 1-chloro- 1-Chloro-2-heptanone	$C_7H_{13}ClO$ 148.63	41055-92-9	83^{16}	4-01-00-03320 0.802^{20}	eth 4; EtOH 4 1.4371^{20}
6514	4-Heptanone, 2,6-dimethyl- Diisobutyl ketone	$C_9H_{18}O$ 142.24	108-83-8 -41.5	169.4	4-01-00-03360 0.8062^{20}	H_2O 1; EtOH 5; eth 5; ctc 3 1.412^{21}
6515	4-Heptanone, 3,5-dimethyl- Di-sec-butyl ketone	$C_9H_{18}O$ 142.24	19549-84-9	162	3-01-00-02893 0.826^{14}	1.4193^{14}
6516	3-Heptanone, 6-(dimethylamino)-4,4-diphenyl-, hydrochloride Methadone hydrochloride	$C_{21}H_{28}ClNO$ 345.91	1095-90-5 235	5852	3-14-00-00279	H_2O 4; EtOH 4
6517	4-Heptanone, 2-hydroxy-	$C_7H_{14}O_2$ 130.19	54862-92-9	95^{12}	4-01-00-04034 0.9296^{19}	1.4357^{19}
6518	2-Heptanone, 4-hydroxy-3-methyl- 4-Heptanol-2-one, 3-methyl	$C_8H_{16}O_2$ 144.21	56072-27-6	110^{16}	3-01-00-03249 0.9238^{20}	1.442^{20}
6519	3-Heptanone, 5-hydroxy-5-methyl- 3-Heptanol-5-one, 3-methyl	$C_8H_{16}O_2$ 144.21	39121-37-4	85^{15}	4-01-00-04045 0.9315^{15}	1.4367^{15}
6520	2-Heptanone, 3-methyl- 3-Methyl-2-heptanone	$C_8H_{16}O$ 128.21	2371-19-9	164	3-01-00-02878 0.8218^{20}	H_2O 2; EtOH 3; eth 3; ace 3 1.4172^{20}
6521	2-Heptanone, 6-methyl-	$C_8H_{16}O$ 128.21	928-68-7	167	4-01-00-03344 0.8151^{20}	H_2O 2; EtOH 4; eth 4; ace 5 1.4162^{20}
6522	3-Heptanone, 2-methyl- Butyl isopropyl ketone	$C_8H_{16}O$ 128.21	13019-20-0	158	4-01-00-03343 0.8163^{20}	H_2O 1; EtOH 3; eth 3; ace 4 1.4115^{20}
6523	3-Heptanone, 6-methyl-	$C_8H_{16}O$ 128.21	624-42-0	164	4-01-00-03344 0.8304^{20}	H_2O 1; EtOH 3; eth 3; bz 3 1.4209^{20}
6524	4-Heptanone, 2-methyl- Isobutyl propyl ketone	$C_8H_{16}O$ 128.21	626-33-5	155	4-01-00-03343 0.813^{22}	H_2O 1; EtOH 3; eth 3
6525	4-Heptanone, 3-methyl- 3-Methyl-4-heptanone	$C_8H_{16}O$ 128.21	15726-15-5	154	4-01-00-03344 0.817^{20}	1.4103^{25}
6526	1-Heptanone, 1-phenyl-	$C_{13}H_{18}O$ 190.29	1671-75-6 16.4	283.3	4-07-00-00780 0.9516^{20}	ace 4; eth 4; EtOH 4 1.5060^{20}
6527	Heptanoyl chloride	$C_7H_{13}ClO$ 148.63	2528-61-2 -83.8	125.2	4-02-00-00963 0.9590^{20}	eth 3; ctc 2; lig 4 1.4345^{18}

No.	Name Synonym	Mol. Form. Mol. Wt.	CAS RN mp/°C	bp/°C	Merck No.	Beil. Ref. den/g cm^{-3}	Solubility n_D
6528	Heptasiloxane, hexadecamethyl- Hexadecamethylheptasiloxane	$C_{16}H_{48}O_6Si_7$ 533.15	541-01-5 -78	270		4-04-00-04129 0.9012[20]	bz 4; lig 4 1.3965[20]
6529	1,3,5-Heptatriene	C_7H_{10} 94.16	2196-23-8 23	113.5; 31[25]		4-01-00-01097 0.764[20]	1.5079
6530	2-Heptenal Butylacrolein	$C_7H_{12}O$ 112.17	2463-63-0	166		4-01-00-03478 0.864[17]	1.4468[17]
6531	3-Heptenal	$C_7H_{12}O$ 112.17	89896-73-1	151		4-01-00-03480 0.851[15]	eth 3; ace 3; bz 3 1.4348[15]
6532	5-Hepten-2-amine, N,6-dimethyl- Isometheptene	$C_9H_{19}N$ 141.26	503-01-5	177	5070	3-04-00-00467	eth 4; EtOH 4
6533	1-Heptene	C_7H_{14} 98.19	592-76-7 -119.7	93.6		4-01-00-00857 0.6970[20]	H_2O 1; EtOH 3; eth 3; ctc 2 1.3998[20]
6534	2-Heptene, (E)-	C_7H_{14} 98.19	14686-13-6 -109.5	98		4-01-00-00860 0.7012[20]	H_2O 1; EtOH 3; eth 3; ace 3 1.4045[20]
6535	2-Heptene, (Z)- cis-2-Heptene	C_7H_{14} 98.19	6443-92-1	98.4		3-01-00-00824 0.708[20]	H_2O 1; EtOH 3; eth 3; ace 3 1.406[20]
6536	3-Heptene, (E)- trans-3-Heptene	C_7H_{14} 98.19	14686-14-7 -136.6	95.7		4-01-00-00861 0.6981[20]	H_2O 1; EtOH 3; eth 3; ace 3 1.4043[20]
6537	3-Heptene, (Z)- cis-3-Heptene	C_7H_{14} 98.19	7642-10-6 -136.6	95.8		4-01-00-00861 0.7030[20]	H_2O 1; EtOH 3; eth 3; ace 3 1.4059[20]
6538	1-Heptene, 1-chloro- 1-Chloro-1-heptene	$C_7H_{13}Cl$ 132.63	2384-75-0	155		3-01-00-00823 0.8948[20]	ace 4; bz 4; eth 4; chl 4 1.4380[20]
6539	1-Heptene, 2-chloro-	$C_7H_{13}Cl$ 132.63	65786-11-0	139		3-01-00-00823 0.8895[20]	eth 4; EtOH 4; chl 4 1.4349[20]
6540	3-Heptene, 4-chloro-	$C_7H_{13}Cl$ 132.63	2431-24-5	139		3-01-00-00827 0.883[14]	eth 4; EtOH 4; chl 4 1.437[14]
6541	2-Heptene, 6-chloro-2-methyl-	$C_8H_{15}Cl$ 146.66	80325-37-7	60-1[15]		2-01-00-00200 0.8931[18]	ace 4; bz 4; eth 4; EtOH 4 1.4458[18]
6542	2-Heptene, 2,3-dimethyl- 2,3-Dimethyl-2-heptene	C_9H_{18} 126.24	3074-64-4 -108.5	145.2		4-01-00-00897 0.731[25]	1.4519[20]
6543	1-Heptene, 2-methyl- 2-Methyl-1-heptene	C_8H_{16} 112.22	15870-10-7 -90	119.3		4-01-00-00881 0.7104[25]	1.4123[20]
6544	1-Heptene, 3-methyl- 3-Methyl-1-heptene	C_8H_{16} 112.22	4810-09-7	111		4-01-00-00883 0.707[25]	1.406[20]
6545	1-Heptene, 4-methyl- 4-Methyl-1-heptene	C_8H_{16} 112.22	13151-05-8	112.8		4-01-00-00884 0.713[25]	1.410[20]
6546	1-Heptene, 5-methyl- 5-Methyl-1-heptene	C_8H_{16} 112.22	13151-04-7	113.3		4-01-00-00884 0.7125[25]	1.4094[20]
6547	1-Heptene, 6-methyl- 6-Methyl-1-heptene	C_8H_{16} 112.22	5026-76-6	113.2		4-01-00-00883 0.7079[25]	1.4070[20]
6548	2-Heptene, 2-methyl- 2-Methyl-2-heptene	C_8H_{16} 112.22	627-97-4	122.6		4-01-00-00882 0.7200[25]	H_2O 1; eth 3; bz 3; ctc 3 1.4170[20]
6549	2-Heptene, 3-methyl-, (E)- 3-Methyl-trans-2-heptene	C_8H_{16} 112.22	22768-20-3	122		4-01-00-00883 1.419[20]	
6550	2-Heptene, 3-methyl-, (Z)- 2-Heptene, 3-methyl (cis)	C_8H_{16} 112.22	22768-19-0	122		3-01-00-00843 0.725[25]	1.419[20]
6551	2-Heptene, 4-methyl-, (E)- 4-Methyl-trans-2-heptene	C_8H_{16} 112.22	66225-17-0	114		4-01-00-00885 0.712[25]	1.410[20]
6552	2-Heptene, 4-methyl-, (Z)- 2-Heptene, 4-methyl (cis)	C_8H_{16} 112.22	66225-16-9	114		4-01-00-00885 0.712[25]	1.408[20]
6553	2-Heptene, 6-methyl- 2-Methyl-5-heptene	C_8H_{16} 112.22	73548-72-8	117		3-01-00-00843 0.714[25]	1.410[20]
6554	3-Heptene, 2-methyl- 2-Methyl-3-heptene	C_8H_{16} 112.22	17618-76-7 -107.5	112		4-01-00-00882 0.702[25]	1.407[20]
6555	3-Heptene, 3-methyl- 2-Ethyl-2-hexene	C_8H_{16} 112.22	7300-03-0	121		3-01-00-00843 0.724[25]	1.418[20]
6556	3-Heptene, 4-methyl- 4-Methyl-3-heptene	C_8H_{16} 112.22	4485-16-9	121		1-01-00-00094 0.721[25]	1.417[20]
6557	3-Heptene, 5-methyl-	C_8H_{16} 112.22	13172-91-3	112		4-01-00-00884 0.709[25]	1.410[20]
6558	3-Heptene, 6-methyl- 2-Methyl-4-heptene	C_8H_{16} 112.22	3404-57-7	115		4-01-00-00882 0.709[25]	1.410[20]
6559	3-Heptene, 4-propyl- 4-Propyl-3-heptene	$C_{10}H_{20}$ 140.27	4485-13-6	160.5		4-01-00-00906 0.7518[17]	1.4302[18]
6560	1-Heptene, 1,1,2,3,3,4,4,5,5,6,6,7,7,7-tetradecafluoro- Perfluoro-1-heptene	C_7F_{14} 350.05	355-63-5	81.0		4-01-00-00859	
6561	2-Heptenoic acid	$C_7H_{12}O_2$ 128.17	18999-28-5 -19	226.5		4-02-00-01577 0.9575[20]	1.4488[20]
6562	4-Heptenoic acid	$C_7H_{12}O_2$ 128.17	35194-37-7	98-100[5]		4-02-00-01578 0.949[15]	1.4418[15]
6563	5-Heptenoic acid	$C_7H_{12}O_2$ 128.17	3593-00-8	223		4-02-00-01578 0.9496[20]	1.4444[20]
6564	6-Heptenoic acid	$C_7H_{12}O_2$ 128.17	1119-60-4 -6.5	226		4-02-00-01576 0.9515[14]	1.4404[14]
6565	2-Heptenoic acid, 6-methyl-	$C_8H_{14}O_2$ 142.20	90112-75-7 3	239.5		1-02-00-00194 0.938[20]	1.4511[10]
6566	1-Hepten-4-ol	$C_7H_{14}O$ 114.19	3521-91-3	152.1		3-01-00-01936 0.8384[22]	1.4347[20]
6567	2-Hepten-1-ol, (E)-	$C_7H_{14}O$ 114.19	33467-76-4	178; 75[10]		3-01-00-01936 0.8516[20]	EtOH 3; ace 3 1.4460[20]

No.	Name / Synonym	Mol. Form. / Mol. Wt.	CAS RN / mp/°C	Merck No. / bp/°C	Beil. Ref. / den/g cm⁻³	Solubility / n_D
6568	2-Hepten-1-ol, (Z)- / cis-2-Hepten-1-ol	$C_7H_{14}O$ / 114.19	55454-22-3	178; 82[20]	4-01-00-02155 / 0.8421[20]	1.4410[20]
6569	2-Hepten-4-ol, (±)-	$C_7H_{14}O$ / 114.19	115113-98-9 /	153	4-01-00-02155 / 0.8445[20]	EtOH 3; eth 3 / 1.4373[20]
6570	3-Hepten-2-ol	$C_7H_{14}O$ / 114.19	98991-54-9	66-7[16]	4-01-00-02156 / 0.834[17]	1.4391[18]
6571	6-Hepten-2-ol	$C_7H_{14}O$ / 114.19	24395-10-6	64-5[13]	2-01-00-00488 / 0.8484[18]	1.4387[20]
6572	6-Hepten-3-ol / 1-Heptene-4-ol	$C_7H_{14}O$ / 114.19	19781-77-2	60-2[11]	4-01-00-02154 / 0.8447[18]	1.4369[18]
6573	1-Hepten-4-ol, 4-methyl- / 1-Heptene-4-ol, 4-methyl	$C_8H_{16}O$ / 128.21	1186-31-8	159.5	3-01-00-01945 / 0.8345[20]	eth 4; EtOH 4 / 1.4479[18]
6574	2-Hepten-4-ol, 5-methyl-	$C_8H_{16}O$ / 128.21	4048-31-1	69-70[18]	2-01-00-00490 / 0.8473[20]	1.4411[20]
6575	2-Hepten-4-ol, 6-methyl- / 6-Methyl-2-hepten-4-ol	$C_8H_{16}O$ / 128.21	4798-62-3	75[21]; 67[11]	2-01-00-00490 / 0.835[15]	1.4378[20]
6576	3-Hepten-2-ol, 2-methyl-	$C_8H_{16}O$ / 128.21	116668-45-2	62-3[14]	2-01-00-00490 / 0.8398[10]	1.4416[10]
6577	4-Hepten-2-ol, 2-methyl-	$C_8H_{16}O$ / 128.21	116668-44-1 / -28	63-4[20]	3-01-00-01944 / 0.8424[14]	1.4407[14]
6578	4-Hepten-3-ol, 3-methyl-	$C_8H_{16}O$ / 128.21	4048-30-0	103[75]; 62[15]	2-01-00-00490 / 0.8477[17]	1.4465[20]
6579	4-Hepten-3-ol, 4-methyl-	$C_8H_{16}O$ / 128.21	81280-12-8	66[17]	4-01-00-02170 / 0.8525[18]	1.4479[18]
6580	5-Hepten-2-ol, 6-methyl-	$C_8H_{16}O$ / 128.21	1569-60-4	175	1-01-00-00230 / 0.8545[20]	1.4505[20]
6581	6-Hepten-1-ol, 3-methyl-	$C_8H_{16}O$ / 128.21	4048-32-2	97-9[22]	2-01-00-00491 / 0.8562[19]	1.4470[14]
6582	6-Hepten-2-ol, 2-methyl- / 1,1-Dimethyl-5-hexen-1-ol	$C_8H_{16}O$ / 128.21	77437-98-0	65-6[16]	3-01-00-01945 / 0.8393[16]	1.4393[16]
6583	1-Hepten-3-one / Butyl vinyl ketone	$C_7H_{12}O$ / 112.17	2918-13-0	42-3[15]	4-01-00-03477 / 0.8434[20]	1.4305[19]
6584	4-Hepten-2-one	$C_7H_{12}O$ / 112.17	24332-22-7	61-2[20]	3-01-00-03004 / 0.8618[21]	1.4290[21]
6585	6-Hepten-2-one / Methyl 4-pentenyl ketone	$C_7H_{12}O$ / 112.17	21889-88-3	41-3[10]	4-01-00-03478 / 0.8673[18]	1.4350[18]
6586	6-Hepten-3-one / 1-Hepten-5-one	$C_7H_{12}O$ / 112.17	2565-39-1	142	4-01-00-03478 / 0.8487[18]	1.4254[18]
6587	3-Hepten-2-one, (E)-	$C_7H_{12}O$ / 112.17	5609-09-6	62[15]	4-01-00-03481 / 0.8496[20]	H_2O 1; EtOH 4; eth 4 / 1.4436[20]
6588	3-Hepten-2-one, (Z)- / 3-Hepten-2-one (cis)	$C_7H_{12}O$ / 112.17	69668-88-8	34-9[11]	4-01-00-03481 / 0.8440[20]	1.4325[20]
6589	5-Hepten-2-one, (E)-	$C_7H_{12}O$ / 112.17	1071-94-9	153	4-01-00-03480 / 0.8445[20]	1.4309[20]
6590	3-Hepten-2-one, 3-ethyl-4-methyl-	$C_{10}H_{18}O$ / 154.25	54244-90-5	190; 83[14]	3-01-00-03028 / 0.8559[20]	1.4535[20]
6591	4-Hepten-2-one, 3-ethyl-4-methyl-	$C_{10}H_{18}O$ / 154.25	53252-08-7	69[11]	3-01-00-03028 / 0.8450[20]	1.4405[20]
6592	4-Hepten-3-one, 5-ethyl-4-methyl-	$C_{10}H_{18}O$ / 154.25	22319-28-4	80[10]	4-01-00-03517 / 0.8622[10]	1.4545[19]
6593	5-Hepten-2-one, 4-ethyl-6-methyl- / 2-Hepten-6-one, 4-ethyl-2-methyl	$C_{10}H_{18}O$ / 154.25	104035-66-7	74-5[13]	4-01-00-03517 / 0.8430[20]	1.4546[20]
6594	5-Hepten-3-one, 5-ethyl-4-methyl-	$C_{10}H_{18}O$ / 154.25	74764-56-0 / 74.5	135-9[3]	4-01-00-03517 / 0.8564[21]	1.4452[21]
6595	1-Hepten-3-one, 5-methyl-	$C_8H_{14}O$ / 126.20	44829-76-7	40-1[7]	3-01-00-03014 / 0.8479[19]	1.4360[19]
6596	2-Hepten-4-one, 5-methyl- / 5-Methyl-2-hepten-4-one	$C_8H_{14}O$ / 126.20	81925-81-7	171	4-01-00-03496 / 0.848[25]	1.4400[25]
6597	2-Hepten-4-one, 6-methyl-	$C_8H_{14}O$ / 126.20	49852-35-9	170	4-01-00-03494 / 0.841[25]	1.4388[25]
6598	3-Hepten-2-one, 6-methyl-	$C_8H_{14}O$ / 126.20	2009-74-7	179	3-01-00-03011 / 0.8443[17]	1.4427[10]
6599	4-Hepten-2-one, 6-methyl- / 2-Methyl-3-hepten-6-one	$C_8H_{14}O$ / 126.20	39273-81-9	163	1-01-00-00385 / 0.8345[20]	1.4315
6600	4-Hepten-3-one, 5-methyl- / Homomesityl oxide	$C_8H_{14}O$ / 126.20	1447-26-3	166.5	3-01-00-03012 / 0.8591[20]	1.4476[20]
6601	5-Hepten-2-one, 3-methyl- / 3-Methyl-5-hepten-2-one	$C_8H_{14}O$ / 126.20	38552-72-6	62-4[20]	4-01-00-03496 / 0.8463[18]	1.4345[18]
6602	5-Hepten-2-one, 5-methyl-	$C_8H_{14}O$ / 126.20	10339-67-0	84[45]	4-01-00-03494 / 0.8647[16]	1.4434[16]
6603	5-Hepten-2-one, 6-methyl-	$C_8H_{14}O$ / 126.20	110-93-0	173.5	4-01-00-03493 / 0.8546[16]	eth 4; EtOH 4 / 1.4445[20]
6604	5-Hepten-3-one, 2-methyl-	$C_8H_{14}O$ / 126.20	77958-21-5	161.5	1-01-00-00385 / 0.842[20]	1.4310[20]
6605	5-Hepten-3-one, 5-methyl- / 5-Methylhept-5-en-3-one	$C_8H_{14}O$ / 126.20	1190-34-7	63[19]	3-01-00-03011 / 0.8524[21]	1.4367[21]
6606	1-Hepten-3-yne	C_7H_{10} / 94.16	2384-73-8	110	4-01-00-01097 / 0.7603[20]	ace 4; bz 4; EtOH 4; peth 4 / 1.4520[25]
6607	1-Hepten-4-yne, 6,6-dimethyl-	C_9H_{14} / 122.21	31508-08-4	125	0.758[20]	ace 4; bz 4; EtOH 4; peth 4 / 1.4312[20]

No.	Name / Synonym	Mol. Form. / Mol. Wt.	CAS RN / mp/°C	Merck No. / bp/°C	Beil. Ref. / den/g cm^{-3} / n_D	Solubility
6608	6-Hepten-4-yn-3-ol, 3-ethyl- / Diethylvinylethynylcarbinol	$C_9H_{14}O$ / 138.21	3142-84-5	62[4]	3-01-00-02034 / 0.8875[20] / 1.4800[20]	eth 4; EtOH 4
6609	1-Heptyne	C_7H_{12} / 96.17	628-71-7 / -81	99.7	4-01-00-01025 / 0.7328[20] / 1.4087[20]	H_2O 2; EtOH 5; eth 5; bz 3
6610	2-Heptyne / 1-Methyl-2-butylacetylene	C_7H_{12} / 96.17	1119-65-9	112	4-01-00-01026 / 0.744[25] / 1.4230[20]	H_2O 1; EtOH 5; eth 5; bz 3
6611	3-Heptyne / 1-Ethyl-2-propylacetylene	C_7H_{12} / 96.17	2586-89-2 / -130.5	107.2	4-01-00-01027 / 0.7336[25] / 1.4189[20]	H_2O 1; EtOH 5; eth 5; bz 3
6612	1-Heptyne, 1-bromo- / 1-Bromo-1-heptyne	$C_7H_{11}Br$ / 175.07	19821-84-2	164; 69[25]	3-01-00-00998 / 1.2120[22] / 1.4678[22]	ace 4; eth 4; EtOH 4; chl 4
6613	2-Heptyne, 2-bromo- / 2-Heptynyl bromide	$C_7H_{11}Br$ / 175.07	18495-26-6	104[55]; 84[20]	4-01-00-01026 / / 1.4878[25]	ace 4; eth 4; EtOH 4
6614	1-Heptyne, 1-chloro- / 1-Chloro-1-heptyne	$C_7H_{11}Cl$ / 130.62	51556-10-6	141	3-01-00-00997 / 0.9250[24] / 1.4411[24]	eth 4; EtOH 4
6615	2-Heptyne, 7-chloro- / 1-Chloro-5-heptyne	$C_7H_{11}Cl$ / 130.62	70396-13-3	166	3-01-00-00998 / / 1.4507[25]	eth 4; EtOH 4
6616	3-Heptyne, 7-chloro- / 4-Heptynyl chloride	$C_7H_{11}Cl$ / 130.62	51575-85-0	164	3-01-00-00999 / / 1.4517[20]	eth 4; EtOH 4
6617	3-Heptyne, 2,6-dimethyl-	C_9H_{16} / 124.23	19549-97-4	130.6	3-01-00-01015 / 0.785[20]	ace 4; eth 4
6618	3-Heptyne, 5,5-dimethyl- / 5,5-Dimethyl-3-heptyne	C_9H_{16} / 124.23	23097-98-5	69[100]	3-01-00-01015 / 0.7610[20] / 1.4360[20]	ace 4; eth 4
6619	3-Heptyne, 5-ethyl-5-methyl- / 5-Ethyl-5-methyl-3-heptyne	$C_{10}H_{18}$ / 138.25	61228-10-2	88[100]	3-01-00-01021 / 0.7714[20] / 1.4386[20]	ace 4; eth 4
6620	1-Heptyne, 1-iodo- / 1-Iodo-1-heptyne	$C_7H_{11}I$ / 222.07	54573-13-6	90-2[17]	3-01-00-00998 / 1.470[19] / 1.5123[19]	
6621	Hexacene	$C_{26}H_{16}$ / 328.41	258-31-1 / 380	sub	4-05-00-02834 / /	H_2O 1; EtOH 1
6622	Hexacosane	$C_{26}H_{54}$ / 366.71	630-01-3 / 56.4	412.2	4-01-00-00583 / 0.7783[60] / 1.4357[60]	bz 4; lig 4; chl 4
6623	Hexacosane, 13-dodecyl-	$C_{38}H_{78}$ / 535.04	55517-73-2 / 13.7	272[1]	4-01-00-00600 / 0.8188[20] / 1.4577[20]	
6624	Hexacosanoic acid / Cerotic acid	$C_{26}H_{52}O_2$ / 396.70	506-46-7 / 88.5		4-02-00-01310 / 0.8198[100] / 1.4301[100]	H_2O 1; EtOH 4; eth 4
6625	Hexacosanoic acid, methyl ester / Methyl hexacosanoate	$C_{27}H_{54}O_2$ / 410.72	5802-82-4 / 63.8	286[15]; 237[2]	4-02-00-01311 / / 1.4301[80]	chl 3
6626	1-Hexacosanol	$C_{26}H_{54}O$ / 382.71	506-52-5 / 80	305 20 dec	4-01-00-01912 / /	H_2O 1; EtOH 3; eth 3
6627	6,10-Hexadecadiyne	$C_{16}H_{26}$ / 218.38	10160-99-3	157[10]	3-01-00-01067 / 0.7907[20] / 1.4523[20]	
6628	Hexadecanal	$C_{16}H_{32}O$ / 240.43	629-80-1 / 35	200[29]	4-01-00-03393 / /	H_2O 1; EtOH 3; eth 3; ace 3
6629	Hexadecanamide	$C_{16}H_{33}NO$ / 255.44	629-54-9 / 107	236[12]	4-02-00-01182 / 1.0000[20] /	H_2O 1; EtOH 2; eth 2; ace 2
6630	Hexadecanamide, N-phenyl-	$C_{22}H_{37}NO$ / 331.54	6832-98-0 / 91.8	283[17]; 132[10]	4-12-00-00400 / /	ace 4; bz 4; EtOH 4; chl 4
6631	1-Hexadecanamine	$C_{16}H_{35}N$ / 241.46	143-27-1 / 46.8	322.5	4-04-00-00818 / 0.8129[20] / 1.4496[20]	H_2O 1; EtOH 4; eth 4; ace 3
6632	Hexadecane / Cetane	$C_{16}H_{34}$ / 226.45	544-76-3 / 18.1	286.8	4-01-00-00537 / 0.7733[20] / 1.4345	H_2O 1; EtOH 2; eth 5; ctc 3
6633	Hexadecane, 1-bromo-	$C_{16}H_{33}Br$ / 305.34	112-82-3 / 18	336	4-01-00-00542 / 0.9991[20] / 1.4618[25]	H_2O 1; eth 3
6634	Hexadecane, 1-chloro-	$C_{16}H_{33}Cl$ / 260.89	4860-03-1 / 17.9	322	4-01-00-00542 / 0.8652[20] / 1.4505[20]	H_2O 1
6635	Hexadecane, 1,16-dibromo- / Hexadecamethylene dibromide	$C_{16}H_{32}Br_2$ / 384.24	45223-18-5 / 56	204[4]	2-01-00-00138 / /	chl 4
6636	Hexadecane, 1,1-dimethoxy- / Palmitaldehyde, dimethyl acetal	$C_{18}H_{38}O_2$ / 286.50	2791-29-9 / 10	144[2]	4-01-00-03393 / 0.8542[20] / 1.4382[25]	ace 4; eth 4; EtOH 4
6637	Hexadecane, 6,11-dipentyl-	$C_{26}H_{54}$ / 366.71	15874-03-0 / -16.2	231[10]	4-01-00-00586 / 0.8072[20] / 1.4502[20]	
6638	Hexadecane, 1-(ethenyloxy)-	$C_{18}H_{36}O$ / 268.48	822-28-6 / 16	160[2]	4-01-00-02057 / 0.821[27] / 1.4444[25]	
6639	Hexadecane, 1-fluoro- / 1-Fluorohexadecane	$C_{16}H_{33}F$ / 244.44	408-38-8 / 18	289	4-01-00-00542 / 0.8321[20] / 1.4317[20]	eth 4; lig 4
6640	Hexadecane, 1-iodo-	$C_{16}H_{33}I$ / 352.34	544-77-4 / 22	212[15]	4-01-00-00543 / 1.123[25] / 1.4806[20]	H_2O 1; EtOH 2; eth 3; ace 3
6641	Hexadecanenitrile	$C_{16}H_{31}N$ / 237.43	629-79-8 / 31	333	4-02-00-01183 / 0.8303[20] / 1.4450[20]	H_2O 1; EtOH 4; eth 4; ace 4
6642	Hexadecane, 1,1'-oxybis-	$C_{32}H_{66}O$ / 466.88	4113-12-6 / 55	270 dec	4-01-00-01878 / 0.978[19] /	H_2O 1; EtOH 3; eth 3; chl 3
6643	1-Hexadecanethiol / Cetyl mercaptan	$C_{16}H_{34}S$ / 258.51	2917-26-2 / 19	123-8[0.5]	4-01-00-01881 / /	H_2O 1; EtOH 2; eth 3; ctc 2
6644	Hexadecanoic acid / Palmitic acid	$C_{16}H_{32}O_2$ / 256.43	57-10-3 / 61.8	6947 / 351.5	4-02-00-01157 / 0.8527[62] / 1.43345[60]	H_2O 1; EtOH 3; eth 5; ace 3
6645	Hexadecanoic acid, aluminum salt	$C_{48}H_{93}AlO_6$ / 793.24	555-35-1	361	4-02-00-01157 / /	H_2O 1; EtOH 3; peth 3; bz 3
6646	Hexadecanoic acid, anhydride	$C_{32}H_{62}O_3$ / 494.84	623-65-4 / 64		4-02-00-01181 / 0.8388[83] / 1.4364[68]	eth 4
6647	Hexadecanoic acid, butyl ester	$C_{20}H_{40}O_2$ / 312.54	111-06-8 / 16.9		4-02-00-01167 / / 1.4312[50]	H_2O 1; EtOH 3; eth 3

No.	Name / Synonym	Mol. Form. / Mol. Wt.	CAS RN / mp/°C	Merck No. / bp/°C	Beil. Ref. / den/g cm⁻³	Solubility / n_D
6648	Hexadecanoic acid, 2-[(dichloroacetyl)amino]-3-hydroxy-3-(4-nitrophenyl)propyl ester, [R-(R*,R*)]- Chloramphenicol palmitate	$C_{27}H_{42}Cl_2N_2O_8$ 561.55	530-43-8 90	2069	4-13-00-02753	bz 4; eth 4; EtOH 4
6649	Hexadecanoic acid, 1,2-ethanediyl ester	$C_{34}H_{66}O_4$ 538.90	624-03-3 72		4-02-00-01169 0.8594[78]	H_2O 1; EtOH 1; eth 3; ace 4
6650	Hexadecanoic acid, ethyl ester	$C_{18}H_{36}O_2$ 284.48	628-97-7 24	191[10]	4-02-00-01165 0.8577[25]	H_2O 1; EtOH 3; eth 3; ace 3 1.4347[34]
6651	Hexadecanoic acid, hexadecyl ester Cetyl palmitate	$C_{32}H_{64}O_2$ 480.86	540-10-3 54	2023	4-02-00-01168 0.989[20]	eth 4; EtOH 4 1.4398[70]
6652	Hexadecanoic acid, 2-hydroxyethyl ester	$C_{18}H_{36}O_3$ 300.48	4219-49-2 51	173-4[3]	4-02-00-01168 0.8768[60]	EtOH 4
6653	Hexadecanoic acid, methyl ester	$C_{17}H_{34}O_2$ 270.46	112-39-0 30	417; 148[2]	4-02-00-01165 0.8247[75]	H_2O 1; EtOH 4; eth 3; ace 4
6654	Hexadecanoic acid, 1-methylethyl ester	$C_{19}H_{38}O_2$ 298.51	142-91-6 13.5	160[2]	4-02-00-01167 0.8404[38]	ace 4; bz 4; eth 4; EtOH 4 1.4364[25]
6655	Hexadecanoic acid, phenylmethyl ester Benzyl palmitate	$C_{23}H_{38}O_2$ 346.55	41755-60-6 36		3-06-00-01481 0.9109[35]	bz 4; eth 4; EtOH 4; chl 4 1.4689[50]
6656	Hexadecanoic acid, 1,2,3-propanetriyl ester Tripalmitin	$C_{51}H_{98}O_6$ 807.34	555-44-2 66.5	9648 315	4-02-00-01176 0.8752[70]	H_2O 1; EtOH 2; eth 4; bz 3 1.4381[80]
6657	Hexadecanoic acid, propyl ester	$C_{19}H_{38}O_2$ 298.51	2239-78-3 20.4	190[12]	4-02-00-01167 0.8455[33]	1.4392[25]
6658	Hexadecanoic acid, 3,7,11,15-tetramethyl- Phytanic acid	$C_{20}H_{40}O_2$ 312.54	14721-66-5 -65			
6659	1-Hexadecanol Cetyl alcohol	$C_{16}H_{34}O$ 242.45	36653-82-4 49.3	2020 334	4-01-00-01876 0.8187[50]	H_2O 1; EtOH 2; eth 4; ace 3 1.4283[79]
6660	2-Hexadecanol	$C_{16}H_{34}O$ 242.45	14852-31-4 44	314	4-01-00-01882 0.8338[20]	1.4479[20]
6661	3-Hexadecanol	$C_{16}H_{34}O$ 242.45	593-03-3 50	152[4]	4-01-00-01882 0.8000[67]	
6662	1-Hexadecanol, acetate	$C_{18}H_{36}O_2$ 284.48	629-70-9 -18.5	220-5[205]	4-02-00-00171 0.8574[25]	H_2O 1; EtOH 2; ctc 3 1.4438[20]
6663	3-Hexadecanone	$C_{16}H_{32}O$ 240.43	18787-64-9 43	184[17]; 140[2]	4-01-00-03394	chl 3
6664	1-Hexadecanone, 1-phenyl-	$C_{22}H_{36}O$ 316.53	6697-12-7 59	251[15]	4-07-00-00860 0.8692[76]	1.4675[76]
6665	Hexadecanoyl chloride	$C_{16}H_{31}ClO$ 274.87	112-67-4 12	199[20]	4-02-00-01182 0.9016[25]	eth 4 1.4514[20]
6666	6,10,14-Hexadecatrienoic acid Hiragonic acid	$C_{16}H_{26}O_2$ 250.38	4444-12-6	180-90[15]	3-02-00-01505 0.9296[20]	eth 4; EtOH 4 1.4850[50]
6667	1-Hexadecene 1-Cetene	$C_{16}H_{32}$ 224.43	629-73-2 4.1	284.9	4-01-00-00927 0.7811[20]	H_2O 1; EtOH 3; eth 3; ctc 3 1.4412[20]
6668	7-Hexadecenoic acid Dinoroleic acid	$C_{16}H_{30}O_2$ 254.41	1191-75-9 33	230[10]	0-02-00-00460	EtOH 4; eth 4
6669	9-Hexadecenoic acid	$C_{16}H_{30}O_2$ 254.41	2091-29-4 0.5	181[1]	4-02-00-01629	bz 4; eth 4; peth 4
6670	9-Hexadecenoic acid, (Z)- Palmitoleic acid	$C_{16}H_{30}O_2$ 254.41	373-49-9 -0.1	131[0.06]	4-02-00-01629	
6671	1-Hexadecen-3-ol, 3,7,11,15-tetramethyl- Isophytol	$C_{20}H_{40}O$ 296.54	505-32-8	5084 107-10[0.01]	2-01-00-00504 0.8519[20]	bz 4; eth 4; EtOH 4 1.4571[20]
6672	2-Hexadecen-1-ol, 3,7,11,15-tetramethyl-, [R-[R*,R*-(E)]]- Phytol	$C_{20}H_{40}O$ 296.54	150-86-7	7362 203[10]	4-01-00-02208 0.8497[25]	1.4595[25]
6673	1-Hexadecyne	$C_{16}H_{30}$ 222.41	629-74-3 15	284	4-01-00-01073 0.7965[20]	bz 4 1.4440[20]
6674	2-Hexadecyne	$C_{16}H_{30}$ 222.41	629-75-4 20	160[15]	3-01-00-01028 0.8039[20]	
6675	2,4-Hexadienal, (E,E)-	C_6H_8O 96.13	142-83-6	174; 76[30]	4-01-00-03545 0.898[20]	1.5384[20]
6676	1,2-Hexadiene Propylallene	C_6H_{10} 82.15	592-44-9	76	4-01-00-01011 0.7149[20]	eth 4; chl 4 1.4282[20]
6677	1,4-Hexadiene	C_6H_{10} 82.15	592-45-0	65	4-01-00-01013 0.7000[20]	H_2O 1; eth 4 1.4150[20]
6678	1,5-Hexadiene Biallyl	C_6H_{10} 82.15	592-42-7 -140.7	59.4	4-01-00-01013 0.6878[25]	H_2O 1; EtOH 3; eth 3; bz 3 1.4042[20]
6679	2,3-Hexadiene Methylethylallene	C_6H_{10} 82.15	592-49-4	78	4-01-00-01016 0.684[15]	1.395[20]
6680	2,4-Hexadiene	C_6H_{10} 82.15	592-46-1 -79	80	3-01-00-00985 0.7196[20]	H_2O 1; EtOH 3; eth 3; chl 3 1.4500[20]
6681	1,3-Hexadiene, (E)-	C_6H_{10} 82.15	20237-34-7 -102.4	73.2	4-01-00-01011 0.6995[20]	1.4406[20]
6682	1,3-Hexadiene, (Z)-	C_6H_{10} 82.15	14596-92-0	73.1	4-01-00-01011 0.7033[25]	1.4379[20]
6683	1,4-Hexadiene, (E)-	C_6H_{10} 82.15	7319-00-8 -138.7	65.0	0.695[25]	1.4104[20]
6684	1,4-Hexadiene, (Z)-	C_6H_{10} 82.15	7318-67-4	66.3	0.695[25]	eth 4 1.4049[20]
6685	2,4-Hexadiene, (E,E)-	C_6H_{10} 82.15	5194-51-4 -44.9	82.2	4-01-00-01016 0.7101[25]	1.4510[20]
6686	2,4-Hexadiene, (E,Z)-	C_6H_{10} 82.15	5194-50-3 -96.1	83.5	4-01-00-01016 0.7185[25]	1.4560[20]

No.	Name Synonym	Mol. Form. Mol. Wt.	CAS RN mp/°C	Merck No. bp/°C	Beil. Ref. den/g cm^{-3}	Solubility n_D
6687	2,4-Hexadiene, (Z,Z)-	C$_6$H$_{10}$ 82.15	6108-61-8	85	0.7298[25]	1.4606[20]
6688	1,3-Hexadiene, 3-chloro-	C$_6$H$_9$Cl 116.59	101870-06-8	68[117]	4-01-00-01012 0.9390[20]	eth 4; chl 4 1.4770[20]
6689	1,5-Hexadiene, 1,1,2,3,3,4,4,5,6,6-decachloro-	C$_6$Cl$_{10}$ 426.59	29030-84-0 49	121[0.03]	4-01-00-01016 1.905[52]	ace 4; eth 4; chl 4 1.6012[51]
6690	2,4-Hexadiene, 1,3-dichloro-	C$_6$H$_8$Cl$_2$ 151.03	73454-83-8	80-2[17]	4-01-00-01017 1.1456[20]	bz 4; chl 4 1.5271[20]
6691	1,5-Hexadiene, 2,5-dimethyl-	C$_8$H$_{14}$ 110.20	627-58-7 -75.6	114.3	4-01-00-01042 0.743[20]	H$_2$O 1; ace 3; chl 3 1.43995[21]
6692	2,4-Hexadiene, 2,5-dimethyl-	C$_8$H$_{14}$ 110.20	764-13-6 14	134.5	4-01-00-01043 0.7577[25]	H$_2$O 1; EtOH 3; eth 3; bz 3 1.4785[20]
6693	1,5-Hexadiene-3,4-diol, (R*,R*)-(±)-	C$_6$H$_{10}$O$_2$ 114.14	19700-97-1 21.7	90-1[8]	1.017[19]	H$_2$O 3; EtOH 3; eth 3; chl 3 1.4790[19]
6694	1,5-Hexadiene, 3-ethyl-2-methyl- 1,5-Hexadiene, 2-methyl-3-ethyl	C$_9$H$_{16}$ 124.23	73398-15-9 -70	145	0-01-00-00260 0.7594[25]	
6695	1,5-Hexadiene, 2-methyl-	C$_7$H$_{12}$ 96.17	4049-81-4 -128.8	88.1	4-01-00-01030 0.7153[25]	1.4183[20]
6696	2,4-Hexadiene, 2-methyl-	C$_7$H$_{12}$ 96.17	28823-41-8 -74.2	111.5	4-01-00-01030 0.7411[25]	
6697	2,4-Hexadiene, 1,3,4,6-tetrachloro-	C$_6$H$_6$Cl$_4$ 219.92	100367-45-1	84-9[2]	3-01-00-00987 1.4013[20]	MeOH 4; chl 4 1.5465[20]
6698	2,4-Hexadienoic acid Sorbic acid	C$_6$H$_8$O$_2$ 112.13	110-44-1 134.5	8677 228 dec; 153[50]	2-02-00-00452 1.204[19]	H$_2$O 3; EtOH 3; eth 4; chl 3
6699	2,4-Hexadienoic acid, ethyl ester	C$_8$H$_{12}$O$_2$ 140.18	110318-09-7	195.5 85[20]	2-02-00-00452 0.9506[20]	H$_2$O 1; EtOH 3; eth 3; chl 3 1.4951[20]
6700	2,4-Hexadienoic acid, ethyl ester, (E,E)-	C$_8$H$_{12}$O$_2$ 140.18	2396-84-1	195.5	4-02-00-01703 0.9506[20]	eth 4; EtOH 4; chl 4 1.4951[20]
6701	2,5-Hexadienoic acid, 3-methoxy-5-methyl-4-oxo- Penicillic acid	C$_8$H$_{10}$O$_4$ 170.17	90-65-3 83	7033	3-03-00-01467	H$_2$O 3; EtOH 4; eth 4; ace 3
6702	2,4-Hexadienoic acid, methyl ester Methyl 2,4-hexadienoate	C$_7$H$_{10}$O$_2$ 126.16	1515-80-6 15	180; 70[20]	2-02-00-00452 0.9777[20]	H$_2$O 1; EtOH 3; eth 3 1.5025[22]
6703	2,4-Hexadienoic acid, methyl ester, (E,E)-	C$_7$H$_{10}$O$_2$ 126.16	689-89-4 15	180	4-02-00-01703 0.9777[20]	eth 4; EtOH 4 1.5025[22]
6704	2,4-Hexadienoic acid, potassium salt, (E,E)-	C$_6$H$_7$KO$_2$ 150.22	24634-61-5 >270 dec	7661	1.361[25]	H$_2$O 4; EtOH 3
6705	1,4-Hexadien-3-ol	C$_6$H$_{10}$O 98.14	1070-14-0	50-2[15]	4-01-00-02238 0.859[23]	1.4502[23]
6706	1,5-Hexadien-3-ol 3-Hydroxy-1,5-hexadiene	C$_6$H$_{10}$O 98.14	924-41-4	133.5	4-01-00-02239 0.8506[25]	1.4464[25]
6707	2,4-Hexadien-1-ol 1-Hydroxy-2,4-hexadiene	C$_6$H$_{10}$O 98.14	111-28-4 30.5	8678 76[12]	4-01-00-02239 0.8967[23]	H$_2$O 1; EtOH 3; eth 3 1.4981[20]
6708	3,5-Hexadien-2-ol 1,3-Hexadien-5-ol	C$_6$H$_{10}$O 98.14	3280-51-1	77-8[26]	4-01-00-02237 0.8678[20]	EtOH 4 1.4896[20]
6709	2,4-Hexadien-1-ol, (E,E)-	C$_6$H$_{10}$O 98.14	17102-64-6 30.5	76[12]	4-01-00-02239 0.8967[20]	eth 4; EtOH 4 1.4981[20]
6710	3,5-Hexadien-2-one, 6-phenyl-	C$_{12}$H$_{12}$O 172.23	4173-44-8 68	171[15]; 155[12]	4-07-00-01183	bz 4; eth 4; EtOH 4; chl 4
6711	2,4-Hexadienoyl chloride, (E,E)-	C$_6$H$_7$ClO 130.57	2614-88-2	78[15]	3-02-00-01460 1.0666[19]	ace 4 1.5545[20]
6712	1,3-Hexadien-5-yne	C$_6$H$_6$ 78.11	10420-90-3 -81	83.5	4-01-00-01120 0.7806[20]	bz 4 1.5095[20]
6713	1,5-Hexadien-3-yne Divinylacetylene	C$_6$H$_6$ 78.11	821-08-9 -88	85	4-01-00-01120 0.7851[20]	H$_2$O 1; bz 3 1.5035[20]
6714	1,5-Hexadien-3-yne, 2,5-dimethyl- Diisopropenylacetylene	C$_8$H$_{10}$ 106.17	3725-05-1	123	4-01-00-01124 0.7863[25]	bz 4; chl 4 1.4845[20]
6715	1,4-Hexadiyne	C$_6$H$_6$ 78.11	10420-91-4 <-80	81.7	3-01-00-01057 0.825[0]	bz 4; chl 4
6716	1,5-Hexadiyne Bipropargyl	C$_6$H$_6$ 78.11	628-16-0 -6	86	4-01-00-01118 0.8049[20]	H$_2$O 1; EtOH 3; eth 3; ace 3 1.4380[23]
6717	2,4-Hexadiyne Dimethyldiacetylene	C$_6$H$_6$ 78.11	2809-69-0 67.8	129.5	4-01-00-01119	EtOH 4; eth 4
6718	Hexanal Caproaldehyde	C$_6$H$_{12}$O 100.16	66-25-1 -56	1761 131	4-01-00-03296 0.8335[20]	H$_2$O 2; EtOH 4; eth 4; ace 3 1.4039[20]
6719	Hexanal, 2-ethyl- 2-Ethylhexanal	C$_8$H$_{16}$O 128.21	123-05-7 <-100	163	4-01-00-03345 0.8540[20]	H$_2$O 1; EtOH 3; eth 3; ctc 2 1.4142[20]
6720	Hexanal, 3-methyl- Caproaldehyde, 3-methyl	C$_7$H$_{14}$O 114.19	19269-28-4	143	2-01-00-00756 0.8203[20]	H$_2$O 1; EtOH 3; eth 3 1.4122[20]
6721	Hexanamide	C$_6$H$_{13}$NO 115.18	628-02-4 101	255	4-02-00-00929 0.999[20]	bz 4; eth 4; EtOH 4; chl 4 1.4200[110]
6722	Hexanamide, N-phenyl-	C$_{12}$H$_{17}$NO 191.27	621-15-8 95		4-12-00-00392 1.112[25]	eth 4; EtOH 4
6723	1-Hexanamine Hexylamine	C$_6$H$_{15}$N 101.19	111-26-2 -22.9	132.8	4-04-00-00709 0.7660[20]	H$_2$O 2; EtOH 5; eth 5; chl 3 1.4180[20]
6724	2-Hexanamine, (±)-	C$_6$H$_{15}$N 101.19	68107-05-1 -19	117.5	4-04-00-00721 0.7533[20]	EtOH 3; eth 3 1.4080[25]

No.	Name Synonym	Mol. Form. Mol. Wt.	CAS RN mp/°C	Merck No. bp/°C	Beil. Ref. den/g cm^{-3}	Solubility n_D
6725	1-Hexanamine, N,N-dihexyl-	C$_{18}$H$_{39}$N 269.51	102-86-3	261.7	4-04-00-00711 0.7976[21]	H$_2$O 1; EtOH 4; eth 4; acid 3
6726	1-Hexanamine, 2-ethyl- 2-Ethyl-1-hexylamine	C$_8$H$_{19}$N 129.25	104-75-6		4-04-00-00766	
6727	1-Hexanamine, N-hexyl-	C$_{12}$H$_{27}$N 185.35	143-16-8 -13.1	236; 75[1]	4-04-00-00711 0.7889[20]	EtOH 3; eth 3 1.4339[20]
6728	1-Hexanamine, 3-methyl- Hexylamine, 3-methyl-	C$_7$H$_{17}$N 115.22	65530-93-0	149; 67[45]	2-04-00-00653 0.772[26]	1.4249[25]
6729	1-Hexanamine, 4-methyl- Hexylamine, 4-methyl-	C$_7$H$_{17}$N 115.22	34263-68-8	153	2-04-00-00654 0.7802[20]	1.4238[20]
6730	2-Hexanamine, 4-methyl- Methylhexaneamine	C$_7$H$_{17}$N 115.22	105-41-9	6000 132.5	4-04-00-00747 0.7655[20]	H$_2$O 2; EtOH 4; eth 4; chl 4 1.4150[25]
6731	Hexane	C$_6$H$_{14}$ 86.18	110-54-3 -95.3	4613 68.7	4-01-00-00338 0.6548[25]	H$_2$O 1; EtOH 4; eth 3; chl 3 1.3749[20]
6732	Hexane, 1-bromo- Hexyl bromide	C$_6$H$_{13}$Br 165.07	111-25-1 -84.7	155.3	4-01-00-00352 1.1744[20]	H$_2$O 1; EtOH 5; eth 5; ace 3 1.4478[20]
6733	Hexane, 2-bromo- 2-Bromohexane	C$_6$H$_{13}$Br 165.07	3377-86-4	144; 78[90]	4-01-00-00353 1.1658[20]	H$_2$O 1; EtOH 4; eth 3; ace 3 1.4832[25]
6734	Hexane, 3-bromo- 3-Bromohexane	C$_6$H$_{13}$Br 165.07	3377-87-5	142	4-01-00-00353 1.1799[20]	ace 4; eth 4; EtOH 4; chl 4 1.4472[20]
6735	Hexane, 1-bromo-6-fluoro- 1-Bromo-6-fluorohexane	C$_6$H$_{12}$BrF 183.06	373-28-4	67-8[11]	4-01-00-00353 1.293[20]	ace 4; eth 4; EtOH 4; chl 4 1.4436[25]
6736	Hexane, 1-bromo-6-methoxy-	C$_7$H$_{15}$BrO 195.10	50592-87-5	112[35]; 98[19]	4-01-00-01704 1.1887[25]	1.4469[25]
6737	Hexane, 1-chloro- Hexyl chloride	C$_6$H$_{13}$Cl 120.62	544-10-5 -94	2144 135	4-01-00-00349 0.8785[20]	H$_2$O 1; EtOH 3; eth 3; ace 3 1.4199[20]
6738	Hexane, 2-chloro- 2-Hexyl chloride	C$_6$H$_{13}$Cl 120.62	638-28-8	122.5	4-01-00-00349 0.8694[21]	ace 4; bz 4; eth 4; EtOH 4 1.4142[22]
6739	Hexane, 3-chloro- 3-Chlorohexane	C$_6$H$_{13}$Cl 120.62	2346-81-8	123	4-01-00-00349 0.8684[20]	ace 4; bz 4; eth 4; EtOH 4 1.4163[20]
6740	Hexane, 2-chloro-2,5-dimethyl- 2-Chloro-2,5-dimethylhexane	C$_8$H$_{17}$Cl 148.68	29342-44-7	86[100]	4-01-00-00434 0.8476[18]	ace 4; bz 4; eth 4; EtOH 4 1.4232[20]
6741	Hexane, 2-chloro-2,3-dimethyl-	C$_8$H$_{17}$Cl 148.68	101654-30-2	41-3[12]	3-01-00-00481 0.8869[20]	EtOH 4 1.4333[25]
6742	Hexane, 1-chloro-6-fluoro- 1-Chloro-6-fluorohexane	C$_6$H$_{12}$ClF 138.61	1550-09-0	168; 62[15]	4-01-00-00350 1.015[20]	eth 4; EtOH 4; chl 4 1.4168[20]
6743	Hexane, 1-chloro-3-methyl-	C$_7$H$_{15}$Cl 134.65	101257-63-0	151	2-01-00-00119 0.8766[20]	eth 4; EtOH 4; chl 4 1.4274[207]
6744	Hexane, 2-chloro-2-methyl-	C$_7$H$_{15}$Cl 134.65	4398-65-6	135 dec; 60[52]	4-01-00-00398 0.8635[20]	eth 4; EtOH 4; chl 4 1.4200[20]
6745	Hexane, 2-chloro-5-methyl- 2-Chloro-5-methylhexane	C$_7$H$_{15}$Cl 134.65	58766-17-9	138 dec	3-01-00-00436 0.863[20]	eth 4; EtOH 4; chl 4
6746	Hexane, 3-chloro-3-methyl-	C$_7$H$_{15}$Cl 134.65	43197-78-0	135	3-01-00-00438 0.8787[20]	eth 4; EtOH 4; chl 4 1.4250[20]
6747	Hexane, 3-chloro-2,2,3-trimethyl-	C$_9$H$_{19}$Cl 162.70	102449-95-6	64-5[13]	3-01-00-00515 0.9010[20]	eth 4; EtOH 4; chl 4 1.4465[20]
6748	Hexanedial	C$_6$H$_{10}$O$_2$ 114.14	1072-21-5 -8	92-4[9]	4-01-00-03686 1.003[19]	bz 4; eth 4; EtOH 4 1.4350[20]
6749	1,6-Hexanediamine Hexamethylenediamine	C$_6$H$_{16}$N$_2$ 116.21	124-09-4 41.5	4614 205	4-04-00-01320	H$_2$O 4; EtOH 3; bz 3
6750	1,6-Hexanediamine, N,N'-dibutyl- N,N'-Dibutylhexamethylenediamine	C$_{14}$H$_{32}$N$_2$ 228.42	4835-11-4	137-40[3-4]	4-04-00-01327	1.4470[25]
6751	2,5-Hexanediamine, 2,5-dimethyl-	C$_8$H$_{20}$N$_2$ 144.26	23578-35-0	63-4[8]	4-04-00-01363 0.832[25]	1.4459[20]
6752	1,6-Hexanediamine, N,N,N',N'-tetramethyl-	C$_{10}$H$_{24}$N$_2$ 172.31	111-18-2	209.5	4-04-00-01322 0.806[25]	1.4359[20]
6753	1,6-Hexanediaminium, N,N'-di-9H-fluoren-9-yl-N,N,N',N'-tetramethyl-dibromide Hexafluorenium bromide	C$_{36}$H$_{42}$Br$_2$N$_2$ 662.55	317-52-2 153.5	4606	4-12-00-03397	
6754	Hexane, 1,2-dibromo- 1,2-Dibromohexane	C$_6$H$_{12}$Br$_2$ 243.97	624-20-4	104[36]; 82[12]	3-01-00-00392 1.5774[20]	bz 4; eth 4; chl 4 1.5024[20]
6755	Hexane, 1,4-dibromo- 1,4-Dibromohexane	C$_6$H$_{12}$Br$_2$ 243.97	25118-28-9	106-8[15]	3-01-00-00392 1.602[15]	1.5084[15]
6756	Hexane, 1,5-dibromo- 1,5-Dibromohexane	C$_6$H$_{12}$Br$_2$ 243.97	627-96-3	153[100]; 103[9]	4-01-00-00354 1.5989[20]	1.5072[15]
6757	Hexane, 1,6-dibromo-	C$_6$H$_{12}$Br$_2$ 243.97	629-03-8 -2.3	245.5	4-01-00-00354 1.5948[15]	H$_2$O 1; eth 3; ace 3; ctc 2 1.5037[20]
6758	Hexane, 2,3-dibromo- 2,3-Dibromohexane	C$_6$H$_{12}$Br$_2$ 243.97	6423-02-5	90[16]	3-01-00-00393 1.5812[20]	1.5025[20]
6759	Hexane, 2,5-dibromo-, (R*,R*)-(±)- Hexane, 2,5-dibromo(DL)	C$_6$H$_{12}$Br$_2$ 243.97	54462-68-9 -44.64	108[30] 94[13]	3-01-00-00393 1.5788[20]	H$_2$O 1; eth 3; ace 3; ctc 3 1.5007[20]
6760	Hexane, 3,4-dibromo-	C$_6$H$_{12}$Br$_2$ 243.97	89583-12-0	80-1[13]	3-01-00-00394 1.6027[20]	1.5043[20]
6761	Hexane, 1,2-dichloro- 1,2-Dichlorohexane	C$_6$H$_{12}$Cl$_2$ 155.07	2162-92-7	173	4-01-00-00350 1.085[15]	eth 4; chl 4
6762	Hexane, 1,6-dichloro- 1,6-Dichlorohexane	C$_6$H$_{12}$Cl$_2$ 155.07	2163-00-0	204	4-01-00-00350 1.0677[20]	H$_2$O 1; eth 3; ctc 3; chl 3 1.4572[20]
6763	Hexane, 2,2-dichloro- 2,2-Dichlorohexane	C$_6$H$_{12}$Cl$_2$ 155.07	42131-89-5	68[49]	3-01-00-00390 1.0150[25]	eth 4; chl 4 1.4353[25]

No.	Name Synonym	Mol. Form. Mol. Wt.	CAS RN	Merck No. mp/°C	bp/°C	Beil. Ref. den/g cm^{-3}	Solubility n_D
6764	Hexane, 2,3-dichloro- 2,3-Dichlorohexane	$C_6H_{12}Cl_2$ 155.07	54305-87-2		163.5	4-01-00-00350 1.0527[11]	eth 4; chl 4
6765	Hexane, 2,5-dichloro-2,5-dimethyl- 2,5-Dimethyl-2,5-dichlorohexane	$C_8H_{16}Cl_2$ 183.12	6223-78-5 67.5			4-01-00-00435 0.9543[20]	bz 4; eth 4; EtOH 4; chl 4
6766	Hexane, 2,5-dichloro-, (R^*,R^*)-(±)- (±)-2,5-Dichlorohexane	$C_6H_{12}Cl_2$ 155.07	41761-12-0 -38		177; 106[91]	4-01-00-00350 1.0474[20]	chl 4 1.4491[20]
6767	Hexane, 3,3-diethyl-	$C_{10}H_{22}$ 142.28	17302-02-2		166.3	4-01-00-00484 0.7575[25]	1.4258[20]
6768	Hexane, 3,4-diethyl- 3,4-Diethylhexane	$C_{10}H_{22}$ 142.28	19398-77-7		163.9	4-01-00-00484 0.7472[25]	1.4190[20]
6769	Hexane, 1,6-diiodo- Hexamethylene diiodide	$C_6H_{12}I_2$ 337.97	629-09-4 9.5		163[17]; 141[10]	4-01-00-00356 2.0416[15]	H_2O 1; EtOH 4; eth 4 1.5899[15]
6770	Hexane, 1,6-diisocyanato- Hexamethylene diisocyanate	$C_8H_{12}N_2O_2$ 168.20	822-06-0		122[10]; 94[1]	4-04-00-01349 1.0528[20]	1.4585[20]
6771	Hexane, 2,2-dimethyl- 2,2-Dimethylhexane	C_8H_{18} 114.23	590-73-8 -121.1		106.8	4-01-00-00432 0.6953[20]	ace 4; bz 4; eth 4; EtOH 4 1.3935[20]
6772	Hexane, 2,3-dimethyl- 2,3-Dimethylhexane	C_8H_{18} 114.23	584-94-1		115.6	4-01-00-00432 0.6912[25]	ace 4; bz 4; EtOH 4; lig 4 1.4011[20]
6773	Hexane, 2,4-dimethyl- 2,4-Dimethylhexane	C_8H_{18} 114.23	589-43-5		109.5	0.6962[25]	1.3929[25]
6774	Hexane, 2,5-dimethyl- 2,5-Dimethylhexane	C_8H_{18} 114.23	592-13-2 -91		109.1	4-01-00-00434 0.6901[25]	H_2O 1; EtOH 5; eth 3; ace 5 1.3925[20]
6775	Hexane, 3,3-dimethyl- 3,3-Dimethylhexane	C_8H_{18} 114.23	563-16-6 -126.1		111.9	4-01-00-00435 0.7100[20]	H_2O 1; EtOH 5; eth 4; ace 4 1.4001[20]
6776	Hexane, 3,4-dimethyl- 3,4-Dimethylhexane	C_8H_{18} 114.23	583-48-2		117.7	4-01-00-00436 0.7151[25]	H_2O 1; EtOH 5; eth 3; ace 5 1.4041[20]
6777	Hexanedinitrile Adiponitrile	$C_6H_8N_2$ 108.14	111-69-3 1	295		4-02-00-01975 0.9676[20]	H_2O 2; EtOH 3; eth 2; chl 3 1.4380[20]
6778	Hexane, 1,6-dinitro- 1,6-Dinitrohexane	$C_6H_{12}N_2O_4$ 176.17	4286-47-9 37.5		100-3[0.3]	4-01-00-00357	H_2O 1; HOAc 3
6779	Hexanedioic acid	$C_6H_{10}O_4$	124-04-9 152		337.5	4-02-00-01956	H_2O 2; EtOH 4; eth 3; HOAc 1
	Adipic acid	146.14	153.2			1.360[25]	
6780	Hexanedioic acid, bis(2-ethylbutyl) ester Adipic acid, di-2-ethylbutyl ester	$C_{18}H_{34}O_4$ 314.47	10022-60-3 -15		200[19]	4-02-00-01964 0.934[25]	ace 4; EtOH 4 1.4434[20]
6781	Hexanedioic acid, bis(2-ethylhexyl) ester Bis(2-ethylhexyl) adipate	$C_{22}H_{42}O_4$ 370.57	103-23-1 -67.8		214[5]	4-02-00-01964 0.922[25]	ace 4; eth 4; EtOH 4 1.4474[20]
6782	Hexanedioic acid, bis(1-methylethyl) ester	$C_{12}H_{22}O_4$ 230.30	6938-94-9 -1.1		120[6.5]	4-02-00-01961 0.9569[20]	ace 4; eth 4; EtOH 4 1.4247[20]
6783	Hexanedioic acid, bis(2-methylpropyl) ester	$C_{14}H_{26}O_4$ 258.36	141-04-8		186-8[15]	4-02-00-01962 0.9543[19]	1.4301[20]
6784	Hexanedioic acid, dibutyl ester	$C_{14}H_{26}O_4$ 258.36	105-99-7 -32.4		165[10]	4-02-00-01961 0.9613[20]	H_2O 1; EtOH 5; eth 5 1.4369[20]
6785	Hexanedioic acid, diethyl ester	$C_{10}H_{18}O_4$ 202.25	141-28-6 -19.8		245	4-02-00-01960 1.0076[20]	H_2O 1; EtOH 3; eth 3 1.4272[20]
6786	Hexanedioic acid, dimethyl ester	$C_8H_{14}O_4$ 174.20	627-93-0 10.3		115[13]	4-02-00-01959 1.0600[20]	H_2O 1; EtOH 3; eth 3; ctc 3 1.4283[20]
6787	Hexanedioic acid, dipropyl ester	$C_{12}H_{22}O_4$ 230.30	106-19-4 -15.7		151[11]	4-02-00-01961 0.9790[20]	eth 4; EtOH 4; chl 4 1.4314[20]
6788	Hexanedioic acid, 2-methyl-	$C_7H_{12}O_4$ 160.17	626-70-0 64		209[13]	4-02-00-02010	H_2O 4; eth 4; EtOH 4; chl 4
6789	Hexanedioic acid, 3-methyl-, (±)- Adipic acid, 3-methyl (DL)	$C_7H_{12}O_4$ 160.17	81177-02-8 97		230[30]	4-02-00-02014	H_2O 4; eth 4; EtOH 4; ace 4
6790	Hexanedioic acid, 3-methyl-, diethyl ester Diethyl 3-methyladipate	$C_{11}H_{20}O_4$ 216.28	55877-01-5		258; 126.5[10]	4-02-00-02015 0.9948[20]	1.4335
6791	Hexanedioic acid, monoethyl ester	$C_8H_{14}O_4$ 174.20	626-86-8 29		285	4-02-00-01960 0.9796[20]	EtOH 3; eth 3; peth 3 1.4311[20]
6792	Hexanedioic acid, monomethyl ester	$C_7H_{12}O_4$ 160.17	627-91-8 9		158[10]	3-02-00-01711 1.0623[20]	EtOH 3 1.4283[20]
6793	1,3-Hexanediol	$C_6H_{14}O_2$ 118.18	21531-91-9		123[13]	4-01-00-02555 0.958[22]	1.4461[22]
6794	1,4-Hexanediol	$C_6H_{14}O_2$ 118.18	16432-53-4		134-5[18.5]	4-01-00-02556 0.982[16]	eth 4; EtOH 4 1.4530[16]
6795	1,5-Hexanediol	$C_6H_{14}O_2$ 118.18	928-40-5		237; 90[0.5]	4-01-00-02556 0.981[25]	1.4511[20]
6796	1,6-Hexanediol Hexamethylene glycol	$C_6H_{14}O_2$ 118.18	629-11-8 42.8	4610	208	4-01-00-02556	H_2O 3; EtOH 3; eth 2; ace 3 1.4579[25]
6797	2,3-Hexanediol	$C_6H_{14}O_2$ 118.18	617-30-1 60		205	4-01-00-02561 0.990[15]	H_2O 4; eth 4; EtOH 4 1.4510[15]
6798	2,4-Hexanediol	$C_6H_{14}O_2$ 118.18	19780-90-6		211; 105[9]	4-01-00-02562 0.9516[21]	1.4418[21]
6799	2,5-Hexanediol Diisopropanol	$C_6H_{14}O_2$ 118.18	2935-44-6 43		218; 86[1]	4-01-00-02562 0.9610[20]	H_2O 3; EtOH 3; eth 3; ctc 2 1.4475[20]
6800	3,4-Hexanediol, 3,4-diethyl- 3,4-Diethyl-3,4-hexanediol	$C_{10}H_{22}O_2$ 174.28	6931-71-1 28		230	4-01-00-02621 0.960[13]	eth 4; EtOH 4 1.467[13]
6801	2,5-Hexanediol, 2,5-dimethyl- 1,1,4,4-Tetramethyl-1,4-butanediol	$C_8H_{18}O_2$ 146.23	110-03-2 92		214	4-01-00-02600 0.898[20]	H_2O 3; EtOH 4; bz 4; chl 4
6802	1,3-Hexanediol, 2-ethyl- Ethohexadiol	$C_8H_{18}O_2$ 146.23	94-96-2 -40	3699	244	4-01-00-02597 0.9325[22]	H_2O 2; EtOH 3; eth 3 1.4497[20]

No.	Name / Synonym	Mol. Form. / Mol. Wt.	CAS RN / mp/°C	Merck No. / bp/°C	Beil. Ref. / den/g cm^{-3}	Solubility / n_D
6803	2,4-Hexanediol, 2-methyl-	$C_7H_{16}O_2$ 132.20	66225-35-2	121[30]	2-01-00-00554 0.9321[18]	1.4407[18]
6804	2,4-Hexanedione Propionylacetone	$C_6H_{10}O_2$ 114.14	3002-24-2	160	4-01-00-03687 0.959[20]	1.4516[20]
6805	2,5-Hexanedione Acetonylacetone	$C_6H_{10}O_2$ 114.14	110-13-4 -5.5	63 194	4-01-00-03688 0.7370[20]	H_2O 4; bz 4; eth 4; EtOH 4 1.4232[20]
6806	3,4-Hexanedione Bipropionyl	$C_6H_{10}O_2$ 114.14	4437-51-8 -10	130	4-01-00-03690 0.941[21]	1.4130[21]
6807	3,4-Hexanedione, 2,5-dimethyl- Biisobutyryl	$C_8H_{14}O_2$ 142.20	4388-87-8	144.5	4-01-00-03713 0.9232[20]	1.4206[20]
6808	2,5-Hexanedione, 3-hydroxy- 3-Hydroxy-2,5-hexanedione	$C_6H_{10}O_3$ 130.14	61892-85-1	62-7[0.5]	3-01-00-03317	bz 4; eth 4 1.4497[25]
6809	2,3-Hexanedione, 5-methyl- 2-Methylhexa-4,5-dione	$C_7H_{12}O_2$ 128.17	13706-86-0	138	4-01-00-03701 0.908[22]	1.4119[20]
6810	2,5-Hexanedione, 3-methyl- 3-Methyl-2,5-hexanedione	$C_7H_{12}O_2$ 128.17	4437-50-7	196; 71[10]	3-01-00-03137 0.9527[20]	eth 4; EtOH 4 1.4260[20]
6811	3,4-Hexanedione, 2,2,5,5-tetramethyl- Pivalil	$C_{10}H_{18}O_2$ 170.25	4388-88-9 -2	169	4-01-00-03726 0.8776[20]	H_2O 1; eth 5 1.4157[20]
6812	1,6-Hexanedithiol	$C_6H_{14}S_2$ 150.31	1191-43-1	118-9[15]	4-01-00-02559 0.983[25]	1.5110[20]
6813	Hexane, 1-(ethenyloxy)- Hexyl vinyl ether	$C_8H_{16}O$ 128.21	5363-64-4	143	4-01-00-02055 0.7966[20]	1.4171[20]
6814	Hexane, 1-ethoxy-	$C_8H_{18}O$ 130.23	5756-43-4	143	4-01-00-01697 0.7722[20]	eth 4; EtOH 4 1.4008[20]
6815	Hexane, 3-ethyl- 3-Ethylhexane	C_8H_{18} 114.23	619-99-8	118.6	4-01-00-00431 0.7136[20]	H_2O 1; EtOH 5; eth 5; ace 5 1.4018[20]
6816	Hexane, 3-ethyl-2,2-dimethyl- 3-Ethyl-2,2-dimethylhexane	$C_{10}H_{22}$ 142.28	20291-91-2	156.1	4-01-00-00484 0.7447[25]	1.4197[20]
6817	Hexane, 3-ethyl-2,3-dimethyl- 3-Ethyl-2,3-dimethylhexane	$C_{10}H_{22}$ 142.28	52897-00-4	166	4-01-00-00484 0.7599[25]	1.4270[207]
6818	Hexane, 3-ethyl-2,4-dimethyl- 3-Ethyl-2,4-dimethylhexane	$C_{10}H_{22}$ 142.28	7220-26-0	162	4-01-00-00484 0.7514[25]	1.4225[20]
6819	Hexane, 3-ethyl-2,5-dimethyl- 3-Ethyl-2,5-dimethylhexane	$C_{10}H_{22}$ 142.28	52897-04-8	154.1	4-01-00-00484 0.7368[25]	1.4157[20]
6820	Hexane, 3-ethyl-3,4-dimethyl- 3-Ethyl-3,4-dimethylhexane	$C_{10}H_{22}$ 142.28	52897-06-0	162.1	4-01-00-00485 0.7596[25]	1.4267[207]
6821	Hexane, 4-ethyl-2,2-dimethyl- Hexane, 2,2-dimethyl-4-ethyl	$C_{10}H_{22}$ 142.28	52896-99-8	147	4-01-00-00485 0.7330[20]	1.4131[20]
6822	Hexane, 4-ethyl-2,3-dimethyl- 4-Ethyl-2,3-dimethylhexane	$C_{10}H_{22}$ 142.28	52897-01-5	162	4-01-00-00485 0.7516[25]	1.4226[20]
6823	Hexane, 4-ethyl-2,4-dimethyl- 4-Ethyl-2,4-dimethylhexane	$C_{10}H_{22}$ 142.28	52897-03-7	161.1	4-01-00-00485 0.7525[25]	1.4235[20]
6824	Hexane, 4-ethyl-3,3-dimethyl- 4-Ethyl-3,3-dimethylhexane	$C_{10}H_{22}$ 142.28	52897-05-9	162.9	4-01-00-00485 0.7598[25]	1.4269[20]
6825	Hexane, 3-ethyl-2-methyl- 3-Ethyl-2-methylhexane	C_9H_{20} 128.26	16789-46-1	138	4-01-00-00459 0.7310[20]	1.4106[20]
6826	Hexane, 3-ethyl-3-methyl- 3-Ethyl-3-methylhexane	C_9H_{20} 128.26	3074-76-8	140.6	4-01-00-00459 0.7371[25]	1.4140[20]
6827	Hexane, 3-ethyl-4-methyl- 4-Ethyl-3-methylhexane	C_9H_{20} 128.26	3074-77-9	140	3-01-00-00514 0.7420[20]	1.4134[20]
6828	Hexane, 4-ethyl-2-methyl- 4-Ethyl-2-methylhexane	C_9H_{20} 128.26	3074-75-7	133.8	4-01-00-00459 0.7195[25]	1.4063[20]
6829	Hexane, 1-fluoro- Hexyl fluoride	$C_6H_{13}F$ 104.17	373-14-8 -103	91.5	4-01-00-00348 0.7995[20]	eth 3; bz 3 1.3738[20]
6830	Hexane, 2-fluoro- 2-Fluorohexane	$C_6H_{13}F$ 104.17	372-54-3	86	4-01-00-00348 0.7916[20]	1.3693[20]
6831	Hexane, 1-iodo- Hexyl iodide	$C_6H_{13}I$ 212.07	638-45-9 -74.2	181	4-01-00-00355 1.4397[20]	H_2O 1 1.4929[20]
6832	Hexane, 2-iodo-, (R)- Hexane, 2-iodo (I)	$C_6H_{13}I$ 212.07	59654-13-6	90[70]; 45[9]	4-01-00-00355 1.4354[17]	H_2O 1; ace 3; chl 3 1.4878[25]
6833	Hexane, 1-methoxy- Hexyl methyl ether	$C_7H_{16}O$ 116.20	4747-07-3	126.1		
6834	Hexane, 2-methyl- 2-Methylhexane	C_7H_{16} 100.20	591-76-4 -118.2	90.0	4-01-00-00397 0.6787[20]	H_2O 1; EtOH 3; eth 5; ace 5 1.3848[20]
6835	Hexane, 3-methyl-, (S)- (+)-3-Methylhexane	C_7H_{16} 100.20	6131-24-4 -119	92	4-01-00-00399 0.6860[20]	H_2O 1; EtOH 3; eth 5; ace 5 1.3887[20]
6836	Hexane, 2-methyl-3-methylene- 2-Isopropyl-1-pentene	C_8H_{16} 112.22	16746-02-4	113	4-01-00-00887 0.721[25]	1.414[20]
6837	Hexane, 2-methyl-4-methylene- 1-Pentene, 2-ethyl-4-methyl-	C_8H_{16} 112.22	3404-80-6	110.3	4-01-00-00888 0.7152[25]	1.4105[20]
6838	Hexane, 3-methyl-4-methylene-	C_8H_{16} 112.22	3404-67-9	112.5	4-01-00-00890 0.725[25]	1.4142[20]
6839	Hexane, 2-methyl-3-(1-methylethyl)- Hexane, 3-isopropyl-2-methyl-	$C_{10}H_{22}$ 142.28	62016-13-1	165	4-01-00-00484 0.7436[25]	1.4195[20]
6840	Hexanenitrile Capronitrile	$C_6H_{11}N$ 97.16	628-73-9 -80.3	163.6	4-02-00-00930 0.8051[20]	H_2O 1; EtOH 3; eth 3; chl 2 1.4068[20]
6841	Hexane, 1-nitro- 1-Nitrohexane	$C_6H_{13}NO_2$ 131.17	646-14-0	193; 84[21]	4-01-00-00356 0.9396[20]	H_2O 1; EtOH 3; eth 3; ace 3 1.4270[20]
6842	Hexane, 1,1'-oxybis-	$C_{12}H_{26}O$ 186.34	112-58-3	223	4-01-00-01698 0.7936[20]	H_2O 1; eth 3; ctc 2 1.4204[20]

No.	Name / Synonym	Mol. Form. / Mol. Wt.	CAS RN / mp/°C	Merck No. / bp/°C	Beil. Ref. / den/g cm^{-3}	Solubility / n_D
6843	Hexane, tetradecafluoro- / Perfluorohexane	C_6F_{14} / 338.04	355-42-0 / -87.1	56.6	4-01-00-00348 / 1.6995[20]	H_2O 1; eth 3; bz 3; chl 3 / 1.2515[20]
6844	Hexane, 2,2,3,3-tetramethyl- / 2,2,3,3-Tetramethylhexane	$C_{10}H_{22}$ / 142.28	13475-81-5 / -54	160.3	4-01-00-00485 / 0.7609[25]	1.4282[20]
6845	Hexane, 2,2,3,4-tetramethyl- / 2,2,3,4-Tetramethylhexane	$C_{10}H_{22}$ / 142.28	52897-08-2 /	157	4-01-00-00485 / 0.7513[25]	1.4216[20]
6846	Hexane, 2,2,3,5-tetramethyl- / 2,2,3,5-Tetramethylhexane	$C_{10}H_{22}$ / 142.28	52897-09-3 /	148.4	4-01-00-00485 / 0.7336[25]	1.4142[20]
6847	Hexane, 2,2,4,4-tetramethyl- / 2,2,4,4-Tetramethylhexane	$C_{10}H_{22}$ / 142.28	51750-65-3 /	153.8	4-01-00-00486 / 0.7424[25]	1.4208[20]
6848	Hexane, 2,2,4,5-tetramethyl- / 2,2,4,5-Tetramethylhexane	$C_{10}H_{22}$ / 142.28	16747-42-5 /	147.9	4-01-00-00486 / 0.7316[25]	1.4132[20]
6849	Hexane, 2,2,5,5-tetramethyl- / 2,2,5,5-Tetramethylhexane	$C_{10}H_{22}$ / 142.28	1071-81-4 / -12.6	137.4	4-01-00-00486 / 0.7148[25]	1.4055[20]
6850	Hexane, 2,3,3,4-tetramethyl- / 2,3,3,4-Tetramethylhexane	$C_{10}H_{22}$ / 142.28	52897-10-6 /	164.6	4-01-00-00486 / 0.7656[25]	1.4298[20]
6851	Hexane, 2,3,3,5-tetramethyl- / 2,3,3,5-Tetramethylhexane	$C_{10}H_{22}$ / 142.28	52897-11-7 /	153.1	4-01-00-00486 / 0.7449[25]	1.4196[20]
6852	Hexane, 2,3,4,4-tetramethyl- / 2,3,4,4-Tetramethylhexane	$C_{10}H_{22}$ / 142.28	52897-12-8 /	161.6	4-01-00-00487 / 0.7586[25]	1.4267[20]
6853	Hexane, 2,3,4,5-tetramethyl- / 2,3,4,5-Tetramethylhexane	$C_{10}H_{22}$ / 142.28	52897-15-1 /	156.2	4-01-00-00487 / 0.7456[25]	1.4204[20]
6854	Hexane, 3,3,4,4-tetramethyl- / 3,3,4,4-Tetramethylhexane	$C_{10}H_{22}$ / 142.28	5171-84-6 /	170.0	3-01-00-00534 / 0.7789[25]	1.4368[20]
6855	Hexane, 1,1'-thiobis-	$C_{12}H_{26}S$ / 202.40	6294-31-1 /	230; 136[20]	4-01-00-01706 / 0.8411[20]	1.4586[20]
6856	1-Hexanethiol / Hexyl mercaptan	$C_6H_{14}S$ / 118.24	111-31-9 / -81	151	4-01-00-01705 / 0.8424[20]	H_2O 1; EtOH 4; eth 4 / 1.4496[20]
6857	2-Hexanethiol	$C_6H_{14}S$ / 118.24	1679-06-7 / -147	142	4-01-00-01711 / 0.8345[20]	H_2O 1; EtOH 3; eth 3; bz 3 / 1.4451[20]
6858	3-Hexanethiol / 3-Mercaptohexane	$C_6H_{14}S$ / 118.24	1633-90-5 /	57[25]	4-01-00-01713 / 0.9206[20]	EtOH 3 / 1.4496[20]
6859	Hexane, 2,2,3-trimethyl- / 2,2,3-Trimethylhexane	C_9H_{20} / 128.26	16747-25-4 /	133.6	4-01-00-00459 / 0.7257[25]	1.4106[20]
6860	Hexane, 2,2,4-trimethyl- / 2,2,4-Trimethylhexane	C_9H_{20} / 128.26	16747-26-5 / -120	126.5	4-01-00-00459 / 0.711[20]	1.4033[20]
6861	Hexane, 2,2,5-trimethyl- / 2,2,5-Trimethylhexane	C_9H_{20} / 128.26	3522-94-9 / -105.7	124.0	4-01-00-00460 / 0.7072[20]	H_2O 1; EtOH 4; eth 4; ace 4 / 1.3997[20]
6862	Hexane, 2,3,3-trimethyl- / 2,3,3-Trimethylhexane	C_9H_{20} / 128.26	16747-28-7 / -116.8	137.7	4-01-00-00461 / 0.7345[25]	1.4141[20]
6863	Hexane, 2,3,4-trimethyl- / 2,3,4-Trimethylhexane	C_9H_{20} / 128.26	921-47-1 /	139.1	4-01-00-00461 / 0.7354[25]	1.4144[20]
6864	Hexane, 2,3,5-trimethyl- / 2,3,5-Trimethylhexane	C_9H_{20} / 128.26	1069-53-0 / -127.9	131.4	4-01-00-00461 / 0.7218[20]	1.4051[20]
6865	Hexane, 2,4,4-trimethyl- / 2,4,4-Trimethylhexane	C_9H_{20} / 128.26	16747-30-1 / -113.4	130.7	4-01-00-00461 / 0.7201[25]	1.4074[20]
6866	Hexane, 3,3,4-trimethyl- / 3,3,4-Trimethylhexane	C_9H_{20} / 128.26	16747-31-2 / -101.2	140.5	4-01-00-00462 / 0.7414[25]	1.4178[20]
6867	1,2,5-Hexanetriol	$C_6H_{14}O_3$ / 134.18	10299-30-6 /	181[10]; 149[3]	4-01-00-02784 / 1.1012[20]	H_2O 4; EtOH 4
6868	1,2,6-Hexanetriol / 1,2,6-Trihydroxyhexane	$C_6H_{14}O_3$ / 134.18	106-69-4 /	170[3]	4-01-00-02784 / 1.03[20]	1.58[20]
6869	2,3,4-Hexanetriol	$C_6H_{14}O_3$ / 134.18	93972-93-1 / 47	256.5	3-01-00-02349	H_2O 4; EtOH 4
6870	Hexanoic acid / Caproic acid	$C_6H_{12}O_2$ / 116.16	142-62-1 / -3	1760 / 205.2	4-02-00-00917 / 0.9274[20]	H_2O 1; EtOH 3; eth 3; chl 3 / 1.4163[20]
6871	Hexanoic acid, 6-(acetylamino)- / ε-Acetamidocaproic acid	$C_8H_{15}NO_3$ / 173.21	57-08-9 / 104.5	38	4-04-00-02700	
6872	Hexanoic acid, 2-acetyl-, ethyl ester	$C_{10}H_{18}O_3$ / 186.25	1540-29-0 /	221.5	4-03-00-01616 / 0.9523[20]	ace 4; eth 4 / 1.4301[20]
6873	Hexanoic acid, 6-amino- / ε-Aminocaproic acid	$C_6H_{13}NO_2$ / 131.17	60-32-2 / 205	442	4-04-00-02695	H_2O 4; EtOH 1; MeOH 2
6874	Hexanoic acid, anhydride	$C_{12}H_{22}O_3$ / 214.30	2051-49-2 / -41	255 dec	4-02-00-00928 / 0.9240[15]	eth 4; EtOH 4 / 1.4297[20]
6875	Hexanoic acid, 2-bromo-, (±)-	$C_6H_{11}BrO_2$ / 195.06	2681-83-6 / 4	240; 140[23]	3-02-00-00736 / 1.2810[33]	EtOH 3; eth 3
6876	Hexanoic acid, 6-bromo-	$C_6H_{11}BrO_2$ / 195.06	4224-70-8 / 35	165-70[20]	4-02-00-00940	peth 4
6877	Hexanoic acid, 2-bromo-, ethyl ester, (±)-	$C_8H_{15}BrO_2$ / 223.11	63927-44-6 /	208.7	4-02-00-00938 / 1.2210[25]	EtOH 3
6878	Hexanoic acid, 5-bromo-, ethyl ester / Caproic acid, 5-bromo, ethyl ester	$C_8H_{15}BrO_2$ / 223.11	90202-08-7 /	121[21]; 70[3]	3-02-00-00737 / 1.1908[25]	1.4525[25]
6879	Hexanoic acid, 6-bromo-, ethyl ester	$C_8H_{15}BrO_2$ / 223.11	25542-62-5 / 33	126-7[21]	4-02-00-00940 / 1.238[23]	1.4566[21]
6880	Hexanoic acid, butyl ester	$C_{10}H_{20}O_2$ / 172.27	626-82-4 / -64.3	208	4-02-00-00922 / 0.8653[20]	H_2O 1; EtOH 3; eth 5 / 1.4152[20]
6881	Hexanoic acid, 2-cyano-, ethyl ester / 2-Cyanocaproic acid, ethyl ester	$C_9H_{15}NO_2$ / 169.22	7391-39-1 /	105[9]	3-02-00-01746 / 0.988[15]	1.4248[20]
6882	Hexanoic acid, 2-ethyl- / 2-Ethylhexanoic acid	$C_8H_{16}O_2$ / 144.21	149-57-5 /	228; 120[13]	3-02-00-00804 / 0.9031[25]	H_2O 3; EtOH 2; eth 3; ctc 3 / 1.4241[20]

No.	Name / Synonym	Mol. Form. / Mol. Wt.	CAS RN / mp/°C	Merck No. / bp/°C	Beil. Ref. / den/g cm^{-3}	Solubility / n_D
6883	Hexanoic acid, 3-ethyl- / 3-Ethylhexanoic acid	$C_8H_{16}O_2$ / 144.21	41065-91-2 /	/ 106[5]	4-02-00-01008 / 0.911[30]	/ 1.4287[25]
6884	Hexanoic acid, ethyl ester / Ethyl hexanoate	$C_8H_{16}O_2$ / 144.21	123-66-0 / -67	3735 / 167	4-02-00-00921 / 0.873[20]	eth 4; EtOH 4 / 1.4073[20]
6885	Hexanoic acid, 2-ethyl-, ethyl ester	$C_{10}H_{20}O_2$ / 172.27	2983-37-1 /	/ 90[28]	4-02-00-01004 / 0.8586[25]	/ 1.4123[25]
6886	Hexanoic acid, 2-ethyl-3-methyl- / 2-Ethyl-3-methylhexanoic acid	$C_9H_{18}O_2$ / 158.24	74581-94-5 /	/ 231	3-02-00-00832 / 0.9060[25]	/ 1.4302[25]
6887	Hexanoic acid, 3-ethyl-3-methyl- / 3-Ethyl-3-methylhexanoic acid	$C_9H_{18}O_2$ / 158.24	50902-82-4 /	/ 139-40[22]	4-02-00-01036 / 0.9205[25]	/ 1.4375[25]
6888	Hexanoic acid, 6-fluoro- / Caproic acid, 6-fluoro	$C_6H_{11}FO_2$ / 134.15	373-05-7 /	/ 138[28]; 67[0.6]	4-02-00-00932 /	eth 4; EtOH 4 / 1.4166[25]
6889	Hexanoic acid, heptyl ester	$C_{13}H_{26}O_2$ / 214.35	6976-72-3 / -34.4	/ 261	4-02-00-00923 / 0.8611[20]	ace 4; bz 4; eth 4; EtOH 4 / 1.4293[15]
6890	Hexanoic acid, hexyl ester	$C_{12}H_{24}O_2$ / 200.32	6378-65-0 / -55	/ 246	4-02-00-00922 / 0.865[18]	ace 4; bz 4; eth 4; EtOH 4 / 1.4264[15]
6891	Hexanoic acid, 2-hydroxy-, (±)-	$C_6H_{12}O_3$ / 132.16	636-36-2 / 61.3	/ 270	4-03-00-00838 /	H$_2$O 4; EtOH 4; eth 3; chl 4
6892	Hexanoic acid, 3-hydroxy-2,2-dimethyl-, β-lactone / Caproic acid, 2,2-dimethyl-3-hydroxy, β-lactone	$C_8H_{14}O_2$ / 142.20	90112-82-6 / 143.5	/ 93-100[14]	/	DMSO 2 / 1.4260[20]
6893	Hexanoic acid, 2-methyl-, (±)- / Caproic acid, 2-methyl (*DL*)	$C_7H_{14}O_2$ / 130.19	22160-12-9 /	/ 215.5	4-02-00-00969 / 0.9612[20]	H$_2$O 5; EtOH 5; eth 5; ace 5 / 1.4195[20]
6894	Hexanoic acid, 3-methyl-	$C_7H_{14}O_2$ / 130.19	3780-58-3 /	/ 213; 116[15]	2-02-00-00298 / 0.9187[20]	/ 1.4222[20]
6895	Hexanoic acid, 4-methyl-, (±)-	$C_7H_{14}O_2$ / 130.19	22160-41-4 / -80	/ 217.5	4-02-00-00974 / 0.9215[20]	H$_2$O 2; EtOH 3; eth 3; ace 3 / 1.4211[20]
6896	Hexanoic acid, 5-methyl-	$C_7H_{14}O_2$ / 130.19	628-46-6 / <-25	/ 216	4-02-00-00970 / 0.9138[21]	ace 4; bz 4; eth 4; EtOH 4 / 1.4220[20]
6897	Hexanoic acid, 3-methylbutyl ester	$C_{11}H_{22}O_2$ / 186.29	2198-61-0 /	/ 225.5	3-02-00-00727 / 0.861[20]	H$_2$O 1; EtOH 3; eth 3
6898	Hexanoic acid, methyl ester / Methyl caproate	$C_7H_{14}O_2$ / 130.19	106-70-7 / -71	/ 149.5	4-02-00-00921 / 0.8846[20]	H$_2$O 1; EtOH 4; eth 4; ace 3 / 1.4049[20]
6899	Hexanoic acid, 3-methyl-, ethyl ester / Ethyl 3-methylhexanoate	$C_9H_{18}O_2$ / 158.24	41692-47-1 /	/ 177	2-02-00-00298 / 0.8679[20]	/ 1.4119[20]
6900	Hexanoic acid, 4-methyl-, ethyl ester / Ethyl 4-methylhexanoate	$C_9H_{18}O_2$ / 158.24	1561-10-0 /	/ 180	4-02-00-00974 / 0.8708[20]	/ 1.4051[20]
6901	Hexanoic acid, 4-methyl-5-oxo-	$C_7H_{12}O_3$ / 144.17	6818-07-1 /	/ 160[15]; 140[5]	4-03-00-01601 / 0.9803[20]	/ 1.4320[20]
6902	Hexanoic acid, 4-methyl-5-oxo-, ethyl ester / Valeric acid, 4-acetyl, ethyl ester	$C_9H_{16}O_3$ / 172.22	53068-88-5 /	/ 114[15]	4-03-00-01601 / 0.9803[20]	/ 1.4282[25]
6903	Hexanoic acid, 4-methyl-5-oxo-, methyl ester / Valeric acid, 4-acetyl, methyl ester	$C_8H_{14}O_3$ / 158.20	36045-56-4 /	/ 113-7[16]	4-03-00-01601 / 1.0988[20]	/ 1.4288[20]
6904	Hexanoic acid, octyl ester	$C_{14}H_{28}O_2$ / 228.38	4887-30-3 / -28	/ 275	3-02-00-00727 / 0.8603[20]	ace 4; bz 4; eth 4; EtOH 4 / 1.4326[15]
6905	Hexanoic acid, 5-oxo-	$C_6H_{10}O_3$ / 130.14	3128-06-1 / 13.5	/ 274.5	4-03-00-01583 / 1.09[25]	H$_2$O 3; EtOH 3; eth 3; ctc 2 / 1.4451[20]
6906	Hexanoic acid, 5-oxo-, ethyl ester / Butyric acid, 4-acetyl, ethyl ester	$C_8H_{14}O_3$ / 158.20	13984-57-1 /	/ 221.5	4-03-00-01583 / 0.989[25]	/ 1.4277[20]
6907	Hexanoic acid, pentyl ester / Amyl caproate	$C_{11}H_{22}O_2$ / 186.29	540-07-8 / -47	640 / 226	4-02-00-00922 / 0.8612[25]	EtOH 3; eth 3; ace 3; ctc 2 / 1.4202[25]
6908	Hexanoic acid, 1,2,3-propanetriyl ester	$C_{21}H_{38}O_6$ / 386.53	621-70-5 / -60	/ >200	4-02-00-00926 / 0.9867[20]	H$_2$O 1; EtOH 5; eth 5; ace 4 / 1.4427[20]
6909	Hexanoic acid, propyl ester	$C_9H_{18}O_2$ / 158.24	626-77-7 / -68.7	/ 187	4-02-00-00922 / 0.8672[20]	eth 4; EtOH 4 / 1.4170[20]
6910	Hexanoic acid, 3,5,6-trichloro-2,2,3,4,4,5,6,6-octafluoro- / Hexanoic acid, 2,2,3,4,4,5,6,6-octafluoro-3,5,6-trichloro	$C_6HCl_3F_8O_2$ / 363.42	2106-54-9 /	/ 105-6[5]	4-02-00-00937 / 1.860[20]	/ 1.3903[20]
6911	1-Hexanol / Caproyl alcohol	$C_6H_{14}O$ / 102.18	111-27-3 / -44.6	4615 / 157.6	4-01-00-01694 / 0.8136[20]	H$_2$O 2; EtOH 3; eth 5; ace 3 / 1.4178[20]
6912	2-Hexanol, (±)-	$C_6H_{14}O$ / 102.18	20281-86-1 /	/ 140	4-01-00-01708 / 0.8159[20]	H$_2$O 2; EtOH 3; eth 3; ctc 2 / 1.4144[20]
6913	2-Hexanol, (*R*)-	$C_6H_{14}O$ / 102.18	26549-24-6 /	/ 137	3-01-00-01663 / 0.8178[18]	H$_2$O 2; EtOH 3; eth 3
6914	3-Hexanol, (±)-	$C_6H_{14}O$ / 102.18	17015-11-1 /	/ 135	4-01-00-01711 / 0.8182[20]	H$_2$O 2; EtOH 5; eth 5; ace 3 / 1.4167[20]
6915	3-Hexanol, (*R*)- / 3-Hexanol (*l*)	$C_6H_{14}O$ / 102.18	13471-42-6 /	/ 135	0.8213[20]	H$_2$O 2; EtOH 3; eth 5; ace 3 / 1.4140[20]
6916	1-Hexanol, 6-chloro-	$C_6H_{13}ClO$ / 136.62	2009-83-8 /	/ 107[12]	4-01-00-01704 / 1.0241[20]	H$_2$O 2; EtOH 4; eth 4 / 1.4550[20]
6917	2-Hexanol, 1-chloro- / 1-Chloro-2-hexanol	$C_6H_{13}ClO$ / 136.62	52802-07-0 /	/ 73-5[12]	4-01-00-01710 / 1.0139[20]	ace 4; eth 4; EtOH 4 / 1.4478[20]
6918	3-Hexanol, 1-chloro- / 1-Chloro-3-hexanol	$C_6H_{13}ClO$ / 136.62	52418-81-2 /	/ 120[35]; 78[6]	4-01-00-01712 / 1.003[25]	ace 4; bz 4; eth 4; EtOH 4 / 1.446[25]
6919	3-Hexanol, 5-chloro-	$C_6H_{13}ClO$ / 136.62	58588-28-6 /	/ 78-9[13]	4-01-00-01712 / 1.0012[15]	ace 4; bz 4; eth 4; EtOH 4 / 1.4433[19]
6920	1-Hexanol, 3,5-dimethoxy-	$C_8H_{18}O_3$ / 162.23	90952-10-6 /	/ 114[13]	4-01-00-02785 / 0.9631[25]	/ 1.4329[25]

No.	Name / Synonym	Mol. Form. / Mol. Wt.	CAS RN / mp/°C	Merck No. / bp/°C	Beil. Ref. / den/g cm⁻³	Solubility / n_D
6921	1-Hexanol, 2,5-dimethyl-	$C_8H_{18}O$ 130.23	6886-16-4	179.5	0-01-00-00422 0.8280[20]	1.5095[17]
6922	2-Hexanol, 2,3-dimethyl- 2,3-Dimethyl-2-hexanol	$C_8H_{18}O$ 130.23	19550-03-9	160; 62[10]	3-01-00-01737 0.8365[20]	1.4335[20]
6923	2-Hexanol, 2,5-dimethyl- 2,5-Dimethyl-2-hexanol	$C_8H_{18}O$ 130.23	3730-60-7	154.5	4-01-00-01792 0.8227[20]	1.4208[20]
6924	2-Hexanol, 3,4-dimethyl-	$C_8H_{18}O$ 130.23	19550-05-1	165.5	3-01-00-01740 0.840[15]	1.4325[15]
6925	3-Hexanol, 2,2-dimethyl- Propyl-tert-butylcarbinol	$C_8H_{18}O$ 130.23	4209-90-9	156	3-01-00-01737 0.8342[20]	1.4275[20]
6926	3-Hexanol, 2,3-dimethyl-	$C_8H_{18}O$ 130.23	4166-46-5	61-2[18]	4-01-00-01791 0.8371[20]	1.4309[20]
6927	3-Hexanol, 2,5-dimethyl- 2,5-Dimethyl-3-hexanol	$C_8H_{18}O$ 130.23	19550-07-3	157.5	4-01-00-01793 0.8212[20]	1.4246[20]
6928	3-Hexanol, 3,5-dimethyl-, (±)- 3-Hexanol, 3,5-dimethyl (DL)	$C_8H_{18}O$ 130.23	19113-78-1	152	4-01-00-01792 0.8373[20]	bz 4; eth 4; EtOH 4 1.4300[20]
6929	3-Hexanol, 4,4-dimethyl-	$C_8H_{18}O$ 130.23	19550-09-5	160.4	4-01-00-01793 0.8341[20]	1.4345[20]
6930	1-Hexanol, 2-ethyl- 2-Ethyl-1-hexanol	$C_8H_{18}O$ 130.23	104-76-7 -70	3764 184.6	4-01-00-01783 0.8319[25]	H_2O 1; EtOH 3; eth 3; ace 3 1.4300[20]
6931	3-Hexanol, 3-ethyl-	$C_8H_{18}O$ 130.23	597-76-2	160	4-01-00-01790 0.8373[20]	H_2O 1; EtOH 3; eth 3; ace 3 1.4300[20]
6932	1-Hexanol, 2-ethyl-3-methyl-	$C_9H_{20}O$ 144.26	66794-04-5	193; 83[10]	4-02-00-02107 0.8486[20]	1.4455[20]
6933	2-Hexanol, 3-ethyl-2-methyl-	$C_9H_{20}O$ 144.26	66794-02-3	71.5[14]	4-01-00-01812 0.8334[25]	1.4331[25]
6934	3-Hexanol, 3-ethyl-2-methyl-	$C_9H_{20}O$ 144.26	66794-03-4	177.5	4-01-00-01811 0.8445[25]	1.4369[25]
6935	3-Hexanol, 3-ethyl-5-methyl-	$C_9H_{20}O$ 144.26	597-77-3	172; 68[18]	3-01-00-01755 0.8396[22]	eth 4; EtOH 4 1.4346[13]
6936	1-Hexanol, 6-fluoro-	$C_6H_{13}FO$ 120.17	373-32-0	85-6[14]	4-01-00-01703 0.975[20]	eth 4; EtOH 4 1.4141[25]
6937	1-Hexanol, 2-methyl-, (±)- 1-Hexanol, 2-methyl (DL)	$C_7H_{16}O$ 116.20	111768-04-8	164; 71[15]	4-01-00-01745 0.8270[20]	eth 4; EtOH 4 1.4226[20]
6938	1-Hexanol, 2-methyl-, (R)- (R)-2-Methyl-1-hexanol	$C_7H_{16}O$ 116.20	66050-98-4	164, 71[15]	0.8313[13]	eth 4; EtOH 4 1.4245[17]
6939	1-Hexanol, 3-methyl-, (±)-	$C_7H_{16}O$ 116.20	111768-08-2	169; 91[13]	4-01-00-01748 0.8258[20]	H_2O 1; EtOH 4; eth 4; ace 3 1.4245[20]
6940	1-Hexanol, 4-methyl-, (±)- 1-Hexanol, 4-methyl (DL)	$C_7H_{16}O$ 116.20	111768-05-9	173	4-01-00-01749 0.8239[20]	ace 4; bz 4; eth 4; EtOH 4 1.4219[20]
6941	1-Hexanol, 5-methyl- 5-Methyl-1-hexanol	C_7H_{16} 100.20	627-98-5	170; 53-5[15]	4-01-00-01748 0.8119[20]	eth 4; EtOH 4 1.4175[20]
6942	2-Hexanol, 2-methyl-	$C_7H_{16}O$ 116.20	625-23-0	143	4-01-00-01745 0.8119[20]	H_2O 2; EtOH 5; eth 5 1.4175[20]
6943	2-Hexanol, 3-methyl- 3-Methyl-2-hexanol	$C_7H_{16}O$ 116.20	2313-65-7	79-81[52]	4-01-00-01746 0.8820[25]	H_2O 1; EtOH 4; eth 4; ace 3 1.4198[18]
6944	2-Hexanol, 5-methyl- 5-Methyl-2-hexanol	$C_7H_{16}O$ 116.20	627-59-8	151; 78[28]	4-01-00-01747 0.814[20]	H_2O 1; EtOH 3; eth 3 1.4180[20]
6945	3-Hexanol, 2-methyl-, (±)- (±)-2-Methyl-3-hexanol	$C_7H_{16}O$ 116.20	100295-82-7	143	4-01-00-01746 0.8407[20]	H_2O 1; EtOH 3; eth 3; ctc 3 1.4149[20]
6946	3-Hexanol, 3-methyl- 2-Ethyl-2-pentanol	$C_7H_{16}O$ 116.20	597-96-6	143	4-01-00-01749 0.8233[20]	H_2O 2; EtOH 3; eth 3; ctc 3 1.4231[20]
6947	3-Hexanol, 5-methyl-, (±)- (±)-5-Methyl-3-hexanol	$C_7H_{16}O$ 116.20	100295-83-8	148	4-01-00-01747 0.827[25]	H_2O 1; EtOH 3; eth 3; ctc 2 1.4128[20]
6948	1-Hexanol, 5-methyl-2-(1-methylethyl)- 2-Isopropyl-5-methyl-1-hexanol	$C_{10}H_{22}O$ 158.28	2051-33-4	211	4-01-00-01833 0.8345[20]	H_2O 1; EtOH 3; eth 4 1.4369[20]
6949	3-Hexanol, 1-phenyl-	$C_{12}H_{18}O$ 178.27	2180-43-0 34	170[50]; 146[16]	4-06-00-03418 0.9525[20]	
6950	3-Hexanol, 2,2,5,5-tetramethyl-	$C_{10}H_{22}O$ 158.28	55073-86-4 52.5	170	3-01-00-01772	ace 4; eth 4; EtOH 4; peth 4
6951	1-Hexanol, 3,5,5-trimethyl-	$C_9H_{20}O$ 144.26	3452-97-9	194	4-01-00-01813 0.8236[25]	1.4300[25]
6952	2-Hexanol, 2,3,4-trimethyl-	$C_9H_{20}O$ 144.26	21102-13-6	57[5]	3-01-00-01756 0.853[15]	ace 4; eth 4; EtOH 4; peth 4 1.4415[15]
6953	3-Hexanol, 2,2,3-trimethyl-	$C_9H_{20}O$ 144.26	5340-41-0	170	3-01-00-01755 0.8474[20]	EtOH 3; eth 3; ace 3 1.4402[20]
6954	3-Hexanol, 2,3,5-trimethyl-	$C_9H_{20}O$ 144.26	65927-60-8	72[21]	4-01-00-01813 0.8256[20]	ace 4; eth 4; EtOH 4 1.4321[20]
6955	3-Hexanol, 2,4,4-trimethyl-	$C_9H_{20}O$ 144.26	66793-92-8	170	3-01-00-01756 0.8489[20]	EtOH 3; eth 3; ace 3 1.4395[20]
6956	3-Hexanol, 2,5,5-trimethyl-	$C_9H_{20}O$ 144.26	66793-72-4	77[32]	3-01-00-01756 0.8250[20]	ace 4; eth 4; EtOH 4 1.4286[20]
6957	3-Hexanol, 3,4,4-trimethyl-	$C_9H_{20}O$ 144.26	66793-74-6	165.5	0-01-00-00425 0.8322[21]	EtOH 3; eth 3; ace 3 1.4341[21]
6958	3-Hexanol, 3,5,5-trimethyl- 3,5,5-Trimethylhexan-3-ol	$C_9H_{20}O$ 144.26	66810-87-5	62[14]	3-01-00-01755 0.8350[20]	ace 4; eth 4; EtOH 4 1.4352[20]
6959	2-Hexanone Butyl methyl ketone	$C_8H_{12}O$ 100.16	591-78-6 -55.5	5955 127.6	4-01-00-03298 0.8113[20]	H_2O 2; EtOH 5; eth 5; ace 3 1.4007[20]
6960	3-Hexanone Ethyl propyl ketone	$C_6H_{12}O$ 100.16	589-38-8 -55.5	123.5	4-01-00-03301 0.8118[20]	H_2O 2; EtOH 5; eth 5; ace 3 1.4004[20]

No.	Name Synonym	Mol. Form. Mol. Wt.	CAS RN mp/°C	Merck No. bp/°C	Beil. Ref. den/g cm^{-3}	Solubility n_D
6961	2-Hexanone, 6-bromo-	C$_6$H$_{11}$BrO 179.06	10226-29-6	216; 136[90]	3-01-00-02829 1.3494[0]	eth 4; EtOH 4
6962	1-Hexanone, 1-(2,4-dihydroxyphenyl)-	C$_{12}$H$_{16}$O$_3$ 208.26	3144-54-5 57	343	4-08-00-01888	H$_2$O 1; EtOH 3; eth 3; ace 3
6963	2-Hexanone, 3,3-dimethyl- 3,3-Dimethyl-2-hexanone	C$_8$H$_{16}$O 128.21	26118-38-7	149	4-01-00-03350 0.838[0]	ace 4; eth 4; EtOH 4 1.4098[20]
6964	2-Hexanone, 3,4-dimethyl- 3,4-Dimethyl-2-hexanone	C$_8$H$_{16}$O 128.21	19550-10-8	158	3-01-00-02882 0.8295[22]	EtOH 3; eth 3; ace 3 1.4193[20]
6965	3-Hexanone, 2,2-dimethyl- tert-Butyl propyl ketone	C$_8$H$_{16}$O 128.21	5405-79-8	147	4-01-00-03347 0.8105[25]	EtOH 3; eth 3; ace 3 1.4119[20]
6966	3-Hexanone, 2,5-dimethyl-	C$_8$H$_{16}$O 128.21	1888-57-9	147.5	4-01-00-03349 0.8269[0]	ace 4; eth 4; EtOH 4 1.4049[20]
6967	3-Hexanone, 4,4-dimethyl-	C$_8$H$_{16}$O 128.21	19550-14-2	148	4-01-00-03350 0.8285[20]	bz 4; EtOH 4; chl 4 1.4208[25]
6968	3-Hexanone, 6-(dimethylamino)-4,4-diphenyl- Normethadone	C$_{20}$H$_{25}$NO 295.42	467-85-6	6628 164-7[3]	4-14-00-00297	
6969	2-Hexanone, 5-hydroxy- 5-Hydroxy-2-hexanone	C$_6$H$_{12}$O$_2$ 116.16	56745-61-0	203[270]; 80[10]	4-01-00-04019 0.9626[25]	H$_2$O 4; eth 4; EtOH 4 1.4312[25]
6970	2-Hexanone, 6-hydroxy-	C$_6$H$_{12}$O$_2$ 116.16	21856-89-3	227	4-01-00-04020 0.9886[15]	eth 4; EtOH 4 1.4494
6971	3-Hexanone, 4-hydroxy-	C$_6$H$_{12}$O$_2$ 116.16	4984-85-4	133[227]; 73[20]	4-01-00-04021 0.956[21]	ace 4; EtOH 4 1.4340[21]
6972	3-Hexanone, 5-hydroxy- 5-Hydroxy-3-hexanone	C$_6$H$_{12}$O$_2$ 116.16	33683-44-2	75-8[12]	4-01-00-04021 0.950[20]	1.4280[25]
6973	3-Hexanone, 2-hydroxy-2-methyl- 2-Hexanol-3-one, 2-methyl	C$_7$H$_{14}$O$_2$ 130.19	18905-91-4	60-2[15]	4-01-00-04035 0.899[18]	1.419[18]
6974	3-Hexanone, 5-hydroxy-2-methyl- 2-Hexanol-4-one, 5-methyl	C$_7$H$_{14}$O$_2$ 130.19	59357-07-2	72-3[9]	4-01-00-04035 0.929[20]	1.4278[20]
6975	3-Hexanone, 4-hydroxy-2,2,5,5-tetramethyl- Pivaloin	C$_{10}$H$_{20}$O$_2$ 172.27	815-66-7 81	80[10]	4-01-00-04060	eth 4
6976	2-Hexanone, 3-methyl- 3-Methyl-2-hexanone	C$_7$H$_{14}$O 114.19	2550-21-2	143.5	3-01-00-02863 0.828[25]	ace 4; bz 4; eth 4; EtOH 4 1.4035[20]
6977	2-Hexanone, 4-methyl- Methyl 2-methylbutyl ketone	C$_7$H$_{14}$O 114.19	105-42-0	144.5	4-01-00-03329 0.8130[20]	H$_2$O 2; EtOH 4; eth 4; ace 4 1.4081[24]
6978	2-Hexanone, 5-methyl- Methyl isopentyl ketone	C$_7$H$_{14}$O 114.19	110-12-3	144	4-01-00-03329 0.888[20]	H$_2$O 2; EtOH 5; eth 5; ace 4 1.4062[20]
6979	3-Hexanone, 2-methyl-	C$_7$H$_{14}$O 114.19	7379-12-6	135	4-01-00-03328 0.8091[20]	EtOH 3; eth 3; ace 4; chl 3 1.4042[20]
6980	3-Hexanone, 4-methyl-	C$_7$H$_{14}$O 114.19	17042-16-9	134.5	4-01-00-03329 0.8162[20]	ace 4; bz 4; eth 4; EtOH 4 1.4069[20]
6981	3-Hexanone, 5-methyl-	C$_7$H$_{14}$O 114.19	623-56-3	135	4-01-00-03328 0.8090[20]	H$_2$O 1; EtOH 5; eth 5; ctc 3 1.4047[20]
6982	2-Hexanone, 5-methyl-, oxime	C$_7$H$_{15}$NO 129.20	624-44-2	195.5	0-01-00-00701 0.8881[20]	chl 2 1.4448[20]
6983	1-Hexanone, 1-phenyl-	C$_{12}$H$_{16}$O 176.26	942-92-7 27	265	4-07-00-00754 0.9576[20]	H$_2$O 2; EtOH 3; eth 3; ace 3 1.5027[25]
6984	2-Hexanone, 3-phenyl-	C$_{12}$H$_{16}$O 176.26	6306-30-5	235.5; 114[13]	3-07-00-01160 0.970[0]	1.5020[20]
6985	3-Hexanone, 2-phenyl- 1-Phenylethyl propyl ketone	C$_{12}$H$_{16}$O 176.26	65248-43-3	241	2-07-00-00258 0.941[25]	1.4988[25]
6986	Hexanoyl chloride Caproyl chloride	C$_6$H$_{11}$ClO 134.61	142-61-0 -87	1763 153	4-02-00-00928 0.9784[20]	eth 3; ace 3 1.4264[20]
6987	Hexanoyl chloride, 2-ethyl-	C$_8$H$_{15}$ClO 162.66	760-67-8	67-8[11]	3-02-00-00805 0.939[25]	1.4335[20]
6988	Hexanoyl chloride, 3-methyl- Caproyl chloride, 3-methyl	C$_7$H$_{13}$ClO 148.63	57323-93-0	163; 82[50]	3-02-00-00777 0.967[20]	bz 4 1.4293[25]
6989	Hexanoyl chloride, 4-methyl- Caproyl chloride, 4-methyl	C$_7$H$_{13}$ClO 148.63	50599-73-0	167	2-02-00-00299 0.9677[20]	eth 4
6990	Hexasiloxane, tetradecamethyl- Tetradecamethylhexasiloxane	C$_{14}$H$_{42}$O$_5$Si$_6$ 459.00	107-52-8 -59	9131 245.5	4-04-00-04127 0.8910[20]	bz 4 1.3948[20]
6991	Hexatriacontane	C$_{36}$H$_{74}$ 506.98	630-06-8 76.5	298.4[3]	4-01-00-00599 0.7803[80]	1.4397[80]
6992	1,3,5-Hexatriene, (E)-	C$_6$H$_8$ 80.13	821-07-8 -12	78.5	4-01-00-01093 0.7369[15]	H$_2$O 1; EtOH 3; ace 3; chl 3 1.5135[20]
6993	1,3,5-Hexatriene, (Z)-	C$_6$H$_8$ 80.13	2612-46-6 -12	78	0.7175[20]	H$_2$O 1; EtOH 3; ace 3; chl 3 1.4577[20]
6994	1,3,5-Hexatriene, 2,5-dimethyl-	C$_8$H$_{12}$ 108.18	4916-63-6 -9	146	4-01-00-01102 0.7822[20]	H$_2$O 1; ace 3; MeOH 3; lig 3 1.5122[20]
6995	Hexazinone 1,3,5-Triazine-2,4(1H,3H)-dione, 3-cyclohexyl- 6-(dimethylamino)-1-methyl-	C$_{12}$H$_{20}$N$_4$O$_2$ 252.32	51235-04-2 99	4617 dec	1.25	
6996	2-Hexenal, (E)-	C$_6$H$_{10}$O 98.14	6728-26-3	146.5; 50[20]	4-01-00-03468 0.8491[20]	1.4480[20]
6997	3-Hexenal, (E)-	C$_6$H$_{10}$O 98.14	69112-21-6	42-3[28]	4-01-00-03469 0.8455[22]	H$_2$O 1; eth 3; ace 3 1.4275[21]
6998	3-Hexenal, (Z)-	C$_6$H$_{10}$O 98.14	6789-80-6	121	4-01-00-03469 0.8533[22]	1.4300[21]
6999	1-Hexene	C$_6$H$_{12}$ 84.16	592-41-6 -139.7	63.4	4-01-00-00828 0.6731[20]	bz 4; eth 4; EtOH 4; peth 4 1.3837[20]

No.	Name Synonym	Mol. Form. Mol. Wt.	CAS RN mp/°C	Merck No. bp/°C	Beil. Ref. den/g cm^{-3}	Solubility n_D
7000	2-Hexene, (E)- trans-2-Hexene	C_6H_{12} 84.16	4050-45-7 -133	67.9	4-01-00-00834 0.6732[25]	H_2O 1; EtOH 3; eth 3; bz 3 1.3936[20]
7001	2-Hexene, (Z)- cis-2-Hexene	C_6H_{12} 84.16	7688-21-3 -141.1	68.8	4-01-00-00833 0.6869[20]	H_2O 1; EtOH 3; eth 3; bz 3 1.3979[20]
7002	3-Hexene, (E)-	C_6H_{12} 84.16	13269-52-8 -115.4	67.1	4-01-00-00837 0.6772[20]	H_2O 1; EtOH 3; eth 3; bz 3 1.3943[20]
7003	3-Hexene, (Z)-	C_6H_{12} 84.16	7642-09-3 -137.8	66.4	4-01-00-00837 0.6796[20]	H_2O 1; EtOH 3; eth 3; bz 3 1.3947[20]
7004	1-Hexene, 1-chloro- 1-Chloro-1-hexene	$C_6H_{11}Cl$ 118.61	22922-67-4 	121	4-01-00-00831 0.8872[22]	ace 4; bz 4; eth 4; chl 4 1.4300[22]
7005	1-Hexene, 2-chloro- 2-Chloro-1-hexene	$C_6H_{11}Cl$ 118.61	10124-73-9 	113	3-01-00-00803 0.8886[25]	ace 4; bz 4; chl 4 1.4278[25]
7006	1-Hexene, 5-chloro- 5-Chloro-1-hexene	$C_6H_{11}Cl$ 118.61	927-54-8 	120.7	4-01-00-00832 0.8891[25]	ace 4; bz 4; eth 4; chl 4 1.4305[20]
7007	2-Hexene, 4-chloro- 4-Chloro-2-hexene	$C_6H_{11}Cl$ 118.61	6734-98-1 	123	4-01-00-00835 0.8934[20]	ace 4; bz 4; eth 4; chl 4 1.4400[20]
7008	3-Hexene, 1-chloro- 3-Hexen-1-yl chloride	$C_6H_{11}Cl$ 118.61	62706-16-5 61[60]		4-01-00-00838 0.900[24]	ace 4; bz 4; eth 4; chl 4 1.435[24]
7009	3-Hexene, 3-chloro-, (Z)- 3-Hexene, 3-chloro (cis)	$C_6H_{11}Cl$ 118.61	17226-34-5 	117	4-01-00-00838 0.9009[20]	ace 4; bz 4; eth 4; chl 4 1.4360[20]
7010	3-Hexene, 1-chloro-4-ethyl-	$C_8H_{15}Cl$ 146.66	82507-04-8 	173	2-01-00-00201 0.9102[20]	bz 4; chl 4 1.4524[20]
7011	1-Hexene, 1,2-dichloro-, (E)-	$C_6H_{10}Cl_2$ 153.05	59697-51-7 63-5[22]		3-01-00-00803 1.1167[25]	bz 4; chl 4 1.4576[25]
7012	1-Hexene, 1,2-dichloro-, (Z)-	$C_6H_{10}Cl_2$ 153.05	59697-55-1 88[30]		3-01-00-00803 1.0812[25]	bz 4; chl 4 1.4631[25]
7013	1-Hexene, 2,3-dimethyl- 2,3-Dimethyl-1-hexene	C_8H_{16} 112.22	16746-86-4 	110.5	3-01-00-00845 0.7172[25]	1.4113[20]
7014	1-Hexene, 2,4-dimethyl- 2,4-Dimethyl-1-hexene	C_8H_{16} 112.22	16746-87-5 	111.2	4-01-00-00887 0.716[25]	1.409[20]
7015	1-Hexene, 2,5-dimethyl- 2,5-Dimethyl-1-hexene	C_8H_{16} 112.22	6975-92-4 	111.6	3-01-00-00846 0.7129[25]	1.4105[20]
7016	1-Hexene, 3,3-dimethyl- 3,3-Dimethyl-1-hexene	C_8H_{16} 112.22	3404-77-1 	104	4-01-00-00889 0.7099[25]	1.4070[20]
7017	1-Hexene, 3,4-dimethyl- 3,4-Dimethyl-1-hexene	C_8H_{16} 112.22	16745-94-1 	112	4-01-00-00889 0.720[25]	1.413[20]
7018	1-Hexene, 3,5-dimethyl-	C_8H_{16} 112.22	7423-69-0 	104	3-01-00-00846 0.712[25]	1.404[20]
7019	1-Hexene, 4,4-dimethyl- 4,4-Dimethyl-1-hexene	C_8H_{16} 112.22	1647-08-1 	107.2	3-01-00-00847 0.7157[25]	1.4102[20]
7020	1-Hexene, 4,5-dimethyl- 4,5-Dimethyl-1-hexene	C_8H_{16} 112.22	16106-59-5 	109	4-01-00-00887 0.724[25]	1.414[20]
7021	1-Hexene, 5,5-dimethyl- 5,5-Dimethyl-1-hexene	C_8H_{16} 112.22	7116-86-1 	104	4-01-00-00886 0.705[25]	1.4049[20]
7022	2-Hexene, 2,3-dimethyl- 2,3-Dimethyl-2-hexene	C_8H_{16} 112.22	7145-20-2 -115.1	121.8	4-01-00-00887 0.7366[25]	1.4268[20]
7023	2-Hexene, 2,4-dimethyl- 2,4-Dimethyl-2-hexene	C_8H_{16} 112.22	14255-23-3 	110.6	4-01-00-00887 0.7171[25]	1.4118[20]
7024	2-Hexene, 2,5-dimethyl- 2,5-Dimethyl-2-hexene	C_8H_{16} 112.22	3404-78-2 	112.2	4-01-00-00888 0.7182[20]	1.4140[20]
7025	2-Hexene, 4,5-dimethyl- 2,3-Dimethyl-4-hexene	C_8H_{16} 112.22	73548-71-7 	110	3-01-00-00846 0.721[25]	1.413[20]
7026	2-Hexene, 5,5-dimethyl-, (E)-	C_8H_{16} 112.22	39782-43-9 	104.1	4-01-00-00886 0.7023[25]	1.4055[20]
7027	2-Hexene, 5,5-dimethyl-, (Z)- cis-5,5-Dimethyl-2-hexene	C_8H_{16} 112.22	39761-61-0 	106.9	4-01-00-00886 0.7125[25]	1.4113[20]
7028	3-Hexene, 2,2-dimethyl-, (E)-	C_8H_{16} 112.22	690-93-7 	100.8	4-01-00-00885 0.6995[25]	1.4063[20]
7029	3-Hexene, 2,2-dimethyl-, (Z)-	C_8H_{16} 112.22	690-92-6 -137.4	105.5	4-01-00-00885 0.7086[25]	1.4099[20]
7030	3-Hexene, 2,4-dimethyl-, (E)- 2,4-Dimethyl-trans-3-hexene	C_8H_{16} 112.22	61847-78-7 	107.6	4-01-00-00887 0.7101[25]	1.4126[20]
7031	3-Hexene, 2,4-dimethyl-, (Z)- 3-Hexene, 2,4-dimethyl (cis)	C_8H_{16} 112.22	37549-89-6 	109	4-01-00-00887 0.7135[25]	1.4140[20]
7032	1-Hexene, 1,1,2,3,3,4,4,5,5,6,6,6- dodecafluoro- Perfluoro-1-hexene	C_6F_{12} 300.05	755-25-9 57.0		4-01-00-00831 	chl 4
7033	1-Hexene, 3-ethyl- 3-Propyl-1-pentene	C_8H_{16} 112.22	3404-58-8 	110.3	4-01-00-00885 0.711[25]	1.407[20]
7034	3-Hexene, 3-ethyl-	C_8H_{16} 112.22	16789-51-8 	116	3-01-00-00845 0.725[25]	1.418[20]
7035	3-Hexene, 1,2,3,4,5,6-hexachloro- 1,2,3,4,5,6-Hexachloro-3-hexene	$C_6H_6Cl_6$ 290.83	1725-74-2 58.5	112[2]	3-01-00-00808 	MeOH 4; chl 4
7036	1-Hexene, 2-methyl- 2-Methyl-1-hexene	C_7H_{14} 98.19	6094-02-6 -102.8	92	4-01-00-00863 0.7000[20]	1.4035[20]
7037	1-Hexene, 3-methyl- 3-Methyl-1-hexene	C_7H_{14} 98.19	3404-61-3 	83.9	4-01-00-00865 0.6871[25]	1.3965[20]
7038	1-Hexene, 4-methyl- 4-Methyl-1-hexene	C_7H_{14} 98.19	3769-23-1 -141.5	86.7	3-01-00-00830 0.6942[25]	1.4000[20]
7039	1-Hexene, 5-methyl- 5-Methyl-1-hexene	C_7H_{14} 98.19	3524-73-0 	85.3	4-01-00-00865 0.6877[25]	1.3967[20]

No.	Name Synonym	Mol. Form. Mol. Wt.	CAS RN mp/°C	Merck No. bp/°C	Beil. Ref. den/g cm⁻³	Solubility n_D
7040	2-Hexene, 2-methyl- 2-Methyl-2-hexene	C_7H_{14} 98.19	2738-19-4 -130.4	95.4	4-01-00-00863 0.7038[25]	1.4106[20]
7041	2-Hexene, 3-methyl-, (Z)-	C_7H_{14} 98.19	10574-36-4 -118.5	95.6	4-01-00-00866 0.712[20]	1.4126[20]
7042	2-Hexene, 4-methyl-, (E)-	C_7H_{14} 98.19	3683-22-5 -125.7	87.6	4-01-00-00867 0.6925[25]	1.4025[20]
7043	2-Hexene, 4-methyl-, (Z)-	C_7H_{14} 98.19	3683-19-0	86.3	4-01-00-00867 0.6952[25]	1.4026[20]
7044	2-Hexene, 5-methyl-, (E)-	C_7H_{14} 98.19	7385-82-2 -124.3	88.1	4-01-00-00865 0.6883[25]	1.4006[20]
7045	2-Hexene, 5-methyl-, (Z)-	C_7H_{14} 98.19	13151-17-2	89.5	4-01-00-00865 0.697[25]	1.404[20]
7046	3-Hexene, 2-methyl-, (E)-	C_7H_{14} 98.19	692-24-0 -141.6	85.9	4-01-00-00864 0.6853[25]	1.4001[20]
7047	3-Hexene, 2-methyl-, (Z)-	C_7H_{14} 98.19	15840-60-5	86	4-01-00-00864 0.690[25]	1.401[20]
7048	1-Hexene, 1,1,2-trichloro-	$C_6H_9Cl_3$ 187.50	53977-99-4	90-3[10]	3-01-00-00803 1.125[25]	eth 4 1.4760[25]
7049	2-Hexenoic acid	$C_6H_{10}O_2$ 114.14	1191-04-4 36.5	216.5	4-02-00-01563 0.965[20]	eth 4 1.4460[40]
7050	3-Hexenoic acid Hydrosorbic acid	$C_6H_{10}O_2$ 114.14	4219-24-3 12	208	4-02-00-01566 0.9640[23]	1.4935[20]
7051	5-Hexenoic acid 5-Hexanoic acid	$C_6H_{10}O_2$ 114.14	1577-22-6 -37	203	4-02-00-01562 0.9610[20]	eth 4; EtOH 4 1.4343[20]
7052	4-Hexenoic acid, 6-(1,3-dihydro-4-hydroxy-6- methoxy-7-methyl-3-oxo-5-isobenzofuranyl)-4- methyl-, (E)- Mycophenolic acid	$C_{17}H_{20}O_6$ 320.34	24280-93-1 141	6238	4-18-00-06513	H_2O 1; EtOH 4; eth 4; bz 2
7053	2-Hexenoic acid, ethyl ester Ethyl 2-hexenoate	$C_8H_{14}O_2$ 142.20	1552-67-6	174.5; 80[22]	3-02-00-01318 0.8986[20]	1.4348[20]
7054	3-Hexenoic acid, ethyl ester Hydrosorbic acid, ethyl ester	$C_8H_{14}O_2$ 142.20	2396-83-0	166.5	4-02-00-01567 0.8957[20]	1.4255[20]
7055	2-Hexenoic acid, 2-methyl-, (E)-	$C_7H_{12}O_2$ 128.17	97961-66-5	205; 118[11]	4-02-00-01581 0.9627[20]	eth 3; ace 3 1.4601[20]
7056	2-Hexenoic acid, 4-methyl- 4-Methyl-2-hexenoic acid	$C_7H_{12}O_2$ 128.17	37549-83-0	125[13]	3-02-00-01336 0.9441[20]	1.4526[20]
7057	2-Hexenoic acid, 5-methyl- 5-Methyl-2-hexenoic acid	$C_7H_{12}O_2$ 128.17	41653-96-7 16.5	226.5	2-02-00-00411 0.942[20]	1.4425[17]
7058	3-Hexenoic acid, 2-methyl- Hydrosorbic acid, 2-methyl-	$C_7H_{12}O_2$ 128.17	73513-50-5	122[24]	4-02-00-01582 0.9353[20]	1.4379[20]
7059	3-Hexenoic acid, 3-methyl-	$C_7H_{12}O_2$ 128.17	35205-71-1	113[10]	3-02-00-01335 0.9549[20]	1.4469[20]
7060	3-Hexenoic acid, 4-methyl- 4-Ethyl-3-pentenoic acid	$C_7H_{12}O_2$ 128.17	55665-79-7	118[12]	3-02-00-01336 0.9644[16]	1.4512[16]
7061	4-Hexenoic acid, 5-methyl-	$C_7H_{12}O_2$ 128.17	5636-65-7 -28	217	4-02-00-01581 0.986[20]	1.4504
7062	3-Hexenoic acid, methyl ester Methyl 3-hexenoate	$C_7H_{12}O_2$ 128.17	2396-78-3	67-8[34]	4-02-00-01566 0.9132[25]	1.4240[23]
7063	2-Hexenoic acid, 2-methyl-, ethyl ester	$C_9H_{16}O_2$ 156.22	26311-33-1	72[10]	3-02-00-01333 0.9031[20]	1.4407[20]
7064	3-Hexenoic acid, 2-methyl-, ethyl ester	$C_9H_{16}O_2$ 156.22	21994-78-5	78[25]	3-02-00-01334 0.8778[20]	1.4237[20]
7065	3-Hexenoic acid, 3-methyl-, ethyl ester	$C_9H_{16}O_2$ 156.22	21994-77-4	84-5[26]	3-02-00-01335 0.8961[20]	1.4309[20]
7066	1-Hexen-3-ol	$C_6H_{12}O$ 100.16	4798-44-1	134	4-01-00-02136 0.834[22]	ace 4; eth 4; EtOH 4 1.4297[18]
7067	4-Hexen-2-ol	$C_6H_{12}O$ 100.16	52387-50-5	137.5	4-01-00-02139 0.8405[18]	1.4392[20]
7068	5-Hexen-2-ol	$C_6H_{12}O$ 100.16	626-94-8	139	3-01-00-01926 0.842[16]	H_2O 2
7069	2-Hexen-1-ol, (E)-	$C_6H_{12}O$ 100.16	928-95-0	157	4-01-00-02138 0.8490[16]	1.4340[20]
7070	2-Hexen-1-ol, (Z)-	$C_6H_{12}O$ 100.16	928-94-9	157	4-01-00-02138 0.8472[20]	H_2O 3; EtOH 4; eth 3; ace 3 1.4397[20]
7071	3-Hexen-1-ol, (Z)-	$C_6H_{12}O$ 100.16	928-96-1	156.5	4-01-00-02141 0.8478[22]	H_2O 3; EtOH 4; eth 3 1.4380[20]
7072	4-Hexen-1-ol, (E)-	$C_6H_{12}O$ 100.16	928-92-7	159	4-01-00-02140 0.8513[20]	1.4402[20]
7073	1-Hexen-3-ol, 3,5-dimethyl-	$C_8H_{16}O$ 128.21	3329-48-4	146.5	4-01-00-02173 0.8382[20]	1.4342[20]
7074	4-Hexen-3-ol, 2,2-dimethyl- 2-Hexen-4-ol, 5,5-dimethyl	$C_8H_{16}O$ 128.21	37674-67-2	75-6[50]	4-01-00-02171 0.8308[25]	1.4369[25]
7075	4-Hexen-3-ol, 2,5-dimethyl- 2,5-Dimethyl-4-hexen-3-ol	$C_8H_{16}O$ 128.21	60703-31-3	162	2-01-00-00491 0.8444[20]	1.4449[20]
7076	4-Hexen-3-ol, 3,5-dimethyl- 2-Hexen-4-ol, 2,4-dimethyl	$C_8H_{16}O$ 128.21	1569-43-3	48-50[3-6]	4-01-00-02172 0.8600[17]	1.4460[17]
7077	1-Hexen-3-ol, 5-methyl-	$C_7H_{14}O$ 114.19	4798-46-3	125; 58[15]	3-01-00-01939 0.8306[15]	1.4263[23]
7078	2-Hexen-1-ol, 5-methyl- 5-Methyl-2-hexenol	$C_7H_{14}O$ 114.19	77053-92-0	169	4-01-00-02158 0.8355[20]	1.4390[20]

No.	Name / Synonym	Mol. Form. / Mol. Wt.	CAS RN mp/°C	Merck No. bp/°C	Beil. Ref. den/g cm⁻³	Solubility n_D
7079	3-Hexen-2-ol, 2-methyl-	$C_7H_{14}O$ 114.19	18812-62-9	49[11]	4-01-00-02157 0.8536[18]	1.443[18]
7080	3-Hexen-2-ol, 3-methyl-	$C_7H_{14}O$ 114.19	110383-31-8	89[55]	0-01-00-00447 0.8678[9]	1.4487[9]
7081	4-Hexen-3-ol, 2-methyl-	$C_7H_{14}O$ 114.19	4798-60-1	139.5	3-01-00-01939 0.8426[20]	1.438[20]
7082	4-Hexen-3-ol, 3-methyl-	$C_7H_{14}O$ 114.19	60111-14-0	72-3[60]	4-01-00-02160 0.8471[25]	1.4268[16]
7083	5-Hexen-2-ol, 2-methyl- Dimethyl(vinylethyl)carbinol	$C_7H_{14}O$ 114.19	16744-89-1 -22	142.5	4-01-00-02158 0.8397[15]	1.4349[17]
7084	5-Hexen-3-ol, 3-methyl- Methylethylallylcarbinol	$C_7H_{14}O$ 114.19	1569-44-4	140; 61[35]	4-01-00-02160 0.8432[20]	1.4370[20]
7085	5-Hexen-3-ol, 4-methyl- 4-Methyl-5-hexen-3-ol	$C_7H_{14}O$ 114.19	1838-77-3	140.5	3-01-00-01940 0.8452[22]	1.4365[22]
7086	3-Hexen-2-one 1-Butenyl methyl ketone	$C_6H_{10}O$ 98.14	763-93-9	140	4-01-00-03468 0.8655[20]	ace 4; eth 4; EtOH 4 1.4418[20]
7087	4-Hexen-2-one 2-Hexen-5-one	$C_6H_{10}O$ 98.14	25659-22-7	127	3-01-00-02994 0.8520[16]	1.4300[16]
7088	4-Hexen-3-one	$C_6H_{10}O$ 98.14	2497-21-4	138.5	4-01-00-03468 0.8559[20]	EtOH 3; eth 3; ace 4 1.4388[20]
7089	5-Hexen-2-one	$C_6H_{10}O$ 98.14	109-49-9	129.5	4-01-00-03467 0.833[27]	1.4178[27]
7090	3-Hexen-2-one, (E)-	$C_6H_{10}O$ 98.14	4376-23-2	140	4-01-00-03469 0.8665[20]	EtOH 3; eth 3; ace 4 1.4418[20]
7091	3-Hexen-2-one, 3,4-dimethyl-	$C_8H_{14}O$ 126.20	1635-02-5	158; 65[20]	4-01-00-03500 0.8585[20]	1.4418[20]
7092	4-Hexen-2-one, 3,4-dimethyl- 2-Hexen-5-one, 3,4-dimethyl-	$C_8H_{14}O$ 126.20	53252-21-4	154	4-01-00-03500 0.8538[19]	1.4377[19]
7093	1-Hexen-3-one, 5-methyl- Isobutyl vinyl ketone	$C_7H_{12}O$ 112.17	2177-32-4	41[22]; 32[10]	4-01-00-03483 0.8400[15]	1.4293[15]
7094	3-Hexen-2-one, 5-methyl- 2-Oxo-5-methylhex-3-ene	$C_7H_{12}O$ 112.17	5166-53-0	77.5[50]	3-01-00-03005 0.8407[22]	1.4395[22]
7095	4-Hexen-2-one, 5-methyl-	$C_7H_{12}O$ 112.17	28332-44-7	76[50]	4-01-00-03482 0.9012[21]	1.4385[20]
7096	4-Hexen-3-one, 2-methyl-	$C_7H_{12}O$ 112.17	53252-19-0	148	4-01-00-03483 0.843[20]	EtOH 3; eth 3; ace 4 1.4345[20]
7097	5-Hexen-2-one, 3-methyl- 1-Methyl-1-allyl acetone	$C_7H_{12}O$ 112.17	2550-22-3	137	4-01-00-03484 0.845[15]	1.4215[25]
7098	5-Hexen-2-one, 4-methyl- 4-Methyl-5-hexen-2-one	$C_7H_{12}O$ 112.17	61675-14-7	137.5	3-01-00-03006 0.8273[25]	1.4193[22]
7099	5-Hexen-2-one, 5-methyl- 5-Hexene-2-one,5-methyl	$C_7H_{12}O$ 112.17	3240-09-3	150	4-01-00-03482 0.8460[20]	ace 4; eth 4; EtOH 4 1.4348[20]
7100	1-Hexen-3-one, 5-methyl-1-phenyl-	$C_{13}H_{16}O$ 188.27	2892-18-4 43	154[25]	4-07-00-01058 0.9509[46]	H_2O 2; EtOH 3; bz 3; chl 3 1.5523[25]
7101	5-Hexen-2-one, 4-phenyl-	$C_{12}H_{14}O$ 174.24	50552-30-2	85-6[1]	3-07-00-01460 0.9848[25]	1.5193[25]
7102	5-Hexen-2-one, 6-phenyl- 6-Phenyl-5-hexen-2-one	$C_{12}H_{14}O$ 174.24	69371-59-1	154[10]; 98[0.3]	3-07-00-01455 0.9967[25]	1.5458[25]
7103	1-Hexen-3-yne Ethylvinylacetylene	C_6H_8 80.13	13721-54-5	85	4-01-00-01091 0.7492[20]	bz 4; eth 4; peth 4; chl 4 1.4522[20]
7104	1-Hexen-4-yne	C_6H_8 80.13	5009-11-0	87	4-01-00-01092 0.767[14]	1.446[14]
7105	1-Hexen-5-yne Hex-5-en-1-yne	C_6H_8 80.13	14548-31-3	70	4-01-00-01092 0.7650[20]	bz 4; eth 4; peth 4; chl 4 1.4318[20]
7106	2-Hexen-4-yne 2-Hexene-4-yne	C_6H_8 80.13	14092-20-7	88.5	4-01-00-01092 0.7710[20]	1.4918[20]
7107	1-Hexen-3-yne, 5-chloro-5-methyl- Dimethylvinylethynylchloromethane	C_7H_9Cl 128.60	819-44-3	48[28]	4-01-00-01098 0.9375[15]	ace 4; bz 4; eth 4; EtOH 4 1.4778[20]
7108	3-Hexen-1-yne, 3-propyl-	C_9H_{14} 122.21	688-52-8	136	3-01-00-01049 0.7799[25]	bz 4; peth 4; chl 4 1.4432[25]
7109	Hexitol, 2,5-anhydro-3,4-dideoxy- Tetrahydro-2,5-furandimethanol	$C_6H_{12}O_3$ 132.16	104-80-3 <-50	9145 265	4-17-00-02052 1.154[20]	H_2O 4; ace 4; bz 4; EtOH 4
7110	D-arabino-Hexose, 2-deoxy- 2-Deoxy-D-glucose	$C_6H_{12}O_5$ 164.16	154-17-6 146.5	2886	4-01-00-04282	
7111	D-ribo-Hexose, 2,6-dideoxy- Digitoxose	$C_6H_{12}O_4$ 148.16	527-52-6 112	3147	4-01-00-04191	H_2O 4; ace 4; py 3; AcOEt 3
7112	ribo-Hexose, 2,6-dideoxy-3-O-methyl- Cymarose	$C_7H_{14}O_4$ 162.19	579-04-4 101	2769	4-01-00-04193	H_2O 4; ace 4; EtOH 4
7113	3-Hexynal	C_6H_8O 96.13	89533-67-5	42-3[17]	4-01-00-03544 0.9036[23]	1.4498[22]
7114	1-Hexyne Butylacetylene	C_6H_{10} 82.15	693-02-7 -131.9	71.3	4-01-00-01006 0.7155[20]	H_2O 1; EtOH 3; eth 3; bz 3 1.3989[20]
7115	2-Hexyne 1-Methyl-2-propylacetylene	C_6H_{10} 82.15	764-35-2 -89.6	84.5	4-01-00-01009 0.7315[20]	H_2O 1; EtOH 5; eth 5; bz 3 1.4138[20]
7116	3-Hexyne Diethylacetylene	C_6H_{10} 82.15	928-49-4 -103	81	4-01-00-01009 0.7231[20]	H_2O 3; EtOH 3; eth 3; bz 3 1.4115[20]
7117	3-Hexyne, 2,5-dichloro-2,5-dimethyl- 3-Hexyne, 2,5-dimethyl-2,5-dichloro-	$C_8H_{12}Cl_2$ 179.09	2431-30-3 29	178; 62[14]	4-01-00-01042 1.0118[20]	chl 4
7118	3-Hexyne-2,5-diol	$C_6H_{10}O_2$ 114.14	3031-66-1	120[11]	3-01-00-02271 1.0180[20]	1.4691[20]

No.	Name Synonym	Mol. Form. Mol. Wt.	CAS RN mp/°C	Merck No. bp/°C	Beil. Ref. den/g cm^{-3}	Solubility n_D
7119	3-Hexyne-2,5-diol, 2,5-dimethyl-	$C_8H_{14}O_2$ 142.20	142-30-3 95	205	4-01-00-02699 0.947[20]	H_2O 3; EtOH 4; eth 4; ace 4
7120	1-Hexyne, 5-methyl- 5-Methyl-1-hexyne	C_7H_{12} 96.17	2203-80-7 -125	92	3-01-00-01000 0.7274[20]	H_2O 1; EtOH 3; eth 3; bz 3 1.4059[20]
7121	2-Hexyne, 5-methyl-	C_7H_{12} 96.17	53566-37-3 -92.9	102.5	4-01-00-01030 0.7378[20]	H_2O 1; EtOH 3; eth 3; ace 3; bz 3 1.4176[20]
7122	3-Hexyne, 2-methyl- 1-Ethyl-2-isopropylacetylene	C_7H_{12} 96.17	36566-80-0 -116.7	95.2	4-01-00-01029 0.7263[20]	bz 4; eth 4; chl 4; peth 4 1.4120[20]
7123	2-Hexynoic acid Propiolic acid, propyl-	$C_6H_8O_2$ 112.13	764-33-0 27	126[24]	4-02-00-01699 0.9820[20]	eth 4; EtOH 4; lig 4
7124	1-Hexyn-3-ol	$C_6H_{10}O$ 98.14	105-31-7 -80	142	4-01-00-02234 0.8704[20]	ctc 3 1.4340[25]
7125	2-Hexyn-1-ol	$C_6H_{10}O$ 98.14	764-60-3	100[53]; 80[21]	4-01-00-02235 0.8472[20]	1.4341[20]
7126	3-Hexyn-1-ol 3-Hexynol	$C_6H_{10}O$ 98.14	1002-28-4	163; 65[12]	4-01-00-02236 0.8982[20]	1.4530[20]
7127	5-Hexyn-2-ol	$C_6H_{10}O$ 98.14	23470-12-4	152	4-01-00-02235 0.899[20]	1.4481[20]
7128	1-Hexyn-3-ol, 3-methyl- Ethynylmethylpropylcarbinol	$C_7H_{12}O$ 112.17	4339-05-3	137	4-01-00-02252 0.8620[20]	H_2O 4; eth 4; EtOH 4 1.4338[20]
7129	3-Hexyn-2-ol, 2-methyl- 2-Methyl-3-hexyn-2-ol	$C_7H_{12}O$ 112.17	5075-33-2 -6	146	4-01-00-02250 0.962[25]	eth 4; EtOH 4 1.4392[25]
7130	3-Hexyn-2-one 3-Hexyne-2-one	C_6H_8O 96.13	1679-36-3	77-8[75]	4-01-00-03544 0.8804[20]	1.4400[20]
7131	5-Hexyn-2-one	C_6H_8O 96.13	2550-28-9	149	4-01-00-03543 0.9065[20]	1.4366[20]
7132	L-Histidine Glyoxaline-5-alanine	$C_6H_9N_3O_2$ 155.16	71-00-1 287 dec	4642	4-25-00-04343	H_2O 3; EtOH 2; eth 1; ace 1
7133	L-Histidine, N-β-alanyl- Carnosine	$C_9H_{14}N_4O_3$ 226.24	305-84-0 260	1857	4-25-00-04381	H_2O 4
7134	Homoserine (S)-2-Amino-4-hydroxybutanoic acid	$C_4H_9NO_3$ 119.12	672-15-1 203 dec	4661	4-04-00-03187	H_2O 4; EtOH 2; eth 1; bz 1
7135	L-Homoserine, O-[(aminoiminomethyl)amino]- Canavanine	$C_5H_{12}N_4O_3$ 176.18	543-38-4	1745		H_2O 4
7136	Hydramethylnon Tetrahydro-5,5-dimethyl-2-(1H)- pyrimidinone[3[4-(trifluoromethyl-styryl)- cinnamylidene- hydrazone	$C_{25}H_{24}F_6N_4$ 494.48	67485-29-4 190	4684		
7137	Hydrazine, 1,2-bis(1-methylethyl)-	$C_6H_{16}N_2$ 116.21	3711-34-0	125	4-04-00-03339 0.7894[20]	ace 4; bz 4; eth 4; EtOH 4 1.4173[20]
7138	Hydrazine, 1,2-bis(3-methylphenyl)-	$C_{14}H_{16}N_2$ 212.29	621-26-1 38	224	4-15-00-01220	H_2O 1; EtOH 3; eth 3; bz 3
7139	Hydrazine, 1,2-bis(4-methylphenyl)-	$C_{14}H_{16}N_2$ 212.29	637-47-8 135		4-15-00-01230 0.957[20]	bz 4; eth 4; EtOH 4
7140	Hydrazine, 1,2-bis(2-methylpropyl)-	$C_8H_{20}N_2$ 144.26	3711-37-3	171; 63[18]	2-04-00-00962 0.8002[20]	ace 4; bz 4; eth 4; EtOH 4 1.4276
7141	Hydrazine, (4-bromophenyl)- (p-Bromophenyl)hydrazine	$C_6H_7BrN_2$ 187.04	589-21-9 108	1417	4-15-00-00282	eth 4; EtOH 4; lig 4
7142	Hydrazinecarbothioamide Thiosemicarbazide	CH_5N_3S 91.14	79-19-6 183	9292	4-03-00-00374	H_2O 4; EtOH 4
7143	Hydrazinecarboxamide, monohydrochloride Semicarbazide hydrochloride	CH_6ClN_3O 111.53	563-41-7 175-7 dec	8396	4-03-00-00177	H_2O 4
7144	Hydrazinecarboxamide, 2-[(5-nitro-2-furanyl)methylene]- Nitrofurazone	$C_6H_6N_4O_4$ 198.14	59-87-0 236-40 dec	6521	5-17-09-00335	H_2O 1; EtOH 2; eth 1; DMSO 2
7145	Hydrazinecarboxamide, 2-phenyl- Phenicarbazide	$C_7H_9N_3O$ 151.17	103-03-7 172	7193	4-15-00-00180	H_2O 2; EtOH 3; eth 2; ace 3
7146	Hydrazinecarboxamide, N-phenyl- 4-Phenylsemicarbazide	$C_7H_9N_3O$ 151.17	537-47-3 128	7283	3-12-00-00822	H_2O 2; EtOH 4; eth 1; chl 4
7147	Hydrazinecarboximidamide Aminoguanidine	CH_6N_4 74.09	79-17-4 dec	453	4-03-00-00236	H_2O 4; EtOH 4
7148	Hydrazinecarboxylic acid, ethyl ester	$C_3H_8N_2O_2$ 104.11	4114-31-2 46	198 dec; 93[9]	4-03-00-00174	EtOH 3; eth 3; chl 2
7149	Hydrazinecarboxylic acid, methyl ester	$C_2H_6N_2O_2$ 90.08	6294-89-9 73	108[12]	2-03-00-00078	H_2O 3; EtOH 3; bz 2; peth 1
7150	1,1-Hydrazinedicarboxylic acid, diethyl ester	$C_6H_{12}N_2O_4$ 176.17	5311-96-6 30.5	138[12]	2-03-00-00079	EtOH 4
7151	1,2-Hydrazinedicarboxylic acid, diethyl ester	$C_6H_{12}N_2O_4$ 176.17	4114-28-7 135	250 dec	4-03-00-00175 1.324[8]	eth 4; EtOH 4
7152	Hydrazine, 1,1-diethyl- 1,1-Diethylhydrazine	$C_4H_{12}N_2$ 88.15	616-40-0	99	4-04-00-03335 0.8804[20]	bz 4; eth 4; EtOH 4; chl 4 1.4214[20]
7153	Hydrazine, 1,2-diethyl- 1,2-Diethylhydrazine	$C_4H_{12}N_2$ 88.15	1615-80-1	85.5	4-04-00-03336 0.797[26]	bz 4; eth 4; EtOH 4 1.4204[20]
7154	Hydrazine, 1,1-dimethyl- 1,1-Dimethylhydrazine	$C_2H_8N_2$ 60.10	57-14-7 -58	3236 63.9	4-04-00-03322 0.791[22]	H_2O 4; EtOH 4; eth 4; MeOH 4 1.4075[22]
7155	Hydrazine, 1,2-dimethyl- 1,2-Dimethylhydrazine	$C_2H_8N_2$ 60.10	540-73-8 81	3237	4-04-00-03323 0.8274[20]	H_2O 5; EtOH 5; eth 5 1.4209[20]
7156	Hydrazine, (2,4-dinitrophenyl)- (2,4-Dinitrophenyl)hydrazine	$C_6H_6N_4O_4$ 198.14	119-26-6 194	3277	4-15-00-00380	H_2O 1; EtOH 3; eth 2; bz 2

No.	Name Synonym	Mol. Form. Mol. Wt.	CAS RN mp/°C	Merck No. bp/°C	Beil. Ref. den/g cm^{-3}	Solubility n_D
7157	Hydrazine, 1,1-diphenyl- 1,1-Diphenylhydrazine	$C_{12}H_{12}N_2$ 184.24	530-50-7 50.5	3326 220[40]	4-15-00-00055 1.190[16]	bz 4; eth 4; EtOH 4; chl 4
7158	Hydrazine, 1,2-diphenyl- N,N'-Diphenylhydrazine	$C_{12}H_{12}N_2$ 184.24	122-66-7 131		4-15-00-00056 1.158[16]	EtOH 4; bz 2; DMSO-d$_6$ 2; HOAc 1
7159	Hydrazine, 1-ethyl-1-phenyl- 1-Phenyl-1-ethylhydrazine	$C_8H_{12}N_2$ 136.20	644-21-3 237		3-15-00-00074 1.0181[21]	ace 4; bz 4; eth 4; EtOH 4 1.5711[21]
7160	Hydrazine, 1-ethyl-2-phenyl- 1-Ethyl-2-phenylhydrazine	$C_8H_{12}N_2$ 136.20	622-82-2 241; 110[14]		4-15-00-00054 1.0150[20]	bz 4; eth 4; EtOH 4; chl 4 1.5676[20]
7161	Hydrazine, methyl- Methylhydrazine	CH_6N_2 46.07	60-34-4 -52.4	6001 87.5	4-04-00-03322	H$_2$O 3; EtOH 5; eth 3; ctc 3 1.4325[20]
7162	Hydrazine, 1-(3-methylbutyl)-1-phenyl- Hydrazine, 1-isopentyl-1-phenyl	$C_{11}H_{18}N_2$ 178.28	636-10-2 236	260	0-15-00-00121 0.9588[15]	
7163	Hydrazine, 1-methyl-2-(3-methylphenyl)- Hydrazine, 1-methyl-2-(3-tolyl)	$C_8H_{12}N_2$ 136.20	116836-06-7 60		2-15-00-00229 1.0265[100]	bz 4; EtOH 4
7164	Hydrazine, (3-methylphenyl)-	$C_7H_{10}N_2$ 122.17	536-89-0 244 dec		4-15-00-01220 1.057[20]	H$_2$O 2; EtOH 4; eth 4; bz 4
7165	Hydrazine, 1-methyl-1-phenyl-	$C_7H_{10}N_2$ 122.17	618-40-6 228; 131[35]		4-15-00-00053 1.0404[20]	H$_2$O 2; EtOH 5; eth 5; bz 5 1.5691[20]
7166	Hydrazine, 1-methyl-2-phenyl- 1-Methyl-2-phenylhydrazine	$C_7H_{10}N_2$ 122.17	622-36-6 231; 112[14]		4-15-00-00054 1.0320[20]	bz 4; eth 4; EtOH 4; chl 4 1.5733[20]
7167	Hydrazine, 1-naphthalenyl-	$C_{10}H_{10}N_2$ 158.20	2243-55-2 117	203[20]	3-15-00-00728	bz 4; eth 4; EtOH 4; chl 4
7168	Hydrazine, (4-nitrophenyl)- (4-Nitrophenyl)hydrazine	$C_6H_7N_3O_2$ 153.14	100-16-3 158 dec	6545	4-15-00-00317	H$_2$O 2; EtOH 3; eth 3; bz 3
7169	Hydrazine, phenyl- Phenylhydrazine	$C_6H_8N_2$ 108.14	100-63-0 19.6	7264 243.5	4-15-00-00050 1.0986[20]	H$_2$O 3; EtOH 5; eth 5; ace 4 1.6084[10]
7170	Hydrazine, (1-phenylethyl)-	$C_8H_{12}N_2$ 136.20	65-64-5 69[0.65]		4-15-00-01267 0.9672[25]	1.5436[25]
7171	Hydrazine, phenyl-, hemihydrate	$C_{12}H_{18}N_4O$ 234.30	6152-31-4 24	120[12]	3-15-00-00067 1.0938[25]	1.6081[20]
7172	Hydrazine, phenyl-, monohydrochloride	$C_6H_9ClN_2$ 144.60	59-88-1 243-6 dec	7265 sub		H$_2$O 4; EtOH 4
7173	Hydrazine, triphenyl- Triphenylhydrazine	$C_{18}H_{16}N_2$ 260.34	606-88-2 142 dec		4-15-00-00058 0.869[70]	EtOH 2; bz 2
7174	Hydrazobenzene, 3-methyl- Hydrazine, 1-phenyl-2-(3-tolyl)	$C_{13}H_{14}N_2$ 198.27	621-25-0 61		2-15-00-00229 1.0265[100]	bz 4; EtOH 4; lig 4
7175	Hydrocinnamic acid, β-ethyl-β-methyl- Valeric acid, 3-methyl-3-phenyl	$C_{12}H_{16}O_2$ 192.26	105401-59-0 284	174[14]	4-09-00-01905 1.050[25]	H$_2$O 3 1.5197[25]
7176	Hydrocyanic acid Hydrogen cyanide	CHN 27.03	74-90-8 -13.4	4722 26	4-02-00-00050 0.6876[20]	H$_2$O 5; EtOH 5; eth 5 1.2614[20]
7177	Hydroperoxide, cyclohexyl Cyclohexyl hydroperoxide	$C_6H_{12}O_2$ 116.16	766-07-4 -20	42[0.1]	4-06-00-00053 1.019[20]	eth 4; EtOH 4; HOAc 4 1.4045[25]
7178	Hydroperoxide, 1,1-dimethylethyl tert-Butyl hydroperoxide	$C_4H_{10}O_2$ 90.12	75-91-2 6	1569 89 dec; 36[17]	4-01-00-01616 0.8960[20]	H$_2$O 3; EtOH 3; eth 3; ctc 3 1.4015[20]
7179	Hydroperoxide, ethyl Ethyl hydrogen peroxide	$C_2H_6O_2$ 62.07	3031-74-1 -100	95	4-01-00-01323 0.9332[20]	H$_2$O 4; bz 4; eth 4; EtOH 4 1.3800[20]
7180	Hydroperoxide, methyl	CH_4O_2 48.04	3031-73-0 38-40[65]		4-01-00-01249 1.9967[15]	H$_2$O 4; bz 4; eth 4; EtOH 4 1.3641[15]
7181	Hydroperoxide, 1-methyl-1-phenylethyl Cumene hydroperoxide	$C_9H_{12}O_2$ 152.19	80-15-9		4-06-00-03221	
7182	Hydroprene 2,4-Dodecadienoic acid, 3,7,11-trimethyl-, ethyl ester, (E,E)-	$C_{17}H_{30}O_2$ 266.42	41096-46-2 174[19]		0.8955[20]	
7183	Hypochlorous acid, 1,1-dimethylethyl ester tert-Butyl hypochlorite	C_4H_9ClO 108.57	507-40-4 77.5	1570	4-01-00-01621 0.9583[18]	H$_2$O 1; eth 4; ace 3; bz 4 1.403[20]
7184	Hypochlorous acid, ethyl ester Ethyl hypochlorite	C_2H_5ClO 80.51	624-85-1 37		4-01-00-01324 1.013[-6]	H$_2$O 1; EtOH 3; eth 5; bz 5
7185	Hypophosphoric acid, tetraethyl ester	$C_8H_{20}O_6P_2$ 274.19	679-37-8 116-7[2]		4-01-00-01358 1.1283[18]	bz 3; ctc 3 1.4284[20]
7186	Ibogamine, 12-methoxy- Ibogaine	$C_{20}H_{26}N_2O$ 310.44	83-74-9 148	4806	5-23-12-00284	chl 3
7187	Imazalil 1H-Imidazole, 1-[2-(2,4-dichlorophenyl)-2-(2- propenyloxy)ethyl]-	$C_{14}H_{14}Cl_2N_2O$ 297.18	35554-44-0 50	3537 dec	1.243[23]	
7188	Imazapyr 2-(4-Isopropyl-4-methyl-5-oxo-2-imidazolin-2- yl)nicotinic acid	$C_{13}H_{15}N_3O_3$ 261.28	81334-34-1 171			
7189	Imazaquin (RS)-2-(4-Isopropyl-4-methyl-5-oxo-2- imidazolin-2-yl)quinoline-3-carboxylic acid	$C_{17}H_{17}N_3O_3$ 311.34	81335-37-7 221	4826		
7190	Imazethapyr (RS)-5-Ethyl-2-(4-isopropyl-4-methyl-5-oxo-2- imidazolin-2-yl)nicotinic acid	$C_{15}H_{19}N_3O_3$ 289.33	81335-77-5 173			
7191	Imidazo[4,5-d]imidazole-2,5(1H,3H)-dione, tetrahydro- Acetyleneurea	$C_4H_6N_4O_2$ 142.12	496-46-8 300 dec	87	2-26-00-00260	H$_2$O 2; EtOH 1; eth 3; HCl 3
7192	1H-Imidazole 1,3-Diazole	$C_3H_4N_2$ 68.08	288-32-4 90.5	4828 257	5-23-04-00191 1.0303[101]	H$_2$O 4; EtOH 4; eth 3; ace 3 1.4801[101]
7193	1H-Imidazole, 1,1'-carbonylbis- N,N'-Carbonyldiimidazole	$C_7H_6N_4O$ 162.15	530-62-1 119	1825	5-23-04-00245	

No.	Name / Synonym	Mol. Form. / Mol. Wt.	CAS RN / mp/°C	Merck No. / bp/°C	Beil. Ref. / den/g cm^{-3}	Solubility / n_D
7194	1H-Imidazole-1-carboxylic acid, 2,3-dihydro-3-methyl-2-thioxo-, ethyl ester Carbimazole	$C_7H_{10}N_2O_2S$ 186.23	22232-54-8 123.5	1803	4-24-00-00064	ace 4; chl 4
7195	1H-Imidazole-4,5-dicarboxylic acid	$C_5H_4N_2O_4$ 156.10	570-22-9 290 dec		4-25-00-01051 1.749[25]	H_2O 2; EtOH 1; eth 1; bz 1
7196	1H-Imidazole, 1,2-diethyl- 1,2-Diethylimidazole	$C_7H_{12}N_2$ 124.19	51807-53-5	219	5-23-05-00165 0.9813[25]	H_2O 4; eth 4; EtOH 4
7197	1H-Imidazole, 4,5-dihydro-2-methyl- Lysidine	$C_4H_8N_2$ 84.12	534-26-9 107	5508 196.5	5-23-03-00385	H_2O 4; EtOH 4; eth 1; chl 3
7198	1H-Imidazole, 4,5-dihydro-2-(phenylmethyl)- 2-Imidazolidine, 2-benzyl	$C_{10}H_{12}N_2$ 160.22	59-98-3	9430	5-23-06-00488	
7199	1H-Imidazole, 4,5-dihydro-2,4,5-triphenyl-, cis-	$C_{21}H_{18}N_2$ 298.39	573-33-1 134.5	198 dec	5-23-10-00053	bz 4; eth 4; EtOH 4
7200	1H-Imidazole, 1,2-dimethyl-	$C_5H_8N_2$ 96.13	1739-84-0	206	5-23-05-00039 1.0051[11]	H_2O 4; eth 4; EtOH 4
7201	1H-Imidazole, 1,4-dimethyl- 1,4-Dimethylimidazole	$C_5H_8N_2$ 96.13	6338-45-0 25.8	199	5-23-05-00094 0.9960[16]	
7202	1H-Imidazole, 1,2-dimethyl-5-nitro- Dimetridazole	$C_5H_7N_3O_2$ 141.13	551-92-8 138.5	3255	5-23-05-00058	eth 4; EtOH 4
7203	1H-Imidazole-4-ethanamine Histamine	$C_5H_9N_3$ 111.15	51-45-6 86	4640 209[18]	4-25-00-02049	H_2O 3; EtOH 3; eth 2; chl 3
7204	1H-Imidazole-4-ethanaminium, α-carboxy-2,3-dihydro-N,N,N-trimethyl-2-thioxo-, hydroxide, in Ergothioneine	$C_9H_{15}N_3O_2S$ 229.30	497-30-3 290 dec	3611	4-25-00-04487	H_2O 4; EtOH 2; eth 1; ace 2
7205	1H-Imidazole-1-ethanol	$C_5H_8N_2O$ 112.13	1615-14-1 39	203[20]	5-23-04-00309	H_2O 4; EtOH 4
7206	1H-Imidazole-1-ethanol, 2-methyl-5-nitro- Metronidazole	$C_6H_9N_3O_3$ 171.16	443-48-1 160.5	6079	5-23-05-00063	
7207	1H-Imidazole, 1-ethyl-	$C_5H_8N_2$ 96.13	7098-07-9	208	5-23-04-00261 0.999[25]	H_2O 5
7208	1H-Imidazole, 1-methyl-	$C_4H_6N_2$ 82.11	616-47-7 -6	195.5	5-23-04-00256 1.0325[20]	H_2O 4; ace 4; eth 4; EtOH 4 1.4970[20]
7209	1H-Imidazole, 2-methyl-	$C_4H_6N_2$ 82.11	693-98-1 144	267	5-23-05-00035	H_2O 4; EtOH 4
7210	1H-Imidazole, 4-methyl-	$C_4H_6N_2$ 82.11	822-36-6 56	263	5-23-05-00089 1.0416[14]	H_2O 4; EtOH 4 1.503[14]
7211	1H-Imidazole, 2-nitro- Azomycin	$C_3H_3N_3O_2$ 113.08	527-73-1 287 dec	934	5-23-04-00471	
7212	1H-Imidazole, 1-phenyl-	$C_9H_8N_2$ 144.18	7164-98-9 13	276	5-23-04-00267 1.1397[15]	H_2O 1; eth 4; ace 4; chl 4 1.6025[25]
7213	1H-Imidazole, 2-phenyl-	$C_9H_8N_2$ 144.18	670-96-2 149.3	340	5-23-07-00194	EtOH 4
7214	2H-Imidazole-2-thione, 1,3-dihydro-1-methyl- Methimazole	$C_4H_6N_2S$ 114.17	60-56-0 146	5892 280 dec	5-24-01-00305	H_2O 4; EtOH 3; eth 2; bz 2
7215	1H-Imidazole, 2,4,5-triphenyl-	$C_{21}H_{16}N_2$ 296.37	484-47-9 275	sub	5-23-10-00107	H_2O 1; EtOH 3; eth 3
7216	2,4-Imidazolidinedione Hydantoin	$C_3H_4N_2O_2$ 100.08	461-72-3 220	4678	5-24-05-00188	H_2O 4; EtOH 3; eth 2; alk 3
7217	2,4-Imidazolidinedione, 1,3-dibromo-5,5-dimethyl- Dibromantine	$C_5H_6Br_2N_2O_2$ 285.92	77-48-5 197-9 dec		5-24-05-00374	
7218	2,4-Imidazolidinedione, 5,5-dimethyl-	$C_5H_8N_2O_2$ 128.13	77-71-4 178	sub	5-24-05-00348	H_2O 4; EtOH 4; eth 4; ace 4
7219	2,4-Imidazolidinedione, 5,5-diphenyl- Phenytoin	$C_{15}H_{12}N_2O_2$ 252.27	57-41-0 286	7293	5-24-08-00376	H_2O 1; EtOH 3; eth 2; ace 3
7220	2,4-Imidazolidinedione, 5-(1-ethylpentyl)-3-[(trichloromethyl)thio]- Chlordantoin	$C_{11}H_{17}Cl_3N_2O_2S$ 347.69	5588-20-5	2080		CS_2 3
7221	2,4-Imidazolidinedione, 3-methyl-5-phenyl- 3-Methyl-5-phenylhydantoin	$C_{10}H_{10}N_2O_2$ 190.20	6846-11-3 164.5	6026	5-24-08-00005	chl 3
7222	2,4-Imidazolidinedione, 1-[[(5-nitro-2-furanyl)methylene]amino]- Nitrofurantoin	$C_8H_6N_4O_5$ 238.16	67-20-9 263	6520	5-24-05-00224	
7223	4-Imidazolidinehexanoic acid, 5-methyl-2-oxo-, (4R-cis)- Desthiobiotin	$C_{10}H_{18}N_2O_3$ 214.26	533-48-2 157	2914	4-25-00-01543	H_2O 3
7224	4-Imidazolidinehexanoic acid, 5-methyl-2-oxo-, methyl ester, (4R-cis)- Desthiobiotin, methyl ester	$C_{11}H_{20}N_2O_3$ 228.29	6020-51-5 69.5	194-7[0.03]	4-25-00-01544	EtOH 4
7225	2-Imidazolidinethione Ethylenethiourea	$C_3H_8N_2S$ 102.16	96-45-7 203	3759	5-24-01-00165	H_2O 4; EtOH 3; eth 1; bz 1
7226	Imidazolidinetrione Parabanic acid	$C_3H_2N_2O_3$ 114.06	120-89-8 244 dec	6970 sub 100	5-24-09-00009	H_2O 3; EtOH 4
7227	2-Imidazolidinone Ethyleneurea	$C_3H_6N_2O$ 86.09	120-93-4 131	4830	5-24-01-00022	H_2O 4; EtOH 4; eth 2; chl 2
7228	4-Imidazolidinone, 5,5-diphenyl- Doxenitoin	$C_{15}H_{14}N_2O$ 238.29	3254-93-1 183	3424	5-24-04-00145	
7229	4-Imidazolidinone, 5-(2-methylpropyl)-3-(2-propenyl)-2-thioxo- Albutoin	$C_{10}H_{16}N_2OS$ 212.32	830-89-7 210.5	210	4-24-00-01139	

No.	Name Synonym	Mol. Form. Mol. Wt.	CAS RN mp/°C	Merck No. bp/°C	Beil. Ref. den/g cm^{-3}	Solubility n_D
7230	4H-Imidazol-4-one, 2-amino-1,5-dihydro-1-methyl- Creatinine	C$_4$H$_7$N$_3$O 113.12	60-27-5 300 dec	2571	4-25-00-03543	H$_2$O 3; EtOH 2; eth 1; ace 1
7231	Imidodicarbonic acid, diethyl ester	C$_6$H$_{11}$NO$_4$ 161.16	19617-44-8 50	226	3-03-00-00050	H$_2$O 4
7232	Imidodicarbonic diamide Biuret	C$_2$H$_5$N$_3$O$_2$ 103.08	108-19-0 190 dec	1319	4-03-00-00141	H$_2$O 2; EtOH 4; eth 1
7233	Imidodicarbonimidic diamide Biguanide	C$_2$H$_7$N$_5$ 101.11	56-03-1 136	1233 142 dec	4-03-00-00162	H$_2$O 4; EtOH 3; bz 1; chl 1
7234	Imidodicarbonimidic diamide, N-phenyl- Phenyl biguanide	C$_8$H$_{11}$N$_5$ 177.21	102-02-3 143	7247	4-12-00-00771	
7235	1H-Indazole 1H-Benzopyrazole	C$_7$H$_6$N$_2$ 118.14	271-44-3 148	4848 269	5-23-06-00156	H$_2$O 3; EtOH 3; eth 3
7236	1H-Indazole, 3-chloro- 3-Chloroindazole	C$_7$H$_5$ClN$_2$ 152.58	29110-74-5 149	sub	5-23-06-00173	H$_2$O 3; EtOH 4; eth 4; bz 4
7237	1H-Indazole, 1,5-dimethyl- 1,5-Dimethylindazole	C$_9$H$_{10}$N$_2$ 146.19	70127-93-4 62.5	260	2-23-00-00157	ace 4; eth 4; EtOH 4; MeOH 4
7238	2H-Indazole, 2,3-dimethyl- 2,3-Dimethylindazole	C$_9$H$_{10}$N$_2$ 146.19	50407-18-6 79.5	286	5-23-06-00317	eth 4; EtOH 4
7239	1H-Indazole, 1-methyl- 1-Methylindazole	C$_8$H$_8$N$_2$ 132.17	13436-48-1 60.5	231	5-23-06-00159 1.0315^{99}	H$_2$O 1; EtOH 3; ace 3; chl 3
7240	1H-Indazole, 3-methyl- 3-Methylindazole	C$_8$H$_8$N$_2$ 132.17	3176-62-3 113	280.5	5-23-06-00312	H$_2$O 3; EtOH 3; eth 3; chl 3
7241	1H-Indazole, 5-methyl- 5-Methylindazole	C$_8$H$_8$N$_2$ 132.17	1776-37-0 117	294	5-23-06-00319	H$_2$O 3; EtOH 3; eth 3; chl 3
7242	2H-Indazole, 2-methyl- 2-Methyl-2H-indazole	C$_8$H$_8$N$_2$ 132.17	4838-00-0 56	261	5-23-06-00165 1.0450^{99}	H$_2$O 1; EtOH 3; eth 3; ace 3
7243	1H-Inden-1-amine, 2,3-dihydro-, (±)- (RS)-1-Aminoindan	C$_9$H$_{11}$N 133.19	61949-83-5 221; 96^8		3-12-00-02798 1.038^{15}	H$_2$O 2; eth 3; ace 3; bz 3 1.5613^{20}
7244	1H-Inden-5-amine, 2,3-dihydro-	C$_9$H$_{11}$N 133.19	24425-40-9 37.5	248; 131^{15}	4-12-00-02929	H$_2$O 2; eth 3; ace 3; bz 3
7245	1H-Indene Indonaphthene	C$_9$H$_8$ 116.16	95-13-6 -1.8	4851 182	4-05-00-01532 0.9960^{25}	H$_2$O 1; EtOH 5; eth 5; ace 3 1.5768^{20}
7246	1H-Indene-1-carboxylic acid 1-Indenecarboxylic acid	C$_{10}$H$_8$O$_2$ 160.17	5020-21-3 74	193-5^{12}	3-09-00-03068	H$_2$O 2; EtOH 4; eth 2; bz 2
7247	1H-Indene, 1,2-dibromo-2,3-dihydro- 1,2-Dibromoindan	C$_9$H$_8$Br$_2$ 275.97	20357-79-3 32	140^{11}	2-05-00-00377 1.7470^{25}	
7248	2H-Indene-2,2-dicarboxylic acid, 1,3-dihydro-1,3-dioxo-, diethyl ester Phthaloylmalonic acid, diethyl ester	C$_{15}$H$_{14}$O$_6$ 290.27	116836-20-5		1-10-00-00441 1.1896^{83}	eth 4; EtOH 4 1.541^{84}
7249	1H-Indene, 1,2-dichloro-2,3-dihydro- 1,2-Dichloroindan	C$_9$H$_8$Cl$_2$ 187.07	74925-48-7 87 90^2		3-05-00-01202 1.264^{25}	1.5715^{23}
7250	1H-Indene, 2,3-dihydro- Indan	C$_9$H$_{10}$ 118.18	496-11-7 -51.4	4844 177.9	4-05-00-01371 0.9639^{20}	H$_2$O 1; EtOH 5; eth 5; chl 2 1.5378^{20}
7251	1H-Indene, 2,3-dihydro-1,1-dimethyl- 1,1-Dimethylindan	C$_{11}$H$_{14}$ 146.23	4912-92-9	191	4-05-00-01415 0.919^{20}	1.5135^{25}
7252	1H-Indene, 2,3-dihydro-1,2-dimethyl- 1,2-Dimethylindan	C$_{11}$H$_{14}$ 146.23	17057-82-8 79-80^{10}		4-05-00-01415 0.927^{20}	1.5186^{20}
7253	1H-Indene, 2,3-dihydro-4,7-dimethyl- 4,7-Dimethylindan	C$_{11}$H$_{14}$ 146.23	6682-71-9 94-7^{10}		4-05-00-01416 0.949^{20}	1.5342^{20}
7254	1H-Indene, 2,3-dihydro-5,6-dimethyl- 5,6-Dimethylindan	C$_{11}$H$_{14}$ 146.23	1075-22-5 94^{10}		4-05-00-01416 0.9449^{20}	1.5360^{20}
7255	1H-Indene, 2,3-dihydro-1-methyl- 1-Methylindan	C$_{10}$H$_{12}$ 132.21	767-58-8 190.6		4-05-00-01397 0.938^{25}	H$_2$O 1 1.5266^{20}
7256	1H-Indene, 2,3-dihydro-2-methyl- 2-Methylindan	C$_{10}$H$_{12}$ 132.21	824-63-5 195		4-05-00-01397 0.940^{25}	H$_2$O 1 1.5220^{20}
7257	1H-Indene, 2,3-dihydro-4-methyl- 4-Methylindan	C$_{10}$H$_{12}$ 132.21	824-22-6 205.3		4-05-00-01397 0.9577^{20}	1.5356^{20}
7258	1H-Indene, 2,3-dihydro-5-methyl- 5-Methylindan	C$_{10}$H$_{12}$ 132.21	874-35-1 202		4-05-00-01398 0.944^{25}	1.5336^{20}
7259	1H-Indene, 2,3-dihydro-2-phenyl- 2-Phenylindan	C$_{15}$H$_{14}$ 194.28	22253-11-8 162-3^{10}		4-05-00-02198 1.082^{16}	1.5955^{15}
7260	1H-Indene, 2,3-dihydro-1,1,4,7-tetramethyl- 1,1,4,7-Tetramethylindan	C$_{13}$H$_{18}$ 174.29	1078-04-2 114.2^{15}		3-05-00-01282 0.934^{25}	1.5216^{25}
7261	1H-Indene, 2,3-dihydro-1,1,5-trimethyl- 1,1,5-Trimethylindan	C$_{12}$H$_{16}$ 160.26	40650-41-7 86^{10}		4-05-00-01433 0.9119^{20}	1.5126^{20}
7262	1H-Indene, 2,3-dihydro-1,4,7-trimethyl- 1,4,7-Trimethylindan	C$_{12}$H$_{16}$ 160.26	54340-87-3 95^{10}		4-05-00-01434 0.938^{20}	1.5252^{20}
7263	1H-Indene, 2,3-dihydro-1,1,3-trimethyl-3-phenyl- Indan, 1-phenyl-1,3,3-trimethyl	C$_{18}$H$_{20}$ 236.36	3910-35-8 52.5	308.5	4-05-00-02246 1.0009^{20}	H$_2$O 1; EtOH 3; bz 3; MeOH 3 1.5681^{20}
7264	1H-Indene-1,3(2H)-dione	C$_9$H$_6$O$_2$ 146.15	606-23-5 131-2 dec		4-07-00-02344 1.37^{21}	H$_2$O 2; EtOH 4; eth 3; bz 3
7265	1H-Indene-1,3(2H)-dione, 2-(4-bromophenyl)- Bromindione	C$_{15}$H$_9$BrO$_2$ 301.14	1146-98-1 138	1381	4-07-00-02572	
7266	1H-Indene-1,3(2H)-dione, 2-(4-chlorophenyl)- Clorindione	C$_{15}$H$_9$ClO$_2$ 256.69	1146-99-2 145.5	2402	4-07-00-02571	bz 4; eth 4; EtOH 4
7267	1H-Indene-1,3(2H)-dione, 2,2-dihydroxy- Ninhydrin	C$_9$H$_6$O$_4$ 178.14	485-47-2 241-3 dec	6470	4-07-00-02786	H$_2$O 4; EtOH 3; eth 2; alk 3

No.	Name / Synonym	Mol. Form. / Mol. Wt.	CAS RN / mp/°C	Merck No. / bp/°C	Beil. Ref. / den/g cm^{-3}	Solubility / n_D
7268	1H-Indene-1,3(2H)-dione, 2-(diphenylacetyl)- / Diphenadione	$C_{23}H_{16}O_3$ / 340.38	82-66-6 / 146.5	3304	4-07-00-02838	ace 4; HOAc 4 / 1.670
7269	1H-Indene-1,3(2H)-dione, 2-(4-methoxyphenyl)- / Anisindione	$C_{16}H_{12}O_3$ / 252.27	117-37-3 / 156.5	698	3-08-00-02931	
7270	1H-Indene-1,3(2H)-dione, 2-phenyl- / Phenindione	$C_{15}H_{10}O_2$ / 222.24	83-12-5 / 150	7196	4-07-00-02570	H_2O 1; EtOH 3; eth 3; ace 3
7271	1H-Indene, 1,3-diphenyl- / 1,3-Diphenylindene	$C_{21}H_{16}$ / 268.36	4467-88-3 / 85	230[15]	3-05-00-02474	ace 4; eth 4
7272	1H-Indene, 2,3-diphenyl- / 2,3-Diphenylindene	$C_{21}H_{16}$ / 268.36	5324-00-5 / 109.3	237[12]	4-05-00-02641	H_2O 1; EtOH 3; eth 3; ace 3
7273	1H-Indene, 1-ethyl-2,3-dihydro- / 1-Ethylindan	$C_{11}H_{14}$ / 146.23	4830-99-3 /	212	4-05-00-01414 / 0.9348[25]	1.5121[25]
7274	1H-Indene, 5-hexyl-2,3-dihydro- / 5-Hexylindan	$C_{15}H_{22}$ / 202.34	54889-55-3 / -34	292.1	4-05-00-01474 / 0.9114[20]	1.5122[20]
7275	1H-Indene, 1-methyl- / 1-Methylindene	$C_{10}H_{10}$ / 130.19	767-59-9 /	199; 82[15]	3-05-00-01371 / 0.970[25]	1.5616[20]
7276	1H-Indene, 2-methyl- / 2-Methylindene	$C_{10}H_{10}$ / 130.19	2177-47-1 / 80	208	4-05-00-01545 / 0.974[25]	H_2O 1; eth 3; ace 3; bz 3 / 1.5652[20]
7277	1H-Indene, 3-methyl- / 3-Methylindene	$C_{10}H_{10}$ / 130.19	767-60-2 /	198	4-05-00-01545 / 0.972[25]	H_2O 1; eth 3; ace 3; bz 3 / 1.5621[20]
7278	1H-Indene, 4-methyl- / 4-Methylindene	$C_{10}H_{10}$ / 130.19	7344-34-5 /	209; 88[13]	2-05-00-00418 / 0.989[25]	1.568[20]
7279	1H-Indene, 6-methyl- / 6-Methylindene	$C_{10}H_{10}$ / 130.19	20232-11-5 /	207; 90[17]	4-05-00-01545 / 0.977[25]	1.566[20]
7280	1H-Indene, 7-methyl- / 7-Methylindene	$C_{10}H_{10}$ / 130.19	7372-92-1 /	209; 90[15]	2-05-00-00418 / 0.989[25]	1.568[20]
7281	1H-Indene, octahydro- / Hexahydroindan	C_9H_{16} / 124.23	496-10-6 / -53	167	4-05-00-00292 / 0.876[25]	1.4702[20]
7282	6H-Indeno[1,2-b]oxirene, 1a,6a-dihydro-	C_9H_8O / 132.16	768-22-9 / 24.5	113[20]; 98[6]	5-17-02-00033 / 1.1255[24]	chl 3
7283	1H-Inden-4-ol, 7-chloro-2,3-dihydro- / Chlorindanol	C_9H_9ClO / 168.62	145-94-8 / 92	2094	4-06-00-03828	
7284	1H-Inden-1-ol, 2,3-dihydro-	$C_9H_{10}O$ / 134.18	6351-10-6 / 54.8	220; 128[12]	4-06-00-03824	bz 4; EtOH 4; chl 4
7285	1H-Inden-5-ol, 2,3-dihydro-	$C_9H_{10}O$ / 134.18	1470-94-6 / 58	253	4-06-00-03829	H_2O 2; EtOH 4; eth 4; sulf 3
7286	1H-Inden-1-one, 2,3-dihydro-	C_9H_8O / 132.16	83-33-0 / 42	243; 129[12]	4-07-00-00999 / 1.0943[40]	H_2O 2; EtOH 4; eth 4; ace 4 / 1.561[25]
7287	2H-Inden-2-one, 1,3-dihydro- / 2-Indanone	C_9H_8O / 132.16	615-13-4 / 59	218 dec	4-07-00-01002 / 1.0712[69]	H_2O 1; EtOH 4; eth 4; ace 4 / 1.538[67]
7288	1H-Inden-1-one, 2,3-dihydro-3,3-dimethyl-	$C_{11}H_{12}O$ / 160.22	26465-81-6 /	130-1[18]	4-07-00-01035 / 1.0320[14]	1.5453[14]
7289	1H-Inden-1-one, 2,3-dihydro-2-methyl-	$C_{10}H_{10}O$ / 146.19	17496-14-9 /	250; 125[18]	4-07-00-01019 / 1.0651[21]	1.553[22]
7290	1H-Inden-1-one, 2,3-dihydro-4-methyl- / 4-Methyl-1-indanone	$C_{10}H_{10}O$ / 146.19	24644-78-8 / 95	144[13]; 120[5]	4-07-00-01020	bz 4; eth 4; EtOH 4
7291	1H-Inden-1-one, 2,3-dihydro-2-phenyl- / 2-Phenyl-1-indanone	$C_{15}H_{12}O$ / 208.26	16619-12-8 / 77.5	344	4-07-00-01673	ace 4; bz 4; EtOH 4
7292	2H-Inden-2-one, octahydro-, cis-	$C_9H_{14}O$ / 138.21	5689-04-3 / 10	225; 113[25]	4-07-00-00174 / 0.9970[20]	EtOH 3; bz 3; lig 3 / 1.4830[20]
7293	2H-Inden-2-one, octahydro-, trans-	$C_9H_{14}O$ / 138.21	16484-17-6 / -12	218	4-07-00-00174 / 0.9807[17]	EtOH 3; bz 3; lig 3 / 1.4769[17]
7294	1H-Indole / 2,3-Benzopyrrole	C_8H_7N / 117.15	120-72-9 / 52.5	4869 / 254	5-20-07-00005 / 1.22[25]	H_2O 3; EtOH 4; eth 4; bz 3
7295	1H-Indole-3-acetic acid / Indoleacetic acid	$C_{10}H_9NO_2$ / 175.19	87-51-4 / 168.5	4870	5-22-03-00065	H_2O 1; EtOH 4; eth 3; ace 3
7296	1H-Indole-3-butanoic acid / Indolebutyric acid	$C_{12}H_{13}NO_2$ / 203.24	133-32-4 / 124.5	4871	5-22-03-00140	bz 4; DMSO 3; peth 1
7297	1H-Indole, 2,3-dihydro-	C_8H_9N / 119.17	496-15-1 /	229	5-20-06-00238 / 1.069[20]	H_2O 2; eth 3; ace 3; bz 3 / 1.5923[20]
7298	1H-Indole, 1,3-dimethyl-	$C_{10}H_{11}N$ / 145.20	875-30-9 / 142	258.5	5-20-07-00071	eth 3
7299	1H-Indole, 2,5-dimethyl- / 2,5-Dimethylindole	$C_{10}H_{11}N$ / 145.20	1196-79-8 / 115.3	188[40]	5-20-07-00111	H_2O 1; EtOH 2; eth 3; bz 3
7300	1H-Indole-2,3-dione / Isatin	$C_8H_5NO_2$ / 147.13	91-56-5 / 203 dec	4985	5-21-10-00221	H_2O 3; EtOH 4; eth 2; ace 3
7301	1H-Indole-5,6-dione, 2,3-dihydro-3-hydroxy-1-methyl- / Adrenochrome	$C_9H_9NO_3$ / 179.18	54-06-8 / 125 dec	162	5-21-13-00079	H_2O 4; EtOH 4; eth 1; bz 1
7302	1H-Indole-3-ethanamine / Tryptamine	$C_{10}H_{12}N_2$ / 160.22	61-54-1 / 118	9705 / 137[0.15]	5-22-10-00045	H_2O 1; EtOH 3; eth 1; ace 3
7303	1H-Indole-3-ethanamine, N,N-dimethyl- / N,N-Dimethyltryptamine	$C_{12}H_{16}N_2$ / 188.27	61-50-7 / 46	3251	5-22-10-00048	
7304	1H-Indole-3-ethanamine, 5-methoxy- / 5-Methoxytryptamine	$C_{11}H_{14}N_2O$ / 190.25	608-07-1 / 121.5	5926	5-22-12-00018	
7305	1H-Indole-3-ethanol / Tryptophol	$C_{10}H_{11}NO$ / 161.20	526-55-6 / 59	9708 / 174[2]	5-21-03-00061	ace 4; eth 4; EtOH 4; chl 4
7306	1H-Indole-3-ethanol, 2-(5-ethenyl-1-azabicyclo[2.2.2]oct-2-yl)- / Cinchonamine	$C_{19}H_{24}N_2O$ / 296.41	482-28-0 / 186	2287	4-23-00-02735	H_2O 1; EtOH 4; eth 4; bz 3

No.	Name / Synonym	Mol. Form. / Mol. Wt.	CAS RN / mp/°C	Merck No. / bp/°C	Beil. Ref. / den/g cm^{-3}	Solubility / n_D
7307	1H-Indole, 1-ethyl-	$C_{10}H_{11}N$ 145.20	10604-59-8 105	252.5; 122[14]	5-20-07-00014 1.2563[15]	
7308	1H-Indole, 3-ethyl-	$C_{10}H_{11}N$ 145.20	1484-19-1 37	279; 138[6]	5-20-07-00099	bz 4; eth 4; EtOH 4
7309	1H-Indole, 3-ethyl-2-methyl-	$C_{11}H_{13}N$ 159.23	35246-18-5 44.5	292; 156[12]	5-20-07-00136 1.0414[0]	eth 4; EtOH 4
7310	1H-Indole-3-methanamine, N,N-dimethyl- Gramine	$C_{11}H_{14}N_2$ 174.25	87-52-5 138.5	4440	5-22-10-00025	H$_2$O 1; EtOH 3; eth 3; chl 3
7311	1H-Indole, 7-methoxy-	C_9H_9NO 147.18	3189-22-8	108-10[0.3]	5-21-03-00024 1.126[25]	1.6120[20]
7312	1H-Indole, 1-methyl-	C_9H_9N 131.18	603-76-9	237	5-20-07-00012 1.0707[25]	H$_2$O 1; EtOH 3; eth 3; bz 3
7313	1H-Indole, 2-methyl-	C_9H_9N 131.18	95-20-5 61	272	5-20-07-00059 1.07[20]	H$_2$O 2; EtOH 4; eth 4; ace 3
7314	1H-Indole, 3-methyl- Skatole	C_9H_9N 131.18	83-34-1 97.5	8503 266	5-20-07-00069	H$_2$O 3; EtOH 3; eth 3; ace 3
7315	1H-Indole, 4-methyl-	C_9H_9N 131.18	16096-32-5 5	267	5-20-07-00079 1.062[20]	
7316	1H-Indole, 5-methyl-	C_9H_9N 131.18	614-96-0 60	267	5-20-07-00080 1.0202[78]	H$_2$O 3; EtOH 3; eth 3; bz 3
7317	1H-Indole, 7-methyl- 7-Methylindole	C_9H_9N 131.18	933-67-5 85	266	5-20-07-00081 1.0202[100]	
7318	1H-Indole, 3-methyl-2-phenyl-	$C_{15}H_{13}N$ 207.27	10257-92-8 91.5	280-90[120]	5-20-08-00274	H$_2$O 1; EtOH 3; bz 3
7319	1H-Indole, octahydro-	$C_8H_{15}N$ 125.21	4375-14-8 143	185.5	5-20-04-00358 0.9472[20]	bz 4; eth 4; EtOH 4 1.4892[20]
7320	1H-Indole, 2-phenyl-	$C_{14}H_{11}N$ 193.25	948-65-2 190.5	250[10]	5-20-08-00232	H$_2$O 2; eth 3; bz 3; chl 3
7321	1H-Indole, 1-(phenylmethyl)-	$C_{15}H_{13}N$ 207.27	3377-71-7 45.8	172[2]	5-20-07-00016	bz 4; eth 4; EtOH 4; lig 4
7322	1H-Indole, 1,2,3-trimethyl- 1,2,3-Trimethylindole	$C_{11}H_{13}N$ 159.23	1971-46-6 18	283	5-20-07-00102	H$_2$O 2; EtOH 2; eth 2; bz 2
7323	1H-Indole, 2,3,5-trimethyl- 2,3,5-Trimethylindole	$C_{11}H_{13}N$ 159.23	21296-92-4 121.5	297	5-20-07-00143	EtOH 4; chl 4; lig 4
7324	Indolizine	C_8H_7N 117.15	274-40-8 75	205	5-20-07-00047	H$_2$O 1; EtOH 3
7325	Indolizine, octahydro-	$C_8H_{15}N$ 125.21	13618-93-4	75[43]	5-20-04-00360 0.9074[10]	eth 4; EtOH 4 1.4748
7326	1H-Indol-5-ol, 3-[2-(dimethylamino)ethyl]- Bufotenine	$C_{12}H_{16}N_2O$ 204.27	487-93-4 146.5	1467 320[0.1]	5-22-12-00026	eth 4; EtOH 4
7327	2H-Indol-2-one, 1,3-dihydro-	C_8H_7NO 133.15	59-48-3 128	227[23]; 195[17]	5-21-08-00007	H$_2$O 3; EtOH 3; eth 3
7328	2H-Indol-2-one, 1,3-dihydro-1,3-dimethyl-	$C_{10}H_{11}NO$ 161.20	24438-17-3 55	273; 137[11]	5-21-08-00077	eth 4; EtOH 4
7329	2H-Indol-2-one, 1,3-dihydro-3,3-dimethyl-	$C_{10}H_{11}NO$ 161.20	19155-24-9 154.3	300	5-21-08-00113	bz 4; eth 4; EtOH 4
7330	2H-Indol-2-one, 1,3-dihydro-4,7-dimethyl- 4,7-Dimethyl-2-indolinone	$C_{10}H_{11}NO$ 161.20	59022-71-8 159	205-10[15]	4-21-00-03676	ace 4; EtOH 4; chl 4
7331	3H-Indol-3-one, 2-(1,3-dihydro-3-oxo-2H-indol-2-ylidene)-1,2-dihydro- Indigo	$C_{16}H_{10}N_2O_2$ 262.27	482-89-3 390 dec	4855 sub 300	5-24-08-00503	
7332	2H-Indol-2-one, 1-ethyl-1,3-dihydro-3-methyl- 2-Indolinone, 1-ethyl-3-methyl-	$C_{11}H_{13}NO$ 175.23	84258-49-1	283; 105[0.5]	2-21-00-00258	eth 4; EtOH 4 1.557[25]
7333	Indolo[2',3':3,4]pyrido[2,1-b]quinazolin-5(7H)-one, 8,13,13b,14-tetrahydro-14-methyl-, (S)- Evodiamine	$C_{19}H_{17}N_3O$ 303.36	518-17-2 28	3865	4-26-00-00525	
7334	Indolo[2',3':3,4]pyrido[2,1-b]quinazolin-5(7H)-one, 8,13-dihydro- Rutecarpine	$C_{18}H_{13}N_3O$ 287.32	84-26-4 259.5	8271	4-26-00-00528	EtOH 2; ace 2; bz 2
7335	Inosine Hypoxanthine riboside	$C_{10}H_{12}N_4O_5$ 268.23	58-63-9 218 dec	4880	4-26-00-02086	EtOH 4
7336	scyllo-Inositol	$C_6H_{12}O_6$ 180.16	488-59-5 353 dec		4-06-00-07920 1.659[19]	H$_2$O 2; EtOH 1; eth 1; bz 1
7337	Iodine cyanide Cyanogen iodide	CIN 152.92	506-78-5 146.7	2702 sub 45	4-03-00-00093 2.84[18]	eth 4; EtOH 4
7338	Iprodione 1-Imidazolidinecarboxamide, 3-(3,5-dichlorophenyl)-N-(1-methylethyl)-2,4-dioxo-	$C_{13}H_{13}Cl_2N_3O_3$ 330.17	36734-19-7 136	4964		
7339	Iron carbonyl Iron pentacarbonyl	C_5FeO_5 195.90	13463-40-6 -20	4980 103	1.5[20]	H$_2$O 1; EtOH 2; bz 3; ace 3 1.453[22]
7340	Isazophos Phosphorothioic acid, O-[2-chloro-l-(l-methylethyl)-lH-imidazol-4-yl] O,O-diethyl ester	$C_9H_{17}ClN_3O_3PS$ 313.74	67329-04-8	4988 170; 100[0.001]	1.22[20]	
7341	5-Isobenzofurancarboxylic acid, 1,3-dihydro-1,3-dioxo- Trimellitic anhydride	$C_9H_4O_5$ 192.13	552-30-7 162	9617 240-3[14]	5-18-08-00562	
7342	1,3-Isobenzofurandione Phthalic anhydride	$C_8H_4O_3$ 148.12	85-44-9 130.8	7346 295	5-17-11-00253 1.527[4]	H$_2$O 2; EtOH 3; eth 2; ace 3

No.	Name Synonym	Mol. Form. Mol. Wt.	CAS RN mp/°C	Merck No. bp/°C	Beil. Ref. den/g cm^{-3}	Solubility n_D
7343	1,3-Isobenzofurandione, 4,5-dichloro- Phthalic anhydride, 3,4-dichloro	C$_8$H$_2$Cl$_2$O$_3$ 217.01	56962-07-3 121	329	4-17-00-06142	EtOH 4; chl 4
7344	1,3-Isobenzofurandione, 5,6-dichloro- 4,5-Dichlorophthalic anhydride	C$_8$H$_2$Cl$_2$O$_3$ 217.01	942-06-3 188	313	5-17-11-00260	eth 4; EtOH 4; tol 4
7345	1,3-Isobenzofurandione, hexahydro- Hexahydrophthalic anhydride	C$_8$H$_{10}$O$_3$ 154.17	85-42-7 32	145^{18}	5-17-11-00076	
7346	1,3-Isobenzofurandione, 4,5,6,7-tetrachloro- Tetrachlorophthalic anhydride	C$_8$Cl$_4$O$_3$ 285.90	117-08-8 254.5	sub	5-17-11-00260 1.49^{275}	eth 2
7347	1,3-Isobenzofurandione, 4,5,6,7-tetrahydro- 3,4,5,6-Tetrahydrophthalic anhydride	C$_8$H$_8$O$_3$ 152.15	2426-02-0 74		5-17-11-00133 1.2^{105}	EtOH 3; eth 4; ace 3; chl 3
7348	1,3-Isobenzofurandione, 4,5,6,7-tetraiodo-	C$_8$I$_4$O$_3$ 651.70	632-80-4 327.5	sub	5-17-11-00266	H$_2$O 1; EtOH 1; bz 1; HOAc 2
7349	1(3H)-Isobenzofuranone	C$_8$H$_6$O$_2$ 134.13	87-41-2 75	290	5-17-10-00007 1.1636^{99}	H$_2$O 3; EtOH 4; eth 4; chl 2 1.536^{99}
7350	1(3H)-Isobenzofuranone, 3,3-bis(3,5-dibromo-4-hydroxyphenyl)- 3',3",5',5"-Tetrabromophenolphthalein	C$_{20}$H$_{10}$Br$_4$O$_4$ 633.91	76-62-0 296	9122	5-18-04-00190	H$_2$O 1; EtOH 2; eth 4; alk 3
7351	1(3H)-Isobenzofuranone, 3,3-bis(4-hydroxy-3,5-diiodophenyl)-	C$_{20}$H$_{10}$I$_4$O$_4$ 821.91	386-17-4 308		4-18-00-01949 2.0201^{22}	H$_2$O 1; EtOH 1; eth 1; chl 2
7352	1(3H)-Isobenzofuranone, 3,3-bis[4-hydroxy-2-methyl-5-(1-methylethyl)phenyl]- Thymolphthalein	C$_{28}$H$_{30}$O$_4$ 430.54	125-20-2 253	9336	5-18-04-00194	H$_2$O 1; EtOH 3; eth 3; ace 3
7353	1(3H)-Isobenzofuranone, 3,3-bis(4-hydroxy-3-methylphenyl)- o-Cresolphthalein	C$_{22}$H$_{18}$O$_4$ 346.38	596-27-0 223	2582	5-18-04-00193	EtOH 4
7354	1(3H)-Isobenzofuranone, 3,3-bis(4-hydroxyphenyl)- Phenolphthalein	C$_{20}$H$_{14}$O$_4$ 318.33	77-09-8 262.5	7208	5-18-04-00188 1.277^{32}	H$_2$O 1; EtOH 4; eth 3; ace 4
7355	1(3H)-Isobenzofuranone, 6,7-dimethoxy- Meconin	C$_{10}$H$_{10}$O$_4$ 194.19	569-31-3 102.5	5665	5-18-03-00068	H$_2$O 2; EtOH 3; eth 3; ace 3
7356	1(3H)-Isobenzofuranone, 6,7-dimethoxy-3-(5,6,7,8-tetrahydro-4-methoxy-6-methyl-1,3-dioxolo[4,5-g]isoquinolin-5 Noscapine	C$_{22}$H$_{23}$NO$_7$ 413.43	128-62-1 176	6638	4-27-00-06838	H$_2$O 1; EtOH 3; eth 2; ace 4
7357	1(3H)-Isobenzofuranone, 6,7-dimethoxy-3-(5,6,7,8-tetrahydro-6-methyl-1,3-dioxolo[4,5-g]isoquinolin-5-yl)-, [S- Hydrastine	C$_{21}$H$_{21}$NO$_6$ 383.40	118-08-1 132	4688	4-27-00-06833	H$_2$O 1; ace 3; bz 3
7358	1(3H)-Isobenzofuranone, 3,3-diphenyl-	C$_{20}$H$_{14}$O$_2$ 286.33	596-29-2 120	235^{15}	5-17-10-00735	chl 2
7359	Isofenphos Benzoic acid, 2-[[ethoxy[(1-methylethyl)amino]phosphinothioyl]oxy]-, 1-methylethyl ester	C$_{15}$H$_{24}$NO$_4$PS 345.40	25311-71-1 <-12	5055 120$^{0.01}$	1.134^{20}	
7360	Isofulminic acid	CHNO 43.03	51060-05-0 stable only at low temp.			
7361	2H-Isoindole-2-acetic acid, 1,3-dihydro-1,3-dioxo-, ethyl ester	C$_{12}$H$_{11}$NO$_4$ 233.22	6974-10-3 112.5	300	5-21-10-00431	bz 4; eth 4; EtOH 4; chl 4
7362	1H-Isoindole-1,3(2H)-dione Phthalimide	C$_8$H$_5$NO$_2$ 147.13	85-41-6 238	7347	5-21-10-00270	bz 4
7363	1H-Isoindole-1,3(2H)-dione, 2-(2,6-dioxo-3-piperidinyl)- Thalidomide	C$_{13}$H$_{10}$N$_2$O$_4$ 258.23	50-35-1 270	9182	5-22-13-00224	py 4; diox 4
7364	1H-Isoindole-1,3(2H)-dione, N-ethyl-	C$_{10}$H$_9$NO$_2$ 175.19	5022-29-7 79	285.5	5-21-10-00274	EtOH 3; eth 3
7365	1H-Isoindole-1,3(2H)-dione, 2-methyl-	C$_9$H$_7$NO$_2$ 161.16	550-44-7 134	286	5-21-10-00273	H$_2$O 1; EtOH 2
7366	1H-Isoindole-1,3(2H)-dione, 2-phenyl-	C$_{14}$H$_9$NO$_2$ 223.23	520-03-6 210	sub	5-21-10-00292	H$_2$O 1; EtOH 2; chl 5
7367	1H-Isoindole-1,3(2H)-dione, 2-(phenylmethyl)-	C$_{15}$H$_{11}$NO$_2$ 237.26	2142-01-0 116		5-21-10-00299 1.343^{18}	EtOH 3; DMSO-d$_6$ 2; HOAc 3
7368	1H-Isoindole-1,3(2H)-dione, 3a,4,7,7a-tetrahydro-2-[(trichloromethyl)thio]- Captan	C$_9$H$_8$Cl$_3$NO$_2$S 300.59	133-06-2 172.5	1771	5-21-10-00136 1.74^{25}	chl 4
7369	1H-Isoindol-1-one, 2,3-dihydro-	C$_8$H$_7$NO 133.15	480-91-1 151	338; 103^{18}	5-21-08-00018	eth 4; EtOH 4; chl 4
7370	L-Isoleucine 2-Amino-3-methylpentanoic acid	C$_6$H$_{13}$NO$_2$ 131.17	73-32-5 285-6 dec	5064	4-04-00-02775	H$_2$O 3; EtOH 1
7371	Isopropalin Benzenamine, 4-(1-methylethyl)-2,6-dinitro-N,N-dipropyl-	C$_{15}$H$_{23}$N$_3$O$_4$ 309.37	33820-53-0	5089		
7372	5-Isoquinolinamine 5-Aminoisoquinoline	C$_9$H$_8$N$_2$ 144.18	1125-60-6 128	196^{18}	5-22-10-00342	chl 2; lig 3
7373	Isoquinoline Benzopyridine	C$_9$H$_7$N 129.16	119-65-3 26.47	5110 243.2	5-20-07-00333 1.0910^{30}	H$_2$O 1; EtOH 4; eth 5; ace 4 1.6148^{20}
7374	Isoquinoline, 4-bromo- 4-Bromoisoquinoline	C$_9$H$_6$BrN 208.06	1532-97-4 41.5	282.5	5-20-07-00363	eth 4
7375	5-Isoquinolinecarbonitrile	C$_{10}$H$_6$N$_2$ 154.17	27655-41-0 137	sub 100	4-22-00-01208	eth 4; EtOH 4

No.	Name Synonym	Mol. Form. Mol. Wt.	CAS RN mp/°C	Merck No. bp/°C	Beil. Ref. den/g cm^{-3}	Solubility n_D
7376	Isoquinoline, 1-chloro- 1-Chloroisoquinoline	C_9H_6ClN 163.61	19493-44-8 37.5	274.5	5-20-07-00361	H_2O 1; bz 3
7377	Isoquinoline, 1-[(3,4-diethoxyphenyl)methyl]-6,7-diethoxy- Isoquinoline, 6,7-diethoxy-1-(3,4-diethoxybenzyl)	$C_{24}H_{29}NO_4$ 395.50	486-47-5 100	3682	5-21-06-00183	H_2O 1; EtOH 3; eth 2; chl 2
7378	Isoquinoline, 1-[(3,4-dimethoxyphenyl)methyl]-6,7-dimethoxy- Papaverine	$C_{20}H_{21}NO_4$ 339.39	58-74-2 147.5	6968 sub 135	5-21-06-00182 1.337[20]	H_2O 2; EtOH 4; eth 2; ace 3 1.625
7379	Isoquinoline, 1-[(3,4-dimethoxyphenyl)methyl]-1,2,3,4-tetrahydro-6,7-dimethoxy-2-methyl-, (S)- Laudanosine	$C_{21}H_{27}NO_4$ 357.45	2688-77-9 89	5249	5-21-06-00049	ace 4; eth 4; EtOH 4; chl 4
7380	Isoquinoline, 7-methoxy- 7-Methoxyisoquinoline	$C_{10}H_9NO$ 159.19	39989-39-4 49	194[50,] 184[34]	5-21-03-00326	H_2O 1; EtOH 3; lig 3
7381	Isoquinoline, 1-methyl- Isoquinaldine	$C_{10}H_9N$ 143.19	1721-93-3 10	248	5-20-07-00407 1.0777[20]	H_2O 2; eth 3; ace 3; bz 3 1.6095[20]
7382	Isoquinoline, 3-methyl- 3-Methylisoquinoline	$C_{10}H_9N$ 143.19	1125-80-0 68	249	5-20-07-00410	H_2O 2; eth 3; ace 3; chl 2
7383	Isoquinoline, 4-methyl- 4-Methylisoquinoline	$C_{10}H_9N$ 143.19	1196-39-0 <-75	256	5-20-07-00413	H_2O 2; eth 3; ace 3; bz 3
7384	Isoquinoline, 6-methyl- 6-Methylisoquinoline	$C_{10}H_9N$ 143.19	42398-73-2 85.5	265.5	5-20-07-00414	ace 4; bz 4; eth 4; EtOH 4
7385	Isoquinoline, 8-methyl- 8-Methylisoquinoline	$C_{10}H_9N$ 143.19	62882-00-2 205	258	5-20-07-00414	ace 4; bz 4; eth 4
7386	Isoquinoline, 5-nitro- 5-Nitroisoquinoline	$C_9H_6N_2O_2$ 174.16	607-32-9 110	sub	5-20-07-00365	H_2O 2; EtOH 4; eth 4; chl 4
7387	Isoquinoline, 3-(phenylmethyl)- Isoquinoline, 3-benzyl	$C_{16}H_{13}N$ 219.29	90210-56-3 104	311[23]	5-20-08-00383	bz 4; chl 4
7388	Isoquinoline, 1,2,3,4-tetrahydro-	$C_9H_{11}N$ 133.19	91-21-4 <-15	232.5	5-20-06-00320 1.0642[24]	H_2O 1; EtOH 3; bz 3; acid 3 1.5668[20]
7389	Isoquinoline, 5,6,7,8-tetrahydro- 5,6,7,8-Tetrahydroisoquinoline	$C_9H_{11}N$ 133.19	36556-06-6 146.5	218; 106[13]	5-20-06-00319 1.0504[10]	H_2O 1; ace 3; bz 3; chl 3 1.5276[10]
7390	Isoquinoline, 1,2,3,4-tetrahydro-6,7-dimethoxy-1,2-dimethyl-, (±)- Carnegine	$C_{13}H_{19}NO_2$ 221.30	490-53-9 170[1]	1854	5-21-04-00501	H_2O 4; eth 4; EtOH 4
7391	Isoquinoline, 1,2,3,4-tetrahydro-6,7-dimethoxy-1,2-dimethyl- Isoquinoline, 6,7-dimethoxy-1,2-dimethyl-1,2,3,4-tetrahydro	$C_{13}H_{19}NO_2$ 221.30	71783-56-7 262.5	170[1]; 100[0.01]	5-21-04-00501	H_2O 3; EtOH 3; eth 3; chl 3
7392	7-Isoquinolinol, 1-[(3,4-dimethoxyphenyl)methyl]-1,2,3,4-tetrahydro-6-methoxy-2-methyl-, (S)- Codamine	$C_{20}H_{25}NO_4$ 343.42	21040-59-5 127	2458	4-21-00-02704	eth 4; EtOH 4; chl 4
7393	8-Isoquinolinol, 1,2,3,4-tetrahydro-6,7-dimethoxy- Anhalamine	$C_{11}H_{15}NO_3$ 209.25	643-60-7 187.5	682	5-21-05-00513	eth 4; EtOH 4
7394	8-Isoquinolinol, 1,2,3,4-tetrahydro-6,7-dimethoxy-1,2-dimethyl-, (S)- Pellotine	$C_{13}H_{19}NO_3$ 237.30	83-14-7 111.5	7018	4-21-00-02525	ace 4; eth 4; EtOH 4; peth 4
7395	8-Isoquinolinol, 1,2,3,4-tetrahydro-6,7-dimethoxy-1-methyl-, (S)- Anhalonidine	$C_{12}H_{17}NO_3$ 223.27	17627-77-9 160.5	683	4-21-00-02524	H_2O 4; EtOH 4
7396	7-Isoquinolinol, 1,2,3,4-tetrahydro-1-[(4-hydroxyphenyl)methyl]-6-methoxy-, (S)- Coclaurine	$C_{17}H_{19}NO_3$ 285.34	486-39-5 220.5	2455	4-21-00-02605	
7397	6-Isoquinolinol, 1,2,3,4-tetrahydro-7-methoxy-1-methyl-, (S)- Salsoline	$C_{11}H_{15}NO_2$ 193.25	89-31-6 221.5	8308	5-21-04-00499	H_2O 2; EtOH 2; eth 1; bz 2
7398	1(2H)-Isoquinolinone, 3,4-dihydro-6,7-dimethoxy- Corydaldine	$C_{11}H_{13}NO_3$ 207.23	493-49-2 175	2542	5-21-13-00086	H_2O 4; bz 4; eth 4; EtOH 4
7399	Isothiocyanic acid	CHNS 59.09	3129-90-6			
7400	DL-Isovaline	$C_5H_{11}NO_2$ 117.15	595-39-1 316.5	sub 300	4-04-00-02655	H_2O 3; EtOH 3; eth 2; bz 1
7401	L-Isovaline 2-Amino-2-methylbutyric acid	$C_5H_{11}NO_2$ 117.15	595-40-4	5124	4-04-00-02655	
7402	Isoxazole 1-Oxa-2-azacyclopentadiene	C_3H_3NO 69.06	288-14-2 95		4-27-00-00959 1.078[20]	H_2O 3 1.4298[17]
7403	Isoxazole, 3,5-dimethyl- 3,5-Dimethylisoxazole	C_5H_7NO 97.12	300-87-8 143		4-27-00-00978 0.99[25]	1.4421[20]
7404	5-Isoxazolemethanol, 3-methyl-	$C_5H_7NO_2$ 113.12	14716-89-3 130[20]		4-27-00-01745 1.133[25]	1.4790[20]
7405	Isoxazole, 5-methyl- 5-Methylisoxazole	C_4H_5NO 83.09	5765-44-6 122		4-27-00-00966 1.023[20]	DMSO 3 1.4386[20]
7406	3-Isoxazolidinone, 4-amino-, (R)- Cycloserine	$C_3H_6N_2O_2$ 102.09	68-41-7 155-6 dec	2758	4-27-00-05549	H_2O 3; MeOH 2
7407	Isoxazolium, 2-ethyl-5-(3-sulfophenyl)-, hydroxide, inner salt Woodward's Reagent K	$C_{11}H_{11}NO_4S$ 253.28	4156-16-5 dec 207	9962		

No.	Name / Synonym	Mol. Form. / Mol. Wt.	CAS RN / mp/°C	Merck No. / bp/°C	Beil. Ref. / den/g cm^{-3}	Solubility / n_D
7408	Kepone 1,3,4-Metheno-2H-cyclobuta[cd]pentalen-2-one, 1,1a,3,3a,4,5,5,5a,5b,6-decachloro-octahydro-	$C_{10}Cl_{10}O$ 490.64	143-50-0 350 dec	2081	1.61[25]	
7409	Ketone, 1-methyl-4-piperidyl phenyl Piperidine, 4-benzoyl-N-methyl	$C_{13}H_{17}NO$ 203.28	92040-00-1 36	190[21]; 133[2]	4-21-00-03689	H_2O 3; EtOH 3; eth 4; ace 3 1.5430[23]
7410	Lactofen (±)-2-Ethoxy-1-ethyl-2-oxoethyl 5-[2-chloro-4-(trifluoromethyl)phenoxy]-2-nitrobenzoate	$C_{19}H_{15}ClF_3NO_7$ 461.78	77501-63-4			
7411	Lanosta-8,24-dien-3-ol, (3β)- Lanosterol	$C_{30}H_{50}O$ 426.73	79-63-0 140.5	5232	4-06-00-04188	eth 4; EtOH 4; chl 4
7412	DL-Leucine	$C_6H_{13}NO_2$ 131.17	328-39-2 293	sub	2-04-00-00870 1.293[18]	H_2O 3; EtOH 2; eth 1
7413	L-Leucine 2-Amino-4-methylpentanoic acid	$C_6H_{13}NO_2$ 131.17	61-90-5 293	5331 sub	4-04-00-02738 1.293[18]	H_2O 2; EtOH 1; eth 1
7414	DL-Leucine, N-glycyl-	$C_8H_{16}N_2O_3$ 188.23	688-14-2 242 dec		4-04-00-02766 1.181[25]	H_2O 4; EtOH 1
7415	Linuron Urea, N'-(3,4-dichlorophenyl)-N-methoxy-N-methyl-	$C_9H_{10}Cl_2N_2O_2$ 249.10	330-55-2 93	5387		
7416	Lup-20(29)-ene-3,28-diol, (3β)- Betulin	$C_{30}H_{50}O_2$ 442.73	473-98-3 250	1212 sub 240	4-06-00-06534	H_2O 1; EtOH 2; eth 3; bz 2
7417	Lup-20(29)-en-3-ol, (3β)- Lupeol	$C_{30}H_{50}O$ 426.73	545-47-1 216	5478	4-06-00-04200 0.9457[218]	H_2O 1; EtOH 4; eth 4; ace 4 1.4910[218]
7418	DL-Lysine 2,6-Diaminohexanoic acid	$C_6H_{14}N_2O_2$ 146.19	70-54-2 224		2-04-00-00858	H_2O 2
7419	L-Lysine 2,6-Diaminohexanoic acid	$C_6H_{14}N_2O_2$ 146.19	56-87-1 224-5 dec	5509	4-04-00-02717	H_2O 4; EtOH 1; eth 1; ace 1
7420	Maneb Manganese, [[1,2-ethanediylbis[carbamodithioato]](2-)]-	$C_4H_6MnN_2S_4$ 265.31	12427-38-2 dec 200	5603		
7421	Manganous dimethyldithiocarbamate Carbamic acid, dimethyldithio-, manganese(2+) salt	$C_6H_{12}MnN_2S_4$ 295.38	15339-36-3			
7422	D-Mannitol Cordycepic acid	$C_6H_{14}O_6$ 182.17	69-65-8 168	5629 295[3.5]	4-01-00-02841 1.489[20]	H_2O 4; EtOH 2; eth 1; py 2 1.3330
7423	D-Mannitol, 1,4:3,6-dianhydro-	$C_6H_{10}O_4$ 146.14	641-74-7 88	274 dec	5-19-03-00201	H_2O 4
7424	D-Mannitol, hexanitrate Mannitol, hexanitrate	$C_6H_8N_6O_{18}$ 452.16	15825-70-4 107	5630 exp	4-01-00-02849 1.8[20]	bz 4; eth 4; EtOH 4
7425	D-Mannose Seminose	$C_6H_{12}O_6$ 180.16	3458-28-4 132 dec	5632	4-01-00-04328 1.539[20]	H_2O 4; EtOH 2; eth 1; bz 1
7426	Matridin-15-one Matrine	$C_{15}H_{24}N_2O$ 248.37	519-02-8	5641 223[6]	5-24-02-00301	H_2O 3; EtOH 4; eth 3; ace 3 1.5286[25]
7427	Mefluidide Acetamide, N-[2,4-dimethyl-5-[[(trifluoromethyl)sulfonyl]amino]phenyl]-	$C_{11}H_{13}F_3N_2O_3S$ 310.30	53780-34-0 184	5684		
7428	m-Menthan-6-ol Cyclohexanol-4-isopropyl-2-methyl	$C_{10}H_{20}O$ 156.27	1490-05-7	119.21[28]	2-06-00-00038 0.9140[20]	eth 4; EtOH 4 1.4666[20]
7429	p-Menth-3-ene, (S)-(-)- Cyclohexene, 1-isopropyl-4-methyl (l)	$C_{10}H_{18}$ 138.25	22564-83-6 167		4-05-00-00301 0.8099[19]	H_2O 1; EtOH 3; eth 3; bz 3 1.4511[20]
7430	Mepiquat chloride Piperidinium, 1,1-dimethyl-, chloride	$C_7H_{16}NCl$ 149.66	24307-26-4 223	5746		
7431	Mercury, (acetato-O)phenyl- Phenylmercuric acetate	$C_8H_8HgO_2$ 336.74	62-38-4 153	7271	4-16-00-01720	chl 3
7432	Mercury, bis(4-methylphenyl)- Bis(p-tolyl)mercury	$C_{14}H_{14}Hg$ 382.86	537-64-4 245.7	3387	4-16-00-01705	
7433	Mercury, dibutyl- Dibutylmercury	$C_8H_{18}Hg$ 314.82	629-35-6 105[10]		4-04-00-04429 1.7779[20]	1.5057[20]
7434	Mercury, diethyl- Diethyl mercury	$C_4H_{10}Hg$ 258.71	627-44-1 159; 57[16]		2.43[20]	eth 3; EtOH 2
7435	Mercury, dimethyl- Dimethyl mercury	C_2H_6Hg 230.66	593-74-8 93	3238	3.17[25]	H_2O 1; EtOH 4; eth 4 1.5452[20]
7436	Mercury, diphenyl- Benzene, mercuridi-	$C_{12}H_{10}Hg$ 354.80	587-85-9 204[10]		4-16-00-01702 2.318[25]	H_2O 1; EtOH 2; eth 2; bz 3
7437	Mercury, [2-methyl-5-nitrophenolato(2-)-C6,O1]- Nitromersol	$C_7H_5HgNO_3$ 351.71	133-58-4	6531		H_2O 1; ace 2; EtOH 2; alk 3
7438	Merphos Phosphorotrithious acid, S,S,S-tributyl ester	$C_{12}H_{27}PS_3$ 298.52	150-50-5 100	137[0.7]; 176[15]	1.02[20]	
7439	Metalaxyl N-(2,6-Dimethylphenyl)-N-(methoxyacetyl)alanine, methyl ester	$C_{15}H_{21}NO_4$ 279.34	57837-19-1 71	5826		
7440	Metaldehyde Acetaldehyde polymer	$(C_2H_4O)_x$	37273-91-9 246	5827 sub 115	4-19-00-05643	H_2O 1; EtOH 2; eth 2; ace 1
7441	Methamidophos Phosphoramidothioic acid, O,S-dimethyl ester	$C_2H_8NO_2PS$ 141.13	10265-92-6 46	5858	1.31[20]	

No.	Name / Synonym	Mol. Form. / Mol. Wt.	CAS RN / mp/°C	Merck No. / bp/°C	Beil. Ref. / den/g cm⁻³	Solubility / n_D
7442	Methanamine Methylamine	CH_5N 31.06	74-89-5 -93.4	5938 -6.3	4-04-00-00118 *0.656[25]	H_2O 4; EtOH 3; eth 5; ace 3
7443	Methanamine, 1,1-diethoxy-N,N-dimethyl-	$C_7H_{17}NO_2$ 147.22	1188-33-6	129	4-04-00-00174 0.859[25]	1.4007[20]
7444	Methanamine, N,N-dimethyl- Trimethylamine	C_3H_9N 59.11	75-50-3 -117.1	9625 2.8	4-04-00-00134 *0.627[25]	H_2O 4; EtOH 3; eth 3; bz 3 1.3631[0]
7445	Methanamine, N,N-dimethyl-, hydrochloride	$C_3H_{10}ClN$ 95.57	593-81-7 277.5	sub 200	4-04-00-00134	H_2O 4; EtOH 4; chl 4
7446	Methanamine, hydrochloride	CH_6ClN 67.52	593-51-1 227.5	5938 227[15]	4-04-00-00118	H_2O 4; EtOH 4
7447	Methanamine, N-hydroxy- Hydroxylamine, N-methyl	CH_5NO 47.06	593-77-1 87.5	62.5[15]	4-04-00-03299 1.0003[20]	H_2O 4; EtOH 4 1.4164[20]
7448	Methanamine, N-methyl- Dimethylamine	C_2H_7N 45.08	124-40-3 -92.2	3217 6.8	4-04-00-00128 0.6804[0]	H_2O 4; EtOH 3; eth 3 1.350[17]
7449	Methanamine, N-methyl-N-nitro-	$C_2H_6N_2O_2$ 90.08	4164-28-7 58	187	4-04-00-03401 1.1090[72]	H_2O 4; ace 4; eth 4; EtOH 4 1.4462[72]
7450	Methanamine, N-methyl-N-nitroso- N-Nitrosodimethylamine	$C_2H_6N_2O$ 74.08	62-75-9 154	6558	4-04-00-03384 1.0059[20]	H_2O 3; EtOH 3; eth 3; chl 3 1.4358[20]
7451	Methanamine, N-nitro-	$CH_4N_2O_2$ 76.06	598-57-2 38	80-5[10]	4-04-00-03400 1.2433[49]	H_2O 4; EtOH 4; eth 3; bz 4 1.4616[49]
7452	Methanamine, N-(phenylmethylene)-	C_8H_9N 119.17	622-29-7 185; 92[34]		4-07-00-00518 0.9671[14]	EtOH 3; eth 3; ace 3; chl 3 1.5526[20]
7453	Methanaminium, N-[4-[bis[4-(dimethylamino)phenyl]methylene]-2,5-cyclohexadien-1-ylidene]-N-methyl-, chloride Crystal violet	$C_{25}H_{30}ClN_3$ 407.99	548-62-9 215 dec	4287		H_2O 4; chl 4
7454	Methanaminium, 1-carboxy-N,N,N-trimethyl-, hydroxide, inner salt Betaine	$C_5H_{11}NO_2$ 117.15	107-43-7 293 dec	1201	4-04-00-02369	H_2O 4; EtOH 3; eth 2; chl 2
7455	Methanaminium, N-[4-[[4-(dimethylamino)phenyl]phenylmethylene]-2,5-cyclohexadien-1-ylidene]-N-methyl-, chlorid Malachite green	$C_{23}H_{25}ClN_2$ 364.92	569-64-2	5581		H_2O 4; EtOH 4; MeOH 4
7456	Methanaminium, N,N,N-trimethyl-, bromide Tetramethylammonium bromide	$C_4H_{12}BrN$ 154.05	64-20-0 230 dec		4-04-00-00145 1.56[25]	H_2O 4; EtOH 2; eth 1; bz 1
7457	Methanaminium, N,N,N trimethyl-, chloride Tetramethylammonium chloride	$C_4H_{12}ClN$ 109.60	75-57-0 420 dec		4-04-00-00145 1.169[20]	H_2O 3; EtOH 2; eth 1; bz 1
7458	Methanaminium, N,N,N-trimethyl-, iodide Tetramethylammonium iodide	$C_4H_{12}IN$ 201.05	75-58-1 >230 dec	9156	4-04-00-00145 1.829[25]	H_2O 2; EtOH 2; eth 1; ace 2
7459	Methane	CH_4 16.04	74-82-8 -182.4	5863 -161.5	4-01-00-00003 0.4228[-162]	H_2O 3; EtOH 3; eth 3; ace 2
7460	Methane, azido- Methyl azide	CH_3N_3 57.06	624-90-8	20.5; exp 500	0.869[15]	
7461	Methane, bis(methylthio)- Bis[methylmercapto]methane	$C_3H_8S_2$ 108.23	1618-26-4	1267 148	4-01-00-03088	
7462	Methane, bromo- Methyl bromide	CH_3Br 94.94	74-83-9 -93.7	5951 3.5	4-01-00-00068 1.6755[20]	H_2O 2; EtOH 5; eth 5; chl 5 1.4218[20]
7463	Methane, bromochloro- Bromochloromethane	CH_2BrCl 129.38	74-97-5 -87.9	68.0	4-01-00-00074 1.9344[20]	H_2O 1; EtOH 3; eth 3; ace 3 1.4838[20]
7464	Methane, bromochlorodifluoro- Bromochlorodifluoromethane	$CBrClF_2$ 165.36	353-59-3 -159.5	-3.7	4-01-00-00075	
7465	Methane, bromochlorodinitro- Bromochlorodinitromethane	$CBrClN_2O_4$ 219.38	33829-48-0 9.3	75-6[15]	2-01-00-00044 2.0393[20]	EtOH 4 1.4793
7466	Methane, bromochlorofluoro- Bromochlorofluoromethane	$CHBrClF$ 147.37	593-98-6 -115	36	4-01-00-00075 1.9771[0]	H_2O 1; eth 3; ace 3; chl 3 1.4144[25]
7467	Methane, bromodichloro- Bromodichloromethane	$CHBrCl_2$ 163.83	75-27-4 -57	90	4-01-00-00076 1.980[20]	H_2O 1; EtOH 4; eth 4; ace 4 1.4964[20]
7468	Methane, bromodichlorofluoro- Bromodichlorofluoromethane	$CBrCl_2F$ 181.82	353-58-2		4-01-00-00076	
7469	Methane, bromodifluoro- Bromodifluoromethane	$CHBrF_2$ 130.92	1511-62-2 -14.5		4-01-00-00072 1.55[16]	H_2O 3; EtOH 4
7470	Methane, bromofluoro- Bromofluoromethane	CH_2BrF 112.93	373-52-4 19		4-01-00-00072	EtOH 3; chl 4
7471	Methane, bromoiodo- Bromoiodomethane	CH_2BrI 220.84	557-68-6 139.5		4-01-00-00095 2.926[17]	chl 4 1.6410[20]
7472	Methane, bromomethoxy-	C_2H_5BrO 124.97	13057-17-5 87		4-01-00-03051 1.5976[20]	1.4562[20]
7473	Methane, bromonitro- Bromonitromethane	CH_2BrNO_2 139.94	563-70-2 149; 71[40]		4-01-00-00106	EtOH 4 1.4880[20]
7474	Methane, bromotrichloro- Bromotrichloromethane	$CBrCl_3$ 198.27	75-62-7 -5.7	105	4-01-00-00077 2.012[25]	eth 4; EtOH 4 1.5065[20]
7475	Methane, bromotrifluoro- Bromotrifluoromethane	$CBrF_3$ 148.91	75-63-8 -172	-57.8	4-01-00-00073 1.580[20]	chl 4
7476	Methane, bromotrinitro- Bromotrinitromethane	$CBrN_3O_6$ 229.93	560-95-2 17.5	56[10]	3-01-00-00116 2.0312[20]	EtOH 4; chl 4 1.4808[20]
7477	Methane, chloro- Methyl chloride	CH_3Cl 50.49	74-87-3 -97.7	5964 -24.0	4-01-00-00028 *0.911[25]	H_2O 3; EtOH 3; eth 5; ace 5 1.3389[20]
7478	Methane, chlorodifluoro- Chlorodifluoromethane	$CHClF_2$ 86.47	75-45-6 -157.4	-40.7	4-01-00-00032 1.4909[-69]	H_2O 4; eth 3; ace 3; chl 3

No.	Name / Synonym	Mol. Form. / Mol. Wt.	CAS RN mp/°C	Merck No. bp/°C	Beil. Ref. den/g cm^{-3}	Solubility n_D
7479	Methane, chlorodiiodo-	CHCII$_2$ 302.28	638-73-3 -4	200 dec; 101^{33}	4-01-00-00097 3.1700^0	ace 4; eth 4; chl 4
7480	Methane, chlorodinitro- Chlorodinitromethane	CHClN$_2$O$_4$ 140.48	921-13-1	34-6^{13}	3-01-00-00115 1.6125^{20}	1.4575^{20}
7481	Methane, chlorofluoro- Chlorofluoromethane	CH$_2$ClF 68.48	593-70-4 -133	-9.1	4-01-00-00032	chl 4
7482	Methane, chloroiodo- Chloroiodomethane	CH$_2$ClI 176.38	593-71-5	109	4-01-00-00094 2.422^{20}	ace 4; bz 4; eth 4; EtOH 4 1.5822^{20}
7483	Methane, chloromethoxy- Chloromethyl methyl ether	C$_2$H$_5$ClO 80.51	107-30-2 -103.5	2146 59.5	4-01-00-03046 1.063^{10}	EtOH 3; eth 3; ace 3; chl 3 1.397^{20}
7484	Methane, chloro(methylthio)-	C$_2$H$_5$ClS 96.58	2373-51-5	105	4-01-00-03079 1.153^{25}	1.4963^{20}
7485	Methane, chloronitro- Chloronitromethane	CH$_2$ClNO$_2$ 95.49	1794-84-9	122.5	4-01-00-00106 1.466^{15}	H$_2$O 4
7486	Methane, chlorotrifluoro- Chlorotrifluoromethane	CClF$_3$ 104.46	75-72-9 -181	-81.4	4-01-00-00034	
7487	Methane, chlorotrinitro-	CClN$_3$O$_6$ 185.48	1943-16-4 4.5	134 dec; 56^{40}	3-01-00-00116 1.6769^{20}	eth 4; EtOH 4; chl 4 1.4500^{20}
7488	Methanediamine, 1-(2-furanyl)-N,N'-bis(2-furanylmethylene)- Hydrofuramide	C$_{15}$H$_{12}$N$_2$O$_3$ 268.27	494-47-3 117	4718	5-17-09-00306	H$_2$O 1; EtOH 4; eth 4
7489	Methanediamine, N,N,N',N'-tetraethyl-	C$_9$H$_{22}$N$_2$ 158.29	102-53-4	165.8	4-04-00-00338 0.8000^{20}	1.4420^{25}
7490	Methanediamine, N,N,N',N'-tetramethyl- Tetramethylmethanediamine	C$_5$H$_{14}$N$_2$ 102.18	51-80-9	83	4-04-00-00153 0.7491^{18}	H$_2$O 3
7491	Methane, diazo- Diazomethane	CH$_2$N$_2$ 42.04	334-88-3 -145	2983 -23	4-01-00-03056	eth 4; diox 4
7492	Methane, dibromo- Methylene bromide	CH$_2$Br$_2$ 173.83	74-95-3 -52.5	5980 97	4-01-00-00078 2.4969^{20}	H$_2$O 2; EtOH 5; eth 5; ace 5 1.5420^{20}
7493	Methane, dibromochloro- Chlorodibromomethane	CHBr$_2$Cl 208.28	124-48-1 -20	120	4-01-00-00081 2.451^{20}	H$_2$O 1; EtOH 3; eth 3; ace 3 1.5482^{20}
7494	Methane, dibromochlorofluoro- Dibromochlorofluoromethane	CBr$_2$ClF 226.27	353-55-9	80.3	4-01-00-00082 2.3173^{22}	1.4570^{20}
7495	Methane, dibromodichloro- Dibromodichloromethane	CBr$_2$Cl$_2$ 242.72	594-18-3 38	150.2	4-01-00-00082 2.42^{25}	H$_2$O 1; EtOH 3; eth 3; ace 3
7496	Methane, dibromodifluoro- Dibromodifluoromethane	CBr$_2$F$_2$ 209.82	75-61-6 -110.1	25	4-01-00-00080	H$_2$O 3; eth 3; ace 3; bz 3
7497	Methane, dibromodinitro- Dibromodinitromethane	CBr$_2$N$_2$O$_4$ 263.83	2973-00-4 5.5	158 dec; 77^{21}	3-01-00-00115 2.4439^{20}	EtOH 4 1.5280^{25}
7498	Methane, dibromofluoro- Dibromofluoromethane	CHBr$_2$F 191.83	1868-53-7 -78	64.9	4-01-00-00080 2.421^{20}	H$_2$O 1; EtOH 3; eth 3; ace 3 1.4685^{20}
7499	Methane, dichloro- Methylene chloride	CH$_2$Cl$_2$ 84.93	75-09-2 -95.1	5982 40	4-01-00-00035 1.3266^{20}	H$_2$O 2; EtOH 5; eth 5; ctc 3 1.4242^{20}
7500	Methane, dichlorodifluoro- Dichlorodifluoromethane	CCl$_2$F$_2$ 120.91	75-71-8 -158	3053 -29.8	4-01-00-00040	H$_2$O 3; EtOH 3; eth 3; HOAc 3
7501	Methane, dichlorodinitro- Dichlorodinitromethane	CCl$_2$N$_2$O$_4$ 174.93	1587-41-3	121.5	3-01-00-00115 1.6124^{20}	bz 4; eth 4; EtOH 4; chl 4 1.4575^{20}
7502	Methane, dichlorofluoro- Dichlorofluoromethane	CHCl$_2$F 102.92	75-43-4 -135	8.9	4-01-00-00039 1.405^9	H$_2$O 1; EtOH 3; eth 3; ctc 3 1.3724^9
7503	Methane, dichloroiodo- Dichloroiodomethane	CHCl$_2$I 210.83	594-04-7	132	4-01-00-00095 2.392^{20}	ace 4; bz 4; eth 4; EtOH 4 1.5840^{20}
7504	Methane, difluoro- Methylene difluoride	CH$_2$F$_2$ 52.02	75-10-5 -136	-51.6	4-01-00-00024 *0.960^{25}	H$_2$O 1; EtOH 3
7505	Methane, difluoroiodo- Difluoroiodomethane	CHF$_2$I 177.92	1493-03-4 -122	21.6	4-01-00-00092 3.238^{-19}	
7506	Methane, diiodo- Methylene iodide	CH$_2$I$_2$ 267.84	75-11-6 6.1	5985 182	4-01-00-00096 3.3212^{20}	H$_2$O 2; EtOH 3; eth 3; bz 3 1.7425^{20}
7507	Methane, dimethoxy- Methylal	C$_3$H$_8$O$_2$ 76.10	109-87-5 -104.8	5936 42	4-01-00-03026 0.8593^{20}	ace 4; bz 4; eth 4; EtOH 4 1.3513^{20}
7508	Methane, dinitro- Dinitromethane	CH$_2$N$_2$O$_4$ 106.04	625-76-3 <-15	exp 100	4-01-00-00107	H$_2$O 1; EtOH 3; eth 3
7509	Methanediol, dibenzoate	C$_{15}$H$_{12}$O$_4$ 256.26	5342-31-4 99	225 dec	4-09-00-00448 1.275^{22}	ace 4; bz 4; eth 4
7510	Methanediol, 2-furanyl-, diacetate	C$_9$H$_{10}$O$_5$ 198.18	613-75-2 53.3	220	5-17-09-00300	bz 4; eth 4; EtOH 4
7511	Methanediol, (3-nitrophenyl)-, diacetate (ester) Benzaldehyde, 3-nitro, diacetate	C$_{11}$H$_{11}$NO$_6$ 253.21	29949-19-7 72		3-07-00-00900 1.393^{25}	ace 4; bz 4; eth 4; EtOH 4
7512	Methanediol, phenyl-, diacetate	C$_{11}$H$_{12}$O$_4$ 208.21	581-55-5 46	220	4-07-00-00516 1.11^{20}	bz 4; eth 4; EtOH 4
7513	Methane, diphenyl-m-tolyl- Triphenylmethane, 3-methyl	C$_{20}$H$_{18}$ 258.36	603-26-9 62	357	3-05-00-02332 1.07^{16}	bz 4; eth 4; chl 4; lig 4
7514	Methanedisulfonic acid Methionic acid	CH$_4$O$_6$S$_2$ 176.17	503-40-2 98	5895	4-01-00-03054	H$_2$O 1; HNO$_3$ 3
7515	Methane, fluoro- Methyl fluoride	CH$_3$F 34.03	593-53-3 -141.8	4103 -78.4	4-01-00-00022 *0.566^{25}	H$_2$O 4; EtOH 4; eth 4; bz 3
7516	Methane, fluorodiiodo-	CHFI$_2$ 285.83	1493-01-2 -34.5	100	4-01-00-00097 3.1969^{22}	eth 4; EtOH 4

No.	Name Synonym	Mol. Form. Mol. Wt.	CAS RN mp/°C	Merck No. bp/°C	Beil. Ref. den/g cm^{-3}	Solubility n_D
7517	Methane, fluoroiodo- Fluoroiodomethane	CH_2FI 159.93	373-53-5 	 53.4	4-01-00-00092 2.366^{20}	ace 4; bz 4; eth 4; chl 4 1.5256^{20}
7518	Methane, iodo- Methyl iodide	CH_3I 141.94	74-88-4 -66.4	6002 42.5	4-01-00-00087 2.279^{20}	H_2O 2; EtOH 5; eth 5; ace 3 1.5380^{20}
7519	Methane, iodomethoxy- (Iodomethyl) methyl ether	C_2H_5IO 171.97	13057-19-7 	 122	4-01-00-03052 2.030^{20}	 1.5472^{20}
7520	Methane, isocyanato- Methyl isocyanate	C_2H_3NO 57.05	624-83-9 -45	6004 39.5	4-04-00-00247 0.9230^{27}	H_2O 4 1.3419^{18}
7521	Methane, isothiocyanato- Methyl isothiocyanate	C_2H_3NS 73.12	556-61-6 36	6005 119	4-04-00-00248 1.0691^{37}	H_2O 2; EtOH 5; eth 4 1.5258
7522	Methane, nitro- Nitromethane	CH_3NO_2 61.04	75-52-5 -28.5	6532 101.1	4-01-00-00100 1.1371^{20}	H_2O 3; EtOH 3; eth 3; ace 3 1.3817^{20}
7523	Methane, oxybis- Dimethyl ether	C_2H_6O 46.07	115-10-6 -141.5	5990 -24.8	4-01-00-01245	H_2O 3; EtOH 3; eth 3; ace 3
7524	Methane, oxybis[bromo-	$C_2H_4Br_2O$ 203.86	4497-29-4 -34	 154.5	4-01-00-03052 2.2013^{20}	
7525	Methane, oxybis[chloro- Bis(chloromethyl) ether	$C_2H_4Cl_2O$ 114.96	542-88-1 -41.5	3058 106	4-01-00-03051 1.323^{15}	EtOH 5; eth 5 1.435^{21}
7526	Methane, selenobis-	C_2H_6Se 109.03	593-79-3 	 55	4-01-00-01288 1.4077^{15}	eth 4; EtOH 4; chl 4
7527	Methanesulfenyl chloride, trichloro- Perchloromethyl mercaptan	CCl_4S 185.89	594-42-3 	 147.5	4-03-00-00290 1.6947^{20}	eth 3 1.5484^{20}
7528	Methanesulfenyl fluoride, trifluoro- Methyl mercaptan, perfluoro	CF_4S 120.07	17742-04-0 	 147.5	 1.6947^{20}	eth 4 1.5484^{20}
7529	Methane, sulfinylbis- Dimethyl sulfoxide	C_2H_6OS 78.14	67-68-5 18.5	3247 189	4-01-00-01277 1.1014^{20}	H_2O 3; EtOH 3; eth 3; ace 3 1.4170^{20}
7530	Methanesulfonic acid Methylsulfonic acid	CH_4O_3S 96.11	75-75-2 20	5865 167^{10}	4-04-00-00010 1.4812^{18}	H_2O 3 1.4317^{18}
7531	Methanesulfonic acid, ethyl ester Ethyl methanesulfonate	$C_3H_8O_3S$ 124.16	62-50-0 	3782 86^{10}	4-04-00-00012	
7532	Methanesulfonic acid, methyl ester Methyl methanesulfonate	$C_2H_6O_3S$ 110.13	66-27-3 20	6010 202.5	4-04-00-00011 1.2943^{20}	 1.4138^{20}
7533	Methanesulfonic acid, trifluoro-, ethyl ester	$C_3H_5F_3O_3S$ 178.13	425-75-2 	 115	4-03-00-00034 1.3740^{0}	eth 3
7534	Methane, sulfonylbis- Dimethyl sulfone	$C_2H_6O_2S$ 94.13	67-71-0 109	3246 238	4-01-00-01279 1.1700^{110}	H_2O 3; EtOH 3; bz 3 1.4226
7535	Methanesulfonyl chloride Methanesulfonic acid chloride	CH_3ClO_2S 114.55	124-63-0 	5866 162; 55^{11}	4-04-00-00027 1.4805^{18}	H_2O 1; EtOH 3; eth 3 1.4573^{20}
7536	Methanesulfonyl chloride, trichloro-	CCl_4O_2S 217.89	2547-61-7 140.5	 170	4-03-00-00036	H_2O 1; EtOH 3; eth 3; CS_2 3
7537	Methanesulfonyl chloride, trifluoro-	$CClF_3O_2S$ 168.52	421-83-0 	 162; 62^{18}	4-03-00-00035	H_2O 1 1.3344^{20}
7538	Methanesulfonyl fluoride Methanesulfonic acid fluoride	CH_3FO_2S 98.10	558-25-8 	 123.5	4-04-00-00027	
7539	Methane, tetrabromo- Carbon tetrabromide	CBr_4 331.63	558-13-4 90.1	 189.5	4-01-00-00085 2.9608^{100}	H_2O 1; EtOH 3; eth 3; chl 3 1.5942^{100}
7540	Methane, tetrachloro- Carbon tetrachloride	CCl_4 153.82	56-23-5 -23	1822 76.8	4-01-00-00056 1.5940^{20}	H_2O 1; EtOH 3; eth 5; ace 3 1.4601^{20}
7541	Methane, tetrafluoro- Carbon tetrafluoride	CF_4 88.00	75-73-0 -183.5	1823 -128.0	4-01-00-00026 3.034^{25}	H_2O 2; bz 3; chl 3
7542	Methane, tetraiodo- Carbon tetraiodide	CI_4 519.63	507-25-5 171	1824 130-40$^{1-2}$	4-01-00-00098 4.23^{20}	py 4; chl 4
7543	Methane, tetramethoxy-	$C_5H_{12}O_4$ 136.15	1850-14-2 -5	 114	4-03-00-00004 1.023^{25}	 1.3845^{20}
7544	Methane, tetranitro- Tetranitromethane	CN_4O_8 196.04	509-14-8 13.8	9164 126.1	4-01-00-00107 1.6380^{20}	H_2O 1; EtOH 3; eth 3 1.4384^{20}
7545	Methane, thiobis- Dimethyl sulfide	C_2H_6S 62.14	75-18-3 -98.3	6042 37.3	4-01-00-01275 0.8483^{20}	H_2O 1; EtOH 3; eth 3 1.4438^{20}
7546	Methanethiol Methyl mercaptan	CH_4S 48.11	74-93-1 -123	5867 5.9	4-01-00-01273 0.8665^{20}	H_2O 2; EtOH 4; eth 4; chl 2
7547	Methanethione, diphenyl-	$C_{13}H_{10}S$ 198.29	1450-31-3 53.5	 174^{14}	4-07-00-01385	EtOH 2; eth 2; bz 4; chl 4
7548	Methane, tribromo- Bromoform	$CHBr_3$ 252.73	75-25-2 8.0	1407 149.1	4-01-00-00082 2.899^{15}	H_2O 2; EtOH 5; eth 5; bz 3 1.6005^{15}
7549	Methane, tribromochloro- Tribromochloromethane	CBr_3Cl 287.18	594-15-0 55	 158.5	4-01-00-00085 2.71^{15}	eth 4
7550	Methane, tribromofluoro- Tribromofluoromethane	CBr_3F 270.72	353-54-8 -73.6	 108		EtOH 3
7551	Methane, tribromonitro-	CBr_3NO_2 297.73	464-10-8 10	 127^{18}	4-01-00-00106 2.811^{12}	H_2O 1; EtOH 3; eth 3; ace 4 1.5790^{20}
7552	Methanetricarboxylic acid, triethyl ester Triethyl methanetricarboxylate	$C_{10}H_{16}O_6$ 232.23	6279-86-3 29	 253	4-02-00-02360 1.1084^{16}	 1.4243^{20}
7553	Methanetricarboxylic acid, trimethyl ester Trimethyl methanetricarboxylate	$C_7H_{10}O_6$ 190.15	1186-73-8 46.5	 242.7; 128^{15}	3-02-00-02023	bz 4; eth 4; EtOH 4; chl 4
7554	Methane, trichloro- Chloroform	$CHCl_3$ 119.38	67-66-3 -63.6	2141 61.1	4-01-00-00042 1.4832^{20}	H_2O 2; EtOH 5; eth 5; ace 3 1.4459^{20}
7555	Methane, trichlorofluoro- Trichlorofluoromethane	CCl_3F 137.37	75-69-4 -111.1	9553 23.7	4-01-00-00054	
7556	Methane, trichloroiodo- Trichloroiodomethane	CCl_3I 245.27	594-22-9 -7.8	 142	4-01-00-00095 2.355^{20}	ace 4; bz 4; eth 4; chl 4 1.5854^{20}

No.	Name / Synonym	Mol. Form. / Mol. Wt.	CAS RN / mp/°C	Merck No. / bp/°C	Beil. Ref. / den/g cm^{-3}	Solubility / n_D
7557	Methane, trichloronitro- / Chloropicrin	CCl$_3$NO$_2$ / 164.37	76-06-2 / -64	2156 / 112	4-01-00-00106 / 1.6558[20]	H$_2$O 3; EtOH 5; ace 5; bz 5 / 1.4611[20]
7558	Methane, trifluoro- / Fluoroform	CHF$_3$ / 70.01	75-46-7 / -155.1	4102 / -82.1	4-01-00-00024 / *0.673[25]	H$_2$O 3; EtOH 4; ace 3; bz 3
7559	Methane, trifluoroiodo- / Trifluoroiodomethane	CF$_3$I / 195.91	2314-97-8 /	/ -22.5	4-01-00-00092 / 2.3607[-32]	1.3790[-32]
7560	Methane, triiodo- / Iodoform	CHI$_3$ / 393.73	75-47-8 / 119	4926 / 218	4-01-00-00097 / 4.008[25]	H$_2$O 1; EtOH 3; eth 3; ace 3
7561	Methane, trimethoxy-	C$_4$H$_{10}$O$_3$ / 106.12	149-73-5 / 15	/ 104	4-02-00-00022 / 0.9676[20]	EtOH 3; eth 3 / 1.3793[20]
7562	Methane, trinitro- / Trinitromethane	CHN$_3$O$_6$ / 151.04	517-25-9 / 15	9642 / exp	4-01-00-00107 / 1.479[20]	ace 4; EtOH 4 / 1.4451[24]
7563	Methanimidamide / Formamidine	CH$_4$N$_2$ / 44.06	463-52-5 / 81	/ dec	4-02-00-00082	H$_2$O 4; EtOH 4
7564	Methanimidamide, N'-(4-chloro-2-methylphenyl)-N,N-dimethyl- / Chlordimeform	C$_{10}$H$_{13}$ClN$_2$ / 196.68	6164-98-3 / 35	2083 / 156-7[0.4]	/ 1.105[25]	bz 4; eth 4; EtOH 4 / 1.5885[25]
7565	Methanimidamide, N,N'-diphenyl-	C$_{13}$H$_{12}$N$_2$ / 196.25	622-15-1 / 142	/ >250	4-12-00-00372	H$_2$O 2; EtOH 3; eth 4; ace 3
7566	Methanimidamide, N-hydroxy-	CH$_4$N$_2$O / 60.06	624-82-8 / 114.5	/ dec	4-02-00-00084	EtOH 4
7567	Methanimidic acid, N-phenyl-, ethyl ester	C$_9$H$_{11}$NO / 149.19	6780-49-0 /	/ 214; 87[10]	4-12-00-00371 / 1.0051[20]	eth 3; bz 3 / 1.5279[20]
7568	1,4-Methanoazulene, decahydro-4,8,8-trimethyl-9-methylene-, [1S-(1α,3aβ,4α,8aβ)]- / Longifolene	C$_{15}$H$_{24}$ / 204.36	475-20-7 /	5446 / 258; 126[15]	4-05-00-01192 / 0.9319[18]	H$_2$O 1; bz 3 / 1.5040[20]
7569	1,4-Methanoazulen-9-ol, decahydro-1,5,5,8a-tetramethyl-, [1R-(1α,3aβ,4α,8aβ,9S*)]	C$_{15}$H$_{26}$O / 222.37	465-24-7 / 112	/ 287 dec	3-06-00-00426 / 1.0441[20]	1.519
7570	1H-3a,7-Methanoazulen-6-ol, octahydro-3,6,8,8-tetramethyl-, / Cedrol	C$_{15}$H$_{26}$O / 222.37	77-53-2 / 86	1919 /	3-06-00-00424 / 0.9479[90]	1.4824[90]
7571	7,14-Methano-2H,6H-dipyrido[1,2-a:1',2'-e][1,5]diazocine, dodecahydro-, / Sparteine	C$_{15}$H$_{26}$N$_2$ / 234.38	90-39-1 / 30.5	8692 / 325; 173[8]	5-23-05-00497 / 1.0196[20]	eth 4; EtOH 4; chl 4 / 1.5312[20]
7572	7,14-Methano-2H,6H-dipyrido[1,2-a:1',2'-e][1,5]diazocin-6-one, dodecahydro-, [7R-(7α,7aβ,14α,14aalp / Aphylline	C$_{15}$H$_{24}$N$_2$O / 248.37	577-37-7 / 54.5	761 / 200 dec	5-24-02-00298	ace 4; bz 4; eth 4; EtOH 4
7573	7,14-Methano-4H,6H-dipyrido[1,2-a:1',2'-e][1,5]diazocin-4-one, 7,7a,8,9,10,11,13,14-octahydro-, / Anagyrine	C$_{15}$H$_{20}$N$_2$O / 244.34	486-89-5 /	660 / 265[12]; 212[4]	5-24-03-00410	H$_2$O 3; EtOH 4; eth 3; bz 3
7574	7,14-Methano-4H,6H-dipyrido[1,2-a:1',2'-e][1,5]diazocin-4-one, dodecahydro-9-hydroxy-, / Hydroxylupanine	C$_{15}$H$_{24}$N$_2$O$_2$ / 264.37	15358-48-2 / 169.5	4760 /	4-25-00-00082	H$_2$O 4; EtOH 4; chl 4
7575	2,5-Methano-2H-furo[3,2-b]pyrrol-6-ol, hexahydro-4-methyl-, (2α,3aβ,5α,6β,6aβ / Scopoline	C$_8$H$_{13}$NO$_2$ / 155.20	487-27-4 / 108.5	8365 / 248	2-27-00-00061 / 1.0891[134]	H$_2$O 3
7576	4,7-Methano-1H-indene, 1,4,5,6,7,8,8-heptachloro-3a,4,7,7a-tetrahydro- / Heptachlor	C$_{10}$H$_5$Cl$_7$ / 373.32	76-44-8 / 95.5	4576 /	/ 1.57[9]	bz 4; eth 4; EtOH 4; lig 4
7577	4,7-Methano-1H-indene, 3a,4,7,7a-tetrahydro-, (3aα,4α,7α,7aα)-	C$_{10}$H$_{12}$ / 132.21	1755-01-7 / 32	/ 170 dec; 65[14]	4-05-00-01399 / 0.9302[35]	eth 4; EtOH 4 / 1.5050[35]
7578	1,5-Methano-8H-pyrido[1,2-a][1,5]diazocin-8-one, 1,2,3,4,5,6-hexahydro-3-methyl-, (1R)- / Caulophylline	C$_{12}$H$_{16}$N$_2$O / 204.27	486-86-2 / 137	1914 /	5-24-02-00538	H$_2$O 4; ace 4; bz 4; EtOH 4
7579	3,6-Methano-8H-1,5,7-trioxacyclopenta[ij]cycloprop[a]azulene-4,8(3H)-dione, hexahydro-2a-hydroxy / Picrotoxin	C$_{30}$H$_{34}$O$_{13}$ / 602.59	124-87-8 / 203.5	7388 /	4-19-00-05245	py 4; EtOH 4
7580	4,7-Methanoisobenzofuran-1,3-dione, 3a,4,7,7a-tetrahydro-, (3aα,4α,7α,7aα)- / Carbic anhydride	C$_9$H$_8$O$_3$ / 164.16	129-64-6 / 164.5	1801 /	5-17-11-00192 / 1.417[25]	ace 4; bz 4; EtOH 4; chl 4
7581	Methanol / Methyl alcohol	CH$_4$O / 32.04	67-56-1 / -97.6	5868 / 64.6	4-01-00-01227 / 0.7914[20]	H$_2$O 5; EtOH 5; eth 5; ace 5 / 1.3288[20]
7582	Methanol, chloro-, acetate	C$_3$H$_5$ClO$_2$ / 108.52	625-56-9 /	/ 116	4-02-00-00280 / 1.194[20]	eth 4; EtOH 4 / 1.409[20]
7583	Methanol, (1,3,5-triazine-2,4,6-triyltriimino)tris- / Trimethylolmelamine	C$_6$H$_{12}$N$_6$O$_3$ / 216.20	1017-56-7 /	9631 /	4-26-00-01273	
7584	Methanol, (1,3,5-triazine-2,4,6-triyltrinitrilo)hexakis- / Hexamethylolmelamine	C$_9$H$_{18}$N$_6$O$_6$ / 306.28	531-18-0 / 137	4611 /	4-26-00-01274	H$_2$O 4
7585	Methanol, trichloro-, carbonate (2:1)	C$_3$Cl$_6$O$_3$ / 296.75	32315-10-9 / 79	/ 203	4-03-00-00033 / 1.6290[80]	
7586	1,6-Methanonaphthalen-1(2H)-ol, octahydro-4,8a,9,9-tetramethyl-, / Patchouli alcohol	C$_{15}$H$_{26}$O / 222.37	5986-55-0 / 56	7001 /	3-06-00-00426 / 0.9906[65]	H$_2$O 1; EtOH 3; eth 3 / 1.5029[65]

No.	Name / Synonym	Mol. Form. / Mol. Wt.	CAS RN / mp/°C	Merck No. / bp/°C	Beil. Ref. / den/g cm^{-3}	Solubility / n_D
7587	Methanone, 1,3,5-benzenetriyltris[phenyl- / 1,3,5-Tribenzoylbenzene	$C_{27}H_{18}O_3$ / 390.44	25871-69-6 / 120	260[1]	4-07-00-02847	chl 3
7588	Methanone, 2-benzofuranylphenyl-	$C_{15}H_{10}O_2$ / 222.24	6272-40-8 / 91	360	5-17-10-00568	H_2O 1
7589	Methanone, bis(3-aminophenyl)-	$C_{13}H_{12}N_2O$ / 212.25	611-79-0 / 173.5	285[11]	4-14-00-00255	eth 4; EtOH 4
7590	Methanone, bis(4-bromophenyl)-	$C_{13}H_8Br_2O$ / 340.01	3988-03-2 / 177	395	4-07-00-01380	bz 4; HOAc 4; chl 4
7591	Methanone, bis(3-chlorophenyl)-	$C_{13}H_8Cl_2O$ / 251.11	7094-34-0 / 124	163[2]	4-07-00-01376	H_2O 1; EtOH 3; eth 3
7592	Methanone, bis(4-chlorophenyl)-	$C_{13}H_8Cl_2O$ / 251.11	90-98-2 / 147.5	353	4-07-00-01376 / 1.4500[20]	H_2O 1; EtOH 3; eth 4; ace 2
7593	Methanone, bis[4-(dimethylamino)phenyl]- / Michler's ketone	$C_{17}H_{20}N_2O$ / 268.36	90-94-8 / 179	6100 / 360 dec	4-14-00-00255	H_2O 1; EtOH 1; eth 1; bz 4
7594	Methanone, bis(2,4-dimethylphenyl)- / 2,2',4,4'-Tetramethylbenzophenone	$C_{17}H_{18}O$ / 238.33	3478-88-4	190[10]	4-07-00-01470 / 1.043[15]	1.5790[25]
7595	Methanone, bis(2-hydroxy-4-methoxyphenyl)- / Benzophenone 6	$C_{15}H_{14}O_5$ / 274.27	131-54-4 / 139.5	1109	4-08-00-03505	
7596	Methanone, bis(2-hydroxyphenyl)-	$C_{13}H_{10}O_3$ / 214.22	835-11-0 / 59.5	333	4-08-00-02446	H_2O 1; EtOH 3; eth 3; chl 3
7597	Methanone, bis(4-hydroxyphenyl)-	$C_{13}H_{10}O_3$ / 214.22	611-99-4 / 210		4-08-00-02452 / 1.133[131]	H_2O 2; EtOH 3; eth 3; ace 3
7598	Methanone, bis(4-iodophenyl)- / 4,4'-Diiodobenzophenone	$C_{13}H_8I_2O$ / 434.01	5630-56-8 / 238.5	281[12]	4-07-00-01381 / 2.2600[20]	
7599	Methanone, bis(4-methylphenyl)-	$C_{15}H_{14}O$ / 210.28	611-97-2 / 96.5	334	4-07-00-01434	ace 4; bz 4; eth 4; EtOH 4
7600	Methanone, (2-bromophenyl)phenyl- / 2-Bromobenzophenone	$C_{13}H_9BrO$ / 261.12	13047-06-8 / 42	345	3-07-00-02079	ace 4; EtOH 4; lig 4
7601	Methanone, (3-bromophenyl)phenyl-	$C_{13}H_9BrO$ / 261.12	1016-77-9 / 81	186[6]	4-07-00-01378	EtOH 4
7602	Methanone, (4-bromophenyl)phenyl-	$C_{13}H_8BrO$ / 261.12	90-90-4 / 82.5	350	4-07-00-01378	H_2O 1; EtOH 2; eth 2; bz 2
7603	Methanone, (2-chlorophenyl)(4-chlorophenyl)-	$C_{13}H_8Cl_2O$ / 251.11	85-29-0 / 67	214-5[22]	4-07-00-01376 / 1.303[14]	EtOH 3; chl 2
7604	Methanone, (4-chlorophenyl)phenyl-	$C_{13}H_9ClO$ / 216.67	134-85-0 / 77.5	332	4-07-00-01375	EtOH 3; eth 3; ace 3; ctc 2
7605	Methanone, cyclobutylphenyl-	$C_{11}H_{12}O$ / 160.22	5407-98-7	260	4-07-00-01027 / 1.0426[25]	1.5472[20]
7606	Methanone, dicyclohexyl-	$C_{13}H_{22}O$ / 194.32	119-60-8	159[20]	4-07-00-00262 / 0.986[0]	eth 3; ace 3; ctc 3
7607	Methanone, dicyclopropyl- / Dicyclopropyl ketone	$C_7H_{10}O$ / 110.16	1121-37-5	161	4-07-00-00139 / 0.977[25]	1.4670[20]
7608	Methanone, (2,6-dihydroxy-4-methoxyphenyl)phenyl- / Cotoin	$C_{14}H_{12}O_4$ / 244.25	479-21-0 / 130.5	2553	4-08-00-03163	ace 4; bz 4; eth 4; EtOH 4
7609	Methanone, (2,4-dihydroxyphenyl)phenyl- / Benzoresorcinol	$C_{13}H_{10}O_3$ / 214.22	131-56-6 / 144	1117	4-08-00-02442	H_2O 1; EtOH 3; eth 4; bz 2
7610	Methanone, (3,4-dihydroxyphenyl)(2,4,6-trihydroxyphenyl)- / Maclurin	$C_{13}H_{10}O_6$ / 262.22	519-34-6 / 222.5	5519	3-08-00-04226	eth 4; EtOH 4
7611	Methanone, (6,7-dimethoxy-1-isoquinolinyl)(3,4-dimethoxyphenyl)- / Papaveraldine	$C_{20}H_{19}NO_5$ / 353.37	522-57-6 / 210.5	6966	5-21-13-00658	H_2O 1; EtOH 2; eth 2; bz 3
7612	Methanone, (2,4-dimethoxyphenyl)phenyl- / 2,4-Dimethoxybenzophenone	$C_{15}H_{14}O_3$ / 242.27	3555-84-8 / 87.5	218[10]	4-08-00-02442	H_2O 1; EtOH 4; eth 2; chl 4
7613	Methanone, [4-(dimethylamino)phenyl]phenyl- / 4-(Dimethylamino)benzophenone	$C_{15}H_{15}NO$ / 225.29	530-44-9 / 92.5	3222	4-14-00-00248	H_2O 1; EtOH 2; eth 4; chl 3
7614	Methanone, (2,4-dimethylphenyl)phenyl-	$C_{15}H_{14}O$ / 210.28	1140-14-3	321.5	4-07-00-01429 / 1.071[20]	bz 4; eth 4; EtOH 4; chl 4
7615	Methanone, (3,4-dimethylphenyl)phenyl- / 3,4-Dimethylbenzophenone	$C_{15}H_{14}O$ / 210.28	2571-39-3 / 47.5	341	4-07-00-01433	bz 4; eth 4; EtOH 4; chl 4
7616	Methanone, diphenyl- / Benzophenone	$C_{13}H_{10}O$ / 182.22	119-61-9 / 47.8	1108 / 305.4	4-07-00-01357	H_2O 1; EtOH 4; eth 4; ace 4 / 1.6077[19]
7617	Methanone, di-2-thienyl- / Di-(2-thienyl) ketone	$C_9H_6OS_2$ / 194.28	704-38-1 / 90	326	5-19-04-00318	ace 4; eth 4
7618	Methanone, 2-furanylphenyl-	$C_{11}H_8O_2$ / 172.18	2689-59-0 / <-15	285	5-17-10-00289 / 1.1732[20]	eth 4; EtOH 4 / 1.6055[20]
7619	Methanone, (2-hydroxy-4-methoxyphenyl)(2-hydroxyphenyl)- / Dioxybenzone	$C_{14}H_{12}O_4$ / 244.25	131-53-3	3298 / 170-5[1]	4-08-00-03163	
7620	Methanone, (2-hydroxy-4-methoxyphenyl)phenyl- / Oxybenzone	$C_{14}H_{12}O_3$ / 228.25	131-57-7 / 65.5	6907	4-08-00-02442	ctc 3
7621	Methanone, [2-hydroxy-4-(octyloxy)phenyl]phenyl- / Octabenzone	$C_{21}H_{26}O_3$ / 326.44	1843-05-6 / 48.5	6662		
7622	Methanone, (2-hydroxyphenyl)phenyl-	$C_{13}H_{10}O_2$ / 198.22	117-99-7 / 40	250[560]	4-08-00-01246	H_2O 1; EtOH 4; eth 4; bz 4
7623	Methanone, (2-methoxyphenyl)phenyl-	$C_{14}H_{12}O_2$ / 212.25	2553-04-0 / 41	210[27], 195[18]	4-08-00-01246	H_2O 1; EtOH 3; bz 3; HOAc 3

No.	Name / Synonym	Mol. Form. / Mol. Wt.	CAS RN / mp/°C	Merck No. / bp/°C	Beil. Ref. / den/g cm^{-3}	Solubility / n_D
7624	Methanone, (3-methoxyphenyl)phenyl-	$C_{14}H_{12}O_2$ 212.25	6136-67-0 44	343; 201[17]	4-08-00-01262	H_2O 1; EtOH 4; bz 4; HOAc 4
7625	Methanone, (4-methoxyphenyl)phenyl-	$C_{14}H_{12}O_2$ 212.25	611-94-9 61.5	355; 168[12]	4-08-00-01263	H_2O 1; EtOH 4; eth 4; ace 3
7626	Methanone, [2-(methylamino)phenyl]phenyl-	$C_{14}H_{13}NO$ 211.26	1859-76-3 69	282	4-14-00-00244	bz 4; EtOH 4; HOAc 4
7627	Methanone, (4-methyl-2-nitrophenyl)phenyl- 4-Methyl-2-nitrobenzophenone	$C_{14}H_{11}NO_3$ 241.25	100224-75-7 126.5	sub	0-07-00-00442	bz 4; EtOH 4; chl 4
7628	Methanone, (2-methylphenyl)(4-methylphenyl)- 2,4'-Dimethylbenzophenone	$C_{15}H_{14}O$ 210.28	1140-16-5	317	4-07-00-01432 1.074[19]	eth 4; EtOH 4; HOAc 4
7629	Methanone, (2-methylphenyl)-1-naphthalenyl-	$C_{18}H_{14}O$ 246.31	68723-25-1 64	365	3-07-00-02639	eth 4; EtOH 4
7630	Methanone, (4-methylphenyl)(4-nitrophenyl)-	$C_{14}H_{11}NO_3$ 241.25	5350-47-0 124	sub	4-07-00-01406	bz 4; eth 4; EtOH 4; chl 4
7631	Methanone, (2-methylphenyl)phenyl-	$C_{14}H_{12}O$ 196.25	131-58-8 <-18	308; 128[12]	4-07-00-01401 1.1098[20]	H_2O 1; EtOH 4; os 3
7632	Methanone, (3-methylphenyl)phenyl-	$C_{14}H_{12}O$ 196.25	643-65-2	317; 170[9]	4-07-00-01403 1.088[17]	H_2O 1; EtOH 3; eth 3; bz 3
7633	Methanone, (4-methylphenyl)phenyl-	$C_{14}H_{12}O$ 196.25	134-84-9 59.5	228[70]	4-07-00-01403 0.9926[0]	H_2O 1; EtOH 2; eth 3; bz 3
7634	Methanone, 1-naphthalenyl-2-naphthalenyl- 1,2'-Dinaphthyl ketone	$C_{21}H_{14}O$ 282.34	605-79-8 136.5	235[0.6]	4-07-00-01886	chl 4
7635	Methanone, (3-nitrophenyl)phenyl-	$C_{13}H_9NO_3$ 227.22	2243-80-3 95	234[18]	4-07-00-01382	H_2O 1; EtOH 3
7636	Methanone, (4-nitrophenyl)phenyl-	$C_{13}H_9NO_3$ 227.22	1144-74-7 138		4-07-00-01382 1.406[9]	bz 4
7637	Methanone, phenyl-1H-pyrrol-2-yl- Pyrrole, 2-benzoyl	$C_{11}H_9NO$ 171.20	7697-46-3 79	305	5-21-08-00461	bz 4; EtOH 4; peth 4; HOAc 4
7638	Methanone, phenyl-2-pyridinyl-	$C_{12}H_9NO$ 183.21	91-02-1 42	317	5-21-08-00566 1.1556[20]	chl 3
7639	Methanone, phenyl-4-pyridinyl-	$C_{12}H_9NO$ 183.21	14548-46-0 72	315; 170[10]	5-21-08-00576	H_2O 2; EtOH 3; eth 3; bz 3
7640	Methanone, phenyl-4-quinolinyl-	$C_{16}H_{11}NO$ 233.27	54885-00-6 60	222[15]	1-21-00-00320	bz 4; eth 4; EtOH 4
7641	Methanone, phenyl-2-thienyl-	$C_{11}H_8OS$ 188.25	135-00-2 56.5	300	5-17-10-00291 1.1890[54]	H_2O 1; EtOH 3; eth 3 1.6181[54]
7642	Methanone, phenyl-3-thienyl-	$C_{11}H_8OS$ 188.25	6453-99-2 63.5	129[3]	5-17-10-00295	H_2O 1; EtOH 3; eth 3
7643	Methazole 1,2,4-Oxadiazolidine-3,5-dione, 2-(3,4-dichlorophenyl)-4-methyl-	$C_9H_6Cl_2N_2O_3$ 261.06	20354-26-1 123	5876	1.24[25]	
7644	1,3,4-Metheno-1H-cyclobuta[cd]pentalene, 1,1a,2,2,3,3a,4,5,5,5a,5b,6-dodecachlorooctahydro- Mirex	$C_{10}Cl_{12}$ 545.54	2385-85-5 485 dec	6126	3-05-00-01237	bz 4; diox 4
7645	Methidathion Phosphorodithioic acid, S-[(5-methoxy-2-oxo-1,3,4-thiadiazol-3(2H)-yl)methyl], O,O-dimethyl ester	$C_6H_{11}N_2O_4PS_3$ 302.34	950-37-8 39	5891		
7646	Methiocarb Phenol, 3,5-dimethyl-4-(methylthio)-, methylcarbamate	$C_{11}H_{15}NO_2S$ 225.31	2032-65-7 120	5893		
7647	DL-Methionine	$C_5H_{11}NO_2S$ 149.21	59-51-8 281 dec		4-04-00-03190 1.340[25]	H_2O 4; EtOH 2; eth 1; bz 1
7648	L-Methionine 2-Amino-4-methylthiobutanoic acid	$C_5H_{11}NO_2S$ 149.21	63-68-3 283 dec	5896	4-04-00-03189	H_2O 3; EtOH 1; eth 1; ace 1
7649	L-Methionine, N-acetyl- Methionamine	$C_7H_{13}NO_3S$ 191.25	65-82-7 105.5	90	4-04-00-03205	
7650	Methomyl Acetimidic acid, N-[(methylcarbamoyl)oxy]thio-, methyl ester	$C_5H_{10}N_2O_2S$ 162.21	16752-77-5 78	5905	1.2946[24]	
7651	Methoprene 2,4-Dodecadienoic acid, 11-methoxy-3,7,11-trimethyl-, 1-methylethyl ester, (E,E)-	$C_{19}H_{34}O_3$ 310.48	40596-69-8 100[0.05]	5906	0.926[20]	
7652	Methylamine, N-(o-chlorobenzylidene)- Toluene, 2-chloro-α-methylimino-	C_8H_8ClN 153.61	17972-08-6 7	78-80[2]	3-07-00-00866	1.5660[20]
7653	Methyl 3,6-dichloro-o-anisate Benzoic acid, 3,6-dichloro-2-methoxy-, methyl ester	$C_9H_8Cl_2O_3$ 235.07	6597-78-0	115[2.4]		
7654	Methylene bispropionate Propanoic acid, methylene ester	$C_7H_{12}O_4$ 160.17	7044-96-4	98[1]; 191[745]	1.053[20]	
7655	Methylisonitrile Methyl isocyanate	C_2H_3N 41.05	593-75-9 -45	59.6 exp	0.756[4]	
7656	Metobromuron Urea, 3-(p-bromophenyl)-1-methoxy-1-methyl-	$C_9H_{11}BrN_2O_2$ 259.10	3060-89-7 95	6061	1.60[20]	
7657	Metolachlor Acetamide, 2-chloro-N-(2-ethyl-6-methylphenyl)-N-(2-methoxy-1-methylethyl)-	$C_{15}H_{22}ClNO_2$ 283.80	51218-45-2 100[0.001]	6067	1.12[20]	

No.	Name Synonym	Mol. Form. Mol. Wt.	CAS RN mp/°C	Merck No. bp/°C	Beil. Ref. den/g cm⁻³	Solubility n_D
7658	Metribuzin 1,2,4-Triazin-5(4H)-one, 4-amino-6-(1,1-dimethylethyl)-3-(methylthio)-	$C_8H_{14}N_4OS$ 214.29	21087-64-9 126	6076	1.31^{20}	
7659	Mevinphos 2-Butenoic acid, 3-[(dimethoxyphosphinyl)oxy]-, methyl ester, (E)-	$C_7H_{13}O_6P$ 224.15	7786-34-7 21 (E), 6.9 (Z)	6089 $101^{0.3}$		
7660	MGK 264 4,7-Methano-1H-isoindole-1,3(2H)-dione, 2-(2-ethylhexyl)-3a,4,7,7a-tetrahydro-	$C_{17}H_{25}NO_2$ 275.39	113-48-4 <-20	157	1.04	
7661	Molinate 1H-Azepine-1-carbothioic acid, hexahydro-, S-ethyl ester	$C_9H_{17}NOS$ 187.31	2212-67-1	202^{10}	1.063^{20}	
7662	Molybdenum carbonyl Molybdenum hexacarbonyl	C_6MoO_6 264.00	13939-06-5 dec 150	subl		os 3
7663	Monocrotophos (E)-Dimethyl 1-methyl-3¹(methylamino)-3-oxo-1-propenyl phosphate	$C_7H_{14}NO_5P$ 223.17	6923-22-4 55	6161 $125^{0.0005}$	1.33^{20}	
7664	Morphinan, 6,7-didehydro-4,5-epoxy-3,6-dimethoxy-17-methyl-, (5α)- Dihydrothebaine	$C_{19}H_{23}NO_3$ 313.40	561-25-1 162.5	3164	2-27-00-00110	H_2O 1; EtOH 3; bz 3; AcOEt 3
7665	Morphinan-3,6-diol, 7,8-didehydro-4,5-epoxy-, (5α,6α)- Normorphine	$C_{16}H_{17}NO_3$ 271.32	466-97-7 273	6630		
7666	Morphinan-3,6-diol, 7,8-didehydro-4,5-epoxy-17-methyl- (5α,6α)- Morphine	$C_{17}H_{19}NO_3$ 285.34	57-27-2 255	6186 sub 190	4-27-00-02223	H_2O 1; EtOH 2; eth 1; ace 1
7667	Morphinan-3,6-diol, 7,8-didehydro-4,5-epoxy-17-methyl- (5α,6α)-, diacetate (ester) Diacetylmorphine	$C_{21}H_{23}NO_5$ 369.42	561-27-3 173	2948 $272-4^{12}$	4-27-00-02236 1.56^{25}	bz 4; chl 4
7668	Morphinan-3,6-diol, 7,8-didehydro-4,5-epoxy-17-methyl- (5α,6α)-, 17-oxide Morphine, N-oxide	$C_{17}H_{19}NO_4$ 301.34	639-46-3 274.5	6192	4-27-00-02249	H_2O 2; EtOH 2; ace 1; bz 1
7669	Morphinan-6-ol, 7,8-didehydro-4,5-epoxy-3-methoxy-17-methyl-, (5α,6α)- Codeine	$C_{18}H_{21}NO_3$ 299.37	76-57-3 157.5	2459 $250^{22·}$ $140^{1.5}$	2-27-00-00136 1.32^{25}	H_2O 3; EtOH 4; eth 3; bz 3
7670	Morphinan-6-ol, 7,8-didehydro-4,5-epoxy-3-methoxy-17-methyl-, (5α,6β)-	$C_{18}H_{21}NO_3$ 299.37	509-64-8 171.5	 dec	2-27-00-00175 1.87^4	 1.675
7671	Morphinan-6-ol, 8,14-didehydro-4,5-epoxy-3-methoxy-17-methyl-, (5α,6α)- Neopine	$C_{18}H_{21}NO_3$ 299.37	467-14-1 127.5	6376		H_2O 3; EtOH 3; eth 3; bz 3
7672	Morphinan-8-ol, 6,7-didehydro-4,5-epoxy-3-methoxy-17-methyl-, (5α,8β)- Pseudocodeine	$C_{18}H_{21}NO_3$ 299.37	466-96-6 181.5	7927	2-27-00-00112 1.290^{80}	 1.574
7673	Morphinan-6-ol, 7,8-didehydro-4,5-epoxy-3-methoxy-17-methyl-, (5α,6α)-, phosphate (1:1) (salt) Codeine, phosphate	$C_{18}H_{24}NO_7P$ 397.36	52-28-8 220-35 dec	2462	4-27-00-02228	EtOH 4; chl 4
7674	Morphinan-6-ol, 4,5-epoxy-3-methoxy-17-methyl-, (5α,6α)- Dihydrocodeine	$C_{18}H_{23}NO_3$ 301.39	125-28-0 112.5	3154 248^{15}	2-27-00-00103	
7675	Morphinan-6-one, 7,8-didehydro-4,5-epoxy-14-hydroxy-3-methoxy-17-methyl-, (5α)- Hydroxycodeinone	$C_{18}H_{19}NO_4$ 313.35	508-54-3 275 dec	4752		
7676	Morphinan-6-one, 7,8-didehydro-4-hydroxy-3,7-dimethoxy-17-methyl-, (9α,13α,14α)- Sinomenine	$C_{19}H_{23}NO_4$ 329.40	115-53-7 162	8496	5-21-13-00516	H_2O 2; EtOH 3; eth 2; ace 3
7677	Morphinan-6-one, 7,8-didehydro-4-hydroxy-3-methoxy-17-methyl- Thebainone	$C_{18}H_{21}NO_3$ 299.37	467-98-1 151.5	9203	5-21-13-00268	H_2O 2; EtOH 2; eth 2; ace 3
7678	Morphinan-6-one, 4,5-epoxy-3-methoxy-17-methyl-, (5α)- Hydrocodone	$C_{18}H_{21}NO_3$ 299.37	125-29-1 198	4708	4-27-00-03580	H_2O 1; EtOH 3
7679	Morphinan, 6,7,8,14-tetradehydro-4,5-epoxy-3,6-dimethoxy-17-methyl-, (5α)- Thebaine	$C_{19}H_{21}NO_3$ 311.38	115-37-7 193	9202 sub 91	4-27-00-02271 1.305^{20}	H_2O 1; EtOH 4; eth 2; bz 3
7680	4-Morpholinamine	$C_4H_{10}N_2O$ 102.14	4319-49-7 166		4-27-00-00623 1.059^{25}	1.4772^{20}
7681	Morpholine Tetrahydro-1,4-oxazine	C_4H_9NO 87.12	110-91-8 -4.9	6194 128	4-27-00-00015 1.0005^{20}	H_2O 5; EtOH 3; eth 3; ace 3 1.4548^{20}
7682	Morpholine, 4-acetyl-	$C_6H_{11}NO_2$ 129.16	1696-20-4 14.5	$152^{50·}$ 118^{12}	4-27-00-00274 1.1145^{20}	H_2O 5; EtOH 3; ace 3; ctc 3 1.4827^{20}
7683	Morpholine, 4-[3-(4-butoxyphenoxy)propyl]- Pramoxine	$C_{17}H_{27}NO_3$ 293.41	140-65-8 196^6	7707	4-27-00-00127	
7684	Morpholine, 4-butyl-	$C_8H_{17}NO$ 143.23	1005-67-0 -57.1	213.5	4-27-00-00026 0.9068^{20}	H_2O 4; ace 4; bz 4; EtOH 4 1.4451^{20}
7685	4-Morpholinecarboxaldehyde	$C_5H_9NO_2$ 115.13	4394-85-8 21	239	4-27-00-00274 1.1520^{20}	1.4845^{20}

No.	Name / Synonym	Mol. Form. / Mol. Wt.	CAS RN / mp/°C	Merck No. / bp/°C	Beil. Ref. / den/g cm^{-3}	Solubility / n_D
7686	4-Morpholinecarboxamide, N,N'-1,2-ethanediylbis[N-butyl- / Dimorpholamine	$C_{20}H_{38}N_4O_4$ / 398.55	119-48-2 / 41.5	3257 / 229[0.4]	4-27-00-00298	H_2O 4
7687	Morpholine, 2,6-dimethyl- / 2,6-Dimethylmorpholine	$C_6H_{13}NO$ / 115.18	141-91-3 / -88	/ 146.6	4-27-00-00674 / 0.9329[20]	H_2O 5; EtOH 5; ace 3; bz 5 / 1.4460[20]
7688	Morpholine, 4,4'-dithiobis- / 4,4'-Dithiodimorpholine	$C_8H_{16}N_2O_2S_2$ / 236.36	103-34-4 / 124.5	3379	4-27-00-00613	chl 3
7689	4-Morpholineethanamine	$C_6H_{14}N_2O$ / 130.19	2038-03-1 / 25.6	/ 205	4-27-00-00370 / 0.9897[20]	H_2O 5; EtOH 5; ace 3; bz 5 / 1.4715[20]
7690	Morpholine, 4,4'-(1,2-ethanediyl)bis-	$C_{10}H_{20}N_2O_2$ / 200.28	1723-94-0 / 75	/ 285; 160[25]	4-27-00-00375	H_2O 4; ace 4; bz 4; EtOH 4
7691	4-Morpholineethanol	$C_6H_{13}NO_2$ / 131.17	622-40-2 / -1.6	/ 227	4-27-00-00056 / 1.0710[20]	H_2O 3; EtOH 3; ctc 2 / 1.4763[20]
7692	4-Morpholineethanol, α-methyl-	$C_7H_{15}NO_2$ / 145.20	2109-66-2	/ 92-4[13]	4-27-00-00119 / 1.0174[20]	H_2O 4; ace 4; bz 4; EtOH 4 / 1.4638[20]
7693	Morpholine, 4-(2-ethoxyethyl)- / Morpholine, N-(2-ethoxyethyl)	$C_8H_{17}NO_2$ / 159.23	622-09-3 / -100	/ 206	4-27-00-00057 / 0.963[20]	H_2O 4; ace 4; bz 4; eth 4
7694	Morpholine, 4-ethyl- / N-Ethylmorpholine	$C_6H_{13}NO$ / 115.18	100-74-3	/ 138.5	4-27-00-00023 / 0.8996[20]	H_2O 5; EtOH 5; eth 5; ace 3 / 1.4400[20]
7695	Morpholine, 4-(2-hydroxybenzoyl)-	$C_{11}H_{10}NO_3$ / 204.21	3202-84-4	8302	4-27-00-00346	DMSO 3
7696	Morpholine, 4-methyl- / N-Methylmorpholine	$C_5H_{11}NO$ / 101.15	109-02-4	/ 116	4-27-00-00022 / 0.9051[20]	H_2O 3; EtOH 3; eth 3 / 1.4332[20]
7697	Morpholine, 3-methyl-2-phenyl- / Phenmetrazine	$C_{11}H_{15}NO$ / 177.25	134-49-6	7200 / 139[12]; 104[1]	4-27-00-01050	
7698	Morpholine, 4-(4-methylphenyl)-	$C_{11}H_{15}NO$ / 177.25	3077-16-5 / 51	/ 167[30]	2-27-00-00004	eth 4; EtOH 4
7699	Morpholine, 4-nitroso- / N-Nitrosomorpholine	$C_4H_8N_2O_2$ / 116.12	59-89-2 / 29	6561 / 225; 140[25]		H_2O 3
7700	Morpholine, 4-(phenylmethyl)-	$C_{11}H_{15}NO$ / 177.25	10316-00-4 / 194	/ 260.5	4-27-00-00039 / 1.0387[20]	ace 4; bz 4 / 1.5302[20]
7701	4-Morpholinepropanamine / 4-(3-Aminopropyl)morpholine	$C_7H_{16}N_2O$ / 144.22	123-00-2 / -15	/ 220; 134[50]	4-27-00-00411 / 0.9854[20]	H_2O 5; EtOH 5; ace 3; bz 5 / 1.4762[20]
7702	Naled / Phosphoric acid, 1,2-dibromo-2,2-dichloroethyl dimethyl ester	$C_4H_7Br_2Cl_2O_4P$ / 380.78	300-76-5 / 27	6272 / 110[0.5]	1.96[20]	
7703	Naphthacene / 2,3-Benzanthracene	$C_{18}H_{12}$ / 228.29	92-24-0 / 357	6288 / sub	4-05-00-02545	H_2O 2; bz 2; con sulf 3; os 2
7704	2-Naphthacenecarboxamide, 7-chloro-4-(dimethylamino)-1,4,4a,5,5a,6,11,12a-octahydro-3,6,10,12,12a-pentahydroxy-6-methyl-1,11-dioxo, [4S-(4α,4aα,5aα,6β,12aα)]- / Chlortetracycline	$C_{22}H_{23}ClN_2O_8$ / 478.89	57-62-5 / 168.5	2193	4-14-00-02629	H_2O 1; EtOH 2; eth 1; ace 2
7705	2-Naphthacenecarboxamide, 4-(dimethylamino)-1,4,4a,5,5a,6,11,12a-octahydro-3,5,6,10,12,12a-hexahydroxy-6-methyl-1,11-dioxo-,[4S-(4α,4aα,5aα,6β,12aα)]- / Oxytetracycline	$C_{22}H_{24}N_2O_9$ / 460.44	79-57-2 / 184.5	6931	4-14-00-02633 / 1.634[20]	
7706	2-Naphthacenecarboxamide, 4-(dimethylamino)-1,4,4a,5,5a,6,11,12a-octahydro-3,6,10,12,12a-pentahydroxy-6-methyl-1,11-dioxo-,[4S-(4α,4aα,5aα,6β,12aα)]- / Tetracycline	$C_{22}H_{24}N_2O_8$ / 444.44	60-54-8 / 170-5 dec	9130	4-14-00-02625	
7707	1,2-Naphthacenedione / 9,10-Naphthacenequinone	$C_{18}H_{10}O_2$ / 258.28	29276-40-2 / 294	/ sub	3-07-00-04273	EtOH 4
7708	1-Naphthalenamine / 1-Naphthylamine	$C_{10}H_9N$ / 143.19	134-32-7 / 49.2	6318 / 300.8	4-12-00-03009 / 1.0228[20]	chl 3 / 1.6140[20]
7709	2-Naphthalenamine / 2-Naphthylamine	$C_{10}H_9N$ / 143.19	91-59-8 / 113	6319 / 366	4-12-00-03122 / 1.6414[98]	H_2O 3; EtOH 3; eth 3 / 1.6493[98]
7710	2-Naphthalenamine, N,N-bis(2-chloroethyl)- / Chlornaphazine	$C_{14}H_{15}Cl_2N$ / 268.19	494-03-1 / 55	2107 / 210[5]	4-12-00-03126	ace 4; bz 4; eth 4; EtOH 4
7711	1-Naphthalenamine, 5-bromo- / 1-Amino-5-bromonaphthalene	$C_{10}H_8BrN$ / 222.08	4766-33-0 / 70	/ sub	4-12-00-03111	bz 4; eth 4; EtOH 4; lig 4
7712	1-Naphthalenamine, decahydro-	$C_{10}H_{19}N$ / 153.27	7250-95-5 / 13.5	/ 227.6	3-12-00-00183 / 0.9508[21]	
7713	1-Naphthalenamine, N,N-diethyl-	$C_{14}H_{17}N$ / 199.30	84-95-7	/ 285	3-12-00-02855 / 1.013[20]	EtOH 3; eth 3; bz 3; ctc 2 / 1.5961[20]
7714	1-Naphthalenamine, 5,8-dihydro- / 1-Naphthylamine, 5,8-dihydro-	$C_{10}H_{11}N$ / 145.20	32666-56-1 / 37.5	/ 247[408]	4-12-00-02995	EtOH 4; chl 4
7715	1-Naphthalenamine, N,N-dimethyl- / N,N-Dimethyl-1-naphthylamine	$C_{12}H_{13}N$ / 171.24	86-56-6	3240 / 250; 140[13]	4-12-00-03013 / 1.0423[20]	H_2O 1; EtOH 3; eth 3; ctc 3 / 1.624[15]
7716	2-Naphthalenamine, N,N-dimethyl-	$C_{12}H_{13}N$ / 171.24	2436-85-3 / 52.5	/ 305	4-12-00-03125 / 1.0279[60]	H_2O 1; EtOH 3; eth 3 / 1.6443[53]
7717	1-Naphthalenamine, N,N-diphenyl- / 1-Naphthylamine, N,N-diphenyl-	$C_{22}H_{17}N$ / 295.38	61231-45-6 / 142	/ 335-40[80]	0-12-00-01225	ace 4; bz 4; eth 4; EtOH 4
7718	1-Naphthalenamine, N-ethyl-	$C_{12}H_{13}N$ / 171.24	118-44-5	/ 305; 191[16]	4-12-00-03013 / 1.060[20]	eth 4; EtOH 4 / 1.6477[15]

No.	Name / Synonym	Mol. Form. / Mol. Wt.	CAS RN / mp/°C	Merck No. / bp/°C	Beil. Ref. / den/g cm^{-3}	Solubility / n_D
7719	2-Naphthalenamine, N-ethyl-	$C_{12}H_{13}N$ / 171.24	2437-03-8 / 236	316.5	4-12-00-03126 / 1.0545[21]	1.6544[21]
7720	1-Naphthalenamine, 4-methyl-	$C_{11}H_{11}N$ / 157.22	4523-45-9 / 51.5	176[12]	4-12-00-03190	eth 4
7721	1-Naphthalenamine, N-methyl-	$C_{11}H_{11}N$ / 157.22	2216-68-4 / 174	294.5	4-12-00-03013	eth 4; EtOH 4 / 1.6722[20]
7722	2-Naphthalenamine, N-methyl-	$C_{11}H_{11}N$ / 157.22	2216-67-3 /	317	4-12-00-03125	eth 4; EtOH 4 / 1.6722[20]
7723	1-Naphthalenamine, N-(2-methylphenyl)- / 1-Naphthyl(2-tolyl)amine	$C_{17}H_{15}N$ / 233.31	634-41-3 / 94.5	200[9]	3-12-00-02857	bz 4; eth 4; EtOH 4
7724	1-Naphthalenamine, N-(4-methylphenyl)-	$C_{17}H_{15}N$ / 233.31	634-43-5 / 79	360[528]; 230[10]	4-12-00-03016	bz 4; eth 4
7725	2-Naphthalenamine, N-(2-methylphenyl)- / 2-Naphthyl(2-tolyl)amine	$C_{17}H_{15}N$ / 233.31	644-15-5 / 95.5	402.5	1-12-00-00535	ace 4; bz 4; eth 4; EtOH 4
7726	1-Naphthalenamine, N-1-naphthalenyl-	$C_{20}H_{15}N$ / 269.35	737-89-3 / 115	311[15]	4-12-00-03018	H_2O 1; EtOH 3; eth 3; ace 3
7727	2-Naphthalenamine, N-2-naphthalenyl- / β,β'-Dinaphthylamine	$C_{20}H_{15}N$ / 269.35	532-18-3 / 172.2	3260 / 471	4-12-00-03131	H_2O 1; EtOH 2; eth 3; bz 2
7728	2-Naphthalenamine, N-phenyl- / N-Phenyl-β-naphthylamine	$C_{16}H_{13}N$ / 219.29	135-88-6 / 108	395.5	4-12-00-03128	H_2O 1; EtOH 3; eth 3; bz 3
7729	2-Naphthalenamine, 1-(phenylazo)- / Yellow AB	$C_{16}H_{13}N_3$ / 247.30	85-84-7 / 103	10003		EtOH 4; HOAc 4
7730	2-Naphthalenamine, N-(phenylmethyl)-	$C_{17}H_{15}N$ / 233.31	13672-18-9 / 68	405	4-12-00-03130	bz 4; lig 4
7731	1-Naphthalenamine, 5,6,7,8-tetrahydro-	$C_{10}H_{13}N$ / 147.22	2217-41-6 / 38	279	4-12-00-02934 / 1.0625[16]	H_2O 2; EtOH 3; eth 3; acid 3 / 1.5900[20]
7732	Naphthalene	$C_{10}H_8$ / 128.17	91-20-3 / 80.2	6289 / 217.9	4-05-00-01640 / 1.0253[20]	H_2O 1; EtOH 3; eth 4; ace 4 / 1.5898[25]
7733	1-Naphthaleneacetamide / Acetic acid, amide, 1-naphthyl	$C_{12}H_{11}NO$ / 185.23	86-86-2 / sub 180		4-09-00-02426	H_2O 1; eth 3; bz 3; CS_2 3
7734	1-Naphthaleneacetic acid / 1-Naphthylacetic acid	$C_{12}H_{10}O_2$ / 186.21	86-87-3 / 135 dec	6290	4-09-00-02424	H_2O 2; EtOH 2; eth 4; ace 4
7735	1-Naphthaleneacetic acid, ethyl ester / Ethyl 1-naphthylacetate	$C_{14}H_{14}O_2$ / 214.26	2122-70-5 / 88.5	222[20]; 118[13]	4-09-00-02425	EtOH 3; eth 3
7736	1-Naphthaleneacetonitrile	$C_{12}H_9N$ / 167.21	132-75-2 / 32.5	192[18]; 163[12]	4-09-00-02428	EtOH 3 / 1.6192[20]
7737	2-Naphthaleneacetonitrile	$C_{12}H_9N$ / 167.21	7498-57-9 / 83	303	3-09-00-03212 / 1.092[25]	
7738	Naphthalene, 1-azido-	$C_{10}H_7N_3$ / 169.19	6921-40-0 / 12	75[0.001]	/ 1.1713[25]	EtOH 4; eth 5; ace 5 / 1.6550[25]
7739	Naphthalene, 1-bromo- / 1-Naphthyl bromide	$C_{10}H_7Br$ / 207.07	90-11-9 / -1.8	1413 / 281	4-05-00-01665 / 1.4785[20]	H_2O 3; EtOH 5; eth 5; ace 3 / 1.658[20]
7740	Naphthalene, 2-bromo- / 2-Bromonaphthalene	$C_{10}H_7Br$ / 207.07	580-13-2 / 59	1414 / 281.5	4-05-00-01667 / 1.605[25]	H_2O 1; EtOH 3; eth 3; bz 3 / 1.6382[60]
7741	Naphthalene, 1-bromo-4-chloro- / 1-Bromo-4-chloronaphthalene	$C_{10}H_8BrCl$ / 241.51	53220-82-9 / 66.5	303	3-05-00-01585	eth 4; HOAc 4
7742	Naphthalene, 1-(bromomethyl)-	$C_{11}H_9Br$ / 221.10	3163-27-7 / 56	183[18]; 167[10]	4-05-00-01693	ace 4; bz 4; eth 4; EtOH 4
7743	Naphthalene, 2-(bromomethyl)-	$C_{11}H_9Br$ / 221.10	939-26-4 / 56	213[100]; 167[14]	4-05-00-01698	EtOH 3; eth 3; chl 3; HOAc 3
7744	Naphthalene, 1-bromo-3-nitro-	$C_{10}H_6BrNO_2$ / 252.07	7499-65-2 / 132.3	sub	4-05-00-01678	eth 4; EtOH 4
7745	Naphthalene, 2-bromo-1-nitro- / 2-Bromo-1-nitronaphthalene	$C_{10}H_6BrNO_2$ / 252.07	4185-62-0 / 104.3	>360	4-05-00-01677	bz 4; eth 4; EtOH 4
7746	Naphthalene, 1-butyl-	$C_{14}H_{16}$ / 184.28	1634-09-9 / -19.8	289.3	4-05-00-01737 / 0.9738[20]	H_2O 1; EtOH 3; eth 3; ace 3 / 1.5819[20]
7747	Naphthalene, 2-butyl- / 2-Butylnaphthalene	$C_{14}H_{18}$ / 186.30	1134-62-9 / -5	292	4-05-00-01738 / 0.9673[20]	ace 4; bz 4; EtOH 4 / 1.5777[20]
7748	1-Naphthalenecarbonitrile	$C_{11}H_7N$ / 153.18	86-53-3 / 37.5	299	4-09-00-02407 / 1.1080[25]	H_2O 1; EtOH 4; eth 4; lig 3 / 1.6298[18]
7749	2-Naphthalenecarbonitrile	$C_{11}H_7N$ / 153.18	613-46-7 / 66	306.5	4-09-00-02419 / 1.0755[60]	H_2O 2; EtOH 3; eth 3; chl 2
7750	2-Naphthalenecarbonyl chloride	$C_{11}H_7ClO$ / 190.63	2243-83-6 / 51	305	4-09-00-02417	bz 4; eth 4; chl 4
7751	1-Naphthalenecarboxaldehyde	$C_{11}H_8O$ / 156.18	66-77-3 / 33.5	292	4-07-00-01286 / 1.1503[20]	H_2O 1; EtOH 3; eth 3; ace 3 / 1.6507[20]
7752	2-Naphthalenecarboxaldehyde	$C_{11}H_8O$ / 156.18	66-99-9 / 62	160[19]	4-07-00-01288 / 1.0775[99]	H_2O 2; EtOH 4; eth 4; ace 3 / 1.621[99]
7753	1-Naphthalenecarboxaldehyde, 2-ethoxy-	$C_{13}H_{12}O_2$ / 200.24	19523-57-0 / 115	185-7[25]	3-08-00-01110	EtOH 3; HOAc 3
7754	1-Naphthalenecarboxaldehyde, 2-hydroxy-	$C_{11}H_8O_2$ / 172.18	708-06-5 / 83	192[27]	4-08-00-01160	H_2O 1; EtOH 3; eth 3; aq alk 3
7755	1-Naphthalenecarboxaldehyde, 4-methoxy-	$C_{12}H_{10}O_2$ / 186.21	15971-29-6 / 34.8	212[40]	4-08-00-01170 / 1.1879[39]	

No.	Name Synonym	Mol. Form. Mol. Wt.	CAS RN mp/°C	Merck No. bp/°C	Beil. Ref. den/g cm^{-3}	Solubility n_D
7756	1-Naphthalenecarboxaldehyde, 4-methyl- 1-Naphthaldehyde, 4-methyl-	$C_{12}H_{10}O$ 170.21	33738-48-6 33.5	175[13]	3-07-00-01972 1.1252[38]	
7757	1-Naphthalenecarboxamide	$C_{11}H_9NO$ 171.20	2243-81-4 205.8	sub	4-09-00-02405	H_2O 2; EtOH 2; con HCl 3; HOAc 3
7758	1-Naphthalenecarboxylic acid 1-Naphthoic acid	$C_{11}H_8O_2$ 172.18	86-55-5 161	6301 >300	4-09-00-02402 1.398[25]	eth 4; EtOH 4; chl 4 1.46
7759	2-Naphthalenecarboxylic acid 2-Naphthoic acid	$C_{11}H_8O_2$ 172.18	93-09-4 185.5	6302 >300	4-09-00-02414 1.077[100]	H_2O 2; EtOH 3; eth 3; chl 3
7760	1-Naphthalenecarboxylic acid, 5-amino-	$C_{11}H_9NO_2$ 187.20	32018-88-5 212.3	sub	4-14-00-01812	EtOH 4; HOAc 4
7761	1-Naphthalenecarboxylic acid, 5-bromo-	$C_{11}H_7BrO_2$ 251.08	16726-67-3 261	sub	3-09-00-03153	H_2O 2; EtOH 2; eth 2; bz 4
7762	1-Naphthalenecarboxylic acid, 5-chloro- 1-Naphthoic acid, 5-chloro-	$C_{11}H_7ClO_2$ 206.63	16650-52-5 245	sub	3-09-00-03149	H_2O 2; EtOH 4; bz 2; HOAc 2
7763	1-Naphthalenecarboxylic acid, 8-chloro-	$C_{11}H_7ClO_2$ 206.63	4537-00-2 171.5	sub	3-09-00-03151	H_2O 3; EtOH 3; ace 4; bz 3
7764	1-Naphthalenecarboxylic acid, 3,4-dihydro- 3,4-Dihydro-1-naphthoic acid	$C_{11}H_{10}O_2$ 174.20	3333-23-1 125	306	4-09-00-02338	H_2O 2; EtOH 3; chl 3; MeOH 3
7765	1-Naphthalenecarboxylic acid, ethyl ester	$C_{13}H_{12}O_2$ 200.24	3007-97-4 	310	4-09-00-02403 1.1264[15]	H_2O 1; EtOH 3 1.5966[15]
7766	1-Naphthalenecarboxylic acid, methyl ester	$C_{12}H_{10}O_2$ 186.21	2459-24-7 59.5	168[20] 101[0.04]	4-09-00-02402 1.1290[20]	bz 4; EtOH 4 1.6086[20]
7767	1-Naphthalenecarboxylic acid, 5-nitro-	$C_{11}H_7NO_4$ 217.18	1975-44-6 241.5	sub	4-09-00-02411	H_2O 1; EtOH 3; eth 2; bz 2
7768	2-Naphthalenecarboxylic acid, 3-amino- 3-Amino-2-naphthoic acid	$C_{11}H_9NO_2$ 187.20	5959-52-4 216.5	464	3-14-00-01341	EtOH 3; eth 3
7769	2-Naphthalenecarboxylic acid, 5-bromo- 5-Bromo-2-naphthoic acid	$C_{11}H_7BrO_2$ 251.08	1013-83-8 270	sub	3-09-00-03196	bz 4; eth 4; EtOH 4; HOAc 4
7770	2-Naphthalenecarboxylic acid, ethyl ester	$C_{13}H_{12}O_2$ 200.24	3007-91-8 32	308.5	4-09-00-02415 1.1143[23]	eth 4; EtOH 4; chl 4; HOAc 4 1.5951[23]
7771	2-Naphthalenecarboxylic acid, 1-hydroxy- 1-Hydroxy-2-naphthoic acid	$C_{11}H_8O_3$ 188.18	86-48-6 195	4766	4-10-00-01181	H_2O 2; EtOH 4; eth 4; bz 3
7772	2-Naphthalenecarboxylic acid, 3-hydroxy- 3-Hydroxy-2-naphthoic acid	$C_{11}H_8O_3$ 188.18	92-70-6 222.5	4767	4-10-00-01184	H_2O 2; EtOH 4; eth 4; bz 3
7773	2-Naphthalenecarboxylic acid, 3-hydroxy-, ethyl ester	$C_{13}H_{12}O_3$ 216.24	7163-25-9 85	291	4-10-00-01186	ace 4; chl 4
7774	2-Naphthalenecarboxylic acid, 3-hydroxy-, hexadecyl ester Hexadecyl 3-hydroxy-2-naphthoate	$C_{27}H_{40}O_3$ 412.61	531-84-0 72.5	4604	3-10-00-01087	bz 4; HOAc 4
7775	2-Naphthalenecarboxylic acid, 3-hydroxy-, methyl ester	$C_{12}H_{10}O_3$ 202.21	883-99-8 75.5	206	4-10-00-01186	H_2O 1; EtOH 3; os 3
7776	2-Naphthalenecarboxylic acid, methyl ester	$C_{12}H_{10}O_2$ 186.21	2459-25-8 77	290	4-09-00-02414	bz 4; eth 4; EtOH 4; chl 4
7777	2-Naphthalenecarboxylic acid, 4,4'- methylenebis[3-hydroxy- Pamoic acid	$C_{23}H_{16}O_6$ 388.38	130-85-8 315	6955	4-10-00-02362	
7778	2-Naphthalenecarboxylic acid, 1,2,3,4- tetrahydro- 1,2,3,4-Tetrahydro-2-naphthoic acid	$C_{11}H_{12}O_2$ 176.22	53440-12-3 97.7	168[15]	3-09-00-02805	bz 4; eth 4; EtOH 4; chl 4
7779	2-Naphthalenecarboxylic acid, 5,6,7,8- tetrahydro-	$C_{11}H_{12}O_2$ 176.22	1131-63-1 155.5	216[14]	4-09-00-02096	EtOH 4; bz 3; HOAc 3; peth 1
7780	Naphthalene, 1-chloro- 1-Naphthyl chloride	$C_{10}H_7Cl$ 162.62	90-13-1 -2.5	2149 259; 106.5[5]	4-05-00-01658 1.1938[20]	H_2O 1; EtOH 3; eth 3; bz 3 1.6326[20]
7781	Naphthalene, 2-chloro- 2-Chloronaphthalene	$C_{10}H_7Cl$ 162.62	91-58-7 61	2150 256	4-05-00-01660 1.1377[71]	H_2O 1; EtOH 3; eth 3; bz 3 1.6079[13]
7782	Naphthalene, 2-chlorodecahydro- Decalin, 2-chloro-	$C_{10}H_{17}Cl$ 172.70	5597-81-9 	112[15]	4-05-00-00313 1.0421[20]	1.5020[20]
7783	Naphthalene, 1-(chloromethyl)-	$C_{11}H_9Cl$ 176.65	86-52-2 32	291.5	4-05-00-01692 1.1813[20]	H_2O 1; EtOH 3; ctc 3; peth 3 1.6380[20]
7784	Naphthalene, 1-chloro-2-methyl- 1-Chloro-2-methylnaphthalene	$C_{11}H_9Cl$ 176.65	5859-45-0 	150-2[10]	3-05-00-01632	ctc 3 1.6265[20]
7785	Naphthalene, 2-(chloromethyl)-	$C_{11}H_9Cl$ 176.65	2506-41-4 48.5	169[20]	4-05-00-01697	H_2O 1; EtOH 3; peth 3
7786	Naphthalene, 1-chloro-5-nitro- 1-Chloro-5-nitronaphthalene	$C_{10}H_6ClNO_2$ 207.62	605-63-0 111	>360	3-05-00-01597	EtOH 4
7787	Naphthalene, 1-chloro-8-nitro- 1-Chloro-8-nitronaphthalene	$C_{10}H_6ClNO_2$ 207.62	602-37-9 95.8	175[2]	4-05-00-01676	EtOH 3; bz 3; chl 2; HOAc 3
7788	Naphthalene, 2-chloro-1-nitro- 2-Chloro-1-nitronaphthalene	$C_{10}H_6ClNO_2$ 207.62	4185-63-1 99.5	>360	4-05-00-01676 1.4854[25]	EtOH 4; eth 4; ace 4; bz 4
7789	Naphthalene, decahydro- Decalin	$C_{10}H_{18}$ 138.25	91-17-8 -43	2839 155.5	3-05-00-00245 0.8965[22]	ace 4; bz 4; eth 4; EtOH 4 1.4810[20]
7790	Naphthalene, decahydro-, cis- cis-Decalin	$C_{10}H_{18}$ 138.25	493-01-6 -42.9	2839 195.8	4-05-00-00310 0.8965[20]	H_2O 1; EtOH 5; eth 4; ace 4 1.4810[20]
7791	Naphthalene, decahydro-, trans- trans-Decalin	$C_{10}H_{18}$ 138.25	493-02-7 -30.3	2839 187.3	4-05-00-00311 0.8699[20]	H_2O 1; EtOH 4; eth 4; ace 4 1.4695[20]

No.	Name / Synonym	Mol. Form. / Mol. Wt.	CAS RN / mp/°C	Merck No. / bp/°C	Beil. Ref. / den/g cm^{-3}	Solubility / n_D
7792	Naphthalene, decahydro-2-methylene-, *trans*- / Decalin, 2-methylene (*trans*)	$C_{11}H_{18}$ / 150.26	7787-72-6 /	/ 201; 82[10]	3-05-00-00397 / 0.8897[20]	/ 1.4841[22]
7793	Naphthalene, 1-decyl- / Decane, 1-(1-naphthyl)-	$C_{20}H_{28}$ / 268.44	26438-27-7 / 15	/ 379	4-05-00-01786 / 0.9322[20]	/ 1.5435[20]
7794	Naphthalene, 2-decyl-	$C_{20}H_{28}$ / 268.44	14188-79-5 / 20	/ 387	4-05-00-01786 / 0.9253[20]	/ 1.5413[20]
7795	1,2-Naphthalenediamine / 1,2-Diaminonaphthalene	$C_{10}H_{10}N_2$ / 158.20	938-25-0 / 98.5	/ 214[13], 150[0.5]	4-13-00-00337 /	H_2O 2; EtOH 4; eth 4; chl 4 /
7796	1,4-Naphthalenediamine / 1,4-Diaminonaphthalene	$C_{10}H_{10}N_2$ / 158.20	2243-61-0 / 120	/	4-13-00-00338 /	ace 4; bz 4; EtOH 4; chl 4 / 1.6441[18]
7797	1,5-Naphthalenediamine	$C_{10}H_{10}N_2$ / 158.20	2243-62-1 / 190	/ sub	4-13-00-00340 / 1.4[25]	H_2O 3; EtOH 3; eth 3; chl 4 /
7798	1,8-Naphthalenediamine / Naphthalene, 1,8-diamino	$C_{10}H_{10}N_2$ / 158.20	479-27-6 / 66.5	6291 / 205[12]	4-13-00-00344 / 1.1265[90]	eth 4; EtOH 4 / 1.6828[99]
7799	2,3-Naphthalenediamine / 2,3-Diaminonaphthalene	$C_{10}H_{10}N_2$ / 158.20	771-97-1 / 199	/	4-13-00-00346 / 1.0968[26]	H_2O 2; EtOH 4; eth 3 / 1.6392[26]
7800	1,4-Naphthalenediamine, 2-methyl- / Vitamin K6	$C_{11}H_{12}N_2$ / 172.23	83-68-1 / 113.5	9937 /	3-13-00-00406 /	
7801	Naphthalene, 1,3-dibromo- / 1,3-Dibromonaphthalene	$C_{10}H_6Br_2$ / 285.97	52358-73-3 / 62.8	/ 290	3-05-00-01586 /	EtOH 4 /
7802	Naphthalene, 1,4-dibromo- / 1,4-Dibromonaphthalene	$C_{10}H_6Br_2$ / 285.97	83-53-4 / 83	/ 310	4-05-00-01669 /	H_2O 1; EtOH 3; eth 3; HOAc 2 /
7803	Naphthalene, 1,5-dibromo- / 1,5-Dibromonaphthalene	$C_{10}H_6Br_2$ / 285.97	7351-74-8 / 131.3	/ 326	4-05-00-01669 /	eth 4; EtOH 4 /
7804	1,8-Naphthalenedicarbonyl dichloride / Naphthaloyl chloride	$C_{12}H_6Cl_2O_2$ / 253.08	6423-29-6 / 85	/ 197[0.2]	3-09-00-04468 /	bz 4; chl 4 /
7805	1,8-Naphthalenedicarboxylic acid / Naphthalic acid	$C_{12}H_8O_4$ / 216.19	518-05-8 / 260	6299 /	4-09-00-03534 /	H_2O 1; EtOH 2; eth 2 /
7806	1,8-Naphthalenedicarboxylic acid, diethyl ester / Naphthalic acid, diethyl ester	$C_{16}H_{16}O_4$ / 272.30	58618-39-6 / 59.5	/ 238[19], 202[11]	4-09-00-03534 / 1.1399[70]	
7807	Naphthalene, 1,2-dichloro- / 1,2-Dichloronaphthalene	$C_{10}H_6Cl_2$ / 197.06	2050-69-3 / 36	/ 296.5	4-05-00-01661 / 1.3147[49]	EtOH 3; eth 3 / 1.5338[49]
7808	Naphthalene, 1,3-dichloro- / 1,3-Dichloronaphthalene	$C_{10}H_6Cl_2$ / 197.06	2198-75-6 / 62.3	/ 291	4-05-00-01661 /	EtOH 3 /
7809	Naphthalene, 1,4-dichloro- / 1,4-Dichloronaphthalene	$C_{10}H_6Cl_2$ / 197.06	1825-31-6 / 67.5	/ 288; 147[12]	4-05-00-01661 / 1.2997[76]	H_2O 1; EtOH 2; eth 3; ace 4 / 1.6228[76]
7810	Naphthalene, 1,5-dichloro- / 1,5-Dichloronaphthalene	$C_{10}H_6Cl_2$ / 197.06	1825-30-5 / 107	/ sub	4-05-00-01662 / 1.4900[20]	H_2O 1; EtOH 2; eth 3 /
7811	Naphthalene, 1,7-dichloro- / 1,7-Dichloronaphthalene	$C_{10}H_6Cl_2$ / 197.06	2050-73-9 / 63.5	/ 285.5	4-05-00-01662 / 1.2611[100]	EtOH 3; eth 3; bz 3; HOAc 3 / 1.6092[100]
7812	Naphthalene, 1,8-dichloro- / 1,8-Dichloronaphthalene	$C_{10}H_6Cl_2$ / 197.06	2050-74-0 / 89	/ sub	4-05-00-01662 / 1.2924[100]	EtOH 3; peth 3 / 1.6236[100]
7813	Naphthalene, 2,6-dichloro- / 2,6-Dichloronaphthalene	$C_{10}H_6Cl_2$ / 197.06	2065-70-5 / 140.5	/ 285	4-05-00-01662 /	EtOH 2; eth 3; bz 3; chl 3 /
7814	Naphthalene, 1,2-dihydro-	$C_{10}H_{10}$ / 130.19	447-53-0 / -8	/ 206.5	4-05-00-01543 / 0.9974[20]	/ 1.5814[20]
7815	Naphthalene, 1,4-dihydro- / Δ 2-Dialin	$C_{10}H_{10}$ / 130.19	612-17-9 / 25	/ 211.5	4-05-00-01544 / 0.9928[33]	/ 1.5577[20]
7816	Naphthalene, 1,2-dihydro-3-methyl-	$C_{11}H_{12}$ / 144.22	2717-44-4 /	/ 105[13]; 64[0.6]	4-05-00-01552 / 0.9819[20]	H_2O 1; eth 3; bz 3 / 1.5751[20]
7817	Naphthalene, 1,2-dihydro-4-methyl-	$C_{11}H_{12}$ / 144.22	4373-13-1 /	/ 112[18]; 84[5]	4-05-00-01551 / 0.9895[20]	H_2O 1; eth 3; bz 3 / 1.5758[20]
7818	Naphthalene, 1,2-dimethyl- / 1,2-Dimethylnaphthalene	$C_{12}H_{12}$ / 156.23	573-98-8 / 1.6	/ 266.5	4-05-00-01708 / 1.0179[20]	H_2O 1; eth 3; bz 3 / 1.6166[20]
7819	Naphthalene, 1,3-dimethyl- / 1,3-Dimethylnaphthalene	$C_{12}H_{12}$ / 156.23	575-41-7 / -6	/ 263	4-05-00-01708 / 1.0144[20]	H_2O 1; eth 3; bz 3 / 1.6140[20]
7820	Naphthalene, 1,4-dimethyl- / 1,4-Dimethylnaphthalene	$C_{12}H_{12}$ / 156.23	571-58-4 / 7.6	/ 268	4-05-00-01709 / 1.0166[20]	H_2O 1; EtOH 4; eth 5; ace 5 / 1.6127[20]
7821	Naphthalene, 1,5-dimethyl- / 1,5-Dimethylnaphthalene	$C_{12}H_{12}$ / 156.23	571-61-9 / 82	/ 265	4-05-00-01710 /	bz 4; eth 4 /
7822	Naphthalene, 1,6-dimethyl- / 1,6-Dimethylnaphthalene	$C_{12}H_{12}$ / 156.23	575-43-9 / -16.9	/ 264	4-05-00-01711 / 1.0021[20]	H_2O 1; eth 3; bz 3 / 1.6166[20]
7823	Naphthalene, 1,7-dimethyl- / 1,7-Dimethylnaphthalene	$C_{12}H_{12}$ / 156.23	575-37-1 / -13.9	/ 263	4-05-00-01711 / 1.0115[20]	H_2O 1; eth 3; bz 3 / 1.6083[20]
7824	Naphthalene, 1,8-dimethyl- / 1,8-Dimethylnaphthalene	$C_{12}H_{12}$ / 156.23	569-41-5 / 65	/ 270	4-05-00-01712 / 1.003[20]	H_2O 1; eth 3; bz 3 /
7825	Naphthalene, 2,3-dimethyl- / Guajen	$C_{12}H_{12}$ / 156.23	581-40-8 / 105	/ 268	4-05-00-01713 / 1.003[20]	bz 4; eth 4 / 1.5060[20]
7826	Naphthalene, 2,6-dimethyl- / 2,6-Dimethylnaphthalene	$C_{12}H_{12}$ / 156.23	581-42-0 / 112	/ 262	4-05-00-01714 / 1.003[20]	
7827	Naphthalene, 2,7-dimethyl- / 2,7-Dimethylnaphthalene	$C_{12}H_{12}$ / 156.23	582-16-1 / 97	/ 265	4-05-00-01715 / 1.003[20]	
7828	Naphthalene, 1-(1,1-dimethylethyl)-	$C_{14}H_{16}$ / 184.28	17085-91-5 / -11.3	/ 279	4-05-00-01739 / 0.9629[20]	/ 1.5726[20]
7829	Naphthalene, 2-(1,1-dimethylethyl)-	$C_{14}H_{16}$ / 184.28	2876-35-9 / -4	/ 276; 125[4]	4-05-00-01740 / 0.9674[20]	ace 4; bz 4; EtOH 4 / 1.5685[20]

No.	Name Synonym	Mol. Form. Mol. Wt.	CAS RN mp/°C	Merck No. bp/°C	Beil. Ref. den/g cm^{-3}	Solubility n_D
7830	Naphthalene, 1,6-dimethyl-4-(1-methylethyl)- Cadalene	$C_{15}H_{18}$ 198.31	483-78-3	1607 294; 149[10]	4-05-00-01758 0.9667[25]	oils 4 1.5785[25]
7831	Naphthalene, 1,3-dinitro- 1,3-Dinitronaphthalene	$C_{10}H_6N_2O_4$ 218.17	606-37-1 148	sub	4-05-00-01680	H_2O 1; EtOH 3; ace 3
7832	Naphthalene, 1,5-dinitro- 1,5-Dinitronaphthalene	$C_{10}H_6N_2O_4$ 218.17	605-71-0 219	sub	4-05-00-01680 1.5860[20]	H_2O 1; EtOH 2; eth 4; ace 2
7833	Naphthalene, 1,8-dinitro- 1,8-Dinitronaphthalene	$C_{10}H_6N_2O_4$ 218.17	602-38-0 173	445 dec	4-05-00-01681	H_2O 1; EtOH 2; ace 3; bz 2
7834	1,3-Naphthalenediol Naphthoresorcinol	$C_{10}H_8O_2$ 160.17	132-86-5 123.5	6316	4-06-00-06543	H_2O 3; EtOH 3; eth 3; ace 2
7835	1,5-Naphthalenediol	$C_{10}H_8O_2$ 160.17	83-56-7 262 dec	sub	4-06-00-06554	H_2O 2; EtOH 2; eth 4; ace 4
7836	1,6-Naphthalenediol	$C_{10}H_8O_2$ 160.17	575-44-0 138	sub	4-06-00-06557	H_2O 2; EtOH 2; eth 3; ace 3
7837	1,7-Naphthalenediol	$C_{10}H_8O_2$ 160.17	575-38-2 180.5	sub	4-06-00-06559	H_2O 2; EtOH 2; eth 4; bz 3
7838	2,6-Naphthalenediol	$C_{10}H_8O_2$ 160.17	581-43-1 220	sub	4-06-00-06566	H_2O 2; EtOH 3; eth 3; ace 3
7839	2,7-Naphthalenediol	$C_{10}H_8O_2$ 160.17	582-17-2 193	sub	4-06-00-06570	H_2O 3; EtOH 3; eth 3; ace 2
7840	1,8-Naphthalenediol, cyclic sulfite 1-Naphthalenesulfonic acid, 8-hydroxy, sulfone	$C_{10}H_6O_3S$ 206.22	21849-97-8 97	>360	2-06-00-00954	bz 4; chl 4
7841	1,4-Naphthalenediol, 2-methyl-, diacetate Menadiol diacetate	$C_{15}H_{14}O_4$ 258.27	573-20-6 113	5710	4-06-00-06581	EtOH 4
7842	1,2-Naphthalenedione 1,2-Naphthoquinone	$C_{10}H_6O_2$ 158.16	524-42-5 146	6314	4-07-00-02417 1.450[25]	H_2O 3; EtOH 3; eth 3; sulf 3
7843	1,4-Naphthalenedione 1,4-Naphthoquinone	$C_{10}H_6O_2$ 158.16	130-15-4 128.5	6315 sub	4-07-00-02422	H_2O 2; EtOH 4; eth 3; bz 3
7844	1,2-Naphthalenedione, 3-bromo- 3-Bromo-1,2-naphthoquinone	$C_{10}H_5BrO_2$ 237.05	7474-83-1 178	sub	4-07-00-02422	bz 4; EtOH 4
7845	1,4-Naphthalenedione, 2-bromo-3-hydroxy-	$C_{10}H_5BrO_3$ 253.05	1203-39-0 202	sub	4-08-00-02364	ace 4
7846	1,4-Naphthalenedione, 2-bromo-3-methyl-	$C_{11}H_7BrO_2$ 251.08	3129-39-3 152.8	sub 100	4-07-00-02431	ace 4; bz 4; eth 4; EtOH 4
7847	1,4-Naphthalenedione, 5-chloro- 5-Chloro-1,4-naphthoquinone	$C_{10}H_5ClO_2$ 192.60	40242-15-7 163	sub	4-07-00-02425	EtOH 4; lig 4
7848	1,4-Naphthalenedione, 2-chloro-3-hydroxy-	$C_{10}H_5ClO_3$ 208.60	1526-73-4 216.5	sub	4-08-00-02363	bz 4; eth 4; EtOH 4
7849	1,2-Naphthalenedione, 3,4-dichloro-	$C_{10}H_4Cl_2O_2$ 227.05	18398-36-2 184	sub	4-07-00-02422	bz 4; chl 4
7850	1,4-Naphthalenedione, 2,3-dichloro- Dichlone	$C_{10}H_4Cl_2O_2$ 227.05	117-80-6 195	3032	4-07-00-02426	H_2O 1; EtOH 2; eth 2; bz 2
7851	1,4-Naphthalenedione, 5,6-dichloro- 1,4-Naphthoquinone, 5,6-dichloro	$C_{10}H_4Cl_2O_2$ 227.05	56961-93-4 181	sub	0-07-00-00730	eth 4
7852	1,4-Naphthalenedione, 2,3-dihydroxy-	$C_{10}H_6O_4$ 190.16	605-37-8 282	sub	4-08-00-02953	H_2O 2; EtOH 2; eth 2; ace 3
7853	1,4-Naphthalenedione, 5,8-dihydroxy-	$C_{10}H_6O_4$ 190.16	475-38-7 232	sub	4-08-00-02946	H_2O 2; EtOH 2; eth 2; HOAc 3
7854	1,4-Naphthalenedione, 5,8-dihydroxy-6-(1-hydroxy-4-methyl-3-pentenyl)-, (S)- Alkannin	$C_{16}H_{16}O_5$ 288.30	23444-65-7 149	243 sub 140	3-08-00-04089	EtOH 4
7855	1,4-Naphthalenedione, 5,8-dihydroxy-6-methoxy-2-methyl-3-(2-oxopropyl)- Javanicin	$C_{15}H_{14}O_6$ 290.27	476-45-9 208 dec	5143	4-08-00-03646	alk 3
7856	1,4-Naphthalenedione, 3,5-dihydroxy-2-methyl-	$C_{11}H_8O_4$ 204.18	478-40-0 181	sub 100	4-08-00-02958	H_2O 3; EtOH 3; eth 3; peth 3
7857	1,4-Naphthalenedione, 2-ethyl-3,5,6,7,8-pentahydroxy- Echinochrome A	$C_{12}H_{10}O_7$ 266.21	517-82-8 220 dec	3470 sub 120	4-08-00-03714	H_2O 2; EtOH 3; eth 4; ace 3
7858	1,4-Naphthalenedione, 6-ethyl-2,3,5,7,8-pentahydroxy-	$C_{12}H_{10}O_7$ 266.21	1471-96-1 220 dec	120[0.0004]	4-08-00-03714	ace 4; eth 4; EtOH 4
7859	1,4-Naphthalenedione, 2 hydroxy- Lawsone	$C_{10}H_8O_3$ 174.16	83-72-7 195-6 dec	5263	4-08-00-02380	EtOH 4; eth 1; bz 1; chl 1
7860	1,4-Naphthalenedione, 5-hydroxy- Juglone	$C_{10}H_8O_3$ 174.16	481-39-0 155	5150 sub	4-08-00-02368	H_2O 1; EtOH 3; eth 3; bz 3
7861	1,4-Naphthalenedione, 2-hydroxy-3-methyl- Phthiocol	$C_{11}H_8O_3$ 188.18	483-55-6 173.5	7352 sub	4-08-00-02375	ace 4; eth 4
7862	1,4-Naphthalenedione, 2-hydroxy-3-(3-methyl-2-butenyl)- Lapachol	$C_{15}H_{14}O_3$ 242.27	84-79-7 139.5	5235	4-08-00-02487	H_2O 1; EtOH 3; eth 3; bz 3
7863	1,4-Naphthalenedione, 5-hydroxy-2-methyl- 1,4-Naphthoquinone, 5-hydroxy-2-methyl	$C_{11}H_8O_3$ 188.18	481-42-5 78.5	7511 sub	4-08-00-02376	ace 4; bz 4; eth 4; EtOH 4
7864	1,4-Naphthalenedione, 2-methyl- Menadione	$C_{11}H_8O_2$ 172.18	58-27-5 107	5714	4-07-00-02430	H_2O 1; EtOH 2; eth 3; bz 3
7865	1,4-Naphthalenedione, 2-methyl-3-(3,7,11,15-tetramethyl-2-hexadecenyl)-, [R-[R*,R*-(E)]]- Vitamin K1	$C_{31}H_{46}O_2$ 450.71	84-80-0 -20	9933 140-5[0.001]	4-07-00-02496 0.964[25]	ace 4; bz 4; eth 4; EtOH 4 1.5250[25]

No.	Name Synonym	Mol. Form. Mol. Wt.	CAS RN mp/°C	Merck No. bp/°C	Beil. Ref. den/g cm^{-3}	Solubility n_D
7866	1,4-Naphthalenedione, 2-(phenylamino)-	$C_{16}H_{11}NO_2$ 249.27	6628-97-3 193	sub	4-14-00-00399	bz 4; eth 4
7867	1,5-Naphthalenedisulfonic acid Armstrong's acid	$C_{10}H_8O_6S_2$ 288.30	81-04-9 240-5 dec	812	4-11-00-00561 1.493[25]	H_2O 4; EtOH 3; eth 1
7868	1,6-Naphthalenedisulfonic acid Eiver-pick acid	$C_{10}H_8O_6S_2$ 288.30	525-37-1 125 dec	6292	4-11-00-00562	H_2O 4; EtOH 3; eth 1
7869	2,7-Naphthalenedisulfonic acid Naphthalene-2,7-disulfonic acid	$C_{10}H_8O_6S_2$ 288.30	92-41-1 199	6294	4-11-00-00563	H_2O 3; con HCl 2
7870	1,3-Naphthalenedisulfonic acid, 7-amino- Amido-G-Acid	$C_{10}H_9NO_6S_2$ 303.32	86-65-7 274	412	4-14-00-02811	H_2O 4; EtOH 4
7871	1,3-Naphthalenedisulfonic acid, 7-hydroxy- 2-Naphthol-6,8-disulfonic acid	$C_{10}H_8O_7S_2$ 304.30	118-32-1	6308	4-11-00-00625	H_2O 3
7872	1,6-Naphthalenedisulfonic acid, 4-amino- 1-Naphthylamine-4,7-disulfonic acid	$C_{10}H_9NO_6S_2$ 303.32	85-75-6	6322	2-14-00-00476	H_2O 4
7873	1,7-Naphthalenedisulfonic acid, 4-amino- 1-Naphthylamine-4,6-disulfonic acid	$C_{10}H_9NO_6S_2$ 303.32	85-74-5	6321	2-14-00-00477	H_2O 4; EtOH 4
7874	2,7-Naphthalenedisulfonic acid, 4-amino-5- hydroxy- 1-Amino-8-naphthol-3,6-disulfonic acid	$C_{10}H_9NO_7S_2$ 319.32	90-20-0	6305	4-14-00-02824	H_2O 2; EtOH 2; eth 2
7875	2,7-Naphthalenedisulfonic acid, 4,5-dihydroxy- Chromotropic acid	$C_{10}H_8O_8S_2$ 320.30	148-25-4	2243	4-11-00-00636	H_2O 3; EtOH 1; eth 1; alk 3
7876	2,7-Naphthalenedisulfonic acid, 3-hydroxy- 2-Naphthol-3,6-disulfonic acid	$C_{10}H_8O_7S_2$ 304.30	148-75-4 dec	6307	4-11-00-00624	H_2O 4; EtOH 4
7877	1,3-Naphthalenedisulfonic acid, 6,6'-[(3,3'- dimethyl[1,1'-biphenyl]-4,4'-diyl)bis(azo)]bis[4- amino-5-hydroxy-, C.I. Direct blue 53	$C_{34}H_{24}N_6Na_4O_{14}S_4$ 960.82	314-13-6	3863		H_2O 3; EtOH 3; acid 3
7878	2,7-Naphthalenedisulfonic acid, 4,5-dihydroxy- 3-[(4-nitrophenyl)azo]-, disodium salt	$C_{16}H_9N_3Na_2O_{10}S_2$ 513.37	548-80-1	2242		
7879	Naphthalene, 1-dodecyl- Dodecane, 1-(1-naphthyl)-	$C_{22}H_{32}$ 296.50	26438-28-8 27	415	4-05-00-01790 0.9240[20]	1.5364[20]
7880	Naphthalene, 2-dodecyl- Dodecane, 1-(2-naphthyl)-	$C_{22}H_{32}$ 296.50	60899-39-0 28	414	4-05-00-01791 0.9177[20]	1.5343[20]
7881	Naphthalene, 1-ethenyl-	$C_{12}H_{10}$ 154.21	826-74-4	124-5[15]	4-05-00-01833 1.036[18]	1.044[20]
7882	Naphthalene, 2-ethenyl- 2-Vinylnaphthalene	$C_{12}H_{10}$ 154.21	827-54-3 66	135[18]; 95[2]	4-05-00-01833	H_2O 1; EtOH 3; ace 3; bz 3
7883	Naphthalene, 1-ethoxy-	$C_{12}H_{12}O$ 172.23	5328-01-8 5.5	280.5	4-06-00-04212 1.060[20]	H_2O 1; EtOH 4; eth 4 1.5953[25]
7884	Naphthalene, 2-ethoxy- 2-Ethoxynaphthalene	$C_{12}H_{12}O$ 172.23	93-18-5 37.5	3709 282	4-06-00-04257 1.0640[20]	H_2O 1; EtOH 3; eth 3; tol 3 1.5975[36]
7885	Naphthalene, 1-ethyl- 1-Ethylnaphthalene	$C_{12}H_{12}$ 156.23	1127-76-0 -13.9	258.6	4-05-00-01705 1.0082[20]	H_2O 1; EtOH 5; eth 5 1.6062[20]
7886	Naphthalene, 2-ethyl- 2-Ethylnaphthalene	$C_{12}H_{12}$ 156.23	939-27-5 -7.4	258	4-05-00-01707 0.9922[20]	H_2O 1; EtOH 5; eth 5; chl 2 1.5999[20]
7887	Naphthalene, 1-ethyl-5-methyl- 1-Methyl-5-ethylnaphthalene	$C_{13}H_{14}$ 170.25	17057-92-0 40	133[10]	3-05-00-01659	1.600[30]
7888	Naphthalene, 1-ethyl-1,2,3,4-tetrahydro- 1-Ethyltetralin	$C_{12}H_{16}$ 160.26	13556-58-6	241	4-05-00-01429 0.9528[20]	1.5318[20]
7889	Naphthalene, 2-ethyl-1,2,3,4-tetrahydro- 2-Ethyltetralin	$C_{12}H_{16}$ 160.26	32367-54-7 -42.8	235	4-05-00-01429 0.9401[15]	1.5250[15]
7890	Naphthalene, 5-ethyl-1,2,3,4-tetrahydro-	$C_{12}H_{16}$ 160.26	42775-75-7 -44.5	245	4-05-00-01428 0.973[20]	1.540[20]
7891	Naphthalene, 6-ethyl-1,2,3,4-tetrahydro- 6-Ethyltetralin	$C_{12}H_{16}$ 160.26	22531-20-0 -70	244	4-05-00-01429 0.9632[17]	1.5414[16]
7892	Naphthalene, 1-ethynyl-	$C_{12}H_8$ 152.20	15727-65-8 2	143-4[25]	3-05-00-01935 1.0513[20]	1.6360[20]
7893	Naphthalene, 1-fluoro- 1-Fluoronaphthalene	$C_{10}H_7F$ 146.16	321-38-0 -9	215; 80[11]	4-05-00-01657 1.1322[20]	H_2O 1; EtOH 3; eth 3; bz 3 1.5939[20]
7894	Naphthalene, 2-fluoro- 2-Fluoronaphthalene	$C_{10}H_7F$ 146.16	323-09-1 61	212; 90[16]	4-05-00-01658	H_2O 1; EtOH 3; eth 3; bz 3
7895	Naphthalene, 1-heptyl- 1-Heptylnaphthalene	$C_{17}H_{22}$ 226.36	2876-52-0 -8	340	4-05-00-01775 0.9491[20]	1.5582[20]
7896	Naphthalene, 2-heptyl- 2-Heptylnaphthalene	$C_{17}H_{22}$ 226.36	2876-45-1 1	341	4-05-00-01775 0.9410[20]	1.5556[20]
7897	Naphthalene, 1,2,3,4,4a,8a-hexahydro- Naphthalene, 1,2,3,4,9,10-hexahydro	$C_{10}H_{14}$ 134.22	62690-62-4 197		4-05-00-01082 0.934[23]	bz 4; eth 4 1.5260[16]
7898	Naphthalene, 1,2,4a,5,8,8a-hexahydro-4,7- dimethyl-1-(1-methylethyl)-, [1S- (1α,4aβ,8aα)]-	$C_{15}H_{24}$ 204.36	523-47-7	274; 136[11]	3-05-00-01086 0.9230[20]	eth 4; lig 4 1.5059[20]
7899	Naphthalene, 1,2,3,5,6,8a-hexahydro-4,7- dimethyl-1-(1-methylethyl)-, (1S-cis)-	$C_{15}H_{24}$ 204.36	483-76-1	125-6[12]	2-05-00-00348 0.9160[15]	1.5089[15]
7900	Naphthalene, 1-hexyl- 1-Hexylnaphthalene	$C_{16}H_{20}$ 212.33	2876-53-1 -18	322	4-05-00-01767 0.9566[20]	1.5647[20]
7901	Naphthalene, 2-hexyl-	$C_{16}H_{20}$ 212.33	2876-46-2 -3	324	4-05-00-01767 0.9479[20]	1.5620[20]
7902	Naphthalene, 1-iodo-	$C_{10}H_7I$ 254.07	90-14-2 4.2	302	4-05-00-01670 1.7399[20]	H_2O 1; EtOH 5; eth 5; bz 5 1.7026[20]

No.	Name / Synonym	Mol. Form. / Mol. Wt.	CAS RN / mp/°C	Merck No. / bp/°C	Beil. Ref. / den/g cm⁻³	Solubility / n_D
7903	Naphthalene, 2-iodo-	$C_{10}H_7I$ 254.07	612-55-5 54.5	308	4-05-00-01671 1.6319[99]	H_2O 1; EtOH 4; eth 4; HOAc 4 1.6662[99]
7904	Naphthalene, 1-isocyanato- 1-Naphthyl isocyanate	$C_{11}H_7NO$ 169.18	86-84-0	6330 269	4-12-00-03094 1.1774[20]	eth 3; bz 3
7905	Naphthalene, 1-isopropyl-2,4,7-trimethyl- Cadalene, 3-methyl	$C_{16}H_{20}$ 212.33	6995-30-8	129-31[1.5]	3-05-00-01689 0.9721[30]	1.5840[30]
7906	Naphthalene, 1-isothiocyanato- 1-Naphthyl isothiocyanate	$C_{11}H_7NS$ 185.25	551-06-4 58	6331	4-12-00-03094	bz 4; eth 4; EtOH 4; chl 4
7907	2-Naphthalenemethanamine	$C_{11}H_{11}N$ 157.22	2018-90-8 59.5	180[24]; 148[12]	4-12-00-03207	eth 4; EtOH 4
7908	1-Naphthalenemethanol	$C_{11}H_{10}O$ 158.20	4780-79-4 64	304; 163[12]	4-06-00-04332 1.1039[80]	H_2O 2; EtOH 4; eth 4
7909	2-Naphthalenemethanol	$C_{11}H_{10}O$ 158.20	1592-38-7 81.3	178[12]	4-06-00-04340	H_2O 2; EtOH 3; eth 3
7910	1-Naphthalenemethanol, α,α-diphenyl- Benzhydrol, α-(1-naphthyl)	$C_{23}H_{18}O$ 310.40	630-95-5 136.5	dec	4-06-00-05134	bz 4; eth 4; EtOH 4
7911	1-Naphthalenemethanol, α-methyl-, (±)- (±)-1-(α-Naphthyl)ethanol	$C_{12}H_{12}O$ 172.23	57605-95-5 66	178[15]	4-06-00-04346 1.1190[14]	ace 4; bz 4; EtOH 4; chl 4 1.6188[25]
7912	1-Naphthalenemethanol, α-phenyl-	$C_{17}H_{14}O$ 234.30	642-28-4 87.8	360	4-06-00-04989	H_2O 1; EtOH 3; eth 3; bz 3
7913	Naphthalene, 1-methoxy-	$C_{11}H_{10}O$ 158.20	2216-69-5 <-10	269	4-06-00-04211 1.0963[14]	H_2O 1; EtOH 3; eth 3; bz 3 1.6940[25]
7914	Naphthalene, 2-methoxy- 2-Methoxynaphthalene	$C_{11}H_{10}O$ 158.20	93-04-9 73.5	5918 274	4-06-00-04257	bz 4; eth 4; chl 4
7915	Naphthalene, 1-methyl- 1-Methylnaphthalene	$C_{11}H_{10}$ 142.20	90-12-0 -30.4	244.7	4-05-00-01687 1.0202[20]	H_2O 1; EtOH 4; eth 4; bz 3 1.6170[20]
7916	Naphthalene, 2-methyl- 2-Methylnaphthalene	$C_{11}H_{10}$ 142.20	91-57-6 34.4	241.1	4-05-00-01693 1.0058[20]	H_2O 1; EtOH 4; eth 4; bz 3 1.6015[40]
7917	Naphthalene, 1-(3-methylbutoxy)-	$C_{15}H_{18}O$ 214.31	20213-30-3	319; 150[3]	3-06-00-02925 1.0069[14]	1.5705[16]
7918	Naphthalene, 2-(3-methylbutoxy)- Isopentyl β-Naphthyl ether	$C_{15}H_{18}O$ 214.31	635-88-1 26.5	325 dec	4-06-00-04259 1.0155[12]	eth 4; EtOH 4 1.5768[12]
7919	Naphthalene, 1,1'-methylenebis-	$C_{21}H_{16}$ 268.36	607-50-1 109	>360; 270[14]	4-05-00-02646	H_2O 2; EtOH 2; eth 3; bz 3
7920	Naphthalene, 1-(1-methylethyl)-	$C_{13}H_{14}$ 170.25	6158-45-8 -16	268	4-05-00-01722 0.9956[20]	1.5952[20]
7921	Naphthalene, 2-(1-methylethyl)-	$C_{13}H_{14}$ 170.25	2027-17-0 14.5	268.2	4-05-00-01723 0.9753[20]	H_2O 1; EtOH 4; eth 4; bz 3 1.5848[20]
7922	Naphthalene, 1-methyl-4-(1-methylethyl)-	$C_{14}H_{16}$ 184.28	1680-53-1	145-8[12]	3-05-00-01671 0.9934[14]	1.5907[14]
7923	Naphthalene, 1-methyl-7-(1-methylethyl)-	$C_{14}H_{16}$ 184.28	490-65-3	140[11]	4-05-00-01741 0.9740[20]	H_2O 1; eth 3; ace 3; bz 3 1.5833[20]
7924	Naphthalene, 7-methyl-1-(1-methylethyl)- Naphthalene, 1-isopropyl-7-methyl	$C_{14}H_{16}$ 184.28	66577-17-1	282	3-05-00-01672 0.9833[20]	1.5884[20]
7925	Naphthalene, 1-methyl-4-nitro- 4-Methyl-1-nitronaphthalene	$C_{11}H_9NO_2$ 187.20	880-93-3 71.5	182[18]	3-05-00-01624	ace 4; eth 4; EtOH 4
7926	Naphthalene, 2-methyl-1-nitro- 2-Methyl-1-nitronaphthalene	$C_{11}H_9NO_2$ 187.20	881-03-8 81.5	188[20]	4-05-00-01698	H_2O 1; EtOH 3; ace 4; os 4
7927	Naphthalene, 1-(1-methylpropyl)-	$C_{14}H_{16}$ 184.28	1680-58-6	273	4-05-00-01738 0.9742[20]	1.5701[20]
7928	Naphthalene, 1-(2-naphthalenyloxy)- 1,2'-Bis-naphthyl ether	$C_{20}H_{14}O$ 270.33	611-49-4 81	264[15]	2-06-00-00600	bz 4; eth 4
7929	Naphthalene, 1-nitro- 1-Nitronaphthalene	$C_{10}H_7NO_2$ 173.17	86-57-7 61	6535 180[14]	4-05-00-01673 1.332[20]	H_2O 1; EtOH 4; eth 4; bz 4
7930	Naphthalene, 2-nitro- 2-Nitronaphthalene	$C_{10}H_7NO_2$ 173.17	581-89-5 79	314; 165[15]	4-05-00-01675	H_2O 1; EtOH 4; eth 4
7931	Naphthalene, 1-nonyl-	$C_{19}H_{26}$ 254.42	26438-26-6 8	366	4-05-00-01784 0.9371[20]	1.5477[20]
7932	Naphthalene, 2-nonyl-	$C_{19}H_{26}$ 254.42	61886-67-7 12	372	4-05-00-01784 0.9298[20]	1.5454[20]
7933	Naphthalene, octachloro- Perchloronaphthalene	$C_{10}Cl_8$ 403.73	2234-13-1 197.5	441[7]; 248[0.5]	4-05-00-01665	EtOH 2; bz 4; chl 4; lig 4
7934	Naphthalene, octadecafluorodecahydro- Perfluorodecalin	$C_{10}F_{18}$ 462.08	306-94-5 -10	142	3-05-00-00249	
7935	Naphthalene, octafluoro- Perfluoronaphthalene	$C_{10}F_8$ 272.10	313-72-4 87.5	209	4-05-00-01658	
7936	Naphthalene, 1,2,3,4,4a,5,6,8a-octahydro-7-methyl-4-methylene-1-(1-methylethyl)-, (1α,4aβ,8aα)- γ-Cadinene	$C_{15}H_{24}$ 204.36	39029-41-9	124-8[12]	3-05-00-01088 0.9182[15]	1.3166[20]
7937	Naphthalene, 1-octyl- 1-Octylnaphthalene	$C_{18}H_{24}$ 240.39	2876-51-9 -2	356	4-05-00-01779 0.9427[20]	1.5526[20]
7938	Naphthalene, 2-octyl-	$C_{18}H_{24}$ 240.39	2876-44-0 12	357	4-05-00-01779 0.9350[20]	1.5501[20]
7939	Naphthalene, 1,1'-oxybis- 1-Naphthyl ether	$C_{20}H_{14}O$ 270.33	607-52-3 110	280-5[22]	4-06-00-04215	bz 4; eth 4
7940	Naphthalene, 2,2'-oxybis-	$C_{20}H_{14}O$ 270.33	613-80-9 105	250[19] dec	4-06-00-04260	H_2O 1; EtOH 2; eth 4; bz 4

No.	Name / Synonym	Mol. Form. / Mol. Wt.	CAS RN / mp/°C	Merck No. / bp/°C	Beil. Ref. / den/g cm^{-3}	Solubility / n_D
7941	Naphthalene, 1-pentyl-	$C_{15}H_{18}$ 198.31	86-89-5 -22	307	4-05-00-01755 0.9656[20]	1.5725[20]
7942	Naphthalene, 2-pentyl- 2-Pentylnaphthalene	$C_{15}H_{18}$ 198.31	93-22-1 -4	310	4-05-00-01755 0.9561[20]	1.5694[20]
7943	Naphthalene, 1-(pentyloxy)- 1-Naphthyl pentyl ether	$C_{15}H_{18}O$ 214.31	108438-40-0 30	322	4-06-00-04213	bz 4; eth 4; EtOH 4
7944	Naphthalene, 2-(pentyloxy)- 2-Naphthyl pentyl ether	$C_{15}H_{18}O$ 214.31	31059-19-5 24.5	335	3-06-00-02973	bz 4; eth 4; EtOH 4; chl 4 1.5587[30]
7945	Naphthalene, 1-phenoxy-	$C_{16}H_{12}O$ 220.27	3402-76-4 55.5	349.5	4-06-00-04214	H_2O 1; EtOH 2; eth 3; ace 3
7946	Naphthalene, 2-phenoxy-	$C_{16}H_{12}O$ 220.27	19420-29-2 46	335	3-06-00-02975	H_2O 1; eth 3; chl 3; HOAc 3
7947	Naphthalene, 1-phenyl- 1-Phenylnaphthalene	$C_{16}H_{12}$ 204.27	605-02-7 45	334	4-05-00-02411 1.096[20]	H_2O 1; EtOH 4; eth 4; bz 4 1.6664[20]
7948	Naphthalene, 2-phenyl-	$C_{16}H_{12}$ 204.27	612-94-2 103.5	345.5	4-05-00-02412 1.2180[20]	EtOH 3; eth 4; bz 3; chl 3
7949	Naphthalene, 1-(phenylmethyl)-	$C_{17}H_{14}$ 218.30	611-45-0 59.5	350	4-05-00-02420 1.166[17]	H_2O 1; EtOH 2; eth 4; bz 3
7950	Naphthalene, 2-(phenylmethyl)- 2-Benzylnaphthalene	$C_{17}H_{14}$ 218.30	613-59-2 58	350	4-05-00-02420 1.176[12]	bz 4; eth 4; chl 4
7951	Naphthalene, 1-(phenylthio)-	$C_{16}H_{12}S$ 236.34	7570-98-1 41.8	255[43]; 222[11]	4-06-00-04242 1.167[15]	eth 4; EtOH 4
7952	1-Naphthalenepropanoic acid	$C_{13}H_{12}O_2$ 200.24	3243-42-3 156.5	179[11]	4-09-00-02435	H_2O 3; EtOH 3; bz 3
7953	2-Naphthalenepropanoic acid, 6-hydroxy- Allenolic acid	$C_{13}H_{12}O_3$ 216.24	553-39-9 180.5	247	3-10-00-01113	py 4; EtOH 4; MeOH 4
7954	Naphthalene, 1-(2-propenyl)-	$C_{13}H_{12}$ 168.24	2489-86-3	266	3-05-00-01805 1.0228[20]	EtOH 3; bz 4; chl 3 1.6140[20]
7955	Naphthalene, 1-propoxy-	$C_{13}H_{14}O$ 186.25	20009-26-1	293.5	4-06-00-04213 1.0447[18]	1.5928[18]
7956	Naphthalene, 2-propoxy-	$C_{13}H_{14}O$ 186.25	19718-45-7 41	305	3-06-00-02973	EtOH 4
7957	Naphthalene, 1-propyl-	$C_{13}H_{14}$ 170.25	2765-18-6 -8.6	274.5	4-05-00-01721 0.9897[20]	1.5923[20]
7958	Naphthalene, 2-propyl-	$C_{13}H_{14}$ 170.25	2027-19-2 -3	273.5	4-05-00-01722 0.9770[20]	1.5872[20]
7959	1-Naphthalenesulfonic acid	$C_{10}H_8O_3S$ 208.24	85-47-2 140	6295	4-11-00-00521	H_2O 3; EtOH 3; eth 2
7960	2-Naphthalenesulfonic acid β-Naphthylsulfonic acid	$C_{10}H_8O_3S$ 208.24	120-18-3 91 dec	6296	4-11-00-00527 1.441[25]	H_2O 4; EtOH 4; eth 3; bz 2
7961	1-Naphthalenesulfonic acid, 2-amino- 2-Naphthylamine-1-sulfonic acid	$C_{10}H_9NO_3S$ 223.25	81-16-3	8328	4-14-00-02792	DMSO 3
7962	1-Naphthalenesulfonic acid, 4-amino- 1-Naphthylamine-4-sulfonic acid	$C_{10}H_9NO_3S$ 223.25	84-86-6 dec	6323	4-14-00-02793 1.6703[25]	H_2O 1; EtOH 2; MeOH 3; py 3
7963	1-Naphthalenesulfonic acid, 5-amino- 1-Naphthylamine-5-sulfonic acid	$C_{10}H_9NO_3S$ 223.25	84-89-9	6324	4-14-00-02800	H_2O 3; eth 1
7964	1-Naphthalenesulfonic acid, 7-amino- Badische acid	$C_{10}H_9NO_3S$ 223.25	86-60-2	952	4-14-00-02802	HOAc 4
7965	1-Naphthalenesulfonic acid, 8-amino- 1-Naphthylamine-8-sulfonic acid	$C_{10}H_9NO_3S$ 223.25	82-75-7	6325	4-14-00-02802	gl HOAc 4
7966	1-Naphthalenesulfonic acid, 4-amino-3-hydroxy- 1-Amino-2-naphthol-4-sulfonic acid	$C_{10}H_9NO_4S$ 239.25	116-63-2	466	4-14-00-02825	
7967	2-Naphthalenesulfonic acid, 8-amino- 1,7-Cleve's acid	$C_{10}H_9NO_3S$ 223.25	119-28-8	2349	4-14-00-02805	
7968	1-Naphthalenesulfonic acid, 3,6-bis(1,1-dimethylethyl)-, ethyl ester Ethyl dibunate	$C_{20}H_{28}O_3S$ 348.51	5560-69-0	3745		chl 3
7969	1-Naphthalenesulfonic acid, 3,3'-[[1,1'-biphenyl]-4,4'-diylbis(azo)]bis[4-amino-, disodium salt Congo red	$C_{32}H_{22}N_6Na_2O_6S_2$ 696.68	573-58-0 >360	2493	1-16-00-00342	H_2O 3; EtOH 2; eth 1
7970	1-Naphthalenesulfonic acid, 3,4-dihydro-3,4-dioxo-, sodium salt 1,2-Naphthoquinone-4-sulfonic acid,sodium salt	$C_{10}H_5NaO_5S$ 260.20	521-24-4 287 dec	8597		
7971	1-Naphthalenesulfonic acid, 4-hydroxy- 1-Naphthol-4-sulfonic acid	$C_{10}H_8O_4S$ 224.24	84-87-7 170 dec	6311	4-11-00-00614	H_2O 4; eth 1
7972	1-Naphthalenesulfonic acid, 7-hydroxy- Croceic acid	$C_{10}H_8O_4S$ 224.24	132-57-0	2591	4-11-00-00623	H_2O 3
7973	2-Naphthalenesulfonic acid, 1-hydroxy- 1-Naphthol-2-sulfonic acid	$C_{10}H_8O_4S$ 224.24	567-18-0 >250	6310	4-11-00-00613	H_2O 2; EtOH 3; eth 1; dil HCl 2
7974	2-Naphthalenesulfonic acid, 6-hydroxy- 2-Naphthol-6-sulfonic acid	$C_{10}H_8O_4S$ 224.24	93-01-6 125	6312	4-11-00-00622	H_2O 4; EtOH 4; eth 1; HOAc 3
7975	1-Naphthalenesulfonyl chloride	$C_{10}H_7ClO_2S$ 226.68	85-46-1 68	209[20]; 147[0.9]	3-11-00-00383	bz 4; eth 4; EtOH 4

No.	Name Synonym	Mol. Form. Mol. Wt.	CAS RN mp/°C	Merck No. bp/°C	Beil. Ref. den/g cm⁻³	Solubility n_D
7976	2-Naphthalenesulfonyl chloride	$C_{10}H_7ClO_2S$ 226.68	93-11-8 81	$201^{13.}$ $148^{0.5}$	4-11-00-00529	H_2O 1; EtOH 3; eth 4; bz 3
7977	1-Naphthalenesulfonyl chloride, 5-(dimethylamino)- Dansyl chloride	$C_{12}H_{12}ClNO_2S$ 269.75	605-65-2 70	2812	4-14-00-02801	
7978	2-Naphthalenesulfonyl chloride, 5,6,7,8-tetrahydro- 6-Tetralinsulfonyl chloride	$C_{10}H_{11}ClO_2S$ 230.71	61551-49-3 58	197^{18}	3-11-00-00375	eth 4
7979	Naphthalene, 1,2,3,4-tetrahydro- Tetralin	$C_{10}H_{12}$ 132.21	119-64-2 -35.7	9152 207.6	4-05-00-01388 0.9660^{25}	H_2O 1; EtOH 4; eth 4; chl 3 1.5413^{20}
7980	Naphthalene, 1,2,3,4-tetrahydro-1,1-dimethyl- 1,1-Dimethyltetralin	$C_{12}H_{16}$ 160.26	1985-59-7	221	4-05-00-01430 0.950^{20}	1.5292^{20}
7981	Naphthalene, 1,2,3,4-tetrahydro-1,2-dimethyl- 1,2,3,4-Tetrahydro-1,2-dimethylnaphthalene	$C_{12}H_{16}$ 160.26	5195-40-4	235	3-05-00-01262 0.9470^{20}	1.5286^{20}
7982	Naphthalene, 1,2,3,4-tetrahydro-1,3-dimethyl- Tetralin, 1,3-dimethyl	$C_{12}H_{16}$ 160.26	5195-37-9	$234; 78^1$	3-05-00-01263 0.940^{20}	1.525^{20}
7983	Naphthalene, 1,2,3,4-tetrahydro-1,4-dimethyl- Tetralin, 1,4-dimethyl	$C_{12}H_{16}$ 160.26	4175-54-6	226	4-05-00-01430 0.940^{20}	1.525^{20}
7984	Naphthalene, 1,2,3,4-tetrahydro-1,5-dimethyl-	$C_{12}H_{16}$ 160.26	21564-91-0	239	3-05-00-01263 0.9410^{20}	1.526^{20}
7985	Naphthalene, 1,2,3,4-tetrahydro-2,2-dimethyl- Tetralin, 2,2-dimethyl	$C_{12}H_{16}$ 160.26	13556-55-3	230	4-05-00-01431 0.935^{20}	1.5200^{20}
7986	Naphthalene, 1,2,3,4-tetrahydro-2,3-dimethyl- Tetralin, 2,3-dimethyl	$C_{12}H_{16}$ 160.26	21564-92-1 -5		3-05-00-01265 0.940^{20}	1.523^{20}
7987	Naphthalene, 1,2,3,4-tetrahydro-2,5-dimethyl- Tetralin, 2,5-dimethyl	$C_{12}H_{16}$ 160.26	25419-37-8	236	3-05-00-01263 0.9460^{20}	1.526^{20}
7988	Naphthalene, 1,2,3,4-tetrahydro-2,6-dimethyl-	$C_{12}H_{16}$ 160.26	7524-63-2 20	$240; 117^{15}$	4-05-00-01432 0.941^{20}	1.526^{20}
7989	Naphthalene, 1,2,3,4-tetrahydro-2,7-dimethyl- Tetralin, 2,7-dimethyl	$C_{12}H_{16}$ 160.26	13065-07-1	237	3-05-00-01265 0.941^{20}	1.526^{20}
7990	Naphthalene, 1,2,3,4-tetrahydro-2,8-dimethyl- Tetralin, 2,8-dimethyl	$C_{12}H_{16}$ 160.26	25419-36-7	236	3-05-00-01264 0.941^{20}	1.526^{20}
7991	Naphthalene, 1,2,3,4-tetrahydro-5,6-dimethyl- Tetralin, 5,6-dimethyl	$C_{12}H_{16}$ 160.26	20027-77-4	252	4-05-00-01430 0.975^{20}	1.552^{20}
7992	Naphthalene, 1,2,3,4-tetrahydro-5,7-dimethyl- Tetralin, 5,7-dimethyl	$C_{12}H_{16}$ 160.26	21693-54-9 -6	253	4-05-00-01430 0.9583^{20}	1.5405^{20}
7993	Naphthalene, 1,2,3,4-tetrahydro-5,8-dimethyl-	$C_{12}H_{16}$ 160.26	14108-88-4 -4.1	254	4-05-00-01430 0.967^{20}	1.547^{20}
7994	Naphthalene, 1,2,3,4-tetrahydro-6,7-dimethyl-	$C_{12}H_{16}$ 160.26	1076-61-5 10	252	4-05-00-01431 0.954^{20}	1.538^{20}
7995	Naphthalene, 1,2,3,4-tetrahydro-1-methyl-	$C_{11}H_{14}$ 146.23	1559-81-5	220.6	4-05-00-01412 0.9583^{20}	1.5353^{20}
7996	Naphthalene, 1,2,3,4-tetrahydro-2-methyl-	$C_{11}H_{14}$ 146.23	3877-19-8 -43.1	221	4-05-00-01414 0.952^{20}	1.531^{20}
7997	Naphthalene, 1,2,3,4-tetrahydro-5-methyl-	$C_{11}H_{14}$ 146.23	2809-64-5 -23	234	4-05-00-01412 0.9720^{20}	1.5439^{20}
7998	Naphthalene, 1,2,3,4-tetrahydro-6-methyl-	$C_{11}H_{14}$ 146.23	1680-51-9 -40	229	4-05-00-01413 0.9537^{20}	1.5357^{20}
7999	Naphthalene, 1,2,3,4-tetrahydro-1,1,2,6-tetramethyl- Irene	$C_{14}H_{20}$ 188.31	1681-22-7	$120-5^{10}$	4-05-00-01460 0.9332^{20}	H_2O 1; ace 3; bz 3 1.5217^{20}
8000	Naphthalene, 1,2,3,4-tetrahydro-1,1,6-trimethyl-	$C_{13}H_{18}$ 174.29	475-03-6	$240; 90^4$	4-05-00-01445 0.9356^{20}	EtOH 3; eth 3; bz 3; chl 3 1.5257^{20}
8001	Naphthalene, 1,2,4,7-tetramethyl-	$C_{14}H_{16}$ 184.28	16020-17-0 -3	$174-5^{12}$	4-05-00-01746 1.011^{20}	1.6032^{20}
8002	Naphthalene, 1,2,5,6-tetramethyl-	$C_{14}H_{16}$ 184.28	2131-43-3 118	152^{12}	4-05-00-01746	H_2O 1; EtOH 3; bz 3; peth 3
8003	Naphthalene, 1,3,6,8-tetramethyl- 1,3,6,8-Tetramethylnaphthalene	$C_{14}H_{16}$ 184.28	14558-14-6 84.5	$115-6^2$	4-05-00-01746	H_2O 1; EtOH 3
8004	Naphthalene, 1,3,6,8-tetranitro-	$C_{10}H_4N_4O_8$ 308.16	28995-89-3 207	exp	4-05-00-01683 1.6400^0	bz 4; HOAc 4
8005	Naphthalene, 1,1'-thiobis-	$C_{20}H_{14}S$ 286.40	607-53-4 110	289^{15}	4-06-00-04243	bz 4; HOAc 4
8006	Naphthalene, 2,2'-thiobis-	$C_{20}H_{14}S$ 286.40	613-81-0 151	296^{15}	4-06-00-04314	EtOH 1; bz 4; CS_2 4; HOAc 4
8007	1-Naphthalenethiol 1-Naphthyl mercaptan	$C_{10}H_8S$ 160.24	529-36-2	6297 285 dec; 161^{20}	4-06-00-04241 1.1607^{20}	H_2O 2; EtOH 4; eth 4; dil alk 2 1.6802^{20}
8008	2-Naphthalenethiol 2-Naphthyl mercaptan	$C_{10}H_8S$ 160.24	91-60-1 81	6298 288	4-06-00-04312 1.550^{25}	H_2O 2; EtOH 4; eth 4; lig 4
8009	Naphthalene, 1,2,4-trimethyl- 1,2,4-Trimethylnaphthalene	$C_{13}H_{14}$ 170.25	2717-42-2 55.5	146^{12}	4-05-00-01725	EtOH 3; eth 3; bz 3; chl 3
8010	Naphthalene, 1,2,5-trimethyl- 1,2,5-Trimethylnaphthalene	$C_{13}H_{14}$ 170.25	641-91-8 33.5	140^{12}	4-05-00-01726 1.0103^{22}	H_2O 1; EtOH 2; eth 4; bz 4 1.6093^{22}
8011	Naphthalene, 1,2,6-trimethyl- 1,2,6-Trimethylnaphthalene	$C_{13}H_{14}$ 170.25	3031-05-8 14	146^{10}	4-05-00-01726	H_2O 1; eth 3; bz 3 1.6010^{20}
8012	Naphthalene, 1,2,7-trimethyl-	$C_{13}H_{14}$ 170.25	486-34-0	$147-8^{16}$	4-05-00-01727 1.0087^{20}	H_2O 1; eth 3; bz 3 1.6097^{20}

No.	Name / Synonym	Mol. Form. / Mol. Wt.	CAS RN / mp/°C	Merck No. / bp/°C	Beil. Ref. / den/g cm^{-3}	Solubility / n_D
8013	Naphthalene, 1,3,7-trimethyl- / 2,4,6-Trimethylnaphthalene	$C_{13}H_{14}$ / 170.25	2131-38-6 / 12	131-3[9]	4-05-00-01728 / 0.9801[21]	1.5972[15]
8014	Naphthalene, 1,2,5-trimethyl-8-(1-methylethyl)- / Cadalene, 5-methyl	$C_{16}H_{20}$ / 212.33	6897-76-3	144-5[1.5]	3-05-00-01689 / 0.9847[30]	1.5926[30]
8015	Naphthalene, 1,3,8-trimethyl-5-(1-methylethyl)- / Cadalene, 8-methyl	$C_{16}H_{20}$ / 212.33	6897-88-7	135-6[1.5]	4-05-00-01770 / 0.9800[30]	1.5876[30]
8016	Naphthalene, 1,3,5-trinitro-	$C_{10}H_5N_3O_6$ / 263.17	2243-94-9 / 124	exp 364	4-05-00-01682	ace 4; EtOH 4; chl 4
8017	1-Naphthalenol / 1-Naphthol	$C_{10}H_8O$ / 144.17	90-15-3 / 95	6303 / 288; 184[40]	4-06-00-04208 / 1.0989[99]	H_2O 1; EtOH 4; eth 4; ace 3 / 1.6224[99]
8018	2-Naphthalenol / 2-Naphthol	$C_{10}H_8O$ / 144.17	135-19-3 / 123	6304 / 285	4-06-00-04253 / 1.28[20]	H_2O 1; EtOH 4; eth 4; bz 3
8019	2-Naphthalenol, 6-amino-	$C_{10}H_9NO$ / 159.19	4363-04-6 / 193	212 dec	3-13-00-01902	H_2O 4; EtOH 4
8020	2-Naphthalenol, 8-amino-	$C_{10}H_9NO$ / 159.19	118-46-7 / 206	sub	3-13-00-01907	H_2O 3; EtOH 4; eth 3; bz 2
8021	2-Naphthalenol, benzoate / 2-Naphthyl benzoate	$C_{17}H_{12}O_2$ / 248.28	93-44-7 / 107	6328	4-09-00-00339	H_2O 1; EtOH 3; eth 2; HOAc 2
8022	2-Naphthalenol, 1-bromo-	$C_{10}H_7BrO$ / 223.07	573-97-7 / 84	130	4-06-00-04301	H_2O 1; EtOH 3; eth 3; bz 3
8023	2-Naphthalenol, 6-chloro-	$C_{10}H_7ClO$ / 178.62	40604-49-7 / 116.5	sub	4-06-00-04294	bz 4; eth 4; EtOH 4; chl 4
8024	2-Naphthalenol, decahydro-	$C_{10}H_{18}O$ / 154.25	825-51-4	109[14]	1-06-00-00044 / 0.996[25]	1.4992[20]
8025	1-Naphthalenol, 2,3-dichloro- / 2,3-Dichloro-1-naphthol	$C_{10}H_6Cl_2O$ / 213.06	71284-96-3 / 101	sub	0-06-00-00612	bz 4; eth 4; EtOH 4; chl 4
8026	1-Naphthalenol, 2,4-dichloro-	$C_{10}H_6Cl_2O$ / 213.06	2050-76-2 / 107.5	180	4-06-00-04233	bz 4; eth 4; EtOH 4
8027	2-Naphthalenol, 1,4-dimethyl- / 2-Naphthol, 1,4-dimethyl-	$C_{12}H_{12}O$ / 172.23	4705-94-6 / 135.5	315.5	4-06-00-04351	eth 4; EtOH 4
8028	2-Naphthalenol, 1-[(2,4-dimethylphenyl)azo]- / C.I. Solvent Orange 7	$C_{18}H_{16}N_2O$ / 276.34	3118-97-6 / 166		4 16 00 00234	eth 4; EtOH 4
8029	2-Naphthalenol, 1-[(2,5-dimethylphenyl)azo]- / 1-(2,5-Xylylazo)-2-naphthol	$C_{18}H_{16}N_2O$ / 276.34	85-82-5 / 153	9997	4-16-00-00234	
8030	1-Naphthalenol, 4-methyl-	$C_{11}H_{10}O$ / 158.20	10240-08-1 / 85.8	166[13]	4-06-00-04330	H_2O 3; EtOH 3; eth 3; ace 3
8031	2-Naphthalenol, 1-methyl-	$C_{11}H_{10}O$ / 158.20	1076-26-2 / 112	100[10]	4-06-00-04329	H_2O 2; EtOH 4; eth 4; ace 4
8032	1-Naphthalenol, methylcarbamate / Carbaryl	$C_{12}H_{11}NO_2$ / 201.22	63-25-2 / 145	1789	4-06-00-04219 / 1.228[25]	ace 4; DMF 4
8033	2-Naphthalenol, 1-[[2-methyl-4-[(2-methylphenyl)azo]phenyl]azo]- / Scarlet red	$C_{24}H_{20}N_4O$ / 380.45	85-83-6 / 186 dec	8349 / 260 dec	4-16-00-00249	chl 4; peth 4
8034	2-Naphthalenol, 1-nitro- / 1-Nitro-2-naphthol	$C_{10}H_7NO_3$ / 189.17	550-60-7 / 104	6536 / 115[0.05]	4-06-00-04307	H_2O 3; EtOH 3; eth 4; chl 2
8035	2-Naphthalenol, 1-nitroso- / 1-Nitroso-2-naphthol	$C_{10}H_7NO_2$ / 173.17	131-91-9 / 109.5	6562	4-07-00-02419	bz 4; eth 4
8036	1-Naphthalenol, 2-(phenylazo)-	$C_{16}H_{12}N_2O$ / 248.28	3375-23-3 / 138	sub	4-16-00-00219	EtOH 4; HOAc 4
8037	2-Naphthalenol, 1-(phenylazo)- / Azo, benzene 1'-naphthalene, 2'-hydroxy	$C_{16}H_{12}N_2O$ / 248.28	842-07-9		4-16-00-00228	
8038	2-Naphthalenol, 1-[[4-(phenylazo)phenyl]azo]- / Sudan III	$C_{22}H_{16}N_4O$ / 352.40	85-86-9 / 195	8858	4-16-00-00248	H_2O 1; EtOH 3; eth 3; ace 3
8039	1-Naphthalenol, 2-(phenylmethyl)-	$C_{17}H_{14}O$ / 234.30	36441-32-4 / 73.5	237[12]	4-06-00-04990	H_2O 1; bz 3; lig 2
8040	2-Naphthalenol, 1-(phenylmethyl)- / 2-Naphthol, 1-benzyl	$C_{17}H_{14}O$ / 234.30	36441-31-3 / 114	247[14]	4-06-00-04988	ace 4; bz 4; eth 4; EtOH 4
8041	1-Naphthalenol, 1,2,3,4-tetrahydro-	$C_{10}H_{12}O$ / 148.20	529-33-9	102-4[2]	3-06-00-02457 / 1.09[25]	1.5638[20]
8042	1-Naphthalenol, 5,6,7,8-tetrahydro-	$C_{10}H_{12}O$ / 148.20	529-35-1 / 70	266; 143[11]	4-06-00-03851 / 1.0556[75]	
8043	2-Naphthalenol, 1,2,3,4-tetrahydro- / 2-Tetralol	$C_{10}H_{12}O$ / 148.20	530-91-6 / 15.5	9153 / 140[12]	0-06-00-00580	
8044	2-Naphthalenol, 5,6,7,8-tetrahydro-	$C_{10}H_{12}O$ / 148.20	1125-78-6 / 57	275.5	4-06-00-03854 / 1.0552[65]	
8045	1(2H)-Naphthalenone, 2-acetyl-3,4-dihydro-	$C_{12}H_{12}O_2$ / 188.23	17216-08-9 / 56	135-42[1]	4-07-00-02358	chl 3
8046	1(2H)-Naphthalenone, 3,4-dihydro-	$C_{10}H_{10}O$ / 146.19	529-34-0 / 8	115[6]	4-07-00-01015 / 1.0988[16]	1.5672[20]
8047	2(1H)-Naphthalenone, 3,4-dihydro-	$C_{10}H_{10}O$ / 146.19	530-93-8 / 18	237	4-07-00-01018 / 1.1055[27]	H_2O 1; eth 3; bz 3 / 1.5598[20]
8048	1(2H)-Naphthalenone, 3,4-dihydro-6-methoxy- / 6-Methoxy-α-tetralone	$C_{11}H_{12}O_2$ / 176.22	1078-19-9 / 78	5924 / 171[11]	4-08-00-00904	
8049	1(2H)-Naphthalenone, 3,4-dihydro-2-methyl-	$C_{11}H_{12}O$ / 160.22	1590-08-5	127-31[12]	4-07-00-01033 / 1.057[25]	1.5535[20]
8050	1(2H)-Naphthalenone, 3,4-dihydro-4-methyl-	$C_{11}H_{12}O$ / 160.22	19832-98-5	133-4[12]	4-07-00-01031 / 1.0779[19]	1.5620[19]

No.	Name Synonym	Mol. Form. Mol. Wt.	CAS RN mp/°C	Merck No. bp/°C	Beil. Ref. den/g cm^{-3}	Solubility n_D
8051	1(2H)-Naphthalenone, 7-ethyl-3,4-dihydro- 1-Tetralone, 7-ethyl	$C_{12}H_{14}O$ 174.24	22531-06-2 	 152-3[12]	4-07-00-01049 1.0556[17]	bz 4; eth 4; EtOH 4 1.5599[17]
8052	2(3H)-Naphthalenone, 4,4a,5,6,7,8- hexahydro-4,4a-dimethyl-6-(1- methylethylidene)-, (4R-cis)-	$C_{15}H_{22}O$ 218.34	15764-04-2 51.5	 144[2]	 1.0035[20]	ace 4 1.5370[20]
8053	1(2H)-Naphthalenone, octahydro-, cis- 1-Decalone (cis)	$C_{10}H_{16}O$ 152.24	32166-40-8 2	 126[20]	4-07-00-00201 1.008[20]	 1.4936[20]
8054	1(2H)-Naphthalenone, octahydro-, trans-	$C_{10}H_{16}O$ 152.24	21370-71-8 33	 122[20]	4-07-00-00201 0.986[20]	 1.4849[21]
8055	2(1H)-Naphthalenone, octahydro-	$C_{10}H_{16}O$ 152.24	4832-17-1 	 96[2.5]	4-07-00-00203 0.979[25]	 1.4900[20]
8056	Naphtho[2,3-b]-1,4-dithiin-2,3-dicarbonitrile, 5,10-dihydro-5,10-dioxo- Dithianone	$C_{14}H_4N_2O_2S_2$ 296.33	3347-22-6 220	3375 	5-19-08-00189 	
8057	Naphtho[1,2-b]furan-2,8(3H,4H)-dione, 3a,5,5a,9b-tetrahydro-3,5a,9-trimethyl-, α-Santonin	$C_{15}H_{18}O_3$ 246.31	481-06-1 175	8325 	5-17-11-00314 1.590[25]	H_2O 2; EtOH 2; eth 2; bz 3
8058	Naphtho[1,2-b]furan-2,8(3H,4H)-dione, 3a,5,5a,9b-tetrahydro-4,5-dihydroxy-9- methoxy-3,5a-dimethyl- Cedrin	$C_{15}H_{18}O_6$ 294.30	6040-62-6 266	1918 	 	EtOH 4; MeOH 4; chl 4
8059	Naphtho[2,3-b]furan-2(3H)-one, 3a,5,6,7,8,8a,9,9a-octahydro-5,8a-dimethyl-3- methylene-, Alantolactone	$C_{15}H_{20}O_2$ 232.32	546-43-0 76	198 275	5-17-10-00097 	bz 4; eth 4; EtOH 4; chl 4
8060	Naphtho[1,2-c]furan-1,3-dione, 4,5-dihydro-	$C_{12}H_8O_3$ 200.19	37845-14-0 126.5	 227-30[23]	5-17-11-00444 	bz 4; MeOH 4
8061	Naphtho[1,2,3,4-def]chrysene Dibenzo[a,e]pyrene	$C_{24}H_{14}$ 302.38	192-65-4 233.5	 	4-05-00-02803 	EtOH 2; ace 2; bz 2; tol 3
8062	Naphtho[2,3-f]quinoline-7,12-dione, 5,6- dihydroxy- Alizarine Blue	$C_{17}H_9NO_4$ 291.26	568-02-5 269	239 	5-21-13-00590 	bz 4; gl HOAc 4
8063	2-Naphthol, 8-chloro-	$C_{10}H_7ClO$ 178.62	29921-50-4 101	 307.5	3-06-00-02992 	bz 4; eth 4; EtOH 4; chl 4
8064	2-Naphthylamine, 1,4-dimethyl- Naphthalene, 2-amino-1,4-dimethyl	$C_{12}H_{13}N$ 171.24	878-93-3 75	 333	0-12-00-01317 	eth 4; EtOH 4
8065	1,5-Naphthyridine 1,5-Diazanaphthalene	$C_8H_6N_2$ 130.15	254-79-5 75	 112[12]	5-23-07-00156 1.2100[20]	
8066	1,8-Naphthyridine-3-carboxylic acid, 1-ethyl- 1,4-dihydro-7-methyl-4-oxo-	$C_{12}H_{12}N_2O_3$ 232.24	389-08-2 229.5	6273 	 	EtOH 2; eth 2; chl 3
8067	Napropamide Propanamide, N,N-diethyl-2-(1- naphthalenyloxy)-	$C_{17}H_{21}NO_2$ 271.36 	15299-99-7 75 	6336 	 	
8068	Nickel carbonyl Nickel tetracarbonyl	C_4NiO_4 170.74	13463-39-3 -19.3	6411 43	 1.31[25]	
8069	Nicosulfuron 2-(4,6-DimethoxypyrimidiN-2- ylcarbamoylsulfamoyl)-N,N- dimethylnicotinamide	$C_{15}H_{18}N_6O_6S$ 410.41 	111991-09-4 172 	 	 	
8070	Nitralin Benzenamine, 4-(methylsulfonyl)-2,6-dinitro- N,N-dipropyl-	$C_{13}H_{19}N_3O_6S$ 345.38 	4726-14-1 150 	6486 	 	
8071	Nitrapyrin Pyridine, 2-chloro-6-(trichloromethyl)-	$C_6H_3Cl_4N$ 230.91	1929-82-4 63	6490 136[11]	 	
8072	Nitric acid, butyl ester Butyl nitrate	$C_4H_9NO_3$ 119.12	928-45-0 	 133	4-01-00-01524 1.0228[30]	H_2O 1; EtOH 3; eth 3; ctc 2 1.4013[23]
8073	Nitric acid, decyl ester Decyl nitrate	$C_{10}H_{21}NO_3$ 203.28	2050-78-4 	 128[11]; 88[1]	4-01-00-01819 0.951[0]	EtOH 3; eth 3
8074	Nitric acid, ethyl ester Ethyl nitrate	$C_2H_5NO_3$ 91.07	625-58-1 -94.6	 87.2	4-01-00-01327 1.1084[20]	H_2O 3; EtOH 5; eth 5 1.3852[20]
8075	Nitric acid, methyl ester Methyl nitrate	CH_3NO_3 77.04	598-58-3 -82.5	6015 64.6 exp	4-01-00-01254 1.2075[20]	H_2O 2; EtOH 3; eth 3 1.3748[20]
8076	Nitric acid, 1-methylethyl ester Isopropyl nitrate	$C_3H_7NO_3$ 105.09	1712-64-7 	 100	4-01-00-01475 1.034[19]	EtOH 3; eth 3 1.3912[16]
8077	Nitric acid, 1-methylpropyl ester	$C_4H_9NO_3$ 119.12	924-52-7 	 124	4-01-00-01573 1.0264[20]	eth 4; EtOH 4 1.4015[20]
8078	Nitric acid, 2-methylpropyl ester Isobutyl nitrate	$C_4H_9NO_3$ 119.12	543-29-3 	5031 123.4	4-01-00-01595 1.0152[20]	 1.4028[20]
8079	Nitric acid, octyl ester	$C_8H_{17}NO_3$ 175.23	629-39-0 	 110-2[20]	4-01-00-01762 0.8409[17]	H_2O 2; EtOH 3; eth 3
8080	Nitric acid, propyl ester Propyl nitrate	$C_3H_7NO_3$ 105.09	627-13-4 	7877 110	4-01-00-01424 1.0538[20]	H_2O 2; EtOH 3; eth 3; ctc 3 1.3973[20]
8081	Nitrofen Benzene, 2,4-dichloro-1-(4-nitrophenoxy)-	$C_{12}H_7Cl_2NO_3$ 284.10	1836-75-5 70	6519 	 	
8082	Nitrofluorfen Benzene, 2-chloro-1-(4-nitrophenoxy)-4- (trifluoromethyl)-	$C_{13}H_7ClF_3NO_3$ 317.65 	42874-01-1 68 	 	 	
8083	Nitrous acid, butyl ester Butyl nitrite	$C_4H_9NO_2$ 103.12	544-16-1 	1581 78	4-01-00-01523 0.9114[25]	EtOH 5; eth 5 1.3762[20]

No.	Name / Synonym	Mol. Form. / Mol. Wt.	CAS RN / mp/°C	Merck No. / bp/°C	Beil. Ref. / den/g cm^{-3}	Solubility / n_D
8084	Nitrous acid, decyl ester / Decyl nitrite	$C_{10}H_{21}NO_2$ / 187.28	1653-57-2	105-8[12]	4-01-00-01819	EtOH 3; eth 3 / 1.4247[20]
8085	Nitrous acid, 1,1-dimethylethyl ester / tert-Butyl nitrite	$C_4H_9NO_2$ / 103.12	540-80-7	1582 / 63	4-01-00-01622 / 0.8670[20]	H_2O 2; EtOH 3; eth 3; chl 3 / 1.368[20]
8086	Nitrous acid, 1,2-ethanediyl ester / Ethylene glycol, dinitrite	$C_2H_4N_2O_4$ / 120.06	629-16-3 / <-15	98	4-01-00-02411 / 1.2156[0]	eth 4; EtOH 4
8087	Nitrous acid, ethyl ester / Ethyl nitrite	$C_2H_5NO_2$ / 75.07	109-95-5	3786 / 18	4-01-00-01327 / 0.899[15]	EtOH 5; eth 5 / 1.3418[10]
8088	Nitrous acid, heptyl ester / Heptyl nitrite	$C_7H_{15}NO_2$ / 145.20	629-43-6	155.8	4-01-00-01735 / 0.8939[0]	eth 4 / 1.4032[20]
8089	Nitrous acid, hexyl ester / Hexyl nitrite	$C_6H_{13}NO_2$ / 131.17	638-51-7	129; 52[44]	4-01-00-01699 / 0.8778[20]	H_2O 1; EtOH 3; eth 3 / 1.3987[20]
8090	Nitrous acid, 3-methylbutyl ester / Isoamyl nitrite	$C_5H_{11}NO_2$ / 117.15	110-46-3	5005 / 99.2	4-01-00-01683 / 0.8828[20]	H_2O 2; EtOH 5; eth 5 / 1.3918[20]
8091	Nitrous acid, methyl ester / Methyl nitrite	CH_3NO_2 / 61.04	624-91-9 / -16	-12	4-01-00-01253 / 0.991[15]	EtOH 3; eth 3
8092	Nitrous acid, 1-methylethyl ester / Isopropyl nitrite	$C_3H_7NO_2$ / 89.09	541-42-4	5104 / 40	4-01-00-01474 / 0.8684[15]	H_2O 1; EtOH 3; eth 3
8093	Nitrous acid, 1-methylpropyl ester	$C_4H_9NO_2$ / 103.12	924-43-6	68.5	4-01-00-01573 / 0.8726[20]	eth 4; EtOH 4; chl 4 / 1.3710[20]
8094	Nitrous acid, 2-methylpropyl ester / Isobutyl nitrite	$C_4H_9NO_2$ / 103.12	542-56-3	5032 / 67	4-01-00-01595 / 0.8699[22]	H_2O 2; EtOH 3; eth 3 / 1.3715[22]
8095	Nitrous acid, octyl ester	$C_8H_{17}NO_2$ / 159.23	629-46-9	174.5	4-01-00-01762 / 0.862[17]	H_2O 2; EtOH 4; eth 4 / 1.4127[20]
8096	Nitrous acid, pentyl ester	$C_5H_{11}NO_2$ / 117.15	463-04-7	104.5	4-01-00-01644 / 0.8817[20]	H_2O 2; EtOH 5; eth 5 / 1.3851[20]
8097	Nitrous acid, phenylmethyl ester / Benzyl nitrite	$C_7H_7NO_2$ / 137.14	935-05-7	80-3[35]	4-06-00-02568 / 1.075[25]	1.4989[25]
8098	Nitrous acid, propyl ester / Propyl nitrite	$C_3H_7NO_2$ / 89.09	543-67-9	7878 / 48	4-01-00-01424 / 0.886[20]	H_2O 2; EtOH 3; eth 3 / 1.3604[20]
8099	Nonacosane / Celidoniol, deoxy-	$C_{29}H_{60}$ / 408.80	630-03-5 / 63.7	440.8	4-01-00-00591 / 0.8083[20]	H_2O 1; EtOH 4; eth 4; ace 4 / 1.4529[20]
8100	1-Nonacosanol / Nonacosanyl alcohol	$C_{29}H_{60}O$ / 424.79	6624-76-6 / 84.5	200[1.0]	4-01-00-01916	EtOH 3
8101	Nonadecane	$C_{19}H_{40}$ / 268.53	629-92-5 / 32.1	329.9	4-01-00-00560 / 0.7855[20]	H_2O 1; EtOH 2; eth 3; ace 3 / 1.4409[20]
8102	1,2,3-Nonadecanetricarboxylic acid, 2-hydroxy- / Agaricic acid	$C_{22}H_{40}O_7$ / 416.56	666-99-9 / 142 dec	174	4-03-00-01284	H_2O 3; EtOH 2; eth 2; bz 1
8103	Nonadecanoic acid	$C_{19}H_{38}O_2$ / 298.51	646-30-0 / 69.4	297[100.]; 228[10]	4-02-00-01256 / 0.8468[70]	H_2O 1; EtOH 4; eth 4; bz 4
8104	1-Nonadecanol	$C_{19}H_{40}O$ / 284.53	1454-84-8 / 63.3	166[0.3]	4-01-00-01898	eth 3; ace 3 / 1.4328[75]
8105	10-Nonadecanone	$C_{19}H_{38}O$ / 282.51	504-57-4 / 65.5	>350; 156[1.1]	4-01-00-03402	H_2O 1; EtOH 2; eth 3; ace 3
8106	2-Nonadecanone	$C_{19}H_{38}O$ / 282.51	629-66-3 / 57	266[110]; 165[2]	4-01-00-03400 / 0.8108[56]	H_2O 1; EtOH 2; eth 4; ace 3
8107	1-Nonadecyne	$C_{19}H_{36}$ / 264.49	26186-01-6 / 37.5	327	3-01-00-01030 / 0.8054[20]	ace 4; bz 4; eth 4 / 1.4488[20]
8108	2,6-Nonadienal / Leaf aldehyde violet	$C_9H_{14}O$ / 138.21	26370-28-5	85-7[11]	4-01-00-03560 / 0.8678[20]	1.4460[20]
8109	2,4-Nonadienal, (E,E)-	$C_9H_{14}O$ / 138.21	5910-87-2	97-8[10]	4-01-00-03560 / 0.862[20]	1.5207[20]
8110	1,8-Nonadiene	C_9H_{16} / 124.23	4900-30-5	142.5	4-01-00-01048 / 0.7511[20]	1.4302[20]
8111	2,7-Nonadiene	C_9H_{16} / 124.23	51333-70-1 / -72.5	151.5	3-01-00-01013 / 0.7499[20]	1.4358[20]
8112	2,6-Nonadien-1-ol	$C_9H_{16}O$ / 140.23	7786-44-9	96-100[11]	4-01-00-02266 / 0.8604[25]	1.4598[25]
8113	6,8-Nonadien-2-one, 8-methyl-5-(1-methylethyl)-, [S-(E)]- / Solanone	$C_{13}H_{22}O$ / 194.32	1937-54-8	8663 / 60[1]	0.870[20]	1.4755[20]
8114	1,4-Nonadiyne	C_9H_{12} / 120.19	6088-94-4	83-4[41]	4-01-00-01125 / 0.8112[25]	1.4518[25]
8115	1,8-Nonadiyne	C_9H_{12} / 120.19	2396-65-8 / -27.3	162	4-01-00-01125 / 0.8158[20]	eth 3; ace 3 / 1.4490[20]
8116	2,7-Nonadiyne	C_9H_{12} / 120.19	31699-35-1 / 4.3	180	3-01-00-01063 / 0.8332[20]	1.4674[20]
8117	Nonanal / Nonaldehyde	$C_9H_{18}O$ / 142.24	124-19-6	191	4-01-00-03352 / 0.8264[22]	eth 3; chl 3 / 1.4273[20]
8118	Nonanamide	$C_9H_{19}NO$ / 157.26	1120-07-6 / 99.5	sub	4-02-00-01023 / 0.8394[110]	H_2O 1; EtOH 2; eth 2 / 1.4248[110]
8119	1-Nonanamine	$C_9H_{21}N$ / 143.27	112-20-9 / -1	202.2	4-04-00-00777 / 0.7886[20]	H_2O 2; EtOH 3; eth 3; chl 2 / 1.4336[20]
8120	Nonane	C_9H_{20} / 128.26	111-84-2 / -53.5	150.8	4-01-00-00447 / 0.7176[20]	H_2O 1; EtOH 4; eth 4; ace 5 / 1.4054[20]
8121	Nonane, 1-bromo-	$C_9H_{19}Br$ / 207.15	693-58-3	88[4]	4-01-00-00451 / 1.0165[20]	1.4533[20]

No.	Name / Synonym	Mol. Form. / Mol. Wt.	CAS RN / mp/°C	Merck No. / bp/°C	Beil. Ref. / den/g cm⁻³	Solubility / n_D
8122	Nonane, 5-butyl-	$C_{13}H_{28}$ 184.37	17312-63-9	217.5	4-01-00-00517 0.7635[18]	1.4273[18]
8123	Nonane, 1-chloro-	$C_9H_{19}Cl$ 162.70	2473-01-0 -39.4	203.5	4-01-00-00450 0.8720[20]	H_2O 1; eth 3; chl 3 1.4345[20]
8124	Nonane, 2-chloro- 2-Chlorononane	$C_9H_{19}Cl$ 162.70	2216-36-6	190	0-01-00-00166 0.8790[20]	chl 4 1.4420[20]
8125	Nonane, 5-chloro- 5-Chlorononane	$C_9H_{19}Cl$ 162.70	28123-70-8	85-7[14]	2-01-00-00128 0.8639[15]	eth 4 1.4314[15]
8126	Nonane, 1-chloro-9-fluoro-	$C_9H_{18}ClF$ 180.69	463-23-0	102[11]	4-01-00-00450 0.966[20]	eth 4; EtOH 4 1.4301[25]
8127	Nonane, 1,2-dibromo- 1,2-Dibromononane	$C_9H_{18}Br_2$ 286.05	73642-91-8	141.5[20]	3-01-00-00506 1.3980[20]	1.4942[20]
8128	Nonane, 1,9-dichloro- 1,9-Dichlorononane	$C_9H_{18}Cl_2$ 197.15	821-99-8	260; 138[17]	4-01-00-00451 1.0173[25]	1.4586[25]
8129	Nonanedinitrile	$C_9H_{14}N_2$ 150.22	1675-69-0	198-9[25]	4-02-00-02059 0.9200[19]	eth 4; EtOH 4; bz 4 1.4518[19]
8130	Nonanedioic acid Azelaic acid	$C_9H_{16}O_4$ 188.22	123-99-9 106.5	921 287[100], 225[10]	4-02-00-02055 1.225[25]	H_2O 2; EtOH 3; eth 2; bz 2 1.4303[111]
8131	Nonanedioic acid, bis(2-ethylbutyl) ester	$C_{21}H_{40}O_4$ 356.55	105-03-3 -45	230[5]	4-02-00-02057 0.928[25]	ace 4; bz 4; EtOH 4 1.443[25]
8132	Nonanedioic acid, bis(2-ethylhexyl) ester	$C_{25}H_{48}O_4$ 412.65	103-24-2 -78	237[5]	3-02-00-01787 0.915[25]	H_2O 1; EtOH 3; ace 3; bz 3 1.446[25]
8133	Nonanedioic acid, diethyl ester	$C_{13}H_{24}O_4$ 244.33	624-17-9 -18.5	291.5	4-02-00-02056 0.9729[20]	H_2O 1; EtOH 3; eth 3 1.4351[20]
8134	Nonanedioic acid, dimethyl ester	$C_{11}H_{20}O_4$ 216.28	1732-10-1	156[20]	4-02-00-02056 1.0082[20]	H_2O 1; EtOH 3; ace 3; bz 3 1.4367[20]
8135	Nonanedioic acid, 3-methyl- Azelaic acid, 3-methyl	$C_{10}H_{18}O_4$ 202.25	76078-85-8 44	168[0.2]	3-02-00-01819	1.4670[19]
8136	1,9-Nonanediol 1,9-Dihydroxynonane	$C_9H_{20}O_2$ 160.26	3937-56-2 45.8	173[20]; 150[3]	4-01-00-02607	H_2O 2; EtOH 4; eth 4; bz 3
8137	2,4-Nonanedione Methyl β-oxoheptyl ketone	$C_9H_{16}O_2$ 156.22	6175-23-1 -18	105-7[22]	4-01-00-03715 0.938[25]	
8138	Nonanedioyl dichloride	$C_9H_{14}Cl_2O_2$ 225.11	123-98-8	166[18]	3-02-00-01789 1.4680[20]	eth 3; bz 4 1.4680[20]
8139	Nonane, icosafluoro- Perfluorononane	C_9F_{20} 488.07	375-96-2	125.3	1.800[25]	
8140	Nonane, 2-methyl- 2-Methylnonane	$C_{10}H_{22}$ 142.28	871-83-0 -74.6	167.1	4-01-00-00473 0.7281[20]	H_2O 1; eth 3; bz 3; chl 3 1.4099[20]
8141	Nonane, 3-methyl- 3-Methylnonane	$C_{10}H_{22}$ 142.28	5911-04-6 -84.8	167.9	2-01-00-00130 0.7354[20]	bz 4; eth 4; chl 4 1.4125[20]
8142	Nonane, 4-methyl- 4-Methylnonane	$C_{10}H_{22}$ 142.28	17301-94-9 -99	165.7	4-01-00-00475 0.7323[20]	bz 4; eth 4; chl 4 1.4123[20]
8143	Nonane, 5-methyl- 5-Methylnonane	$C_{10}H_{22}$ 142.28	15869-85-9 -87.7	165.1	4-01-00-00475 0.7326[20]	H_2O 1; eth 3; bz 3; chl 3 1.4116[20]
8144	Nonanenitrile	$C_9H_{17}N$ 139.24	2243-27-8 -34.2	224.4	4-02-00-01024 0.8178[20]	H_2O 1; EtOH 3; eth 3; ctc 2 1.4255[20]
8145	1-Nonanethiol Nonyl mercaptan	$C_9H_{20}S$ 160.32	1455-21-6 -20.1	220	4-01-00-01802 0.842[25]	1.4548[20]
8146	Nonanoic acid Pelargonic acid	$C_9H_{18}O_2$ 158.24	112-05-0 12.3	7013 254.5	4-02-00-01018 0.9052[20]	H_2O 1; EtOH 3; eth 3; chl 3 1.4343[19]
8147	Nonanoic acid, butyl ester Pelargonic acid, butyl ester	$C_{13}H_{26}O_2$ 214.35	50623-57-9 -38	122-4[20]	3-02-00-00820 0.8520[25]	1.4262[25]
8148	Nonanoic acid, 9-cyano-, methyl ester Sebacic acid, mononitrile, methyl ester	$C_{11}H_{19}NO_2$ 197.28	53663-26-6 3.5	178[16]	4-02-00-02089 0.9340[20]	eth 4 1.4398[25]
8149	Nonanoic acid, ethenyl ester	$C_{11}H_{20}O_2$ 184.28	6280-03-1	133.5[50]	4-02-00-01020 0.8689[30]	1.4447[30]
8150	Nonanoic acid, ethyl ester Ethyl nonanoate	$C_{11}H_{22}O_2$ 186.29	123-29-5 -36.7	3793 227.0	4-02-00-01019 0.8657[20]	H_2O 1; EtOH 3; eth 3; ace 3 1.4220[20]
8151	Nonanoic acid, 9-fluoro- Pelargonic acid, 9-fluoro	$C_9H_{17}FO_2$ 176.23	463-16-1 18	89[0.2]	4-02-00-01024	eth 4; EtOH 4 1.4289[25]
8152	Nonanoic acid, heptyl ester Pelargonic acid, heptyl ester	$C_{16}H_{32}O_2$ 256.43	71605-85-1 -15.5	210[75]	3-02-00-00821 0.8553[25]	1.4350[20]
8153	Nonanoic acid, 3-methyl- Pelargonic acid, 3-methyl	$C_{10}H_{20}O_2$ 172.27	35205-79-9	147-8[12]	4-02-00-01057 0.9012[20]	1.4342[20]
8154	Nonanoic acid, methyl ester	$C_{10}H_{20}O_2$ 172.27	1731-84-6	213.5	4-02-00-01019 0.8799[15]	H_2O 1; EtOH 3; eth 3; ctc 2 1.4214[20]
8155	Nonanoic acid, 3-methyl-, ethyl ester Pelargonic acid 3-methyl, ethyl ester	$C_{12}H_{24}O_2$ 200.32	86051-37-8	115[13]	3-02-00-00849 0.8653[20]	1.4240[20]
8156	Nonanoic acid, pentyl ester Pelargonic acid, pentyl ester	$C_{14}H_{28}O_2$ 228.38	61531-45-1 -27	130-2[20]	3-02-00-00820 0.8506[25]	1.4318[20]
8157	1-Nonanol Nonyl alcohol	$C_9H_{20}O$ 144.26	143-08-8 -5	6598 213.3	4-01-00-01798 0.8273[20]	H_2O 1; EtOH 3; eth 3; ctc 2 1.4333[20]
8158	2-Nonanol, (±)-	$C_9H_{20}O$ 144.26	74683-66-2 -35	193.5	4-01-00-01803 0.8471[20]	eth 4; EtOH 4 1.4353[20]
8159	3-Nonanol, (±)-	$C_9H_{20}O$ 144.26	74742-08-8 22	195; 93[18]	4-01-00-01803 0.8250[20]	H_2O 1; EtOH 3; eth 3 1.4289[20]
8160	4-Nonanol, (S)-	$C_9H_{20}O$ 144.26	52708-03-9	192.5; 94[18]	4-01-00-01747 0.8282[20]	H_2O 1; EtOH 3; eth 3 1.4197[20]
8161	5-Nonanol Dibutylcarbinol	$C_9H_{20}O$ 144.26	623-93-8	193; 97[20]	4-01-00-01804 0.8356[20]	H_2O 1; EtOH 3 1.4289[20]

No.	Name / Synonym	Mol. Form. / Mol. Wt.	CAS RN / mp/°C	Merck No. / bp/°C	Beil. Ref. / den/g cm^{-3}	Solubility / n_D
8162	5-Nonanol, 5-butyl-	$C_{13}H_{28}O$ / 200.36	597-93-3 / 20	/ 232 dec	4-01-00-01863 / 0.8408[20]	EtOH 4 / 1.4445[20]
8163	1-Nonanol, 9-chloro- / 9-Chlorononan-1-ol	$C_9H_{19}ClO$ / 178.70	51308-99-7 / 28	/ 147[14]	4-01-00-01802 /	eth 4; EtOH 4 / 1.4575[20]
8164	5-Nonanol, 2,8-dimethyl- / 2,8-Dimethylnonan-5-ol	$C_{11}H_{24}O$ / 172.31	19780-96-2 /	/ 105[9]	0-01-00-00428 / 0.8305[12]	/ 1.4380[12]
8165	1-Nonanol, 9-fluoro-	$C_9H_{19}FO$ / 162.25	463-24-1 /	/ 125-6[15]	4-01-00-01801 / 0.928[20]	eth 4; EtOH 4 / 1.4279[25]
8166	1-Nonanol, 3-methyl- / 3-Hexylbutanol-1	$C_{10}H_{22}O$ / 158.28	22663-64-5 /	/ 114[14]; 103[9]	4-01-00-01825 / 0.8342[20]	/ 1.4361[20]
8167	1-Nonanol, 4-methyl-	$C_{10}H_{22}O$ / 158.28	1489-47-0 /	/ 120[17]	4-01-00-01826 / 0.826[27]	/ 1.4364[20]
8168	1-Nonanol, 8-methyl / 8-Methyl-1-nonanol	$C_{10}H_{22}O$ / 158.28	55505-26-5 /	/ 108[10]		
8169	3-Nonanol, 2-methyl-	$C_{10}H_{22}O$ / 158.28	26533-33-5 /	/ 115[28]	4-01-00-01825 / 0.8290[20]	/ 1.4346[20]
8170	4-Nonanol, 4-methyl- / 1-Methyl-1-propylhexanol	$C_{10}H_{22}O$ / 158.28	23418-38-4 /	/ 92-3[15]	4-01-00-01826 / 0.8245[25]	/ 1.4338[20]
8171	5-Nonanol, 5-methyl- / 2-Butyl-2-hexanol	$C_{10}H_{22}O$ / 158.28	33933-78-7 /	/ 84-5[10]	4-01-00-01827 / 0.8290[20]	/ 1.4341[20]
8172	2-Nonanone / Heptyl methyl ketone	$C_9H_{18}O$ / 142.24	821-55-6 / -7.5	/ 195.3	4-01-00-03353 / 0.8208[20]	H_2O 1; EtOH 3; eth 3; ace 4 / 1.4210[20]
8173	3-Nonanone / Ethyl hexyl ketone	$C_9H_{18}O$ / 142.24	925-78-0 / -8	/ 190; 86[20]	4-01-00-03354 / 0.8241[20]	H_2O 1; EtOH 3; eth 3; ace 4 / 1.4208[20]
8174	4-Nonanone / Propyl amyl ketone	$C_9H_{18}O$ / 142.24	4485-09-0 /	/ 187.5	4-01-00-03354 / 0.8190[25]	H_2O 1; EtOH 3; eth 3; ace 4 / 1.4189[20]
8175	5-Nonanone / Dibutyl ketone	$C_9H_{18}O$ / 142.24	502-56-7 / -5.9	/ 188.4	4-01-00-03355 / 0.8217[20]	H_2O 1; EtOH 3; eth 4; chl 4 / 1.4195[20]
8176	5-Nonanone, 2-methyl- / 2-Methyl-5-nonanone	$C_{10}H_{20}O$ / 156.27	22287-02-1 /	/ 203.5	3-01-00-02900 / 0.8213[20]	/ 1.4239[20]
8177	Nonanoyl chloride	$C_9H_{17}ClO$ / 176.69	764-85-2 / -60.5	/ 215.3	4-02-00-01023 / 0.9463[15]	eth 3; ace 3
8178	Nonasiloxane, eicosamethyl- / Eicosamethylnonasiloxane	$C_{20}H_{60}O_8Si_9$ / 681.46	2652-13-3 /	3494 / 307.5; 198[16]	4-04-00-04130 / 0.9173[20]	bz 4 / 1.3980[20]
8179	2-Nonenal / 2-Nonen-1-al	$C_9H_{16}O$ / 140.23	2463-53-8 /	/ 56-8[0.1]	4-01-00-03501 / 0.8418[25]	/ 1.4502[25]
8180	2-Nonenal, (E)-	$C_9H_{16}O$ / 140.23	18829-56-6 /	6595 /	4-01-00-03502 /	
8181	6-Nonenamide, N-[(4-hydroxy-3-methoxyphenyl)methyl]-8-methyl-, (E)- / Capsaicin	$C_{18}H_{27}NO_3$ / 305.42	404-86-4 / 65	1767 / 210 20[0.01]	4-13-00-02588 /	H_2O 1; EtOH 4; eth 3; bz 3
8182	1-Nonene	C_9H_{18} / 126.24	124-11-8 / -81.3	/ 146.9	4-01-00-00894 / 0.7253[26]	/ 1.4257[20]
8183	3-Nonene, (E)-	C_9H_{18} / 126.24	20063-92-7 /	/ 147.5	3-01-00-00853 / 0.732[21]	H_2O 1; eth 3; bz 3; ctc 3 / 1.4181[21]
8184	4-Nonene, 5-butyl- / 5-Butyl-4-nonene	$C_{13}H_{26}$ / 182.35	7367-38-6 /	/ 215.5	4-01-00-00922 / 0.7745[20]	/ 1.4375[20]
8185	1-Nonene, 2-methyl- / 2-Methylnon-1-ene	$C_{10}H_{20}$ / 140.27	2980-71-4 / -64.2	/ 168.4	4-01-00-00903 / 0.7412[25]	/ 1.4241[20]
8186	3-Nonene, 2-methyl-	$C_{10}H_{20}$ / 140.27	53966-53-3 /	/ 161.0	4-01-00-00903 / 0.7340[20]	/ 1.4202[20]
8187	3-Nonenoic acid	$C_9H_{16}O_2$ / 156.22	4124-88-3 /	/ 106[1]	4-02-00-01598 / 0.9254[20]	/ 1.4454[25]
8188	8-Nonenoic acid	$C_9H_{16}O_2$ / 156.22	31642-67-8 / 5	/ 116-8[1]	4-02-00-01597 / 0.9146[16]	/ 1.4492[15]
8189	1-Nonen-3-ol / 1-Vinylheptanol	$C_9H_{18}O$ / 142.24	21964-44-3 /	/ 193.5	3-01-00-01947 / 0.824[21]	/ 1.4382[15]
8190	8-Nonen-1-ol	$C_9H_{18}O$ / 142.24	13038-21-6 /	/ 135[20]	3-01-00-01947 / 0.8394[25]	/ 1.4450[23]
8191	5-Nonen-4-one, 6-methyl-	$C_{10}H_{18}O$ / 154.25	7036-98-8 /	/ 90-2[16]	3-01-00-03022 / 0.8608[20]	/ 1.4518[20]
8192	6-Nonen-4-one, 6-methyl-	$C_{10}H_{18}O$ / 154.25	53252-20-3 /	/ 94[18]	3-01-00-03022 / 0.8413[21]	/ 1.4429[21]
8193	1-Nonen-3-yne	C_9H_{14} / 122.21	57223-18-4 /	/ 27-8[4]	4-01-00-01103 / 0.7602[25]	ace 4; eth 4 / 1.4487[25]
8194	1-Nonen-4-yne	C_9H_{14} / 122.21	31508-12-0 /	/ 58[22]	4-01-00-01103 / 0.777[25]	ace 4; eth 4 / 1.4418[25]
8195	2-Nonen-4-yne	C_9H_{14} / 122.21	90644-60-3 /	/ 70[20]	3-01-00-01048 / 0.7832[25]	ace 4; eth 4 / 1.4590[25]
8196	1-Nonyne / Heptylacetylene	C_9H_{16} / 124.23	3452-09-3 / -50	/ 150.8	4-01-00-01047 / 0.7658[20]	H_2O 1; eth 3; bz 3; ctc 3 / 1.4217[20]
8197	2-Nonyne	C_9H_{16} / 124.23	19447-29-1 /	/ 161.9	3-01-00-01013 / 0.7690[20]	H_2O 1; eth 3; lig 3 / 1.4337[20]
8198	3-Nonyne	C_9H_{16} / 124.23	20184-89-8 /	/ 157.1	3-01-00-01013 / 0.7616[20]	H_2O 1; eth 3; lig 3 / 1.4299[20]
8199	4-Nonyne	C_9H_{16} / 124.23	20184-91-2 /	/ 152	4-01-00-01047 / 0.757[25]	H_2O 1; eth 3; ace 3 / 1.4296[25]
8200	1-Nonyne, 1-chloro-	$C_9H_{15}Cl$ / 158.67	90722-14-8 /	/ 75-7[15]	3-01-00-01012 / 0.906[20]	eth 4 / 1.450[20]
8201	4-Nonyne, 3,3-dimethyl- / 3,3-Dimethyl-4-nonyne	$C_{11}H_{20}$ / 152.28	29022-31-9 /	/ 82[40]	3-01-00-01024 / 0.7667[20]	bz 4; eth 4 / 1.4317[20]

No.	Name Synonym	Mol. Form. Mol. Wt.	CAS RN mp/°C	Merck No. bp/°C	Beil. Ref. den/g cm^{-3}	Solubility n_D
8202	4-Nonyne, 8-methyl- 8-Methyl-4-nonyne	$C_{10}H_{18}$ 138.25	70732-43-3	104.5[97]	3-01-00-01018 0.7681[20]	bz 4; eth 4 1.4311[20]
8203	2-Nonynoic acid	$C_9H_{14}O_2$ 154.21	1846-70-4 -8	154-6[18]	4-02-00-01721 0.9525[12]	1.4605[25]
8204	3-Nonynoic acid	$C_9H_{14}O_2$ 154.21	56630-33-2 17	118[2]	4-02-00-01721	1.4603[25]
8205	5-Nonynoic acid	$C_9H_{14}O_2$ 154.21	56630-34-3 -3	115[1]	4-02-00-01723	1.4558[25]
8206	6-Nonynoic acid	$C_9H_{14}O_2$ 154.21	56630-31-0 14	130[3]	4-02-00-01723	1.4578[25]
8207	8-Nonynoic acid	$C_9H_{14}O_2$ 154.21	30964-01-3 19	100[6]	4-02-00-01721	1.4524[25]
8208	1-Nonyn-3-ol	$C_9H_{16}O$ 140.23	7383-20-2	85[10]; 63[2]	4-01-00-02265 0.8627[20]	1.4440[20]
8209	2-Nonyn-1-ol	$C_9H_{16}O$ 140.23	5921-73-3	109[11]	4-01-00-02265 0.878[16]	1.4576[16]
8210	24-Norcholan-23-oic acid, (5β)- Norcholanic acid	$C_{23}H_{38}O_2$ 346.55	511-18-2 177	6606	3-09-00-02652	
8211	29-Nordammara-1,17(20),24-trien-21-oic acid, 6,16-bis(acetyloxy)-3,7-dioxo-, Helvolic acid	$C_{33}H_{44}O_8$ 568.71	29400-42-8 212 dec	4552	4-10-00-04083	H_2O 2; EtOH 2; eth 3; ace 3
8212	Norea Urea, N,N-dimethyl-N'-(octahydro-4,7- methano-1H-inde N-5-yl)-, (3aα,4α, 5α,7α,7aα)-	$C_{13}H_{22}N_2O$ 222.33	18530-56-8 177	6611		
8213	Norflurazon 3(2H)-Pyridazinone, 4-chloro-5- (methylamino)-2-[3-(trifluoromethyl)phenyl]-	$C_{12}H_9ClF_3N_3O$ 303.67	27314-13-2 184	6618		
8214	Norleucine 2-Aminohexanoic acid	$C_6H_{13}NO_2$ 131.17	327-57-1 298.5	6624	4-04-00-02686	H_2O 4
8215	DL-Norleucine Hexanoic acid, 2-amino(DL)	$C_6H_{13}NO_2$ 131.17	616-06-8 327 dec	6624	3-04-00-01386 1.172[25]	H_2O 3; EtOH 2; eth 1
8216	19-Norpregna-1,3,5(10)-trien-20-yne-3,17-diol, (17α)- 17α-Ethinyl-17β-estradiol	$C_{20}H_{24}O_2$ 296.41	57-63-6	3689	4-06-00-06877	chl 2
8217	DL-Norvaline	$C_5H_{11}NO_2$ 117.15	760-78-1 303 sub	6636	4-04-00-02629	H_2O 3; EtOH 1; eth 1; chl 1
8218	L-Norvaline (S)-2-Aminopentanoic acid	$C_5H_{11}NO_2$ 117.15	6600-40-4 307	6636	4-04-00-02629	H_2O 4
8219	Nuarimol 5-Pyrimidinemethanol, α-(2-chlorophenyl)-α- (4-fluorophenyl)-	$C_{17}H_{12}ClFN_2O$ 314.75	63284-71-9 126			
8220	Octacosane	$C_{28}H_{58}$ 394.77	630-02-4 64.5	431.6	4-01-00-00588 0.8067[20]	H_2O 1; ace 5; bz 3; chl 3 1.4330[70]
8221	Octacosanoic acid Montanic acid	$C_{28}H_{56}O_2$ 424.75	506-48-9 90.9		4-02-00-01318 0.8191[100]	bz 4; chl 4 1.4313[100]
8222	1-Octacosanol Montanyl alcohol	$C_{28}H_{58}O$ 410.77	557-61-9 83.3	6664 200[1]	4-01-00-01915	
8223	9,12-Octadecadienoic acid (Z,Z)- Linoleic acid	$C_{18}H_{32}O_2$ 280.45	60-33-3 -8.5	5382 229-30[16]	4-02-00-01754 0.9022[20]	ace 4; bz 4; eth 4; EtOH 4 1.4699[20]
8224	10,12-Octadecadienoic acid, (E,E)-	$C_{18}H_{32}O_2$ 280.45	1072-36-2 56.5		4-02-00-01752 0.8686[70]	1.4689[40]
8225	9,12-Octadecadienoic acid	$C_{18}H_{32}I_2$ 502.26	2197-37-7 -5	229-30[16]	4-02-00-01754 0.9022[20]	ace 4; bz 4; eth 4; EtOH 4 1.4699[20]
8226	9,12-Octadecadienoic acid (Z,Z)-, ethyl ester Ethyl linoleate	$C_{20}H_{36}O_2$ 308.50	544-35-4	3776 272[180]; 212[12]	4-02-00-01757 0.8865[20]	eth 4; EtOH 4
8227	9,12-Octadecadienoic acid, ethyl ester	$C_{20}H_{36}O_2$ 308.50	7619-08-1	272[180] 212[12]	2-02-00-00461 0.8865[20]	H_2O 1; EtOH 3; eth 3
8228	9,12-Octadecadienoic acid (Z,Z)-, methyl ester Methyl linoleate	$C_{19}H_{34}O_2$ 294.48	112-63-0 -35	6008 215[20]	4-02-00-01756 0.8886[10]	eth 4; EtOH 4 1.4638[20]
8229	9,12-Octadecadienoic acid, methyl ester Methyl 9,12-octadecadienoate	$C_{19}H_{34}O_2$ 294.48	2462-85-3 -35	215[20] 169[1]	2-02-00-00461 0.8886[18]	H_2O 1; EtOH 3; eth 3; ctc 2 1.4638[20]
8230	7,11-Octadecadiyne	$C_{18}H_{30}$ 246.44	103697-14-9	167-8[7]	3-01-00-01068 0.841[19]	1.4698[19]
8231	Octadecanamide	$C_{18}H_{37}NO$ 283.50	124-26-5 109	250[12]	4-02-00-01240	eth 4; chl 4
8232	Octadecanamide, N-(4-hydroxyphenyl)-	$C_{24}H_{41}NO_2$ 375.60	103-99-1 133.8	239.5[10]	4-13-00-01113	H_2O 1; eth 2; ace 3; bz 2
8233	1-Octadecanamine	$C_{18}H_{39}N$ 269.51	124-30-1 52.9	346.8	4-04-00-00825 0.8618[20]	H_2O 1; EtOH 3; eth 3; ace 2 1.4522[20]
8234	Octadecane	$C_{18}H_{38}$ 254.50	593-45-3 28.2	316.3	4-01-00-00553 0.7768[28]	H_2O 1; EtOH 2; eth 3; ace 3 1.4390[20]
8235	Octadecane, 1-bromo-	$C_{18}H_{37}Br$ 333.40	112-89-0 28.2	210[10]	4-01-00-00555 0.9848[20]	H_2O 1; EtOH 3; eth 3; ctc 2 1.4631[20]
8236	Octadecane, 1-chloro-	$C_{18}H_{37}Cl$ 288.94	3386-33-2 28.6	348	4-01-00-00554 0.8641[20]	H_2O 1; ctc 2 1.4531[20]
8237	Octadecane, 1,1-dimethoxy- Octadecanal, dimethyl acetal	$C_{20}H_{42}O_2$ 314.55	14620-55-4	168-70[3]	4-01-00-03397	eth 4; EtOH 4 1.4410[25]

No.	Name / Synonym	Mol. Form. / Mol. Wt.	CAS RN / mp/°C	Merck No. / bp/°C	Beil. Ref. / den/g cm^{-3}	Solubility / n_D
8238	Octadecanedioic acid, diethyl ester	$C_{22}H_{42}O_4$ 370.57	1472-90-8 54.5	240[12]	4-02-00-02176	eth 4; EtOH 4
8239	Octadecanedioic acid, 9,10-dihydroxy-, (R*,R*)-(±)- Phloionic acid	$C_{18}H_{34}O_6$ 346.46	23843-52-9 126	7297	4-03-00-01250	
8240	Octadecane, 1-(ethenyloxy)-	$C_{20}H_{40}O$ 296.54	930-02-9 30	182[3]	4-01-00-02057 0.8138[40]	chl 2
8241	Octadecane, 3-ethyl-5-(2-ethylbutyl)-	$C_{26}H_{54}$ 366.71	55282-12-7	229.5[10]	4-01-00-00586 0.8115[20]	1.45239[20]
8242	Octadecane, 1-iodo-	$C_{18}H_{37}I$ 380.40	629-93-6 34	383	4-01-00-00556 1.0994[20]	H$_2$O 1; EtOH 2; eth 2 1.4810[20]
8243	Octadecanenitrile	$C_{18}H_{35}N$ 265.48	638-65-3 41	362	4-02-00-01242 0.8325[20]	H$_2$O 1; EtOH 3; eth 4; ace 4 1.4389[45]
8244	1-Octadecanethiol Stearyl mercaptan	$C_{18}H_{38}S$ 286.57	2885-00-9 30	204-10[11]	4-01-00-01894 0.8475[20]	eth 4 1.4645[20]
8245	Octadecane, 9-p-tolyl- Toluene, p-(1-octyldecyl)-	$C_{25}H_{44}$ 344.62	4445-08-3	185[10]	4-05-00-01221 0.8549[20]	1.4811[20]
8246	9,11,13-Octadecanetrienoic acid (Z,Z,Z) Eleostearic acid	$C_{18}H_{30}O_2$ 278.44	3884-88-6 48.5			
8247	Octadecanoic acid Stearic acid	$C_{18}H_{36}O_2$ 284.48	57-11-4 68.8	8761 350 dec; 232[15]	4-02-00-01206 0.9408[20]	H$_2$O 1; EtOH 2; eth 4; ace 3 1.4299[80]
8248	Octadecanoic acid, aluminum salt	$C_{54}H_{105}AlO_6$ 877.41	637-12-7 118	370	4-02-00-01206	H$_2$O 1; EtOH 3; peth 3
8249	Octadecanoic acid, anhydride	$C_{36}H_{70}O_3$ 550.95	638-08-4 72		4-02-00-01239 0.8365[82]	H$_2$O 1; EtOH 1; eth 2; bz 2 1.4362[80]
8250	Octadecanoic acid, 18-bromo- Stearic acid, 18-bromo	$C_{18}H_{35}BrO_2$ 363.38	2536-38-1 75.5	240[4]	2-02-00-00361	bz 4; eth 4; EtOH 4
8251	Octadecanoic acid, butyl ester Butyl stearate	$C_{22}H_{44}O_2$ 340.59	123-95-5 27	1589 343	4-02-00-01219 0.854[25]	H$_2$O 1; EtOH 3; ace 4 1.4328[50]
8252	Octadecanoic acid, calcium salt	$C_{36}H_{70}CaO_4$ 607.03	1592-23-0 179.5	1710	4-02-00-01206	H$_2$O 1; EtOH 1; eth 1
8253	Octadecanoic acid, cyclohexyl ester Stearic acid, cyclohexyl ester	$C_{24}H_{46}O_2$ 366.63	104-07-4 44		4-06-00-00038 0.889[15]	eth 4
8254	Octadecanoic acid, 9,10-dihydroxy 9,10-Dihydroxystearic acid	$C_{18}H_{36}O_4$ 316.48	120-87-6 90	3171	4-03-00-01092	H$_2$O 1; EtOH 2; eth 2
8255	Octadecanoic acid, 2,3-dihydroxypropyl ester, (±)-	$C_{21}H_{42}O_4$ 358.56	22610-63-5 74		4-02-00-01225 0.9841[20]	H$_2$O 1; EtOH 2; eth 2; lig 3 1.4400[86]
8256	Octadecanoic acid, 1,2-ethanediyl ester	$C_{38}H_{74}O_4$ 595.00	627-83-8 79	241[20]	4-02-00-01223 0.8581[78]	H$_2$O 1; EtOH 1; eth 4; ace 4
8257	Octadecanoic acid, ethyl ester	$C_{20}H_{40}O_2$ 312.54	111-01-5 33	199[10]	4-02-00-01218 1.057[20]	H$_2$O 1; EtOH 3; eth 3; ace 4 1.4349[40]
8258	Octadecanoic acid, hexadecyl ester	$C_{34}H_{68}O_2$ 508.91	1190-63-2 57		4-02-00-01220	ace 4; eth 4; chl 4 1.4410[70]
8259	Octadecanoic acid, 2-[2-[2-(2-hydroxyethoxy)ethoxy]ethoxy]ethyl ester	$C_{26}H_{52}O_6$ 460.70	106-07-0 40	328	1.1285[15]	1.4593[20]
8260	Octadecanoic acid, 2-hydroxyethyl ester	$C_{20}H_{40}O_3$ 328.54	111-60-4 60.5	189-91[3]	4-02-00-01222 0.8780[60]	EtOH 2; eth 3 1.4310[60]
8261	Octadecanoic acid, lead (II) salt Lead stearate	$C_{36}H_{70}O_4Pb$ 774.15	7428-48-0 125		4-02-00-01206 1.4	H$_2$O 1; Hot EtOH 3; eth 1
8262	Octadecanoic acid, 14-methyl- Stearic acid, 14-methyl	$C_{19}H_{38}O_2$ 298.51	94434-64-7 37.5	182[0.4]	4-02-00-01265 0.9400[20]	
8263	Octadecanoic acid, 17-methyl-	$C_{19}H_{38}O_2$ 298.51	2724-59-6 67.5	180[0.3]	4-02-00-01260 0.8420[70]	1.4336[70]
8264	Octadecanoic acid, 9-methyl- Stearic acid, 9-methyl	$C_{19}H_{38}O_2$ 298.51	86073-38-3 40	171[0.1]	4-02-00-01271 0.9980[20]	
8265	Octadecanoic acid, 3-methylbutyl ester Stearic acid, isopentyl ester	$C_{23}H_{46}O_2$ 354.62	627-88-3 25.5	192[2]	2-02-00-00353 0.855[20]	H$_2$O 1; EtOH 2; eth 3; ace 3 1.433[50]
8266	Octadecanoic acid, methyl ester	$C_{19}H_{38}O_2$ 298.51	112-61-8 39.1	443; 215[15]	4-02-00-01216 0.8498[40]	eth 4; chl 4 1.4367[40]
8267	Octadecanoic acid, 1-methylethyl ester	$C_{21}H_{42}O_2$ 326.56	112-10-7 28	207[6]	4-02-00-01219 0.8403[38]	ace 4; eth 4; EtOH 4; chl 4
8268	Octadecanoic acid, 2-methylpropyl ester Isobutyl stearate	$C_{22}H_{44}O_2$ 340.59	646-13-9 28.9	5034 223[15]	3-02-00-01017 0.8498[20]	eth 4
8269	Octadecanoic acid, 12-oxo-, ethyl ester Stearic acid, 12-oxo, ethyl ester	$C_{20}H_{38}O_3$ 326.52	88472-61-1 38	199[3]	3-03-00-01294	EtOH 4
8270	Octadecanoic acid, pentyl ester Stearic acid, pentyl ester	$C_{23}H_{46}O_2$ 354.62	6382-13-4 30		4-02-00-01220	eth 4; EtOH 4 1.4342[50]
8271	Octadecanoic acid, phenyl ester Stearic acid, phenyl ester	$C_{24}H_{40}O_2$ 360.58	637-55-8 52	267[15]	4-06-00-00618	H$_2$O 1; EtOH 3; eth 3
8272	Octadecanoic acid, 1,2,3-propanetriyl ester Tristearin	$C_{57}H_{110}O_6$ 891.50	555-43-1	9669	4-02-00-01233 0.8559[90]	H$_2$O 1; EtOH 1; ace 3; bz 2 1.4395[80]
8273	Octadecanoic acid, propyl ester	$C_{21}H_{42}O_2$ 326.56	3634-92-2 28.9	186.8[2]	4-02-00-01219 0.8452[38]	ace 4; eth 4; EtOH 4 1.4400[30]
8274	Octadecanoic acid, 9,10,12,13-tetrabromo-, methyl ester Stearic acid, 9,10,12,13-tetrabromo, methyl ester	$C_{19}H_{34}Br_4O_2$ 614.09	62080-86-8 63	215[15]	3-02-00-01049	eth 4; EtOH 4; chl 4 1.4346[45]

No.	Name / Synonym	Mol. Form. / Mol. Wt.	CAS RN / mp/°C	Merck No. / bp/°C	Beil. Ref. / den/g cm^{-3}	Solubility / n_D
8275	Octadecanoic acid, (tetrahydro-2-furanyl)methyl ester	$C_{23}H_{44}O_3$	6940-09-6		4-02-00-01248	eth 4; EtOH 4
	Stearic acid, tetrahydrofurfuryl ester	368.60	22		0.914^{25}	
8276	Octadecanoic acid, 9,10,18-trihydroxy-, (R*,R*)-	$C_{18}H_{36}O_5$	583-86-8	7298	4-03-00-01119	
	Phloionolic acid	332.48	101.5			
8277	Octadecanoic acid, zinc salt	$C_{36}H_{70}O_4Zn$	557-05-1	10063	4-02-00-01206	H_2O 1; EtOH 1; eth 1
		632.34	130			
8278	1-Octadecanol	$C_{18}H_{38}O$	112-92-5		4-01-00-01888	H_2O 1; EtOH 3; eth 3; ace 2
	Stearyl alcohol	270.50	59.5	210.5^{15}	0.8124^{59}	
8279	3-Octadecanol, (±)-	$C_{18}H_{38}O$	111897-18-8		4-01-00-01896	
	3-Octadecanol (DL)	270.50	51	202^{13}	0.7858^{80}	$1 429 0^{80}$
8280	3-Octadecanone	$C_{18}H_{36}O$	18261-92-2		4-01-00-03398	ctc 3
		268.48	51	197^{11}		
8281	Octadecanoyl chloride	$C_{18}H_{35}ClO$	112-76-5		4-02-00-01240	EtOH 2
		302.93	23	215^{15}	0.8969^0	1.4523^{24}
8282	9,11,13-Octadecatrienoic acid, (E,E,E)-	$C_{18}H_{30}O_2$	544-73-0		4-02-00-01787	EtOH 4
		278.44	71.5	188^1	0.8839^{80}	1.5000^{80}
8283	9,12,15-Octadecatrienoic acid, ethyl ester, (Z,Z,Z)-	$C_{20}H_{34}O_2$	1191-41-9		4-02-00-01782	eth 4; EtOH 4
		306.49		218^{15}	0.8919^{20}	1.4694^{20}
8284	9,11,13-Octadecatrienoic acid, (E,Z,E)-	$C_{18}H_{30}O_2$	506-23-0		4-02-00-01787	eth 4; EtOH 4
		278.44	49	235^{12} dec; 170^1	0.9028^{50}	1.5112^{50}
8285	9,12,15-Octadecatrienoic acid, methyl ester, (Z,Z,Z)-	$C_{19}H_{32}O_2$	301-00-8		4-02-00-01782	
		292.46		182^3	0.896^{25}	1.4709^{20}
8286	9,11,13-Octadecatrienoic acid, (Z,Z,E)-	$C_{18}H_{30}O_2$	544-72-9		4-02-00-01787	EtOH 4; peth 4
		278.44	62		0.9027^{20}	1.5114^{50}
8287	9,12,15-Octadecatrienoic acid, (Z,Z,Z)-	$C_{18}H_{30}O_2$	463-40-1	5383	4-02-00-01781	H_2O 1; EtOH 3; eth 3; bz 2
	Linolenic acid	278.44	-16.5	231^{17}; $129^{0.05}$	0.9164^{20}	1.4800^{20}
8288	9-Octadecenal	$C_{18}H_{34}O$	5090-41-5		4-01-00-03533	
	Octadecenyl aldehyde	266.47		$168-9^3$	0.8509^{20}	1.4558^{20}
8289	9-Octadecenal, (Z)-	$C_{18}H_{34}O$	2423-10-1		4-01-00-03533	
		266.47		$168-9^3$	0.8509^{20}	1.4558^{20}
8290	9-Octadecenamide, N-phenyl-, (Z)-	$C_{24}H_{39}NO$	5429-85-6		4-12-00-00411	bz 4; eth 4; HOAc 4
		357.58	41	143.5^{10}		
8291	1-Octadecene	$C_{18}H_{36}$	112-88-9		4-01-00-00930	H_2O 1; ace 3; ctc 3
		252.48	17.5	179^{15}; 145^8	0.7891^{20}	1.4448^{20}
8292	9-Octadecene	$C_{18}H_{36}$	5557-31-3		4-01-00-00932	
		252.48	-30.5	162^9	0.7916^{20}	1.4470^{20}
8293	9-Octadecenenitrile, (Z)-	$C_{18}H_{33}N$	112-91-4		4-02-00-01668	EtOH 4
		263.47	-1	332 dec	0.847^{17}	1.4566^{20}
8294	9-Octadecenoic acid (Z)-	$C_{18}H_{34}O_2$	112-80-1	6788	4-02-00-01641	H_2O 1; EtOH 5; eth 5; ace 5
	Oleic acid	282.47	13.4	286^{100}	0.8935^{20}	1.4582^{20}
8295	11-Octadecenoic acid, (E)-	$C_{18}H_{34}O_2$	693-72-1	9810	4-02-00-01640	ace 3
	Vaccenic acid	282.47	44			1.4499^{60}
8296	2-Octadecenoic acid, (E)-	$C_{18}H_{34}O_2$	2825-79-8		4-02-00-01632	H_2O 1; EtOH 2; eth 3; bz 3
		282.47	58.5		0.8484^{90}	
8297	9-Octadecenoic acid, (E)-	$C_{18}H_{34}O_2$	112-79-8	3497	4-02-00-01647	H_2O 1; EtOH 3; eth 3; bz 3
	Elaidic acid	282.47	45	288^{100}; 234^{15}	0.8734^{45}	1.4499^{45}
8298	9-Octadecenoic acid (Z)-, butyl ester	$C_{22}H_{42}O_2$	142-77-8		4-02-00-01653	EtOH 4
		338.57	-26.4	$227-8^{15}$	0.8704^{15}	1.4480^{25}
8299	9-Octadecenoic acid (Z)-, 2,3-dihydroxypropyl ester	$C_{21}H_{40}O_4$	111-03-5		2-02-00-00439	H_2O 1; EtOH 3; eth 3; chl 3
		356.55	35	$238-40^3$	0.9420^{20}	1.4626^{20}
8300	6-Octadecenoic acid, 6,7-diiodo-	$C_{18}H_{32}I_2O_2$	533-86-8		4-02-00-01638	bz 4; eth 4; chl 4
	Diiodotariric acid	534.26	50.3	dec		
8301	9-Octadecenoic acid (Z)-, ethyl ester	$C_{20}H_{38}O_2$	111-62-6		4-02-00-01651	eth 4; EtOH 4
		310.52		207^{13}	0.8720^{20}	1.4515^{20}
8302	9-Octadecenoic acid, ethyl ester, (E)-	$C_{20}H_{38}O_2$	6114-18-7		4-02-00-01652	eth 4; EtOH 4
		310.52	5.8	$217-9^{15}$	0.8664^{25}	1.4480^{25}
8303	9-Octadecenoic acid, 12-hydroxy-, [R-(Z)]-	$C_{18}H_{34}O_3$	141-22-0	8213	4-03-00-01026	eth 4; EtOH 4
	Ricinoleic acid	298.47		227^{10}	0.9450^{21}	1.4716^{21}
8304	9-Octadecenoic acid, 12-hydroxy-, butyl ester, [R-(Z)]-	$C_{22}H_{42}O_3$	151-13-3		0-03-00-00388	eth 4
		354.57		278^{12}	0.9058^{22}	1.4566^{22}
8305	9-Octadecenoic acid, 12-hydroxy-, ethyl ester, [R-(Z)]-	$C_{20}H_{38}O_3$	55066-53-0		4-03-00-01029	
		326.52		258^{13}	0.9045^{22}	1.4618^{22}
8306	9-Octadecenoic acid (Z)-, 3-methylbutyl ester	$C_{23}H_{44}O_2$	627-89-4		3-02-00-01414	eth 4; EtOH
	Oleic acid, isopentyl ester	352.60		$223-4^{10}$	0.897^{15}	
8307	9-Octadecenoic acid (Z)-, methyl ester	$C_{19}H_{36}O_2$	112-62-9		4-02-00-01649	H_2O 1; EtOH 5; eth 5; chl 3
	Methyl oleate	296.49	-19.9	218.5^{20}	0.8739^{20}	1.4522^{20}
8308	9-Octadecenoic acid, methyl ester, (E)-	$C_{19}H_{36}O_2$	1937-62-8		4-02-00-01651	eth 4; EtOH 4
		296.49		$213-5^{15}$	0.8730^{20}	1.4513^{20}
8309	9-Octadecenoic acid (Z)-, phenylmethyl ester	$C_{25}H_{40}O_2$	55130-16-0		4-06-00-02269	eth 4; EtOH 4
	Oleic acid, benzyl ester	372.59		237^1	0.9302^{25}	1.4875^{25}

No.	Name Synonym	Mol. Form. Mol. Wt.	CAS RN mp/°C	Merck No. bp/°C	Beil. Ref. den/g cm⁻³	Solubility n_D
8310	9-Octadecenoic acid (Z)-, 1,2,3-propanetriyl ester Triolein	$C_{57}H_{104}O_6$ 885.45	122-32-7 -32	9644 235-40[18]	4-02-00-01664 0.8988[40]	H_2O 1; EtOH 2; eth 4; chl 3 1.4621[40]
8311	9-Octadecen-1-ol, (E)-	$C_{18}H_{36}O$ 268.48	506-42-3 36.5	333	4-01-00-02204 0.8338[40]	ace 4; eth 4; EtOH 4 1.4552[40]
8312	9-Octadecen-1-ol, (Z)- Oleyl alcohol	$C_{18}H_{36}O$ 268.48	143-28-2 6.5	6791 205-10[15]	4-01-00-02204 0.8489[20]	H_2O 1; EtOH 3; eth 3; ctc 2 1.4606[20]
8313	1-Octadecyne	$C_{18}H_{34}$ 250.47	629-89-0 22.5	313	4-01-00-01075 0.8025[20]	1.4774[20]
8314	2-Octadecyne	$C_{18}H_{34}$ 250.47	61847-97-0 30	184[15]	4-01-00-01075 0.8016[30]	
8315	9-Octadecyne	$C_{18}H_{34}$ 250.47	35365-59-4 3	163-4[7]	4-01-00-01076 0.8012[20]	1.4488[25]
8316	9-Octadecynoic acid 9-Stearolic acid	$C_{18}H_{32}O_2$ 280.45	506-24-1 48	189[1.8]	4-02-00-01751 	H_2O 1; EtOH 3; eth 4 1.4510[54]
8317	2,6-Octadienal, 3,7-dimethyl-, (E)-	$C_{10}H_{16}O$ 152.24	141-27-5 	229	4-01-00-03569 0.8888[20]	H_2O 1; EtOH 5; eth 5 1.4898[20]
8318	2,6-Octadienal, 3,7-dimethyl-, (Z)-	$C_{10}H_{16}O$ 152.24	106-26-3 	120[20]	4-01-00-03569 0.8869[20]	H_2O 1; EtOH 5; eth 5 1.4869[20]
8319	1,7-Octadiene α,ω-Octadiene	C_8H_{14} 110.20	3710-30-3 	115.5	4-01-00-01038 0.734[20]	1.4245[20]
8320	2,6-Octadiene 1,6-Dimethyldiallyl	C_8H_{14} 110.20	4974-27-0 -76	124.5	4-01-00-01039 0.7406[25]	1.4292[17]
8321	2,6-Octadiene, 1-bromo-3,7-dimethyl-, (E)- trans-Geranyl bromide	$C_{10}H_{17}Br$ 217.15	6138-90-5 	101[12]; 47[0.005]	4-01-00-01059 1.0940[22]	1.5027[20]
8322	2,4-Octadiene, 3,7-dimethyl- 3,7-Dimethylocta-2,4-diene	$C_{10}H_{18}$ 138.25	56523-26-3 	165.5	3-01-00-01020 0.7933[20]	1.456[20]
8323	2,5-Octadiene, 2,6-dimethyl-	$C_{10}H_{18}$ 138.25	116668-48-5 	59.5[12]	3-01-00-01018 0.7773[20]	1.4500[20]
8324	2,6-Octadiene, 2,6-dimethyl-, (Z)- cis-2,6-Dimethyl-2,6-octadiene	$C_{10}H_{18}$ 138.25	2492-22-0 	168	4-01-00-01058 0.775[21]	ace 4; eth 4; EtOH 4 1.4498[20]
8325	2,6-Octadiene, 2,7-dimethyl 2,7-Dimethyl-2,6-octadiene	$C_{10}H_{18}$ 138.25	16736-42-8 -74.4	166	4-01-00-01061 0.7849[20]	1.4481[20]
8326	2,6-Octadiene, 3,6-dimethyl-	$C_{10}H_{18}$ 138.25	116668-49-6 	154	4-01-00-01061 0.7767[20]	1.4445[20]
8327	2,6-Octadiene, 4,5-dimethyl- 4,5-Dimethyl-2,6-octadiene	$C_{10}H_{18}$ 138.25	18476-57-8 -64.8	153.5	3-01-00-01021 0.7611[25]	1.4375[25]
8328	2,4-Octadiene, 7-methyl-	C_9H_{16} 124.23	2216-70-8 	149	2-01-00-00239 0.7521[18]	1.4543[18]
8329	3,5-Octadiene, 4-methyl- 4-Methyl-3,5-octadiene	C_9H_{16} 124.23	36903-95-4 	149.5	2-01-00-00239 0.7640[25]	1.4628[25]
8330	1,6-Octadiene, 7-methyl-3-methylene- β-Myrcene	$C_{10}H_{16}$ 136.24	123-35-3 	167	4-01-00-01108 0.8013[15]	H_2O 1; EtOH 3; eth 3; bz 3 1.4722[20]
8331	2,4-Octadienoic acid, 3,7-dimethyl-	$C_{10}H_{16}O_2$ 168.24	101715-60-0 	151-4[14]	4-02-00-01736 0.959[17]	1.4919
8332	1,6-Octadien-3-ol, 3,7-dimethyl-, (±)- Linalol	$C_{10}H_{18}O$ 154.25	22564-99-4 	198; 86[13]	4-01-00-02278 0.870[15]	1.4627
8333	2,6-Octadien-1-ol, 3,7-dimethyl-, (E)- trans-Geraniol	$C_{10}H_{18}O$ 154.25	106-24-1 <-15	4298 230	4-01-00-02277 0.8894[20]	H_2O 1; EtOH 3; eth 3; ace 3 1.4766[20]
8334	2,6-Octadien-1-ol, 3,7-dimethyl-, (Z)- Nerol	$C_{10}H_{18}O$ 154.25	106-25-2 <-15	6387 225; 125[25]	4-01-00-02276 0.8756[20]	EtOH 4 1.4746[20]
8335	1,6-Octadien-3-ol, 3,7-dimethyl-, acetate, (R)-	$C_{12}H_{20}O_2$ 196.29	16509-46-9 	220; 115[25]	4-02-00-00204 0.8951[20]	eth 4; EtOH 4 1.4544[21]
8336	2,6-Octadien-1-ol, 3,7-dimethyl-, acetate, (Z)-	$C_{12}H_{20}O_2$ 196.29	141-12-8 	134[25]; 93[3]	2-02-00-00153 0.905[15]	1.452[20]
8337	1,6-Octadien-3-ol, 3,7-dimethyl-, formate Linalyl formate	$C_{11}H_{18}O_2$ 182.26	115-99-1 	100-3[10]	4-02-00-00035 0.915[25]	EtOH 4 1.456[20]
8338	2,6-Octadien-1-ol, 3,7-dimethyl-, formate, (E)-	$C_{11}H_{18}O_2$ 182.26	105-86-2 	229 dec; 113[25]	4-02-00-00035 0.9086[25]	H_2O 1; EtOH 4; eth 3; ace 3 1.4659[20]
8339	2,6-Octadien-4-yne, 3,6-diethyl-	$C_{12}H_{18}$ 162.27	100319-48-0 	170	3-01-00-01066 0.8196[20]	ace 4; eth 4 1.4965[20]
8340	2,6-Octadien-4-yne, 3,6-dimethyl- 3,6-Dimethyl-2,6-octadiene-4-yne	$C_{10}H_{14}$ 134.22	3725-07-3 -45	170	4-01-00-01129 0.8071[22]	ace 4; eth 4 1.4998[20]
8341	1,5-Octadien-3-yne, 5-propyl- 5-Propyl-1,5-octadien-3-yne	$C_{11}H_{16}$ 148.25	70732-44-4 	57-8[6]	4-01-00-01129 0.8047[20]	ace 4; eth 4 1.4949[20]
8342	1,7-Octadiyne	C_8H_{10} 106.17	871-84-1 	135.5; 59[35]	4-01-00-01122 0.8169[21]	eth 3 1.4521[18]
8343	2,6-Octadiyne	C_8H_{10} 106.17	764-73-8 27	62[19]	4-01-00-01123 0.8288[80]	eth 4 1.4658[30]
8344	3,5-Octadiyne	C_8H_{10} 106.17	16387-70-5 	163.5	4-01-00-01123 0.826[0]	eth 3 1.4968[0]
8345	3,5-Octadiyne, 2,7-dimethyl- Diisopropylbutadiyne	$C_{10}H_{14}$ 134.22	14813-68-4 	74[12]	1-01-00-00128 0.8090[20]	eth 4
8346	Octanal Caprylic aldehyde	$C_8H_{16}O$ 128.21	124-13-0 	1766 171	4-01-00-03337 0.8211[20]	ace 4; bz 4; eth 4; EtOH 4 1.4217[20]
8347	Octanal, 7-hydroxy-3,7-dimethyl-	$C_{10}H_{20}O_2$ 172.27	107-75-5 	103[3]	4-01-00-04058 0.9220[20]	H_2O 2; EtOH 3; ace 3 1.4494[20]
8348	Octanal, oxime	$C_8H_{17}NO$ 143.23	929-55-5 60	112[9]	2-01-00-00758 	ace 4; EtOH 4

No.	Name Synonym	Mol. Form. Mol. Wt.	CAS RN mp/°C	Merck No. bp/°C	Beil. Ref. den/g cm^{-3}	Solubility n_D
8349	Octanamide	$C_8H_{17}NO$ 143.23	629-01-6 108	239	4-02-00-00992 0.8450[110]	H_2O 2; EtOH 4; eth 3; ace 3
8350	1-Octanamine	$C_8H_{19}N$ 129.25	111-86-4 0	179.6	4-04-00-00751 0.7826[20]	H_2O 2; EtOH 4; eth 4; ctc 3 1.4292[20]
8351	2-Octanamine, (±)-	$C_8H_{19}N$ 129.25	44855-57-4 97	164	4-04-00-00762 0.7744[20]	eth 4; EtOH 4 1.4232[25]
8352	2-Octanamine, N-(1-methylheptyl)-	$C_{16}H_{35}N$ 241.46	5412-92-0	283	4-04-00-00764 0.7948[20]	H_2O 1; EtOH 4; eth 4
8353	1-Octanamine, N-methyl-N-octyl-	$C_{17}H_{37}N$ 255.49	4455-26-9 -30.1	158[10]	4-04-00-00753	1.4424[20]
8354	1-Octanamine, N-octyl-	$C_{16}H_{35}N$ 241.46	1120-48-5 35.5	297.5	4-04-00-00753 0.7963[26]	eth 4; EtOH 4 1.4415[26]
8355	Octane	C_8H_{18} 114.23	111-65-9 -56.8	6672 125.6	4-01-00-00412 0.6986[25]	H_2O 1; EtOH 5; eth 3; ace 5 1.3974[20]
8356	Octane, 1-bromo- Octyl bromide	$C_8H_{17}Br$ 193.13	111-83-1 -55	6684 200	4-01-00-00422 1.108[25]	H_2O 1; EtOH 5; eth 5; ctc 2 1.4524[20]
8357	Octane, 2-bromo-, (±)- Octane, 2-bromo (DL)	$C_8H_{17}Br$ 193.13	60251-57-2 188.5		4-01-00-00423 1.0878[25]	H_2O 1; EtOH 5; eth 5 1.4442[25]
8358	Octane, 1-bromo-8-fluoro-	$C_8H_{16}BrF$ 211.12	593-12-4	118-20[22.5]	4-01-00-00424	eth 4; EtOH 4 1.4500[20]
8359	Octane, 1-chloro- Octyl chloride	$C_8H_{17}Cl$ 148.68	111-85-3 -57.8	181.5	4-01-00-00419 0.8738[20]	H_2O 1; EtOH 4; eth 4; ctc 2 1.4305[20]
8360	Octane, 2-chloro-	$C_8H_{17}Cl$ 148.68	628-61-5	172; 75[28]	1-01-00-00060 0.8658[17]	H_2O 1; EtOH 4; eth 3 1.4273[21]
8361	Octane, 1-chloro-8-fluoro-	$C_8H_{16}ClF$ 166.67	593-14-6	87[10]	4-01-00-00420 0.978[20]	ace 4; eth 4; EtOH 4 1.4266[25]
8362	Octane, 3-chloro-3-methyl-	$C_9H_{19}Cl$ 162.70	28320-88-9	73-4[15]	3-01-00-00508 0.8680[25]	ace 4; bz 4; eth 4; chl 4 1.4351[20]
8363	Octane, 4-chloro-4-methyl- 4-Chloro-4-methyloctane	$C_9H_{19}Cl$ 162.70	36903-89-6	71[14.5]	3-01-00-00509 0.8723[20]	eth 4; chl 4 1.4360[20]
8364	Octanedial	$C_8H_{14}O_2$ 142.20	638-54-0	235 dec; 97[3]	4-01-00-03706	H_2O 4; EtOH 4 1.4439[20]
8365	1,8-Octanediamine	$C_8H_{20}N_2$ 144.26	373-44-4 52	225	3-04-00-00612	H_2O 4; eth 4; EtOH 4
8366	Octane, 1,2-dibromo- 1,2-Dibromooctane	$C_8H_{16}Br_2$ 272.02	6269-92-7 241		3-01-00-00468 1.4580[20]	1.4970[20]
8367	Octane, 1,4-dibromo-	$C_8H_{16}Br_2$ 272.02	70690-24-3	125-6[11]	4-01-00-00424 1.466[11]	1.5003[11]
8368	Octane, 1,5-dibromo- 1,5-Dibromooctane	$C_8H_{16}Br_2$ 272.02	17912-17-3	127-8[11]	4-01-00-00424 1.447[16]	1.4968[16]
8369	Octane, 1,8-dibromo- Octamethylene dibromide	$C_8H_{16}Br_2$ 272.02	4549-32-0 15.5	271	4-01-00-00424 1.4594[25]	H_2O 1; eth 3; ctc 3; chl 3 1.4971[25]
8370	Octane, 4,5-dibromo-	$C_8H_{16}Br_2$ 272.02	61539-75-1	84[4.3]	3-01-00-00469 1.4569[20]	1.4981[20]
8371	Octane, 1,8-dichloro- 1,8-Dichlorooctane	$C_8H_{16}Cl_2$ 183.12	2162-99-4	241	4-01-00-00420 1.0248[25]	1.4572[25]
8372	Octane, 2,3-dichloro-	$C_8H_{16}Cl_2$ 183.12	21948-47-0 -55.1	206	4-01-00-00425 0.9712[20]	1.4523[20]
8373	Octane, 2,2-dimethyl- 2,2-Dimethyloctane	$C_{10}H_{22}$ 142.28	15869-87-1	155	4-01-00-00476 0.7208[25]	1.4082[20]
8374	Octane, 2,3-dimethyl- 2,3-Dimethyloctane	$C_{10}H_{22}$ 142.28	7146-60-3	164.3	4-01-00-00476 0.7377[20]	1.4146[20]
8375	Octane, 2,4-dimethyl- 2,4-Dimethyloctane	$C_{10}H_{22}$ 142.28	4032-94-4	156	4-01-00-00477 0.7226[25]	1.4091[20]
8376	Octane, 2,5-dimethyl- 2,5-Dimethyloctane	$C_{10}H_{22}$ 142.28	15869-89-3	158.5	4-01-00-00477 0.7264[25]	1.4112[20]
8377	Octane, 2,6-dimethyl- 2,6-Dimethyloctane	$C_{10}H_{22}$ 142.28	2051-30-1	160.4	2-01-00-00131 0.7313[20]	1.4097[20]
8378	Octane, 2,7-dimethyl-	$C_{10}H_{22}$ 142.28	1072-16-8 -54.9	159.9	4-01-00-00478 0.7202[25]	eth 3; HOAc 3 1.4086[20]
8379	Octane, 3,3-dimethyl- 3,3-Dimethyloctane	$C_{10}H_{22}$ 142.28	4110-44-5	161.2	4-01-00-00479 0.7351[25]	1.4165[20]
8380	Octane, 3,4-dimethyl- 3,4-Dimethyloctane	$C_{10}H_{22}$ 142.28	15869-92-8	163.4	4-01-00-00479 0.7410[25]	1.4182[20]
8381	Octane, 3,5-dimethyl- 3,5-Dimethyloctane	$C_{10}H_{22}$ 142.28	15869-93-9	159.4	4-01-00-00479 0.7329[25]	1.4139[20]
8382	Octane, 3,6-dimethyl- 3,6-Dimethyloctane	$C_{10}H_{22}$ 142.28	15869-94-0	160.8	4-01-00-00479 0.7324[25]	1.4139[20]
8383	Octane, 4,4-dimethyl- 4,4-Dimethyloctane	$C_{10}H_{22}$ 142.28	15869-95-1	157.5	4-01-00-00480 0.7312[25]	1.4144[20]
8384	Octane, 4,5-dimethyl- 4,5-Dimethyloctane	$C_{10}H_{22}$ 142.28	15869-96-2	162.15	4-01-00-00480 0.7432[25]	1.4190[20]
8385	Octanedinitrile Suberonitrile	$C_8H_{12}N_2$ 136.20	629-40-3 -3.5	185[15]	3-02-00-01768 0.954[25]	1.4436[20]
8386	Octanedioic acid Suberic acid	$C_8H_{14}O_4$ 174.20	505-48-6 144	8833 219[20]	4-02-00-02028	H_2O 1; eth 5; bz 5
8387	Octanedioic acid, diethyl ester	$C_{12}H_{22}O_4$ 230.30	2050-23-9 5.9	282.6	4-02-00-02029 0.9811[20]	H_2O 1; EtOH 3; eth 3; ctc 2 1.4328[20]
8388	Octanedioic acid, dimethyl ester	$C_{10}H_{18}O_4$ 202.25	1732-09-8 -3.1	268	4-02-00-02029 1.0217[20]	H_2O 1; EtOH 3; eth 3; ace 3 1.4341[20]

No.	Name Synonym	Mol. Form. Mol. Wt.	CAS RN mp/°C	Merck No. bp/°C	Beil. Ref. den/g cm^{-3}	Solubility n_D
8389	1,8-Octanediol	$C_8H_{18}O_2$ 146.23	629-41-4 63	172[20]	4-01-00-02592	H_2O 2; EtOH 4; eth 2; bz 3
8390	2,4-Octanediol	$C_8H_{18}O_2$ 146.23	90162-24-6 	117-8[8]	3-01-00-02219 0.918[25]	1.4422[25]
8391	4,5-Octanediol, (±)-	$C_8H_{18}O_2$ 146.23	22520-40-7 30	110[8]	4-01-00-02593	ctc 2 1.4419[25]
8392	1,7-Octanediol, 3,7-dimethyl-	$C_{10}H_{22}O_2$ 174.28	107-74-4 	265	3-01-00-02233 0.937[20]	bz 2; tol 2 1.4599[20]
8393	2,4-Octanedione Valerylacetone	$C_8H_{14}O_2$ 142.20	14090-87-0 	79-83[20]	4-01-00-03706 0.9233[20]	1.4559[20]
8394	2,7-Octanedione	$C_8H_{14}O_2$ 142.20	1626-09-1 44	110[13]	4-01-00-03707	EtOH 4
8395	3,6-Octanedione	$C_8H_{14}O_2$ 142.20	2955-65-9 35.5	98[14]	4-01-00-03708	EtOH 4
8396	4,5-Octanedione	$C_8H_{14}O_2$ 142.20	5455-24-3 	168	4-01-00-03708 0.934[0]	EtOH 4; eth 4; ace 4
8397	4,5-Octanedione, dioxime Dipropylglyoxime	$C_8H_{16}N_2O_2$ 172.23	61050-68-8 186.5	sub	2-01-00-00846	eth 4; EtOH 4
8398	3,5-Octanedione, 6,6,7,7,8,8,8-heptafluoro-2,2-dimethyl-	$C_{10}H_{11}F_7O_2$ 296.18	17587-22-3 38	46-7[5]	1.273[25]	1.3766[20]
8399	2,3-Octanedione, 3-oxime 3-Oximino-2-octanone	$C_8H_{15}NO_2$ 157.21	584-92-9 59	133[13]	0-01-00-00795	eth 4
8400	3,6-Octanedione, 2,2,7,7-tetramethyl- 2,2,7,7-Tetramethyl-3,6-octanedione	$C_{12}H_{22}O_2$ 198.31	27610-88-4 2.5	116[17]; 57[0.5]	4-01-00-03734 0.900[27]	EtOH 4 1.4400[20]
8401	Octane, 1-(ethenyloxy)-	$C_{10}H_{20}O$ 156.27	929-62-4 	58[4]	4-01-00-02056 0.8020[20]	1.4268[20]
8402	Octane, 1-ethoxy-	$C_{10}H_{22}O$ 158.28	929-61-3 12.5	186.3	4-01-00-01759 0.7847[20]	EtOH 4 1.4127[20]
8403	Octane, 3-ethyl- 3-Ethyloctane	$C_{10}H_{22}$ 142.28	5881-17-4 	166.5	4-01-00-00476 0.7359[25]	1.4156[20]
8404	Octane, 4-ethyl- 4-Ethyloctane	$C_{10}H_{22}$ 142.28	15869-86-0 	163.7	4-01-00-00476 0.7343[25]	1.4151[20]
8405	Octane, 1-fluoro- Octyl fluoride	$C_8H_{17}F$ 132.22	463-11-6 	142.4	4-01-00-00418 0.8108[20]	1.3935[20]
8406	Octane, 1-iodo-	$C_8H_{17}I$ 240.13	629-27-6 -45.7	225.5	4-01-00-00425 1.3297[20]	EtOH 3; eth 3 1.4889[20]
8407	Octane, 2-iodo-, (±)- (±)-2-Octyl iodide	$C_8H_{17}I$ 240.13	36049-78-2 	210; 95[16]	4-01-00-00425 1.3251[20]	H_2O 1; EtOH 3; eth 3; lig 3 1.4896[20]
8408	Octane, 2-methyl- 2-Methyloctane	C_9H_{20} 128.26	3221-61-2 -80.3	143.2	4-01-00-00454 0.7095[25]	H_2O 1; EtOH 3; eth 3; ctc 2 1.4031[20]
8409	Octane, 3-methyl- 3-Methyloctane	C_9H_{20} 128.26	2216-33-3 -107.6	144.2	4-01-00-00454 0.7177[25]	1.4040[25]
8410	Octane, 4-methyl- 4-Methyloctane	C_9H_{20} 128.26	2216-34-4 -113.3	142.4	4-01-00-00454 0.716[25]	1.4039[25]
8411	Octanenitrile Caprylnitrile	$C_8H_{15}N$ 125.21	124-12-9 -45.6	205.2	4-02-00-00993 0.8136[20]	eth 4 1.4203[20]
8412	Octane, 1-nitro- 1-Nitrooctane	$C_8H_{17}NO_2$ 159.23	629-37-8 15	208.5	4-01-00-00426 0.9346[20]	1.4322[20]
8413	Octane, 2-nitro-, (+)-	$C_8H_{17}NO_2$ 159.23	116836-12-5 	102-5[23]	4-01-00-00426 0.9166[20]	1.4280[20]
8414	Octane, 3-nitro-	$C_8H_7NO_2$ 149.15	4609-92-1 	91[25]	4-01-00-00427 0.917[25]	1.4210[25]
8415	Octane, octadecafluoro- Perfluorooctane	C_8F_{18} 438.06	307-34-6 	105.9	4-01-00-00418 1.73[20]	1.282[20]
8416	Octane, 1,1'-oxybis-	$C_{16}H_{34}O$ 242.45	629-82-3 -7.6	283	4-01-00-01760 0.8063[20]	H_2O 2; EtOH 3; eth 3; ctc 3 1.4327[20]
8417	Octane, 1,1'-thiobis-	$C_{16}H_{34}S$ 258.51	2690-08-6 	180[10]	4-01-00-01768 0.842[25]	1.4610[20]
8418	1-Octanethiol Octyl mercaptan	$C_8H_{18}S$ 146.30	111-88-6 -49.2	199.1	4-01-00-01767 0.8433[20]	EtOH 3; ctc 2 1.4540[20]
8419	2-Octanethiol, (±)- 2-Octanethiol (DL)	$C_8H_{18}S$ 146.30	10435-81-1 -79	186.4	4-01-00-01777 0.8366[20]	bz 4; eth 4; EtOH 4 1.4504[20]
8420	Octane, 2,2,4-trimethyl- Octane, 2,2,4-trimethyl	$C_{11}H_{24}$ 156.31	18932-14-4 	168.5	4-01-00-00494 0.7381[20]	1.4155[20]
8421	Octane, 2,4,7-trimethyl 2,4,7-Trimethyloctane	$C_{11}H_{24}$ 156.31	62016-38-0 	168.1		
8422	Octanoic acid Caprylic acid	$C_8H_{16}O_2$ 144.21	124-07-2 16.3	1765 239	4-02-00-00982 0.9106[20]	H_2O 2; EtOH 5; chl 5; CH_3CN 5 1.4285[20]
8423	Octanoic acid, 2-amino-, (±)-	$C_8H_{17}NO_2$ 159.23	644-90-6 270	sub	4-04-00-02799	H_2O 2; EtOH 2; eth 2; bz 2
8424	Octanoic acid, anhydride	$C_{16}H_{30}O_3$ 270.41	623-66-5 -1	282.5	3-02-00-00796 0.9065[18]	ace 4; eth 4; EtOH 4 1.4358[18]
8425	Octanoic acid, 2-bromo-	$C_8H_{15}BrO_2$ 223.11	2623-82-7 	140[5]; 118[0.5]	4-02-00-01000 1.2785[24]	1.4613[24]
8426	Octanoic acid, 8-bromo-	$C_8H_{15}BrO_2$ 223.11	17696-11-6 38.5	147[2]	4-02-00-01000	bz 4; eth 4; EtOH 4
8427	Octanoic acid, butyl ester	$C_{12}H_{24}O_2$ 200.32	589-75-3 -42.9	240.5	4-02-00-00987 0.8628[20]	ace 4; eth 4; EtOH 4 1.4232[25]

No.	Name / Synonym	Mol. Form. / Mol. Wt.	CAS RN / mp/°C	Merck No. / bp/°C	Beil. Ref. / den/g cm^{-3}	Solubility / n_D
8428	Octanoic acid, ethyl ester / Ethyl octanoate	$C_{10}H_{20}O_2$ / 172.27	106-32-1 / -43.1	3736 / 208.5	4-02-00-00987 / 0.866[18]	H$_2$O 1; EtOH 4; eth 4; ctc 2 / 1.4178[20]
8429	Octanoic acid, 8-fluoro- / Caprylic acid, 8-fluoro	$C_8H_{15}FO_2$ / 162.20	353-25-3 / 35	132[4]	4-02-00-00994	eth 4; EtOH 4
8430	Octanoic acid, heptyl ester	$C_{15}H_{30}O_2$ / 242.40	4265-97-8 / -10.6	290.5	3-02-00-00794 / 0.8596[20]	ace 4; eth 4; EtOH 4 / 1.4340[20]
8431	Octanoic acid, hexyl ester	$C_{14}H_{28}O_2$ / 228.38	1117-55-1 / -30.6	277.4	3-02-00-00794 / 0.8603[20]	H$_2$O 1; EtOH 3; eth 3; ace 3 / 1.4323[25]
8432	Octanoic acid, 2-hydroxy-	$C_8H_{16}O_3$ / 160.21	617-73-2 / 70	162[10]	3-03-00-00642	H$_2$O 2; EtOH 4; eth 4; chl 2
8433	Octanoic acid, 3-methyl- / Caprylic acid, 3-methyl	$C_9H_{18}O_2$ / 158.24	6061-10-5	135[16]	3-02-00-00826 / 0.899[23]	1.4298[25]
8434	Octanoic acid, methyl ester / Methyl caprylate	$C_9H_{18}O_2$ / 158.24	111-11-5 / -40	192.9	4-02-00-00986 / 0.8775[20]	H$_2$O 1; EtOH 4; eth 4; ctc 2 / 1.4170[20]
8435	Octanoic acid, 1-methylethyl ester	$C_{11}H_{22}O_2$ / 186.29	5458-59-3	93.8[10]	4-02-00-00987 / 0.8555[20]	1.4147[25]
8436	Octanoic acid, 2-methyl-3-oxo-, ethyl ester / Caprylic acid, 2-methyl-3-oxo, ethyl ester	$C_{11}H_{20}O_3$ / 200.28	10488-94-5	128-9[12]	4-03-00-01629 / 0.963[0]	EtOH 4
8437	Octanoic acid, octyl ester	$C_{16}H_{32}O_2$ / 256.43	2306-88-9 / -18.1	306.8	4-02-00-00988 / 0.8554[20]	ace 4; eth 4; EtOH 4 / 1.4352[20]
8438	Octanoic acid, pentadecafluoro-, methyl ester	$C_9H_3F_{15}O_2$ / 428.10	376-27-2	158	4-02-00-00995 / 1.684[20]	1.304[27]
8439	Octanoic acid, pentyl ester	$C_{13}H_{26}O_2$ / 214.35	638-25-5 / -34.8	260.2	3-02-00-00794 / 0.8613[20]	H$_2$O 1; EtOH 3; eth 3; ace 3 / 1.4262[25]
8440	Octanoic acid, 1,2,3-propanetriyl ester	$C_{27}H_{50}O_6$ / 470.69	538-23-8 / 10	233	4-02-00-00991 / 0.9540[20]	H$_2$O 1; EtOH 5; eth 4; bz 4 / 1.4482[20]
8441	Octanoic acid, propyl ester	$C_{11}H_{22}O_2$ / 186.29	624-13-5 / -46.2	226.4	4-02-00-00987 / 0.8659[20]	ace 4; eth 4; EtOH 4 / 1.4191[25]
8442	Octanoic acid, 3,5,7,8-tetrachloro-2,2,3,4,4,5,6,6,7,8,8-undecafluoro- / Octanoic acid, 3,5,7,8-tetrachloro-2,2,3,4,4,5,6,6,7,8,8-undecafluoro	$C_8HCl_4F_{11}O_2$ / 479.89	2923-68-4	154.0-6.5[10]	4-02-00-00998 / 1.899[20]	1.3980[20]
8443	1-Octanol / Caprylic alcohol	$C_8H_{18}O$ / 130.23	111-87-5 / -15.5	6674 / 195.1	4-01-00-01756 / 0.8262[25]	H$_2$O 1; EtOH 5; eth 5; ctc 3 / 1.4295[20]
8444	2-Octanol, (±)-	$C_8H_{18}O$ / 130.23	4128-31-8 / -31.6	6675 / 180	4-01-00-01770 / 0.8193[20]	H$_2$O 2; EtOH 3; eth 3; ace 3 / 1.4203[20]
8445	3-Octanol / Ethylamylcarbinol	$C_8H_{18}O$ / 130.23	589-98-0 / -45	171	3-01-00-01723 / 0.8258[20]	
8446	4-Octanol, (±)-	$C_8H_{18}O$ / 130.23	74778-22-6 / -40.7	176.3	4-01-00-01779 / 0.8186[20]	H$_2$O 2; EtOH 3; ctc 2 / 1.4248[20]
8447	2-Octanol, acetate, (±)- / Acetic acid, 2-octyl ester(DL)	$C_{10}H_{20}O_2$ / 172.27	74112-36-0	195	4-02-00-00165 / 0.8626[14]	eth 4; EtOH 4 / 1.4146[20]
8448	1-Octanol, 8-chloro-	$C_8H_{17}ClO$ / 164.68	23144-52-7	139[19]	4-01-00-01766	eth 4; EtOH 4 / 1.4563[25]
8449	1-Octanol, 3,7-dimethyl-, (±)- / Geraniol, tetrahydro (DL)	$C_{10}H_{22}O$ / 158.28	59204-02-3	212.5; 106[13]	4-01-00-01830 / 0.8308[10]	H$_2$O 1; EtOH 3; eth 3; ace 3 / 1.4367[20]
8450	2-Octanol, 2,6-dimethyl- / Tetrahydromyrcenol	$C_{10}H_{22}O$ / 158.28	18479-57-7	80.5[10]	4-01-00-01829 / 0.8023[25]	1.4220[25]
8451	3-Octanol, 2,3-dimethyl- / Amylisopropylmethylcarbinol	$C_{10}H_{22}O$ / 158.28	19781-10-3	69-70[5]	4-01-00-01828 / 0.8401[20]	1.4380[20]
8452	3-Octanol, 2,7-dimethyl-	$C_{10}H_{22}O$ / 158.28	66719-55-9	193.5	1-01-00-00214 / 0.8152[20]	1.4302[20]
8453	3-Octanol, 3,6-dimethyl-	$C_{10}H_{22}O$ / 158.28	151-19-9 / -67.5	202.2	1-01-00-00214 / 0.8347[22]	1.4370[20]
8454	3-Octanol, 3,6-dimethyl-(+-)- / Linalool, tetrahydro (DL)	$C_{10}H_{22}O$ / 158.28	57706-88-4 / 31.5	196.5	3-01-00-01767 / 0.8280[20]	EtOH 4 / 1.4335[20]
8455	4-Octanol, 2,4-dimethyl-	$C_{10}H_{22}O$ / 158.28	33933-79-8	97[27]	4-01-00-01828 / 0.8238[20]	1.4338[20]
8456	4-Octanol, 2,5-dimethyl-	$C_{10}H_{22}O$ / 158.28	66719-53-7	102-4[34]	1-01-00-00213 / 0.8125[25]	1.4260[25]
8457	4-Octanol, 2,6-dimethyl-	$C_{10}H_{22}O$ / 158.28	66719-54-8	195	2-01-00-00460 / 0.8114[20]	1.4270[20]
8458	4-Octanol, 2,7-dimethyl- / 2,7-Dimethyloctan-4-ol	$C_{10}H_{22}O$ / 158.28	19781-11-4	96[18]	3-01-00-01769 / 0.8139[20]	1.4280[20]
8459	3-Octanol, 3-ethyl-	$C_{10}H_{22}O$ / 158.28	2051-32-3	199	3-01-00-01766 / 0.8361[25]	EtOH 4 / 1.4390[20]
8460	1-Octanol, 8-fluoro-	$C_8H_{17}FO$ / 148.22	408-27-5	106-7[10]	4-01-00-01766 / 0.945[20]	eth 4; EtOH 4 / 1.4248[25]
8461	1-Octanol, 3-methyl- / 3-Methyl-1-octanol	$C_9H_{20}O$ / 144.26	38514-02-2	110[25]	4-01-00-01806 / 0.827[4]	1.4328[25]
8462	1-Octanol, 4-methyl- / 4-Methyl-1-octanol	$C_9H_{20}O$ / 144.26	38514-03-3	106[17]	4-01-00-01807 / 0.820[27]	1.4335[25]
8463	1-Octanol, 7-methyl-	$C_9H_{20}O$ / 144.26	2430-22-0 / 64.5	206	4-01-00-01806 / 0.826[20]	1.4316[25]
8464	2-Octanol, 2-methyl- / 2-Methyl-2-octanol	$C_9H_{20}O$ / 144.26	628-44-4	178	4-01-00-01805 / 0.8210[20]	H$_2$O 1; EtOH 3; eth 3 / 1.4280[20]
8465	2-Octanol, 3-methyl-	$C_9H_{20}O$ / 144.26	27644-49-1	75[15]	3-01-00-01748 / 0.831[27]	1.437[27]
8466	3-Octanol, 3-methyl- / Amylethylmethylcarbinol	$C_9H_{20}O$ / 144.26	5340-36-3	97.5[50]; 36[3]	4-01-00-01807 / 0.8258[25]	1.4257[25]

No.	Name Synonym	Mol. Form. Mol. Wt.	CAS RN mp/°C	Merck No. bp/°C	Beil. Ref. den/g cm⁻³	Solubility n_D
8467	3-Octanol, 4-methyl-	$C_9H_{20}O$ 144.26	66793-80-4	132-3[20]	3-01-00-01750 0.8437[25]	1.4372[20]
8468	3-Octanol, 6-methyl- 6-Methyl-3-octanol	$C_9H_{20}O$ 144.26	40225-75-0	81-3[15]	3-01-00-01749 0.8320[28]	1.4372[20]
8469	4-Octanol, 4-methyl- 4-Methyl-4-octanol	$C_9H_{20}O$ 144.26	23418-37-3	180; 78[15]	4-01-00-01807 0.8267[20]	1.4327[20]
8470	4-Octanol, 5-methyl-	$C_9H_{20}O$ 144.26	59734-23-5	74-6[9]	1-01-00-00211 0.8156[25]	1.4262[25]
8471	2-Octanone Hexyl methyl ketone	$C_8H_{16}O$ 128.21	111-13-7 -16	4632 172.5	4-01-00-03339 0.820[20]	H_2O 2; EtOH 5; eth 5 1.4151[20]
8472	3-Octanone Ethyl amyl ketone	$C_8H_{16}O$ 128.21	106-68-3	3721 167.5	4-01-00-03341 0.822[20]	H_2O 1; EtOH 5; eth 5 1.4150[20]
8473	4-Octanone Butyl propyl ketone	$C_8H_{16}O$ 128.21	589-63-9	163	4-01-00-03342 0.8146[25]	H_2O 1; EtOH 5; eth 5; ctc 3 1.4173[14]
8474	4-Octanone, 5-hydroxy- Butyroin	$C_8H_{16}O_2$ 144.21	496-77-5 -10	1595 185	4-01-00-04042 0.9107[16]	1.4345[16]
8475	2-Octanone, 3-methyl- 3-Methyl-2-octanone	$C_9H_{18}O$ 142.24	6137-08-2	64-5[18]	4-01-00-03356 0.832[27]	1.424[27]
8476	3-Octanone, 4-methyl- 4-Methyl-3-octanone	$C_9H_{18}O$ 142.24	6137-15-1	62-4[10]	4-01-00-03357 0.820[25]	1.4186[25]
8477	3-Octanone, 7-methyl- 7-Methyl-3-octanone	$C_9H_{18}O$ 142.24	5408-57-1	182.5	3-01-00-02889 0.8353[25]	1.4748
8478	4-Octanone, 3-methyl-	$C_9H_{18}O$ 142.24	20754-04-5	174	4-01-00-03357 0.829[14]	1.4200[14]
8479	4-Octanone, 7-methyl- 7-Methyl-4-octanone	$C_9H_{18}O$ 142.24	20809-46-5	178	4-01-00-03356 0.8239[20]	H_2O 1; EtOH 3; eth 3 1.4210[20]
8480	1-Octanone, 1-phenyl-	$C_{14}H_{20}O$ 204.31	1674-37-9 22.8	285; 164[15]	4-07-00-00798 0.9360[30]	EtOH 3; eth 3
8481	Octanoyl chloride	$C_8H_{15}ClO$ 162.66	111-64-8 -63	195.6	4-02-00-00992 0.9535[15]	eth 3 1.4335[20]
8482	Octasiloxane, octadecamethyl- Octadecamethyloctasiloxane	$C_{18}H_{54}O_7Si_8$ 607.31	556-69-4	153[5.1]	4-04-00-04130 0.913[25]	bz 4; peth 4; lig 4 1.3970[20]
8483	1,3,5,7-Octatetraene Octatetraene	C_8H_{10} 106.17	1482-91-3 50	sub	4-01-00-01124	peth 3; HOAc 3
8484	2,4,6-Octatrienal Octatrienal	$C_8H_{10}O$ 122.17	17609-31-3 56.8	75[3]	4-01-00-03591 0.8891[60]	eth 4; EtOH 4
8485	1,3,6-Octatriene, 3,7-dimethyl- Ocimene	$C_{10}H_{16}$ 136.24	13877-91-3	6661 177 dec; 73[21]	3-01-00-01052 0.800[20]	bz 4; eth 4; EtOH 4; chl 4 1.4862[20]
8486	1,3,7-Octatriene, 3,7-dimethyl-	$C_{10}H_{16}$ 136.24	502-99-8	177 dec	3-01-00-01050 0.8000[20]	H_2O 1; EtOH 3; eth 3; chl 3 1.4862[20]
8487	2,4,6-Octatriene, 2,6-dimethyl-, (E,E)-	$C_{10}H_{16}$ 136.24	3016-19-1 -35.4	188; 91[20]	4-01-00-01106 0.8118[20]	1.5446[20]
8488	2,4,6-Octatriene, 2,6-dimethyl-, (E,Z)-	$C_{10}H_{16}$ 136.24	7216-56-0 -20.6	89[20]	4-01-00-01106 0.8060[20]	1.5446[20]
8489	2,4,6-Octatriene, (E,E,E)-	C_8H_{12} 108.18	15192-80-0 52	147.5	4-01-00-01101 0.7961[23]	EtOH 3; chl 3; lig 3 1.5131[27]
8490	6-Octenal, 3,7-dimethyl- Resorcinol, 4-acetyl	$C_{10}H_{18}O$ 154.25	106-23-0	2331 47[1]	1-01-00-00387 0.85[20]	H_2O 2; EtOH 3
8491	1-Octene Caprylene	C_8H_{16} 112.22	111-66-0 -101.7	1764 121.2	4-01-00-00874 0.7149[20]	H_2O 1; EtOH 5; eth 3; ace 3 1.4087[20]
8492	2-Octene, (E)-	C_8H_{16} 112.22	13389-42-9 -87.7	125	4-01-00-00879 0.7199[20]	H_2O 1; EtOH 3; eth 3; ace 3 1.4132[20]
8493	2-Octene, (Z)-	C_8H_{16} 112.22	7642-04-8 -100.1	125.6	4-01-00-00878 0.7243[20]	H_2O 1; EtOH 3; eth 3; ace 3 1.4150[20]
8494	3-Octene, (E)-	C_8H_{16} 112.22	14919-01-8 -110	123.3	4-01-00-00880 0.7152[20]	H_2O 1; EtOH 3; eth 3; ace 3 1.4126[20]
8495	3-Octene, (Z)-	C_8H_{16} 112.22	14850-22-7 -126	122.9	4-01-00-00880 0.7159[20]	ace 4; bz 4; eth 4; EtOH 4 1.4135[20]
8496	4-Octene, (E)-	C_8H_{16} 112.22	14850-23-8 -93.8	122.3	4-01-00-00880 0.7141[20]	H_2O 1; EtOH 3; eth 3; ace 3 1.4114[20]
8497	4-Octene, (Z)-	C_8H_{16} 112.22	7642-15-1 -118.7	122.5	4-01-00-00880 0.7212[20]	ace 4; bz 4; eth 4; EtOH 4 1.4148[20]
8498	1-Octene, 2-chloro-	$C_8H_{15}Cl$ 146.66	31283-43-9	169	3-01-00-00840 0.9273[0]	ace 4; bz 4; eth 4
8499	2-Octene, 2-chloro-	$C_8H_{15}Cl$ 146.66	90202-21-4	167.5; 82[50]	4-01-00-00879 0.8914[16]	ace 4; bz 4; eth 4; EtOH 4 1.4442[16]
8500	4-Octene, 4-chloro-, (Z)- 4-Octene, 4-chloro (cis)	$C_8H_{15}Cl$ 146.66	7321-48-4	165.3; 75[50]	4-01-00-00881 0.8912[20]	eth 4 1.4447[20]
8501	2-Octene, 4-chloro-	$C_8H_{15}Cl$ 146.66	116668-50-9	153	3-01-00-00840 0.8924[20]	ace 4; bz 4; eth 4; chl 4 1.4952[20]
8502	1-Octene, 3,7-dimethyl- 2,6-Dimethyl-7-octene	$C_{10}H_{20}$ 140.27	4984-01-4	154	3-01-00-00862 0.7396[20]	1.4212[20]
8503	2-Octene, 2,6-dimethyl- 2,6-Dimethyl-2-octene	$C_{10}H_{20}$ 140.27	4057-42-5	163	3-01-00-00861 0.746[22]	1.425[22]
8504	4-Octene-2,7-dione	$C_8H_{12}O_2$ 140.18	2130-23-6 30.5	72[0.03]	4-01-00-03757	eth 4; EtOH 4
8505	1-Octene, 2-methyl- 2-Methyl-1-octene	C_9H_{18} 126.24	4588-18-5 -77.8	144.8	4-01-00-00896 0.7343[20]	1.4184[20]
8506	2-Octene, 3-methyl-	C_9H_{18} 126.24	113426-22-5	145	3-01-00-00854 0.7409[25]	1.4247[20]

No.	Name Synonym	Mol. Form. Mol. Wt.	CAS RN mp/°C	Merck No. bp/°C	Beil. Ref. den/g cm^{-3}	Solubility n_D
8507	3-Octene, 7-methyl-	C_9H_{18} 126.24	86668-33-9	142	3-01-00-00854 0.7278[20]	1.4168[20]
8508	4-Octene, 2-methyl-	C_9H_{18} 126.24	64501-77-5	139	3-01-00-00854 0.7379[20]	1.4181[20]
8509	1-Octen-4-ol	$C_8H_{12}O$ 124.18	40575-42-6	172; 68[10]	3-01-00-01942 0.8373[20]	1.4383[20]
8510	2-Octen-1-ol	$C_8H_{16}O$ 128.21	22104-78-5	87-9[11]	4-01-00-02164 0.850[20]	1.4470[20]
8511	2-Octen-4-ol	$C_8H_{16}O$ 128.21	4798-61-2	175	3-01-00-01943 0.8393[20]	1.4395
8512	4-Octen-1-ol	$C_8H_{16}O$ 128.21	67700-26-9	95[17]	4-01-00-02166 0.842[20]	1.4462[20]
8513	5-Octen-1-ol	$C_8H_{16}O$ 128.21	90200-83-2	91.5[14]	4-01-00-02166 0.8492[20]	1.4478[21]
8514	6-Octen-1-ol, 3,7-dimethyl-, (±)- (±)-Citronellol	$C_{10}H_{20}O$ 156.27	26489-01-0	2332 224; 99[10]	4-01-00-02188 0.8560[20]	H$_2$O 2; EtOH 5; eth 5 1.4543[20]
8515	6-Octen-3-ol, 3,7-dimethyl-	$C_{10}H_{20}O$ 156.27	18479-51-1	94[14]	2-01-00-00494 0.8695[15]	1.4569[15]
8516	7-Octen-1-ol, 3,7-dimethyl-, (S)- (-)-α-Citronellol	$C_{10}H_{20}O$ 156.27	6812-78-8	8186 114-5[12]	2-01-00-00497 0.8549[20]	eth 4; EtOH 4 1.4556[20]
8517	1-Octen-4-ol, 4-methyl-	$C_9H_{18}O$ 142.24	62108-06-9	178	2-01-00-00492 0.8440[20]	1.4399
8518	2-Octen-4-ol, 7-methyl-	$C_9H_{18}O$ 142.24	4798-64-5	85[12]	2-01-00-00492 0.8402[14]	1.4449[14]
8519	3-Octen-2-ol, 2-methyl-, (Z)- 3-Octen-2-ol, 2-methyl (cis)	$C_9H_{18}O$ 142.24	18521-07-8	177	4-01-00-02177 0.8378[20]	1.4426[20]
8520	4-Octen-3-ol, 3-methyl-	$C_9H_{18}O$ 142.24	90676-55-4	69-70[10]	4-01-00-02178 0.812[14]	1.4414[14]
8521	5-Octen-4-ol, 5-methyl- 3-Octen-5-ol, 4-methyl	$C_9H_{18}O$ 142.24	36903-92-1	89[16]	2-01-00-00492 0.8468[25]	1.4446[25]
8522	1-Octen-3-one, 5-methyl-	$C_9H_{16}O$ 140.23	116836-13-6	72-3[16]	3-01-00-03018 0.8450[19]	1.4380[19]
8523	5-Octen-4-one, 2-methyl- 2-Methyl-5-octen-4-one	$C_9H_{16}O$ 140.23	17577-93-4	183	4-01-00-03504 0.839[25]	1.4387[25]
8524	5-Octen-4-one, 7-methyl-	$C_9H_{16}O$ 140.23	32064-78-1	68-78[24]	3-01-00-03018 0.9011[12]	1.4748
8525	7-Octen-4-one, 2-methyl- 1-Octen-5-one, 7-methyl	$C_9H_{16}O$ 140.23	54298-97-4	62-3[14]	2-01-00-00801 0.8361[12]	1.4288[12]
8526	1-Octen-3-yne	C_8H_{12} 108.18	17679-92-4	62[60]	4-01-00-01100 0.7830[20]	eth 4 1.4592[20]
8527	7-Octen-5-yn-4-ol, 4-methyl- 1-Octen-3-yne-5-ol, 5-methyl	$C_9H_{14}O$ 138.21	39118-35-9	80[13]	3-01-00-02034 0.8851[15]	EtOH 4 1.4785[20]
8528	Octhilinone 3(2H)-Isothiazolone, 2-octyl-	$C_{11}H_{19}NOS$ 213.34	26530-20-1	6677 120[0.01]		
8529	1-Octyne Hexylacetylene	C_8H_{14} 110.20	629-05-0 -79.3	126.3	4-01-00-01034 0.7461[20]	H$_2$O 1; EtOH 3; eth 3 1.4159[20]
8530	2-Octyne Methylpentylacetylene	C_8H_{14} 110.20	2809-67-8 -61.6	137.6	4-01-00-01035 0.7596[20]	H$_2$O 1; EtOH 3; eth 3 1.4278[20]
8531	3-Octyne	C_8H_{14} 110.20	15232-76-5 -103.9	133.1	4-01-00-01036 0.7529[20]	H$_2$O 1; EtOH 3; eth 3 1.4250[20]
8532	4-Octyne Dipropylacetylene	C_8H_{14} 110.20	1942-45-6 -101	131.6	4-01-00-01037 0.7509[20]	H$_2$O 1; EtOH 3; eth 3 1.4248[20]
8533	1-Octyne, 1-chloro- 1-Chloro-1-octyne	$C_8H_{13}Cl$ 144.64	64531-26-6	61-2[17]	3-01-00-01005 0.912[20]	eth 4; EtOH 4 1.445[20]
8534	3-Octyne, 2-chloro-2-methyl- 2-Chloro-2-methyloct-3-yne	$C_9H_{15}Cl$ 158.67	20599-21-7	68[15]	4-01-00-01048 0.8929[20]	H$_2$O 1; eth 4 1.4480[20]
8535	3-Octyne, 2,2-dimethyl- 2,2-Dimethyl-3-octyne	$C_{10}H_{18}$ 138.25	19482-57-6	79[60]	3-01-00-01018 0.7491[20]	eth 4 1.4270[20]
8536	2-Octyne, 1-methoxy- Methyl (2-octyn-1-yl) ether	$C_9H_{16}O$ 140.23	18495-23-3	77[19]	3-01-00-01996 0.8370[25]	eth 4 1.4380[20]
8537	3-Octyne, 7-methyl- 7-Methyl-3-octyne	C_9H_{16} 124.23	37050-06-9	87[99]	3-01-00-01014 0.7599[20]	eth 4 1.4280[20]
8538	2-Octynoic acid Pentylpropiolic acid	$C_8H_{12}O_2$ 140.18	5663-96-7 4	148[19]; 114[2]	4-02-00-01714 0.9623[13]	1.4595[20]
8539	3-Octynoic acid	$C_8H_{12}O_2$ 140.18	57074-96-1 18	110[2]	4-02-00-01715	1.4577[25]
8540	5-Octynoic acid	$C_8H_{12}O_2$ 140.18	76469-08-4 8	111[2]	4-02-00-01715 0.9787[27]	1.4540[25]
8541	7-Octynoic acid	$C_8H_{12}O_2$ 140.18	10297-09-3 19	123[2]	4-02-00-01714	1.4502[25]
8542	2-Octynoic acid, methyl ester	$C_9H_{14}O_2$ 154.21	111-12-6	107[20]; 94[10]	4-02-00-01715 0.926[20]	1.4464[20]
8543	2-Octyn-1-ol 2-Octynol	$C_8H_{14}O$ 126.20	20739-58-6 -18	98-9[15]	4-01-00-02256 0.8805[20]	eth 4 1.4556[20]
8544	1-Octyn-3-ol, 3-methyl- 1-Octyne-3-ol, 3-methyl	$C_9H_{16}O$ 140.23	23580-51-0	75[10]	4-01-00-02268 0.863[10]	1.443[10]
8545	Olean-12-en-28-oic acid, 3,23-dihydroxy-, (3β,4α)- Hederagenin	$C_{30}H_{48}O_4$ 472.71	465-99-6 333	4540		

No.	Name / Synonym	Mol. Form. / Mol. Wt.	CAS RN / mp/°C	Merck No. / bp/°C	Beil. Ref. / den/g cm^{-3}	Solubility / n_D
8546	Olean-12-en-28-oic acid, 3,16-dihydroxy-23-oxo-, (3β,4α,16α)- Quillaic acid	$C_{30}H_{46}O_5$ 486.69	631-01-6 294	8049	3-10-00-04652	ace 4; eth 4; py 4; EtOH 4
8547	Olean-12-en-28-oic acid, 3-hydroxy-, (3β)- Oleanolic acid	$C_{30}H_{48}O_3$ 456.71	508-02-1 310 dec	6787 sub 280	4-10-00-01164	H_2O 1; EtOH 2; eth 2; ace 2
8548	Olean-12-en-3-ol, (3β)- β-Amyrin	$C_{30}H_{50}O$ 426.73	559-70-6 197	654 260[05]	4-06-00-04195	H_2O 1; EtOH 2; eth 3; bz 3
8549	Orcein	$C_{28}H_{24}N_2O_7$ 500.51	1400-62-0	6818	2-06-00-00876	
8550	L-Ornithine (S)-2,5-Diaminopentanoic acid	$C_5H_{12}N_2O_2$ 132.16	70-26-8 140	6826	4-04-00-02644	H_2O 4; EtOH 4
8551	L-Ornithine, N5-(aminocarbonyl)- Citrulline	$C_6H_{13}N_3O_3$ 175.19	372-75-8 235.5	2333	4-04-00-02647	H_2O 3; EtOH 1; MeOH 1
8552	Orthoformic acid, triisopentyl ester Isopentyl alcohol, orthoformate (3:1)	$C_{16}H_{34}O_3$ 274.44	5337-70-2	 268 dec	4-02-00-00030 0.8628[20]	eth 4; EtOH 4 1.4238[20]
8553	Oryzalin Benzenesulfonamide, 4-(dipropylamino)-3,5-dinitro-	$C_{12}H_{18}N_4O_6S$ 346.36	19044-88-3 141	6840		
8554	7-Oxabicyclo[4.1.0]heptane	$C_6H_{10}O$ 98.14	286-20-4 <-10	 131.5	5-17-01-00203 0.9663[20]	H_2O 1; EtOH 4; eth 4; ace 4 1.4519[20]
8555	7-Oxabicyclo[4.1.0]heptane, 3-ethenyl- Cyclohexane, 1,2-epoxy-4-vinyl	$C_8H_{12}O$ 124.18	106-86-5 <-100	 169; 70[20]	4-17-00-00314 0.9581[20]	1.4700[20]
8556	7-Oxabicyclo[2.2.1]heptane, 1-methyl-4-(1-methylethyl)-	$C_{10}H_{18}O$ 154.25	470-67-7 1	 173.5	5-17-01-00273 0.8997[20]	H_2O 2; EtOH 5; eth 5; bz 3 1.4562[20]
8557	7-Oxabicyclo[4.1.0]heptane, 3-oxiranyl- 4-Vinyl-1-cyclohexene dioxide	$C_8H_{12}O_2$ 140.18	106-87-6 <-55	 227	5-19-01-00295 1.0966[20]	H_2O 4 1.4787[20]
8558	6-Oxabicyclo[3.1.0]hexane	C_5H_8O 84.12	285-67-6	 102	5-17-01-00190 0.9642[5]	1.4336[20]
8559	3-Oxabicyclo[3.2.1]octane-2,4-dione, 1,8,8-trimethyl-, (±)-	$C_{10}H_{14}O_3$ 182.22	595-30-2 221	 270	5-17-11-00089 1.194[20]	bz 4
8560	2-Oxabicyclo[2.2.2]octane, 1,3,3-trimethyl- Eucalyptol	$C_{10}H_{18}O$ 154.25	470-82-6 1.5	3851 176 4	5-17-01-00273 0.9267[20]	H_2O 1; EtOH 3; eth 3; ctc 2 1.4586[20]
8561	6-Oxabicyclo[3.2.1]oct-3-ene, 4,7,7-trimethyl-, (±)- (±)-Pinol	$C_{10}H_{16}O$ 152.24	60761-04-8 183.5		4-17-00-00327 0.9515[20]	eth 4; EtOH 4 1.4695[20]
8562	Oxacyclohexadecan-2-one Exaltolide	$C_{15}H_{28}O_2$ 240.39	106-02-5	3867	5-17-09-00106	
8563	1,3,4-Oxadiazole 1-Oxa-3,4-diazacyclopentadiene	$C_2H_2N_2O$ 70.05	288-99-3 150	6859		1.4300[25]
8564	1,2,4-Oxadiazole, 3,5-diphenyl- 3,5-Diphenyl-1,2,4-oxadiazole	$C_{14}H_{10}N_2O$ 222.25	888-71-1 109	 290; 210[17]	4-27-00-07210	bz 4; eth 4; EtOH 4
8565	Oxadiazon 1,2,4-Oxadiazolidin-3,5-dione, 2-tert-butyl-4-(2,4-dichloro-5-isopropoxyphenyl)-	$C_{15}H_{18}Cl_2N_2O_3$ 345.23	19666-30-9 90	6860		
8566	Oxadixyl N-(2,6-Dimethylphenyl)-2-methoxy-N-(2-oxo-3-oxazolidinyl)acetamide	$C_{14}H_{18}N_2O_4$ 278.31	77732-09-3 104			
8567	16(15H)-Oxaerythrinan-15-one, 1,2,6,7-tetradehydro-14,17-dihydro-3-methoxy-, (3β)- β-Erythroidine	$C_{16}H_{19}NO_3$ 273.33	466-81-9 99.5	3625	4-27-00-03568	H_2O 3; EtOH 4; eth 3; bz 4
8568	Oxamyl Ethanimidothioic acid, 2-(dimethylamino)-N-[[(methylamino)carbonyl]oxy]-2-oxo-, methyl ester	$C_7H_{13}N_3O_3S$ 219.26	23135-22-0 109	6873 dec	0.97[25]	
8569	1,4-Oxathiane	C_4H_8OS 104.17	15980-15-1 147		5-19-01-00033 1.1174[20]	H_2O 2
8570	1,2-Oxathiolane, 2,2-dioxide Propane sultone	$C_3H_6O_3S$ 122.14	1120-71-4		5-19-01-00005	chl 3
8571	Oxayohimban-16-carboxylic acid, 16,17-didehydro-19-methyl-, methyl ester, (19α)- Raubasine	$C_{21}H_{24}N_2O_3$ 352.43	483-04-5 258 dec	8129	4-27-00-07927	H_2O 1; MeOH 3
8572	Oxayohimbanium, 3,4,5,6,16,17-hexadehydro-16-(methoxycarbonyl)-19-methyl-, (19α)- Serpentine	$C_{21}H_{20}N_2O_3$ 348.40	18786-24-8 175	8415		H_2O 1; EtOH 3; eth 3; ace 3
8573	2H-1,3-Oxazine-2,4(3H)-dione, 5,5-diethyldihydro- Diethadione	$C_8H_{13}NO_3$ 171.20	702-54-5 97.5	3096	4-27-00-03263	
8574	2-Oxazolamine, 4,5-dihydro-5-phenyl- Aminorex	$C_9H_{10}N_2O$ 162.19	2207-50-3 137	490		
8575	Oxazole 1,3-Oxazole	C_3H_3NO 69.06	288-42-6 69.5		4-27-00-00960	1.4285[17]
8576	Oxazole, 4,5-dihydro-2-methyl-	C_4H_7NO 85.11	1120-64-5 111		4-27-00-00921 1.005[25]	1.4340[20]
8577	Oxazole, 2,4-dimethyl- 2,4-Dimethyloxazole	C_5H_7NO 97.12	7208-05-1 108		4-27-00-00980 0.9352[15]	H_2O 4; eth 4; EtOH 4 1.4166[15]
8578	Oxazole, 2,5-dimethyl- 2,5-Dimethyloxazole	C_5H_7NO 97.12	23012-11-5 117.5		4-27-00-00985 0.9958[21]	H_2O 4 1.4385[21]

No.	Name Synonym	Mol. Form. Mol. Wt.	CAS RN mp/°C	Merck No. bp/°C	Beil. Ref. den/g cm^{-3}	Solubility n_D
8579	Oxazole, 2,4-diphenyl- 2,4-Diphenyloxazole	C$_{15}$H$_{11}$NO 221.26	838-41-5 103	339	4-27-00-01435	EtOH 4; eth 4; bz 4
8580	Oxazole, 2,5-diphenyl-	C$_{15}$H$_{11}$NO 221.26	92-71-7 74	360	4-27-00-01435 1.0940^{100}	H$_2$O 1; EtOH 4; eth 4; chl 2 1.6231^{100}
8581	Oxazole, 4,5-diphenyl- 4,5-Diphenyloxazole	C$_{15}$H$_{11}$NO 221.26	4675-18-7 44	193$^{15.}$ 132$^{0.05}$	4-27-00-01438	H$_2$O 2; EtOH 4; eth 4; con acid 3 1.6283^{100}
8582	Oxazole, 5-methoxy-2-phenyl- 5-Methoxy-2-phenyloxazole	C$_{10}$H$_9$NO$_2$ 175.19	40527-16-0	140^{10}	4-27-00-01958	ctc 3 1.584^{23}
8583	Oxazole, 4-methyl- 4-Methyloxazole	C$_4$H$_5$NO 83.09	693-93-6	88	4-27-00-00968 1.015^{25}	1.4317^{20}
8584	2,4-Oxazolidinedione, 5,5-dimethyl- Dimethadione	C$_5$H$_7$NO$_3$ 129.12	695-53-4 76.5	3201	4-27-00-03237	
8585	2,4-Oxazolidinedione, 5,5-dipropyl- 5,5-Dipropyl-2,4-oxazolidinedione	C$_9$H$_{15}$NO$_3$ 185.22	512-12-9 42.5	3352 148-50^3	4-27-00-03269	
8586	2,4-Oxazolidinedione, 5-methyl-3-(2-propenyl)- Aloxidone	C$_7$H$_9$NO$_3$ 155.15	526-35-2	305 138^{35}; 86$^{0.5}$	4-27-00-03214	1.4688^{25}
8587	2,4-Oxazolidinedione, 3,5,5-trimethyl- Trimethadione	C$_6$H$_9$NO$_3$ 143.14	127-48-0 46	9620 78-80^5	4-27-00-03237	H$_2$O 3; EtOH 4; eth 4; ace 4
8588	2-Oxazolidinone, 5-[(2-methoxyphenoxy)methyl]- Mephenoxalone	C$_{11}$H$_{13}$NO$_4$ 223.23	70-07-5 144	5739		
8589	2-Oxazolidinone, 3-[[(5-nitro-2-furanyl)methylene]amino]- Furazolidone	C$_8$H$_7$N$_3$O$_5$ 225.16	67-45-8 255	4210	4-27-00-02530	
8590	Oxepane	C$_6$H$_{12}$O 100.16	592-90-5	119	5-17-01-00090 0.89^{25}	1.4400^{20}
8591	2-Oxepanone Caprolactone	C$_6$H$_{10}$O$_2$ 114.14	502-44-3 -18	215	5-17-09-00034 1.0693^{20}	EtOH 3; eth 3; ace 3 1.4611^{20}
8592	Oxetane Trimethylene oxide	C$_3$H$_6$O 58.08	503-30-0 -97	9630 47.6	5-17-01-00011 0.8930^{25}	H$_2$O 5; EtOH 5; eth 3; ace 4 1.3961^{20}
8593	Oxetane, 3,3-dimethyl- 3,3-Dimethyloxetane	C$_5$H$_{10}$O 86.13	6921-35-3	80.6	5-17-01-00085 0.834^{25}	1.3965^{20}
8594	Oxetane, 2-methyl-	C$_4$H$_8$O 72.11	2167-39-7	59	5-17-01-00054 0.841^{25}	1.3885^{20}
8595	2-Oxetanone β-Propiolactone	C$_3$H$_4$O$_2$ 72.06	57-57-8 -33.4	7832 162	5-17-09-00003 1.1460^{20}	eth 5; chl 3 1.4105^{20}
8596	3-Oxetanone	C$_3$H$_4$O$_2$ 72.06	6704-31-0			
8597	2-Oxetanone, 3,3-dimethyl-4-(1-methylethylidene)-	C$_8$H$_{12}$O$_2$ 140.18	3173-79-3 -18	170	5-17-09-00181 0.947^{25}	1.4380^{20}
8598	2-Oxetanone, 4-methylene- Diketene	C$_4$H$_4$O$_2$ 84.07	674-82-8 -6.5	126.1	5-17-09-00115 1.0877^{20}	1.4379^{20}
8599	Oxirane Ethylene oxide	C$_2$H$_4$O 44.05	75-21-8 -111.7	3758 10.6	5-17-01-00003 0.8821^{10}	H$_2$O 3; EtOH 3; eth 3; ace 3 1.3579^7
8600	Oxirane, (bromomethyl)-, (±)- Propane, 3-bromo-1,2-epoxy (DL)	C$_3$H$_5$BrO 136.98	82584-73-4 -40	137	5-17-01-00023 1.615^{14}	H$_2$O 1; EtOH 3; eth 3; bz 3 1.4841^{20}
8601	Oxirane, 2,2'-[1,4-butanediylbis(oxymethylene)]bis-	C$_{10}$H$_{18}$O$_4$ 202.25	2425-79-8	266; 155^{11}	5-17-03-00023 1.1^{25}	1.4611^{20}
8602	Oxirane, (butoxymethyl)- Butyl glycidyl ether	C$_7$H$_{14}$O$_2$ 130.19	2426-08-6	169; 75^{26}	5-17-03-00011 0.918^{20}	
8603	Oxiranecarboxaldehyde Glycidaldehyde	C$_3$H$_4$O$_2$ 72.06	765-34-4 -62	112.5	5-17-09-00006	1.4265^{20}
8604	Oxiranecarboxylic acid, 3-methyl-3-phenyl-, ethyl ester	C$_{12}$H$_{14}$O$_3$ 206.24	77-83-8	273.5	5-18-06-00385 1.044^{20}	1.5182^{20}
8605	Oxirane, (chloromethyl)-, (±)- DL-α-Epichlorohydrin	C$_3$H$_5$ClO 92.52	13403-37-7 -26	3563 117; 62^{100}	5-17-01-00020 1.1812^{20}	H$_2$O 1; EtOH 5; eth 5; bz 3 1.4361^{20}
8606	Oxirane, 2-(chloromethyl)-2-methyl-	C$_4$H$_7$ClO 106.55	598-09-4	122	5-17-01-00059 1.1011^{20}	H$_2$O 4; eth 4 1.4310^{20}
8607	Oxirane, 2,2-dimethyl-	C$_4$H$_8$O 72.11	558-30-5	52	5-17-01-00058 0.8112^{20}	EtOH 3; eth 3 1.3712^{22}
8608	Oxirane, 2,3-dimethyl-, cis-	C$_4$H$_8$O 72.11	1758-33-4 -80	60	5-17-01-00061 0.8226^{25}	eth 4; ace 4; bz 4; os 4 1.3802^{20}
8609	Oxirane, 2,3-dimethyl-, trans-	C$_4$H$_8$O 72.11	21490-63-1 -85	56.5	5-17-01-00061 0.8010^{25}	eth 4; ace 4; bz 4; os 4 1.3736^{20}
8610	Oxirane, [(1,1-dimethylethoxy)methyl]-	C$_7$H$_{14}$O$_2$ 130.19	7665-72-7 -70	152	5-17-03-00011 0.898^{20}	
8611	Oxirane, [[4-(1,1-dimethylethyl)phenoxy]methyl]-	C$_{13}$H$_{18}$O$_2$ 206.28	3101-60-8	165-70^{14}	5-17-03-00017 1.038^{25}	1.5145^{20}
8612	Oxirane, 2-(2,2-dimethylpropyl)-2-methyl-	C$_8$H$_{16}$O 128.21	107-48-2 -64	140.9	4-17-00-00112 0.8272^{20}	bz 4; eth 4 1.4097^{20}
8613	Oxirane, ethenyl-	C$_4$H$_6$O 70.09	930-22-3	68	5-17-01-00180 0.9006^{25}	EtOH 3; eth 3; bz 3; os 3 1.4168^{20}
8614	Oxirane, (ethoxymethyl)- 2,3-Epoxypropyl ethyl ether	C$_5$H$_{10}$O$_2$ 102.13	4016-11-9	128	5-17-03-00010 0.9700^{20}	H$_2$O 3; EtOH 3; eth 3; ctc 2 1.4320^{20}

No.	Name Synonym	Mol. Form. Mol. Wt.	CAS RN mp/°C	Merck No. bp/°C	Beil. Ref. den/g cm^{-3}	Solubility n_D
8615	Oxirane, ethyl- 1,2-Epoxybutane	$C_4H_8O_2$ 88.11	106-88-7 	 63.3	5-17-01-00056 0.8297[20]	EtOH 4; eth 5; ace 4; os 4 1.3851[20]
8616	Oxirane, (fluoromethyl)-	C_3H_5FO 76.07	503-09-3 	 85.5	5-17-01-00019 1.067[25]	 1.3715[20]
8617	Oxirane, (iodomethyl)-	C_3H_5IO 183.98	624-57-7 	 161; 62[24]	5-17-01-00024 1.982[24]	eth 4; EtOH 4
8618	Oxiranemethanol Glycidol	$C_3H_6O_2$ 74.08	556-52-5 	4385 167 dec; 66[2.5]	5-17-03-00009 1.1143[25]	H_2O 5; EtOH 5; eth 3; ace 3 1.4287[20]
8619	Oxiranemethanol, 3-phenyl-	$C_9H_{10}O_2$ 150.18	21915-53-7 26.5	134[4]	4-17-00-01349 1.1512[27]	 1.5427[27]
8620	Oxirane, (methoxymethyl)-	$C_4H_8O_2$ 88.11	930-37-0 	 113	5-17-03-00010 0.9890[20]	H_2O 4; ace 4; eth 4; EtOH 4 1.4320[20]
8621	Oxirane, methyl- Propylene oxide	C_3H_6O 58.08	75-56-9 -111.9	7869 35	5-17-01-00017 0.859[0]	H_2O 4; eth 4; EtOH 4 1.3670[20]
8622	Oxirane, 2-methyl-2-phenyl-	$C_9H_{10}O$ 134.18	2085-88-3 	 83-5[15]	5-17-01-00591 1.0280[20]	 1.5232[20]
8623	Oxirane, (phenoxymethyl)- Phenyl glycidyl ether	$C_9H_{10}O_2$ 150.18	122-60-1 	 243	5-17-03-00013 1.1109[21]	 1.5307[21]
8624	Oxirane, phenyl- Phenylethylene oxide	C_8H_8O 120.15	96-09-3 -35.6	 194.1	5-17-01-00577 1.0523[16]	H_2O 1; EtOH 3; eth 3; chl 3 1.5342[20]
8625	Oxirane, 2,2'-[1,3- phenylenebis(oxymethylene)]bis- Diglycidyl resorcinol ether	$C_{12}H_{14}O_4$ 222.24	101-90-6 42.5	143-51[0.4]	5-17-03-00024 1.2183[30]	 1.5408[20]
8626	Oxirane, [(2-propenyloxy)methyl]- Allyl glycidyl ether	$C_6H_{10}O_2$ 114.14	106-92-3 	 154	5-17-03-00012 0.9698[20]	 1.4332[20]
8627	Oxirane, tetramethyl-	$C_6H_{12}O$ 100.16	5076-20-0 	 90-4[74.5]	5-17-01-00108 0.8156[16]	H_2O 3 1.3984[16]
8628	Oxirane, (trichloromethyl)-	$C_3H_3Cl_3O$ 161.41	3083-23-6 	 149; 44[13]	5-17-01-00023 1.495[20]	eth 4; chl 3 1.4737[25]
8629	Oxirane, trifluoro(trifluoromethyl)- Perfluoropropylene oxide	C_3F_6O 166.02	428-59-1 		5-17-01-00020	
8630	Oxyacanthan, 6',12'-dimethoxy-2,2'-dimethyl- 6,7-[methylenebis(oxy)]- Cepharanthine	$C_{37}H_{38}N_2O_6$ 606.72	481-49-2 150	1981	4-27-00-09061	
8631	Oxyacanthan-12'-ol, 6,6',7-trimethoxy-2,2'- dimethyl- Oxyacanthine	$C_{37}H_{40}N_2O_6$ 608.73	548-40-3 216.5	6906	2-27-00-00894	H_2O 1; EtOH 3; eth 3; bz 3
8632	Oxydemeton-methyl Phosphorothioic acid, S-[2-(ethylsulfinyl)ethyl] O,O-dimethyl ester	$C_6H_{15}O_4PS_2$ 246.29	301-12-2 <-20	106[0.01]	 1.289[20]	
8633	Oxyfluorfen Benzene, 2-chloro-1-(3-ethoxy-4- nitrophenoxy)-4-(trifluoromethyl)-	$C_{15}H_{11}ClF_3NO_4$ 361.70	42874-03-3 84	6916 358.2 dec	 1.35[73]	
8634	Paraformaldehyde Formaldehyde polymer	$(CH_2O)_x$ 	30525-89-4 164 dec	6974	5-19-09-00103	H_2O 2; EtOH 1; eth 1; alk 3
8635	Pendimethalin Benzenamine, N-(1-ethylpropyl)-3,4-dimethyl- 2,6-dinitro-	$C_{13}H_{19}N_3O_4$ 281.31	40487-42-1 56	7026 dec	 1.19[25]	
8636	Pentacene Benzo[b]naphthacene	$C_{22}H_{14}$ 278.35	135-48-8 257	7057	4-05-00-02721	H_2O 1; bz 2; $PhNO_2$ 3; os 2
8637	Pentacosane	$C_{25}H_{52}$ 352.69	629-99-2 54	 401.9; 282[40]	4-01-00-00582 0.8012[20]	bz 3; chl 3 1.4491[20]
8638	Pentacosane, 13-phenyl- Benzene, (1-dodecyltridecyl)-	$C_{31}H_{56}$ 428.79	6006-90-2 31.7	 235[10]	4-05-00-01241 0.8537[20]	 1.4787[20]
8639	Pentacosane, 13-undecyl-	$C_{36}H_{74}$ 506.98	55517-89-0 9.7	 307[10]	4-01-00-00600 0.8168[20]	 1.4567[20]
8640	Pentadecanal Pentadecaldehyde	$C_{15}H_{30}O$ 226.40	2765-11-9 24.5	 185[25]	4-01-00-03391	ace 4; eth 4; EtOH 4
8641	1-Pentadecanamine	$C_{15}H_{33}N$ 227.43	2570-26-5 37.3	 307.6	4-04-00-00817 0.8104[20]	eth 4; EtOH 4 1.4480[20]
8642	Pentadecane	$C_{15}H_{32}$ 212.42	629-62-9 9.9	 270.6	4-01-00-00529 0.7685[20]	H_2O 1; EtOH 4; eth 4 1.4315[20]
8643	Pentadecane, 1-bromo-	$C_{15}H_{31}Br$ 291.32	629-72-1 19	 322	4-01-00-00531 1.0675[20]	H_2O 1; ace 3; chl 4 1.4611[20]
8644	Pentadecane, 2,6,10,14-tetramethyl- Pristane	$C_{19}H_{40}$ 268.53	1921-70-6 	7757 296	3-01-00-00570 0.783[20]	bz 4; eth 4; chl 4; peth 4 1.4379[20]
8645	Pentadecanoic acid Pentadecylic acid	$C_{15}H_{30}O_2$ 242.40	1002-84-2 52.3	 257[100]; 158[1]	4-02-00-01147 0.8423[80]	H_2O 1; EtOH 4; eth 3; ace 4 1.4254[80]
8646	Pentadecanoic acid, methyl ester Methyl pentadecanoate	$C_{16}H_{32}O_2$ 256.43	7132-64-1 18.5	 153.5	3-02-00-00936 0.8618[25]	EtOH 3; eth 3; os 3 1.4390[25]
8647	2-Pentadecanol sec-Pentadecyl alcohol	$C_{15}H_{32}O$ 228.42	1653-34-5 35	 299	4-01-00-01871 0.8328[20]	 1.4463[20]
8648	3-Pentadecanol	$C_{15}H_{32}O$ 228.42	53346-71-7 38.8	 388; 163[12]	4-01-00-01871	 1.4227[80]
8649	1-Pentadecanol, 2-methyl- 2-Methyl-1-pentadecanol	$C_{16}H_{34}O$ 242.45	25354-98-7 12.5	 185-7[15]	3-01-00-01827 0.8320[25]	 1.4453[25]
8650	6-Pentadecanol, 6-methyl-	$C_{16}H_{34}O$ 242.45	108836-86-8 	 199-200[50]	3-01-00-01828 0.8316[25]	 1.4446[25]

No.	Name Synonym	Mol. Form. Mol. Wt.	CAS RN mp/°C	Merck No. bp/°C	Beil. Ref. den/g cm^{-3}	Solubility n_D
8651	2-Pentadecanone	$C_{15}H_{30}O$ 226.40	2345-28-0 39.5	294	4-01-00-03391 0.8182[39]	
8652	8-Pentadecanone	$C_{15}H_{30}O$ 226.40	818-23-5 43	291	4-01-00-03392 0.8180[39]	EtOH 3; eth 3; bz 3; ctc 3
8653	1-Pentadecene	$C_{15}H_{30}$ 210.40	13360-61-7 -2.8	268.2	4-01-00-00926 0.7764[20]	H_2O 1; ace 3 1.4389[20]
8654	7-Pentadecene	$C_{15}H_{30}$ 210.40	15430-98-5	114[3.2]	3-01-00-00875 0.7765[20]	1.4420[20]
8655	1-Pentadecen-3-ol	$C_{15}H_{30}O$ 226.40	99814-65-0 27.5	150-2[6]	3-01-00-01961 0.834[32]	1.4481[32]
8656	1-Pentadecyne	$C_{15}H_{28}$ 208.39	765-13-9 10	268	4-01-00-01072 0.7928[20]	ace 4 1.4419[20]
8657	2,4-Pentadienal, 5-phenyl- 5-Phenyl-2,4-pentadienal	$C_{11}H_{10}O$ 158.20	13466-40-5 42.5	160[3]; 133[1.0]	4-07-00-01179	H_2O 1; EtOH 5; eth 4; bz 5
8658	1,2-Pentadiene Ethylallene	C_5H_8 68.12	591-95-7 -137.3	44.9	4-01-00-00993 0.6926[20]	EtOH 5; eth 5; ace 5; bz 5 1.4209[20]
8659	1,4-Pentadiene	C_5H_8 68.12	591-93-5 -148.8	26	4-01-00-00998 0.6608[20]	H_2O 1; EtOH 4; eth 4; ace 4 1.3888[20]
8660	2,3-Pentadiene 1,3-Dimethylallene	C_5H_8 68.12	591-96-8 -125.6	48.2	4-01-00-00999 0.6950[20]	H_2O 1; EtOH 5; eth 5; ace 5 1.4284[20]
8661	1,3-Pentadiene, (E)- trans-Piperylene	C_5H_8 68.12	2004-70-8 -87.4	42	4-01-00-00994 0.6710[25]	1.4301[20]
8662	1,3-Pentadiene, (Z)- cis-Piperylene	C_5H_8 68.12	1574-41-0 -140.8	44.1	4-01-00-00994 0.6910[20]	EtOH 5; eth 5; ace 5; bz 5 1.4363[20]
8663	1,3-Pentadiene, 3-chloro- 3-Chloro-1,3-pentadiene	C_5H_7Cl 102.56	37710-49-9	100	3-01-00-00962 0.9576[20]	ace 4; bz 4; eth 4 1.4785[20]
8664	1,2-Pentadiene, 1-chloro-3-methyl-	C_6H_9Cl 116.59	32337-74-9	68-70[100]	3-01-00-00990 0.9562[20]	eth 4
8665	1,3-Pentadiene, 2,4-dimethyl-	C_7H_{12} 96.17	1000-86-8 -114	93.2	4-01-00-01034 0.7343[23]	1.4390[23]
8666	1,2-Pentadiene, 3-methyl- 1-Ethyl-1-methylallene	C_6H_{10} 82.15	7417-48-3	70	4-01-00-01021 0.710[25]	1.425[20]
8667	1,2-Pentadiene, 4-methyl-	C_6H_{10} 82.15	13643-05-5	70	2-01-00-00232 0.703[25]	1.424[20]
8668	1,3-Pentadiene, 2-methyl-, (Z)- cis-2-Methyl-1,3-pentadiene	C_6H_{10} 82.15	1501-60-6 -117.6	76	4-01-00-01019 0.714[25]	1.446[20]
8669	1,3-Pentadiene, 3-methyl-	C_6H_{10} 82.15	4549-74-0	77	4-01-00-01021 0.730[25]	1.452[20]
8670	1,3-Pentadiene, 4-methyl- 1,1-Dimethylbutadiene	C_6H_{10} 82.15	926-56-7	76.5	4-01-00-01020 0.7181[20]	1.4532[20]
8671	1,4-Pentadiene, 2-methyl- 2-Methyl-1,4-pentadiene	C_6H_{10} 82.15	763-30-4	56	4-01-00-01020 0.689[25]	1.405[20]
8672	2,3-Pentadiene, 2-methyl-	C_6H_{10} 82.15	3043-33-2	72	4-01-00-01020 0.706[25]	1.425[20]
8673	2,4-Pentadienenitrile	C_5H_5N 79.10	1615-70-9 -60	136.5	4-02-00-01692 0.8444[20]	1.4880
8674	2,4-Pentadienenitrile, (E)-	C_5H_5N 79.10	2180-68-9 -43	41[13]	4-02-00-01695 0.8576[20]	eth 3; ace 3 1.4986[20]
8675	2,4-Pentadienenitrile, (Z)-	C_5H_5N 79.10	2180-69-0 -64	49[32]; 32[13]	4-02-00-01695 0.8541[26]	eth 3; ace 3 1.4855[20]
8676	2,4-Pentadienoic acid	$C_5H_6O_2$ 98.10	626-99-3 79	112 dec	3-02-00-01451	H_2O 4; EtOH 4; eth 4; bz 4
8677	2,4-Pentadienoic acid, 5-(1,3-benzodioxol-5-yl)- 2,4-Pentadienoic acid, 5(3,4-metylenedioxyphenyl)	$C_{12}H_{10}O_4$ 218.21	5285-18-7 215	7437	2-19-00-00300	H_2O 1; EtOH 3; eth 2; bz 2
8678	2,4-Pentadienoic acid, 5-(1,3-benzodioxol-5-yl)-, (E,E)-	$C_{12}H_{10}O_4$ 218.21	136-72-1 215.8	sub	5-19-07-00382	EtOH 4
8679	2,4-Pentadienoic acid, 5-(1-hydroxy-2,6,6-trimethyl-4-oxo-2-cyclohexen-1-yl)-3-methyl-, [S-(Z,E)]- Abscisic acid	$C_{15}H_{20}O_4$ 264.32	21293-29-8 160	6 sub 120		ace 4; eth 4; chl 4
8680	2,4-Pentadienoic acid, 5-phenyl-, ethyl ester Ethyl cinnamylideneacetate	$C_{13}H_{14}O_2$ 202.25	1552-95-0 26	149-50[4]	2-09-00-00441 1.0467[20]	eth 4; EtOH 4 1.5768[80]
8681	1,4-Pentadien-3-ol	C_5H_8O 84.12	922-65-6	115.5	4-01-00-02229 0.860[23]	1.4400[17]
8682	1,4-Pentadien-3-one, 1,5-di-2-furanyl-	$C_{13}H_{10}O_3$ 214.22	886-77-1 60.5	181[4]	5-19-04-00459	eth 4; EtOH 4; chl 4
8683	1,4-Pentadien-3-one, 1,5-diphenyl- Dibenzalacetone	$C_{17}H_{14}O$ 234.30	538-58-9 113 dec	2988 dec	4-07-00-01747	H_2O 1; EtOH 2; eth 2; ace 3
8684	1,3-Pentadiyne Methyldiacetylene	C_5H_4 64.09	4911-55-1 -38.5	55	4-01-00-01117 0.7909[20]	H_2O 1; eth 3; bz 3; chl 3 1.4431[21]
8685	1,5-Pentalenedione, hexahydro- Bicyclo[3.3.0]octane-2,6-dione	$C_8H_{10}O_2$ 138.17	62353-69-9 45	86-8[0.2]	3-07-00-03280 1.1290[60]	1.4877[54]
8686	Pentalene, octahydro-, cis-	C_8H_{14} 110.20	1755-05-1 <-80	139	4-05-00-00277 0.8638[25]	H_2O 1; EtOH 3 1.4595[25]
8687	Pentalene, octahydro-, trans-	C_8H_{14} 110.20	5597-89-7 -30	132	4-05-00-00277 0.8624[18]	H_2O 1; EtOH 3 1.4625[18]
8688	2(1H)-Pentalenone, hexahydro- Bicyclo[3.3.0]octane-2-one	$C_8H_{12}O$ 124.18	56180-61-1	72[13]	4-07-00-00156 1.0097[20]	ace 4; EtOH 4 1.4790[20]

No.	Name Synonym	Mol. Form. Mol. Wt.	CAS RN mp/°C	Merck No. bp/°C	Beil. Ref. den/g cm⁻³	Solubility n_D
8689	2(1H)-Pentalenone, hexahydro-, cis- cis-Bicyclo[3.3.0]octan-3-one	$C_8H_{12}O$ 124.18	19915-11-8 	$50^{2.3}$	4-07-00-00156 1.0660^{25}	1.4766^{25}
8690	Pentanal Valeraldehyde	$C_5H_{10}O$ 86.13	110-62-3 -91.5	9813 103	4-01-00-03268 0.8095^{20}	H$_2$O 2; EtOH 3; eth 3 1.3944^{20}
8691	Pentanal, 3-hydroxy-2-methyl-	$C_6H_{12}O_2$ 116.16	615-30-5 	$94-6^{23}$	4-01-00-04022 0.986^{25}	H$_2$O 4; ace 4; eth 4; EtOH 4 1.4502^{20}
8692	Pentanal, 3-hydroxy-2,2,4-trimethyl-	$C_8H_{16}O_2$ 144.21	918-79-6 	$118-20^{14}$	4-01-00-04049 0.9482^{20}	1.4501^{20}
8693	Pentanal, 2-methyl- 2-Methylvaleraldehyde	$C_6H_{12}O$ 100.16	123-15-9 	117	4-01-00-03304 	H$_2$O 3; eth 3; ace 3; ctc 2
8694	Pentanal, 3-methyl-	$C_6H_{12}O$ 100.16	15877-57-3 	121	3-01-00-02836 0.811^{25}	1.4026^{20}
8695	Pentanal, 2-oxo-	$C_5H_8O_2$ 100.12	7332-93-6 	112	4-01-00-03658 	H$_2$O 4; eth 4; EtOH 4 1.4043^{25}
8696	Pentanal, 4-oxo-	$C_5H_8O_2$ 100.12	626-96-0 <-21	187 dec	4-01-00-03659 1.0134^{21}	H$_2$O 4; ace 4; eth 4; EtOH 4 1.4257^{22}
8697	Pentanal, 2-propyl- Valeraldehyde, 2-propyl-	$C_8H_{16}O$ 128.21	18295-59-5 	160	3-01-00-02880 0.8347^{15}	1.4142^{15}
8698	Pentanamide	$C_5H_{11}NO$ 101.15	626-97-1 106	225	4-02-00-00874 0.8735^{110}	H$_2$O 4; EtOH 4; eth 4; chl 2 1.4183^{110}
8699	Pentanamide, N,N-dimethyl-	$C_7H_{15}NO$ 129.20	6225-06-5 -51	141^{100}	3-04-00-00127 0.8962^{25}	H$_2$O 4; eth 4; EtOH 4 1.4419^{25}
8700	1-Pentanamine Amylamine	$C_5H_{13}N$ 87.16	110-58-7 -55	630 104.3	4-04-00-00674 0.7544^{20}	H$_2$O 5; EtOH 5; eth 5; ace 4 1.448^{20}
8701	2-Pentanamine	$C_5H_{13}N$ 87.16	625-30-9 	92	4-04-00-00689 0.7384^{20}	ace 4; bz 4; eth 4; EtOH 4 1.4027^{20}
8702	3-Pentanamine	$C_5H_{13}N$ 87.16	616-24-0 	89	4-04-00-00692 0.7487^{20}	EtOH 3; chl 2 1.4063^{20}
8703	1-Pentanamine, N,N-dimethyl-	$C_7H_{17}N$ 115.22	26153-88-8 	123	4-04-00-00675 0.743^{20}	eth 4 1.4083^{20}
8704	2-Pentanamine, 2,4-dimethyl-	$C_7H_{17}N$ 115.22	64379-30-2 	122	4-04-00-00749 0.7719^{20}	1.4009^{20}
8705	1-Pentanamine, N,N-dipentyl-	$C_{15}H_{33}N$ 227.43	621-77-2 	242.5	4-04-00-00676 0.7907^{20}	H$_2$O 1; EtOH 3; eth 3; acid 3 1.4366^{20}
8706	1-Pentanamine, 3-methyl- 3-Methylpentylamine	$C_6H_{15}N$ 101.19	42245-37-4 	122	3-04-00-00366 0.767^{20}	1.4190^{25}
8707	1-Pentanamine, 4-methyl-	$C_6H_{15}N$ 101.19	5344-20-7 -94.4	125	4-04-00-00727 0.758^{52}	
8708	1-Pentanamine, N-pentyl- Diamylamine	$C_{10}H_{23}N$ 157.30	2050-92-2 	202.5	4-04-00-00676 0.7771^{20}	H$_2$O 2; EtOH 4; eth 5; ace 3 1.4272^{20}
8709	Pentane	C_5H_{12} 72.15	109-66-0 -129.7	7072 36.0	4-01-00-00303 0.6262^{20}	H$_2$O 2; EtOH 5; eth 5; ace 5 1.3575^{20}
8710	Pentane, 1-bromo- Amyl bromide	$C_5H_{11}Br$ 151.05	110-53-2 -95	637 129.8	4-01-00-00312 1.2182^{20}	H$_2$O 1; EtOH 3; eth 5; bz 3 1.4447^{20}
8711	Pentane, 2-bromo- 2-Pentyl bromide	$C_5H_{11}Br$ 151.05	107-81-3 -95.5	117.4	4-01-00-00312 1.2075^{20}	bz 4; eth 4; EtOH 4; chl 4 1.4413^{20}
8712	Pentane, 3-bromo- 3-Pentyl bromide	$C_5H_{11}Br$ 151.05	1809-10-5 -126.2	118.6	4-01-00-00313 1.214^{20}	H$_2$O 1; EtOH 3; eth 3; bz 3 1.4441^{20}
8713	Pentane, 1-bromo-5-fluoro- 1-Bromo-5-fluoropentane	$C_5H_{10}BrF$ 169.04	407-97-6 	162	4-01-00-00313 1.3604^{25}	eth 4; EtOH 4 1.4406^{25}
8714	Pentane, 1-bromo-2-methyl- 2-Methylpentyl bromide	$C_6H_{13}Br$ 165.07	25346-33-2 	141	4-01-00-00361 1.1624^{20}	eth 4; chl 4 1.4495^{20}
8715	Pentane, 1-bromo-3-methyl- 3-Methylpentyl bromide	$C_6H_{13}Br$ 165.07	51116-73-5 	148	4-01-00-00365 1.1829^{20}	H$_2$O 1; eth 3; chl 4 1.4496^{20}
8716	Pentane, 1-bromo-4-methyl-	$C_6H_{13}Br$ 165.07	626-88-0 	145	4-01-00-00361 1.1683^{20}	eth 4; chl 4 1.4490
8717	Pentane, 2-bromo-2-methyl- 1-Bromo-1,1-dimethylbutane	$C_6H_{13}Br$ 165.07	4283-80-1 	142.5	4-01-00-00361 	eth 4; chl 4 1.442^{23}
8718	Pentane, 3-bromo-3-methyl- 3-Bromo-3-methylpentane	$C_6H_{13}Br$ 165.07	25346-31-0 	$130; 76^{100}$	4-01-00-00365 1.1835^{20}	eth 4; chl 4 1.4525^{20}
8719	Pentane, 1-chloro- Amyl chloride	$C_5H_{11}Cl$ 106.60	543-59-9 -99	642 107.8	4-01-00-00309 0.8820^{20}	H$_2$O 1; EtOH 5; eth 5; bz 3 1.4127^{20}
8720	Pentane, 2-chloro-, (+)- Pentane, 2-chloro (d)	$C_5H_{11}Cl$ 106.60	29882-57-3 -137	96.9	4-01-00-00309 0.8698^{20}	H$_2$O 1; EtOH 3; eth 3; bz 3 1.4069^{20}
8721	Pentane, 3-chloro- 1-Ethylpropyl chloride	$C_5H_{11}Cl$ 106.60	616-20-6 -105	97.5	4-01-00-00309 0.8731^{20}	H$_2$O 1; EtOH 3; eth 3; ace 2 1.4082^{20}
8722	Pentane, 2-chloro-2,3-dimethyl-	$C_7H_{15}Cl$ 134.65	59889-45-1 	$38-9^{20}$	4-01-00-00406 	bz 4; eth 4 1.4264^{20}
8723	Pentane, 2-chloro-2,4-dimethyl-	$C_7H_{15}Cl$ 134.65	35951-33-8 	$129; 33^{20}$	4-01-00-00408 0.861^{20}	eth 4 1.4180^{20}
8724	Pentane, 3-chloro-2,3-dimethyl-	$C_7H_{15}Cl$ 134.65	595-38-0 	136 dec; 41^{20}	4-01-00-00406 0.882^{22}	eth 4; chl 4 1.4318^{20}
8725	Pentane, 4-chloro-2,2-dimethyl-	$C_7H_{15}Cl$ 134.65	33429-72-0 	93^{250}	4-01-00-00404 0.855^{20}	eth 4; chl 4 1.4180^{20}
8726	Pentane, 3-chloro-3-ethyl-	$C_7H_{15}Cl$ 134.65	994-25-2 	143.5	4-01-00-00403 0.8856^{20}	eth 4 1.4400^{20}
8727	Pentane, 3-chloro-3-ethyl-2,2-dimethyl- Pentane 3-chloro-2,2-dimethyl-3-ethyl	$C_9H_{19}Cl$ 162.70	86661-53-2 	53^6 dec	4-01-00-00462 	eth 4; chl 4 1.4528^{25}
8728	Pentane, 1-chloro-5-fluoro-	$C_5H_{10}ClF$ 124.59	407-98-7 	143.2	4-01-00-00309 1.0325^{25}	eth 4; EtOH 4 1.4120^{23}

No.	Name Synonym	Mol. Form. Mol. Wt.	CAS RN mp/°C	Merck No. bp/°C	Beil. Ref. den/g cm^{-3}	Solubility n_D
8729	Pentane, 2-chloro-2-methyl-	$C_6H_{13}Cl$ 120.62	4325-48-8	110 dec; 36[15]	4-01-00-00360 0.863[20]	eth 4 1.4126[20]
8730	Pentane, 2-chloro-4-methyl- 2-Chloro-4-methylpentane	$C_6H_{13}Cl$ 120.62	25346-32-1	113	3-01-00-00399 0.8610[20]	eth 4 1.4113[20]
8731	Pentane, 3-(chloromethyl)-	$C_6H_{13}Cl$ 120.62	4737-41-1	126; 83[202]	3-01-00-00403 0.8914[20]	bz 4; eth 4; chl 4 1.4222[20]
8732	Pentane, 3-chloro-3-methyl-	$C_6H_{13}Cl$ 120.62	918-84-3	116	4-01-00-00365 0.8900[20]	bz 4; eth 4; chl 4 1.4210[20]
8733	Pentane, 2-chloro-2,4,4-trimethyl-	$C_8H_{17}Cl$ 148.68	6111-88-2 -26	147 dec; 44[16]	4-01-00-00444 0.8746[20]	EtOH 4 1.4308[20]
8734	Pentanedial Glutaraldehyde	$C_5H_8O_2$ 100.12	111-30-8	4366 188 dec	4-01-00-03659	H_2O 5; EtOH 5; bz 3
8735	1,5-Pentanediamine Cadaverine	$C_5H_{14}N_2$ 102.18	462-94-2 9	1608 179	4-04-00-01310 0.873[25]	H_2O 3; EtOH 3; eth 2 1.463[20]
8736	1,4-Pentanediamine, N1,N1-diethyl- Novoldiamine	$C_9H_{22}N_2$ 158.29	140-80-7	6642 201	4-04-00-01308 0.814[20]	1.4429[20]
8737	1,5-Pentanediamine, N-(6-methoxy-8-quinolinyl)-N'-(1-methylethyl)-	$C_{18}H_{27}N_3O$ 301.43	86-78-2	165-70[0.02]	5-22-12-00126	dil HCl 4 1.5785[25]
8738	1,5-Pentanediaminium, N,N,N,N',N',N'-hexamethyl-, dibromide Ammonium, pentamethylenebis trimethyl, bromide	$C_{11}H_{28}Br_2N_2$ 348.16	541-20-8 301	7070	4-04-00-01311	H_2O 2
8739	Pentane, 1,3-dibromo- 1,3-Dibromopentane	$C_5H_{10}Br_2$ 229.94	42474-20-4	193; 64[10]	4-01-00-00314 1.6721[20]	1.5048[207]
8740	Pentane, 1,4-dibromo- 1,4-Dibromopentane	$C_5H_{10}Br_2$ 229.94	626-87-9 -34.4	146[150]; 99[14]	4-01-00-00314 1.6222[20]	1.5086[20]
8741	Pentane, 1,5-dibromo-	$C_5H_{10}Br_2$ 229.94	111-24-0 -39.5	222.3	4-01-00-00314 1.7018[20]	H_2O 1; bz 3; ctc 2; chl 3 1.5126[20]
8742	Pentane, 2,3-dibromo- 2,3-Dibromopentane	$C_5H_{10}Br_2$ 229.94	5398-25-4 -56	178; 91[50]	3-01-00-00347 1.6865[14]	1.5087[20]
8743	Pentane, 2,4-dibromo- 2,4-Dibromopentane	$C_5H_{10}Br_2$ 229.94	19398-53-9	60[12]	3-01-00-00348 1.6659[20]	1.4987[20]
8744	Pentane, 1,2-dichloro- 1,2-Dichloropentane	$C_5H_{10}Cl_2$ 141.04	1674-33-5	148.3	3-01-00-00341 1.0872[20]	H_2O 1; EtOH 3; chl 4 1.4485[20]
8745	Pentane, 1,3-dichloro- 1,3-Dichloropentane	$C_5H_{10}Cl_2$ 141.04	30122-12-4	80.4[60]	3-01-00-00342 1.0834[20]	ace 4; chl 4 1.4485[20]
8746	Pentane, 1,4-dichloro- 1,4-Dichloropentane	$C_5H_{10}Cl_2$ 141.04	626-92-6	162; 88[60]	4-01-00-00309 1.0840[20]	ace 3; chl 3 1.4503[20]
8747	Pentane, 1,5-dichloro- 1,5-Dichloropentane	$C_5H_{10}Cl_2$ 141.04	628-76-2 -72.8	179	4-01-00-00310 1.1006[22]	H_2O 1; EtOH 3; eth 3; bz 3 1.4564[20]
8748	Pentane, 2,2-dichloro- 2,2-Dichloropentane	$C_5H_{10}Cl_2$ 141.04	34887-14-4	128.5	2-01-00-00095 1.040[20]	H_2O 1; eth 3; bz 3; chl 3 1.434[20]
8749	Pentane, 2,3-dichloro- 2,3-Dichloropentane	$C_5H_{10}Cl_2$ 141.04	600-11-3 -77.3	139	4-01-00-00310 1.0789[20]	1.4464[20]
8750	Pentane, 2,4-dichloro- 2,4-Dichloropentane	$C_5H_{10}Cl_2$ 141.04	625-67-2	145	3-01-00-00343 1.0634[15]	H_2O 1; eth 4; bz 3; chl 4 1.447[18]
8751	Pentane, 3,3-dichloro- 3,3-Dichloropentane	$C_5H_{10}Cl_2$ 141.04	21571-91-5	132; 32[14]	3-01-00-00343 1.053[20]	bz 4; chl 4 1.442[20]
8752	Pentane, 1,2-dichloro-4,4-dimethyl- 1,2-Dichloro-4,4-dimethylpentane	$C_7H_{14}Cl_2$ 169.09	6065-90-3	175; 58[12]	3-01-00-00445 1.0259[20]	bz 4; chl 4 1.4489[20]
8753	Pentane, 1,5-dichloro-3,3-dimethyl- 1,5-Dichloro-3,3-dimethylpentane	$C_7H_{14}Cl_2$ 169.09	62496-53-1	135[80]; 58[8]	4-01-00-00409 1.0563[20]	chl 4 1.4652[20]
8754	Pentane, 2,4-dichloro-2,4-dimethyl- 2,4-Dichloro-2,4-dimethylpentane	$C_7H_{14}Cl_2$ 169.09	33553-93-4 23.5	51-7[8]	4-01-00-00408 1.0292[20]	1.4537[20]
8755	Pentane, 1,1-diethoxy-	$C_9H_{20}O_2$ 160.26	3658-79-5	59[12]	4-01-00-03269 0.829[22]	1.4029[22]
8756	Pentane, 3,3-diethyl- 3,3-Diethylpentane	C_9H_{20} 128.26	1067-20-5 -33.1	146.3	4-01-00-00462 0.7536[20]	H_2O 1; eth 3; bz 3; os 3 1.4206[20]
8757	Pentane, 3,3-diethyl-2-methyl- 3,3-Diethyl-2-methylpentane	$C_{10}H_{22}$ 142.28	52897-16-2	172	4-01-00-00487 0.7755[25]	1.4343[20]
8758	Pentane, 1,5-diiodo- Pentamethylene diiodide	$C_5H_{10}I_2$ 323.94	628-77-3 9	149[20]; 101[3]	4-01-00-00317 2.1903[15]	H_2O 1; eth 3; chl 3 1.6046[15]
8759	Pentane, 2,4-diiodo-	$C_5H_{10}I_2$ 323.94	66719-29-7	147[34]	0-01-00-00133 2.195[20]	1.600[20]
8760	Pentane, 2,2-dimethyl- 2,2-Dimethylpentane	C_7H_{16} 100.20	590-35-2 -123.8	79.2	4-01-00-00403 0.6739[20]	H_2O 1; EtOH 3; eth 3; ace 5 1.3822[20]
8761	Pentane, 2,3-dimethyl- 2,3-Dimethylpentane	C_7H_{16} 100.20	565-59-3	89.7	3-01-00-00445 0.6951[20]	H_2O 1; EtOH 3; eth 3; ace 5 1.3919[20]
8762	Pentane, 2,4-dimethyl- 2,4-Dimethylpentane	C_7H_{16} 100.20	108-08-7 -119.9	80.4	4-01-00-00406 0.6727[20]	H_2O 1; EtOH 3; eth 3; ace 5 1.3815[20]
8763	Pentane, 3,3-dimethyl- 3,3-Dimethylpentane	C_7H_{16} 100.20	562-49-2 -134.9	86.0	4-01-00-00409 0.6936[20]	H_2O 1; EtOH 3; eth 3; ace 5 1.3909[20]
8764	Pentane, 2,2-dimethyl-3-methylene-	C_8H_{16} 112.22	18231-53-3	110	3-01-00-00848 0.724[25]	1.4159[20]
8765	Pentane, 2,4-dimethyl-3-methylene- 1-Butene, 2-isopropyl-3-methyl-	C_8H_{16} 112.22	111823-35-9	104	3-01-00-00851 0.718[25]	1.4086[20]

No.	Name / Synonym	Mol. Form. / Mol. Wt.	CAS RN / mp/°C	Merck No. / bp/°C	Beil. Ref. / den/g cm⁻³	Solubility / n_D
8766	Pentane, 2,4-dimethyl-3-(1-methylethyl)-	$C_{10}H_{22}$ 142.28	13475-79-1 -81.7	157.1	3-01-00-00534 0.7545[25]	1.4246[20]
8767	Pentanedinitrile Glutaronitrile	$C_5H_6N_2$ 94.12	544-13-8 -29	4368 286	4-02-00-01941 0.9911[15]	EtOH 4; chl 4 1.4295[20]
8768	Pentanedinitrile, 2-bromo-2-(bromomethyl)- 1,2-Dibromo-2,4-dicyanobutane	$C_6H_6Br_2N_2$ 265.94	35691-65-7 52	3004		H_2O 1; ace 4; bz 4; DMF 4
8769	Pentane, 1,5-dinitro- 1,5-Dinitropentane	$C_5H_{10}N_2O_4$ 162.15	6848-84-6	134[1.2]	4-01-00-00319	H_2O 1; bz 3 1.461[20]
8770	Pentanedioic acid Glutaric acid	$C_5H_8O_4$ 132.12	110-94-1 97.8	4367 303 dec	4-02-00-01934 1.429[15]	H_2O 4; EtOH 4; eth 4; bz 1 1.4188[106]
8771	Pentanedioic acid, 2-acetyl-, diethyl ester	$C_{11}H_{18}O_5$ 230.26	1501-06-0	271 dec	4-03-00-01835 1.0712[20]	H_2O 2; EtOH 3; eth 3; ctc 2 1.4420[15]
8772	Pentanedioic acid, diethyl ester	$C_9H_{16}O_4$ 188.22	818-38-2 -24.1	236.5	4-02-00-01937 1.0220[20]	eth 4 1.4241[20]
8773	Pentanedioic acid, 3,3-dimethyl-	$C_7H_{12}O_4$ 160.17	4839-46-7 103.5	126[415]; 89[2]	4-02-00-02023 1.4278[20]	H_2O 4; EtOH 4; eth 4; bz 2
8774	Pentanedioic acid, 3,3-dimethyl-, dimethyl ester	$C_9H_{16}O_4$ 188.22	19184-67-9	103-4[15]	3-02-00-01758 1.0366[20]	ctc 3
8775	Pentanedioic acid, dimethyl ester	$C_7H_{12}O_4$ 160.17	1119-40-0 -42.5	214; 109[21]	4-02-00-01937 1.0876[20]	EtOH 4; eth 4; chl 3 1.4242[20]
8776	Pentanedioic acid, diphenyl ester Diphenyl glutarate	$C_{17}H_{16}O_4$ 284.31	47172-89-4 54	236.5[15]	4-06-00-00626	ace 4; EtOH 4; lig 4
8777	Pentanedioic acid, 3-ethyl-3-methyl- 3-Ethyl-3-methylglutaric acid	$C_8H_{14}O_4$ 174.20	5345-01-7 87	261	3-02-00-01781	H_2O 3; EtOH 3; eth 3; bz 3
8778	Pentanedioic acid, 3-hydroxy-, diethyl ester Diethyl 3-hydroxyglutarate	$C_9H_{18}O_5$ 204.22	32328-03-3	156-7[23]	4-03-00-01149 1.103[25]	1.4368[20]
8779	Pentanedioic acid, 3-methyl- Glutaric acid, 3-methyl	$C_6H_{10}O_4$ 146.14	626-51-7 87	165-7[0.5]	4-02-00-01992	H_2O 3; EtOH 3; eth 3; bz 2
8780	Pentanedioic acid, monomethyl ester	$C_6H_{10}O_4$ 146.14	1501-27-5	150-1[10]	4-02-00-01937 1.169[25]	1.4381[20]
8781	Pentanedioic acid, 2-oxo- α-Ketoglutaric acid	$C_5H_6O_5$ 146.10	328-50-7 115.5	5182	4-03-00-01813	H_2O 4; EtOH 4; eth 4; ace 3
8782	Pentanedioic acid, 3-oxo- Acetonedicarboxylic acid	$C_5H_6O_5$ 146.10	542-05-2 138 dec	60	4-03-00-01816	H_2O 3; EtOH 3; eth 2; bz 1
8783	Pentanedioic acid, 3-oxo-, diethyl ester	$C_9H_{14}O_5$ 202.21	105-50-0	250	4-03-00-01817 1.113[20]	H_2O 2; EtOH 5
8784	Pentanedioic acid, 3-oxo-, dimethyl ester	$C_7H_{10}O_5$ 174.15	1830-54-2	150[25]	4-03-00-01817 1.185[25]	1.4434[20]
8785	1,3-Pentanediol	$C_5H_{12}O_2$ 104.15	3174-67-2	120-2[16]	4-01-00-02538 0.9863[20]	1.4659[20]
8786	1,4-Pentanediol 4-Hydroxy-1-pentanol	$C_5H_{12}O_2$ 104.15	626-95-9	202, 125[10]	3-01-00-02191 0.8883[20]	H_2O 4; EtOH 4; chl 4 1.4452[23]
8787	1,5-Pentanediol Pentamethylene glycol	$C_5H_{12}O_2$ 104.15	111-29-5 -18	7073 239	4-01-00-02540 0.9914[20]	H_2O 3; EtOH 3; eth 2; bz 2 1.4494[20]
8788	2,3-Pentanediol	$C_5H_{12}O_2$ 104.15	42027-23-6	187.5; 100[17]	4-01-00-02543 0.9798[19]	H_2O 3; EtOH 3; eth 2 1.4412[25]
8789	2,4-Pentanediol 2,4-Amylene glycol	$C_5H_{12}O_2$ 104.15	625-69-4	199; 97-8[13]	3-01-00-02195 0.9635[20]	H_2O 4; EtOH 4 1.4349[20]
8790	1,2-Pentanediol, (±)- 1,2-Pentanediol (DL)	$C_5H_{12}O_2$ 104.15	91049-43-3	211	4-01-00-02538 0.9784[20]	1.4412[19]
8791	1,5-Pentanediol, diacetate Pentamethylene acetate	$C_9H_{16}O_4$ 188.22	6963-44-6	122-3[3]	4-02-00-00226 1.0296[20]	1.4261[19]
8792	1,3-Pentanediol, 2,2-dimethyl- 2,2-Dimethyl-1,3-dihydroxypentane	$C_7H_{16}O_2$ 132.20	2157-31-5 62.3	213	0-01-00-00490	EtOH 4
8793	2,4-Pentanediol, 2-methyl- Hexylene glycol	$C_8H_{14}O_2$ 118.18	107-41-5 -50	4631 197.1	4-01-00-02565 0.923[15]	H_2O 3; EtOH 3; eth 3; ctc 2 1.4276[20]
8794	2,4-Pentanediol, 3-methyl- 3-Methyl-2,4-pentanediol	$C_6H_{14}O_2$ 118.18	5683-44-3	211.5	4-01-00-02572 0.9640[20]	H_2O 3; EtOH 3 1.4433[20]
8795	1,3-Pentanediol, 2,2,4-trimethyl- 2,2,4-Trimethyl-1,3-pentanediol	$C_8H_{18}O_2$ 146.23	144-19-4 51.5	235; 81-2[1]	4-01-00-02604 0.936[15]	H_2O 2; EtOH 4; eth 4; bz 3 1.4513[15]
8796	1,4-Pentanediol, 2,2,4-trimethyl-	$C_8H_{18}O_2$ 146.23	80864-10-4 86	210	0-01-00-00493	eth 4; EtOH 4
8797	2,3-Pentanedione Acetylpropionyl	$C_5H_8O_2$ 100.12	600-14-6	108	4-01-00-03660 0.9565[19]	H_2O 3; EtOH 5; eth 5; ace 5 1.4014[19]
8798	2,4-Pentanedione Acetylacetone	$C_5H_8O_2$ 100.12	123-54-6 -23	75 138	4-01-00-03662 0.9721[25]	H_2O 4; EtOH 5; eth 5; ace 5 1.4494[20]
8799	2,4-Pentanedione, 3,3-dimethyl- 3,3-Dimethylacetoacetone	$C_7H_{12}O_2$ 128.17	3142-58-3 19	173	4-01-00-03705 0.9575[20]	eth 4 1.4306[20]
8800	2,3-Pentanedione, dioxime	$C_5H_{10}N_2O_2$ 130.15	4775-86-4 172.5	sub	4-01-00-03661	EtOH 4
8801	2,4-Pentanedione, 3-ethyl-	$C_7H_{12}O_2$ 128.17	1540-34-7	178.5	4-01-00-03703 0.9531[19]	eth 4; EtOH 4; chl 4 1.4408[19]
8802	2,4-Pentanedione, 1,1,1,5,5,5-hexafluoro- Acetylacetone, hexafluoro	$C_5H_2F_6O_2$ 208.06	1522-22-1	54.1	4-01-00-03681 1.485[20]	1.3333[20]
8803	2,3-Pentanedione, 3-oxime 3-Oximino-2-pentanone	$C_5H_9NO_2$ 115.13	609-29-0 58.5	183	4-01-00-03661	eth 4; EtOH 4; chl 4
8804	1,4-Pentanedione, 1-phenyl-	$C_{11}H_{12}O_2$ 176.22	583-05-1	162[12]	4-07-00-02160	ace 4 1.5250[30]
8805	2,4-Pentanedione, 3-(2-propenyl)-	$C_8H_{12}O_2$ 140.18	3508-78-9	199; 92[16]	4-01-00-03759 0.974[15]	1.4698[14]

No.	Name / Synonym	Mol. Form. / Mol. Wt.	CAS RN / mp/°C	Merck No. / bp/°C	Beil. Ref. / den/g cm^{-3}	Solubility / n_D
8806	Pentanedioyl dichloride	$C_5H_6Cl_2O_2$ 169.01	2873-74-7	217	4-02-00-01939 1.324[20]	eth 3; chl 2 1.4728[20]
8807	Pentane, dodecafluoro- Perfluoropentane	C_5F_{12} 288.04	678-26-2 -10	29.2	4-01-00-00308	
8808	Pentane, 1-ethoxy- Ethyl pentyl ether	$C_7H_{16}O$ 116.20	17952-11-3	117.6	3-01-00-01602 0.7622[20]	eth 4; EtOH 4 1.3927[20]
8809	Pentane, 3-ethyl- 3-Ethylpentane	C_7H_{16} 100.20	617-78-7 -118.6	93.5	4-01-00-00402 0.6982[20]	H_2O 1; EtOH 3; eth 3; ace 5 1.3934[20]
8810	Pentane, 3-ethyl-2,2-dimethyl-	C_9H_{20} 128.26	16747-32-3 -99.3	133.8	4-01-00-00462 0.7438[20]	1.4123[20]
8811	Pentane, 3-ethyl-2,3-dimethyl-	C_9H_{20} 128.26	16747-33-4	144.7	4-01-00-00462 0.7508[25]	1.4221[20]
8812	Pentane, 3-ethyl-2,4-dimethyl-	C_9H_{20} 128.26	1068-87-7 -122.4	136.7	4-01-00-00462 0.7365[20]	1.4131[20]
8813	Pentane, 3-ethyl-2-methyl- 3-Ethyl-2-methylpentane	C_8H_{18} 114.23	609-26-7 -114.9	115.6	4-01-00-00437 0.7193[20]	H_2O 1; EtOH 5; eth 3; ace 5 1.4040[20]
8814	Pentane, 3-ethyl-3-methyl- 3-Ethyl-3-methylpentane	C_8H_{18} 114.23	1067-08-9 -90.9	118.2	4-01-00-00438 0.7274[20]	H_2O 1; EtOH 5; eth 3; ace 5 1.4078[20]
8815	Pentane, 3-ethyl-2,2,3-trimethyl- 3-Ethyl-2,2,3-trimethylpentane	$C_{10}H_{22}$ 142.28	52897-17-3 -42.5	169.5	4-01-00-00487 0.7780[25]	1.4420[20]
8816	Pentane, 3-ethyl-2,3,4-trimethyl- 3-Ethyl-2,3,4-trimethylpentane	$C_{10}H_{22}$ 142.28	52897-19-5	169.47	4-01-00-00487 0.7735[25]	1.4333[20]
8817	Pentane, 1-fluoro- Pentyl fluoride	$C_5H_{11}F$ 90.14	592-50-7 -120	62.8	4-01-00-00308 0.7907[20]	eth 4; EtOH 4 1.3591[2-]
8818	Pentane, 1-iodo- Amyl iodide	$C_5H_{11}I$ 198.05	628-17-1 -85.6	155	4-01-00-00315 1.5161[20]	chl 3 1.4959[20]
8819	Pentane, 2-iodo-, (±)- Pentane, 2-iodo (DL)	$C_5H_{11}I$ 198.05	52152-72-4	141	4-01-00-00316 1.5096[20]	H_2O 1; eth 3; ace 3; bz 3 1.4961[20]
8820	Pentane, 3-iodo- 3-Iodopentane	$C_5H_{11}I$ 198.05	1809-05-8	145.5	3-01-00-00349 1.5176[20]	ace 4; bz 4; eth 4 1.4974[20]
8821	Pentane, 1-isocyano- Pentyl isocyanide	$C_6H_{11}N$ 97.16	18971-59-0 -51.1	155.5	3-04-00-00331 0.806[20]	H_2O 1; EtOH 3
8822	Pentane, 1-methoxy- Methyl pentyl ether	$C_6H_{14}O$ 102.18	628-80-8	99	4-01-00-01643 0.759[22]	ace 4; eth 4; EtOH 4 1.3862[22]
8823	Pentane, 2-methyl- Isohexane	C_6H_{14} 86.18	107-83-5 -153.7	60.2	4-01-00-00358 0.650[25]	H_2O 1; EtOH 3; eth 3; ace 5 1.3715[20]
8824	Pentane, 3-methyl- 3-Methylpentane	C_6H_{14} 86.18	96-14-0 -162.9	63.2	4-01-00-00363 0.6598[25]	H_2O 1; EtOH 5; eth 5; ace 5 1.3765[20]
8825	Pentane, 3-methylene- 2-Ethyl-1-butene	C_6H_{12} 84.16	760-21-4 -131.5	64.7	4-01-00-00850 0.6894[20]	H_2O 1; eth 3; ace 3; bz 3 1.3969[20]
8826	Pentane, 1-(methylthio)- Methyl pentyl sulfide	$C_6H_{14}S$ 118.24	1741-83-9 -94	145.1	3-01-00-01608 0.8431[20]	EtOH 3; eth 3; ace 3; bz 3 1.4506[20]
8827	Pentanenitrile Valeronitrile	C_5H_9N 83.13	110-59-8 -96.2	141.3	4-02-00-00875 0.8008[20]	eth 3; ace 3; bz 3; ctc 2 1.3971[20]
8828	Pentanenitrile, 5-bromo-	C_5H_8BrN 162.03	5414-21-1	111[12]	4-02-00-00884 1.386[25]	1.4780[20]
8829	Pentanenitrile, 2-hydroxy-2-propyl-	$C_8H_{15}NO$ 141.21	5699-74-1	119-20[21]	4-03-00-00880 0.9077[18]	EtOH 4 1.4337[18]
8830	Pentanenitrile, 3-imino-2-methyl- Valeronitrile, 3-imino-2-methyl	$C_6H_{10}N_2$ 110.16	95642-49-2 46.3	257.5	0.9525[19]	eth 4; EtOH 4
8831	Pentanenitrile, 4-methyl- Isoamyl cyanide	$C_6H_{11}N$ 97.16	542-54-1 -51	156.5	4-02-00-00946 0.8030[20]	H_2O 1; EtOH 3; eth 5; ctc 2 1.4059[20]
8832	Pentane, 1-nitro- 1-Nitropentane	$C_5H_{11}NO_2$ 117.15	628-05-7	172.5	4-01-00-00317 0.9525[20]	EtOH 3; eth 3; bz 3 1.4175[20]
8833	Pentane, 2-nitro- 2-Nitropentane	$C_5H_{11}NO_2$ 117.15	4609-89-6	149; 88[100]	4-01-00-00318 0.938[25]	1.4092[25]
8834	Pentane, 3-nitro- 3-Nitropentane	$C_5H_{11}NO_2$ 117.15	551-88-2	154	4-01-00-00318 0.957[0]	ace 4; eth 4; EtOH 4
8835	Pentane, 1,1'-oxybis- Amyl ether	$C_{10}H_{22}O$ 158.28	693-65-2 -69	646 190	4-01-00-01643 0.7833[20]	H_2O 1; EtOH 5; eth 5; chl 3 1.4119[20]
8836	Pentane, 2,2,3,3,4-pentamethyl- 2,2,3,3,4-Pentamethylpentane	$C_{10}H_{22}$ 142.28	16747-44-7 -36.4	166.1	4-01-00-00487 0.7767[25]	1.4361[20]
8837	Pentane, 2,2,3,4,4-pentamethyl- 2,2,3,4,4-Pentamethylpentane	$C_{10}H_{22}$ 142.28	16747-45-8 -38.7	159.3	4-01-00-00487 0.7636[25]	1.4307[20]
8838	Pentane, 1,1,1,5-tetrachloro- 1,1,1,5-Tetrachloropentane	$C_5H_8Cl_4$ 209.93	2467-10-9	112[24]	4-01-00-00311 1.3416[25]	1.4859[25]
8839	Pentane, 2,2,3,3-tetramethyl- 2,2,3,3-Tetramethylpentane	C_9H_{20} 128.26	7154-79-2 -9.8	140.2	4-01-00-00463 0.7530[25]	1.4236[20]
8840	Pentane, 2,2,3,4-tetramethyl- 2,2,3,4-Tetramethylpentane	C_9H_{20} 128.26	1186-53-4 -121.0	133.0	4-01-00-00463 0.7389[20]	1.4147[20]
8841	Pentane, 2,2,4,4-tetramethyl- Di-tert-butylmethane	C_9H_{20} 128.26	1070-87-7 -66.5	122.2	4-01-00-00464 0.7195[20]	H_2O 1; EtOH 4; bz 4; os 4 1.4069[20]
8842	Pentane, 2,3,3,4-tetramethyl- 2,3,3,4-Tetramethylpentane	C_9H_{20} 128.26	16747-38-9 -102.1	141.5	4-01-00-00464 0.7547[20]	1.4222[20]
8843	Pentane, 1,1'-thiobis-	$C_{10}H_{22}S$ 174.35	872-10-6 -51.3	86[3.7]	4-01-00-01654 0.8407[20]	H_2O 1; eth 3 1.4561[20]
8844	1-Pentanethiol Pentyl mercaptan	$C_5H_{12}S$ 104.22	110-66-7 -75.7	648 126.6	4-01-00-01653 0.850[20]	H_2O 1; EtOH 5; eth 5 1.4469[20]
8845	2-Pentanethiol sec-Amylmercaptan	$C_5H_{12}S$ 104.22	2084-19-7 -169	112.9	4-01-00-01662 0.8327[20]	EtOH 3; lig 3 1.4412[20]

No.	Name / Synonym	Mol. Form. / Mol. Wt.	CAS RN / mp/°C	Merck No. / bp/°C	Beil. Ref. / den/g cm^{-3}	Solubility / n_D
8846	3-Pentanethiol / 3-Pentyl mercaptan	$C_5H_{12}S$ / 104.22	616-31-9 / -110.8	/ 105	4-01-00-01665 / 0.8410[20]	EtOH 3; DMSO-d$_6$ 2 / 1.4447[20]
8847	Pentane, 2,2,3-trimethyl- / 2-tert-Butylbutane	C_8H_{18} / 114.23	564-02-3 / -112.2	/ 110	4-01-00-00438 / 0.7161[20]	H$_2$O 1; EtOH 5; eth 5; acc 5 / 1.4030[20]
8848	Pentane, 2,2,4-trimethyl- / Isooctane	C_8H_{18} / 114.23	540-84-1 / -107.3	5079 / 99.2	4-01-00-00439 / 0.6877[25]	H$_2$O 1; EtOH 3; eth 3; ace 5 / 1.3915[20]
8849	Pentane, 2,3,3-trimethyl- / 2,3,3-Trimethylpentane	C_8H_{18} / 114.23	560-21-4 / -100.9	/ 114.8	4-01-00-00445 / 0.7262[20]	H$_2$O 1; EtOH 4; eth 5; ace 5 / 1.4075[20]
8850	Pentane, 2,3,4-trimethyl- / 2,3,4-Trimethylpentane	C_8H_{18} / 114.23	565-75-3 / -109.2	/ 113.5	4-01-00-00446 / 0.7191[20]	H$_2$O 1; EtOH 4; eth 5; ace 5 / 1.4042[20]
8851	1,2,5-Pentanetriol / 1,4,5-Pentanetriol	$C_5H_{12}O_3$ / 120.15	14697-46-2 /	/ 190-1[13]	4-01-00-02779 / 1.136[20]	H$_2$O 4; EtOH 4 / 1.4730[20]
8852	1,3,5-Pentanetriol	$C_5H_{12}O_3$ / 120.15	4328-94-3 /	/ 188[11]; 139[0.1]	4-01-00-02779 / 1.1291[20]	H$_2$O 4; EtOH 4; eth 3; ace 3 / 1.4785[20]
8853	Pentane, 1,1,1,2,2,3,4,4,5,5,5-undecafluoro-3-(trifluoromethyl) / Perfluoro-3-methylpentane	C_6F_{14} / 338.04	865-71-4 / -115	/ 58.4	/	bz 3
8854	Pentane, 1,1,1,2,2,3,3,4,5,5,5-undecafluoro-4-(trifluoromethyl)- / Perfluoro(2-methylpentane)	C_6F_{14} / 338.04	355-04-4 /	/ 57.6	4-01-00-00360 / 1.7326[20]	H$_2$O 1; bz 3 / 1.2564[22]
8855	Pentanoic acid / Valeric acid	$C_5H_{10}O_2$ / 102.13	109-52-4 / -34	9815 / 186.1	4-02-00-00868 / 0.9391[20]	H$_2$O 3; EtOH 3; eth 3; ctc 2 / 1.4085[20]
8856	Pentanoic acid, 2-acetyl-, ethyl ester	$C_9H_{16}O_3$ / 172.22	1540-28-9 /	/ 224; 90[15]	4-03-00-01602 / 0.9661[20]	eth 4 / 1.4255[20]
8857	Pentanoic acid, 5-amino-	$C_5H_{11}NO_2$ / 117.15	660-88-8 / 157-8 dec	dec	4-04-00-02636 /	H$_2$O 3; EtOH 2; eth 1; bz 1
8858	Pentanoic acid, anhydride	$C_{10}H_{18}O_3$ / 186.25	2082-59-9 / -56.1	/ 227	4-02-00-00874 / 0.924[20]	eth 4; EtOH 4 / 1.4171[26]
8859	Pentanoic acid, 2-bromo-, (±)-	$C_5H_9BrO_2$ / 181.03	2681-92-7 /	/ 118[12]	4-02-00-00883 / 1.381[20]	eth 4; EtOH 4
8860	Pentanoic acid, 2-bromo-, ethyl ester	$C_7H_{13}BrO_2$ / 209.08	615-83-8 /	/ 191	3-02-00-00681 / 1.226[18]	H$_2$O 1; EtOH 3; eth 3 / 1.4496[20]
8861	Pentanoic acid, 5-bromo-, ethyl ester	$C_7H_{13}BrO_2$ / 209.08	14660-52-7 /	/ 107-8[20]	4-02-00-00884 / 1.2804[20]	ctc 2 / 1.4643[20]
8862	Pentanoic acid, butyl ester	$C_9H_{18}O_2$ / 158.24	591-68-4 / -92.8	/ 185.8	3-02-00-00671 / 0.8710[15]	H$_2$O 2; EtOH 3; eth 3 / 1.4128[20]
8863	Pentanoic acid, 2-chloro-, (±)-	$C_5H_9ClO_2$ / 136.58	94347-45-2 / -15	/ 222	4-02-00-00878 / 1.141[13]	eth 4; EtOH 4 / 1.4481[11]
8864	Pentanoic acid, 3-chloro- / Valeric acid, 3-chloro	$C_5H_9ClO_2$ / 136.58	51637-47-9 / 33	/ 112[10]	4-02-00-00879 / 1.1484[20]	eth 4; EtOH 4 / 1.4462[20]
8865	Pentanoic acid, 4-chloro- / 4-Chlorovaleric acid	$C_5H_9ClO_2$ / 136.58	32607-54-8 /	/ 116[10]	4-02-00-00879 / 1.1514[20]	eth 4, EtOH 4 / 1.4458[20]
8866	Pentanoic acid, 5-chloro-	$C_5H_9ClO_2$ / 136.58	1119-46-6 / 18	/ 230	4-02-00-00880 / 1.3416[25]	eth 4; EtOH 4 / 1.4555[20]
8867	Pentanoic acid, 4-chloro-, ethyl ester / Ethyl 4-chlorovalerate	$C_7H_{13}ClO_2$ / 164.63	70786-82-2 /	/ 196	4-02-00-00879 / 1.0393[20]	eth 4; EtOH 4 / 1.4310[20]
8868	Pentanoic acid, 5-chloro-, ethyl ester	$C_7H_{13}ClO_2$ / 164.63	2323-81-1 /	/ 205.5	4-02-00-00880 / 1.0561[20]	H$_2$O 1; EtOH 3; eth 3; chl 2 / 1.4355[20]
8869	Pentanoic acid, 2-chloroethyl ester	$C_7H_{13}ClO_2$ / 164.63	7735-33-3 /	/ 185	3-02-00-00670 / 1.040[12]	eth 4 / 1.4307[11]
8870	Pentanoic acid, 2,3-dibromo- / Valeric acid, 2,3-dibromo-	$C_5H_8Br_2O_2$ / 259.93	79912-57-5 / 57	/ 104-6[13]	3-02-00-00683 /	bz 4; CS$_2$ 4
8871	Pentanoic acid, 2,5-dibromo- / Valeric acid, 2,5-dibromo-	$C_5H_8Br_2O_2$ / 259.93	1450-81-3 /	/ 173[15]; 151[5]	3-02-00-00683 / 1.8629[25]	1.5347[25]
8872	Pentanoic acid, 2,3-dibromo-, ethyl ester / Valeric acid, 2,3-dibromo-, ethyl ester	$C_7H_{12}Br_2O_2$ / 287.98	79912-55-3 /	/ 117[14]	2-02-00-00269 / 1.6199[15]	1.4953[15]
8873	Pentanoic acid, 2,5-dibromo-, ethyl ester / Valeric acid, 2,5-dibromo-, ethyl ester	$C_7H_{12}Br_2O_2$ / 287.98	29823-16-3 /	/ 133-5[14]	3-02-00-00683 / 1.6289[25]	1.4947[25]
8874	Pentanoic acid, 2,2-dimethyl-4-oxo-	$C_7H_{12}O_3$ / 144.17	470-49-5 / 76.8	/ 138[15]	3-03-00-01247 /	H$_2$O 4; bz 4; eth 4; EtOH 4
8875	Pentanoic acid, 2,4-dioxo-	$C_5H_6O_4$ / 130.10	5699-58-1 / 99.5	/ 130[37]	3-03-00-01331 /	H$_2$O 3; EtOH 3; eth 3; ace 3
8876	Pentanoic acid, 2,4-dioxo-, ethyl ester	$C_7H_{10}O_4$ / 158.15	615-79-2 / 18	/ 214	4-03-00-01777 / 1.1251[20]	eth 4; EtOH 4 / 1.4757[17]
8877	Pentanoic acid, 2-ethyl-, (±)- / Valeric acid, 2-ethyl (DL)	$C_7H_{14}O_2$ / 130.19	116908-83-9 /	/ 209.2	4-02-00-00975 / 0.9311[33]	eth 4; EtOH 4
8878	Pentanoic acid, ethyl ester / Ethyl valerate	$C_7H_{14}O_2$ / 130.19	539-82-2 / -91.2	/ 146.1	4-02-00-00872 / 0.8770[20]	H$_2$O 1; EtOH 5; eth 5; ctc 2 / 1.4120[20]
8879	Pentanoic acid, 5-fluoro- / Valeric acid, 5-fluoro-	$C_5H_9FO_2$ / 120.12	407-75-0 /	/ 83[2]	4-02-00-00876 /	eth 4; EtOH 4 / 1.4080[25]
8880	Pentanoic acid, 2-furanylmethyl ester / Furfuryl valerate	$C_{10}H_{14}O_3$ / 182.22	36701-01-6 /	/ 228; 82[1]	2-17-00-00115 / 1.0284[20]	eth 4; EtOH 4
8881	Pentanoic acid, heptyl ester	$C_{12}H_{24}O_2$ / 200.32	5451-80-9 / -46.4	/ 245.2	4-02-00-00872 / 0.8623[20]	ace 4; eth 4; EtOH 4 / 1.4254[15]
8882	Pentanoic acid, hexyl ester	$C_{11}H_{22}O_2$ / 186.29	1117-59-5 / -63.1	/ 226.3	3-02-00-00671 / 0.8635[20]	ace 4; eth 4; EtOH 4 / 1.4228[15]
8883	Pentanoic acid, 2-hydroxy-	$C_5H_{10}O_3$ / 118.13	617-31-2 / 34	sub	4-03-00-00807 /	H$_2$O 3; EtOH 3; eth 3
8884	Pentanoic acid, 4-hydroxy-, ethyl ester / Valeric acid, 4-hydroxy-, ethyl ester	$C_7H_{14}O_3$ / 146.19	6149-46-8 /	/ 85-6[2]	3-03-00-00612 / 0.9504[25]	eth 4; EtOH 4 / 1.4265[25]

No.	Name / Synonym	Mol. Form. / Mol. Wt.	CAS RN / mp/°C	Merck No. / bp/°C	Beil. Ref. / den/g cm^{-3}	Solubility / n_D
8885	Pentanoic acid, 2-hydroxyethyl ester Valeric acid, 2-hydroxyethyl ester	$C_7H_{14}O_3$ 146.19	16179-36-5 	 190	3-02-00-00672 0.9839[20]	 1.4220[20]
8886	Pentanoic acid, 2-methoxyphenyl ester Guaiacol valerate	$C_{12}H_{16}O_3$ 208.26	531-39-5 	4461 265	 1.05[25]	bz 4; eth 4; EtOH 4
8887	Pentanoic acid, 2-methyl-, (±)-	$C_6H_{12}O_2$ 116.16	22160-39-0 	 195.6	4-02-00-00942 0.9230[20]	H_2O 3; EtOH 3; eth 3; ctc 2 1.413[20]
8888	Pentanoic acid, 3-methyl-, (±)- Valeric acid, 3-methyl (DL)	$C_6H_{12}O_2$ 116.16	22160-40-3 -41.6	 197.5	4-02-00-00948 0.9262[20]	eth 4; EtOH 4 1.4159[20]
8889	Pentanoic acid, 4-methyl-	$C_6H_{12}O_2$ 116.16	646-07-1 -33	 200.5	4-02-00-00944 0.9225[20]	H_2O 2; EtOH 3; eth 3; chl 3 1.4144[20]
8890	Pentanoic acid, 2-methylene- Valeric acid, 2-methylene	$C_6H_{10}O_2$ 114.14	5650-75-9 24.4	 214	0-02-00-00437 0.9783[25]	bz 4; eth 4; chl 4
8891	Pentanoic acid, methyl ester Methyl valerate	$C_6H_{12}O_2$ 116.16	624-24-8 	 127.4	4-02-00-00871 0.8947[20]	H_2O 2; EtOH 5; eth 5; ace 3 1.4003[20]
8892	Pentanoic acid, 4-methyl-, ethyl ester	$C_8H_{16}O_2$ 144.21	25415-67-2 	163; 52[10]	4-02-00-00945 0.8705[20]	 1.4050[20]
8893	Pentanoic acid, 1-methylethyl ester	$C_8H_{16}O_2$ 144.21	18362-97-5 	 	3-02-00-00671 0.8579[20]	H_2O 1; EtOH 3; eth 3; ace 3 1.4061[20]
8894	Pentanoic acid, 4-methyl-, 2-methylpropyl ester	$C_{10}H_{20}O_2$ 172.27	25415-70-7 	172; 76[10]	4-02-00-00945 0.8536[20]	 1.4135[20]
8895	Pentanoic acid, 2-methyl-4-oxo-	$C_6H_{10}O_3$ 130.14	6641-83-4 	165[40]	4-03-00-01587 1.11[25]	 1.4431[20]
8896	Pentanoic acid, 3-methyl-4-oxo-	$C_6H_{10}O_3$ 130.14	6628-79-1 31.5	 241.5	3-03-00-01238 1.0932[17]	H_2O 4; EtOH 4 1.4443[17]
8897	Pentanoic acid, 4-methyl-3-oxo-, ethyl ester	$C_8H_{14}O_3$ 158.20	7152-15-0 -9	 173	4-03-00-01589 0.98[25]	 1.250[20]
8898	Pentanoic acid, 3-methyl-4-oxo-, methyl ester Levulinic acid, 3-methyl-, methyl ester	$C_7H_{12}O_3$ 144.17	25234-83-7 	200; 82[10]	1-03-00-00240 1.022[20]	 1.4052[73]
8899	Pentanoic acid, 1-methylpropyl ester, (+)- Valeric acid, sec-butyl ester (d)	$C_9H_{18}O_2$ 158.24	116836-32-9 	 174.5	3-02-00-00671 0.8605[20]	bz 4; eth 4; py 4; EtOH 4 1.4070[20]
8900	Pentanoic acid, 2-methylpropyl ester	$C_9H_{18}O_2$ 158.24	10588-10-0 	 179	3-02-00-00671 0.8625[25]	H_2O 1; EtOH 5; eth 3; ace 3 1.4046[20]
8901	Pentanoic acid, octyl ester Valeric acid, octyl ester	$C_{13}H_{28}O_2$ 214.35	5451-85-4 -42.3	 261.6	3-02-00-00672 0.8615[20]	ace 4; eth 4; EtOH 4 1.4273[15]
8902	Pentanoic acid, 2-oxo-	$C_5H_8O_3$ 116.12	1821-02-9 6.5	 179	4-03-00-01558 1.0970[14]	H_2O 2; eth 3; bz 3; chl 3
8903	Pentanoic acid, 4-oxo- Levulinic acid	$C_5H_8O_3$ 116.12	123-76-2 33	5352 245 dec	4-03-00-01560 1.1335[20]	H_2O 4; EtOH 4; eth 4; chl 3 1.4396[20]
8904	Pentanoic acid, 4-oxo-, butyl ester	$C_9H_{16}O_3$ 172.22	2052-15-5 	 237.5	4-03-00-01563 0.9735[20]	chl 2 1.4290[20]
8905	Pentanoic acid, 2-oxo-, ethyl ester Valeric acid, 2-oxo-, ethyl ester	$C_7H_{12}O_3$ 144.17	50461-74-0 	182.5; 70[11]	4-03-00-01558 0.9985[18]	 1.4170[18]
8906	Pentanoic acid, 3-oxo-, ethyl ester	$C_7H_{12}O_3$ 144.17	4949-44-4 	 191	4-03-00-01559 1.0120[20]	bz 4; eth 4; EtOH 4 1.4230[20]
8907	Pentanoic acid, 4-oxo-, ethyl ester Ethyl levulinate	$C_7H_{12}O_3$ 144.17	539-88-8 	3775 205.8	4-03-00-01562 1.0111[20]	H_2O 4; EtOH 4 1.4229[20]
8908	Pentanoic acid, 4-oxo-, methyl ester	$C_6H_{10}O_3$ 130.14	624-45-3 	 196	4-03-00-01562 1.0511[20]	H_2O 2; EtOH 3; eth 5; ace 3 1.4233[20]
8909	Pentanoic acid, 4-oxo-, 1-methylethyl ester Levulinic acid, isopropyl ester	$C_8H_{14}O_3$ 158.20	21884-26-4 	 209.3	4-03-00-01563 0.9842[20]	ace 4; bz 4; eth 4; EtOH 4 1.4420[20]
8910	Pentanoic acid, 4-oxo-, 2-methylpropyl ester	$C_9H_{16}O_3$ 172.22	3757-32-2 	 228	4-03-00-01563 0.9669[20]	ace 4; eth 4; EtOH 4 1.4249[20]
8911	Pentanoic acid, 4-oxo-, phenylmethyl ester Levulinic acid, benzyl ester	$C_{12}H_{14}O_3$ 206.24	6939-75-9 	132-4[2]	4-06-00-02481 1.0935[20]	tol 4 1.5090[20]
8912	Pentanoic acid, 4-oxo-, propyl ester Levulinic acid, propyl ester	$C_8H_{14}O_3$ 158.20	645-67-0 	 218	4-03-00-01563 0.9896[20]	H_2O 2; EtOH 3; eth 3; ace 3 1.4258[20]
8913	Pentanoic acid, pentyl ester	$C_{10}H_{20}O_2$ 172.27	2173-56-0 -78.8	 203.7	4-02-00-00872 0.8638[20]	H_2O 2; EtOH 5; eth 5 1.4164[20]
8914	Pentanoic acid, propyl ester	$C_8H_{16}O_2$ 144.21	141-06-0 -70.7	 167.5	4-02-00-00872 0.8699[20]	H_2O 1; EtOH 3; eth 3; chl 3 1.4065[20]
8915	1-Pentanol Amyl alcohol	$C_5H_{12}O$ 88.15	71-41-0 -78.9	7074 137.9	4-01-00-01640 0.8144[20]	H_2O 1; EtOH 5; eth 5; ace 3 1.4101[20]
8916	2-Pentanol sec-Amyl alcohol	$C_5H_{12}O$ 88.15	6032-29-7 -73	7075 119.3	3-01-00-01609 0.8094[20]	H_2O 4; EtOH 3; eth 3; ctc 3 1.4053[20]
8917	3-Pentanol Diethyl carbinol	$C_5H_{12}O$ 88.15	584-02-1 -69	7076 116.2	4-01-00-01662 0.8203[20]	H_2O 2; EtOH 3; eth 3; ace 3 1.4104[20]
8918	2-Pentanol, acetate, (R)- sec-Amyl acetate (R)	$C_7H_{14}O_2$ 130.19	54638-10-7 	 142	0-02-00-00132 0.8803[18]	eth 4; EtOH 4 1.4012[20]
8919	3-Pentanol, acetate	$C_7H_{14}O_2$ 130.19	620-11-1 	 133	4-02-00-00155 0.8712[20]	eth 4; EtOH 4 1.4005[20]
8920	1-Pentanol, 5-amino-	$C_5H_{13}NO$ 103.16	2508-29-4 38.5	 221.5	4-04-00-01750 0.9488[17]	H_2O 5; EtOH 5; ace 5 1.4618[17]
8921	3-Pentanol, 2-amino-	$C_5H_{13}NO$ 103.16	116836-16-9 	 171	1-04-00-00442 0.9289[23]	 1.4458
8922	1-Pentanol, 2-amino-4-methyl-, (±)-	$C_6H_{15}NO$ 117.19	16369-17-8 	 199	2-04-00-00748 0.9173[13]	H_2O 4; EtOH 4
8923	1-Pentanol, 5-chloro-	$C_5H_{11}ClO$ 122.59	5259-98-3 	112[12]	4-01-00-01650 	eth 4; EtOH 4 1.4518[20]
8924	2-Pentanol, 1-chloro-	$C_5H_{11}ClO$ 122.59	17658-32-1 	160; 60[14]	3-01-00-01613 1.035[20]	eth 4; EtOH 4 1.4404[25]

No.	Name Synonym	Mol. Form. Mol. Wt.	CAS RN mp/°C	Merck No. bp/°C	Beil. Ref. den/g cm^{-3}	Solubility n_D
8925	3-Pentanol, 1-chloro- 1-Chloro-3-hydroxypentane	$C_5H_{11}ClO$ 122.59	32541-33-6 	 173	4-01-00-01665 1.0327[25]	eth 4; EtOH 4 1.448[25]
8926	1-Pentanol, 2-(diethylamino)-4-methyl-, 4-aminobenzoate (ester), monomethanesulfonate (salt) Panthesin	$C_{18}H_{32}N_2O_5S$ 388.53	135-44-4 158	5332 	4-14-00-01143 	H_2O 4; EtOH 4
8927	1-Pentanol, 2,4-dimethyl-, (±)- 1-Pentanol, 2,4-dimethyl (DL)	$C_7H_{16}O$ 116.20	111768-02-6 	 159	4-01-00-01753 0.793[20]	eth 4; EtOH 4 1.427[20]
8928	2-Pentanol, 2,4-dimethyl- Isobutyldimethylcarbinol	$C_7H_{16}O$ 116.20	625-06-9 <-20	 133.1	4-01-00-01753 0.8103[20]	H_2O 1; EtOH 3; eth 3; ctc 3 1.4172[20]
8929	3-Pentanol, 2,2-dimethyl- 2,2-Dimethyl-3-pentanol	$C_7H_{16}O$ 116.20	3970-62-5 -5	 135	3-01-00-01697 0.8253[20]	H_2O 1; EtOH 3; eth 3 1.4223[20]
8930	3-Pentanol, 2,3-dimethyl- 2,3-Dimethyl-3-pentanol	$C_7H_{16}O$ 116.20	595-41-5 <-30	 139.7	4-01-00-01752 0.8332[20]	H_2O 1; EtOH 3; eth 3; bz 2 1.4287[20]
8931	3-Pentanol, 2,4-dimethyl-	$C_7H_{16}O$ 116.20	600-36-2 <-70	 138.7	4-01-00-01754 0.8288[20]	H_2O 2; EtOH 3; eth 3 1.4250[20]
8932	3-Pentanol, 3-ethyl-	$C_7H_{16}O$ 116.20	597-49-9 -12.5	 142	4-01-00-01750 0.8407[22]	H_2O 1; EtOH 3; eth 3 1.4294[20]
8933	2-Pentanol, 3-ethyl-2-methyl-	$C_8H_{18}O$ 130.23	19780-63-3 	 156	4-01-00-01794 0.8382[20]	bz 4; eth 4; EtOH 4; lig 4 1.4325[20]
8934	3-Pentanol, 3-ethyl-2-methyl- Diethylisopropylcarbinol	$C_8H_{18}O$ 130.23	597-05-7 	160; 56[48]	4-01-00-01794 0.8280[20]	H_2O 1; EtOH 3; eth 3 1.4372[10]
8935	1-Pentanol, 5-fluoro-	$C_5H_{11}FO$ 106.14	592-80-3 	70-1[11]	4-01-00-01647 	eth 4; EtOH 4 1.4057[25]
8936	2-Pentanol, 1,1'-iminobis[2-methyl- Dipentylamine, 2,2'-dimethyl-2,2'-dihydroxy	$C_{12}H_{27}NO_2$ 217.35	85733-97-7 	165-70[15]	 0.9264[20]	ace 4; eth 4; EtOH 4 1.4584[20]
8937	1-Pentanol, 2-methyl- 2-Methyl-1-pentanol	$C_6H_{14}O$ 102.18	105-30-6 	 149	3-01-00-01665 0.8263[20]	EtOH 3; eth 3; ace 3; ctc 3 1.4182[20]
8938	1-Pentanol, 3-methyl-, (±)- 1-Pentanol, 3-methyl (DL)	$C_6H_{14}O$ 102.18	20281-83-8 	 153	4-01-00-01722 0.8242[20]	H_2O 1; EtOH 3; eth 3 1.4112[23]
8939	1-Pentanol, 4-methyl- Isohexyl alcohol	$C_6H_{14}O$ 102.18	626-89-1 	 151.9	4-01-00-01721 0.8131[20]	H_2O 1; EtOH 3; eth 3 1.4134[25]
8940	2-Pentanol, 2-methyl- 2-Methyl-2-pontanol	$C_6H_{14}O$ 102.18	590-36-3 -103	 121.1	4-01-00-01714 0.8350[16]	H_2O 2; EtOH 3; eth 3 1.4100[20]
8941	2-Pentanol, 3-methyl- 3-Methyl-2-pentanol	$C_6H_{14}O$ 102.18	565-60-6 	 134.3	3-01-00-01672 0.8307[20]	H_2O 2; EtOH 3; oth 3 1.4182[20]
8942	2-Pentanol, 4-methyl- 4-Methyl-2-pentanol	$C_6H_{14}O$ 102.18	108-11-2 -90	 131.6	4-01-00-01717 0.8075[20]	H_2O 2; EtOH 3; eth 3; ctc 2 1.4100[20]
8943	3-Pentanol, 2-methyl- 2-Methyl-3-pentanol	$C_6H_{14}O$ 102.18	565-67-3 	 126.5	4-01-00-01716 0.8243[20]	H_2O 2; EtOH 5; eth 5 1.4175[20]
8944	3-Pentanol, 3-methyl-	$C_6H_{14}O$ 102.18	77-74-7 -23.6	 122.4	4-01-00-01723 0.8286[20]	H_2O 2; EtOH 5; eth 5; ctc 2 1.4186[20]
8945	1-Pentanol, 2-methyl-, acetate Acetic acid, (2-methylpentyl) ester	$C_8H_{16}O_2$ 144.21	7789-99-3 	 163	4-02-00-00160 0.8691[25]	eth 4; EtOH 4
8946	2-Pentanol, 2-methyl-, acetate 1,1-Dimethylbutyl acetate	$C_8H_{16}O_2$ 144.21	34859-98-8 	153; 34[10]	3-02-00-00257 0.8798[18]	eth 4; EtOH 4 1.4068[10]
8947	2-Pentanol, 4-methyl-, acetate 1,3-Dimethylbutyl acetate	$C_8H_{16}O_2$ 144.21	108-84-9 	 147.5	4-02-00-00161 0.8805[25]	eth 4; EtOH 4 1.3980[20]
8948	3-Pentanol, 2-methyl-, acetate Acetic acid, (2-methyl-3-pentyl) ester	$C_8H_{16}O_2$ 144.21	35897-16-6 	 149	4-02-00-00160 0.8688[20]	eth 4; EtOH 4
8949	3-Pentanol, 3-methyl-, acetate Acetic acid, (3-methyl-3-pentyl) ester	$C_8H_{16}O_2$ 144.21	10250-47-2 	 148	4-02-00-00161 0.8818[20]	eth 4; EtOH 4 1.4169[18]
8950	1-Pentanol, 2-nitro- 2-Nitro-1-pentanol	$C_5H_{11}NO_3$ 133.15	2899-90-3 	133[28]; 117[10]	3-01-00-01606 1.0818[25]	 1.4405[25]
8951	2-Pentanol, 1-nitro-	$C_5H_{11}NO_3$ 133.15	2224-37-5 	87-8[1]	4-01-00-01661 1.0968[20]	 1.4493[20]
8952	2-Pentanol, 3-nitro- 3-Nitro-2-pentanol	$C_5H_{11}NO_3$ 133.15	5447-99-4 	60[0.5]	3-01-00-01614 1.075[25]	 1.4430[20]
8953	1-Pentanol, 2,2,3,3,4,4,5,5-octafluoro- 2,2,3,3,4,4,5,5-Octafluoro-1-pentanol	$C_5H_4F_8O$ 232.07	355-80-6 	6667 140.5	4-01-00-01648 1.6647[20]	 1.3178[20]
8954	2-Pentanol, 2,4,4-trimethyl- 2,4,4-Trimethyl-2-pentanol	$C_8H_{18}O$ 130.23	690-37-9 -20	 147.5	4-01-00-01796 0.8225[20]	H_2O 1; EtOH 2; eth 3 1.4284[20]
8955	3-Pentanol, 2,2,4-trimethyl- Isopropyl-tert-butylcarbinol	$C_8H_{18}O$ 130.23	5162-48-1 -13	 150.5	4-01-00-01795 0.8297[20]	 1.4288[20]
8956	3-Pentanol, 2,3,4-trimethyl-	$C_8H_{18}O$ 130.23	3054-92-0 	 156.5	4-01-00-01798 0.8492[20]	 1.4353[20]
8957	2-Pentanone Methyl propyl ketone	$C_5H_{10}O$ 86.13	107-87-9 -76.9	6032 102.2	4-01-00-03271 0.809[20]	H_2O 2; EtOH 5; eth 5; ctc 2 1.3895[20]
8958	3-Pentanone Diethyl ketone	$C_5H_{10}O$ 86.13	96-22-0 -39	3111 101.9	4-01-00-03279 0.816[19]	H_2O 4; EtOH 5; eth 5; ctc 3 1.3905[25]
8959	2-Pentanone, 4-amino-4-methyl- Diacetonamine	$C_6H_{13}NO$ 115.18	625-04-7 	2942 25[0.14]	3-04-00-00894 	H_2O 3; EtOH 5; eth 5
8960	1-Pentanone, 1-(4-aminophenyl)-	$C_{11}H_{15}NO$ 177.25	38237-74-0 74.5	160-3[3]	4-14-00-00178 	H_2O 1; EtOH 3; eth 3
8961	2-Pentanone, 5-chloro-	C_5H_9ClO 120.58	5891-21-4 	76[34]	4-01-00-03277 1.0523[20]	eth 3; ace 3; ctc 2 1.4375[20]
8962	3-Pentanone, 1-chloro-	C_5H_9ClO 120.58	32830-97-0 	68[20]	4-01-00-03284 	eth 4; EtOH 4 1.4361[20]
8963	2-Pentanone, 3,5-dichloro- 3,5-Dichloro-2-pentanone	$C_5H_8Cl_2O$ 155.02	58371-98-5 	99[15]; 58[3]	4-01-00-03277 1.239[15]	 1.4632[20]

No.	Name / Synonym	Mol. Form. / Mol. Wt.	CAS RN / mp/°C	Merck No. / bp/°C	Beil. Ref. / den/g cm^{-3}	Solubility / n_D
8964	2-Pentanone, 5-(diethylamino)-	$C_9H_{19}NO$ 157.26	105-14-6	83-5[15]	4-04-00-01935 0.861[25]	1.4350[20]
8965	2-Pentanone, 4,4-dimethyl-	$C_7H_{14}O$ 114.19	590-50-1 -64	126	4-01-00-03331 0.809[25]	1.4036[20]
8966	3-Pentanone, 2,2-dimethyl- 2,2-Dimethyl-3-pentanone	$C_7H_{14}O$ 114.19	564-04-5 -45	125.6	4-01-00-03331 0.8125[20]	H_2O 2; EtOH 3; eth 3; ace 3 1.4065[20]
8967	3-Pentanone, 2,4-dimethyl- Diisopropyl ketone	$C_7H_{14}O$ 114.19	565-80-0 -69	125.4	4-01-00-03334 0.8108[20]	H_2O 1; EtOH 5; eth 5; bz 3 1.3999[20]
8968	3-Pentanone, 2,4-dimethyl-, oxime Diisopropyl ketone oxime	$C_7H_{15}NO$ 129.20	1113-74-2 34	181; 91.5[21]	4-01-00-03335 0.9022[17]	
8969	3-Pentanone, 1,5-diphenyl- Diphenethyl ketone	$C_{17}H_{18}O$ 238.33	5396-91-8 13.5	352	4-07-00-01462 1.0356[0]	
8970	2-Pentanone, 3-ethyl-4-methyl-	$C_8H_{16}O$ 128.21	71172-57-1	154.5	3-01-00-02883 0.812[20]	H_2O 3; EtOH 5; eth 5; bz 5 1.4105[20]
8971	2-Pentanone, 1-hydroxy- 1-Pentanol-2-one	$C_5H_{10}O_2$ 102.13	64502-89-2	152	4-01-00-04004 0.9860[20]	H_2O 4; eth 4; EtOH 4 1.4234[12]
8972	2-Pentanone, 3-hydroxy-	$C_5H_{10}O_2$ 102.13	3142-66-3	147.5	3-01-00-03220 0.9500[20]	H_2O 4; ace 4; bz 4; eth 4 1.4350[10]
8973	2-Pentanone, 4-hydroxy-	$C_5H_{10}O_2$ 102.13	4161-60-8	177	4-01-00-04005 1.0071[20]	H_2O 3; EtOH 3; eth 3 1.4415[18]
8974	2-Pentanone, 5-hydroxy-	$C_5H_{10}O_2$ 102.13	1071-73-4	209; 117[33]	4-01-00-04006 1.0071[20]	H_2O 5; EtOH 3; eth 3 1.4390[20]
8975	3-Pentanone, 2-hydroxy- 2-Pentanol-3-one	$C_5H_{10}O_2$ 102.13	5704-20-1	152.5	4-01-00-04008 0.9742[20]	eth 4; EtOH 4 1.4128[20]
8976	2-Pentanone, 4-hydroxy-4-methyl- Diacetone alcohol	$C_6H_{12}O_2$ 116.16	123-42-2 -44	2944 167.9	4-01-00-04023 0.9387[20]	H_2O 5; EtOH 5; eth 5; chl 3 1.4213[20]
8977	2-Pentanone, 4-imino-	C_5H_9NO 99.13	870-74-6 43	209	4-01-00-03678 0.9427[0]	H_2O 4; eth 4
8978	2-Pentanone, 4-methoxy-4-methyl- Pentoxone	$C_7H_{14}O_2$ 130.19	107-70-0	160	4-01-00-04024 0.8980[25]	1.418[20]
8979	1-Pentanone, 1-(3-methoxyphenyl)-	$C_{12}H_{16}O_2$ 192.26	20359-55-1	196[30]	4-08-00-00533 1.012[20]	1.5242[31]
8980	2-Pentanone, 3-methyl-, (±)- 2-Pentanone, 3-methyl (DL)	$C_6H_{12}O$ 100.16	55156-16-6	118	4-01-00-03309 0.8130[20]	H_2O 2; EtOH 5; eth 5; chl 3 1.4002[20]
8981	2-Pentanone, 4-methyl- Isobutyl methyl ketone	$C_6H_{12}O$ 100.16	108-10-1 -84	5095 116.5	4-01-00-03305 0.7978[20]	H_2O 2; EtOH 5; eth 5; ace 5 1.3962[20]
8982	3-Pentanone, 2-methyl- Ethyl isopropyl ketone	$C_6H_{12}O$ 100.16	565-69-5	113.5	4-01-00-03304 0.814[18]	H_2O 2; EtOH 4; eth 5; ace 5 1.3975[20]
8983	1-Pentanone, 4-methyl-1-phenyl-	$C_{12}H_{16}O$ 176.26	2050-07-9 -2	255.5	3-07-00-01158 0.9623[15]	ace 4; bz 4; eth 4; EtOH 4 1.533[20]
8984	2-Pentanone, 3-methyl-1-phenyl- Benzyl sec-butyl ketone	$C_{12}H_{16}O$ 176.26	27993-42-6	127-8[17]	4-07-00-00758 0.964[20]	1.5019[20]
8985	2-Pentanone, oxime Methyl propyl ketone oxime	$C_5H_{11}NO$ 101.15	623-40-5	168	4-01-00-03274 0.9095[20]	H_2O 4; eth 4; EtOH 4 1.4450[20]
8986	1-Pentanone, 1-phenyl-	$C_{11}H_{14}O$ 162.23	1009-14-9 -9.4	245	4-07-00-00733 0.986[20]	H_2O 1; EtOH 4; eth 4; ctc 2 1.5158[20]
8987	2-Pentanone, 3-(phenylmethylene)- Methyl α-ethylstyryl ketone	$C_{12}H_{14}O$ 174.24	3437-89-6	136-8[12]	3-07-00-01461 1.0005[22]	1.5650[22]
8988	3-Pentanone, 2,2,4,4-tetramethyl-	$C_9H_{18}O$ 142.24	815-24-7 -25.2	152	4-01-00-03365 0.8240[18]	H_2O 1; EtOH 3; eth 3; ace 3 1.4194[20]
8989	3-Pentanone, 2,2,4-trimethyl- tert-Butyl isopropyl ketone	$C_8H_{16}O$ 128.21	5857-36-3	135.1	4-01-00-03351 0.8065[20]	H_2O 1; eth 3; ace 3 1.4060
8990	Pentanoyl chloride Valeroyl chloride	C_5H_9ClO 120.58	638-29-9 -110	109	4-02-00-00874 1.0155[15]	1.4200[20]
8991	Pentanoyl chloride, 2-chloro-	$C_5H_8Cl_2O$ 155.02	61589-68-2	156	4-02-00-00879 1.1765[20]	eth 4 1.4465[20]
8992	Pentanoyl chloride, 5-chloro-	$C_5H_8Cl_2O$ 155.02	1575-61-7	83[12]	4-02-00-00881 1.210[18]	eth 4 1.4639[20]
8993	Pentanoyl chloride, 2-methyl-, (±)- Valeryl chloride, 2-methyl (DL)	$C_6H_{11}ClO$ 134.61	116908-84-0	141	3-02-00-00741 0.9781[20]	eth 4 1.4330[27]
8994	Pentanoyl chloride, 3-methyl-, (±)- Valeryl chloride, 3-methyl (DL)	$C_6H_{11}ClO$ 134.61	116908-85-1	141	4-02-00-00949 0.9781[20]	eth 4
8995	2,5,8,11,14-Pentaoxapentadecane Tetraethylene glycol dimethyl ether	$C_{10}H_{22}O_5$ 222.28	143-24-8	9141 275.3	4-01-00-02404 1.0114[20]	H_2O 5; EtOH 3; eth 3; ctc 3
8996	Pentaphene 2,3:6,7-Dibenzphenanthrene	$C_{22}H_{14}$ 278.35	222-93-5 257	2992	4-05-00-02721	H_2O 1; EtOH 2; eth 2; bz 3
8997	Pentasiloxane, dodecamethyl- Dodecamethylpentasiloxane	$C_{12}H_{36}O_4Si_5$ 384.84	141-63-9 -80	3401 232; 105[12]	4-04-00-04124 0.8755[20]	ctc 3; CS_2 3 1.3925[20]
8998	1,2,3,5,6-Pentathiepane Lenthionine	$C_2H_4S_5$ 188.38	292-46-6 60.5	5321	5-19-12-00251	
8999	Pentatriacontane	$C_{35}H_{72}$ 492.96	630-07-9 75	490	4-01-00-00598 0.8157[20]	H_2O 1; eth 2; ace 3 1.4568[20]
9000	18-Pentatriacontanone	$C_{35}H_{70}O$ 506.94	504-53-0 89.0	270[0.1]	4-01-00-03413 0.793[95]	H_2O 1; EtOH 2; eth 2; ace 2
9001	4-Pentenal Pent-4-en-1-al	C_5H_8O 84.12	2100-17-6	99	4-01-00-03459 0.852[20]	H_2O 1; eth 3; ace 3 1.4191[20]
9002	2-Pentenal, 2-methyl-	$C_6H_{10}O$ 98.14	623-36-9	136.5	4-01-00-03471 0.8581[20]	H_2O 3; EtOH 3; eth 3; bz 3 1.4488[20]
9003	4-Pentenamide, 2,2-diethyl- Novonal	$C_9H_{17}NO$ 155.24	512-48-1 75.5	6643	3-02-00-01351	eth 4; EtOH 4

No.	Name / Synonym	Mol. Form. / Mol. Wt.	CAS RN / mp/°C	Merck No. / bp/°C	Beil. Ref. / den/g cm^{-3}	Solubility / n_D
9004	1-Pentene / α-Amylene	C_5H_{10} / 70.13	109-67-1 / -165.2	7079 / 29.9	4-01-00-00808 / 0.6405[20]	H_2O 1; EtOH 5; eth 5; bz 3 / 1.3715[20]
9005	2-Pentene, (E)- / trans-2-Pentene	C_5H_{10} / 70.13	646-04-8 / -140.2	7080 / 36.3	4-01-00-00814 / 0.6431[25]	H_2O 1; EtOH 5; eth 5; bz 3 / 1.3793[20]
9006	2-Pentene, (Z)- / cis-2-Pentene	C_5H_{10} / 70.13	627-20-3 / -151.4	7080 / 36.9	4-01-00-00814 / 0.6556[20]	H_2O 1; EtOH 5; eth 5; bz 3 / 1.3830[20]
9007	1-Pentene, 1-bromo- / 1-Bromo-1-pentene	C_5H_9Br / 149.03	60468-23-7 /	/ 119	3-01-00-00775 / 1.2606[20]	bz 4; eth 4; chl 4 / 1.4572[20]
9008	1-Pentene, 2-bromo- / 2-Bromo-1-pentene	C_5H_9Br / 149.03	31844-95-8 /	/ 107.5	3-01-00-00775 / 1.228[20]	bz 4; eth 4; lig 4 / 1.4535[20]
9009	1-Pentene, 3-bromo- / 1-Ethylallyl bromide	C_5H_9Br / 149.03	53045-71-9 /	/ 30.5[30]	2-01-00-00183 / 1.2417[25]	ace 4; bz 4; chl 4 / 1.4626[25]
9010	1-Pentene, 5-bromo-	C_5H_9Br / 149.03	1119-51-3 /	/ 125.5	3-01-00-00775 / 1.2581[20]	/ 1.4640[20]
9011	2-Pentene, 1-bromo-	C_5H_9Br / 149.03	20599-27-3 /	/ 123.5	3-01-00-00783 / 1.2545[20]	ace 4; bz 4; chl 4 / 1.4731[20]
9012	2-Pentene, 2-bromo- / 2-Bromo-2-pentene	C_5H_9Br / 149.03	80204-19-9 /	/ 111	3-01-00-00784 / 1.275[20]	ace 4; bz 4; chl 4 / 1.4580[20]
9013	2-Pentene, 3-bromo- / 3-Bromo-2-pentene	C_5H_9Br / 149.03	21964-23-8 /	/ 116	3-01-00-00784 / 1.271[20]	ace 4; bz 4; chl 4 / 1.4628[20]
9014	2-Pentene, 4-bromo-	C_5H_9Br / 149.03	1809-26-3 /	/ 117 dec; 22[9]	4-01-00-00816 / 1.2312[21]	ace 4; bz 4; chl 4 / 1.4752[21]
9015	2-Pentene, 5-bromo- / 3-Pentenyl bromide	C_5H_9Br / 149.03	51952-42-2 /	/ 121.7[621]	4-01-00-00816 / 1.2715[20]	ace 4; bz 4; chl 4 / 1.4695[20]
9016	1-Pentene, 2-chloro- / 2-Chloro-1-pentene	C_5H_9Cl / 104.58	42131-85-1 /	/ 96	3-01-00-00774 / 0.872[5]	eth 4; EtOH 4 /
9017	1-Pentene, 3-chloro- / 1-Ethylallyl chloride	C_5H_9Cl / 104.58	24356-00-1 /	/ 93	3-01-00-00774 / 0.8978[20]	ace 4; eth 4; EtOH 4 / 1.4254[20]
9018	1-Pentene, 4-chloro- / 4-Chloro-1-pentene	C_5H_9Cl / 104.58	10524-08-0 /	/ 97	2-01-00-00183 / 0.9341[15]	H_2O 1; eth 4; chl 4 / 1.417[15]
9019	1-Pentene, 5-chloro-	C_5H_9Cl / 104.58	928-50-7 /	/ 105	4-01-00-00810 / 0.9125[20]	ace 4; eth 4 / 1.4297[20]
9020	2-Pentene, 1-chloro- / γ-Ethylallyl chloride	C_5H_9Cl / 104.58	10071-60-0 /	/ 108	3-01-00-00782 / 0.908[22]	ace 4; eth 4; EtOH 4; chl 4 / 1.4352[22]
9021	2-Pentene, 2-chloro- / 2-Chloro-2-pentene	C_5H_9Cl / 104.58	67747-70-0 /	/ 96	3-01-00-00782 / 0.9067[20]	ace 4; eth 4 / 1.4261[20]
9022	2-Pentene, 3-chloro- / 3-Chloro-2-pentene	C_5H_9Cl / 104.58	34238-52-3 /	/ 91	3-01-00-00782 / 0.9125[20]	ace 4; eth 4 / 1.423[24]
9023	2-Pentene, 4-chloro-	C_5H_9Cl / 104.58	1458-99-7 /	/ 103; 47[25]	4-01-00-00816 / 0.8988[20]	ace 4; eth 4; chl 4 / 1.4322[20]
9024	2-Pentene, 5-chloro-	C_5H_9Cl / 104.58	16435-50-0 /	/ 108	4-01-00-00816 / 0.9043[20]	ace 4; bz 4; chl 4 / 1.4310[20]
9025	1-Pentene, 3-chloro-2-methyl-	$C_6H_{11}Cl$ / 118.61	4104-01-2 /	/ 122.5	3-01-00-00809 /	bz 4; eth 4; chl 4 / 1.4422[20]
9026	2-Pentene, 5-chloro-2-methyl- / 5-Chloro-2-methyl-2-pentene	$C_6H_{11}Cl$ / 118.61	7712-60-9 /	/ 133	4-01-00-00843 / 0.9135[20]	ace 4; chl 4 /
9027	1-Pentene, 1,1,2,3,3,4,4,5,5,5-decafluoro-	C_5F_{10} / 250.04	376-87-4 /	/ 30	4-01-00-00810 /	chl 4 / 1.2571[25]
9028	2-Pentene, 2,5-dichloro- / 1,4-Dichloro-3-pentene	$C_5H_8Cl_2$ / 139.02	20177-02-0 /	/ 40-1[8]	4-01-00-00816 / 1.1182[15]	chl 4 /
9029	1-Pentene, 2,3-dimethyl- / 2,3-Dimethyl-1-pentene	C_7H_{14} / 98.19	3404-72-6 / -134.3	/ 84.3	3-01-00-00832 / 0.7051[20]	H_2O 1; EtOH 5; eth 5; dil sulf 4 / 1.4033[20]
9030	1-Pentene, 2,4-dimethyl- / 2,4-Dimethyl-1-pentene	C_7H_{14} / 98.19	2213-32-3 / -124.1	/ 81.6	4-01-00-00871 / 0.6943[20]	H_2O 1; EtOH 5; eth 5; bz 3 / 1.3986[20]
9031	1-Pentene, 3,3-dimethyl- / 3,3-Dimethyl-1-pentene	C_7H_{14} / 98.19	3404-73-7 / -134.3	/ 77.5	4-01-00-00873 / 0.6974[20]	H_2O 1; EtOH 5; eth 5; bz 3 / 1.3984[20]
9032	1-Pentene, 3,4-dimethyl- / 3,4-Dimethyl-1-pentene	C_7H_{14} / 98.19	7385-78-6 /	/ 80.8	4-01-00-00871 / 0.6934[25]	/ 1.3992[20]
9033	1-Pentene, 4,4-dimethyl-	C_7H_{14} / 98.19	762-62-9 / -136.6	/ 72.5	4-01-00-00869 / 0.6827[20]	H_2O 1; EtOH 5; eth 5; bz 3 / 1.3818[20]
9034	2-Pentene, 2,3-dimethyl- / Ethyltrimethylethylene	C_7H_{14} / 98.19	10574-37-5 / -118.3	/ 97.5	4-01-00-00870 / 0.7277[20]	H_2O 1; EtOH 3; eth 3; bz 3 / 1.4208[20]
9035	2-Pentene, 2,4-dimethyl- / 2,4-Dimethyl-2-pentene	C_7H_{14} / 98.19	625-65-0 / -127.7	/ 83.4	4-01-00-00872 / 0.6954[20]	H_2O 1; EtOH 3; eth 3; bz 3 / 1.4040[20]
9036	2-Pentene, 3,4-dimethyl- / 3,4-Dimethyl-2-pentene	C_7H_{14} / 98.19	24910-63-2 / -124.2	/ 89	3-01-00-00833 / 0.7126[20]	H_2O 1; EtOH 3; eth 3; bz 3 / 1.4070[20]
9037	2-Pentene, 3,4-dimethyl-, (E)-	C_7H_{14} / 98.19	4914-92-5 / -124.2	/ 91.5	4-01-00-00871 / 0.7124[25]	/ 1.4128[20]
9038	2-Pentene, 3,4-dimethyl-, (Z)-	C_7H_{14} / 98.19	4914-91-4 / -113.4	/ 89.3	4-01-00-00871 / 0.7092[25]	/ 1.4104[20]
9039	2-Pentene, 4,4-dimethyl-, (E)-	C_7H_{14} / 98.19	690-08-4 / -115.2	/ 76.7	4-01-00-00868 / 0.6889[20]	H_2O 1; EtOH 3; eth 3; bz 3 / 1.3982[20]
9040	2-Pentene, 4,4-dimethyl-, (Z)-	C_7H_{14} / 98.19	762-63-0 / -135.4	/ 80.4	4-01-00-00868 / 0.6951[25]	/ 1.4026[20]
9041	2-Pentenedioic acid, diethyl ester / Diethyl glutaconate	$C_9H_{14}O_4$ / 186.21	2049-67-4 /	/ 237	/ 1.0496[20]	eth 4; EtOH 4 / 1.4411[20]
9042	2-Pentenedioic acid, diethyl ester, (E)- / trans-2-Pentenedioic acid diethyl ester	$C_9H_{14}O_4$ / 186.21	73178-43-5 /	/ 237; 126[14]	4-02-00-02227 / 1.0496[20]	eth 4; EtOH 4 / 1.4411[20]

No.	Name Synonym	Mol. Form. Mol. Wt.	CAS RN mp/°C	Merck No. bp/°C	Beil. Ref. den/g cm^{-3}	Solubility n_D
9043	1-Pentene, 2-ethyl- 2-Ethyl-1-pentene	C_7H_{14} 98.19	3404-71-5	94	4-01-00-00867 0.7079[20]	bz 4; eth 4; EtOH 4 1.405[20]
9044	1-Pentene, 3-ethyl- 3-Ethyl-1-pentene	C_7H_{14} 98.19	4038-04-4 -127.5	84.1	4-01-00-00867 0.6917[25]	1.3982[20]
9045	2-Pentene, 3-ethyl- 3-Ethyl-2-pentene	C_7H_{14} 98.19	816-79-5	96	4-01-00-00868 0.7204[20]	H_2O 1; EtOH 3; eth 3; bz 3 1.4148[20]
9046	1-Pentene, 3-ethyl-2-methyl-	C_8H_{16} 112.22	19780-66-6 -112.9	109.5	4-01-00-00891 0.7262[20]	1.4140[20]
9047	1-Pentene, 3-ethyl-3-methyl- 3-Ethyl-3-methyl-1-pentene	C_8H_{16} 112.22	6196-60-7	112	4-01-00-00891 0.7264[20]	1.418[20]
9048	1-Pentene, 3-ethyl-4-methyl- 1-Pentene, 4-methyl-3-ethyl	C_8H_{16} 112.22	61847-80-1	107.5	4-01-00-00890 0.7158[25]	1.4097[20]
9049	2-Pentene, 3-ethyl-2-methyl- 3-Ethyl-2-methyl-2-pentene	C_8H_{16} 112.22	19780-67-7	115	4-01-00-00890 0.735[25]	1.4247[20]
9050	2-Pentene, 3-ethyl-4-methyl-, (E)- 2-Pentene, 4-methyl-3-ethyl (trans)	C_8H_{16} 112.22	42067-49-2	114.3	4-01-00-00890 0.7308[25]	1.4210[20]
9051	2-Pentene, 3-ethyl-4-methyl-, (Z)- 2-Pentene, 4-methyl-3-ethyl (cis)	C_8H_{16} 112.22	42067-48-1	116	4-01-00-00890 0.735[25]	1.424[20]
9052	1-Pentene, 2-methyl- 2-Methyl-1-pentene	C_6H_{12} 84.16	763-29-1 -135.7	62.1	4-01-00-00841 0.6799[20]	H_2O 1; EtOH 3; bz 3; ctc 2 1.3920[20]
9053	1-Pentene, 3-methyl- sec-Butylethene	C_6H_{12} 84.16	760-20-3 -153	54.2	3-01-00-00812 0.6675[20]	H_2O 1; EtOH 3; bz 3; chl 3 1.3841[20]
9054	1-Pentene, 4-methyl- Isobutylethene	C_6H_{12} 84.16	691-37-2 -153.6	53.9	4-01-00-00846 0.6642[20]	H_2O 1; EtOH 3; bz 3; chl 3 1.3828[20]
9055	2-Pentene, 2-methyl- 2-Methyl-2-pentene	C_6H_{12} 84.16	625-27-4 -135	67.3	4-01-00-00842 0.6863[20]	H_2O 1; EtOH 3; bz 3; ctc 3 1.4004[20]
9056	2-Pentene, 3-methyl-, (E)-	C_6H_{12} 84.16	616-12-6 -138.5	70.4	4-01-00-00848 0.6930[25]	H_2O 1; EtOH 3; bz 3; ctc 3 1.4045[20]
9057	2-Pentene, 3-methyl-, (Z)- (cis)-3-Methyl-2-pentene	C_6H_{12} 84.16	922-62-3 -134.8	67.7	4-01-00-00848 0.6886[25]	H_2O 1; EtOH 3; bz 3; chl 3 1.4016[20]
9058	2-Pentene, 4-methyl-, (E)- 4-Methyl-trans-2-pentene	C_6H_{12} 84.16	674-76-0 -140.8	58.6	4-01-00-00844 0.6686[20]	H_2O 1; EtOH 3; bz 3; ctc 2 1.3889[20]
9059	2-Pentene, 4-methyl-, (Z)- cis-4-Methyl-2-pentene	C_6H_{12} 84.16	691-38-3 -134.8	56.3	4-01-00-00844 0.6690[20]	H_2O 1; EtOH 3; bz 3; chl 3 1.3800[20]
9060	1-Pentene, 3-methylene-	C_6H_{10} 82.15	3404-63-5	75	4-01-00-01022 0.712[25]	1.445[20]
9061	3-Pentenenitrile, (E) trans-3-Pentenenitrile	C_5H_7N 81.12	16545-78-1	142.6		
9062	4-Pentenenitrile	C_5H_7N 81.12	592-51-8	140	4-02-00-01543 0.8239[24]	H_2O 1; EtOH 5; eth 5 1.4213[14]
9063	3-Pentenenitrile, 2-hydroxy-, (Z)-	C_5H_7NO 97.12	116908-78-2	139[70]	4-03-00-01001 0.9675[15]	bz 4; eth 4; EtOH 4; chl 4 1.4460[21]
9064	Pentene, 2,4,4-trimethyl- Diisobutylene	C_8H_{16} 112.22	25167-70-8	101	4-01-00-00892	
9065	1-Pentene, 2,3,3-trimethyl- 2,3,3-Trimethyl-1-pentene	C_8H_{16} 112.22	560-23-6 -69	108.3	4-01-00-00894 0.7308[25]	1.4174[20]
9066	1-Pentene, 2,3,4-trimethyl- 2,3,4-Trimethyl-1-pentene	C_8H_{16} 112.22	565-76-4	108	4-01-00-00894 0.725[25]	1.415[20]
9067	1-Pentene, 2,4,4-trimethyl- 2,2,4-Trimethyl-4-pentene	C_8H_{16} 112.22	107-39-1 -93.5	101.4	4-01-00-00892 0.7150[20]	H_2O 1; eth 3; bz 3; ctc 3 1.4086[20]
9068	1-Pentene, 3,3,4-trimethyl-	C_8H_{16} 112.22	560-22-5	105	3-01-00-00851 0.725[25]	1.4144[20]
9069	1-Pentene, 3,4,4-trimethyl- 3,4,4-Trimethyl-1-pentene	C_8H_{16} 112.22	564-03-4	104	4-01-00-00891 0.715[25]	1.412[20]
9070	2-Pentene, 2,3,4-trimethyl- 2,3,4-Trimethyl-2-pentene	C_8H_{16} 112.22	565-77-5 -113.4	116.5	4-01-00-00894 0.7434[20]	1.4274[20]
9071	2-Pentene, 2,4,4-trimethyl- 2,4,4-Trimethyl-2-pentene	C_8H_{16} 112.22	107-40-4 -106.3	104.9	4-01-00-00891 0.7218[20]	H_2O 1; eth 3; bz 3; ctc 3 1.4160[20]
9072	2-Pentene, 3,4,4-trimethyl- 3,4,4-Trimethyl-2-pentene	C_8H_{16} 112.22	598-96-9	112	4-01-00-00891 0.735[25]	EtOH 4 1.423[20]
9073	4-Pentenoic acid Allylacetic acid	$C_5H_8O_2$ 100.12	591-80-0 -22.5	188.5	4-02-00-01542 0.9809[20]	H_2O 2; EtOH 4; eth 4 1.4281[20]
9074	4-Pentenoic acid, 2-acetyl, ethyl ester	$C_9H_{14}O_3$ 170.21	610-89-9	208	3-03-00-01314 0.9898[20]	EtOH 5; eth 5; bz 5 1.4388[18]
9075	4-Pentenoic acid, 2,2-dimethyl- 2,2-Dimethylpent-4-enoic acid	$C_7H_{12}O_2$ 128.17	16386-93-9	116[30]	4-02-00-01585 0.933[20]	1.4338[20]
9076	2-Pentenoic acid, 2-methyl-, (E)- 2-Pentenoic acid, 2-methyl (trans)	$C_6H_{10}O_2$ 114.14	16957-70-3 24.4	214; 112[12]	4-02-00-01568 0.9751[20]	H_2O 2; eth 3; chl 3; CS_2 3 1.4513[20]
9077	2-Pentenoic acid, 3-methyl-, (Z)- 2-Pentenoic acid, 3-methyl (cis)	$C_6H_{10}O_2$ 114.14	19866-51-4 12	96[5]	4-02-00-01571 0.9830[20]	1.4650
9078	2-Pentenoic acid, 4-methyl- 4,4-Dimethyl-2-butenoic acid	$C_6H_{10}O_2$ 114.14	10321-71-8 35	217	3-02-00-01324 0.9529[21]	ace 4; eth 4; EtOH 4 1.4489[21]
9079	3-Pentenoic acid, 2-methyl-	$C_6H_{10}O_2$ 114.14	37674-63-8	199	3-02-00-01324 0.966[15]	1.4402[25]
9080	2-Pentenoic acid, 4-methyl-, ethyl ester Ethyl 4,4-dimethylcrotonate	$C_8H_{14}O_2$ 142.20	2351-97-5	174; 76-7[25]	3-02-00-01324 0.9048[21]	1.4328[20]
9081	3-Pentenoic acid, 3-methyl-, ethyl ester Ethyl 3-methyl-3-pentenoate	$C_8H_{14}O_2$ 142.20	86623-80-5	61[13]	3-02-00-01328 0.9183[20]	1.4298[20]
9082	3-Pentenoic acid, 4-methyl-, ethyl ester Ethyl 4-methyl-3-pentenoate	$C_8H_{14}O_2$ 142.20	6849-18-9	58[11]	3-02-00-01324 0.9134[17]	1.4329[17]

No.	Name Synonym	Mol. Form. Mol. Wt.	CAS RN	mp/°C	bp/°C	Merck No.	Beil. Ref. den/g cm^{-3}	Solubility n_D
9083	2-Pentenoic acid, 2-methyl-, methyl ester	$C_7H_{12}O_2$ 128.17	10478-12-3		51[11]		4-02-00-01568 0.920[25]	1.4336
9084	2-Pentenoic acid, 3-methyl-, methyl ester, (E)-	$C_7H_{12}O_2$ 128.17	17447-01-7		79[50]		4-02-00-01571 0.9302[20]	1.4446[20]
9085	2-Pentenoic acid, 3-methyl-, methyl ester, (Z)- cis-3-Methyl-2-pentenoic acid methyl ester	$C_7H_{12}O_2$ 128.17	17447-00-6		74[50]		4-02-00-01571 0.9279[20]	1.4420[20]
9086	3-Pentenoic acid, 3-methyl-, methyl ester Methyl 3-methyl-3-pentenoate	$C_7H_{12}O_2$ 128.17	2258-58-4		74[50]		4-02-00-01573 0.9949[20]	1.4306[20]
9087	1-Penten-3-ol	$C_5H_{10}O$ 86.13	616-25-1		115		1-01-00-00227 0.839[20]	H_2O 2; EtOH 5; eth 5 1.4239[20]
9088	4-Penten-1-ol	$C_5H_{10}O$ 86.13	821-09-0		141		4-01-00-02119 0.8457[20]	H_2O 2; eth 3; ctc 2 1.4309[20]
9089	4-Penten-2-ol	$C_5H_{10}O$ 86.13	625-31-0		116		1-01-00-00227 0.8367[20]	H_2O 4; EtOH 5; eth 5 1.4225[20]
9090	2-Penten-1-ol, (E)-	$C_5H_{10}O$ 86.13	1576-96-1		138		4-01-00-02121 0.8471[20]	EtOH 3; eth 3; ace 3 1.4341[20]
9091	2-Penten-1-ol, (Z)-	$C_5H_{10}O$ 86.13	1576-95-0		138		4-01-00-02121 0.8529[20]	EtOH 3; eth 3; ace 3 1.4354[20]
9092	3-Penten-2-ol, (±)-	$C_5H_{10}O$ 86.13	42569-16-4		121.6; 65[70]		4-01-00-02122 0.8328[25]	ace 4; eth 4; EtOH 4 1.4280[20]
9093	2-Penten-1-ol, 5-(2,3-dimethyltricyclo[2.2.1.02,6]hept-3-yl)-2-methyl-, stereoisomer α-Santalol	$C_{15}H_{24}O$ 220.35	115-71-9		8321 301.5		4-06-00-03521 0.9679[20]	H_2O 1; EtOH 3 1.5023[20]
9094	2-Penten-1-ol, 2-methyl-	$C_6H_{12}O$ 100.16	1610-29-3		167.5		4-01-00-02144 0.8501[24]	1.444[24]
9095	2-Penten-1-ol, 4-methyl-	$C_6H_{12}O$ 100.16	5362-55-0		159		4-01-00-02146 0.8489[16]	1.4403[16]
9096	3-Penten-1-ol, 4-methyl- 4-Methyl-3-pentenol	$C_6H_{12}O$ 100.16	763-89-3		157		4-01-00-02145 0.8600[20]	1.4432[21]
9097	3-Penten-2-ol, 2-methyl- 2-Methyl-3-penten-2-ol	$C_6H_{12}O$ 100.16	63468-05-3		112; 65[72]		4-01-00-02145 0.8347[20]	1.4302[20]
9098	3-Penten-2-ol, 3-methyl- 3-Methyl-3-penten-2-ol	$C_6H_{12}O$ 100.16	2747-53-7		140		4-01-00-02149 0.8792[0]	1.4428[17]
9099	3-Penten-2-ol, 4-methyl-	$C_6H_{12}O$ 100.16	4325-82-0		134		4-01-00-02144 0.840[15]	1 937[15]
9100	4-Penten-2-ol, 2-methyl-	$C_6H_{12}O$ 100.16	624-97-5	-73	119.5		4-01-00-02146 0.8300[20]	H_2O 2; EtOH 3; eth 3; ctc 2 1.4268[20]
9101	4-Penten-2-ol, 3-methyl-	$C_6H_{12}O$ 100.16	1569-59-1		125.5		3-01-00-01933 0.8429[22]	1.4326[22]
9102	4-Penten-2-ol, 4-methyl-	$C_6H_{12}O$ 100.16	2004-67-3		129		4-01-00-02143 0.8436[20]	1.4297[17]
9103	2-Penten-1-ol, 2-methyl-5-(2-methyl-3-methylenebicyclo[2.2.1]hept-2-yl)-, [1S-[1α,2α(Z),4α]- β-Santalol	$C_{15}H_{24}O$ 220.35	77-42-9		8322 167-8[10]		4-06-00-03519 0.9750[20]	1.5115[20]
9104	1-Penten-3-one Ethyl vinyl ketone	C_5H_8O 84.12	1629-58-9		103; 44[90]		4-01-00-03457 0.8468[20]	H_2O 1; EtOH 3; eth 3; ace 3 1.4195[20]
9105	3-Penten-2-one, (E)-	C_5H_8O 84.12	3102-33-8		122		4-01-00-03460 0.8624[20]	H_2O 3; eth 3; ace 3; ctc 3 1.4350[20]
9106	1-Penten-3-one, 4,4-dimethyl-1-phenyl-	$C_{13}H_{16}O$ 188.27	538-44-3	43	154[25]		3-07-00-01490 0.9508[46]	1.5523[25]
9107	1-Penten-3-one, 2-methyl-	$C_6H_{10}O$ 98.14	25044-01-3	-69.5	118.5		4-01-00-03470 1.8530[20]	ace 4; EtOH 4 1.4289[20]
9108	3-Penten-2-one, 4-methyl- Mesityl oxide	$C_6H_{10}O$ 98.14	141-79-7	-59	130	5811	4-01-00-03471 0.8653[20]	H_2O 3; EtOH 5; eth 5; ace 3 1.4440[20]
9109	4-Penten-2-one, 4-methyl-	$C_6H_{10}O$ 98.14	3744-02-3	-72.6	124.2		4-01-00-03470 0.8411[20]	
9110	1-Penten-3-one, 1-phenyl-	$C_{11}H_{12}O$ 160.22	3152-68-9	38.5	142[12]		2-07-00-00298 0.8697[20]	H_2O 2; EtOH 4; eth 4; bz 4 1.5684[20]
9111	1-Penten-3-yne Methylvinylacetylene	C_5H_6 66.10	646-05-9		59.5		4-01-00-01087 0.7401[20]	bz 4; eth 4 1.4496[20]
9112	1-Penten-4-yne	C_5H_6 66.10	871-28-3		42.5		4-01-00-01088 0.738[16]	H_2O 1; eth 3; bz 3 1.4125[16]
9113	3-Penten-1-yne, (E)- trans-3-Penten-1-yne	C_5H_6 66.10	2004-69-5		52.2		4-01-00-01085	
9114	3-Penten-1-yne, (Z)- cis-3-Penten-1-yne	C_5H_6 66.10	1574-40-9		44.6			
9115	3-Penten-1-yne, 3-ethyl-	C_7H_{10} 94.16	14272-54-9		96.5		4-01-00-01099 0.7886[25]	bz 4; eth 4 1.4338[25]
9116	1-Penten-3-yne, 2-methyl- Methylisopropenylacetylene	C_6H_8 80.13	926-55-6		81-2[100]		4-01-00-01095	bz 4; eth 4 1.4002[20]
9117	3-Penten-1-yne, 3-methyl- 3-Methyl-3-penten-1-yne	C_6H_8 80.13	1574-33-0		66.5		4-01-00-01096 0.739[20]	eth 3; bz 3 1.4332[20]
9118	4-Penten-2-yn-1-ol 1-Hydroxypent-2-yn-4-ene	C_5H_6O 82.10	2919-05-3		57[10]; 44[2.5]		4-01-00-02303 0.9581[20]	1.4986[20]
9119	Pentitol, 1,2-dideoxy- 1,2,3-Pentanetriol	$C_5H_{12}O_3$ 120.15	5371-48-2		192[63]		4-01-00-02778 1.0849[34]	H_2O 4; EtOH 4; eth 4
9120	Pentonic acid, 5-C-[3,5-bis(1-methylpropyl)-1-cyclopenten-1-yl]-4-deoxy-	$C_{18}H_{32}O_5$ 328.45	491-14-5		196		3-10-00-02045 1.292[14]	EtOH 4

No.	Name Synonym	Mol. Form. Mol. Wt.	CAS RN	Merck No. mp/°C bp/°C	Beil. Ref. den/g cm^{-3}	Solubility n_D
9121	1-Pentyn-3-amine, 3-ethyl-	$C_7H_{13}N$ 111.19	3234-64-8	71-2[90]	4-04-00-01154 0.828[25]	1.4409[20]
9122	1-Pentyne Propylacetylene	C_5H_8 68.12	627-19-0 -90	40.1	4-01-00-00990 0.6901[20]	H_2O 1; EtOH 4; eth 5; bz 3 1.3852[20]
9123	2-Pentyne	C_5H_8 68.12	627-21-4 -109.3	56.1	4-01-00-00992 0.7058[25]	H_2O 1; EtOH 4; eth 5; bz 3 1.4039[20]
9124	1-Pentyne, 1-bromo- 1-Pentynyl bromide	C_5H_7Br 147.01	14752-60-4	44-6[52]	3-01-00-00958 1.281[13]	1.4579[13]
9125	1-Pentyne, 3-chloro-3-ethyl- 3-Chloro-3-ethyl-1-pentyne	$C_7H_{11}Cl$ 130.62	6080-79-1	73-6[100]	4-01-00-01032 0.9230[19]	bz 4; eth 4; chl 4 1.4437[19]
9126	1-Pentyne, 3-chloro-3-methyl- 3-Chloro-3-methyl-1-pentyne	C_6H_9Cl 116.59	14179-94-3	102.3	4-01-00-01021 0.9163[20]	H_2O 1; eth 3; bz 3; chl 3 1.4330[20]
9127	2-Pentyne, 4-chloro-4-methyl-	C_6H_9Cl 116.59	999-79-1	55[70]	4-01-00-01018	ace 4; bz 4; chl 4 1.4143[20]
9128	1-Pentyne, 3,3-dimethyl- 3,3-Dimethyl-1-pentyne	C_7H_{12} 96.17	918-82-1	70	4-01-00-01034 0.7032[25]	1.3934[20]
9129	1-Pentyne, 4,4-dimethyl- 4,4-Dimethyl-1-pentyne	C_7H_{12} 96.17	13361-63-2 -75.7	76.1	4-01-00-01033 0.7142[20]	bz 4; eth 4; chl 4 1.3983[20]
9130	2-Pentyne, 4,4-dimethyl-	C_7H_{12} 96.17	999-78-0 -82.4	83	4-01-00-01032 0.7176[20]	H_2O 1; eth 3; bz 3; ctc 2 1.4071[20]
9131	1-Pentyne, 3-ethyl- 3-Ethyl-1-pentyne	C_7H_{12} 96.17	21020-26-8	84	3-01-00-01002 0.7246[25]	bz 4; eth 4; chl 4 1.4043[25]
9132	1-Pentyne, 3-ethyl-3-methyl- 3-Ethyl-3-methyl-1-pentyne	C_8H_{14} 110.20	919-12-0	101.5	4-01-00-01046 0.7422[20]	bz 4; eth 4; chl 4 1.4110[20]
9133	1-Pentyne, 1-iodo- 1-Iodo-1-pentyne	C_5H_7I 194.02	14752-61-5	54[23]	4-01-00-00992 1.6127[19]	1.5148[19]
9134	1-Pentyne, 3-methyl- 3-Methyl-1-pentyne	C_6H_{10} 82.15	922-59-8	57.7	4-01-00-01021 0.6992[25]	1.3916[20]
9135	1-Pentyne, 4-methyl-	C_6H_{10} 82.15	7154-75-8 -104.6	61.2	4-01-00-01019 0.7000[25]	H_2O 1; bz 3; chl 3 1.3936[20]
9136	2-Pentyne, 4-methyl-	C_6H_{10} 82.15	21020-27-9 -110.3	73.1	4-01-00-01018 0.7112[25]	bz 4; chl 4 1.4057[20]
9137	2-Pentynoic acid Ethylpropiolic acid	$C_5H_6O_2$ 98.10	5963-77-9 50	122[10]; 81[3]	4-02-00-01693 0.978[20]	H_2O 4 1.4619[20]
9138	4-Pentynoic acid Propargylacetic acid	$C_5H_6O_2$ 98.10	6089-09-4 57.7	110[30]; 102[17]	4-02-00-01693	eth 4; EtOH 4
9139	2-Pentynoic acid, methyl ester Methyl 2-pentynoate	$C_6H_8O_2$ 112.13	24342-04-9	72[10]	4-02-00-01694 0.937[20]	1.4455[20]
9140	1-Pentyn-3-ol Ethylethynylcarbinol	C_5H_8O 84.12	4187-86-4	125	4-01-00-02224 0.8926[15]	1.4347[15]
9141	2-Pentyn-1-ol	C_5H_8O 84.12	6261-22-9	61-2[15]	4-01-00-02226 0.913[17]	1.4518[17]
9142	3-Pentyn-1-ol	C_5H_8O 84.12	10229-10-4	154	4-01-00-02227 0.9002[20]	1.4454[20]
9143	4-Pentyn-1-ol	C_5H_8O 84.12	5390-04-5	154	4-01-00-02225 0.913[20]	1.4414[20]
9144	1-Pentyn-3-ol, 3,4-dimethyl-	$C_7H_{12}O$ 112.17	1482-15-1	134	4-01-00-02255 0.8691[20]	H_2O 4; eth 4; EtOH 4 1.4372[20]
9145	1-Pentyn-3-ol, 3-ethyl-	$C_7H_{12}O$ 112.17	6285-06-9	138	4-01-00-02254 0.8691[20]	H_2O 2
9146	1-Pentyn-3-ol, 3-methyl- Meparfynol	$C_6H_{10}O$ 98.14	77-75-8 30.5	5731 120.5	2-01-00-00506 0.8688[20]	1.4310[20]
9147	2-Pentynoyl chloride, 4,4-dimethyl- 4,4-Dimethyl-2-pentynoyl chloride	C_7H_9ClO 144.60	52324-03-5	52[20]	2-02-00-00454 0.9743[20]	1.4443[20]
9148	Perchloric acid, trichloromethyl ester Trichloro perchlorate	CCl_4O_4 217.82	67632-66-0 -55	exp 40	3-03-00-00037	eth 4
9149	Perfluidone o-Toluidine, 4-(phenylsulfonyl)-N-[(trifluoromethyl)sulfono]-	$C_{14}H_{12}F_3NO_4S_2$ 379.38	37924-13-3 143	7113		
9150	Permethrin Cyclopropanecarboxylic acid, 3-(2,2-dichloroethenyl)-2,2-dimethyl-, (3-phenoxyphenyl), methyl ester	$C_{21}H_{20}Cl_2O_3$ 391.29	52645-53-1 34	7132 200[0.01]	1.23[20]	
9151	Peroxide, acetyl benzoyl Acetozone	$C_9H_8O_4$ 180.16	644-31-5 37	71 130[19]	4-09-00-00717	eth 4
9152	Peroxide, bis(2,4-dichlorobenzoyl)	$C_{14}H_8Cl_4O_4$ 380.01	133-14-2 106		4-09-00-01001	
9153	Peroxide, bis(1,1-dimethylethyl) DTBP	$C_8H_{18}O_2$ 146.23	110-05-4 -40	3446 111	4-01-00-01619 0.704[20]	H_2O 1; ace 5; ctc 3; os 3 1.3890[20]
9154	Peroxide, bis(1,1-dimethylpropyl)	$C_{10}H_{22}O_2$ 174.28	10508-09-5	58.5[14]; 44[10]	4-01-00-01671 0.821[20]	1.4095[20]
9155	Peroxide, bis(1-oxododecyl) Lauroyl peroxide	$C_{24}H_{46}O_4$ 398.63	105-74-8 49		4-02-00-01102	H_2O 1; chl 3
9156	Peroxide, diacetyl Acetyl peroxide	$C_4H_6O_4$ 118.09	110-22-5 30	63[21]	4-02-00-00392	eth 4; EtOH 4
9157	Peroxide, dibenzoyl Benzoyl peroxide	$C_{14}H_{10}O_4$ 242.23	94-36-0 105	1128 exp	4-09-00-00717	H_2O 2; EtOH 3; eth 3; ace 3 1.543
9158	Peroxide, diethyl	$C_4H_{10}O_2$ 90.12	628-37-5 -70	65	4-01-00-01323 0.8240[19]	H_2O 2; EtOH 5; eth 5 1.3715[17]

No.	Name Synonym	Mol. Form. Mol. Wt. mp/°C	CAS RN	Merck No. bp/°C	Beil. Ref. den/g cm^{-3}	Solubility n_D
9159	Peroxide, dimethyl	$C_2H_6O_2$ 62.07	690-02-8	14	4-01-00-01249 0.8677^0	EtOH 2; eth; tol 3; HOAc 3 1.3503^0
9160	Perthane Ethane, 1,1-dichloro-2,2-bis(p-ethylphenyl)-	$C_{18}H_{20}Cl_2$ 307.26	72-56-0 56	3050		
9161	Perylene Dibenz[de,kl]anthracene	$C_{20}H_{12}$ 252.32	198-55-0 274	7137	4-05-00-02689 1.35^{25}	H_2O 1; EtOH 2; eth; ace 4
9162	9-Phenanthrenamine	$C_{14}H_{11}N$ 193.25	947-73-9 138.3	sub	4-12-00-03443	eth 2; bz 2; chl 2
9163	Phenanthrene	$C_{14}H_{10}$ 178.23	85-01-8 99.2	7167 340	4-05-00-02297 0.9800^4	H_2O 1; EtOH 3; eth 3; ace 3 1.5943
9164	Phenanthrene, 9-bromo- 9-Phenanthryl bromide	$C_{14}H_9Br$ 257.13	573-17-1 64.5	>360	4-05-00-02303 1.4093^{10}	H_2O 1; EtOH 3; eth 3; chl 2
9165	9-Phenanthrenecarboxylic acid 9-Phenanthroic acid	$C_{15}H_{10}O_2$ 222.24	837-45-6 257.3	sub	4-09-00-02674	H_2O 1; EtOH 3; eth 3; bz 3
9166	1-Phenanthrenecarboxylic acid, 1,2,3,4,4a,5,6,9,10,10a-decahydro-1,4a- dimethyl-7-(1-methylethyl)-, Palustric acid	$C_{20}H_{30}O_2$ 302.46	1945-53-5 164.5	6950	4-09-00-02174	
9167	1-Phenanthrenecarboxylic acid, 1,2,3,4,4a,4b,5,6,10,10a-decahydro-1,4a- dimethyl-7-(1-methylethyl)-, Abietic acid	$C_{20}H_{30}O_2$ 302.46	514-10-3 173.5	2 250^9	4-09-00-02175	ace 4; bz 4; eth 4; EtOH 4
9168	1-Phenanthrenecarboxylic acid, 1,2,3,4,4a,4b,5,6,10,10a-decahydro-1,4a- dimethyl-7-(1-methylethyl)-, methyl ester, [1R- (1α,4aβ,4bα,10aα)]- Methyl abietate	$C_{21}H_{32}O_2$ 316.48	127-25-3 225-6^{16}	5930	1.049^{20}	H_2O 1; EtOH 3; HOAc 3 1.5344
9169	1-Phenanthrenecarboxylic acid, 1,2,3,4,4a,4b,5,9,10,10a-decahydro-1,4a- dimethyl-7-(1-methylethyl)-, Levopimaric acid	$C_{20}H_{30}O_2$ 302.46	79-54-9 150	5348	4-09-00-02173	
9170	1-Phenanthrenecarboxylic acid, 7-ethenyl- 1,2,3,4,4a,4b,5,6,7,9,10,10a-dodecahydro- 1,4a,7-trimethyl-, [1R-(1α,4aβ,4bα,7β,10aα)]- Pimaric acid	$C_{20}H_{30}O_2$ 302.46	127-27-5 218.5	7398 282^{18}	4-09-00-02179	eth 4; py 4; EtOH 4
9171	Phenanthrene, 9,10-dihydro- 9,10-Dihydrophenanthrene	$C_{14}H_{12}$ 180.25	776-35-2 34.5	168-9^{15}	4-05-00-02183 1.0757^{40}	chl 3 1.6415^{20}
9172	Phenanthrene, 9,10-dimethyl- 9,10-Phenanthraquinodimethane	$C_{16}H_{14}$ 206.29	604-83-1 144	sub	4-05-00-02335	EtOH 2; bz 3; chl 3; HOAc 3
9173	3,4-Phenanthrenediol	$C_{14}H_{10}O_2$ 210.23	478-71-7 143	sub 130	3-06-00-05689	eth 4; EtOH 4
9174	9,10-Phenanthrenedione Phenanthrenequinone	$C_{14}H_8O_2$ 208.22	84-11-7 209	7168	4-07-00-02565 1.405^{22}	H_2O 1; EtOH 2; eth 3; bz 2
9175	9,10-Phenanthrenedione, 2-hydroxy-	$C_{14}H_8O_3$ 224.22	4088-81-7 284	sub	3-08-00-02929	H_2O 4
9176	9,10-Phenanthrenedione, 3-hydroxy- 9,10-Phenanthraquinone, 3-hydroxy	$C_{14}H_8O_3$ 224.22	57404-54-3 330 dec	sub	2-08-00-00398	H_2O 4; EtOH 4
9177	9,10-Phenanthrenedione, 1-methyl-7-(1-methylethyl)-	$C_{18}H_{16}O_2$ 264.32	5398-75-4 197.5	sub	4-07-00-02595	eth 4; EtOH 4
9178	Phenanthrene, dodecahydro- Dodecahydrophenanthrene	$C_{14}H_{22}$ 190.33	1322-67-4 	81-2$^{1.5}$	3-05-00-01074 0.9674^{20}	ace 4; bz 4 1.5102^{20}
9179	Phenanthrene, 9-ethyl- 9-Ethylphenanthrene	$C_{16}H_{14}$ 206.29	3674-75-7 62.5	199-200^1	4-05-00-02332 1.0603^{78}	H_2O 1; EtOH 3; bz 3; peth 2 1.6582^{78}
9180	Phenanthrene, 3-methyl- 3-Methylphenanthrene	$C_{15}H_{12}$ 192.26	832-71-3 65	350; 145^6	4-05-00-02314	H_2O 1; EtOH 3; ace 3; chl 2
9181	Phenanthrene, 4-methyl- 4-Methylphenanthrene	$C_{15}H_{12}$ 192.26	832-64-4 53.5	175-80^{10}	3-05-00-02154	H_2O 1; EtOH 3; ctc 3
9182	Phenanthrene, 9-(1-methylethenyl)- 9-Isopropenylphenanthrene	$C_{17}H_{14}$ 218.30	58873-44-2 38	163^{20}	3-05-00-02240	1.6765^{22}
9183	Phenanthrene, 1-methyl-7-(1-methylethyl)- Retene	$C_{18}H_{18}$ 234.34	483-65-8 101	8162 390	4-05-00-02368 1.035^{25}	H_2O 1; EtOH 3; eth 3; bz 3
9184	Phenanthrene, 1,2,3,4,4a,9,10,10a-octahydro- 1,2,3,4,4a,9,10,10a-Octahydrophenanthrene	$C_{14}H_{18}$ 186.30	16306-39-1 	159^{15}	4-05-00-01585 0.9973^{32}	1.5527^{19}
9185	Phenanthrene, 1,2,3,4,5,6,7,8-octahydro- 1,2,3,4,5,6,7,8-Octahydrophenanthrene	$C_{14}H_{18}$ 186.30	5325-97-3 16.7	295	4-05-00-01585 1.026^{20}	H_2O 1; ace 3; bz 3; CS_2 3 1.5569^{17}
9186	Phenanthrene, tetradecahydro-	$C_{14}H_{24}$ 192.34	5743-97-5 	86-9^2	4-05-00-00492 0.9447^{20}	H_2O 1; eth 3; ace 3; bz 3 1.5011^{20}
9187	Phenanthrene, 1,2,3,4-tetrahydro- 1,2,3,4-Tetrahydrophenanthrene	$C_{14}H_{14}$ 182.27	1013-08-7 33.5	173^{11}	4-05-00-01909 1.0601^{40}	H_2O 1; EtOH 3; eth 3; ace 3
9188	Phenanthridine	$C_{13}H_9N$ 179.22	229-87-8 107.4	349	5-20-08-00223	H_2O 2; EtOH 4; eth 4; ace 3
9189	1,10-Phenanthroline o-Phenanthroline	$C_{12}H_8N_2$ 180.21	66-71-7 117	7169 >300	5-23-08-00419	H_2O 4; EtOH 3; ace 3; bz 3
9190	1,7-Phenanthroline	$C_{12}H_8N_2$ 180.21	230-46-6 78	360	5-23-08-00417	H_2O 3; EtOH 4; eth 1; bz 1
9191	4,7-Phenanthroline	$C_{12}H_8N_2$ 180.21	230-07-9 177	sub 100	5-23-08-00430	H_2O 3; EtOH 4; eth 2; bz 2
9192	1,10-Phenanthroline, 2,9-dimethyl- Neocuproine	$C_{14}H_{12}N_2$ 208.26	484-11-7 159.5	6364	5-23-08-00527	
9193	Phenazine Dibenzopyrazine	$C_{12}H_8N_2$ 180.21	92-82-0 176.5	7173	5-23-08-00389	H_2O 2; EtOH 3; eth 2; bz 3

No.	Name / Synonym	Mol. Form. / Mol. Wt.	CAS RN / mp/°C	Merck No. / bp/°C	Beil. Ref. / den/g cm^{-3}	Solubility / n_D
9194	2,3-Phenazinediamine / 2,3-Diaminophenazine	$C_{12}H_{10}N_4$ / 210.24	655-86-7 / 264	2960 / sub	4-25-00-03028	bz 4; EtOH 4
9195	Phenazine, 2-methyl- / Tolazine	$C_{13}H_{10}N_2$ / 194.24	1016-94-0 / 118.5	/ 350	5-23-08-00475	eth 4; EtOH 4; chl 4
9196	Phenazinium, 1-hydroxy-5-methyl-, hydroxide, inner salt / Pyocyanine	$C_{13}H_{10}N_2O$ / 210.24	85-66-5 / 133 dec	7965	5-24-04-00100	H_2O 2; EtOH 3; eth 1; ace 3
9197	1-Phenazinol / Hemipyocyanine	$C_{12}H_8N_2O$ / 196.21	528-71-2 / 158	4564 / sub	5-23-12-00297	H_2O 2; EtOH 2; bz 3; os 4
9198	Phenmedipham / Carbamic acid, (3-methylphenyl)-, 3-[(methoxycarbonyl)amino]phenyl ester	$C_{16}H_{16}N_2O_4$ / 300.31	13684-63-4 / 143	7199		
9199	Phenol / Hydroxybenzene	C_6H_6O / 94.11	108-95-2 / 40.9	7206 / 181.8	4-06-00-00531 / 1.0545[45]	H_2O 3; EtOH 3; eth 4; ace 5 / 1.5408[41]
9200	Phenol, amino- / Hydroxyphenylamine	C_6H_7NO / 109.13	27598-85-2 / 83.5		2-15-00-00004	H_2O 3; EtOH 4; eth 4; bz 4
9201	Phenol, 2-amino- / o-Aminophenol	C_6H_7NO / 109.13	95-55-6 / 174	473 / sub 153	4-13-00-00805 / 1.328[25]	H_2O 3; EtOH 4; eth 3; bz 2
9202	Phenol, 3-amino- / m-Aminophenol	C_6H_7NO / 109.13	591-27-5 / 123	472 / 164[11]	4-13-00-00952	H_2O 3; EtOH 4; eth 4; bz 2
9203	Phenol, 4-amino- / p-Aminophenol	C_6H_7NO / 109.13	123-30-8 / 187.5	474 / 110[0.3]	4-13-00-01014	H_2O 2; EtOH 4; bz 1; chl 1
9204	Phenol, 2-amino-4-chloro- / 2-Hydroxy-5-chloroaniline	C_6H_6ClNO / 143.57	95-85-2 / 140		4-13-00-00878	DMSO 2
9205	Phenol, 2-amino-4,6-dichloro- / 2-Amino-4,6-dichlorophenol	$C_6H_5Cl_2NO$ / 178.02	527-62-8 / 95.5	448 / sub 70	4-13-00-00889	
9206	Phenol, 4-amino-2,6-dichloro-	$C_6H_5Cl_2NO$ / 178.02	5930-28-9 / 168	/ sub	2-13-00-00274	H_2O 1; EtOH 4; eth 4; ace 3
9207	Phenol, 2-amino-4,6-dinitro- / Picramic acid	$C_6H_5N_3O_5$ / 199.12	96-91-3 / 169	7379	4-13-00-00909	bz 4; EtOH 4
9208	Phenol, 4-(2-aminoethyl)- / Tyramine	$C_8H_{11}NO$ / 137.18	51-67-2 / 164.5	9743 / 205-7[25]	4-13-00-01788	H_2O 2; EtOH 3; bz 2; xyl 3
9209	Phenol, 2-amino-3-methyl-	C_7H_9NO / 123.15	2835-97-4 / 150	/ sub	2-13-00-00324	eth 4
9210	Phenol, 2-amino-4-methyl-	C_7H_9NO / 123.15	95-84-1 / 136	/ sub	4-13-00-01712	H_2O 2; EtOH 3; eth 3; bz 2
9211	Phenol, 3-amino-4-methyl-	C_7H_9NO / 123.15	2836-00-2 / 157.8	/ sub	4-13-00-01710	eth 4
9212	Phenol, 3-amino-5-methyl- / m-Cresol, 5-amino-	C_7H_9NO / 123.15	76619-89-1 / 139	/ 245	4-13-00-01697	EtOH 4
9213	Phenol, 4-amino-2-methyl-	C_7H_9NO / 123.15	2835-96-3 / 176.5	/ sub	4-13-00-01672	H_2O 2; EtOH 3; eth 3; bz 2
9214	Phenol, 2-amino-3-nitro-	$C_6H_6N_2O_3$ / 154.13	603-85-0 / 216.5	/ sub	4-13-00-00895	H_2O 4
9215	Phenol, 4-(2-aminopropyl)-, (±)- / Hydroxyamphetamine	$C_9H_{13}NO$ / 151.21	1518-86-1 / 125.5	4740	4-13-00-01871	H_2O 3; EtOH 3; bz 3; chl 3
9216	Phenol, 2,2'-azobis-	$C_{12}H_{10}N_2O_2$ / 214.22	2050-14-8 / 173	/ 140[0.001]	2-16-00-00033	H_2O 1; EtOH 2; eth 4; bz 2
9217	Phenol, 2-(2-benzothiazolyl)-	$C_{13}H_9NOS$ / 227.29	3411-95-8 / 131	/ 175-93[3]	4-27-00-02081	EtOH 3
9218	Phenol, 2-(2-benzoxazolyl)-	$C_{13}H_9NO_2$ / 211.22	835-64-3 / 123.5	/ 338	4-27-00-02080	H_2O 2; EtOH 4; eth 3; ace 3
9219	Phenol, 2,4-bis(1,1-dimethylethyl)-	$C_{14}H_{22}O$ / 206.33	96-76-4 / 56.5	/ 263.5	4-06-00-03493	ctc 2; alk 1 / 1.5080[20]
9220	Phenol, 2,6-bis(1,1-dimethylethyl)-	$C_{14}H_{22}O$ / 206.33	128-39-2 / 39	/ 161[50]; 133[20]	4-06-00-03492	EtOH 2; ctc 3; alk 1 / 1.5001[20]
9221	Phenol, 2,6-bis(1,1-dimethylethyl)-4-ethyl-	$C_{16}H_{26}O$ / 234.38	4130-42-1 / 44	/ 272	4-06-00-03529	alk 1
9222	Phenol, 2,4-bis(1,1-dimethylethyl)-5-methyl- / DBMC	$C_{15}H_{24}O$ / 220.35	497-39-2 / 62.1	2829 / 282	4-06-00-03510 / 0.912[80]	H_2O 1; EtOH 3; eth 3; ace 3
9223	Phenol, 2,4-bis(1,1-dimethylethyl)-6-methyl-	$C_{15}H_{24}O$ / 220.35	616-55-7 / 51	/ 269	4-06-00-03510 / 0.891[80]	alk 1
9224	Phenol, 2,6-bis(1,1-dimethylethyl)-4-methyl- / 2,6-Di-tert-butyl-4-methylphenol	$C_{15}H_{24}O$ / 220.35	128-37-0 / 71	1548 / 265	4-06-00-03511 / 0.8937[75]	H_2O 1; EtOH 3; ace 3; bz 3 / 1.4859[75]
9225	Phenol, 2,6-bis(1,1-dimethylpropyl)-4-methyl- / p-Cresol, 2,6-di-tert-pentyl-	$C_{17}H_{28}O$ / 248.41	56103-67-4 /	/ 283	4-06-00-03533 / 0.931[25]	/ 1.4950[20]
9226	Phenol, 2,6-bis(1-methylethyl)- / Propofol	$C_{12}H_{18}O$ / 178.27	2078-54-8 / 18	7847 / 256	4-06-00-03435 / 1.5140[20]	/ 1.5140[20]
9227	Phenol, 2,6-bis(1-methylpropyl)- / 2,6-Di-sec-butylphenol	$C_{14}H_{22}O$ / 206.33	5510-99-6 / -42	/ 257.5	4-06-00-03490	/ 1.5080[20]
9228	Phenol, 2-bromo-	C_6H_5BrO / 173.01	95-56-7 / 5.6	/ 194.5	4-06-00-01037 / 1.4924[20]	H_2O 2; EtOH 3; eth 3; chl 2 / 1.589[20]
9229	Phenol, 3-bromo-	C_6H_5BrO / 173.01	591-20-8 / 33	/ 236.5	4-06-00-01042	H_2O 2; EtOH 4; eth 4; ctc 2
9230	Phenol, 4-bromo-	C_6H_5BrO / 173.01	106-41-2 / 66.4	/ 238	4-06-00-01043 / 1.840[15]	H_2O 3; EtOH 4; eth 4; chl 3
9231	Phenol, 3-bromo-5-chloro-	C_6H_4BrClO / 207.45	56962-04-0 / 70	/ 256	2-06-00-00187	EtOH 4; HOAc 4
9232	Phenol, 4-bromo-2-chloro- / 4-Bromo-2-chlorophenol	C_6H_4BrClO / 207.45	3964-56-5 / 50.5	/ 233.5	3-06-00-00750 / 1.6170[20]	EtOH 4; eth 4; ace 4; bz 4 / 1.5859[20]

No.	Name Synonym	Mol. Form. Mol. Wt.	CAS RN mp/°C	Merck No. bp/°C	Beil. Ref. den/g cm^{-3}	Solubility n_D
9233	Phenol, 2-bromo-4,6-dichloro-	C$_6$H$_3$BrCl$_2$O 241.90	4524-77-0 68	220[200]	3-06-00-00751	bz 4; eth 4; chl 4
9234	Phenol, 2-bromo-4,6-dinitro-	C$_6$H$_3$BrN$_2$O$_5$ 263.00	2316-50-9 119.3	sub	3-00-00-00071	H$_2$O 2; EtOH 3; eth 4; bz 4
9235	Phenol, 4-bromo-2,6-dinitro-	C$_6$H$_3$BrN$_2$O$_5$ 263.00	40466-95-3 78	sub	3-06-00-00872	bz 4; eth 4; EtOH 4; chl 4
9236	Phenol, 2-bromo-4-methyl-	C$_7$H$_7$BrO 187.04	6627-55-0 56.5	213.5	4-06-00-02143 1.5422[25]	H$_2$O 2; EtOH 3; bz 3; chl 3 1.5772[20]
9237	Phenol, 2-bromo-5-methyl- 6-Bromo-m-cresol	C$_7$H$_7$BrO 187.04	14847-51-9 38.8	209	3-06-00-01320	ace 4; eth 4; EtOH 4
9238	Phenol, 3-bromo-2-methyl-	C$_7$H$_7$BrO 187.04	7766-23-6 95	55[4]	0-06-00-00360	ace 4; bz 4; eth 4; EtOH 4
9239	Phenol, 3-bromo-4-methyl-	C$_7$H$_7$BrO 187.04	60710-39-6 56	246	4-06-00-02143	ace 4; bz 4; eth 4; EtOH 4
9240	Phenol, 3-bromo-5-methyl-	C$_7$H$_7$BrO 187.04	74204-00-5 56.5	161[28]	2-06-00-00357	eth 4; EtOH 4
9241	Phenol, 4-bromo-2-methyl-	C$_7$H$_7$BrO 187.04	2362-12-1 64.7	235	4-06-00-02006	ace 4; eth 4; EtOH 4
9242	Phenol, 4-bromo-3-methyl-	C$_7$H$_7$BrO 187.04	14472-14-1 63.5	142[23]	4-06-00-02072	eth 4; py 4
9243	Phenol, 2-bromo-3-nitro-	C$_6$H$_4$BrNO$_3$ 218.01	101935-40-4 147.5	sub	3-06-00-00844	eth 4; EtOH 4
9244	Phenol, 4-bromo-2-nitro- 4-Bromo-2-nitrophenol	C$_6$H$_4$BrNO$_3$ 218.01	7693-52-9 92	sub	4-06-00-01363	bz 4; eth 4; EtOH 4; chl 4
9245	Phenol, 2-butoxy-	C$_{10}$H$_{14}$O$_2$ 166.22	39075-90-6	232.5	4-06-00-05566 1.026[25]	1.5113[25]
9246	Phenol, 2-butyl-	C$_{10}$H$_{14}$O 150.22	3180-09-4 -20	235	4-06-00-03267 0.975[20]	H$_2$O 1; EtOH 3; eth 3; alk 3 1.5180[25]
9247	Phenol, 3-butyl-	C$_{10}$H$_{14}$O 150.22	4074-43-5	248	4-06-00-03269 0.974[20]	eth 4; EtOH 4
9248	Phenol, 4-butyl-	C$_{10}$H$_{14}$O 150.22	1638-22-8 22	248	4-06-00-03269 0.976[22]	H$_2$O 1; EtOH 3; eth 3; ctc 2 1.5165[25]
9249	Phenol, 4,4'-butylidenebis-	C$_{16}$H$_{18}$O$_2$ 242.32	4731-84-4 137	270[12]	4-06-00-06736	bz 4; EtOH 4; tol 4
9250	Phenol, 2-butyl-4-methyl-	C$_{11}$H$_{16}$O 164.25	6891-45-8 19	228; 124[15]	4-00-00-03392 0.9680[20]	
9251	Phenol, 2-chloro- o-Chlorophenol	C$_6$H$_5$ClO 128.56	95-57-8 9.8	2154 174.9	4-06-00-00782 1.2634[20]	H$_2$O 2; EtOH 3; eth 3; bz 4 1.5524[20]
9252	Phenol, 3-chloro- m-Chlorophenol	C$_6$H$_5$ClO 128.56	108-43-0 32.6	2154 214	4-06-00-00810 1.245[45]	H$_2$O 2; EtOH 3; eth 3; bz 4 1.5565[40]
9253	Phenol, 4-chloro- p-Chlorophenol	C$_6$H$_5$ClO 128.56	106-48-9 42.7	2154 220	4-06-00-00820 1.2651[40]	H$_2$O 1; EtOH 4; eth 4; bz 4 1.5579[40]
9254	Phenol, 2-chloro-4-cyclohexyl- 2-Chloro-4-cyclohexylphenol	C$_{12}$H$_{15}$ClO 210.70	3964-61-2 40.5	160[12]; 122[3]	3-06-00-02508 1.160[15]	ctc 3; CS$_2$ 3
9255	Phenol, 4-chloro-2-cyclopentyl- Dowicide 9	C$_{11}$H$_{13}$ClO 196.68	13347-42-7	3420 181-5[18]	3-06-00-02478	
9256	Phenol, 2-chloro-3,4-dimethyl-	C$_8$H$_9$ClO 156.61	10283-15-5 27	188	2-06-00-00456 1.5538[20]	peth 4
9257	Phenol, 4-chloro-3,5-dimethyl- Chloroxylenol	C$_8$H$_9$ClO 156.61	88-04-0 115	2176 246	4-06-00-03152	H$_2$O 2; EtOH 3; eth 3; bz 2
9258	Phenol, 4-chloro-2,6-dinitro- 4-Chloro-2,6-dinitrophenol	C$_6$H$_3$ClN$_2$O$_5$ 218.55	88-87-9 81		4-06-00-01386 1.74[22]	eth 4; EtOH 4; chl 4
9259	Phenol, 5-chloro-2,4-dinitro- 3-Chloro-4,6-dinitrophenol	C$_6$H$_3$ClN$_2$O$_5$ 218.55	54715-57-0 92		4-06-00-01385 1.74[22]	eth 4; EtOH 4; chl 4; peth 4
9260	Phenol, 2-chloro-3-methyl-	C$_7$H$_7$ClO 142.58	608-26-4 56	230	4-06-00-02062	H$_2$O 2; HOAc 4; peth 3
9261	Phenol, 2-chloro-4-methyl-	C$_7$H$_7$ClO 142.58	6640-27-3	195.5	4-06-00-02135 1.1785[27]	bz 4; eth 4; EtOH 4 1.5200[27]
9262	Phenol, 2-chloro-5-methyl- 6-Chloro-m-cresol	C$_7$H$_7$ClO 142.58	615-74-7 45.5	196	3-06-00-01315 1.215[15]	H$_2$O 4; EtOH 4
9263	Phenol, 2-chloro-6-methyl-	C$_7$H$_7$ClO 142.58	87-64-9	189; 80[20]	4-06-00-01984 1.5449[20]	eth 3
9264	Phenol, 3-chloro-2-methyl-	C$_7$H$_7$ClO 142.58	3260-87-5 86	225	4-06-00-02000	bz 4; eth 4; EtOH 4
9265	Phenol, 3-chloro-4-methyl-	C$_7$H$_7$ClO 142.58	615-62-3 55.5	228	3-06-00-01374	bz 4; eth 4; EtOH 4
9266	Phenol, 4-chloro-2-methyl- 4-Chloro-o-cresol	C$_7$H$_7$ClO 142.58	1570-64-5 51	223	4-06-00-01987	H$_2$O 2; peth 3
9267	Phenol, 4-chloro-3-methyl- 4-Chloro-m-cresol	C$_7$H$_7$ClO 142.58	59-50-7 67	2133 235	4-06-00-02064	H$_2$O 2; EtOH 3; eth 3; chl 2
9268	Phenol, 4-chloro-5-methyl-2-(1-methylethyl)- Chlorothymol	C$_{10}$H$_{13}$ClO 184.67	89-68-9 63	2171 258.5	4-06-00-03344	H$_2$O 4; EtOH 3; eth 3; bz 3
9269	Phenol, 5-chloro-2-nitro- 5-Chloro-2-nitrophenol	C$_6$H$_4$ClNO$_3$ 173.56	611-07-4 41	sub	3-06-00-00836	H$_2$O 2; EtOH 3; eth 3; HOAc 3
9270	Phenol, 4-chloro-2-(phenylmethyl)- Clorophene	C$_{13}$H$_{11}$ClO 218.68	120-32-1 48.5	2403 160-2[3.5]	4-06-00-04636 1.185[58]	ctc 3; CS$_2$ 3
9271	Phenol, 4-chloro-2-(2-propenyl)-	C$_9$H$_9$ClO 168.62	13997-73-4 48	130-2[15]	4-06-00-03813 1.171[15]	bz 4; EtOH 4
9272	Phenol, 4-cyclohexyl-	C$_{12}$H$_{16}$O 176.26	1131-60-8 133	294; 133[4]	4-06-00-03904	H$_2$O 1; EtOH 4; eth 4; bz 3

No.	Name Synonym	Mol. Form. Mol. Wt.	CAS RN mp/°C	Merck No. bp/°C	Beil. Ref. den/g cm^{-3}	Solubility n_D
9273	Phenol, 2-cyclohexyl-4,6-dinitro- 4,6-Dinitro-o-cyclohexylphenol	$C_{12}H_{14}N_2O_5$ 266.25	131-89-5	2739	4-06-00-03903	
9274	Phenol, 2,4-diamino- 2,4-Diaminophenol	$C_6H_8N_2O$ 124.14	95-86-3 78-80 dec	2961	4-13-00-01425	H_2O 4; ace 4; EtOH 4
9275	Phenol, 4,4'-(1,2-diarsenediyl)bis[2-amino-, dihydrochloride Salvarsan	$C_{12}H_{20}As_2Cl_2O_5$ 465.04	139-93-5 185-95 dec	839	2-16-00-00560	H_2O 4
9276	Phenol, 2,4-dibromo- 2,4-Dibromophenol	$C_6H_4Br_2O$ 251.91	615-58-7 38	238.5	4-06-00-01061 2.0700[20]	H_2O 2; EtOH 4; eth 4; bz 4
9277	Phenol, 2,6-dibromo- 2,6-Dibromophenol	$C_6H_4Br_2O$ 251.91	608-33-3 56.5	255; 162[21]	4-06-00-01064	H_2O 3; EtOH 4; eth 4
9278	Phenol, 3,5-dibromo- 3,5-Dibromophenol	$C_6H_4Br_2O$ 251.91	626-41-5 81	274; 121[3]	4-06-00-01065	H_2O 2; EtOH 3; eth 3; lig 2
9279	Phenol, 2,4-dibromo-6-methyl-	$C_7H_6Br_2O$ 265.93	609-22-3 58	265 dec; 105[4]	3-06-00-01271	chl 3
9280	Phenol, 3,5-dibromo-2-methyl-	$C_7H_6Br_2O$ 265.93	14122-00-0 100.3	285	2-06-00-00334	peth 4
9281	Phenol, 2,4-dibromo-3-methyl-6-(1- methylethyl)- Thymol, 2,6-dibromo	$C_{10}H_{12}Br_2O$ 308.01	52262-38-1 3.7	160-1[16]	2-06-00-00500 1.6568[17]	
9282	Phenol, 2,3-dichloro- 2,3-Dichlorophenol	$C_6H_4Cl_2O$ 163.00	576-24-9 58		4-06-00-00883	EtOH 3; eth 3; bz 3; lig 3
9283	Phenol, 2,4-dichloro- 2,4-Dichlorophenol	$C_6H_4Cl_2O$ 163.00	120-83-2 45	3061 210	4-06-00-00885	H_2O 2; EtOH 3; eth 3; bz 3
9284	Phenol, 2,5-dichloro- 2,5-Dichlorophenol	$C_6H_4Cl_2O$ 163.00	583-78-8 59	211	4-06-00-00942	H_2O 2; EtOH 4; eth 4; bz 3
9285	Phenol, 2,6-dichloro- 2,6-Dichlorophenol	$C_6H_4Cl_2O$ 163.00	87-65-0 68.5	3062 220; 82[4]	4-06-00-00949 1.653[20]	EtOH 4; eth 4; bz 3; peth 3
9286	Phenol, 3,4-dichloro- 3,4-Dichlorophenol	$C_6H_4Cl_2O$ 163.00	95-77-2 68	253	4-06-00-00952	H_2O 2; EtOH 4; eth 4; bz 3
9287	Phenol, 3,5-dichloro- 3,5-Dichlorophenol	$C_6H_4Cl_2O$ 163.00	591-35-5 68	233	4-06-00-00957	H_2O 2; EtOH 4; eth 4; peth 3
9288	Phenol, 2,4-dichloro-, acetate 2,4-Dichlorophenyl acetate	$C_8H_6Cl_2O_2$ 205.04	6341-97-5 244.5		3-06-00-00704	EtOH 3; eth 3; chl 3
9289	Phenol, 2,4-dichloro-, benzenesulfonate Genite	$C_{12}H_8Cl_2O_3S$ 303.17	97-16-5 45.5	4283	4-11-00-00034	ctc 3; CS_2 3
9290	Phenol, 2,4-dichloro-3,5-dimethyl- Dichloroxylenol	$C_8H_8Cl_2O$ 191.06	133-53-9 83	3066	4-06-00-03156	eth 4
9291	Phenol, 2,6-dichloro-3,5-dimethyl-	$C_8H_8Cl_2O$ 191.06	1943-54-0 87.5	105-10[1]	4-06-00-03157	eth 4; chl 4
9292	Phenol, 2,4-dichloro-3-methyl- m-Cresol, 2,4-dichloro	$C_7H_6Cl_2O$ 177.03	17788-00-0 58	236; 77[4]	3-06-00-01319	eth 4; chl 4
9293	Phenol, 2,4-dichloro-5-methyl-	$C_7H_6Cl_2O$ 177.03	1124-07-8 73	234	4-06-00-02069	chl 4; peth 3 1.572[20]
9294	Phenol, 2,6-dichloro-3-methyl- 2,6-Dichloro-m-cresol	$C_7H_6Cl_2O$ 177.03	13481-70-4 28	237; 77[4]	3-06-00-01319	eth 4; chl 4
9295	Phenol, 2,6-dichloro-4-methyl-	$C_7H_6Cl_2O$ 177.03	2432-12-4 39	231; 138[28]	4-06-00-02141	eth 4; EtOH 4; HOAc 4
9296	Phenol, 2,6-dichloro-4-nitro-	$C_6H_3ClNO_3$ 172.55	618-80-4 127 exp		4-06-00-01361 1.822[25]	eth 4; chl 4
9297	Phenol, 3-(diethylamino)-	$C_{10}H_{15}NO$ 165.24	91-68-9 78	276; 170[15]	4-13-00-00969	H_2O 3; EtOH 3; eth 3; CS_2 3
9298	Phenol, 4,4'-(1,2-diethyl-1,2-ethanediyl)bis-, (R*,S*)- Hexestrol	$C_{18}H_{22}O_2$ 270.37	84-16-2 186.5	4621	4-06-00-06761	ace 4; eth 4; EtOH 4
9299	Phenol, 4,4'-(1,2-diethyl-1,2-ethanediyl)bis[2- methyl- Methestrol	$C_{20}H_{26}O_2$ 298.43	130-73-4 145	5888	4-06-00-06786	
9300	Phenol, 4,4'-(1,2-diethyl-1,2-ethenediyl)bis-, (E)- Diethylstilbestrol	$C_{18}H_{20}O_2$ 268.36	56-53-1 170.5	3118	4-06-00-06856	eth 4; EtOH 4; chl 4
9301	Phenol, 4,4'-(1,2-diethyl-1,2-ethenediyl)bis-, dipropanoate, (E)- Diethylstilbestrol dipropionate	$C_{24}H_{28}O_4$ 380.48	130-80-3 104	3119	4-06-00-06857	bz 4; eth 4; EtOH 4
9302	Phenol, 4,4'-(1,2-diethylidene-1,2- ethanediyl)bis- Dienestrol	$C_{18}H_{18}O_2$ 266.34	84-17-3 227.5	3094 sub 130	4-06-00-06916	ace 4; eth 4; EtOH 4
9303	Phenol, 4,4'-(1,2-diethyl-3-methyl-1,3- propanediyl)bis- Benzestrol	$C_{20}H_{26}O_2$ 298.43	85-95-0 164	1082	3-06-00-05538	ace 4; eth 4; EtOH 4; HOAc 4
9304	Phenol, 2,6-diiodo-4-nitro- Disophenol	$C_6H_3I_2NO_3$ 390.90	305-85-1 157	3365	4-06-00-01368	EtOH 4
9305	Phenol, 2,6-dimethoxy-	$C_8H_{10}O_3$ 154.17	91-10-1 56.5	261	4-06-00-07329	eth 4; EtOH 4
9306	Phenol, 3,5-dimethoxy-	$C_8H_{10}O_3$ 154.17	500-99-2 37	199[35]; 170[10]	4-06-00-07362	eth 3; bz 3; lig 2
9307	Phenol, 2,3-dimethyl- 2,3-Xylenol	$C_8H_{10}O$ 122.17	526-75-0 72.8	9989 216.9	4-06-00-03096	H_2O 2; EtOH 3; eth 3 1.5420[20]

No.	Name / Synonym	Mol. Form. / Mol. Wt.	CAS RN / mp/°C	Merck No. / bp/°C	Beil. Ref. / den/g cm^{-3}	Solubility / n_D
9308	Phenol, 2,4-dimethyl- / 2,4-Xylenol	$C_8H_{10}O$ / 122.17	105-67-9 / 24.5	9989 / 210.9	4-06-00-03126 / 0.9650[20]	H_2O 2; EtOH 5; eth 5; ctc 3 / 1.5420[14]
9309	Phenol, 2,5-dimethyl- / 2,5-Xylenol	$C_8H_{10}O$ / 122.17	95-87-4 / 74.8	9989 / 211.1	4-06-00-03164	H_2O 3; EtOH 3; eth 4; chl 2
9310	Phenol, 2,6-dimethyl- / 2,6-Xylenol	$C_8H_{10}O$ / 122.17	576-26-1 / 45.7	9989 / 201.0	4-06-00-03112	H_2O 3; EtOH 3; eth 3; ctc 3
9311	Phenol, 3,4-dimethyl- / 3,4-Xylenol	$C_8H_{10}O$ / 122.17	95-65-8 / 60.8	9989 / 227	4-06-00-03099 / 0.9830[20]	H_2O 2; EtOH 3; eth 5; ctc 3
9312	Phenol, 3,5-dimethyl- / 3,5-Xylenol	$C_8H_{10}O$ / 122.17	108-68-9 / 63.6	9989 / 221.7	4-06-00-03141 / 0.9680[20]	H_2O 3; EtOH 3; ctc 3
9313	Phenol, 2,4-dimethyl-, acetate	$C_{10}H_{12}O_2$ / 164.20	877-53-2 /	/ 226	2-06-00-00459 / 1.0298[15]	eth 4; EtOH 4 / 1.4990[15]
9314	Phenol, 2,5-dimethyl-, acetate	$C_{10}H_{12}O_2$ / 164.20	877-48-5 / <-20	/ 237	2-06-00-00467 / 1.0624[15]	eth 4; EtOH 4
9315	Phenol, 3,4-dimethyl-, acetate	$C_{10}H_{12}O_2$ / 164.20	22618-23-1 / 22	/ 235	4-06-00-03102	bz 4; eth 4; EtOH 4
9316	Phenol, 3,5-dimethyl-, acetate	$C_{10}H_{12}O_2$ / 164.20	877-82-7 / <-20	/ 237	3-06-00-01757 / 1.0624[15]	eth 4; EtOH 4
9317	Phenol, 2-(dimethylamino)-	$C_8H_{11}NO$ / 137.18	3743-22-4 / 45	/ 199.5	4-13-00-00815	eth 4; EtOH 4; HOAc 4
9318	Phenol, 3-(dimethylamino)-	$C_8H_{11}NO$ / 137.18	99-07-0 / 86	/ 266.5	4-13-00-00959	H_2O 1; EtOH 3; eth 3; ace 3 / 1.5895[26]
9319	Phenol, 4-(dimethylamino)-	$C_8H_{11}NO$ / 137.18	619-60-3 / 77	/ 165[30]	4-13-00-01040	H_2O 2; EtOH 3; eth 3
9320	Phenol, 4-[2-(dimethylamino)ethyl]- / Hordenine	$C_{10}H_{15}NO$ / 165.24	539-15-1 / 117.5	4666 / 173[11]	4-13-00-01790	eth 4; EtOH 4; chl 4
9321	Phenol, 2-(1,1-dimethylethyl)-	$C_{10}H_{14}O$ / 150.22	88-18-6 / -6.8	/ 223	4-06-00-03292 / 0.9783[20]	EtOH 3; eth 4; ctc 3; alk 3 / 1.5160[20]
9322	Phenol, 3-(1,1-dimethylethyl)-	$C_{10}H_{14}O$ / 150.22	585-34-2 / 42.3	/ 240	4-06-00-03294	EtOH 3; eth 4; alk 3
9323	Phenol, 4-(1,1-dimethylethyl)- / p-tert-Butylphenol	$C_{10}H_{14}O$ / 150.22	98-54-4 / 98	1584 / 237	4-06-00-03296 / 0.908[80]	H_2O 3; EtOH 3; eth 3; chl 3 / 1.4787[114]
9324	Phenol, 2-(1,1-dimethylethyl)-4,5-dimethyl- / 3,4-Xylenol, 6-tert-butyl-	$C_{12}H_{18}O$ / 178.27	1445-23-4 / 46	/ 258.5	3-06-00-02019 / 0.920[80]	/ 1.5222[20]
9325	Phenol, 2-(1,1-dimethylethyl)-4,6-dimethyl-	$C_{12}H_{18}O$ / 178.27	1879-09-0 / 22.3	/ 249	4-06-00-03442 / 0.9178[0]	alk 1 / 1.5183[20]
9326	Phenol, 4-(1,1-dimethylethyl)-2,5-dimethyl-	$C_{12}H_{18}O$ / 178.27	17696-37-6 / 71.2	/ 264	3-06-00-02019 / 0.9398[0]	alk 3 / 1.5311[20]
9327	Phenol, 4-(1,1-dimethylethyl)-2,6-dimethyl-	$C_{12}H_{18}O$ / 178.27	879-97-0 / 82.4	/ 248	3-06-00-02019 / 0.916[80]	alk 3
9328	Phenol, 2-(1,1-dimethylethyl)-4,6-dinitro- / 2-tert-Butyl-4,6-dinitrophenol	$C_{10}H_{12}N_2O_5$ / 240.22	1420-07-1 /		4-06-00-03293	
9329	Phenol, (1,1-dimethylethyl)-4-methoxy- / Butylated hydroxyanisole	$C_{11}H_{16}O_2$ / 180.25	25013-16-5 / 51	1547 / 268		H_2O 1; peth 3; EtOH 3
9330	Phenol, 2-(1,1-dimethylethyl)-4-methyl-	$C_{11}H_{16}O$ / 164.25	2409-55-4 / 51.5	/ 237	4-06-00-03397 / 0.9247[75]	H_2O 2; ace 3; bz 3; chl 3 / 1.4969[75]
9331	Phenol, 2-(1,1-dimethylethyl)-5-methyl-	$C_{11}H_{16}O$ / 164.25	88-60-8 / 46.5	/ 127[11]	4-06-00-03400 / 0.922[80]	H_2O 1; EtOH 3; eth 3; ace 3 / 1.5250[20]
9332	Phenol, 2-(1,1-dimethylethyl)-6-methyl-	$C_{11}H_{16}O$ / 164.25	2219-82-1 / 31	/ 230	4-06-00-03396 / 0.9240[80]	/ 1.5195[20]
9333	Phenol, 4-(1,1-dimethylethyl)-2-methyl-	$C_{11}H_{16}O$ / 164.25	98-27-1 / 27.5	/ 237; 132[20]	4-06-00-03398 / 0.965[20]	H_2O 1; eth 3; ace 3; bz 3 / 1.5230[20]
9334	Phenol, 3,6-dimethyl-2-nitro-	$C_8H_9NO_3$ / 167.16	71608-10-1 / 34.5	/ 236 dec; 150[15]	4-06-00-03169	EtOH 4; eth 3; ace 3; bz 3
9335	Phenol, 2,4-dimethyl-, phosphate (3:1)	$C_{24}H_{27}O_4P$ / 410.45	3862-12-2 /	/ 233.5	0-06-00-00488 / 1.142[38]	H_2O 1; bz 3; chl 3; hx 3 / 1.5550[20]
9336	Phenol, 2,5-dimethyl-, phosphate (3:1)	$C_{24}H_{27}O_4P$ / 410.45	19074-59-0 / 79.8	/ 260-5[8]	3-06-00-01773 / 1.197[25]	H_2O 1; EtOH 2; eth 3; bz 3
9337	Phenol, 2,6-dimethyl-, phosphate (3:1)	$C_{24}H_{27}O_4P$ / 410.45	121-06-2 / 137.8	/ 262-4[6]	4-06-00-03121	H_2O 1; EtOH 2; bz 3; hx 2
9338	Phenol, 3,4-dimethyl-, phosphate (3:1)	$C_{24}H_{27}O_4P$ / 410.45	3862-11-1 / 72	/ 260-3[7]	0-06-00-00482	H_2O 1; EtOH 2; bz 3; chl 2
9339	Phenol, 4-(1,1-dimethylpropyl)- / p-tert-Pentylphenol	$C_{11}H_{16}O$ / 164.25	80-46-6 / 95	7098 / 262.5	4-06-00-03383	
9340	Phenol, 4-(1,1-dimethylpropyl)-2-methyl-	$C_{12}H_{18}O$ / 178.27	71745-63-6 / 15	/ 260	3-06-00-02010 / 0.969[20]	chl 2 / 1.5240[20]
9341	Phenol, 2,3-dinitro- / 2,3-Dinitrophenol	$C_6H_4N_2O_5$ / 184.11	66-56-8 / 144.5		4-06-00-01369 / 1.681[20]	H_2O 4; EtOH 4; eth 4; bz 3
9342	Phenol, 2,4-dinitro- / 2,4-Dinitrophenol	$C_6H_4N_2O_5$ / 184.11	51-28-5 / 115.5	3274 / sub	4-06-00-01369 / 1.683[24]	H_2O 2; EtOH 3; eth 3; ace 3
9343	Phenol, 2,5-dinitro- / 2,5-Dinitrophenol	$C_6H_4N_2O_5$ / 184.11	329-71-5 / 108	3275	4-06-00-01383	bz 4; eth 4
9344	Phenol, 2,6-dinitro- / 2,6-Dinitrophenol	$C_6H_4N_2O_5$ / 184.11	573-56-8 / 63.5	3276	4-06-00-01383	H_2O 1; EtOH 4; eth 4; ace 4
9345	Phenol, 3,4-dinitro-	$C_6H_4N_2O_5$ / 184.11	577-71-9 / 134		4-06-00-01384 / 1.672[25]	bz 4; eth 4; EtOH 4
9346	Phenol, 3,5-dinitro- / 3,5-Dinitrophenol	$C_6H_4N_2O_5$ / 184.11	586-11-8 / 125.1		4-06-00-01385 / 1.702[25]	bz 4; eth 4; EtOH 4; chl 4
9347	Phenol, 2,6-dipropyl- / 2,6-Dipropylphenol	$C_{12}H_{18}O$ / 178.27	6626-32-0 / 27	/ 256	3-06-00-02013 / 1.6450[0]	eth 4

No.	Name / Synonym	Mol. Form. / Mol. Wt.	CAS RN / mp/°C	Merck No. / bp/°C	Beil. Ref. / den/g cm^{-3}	Solubility / n_D
9348	Phenol, 2-ethenyl-	C_8H_8O 120.15	695-84-1 29.5	101[14]	4-06-00-03771 1.0609[18]	eth 4; EtOH 4 1.5851[20]
9349	Phenol, 3-ethenyl-	C_8H_8O 120.15	620-18-8	114-6[16]	4-06-00-03773 1.0353[21]	1.5804[21]
9350	Phenol, 2-ethoxy-	$C_8H_{10}O_2$ 138.17	94-71-3 29	217	4-06-00-05565 1.0903[25]	H_2O 2; EtOH 5; eth 5; ctc 2
9351	Phenol, 3-ethoxy-	$C_8H_{10}O_2$ 138.17	621-34-1	131[10]	4-06-00-05664 1.0705[4]	H_2O 1; EtOH 3; eth 3; bz 3
9352	Phenol, 4-ethoxy-	$C_8H_{10}O_2$ 138.17	622-62-8 66.5	246.5	4-06-00-05719	H_2O 2; EtOH 4; eth 4; chl 3
9353	Phenol, 2-ethyl- o-Ethylphenol	$C_8H_{10}O$ 122.17	90-00-6 18	7302 204.5	4-06-00-03011 1.0146[25]	ace 4; bz 4; eth 4; EtOH 4 1.5367[20]
9354	Phenol, 3-ethyl- m-Ethylphenol	$C_8H_{10}O$ 122.17	620-17-7 -4	218.4	4-06-00-03016 1.0283[20]	H_2O 2; EtOH 4; eth 4; chl 2
9355	Phenol, 4-ethyl- p-Ethylphenol	$C_8H_{10}O$ 122.17	123-07-9 45.0	217.9	4-06-00-03020	H_2O 2; EtOH 4; eth 4; ace 3 1.5239[25]
9356	Phenol, 3-(ethylamino)-	$C_8H_{11}NO$ 137.18	621-31-8 62	176[12]	0-13-00-00408	bz 4; eth 4; EtOH 4
9357	Phenol, 4-ethyl-2-methoxy-	$C_9H_{12}O_2$ 152.19	2785-89-9 -7	236.5	4-06-00-05927 1.0931[18]	
9358	Phenol, 4-[1-ethyl-2-(4-methoxyphenyl)-1-butenyl]- Mestilbol	$C_{19}H_{22}O_2$ 282.38	7773-60-6 117.5	5818 185-95[0.3]	4-06-00-06856	ace 4; eth 4; EtOH 4
9359	Phenol, 4-[1-ethyl-2-(4-methoxyphenyl)-1-butenyl]-, (E)-	$C_{19}H_{22}O_2$ 282.38	18839-90-2 117	185-95[0.3]	4-06-00-06856	EtOH 4
9360	Phenol, 2-fluoro-	C_6H_5FO 112.10	367-12-4 16.1	151.5	4-06-00-00770 1.120[25]	H_2O 3 1.5144[20]
9361	Phenol, 3-fluoro-	C_6H_5FO 112.10	372-20-3 13.7	178	4-06-00-00772 1.238[25]	1.5140[20]
9362	Phenol, 4-fluoro-	C_6H_5FO 112.10	371-41-5 48	185.5	4-06-00-00773 1.1889[56]	H_2O 2; ace 3; peth 3
9363	Phenol, 4,4'-(3H-2,1-benzoxathiol-3-ylidene)bis-, S,S-dioxide Phenolsulfonphthalein	$C_{19}H_{14}O_5S$ 354.38	143-74-8 >300	7213	5-19-03-00457	H_2O 2; EtOH 2; eth 1; ace 2
9364	Phenol, 4,4'-(3H-2,1-benzoxathiol-3-ylidene)bis[2-bromo-6-methyl-, S,S-dioxide Bromcresol purple	$C_{21}H_{16}Br_2O_5S$ 540.23	115-40-2 241.5	1376	5-19-03-00460	
9365	Phenol, 4,4'-(3H-2,1-benzoxathiol-3-ylidene)bis[2-bromo-3-methyl-6-(1-methylethyl)-, S,S-dioxide Bromthymol blue	$C_{27}H_{28}Br_2O_5S$ 624.39	76-59-5 201	1435	5-19-03-00461	eth 4; EtOH 4
9366	Phenol, 4,4'-(3H-2,1-benzoxathiol-3-ylidene)bis[2,6-dibromo-, S,S-dioxide Bromphenol blue	$C_{19}H_{10}Br_4O_5S$ 669.97	115-39-9 279 dec	1434	5-19-03-00458	H_2O 2; EtOH 3; bz 3; HOAc 3
9367	Phenol, 4,4'-(3H-2,1-benzoxathiol-3-ylidene)bis[2,6-dibromo-3-methyl-, S,S-dioxide Bromcresol green	$C_{21}H_{14}Br_4O_5S$ 698.02	76-60-8 218.5	1375	5-19-03-00460	H_2O 2; EtOH 4; eth 4; bz 3
9368	Phenol, 4,4'-(3H-2,1-benzoxathiol-3-ylidene)bis[2-methyl-, S,S-dioxide Cresol red	$C_{21}H_{18}O_5S$ 382.44	1733-12-6 >300	2583	5-19-03-00460	H_2O 4; EtOH 4
9369	Phenol, 4,4'-(3H-2,1-benzoxathiol-3-ylidene)bis[5-methyl-2-(1-methylethyl)-, S,S-dioxide Thymol blue	$C_{27}H_{30}O_5S$ 466.60	76-61-9 221-4 dec	9334	5-19-03-00462	H_2O 2; EtOH 3; ace 2; bz 2
9370	Phenol, 4-(3-hydroxy-1-propenyl)-2-methoxy- Coniferyl alcohol	$C_{10}H_{12}O_3$ 180.20	458-35-5 74	2499 163-5[3]	2-06-00-01093	H_2O 1; EtOH 3; eth 4; alk 3
9371	Phenol, 2-iodo- o-Iodophenol	C_6H_5IO 220.01	533-58-4 43	4928 186[160]; 91[2]	4-06-00-01070 1.8757[80]	H_2O 3; EtOH 4; eth 4; CS_2 4
9372	Phenol, 3-iodo-	C_6H_5IO 220.01	626-02-8 118	186[100]	4-06-00-01073	H_2O 2; EtOH 3; eth 3
9373	Phenol, 4-iodo- p-Iodophenol	C_6H_5IO 220.01	540-38-5 93.5	4929 139[5] dec	4-06-00-01074 1.8573[112]	H_2O 2; EtOH 4; eth 4
9374	Phenol, 2-iodo-4-methyl- p-Cresol, 2-iodo-	C_7H_7IO 234.04	16188-57-1 35	117[12]	4-06-00-02147 1.684[25]	1.5331[25]
9375	Phenol, 2-iodo-6-methyl-	C_7H_7IO 234.04	24885-45-8 15.5	70-5[2]	4-06-00-02008 1.661[25]	1.610[25]
9376	Phenol, 2-mercapto-	C_6H_6OS 126.18	1121-24-0 5.5	217; 89[8]	4-06-00-05633 1.2371[0]	bz 4; eth 4; EtOH 4
9377	Phenol, 3-mercapto-	C_6H_6OS 126.18	40248-84-8 16.5	168[35]	2-06-00-00827	bz 4; EtOH 4
9378	Phenol, 4-mercapto-	C_6H_6OS 126.18	637-89-8 29.5	167[45]; 135[11]	4-06-00-05790 1.1285[25]	H_2O 3; EtOH 3; alk 3; con sulf 3 1.5101[25]
9379	Phenol, 2-methoxy- Guaiacol	$C_7H_8O_2$ 124.14	90-05-1 32	4457 205	4-06-00-05563 1.1287[21]	H_2O 2; EtOH 3; eth 3; ctc 3 1.5429[20]
9380	Phenol, 3-methoxy-	$C_7H_8O_2$ 124.14	150-19-6 <-17	114[5]	4-06-00-05662 1.131[25]	H_2O 2; EtOH 5; eth 5; chl 2 1.5510[20]

No.	Name Synonym	Mol. Form. Mol. Wt.	CAS RN mp/°C	Merck No. bp/°C	Beil. Ref. den/g cm^{-3}	Solubility n_D
9381	Phenol, 4-methoxy- 4-Methoxyphenol	$C_7H_8O_2$ 124.14	150-76-5 57	243	4-06-00-05717	H_2O 3; EtOH 4; eth 4; bz 3
9382	Phenol, 2-methoxy-, acetate	$C_9H_{10}O_3$ 166.18	613-70-7 31.5	123-4[13]	4-06-00-05581 1.1285[25]	H_2O 1; EtOH 3; eth 3 1.5101[25]
9383	Phenol, 2-methoxy-, benzoate Guaiacol benzoate	$C_{14}H_{12}O_3$ 228.25	531-37-3 57.5	4458	4-09-00-00371	eth 4; chl 4
9384	Phenol, 2-methoxy-, carbonate (2:1) Guaiacol carbonate	$C_{15}H_{14}O_5$ 274.27	553-17-3 89	4459	3-06-00-04233	H_2O 1; EtOH 2; eth 3; chl 4
9385	Phenol, 2-methoxy-4-methyl- Creosol	$C_8H_{10}O_2$ 138.17	93-51-6 5.5	2573 221	4-06-00-05878 1.098[20]	eth 4; EtOH 4 1.5353[25]
9386	Phenol, 2-methoxy-3-nitro-	$C_7H_7NO_4$ 169.14	20734-71-8 70	sub	3-06-00-04263	H_2O 2; EtOH 4; chl 3; peth 3
9387	Phenol, 2-methoxy-, phosphate (3:1) Guaiacol phosphate	$C_{21}H_{21}O_7P$ 416.37	563-03-1 91	4460 275-80[3]	3-06-00-04246	ace 4; tol 4; chl 4
9388	Phenol, 2-methoxy-4-(1-propenyl)- Isoeugenol	$C_{10}H_{12}O_2$ 164.20	97-54-1 266	5054	3-06-00-04993 1.080[25]	eth 4; EtOH 4 1.5739[19]
9389	Phenol, 2-methoxy-4-(2-propenyl)- Eugenol	$C_{10}H_{12}O_2$ 164.20	97-53-0 -7.5	3855 253.2	4-06-00-06337 1.0652[20]	H_2O 1; EtOH 5; eth 5; chl 3 1.5405[20]
9390	Phenol, 2-methoxy-5-(2-propenyl)-	$C_{10}H_{12}O_2$ 164.20	501-19-9 8.5	253.5	3-06-00-05024 1.0613[25]	eth 4; EtOH 4 1.5413[20]
9391	Phenol, 2-methoxy-6-(1-propenyl)-	$C_{10}H_{12}O_2$ 164.20	1076-55-7 81	267.5	3-06-00-04991	H_2O 2; EtOH 3; eth 3; ace 3
9392	Phenol, 2-methoxy-6-(2-propenyl)-	$C_{10}H_{12}O_2$ 164.20	579-60-2 250.5		4-06-00-06334 1.062[20]	ace 3; bz 3 1.5545[20]
9393	Phenol, 2-methoxy-4-(1-propenyl)-, acetate, (E)-	$C_{12}H_{14}O_3$ 206.24	5912-87-8 80.5	282.5	4-06-00-06330 1.0251[100]	eth 4 1.5052[100]
9394	Phenol, 2-methoxy-4-(1-propenyl)-, acetate, (Z)- Isoeugenol, acetate (cis)	$C_{12}H_{14}O_3$ 206.24	97412-23-2 160-2[13]		3-06-00-05007 1.0947[19]	H_2O 1; eth 3 1.5418[20]
9395	Phenol, 2-methoxy-4-(2-propenyl)-, acetate	$C_{12}H_{14}O_3$ 206.24	93-28-7 30.5	281; 127-8[6]	3-06-00-05029 1.0806[20]	H_2O 1; EtOH 3; ctc 2 1.5205[20]
9396	Phenol, 2-methoxy-4-(1-propenyl)-, (E)-	$C_{10}H_{12}O_2$ 164.20	5932-68-3 33.5	141-2[13]	4-06-00-06324 1.0852[20]	H_2O 2; EtOH 3; eth 3; chl 3 1.5784[20]
9397	Phenol, 2-methoxy-4-(1-propenyl)-, (Z)-	$C_{10}H_{12}O_2$ 164.20	5912-86-7 133[13]; 80[0.5]		4-06 00 06324 1.0837[20]	H_2O 2; EtOH 3; eth 3 1.5726[20]
9398	Phenol, 2-methoxy-5-[(1,2,3,4-tetrahydro-6,7-dimethoxy-2-methyl-1-isoquinolinyl)methyl]-, (±)- Laudanine	$C_{20}H_{25}NO_4$ 343.42	85-64-3 167	5248	5-21-06-00048 1.26[20]	H_2O 2; EtOH 2; eth 2; bz 3
9399	Phenol, 2-methoxy-5-[(1,2,3,4-tetrahydro-6,7-dimethoxy-2-methyl-1-isoquinolinyl)methyl]-, (R)- Laudanidine	$C_{20}H_{25}NO_4$ 343.42	301-21-3 184.5	5247	5-21-06-00048	H_2O 4; bz 4
9400	Phenol, 2-methyl- o-Cresol	C_7H_8O 108.14	95-48-7 29.8	2580 191.0	4-06-00-01940 1.135[25]	H_2O 3; EtOH 4; eth 4; ace 5 1.5361[20]
9401	Phenol, 3-methyl- m-Cresol	C_7H_8O 108.14	108-39-4 11.8	2579 202.2	4-06-00-02035 1.0341[20]	H_2O 2; EtOH 5; eth 5; ace 5 1.5438[20]
9402	Phenol, 4-methyl- p-Cresol	C_7H_8O 108.14	106-44-5 35.5	2581 201.9	4-06-00-02093 1.0185[40]	H_2O 2; EtOH 5; eth 5; ace 5 1.5312[20]
9403	Phenol, 4-[2-(methylamino)propyl]- Pholedrine	$C_{10}H_{15}NO$ 165.24	370-14-9 161	7304	3-13-00-01710	eth 4; EtOH 4
9404	Phenol, 4-(3-methylbutyl)-	$C_{11}H_{16}O$ 164.25	1805-61-4 29	255; 128[7]	4-06-00-03378 0.9562[23]	1.505[27]
9405	Phenol, 2-methyl-4,6-dinitro- Dinitrocresol	$C_7H_6N_2O_5$ 198.14	534-52-1 86.5	3272	4-06-00-02014	H_2O 2; EtOH 3; eth 3; ace 3
9406	Phenol, 3,3'-methylenebis- Diphenylmethane, 3,3'-dihydroxy-	$C_{13}H_{12}O_2$ 200.24	10193-50-7 102.5	235[3]	3-06-00-05411	eth 4; EtOH 4; HOAc 4
9407	Phenol, 4,4'-methylenebis-	$C_{13}H_{12}O_2$ 200.24	620-92-8 162.5	314[3] sub	4-06-00-06664	EtOH 3; eth 3; chl 3; DMSO-d_6 2
9408	Phenol, 2,2'-methylenebis[4-chloro- Dichlorophene	$C_{13}H_{10}Cl_2O_2$ 269.13	97-23-4 177.5	3059	4-06-00-06658	H_2O 1; EtOH 3; ace 3
9409	Phenol, 2,2'-methylenebis[3,4,6-trichloro- Hexachlorophene	$C_{13}H_6Cl_6O_2$ 406.91	70-30-4 166.5	4602	4-06-00-06659	H_2O 1; EtOH 3; eth 3; ace 3
9410	Phenol, 2-(1-methylethyl)-	$C_9H_{12}O$ 136.19	88-69-7 15.5	213.5	4-06-00-03210 1.012[20]	H_2O 2; EtOH 3; eth 3; bz 3 1.5315[20]
9411	Phenol, 3-(1-methylethyl)-	$C_9H_{12}O$ 136.19	618-45-1 26	228	4-06-00-03214	eth 4 1.5261[20]
9412	Phenol, 4-(1-methylethyl)-	$C_9H_{12}O$ 136.19	99-89-8 62.3	230; 110[10]	4-06-00-03215 0.990[20]	H_2O 2; EtOH 3; chl 3 1.5228[20]
9413	Phenol, 4,4'-(1-methylethylidene)bis- Bisphenol A	$C_{15}H_{18}O_2$ 228.29	80-05-7 153	1311 220[4]; 222[3]	4-06-00-06717	H_2O 1; EtOH 4; eth 4; bz 4
9414	Phenol, 2-methyl-4-(1-methylethyl)-	$C_{10}H_{14}O$ 150.22	1740-97-2 8.6	230; 83[3]	4-06-00-03330 0.9793[25]	H_2O 3; EtOH 3; bz 3; chl 3 1.5253[20]
9415	Phenol, 2-methyl-5-(1-methylethyl)- Carvacrol	$C_{10}H_{14}O$ 150.22	499-75-2 1	1881 237.7	4-06-00-03331 0.9772[20]	H_2O 2; EtOH 3; eth 3; ace 4 1.5230[20]
9416	Phenol, 2-methyl-6-(1-methylethyl)-	$C_{10}H_{14}O$ 150.22	3228-04-4 -14.5	225.5	3-06-00-01882 0.9782[25]	EtOH 3; bz 3; chl 3; os 3 1.5239[15]
9417	Phenol, 4-methyl-2-(1-methylethyl)-	$C_{10}H_{14}O$ 150.22	4427-56-9 36.5	228.5	4-06-00-03329 0.9910[20]	H_2O 3; EtOH 3; bz 3; chl 3 1.5275[20]

No.	Name / Synonym	Mol. Form. / Mol. Wt.	CAS RN / mp/°C	Merck No. / bp/°C	Beil. Ref. / den/g cm^{-3}	Solubility / n_D
9418	Phenol, 5-methyl-2-(1-methylethyl)- / Thymol	$C_{10}H_{14}O$ / 150.22	89-83-8 / 51.5	9333 / 232.5	4-06-00-03334 / 0.970[25]	H_2O 1; EtOH 4; eth 4; chl 4 / 1.5227[20]
9419	Phenol, 2-methyl-5-(1-methylethyl)-, acetate / Carvacrol, acetate	$C_{12}H_{16}O_2$ / 192.26	6380-28-5 /	/ 246.5	2-06-00-00494 / 0.9896[25]	eth 4; EtOH 4 / 1.4913[28]
9420	Phenol, 5-methyl-2-(1-methylethyl)-, acetate / Thymol, acetate	$C_{12}H_{16}O_2$ / 192.26	528-79-0 /	/ 245	4-06-00-03337 / 1.009[9]	bz 4; eth 4; EtOH 4; chl 4
9421	Phenol, 5-methyl-2-(1-methylpropyl)- / 6-sec-Butyl-m-cresol	$C_{11}H_{16}O$ / 164.25	29472-95-5 /	/ 89-90[2]	4-06-00-03394 / 0.9612[25]	/ 1.5188[20]
9422	Phenol, 2-methyl-4-nitro-	$C_7H_7NO_3$ / 153.14	99-53-6 / 96	/ 188[9]; 132[0.2]	4-06-00-02011	H_2O 2; EtOH 4; eth 4; bz 4
9423	Phenol, 2-methyl-6-nitro-	$C_7H_7NO_3$ / 153.14	13073-29-5 / 70	/ 255 dec; 102[9]	4-06-00-02009	H_2O 1; EtOH 4; eth 4; peth 3
9424	Phenol, 4-methyl-2-nitro-	$C_7H_7NO_3$ / 153.14	119-33-5 / 36.5	/ 125[22]	4-06-00-02149 / 1.2399[20]	ace 4; bz 4; eth 4; EtOH 4 / 1.5744[40]
9425	Phenol, 5-methyl-2-pentyl- / 6-n-Amyl-m-cresol	$C_{12}H_{18}O$ / 178.27	1300-94-3 / 24	643 / 137-9[15]	3-06-00-02005	ace 4; eth 4; EtOH 4
9426	Phenol, 4-methyl-2-(phenylazo)-	$C_{13}H_{12}N_2O$ / 212.25	952-47-6 / 108.5	/ sub	4-16-00-00197	EtOH 3; eth 3; bz 3; chl 3
9427	Phenol, 4-methyl-6-(2-propenyl)-	$C_{10}H_{12}O$ / 148.20	3354-58-3 /	/ 232	4-06-00-03844 / 0.992[25]	/ 1.5381[20]
9428	Phenol, 4-methyl-2-(1-propenyl)-	$C_{10}H_{12}O$ / 148.20	53889-94-4 /	/ 126-8[14]	4-06-00-03844 / 1.019[20]	/ 1.5727[20]
9429	Phenol, 4-methyl-2-(2-propenyl)-	$C_{10}H_{12}O$ / 148.20	6628-06-4 /	/ 237	4-06-00-03845 / 0.96[25]	/ 1.5385[20]
9430	Phenol, 2-(1-methylpropyl)- / o-sec-Butylphenol	$C_{10}H_{14}O$ / 150.22	89-72-5 / 16	/ 228; 116[21]	3-06-00-01852 / 0.9804[25]	/ 1.5200[25]
9431	Phenol, 4-(1-methylpropyl)-	$C_{10}H_{14}O$ / 150.22	99-71-8 / 61.5	/ 241	4-06-00-03279 / 0.986[20]	H_2O 1; EtOH 3; eth 4; alk 3 / 1.5182[21]
9432	Phenol, 4-(2-methylpropyl)-	$C_{10}H_{14}O$ / 150.22	4167-74-2 / 51.5	/ 237	4-06-00-03288 / 0.9778[20]	ace 4; eth 4; EtOH 4 / 1.5319[25]
9433	Phenol, 4,4'-(1-methylpropylidene)bis- / Bisphenol B	$C_{16}H_{18}O_2$ / 242.32	77-40-7 / 120.5	1312	4-06-00-06738	ace 4; MeOH 4
9434	Phenol, 3-methyl-2,4,6-trinitro- / 3-Methylpicric acid	$C_7H_5N_3O_7$ / 243.13	602-99-3 / 109.5	/ exp 150	4-06-00-02079	H_2O 2; EtOH 3; eth 3; ace 3
9435	Phenol, 2-nitro- / o-Nitrophenol	$C_6H_5NO_3$ / 139.11	88-75-5 / 44.8	6541 / 216	4-06-00-01246 / 1.294[40]	H_2O 2; EtOH 4; eth 4; ace 4 / 1.5723[50]
9436	Phenol, 3-nitro- / m-Nitrophenol	$C_6H_5NO_3$ / 139.11	554-84-7 / 96.8	6540 / 194[70]	4-06-00-01269 / 1.2797[100]	H_2O 2; EtOH 4; eth 4; ace 4
9437	Phenol, 4-nitro- / p-Nitrophenol	$C_6H_5NO_3$ / 139.11	100-02-7 / 113.8	6542	4-06-00-01279 / 1.479[20]	H_2O 2; EtOH 4; eth 4; ace 4
9438	Phenol, 4-nitroso- / 4-Nitrosophenol	$C_6H_5NO_2$ / 123.11	104-91-6 / 144 dec	6563	4-07-00-02073	H_2O 2; EtOH 3; eth 3; ace 3
9439	Phenol, pentabromo- / Pentabromophenol	C_6HBr_5O / 488.59	608-71-9 / 229.5	/ sub	3-06-00-00766	H_2O 1; EtOH 3; eth 2; bz 3
9440	Phenol, pentachloro- / Pentachlorophenol	C_6HCl_5O / 266.34	87-86-5 / 174	7059 / 310 dec	4-06-00-01025 / 1.978[22]	H_2O 2; EtOH 4; eth 4; bz 3
9441	Phenol, 3-pentadecyl-	$C_{21}H_{36}O$ / 304.52	501-24-6 / 53.5	230[8]; 197[1.5]	4-06-00-03556	ace 4; bz 4; EtOH 4
9442	Phenol, pentafluoro- / Pentafluorophenol	C_6HF_5O / 184.07	771-61-9 / 32.8	/ 145.6	4-06-00-00782	/ 1.4263[20]
9443	Phenol, pentamethyl- / Pentamethylphenol	$C_{11}H_{16}O$ / 164.25	2819-86-5 / 128	/ 267	4-06-00-03412	H_2O 1; EtOH 3
9444	Phenol, 4-pentyl-	$C_{11}H_{16}O$ / 164.25	14938-35-3 / 23	/ 250.5	4-06-00-03370 / 0.960[20]	eth 4; EtOH 4 / 1.5272[25]
9445	Phenol, 2-(phenylamino)-	$C_{12}H_{11}NO$ / 185.23	644-71-3 / 69.5	/ 185[20]	4-13-00-00822	H_2O 2; EtOH 3; eth 3; bz 2
9446	Phenol, 3-(phenylamino)-	$C_{12}H_{11}NO$ / 185.23	101-18-8 / 81.5	/ 340	3-13-00-00944	H_2O 2; EtOH 4; eth 4; ace 4
9447	Phenol, 4-(phenylamino)-	$C_{12}H_{11}NO$ / 185.23	122-37-2 / 73	/ 330	4-13-00-01052	H_2O 2; EtOH 4; eth 4; bz 4
9448	Phenol, 4-(phenylazo)-	$C_{12}H_{10}N_2O$ / 198.22	1689-82-3 / 155	/ 225[20] dec	2-16-00-00038	H_2O 1; EtOH 4; eth 4; bz 3
9449	Phenol, 4-(phenylazo)-, acetate (ester)	$C_{14}H_{12}N_2O_2$ / 240.26	13102-31-3 / 89	/ 360 dec	1-16-00-00236	EtOH 4; lig 4
9450	Phenol, 2-[(phenylimino)methyl]-	$C_{13}H_{11}NO$ / 197.24	779-84-0 / 49.5		4-12-00-00339 / 1.087[25]	H_2O 1; EtOH 3
9451	Phenol, 2-(phenylmethoxy)-	$C_{13}H_{12}O_2$ / 200.24	6272-38-4 /	/ 173-4[13]	4-06-00-05572 / 1.154[22]	eth 4; EtOH 4 / 1.5906[18]
9452	Phenol, 3-(phenylmethoxy)-	$C_{13}H_{12}O_2$ / 200.24	3769-41-3 / 69.2	/ 200[5]	4-06-00-05669	H_2O 4
9453	Phenol, 4-(phenylmethoxy)- / Monobenzone	$C_{13}H_{12}O_2$ / 200.24	103-16-2 / 122	6159	4-06-00-05728	H_2O 2; EtOH 4; eth 4; ace 3
9454	Phenol, 2-(phenylmethyl)- / o-Benzylphenol	$C_{13}H_{12}O$ / 184.24	28994-41-4 / 21	1158 / 312	4-06-00-04628	ace 4; bz 4; EtOH 4 / 1.5994[20]
9455	Phenol, 4-(phenylmethyl)- / p-Benzylphenol	$C_{13}H_{12}O$ / 184.24	101-53-1 / 84	1159 / 322	4-06-00-04640	H_2O 3; EtOH 3; eth 3; bz 3
9456	Phenol, 2-(1-propenyl)-	$C_9H_{10}O$ / 134.18	6380-21-8 / -6	/ 220	3-06-00-02393 / 1.0246[15]	eth 4 / 1.5181[20]
9457	Phenol, 2-(2-propenyl)-	$C_9H_{10}O$ / 134.18	1745-81-9 / -6	/ 220	4-06-00-03807 / 1.0246[15]	eth 4 / 1.5181[20]

No.	Name Synonym	Mol. Form. Mol. Wt.	CAS RN mp/°C	Merck No. bp/°C	Beil. Ref. den/g cm^{-3}	Solubility n_D
9458	Phenol, 4-(2-propenyl)- Chavicol	$C_9H_{10}O$ 134.18	501-92-8 15.8	2039 238	4-06-00-03817 1.0203[15]	eth 4; EtOH 4; chl 4 1.5441[18]
9459	Phenol, 2-(1-propenyl)-, (E)-	$C_9H_{10}O$ 134.18	23019-59-2 37.5	 230.5	4-06-00-03793 1.0441[16]	eth 4; lig 4 1.5823[21]
9460	Phenol, 2-propoxy-	$C_9H_{12}O_2$ 152.19	6280-96-2 	 227	4-06-00-05566 1.0523[25]	EtOH 4 1.5176[25]
9461	Phenol, 2-propyl-	$C_9H_{12}O$ 136.19	644-35-9 7	 220	4-06-00-03176 1.015[20]	eth 4; EtOH 4
9462	Phenol, 3-propyl-	$C_9H_{12}O$ 136.19	621-27-2 26	 228	4-06-00-03180 0.987[20]	H_2O 2; EtOH 3; eth 3 1.5223[20]
9463	Phenol, 4-propyl-	$C_9H_{12}O$ 136.19	645-56-7 22	 232.6	4-06-00-03181 1.009[20]	H_2O 2; EtOH 3; ctc 2 1.5379[25]
9464	Phenol, 4,4'-propylidenebis-	$C_{15}H_{16}O_2$ 228.29	1576-13-2 132	 275[20]	3-06-00-05457	eth 4; EtOH 4
9465	Phenol, 4,4'-(2-pyridinylmethylene)bis-, diacetate (ester) Bisacodyl	$C_{22}H_{19}NO_4$ 361.40	603-50-9 133.5	1253	5-21-05-00415	
9466	Phenol, 4,4'-sulfonylbis-	$C_{12}H_{10}O_4S$ 250.28	80-09-1 240.5		4-06-00-05809 1.3663[15]	H_2O 1; EtOH 3; eth 3; bz 2
9467	Phenol, 2,3,4,6-tetrabromo- 2,3,4,6-Tetrabromophenol	$C_6H_2Br_4O$ 409.70	14400-94-3 113.5	 sub	3-06-00-00766	bz 4; EtOH 4
9468	Phenol, 2,3,4,5-tetrabromo-6-methyl- 3,4,5,6-Tetrabromo-o-cresol	$C_7H_4Br_4O$ 423.72	576-55-6 208	9120 dec	3-06-00-01272	H_2O 1; EtOH 3; eth 3; bz 3
9469	Phenol, 2,3,4,5-tetrachloro- 2,3,4,5-Tetrachlorophenol	$C_6H_2Cl_4O$ 231.89	4901-51-3 116.5	 sub	4-06-00-01020	EtOH 4
9470	Phenol, 2,3,4,6-tetrachloro- 2,3,4,6-Tetrachlorophenate	$C_6H_2Cl_4O$ 231.89	58-90-2 70	 150[15]	4-06-00-01021	H_2O 1; EtOH 3; bz 3; chl 3
9471	Phenol, 2,3,5,6-tetrachloro- 2,3,5,6-Tetrachlorophenol	$C_6H_2Cl_4O$ 231.89	935-95-5 115		4-06-00-01025	H_2O 2; bz 4; lig 3
9472	Phenol, 2,3,5,6-tetrachloro-4-methoxy- Drosophilin A	$C_7H_4Cl_4O_2$ 261.92	484-67-3 116	3441	3-06-00-04436	
9473	Phenol, 2,3,4,5-tetramethyl- Prehnitenol	$C_{10}H_{14}O$ 150.22	488-70-0 85.3	 266	4-06-00-03359	H_2O 2; EtOH 4; eth 4; lig 2
9474	Phenol, 2,3,4,6-tetramethyl-	$C_{10}H_{14}O$ 150.22	3238-38-8 80.5	 240	4-06-00-03360	EtOH 3
9475	Phenol, 2,3,5,6-tetramethyl-	$C_{10}H_{14}O$ 150.22	527-35-5 118.5	 247	4-06-00-03363	chl 3; peth 3; HOAc 3
9476	Phenol, 2,3,4,6-tetranitro- 2,3,4,6-Tetranitrophenol	$C_6H_2N_4O_9$ 274.10	641-16-7 140 dec	 exp	2-06-00-00284	H_2O 4; chl 3
9477	Phenol, 2,2'-thiobis[4-chloro- Fenticlor	$C_{12}H_8Cl_2O_2S$ 287.17	97-24-5 174	3947	4-06-00-05645	H_2O 1; EtOH 3; eth 3; gl HOAc 3
9478	Phenol, 2,2'-thiobis[4,6-dichloro- Bithionol	$C_{12}H_6Cl_4O_2S$ 356.06	97-18-7 188	1316	4-06-00-05646 1.73[25]	ace 4
9479	Phenol, 2,4,6-tribromo- 2,4,6-Tribromophenol	$C_6H_3Br_3O$ 330.80	118-79-6 95.5	9526 286	4-06-00-01067 2.55[20]	H_2O 2; EtOH 4; eth 3; bz 3
9480	Phenol, 2,4,6-tribromo-3-methyl- 2,4,6-Tribromo-m-cresol	$C_7H_5Br_3O$ 344.83	4619-74-3 84	9524	3-06-00-01324	EtOH 3; eth 3; bz 3; chl 2
9481	Phenol, 2,3,4-trichloro- 2,3,4-Trichlorophenol	$C_6H_3Cl_3O$ 197.45	15950-66-0 83.5	 sub	3-06-00-00716	EtOH 3; eth 3; bz 3; alk 3
9482	Phenol, 2,3,5-trichloro- 2,3,5-Trichlorophenol	$C_6H_3Cl_3O$ 197.45	933-78-8 62	 248-9[250]	3-06-00-00716	eth 4; EtOH 4
9483	Phenol, 2,3,6-trichloro- 2,3,6-Trichlorophenol	$C_6H_3Cl_3O$ 197.45	933-75-5 58		4-06-00-00962	H_2O 2; EtOH 4; eth 4; bz 4
9484	Phenol, 2,4,5-trichloro- 2,4,5-Trichlorophenol	$C_6H_3Cl_3O$ 197.45	95-95-4 69	9555 247	4-06-00-00962	H_2O 2; EtOH 4; eth 4; bz 4
9485	Phenol, 2,4,6-trichloro- 2,4,6-Trichlorophenol	$C_6H_3Cl_3O$ 197.45	88-06-2 69	9556 246	4-06-00-01005 1.490[75]	H_2O 2; EtOH 3; eth 3; HOAc 3
9486	Phenol, 3,4,5-trichloro- 3,4,5-Trichlorophenol	$C_6H_3Cl_3O$ 197.45	609-19-8 101	 275	3-06-00-00729	H_2O 2; eth 3; lig 2
9487	Phenol, 2,3,4-trichloro-6-methyl- 4,5,6-Trichloro-o-cresol	$C_7H_5Cl_3O$ 211.47	551-78-0 77	9548	3-06-00-01268	
9488	Phenol, 2,3,6-trichloro-4-methyl- 2,3,6-Trichloro-p-cresol	$C_7H_5Cl_3O$ 211.47	551-77-9 66.5	9546	0-06-00-00404	EtOH 4
9489	Phenol, 2,4,6-trichloro-3-methyl- 2,4,6-Trichloro-m-cresol	$C_7H_5Cl_3O$ 211.47	551-76-8 46	9547 265	4-06-00-02070	EtOH 4; MeOH 4; chl 4
9490	Phenol, 3,4,6-trichloro-2-nitro- 3,4,6-Trichloro-2-nitrophenol	$C_6H_2Cl_3NO_3$ 242.44	82-62-2 92.5	9554	3-06-00-00842	
9491	Phenol, 3-(trifluoromethyl)-	$C_7H_5F_3O$ 162.11	98-17-9 -1.8	 178	4-06-00-02060 1.3418[25]	
9492	Phenol, 2,4,6-triiodo- 2,4,6-Triiodophenol	$C_6H_3I_3O$ 471.80	609-23-4 159.8	 sub	4-06-00-01085	H_2O 1; EtOH 2; eth 3; ace 3
9493	Phenol, 2,3,4-trimethyl-	$C_9H_{12}O$ 136.19	526-85-2 81	 236	4-06-00-03245	bz 4; eth 4; EtOH 4
9494	Phenol, 2,4,5-trimethyl-	$C_9H_{12}O$ 136.19	496-78-6 72	 232	4-06-00-03247	H_2O 1; EtOH 4; eth 4
9495	Phenol, 2,4,6-trimethyl-	$C_9H_{12}O$ 136.19	527-60-6 73	 220	4-06-00-03253	eth 4; EtOH 4
9496	Phenol, 2,4,5-trimethyl-, acetate 2,4,5-Trimethylphenyl acetate	$C_{11}H_{14}O_2$ 178.23	69305-42-6 34	 245.5	2-06-00-00482	eth 4; EtOH 4

No.	Name Synonym	Mol. Form. Mol. Wt.	CAS RN mp/°C	Merck No. bp/°C	Beil. Ref. den/g cm^{-3}	Solubility n_D
9497	Phenol, 2,4,6-trinitro- Picric acid	$C_6H_3N_3O_7$ 229.11	88-89-1 122.5	7380 exp 300	4-06-00-01388	H_2O 2; EtOH 3; eth 3; ace 4 1.763
9498	Phenol, 2,4,6-tris(1,1-dimethylethyl)-	$C_{18}H_{30}O$ 262.44	732-26-3 131	278	4-06-00-03539 0.864^{27}	H_2O 1; EtOH 3; ace 3; ctc 3
9499	10H-Phenothiazine Thiodiphenylamine	$C_{12}H_9NS$ 199.28	92-84-2 187.5	7220 371	4-27-00-01214	ace 4; bz 4; eth 4; EtOH 4
9500	10H-Phenothiazine-2-acetic acid, 10-methyl- Metiazinic acid	$C_{15}H_{13}NO_2S$ 271.34	13993-65-2 144	6057		chl 3
9501	10H-Phenothiazine, 2-chloro-10-[3-(4-methyl-1-piperazinyl)propyl]- Phenothiazine, 2-chloro-10(3(4-methyl-1-piperazinyl)propyl)	$C_{20}H_{24}ClN_3S$ 373.95	58-38-8 228	7768	4-27-00-01307	
9502	10H-Phenothiazine, 10-[(dimethylamino)acetyl]- Ahistan	$C_{16}H_{16}N_2OS$ 284.38	518-61-6 144.5	179	4-27-00-01272	
9503	10H-Phenothiazine-10-ethanamine, N,N-diethyl- Diethazine	$C_{18}H_{22}N_2S$ 298.45	60-91-3	3098 167$^{0.5}$	4-27-00-01240	H_2O 1; dil HCl 3
9504	10H-Phenothiazine-10-ethanamine, N,N,α-trimethyl- Promethazine	$C_{17}H_{20}N_2S$ 284.43	60-87-7 60	7797 190-3$^{0.5}$	4-27-00-01253	H_2O 1; dil HCl 4
9505	10H-Phenothiazine-10-propanamine, 2-chloro-N,N-dimethyl- Chlorpromazine	$C_{17}H_{19}ClN_2S$ 318.87	50-53-3 200-5$^{0.8}$	2186	4-27-00-01300	H_2O 1; EtOH 4; eth 4; bz 4
9506	Phenothiazin-5-ium, 3,7-bis(dimethylamino)-, chloride Methylene blue	$C_{16}H_{18}ClN_3S$ 319.86	61-73-4	5979	4-27-00-05152	H_2O 3; EtOH 3; eth 1; chl 3
9507	10H-Phenoxazine Phenoxazine	$C_{12}H_9NO$ 183.21	135-67-1 156	7222 dec	4-27-00-01209	bz 4; eth 4; EtOH 4
9508	3H-Phenoxazin-3-one, 7-hydroxy-, 10-oxide Resazurin	$C_{12}H_7NO_4$ 229.19	550-82-3 sub	8144	4-27-00-03594	H_2O 1; EtOH 2; eth 1; alk 3
9509	DL-Phenylalanine	$C_9H_{11}NO_2$ 165.19	150-30-1 284-8 dec	7242 sub	4-14-00-01553	H_2O 4; EtOH 1; eth 1; bz 1
9510	L-Phenylalanine (S)-α-Aminobenzenepropanoic acid	$C_9H_{11}NO_2$ 165.19	63-91-2 283 dec	7242	4-14-00-01552	H_2O 4; EtOH 1; eth 1; bz 1
9511	L-Phenylalanine, N-L-α-aspartyl-, 1-methyl ester Aspartame	$C_{14}H_{18}N_2O_5$ 294.31	22839-47-0 246.5	861		
9512	L-Phenylalanine, ethyl ester	$C_{11}H_{15}NO_2$ 193.25	3081-24-1 148^{13}		4-14-00-01556 1.065^{15}	H_2O 2
9513	Phorate Phosphorodithioic acid, O,O-diethyl S-[(ethylthio)methyl] ester	$C_7H_{17}O_2PS_3$ 260.38	298-02-2 <-15 119$^{0.8}$	7305	1.16^{25}	
9514	Phosalone Phosphorodithioic acid, S-[(6-chloro-2-oxo-3(2H)-benzoxazolyl)methyl]O,O-diethyl ester	$C_{12}H_{15}ClNO_4PS_2$ 367.81	2310-17-0 46	7308		
9515	Phosmet Phosphorodithioic acid, S-[(1,3-dihydro-1,3-dioxo-2H-isoindol-2-yl)methyl] O,O-dimethyl ester	$C_{11}H_{12}NO_4PS_2$ 317.33	732-11-6 72 dec	7311		
9516	Phosphine, dicyclohexyl- Dicyclohexylphosphine	$C_{12}H_{23}P$ 198.29	829-84-5 281; 129^8		4-16-00-00947 0.904^{25}	1.5163^{20}
9517	Phosphine oxide, triethyl-	$C_6H_{15}OP$ 134.16	597-50-2 48	243	4-04-00-03463	H_2O 4; eth 4; EtOH 4
9518	Phosphine oxide, triphenyl-	$C_{18}H_{15}OP$ 278.29	791-28-6 156.5	>360	4-16-00-01011 1.2124^{23}	H_2O 2; EtOH 4; eth 2; bz 4
9519	Phosphine, phenyl- Monophenylphosphine	C_6H_7P 110.10	638-21-1 160.5		4-16-00-00948 1.001^{15}	1.5796^{20}
9520	Phosphine, tributyl- Tributylphosphine	$C_{12}H_{27}P$ 202.32	998-40-3 150^{50}		4-04-00-03436 0.812^{25}	1.4619^{20}
9521	Phosphine, triethyl- Triethyl phosphine	$C_6H_{15}P$ 118.16	554-70-1 -88 129	9590	4-04-00-03431 0.8006^{19}	H_2O 1; EtOH 5; eth 5 1.458^{15}
9522	Phosphine, trimethyl- Trimethylphosphine	C_3H_9P 76.08	594-09-2 -85 37.5		4-04-00-03429	H_2O 1; eth 3
9523	Phosphine, triphenyl- Triphenylphosphine	$C_{18}H_{15}P$ 262.29	603-35-0 80	9657 188^1	4-16-00-00951 1.0749^{80}	H_2O 1; EtOH 3; eth 4; bz 3 1.6358^{80}
9524	Phosphine, tris(chloromethyl)- Tris(chloromethyl)phosphine	$C_3H_6Cl_3P$ 179.41	13482-62-7 100^7		3-01-00-02608 1.414^{20}	ace 4; bz 4; eth 4; EtOH 4
9525	Phosphinic acid, diethyl- Diethylphosphinic acid	$C_4H_{11}O_2P$ 122.10	813-76-3 18.5	320	4-04-00-03474	H_2O 4; EtOH 4; eth 4
9526	Phosphinic acid, diethyl-, anhydride	$C_8H_{20}O_3P_2$ 226.19	7495-97-8 188^{14}		4-04-00-03477 1.1053^{20}	1.4647^{20}
9527	Phosphinic acid, diethyl-, ethyl ester Ethyl diethylphosphinate	$C_6H_{15}O_2P$ 150.16	4775-09-1 95^{14}		4-04-00-03475 0.9908^{20}	eth 4; EtOH 4 1.4337^{20}
9528	Phosphinic acid, dimethyl- Dimethylphosphinic acid	$C_2H_7O_2P$ 94.05	3283-12-3 92	377	4-04-00-03470	H_2O 4; EtOH 4; eth 4; bz 3
9529	Phosphinic chloride, diethyl-	$C_4H_{10}ClOP$ 140.55	1112-37-4 104^{15}		4-04-00-03476 1.1394^{20}	1.4647^{20}
9530	Phosphonic acid, acetyl-, diethyl ester O,O-Diethyl acetylphosphonate	$C_6H_{13}O_4P$ 180.14	919-19-7 114-5^{20}		4-02-00-00440 1.1005^{20}	1.4200^{26}

No.	Name / Synonym	Mol. Form. / Mol. Wt.	CAS RN / mp/°C	Merck No. / bp/°C	Beil. Ref. / den/g cm^{-3}	Solubility / n_D
9531	Phosphonic acid, bis(1,1-dimethylethyl) ester	$C_8H_{19}O_3P$ 194.21	13086-84-5	70-2[10]	4-01-00-01623 0.975[25]	1.4168[25]
9532	Phosphonic acid, dibutyl ester	$C_8H_{19}O_3P$ 194.21	1809-19-4	131[19]; 77[1.5]	4-01-00-01525 0.9485[20]	1.4220[20]
9533	Phosphonic acid, (1,1-dimethylethyl)-	$C_4H_{11}O_3P$ 138.10	4923-84-6 192	dec	4-04-00-03557	H_2O 4; EtOH 4
9534	Phosphonic acid, diphenyl ester	$C_{12}H_{11}O_3P$ 234.19	4712-55-4 12	218-9[26]	4-06-00-00703 1.223[25]	1.5575[20]
9535	Phosphonic acid, ethenyl-, diethyl ester	$C_6H_{13}O_3P$ 164.14	682-30-4	110[2]	4-04-00-03569 1.068[25]	1.4290[20]
9536	Phosphonic acid, ethyl-, diethyl ester Diethyl ethylphosphonate	$C_6H_{15}O_3P$ 166.16	78-38-6	198; 90[16]	4-04-00-03524 1.0259[20]	H_2O 2; EtOH 3; eth 3 1.4163[20]
9537	Phosphonic acid, ethyl-, dimethyl ester	$C_4H_{11}O_3P$ 138.10	6163-75-3	82[18]	4-04-00-03523 1.1029[30]	H_2O 4; bz 4; EtOH 4 1.4128[30]
9538	Phosphonic acid, methyl-	CH_5O_3P 96.02	993-13-5 108.5	dec	4-04-00-03498	H_2O 4; EtOH 4; eth 4; bz 1
9539	Phosphonic acid, methyl-, diethyl ester	$C_5H_{13}O_3P$ 152.13	683-08-9	194	4-04-00-03500 1.0406[30]	H_2O 3; EtOH 3; eth 3; bz 1 1.4101[30]
9540	Phosphonic acid, methyl-, dimethyl ester	$C_3H_9O_3P$ 124.08	756-79-6	181; 79.5[20]	4-04-00-03499 1.1507[20]	H_2O 3; EtOH 3; eth 3 1.4099[30]
9541	Phosphonic acid, methyl-, diphenyl ester Diphenyl methylphosphonate	$C_{13}H_{13}O_3P$ 248.22	7526-26-3 35	205[13]	4-06-00-00703 1.2051[20]	H_2O 1
9542	Phosphonic acid, (1-methylethyl)- 2-Propanephosphonic acid	$C_3H_9O_3P$ 124.08	4721-37-3 74.5	dec	4-04-00-03548	H_2O 4; eth 4; EtOH 4
9543	Phosphonic acid, methyl-, monoethyl ester	$C_3H_9O_3P$ 124.08	1832-53-7	106-7[0.1]	4-04-00-03499 1.1800[20]	1.4258[20]
9544	Phosphonic acid, (2-methylpropyl)-	$C_4H_{11}O_3P$ 138.10	4721-34-0 119	dec	4-04-00-03556	H_2O 4; eth 4; EtOH 4; xyl 4
9545	Phosphonic acid, (phenylmethyl)-, diethyl ester	$C_{11}H_{17}O_3P$ 228.23	1080-32-6	110[2]	4-16-00-01093	ctc 3 1.4930[20]
9546	Phosphonic acid, 1,3-propanediylbis-, tetraethyl ester	$C_{11}H_{26}O_6P_2$ 316.27	22401-25-8	185-6[3]	4-04-00-03582 1.1227[20]	1.4500[20]
9547	Phosphonic acid, propyl-	$C_3H_9O_3P$ 124.08	4672-38-2 73.8	dec	4-04-00-03542	H_2O 4; eth 4; EtOH 4
9548	Phosphonic dichloride, phenyl-	$C_6H_5Cl_2OP$ 194.98	824-72-6 1	258	4-16-00-01074 1.197[25]	DMSO-d_6 2 1.5581[25]
9549	Phosphonothioic acid, methyl-, O,O-bis(1-methylethyl) ester Phosphonic acid, thiono, methyl, diisopropyl ester	$C_7H_{17}O_2PS$ 196.25	66295-45-2	42[1]	4-04-00-03516 0.9885[25]	1.4512[25]
9550	Phosphonothioic acid, methyl-, O,O-dimethyl ester	$C_3H_9O_2PS$ 140.14	681-06-1	43[4]	4-04-00-03514 1.1375[25]	1.4738[25]
9551	Phosphonothioic acid, phenyl-, O-ethyl O-(4-nitrophenyl) ester O-Ethyl O-p-nitrophenyl benzenethiophosphonate	$C_{14}H_{14}NO_4PS$ 323.31	2104-64-5 36	3576	4-16-00-01089 1.27[25]	bz 4; eth 4; EtOH 4 1.5978[30]
9552	Phosphonothioic dichloride, phenyl- Dichlorophenylphosphine sulfide	$C_6H_5Cl_2PS$ 211.05	3497-00-5	205[130]	4-16-00-01090 1.376[13]	
9553	Phosphonous dichloride, (2,5-dimethylphenyl)- Phosphonous dichloride, 2,5-xylyl-	$C_8H_9Cl_2P$ 207.04	57150-66-0 -30	253.5	3-16-00-00848 1.25[18]	
9554	Phosphonous dichloride, (4-ethylphenyl)-	$C_8H_9Cl_2P$ 207.04	5274-50-0	251	4-16-00-00974 1.237[20]	1.584[20]
9555	Phosphonous dichloride, [4-(1-methylethyl)phenyl]- Benzene, 1-chlorophosphino-4-isopropyl	$C_9H_{11}Cl_2P$ 221.07	82906-78-3	269	3-16-00-00848 1.190[12]	ace 4 1.5677[25]
9556	Phosphonous dichloride, phenyl- Dichlorophenylphosphine	$C_6H_5Cl_2P$ 178.98	644-97-3 -51	225; 142[57]	4-16-00-00972 1.356[20]	bz 4 1.6030[20]
9557	Phosphoramidothioic dichloride, dimethyl- Phosphoric acid, amide,thiono,dichloride,N,N-dimethyl	$C_2H_6Cl_2NPS$ 178.02	1498-65-3	77.5[10]	4-04-00-00293 1.3731[20]	1.5390[20]
9558	Phosphoric acid, diethyl ester	$C_4H_{11}O_4P$ 154.10	598-02-7	203 dec	4-01-00-01339 1.186[25]	eth 4 1.4170[20]
9559	Phosphoric acid, dimethyl ester	$C_2H_7O_4P$ 126.05	813-78-5	174 dec	4-01-00-01259 1.335[25]	H_2O 4; ace 4; EtOH 4 1.408[25]
9560	Phosphoric acid, 2-ethylhexyl diphenyl ester	$C_{20}H_{27}O_4P$ 362.41	1241-94-7	232[5]	4-06-00-00718 1.090[25]	1.510[25]
9561	Phosphoric acid, monoethyl ester	$C_2H_7O_4P$ 126.05	1623-14-9	dec	4-01-00-01338 1.430[25]	H_2O 4; ace 4; eth 4; EtOH 4 1.427
9562	Phosphoric acid tributyl ester Tributyl phosphate	$C_{12}H_{27}O_4P$ 266.32	126-73-8	289	4-01-00-01531 0.9727[25]	H_2O 3; EtOH 5; eth 3; bz 3 1.4224[25]
9563	Phosphoric acid, triethyl ester Ethyl phosphate	$C_6H_{15}O_4P$ 182.16	78-40-0 -56.4	9589 215.5	4-01-00-01339 1.0695[20]	H_2O 3; EtOH 4; eth 3; bz 3 1.4053[20]
9564	Phosphoric acid, trimethyl ester	$C_3H_9O_4P$ 140.08	512-56-1 -46	197.2	4-01-00-01259 1.2144[20]	H_2O 4; EtOH 2; eth 3 1.3967[20]
9565	Phosphoric acid, tripentyl ester	$C_{15}H_{33}O_4P$ 308.40	2528-38-3	225[50]; 167[5]	4-01-00-01645 0.9608[20]	H_2O 1; EtOH 3; eth 3; tol 3 1.4319[20]
9566	Phosphoric acid, triphenyl ester Triphenyl phosphate	$C_{18}H_{15}O_4P$ 326.29	115-86-6 50.5	9656 245[11]	4-06-00-00720 1.2055[50]	H_2O 1; EtOH 3; eth 4; bz 4

No.	Name / Synonym	Mol. Form. / Mol. Wt.	CAS RN / mp/°C	Merck No. / bp/°C	Beil. Ref. / den/g cm^{-3}	Solubility / n_D
9567	Phosphoric acid, tripropyl ester / Tripropyl phosphate	$C_9H_{21}O_4P$ / 224.24	513-08-6 /	/ 252	4-01-00-01428 / 1.0121[20]	H_2O 2; EtOH 3; eth 3; chl 2 / 1.4165[20]
9568	Phosphoric acid, tris(1-methylethyl) ester	$C_9H_{21}O_4P$ / 224.24	513-02-0 /	/ 219	4-01-00-01478 / 0.9867[20]	EtOH 4 / 1.4057[20]
9569	Phosphoric acid, tris(2-methylphenyl) ester / Tri-o-tolyl phosphate	$C_{21}H_{21}O_4P$ / 368.37	78-30-8 / 11	9676 / 410	4-06-00-01979 / 1.1955[20]	H_2O 1; EtOH 4; eth 4; ctc 4 / 1.5575[20]
9570	Phosphoric acid, tris(3-methylphenyl) ester	$C_{21}H_{21}O_4P$ / 368.37	563-04-2 / 25.5	/ 260[15]	4-06-00-02057 / 1.150[25]	H_2O 1; EtOH 2; eth 3; ctc 4 / 1.5575[20]
9571	Phosphoric acid, tris(4-methylphenyl) ester	$C_{21}H_{21}O_4P$ / 368.37	78-32-0 / 77.5	/ 224[35]	4-06-00-02130 / 1.247[25]	EtOH 3; eth 3; bz 3; chl 3
9572	Phosphoric acid, tris(2-methylpropyl) ester / Triisobutyl phosphate	$C_{12}H_{27}O_4P$ / 266.32	126-71-6 /	/ 264	4-01-00-01598 / 0.9681[20]	H_2O 4; bz 4; eth 4; EtOH 4 / 1.4193[20]
9573	Phosphoric triamide, hexamethyl- / Tris(dimethylamino)phosphine oxide	$C_6H_{18}N_3OP$ / 179.20	680-31-9 /	4568 / 232.5	4-04-00-00284 / 1.03[20]	EtOH 3; eth 3 / 1.4579[20]
9574	Phosphorochloridic acid, diethyl ester / Diethyl chlorophosphate	$C_4H_{10}ClO_3P$ / 172.55	814-49-3 /	/ 93.5	4-01-00-01345 / 1.205[19]	/ 1.4170[20]
9575	Phosphorochloridic acid, diphenyl ester	$C_{12}H_{10}ClO_3P$ / 268.64	2524-64-3 /	/ 314-6[272]	4-06-00-00737 / 1.296[25]	tfa 3 / 1.5500[20]
9576	Phosphorochloridothioic acid, O,O-diethyl ester / Diethyl chlorothiophosphate	$C_4H_{10}ClO_2PS$ / 188.61	2524-04-1 /	/ 45[3]	4-01-00-01352 /	ctc 3
9577	Phosphorodichloridic acid, ethyl ester / Ethyl dichlorophosphate	$C_2H_5Cl_2O_2P$ / 162.94	1498-51-7 /	/ 60-5[10]	4-01-00-01345 /	/ 1.4338[20]
9578	Phosphorodichloridothioic acid, O-(2,4,5-trichlorophenyl) ester / Phenol, 2,4,5-trichloro-, O-phosphorodichloridothioate	$C_6H_2Cl_5OPS$ / 330.38	16805-78-0 /	/ 110[1]	4-06-00-00996 / 1.6653[20]	/ 1.6084[20]
9579	Phosphorodithioic acid, O,O,S-triethyl ester	$C_6H_{15}O_2PS_2$ / 214.29	2524-09-6 /	/ 128[20]	4-01-00-01402 / 1.1336[0]	/ 1.5043
9580	Phosphorofluoridic acid, bis(1-methylethyl) ester / Isoflurophate	$C_6H_{14}FO_3P$ / 184.15	55-91-4 /	5060 / 62[9]	4-01-00-01480 / 1.055[25]	H_2O 2; eth 3; oils 4; lig 2 / 1.3830[25]
9581	Phosphorothioic acid, O-(2,4-dichlorophenyl) O,O-diethyl ester / Dichlofenthion	$C_{10}H_{13}Cl_2O_3PS$ / 315.16	97-17-6 /	3030 /	4-06-00-00939 /	ctc 3; CS_2 3
9582	Phosphorothioic acid, O,O-diethyl O-[6-methyl-2-(1-methylethyl)-4-pyrimidinyl] ester / Diazinon	$C_{12}H_{21}N_2O_3PS$ / 304.35	333-41-5 /	2978 / 85-90[0.05]	5-23-11-00187 / 1.1088[20]	/ 1.4922[20]
9583	Phosphorothioic acid, O,O-diethyl O-(4-nitrophenyl) ester / Parathion	$C_{10}H_{14}NO_5PS$ / 291.26	56-38-2 / 6.1	6983 / 375	4-06-00-01337 / 1.2681[20]	H_2O 1; EtOH 5; eth 3; ace 3 / 1.5370[25]
9584	Phosphorothioic acid, O,O-dimethyl O-(2,4,5-trichlorophenyl) ester / Ronnel	$C_8H_8Cl_3O_3PS$ / 321.55	299-84-3 / 41	8239 / 151-4[0.4]	4-06-00-00994 / 1.44[32]	/ 1.5335[35]
9585	Phosphorothioic acid, O,O,O-triethyl ester	$C_6H_{15}O_3PS$ / 198.22	126-68-1 /	/ 100[16]	4-01-00-01348 / 1.082[25]	/ 1.4480[20]
9586	Phosphorous acid, bis(phenylmethyl) ester / Dibenzyl phosphite	$C_{14}H_{15}O_3P$ / 262.25	17176-77-1 / -5	2996 / 162[0.1]	4-06-00-02571 /	/ 1.5521[18]
9587	Phosphorous acid, diethyl ester	$C_4H_{11}O_3P$ / 138.10	123-22-8 /	/ 87[20]	/ 1.0720[20]	EtOH 3; eth 3 / 1.4101[20]
9588	Phosphorous acid, dimethyl ester	$C_2H_7O_3P$ / 110.05	868-85-9 /	/ 170.5	/ 1.2002[20]	EtOH 3; ctc 2; py 3 / 1.4036[20]
9589	Phosphorous acid, 2-ethylhexyl diphenyl ester	$C_{20}H_{27}O_3P$ / 346.41	15647-08-2 /	/ 148-56[0.15]	4-06-00-00695 / 1.054[20]	/ 1.5207[27]
9590	Phosphorous acid, tributyl ester	$C_{12}H_{27}O_3P$ / 250.32	102-85-2 /	/ 122[12]	4-01-00-01527 / 0.9259[20]	EtOH 3; eth 4; ctc 2 / 1.4321[19]
9591	Phosphorous acid, triethyl ester / Triethyl phosphite	$C_6H_{15}O_3P$ / 166.16	122-52-1 /	/ 157.9	4-01-00-01333 / 0.9629[20]	H_2O 1; EtOH 4; eth 4 / 1.4127[20]
9592	Phosphorous acid, trimethyl ester / Trimethyl phosphite	$C_3H_9O_3P$ / 124.08	121-45-9 /	/ 111.5	4-01-00-01256 / 1.0518[20]	EtOH 4; eth 4; ctc 2 / 1.4095[20]
9593	Phosphorous acid, triphenyl ester	$C_{18}H_{15}O_3P$ / 310.29	101-02-0 / 25	/ 360	4-06-00-00695 / 1.1842[20]	H_2O 1; EtOH 4; os 4 / 1.5900[20]
9594	Phosphorous acid, tripropyl ester / Propyl phosphite	$C_9H_{21}O_3P$ / 208.24	923-99-9 /	/ 206.5	4-01-00-01426 / 0.9417[20]	eth 4; EtOH 4 / 1.4282[20]
9595	Phosphorous acid, tris(1-methylethyl) ester	$C_9H_{21}O_3P$ / 208.24	116-17-6 /	/ 60-1[10]	4-01-00-01476 / 0.9685[18]	EtOH 3; eth 3; chl 3 / 1.4085[25]
9596	Phosphorous acid, tris(2-methylphenyl) ester	$C_{21}H_{21}O_3P$ / 352.37	2622-08-4 /	/ 238[11] 198[2]	4-01-00-01977 / 1.1423[20]	eth 3; chl 2 / 1.5740[28]
9597	Phosphorous acid, tris(4-methylphenyl) ester	$C_{21}H_{21}O_3P$ / 352.37	620-42-8 /	/ 250-5[10]	4-06-00-02128 / 1.1280[25]	eth 4 / 1.5703[28]
9598	p-Phthalaldehydic acid, 3-hydroxy- / 3-Hydroxybenzoic acid, 4-formyl	$C_8H_6O_4$ / 166.13	619-12-5 / 234	/ sub	4-10-00-03623 /	eth 4; EtOH 4
9599	Phthalazine / 2,3-Benzodiazine	$C_8H_6N_2$ / 130.15	253-52-1 / 90.5	7344 / 316	5-23-07-00149 /	H_2O 3; EtOH 3; eth 2; bz 3
9600	1,4-Phthalazinedione, 5-amino-2,3-dihydro- / Luminol	$C_8H_7N_3O_2$ / 177.16	521-31-3 / 330.5	5470 /	4-25-00-04192 /	H_2O 1; EtOH 2; eth 2; alk 4
9601	Picene / Benzo[a]chrysene	$C_{22}H_{14}$ / 278.35	213-46-7 / 368	7368 / 519	4-05-00-02724 /	H_2O 1; EtOH 2; bz 2; chl 2
9602	Piperazine / Diethylenediamine	$C_4H_{10}N_2$ / 86.14	110-85-0 / 106	7431 / 146	5-23-01-00030 /	H_2O 4; EtOH 3; eth 1; chl 3 / 1.446[113]

No.	Name / Synonym	Mol. Form. / Mol. Wt.	CAS RN / mp/°C	Merck No. / bp/°C	Beil. Ref. / den/g cm^{-3}	Solubility / n_D
9603	1-Piperazinecarboxamide, N,N-diethyl-4-methyl- / Diethylcarbamazine	$C_{10}H_{21}N_3O$ / 199.30	90-89-1 / 48	3106 / 109-11[3]	4-23-00-00225	
9604	1-Piperazinecarboxylic acid, ethyl ester	$C_7H_{14}N_2O_2$ / 158.20	120-43-4 /	/ 237	5-23-01-00055	H_2O 4; eth 4; EtOH 4 / 1.4760[25]
9605	Piperazine, 1-[(4-chlorophenyl)phenylmethyl]-4-[(3-methylphenyl)methyl]- / Meclizine	$C_{25}H_{27}ClN_2$ / 390.96	569-65-3 /	5657 / 230	5-23-01-00235	CS_2 3
9606	1,4-Piperazinedicarboxylic acid, diethyl ester / N,N'-Piperazinedicarboxylic acid, diethyl ester	$C_{10}H_{18}N_2O_4$ / 230.26	5470-28-0 / 47	/ 315	4-23-00-00254	eth 4; EtOH 4
9607	Piperazine, dihydrochloride / Diethylenediamine dihydrochloride	$C_4H_{12}Cl_2N_2$ / 159.06	142-64-3 /		5-23-01-00030	H_2O 2; EtOH 1
9608	Piperazine, 1,4-dimethyl-	$C_6H_{14}N_2$ / 114.19	106-58-1 /	/ 131	5-23-01-00128 / 0.8600[20]	H_2O 4; EtOH 4; eth 4 / 1.4474[20]
9609	Piperazine, 2,5-dimethyl-, cis- / cis-2,5-Dimethylpiperazine	$C_6H_{14}N_2$ / 114.19	6284-84-0 / 114	/ 162	5-23-03-00304	H_2O 4; EtOH 4; eth 2; bz 2 / 1.4720[20]
9610	Piperazine, 2,5-dimethyl-, trans- / trans-2,5-Dimethylpiperazine	$C_6H_{14}N_2$ / 114.19	2815-34-1 / 118.5	/ 162	5-23-03-00304	H_2O 4; EtOH 4; eth 2; bz 2
9611	Piperazine, 2,6-dimethyl-, cis- / cis-2,6-Dimethylpiperazine	$C_6H_{14}N_2$ / 114.19	21655-48-1 / 114	/ 160	5-23-03-00326	H_2O 3; EtOH 3; eth 1; bz 2
9612	2,5-Piperazinedione / Glycine anhydride	$C_4H_8N_2O_2$ / 114.10	106-57-0 / 312 dec	7434 / sub 260	5-24-05-00294	H_2O 2; EtOH 2; HCl 3
9613	Piperazine, 1,4-diphenyl-	$C_{16}H_{18}N_2$ / 238.33	613-39-8 / 165	/ 300 dec; 232[12]	5-23-01-00173	H_2O 1; EtOH 2
9614	1,4-Piperazinedipropanamine / 1,4-Bis(3-aminopropyl)piperazine	$C_{10}H_{24}N_4$ / 200.33	7209-38-3 / 15	/ 150-2[2]	5-23-01-00356 / 0.973[25]	1.5015[20]
9615	1-Piperazineethanamine / 1-(2-Aminoethyl)piperazine	$C_6H_{15}N_3$ / 129.21	140-31-8 /	/ 220	5-23-01-00257 / 0.985[25]	1.4983[20]
9616	1-Piperazineethanol	$C_6H_{14}N_2O$ / 130.19	103-76-4 /	/ 246	5-23-01-00406 / 1.061[25]	1.5065[20]
9617	1-Piperazineethanol, 4-[3-(2-chloro-10H-phenothiazin-10-yl)propyl]- / Perphenazine	$C_{21}H_{26}ClN_3OS$ / 403.98	58-39-9 / 97	7135 /	4-27-00-01307	
9618	Piperazine, 1-methyl-	$C_5H_{12}N_2$ / 100.16	109-01-3 /	/ 138	5-23-01-00082	H_2O 4; eth 4; EtOH 4 / 1.4378[20]
9619	Piperazine, 2-methyl- / 2-Methylpiperazine	$C_5H_{12}N_2$ / 100.16	109-07-9 / 62	/ 153	5-23-03-00267	H_2O 4; EtOH 3; eth 3; bz 3
9620	Piperazine, 1-phenyl-	$C_{10}H_{14}N_2$ / 162.23	92-54-6 /	/ 286.5; 161[15]	5-23-01-00160 / 1.0621[20]	H_2O 1; EtOH 5; eth 5; chl 3 / 1.5875[20]
9621	Piperazine, 1-(phenylmethyl)-	$C_{11}H_{16}N_2$ / 176.26	2759-28-6 /	/ 145-7[12]	5-23-01-00193	H_2O 3; EtOH 3; eth 3; chl 2 / 1.5430[28]
9622	Piperazine, 1,2,4-trimethyl- / 1,2,4-Trimethylpiperazine	$C_7H_{16}N_2$ / 128.22	120-85-4 /	/ 149.5	5-23-03-00271	ctc 3 / 1.4433[20]
9623	Piperazinium, 4-(2-cyclohexyl-2-hydroxy-2-phenylethyl)-1,1-dimethyl-, methyl sulfate (salt) / Hexocyclium methyl sulfate	$C_{21}H_{36}N_2O_5S$ / 428.59	115-63-9 /	4627 /		
9624	1-Piperidinamine	$C_5H_{12}N_2$ / 100.16	2213-43-6 /	/ 147	5-20-03-00516 / 0.928[25]	1.4750[20]
9625	1-Piperidinamine, 2,6-dimethyl-	$C_7H_{16}N_2$ / 128.22	39135-39-2 /	/ 65-80[30]	5-20-04-00187 / 0.865[25]	1.4650[20]
9626	3-Piperidinamine, 1-ethyl-	$C_7H_{16}N_2$ / 128.22	6789-94-2 /	/ 155; 82[40]	5-22-08-00026 / 0.923[25]	1.4715[20]
9627	4-Piperidinamine, 1-methyl-N-phenyl-N-(phenylmethyl)- / Bamipine	$C_{19}H_{24}N_2$ / 280.41	4945-47-5 / 115	966 /	5-22-08-00036	
9628	4-Piperidinamine, 2,2,6,6-tetramethyl-	$C_9H_{20}N_2$ / 156.27	36768-62-4 / 17	/ 188.5	5-22-08-00162 / 0.912[25]	1.4706[20]
9629	Piperidine / Azacyclohexane	$C_5H_{11}N$ / 85.15	110-89-4 / -11.03	7438 / 106.2	5-20-02-00003 / 0.8606[20]	H_2O 5; EtOH 5; eth 3; ace 3 / 1.4530[20]
9630	Piperidine, 1-acetyl-	$C_7H_{13}NO$ / 127.19	618-42-8 / -13.4	/ 226.5	5-20-02-00440 / 1.011[9]	H_2O 4; EtOH 4 / 1.4790[25]
9631	Piperidine, 1-acetyl-3-methyl-	$C_8H_{15}NO$ / 141.21	4593-16-2 / -13.6	/ 239	5-20-04-00107 / 0.9684[25]	H_2O 4 / 1.4731[25]
9632	Piperidine, 1-[5-(1,3-benzodioxol-5-yl)-1-oxo-2,4-pentadienyl]-, (E,E)- / Piperine	$C_{17}H_{19}NO_3$ / 285.34	94-62-2 / 131.5	7442 /	5-20-03-00469	H_2O 1; EtOH 3; eth 2; bz 3
9633	Piperidine, 1-[5-(1,3-benzodioxol-5-yl)-1-oxo-2,4-pentadienyl]-, (Z,Z)- / Chavicine	$C_{17}H_{19}NO_3$ / 285.34	495-91-0 /	2038 /	5-20-03-00469	eth 4; EtOH 4; peth 4
9634	Piperidine, 1-benzoyl-	$C_{12}H_{15}NO$ / 189.26	776-75-0 / 49	/ 320.5	5-20-02-00450	H_2O 1; EtOH 3; eth 3; ctc 2
9635	Piperidine, 1-butyl-	$C_9H_{19}N$ / 141.26	4945-48-6 /	/ 176	5-20-02-00029 / 0.8245[20]	1.4467[20]
9636	Piperidine, 3-butyl- / 3-Butylpiperidine	$C_9H_{19}N$ / 141.26	13603-21-9 /	/ 196.5	5-20-04-00230 / 0.8620[18]	eth 4; EtOH 4; chl 4
9637	Piperidine, 4-butyl- / 4-n-Butylpiperidine	$C_9H_{19}N$ / 141.26	24152-39-4 /	/ 194	4-20-00-01635 / 0.879[23]	1.4472[23]
9638	1-Piperidinecarboxaldehyde	$C_6H_{11}NO$ / 113.16	2591-86-8 / -30.8	/ 222.5	5-20-02-00432 / 1.0158[25]	H_2O 5; EtOH 5; eth 5; bz 5 / 1.4805[25]

No.	Name / Synonym	Mol. Form. / Mol. Wt.	CAS RN / mp/°C	Merck No. / bp/°C	Beil. Ref. / den/g cm^{-3}	Solubility / n_D
9639	2-Piperidinecarboxamide, N-(2,6-dimethylphenyl)-1-methyl- / Mepivacaine	$C_{15}H_{22}N_2O$ / 246.35	96-88-8	5748	5-22-01-00223	CS_2 3
9640	2-Piperidinecarboxylic acid / Pipecolic acid	$C_6H_{11}NO_2$ / 129.16	535-75-1 / 264	7425	5-22-01-00220	H_2O 3; EtOH 2
9641	3-Piperidinecarboxylic acid / Nipecotic acid	$C_6H_{11}NO_2$ / 129.16	498-95-3 / 261 dec	6478	5-22-01-00235	H_2O 4
9642	4-Piperidinecarboxylic acid / Isonipecotic acid	$C_6H_{11}NO_2$ / 129.16	498-94-2 / 336	5075	5-22-01-00244	
9643	4-Piperidinecarboxylic acid, ethyl ester	$C_8H_{15}NO_2$ / 157.21	1126-09-6 /	100-1[10]	4-22-00-00128	H_2O 4; bz 4; eth 4; EtOH 4 / 1.4591[20]
9644	4-Piperidinecarboxylic acid, methyl ester / Methyl isonipecotate	$C_7H_{13}NO_2$ / 143.19	2971-79-1 /	107-10[22]	4-22-00-00128	H_2O 4; bz 4; eth 4; EtOH 4 / 1.4635[25]
9645	4-Piperidinecarboxylic acid, 1-methyl-, methyl ester / Methyl N-methylisonipecotinate	$C_8H_{15}NO_2$ / 157.21	1690-75-1 /	98[20]; 84[10]	5-22-01-00245	H_2O 3; EtOH 3; eth 3; bz 3 / 1.4539[24]
9646	2,6-Piperidinediethanol, 1-methyl-α,α'-diphenyl-, [2α(R^*),6α(S^*)]- / Lobelanidine	$C_{22}H_{29}NO_2$ / 339.48	552-72-7 / 150	5429	4-21-00-02386	H_2O 1; EtOH 3; eth 2; ace 4
9647	Piperidine, 2,5-diethyl-	$C_9H_{19}N$ / 141.26	116836-21-6 /	190	0-20-00-00128 / 0.8722[25]	eth 4; chl 4
9648	Piperidine, 1,2-dimethyl-, (±)- / Piperidine, N,2-dimethyl (DL)	$C_7H_{15}N$ / 113.20	2512-81-4 /	127.5	5-20-04-00072 / 0.824[15]	H_2O 4; eth 4; EtOH 4 / 1.4395[20]
9649	Piperidine, 2,5-dimethyl- / 2,5-Dimethylpiperidine	$C_7H_{15}N$ / 113.20	34893-50-0 /	137	5-20-04-00179 / 0.8317[20]	EtOH 4 / 1.4452[25]
9650	Piperidine, 2,6-dimethyl-	$C_7H_{15}N$ / 113.20	504-03-0 /	127	5-20-04-00180 / 0.8158[25]	H_2O 5; EtOH 5; eth 5; ctc 2 / 1.4377[20]
9651	Piperidine, 3,3-dimethyl- / 3,3-Dimethylpiperidine	$C_7H_{15}N$ / 113.20	1193-12-0 /	137	5-20-04-00189	EtOH 4 / 1.4452[25]
9652	Piperidine, 3,5-dimethyl- / 3,5-Lupetidine	$C_7H_{15}N$ / 113.20	35794-11-7 /	144	5-20-04-00191 / 0.853[25]	1.4454[20]
9653	Piperidine, 1-(1,1-dimethylethyl)- / Piperidine, N-$tert$-butyl	$C_9H_{19}N$ / 141.26	14446-69-6 /	166	5-20-02-00031 / 0.8465[20]	eth 4; EtOH 4 / 1.4532[20]
9654	2,4-Piperidinedione, 3,3-diethyl- / Piperidione	$C_9H_{15}NO_2$ / 169.22	77-03-2 / 104	7439	5-21-09-00609	H_2O 4; EtOH 4; chl 4; MeOH 4
9655	2,6-Piperidinedione, 4-[2-(3,5-dimethyl-2-oxocyclohexyl)-2-hydroxyethyl]-, / Cycloheximide	$C_{15}H_{23}NO_4$ / 281.35	66-81-9 / 119	2734	5-21-13-00434	EtOH 4
9656	2,6-Piperidinedione, 4-ethyl-4-methyl- / Bemegride	$C_8H_{13}NO_2$ / 155.20	64-65-3 / 126.5	1036 / sub 100	5-21-09-00601	chl 3
9657	Piperidine, 1-[2-[3-(2,2-diphenylethyl)-1,2,4-oxadiazol-5-yl]ethyl]-, monohydrochloride / Prenoxdiazine hydrochloride	$C_{23}H_{28}ClN_3O$ / 397.95	982-43-4 / 186.5	7743		
9658	Piperidine, 1-(3,3-diphenylpropyl)- / Fenpiprane	$C_{20}H_{25}N$ / 279.43	3540-95-2 / 41.5	3934 / 210-20[8]	5-20-02-00091	
9659	Piperidine, 1,1'-dithiobis-	$C_{10}H_{20}N_2S_2$ / 232.41	10220-20-9 / 64.2	/ 160.1[11]	5-20-03-00487	acid 1
9660	Piperidine, 1-dodecyl-	$C_{17}H_{35}N$ / 253.47	5917-47-5 /	161[5]; 115[0.6]	4-20-00-00321 / 0.8378[20]	1.4588[20]
9661	Piperidine, 1,1'-(1,2-ethanediyl)bis-	$C_{12}H_{24}N_2$ / 196.34	1932-04-3 / -0.5	265	5-20-03-00148 / 0.9160[25]	1.4853[25]
9662	1-Piperidineethanol	$C_7H_{15}NO$ / 129.20	3040-44-6 / 17.9	202; 90[12]	5-20-02-00101 / 0.9703[25]	H_2O 5; EtOH 4 / 1.4749[20]
9663	2-Piperidineethanol / 2-(2-Hydroxyethyl)piperidine	$C_7H_{15}NO$ / 129.20	1484-84-0 / 69	202; 145[36]	5-21-01-00121 / 1.01[27]	H_2O 4
9664	4-Piperidineethanol / 4-(2-Hydroxyethyl)piperidine	$C_7H_{15}NO$ / 129.20	622-26-4 / 132.5	227.5	4-21-00-00093 / 1.0059[15]	H_2O 4; eth 4; EtOH 4 / 1.4907[20]
9665	1-Piperidineethanol, 2-(diphenylmethyl)- / Diphemethoxidine	$C_{20}H_{25}NO$ / 295.42	13862-07-2 / 106.5	3303 / 180-1[0.1]	5-20-08-00173	
9666	Piperidine, 3-ethenyl- / Piperidine, 3-vinyl-	$C_7H_{13}N$ / 111.19	57502-49-5 /	153.5	4-20-00-01946 / 0.9274[25]	1.4731[25]
9667	Piperidine, 1-ethyl- / N-Ethylpiperidine	$C_7H_{15}N$ / 113.20	766-09-6 /	130.8	5-20-02-00025 / 0.8237[20]	1.4480[20]
9668	Piperidine, 2-ethyl-, (±)- / Piperidine, 2-ethyl (DL)	$C_7H_{15}N$ / 113.20	78738-37-1 /	142.5	5-20-04-00169 / 0.8649[0]	1.4494[21]
9669	Piperidine, 3-ethyl-, (±)- / Piperidine, 3-ethyl (DL)	$C_7H_{15}N$ / 113.20	59433-08-8 /	152.6	5-20-04-00173 / 0.8565[23]	ace 4 / 1.4531[20]
9670	Piperidine, 4-ethyl- / 4-Ethylpiperidine	$C_7H_{15}N$ / 113.20	3230-23-7 /	154	5-20-04-00174 / 0.8759[25]	1.4503[25]
9671	Piperidine, 1-(1-ethyl-1-methylpropyl)- / Piperidine, N-(3-methyl-3-pentyl)	$C_{11}H_{23}N$ / 169.31	14045-26-2 /	214	5-20-02-00033 / 0.8614[20]	1.4637[20]
9672	Piperidine, 1-hexyl- / 1-Hexylpiperidine	$C_{11}H_{23}N$ / 169.31	7335-01-5 /	219.2	5-20-02-00032 / 0.8292[20]	1.4522[20]
9673	4-Piperidinemethanamine / 4-(Aminomethyl)piperidine	$C_6H_{14}N_2$ / 114.19	7144-05-0 / 25	200; 31[10]	5-22-08-00128	1.4900[20]
9674	2-Piperidinemethanol	$C_6H_{13}NO$ / 115.18	3433-37-2 / 69	104[10]; 80[1]	5-21-01-00098	chl 2
9675	3-Piperidinemethanol	$C_6H_{13}NO$ / 115.18	4606-65-9 / 61	106-7[3.5]	5-21-01-00107 / 1.0263[20]	chl 2 / 1.4964[20]

No.	Name / Synonym	Mol. Form. / Mol. Wt.	CAS RN / mp/°C	Merck No. / bp/°C	Beil. Ref. / den/g cm^{-3}	Solubility / n_D
9676	2-Piperidinemethanol, α,α-diphenyl- / Pipradrol	$C_{18}H_{21}NO$ / 267.37	467-60-7	7455	5-21-04-00070	
9677	2-Piperidinemethanol, α-ethyl-, (R*,S*)-(±)- / Conhydrine (DL) (lower m.p.)	$C_8H_{17}NO$ / 143.23	3238-62-8 / 121	2495 / 226	5-21-01-00138	H_2O 2; bz 4; eth 4; EtOH 4
9678	3-Piperidinemethanol, 1-methyl-	$C_7H_{15}NO$ / 129.20	7583-53-1	/ 142.5; 114[20]	5-21-01-00108 / 1.013[25]	1.4772[20]
9679	Piperidine, 1-methyl- / N-Methylpiperidine	$C_6H_{13}N$ / 99.18	626-67-5	/ 107	5-20-02-00021 / 0.8159[20]	H_2O 4; EtOH 5; eth 5; ctc 3 / 1.4355[20]
9680	Piperidine, 2-methyl-, (±)-	$C_6H_{13}N$ / 99.18	3000-79-1 / -4.9	/ 118	5-20-04-00071 / 0.8436[24]	H_2O 4; EtOH 3; eth 3; chl 2 / 1.4459[20]
9681	Piperidine, 3-methyl-, (±)- / Piperidine, 3-methyl (DL)	$C_6H_{13}N$ / 99.18	53152-98-0 / -24	/ 125.5	5-20-04-00099 / 0.8446[26]	H_2O 4; chl 2 / 1.4470[20]
9682	Piperidine, 4-methyl-	$C_6H_{13}N$ / 99.18	626-58-4	/ 130	5-20-04-00116 / 0.8674[25]	H_2O 4; chl 2 / 1.4458[20]
9683	Piperidine, 1,1'-methylenebis-	$C_{11}H_{22}N_2$ / 182.31	880-09-1	/ 122-3[15]	5-20-02-00293 / 0.915[25]	1.4820[20]
9684	Piperidine, 1-(1-methylethyl)-	$C_8H_{17}N$ / 127.23	766-79-0	/ 149.5	5-20-02-00028 / 0.8389[20]	H_2O 4 / 1.4491[20]
9685	Piperidine, 1-(2-methylpropyl)-	$C_9H_{19}N$ / 141.26	10315-89-6	/ 160.5	5-20-02-00030 / 0.8161[25]	H_2O 4 / 1.4382[25]
9686	Piperidine, 1-methyl-2-propyl-, (S)- / Methylconiine	$C_9H_{19}N$ / 141.26	35305-13-6	5968 / 174	5-20-04-00206 / 0.8326[22]	ace 4; EtOH 4 / 1.4538[12]
9687	Piperidine, 1-nitro-	$C_5H_{10}N_2O_2$ / 130.15	7119-94-0 / -5.5	/ 245; 121[20]	5-20-03-00531 / 1.1519[26]	1.4954[26]
9688	Piperidine, 1-nitroso- / N-Nitrosopiperidine	$C_5H_{10}N_2O$ / 114.15	100-75-4	/ 219; 109[20]	5-20-03-00530 / 1.0631[18]	H_2O 3; HCl 3 / 1.4933[18]
9689	Piperidine, 1-nonyl-	$C_{14}H_{29}N$ / 211.39	30538-80-8	/ 135-7[11]	4-20-00-00319 / 0.8313[25]	1.4538[25]
9690	Piperidine, 1-octyl-	$C_{13}H_{27}N$ / 197.36	7335-02-6	/ 137[13]; 89[1]	5-20-02-00034 / 0.8324[20]	1.4544[20]
9691	Piperidine, 1,2,2,6,6-pentamethyl- / Pempidine	$C_{10}H_{21}N$ / 155.28	79-55-0	7022 / 147	5-20-04-00238 / 0.8580[0]	1.4550[21]
9692	Piperidine, 1-pentyl-	$C_{10}H_{21}N$ / 155.28	10324-58-0	/ 198.2	5-20-02-00031 / 0.8282[20]	1.4498[20]
9693	Piperidine, 1-phenyl-	$C_{11}H_{15}N$ / 161.25	4096-20-2 / 4.7	/ 258	5-20-02-00057 / 0.9944[25]	EtOH 4; eth 4; bz 4; chl 4 / 1.5598[25]
9694	Piperidine, 3-phenyl-	$C_{11}H_{15}N$ / 161.25	3973-62-4 / 14.5	/ 257; 140[19]	5-20-06-00405 / 1.0010[25]	1.5473[25]
9695	Piperidine, 4-phenyl- / 4-Phenylpiperidine	$C_{11}H_{15}N$ / 161.25	771-99-3 / 60.5	/ 257	5-20-06-00408 / 0.9996[16]	chl 3
9696	Piperidine, 1-(1-phenylcyclohexyl)- / Phencyclidine	$C_{17}H_{25}N$ / 243.39	77-10-1 / 46.5	7179 / 135-7[1.0]	5-20-02-00078	
9697	Piperidine, 1-(1-phenylethyl)-	$C_{13}H_{19}N$ / 189.30	7529-63-7 / -13	/ 272	5-20-02-00064 / 0.945[25]	1.5218[25]
9698	Piperidine, 1-(phenylmethyl)-	$C_{12}H_{17}N$ / 175.27	2905-56-8	/ 245	5-20-02-00062 / 0.9625[16]	1.5227[20]
9699	Piperidine, 2-(phenylmethyl)- / 2-Benzylpiperidine	$C_{12}H_{17}N$ / 175.27	32838-55-4 / 32	/ 267.5	5-20-06-00438 / 0.9660[25]	eth 4; EtOH 4
9700	Piperidine, 4-(phenylmethyl)- / 4-Benzylpiperidine	$C_{12}H_{17}N$ / 175.27	31252-42-3 / 16.8	/ 270; 150[17]	5-20-06-00439 / 0.9970[20]	H_2O 1; EtOH 3; eth 3 / 1.5337[25]
9701	Piperidine, 4,4'-(1,3-propanediyl)bis-	$C_{13}H_{26}N_2$ / 210.36	16898-52-5 / 67.1	/ 329	5-23-04-00101	H_2O 4
9702	Piperidine, 4,4'-(1,3-propanediyl)bis[1-methyl- / Propane, 1,3-bis(1-methyl-4-piperidyl)	$C_{15}H_{30}N_2$ / 238.42	64168-11-2 / 13.7	/ 215[50]	4-23-00-00536 / 0.896[25]	1.4804[25]
9703	1-Piperidinepropanenitrile	$C_8H_{14}N_2$ / 138.21	3088-41-3 / -6.8	/ 145[50]	5-20-03-00047 / 0.9403[25]	1.4676[25]
9704	1-Piperidinepropanoic acid, ethyl ester / Piperidine, N-carboxyethyl, ethyl ester	$C_{10}H_{19}NO_2$ / 185.27	19653-33-9	/ 139[50]	5-20-03-00042 / 0.9627[25]	H_2O 4 / 1.4525[25]
9705	1-Piperidinepropanol, α-cyclohexyl-α-phenyl-, hydrochloride / Trihexyphenidyl hydrochloride	$C_{20}H_{32}ClNO$ / 337.93	52-49-3 / 258.5	9607		
9706	Piperidine, 2-(1-propenyl)-, (±)-	$C_8H_{15}N$ / 125.21	5913-85-9 / 6	/ 168	5-20-04-00346 / 0.8716[15]	eth 4; EtOH 4
9707	Piperidine, 1-propyl-	$C_8H_{17}N$ / 127.23	5470-02-0	/ 151.5	5-20-02-00027 / 0.8231[20]	H_2O 4; eth 4; EtOH 4 / 1.4446[20]
9708	Piperidine, 2-propyl-, (S)- / Coniine	$C_8H_{17}N$ / 127.23	458-88-8 / -2	2500 / 166.5	5-20-04-00205 / 0.8440[20]	H_2O 2; EtOH 5; eth 4; bz 3 / 1.4512[22]
9709	Piperidine, 4-propyl- / 4-Propylpiperidine	$C_8H_{17}N$ / 127.23	22398-09-0	/ 173	5-20-04-00208 / 0.864[22]	EtOH 3; chl 2 / 1.4465[23]
9710	Piperidine, 2,2,4,6-tetramethyl- / 2,2,4,6-Tetramethylpiperidine	$C_9H_{19}N$ / 141.26	6292-82-6	/ 156	4-20-00-01638 / 0.816[25]	1.4374[25]
9711	Piperidine, 2,2,6,6-tetramethyl- / Norpempidine	$C_9H_{19}N$ / 141.26	768-66-1 / 28	/ 156	5-20-04-00237 / 0.8367[16]	eth 4 / 1.4455[20]
9712	Piperidine, 2,2,4-trimethyl-	$C_8H_{17}N$ / 127.23	101257-71-0	/ 148	4-20-00-01624 / 0.832[15]	eth 4; EtOH 4 / 1.4458[20]
9713	Piperidine, 2,3,6-trimethyl- / 2,3,6-Trimethylpiperidine	$C_8H_{17}N$ / 127.23	50402-72-7	36[5]	4-20-00-01625 / 0.8302[20]	1.4434[20]
9714	Piperidine, 2,4,6-trimethyl- / 2,4,6-Trimethylpiperidine	$C_8H_{17}N$ / 127.23	21974-48-1 / <20	/ 145	5-20-04-00220 / 0.8315[19]	H_2O 2 / 1.4412[20]
9715	3-Piperidinol, 1-ethyl- / 1-Ethyl-3-piperidinol	$C_7H_{15}NO$ / 129.20	13444-24-1	3799 / 93-5[15]	5-21-01-00039	1.4777[14]

No.	Name / Synonym	Mol. Form. / Mol. Wt.	CAS RN / mp/°C	Merck No. / bp/°C	Beil. Ref. / den/g cm⁻³	Solubility / n_D
9716	3-Piperidinol, 1-methyl-	$C_6H_{13}NO$ 115.18	3554-74-3	$76-8^{11}$	5-21-01-00029 0.999[25]	1.4735[20]
9717	4-Piperidinol, 1-methyl-	$C_6H_{13}NO$ 115.18	106-52-5 29	200	5-21-01-00052	1.4775[20]
9718	3-Piperidinol, 6-propyl-, (3S-trans)- Pseudoconhydrine	$C_8H_{17}NO$ 143.23	140-55-6 106	7928 236	5-21-01-00138	H_2O 4; eth 4; EtOH 4
9719	2-Piperidinone	C_5H_9NO 99.13	675-20-7 39.5	256	5-21-06-00396	H_2O 4; EtOH 4; eth 4; dil acid 3
9720	4-Piperidinone, 1-acetyl-	$C_7H_{11}NO_2$ 141.17	32161-06-1	218; $124^{0.2}$	5-21-06-00426 1.146[25]	1.5026[20]
9721	2-Piperidinone, 1-methyl-	$C_6H_{11}NO$ 113.16	931-20-4	105^{12}; 94^9	5-21-06-00398 1.0263[25]	1.4820[20]
9722	2-Piperidinone, 3-methyl-	$C_6H_{11}NO$ 113.16	3768-43-2 55.3	249.5	5-21-06-00479	bz 4; eth 4; EtOH 4
9723	3-Piperidinone, 1-methyl-	$C_6H_{11}NO$ 113.16	5519-50-6	$63-4^{13}$	5-21-06-00417 0.9655[25]	1.4559[25]
9724	4-Piperidinone, 1-methyl-	$C_6H_{11}NO$ 113.16	1445-73-4	$56-8^{11}$	5-21-06-00419 0.9696[25]	1.4580[25]
9725	4-Piperidinone, 1,2,3,6-tetramethyl- 1,2,5,6-Tetramethyl-4-piperidinone	$C_9H_{17}NO$ 155.24	53381-90-1	$86-8^{7.5}$	4-21-00-03272 0.9499[20]	1.4680[20]
9726	4-Piperidinone, 2,2,6,6-tetramethyl-	$C_9H_{17}NO$ 155.24	826-36-8 36	205	5-21-06-00538	H_2O 3; EtOH 3; eth 3; chl 2
9727	Piperonyl butoxide 1,3-Benzodioxole, 5-[[2-(2-butoxyethoxy)ethoxy]methyl]-6-propyl-	$C_{19}H_{30}O_5$ 338.44	51-03-6	7446 180^1	1.05[25]	
9728	Pirimicarb Carbamic acid, dimethyl-, 2-(dimethylamino)-5,6-dimethyl-4-pyrimidinyl ester	$C_{11}H_{18}N_4O_2$ 238.29	23103-98-2 90.5	7468		
9729	Pirimiphos-ethyl Phosphorothioic acid, O-[2-(diethylamino)-6-methyl-4-pyrimidinyl] O,O-diethyl ester	$C_{13}H_{24}N_3O_3PS$ 333.39	23505-41-1	7469 dec >130	1.14[20]	
9730	Pirimiphos-methyl Phosphorothioic acid, O-[2-(diethylamino)-6-methyl-4-pyrimidinyl] O,O-dimethyl ester	$C_{11}H_{20}N_3O_3PS$ 305.34	29232-93-7 15	dec	1.17[20]	
9731	Plumbane, tetraethyl- Tetraethyl lead	$C_8H_{20}Pb$ 323.45	78-00-2	9136 200 dec	1.653[20]	H_2O 1; bz 3; EtOH 2 1.5198[20]
9732	Plumbane, tetramethyl- Tetramethyl lead	$C_4H_{12}Pb$ 267.34	75-74-1 -27.5	110	1.995[20]	
9733	Plumbane, tetraphenyl-	$C_{24}H_{20}Pb$ 515.62	595-89-1 228.3	126^{13}	4-16-00-01615 1.5298[20]	chl 3
9734	21H,23H-Porphine	$C_{20}H_{14}N_4$ 310.36	101-60-0 360	7574 sub 300	4-26-00-01900 1.336[25]	H_2O 1; EtOH 2; eth 1; ace 1
9735	21H,23H-Porphine-2,18-dipropanoic acid, 7,12-bis(1-hydroxyethyl)-3,8,13,17-tetramethyl- Hematoporphyrin	$C_{34}H_{38}N_4O_6$ 598.70	14459-29-1 172.5	4555	4-26-00-03157	H_2O 1; EtOH 3; eth 2; chl 2
9736	Pregna-4,6-diene-21-carboxylic acid, 17-hydroxy-3-oxo-, γ-lactone, (17α)- Canrenone	$C_{22}H_{28}O_3$ 340.46	976-71-6 150	1753	5-17-11-00476	
9737	Pregna-1,4-diene-3,20-dione, 9-fluoro-11,17,21-trihydroxy-16-methyl-, (11β,16α)- Dexamethasone	$C_{22}H_{29}FO_5$ 392.47	50-02-2 262	2922		
9738	Pregna-1,4-diene-3,20-dione, 11,17,21-trihydroxy-, (11β)- Prednisolone	$C_{21}H_{28}O_5$ 360.45	50-24-8 235	7719	4-08-00-03467	
9739	Pregna-1,4-diene-3,11,20-trione, 17,21-dihydroxy- Prednisone	$C_{21}H_{26}O_5$ 358.43	53-03-2 233-5 dec	7727	4-08-00-03531	
9740	Pregnane, (5α)- Allopregnane	$C_{21}H_{36}$ 288.52	641-85-0 84.5	257	4-05-00-01215	
9741	Pregnane, (5β)- 17β-Ethyletiocholane	$C_{21}H_{36}$ 288.52	481-26-5 83.5	7731	4-05-00-01215 1.032[15]	H_2O 1; chl 3; MeOH 3
9742	Pregnane-3,20-diol, (3α,5β,20S)- Pregnanediol	$C_{21}H_{36}O_2$ 320.52	80-92-2 243.5	7732	4-06-00-06111 1.15[25]	EtOH 2; eth 2; ace 3; os 2
9743	Pregnane-3,20-dione, (5β)- 5β-Pregnane-3,20-dione	$C_{21}H_{32}O_2$ 316.48	128-23-4 123	7733	4-07-00-02198	H_2O 1; EtOH 4; eth 3; ace 3
9744	Pregnane-11,20-dione, 3,17,21-trihydroxy-, (3α,5β)- Tetrahydrocortisone	$C_{21}H_{32}O_5$ 364.48	53-05-4 190	9143	4-08-00-03395	
9745	Pregnan-20-one, 3-hydroxy-, (3α,5β)- Pregnan-3α-ol-20-one	$C_{21}H_{34}O_2$ 318.50	128-20-1 149.5	7734		EtOH 4
9746	Pregn-4-en-18-al, 11,21-dihydroxy-3,20-dioxo-, (11β)- Aldosterone	$C_{21}H_{28}O_5$ 360.45	52-39-1 166.5	218	4-08-00-03491	
9747	Pregn-4-ene-3,20-dione, 21-(acetyloxy)-11,17-dihydroxy-, (11β)- Hydrocortisone acetate	$C_{23}H_{23}O_6$ 395.43	50-03-3 223 dec	4711	4-08-00-03424 1.289[20]	
9748	Pregn-4-ene-3,20-dione, 11,21-dihydroxy-, (11β)- Corticosterone	$C_{21}H_{30}O_4$ 346.47	50-22-6 181	2532 $190^{0.01}$	4-08-00-02907	H_2O 1; EtOH 3; eth 3; ace 3

No.	Name Synonym	Mol. Form. Mol. Wt.	CAS RN mp/°C	Merck No. bp/°C	Beil. Ref. den/g cm^{-3}	Solubility n_D
9749	Pregn-4-ene-3,20-dione, 17,21-dihydroxy- 11-Deoxy-17-hydroxycorticosterone	$C_{21}H_{30}O_4$ 346.47	152-58-9 215	2912	4-08-00-02913	ace 4; EtOH 4; chl 4
9750	Pregn-4-ene-3,20-dione, 17-hydroxy- 17α-Hydroxyprogesterone	$C_{21}H_{30}O_3$ 330.47	68-96-2	4773	4-08-00-02189	chl 2
9751	Pregn-4-ene-3,20-dione, 21-hydroxy- Deoxycorticosterone	$C_{21}H_{30}O_3$ 330.47	64-85-7 141.5	2882	4-08-00-02195	H_2O 2; EtOH 4; eth 2; ace 4
9752	Pregn-4-ene-3,20-dione, 17-hydroxy-6- methyl-, (6α)- Medroxyprogesterone	$C_{22}H_{32}O_3$ 344.49	520-85-4 214.5	5677	4-08-00-02211	chl 4
9753	Pregn-4-ene-3,20-dione, 11,17,21-trihydroxy-, (11β)- Hydrocortisone	$C_{21}H_{30}O_5$ 362.47	50-23-7 220	4710	4-08-00-03422	H_2O 2; EtOH 3; diox 3; HOAc 3
9754	Pregn-4-ene-3,11,20-trione, 17,21-dihydroxy- Cortisone	$C_{21}H_{28}O_5$ 360.45	53-06-5 222	2533	4-08-00-03480	H_2O 2; EtOH 3; eth 2; ace 3
9755	Pregn-4-ene-3,11,20-trione, 21-hydroxy- 11-Dehydrocorticosterone	$C_{21}H_{28}O_4$ 344.45	72-23-1 183.5	2859	4-08-00-02998	H_2O 1; EtOH 3; ace 3; bz 3
9756	Pregn-5-en-20-one, 3-hydroxy-, (3β)- Pregnenolone	$C_{21}H_{32}O_2$ 316.48	145-13-1 192	7739	4-08-00-01019	
9757	Pregn-4-en-3-one, 11,17,20,21-tetrahydroxy-, (11β,20R)- 4-Pregnene-11β,17α,20β,21-tetrol-3-one	$C_{21}H_{32}O_5$ 364.48	116-58-5 125 dec	7736	4-08-00-03384	ace 4; EtOH 4
9758	Pregn-4-en-20-yn-3-one, 17-hydroxy-, (17α)- Ethisterone	$C_{21}H_{28}O_2$ 312.45	434-03-7 272	3696	4-08-00-01225	
9759	Procyazine Propanenitrile, 2-[[4-chloro-6- (cyclopropylamino)-1,3,5-triazin-2-yl]amino]-2- methyl-	$C_{10}H_{13}ClN_6$ 252.71	32889-48-8 170			
9760	Procymidone N-(3,5-Dichlorophenyl)-1,2- dimethylcyclopropane-1,2-dicarboximide	$C_{13}H_{11}Cl_2NO_2$ 284.14	32809-16-8 166		1.452^{25}	
9761	Prodiamine 1,3-Benzenediamine, 2,4-dinitro-N3,N3- dipropyl-6-(trifluoromethyl)-	$C_{13}H_{17}F_3N_4O_4$ 350.30	29091-21-2 124	7773	1.47^{25}	
9762	Profenofos Phosphorothioic acid, O-(4-bromo-2- chlorophenyl) O-ethyl S-propyl ester	$C_{11}H_{15}BrClO_3PS$ 373.63	41198-08-7 110$^{0.001}$		1.455^{20}	
9763	Profluralin Benzenamine, N-(cyclopropylmethyl)-2,6- dinitro-N-propyl-4-(trifluoromethyl)-	$C_{14}H_{16}F_3N_3O_4$ 347.29	26399-36-0 34	7781		
9764	L-Proline 2-Pyrrolidinecarboxylic acid	$C_5H_9NO_2$ 115.13	147-85-3 220-2 dec	7790	5-22-01-00031	H_2O 4; EtOH 2; eth 1; ace 2
9765	DL-Proline	$C_5H_9NO_2$ 115.13	609-36-9 205 dec		5-22-01-00033	H_2O 4; EtOH 4
9766	L-Proline, 1-acetyl-4-hydroxy-, trans- Oxaceprol	$C_7H_{11}NO_4$ 173.17	33996-33-7 132	6857	5-22-05-00013	H_2O 4; MeOH 4
9767	L-Proline, 4-hydroxy-, trans- 4-Hydroxy-L-proline	$C_5H_9NO_3$ 131.13	51-35-4 274	4775	5-22-05-00007	H_2O 4; EtOH 2
9768	L-Proline, 5-oxo- L-Pyroglutamic acid	$C_5H_7NO_3$ 129.12	98-79-3 184.7	8012	5-22-06-00007	DMSO 3
9769	Promecarb Phenol, 3-methyl-5-(1-methylethyl)-, methylcarbamate	$C_{12}H_{17}NO_2$ 207.27	2631-37-0 87	7794 117$^{0.01}$		
9770	Prometryn 1,3,5-Triazine-2,4-diamine, N,N'-bis(1- methylethyl)-6-(methylthio)-	$C_{10}H_{19}N_5S$ 241.36	7287-19-6 119	7800	1.157^{20}	
9771	Propachlor Acetamide, 2-chloro-N-(1-methylethyl)-N- phenyl-	$C_{11}H_{14}ClNO$ 211.69	1918-16-7 77	7805 110$^{0.03}$	1.242^{25}	
9772	1,2-Propadiene Allene	C_3H_4 40.06	463-49-0 -136.2	-34.4	4-01-00-00966 *0.584^{25}	bz 4; peth 4 1.4168
9773	1,2-Propadiene-1,3-dione Carbon suboxide	C_3O_2 68.03	504-64-3 -107	1821 6.8	4-01-00-03764 1.114^0	eth 3; bz 3; CS_2 3 1.4538^0
9774	Propanal Propionaldehyde	C_3H_6O 58.08	123-38-6 -80	7835 48	4-01-00-03165 0.865^{25}	H_2O 3; EtOH 5; eth 5 1.3636^{20}
9775	Propanal, 2-bromo-	C_3H_5BrO 136.98	19967-57-8 109.5		4-01-00-03177 1.592^{20}	eth 4 1.4813^{20}
9776	Propanal, 2-bromo-2-methyl-	C_4H_7BrO 151.00	13206-46-7 113		4-01-00-03267 1.383^{25}	1.4518^{25}
9777	Propanal, 2-chloro-	C_3H_5ClO 92.52	683-50-1 86		3-01-00-02691 1.182^{15}	bz 4; eth 4 1.431^{17}
9778	Propanal, 3-chloro-	C_3H_5ClO 92.52	19434-65-2 <20	130; 45^{10}	4-01-00-03174 1.268^{15}	eth 4; EtOH 4 1.475^{25}
9779	Propanal, 2-chloro-2-methyl-	C_4H_7ClO 106.55	917-93-1 90		4-01-00-03267 1.053^{15}	eth 4; EtOH 4 1.4160^{16}
9780	Propanal, 2,3-dibromo-	$C_3H_4Br_2O$ 215.87	5221-17-0 73-5^{10}		4-01-00-03178 2.198^{15}	eth 4 1.5082^{20}
9781	Propanal, 2,3-dichloro-	$C_3H_4Cl_2O$ 126.97	10140-89-3 73^{50}; 48^{14}		4-01-00-03174 1.400^{20}	1.4762^{20}
9782	Propanal, 2,3-dihydroxy-, (±)-	$C_3H_6O_3$ 90.08	56-82-6 145	140-50$^{0.8}$	3-01-00-03282 1.453^{18}	H_2O 3; EtOH 2; eth 2; bz 1

No.	Name / Synonym	Mol. Form. / Mol. Wt.	CAS RN / mp/°C	Merck No. / bp/°C	Beil. Ref. / den/g cm^{-3}	Solubility / n_D
9783	Propanal, 2,2-dimethyl- / Pivaldehyde	$C_5H_{10}O$ / 86.13	630-19-3 / 6	/ 77.5	4-01-00-03295 / 0.7923[17]	EtOH 3; eth 3 / 1.3791[20]
9784	Propanal, 2-(hydroxyimino)-, oxime	$C_3H_6N_2O_2$ / 102.09	1804-15-5 / 157	/ sub	4-01-00-03633	H_2O 2; EtOH 3; eth 3; bz 3
9785	Propanal, 2-methyl- / Isobutanal	C_4H_8O / 72.11	78-84-2 / -65.9	5038 / 64.5	4-01-00-03262 / 0.7891[20]	H_2O 3; eth 3; ace 3; ctc 2 / 1.3730[20]
9786	Propanal, 3-(methylthio)- / 3-(Methylthio)propionaldehyde	C_4H_8OS / 104.17	3268-49-3 /	/ 62[11]	4-01-00-03974	
9787	Propanal, oxime	C_3H_7NO / 73.09	627-39-4 / 40	/ 131.5	4-01-00-03170 / 0.9258[20]	1.4287[20]
9788	Propanal, 2-oxo- / Pyruvaldehyde	$C_3H_4O_2$ / 72.06	78-98-8 /	8030 / 72	4-01-00-03631 / 1.0455[20]	EtOH 3; eth 3; bz 3 / 1.4002[18]
9789	Propanal, 2-oxo-, 1-oxime / Isonitrosoacetone	$C_3H_5NO_2$ / 87.08	306-44-5 / 69	5076 / sub	3-01-00-03092 / 1.0744[67]	H_2O 3; eth 3; bz 2; ctc 2
9790	Propanal, 2,2,3-trichloro-	$C_3H_3Cl_3O$ / 161.41	7789-90-4 /	/ 63-5[45]	3-01-00-02693 / 1.470[25]	eth 4 / 1.473[25]
9791	Propanamide / Propionamide	C_3H_7NO / 73.09	79-05-0 / 81.3	7836 / 213	4-02-00-00725 / 0.9262[110]	H_2O 4; EtOH 4; eth 4; chl 4 / 1.4180[110]
9792	Propanamide, 2-bromo-2-methyl- / Isobutyramide, 2-bromo	C_4H_8BrNO / 166.02	7462-74-0 / 148	/ 145[17]	4-02-00-00863	EtOH 4; chl 4
9793	Propanamide, N,N-diethyl-	$C_7H_{15}NO$ / 129.20	1114-51-8 /	/ 191	4-04-00-00353 / 0.8972[20]	EtOH 4 / 1.4425[20]
9794	Propanamide, N,N-diethyl-2,2-dimethyl- / Pivalamide, N,N-diethyl	$C_9H_{19}NO$ / 157.26	24331-72-4 /	/ 206	3-04-00-00211 / 0.891[15]	eth 4; EtOH 4
9795	Propanamide, N-(2,3-dihydro-1,5-dimethyl-3-oxo-2-phenyl-1H-pyrazol-4-yl)-2-(dimethylamino)- / Aminopropylon	$C_{16}H_{22}N_4O_2$ / 302.38	3690-04-8 / 181	484	4-25-00-03595	H_2O 4
9796	Propanamide, N-(4-ethoxyphenyl)-2-hydroxy- / p-Lactophenetide	$C_{11}H_{15}NO_3$ / 209.25	539-08-2 / 118	5220	3-13-00-01135	H_2O 3; EtOH 4; eth 2; bz 2
9797	Propanamide, 2-hydroxy-, (±)-	$C_3H_7NO_2$ / 89.09	65144-02-7 / 75.5		4-03-00-00674 / 1.1381[80]	H_2O 4; EtOH 4
9798	Propanamide, 2-methyl-	C_4H_9NO / 87.12	563-83-7 / 128	/ 218	4-02-00-00852	chl 3
9799	Propanamide, N-methyl-	C_4H_9NO / 87.12	1187-58-2 / -30.9	/ 148	4-04-00-00183 / 0.9305[25]	1.4345[25]
9800	Propanamide, N-phenyl-	$C_9H_{11}NO$ / 149.19	620-71-3 / 105.5	/ 222.2	4-12-00-00385 / 1.175[25]	H_2O 2; EtOH 4; eth 4
9801	Propanamide, N-phenyl-N-[1-(2-phenylethyl)-4-piperidinyl]- / Fentanyl	$C_{22}H_{28}N_2O$ / 336.48	437-38-7 / 87.5	3944	5-22-08-00049	
9802	Propanamide, N-propyl-	$C_6H_{13}NO$ / 115.18	3217-86-5 / 154	/ 215; 108[9]	4-04-00-00477 / 0.8985[25]	H_2O 2; eth 2
9803	1-Propanamine / Propylamine	C_3H_9N / 59.11	107-10-8 / -83	7855 / 47.2	4-04-00-00464 / 0.7173[20]	H_2O 3; EtOH 4; eth 4; ace 4 / 1.3870[20]
9804	2-Propanamine / Isopropylamine	C_3H_9N / 59.11	75-31-0 / -95.1	5097 / 31.7	4-04-00-00504 / 0.6891[20]	H_2O 5; EtOH 5; eth 5; ace 4 / 1.3742[20]
9805	1-Propanamine, N,N-dichloro- / N,N-Dichloropropylamine	$C_3H_7Cl_2N$ / 128.00	69947-01-9 /	/ 117	0-04-00-00145 / 1.1454[23]	1.4525[23]
9806	1-Propanamine, 2,2-dimethyl-	$C_5H_{13}N$ / 87.16	5813-64-9 /	/ 82	4-04-00-00707 / 0.7455[20]	eth 4 / 1.4023[20]
9807	1-Propanamine, N,N-dimethyl- / Dimethylpropylamine	$C_5H_{13}N$ / 87.16	926-63-6 /	/ 66	4-04-00-00467 / 0.7152[20]	bz 4; eth 4; EtOH 4 / 1.3860[20]
9808	1-Propanamine, N,N-dipropyl- / Tripropylamine	$C_9H_{21}N$ / 143.27	102-69-2 / -93.5	/ 156	4-04-00-00470 / 0.7558[20]	eth 4; EtOH 4 / 1.4181[20]
9809	1-Propanamine, N-ethyl-	$C_5H_{13}N$ / 87.16	20193-20-8 /	/ 81	4-04-00-00468 / 0.7204[17]	H_2O 2; ace 4; EtOH 4 / 1.3858[25]
9810	2-Propanamine, N-ethyl- / Ethylisopropylamine	$C_5H_{13}N$ / 87.16	19961-27-4 /	/ 69.6	4-04-00-00508	1.3872[25]
9811	2-Propanamine, N-ethyl-N-(1-methylethyl)-	$C_8H_{19}N$ / 129.25	7087-68-5 /	/ 126.5	4-04-00-00511 / 0.742[25]	ctc 3 / 1.4138[20]
9812	1-Propanamine, N-ethyl-N-propyl-	$C_8H_{19}N$ / 129.25	20634-92-8 /	/ 138	2-04-00-00622 / 0.807[24]	ace 4; bz 4; eth 4
9813	1-Propanamine, 1,1,2,2,3,3,3-heptafluoro-N,N-bis(heptafluoropropyl)-	$C_9F_{21}N$ / 521.07	338-83-0 /	/ 130	4-02-00-00742 / 1.822[4]	1.279[25]
9814	1-Propanamine, N-hydroxy-N-propyl- / N,N-Dipropylhydroxylamine	$C_6H_{15}NO$ / 117.19	7446-43-7 / 29	/ 154.5	4-04-00-03308	bz 4; eth 4; EtOH 4; chl 4
9815	2-Propanamine, N,N'-methanetetraylbis-	$C_7H_{14}N_2$ / 126.20	693-13-0 /	/ 147	4-04-00-00531 / 0.806[25]	1.4320[20]
9816	1-Propanamine, 3-methoxy-	$C_4H_{11}NO$ / 89.14	5332-73-0 /	/ 117.5	4-04-00-01623 / 0.8727[20]	H_2O 3; ace 3; bz 3; ctc 3 / 1.4391[20]
9817	1-Propanamine, 2-methyl- / Isobutylamine	$C_4H_{11}N$ / 73.14	78-81-9 / -86.7	5016 / 67.7	4-04-00-00625 / 0.724[25]	1.3988[19]
9818	2-Propanamine, 2-methyl- / $tert$-Butylamine	$C_4H_{11}N$ / 73.14	75-64-9 / -66.9	1545 / 44.0	4-04-00-00657 / 0.6958[20]	H_2O 5; EtOH 5; eth 5; chl 3 / 1.3784[20]
9819	2-Propanamine, N-methyl- / Isopropylmethylamine	$C_4H_{11}N$ / 73.14	4747-21-1 /	/ 50.4		
9820	1-Propanamine, 2-methyl-N,N-bis(2-methylpropyl)- / Triisobutylamine	$C_{12}H_{27}N$ / 185.35	1116-40-1 / -21.8	/ 191.5	4-04-00-00631 / 0.7684[20]	eth 4; EtOH 4 / 1.4252[17]

No.	Name / Synonym	Mol. Form. / Mol. Wt.	CAS RN / mp/°C	Merck No. / bp/°C	Beil. Ref. / den/g cm^{-3}	Solubility / n_D
9821	2-Propanamine, N-(1-methylethyl)- / Diisopropylamine	$C_6H_{15}N$ / 101.19	108-18-9 / -61	3181 / 83.9	4-04-00-00510 / 0.7153[20]	ace 4; bz 4; eth; EtOH 4 / 1.3924[20]
9822	2-Propanamine, N-(1-methylethyl)-N-nitroso-	$C_6H_{14}N_2O$ / 130.19	601-77-4 / 48	/ 194.5	4-04-00-03388 / 0.9422[20]	H_2O 2; EtOH 3; eth 3; bz 3
9823	1-Propanamine, 2-methyl-N-(2-methylpropyl)- / Diisobutylamine	$C_8H_{19}N$ / 129.25	110-96-3 / -73.5	/ 139.6	4-04-00-00630 /	H_2O 2; EtOH 3; eth 3; ace 3 / 1.4090[20]
9824	1-Propanamine, N-nitro-	$C_3H_8N_2O_2$ / 104.11	627-07-6 / -21	/ 128-9[40]	4-04-00-03405 / 1.1046[15]	eth 4; EtOH 4 / 1.4610[20]
9825	1-Propanamine, N-nitroso-N-propyl- / Dipropylnitrosamine	$C_6H_{14}N_2O$ / 130.19	621-64-7 /	/ 206; 113[40]	4-04-00-03388 / 0.9163[20]	H_2O 2; EtOH 5; eth 5 / 1.4437[20]
9826	1-Propanamine, N-propyl- / Dipropylamine	$C_6H_{15}N$ / 101.19	142-84-7 / -63	3350 / 109.3	4-04-00-00469 / 0.7400[20]	H_2O 3; EtOH 3; eth 5; ace 4 / 1.4050[20]
9827	2-Propanamine, N-propyl / Isopropylpropylamine	$C_6H_{15}N$ / 101.19	21968-17-2 /	/ 96.9		
9828	1-Propanamine, 3-(triethoxysilyl)-	$C_9H_{23}NO_3Si$ / 221.37	919-30-2 /	/ 122-3[31]	4-04-00-04273 / 0.9506[20]	/ 1.4225[20]
9829	1-Propanamine, N,N,2-trimethyl-	$C_6H_{15}N$ / 101.19	7239-24-9 /	/ 80.5	4-04-00-00627 / 0.7097[20]	H_2O 4 / 1.3907[20]
9830	1-Propanaminium, 2-(acetyloxy)-N,N,N-trimethyl-, chloride / Methacholine chloride	$C_8H_{18}ClNO_2$ / 195.69	62-51-1 / 172	5847 /		H_2O 4; EtOH 4; chl 4
9831	1-Propanaminium, 2-hydroxy-N,N,N-trimethyl-, chloride / (2-Hydroxypropyl)trimethyl ammonium chloride	$C_6H_{16}ClNO$ / 153.65	2382-43-6 / 165	/ dec	3-04-00-00754 /	H_2O 4; EtOH 4
9832	2-Propanaminium, N-methyl-N-(1-methylethyl)-N-[2-[(9H-xanthen-9-ylcarbonyl)oxy]ethyl]-, bromide / Propantheline bromide	$C_{23}H_{30}BrNO_3$ / 448.40	50-34-0 / 159.5	7816 /	5-18-06-00590 /	
9833	1-Propanaminium, N,N,N-tripropyl-, iodide / Tetrapropylammonium iodide	$C_{12}H_{28}IN$ / 313.27	631-40-3 / 280 dec		4-04-00-00471 / 1.3138[25]	H_2O 4; EtOH 3; eth 2; chl 4
9834	Propane	C_3H_8 / 44.10	74-98-6 / -187.6	7809 / -42.1	4-01-00-00176 / *0.493[25]	H_2O 3; EtOH 4; eth 4; ace 2
9835	Propane, 2,2-bis(ethylsulfonyl)- / Sulfonmethane	$C_7H_{16}O_4S_2$ / 228.33	115-24-2 / 125.8	8939 / 300 dec	3-01-00-02754 /	bz 4; EtOH 4; chl 4
9836	Propane, 1-bromo- / Propyl bromide	C_3H_7Br / 122.99	106-94-5 / -110	7857 / 71.1	4-01-00-00205 / 1.3537[20]	H_2O 2; EtOH 3; eth 3; ace 3 / 1.4343[20]
9837	Propane, 2-bromo- / Isopropyl bromide	C_3H_7Br / 122.99	75-26-3 / -89	5098 / 59.5	4-01-00-00208 / 1.3140[20]	H_2O 2; EtOH 5; eth 5; ace 3 / 1.4251[20]
9838	Propane, 1-bromo-2-chloro- / 1-Bromo-2-chloropropane	C_3H_6BrCl / 157.44	3017-96-7 /	/ 118	4-01-00-00212 / 1.531[20]	ace 4; bz 4; eth 4; EtOH 4 / 1.4745[20]
9839	Propane, 1-bromo-3-chloro- / 1-Bromo-3-chloropropane	C_3H_6BrCl / 157.44	109-70-6 / -58.9	/ 143.3	4-01-00-00212 / 1.5960[20]	H_2O 1; EtOH 4; eth 4; chl 4 / 1.4864[20]
9840	Propane, 2-bromo-1-chloro- / 2-Bromo-1-chloropropane	C_3H_6BrCl / 157.44	3017-95-6 /	/ 118	4-01-00-00212 / 1.537[20]	H_2O 1; EtOH 4; eth 4; ace 3 / 1.4795[20]
9841	Propane, 2-bromo-2-chloro- / 2-Bromo-2-chloropropane	C_3H_6BrCl / 157.44	2310-98-7 /	/ 95	4-01-00-00213 / 1.474[22]	ace 4; bz 4; eth 4; EtOH 4 / 1.4575[20]
9842	Propane, 1-bromo-3-chloro-2-methyl-	C_4H_8BrCl / 171.46	6974-77-2 /	/ 155	4-01-00-00298 / 1.4839[20]	/ 1.4796[25]
9843	Propane, 1-bromo-2,2-dimethyl-	$C_5H_{11}Br$ / 151.05	630-17-1 /	/ 106	4-01-00-00337 / 1.1997[20]	H_2O 1; EtOH 3; eth 3; ace 3 / 1.4370[20]
9844	Propane, 1-bromo-3-fluoro-	C_3H_6BrF / 140.98	352-91-0 /	/ 101.4	4-01-00-00210 / 1.542[25]	bz 4; eth 4; EtOH 4; chl 4 / 1.4290[25]
9845	Propane, 1-bromo-2-methyl- / Isobutyl bromide	C_4H_9Br / 137.02	78-77-3 / -119	5019 / 91.1	4-01-00-00294 / 1.272[15]	EtOH 4; eth 4; ace 4; bz 4 / 1.4348[20]
9846	Propane, 2-bromo-2-methyl- / tert-Butyl bromide	C_4H_9Br / 137.02	507-19-7 / -16.2	1555 / 73.3	4-01-00-00295 / 1.4278[20]	H_2O 1; ctc 2 / 1.4278[20]
9847	Propane, 2-bromo-2-nitro-	$C_3H_6BrNO_2$ / 167.99	5447-97-2 /	/ 153; 73-5[50]	4-01-00-00233 / 1.6562[25]	eth 4; EtOH 4
9848	Propane, 1-chloro- / Propyl chloride	C_3H_7Cl / 78.54	540-54-5 / -122.8	7859 / 46.5	4-01-00-00189 / 0.8899[20]	H_2O 2; EtOH 5; eth 5; bz 3 / 1.3879[20]
9849	Propane, 2-chloro- / Isopropyl chloride	C_3H_7Cl / 78.54	75-29-6 / -117.2	5099 / 35.7	4-01-00-00191 / 0.8617[20]	H_2O 2; EtOH 5; eth 5; bz 3 / 1.3777[20]
9850	Propane, 3-chloro-1,1-diethoxy-	$C_7H_{15}ClO_2$ / 166.65	35573-93-4 /	/ 84[25]	3-01-00-02592 / 0.9951[19]	ace 4; bz 4 / 1.4268[20]
9851	Propane, 1-chloro-2,2-difluoro- / 2,2-Difluoropropyl chloride	$C_3H_5ClF_2$ / 114.52	420-99-5 / -56.2	/ 55	4-01-00-00193 / 1.2001[20]	eth 3; bz 3; chl 3 / 1.3520[20]
9852	Propane, 3-chloro-1,1-dimethoxy-	$C_5H_{11}ClO_2$ / 138.59	35502-06-8 /	/ 86[100]; 51[19]	3-01-00-02692 / 1.064[15]	/ 1.4163[20]
9853	Propane, 1-chloro-2,2-dimethyl-	$C_5H_{11}Cl$ / 106.60	753-89-9 / -20	/ 84.3	4-01-00-00336 / 0.8660[20]	bz 4; eth 4; EtOH 4; chl 4 / 1.4044[20]
9854	Propane, 1-(1-chloroethoxy)- / 1-Chloroethyl propyl ether	$C_5H_{11}ClO$ / 122.59	692-35-3 /	/ 114	3-01-00-02655 / 0.9322[20]	eth 4 / 1.4013[20]
9855	Propane, 2-[(2-chloroethyl)thio]-2-methyl-	$C_6H_{13}ClS$ / 152.69	4303-44-0 / -49	/ 81-2[30]	3-01-00-01592 / 1.0001[25]	
9856	Propane, 1-chloro-3-fluoro- / 1-Chloro-3-fluoropropane	C_3H_6ClF / 96.53	462-38-4 /	/ 80	4-01-00-00193 /	bz 4; eth 4; EtOH 4; chl 4 / 1.3871[25]
9857	Propane, 1-chloro-3-iodo-	C_3H_6ClI / 204.44	6940-76-7 /	/ 171	4-01-00-00226 / 1.904[20]	H_2O 1; eth 3; bz 3; ctc 2 / 1.5472[20]
9858	Propane, 1-(chloromethoxy)-	C_4H_9ClO / 108.57	3587-57-3 /	/ 109	3-01-00-02589 / 0.9884[20]	eth 4; EtOH 4 / 1.4125[20]

No.	Name Synonym	Mol. Form. Mol. Wt.	CAS RN mp/°C	Merck No. bp/°C	Beil. Ref. den/g cm^{-3}	Solubility n_D
9859	Propane, 1-chloro-2-methoxy-	C_4H_9ClO 108.57	5390-72-7	103.5	3-01-00-01470 1.009[20]	1.4137[20]
9860	Propane, 1-chloro-3-methoxy-	C_4H_9ClO 108.57	36215-07-3	111	4-01-00-01441 1.0013[20]	1.4131[20]
9861	Propane, 2-chloro-1-methoxy-	C_4H_9ClO 108.57	5390-71-6	99	3-01-00-01424 0.9946[20]	1.4075[20]
9862	Propane, 1-chloro-2-methyl- Isobutyl chloride	C_4H_9Cl 92.57	513-36-0 -130.3	5022 68.5	4-01-00-00287 0.8773[20]	eth 3; ace 3; ctc 2; chl 3 1.3984[20]
9863	Propane, 2-chloro-2-methyl- tert-Butyl chloride	C_4H_9Cl 92.57	507-20-0 -26	1562 50.9	4-01-00-00288 0.8420[20]	H_2O 2; EtOH 5; eth 5; bz 3 1.3857[20]
9864	Propane, 1-chloro-3-(methylthio)- 3-Chloropropyl methyl sulfide	C_4H_9ClS 124.63	13012-59-4	71-2[29]	4-01-00-01458 1.0844[20]	1.4833[20]
9865	Propane, 2-chloro-1-(methylthio)- 2-Chloropropyl methyl sulfide	C_4H_9ClS 124.63	19987-13-4	67[37]	4-01-00-01457 1.076[20]	1.4905[20]
9866	Propane, 1-chloro-1-nitro- 1-Chloro-1-nitropropane	$C_3H_6ClNO_2$ 123.54	600-25-9	142	4-01-00-00232 1.207[20]	H_2O 2; EtOH 3; eth 3; chl 2 1.4251[20]
9867	Propane, 1-chloro-2-nitro-	$C_3H_6ClNO_2$ 123.54	2425-66-3	172.5	4-01-00-00233 1.245[22]	eth 4; EtOH 4; chl 4 1.4432[25]
9868	Propane, 1-chloro-3-nitro- 1-Chloro-3-nitropropane	$C_3H_6ClNO_2$ 123.54	16694-52-3	197 dec; 115[40]	3-01-00-00259 1.267[20]	eth 4; EtOH 4
9869	Propane, 2-chloro-1-nitro-	$C_3H_6ClNO_2$ 123.54	503-76-4	172	4-01-00-00232 1.2361[15]	eth 4; EtOH 4 1.4447[20]
9870	Propane, 2-chloro-2-nitro- 2-Chloro-2-nitropropane	$C_3H_6ClNO_2$ 123.54	594-71-8	134 dec; 57[50]	4-01-00-00233 1.230[19]	H_2O 2; EtOH 3; eth 3; ctc 3 1.4378[19]
9871	Propane, 3-chloro-1,1,1-trifluoro-	$C_3H_4ClF_3$ 132.51	460-35-5 -106.5	45.1	4-01-00-00193 1.3253[20]	1.3350[20]
9872	1,2-Propanediamine Propylenediamine	$C_3H_{10}N_2$ 74.13	78-90-0	7865 119.5	2-04-00-00698 0.878[15]	H_2O 4; eth 1 1.4460[20]
9873	1,3-Propanediamine 1,3-Diaminopropane	$C_3H_{10}N_2$ 74.13	109-76-2	139.8	4-04-00-01258 0.884[25]	H_2O 3; EtOH 5; eth 5 1.4600[20]
9874	1,3-Propanediamine, N-(3-aminopropyl)- Bis(3-aminopropyl)amine	$C_6H_{17}N_3$ 131.22	56-18-8 -14	151[50]	4-04-00-01278 0.938[25]	chl 3 1.4810[20]
9875	1,3-Propanediamine, N-(3-aminopropyl)-N-methyl-	$C_7H_{19}N_3$ 145.25	105-83-9	232.5; 112[6]	4-04-00-01279 0.9023[20]	1.4705[25]
9876	1,3-Propanediamine, N,N-diethyl- 3-(Diethylamino)propylamine	$C_7H_{18}N_2$ 130.23	104-78-9	168.5	4-04-00-01260 0.822[20]	1.443[20]
9877	1,3-Propanediamine, N,N-diethyl-N'-(6-methoxy-8-quinolinyl)- Plasmocid	$C_{17}H_{25}N_3O$ 287.41	551-01-9	182[1.0]	5-22-12-00126 1.0569[24]	1.5855[24]
9878	1,2-Propanediamine, 2-methyl-	$C_4H_{12}N_2$ 88.15	811-93-8	123	4-04-00-01306 0.841[25]	ctc 3 1.4410[20]
9879	1,3-Propanediamine, N,N,N',N'-tetramethyl-	$C_7H_{18}N_2$ 130.23	110-95-2	144	4-04-00-01259 0.7837[18]	H_2O 5; EtOH 5; eth 5
9880	Propane, 1,1-dibromo- 1,1-Dibromopropane	$C_3H_6Br_2$ 201.89	598-17-4	133.5	4-01-00-00215 1.982[20]	eth 4; EtOH 4; chl 4 1.5100[20]
9881	Propane, 1,2-dibromo- Propylene dibromide	$C_3H_6Br_2$ 201.89	78-75-1 -55.2	7866 141.9	4-01-00-00215 1.9324[20]	EtOH 3; eth 3; ctc 2; chl 3 1.5201[20]
9882	Propane, 1,3-dibromo- 1,3-Dibromopropane	$C_3H_6Br_2$ 201.89	109-64-8 -34.2	9628 167.3	4-01-00-00216 1.9822[20]	EtOH 3; eth 3; ctc 2; chl 3 1.5232[20]
9883	Propane, 2,2-dibromo- 2,2-Dibromopropane	$C_3H_6Br_2$ 201.89	594-16-1	115	4-01-00-00217 1.7825[20]	eth 4; EtOH 4; chl 4
9884	Propane, 1,3-dibromo-2,2-bis(bromomethyl)-	$C_5H_8Br_4$ 387.73	3229-00-3 163	305.5	4-01-00-00337 2.596[15]	EtOH 3; eth 2; bz 3; chl 2
9885	Propane, 1,2-dibromo-3-chloro- Dibromochloropropane	$C_3H_5Br_2Cl$ 236.33	96-12-8	3003 196	3-01-00-00250 2.093[14]	1.553[14]
9886	Propane, 1,1-dibromo-2,2-dimethyl- Neopentane, 1,1-dibromo	$C_5H_{10}Br_2$ 229.94	2443-91-6	64-5[43]	0-01-00-00141 1.7880[0]	1.5085
9887	Propane, 1,3-dibromo-2,2-dimethyl-	$C_5H_{10}Br_2$ 229.94	5434-27-5	187 dec; 80[26]	4-01-00-00337 1.7049[0]	1.5090
9888	Propane, 1,2-dibromo-3-methoxy-	$C_4H_8Br_2O$ 231.91	5836-66-8	185	4-01-00-01447 1.8320[12]	eth 4 1.5123[20]
9889	Propane, 1,2-dibromo-2-methyl-	$C_4H_8Br_2$ 215.92	594-34-3 10.5	150	4-01-00-00298 1.7827[20]	EtOH 3; eth 3; chl 3 1.5119[20]
9890	Propane, 1,3-dibromo-2-methyl-	$C_4H_8Br_2$ 215.92	28148-04-1	177.5; 62[13]	4-01-00-00299 1.8204[20]	1.5068[20]
9891	Propane, 1,1-dichloro- Propylidene chloride	$C_3H_6Cl_2$ 112.99	78-99-9	7874 88.1	4-01-00-00195 1.1321[20]	EtOH 3; eth 3; bz 3; chl 3 1.4289[20]
9892	Propane, 1,2-dichloro-, (±)- Propane, 1,2-dichloro (DL)	$C_3H_6Cl_2$ 112.99	26198-63-0 -100.4	7867 96.4	4-01-00-00195 1.1560[20]	H_2O 2; EtOH 3; eth 3; bz 3 1.4394[20]
9893	Propane, 1,3-dichloro- Trimethylene dichloride	$C_3H_6Cl_2$ 112.99	142-28-9 -99.5	120.9	4-01-00-00196 1.1876[20]	H_2O 2; EtOH 4; eth 4; bz 3 1.4487[20]
9894	Propane, 2,2-dichloro-	$C_3H_6Cl_2$ 112.99	594-20-7 -33.8	69.3	4-01-00-00196 1.1136[20]	H_2O 1; EtOH 3; eth 5; bz 3 1.4148[20]
9895	Propane, 1,3-dichloro-2,2-bis(chloromethyl)-	$C_5H_8Cl_4$ 209.93	3228-99-7 97	110[12]	4-01-00-00336	H_2O 1; eth 3; ctc 3; chl 3

No.	Name / Synonym	Mol. Form. / Mol. Wt.	CAS RN / mp/°C	Merck No. / bp/°C	Beil. Ref. / den/g cm^{-3}	Solubility / n_D
9896	Propane, 1,3-dichloro-2-(ethoxymethoxy)- Formaldehyde, (2,2'-dichloroisopropyl) ethyl acetal	$C_6H_{12}Cl_2O_2$ 187.07	89583-61-9	96-8[16]	3-01-00-02574 1.181[17]	eth 4 1.4491[17]
9897	Propane, 1,2-dichloro-2-fluoro-	$C_3H_5Cl_2F$ 130.98	420-97-3 -91.7	88.6	4-01-00-00197 1.2624[20]	H_2O 2; ace 3; bz 3 1.4099[20]
9898	Propane, 1,2-dichloro-1,1,2,3,3,3-hexafluoro- 1,2-Dichlorohexafluoropropane	$C_3Cl_2F_6$ 220.93	661-97-2	34.1		
9899	Propane, 1,1-dichloro-2-methyl- Isobutane, 1,1-dichloro	$C_4H_8Cl_2$ 127.01	598-76-5	105.5	4-01-00-00292 1.011[12]	bz 4; eth 4; EtOH 4; chl 4 1.4330[25]
9900	Propane, 1,2-dichloro-2-methyl- Isobutane, 1,2-dichloro	$C_4H_8Cl_2$ 127.01	594-37-6	106.5	4-01-00-00292 1.093[20]	H_2O 1; EtOH 5; eth 5; ace 5 1.4370[20]
9901	Propane, 1,3-dichloro-2-methyl-	$C_4H_8Cl_2$ 127.01	616-19-3	136.4	3-01-00-00319 1.131[20]	H_2O 1; EtOH 3; eth 4; bz 4 1.4488[25]
9902	Propane, 1,1-dichloro-1-nitro- 1,1-Dichloro-1-nitropropane	$C_3H_5Cl_2NO_2$ 157.98	595-44-8	145	3-01-00-00259 1.312[20]	ctc 3
9903	Propane, 1,1-diethoxy-	$C_7H_{16}O_2$ 132.20	4744-08-5	123	4-01-00-03168 0.8232[20]	H_2O 3; EtOH 4; eth 4; ace 3 1.3924[19]
9904	Propane, 2,2-diethoxy-	$C_7H_{16}O_2$ 132.20	126-84-1	114	4-01-00-03200 0.8200[21]	EtOH 3; eth 4; ace 3; bz 3 1.3891[20]
9905	Propane, 1,3-difluoro- 1,3-Difluoropropane	$C_3H_6F_2$ 80.08	462-39-5	41.6	3-01-00-00218 1.0057[25]	bz 4 1.3190[26]
9906	Propane, 2,2-difluoro- Dimethyldifluoromethane	$C_3H_6F_2$ 80.08	420-45-1 -104.8	-0.4	4-01-00-00187 0.9205[20]	
9907	Propane, 1,3-diiodo- Trimethylene diiodide	$C_3H_6I_2$ 295.89	627-31-6 -20	227 dec; 110[19]	4-01-00-00228 2.5754[20]	H_2O 1; eth 3; ctc 3; chl 3 1.6423[20]
9908	Propane, 2,2-diiodo- 2,2-Diiodopropane	$C_3H_6I_2$ 295.89	630-13-7	173; 156[500]	3-01-00-00255 2.5754[20]	eth 4; chl 4 1.651[20]
9909	Propane, 2,2-dimethoxy-	$C_5H_{12}O_2$ 104.15	77-76-9 -47	83	4-01-00-03199 0.847[25]	1.3780[20]
9910	Propane, 2,2-dimethyl- Neopentane	C_5H_{12} 72.15	463-82-1 -16.6	6372 9.4	4-01-00-00333 *0.5852[25]	H_2O 1; EtOH 3; eth 3; ctc 3 1.3476[6]
9911	Propanedinitrile Malononitrile	$C_3H_2N_2$ 66.06	109-77-3 32	5592 218.5	4-02-00-01892 1.1910[20]	H_2O 3; EtOH 4; eth 4; ace 3 1.4146[34]
9912	Propanedinitrile, dichloro-	$C_3Cl_2N_2$ 134.95	13063-43-9	97	3-02-00-01638 1.312[25]	1.4312[20]
9913	Propanedinitrile, dimethyl-	$C_5H_6N_2$ 94.12	7321-55-3 31.5	169.5; 62[22]	3-02-00-01704	H_2O 3; EtOH 2
9914	Propanedinitrile, (ethoxymethylene)-	$C_6H_8N_2O$ 122.13	123-06-8 66	160[12]	4-03-00-01194	EtOH 3; eth 3; chl 2
9915	Propanedinitrile, oxo- Carbonyl cyanide	C_3N_2O 80.05	1115-12-4 -36	65.5	4-03-00-01806 1.124[20]	eth 3; ace 3; ctc 3; chl 3 1.3919[20]
9916	Propanedinitrile, phenyl- Phenylmalononitrile	$C_9H_6N_2$ 142.16	3041-40-5 70.5	152[21]	4-09-00-03341	EtOH 4
9917	Propanedinitrile, (phenylmethyl)-	$C_{10}H_8N_2$ 156.19	1867-37-4 91	174[23]	4-09-00-03359	bz 4; eth 4; EtOH 4
9918	Propane, 1,1-dinitro- 1,1-Dinitropropane	$C_3H_6N_2O_4$ 134.09	601-76-3 -42	184	4-01-00-00234 1.2610[25]	alk 3 1.4339[20]
9919	Propane, 1,3-dinitro- 1,3-Dinitropropane	$C_3H_6N_2O_4$ 134.09	6125-21-9 -21.4	103[1]	4-01-00-00234 1.353[26]	H_2O 1; eth 3 1.4654[20]
9920	Propane, 2,2-dinitro- 2,2-Dinitropropane	$C_3H_6N_2O_4$ 134.09	595-49-3 53	185.5	4-01-00-00234 1.30[25]	H_2O 2
9921	Propanedioic acid Malonic acid	$C_3H_4O_4$ 104.06	141-82-2 135 dec	5591 sub	4-02-00-01874 1.619[10]	H_2O 4; EtOH 3; eth 3; bz 1
9922	Propanedioic acid, (acetylamino)-, diethyl ester	$C_9H_{15}NO_5$ 217.22	1068-90-2 96.3	185[20]	4-04-00-02993	H_2O 2; EtOH 3; eth 2; tfa 3
9923	Propanedioic acid, acetyl-, diethyl ester	$C_9H_{14}O_5$ 202.21	570-08-1	232	4-03-00-01819 1.0834[26]	H_2O 1; ace 3 1.4435[25]
9924	Propanedioic acid, amino-, diethyl ester Diethyl aminomalonate	$C_7H_{13}NO_4$ 175.18	6829-40-9	122-3[16]	4-04-00-02991 1.100[16]	H_2O 4; EtOH 4; eth 4; ace 3 1.4353[16]
9925	Propanedioic acid, bis(1,1-dimethylethyl) ester Di-*tert*-butyl malonate	$C_{11}H_{20}O_4$ 216.28	541-16-2 -6	3021 113[31]; 66[2]	4-02-00-01884 1.4184[20]	ace 3; chl 3 1.4184[29]
9926	Propanedioic acid, bis(phenylmethyl)-, diethyl ester Diethyl dibenzylmalonate	$C_{21}H_{24}O_4$ 340.42	597-55-7 14	234-5[23]	3-09-00-04552 1.093[20]	eth 4; EtOH 4
9927	Propanedioic acid, bis(phenylmethyl) ester	$C_{17}H_{16}O_4$ 284.31	15014-25-2	188[0.2]	4-06-00-02270 1.137[25]	1.5447[20]
9928	Propanedioic acid, bromo-, diethyl ester	$C_7H_{11}BrO_4$ 239.07	685-87-0 -54	254 dec	4-02-00-01904 1.4022[25]	H_2O 1; EtOH 5; eth 5; ace 3 1.4521[20]
9929	Propanedioic acid, butyl- *n*-Butylmalonic acid	$C_7H_{12}O_4$ 160.17	534-59-8 104.5	1574	4-02-00-02011	H_2O 4; EtOH 3; eth 3
9930	Propanedioic acid, butyl-, diethyl ester	$C_{11}H_{20}O_4$ 216.28	133-08-4	237.5	4-02-00-02011	EtOH 4; eth 4 1.4250[20]
9931	Propanedioic acid, chloro-, diethyl ester	$C_7H_{11}ClO_4$ 194.61	14064-10-9	222	4-02-00-01903 1.2040[20]	H_2O 1; EtOH 5; eth 5; chl 5 1.4327[20]
9932	Propanedioic acid, (3-chloropropyl)-, diethyl ester	$C_{10}H_{17}ClO_4$ 236.70	18719-43-2	147-9[10]	4-02-00-01992	H_2O 1; EtOH 3; eth 3; ctc 2 1.4429[20]

No.	Name Synonym	Mol. Form. Mol. Wt.	CAS RN mp/°C	Merck No. bp/°C	Beil. Ref. den/g cm^{-3}	Solubility n_D
9933	Propanedioic acid, (2-cyanoethyl)-, diethyl ester	$C_{10}H_{15}NO_4$ 213.23	17216-62-5 115-6[2]		4-02-00-02365 1.078[25]	1.4368[20]
9934	Propanedioic acid, cyclohexyl-, diethyl ester	$C_{13}H_{22}O_4$ 242.32	2163-44-2 163-5[20]		3-09-00-03834 1.0281[19]	ace 4; bz 4; eth 4; EtOH 4 1.4478[25]
9935	Propanedioic acid, 2-cyclopenten-1-yl-, diethyl ester Malonic acid, 2-cyclopentenyl, diethyl ester	$C_{12}H_{18}O_4$ 226.27	53608-93-8 141[10]		3-09-00-03951 1.0507[20]	H_2O 1; eth 3; ace 3 1.4536[20]
9936	Propanedioic acid, cyclopentylidene-, diethyl ester Malonic acid, cyclopentylidene, diethyl ester	$C_{12}H_{18}O_4$ 226.27	41589-42-8 140[10]		3-09-00-03952 1.0616[20]	ace 4; eth 4 1.4724[20]
9937	Propanedioic acid, dibutyl-, diethyl ester Diethyl dibutylmalonate	$C_{15}H_{28}O_4$ 272.38	596-75-8 153-4[14]		4-02-00-02119 0.9457[20]	H_2O 1; EtOH 3; eth 3; ctc 3 1.4341[20]
9938	Propanedioic acid, dibutyl ester	$C_{11}H_{20}O_4$ 216.28	1190-39-2 -83	251.5	4-02-00-01884 0.9824[20]	H_2O 1; EtOH 3; eth 3; ace 3 1.4262[20]
9939	Propanedioic acid, diethyl- Diethylmalonic acid	$C_7H_{12}O_4$ 160.17	510-20-3 127 dec	3114	4-02-00-02026	H_2O 4; EtOH 4; eth 4; bz 2
9940	Propanedioic acid, diethyl-, diethyl ester Diethyl 2,2-diethylmalonate	$C_{11}H_{20}O_4$ 216.28	77-25-8 230	3746	4-02-00-02026 0.9643[30]	H_2O 1; EtOH 5; eth 5; ctc 3 1.4240[20]
9941	Propanedioic acid, diethyl ester Diethyl malonate	$C_7H_{12}O_4$ 160.17	105-53-3 -50	3779 200	4-02-00-01881 1.0551[20]	H_2O 2; EtOH 5; eth 5; ace 4 1.4139[20]
9942	Propanedioic acid, dimethyl-, diethyl ester Diethyl dimethylmalonate	$C_9H_{16}O_4$ 188.22	1619-62-1 -30.4	197	4-02-00-01955 0.9964[20]	H_2O 1; EtOH 5; eth 5; ctc 2 1.4129[20]
9943	Propanedioic acid, dimethyl ester Methyl malonate	$C_5H_8O_4$ 132.12	108-59-8 -61.9	6009 181.4	4-02-00-01880 1.528[20]	H_2O 2; EtOH 5; eth 5; ace 4 1.4135[20]
9944	Propanedioic acid, di-2-propenyl-, diethyl ester	$C_{13}H_{20}O_4$ 240.30	3195-24-2 243.5		3-02-00-02003 0.9943[20]	EtOH 3; eth 3; ace 3; ctc 2 1.4445[22]
9945	Propanedioic acid, dipropyl ester Dipropyl malonate	$C_9H_{16}O_4$ 188.22	1117-19-7 -77.1	229	3-02-00-01619 1.0097[20]	ace 4; bz 4; eth 4; EtOH 4 1.4206[20]
9946	Propanedioic acid, di-2-propynyl-, diethyl ester Diethyl dipropargylmalonate	$C_{13}H_{16}O_4$ 236.27	2689-88-5 45.5	129[11]	4-02-00-02340	chl 3
9947	Propanedioic acid, (ethoxymethylene)-, diethyl ester	$C_{10}H_{16}O_5$ 216.23	87-13-8	280 dec; 165[19]	4-03-00-01192	H_2O 1; EtOH 3; eth 3; chl 2 1.4600[20]
9948	Propanedioic acid, ethyl-	$C_5H_8O_4$ 132.12	601-75-2 114	180[0.05]	4-02-00-01952	H_2O 4; EtOH 3; eth 3; ace 1
9949	Propanedioic acid, ethyl-, diethyl ester	$C_9H_{16}O_4$ 188.22	133-13-1 208; 98[12]		4-02-00-01953 1.0047[20]	H_2O 2; EtOH 4; eth 4; ace 3 1.4166[20]
9950	Propanedioic acid, ethylidene-, diethyl ester	$C_9H_{14}O_4$ 186.21	1462-12-0 115-8[17]		4-02-00-02234 1.0194[17]	eth 4; EtOH 4 1.4308[17]
9951	Propanedioic acid, ethyl(1-methylethyl)-, diethyl ester Malonic acid, ethylisopropyl-, diethyl ester	$C_{12}H_{22}O_4$ 230.30	2049-66-3 234; 109[11]		4-02-00-02052	ace 4; eth 4; EtOH 4 1.4280[25]
9952	Propanedioic acid, ethylphenyl-, diethyl ester	$C_{15}H_{20}O_4$ 264.32	76-67-5 170[19]		4-09-00-03385 1.071[20]	H_2O 1; EtOH 3; eth 3; chl 2 1.4896[25]
9953	Propanedioic acid, (formylamino)-, diethyl ester	$C_8H_{13}NO_5$ 203.19	6326-44-9 48.5	173[11]	4-04-00-02992	EtOH 4
9954	Propanedioic acid, hexadecyl-, diethyl ester	$C_{23}H_{44}O_4$ 384.60	41433-81-2 238[14]		4-02-00-02183 0.9118[20]	EtOH 4 1.4433[20]
9955	Propanedioic acid, hexyl-, diethyl ester	$C_{13}H_{24}O_4$ 244.33	5398-10-7 269; 143[15]		3-02-00-01791 0.9577[21]	ace 4; bz 4; eth 4; EtOH 4 1.4278[21]
9956	Propanedioic acid, hydroxy- Tartronic acid	$C_3H_4O_5$ 120.06	80-69-3 157 sub	9042	4-03-00-01120	H_2O 3; EtOH 3; eth 2
9957	Propanedioic acid, hydroxy-, diethyl ester	$C_7H_{12}O_5$ 176.17	13937-08-1 -2.5	221	3-03-00-00905 1.1520[15]	ace 4; bz 4; eth 4; EtOH 4
9958	Propanedioic acid, (hydroxyimino)-, diethyl ester	$C_7H_{11}NO_5$ 189.17	6829-41-0 172[12]		4-03-00-01805 1.1821[18]	ace 4; bz 4; eth 4; EtOH 4 1.4544[18]
9959	Propanedioic acid, (hydroxymethyl)- β-Isomalic acid	$C_4H_6O_5$ 134.09	4360-96-7 142 dec	170 dec	0-03-00-00441	H_2O 4; EtOH 4; eth 4
9960	Propanedioic acid, methyl-	$C_4H_6O_4$ 118.09	516-05-2 135 dec		4-02-00-01932 1.455[20]	H_2O 4; EtOH 4; eth 4; bz 4
9961	Propanedioic acid, (3-methylbutyl)-, diethyl ester	$C_{12}H_{22}O_4$ 230.30	5398-08-3 241		3-02-00-01777 0.9580[25]	ace 4; eth 4; EtOH 4 1.4255[25]
9962	Propanedioic acid, methyl-, diethyl ester	$C_8H_{14}O_4$ 174.20	609-08-5 201		4-02-00-01932 1.0225[20]	H_2O 2; EtOH 4; eth 4; ace 3 1.4126[20]
9963	Propanedioic acid, methyl-, dimethyl ester	$C_6H_{10}O_4$ 146.14	609-02-9 174		3-02-00-01681 1.0977[20]	ace 4; eth 4; EtOH 4; chl 4 1.4128[20]
9964	Propanedioic acid, (1-methylethyl)-, diethyl ester	$C_{10}H_{18}O_4$ 202.25	759-36-4 215		4-02-00-02001 0.9961[20]	H_2O 2; EtOH 4; eth 4; ctc 2 1.4188[21]
9965	Propanedioic acid, (1-methylethylidene)-	$C_6H_8O_4$ 144.13	4441-90-1 170.5 dec		3-02-00-01948	EtOH 4
9966	Propanedioic acid, (1-methylethylidene)-, diethyl ester Diethyl isopropylidenemalonate	$C_{10}H_{16}O_4$ 200.23	6802-75-1 176.5; 116[14]		4-02-00-02244 1.0282[18]	ace 4; EtOH 4 1.4486[17]

No.	Name Synonym	Mol. Form. Mol. Wt.	CAS RN mp/°C	Merck No. bp/°C	Beil. Ref. den/g cm^{-3}	Solubility n_D
9967	Propanedioic acid, (1-methylpropyl)-, diethyl ester	$C_{11}H_{20}O_4$ 216.28	83-27-2		4-02-00-02019 0.988[15]	H$_2$O 1; EtOH 4; eth 4; ctc 2 1.4248[20]
9968	Propanedioic acid, (2-methylpropyl)-	$C_7H_{12}O_4$ 160.17	4361-06-2 112 dec	dec	4-02-00-02023	eth 4; EtOH 4
9969	Propanedioic acid, (2-methylpropyl)-, diethyl ester Diethyl isobutylmalonate	$C_{11}H_{20}O_4$ 216.28	10203-58-4		3-02-00-01756 0.9804[20]	H$_2$O 1; EtOH 4; eth 4; chl 3 1.4236[20]
9970	Propanedioic acid, 1-naphthalenyl-, diethyl ester	$C_{17}H_{18}O_4$ 286.33	6341-60-2 61.3	176[1.0]	4-09-00-03536	EtOH 3; bz 3; chl 2
9971	Propanedioic acid, octyl-	$C_{11}H_{20}O_4$ 216.28	760-55-4 116		4-02-00-02114 1.173[17]	ace 4; EtOH 4
9972	Propanedioic acid, (1-oxobutyl)-, diethyl ester	$C_{11}H_{18}O_5$ 230.26	21633-79-4	249.5	4-03-00-01833 1.056[20]	1.4451[20]
9973	Propanedioic acid, oxo-, diethyl ester	$C_7H_{10}O_5$ 174.15	609-09-6 -30	210; 105[19]	3-03-00-01356 1.1419[16]	H$_2$O 4; EtOH 3; eth 3; chl 3 1.4310[22]
9974	Propanedioic acid, pentyl-, diethyl ester	$C_{12}H_{22}O_4$ 230.30	6065-59-4	134-6[14]	4-02-00-02034 0.9652[20]	H$_2$O 1; EtOH 4; eth 4; ctc 2 1.4253[20]
9975	Propanedioic acid, phenyl-, diethyl ester	$C_{13}H_{16}O_4$ 236.27	83-13-6 16.5	205 dec; 168[12]	4-09-00-03339 1.0950[20]	ace 4; EtOH 4 1.4977[20]
9976	Propanedioic acid, phenyl-, dimethyl ester	$C_{11}H_{12}O_4$ 208.21	37434-59-6 51.6	147[13]	4-09-00-03339	eth 4; EtOH 4
9977	Propanedioic acid, (phenylmethyl)-, diethyl ester	$C_{14}H_{18}O_4$ 250.29	607-81-8	300	4-09-00-03357 1.076[15]	H$_2$O 1; chl 2 1.4872[20]
9978	Propanedioic acid, (phenylmethylene)-, diethyl ester	$C_{14}H_{16}O_4$ 248.28	5292-53-5 32	216[30]; 196[14]	4-09-00-03461 1.1045[20]	H$_2$O 1; EtOH 3; eth 3; ace 3 1.5389[20]
9979	Propanedioic acid, 2-propenyl-	$C_6H_8O_4$ 144.13	2583-25-7 105	180 dec	3-02-00-01945	H$_2$O 4; eth 4; EtOH 4
9980	Propanedioic acid, 2-propenyl-, diethyl ester	$C_{10}H_{16}O_4$ 200.23	2049-80-1	222.5; 93[6]	4-02-00-02240 1.0098[20]	H$_2$O 1; EtOH 4; eth 4; ctc 3 1.4305[20]
9981	Propanedioic acid, propyl-, diethyl ester	$C_{10}H_{18}O_4$ 202.25	2163-48-6	221; 114[22]	4-02-00-01991 0.9873[20]	H$_2$O 2; EtOH 4; eth 4 1.4197[20]
9982	1,2-Propanediol 1,2-Propylene glycol	$C_3H_8O_2$ 76.10	57-55-6 -60	7868 187.6	3-01-00-02142 1.0361[20]	H$_2$O 5; EtOH 5; eth 3; bz 3 1.4324[20]
9983	1,3-Propanediol Trimethylene glycol	$C_3H_8O_2$ 76.10	504-63-2 -26.7	9629 214.4	4-01-00-02493 1.0538[20]	H$_2$O 5; EtOH 5; eth 4; bz 2 1.4398[20]
9984	1,2-Propanediol, 3-amino-, (±)- 1,2-Propanediol, 3-amino(DL)	$C_3H_9NO_2$ 91.11	13552-31-3	265 dec; 145[9]	3-04-00-00839 1.1752[20]	H$_2$O 3; EtOH 3; eth 1; bz 1 1.4910[25]
9985	1,3-Propanediol, 2-amino-2-ethyl- 2-Amino-2-ethyl-1,3-propanediol	$C_5H_{13}NO_2$ 119.16	115-70-8 37.5	451 152-3[10]	3-04-00-00850 1.099[20]	H$_2$O 5 1.490[20]
9986	1,3-Propanediol, 2-amino-2-(hydroxymethyl)- Tromethamine	$C_4H_{11}NO_3$ 121.14	77-86-1 171.5	9684 219-20[10]	4-04-00-01903	H$_2$O 4; MeOH 3
9987	1,3-Propanediol, 2-amino-2-methyl- 2-Amino-2-methyl-1,3-propanediol	$C_4H_{11}NO_2$ 105.14	115-69-5 110	460 151-2[10]	4-04-00-01881	H$_2$O 4; EtOH 3
9988	1,3-Propanediol, 2,2-bis[(acetyloxy)methyl]-, diacetate Pentaerythrityl tetraacetate	$C_{13}H_{20}O_8$ 304.30	597-71-7 83.5	7065	4-02-00-00264 1.273[18]	H$_2$O 3; EtOH 4; eth 4
9989	1,3-Propanediol, 2,2-bis(hydroxymethyl)- Pentaerythritol	$C_5H_{12}O_4$ 136.15	115-77-5 260	7062 sub	4-01-00-02812 1.548	H$_2$O 3; eth 1; bz 1
9990	1,3-Propanediol, 2,2-bis[(nitrooxy)methyl]-, dinitrate (ester) Pentaerythrityl tetranitrate	$C_5H_8N_4O_{12}$ 316.14	78-11-5 140.5	7066	4-01-00-02816 1.773[20]	H$_2$O 2; EtOH 2; eth 2; ace 4
9991	1,3-Propanediol, 2-bromo-2-nitro- Bronopol	$C_3H_6BrNO_4$ 199.99	52-51-7 131.5	1437	4-01-00-02501	
9992	1,3-Propanediol, 2-butyl-2-ethyl-	$C_9H_{20}O_2$ 160.26	115-84-4 43.8	262	4-01-00-02611 0.927[50]	H$_2$O 2; EtOH 3; ace 2 1.4587[25]
9993	1,3-Propanediol, 2-butyl-2-methyl-	$C_8H_{18}O_2$ 146.23	3121-83-3 43.8	262	4-01-00-02599 0.927[50]	EtOH 4 1.4587[25]
9994	1,2-Propanediol, 3-chloro- α-Chlorohydrin	$C_3H_7ClO_2$ 110.54	96-24-2	2145 213 dec; 116[11]	4-01-00-02484 1.325[18]	H$_2$O 3; EtOH 3; eth 3 1.4809[20]
9995	1,3-Propanediol, 2-chloro- Glycerol β-chlorohydrin	$C_3H_7ClO_2$ 110.54	497-04-1	146[18]	4-01-00-02499 1.3219[20]	H$_2$O 4; ace 4; EtOH 4 1.4831[20]
9996	1,2-Propanediol, 3-chloro-, diacetate	$C_7H_{11}ClO_4$ 194.61	869-50-1	245	3-02-00-00313 1.199[25]	eth 4; EtOH 4 1.4407[20]
9997	1,2-Propanediol, 3-chloro-, dinitrate Clonitrate	$C_3H_5ClN_2O_6$ 200.54	2612-33-1 192.5	2390	4-01-00-02487 1.5112[9]	ace 4; EtOH 4; chl 4
9998	1,2-Propanediol, 3-chloro-2-methyl- Isobutylene glycol, 3-chloro	$C_4H_9ClO_2$ 124.57	597-33-1	114-7[20]	4-01-00-02536 1.2362[20]	H$_2$O 4; eth 4; EtOH 4 1.4748[20]
9999	1,2-Propanediol, 3-(2-chlorophenoxy)-	$C_9H_{11}ClO_3$ 202.64	5112-21-0 71.5	250[19]	4-06-00-00790	H$_2$O 3; eth 4; bz 3
10000	1,2-Propanediol, 3-(4-chlorophenoxy)- Chlorphenesin	$C_9H_{11}ClO_3$ 202.64	104-29-0 78	2178 214-5[19]	4-06-00-00831	H$_2$O 1; EtOH 4; eth 4; bz 3

No.	Name / Synonym	Mol. Form. / Mol. Wt.	CAS RN / mp/°C	Merck No. / bp/°C	Beil. Ref. / den/g cm^{-3}	Solubility / n_D
10001	1,2-Propanediol, 3-(4-chlorophenoxy)-, 1-carbamate Chlorphenesin carbamate	$C_{10}H_{12}ClNO_4$ 245.66	886-74-8 90	2179		ace 4; EtOH 4; diox 4
10002	1,2-Propanediol, diacetate	$C_7H_{12}O_4$ 160.17	623-84-7 	190.5	4-02-00-00220 1.059[20]	H_2O 4; EtOH 3; eth 3 1.4173[20]
10003	1,3-Propanediol, diacetate	$C_7H_{12}O_4$ 160.17	628-66-0 	209.5	4-02-00-00221 1.070[14]	H_2O 4; EtOH 3 1.4192
10004	1,3-Propanediol, 2,2-diethyl-	$C_7H_{16}O_2$ 132.20	115-76-4 61.5	240.5	4-01-00-02589 1.050[20]	H_2O 4; EtOH 4; eth 4; chl 3 1.4574[25]
10005	1,3-Propanediol, 2,2-dimethyl- Neopentyl glycol	$C_5H_{12}O_2$ 104.15	126-30-7 130	6374 208	4-01-00-02551	H_2O 3; EtOH 4; eth 4; bz 3
10006	1,3-Propanediol, 2-ethyl-2-(hydroxymethyl)- Trimethylolpropane	$C_6H_{14}O_3$ 134.18	77-99-6 58	160[5]	4-01-00-02786	H_2O 4; EtOH 4
10007	1,3-Propanediol, 2-ethyl-2-methyl-	$C_6H_{14}O_2$ 118.18	77-84-9 40.3	225.8	4-01-00-02574	chl 3
10008	1,3-Propanediol, 2-ethyl-2-nitro- 2-Nitro-2-ethyl-1,3-propanediol	$C_5H_{11}NO_4$ 149.15	597-09-1 57.5	dec	4-01-00-02550	H_2O 4; eth 4; EtOH 4
10009	1,2-Propanediol, 3-(hexadecyloxy)-	$C_{19}H_{40}O_3$ 316.52	53584-29-5 64	205[1.0]	3-01-00-02322	ace 4; chl 4; peth 4
10010	1,2-Propanediol, 3-(hexadecyloxy)-, (S)- Chimyl alcohol	$C_{19}H_{40}O_3$ 316.52	506-03-6 64	2049 120[0.005]	3-01-00-02322	ace 4; peth 4; chl 4
10011	1,3-Propanediol, 2-(hydroxymethyl)-2-methyl-	$C_5H_{12}O_3$ 120.15	77-85-0 204	135-7[15]	4-01-00-02780	H_2O 5; EtOH 5; eth 1; bz 1
10012	1,3-Propanediol, 2-(hydroxymethyl)-2-nitro- Tris(hydroxymethyl)nitromethane	$C_4H_9NO_5$ 151.12	126-11-4 165	9667 dec	4-01-00-02777	H_2O 4; eth 4; EtOH 4
10013	1,2-Propanediol, 3-mercapto- Thioglycerol	$C_3H_8O_2S$ 108.16	96-27-5 	9264 100-1[1]	3-01-00-02339 1.2455[20]	H_2O 2; EtOH 5; eth 2; ace 4 1.5268[20]
10014	1,2-Propanediol, 3-methoxy-	$C_4H_{10}O_3$ 106.12	623-39-2 	220	4-01-00-02755 1.114[20]	H_2O 4; EtOH 4; eth 3; ace 4 1.442[25]
10015	1,3-Propanediol, 2-methoxy- Glycerol, 2-methyl ether	$C_4H_{10}O_3$ 106.12	761-06-8 	232	3-01-00-02317 1.124[25]	H_2O 4; ace 4; eth 4; EtOH 4 1.4505[12]
10016	1,2-Propanediol, 3-(2-methoxyphenoxy)- Guaifenesin	$C_{10}H_{14}O_4$ 198.22	93-14-1 78.5	4465 215[19.] 127[0.2]	4-06-00-05576	H_2O 3; EtOH 4; bz 3; chl 3
10017	1,2-Propanediol, 2-methyl-	$C_4H_{10}O_2$ 90.12	558-43-0 	176	4-01-00-02533 1.0024[20]	H_2O 4; eth 4; EtOH 4 1.4350[20]
10018	1,3-Propanediol, 2-methyl-2-nitro-	$C_4H_9NO_4$ 135.12	77-49-6 150.1	dec	4-01-00-02537	H_2O 4; EtOH 4; DMSO-d$_6$ 2
10019	1,2-Propanediol, 3-(2-methylphenoxy)- Mephenesin	$C_{10}H_{14}O_3$ 182.22	59-47-2 70 dec	5737	4-06-00-01952	H_2O 2; EtOH 3; eth 2
10020	1,3-Propanediol, 2-methyl-2-propyl-	$C_7H_{16}O_2$ 132.20	78-26-2 62.5	234; 121[10]	4-01-00-02585	H_2O 3; chl 2; os 3; hx 3
10021	1,3-Propanediol, 2-methyl-2-propyl-, dicarbamate Meprobamate	$C_9H_{18}N_2O_4$ 218.25	57-53-4 105	5751	4-03-00-00073	bz 4; eth 4; EtOH 4
10022	1,2-Propanediol, 1-nitrate	$C_3H_7NO_4$ 121.09	20266-65-3 	148-9[4]	4-01-00-02476 1.2417[20]	1.4368[20]
10023	1,2-Propanediol, 3-(octadecyloxy)- Batyl alcohol	$C_{21}H_{44}O_3$ 344.58	544-62-7 70.5	1021 217[2]	4-01-00-02758	eth 4
10024	Propanediol, oxybis-	$C_6H_{14}O_5$ 166.17	59113-36-9 	267[15]; 225[3]	3-01-00-02327 1.2751[20]	H_2O 4; EtOH 4 1.4890[20]
10025	1,2-Propanediol, 3,3'-oxybis-	$C_6H_{14}O_5$ 166.17	627-82-7 	267[15]; 225[3]	3-01-00-02327 1.2751[20]	H_2O 4; EtOH 4 1.4890[20]
10026	1,2-Propanediol, 3-phenoxy- Phenylglyceryl ether	$C_9H_{12}O_3$ 168.19	538-43-2 67.5	7261 200[22]	4-06-00-00589 1.225[20]	H_2O 4; bz 4; eth 4; EtOH 4
10027	1,3-Propanedione, 1,3-diphenyl- Dibenzoylmethane	$C_{15}H_{12}O_2$ 224.26	120-46-7 70.5	2991	4-07-00-02512	EtOH 3; eth 3; chl 3; dil NaOH 3
10028	1,2-Propanedione, 1-phenyl-	$C_9H_8O_2$ 148.16	579-07-7 <20	222; 102[12]	4-07-00-02143 1.1006[20]	H_2O 3; EtOH 3; eth 3 1.537[10]
10029	Propanedioyl dichloride	$C_3H_2Cl_2O_2$ 140.95	1663-67-8 	58[26]	4-02-00-01887 1.4509[20]	eth 3; AcOEt 3 1.4639[20]
10030	1,2-Propanedithiol	$C_3H_8S_2$ 108.23	814-67-5 	152	4-01-00-02492 1.08[20]	chl 3 1.532[20]
10031	1,3-Propanedithiol Trimethylene dimercaptan	$C_3H_8S_2$ 108.23	109-80-8 -79	7811 172.9	4-01-00-02503 1.0772[20]	H_2O 2; EtOH 5; eth 5; bz 5 1.5392[20]
10032	Propane, 1,1'-[1,2-ethanediylbis(oxy)]bis[2-methyl- Ethylene glycol diisobutyl ether	$C_{10}H_{22}O_2$ 174.28	54922-86-0 	171.3	4-01-00-02382 0.8200[20]	1.4021[20]
10033	Propane, 1-(ethenyloxy)-	$C_5H_{10}O$ 86.13	764-47-6 	65	4-01-00-02052 0.7674[20]	1.3908[20]
10034	Propane, 2-(ethenyloxy)-	$C_5H_{10}O$ 86.13	926-65-8 -140	55.5	4-01-00-02052 0.7534[20]	ace 4; bz 4; eth 4; EtOH 4 1.3840[20]
10035	Propane, 1-(ethenyloxy)-2-methyl- Isobutyl vinyl ether	$C_6H_{12}O$ 100.16	109-53-5 -112	83	4-01-00-02054 0.7645[20]	H_2O 2; EtOH 4; eth 5; ace 4 1.3966[20]
10036	Propane, 2-(ethenyloxy)-2-methyl-	$C_6H_{12}O$ 100.16	926-02-3 -112	75	4-01-00-02055 0.7691[20]	1.3922[20]
10037	Propane, 1-ethoxy- Ethyl propyl ether	$C_5H_{12}O$ 88.15	628-32-0 -127.5	63.2	4-01-00-01421 0.7386[20]	eth 4; EtOH 4; HOAc 4 1.3695[20]
10038	Propane, 2-ethoxy- Ethyl isopropyl ether	$C_5H_{12}O$ 88.15	625-54-7 	54.1	4-01-00-01471 0.720[25]	H_2O 3; EtOH 5; eth 5; ace 3 1.3698[25]

No.	Name / Synonym	Mol. Form. / Mol. Wt.	CAS RN / mp/°C	Merck No. / bp/°C	Beil. Ref. / den/g cm^{-3}	Solubility / n_D
10039	Propane, 1-ethoxy-2-methyl-	$C_6H_{14}O$ 102.18	627-02-1	81	4-01-00-01593 0.751[20]	H_2O 1; EtOH 5; eth 5; ace 3 1.3739[25]
10040	Propane, 2-ethoxy-2-methyl- Ethyl tert-butyl ether	$C_6H_{14}O$ 102.18	637-92-3 -94	3732 73.1	4-01-00-01615 0.7519[25]	H_2O 1; EtOH 4; eth 4 1.3794[20]
10041	Propane, 1,1'-[ethylidenebis(oxy)]bis[2-methyl-	$C_{10}H_{22}O_2$ 174.28	5669-09-0	171.3	4-01-00-03107 0.8200[20]	1.4021[20]
10042	Propane, 1-(ethylthio)- Ethyl propyl sulfide	$C_5H_{12}S$ 104.22	4110-50-3 -117	118.6	4-01-00-01451 0.8370[20]	EtOH 3 1.4462[20]
10043	Propane, 2-(ethylthio)- Ethyl isopropyl sulfide	$C_5H_{12}S$ 104.22	5145-99-3 -122.2	107.5	4-01-00-01501 0.8246[20]	
10044	Propane, 1-(ethylthio)-2-methyl- Ethyl isobutyl sulfide	$C_6H_{14}S$ 118.24	1613-45-2	134.2	4-01-00-01606 0.8306[20]	chl 3 1.4450[20]
10045	Propane, 1-(ethynyloxy)-	C_5H_8O 84.12	42842-08-0	75	4-01-00-02213 0.8080[20]	eth 4; EtOH 4 1.3935[20]
10046	Propane, 1-fluoro- Propyl fluoride	C_3H_7F 62.09	460-13-9 -159	2.5	4-01-00-00187 0.7956[20]	H_2O 2; EtOH 4; eth 4 1.3115[20]
10047	Propane, 2-fluoro- Isopropyl fluoride	C_3H_7F 62.09	420-26-8		4-01-00-00187	
10048	Propane, 1,1,1,2,2,3,3-heptachloro- 1,1,1,2,2,3,3-Heptachloropropane	C_3HCl_7 285.21	594-89-8 29.4	247.5	4-01-00-00205 1.8048[34]	ctc 2; chl 3
10049	Propane, 1,1,1,2,3,3,3-heptachloro- 1,1,1,2,3,3,3-Heptachloropropane	C_3HCl_7 285.21	3849-33-0 11	249	4-01-00-00205 1.7921[34]	chl 4 1.5427[21]
10050	Propane, 1,1,1,2,3,3,3-heptafluoro-2-(trifluoromethyl)- Perfluoroisobutane	C_4F_{10} 238.03	354-92-7	0		
10051	Propane, 1,1,1,2,2,3-hexachloro- 1,1,1,2,2,3-Hexachloropropane	$C_3H_2Cl_6$ 250.77	24425-97-6 26	216	4-01-00-00204 1.7187[25]	1.5282[25]
10052	Propane, 1,1,1,3,3,3-hexachloro- 1,1,1,3,3,3-Hexachloropropane	$C_3H_2Cl_6$ 250.77	3607-78-1 -27	206	4-01-00-00205 1.6800[20]	1.5179[20]
10053	Propane, 1,1,2,2,3,3-hexachloro- 1,1,2,2,3,3-Hexachloropropane	$C_3H_2Cl_6$ 250.77	15600-01-8	218.5	4-01-00-00205 1.7137[34]	1.5262[18]
10054	Propane, 1-iodo- Propyl iodide	C_3H_7I 169.99	107-08-4 -101.3	7875 102.6	4-01-00-00222 1.7489[20]	H_2O 2; EtOH 5; eth 5; ctc 2 1.5058[20]
10055	Propane, 2-iodo- Isopropyl iodide	C_3H_7I 169.99	75-30-9 -90	5102 89.5	4-01-00-00223 1.7042[20]	H_2O 2; EtOH 5; eth 5; bz 5 1.5028[20]
10056	Propane, 1-iodo-2,2-dimethyl-	$C_5H_{11}I$ 198.05	15501-33-4	128 dec	4-01-00-00338 1.4940[20]	H_2O 1; EtOH 3; eth 3 1.4890[20]
10057	Propane, 1-iodo-2-methyl- Isobutyl iodide	C_4H_9I 184.02	513-38-2	5027 121.1	4-01-00-00299 1.6035[20]	1.4959[20]
10058	Propane, 2-iodo-2-methyl- tert-Butyl iodide	C_4H_9I 184.02	558-17-8 -38.2	100.1	4-01-00-00300 1.571[25]	EtOH 5; eth 5 1.4910[20]
10059	Propane, 1-isocyanato- Propyl isocyanate	C_4H_7NO 85.11	110-78-1	83.5	4-04-00-00491 0.908[25]	1.3970[20]
10060	Propane, 2-isocyanato- Isopropyl isocyanate	C_4H_7NO 85.11	1795-48-8	74.5	4-04-00-00530 0.866[25]	1.3825[20]
10061	Propane, 1-isocyanato-2-methyl- Isobutyl isocyanate	C_5H_9NO 99.13	1873-29-6	106	4-04-00-00653	
10062	Propane, 2-isocyanato-2-methyl- tert-Butyl isocyanate	C_5H_9NO 99.13	1609-86-5 85.5		4-04-00-00669 0.8670[7]	1.4061[20]
10063	Propane, 2-isocyano- Isopropyl isocyanide	C_4H_7N 69.11	598-45-8	87	0-04-00-00154 0.7596[25]	H_2O 1; EtOH 5; eth 5
10064	Propane, 1-isothiocyanato-	C_4H_7NS 101.17	628-30-8	153	4-04-00-00491 0.9781[16]	H_2O 2; EtOH 5; eth 5 1.5085[16]
10065	Propane, 1-isothiocyanato-2-methyl-	C_5H_9NS 115.20	591-82-2	160	4-04-00-00653 0.9631[14]	1.5005[14]
10066	Propane, 2-isothiocyanato-2-methyl- tert-Butyl isothiocyanate	C_5H_9NS 115.20	590-42-1 10.5	140	4-04-00-00669 0.9187[10]	
10067	Propane, 1-isothiocyanato-3-(methylsulfonyl)- Cheirolin	$C_5H_9NO_2S_2$ 179.26	505-34-0 47.5	2040 200[3]	4-04-00-01664	EtOH 4; chl 4
10068	Propane, 1,1',1'',1'''-[methanetetrayltetrakis(oxy)]tetrakis-	$C_{13}H_{28}O_4$ 248.36	597-72-8	224.2	4-03-00-00007 0.897[20]	EtOH 4; eth 4 1.4100[20]
10069	Propane, 1-methoxy- Methyl propyl ether	$C_4H_{10}O$ 74.12	557-17-5	6031 39.1	4-01-00-01421 0.7356[13]	H_2O 3; EtOH 5; eth 5; ace 3 1.3579[25]
10070	Propane, 2-methoxy- Isopropyl methyl ether	$C_4H_{10}O$ 74.12	598-53-8	30.7	4-01-00-01471 0.7237[15]	H_2O 2; EtOH 5; eth 5 1.3576[20]
10071	Propane, 1-methoxy-2-methyl- Isobutyl methyl ether	$C_5H_{12}O$ 88.15	625-44-5	58.6	4-01-00-01593 0.7311[20]	eth 4; EtOH 4
10072	Propane, 2-methoxy-2-methyl- tert-Butyl methyl ether	$C_5H_{12}O$ 88.15	1634-04-4 -108.6	5954 55.2	4-01-00-01615 0.7405[20]	H_2O 3; EtOH 4; eth 4 1.3690[20]
10073	Propane, 2-methyl- Isobutane	C_4H_{10} 58.12	75-28-5 -138.3		4-01-00-00282 *0.5510[25]	H_2O 2; EtOH 3; eth 3; chl 3 1.3518[-25]
10074	Propane, 1,1'-[methylenebis(oxy)]bis-	$C_7H_{16}O_2$ 132.20	505-84-0 -97.3	140.5	4-01-00-03029 0.8345[20]	ace 4; bz 4; eth 4; EtOH 4 1.3939[19]
10075	Propane, 1,1'-[methylenebis(oxy)]bis[2-methyl-	$C_9H_{20}O_2$ 160.26	2568-91-4	165.5	4-01-00-03030 0.8268[20]	1.4108[17]
10076	Propane, 1-(1-methylethoxy)-	$C_6H_{14}O$ 102.18	627-08-7	83	4-01-00-01471 0.7370[20]	H_2O 2; EtOH 4; eth 3; ace 3 1.376[21]
10077	Propane, 1-[(1-methylethyl)thio]- Isopropyl propyl sulfide	$C_6H_{14}S$ 118.24	5008-73-1	132.1	4-01-00-01501 0.8269[20]	

No.	Name / Synonym	Mol. Form. / Mol. Wt.	CAS RN / mp/°C	Merck No. / bp/°C	Beil. Ref. / den/g cm^{-3}	Solubility / n_D
10078	Propane, 1,1',1"-[methylidynetris(oxy)]tris-	$C_{10}H_{22}O_3$ 190.28	621-76-1	191; 93[30]	4-02-00-00027 0.8805[20]	eth 4; EtOH 4 1.4072[20]
10079	Propane, 2,2',2"-[methylidynetris(oxy)]tris-	$C_{10}H_{22}O_3$ 190.28	4447-60-3	167	4-02-00-00028 0.8621[20]	eth 4; EtOH 4 1.4000[20]
10080	Propane, 1,1',1"-[methylidynetris(oxy)]tris[2-methyl- Orthoformic acid, triisobutyl ester	$C_{13}H_{28}O_3$ 232.36	16754-49-7	223	4-02-00-00029 0.8582[20]	eth 4; EtOH 4 1.4120[20]
10081	Propane, 2-methyl-2-(1-methylethoxy)-	$C_7H_{16}O$ 116.20	17348-59-3 -88	87.6	4-01-00-01616 0.7365[25]	chl 3
10082	Propane, 2-methyl-1-(methylthio)-	$C_5H_{12}S$ 104.22	5008-69-5	112.5	4-01-00-01606 0.8335[20]	EtOH 3; eth 3; ace 3 1.4433[20]
10083	Propane, 2-methyl-2-(methylthio)- tert-Butyl methyl sulfide	$C_5H_{12}S$ 104.22	6163-64-0		4-01-00-01636	
10084	Propane, 2-methyl-1-nitro-	$C_4H_9NO_2$ 103.12	625-74-1	140.5	4-01-00-00301 0.9597[25]	H_2O 2; EtOH 5; eth 5 1.4066[25]
10085	Propane, 2-methyl-2-nitro-	$C_4H_9NO_2$ 103.12	594-70-7 26.23	127.16	4-01-00-00301 0.9501[28]	EtOH 5; eth 5; ace 5; bz 5 1.4015[20]
10086	Propane, 2-methyl-1-propoxy-	$C_7H_{16}O$ 116.20	15268-49-2	107	4-01-00-01594 0.7549[20]	eth 4; EtOH 4 1.3852[25]
10087	Propane, 2-methyl-2-propoxy-	$C_7H_{16}O$ 116.20	29072-93-3	100	3-01-00-01578 0.7472[25]	1.3830[25]
10088	Propane, 1-(methylthio)- Methyl propyl sulfide	$C_4H_{10}S$ 90.19	3877-15-4 -113	95.6	4-01-00-01450 0.8424[20]	H_2O 3; EtOH 3; eth 3; ace 3 1.4442[20]
10089	Propane, 2-(methylthio)- Isopropyl methyl sulfide	$C_4H_{10}S$ 90.19	1551-21-9 -101.5	84.8	4-01-00-01500 0.8291[20]	EtOH 3; eth 3; ace 3 1.4932[20]
10090	Propanenitrile Ethyl cyanide	C_3H_5N 55.08	107-12-0 -92.8	7839 97.1	4-02-00-00728 0.7818[20]	H_2O 4; EtOH 3; eth 3; ace 3 1.3655[20]
10091	Propanenitrile, 2-(acetyloxy)-	$C_5H_7NO_2$ 113.12	15657-96-2 -75	172.5	4-03-00-00676 1.0278[20]	H_2O 4 1.4027[20]
10092	Propanenitrile, 3-amino- 3-Aminopropionitrile	$C_3H_6N_2$ 70.09	151-18-8	481 185; 88[20]	4-04-00-02530 0.9584[20]	1.4396[20]
10093	Propanenitrile, 2,2'-azobis[2-methyl- 2,2'-Azobis(isobutyronitrile)	$C_8H_{12}N_4$ 164.21	78-67-1	931	4-04-00-03377	H_2O 1; EtOH 2; eth 2
10094	Propanenitrile, 2-bromo-	C_3H_4BrN 133.98	19481-82-4	59[24]	4-02-00-00764 1.5505[20]	ace 4; eth 4 1.4585[20]
10095	Propanenitrile, 3-bromo-	C_3H_4BrN 133.98	2417-90-5	92[25]; 69[7]	4-02-00-00766 1.6152[20]	EtOH 4; eth 4; ctc 2 1.4800[20]
10096	Propanenitrile, 2-bromo-2-methyl-	C_4H_6BrN 148.00	41658-69-9	139.5	4-02-00-00863 1.4796[15]	ace 4; eth 4; EtOH 4 1.4379[15]
10097	Propanenitrile, 3-chloro- β-Chloropropionitrile	C_3H_4ClN 89.52	542-76-7 -51	2160 175.5	4-02-00-00751 1.1573[20]	ctc 2 1.4360[20]
10098	Propanenitrile, 3-chloro-2-hydroxy-2-methyl- Lactonitrile, 3-chloro-2-methyl-	C_4H_6ClNO 119.55	33401-05-7	110[27]	3-03-00-00599 1.2027[15]	H_2O 4; EtOH 4; ace 4 1.4356[11]
10099	Propanenitrile, 2,3-dichloro-	$C_3H_3Cl_2N$ 123.97	2601-89-0	58-9[8]	4-02-00-00757 1.3500[20]	1.4640[20]
10100	Propanenitrile, 2,2-dimethyl- tert-Butyl cyanide	C_5H_9N 83.13	630-18-2 15	106.1	4-02-00-00913 0.7586[25]	1.3774[20]
10101	Propanenitrile, 3-(dimethylamino)- 3-(Dimethylamino)propionitrile	$C_5H_{10}N_2$ 98.15	1738-25-6	173	4-04-00-02533 0.8705[20]	
10102	Propanenitrile, 2-ethoxy- 2-Ethoxypropionitrile	C_5H_9NO 99.13	14631-45-9	136	4-03-00-00675 0.8743[20]	eth 4; EtOH 4 1.3890[22]
10103	Propanenitrile, 3-ethoxy-	C_5H_9NO 99.13	2141-62-0	171	4-03-00-00708 0.9285[15]	eth 4; EtOH 4 1.4068[20]
10104	Propanenitrile, 2-(β-D-glucopyranosyloxy)-2-methyl- Linamarin	$C_{10}H_{17}NO_6$ 247.25	554-35-8 145	5375	5-17-07-00397	ace 4
10105	Propanenitrile, 2-hydroxy- Acetaldehyde, cyanohydrin	C_3H_5NO 71.08	78-97-7 -40	183	3-03-00-00451 0.9877[20]	H_2O 5; EtOH 5; eth 3; chl 3 1.4058[18]
10106	Propanenitrile, 3-hydroxy- Hydracrylonitrile	C_3H_5NO 71.08	109-78-4 -46	3751 221	4-03-00-00708 1.0404[25]	H_2O 5; EtOH 5; eth 2; chl 3 1.4248[20]
10107	Propanenitrile, 3-[2-(2-hydroxyethoxy)ethoxy]-	$C_7H_{13}NO_3$ 159.19	10143-54-1	186[9]	4-03-00-00713 1.089[32]	1.4452[22]
10108	Propanenitrile, 2-hydroxy-2-methyl- Acetone cyanohydrin	C_4H_7NO 85.11	75-86-5 -19	59 82[23]	4-03-00-00785 0.932[19]	H_2O 4; EtOH 4; eth 4; ace 3 1.3992[20]
10109	Propanenitrile, 2-methoxy- 2-Methoxypropionitrile	C_4H_7NO 85.11	33695-59-9	119	4-03-00-00675 0.8928[20]	EtOH 4 1.3818[20]
10110	Propanenitrile, 3-methoxy-	C_4H_7NO 85.11	110-67-8	163	4-03-00-00708 0.9379[20]	EtOH 3; eth 3; chl 3 1.4043[20]
10111	Propanenitrile, 2-methyl- Isobutyronitrile	C_4H_7N 69.11	78-82-0 -71.5	103.9	4-02-00-00853 0.7704[20]	H_2O 2; EtOH 4; eth 4; ace 3 1.3720[20]
10112	Propanenitrile, 3-(methylamino)-	$C_4H_8N_2$ 84.12	693-05-0	101-4[49]	3-04-00-01264 0.8992[20]	H_2O 3; ace 3; bz 3; chl 3 1.4320[20]
10113	Propanenitrile, 2-oxo-	C_3H_3NO 69.06	631-57-2	92.3	4-03-00-01515 0.9745[20]	eth 3; ace 3; CH_3CN 3 1.3764[20]
10114	Propanenitrile, 3,3'-oxybis-	$C_6H_8N_2O$ 124.14	1656-48-0	110-12[0.5]	4-03-00-00720 1.043[25]	1.4405[20]
10115	Propanenitrile, 3,3,3-trichloro-2-hydroxy- Chlorocyanohydrin	$C_3H_2Cl_3NO$ 174.41	513-96-2 61	2134 217 dec	4-03-00-00680	H_2O 4; eth 4; EtOH 4
10116	Propane, 1-nitro- 1-Nitropropane	$C_3H_7NO_2$ 89.09	108-03-2 -108	6548 131.1	4-01-00-00229 0.9961[25]	H_2O 2; EtOH 5; eth 5; chl 3 1.4018[20]

No.	Name / Synonym	Mol. Form. / Mol. Wt.	CAS RN / mp/°C	Merck No. / bp/°C	Beil. Ref. / den/g cm⁻³	Solubility / n_D
10117	Propane, 2-nitro- / 2-Nitropropane	$C_3H_7NO_2$ / 89.09	79-46-9 / -91.3	6549 / 120.2	4-01-00-00230 / 0.9821[25]	H_2O 2; chl 3 / 1.3944[20]
10118	Propane, octafluoro- / Perfluoropropane	C_3F_8 / 188.02	76-19-7 / -147.6	/ -36.6	4-01-00-00189 /	
10119	Propane, 1,1'-oxybis- / Dipropyl ether	$C_6H_{14}O$ / 102.18	111-43-3 / -126.1	7870 / 90.0	4-01-00-01422 / 0.7466[20]	eth 4; EtOH 4 / 1.3809[20]
10120	Propane, 2,2'-oxybis- / Diisopropyl ether	$C_6H_{14}O$ / 102.18	108-20-3 / -86.8	5100 / 68.5	4-01-00-01471 / 0.7241[20]	H_2O 2; EtOH 5; eth 5; ace 3 / 1.3679[20]
10121	Propane, 1,1'-oxybis[2-chloro-	$C_6H_{12}Cl_2O$ / 171.07	54460-96-7 /	/ 188	2-01-00-00370 / 1.109[20]	EtOH 3; eth 3 / 1.4467[20]
10122	Propane, 1,1'-oxybis[3-chloro-	$C_6H_{12}Cl_2O$ / 171.07	629-36-7 /	/ 216; 90.5[11]	4-01-00-01442 / 1.138[20]	EtOH 3; eth 3 / 1.4158[20]
10123	Propane, 2,2'-oxybis[1-chloro- / 2,2'-Dichlorodiisopropyl ether	$C_6H_{12}Cl_2O$ / 171.07	108-60-1 /	/ 187	3-01-00-01470 / 1.103[20]	H_2O 1; EtOH 5; eth 5; ace 5 / 1.4505[20]
10124	Propane, 2,2'-oxybis[2-methyl- / Di-tert-Butyl ether	$C_8H_{18}O$ / 130.23	6163-66-2 /	3020 / 107.2	4-01-00-01616 / 0.7658[20]	/ 1.3949[20]
10125	Propane, 1,1,1,2,3-pentachloro- / 1,1,1,2,3-Pentachloropropane	$C_3H_3Cl_5$ / 216.32	21700-31-2 /	/ 179.5	4-01-00-00203 /	eth 4; EtOH 4 / 1.5130[20]
10126	Propane, 1,1,2,3,3-pentachloro- / 1,1,2,3,3-Pentachloropropane	$C_3H_3Cl_5$ / 216.32	15104-61-7 /	/ 199; 99[20]	4-01-00-00204 / 1.6086[39]	H_2O I; EtOH 4; eth 3; chl 3 / 1.5131[17]
10127	Propane, 1,1,1,2,3-pentachloro-2-methyl- / Isobutane, 1,1,1,2,3-pentachloro	$C_4H_5Cl_5$ / 230.35	4749-31-9 / 73.5	/ 215	4-01-00-00294 / 1.5686[25]	chl 4 / 1.5165[20]
10128	Propane, 1,1,1,2,2-pentafluoro- / 1,1,1,2,2-Pentafluoropropane	$C_3H_3F_5$ / 134.05	1814-88-6 /	/ -17.4		
10129	Propane, 1,1'-sulfinylbis-	$C_6H_{14}OS$ / 134.24	4253-91-2 / 22.5	/ 80[2]	4-01-00-01453 / 0.9654[20]	eth 4; EtOH 4 / 1.4663[20]
10130	2-Propanesulfonic acid / Isopropanesulfonic acid	$C_3H_8O_3S$ / 124.16	14159-48-9 / -37	/ 159[1.4]	4-04-00-00042 / 1.187[25]	H_2O 4 / 1.4332[20]
10131	Propane, 1,1'-sulfonylbis-	$C_6H_{14}O_2S$ / 150.24	598-03-8 / 29.5		4-01-00-01453 / 1.0278[50]	H_2O 2; EtOH 3; eth 3 / 1.4456[30]
10132	Propane, 1,1'-sulfonylbis[2-methyl-	$C_8H_{18}O_2S$ / 178.30	10495-45-1 / 17	/ 265	4-01-00-01607 / 1.0056[18]	
10133	1-Propanesulfonyl chloride	$C_3H_7ClO_2S$ / 142.61	10147-36-1 /	/ 180 dec, 77[12]	4-04-00-00039 / 1.2826[15]	/ 1.452[20]
10134	1-Propanesulfonyl chloride, 2-methyl-	$C_4H_9ClO_2S$ / 156.63	35432-36-1 /	/ 190	4-04-00-00049 / 1.2014[20]	/ 1.4520[25]
10135	Propane, 1,1,1,2-tetrabromo-	$C_3H_4Br_4$ / 359.68	62127-49-5 /	/ 122[15]	3-01-00-00251 / 2.679[20]	/ 1.6187[20]
10136	Propane, 1,1,2,2-tetrabromo- / 1,1,2,2-Tetrabromopropane	$C_3H_4Br_4$ / 359.68	34570-59-7 /	/ 105-7[9]	4-01-00-00222 / 2.687[20]	/ 1.6166[20]
10137	Propane, 1,1,3,3-tetrabromo-	$C_3H_4Br_4$ / 359.68	51525-07-4 /	/ 154-6[19]	2-01-00-00077 / 2.702[21]	/ 1.6225[21]
10138	Propane, 1,2,2,3-tetrabromo- / 1,2,2,3-Tetrabromopropane	$C_3H_4Br_4$ / 359.68	54268-02-9 / 10.5	/ 169[80]; 115[9]	3-01-00 00251 / 2.703[20]	/ 1.6200[20]
10139	1,1,2,3-Propanetetracarboxylic acid, tetraethyl ester / Tetraethyl 1,1,2,3-propanetetracarboxylate	$C_{15}H_{24}O_8$ / 332.35	635-03-0 /	/ 203-4[18]	3-02-00-02077 / 1.1184[20]	EtOH 4 / 1.4395[20]
10140	1,1,3,3-Propanetetracarboxylic acid, tetraethyl ester / 1,1,3,3-Tetracarbethoxypropane	$C_{15}H_{24}O_8$ / 332.35	2121-66-6 / -30	/ 305 dec; 195[8]	4-02-00-02417 / 1.116[20]	EtOH 3 / 1.4398[20]
10141	Propane, 1,1,1,2-tetrachloro- / 1,1,1,2-Tetrachloropropane	$C_3H_4Cl_4$ / 181.88	812-03-3 / -64	/ 152.5	4-01-00-00201 / 1.473[20]	H_2O 1; EtOH 4; eth 3; chl 3 / 1.4867[20]
10142	Propane, 1,1,1,3-tetrachloro- / 1,1,1,3-Tetrachloropropane	$C_3H_4Cl_4$ / 181.88	1070-78-6 /	/ 157	4-01-00-00201 / 1.4509[20]	H_2O 1; EtOH 4; eth 4; bz 4 / 1.4825[20]
10143	Propane, 1,1,2,2-tetrachloro- / 1,1,2,2-Tetrachloropropane	$C_3H_4Cl_4$ / 181.88	13116-60-4 /	/ 153	4-01-00-00201 / 1.471[2]	H_2O 1; EtOH 5; eth 5; chl 3 / 1.4850[25]
10144	Propane, 1,1,2,3-tetrachloro- / 1,1,2,3-Tetrachloropropane	$C_3H_4Cl_4$ / 181.88	18495-30-2 /	/ 179.5	4-01-00-00202 / 1.513[17]	H_2O 1; EtOH 4; eth 4; chl 3 / 1.5037[17]
10145	Propane, 1,2,2,3-tetrachloro- / 1,2,2,3-Tetrachloropropane	$C_3H_4Cl_4$ / 181.88	13116-53-5 /	/ 165	4-01-00-00202 / 1.500[18]	H_2O 1; EtOH 4; eth 4; chl 3 / 1.4940[18]
10146	Propane, 1,1,2,3-tetrachloro-2-(chloromethyl)- / 1,1,2,3-Tetrachloro-2-(chloromethyl)propane	$C_4H_5Cl_5$ / 230.35	66997-45-3 /	/ 227; 95[9]	3-01-00-00321 / 1.5686[25]	bz 4; EtOH 4; chl 4 / 1.5165[25]
10147	Propane, 1,1,1,2-tetrachloro-2-methyl- / Isobutane, 1,1,1,2-tetrachloro	$C_4H_6Cl_4$ / 195.90	7086-07-9 / 178.5	/ 192[117]	4-01-00-00293 /	eth 4; EtOH 4; chl 4
10148	Propane, 1,1,2,3-tetrachloro-2-methyl-	$C_4H_6Cl_4$ / 195.90	18963-01-4 / -46	/ 190.5	4-01-00-00294 / 1.4393[25]	bz 4; eth 4; chl 4 / 1.4963[20]
10149	Propane, 1,2,2,3-tetrachloro-1,1,3,3-tetrafluoro- / Propane, 1,2,2,3-tetrachloro-tetrafluoro	$C_3Cl_4F_4$ / 253.84	677-68-9 / -42.9	/ 112	4-01-00-00203 / 1.7199[20]	/ 1.3958[20]
10150	Propane, 1,1,3,3-tetramethoxy-	$C_7H_{16}O_4$ / 164.20	102-52-3 /	/ 183; 66[12]	4-01-00-03635 / 0.997[25]	/ 1.4081[20]
10151	Propane, 1,1'-thiobis- / Dipropyl sulfide	$C_6H_{14}S$ / 118.24	111-47-7 / -102.5	7881 / 142.9	4-01-00-01452 / 0.814[17]	H_2O 1; EtOH 3; eth 3 / 1.4487[20]
10152	Propane, 2,2'-thiobis- / Diisopropyl sulfide	$C_6H_{14}S$ / 118.24	625-80-9 / -78.1	/ 120.1	4-01-00-01502 / 0.8142[20]	H_2O 1; EtOH 3; eth 3 / 1.4438[20]
10153	Propane, 1,1'-thiobis[3-chloro- / Bis(3-chloropropyl) sulfide	$C_6H_{12}Cl_2S$ / 187.13	55882-21-8 /	/ 162[43]	3-01-00-01438 / 1.1774[25]	eth 4; EtOH 4; tol 4 / 1.5075[20]
10154	Propane, 1,1'-thiobis[2-methyl- / Diisobutyl sulfide	$C_8H_{18}S$ / 146.30	592-65-4 / -105.5	5035 / 171	4-01-00-01607 / 0.8363[10]	

No.	Name / Synonym	Mol. Form. / Mol. Wt.	CAS RN / mp/°C	Merck No. / bp/°C	Beil. Ref. / den/g cm^{-3}	Solubility / n_D
10155	Propane, 2,2'-thiobis[2-methyl- Di-*tert*-butyl sulfide	$C_8H_{18}S$ 146.30	107-47-1 -9.0	 149.1	4-01-00-01637 0.815[25]	 1.4506[20]
10156	1-Propanethiol Propyl mercaptan	C_3H_8S 76.16	107-03-9 -113.3	 67.8	4-01-00-01449 0.8411[20]	H_2O 2; EtOH 3; eth 3; ace 3 1.4380[20]
10157	2-Propanethiol Isopropyl mercaptan	C_3H_8S 76.16	75-33-2 -130.5	 52.6	4-01-00-01498 0.8143[20]	H_2O 2; EtOH 5; eth 5; ace 4 1.4255[20]
10158	1-Propanethiol, 2-methyl- Isobutyl mercaptan	$C_4H_{10}S$ 90.19	513-44-0 <-70	5030 88.5	4-01-00-01605 0.8357[20]	H_2O 2; EtOH 4; eth 4; ace 4 1.4387[20]
10159	2-Propanethiol, 2-methyl- *tert*-Butyl mercaptan	$C_4H_{10}S$ 90.19	75-66-1 -0.5	1577 64.3	4-01-00-01634 0.7943[25]	H_2O 1; ctc 3; hp 3 1.4232[20]
10160	Propane, 1,1,1-tribromo-	$C_3H_5Br_3$ 280.78	62127-61-1 	 77-83[25]	4-01-00-00221 2.35[25]	 1.5606[25]
10161	Propane, 1,1,2-tribromo-	$C_3H_5Br_3$ 280.78	14602-62-1 	 200.5	3-01-00-00251 2.3547[20]	H_2O 1; EtOH 3; eth 4; chl 3 1.5790[20]
10162	Propane, 1,2,2-tribromo-	$C_3H_5Br_3$ 280.78	14476-30-3 	 190.5	2-01-00-00077 2.2984[20]	eth 4; EtOH 4; chl 4 1.5670[20]
10163	Propane, 1,2,3-tribromo- 1,2,3-Tribromopropane	$C_3H_5Br_3$ 280.78	96-11-7 16.9	9527 222.1	4-01-00-00221 2.4208[20]	H_2O 1; EtOH 4; eth 4; ctc 2 1.5862[20]
10164	Propane, 1,2,3-tribromo-2-methyl-	$C_4H_7Br_3$ 294.81	631-28-7 	178[235]; 108[18]	4-01-00-00299 2.2105[14]	H_2O 1; eth 3; chl 3 1.5701[21]
10165	1,2,3-Propanetricarboxylic acid Tricarballylic acid	$C_6H_8O_6$ 176.13	99-14-9 166	9534	4-02-00-02366	H_2O 4; EtOH 4; eth 2
10166	1,2,3-Propanetricarboxylic acid, 2-(acetyloxy)-, triethyl ester 	$C_{14}H_{22}O_8$ 318.32	77-89-4 	 214[40]	3-03-00-01105 1.135[25]	 1.4380
10167	1,2,3-Propanetricarboxylic acid, 2-hydroxy- Citric acid	$C_6H_8O_7$ 192.13	77-92-9 153	2328 dec	4-03-00-01272 1.665[20]	H_2O 4; EtOH 4; eth 3; bz 1
10168	1,2,3-Propanetricarboxylic acid, 2-hydroxy-, tributyl ester Butyl citrate	$C_{18}H_{32}O_7$ 360.45	77-94-1 -20	1564 233[22]	4-03-00-01276 1.043[20]	 1.4460[20]
10169	1,2,3-Propanetricarboxylic acid, 2-hydroxy-, triethyl ester 	$C_{12}H_{20}O_7$ 276.29	77-93-0 	 294	4-03-00-01276 1.1369[20]	H_2O 1; EtOH 3; eth 3; ctc 2 1.4455[20]
10170	1,2,3-Propanetricarboxylic acid, 2-hydroxy-, trimethyl ester Trimethyl citrate	$C_9H_{14}O_7$ 234.21	1587-20-8 79.3	 285; 176[16]	4-03-00-01276	eth 4; EtOH 4
10171	Propane, 1,1,1-trichloro- 1,1,1-Trichloropropane	$C_3H_5Cl_3$ 147.43	7789-89-1 	 108	4-01-00-00198 1.287[23]	H_2O 1; EtOH 3; eth 4; chl 4
10172	Propane, 1,1,2-trichloro- 1,1,2-Trichloropropane	$C_3H_5Cl_3$ 147.43	598-77-6 	132.0; 117[500]	4-01-00-00199 1.372[15]	H_2O 1; EtOH 3; eth 4; ctc 2
10173	Propane, 1,1,3-trichloro- 1,1,3-Trichloropropane	$C_3H_5Cl_3$ 147.43	20395-25-9 -59	 145.5	4-01-00-00199 1.3557[20]	eth 4; EtOH 4; chl 4 1.4718[20]
10174	Propane, 1,2,2-trichloro- 1,2,2-Trichloropropane	$C_3H_5Cl_3$ 147.43	3175-23-3 	 124	4-01-00-00199 1.318[25]	H_2O 1; EtOH 3; eth 3; chl 4 1.4609[20]
10175	Propane, 1,2,3-trichloro- 1,2,3-Trichloropropane	$C_3H_5Cl_3$ 147.43	96-18-4 -14.7	 157	4-01-00-00199 1.3889[20]	H_2O 2; EtOH 3; eth 3; ctc 2 1.4852[20]
10176	Propane, 1,2,3-trichloro-2-(chloromethyl)-	$C_4H_6Cl_4$ 195.90	18963-00-3 	 211; 87[9]	3-01-00-00321 1.5036[25]	chl 4 1.508[20]
10177	Propane, 1,1,2-trichloro-2-methyl- Isobutane, 1,1,2-trichloro	$C_4H_7Cl_3$ 161.46	29559-52-2 -6	 163	4-01-00-00293 1.2588[20]	eth 4; chl 4 1.4666[20]
10178	Propane, 1,2,3-trichloro-2-methyl-	$C_4H_7Cl_3$ 161.46	1871-58-5 	 162.5	4-01-00-00293 1.3012[25]	chl 4 1.4765[20]
10179	Propane, 1,1,1-trichloro-2,2,3,3,3-pentafluoro-	$C_3Cl_3F_5$ 237.38	4259-43-2 -80	 74	4-01-00-00200 1.643[25]	
10180	Propane, 1,2,2-trichloro-1,1,3,3,3-pentafluoro-	$C_3Cl_3F_5$ 237.38	1599-41-3 -4.3	 72.0	3-01-00-00233 1.6681[20]	 1.3519[20]
10181	Propane, 1,2,3-trichloro-1,1,2,3,3-pentafluoro-	$C_3Cl_3F_5$ 237.38	76-17-5 -72	 73.7	4-01-00-00200 1.6631[20]	 1.3512[20]
10182	Propane, 1,1,1-triethoxy-	$C_9H_{20}O_3$ 176.26	115-80-0 	 171	4-02-00-00707 	eth 4; EtOH 4 1.4000[25]
10183	Propane, 1,1,3-triethoxy-	$C_9H_{20}O_3$ 176.26	7789-92-6 	185 dec; 78[14]	4-01-00-03971 0.898[19]	H_2O 2; EtOH 3; ctc 3 1.4067[20]
10184	Propane, 1,1,1-trifluoro-3-nitro- Propane, 1-nitro-3,3,3-trifluoro	$C_3H_4F_3NO_2$ 143.07	461-35-8 	 135	4-01-00-00232 1.4178[20]	eth 4 1.3549[20]
10185	Propane, 1,2,3-trimethoxy- Glycerol trimethyl ether	$C_6H_{14}O_3$ 134.18	20637-49-4 	 148	4-01-00-02755 0.9460[15]	H_2O 5; eth 3; ace 4; bz 3 1.4055[15]
10186	1,2,3-Propanetriol Glycerol	$C_3H_8O_3$ 92.09	56-81-5 18.2	4379 290	4-01-00-02751 1.2613[20]	H_2O 5; EtOH 5; eth 2; bz 1 1.4746[20]
10187	1,2,3-Propanetriol, 1-acetate, (±)- Glycerol, 1-monoacetate(*DL*)	$C_5H_{10}O_4$ 134.13	93713-40-7 	158[165] 130[3]	4-02-00-00251 1.2060[20]	H_2O 3; EtOH 3; eth 2; bz 1 1.4157[20]
10188	1,2,3-Propanetriol, 1-(4-aminobenzoate) Glyceryl *p*-aminobenzoate	$C_{10}H_{13}NO_4$ 211.22	136-44-7 	4382		H_2O 1; EtOH 3
10189	1,2,3-Propanetriol, diacetate Diacetin	$C_7H_{12}O_5$ 176.17	25395-31-7 	2941 259	4-02-00-00252 1.184[16]	H_2O 4; eth 4; EtOH 4 1.44[20]
10190	1,2,3-Propanetriol, 1,3-diacetate 1,3-Diacetin	$C_7H_{12}O_5$ 176.17	105-70-4 	2941 173[40]	4-02-00-00252 1.179[15]	H_2O 4; EtOH 4; eth 2; CS_2 1 1.4395[20]

No.	Name / Synonym	Mol. Form. / Mol. Wt.	CAS RN / mp/°C	Merck No. / bp/°C	Beil. Ref. / den/g cm^{-3}	Solubility / n_D
10191	1,2,3-Propanetriol, 1,3-dinitrate	$C_3H_6N_2O_7$ 182.09	623-87-0 26	148[15]; 116[0.6]	2-01-00-00591 1.523[20]	H_2O 4; eth 4; EtOH 4 1.4715[20]
10192	1,2,3-Propanetriol, 1-nitrate	$C_3H_7NO_5$ 137.09	624-43-1 61	157.5	4-01-00-02761 1.4164[20]	H_2O 4; EtOH 4; eth 2 1.4698[20]
10193	1,2,3-Propanetriol, 2-nitrate	$C_3H_7NO_5$ 137.09	620-12-2 54	157.5	2-01-00-00591 1.40[22]	H_2O 4; eth 4; EtOH 4
10194	1,2,3-Propanetriol, 1-phenyl- Glycerol, 1-phenyl	$C_9H_{12}O_3$ 168.19	63157-81-3 100.5	185	4-06-00-07407 1.2213[13]	1.5606[15]
10195	1,2,3-Propanetriol, triacetate Triacetin	$C_9H_{14}O_6$ 218.21	102-76-1 -78	9504 259	4-02-00-00253 1.1583[20]	H_2O 2; EtOH 5; eth 5; ace 4 1.4301[20]
10196	1,2,3-Propanetriol, tribenzoate	$C_{24}H_{20}O_6$ 404.42	614-33-5 76		3-09-00-00666 1.228[12]	H_2O 1; EtOH 3; eth 4; ace 4
10197	1,2,3-Propanetriol, trinitrate Nitroglycerin	$C_3H_5N_3O_9$ 227.09	55-63-0 13.5	6528 93[0.31]; exp 218	4-01-00-02762 1.5931[20]	H_2O 2; EtOH 3; eth 5; ace 4 1.4786[12]
10198	1,2,3-Propanetriol, tripropanoate	$C_{12}H_{20}O_6$ 260.29	139-45-7	175-6[20]	4-02-00-00717 1.099[20]	H_2O 1; EtOH 3; eth 4; chl 3 1.4318[19]
10199	Propanetrione, diphenyl-	$C_{15}H_{10}O_3$ 238.24	643-75-4 70	289[175]; 179[4]	4-07-00-02822	eth 4
10200	Propanil Propanamide, N-(3,4-dichlorophenyl)-	$C_9H_9Cl_2NO$ 218.08	709-98-8 92	7814	1.25[25]	
10201	Propanoic acid Propionic acid	$C_3H_6O_2$ 74.08	79-09-4 -20.7	7837 141.1	4-02-00-00695 0.9930[20]	H_2O 5; EtOH 5; eth 3; chl 2 1.3809[20]
10202	Propanoic acid, 2-(acetyloxy)-, (±)-	$C_5H_8O_4$ 132.12	3853-80-3	167-70[78]	4-03-00-00638 1.1758[20]	bz 4; EtOH 4 1.4240[20]
10203	Propanoic acid, anhydride Propionic anhydride	$C_6H_{10}O_3$ 130.14	123-62-6 -45	7838 170; 67.5[18]	4-02-00-00722 1.0110[20]	eth 5; ctc 2 1.4038[20]
10204	Propanoic acid, 2-bromo-, (±)-	$C_3H_5BrO_2$ 152.98	10327-08-9 25.7	203.5	4-02-00-00764 1.7000[20]	H_2O 4; EtOH 4; eth 4; chl 3 1.4753[20]
10205	Propanoic acid, 3-bromo- β-Bromopropionic acid	$C_3H_5BrO_2$ 152.98	590-92-1 62.5	1421 140-2[45]	4-02-00-00764 1.48[25]	H_2O 3; EtOH 3; eth 3; bz 3
10206	Propanoic acid, 3-bromo-, butyl ester	$C_7H_{13}BrO_2$ 209.08	6973-79-1	130[26]	4-02-00-00765 1.4549[20]	ace 4; eth 4; EtOH 4 1.3051[20]
10207	Propanoic acid, 2-bromo-, ethyl ester Ethyl α-bromopropionate	$C_5H_9BrO_2$ 181.03	535-11-5	3731 160 dec; 71[26]	4-02-00-00762 1.4135[20]	H_2O 4; EtOH 5; eth 5; chl 3 1.4490[20]
10208	Propanoic acid, 3-bromo-, ethyl ester	$C_5H_9BrO_2$ 181.03	539-74-2	179; 65[15]	4-02-00-00765 1.4123[18]	EtOH 3; eth 3; ace 3; ctc 2 1.4516[20]
10209	Propanoic acid, 2-bromo-2-methyl- α-Bromoisobutyric acid	$C_4H_7BrO_2$ 167.00	2052-01-9 48.5	1408 199; 115[24]	4-02-00-00862 1.4969[60]	
10210	Propanoic acid, 3-bromo-, 3-methylbutyl ester Isopentyl 3-bromopropionate	$C_8H_{15}BrO_2$ 223.11	100983-12-8	110-1[11]	2-02-00-00231 1.2217[15]	eth 4; EtOH 4 1.4556[9]
10211	Propanoic acid, 2-bromo-, methyl ester, (±)-	$C_4H_7BrO_2$ 167.00	57885-43-5	144	4-02-00-00762 1.4966[25]	eth 4; EtOH 4 1.4451[22]
10212	Propanoic acid, 3-bromo-, methyl ester	$C_4H_7BrO_2$ 167.00	3395-91-3	105.5[60]	4-02-00-00765 1.4897[15]	EtOH 3; eth 3; ace 3; os 4 1.4542[20]
10213	Propanoic acid, 2-bromo-2-methyl-, ethyl ester	$C_6H_{11}BrO_2$ 195.06	600-00-0	163	4-02-00-00862 1.3263[20]	H_2O 1; EtOH 3; eth 5 1.4446[20]
10214	Propanoic acid, butyl ester Butyl propionate	$C_7H_{14}O_2$ 130.19	590-01-2 -89	1587 146.8	4-02-00-00708 0.8754[20]	H_2O 2; EtOH 5; eth 5; ctc 2 1.4014[20]
10215	Propanoic acid, 2-chloro-	$C_3H_5ClO_2$ 108.52	598-78-7	185	4-02-00-00745 1.2585[20]	H_2O 5; EtOH 5; eth 5; ace 3 1.4380[20]
10216	Propanoic acid, 3-chloro- β-Chloropropionic acid	$C_3H_5ClO_2$ 108.52	107-94-8 41	2159 204 dec	4-02-00-00748	H_2O 3; EtOH 3; eth 5; chl 3
10217	Propanoic acid, 2-chloro-, butyl ester	$C_7H_{13}ClO_2$ 164.63	54819-86-2	184	4-02-00-00746 1.0253[20]	eth 4 1.4263[20]
10218	Propanoic acid, 3-chloro-, butyl ester	$C_7H_{13}ClO_2$ 164.63	27387-79-7	104[22]	4-02-00-00749 1.0370[20]	H_2O 4; eth 4 1.4321[20]
10219	Propanoic acid, 3-chloro-2,2-dimethyl-	$C_5H_9ClO_2$ 136.58	13511-38-1 41.5	110[10]	4-02-00-00914	ctc 4
10220	Propanoic acid, 2-chloro-, ethyl ester Ethyl α-chloropropionate	$C_5H_9ClO_2$ 136.58	535-13-7	3743 147	4-02-00-00746 1.0793[20]	H_2O 1; EtOH 5; eth 5; ctc 2 1.4178[20]
10221	Propanoic acid, 3-chloro-, ethyl ester	$C_5H_9ClO_2$ 136.58	623-71-2	162	4-02-00-00749 1.1086[20]	H_2O 2; EtOH 5; eth 5; os 3 1.4254[20]
10222	Propanoic acid, 2-chloro-, methyl ester Methyl 2-chloropropionate	$C_4H_7ClO_2$ 122.55	17639-93-9	132.5	4-02-00-00746 1.0750[25]	
10223	Propanoic acid, 2-chloro-, 1-methylethyl ester Isopropyl 2-chloropropionate	$C_6H_{11}ClO_2$ 150.60	40058-87-5	151.5	4-02-00-00746 1.0315[20]	H_2O 1; EtOH 3; eth 3 1.4149[20]
10224	Propanoic acid, 2-chloro-, 2-methylpropyl ester, (R)- Propionic acid, 2-chloro, isobutyl ester (d)	$C_7H_{13}ClO_2$ 164.63	114489-96-2	176	0-02-00-00248 1.0312[20]	1.4247[20]
10225	Propanoic acid, 3-chloro-, methyl ester	$C_4H_7ClO_2$ 122.55	6001-87-2	156	4-02-00-00748 1.1861[15]	EtOH 3; ctc 3 1.4263[20]
10226	Propanoic acid, 3-chloro-, 3-methylbutyl ester	$C_8H_{15}ClO_2$ 178.66	62108-70-7	208; 87[12]	4-02-00-00749 1.0171[20]	eth 4; EtOH 4 1.4343[20]
10227	Propanoic acid, 3-chloro-, 2-methylpropyl ester	$C_7H_{13}ClO_2$ 164.63	62108-68-3	191.3	4-02-00-00749 1.0323[20]	eth 4; EtOH 4 1.4295[20]

No.	Name / Synonym	Mol. Form. / Mol. Wt.	CAS RN / mp/°C	Merck No. / bp/°C	Beil. Ref. / den/g cm^{-3}	Solubility / n_D
10228	Propanoic acid, 2-chloro-2-methyl-	$C_4H_7ClO_2$ 122.55	594-58-1 31	118[50]; 93[20]	4-02-00-00858 1.1790[20]	EtOH 4 1.450[20]
10229	Propanoic acid, 2-chloro-2-methyl-, methyl ester	$C_5H_9ClO_2$ 136.58	22421-97-2	135	4-02-00-00859 1.0883[15]	eth 4 1.4122[21]
10230	Propanoic acid, 3-chloro-2-methyl-	$C_4H_7ClO_2$ 122.55	16674-04-7	128-33[50]	4-02-00-00860 1.0153[20]	eth 4; EtOH 4 1.4310[20]
10231	Propanoic acid, 2-(4-chlorophenoxy)-2-methyl-, ethyl ester Clofibrate	$C_{12}H_{15}ClO_3$ 242.70	637-07-0	2374 148-50[20]	4-06-00-00851	
10232	Propanoic acid, 3-chloro-, propyl ester	$C_6H_{11}ClO_2$ 150.60	62108-66-1	180	4-02-00-00749 1.0656[20]	eth 4; EtOH 4 1.4290[20]
10233	Propanoic acid, 2-cyano-	$C_4H_5NO_2$ 99.09	632-07-5 35	143[11]	3-02-00-01683 1.1400[20]	
10234	Propanoic acid, 3-cyano-, methyl ester	$C_5H_7NO_2$ 113.12	4107-62-4	211	3-02-00-01672	chl 3 1.4250[20]
10235	Propanoic acid, cyclohexyl ester	$C_9H_{16}O_2$ 156.22	6222-35-1	193; 93[35]	4-06-00-00037 0.9359[20]	H_2O 1; EtOH 3; eth 3; ace 3 1.4403[20]
10236	Propanoic acid, 2-diazo-, ethyl ester Ethyl 2-diazopropionate	$C_5H_8N_2O_2$ 128.13	6111-99-5	65-8[41]	2-03-00-00403 1.086[12]	1.4472[18]
10237	Propanoic acid, 2-diazo-, methyl ester Methyl 2-diazopropionate	$C_4H_6N_2O_2$ 114.10	34757-14-7	43-5[11]	2-03-00-00403 1.1011[13]	1.4487[20]
10238	Propanoic acid, 2,3-dibromo-	$C_3H_4Br_2O_2$ 231.87	600-05-5 66.5	160[20]; 138[12]	4-02-00-00767	bz 4; eth 4; EtOH 4
10239	Propanoic acid, 2,3-dibromo-, butyl ester Butyl 2,3-dibromopropionate	$C_7H_{12}Br_2O_2$ 287.98	21179-48-6	237.5	4-02-00-00768 1.6107[20]	1.4890[20]
10240	Propanoic acid, 2,3-dibromo-, ethyl ester	$C_5H_8Br_2O_2$ 259.93	3674-13-3	214.5	4-02-00-00767 1.7966[20]	EtOH 3; eth 3 1.5007[20]
10241	Propanoic acid, 2,3-dibromo-, methyl ester	$C_4H_6Br_2O_2$ 245.90	1729-67-5	206	4-02-00-00767 1.9333[20]	EtOH 3 1.5127[20]
10242	Propanoic acid, 2,3-dibromo-2-methyl-, methyl ester	$C_5H_8Br_2O_2$ 259.93	3673-79-8	203	4-02-00-00864 1.805[20]	1.505[20]
10243	Propanoic acid, 2,3-dibromo-, propyl ester Propyl 2,3-dibromopropionate	$C_6H_{10}Br_2O_2$ 273.95	79762-76-8	222.5	3-02-00-00572 1.6799[20]	1.4950[20]
10244	Propanoic acid, 2,2-dichloro- Dalapon	$C_3H_4Cl_2O_2$ 142.97	75-99-0	2806 187.5; 92[14]	4-02-00-00753 1.389[12]	H_2O 4; EtOH 4; eth 3; ctc 3
10245	Propanoic acid, 2,3-dichloro-	$C_3H_4Cl_2O_2$ 142.97	565-64-0 50	210; 130[26]	4-02-00-00756	H_2O 3; EtOH 3; eth 3; peth 2 1.4650[20]
10246	Propanoic acid, 2,3-dichloro-, ethyl ester	$C_5H_8Cl_2O_2$ 171.02	6628-21-3	183.5	4-02-00-00756 1.2401[20]	eth 4; EtOH 4 1.4482[20]
10247	Propanoic acid, 3,3-dichloro-2-hydroxy-2-methyl-	$C_4H_6Cl_2O_3$ 173.00	31340-44-0 82.5	dec	2-03-00-00224	H_2O 3; EtOH 3; eth 3
10248	Propanoic acid, 2,3-dichloro-, methyl ester	$C_4H_6Cl_2O_2$ 157.00	3674-09-7	92[50]	4-02-00-00756 1.3282[20]	ace 4; eth 4; EtOH 4
10249	Propanoic acid, 2,3-dichloro-, 1-methylethyl ester	$C_6H_{10}Cl_2O_2$ 185.05	54774-99-1	61-2[5]	4-02-00-00756 1.2010[25]	1.4470
10250	Propanoic acid, 2-(2,4-dichlorophenoxy)- Dichlorprop	$C_9H_8Cl_2O_3$ 235.07	120-36-5 117.5	3068	3-06-00-00707	H_2O 2; EtOH 3; eth 3; lig 2
10251	Propanoic acid, 2,2-dichloro-3,3,3-trifluoro-, methyl ester Propionic acid, 2,2-dichloro-3,3,3-trifluoro-, methyl ester	$C_4H_3Cl_2F_3O_2$ 210.97	378-68-7	116-7[625]	4-02-00-00758 1.5092[20]	1.3806[20]
10252	Propanoic acid, 2,3-dihydroxy- Glyceric acid	$C_3H_6O_4$ 106.08	473-81-4	4378 dec	4-03-00-01050	H_2O 4; ace 4; EtOH 4
10253	Propanoic acid, 2,3-dihydroxy-, ethyl ester, (±)- Glyceric acid, ethyl ester (DL)	$C_5H_{10}O_4$ 134.13	89300-69-6	235; 120[14]	4-03-00-01052 1.1897[15]	H_2O 4; eth 4; EtOH 4
10254	Propanoic acid, 2,3-dihydroxy-, methyl ester, (±)- Methyl DL-glycerate	$C_4H_8O_4$ 120.11	15909-76-9	241.5; 119[14]	4-03-00-01052 1.2802[15]	H_2O 4; EtOH 4 1.4502[20]
10255	Propanoic acid, 2,3-dihydroxy-2-methyl-, ethyl ester Glyceric acid, 2-methyl, ethyl ester	$C_6H_{12}O_4$ 148.16	67535-07-3	95[10]	3-03-00-00860 1.114[25]	1.4370[25]
10256	Propanoic acid, 2,3-dihydroxy-2-methyl-, methyl ester Glyceric acid, 2-methyl-, methyl ester	$C_5H_{10}O_4$ 134.13	19860-56-1	89[10]	4-03-00-01060 1.185[25]	1.4438[25]
10257	Propanoic acid, 2,3-dihydroxy-, propyl ester	$C_6H_{12}O_4$ 148.16	116435-95-1	132-4[3]	3-03-00-00854 1.1537[20]	1.4503[20]
10258	Propanoic acid, 2,2-dimethyl- Pivalic acid	$C_5H_{10}O_2$ 102.13	75-98-9 35	7482 164	4-02-00-00908 0.905[50]	H_2O 2; EtOH 4; eth 4 1.3931[30]
10259	Propanoic acid, 2,2-dimethyl-, ethyl ester Ethyl 2,2-dimethylpropionate	$C_7H_{14}O_2$ 130.19	3938-95-2 -89.5	118.4	4-02-00-00910 0.856[20]	EtOH 3; eth 3 1.3906[20]
10260	Propanoic acid, 2,2-dimethyl-, methyl ester Methyl 2,2-dimethylpropionate	$C_6H_{12}O_2$ 116.16	598-98-1	101.1	4-02-00-00909 0.891[0]	eth 4; EtOH 4 1.3905[20]
10261	Propanoic acid, ethenyl ester Vinyl propionate	$C_5H_8O_2$ 100.12	105-38-4	91.2	4-02-00-00711	

No.	Name / Synonym	Mol. Form. / Mol. Wt.	CAS RN / mp/°C	Merck No. / bp/°C	Beil. Ref. / den/g cm^{-3}	Solubility / n_D
10262	Propanoic acid, 3-ethoxy-, ethyl ester	$C_7H_{14}O_3$ 146.19	763-69-9	48-9[5]	4-03-00-00697 0.92[25]	1.4065[20]
10263	Propanoic acid, ethyl ester Ethyl propionate	$C_5H_{10}O_2$ 102.13	105-37-3 -73.9	3801 99.1	4-02-00-00705 0.8917[20]	H_2O 2; EtOH 5; eth 5; ace 3 1.3839[20]
10264	Propanoic acid, heptyl ester	$C_{10}H_{20}O_2$ 172.27	2216-81-1 -50.9	210	4-02-00-00710 0.8679[20]	H_2O 1; EtOH 3; eth 3; ace 3 1.4201[15]
10265	Propanoic acid, hexyl ester	$C_9H_{18}O_2$ 158.24	2445-76-3 -57.5	190	4-02-00-00709 0.8698[20]	H_2O 1; EtOH 3; eth 3; ace 3 1.4162[15]
10266	Propanoic acid, 2-hydroxy-, (±)- Lactic acid (DL)	$C_3H_6O_2$ 90.08	598-82-3 18	5215 122[15]	2-03-00-00192 1.2060[21]	H_2O 4; eth 2; EtOH 4 1.4392[20]
10267	Propanoic acid, 3-hydroxy- Hydracrylic acid	$C_3H_6O_3$ 90.08	503-66-2	4681 dec	4-03-00-00689	H_2O 4; EtOH 3; eth 5 1.4489[20]
10268	Propanoic acid, 2-hydroxy-, butyl ester, (R)- Lactic acid, butyl ester (d)	$C_7H_{14}O_3$ 146.19	34451-18-8	77[10]	3-03-00-00450 0.9744[27]	eth 4; EtOH 4
10269	Propanoic acid, 2-hydroxy-, ethyl ester, (±)- Ethyl lactate	$C_5H_{10}O_3$ 118.13	2676-33-7 -26	3773 154.5	1.0328[20]	H_2O 4; eth 4; EtOH 4 1.4124[20]
10270	Propanoic acid, 3-hydroxy-, ethyl ester	$C_5H_{10}O_3$ 118.13	623-72-3	187.5	2-03-00-00213 1.059[20]	1.4271[23]
10271	Propanoic acid, 2-hydroxy-, hexadecyl ester Cetyl lactate	$C_{19}H_{38}O_3$ 314.51	35274-05-6 41	2022 219[10]; 170[1]	4-03-00-00660	1.4410[40]
10272	Propanoic acid, 3-hydroxy-2-(hydroxymethyl)-2-methyl- Dimethylolpropionic acid	$C_5H_{10}O_4$ 134.13	4767-03-7 190	3241	4-03-00-01066	
10273	Propanoic acid, 2-hydroxy-, methyl ester, (±)- Lactic acid, methyl ester (DL)	$C_4H_8O_3$ 104.11	2155-30-8	144.8	4-03-00-00640 1.0928[20]	H_2O 4; eth 4; EtOH 4 1.4141[20]
10274	Propanoic acid, 2-hydroxy-, 3-methylbutyl ester	$C_8H_{16}O_3$ 160.21	19329-89-6	202.4	3-03-00-00483 0.9589[25]	eth 4; EtOH 4 1.4240[25]
10275	Propanoic acid, 2-hydroxy-, 1-methylethyl ester Isopropyl lactate	$C_6H_{12}O_3$ 132.16	617-51-6	167	3-03-00-00479 0.9980[20]	H_2O 4; bz 4; eth 4; EtOH 4 1.4082[25]
10276	Propanoic acid, 3-hydroxy-, methyl ester	$C_4H_8O_3$ 104.11	6149-41-3	179	3-03-00-00525 1.105[16]	H_2O 5; EtOH 5; eth 5 1.43[23]
10277	Propanoic acid, 3-hydroxy-, 1-methylethyl ester Hydracrylic acid, isopropyl ester	$C_6H_{12}O_3$ 132.16	84098-45-3	128.5[82]; 95[12]	1-03-00 00113 1.058[25]	1.4303[23]
10278	Propanoic acid, 2-hydroxy-2-methyl-	$C_4H_8O_3$ 104.11	594-61-6 82.5	212	4-03-00-00782	H_2O 4; EtOH 4; eth 4; bz 2
10279	Propanoic acid, 2-hydroxy-2-methyl-, ethyl ester	$C_6H_{12}O_3$ 132.16	80-55-7	150	4-03-00-00783 0.987[20]	H_2O 5; EtOH 5; ctc 3 1.4080[20]
10280	Propanoic acid, 2-hydroxy-2-methyl-, methyl ester	$C_5H_{10}O_3$ 118.13	2110-78-3	137	4-03-00-00783	H_2O 4; EtOH 4 1.4056[20]
10281	Propanoic acid, 2-hydroxy-, 2-propenyl ester	$C_6H_{10}O_3$ 130.14	5349-55-3	56-60[8]	3-03-00-00486 1.0452[20]	py 3 1.4369[20]
10282	Propanoic acid, 2-iodo-, (±)-	$C_3H_5IO_2$ 199.98	18791-49-6 46	93-6[0.2]	3-02-00-00573 2.073[18]	H_2O 2; EtOH 4; eth 4
10283	Propanoic acid, 3-iodo-, ethyl ester	$C_5H_9IO_2$ 228.03	6414-69-3	202	4-02-00-00770 1.6226[30]	1.4961[25]
10284	Propanoic acid, 3-iodo-, methyl ester	$C_4H_7IO_2$ 214.00	5029-66-3	188	3-02-00-00574 1.8408[7]	EtOH 4
10285	Propanoic acid, 2-mercapto- Thiolactic acid	$C_3H_6O_2S$ 106.15	79-42-5 12	9269 106-7[15]	4-03-00-00682 1.1938[20]	H_2O 4; eth 4; EtOH 4 1.4810[20]
10286	Propanoic acid, 3-mercapto-	$C_3H_6O_2S$ 106.15	107-96-0 18	111[15]; 86[3]	4-03-00-00726 1.218[21]	H_2O 3; EtOH 3; eth 3; ctc 3 1.494[20]
10287	Propanoic acid, 3-mercapto-, methyl ester	$C_4H_8O_2S$ 120.17	2935-90-2	54-5[14]	3-03-00-00553 1.085[25]	1.4640[20]
10288	Propanoic acid, 2-mercapto-2-methyl-	$C_4H_8O_2S$ 120.17	4695-31-2 47	101[15]	4-03-00-00797	H_2O 4
10289	Propanoic acid, 3-methoxy-, methyl ester	$C_5H_{10}O_3$ 118.13	3852-09-3	142.8	4-03-00-00693 1.0139[15]	1.4030[20]
10290	Propanoic acid, 2-methyl- Isobutyric acid	$C_4H_8O_2$ 88.11	79-31-2 -46	5039 154.4	4-02-00-00843 0.9681[20]	H_2O 4; EtOH 5; eth 5; ctc 2 1.3930[20]
10291	Propanoic acid, 2-methyl-, anhydride Isobutyric anhydride	$C_8H_{14}O_3$ 158.20	97-72-3 -53.5	183; 89[32]	4-02-00-00851 0.9535[20]	eth 5; chl 3 1.4061[19]
10292	Propanoic acid, 2-methyl-, 2-chloroethyl ester 2-Chloroethyl isobutyrate	$C_6H_{11}ClO_2$ 150.60	33662-96-3	170	4-02-00-00847 1.062[25]	eth 4; EtOH 4 1.4109[16]
10293	Propanoic acid, 2-methyl-, cyclohexyl ester Isobutyric acid, cyclohexyl ester	$C_{10}H_{18}O_2$ 170.25	1129-47-1	204	3-06-00-00024 0.9489[0]	eth 4; EtOH 4
10294	Propanoic acid, 2-methyl-, 1,1-dimethylethyl ester	$C_8H_{16}O_2$ 144.21	16889-72-8	126.7	3-02-00-00648	ace 4; eth 4; EtOH 4 1.3921[20]
10295	Propanoic acid, 2-methyl-, 3,7-dimethyl-2,6-octadienyl ester, (E)-	$C_{14}H_{24}O_2$ 224.34	2345-26-8	135-7[13]	2-02-00-00261 0.8997[15]	1.4576[20]
10296	Propanoic acid, methyl ester Methyl propionate	$C_4H_8O_2$ 88.11	554-12-1 -87.5	6030 79.8	4-02-00-00704 0.9150[20]	H_2O 2; EtOH 5; eth 5; ace 3 1.3775[20]
10297	Propanoic acid, 2-methyl-, ethenyl ester	$C_6H_{10}O_2$ 114.14	2424-98-8	104.5	4-02-00-00848 0.8932[20]	1.4061[20]

No.	Name / Synonym	Mol. Form. / Mol. Wt.	CAS RN / mp/°C	Merck No. / bp/°C	Beil. Ref. / den/g cm^{-3}	Solubility / n_D
10298	Propanoic acid, 2-methyl-, ethyl ester / Ethyl isobutanoate	$C_6H_{12}O_2$ / 116.16	97-62-1 / -88.2	3770 / 110.1	4-02-00-00846 / 0.868[20]	H_2O 2; EtOH 5; eth 5; ace 3 / 1.3869[18]
10299	Propanoic acid, 1-methylethyl ester / Isopropyl propionate	$C_6H_{12}O_2$ / 116.16	637-78-5 /	/ 109.5	4-02-00-00708 / 0.8660[20]	H_2O 2; EtOH 5; eth 5 / 1.3872[20]
10300	Propanoic acid, 2-methyl-, 2-furanylmethyl ester	$C_9H_{12}O_3$ / 168.19	6270-55-9 /	/ 85-6[15]	2-17-00-00115 / 1.0313[20]	eth 4; EtOH 4
10301	Propanoic acid, 2-methyl-, methyl ester / Methyl isobutyrate	$C_5H_{10}O_2$ / 102.13	547-63-7 / -84.7	6003 / 92.5	4-02-00-00846 / 0.8906[20]	H_2O 2; EtOH 5; eth 5; ace 3 / 1.3840[20]
10302	Propanoic acid, 2-methyl-, 3-methylbutyl ester	$C_9H_{18}O_2$ / 158.24	2050-01-3 /	/ 168.5	3-02-00-00649 / 0.8627[20]	H_2O 2; EtOH 3; eth 3; ace 3
10303	Propanoic acid, 2-methyl-, 1-methylethyl ester / Isopropyl isobutyrate	$C_7H_{14}O_2$ / 130.19	617-50-5 /	/ 120.7	4-02-00-00847 / 0.8471[21]	H_2O 1; EtOH 3; eth 3; ace 3
10304	Propanoic acid, 2-methyl-, 2-methylpropyl ester / Isobutyl isobutanoate	$C_8H_{16}O_2$ / 144.21	97-85-8 / -80.7	5028 / 148.6	4-02-00-00847 / 0.8542[20]	H_2O 2; EtOH 3; eth 5; ace 3 / 1.3999[20]
10305	Propanoic acid, 2-methyl-, 2-nitropentyl ester / Isobutyric acid, (2-nitropentyl) ester	$C_9H_{17}NO_4$ / 203.24	116530-63-3 /	/ 249.5; 121[10]	3-02-00-00648 / 1.0310[20]	ace 4; eth 4 / 1.4315[20]
10306	Propanoic acid, 2-methyl-, 2-phenylethyl ester	$C_{12}H_{16}O_2$ / 192.26	103-48-0 /	/ 122-4[15]	4-06-00-03073 / 0.9950[15]	1.4871[20]
10307	Propanoic acid, 2-methyl-, phenylmethyl ester	$C_{11}H_{14}O_2$ / 178.23	103-28-6 /	/ 114-5[20]	4-06-00-02266 / 1.0075[15]	1.4883[20]
10308	Propanoic acid, 2-methyl-, propyl ester / Propyl isobutyrate	$C_7H_{14}O_2$ / 130.19	644-49-5 /	/ 135.4	4-02-00-00847 / 0.8843[0]	H_2O 2; EtOH 3; eth 4; ace 3 / 1.3955[20]
10309	Propanoic acid, 1-methylpropyl ester	$C_7H_{14}O_2$ / 130.19	591-34-4 /	/ 133	4-02-00-00709 / 0.8657[20]	EtOH 3; eth 3 / 1.3952[20]
10310	Propanoic acid, 2-methylpropyl ester / Isobutyl propionate	$C_7H_{14}O_2$ / 130.19	540-42-1 / -71.4	5033 / 137	4-02-00-00709 / 0.888[0]	H_2O 2; EtOH 4; eth 4; ace 3 / 1.3973[20]
10311	Propanoic acid, 3-(methylthio)-, methyl ester	$C_5H_{10}O_2S$ / 134.20	13532-18-8 /	/ 74-5[13]	4-03-00-00737 / 1.077[25]	1.4650[20]
10312	Propanoic acid, 2-nitro-, ethyl ester	$C_5H_9NO_4$ / 147.13	2531-80-8 /	/ 190.5	4-02-00-00771 / 1.4210[20]	bz 4; eth 4; EtOH 4
10313	Propanoic acid, octyl ester	$C_{11}H_{22}O_2$ / 186.29	142-60-9 / -42.6	/ 228	4-02-00-00710 / 0.8663[20]	H_2O 1; EtOH 3; eth 3; bz 3 / 1.4221[15]
10314	Propanoic acid, 2-oxo- / Pyruvic acid	$C_3H_4O_3$ / 88.06	127-17-3 / 13.8	8032 / 165 dec; 54[10]	4-03-00-01505 / 1.2272[20]	H_2O 5; EtOH 5; eth 5; ace 3 / 1.4280[20]
10315	Propanoic acid, 2-oxo-, ethyl ester	$C_5H_8O_3$ / 116.12	617-35-6 / -50	/ 155	4-03-00-01513 / 1.0596[15]	H_2O 2; EtOH 5; eth 5; ace 3 / 1.4052[20]
10316	Propanoic acid, 2-oxo-, methyl ester	$C_4H_6O_3$ / 102.09	600-22-6 /	/ 135.5	4-01-00-04146 / 1.154[0]	H_2O 2; EtOH 5; eth 5; ace 3 / 1.4046[25]
10317	Propanoic acid, pentafluoro-, methyl ester / Pentafluoropropionic acid, methyl ester	$C_4H_3F_5O_2$ / 178.06	378-75-6 /	/ 59.5	4-02-00-00740 / 1.390[25]	1.2869[25]
10318	Propanoic acid, pentyl ester	$C_8H_{16}O_2$ / 144.21	624-54-4 / -73.1	/ 168.6	4-02-00-00709 / 0.8761[25]	H_2O 1; EtOH 5; eth 5; bz 3 / 1.4096[15]
10319	Propanoic acid, 2-phenoxy-, (±)-	$C_9H_{10}O_3$ / 166.18	1912-21-6 / 115.5	/ 265.5	4-06-00-00642 / 1.1865[20]	eth 4; EtOH 4 / 1.5184[20]
10320	Propanoic acid, 3-phenoxy-	$C_9H_{10}O_3$ / 166.18	7170-38-9 / 97.5	/ 240; 188[26]	4-06-00-00643 /	H_2O 3; chl 2; lig 3
10321	Propanoic acid, 2-phenoxy-, ethyl ester	$C_{11}H_{14}O_3$ / 194.23	42412-84-0 /	/ 243.5	4-06-00-00643 / 1.360[17]	H_2O 1; EtOH 3; eth 3 / 1.360[17]
10322	Propanoic acid, 3-phenoxy-, ethyl ester	$C_{11}H_{14}O_3$ / 194.23	22409-91-2 / 24	/ 142[11]	4-06-00-00644 / 1.0745[20]	1.5007[18]
10323	Propanoic acid, phenyl ester	$C_9H_{10}O_2$ / 150.18	637-27-4 / 20	/ 211	4-06-00-00615 / 1.0436[25]	H_2O 1; EtOH 4; eth 4; bz 3 / 1.4980[20]
10324	Propanoic acid, 2-phenylethyl ester / 2-Phenethyl propionate	$C_{11}H_{14}O_2$ / 178.23	122-70-3 /	/	3-06-00-01709 /	
10325	Propanoic acid, 2-propenyl ester	$C_6H_{10}O_2$ / 114.14	2408-20-0 /	/ 124	4-02-00-00711 / 0.9140[20]	EtOH 3; eth 3; ace 3 / 1.4105[20]
10326	Propanoic acid, propyl ester / Propyl propionate	$C_6H_{12}O_2$ / 116.16	106-36-5 / -75.9	7880 / 122.5	4-02-00-00707 / 0.8809[20]	H_2O 2; EtOH 5; eth 5; ace 3 / 1.3935[20]
10327	Propanoic acid, sodium salt	$C_3H_5NaO_2$ / 96.06	137-40-6 /	8623 /	4-02-00-00695 /	H_2O 2
10328	Propanoic acid, 2,3,3,3-tetrafluoro-, ethyl ester / Ethyl 2,3,3,3-tetrafluoropropionate	$C_5H_6F_4O_2$ / 174.10	399-92-8 /	/ 108.5	4-02-00-00737 / 1.289[20]	1.3260[20]
10329	Propanoic acid, 3,3'-thiobis- / 3,3'-Thiodipropionic acid	$C_6H_{10}O_4S$ / 178.21	111-17-1 / 129	9261 /	4-03-00-00735 /	H_2O 4; EtOH 4
10330	Propanoic acid, 3,3'-thiobis-, diethyl ester	$C_{10}H_{18}O_4S$ / 234.32	673-79-0 /	/ 121[2]	4-03-00-00741 / 1.095[25]	1.4655[20]
10331	Propanoic acid, 3,3'-thiobis-, dimethyl ester	$C_8H_{14}O_4S$ / 206.26	4131-74-2 /	/ 148[18]	4-03-00-00738 / 1.198[25]	1.4740[20]
10332	Propanoic acid, 2,2,2-trichloro-	$C_3H_3Cl_3O_2$ / 177.41	3278-46-4 / 65.5	/ 140[40]	4-02-00-00759 /	H_2O 3; EtOH 3; bz 3; CS_2 3
10333	Propanoic acid, 3,3,3-trichloro-2-hydroxy- / Lactic acid, 3,3,3-trichloro-	$C_3H_3Cl_3O_3$ / 193.41	599-01-9 / 125	/ 140[45]	3-03-00-00501 /	H_2O 4; eth 4; EtOH 4; chl 4
10334	1-Propanol / Propyl alcohol	C_3H_8O / 60.10	71-23-8 / -126.1	7854 / 97.2	4-01-00-01413 / 0.8035[20]	H_2O 5; EtOH 5; eth 5; ace 3 / 1.3850[20]
10335	2-Propanol / Isopropyl alcohol	C_3H_8O / 60.10	67-63-0 / -89.5	5096 / 82.3	4-01-00-01461 / 0.7855[20]	H_2O 5; EtOH 5; eth 5; ace 3 / 1.3776[20]

No.	Name / Synonym	Mol. Form. / Mol. Wt.	CAS RN / mp/°C	Merck No. / bp/°C	Beil. Ref. / den/g cm^{-3}	Solubility / n_D
10336	1-Propanol, 2-amino-, (±)-	C_3H_9NO 75.11	6168-72-5	174.5	3-04-00-00737	H_2O 4; EtOH 4; eth 4; chl 2 1.4502[20]
10337	1-Propanol, 3-amino- Propanolamine	C_3H_9NO 75.11	156-87-6 11	187.5	4-04-00-01623 0.9824[26]	H_2O 3; EtOH 3; eth 3 1.4617[20]
10338	2-Propanol, 1-amino-, (±)- DL-1-Amino-2-propanol	C_3H_9NO 75.11	1674-56-2 1.7	159.4	4-04-00-01665 0.9611[20]	H_2O 5; EtOH 5; eth 5; ace 5 1.4479[20]
10339	2-Propanol, 1-amino-3-(diethylamino)- 3-(Diethylamino)-2-hydroxypropylamine	$C_7H_{18}N_2O$ 146.23	6322-01-6	223	3-04-00-00767 0.931[20]	EtOH 3; eth 3; chl 2 1.465[20]
10340	2-Propanol, 1-[(2-aminoethyl)amino]- N-(2-Hydroxypropyl)ethylenediamine	$C_5H_{14}N_2O$ 118.18	123-84-2	450 94[3]	4-04-00-01684 0.9837[25]	1.4738[20]
10341	1-Propanol, 2-amino-2-methyl- 2-Aminoisobutanol	$C_4H_{11}NO$ 89.14	124-68-5 25.5	461 165.5	4-04-00-01740 0.934[20]	H_2O 5; ctc 3 1.449[20]
10342	1-Propanol, 2-[(2-amino-2-methylpropyl)amino]-2-methyl- 1-Propanol, 2-[2-(2-amino-2-methylpropyl)amino]-2-methyl-	$C_8H_{20}N_2O$ 160.26	72622-74-3 46	231	3-04-00-00788 0.9360[20]	chl 2 1.4635[25]
10343	2-Propanol, 1,3-bis(dimethylamino)-	$C_7H_{18}N_2O$ 146.23	5966-51-8	181.5	4-04-00-01695 0.8788[20]	H_2O 4 1.4418[20]
10344	1-Propanol, 3-bromo-	C_3H_7BrO 138.99	627-18-9	105[185]; 62[5]	4-01-00-01446 1.5374[20]	H_2O 3; EtOH 5; eth 5 1.4834[25]
10345	2-Propanol, 1-bromo-	C_3H_7BrO 138.99	19686-73-8	146.5	3-01-00-01474 1.5585[30]	H_2O 3; EtOH 4; eth 4 1.4801[20]
10346	1-Propanol, 1-butoxy-	$C_7H_{16}O_2$ 132.20	5131-66-8	171.5; 71[20]	4-01-00-02471 0.882[20]	EtOH 3; eth 3; bz 3; ctc 3 1.4168[20]
10347	1-Propanol, 2-chloro- Propylene chlorohydrin	C_3H_7ClO 94.54	78-89-7	7863 133.5	4-01-00-01440 1.103[20]	H_2O 4; eth 4; EtOH 4 1.4390[20]
10348	1-Propanol, 3-chloro- 3-Chloro-1-propanol	C_3H_7ClO 94.54	627-30-5	165	4-01-00-01441 1.1309[20]	H_2O 4; EtOH 3; eth 3; ctc 2 1.4459[20]
10349	2-Propanol, 1-chloro- sec-Propylene chlorohydrin	C_3H_7ClO 94.54	127-00-4	7864 127	4-01-00-01490 1.113[20]	H_2O 5; EtOH 5; eth 5; ctc 2 1.4392[20]
10350	1-Propanol, 3-chloro-, acetate	$C_5H_9ClO_2$ 136.58	628-09-1	164; 62[10]	4-02-00-00140 1.250[19]	H_2O 1; EtOH 3; eth 3; ace 3 1.431[20]
10351	2-Propanol, 1-chloro-, acetate	$C_5H_9ClO_2$ 136.58	623-60-9	149.5	4-02-00-00142 1.0788[20]	eth 4; EtOH 4 1.4223[20]
10352	1-Propanol, 2-chloro-2-methyl- Isobutyl alcohol, 2-chloro	C_4H_9ClO 108.57	558-38-3	132 dec; 60[50]	4-01-00-01603 1.0472[20]	1.4388[20]
10353	1-Propanol, 3-chloro-2-methyl- Isobutyl alcohol, 3-chloro	C_4H_9ClO 108.57	10317-10-9	76-8[11]	3-01-00-01564 1.083[25]	eth 4; EtOH 4 1.4460[25]
10354	2-Propanol, 1-chloro-2-methyl- Isobutylene chlorohydrin	C_4H_9ClO 108.57	558-42-9 -20	128.5	4-01-00-01628 1.0628[20]	H_2O 4; EtOH 4 1.4380[24]
10355	2-Propanol, 1-chloro-3-(1-methylethoxy)- 2-Propanol, 1-chloro-3-isopropoxy-	$C_6H_{13}ClO_2$ 152.62	4288-84-0	87[20]	4-01-00-02485 1.0530[25]	EtOH 3; eth 3 1.4370[25]
10356	2-Propanol, 1-chloro-3-propoxy-	$C_6H_{13}ClO_2$ 152.62	6943-58-4	92-5[15]	3-01-00-02153 1.0526[25]	ace 4; eth 4 1.4378[25]
10357	2-Propanol, 1,3-dibromo-	$C_3H_6Br_2O$ 217.89	96-21-9	219 dec; 105[16]	4-01-00-01496 2.1201[25]	ace 4; eth 4; EtOH 4 1.5495[25]
10358	1-Propanol, 2,3-dibromo-, phosphate (3:1) Tris(2,3-dibromopropyl) phosphate	$C_9H_{15}Br_6O_4P$ 697.61	126-72-7	9665		chl 3
10359	1-Propanol, 2,3-dichloro-	$C_3H_6Cl_2O$ 128.99	616-23-9	184	4-01-00-01442 1.3607[20]	H_2O 2; EtOH 5; eth 5; ace 5 1.4819[20]
10360	1-Propanol, 3,3-dichloro-	$C_3H_6Cl_2O$ 128.99	83682-72-8	82-3[20]	3-01-00-01427 1.316[25]	1.467[20]
10361	2-Propanol, 1,1-dichloro-	$C_3H_6Cl_2O$ 128.99	53894-19-2	147	2-01-00-00383 1.3334[22]	ace 4; eth 4; EtOH 4
10362	2-Propanol, 1,3-dichloro- 1,3-Dichloro-2-propanol	$C_3H_6Cl_2O$ 128.99	96-23-1	3063 176	4-01-00-01491 1.3506[17]	H_2O 4; EtOH 4; eth 5; ace 3 1.4837[20]
10363	2-Propanol, 1,3-dichloro-, acetate	$C_5H_8Cl_2O_2$ 171.02	3674-10-0	205	3-02-00-00233 1.281[20]	eth 4; EtOH 4; chl 4 1.4542[20]
10364	2-Propanol, 1,1-dichloro-2-methyl- 2,2-Dichloro-1,1-dimethylethanol	$C_4H_8Cl_2O$ 143.01	4773-53-9 8	152.5	4-01-00-01629 1.2363[19]	bz 4; EtOH 4 1.4598[20]
10365	2-Propanol, 1,3-dichloro-2-methyl- β-Methylglycerol α,γ-dichlorohydrin	$C_4H_8Cl_2O$ 143.01	597-32-0	174.5	4-01-00-01629 1.2745[20]	H_2O 3; EtOH 4; ace 3 1.4744[21]
10366	1-Propanol, 1-(diethylamino)- 2-Diethylamino-1-propanol	$C_7H_{17}NO$ 131.22	611-12-1	168; 78[12]	4-04-00-01618 0.8649[27]	EtOH 3; eth 3; ace 3; bz 3 1.4332[20]
10367	1-Propanol, 3-(diethylamino)-	$C_7H_{17}NO$ 131.22	622-93-5	189.5	4-04-00-01633 0.8600[20]	EtOH 3; eth 3; ace 3; bz 3 1.4439[20]
10368	2-Propanol, 1-(diethylamino)-, (±)- 2-Propanol, 1-diethylamino (DL)	$C_7H_{17}NO$ 131.22	78738-36-0 13.5	159; 63[22]	3-04-00-00759 0.8511[20]	H_2O 4; ace 4; EtOH 4 1.4255[20]
10369	2-Propanol, 1,3-difluoro- 1,3-Difluoro-2-propanol	$C_3H_8F_2O$ 96.08	453-13-4	54-5[34]	4-01-00-01489 1.24[25]	1.3725[20]
10370	1-Propanol, 2,3-dimercapto- Dimercaprol	$C_3H_8OS_2$ 124.23	59-52-9	3196 82-4[0.8]	4-01-00-02770 1.2463[20]	EtOH 3; eth 3; chl 2; oils 3 1.5749[20]
10371	2-Propanol, 1,3-dimercapto-	$C_3H_8OS_2$ 124.23	584-04-3	94[12]	4-01-00-02773 1.2386[20]	1.5700[20]
10372	2-Propanol, 1,3-dimethoxy- Glycerol 1,3-dimethyl ether	$C_5H_{12}O_3$ 120.15	623-69-8	169	3-01-00-02318 1.0085[20]	H_2O 5; EtOH 4; eth 3 1.4192[20]
10373	1-Propanol, 2,2-dimethyl- Neopentyl alcohol	$C_5H_{12}O$ 88.15	75-84-3 52.5	6373 113.5	4-01-00-01690 0.812[20]	H_2O 2; EtOH 4; eth 4; ctc 3

No.	Name / Synonym	Mol. Form. / Mol. Wt.	CAS RN / mp/°C	Merck No. / bp/°C	Beil. Ref. / den/g cm^{-3}	Solubility / n_D
10374	1-Propanol, 2-(dimethylamino)- / 2-Dimethylamino-1-propanol	$C_5H_{13}NO$ / 103.16	15521-18-3 /	/ 150.3	4-04-00-01617 / 0.8820[26]	H_2O 3
10375	1-Propanol, 3-(dimethylamino)-	$C_5H_{13}NO$ / 103.16	3179-63-3 /	/ 163.5	4-04-00-01628 / 0.872[25]	ctc 3 / 1.4360[20]
10376	2-Propanol, 1-(dimethylamino)-	$C_5H_{13}NO$ / 103.16	108-16-7 /	/ 124.5	3-04-00-00756 / 0.837[25]	ctc 3 / 1.4193[20]
10377	2-Propanol, 1,3-diphenoxy-	$C_{15}H_{16}O_3$ / 244.29	622-04-8 / 81.5	224[17]; 175[2]	4-06-00-00590 / 1.179[24]	bz 4; eth 4; EtOH 4; chl 4
10378	2-Propanol, 1,3-diphenoxy-, acetate / Glycerol, 1,3-diphenyl ether-2-acetate	$C_{17}H_{18}O_4$ / 286.33	71159-31-4 / 33	/ 190[160]	3-06-00-00583 /	bz 4; eth 4; EtOH 4; chl 4
10379	2-Propanol, 1,1',1'',1'''-(1,2-ethanediyldinitrilo)tetrakis- / ENTPROL	$C_{14}H_{32}N_2O_4$ / 292.42	102-60-3 /	3551	4-04-00-01685 / 1.030[25]	chl 2 / 1.478[25]
10380	1-Propanol, 2-ethoxy- / Propylene glycol β-monoethyl ether	$C_5H_{12}O_2$ / 104.15	19089-47-5 /	/ 140.5	3-01-00-02147 / 0.9044[20]	H_2O 3; EtOH 3; eth 3 / 1.4122[20]
10381	2-Propanol, 1-ethoxy- / 1-Ethoxy-2-propanol	$C_5H_{12}O_2$ / 104.15	1569-02-4 /	/ 131	4-01-00-02471 / 0.9028[20]	H_2O 3; EtOH 3; eth 3 / 1.4075[20]
10382	1-Propanol, 3-fluoro- / 3-Hydroxypropyl fluoride	C_3H_7FO / 78.09	462-43-1 /	/ 127.8	4-01-00-01437 / 1.039[25]	H_2O 4; eth 4; EtOH 4 / 1.3771[25]
10383	2-Propanol, 1-[(2-hydroxyethyl)amino]-	$C_5H_{13}NO_2$ / 119.16	6579-55-1 /	120-3[1]	4-04-00-01678 / 1.0455[20]	H_2O 3; EtOH 3; ace 3 / 1.4695[20]
10384	2-Propanol, 1,1'-iminobis- / Diisopropanolamine	$C_6H_{15}NO_2$ / 133.19	110-97-4 / 44.5	250; 151[23]	3-04-00-00761 / 0.989[20]	H_2O 3; EtOH 3; eth 2
10385	1-Propanol, 3-iodo-	C_3H_7IO / 185.99	627-32-7 /	115[38]; 88[4]	4-01-00-01448 / 1.9976[20]	1.5585[20]
10386	1-Propanol, 2-methoxy- / 2-Methoxy-1-propanol	$C_4H_{10}O_2$ / 90.12	1589-47-5 /	/ 130	4-01-00-02471 / 0.938[20]	1.4070[20]
10387	2-Propanol, 1-methoxy- / 1-Methoxy-2-propanol	$C_4H_{10}O_2$ / 90.12	107-98-2 /	/ 119	3-01-00-02146 / 0.9620[20]	1.4034[20]
10388	1-Propanol, 2-methyl- / Isobutyl alcohol	$C_4H_{10}O$ / 74.12	78-83-1 / -108	5015 / 107.8	4-01-00-01588 / 0.8018[20]	H_2O 2; EtOH 3; eth 3; ace 3 / 1.3955[20]
10389	2-Propanol, 2-methyl- / tert-Butyl alcohol	$C_4H_{10}O$ / 74.12	75-65-0 / 25.4	1542 / 82.4	4-01-00-01609 / 0.7887[20]	H_2O 5; EtOH 5; eth 5; chl 3 / 1.3878[20]
10390	2-Propanol, 1-(1-methylethoxy)- / 1-Isopropoxy-2-propanol	$C_6H_{14}O_2$ / 118.18	3944-36-3 /	/ 137.5	4-01-00-02471 / 0.879[20]	1.4070[20]
10391	1-Propanol, 2-methyl-2-nitro-	$C_4H_9NO_3$ / 119.12	76-39-1 / 89.5	94[10]	4-01-00-01604 /	H_2O 2; EtOH 4; eth 4; chl 3
10392	1-Propanol, 3-[methyl(phenylmethyl)amino]-	$C_{11}H_{17}NO$ / 179.26	5814-42-6 /	132-5[4]	4-12-00-02208 / 1.001[25]	ctc 2; CS_2 3 / 1.5230[20]
10393	2-Propanol, 1,1',1''-nitrilotris-	$C_9H_{21}NO_3$ / 191.27	122-20-3 / 45	170-80[10]	4-04-00-01680 / 1.0[20]	H_2O 3; EtOH 3; chl 2
10394	1-Propanol, 2-nitro-	$C_3H_7NO_3$ / 105.09	2902-96-7 /	100[12]	4-01-00-01448 / 1.1841[25]	H_2O 3; EtOH 3; eth 3; chl 2 / 1.4379[20]
10395	1-Propanol, 3-nitro-	$C_3H_7NO_3$ / 105.09	25182-84-7 /	138-40[32]	3-01-00-01429 / 1.173[13]	H_2O 4; eth 4; EtOH 4
10396	2-Propanol, 1-nitro- / 1-Nitro-2-propanol	$C_3H_7NO_3$ / 105.09	3156-73-8 /	112[13]; 68[1]	4-01-00-01496 / 1.1906[20]	H_2O 3; EtOH 3; eth 3; bz 3 / 1.4383[20]
10397	Propanol, oxybis- / Dipropylene glycol	$C_6H_{14}O_3$ / 134.18	25265-71-8 /	/ 230.5	4-01-00-02473 / 1.0206[20]	H_2O 5; EtOH 3
10398	1-Propanol, 2-phenoxy-	$C_9H_{12}O_2$ / 152.19	4169-04-4 /	/ 244	4-06-00-00582 / 0.9801[25]	EtOH 3; eth 3 / 1.4760[25]
10399	2-Propanol, 1-phenoxy-	$C_9H_{12}O_2$ / 152.19	770-35-4 /	134.5[20]	4-06-00-00582 / 1.0622[20]	1.5232[20]
10400	1-Propanol, 3-(phenylmethoxy)-	$C_{10}H_{14}O_2$ / 166.22	4799-68-2 /	172[43]	4-06-00-02243 / 1.0474[20]	EtOH 3; eth 3
10401	2-Propanol, 1-propoxy- / 1,2-Propylene glycol 1-propyl ether	$C_6H_{14}O_2$ / 118.18	1569-01-3 /	/ 150	3-01-00-02147 / 0.8886[20]	1.4130[20]
10402	2-Propanol, 1,1,3,3-tetrachloro- / 1,1,3,3-Tetrachloro-2-propanol	$C_3H_4Cl_4O$ / 197.87	18992-39-7 /	80-90[14]	3-01-00-01474 / 1.612[20]	eth 3 / 1.5133[20]
10403	2-Propanol, 1,1,1,3-tetrachloro-	$C_3H_4Cl_4O$ / 197.87	88947-16-4 /	95-6[17]	3-01-00-01474 / 1.610[20]	eth 4 / 1.5145[20]
10404	1-Propanol, 2,2,3,3-tetrafluoro- / 2,2,3,3-Tetrafluoro-1-propanol	$C_3H_4F_4O$ / 132.06	76-37-9 / -15	9139 / 109.5	4-01-00-01438 / 1.4853[20]	EtOH 3; ace 3; chl 3 / 1.3197[20]
10405	2-Propanol, 1,1,1-tribromo-2-methyl- / Tribromo-tert-butyl alcohol	$C_4H_7Br_3O$ / 310.81	76-08-4 / 169	9523 / sub	3-01-00-01588 /	H_2O 2; EtOH 3; eth 3; chl 2
10406	2-Propanol, 1,1,1-trichloro- / 1,1,1-Trichloro-2-propanol	$C_3H_5Cl_3O$ / 163.43	76-00-6 / 50.5	9557 / 163; 53-5[12]	3-01-00-01474 /	ace 4; bz 4; eth 4; EtOH 4
10407	2-Propanol, 1,1,1-trichloro-2-methyl- / Chlorobutanol	$C_4H_7Cl_3O$ / 177.46	57-15-8 / 97	2129 / 167	4-01-00-01629 /	H_2O 1; EtOH 3; eth 3; ace 3
10408	2-Propanol, 1,1,1-trichloro-3-nitro- / 2-Propanol, 1-nitro-3,3,3-trichloro	$C_3H_4Cl_3NO_3$ / 208.43	760-40-7 / 45.5	105[3.5]	4-01-00-01497 /	eth 4; EtOH 4
10409	2-Propanol, 1,1,1-trifluoro-, (±)- / 2-Propanol, 1,1,1-trifluoro(DL)	$C_3H_5F_3O$ / 114.07	17556-48-8 / -52	/ 78	4-01-00-01489 / 1.2632[25]	EtOH 4; eth 4; ace 3; bz 3 / 1.3130[25]
10410	1-Propanol, 3-(trimethylsilyl)-	$C_6H_{16}OSi$ / 132.28	2917-47-7 /	141; 82[24]	4-04-00-03953 / 0.822[25]	1.4298[20]
10411	2-Propanone / Acetone	C_3H_6O / 58.08	67-64-1 / -94.8	58 / 56.0	4-01-00-03180 / 0.7899[20]	H_2O 5; EtOH 5; eth 5; ace 5 / 1.3588[20]
10412	2-Propanone, 1-(acetyloxy)-	$C_5H_8O_3$ / 116.12	592-20-1 /	171; 63[11]	4-02-00-00297 / 1.0757[20]	H_2O 4; eth 4; EtOH 4 / 1.4141[20]
10413	1-Propanone, 1-(4-aminophenyl)- / p-Aminopropiophenone	$C_9H_{11}NO$ / 149.19	70-69-9 / 140	482	4-14-00-00139 /	DMSO 3

No.	Name Synonym	Mol. Form. Mol. Wt.	CAS RN mp/°C	Merck No. bp/°C	Beil. Ref. den/g cm^{-3}	Solubility n_D
10414	1-Propanone, 2-amino-1-phenyl-	C$_9$H$_{11}$NO 149.19	5265-18-9 46.5	93[0.8]	3-14-00-00147	ace 4; eth 4; EtOH 4
10415	2-Propanone, 1,3-bis(1-methyl-2-pyrrolidinyl)-, (R*,S*)- Cuscohygrine	C$_{13}$H$_{24}$N$_2$O 224.35	454-14-8	2685 169[23]; 122[2]	5-24-01-00458 0.9733[20]	H$_2$O 4; bz 4; eth 4; EtOH 4 1.4832[20]
10416	2-Propanone, 1-bromo- Bromoacetone	C$_3$H$_5$BrO 136.98	598-31-2 -36.5	1389 138; 31.5[8]	4-01-00-03223 1.634[23]	H$_2$O 2; EtOH 3; eth 3; ace 3 1.4697[15]
10417	2-Propanone, 1-bromo-3-chloro- 1-Bromo-3-chloroacetone	C$_3$H$_4$BrClO 171.42	53535-68-5 34.5	179	4-01-00-03224	eth 4; EtOH 4
10418	1-Propanone, 1-(4-bromophenyl)-	C$_9$H$_9$BrO 213.07	10342-83-3 48	169[15]; 137[2]	4-07-00-00684	H$_2$O 1; EtOH 3; eth 3; ace 3
10419	1-Propanone, 2-bromo-1-phenyl-	C$_9$H$_9$BrO 213.07	2114-00-3	247.5	4-07-00-00684 1.4298[20]	H$_2$O 1; EtOH 3; eth 3; ace 3 1.5720[20]
10420	2-Propanone, 1-chloro- Chloroacetone	C$_3$H$_5$ClO 92.52	78-95-5 -44.5	2113 119	4-01-00-03215 1.15[20]	H$_2$O 3; EtOH 3; eth 3; chl 3
10421	2-Propanone, 1-chloro-1,1,3,3,3-pentafluoro- Chloropentafluoroacetone	C$_3$ClF$_5$O 182.48	79-53-8 -133	8		
10422	1-Propanone, 1-(4-chlorophenyl)-	C$_9$H$_9$ClO 168.62	6285-05-8 37.3	135[31]; 114[2]	4-07-00-00683	H$_2$O 1; EtOH 3; chl 2; CS$_2$ 3
10423	1-Propanone, 1-cyclohexyl- Cyclohexyl ethyl ketone	C$_9$H$_{16}$O 140.23	1123-86-0	196	4-07-00-00064 0.9105[20]	1.4530[20]
10424	2-Propanone, 1,3-dibromo-	C$_3$H$_4$Br$_2$O 215.87	816-39-7 26	97[22]	4-01-00-03225 2.1670[18]	eth 4; CS$_2$ 4
10425	2-Propanone, 1,1-dichloro- 1,1-Dichloroacetone	C$_3$H$_4$Cl$_2$O 126.97	513-88-2	3038 120	4-01-00-03218 1.304[18]	H$_2$O 2; EtOH 3; eth 5
10426	2-Propanone, 1,3-dichloro- 1,3-Dichloroacetone	C$_3$H$_4$Cl$_2$O 126.97	534-07-6 45	3039 173.4	4-01-00-03219 1.3826[46]	H$_2$O 3; EtOH 3; eth 3 1.4716[40]
10427	2-Propanone, 1-(diethylamino)-	C$_7$H$_{15}$NO 129.20	1620-14-0	157 dec	4-04-00-01925 0.8620[20]	H$_2$O 5; EtOH 5; eth 5; ctc 3 1.4249[20]
10428	1-Propanone, 2-(diethylamino)-1-phenyl- Diethylpropion	C$_{13}$H$_{19}$NO 205.30	90-84-6	3116	4-14-00-00144	
10429	2-Propanone, 1,3-dihydroxy- Dihydroxyacetone	C$_3$H$_6$O$_3$ 90.08	96-26-4 90	3166	4-01-00-04119	H$_2$O 3; EtOH 3; eth 3; ace 3
10430	1-Propanone, 1-(2,4-dihydroxyphenyl)-	C$_9$H$_{10}$O$_3$ 166.18	5792-36-9 99	176[6]	4-08-00-01837	bz 4; eth 4; EtOH 4
10431	1-Propanone, 2,2-dimethyl-1-phenyl-	C$_{11}$H$_{14}$O 162.23	938-16-9	220	4-07-00-00740 0.963[26]	ace 3 1.5086[19]
10432	2-Propanone, 1,1-diphenyl-	C$_{15}$H$_{14}$O 210.28	781-35-1 46	307; 174[10]	4-07-00-01426	EtOH 3; eth 3; bz 3; chl 3 1.5361[16]
10433	2-Propanone, 1,3-diphenyl-	C$_{15}$H$_{14}$O 210.28	102-04-5 35	331	4-07-00-01420 1.195[0]	H$_2$O 1; EtOH 3; eth 3; peth 3
10434	1-Propanone, 3-(3-ethenyl-4-piperidinyl)-1-(6-methoxy-4-quinolinyl)-, (3R-cis)- Viquidil	C$_{20}$H$_{24}$N$_2$O$_2$ 324.42	84-55-9 60	9908	4-25-00-00181	eth 4; EtOH 4; chl 4
10435	1-Propanone, 3-(3-ethenyl-4-piperidinyl)-1-(4-quinolinyl)-, (3R-cis)- Cinchotoxine	C$_{19}$H$_{22}$N$_2$O 294.40	69-24-9 59	2291	4-24-00-00651	H$_2$O 1; EtOH 4; eth 4; ace 4
10436	1-Propanone, 3-ethoxy-1-phenyl-	C$_{11}$H$_{14}$O$_2$ 178.23	34008-71-4 11.5	135[18]	4-08-00-00453 1.0386[20]	1.5190[20]
10437	2-Propanone, 1-fluoro-	C$_3$H$_5$FO 76.07	430-51-3	77	4-01-00-03213 1.0288[20]	1.3700[20]
10438	1-Propanone, 1-(2-furanyl)-	C$_7$H$_8$O$_2$ 124.14	3194-15-8 28	88[14]	5-17-09-00415 1.0626[28]	eth 3; ctc 2 1.4922[25]
10439	2-Propanone, 1-(2-furanyl)-	C$_7$H$_8$O$_2$ 124.14	6975-60-6 29	179.5	5-17-09-00418 1.104[20]	1.5035[20]
10440	2-Propanone, 1,1,1,3,3,3-hexachloro- Hexachloroacetone	C$_3$Cl$_6$O 264.75	116-16-5 -2	203	4-01-00-03223 1.7434[12]	H$_2$O 2; ace 3 1.5112[20]
10441	2-Propanone, 1,1,1,3,3,3-hexafluoro- Perfluoroacetone	C$_3$F$_6$O 166.02	684-16-2 -125	-27.4	4-01-00-03215	
10442	2-Propanone, 1-(1H-indol-3-yl)- 3-Indolylacetone	C$_{11}$H$_{11}$NO 173.21	1201-26-9 116	4873	5-21-08-00327	
10443	2-Propanone, 1-hydroxy-	C$_3$H$_6$O$_2$ 74.08	116-09-6 -17	145.5	4-01-00-03977 1.0805[20]	H$_2$O 4; EtOH 4; eth 4 1.4295[20]
10444	1-Propanone, 1-(2-hydroxy-3-methylphenyl)- 2-Methyl-6-propionylphenol	C$_{10}$H$_{12}$O$_2$ 164.20	3338-15-6 22.5	124[13]	4-08-00-00517 1.0900[20]	ctc 3
10445	1-Propanone, 1-(2-hydroxy-4-methylphenyl)- 2'-Hydroxy-4'-methylpropiophenone	C$_{10}$H$_{12}$O$_2$ 164.20	2886-52-4 44	115[10]	4-08-00-00519	chl 3
10446	1-Propanone, 1-(2-hydroxy-5-methylphenyl)-	C$_{10}$H$_{12}$O$_2$ 164.20	938-45-4 2	129-30[16.5]	4-08-00-00517 1.084[14]	chl 3 1.549[13]
10447	1-Propanone, 1-(2-hydroxy-5-methylphenyl)-2-methyl-	C$_{11}$H$_{14}$O$_2$ 178.23	64207-03-0	251; 125[11]	4-08-00-00545 1.0460[16]	1.538[16]
10448	1-Propanone, 1-(4-hydroxy-3-methylphenyl)-2-methyl- o-Cresol, 4-isobutyryl	C$_{11}$H$_{14}$O$_2$ 178.23	73206-57-2 122	182[12]	2-08-00-00130	eth 4; EtOH 4
10449	1-Propanone, 1-(2-hydroxyphenyl)-	C$_9$H$_{10}$O$_2$ 150.18	610-99-1	150[80]; 115[15]	4-08-00-00436	H$_2$O 2; EtOH 3; eth 3; ctc 3 1.5501[20]
10450	1-Propanone, 1-(4-hydroxyphenyl)- Paroxypropione	C$_9$H$_{10}$O$_2$ 150.18	70-70-2 149	6992	4-08-00-00441	H$_2$O 2; EtOH 3; eth 3; ace 2

No.	Name Synonym	Mol. Form. Mol. Wt.	CAS RN mp/°C	Merck No. bp/°C	Beil. Ref. den/g cm^{-3}	Solubility n_D
10451	1-Propanone, 2-hydroxy-1-phenyl-	$C_9H_{10}O_2$ 150.18	5650-40-8	251	4-08-00-00450 1.1085[18]	1.536[23]
10452	1-Propanone, 3-hydroxy-1-phenyl-	$C_9H_{10}O_2$ 150.18	5650-41-9	98[0.2]	4-08-00-00453 1.1170[20]	1.5450[20]
10453	1-Propanone, 3-(4-hydroxyphenyl)-1-(2,4,6-trihydroxyphenyl)- Phloretin	$C_{15}H_{14}O_5$ 274.27	60-82-2 263-4 dec	7299	4-08-00-03518	H_2O 2; EtOH 5; eth 1; ace 3
10454	2-Propanone, 1-iodo-	C_3H_5IO 183.98	3019-04-3	62[12]	4-01-00-03226 2.17[15]	EtOH 3
10455	2-Propanone, 1-methoxy-	$C_4H_8O_2$ 88.11	5878-19-3	116	4-01-00-03977 0.957[25]	1.3970[20]
10456	1-Propanone, 1-(3-methoxyphenyl)- 3-Methoxypropiophenone	$C_{10}H_{12}O_2$ 164.20	37951-49-8	259	4-08-00-00441 1.0812[25]	eth 4; EtOH 4 1.5230[25]
10457	1-Propanone, 2-methoxy-1-phenyl- Propiophenone, 2-methoxy-	$C_{10}H_{12}O_2$ 164.20	6493-83-0	79-80[1]	4-08-00-00451 1.0704[20]	1.5225[20]
10458	1-Propanone, 3-methoxy-1-phenyl-	$C_{10}H_{12}O_2$ 164.20	55563-72-9	125-6[16]	4-08-00-00453 1.0602[20]	1.5250[20]
10459	2-Propanone, 1-(methoxyphenyl)- 2-Propanone, (methoxyphenyl)-	$C_{10}H_{12}O_2$ 164.20	31116-84-4 <-15	267-9 142[14]	2-08-00-00104 1.0670[18]	H_2O 2; EtOH 3; eth 3 1.5253[20]
10460	2-Propanone, 1-(3-methoxyphenyl)-	$C_{10}H_{12}O_2$ 164.20	3027-13-2	259	4-08-00-00459 1.0812[6]	EtOH 3; eth 3 1.5230[25]
10461	2-Propanone, 1-(4-methoxyphenyl)-	$C_{10}H_{12}O_2$ 164.20	122-84-9 <-15	268	4-08-00-00460 1.0694[17]	eth 4; EtOH 4 1.5253[20]
10462	1-Propanone, 2-methyl-1,2-di-3-pyridinyl- Metyrapone	$C_{14}H_{14}N_2O$ 226.28	54-36-4 50.5	6083	5-24-04-00048	
10463	2-Propanone, (1-methylethylidene)hydrazone	$C_6H_{12}N_2$ 112.17	627-70-3 -12.5	133	4-01-00-03207 0.8390[20]	H_2O 5; EtOH 5; eth 5; ace 3 1.4535[20]
10464	1-Propanone, 2-methyl-1-(1-naphthalenyl)- Isopropyl 1-naphthyl ketone	$C_{14}H_{14}O$ 198.26	61838-78-6	307	3-07-00-01985 1.0761[25]	bz 4; eth 4; EtOH 4 1.5948[20]
10465	1-Propanone, 2-methyl-1-(2-naphthalenyl)- 2-Isobutyronaphthone	$C_{14}H_{14}O$ 198.26	107574-57-2	313	3-07-00-01985 1.0617[25]	bz 4; eth 4; EtOH 4
10466	1-Propanone, 1-(2-methylphenyl)-	$C_{10}H_{12}O$ 148.20	2040-14-4 -27.6	219.5	4-07-00-00720 1.0119[0]	
10467	1-Propanone, 1-(3-methylphenyl)-	$C_{10}H_{12}O$ 148.20	51772-30-6	235; 132[31]	4-07-00-00721 1.0059[0]	H_2O 1; EtOH 3; eth 3; ace 3
10468	1-Propanone, 1-(4-methylphenyl)-	$C_{10}H_{12}O$ 148.20	5337-93-9 7.2	236	4-07-00-00722 0.9926[20]	H_2O 1; EtOH 3; eth 3; ace 3 1.5278[20]
10469	1-Propanone, 2-methyl-1-phenyl-	$C_{10}H_{12}O$ 148.20	611-70-1 -1.3	220	4-07-00-00716 0.9863[11]	eth 4; EtOH 4 1.5172[20]
10470	2-Propanone, 1-(1-methyl-2-piperidinyl)-, (±)-	$C_9H_{17}NO$ 155.24	18747-42-7	96-8[13]	5-21-06-00523 0.9478[20]	H_2O 4; lig 4 1.4674[20]
10471	2-Propanone, 1-(1-methyl-2-pyrrolidinyl)-, (R)- Hygrine	$C_8H_{15}NO$ 141.21	496-49-1 76.5[11]	4788	5-21-06-00511	EtOH 4; chl 4 1.4555[20]
10472	1-Propanone, 2-methyl-1-(2-thienyl)- Isopropyl 2-thienyl ketone	$C_8H_{10}OS$ 154.23	36448-60-9	229	5-17-09-00439 1.0894[20]	1.5405[20]
10473	1-Propanone, 1-(1-naphthalenyl)-	$C_{13}H_{12}O$ 184.24	2876-63-3	306	4-07-00-01298 1.0971[20]	1.6108[20]
10474	1-Propanone, 1-(2-naphthalenyl)-	$C_{13}H_{12}O$ 184.24	6315-96-4 60	313	4-07-00-01299	EtOH 4; chl 4
10475	2-Propanone, 1-nitro-	$C_3H_5NO_3$ 103.08	10230-68-9 50.3	103[24]	4-01-00-03226	bz 4; eth 4; EtOH 4
10476	2-Propanone, oxime Acetoxime	C_3H_7NO 73.09	127-06-0 61	68 136; 61[20]	4-01-00-03202 0.9113[62]	H_2O 3; EtOH 3; eth 3; chl 3 1.4156[20]
10477	2-Propanone, 1,1,1,3,3-pentabromo- Pentabromoacetone	C_3HBr_5O 452.56	79-49-2 79.5	7056 sub	4-01-00-03226	H_2O 1; EtOH 4; eth 4; ace 4
10478	2-Propanone, 1,1,1,3,3-pentachloro-	C_3HCl_5O 230.30	1768-31-6 2.1	192[253]; 98[40]	4-01-00-03222 1.69[15]	
10479	2-Propanone, 1-phenoxy-	$C_9H_{10}O_2$ 150.18	621-87-4	229.5	4-06-00-00604 1.0903[20]	eth 3; ace 3 1.5228[20]
10480	1-Propanone, 1-phenyl- Propiophenone	$C_9H_{10}O$ 134.18	93-55-0 18.6	7842 217.5	4-07-00-00680 1.0096[20]	H_2O 1; EtOH 3; eth 3; chl 3 1.5269[20]
10481	2-Propanone, 1-phenyl- Phenylacetone	$C_9H_{10}O$ 134.18	103-79-7 -15	7240 216.5	4-07-00-00687 1.0157[20]	H_2O 1; EtOH 4; eth 4; bz 5 1.5168[20]
10482	2-Propanone, phenylhydrazone Acetone, phenylhydrazone	$C_9H_{12}N_2$ 148.21	103-02-6 42	163[50]	4-15-00-00060	EtOH 3; eth 3; dil acid 3
10483	1-Propanone, 1-phenyl-, oxime	$C_9H_{11}NO$ 149.19	2157-50-8 54.8	245; 165[38]	4-07-00-00682 1.0639[18]	H_2O 1; EtOH 3; eth 3; chl 2
10484	1-Propanone, 1-(2-piperidinyl)- Piperidine, 2(1-propionyl)-	$C_8H_{15}NO$ 141.21	97073-23-9	94[18]	5-21-06-00522 0.9380[20]	1.4621[20]
10485	2-Propanone, 1-(2-piperidinyl)-, (R)- (-)-Pelletierine	$C_8H_{15}NO$ 141.21	2858-66-4	7015 195; 125[100]	5-21-06-00522 0.988[20]	eth 4; EtOH 4; chl 4 1.4683[20]
10486	2-Propanone, 1,1,1,3-tetrachloro-	$C_3H_2Cl_4O$ 195.86	16995-35-0 65	183	4-01-00-03222 1.624[25]	H_2O 2; eth 4; ace 4 1.497[18]
10487	2-Propanone, 1,1,3,3-tetrachloro-	$C_3H_2Cl_4O$ 195.86	632-21-3	183	4-01-00-03222	H_2O 4; EtOH 4; eth 4; ace 4 1.4944[20]
10488	2-Propanone, 1,1,1-trichloro-	$C_3H_3Cl_3O$ 161.41	918-00-3	149; 28[10]	4-01-00-03221 1.435[20]	H_2O 1; EtOH 4; eth 4 1.4635[17]
10489	1-Propanone, 1-(2,4,6-trihydroxyphenyl)- Flopropione	$C_9H_{10}O_4$ 182.18	2295-58-1 175.5	4039	3-08-00-03413	eth 4; EtOH 4

No.	Name / Synonym	Mol. Form. / Mol. Wt.	CAS RN / mp/°C	Merck No. / bp/°C	Beil. Ref. / den/g cm^{-3}	Solubility / n_D
10490	Propanoyl bromide / Propionic acid bromide	C_3H_5BrO / 136.98	598-22-1	102	4-02-00-00724 / 1.5210[16]	eth 4 / 1.4578[16]
10491	Propanoyl bromide, 2-bromo-2-methyl-	$C_4H_6Br_2O$ / 229.90	20769-85-1	163	3-02-00-00661 / 1.4067[14]	ace 4; CS_2 4
10492	Propanoyl bromide, 2-methyl-	C_4H_7BrO / 151.00	2736-37-0	117	4-02-00-00852 / 1.4067[15]	1.4552[25]
10493	Propanoyl chloride / Propionyl chloride	C_3H_5ClO / 92.52	79-03-8 / -94	7840 / 80	4-02-00-00724 / 1.0646[20]	eth 3 / 1.4032[20]
10494	Propanoyl chloride, 2-(acetyloxy)-, (±)- / Lactyl chloride, acetate (DL)	$C_5H_7ClO_3$ / 150.56	55057-45-9	150 dec; 56[11]	4-03-00-00674 / 1.1920[17]	1.4241[17]
10495	Propanoyl chloride, 2-bromo-	C_3H_4BrClO / 171.42	7148-74-5	132	4-02-00-00764 / 1.697[11]	eth 3; ctc 2; chl 3 / 1.4780[20]
10496	Propanoyl chloride, 2-chloro-, (S)- / Propanoic acid, chloride, 2-chloro(d)	$C_3H_4Cl_2O$ / 126.97	70110-24-6	110	4-02-00-00747 / 1.2394[7]	1.4400[20]
10497	Propanoyl chloride, 3-chloro-	$C_3H_4Cl_2O$ / 126.97	625-36-5	144	4-02-00-00750 / 1.3307[13]	H_2O 2; EtOH 4; eth 4; chl 4 / 1.4549[20]
10498	Propanoyl chloride, 2-chloro-2-methyl-	$C_4H_6Cl_2O$ / 141.00	13222-26-9	126.5	4-02-00-00859	eth 4 / 1.4369[20]
10499	Propanoyl chloride, 3-chloro-2-methyl- / Isobutyryl chloride, 3-chloro	$C_4H_6Cl_2O$ / 141.00	7623-10-1 / -90	92	4-02-00-00860 / 1.0174[20]	eth 4 / 1.4079[20]
10500	Propanoyl chloride, 2,2-dichloro-	$C_3H_3Cl_3O$ / 161.41	26073-26-7	118; 70[89]	4-02-00-00755 / 1.4062[20]	1.4524[20]
10501	Propanoyl chloride, 2,3-dichloro-	$C_3H_3Cl_3O$ / 161.41	7623-13-4	52-4[16]	4-02-00-00757 / 1.4757[20]	1.4764[20]
10502	Propanoyl chloride, 3,3-dichloro- / 3,3-Dichloropropionyl chloride	$C_3H_3Cl_3O$ / 161.41	17880-86-3	43-4[10]	3-02-00-00562 / 1.4557[20]	eth 4; Diox 4 / 1.4738[20]
10503	Propanoyl chloride, 2,2-dimethyl- / Pivalic acid chloride	C_5H_9ClO / 120.58	3282-30-2	107	4-02-00-00912 / 1.003[20]	eth 4 / 1.4139[20]
10504	Propanoyl chloride, 2-methyl- / Isobutyric acid chloride	C_4H_7ClO / 106.55	79-30-1 / -90	92	4-02-00-00852 / 1.0174[20]	eth 3 / 1.4079[20]
10505	Propanoyl chloride, 2-phenoxy-	$C_9H_9ClO_2$ / 184.62	122-35-0	147; 116[10]	4-06-00-00643 / 1.1865[20]	eth 3 / 1.5178[20]
10506	Propanoyl fluoride / Propionyl fluoride	C_3H_5FO / 76.07	430-71-7	44	4-02-00-00724 / 0.972[15]	1.329[13]
10507	Propargite / Sulfurous acid, 2-[4-(1,1-dimethylethyl)phenoxy]cyclohexyl 2-propynyl ester	$C_{19}H_{26}O_4S$ / 350.48	2312-35-8	7818	1.10[25]	
10508	Propazine / 1,3,5-Triazine-2,4-diamine, 6-chloro-N,N'-bis(1-methylethyl)-	$C_9H_{16}ClN_5$ / 229.71	139-40-2 / 213	7822	1.162[20]	
10509	2-Propenal / Acrolein	C_3H_4O / 56.06	107-02-8 / -87.7	122 / 52.6	4-01-00-03435 / 0.840[20]	H_2O 4; EtOH 3; eth 3; ace 3 / 1.4017[20]
10510	2-Propenal, 3-bromo-3-phenyl-	C_9H_7BrO / 211.06	14804-59-2	144-6[12]	3-07-00-01383 / 1.492[20]	eth 4
10511	2-Propenal, 2-chloro-	C_3H_3ClO / 90.51	683-51-2	40[30]	4-01-00-03440 / 1.199[20]	eth 4; EtOH 4 / 1.463[20]
10512	2-Propenal, 3-(2-furanyl)-	$C_7H_6O_2$ / 122.12	623-30-3 / 54	135[14]	5-17-09-00570	H_2O 1; EtOH 5; eth 3; chl 2
10513	2-Propenal, 3-(2-furanyl)-, (E)-	$C_7H_6O_2$ / 122.12	39511-08-5 / 50	143[37]; 100[12]	5-17-09-00570	bz 4; eth 4; EtOH 4 / 1.5286[20]
10514	2-Propenal, 3-(4-hydroxy-3-methoxyphenyl)-	$C_{10}H_{10}O_3$ / 178.19	458-36-6 / 84		4-08-00-01979 / 1.1562[102]	bz 4; eth 4; EtOH 4
10515	2-Propenal, 2-methyl- / Methacrolein	C_4H_6O / 70.09	78-85-3	68.4	4-01-00-03455 / 0.840[25]	H_2O 5; EtOH 5; eth 5 / 1.4144[20]
10516	2-Propenal, 2-methyl-3-phenyl-	$C_{10}H_{10}O$ / 146.19	101-39-3	150[100]	2-07-00-00291 / 1.0407[17]	1.6057[17]
10517	2-Propenal, 3-(4-methylphenyl)-	$C_{10}H_{10}O$ / 146.19	1504-75-2 / 41.5	154[25]	4-07-00-01013 / 0.9670[23]	EtOH 4
10518	2-Propenal, 3-phenyl-, (E)- / trans-Cinnamaldehyde	C_9H_8O / 132.16	14371-10-9 / -7.5	253 dec; 127[16]	4-07-00-00984 / 1.0497[20]	H_2O 2; EtOH 3; eth 3; chl 3 / 1.6195[20]
10519	2-Propenamide / Acrylamide	C_3H_5NO / 71.08	79-06-1 / 84.5	123 / 192.6	4-02-00-01471	H_2O 4; EtOH 3; eth 3; ace 3
10520	2-Propenamide, N-(3,4-dichlorophenyl)-2-methyl- / Dicryl	$C_{10}H_9Cl_2NO$ / 230.09	2164-09-2 / 128	3078		ace 4; EtOH 4
10521	2-Propenamide, N-(1,1-dimethyl-3-oxobutyl)- / Diacetone acrylamide	$C_9H_{15}NO_2$ / 169.22	2873-97-4	2943		chl 3
10522	2-Propen-1-amine / Allylamine	C_3H_7N / 57.10	107-11-9 / -88.2	285 / 53.3	4-04-00-01057 / 0.758[20]	H_2O 5; EtOH 5; eth 5; chl 3 / 1.4205[20]
10523	2-Propen-1-amine, N,N-di-2-propenyl- / Triallylamine	$C_9H_{15}N$ / 137.22	102-70-5 / 94	155.5	4-04-00-01061 / 0.809[20]	EtOH 3; eth 3; ace 3; bz 3 / 1.4502[20]
10524	2-Propen-1-amine, N-methyl-	C_4H_9N / 71.12	627-37-2	65	4-04-00-01058	H_2O 4; ace 4; eth 4; EtOH 4 / 1.4065[20]
10525	2-Propen-1-amine, N-(1-methylethyl)-	$C_6H_{13}N$ / 99.18	35000-22-7	96.5	4-04-00-01059 / 0.7400[30]	1.4140[25]
10526	2-Propen-1-amine, N-2-propenyl- / Diallylamine	$C_6H_{11}N$ / 97.16	124-02-7	2951 / 111	4-04-00-01060	EtOH 3; eth 3 / 1.4387[20]

No.	Name / Synonym	Mol. Form. / Mol. Wt.	CAS RN / mp/°C	Merck No. / bp/°C	Beil. Ref. / den/g cm⁻³	Solubility / n_D
10527	2-Propen-1-amine, *N*-propyl-	$C_6H_{13}N$ 99.18	5666-21-7	112	0-04-00-00207 0.7708[18]	H_2O 4 1.4140[25]
10528	1-Propene Propylene	C_3H_6 42.08	115-07-1 -185.2	7862 -47.6	4-01-00-00725 *0.505[25]	H_2O 4; EtOH 4; HOAc 4 1.3567[-70]
10529	1-Propene, 1-bromo-, (*Z*)-	C_3H_5Br 120.98	590-13-6 -113	57.8	4-01-00-00754 1.4291[20]	H_2O 1; eth 3; ace 3; chl 3 1.4560[20]
10530	1-Propene, 2-bromo-	C_3H_5Br 120.98	557-93-7 -126	48.4	4-01-00-00754 1.3965[16]	H_2O 1; eth 3; ace 3; chl 3 1.4467[16]
10531	1-Propene, 3-bromo- Allyl bromide	C_3H_5Br 120.98	106-95-6 -119	286 70.1	4-01-00-00754 1.398[20]	H_2O 1; EtOH 5; eth 5; ctc 3 1.4697[20]
10532	1-Propene, 3-bromo-3,3-difluoro- Propylene, 3-bromo-3,3-difluoro	$C_3H_3BrF_2$ 156.96	420-90-6	42	4-01-00-00756 1.543[25]	1.3773[25]
10533	1-Propene, 1-chloro-, (*E*)- *trans*-1-Chloropropene	C_3H_5Cl 76.53	16136-85-9 -99	37.4	4-01-00-00737 0.9349[20]	H_2O 1; eth 3; ace 3; bz 3 1.4054[20]
10534	1-Propene, 1-chloro-, (*Z*)- *cis*-1-Chloropropene	C_3H_5Cl 76.53	16136-84-8 -134.8	32.8	4-01-00-00737 0.9347[20]	H_2O 1; eth 3; ace 3; bz 3 1.4055[20]
10535	1-Propene, 2-chloro- Isopropenyl chloride	C_3H_5Cl 76.53	557-98-2 -137.4	22.6	4-01-00-00737 0.9017[20]	H_2O 1; eth 3; ace 3; bz 3 1.3973[20]
10536	1-Propene, 3-chloro- Allyl chloride	C_3H_5Cl 76.53	107-05-1 -134.5	287 45.1	4-01-00-00738 0.9376[20]	H_2O 1; EtOH 5; eth 5; ace 5 1.4157[20]
10537	1-Propene, 3-chloro-2-(chloromethyl)-	$C_4H_6Cl_2$ 125.00	1871-57-4 -14	138	4-01-00-00805 1.1782[20]	EtOH 4; chl 4 1.4753
10538	1-Propene, 2-chloro-3-isothiocyanato-	C_4H_4ClNS 133.60	14214-31-4	182	0-04-00-00219 1.27[12]	EtOH 2
10539	1-Propene, 1-chloro-2-methyl- 1-Chloro-2-methylpropene	C_4H_7Cl 90.55	513-37-1	2147 68	4-01-00-00803 0.9186[20]	chl 3 1.4221[20]
10540	1-Propene, 3-chloro-2-methyl- 3-Chloro-2-methylpropene	C_4H_7Cl 90.55	563-47-3	2148 71.5	4-01-00-00803 0.9165[20]	EtOH 5; eth 5; ace 3; chl 4 1.4291[20]
10541	1-Propene, 2-chloro-1,1,3,3,3-pentafluoro-	C_3ClF_5 166.48	2804-50-4 -130	6.8	4-01-00-00741 1.5920[20]	
10542	1-Propene, 1,1-dibromo-	$C_3H_4Br_2$ 199.87	13195-80-7	125	4-01-00-00759 1.9767[20]	H_2O 2; bz 3; ctc 3; chl 3 1.5260[20]
10543	1-Propene, 1,3-dibromo-, (*E*)- Propylene, 1,3-dibromo (*trans*)	$C_3H_4Br_2$ 199.87	32121-07-6	66[25]	4-01-00-00760 1.9791[20]	1.5495[20]
10544	1-Propene, 1,3-dibromo-, (*Z*)- Propylene, 1,3-dibromo (*cis*)	$C_3H_4Br_2$ 199.87	32121-06-5	60[25]	4-01-00-00760 2.0599[20]	1.5550[20]
10545	1-Propene, 2,3-dibromo- 2,3-Dibromopropene	$C_3H_4Br_2$ 199.87	513-31-5	3009 141; 37.7[11]	4-01-00-00760 2.0345[25]	H_2O 1; eth 3; ace 3; chl 3 1.5416[25]
10546	1-Propene, 1,3-dibromo-2-methyl- Isobutylene, 1,3-dibromo	$C_4H_6Br_2$ 213.90	35911-17-2	102[100]	3-01-00-00769 1.8939[4]	1.5396[15]
10547	1-Propene, 1,1-dichloro-	$C_3H_4Cl_2$ 110.97	563-58-6	76.5	4-01-00-00742 1.1864[25]	H_2O 1; eth 3; ace 3; chl 3 1.4430[25]
10548	1-Propene, 1,2-dichloro-, (*E*)-	$C_3H_4Cl_2$ 110.97	7069-38-7	77	4-01-00-00742 1.1818[20]	H_2O 1; EtOH 4; ctc 4; MeOH 4 1.4471[20]
10549	1-Propene, 1,2-dichloro-, (*Z*)-	$C_3H_4Cl_2$ 110.97	6923-20-2	93	4-01-00-00742	H_2O 1; ace 3; bz 3; chl 3 1.4549[20]
10550	1-Propene, 1,3-dichloro-, (*E*)- *trans*-1,3-Dichloropropylene	$C_3H_4Cl_2$ 110.97	10061-02-6	3064 112	4-01-00-00744 1.217[20]	H_2O 1; eth 3; bz 3; chl 3 1.4730[20]
10551	1-Propene, 1,3-dichloro-, (*Z*)- *cis*-1,3-Dichloropropylene	$C_3H_4Cl_2$ 110.97	10061-01-5	3064 104.3	4-01-00-00743 1.224[20]	H_2O 1; eth 3; bz 3; chl 3 1.4682[20]
10552	1-Propene, 2,3-dichloro- 2,3-Dichloropropene	$C_3H_4Cl_2$ 110.97	78-88-6 10	94	4-01-00-00744 1.211[20]	H_2O 1; EtOH 5; eth 3; bz 3 1.4603[20]
10553	1-Propene, 3,3-dichloro-	$C_3H_4Cl_2$ 110.97	563-57-5	84.4	4-01-00-00745 1.175[20]	bz 4; eth 4; EtOH 4 1.4510[20]
10554	1-Propene, 1,1-dichloro-2-fluoro- Propylene, 1,1-dichloro-2-fluoro	$C_3H_3Cl_2F$ 128.96	430-95-5	78	4-01-00-00745 1.3026[25]	1.4196[25]
10555	1-Propene, 1,1-dichloro-2-methyl-	$C_4H_6Cl_2$ 125.00	6065-93-6 -43	108.5	4-01-00-00804 1.1447[20]	bz 4; eth 4; chl 4 1.4580[20]
10556	1-Propene, 3,3-dichloro-2-methyl-	$C_4H_6Cl_2$ 125.00	22227-75-4	108	4-01-00-00805 1.136[24]	bz 4; eth 4; chl 4 1.4523[24]
10557	1-Propene, 1,2-dichloro-1,3,3,3-tetrafluoro-	$C_3Cl_2F_4$ 182.93	431-53-8 -137	47.3	4-01-00-00747 1.5468[25]	1.3511[20]
10558	Propene, 1,2-dichloro-3,3,3-trifluoro- Propylene, 1,2-dichloro-3,3,3-trifluoro	$C_3HCl_2F_3$ 164.94	431-27-6 -109.2	53.7	4-01-00-00746 1.4653[20]	1.3670[20]
10559	1-Propene, 1,1-diethoxy-	$C_7H_{14}O_2$ 130.19	21504-43-8	133.5	4-01-00-03434 0.8628[25]	1.4083[25]
10560	1-Propene, 3,3-diethoxy- Acrolein, diethyl acetal	$C_7H_{14}O_2$ 130.19	3054-95-3	123.5	4-01-00-03437 0.8543[15]	H_2O 2; EtOH 5; eth 5 1.4000[20]
10561	1-Propene, 1,1-difluoro- 1,1-Difluoro-1-propene	$C_3H_4F_2$ 78.06	430-63-7		4-01-00-00741	
10562	1-Propene, 3,3-dimethoxy-	$C_5H_{10}O_2$ 102.13	6044-68-4	88	4-01-00-03437 0.862[25]	1.3954[20]
10563	2-Propene-1,1-diol, diacetate	$C_7H_{10}O_4$ 158.15	869-29-4 -37.6	180	4-02-00-00291 1.0760[20]	ace 4; bz 4; eth 4; EtOH 4 1.4193[20]
10564	2-Propene-1,1-diol, 2-methyl-, diacetate Methacrolein diacetate	$C_8H_{12}O_4$ 172.18	10476-95-6	191	4-02-00-00292	1.4241[20]
10565	1-Propene, 3-(ethenyloxy)-	C_5H_8O 84.12	3917-15-5	66	4-01-00-02085 0.8050[20]	H_2O 1; eth 3; ace 3; chl 3 1.4062[20]

No.	Name / Synonym	Mol. Form. / Mol. Wt.	CAS RN / mp/°C	Merck No. / bp/°C	Beil. Ref. / den/g cm⁻³	Solubility / n_D
10566	1-Propene, 3-ethoxy- / Allyl ethyl ether	$C_5H_{10}O$ / 86.13	557-31-3	291 / 67.6	4-01-00-02083 / 0.7651^{20}	H₂O 1; EtOH 5; eth 5; ace 3 / 1.3881^{20}
10567	1-Propene, 1-fluoro- (Z) / cis-1-Fluoropropene	C_3H_5F / 60.07	19184-10-2			
10568	1-Propene, 1-fluoro- (E) / trans-1-Fluoropropene	C_3H_5F / 60.07	20327-65-5			
10569	1-Propene, 2-fluoro- / 2-Fluoropropene	C_3H_5F / 60.07	1184-60-7	/ -24	4-01-00-00733	
10570	1-Propene, 3-fluoro- / 3-Fluoropropene	C_3H_5F / 60.07	818-92-8	/ -3	4-01-00-00733	H₂O 2; EtOH 4; eth 4; chl 3
10571	1-Propene, 1,1,2,3,3,3-hexachloro- / Hexachloropropylene	C_3Cl_6 / 248.75	1888-71-7 / -72.9	209.5	4-01-00-00753 / 1.7632^{20}	H₂O 1; ctc 3; chl 3 / 1.5091^{20}
10572	1-Propene, 1,1,2,3,3,3-hexafluoro- / Perfluoropropene	C_3F_6 / 150.02	116-15-4 / -156.5	-29.6	4-01-00-00735	/ 1.583^{-40}
10573	1-Propene, 3-iodo- / Allyl iodide	C_3H_5I / 167.98	556-56-9 / -99.3	292 / 103	4-01-00-00761 / 1.848^{12}	H₂O 1; EtOH 3; eth 3; chl 3 / 1.5540^{21}
10574	1-Propene, 3-isocyano- / Allyl isocyanide	C_4H_5N / 67.09	2835-21-4	/ 98	0-04-00-00208 / 0.794^{17}	eth 4; EtOH 4
10575	1-Propene, 3-isothiocyanato- / Allyl isothiocyanate	C_4H_5NS / 99.16	57-06-7 / -80	293 / 152	4-04-00-01081 / 1.0126^{20}	bz 4; eth 4; EtOH 4 / 1.5306^{20}
10576	1-Propene, 3-methoxy-	C_4H_8O / 72.11	627-40-7	/ 44	4-01-00-02083 / 0.77^{11}	H₂O 1; EtOH 5; eth 5; ace 3 / 1.3778^{20}
10577	1-Propene, 3-methoxy-2-methyl-	$C_5H_{10}O$ / 86.13	22418-49-1	/ 68	3-01-00-01904 / 0.7698^{20}	/ 1.3964^{20}
10578	1-Propene, 2-methyl- / Isobutene	C_4H_8 / 56.11	115-11-7 / -140.4	5024 / -6.9	4-01-00-00796 / $*0.589^{25}$	H₂O 1; EtOH 4; eth 4; bz 3 / 1.3926^{-25}
10579	1-Propene, 3-(1-methylethoxy)-	$C_6H_{12}O$ / 100.16	6140-80-3	/ 83.5	4-01-00-02083 / 0.7764^{20}	ace 4; eth 4; EtOH 4 / 1.3946^{20}
10580	1-Propene, 2-methyl-, tetramer / Propene, 2-methyl(tetramer)	$C_{16}H_{32}$ / 224.43	15220-85-6 / -98	244; 109^{15}	3-01-00-00763 / 0.7944^{20}	/ 1.4482^{20}
10581	1-Propene, 3-(methylthio)-	C_4H_8S / 88.17	10152-76-8	/ 92	4-01-00-02095 / 0.8767^{20}	/ 1.4714^{20}
10582	1-Propene, 2-methyl-, trimer / Isobutylene (trimer)	$C_{12}H_{24}$ / 168.32	7756-94-7 / -76	180	3-01-00-00763 / 0.7590^{20}	/ 1.4314^{20}
10583	2-Propenenitrile / Acrylonitrile	C_3H_3N / 53.06	107-13-1 / -83.5	125 / 77.3	4-02-00-01473 / 0.8060^{20}	ace 4; bz 4; eth 4; EtOH 4 / 1.3911^{20}
10584	2-Propenenitrile, 2-chloro-	C_3H_2ClN / 87.51	920-37-6 / -65	88.5	4-02-00-01484 / 1.096^{25}	/ 1.4290^{20}
10585	2-Propenenitrile, 3-(2-furanyl)- / 2-Furanacrylonitrile	C_7H_5NO / 119.12	7187-01-1 / 38	4208 / 96	4-18-00-04152	tol 4 / 1.5824^{25}
10586	2-Propenenitrile, 2-methyl- / Methylacrylonitrile	C_4H_5N / 67.09	126-98-7 / -35.8	5850 / 90.3	4-02-00-01539 / 0.8001^{20}	H₂O 1; EtOH 5; eth 5; ace 5 / 1.4003^{20}
10587	2-Propenenitrile, 3-phenyl-, (E)-	C_9H_7N / 129.16	1885-38-7 / 22	263.8	4-09-00-02027 / 1.0304^{20}	H₂O 1; EtOH 3; ace 3; ctc 3 / 1.6013^{20}
10588	2-Propenenitrile, 3-phenyl-, (Z)-	C_9H_7N / 129.16	24840-05-9 / -4.4	249; 139^{30}	4-09-00-02027 / 1.0289^{20}	H₂O 1; EtOH 3; bz 4 / 1.5843^{20}
10589	1-Propene, 1-nitro-	$C_3H_5NO_2$ / 87.08	3156-70-5	60^{34}; 37^{10}	3-01-00-00715 / 1.0661^{20}	eth 3; ace 3; chl 3 / 1.4527^{20}
10590	1-Propene, 2-nitro-	$C_3H_5NO_2$ / 87.08	4749-28-4	57^{100}; 32^{30}	4-01-00-00764 / 1.0643^{20}	eth 3; ace 3; chl 3 / 1.4358^{20}
10591	1-Propene, 3-nitro-	$C_3H_5NO_2$ / 87.08	625-46-7	/ 127.5	4-01-00-00764 / 1.051^{22}	eth 4; EtOH 4
10592	1-Propene, 3,3'-oxybis- / Allyl ether	$C_6H_{10}O$ / 98.14	557-40-4 / -6	290 / 94	4-01-00-02086 / 0.8260^{20}	H₂O 1; EtOH 5; eth 5; ace 4 / 1.4163^{20}
10593	1-Propene, 1,1,2,3,3-pentachloro-	C_3HCl_5 / 214.30	1600-37-9	185	4-01-00-00753 / 1.6317^{34}	eth 4 / 1.5313^{20}
10594	1-Propene, 3-propoxy-	$C_6H_{12}O$ / 100.16	1471-03-0	/ 91	4-01-00-02083 / 0.7764^{20}	ace 4; eth 4; EtOH 4 / 1.3919^{20}
10595	2-Propene-1-sulfinothioic acid, S-2-propenyl ester / Allicin	$C_6H_{10}OS_2$ / 162.28	539-86-6 / dec	249	4-04-00-00007 / 1.112^{20}	H₂O 4 / 1.561^{20}
10596	1-Propene, 3,3'-sulfinylbis-	$C_6H_{10}OS$ / 130.21	14180-63-3 / 23.5	114^{12}; $89^{2.6}$	4-01-00-02097 / 1.0261^{20}	EtOH 3; eth 3; ace 3 / 1.5115^{20}
10597	1-Propene, 3,3'-sulfonylbis-	$C_6H_{10}O_2S$ / 146.21	16841-48-8	109^3	4-01-00-02098 / 1.1213^{20}	/ 1.4893^{20}
10598	1-Propene, 1,2,3,3-tetrachloro-	$C_3H_2Cl_4$ / 179.86	20589-85-9	162	3-01-00-00708 / 1.537^{19}	bz 4; chl 4 / 1.5121^{19}
10599	1-Propene, 3,3'-thiobis- / Diallyl sulfide	$C_6H_{10}S$ / 114.21	592-88-1 / -85	295 / 138.6	4-01-00-02097 / 0.8877^{27}	eth 4; EtOH 4 / 1.4870^{25}
10600	2-Propene-1-thiol	C_3H_6S / 74.15	870-23-5	/ 65	4-01-00-02095 / 0.925^{23}	H₂O 1; EtOH 5; eth 5; chl 3 / 1.4832^{20}
10601	1-Propene-1,2,3-tricarboxylic acid, triethyl ester, (E)- / Aconitic acid, triethyl ester (trans)	$C_{12}H_{18}O_6$ / 258.27	68077-28-1	275 dec; 159^9	4-02-00-02405 / 1.1064^{20}	eth 4; EtOH 4
10602	1-Propene, 1,1,2-trichloro-	$C_3H_3Cl_3$ / 145.41	21400-25-9	118	4-01-00-00747 / 1.382^{20}	H₂O 1; EtOH 3; eth 3; bz 3 / 1.4827^{20}
10603	1-Propene, 1,2,3-trichloro-	$C_3H_3Cl_3$ / 145.41	96-19-5	142	4-01-00-00748 / 1.412^{20}	H₂O 1; EtOH 4; eth 4; bz 3 / 1.5030^{20}
10604	1-Propene, 3,3,3-trichloro-	$C_3H_3Cl_3$ / 145.41	2233-00-3 / -30	114.5	4-01-00-00749 / 1.367^{20}	H₂O 1; EtOH 3; eth 3; bz 3 / 1.4827^{20}

No.	Name / Synonym	Mol. Form. / Mol. Wt.	CAS RN / mp/°C	Merck No. / bp/°C	Beil. Ref. / den/g cm⁻³	Solubility / n_D
10605	1-Propene, 1,1,3-trichloro-2-methyl-	$C_4H_5Cl_3$ 159.44	31702-33-7	156	4-01-00-00805 1.346[20]	ace 4; bz 4; chl 4 1.4990[20]
10606	1-Propene, 3,3,3-trichloro-2-methyl-	$C_4H_5Cl_3$ 159.44	4749-27-3	133	4-01-00-00805 1.293[20]	H_2O 1; ace 3; bz 3; chl 4 1.4770[20]
10607	1-Propene, 1,1,2-trichloro-3,3,3-trifluoro-	$C_3Cl_3F_3$ 199.39	431-52-7 -114.6	88.1	4-01-00-00751 1.617[20]	
10608	1-Propene, 3,3,3-trifluoro- 3,3,3-Trifluoropropene	$C_3H_3F_3$ 96.05	677-21-4	-17	4-01-00-00734 0	
10609	2-Propenoic acid Acrylic acid	$C_3H_4O_2$ 72.06	79-10-7 12.3	124 141	4-02-00-01455 1.0511[20]	H_2O 5; EtOH 5; eth 5; ace 3 1.4224[20]
10610	2-Propenoic acid, 2-bromo-	$C_3H_3BrO_2$ 150.96	10443-65-9 71.5	dec	4-02-00-01486	H_2O 4; EtOH 4
10611	2-Propenoic acid, 2-bromoethyl ester	$C_5H_7BrO_2$ 179.01	4823-47-6	156.5; 53[5]	4-02-00-01462 1.4581[25]	1.4660[25]
10612	2-Propenoic acid, 3-bromo-2-methyl- Methacrylic acid, 3-bromo	$C_4H_5BrO_2$ 164.99	89123-63-7 64	229	0-02-00-00424	H_2O 4
10613	2-Propenoic acid, 2-bromo-3-phenyl-, ethyl ester	$C_{11}H_{11}BrO_2$ 255.11	57296-00-1	295	4-09-00-02042 1.3885[25]	1.5845[25]
10614	2-Propenoic acid, 3-bromo-3-phenyl-, ethyl ester, (Z)- Cinnamic acid, β-bromo-, ethyl ester	$C_{11}H_{11}BrO_2$ 255.11	1504-70-7	167[12]	3-09-00-02733 1.4044[10]	1.5841[10]
10615	2-Propenoic acid, 2-bromo-3-phenyl-, methyl ester Methyl α-bromocinnamate	$C_{10}H_9BrO_2$ 241.08	24127-62-6 23	120-1[0.6]	4-09-00-02042 1.4975[20]	
10616	2-Propenoic acid, butyl ester Butyl acrylate	$C_7H_{12}O_2$ 128.17	141-32-2 -64.6	1539 145	4-02-00-01463 0.8898[20]	H_2O 1; EtOH 3; eth 3; ace 3 1.4185[20]
10617	2-Propenoic acid, 3-chloro-, (E)-	$C_3H_3ClO_2$ 106.51	2345-61-1 86	94[18]	4-02-00-01481	EtOH 3; eth 3; chl 2
10618	2-Propenoic acid, 3-chloro-, (Z)-	$C_3H_3ClO_2$ 106.51	1609-93-4 63.5	106[17]	4-02-00-01481	EtOH 3; eth 3
10619	2-Propenoic acid, chloro-, ethyl ester Acrylic acid, chloro-, ethyl ester	$C_5H_7ClO_2$ 134.56	30600-19-2	51-3[18]	3-02-00-01245 1.1404[20]	EtOH 4; eth 4 1.4384[20]
10620	2-Propenoic acid, 2-chloroethyl ester	$C_5H_7ClO_2$ 134.56	2206-89-5	51-3[18]	3-02-00-01225 1.1604[20]	eth 4; EtOH 4 1.4384[20]
10621	2-Propenoic acid, 2-chloro-, methyl ester Methyl 2-chloroacrylate	$C_4H_5ClO_2$ 120.54	80-63-7	52[51]	4-02-00-01482 1.189[20]	eth 4 1.4420[20]
10622	2-Propenoic acid, 2-chloro-3-phenyl-, methyl ester, (E)- Cinnamic acid, α-chloro-, methyl ester, (E)-	$C_{10}H_9ClO_2$ 196.63	14737-94-1 33	108-9[0.5]	4-09-00-02035	ace 4
10623	2-Propenoic acid, 3-chloro-3-phenyl-, methyl ester Cinnamic acid, β-chloro-, methyl ester	$C_{10}H_9ClO_2$ 196.63	87541-87-5 29	113-4[0.5]	1.2248[21]	ace 4 1.5781[21]
10624	2-Propenoic acid, 2-cyano-, methyl ester Mecrylate	$C_5H_5NO_2$ 111.10	137-05-3	5667 47-8[2]	4-02-00-02226 1.1012[20]	1.4430
10625	2-Propenoic acid, 2-cyano-3-phenyl-, ethyl ester	$C_{12}H_{11}NO_2$ 201.22	2025-40-3 51	188[15]	2-09-00-00640 1.1076[25]	ace 4; chl 4 1.5033
10626	2-Propenoic acid, cyclohexyl ester	$C_9H_{14}O_2$ 154.21	3066-71-5	183; 88[20]	4-06-00-00038 1.0275[20]	H_2O 1; EtOH 5; eth 5; chl 3 1.4673[20]
10627	2-Propenoic acid, 2,3-dibromo-, (Z)-	$C_3H_2Br_2O_2$ 229.86	24557-10-6 85.5	148[6]	3-02-00-01249	H_2O 2; EtOH 3; eth 3; bz 2
10628	2-Propenoic acid, 3,3-dibromo- Acrylic acid, 3,3-dibromo-	$C_3H_2Br_2O_2$ 229.86	1578-21-8 86.3	245	4-02-00-01487	eth 4; EtOH 4
10629	2-Propenoic acid, 2,3-dibromo-3-phenyl-, (Z)-	$C_9H_6Br_2O_2$ 305.95	708-82-7 100	124[0.5]	4-09-00-02043	H_2O 1; EtOH 3; eth 3; chl 3
10630	2-Propenoic acid, 3,3-dichloro- Acrylic acid, 3,3-dichloro-	$C_3H_2Cl_2O_2$ 140.95	1561-20-2 77.5	sub	4-02-00-01484	eth 4; chl 4
10631	2-Propenoic acid, 2-(diethylamino)ethyl ester	$C_9H_{17}NO_2$ 171.24	2426-54-2 <-60	81[10]	4-04-00-01477 0.937[20]	1.4376[25]
10632	2-Propenoic acid, 3-(3,4-dihydroxyphenyl)- Caffeic acid	$C_9H_8O_4$ 180.16	331-39-5 225 dec	1634	2-10-00-00294	EtOH 4
10633	2-Propenoic acid, 2-(dimethylamino)ethyl ester	$C_7H_{13}NO_2$ 143.19	2439-35-2 <-60	95[50]	4-04-00-01431 0.938[20]	
10634	2-Propenoic acid, 2-ethoxyethyl ester	$C_7H_{12}O_3$ 144.17	106-74-1 -47	174	4-02-00-01469 0.983[20]	1.4274[20]
10635	2-Propenoic acid, ethyl ester Ethyl acrylate	$C_5H_8O_2$ 100.12	140-88-5 -71.2	3715 99.4	4-02-00-01460 0.9234[20]	H_2O 2; EtOH 5; eth 5; chl 3 1.4068[20]
10636	2-Propenoic acid, 2-ethylhexyl ester 2-Ethylhexyl acrylate	$C_{11}H_{20}O_2$ 184.28	103-11-7 -90	123-7[60]	4-02-00-01467 0.880[25]	1.4332[25]
10637	2-Propenoic acid, 2-fluoro-3-phenyl- Cinnamic acid, α-fluoro-	$C_9H_7FO_2$ 166.15	350-90-3 157.6	290	1-09-00-00237	EtOH 3; eth 3
10638	2-Propenoic acid, 3-(2-furanyl)- 2-Furanacrylic acid	$C_7H_6O_3$ 138.12	539-47-9 141	4207 286	5-18-06-00306	eth 4; EtOH 4
10639	2-Propenoic acid, 3-(2-furanyl)-, ethyl ester	$C_9H_{10}O_3$ 166.18	623-20-1 24.5	232	5-18-06-00307 1.0891[25]	H_2O 1; EtOH 5; eth 5; chl 2 1.5286[20]
10640	2-Propenoic acid, 3-(2-furanyl)-, methyl ester	$C_8H_8O_3$ 152.15	623-18-7 36	227; 117[23]	4-18-00-04144 1.4447[20]	bz 4; eth 4; EtOH 4

No.	Name / Synonym	Mol. Form. / Mol. Wt.	CAS RN / mp/°C	Merck No. / bp/°C	Beil. Ref. / den/g cm^{-3}	Solubility / n_D
10641	2-Propenoic acid, 3-(2-furanyl)-, pentyl ester Acrylic acid, 3-(2-furyl), pentyl ester	$C_{12}H_{16}O_3$ 208.26	2438-19-9	116-8[2]	4-18-00-04146 1.0322[20]	eth 4; EtOH 4 1.5289[14]
10642	2-Propenoic acid, 3-(2-furanyl)-, propyl ester Acrylic acid, 3-(2-furyl), propyl ester	$C_{10}H_{12}O_3$ 180.20	623-22-3	91-4[3]	4-18-00-04146 1.0744[20]	bz 4; eth 4; EtOH 4 1.5392[24]
10643	2-Propenoic acid, hexyl ester	$C_9H_{16}O_2$ 156.22	2499-95-8 -45	40[1]	4-02-00-01466 0.878[20]	
10644	2-Propenoic acid, 3-(4-hydroxyphenyl)- p-Coumaric acid	$C_9H_8O_3$ 164.16	7400-08-0 211.5	2561	4-10-00-01005	eth 4; EtOH 4
10645	2-Propenoic acid, 2-isocyano-3-phenyl-, ethyl ester	$C_{12}H_{11}NO_2$	52744-85-1			H_2O 1; EtOH 2; eth 3; bz 4
	Malonononitrile, benzylidene,monoethyl ester	201.22	51	188[15.] 118[0.4]	1.0762[25]	1.5033
10646	2-Propenoic acid, isodecyl ester	$C_{13}H_{24}O_2$ 212.33	1330-61-6 -100	158[50]	0.885[20]	1.4416[20]
10647	2-Propenoic acid, 3-(2-methoxyphenyl)-, methyl ester	$C_{11}H_{12}O_3$ 192.21	15854-58-7	304; 161-3[3]	3-10-00-00835 1.1366[17]	1.5854[16]
10648	2-Propenoic acid, 2-methyl- Methacrylic acid	$C_4H_6O_2$ 86.09	79-41-4 16	5849 162.5	4-02-00-01518 1.0153[20]	H_2O 3; EtOH 5; eth 5; chl 3 1.4314[20]
10649	2-Propenoic acid, 2-methyl-, anhydride Methacrylic acid anhydride	$C_8H_{10}O_3$ 154.17	760-93-0	89[5]	4-02-00-01537	EtOH 5; eth 5 1.4540[20]
10650	2-Propenoic acid, 2-methyl-, butyl ester Butyl methacrylate	$C_8H_{14}O_2$ 142.20	97-88-1	160	4-02-00-01525 0.8936[20]	eth 4; EtOH 4 1.4240[20]
10651	2-Propenoic acid, methylbutyl ester Propenoic acid, 2-methyl,butyl ester	$C_8H_{14}O_2$ 142.20	90638-48-5	160; 71[37]	3-02-00-01286 0.8936[20]	H_2O 1; EtOH 5; eth 5; ctc 2 1.4240[20]
10652	2-Propenoic acid, 2-methylbutyl ester Acrylic acid, (2-methylbutyl) ester	$C_8H_{14}O_2$ 142.20	44914-03-6	160; 45[10]	3-02-00-01228 0.8936[20]	eth 4; EtOH 4 1.4240[20]
10653	2-Propenoic acid, 2-methyl-, cyclohexyl ester	$C_{10}H_{16}O_2$ 168.24	101-43-9	210	4-06-00-00039 0.9626[20]	1.4578[20]
10654	2-Propenoic acid, 2-methyl-, 2-(dimethylamino)ethyl ester β-Dimethylaminoethyl methacrylate	$C_8H_{15}NO_2$ 157.21	2867-47-2	62-5[6]	4-04-00-01432	
10655	2-Propenoic acid, methyl ester Methyl acrylate	$C_4H_6O_2$ 86.09	96-33-3 <-75	5935 80.7	4-02-00-01457 0.9535[20]	H_2O 2; EtOH 3; eth 3; ace 3 1.4040[20]
10056	2-Propenoic acid, 2-methyl-, 1,2-ethanediyl ester	$C_{10}H_{14}O_4$ 198.22	97-90-5 -40	260	4-02-00-01532 1.053[20]	bz 4; EtOH 4; lig 4 1.4532[25]
10657	2-Propenoic acid, 2-methyl-, 1,2-ethanediylbis(oxy-2,1-ethanediyl) ester	$C_{14}H_{22}O_6$ 286.33	109-16-0		4-02-00-01531 1.072[25]	ace 4; eth 4; EtOH 4; peth 4 1.4595[25]
10658	2-Propenoic acid, 2-methyl-, ethyl ester Ethyl methacrylate	$C_6H_{10}O_2$ 114.14	97-63-2	117	4-02-00-01523 0.9135[20]	H_2O 2; EtOH 5; cth 5; chl 2 1.4147[20]
10659	2-Propenoic acid, 2-methyl-, 2-ethylhexyl ester 2-Ethylhexyl methacrylate	$C_{12}H_{22}O_2$ 198.31	688-84-6	110[14]	4-02-00-01528 0.880[25]	1.436[25]
10660	2-Propenoic acid, 2-methyl-, hexyl ester	$C_{10}H_{18}O_2$ 170.25	142-09-6	162; 86[17]	4-02-00-01527 0.880[25]	ace 4; bz 4; eth 4; EtOH 4 1.429[25]
10661	2-Propenoic acid, 2-methyl-, 2-hydroxyethyl ester	$C_6H_{10}O_3$ 130.14	868-77-9	67[3.5]	4-02-00-01530 1.034[25]	1.4515[20]
10662	2-Propenoic acid, 2-methyl-, 2-hydroxypropyl ester	$C_7H_{12}O_3$ 144.17	923-26-2	90[9]	4-02-00-01532 1.066[25]	1.4458[20]
10663	2-Propenoic acid, 2-methyl-, methyl ester Methyl methacrylate	$C_5H_8O_2$ 100.12	80-62-6 -48	5849 100.5	4-02-00-01519 0.9440[20]	H_2O 2; EtOH 5; eth 5; ace 5 1.4142[20]
10664	2-Propenoic acid, 2-methyl-, 1-methylethyl ester	$C_7H_{12}O_2$ 128.17	4655-34-9	125	4-02-00-01525 0.8847[20]	ace 4; bz 4; eth 4; EtOH 4 1.4122[20]
10665	2-Propenoic acid, 2-methyl-, 1-methylpropyl ester	$C_8H_{14}O_2$ 142.20	2998-18-7	147; 35[4]	4-02-00-01526 0.886[20]	1.4179[20]
10666	2-Propenoic acid, 2-methyl-, 2-methylpropyl ester Isobutyl methacrylate	$C_8H_{14}O_2$ 142.20	97-86-9	155	4-02-00-01526 0.8858[20]	H_2O 1; EtOH 5; eth 5 1.4199[20]
10667	2-Propenoic acid, 2-methyl-, octadecyl ester	$C_{22}H_{42}O_2$ 338.57	32360-05-7	195[6]	4-02-00-01529 0.880[25]	1.429[25]
10668	2-Propenoic acid, 2-methyl-, oxiranylmethyl ester	$C_7H_{10}O_3$ 142.15	106-91-2	75[10]	5-17-03-00035 1.07[25]	bz 4; eth 4; EtOH 4 1.448[25]
10669	2-Propenoic acid, 2-methyl-, oxydi-2,1-ethanediyl ester	$C_{12}H_{18}O_5$ 242.27	2358-84-1	>200	4-02-00-01531 1.056[20]	1.4571[25]
10670	2-Propenoic acid, 2-methyl-3-phenyl-, (Z)-	$C_{10}H_{10}O_2$ 162.19	15250-29-0 94	288	4-09-00-02063	H_2O 2; EtOH 4; eth 4; bz 4
10671	2-Propenoic acid, 2-methyl-, 2-propenyl ester	$C_7H_{10}O_2$ 126.16	96-05-9	55[30]; 60[43]	4-02-00-01529 0.927[20]	1.4360[20]
10672	2-Propenoic acid, 2-methyl-, propyl ester Propyl methacrylate	$C_7H_{12}O_2$ 128.17	2210-28-8	141	4-02-00-01524 0.9022[20]	H_2O 1; EtOH 5; eth 5 1.4190[20]
10673	2-Propenoic acid, 2-methylpropyl ester Isobutyl acrylate	$C_7H_{12}O_2$ 128.17	106-63-8 -61	132	4-02-00-01465 0.8896[20]	H_2O 2; EtOH 3; eth 3; MeOH 3 1.4150[20]

No.	Name Synonym	Mol. Form. Mol. Wt.	CAS RN mp/°C	Merck No. bp/°C	Beil. Ref. den/g cm^{-3}	Solubility n_D
10674	2-Propenoic acid, 2-methyl-, (tetrahydro-2-furanyl)methyl ester	$C_9H_{14}O_3$ 170.21	2455-24-5 	 265; 81[4]	4-17-00-01105 1.040[25]	 1.4554[25]
10675	2-Propenoic acid, 3-(1-naphthalenyl)-, (E)- Acrylic acid, 3-(1-naphtyl) (trans)	$C_{13}H_{10}O_2$ 198.22	2006-14-6 211.5	 sub	4-09-00-02486 	H$_2$O 2; EtOH 3; eth 3; chl 3
10676	2-Propenoic acid, 3-(2-nitrophenyl)-, (E)-	$C_9H_7NO_4$ 193.16	1013-96-3 244.3	 sub	4-09-00-02043 	H$_2$O 1; EtOH 3; bz 1
10677	2-Propenoic acid, 3-(2-nitrophenyl)-, methyl ester, (E)- Cinnamic acid, 2-nitro, methyl ester (trans)	$C_{10}H_9NO_4$ 207.19	39228-29-0 73	 187[15]	4-09-00-02044 	EtOH 4
10678	2-Propenoic acid, 3-(3-nitrophenyl)-, methyl ester, (E)- Cinnamic acid, 3-nitro, methyl ester (trans)	$C_{10}H_9NO_4$ 207.19	659-04-1 125.3	 dec	3-09-00-02742 	bz 4; eth 4; chl 4
10679	2-Propenoic acid, 3-(4-nitrophenyl)-, methyl ester, (E)-	$C_{10}H_9NO_4$ 207.19	637-57-0 162	 283.5	4-09-00-02046 	EtOH 4
10680	2-Propenoic acid, oxiranylmethyl ester	$C_6H_8O_3$ 128.13	106-90-1 	 53[10]	4-17-00-01005 1.1[20]	bz 4 1.4490[20]
10681	2-Propenoic acid, 2-phenoxyethyl ester	$C_{11}H_{12}O_3$ 192.21	48145-04-6 	 110[2]	3-06-00-00572 1.090[25]	ace 4; eth 4; chl 4
10682	2-Propenoic acid, 3-phenyl-, (E)- trans-Cinnamic acid	$C_9H_8O_2$ 148.16	140-10-3 133	2300 300	4-09-00-02002 1.2475[4]	H$_2$O 1; EtOH 4; eth 3; ace 3
10683	2-Propenoic acid, 3-phenyl-, (Z)-	$C_9H_8O_2$ 148.16	102-94-3 42		4-09-00-02001 	EtOH 4; HOAc 4; lig 4
10684	2-Propenoic acid, 3-phenyl-, anhydride Cinnamic anhydride	$C_{18}H_{14}O_3$ 278.31	538-56-7 136	2300	 	bz 4
10685	2-Propenoic acid, 3-phenyl-, ethyl ester, (E)-	$C_{11}H_{12}O_2$ 176.22	4192-77-2 10	2300 271.5	4-09-00-02006 1.0491[20]	H$_2$O 1; EtOH 4; eth 4; ace 4 1.5598[20]
10686	2-Propenoic acid, 3-phenyl-, methyl ester, (E)-	$C_{10}H_{10}O_2$ 162.19	1754-62-7 36.5	2300 261.9	4-09-00-02005 1.042[36]	H$_2$O 1; EtOH 4; eth 4; ace 3 1.5766[22]
10687	2-Propenoic acid, 3-phenyl-, 1-methylethyl ester, (E)- Cinnamic acid, isopropyl ester (trans)	$C_{12}H_{14}O_2$ 190.24	60512-85-8 	 269	4-09-00-02008 1.0320[20]	H$_2$O 1; EtOH 3; eth 3; ace 3 1.5455[20]
10688	2-Propenoic acid, phenylmethyl ester	$C_{10}H_{10}O_2$ 162.19	2495-35-4 	 228	3-06-00-01481 1.0573[20]	H$_2$O 1; EtOH 3; eth 3; ace 3 1.5143[20]
10689	2-Propenoic acid, 3-phenyl-, phenyl ester, (E)- Cinnamic acid, phenyl ester (trans)	$C_{15}H_{12}O_2$ 224.26	25695-77-6 74.3	 205[15]	4-09-00-02011 1.0650[100]	bz 4; eth 4
10690	2-Propenoic acid, 3-phenyl-, phenylmethyl ester, (E)- Cinnamic acid, benzyl ester (trans)	$C_{16}H_{14}O_2$ 238.29	78277-23-3 39	 350 dec; 244[5]	4-09-00-02012 1.109[15]	H$_2$O 1; EtOH 3; eth 3; bz 2
10691	2-Propenoic acid, 3-phenyl-, 3-phenyl-2-propenyl ester Cinnamyl cinnamate	$C_{18}H_{16}O_2$ 264.32	122-69-0 44	2307	4-09-00-02013 1.1565[4]	H$_2$O 1; EtOH 3; eth 4; chl 3
10692	2-Propenoic acid, 3-phenyl-, 2-propenyl ester Cinnamic acid, allyl ester	$C_{12}H_{12}O_2$ 188.23	1866-31-5 	 268 dec; 163[17]	4-09-00-02010 1.048[23]	H$_2$O 1; EtOH 4; eth 5; ctc 2 1.530[20]
10693	2-Propenoic acid, 3-phenyl-, propyl ester, (E)- Cinnamic acid, propyl ester (trans)	$C_{12}H_{14}O_2$ 190.24	74513-58-9 	 285	4-09-00-02007 1.0433[0]	H$_2$O 1
10694	2-Propenoic acid, 2-propenyl ester	$C_6H_8O_2$ 112.13	999-55-3 	 121	4-02-00-01468 0.9441[20]	H$_2$O 2; EtOH 3; eth 3; acid 3 1.4320[20]
10695	2-Propenoic acid, (tetrahydro-2-furanyl)methyl ester	$C_8H_{12}O_3$ 156.18	2399-48-6 <-60	 96[6]	4-17-00-01104 1.061[20]	
10696	2-Propenoic acid, 2,3,3-trichloro-	$C_3HCl_3O_2$ 175.40	2257-35-4 76	 222	4-02-00-01486 	H$_2$O 3; EtOH 3; eth 3; ctc 3 1.5271[18]
10697	2-Propen-1-ol Allyl alcohol	C_3H_6O 58.08	107-18-6 -129	284 97.0	4-01-00-02079 0.8540[20]	H$_2$O 5; EtOH 5; eth 5; chl 3 1.4135[20]
10698	1-Propen-2-ol, acetate Isopropenyl acetate	$C_5H_8O_2$ 100.12	108-22-5 -92.9	5091 94	4-02-00-00179 0.9090[20]	H$_2$O 2; EtOH 3; eth 4; ace 3 1.4033[20]
10699	2-Propen-1-ol, 2-bromo-	C_3H_5BrO 136.98	598-19-6 	 154; 62[11]	4-01-00-02094 1.621[18]	eth 4; chl 4 1.500[18]
10700	2-Propen-1-ol, 2-chloro-	C_3H_5ClO 92.52	5976-47-6 	 135	4-01-00-02091 1.1618[20]	EtOH 3; ctc 2 1.4588[20]
10701	2-Propen-1-ol, 3-chloro- Allyl alcohol, 3-chloro	C_3H_5ClO 92.52	29560-84-7 	 147	4-01-00-02090 1.1769[20]	 1.4738[20]
10702	2-Propen-1-ol, 2-methyl- Methallyl alcohol	C_4H_8O 72.11	513-42-8 	 114.5	4-01-00-02114 0.8515[20]	H$_2$O 4; EtOH 5; eth 5 1.4255[20]
10703	2-Propen-1-ol, 2-methyl-, carbonate (2:1) Bis-(2-methallyl)carbonate	$C_9H_{14}O_3$ 170.21	64057-79-0 201.3	 66[3]	4-03-00-00011 0.943[25]	 1.4371[20]
10704	2-Propen-1-ol, 3-phenyl-, (E)-	$C_9H_{10}O$ 134.18	4407-36-7 34	 257.5	4-06-00-03799 1.0440[20]	H$_2$O 2; EtOH 4; eth 4; chl 2 1.5819[20]
10705	2-Propen-1-ol, 3-phenyl-, (Z)-	$C_9H_{10}O$ 134.18	4510-34-3 34	 257.5	3-06-00-02401 1.0440[20]	eth 4; EtOH 4 1.5819[20]
10706	2-Propen-1-ol, 3-phenyl-, acetate, (E)-	$C_{11}H_{12}O_2$ 176.22	21040-45-9 	 145[15]; 114[1]	4-06-00-03801 1.0567[20]	H$_2$O 1; EtOH 3; eth 3; ace 3 1.5425[20]
10707	2-Propen-1-ol, 3-phenyl-, formate	$C_{10}H_{10}O_2$ 162.19	104-65-4 0	 252	0-06-00-00571 1.086[25]	
10708	2-Propen-1-one, 1,3-diphenyl- Chalcone	$C_{15}H_{12}O$ 208.26	94-41-7 59	2028 346 dec	4-07-00-01658 1.0712[62]	bz 4; EtOH 4; chl 4

No.	Name Synonym	Mol. Form. Mol. Wt.	CAS RN mp/°C	Merck No. bp/°C	Beil. Ref. den/g cm^{-3}	Solubility n_D
10709	2-Propen-1-one, 1,3-diphenyl-, (E)-	C$_{15}$H$_{12}$O 208.26	614-47-1 59	346 dec	4-07-00-01658 1.0712[62]	H$_2$O 1; EtOH 2; eth 3; bz 3
10710	2-Propen-1-one, 3-(2-furanyl)-1-phenyl-	C$_{13}$H$_{10}$O$_2$ 198.22	717-21-5 47	317	5-17-10-00371 1.1140[20]	EtOH 3; eth 3
10711	2-Propen-1-one, 3-hydroxy-1,3-diphenyl-	C$_{15}$H$_{12}$O$_2$ 224.26	1704-15-0 79.8	220[18.] 140[1.0]	4-07-00-02512	eth 4; EtOH 4; chl 4
10712	2-Propen-1-one, 3-(3-methoxyphenyl)-1-phenyl-	C$_{16}$H$_{14}$O$_2$ 238.29	5470-91-7 65	247[12]	1-08-00-00579	ace 4; bz 4; eth 4; EtOH 4
10713	2-Propen-1-one, 3-(4-methoxyphenyl)-1-phenyl-	C$_{16}$H$_{14}$O$_2$ 238.29	959-33-1 79	187[19]	2-08-00-00218	H$_2$O 1; EtOH 4; eth 3; ctc 3
10714	2-Propenoyl chloride Acrylic acid chloride	C$_3$H$_3$ClO 90.51	814-68-6	75.5	4-02-00-01471 1.1136[20]	chl 4 1.4343[20]
10715	2-Propenoyl chloride, 2-methyl- Methacrylic acid chloride	C$_4$H$_5$ClO 104.54	920-46-7 -60	96	4-02-00-01537 1.0871[20]	eth 3; ace 3; chl 3 1.4435[20]
10716	2-Propenoyl chloride, 3-phenyl- Cinnamoyl chloride	C$_9$H$_7$ClO 166.61	102-92-1 36	2303 257	2-09-00-00395 1.1617[45]	lig 4 1.614[42]
10717	2-Propenoyl chloride, 3-phenyl-, (E)-	C$_9$H$_7$ClO 166.61	17082-09-6 37.5	257.5	2-09-00-00390 1.1617[45]	H$_2$O 1; EtOH 3; ctc 3; lig 3 1.614[42]
10718	2-Propenoyl chloride, 2,3,3-trichloro-	C$_3$Cl$_4$O 193.84	815-58-7	158	4-02-00-01486	bz 4 1.5271[18]
10719	Propetamphos 2-Butenoic acid, 3-[[(ethylamino)methoxyphosphinothioyl]oxy]-, 1-methylethyl ester, (E)-	C$_{10}$H$_{20}$NO$_4$PS 281.31	31218-83-4	7827 88[0.005]	1.1294[20]	
10720	Propiconazole 1-[[2-(2,4-Dichlorophenyl)-4-propyl-1,3-dioxolan-2-yl]methyl]-1H-1,2,4-triazole	C$_{15}$H$_{17}$Cl$_2$N$_3$O$_2$ 342.22	60207-90-1	180[0.1]	1.27[20]	
10721	Propionic acid, 2,2-dichloro-3,3,3-trifluoro-, propyl ester Propanoic acid, 2,2-dichloro-3,3,3-trifluoro-, propyl ester	C$_6$H$_7$Cl$_2$F$_3$O$_2$ 239.02	357-49-3	144.5[625]	4-02-00-00758 1.3531[20]	1.3888[20]
10722	Propoxur Phenol, 2-(1-methylethoxy)-, methylcarbamate	C$_{11}$H$_{15}$NO$_3$ 209.25	114-26-1 87	7849 dec	1.12[20]	
10723	2-Propynal Propargyl aldehyde	C$_3$H$_2$O 54.05	624-67-9	60	4-01-00-03537 0.9152[20]	H$_2$O 5; EtOH 3; eth 3; ace 3 1.4033[25]
10724	2-Propynal, 3-phenyl-	C$_9$H$_6$O 130.15	2579-22-8	127[28]; 104[11]	4-07-00-01173 1.0680[12]	1.6079[12]
10725	2-Propyn-1-amine	C$_3$H$_5$N 55.08	2450-71-7	83	4-04-00-01135 0.803[25]	1.4480[20]
10726	2-Propyn-1-amine, N-methyl-	C$_4$H$_7$N 69.11	35161-71-8	83	0-04-00-00228 0.819[25]	1.4332[20]
10727	1-Propyne Methylacetylene	C$_3$H$_4$ 40.06	74-99-7 -102.7	-23.2	4-01-00-00958 *0.607[25]	H$_2$O 2; EtOH 4; bz 3; chl 3 1.3863[-40]
10728	1-Propyne, 3-bromo- Propargyl bromide	C$_3$H$_3$Br 118.96	106-96-7	89	4-01-00-00964 1.579[19]	EtOH 3; eth 3; bz 3; ctc 3 1.4922[20]
10729	1-Propyne, 3-chloro- Propargyl chloride	C$_3$H$_3$Cl 74.51	624-65-7	7820 58	4-01-00-00963 1.030[25]	H$_2$O 1; EtOH 5; eth 5; bz 3 1.4349[20]
10730	1-Propyne, 1,3-dibromo- Propyne, 1,3-dibromo-	C$_3$H$_2$Br$_2$ 197.86	627-16-7	73-4[30]	4-01-00-00965 2.1894[20]	eth 4; chl 4 1.5690[20]
10731	1-Propyne, 3,3-diethoxy-	C$_7$H$_{12}$O$_2$ 128.17	10160-87-9	139	4-01-00-03538 0.8942[22]	ace 4; eth 4; EtOH 4; chl 4 1.4140[20]
10732	1-Propyne, 1-ethoxy-	C$_5$H$_8$O 84.12	14273-06-4	84	3-01-00-01969 0.8276[20]	eth 4; EtOH 4 1.4039[20]
10733	1-Propyne, 3-ethoxy-	C$_5$H$_8$O 84.12	628-33-1	82	4-01-00-02215 0.8326[20]	eth 4; EtOH 4 1.4039[20]
10734	1-Propyne, 1-iodo-	C$_3$H$_3$I 165.96	624-66-8	110	4-01-00-00965 2.08[22]	eth 4; EtOH 4
10735	1-Propyne, 3-iodo-	C$_3$H$_3$I 165.96	659-86-9	115	4-01-00-00965 2.0177[25]	eth 3
10736	1-Propyne, 3-methoxy-	C$_4$H$_6$O 70.09	627-41-8	63	4-01-00-02215 0.83[12]	eth 4; EtOH 4 1.5035[20]
10737	2-Propynenitrile Cyanoacetylene	C$_3$HN 51.05	1070-71-9 5	42.5	4-02-00-01689 0.8167[17]	H$_2$O 2; EtOH 3 1.3868[25]
10738	2-Propynoic acid Propiolic acid	C$_3$H$_2$O$_2$ 70.05	471-25-0 9	7833 144 dec; 72[50]	4-02-00-01687 1.1380[20]	H$_2$O 4; eth 4; EtOH 4; chl 4 1.4306[20]
10739	2-Propynoic acid, ethyl ester	C$_5$H$_8$O$_2$ 98.10	623-47-2	120	4-02-00-01688 0.9560[25]	H$_2$O 1; EtOH 4; eth 4; chl 4 1.4105[20]
10740	2-Propynoic acid, 3-(2-nitrophenyl)- o-Nitrophenylpropiolic acid	C$_9$H$_5$NO$_4$ 191.14	530-85-8 157 dec	6546 exp	4-09-00-02330	eth 4; EtOH 4
10741	2-Propynoic acid, 3-phenyl-, ethyl ester	C$_{11}$H$_{10}$O$_2$ 174.20	2216-94-6	265; 128[1.6]	4-09-00-02328 1.055[25]	eth 3 1.5520[20]
10742	2-Propynoic acid, 3-phenyl-, methyl ester	C$_{10}$H$_8$O$_2$ 160.17	4891-38-7 26	158[48]; 132[16]	4-09-00-02328 1.0830[25]	1.5618[25]

No.	Name / Synonym	Mol. Form. / Mol. Wt.	CAS RN / mp/°C	Merck No. / bp/°C	Beil. Ref. / den/g cm^{-3}	Solubility / n_D
10743	2-Propyn-1-ol / Propargyl alcohol	C_3H_4O / 56.06	107-19-7 / -51.8	7819 / 113.6	4-01-00-02214 / 0.9478[20]	H_2O 3; EtOH 5; eth 5; chl 3 / 1.4322[20]
10744	2-Propyn-1-ol, acetate / Propargyl acetate	$C_5H_6O_2$ / 98.10	627-09-8 /	/ 121.5	4-02-00-00197 / 0.9982[20]	H_2O 2; EtOH 3; eth 3 / 1.4187[20]
10745	2-Propyn-1-ol, 3-phenyl-	C_9H_8O / 132.16	1504-58-1 /	/ 137-8[15]	4-06-00-04065 / 1.07[18]	eth 3; ace 3; bz 3 / 1.5873[28]
10746	Propyzamide / Benzamide, 3,5-dichloro-N-(1,1-dimethyl-2-propynyl)-	$C_{12}H_{11}Cl_2NO$ / 256.13	23950-58-5 / 155	7886 /		
10747	Pteridine / Pyrazino[2,3-d]pyrimidine	$C_6H_4N_4$ / 132.12	91-18-9 / 139.5	7947 / sub 125	4-26-00-01770	H_2O 4; EtOH 3; eth 2; bz 2
10748	2,4(1H,3H)-Pteridinedione / Lumazine	$C_6H_4N_4O_2$ / 164.12	487-21-8 / 348.5	5467 /	4-26-00-02489	HOAc 4
10749	4,6-Pteridinedione, 2-amino-1,5-dihydro- / Xanthopterin	$C_6H_5N_5O_2$ / 179.14	119-44-8 / >410 dec	9973 / 98-100[18]	4-26-00-04000 / 1.559[25]	H_2O 1; EtOH 2; eth 2; acid 4
10750	2,4,7-Pteridinetriamine, 6-phenyl- / Triamterene	$C_{12}H_{11}N_7$ / 253.27	396-01-0 /	9515 /	4-26-00-03837	
10751	1H-Purin-6-amine / Adenine	$C_5H_5N_5$ / 135.13	73-24-5 / 360 dec	141 / sub 220	4-26-00-03561	H_2O 3; EtOH 3; eth 1; chl 1
10752	1H-Purine / 7H-Imidazo[4,5-d]pyrimidine	$C_5H_4N_4$ / 120.11	120-73-0 / 216.5	7959 /	4-26-00-01736	H_2O 4; EtOH 4; eth 2; ace 3
10753	1H-Purine, 6-chloro- / 6-Chloropurine	$C_5H_3ClN_4$ / 154.56	87-42-3 / 175-7 dec	2161 /	4-26-00-01742	
10754	1H-Purine-2,6-diamine / 2,6-Diaminopurine	$C_5H_6N_6$ / 150.14	1904-98-9 / 302	2963 /	4-26-00-03785	
10755	1H-Purine-2,6-dione, 3,7-dihydro- / Xanthine	$C_5H_4N_4O_2$ / 152.11	69-89-6 / dec	9968 / sub	4-26-00-02327	
10756	1H-Purine-2,6-dione, 3,7-dihydro-1,3-dimethyl- / Theophylline	$C_7H_8N_4O_2$ / 180.17	58-55-9 / 273	9212 /	4-26-00-02331	H_2O 4; EtOH 2; eth 2; chl 2
10757	1H-Purine-2,6-dione, 3,7-dihydro-3,7-dimethyl- / Theobromine	$C_7H_8N_4O_2$ / 180.17	83-67-0 / 357	9209 / sub 290	4-26-00-02336	H_2O 2; EtOH 1; eth 1; bz 1
10758	1H-Purine-2,6-dione, 3,7-dihydro-7-(2-hydroxyethyl)-1,3-dimethyl- / Etofylline	$C_9H_{12}N_4O_3$ / 224.22	519-37-9 / 158	3835 /	4-26-00-02357	H_2O 4; EtOH 3; eth 2; bz 2
10759	1H-Purine-2,6-dione, 3,7-dihydro-8-methoxy-1,3,7-trimethyl-	$C_9H_{12}N_4O_3$ / 224.22	569-34-6 / 176		4-26-00-02727 / 1.399[25]	H_2O 3; EtOH 4; bz 3
10760	1H-Purine-2,6-dione, 3,7-dihydro-1,3,7-trimethyl- / Caffeine	$C_8H_{10}N_4O_2$ / 194.19	58-08-2 / 238	1635 / sub 90	4-26-00-02338 / 1.23[19]	H_2O 2; EtOH 2; eth 1; ace 2
10761	1H-Purine-2,6-dione, 7-(2,3-dihydroxypropyl)-3,7-dihydro-1,3-dimethyl- / Dyphylline	$C_{10}H_{14}N_4O_4$ / 254.25	479-18-5 / 161.5	3459 /	4-26-00-02370	
10762	6H-Purine-6-thione, 2-amino-1,7-dihydro- / Thioguanine	$C_5H_5N_5S$ / 167.19	154-42-7 / >360	9266 /	4-26-00-03926	
10763	6H-Purine-6-thione, 1,7-dihydro- / 6-Mercaptopurine	$C_5H_4N_4S$ / 152.18	50-44-2 / 313 dec	5762 /		H_2O 1; alk 3
10764	1H-Purine-2,6,8(3H)-trione, 7,9-dihydro- / Uric acid	$C_5H_4N_4O_3$ / 168.11	69-93-2 / dec	9790 / dec	4-26-00-02619 / 1.89[25]	H_2O 1; EtOH 1; eth 1; alk 4
10765	1H-Purine-2,6,8(3H)-trione, 4,9-dihydro-3-methyl- / 3-Methyluric acid	$C_6H_6N_4O_3$ / 182.14	605-99-2 / >350		4-26-00-02621 / 1.6104[25]	H_2O 2; EtOH 2; alk 3 / 1.6334[25]
10766	1H-Purine-2,6,8(3H)-trione, 7,9-dihydro-7-methyl- / 7-Methyluric acid	$C_6H_6N_4O_3$ / 182.14	612-37-3 / 370-80 dec		4-26-00-02622 / 1.706[25]	H_2O 2; NaOH 3
10767	2H-Purin-2-one, 6-amino-1,3-dihydro- / Isoguanine	$C_5H_5N_5O$ / 151.13	3373-53-3 / >360			
10768	6H-Purin-6-one, 2-amino-1,7-dihydro- / Guanine	$C_5H_5N_5O$ / 151.13	73-40-5 / 360 dec	4477 / sub	4-26-00-03890	H_2O 1; EtOH 2; eth 2; alk 3
10769	6H-Purin-6-one, 2-amino-1,7-dihydro-7-methyl / 7-Methylguanine	$C_6H_7N_5O$ / 165.15	578-76-7 / 370			
10770	6H-Purin-6-one, 1,7-dihydro- / Hypoxanthine	$C_5H_4N_4O$ / 136.11	68-94-0 / 150 dec	4805 /	4-26-00-02081	H_2O 2; alk 3; dil acid 3
10771	Pyran	C_5H_6O / 82.10	33941-07-0 /	/ 80	0-17-00-00036	EtOH 3; eth 3; bz 3 / 1.4559[20]
10772	2H-Pyran-5-carboxylic acid, 3,4-dihydro-3-(hydroxymethyl)-2-methyl-4-[(9-methyl-9H-pyrido[3,4-b]indol-1-yl)methyl]-, methyl ester, [2S-(2α,3β,4β)]- / Alstonidine	$C_{22}H_{24}N_2O_4$ / 380.44	25394-75-6 / 189	313 /	4-27-00-07962	ace 4; EtOH 4
10773	2H-Pyran-5-carboxylic acid, 2-oxo- / Coumalic acid	$C_6H_4O_4$ / 140.10	500-05-0 / 205-10 dec	2558 / 218[120]	5-18-08-00120	H_2O 2; EtOH 3; eth 2; ace 2
10774	2H-Pyran-4-carboxylic acid, tetrahydro- / Pyran-4-carboxylic acid, tetrahydro-	$C_6H_{10}O_3$ / 130.14	5337-03-1 / 88	/ 144[11]	5-18-06-00023	H_2O 3; peth 2
10775	4H-Pyran-2,6-dicarboxylic acid, 3-hydroxy-4-oxo- / Meconic acid	$C_7H_4O_7$ / 200.10	497-59-6 / 120 dec	5664 / 120 dec	5-18-09-00319	H_2O 2; EtOH 3; eth 2; ace 2
10776	4H-Pyran-2,6-dicarboxylic acid, 4-oxo- / Chelidonic acid	$C_7H_4O_6$ / 184.11	99-32-1 / 262	2042 /	5-18-08-00646	H_2O 3; EtOH 2

No.	Name Synonym	Mol. Form. Mol. Wt.	CAS RN mp/°C	Merck No. bp/°C	Beil. Ref. den/g cm^{-3}	Solubility n_D
10777	2H-Pyran, 3,4-dihydro- 3,4-Dihydro-2H-pyran	C_5H_8O 84.12	110-87-2	86	5-17-01-00181 0.921[19]	H_2O 3; EtOH 3; chl 2 1.4402[19]
10778	2H-Pyran, 3,6-dihydro-4-methyl-	$C_6H_{10}O$ 98.14	16302-35-5	117.5	5-17-01-00194 0.912[25]	1.4495[20]
10779	2H-Pyran-2,4(3H)-dione, 3-acetyl-6-methyl- Dehydroacetic acid	$C_8H_8O_4$ 168.15	520-45-6 109	2855 270	4-17-00-06699	H_2O 4; EtOH 2; eth 4; chl 2
10780	2H-Pyran-2,6(3H)-dione, dihydro-	$C_5H_6O_3$ 114.10	108-55-4 56.3	158[15]	5-17-11-00009 1.4110[20]	
10781	2H-Pyran-2,6(3H)-dione, dihydro-3-phenyl-	$C_{11}H_{10}O_3$ 190.20	2959-96-8 95	224[13]	5-17-11-00279	EtOH 4
10782	2H-Pyran-2,6(3H)-dione, dihydro-4-phenyl-	$C_{11}H_{10}O_3$ 190.20	4160-80-9 105.6	217[15]	5-17-11-00279	bz 4; eth 4; chl 4
10783	2H-Pyran, 2-ethoxy-3,4-dihydro-	$C_7H_{12}O_2$ 128.17	103-75-3	42[16]	5-17-03-00196 0.957[25]	1.4394[20]
10784	2H-Pyran-2-methanol, tetrahydro-	$C_6H_{12}O_2$ 116.16	100-72-1	185	5-17-03-00136 1.027[25]	1.458[20]
10785	4H,5H-Pyrano[4,3-b]pyran-4,5-dione, 2,3-dihydro-3-hydroxy-2-methyl-7-(1-propenyl)-, [2S-[2α,3β,7(E)]]- Radicinin	$C_{12}H_{12}O_5$ 236.22	10088-95-6 221.5	8117	4-19-00-02746	chl 2
10786	Pyrano[3,2-c][2]benzopyran-6(2H)-one, 3,4,4a,10b-tetrahydro-3,4,8,10-tetrahydroxy-2-(hydroxymethyl)-9-methoxy-[2R-(2α,3β,4α,4aα,10bβ)]- Bergenin	$C_{14}H_{16}O_9$ 328.28	477-90-7 238	1174	5-19-07-00132	H_2O 4; EtOH 4
10787	2H-Pyran-4-ol, tetrahydro-	$C_5H_{10}O_2$ 102.13	2081-44-9	87[15]	5-17-03-00111 1.071[25]	1.4600[20]
10788	2H-Pyran-2-one	$C_5H_4O_2$ 96.09	504-31-4 8.5	207.5	5-17-09-00288 1.200[20]	H_2O 5; ace 4 1.5270[25]
10789	4H-Pyran-4-one	$C_5H_4O_2$ 96.09	108-97-4 32.5	212.5	5-17-09-00290 1.190[25]	H_2O 4; EtOH 3; eth 4; bz 3 1.5238
10790	2H-Pyran-2-one, 6-[2-(1,3-benzodioxol-5-yl)ethenyl]-5,6-dihydro-4-methoxy-, [R-(E)]- Methysticin	$C_{15}H_{14}O_5$ 274.27	495-85-2 137	6056	5-19-10-00456	
10791	2H-Pyran-2-one, 5,6-dihydro-6-methyl-, (S)- Parasorbic acid	$C_6H_8O_2$ 112.13	10048-32-5	6981 100[15]	5-17-09-00130 1.079[18]	H_2O 4; eth 4; EtOH 4 1.4736[20]
10792	2H-Pyran-2-one, 4,6-dimethyl-	$C_7H_8O_2$ 124.14	675-09-2 51.5	245	5-17-09-00409	H_2O 4; eth 4; EtOH
10793	4H-Pyran-4-one, 2,6-dimethyl-	$C_7H_8O_2$ 124.14	1004-36-0 132	251; 140[25]	5-17-09-00410 0.9953[137]	H_2O 3; EtOH 3; eth 3; ace 3
10794	2H-Pyran-2-one, 3-hydroxy-	$C_5H_4O_3$ 112.08	496-64-0 93.8	112[20]. 102[15]	5-18-01-00095	H_2O 3, EtOH 3; eth 3; bz 3
10795	4H-Pyran-4-one, 5-hydroxy-2-(hydroxymethyl)- Kojic acid	$C_6H_6O_4$ 142.11	501-30-4 153.5	5197	5-18-02-00516	H_2O 2; EtOH 3; eth 3; ace 3
10796	4H-Pyran-4-one, 3-hydroxy-2-methyl- Maltol	$C_6H_6O_3$ 126.11	118-71-8 161.5	5594 sub 93	5-18-01-00114	H_2O 2; eth 2; bz 2; chl 4
10797	4H-Pyran-4-one, 2-methoxy-3,5-dimethyl-6-[tetrahydro-4-[2-methyl-3-(4-nitrophenyl)-2-propenylidene]-2-furanyl] Aureothin	$C_{22}H_{23}NO_6$ 397.43	2825-00-5 158	897	5-19-06-00099	ace 4; EtOH 4; chl 4
10798	2H-Pyran-2-one, tetrahydro-	$C_5H_8O_2$ 100.12	542-28-9 -12.5	219	5-17-09-00017 1.0794[20]	H_2O 3; EtOH 5; eth 5; ctc 2 1.4503[20]
10799	4H-Pyran-4-one, tetrahydro-	$C_5H_8O_2$ 100.12	29943-42-8	166.5	5-17-09-00021 1.084[25]	1.4520[20]
10800	2H-Pyran, tetrahydro- Oxane	$C_5H_{10}O$ 86.13	142-68-7 -45	9149 88	5-17-01-00064 0.8814[20]	EtOH 3; eth 3; bz 3; ctc 3 1.4200[20]
10801	Pyrazine 1,4-Diazine	$C_4H_4N_2$ 80.09	290-37-9 55	7971 115	5-23-05-00351 1.0311[61]	H_2O 3; EtOH 3; eth 3; ace 3 1.4953[61]
10802	Pyrazinecarboxamide Pyrazinamide	$C_5H_5N_3O$ 123.11	98-96-4 192	7970 sub	4-25-00-00772	H_2O 3; EtOH 3
10803	Pyrazinecarboxylic acid Pyrazinoic acid	$C_5H_4N_2O_2$ 124.10	98-97-5 225 dec	7973 sub	4-25-00-00771	
10804	2,3-Pyrazinedicarboxylic acid 2,3-Dicarboxypyrazine	$C_6H_4N_2O_4$ 168.11	89-01-0 193 dec	7972	4-25-00-01064	H_2O 4; EtOH 2; eth 2; ace 3
10805	Pyrazine, 2,3-dimethyl- 2,3-Dimethylpyrazine	$C_6H_8N_2$ 108.14	5910-89-4	156	5-23-05-00402 1.0281[0]	H_2O 3; EtOH 3; eth 3
10806	Pyrazine, 2,5-dimethyl- 2,5-Dimethylpyrazine	$C_6H_8N_2$ 108.14	123-32-0 15	155	5-23-05-00403 0.9887[20]	H_2O 5; EtOH 5; eth 5; ace 3 1.4980[20]
10807	Pyrazine, 2,6-dimethyl- 2,6-Dimethylpyrazine	$C_6H_8N_2$ 108.14	108-50-9 47.5	155.6	5-23-05-00406 0.9647[50]	H_2O 3; EtOH 3; eth 3; ctc 2
10808	Pyrazineethanol	$C_6H_8N_2O$ 124.14	6705-31-3	129[10]	5-23-11-00173 1.163[20]	1.5378[20]
10809	Pyrazine, 3-ethyl-2,5-dimethyl-	$C_8H_{12}N_2$ 136.20	13360-65-1	180.5	5-23-05-00440 0.965[24]	H_2O 2; EtOH 2; eth 2 1.5014[24]
10810	Pyrazine, methyl- 2-Methylpyrazine	$C_5H_6N_2$ 94.12	109-08-0 -29	137	5-23-05-00386 1.03[20]	H_2O 5; EtOH 5; eth 5; ace 3 1.5042[20]
10811	2(1H)-Pyrazinone, 1-hydroxy-6-(1-methylpropyl)-3-(2-methylpropyl)-, (+)- Aspergillic acid	$C_{12}H_{20}N_2O_2$ 224.30	490-02-8 98	863	4-24-00-00235	bz 4; eth 4; EtOH 4

No.	Name / Synonym	Mol. Form. / Mol. Wt.	CAS RN / mp/°C	Merck No. / bp/°C	Beil. Ref. / den/g cm^{-3}	Solubility / n_D
10812	1H-Pyrazol-4-amine / 4-Aminopyrazole	C$_3$H$_5$N$_3$ / 83.09	28466-26-4 / 81	/ sub	4-25-00-02029	H$_2$O 4; EtOH 4; chl 4
10813	1H-Pyrazol-4-amine, 3-methyl-1-phenyl- / Pyrazole, 4-amino-3-methyl-1-phenyl-	C$_{10}$H$_{11}$N$_3$ / 173.22	103095-51-8 / 88.8	/ 314.5	4-25-00-02033	bz 4; eth 4; EtOH 4; chl 4
10814	1H-Pyrazol-5-amine, 3-methyl-1-phenyl-	C$_{10}$H$_{11}$N$_3$ / 173.22	1131-18-6 / 116	/ 333	4-25-00-02034	H$_2$O 3; EtOH 3; bz 2; chl 3
10815	1H-Pyrazole / 1,2-Diazole	C$_3$H$_4$N$_2$ / 68.08	288-13-1 / 68	7974 / 187	5-23-04-00122	H$_2$O 3; EtOH 3; eth 3; bz 3 / 1.4203
10816	1H-Pyrazole, 4-bromo-1,3-dimethyl- / 4-Bromo-1,3-dimethylpyrazole	C$_5$H$_7$BrN$_2$ / 175.03	5775-82-6 /	/ 76-7[10]	5-23-05-00022 / 1.4975[15]	ace 4; eth 4; EtOH 4 / 1.5214[15]
10817	1H-Pyrazole, 4-bromo-3-methyl- / 4-Bromo-3-methylpyrazole	C$_4$H$_5$BrN$_2$ / 161.00	13808-64-5 / 76.5		5-23-05-00022 / 1.5638[100]	ace 4; eth 4; EtOH 4 / 1.5182[100]
10818	1H-Pyrazole, 3-chloro-1,5-dimethyl- / 3-Chloro-1,5-dimethylpyrazole	C$_5$H$_7$ClN$_2$ / 130.58	51500-32-4 / 47.5	/ 211	2-23-00-00049 / 1.0823[100]	ace 4; bz 4; eth 4; EtOH 4 / 1.4648[100]
10819	1H-Pyrazole, 4-chloro-3,5-dimethyl-	C$_5$H$_7$ClN$_2$ / 130.58	15953-73-8 / 117.5	/ 221	5-23-05-00151	ace 4; bz 4; eth 4; EtOH 4
10820	1H-Pyrazole, 5-chloro-1,3-dimethyl- / 5-Chloro-1,3-dimethylpyrazole	C$_5$H$_7$ClN$_2$ / 130.58	54454-10-3 /	/ 157.5	5-23-05-00019 / 1.1367[18]	H$_2$O 4 / 1.4877[18]
10821	1H-Pyrazole, 3-chloro-5-methyl-	C$_4$H$_5$ClN$_2$ / 116.55	15953-45-4 / 118.5	/ 258; 138[15]	5-23-05-00019	H$_2$O 4
10822	1H-Pyrazole, 4,5-dihydro- / 2-Pyrazoline	C$_3$H$_6$N$_2$ / 70.09	109-98-8 /	7975 / 144	5-23-03-00361 / 1.0200[17]	H$_2$O 4; eth 4; EtOH 4 / 1.4796[17]
10823	1H-Pyrazole, 4,5-dihydro-1-phenyl- / 1-Phenyl-2-pyrazoline	C$_9$H$_{10}$N$_2$ / 146.19	936-53-8 / 52	/ 273; 151[17]	5-23-03-00363 / 1.0689[58]	H$_2$O 1; EtOH 3; eth 3; bz 3 / 1.6015[58]
10824	1H-Pyrazole, 4,5-dihydro-3-phenyl- / 2-Pyrazoline, 3-phenyl	C$_9$H$_{10}$N$_2$ / 146.19	936-48-1 / 45.8	/ 164[17]	5-23-06-00396 / 1.0892[54]	
10825	1H-Pyrazole, 1,3-dimethyl-	C$_5$H$_8$N$_2$ / 96.13	694-48-4 /	/ 137	5-23-05-00012 / 0.9561[17]	H$_2$O 4 / 1.4734[15]
10826	1H-Pyrazole, 1,5-dimethyl-	C$_5$H$_8$N$_2$ / 96.13	694-31-5 /	/ 149	5-23-05-00006 / 0.9813[17]	H$_2$O 4; EtOH 4; eth 4 / 1.4782[16]
10827	1H-Pyrazole, 3,4-dimethyl-	C$_5$H$_8$N$_2$ / 96.13	2820-37-3 / 58	/ 110[11]	5-23-05-00108 / 0.9325[99]	EtOH 3; eth 3; ace 3; bz 3
10828	1H-Pyrazole, 3,5-dimethyl-	C$_5$H$_8$N$_2$ / 96.13	67-51-6 / 107.5	/ 218	5-23-05-00110 / 0.883[16]	H$_2$O 3; EtOH 4; eth 4; ace 3
10829	1H-Pyrazole, 3,5-dimethyl-1-phenyl-	C$_{11}$H$_{12}$N$_2$ / 172.23	1131-16-4 /	/ 273; 145[12.5]	5-23-05-00127 / 1.0566[20]	eth 4; EtOH 4; chl 4 / 1.5738[19]
10830	1H-Pyrazole, 1,3-diphenyl- / 1,3-Diphenylpyrazole	C$_{15}$H$_{12}$N$_2$ / 220.27	4492-01-7 / 85.8	/ 341[270]	5-23-07-00178 / 1.0794[101]	
10831	1H-Pyrazole, 1,5-diphenyl- / 1,5-Diphenylpyrazole	C$_{15}$H$_{12}$N$_2$ / 220.27	6831-89-6 / 55.5	/ 338; 167[12]	5-23-07-00176 / 1.0696[100]	bz 4; eth 4; EtOH 4
10832	1H-Pyrazole, 1-ethyl-	C$_5$H$_8$N$_2$ / 96.13	2817-71-2 /	/ 139	5-23-04-00135 / 0.9537[20]	ace 4; bz 4; eth 4; EtOH 4 / 1.4700[20]
10833	1H-Pyrazole, 1-ethyl-3-methyl- / 1-Ethyl-3-methylpyrazole	C$_6$H$_{10}$N$_2$ / 110.16	30433-57-9 / 115	/ 152	5-23-05-00013 / 0.936[20]	
10834	1H-Pyrazole, 1-methyl- / 1-Methylpyrazole	C$_4$H$_6$N$_2$ / 82.11	930-36-9 /	/ 127	5-23-04-00134 / 0.992[13]	1.4787[13]
10835	1H-Pyrazole, 3-methyl-	C$_4$H$_6$N$_2$ / 82.11	1453-58-3 / 36.5	/ 204; 108[25]	5-23-05-00004 / 1.0203[16]	H$_2$O 5; EtOH 5; eth 5 / 1.4915[20]
10836	1H-Pyrazole, 4-methyl- / 4-Methylpyrazole	C$_4$H$_6$N$_2$ / 82.11	7554-65-6 /	10146 / 206; 95[13]	5-23-05-00031 / 1.015[20]	
10837	1H-Pyrazole, 1-methyl-3-phenyl-	C$_{10}$H$_{10}$N$_2$ / 158.20	3463-26-1 / 56	/ 280.5	5-23-07-00178 / 1.0232[99]	
10838	1H-Pyrazole, 3-methyl-1-phenyl-	C$_{10}$H$_{10}$N$_2$ / 158.20	1128-54-7 / 37	/ 255	5-23-05-00013 / 1.076[20]	
10839	1H-Pyrazole, 1,3,4-trimethyl- / 1,3,4-Trimethylpyrazole	C$_6$H$_{10}$N$_2$ / 110.16	15802-99-0 /	/ 160	5-23-05-00109 / 0.9567[17]	1.4866[17]
10840	1H-Pyrazole, 1,3,5-trimethyl- / 1,3,5-Trimethylpyrazole	C$_6$H$_{10}$N$_2$ / 110.16	1072-91-9 /	/ 170	5-23-05-00125 / 0.9130[57]	1.4589[57]
10841	1H-Pyrazole, 1,4,5-trimethyl- / 1,4,5-Trimethylpyrazole	C$_6$H$_{10}$N$_2$ / 110.16	15802-97-8 /	/ 176.5	5-23-05-00108 / 0.9685[17]	1.4849[17]
10842	1H-Pyrazole, 3,4,5-trimethyl-	C$_6$H$_{10}$N$_2$ / 110.16	5519-42-6 / 138.5	/ 233	5-23-05-00198	bz 4; eth 4; EtOH 4
10843	3,5-Pyrazolidinedione, 1,4-diphenyl- / Phenopyrazone	C$_{15}$H$_{12}$N$_2$O$_2$ / 252.27	3426-01-5 / 233.5	7217	5-24-08-00003	
10844	3-Pyrazolidinone, 1-phenyl- / 1-Phenyl-3-pyrazolidinone	C$_9$H$_{10}$N$_2$O / 162.19	92-43-3 / 126	7281	5-24-01-00011	eth 1; lig 1
10845	3H-Pyrazol-3-one, 4-amino-1,2-dihydro-1,5-dimethyl-2-phenyl- / Ampyrone	C$_{11}$H$_{13}$N$_3$O / 203.24	83-07-8 / 109	625	4-25-00-03554	H$_2$O 3; EtOH 3; eth 2; bz 3
10846	3H-Pyrazol-3-one, 4-bromo-1,2-dihydro-1,5-dimethyl-2-phenyl- / 4-Bromoantipyrine	C$_{11}$H$_{11}$BrN$_2$O / 267.13	5426-65-3 / 122	/ 300[9]	5-24-01-00343 / 1.5900[20]	H$_2$O 3; EtOH 4; eth 3; chl 4
10847	3H-Pyrazol-3-one, 1,2-dihydro-1,5-dimethyl-2-phenyl- / Antipyrine	C$_{11}$H$_{12}$N$_2$O / 188.23	60-80-0 / 114	748 / 319	5-24-01-00330	H$_2$O 4; EtOH 4
10848	3H-Pyrazol-3-one, 2,4-dihydro-5-methyl-4-nitro-2-(4-nitrophenyl)- / Picrolonic acid	C$_{10}$H$_8$N$_4$O$_5$ / 264.20	550-74-3 / 116	7383 / dec		H$_2$O 2; EtOH 3; eth 3; MeOH 3

No.	Name / Synonym	Mol. Form. / Mol. Wt.	CAS RN / mp/°C	Merck No. / bp/°C	Beil. Ref. / den/g cm^{-3}	Solubility / n_D
10849	3H-Pyrazol-3-one, 1,2-dihydro-5-methyl-2-phenyl-	$C_{10}H_{10}N_2O$	19735-89-8		5-24-01-00317	H_2O 3; EtOH 3; bz 2; peth 1
		174.20	128	287[105]; 191[17]	1.2600[20]	1.637
10850	3H-Pyrazol-3-one, 1,2-dihydro-2-(1-methyl-4-piperidinyl)-5-phenyl-4-(phenylmethyl)- Benzpiperylon	$C_{22}H_{25}N_3O$	53-89-4	1131	5-24-04-00398	
		347.46	181-3 dec			
10851	3H-Pyrazol-3-one, 4-(dimethylamino)-1,2-dihydro-1,5-dimethyl-2-phenyl- Aminopyrine	$C_{13}H_{17}N_3O$	58-15-1	488	4-24-00-00096	H_2O 4; bz 4; EtOH 4
		231.30	134.5			
10852	4H-Pyrazolo[3,4-d]pyrimidin-4-one, 1,5-dihydro- Allopurinol	$C_5H_4N_4O$	315-30-0	278	4-26-00-02076	
		136.11	350			
10853	Pyrene Benzo[def]phenanthrene	$C_{16}H_{10}$ 202.26	129-00-0 151.2	7977 404	4-05-00-02467 1.271[23]	H_2O 1; EtOH 3; eth 3; bz 3
10854	1-Pyrenecarboxylic acid 3-Pyrenecarboxylic acid	$C_{17}H_{10}O_2$ 246.27	19694-02-1 274	sub	3-09-00-03575	eth 4; chl 4
10855	Pyridazine 1,2-Diazabenzene	$C_4H_4N_2$ 80.09	289-80-5 -8	7982 208	5-23-05-00321 1.1035[23]	H_2O 5; EtOH 5; eth 4; ace 4 1.5218[20]
10856	Pyridazine, 3-methyl- 3-Methyl-1,2-diazine	$C_5H_6N_2$ 94.12	1632-76-4 184	214	5-23-05-00371 1.0450[26]	1.5145[20]
10857	2-Pyridinamine 2-Aminopyridine	$C_5H_6N_2$ 94.12	504-29-0 57.5	486 105[20]	5-22-08-00280	EtOH 3; eth 3; ace 3; bz 3
10858	3-Pyridinamine β-Aminopyridine	$C_5H_6N_2$ 94.12	462-08-8 64.5	487 252	5-22-09-00003	H_2O 3; EtOH 3; eth 3; lig 2
10859	4-Pyridinamine 4-Aminopyridine	$C_5H_6N_2$ 94.12	504-24-5 158.5	273	5-22-09-00106	H_2O 3; EtOH 4; eth 3; bz 3
10860	3-Pyridinamine, 6-butoxy-	$C_9H_{14}N_2O$	539-23-1		4-22-00-05574	EtOH 3; eth 3; chl 2; dil acid 3
		166.22		148-50[12]	1.037[25]	1.5373[20]
10861	2-Pyridinamine, 5-chloro-	$C_5H_5ClN_2$ 128.56	1072-98-6 137	127[11]	5-22-08-00451	H_2O 3; EtOH 3; peth 1; lig 1
10862	3-Pyridinamine, 2,6-dichloro- Pyridine, 3-amino-2,6-dichloro	$C_5H_4Cl_2N_2$ 163.01	62476-56-6 119	110[0.3]	4-22-00-04093	eth 4; EtOH 4
10863	2-Pyridinamine, N,N-dimethyl-	$C_7H_{10}N_2$ 122.17	5683-33-0 182	196	5-22-08-00287 1.0149[14]	EtOH 3; eth 3; bz 3 1.5663[20]
10864	4-Pyridinamine, 2,6-dimethyl-	$C_7H_{10}N_2$ 122.17	3512-80-9 192.5	246	5-22-09-00417	H_2O 3; EtOH 3; eth 2; ace 3
10865	3-Pyridinamine, 6-methoxy-	$C_6H_8N_2O$ 124.14	6628-77-9 30	125[10]; 87[1]	5-22-11-00408	1.5745[20]
10866	2-Pyridinamine, 3-methyl-	$C_6H_8N_2$ 108.14	1603-40-3 33.5	222; 95[8]	5-22-09-00272	H_2O 4; EtOH 3; eth 3; ace 3
10867	2-Pyridinamine, 4-methyl- 2-Amino-4-picoline	$C_6H_8N_2$ 108.14	695-34-1 100	478 115-7[11]	5-22-09-00325	H_2O 4; EtOH 3; eth 3; ace 3
10868	2-Pyridinamine, 6-methyl-	$C_6H_8N_2$ 108.14	1824-81-3 41	208.5	5-22-09-00210	H_2O 4; EtOH 3; eth 3; ace 3
10869	2-Pyridinamine, N-methyl-	$C_6H_8N_2$ 108.14	4597-87-9 15	200.5	5-22-08-00286 1.048[29]	H_2O 3; EtOH 4; eth 4; bz 3
10870	3-Pyridinamine, 4-methyl-	$C_6H_8N_2$ 108.14	3430-27-1 106	255	5-22-09-00344	H_2O 2; EtOH 2; eth 2; ace 2
10871	2-Pyridinamine, N-2-pyridinyl-	$C_{10}H_9N_3$ 171.20	1202-34-2 90.5	307.5	5-22-08-00415	H_2O 2; EtOH 4; eth 4; ace 4
10872	Pyridine Azine	C_5H_5N 79.10	110-86-1 -41.6	7983 115.2	5-20-05-00160 0.9819[20]	H_2O 5; EtOH 5; eth 5; ace 5 1.5095[20]
10873	1(4H)-Pyridineacetic acid, 3,5-diiodo-4-oxo-, propyl ester Propyliodone	$C_{10}H_{11}I_2NO_3$ 447.01	587-61-1 186	7876	5-21-07-00164	
10874	Pyridine, 2,6-bis(1,1-dimethylethyl)- 2,6-Di-tert-butylpyridine	$C_{13}H_{21}N$ 191.32	585-48-8	3022 120-1[20]	5-20-06-00181	
10875	Pyridine, 2-bromo-	C_5H_4BrN 158.00	109-04-6	193; 75[13]	5-20-05-00422 1.657[15]	H_2O 2; EtOH 3; eth 3; ctc 3 1.5734[20]
10876	Pyridine, 3-bromo- 3-Bromopyridine	C_5H_4BrN 158.00	626-55-1 -27.3	173; 69[18]	5-20-05-00426 1.645[0]	H_2O 3; EtOH 4; eth 4 1.5694[20]
10877	Pyridine, 4-bromo- 4-Bromopyridine	C_5H_4BrN 158.00	1120-87-2 0.5	28-30[0.4]	5-20-05-00431 1.6450[0]	ace 3; bz 3 1.5694[20]
10878	Pyridine, 2-butyl- 2-Butylpyridine	$C_9H_{13}N$ 135.21	5058-19-5 97	192	5-20-06-00118 0.9135[15]	
10879	Pyridine, 4-butyl-	$C_9H_{13}N$ 135.21	5335-75-1	209; 98[20]	5-20-06-00120 0.9157[20]	1.4946[20]
10880	Pyridine, 4-(1-butylpentyl)-	$C_{14}H_{23}N$ 205.34	2961-47-9	181[50]	5-20-06-00189 0.8878[25]	1.4846[25]
10881	2-Pyridinecarbonitrile	$C_6H_4N_2$ 104.11	100-70-9 29	224.5	5-22-02-00019 1.0810[25]	H_2O 3; EtOH 4; eth 4; bz 4 1.5242[25]
10882	3-Pyridinecarbonitrile	$C_6H_4N_2$ 104.11	100-54-9 51	206.9; 170[300]	5-22-02-00115 1.1590[25]	H_2O 4; EtOH 4; eth 4; bz 4
10883	3-Pyridinecarbonitrile, 1,2-dihydro-4-methoxy-1-methyl-2-oxo- Ricinine	$C_8H_8N_2O_2$ 164.16	524-40-3 201.5	8212 sub 170	5-22-07-00318	H_2O 3; EtOH 2; bz 2; chl 3

No.	Name / Synonym	Mol. Form. / Mol. Wt.	CAS RN / mp/°C	Merck No. / bp/°C	Beil. Ref. / den/g cm^{-3}	Solubility / n_D
10884	3-Pyridinecarbonitrile, 1,2-dihydro-1-methyl-2-oxo-	$C_7H_8N_2O$	767-88-4		5-22-06-00129	EtOH 4
		134.14	140	243[28]		
10885	4-Pyridinecarbothioamide, 2-ethyl- Ethionamide	$C_8H_{10}N_2S$ 166.25	536-33-4 163	3692	5-22-02-00360	
10886	4-Pyridinecarbothioamide, 2-propyl- Protionamide	$C_9H_{12}N_2S$ 180.27	14222-60-7 136.7	7905	5-22-02-00376	
10887	2-Pyridinecarboxaldehyde	C_6H_5NO 107.11	1121-60-4	180; 62[13]	5-21-07-00293 1.1255[18]	H_2O 3; EtOH 3; eth 3; ctc 2 1.5389[18]
10888	3-Pyridinecarboxaldehyde	C_6H_5NO 107.11	500-22-1	89.5[14]	5-21-07-00334 1.1394[25]	H_2O 3; EtOH 3; eth 2; ace 3
10889	4-Pyridinecarboxaldehyde	C_6H_5NO 107.11	872-85-5	77-8[12]	5-21-07-00351	H_2O 3; eth 3; ctc 3 1.5423[20]
10890	3-Pyridinecarboxamide Niacinamide	$C_6H_6N_2O$ 122.13	98-92-0 130	6398 157[0.0005]	5-22-02-00080 1.400[25]	H_2O 4; EtOH 4; chl 2; glycerol 4 1.466
10891	3-Pyridinecarboxamide, N,N-diethyl- Nikethamide	$C_{10}H_{14}N_2O$ 178.23	59-26-7 25	6459 280 dec; 175[25]	5-22-02-00084 1.060[25]	DMSO-d_6 2 1.525[20]
10892	4-Pyridinecarboxamide, N,N-diethyl- Isonicotinic acid diethylamide	$C_{10}H_{14}N_2O$ 178.23	530-40-5	5074 119-20[1]	5-22-02-00196	H_2O 4; ace 4; eth 4; EtOH 4 1.525[20]
10893	2-Pyridinecarboxylic acid Picolinic acid	$C_6H_5NO_2$ 123.11	98-98-6 136.5	7375 sub	5-22-02-00003	H_2O 2; EtOH 3; eth 1; bz 2
10894	3-Pyridinecarboxylic acid Nicotinic acid	$C_6H_5NO_2$ 123.11	59-67-6 236.6	6435 sub	5-22-02-00057 1.473[25]	H_2O 2; EtOH 2; eth 2
10895	4-Pyridinecarboxylic acid Isonicotinic acid	$C_6H_5NO_2$ 123.11	55-22-1 315	5073 sub 260	5-22-02-00188	H_2O 2; EtOH 2; eth 2; bz 2
10896	2-Pyridinecarboxylic acid, 4-amino-3,5,6-trichloro- Picloram	$C_6H_3Cl_3N_2O_2$ 241.46	1918-02-1 218.5	7370	5-22-13-00585	
10897	3-Pyridinecarboxylic acid, 6-amino- 6-Aminonicotinic acid	$C_6H_6N_2O_2$ 138.13	3167-49-5 312	468	4-22-00-06726	
10898	2-Pyridinecarboxylic acid, 5-butyl- Fusaric acid	$C_{10}H_{13}NO_2$ 179.22	536-69-6 97	4227	5-22-02-00384	
10899	3-Pyridinecarboxylic acid, 1,6-dihydro-6-oxo-	$C_6H_5NO_3$ 139.11	5006-66-6 310 dec	sub	5-22-06-00119	H_2O 2; EtOH 1; eth 1; bz 1
10900	4-Pyridinecarboxylic acid, 1,2-dihydro-6-hydroxy-2-oxo- Citrazinic acid	$C_6H_5NO_4$ 155.11	99-11-6 >300 dec	2327	5-22-07-00024	H_2O 3; alk 3; HCl 2
10901	2-Pyridinecarboxylic acid, ethyl ester	$C_8H_9NO_2$ 151.17	2524-52-9 1	243	5-22-02-00004 1.1194[20]	H_2O 4; eth 4; EtOH 4 1.5104[20]
10902	3-Pyridinecarboxylic acid, ethyl ester	$C_8H_9NO_2$ 151.17	614-18-6 8.5	224	5-22-02-00060 1.1070[20]	H_2O 4; EtOH 4; eth 4; bz 4 1.5024[20]
10903	4-Pyridinecarboxylic acid, ethyl ester	$C_8H_9NO_2$ 151.17	1570-45-2 23	219.5	5-22-02-00190 1.0091[15]	H_2O 2; EtOH 3; eth 4; bz 3 1.5017[20]
10904	4-Pyridinecarboxylic acid, [1-(2-furanyl)ethylidene]hydrazide Furonazide	$C_{12}H_{11}N_3O_2$ 229.24	3460-67-1 202.3	4220	5-22-02-00264	
10905	4-Pyridinecarboxylic acid, hydrazide Isoniazid	$C_6H_7N_3O$ 137.14	54-85-3 171.4	5071	5-22-02-00219	H_2O 4; EtOH 4
10906	3-Pyridinecarboxylic acid, methyl ester Methyl nicotinate	$C_7H_7NO_2$ 137.14	93-60-7 42.5	6014 204	5-22-02-00059	H_2O 3; EtOH 3; bz 3
10907	3-Pyridinecarboxylic acid, 1-methylethyl ester	$C_9H_{11}NO_2$ 165.19	553-60-6	126[30]; 92.5[5]	5-22-02-00060 1.0624[20]	1.4926[20]
10908	4-Pyridinecarboxylic acid, methyl ester	$C_7H_7NO_2$ 137.14	2459-09-8 16.1	208	5-22-02-00189 1.1599[20]	H_2O 2; EtOH 3; eth 3; bz 3 1.5135[20]
10909	3-Pyridinecarboxylic acid, 1-oxide Oxiniacic acid	$C_6H_5NO_3$ 139.11	2398-81-4 254-5 dec	6895	5-22-02-00140	H_2O 4; MeOH 4
10910	4-Pyridinecarboxylic acid, 2-[3-oxo-3-[(phenylmethyl)amino]propyl]hydrazide Nialamide	$C_{16}H_{18}N_4O_2$ 298.34	51-12-7 151.6	6399	5-22-02-00251	
10911	3-Pyridinecarboxylic acid, phenylmethyl ester Nicotinic acid benzyl ester	$C_{13}H_{11}NO_2$ 213.24	94-44-0	6436 170[3]	5-22-02-00062	
10912	3-Pyridinecarboxylic acid, 1,2,5,6-tetrahydro- Guvacine	$C_6H_9NO_2$ 127.14	498-96-4 295 dec	4495	5-22-01-00322	H_2O 4
10913	3-Pyridinecarboxylic acid, 1,2,5,6-tetrahydro-1-methyl- Arecaidine	$C_7H_{11}NO_2$ 141.17	499-04-7 232 dec	801	5-22-01-00322	H_2O 4; EtOH 1; eth 1; bz 1
10914	3-Pyridinecarboxylic acid, 1,2,5,6-tetrahydro-1-methyl-, methyl ester Arecoline	$C_8H_{13}NO_2$ 155.20	63-75-2 209	802	5-22-01-00322 1.0485[20]	H_2O 5; EtOH 5; eth 5; chl 3 1.486[20]
10915	Pyridine, 2-chloro- 2-Chloropyridine	C_5H_4ClN 113.55	109-09-1 170		5-20-05-00402 1.205[15]	H_2O 2; EtOH 3; eth 3 1.5320[20]
10916	Pyridine, 3-chloro- 3-Chloropyridine	C_5H_4ClN 113.55	626-60-8	148; 86[100]	5-20-05-00406	H_2O 2 1.5304[20]
10917	Pyridine, 4-chloro- 4-Chloropyridine	C_5H_4ClN 113.55	626-61-9 -43.5	147.5	5-20-05-00410 1.2000[25]	H_2O 3; EtOH 5
10918	Pyridine, 2-chloro-6-methoxy-	C_6H_6ClNO 143.57	17228-64-7	185.5	5-21-02-00018 1.207[25]	1.5263[20]
10919	Pyridine, 2-chloro-4-methyl-	C_6H_6ClN 127.57	3678-62-4 115	192	5-20-05-00579 1.1459[20]	1.5293[8]

No.	Name Synonym	Mol. Form. Mol. Wt.	CAS RN mp/°C	Merck No. bp/°C	Beil. Ref. den/g cm^{-3}	Solubility n_D
10920	Pyridine, 2-chloro-6-methyl-	C_6H_6ClN 127.57	18368-63-3	183.5; 64^{10}	5-20-05-00498 1.167^{25}	1.5270^{20}
10921	Pyridine, 2-[(4-chlorophenyl)methyl]-	$C_{12}H_{10}ClN$ 203.67	4350-41-8	181-3^{20}	5-20-07-00558 1.39^{25}	1.5868^{20}
10922	Pyridine, 4-(3-cyclohexen-1-yl)- 4-(Cyclohex-3-enyl)pyridine	$C_{11}H_{13}N$ 159.23	70644-46-1 22.1	226	4-20-00-03239 1.0222^{25}	1.5466^{25}
10923	2,3-Pyridinediamine	$C_5H_7N_3$ 109.13	452-58-4 120.8	148-50^5	5-22-11-00241	H_2O 3; EtOH 3; bz 3
10924	2,5-Pyridinediamine 2,5-Diaminopyridine	$C_5H_7N_3$ 109.13	4318-76-7 110.3	182^{12}	5-22-11-00252	H_2O 4; EtOH 4
10925	2,6-Pyridinediamine	$C_5H_7N_3$ 109.13	141-86-6 121.5	285; 148^5	5-22-11-00255	H_2O 2; ace 2
10926	Pyridine, 3,5-dibromo- 3,5-Dibromopyridine	$C_5H_3Br_2N$ 236.89	625-92-3 112	222	5-20-05-00436	H_2O 2; EtOH 3; eth 3
10927	2,3-Pyridinedicarboxylic acid Quinolinic acid	$C_7H_5NO_4$ 167.12	89-00-9 228.5	8102	5-22-04-00115	H_2O 2; EtOH 1; eth 1; bz 1
10928	2,4-Pyridinedicarboxylic acid	$C_7H_5NO_4$ 167.12	499-80-9 249		5-22-04-00119 0.942^{25}	H_2O 2; EtOH 1; eth 1; bz 1
10929	2,5-Pyridinedicarboxylic acid Isocinchomeronic acid	$C_7H_5NO_4$ 167.12	100-26-5 254	5043	5-22-04-00124	H_2O 3; EtOH 1; eth 1; bz 1
10930	3,4-Pyridinedicarboxylic acid Cinchomeronic acid	$C_7H_5NO_4$ 167.12	490-11-9 256	2285 sub	5-22-04-00135	H_2O 2; EtOH 1; eth 1; bz 2
10931	3,5-Pyridinedicarboxylic acid	$C_7H_5NO_4$ 167.12	499-81-0 324	sub	5-22-04-00139	H_2O 1; eth 2; HCl 3; HOAc 2
10932	3,5-Pyridinedicarboxylic acid, 2,6-dimethyl-, diethyl ester	$C_{13}H_{17}NO_4$ 251.28	1149-24-2 71	301; 208^{40}	5-22-04-00162	H_2O 1; EtOH 3; eth 3; bz 3
10933	Pyridine, 2,4-diethyl- 2,4-Diethylpyridine	$C_9H_{13}N$ 135.21	626-21-1 112	187.5	4-20-00-02825 0.9338^{25}	
10934	3,4-Pyridinedimethanol, 5-hydroxy-6-methyl-, hydrochloride Pyridoxine hydrochloride	$C_8H_{12}ClNO_3$ 205.64	58-56-0 207	7995 sub	5-21-05-00492	H_2O 4
10035	Pyridine, 2,6-dimethoxy- 2,6-Dimethoxypyridine	$C_7H_9NO_2$ 139.15	6231-18-1	179	5-21-04-00416 1.053^{25}	1.5029^{20}
10936	Pyridine, 2,3-dimethyl- 2,3-Lutidine	C_7H_9N 107.16	583-61-9	161.2	5-20-06-00015 0.9319^{25}	H_2O 3; EtOH 3; eth 3 1.5057^{20}
10937	Pyridine, 2,4-dimethyl- 2,4-Lutidine	C_7H_9N 107.16	108-47-4 -64	158.5	5-20-06-00019 0.9309^{20}	H_2O 4; EtOH 4; eth 4; ace 3 1.5010^{20}
10938	Pyridine, 2,5-dimethyl- 2,5-Lutidine	C_7H_9N 107.16	589-93-5 -16	157.0	5-20-06-00027 0.9297^{20}	H_2O 2; EtOH 4; eth 5; ace 3 1.5006^{20}
10939	Pyridine, 2,6-dimethyl- 2,6-Lutidine	C_7H_9N 107.16	108-48-5 -6.1	5485 144.1	5-20-06-00032 0.9226^{20}	H_2O 5; EtOH 2; eth 3; ace 3 1.4953^{20}
10940	Pyridine, 3,4-dimethyl- 3,4-Lutidine	C_7H_9N 107.16	583-58-4 -11	179.1	5-20-06-00054 0.9281^{20}	H_2O 2; EtOH 3; eth 3; ace 3 1.5090^{20}
10941	Pyridine, 3,5-dimethyl- 3,5-Lutidine	C_7H_9N 107.16	591-22-0 -6.6	172	5-20-06-00060 0.9419^{20}	H_2O 3; EtOH 3; eth 3; ace 3 1.5061^{20}
10942	Pyridine, 2-(1,1-dimethylethyl)- 2-tert-Butylpyridine	$C_9H_{13}N$ 135.21	5944-41-2 -33	170	5-20-06-00121	1.4891^{20}
10943	Pyridine, 3-(1,1-dimethylethyl)- 3-tert-Butylpyridine	$C_9H_{13}N$ 135.21	38031-78-6 -44	194.8	5-20-06-00122	1.4965^{20}
10944	Pyridine, 4-(1,1-dimethylethyl)-	$C_9H_{13}N$ 135.21	3978-81-2 -41	196.5	5-20-06-00123 0.915^{25}	ctc 3; CS_2 3 1.4958^{20}
10945	Pyridine, 2,6-dimethyl-, 1-oxide	C_7H_9NO 123.15	1073-23-0	115-9^{18}	5-20-06-00040 1.073^{25}	1.5706^{20}
10946	2,4(1H,3H)-Pyridinedione, 3,3-diethyl- Pyrithyldione	$C_9H_{13}NO_2$ 167.21	77-04-3 90.7	8005	5-21-10-00086	
10947	Pyridine, 2,6-diphenyl- 2,6-Diphenylpyridine	$C_{17}H_{13}N$ 231.30	3558-69-8 82	397	5-20-08-00459	eth 4; EtOH 4
10948	2-Pyridineethanamine	$C_7H_{10}N_2$ 122.17	2706-56-1	131.5^{50}	5-22-09-00368 1.0220^{25}	1.5335^{25}
10949	4-Pyridineethanamine	$C_7H_{10}N_2$ 122.17	13258-63-4	121^{10}	4-22-00-04201 1.0302^{25}	H_2O 4 1.5381^{25}
10950	2-Pyridineethanol	C_7H_9NO 123.15	103-74-2 -7.8	190^{200} 170^{100}	5-21-02-00206 1.091^{25}	H_2O 4; EtOH 4; eth 2; chl 4 1.5366^{20}
10951	2-Pyridineethanol, α-methyl- 2-(2-Hydroxypropyl)pyridine	$C_8H_{11}NO$ 137.18	5307-19-7 33	124^{20} 110^{10}	5-21-02-00234	H_2O 4; EtOH 4; chl 4
10952	Pyridine, 2-ethenyl-	C_7H_7N 105.14	100-69-6	159.5	5-20-06-00211 0.9983^{20}	H_2O 2; EtOH 4; eth 4; ace 4 1.5495^{20}
10953	Pyridine, 3-ethenyl-	C_7H_7N 105.14	1121-55-7	162	5-20-06-00212	H_2O 2; EtOH 3; eth 3 1.5530^{20}
10954	Pyridine, 4-ethenyl-	C_7H_7N 105.14	100-43-6	121^{150} 65^{15}	5-20-06-00213 0.9800^{20}	H_2O 3; EtOH 3; eth 2; chl 3 1.5449^{20}
10955	Pyridine, 5-ethenyl-2-methyl- 5-Vinyl-2-picoline	C_8H_9N 119.17	140-76-1		5-20-06-00218	
10956	Pyridine, 2-ethyl- 2-Ethylpyridine	C_7H_9N 107.16	100-71-0 -63.1	148.6	5-20-06-00003 0.9502^{25}	H_2O 3; EtOH 5; eth 4; ace 4 1.4964^{20}
10957	Pyridine, 3-ethyl- 3-Ethylpyridine	C_7H_9N 107.16	536-78-7 -76.9	3802 165	5-20-06-00007 0.9539^{25}	H_2O 3; EtOH 3; eth 3; ace 4 1.5021^{20}

No.	Name / Synonym	Mol. Form. / Mol. Wt.	CAS RN / mp/°C	Merck No. / bp/°C	Beil. Ref. / den/g cm⁻³	Solubility / n_D
10958	Pyridine, 4-ethyl- / 4-Ethylpyridine	C_7H_9N / 107.16	536-75-4 / -90.5	3803 / 168.3	5-20-06-00010 / 0.9417[20]	H_2O 3; EtOH 3; eth 3; ace 4 / 1.5009[20]
10959	Pyridine, 2-ethyl-3,5-dimethyl-	$C_9H_{13}N$ / 135.21	1123-96-2 /	/ 198; 86[15]	5-20-06-00129 / 0.9338[25]	H_2O 3; EtOH 4; eth 4; ace 4
10960	Pyridine, 3-ethyl-2,6-dimethyl-	$C_9H_{13}N$ / 135.21	23580-52-1 /	/ 83-5[23]	5-20-06-00129 / 0.9120[18]	H_2O 2
10961	Pyridine, 4-ethyl-2,6-dimethyl-	$C_9H_{13}N$ / 135.21	36917-36-9 / 97	/ 187.5	5-20-06-00130 / 0.9089[25]	/ 1.4964[25]
10962	Pyridine, 2-ethyl-4-methyl-	$C_8H_{11}N$ / 121.18	2150-18-7 /	/ 175	5-20-06-00086 / 0.9238[20]	ace 4; eth 4; EtOH 4
10963	Pyridine, 2-ethyl-6-methyl-	$C_8H_{11}N$ / 121.18	1122-69-6 /	/ 160.5	5-20-06-00089 / 0.9207[25]	H_2O 2; EtOH 3; eth 3; ace 4 / 1.4920[25]
10964	Pyridine, 3-ethyl-4-methyl- / 3-Ethyl-4-picoline	$C_8H_{11}N$ / 121.18	529-21-5 /	3796 / 198	5-20-06-00090 / 0.9286[17]	H_2O 2; EtOH 3; eth 3; ace 4
10965	Pyridine, 4-ethyl-2-methyl- / 4-Ethyl-2-picoline	$C_8H_{11}N$ / 121.18	536-88-9 /	3797 / 179	5-20-06-00085 / 0.9130[25]	ace 4; bz 4; eth 4; EtOH 4
10966	Pyridine, 5-ethyl-2-methyl- / 5-Ethyl-2-picoline	$C_8H_{11}N$ / 121.18	104-90-5 /	3798 / 178.3	5-20-06-00086 / 0.9202[20]	H_2O 2; EtOH 3; eth 3; ace 4 / 1.4971[20]
10967	Pyridine, 2-(1-ethylpropyl)-	$C_{10}H_{15}N$ / 149.24	7399-50-0 /	/ 195.4	5-20-06-00145 / 0.8981[20]	/ 1.4850[25]
10968	Pyridine, 4-(1-ethylpropyl)- / 4-(3-Pentyl)pyridine	$C_{10}H_{15}N$ / 149.24	35182-51-5 / 125.5	/ 217; 80[12]	5-20-06-00145 / 0.9085[25]	/ 1.40905[25]
10969	Pyridine, hydrochloride / Pyridinium chloride	C_5H_6ClN / 115.56	628-13-7 / 146	/ 222	5-20-05-00160 /	H_2O 4; EtOH 4; chl 4
10970	Pyridine, 2-iodo- / 2-Iodopyridine	C_5H_4IN / 205.00	5029-67-4 /	/ 93[13]	5-20-05-00438 / 1.9735[20]	EtOH 3; eth 3; ace 3; bz 3 / 1.6366[20]
10971	2-Pyridinemethanamine	$C_6H_8N_2$ / 108.14	3731-51-9 /	/ 203; 91[17]	5-22-09-00232 / 1.0525[25]	H_2O 4 / 1.5431[25]
10972	3-Pyridinemethanamine	$C_6H_8N_2$ / 108.14	3731-52-0 / -21.1	/ 226	5-22-09-00301 / 1.064[20]	H_2O 4; eth 4; EtOH 4 / 1.552[20]
10973	4-Pyridinemethanamine	$C_6H_8N_2$ / 108.14	3731-53-1 / -7.6	/ 230; 103[11]	5-22-09-00345 / 1.072[20]	H_2O 4 / 1.5495[25]
10974	2-Pyridinemethanamine, N-(2-pyridinylmethyl)-	$C_{12}H_{13}N_3$ / 199.26	1539-42-0 /	/ 200[10]	5-22-09-00267 / 1.1074[25]	/ 1.5757[25]
10975	2-Pyridinemethanol	C_6H_7NO / 109.13	586-98-1 /	/ 112[16]; 102.5[8]	5-21-02-00150 / 1.1317[20]	H_2O 5; EtOH 4; eth 4; ace 4 / 1.5444[20]
10976	3-Pyridinemethanol / Nicotinyl alcohol	C_6H_7NO / 109.13	100-55-0 / -6.5	6438 / 266	5-21-02-00172 / 1.131[20]	H_2O 4; eth 4 / 1.5455[20]
10977	4-Pyridinemethanol / 4-Picolyl alcohol	C_6H_7NO / 109.13	586-95-8 / 53	/ 140-2[12]	5-21-02-00191 /	chl 3
10978	3-Pyridinemethanol, 4-(aminomethyl)-5-hydroxy-6-methyl-, dihydrochloride / Pyridoxamine dihydrochloride	$C_8H_{14}Cl_2N_2O_2$ / 241.12	524-36-7 / 226-7 dec	7993 /	5-22-12-00324 /	H_2O 4; EtOH 2
10979	Pyridine, 2-(2-methoxyethyl)- / Metyridine	$C_8H_{11}NO$ / 137.18	114-91-0 /	6084 / 203; 96[17]	5-21-02-00206 / 0.988[20]	H_2O 4; EtOH 4 / 1.4975[20]
10980	Pyridine, 2-methyl- / 2-Picoline	C_6H_7N / 93.13	109-06-8 / -66.7	7372 / 129.3	5-20-05-00464 / 0.9443[20]	H_2O 4; EtOH 5; eth 5; ace 4 / 1.4957[20]
10981	Pyridine, 3-methyl- / 3-Picoline	C_6H_7N / 93.13	108-99-6 / -18.1	7373 / 144.1	5-20-05-00506 / 0.9566[20]	H_2O 5; EtOH 5; eth 5; ace 4 / 1.5040[20]
10982	Pyridine, 4-methyl- / 4-Picoline	C_6H_7N / 93.13	108-89-4 / 3.66	7374 / 145.3	5-20-05-00543 / 0.9548[20]	H_2O 5; EtOH 5; eth 5; ace 3 / 1.5037[20]
10983	Pyridine, 2-[2-(1-methylethoxy)ethyl]- / Pyridine, 2-(2-isopropoxyethyl)	$C_{10}H_{15}NO$ / 165.24	70715-19-4 /	/ 133[50]	/ 0.9502[25]	H_2O 4 / 1.4820[25]
10984	Pyridine, 2-(1-methylethyl)-	$C_8H_{11}N$ / 121.18	644-98-4 / -141	/ 159.8	5-20-06-00080 / 0.9342[25]	H_2O 2; EtOH 5; eth 5; ace 4 / 1.4915[20]
10985	Pyridine, 3-(1-methylethyl)- / 3-Isopropylpyridine	$C_8H_{11}N$ / 121.18	6304-18-3 / -45.8	/ 177.5	5-20-06-00081 / 0.9217[16]	
10986	Pyridine, 4-(1-methylethyl)-	$C_8H_{11}N$ / 121.18	696-30-0 / -54.9	/ 178	5-20-06-00082 / 0.9382[25]	H_2O 2; EtOH 5; eth 5; ace 4 / 1.4962[20]
10987	Pyridine, 3-(1-methyl-1H-pyrrol-2-yl)-	$C_{10}H_{10}N_2$ / 158.20	487-19-4 /	/ 281; 150[16]	5-23-07-00227 / 1.241[20]	EtOH 4 / 1.6057[20]
10988	Pyridine, 3-methyl-, 1-oxide	C_6H_7NO / 109.13	1003-73-2 / 39	/ 148[15]	5-20-05-00517 /	chl 3
10989	Pyridine, 2-(4-methylphenyl)-	$C_{12}H_{11}N$ / 169.23	4467-06-5 /	/ 170-80[20]	5-20-07-00569 / 0.999[25]	/ 1.6125[20]
10990	Pyridine, 2-methyl-5-phenyl-	$C_{12}H_{11}N$ / 169.23	3256-88-0 /	/ 189[50]	5-20-07-00573 / 1.0590[25]	/ 1.6055[25]
10991	Pyridine, 3-methyl-2-phenyl- / 3-Picoline, 2-phenyl-	$C_{12}H_{11}N$ / 169.23	10273-90-2 /	/ 148-9[16]	5-20-07-00571 / 1.065[25]	/ 1.6026[20]
10992	Pyridine, 3-(1-methyl-2-pyrrolidinyl)-, (S)- / Nicotine	$C_{10}H_{14}N_2$ / 162.23	54-11-5 / -79	6434 / 247; 125[18]	5-23-06-00064 / 1.0097[20]	H_2O 5; EtOH 4; eth 4; chl 4 / 1.5282[20]
10993	Pyridine, 4-nitro-, 1-oxide / 4-Nitropyridine N-oxide	$C_5H_4N_2O_3$ / 140.10	1124-33-0 / 160.5	/	5-20-05-00448 /	
10994	Pyridine, 1-oxide / Pyridine N-oxide	C_5H_5NO / 95.10	694-59-7 / 65.5	7984 / 146-7[13]	5-20-05-00217 /	
10995	Pyridine, pentachloro-	C_5Cl_5N / 251.33	2176-62-7 / 125.5	/ 280	5-20-05-00422 /	bz 4; EtOH 4; lig 4
10996	Pyridine, 2-(3-pentenyl)-	$C_{10}H_{13}N$ / 147.22	2057-43-4 /	/ 216; 93[12]	5-20-06-00355 / 0.9234[25]	/ 1.5076[25]

No.	Name Synonym	Mol. Form. Mol. Wt.	CAS RN mp/°C	Merck No. bp/°C	Beil. Ref. den/g cm^{-3}	Solubility n_D
10997	Pyridine, 2-phenyl- 2-Phenylpyridine	$C_{11}H_9N$ 155.20	1008-89-5 	 271	5-20-07-00545 1.0833[25]	H_2O 2; EtOH 5; eth 5 1.6210[20]
10998	Pyridine, 3-phenyl- 3-Phenylpyridine	$C_{11}H_9N$ 155.20	1008-88-4 164	 272	5-20-07-00548 	H_2O 2; EtOH 3; eth 3 1.6123[25]
10999	Pyridine, 4-phenyl- 4-Phenylpyridine	$C_{11}H_9N$ 155.20	939-23-1 77.5	 281	5-20-07-00549 	H_2O 3; EtOH 3; eth 3
11000	Pyridine, 2-(2-phenylethenyl)-, (E)-	$C_{13}H_{11}N$ 181.24	538-49-8 91.5	 325; 194[14]	5-20-08-00057 1.0147[100]	H_2O 1; EtOH 3; eth 3; bz 3
11001	Pyridine, 4-(2-phenylethenyl)-, (E)-	$C_{13}H_{11}N$ 181.24	5097-93-8 131	 208[33]	5-20-08-00063 	H_2O 1; EtOH 3; eth 3; chl 3
11002	Pyridine, 2-(2-phenylethenyl)-, (Z)-	$C_{13}H_{11}N$ 181.24	1519-59-1 -50	 141[10]	4-20-00-03874 	EtOH 3; eth 3; bz 3
11003	Pyridine, 2-(2-phenylethyl)-	$C_{13}H_{13}N$ 183.25	2116-62-3 -3	 289	5-20-07-00588 1.0465[0]	
11004	Pyridine, 2-(phenylmethyl)- 2-Benzylpyridine	$C_{12}H_{11}N$ 169.23	101-82-6 12.5	 277; 149[16]	5-20-07-00556 1.067[0]	H_2O 1; EtOH 3; eth 3; chl 3 1.5785[20]
11005	Pyridine, 3-(phenylmethyl)- 3-Benzylpyridine	$C_{12}H_{11}N$ 169.23	620-95-1 34	 287	5-20-07-00560 1.061[25]	eth 4; EtOH 4
11006	Pyridine, 4-(phenylmethyl)-	$C_{12}H_{11}N$ 169.23	2116-65-6 12.4	 288; 180[31]	5-20-07-00561 1.0612[20]	H_2O 1; EtOH 3; eth 4; ctc 3 1.5818[20]
11007	Pyridine, 4-(3-phenylpropyl)-	$C_{14}H_{15}N$ 197.28	2057-49-0 	 322; 150[5]	5-20-07-00607 1.024[25]	bz 4; eth 4; py 4; EtOH 4 1.5616[25]
11008	Pyridine, 3-(2-piperidinyl)-, (S)- Anabasine	$C_{10}H_{14}N_2$ 162.23	494-52-0 9	655 276	5-23-06-00096 1.0436[20]	H_2O 5; EtOH 3; eth 3; bz 3 1.5430[20]
11009	Pyridine, 4-(4-piperidinyl)-	$C_{10}H_{14}N_2$ 162.23	581-45-3 79	 292	5-23-06-00115 	H_2O 2; EtOH 4; eth 3; bz 3
11010	2-Pyridinepropanamine, N,N-dimethyl-γ-phenyl- Pheniramine	$C_{16}H_{20}N_2$ 240.35	86-21-5 	7198 181[13]; 135[0.5]	5-22-10-00487 1.0081[25]	bz 4; eth 4; EtOH 4; chl 4 1.5519[25]
11011	1(4H)-Pyridinepropanoic acid, α-amino-3-hydroxy-4-oxo-, (S)- Mimosine	$C_8H_{10}N_2O_4$ 198.18	500-44-7 228 dec	6118 	5-21-12-00107 	H_2O 2; EtOH 1; eth 1; ace 1
11012	2-Pyridinepropanol 2-(3-Hydroxypropyl)pyridine	$C_8H_{11}NO$ 137.18	2859-68-9 34	 260.2; 116[4]	5-21-02-00236 1.060[25]	H_2O 4 1.5298[20]
11013	3-Pyridinepropanol	$C_8H_{11}NO$ 137.18	2859-67-8 	 284; 130[3]	5-21-02-00237 1.063[25]	H_2O 4 1.5313[20]
11014	Pyridine, 2-(2-propenyl)-	C_8H_9N 119.17	2835-33-8 	 190; 63[15]	5-20-06-00216 0.959[20]	eth 4; EtOH 4
11015	Pyridine, 2-propyl-	$C_8H_{11}N$ 121.18	622-39-9 2	 167	5-20-06-00076 0.9119[20]	H_2O 2; EtOH 5; eth 5; ace 4 1.4925[20]
11016	Pyridine, 4-propyl-	$C_8H_{11}N$ 121.18	1122-81-2 	 185	5-20-06-00078 0.9381[15]	eth 4; EtOH 4 1.4966[20]
11017	Pyridine, 3-(2-pyrrolidinyl)-, (S)- Nornicotine	$C_9H_{12}N_2$ 148.21	494-97-3 	6631 270	4-23-00-00998 1.0737[19]	H_2O 4; ace 4; eth 4; EtOH 4 1.5378[18]
11018	3-Pyridinesulfonic acid 3-Pyridylsulfonic acid	$C_5H_5NO_3S$ 159.17	636-73-7 357 dec	 	5-22-07-00552 1.713[25]	H_2O 4; EtOH 2; eth 1
11019	Pyridine, 2,3,5,6-tetrachloro- 2,3,5,6-Tetrachloropyridine	C_5HCl_4N 216.88	2402-79-1 90.5	 250.5	5-20-05-00421 	eth 4; EtOH 4; peth 4
11020	Pyridine, 1,2,3,6-tetrahydro- Δ 3-Piperidine	C_5H_9N 83.13	694-05-3 -48	 108	5-20-04-00282 0.911[25]	chl 3 1.4800[20]
11021	Pyridine, 2,3,4,5-tetrahydro-6-propyl- γ-Coniceine	$C_8H_{15}N$ 125.21	1604-01-9 	2497 174	5-20-04-00345 0.8753[15]	 1.4661[16]
11022	Pyridine, 2,3,4,6-tetramethyl- 2,3,4,6-Tetramethylpyridine	$C_9H_{13}N$ 135.21	20820-82-0 	 203	5-20-06-00132 0.9322[25]	eth 4; EtOH 4 1.5087[25]
11023	Pyridine, 2-(trichloromethyl)- 2-(Trichloromethyl)pyridine	$C_6H_4Cl_3N$ 196.46	4377-37-1 -10	 125-6[25]	5-20-05-00499 1.4526[25]	 1.5596[25]
11024	Pyridine, 2,3,4-trimethyl- 2,3,4-Collidine	$C_8H_{11}N$ 121.18	2233-29-6 	 192.5	5-20-06-00091 0.9127[15]	H_2O 3; EtOH 3; eth 3 1.5150[20]
11025	Pyridine, 2,3,5-trimethyl- 2,3,5-Collidine	$C_8H_{11}N$ 121.18	695-98-7 	 184	5-20-06-00092 0.9352[19]	H_2O 2; EtOH 3; eth 3; ace 3 1.5057[25]
11026	Pyridine, 2,3,6-trimethyl- 2,3,6-Collidine	$C_8H_{11}N$ 121.18	1462-84-6 	 171.6	5-20-06-00092 0.9220[25]	H_2O 3; EtOH 3; eth 3; ace 3 1.5053[20]
11027	Pyridine, 2,4,5-trimethyl- 2,4,5-Collidine	$C_8H_{11}N$ 121.18	1122-39-0 	 188	5-20-06-00093 0.9330[25]	ace 4; bz 4; eth 4; EtOH 4 1.5054[25]
11028	Pyridine, 2,4,6-trimethyl- 2,4,6-Collidine	$C_8H_{11}N$ 121.18	108-75-8 -46	9632 170.6	5-20-06-00093 0.9166[22]	H_2O 3; EtOH 3; eth 3; ace 3 1.4959[25]
11029	Pyridinium, 3-carboxy-1-methyl-, hydroxide, inner salt Trigonelline	$C_7H_7NO_2$ 137.14	535-83-1 	9606 	5-22-02-00143 	H_2O 4
11030	Pyridinium, 1-hexadecyl-, chloride Hexadecyl pyridinium chloride	$C_{21}H_{38}ClN$ 339.99	123-03-5 80	2024 	5-20-05-00233 	H_2O 4; chl 4
11031	2-Pyridinol	C_5H_5NO 95.10	72762-00-6 107.8	 	5-21-07-00106 1.3910[20]	H_2O 4; bz 4; EtOH 4
11032	4-Pyridinol	C_5H_5NO 95.10	626-64-2 149.8	 >350; 257[10]	5-21-07-00152 	H_2O 3; EtOH 3; eth 1; bz 1
11033	2(1H)-Pyridinone	C_5H_5NO 95.10	142-08-5 107.8	 280	5-21-07-00106 1.3910[20]	H_2O 3; EtOH 3; eth 2; bz 3
11034	2(1H)-Pyridinone, hydrazone	$C_5H_7N_3$ 109.13	4930-98-7 46.6	 185[140]; 90[1]	5-22-14-00486 	chl 3

No.	Name / Synonym	Mol. Form. / Mol. Wt.	CAS RN / mp/°C	Merck No. / bp/°C	Beil. Ref. / den/g cm⁻³ / n_D	Solubility
11035	2(1H)-Pyridinone, 1-methyl-	C_6H_7NO / 109.13	694-85-9 / 31	250	5-21-07-00109 / 1.1120^{20}	H_2O 5; peth 2; lig 2
11036	2(1H)-Pyridinone, 3-methyl-	C_6H_7NO / 109.13	1003-56-1 / 141.5	289	5-21-07-00199	H_2O 4; EtOH 4; chl 4
11037	2(1H)-Pyridinone, 4-methyl-	C_6H_7NO / 109.13	13466-41-6	308	5-21-07-00201	H_2O 3; EtOH 3; eth 2; bz 2
11038	3H-Pyrido[3,4-b]indole, 4,9-dihydro-7-methoxy-1-methyl- / Harmaline	$C_{13}H_{14}N_2O_{13}$ / 406.26	304-21-2 / 250 dec	4528	5-23-12-00148	H_2O 2; EtOH 2; eth 2; chl 3
11039	9H-Pyrido[3,4-b]indole, 7-methoxy-1-methyl- / Harmine	$C_{13}H_{12}N_2O$ / 212.25	442-51-3 / 273	4531 sub	5-23-12-00237	H_2O 2; EtOH 2; eth 2; chl 2
11040	9H-Pyrido[3,4-b]indole, 1-methyl- / Harman	$C_{12}H_{10}N_2$ / 182.22	486-84-0 / 236.5	4530	5-23-08-00261	
11041	9H-Pyrido[3,4-b]indole, 1-[(2-methylphenyl)methyl]-	$C_{19}H_{16}N_2$ / 272.35	525-15-5 / 217.8	$150^{0.01}$	5-23-09-00536	EtOH 3; eth 3; bz 3; chl 3
11042	2-Pyrimidinamine	$C_4H_5N_3$ / 95.10	109-12-6 / 127.5	sub	4-25-00-02071	H_2O 3; chl 2
11043	4-Pyrimidinamine, 2-chloro-N,N,6-trimethyl- / Crimidine	$C_7H_{10}ClN_3$ / 171.63	535-89-7 / 87	2590 $140-7^4$	4-25-00-02184	EtOH 4
11044	2-Pyrimidinamine, 4,5-dimethyl- / 2-Amino-4,5-dimethylpyrimidine	$C_6H_9N_3$ / 123.16	1193-74-4 / 215.8	sub	4-25-00-02203	ace 4; bz 4; EtOH 4; chl 4
11045	4-Pyrimidinamine, 2,6-dimethyl- / Kyanmethin	$C_6H_9N_3$ / 123.16	461-98-3 / 183	5205 sub	4-25-00-02192	H_2O 2; EtOH 2; bz 2; chl 2
11046	2-Pyrimidinamine, 4-methyl-	$C_5H_7N_3$ / 109.13	108-52-1 / 160.3	sub	4-25-00-02152	H_2O 3; EtOH 3; chl 2
11047	2-Pyrimidinamine, 5-methyl-	$C_5H_7N_3$ / 109.13	50840-23-8 / 193.5	sub	5-25-10-00191	EtOH 4
11048	4-Pyrimidinamine, 6-methyl-	$C_5H_7N_3$ / 109.13	3435-28-7 / 197	sub	4-25-00-02182	H_2O 3; EtOH 3
11049	Pyrimidine / 1,3-Diazine	$C_4H_4N_2$ / 80.09	289-95-2 / 22	7998 123.8	5-23-05-00334 / 1.4998^{20}	H_2O 5; EtOH 3
11050	4-Pyrimidinecarboxylic acid, 1,2,3,6-tetrahydro-2,6-dioxo- / Orotic acid	$C_5H_4N_2O_4$ / 156.10	65-86-1 / 345.5	6828	4-25-00-01759	H_2O 2; os 1
11051	2,4-Pyrimidinediamine, 5-(4-chlorophenyl)-6-ethyl- / Pyrimethamine	$C_{12}H_{13}ClN_4$ / 248.71	58-14-0 / 233.5	7997	4-25-00-03014	
11052	Pyrimidine, 2,4-dichloro-5-methyl- / 2,4-Dichloro-5-methylpyrimidine	$C_5H_4Cl_2N_2$ / 163.01	1780-31-0 / 26	235	5-23-05-00385	H_2O 2; EtOH 4; eth 4; bz 4
11053	Pyrimidine, 2,4-dichloro-6-methyl- / 2,6-Dichloro-4-methylpyrimidine	$C_5H_4Cl_2N_2$ / 163.01	5424-21-5 / 46.5	219	5-23-05-00381	bz 4; eth 4; EtOH 4; chl 4
11054	Pyrimidine, 2,4-dichloro-5-nitro- / 2,4-Dichloro-5-nitropyrimidine	$C_4HCl_2N_3O_2$ / 193.98	49845-33-2 / 29.3	154^{58}, 135^{17}	5-23-05-00350	EtOH 2; eth 2
11055	Pyrimidine, 2,4-dimethyl- / 2,4-Dimethylpyrimidine	$C_6H_8N_2$ / 108.14	14331-54-5 / -4	150.5	5-23-05-00397 / 1.168^{14}	1.4880^{25}
11056	Pyrimidine, 2,5-dimethyl- / 2,5-Dimethylpyrimidine	$C_6H_8N_2$ / 108.14	22868-76-4 / 19	59^{13}	5-23-05-00399	H_2O 4
11057	Pyrimidine, 4,5-dimethyl- / 4,5-Dimethylpyrimidine	$C_6H_8N_2$ / 108.14	694-81-5 / 3	177	5-23-05-00399	H_2O 4
11058	Pyrimidine, 4,6-dimethyl- / 4,6-Dimethylpyrimidine	$C_6H_8N_2$ / 108.14	1558-17-4 / 25	159	5-23-05-00400	H_2O 4 / 1.4880^{20}
11059	2,4(1H,3H)-Pyrimidinedione / Uracil	$C_4H_4N_2O_2$ / 112.09	66-22-8 / 338	9761	5-24-06-00043	H_2O 2; EtOH 4; eth 4
11060	2,4(1H,3H)-Pyrimidinedione, 6-amino-3-ethyl-1-(2-propenyl)- / Aminometradine	$C_9H_{13}N_3O_2$ / 195.22	642-44-4 / 143	463	4-25-00-04113	
11061	2,4(1H,3H)-Pyrimidinedione, 5-[bis(2-chloroethyl)amino]- / Uracil mustard	$C_8H_{11}Cl_2N_3O_2$ / 252.10	66-75-1 / 206 dec	9762	4-25-00-04127	H_2O 2
11062	2,4(1H,3H)-Pyrimidinedione, 5-bromo- / 5-Bromouracil	$C_4H_3BrN_2O_2$ / 190.98	51-20-7 / 310	1430	5-24-06-00322	
11063	4,6(1H,5H)-Pyrimidinedione, dihydro-5-(2-methylpropyl)-5-(2-propenyl)-2-thioxo-, monosodium salt / Barbituric acid, 5-allyl-5-isobutyl-2-thio, sodium salt	$C_{11}H_{15}N_2NaO_2S$ / 262.31	510-90-7	1518		H_2O 4; EtOH 2; eth 1; bz 1
11064	4,6(1H,5H)-Pyrimidinedione, 5-ethyldihydro-5-(2-methyl-2-propenyl)-2-thioxo- / Methallatal	$C_{10}H_{14}N_2O_2S$ / 226.30	115-56-0 / 160.5	5855	5-24-09-00224	
11065	4,6(1H,5H)-Pyrimidinedione, 5-ethyldihydro-5-(1-methylpropyl)-2-thioxo- / Barbituric acid, 5-sec-butyl-5-ethyl-2-thio-	$C_{10}H_{16}N_2O_2S$ / 228.32	2095-57-0 / 169	9250	5-24-09-00164	
11066	4,6(1H,5H)-Pyrimidinedione, 5-ethyldihydro-5-phenyl- / Primidone	$C_{12}H_{14}N_2O_2$ / 218.26	125-33-7 / 281.5	7753	5-24-08-00102	
11067	2,4(1H,3H)-Pyrimidinedione, 5-fluoro- / 5-Fluorouracil	$C_4H_3FN_2O_2$ / 130.08	51-21-8 / 283	4109 sub 190	5-24-06-00271	
11068	2,4(1H,3H)-Pyrimidinedione, 5-methyl- / Thymine	$C_5H_6N_2O_2$ / 126.11	65-71-4 / 316	9332	5-24-07-00055	H_2O 2; EtOH 2; eth 2

No.	Name Synonym	Mol. Form. Mol. Wt.	CAS RN mp/°C	Merck No. bp/°C	Beil. Ref. den/g cm⁻³	Solubility n_D
11069	2,4(1H,3H)-Pyrimidinedione, 6-methyl- 6-Methyluracil	$C_5H_6N_2O_2$ 126.11	626-48-2 270-80 dec	6051	5-24-07-00006	H_2O 3; EtOH 3; eth 2; tfa 2
11070	Pyrimidine, 2-methyl- 2-Methyl-1,3-diazine	$C_5H_6N_2$ 94.12	5053-43-0 -4	138	5-23-05-00376	H_2O 5
11071	Pyrimidine, 4-methyl- 4-Methyl-1,3-diazine	$C_5H_6N_2$ 94.12	3438-46-8 32	142	5-23-05-00379 1.030¹⁶	H_2O 5 1.500²⁰
11072	Pyrimidine, 5-methyl- 5-Methyl-1,3-diazine	$C_5H_6N_2$ 94.12	2036-41-1 30.5	153	5-23-05-00383	H_2O 4
11073	2,4,5,6(1H,3H)-Pyrimidinetetrone Alloxan	$C_4H_2N_2O_4$ 142.07	50-71-5 256 dec	281 sub	5-24-09-00382	H_2O 4; EtOH 3; ace 3; bz 3
11074	2,4,5,6(1H,3H)-Pyrimidinetetrone, 5-oxime Violuric acid	$C_4H_3N_3O_4$ 157.09	87-39-8 203-4 dec	9904	5-24-09-00386	H_2O 2; EtOH 3
11075	Pyrimidine, 2,4,6-trichloro- 2,4,6-Trichloropyrimidine	$C_4HCl_3N_2$ 183.42	3764-01-0 22.5	212.5	5-23-05-00347	1.5700²⁰
11076	2,4,6(1H,3H,5H)-Pyrimidinetrione Barbituric acid	$C_4H_4N_2O_3$ 128.09	67-52-7 248	973 260 dec	5-24-09-00078	H_2O 3; EtOH 2; eth 3
11077	2,4,6(1H,3H,5H)-Pyrimidinetrione, 5-amino- Uramil	$C_4H_5N_3O_3$ 143.10	118-78-5 >400	9763	4-25-00-04228	H_2O 3; eth 1; bz 1; chl 3
11078	2,4,6(1H,3H,5H)-Pyrimidinetrione, 5-(2-bromo-2-propenyl)-5-(1-methylethyl)- Propallylonal	$C_{10}H_{13}BrN_2O_3$ 289.13	545-93-7 181	7807	4-24-00-01994	H_2O 2; EtOH 4; eth 2; ace 4
11079	2,4,6(1H,3H,5H)-Pyrimidinetrione, 5-(2-bromo-2-propenyl)-1-methyl-5-(1-methylethyl)- Narcobarbital	$C_{11}H_{15}BrN_2O_3$ 303.16	125-55-3 115	6342	4-24-00-01995	H_2O 2; EtOH 3; py 3
11080	2,4,6(1H,3H,5H)-Pyrimidinetrione, 5-(2-bromo-2-propenyl)-5-(1-methylpropyl)- Butallylonal	$C_{11}H_{15}BrN_2O_3$ 303.16	1142-70-7 131.5	1503	5-24-09-00231	eth 4; EtOH 4
11081	2,4,6(1H,3H,5H)-Pyrimidinetrione, 5-butyl-1-cyclohexyl- Bucolome	$C_{14}H_{22}N_2O_3$ 266.34	841-73-6 84	1452 185-7⁰·⁸	5-24-09-00132	
11082	2,4,6(1H,3H,5H)-Pyrimidinetrione, 5-butyl-5-ethyl- Butethal	$C_{10}H_{16}N_2O_3$ 212.25	77-28-1 128.5	1515	5-24-09-00161	
11083	2,4,6(1H,3H,5H)-Pyrimidinetrione, 5-(1-cyclohexen-1-yl)-1,5-dimethyl- Hexobarbital	$C_{12}H_{16}N_2O_3$ 236.27	56-29-1 146.5	4625	5-24-09-00253	
11084	2,4,6(1H,3H,5H)-Pyrimidinetrione, 5-(1-cyclohexen-1-yl)-5-ethyl- Cyclobarbital	$C_{12}H_{16}N_2O_3$ 236.27	52-31-3 173	2717	5-24-09-00258	H_2O 1; EtOH 4; eth 3; dil alk 3
11085	2,4,6(1H,3H,5H)-Pyrimidinetrione, 5-(2-cyclopenten-1-yl)-5-(2-propenyl)- Cyclopentobarbital	$C_{12}H_{14}N_2O_3$ 234.25	76-68-6 139.5	2751	5-24-09-00269	H_2O 2; EtOH 4
11086	2,4,6(1H,3H,5H)-Pyrimidinetrione, 5,5-dibromo- 	$C_4H_2Br_2N_2O_3$ 285.88	511-67-1 235	dec	4-24-00-01881	eth 4; EtOH 4
11087	2,4,6(1H,3H,5H)-Pyrimidinetrione, 5,5-diethyl- Barbital	$C_8H_{12}N_2O_3$ 184.19	57-44-3 190	972	5-24-09-00137 1.220²⁵	H_2O 2; EtOH 3; eth 3; ace 3
11088	2,4,6(1H,3H,5H)-Pyrimidinetrione, 5,5-diethyl-1-methyl- Metharbital	$C_9H_{14}N_2O_3$ 198.22	50-11-3 150.5	5873	5-24-09-00144	H_2O 3; chl 2
11089	2,4,6(1H,3H,5H)-Pyrimidinetrione, 5,5-diethyl-1-phenyl- Phenetharbital	$C_{14}H_{16}N_2O_3$ 260.29	357-67-5 178	7183	4-24-00-01908	EtOH 4
11090	2,4,6(1H,3H,5H)-Pyrimidinetrione, 5-(2,2-dimethylpropyl)-5-(2-propenyl)- Nealbarbital	$C_{12}H_{18}N_2O_3$ 238.29	561-83-1 156	6350	5-24-09-00238	ace 4; eth 4; EtOH 4
11091	2,4,6(1H,3H,5H)-Pyrimidinetrione, 5,5-di-2-propenyl- Allobarbital	$C_{10}H_{12}N_2O_3$ 208.22	52-43-7 172	252	5-24-09-00245	H_2O 2; EtOH 3; eth 3; bz 3
11092	2,4,6(1H,3H,5H)-Pyrimidinetrione, 5-ethyl-5-(3-methylbutyl)- Amobarbital	$C_{11}H_{18}N_2O_3$ 226.28	57-43-2 157	601	5-24-09-00173	bz 4; EtOH 4; chl 4
11093	2,4,6(1H,3H,5H)-Pyrimidinetrione, 5-ethyl-5-(1-methylethyl)- Probarbital	$C_9H_{14}N_2O_3$ 198.22	76-76-6 203	7759	5-24-09-00159	H_2O 2; EtOH 1; eth 3
11094	2,4,6(1H,3H,5H)-Pyrimidinetrione, 5-ethyl-1-methyl-5-phenyl- Mephobarbital	$C_{13}H_{14}N_2O_3$ 246.27	115-38-8 176	5742	5-24-09-00294	H_2O 2; EtOH 4; eth 2; chl 3
11095	2,4,6(1H,3H,5H)-Pyrimidinetrione, 5-ethyl-5-phenyl- Phenobarbital	$C_{12}H_{12}N_2O_3$ 232.24	50-06-6 174	7201	5-24-09-00286	H_2O 1; EtOH 3; eth 3; bz 1
11096	2,4,6(1H,3H,5H)-Pyrimidinetrione, 5-ethyl-5-(1-piperidinyl)- 5-Ethyl-5-(1-piperidyl)barbituric acid	$C_{11}H_{17}N_3O_3$ 239.27	509-87-5 215	3800	4-25-00-04239	ace 4; eth 4; EtOH 4
11097	2,4,6(1H,3H,5H)-Pyrimidinetrione, 5-[(hexahydro-2,4,6-trioxo-5-pyrimidinyl)imino]-, monoammonium salt Murexide	$C_8H_{10}N_6O_7$ 302.20	3051-09-0	6215	4-25-00-04236	H_2O 2; EtOH 1; eth 1; alk 3
11098	2,4,6(1H,3H,5H)-Pyrimidinetrione, 5-(1-methylethyl)-5-(2-propenyl)- Aprobarbital	$C_{10}H_{14}N_2O_3$ 210.23	77-02-1 141	782	5-24-09-00224	ace 4; eth 4; EtOH 4; chl 4

No.	Name / Synonym	Mol. Form. / Mol. Wt.	CAS RN / mp/°C	Merck No. / bp/°C	Beil. Ref. / den/g cm^{-3}	Solubility / n_D
11099	2,4,6(1H,3H,5H)-Pyrimidinetrione, 1-methyl-5-(1-methylethyl)-5-(2-propenyl)- Enallylpropymal	C$_{11}$H$_{16}$N$_2$O$_3$ 224.26	1861-21-8 56.5	3523 176-8[12]	5-24-09-00225	bz 4; eth 4; EtOH 4; chl 4
11100	2,4,6(1H,3H,5H)-Pyrimidinetrione, 5-methyl-5-phenyl- Phenylmethylbarbituric acid	C$_{11}$H$_{10}$N$_2$O$_3$ 218.21	76-94-8 220	7275	5-24-09-00283	H$_2$O 1; EtOH 3; eth 3; alk 3
11101	2,4,6(1H,3H,5H)-Pyrimidinetrione, 5-(2-methylpropyl)-5-(2-propenyl)- Butalbital	C$_{11}$H$_{16}$N$_2$O$_3$ 224.26	77-26-9 138.5	1502	5-24-09-00231	H$_2$O 2; EtOH 3; eth 3; ace 3
11102	2,4,6(1H,3H,5H)-Pyrimidinetrione, 5-nitro- 5-Nitrobarbituric acid	C$_4$H$_3$N$_3$O$_5$ 173.09	480-68-2 180.5	6507	5-24-09-00105	H$_2$O 3; EtOH 3; eth 1
11103	2,4,6(1H,3H,5H)-Pyrimidinetrione, 5-phenyl-5-(2-propenyl)- Phenallymal	C$_{13}$H$_{12}$N$_2$O$_3$ 244.25	115-43-5 156.5	7162	5-24-09-00334	H$_2$O 2; EtOH 4; eth 4; bz 2
11104	2(1H)-Pyrimidinone, 4-amino- Cytosine	C$_4$H$_5$N$_3$O 111.10	71-30-7 320-5 dec	2801	4-25-00-03654	H$_2$O 3; EtOH 2; eth 1; chl 2
11105	2(1H)-Pyrimidinone, 4-amino-5-(hydroxymethyl)- 5-Hydroxymethylcytosine	C$_5$H$_7$N$_3$O$_2$ 141.13	1123-95-1 >300 dec			
11106	2(1H)-Pyrimidinone, 4-amino-5-methyl- 5-Methylcytosine	C$_5$H$_7$N$_3$O 125.13	554-01-8 270 dec	5970	4-25-00-03727	H$_2$O 3; EtOH 2; eth 1; acid 3
11107	4(1H)-Pyrimidinone, 2,6-diamino-5-(β-D-glucopyranosyloxy)- Vicine	C$_{10}$H$_{16}$N$_4$O$_7$ 304.26	152-93-2 239-42 dec	9880	4-25-00-04285	H$_2$O 2; EtOH 2; acid 4; alk 4
11108	4(1H)-Pyrimidinone, 2,3-dihydro-6-methyl-2-thioxo- Methylthiouracil	C$_5$H$_6$N$_2$OS 142.18	56-04-2 330 dec	6048 sub	5-24-07-00049	H$_2$O 1; EtOH 2; eth 2; bz 2
11109	4(1H)-Pyrimidinone, 2,3-dihydro-6-propyl-2-thioxo- Propylthiouracil	C$_7$H$_{10}$N$_2$OS 170.24	51-52-5 219	7882	5-24-07-00257	H$_2$O 2; EtOH 2; eth 1; bz 1
11110	4(1H)-Pyrimidinone, 2,3-dihydro-2-thioxo- 2-Thiouracil	C$_4$H$_4$N$_2$OS 128.15	141-90-2 >340 dec	9298	5-24-06-00361	H$_2$O 2; EtOH 2; anh HF 3
11111	5H-Pyrimido[4,5-d]thiazolo[3,2-a]pyrimidine-8-ethanol, 2,7-dimethyl- Thiochrome	C$_{12}$H$_{14}$N$_4$OS 262.34	92-35-3 228.8	sub	4-27-00-09599	H$_2$O 3; EtOH 2; eth 2; ace 2
11112	1H-Pyrrole Azole	C$_4$H$_5$N 67.09	109-97-7 -23.4	8025 129.7	5-20-05-00003 0.9698[20]	H$_2$O 2; EtOH 3; eth 3; ace 3 1.5085[20]
11113	1H-Pyrrole, 1-butyl-	C$_8$H$_{13}$N 123.20	589-33-3	170.5	5-20-05-00010 0.8747[20]	1.4727[20]
11114	1H-Pyrrole-2-carboxaldehyde	C$_5$H$_5$NO 95.10	1003-29-8 46.5	218	5-21-07-00169	chl 2; os 4; lig 2 1.5939[16]
11115	1H-Pyrrole-3-carboxylic acid, 2,4-dimethyl-, ethyl ester	C$_9$H$_{13}$NO$_2$ 167.21	2199-51-1 78.5	291	5-22-01-00427	eth 4; EtOH 4
11116	1H-Pyrrole-3-carboxylic acid, 2,5-dimethyl-, ethyl ester	C$_9$H$_{13}$NO$_2$ 167.21	2199-52-2 117.5	291; 130[15]	5-22-01-00432	EtOH 4
11117	1H-Pyrrole, 3,4-diethyl-2-methyl- 3,4-Diethyl-2-methylpyrrole	C$_9$H$_{15}$N 137.22	34874-30-1 202.5		4-20-00-02171 0.9100[17]	1.4988[17]
11118	1H-Pyrrole, 2,5-dihydro- 3-Pyrroline	C$_4$H$_7$N 69.11	109-96-6 90.5	8028	5-20-04-00274 0.9097[20]	H$_2$O 4; ace 4; eth 4; EtOH 4 1.4664[20]
11119	1H-Pyrrole, 2,4-dimethyl- 2,4-Dimethylpyrrole	C$_6$H$_9$N 95.14	625-82-1 168		5-20-05-00063 0.9236[20]	H$_2$O 2; EtOH 4; eth 4; bz 4 1.5048[20]
11120	1H-Pyrrole, 2,5-dimethyl-	C$_6$H$_9$N 95.14	625-84-3 171; 51[8]		5-20-05-00064 0.9353[20]	H$_2$O 1; EtOH 4; eth 4 1.5036[20]
11121	1H-Pyrrole-2,5-dione	C$_4$H$_3$NO$_2$ 97.07	541-59-3 94	sub	5-21-10-00003 1.2493[106]	H$_2$O 3; EtOH 3; eth 3
11122	1H-Pyrrole-2,5-dione, 1-ethyl- N-Ethylmaleimide	C$_6$H$_7$NO$_2$ 125.13	128-53-0 45.5	3778	5-21-10-00006	H$_2$O 2; EtOH 4; eth 4; chl 3
11123	1H-Pyrrole-2,5-dione, 1-phenyl- N-Phenylmaleimide	C$_{10}$H$_7$NO$_2$ 173.17	941-69-5 90.5	7270 162[12]	5-21-10-00009	bz 4; eth 4; EtOH 4
11124	1H-Pyrrole, 1-ethyl-	C$_6$H$_9$N 95.14	617-92-5 129.5		5-20-05-00010 0.9009[20]	EtOH 4 1.4841[20]
11125	1H-Pyrrole, 2-ethyl-	C$_6$H$_9$N 95.14	1551-06-0 164		5-20-05-00061 0.9042[20]	H$_2$O 1; EtOH 3; chl 3 1.4942[20]
11126	1H-Pyrrole, 3-ethyl-2,4-dimethyl-	C$_8$H$_{13}$N 123.20	517-22-6 0	199; 96[16]	5-20-05-00099 0.913[20]	H$_2$O 2; EtOH 3; eth 3; bz 3 1.4961[20]
11127	1H-Pyrrole, 4-ethyl-2,3-dimethyl-	C$_8$H$_{13}$N 123.20	491-18-9 16.5	200; 113[16]	4-20-00-02152 0.915[20]	H$_2$O 4
11128	1H-Pyrrole, 3-ethyl-4-methyl-	C$_7$H$_{11}$N 109.17	488-92-6 3	70[11]	5-20-05-00083 0.9059[20]	eth 4; EtOH 4 1.4913[20]
11129	Pyrrole, 2-isopropyl-	C$_7$H$_{11}$N 109.17	7696-51-7	172	5-20-05-00081 0.908[25]	eth 4; EtOH 4 1.491[25]
11130	1H-Pyrrole, 1-methyl-	C$_5$H$_7$N 81.12	96-54-8	115	5-20-05-00008 0.9145[15]	H$_2$O 1; EtOH 5; eth 5 1.4875[20]
11131	1H-Pyrrole, 2-methyl-	C$_5$H$_7$N 81.12	636-41-9 -35.6	147	5-20-05-00053 0.9446[15]	H$_2$O 1; EtOH 5; eth 5 1.5035[16]
11132	1H-Pyrrole, 3-methyl- 3-Methylpyrrole	C$_5$H$_7$N 81.12	616-43-3 -48.4	143; 45[11]	5-20-05-00056	EtOH 5; eth 5 1.4970[20]
11133	1H-Pyrrole, 2,2'-methylenebis-	C$_9$H$_{10}$N$_2$ 146.19	21211-65-4 73	164[12]	5-23-06-00468	bz 4; eth 4; EtOH 4

No.	Name / Synonym	Mol. Form. / Mol. Wt.	CAS RN / mp/°C	Merck No. / bp/°C	Beil. Ref. / den/g cm⁻³	Solubility / n_D
11134	1H-Pyrrole, 1-phenyl-	$C_{10}H_9N$ 143.19	635-90-5 62	234	5-20-05-00012	H_2O 1; EtOH 3; eth 3; ace 3
11135	1H-Pyrrole, 2-phenyl- 2-Phenylpyrrole	$C_{10}H_9N$ 143.19	3042-22-6 129	272	5-20-07-00369	H_2O 1; EtOH 4; eth 4; bz 4
11136	1H-Pyrrole, 1-(phenylmethyl)-	$C_{11}H_{11}N$ 157.22	2051-97-0 15	247	5-20-05-00013 1.0183[20]	H_2O 1; EtOH 4; eth 4 1.5655[24]
11137	1H-Pyrrole-1-propanenitrile	$C_7H_8N_2$ 120.15	43036-06-2	132-3[10]	5-20-05-00023 1.048[25]	1.5103[20]
11138	1H-Pyrrole, 1-propyl-	$C_7H_{11}N$ 109.17	5145-64-2	145.5	5-20-05-00010 0.8833[20]	eth 4; EtOH 4
11139	1H-Pyrrole, 3-propyl- 3-Propylpyrrole	$C_7H_{11}N$ 109.17	1551-09-3	90[30]; 49[2]	5-20-05-00081	eth 3; ace 3; chl 2 1.4900[25]
11140	1H-Pyrrole, 2,3,4,5-tetraiodo- Iodopyrrole	C_4HI_4N 570.68	87-58-1 150 dec	4934	5-20-05-00045	ace 4; eth 4; chl 4
11141	1H-Pyrrole, 2,3,4,5-tetramethyl-	$C_8H_{13}N$ 123.20	1003-90-3 111	130[7]	5-20-05-00099	EtOH 5; eth 5; ace 5; bz 4
11142	1H-Pyrrole, 1,2,5-trimethyl- 1,2,5-Trimethylpyrrole	$C_7H_{11}N$ 109.17	930-87-0	171	5-20-05-00065 0.807[25]	1.4969[20]
11143	Pyrrolidine Azacyclopentane	C_4H_9N 71.12	123-75-1 -57.8	8026 86.5	5-20-01-00162 0.8586[20]	H_2O 5; EtOH 3; eth 3; bz 2 1.4431[20]
11144	3-Pyrrolidineacetic acid, 2-carboxy-4-(1-methylethenyl)-, [2S-(2α,3β,4β)]- Kainic acid	$C_{10}H_{15}NO_4$ 213.23	487-79-6	5157	4-22-00-01523	
11145	1-Pyrrolidineacetic acid, 2,5-dioxo-, ethyl ester Aminoacetic acid, N-succinyl, ethyl ester	$C_8H_{11}NO_4$ 185.18	14181-05-6 67	198[32]	5-21-09-00474	H_2O 4; EtOH 4
11146	Pyrrolidine, 1-butyl-	$C_8H_{17}N$ 127.23	767-10-2	154.5	5-20-01-00170 0.816[25]	EtOH 4 1.4373[25]
11147	Pyrrolidine, 2-butyl- 2-Butylpyrrolidine	$C_8H_{17}N$ 127.23	3446-98-8	174; 67[18]	5-20-04-00222 0.8277[20]	H_2O 4; EtOH 4 1.4490[20]
11148	1,2-Pyrrolidinedicarboxylic acid, 1-(phenylmethyl) ester, (S)-	$C_{13}H_{15}NO_4$ 249.27	1148-11-4 77		5-22-01-00084	chl 2 1.5310[20]
11149	Pyrrolidine, 2,2-dimethyl- 2,2-Dimethylpyrrolidine	$C_8H_{13}N$ 99.18	35018-15-6	108	4-20-00-01529 0.8211[25]	1.4304[25]
11150	Pyrrolidine, 2,4-dimethyl-	$C_8H_{13}N$ 99.18	13603-04-8	116	4-20-00-01537 0.8297[20]	H_2O 4; eth 4; EtOH 4 1.4325[20]
11151	Pyrrolidine, 2,5-dimethyl-, cis- cis-2,5-Dimethylpyrrolidine	$C_6H_{13}N$ 99.18	39713-71-8	106.5	5-20-04-00146 0.8205[20]	H_2O 4; eth 4; EtOH 4 1.4299[20]
11152	2,5-Pyrrolidinedione Succinimide	$C_4H_5NO_2$ 99.09	123-56-8 126.5	8842 287 dec	5-21-09-00438 1.418[25]	H_2O 3; EtOH 2; eth 2; ace 2
11153	2,5-Pyrrolidinedione, 1-bromo- N-Bromosuccinimide	$C_4H_4BrNO_2$ 177.99	128-08-5 174	1428	5-21-09-00543 2.098[25]	H_2O 2; eth 2; ace 4
11154	2,5-Pyrrolidinedione, 1-chloro- N-Chlorosuccinimide	$C_4H_4ClNO_2$ 133.53	128-09-6 150	2165	5-21-09-00543 1.65[25]	H_2O 2; EtOH 2; ace 3; bz 2
11155	2,5-Pyrrolidinedione, 1-(chloromethyl)-	$C_5H_6ClNO_2$ 147.56	54553-14-9 58	158[12]	5-21-09-00460	ace 4; bz 4
11156	2,5-Pyrrolidinedione, 1,3-dimethyl-3-phenyl- Methsuximide	$C_{12}H_{13}NO_2$ 203.24	77-41-8 52.5	5928 121-2[0.1]	5-21-11-00209	
11157	2,5-Pyrrolidinedione, 1-ethyl-	$C_6H_9NO_2$ 127.14	2314-78-5 26	236	5-21-09-00442	eth 4; EtOH 4
11158	2,5-Pyrrolidinedione, 3-ethyl-3-methyl- Ethosuximide	$C_7H_{11}NO_2$ 141.17	77-67-8 64.5	3704	5-21-09-00595	H_2O 4
11159	2,5-Pyrrolidinedione, 1-iodo- N-Iodosuccinimide	$C_4H_4INO_2$ 224.99	516-12-1 200.5	4936	5-21-09-00544 2.245[25]	H_2O 4; EtOH 3; eth 2; ace 3
11160	2,5-Pyrrolidinedione, 1-methyl-	$C_5H_7NO_2$ 113.12	1121-07-9 71	234	5-21-09-00441	H_2O 3; EtOH 3; eth 4
11161	2,5-Pyrrolidinedione, 1-methyl-3-phenyl- Phensuximide	$C_{11}H_{11}NO_2$ 189.21	86-34-0 72	7231	5-21-11-00188	EtOH 4; MeOH 4
11162	2,5-Pyrrolidinedione, 1-phenyl- Succinanil	$C_{10}H_9NO_2$ 175.19	83-25-0 156	8838 400	5-21-09-00446 1.356[25]	H_2O 1; EtOH 3; eth 3
11163	2,5-Pyrrolidinedione, 1-(phenylmethyl)- N-Benzylsuccinimide	$C_{11}H_{11}NO_2$ 189.21	2142-06-5 103	393	5-21-09-00449	bz 4; eth 4; EtOH 4; chl 4
11164	1-Pyrrolidineethanamine	$C_6H_{14}N_2$ 114.19	7154-73-6	66-70[23]	5-20-01-00405 0.901[25]	1.4687[20]
11165	1-Pyrrolidineethanol	$C_6H_{13}NO$ 115.18	2955-88-6	79-81[13]	4-20-00-00092 0.985[25]	1.4713[20]
11166	Pyrrolidine, 1-ethyl-	$C_6H_{13}N$ 99.18	7335-06-0	104	5-20-01-00168 0.8156[20]	1.4113[15]
11167	2-Pyrrolidinemethanamine, 1-ethyl-	$C_7H_{16}N_2$ 128.22	26116-12-1	58-60[16]	5-22-08-00096 0.887[25]	1.4665[20]
11168	Pyrrolidine, 1-methyl- N-Methylpyrrolidine	$C_5H_{11}N$ 85.15	120-94-5	81	5-20-01-00166 0.8188[20]	H_2O 4; eth 4 1.4247[20]
11169	Pyrrolidine, 1-[2-(3-methylbutoxy)-2-phenylethyl]- Amixetrine	$C_{17}H_{27}NO$ 261.41	24622-72-8 507	121[2]		1.4978[22]
11170	Pyrrolidine, 1-nitroso- N-Nitrosopyrrolidine	$C_4H_8N_2O$ 100.12	930-55-2	6564 214	5-20-01-00521 1.085[25]	1.4880[25]
11171	Pyrrolidine, 1-phenyl-	$C_{10}H_{13}N$ 147.22	4096-21-3	119-20[12]	5-20-01-00182 1.0260[25]	eth 3 1.5813[20]

No.	Name / Synonym	Mol. Form. / Mol. Wt.	CAS RN / mp/°C	Merck No. / bp/°C	Beil. Ref. / den/g cm⁻³	Solubility / n_D
11172	Pyrrolidine, 1-(phenylmethyl)-	$C_{11}H_{15}N$ 161.25	29897-82-3	45[0.7]	5-20-01-00184 0.965[25]	1.5270[20]
11173	Pyrrolidine, 1-propyl-	$C_7H_{15}N$ 113.20	7335-07-1 105	132.5	5-20-01-00169 0.8171[20]	1.4389[20]
11174	Pyrrolidine, 3-propyl-	$C_7H_{15}N$ 113.20	116632-47-4	159	4-20-00-01598 0.8450[20]	1.4469[20]
11175	Pyrrolidine, 2,2,4-trimethyl- 2,2,4-Trimethylpyrrolidine	$C_7H_{15}N$ 113.20	35018-28-1	119	5-20-04-00198 0.8063[25]	1.4259[25]
11176	Pyrrolidine, 2,2,5-trimethyl-	$C_7H_{15}N$ 113.20	6496-48-6	112	4-20-00-01603 0.7980[25]	1.4223[25]
11177	Pyrrolidinium, 2-carboxy-1,1-dimethyl-, hydroxide, inner salt, (S)- Stachydrine	$C_7H_{13}NO_2$ 143.19	471-87-4 235	8729		H_2O 4; EtOH 4; EtOH 4
11178	Pyrrolidinium, 2-carboxy-4-hydroxy-1,1-dimethyl-, hydroxide, inner salt, (2S-trans)- Betonicine	$C_7H_{13}NO_3$ 159.19	515-25-3 252 dec	1209		EtOH 4
11179	Pyrrolidinium, 3-[(cyclopentylhydroxyphenylacetyl)oxy]-1,1-dimethyl-, bromide Glycopyrrolate	$C_{19}H_{28}BrNO_3$ 398.34	596-51-0 192.5	4397	5-21-01-00015	
11180	2-Pyrrolidinone 2-Pyrrolidone	C_4H_7NO 85.11	616-45-5 25	8027 251; 133[12]	5-21-06-00317 1.120[20]	H_2O 4; EtOH 4; eth 4; bz 4 1.4806[30]
11181	2-Pyrrolidinone, 1,5-dimethyl-	$C_6H_{11}NO$ 113.16	5075-92-3	217; 87.5[10]	5-21-06-00431	H_2O 4; eth 4 1.4650[20]
11182	2-Pyrrolidinone, 3,3-dimethyl- 3,3-Dimethyl-2-pyrrolidone	$C_6H_{11}NO$ 113.16	4831-43-0 66	237	0-21-00-00242	H_2O 4; ace 4; eth 4; EtOH 4
11183	2-Pyrrolidinone, 1-ethenyl-	C_6H_9NO 111.14	88-12-0 13.5	193[400]; 93[11]	5-21-06-00330 1.04[20]	
11184	2-Pyrrolidinone, 1-(2-hydroxyethyl)-	$C_6H_{11}NO_2$ 129.16	3445-11-2 20	295	5-21-06-00341 1.1435[20]	
11185	2-Pyrrolidinone, 1-methyl- N-Methyl-2-pyrrolidone	C_5H_9NO 99.13	872-50-4 -24	202	5-21-06-00321 1.0230[25]	H_2O 4; eth 3; ace 3; chl 3 1.4684[20]
11186	2-Pyrrolidinone, 5-methyl-	C_5H_9NO 99.13	108-27-0 43	248	5-21-06-00431 1.0458[20]	
11187	3-Pyrrolidinone, 1-methyl-	C_5H_9NO 99.13	68165-06-0	46-7[18]	5-21-06-00389 0.9646[25]	1.4431[25]
11188	2-Pyrrolidinone, 1-[4-(1-pyrrolidinyl)-2-butynyl]- Oxotremorine	$C_{12}H_{18}N_2O$ 206.29	70-22-4	6904 124[0.1]	5-21-06-00370 0.991[25]	1.5160[20]
11189	2-Pyrroline, 2-methyl-	C_5H_9N 83.13	100791-95-5	96	0-20-00-00135 0.8995[22]	H_2O 4; eth 4; EtOH 4
11190	Pyrrolo[1,2-a]pyrimidine, 2,3,4,6,7,8-hexahydro-	$C_7H_{12}N_2$ 124.19	3001-72-7	95-8[7]	5-23-05-00239 1.005[25]	1.5190[20]
11191	Pyrrolo[2,3-b]indol-5-ol, 1,2,3,3a,8,8a-hexahydro-1,3a,8-trimethyl-, methylcarbamate (ester), (3aS-cis)- Physostigmine	$C_{15}H_{21}N_3O_2$ 275.35	57-47-6 105.5	7357	5-23-11-00401	H_2O 2; EtOH 3; eth 3; bz 3
11192	Pyrrolo[2,1-b]quinazolin-3-ol, 1,2,3,9-tetrahydro-, (R)- Vasicine	$C_{11}H_{12}N_2O$ 188.23	6159-55-3 211.5	9842	5-23-12-00040	H_2O 2; EtOH 3; eth 2; ace 3
11193	1,1':2',1'':2'',1'''-Quaterphenyl	$C_{24}H_{18}$ 306.41	641-96-3 118.5	420; 193[1.0]	4-05-00-02729	H_2O 1; EtOH 2; eth 3; ace 3
11194	1,1':4',1'':4'',1'''-Quaterphenyl	$C_{24}H_{18}$ 306.41	135-70-6 320	428[18]	4-05-00-02730	H_2O 1; EtOH 1; eth 1; bz 3
11195	Quinazoline 1,3-Benzodiazine	$C_8H_6N_2$ 130.15	253-82-7 48	8062 241	5-23-07-00129	H_2O 4; EtOH 3; eth 3; ace 3
11196	Quinazoline, 3,4-dihydro-3-phenyl- Orexine	$C_{14}H_{12}N_2$ 208.26	612-97-5 96.5	dec	4-23-00-01108 1.290[4]	bz 4; eth 4; EtOH 4; chl 4
11197	6-Quinazolinesulfonamide, 7-chloro-2-ethyl-1,2,3,4-tetrahydro-4-oxo- Quinethazone	$C_{10}H_{12}ClN_3O_3S$ 289.74	73-49-4	8067		tfa 3
11198	4(1H)-Quinazolinone, 2,3-dihydro-1-(4-morpholinylacetyl)-3-phenyl- Moquizone	$C_{20}H_{21}N_3O_3$ 351.41	19395-58-5 136	6174		chl 3
11199	4(3H)-Quinazolinone, 2-methyl-3-(2-methylphenyl)- Methaqualone	$C_{16}H_{14}N_2O$ 250.30	72-44-6 120	5872	5-24-03-00132	eth 4; EtOH 4; chl 4
11200	4(1H)-Quinazolinone, 1-methyl-2-(phenylmethyl)- Glycosine	$C_{16}H_{14}N_2O$ 250.30	6873-15-0 161.5	4398	5-24-04-00363	
11201	Quinclorac 3,7-Dichloroquinoline-8-carboxylic acid	$C_{10}H_5Cl_2NO_2$ 242.06	84087-01-4 274		1.75	
11202	2-Quinolinamine 2-Aminoquinoline	$C_9H_8N_2$ 144.18	580-22-3 131.5	sub	5-22-10-00220	H_2O 4; EtOH 3; eth 3; ace 3
11203	4-Quinolinamine	$C_9H_8N_2$ 144.18	578-68-7 154.8	180[12]	5-22-10-00241	H_2O 3; EtOH 4; eth 4; bz 3
11204	5-Quinolinamine	$C_9H_8N_2$ 144.18	611-34-7 110	310; 184[10]	5-22-10-00297	H_2O 2; EtOH 4; eth 4; bz 3
11205	6-Quinolinamine 6-Aminoquinoline	$C_9H_8N_2$ 144.18	580-15-4 114	187[12]	5-22-10-00303	H_2O 2; EtOH 3; eth 2; NH_3 3

No.	Name Synonym	Mol. Form. Mol. Wt.	CAS RN mp/°C	Merck No. bp/°C	Beil. Ref. den/g cm^{-3}	Solubility n_D
11206	8-Quinolinamine	$C_9H_8N_2$ 144.18	578-66-5 70	157[19]	5-22-10-00316	H_2O 4; EtOH 4
11207	4-Quinolinamine, 3-bromo- 4-Amino-3-bromoquinoline	$C_9H_7BrN_2$ 223.07	36825-36-2 203	sub	4-22-00-04662	eth 4; EtOH 4
11208	2-Quinolinamine, 4-methyl- Lepidine, 2-amino-	$C_{10}H_{10}N_2$ 158.20	27063-27-0 133	320	5-22-10-00375	eth 4; EtOH 4; chl 4
11209	3-Quinolinamine, 2-methyl-	$C_{10}H_{10}N_2$ 158.20	21352-22-7 160	278; 198[16]	5-22-10-00347	bz 4; eth 4; EtOH 4
11210	4-Quinolinamine, 2-methyl-	$C_{10}H_{10}N_2$ 158.20	6628-04-2 169	333	5-22-10-00347	H_2O 2; EtOH 4; eth 4; ace 4
11211	7-Quinolinamine, 2-methyl- 7-Amino-2-methylquinoline	$C_{10}H_{10}N_2$ 158.20	64334-96-9 148	304	5-22-10-00364	ace 4; bz 4; eth 4; EtOH 4
11212	8-Quinolinamine, 6-methyl- 8-Amino-6-methylquinoline	$C_{10}H_{10}N_2$ 158.20	68420-93-9 73	sub	4-22-00-04822	ace 4; bz 4; eth 4; EtOH 4
11213	6-Quinolinamine, N,N,2-trimethyl-	$C_{12}H_{14}N_2$ 186.26	92-99-9 101	319	2-22-00-00361	ctc 3; CS_2 3
11214	Quinoline 1-Azanaphthalene	C_9H_7N 129.16	91-22-5 -14.78	8097 237.1	5-20-07-00276 1.0977[15]	H_2O 1; EtOH 5; eth 5; ace 5 1.6268[20]
11215	Quinoline, 6-amino-3-bromo-	$C_9H_7BrN_2$ 223.07	7101-96-4 106	sub	5-22-10-00312	EtOH 4
11216	Quinoline, 2-[2-(1,3-benzodioxol-5-yl)ethyl]-4-methoxy- Cusparine	$C_{19}H_{17}NO_3$ 307.35	529-92-0 92(form a); 116(form b)	2687	2-27-00-00545	ace 4; bz 4; eth 4; EtOH 4
11217	Quinoline, 3-bromo- 3-Bromoquinoline	C_9H_6BrN 208.06	5332-24-1 13.3	275	5-20-07-00318	chl 3; HOAc 4 1.6641[20]
11218	Quinoline, 5-bromo- 5-Bromoquinoline	C_9H_6BrN 208.06	4964-71-0 52	280; 106[12]	5-20-07-00319	acid 3
11219	Quinoline, 6-bromo- 6-Bromoquinoline	C_9H_6BrN 208.06	5332-25-2 24	281	5-20-07-00319	EtOH 3; eth 3; acid 3
11220	Quinoline, 7-bromo- 7-Bromoquinoline	C_9H_6BrN 208.06	4965-36-0 34	290	4-20-00-03390	EtOH 3; eth 3; acid 4
11221	Quinoline, 6-bromo-5-nitro- 6-Bromo-5-nitroquinoline	$C_9H_5BrN_2O_2$ 253.05	98203-04-4 130	sub	0-20-00-00376	eth 4; EtOH 4
11222	Quinoline, 2-butyl- 2-Butylquinoline	$C_{13}H_{15}N$ 185.27	7661-39-4	153[14]	5-20-07-00482 1.003[20]	1.5699[20]
11223	2-Quinolinecarbonitrile 2-Cyanoquinoline	$C_{10}H_6N_2$ 154.17	1436-43-7 95	160[23]	5-22-03-00190	H_2O 3; EtOH 4; eth 4; bz 4
11224	4-Quinolinecarbonitrile	$C_{10}H_6N_2$ 154.17	2973-27-5 103.5	242.5	5-22-03-00206	ace 4; bz 4; eth 4; EtOH 4
11225	4-Quinolinecarboxaldehyde Cinchoninaldehyde	$C_{10}H_7NO$ 157.17	4363-93-3 51	122[4]	5-21-08-00443	eth 4; tol 4
11226	4-Quinolinecarboxamide, 2-butoxy-N-[2-(diethylamino)ethyl]-, monohydrochloride Dibucaine hydrochloride	$C_{20}H_{30}ClN_3O_2$ 379.93	61-12-1 94 dec	3016	5-22-05-00278	chl 3
11227	2-Quinolinecarboxylic acid Quinaldic acid	$C_{10}H_7NO_2$ 173.17	93-10-7 156	8054	5-22-03-00183	H_2O 3; bz 4
11228	7-Quinolinecarboxylic acid	$C_{10}H_7NO_2$ 173.17	1078-30-4 249.5	sub	5-22-03-00217	EtOH 4
11229	8-Quinolinecarboxylic acid 8-Carboxyquinoline	$C_{10}H_7NO_2$ 173.17	86-59-9 187	8099 sub	5-22-03-00217	EtOH 4
11230	2-Quinolinecarboxylic acid, 4,8-dihydroxy- Xanthurenic acid	$C_{10}H_7NO_4$ 205.17	59-00-7 289	9977	4-22-00-02513	H_2O 1; EtOH 3; eth 2; bz 2
11231	1(2H)-Quinolinecarboxylic acid, 2-ethoxy-, ethyl ester EEDQ	$C_{14}H_{17}NO_3$ 247.29	16357-59-8 56.5	3488 125-8[0.1]	5-21-03-00028	chl 3
11232	2-Quinolinecarboxylic acid, 4-hydroxy- Kynurenic acid	$C_{10}H_7NO_3$ 189.17	492-27-3 282.5	5206	5-22-06-00280	H_2O 2; EtOH 3; eth 1; os 2
11233	4-Quinolinecarboxylic acid, 3-hydroxy-2-phenyl- Oxycinchophen	$C_{16}H_{11}NO_3$ 265.27	485-89-2 206-7 dec	6910	4-22-00-02383	bz 4; EtOH 4; HOAc 4
11234	4-Quinolinecarboxylic acid, 6-methoxy- Quininic acid	$C_{11}H_9NO_3$ 203.20	86-68-0 285 dec	8092 sub	4-22-00-02297	H_2O 2; EtOH 3; eth 2; bz 2
11235	4-Quinolinecarboxylic acid, 2-phenyl- Cinchophen	$C_{16}H_{11}NO_2$ 249.27	132-60-5 214.5	2290	5-22-03-00484	H_2O 1; EtOH 3; eth 3; ace 2
11236	4-Quinolinecarboxylic acid, 2-phenyl-, 2-propenyl ester	$C_{19}H_{15}NO_2$ 289.33	524-34-5 31.5	262[15]	4-22-00-01361	ace 4; eth 4; EtOH 4
11237	Quinoline, 2-chloro- 2-Chloroquinoline	C_9H_6ClN 163.61	612-62-4 38	266; 153[22]	5-20-07-00312 1.2464[25]	H_2O 1; EtOH 4; eth 4; bz 3 1.6342[25]
11238	Quinoline, 4-chloro- 4-Chloroquinoline	C_9H_6ClN 163.61	611-35-8 34.5	262; 130[15]	5-20-07-00314 1.251[25]	H_2O 2; EtOH 4; eth 4; dil HCl 3
11239	Quinoline, 5-chloro- 5-Chloroquinoline	C_9H_6ClN 163.61	635-27-8 45	256	5-20-07-00314	EtOH 3
11240	Quinoline, 6-chloro- 6-Chloroquinoline	C_9H_6ClN 163.61	612-57-7 43.8	263	5-20-07-00315	1.6110[56]
11241	Quinoline, 7-chloro- 7-Chloroquinoline	C_9H_6ClN 163.61	612-61-3 31.5	267.5	5-20-07-00315 1.2158[58]	H_2O 2; EtOH 4; eth 4; ace 4 1.6108[58]
11242	Quinoline, 8-chloro- 8-Chloroquinoline	C_9H_6ClN 163.61	611-33-6 -20	288.5	5-20-07-00315 1.2834[14]	H_2O 3; EtOH 4; eth 4; ace 4 1.6408[14]

No.	Name / Synonym	Mol. Form. / Mol. Wt.	CAS RN / mp/°C	Merck No. / bp/°C	Beil. Ref. / den/g cm^{-3}	Solubility / n_D
11243	Quinoline, 2-chloro-4-methyl-	$C_{10}H_8ClN$ 177.63	634-47-9 59	296	5-20-07-00396	H_2O 1; EtOH 4; eth 4; chl 4
11244	Quinoline, 2-chloro-8-methyl- 2-Chloro-8-methylquinoline	$C_{10}H_8ClN$ 177.63	4225-85-8 61	287	5-20-07-00407	bz 4; eth 4; EtOH 4; chl 4
11245	Quinoline, 4-chloro-2-methyl-	$C_{10}H_8ClN$ 177.63	4295-06-1 42.5	269.5	5-20-07-00383 0.881[25]	H_2O 2; EtOH 3; eth 3; bz 3 1.6224[20]
11246	Quinoline, decahydro-, cis- cis-Decahydroquinoline	$C_9H_{17}N$ 139.24	10343-99-4 -40	206; 90[20]	5-20-04-00405 0.9426[20]	H_2O 2; EtOH 3; eth 3 1.4926[20]
11247	Quinoline, decahydro-, trans-(±)- Quinoline, decahydro (trans, DL)	$C_9H_{17}N$ 139.24	105728-23-2 48	204	5-20-04-00405 0.9610[22]	ace 4; bz 4; eth 4; EtOH 4 1.4692[56]
11248	Quinoline, 2,4-dichloro- 2,4-Dichloroquinoline	$C_9H_5Cl_2N$ 198.05	703-61-7 67.5	281	4-20-00-03382	bz 4; eth 4; EtOH 4; chl 4
11249	Quinoline, 2,7-dichloro- 2,7-Dichloroquinoline	$C_9H_5Cl_2N$ 198.05	613-77-4 120	100[2]	4-20-00-03383	eth 4; EtOH 4
11250	Quinoline, 4,5-dichloro-	$C_9H_5Cl_2N$ 198.05	21617-18-5 118	134[5.5]	4-20-00-03383	eth 4; EtOH 4
11251	Quinoline, 4,7-dichloro- 4,7-Dichloroquinoline	$C_9H_5Cl_2N$ 198.05	86-98-6 93	148[10]	5-20-07-00316	chl 2
11252	Quinoline, 5,8-dichloro-	$C_9H_5Cl_2N$ 198.05	703-32-2 97.5	sub	5-20-07-00317	eth 4; EtOH 4
11253	Quinoline, 2-[2-(3,4-dimethoxyphenyl)ethyl]-4-methoxy- Galipine	$C_{20}H_{21}NO_3$ 323.39	525-68-8 115.5	4247	4-21-00-02656	ace 4; bz 4; eth 4; EtOH 4
11254	Quinoline, 2,3-dimethyl- 2,3-Dimethylquinoline	$C_{11}H_{11}N$ 157.22	1721-89-7 68.5	263	5-20-07-00431 1.1013[25]	eth 4; EtOH 4; lig 4
11255	Quinoline, 2,4-dimethyl- 4-Methylquinaldine	$C_{11}H_{11}N$ 157.22	1198-37-4	265	5-20-07-00431 1.0611[15]	H_2O 2; EtOH 4; eth 4; chl 2 1.6075[20]
11256	Quinoline, 2,5-dimethyl-	$C_{11}H_{11}N$ 157.22	26190-82-9 61	264.5	4-20-00-03524	bz 4; eth 4; EtOH 4
11257	Quinoline, 2,6-dimethyl-	$C_{11}H_{11}N$ 157.22	877-43-0 60	266.5	5-20-07-00433	H_2O 2; EtOH 2; eth 2; bz 4
11258	Quinoline, 2,7-dimethyl- m-Toluquinaldine	$C_{11}H_{11}N$ 157.22	93-37-8 61	264.5	5-20-07-00435	H_2O 2; EtOH 3; eth 3; chl 3
11259	Quinoline, 2,8-dimethyl-	$C_{11}H_{11}N$ 157.22	1463-17-8 27	255.3	5-20-07-00436 1.0394[20]	H_2O 2; EtOH 4; eth 4; ctc 3 1.6022[20]
11260	Quinoline, 3,4-dimethyl- 3,4-Dimethylquinoline	$C_{11}H_{11}N$ 157.22	2436-92-2 73.5	291; 114[4]	5-20-07-00437	eth 4; EtOH 4
11261	Quinoline, 3,8-dimethyl- 3,8-Dimethylquinoline	$C_{11}H_{11}N$ 157.22	20668-29-5	270	5-20-07-00437 1.049[20]	1.6063[20]
11262	Quinoline, 4,6-dimethyl- 6-Methyllepidine	$C_{11}H_{11}N$ 157.22	826-77-7 22	273.5; 140[12]	5-20-07-00437 1.0577[20]	eth 4; EtOH 4 1.6100[20]
11263	Quinoline, 5,8-dimethyl- 5,8-Dimethylquinoline	$C_{11}H_{11}N$ 157.22	2623-50-9 4.5	266	5-20-07-00438 1.070[21]	eth 4; EtOH 4
11264	Quinoline, 6,8-dimethyl- 6,8-Dimethylquinoline	$C_{11}H_{11}N$ 157.22	2436-93-3	268.5	5-20-07-00439 1.0665[4]	eth 4; EtOH 4
11265	Quinoline, 6-ethoxy-1,2-dihydro-2,2,4-trimethyl- Ethoxyquin	$C_{14}H_{19}NO$ 217.31	91-53-2	3710 123-5[2]	5-21-03-00095 1.026[25]	1.569[25]
11266	Quinoline, 2-ethyl- 2-Ethylquinoline	$C_{11}H_{11}N$ 157.22	1613-34-9	245.5; 134[16]	5-20-07-00428 1.050[17]	eth 4; EtOH 4; chl 4 1.5979[23]
11267	Quinoline, 3-ethyl- 3-Ethylquinoline	$C_{11}H_{11}N$ 157.22	1873-54-7	266	5-20-07-00428 1.0508[20]	1.6160[20]
11268	Quinoline, 6-ethyl- 6-Ethylquinoline	$C_{11}H_{11}N$ 157.22	19655-60-8	135-6[14]	5-20-07-00429 1.043[25]	1.6009[25]
11269	Quinoline, 2-ethyl-3-methyl- 2-Ethyl-3-methylquinoline	$C_{12}H_{13}N$ 171.24	27356-52-1 57	269; 140[12]	5-20-07-00456	bz 4; eth 4; EtOH 4
11270	Quinoline, 2-ethyl-4-methyl- Lepidine, 2-ethyl-	$C_{12}H_{13}N$ 171.24	33357-44-7	272.5	5-20-07-00456 1.025[20]	1.5941[20]
11271	Quinoline, 3-ethyl-6-methyl-	$C_{12}H_{13}N$ 171.24	105688-74-2	284.5	4-20-00-03555 1.0280[20]	1.5955[20]
11272	Quinoline, 4-iodo- 4-Iodoquinoline	C_9H_6IN 255.06	16560-43-3 98.5	sub	5-20-07-00322	eth 4; EtOH 4
11273	Quinoline, 6-iodo-	C_9H_6IN 255.06	13327-31-6 91	120[1.0]	5-20-07-00322	eth 4; EtOH 4
11274	Quinoline, 4-methoxy- 4-Methoxyquinoline	$C_{10}H_9NO$ 159.19	607-31-8 41	245	5-21-03-00229	H_2O 1; EtOH 3; eth 3; bz 3
11275	Quinoline, 6-methoxy- 6-Methoxyquinoline	$C_{10}H_9NO$ 159.19	5263-87-6 26.5	306; 153[12]	5-21-03-00244 1.152[20]	EtOH 3; eth 3; chl 3; dil HCl 3
11276	Quinoline, 8-methoxy- 8-Methoxyquinoline	$C_{10}H_9NO$ 159.19	938-33-0 49.5	283; 164[14]	5-21-03-00254 1.034[29]	EtOH 3; eth 3; bz 3; peth 3
11277	Quinoline, 2-methyl- Quinaldine	$C_{10}H_9N$ 143.19	91-63-4 -1.5	8055 246.5	5-20-07-00375 1.06[25]	H_2O 2; EtOH 3; eth 3; ace 3 1.6116[20]
11278	Quinoline, 3-methyl- 3-Methylquinoline	$C_{10}H_9N$ 143.19	612-58-8 16.5	259.8	5-20-07-00388 1.0673[20]	ace 4; eth 4; EtOH 4 1.6171[20]
11279	Quinoline, 4-methyl- Lepidine	$C_{10}H_9N$ 143.19	491-35-0 9.5	5324 262	5-20-07-00389 1.083[20]	H_2O 2; EtOH 3; eth 3; ace 3 1.6200[20]
11280	Quinoline, 5-methyl- 5-Methylquinoline	$C_{10}H_9N$ 143.19	7661-55-4 19	262.7	5-20-07-00399 1.0832[20]	H_2O 2; EtOH 5; eth 5; ace 3 1.6219[20]
11281	Quinoline, 6-methyl- 6-Methylquinoline	$C_{10}H_9N$ 143.19	91-62-3 -22	258.6	5-20-07-00400 1.0654[20]	H_2O 2; EtOH 3; eth 3; ace 3 1.6157[20]

No.	Name Synonym	Mol. Form. Mol. Wt.	CAS RN mp/°C	Merck No. bp/°C	Beil. Ref. den/g cm^{-3}	Solubility n_D
11282	Quinoline, 7-methyl- m-Toluquinoline	$C_{10}H_9N$ 143.19	612-60-2 39	257.6	5-20-07-00402 1.0609[20]	H_2O 2; EtOH 3; eth 3; ace 3 1.6150[20]
11283	Quinoline, 8-methyl- 8-Methylquinoline	$C_{10}H_9N$ 143.19	611-32-5 -80	247.5	5-20-07-00405 1.0719[20]	H_2O 2; EtOH 5; eth 5; ace 3 1.6164[20]
11284	Quinoline, 5-nitro- 5-Nitroquinoline	$C_9H_6N_2O_2$ 174.16	607-34-1 74	sub	5-20-07-00325	H_2O 2; EtOH 3; bz 3; chl 2
11285	Quinoline, 6-nitro- 6-Nitroquinoline	$C_9H_6N_2O_2$ 174.16	613-50-3 153.5	170[0.2]	5-20-07-00326	H_2O 3; EtOH 3; eth 2; bz 4
11286	Quinoline, 7-nitro- 7-Nitroquinoline	$C_9H_6N_2O_2$ 174.16	613-51-4 132.5	sub	5-20-07-00327	H_2O 3; EtOH 3; eth 3; chl 4
11287	Quinoline, 4-nitro-, 1-oxide 4-Nitroquinoline oxide	$C_9H_6N_2O_3$ 190.16	56-57-5 154		5-20-07-00324	
11288	Quinoline, 2-phenyl- 2-Phenylquinoline	$C_{15}H_{11}N$ 205.26	612-96-4 86	363; 194[6]	5-20-08-00355	H_2O 2; EtOH 4; eth 4; ace 4
11289	Quinoline, 3-phenyl- 3-Phenylquinoline	$C_{15}H_{11}N$ 205.26	1666-96-2 52	205[12]	5-20-08-00359	H_2O 2; EtOH 3; eth 3; ace 4
11290	Quinoline, 4-phenyl- 4-Phenylquinoline	$C_{15}H_{11}N$ 205.26	605-03-8 61.5	260[77.] 156[0.3]	5-20-08-00360	H_2O 1; EtOH 4; eth 4; ace 4
11291	Quinoline, 6-phenyl- 6-Phenylquinoline	$C_{15}H_{11}N$ 205.26	612-95-3 110	260[77]	5-20-08-00361 1.1945[20]	H_2O 2; EtOH 4; eth 2; ace 4
11292	Quinoline, 2-propyl- 2-Propylquinoline	$C_{12}H_{13}N$ 171.24	1613-32-7	130-1[10]	5-20-07-00453 1.038[17]	1.5886[23]
11293	5-Quinolinesulfonic acid, 8-ethoxy- Actinoquinol	$C_{11}H_{11}NO_4S$ 253.28	15301-40-3 286 dec	134	5-22-07-00581	alk 3
11294	5-Quinolinesulfonic acid, 8-hydroxy- 8-Hydroxy-5-quinolinesulfonic acid	$C_9H_7NO_4S$ 225.22	84-88-8 322.5	4780	5-22-07-00581	H_2O 2
11295	5-Quinolinesulfonic acid, 8-hydroxy-7-iodo- Ferron	$C_9H_6INO_4S$ 351.12	547-91-1 260 dec	4757	5-22-07-00582	H_2O 2; EtOH 2; eth 1; bz 1
11296	Quinoline, 1,2,3,4-tetrahydro-	$C_9H_{11}N$ 133.19	635-46-1 20	251	5-20-06-00293 1.0588[20]	H_2O 3; EtOH 5; eth 5; chl 3 1.6062[19]
11297	Quinoline, 5,6,7,8-tetrahydro- 2,3-Cyclohexenopyridine	$C_9H_{11}N$ 133.19	10500-57-9	222	5-20-06-00292 1.0304[13]	H_2O 2; EtOH 3; eth 3; ace 3 1.5435[20]
11298	Quinoline, 1,2,3,4-tetrahydro-6-methoxy-	$C_{10}H_{13}NO$ 163.22	120-15-0 42.5	284; 128[1]	5-21-02-00314	chl 3 1.5718[20]
11299	Quinoline, 1,2,3,4-tetrahydro-1-methyl-	$C_{10}H_{13}N$ 147.22	491-34-9	246	5-20-06-00294 1.022[20]	EtOH 4 1.5802[23]
11300	Quinoline, 2,3,6-trimethyl-	$C_{12}H_{13}N$ 171.24	2437-73-2 86.5	285	4-20-00-03557	eth 4; EtOH 4
11301	Quinoline, 2,3,8-trimethyl-	$C_{12}H_{13}N$ 171.24	4945-28-2	283	4-20-00-03557 1.0069[20]	H_2O 1; EtOH 3; eth 3; bz 2
11302	Quinoline, 2,4,6-trimethyl- 4,6-Dimethylquinaldine	$C_{12}H_{13}N$ 171.24	2243-89-2 68	280	5-20-07-00458	ace 4; bz 4; eth 4; EtOH 4
11303	Quinoline, 2,4,7-trimethyl- 2,4,7-Trimethylquinoline	$C_{12}H_{13}N$ 171.24	71633-43-7 63.5	280.5	5-20-07-00458 1.0337[20]	H_2O 2; EtOH 4; eth 4; ace 4 1.5973[24]
11304	Quinoline, 2,5,7-trimethyl-	$C_{12}H_{13}N$ 171.24	102871-67-0	287; 107[1.2]	4-20-00-03559	H_2O 1; EtOH 4; eth 4 1.5980[20]
11305	5-Quinolinol 5-Hydroxyquinoline	C_9H_7NO 145.16	578-67-6 226 dec	sub	5-21-03-00240	H_2O 3; EtOH 2; eth 2; bz 3
11306	6-Quinolinol	C_9H_7NO 145.16	580-16-5 195	360	5-21-03-00244	H_2O 1; EtOH 2; eth 2; bz 1
11307	7-Quinolinol 7-Hydroxyquinoline	C_9H_7NO 145.16	580-20-1 239	sub	5-21-03-00249	EtOH 4
11308	8-Quinolinol 8-Hydroxyquinoline	C_9H_7NO 145.16	148-24-3 75.5	4778 267	5-21-03-00252 1.034[20]	H_2O 1; EtOH 4; eth 1; ace 3
11309	8-Quinolinol, benzoate (ester) Benzoxiquine	$C_{16}H_{11}NO_2$ 249.27	86-75-9	1122	5-21-03-00259	chl 2
11310	8-Quinolinol, 5-bromo-	C_9H_6BrNO 224.06	1198-14-7 124	sub	5-21-03-00288	H_2O 4; EtOH 4; bz 4; chl 4
11311	8-Quinolinol, 5-chloro- Cloxyquin	C_9H_6ClNO 179.61	130-16-5 130	2416	5-21-03-00279	
11312	8-Quinolinol, 5-chloro-7-iodo- Iodochlorhydroxyquin	C_9H_5ClINO 305.50	130-26-7 178.5	4924	5-21-03-00294	EtOH 2; HOAc 3
11313	8-Quinolinol, 5,7-dibromo- Broxyquinoline	$C_9H_5Br_2NO$ 302.95	521-74-4 196	1441 sub	5-21-03-00290	H_2O 1; EtOH 3; eth 2; ace 3
11314	8-Quinolinol, 5,7-dichloro- Chloroxine	$C_9H_5Cl_2NO$ 214.05	773-76-2 179.5	2175	5-21-03-00286	EtOH 2; ace 2; bz 3; chl 2
11315	8-Quinolinol, 5,7-diiodo- Iodoquinol	$C_9H_5I_2NO$ 396.95	83-73-8 210	4935	5-21-03-00296	H_2O 2; EtOH 4; eth 2; bz 2
11316	4-Quinolinol, 2,8-dimethyl-	$C_{11}H_{11}NO$ 173.21	15644-80-1 262.3	sub	5-21-08-00321	EtOH 4
11317	6-Quinolinol, 2,4-dimethyl- 2,4-Dimethyl-6-hydroxyquinoline	$C_{11}H_{11}NO$ 173.21	64165-34-0 214	360 dec	2-21-00-00068	ace 4; EtOH 4
11318	8-Quinolinol, 2,4-dimethyl- Quinoline, 2,4-dimethyl-8-hydroxy	$C_{11}H_{11}NO$ 173.21	115310-98-0 65	281	4-21-00-01292	ace 4; bz 4; eth 4; EtOH 4
11319	6-Quinolinol, 2-methyl-	$C_{10}H_9NO$ 159.19	613-21-8 213	304.5	5-21-03-00339 1.1665[25]	eth 4; EtOH 4
11320	7-Quinolinol, 6-methyl- Quinoline, 7-hydroxy-6-methyl	$C_{10}H_9NO$ 159.19	84583-53-9 244	sub	0-21-00-00111	bz 4; eth 4; EtOH 4
11321	8-Quinolinol, 2-methyl-	$C_{10}H_9NO$ 159.19	826-81-3 73.8	267	5-21-03-00341	H_2O 1; EtOH 3; eth 3; bz 3

No.	Name / Synonym	Mol. Form. / Mol. Wt.	CAS RN / mp/°C	Merck No. / bp/°C	Beil. Ref. / den/g cm^{-3}	Solubility / n_D
11322	8-Quinolinol, 6-methyl-	$C_{10}H_9NO$ 159.19	20984-33-2 95.5	sub	4-21-00-01275	EtOH 4
11323	8-Quinolinol, 5-nitro- Nitroxoline	$C_9H_6N_2O_3$ 190.16	4008-48-4 180	6577	5-21-03-00297	
11324	8-Quinolinol, sulfate (2:1) (salt) 8-Hydroxyquinoline sulfate	$C_9H_8NO_3S$ 210.23	134-31-6 177.5	4779	5-21-03-00252	H_2O 4; EtOH 3; eth 1
11325	2(1H)-Quinolinone Carbostyril	C_9H_7NO 145.16	59-31-4 199.5	1831 sub	5-21-08-00217	H_2O 2; EtOH 4; eth 4
11326	2(1H)-Quinolinone, 4-bromo-	C_9H_6BrNO 224.06	938-39-6 266.5	sub	5-21-08-00229	EtOH 4
11327	2(1H)-Quinolinone, 3,4-dihydro- Hydrocarbostyril	C_9H_9NO 147.18	553-03-7 163.5	4702 201[45]	5-21-08-00057	eth 4; EtOH 4
11328	2(1H)-Quinolinone, 1-methyl-	$C_{10}H_9NO$ 159.19	606-43-9 74	325	5-21-08-00218	H_2O 2; EtOH 3; eth 3; ace 3
11329	2(1H)-Quinolinone, 6-methyl-	$C_{10}H_9NO$ 159.19	4053-34-3 237	240[12]	5-21-08-00289	ace 4; bz 4; eth 4; EtOH 4
11330	4(1H)-Quinolinone, 1-methyl- Echinopsine	$C_{10}H_9NO$ 159.19	83-54-5	3472	5-21-08-00210	H_2O 3; EtOH 4; eth 2; bz 4
11331	2H-Quinolizine-1-methanol, octahydro-, (1R-trans)- Lupinine	$C_{10}H_{19}NO$ 169.27	486-70-4 70	5479 270	5-21-01-00338	H_2O 3; EtOH 3; eth 3; bz 3
11332	6-Quinoxalinamine 6-Aminoquinoxaline	$C_8H_7N_3$ 145.16	6298-37-9 159	sub	4-25-00-02611	H_2O 4; EtOH 4; eth 3; bz 2
11333	Quinoxaline 1,4-Benzodiazine	$C_8H_6N_2$ 130.15	91-19-0 28	8107 229.5	5-23-07-00135 1.1334[48]	H_2O 3; EtOH 5; eth 5; ace 5 1.6231[48]
11334	Quinoxaline, 6-chloro- 6-Chloroquinoxaline	$C_8H_5ClN_2$ 164.59	5448-43-1 64	117[10]	5-23-07-00144	H_2O 2
11335	Quinoxaline, 2,6-dimethyl-	$C_{10}H_{10}N_2$ 158.20	60814-29-1 54	268	0-23-00-00192	H_2O 4; ace 4; bz 4; EtOH 4
11336	Quinoxaline, 6-methoxy- 6-Methoxyquinoxaline	$C_9H_8N_2O$ 160.18	6639-82-3 58.9	128[7]	5-23-11-00461	ace 4; eth 4
11337	Quinoxaline, 2-methyl- 2-Methylquinoxaline	$C_9H_8N_2$ 144.18	7251-61-8 180.5	244	5-23-07-00219	H_2O 5; EtOH 4; eth 5; ace 5
11338	Quinoxaline, 6-methyl- 6-Methylquinoxaline	$C_9H_8N_2$ 144.18	6344-72-5 218.5	249; 141.5[29]	5-23-07-00223 1.1164[20]	H_2O 5; EtOH 5; eth 5; ace 5 1.6211[18]
11339	Quinoxaline, 1,2,3,4-tetrahydro- 1,2,3,4-Tetrahydroquinoxaline	$C_8H_{10}N_2$ 134.18	3476-89-9 99	289	5-23-06-00024	H_2O 3; EtOH 4; eth 4; bz 4
11340	(-)-cis-Resmethrin Cyclopropanecarboxylic acid, 2,2-dimethyl-3-(2-methyl-1-propenyl)-, [5-(phenylmethyl)-3-furanyl] methyl ester	$C_{22}H_{26}O_3$ 338.45	10453-86-8 75			
11341	Retinol Vitamin A	$C_{20}H_{30}O$ 286.46	68-26-8 63.5	9918 137-8[10-6]	4-06-00-04133	H_2O 1; EtOH 3; eth 3; ace 3
11342	Rheadan, 8-methoxy-16-methyl-2,3:10,11-bis[methylenebis(oxy)]-, (8β)- Rheadine	$C_{21}H_{21}NO_6$ 383.40	2718-25-4 257	8174 sub	4-27-00-06897	
11343	Ribitol Adonitol	$C_5H_{12}O_5$ 152.15	488-81-3 104	157	4-01-00-02832	H_2O 3; EtOH 3; eth 1; lig 1
11344	Riboflavin Riboflavine	$C_{17}H_{20}N_4O_6$ 376.37	83-88-5 280 dec	8201	4-26-00-02542	H_2O 1; EtOH 2; eth 1; ace 1
11345	D-Ribose	$C_5H_{10}O_5$ 150.13	50-69-1 95	8205	4-01-00-04211	H_2O 3; EtOH 2
11346	Sarpagan-17-al Vellosimine	$C_{19}H_{20}N_2O$ 292.38	6874-98-2 305.5	9846 sub 180		
11347	Sarpagan-10,17-diol Sarpagine	$C_{19}H_{22}N_2O_2$ 310.40	482-68-8 320	8336	5-23-13-00401	H_2O 1; EtOH 3
11348	9,10-Secocholesta-5,7,10(19)-trien-3-ol, (3β,5Z,7E)- Vitamin D3	$C_{27}H_{44}O$ 384.65	67-97-0 84.5	9929	4-06-00-04149	H_2O 1; os 3
11349	9,10-Secoergosta-5(10),6,8,22-tetraen-3-ol, (3β,6E,22E)- Tachysterol	$C_{28}H_{44}O$ 396.66	115-61-7	9002	4-06-00-04404	H_2O 1; EtOH 3; eth 3; ace 3
11350	9,10-Secoergosta-5,7,10(19),22-tetraen-3-ol, (3β,5Z,7E,22E)- Vitamin D2	$C_{28}H_{44}O$ 396.66	50-14-6 116.5	9928 sub	4-06-00-04404	H_2O 1; EtOH 3; eth 3; ace 3
11351	16,19-Secostrychnidine-10,16-dione, 4-hydroxy-19-methyl- Vomicine	$C_{22}H_{24}N_2O_4$ 380.44	125-15-5 282	9946	4-27-00-07873	EtOH 2; eth 2; ace 2; chl 4
11352	9,10-Secostrychnidin-10-oic acid, 2,3-dihydro-4-nitro-2,3-dioxo- Cacotheline	$C_{21}H_{21}N_3O_7$ 427.41	561-20-6 >300	1604	4-27-00-08014	H_2O 2
11353	Selenanthrene Selanthrene	$C_{12}H_8Se_2$ 310.12	262-30-6 181	223[11]	5-19-02-00057	EtOH 2; eth 2; CS_2 3
11354	Selenourea Carbamimidoselenoic acid	CH_4N_2Se 123.02	630-10-4	200 dec	4-03-00-00435	H_2O 4
11355	D-Serine	$C_3H_7NO_3$ 105.09	312-84-5 229 dec	dec	4-04-00-03118	H_2O 4; EtOH 1; eth 1; bz 1
11356	DL-Serine	$C_3H_7NO_3$ 105.09	302-84-1 246 dec		4-04-00-03119 1.603[22]	H_2O 3; EtOH 1; eth 1; bz 1
11357	L-Serine (S)-2-Amino-3-hydroxypropanoic acid	$C_3H_7NO_3$ 105.09	56-45-1 228 dec	8411 sub 150	4-04-00-03118 1.6[22]	H_2O 4; EtOH 1; eth 1; bz 1

No.	Name / Synonym	Mol. Form. / Mol. Wt.	CAS RN / mp/°C	Merck No. / bp/°C	Beil. Ref. / den/g cm^{-3}	Solubility / n_D
11358	Sesone / Ethanol, 2-(2,4-dichlorophenoxy)-, hydrogen sulfate, sodium salt	$C_8H_8Cl_2NaO_5S$ / 310.11	136-78-7 / 245 dec	3372		
11359	Sethoxydim / (±)-2-[1-(Ethoxyimino)butyl]-5-[2-(ethylthio)propyl]-3-hydroxy-2-cyclohexen-1-one	$C_{17}H_{29}NO_3S$ / 327.49	74051-80-2	8424 / >90[0.00003]	1.043[25]	
11360	Silacyclohexane, 1,1-diphenyl- / 1,1-Diphenylsilacyclohexane	$C_{17}H_{20}Si$ / 252.43	18002-79-4	/ 193-8[5]	4-27-00-09833 / 1.0319[25]	1.5779[20]
11361	Silanamine, pentamethyl-	$C_5H_{15}NSi$ / 117.27	2083-91-2	/ 86	4-04-00-04010 / 0.7400[20]	1.4379[24]
11362	Silanamine, 1,1,1-trimethyl-N-phenyl-N-(trimethylsilyl)-	$C_{12}H_{23}NSi_2$ / 237.49 / 16.5	4147-89-1	98-9[11]	0.8951[23]	1.4855[20]
11363	Silanamine, 1,1,1-trimethyl-N-(trimethylsilyl)- / Hexamethyldisilazane	$C_6H_{19}NSi_2$ / 161.39	999-97-3 / 125	10158	4-04-00-04014 / 0.7741[25]	1.4090[20]
11364	Silane, [1,1'-biphenyl]-4-yltrichloro-	$C_{12}H_9Cl_3Si$ / 287.65	18030-61-0	/ 209-11[30]	4-16-00-01572 / 1.3190[20]	ctc 3; CS_2 3
11365	Silane, (bromomethyl)chlorodimethyl- / (Bromomethyl)dimethylchlorosilane	$C_3H_8BrClSi$ / 187.54	16532-02-8 / 131		4-04-00-04024 / 1.375[25]	1.4630[25]
11366	Silane, (bromomethyl)trimethyl- / Bromomethyltrimethylsilane	$C_4H_{11}BrSi$ / 167.12	18243-41-9 / 116.5		4-04-00-03878 / 1.170[25]	1.4460[20]
11367	Silane, (4-bromophenoxy)trimethyl-	$C_9H_{13}BrOSi$ / 245.19	17878-44-3 / 126[25]		4-06-00-01056 / 1.2619[20]	1.5145[20]
11368	Silane, bromotriethyl-	$C_6H_{15}BrSi$ / 195.17	1112-48-7 / 66-7[24]		4-04-00-04051 / 1.1403[20]	1.4561[20]
11369	Silane, butylchlorodimethyl-	$C_6H_{15}ClSi$ / 150.72	1000-50-6 / 139		4-04-00-04068 / 0.8751[20]	1.5145[20]
11370	Silane, butyltrichloro- / Trichlorobutylsilane	$C_4H_9Cl_3Si$ / 191.56	7521-80-4 / 148.5		4-04-00-04246 / 1.1606[20]	eth 3; bz 3; tol 3; AcOEt 3 / 1.4363[20]
11371	Silane, butyltrimethyl- / Trimethylbutylsilane	$C_7H_{18}Si$ / 130.31	1000-49-3 / 115		4-04-00-03907 / 0.7353[0]	H_2O 1
11372	Silane, chloro(chloromethyl)dimethyl-	$C_3H_8Cl_2Si$ / 143.09	1719-57-9 / 115.5		4-04-00-04024 / 1.0865[20]	1.4360[20]
11373	Silane, chloro(dichloromethyl)dimethyl- / (Dichloromethyl)dimethylchlorosilane	$C_3H_7Cl_3Si$ / 177.53 / -48	18171-59-0 / 149		4-04-00-04031 / 1.2369[20]	1.461[20]
11374	Silane, chlorodimethylphenyl-	$C_8H_{11}ClSi$ / 170.71	768-33-2 / 80-4[16]		4-16-00-01475 / 1.032[20]	1.5082[20]
11375	Silane, chlorodimethyl-2-propenyl-	$C_5H_{11}ClSi$ / 134.68	4028-23-3 / 111		4-04-00-04080 / 0.8964[20]	1.4195[20]
11376	Silane, chloroethenyldimethyl-	C_4H_9ClSi / 120.65	1719-58-0 / 83.5		4-04-00-04080 / 0.8744[20]	1.4141[20]
11377	Silane, chloroethyldimethyl-	$C_4H_{11}ClSi$ / 122.67	6917-76-6 / 89.5		4-04-00-04035 / 0.8675[20]	1.4105[20]
11378	Silane, chloromethyl / Chloromethylsilane	CH_5ClSi / 80.59	993-00-0 / -135	/ 7; -45[63]		
11379	Silane, (chloromethyl)dimethylphenyl-	$C_9H_{13}ClSi$ / 184.74	1833-51-8 / 225		4-16-00-01376 / 1.0240[25]	ctc 3; CS_2 3
11380	Silane, chloromethyldiphenyl-	$C_{13}H_{13}ClSi$ / 232.78	144-79-6 / 295		4-16-00-01479 / 1.1277[20]	1.5742[20]
11381	Silane, chloromethylphenyl-	C_7H_9ClSi / 156.69	1631-82-9 / 113[100]		4-16-00-01473 / 1.0540[20]	1.5171[20]
11382	Silane, (chloromethyl)trimethyl-	$C_4H_{11}ClSi$ / 122.67	2344-80-1 / 98.5		4-04-00-03877 / 0.879[25]	1.4175[20]
11383	Silane, (4-chlorophenoxy)trimethyl-	$C_9H_{13}ClOSi$ / 200.74	17005-59-3 / 214; 101[14]		4-06-00-00878 / 1.0320[20]	1.4930[20]
11384	Silane, chlorophenyl- / Phenylchlorosilane	C_6H_7ClSi / 142.66	4206-75-1 / 162.5		4-16-00-01472 / 1.0683[20]	1.5340[20]
11385	Silane, (3-chlorophenyl)trimethyl-	$C_9H_{13}ClSi$ / 184.74	4405-42-9 / 206.5		4-16-00-01386 / 1.0071[20]	1.5108[20]
11386	Silane, (3-chloropropyl)trimethyl-	$C_6H_{15}ClSi$ / 150.72	2344-83-4 / 151		4-04-00-03905 / 0.8789[20]	1.4319[20]
11387	Silane, chlorotriethoxy-	$C_6H_{15}ClO_3Si$ / 198.72	4667-99-6 / -51	/ 156	4-01-00-01363 / 1.030[20]	EtOH 4 / 1.3999[20]
11388	Silane, chlorotriethyl-	$C_6H_{15}ClSi$ / 150.72	994-30-9 / 144.5		4-04-00-04051 / 0.8967[20]	1.4314[20]
11389	Silane, chlorotrimethyl- / Trimethylchlorosilane	C_3H_9ClSi / 108.64	75-77-4 / -40	/ 60	4-04-00-04007 / 0.856[25]	1.3870[20]
11390	Silane, dichloro(chloromethyl)methyl- / Dichloro(chloromethyl)methylsilane	$C_2H_5Cl_3Si$ / 163.51	1558-33-4 / 121.5		4-04-00-04135 / 1.2858[20]	1.4500[20]
11391	Silane, dichloro(dichloromethyl)methyl- / Dichloromethyl methyl dichlorosilane	$C_2H_4Cl_4Si$ / 197.95	1558-31-2 / 149		4-04-00-04144 / 1.4116[20]	1.4700[20]
11392	Silane, dichlorodiethoxy-	$C_4H_{10}Cl_2O_2Si$ / 189.11	4667-38-3 / -130	/ 135.9	4-01-00-01364 / 1.1290[20]	EtOH 4
11393	Silane, dichlorodiethyl- / Dichlorodiethylsilane	$C_4H_{10}Cl_2Si$ / 157.11	1719-53-5 / -96.5	/ 129 dec	4-04-00-04158 / 1.0504[20]	1.4309[20]
11394	Silane, dichlorodimethyl- / Dichlorodimethylsilane	$C_2H_6Cl_2Si$ / 129.06	75-78-5 / -16	/ 70.3	4-04-00-04111 / 1.064[25]	H_2O 6; EtOH 6 / 1.4038[20]
11395	Silane, dichlorodiphenyl- / Dichlorodiphenylsilane	$C_{12}H_{10}Cl_2Si$ / 253.20	80-10-4 / 305		4-16-00-01526 / 1.204[25]	EtOH 3; eth 3; ace 3; bz 3 / 1.5800[20]

No.	Name / Synonym	Mol. Form. / Mol. Wt.	CAS RN / mp/°C	Merck No. / bp/°C	Beil. Ref. / den/g cm^{-3}	Solubility / n_D
11396	Silane, dichloroethenylmethyl-	$C_3H_8Cl_2Si$ 141.07	124-70-9	92.5	4-04-00-04184 1.0868[20]	H_2O 6 1.4270[20]
11397	Silane, dichloroethylmethyl- Methylethyldichlorosilane	$C_3H_8Cl_2Si$ 143.09	4525-44-4	101	4-04-00-04152 1.0630[20]	1.4197[20]
11398	Silane, dichloromethyl- Dichloromethylsilane	CH_4Cl_2Si 115.03	75-54-7 -93	41	4-04-00-04096 1.105[25]	
11399	Silane, dichloromethyl(1-methylethyl)-	$C_4H_{10}Cl_2Si$ 157.11	18236-89-0	121.5	4-04-00-04171 1.0385[20]	1.4270[20]
11400	Silane, dichloromethyl(4-methylphenyl)-	$C_8H_{10}Cl_2Si$ 205.16	18236-57-2 57		1.1619[20]	1.5170[20]
11401	Silane, dichloromethylphenyl- Dichloromethylphenylsilane	$C_7H_8Cl_2Si$ 191.13	149-74-6	206.5	4-16-00-01517 1.1866[20]	1.5180[20]
11402	Silane, dichloromethyl-2-propenyl-	$C_4H_8Cl_2Si$ 155.10	1873-92-3	119.5	4-04-00-04188 1.0758[20]	1.4419[20]
11403	Silane, dichlorophenyl- Phenyldichlorosilane	$C_6H_6Cl_2Si$ 177.10	1631-84-1	181	4-16-00-01514 1.221[25]	H_2O 6
11404	Silane, diethenyldiphenyl-	$C_{16}H_{16}Si$ 236.39	17937-68-7	130-1[0.05]	4-16-00-01368 1.0092[25]	1.5350[25]
11405	Silane, diethoxydimethyl- Dimethyldiethoxysilane	$C_6H_{16}O_2Si$ 148.28	78-62-6 -87	114	4-04-00-04101 0.865[25]	ctc 3 1.3811[20]
11406	Silane, diethoxydiphenyl- Diphenyldiethoxysilane	$C_{16}H_{20}O_2Si$ 272.42	2553-19-7	167[15]	4-16-00-01524 1.0329[20]	1.5269[20]
11407	Silane, diethoxymethylphenyl-	$C_{11}H_{18}O_2Si$ 210.35	775-56-4	218	4-16-00-01515 0.9627[20]	1.4690[20]
11408	Silane, diethoxymethyl-2-propenyl- Silane, allyldiethoxymethyl-	$C_8H_{18}O_2Si$ 174.32	18388-45-9	155	4-04-00-04187 0.8572[25]	1.4104[20]
11409	Silane, diethyl- 3-Silapentane	$C_4H_{12}Si$ 88.22	542-91-6 -134.3	3117 57	4-04-00-03894 0.6843[20]	H_2O 1 1.3921[20]
11410	Silane, diethyldifluoro- Diethyldifluorosilane	$C_4H_{10}F_2Si$ 124.21	358-06-5 -78.7	61	4-04-00-04157 0.9348[20]	1.3385[20]
11411	Silane, difluorodiphenyl-	$C_{12}H_{10}F_2Si$ 220.29	312-40-3	157[50]	4-16-00-01525 1.145[17]	1.5221[25]
11412	Silane, dimethoxydimethyl-	$C_4H_{12}O_2Si$ 120.22	1112-39-6	82	4-04-00-04100 0.8646[20]	H_2O 6 1.3708[20]
11413	Silane, dimethoxydiphenyl-	$C_{14}H_{16}O_2Si$ 244.37	6843-66-9	161[15]	4-16-00-01524 1.0771[20]	1.5447[20]
11414	Silane, dimethyl- 2-Silapropane	C_2H_8Si 60.17	1111-74-6 -150	-20	4-04-00-03874 0.68[-80]	
11415	Silane, dimethyldiphenoxy- Diphenoxydimethylsilane	$C_{14}H_{16}O_2Si$ 244.37	3440-02-6 -23	130-2[5]	4-06-00-00764 1.0599[25]	1.5330[20]
11416	Silane, dimethyldiphenyl- Diphenyl(dimethyl)silane	$C_{14}H_{16}Si$ 212.37	778-24-5	277; 173[45]	4-16-00-01366 0.9867[20]	1.5644[20]
11417	Silane, dimethyldi-2-propenyl-	$C_8H_{16}Si$ 140.30	1113-12-8	68[50]	4-04-00-03929 0.7679[20]	1.4420[20]
11418	Silane, dimethylphenyl-	$C_8H_{12}Si$ 136.27	766-77-8	156.5	4-16-00-01360 0.8891[20]	H_2O 1 1.4995[20]
11419	Silanediol, dimethyl-, diacetate	$C_8H_{12}O_4Si$ 176.24	2182-66-3 -12.5	165	4-04-00-04108 1.0540[20]	1.4030[20]
11420	Silane, diphenyl- Diphenylsilane	$C_{12}H_{12}Si$ 184.31	775-12-2	95-7[13]	4-16-00-01366 0.9969[20]	ctc 3; CS_2 3 1.5800[20]
11421	Silane, 1,2-ethenediylbis[trimethyl-, (E)-	$C_8H_{20}Si_2$ 172.42	18178-59-1 -18.5	145.5	4-04-00-03948 0.7589[20]	H_2O 1 1.4310[20]
11422	Silane, ethenyldiethoxymethyl-	$C_7H_{16}O_2Si$ 160.29	5507-44-8	133	4-04-00-04183 0.8620[20]	1.4001[20]
11423	Silane, ethenylethoxydimethyl-	$C_6H_{14}OSi$ 130.26	5356-83-2	99	4-04-00-04079 0.790[20]	1.3983[20]
11424	Silane, ethenyltriethoxy-	$C_8H_{18}O_3Si$ 190.31	78-08-0	148[535] 62[20]	4-04-00-04256 0.9027[20]	chl 3 1.3960[25]
11425	Silane, ethenyltrimethyl-	$C_5H_{12}Si$ 100.24	754-05-2	55	4-04-00-03922 0.6910[20]	H_2O 1 1.3914[20]
11426	Silane, ethenyltris(1-methylethoxy)-	$C_{11}H_{24}O_3Si$ 232.40	18023-33-1	179.5; 77[20]	4-04-00-04257 0.8627[25]	ctc 3 1.3981[20]
11427	Silane, ethenyltris(2-propenyloxy)- Tris(allyloxy)vinylsilane	$C_{11}H_{18}O_3Si$ 226.35	17988-31-7	90.3[7.5]	4-04-00-04257 0.9435[25]	1.4380[20]
11428	Silane, ethoxytriethyl-	$C_8H_{20}OSi$ 160.33	597-67-1	154.5	4-04-00-04042 0.8160[20]	eth 4; EtOH 4 1.4140[20]
11429	Silane, ethoxytrifluoro- Ethoxytrifluorosilicon	$C_2H_5F_3OSi$ 130.14	460-55-9 -122	-7	3-01-00-01336 1.3200[-63]	
11430	Silane, ethoxytrimethyl-	$C_5H_{14}OSi$ 118.25	1825-62-3	76	4-04-00-03994 0.7573[20]	H_2O 1; EtOH 3; eth 3; ace 3 1.3741[20]
11431	Silane, ethoxytriphenyl-	$C_{20}H_{20}OSi$ 304.46	1516-80-9 65	344	4-16-00-01481	chl 3
11432	Silane, ethyltrifluoro- Trifluoroethylsilane	$C_2H_5F_3Si$ 114.14	353-89-9 -105	-4.4	4-04-00-04226 1.227[-76]	
11433	Silane, ethyltrimethoxy- Ethyltrimethoxysilane	$C_5H_{14}O_3Si$ 150.25	5314-55-6	124.3	4-04-00-04223 0.9488[20]	EtOH 4 1.3838[20]
11434	Silane, 1,2-ethynediylbis[trimethyl-	$C_8H_{18}Si_2$ 170.40	14630-40-1 26	134	4-04-00-03950 0.770[20]	1.413[20]
11435	Silane, ethynyl Ethynylsilane	C_2H_4Si 56.14	1066-27-9	-22.5		

No.	Name / Synonym	Mol. Form. / Mol. Wt.	CAS RN / mp/°C	Merck No. / bp/°C	Beil. Ref. / den/g cm^{-3}	Solubility / n_D
11436	Silane, methoxy / Methyl silyl ether	CH_6OSi / 62.14	2171-96-2			
11437	Silane, methyl- / Methylsilane	CH_6Si / 46.14	992-94-9 / -156.5	-57.5	4-04-00-03873	
11438	Silane, methyldiphenyl-	$C_{13}H_{14}Si$ / 198.34	776-76-1	93.5[1]	4-16-00-01366 / 0.9973[20]	ctc 3 / 1.5694[20]
11439	Silane, methylenebis-	CH_8Si_2 / 76.25	1759-88-2	15	4-01-00-03072 / 0.6979[4]	1.4115[4]
11440	Silane, methylenebis[trichloro-	$CH_2Cl_6Si_2$ / 282.91	4142-85-2	183; 64[10]	4-01-00-03077 / 1.5567[20]	eth 4; EtOH 4 / 1.4740[20]
11441	Silane, (2-methylphenoxy)triphenyl-	$C_{25}H_{22}OSi$ / 366.53	18858-65-6	192	4-16-00-01482 / 0.9287[20]	1.4830[20]
11442	Silane, methylphenyl-	$C_7H_{10}Si$ / 122.24	766-08-5	140	4-16-00-01360 / 0.8895[20]	1.5058[20]
11443	Silane, methyltriphenoxy- / Methyltriphenoxysilane	$C_{19}H_{18}O_3Si$ / 322.44	3439-97-2	269[100]; 210[12]	4-06-00-00765 / 1.135[20]	1.5599[20]
11444	Silane, methyltriphenyl- / Triphenylmethylsilane	$C_{19}H_{18}Si$ / 274.44	791-29-7 / 69.5	198[9]	4-16-00-01370 / 1.0880[20]	ctc 3; CS$_2$ 3
11445	Silane, methyltri-p-tolyl- / Silane, methyl tri-(4-tolyl)	$C_{22}H_{24}Si$ / 316.52	18752-92-6 / 92	198[4]	4-16-00-01399	chl 2
11446	Silane, phenyl-	C_6H_8Si / 108.22	694-53-1	119	4-16-00-01360 / 0.8681[20]	H$_2$O 1 / 1.5125[20]
11447	Silane, [1,3-phenylenebis(oxy)]bis[trimethyl-	$C_{12}H_{22}O_2Si_2$ / 254.48	4520-29-0	240	4-06-00-05683 / 0.950[20]	1.4748[20]
11448	Silane, phenyltripropyl- / Phenyltripropylsilane	$C_{15}H_{26}Si$ / 234.46	78938-11-1	146-7[18]	4-16-00-01362 / 0.8799[20]	1.4950[20]
11449	Silane, tetraethenyl- / Tetravinylsilane	$C_8H_{12}Si$ / 136.27	1112-55-6	130.2	4-04-00-03925 / 0.7999[20]	1.4625[20]
11450	Silane, tetraethyl- / Tetraethylsilane	$C_8H_{20}Si$ / 144.33	631-36-7	154.7	4-04-00-03895 / 0.7658[20]	H$_2$O 1 / 1.4268[20]
11451	Silane, tetramethyl- / Tetramethylsilane	$C_4H_{12}Si$ / 88.22	75-76-3 / -99.0	26.6	4-04-00-03875 / 0.648[19]	H$_2$O 1; EtOH 4; eth 4; sulf 1 / 1.3587[20]
11452	Silane, tetraphenyl- / Tetraphenylsilane	$C_{24}H_{20}Si$ / 336.51	1048-08-4 / 236.5	228[3]	4-16-00-01372 / 1.078[20]	ctc 3; CS$_2$ 3
11453	Silane, tribromomethyl-	CH_3Br_3Si / 282.83	4095-09-4 / -28.4	132	4-04-00-04213 / 2.2130[25]	1.5152[25]
11454	Silane, tributyl- / Tributylsilane	$C_{12}H_{28}Si$ / 200.44	998-41-4	221	4-04-00-03908 / 0.7794[20]	1.4380[20]
11455	Silane, tributylphenyl- / Silane, phenyl-tributyl	$C_{18}H_{32}Si$ / 276.54	18510-29-7	140-2[4]	4-16-00-01362 / 0.8753[20]	1.4915[20]
11456	Silane, trichloro(chloromethyl)- / (Chloromethyl)trichlorosilane	CH_2Cl_4Si / 183.92	1550-25-4	118	4-01-00-03074 / 1.4650[20]	1.4555[20]
11457	Silane, trichloro(4-chlorophenyl)-	$C_6H_4Cl_4Si$ / 245.99	825-94-5	115-7[20]	4-16-00-01563 / 1.4316[20]	1.5418[20]
11458	Silane, trichloro(3-chloropropyl)-	$C_3H_6Cl_4Si$ / 211.98	2550-06-3	181.5	4-04-00-04243 / 1.3590[20]	1.4668[20]
11459	Silane, trichloro(dichloromethyl)- / (Dichloromethyl)trichlorosilane	$CHCl_5Si$ / 218.37	1558-24-3	145	4-02-00-00088 / 1.5518[20]	1.4714[20]
11460	Silane, trichlorododecyl- / Dodecyltrichlorosilane	$C_{12}H_{25}Cl_3Si$ / 303.77	4484-72-4	155[10]	4-04-00-04255	1.4581[20]
11461	Silane, trichloroethenyl- / Vinyltrichlorosilane	$C_2H_3Cl_3Si$ / 161.49	75-94-5 / -95	91.5	4-04-00-04258 / 1.2426[20]	chl 4 / 1.4295[20]
11462	Silane, trichloroethoxy-	$C_2H_5Cl_3OSi$ / 179.50	1825-82-7 / -135	101.9	4-01-00-01364 / 1.2274[20]	EtOH 4 / 1.4045[20]
11463	Silane, trichloroethyl- / Ethyltrichlorosilane	$C_2H_5Cl_3Si$ / 163.51	115-21-9 / -105.6	100.5	4-04-00-04227 / 1.2373[20]	ctc 3 / 1.4256[20]
11464	Silane, trichlorohexyl- / Hexyltrichlorosilane	$C_6H_{13}Cl_3Si$ / 219.61	928-65-4	190	4-04-00-04251 / 1.1100[20]	H$_2$O 6
11465	Silane, trichloromethyl- / Methyltrichlorosilane	CH_3Cl_3Si / 149.48	75-79-6 / -90	65.6	4-04-00-04212 / 1.273[20]	H$_2$O 6; EtOH 6 / 1.4106[20]
11466	Silane, trichloro(1-methylethyl)-	$C_3H_7Cl_3Si$ / 177.53	4170-46-1 / -87.7	121	4-04-00-04244 / 1.1934[20]	H$_2$O 6 / 1.4319[20]
11467	Silane, trichloro(2-methylphenyl)-	$C_7H_7Cl_3Si$ / 225.58	13835-81-9 / -25.5	226.7	4-16-00-01565 / 1.306[25]	1.5336[25]
11468	Silane, trichloro(3-methylphenyl)-	$C_7H_7Cl_3Si$ / 225.58	13688-75-0 / -26	226.5	4-16-00-01565 / 1.306[25]	1.5336[25]
11469	Silane, trichloro(2-methylpropyl)-	$C_4H_9Cl_3Si$ / 191.56	18169-57-8	143.3	4-04-00-04248 / 1.154[20]	H$_2$O 6
11470	Silane, trichlorooctadecyl- / Octadecyltrichlorosilane	$C_{18}H_{37}Cl_3Si$ / 387.94	112-04-9	223[10]	4-04-00-04256 / 0.984[25]	1.4602[20]
11471	Silane, trichlorooctyl- / Octyltrichlorosilane	$C_8H_{17}Cl_3Si$ / 247.67	5283-66-9	232	4-04-00-04253	H$_2$O 6; EtOH 6; ctc 3 / 1.4480[20]
11472	Silane, trichloropentyl- / Amyltrichlorosilane	$C_5H_{11}Cl_3Si$ / 205.59	107-72-2	172; 60.5[15]	4-04-00-04249 / 1.1330[20]	1.4503[20]
11473	Silane, trichlorophenyl- / Phenyltrichlorosilane	$C_6H_5Cl_3Si$ / 211.55	98-13-5	201	4-16-00-01560 / 1.321[20]	ctc 3; chl 3; CS$_2$ 3 / 1.5230[20]
11474	Silane, trichloro(2-phenylethyl)-	$C_8H_9Cl_3Si$ / 239.60	940-41-0	242; 98-9[5]	4-16-00-01568 / 1.2397[20]	1.5185[20]
11475	Silane, trichloro-2-propenyl- / Allyltrichlorosilane	$C_3H_5Cl_3Si$ / 175.52	107-37-9 / 35	117.5	4-04-00-04261 / 1.2011[20]	1.4460[20]

No.	Name / Synonym	Mol. Form. / Mol. Wt.	CAS RN / mp/°C	Merck No. / bp/°C	Beil. Ref. / den/g cm^{-3}	Solubility / n_D
11476	Silane, trichloropropyl- / Propyltrichlorosilane	$C_3H_7Cl_3Si$ / 177.53	141-57-1 /	/ 123.5	4-04-00-04239 / 1.195[20]	/ 1.4310[20]
11477	Silane, triethoxy- / Triethoxysilane	$C_6H_{16}O_3Si$ / 164.28	998-30-1 /	/ 133.5	4-04-00-01359 / 0.8745[20]	
11478	Silane, triethoxyethyl-	$C_8H_{20}O_3Si$ / 192.33	78-07-9 /	/ 158.5	4-04-00-04223 / 0.8963[20]	H_2O 1; EtOH 5; eth 5; chl 3 / 1.3955[20]
11479	Silane, triethoxymethyl-	$C_7H_{18}O_3Si$ / 178.30	2031-67-6 /	/ 142	4-04-00-04204 / 0.8948[25]	/ 1.3832[20]
11480	Silane, triethoxypentyl-	$C_{11}H_{26}O_3Si$ / 234.41	2761-24-2 /	/ 95[13]	4-04-00-04248 / 0.8862[20]	/ 1.4059[20]
11481	Silane, triethoxyphenyl-	$C_{12}H_{20}O_3Si$ / 240.37	780-69-8 /	/ 112-3[10]	4-16-00-01556 / 0.9962[5]	/ 1.4604[20]
11482	Silane, triethoxy-2-propenyl-	$C_9H_{20}O_3Si$ / 204.34	2550-04-1 /	/ 100[50]	4-04-00-04261 / 0.9030[20]	/ 1.4072[20]
11483	Silane, triethyl- / Triethylsilane	$C_6H_{16}Si$ / 116.28	617-86-7 /	/ 109	4-04-00-03895 / 0.7302[20]	H_2O 1; sulf 1 / 1.447[20]
11484	Silane, triethylfluoro-	$C_6H_{15}FSi$ / 134.27	358-43-0 /	/ 110	4-04-00-04051 / 0.8354[25]	peth 4 / 1.3900[25]
11485	Silane, triethylphenyl-	$C_{12}H_{20}Si$ / 192.38	2987-77-1 /	/ 236	4-16-00-01362 / 0.8915[20]	/ 1.4999[20]
11486	Silane, trifluorophenyl-	$C_6H_5F_3Si$ / 162.19	368-47-8 / -18	/ 101.5	4-16-00-01559 / 1.2169[20]	bz 4; EtOH 4 / 1.4110[20]
11487	Silane, trimethoxymethyl-	$C_4H_{12}O_3Si$ / 136.22	1185-55-3 /	/ 102.5	4-04-00-04203 / 0.9548[20]	chl 3 / 1.3696[20]
11488	Silane, trimethoxyphenyl-	$C_9H_{14}O_3Si$ / 198.29	2996-92-1 /	/ 130[45]	4-16-00-01556 / 1.064[20]	ctc 3; CS_2 3 / 1.4734[20]
11489	Silane, trimethyl- / Trimethylsilane	$C_3H_{10}Si$ / 74.20	993-07-7 / -135.9	/ 6.7		
11490	Silane, trimethyl(4-methylphenyl)-	$C_{10}H_{16}Si$ / 164.32	3728-43-6 /	/ 192; 73[10]	4-16-00-01396 / 0.860[25]	/ 1.4900[20]
11491	Silane, trimethyl(2-methylpropyl)-	$C_7H_{18}Si$ / 130.31	1118-09-8 /	/ 108.5	1-04-00-00580 / 0.7330[0]	
11492	Silane, trimethylphenoxy-	$C_9H_{14}OSi$ / 166.30	1529-17-5 / -55	/ 119	4-06-00-00762 / 0.8681[20]	/ 1.5125[20]
11493	Silane, trimethylphenyl-	$C_9H_{14}Si$ / 150.30	768-32-1 /	/ 169.5	4-16-00-01361 / 0.8722[20]	ctc 3; CS_2 3 / 1.4907[20]
11494	Silane, trimethyl(phenylmethyl)-	$C_{10}H_{16}Si$ / 164.32	770-09-2 /	/ 190.5	4-16-00-01401 / 0.8933[20]	/ 1.4941[20]
11495	Silane, trimethyl-2-propenyl-	$C_6H_{14}Si$ / 114.26	762-72-1 /	/ 86	4-04-00-03927 / 0.7158[20]	H_2O 1 / 1.4074[20]
11496	Silane, trimethylpropyl- / Trimethylpropylsilane	$C_6H_{16}Si$ / 116.28	3510-70-1 /	/ 89	4-04-00-03901 / 0.7196[0]	/ 1.3929[20]
11497	Silane, trimethyl[4-[(trimethylsilyl)oxy]phenyl]- / Silane, (oxy-p-phenylene)bis[trimethyl-	$C_{12}H_{22}OSi_2$ / 238.48	18036-81-2 /	/ 132[25]	4-16-00-01441 / 0.900[25]	/ 1.4794[25]
11498	Silanetriol, ethenyl-, triacetate	$C_8H_{12}O_6Si$ / 232.27	4130-08-9 /	/ 115[10]	4-04-00-04258 / 1.167[20]	/ 1.4226[20]
11499	Silanetriol, methyl-, triacetate	$C_7H_{12}O_6Si$ / 220.25	4253-34-3 / 40.5	/ 110-12[17]	4-04-00-04208 / 1.1750[20]	/ 1.4083[20]
11500	Silane, tripropyl- / Tripropylsilane	$C_9H_{22}Si$ / 158.36	998-29-8 /	/ 172	4-04-00-03902 / 0.7723[0]	H_2O 1 / 1.4280[20]
11501	Silanol, ethyldimethyl- / Silanol, dimethyl ethyl	$C_4H_{12}OSi$ / 104.22	5906-73-0 /	/ 120; 58[50]	4-04-00-04034 / 0.8332[20]	/ 1.4070[20]
11502	Silanol, methyldiphenyl-	$C_{13}H_{14}OSi$ / 214.34	778-25-6 / 167	/ 184[24]; 148[3]	4-16-00-01478 / 1.0840[25]	ctc 3; CS_2 3
11503	Silanol, triethyl-	$C_6H_{16}OSi$ / 132.28	597-52-4 /	/ 154	4-04-00-04040 / 0.8647[20]	H_2O 1; EtOH 5; eth 5 / 1.4329[20]
11504	Silanol, triphenyl-	$C_{18}H_{16}OSi$ / 276.41	791-31-1 / 154.8		4-16-00-01480 / 1.1777[20]	ctc 3; CS_2 3
11505	Silicic acid (H4SiO4), tetrabutyl ester	$C_{16}H_{36}O_4Si$ / 320.54	4766-57-8 /	/ 120[3]	4-01-00-01541 / 0.8990[20]	/ 1.4128[20]
11506	Silicic acid (H4SiO4), tetraethyl ester / Ethyl silicate	$C_8H_{20}O_4Si$ / 208.33	78-10-4 / -82.5	3805 / 168.8	4-01-00-01360 / 0.9320[20]	H_2O 6 / 1.3928[20]
11507	Silicic acid (H4SiO4), tetrakis(2-ethylbutyl) ester	$C_{24}H_{52}O_4Si$ / 432.76	78-13-7 /		4-01-00-01725 / 0.8920[20]	H_2O 1; EtOH 2; eth 3; bz 3 / 1.4307[20]
11508	Silicic acid (H4SiO4), tetramethyl ester / Methyl silicate	$C_4H_{12}O_4Si$ / 152.22	681-84-5 / -2	/ 121	4-01-00-01266 / 1.0232[20]	EtOH 4 / 1.3683[20]
11509	Silicic acid (H4SiO4), tetraphenyl ester	$C_{24}H_{20}O_4Si$ / 400.51	1174-72-7 / 49	/ 417; 236[1]	4-06-00-00766 / 1.1412[60]	
11510	Silicic acid (H4SiO4), tetrapropyl ester	$C_{12}H_{28}O_4Si$ / 264.44	682-01-9 /	/ 226	4-01-00-01435 / 0.9158[20]	ctc 3; CS_2 3 / 1.4012[20]
11511	Silicic acid (H4SiO4), triethyl phenyl ester	$C_{12}H_{20}O_4Si$ / 256.37	18023-36-4 /	/ 120-2[10]	4-06-00-00765 / 1.0283[20]	/ 1.4525[20]
11512	Silvex / Propanoic acid, 2-(2,4,5-trichlorophenoxy)-	$C_9H_7Cl_3O_3$ / 269.51	93-72-1 / 181.6	8483		
11513	Simazine / 1,3,5-Triazine-2,4-diamine, 6-chloro-N,N'-diethyl-	$C_7H_{12}ClN_5$ / 201.66	122-34-9 / 226	8485	1.302[20]	
11514	Sodium dimethyldithiocarbamate / Carbamodithioic acid, dimethyl-, sodium salt	$C_3H_7NNaS_2$ / 144.22	128-04-1 /			

No.	Name / Synonym	Mol. Form. / Mol. Wt.	CAS RN / mp/°C	Merck No. / bp/°C	Beil. Ref. / den/g cm⁻³	Solubility / n_D
11515	Solanid-5-ene-3,12-diol, (3β,12α)- / Rubijervine	$C_{27}H_{43}NO_2$ / 413.64	79-58-3 / 242	8264	5-21-05-00208	bz 4; EtOH 4; chl 4
11516	Solanid-5-ene-3,18-diol, (3β)- / Isorubijervine	$C_{27}H_{43}NO_2$ / 413.64	468-45-1 / 242.5	5111	5-21-05-00208	bz 4; chl 4
11517	Sonar / 2-Naphthalenol, 1,1'-methylenebis-, monosodium salt	$C_{21}H_{16}NaO_2$ / 323.35	38827-66-6 / 132			
11518	DL-Sorbose	$C_6H_{12}O_6$ / 180.16	65732-90-3 / 162.5	8681	4-01-00-04410 / 1.638[17]	H_2O 4; MeOH 4
11519	L-Sorbose / L-Sorbinose	$C_6H_{12}O_6$ / 180.16	87-79-6 / 165	8681	4-01-00-04412 / 1.612[17]	H_2O 3; EtOH 2; eth 2; MeOH 2
11520	Spiro[benzofuran-2(3H),1'-[2]cyclohexene]-3,4'-dione, 7-chloro-2',4,6-trimethoxy- / Griseofulvin, (+)-	$C_{17}H_{17}ClO_6$ / 352.77	126-07-8 / 220	4453	5-18-05-00150	H_2O 1; EtOH 2; eth 2; ace 2
11521	Spiro[4.5]dec-6-en-8-one, 6,10-dimethyl-2-(1-methylethylidene)-, (5R-cis)-	$C_{15}H_{22}O$ / 218.34	18444-79-6 / 44.5		141-2[2]	4-07-00-00822 / 1.0001[20]	ace 3 / 1.5309[20]
11522	Spiro[isobenzofuran-1(3H),9'-[9H]xanthen]-3-one, 3',6'-dihydroxy- / Fluorescein	$C_{20}H_{12}O_5$ / 332.31	2321-07-5 / 315 dec	4085	5-19-06-00456	H_2O 2; EtOH 2; eth 2; ace 4
11523	Spiro[isobenzofuran-1(3H),9'-[9H]xanthen]-3-one, 3',6'-dihydroxy-4',5'-diiodo- / Fluorescein, 4',5'-diiodo-	$C_{20}H_{10}I_2O_5$ / 584.11	38577-97-8	3174	5-19-06-00466	
11524	Spiro[isobenzofuran-1(3H),9'-[9H]xanthen]-3-one, 3',6'-dihydroxy-2',4',5',7'-tetraiodo-, disodium salt / Erythrosin	$C_{20}H_8I_4O_5$ / 835.90	16423-68-0	3640	4-19-00-02923	eth 4; EtOH 4
11525	Spiro[isobenzofuran-1(3H),9'-[9H]xanthen]-3-one, 2',4',5',7'-tetrabromo-3',6'-dihydroxy-, disodium salt / Eosin	$C_{20}H_8Br_4Na_2O_5$ / 691.86	17372-87-1 / 295.5	3555		EtOH 4
11526	Spiro[isobenzofuran-1(3H),9'-[9H]xanthen]-3-one, 3',4',5',6'-tetrahydroxy- / Gallein	$C_{20}H_{12}O_5$ / 332.31	2103-64-2 / >300	4250	5-19-07-00083	ace 4; EtOH 4
11527	Spiropentane / Spiro[2.2]pentane	C_5H_8 / 68.12	157-40-4 / -134.6	39	4-05-00-00218 / 0.7266[20]	1.4120[20]
11528	Spirosolan-3-ol, (3β,5α,22β,25S)- / Tomatidine	$C_{27}H_{45}NO_2$ / 415.66	77-59-8 / 210.5	9470		EtOH 3; eth 3
11529	Spirosol-5-en-3-ol, (3β,22α,25R)- / Solasodine	$C_{27}H_{43}NO_2$ / 413.64	126-17-0 / 202	8665	4-27-00-02000	EtOH 3; eth 2; ace 3; bz 3
11530	Spirostan-2,3-diol, (2α,3β,5α,25R)- / Gitogenin	$C_{27}H_{44}O_4$ / 432.64	511-06-6 / 271.5	4325	5-19-03-00338	H_2O 1; EtOH 3; eth 2; chl 3
11531	Spirostan-3-ol, (3β,5β,25S)- / Sarsasapogenin	$C_{27}H_{44}O_3$ / 416.64	126-19-2 / 200.5	8338	5-19-02-00672	EtOH 3; ace 3; bz 3; chl 3
11532	Spirostan-3-ol, (3β,5α,25R)- / Tigogenin	$C_{27}H_{44}O_3$ / 416.64	77-60-1 / 205.5	9367		EtOH 3; eth 3; ace 3; ctc 3
11533	Spirostan-3-ol, (3β,5β,25R)- / Smilagenin	$C_{27}H_{44}O_3$ / 416.64	126-18-1 / 185	8508	5-19-02-00673	ace 4; bz 4; EtOH 4
11534	Spirostan-12-one, 3-hydroxy-, (3β,5α,25R)- / Hecogenin	$C_{27}H_{42}O_4$ / 430.63	467-55-0 / 266.5	4537	5-19-05-00592	ace 4; eth 4; EtOH 4
11535	Spirostan-2,3,15-triol, (2α,3β,5α,15β,25R)- / Digitogenin	$C_{27}H_{44}O_5$ / 448.64	511-34-2 / 281.5	3143	5-19-03-00603	chl 4
11536	Spirost-5-en-3-ol, (3β,25R)- / Diosgenin	$C_{27}H_{42}O_3$ / 414.63	512-04-9 / 205.5	3289	5-19-03-00030	EtOH 4
11537	Spiro[5.5]undecane / Spir[5.]undecane	$C_{11}H_{20}$ / 152.28	180-43-8	207	4-05-00-00328 / 0.8783[20]	1.4731
11538	Stannane, (acetyloxy)tributyl- / Acetoxytributylstannane	$C_{14}H_{30}O_2Sn$ / 349.10	56-36-0 / 84.7			
11539	Stannane, (acetyloxy)triphenyl- / Triphenyltin acetate	$C_{20}H_{18}O_2Sn$ / 409.07	900-95-8 / 121.5			
11540	Stannane, chlorotrimethyl-	C_3H_9ClSn / 199.27	1066-45-1 / 38.5	148		H_2O 3; chl 3; os 3
11541	Stannane, chlorotriphenyl- / Triphenyltin chloride	$C_{18}H_{15}ClSn$ / 385.48	639-58-7 / 103.5			chl 3
11542	Stannane, chlorotripropyl-	$C_9H_{21}ClSn$ / 283.43	2279-76-7 / -23.5	123[13]	1.2678[28]	ctc 3; os 3 / 1.49102[28]
11543	Stannane, tetraethenyl-	$C_8H_{12}Sn$ / 226.89	1112-56-7	55-7[17]	4-04-00-04316 / 1.2651[25]	ctc 3 / 1.4993[25]
11544	Stannane, tetramethyl- / Tetramethylstannane	$C_4H_{12}Sn$ / 178.85	594-27-4 / -54.8	78	4-04-00-04307 / 1.314[25]	H_2O 1; ctc 3; CS_2 3; os 3 / 1.4386
11545	Stannane, tetraphenyl-	$C_{24}H_{20}Sn$ / 427.13	595-90-4 / 228	420	4-16-00-01592	chl 2
11546	Stannane, tetrapropyl-	$C_{12}H_{28}Sn$ / 291.06	2176-98-9 / -109.1	228	4-04-00-04310 / 1.1065[20]	1.4745[20]
11547	Stibine, dichlorophenyl-	$C_6H_5Cl_2Sb$ / 269.77	5035-52-9 / 69.5	110-5[10]	4-16-00-01205	ace 4; EtOH 4; HOAc 4
11548	Stibine oxide, dihydroxyphenyl- / Benzenestibonic acid	$C_6H_7O_3Sb$ / 248.88	535-46-6 / 139	1077	3-16-00-01178	

No.	Name Synonym	Mol. Form. Mol. Wt.	CAS RN mp/°C	Merck No. bp/°C	Beil. Ref. den/g cm^{-3}	Solubility n_D
11549	Stibine, triethyl- Triethylstibine	$C_6H_{15}Sb$ 208.94	617-85-6 -98	161.4	4-04-00-03685 1.3224[15]	H_2O 1; EtOH 3; eth 3
11550	Stibine, trimethyl- Trimethylstibine	C_3H_9Sb 166.86	594-10-5 -62	80.6	4-04-00-03684 1.523[15]	H_2O 1; EtOH 3; eth 3; CS_2 3 1.42[15]
11551	Stibine, triphenyl-	$C_{18}H_{15}Sb$ 353.07	603-36-1 53.5	>360	4-16-00-01198 1.4343[25]	H_2O 1; EtOH 3; eth 4; ace 4 1.6948[42]
11552	Stibine, tris(3-methylphenyl)-	$C_{21}H_{21}Sb$ 395.15	35569-54-1 72		4-16-00-01199 1.3957[16]	bz 4; eth 4; EtOH 4; chl 4
11553	Stibine, tris(4-methylphenyl)-	$C_{21}H_{21}Sb$ 395.15	5395-43-7 127.5	218[3]	4-16-00-01199 1.3595[16]	bz 4; eth 4; chl 4
11554	Stigmasta-5,7-dien-3-ol, (3β)- 7-Dehydrositosterol	$C_{29}H_{48}O$ 412.70	521-04-0 144.5	2863	3-06-00-02855	bz 4; eth 4; EtOH 4
11555	Stigmasta-5,22-dien-3-ol, (3β,22E)- Stigmasterol	$C_{29}H_{48}O$ 412.70	83-48-7 170	8771	4-06-00-04170	bz 4; eth 4; EtOH 4
11556	Stigmasta-7,24(28)-dien-3-ol, 4-methyl-, (3β,4α,5α,24Z)- α1-Sitosterol	$C_{30}H_{50}O$ 426.73	474-40-8 166	8499	3-06-00-02876	EtOH 4; chl 4
11557	Stigmastan-3-ol, (3β,5α)- Stigmastanol	$C_{29}H_{52}O$ 416.73	83-45-4 144	8770	3-06-00-02172	
11558	Stigmast-5-en-3-ol, (3β)- β-Sitosterol	$C_{29}H_{50}O$ 414.72	83-46-5 140	8500	4-06-00-04037	EtOH 3; eth 3; HOAc 3
11559	Stigmast-5-en-3-ol, (3β,24S)- γ-Sitosterol	$C_{29}H_{50}O$ 414.72	83-47-6	8501		
11560	Stilbene, α-ethyl- 1-Butene, 1,2-diphenyl	$C_{16}H_{16}$ 208.30	22692-70-2 57	296.5	4-05-00-02206 1.0124[18]	1.593[18]
11561	D-Streptamine, O-2-deoxy-2-(methylamino)-α- L-glucopyranosyl-(12)-O-5-deoxy-3-C-formyl- α-L-lyxofuranosyl-(1-4)-N,N ' 55- bis(aminoiminomethyl)-, sulfate (2:3)(salt) Streptomycin, sulfate	$C_{42}H_{84}N_{14}O_{36}S_3$ 1457.40	3810-74-0		5-18-11-00082	
11562	Streptomycin N-Methyl-L- glucosamidinostreptosidostreptidine	$C_{21}H_{39}N_7O_{12}$ 581.58	57-92-1	8786		
11563	Strychnidin-10-one Strychnine	$C_{21}H_{22}N_2O_2$ 334.42	57-24-9 287	8822 270[5]	4-27-00-07537 1.36[20]	H_2O 2; EtOH 2; eth 1; ace 2
11564	Strychnidin-10-one, 2,3-dimethoxy- Brucine	$C_{23}H_{26}N_2O_4$ 394.47	357-57-3 178	1443	4-27-00-07875	H_2O 2; EtOH 4; eth 2; bz 2
11565	Strychnidin-10-one, mononitrate Strychnine nitrate	$C_{21}H_{23}N_3O_5$ 397.43	66-32-0 295	8823	2-27-00-00723 1.627[25]	H_2O 4; MeOH 4
11566	Strychnidin-10-one, sulfate (2:1) Strychnine, sulfate	$C_{42}H_{46}N_4O_8S$ 766.92	60-41-3 200 dec	8826		H_2O 3; EtOH 3; eth 1; chl 2
11567	Succinimide, N-(ethoxymethyl)-	$C_7H_{11}NO_3$ 157.17	98431-97-1 31.5	262	2-21-00-00304	H_2O 4; EtOH 4
11568	Sulfamic acid, cyclohexyl- Cyclamic acid	$C_6H_{13}NO_3S$ 179.24	100-88-9 169.5	2707	4-12-00-00102	alk 4
11569	Sulfamide, tetramethyl-	$C_4H_{12}N_2O_2S$ 152.22	3768-63-6 73	225	4-04-00-00270	EtOH 4
11570	Sulfide, m-tolyl p-tolyl 3,4'-Ditolyl sulfide	$C_{14}H_{14}S$ 214.33	107770-92-3 28	179[11]	4-06-00-02173	eth 4; EtOH 4
11571	Sulfoxylic acid, diethyl ester	$C_4H_{10}O_2S$ 122.19	10297-38-8	118	3-01-00-01314 0.9940[20]	H_2O 3; EtOH 4; eth 4; ace 4 1.4234[20]
11572	Sulfur cyanide	C_2N_2S 84.10	627-52-1 63.5	sub 30	4-03-00-00339	eth 4; EtOH 4
11573	Sulfuric acid, diethyl ester Diethyl sulfate	$C_4H_{10}O_4S$ 154.19	64-67-5 -24	3120 208	4-01-00-01326 1.172[25]	H_2O 1; EtOH 5; eth 5 1.3989[20]
11574	Sulfuric acid, dimethyl ester Dimethyl sulfate	$C_2H_6O_4S$ 126.13	77-78-1 -27	3244 188 dec; 76[15]	4-01-00-01251 1.3322[20]	H_2O 3; EtOH 5; eth 3; bz 3 1.3874[20]
11575	Sulfuric acid, dipropyl ester	$C_6H_{14}O_4S$ 182.24	598-05-0	121[30]	4-01-00-01423 1.1064[20]	peth 4 1.4135[20]
11576	Sulfuric acid, monoethyl ester Ethyl sulfate	$C_2H_6O_4S$ 126.13	540-82-9	3808 280 dec	4-01-00-01324 1.3657[20]	H_2O 4 1.4105[20]
11577	Sulfuric acid, monomethyl ester Methyl sulfate	CH_4O_4S 112.11	75-93-4 <-30	6041 135 dec	4-01-00-01250	H_2O 4; eth 4; EtOH 4
11578	Sulfurous acid, bis(2-methylpropyl) ester	$C_8H_{18}O_3S$ 194.30	18748-27-1	210; 92-4[13]	4-01-00-01595 0.9862[20]	1.4268[20]
11579	Sulfurous acid, 2-chloroethyl 2-[4-(1,1- dimethylethyl)phenoxy]-1-methylethyl ester Aramite	$C_{15}H_{23}ClO_4S$ 334.86 -37.3	140-57-8 794 195[2]	1.143[20]	4-06-00-03301 1.5100[20]	ace 4; bz 4; eth 4; EtOH 4
11580	Sulfurous acid, dibutyl ester Butyl sulfite	$C_8H_{18}O_3S$ 194.30	626-85-7 230		4-01-00-01522 0.9957[20]	EtOH 3; eth 3 1.4310[20]
11581	Sulfurous acid, diethyl ester Ethyl sulfite	$C_4H_{10}O_3S$ 138.19	623-81-4 158; 51[13]		4-01-00-01324 1.1[20]	EtOH 3; eth 3 1.4310[20]
11582	Sulfurous acid, dimethyl ester	$C_2H_6O_3S$ 110.13	616-42-2 126		4-01-00-01250 1.2129[20]	H_2O 3; EtOH 3; eth 3 1.4083[20]
11583	Sulfurous diamide, tetramethyl-	$C_4H_{12}N_2OS$ 136.22	3768-60-3 31	209; 74[13]	4-04-00-00269	eth 4
11584	Sulfur, pentafluoro(trifluoromethyl)-, (OC-6-21)-	CF_8S 196.06	373-80-8 -87	-20	4-03-00-00035	H_2O 1; EtOH 1; CS_2 3

No.	Name / Synonym	Mol. Form. / Mol. Wt.	CAS RN / mp/°C	Merck No. / bp/°C	Beil. Ref. / den/g cm^{-3}	Solubility / n_D
11585	Sulfuryl chloride isocyanate	CClNO$_3$S 141.53	1189-71-5 -44	107	4-03-00-00086 1.626[25]	1.4467[20]
11586	Sutan Carbamothioic acid, bis(2-methylpropyl)-, S-ethyl ester	C$_{11}$H$_{23}$NOS 217.38	2008-41-5	1546 138[21]	0.9402[25]	
11587	D-Tagatose Tagatose, D-	C$_6$H$_{12}$O$_6$ 180.16	87-81-0 134.5	9005	4-01-00-04414	H$_2$O 4
11588	Tartar emetic Antimonate(2-), bis[μ-[2,3-dihydroxybutanedioato(4-)-O1,O2:O3,O4]]di-, dipotassium, trihydrate, stereoisomer	C$_8$H$_{10}$K$_2$O$_{15}$Sb$_2$ 667.87	28300-74-5	732	2.6	
11589	Tebuconazole (RS)-1-p-Chlorophenyl-4,4-dimethyl-3-(1H-1,2,4-triazol-1-ylmethyl)pentan-3-ol	C$_{16}$H$_{23}$ClN$_3$O 308.83	107534-96-3 102.4			
11590	Tebuthiuron Urea, N-[5-(1,1-dimethylethyl)-1,3,4-thiadiazol-2-yl]-N,N'-dimethyl-	C$_9$H$_{16}$N$_4$OS 228.32	34014-18-1 163 dec	9053		
11591	Terbacil 2,4(1H,3H)-Pyrimidinedione, 5-chloro-3-(1,1-dimethylethyl)-6-methyl-	C$_9$H$_{13}$ClN$_2$O$_2$ 216.67	5902-51-2 176	9085 sub 175	1.34[25]	
11592	Terbufos Phosphorodithioic acid, S-[[(1,1-dimethylethyl)thio]methyl] O,O-diethyl ester	C$_9$H$_{21}$O$_2$PS$_3$ 288.44	13071-79-9 -29.2	9088 69[0.01]	1.105[24]	
11593	Terbuthylazine 1,3,5-Triazine-2,4-diamine, 6-chloro-N-(1,1-dimethylethyl)-N'-ethyl-	C$_9$H$_{16}$ClN$_5$ 229.71	5915-41-3 178		1.188[20]	
11594	Terbutryn 1,3,5-Triazine-2,4-diamine, N-(1,1-dimethylethyl)-N'-ethyl-6-(methylthio)-	C$_{10}$H$_{19}$N$_5$S 241.36	886-50-0 104	157[0.06]	1.115[20]	
11595	1,1':2',1''-Terphenyl o-Terphenyl	C$_{18}$H$_{14}$ 230.31	84-15-1 56.2	332	4-05-00-02478	H$_2$O 1; ace 3; bz 3; chl 3
11596	1,1':3',1''-Terphenyl m-Terphenyl	C$_{18}$H$_{14}$ 230.31	92-06-8 87	363	4-05-00-02480 1.199[20]	H$_2$O 1; EtOH 3; eth 3; bz 3
11597	1,1':4',1''-Terphenyl p-Terphenyl	C$_{18}$H$_{14}$ 230.31	92-94-4 210.1	376	4-05-00-02483	H$_2$O 1; EtOH 2; eth 3; bz 3
11598	1,1':3',1''-Terphenyl, 5'-phenyl-	C$_{24}$H$_{18}$ 306.41	612-71-5 176	462	4-05-00-02732 1.199[30]	H$_2$O 1; EtOH 3; eth 3; bz 4
11599	Terrazole 1,2,4-Thiadiazole, 5-ethoxy-3-(trichloromethyl)-	C$_5$H$_5$Cl$_3$N$_2$OS 247.53	2593-15-9 19.9	95[1]	1.503[25]	
11600	2,2':5',2''-Terthiophene α-Terthienyl	C$_{12}$H$_8$S$_3$ 240.39	1081-34-1	9106	5-19-09-00226	
11601	1,3,5,7-Tetraazatricyclo[3.3.1.1^3,7]decane Methenamine	C$_6$H$_{12}$N$_4$ 140.19	100-97-0 >250	5879 sub	4-26-00-01680 1.331[-5]	H$_2$O 4; EtOH 3; eth 2; ace 3
11602	2,6,10,14,18,22-Tetracosahexaene, 2,6,10,15,19,23-hexamethyl-, (all-E)- Squalene	C$_{30}$H$_{50}$ 410.73	111-02-4 <-20	8727 280[17]; 241[4]	3-01-00-01071 0.8584[20]	H$_2$O 1; EtOH 2; eth 3; ace 3 1.4990[20]
11603	Tetracosane	C$_{24}$H$_{50}$ 338.66	646-31-1 54	391.3	4-01-00-00578 0.7991[20]	H$_2$O 1; EtOH 2; eth 4 1.4283[70]
11604	Tetracosane, 11-decyl-	C$_{34}$H$_{70}$ 478.93	55429-84-0 10.8	301[10]	4-01-00-00598 0.8161[20]	1.4556[20]
11605	Tetracosane, 3-ethyl-	C$_{26}$H$_{54}$ 366.71	55282-17-2 30.1	255.5[10]	4-01-00-00584 0.7949[40]	1.4436[40]
11606	Tetracosane, 2,6,10,15,19,23-hexamethyl- Squalane	C$_{30}$H$_{62}$ 422.82	111-01-3 -38	8726 350	4-01-00-00593 0.8115[15]	H$_2$O 1; EtOH 2; eth 3; ace 2 1.4530[15]
11607	Tetracosanoic acid Lignoceric acid	C$_{24}$H$_{48}$O$_2$ 368.64	557-59-5 84.2	5364 272[10]	4-02-00-01301 0.8207[100]	bz 4; eth 4 1.4287[100]
11608	15-Tetracosenoic acid	C$_{24}$H$_{46}$O$_2$ 366.63	14490-79-0 61		4-02-00-01681	EtOH 3; chl 3; ace 3
11609	5,9-Tetradecadien-7-yne, 6,9-dimethyl-	C$_{16}$H$_{26}$ 218.38	101427-54-7	95-8[0.3]	4-01-00-01133 0.8241[20]	bz 4; eth 4 1.4866[20]
11610	6,8-Tetradecadiyne	C$_{14}$H$_{22}$ 190.33	16387-72-7 2	118-9[4]	3-01-00-01067 0.8699[16]	eth 4
11611	Tetradecanamide	C$_{14}$H$_{29}$NO 227.39	638-58-4 104	217[12]	4-02-00-01138	EtOH 4
11612	1-Tetradecanamine	C$_{14}$H$_{31}$N 213.41	2016-42-4 83.1	291.2	4-04-00-00812 0.8079[20]	H$_2$O 1; EtOH 4; eth 4; ace 3 1.4463[20]
11613	Tetradecane	C$_{14}$H$_{30}$ 198.39	629-59-4 5.8	253.5	4-01-00-00520 0.7628[20]	H$_2$O 1; EtOH 4; eth 4; ctc 3 1.4290[20]
11614	Tetradecane, 1-bromo-	C$_{14}$H$_{29}$Br 277.29	112-71-0 5.6	307	4-01-00-00523 1.0170[20]	ace 4; bz 4; EtOH 4 1.4603[20]
11615	Tetradecane, 1-chloro-	C$_{14}$H$_{29}$Cl 232.84	2425-54-9 4.9	292	3-01-00-00550 0.8665[20]	H$_2$O 1; EtOH 3; ace 4; bz 4 1.4473[20]
11616	Tetradecane, 1,14-dibromo- Tetradecamethylene dibromide	C$_{14}$H$_{28}$Br$_2$ 356.18	37688-96-3 50.4	190[8]	4-01-00-00523	eth 4; EtOH 4; chl 4
11617	Tetradecane, 1,1-dimethoxy- Myristaldehyde, dimethyl acetal	C$_{16}$H$_{34}$O$_2$ 258.44	14620-53-2	134-6[4]	4-01-00-03389	eth 4; EtOH 4 1.4342[25]
11618	1,14-Tetradecanediol	C$_{14}$H$_{30}$O$_2$ 230.39	19812-64-7 85.8	200[9]	4-01-00-02631	eth 4; EtOH 4

No.	Name / Synonym	Mol. Form. / Mol. Wt.	CAS RN / mp/°C	Merck No. / bp/°C	Beil. Ref. / den/g cm^{-3}	Solubility / n_D
11619	Tetradecanenitrile / Myristonitrile	$C_{14}H_{27}N$ / 209.38	629-63-0 / 19	/ 226[100]; 119[1]	4-02-00-01139 / 0.8281[19]	H_2O 1; EtOH 5; eth 5; ace 5 / 1.4392[23]
11620	1-Tetradecanesulfonic acid / Tetradecylsulfonic acid	$C_{14}H_{30}O_3S$ / 278.46	7314-37-6 / 65		4-04-00-00066 / 0.9996[25]	H_2O 4
11621	1-Tetradecanethiol	$C_{14}H_{30}S$ / 230.46	2079-95-0 /	/ 176-80[22]	4-01-00-01867 / 0.8469[20]	H_2O 1; EtOH 3; eth 3; ctc 3 / 1.4597[20]
11622	Tetradecanoic acid / Myristic acid	$C_{14}H_{28}O_2$ / 228.38	544-63-8 / 53.9	6246 / 250[100]	4-02-00-01126 / 0.8622[54]	H_2O 1; EtOH 3; eth 2; ace 3 / 1.4723[70]
11623	Tetradecanoic acid, anhydride	$C_{28}H_{54}O_3$ / 438.73	626-29-9 / 53.4		4-02-00-01138 / 0.8502[70]	eth 4; EtOH 4 / 1.4335[70]
11624	Tetradecanoic acid, ethyl ester	$C_{16}H_{32}O_2$ / 256.42	124-06-1 / 12.3	/ 295	4-02-00-01131 / 0.8573[25]	H_2O 1; EtOH 3; eth 2; ctc 3 / 1.4362[20]
11625	Tetradecanoic acid, methyl ester	$C_{15}H_{30}O_2$ / 242.40	124-10-7 / 19	/ 295; 155[7]	4-02-00-01131 / 0.8671[20]	H_2O 1; EtOH 5; eth 5; ace 5 / 1.425[45]
11626	Tetradecanoic acid, 1-methylethyl ester / Isopropyl myristate	$C_{17}H_{34}O_2$ / 270.46	110-27-0 /	5103 / 193[20]; 140[2]	4-02-00-01132 / 0.8532[20]	H_2O 1; EtOH 3; eth 3; ace 4 / 1.4325[25]
11627	Tetradecanoic acid, phenylmethyl ester	$C_{21}H_{34}O_2$ / 318.50	31161-71-4 / 20.5	/ 229.3[11]	2-06-00-00417 / 0.9293[25]	bz 4; eth 4; EtOH 4; chl 4
11628	Tetradecanoic acid, 1,2,3-propanetriyl ester / Trimyristin	$C_{45}H_{86}O_6$ / 723.17	555-45-3 / 56.5	9638 / 311	4-02-00-01135 / 0.8848[60]	H_2O 1; EtOH 3; eth 3; ace 3 / 1.4428[60]
11629	Tetradecanoic acid, propyl ester	$C_{17}H_{34}O_2$ / 270.46	14303-70-9 /	/ 147[2]	4-02-00-01132 / 0.8592[20]	ace 4; bz 4; eth 4; EtOH 4 / 1.4356[25]
11630	1-Tetradecanol / Tetradecyl alcohol	$C_{14}H_{30}O$ / 214.39	112-72-1 / 39.5	6248 / 289	4-01-00-01864 / 0.8236[38]	H_2O 1; EtOH 4; eth 4; ace 4
11631	2-Tetradecanol	$C_{14}H_{30}O$ / 214.39	4706-81-4 / 34	/ 284	4-01-00-01867 / 0.8315[20]	/ 1.4444[20]
11632	3-Tetradecanol	$C_{14}H_{30}O$ / 214.39	1653-32-3 / 31.5	/ 173[25]; 146[10]	4-01-00-01868 / 0.8098[53]	bz 4; eth 4; EtOH 4 / 1.4340[45]
11633	2-Tetradecanone / Dodecylmethylketone	$C_{14}H_{28}O$ / 212.38	2345-27-9 / 33.5	/ 205[100]; 134[13]	4-01-00-03389	H_2O 1; EtOH 3; ace 3; os 3
11634	3-Tetradecanone	$C_{14}H_{28}O$ / 212.38	629-23-2 / 34	/ 152[16]	4-01-00-03389	H_2O 1; EtOH 3; ace 3; os 3
11635	Tetradecanoyl chloride / Myristoyl chloride	$C_{14}H_{27}ClO$ / 246.82	112-64-1 / -1	/ 171[16]	4-02-00-01138 / 0.9078[25]	eth 3
11636	1-Tetradecene	$C_{14}H_{28}$ / 196.38	1120-36-1 / -12	/ 233	4-01-00-00924 / 0.7745[25]	H_2O 1; EtOH 4; eth 4; bz 3 / 1.4351[20]
11637	4-Tetradecenoic acid / Tuduic acid	$C_{14}H_{26}O_2$ / 226.36	544-65-0 / 18.5	/ 185-8[13]	4-02-00-01626 / 0.9024[20]	bz 4; peth 4 / 1.4559[20]
11638	5-Tetradecenoic acid / Physoteric acid	$C_{14}H_{26}O_2$ / 226.36	544-66-1 / 20	/ 190-5[15]	3-02-00-01373 / 0.9046[20]	/ 1.4552[20]
11639	9-Tetradecenoic acid	$C_{14}H_{26}O_2$ / 226.36	13147-06-3 / -4	/ 144[0.6]	4-02-00-01626 / 0.9018[20]	/ 1.4519[20]
11640	2-Tetradecyne	$C_{14}H_{26}$ / 194.36	638-60-8 / 6.5	/ 252.5	0-01-00-00262 / 0.8000[20]	eth 4; EtOH 4
11641	7-Tetradecyne	$C_{14}H_{26}$ / 194.36	35216-11-6 /	/ 144[30]	3-01-00-01027 / 0.7991[20]	eth 4; EtOH 4 / 1.4330[25]
11642	2,5,8,11-Tetraoxadodecane / Triglyme	$C_8H_{18}O_4$ / 178.23	112-49-2 / -45	9604 / 216	4-01-00-02401 / 0.986[20]	H_2O 4; bz 4 / 1.4224[20]
11643	2,4,8,10-Tetraoxaspiro[5.5]undecane	$C_7H_{12}O_4$ / 160.17	126-54-5 / 48.3	/ 147[53]; 68[1]	5-19-11-00342	H_2O 4; ace 4; eth 4; EtOH 4
11644	Tetraphenylene / Tetrabenzocyclooctatetraene	$C_{24}H_{16}$ / 304.39	212-74-8 / 233	/ sub 200	4-05-00-02773	EtOH 3; eth 2; AcOEt 3; $PhNO_2$ 3
11645	Tetraphosphoric acid, hexaethyl ester / Ethyl tetraphosphate	$C_{12}H_{30}O_{13}P_4$ / 506.26	757-58-4 / -40	/ 150 dec	1.2917[27]	ace 4; bz 4; EtOH 4 / 1.4273[27]
11646	Tetrasiloxane, decamethyl- / Decamethyltetrasiloxane	$C_{10}H_{30}O_3Si_4$ / 310.69	141-62-8 / -76	2843 / 194	4-04-00-04119 / 0.8536[25]	H_2O 1; EtOH 2; bz 3; peth 3 / 1.3895[20]
11647	Tetrasiloxane, 1,1,1,3,5,7,7,7-octamethyl- / 1,1,1,3,5,7,7,7-Octamethyltetrasiloxane	$C_8H_{26}O_3Si_4$ / 282.63	16066-09-4 /	/ 170	4-04-00-04098 / 0.8559[20]	/ 1.3854[20]
11648	Tetrasul / p-Chlorophenyl 2,4,5-trichlorophenyl sulfide	$C_{12}H_6Cl_4S$ / 324.06	2227-13-6			
11649	Tetrasulfide, bis(1,1-dimethylethyl) / Di-tert-Butyl tetrasulfide	$C_8H_{18}S_4$ / 242.49	5943-35-1 / 2.3	/ 70[0.2]	4-01-00-01638 / 1.0690[20]	/ 1.5660[20]
11650	Tetratriacontane	$C_{34}H_{70}$ / 478.93	14167-59-0 / 72.6	/ 285.4[3]	4-01-00-00597 / 0.7728[90]	/ 1.4296[90]
11651	1,2,4,5-Tetrazine / sym-Tetrazine	$C_2H_2N_4$ / 82.06	290-96-0 / 99	/ sub	4-26-00-01710	H_2O 3; EtOH 3; eth 3; sulf 3
11652	1H-Tetrazole	CH_2N_4 / 70.05	288-94-8 / 157	/ sub	4-26-00-01652 / 1.4060[20]	H_2O 2
11653	2H-Tetrazolium, 2,3,5-triphenyl-, chloride / Triphenyltetrazolium chloride	$C_{19}H_{15}ClN_4$ / 334.81	298-96-4 / 243 dec	9658	4-26-00-01774	H_2O 3; EtOH 3; eth 1; ace 3
11654	5H-Tetrazolo[1,5-a]azepine, 6,7,8,9-tetrahydro- / Pentylenetetrazole	$C_6H_{10}N_4$ / 138.17	54-95-5 / 59.5	7097 / 194[12]	4-26-00-01712	H_2O 4; EtOH 4; eth 3; ace 4
11655	4-Thia-1-azabicyclo[3.2.0]heptane-2-carboxylic acid, 3,3-dimethyl-7-oxo-6-[(phenylacetyl)amino]- / Penicillin G, sodium salt	$C_{16}H_{17}N_2NaO_4S$ / 356.38	69-57-8	1157	4-27-00-05861	

No.	Name / Synonym	Mol. Form. / Mol. Wt.	CAS RN mp/°C	Merck No. bp/°C	Beil. Ref. den/g cm⁻³	Solubility n_D
11656	Thiabendazole 1H-Benzimidazole, 2-(4-thiazolyl)-	$C_{10}H_7N_3S$ 201.25	148-79-8	9217 sub 305		
11657	1,2,4-Thiadiazole 1-Thia-2,4-diazacyclopentadiene	$C_2H_2N_2S$ 86.12	288-92-6 -33	121	4-27-00-07093 1.3298[20]	
11658	1,2,5-Thiadiazole Piazthiole	$C_2H_2N_2S$ 86.12	288-39-1 -50.1	94	1.268[25]	1.5150[25]
11659	1,3,4-Thiadiazole, 2,5-dimethyl- 2,5-Dimethyl-1,3,4-thiadiazole	$C_4H_6N_2S$ 114.17	27464-82-0 65	202.5	4-27-00-07097	H_2O 2; EtOH 2; eth 2
11660	Thianthrene	$C_{12}H_8S_2$ 216.33	92-85-3 159.3	365	5-19-02-00049 1.4420[20]	H_2O 1; EtOH 2; eth 3; bz 3
11661	Thianthrene, 2,7-dimethyl- Mesulphen	$C_{14}H_{12}S_2$ 244.38	135-58-0 123	5821 184[3]	5-19-02-00072	ace 4; eth 4; peth 4; chl 4
11662	2-Thiazolamine 2-Aminothiazole	$C_3H_4N_2S$ 100.14	96-50-4 93	494 140[11]	4-27-00-04574	H_2O 2; EtOH 2; eth 2; chl 2
11663	2-Thiazolamine, 5-[(4-aminophenyl)sulfonyl]- Thiazolsulfone	$C_9H_9N_3O_2S_2$ 255.32	473-30-3 219-21 dec	9237	4-27-00-05401	ace 4; eth 4; EtOH 4; diox 4
11664	2-Thiazolamine, 4,5-dihydro-	$C_3H_6N_2S$ 102.16	1779-81-3 85.3	dec	4-27-00-04555	H_2O 4; EtOH 4; bz 4; chl 4
11665	2-Thiazolamine, 4-methyl- 2-Amino-4-methylthiazole	$C_4H_6N_2S$ 114.17	1603-91-4 45.5	462 125[20]; 70[0.4]	4-27-00-04687	H_2O 4; EtOH 4; eth 4
11666	2-Thiazolamine, 5-nitro- 2-Amino-5-nitrothiazole	$C_3H_3N_3O_2S$ 145.14	121-66-4 202 dec	469	4-27-00-04675	
11667	Thiazole	C_3H_3NS 85.13	288-47-1 118	9235	4-27-00-00960 1.1998[17]	H_2O 2; EtOH 3; eth 3; ace 3 1.5969[20]
11668	Thiazole, 2-bromo- 2-Bromo-1,3-thiazole	C_3H_2BrNS 164.03	3034-53-5 171		4-27-00-00962 1.82[25]	1.5927[20]
11669	5-Thiazolecarboxylic acid, 4-methyl- 4-Methyl-5-thiazolecarboxylic acid	$C_5H_5NO_2S$ 143.17	20485-41-0 280 dec	sub 250	4-27-00-04008	H_2O 2; EtOH 3; eth 3; bz 2
11670	Thiazole, 5-(2-chloroethyl)-4-methyl- Clomethiazole	C_6H_8ClNS 161.65	533-45-9 92[7]	2382	4-27-00-00990 1.233[25]	
11671	2,4-Thiazolediamine, 5-phenyl- Amiphenazole	$C_9H_9N_3S$ 191.26	490-55-1 163-4 dec	498	4-27-00-05139	
11672	Thiazole, 4,5-dihydro-2-methyl-	C_4H_7NS 101.17	2346-00-1 -101	145	4-27-00-00921 1.067[25]	1.5200[20]
11673	Thiazole, 4,5-dihydro-2,4,4-trimethyl-	$C_6H_{11}NS$ 129.23	4145-94-2 147		4-27-00-00931 0.969[25]	1.4825[25]
11674	Thiazole, 2,4-dimethyl- 2,4-Dimethylthiazole	C_5H_7NS 113.18	541-58-2 146; 71[50]	3249	4-27-00-00981 1.0562[15]	H_2O 2; EtOH 3; eth 3; chl 3 1.5091[20]
11675	Thiazole, 4,5-dimethyl- 4,5-Dimethylthiazole	C_5H_7NS 113.18	3581-91-7 83.5	158	4-27-00-00986 1.0699[20]	eth 4; EtOH 4
11676	Thiazole, 2,4-diphenyl- 2,4-Diphenylthiazole	$C_{15}H_{11}NS$ 237.33	1826-14-8 92.5	>380	4-27-00-01435 1.1554[08]	eth 4; EtOH 4
11677	5-Thiazoleethanol, 4-methyl- 4-Methyl-5-thiazoleethanol	C_6H_9NOS 143.21	137-00-8 135[7]	6046	4-27-00-01754 1.196[24]	H_2O 4; EtOH 3; eth 3; bz 3
11678	Thiazole, 2-methyl- 2-Methylthiazole	C_4H_5NS 99.16	3581-87-1 128.5		4-27-00-00967	H_2O 5; EtOH 3; ace 3 1.510
11679	Thiazole, 4-methyl- 4-Methylthiazole	C_4H_5NS 99.16	693-95-8 133.3		4-27-00-00969 1.112[25]	H_2O 3; EtOH 3; eth 3
11680	2(3H)-Thiazolethione, 4-methyl-	$C_4H_5NS_2$ 131.22	5685-06-3 89.3	188[3]	4-27-00-02584	EtOH 4
11681	Thiazolidine	C_3H_7NS 89.16	504-78-9 164.5		4-27-00-00009 1.131[25]	H_2O 5; EtOH 3; eth 4; ace 4 1.551[20]
11682	4-Thiazolidinecarboxylic acid Timonacic	$C_4H_7NO_2S$ 133.17	444-27-9 196.5	9375		H_2O 4
11683	2,4-Thiazolidinedione	$C_3H_3NO_2S$ 117.13	2295-31-0 128	179[19]	4-27-00-03144	eth 4
11684	4-Thiazolidinone, 5-[[4-(dimethylamino)phenyl]methylene]-2-thioxo- p-(Dimethylamino)benzal-5-rhodanine	$C_{12}H_{12}N_2OS_2$ 264.37	536-17-4 270 dec	3220	4-27-00-05697	H_2O 1; EtOH 2; eth 4; ace 3
11685	4-Thiazolidinone, 2-thioxo- Rhodanine	$C_3H_3NOS_2$ 133.19	141-84-4 170	8184	4-27-00-03188 0.868[25]	H_2O 2; EtOH 4; eth 4
11686	Thiazolium, 3-[(4-amino-2-methyl-5-pyrimidinyl)methyl]-5-(2-hydroxyethyl)-4-methyl- chloride, monohydrochlorid Thiamine hydrochloride	$C_{12}H_{18}Cl_2N_4OS$ 337.27	67-03-8 248 dec	9222	4-27-00-01764	H_2O 4; EtOH 2; eth 1; ace 2
11687	Thidiazuron Urea, N-phenyl-N'-1,2,3-thiadiazol-5-yl-	$C_9H_8N_4OS$ 220.25	51707-55-2 211 dec			
11688	Thieno[2,3-b]thiophene	$C_6H_4S_2$ 140.23	250-84-0 6.5	225	5-19-01-00431	eth 4
11689	1H-Thieno[3,4-d]imidazole-4-pentanoic acid, hexahydro-2-oxo-, [3aS-(3aα,4β,6aα)]- Biotin	$C_{10}H_{16}N_2O_3S$ 244.31	58-85-5 232 dec	1244	4-27-00-07979	H_2O 3; EtOH 3; eth 2; chl 2
11690	Thiepane	$C_6H_{12}S$ 116.23	4753-80-4 0.5	173.5	5-17-01-00090 0.9883[20]	H_2O 1; eth 3; ace 3; chl 3 1.5044[18]
11691	Thietane Trimethylene sulfide	C_3H_6S 74.15	287-27-4 -73.2	95	5-17-01-00014 1.0200[20]	H_2O 1; EtOH 4; ace 3; bz 4 1.5102[20]
11692	Thietane, 1,1-dioxide Trimethylene sulfone	$C_3H_6O_2S$ 106.15	5687-92-3 75.5	91.2[14]	5-17-01-00015	H_2O 3; EtOH 2; eth 2; peth 2 1.5156[20]
11693	Thietane, 2-methyl-	C_4H_8S 88.17	17837-41-1 107		5-17-01-00055 0.9571[20]	H_2O 1; EtOH 3; eth 3; ace 3 1.4852[20]

No.	Name Synonym	Mol. Form. Mol. Wt.	CAS RN mp/°C	Merck No. bp/°C	Beil. Ref. den/g cm^{-3}	Solubility n_D
11694	Thietane, 1-oxide 1,3-Propanesulfone	$C_3H_6O_3S$ 122.14	13153-11-2 31.5	180[30]	5-17-01-00015 1.392[25]	
11695	Thiirane	C_2H_4S 60.12	420-12-2 57 dec		5-17-01-00009 1.0368[0]	EtOH 2; eth 2; ace 3; chl 3 1.4935[20]
11696	Thiirane, methyl-	C_3H_6S 74.15	1072-43-1 -91	72.5	5-17-01-00025 0.941[20]	chl 3 1.472[20]
11697	Thiobencarb Carbamothioic acid, diethyl-, S-[(4-chlorophenyl)methyl] ester	$C_{12}H_{16}CINOS$ 257.78	28249-77-6 3.3	127[0.008]	1.16[20]	
11698	Thioboric acid, trimethyl ester	$C_3H_9BS_3$ 152.11	997-49-9 5	218.2	4-01-00-01287 1.126[20]	ace 4; eth 4 1.5788[20]
11699	Thiocyanic acid, butyl ester Butyl thiocyanate	C_5H_9NS 115.20	628-83-1 	186	4-03-00-00329 0.9563[15]	H_2O 1; EtOH 3; eth 3 1.4360[20]
11700	Thiocyanic acid, chloromethyl ester	C_2H_2CINS 107.56	3268-79-9 	185	2-03-00-00124 1.37[15]	os 5
11701	Thiocyanic acid, 1,1-dimethylethyl ester tert-Butyl thiocyanate	C_5H_9NS 115.20	37985-18-5 10.5	140 dec; 40[10]	4-03-00-00330 0.9187[10]	
11702	Thiocyanic acid, 1,2-ethanediyl ester	$C_4H_4N_2S_2$ 144.22	629-17-4 90	dec	4-03-00-00333 1.4200[0]	H_2O 2; EtOH 3; eth 3; ace 4
11703	Thiocyanic acid, ethyl ester Ethyl thiocyanate	C_3H_5NS 87.15	542-90-5 -85.5	146	4-03-00-00328 1.007[23]	H_2O 1; EtOH 5; eth 5; chl 3 1.4684[15]
11704	Thiocyanic acid, heptyl ester Heptyl thiocyanate	$C_8H_{15}NS$ 157.28	5416-94-4 	235	4-03-00-00331 0.92[20]	eth 4; EtOH 4
11705	Thiocyanic acid, methyl ester Methyl thiocyanate	C_2H_3NS 73.12	556-64-9 -5	6047 132.9	4-03-00-00327 1.0678[25]	H_2O 2; EtOH 5; eth 5; ctc 3 1.4669[25]
11706	Thiocyanic acid, 1-methylethyl ester Isopropyl thiocyanate	C_4H_7NS 101.17	625-59-2 	153	4-03-00-00329 0.9784[20]	H_2O 1; EtOH 5; eth 5
11707	Thiocyanic acid, 2-methylpropyl ester Isobutyl thiocyanate	C_5H_9NS 115.20	591-84-4 -59	5036 175.4	4-03-00-00330	eth 4; EtOH 4
11708	Thiocyanic acid, octyl ester	$C_9H_{17}NS$ 171.31	19942-78-0 105	141-2[19]	4-03-00-00331 0.9149[25]	H_2O 1; EtOH 4; eth 4; bz 3 1.4649[20]
11709	Thiocyanic acid, phenyl ester Phenyl thiocyanate	C_7H_5NS 135.19	5285-87-0 	232.5	4-06-00-01536 1.153[18]	H_2O 1; EtOH 3; eth 3
11710	Thiocyanic acid, phenylmethyl ester	C_8H_7NS 149.22	3012-37-1 43	232	4-06-00-02680	H_2O 1; EtOH 3; eth 3; chl 3
11711	Thiocyanic acid, potassium salt KSCN	CKNS 97.18	333-20-0	7687	4-03-00-00299	
11712	Thiocyanic acid, 2-propenyl ester	C_4H_5NS 99.16	764-49-8 	161	4-03-00-00332 1.056[15]	eth 4; EtOH 4
11713	Thiodicarb Dimethyl N,N'-[thiobis[[(methylimino)carbonyl]oxyl]]bis[ethanimidothioate]	$C_{10}H_{18}N_4O_4S_3$ 354.48	59669-26-0 173	9258	1.4[20]	
11714	Thiodicarbonic diamide, tetraethyl- Sulfiram	$C_{10}H_{20}N_2S_3$ 264.48	95-05-6 	8928 225-40[3]	4-04-00-00397 1.12[20]	chl 3
11715	Thiodicarbonic diamide, tetramethyl-	$C_6H_{12}N_2S_3$ 208.37	97-74-5 109.5		4-04-00-00238 1.37[25]	H_2O 1; EtOH 3; eth 2; ace 3
11716	Thiodiphosphoric acid, tetraethyl ester Sulfotep	$C_8H_{20}O_5P_2S_2$ 322.32	3689-24-5 	8945 136-9[2]	4-01-00-01351 1.196[25]	H_2O 1; EtOH 3 1.4753[25]
11717	Thiofanox 2-Butanone, 3,3-dimethyl-1-(methylthio)-, O-[(methylamino)carbonyl]oxime	$C_9H_{18}N_2O_2S$ 218.32	39196-18-4 57			
11718	Thioimidodicarbonic diamide 2,4-Dithiobiuret	$C_2H_5N_3S_2$ 135.21	541-53-7 181 dec	3378	4-03-00-00356	ace 4
11719	Thiomorpholine Thiamorpholine	C_4H_9NS 103.19	123-90-0 	9229 175; 110[100]	4-27-00-00636 1.0882[20]	H_2O 4; ace 4; eth 4; EtOH 4 1.5386[20]
11720	Thionazin Phosphorothioic acid, O,O-diethyl O-pyrazinyl ester	$C_8H_{13}N_2O_3PS$ 248.24	297-97-2 -1.7	9275 80		
11721	Thioperoxydicarbonic diamide ([(H2N)C(S)]2S2), tetraethyl- Disulfiram	$C_{10}H_{20}N_2S_4$ 296.55	97-77-8 71.5	3370 117[17]	4-04-00-00398	H_2O 1; EtOH 3; eth 2; chl 4
11722	Thioperoxydicarbonic diamide ([(H2N)C(S)]2S2), tetramethyl- Thiram	$C_6H_{12}N_2S_4$ 240.44	137-26-8 155.6	9304 129[20]	4-04-00-00242	chl 4
11723	Thiophanate-methyl Carbamic acid, [1,2-phenylenebis(iminocarbonothioyl)]bis-, dimethyl ester	$C_{12}H_{14}N_4O_4S_2$ 342.40	23564-05-8 172 dec			
11724	Thiophene Thiofuran	C_4H_4S 84.14	110-02-1 -39.4	9283 84.0	5-17-01-00297 1.0649[20]	EtOH 5; eth 5; ace 5; bz 5 1.5289[20]
11725	2-Thiopheneacetonitrile	C_6H_5NS 123.18	20893-30-5 	115-20[22]	5-18-06-00208 1.157[25]	1.5425[20]
11726	Thiophene, 2-bromo- 2-Thienyl bromide	C_4H_3BrS 163.04	1003-09-4 	150	5-17-01-00305 1.684[20]	H_2O 1; eth 4; ace 4; ctc 3 1.5868[20]
11727	Thiophene, 3-bromo-	C_4H_3BrS 163.04	872-31-1 	159.5	5-17-01-00306 1.735[20]	H_2O 1; ace 3; bz 3; chl 2 1.5919[20]
11728	Thiophene, 2-bromo-5-chloro- 2-Bromo-5-chlorothiophene	C_4H_2BrCIS 197.48	2873-18-9 -20	70[18]	5-17-01-00307 1.803[25]	eth 4 1.5925[25]

No.	Name / Synonym	Mol. Form. / Mol. Wt.	CAS RN / mp/°C	Merck No. / bp/°C	Beil. Ref. / den/g cm⁻³	Solubility / n_D
11729	Thiophene, 2-bromo-3-methyl- / 2-Bromo-3-methylthiophene	C_5H_5BrS / 177.06	14282-76-9	176; 27[1.8]	5-17-01-00333 / 1.5844[18]	bz 4; eth 4 / 1.5714[20]
11730	Thiophene, 2-bromo-5-methyl-	C_5H_5BrS / 177.06	765-58-2	178; 29[1.8]	5-17-01-00327 / 1.5529[20]	eth 3; bz 3 / 1.5673[20]
11731	Thiophene, 2-butyl- / 2-Butylthiophene	$C_8H_{12}S$ / 140.25	1455-20-5	181.5	5-17-01-00369 / 0.9537[20]	1.5090[20]
11732	2-Thiophenecarbonitrile / 2-Cyanothiophene	C_5H_3NS / 109.15	1003-31-2	192	5-18-06-00171 / 1.172[25]	chl 3 / 1.5629[20]
11733	2-Thiophenecarboxaldehyde	C_5H_4OS / 112.15	98-03-3	197; 85[16]	5-17-09-00349 / 1.2127[21]	H_2O 1; EtOH 4; eth 3; chl 2 / 1.5920[20]
11734	3-Thiophenecarboxaldehyde / 3-Formylthiophene	C_5H_4OS / 112.15	498-62-4	86.7[20]	5-17-09-00373	H_2O 1; EtOH 4; eth 3 / 1.5855[20]
11735	2-Thiophenecarboxaldehyde, 5-chloro-	C_5H_4ClOS / 147.60	7283-96-7	77.5[5]	5-17-09-00360	chl 2 / 1.6036[25]
11736	2-Thiophenecarboxaldehyde, 5-methyl-	C_6H_6OS / 126.18	13679-70-4	114[25]	5-17-09-00406	chl 3 / 1.5825[20]
11737	2-Thiophenecarboxylic acid / 2-Carboxythiophene	$C_5H_4O_2S$ / 128.15	527-72-0 / 129.5	9284 / 260 dec	5-18-06-00158	H_2O 4; EtOH 4; eth 4; chl 3
11738	3-Thiophenecarboxylic acid / 3-Thenoic acid	$C_5H_4O_2S$ / 128.15	88-13-1 / 138	9206	5-18-06-00199	H_2O 3
11739	2-Thiophenecarboxylic acid, ethyl ester	$C_7H_8O_2S$ / 156.21	2810-04-0	218	5-18-06-00159 / 1.1623[16]	EtOH 3; ace 3; ctc 2; os 3 / 1.5248[20]
11740	Thiophene, 2-chloro- / 2-Thienyl chloride	C_4H_3ClS / 118.59	96-43-5 / -71.9	128.3	5-17-01-00303 / 1.2863[20]	H_2O 1; EtOH 5; eth 5; chl 2 / 1.5487[20]
11741	Thiophene, 2-chloro-5-methyl- / 2-Chloro-5-methylthiophene	C_5H_5ClS / 132.61	17249-82-0	155; 55[19]	5-17-01-00326 / 1.2147[25]	ace 4; bz 4; eth 4; EtOH 4 / 1.5372[20]
11742	Thiophene, 2,3-dibromo- / 2,3-Dibromothiophene	$C_4H_2Br_2S$ / 241.93	3140-93-0 / -17.5	218.5; 89[13]	5-17-01-00308	1.6304[22]
11743	Thiophene, 2,4-dibromo- / 2,4-Dibromothiophene	$C_4H_2Br_2S$ / 241.93	3140-92-9 / -26.3	210	5-17-01-00308 / 2.1488[20]	
11744	Thiophene, 2,5-dibromo- / 2,5-Dibromothiophene	$C_4H_2Br_2S$ / 241.93	3141-27-3 / -6	210.3	5-17-01-00308 / 2.142[23]	H_2O 4; EtOH 4; ctc 3 / 1.6288[20]
11745	2,5-Thiophenedicarboxylic acid / 2,5-Dicarboxythiophene	$C_6H_4O_4S$ / 172.16	4282-31-9 / 359	sub 150	5-18-07-00049	H_2O 2; EtOH 3; eth 3
11746	Thiophene, 2,3-dichloro- / 2,3-Dichlorothiophene	$C_4H_2Cl_2S$ / 153.03	17249-79-5 / -26.2	173.5	5-17-01-00304 / 1.4605[20]	H_2O 1; EtOH 5; eth 3; ace 3 / 1.5651[20]
11747	Thiophene, 2,4-dichloro- / 2,4-Dichlorothiophene	$C_4H_2Cl_2S$ / 153.03	17249-75-1 / -34	167.6	5-17-01-00304 / 1.4553[20]	H_2O 1; EtOH 4; eth 3; ace 3 / 1.5660[20]
11748	Thiophene, 2,5-dichloro- / 5-Chloro-2-thienyl chloride	$C_4H_2Cl_2S$ / 153.03	3172-52-9 / -40.5	162	5-17-01-00304 / 1.4422[20]	H_2O 1; EtOH 5; eth 5; ctc 3 / 1.5626[20]
11749	Thiophene, 3,4-dichloro- / 3,4-Dichlorothiophene	$C_4H_2Cl_2S$ / 153.03	17249-76-2 / 1	185	5-17-01-00305 / 1.4867[20]	H_2O 1; EtOH 4; eth 4; ace 3 / 1.5821[19]
11750	Thiophene, 2,3-dihydro- / 2,3-Dihydrothiophene	C_4H_6S / 86.16	1120-59-8	112.1		
11751	Thiophene, 2,5-dihydro- / 2,5-Dihydrothiophene	C_4H_6S / 86.16	1708-32-3	122.4		
11752	Thiophene, 2,5-dihydro-, 1,1-dioxide / 3-Sulfolene	$C_4H_6O_2S$ / 118.16	77-79-2 / 64.5	8935	5-17-01-00177	chl 3
11753	Thiophene, 2,5-diiodo- / 2,5-Diiodothiophene	$C_4H_2I_2S$ / 335.93	625-88-7 / 41.5	139[15]	5-17-01-00313	H_2O 1; EtOH 4
11754	Thiophene, 2,3-dimethyl- / 2,3-Dimethylthiophene	C_6H_8S / 112.20	632-16-6 / -49	141.6	5-17-01-00342 / 1.0021[20]	H_2O 1; EtOH 4; eth 4; bz 3 / 1.5192[20]
11755	Thiophene, 2,4-dimethyl- / 2,4-Dimethylthiophene	C_6H_8S / 112.20	638-00-6	140.7	5-17-01-00344 / 0.9938[20]	H_2O 1; EtOH 3; eth 3; bz 3 / 1.5104[20]
11756	Thiophene, 2,5-dimethyl- / 2,5-Dimethylthiophene	C_6H_8S / 112.20	638-02-8 / -62.6	136.5	5-17-01-00345 / 0.9850[20]	H_2O 1; EtOH 3; eth 3; bz 3 / 1.5129[20]
11757	Thiophene, 3,4-dimethyl- / 3,4-Dimethylthiophene	C_6H_8S / 112.20	632-15-5	145	5-17-01-00347 / 0.993[25]	H_2O 1; EtOH 3; eth 4 / 1.5206[20]
11758	Thiophene, 2,5-dinitro- / 2,5-Dinitrothiophene	$C_4H_2N_2O_4S$ / 174.14	59434-05-8 / 81	290	5-17-01-00315	eth 4; EtOH 4
11759	Thiophene, 2-ethyl- / 2-Ethylthiophene	C_6H_8S / 112.20	872-55-9	134	5-17-01-00339 / 0.9930[20]	H_2O 1; EtOH 4; eth 4 / 1.5122[20]
11760	Thiophene, 3-ethyl- / 3-Ethylthiophene	C_6H_8S / 112.20	1795-01-3 / -89.1	136	5-17-01-00341 / 0.9980[20]	H_2O 1; EtOH 3; eth 4 / 1.5146[20]
11761	Thiophene, 2-ethyl-3-methyl- / 2-Ethyl-3-methylthiophene	$C_7H_{10}S$ / 126.22	31805-48-8	161	5-17-01-00357 / 0.9815[20]	1.5105[20]
11762	Thiophene, 2-ethyl-5-methyl- / 2-Ethyl-5-methylthiophene	$C_7H_{10}S$ / 126.22	40323-88-4 / -68.4	160	5-17-01-00357 / 0.9663[20]	1.5073[20]
11763	Thiophene, 3-ethyl-4-methyl-	$C_7H_{10}S$ / 126.22	66577-03-5	165.5	4-17-00-00301 / 0.9976[20]	1.5186[20]
11764	Thiophene, 4-ethyl-2-methyl- / 5-Methyl-3-ethylthiophene	$C_7H_{10}S$ / 126.22	13678-54-1 / -60	163	4-17-00-00299 / 0.9742[20]	1.5098[20]
11765	Thiophene, 2-ethyltetrahydro-	$C_6H_{12}S$ / 116.23	1551-32-2	157	5-17-01-00096 / 0.944[20]	ace 4; bz 4; eth 4; EtOH 4 / 1.490[20]
11766	Thiophene, 3-ethyltetrahydro- / 3-Ethyltetrahydrothiophene	$C_6H_{12}S$ / 116.23	62184-67-2	165	5-17-01-00096 / 0.950[20]	1.491[20]
11767	Thiophene, 2-iodo-	C_4H_3IS / 210.04	3437-95-4 / -40	181	5-17-01-00309 / 2.0595[25]	EtOH 4; eth 4; chl 2 / 1.6465[25]
11768	2-Thiophenemethanol	C_5H_6OS / 114.17	636-72-6	207; 86[10]	5-17-03-00358 / 1.2053[16]	EtOH 3; ace 3 / 1.5280[20]

No.	Name / Synonym	Mol. Form. / Mol. Wt.	CAS RN / mp/°C	Merck No. / bp/°C	Beil. Ref. / den/g cm^{-3}	Solubility / n_D
11769	Thiophene, 2-methyl- / 2-Methylthiophene	C_5H_6S / 98.17	554-14-3 / -63.4	112.6	5-17-01-00324 / 1.0193[20]	H_2O 1; EtOH 5; eth 5; ace 5 / 1.5203[20]
11770	Thiophene, 3-methyl- / 3-Thiotolene	C_5H_6S / 98.17	616-44-4 / -69	115.5	5-17-01-00331 / 1.0218[20]	H_2O 1; EtOH 5; eth 5; ace 5 / 1.5204[20]
11771	Thiophene, 2-(3-methylbutyl)- / 2-Isopentylthiophene	$C_9H_{14}S$ / 154.28	26963-33-7 /	74-5[11]	4-17-00-00316 / 0.9481[12]	1.5014[12]
11772	Thiophene, 2-(1-methylethyl)- / 2-Isopropylthiophene	$C_7H_{10}S$ / 126.22	4095-22-1 /	153	5-17-01-00356 / 0.9678[20]	1.5038[20]
11773	Thiophene, 3-(1-methylethyl)- / 3-Isopropylthiophene	$C_7H_{10}S$ / 126.22	29488-27-5 /	157	5-17-01-00356 / 0.9733[20]	1.5052[20]
11774	Thiophene, 2-methyl-5-phenyl- / 2-Methyl-5-phenylthiophene	$C_{11}H_{10}S$ / 174.27	5069-26-1 / 51	271	5-17-02-00153	EtOH 4; eth 4; lig 4
11775	Thiophene, 2-nitro- / 2-Nitrothiophene	$C_4H_3NO_2S$ / 129.14	609-40-5 / 46.5	224.5	5-17-01-00313 / 1.3644[43]	H_2O 1; EtOH 4; alk 3; peth 2
11776	Thiophene, 3-nitro- / 3-Nitrothiophene	$C_4H_3NO_2S$ / 129.14	822-84-4 / 78.5	225	5-17-01-00313	H_2O 1; EtOH 4; ctc 2; peth 3
11777	Thiophene-2-ol	C_4H_4OS / 100.14	17236-58-7 / 9	218	1.255[20]	H_2O 3; ace 3; chl 3 / 1.5644[20]
11778	Thiophene, 2-propyl- / 2-Propylthiophene	$C_7H_{10}S$ / 126.22	1551-27-5 /	158	5-17-01-00355 / 0.9683[20]	H_2O 1; EtOH 3; eth 3; bz 3 / 1.5076[20]
11779	Thiophene, 3-propyl- / 3-Propylthiophene	$C_7H_{10}S$ / 126.22	1518-75-8 /	163.2	5-17-01-00355 / 0.9738[20]	H_2O 1; EtOH 3; eth 3; bz 3 / 1.5057[20]
11780	2-Thiophenesulfonyl chloride	$C_4H_3ClO_2S_2$ / 182.65	16629-19-9 / 28	100[6]	5-18-09-00426	eth 3
11781	Thiophene, tetrabromo- / 2,3,4,5-Tetrabromothiophene	C_4Br_4S / 399.73	3958-03-0 / 117.5	326	5-17-01-00309	H_2O 1; EtOH 3; eth 4
11782	Thiophene, tetrachloro-	C_4Cl_4S / 221.92	6012-97-1 / 30.5	233.4	5-17-01-00305 / 1.7036[30]	H_2O 1; EtOH 4; eth 5 / 1.5915[30]
11783	Thiophene, tetrahydro- / Thiacyclopentane	C_4H_8S / 88.17	110-01-0 / -96.1	121.0	5-17-01-00036 / 0.9987[20]	H_2O 1; EtOH 5; eth 5; ace 5 / 1.4871[18]
11784	Thiophene, tetrahydro-2,4-dimethyl-, 1,1-dioxide / 2,4-Dimethylsulfolane	$C_6H_{12}O_2S$ / 148.23	1003-78-7 / -3	3245 / 281	5-17-01-00097 / 1.1362[20]	lig 4 / 1.4732[20]
11785	Thiophene, tetrahydro-2,5-dimethyl-, cis-	$C_6H_{12}S$ / 116.23	5161-13-7 / -89	142.3	5-17-01-00098 / 0.9222[20]	ace 4; bz 4; eth 4; EtOH 4 / 1.4799[20]
11786	Thiophene, tetrahydro-, 1,1-dioxide / Sulfolane	$C_4H_8O_2S$ / 120.17	126-33-0 / 27.6	8934 / 287.3	5-17-01-00039 / 1.2723[18]	chl 3 / 1.4833[18]
11787	Thiophene, tetrahydro-2-methyl-	$C_5H_{10}S$ / 102.20	1795-09-1 / -100.7	134	5-17-01-00080 / 0.9564[18]	H_2O 1; EtOH 4; eth 4; ace 4 / 1.4886[15]
11788	Thiophene, tetrahydro-3-methyl-	$C_5H_{10}S$ / 102.20	4740-00-5 /	139	5-17-01-00082 / 0.9596[18]	H_2O 1; EtOH 4; eth 4; ace 4 / 1.4886[18]
11789	Thiophene, tetrahydro-3-methyl-, 1,1-dioxide / 3-Methylsulfolane	$C_5H_{10}O_2S$ / 134.20	872-93-5 / 1	276	5-17-01-00082 / 1.188[25]	1.4772[20]
11790	Thiophene, 2,3,5-tribromo- / 2,3,5-Tribromothiophene	C_4HBr_3S / 320.83	3141-24-0 / 29	260	5-17-01-00309	chl 3
11791	Thiophene, 2,3,5-trichloro- / 2,3,5-Trichlorothiophene	C_4HCl_3S / 187.48	17249-77-3 / -16.1	198.7	5-17-01-00305 / 1.5856[20]	H_2O 1; EtOH 5; eth 5 / 1.5791[20]
11792	Thiophene, 2,3,4-trimethyl- / 2,3,4-Trimethylthiophene	$C_7H_{10}S$ / 126.22	1795-04-6 / -26.2	162; 75[30]	5-17-01-00359 / 0.995[20]	1.5208[20]
11793	Thiophene, 2,3,5-trimethyl- / 2,3,5-Trimethylthiophene	$C_7H_{10}S$ / 126.22	1795-05-7 /	164.5	5-17-01-00359 / 0.9753[20]	H_2O 1; EtOH 4; eth 4; ace 4 / 1.5112[20]
11794	2(3H)-Thiophenone, dihydro-	C_4H_6OS / 102.16	1003-10-7 /	39-40[1]	5-17-09-00012 / 1.18[25]	1.5230[20]
11795	2H-Thiopyran, tetrahydro- / Thiacyclohexane	$C_5H_{10}S$ / 102.20	1613-51-0 / 19	141.8	5-17-01-00068 / 0.9861[20]	H_2O 1; EtOH 3; eth 3; ace 3 / 1.5067[20]
11796	2H-Thiopyran, tetrahydro-2-methyl-	$C_6H_{12}S$ / 116.23	5161-16-0 / -58	153	5-17-01-00093 / 0.9428[20]	1.4905[20]
11797	2H-Thiopyran, tetrahydro-3-methyl-	$C_6H_{12}S$ / 116.23	5258-50-4 / -60	158	5-17-01-00094 / 0.9473[20]	1.4922[20]
11798	2H-Thiopyran, tetrahydro-4-methyl-	$C_6H_{12}S$ / 116.23	5161-17-1 / -28	158.6	5-17-01-00094 / 0.9471[20]	1.4923[20]
11799	Thiourea / Thiocarbamide	CH_4N_2S / 76.12	62-56-6 / 182	9299	4-03-00-00342 / 1.405[25]	H_2O 3; EtOH 3; eth 1
11800	Thiourea, N,N'-bis[4-(3-methylbutoxy)phenyl]- / Tiocarlide	$C_{23}H_{32}N_2O_2S$ / 400.59	910-86-1 / 146	9382	4-13-00-01187	
11801	Thiourea, (2-chlorophenyl)- / 1-(2-Chlorophenyl)thiourea	$C_7H_7ClN_2S$ / 186.66	5344-82-1 / 146		3-12-00-01296	bz 4; EtOH 4
11802	Thiourea, N-(3,5-dichlorophenyl)-N'-(4-fluorophenyl)- / Loflucarban	$C_{13}H_9Cl_2FN_2S$ / 315.20	790-69-2 / 163.5	5440	4-12-00-01275	
11803	Thiourea, N,N'-diethyl-	$C_5H_{12}N_2S$ / 132.23	105-55-5 / 78	dec	4-04-00-00375	H_2O 3; EtOH 3; eth 4; ctc 2
11804	Thiourea, N,N'-dimethyl- / N,N'-Dimethylthiourea	$C_3H_8N_2S$ / 104.18	534-13-4 / 62	3250	4-04-00-00217	H_2O 4; EtOH 4; eth 2; ace 4
11805	Thiourea, N,N'-diphenyl- / sym-Diphenylthiourea	$C_{13}H_{12}N_2S$ / 228.32	102-08-9 / 154.5	3337	4-12-00-00810 / 1.32[25]	H_2O 2; EtOH 4; eth 4; chl 4
11806	Thiourea, (2-methylphenyl)- / o-Tolylthiourea	$C_8H_{10}N_2S$ / 166.25	614-78-8 / 162		4-12-00-01766	H_2O 4; EtOH 4; eth 2

No.	Name / Synonym	Mol. Form. / Mol. Wt.	CAS RN / mp/°C	Merck No. / bp/°C	Beil. Ref. / den/g cm⁻³	Solubility / n_D
11807	Thiourea, 1-naphthalenyl- / ANTU	$C_{11}H_{10}N_2S$ / 202.28	86-88-4 / 198	755	4-12-00-03086	H_2O 1; EtOH 2; eth 2; ace 2
11808	Thiourea, phenyl- / Phenylthiourea	$C_7H_8N_2S$ / 152.22	103-85-5 / 154	7286	4-12-00-00804	H_2O 2; EtOH 3; NaOH 3
11809	Thiourea, 2-propenyl- / Thiosinamine	$C_4H_8N_2S$ / 116.19	109-57-9 / 78	9293	4-04-00-01072 1.217^{20}	H_2O 3; EtOH 3; eth 2; bz 1 1.5936^{78}
11810	Thiourea, tetramethyl-	$C_5H_{12}N_2S$ / 132.23	2782-91-4 / 79.3	245	4-04-00-00232	H_2O 3; EtOH 3; eth 2; chl 3
11811	9H-Thioxanthene / Dibenzothiapyran	$C_{13}H_{10}S$ / 198.29	261-31-4 / 128.5	9300 / 341	5-17-02-00254	chl 3
11812	9H-Thioxanthen-9-one / Thioxanthone	$C_{13}H_8OS$ / 212.27	492-22-8 / 209	9301 / 373	5-17-10-00437	H_2O 1; EtOH 2; bz 3; chl 3
11813	L-Threonine / [R-(R*,S*)]-2-Amino-3-hydroxybutanoic acid	$C_4H_9NO_3$ / 119.12	72-19-5 / 256 dec	9316	4-04-00-03171	H_2O 3; EtOH 1; eth 1; chl 1
11814	Thujane, (1S,4R,5S)-(+)- / Sabinane (d)	$C_{10}H_{18}$ / 138.25	7712-66-5	157	4-05-00-00318 0.8139^{20}	1.4376^{20}
11815	4(10)-Thujene, (+)- / 4(10)-Thujene (d)	$C_{10}H_{16}$ / 136.24	2514-91-2	164	4-05-00-00451 0.842^{20}	ace 4; bz 4; eth 4; EtOH 4 1.4678^{20}
11816	Thymidine / Thymine 2-desoxyriboside	$C_{10}H_{14}N_2O_5$ / 242.23	50-89-5 / 186.5	9331	5-24-07-00109	H_2O 3; EtOH 3; ace 3; chl 2
11817	Toluene, 3-chloro-2-iodo-	C_7H_6ClI / 252.48	5100-98-1 / 27.5	123^{14}	3-05-00-00726	1.608^{23}
11818	o-Toluenesulfonic acid, thio-, S-phenyl ester / Benzenesulfonic acid, thiolo,2-methyl,phenyl ester	$C_{13}H_{12}O_2S_2$ / 264.37	96097-69-7 / 59.5	241.6	4-11-00-00237	H_2O 2; EtOH 4; eth 3; chl 2 1.53412^{20}
11819	Triacontane	$C_{30}H_{62}$ / 422.82	638-68-6 / 65.8	449.7	4-01-00-00592 0.8097^{20}	H_2O 1; EtOH 2; eth 3; bz 4 1.4352^{70}
11820	Triacontanoic acid	$C_{30}H_{60}O_2$ / 452.81	506-50-3 / 93.6		4-02-00-01321	bz 4; CS_2 4; chl 4 1.4323^{100}
11821	1-Triacontanol / Myricyl alcohol	$C_{30}H_{62}O$ / 438.82	593-50-0 / 88	9506	4-01-00-01918 0.777^{95}	bz 4; eth 4; EtOH 4
11822	Triadimenol / Mercury, chloro(2-methoxyethyl)-	C_3H_7ClHgO / 295.13	123-88-6			
11823	Triallate / Carbamothioic acid, bis(1-methylethyl)-, S-(2,3,3-trichloro-2-propenyl) ester	$C_{10}H_{16}Cl_3NOS$ / 304.67	2303-17-5 / 29	9510 / $117^{0.0003}$	1.273^{25}	
11824	Triasulfuron / 2-(2-Chloroethoxy)-N-[[(4-methoxy-6-methyl-1,3,5-triaziN-2-yl)amino]carbonyl]benzene sulfonamide	$C_{14}H_{16}ClN_5O_5S$ / 401.83	82097-50-5 / 186			
11825	3,5,7-Triaza-1-azoniatricyclo[3.3.1.1 13,7]decane, 1-(2-propenyl)-, iodide / Hexamethylenetetramine, allyl iodide	$C_9H_{17}IN_4$ / 308.17	36895-62-2	5880		
11826	1-Triazene, 1,3-diphenyl- / Diazoaminobenzene	$C_{12}H_{11}N_3$ / 197.24	136-35-6 / 98	2981	4-16-00-00904	H_2O 1; EtOH 4; eth 4; bz 4
11827	1,3,5-Triazine	$C_3H_3N_3$ / 81.08	290-87-9 / 86	114	4-26-00-00063 1.38^{25}	EtOH 3; eth 3
11828	1,3,5-Triazine-2,4-diamine, 6-chloro-N,N,N',N'-tetraethyl-	$C_{11}H_{20}ClN_5$ / 257.77	580-48-3 / 27	$154-6^9$	4-26-00-01208 1.0956^{20}	bz 4; chl 4; EtOH 4; lig 4 1.5320^{20}
11829	1,3,5-Triazine-2,4-diamine, 6-phenyl- / Benzoguanamine	$C_9H_9N_5$ / 187.20	91-76-9 / 226.5	1099	4-26-00-01244	EtOH 3; eth 3; tfa 2
11830	1,3,5-Triazine-2,4-diamine, N-phenyl- / Amanozine	$C_9H_9N_5$ / 187.20	537-17-7 / 235.5	379	4-26-00-01195	
11831	1,2,4-Triazine-3,5(2H,4H)-dione, 6-methyl- / 6-Azathymine	$C_4H_5N_3O_2$ / 127.10	932-53-6 / 211	919	4-26-00-00556	H_2O 3; EtOH 3; ace 3
11832	1,2,4-Triazine-3,5(2H,4H)-dione, 2-β-D-ribofuranosyl- / 6-Azauridine	$C_8H_{11}N_3O_6$ / 245.19	54-25-1 / 158	920	4-26-00-00554	H_2O 3
11833	1,3,5-Triazine, hexahydro-1,3,5-trinitro- / Cyclonite	$C_3H_6N_6O_6$ / 222.12	121-82-4 / 205.5	2742	4-26-00-00022 1.82^{20}	H_2O 1; EtOH 1; eth 2; ace 3
11834	1,3,5-Triazine, hexahydro-1,3,5-triphenyl-	$C_{21}H_{21}N_3$ / 315.42	91-78-1 / 144	$185; 60^{29}$	4-26-00-00007	H_2O 1; EtOH 2; eth 3; ace 3
11835	1,3,5-Triazine-2,4,6-triamine / Melamine	$C_3H_6N_6$ / 126.12	108-78-1 / 345 dec	5691 / sub	4-26-00-01253 1.573^{16}	H_2O 2; EtOH 2; eth 1 1.872^{20}
11836	1,3,5-Triazine-2,4,6-tricarbonitrile	C_6N_6 / 156.11	7615-57-8 / 119	$262; 119^{1.0}$	1-26-00-00091	bz 4
11837	1,3,5-Triazine, 2,4,6-trichloro- / Cyanuric acid trichloride	$C_3Cl_3N_3$ / 184.41	108-77-0 / 154	192	4-26-00-00066	EtOH 4
11838	1,3,5-Triazine-2,4,6(1H,3H,5H)-trione, 1,3-dichloro- / Dichlorocyanuric acid	$C_3HCl_2N_3O_3$ / 197.96	2782-57-2			
11839	1,3,5-Triazine-2,4,6(1H,3H,5H)-trione, dihydrate / Cyanuric acid, dihydrate	$C_3H_7N_3O_5$ / 165.11	6202-04-6 / >360 dec	dec	4-26-00-00632 2.50^{20}	H_2O 3; EtOH 2; eth 1; ace 1 2.500^{20}
11840	1,3,5-Triazine-2,4,6(1H,3H,5H)-trione, 1,3,5-trichloro- / Symclosene	$C_3Cl_3N_3O_3$ / 232.41	87-90-1 / 246.7 dec	8993	4-26-00-00642	

No.	Name / Synonym	Mol. Form. / Mol. Wt.	CAS RN / mp/°C	Merck No. / bp/°C	Beil. Ref. / den/g cm^{-3}	Solubility / n_D
11841	1,3,5-Triazine-2,4,6(1H,3H,5H)-trione, 1,3,5-trimethyl-	$C_6H_9N_3O_3$ 171.16	827-16-7 176.5	274	4-26-00-00635	H_2O 1; EtOH 3
11842	1,3,5-Triazine-2,4,6(1H,3H,5H)-trione, 1,3,5-tri-2-propenyl-	$C_{12}H_{15}N_3O_3$ 249.27	1025-15-6 20.5	149[4]; 105[0.5]	4-26-00-00637 1.1590[20]	
11843	1,3,5-Triazine, 2,4,6-tri-2-pyridinyl- 2,4,6-Tripyridyl-s-triazine	$C_{18}H_{12}N_6$ 312.33	3682-35-7 210	9664	4-26-00-04192	
11844	1,3,5-Triazine, 2,4,6-tris(1-aziridinyl)- Triethylenemelamine	$C_9H_{12}N_6$ 204.23	51-18-3	9586	4-26-00-01268	
11845	1H-1,2,4-Triazol-3-amine Amitrole	$C_2H_4N_4$ 84.08	61-82-5 159	506	4-26-00-01073	H_2O 4; EtOH 4; eth 1; ace 1
11846	1H-1,2,3-Triazole	$C_2H_3N_3$ 69.07	288-36-8 23	204	4-26-00-00029 1.1861[25]	H_2O 3; eth 3; ace 3; os 3 1.4854[25]
11847	1H-1,2,4-Triazole, 1-butyl 1-Butyl-s-triazole	$C_6H_{11}N_3$ 125.17	6086-22-2			
11848	4H-1,2,4-Triazole, 4-cyclohexyl-3-ethyl- Hexazole	$C_{10}H_{17}N_3$ 179.27	4671-03-8 89.5	4618 227[10]	4-26-00-00049	H_2O 4; bz 4; chl 4
11849	1H-1,2,3-Triazole, 1,5-dimethyl- 1,5-Dimethyl-1,2,3-triazole	$C_4H_7N_3$ 97.12	15922-53-9 -4	255	1-26-00-00005	H_2O 4
11850	1,2,4-Triazolidine-3,5-dione Urazole	$C_2H_3N_3O_2$ 101.06	3232-84-6 249-50 dec	9780	4-26-00-00538	
11851	7H-1,2,3-Triazolo[4,5-d]pyrimidin-7-one, 5-amino-1,4-dihydro- 8-Azaguanine	$C_4H_4N_6O$ 152.12	134-58-7 300	912	4-26-00-04171	
11852	S,S,S-Tributyl phosphorotrithioate S,S,S-Tributyl trithiophosphate	$C_{12}H_{27}OPS_3$ 314.52	78-48-8 <-25	150[0.3]	1.057[20]	
11853	Trichlorfon Phosphonic acid, (2,2,2-trichloro-1-hydroxyethyl)-, dimethyl ester	$C_4H_8Cl_3O_4P$ 257.44	52-68-6 77	9536 100[0.1]	1.73[20]	
11854	Triclopyr Acetic acid, [(3,5,6-trichloro-2-pyridinyl)oxy]-	$C_7H_4Cl_3NO_3$ 256.47	55335-06-3 149	9572 dec 290		
11855	Tricosane	$C_{23}H_{48}$ 324.63	638-67-5 47.6	380	4-01-00-00576 0.7785[48]	H_2O 1; EtOH 2; eth 3; ctc 3 1.4468[20]
11856	Tricosane, 2-methyl- 2-Methyltricosane	$C_{24}H_{50}$ 338.66	1928-30-9 37.6	205[3]	3-01-00-00576 0.7539[90]	1.4201[90]
11857	12-Tricosanone Diundecyl ketone	$C_{23}H_{46}O$ 338.62	540-09-0 70.2		4-01-00-03408 0.8086[69]	bz 4; eth 4; chl 4 1.4283[80]
11858	Tricyclazole 1,2,4-Triazolo[3,4-b]benzothiazole, 5-methyl-	$C_9H_7N_3S$ 189.24	41814-78-2 187			
11859	Tricyclo[3.3.1.13,7]decan-1-amine Amantadine	$C_{10}H_{17}N$ 151.25	768-94-5 180	380		H_2O 2
11860	Tricyclo[3.3.1.13,7]decane Adamantane	$C_{10}H_{16}$ 136.24	281-23-2 268	140 sub	4-05-00-00469 1.07[25]	bz 3; ctc 3 1.568
11861	Tricyclo[4.4.0.02,7]dec-3-ene, 1,3-dimethyl-8-(1-methylethyl)-, stereoisomer Copaene	$C_{15}H_{24}$ 204.36	3856-25-5	2510 248.5	4-05-00-01189 0.8996[20]	H_2O 1; eth 3; ace 3; HOAc 3 1.4894[20]
11862	Tricyclo[2.2.1.02,6]heptane-1-carboxylic acid, 7,7-dimethyl- Tricyclenic acid	$C_{10}H_{14}O_2$ 166.22	512-60-7 151.3	263	4-09-00-00246	bz 4; EtOH 4; peth 4
11863	Tricyclo[2.2.1.02,6]heptane-1-carboxylic acid, 7,7-dimethyl-, ethyl ester Tricyclenic acid, ethyl ester	$C_{12}H_{18}O_2$ 194.27	56694-98-5	100-1[10]	2-09-00-00064 1.0143[20]	1.4230[20]
11864	Tricyclo[2.2.1.02,6]heptane-1-carboxylic acid, 7,7-dimethyl-, methyl ester Tricyclenic acid, methyl ester	$C_{11}H_{18}O_2$ 180.25	56694-97-4 45.5	99[14]	2-09-00-00064 1.0255[42]	1.4695[42]
11865	Tricyclo[2.2.1.02,6]heptane, 3,3-dimethyl-	C_9H_{14} 122.21	473-02-9 42.5	138	4-05-00-00430 0.8710[40]	bz 4; eth 4; EtOH 4 1.4514[40]
11866	Tricyclo[2.2.1.02,6]heptane, 1,3,3-trimethyl- Cyclofenchene	$C_{10}H_{16}$ 136.24	488-97-1	145	3-05-00-00392 0.859[20]	1.4503[22]
11867	Tricyclo[2.2.1.02,6]heptane, 1,7,7-trimethyl-	$C_{10}H_{16}$ 136.24	508-32-7 67.5	152.5	4-05-00-00468 0.8668[80]	1.4296[80]
11868	1,12-Tridecadiyne	$C_{13}H_{20}$ 176.30	38628-39-6 -3	115.5[12]	2-01-00-00248 0.8262[21]	1.454[20]
11869	Tridecanal	$C_{13}H_{26}O$ 198.35	10486-19-8 14	156[13]	4-01-00-03386 0.8356[18]	H_2O 1; EtOH 3; os 4 1.4384[18]
11870	1-Tridecanamine Tridecane, 1-amino	$C_{13}H_{29}N$ 199.38	2869-34-3 27.4	275.8	4-04-00-00810 0.8049[20]	H_2O 2; EtOH 3; eth 3 1.4443[20]
11871	Tridecane	$C_{13}H_{28}$ 184.37	629-50-5 -5.3	235.4	4-01-00-00512 0.7564[20]	H_2O 1; EtOH 4; eth 4; ctc 3 1.4256[20]
11872	Tridecane, 1-bromo-	$C_{13}H_{27}Br$ 263.26	765-09-3 6.2	296	4-01-00-00514 1.0177[20]	H_2O 1; chl 4 1.4593[20]
11873	Tridecane, 1-bromo-12-methyl- Tridecane, 2-methyl-13-bromo	$C_{14}H_{29}Br$ 277.29	111772-90-8	120-2[3]	3-01-00-00551 1.0241[20]	1.4598[20]
11874	Tridecane, 1,13-dibromo- Tridecamethylene dibromide	$C_{13}H_{26}Br_2$ 342.16	31772-05-1 9	188-92[12]	3-01-00-00548 1.276[15]	eth 4; chl 4 1.4880[27]
11875	1,12-Tridecanediol	$C_{13}H_{28}O_2$ 216.36	99706-68-0 60.5	188[8]	2-01-00-00563	EtOH 4
11876	1,13-Tridecanediol	$C_{13}H_{28}O_2$ 216.36	13362-52-2 76.5	195[10]	4-01-00-02630	EtOH 4

No.	Name Synonym	Mol. Form. Mol. Wt.	CAS RN mp/°C	Merck No. bp/°C	Beil. Ref. den/g cm^{-3}	Solubility n_D
11877	Tridecane, 2,2,4,10,12,12-hexamethyl-7-(3,5,5-trimethylhexyl)-	$C_{28}H_{58}$ 394.77	3035-75-4 	 166[0.5]	4-01-00-00590 0.8137[20]	 1.4558[20]
11878	Tridecane, 7-hexyl-	$C_{19}H_{40}$ 268.53	7225-66-3 -28.3	 170.5[10]	4-01-00-00562 0.7877[20]	 1.4409[20]
11879	Tridecane, 7-methyl- 7-Methyltridecane	$C_{14}H_{30}$ 198.39	26730-14-3 -37.2	 115.5[10]	4-01-00-00525 0.7634[20]	 1.4291[20]
11880	Tridecanenitrile	$C_{13}H_{25}N$ 195.35	629-60-7 9.7	 293	3-02-00-00906 0.8257[20]	EtOH 4; eth 4 1.4378[20]
11881	Tridecanoic acid	$C_{13}H_{26}O_2$ 214.35	638-53-9 44.5	 236[100]; 140[1]	4-02-00-01117 0.8458[80]	H_2O 1; EtOH 4; eth 4; ace 3 1.4286[60]
11882	Tridecanoic acid, methyl ester Methyl tridecanoate	$C_{14}H_{28}O_2$ 228.38	1731-88-0 6.5	 92[1]	4-02-00-01118 	EtOH 5; ctc 3 1.4405[20]
11883	1-Tridecanol Tridecyl alcohol	$C_{13}H_{28}O$ 200.36	112-70-9 32.5	 152[14]	4-01-00-01860 0.8223[31]	H_2O 1; EtOH 3; eth 3
11884	2-Tridecanol 2-Hydroxytridecane	$C_{13}H_{28}O$ 200.36	1653-31-2 23	 161[30]; 145[12]	4-01-00-01861 	bz 4; eth 4; EtOH 4 1.4188[70]
11885	3-Tridecanol	$C_{13}H_{28}O$ 200.36	10289-68-6 32	 139[12]	4-01-00-01861 0.8139[46]	
11886	1-Tridecanol, 12-methyl- 12-Methyl-1-tridecanol	$C_{14}H_{30}O$ 214.39	21987-21-3 10.5	 145-50[8]	4-01-00-01869 0.8414[20]	 1.4437[20]
11887	3-Tridecanol, 2-methyl-	$C_{14}H_{30}O$ 214.39	98930-89-3 20	 274	1-01-00-00219 0.8390[20]	 1.4460[20]
11888	4-Tridecanol, 2-methyl Acetic acid, trifluoro,(1,2-dimethyl-1-ethylbutyl) ester	$C_{14}H_{30}O$ 214.39	36691-77-7 	 147-8[14]	4-01-00-01869 	chl 2 1.4404[25]
11889	2-Tridecanone Methyl undecyl ketone	$C_{13}H_{26}O$ 198.35	593-08-8 30.5	 263	4-01-00-03386 0.8217[30]	H_2O 1; EtOH 4; eth 4; ace 4 1.4318[20]
11890	3-Tridecanone	$C_{13}H_{26}O$ 198.35	1534-26-5 31	 140[17]	4-01-00-03387 	H_2O 1; ace 3
11891	7-Tridecanone Dihexyl ketone	$C_{13}H_{26}O$ 198.35	462-18-0 33	 261	4-01-00-03315 0.825[30]	EtOH 3; eth 4; chl 3; lig 3
11892	3-Tridecanone, 2-methyl- Decyl isopropyl ketone	$C_{14}H_{28}O$ 212.38	40239-35-8 47	 267; 142[17]	4-01-00-03390 0.8314[20]	
11893	1-Tridecene	$C_{13}H_{26}$ 182.35	2437-56-1 -13	 232.8	4-01-00-00921 0.7658[20]	H_2O 1; EtOH 4; eth 4; bz 3 1.4340[20]
11894	2-Tridecenoic acid	$C_{13}H_{24}O_2$ 212.33	6969-16-0 	 167-71[2]	4-02-00-01622 0.8995[30]	 1.4612[20]
11895	12-Tridecenoic acid (10-Undecenyl)acetic acid	$C_{13}H_{24}O_2$ 212.33	6006-06-0 38.5	 192[20]	2-02-00-00422 	bz 4; eth 4; EtOH 4; peth 4
11896	12-Tridecenoic acid, methyl ester	$C_{14}H_{26}O_2$ 226.36	29780-00-5 	 143[8]	4-02-00-01622 0.8803[20]	 1.4438[20]
11897	1-Tridecyne	$C_{13}H_{24}$ 180.33	26186-02-7 	 94.5[25]	4-01-00-01069 0.7729[20]	bz 4; eth 4 1.4309[20]
11898	Tridemorph Morpholine, 2,6-dimethyl-4-tridecyl-	$C_{19}H_{39}NO$ 297.52	24602-86-6 	9576 141[1.3]	 0.86	
11899	Tridiphane (±)-2-(3,5-Dichlorophenyl)-2-(2,2,2-trichloroethyl)oxirane	$C_{10}H_7Cl_5O$ 320.43	58138-08-2 42.8			
11900	Triflumizole (E)-1-[1-[[4-Chloro-2-(trifluoromethyl)phenyl]imino]-2-propoxyethyl]-1H-imidazole	$C_{15}H_{15}ClF_3N_3O$ 345.75	68694-11-1 63.5			
11901	Trifluralin Benzenamine, 2,6-dinitro-N,N-dipropyl-4-(trifluoromethyl)-	$C_{13}H_{16}F_3N_3O_4$ 335.28	1582-09-8 49	9598 140[4.2]		
11902	Triforine Formamide, N,N'-[1,4-piperazinediylbis(2,2,2-trichloroethylidene)]bis-	$C_{10}H_{14}Cl_6N_4O_2$ 434.96	26644-46-2 155 dec	9602		
11903	2,5,7-Trioxabicyclo[2.2.1]heptane, 1,3,3,4,6,6-hexamethyl- Bicyclo[2.2.1]heptane, 1,3,3,4,6,6-hexamethyl-2,5,7-trioxa	$C_{10}H_{18}O_3$ 186.25	7045-89-8 	 166	5-19-09-00139 0.9702[17]	H_2O 1
11904	2H,4aH-1,4,5-Trioxadicyclopent[a,hi]indene-7-carboxylic acid, 3-ethylidene-3,3a,7a,9b-tetrahydro-2-oxo-, methyl ester, [3aS-(3E,3aα,4aβ,7aβ,9aR*,9bβ)]- Plumericin	$C_{15}H_{14}O_6$ 290.27	77-16-7 	7512	5-19-11-00097 	chl 3
11905	1,3,5-Trioxane Formaldehyde, trimer	$C_3H_6O_3$ 90.08	110-88-3 60.2	9646 114.5	5-19-09-00103 1.17[65]	H_2O 4; EtOH 3; eth 3; bz 3
11906	1,3,5-Trioxane, 2,4,6-triethenyl- Metacrolein	$C_9H_{12}O_3$ 168.19	588-01-2 50	 170	5-19-09-00165 	eth 4; EtOH 4
11907	1,3,5-Trioxane-2,4,6-triimine Cyamelide	$C_3H_3N_3O_3$ 129.08	462-02-2 dec	2689 dec	3-03-00-00030 1.127[15]	eth 4; EtOH 4
11908	1,3,5-Trioxane, 2,4,6-trimethyl- Paraldehyde	$C_6H_{12}O_3$ 132.16	123-63-7 12.6	6975 124.3	5-19-09-00112 0.9943[20]	H_2O 2; EtOH 5; eth 5; chl 5 1.4049[20]

No.	Name / Synonym	Mol. Form. / Mol. Wt.	CAS RN / mp/°C	Merck No. / bp/°C	Beil. Ref. / den/g cm^{-3}	Solubility / n_D
11909	2,6,7-Trioxa-1-phosphabicyclo[2.2.2]octane, 4-ethyl- / Trimethylolpropane phosphite	$C_6H_{11}O_3P$ / 162.13	824-11-3 / 53.7			chl 3
11910	Triphenylene / 9,10-Benzophenanthrene	$C_{18}H_{12}$ / 228.29	217-59-4 / 199	9654 / 425	4-05-00-02556	H_2O 1; EtOH 3; bz 4; chl 4
11911	Triphenyltin hydroxide / Stannane, hydroxytriphenyl-	$C_{18}H_{16}OSn$ / 367.03	76-87-9 / 119	9659	1.54[20]	
11912	Tripropylamine, eicosafluoro-2-(trifluoromethyl)- / Dipropylisobutylamine, perfluoro	$C_{10}F_{23}N$ / 571.08	307-11-9 / 147		4-02-00-00858 / 1.84[25]	eth 4; EtOH 4 / 1.283[25]
11913	Trisiloxane, 1,5-dichloro-1,1,3,3,5,5-hexamethyl-	$C_6H_{18}Cl_2O_2Si_3$ / 277.37	3582-71-6 / -53	184	4-04-00-04123 / 1.018[20]	H_2O 6
11914	Trisiloxane, octamethyl- / Octamethyltrisiloxane	$C_8H_{24}O_2Si_3$ / 236.53	107-51-7 / -80	6669 / 153; 50-2[17]	4-04-00-04115 / 0.8200[20]	EtOH 2; bz 3; peth 3 / 1.3840[20]
11915	Trisiloxane, 1,1,3,5,5-pentamethyl-1,3,5-triphenyl- / 1,1,3,5,5-Pentamethyl-1,3,5-triphenyltrisiloxane	$C_{23}H_{30}O_2Si_3$ / 422.75	80-14-8 /	169[0.7]	3-16-00-01210 / 1.0227[20]	bz 4; lig 4 / 1.5280[20]
11916	Trisulfide, diethyl	$C_4H_{10}S_3$ / 154.32	3600-24-6 /	96-7[26]	4-01-00-01398 / 1.114[20]	1.5689[13]
11917	Trisulfide, di-2-propenyl	$C_6H_{10}S_3$ / 178.34	2050-87-5 /	112-22[16]	0-01-00-00441 / 1.0845[15]	eth 4
11918	Tritetracontane	$C_{43}H_{88}$ / 605.17	7098-21-7 / 85.5	332[3]	3-01-00-00593 / 0.7812[90]	1.4340[90]
11919	1,3,5-Trithiane	$C_3H_6S_3$ / 138.28	291-21-4 / 220	sub	5-19-00-00105 / 1.6374[24]	H_2O 2; EtOH 2; eth 2; bz 3
11920	1,3,5-Trithiane, 2,4,6-trimethyl- / Thioacetaldehyde trimer	$C_6H_{12}S_3$ / 180.36	2765-04-0 /	9245	4-19-00-04718	
11921	1,3,5-Trithiane, 2,4,6-trimethyl-, (2α,4α,6β)-	$C_6H_{12}S_3$ / 180.36	23769-39-3 / 101	246.5	2-19-00-00399	H_2O 1; EtOH 3; eth 3; ace 3
11922	L-Tryptophan / l-α-Aminoindole-3-propionic acid	$C_{11}H_{12}N_2O_2$ / 204.23	73-22-3 / 290-2 dec	9707	5-22-14-00014	H_2O 2; EtOH 3; eth 1; chl 1
11923	L-Tryptophan, N-methyl- / L-Abrine	$C_{12}H_{14}N_2O_2$ / 218.26	526-31-8 / 295 dec	5	5-22-14-00040	H_2O 2; MeOH 2; eth 1; alk 3
11924	Tubocuraran-7',12'-diol, 6,6'-dimethoxy-2,2'-dimethyl-, (1'α)- / Bebeerine	$C_{36}H_{38}N_2O_6$ / 594.71	477-60-1 / 221	1024	2-27-00-00896	EtOH 3; eth 3; ace 4; chl 4
11925	Tubocuraran-7',12'-diol, 6,6'-dimethoxy-2,2'-dimethyl-, (1β)- / Curine	$C_{36}H_{38}N_2O_6$ / 594.71	436-05-5 / 221	2682	4-27-00-08725	ace 4; bz 4; py 4
11926	L-Tyrosine / L-Phenylalanine, 4-hydroxy-	$C_9H_{11}NO_3$ / 181.19	60-18-4 / 343 dec	9747 / sub	4-14-00-02264	H_2O 2; EtOH 1; eth 1; HOAc 2
11927	L-Tyrosine, 3,5-dibromo- / Tyrosine, 3,5-dibromo(L)	$C_9H_9Br_2NO_3$ / 338.98	300-38-9 / 245	3014	4-14-00-02365	H_2O 2; EtOH 2; eth 1; alk 3
11928	Tyrosine, 3,5-diiodo- / 3,5-Diiodotyrosine	$C_9H_9I_2NO_3$ / 432.98	66-02-4 / 213 dec	3177	2-14-00-00384	H_2O 2; EtOH 1; eth 1; bz 1
11929	L-Tyrosine, 3-hydroxy- / Levodopa	$C_9H_{11}NO_4$ / 197.19	59-92-7 / 285 dec	5344	4-14-00-02492	H_2O 3; EtOH 1; eth 1; ace 1
11930	L-Tyrosine, O-(4-hydroxy-3,5-diiodophenyl)-3,5-diiodo- / Thyroxine	$C_{15}H_{11}I_4NO_4$ / 776.87	51-48-9 / 235.5	9348	4-14-00-02373	H_2O 2; EtOH 1; bz 1
11931	L-Tyrosine, 3-hydroxy-α-methyl- / Methyldopa	$C_{10}H_{13}NO_4$ / 211.22	555-30-6 / 300 dec	5974		
11932	L-Tyrosine, 3-iodo- / 3-Iodotyrosine	$C_9H_{10}INO_3$ / 307.09	70-78-0 / 204-6 dec	4937	4-14-00-02369	
11933	L-Tyrosine, N-methyl- / Surinamine	$C_{10}H_{13}NO_3$ / 195.22	537-49-5 / 293	8988	3-14-00-01513	
11934	1,10-Undecadiyne	$C_{11}H_{16}$ / 148.25	4117-15-1 / -17	83[12]	2-01-00-00248 / 0.8182[21]	ace 4; bz 4 / 1.453[21]
11935	Undecanal	$C_{11}H_{22}O$ / 170.30	112-44-7 / -4	117[18]	4-01-00-03374 / 0.8251[23]	H_2O 1; EtOH 3; eth 3 / 1.4520[20]
11936	Undecanal, 2-methyl-	$C_{12}H_{24}O$ / 184.32	110-41-8 /	114[10]	4-01-00-03383 / 0.830[15]	H_2O 2; EtOH 3; eth 3 / 1.4321[20]
11937	1-Undecanamine	$C_{11}H_{25}N$ / 171.33	7307-55-3 / 17	242	4-04-00-00792 / 0.7979[20]	H_2O 3; EtOH 3; eth 1; ctc 2 / 1.4398[20]
11938	Undecane / Hendecane	$C_{11}H_{24}$ / 156.31	1120-21-4 / -25.6	195.9	4-01-00-00487 / 0.7402[20]	H_2O 1; EtOH 5; eth 5 / 1.4398[20]
11939	Undecane, 1-bromo-11-fluoro- / 1-Bromo-11-fluoroundecane	$C_{11}H_{22}BrF$ / 253.20	463-33-2 /	95[0.6]	4-01-00-00490	eth 4; EtOH 4 / 1.4518[25]
11940	Undecane, 1,11-dibromo- / 1,11-Dibromoundecane	$C_{11}H_{22}Br_2$ / 314.10	16696-65-4 / 166	170-5[15]	3-01-00-00536 / 1.332[15]	
11941	Undecanenitrile / Decyl cyanide	$C_{11}H_{21}N$ / 167.29	2244-07-7 /	253	3-02-00-00860 / 0.8254[30]	H_2O 1; EtOH 3; eth 3; ctc 3 / 1.4293[30]
11942	1-Undecanethiol / Undecyl mercaptan	$C_{11}H_{24}S$ / 188.38	5332-52-5 / -3	257.4	4-01-00-01838 / 0.8448[20]	1.4585[20]
11943	Undecanoic acid	$C_{11}H_{22}O_2$ / 186.29	112-37-8 / 28.6	280	4-02-00-01068 / 0.8907[20]	H_2O 1; EtOH 4; eth 3; ace 4 / 1.4294[45]

No.	Name / Synonym	Mol. Form. / Mol. Wt.	CAS RN	mp/°C	Merck No.	bp/°C	Beil. Ref.	den/g cm^{-3}	Solubility	n_D
11944	Undecanoic acid, 11-bromo- / 11-Bromoundecanoic acid	$C_{11}H_{21}BrO_2$ / 265.19	2834-05-1	57		188^{18}	4-02-00-01073		ace 4; bz 4; eth 4; EtOH 4	
11945	Undecanoic acid, 2-bromo- / 2-Bromoundecanoic acid	$C_{11}H_{21}BrO_2$ / 265.19	2623-84-9	10		180^{14}	4-02-00-01073	1.1586^{24}		
11946	Undecanoic acid, ethyl ester / Ethyl undecylate	$C_{13}H_{26}O_2$ / 214.35	627-90-7	-15		131^{14}	3-02-00-00858	0.8663^{20}	H_2O 1; EtOH 3; eth 3; ace 3	1.4285^{20}
11947	Undecanoic acid, 11-fluoro-	$C_{11}H_{21}FO_2$ / 204.28	463-17-2	36		$113^{0.25}$	4-02-00-01071		eth 4; EtOH 4; lig 4	
11948	Undecanoic acid, 11-hydroxy- / 11-Hydroxyundecanoic acid	$C_{11}H_{22}O_3$ / 202.29	3669-80-5	65.5		$148-50^4$	4-03-00-00903		eth 4; EtOH 4	
11949	Undecanoic acid, 11-hydroxy-, methyl ester / Methyl 11-hydroxyhendecanoate	$C_{12}H_{24}O_3$ / 216.32	24724-07-0	27.5		168^8	2-03-00-00243	0.9542^{20}	bz 4; eth 4; EtOH 4	
11950	1-Undecanol	$C_{11}H_{24}O$ / 172.31	112-42-5	19		243	4-01-00-01835	0.8298^{20}	H_2O 1; EtOH 3; eth 4	1.4392^{20}
11951	2-Undecanol, (±)-	$C_{11}H_{24}O$ / 172.31	113666-64-1	12		228	4-01-00-01838	0.8268^{19}	H_2O 1; EtOH 3; eth 3	1.4369^{20}
11952	3-Undecanol, (R)-	$C_{11}H_{24}O$ / 172.31	107494-37-1	17		229	2-01-00-00462	0.8295^{20}	bz 4; eth 4; EtOH 4	1.4367^{20}
11953	5-Undecanol / 1-Butylheptyl alcohol	$C_{11}H_{24}O$ / 172.31	37493-70-2	-3.5		229	4-01-00-01839	0.8292^{20}	eth 4; EtOH 4	1.4354^{24}
11954	6-Undecanol	$C_{11}H_{24}O$ / 172.31	23708-56-7	25		228	4-01-00-01839	0.8334^{20}	H_2O 1; EtOH 3; ace 3	1.4374^{20}
11955	1-Undecanol, 2-methyl- / 2-Methyl-1-undecanol	$C_{12}H_{26}O$ / 186.34	10522-26-6			$129-31^{12}$	4-01-00-01855	0.8300^{15}	eth 4; EtOH 4	1.4382^{20}
11956	3-Undecanol, 2-methyl- / 3-Hydroxy-2-methylundecane	$C_{12}H_{26}O$ / 186.34	60671-36-5			135^{26}	1-01-00-00217	0.8327^{20}		1.4405^{20}
11957	5-Undecanol, 2-methyl-	$C_{12}H_{26}O$ / 186.34	33978-71-1			$130-3^{20}$	4-01-00-01855	0.8235^{20}	eth 4; EtOH 4	1.4346^{20}
11958	2-Undecanone / Methyl nonyl ketone	$C_{11}H_{22}O$ / 170.30	112-12-9	15		231.5	4-01-00-03374	0.8250^{20}	H_2O 1; EtOH 3; eth 3; ace 3	1.4291^{20}
11959	3-Undecanone / Ethyl octyl ketone	$C_{11}H_{22}O$ / 170.30	2216-87-7	12		227	4-01-00-03375	0.8272^{20}	H_2O 1; EtOH 3; eth 3; ctc 2	1.4296^{20}
11960	4-Undecanone / Heptyl propyl ketone	$C_{11}H_{22}O$ / 170.30	14476-37-0	4.5		106^{13}	4-01-00-03375	0.8274^{25}	eth 4; EtOH 4	1.4248^{24}
11961	5-Undecanone / Butyl hexyl ketone	$C_{11}H_{22}O$ / 170.30	33083-83-9	2		227	4-01-00-03375	0.8278^{19}	H_2O 1; EtOH 3; eth 3; ctc 3	1.4275^{18}
11962	6-Undecanone / Diamyl ketone	$C_{11}H_{22}O$ / 170.30	927-49-1	14.5		228	4-01-00-03376	0.8308^{20}	H_2O 1; EtOH 4; eth 4	1.4270^{20}
11963	4-Undecanone, 7-ethyl-2-methyl-	$C_{14}H_{28}O$ / 212.38	6976-00-7			252.5	3-01-00-02920	0.8362^{20}	H_2O 1; EtOH 4; eth 4; ace 4	1.4370^{20}
11964	Undecasiloxane, tetracosamethyl- / Tetracosamethylhendecasiloxane	$C_{24}H_{72}O_{10}Si_{11}$ / 829.77	107-53-9		9128	322.8; 202^{47}	3-04-00-01881	0.9247^{25}	bz 4	1.3994^{20}
11965	3,5,9-Undecatrien-2-one, 6,10-dimethyl- / Pseudoionone	$C_{13}H_{20}O$ / 192.30	141-10-6		7931	$143-5^{12}$	3-01-00-03067	0.8984^{20}	EtOH 3; eth 3; chl 3; MeOH 3	1.5335^{20}
11966	1-Undecene	$C_{11}H_{22}$ / 154.30	821-95-4	-49.2		192.7	4-01-00-00910	0.7503^{20}	H_2O 1; eth 3; chl 3; lig 3	1.4261^{20}
11967	3-Undecene	$C_{11}H_{22}$ / 154.30	60669-40-1	-62		193.5		0.7516^{20}		1.4290^{20}
11968	2-Undecene, (E)-	$C_{11}H_{22}$ / 154.30	693-61-8	-48.3		192.5	4-01-00-00911	0.7528^{20}		1.4292^{20}
11969	2-Undecene, (Z)-	$C_{11}H_{22}$ / 154.30	821-96-5	-66.5		196.1		0.7576^{20}		
11970	4-Undecene, (E)-	$C_{11}H_{22}$ / 154.30	693-62-9	-63.7		193		0.7508^{20}		1.4285^{20}
11971	4-Undecene, (Z)-	$C_{11}H_{22}$ / 154.30	821-98-7	-97		192.6		0.7541^{20}		1.4302^{20}
11972	5-Undecene, (E)-	$C_{11}H_{22}$ / 154.30	764-97-6	-61.1		192	4-01-00-00911	0.7497^{20}	eth 4; chl 4; lig 4	1.4285^{20}
11973	5-Undecene, (Z)-	$C_{11}H_{22}$ / 154.30	764-96-5	-106.5		192.3	4-01-00-00911	0.7537^{20}		1.4302^{20}
11974	10-Undecenoic acid / Undecylenic acid	$C_{11}H_{20}O_2$ / 184.28	112-38-9	24.5	9760	275	4-02-00-01612	0.9072^{24}	H_2O 1; EtOH 3; eth 3; ctc 2	1.4486^{24}
11975	9-Undecenoic acid, (E)- / trans-9-Undecenoic acid	$C_{11}H_{20}O_2$ / 184.28	37973-84-5	19		273	4-02-00-01616	0.9118^{25}	eth 4	1.4519^{20}
11976	9-Undecenoic acid, (Z)-	$C_{11}H_{20}O_2$ / 184.28	116836-30-7	2.5		156^4; $119^{0.8}$	4-02-00-01616	0.9150^{20}		1.4530^{20}
11977	10-Undecenoic acid, ethyl ester	$C_{13}H_{24}O_2$ / 212.33	692-86-4	-38		264.5	4-02-00-01614	0.8827^{15}	H_2O 1; EtOH 3; eth 3; ctc 2	1.4449^{25}
11978	10-Undecenoic acid, methyl ester	$C_{12}H_{22}O_2$ / 198.31	111-81-9	-27.5		248	4-02-00-01613	0.889^{15}	H_2O 1; EtOH 3; eth 3; ctc 2	1.4393^{20}
11979	10-Undecen-1-ol	$C_{11}H_{22}O$ / 170.30	112-43-6	-2		250	4-01-00-02194	0.8495^{15}	H_2O 1; EtOH 3; eth 3; ctc 2	1.4500^{20}
11980	2-Undecen-4-ol	$C_{11}H_{22}O$ / 170.30	22381-86-8			$114-5^{13}$	4-01-00-02194	0.843^{22}		1.4485^{22}
11981	9-Undecen-1-ol	$C_{11}H_{22}O$ / 170.30	112-46-9	6		248.5	3-01-00-01957	0.8507^{15}		1.4535^{19}
11982	10-Undecenoyl chloride	$C_{11}H_{19}ClO$ / 202.72	38460-95-6			$120-2^{10}$	4-02-00-01615	0.944^{20}		1.454^{20}
11983	1-Undecen-3-yne	$C_{11}H_{18}$ / 150.26	74744-28-8			74^9	3-01-00-01054	0.7962^{20}		1.4606^{20}

No.	Name Synonym	Mol. Form. Mol. Wt.	CAS RN mp/°C	Merck No. bp/°C	Beil. Ref. den/g cm^{-3}	Solubility n_D
11984	1-Undecyne	$C_{11}H_{20}$ 152.28	2243-98-3 -25	196	4-01-00-01064 0.7728[20]	ace 4; bz 4; eth 4; EtOH 4 1.4306[20]
11985	2-Undecyne	$C_{11}H_{20}$ 152.28	60212-29-5 -30.1	204.2	0-01-00-00261 0.7827[20]	1.4391[20]
11986	4-Undecyne	$C_{11}H_{20}$ 152.28	60212-31-9 -74.7	198.5	3-01-00-01022 0.7752[20]	1.4369
11987	5-Undecyne	$C_{11}H_{20}$ 152.28	2294-72-6 -74.1	198	4-01-00-01064 0.7753[20]	1.4369[20]
11988	6-Undecynoic acid	$C_{11}H_{18}O_2$ 182.26	55182-83-7	124-5[0.17]	4-02-00-01739 0.9537[25]	1.4566[25]
11989	9-Undecynoic acid	$C_{11}H_{18}O_2$ 182.26	22202-65-9 61	170[15], 132[0.1]	3-02-00-01471	H$_2$O 4; eth 4; EtOH 4
11990	9-Undecynoic acid, methyl ester Methyl 9-undecynoate	$C_{12}H_{20}O_2$ 196.29	18937-76-3	93-4[1]	3-02-00-01472 0.9245[17]	1.4732[19]
11991	10-Undecynoic acid, methyl ester Methyl 10-undecynoate	$C_{12}H_{20}O_2$ 196.29	2777-66-4	121[15]	4-02-00-01738 0.9177[20]	1.4421[21]
11992	Urea Carbamide	CH_4N_2O 60.06	57-13-6 132.7	9781 dec	4-03-00-00094 1.3230[20]	H$_2$O 4; EtOH 4; eth 1; bz 1 1.484
11993	Urea, (aminoiminomethyl)-	$C_2H_6N_4O$ 102.10	141-83-3 105	160 dec	4-03-00-00155	H$_2$O 3; EtOH 2; eth 1; bz 1
11994	Urea, N,N'-bis(2-chloroethyl)-N-nitroso- Carmustine	$C_5H_9Cl_2N_3O_2$ 214.05	154-93-8 31	1852		H$_2$O 4; EtOH 4
11995	Urea, N,N-bis(hydroxymethyl)- Oxymethurea	$C_3H_8N_2O_3$ 120.11	140-95-4 126	6921 149[25]	4-03-00-00107	H$_2$O 3; EtOH 3; eth 1; MeOH 3
11996	Urea, N,N-bis(2-methylpropyl)- 1,1-Diisobutylurea	$C_9H_{20}N_2O$ 172.27	77464-06-3 73	180[25]	0-04-00-00170	eth 4; EtOH 4
11997	Urea, N,N'-bis(4-nitrophenyl)- 4,4'-Dinitrocarbanilide	$C_{13}H_{10}N_4O_5$ 302.25	587-90-6 312 dec	3271	4-12-00-01646	
11998	Urea, (4-bromophenyl)-	$C_7H_7BrN_2O$ 215.05	1967-25-5 226	260	4-12-00-01511	H$_2$O 2; EtOH 4; eth 4; bz 4
11999	Urea, N'-(4-chlorophenyl)-N,N-dimethyl- Monuron	$C_9H_{11}ClN_2O$ 198.65	150-68-5 170.5	6169	4-12-00-01191	H$_2$O 1; EtOH 2; ace 2
12000	Urea, N'-(3,4-dichlorophenyl)-N,N-dimethyl- Diuron	$C_9H_{10}Cl_2N_2O$ 233.10	330-54-1 158	3388	4-12-00-01263	
12001	Urea, N,N'-diethyl-	$C_5H_{12}N_2O$ 116.16	623-76-7 112.5	263	4-04-00-00370 1.0415[25]	H$_2$O 4; EtOH 4; eth 4 1.4616[40]
12002	Urea, N,N-diethyl-	$C_5H_{12}N_2O$ 116.16	634-95-7 75	94-6[0.02]	4-04-00-00380	H$_2$O 4; EtOH 4; eth 3; bz 4
12003	Urea, N,N'-diethyl-N,N'-diphenyl- N,N'-Diethylcarbanilide	$C_{17}H_{20}N_2O$ 268.36	85-98-3 79	3107	4-12-00-00844	H$_2$O 1; EtOH 4; chl 3
12004	Urea, N,N'-dimethyl-	$C_3H_8N_2O$ 88.11	96-31-1 108	269	4-04-00-00207 1.142[25]	H$_2$O 4; EtOH 4; eth 1; chl 2
12005	Urea, N,N-dimethyl-	$C_3H_8N_2O$ 88.11	598-94-7 183.5		4-04-00-00224 1.2555[25]	H$_2$O 3; EtOH 2; eth 1; tfa 2
12006	Urea, N,N'-dimethyl-N,N'-diphenyl-	$C_{15}H_{16}N_2O$ 240.30	611-92-7 122	350	4-12-00-00839	H$_2$O 4; EtOH 4; eth 2; ace 4
12007	Urea, N,N-dimethyl-N'-[3-(trifluoromethyl)phenyl]- Fluometuron	$C_{10}H_{11}F_3N_2O$ 232.21	2164-17-2 164	4080		ace 4; EtOH 4
12008	Urea, N,N'-di-1-naphthalenyl-	$C_{21}H_{16}N_2O$ 312.37	607-56-7 296	sub	4-12-00-03078	py 4
12009	Urea, (2,5-dioxo-4-imidazolidinyl)- Allantoin	$C_4H_6N_4O_3$ 158.12	97-59-6 239	246	4-25-00-04071	H$_2$O 2; EtOH 3; eth 1; MeOH 1
12010	Urea, N,N'-diphenyl- Carbanilide	$C_{13}H_{12}N_2O$ 212.25	102-07-8 236	1787 262	4-12-00-00741 1.239[25]	H$_2$O 2; EtOH 2; eth 3; ace 2
12011	Urea, N,N-diphenyl- 1,1-Diphenylurea	$C_{13}H_{12}N_2O$ 212.25	603-54-3 189	dec	4-12-00-00852 1.276[25]	H$_2$O 2; EtOH 3; eth 3; chl 3
12012	Urea, N,N'-dipropyl-	$C_7H_{16}N_2O$ 144.22	623-95-0 105	255	4-04-00-00482	eth 4; EtOH 4
12013	Urea, (4-ethoxyphenyl)- Dulcin	$C_9H_{12}N_2O_2$ 180.21	150-69-6 173.5	3448 dec	4-13-00-01154	H$_2$O 2; EtOH 3; AcOEt 4
12014	Urea, ethyl-	$C_3H_8N_2O$ 88.11	625-52-5 92.5	dec	4-04-00-00369 1.2130[18]	H$_2$O 4; EtOH 4; eth 3; bz 4
12015	Urea, N-ethyl-N-nitroso- N-Ethyl-N-nitrosourea	$C_3H_7N_3O_2$ 117.11	759-73-9 98-100 dec	3788	3-04-00-00233	chl 3
12016	Urea, hydroxy- Hydroxyurea	$CH_4N_2O_2$ 76.06	127-07-1 141	4785 dec	4-03-00-00170	H$_2$O 4
12017	Urea, methyl-	$C_2H_6N_2O$ 74.08	598-50-5 103	dec	4-04-00-00205 1.2040[0]	H$_2$O 4; EtOH 4; eth 1; bz 1
12018	Urea, N-methyl-N,N'-diphenyl-	$C_{14}H_{14}N_2O$ 226.28	612-01-1 106	203	4-12-00-00839	H$_2$O 2; EtOH 2; eth 4; bz 3
12019	Urea, N-methyl-N-nitroso- Methylnitrosourea	$C_2H_5N_3O_2$ 103.08	684-93-5 123-4 dec		4-04-00-03385	H$_2$O 2; EtOH 2; eth 2
12020	Urea, monohydrochloride Urea hydrochloride	CH_5ClN_2O 96.52	506-89-8 145 dec	9782	4-03-00-00094	H$_2$O 3
12021	Urea, mononitrate Urea nitrate	$CH_5N_3O_4$ 123.07	124-47-0 152 dec	9784	4-03-00-00094 1.690[20]	EtOH 4

No.	Name / Synonym	Mol. Form. / Mol. Wt.	CAS RN / mp/°C	Merck No. / bp/°C	Beil. Ref. / den/g cm⁻³	Solubility / n_D
12022	Urea, 1-naphthalenyl- / 1-Naphthylurea	$C_{11}H_{10}N_2O$ / 186.21	6950-84-1 / 219.5		4-12-00-03076	H_2O 1; EtOH 4; eth 3
12023	Urea, 2-naphthalenyl- / 2-Naphthylurea	$C_{11}H_{10}N_2O$ / 186.21	13114-62-0 / 219		4-12-00-03149	EtOH 4
12024	Urea, nitro- / Nitrourea	$CH_3N_3O_3$ / 105.05	556-89-8 / 158-9 dec	6573	4-03-00-00248	ace 4; EtOH 4
12025	Urea, (4-nitrophenyl)- / 4-Nitrophenylurea	$C_7H_7N_3O_3$ / 181.15	556-10-5 / 238	6547	4-12-00-01645	H_2O 4; EtOH 4
12026	Urea, phenyl- / Phenylurea	$C_7H_8N_2O$ / 136.15	64-10-8 / 147	7290 / 238	4-12-00-00734 / 1.302^{25}	H_2O 2; EtOH 3; eth 2; AcOEt 3
12027	Urea, (phenylmethyl)- / Benzylurea	$C_8H_{10}N_2O$ / 150.18	538-32-9 / 148	1165 / 200 dec	4-12-00-02253	ace 4; EtOH 4
12028	Urea, 2-propenyl- / Allylurea	$C_4H_8N_2O$ / 100.12	557-11-9 / 85	296	4-04-00-01070	H_2O 5; EtOH 5; eth 2; chl 2
12029	Urea, tetraethyl- / Urea, 1,1,3,3-tetraethyl-	$C_9H_{20}N_2O$ / 172.27	1187-03-7 /	/ 209	4-04-00-00380 / 0.919^{20}	H_2O 1; alk 1; acid 1 / 1.4474^{20}
12030	Urea, tetramethyl- / Tetramethylurea	$C_5H_{12}N_2O$ / 116.16	632-22-4 / -1.2	9160 / 176.5	4-04-00-00225 / 0.9687^{20}	EtOH 2; eth 2; ctc 2 / 1.4496^{23}
12031	Urea, tetraphenyl- / Tetraphenylurea	$C_{25}H_{20}N_2O$ / 364.45	632-89-3 / 183		4-12-00-00853 / 1.222^{25}	H_2O 1; EtOH 2
12032	Urea, trimethyl- / Trimethylurea	$C_4H_{10}N_2O$ / 102.14	632-14-4 / 75.5	/ 232.5	4-04-00-00224 / 1.1900^{20}	H_2O 3; EtOH 3; eth 2; bz 2
12033	Uridine / 1-β-D-Ribofuranosyluracil	$C_9H_{12}N_2O_6$ / 244.20	58-96-8 / 165	9792	4-24-06-00132	H_2O 3; EtOH 3; py 3
12034	Uridine, 2'-deoxy-5-fluoro- / Floxuridine	$C_9H_{11}FN_2O_5$ / 246.20	50-91-9	4045	5-24-06-00291	
12035	5'-Uridylic acid / Uridine 5'-phosphoric acid	$C_9H_{13}N_2O_9P$ / 324.18	58-97-9 / 202 dec	9796	5-24-06-00173	H_2O 4; MeOH 3
12036	Urs-12-ene-27,28-dioic acid, 3-hydroxy-, (3β)- / Quinovic acid	$C_{30}H_{46}O_5$ / 486.69	465-74-7 / 298 dec	8104	4-10-00-02197	
12037	Urs-12-en-28-oic acid, 3-hydroxy-, (3β)- / Ursolic acid	$C_{30}H_{48}O_3$ / 456.71	77-52-1 / 284	9802	4-10-00-01160	ace 4; eth 4; chl 4
12038	Urs-12-en-3-ol, (3β)- / α-Amyrin	$C_{30}H_{50}O$ / 426.73	638-95-9 / 186	653 / $243^{0.5}$	4-06-00-04191	EtOH 3; eth 3; bz 3; chl 3
12039	Valeric acid, 3-chloro-, ethyl ester	$C_7H_{13}ClO_2$ / 164.63	6513-13-9 /	/ 189	3-02-00-00678 / 1.0330^{20}	eth 4; EtOH 4 / 1.4278^{20}
12040	DL-Valine	$C_5H_{11}NO_2$ / 117.15	516-06-3 / 298 dec	9818 / sub	3-04-00-01370 / 1.316^{25}	H_2O 3; EtOH 2; eth 2; bz 2
12041	L-Valine / 2-Aminoisovaleric acid	$C_5H_{11}NO_2$ / 117.15	72-18-4 / 315	9818 / sub	4-04-00-02659 / 1.23^{25}	H_2O 4
12042	D-Valine, 3-[(2-amino-2-carboxyethyl)dithio]-, (R)- / Penicillamine cysteine disulfide	$C_8H_{16}N_2O_4S_2$ / 268.36	18840-45-4 / 195	7030	4-04-00-03231	
12043	D-Valine, 3-mercapto- / Penicillamine	$C_5H_{11}NO_2S$ / 149.21	52-67-5 / 198.5	7029	4-04-00-03228	
12044	Veratraman-3,23-diol, 14,15,16,17-tetradehydro-, (3β,23β)- / Veratramine	$C_{27}H_{39}NO_2$ / 409.61	60-70-8 / 206	9853	5-21-05-00297	EtOH 3; bz 3; chl 3; dil acid 3
12045	Veratraman-11-one, 17,23-epoxy-3-(β-D-glucopyranosyloxy)-, (3β,23β)- / Pseudojervine	$C_{33}H_{49}NO_8$ / 587.75	36069-05-3 / 304-5 dec	7932	4-27-00-03592	H_2O 1; EtOH 3; eth 1; bz 1
12046	Veratraman-11-one, 17,23-epoxy-3-hydroxy-, (3β,23β)- / Jervine	$C_{27}H_{39}NO_3$ / 425.61	469-59-0 / 243 dec	5145	4-27-00-03590	H_2O 1; EtOH 3; eth 2; ace 3
12047	Vernolate / Carbamothioic acid, dipropyl-, S-propyl ester	$C_{10}H_{21}NOS$ / 203.35	1929-77-7 / 150^{30}	9866	0.952^{20}	
12048	Vinclozolin / 2,4-Oxazolidinedione, 3-(3,5-dichlorophenyl)-5-ethenyl-5-methyl-	$C_{12}H_9Cl_2NO_3$ / 286.11	50471-44-8 / 108	9890 / $131^{0.05}$	1.51	
12049	9H-Xanthene / 10H-9-Oxaanthracene	$C_{13}H_{10}O$ / 182.22	92-83-1 / 100.5	/ 311	5-17-02-00252	H_2O 1; EtOH 2; eth 3; bz 3
12050	9H-Xanthen-9-one / Xanthone	$C_{13}H_8O_2$ / 196.21	90-47-1 / 174	9971 / 351; 146^3	5-17-10-00430	H_2O 1; EtOH 3; eth 3; bz 3
12051	9H-Xanthen-9-one, 1,3-dihydroxy-	$C_{13}H_8O_4$ / 228.20	3875-68-1 / 259	/ sub	5-18-04-00004	EtOH 4
12052	9H-Xanthen-9-one, 2,7-dihydroxy- / Xanthone, 2,7-dihydroxy	$C_{13}H_8O_4$ / 228.20	64632-72-0 / 330	/ sub	1-18-00-00357	eth 4; EtOH 4
12053	9H-Xanthen-9-one, 3,6-dihydroxy-	$C_{13}H_8O_4$ / 228.20	1214-24-0 / 350 dec	/ sub	5-18-04-00010	EtOH 4; HOAc 4
12054	9H-Xanthen-9-one, 1,7-dihydroxy-3-methoxy- / Gentisin	$C_{14}H_{10}O_5$ / 258.23	437-50-3 / 266.5	4291	5-18-04-00497	H_2O 1; EtOH 4; ace 1; py 3
12055	peri-Xanthenoxanthene / Dinaphthalene dioxide	$C_{20}H_{10}O_2$ / 282.30	191-28-6 / 243.5	/ 400^{20-25}	5-19-02-00242	bz 3; chl 3
12056	Xanthylium, 9-(2-carboxyphenyl)-3,6-bis(diethylamino)-, chloride / Rhodamine, B	$C_{28}H_{32}ClN_2O_3$ / 480.03	81-88-9 / 165	8182	5-19-08-00669	H_2O 3; EtOH 3; eth 3; bz 3
12057	o-Xylene, 4-propyl-	$C_{11}H_{16}$ / 148.25	3982-66-9 / -20	/ 208.9	3-05-00-01005 / 0.8715^{20}	ace 4; bz 4; eth 4; EtOH 4 / 1.5000^{20}

No.	Name Synonym	Mol. Form. Mol. Wt.	CAS RN mp/°C	Merck No. bp/°C	Beil. Ref. den/g cm^{-3}	Solubility n_D
12058	Xylitol Xylite	$C_5H_{12}O_5$ 152.15	87-99-0 93.5	9992 216	4-01-00-02832	H_2O 4; py 4; EtOH 4
12059	D-Xylose	$C_5H_{10}O_5$ 150.13	58-86-6 90.5	9995	4-01-00-04223 1.525[20]	H_2O 4; EtOH 3; eth 2
12060	Yohimban-16-carboxylic acid, 11,17-dimethoxy-18-[[1-oxo-3-(3,4,5-trimethoxyphenyl)-2-propenyl]oxy]-, methyl ester, (3β,16β,17α,18β,20α)- Rescinnamine	$C_{35}H_{42}N_2O_9$ 634.73	24815-24-5 238.5	8146	4-25-00-01323	H_2O 1; EtOH 2; ace 3; chl 3
12061	Yohimban-16-carboxylic acid, 11,17-dimethoxy-18-[(3,4,5-trimethoxybenzoyl)oxy]-, methyl ester, Reserpine	$C_{33}H_{40}N_2O_9$ 608.69	50-55-5 264.5	8149	4-25-00-01319	H_2O 2; EtOH 3; eth 2; ace 2
12062	Yohimban-16-carboxylic acid, 17-hydroxy-, methyl ester, (16α,17α)- Yohimbine	$C_{21}H_{26}N_2O_3$ 354.45	146-48-5 241	10011 sub 160	4-25-00-01237	H_2O 2; EtOH 3; eth 3; bz 2
12063	Yohimban-16-carboxylic acid, 18-hydroxy-11,17-dimethoxy-, (3β,16β,17α,18β,20 Reserpic acid	$C_{22}H_{28}N_2O_5$ 400.47	83-60-3 242	8147	4-25-00-01305	
12064	Yohimban-16-carboxylic acid, 17-hydroxy-18-[(3,4,5-trimethoxybenzoyl)oxy]-, methyl ester, (3β,16β Raunescine	$C_{31}H_{36}N_2O_8$ 564.64	117-73-7 165	8130	4-25-00-01281	H_2O 1; EtOH 3; chl 3; HOAc 3
12065	Yohimban-16-carboxylic acid, 17-methoxy-18-[(3,4,5-trimethoxybenzoyl)oxy]-, methyl ester, (3β,16β Deserpidine	$C_{32}H_{38}N_2O_8$ 578.66	131-01-1 230.5	2901	4-25-00-01282	H_2O 1; EtOH 3; chl 3
12066	Zinc, dimethyl Dimethyl zinc	C_2H_6Zn 95.46	544-97-8 -40	3252 46	1.386[10]	eth 3; peth 5
12067	Zinc, [[1,2-ethanediylbis[carbamodithioato]](2-)]- Zineb	$C_4H_6N_2S_4Zn$ 275.76	12122-67-7 157 dec	10071		
12068	Ziram Zinc, bis(dimethylcarbamodithioato-S,S')-, (T-4)-	$C_6H_{12}N_2S_4Zn$ 305.83	137-30-4			

STRUCTURAL FORMULAS OF ORGANIC COMPOUNDS

In numeric order as they occur in the Table of Physical Constants of Organic Compounds

1   **12**   **22**   **30**   **39**

14   **23**   **31**   **40**

3   **15**

4   **16**   **24**   **32**   **41**

5   **17**   **25**   **33**   **42**

6   **18**   **26**   **34**   **43**

8   **19**   **35**   **44**

9   **27**   **20**

10   **37**   **45**

28   **46**

21   **11**

29   **38**

47

48

49

50

51

52

53

54

55

56

57

58

59

60

61

62

63

64

65

66

67

68

69

70

71

72

73

74

75

76

77

78

79

80

81

82

83

84

85

86

87

88

89

90

91

92

93

95

96

97

98

99

100

101

102

103

104

105

107

108

109

111

112

113

114

115

116

117

118

119

120

121

123

124

125

126

127

128

129

130

131

132

141

150

161

169

133

142

151

162

170

134

143

152

163

171

135

144

153

164

172

136

145

154

165

173

137

146

157

166

174

138

147

158

167

175

139

148

159

168

176

140

149

160

177

178

179

180

181

182

183

184

185

186

187

188

189

190

191

192

193

194

195

196

197

198

199

200

201

202

203

204

205

206

207

208

209

211

212

213

214

215

216

217

218

219

220

221

222

223

224

225

226

236

246

255

263

227

238

247

256

264

228

239

248

257

265

229

240

249

258

266

230

241

250

259

267

231

242

252

260

268

232

243

253

261

269

233

244

254

262

270

235

245

271

272

273

274

277

278

279

280

281

282

283

284

285

286

287

288

289

290

291

292

293

294

295

296

297

298

299

300

301

302

303

304

306

312

313

314

315

316

317

318

319

320

321

322

323

324

325

326

327

329

331

332

333

334

335

336

337

338

339

340

341

342

343

344

345

346

347

348

349

350

351

352

353

354

355

356

357

358

359

360

362

364

365

366

367

368

369

370

371

372

373

374

375

376

377

379

380

381

382

383

384

385

386

387

388

389

390

391

392

393

394

395

396

397

398

399

400

401

402

403

404

405

406

414

423

432

441

407

415

424

433

442

408

416

425

434

443

409

417

426

435

444

410

418

427

436

445

411

419

428

437

446

412

420

429

438

447

413

421

430

439

448

422

431

440

449

450

451

452

453

454

455

456

457

458

459

460

461

462

463

464

465

466

467

468

470

471

472

473

474

475

476

477

478

479

480

481

482

483

484

485

486

487

488

489

490

491

492

493

494

495

496

497

498

499

500

501

502

503

504

505

506

507

509

510

511

512

513

514

515

516

517

518

519

520

521

522

523

524

525

526

527

528

529

530

531

532

533

535

536

537

538

539

540

541

542

543

544

545

546

547

548

549

550

551

552

553

554

555

556

557

558

559

560

561

562

563

564

565

566

567

568

569

570

571

572

573

574

575

STRUCTURAL FORMULAS OF ORGANIC COMPOUNDS (continued)

In numeric order as they occur in the Table of Physical Constants of Organic Compounds

576

577

578

579

580

581

582

583

584

585

586

587

588

589

590

591

592

593

594

595

596

597

598

599

600

601

603

604

605

606

607

608

609

610

611

612

613

614

615

616

617

618

619

620

621

622

623

624

625

626

627

628

629

630

631

632

633

635

636

637

638

639

640

641

642

643

644

645

646

647

648

649

650

651

652

653

654

655

656

657

658

659

660

661

662

663

666

667

668

669

670

671

672

673

674

675

676

677

678

679

680

681

682

683

684

685

686

687

688

689

690

692

693

694

695

696

697

698

699

700

701

709

718

725

732

702

710

719

726

733

703

711

720

727

734

704

712

721

728

735

705

713

722

729

736

706

714

723

730

737

707

715

724

731

738

708

716

740

741

742

743

744

745

746

747

748

749

750

751

752

753

754

755

756

757

758

759

760

761

762

763

764

765

766

767

768

769

770

771

772

773

775

776

777

778

779

780

781

782

783

784

785

786

787

788

789

790

791

793

794

795

796

797

798

799

800

801

802

803

804

805

806

807

808

809

810

811

812

813

814

815

816

817

818

819

820

822

823

824

825

826

828

829

830

831

832

833

834

835

836

837

838

839

840

841

842

843

844

845

846

847

848

849

850

851

852

853

854

855

856

857

858

859

860

861

862

863

864

867

868

869

870

871

872

873

874

875

877

878

879

880

881

882

883

884

886

887

888

890

891

892

893

894

895

STRUCTURAL FORMULAS OF ORGANIC COMPOUNDS (continued)

In numeric order as they occur in the Table of Physical Constants of Organic Compounds

896

897

898

899

900

901

902

903

904

905

906

907

908

909

910

911

912

913

914

915

916

917

918

919

920

921

923

924

925

926

927

928

930

931

932

933

934

935

937

938

939

940

941

942

944

945

946

947

948

949

951

952

953

954

956

957

959

960

961

962

963

964

965

966

967

968

969

970

971

972

973

974

975

976

977

978

979

980

981

982

983

984

985

986

987

988

989

990

991

992

993

994

995

996

997

998

999

1000

1001

1002

1003

1004

1005

1006

1007

1008

1010

1011

1012

1013

1014

1015

1016

1017

1018

1019

1020

1021

1022

1023

1024

1025

1026

1027

1028

1029

1030

1031

1032

1033

1034

1035

1036

1037

1038

1039

1040

1041

1042

1043

1044

1045

1046

1047

1048

1049

1050

1051

1052

1053

1054

1055

1056

1057

1058

1059

1060

1061

1062

1063

1064

1065

1066

1067

1068

1069

1070

1071

1072

1073

1074

1075

1076

1077

1078

1079

1080

1081

1082

1083

1084

1085

1086

1087

1088

1089

1090

1097

1105

1115

1125

1091

1098

1106

1116

1126

1092

1099

1108

1117

1127

1093

1100

1109

1118

1128

1094

1101

1110

1119

1129

1095

1102

1111

1120

1130

1096

1103

1112

1121

1131

1104

1113

1122

1132

1114

1123

1133

1124

1134

1135

1136

1137

1138

1139

1140

1141

1142

1143

1144

1145

1146

1147

1148

1149

1150

1151

1153

1154

1155

1156

1157

1158

1159

1160

1161

1162

1163

1164

1166

1167

1168

1169

1170

1171

1172

1173

1175

1176

1177

1178

1179

1180

1181

1191

1200

1208

1216

1182

1192

1201

1209

1217

1183

1193

1202

1210

1218

1184

1194

1203

1211

1219

1185

1195

1204

1212

1220

1186

1196

1205

1213

1221

1187

1197

1206

1214

1222

1189

1198

1207

1215

1223

1190

1199

1224

1225

1226

1227

1228

1229

1231

1232

1234

1235

1236

1237

1238

1239

1240

1241

1242

1243

1244

1245

1246

1247

1248

1249

1250

1251

1252

1253

1254

1255

1256

1257

1258

1259

1260

1261

1262

1263

1264

1265

1266

1267

1275

1281

1288

1297

1268

1276

1298

1282

1289

1270

1277

1283

1290

1299

1271

1284

1291

1300

1278

1272

1279

1285

1292

1301

1273

1293

1302

1286

1280

1294

1274

1287

1295

1303

STRUCTURAL FORMULAS OF ORGANIC COMPOUNDS (continued)

In numeric order as they occur in the Table of Physical Constants of Organic Compounds

1304

1305

1306

1307

1308

1309

1310

1311

1312

1313

1314

1315

1316

1317

1318

1319

1320

1321

1322

1323

1324

1325

1327

1328

1329

1330

1331

1332

1333

1334

1335

1336

1337

1338

1339

1340

1341

1342

1343

1351

1359

1368

1377

1344

1352

1361

1369

1378

1345

1353

1362

1370

1379

1346

1354

1363

1371

1380

1347

1355

1364

1372

1381

1348

1356

1365

1373

1382

1349

1357

1366

1374

1383

1350

1358

1367

1375

1376

1384

3-363

1385

1386

1387

1388

1389

1390

1391

1392

1393

1394

1395

1396

1397

1398

1399

1400

1401

1402

1403

1404

1405

1406

1407

1408

1409

1410

1411

1412

1413

1414

1415

1416

1417

1418

1419

1420

1421

1422

1423

1424

1425

1426

1427

1428

1429

1430

1431

1432

1433

1434

1435

1436

1437

1438

1439

1440

1441

1442

1443

1444

1445

1447

1448

1449

1450

1451

1452

1453

1454

1455

1456

1457

1458

1459

1460

1461

1462

1463

1464

1465

1466

1467

1468

1469

1470

1471

1472

1473

1474

1475

1476

1477

1478

1479

1480

1481

1482

1483

1484

1486

1487

1488

1489

1490

1491

1492

1493

1494

1495

1496

1497

1498

1499

1500

1501

1502

1503

1504

1505

1506

1507

1508

1509

1510

1511

1512

1513

1514

1515

1516

1517

1518

1519

1520

1527

1535

1543

1551

1521

1528

1536

1544

1552

1522

1529

1537

1545

1553

1523

1530

1538

1546

1554

1524

1531

1539

1547

1555

1525

1532

1540

1548

1556

1526

1533

1541

1549

1557

1534

1542

1550

1558

1567

1574

1581

1590

1560

1568

1575

1583

1591

1561

1569

1576

1584

1592

1562

1570

1577

1585

1593

1563

1571

1578

1586

1594

1564

1572

1579

1587

1595

1565

1573

1580

1588

1596

1566

1589

1597

1598

1599

1600

1601

1602

1603

1604

1606

1607

1608

1609

1610

1612

1613

1614

1615

1616

1617

1618

1619

1620

1621

1622

1623

1624

1626

1627

1628

1629

1630

1631

1632

1633

1634

1635

1636

1637

1638

1639

In numeric order as they occur in the Table of Physical Constants of Organic Compounds

1640

1641

1642

1643

1644

1645

1646

1647

1648

1649

1650

1651

1652

1653

1654

1655

1656

1657

1658

1659

1660

1661

1662

1663

1664

1665

1666

1667

1668

1669

1670

1671

1672

1673

1674

1675

1676

1677

1678

1679

1680

1681

1682

1683

1684

1693

1702

1710

1719

1685

1694

1703

1711

1720

1686

1695

1704

1712

1721

1687

1696

1705

1713

1722

1689

1697

1706

1714

1723

1690

1698

1707

1715

1724

1691

1699

1708

1716

1725

1692

1700

1709

1717

1726

1701

1718

1727

1728

1729

1730

1731

1732

1733

1734

1735

1736

1737

1738

1739

1740

1741

1742

1743

1744

1745

1746

1747

1748

1749

1750

1751

1752

1753

1754

1755

1756

1757

1758

1759

1760

1761

1762

1763

1764

1765

1766

1767

1768

1777

1786

1794

1802

1769

1778

1787

1795

1803

1770

1779

1788

1796

1804

1771

1780

1789

1797

1805

1772

1781

1790

1798

1806

1773

1782

1791

1799

1807

1774

1783

1792

1800

1808

1775

1784

1793

1801

1809

1776

1785

In numeric order as they occur in the Table of Physical Constants of Organic Compounds

1810

1811

1812

1813

1814

1815

1816

1817

1818

1819

1820

1821

1822

1823

1826

1827

1828

1829

1830

1831

1832

1833

1834

1835

1836

1837

1838

1839

1840

1841

1842

1843

1844

1845

1846

1847

1848

1849

1850

1851

1852

1853

1854

1855

1856

1857

1858

1859

1860

1861

1862

1863

1864

1865

1866

1867

1868

1869

1870

1871

1872

1873

1874

1875

1876

1877

1878

1879

1880

1881

1882

1883

1884

1885

1886

1887

1888

1889

1890

1891

1892

1893

1894

1895

1896

1897

1898

1899

1900

1901

1902

1903

1904

1905

1906

1907

1908

1909

1910

1911

1912

1913

1914

1915

1916

1917

1918

1919

1920

1921

1923

1924

1925

1926

1927

1928

1929

1930

1931

1932

1933

1934

1935

1936

1937

1938

1939

1940

1941

1942

1943

1944

1945

1946

1947

1948

1949

1950

1951

1952

1953

1954

1955

1956

1957

1958

1959

1960

1961

1962

1963

1964

1965

1966

1967

1968

1969

1970

1971

1972

1973

1974

1975

1976

1977

1978

1979

1980

1981

1982

1983

1984

1985

1986

1987

1988

1989

1990

1991

1992

1993

1995

1996

1997

1998

1999

2000

2001

2002

2003

2004

2005

2006

2007

2009

2010

2011

2012

2013

2014

2015

2016

2017

2018

2019

2021

2022

2023

2024

2025

2026

2027

2028

2029

2030

2031

2033

2034

2035

2036

2037

2038

2039

2040

2041

2042

2044

2045

2047

2048

2049

2050

2051

2052

2053

2054

2055

2056

2057

2058

2059

2060

2061

2062

2063

2064

2065

2066

2067

2068

2069

2070

2071

2072

2073

2074

2075

2076

2077

2078

2079

2080

2081

2083

2084

2085

2086

2087

2088

2089

2090

2091

2092

2093

2094

2095

2097

2098

2099

2100

2101

2102

2111

2122

2131

2141

2123

2103

2113

2132

2142

2104

2124

2133

2114

2143

2105

2125

2134

2116

2144

2106

2126

2137

2145

2117

2107

2127

2146

2138

2118

2108

2129

2139

2147

2119

2110

2130

2140

2148

2121

STRUCTURAL FORMULAS OF ORGANIC COMPOUNDS (continued)

In numeric order as they occur in the Table of Physical Constants of Organic Compounds

2149

2150

2151

2152

2153

2154

2155

2156

2157

2158

2159

2160

2161

2162

2163

2164

2165

2166

2167

2168

2169

2170

2171

2172

2173

2174

2175

2176

2177

2178

2179

2181

2182

2183

2184

2185

2186

2187

2188

2189

STRUCTURAL FORMULAS OF ORGANIC COMPOUNDS (continued)

In numeric order as they occur in the Table of Physical Constants of Organic Compounds

2190

2191

2192

2193

2194

2195

2196

2197

2198

2199

2200

2202

2203

2204

2205

2206

2207

2208

2210

2211

2212

2213

2214

2215

2216

2217

2220

2221

2223

2224

2225

2226

2227

2228

2229

2230

2231

2232

2233

2234

2235

2236

2237

2238

2239

2240

2242

2243

2244

2245

2246

2247

2248

2249

2250

2251

2252

2253

2254

2255

2256

2257

2258

2259

2260

2261

2262

2264

2265

2266

2267

2268

2269

2270

2271

2272

2276

2277

2278

2279

2280

2281

2283

2284

2285

2286

2287

2289

2290

2291

2292

2293

2294

2295

2296

2297

2299

2300

2301

2302

2303

2304

2305

2306

2308

2309

2310

2311

2312

2313

2314

2316

2324

2331

2339

2347

2317

2325

2332

2340

2348

2318

2326

2333

2341

2352

2319

2327

2334

2342

2354

2321

2328

2335

2343

2355

2322

2329

2336

2344

2356

2323

2330

2337

2345

2357

2346

2358

2359

STRUCTURAL FORMULAS OF ORGANIC COMPOUNDS (continued)

In numeric order as they occur in the Table of Physical Constants of Organic Compounds

2360

2367

2375

2383

2391

2361

2368

2376

2384

2392

2362

2369

2377

2385

2393

2363

2370

2378

2386

2394

2364

2371

2379

2387

2395

2365

2372

2380

2388

2396

2366

2373

2381

2389

2397

2374

2382

2390

2398

2399

2400

2401

2402

2403

2404

2405

2406

2407

2408

2409

2410

2411

2412

2413

2414

2415

2416

2417

2418

2419

2420

2421

2422

2423

2424

2425

2426

2427

2428

2429

2430

2431

2432

2434

2435

2436

2437

2438

2439

2440

2441

2442

2443

2444

2445

2446

2447

2448

2449

2451

2452

2453

2454

2455

2456

2457

2458

2459

2460

2461

2462

2463

2464

2465

2466

2468

2469

2470

2471

2472

2474

2475

2476

2477

2478

2479

2480

2481

2482

2483

2484

2485

2486

2487

2488

2489

2490

2491

2492

2493

2494

2495

2496

2497

2498

2499

2500

2501

2502

2503

2504

2505

2506

2507

2508

2509

2510

2511

2512

2513

2514

2515

2516

2517

2518

2519

2520

2521

2522

2523

2524

2525

2526

2527

2528

2529

2530

2532

2533

2534

2535

2536

2537

2538

2539

2540

2541

2542

2543

2544

2545

2546

2547

2548

2549

2550

2551

2552

2553

2554

2555

2556

2557

2558

2559

2560

2561

2562

2563

2564

2565

2566

2567

2568

2569

2570

2571

2572

2573

2574

2575

2576

2577

2578

2579

2580

2581

2582

2583

2584

2585

2586

2587

2588

2589

2590

2591

2592

2593

2594

2595

2596

2597

2598

2599

2600

2601

2602

2603

2604

2605

2606

2613

2620

2628

2635

2607

2614

2621

2629

2636

2608

2615

2622

2630

2637

2609

2616

2624

2631

2638

2610

2617

2625

2632

2639

2611

2618

2626

2633

2640

2612

2619

2627

2634

2641

In numeric order as they occur in the Table of Physical Constants of Organic Compounds

2642

2643

2644

2645

2646

2647

2648

2649

2650

2651

2652

2653

2654

2655

2656

2657

2658

2659

2660

2661

2662

2663

2664

2665

2666

2667

2668

2669

2670

2671

2672

2673

2674

2675

2676

2677

2678

2686

2693

2700

2707

2679

2687

2694

2708

2701

2680

2688

2695

2709

2702

2681

2689

2696

2710

2703

2682

2690

2697

2711

2704

2683

2691

2698

2712

2705

2684

2692

2699

2706

2713

2685

2714

2715

2716

2717

2718

2719

2720

2721

2722

2723

2724

2726

2727

2728

2729

2730

2731

2732

2733

2734

2735

2736

2737

2738

2739

2740

2741

2742

2744

2745

2746

2747

2748

2749

2750

2751

2752

2759

2766

2773

2783

2753

2760

2767

2774

2784

2754

2761

2768

2775

2785

2755

2762

2769

2776

2787

2756

2763

2770

2777

2788

2757

2764

2771

2778

2789

2758

2765

2772

2780

2790

2781

2782

2791

2792

2803

2811

2819

2830

2793

2804

2812

2820

2831

2795

2805

2813

2822

2832

2797

2806

2814

2825

2833

2798

2807

2815

2826

2834

2799

2808

2816

2827

2835

2800

2809

2817

2828

2836

2801

2810

2818

2829

2837

2802

2838

2847

2856

2865

2873

2839

2849

2857

2866

2874

2840

2850

2858

2867

2875

2841

2851

2859

2868

2876

2842

2852

2860

2869

2877

2843

2853

2861

2870

2878

2844

2854

2862

2871

2879

2845

2855

2863

2872

2880

2864

2881

2882

2883

2884

2885

2886

2887

2888

2889

2890

2891

2892

2893

2894

2895

2896

2897

2898

2899

2900

2901

2902

2903

2904

2905

2906

2907

2908

2909

2910

2911

2912

2913

2914

2915

2916

2918

2919

2920

2921

2922

2923

2925

2926

2927

2928

2929

2930

2932

2933

2934

2935

2936

2937

2939

2941

2944

2945

2946

2947

2948

2950

2951

2952

2953

2954

2955

2957

2958

2959

2960

2961

2962

2963

2964

2966

2967

2968

2969

2977

2986

2995

3003

2970

2978

2987

2996

3004

2971

2980

2988

2997

3005

2972

2981

2989

2998

3007

2973

2982

2990

2999

3008

2974

2983

2991

3000

3009

2975

2984

2992

3001

3010

2976

2985

2993

3002

3011

2994

3013

3014

3015

3016

3017

3018

3019

3020

3021

3022

3023

3024

3025

3026

3027

3028

3029

3030

3031

3032

3033

3034

3035

3037

3038

3039

3040

3041

3042

3043

3044

3045

3046

3050

3051

3052

3053

3054

3055

3056

3057

3058

3066

3078

3086

3095

3059

3070

3079

3087

3096

3060

3071

3080

3089

3098

3061

3072

3081

3090

3099

3062

3073

3082

3091

3100

3063

3074

3083

3092

3101

3064

3075

3084

3093

3102

3065

3076

3085

3094

3077

3103

3104

3105

3108

3109

3111

3112

3113

3114

3115

3116

3117

3118

3119

3120

3121

3122

3123

3124

3125

3126

3127

3128

3129

3130

3131

3132

3134

3135

3136

3137

3138

3139

3140

3141

3143

3144

3145

3146

3147

3148

3150

3151

3152

3153

3154

3155

3156

3157

3158

3159

3160

3161

3162

3163

3164

3166

3167

3168

3169

3170

3171

3172

3173

3174

3175

3176

3177

3178

3179

3180

3181

3182

3183

3184

3185

3186

3188

3189

3190

3191

3192

3194

3195

3196

3197

3198

3199

3200

3201

3202

3203

3205

3206

3207

3208

3209

3210

3211

3212

3213

3214

3215

3216

3217

3218

3220

3221

3222

3223

3224

3225

3226

3227

3228

3229

3230

STRUCTURAL FORMULAS OF ORGANIC COMPOUNDS (continued)

In numeric order as they occur in the Table of Physical Constants of Organic Compounds

3231

3232

3233

3234

3235

3236

3237

3238

3239

3240

3241

3242

3243

3244

3245

3246

3247

3248

3249

3250

3251

3252

3253

3254

3255

3256

3257

3258

3259

3262

3263

3264

3265

3266

3267

3268

3269

3270

3271

3272

3273

3274

3275

3276

3284

3292

3301

3311

3277

3285

3293

3302

3312

3278

3286

3294

3303

3313

3279

3287

3295

3304

3316

3280

3288

3296

3305

3317

3281

3289

3297

3307

3318

3282

3290

3298

3308

3319

3283

3291

3299

3309

3320

3300

3310

3321

3322

3323

3324

3325

3326

3327

3328

3329

3330

3331

3332

3333

3334

3336

3339

3340

3341

3342

3343

3344

3345

3346

3347

3348

3349

3350

3351

3352

3353

3354

3355

3356

3357

3359

3360

3361

3362

3363

3364

3365

3366

3367

3368

3369

3370

3371

3372

3373

3374

3375

3376

3377

3378

3379

3381

3382

3383

3384

3385

3387

3388

3390

3391

3392

3394

3395

3396

3397

3398

3399

3400

3401

3402

3403

3404

3405

3406

3407

3408

3409

3410

3411

STRUCTURAL FORMULAS OF ORGANIC COMPOUNDS (continued)

In numeric order as they occur in the Table of Physical Constants of Organic Compounds

3412

3414

3415

3416

3417

3418

3419

3420

3421

3422

3423

3426

3427

3428

3429

3430

3431

3433

3434

3435

3436

3437

3438

3439

3440

3441

3442

3443

3444

3445

3446

3447

3448

3449

3450

3451

3452

3453

3454

3455

3456

3457

3458

3459

3470

3479

3488

3498

3460

3471

3480

3489

3499

3461

3472

3481

3490

3500

3462

3473

3482

3492

3502

3464

3474

3483

3493

3503

3465

3475

3484

3494

3504

3467

3476

3485

3495

3505

3468

3477

3486

3497

3506

3469

3478

3487

STRUCTURAL FORMULAS OF ORGANIC COMPOUNDS (continued)

In numeric order as they occur in the Table of Physical Constants of Organic Compounds

3507

3508

3509

3510

3511

3512

3513

3514

3515

3516

3517

3518

3519

3520

3521

3522

3523

3524

3525

3526

3527

3528

3529

3530

3531

3532

3533

3534

3535

3537

3538

3539

3540

3541

3542

3543

3544

3545

3546

3547

3548

3549

3551

3552

3553

3563

3573

3582

3594

3555

3565

3574

3584

3595

3556

3566

3575

3585

3596

3567

3576

3557

3586

3597

3558

3568

3577

3587

3598

3569

3578

3588

3599

3560

3570

3590

3561

3571

3579

3592

3601

3580

3562

3572

3593

3604

3581

STRUCTURAL FORMULAS OF ORGANIC COMPOUNDS (continued)

In numeric order as they occur in the Table of Physical Constants of Organic Compounds

3605

3606

3608

3609

3612

3613

3614

3615

3616

3617

3618

3619

3620

3621

3622

3623

3624

3625

3626

3627

3628

3629

3630

3631

3632

3633

3634

3635

3636

3637

3638

3639

3640

3641

3642

3643

3644

3645

3646

3647

3648

STRUCTURAL FORMULAS OF ORGANIC COMPOUNDS (continued)

In numeric order as they occur in the Table of Physical Constants of Organic Compounds

3649

3658

3666

3675

3683

3650

3659

3667

3676

3684

3651

3660

3668

3677

3685

3653

3661

3669

3678

3688

3654

3662

3670

3679

3690

3655

3663

3671

3680

3691

3656

3664

3673

3681

3692

3657

3665

3682

3693

3695

3696

3697

3698

3699

3700

3702

3704

3705

3706

3707

3708

3709

3710

3711

3712

3713

3714

3715

3716

3717

3718

3720

3721

3722

3723

3724

3725

3726

3727

3728

3729

3730

3731

3732

3733

3734

3735

3736

3737

3738

3739

3740

3741

3742

3743

3744

3745

3746

3747

3748

3749

3750

3751

3752

3753

3754

3755

3756

3757

3758

3759

3760

3761

3762

3763

3764

3765

3766

3767

3769

3770

3771

3772

3773

3774

3775

3776

3777

3778

3779

3781

3782

STRUCTURAL FORMULAS OF ORGANIC COMPOUNDS (continued)

In numeric order as they occur in the Table of Physical Constants of Organic Compounds

3783

3784

3785

3786

3787

3788

3789

3790

3791

3792

3793

3794

3795

3796

3797

3798

3799

3800

3801

3802

3803

3804

3805

3806

3807

3808

3809

3810

3812

3813

3814

3815

3816

3817

3819

3820

3821

3822

3823

3824

3825

3826

3827

3836

3844

3852

3865

3828

3866

3837

3856

3845

3829

3857

3868

3830

3838

3846

3858

3869

3831

3839

3847

3859

3870

3832

3840

3848

3860

3871

3833

3841

3849

3861

3842

3834

3850

3862

3873

3843

3851

3864

3835

3874

3875

3877

3878

3879

3883

3884

3885

3886

3887

3888

3889

3890

3891

3892

3893

3894

3895

3896

3897

3898

3899

3900

3901

3902

3904

3905

3907

3908

3909

3910

3911

3912

3913

3914

3915

3916

3917

3918

3919

3920

3921

3922

3923

3925

3926

3927

3929

3930

3931

3932

3933

3934

3935

3936

3937

3938

3939

3940

3941

3942

3943

3944

3947

3948

3949

3950

3953

3954

3955

3956

3957

3959

3960

3961

3962

3963

3964

3966

3969

3970

3971

3972

3973

3976

3978

3979

3980

3982

3989

3990

3991

3992

3993

3994

3995

3997

3998

3999

4000

4002

4003

4005

4006

4007

4008

4009

4012

4013

4014

4015

4016

4017

4018

4019

4020

4021

4022

4023

4024

4025

4028

4029

4030

4031

4032

4033

4034

4035

4036

4037

4038

4039

4040

4043

4044

4046

4047

4048

4049

4050

4065

4066

4067

4068

4069

4071

4072

4075

4076

4077

4078

4079

4080

4081

4082

4083

4084

4085

4086

4087

4088

4089

4090

4091

4092

4093

4094

4095

4096

4097

4098

4099

4100

4101

4102

4103

4106

4107

4108

4110

4111

4113

4116

4118

4119

4120

4121

4122

4123

4124

4125

4126

4127

4128

4130

4133

4134

4137

4138

4139

4140

4141

4142

4143

4144

4146

4147

4150

4151

4152

4153

4154

4155

4156

4157

4159

4161

4162

4163

4164

4165

4166

4167

4168

4169

4170

4171

4172

4173

4174

4175

4176

4177

4178

4179

4180

STRUCTURAL FORMULAS OF ORGANIC COMPOUNDS (continued)

In numeric order as they occur in the Table of Physical Constants of Organic Compounds

4181

4182

4183

4184

4185

4186

4187

4188

4189

4190

4191

4192

4193

4194

4195

4196

4197

4198

4199

4200

4201

4202

4203

4204

4205

4206

4207

4208

4209

4210

4211

4212

4213

4214

4215

4216

4217

4218

4219

4220

4221

4222

4223

4224

4225

4226

4227

4228

4229

4230

4231

4233

4234

4235

4236

4237

4238

4239

4240

4241

4242

4243

4244

4245

4246

4247

4248

4249

4250

4251

4252

4253

4254

4255

4256

4257

4258

4259

4260

4261

4262

4263

4264

4265

4266

4267

4268

4269

4270

4271

4273

4274

4275

4276

4277

4278

4279

4280

4281

4282

4283

4284

4285

4286

4287

4288

4289

4290

4291

4292

4293

4294

4295

4296

4297

4298

4299

4300

4301

4302

4303

4304

4305

4306

4307

4308

4309

4310

4311

4312

4313

4314

STRUCTURAL FORMULAS OF ORGANIC COMPOUNDS (continued)

In numeric order as they occur in the Table of Physical Constants of Organic Compounds

4315

4316

4317

4318

4319

4320

4322

4323

4324

4325

4326

4327

4328

4329

4330

4331

4332

4333

4334

4335

4336

4337

4339

4340

4341

4342

4343

4344

4345

4346

4347

4348

4349

4350

4351

4352

4353

4354

4356

4357

4358

4359

4360

STRUCTURAL FORMULAS OF ORGANIC COMPOUNDS (continued)

In numeric order as they occur in the Table of Physical Constants of Organic Compounds

4361

4362

4364

4365

4366

4367

4368

4369

4370

4371

4372

4373

4374

4375

4376

4377

4378

4379

4380

4381

4382

4383

4384

4385

4386

4387

4388

4389

4390

4391

4392

4393

4394

4395

4396

4397

4398

4399

4400

4401

4402

4403

4404

4405

4406

4407

4418

4427

4437

4446

4408

4419

4428

4438

4447

4409

4420

4429

4439

4448

4410

4421

4430

4440

4449

4411

4431

4450

4413

4422

4432

4441

4451

4414

4423

4433

4442

4452

4415

4424

4434

4443

4453

4416

4425

4435

4444

4454

4417

4426

4436

4445

4455

4456

4457

4458

4459

4460

4461

4462

4463

4464

4465

4466

4467

4468

4469

4470

4471

4472

4473

4474

4475

4476

4477

4478

4479

4480

4481

4482

4483

4484

4485

4486

4487

4488

4489

4490

4491

4492

4493

4494

4495

4496

4497

4498

4499

4500

4501

4502

4503

4504

4505

4507

4508

4509

4510

4511

4513

4514

4515

4516

4517

4518

4519

4521

4522

4523

4525

4526

4527

4528

4529

4530

4531

4532

4533

4534

4535

4536

4537

4538

4539

4540

4541

4542

4543

4544

4545

4546

4547

4548

4549

4550

4551

4552

4553

4554

4555

4556

4557

4558

4559

4560

4562

4563

4564

4565

4566

4568

4569

4570

4571

4572

4573

4574

4575

4576

4577

4578

4579

4580

4581

4582

4583

4585

4586

4587

4588

4589

4590

4591

4592

4593

4594

4595

4596

4597

4598

4599

4600

4601

4602

4603

4604

4605

4606

4607

4608

4609

4610

4611

4612

4613

4614

4615

4616

4617

4618

4619

4620

4621

4622

4623

4624

4625

4626

4627

4628

4629

4630

4631

4632

4633

4634

4635

4636

4637

4638

4639

4640

4642

4643

4644

4645

4646

4647

4648

4649

4650

4651

4652

4653

4654

4655

4656

4657

4658

4659

4660

4661

4662

4663

4664

4665

4666

4667

4668

4669

4670

4671

4672

4673

4674

4675

4676

4677

4678

4679

4680

4681

4682

4683

4684

4685

4686

4687

4688

4689

4690

STRUCTURAL FORMULAS OF ORGANIC COMPOUNDS (continued)

In numeric order as they occur in the Table of Physical Constants of Organic Compounds

4691

4692

4694

4695

4696

4697

4698

4699

4700

4701

4702

4703

4704

4705

4706

4707

4708

4709

4710

4711

4712

4713

4714

4715

4716

4718

4719

4720

4721

4722

4723

4724

4725

4726

4727

4728

4729

4731

4732

4733

4734

4735

4736

4737

4738

4739

4740

3-439

4741

4742

4743

4744

4745

4746

4747

4748

4749

4750

4751

4752

4753

4754

4755

4756

4757

4758

4759

4760

4761

4762

4763

4764

4765

4766

4767

4768

4770

4771

4772

4773

4774

4775

4776

4777

4778

4779

4780

4781

4782

4783

4784

4785

4786

4789

4790

4791

4792

4793

4794

4796

4797

4798

4799

4800

4801

4802

4803

4804

4805

4806

4807

4808

4809

4810

4811

4812

4813

4814

4815

4816

4817

4818

4819

4820

4821

4822

4823

4824

4825

4826

4827

4828

4829

4830

4831

4832

4833

4834

4835

4836

4837

4838

4839

4840

4841

4842

4843

4844

4845

4846

4847

4848

4849

4850

4851

4852

4853

4854

4855

4856

4857

4858

4859

4860

4861

4862

4863

4864

4865

4866

4867

4868

4869

4870

4871

4872

4873

4874

4875

4876

4877

4878

4879

4880

4881

4882

4883

4884

4885

4886

4887

4888

4889

4890

4891

4892

4893

4894

4895

4896

4897

4898

4899

4900

4901

4902

4903

4904

4905

4906

4907

4908

4909

4910

4911

4912

4913

4914

4915

4916

4917

4918

4919

4920

4921

4922

4923

4924

4925

4926

4927

4928

4929

4930

4931

4932

4933

4934

4935

4936

4937

4938

4939

4940

4941

4942

4943

4944

4945

4946

4947

4948

4949

4950

4951

4952

4953

4954

4955

4956

4957

4958

4959

4960

4961

4962

4963

4964

4966

4967

4968

4969

4970

4971

4972

4973

4974

4975

4976

4977

4978

4979

4980

4981

4982

4983

4984

4985

4986

4987

4988

4989

4990

4991

4992

4993

4994

4995

4996

4997

4998

4999

5000

5001

5002

5003

5004

5005

5006

5008

5009

5010

5011

5012

5013

5014

5015

5016

5017

5018

5020

5021

5022

5023

5024

5026

5027

5028

5029

5030

5032

5033

5034

5035

5036

5037

5038

5039

5040

5041

5042

5043

5044

5045

5046

5047

5048

5049

5050

5051

5052

5053

5054

5055

5056

5057

5058

5059

5060

5061

5062

5063

5064

5065

5066

5067

5068

5069

5070

5071

5072

5073

5074

5075

5076

5077

5078

5080

5081

5082

5083

5084

5085

5086

5087

5088

5089

5090

5091

5092

5093

5094

5095

5097

5098

5099

5100

5101

5102

5103

5104

5105

5106

5109

5110

5111

5112

5113

5114

5115

5116

5118

5120

5121

5122

5123

5124

5125

5126

5127

5128

5129

5130

5131

5132

5133

5134

5135

5136

5137

5140

5141

5142

5143

5144

5145

5146

5147

5148

5149

5150

5151

5152

5153

5154

5155

5156

5157

5158

5159

5160

5161

5162

5163

5164

5165

5166

5167

5168

5169

5170

5171

5172

5173

5174

5175

5176

5177

5178

5179

5180

5181

5182

5183

5184

5186

5187

5188

5189

5190

5192

5193

5194

5195

5196

5197

5200

5201

5202

5203

5204

5205

5206

5207

5208

5209

5210

5211

5212

5213

5214

5215

5217

5219

5220

5221

5222

5224

5225

5226

5227

5228

5229

5230

5232

5233

5234

5235

5236

5237

5238

5239

5240

5241

5242

5243

5244

5245

5246

5247

5248

5249

5250

5251

5252

5253

5254

5257

5258

5265

5275

5285

5296

5259

5266

5276

5286

5297

5260

5277

5287

5298

5267

5261

5278

5288

5299

5268

5279

5289

5300

5262

5269

5280

5290

5301

5263

5270

5281

5291

5302

5271

5282

5292

5303

5264

5272

5283

5293

5304

5274

5284

5294

5305

5295

5307

5308

5319

5329

5339

5348

5309

5320

5330

5340

5349

5310

5321

5331

5341

5350

5311

5322

5332

5342

5351

5312

5323

5333

5343

5352

5314

5324

5334

5344

5354

5315

5325

5335

5345

5355

5316

5326

5337

5346

5356

5317

5327

5338

5347

5318

5328

5357

5358

5359

5361

5362

5363

5364

5365

5366

5367

5368

5369

5370

5371

5372

5373

5374

5376

5379

5383

5384

5385

5386

5387

5391

5393

5394

5396

5397

5398

5399

5400

5403

5404

5405

5406

5409

5410

5411

5412

5413

5414

5415

5416

5417

5419

5420

5421

5422

5423

5424

5426

5427

5429

5430

5431

5432

5436

5441

5442

5445

5448

5449

5450

5451

5452

5453

5454

5455

5456

5457

5458

5459

5460

5461

5462

5463

5464

5465

5466

5467

5468

5469

5470

5471

5472

5473

5474

5475

5476

5477

5478

5479

5480

5481

5482

5483

5484

5485

5486

5487

5488

5489

5490

5491

5492

5493

5495

5496

5497

5498

5499

5500

5501

5502

5503

5505

5506

5507

5508

5509

5510

5511

5512

5513

5514

5515

5516

5517

5518

5519

5520

5521

5522

5523

5524

5525

5526

5527

5528

5529

5530

5531

5532

5533

5534

5535

5536

5537

5538

5539

5540

5541

5542

5543

5544

5545

5546

5547

5548

5549

5550

5551

5552

5553

5554

5555

5556

5557

5558

5559

5560

5562

5563

5564

5565

5567

5569

5570

5571

5572

5573

5574

5576

5577

5578

5579

5580

5582

5583

5584

5585

5586

5588

5589

5590

5591

5592

5593

5595

5596

5597

5598

5599

5600

5601

5602

5603

5604

5605

5606

5607

5608

5609

5610

5611

5612

5613

5614

5615

5616

5617

5618

5619

5620

5621

5622

5626

5627

5628

5629

5630

5631

5632

5633

5634

5635

5636

5637

5638

5639

5640

5641

5642

5643

5653

5660

5667

5678

5644

5654

5661

5668

5679

5645

5655

5662

5669

5680

5646

5656

5663

5672

5681

5647

5657

5664

5673

5682

5648

5658

5665

5674

5683

5650

5659

5666

5675

5684

5651

5676

5652

5677

5685

5686

5696

5704

5712

5723

5687

5697

5705

5713

5724

5688

5698

5706

5714

5726

5689

5699

5707

5715

5727

5690

5700

5708

5716

5728

5691

5701

5717

5709

5729

5692

5702

5718

5693

5710

5719

5730

5694

5703

5711

5720

5731

5695

5721

5722

5732

5733

5734

5735

5736

5737

5738

5739

5740

5741

5742

5743

5744

5745

5746

5747

5748

5749

5750

5751

5752

5753

5754

5755

5756

5757

5758

5759

5760

5761

5762

5763

5764

5765

5766

5767

5768

5769

5770

5771

5772

5773

5774

5775

5776

5777

5779

5780

5781

5782

5783

5784

5786

5787

5788

5789

5790

5791

5792

5793

5794

5795

5796

5797

5798

5800

5801

5802

5803

5805

5806

5807

5808

5809

5810

5811

5812

5813

5814

5815

5816

5817

5818

5819

5820

5821

5822

5823

5824

5826

5827

5828

5829

5830

5831

5832

5833

5834

5835

5836

5837

5838

5839

5840

5841

5842

5843

5844

5845

5846

5847

5848

5849

5850

5851

5852

5853

5855

5856

5857

5858

5859

5860

5861

5862

5863

5864

5865

5866

5867

5868

5869

5870

5871

5872

5873

5874

5875

5876

5877

5879

5880

5881

5882

5883

5884

5885

5886

5887

5888

5889

5890

5891

5892

5893

5894

5895

5896

5897

5898

5899

5900

5901

5902

5903

5904

5905

5906

5907

5908

5909

5910

5917

5924

5932

5941

5911

5918

5925

5942

5933

5943

5912

5919

5926

5934

5944

5913

5945

5920

5927

5935

5914

5921

5929

5936

5946

5915

5922

5930

5937

5947

5938

5916

5923

5939

5948

5931

5940

5950

5960

5970

5981

5989

5951

5961

5971

5982

5990

5952

5962

5973

5984

5991

5953

5963

5974

5985

5992

5954

5964

5975

5986

5993

5955

5965

5977

5987

5994

5956

5967

5978

5995

5957

5968

5979

5988

5996

5980

STRUCTURAL FORMULAS OF ORGANIC COMPOUNDS (continued)

In numeric order as they occur in the Table of Physical Constants of Organic Compounds

5997

5999

6000

6001

6002

6004

6005

6006

6007

6008

6009

6010

6011

6012

6013

6014

6015

6016

6017

6018

6019

6020

6021

6024

6025

6026

6027

6028

6029

6031

6032

6033

6034

6035

6036

6037

6039

6040

6041

6042

6043

6044

6045

6046

6047

6049

6050

6051

6052

6053

6055

6056

6057

6058

6059

6060

6061

6062

6063

6064

6066

6067

6068

6069

6070

6071

6072

6073

6074

6075

6076

6077

6081

6082

6083

6084

6085

6086

6087

6088

6089

6090

6091

STRUCTURAL FORMULAS OF ORGANIC COMPOUNDS (continued)

In numeric order as they occur in the Table of Physical Constants of Organic Compounds

6092

6093

6094

6095

6096

6097

6098

6099

6101

6102

6103

6104

6105

6106

6107

6108

6109

6110

6111

6112

6113

6114

6115

6116

6117

6118

6119

6120

6121

6122

6123

6124

6125

6126

6127

6128

6129

6130

6131

6132

6133

6134

STRUCTURAL FORMULAS OF ORGANIC COMPOUNDS (continued)

In numeric order as they occur in the Table of Physical Constants of Organic Compounds

6135

6137

6138

6139

6140

6141

6142

6143

6145

6146

6147

6148

6149

6150

6151

6152

6153

6154

6155

6156

6157

6158

6159

6160

6163

6164

6165

6166

6167

6168

6169

6170

6171

6172

6174

6175

6176

6180

6181

6183

6184

6185

6186

6187

6188

6190

6191

6192

6194

6206

6221

6240

6197

6208

6241

6262

6199

6212

6222

6243

6263

6200

6214

6227

6249

6265

6201

6216

6228

6250

6266

6202

6217

6229

6251

6267

6203

6218

6238

6252

6268

6204

6220

6239

6254

6272

6255

6275

6276

6277

6278

6279

6280

6282

6284

6287

6288

6289

6291

6292

6294

6295

6296

6297

6298

6299

6300

6301

6302

6304

6305

6306

6307

6308

6309

6310

6311

6312

6313

6314

6315

6316

6317

6318

6319

6320

6321

6322

6323

6324

6325

6326

6327

6328

6329

6330

6331

6332

6333

6334

6335

6336

6337

6338

6339

6340

6341

6342

6343

6344

6345

6346

6347

6348

6349

6350

6351

6352

6353

6354

6355

6356

6357

6358

6359

6360

6361

6362

6363

6364

6365

6366

6367

6368

6369

6370

6371

6372

6373

6374

6375

6376

6377

6378

6379

6380

6381

6383

6392

6402

6411

6420

6384

6393

6403

6412

6421

6385

6394

6404

6413

6422

6386

6395

6405

6414

6423

6396

6387

6406

6424

6397

6415

6388

6407

6425

6398

6416

6389

6408

6426

6399

6417

6390

6409

6418

6427

6400

6410

6428

6391

6401

6419

6429

6430

6431

6432

6433

6434

6435

6436

6437

6438

6439

6440

6441

6442

6443

6444

6445

6446

6447

6448

6449

6450

6451

6452

6453

6454

6455

6456

6457

6459

6461

6462

6463

6464

6465

6466

6467

6468

6469

6470

6471

6472

6473

6474

6475

6476

6477

6478

6479

STRUCTURAL FORMULAS OF ORGANIC COMPOUNDS (continued)

In numeric order as they occur in the Table of Physical Constants of Organic Compounds

6480

6481

6482

6483

6484

6485

6486

6487

6488

6489

6490

6491

6492

6493

6494

6495

6496

6497

6498

6499

6500

6501

6502

6503

6504

6505

6506

6508

6509

6510

6511

6512

6513

6514

6515

6517

6518

6519

6520

6521

6522

6523

6524

6525

STRUCTURAL FORMULAS OF ORGANIC COMPOUNDS (continued)

In numeric order as they occur in the Table of Physical Constants of Organic Compounds

6526	6537	6547	6556	6565
6527	6538	6548	6557	6566
6528	6539	6549	6558	6567
6529	6540	6550	6559	6568
6530	6541	6551	6560	6570
6531	6542	6552	6561	6572
6532	6543	6553		6573
6533	6544	6554	6562	6574
6534	6545	6555	6563	6575
6535	6546		6564	6578
6536				

6579

6580

6581

6582

6583

6584

6585

6586

6587

6588

6589

6590

6591

6592

6593

6594

6595

6596

6597

6598

6599

6600

6601

6602

6603

6604

6605

6606

6607

6608

6609

6610

6611

6612

6613

6614

6615

6616

6617

6618

6619

6620

6621

6622

6623

6625

6626

6627

6628

6629

6630

6631

6632

6633

6634

6635

6636

6637

6638

6639

6640

6641

6642

6643

6644

6647

6649

6650

6651

6652

6653

6654

6655

6656

6657

6658

6659

6660

6661

6662

6663

6664

6665

6666

6667

6668

6669

6670

6671

6672

6673

6674

6675

6676

6677

6678

6679

6680

6681

6690

6698

6709

6719

6682

6691

6699

6710

6720

6683

6692

6700

6711

6721

6684

6693

6701

6712

6722

6685

6702

6713

6694

6703

6714

6723

6686

6695

6705

6715

6724

6687

6696

6706

6716

6725

6688

6707

6717

6689

6697

6708

6718

6726

6727

6737

6746

6757

6767

6728

6738

6747

6758

6768

6729

6739

6748

6759

6769

6730

6740

6749

6760

6770

6731

6741

6750

6761

6771

6732

6751

6762

6772

6733

6742

6752

6763

6773

6743

6734

6753

6764

6744

6735

6754

6765

6774

6745

6755

6736

6756

6766

6775

6776

6787

6797

6805

6814

6777

6788

6798

6806

6815

6778

6789

6799

6807

6816

6779

6790

6800

6808

6817

6781

6791

6801

6809

6818

6782

6792

6802

6810

6819

6783

6793

6803

6811

6820

6784

6794

6804

6812

6821

6785

6795

6813

6822

6786

6796

6823

6824

6825

6826

6827

6828

6829

6830

6831

6832

6833

6834

6835

6836

6837

6838

6839

6840

6841

6842

6843

6844

6845

6846

6847

6848

6849

6850

6851

6852

6853

6854

6855

6856

6857

6858

6859

6860

6861

6862

6863

6864

6865

6866

6868

6869

6870

6872

6873

STRUCTURAL FORMULAS OF ORGANIC COMPOUNDS (continued)

In numeric order as they occur in the Table of Physical Constants of Organic Compounds

6874

6883

6892

6904

6914

6875

6884

6893

6905

6915

6876

6885

6894

6906

6916

6877

6886

6895

6907

6917

6878

6887

6896

6908

6918

6879

6888

6897

6909

6919

6880

6889

6898

6910

6921

6881

6890

6899

6911

6922

6882

6891

6900

6912

6923

6901

6913

6924

6932

6941

6949

6957

6925

6933

6942

6950

6958

6926

6934

6943

6951

6959

6927

6935

6944

6952

6960

6928

6936

6945

6953

6961

6929

6937

6946

6954

6962

6930

6938

6947

6955

6963

6931

6939

6948

6956

6964

6940

6965

6973

6981

6989

6998

6966

6974

6982

6990

6999

6967

6975

6983

6991

7000

6968

6976

6984

6992

7001

6969

6977

6985

6993

7002

6970

6978

6986

6994

7003

6971

6979

6987

6995

7004

6972

6980

6988

6996

6997

7005

7007

STRUCTURAL FORMULAS OF ORGANIC COMPOUNDS (continued)

In numeric order as they occur in the Table of Physical Constants of Organic Compounds

7008

7009

7010

7011

7012

7013

7014

7015

7016

7017

7018

7019

7020

7021

7022

7023

7024

7025

7026

7027

7028

7029

7030

7031

7032

7033

7034

7035

7036

7037

7038

7039

7040

7041

7042

7043

7044

7045

7046

7047

7048

7049

7050

7051

7053

7054

7055

7056

7057

7058

STRUCTURAL FORMULAS OF ORGANIC COMPOUNDS (continued)

In numeric order as they occur in the Table of Physical Constants of Organic Compounds

7060

7061

7062

7063

7064

7065

7066

7067

7068

7069

7070

7071

7072

7073

7074

7075

7076

7077

7078

7079

7080

7081

7082

7083

7084

7085

7086

7087

7088

7089

7090

7091

7092

7093

7094

7095

7096

7097

7098

7099

7100

7101

7102

7103

7104

7105

7106

7107

7108

7109

7113

7114

7115

7116

7117

7118

7119

7120

7121

7122

7123

7124

7125

7126

7127

7128

7129

7130

7131

7132

7136

7137

7138

7140

7142

7144

7145

7146

7147

7148

7149

7150

7151

7152

7153

7154

7155

7156

7158

7159

7160

7161

7164

7165

7167

7168

7170

7172

7175

7176

7177

7178

7179

7180

7181

7182

7183

7184

7185

7186

7187

7188

7189

7190

7192

7193

7194

7195

7196

7197

7198

7199

7200

7201

7202

7203

7205

7206

7207

7208

7209

7210

7211

7212

7213

7214

7215

7216

7217

7218

7219

7222

7224

7225

7226

7227

7228

7230

7231

7232

7233

7235

7236

7237

7238

7239

7240

7241

7242

7243

7244

7245

7246

7247

7249

7250

7251

7252

7253

7254

7255

7257

7258

7259

7260

7262

7263

7264

7267

7268

7269

7271

7272

7273

7274

7276

7277

7278

7279

7281

7282

7284

7285

7286

7287

7289

7290

7291

7292

7293

7294

7295

7296

7297

7298

7299

7300

7302

7303

7304

7305

7307

7308

7309

7310

7319

7327

7337

7346

7312

7320

7328

7338

7347

7313

7321

7329

7339

7350

7314

7322

7330

7340

7351

7315

7323

7331

7342

7354

7316

7324

7335

7343

7355

7317

7325

7336

7344

7356

7326

7345

7357

7367

7375

7386

7403

7387

7404

7359

7368

7376

7388

7405

7361

7369

7378

7389

7406

7362

7370

7380

7390

7408

7381

7363

7371

7398

7409

7372

7383

7399

7410

7384

7364

7402

7411

7365

7374

7385

STRUCTURAL FORMULAS OF ORGANIC COMPOUNDS (continued)

In numeric order as they occur in the Table of Physical Constants of Organic Compounds

7412

7421

7432

7442

7450

7413

7422

7433

7443

7451

7415

7423

7434

7444

7452

7416

7425

7435

7446

7454

7417

7426

7436

7447

7455

7418

7427

7438

7448

7459

7419

7430

7439

7449

7460

7420

7431

7441

7461

7462

7463

7464

7465

7466

7467

7468

7469

7470

7471

7472

7473

7474

7475

7476

7477

7478

7479

7480

7481

7482

7483

7484

7485

7486

7487

7489

7490

7491

7492

7493

7494

7495

7496

7497

7498

7499

7500

7501

7509

7519

7528

7535

7502

7510

7520

7529

7536

7503

7512

7521

7530

7537

7504

7513

7522

7531

7538

7505

7515

7523

7532

7539

7506

7516

7524

7533

7540

7507

7517

7525

7534

7541

7508

7518

7526

7527

7542

7543

7544

7545

7546

7547

7548

7549

7550

7551

7552

7553

7554

7555

7556

7557

7558

7559

7560

7561

7562

7564

7565

7567

7568

7569

7570

7571

7572

7573

7574

7575

7576

7577

7578

7579

7580

STRUCTURAL FORMULAS OF ORGANIC COMPOUNDS (continued)

In numeric order as they occur in the Table of Physical Constants of Organic Compounds

7581

7582

7585

7586

7587

7588

7590

7591

7592

7593

7594

7595

7597

7598

7599

7600

7601

7602

7604

7605

7606

7607

7609

7612

7614

7615

7616

7617

7618

7619

7621

7622

7623

7624

7625

7626

7629

7631

7632

7633

7635

7636

7639

7640

7641

7642

7643

7650

7659

7667

7675

7644

7651

7660

7669

7678

7645

7653

7661

7670

7679

7646

7654

7662

7671

7680

7655

7663

7647

7656

7665

7672

7681

7648

7657

7666

7674

7682

7649

7658

7683

7684

7685

7686

7687

7689

7690

7691

7692

7693

7694

7696

7697

7698

7699

7700

7701

7702

7703

7704

7705

7706

7708

7709

7710

7712

7713

7714

7715

7716

7717

7718

7720

7721

7722

7723

7724

7725

7727

7728

7730

7731

7732

7733

7734

7735

7736

7737

7738

7739

7740

7742

7743

7746

7747

7749

7750

7751

7752

7753

7754

7756

7758

7759

7764

7765

7766

7770

7771

7772

7773

7775

7776

7778

7779

7780

7781

7782

7783

7784

7785

7789

7790

7791

7792

7793

7794

7795

7796

7797

7798

7799

7801

7802

7803

7804

7806

7807

7808

7809

7810

7811

7812

7813

7814

7815

7816

7817

7818

7819

7820

7821

7822

7823

7824

7825

7826

7827

7828

7829

7830

7831

7832

7833

7834

7835

7836

7838

7839

7842

7843

7846

7848

7850

7852

7853

7859

7860

7861

7862

7863

7864

7865

7879

7880

7881

7882

7884

7885

7886

7887

7889

7890

7891

7892

7894

7895

7896

7897

7898

7899

7900

7901

7902

7903

7904

7906

7907

7908

7909

7911

7913

7915

7916

7917

7918

7920

7921

7922

7923

7924

7925

7926

7927

7928

7929

7931

7932

7933

7934

7935

7936

7937

7938

7939

7941

7942

7943

7944

7945

7946

7948

7949

7950

7951

7952

7954

7955

7956

7957

7958

7959

7966

7975

7976

7977

7978

7979

7980

7981

7982

7983

7984

7985

7986

7987

7988

7989

7990

7991

7992

7993

7994

7995

7996

7997

7998

7999

8000

8001

8002

8003

8005

8017

8030

8039

8052

8006

8018

8031

8041

8054

8007

8020

8032

8043

8055

8008

8022

8033

8044

8056

8009

8024

8034

8045

8057

8010

8026

8035

8046

8059

8011

8027

8037

8047

8061

8012

8028

8038

8048

8062

8013

8015

8050

8063

8064

8065

8066

8067

8068

8069

8070

8071

8072

8073

8074

8075

8076

8077

8078

8079

8080

8081

8082

8083

8084

8085

8086

8087

8088

8089

8090

8091

8092

8093

8094

8095

8096

8097

8098

8099

8101

8103

8104

8105

8106

8107

8108

8109

8110

8111

8112

8113

8114

STRUCTURAL FORMULAS OF ORGANIC COMPOUNDS (continued)

In numeric order as they occur in the Table of Physical Constants of Organic Compounds

8115

8116

8117

8118

8119

8120

8121

8122

8123

8124

8125

8126

8127

8128

8129

8130

8131

8132

8133

8134

8135

8136

8137

8138

8139

8140

8141

8142

8143

8144

8145

8146

8147

8148

8149

8150

8151

8152

8154

8155

8156

8157

8158

8159

8160

8161

8162

8163

8164

8165

8166

8167

8168

8169

8170

8171

8172

8173

8174

8175

8176

8177

8178

8179

8180

8181

8182

8183

8184

8185

8186

8187

8188

8189

8190

8191

8192

8193

8194

8195

8196

8197

8198

8199

8200

8201

8202

8203

8204

8205

8206

8207

8208

8209

8212

8213

8214

8215

8216

8218

8219

8220

8223

8237

8260

8281

8293

8226

8238

8262

8282

8294

8227

8263

8283

8240

8228

8264

8241

8297

8265

8284

8229

8298

8242

8266

8230

8285

8243

8299

8267

8231

8244

8287

8268

8301

8288

8232

8245

8269

8289

8246

8302

8271

8233

8290

8247

8303

8273

8234

8250

8278

8291

8235

8279

8251

8304

8236

8292

8257

8280

8305

8306

8307

8308

8309

8310

8311

8312

8313

8314

8315

8316

8317

8318

8319

8320

8321

8322

8324

8325

8327

8328

8329

8330

8331

8332

8333

8334

8335

8336

8337

8338

8339

8340

8341

8342

8343

8344

8345

8346

8347

8348

8349

8350

8351

8352

8353

8354

8355

8356

8357

8358

8359

8360

8361

8363

8364

8365

8366

8367

8368

8369

8370

8371

8372

8373

8374

8375

8376

8377

8378

8379

8380

8381

8382

8383

8384

8385

8386

8387

8388

8389

8390

8391

8392

8393

8394

8395

8396

8398

8399

8400

8401

8402

8403

8404

8405

8406

8407

8408

8409

8410

8411

8412

8415

8416

8417

8418

8419

8420

8421

8422

8424

8425

8436

8447

8457

8468

8426

8437

8449

8458

8469

8427

8438

8450

8459

8470

8428

8439

8451

8460

8471

8429

8440

8452

8461

8472

8430

8441

8453

8462

8473

8431

8442

8454

8464

8474

8432

8443

8455

8465

8475

8433

8444

8456

8466

8476

8434

8445

8477

8435

8446

STRUCTURAL FORMULAS OF ORGANIC COMPOUNDS (continued)

In numeric order as they occur in the Table of Physical Constants of Organic Compounds

8478

8479

8480

8481

8482

8484

8485

8486

8487

8488

8489

8490

8491

8492

8493

8494

8495

8496

8497

8498

8499

8500

8502

8503

8504

8505

8507

8508

8509

8510

8511

8512

8513

8514

8515

8516

8517

8518

8519

8520

8521

8523

8524

8525

8526

8527

8528

8529

8530

8531

8532

8533

8534

8544

8557

8569

8579

8535

8545

8558

8570

8580

8536

8559

8571

8581

8537

8547

8560

8574

8582

8538

8548

8562

8575

8583

8539

8552

8563

8576

8584

8540

8553

8564

8577

8585

8541

8565

8578

8586

8542

8554

8566

8543

8555

8568

8587

8556

8588

8598

8599

8609

8610

8619

8620

8621

8629

8632

8589

8600

8611

8590

8601

8612

8622

8633

8591

8602

8613

8623

8635

8592

8603

8614

8624

8636

8593

8604

8615

8625

8637

8594

8605

8616

8626

8638

8595

8607

8617

8627

8639

8596

8608

8618

8628

8640

8597

8641

8642

8643

8644

8645

8646

8647

8648

8649

8650

8651

8652

8653

8654

8655

8656

8657

8658

8659

8660

8661

8662

8663

8664

8665

8666

8667

8668

8669

8670

8671

8672

8673

8674

8675

8676

8679

8680

8681

8682

8683

8684

8685

8686

8687

8688

8689

8690

8691

8692

8693

8694

8695

8696

8697

STRUCTURAL FORMULAS OF ORGANIC COMPOUNDS (continued)

In numeric order as they occur in the Table of Physical Constants of Organic Compounds

8698

8699

8700

8701

8702

8703

8704

8705

8706

8707

8708

8709

8710

8711

8712

8713

8714

8715

8716

8717

8718

8719

8720

8721

8722

8723

8726

8727

8728

8729

8730

8731

8732

8733

8734

8735

8736

8737

8739

8740

8741

8742

8743

8744

8753

8764

8774

8786

8765

8745

8754

8775

8787

8766

8746

8755

8776

8788

8767

8747

8777

8789

8768

8756

8748

8778

8790

8757

8769

8780

8749

8758

8791

8770

8781

8750

8760

8792

8771

8783

8751

8761

8793

8772

8784

8762

8752

8773

8785

8794

8763

8795

8796

8797

8798

8799

8801

8802

8803

8804

8805

8806

8807

8808

8809

8810

8811

8812

8813

8814

8815

8816

8817

8818

8819

8820

8821

8822

8823

8824

8825

8826

8827

8828

8829

8830

8831

8832

8833

8834

8835

8836

8837

8838

8839

8840

8841

8842

8843

8844

8845

8846

8847

8848

8849

8850

8851

8852

8853

8854

8855

8856

8857

8858

8859

8860

8861

8862

8865

8866

8867

8868

8869

8870

8871

8872

8873

8874

8875

8876

8878

8879

8881

8882

8884

8885

8886

8887

8888

8889

8891

8892

8893

8894

8895

8896

8897

8898

8900

8903

8904

In numeric order as they occur in the Table of Physical Constants of Organic Compounds

8905

8906

8907

8908

8909

8910

8911

8912

8913

8914

8915

8916

8917

8918

8919

8920

8922

8923

8924

8925

8927

8928

8929

8930

8931

8932

8933

8934

8935

8936

8937

8938

8939

8940

8941

8942

8943

8944

8945

8946

8947

8950

8952

8953

8954

8955

8964

8972

8980

8989

8956

8965

8973

8981

8990

8957

8966

8974

8982

8991

8958

8967

8975

8983

8992

8959

8968

8976

8984

8995

8985

8996

8961

8969

8977

8986

8997

8962

8970

8978

8998

8987

8999

8963

8979

8988

9000

8971

9001

STRUCTURAL FORMULAS OF ORGANIC COMPOUNDS (continued)

In numeric order as they occur in the Table of Physical Constants of Organic Compounds

9002

9004

9005

9006

9007

9008

9009

9010

9011

9012

9013

9014

9015

9016

9017

9018

9019

9020

9021

9022

9023

9024

9025

9026

9027

9028

9029

9030

9031

9032

9033

9034

9035

9036

9037

9038

9039

9040

9041

9042

9043

9044

9045

9046

9047

9048

9049

9050

9051

9052

9053

9054

9055

9056

9057

STRUCTURAL FORMULAS OF ORGANIC COMPOUNDS (continued)

In numeric order as they occur in the Table of Physical Constants of Organic Compounds

9058

9059

9060

9061

9062

9065

9066

9067

9068

9069

9070

9071

9072

9073

9074

9075

9076

9077

9078

9079

9080

9081

9082

9083

9084

9085

9086

9087

9088

9089

9090

9091

9092

9093

9094

9095

9096

9097

9098

9099

9100

9101

9102

9103

9104

9105

9106

9107

9108

9117

9126

9135

9143

9109

9118

9128

9136

9144

9110

9119

9129

9137

9145

9111

9121

9130

9138

9146

9112

9122

9131

9139

9147

9113

9123

9132

9140

9149

9114

9124

9133

9141

9150

9115

9125

9134

9142

9151

9116

9153

9167

9182

9192

9206

9154

9168

9183

9193

9208

9156

9170

9184

9195

9210

9158

9171

9185

9198

9211

9159

9172

9186

9199

9212

9160

9173

9187

9201

9213

9161

9174

9188

9202

9216

9162

9179

9189

9204

9164

9180

9190

9205

9217

9181

9191

9218

9219

9220

9221

9223

9224

9225

9226

9227

9228

9230

9231

9232

9233

9249

9250

9251

9253

9254

9255

9256

9257

9260

9262

9263

9264

9265

9266

9267

9270

9271

9272

9273

9276

9277

9279

9280

9282

9283

9284

9285

9287

9288

9289

9290

9291

9292

9294

9295

9296

9297

9298

9300

9305

9306

9307

9308

9309

9310

9312

9313

9314

9315

9316

9317

9319

9320

9321

9322

9323

9324

9326

9327

9328

9330

9331

9333

9334

9335

9336

9337

9338

9340

9341

9342

9343

9344

9345

9348

9349

9350

9351

9352

9353

9355

9356

9357

9358

9359

9360

9362

9364

9365

9370

9371

9372

9374

9375

9376

9377

9378

9379

9381

9382

9385

9387

9388

9389

9391

9392

9394

9395

9396

9397

9401

9402

9403

9404

9405

9407

9409

9410

9411

9412

9413

9414

9416

9417

9418

9419

9420

9421

9423

9424

9425

9426

9427

9428

9430

9431

9432

9434

9435

9436

9438

9439

9440

9441

9442

9443

9445

9446

9447

9448

9451

9452

9454

9455

9456

9457

9458

9459

9461

9462

9463

9465

9466

9469

9471

9472

9473

9474

9475

9479

9482

9483

9484

9485

9486

9489

9492

9493

9494

9495

9496

9497

9499

9501

9503

9504

9505

9507

9510

9512

9513

9514

9515

9517

9518

9519

9520

9521

9522

9523

9524

9525

9527

9528

9529

9530

9532

9535

9536

9537

9539

9540

9541

9543

9545

9555

9564

9572

9581

9546

9556

9565

9573

9582

9548

9557

9566

9574

9583

9549

9558

9567

9575

9584

9550

9559

9568

9576

9585

9551

9560

9569

9577

9586

9553

9562

9570

9578

9587

9554

9563

9571

9579

9588

9580

9589

9590

9591

9592

9593

9594

9595

9596

9597

9599

9600

9601

9602

9603

9604

9605

9606

9608

9609

9610

9611

9612

9613

9614

9615

9616

9617

9618

9619

9620

9621

9622

9624

9625

9626

9627

9628

9629

9630

9631

9632

9634

9644

9656

9666

9675

9635

9645

9658

9667

9676

9636

9648

9659

9668

9678

9637

9649

9660

9669

9679

9638

9650

9661

9670

9680

9639

9651

9662

9671

9681

9642

9652

9663

9673

9682

9643

9655

9664

9674

9683

9684

9685

9686

9687

9688

9689

9690

9691

9692

9693

9695

9696

9697

9698

9699

9700

9701

9702

9703

9704

9705

9706

9707

9708

9709

9710

9711

9712

9713

9714

9715

9716

9717

9718

9719

9720

9721

9722

9723

9724

9726

9727

9728

9729

9730

9731

9732

9733

9734

9736

9738

9739

9740

9741

9743

9745

9746

9748

9749

9750

9751

9752

9753

9754

9756

9758

9759

9760

9761

9762

9763

9764

9765

9766

9767

9768

9769

9770

9771

9772

9773

9774

9775

9784

9798

9808

9817

9776

9785

9799

9809

9818

9786

9810

9777

9800

9819

9787

9801

9811

9820

9778

9788

9803

9812

9821

9779

9790

9804

9813

9782

9822

9780

9791

9805

9814

9823

9781

9792

9806

9815

9824

9782

9793

9807

9816

9825

9783

9826

9827

9828

9829

9832

9834

9835

9836

9837

9838

9839

9840

9841

9842

9843

9844

9845

9846

9847

9848

9849

9850

9851

9852

9853

9854

9855

9856

9857

9858

9859

9860

9861

9862

9863

9864

9865

9866

9867

9868

9869

9870

9871

9872

9873

9874

9875

STRUCTURAL FORMULAS OF ORGANIC COMPOUNDS (continued)

In numeric order as they occur in the Table of Physical Constants of Organic Compounds

9876

9877

9878

9879

9880

9881

9882

9883

9884

9885

9886

9887

9888

9889

9890

9891

9892

9893

9894

9895

9896

9897

9898

9899

9900

9901

9902

9903

9904

9905

9906

9907

9908

9909

9910

9911

9912

9913

9914

9915

9916

9917

9918

9919

9920

STRUCTURAL FORMULAS OF ORGANIC COMPOUNDS (continued)

In numeric order as they occur in the Table of Physical Constants of Organic Compounds

9921

9922

9923

9924

9925

9926

9927

9928

9930

9931

9932

9933

9934

9935

9936

9937

9938

9939

9940

9941

9942

9943

9944

9945

9946

9947

9948

9949

9950

9951

9952

9953

9954

9955

9957

9958

9959

9960

9961

9962

9963

9964

9966

9967

9978

9989

9999

10007

9969

9980

9990

10000

10008

9970

9981

9991

10001

10009

9972

9982

9992

10002

10010

9973

9983

9994

10003

10011

9974

9985

9995

10004

10012

9975

9986

9997

10005

10013

9976

9987

9998

10006

10014

9977

10016

10017

10027

10037

10047

10056

10018

10028

10038

10048

10057

10019

10029

10039

10049

10058

10020

10030

10040

10050

10059

10021

10031

10041

10051

10060

10022

10032

10042

10052

10061

10023

10033

10043

10053

10062

10024

10034

10044

10054

10063

10025

10035

10045

10055

10064

10026

10036

10046

10065

10066

10067

10068

10069

10070

10071

10072

10073

10074

10075

10076

10077

10078

10079

10080

10081

10082

10083

10084

10085

10086

10087

10088

10089

10090

10091

10092

10093

10094

10095

10096

10097

10098

10099

10100

10101

10102

10103

10105

10106

10107

10108

10109

10110

10111

10112

10113

10114

10115

10116

10117

10118

10119

10120

10121

10122

10123

10124

10125

10126

10127

10128

10129

10130

10131

10132

10133

10134

10135

10136

10138

10139

10140

10141

10142

10143

10144

10145

10146

10147

10148

10149

10150

10151

10152

10153

10154

10155

10156

10157

10158

10159

10160

10161

10162

10163

10164

10166

10167

10168

10169

10170

10171

10172

10173

10174

10175

10176

10177

10178

10179

10180

10181

10182

10183

10184

10185

10186

10187

10188

10189

10190

10191

10192

10193

10194

10195

10196

10197

10198

STRUCTURAL FORMULAS OF ORGANIC COMPOUNDS (continued)

In numeric order as they occur in the Table of Physical Constants of Organic Compounds

10199

10209

10218

10228

10236

10200

10210

10219

10229

10237

10201

10211

10220

10230

10238

10202

10212

10221

10231

10239

10203

10213

10222

10232

10240

10204

10214

10223

10233

10241

10205

10215

10225

10234

10242

10207

10216

10226

10235

10243

10208

10217

10227

10244

10254

10263

10275

10283

10245

10255

10264

10276

10285

10246

10256

10265

10277

10286

10248

10258

10268

10278

10287

10249

10259

10269

10279

10288

10250

10260

10270

10280

10289

10251

10261

10273

10281

10290

10253

10262

10274

10282

10291

10292

10293

10294

10295

10296

10297

10298

10299

10300

10301

10302

10303

10304

10306

10307

10308

10309

10310

10311

10312

10313

10314

10315

10316

10317

10318

10319

10320

10321

10322

10323

10324

10325

10326

10328

10329

10330

10331

10332

10333

10334

10335

10336

10337

10338

10339

10340

10341

10342

10343

10344

10345

10346

10347

10348

10349

10350

10351

10352

10353

10354

10355

10356

10357

10358

10359

10360

10361

10362

10363

10364

10365

10366

10367

10368

10369

10370

10371

10372

10373

10374

10375

10376

10378

10380

10381

10382

10383

10384

10385

10386

10387

10388

10389

10390

10391

10393

10394

10395

10396

10398

10399

10401

10402

10404

10405

10406

10407

10408

10409

10410

10411

10412

10413

10414

10415

10416

10417

10418

10419

10420

10421

10422

10423

10424

10425

10426

10427

10428

10429

10430

10431

10432

10433

10436

10437

10438

10439

10440

10441

10442

10443

10444

10445

10446

10447

10448

10449

10450

10451

10452

10454

10455

10457

10458

10460

10461

10462

10463

10464

10465

10466

10467

10468

10469

10470

10471

10472

10473

10474

10475

10476

10478

10479

10480

10481

10482

10483

10484

10485

10486

10487

10488

10489

10490

10491

10492

10493

10494

10495

10496

10497

10498

10500

10501

10502

10503

10504

10505

10506

10507

10508

10509

10510

10511

10512

10514

10515

10516

10517

10518

10519

10520

10521

10522

10523

10524

10525

10526

10527

10528

10529

10530

10531

10532

10533

10534

10535

10536

10537

10538

10539

10540

10541

10542

10543

10544

10545

10546

10547

10548

10549

10550

10551

10552

10553

10554

10555

10556

10557

10558

10559

10560

10561

10562

10563

10564

10565

10566

10567

10568

10569

10570

10571

10572

10573

10574

10575

10576

10577

10578

10579

10580

10581

10582

10583

10584

10585

10586

10587

10588

10589

10590

10591

10592

10593

10594

10596

10597

10598

10599

10600

10601

10602

10603

10604

10605

10606

STRUCTURAL FORMULAS OF ORGANIC COMPOUNDS (continued)

In numeric order as they occur in the Table of Physical Constants of Organic Compounds

10607

10618

10628

10638

10649

10608

10620

10629

10639

10650

10609

10621

10631

10640

10652

10611

10622

10632

10643

10653

10613

10624

10633

10644

10654

10614

10625

10634

10645

10655

10616

10626

10635

10647

10656

10617

10627

10636

10648

10657

10637

10658

10659

10668

10680

10692

10702

10660

10670

10681

10693

10703

10661

10671

10683

10694

10704

10662

10672

10685

10695

10706

10663

10673

10686

10696

10707

10664

10674

10687

10697

10708

10665

10677

10688

10698

10709

10666

10679

10689

10699

10710

10691

10700

10711

10713

10714

10715

10716

10717

10718

10719

10720

10721

10722

10723

10724

10725

10726

10727

10728

10729

10730

10731

10732

10733

10734

10735

10736

10737

10738

10739

10743

10744

10745

10746

10747

10748

10749

10750

10751

10752

10753

10755

10756

10757

10758

10759

10760

10761

10763

10764

10767

10768

10769

10770

10773

10774

10775

10777

10778

10779

10780

10781

10782

10783

10784

10787

10788

10789

10790

10791

10792

10793

10794

10795

10796

10798

10799

10800

10801

10802

10803

10805

10806

10807

10809

10810

10811

10813

10814

10815

10816

10818

10819

10820

10821

10822

10823

10824

10825

10826

10827

10828

10829

10830

10831

10832

10833

10834

10835

10836

10837

10838

10839

10840

10841

10842

10844

10845

10846

10847

10849

10851

10852

10853

10855

10864

10873

10883

10892

10856

10875

10884

10893

10865

10858

10876

10894

10866

10886

10859

10877

10887

10895

10868

10860

10878

10888

10896

10869

10861

10879

10890

10897

10870

10862

10880

10891

10871

10863

10882

10898

10872

STRUCTURAL FORMULAS OF ORGANIC COMPOUNDS (continued)

In numeric order as they occur in the Table of Physical Constants of Organic Compounds

10899

10909

10921

10932

10941

10901

10911

10922

10933

10942

10903

10914

10923

10935

10943

10905

10915

10924

10936

10944

10906

10916

10925

10937

10945

10907

10917

10926

10938

10946

10908

10918

10927

10939

10947

10920

10929

10940

10948

10949

10959

10968

10977

10988

10951

10960

10969

10980

10989

10952

10961

10970

10981

10990

10953

10962

10972

10982

10991

10954

10963

10973

10983

10992

10965

10984

10955

10974

10994

10966

10985

10956

10975

10995

10987

10958

10967

10976

10996

10997

11006

11014

11024

11033

10998

11007

11016

11025

11034

10999

11008

11017

11026

11035

11001

11009

11018

11027

11036

11002

11010

11019

11028

11038

11003

11012

11020

11029

11039

11004

11013

11021

11031

11040

11005

11023

11032

11041

11042

11055

11067

11080

11090

11043

11056

11068

11081

11091

11045

11057

11069

11082

11092

11049

11058

11071

11083

11093

11050

11059

11072

11084

11094

11051

11062

11075

11085

11095

11052

11064

11076

11087

11098

11054

11066

11078

11088

11099

11100

11101

11103

11104

11105

11106

11109

11110

11112

11113

11114

11115

11116

11117

11118

11119

11120

11121

11122

11124

11125

11126

11127

11128

11129

11130

11131

11132

11133

11134

11135

11136

11137

11138

11139

11141

11142

11143

11146

11147

11148

11149

11150

11151

11152

11153

11154

11155

11156

11157

11158

11160

11161

11163

11164

11165

11166

11167

11168

11169

11170

11171

11172

11173

11175

11176

11180

11181

11182

11183

11184

11185

11186

11187

11188

11189

11190

11191

11193

11194

11195

11208

11219

11231

11242

11199

11209

11220

11232

11243

11201

11210

11222

11236

11244

11202

11211

11223

11237

11245

11203

11213

11224

11238

11246

11204

11214

11225

11239

11247

11217

11227

11240

11248

11206

11218

11230

11241

11250

11251

11262

11270

11280

11288

11254

11263

11271

11281

11289

11255

11264

11274

11282

11290

11257

11265

11275

11283

11291

11258

11266

11276

11284

11292

11259

11267

11277

11285

11296

11260

11268

11278

11286

11297

11261

11269

11279

11287

11298

11299

11300

11301

11302

11303

11304

11305

11306

11307

11308

11311

11312

11313

11314

11315

11317

11318

11319

11320

11321

11325

11327

11328

11329

11330

11331

11333

11334

11335

11337

11338

11339

11340

11341

11342

11343

11344

11345

11348

11350

11351

11353

11354

11356

11357

11358

11359

11360

11361

11362

11363

11364

11365

11366

11367

11368

11369

11370

11371

11372

11373

11374

11375

11376

11377

11378

11380

11381

11382

11383

11384

11385

11386

11387

11388

11389

11390

11391

11392

11393

11401

11409

11418

11426

11394

11402

11419

11427

11410

11395

11403

11420

11428

11411

11396

11404

11412

11421

11429

11397

11405

11413

11422

11430

11398

11406

11414

11423

11431

11399

11407

11415

11424

11432

11400

11408

11417

11425

11433

STRUCTURAL FORMULAS OF ORGANIC COMPOUNDS (continued)

In numeric order as they occur in the Table of Physical Constants of Organic Compounds

11434

11435

11436

11437

11438

11439

11440

11443

11444

11445

11446

11447

11448

11449

11450

11451

11452

11453

11454

11455

11456

11457

11458

11459

11460

11461

11462

11463

11464

11465

11466

11467

11468

11469

11470

11471

11472

11474

11475

11476

11477

STRUCTURAL FORMULAS OF ORGANIC COMPOUNDS (continued)

In numeric order as they occur in the Table of Physical Constants of Organic Compounds

11478

11479

11480

11481

11482

11483

11484

11485

11486

11487

11488

11489

11490

11491

11492

11493

11494

11495

11496

11497

11498

11499

11500

11501

11502

11503

11504

11505

11506

11507

11508

11509

11510

11511

11512

11513

11514

11517

11519

11520

11521

11527

11528

11529

11531

11532

11536

11537

11539

11540

11541

11542

11543

11544

11545

11546

11549

11550

11551

11555

11556

11558

11559

11560

11562

11563

11564

11567

11568

11569

11570

11571

11572

11573

11574

11575

11576

11577

11578

11579

11580

11581

11582

11583

11584

11585

11586

11588

11590

11591

11592

11593

11594

11595

11596

11598

11599

11600

11601

11602

11603

11604

11605

11606

11607

11609

11610

11611

11612

11613

11614

11615

11616

11617

11618

11619

11621

11622

11624

11625

11626

11627

11628

11629

11630

11631

11632

11633

11634

11635

11636

11637

11638

11640

11641

11642

11643

11644

11646

11647

11648

11649

11650

11651

11652

11654

11656

11657

11658

11659

11660

11661

11662

11664

11665

11666

11667

11668

11670

11671

11672

11673

11674

11675

11676

11677

11678

11679

11680

11681

11682

11683

11684

11685

11687

11688

11689

11690

11691

11692

11693

11694

11695

11696

11697

11698

11699

11700

11701

11703

11704

11705

11706

11707

11708

11709

11710

11712

11713

11714

11715

11716

11717

11719

11720

11721

11722

11723

11724

11725

11726

11727

11728

11729

11730

11731

11732

11733

11734

11735

11736

11737

11738

11739

11748

11756

11764

11773

11740

11749

11757

11765

11774

11741

11750

11758

11767

11775

11742

11751

11759

11768

11776

11743

11752

11760

11769

11777

11744

11753

11761

11770

11778

11746

11754

11762

11771

11779

11747

11755

11763

11772

11780

11781

11789

11797

11807

11819

11782

11790

11798

11808

11822

11783

11791

11799

11809

11823

11784

11792

11800

11810

11824

11785

11793

11801

11812

11826

11786

11794

11803

11813

11827

11787

11795

11804

11816

11828

11788

11796

11805

11817

11829

11806

In numeric order as they occur in the Table of Physical Constants of Organic Compounds

11831

11833

11834

11835

11836

11837

11840

11841

11842

11845

11846

11847

11848

11849

11852

11853

11854

11855

11856

11857

11858

11859

11860

11861

11862

11863

11865

11866

11867

11868

11869

11870

11871

11872

11873

11874

11875

11877

11878

11879

11880

11881

11882

11883

11884

11885

11886

11887

11888

11889

STRUCTURAL FORMULAS OF ORGANIC COMPOUNDS (continued)

In numeric order as they occur in the Table of Physical Constants of Organic Compounds

11890

11891

11892

11893

11894

11895

11897

11898

11899

11900

11901

11902

11903

11904

11905

11906

11908

11909

11910

11911

11912

11913

11914

11915

11916

11917

11919

11920

11921

11922

11926

11929

11931

11934

11935

11936

11937

11938

11939

11940

11941

11942

11943

11944

11945

11946

11947

11948

11949

11950

11951

11953

11954

11955

11956

11957

11958

11959

11960

11961

11962

11963

11964

11965

11966

11967

11968

11969

11970

11971

11972

11973

11974

11975

11977

11978

11979

11980

11981

11982

11983

11984

11985

11986

11987

11988

11990

11991

11992

11993

11994

11995

11996

11999

12000

12001

12002

12004

12005

12006

12007

12009

12018

12032

12043

12058

12019

12010

12033

12044

12059

12012

12026

12034

12047

12061

12013

12037

12048

12062

12027

12014

12038

12049

12065

12028

12015

12039

12050

12066

12029

12016

12040

12055

12067

12030

12017

12041

12056

12031

12057

12068

Adiponitrile	: 6777	Aluminum lactate	: 340	2-Aminoglutaric acid	: 6241
Adol	: 4900	Amanozine	: 11830	Aminoglycol	: 9987
Adonitol	: 11343	Amantadine	: 11859	Aminoguanidine	: 7147
Adrenalone	: 5804	Amediol	: 9987	2-Aminohexanoic acid	: 8214
Adrenochrome	: 7301	Amethocaine monohydrochloride	: 2531	p-Aminohippuric acid	: 6252
Adrenosterone	: 353	Amidazophen	: 10851	1-Amino-2-hydroxybenzene	: 9201
Aflatoxin B2	: 4757	N-Amidinoglycine	: 6253	1-Amino-3-hydroxybenzene	: 9202
Aflatoxin G1	: 6184	Amidocyanogen	: 4155	4-Amino-1-hydroxybenzene	: 9203
Agaricic acid	: 8102	Amido-G-Acid	: 7870	(S)-2-Amino-4-hydroxybutanoic acid	: 7134
Agaricin	: 8102	Amikapron	: 4333	[R-(R*,S*)]-2-Amino-3-	: 11813
Agaritine	: 6247	Aminacrine	: 313	hydroxybutanoic acid	
Agifutol S	: 6264	Aminate base	: 7147	(S)-2-Amino-3-hydroxypropanoic	: 11357
Agroxone	: 141	Aminetrimethylboron	: 3120	acid	
Ahistan	: 9502	Aminic acid	: 6002	2-Amino-6-hydroxypurine	: 10768
Ajmaline	: 321	Aminitrozole	: 85	2-Aminohypoxanthine	: 10768
Akalbir	: 2842	2-(Aminoacetamido)acetic acid	: 6265	(RS)-1-Aminoindan	: 7243
Alanine, 3-[(carboxymethyl)thio]-, L-	: 5017	p-Aminoacetanilide	: 29	l-α-Aminoindole-3-propionic acid	: 11922
Alant camphor	: 8059	2-Aminoacetic acid	: 6250	2-Aminoisobutanol	: 10341
Alantolactone	: 8059	Aminoacetic acid	: 6250	α-Aminoisobutyric acid	: 331
Alar	: 3334	Aminoacetic acid, N,N-	: 6258	5-Aminoisoquinoline	: 7372
Albutoin	: 7229	(dicarboethoxy), ethyl ester		2-Aminoisovaleric acid	: 12041
Alcanfor	: 2961	Aminoacetic acid hydrazide	: 6266	(R)-2-Amino-3-mercaptopropanoic	: 5014
Aldocortene	: 9746	Aminoacetic acid, N-succinyl, ethyl	: 11145	acid	
Aldocortin	: 9746	ester		Aminomethanamidine	: 6278
Aldol	: 3184	Aminoacetonitrile	: 272	1-Amino-3-methylbutane	: 3220
Aldosterone	: 9746	4-Amino-2-anilino-1,3,5-triazine	: 11830	2-Amino-2-methylbutyric acid	: 7401
Aldrin	: 5177	4-Aminoanthranilic acid	: 2553	trans-4-(Aminomethyl)-1-	: 4333
Aldrin epoxide	: 5178	1-Aminoanthraquinone	: 380	cyclohexanecarboxylic acid	
Alizarin	: 391	2-Aminoanthraquinone	: 381	2-Amino-3-methylpentanoic acid	: 7370
Alizarine Blue	: 8062	p-Aminoazobenzene	: 837	2-Amino-4-methylpentanoic acid	: 7413
Alizarine Bordeaux B	: 413	1,2-Aminoazophenylene	: 2871	4-Amino-4-methyl-2-pentanone	: 8959
Alizarine Brown R	: 415	(S)-α-Aminobenzenepropanoic acid	: 9510	3-Amino-5-methylphenol	: 9212
Alizarine Maroon	: 382	3-(p-Aminobenzenesulfamido)-6-	: 2197	4-(Aminomethyl)piperidine	: 9673
Alizarine Orange	: 399	methoxypyridazine		2-Amino-2-methyl-1,3-propanediol	: 9987
Alizarine Yellow C	: 5873	p-Aminobenzenesulfonamide	: 2187	7-Amino-2-methylquinoline	: 11211
Alizarine Yellow R	: 2648	4-(p-Aminobenzenesulfonamido)-	: 2194	8-Amino-6-methylquinoline	: 11212
Alkannin	: 7854	2,6-dimethoxypyrimidine		2-Amino-4-methylthiazole	: 11665
Allacil	: 11000	2-(4-Aminobenzenesulfonamido)-	: 2196	2-Amino-4-methylthiobutanoic acid	: 7648
Allantoin	: 12009	4,6-dimethylpyrimidine		3-(Aminomethyl)-3,5,5-	: 4502
Allene	: 9772	2-(4-Aminobenzenesulfonamido)-4-	: 2199	trimethylcyclohexanol	
Allenolic acid	: 7953	methylpyrimidine		Aminometradine	: 11060
Allergosil	: 5587	2-(p-Aminobenzenesulfonamido)-4-	: 2200	2-Aminonaphthalene-6,8-disulfonic	: 7870
Allicin	: 10595	methylthiazole		acid	
Allidochlor	: 44	N'-(4-Aminobenzenesulfonyl)-N-	: 2192	3-Amino-2-naphthoic acid	: 7768
Alliin	: 333	butylurea		1-Amino-8-naphthol-3,6-disulfonic	: 7874
Allobarbital	: 11091	3-Aminobenzoic acid, 6-chloro	: 2490	acid	
Allocholesterol	: 4111	4-Aminobenzoic acid, isobutyl ester	: 2505	1-Amino-2-naphthol-4-sulfonic acid	: 7966
Allocyclogeranic acid	: 4307	2-Aminobenzothiazole	: 2859	6-Aminonicotinic acid	: 10897
Allomaleic acid	: 3764	6-Aminobenzothiazole	: 2860	2-Amino-5-nitrothiazole	: 11666
Allomethadione	: 8586	N-(p-Aminobenzoyl)-L-(+)-glutamic	: 6242	(S)-2-Aminopentanedioic acid	: 6241
Allopregnane	: 9740	acid		(S)-2-Aminopentanoic acid	: 8218
Allopurinol	: 10852	N-(p-Aminobenzoyl)glycine	: 6252	o-Aminophenol	: 9201
Alloxan	: 11073	p-Aminobiphenyl	: 3011	m-Aminophenol	: 9202
Alloxantin	: 3106	1-Amino-5-bromonaphthalene	: 7711	p-Aminophenol	: 9203
Allyl acetate	: 235	4-Amino-3-bromoquinoline	: 11207	p-Aminophenylacetic acid	: 886
Allylacetic acid	: 9073	1-Aminobutane	: 3209	3-Amino-1-phenylbutane	: 2119
Allyl alcohol	: 10697	γ-Aminobutyric acid	: 3464	4-Amino-3-phenylbutyric acid	: 2128
Allyl alcohol, 3-chloro	: 10701	ε-Aminocaproic acid	: 6873	5-(4-Aminophenylsulfonamido)-3,4-	: 2195
Allyl alcohol oxide	: 8618	(L)-2-Amino-3-	: 5017	dimethylisoxazole	
Allyl aldehyde	: 10509	(carboxymethylthio)propionic acid		2-Aminophenylsulfonic acid	: 2215
Allylamine	: 10522	5-Amino-2-chlorobenzoic acid	: 2490	3-Aminophenylsulfonic acid	: 2216
Allyl bromide	: 10531	2-Amino-5-chlorobenzoxazole	: 2876	4-Aminophenylsulfonic acid	: 2217
Allylcarbamide	: 12028	4-Amino-4'-chlorodiphenyl	: 3012	p-Aminophenylsulfonylthiourea	: 2189
Allylcatechol methylene ether	: 2429	6-Aminochrysene	: 4113	2-Amino-3-picoline	: 10866
Allyl chloride	: 10536	Aminocyclobutane	: 4169	2-Amino-4-picoline	: 10867
Allyl cinnamate	: 10692	Aminocyclooctane	: 4735	2-Amino-6-picoline	: 10868
Allyl crotonate	: 3856	1-Aminocyclopentanecarboxylic acid	: 4782	3-Amino-4-picoline	: 10870
Allyl cyanide	: 3803	1-Amino-1-deoxysorbitol	: 6207	1-Aminopropane-1,3-dicarboxylic	: 6241
5-Allyl-4,7-dimethoxy-1,3-	: 2423	[(4-Amino-3,5-dichloro-6-fluoro-2-	: 5982	acid	
benzodioxole		pyridyl)oxy]acetic acid		3-Aminopropanoic acid	: 323
Allyl ether	: 10592	2-Amino-4,6-dichlorophenol	: 9205	DL-2-Aminopropanoic acid	: 324
Allyl ethyl ether	: 10566	1-Amino-3-diethylamino-2-propanol	: 10339	(S)-2-Aminopropanoic acid	: 325
Allyl fluoride	: 10570	2-Amino-1,9-dihydro-9-β-D-	: 6289	DL-1-Amino-2-propanol	: 10338
Allyl formate	: 6017	ribofuranosyl-6H-purin-6-one		3-Aminopropionitrile	: 10092
Allyl glycidyl ether	: 8626	2-Amino-2,2-dimethylethanol	: 10341	β-Aminopropionitrile	: 10092
Allyl iodide	: 10573	2-Amino-4,5-dimethylpyrimidine	: 11044	4-Aminopropiophenone	: 10413
Allyl isocyanide	: 10574	p-Aminodiphenylamine	: 1283	p-Aminopropiophenone	: 10413
Allyl isothiocyanate	: 10575	2-Amino-6,8-disulfonaphthalene	: 7870	α-(α-Aminopropyl)benzyl alcohol	: 1840
4-Allyl-1,2-(methylenedioxy)benzene	: 2429	Aminoethanoic acid	: 6250	N-(3-Aminopropyl)diethanolamine	: 5679
m-Allylpyrocatechin methylene ether	: 2429	1-Aminoethanol	: 5674	Aminopropylin	: 9795
Allyl sulfide	: 10599	2-Aminoethanol	: 5675	4-(3-Aminopropyl)morpholine	: 7701
Allyltrichlorosilane	: 11475	N-(Aminoethyl)ethanolamine	: 5676	Aminopropylon	: 9795
Allylurea	: 12028	1-(2-Aminoethyl)piperazine	: 9615	6-Aminopurine	: 10751
Aloe-emodol	: 397	2-Amino-2-ethyl-1,3-propanediol	: 9985	4-Aminopyrazole	: 10812
Aloperidin	: 3651	2-Aminoethylsulfonic acid	: 5623	2-Aminopyridine	: 10857
Aloxidone	: 8586	Aminoformamidine	: 6278	α-Aminopyridine	: 10857
Alstonidine	: 10772	2-Aminoglutaramic acid	: 6249		

β-Aminopyridine	: 10858
4-Aminopyridine	: 10859
Aminopyrine	: 10851
2-Aminoquinoline	: 11202
4-Aminoquinoline	: 11203
5-Aminoquinoline	: 11204
6-Aminoquinoline	: 11205
8-Aminoquinoline	: 11206
6-Aminoquinoxaline	: 11332
Aminorex	: 8574
4-Amino-1-β-D-ribofuranosyl-2(1H)-pyrimidinone	: 5021
p-Aminosalicylic acid	: 2498
5-Aminosalicylic acid	: 2499
p-Aminosalicylic acid hydrazide	: 2500
α-Aminosuccinamic acid	: 468
L-Aminosuccinic acid	: 471
2-Amino-1,3-thiazole	: 11662
2-Aminothiazole	: 11662
Aminoxafen	: 8574
Aminozide	: 3334
Amiphenazole	: 11671
Amitrole	: 11845
Amixetrine	: 11169
Amizepin	: 5144
Ammonium, acetate	: 106
Ammonium, (3-carbamoyl-3,3-diphenylpropyl) diisopropyl methyl,iodide	: 2120
Ammonium, diethyl(2-hydroxyethyl)methyl-, bromide, xanthene-9-carboxylate	: 5445
Ammonium, dimethyl ethyl 3-hydroxyphenyl,chloride	: 865
Ammonium formate	: 6003
Ammonium, pentamethylenebis trimethyl, bromide	: 8738
Ammonium, phenyl trimethyl,iodide	: 866
Ammonium, tetraethyl,bromide	: 5447
Ammonium, tetramethyl,bromide	: 7456
Ammonium, tetramethyl,chloride	: 7457
Ammonium, tetramethyl,iodide	: 7458
Ammonium, tetrapropyl,iodide	: 9833
Amobarbital	: 11092
Amolanone	: 2463
Amphecloral	: 1583
Amphetamine	: 1578
Ampyrone	: 10845
Amudane	: 11520
Amygdalin	: 954
Amyl acetate	: 227
sec-Amyl acetate (R)	: 8918
Amylacetylene	: 6609
tert-Amyl alcohol	: 3628
Amyl alcohol	: 8915
sec-Amyl alcohol	: 8916
iso-Amylamine	: 3223
Amylamine	: 8700
Amylbenzene	: 2107
d-Amyl bromide	: 3238
tert-Amyl bromide	: 3240
Amyl bromide	: 8710
Amyl butyrate	: 3579
Amyl caproate	: 6907
tert-Amyl carbamate	: 3634
Amyl chloride	: 8719
6-n-Amyl-m-cresol	: 9425
Amylene	: 3801
α-Amylene	: 9004
Amylene dichloride	: 3286
2,4-Amylene glycol	: 8789
Amyl ether	: 8835
Amyl ethyl ketone	: 8472
Amylethylmethylcarbinol	: 8466
Amyl formate	: 6015
Amyl heptanoate	: 6470
Amyl hexanoate	: 6907
Amyl hexyl ketone	: 5326
Amyl iodide	: 8818
Amylisopropylmethylcarbinol	: 8451
tert-Amyl isovalerate	: 3546
Amyl ketone	: 11962
Amyl mercaptan	: 8844
sec-Amylmercaptan	: 8845
Amyl nitrite	: 8090
	: 8096
Amylol	: 8915

4-Amylresorcinol	: 1544
Amyl salicylate	: 2649
Amyl stearate	: 8270
Amyl sulfide	: 8843
Amyltrichlorosilane	: 11472
Amyl valerate	: 8913
β-Amyrin	: 8548
α-Amyrin	: 12038
Anabasine	: 11008
Anagyrine	: 7573
Analeptin	: 1887
Anastil	: 9379
Anatabine	: 3105
Anchovyxanthin, all-trans-	: 4057
Androfluorone	: 355
4-Androstene-3,17-dione	: 352
Androsterone	: 348
Anemonin	: 5185
Anesthesin	: 2497
Anethole	: 1942
Angelic acid	: 3845
Anhalamine	: 7393
Anhalonidine	: 7395
Anhydroecgonine	: 491
Anhydron	: 2851
Aniline	: 633
Aniline, 2-bromo	: 638
Aniline, 3-bromo	: 639
Aniline, 4-bromo	: 640
Aniline, 2-bromo-6-nitro-	: 653
Aniline, 4-butyl	: 655
Aniline, N-butyl-	: 656
Aniline, 2-tert-butyl	: 714
Aniline, 4-tert-butyl	: 715
Aniline, N-tert-butyl-	: 716
Aniline-4-sec-butyl	: 813
Aniline, N-sec-butyl-	: 816
Aniline-3-carboxylic acid	: 2485
Aniline-4-carboxylic acid	: 2486
Aniline, 2-chloro	: 658
Aniline, 3-chloro	: 659
Aniline, 4-chloro	: 660
Aniline, 2-chloro, hydrochloride	: 665
Aniline, 2,4-dibromo-	: 681
Aniline, 2,6-dibromo-	: 682
Aniline, 3,4-dibromo-	: 683
Aniline, 2,3-dichloro-	: 685
Aniline, 2,4-dichloro-	: 686
Aniline, 2,5-dichloro-	: 687
Aniline, 2,6-dichloro-	: 688
Aniline, 3,4-dichloro-	: 689
Aniline, 3,5-dichloro-	: 690
Aniline, 2,6-dichloro-4-ethoxy	: 691
Aniline, 3,4-diethoxy-	: 693
Aniline, N,N-diethyl-4-nitroso	: 699
Aniline, 2,4-difluoro-	: 700
Aniline, 2,4-diiodo-	: 701
Aniline, 2,6-diisopropyl-	: 636
Aniline, 2,6-dimethoxy-	: 703
Aniline, N,N-dimethyl-p-[(α,α,α-trifluoro-m-tolyl)azo]-	: 723
Aniline, 2,3-dinitro-	: 724
Aniline, 2,6-dinitro-	: 726
Aniline, 3,5-dinitro-	: 727
Aniline, hydrochloride	: 756
Aniline, 2-methylthio	: 819
Aniline nitrate	: 821
Aniline, 5-nitro-2-propoxy	: 827
p-Anilinesulfonamide	: 2187
2-Anilinesulfonic acid	: 2215
3-Anilinesulfonic acid	: 2216
Aniline-4-sulfonic acid	: 2217
Aniline, 2,3,4,5-tetramethyl	: 847
Aniline, 2,3,4-trichloro-	: 852
p-Anilinobenzenesulfonic acid	: 2234
o-Anilinobenzoic acid	: 2715
Anilotic acid	: 2647
p-Anisaldehyde	: 585
Anise alcohol	: 1891
p-Anisic acid	: 2666
m-Anisic acid, 2-(methylamino)-, methyl ester	: 2672
o-Anisidine	: 766
m-Anisidine	: 767
p-Anisidine	: 768
m-Anisidine, 4-nitro-	: 776
Anisindione	: 7269

Anisole	: 1915
Anisole, 3,5-dinitro-	: 1924
Anisole, 2,5-dinitro-	: 1926
Anisole, 2,3-dinitro-	: 1922
Anisole, 4-nitro-	: 1937
Anisole, 2,3,4,6-tetrabromo	: 2256
Anisole, 3,4,5-tribromo	: 2310
Anisole, 2,4,5-tribromo	: 2311
Anisole, 2,3,5-tribromo	: 2312
o-(p-Anisoyl)benzoic acid	: 2667
p-Anisoyl chloride	: 2901
p-Anisyl alcohol	: 1891
o-Anisyl benzoate	: 9383
1-(p-Anisyl)ethanol	: 1895
[6]Annulene	: 867
[8]Annulene	: 4742
Anthorine	: 475
Anthracene, 9,10-dibenzyl-	: 367
Anthracene, 9,3-dimethyl	: 378
Anthragallol	: 415
Anthralin	: 430
Anthranil	: 2390
o-Anthranilic acid	: 2484
Anthranilic acid, N-methyl-, ethyl ester	: 2680
Anthranilic acid, 3-nitro-	: 2506
Anthranilic acid, N-(α,α,α-trifluoro-m-tolyl)-	: 2729
Anthranol	: 432
Anthraquinone	: 379
Anthraquinone, 2-amino-1-hydroxy	: 383
Anthraquinone, 2,7-dibromo-	: 390
9,10-Anthraquinone, 1,7-dihydroxy	: 394
Anthrarobin	: 429
Anthrarufin	: 393
1-Anthroic acid	: 368
2-Anthroic acid	: 369
9-Anthroic acid	: 370
Anthrone	: 433
Anthropodeoxycholic acid	: 4097
Anticanitic vitamin	: 2486
Antifebrin	: 86
Antimonate(2-), bis[μ-[2,3-dihydroxybutanedioato(4-)-O1,O2:O3,O4]]di-, dipotassium, trihydrate, stereoisomer	: 11588
Antipyrine	: 10847
Antipyrine, 4-bromo-	: 10846
ANTU	: 11807
Apacil	: 2498
Apacizin	: 2500
Aphylline	: 7572
Apigenin	: 2806
Apigenin 4'-methyl ether	: 2809
Apiole	: 2423
Apoatropine	: 928
Apocodeine	: 5160
Apocynin	: 5824
Apomorphine	: 5155
6aα-Aporphine-2,9-diol, 1,10-dimethoxy-	: 5156
6aα-Aporphine, 1,2,9,10-tetramethoxy-	: 5157
6aα-Aporphin-11-ol, 10-methoxy-1,2-(methylenedioxy)-	: 2470
6aα-Aporphin-11-ol, 1,2,10-trimethoxy-	: 5161
Aprobarbital	: 11098
Arabinose (D)(β)	: 436
Arachic alcohol	: 5358
Arachidic acid	: 5355
Arachidonic acid	: 5361
Aramite	: 11579
Arborine	: 11200
Arbutin	: 6224
Arecaidine	: 10913
Arecoline	: 10914
Argobyl	: 10489
Armstrong's acid	: 7867
Arsanilic acid	: 456
Arsine, difluoromethyl-	: 463
Arsine, oxybis[dimethyl-	: 452
Ascaridole	: 5184
Ascensil	: 10867
Asculetine	: 2804
Asmacoril	: 2840
Aspartame	: 9511

Aspergillic acid	: 10811
Aspidospermine	: 473
Aspirin	: 2480
Atrolactic acid	: 914
Atropic acid	: 927
Atropine	: 902
Aureothin	: 10797
Aurin	: 4278
12-Azabenz[a]anthracene	: 630
Azacycloheptane	: 492
Azacyclohexane	: 9629
Azacyclopentane	: 11143
7-Azadibenz[a,j]anthracene	: 5143
8-Azaguanine	: 11851
7-Aza-7H-benzo[c]fluorene	: 2405
2-Azanaphthalene	: 7373
1-Azanaphthalene	: 11214
6-Azathymine	: 11831
6-Azauridine	: 11832
Azelaic acid	: 8130
Azelaic acid, 3-methyl	: 8135
1H-Azepine-1-carbothioic acid, hexahydro-, S-ethyl ester	: 7661
Aziminobenzene	: 2871
Azindole	: 2376
Azine	: 10872
trans-Azobenzene	: 5129
cis-Azobenzene	: 5130
Azobenzene, 3,3'-bis(dimethylamino)	: 361
3-Azobenzenecarboxylic acid, 4-hydroxy-4'-nitro	: 2648
Azobenzene, 2,2'-diethoxy-	: 5112
Azobenzene, 4-dimethylamino-3'-trifluoromethyl	: 723
Azobenzene, 4-methyl-	: 5139
Azobenzene, 3-methyl	: 5138
Azo, benzene 1'-naphthalene,2'-hydroxy	: 8037
2,2'-Azobis(isobutyronitrile)	: 10093
1,1'-Azobisnaphthalene	: 5127
Azobutane	: 5123
Azodicarbonamide	: 5124
Azodicarboxylic acid diamide	: 5124
Azoformic acid, diamide	: 5124
Azole	: 11112
Azomethane	: 5125
Azomycin	: 7211
1,1'-Azonaphthalene	: 5127
2,2'-Azonaphthalene	: 5128
Azopropane	: 5132
Azopyrin	: 2654
cis-Azoxybenzene	: 5131
Azoxybenzene, 4-bromo-	: 5120
Azoxybenzene, 2,2'-dimethyl (trans)	: 5116
Azoxybenzene, 3,3'-dimethyl (trans)	: 5118
Azoxybenzene, 2,2'-dimethyl(trans)	: 5117
Azoxybenzene, 3,3'-dimethyl(trans)	: 5119
Badische acid	: 7964
Baicalein	: 2844
Bamipine	: 9627
Barbital	: 11087
Barbituric acid	: 11076
Barbituric acid, 5-allyl-5-isobutyl-2-thio, sodium salt	: 11063
Barbituric acid, 5-sec-butyl-5-ethyl-2-thio-	: 11065
Basedol	: 11662
Batyl alcohol	: 10023
BCPE	: 1850
Bebeerine	: 11924
Behenic acid	: 5276
Bemegride	: 9656
Bendazol	: 2385
Benemid	: 2595
Benovocylin	: 5403
Benzalacetone	: 3881
β-Benzalbutyramide	: 3703
Benzal chloride	: 1382
Benzaldehyde, 3-nitro, diacetate	: 7511
Benzalhydrazone	: 563
Benzamide, N-[[(4-chlorophenyl)amino]carbonyl]-2,6-difluoro-	: 5176
Benzamide, 3,5-dichloro-N-(1,1-dimethyl-2-propynyl)-	: 10746
Benzamide, N-(3-tolyl)	: 622

Benzanilide	: 626
1,2-Benzanthracene	: 519
2,3-Benzanthracene	: 7703
1,2-Benzanthracene, 3-methyl	: 522
Benzanthrone	: 6292
Benzathine	: 5500
Benzazide	: 2884
2-Benzazine	: 7373
Benz[b]anthracene	: 7703
Benzcurine iodide	: 5441
Benzenamine, N-butyl-N-ethyl-2,6-dinitro-4-(trifluoromethyl)-	: 509
Benzenamine, N-(2-chloroethyl)-2,6-dinitro-N-propyl-4-(trifluoromethyl)-	: 5964
Benzenamine, N-(cyclopropylmethyl)-2,6-dinitro-N-propyl-4-(trifluoromethyl)-	: 9763
Benzenamine, 4-(1,1-dimethylethyl)-N-(1-methylpropyl)-2,6-dinitro-	: 3893
Benzenamine, 2,6-dinitro-N,N-dipropyl-4-(trifluoromethyl)-	: 11901
Benzenamine, N,N'-1,2-ethanediylbis-	: 5508
Benzenamine, N-ethyl-N-(2-methyl-2-propenyl)-2,6-dinitro-4-(trifluoromethyl)-	: 5409
Benzenamine, N-(1-ethylpropyl)-3,4-dimethyl-2,6-dinitro-	: 8635
Benzenamine, 4-(1-methylethyl)-2,6-dinitro-N,N-dipropyl-	: 7371
Benzenamine, 4-(methylsulfonyl)-2,6-dinitro-N,N-dipropyl-	: 8070
Benzeneacetamide, N,N-dimethyl-α-phenyl-	: 5219
Benzeneacetic acid, 4-bromo-α-(4-bromophenyl)-α-hydroxy-, 1-methylethyl ester	: 3140
Benzeneacetic acid, 4-chloro-α-(4-chlorophenyl)-α-hydroxy-, ethyl ester	: 4082
Benzeneacetic acid, 4-chloro-α-(1-methylethyl)-, cyano(3-phenoxyphenyl)methyl ester	: 5956
Benzeneamine, N-(1-methylpropyl)-2,6-dinitro	: 3894
Benzene, 3-aminobutyl	: 2119
Benzenearsonic acid	: 460
Benzene, 1,3-bis(diethylamino)	: 1285
Benzeneboronic acid	: 3135
Benzene, (1-bromopropyl) (DL)	: 1086
Benzenecarbonyl chloride	: 2886
Benzenecarboxaldehyde	: 525
Benzenecarboxylic acid	: 2475
Benzene, 1(3-chloroallyl)-4-methoxy	: 1231
Benzene, 1-chloro-4-[[(4-chlorophenyl)methyl]thio]-	: 4075
Benzene, 2-chloro-1-(3-ethoxy-4-nitrophenoxy)-4-(trifluoromethyl)-	: 8633
Benzene, 2-chloro-1-(4-nitrophenoxy)-4-(trifluoromethyl)-	: 8082
Benzene, 1-chlorophosphino-4-isopropyl	: 9555
Benzene, (1-decylundecyl)-	: 6297
1,3-Benzenediamine, N3,N3-diethyl-2,4-dinitro-6-(trifluoromethyl)-	: 5181
1,3-Benzenediamine, 2,4-dinitro-N3,N3-dipropyl-6-(trifluoromethyl)-	: 9761
Benzene, 2,3-dibromo-1-chloro	: 1292
Benzene, 3,4-dibromo-1-chloro	: 1293
Benzene, 2,6-dibromo-1-chloro	: 1294
Benzene, 2,5-dibromo-1-chloro	: 1295
Benzene, 1,4-di-tert-butyl	: 990
1,4-Benzenedicarboxylic acid, 2,3,5,6-tetrachloro-, dimethyl ester	: 5180
Benzene, 1,4-dichloro-2,5-dimethoxy-	: 4084
Benzene, 2,5-dichloro-1-ethyl	: 1377
Benzene, 2,4-dichloro-1-(4-nitrophenoxy)-	: 8081
Benzene, 1,3-diisopropyl-4-methyl-	: 1948
Benzene, 1,4-dinitro-2-methoxy	: 1926
1,3-Benzenedisulfonic acid, diamide,4,5-dichloro	: 1559
Benzene, (1-dodecyltridecyl)-	: 8638
Benzene, 1-ethyl-2-isopropyl	: 1688
Benzene, 1-ethyl-3-isopropyl	: 1689
Benzene, 1-ethyl-4-isopropyl	: 1690

Benzene, 4-ethyl-1,2,3-trimethyl	: 1708
Benzene, 6-ethyl-1,2,4-trimethyl	: 1709
Benzene, 5-ethyl-1,2,4-trimethyl	: 1710
Benzene, 3-ethyl-1,2,4-trimethyl	: 1711
Benzene hexachloride	: 4409
Benzene, 1-hydroxy-2-methoxy-4-propenyl,acetate(cis)	: 9394
Benzene, 1-isobutoxy-2-nitro	: 2034
Benzene, mercuriodi-	: 7436
Benzenenitrile	: 2740
Benzene, 2-nitro-1-(4-nitrophenoxy)-4-(trifluoromethyl)-	: 5980
Benzene, 1-nitro-3,4,5-tribromo	: 2315
Benzeneselenic acid	: 2174
Benzenestibonic acid	: 11548
Benzenesulfonamide, 4-(acetylamino)-	: 31
Benzenesulfonamide, 4-(dipropylamino)-3,5-dinitro-	: 8553
Benzenesulfonic acid, amide,4(4-aminophenylsulfonamido)	: 2188
Benzenesulfonic acid, amide,4(benzylamino)	: 2213
Benzenesulfonic acid, chloride	: 2219
Benzenesulfonic acid, p-chlorothio-, S-phenyl ester	: 2236
Benzenesulfonic acid, (2,4-dichlorophenyl) ester	: 9289
Benzenesulfonic acid, fluoride	: 2251
Benzenesulfonic acid, guanidine,4-amino	: 2191
Benzenesulfonic acid, thiolo, 4-chloro, S-phenyl ester	: 2236
Benzenesulfonic acid, thiolo,2-methyl,phenyl ester	: 11818
Benzene, 1,2,4,5-tetraisopropyl-	: 2276
Benzene, 1,2,3-triamino	: 2305
Benzene, 1,3,5-tribenzoyl-	: 7587
Benzene, 1,3,5-trichloro-2,4,6-trihydroxy	: 2371
Benzene, 1,2,4-triisopropyl-	: 2372
Benzene, 1,2,3-tris(2(diethylamino)ethoxy),tri(ethyl iodide)	: 5441
Benzene, uneicosyl	: 1741
Benzestrol	: 9303
Benzhydramine	: 5422
Benzhydrol, α-(1-naphthyl)	: 7910
Benzhydrylamine	: 1826
Benzhydryl bromide	: 1058
2-(Benzhydryloxy)-N,N-dimethylethylamine	: 5422
p-Benzidine	: 3031
Benzidine	: 3031
Benzidine, 2,2'-dichloro-	: 3032
Benzil	: 5584
Benzilic acid	: 918
Benzimidazole, 1,5-dimethyl-	: 2378
Benzimidazole, 1-trifluoromethyl-	: 2381
1H-Benzimidazole, 2-(4-thiazolyl)	: 11656
2-Benzimidazolethiol	: 2386
Benzo[a]chrysene	: 9601
1H-Benzo[a]fluorene, veratraman-3,23-diol deriv.	: 12044
2H,8H-Benzo[1,2-b:3,4-b']dipyran, butanoic acid deriv.	: 3540
Benzo[b]fluoranthene	: 631
Benzo[b]naphthacene	: 8636
Benzo[b]quinoline	: 314
3,4-Benzocarbazole	: 2405
Benzo[def]phenanthrene	: 10853
1,2-Benzodiazine	: 4136
2,3-Benzodiazine	: 9599
1,3-Benzodiazine	: 11195
1,4-Benzodiazine	: 11333
1,3-Benzodiazole	: 2376
Benzodihydrofuran	: 2442
1,3-Benzodioxole, 5-[[2-(2-butoxyethoxy)ethoxy]methyl]-6-propyl-	: 9727
1,3-Benzodioxol-4-ol, 2,2-dimethyl-, methylcarbamate	: 514
Benzo[d,e,f]chrysene	: 2395
Benzo[7,8]fluoreno[2,1-b]quinolizine, cevane-3,4,12,14,16,17,20-heptol deriv.	: 4067
Benzofuran, 2-hydroxymethyl	: 2455
Benzofuran, 7-isopropyl-4-methyl	: 2460

2,3,1',8'-Binaphthylene	2739
Bineopentyl	6849
Bineophyl	2281
Biochanin B	2830
Biotin	11689
Biphenyl	3008
p-Biphenylacetic acid	3009
p-Biphenylamine	3011
4-Biphenylamine, 4'-methyl-	3015
Biphenyl, 2-amino-6-methyl	3013
Biphenyl, 4-amino-4'-methyl	3015
2,4'-Biphenyldiamine	3029
2,2'-Biphenyldicarboxylic acid	3038
Biphenyl, 2,2'-dimethoxy-	3045
Biphenyl, 3,3'-dimethoxy-	3046
Biphenyl, 2,2'-dimethoxy-5,5'-dimethyl	3048
Biphenyl, 2,5'-dimethoxy-2',5-dimethyl	3049
Biphenyl, 2,3-dimethyl-	3052
Biphenyl, 2,4-dimethyl-	3054
Biphenyl, 3,4-dimethyl-	3059
Biphenyl, 3,5-dimethyl-	3060
4,4'-Biphenyldisulfonic acid	3070
o-Biphenylenemethane	5967
2,2'-Biphenylene oxide	5163
Biphenyl, 3-ethoxy	3071
2-Biphenylol	3088
4-Biphenylol	3090
2-Biphenylol, 3-chloro-	3092
2,3'-Bipiperidyl	3098
Bipropargyl	6716
Bipropionyl	6806
3,3'-Bipyridyl	3102
3,4'-Bipyridyl	3103
4,4'-Bipyridyl	3104
2,3'-Bipyridyl	3100
2,3'-Riquinolyl	3107
5,5'-Biresorcinol	3097
Birodan	2136
Bis(4-acetamidophenyl) sulfone	91
N,N'-Bis(acetoacetyl)-3,3'-dimethylbenzidine	3204
Bisacodyl	9465
Bis(β-amino-β-carboxyethyl) disulfide	5020
Bis(p-aminophenyl) ether	832
Bis(4-aminophenyl) sulfide	850
Bis(3-aminopropyl)amine	9874
1,4-Bis(3-aminopropyl)piperazine	9614
N,N'-Bis(3-aminopropyl)-1,4-tetramethylenediamine	3259
1,3-Bis(bromomethyl)tetramethyl-disiloxane	5229
1,2-Bis(chlorocarbonyl)ethane	3366
Bis(2-chloroethyl) carbonate	5697
Bis(2-chloroethyl) ether	5611
Bis(2-chloroethyl) sulfide	5645
Bis(chloromethyl) ether	7525
Bis(p-chlorophenoxy)methane	1965
1,1-Bis(4-chlorophenyl)ethane	1678
1,1-Bis(4-chlorophenyl)ethanol	1850
3,6-Bis(2-chlorophenyl)-1,2,4,5-tetrazine	4139
1,1-Bis(4-chlorophenyl)-2,2,2-trichloroethanol	1851
1,1-Bis(p-chlorophenyl)-2,2,2-trichloroethanol	1851
Bis(3-chloropropyl) sulfide	10153
Bis(cyclohexyl)carbodiimide	4291
Bis(cyclopentadienyl)iron	5960
1,3-Bis-(dichloromethyl)tetramethyl disiloxane	5231
Bis(2,2-diethoxyethyl) ether	5612
Bis-O,O-diethylphosphoric anhydride	5221
Bis(diethylthiocarbamoyl) sulfide	11714
Bis(dimethylamino)phosphonous anhydride	5220
2,3-Bis(ethoxycarbonyl)furan	6082
Bis-(2-ethoxyethyl) amine	5425
Bis(ethoxymethyl)ether	5614
Bis(2-ethylhexyl) adipate	6781
Bis(2-ethylhexyl) decanedioate	5043
Bis(2-ethylhexyl) phthalate	1334
Bis(2-ethylhexyl) sebacate	5043
2,2-Bis(ethylsulfonyl)propane	9835

Bis(2-furyl)glyoxime	5583
Bis-(3-hydroxy-2-butyl)amine	3624
Bishydroxycoumarin	2839
Bis(hydroxymethyl)acetylene	3911
Bis-(2-methallyl)carbonate	10703
Bis(methoxydimethylsilyl) oxide	5234
Bis(2-methoxyethyl) carbonate	5736
Bis(2-methylbutyl) sulfide	3435
Bis[methylmercapto]methane	7461
Bis(methylthio)methane	7461
1,2'-Bis-naphthyl ether	7928
Bis(m-nitrophenyl) disulfide	5247
Bisphenol A	9413
Bisphenol B	9433
1,2-Bis(phenylamino)ethane	5508
Bis-(1-phenylethyl)ether (DL)	2081
Bis(1,2,2,2-tetrachloroethyl) ether	5615
1,5-Bis(2-tetrahydrofuryl)-3-pentanol	6166
Bis(p-tolyl)mercury	7432
Bis(tributyltin) oxide	5241
Bithionol	9478
Bitoscanate	1432
Δ 1,1'-Biurea	5124
Biuret	7232
Bivinyl	3146
Bixin	5108
Boldenone	343
Boldine	5156
Boldine dimethyl ether	5157
Bolusanthin	2801
Borane, triisopentyl-	3122
Borane-trimethylamine (1:1)	3132
Borine, dimethoxy-	3117
Borine, triisopentyl-	3122
Bornane, 2-chloro-	2919
10-Bornanesulfonic acid, 2-oxo-, (1S,4R)-(+)-	2929
3-Bornanol, acetate, (1S,3R,4S)-(-)-	2947
(+)-2-Bornanone	2961
2-Bornanone, 3-acetyl-17O-	2950
Borneol, acetate	2948
Bornyl chloride	2920
Bornyl formate (d)	2949
Bornyl isovalerate (d)	3565
d-Bornyl isovalerate	3565
Boron fluoride-diethyl ether compound	3137
Boron trifluoride compd. with methyl ether	3138
Boron trifluoride etherate	3137
Boron trimethoxide	3126
Brassidic acid	5282
Brilliant green	5444
Bristurin	2858
Bromacetocarbamide	3195
ω-Bromacetophenone	5782
Bromal	20
Bromcarbamide	3196
Bromcholitin	5157
Bromcresol green	9367
Bromcresol purple	9364
Bromdiethylacetylurea	3195
Bromindione	7265
Bromine cyanide	4165
Bromisovalum	3196
5-Bromoacenaphthene	4
N-Bromoacetamide	36
p-Bromoacetanilide	37
Bromoacetic acid	111
Bromoacetic acid, propyl ester	120
Bromoacetone	10416
p-Bromoacetophenone	5781
Bromoacetyl bromide	292
Bromoacetylene	5937
5-Bromoacetylsalicylic acid	2481
o-Bromoaniline	638
m-Bromoaniline	639
p-Bromoaniline	640
5-Bromoanthranilic acid	2487
4-Bromoantipyrine	10846
Bromoaspirin	2481
Bromobenzene	998
p-Bromobenzenesulfonyl chloride	2244
p-Bromobenzoic acid	2520
2-Bromobenzophenone	7600
p-Bromobenzyl bromide	1000
p-Bromobenzyl chloride	1008

α-Bromobenzyl cyanide	947
4-Bromo-1,2-butadiene	3147
1-Bromo-1,3-butadiene	3148
2-Bromo-1-butene	3709
3-Bromo-1-butene	3710
(±)-2-Bromobutyric acid	3468
DL-α-Bromobutyric acid	3468
(RS)-α-Bromobutyric acid	3468
Bromochloroacetic acid	112
Bromochloroacetic acid, ethyl ester	113
1-Bromo-3-chloroacetone	10417
2-Bromo-1-chlorobutane	3234
Bromochlorodifluoromethane	7464
Bromochlorodinitromethane	7465
1-Bromo-1-chloroethane	5455
Bromochlorofluoromethane	7466
Bromochloromethane	7463
1-Bromo-4-chloronaphthalene	7741
4-Bromo-2-chlorophenol	9232
1-Bromo-2-chloropropane	9838
1-Bromo-3-chloropropane	9839
2-Bromo-1-chloropropane	9840
2-Bromo-2-chloropropane	9841
2-Bromo-5-chlorothiophene	11728
4-Bromo-3-chloro-3,4,4-trifluoro-1-butene	3715
1-Bromo-2-chloro-1,1,2-trifluoroethane	5457
6-Bromo-m-cresol	9237
Bromocycloheptane	4222
2-Bromo-1-cyclohexanol	4503
Bromocyclopentane	4777
2-Bromo-1-decene	5086
1-Bromo-2-decene	5087
2-Bromodibenzofuran	5164
3-Bromodibenzofuran	5165
Bromodichlorofluoromethane	7468
Bromodichloromethane	7467
2-Bromo-2,2-diethylacetamide	3198
1-Bromo-1,2-difluoroethylene	5883
Bromodifluoromethane	7469
2-Bromo-2,3-dimethylbutane	3236
1-Bromo-1,1-dimethylbutane	8717
4-Bromo-1,3-dimethylpyrazole	10816
1-Bromo-2,3-dinitrobenzene	1024
Bromodiphenylmethane	1058
2-Bromoethanesulfonyl chloride	5628
2-Bromoethanol	5685
2-Bromoethyl ethyl ether	5461
2-Bromo-2-ethylisovaleramide	3199
2-Bromofluorene	5968
Bromofluoroacetic acid	116
1-Bromo-4-fluorobutane	3237
1-Bromo-10-fluorodecane	5035
1-Bromo-12-fluorododecane	5294
1-Bromo-6-fluorohexane	6735
Bromofluoromethane	7470
1-Bromo-5-fluoropentane	8713
1-Bromo-11-fluoroundecane	11939
Bromoform	7548
2-Bromofumaric acid	3770
3-Bromofuran	6034
1-Bromoheptadecane	6307
2-Bromoheptane	6357
4-Bromoheptane	6358
2-Bromoheptanoic acid	6457
1-Bromo-1-heptyne	6612
1-Bromo-2-heptyne	6613
2-Bromohexane	6733
3-Bromohexane	6734
1-Bromo-3-iodobenzene	1047
m-Bromoiodobenzene	1047
1-Bromo-2-iodoethane	5463
Bromoiodomethane	7471
α-Bromoisobutyric acid	10209
4-Bromoisoquinoline	7374
β-Bromoisovaleric acid	3472
2-Bromoisovalerylurea	3196
Bromomethane	7462
Bromomethoxymethane	7472
2-Bromo-p-methylacetophenone	5779
2-Bromo-5-methylaniline	648
(+)-1-Bromo-2-methylbutane	3238
(S)-1-Bromo-2-methylbutane	3238
2-Bromo-3-methyl-1-butene	3716
(Bromomethyl)chlorodimethylsilane	11365
(Bromomethyl)dimethylchlorosilane	11365

11-Butyldocosane	: 5272
7-Butyldocosane	: 5270
9-Butyldocosane	: 5271
Butyl enanthate	: 6459
1-Butylene	: 3704
α-Butylene chlorohydrin	: 3605
β-Butylene chlorohydrin	: 3606
1,4-Butylenediamine	: 3258
α-Butylene dibromide	: 3265
1,3-Butylene glycol	: 3345
1,4-Butylene glycol	: 3346
Butylene glycol	: 3346
2,4-D Butyl ester	: 171
Butyl ethanoate	: 121
sec-Butylethene	: 9053
Butyl ether	: 3415
sec-Butyl ether, (±)-	: 3416
Butylethylamine	: 3218
sec-Butyl ethyl amine (DL)	: 3219
Butyl ethyl ether	: 3372
sec-Butyl ethyl ether	: 3373
(RS)-S-sec-Butyl O-ethyl 2-oxo-1,3-thiazolidin-3-ylphosphonothioate	: 6021
Butyl ethyl sulfide	: 3376
Butyl fluoride	: 3378
sec-Butyl fluoride	: 3379
Butyl formate	: 6004
2-tert-Butylfuran	: 6090
Butyl glycidyl ether	: 8602
Butyl heptanoate	: 6459
1-Butylheptyl alcohol	: 11953
2-Butyl-2-hexanol	: 8171
Butyl hexyl ketone	: 11961
tert-Butyl hydroperoxide	: 7178
Butyl 4-hydroxybenzoate	: 2615
Butyl 2-hydroxypropionate	: 10268
tert-Butyl hypochlorite	: 7183
Butylidene chloride	: 3276
2-Butylidenecyclohexanone	: 4576
Butyl iodide	: 3381
tert-Butyl iodide	: 10058
Butyl isobutyl ether	: 3401
tert-Butyl isocyanate	: 10062
Butyl isocyanide	: 3387
Butyl isopropyl ether	: 3399
Butyl isopropyl ketone	: 6522
tert-Butyl isopropyl ketone	: 8989
Butylisopropylmethylcarbinol	: 6480
tert-Butyl isothiocyanate	: 10066
sec-Butyl isothiocyanate (DL)	: 3389
Butyl isovalerate	: 3543
Butyl ketone	: 8175
n-Butylmalonic acid	: 9929
Butyl mercaptan	: 3437
tert-Butyl mercaptan	: 10159
Butyl methacrylate	: 10650
Butylmethylamine	: 3221
Butyl m-methylbenzoate	: 2685
Butyl methyl ether	: 3391
sec-Butyl methyl ether (d)	: 3392
tert-Butyl methyl ether	: 10072
sec-Butyl methyl ether (DL)	: 3393
tert-Butyl methyl ketone	: 3657
Butyl methyl ketone	: 6959
1-Butyl-3-(p-methylphenylsulfonyl)urea	: 2207
Butyl methyl sulfide	: 3402
tert-Butyl methyl sulfide	: 10083
2-Butylnaphthalene	: 7747
Butyl nitrate	: 8072
Butyl nitrite	: 8083
tert-Butyl nitrite	: 8085
5-Butylnonane	: 8122
Butyl nonanoate	: 8147
5-Butyl-4-nonene	: 8184
Butyl oxide	: 3415
Butylparaben	: 2615
Butyl parasept	: 2615
p-tert-Butylphenol	: 9323
o-sec-Butylphenol	: 9430
5-Butylpicolinic acid	: 10898
3-Butylpiperidine	: 9636
4-n-Butylpiperidine	: 9637
Butyl propionate	: 10214
Butyl propyl ether	: 3421
tert-Butyl propyl ketone	: 6965

Butyl propyl ketone	: 8473
2-Butylpyridine	: 10878
2-tert-Butylpyridine	: 10942
3-tert-Butylpyridine	: 10943
5-Butyl-2-pyridinecarboxylic acid	: 10898
2-Butylpyrrolidine	: 11147
2-Butylquinoline	: 11222
Butyl stearate	: 8251
1-Butyl-3-sulfanilylurea	: 2192
Butyl sulfide, 2,2'-dimethyl	: 3435
Butyl sulfite	: 11580
tert-Butyl tetrasulfide	: 11649
Butyl thiocyanate	: 11699
tert-Butyl thiocyanate	: 11701
Butylthiol	: 3437
2-Butylthiophene	: 11731
Butyl m-toluate	: 2685
p-tert-Butyltoluene	: 1466
1-Butyl-s-triazole	: 11847
Butyltrichlorosilane	: 11370
Butyl trifluoroacetate	: 265
2-Butyl-1,1,3-trimethylcyclohexane	: 4327
Butyltrimethylsilane	: 11371
Butyl vinyl ether	: 3369
Butyl vinyl ketone	: 6583
2-Butyne-1-al	: 3895
1-Butyne-3,4-diol	: 3912
Butyraldehyde	: 3178
Butyraldehyde, 3-chloro-, diethyl acetal	: 3243
Butyraldehyde, 3-chloro-, dimethyl acetal	: 3244
Butyraldehyde, 3-methyl, oxime	: 3187
Butyraldehyde, oxime	: 3188
Butyraldehyde, phenylhydrazone	: 3189
Butyramide	: 3194
p-Butyramidophenol	: 3205
Butyrhydrazide	: 3529
Butyric acid	: 3460
Butyric acid, 4-acetyl, ethyl ester	: 6906
Butyric acid, 3-bromo-3-methyl-	: 3472
Butyric acid, sec-butyl ester (d)	: 3564
Butyric acid chloride	: 3681
Butyric acid, 2-chloro-2-methyl-	: 3486
Butyric acid, 2-chloro-3-methyl, ethyl ester	: 3490
Butyric acid, 2-chloro-2-methyl, ethyl ester	: 3491
Butyric acid, 2-chloro-2-methyl-3-oxo, ethyl ester	: 3492
Butyric acid, 2-cyano-2-ethyl-, ethyl ester	: 3495
Butyric acid, 2,2-dichloro-	: 3500
Butyric acid, 3,4-dichloro, butyl ester	: 3501
Butyric acid, 2,3-dimethyl	: 3507
Butyric acid, heptafluoro-, isopropyl ester	: 3526
Butyric acid, 2-hydroxy, ethyl ester (DL)	: 3532
Butyric acid, 2-hydroxy, isobutyl ester	: 3536
Butyric acid, 3-methylamino-, ethyl ester	: 3541
Butyric acid, (2-methylbutyl) ester	: 3545
Butyric acid, 3-methyl, isopropyl ester	: 3553
Butyric acid (1-methylpentyl) ester (d)	: 3559
Butyric acid, 4-phenoxy-, ethyl ester	: 3581
Butyric acid, phenyl ester	: 3584
Butyric acid, 3-thioxo, ethyl ester	: 3589
Butyric acid, 2,2,3-trichloro-	: 3590
Butyric acid, 2,2,3-trichloro, ethyl ester	: 3591
Butyric aldehyde	: 3178
Butyric anhydride	: 3467
Butyroin	: 8474
γ-Butyrolactone	: 6142
1-Butyronaphthone	: 3674
Butyronitrile	: 3403
Butyrophenone, 4-[4-(p-chlorophenyl)-4-hydroxypiperidino]-4'-fluoro-	: 3651
Butyrophenone, 2-hydroxy-	: 3666
Butyrophenone, 4-methyl	: 3672
p-(Butyrylamino)phenol	: 3205
N-Butyryl-p-aminophenol	: 3205

Butyryl bromide	: 3680
Butyryl chloride	: 3681
Butyryl chloride, 3-chloro-	: 3684
Butyryl chloride, 2-chloro-3-methyl	: 3686
Butyryl chloride, 4-chloro-3-oxo	: 3687
Butyryl chloride, 2,3-dimethyl	: 3689
Butyryl chloride, 3-methyl	: 3693
Butyryl chloride, 2-methyl (DL)	: 3694
Butyryl oxide	: 3467
Butyryl triglyceride	: 3586
C.I. Direct blue 53	: 7877
C.I. Solvent Orange 7	: 8028
C.I. Solvent Orange 35	: 384
Cacodyl	: 5110
Cacodylic acid	: 451
Cacodyl oxide	: 452
Cacotheline	: 11352
Cadalene	: 7830
Cadalene, 3-methyl	: 7905
Cadalene, 8-methyl	: 8015
Cadalene, 5-methyl	: 8014
Cadaverine	: 8735
γ-Cadinene	: 7936
Caffeic acid	: 10632
Caffeine	: 10760
Calcyan	: 3930
Calusterone	: 357
Campesterol	: 5386
Camphane-3-carboxylic acid	: 3131
d-Camphocarboxylic acid	: 2917
Camphocean	: 4855
Campholic acid	: 4786
α-Campholytic acid	: 4913
(+)-Camphor	: 2961
Camphor, 3-acetyl	: 2950
d-Camphor, 10-bromo-	: 2951
Camphor, 3,3-dibromo(d)	: 2953
Camphoric acid, dimethyl ester, (+)-	: 4795
(-)-Camphor oxime	: 2982
D-Camphorsulfonic acid	: 2929
α-Camphylamine	: 2927
Canavanine	: 7135
Cannabidiol	: 1537
Cannabinol	: 5154
Canrenone	: 9736
Cantharidin	: 5371
Capraldehyde	: 5030
Capramide	: 5031
Capric acid	: 5059
Capric acid, 10-fluoro	: 5064
Capric alcohol	: 5070
Caprinitrile	: 5057
Caprinoyl chloride	: 5078
Caproaldehyde	: 6718
Caproaldehyde, 3-methyl	: 6720
Caproic acid	: 6870
Caproic acid, 5-bromo,ethyl ester	: 6878
Caproic acid, 6-cyclohexyl	: 4412
Caproic acid, 2,2-dimethyl-3-hydroxy, β-lactone	: 6892
Caproic acid, 6-fluoro	: 6888
Caproic acid, 2-methyl (DL)	: 6893
Caprolactam	: 493
Caprolactone	: 8591
Caproleic acid	: 5091
Capronitrile	: 6840
Caproyl alcohol	: 6911
Caproyl chloride	: 6986
Caproyl chloride, 3-methyl	: 6988
Caproyl chloride, 4-methyl	: 6989
Caprylene	: 8491
Caprylic acid	: 8422
Caprylic acid, 8-fluoro	: 8429
Caprylic acid, 3-methyl	: 8433
Caprylic acid, 2-methyl-3-oxo, ethyl ester	: 8436
Caprylic alcohol	: 8443
Caprylic aldehyde	: 8346
Caprylnitrile	: 8411
Capsaicin	: 8181
Capsanthin	: 4063
Captan	: 7368
Carbamamidine	: 6278
Carbamazepine	: 5144
Carbamic acid, 1H-benzimidazol-2-yl-, methyl ester	: 3990

Carbamic acid, [1-[(butylamino)carbonyl]-1H-benzimidazol-2-yl]-, methyl ester	: 515
Carbamic acid, sec-butyl, ethyl ester	: 3960
Carbamic acid, butyl-, 2-(hydroxymethyl)-2-methylpentyl ester, carbamate	: 3933
Carbamic acid, N-sec-butyl, methyl ester	: 3962
Carbamic acid, (3-chlorophenyl)-, 4-chloro-2-butynyl ester	: 510
Carbamic acid, N,N-diethyl	: 3941
Carbamic acid, dimethyl-, 2-(dimethylamino)-5,6-dimethyl-4-pyrimidinyl ester	: 9728
Carbamic acid, dimethyldithio-, manganese(2+) salt	: 7421
Carbamic acid, N-ethyl, butyl ester	: 3947
Carbamic acid, N-ethyl-N-nitro, butyl ester	: 3951
Carbamic acid, N-ethyl-N-nitro, ethyl ester	: 3950
Carbamic acid, isobutyl ester	: 3959
Carbamic acid, isopentyl-, ethyl ester	: 3952
Carbamic acid, isopropyl ester	: 3955
Carbamic acid, N-isopropyl-N-nitro, ethyl ester	: 3957
Carbamic acid, (3-methylphenyl)-, 3-[(methoxycarbonyl)amino]phenyl ester	: 9198
Carbamic acid, [3-[[(phenylamino)-carbonyl]oxy]phenyl]-, ethyl ester	: 5104
Carbamic acid, [1,2-phenylenebis-(iminocarbonothioyl)]bis-, dimethyl ester	: 11723
Carbamic acid, 3-phenylpropyl ester	: 2149
Carbamic acid, 1-phenyl-2-propynyl ester	: 1877
Carbamic acid, N-propyl-N-nitro, ethyl ester	: 3964
Carbamide	: 11992
Carbamimidoselenoic acid	: 11354
Carbamodithioic acid, diethyl-, 2-chloro-2-propenyl ester	: 4065
Carbamodithioic acid, dimethyl-, sodium salt	: 11514
Carbamonitrile	: 4155
Carbamothioic acid, bis(2-methylpropyl)-, S-ethyl ester	: 11586
Carbamothioic acid, bis(1-methylethyl)-, S-(2,3,3-trichloro-2-propenyl) ester	: 11823
Carbamothioic acid, bis(1-methylethyl)-, S-(2,3-dichloro-2-propenyl) ester	: 5106
Carbamothioic acid, cyclohexyl-ethyl-, S-ethyl ester	: 4168
Carbamothioic acid, diethyl-, S-[(4-chlorophenyl)methyl] ester	: 11697
Carbamothioic acid, diethyl-, S-ethyl ester	: 5930
Carbamothioic acid, dipropyl-, S-ethyl ester	: 5372
Carbamothioic acid, dipropyl-, S-propyl ester	: 12047
Carbamoyl chloride, N,N-diethyl	: 3972
Carbamoylcholine chloride	: 5440
5-Carbamoyl-5H-dibenz[b,f]azepine	: 5144
Carbamyl chloride	: 3971
Carbanilaldehyde	: 5999
Carbanilide	: 12010
Carbarsone	: 455
Carbaryl	: 8032
Carbazide	: 4009
(N-Carbazolyl)acetic acid	: 3982
Carbazone, 1,5-diphenyl	: 5122
Carbic anhydride	: 7580
Carbimazole	: 7194
Carbitol	: 5720
Carbitol acetate	: 5721
Carbobenzoxy chloride	: 4027
Carbocysteine	: 5017
Carbodiimide, dicyclohexyl-	: 4291
Carbohydrazide	: 4009
Carbomethene	: 5927

o-Carbomethoxyaniline	: 2503
2-Carbomethoxyphenol	: 2634
Carbon bisulfide	: 3993
Carbon dibromide dichloride	: 7495
Carbon diselenide	: 4036
Carbonic acid, bis(2-chloroethyl) ester	: 5697
Carbonic acid, bis(2-ethoxyethyl) ester	: 5719
Carbonic acid, bis(2-methoxyethyl) ester	: 5736
Carbonic acid, bis(2-methoxyphenyl) ester	: 9384
Carbonic acid, bis(2-methylallyl) ester	: 10703
Carbonic acid, dithio-, cyclic S,S-(6-methyl-2,3-quinoxalinediyl) ester	: 4072
Carbonic acid, monochloride,monoethyl ester	: 4018
Carbonic anhydride	: 3992
Carbon oxide	: 4011
Carbon oxysulfide	: 4035
Carbon suboxide	: 9773
Carbon tetrabromide	: 7539
Carbon tetrachloride	: 7540
Carbon tetrafluoride	: 7541
Carbon tetraiodide	: 7542
Carbonyl bromide	: 4006
Carbonyl chloride	: 4007
Carbonyl chloride fluoride	: 4005
Carbonyl cyanide	: 9915
N,N'-Carbonyldiimidazole	: 7193
Carbonyl fluoride	: 4008
Carbonyl sulfide	: 4035
Carbostyril	: 11325
3-Carboxyaniline	: 2485
4-Carboxyaniline	: 2486
9-Carboxyanthracene	: 370
4-Carboxybenzenesulfonamide	: 2508
3-Carboxycoumarin	: 2775
4-Carboxy-1,3-dimethylbenzene	: 2580
2-Carboxyethylamine	: 323
2-Carboxyfuran	: 6050
4-Carboxymethylbiphenyl	: 3009
S-(Carboxymethyl)-L-cysteine	: 5017
3-[(Carboxymethyl)thio]-L-alanine	: 5017
m-Carboxynitrobenzene	: 2704
p-Carboxynitrobenzene	: 2705
4-Carboxyphenol	: 2613
4'-Carboxyphenylmethane-sulfonanilide	: 2721
4-Carboxyphthalic acid	: 2319
3-Carboxypropylamine	: 3464
8-Carboxyquinoline	: 11229
2-Carboxythiophene	: 11737
Carbromal	: 3195
Carbromide	: 3198
Carbutamide	: 2192
Carbyloxime	: 6024
Cardilate	: 3432
Cardine	: 3540
Carditoxin	: 4043
Cardrase	: 2865
Δ4-Carene (l)	: 4050
Carfimate	: 1877
Carinamide	: 2721
Carminic acid	: 373
Carmustine	: 11994
Carnegine	: 7390
Carnosine	: 7133
γ-Carotene	: 4052
α-Carotene	: 4054
β-Carotene, all-trans-	: 4051
α-Carotene, 6',7'-didehydro-5,6-epoxy-4',5,5',6,7,8-hexahydro-3,3'	: 4055
Carvacrol	: 9415
Carvacrol, acetate	: 9419
Carvacryl acetate	: 9419
Carvenol	: 4698
Carvenone (DL)	: 4717
(+)-Carvone	: 4715
(-)-β-Caryophyllene	: 2996
β-Caryophyllene	: 2996
trans-Caryophyllene	: 2996
Caryophyllene	: 2996
Carzenide	: 2508
Casimiroin	: 5218

Catapyrin	: 11060
Catechol, 5-iodo	: 1523
Caulophylline	: 7578
CDT	: 4212
Cedrin	: 8058
Cedrol	: 7570
Celidoniol, deoxy-	: 8099
Cellocidin	: 3902
Cellosolve	: 5717
Cephaeline	: 5367
Cepharanthine	: 8630
Cerotic acid	: 6624
Cetane	: 6632
1-Cetene	: 6667
Cetyl alcohol	: 6659
Cetylamine	: 6631
Cetyl bromide	: 6633
Cetyl chloride	: 6634
α-Cetylcitric acid	: 8102
Cetyl iodide	: 6640
Cetyl lactate	: 10271
Cetyl mercaptan	: 6643
Cetyl palmitate	: 6651
Cetyl stearate	: 8258
Cevadic acid	: 3844
Cevadine	: 4069
CFC 11	: 7555
CFC 12	: 7500
CFC 13	: 7486
CFC 13B1	: 7475
CFC 21	: 7502
CFC 22	: 7478
CFC 113	: 5663
CFC 114	: 5538
CFC 115	: 5484
CFC 123	: 5540
CFC 123a	: 5539
C16 Guerbet fatty acid	: 5065
Chalcone	: 10708
Champacol	: 507
Chaulmoogric acid	: 4936
Chavicine	: 9633
Chavicol	: 9458
Cheirolin	: 10067
Chelerythrine	: 2433
Chelidonic acid	: 10776
Chenic acid	: 4097
Chenodeoxycholic acid	: 4097
Chenodiol	: 4097
Chimyl alcohol	: 10010
Chitosamine	: 6233
Chloral	: 21
Chloral alcoholate	: 5769
Chloral ammonia	: 5680
Chloral, diethyl acetal	: 5657
Chloral formamide	: 6000
Chloral hydrate	: 5582
Chloralide	: 5212
Chloramben	: 2492
Chlorambucil	: 1101
Chloraminophenamide	: 1558
Chloramphenicol	: 52
Chloramphenicol palmitate	: 6648
Chloranil	: 4268
Chloranilic acid	: 4258
Chlorocyanohydrin	: 10115
Chlordantoin	: 7220
Chlordimeform	: 7564
Chlorfenac	: 944
Chlorfenethol	: 1850
Chlorfenson	: 2221
Chlorindanol	: 7283
Chlorine cyanide	: 4166
Chlornaphazine	: 7710
5-Chloroacenaphthene	: 5
2-Chloroacetaldehyde	: 12
Chloroacetaldehyde, trimer	: 13
Chloroacetamide	: 41
Chloroacetamide, N-3-chloroallyl	: 43
Chloroacetic acid (β)	: 123
Chloroacetic acid, methyl ester	: 138
Chloroacetic acid, phenyl ester	: 147
Chloroacetic acid, propyl ester	: 149
Chloroacetic anhydride	: 124
Chloroacetone	: 10420
m-Chloroacetophenone	: 5788
p-Chloroacetophenone	: 5789

ω-Chloroacetophenone	: 5790
Chloroacetyl chloride	: 295
Chloroacetylene	: 5938
9-Chloroacridine	: 315
o-Chloroaniline	: 658
m-Chloroaniline	: 659
p-Chloroaniline	: 660
4-Chloro-o-anisidine	: 667
1-Chloroanthracene	: 374
Chlorobenzene	: 1131
p-Chlorobenzenesulfonic acid	: 2220
1-(p-Chlorobenzenesulfonyl)-3-propylurea	: 2208
2-Chloro-1,3,5-benzenetricarboxylic acid	: 2320
N-(p-Chlorobenzhydryl)-N'-methylhomopiperazine	: 5140
o-Chlorobenzoic acid	: 2534
m-Chlorobenzoic acid	: 2535
p-Chlorobenzoic acid	: 2536
2-Chlorobenzothiazole	: 2862
2-Chlorobenzoxazole	: 2878
m-Chlorobenzoyl chloride	: 2891
4-Chlorobenzyl chloride	: 1134
2-(2-Chlorobenzyl)-4,4-dimethyl-1,2-oxazolidin-3-one	: 4140
4'-Chloro-4-biphenylamine	: 3012
Chlorobromoacetic acid	: 112
Chlorobromodifluoromethane	: 7464
Chlorobromomethane	: 7463
2-Chloro-4-bromophenol	: 9232
2-Chloro-1-bromopropane	: 9838
1-Chloro-2-bromopropane	: 9840
2-Chloro-1,3-butadiene	: 3152
1-Chlorobutane	: 3241
1-Chloro-2-butanol	: 3605
3-Chloro-2-butanol	: 3606
Chlorobutanol	: 10407
1-Chloro-2-butanone	: 3646
1-Chloro-3-butanone	: 3648
trans-1-Chloro-1-butene	: 3720
cis-1-Chloro-1-butene	: 3721
2-Chloro-1-butene	: 3722
3-Chloro-1-butene	: 3723
3-Chloro-2-buten-1-ol	: 3864
4-Chloro-2-buten-1-ol	: 3865
1-Chloro-3-buten-2-ol	: 3866
3-Chloro-3-buten-2-ol	: 3868
Chloro-tert-butyl alcohol	: 10354
3-Chlorobutyryl chloride	: 3684
m-Chlorocarbanilic acid isopropyl ester	: 3939
3-Chloro-2-(chloromethyl)-1-butene	: 3729
4-Chloro-o-cresol	: 9266
4-Chloro-m-cresol	: 9267
6-Chloro-m-cresol	: 9262
α-Chlorocrotonaldehyde	: 3697
γ-Chlorocrotonic acid	: 3822
2-Chlorocrotonic acid	: 4150
3-Chlorocrotonic acid ethyl ester	: 3826
γ-Chlorocrotyl alcohol	: 3864
trans-2-Chlorocyclohexanol	: 4505
trans-4-Chlorocyclohexanol	: 4507
2-Chlorocyclohexanone	: 4577
4-Chlorocyclohexanone	: 4578
1-Chlorocyclohexene	: 4642
2-Chloro-4-cyclohexylphenol	: 9254
Chlorocyclooctatetraene	: 4745
Chlorocyclopentane	: 4789
Chlorocyclopropylmethane	: 4966
2-Chloro-N,N-diallylacetamide	: 44
Chlorodiazepoxide	: 2409
Chlorodibromomethane	: 7493
N-Chloro-2,6-dibromoquinoneimine	: 4279
Chloro(dichloromethyl)dimethyl-silane	: 11373
Chlorodifluoroacetic acid	: 127
1-Chloro-1,1-difluoroethane	: 5470
1-Chloro-2,2-difluoroethane	: 5471
2-Chloro-1,1-difluoroethylene	: 5885
Chlorodifluoromethane	: 7478
Chlorodiiodomethane	: 7479
1-Chloro-2,3-dimethylbutane	: 3245
1-Chloro-2,3-dimethyl-1-butene	: 3731
1-Chloro-2,3-dimethyl-2-butanol	: 3608
1-Chloro-3,3-dimethyl-2-butanol	: 3609

2-Chloro-2,5-dimethylhexane	: 6740
1-Chloro-3,3-dimethylpropane	: 3252
3-Chloro-1,5-dimethylpyrazole	: 10818
5-Chloro-1,3-dimethylpyrazole	: 10820
Chlorodimethyl(vinylethynyl)-methane	: 7107
1-Chloro-2,4-dinitrobenzene	: 1144
2-Chloro-1,3-dinitrobenzene	: 1145
Chlorodinitromethane	: 7480
4-Chloro-2,6-dinitrophenol	: 9258
3-Chloro-4,6-dinitrophenol	: 9259
Chlorodiphenylarsine	: 454
2-Chloroethanal	: 12
2-Chloroethanethiol	: 5651
2-(2-Chloroethoxy)-N-[[(4-methoxy-6-methyl-1,3,5-triaziN-2-yl)amino]carbonyl]benzene sulfonamide	: 11824
β-Chloroethyl acetate	: 5695
1-Chloroethyl 2-chloroethyl ether	: 5468
2-Chloroethyl chloroformate	: 4013
N-(2-Chloroethyl)dibenzylamine hydrochloride	: 1801
2-Chloroethyl isobutyrate	: 10292
6-Chloro-N-ethyl-N'-(1-methylethyl)-1,3,5-triazine-2,4-diamine	: 476
3-Chloro-3-ethyl-1-pentyne	: 9125
1-Chloroethyl propyl ether	: 9854
2-Chloroethyl vinyl ether	: 5887
Chlorofluoroacetic acid	: 131
Chlorofluoroacetic acid, butyl ester	: 132
Chlorofluoroacetic acid, methyl ester	: 134
Chlorofluoroacetic acid, propyl ester	: 135
Chlorofluoroacetyl chloride	: 296
1-Chloro-4-fluorobutane	: 3250
1-Chloro-10-fluorodecane	: 5037
1-Chloro-1-fluoroethane	: 5475
1-Chlorofluoroethane	: 5475
1-Chloro-6-fluorohexane	: 6742
Chlorofluoromethane	: 7481
1-Chloro-3-fluoropropane	: 9856
Chloroflurazole	: 2377
Chloroform	: 7554
Chloroformic acid, isopropyl ester	: 4022
Chloroformic digitalin	: 4041
2-Chlorofuran	: 6073
3-Chlorofuran	: 6074
Chloroheptafluorocyclobutane	: 4177
2-Chloroheptane	: 6361
3-Chloroheptane	: 6362
4-Chloroheptane	: 6363
7-Chloro-1-heptanol	: 6477
1-Chloro-2-heptanol	: 6478
1-Chloro-2-heptanone	: 6513
1-Chloro-1-heptene	: 6538
1-Chloro-1-heptyne	: 6614
1-Chloro-5-heptyne	: 6615
1-Chloro-4-heptyne	: 6616
2-Chlorohexane	: 6738
3-Chlorohexane	: 6739
1-Chloro-2-hexanol	: 6917
1-Chloro-3-hexanol	: 6918
1-Chloro-1-hexene	: 7004
2-Chloro-1-hexene	: 7005
5-Chloro-1-hexene	: 7006
4-Chloro-2-hexene	: 7007
α-Chloro-N-hexylacetamide	: 45
α-Chlorohydrin	: 9994
1-Chloro-3-hydroxypentane	: 8925
3-Chloroindazole	: 7236
Chloroiodomethane	: 7482
4-Chloroisoprene	: 3154
2-Chloroisopropyl ether	: 10123
1-Chloroisoquinoline	: 7376
Chloromethane	: 7477
Chloromethoxymethane	: 7483
2-Chloro-N-[[(4-methoxy-6-methyl-1,3,5-triaziN-2-yl)amino]carbonyl]benzene-sulfonamide	: 4091
α-Chloro-N-(3-methoxypropyl)acetamide	: 46
2-(4-Chloro-6-methoxypyrimidin-2-ylcarbamoylsulfamoyl) benzoic acid, ethyl ester	: 4081
2-Chloro-2'-methylbiphenyl	: 3026

Chloromethylbutadiene	: 3153
3-Chloro-3-methyl-2-butanone	: 3649
1-Chloro-3-methyl-1-butene	: 3733
1-Chloro-2-methyl-2-butene	: 3735
1-Chloro-2-methyl-1-butene	: 3732
3-Chloro-2-methyl-1-butene	: 3734
2-Chloro-3-methyl-2-butene	: 3737
Chloromethyl chloroformate	: 4014
Chloromethyl cyanide	: 274
2-(Chloromethyl)-2,3-dihydrobenzofuran	: 2441
Chloromethyl ethyl ether	: 5478
Chloromethyl ethyl ketone	: 3646
2-Chloro-2-methylheptane	: 6366
2-Chloro-6-methylheptane	: 6367
3-Chloro-3-methylheptane	: 6369
4-Chloro-4-methylheptane	: 6370
2-Chloro-5-methylhexane	: 6745
Chloromethyl iodide	: 7482
Chloromethyl methyl dichloro silane	: 11390
Chloromethyl methyl ether	: 7483
1-Chloro-2-methylnaphthalene	: 7784
4-Chloro-4-methyloctane	: 8363
2-Chloro-2-methyloct-3-yne	: 8534
2-Chloro-4-methylpentane	: 8730
5-Chloro-2-methyl-2-pentene	: 9026
3-Chloro-3-methyl-1-pentyne	: 9126
4-(4-Chloro-2-methylphenoxy)butyric acid	: 3493
2-Chloromethylpropanol	: 10353
1-Chloro-2-methylpropene	: 10539
3-Chloro-2-methylpropene	: 10540
2-Chloro-8-methylquinoline	: 11244
Chloromethylsilane	: 11378
2-Chloro-5-methylthiophene	: 11741
cis,trans-3-Chloro-4-[4-methyl-2-(1H-1,2,4-triazol-1-ylmethyl)-1,3-dioxolan-2-yl]phenyl 4-chlorophenyl ether	: 5174
(Chloromethyl)trichlorosilane	: 11456
2-Chloronaphthalene	: 7781
5-Chloro-1-naphthalenecarboxylic acid	: 7762
5-Chloro-1,4-naphthoquinone	: 7847
o-Chloronitrobenzene	: 1214
m-Chloronitrobenzene	: 1215
p-Chloronitrobenzene	: 1216
2-Chloro-3-nitrobenzoic acid	: 2544
3-Chloro-2-nitrobenzoic acid	: 2546
1-Chloro-1-nitroethane	: 5482
2-Chloro-1-nitroethane	: 5483
Chloronitromethane	: 7485
1-Chloro-5-nitronaphthalene	: 7786
1-Chloro-8-nitronaphthalene	: 7787
2-Chloro-1-nitronaphthalene	: 7788
5-Chloro-2-nitrophenol	: 9269
1-Chloro-1-nitropropane	: 9866
3-Chloro-1-nitropropane	: 9868
1-Chloro-3-nitropropane	: 9868
2-Chloro-2-nitropropane	: 9870
3-Chloro-5-nitrotoluene	: 1207
2-Chlorononane	: 8124
5-Chlorononane	: 8125
9-Chlorononan-1-ol	: 8163
(Z)-4-Chloro-4-octene	: 8500
1-Chloro-1-octyne	: 8533
3-Chloro-1,3-pentadiene	: 8663
Chloropentafluoroacetone	: 10421
Chloropentafluorobenzene	: 1220
Chloropentafluoroethane	: 5484
3-Chloropentane	: 8721
3-Chloropentanoic acid	: 8864
1-Chloro-3-pentanol	: 8925
2-Chloro-1-pentene	: 9016
3-Chloro-1-pentene	: 9017
4-Chloro-1-pentene	: 9018
1-Chloro-2-pentene	: 9020
2-Chloro-2-pentene	: 9021
3-Chloro-2-pentene	: 9022
3-Chloroperoxybenzoic acid	: 1127
p-Chlorophenacyl bromide	: 5778
o-Chlorophenol	: 9251
m-Chlorophenol	: 9252
p-Chlorophenol	: 9253
(p-Chlorophenoxy)acetic acid	: 146
p'-Chloro-p-phenylaniline	: 3012

Coumarin-3-carboxylic acid	: 2775
Coumarin, 3-chloro-7-hydroxy-4-methyl, O-ester with O,O-diethylphosphorothioate	: 4147
Coumarin, 3-ethyl-	: 2820
Coumarin, 7-hydroxy-	: 2826
Coumarin, 7-hydroxy-6-methoxy-	: 2829
cis-o-Coumarinic acid lactone	: 2792
Coumarin, 3-methyl-	: 2833
Coumarin, 5-methyl-	: 2834
Coumarin, 7-methyl-	: 2836
Coumarin, 8-methyl-	: 2837
Coumarone	: 2438
Coumarone-2-carboxylic acid	: 2439
Coumestrol	: 2466
Creatine	: 6254
Creatinine	: 7230
Creosol	: 9385
o-Cresol	: 9400
m-Cresol	: 9401
p-Cresol	: 9402
m-Cresol, 5-amino-	: 9212
m-Cresol, 6-bromo-	: 9237
m-Cresol, 6-sec-butyl-	: 9421
m-Cresol, 2,4-dichloro	: 9292
m-Cresol, 2,6-dichloro-	: 9294
p-Cresol, 2,6-di-tert-pentyl-	: 9225
p-Cresol, 2-iodo-	: 9374
o-Cresol, 4-isobutyryl	: 10448
o-Cresol, 3-isopropyl, methyl ether	: 363
m-Cresol, 6-pentyl-	: 9425
o-Cresol, α-phenyl-	: 9454
p-Cresol, α-phenyl-	: 9455
o-Cresolphthalein	: 7353
Cresol red	: 9368
o-Cresol, 4,5,6-trichloro-	: 9487
p-Cresol, 2,3,6-trichloro-	: 9488
m-Cresol, 2,4,6-trichloro-	: 9489
o-Cresotic acid	: 2629
m-Cresotic acid	: 2630
p-Cresotic acid	: 2631
o-Cresyl acetate	: 211
m-Cresyl acetate	: 212
Crimidine	: 11043
Crithminic acid	: 2678
Croceic acid	: 7972
Crocetin (trans)	: 5107
Croconic acid	: 4939
Crodimyl	: 2838
Cromaril	: 2840
trans-Crotonaldehyde	: 3695
Crotonaldehyde, 2-chloro-	: 3697
Crotonaldehyde-2-chloro, dimethylacetal	: 3730
Crotonaldehyde, diethyl acetal	: 3758
Crotonchloral	: 3191
trans-Crotonic acid	: 3814
cis-Crotonic acid	: 3815
Crotonic acid, allyl ester	: 3856
Crotonic acid anhydride	: 3818
Crotonic acid, 4-chloro-	: 3822
Crotonic acid, 2-chloro-, ethyl ester	: 3823
Crotonic acid, 3-chloro-, ethyl ester	: 3826
Crotonic acid, 4-chloro-, ethyl ester	: 3829
Crotonic acid, 4-chloro, methyl ester	: 3833
Crotonic acid, 2,3-dimethyl-, ethyl ester	: 3836
Crotonic acid, 2-ethyl-, stereoisomer	: 3837
trans-Crotyl alcohol	: 3859
cis-Crotyl alcohol	: 3860
Cryptocavine	: 2435
Cryptopine	: 2435
all-trans-β-Cryptoxanthin	: 4061
Cryptoxanthin	: 4061
Cryptoxanthol, all-trans-	: 4061
Crystal violet	: 7453
Cubebin	: 6136
Cumene	: 1975
Cumene hydroperoxide	: 7181
3-p-Cumenyl-2-methylpropionaldehyde	: 2117
Cumic acid	: 2689
Cumic alcohol	: 1902
Cumidine	: 789
Cuminaldehyde	: 590

Cuminic acid	: 2689
Cuminol	: 1902
Cumotocopherol	: 2789
α-Cumyl alcohol	: 1857
Cupferron	: 759
Cupreine	: 4119
Cupreol	: 11558
ar-Curcumene	: 1468
Curcumin	: 6334
Curine	: 11925
Curling factor	: 11520
Cuscohygrine	: 10415
Cusparine	: 11216
Cyacetacide	: 153
Cyamelide	: 11907
Cyanidelonon 1522	: 2817
Cyanoacetamide	: 49
Cyanoacetic acid	: 150
Cyanoacetic acid, diethyl, ethyl ester	: 3495
2-Cyanoacetophenone	: 2124
Cyanoacetylene	: 10737
N-Cyanoamine	: 4155
Cyanobenzene	: 2740
2-Cyanocaproic acid, ethyl ester	: 6881
Cyanocyclobutane	: 4173
1-Cyanocyclohexene	: 4630
Cyanocyclopentane	: 4779
1-Cyanocyclopentene	: 4907
4-Cyano-2,6-dibromophenol	: 2751
2-Cyanoethylamine	: 10092
(2-Cyanoethyl)benzene	: 2123
Cyanoethylene	: 10583
Cyano(4-fluoro-3-phenoxyphenyl)methyl 3-(2,2-dichloroethenyl)-2,2-dimethylcyclopropanecarboxylate	: 5008
1-Cyano-2-(2-furyl)ethene	: 10585
Cyanogen	: 5550
Cyanogenamide	: 4155
Cyanogen iodide	: 7337
Cyanogen nitride	: 4155
Cyanoiodomethane	: 283
Cyanomethane	: 271
Cyanomethyl methyl sulfide	: 286
Cyanomethyl phenyl ketone	: 2124
Cyano(3-phenoxyphenyl)methyl-3-(2,2-dibromoethenyl)-2,2-dimethylcyclopropanecarboxylate	: 5102
Cyano(3-phenoxyphenyl)methyl 2,2,3,3-tetramethylcyclo-propanecarboxylate	: 5953
1-Cyano-2-phenylethane	: 2123
1-Cyanopropane	: 3403
2-Cyano-2-propanol	: 10108
3-Cyano-1-propene	: 3803
2-Cyanoquinoline	: 11223
2-Cyanothiophene	: 11732
2-Cyanotoluene	: 2763
m-Cyanotoluene	: 2764
p-Cyanotoluene	: 2765
1-Cyanoundecane	: 5303
Cyanuric acid, dihydrate	: 11839
Cyanuric acid trichloride	: 11837
Cyclamal	: 2117
Cyclamic acid	: 11568
Cyclobarbital	: 11084
Cyclobutaneacetic acid, 3-carboxy-2,2-dimethyl-, diethyl ester	: 4171
Cyclobutane-1-carboxylic acid	: 4174
Cyclobutane-1,1-dicarboxylic acid, diethyl ester	: 4180
Cyclobutane-1,2-dicarboxylic acid, dimethyl ester (trans)	: 4181
Cyclobutane-1,3-dicarboxylic acid (cis), 2,2-dimethyl	: 4183
Cyclobutane-1,3-dicarboxylic acid, 2,2-dimethyl, dimethyl ester	: 4184
Cyclobutane, 1,2-divinyl-, trans-	: 4187
Cyclobutylamine	: 4169
Cyclobutylcarboxylic acid	: 4174
Cyclobutyl hydroxide	: 4196
Cyclobutyl methyl ketone	: 5792
Cyclobutyrol	: 4309
Cyclocapron	: 4333
β-Cyclocitral	: 4636

1,5,9-Cyclododecatriene, (cis, trans, trans)	: 4213
cis,cis,cis-Cyclododeca-1,5,9-triene	: 4214
Cyclodoecanone	: 4211
Cyclofenchene	: 11866
Cycloform	: 2505
β-Cyclogeraniol	: 4666
α-Cyclogeraniol	: 4667
Cyclogeraniolane	: 4486
α-Cyclogeraniolene	: 4691
Cyclogeranyl alcohol	: 4666
Cycloglycylglycine	: 9612
9-Cycloheptadecene-1-one	: 4217
Cycloheptyl bromide	: 4222
Cycloheptylcarboxylic acid	: 4223
1,3-Cyclohexadiene, 3,5-dimethyl	: 4256
1,3-Cyclohexadiene, 5-methyl (DL)	: 4272
Cyclohexanacetic acid, 2-methyl	: 4311
Cyclohexanacetic acid, 3-methyl	: 4312
Cyclohexane, 1,3-bis(aminomethyl)	: 4362
Cyclohexane, 1,3-bis-(hydroxymethyl)	: 4363
Cyclohexane, sec-butyl-	: 4448
Cyclohexane, decafluoro-1,2-bis(trifluoromethyl)-	: 4342
Cyclohexane-1,2-dicarboxylic acid, diethyl ester (trans, DL)	: 4350
Cyclohexane-1,3-dicarboxylic acid, diethyl ester (DL)	: 4351
Cyclohexane-1,3-dicarboxylic acid, dimethyl ester (cis)	: 4352
Cyclohexane-1,3-dicarboxylic acid, dimethyl ester (trans)	: 4353
Cyclohexane-1,4-dicarboxylic acid, diethyl ester (trans)	: 4356
Cyclohexane-1,4-dicarboxylic acid, dimethyl ester (cis)	: 4357
Cyclohexane-1,4-dicarboxylic acid, diethyl ester (cis)	: 4355
Cyclohexane, 2,4-diisopropenyl-1-methyl-1-vinyl-, stereoisomer	: 4389
Cyclohexane, 1,2-dimethyl, perfluoro	: 4342
Cyclohexane, 1,2-epoxy-4-vinyl	: 8555
Cyclohexane, 1,2,3,4,5,6-hexachloro, (α, DL)	: 4408
Cyclohexane hydroperoxide	: 7177
Cyclohexane, 1-hydroxymethyl-2-methyl (cis)	: 4422
Cyclohexane, 1-hydroxymethyl-2-methyl (trans)	: 4423
Cyclohexane, 1-isopropenyl-3-methyl	: 4435
Cyclohexane, isopropyl-	: 4431
Cyclohexane, 1-methylamino-4-methyl	: 4723
Cyclohexane, 1-methyl-2-ethyl (trans)	: 4396
Cyclohexane, 2-methyl-1-ethylidene	: 4393
Cyclohexane, 2-nitrophenyl	: 1251
1-L-1,3,4/2,5-Cyclohexanepentol	: 4074
Cyclohexanol, acetate	: 155
Cyclohexanol, 2-allyl-, trans-	: 4556
Cyclohexanol, 2-amino (trans, DL)	: 4501
Cyclohexanol, 1,2-dimethyl-, stereoisomer	: 4508
	: 4509
Cyclohexanol, 2-ethyl (trans, DL)	: 4523
Cyclohexanol, 2-isopropyl-, cis-	: 4539
Cyclohexanol, 4-isopropyl-, cis-	: 4540
Cyclohexanol, 4-isopropyl-, trans-	: 4541
Cyclohexanol-4-isopropyl-2-methyl	: 7428
Cyclohexanol, 2-methyl (cis, DL)	: 4531
Cyclohexanol, 2-phenyl (cis, DL)	: 4553
Cyclohexanol, 2-phenyl (trans, DL)	: 4554
Cyclohexanone-4-carboxylic acid, 2-methyl	: 4336
Cyclohexanone, 2,4-dimethyl (trans, d)	: 4583
Cyclohexanone, 2-isopropyl-	: 4599
Cyclohexanone, 4-isopropyl-	: 4600
Cyclohexanone, 2-isopropyl-5-methyl (l)	: 4606
1-Cyclohexene-1-carboxaldehyde, 4-isopropenyl-	: 4633
1-Cyclohexene-1-carboxaldehyde, 4-isopropyl-	: 4635

Dexpanthenol	: 3202	Diazomethane	: 7491	1,2-Dibromo-2,3-dimethylbutane	: 3269
Dextroamphetamine sulfate	: 1582	Dibenzalacetone	: 8683	1,1-Dibromo-2,2-dimethylpropane	: 9886
Dextrotartaric acid	: 3309	1,2:5,6-Dibenzanthracene	: 5142	Dibromodinitromethane	: 7497
4,4'-Diacetamidodiphenyl sulfone	: 91	2,3:6,7-Dibenzazepine-1-carboxylic	: 5144	1,1 Dibromoethane	: 5517
Diacetic acid	: 3568	acid,amide		1,1-Dibromoethylene	: 5890
Diacetin	: 10189	Dibenz[de,kl]anthracene	: 9161	trans-1,2-Dibromoethylene	: 5891
1,3-Diacetin	: 10190	Dibenzo[a,e]pyrene	: 8061	cis-1,2-Dibromoethylene	: 5892
Diacetonamine	: 8959	Dibenzo[a,i]phenanthrene	: 9601	1,2-Dibromoethyl methyl ketone	: 3653
Diacetone acrylamide	: 10521	Dibenzo[a,i]pyrene	: 2850	Dibromofluoroacetic acid, ethyl	: 161
Diacetone alcohol	: 8976	Dibenzo[a,jk]fluorene	: 2738	ester	
Diacetonyl	: 6805	2,3:6,7-Dibenzoanthracene	: 8636	Dibromofluoromethane	: 7498
1,4-Diacetoxy-2-butyne	: 3913	Dibenzo[b,e]pyridine	: 314	2,3-Dibromofuran	: 6077
1,2-Diacetoxyethane	: 5571	Dibenzo[b,h]pyrene	: 2850	2,5-Dibromofuran	: 6078
Diacetyl	: 3359	Dibenzo[b,d]pyrrole	: 3981	Dibromogallic acid	: 2560
Diacetyldioxime	: 3360	1,2:6,7-Dibenzo-9-fluorenone	: 5152	1,2-Dibromoheptane	: 6372
Diacetyl disulfide	: 5250	Dibenzopyrazine	: 9193	1,5-Dibromoheptane	: 6373
1,2-Diacetylethane	: 6805	1,2:7,8-Dibenzopyrene	: 2850	1,7-Dibromoheptane	: 6374
Diacetylmethane	: 8798	1,2:4,5-Dibenzopyrene	: 8061	2,3-Dibromoheptane	: 6375
Diacetylmorphine	: 7667	Dibenzothiapyran	: 11811	3,4-Dibromoheptane	: 6376
Diacetylperoxide	: 9156	Dibenzothiazine	: 9499	1,16-Dibromohexadecane	: 6635
Diadol	: 11091	2,3,:5,6-Dibenzo-1,4-thiazine,	: 9502	1,2-Dibromohexane	: 6754
Δ 2-Dialin	: 7815	10(dimethyaminacetyl)		1,4-Dibromohexane	: 6755
Diallylamine	: 10526	Dibenzoylmethane	: 10027	1,5-Dibromohexane	: 6756
5,5-Diallylbarbituric acid	: 11091	2,3:6,7-Dibenzphenanthrene	: 8996	2,3-Dibromohexane	: 6758
N,N-Diallyl-2-chloroacetamide	: 44	Dibenzyl	: 1587	1,2-Dibromoindan	: 7247
Diallylcyanamide	: 4159	Dibenzylamine	: 1828	Dibromomethane	: 7492
Diallyl fumarate	: 3781	9,10-Dibenzylanthracene	: 367	1,2-Dibromo-2-methylbutane	: 3272
Diallyl maleate	: 3782	Dibenzyl disulfide	: 5248	2,3-Dibromo-2-methylbutane	: 3274
Diallyl oxalate	: 5564	Dibenzyl phosphite	: 9586	1,3-Dibromo-2-methylpropene	: 10546
Diallyl sulfide	: 10599	Diborane, methylthio	: 5168	Dibromomonochloromethane	: 7493
p-Diaminoazobenzene	: 635	Dibromantine	: 7217	1,3-Dibromonaphthalene	: 7801
2,3-Diaminobenzoic acid	: 2552	Dibromoacetic acid	: 159	1,4-Dibromonaphthalene	: 7802
2,4-Diaminobenzoic acid	: 2553	Dibromoacetic acid ethyl ester	: 160	1,5-Dibromonaphthalene	: 7803
3,5-Diaminobenzoic acid	: 2554	Dibromoacetonitrile	: 277	3,4-Dibromonitrobenzene	: 1307
o,p'-Diaminobiphenyl	: 3029	β,β-Dibromoacrylic acid	: 10628	1,2-Dibromo-4-nitrobenzene	: 1307
4,4'-Diamino-1,1'-biphenyl	: 3031	3,4-Dibromoaniline	: 683	2,6-Dibromonitrobenzene	: 1308
1,4-Diaminobutane	: 3258	9,10-Dibromoanthracene	: 375	1,3-Dibromo-2-nitrobenzene	: 1308
1,8-Diamino-3,6-diazaoctane	: 5490	2,7-Dibromoanthraquinone	: 390	1,2-Dibromononane	: 8127
β,β'-Diamino-β,β'-dicarboxydiethyl	: 5020	o-Dibromobenzene	: 1288	1,4-Dibromooctafluorobutane	: 3275
disulfide		m-Dibromobenzene	: 1289	1,2-Dibromooctane	: 8366
4,4'-Diamino-3,3'-dimethylbiphenyl	: 3034	p-Dibromobenzene	: 1290	1,5-Dibromooctane	: 8368
p-Diaminodiphenyl	: 3031	2,4-Dibromobenzoic acid	: 2555	1,8-Dibromooctane	: 8369
4,4'-Diaminodiphenylamine	: 1263	2,5-Dibromobenzoic acid	: 2556	1,3-Dibromopentane	: 8739
p,p'-Diaminodiphenylmethane	: 784	2,6-Dibromobenzoic acid	: 2557	1,4-Dibromopentane	: 8740
3,3'-Diaminodipropylamine	: 9874	3,5-Dibromobonzyl bromide	: 1291	2,3-Dibromopentane	: 8742
2,7-Diaminofluorene	: 5971	d-3,3-Dibromo-d-bornanone	: 2953	2,4-Dibromopentane	: 8743
2,6-Diaminohexanoic acid	: 7418	1,1-Dibromobutane	: 3264	2,3-Dibromopentanoic acid	: 8870
	: 7419	1,2-Dibromobutane	: 3265	2,4-Dibromophenol	: 9276
2,6-Diamino-1-methylbenzene	: 1277	1,3-Dibromobutane	: 3266	2,6-Dibromophenol	: 9277
1,2-Diaminonaphthalene	: 7795	1,4-Dibromobutane	: 3267	3,5-Dibromophenol	: 9278
1,4-Diaminonaphthalene	: 7796	2,3-Dibromobutane	: 3268	1,1-Dibromopropane	: 9880
2,3-Diaminonaphthalene	: 7799	2,3-Dibromo-1-butanol	: 3612	1,2-Dibromopropane	: 9881
(S)-2,5-Diamino-5-oxopentanoic	: 6249	3,4-Dibromobutanol	: 3613	1,3-Dibromopropane	: 9882
acid		1,4-Dibromo-2-butanol	: 3614	2,2-Dibromopropane	: 9883
(S)-2,5-Diaminopentanoic acid	: 8550	3,4-Dibromo-2-butanone	: 3653	2,3-Dibromopropane	: 10545
2,3-Diaminophenazine	: 9194	1,1-Dibromo-1-butene	: 3738	2,3-Dibromopropylene	: 10545
2,4-Diaminophenol	: 9274	3,4-Dibromo-1-butene	: 3740	3,5-Dibromopyridine	: 10926
2,4-Diamino-5-phenylthiazole	: 11671	1,2-Dibromo-3-butene	: 3740	2,6-Dibromoquinone-4-chlorimide	: 4279
1,3-Diaminopropane	: 9873	2,4-Dibromo-2-butene	: 3741	3,5-Dibromosalicylaldehyde	: 534
2,3-Diaminopropionic acid	: 326	1,3-Dibromo-2-butene	: 3741	3,5-Dibromosalicylic acid	: 2558
2,6-Diaminopurine	: 10754	1,2-dibromo-1-butyne	: 3739	1,14-Dibromotetradecane	: 11616
2,5-Diaminopyridine	: 10924	1,4-Dibromo-2-butyne	: 3904	1,2-Dibromotetrafluoroethane	: 5525
1,3-Diamino-2-thiourea	: 4033	(+)-3,3-Dibromocamphor	: 2953	1,2-Dibromo-1,1,2,2-	: 3271
2,6-Diaminotoluene	: 1277	α,α'-Dibromo-d-camphor	: 2953	tetramethylethane	
Diamylamine	: 8708	1,2-Dibromo-3-chlorobenzene	: 1292	2,3-Dibromothiophene	: 11742
Diamyl ketone	: 11962	1,2-Dibromo-4-chlorobenzene	: 1293	2,4-Dibromothiophene	: 11743
Dianat	: 2567	1,3-Dibromo-2-chlorobenzene	: 1294	2,5-Dibromothiophene	: 11744
1,2,3,4-Dianhydroerythritol	: 3007	1,4-Dibromo-2-chlorobenzene	: 1295	2,3-Dibromotoluene	: 1302
o,p'-Dianiline	: 3029	Dibromochlorofluoromethane	: 7494	3,4-Dibromotoluene	: 1303
1,2-Dianilinoethane	: 5508	Dibromochloromethane	: 7493	2,6-Dibromotoluene	: 1304
o-Dianisidine	: 3033	Dibromochloropropane	: 9885	2,5-Dibromotoluene	: 1305
7,12-Diazabenz[a]anthracene	: 2394	1,2-Dibromo-1-chloro-1,2,2-	: 5519	1,2-Dibromo-1,1,2-trifluoroethane	: 5526
1,2-Diazabenzene	: 10855	trifluoroethane		1,2-Dibromo-1,1,2-trimethylethane	: 3274
1,4-Diazabicyclooctane	: 5111	1,1-Dibromocyclobutane	: 4178	1,11-Dibromoundecane	: 11940
4,9-Diazadodecane-1,12-diamine	: 3259	1,2-Dibromocyclobutane	: 4179	3,4-Dibromo-o-xylene	: 1297
2,3-Diazaindole	: 2871	1,1-Dibromocyclohexane	: 4345	Dibucaine hydrochloride	: 11226
Diazald	: 2212	cis-1,2-Dibromocyclohexane	: 4346	Dibutylacetylene	: 5100
1,5-Diazanaphthalene	: 8065	trans-1,3-Dibromocyclohexane	: 4348	Di-tert-Butylacyloin	: 6975
3,6-Diazaoctane-1,8-diamine	: 5499	cis-1,4-Dibromocyclohexane	: 4349	Dibutylamine	: 3211
Diazepam	: 2410	1,10-Dibromodecane	: 5038	Dibutylbutadiyne	: 5287
1,4-Diazine	: 10801	5,6-Dibromodecane	: 5039	Dibutylcarbinol	: 8161
1,3-Diazine	: 11049	Dibromodichloromethane	: 7495	Dibutyl carbonate	: 3998
Diazinon	: 9582	1,2-Dibromo-2,4-dicyanobutane	: 8768	Dibutyl ether	: 3415
Diazoacetic ester	: 158	1,2-Dibromo-1,1-difluoroethane	: 5522	Di-tert-Butyl ether	: 10124
Diazoaminobenzene	: 11826	Dibromodifluoromethane	: 7496	N,N'-Dibutylhexamethylenediamine	: 6750
1,3-Diazole	: 7192	1,4-Dibromo-2,3-dimethylbutane	: 3270	Di-sec-butyl ketone	: 6515
1,2-Diazole	: 10815	2,3-Dibromo-2,3-dimethylbutane	: 3271	Dibutyl ketone	: 8175

2,5-Dichlorothiophene	: 11748
3,4-Dichlorothiophene	: 11749
2,3-Dichlorotoluene	: 1383
3,4-Dichlorotoluene	: 1384
2,6-Dichlorotoluene	: 1385
3,5-Dichlorotoluene	: 1386
2,5-Dichlorotoluene	: 1387
2,4-Dichlorotoluene	: 1388
N,N-Dichloro-p-toluenesulfonamide	: 2210
2,2-Dichloro-1,1,1-trifluoroethane	: 5540
Dichloroxylenol	: 9290
Dichlorprop	: 10250
Dicopur M	: 141
Dicryl	: 10520
Dictamnine	: 6180
Dicumarol	: 2839
Dicurone	: 6239
1,3-Dicyanobenzene	: 1315
Dicyanodiamide	: 6280
1,2-Dicyanoethane	: 3291
1,6-Dicyanohexane	: 8385
Dicyanomethane	: 9911
1,3-Dicyanopropane	: 8767
Dicyclohexylamine	: 4284
1,4-Dicyclohexylbutane	: 4323
Dicyclohexylcarbodiimide	: 4291
1,2-Dicyclohexylethane	: 4386
1,1-Dicyclohexylethane	: 4392
Dicyclohexylmethane	: 4429
1,3-Dicyclohexyl-2-methylpropane	: 4447
Dicyclohexyl oxalate	: 5559
Dicyclohexylphosphine	: 9516
1,2-Dicyclohexylpropane	: 4430
1,3-Dicyclohexylpropane	: 4466
1,1-Dicyclohexyltetradecane	: 4476
Dicyclopropyl ketone	: 7607
Dicysteine	: 5020
Dieldrin	: 5178
Dienestrol	: 9302
meso-Diepoxybutane	: 3007
Diethadione	: 8573
Diethanolamine	: 5730
Diethazine	: 9503
3,4-Diethoxyaniline	: 693
3,4-Diethoxybenzaldehyde	: 638
Diethoxydimethylsilane	: 11405
Diethoxydiphenylsilane	: 11406
1,1-Diethoxyethane	: 5541
Diethoxymethane	: 5603
3,3-Diethoxy-1-propene	: 10560
2,5-Diethoxytoluene	: 1410
Diethylacetaldehyde	: 3183
Diethylacetic acid	: 3514
Diethylacetylene	: 7116
O,O-Diethyl acetylphosphonate	: 9530
Diethylamine	: 5427
2-(Diethylamino)ethanol	: 5706
2-(Diethylamino)ethyl p-aminobenzoate	: 2493
S-[2-(Diethylamino)ethyl] diphenylthiocarbamate	: 3977
3-(Diethylamino)-2-hydroxypropylamine	: 10339
Diethyl aminomalonate	: 9924
2-Diethylamino-1-propanol	: 10366
3-(Diethylamino)propylamine	: 9876
N,N-Diethylaniline	: 695
Diethylarsine	: 441
o-Diethylbenzene	: 1413
m-Diethylbenzene	: 1414
p-Diethylbenzene	: 1415
N,N-Diethylbenzhydrylamine	: 1804
Diethylbromoacetamide	: 3198
Diethyl butanedioate	: 3303
Diethylcarbamazine	: 9603
N,N'-Diethylcarbanilide	: 12003
Diethyl carbinol	: 8917
Diethyl Carbitol	: 5612
Diethyl carbonate	: 3999
Diethyl chloromaleate	: 3771
Diethyl chlorophosphate	: 9574
Diethyl chlorothiophosphate	: 9576
Diethyl 1,1-cyclobutanedicarboxylate	: 4180
Diethyl 1-cyclohexene-1,3-dicarboxylate	: 4645
1,1-Diethylcyclopentane	: 4796
cis-1,2-Diethylcyclopentane	: 4797

trans-1,2-Diethylcyclopentane	: 4798
1,3-Diethylcyclopentane	: 4799
Diethyl 1-cyclopentene-1,2-dicarboxylate	: 4915
Diethyl dibenzylmalonate	: 9926
Diethyl 2,5-dibromovalerate	: 8873
Diethyl dibutylmalonate	: 9937
Diethyl dicarbonate	: 5169
Diethyl 2,2-diethylmalonate	: 9940
Diethyldifluorosilane	: 11410
Diethyl dimethylmalonate	: 9942
Diethyl dipropargylmalonate	: 9946
Diethyl disulfide	: 5253
Diethylenediamine	: 9602
Diethylenediamine dihydrochloride	: 9607
Diethylene glycol	: 5753
Diethylene glycol, diacetate	: 5754
Diethylene glycol, dibenzoate	: 5755
Diethylene glycol diethyl ether	: 5612
Diethylene glycol dimethyl ether	: 5613
Diethylene glycol, dodecanoate	: 5313
Diethylene glycol, monoacetate	: 5673
Diethylene glycol monobutyl ether	: 5688
Diethylene glycol monoethyl ether	: 5720
Diethylene glycol monoethyl ether acetate	: 5721
Diethylene glycol monolaurate	: 5313
Diethylenetriamine	: 5497
Diethyl ether	: 5608
Diethyl ether-trifluoroborane (1:1)	: 3137
N,N-Diethylethylenediamine	: 5502
Diethyl ethylphosphonate	: 9536
Diethyl 3,4-furandicarboxylate	: 6082
Diethyl glutaconate	: 9041
3,4-Diethylhexane	: 6768
3,4-Diethyl-3,4-hexanediol	: 6800
Diethylhexyl adipate	: 6781
1,1-Diethylhydrazine	: 7152
1,2-Diethylhydrazine	: 7153
Diethyl(2-hydroxyethyl)methylammonium bromide, xanthene-9-carboxylate	: 5445
Diethyl 3-hydroxyglutarate	: 8778
1,2-Diethylimidazole	: 7196
Diethyl isobutylmalonate	: 9989
Diethyl isophthalate	: 1346
Diethylisopropylcarbinol	: 8934
Diethyl isopropylidenemalonate	: 9966
N,N-Diethylisovaleramide	: 3201
Diethylketene	: 3877
Diethyl ketone	: 8958
Diethyl maleate	: 3776
Diethyl malonate	: 9941
Diethylmalonic acid	: 9939
Diethyl mercury	: 7434
Diethyl mesaconate	: 3787
Diethyl 3-methyladipate	: 6790
N,N-Diethyl-3-methylbenzamide	: 612
3,3-Diethyl-2-methylpentane	: 8757
3,4-Diethyl-2-methylpyrrole	: 11117
Diethyl oxalacetate	: 3340
Diethyl oxalate	: 5560
Diethyl oxydiformate	: 5169
3,3-Diethylpentane	: 8756
Diethylphosphinic acid	: 9525
Diethyl phthalate	: 1342
Diethyl pimelate	: 6394
N,N-Diethyl-p-phenylenediamine	: 1266
Diethylpropion	: 10428
2,4-Diethylpyridine	: 10933
Diethyl pyrocarbonate	: 5169
Diethylsilane	: 11409
Diethylstilbestrol	: 9300
Diethylstilbestrol dipropionate	: 9301
Diethyl succinate	: 3303
Diethyl sulfate	: 11573
Diethyl sulfide	: 5644
Diethyl sulfite	: 11581
Diethyl DL-tartrate	: 3307
Diethyl terephthalate	: 1350
2,2-Diethyltetrahydrofuran	: 6085
Diethylvinylethynylcarbinol	: 6608
Difluoroacetic acid	: 175
Difluoroacetic acid, ethyl ester	: 176
2,4-Difluoroaniline	: 700
o-Difluorobenzene	: 1422
m-Difluorobenzene	: 1423

p-Difluorobenzene	: 1424
Difluorochlorobromomethane	: 7464
1,1-Difluoro-2-chloroethylene	: 5885
Difluorochloromethane	: 7478
Difluorodibromomethane	: 7496
1,1-Difluoro-1,2-dichloroethane	: 5531
1,2-Difluoro-1,2-dichloroethane	: 5532
1,1-Difluoro-2,2-dichloroethane	: 5897
1,1-Difluoro-2,2-dichloroethylene	: 5897
Difluorodichloromethane	: 7500
4,4'-Difluorodiphenyl	: 3044
1,1-Difluoroethane	: 5544
1,2-Difluoroethane	: 5545
2,2-Difluoroethanol	: 5708
1,1-Difluoroethylene	: 5902
cis-1,2-Difluoroethylene	: 5903
Difluoroiodomethane	: 7505
Difluoromethane	: 7504
Difluoromethylarsine	: 463
1,1-Difluoroperchloroethane	: 5636
1,3-Difluoropropane	: 9905
2,2-Difluoropropane	: 9906
1,3-Difluoro-2-propanol	: 10369
1,1-Difluoro-1-propene	: 10561
2,2-Difluoropropyl chloride	: 9851
2,2-Difluoro-1,1,1,2-tetrachloroethane	: 5636
1,2-Difluoro-1,1,2,2-tetrachloroethane	: 5637
1,2-Difluorotetrachloroethane	: 5637
1,2-Difluoro-1,1,2,2-tetramethyldisilane	: 5227
1,2-Difluorotetramethyldisilane	: 5227
2,2-Difluoro-1,1,2-trichloroethane	: 5659
1,1-Difluoro-1,2,2-trichloroethane	: 5659
Difurfurylamine	: 6114
Digallic acid	: 2575
Digin	: 11530
Digitalose	: 6108
Digitoflavone	: 2814
Digitogenin	: 11535
Digitonin	: 6195
Digitoxigenin	: 4044
Digitoxin	: 4043
Digitoxose	: 7111
Diglycidyl resorcinol ether	: 8625
Diglycine	: 6265
Diglycol laurate	: 5313
Digoxigenin	: 4047
Digoxin	: 4041
Diheptylamine	: 6352
Dihexyl ketone	: 11891
1,2-Dihydroacenaphthylene	: 6
9,10-Dihydroanthracene	: 376
1,4-Dihydrobenzene	: 4249
2,3-Dihydrobenzofuran	: 2442
2,3-Dihydro-2-chloromethylbenzofuran	: 2441
Dihydrocholesterol	: 4107
(+)-Dihydrocinchonine	: 4122
Dihydrocinnamaldehyde	: 2114
Dihydrocodeine	: 7674
Dihydro-p-coumaric acid	: 2134
2,5-Dihydro-2,5-dimethoxyfuran	: 6087
2,3-Dihydro-2,2-dimethyl-7-benzofuranyl (dibutylamino)thio methylcarbamate	: 4038
2,3-Dihydro-5,6-dimethyl-1,4-dithiin, 1,1,4,4-tetraoxide	: 5179
α-Dihydrofucosterol	: 11558
Dihydrofumaric acid	: 3294
2,3-Dihydrofuran	: 6086
3,4-Dihydro-2H-pyran	: 10777
4,6-Dihydro-8-hydroxy-3,4,5-trimethyl-6-oxo-3H-2-benzopyran-7-carboxylic acid	: 2774
3,6-Dihydro-6-iminopurine	: 10751
1,2-Dihydro-2-iminopyridine	: 10857
Dihydroisophorone	: 4622
2,3-Dihydro-5-methylbenzofuran	: 2444
1,4-Dihydronaphthalene	: 7815
3,4-Dihydro-1-naphthoic acid	: 7764
2,3-Dihydro-3-oxobenzisosulfonazole	: 2388
9,10-Dihydrophenanthrene	: 9171
Dihydrosafrole	: 2431
Dihydrothebaine	: 7664

2,3-Dihydrothiophene	: 11750
2,5-Dihydrothiophene	: 11751
4,5-Dihydrotoluene	: 4271
Dihydroxyacetone	: 10429
1,3-Dihydroxy-2-amino-2-methylpropane	: 9987
2,5-Dihydroxybenzyl alcohol	: 1520
1,3-Dihydroxybutane	: 3345
1,4-Dihydroxybutane	: 3346
2,3-Dihydroxybutanedioic acid	: 3309
1,4-Dihydroxy-2-butyne	: 3911
3α,12α-Dihydroxycholanic acid	: 4095
3α,6α-Dihydroxy-5β-cholanic acid	: 4096
3α,7α-Dihydroxy-5β-cholan-24-oic acid	: 4097
3α,7β-Dihydroxy-5β-cholan-24-oic acid	: 4098
6,7-Dihydroxycoumarin	: 2804
6,7-Dihydroxycoumarin 6-glucoside	: 2822
trans-1,4-Dihydroxycyclohexane	: 4374
(+)-2,9-Dihydroxy-1,10-dimethoxyaporphine	: 5156
3,3'-Dihydroxy-5,5'-dimethylbiquinone	: 2979
2,3-Dihydroxy-2,3-dimethylbutane	: 3350
4,4'-Dihydroxydiphenylmethane-3,3'-dicarboxylic acid	: 2686
1,12-Dihydroxydodecane	: 5301
3,17-Dihydroxyestratriene	: 5400
Dihydroxyestrin	: 5399
1,2-Dihydroxyethylbenzene	: 5579
2,2'-Dihydroxyethyl ether monododecanoate	: 5313
4,4'-Dihydroxyfuchsone	: 4278
5',7-Dihydroxy-4'-glucosyloxyisoflavone	: 2824
2,5-Dihydroxyhexane	: 6799
1,3-Dihydroxy-5-iodobenzene	: 1524
4',7-Dihydroxyisoflavone	: 2827
Dihydroxymaleic acid	: 3777
5,7-Dihydroxy-4'-methoxyflavone	: 2809
5,7-Dihydroxy-6-methoxyflavone	: 2810
3,6-Dihydroxy-4-methoxy-2,5-toluquinone	: 4260
1,7-Dihydroxy-3-methoxyxanthone	: 12054
2,3-Dihydroxy-2-methylbutane	: 3355
1,4-Dihydroxy-1-methylbutane	: 8786
1,1-Di(hydroxymethyl)ethylamine	: 9987
2,5-Dihydroxy-3-methyl-6-methoxy-1,4-benzoquinone	: 4260
1,9-Dihydroxynonane	: 8136
9,10-Dihydroxyoctadecanoic acid	: 8254
2,4-Dihydroxypentane	: 8789
2,3-Dihydroxyphenol	: 2366
2,5-Dihydroxyphenol	: 2367
Dihydroxyphenylborane	: 3135
2,3-Dihydroxypropionic acid	: 10252
(±)-2,3-Dihydroxypropyl acetate	: 10187
3,6-Dihydroxyspiro[xanthene-9,3'-phthalide]	: 11522
9,10-Dihydroxystearic acid	: 8254
Dihydroxytartaric acid	: 3342
4,6-Dihydroxy-o-toluic acid	: 2573
Diiodoacetylene	: 5941
2,4-Diiodoaniline	: 701
4,4'-Diiodobenzophenone	: 7598
Diiodochloromethane	: 7479
1,1-Diiodoethane	: 5546
Diiodofluoromethane	: 7516
1,6-Diiodohexane	: 6769
Diiodomethane	: 7506
1,5-Diiodopentane	: 8758
3,5-Diiodo-α-phenylphloretic acid	: 2135
1,3-Diiodopropane	: 9907
2,2-Diiodopropane	: 9908
3,5-Diiodosalicylic acid	: 2617
Diiodotariric acid	: 8300
2,5-Diiodothiophene	: 11753
3,5-Diiodotyrosine	: 11928
Diisoamylamine	: 3226
Diisoamyl ether	: 3418
Diisoamyl phthalate	: 1335
Diisobutylamine	: 9823
Diisobutylcarbinol	: 6485
Diisobutylene	: 9064
Diisobutyl fumarate	: 3767

Diisobutyl ketone	: 6514
Diisobutyl oxalate	: 5557
Diisobutyl sulfide	: 10154
1,1-Diisobutylurea	: 11996
2,3-Diisonitrosobutane	: 3360
Diisopentyl ether	: 3418
Diisopentyl 2-hydroxyethylamine	: 5682
Diisopentyl phthalate	: 1335
Diisopentyl sulfide	: 3436
Diisopropanol	: 6799
Diisopropanolamine	: 10384
Diisopropenyl	: 3159
Diisopropenylacetylene	: 6714
Diisopropylamine	: 9821
N,N-Diisopropyl-2-aminoethanol	: 5683
Diisopropylbutadiyne	: 8345
Diisopropyl ether	: 10120
Diisopropyl fumarate	: 3766
Diisopropyl ketone	: 8967
Diisopropyl ketone, cyanohydrin	: 3408
Diisopropyl ketone oxime	: 8968
Diisopropyl sulfide	: 10152
Diketene	: 8598
Dilauroyl peroxide	: 9155
Diluran	: 32
Dimedazole	: 2380
Dimedone	: 4378
Dimefline	: 2819
Dimercaprol	: 10370
1,4-Dimercaptobutane	: 3368
1,2-Dimercaptoethane	: 5590
Dimethadione	: 8584
1,4:5,8-Dimethanonaphthalene, 1,2,3,4,10,10-hexachloro-6,7-epoxy-1,4,4a,5,6,7,8,8a-octahydro-, endo,exo-	: 5178
Dimethisoquin	: 5411
2,6-Dimethoxyaniline	: 703
2,4-Dimethoxybenzaldehyde	: 543
2,5-Dimethoxybenzaldehyde	: 544
3,5-Dimethoxybenzaldehyde	: 546
3,5-Dimethoxybenzoic acid	: 2577
2,6-Dimethoxybenzonitrile	: 2753
2,4-Dimethoxybenzophenone	: 7612
2,6-Dimethoxy-p-benzoquinone	: 4262
2,2'-Dimethoxybiphenyl	: 3045
3,3'-Dimethoxybiphenyl	: 3046
Dimethoxyborane	: 3117
5,7-Dimethoxycoumarin	: 2818
3,3'-Dimethoxy-4,4'-diaminodiphenyl	: 3033
2,5-Dimethoxy-2,5-dihydrofuran	: 6087
2,4-Dimethoxy-1-nitrobenzene	: 1448
1,3-Dimethoxy-4-nitrobenzene	: 1448
[(Dimethoxyphosphinothioyl)thio]butanedioic acid diethyl ester	: 3316
5,6-Dimethoxyphthalaldehydic acid	: 2608
1,10-Dimethoxy-6aα-aporphine-2,9-diol	: 5156
1,3-Dimethoxy-2-propanol	: 10372
2,6-Dimethoxypyridine	: 10935
2-[[[[[[(4,6-Dimethoxypyrimidin-2-yl)amino]carbonyl]amino]sulfonyl]methyl]benzoic acid, methyl ester	: 516
N1-(2,6-Dimethoxy-4-pyrimidinyl)sulfanilamide	: 2194
2-(4,6-DimethoxypyrimidiN-2-ylcarbamoylsulfamoyl)-N,N-dimethylnicotinamide	: 8069
2,6-Dimethoxy-p-quinone	: 4262
Dimethyl acetal	: 5548
N,N-Dimethylacetamide	: 55
3,3-Dimethylacetylacetone	: 8799
Dimethylacetylene	: 3898
Dimethylacetylenecarbinol	: 3922
3,3-Dimethylacrolein	: 3700
trans-2,3-Dimethylacrylic acid	: 3844
1,3-Dimethylallene	: 8660
7,8-Dimethylalloxazine	: 2472
Dimethylamide acetate	: 55
Dimethylamine	: 7448
4-(Dimethylamino)azobenzene	: 721
2',3-Dimethyl-4-aminoazobenzene	: 798
p-(Dimethylamino)azobenzene-o-carboxylic acid	: 2587
p-(Dimethylamino)benzal-5-rhodanine	: 11684

p-(Dimethylamino)benzoic acid	: 2585
4-(Dimethylamino)benzophenone	: 7613
3-(Dimethylamino)benzophenone	: 2771
4,4'-Dimethylaminobenzo-phenonimide	: 657
2-(Dimethylamino)ethanol	: 5709
β-Dimethylaminoethanol diphenylmethyl ether	: 5422
β-Dimethylaminoethyl methacrylate	: 10654
2-[[(2-Dimethylaminoethyl)(p-methoxybenzyl)amino]pyridine	: 5511
2-[(Dimethylamino)methyl]-cyclohexanone	: 4591
3-[[(Dimethylamino)methyl]indole	: 7310
4-Dimethylaminophenazone	: 10851
2-Dimethylamino-1-propanol	: 10374
4-[2-(Dimethylamino)propionamido]-antipyrine	: 9795
3-(Dimethylamino)propionitrile	: 10101
N,N-Dimethylaniline	: 713
1,3-Dimethylanthracene	: 377
Dimethylarsine	: 442
2,4-Dimethylbenzaldehyde	: 547
2,5-Dimethylbenzaldehyde	: 548
9,10-Dimethyl-1,2-benzanthracene	: 520
2,4-Dimethylbenzene-carboxaldehyde	: 547
3,3'-Dimethylbenzidine	: 3034
1,5-Dimethylbenzimidazole	: 2378
2,5-Dimethylbenzofuran	: 2445
2,6-Dimethylbenzofuran	: 2446
3,5-Dimethylbenzofuran	: 2448
4,7-Dimethylbenzofuran	: 2451
5,6-Dimethylbenzofuran	: 2452
5,7-Dimethylbenzofuran	: 2453
2,4-Dimethylbenzoic acid	: 2580
2,6-Dimethylbenzoic acid	: 2582
3,5-Dimethylbenzoic acid	: 2583
3,4-Dimethylbenzophenone	: 7615
2,4'-Dimethylbenzophenone	: 7628
2,5-Dimethylbenzoxazole	: 2879
α,α-Dimethylbenzyl alcohol	: 1857
m-α-Dimethylbenzyl alcohol	: 1858
p-α-Dimethylbenzyl alcohol	: 1859
Dimethylbenzylamine	: 1807
3,5-Dimethylbenzyl bromide	: 1057
2,3-Dimethylbiphenyl	: 3052
2,4-Dimethylbiphenyl	: 3054
3,4-Dimethylbiphenyl	: 3059
3,5-Dimethylbiphenyl	: 3060
2,3-Dimethylbutadiene	: 3159
1,1-Dimethylbutadiene	: 8670
Dimethylbutadiyne	: 6717
2,2-Dimethylbutane	: 3289
2,3-Dimethylbutane	: 3290
2,2-Dimethyl-1-butanol	: 3615
3,3-Dimethylbutan-2-ol (DL-)	: 3619
2,3-Dimethyl-1-butene	: 3759
2,3-Dimethyl-2-butene	: 3761
2,2-Dimethyl-3-butenoic acid	: 3835
4,4-Dimethyl-2-butenoic acid	: 9078
1,1-Dimethylbutyl acetate	: 8946
1,3-Dimethylbutyl acetate	: 8947
3,4'-Dimethylbutyrophenone	: 3670
Dimethyl cadmium	: 3929
Dimethylcarbamoyl chloride	: 3973
Dimethylchloroarsine	: 453
Dimethyl chloromaleate	: 3773
5,6-Dimethylchrysene	: 4115
trans-1,2-Dimethylcyclobutane	: 4188
cis-1,3-Dimethylcyclobutane	: 4189
trans-1,3-Dimethylcyclobutane	: 4190
Dimethyl trans-1,2-cyclobutanedicarboxylate	: 4181
1,3-Dimethyl-1,3-cyclohexadiene	: 4252
1,4-Dimethyl-1,3-cyclohexadiene	: 4253
1,5-Dimethyl-1,3-cyclohexadiene	: 4254
2,5-Dimethyl-1,3-cyclohexadiene	: 4255
1,1-Dimethylcyclohexane	: 4364
gem-Dimethylcyclohexane	: 4364
cis-1,2-Dimethylcyclohexane	: 4365
trans-1,2-Dimethylcyclohexane	: 4366
cis-1,3-Dimethylcyclohexane	: 4367
cis-1,4-Dimethylcyclohexane	: 4368
trans-1,3-Dimethylcyclohexane	: 4370
trans-1,4-Dimethylcyclohexane	: 4371

1,3-Dithiacyclopentane	: 5266
Dithianone	: 8056
1,3,5-Dithiazine, 4,5-dihydro-5-methyl	: 5264
Di-(2-thienyl) ketone	: 7617
3,3'-Dithiobis(2-aminopropanoic acid)	: 5020
2,4-Dithiobiuret	: 11718
4,4'-Dithiodimorpholine	: 7688
Dithioglycol	: 5590
Dithiooxamide	: 5588
Dithiosalicylic acid	: 1125
Dithizone	: 5121
1,2-Di-p-tolylethane	: 1590
2,4'-Ditolyl sulfide	: 1993
3,4'-Ditolyl sulfide	: 11570
Diundecyl ketone	: 11857
Diurea glyoxalate	: 7191
Diurite	: 2852
Diuron	: 12000
Diuroral	: 2854
Divinyl	: 3146
Divinylacetylene	: 6713
o-Divinylbenzene	: 1403
m-Divinylbenzene	: 1404
p-Divinylbenzene	: 1405
trans-1,2-Divinylcyclobutane	: 4187
Divinyl ether	: 5912
Djenkolic acid	: 5018
n-Docosane	: 5268
Docosane, 1-cyclohexyl-	: 4380
Docosane, 11-decyl	: 5273
Docosane, 1-phenyl-	: 1565
cis-13-Docosenol	: 5284
Docosylcyclopentane	: 4808
2,4-Dodecadienoic acid, 11-methoxy-3,7,11-trimethyl-, 1-methylethyl ester, (E,E)-	: 7651
2,4-Dodecadienoic acid, 3,7,11-trimethyl-, ethyl ester, (E,E)-	: 7182
Dodecahydrophenanthrene	: 9178
Dodecamethylcyclohexasiloxane	: 4624
Dodecamethylpentasiloxane	: 8997
Dodecanal dimethyl acetal	: 5297
1,12-Dodecanediol, dimethyl ether	: 5298
Dodecane, 1,1-di-(4-tolyl)	: 1567
Dodecane, 1-(1-naphthyl)-	: 7879
Dodecane, 1-(2-naphthyl)-	: 7880
N-Dodecanoyl-N-methylglycine sodium salt	: 6269
Dodecanoyl peroxide	: 9155
Dodecylmethylketone	: 11633
2-Dodecylnaphthalene	: 7880
Dodecyltrichlorosilane	: 11460
Donaxine	: 7310
Dormin	: 8679
2,6,8,10,12,14,16,18,20,22,24,26,30-Dotriacontatridecaene, 2,6,10,14,19,23,27,31-octamethyl-, (all-E)-	: 4053
Dotriacontylcyclohexane	: 4383
Dowicide 9	: 9255
Doxenitoin	: 7228
Draconic acid	: 2666
Dromostanolone propionate	: 351
Drosophilin A	: 9472
DTBP	: 9153
Dulcin	: 12013
Dulcitol	: 6194
Dulcose	: 6194
Duotal	: 9384
Duracaine	: 2493
Durene	: 2280
Durene aldehyde	: 598
Durohydroquinone	: 1552
Duroquinone	: 4270
1,2-Dyclodecanedione	: 4203
Dyphylline	: 10761
Dypnone	: 3876
Ecgonidine	: 491
Ecgonine	: 483
Echinochrome A	: 7857
Echinopsine	: 11330
Echitamine	: 4730
Edetic acid	: 6262
EEDQ	: 11231
Efloxate	: 225

Egonol	: 2464
Ehrlich's reagent	: 550
Eicosamethylnonasiloxane	: 8178
Eicosane, 1-cyclohexyl-	: 4384
Eicosane, 9-octyl	: 5354
Eicosane, 5-phenyl-	: 1119
Eicosane, 3-phenyl-	: 1695
Eicosane, 2-phenyl-	: 2007
Eicosane, 9-phenyl-	: 2068
Eicosane, 4-phenyl-	: 2170
cis-5,8,11,14-Eicosatetraenoic acid	: 5361
Δ 11-Eicosenoic acid	: 5363
5,8,11,14-Eicosotetraenoic acid	: 5360
Eiver-pick acid	: 7868
Elaidic acid	: 8297
Electrocortin	: 9746
β-Elemene	: 4389
Eleostearic acid	: 8246
Embelin	: 4261
Emetine	: 5368
Emodin	: 419
Enallylpropymal	: 11099
Enanthaldoxime	: 6346
Enanthylic acid	: 6455
Enheptin	: 11666
Enidin	: 2847
ENTPROL	: 10379
Eosin	: 11525
Ephdine	: 2849
(±)-Ephedrine	: 1900
Epiandrosterone	: 347
Epiborneol, acetate	: 2947
DL-α-Epichlorohydrin	: 8605
Epicholestanol	: 4108
Epicholesterol	: 4112
3,17-Epidihydroxyestratriene	: 5399
Epiestradiol	: 5400
16-Epiestriol	: 5405
Epihydrinaldehyde	: 8603
Epinephrine	: 1521
Epiquinidine	: 4125
EPN	: 9551
1,2-Epoxybutane	: 8615
2,3-Epoxypropanal	: 8603
1,2-Epoxypropane	: 8621
2,3 Epoxy 1-propanol	: 8618
2,3-Epoxypropyl ethyl ether	: 8614
Epsilan	: 2787
Epsilcapramin	: 6873
Equilenin	: 5397
Equilin	: 5398
Equol	: 2786
Ergobasinine	: 5373
Ergocornine	: 5387
Ergocorninine	: 5388
Ergocristine	: 5391
Ergocristinine	: 5392
Ergocryptine	: 5389
Ergocryptinine	: 5390
Ergoline-8α-carboxamide, 9,10-didehydro-N-((S)-2-hydroxy-1-methylethyl)-6-methyl-	: 5373
Ergometrinine	: 5373
Ergosine	: 5393
Ergostanol	: 5379
γ-Ergostenol	: 5384
α-Ergostenol	: 5385
Ergosterol	: 5383
Ergotamine	: 5394
Ergotaminine	: 5395
Ergothioneine	: 7204
Eriodictyol	: 2812
Eriodictyol 4'-monomethyl ether	: 2800
Erucic acid	: 5283
Erucyl alcohol	: 5284
Erysimupicrone	: 4049
Erythraric acid	: 3304
meso-Erythritol	: 3431
Erythritol	: 3431
Erythritol anhydride	: 3007
Erythrityl tetranitrate	: 3432
β-Erythroidine	: 8567
D-Erythrose	: 3192
Erythrosin	: 11524
L-Erythrulose	: 3679
Erytrin	: 31
Esculetin	: 2804

Esculin	: 2822
Esculoside	: 2822
d-3,17β-Estradiol	: 5399
Estradiol	: 5399
α-Estradiol	: 5400
17β-Estradiol 3-benzoate	: 5403
Estradiol benzoate	: 5403
Estragole	: 1943
Estra-1,3,5(10)-triene-3,17β-diol	: 5399
Estra-1,3,5(10)-triene-3,17α-diol	: 5400
Estra-1,3,5(10)-triene-3,16α,17β-triol	: 5404
Estratriol	: 5404
Estriol	: 5404
Estrone	: 5406
Ethacridine	: 317
Ethacrynic acid	: 168
Ethamivan	: 611
Ethanal	: 11
Ethanamide	: 22
Ethanamidine hydrochloride	: 5671
Ethanearsonic acid	: 457
Ethane, 1,2-bis(di-2-hydroxypropylamino)	: 10379
Ethane, 1-chloro-2-iodo-1,1,2-trifluoro	: 5490
1,2-Ethanedicarboxylic acid	: 3294
Ethane, 1,1-dichloro-2,2-bis(p-ethylphenyl)-	: 9160
Ethane, 1,2-dicyclohexyl-	: 4386
Ethane, 1,1-dicyclohexyl-	: 4392
Ethane, 1,2-dinitro-1,1,2,2-tetrafluoro	: 5642
Ethanedione	: 5491
1,2-Ethanediylbis(chloroformate)	: 3366
Ethane, 1-ethoxy-2-methylamino	: 5426
Ethane, 1-ethoxy-2-phenoxy-	: 1652
Ethanenitrile	: 271
Ethane, perbromo	: 5594
Ethanephosphonic acid, diethyl ester	: 9536
Ethane, 1,1,1,2-tetraphenyl-	: 1618
Ethanethiol, 1-phenyl (1)	: 1834
Ethanimidothioic acid, 2-(dimethylamino)-N-[[(methylamino)carbonyl]oxy]-2-oxo-, methyl ester	: 8568
3,5-Ethano-3H-pyrrolo[2,3 d]carbazole, curan-17-ol deriv.	: 4154
Ethanoic acid	: 100
Ethanoic anhydride	: 107
Ethanolamine	: 5675
Ethanolamine, N,N-diisobutyl	: 5684
Ethanolamine, N-isobutyl	: 5747
Ethanolamine, N-(β-naphthyl)	: 5750
Ethanol, 2-(2,4-dichlorophenoxy)-, hydrogen sulfate, sodium salt	: 11358
Ethanol, 1(3,4-dihydroxyphenyl)-2(isopropylamino)	: 1522
Ethanol, 2-(diisobutylamino)-	: 5684
Ethanol, 2-(diisopentylamino)-	: 5682
Ethanol, 2,2-diphenyl-	: 1615
Ethanol, 2-(isobutylamino)-	: 5747
Ethanol, 2,2'-oxybis-, diethyl ether	: 5612
Ethanol, 1(3-tolyl)	: 1858
Ethanoylaminoethanoic acid	: 6251
Ethanoyl chloride	: 294
11aH-12,3,6a-Ethanylylidene-7,9-methanonaphth[2,3-b]azocine, aconitane-3,8,13,14,15-pentol deriv.	: 305
Ethenylmethylphenylcarbinol	: 1869
Ethenzamide	: 615
Ether, allyl p-chlorophenyl	: 1232
Ether, allyl ethenyl	: 10565
Ether, allyl ethyl	: 10566
Ether, allyl isopropyl	: 10579
Ether, allyl methyl	: 10576
Ether, allyl phenyl	: 2167
Ether, p-allylphenyl methyl	: 1943
Ether, allyl propyl	: 10594
Ether, allyl o-tolyl	: 2026
Ether, allyl m-tolyl	: 2027
Ether, allyl p-tolyl	: 2028
Ether, allyl vinyl	: 10565
Ether, benzyl butyl	: 1115
Ether, benzyl chloromethyl	: 1181
Ether, benzyl ethyl	: 1658
Ether, benzyl isobutyl	: 2033

Ether, benzyl isopentyl	: 1951
Ether, benzyl methyl	: 1930
Ether, benzyl (3-methylbutyl)	: 1951
Ether, bis(2-bromoethyl)	: 5609
Ether, bis(bromomethyl)	: 7524
Ether, bis(p-bromophenyl)	: 2072
Ether, bis(2-butoxyethyl)	: 3417
Ether, bis(1-chloroethyl)	: 5610
Ether, bis(chloroethyl)	: 5611
Ether, bis(2-chloroethyl)	: 5611
Ether, bis(chloromethyl)	: 7525
Ether, bis(2-chloro-1-methylethyl)	: 10123
Ether, bis(p-chlorophenyl)	: 2073
Ether, bis(2-chloropropyl)	: 10121
Ether, bis(3-chloropropyl)	: 10122
Ether, bis(2-ethoxyethyl)	: 5612
Ether, bis[2-(2-methoxyethoxy)ethyl]	: 8995
Ether, bis(2-methoxyethyl)	: 5613
Ether, bis(o-methoxyphenyl)	: 2075
Ether, 2-bromethyl phenyl	: 1033
Ether, 4-bromobutyl phenyl	: 1003
Ether, 2-bromocyclohexyl methyl	: 4317
Ether, 2-bromoethyl ethyl	: 5461
Ether, 2-bromoethyl methyl	: 5465
Ether, 6-bromohexyl methyl	: 6736
Ether, bromomethyl ethyl	: 5464
Ether, bromomethyl methyl	: 7472
Ether, 4-bromophenyl hexyl	: 1045
Ether, p-bromophenyl hexyl	: 1045
Ether, p-bromophenyl phenyl	: 1079
Ether, 3-bromopropyl phenyl	: 1085
Ether, 1,3-butadienyl ethyl	: 3160
Ether, 2,3-butadienyl methyl	: 3167
Ether, 1,3-butadienyl methyl	: 3168
Ether, 1-buten-3-ynyl methyl	: 3891
Ether, butyl 2-chloroethyl	: 3249
Ether, butyl ethyl	: 3372
Ether, sec-butyl ethyl	: 3373
Ether, tert-butyl ethyl	: 10040
Ether, butyl ethynyl	: 3377
Ether, butyl isobutyl	: 3401
Ether, butyl isopropyl	: 3399
Ether, tert-butyl isopropyl	: 10081
Ether, butyl methyl	: 3391
Ether, tert-butyl methyl	: 10072
Ether, butyl phenyl	: 1114
Ether, tert-butyl phenyl	: 1458
Ether, sec-butyl phenyl	: 2030
Ether, butyl propyl	: 3421
Ether, tert-butyl propyl	: 10087
Ether, butyl m-tolyl	: 1116
Ether, butyl p-tolyl	: 1117
Ether, butyl vinyl	: 3369
Ether, sec-butyl vinyl	: 3370
Ether, tert-butyl vinyl	: 10036
Ether, 2-butynyl methyl	: 3915
Ether chloratus	: 5467
Ether, 1-chloroethyl 2-chloroethyl	: 5468
Ether, 1-chloroethyl ethyl	: 5473
Ether, 2-chloroethyl ethyl	: 5474
Ether, 1-chloroethyl methyl	: 5479
Ether, 2-chloroethyl methyl	: 5480
Ether, 1-chloroethyl propyl	: 9854
Ether, 2-chloroethyl vinyl	: 5887
Ether, chloromethyl ethyl	: 5478
Ether, 2-chloro-1-methylethyl methyl	: 9859
Ether, chloromethyl methyl	: 7483
Ether, chloromethyl propyl	: 9858
Ether, p-chlorophenyl phenyl	: 1221
Ether, 3-chloropropyl methyl	: 9860
Ether, 2-chloropropyl methyl	: 9861
Ether, 2-chloro-α,α,α-trifluoro-p-tolyl 3-ethoxy-4-nitrophenyl	: 8633
Ether, 2-chloro-α,α,α-trifluoro-p-tolyl p-nitrophenyl	: 8082
Ether, 2-chlorovinyl ethyl	: 5886
Ether, cinnamyl ethyl	: 1666
Ether cyanatus	: 10090
Ether, cyclohexyl methyl	: 4425
Ether, cyclohexyl phenyl	: 1253
Ether, cyclopropyl methyl	: 4980
Ether, decyl vinyl	: 5051
Ether, diallyl	: 10592
Ether, dibenzyl	: 2079
Ether, 1,2-dibromoethyl ethyl	: 5523
Ether, 2,3-dibromopropyl methyl	: 9888
Ether, dibutyl	: 3415

Ether, di-tert-butyl	: 10124
Ether, 1,2-dichloroethyl ethyl	: 5533
Ether, 2,4-dichlorophenyl p-nitrophenyl	: 8081
Ether, 1,2-dichlorovinyl ethyl	: 5899
Ether, dicyclohexyl	: 4461
Ether, diethyl	: 5608
Ether, diethyl,2-amino	: 5423
Ether, diethyl,2-bromo	: 5461
Ether, diethyl,1-chloro	: 5473
Ether, diethyl,2-chloro	: 5474
Ether, diethyl,2,2'-dibromo	: 5609
Ether, diethyl,1,2,-dichloro	: 5533
Ether, diethyl,2,2,'-dichloro	: 5611
Ether, diethyl,2,2'-dicyano	: 10114
Ether, diethyl,2,2'-diphenyl	: 2074
Ether, diethyl,1,1'-diphenyl(DL)	: 2081
Ether, difurfuryl	: 6156
Ether, dihexdecyl	: 6642
Ether, dihexyl	: 6842
Ether, dihexyl,2,2'-diethyl	: 6431
Ether, diisopropyl	: 10120
Ether, diisopropyl,β,β'-dichloro	: 10123
Ether, dimethyl	: 7523
Ether, dimethyl,chloro	: 7483
Ether, 2,2'-dinaphthyl	: 7940
Ether, 2,4-dinitrophenyl phenyl	: 1495
Ether, 2,4'-dinitro-4-trifluoromethyldiphenyl	: 5980
Ether, dioctyl	: 8416
Ether, dipentyl	: 8835
Ether, diphenyl	: 2071
Ether, diphenyl,4-bromo	: 1079
Ether, diphenyl,4-chloro	: 1221
Ether, diphenyl,4,4'-diamino	: 832
Ether, diphenyl,4,4'-dibromo	: 2072
Ether, diphenyl,4,4'-dichloro	: 2073
Ether, diphenyl,2,2'-dimethoxy	: 2075
Ether, diphenyl,2,2'-dimethyl	: 2076
Ether, diphenyl,3,3'-dimethyl	: 2077
Ether, diphenyl,4,4'-dimethyl	: 2078
Ether, diphenyl,2-methoxy	: 1938
Ether, diphenyl,3-methyl	: 2013
Ether, diphenyl,2-nitro	: 2053
Ether, diphenyl,4-nitro	: 2054
Ether, dipropyl	: 10119
Ether, dipropyl,2,2'-dichloro	: 10121
Ether, dipropyl,3,3'-dichloro	: 10122
Ether, divinyl	: 5912
Ether, 2,3-epoxypropyl ethyl	: 8614
Ether, 2,3-epoxypropyl phenyl	: 8623
Ether, ethyl ethynyl	: 5942
Ether, ethyl furfuryl	: 6104
Ether, ethyl heptyl	: 6406
Ether, ethyl hexyl	: 6814
Ether, 2-ethylhexyl vinyl	: 6405
Ether, ethyl isobutyl	: 10039
Ether, ethyl isopentyl	: 3374
Ether, ethyl isopropyl	: 10038
Ether, ethyl methyl	: 5602
Ether, ethyl 1-methyleneallyl	: 3161
Ether, ethyl octyl	: 8402
Ether, ethyl tert-pentyl	: 3375
Ether, ethyl pentyl	: 8808
Ether, ethyl propyl	: 10037
Ether, ethyl 1-propynyl	: 10732
Ether, ethyl 2-propynyl	: 10733
Ether, ethyl styryl	: 1651
Ether, ethyl vinyl	: 5905
Ether, ethynyl methyl	: 5944
Ether, ethynyl phenyl	: 1716
Ether, ethynyl propyl	: 10045
Ether, furfuryl methyl	: 6128
Ether, 2-furyl methyl	: 6157
Ether, heptyl methyl	: 6421
Ether, heptyl phenyl	: 1747
Ether, hexadecyl phenyl	: 1752
Ether, hexadecyl vinyl	: 6638
Ether, hexyl phenyl	: 1765
Ether hydrochloric	: 5467
Ether, iodomethyl methyl	: 7519
Ether, isobutyl methyl	: 10071
Ether, isobutyl phenyl	: 2031
Ether, isobutyl propyl	: 10086
Ether, isobutyl vinyl	: 10035
Ether, isopentyl methyl	: 3395
Ether, isopentyl vinyl	: 3371

Ether, isopropyl methyl	: 10070
Ether, isopropyl phenyl	: 1974
Ether, isopropyl propyl	: 10076
Ether, isopropyl m-tolyl	: 1987
Ether, isopropyl vinyl	: 10034
Ether, m-methoxyphenyl o-methoxyphenyl	: 1929
Ether, methyl 2-methylallyl	: 10577
Ether, methyl 2-methylbutyl	: 3394
Ether, methyl 1-methyleneallyl	: 3169
Ether, methyl 2-octynyl	: 8536
Ether, methyl tert-pentyl	: 3396
Ether, methyl pentyl	: 8822
Ether, methyl propyl	: 10069
Ether, methyl 2-propynyl	: 10736
Ether, methyl styryl	: 1928
Ether, methyl vinyl	: 5908
Ether muriatic	: 5467
Ether, 1-naphthyl 2-naphthyl	: 7928
Ether, 1-naphthyl phenyl	: 7945
Ether, 2-naphthyl phenyl	: 7946
Ether, o-nitrophenyl phenyl	: 2053
Ether, p-nitrophenyl phenyl	: 2054
Ether, p-nitrophenyl p-tolyl	: 2006
Ether, p-nitrophenyl α,α,α-trifluoro-2-nitro-p-tolyl	: 5980
Ether, octadecyl vinyl	: 8240
Ether, octyl phenyl	: 2069
Ether, octyl vinyl	: 8401
Ether, phenyl m-tolyl	: 2013
Ether, phenyl propyl	: 2168
Ether, phenyl o-tolyl	: 2012
Ether, phenyl vinyl	: 1646
Ether, propyl vinyl	: 10033
Ethide	: 5536
17α-Ethinyl-17β-estradiol	: 8216
Ethionamide	: 10885
Ethionine	: 991
Ethisterone	: 9758
Ethohexadiol	: 6802
Ethosuximide	: 11158
α-Ethoxyacetaldehyde	: 14
2-Ethoxyacetic acid	: 180
Ethoxyacetic acid methyl ester	: 182
2-Ethoxyacetophenone	: 5814
o-Ethoxybenzonitrile	: 2754
6-Ethoxy-2-benzothiazolesulfonamide	: 2865
2-Ethoxybenzylamine	: 1808
3-Ethoxybiphenyl	: 3071
N-(Ethoxycarbonyl)aniline	: 3965
Ethoxycarbonyl chloride	: 4018
Ethoxydiethylphosphine oxide	: 9527
2-Ethoxy-2,3-dihydro-3,3-dimethyl-5-benzofuranyl methanesulfonate	: 5933
2-Ethoxy-1,3-dinitrobenzene	: 1650
2-Ethoxyethyl acetate	: 5718
(±)-2-Ethoxy-1-ethyl-2-oxoethyl 5-[2-chloro-4-(trifluoromethyl)phenoxy]-2-nitrobenzoate	: 7410
5-Ethoxy-2-hydroxybenzaldehyde	: 556
(±)-2-[1-(Ethoxyimino)butyl]-5-[2-(ethylthio)propyl]-3-hydroxy-2-cyclohexen-1-one	: 11359
1-Ethoxy-2-methoxyethane	: 5591
o-Ethoxy-β-methylstyrene	: 1665
2-Ethoxynaphthalene	: 7884
2-Ethoxy-1-propanol	: 10380
1-Ethoxy-2-propanol	: 10381
2-Ethoxypropionitrile	: 10102
Ethoxyquin	: 11265
8-Ethoxy-5-quinolinesulfonic acid	: 11293
Ethoxytrifluorosilicon	: 11429
Ethoxzolamide	: 2865
N-Ethylacetamide	: 61
Ethyl acetate	: 186
Ethylacetic acid	: 3460
Ethyl acetoacetate	: 3571
Ethyl acetylacetate	: 3571
Ethylacetylene	: 3897
Ethyl acrylate	: 10635
Ethyl adipate	: 6785
Ethyl alcohol	: 5672
Ethylallene	: 8658
3-Ethyl-1-allyl-6-aminouracil	: 11060
1-Ethylallyl bromide	: 9009
1-Ethylallyl chloride	: 9017

γ-Ethylallyl chloride	: 9020
Ethyl allyl ether	: 10566
Ethylamine	: 5410
Ethyl aminoacetate	: 6263
Ethyl aminobenzoate	: 2497
Ethylammonium chloride	: 5433
N-Ethylamphetamine	: 1576
Ethylamylcarbinol	: 8445
Ethyl tert-amyl ether	: 3375
Ethyl amyl ether	: 8808
Ethyl amyl ketone	: 8472
2-Ethylaniline	: 740
p-Ethylaniline	: 742
N-Ethylaniline	: 743
Ethyl anisate	: 2671
9-Ethylanthracene	: 420
Ethyl anthranilate	: 2495
Ethyl arachate	: 5356
Ethyl arachidate	: 5356
Ethyl arsenate	: 439
Ethyl arsenite	: 440
Ethylarsine	: 444
Ethylarsonic acid	: 457
Ethyl behenate	: 5277
Ethylbenzene	: 1668
Ethyl benzeneacetate	: 899
Ethyl benzenesulfonate	: 2223
Ethyl benzoate	: 2602
Ethyl benzoin ether	: 5813
Ethyl benzoylacetate	: 2141
α-Ethylbenzyl alcohol	: 1870
Ethylbenzylamine	: 1810
Ethylbenzylaniline	: 1811
Ethyl benzyl ether	: 1658
Ethyl benzyl ketone	: 3677
Ethyl bromide	: 5454
Ethyl bromoacetate	: 115
2-Ethyl-2-bromobutyramide	: 3198
Ethyl bromochloroacetate	: 113
2-Ethyl-2-bromo-3-methylbutyramide	: 3199
Ethyl α-bromopropionate	: 10207
Ethyl butanoate	: 3515
2-Ethyl-1-butanol	: 3620
2 Ethyl 1 butene	: 8825
Ethyl 3-butenoate	: 3841
2-Ethylbutyl acetate	: 185
Ethylbutylamine	: 3218
Ethyl butyl ether	: 3372
Ethyl sec-butyl ether	: 3373
Ethyl tert-butyl ether	: 10040
Ethyl butyl ketone	: 6511
Ethyl sec-butyl ketone	: 6980
Ethyl tert-butyl ketone	: 8966
Ethyl butyl sulfide	: 3376
Ethyl butyrate	: 3515
2-Ethylbutyric acid	: 3514
α-Ethylbutyric acid	: 3514
Ethyl caprate	: 5063
Ethyl caproate	: 6884
Ethyl carbanilate	: 3965
Ethyl carbitol	: 5720
Ethyl carbonate	: 3999
Ethyl Cellosolve	: 5717
Ethyl cellosolve acetate	: 5718
Ethyl chloride	: 5467
Ethyl chloroacetate	: 130
Ethyl chlorocarbonate	: 4018
Ethyl β-chlorocrotonate	: 3823
Ethyl chloroethanoate	: 130
Ethyl chloroformate	: 4018
Ethyl 5-chloro-2-furancarboxylate	: 6053
Ethyl 2-chloro-3-methylbutanoate	: 3490
Ethyl chloromethyl ether	: 5478
Ethyl α-chloropropionate	: 10220
Ethyl β-chloropropionate	: 10221
Ethyl chlorosilicate	: 11462
Ethyl chlorosulfate	: 4085
Ethyl 4-chlorovalerate	: 8867
Ethyl 5-chlorovalerate	: 8868
Ethyl trans-cinnamate	: 10685
Ethyl cinnamylideneacetate	: 8680
Ethyl crotonate	: 3838
Ethyl trans-crotonate	: 3839
cis-Ethyl crotonate	: 3840
Ethyl cyanate	: 4163
Ethyl cyanide	: 10090

Ethyl cyanoacetate	: 152
Ethylcyclobutane	: 4191
Ethyl cyclobutanecarboxylate	: 4175
Ethyl cycloheptanecarboxylate	: 4224
Ethylcyclohexane	: 4390
1-Ethylcyclohexanol	: 4522
4-Ethylcyclohexanol	: 4525
4-Ethylcyclohexanone	: 4593
3-Ethyl-1-cyclohexene	: 4661
4-Ethyl-1-cyclohexene	: 4662
2-Ethylcyclohexylamine	: 4287
Ethyl cyclohexyl ketone	: 10423
Ethylcyclopentane	: 4812
4-Ethylcyclopentene	: 4926
Ethylcyclopropane	: 4974
Ethyl cyclopropanecarboxylate	: 4961
Ethyl cyclopropylcarboxylate	: 4961
Ethyl 2-diazopropionate	: 10236
Ethyl dibromoacetate	: 160
Ethyl dibromofluoroacetate	: 161
Ethyl 2,3-dibromopentanoate	: 8872
Ethyl dibunate	: 7968
Ethyl dichloroacetate	: 164
Ethyl dichlorophosphate	: 9577
Ethyl 3-(diethylamino)propionate	: 328
Ethyl diethylphosphinate	: 9527
Ethyl difluoroacetate	: 176
9-Ethyl-9,10-dihydroanthracene	: 421
Ethyl 2,3-dimethyl-2-butenoate	: 3836
Ethyl 4,4-dimethylcrotonate	: 9080
3-Ethyl-2,2-dimethylhexane	: 6816
3-Ethyl-2,3-dimethylhexane	: 6817
3-Ethyl-2,4-dimethylhexane	: 6818
3-Ethyl-2,5-dimethylhexane	: 6819
3-Ethyl-3,4-dimethylhexane	: 6820
4-Ethyl-2,3-dimethylhexane	: 6822
4-Ethyl-2,4-dimethylhexane	: 6823
4-Ethyl-3,3-dimethylhexane	: 6824
Ethyl 2,2-dimethylpropionate	: 10259
Ethyl 3,5-dinitrobenzoate	: 2593
Ethylene	: 5879
Ethylene, 1-bromo-2-chloro-	: 5881
Ethylene, 1-bromo-1,2-dichloro (dis)	: 5935
Ethylene, 1-bromo-1,2-difluoro-	: 5883
Ethylene bromohydrin	: 5685
Ethylenecarboxamide	: 10519
Ethylenecarboxylic acid	: 10609
Ethylene chlorohydrin	: 5693
Ethylenediamine	: 5496
Ethylenediamine, N-benzyl-N',N'-dimethyl-N-phenyl-	: 5506
Ethylenediamine, N,N-dimethyl-N'-ethyl	: 5510
Ethylene dibromide	: 5518
Ethylene, 1,1-dibromo-2-ethoxy	: 5893
Ethylene dichloride	: 5529
Ethylene dicyanide	: 3291
Ethylene diethanoate	: 5571
Ethylenedimercaptan	: 5590
Ethylenedisulfonic acid	: 5587
Ethylene fluorohydrin	: 5727
Ethylene glycol	: 5569
Ethylene glycol carbonate	: 5211
Ethylene glycol diacetate	: 5571
Ethylene glycol dibutyl ether	: 5527
Ethylene glycol didodecanoate	: 5311
Ethylene glycol diethyl ether	: 5542
Ethylene glycol diisobutyl ether	: 10032
Ethylene glycol dilaurate	: 5311
Ethylene glycol dimethyl ether	: 5549
Ethylene glycol dinitrate	: 5574
Ethylene glycol, dinitrite	: 8086
Ethylene glycol, methyl ethyl ether	: 5591
Ethylene glycol, monoacetate	: 5577
Ethylene glycol monoallyl ether	: 5763
Ethylene glycol monobenzyl ether	: 5761
Ethylene glycol monobutyl ether	: 5687
Ethylene glycol monoethyl ether	: 5717
Ethylene glycol monoethyl ether acetate	: 5718
Ethylene glycol monomethyl ether	: 5733
Ethylene glycol monopentyl ether	: 5757
Ethylene glycol monosalicylate	: 2622
Ethylene glycol, monovalerate	: 8885
Ethyleneimine	: 498
Ethylene iodohydrin	: 5731
1,8-Ethylenenaphthalene	: 6

Ethylene nitrite	: 8086
Ethylene oxide	: 8599
1,4-Ethylenepiperazine	: 5111
Ethylenethiourea	: 7225
Ethyleneurea	: 7227
Ethyl 2-ethoxybenzoate	: 2598
Ethyl 4-ethoxybenzoate	: 2599
Ethylethynylcarbinol	: 9140
17β-Ethyletiocholane	: 9741
Ethyl fluoride	: 5593
N-Ethylformamide	: 5994
Ethyl formate	: 6007
2-Ethylfuran	: 6105
Ethyl 2-furyl ether	: 6103
γ-Ethyl L-glutamate	: 6245
Ethyl glycidyl ether	: 8614
3-Ethylheptane	: 6407
4-Ethylheptane	: 6408
Ethyl heptyl ketone	: 5076
2-Ethylhexanal	: 6719
3-Ethylhexane	: 6815
Ethyl hexanoate	: 6884
2-Ethylhexanoic acid	: 6882
3-Ethylhexanoic acid	: 6883
2-Ethyl-2-hexanol	: 6500
2-Ethyl-1-hexanol	: 6930
2-Ethyl-2-hexene	: 6555
3-Ethyl-1-hexene	: 7033
Ethyl 2-hexenoate	: 7053
Ethyl 3-hexenoate	: 7054
2-Ethylhexyl acetate	: 187
2-Ethylhexyl acrylate	: 10636
2-Ethyl-1-hexylamine	: 6726
Ethyl hexyl ketone	: 8173
2-Ethylhexyl methacrylate	: 10659
Ethyl hydrogen peroxide	: 7179
Ethyl hydrogen phosphonate, aluminum salt	: 6020
Ethyl hydrogen sulfate	: 11576
Ethyl 4-hydroxybenzoate	: 2621
Ethyl 2-hydroxy-3-butenoate	: 3843
N-Ethylhydroxylamine	: 5434
Ethyl 4-hydroxypentanoate	: 8884
Ethyl hypochlorite	: 7184
Ethylidenecyclohexane	: 4391
2-Ethylidenecyclohexanone	: 4594
Ethylidenecyclopentane	: 4813
Ethylidene diacetate	: 5570
Ethylidene dibromide	: 5517
Ethylidene dichloride	: 5528
Ethylidene diethyl ether	: 5541
1-Ethylindan	: 7273
Ethyl iodide	: 5597
o-Ethyliodobenzene	: 1680
Ethyl isobutanoate	: 10298
Ethyl isobutyl sulfide	: 10044
Ethyl isocyanate	: 5598
Ethyl isopentyl ether	: 3374
1-Ethyl-2-isopropylacetylene	: 7122
Ethylisopropylamine	: 9810
Ethyl isopropyl ether	: 10038
Ethyl isopropyl ketone	: 8982
(RS)-5-Ethyl-2-(4-isopropyl-4-methyl-5-oxo-2-imidazolin-2-yl)nicotinic acid	: 7190
Ethyl isopropyl sulfide	: 10043
Ethyl isothiocyanate	: 5600
Ethyl isovalerate	: 3548
Ethyl laurate	: 5312
Ethyl levulinate	: 8907
Ethyl linoleate	: 8226
N-Ethylmaleimide	: 11122
Ethyl mercaptan	: 5648
Ethyl methacrylate	: 10658
Ethyl methanesulfonate	: 7531
Ethyl methoxyacetate	: 206
1-Ethyl-1-methylallene	: 8666
Ethyl methyl amine, hydrochloride	: 5435
Ethyl 2-(methylamino)benzoate	: 2680
1-Ethyl-2-methylbenzimidazole	: 2381
2-Ethyl-3-methylbut-1-ene	: 3797
Ethyl methyl carbonate	: 4003
cis-1-Ethyl-3-methylcyclohexane	: 4397
cis-1-Ethyl-4-methylcyclohexane	: 4399
trans-1-Ethyl-4-methylcyclohexane	: 4400
1-Ethyl-2-methylcyclohexanol	: 4526
1-Ethyl-3-methylcyclohexanol	: 4527

2-Ethyl-2-methylcyclohexanone	: 4595
1-Ethyl-4-methylcyclohexene	: 4663
1-Ethyl-1-methylcyclopentane	: 4814
cis-1-Ethyl-2-methylcyclopentane	: 4815
cis-1-Ethyl-3-methylcyclopentane	: 4817
trans-1-Ethyl-3-methylcyclopentane	: 4818
2-Ethyl-5-methylcyclopentanone	: 4883
cis-1-Ethyl-2-methylcyclopropane	: 4976
Ethyl methyl ether	: 5602
N-Ethyl-N '-(1-methylethyl)-6-(methylthio)1,3,5-triazine-2,4-diamine	: 341
2-Ethyl-5-methylfuran	: 6106
3-Ethyl-3-methylglutaric acid	: 8777
3-Ethyl-4-methylheptane	: 6411
3-Ethyl-5-methylheptane	: 6412
4-Ethyl-2-methylheptane	: 6413
4-Ethyl-3-methylheptane	: 6414
4-Ethyl-4-methylheptane	: 6415
5-Ethyl-5-methyl-3-heptyne	: 6619
3-Ethyl-2-methylhexane	: 6825
3-Ethyl-3-methylhexane	: 6826
4-Ethyl-3-methylhexane	: 6827
3-Ethyl-4-methylhexane	: 6827
4-Ethyl-2-methylhexane	: 6828
Ethyl 3-methylhexanoate	: 6899
Ethyl 4-methylhexanoate	: 6900
2-Ethyl-3-methylhexanoic acid	: 6886
3-Ethyl-3-methylhexanoic acid	: 6887
Ethyl 4-methyl-5-oxohexanoate	: 6902
3-Ethyl-2-methylpentane	: 8813
3-Ethyl-3-methylpentane	: 8814
3-Ethyl-3-methyl-1-pentene	: 9047
3-Ethyl-2-methyl-2-pentene	: 9049
Ethyl 4-methyl-2-pentenoate	: 9080
Ethyl 3-methyl-3-pentenoate	: 9081
Ethyl 4-methyl-3-pentenoate	: 9082
3-Ethyl-3-methyl-1-pentyne	: 9132
1-Ethyl-1-methylpropylbenzene	: 1691
1-Ethyl-3-methylpyrazole	: 10833
2-Ethyl-3-methylquinoline	: 11269
2-Ethyl-4-methylquinoline	: 11270
Ethyl methyl sulfide	: 5606
2-Ethyl-3-methylthiophene	: 11761
2-Ethyl-5-methylthiophene	: 11762
N-Ethylmorpholine	: 7694
1-Ethylnaphthalene	: 7885
2-Ethylnaphthalene	: 7886
Ethyl 1-naphthylacetate	: 7735
Ethyl nitrate	: 8074
Ethyl nitrite	: 8087
Ethyl nitroacetate	: 218
O-Ethyl O-p-nitrophenyl benzenethiophosphonate	: 9551
2-Ethyl-2-nitro-1,3-propanediol	: 10008
N-Ethyl-N-nitrosoethanamine	: 5432
N-Ethyl-N-nitrosourea	: 12015
Ethyl nonanoate	: 8150
N-Ethylnorephedrine	: 1871
3-Ethyloctane	: 8403
4-Ethyloctane	: 8404
Ethyl octanoate	: 8428
Ethyl octyl ketone	: 11959
Ethyl oenanthate	: 6462
Ethyl orthoformate	: 5604
Ethyl 3-oxo-3-phenylpropanoate	: 2141
Ethylparaben	: 2621
3-Ethylpentane	: 8809
2-Ethyl-2-pentanol	: 6946
2-Ethyl-1-pentene	: 9043
3-Ethyl-1-pentene	: 9044
3-Ethyl-2-pentene	: 9045
4-Ethyl-3-pentenoic acid	: 7060
Ethyl pentyl ether	: 8808
3-Ethyl-1-pentyne	: 9131
9-Ethylphenanthrene	: 9179
o-Ethylphenol	: 9353
m-Ethylphenol	: 9354
p-Ethylphenol	: 9355
Ethyl phenoxyacetate	: 229
Ethyl 4-phenoxybutyrate	: 3581
Ethyl 2-(4-phenoxyphenoxy)ethylcarbamate	: 5952
Ethyl phenylacetate	: 899
Ethyl N-phenylcarbamate	: 3965
1-Ethyl-1-phenylhydrazine	: 7159
1-Ethyl-2-phenylhydrazine	: 7160

Ethyl 5-phenyl-2,4-pentadienoate	: 8680
Ethyl N-phenylurethane	: 3965
Ethyl phosphate	: 9563
3-Ethyl-4-picoline	: 10964
4-Ethyl-2-picoline	: 10965
5-Ethyl-2-picoline	: 10966
N-Ethylpiperidine	: 9667
4-Ethylpiperidine	: 9670
1-Ethyl-3-piperidinol	: 9715
5-Ethyl-5-(1-piperidyl)barbituric acid	: 11096
Ethyl 2-propenyl ether	: 10566
Ethylpropiolic acid	: 9137
Ethyl propionate	: 10263
1-Ethyl-2-propylacetylene	: 6611
1-Ethyl-4-propylbenzene	: 1703
1-Ethylpropyl chloride	: 8721
Ethyl propyl ether	: 10037
Ethyl propyl ketone	: 6960
Ethyl propyl sulfide	: 10042
2-Ethylpyridine	: 10956
3-Ethylpyridine	: 10957
4-Ethylpyridine	: 10958
2-Ethylquinoline	: 11266
3-Ethylquinoline	: 11267
6-Ethylquinoline	: 11268
Ethyl salicylate	: 2619
Ethyl silicate	: 11506
Ethyl succinate	: 3303
Ethyl sulfate	: 11576
Ethyl sulfite	: 11581
Ethyl sulfocyanate	: 11703
Ethylsulfonal	: 3230
Ethylsulfonic acid	: 5622
2-(Ethylsulfonyl)ethanol	: 5725
Ethylsulfonylethyl alcohol	: 5725
Ethyl 2,3,3,3-tetrafluoropropionate	: 10328
2-Ethyltetrahydrofuran	: 6107
3-Ethyltetrahydrothiophene	: 11766
1-Ethyltetralin	: 7888
2-Ethyltetralin	: 7889
6-Ethyltetralin	: 7891
Ethyl tetraphosphate	: 11645
Ethyl thiocyanate	: 11703
2-(Ethylthio)ethanol	: 5726
2-Ethylthiophene	: 11759
3-Ethylthiophene	: 11760
3-Ethylthiophenol	: 2297
o-Ethyltoluene	: 1685
m-Ethyltoluene	: 1686
p-Ethyltoluene	: 1687
Ethyl p-toluenesulfonate	: 2231
N-Ethyl-4-toluidine	: 748
Ethyl m-tolyl sulfide	: 1706
Ethyl tosylate	: 2231
Ethyl tribromoacetate	: 242
Ethyl trichloroacetate	: 248
Ethyl(trichloromethyl)carbinol	: 3643
Ethyltrichlorosilane	: 11463
Ethyltrifluorosilane	: 11432
Ethyltrimethoxysilane	: 11433
1-Ethyl-2,3,5-trimethylbenzene	: 1709
1-Ethyl-2,4,5-trimethylbenzene	: 1710
5-Ethyl-1,2,3-trimethylbenzene	: 1713
Ethyltrimethylethylene	: 9034
3-Ethyl-2,2,3-trimethylpentane	: 8815
3-Ethyl-2,3,4-trimethylpentane	: 8816
Ethyl undecanoate	: 11946
Ethyl undecylate	: 11946
Ethyl urethane	: 3948
O-Ethylurethane	: 3948
Ethyl valerate	: 8878
Ethylvanillin	: 555
Ethylvinylacetylene	: 7103
Ethyl vinyl ether	: 5905
Ethyl vinyl ketone	: 9104
3-Ethyl-o-xylene	: 1670
Ethynylbenzene	: 1715
α-Ethynylbenzyl alcohol	: 1876
α-Ethynylbenzyl carbamate	: 1877
1-Ethynyl-1-cyclopentanol	: 4866
Ethynyldimethylcarbinol	: 3922
Ethynylestradiol	: 8216
Ethynyl methyl ketone	: 3923
Ethynylmethylpropylcarbinol	: 7128
Ethynylsilane	: 11435
Etiocholanic acid	: 346
Etofylline	: 10758

Eucalyptol	: 8560
Eucarvone	: 4220
Eucazulen	: 505
4αH-Eudesma-5,11(13)-dien-12-oic acid, 8β-hydroxy-, γ-lactone	: 8059
Eugenol	: 9389
Euonymit	: 6194
Euparin	: 5825
Euprax	: 7229
Evasprin	: 1913
Evodiamine	: 7333
Exaltol	: 4760
Exaltolide	: 8562
Fagarine	: 6179
(E,E)-α-Farnesene	: 5329
α-Farnesene	: 5329
β-Farnesene	: 5331
Felbinac	: 3009
Fenamic acid	: 2715
Fenamisal	: 2502
Fenchane(d)	: 2931
α-Fenchene (DL)	: 2923
l-α-Fenchyl alcohol	: 2941
Fenibut	: 2128
Fenicol	: 1870
Fenpiprane	: 9658
Fentanyl	: 9801
Fenticlor	: 9477
Fenyramidol	: 1913
Ferrocene, 1,1'-bis(3-phenyl-1,1,3,3-tetramethyldisiloxanyl)	: 5961
Ferron	: 11295
Fisetin	: 2813
Flacitran	: 2814
Flavan	: 2782
Flavanone, 4'-methoxy-3',5,7-trihydroxy	: 2801
Flavanone, 3',5,7-trihydroxy-4'-methoxy-, 7-(6-O-α-L-rhamnosyl-D-glucoside)	: 2794
Flavone	: 2840
Flexin	: 2876
Flopropione	: 10489
Floxuridine	: 12034
Flufenamic acid	: 2729
Fluimucetin	: 5015
Fluometuron	: 12007
Fluonilid	: 11802
Fluorakil 100	: 63
3',6'-Fluorandiol	: 11522
Fluorene	: 5967
Fluorene, 2-bromo-	: 5968
9H-Fluorene-9-carboxylic acid, 2-chloro-9-hydroxy-	: 4079
Fluorene, 9-(2-chlorobenzylidene)	: 5970
Fluorene, 9-methyl-	: 5974
N-2-Fluorenylacetamide	: 64
Fluorescein	: 11522
Fluorescein, 4',5'-diiodo	: 11523
Fluorescin	: 2572
Fluorine cyanide	: 4167
Fluoroacetic acid amide	: 63
Fluoroacetic acid, methyl ester	: 192
Fluoroacetylene	: 5943
o-Fluoroaniline	: 752
m-Fluoroaniline	: 753
p-Fluoroaniline	: 754
Fluorobenzene	: 1717
4-Fluorobenzeneacetic acid	: 901
p-Fluorobenzoic acid	: 2605
Fluorobromomethane	: 7470
2-Fluoro-1,3-butadiene	: 3162
1-Fluorobutane	: 3378
2-Fluorobutane	: 3379
4-Fluoro-1-butanol	: 3621
4-Fluorobutyric acid	: 3927
Fluorochlorobromomethane	: 7466
α-Fluorocinnamic acid	: 10637
Fluorocyclohexane	: 4403
1-Fluorodecane	: 5052
10-Fluorodecanoic acid	: 5064
10-Fluoro-1-decanol	: 5074
Fluorodichloroacetonitrile	: 279
1-Fluoro-1,1-dichloroethane	: 5534
Fluorodichloromethane	: 7502
Fluorodiiodomethane	: 7516
Fluoroethanoic acid	: 189

Glyceric acid, ethyl ester (*DL*)	: 10253
Glyceric acid, methyl ester, *DL-*	: 10254
Glyceric acid, 2-methyl, ethyl ester	: 10255
Glyceric acid, 2-methyl-, methyl ester	: 10256
Glycerin	: 10186
Glycerin tributyrate	: 3586
Glycerol	: 10186
Glycerol acetonide	: 5207
Glycerol β-chlorohydrin	: 9995
Glycerol diacetate	: 10189
Glycerol, 1,3-dimethyl ether	: 10372
Glycerol, 1,3-diphenyl ether-2-acetate	: 10378
Glycerol guaiacolate	: 10016
Glycerol, 1-laurate (*DL*)	: 5310
Glycerol, 1-(2-methoxyphenyl)ether	: 10016
Glycerol, 2-methyl ether	: 10015
Glycerol, 1-monoacetate(*DL*)	: 10187
Glycerol, 1-monododecanoate(*DL*)	: 5310
Glycerol, 1-phenyl	: 10194
Glycerol tributanoate	: 3586
Glycerol trimethyl ether	: 10185
Glyceryl *p*-aminobenzoate	: 10188
Glycidaldehyde	: 8603
Glycidol	: 8618
Glycidyl allyl ether	: 8626
Glycine anhydride	: 9612
Glycine, *N*,*N*-bis(phosphonomethyl)-	: 6277
Glycine, *N*-(chloroacetyl)-*N*-(2,6-diethylphenyl)-, ethyl ester	: 5173
Glycine cholate	: 6273
Glycine, *N*-(phosphonomethyl)-	: 6276
Glycinol	: 5675
Glycinonitrile, *N*,*N*-dimethyl-	: 281
Glycocholic acid	: 6273
Glycocyamine	: 6253
Glycolaldehyde	: 15
Glycol diacetate	: 5571
Glycolic acid	: 196
Glycolic aldehyde	: 15
Glycol salicylate	: 2622
Glyconitrile	: 282
Glycopyrrolate	: 11179
Glycopyrronium bromide	: 11179
Glycosine	: 11200
Glycosthene	: 6250
Glycurone	: 6239
Glycyl alcohol	: 10186
α-Glycylglycine	: 6265
Glycyrrhetinic acid glycoside	: 6230
Glycyrrhizic acid	: 6230
Glycyrrhizin	: 6230
Glyoxal	: 5491
Glyoxaldiureine	: 7191
Glyoxaline-5-alanine	: 7132
Glyoxylaldehyde	: 5491
Glyoxylic acid	: 223
Glyoxylic diureide	: 12009
Gramine	: 7310
Grasex	: 21
Griseofulvin, (+)-	: 11520
Guaiacol	: 9379
Guaiacol benzoate	: 9383
Guaiacol carbonate	: 9384
Guaiacol glycerol ether	: 10016
Guaiacol phosphate	: 9387
Guaiacol valerate	: 8886
Guaiazulene	: 505
Guai-1(5)-en-11-ol	: 507
Guaifenesin	: 10016
Guaiol	: 507
Guajen	: 7825
Guanidine, dodecyl-, monoacetate	: 5347
Guanidinoacetic acid	: 6253
Guanine	: 10768
Guanyl glycine	: 6253
Guanyl hydrazine	: 7147
Gulose, *D*-	: 6290
Guvacine	: 10912
Gynecormone	: 5403
Halazone	: 2565
Halon 2302	: 5526
Haloperidol	: 3651
Halotestin	: 355
Halothane	: 5458
Harmaline	: 11038

Harman	: 11040
Harmine	: 11039
Hecogenin	: 11534
Hederagenin	: 8545
Helenin	: 8059
Helicin	: 562
Heliotropic acid	: 2418
Heliotropin	: 2417
Helvolic acid	: 8211
Hematein	: 628
Hematin	: 5959
Hematoidin	: 3000
Hematoporphyrin	: 9735
Hematoxylin	: 629
Hemimellitic acid	: 2318
Hemin	: 5958
Hemipyocyanine	: 9197
Hendecane	: 11938
Heneicosylbenzene	: 1741
Hepin	: 6873
Heptachlor	: 7576
1,1,1,2,2,3,3-Heptachloropropane	: 10048
1,1,1,2,3,3,3-Heptachloropropane	: 10049
Heptacosylbenzene	: 1742
Heptadecane, 1-cyclohexyl-	: 4404
Heptadecane, 9-phenethyl-	: 2070
Heptadecane, 1-phenyl-	: 1743
Heptadecyl alcohol	: 6317
Heptadecylbenzene	: 1743
Heptadecylcyclohexane	: 4404
1,6-Heptadiene-4-ol, 4-methyl	: 6338
1,6-Heptadiene-3-yne	: 6341
Heptaldehyde	: 6344
Heptamethylcyclotetrasiloxane	: 4992
Heptamethylene chlorohydrin	: 6477
Heptamethylene dibromide	: 6374
Heptane, 1-cyclopentyl-	: 4820
Heptane, 1,1-diphenyl-	: 1746
Heptane, 4-isopropyl-	: 6427
Heptane, 2-methyl-3-ethyl	: 6409
4-Heptanol, 3-methyl(*DL*)	: 6504
4-Heptanol-2-one, 3-methyl	: 6518
3-Heptanol-5-one, 3-methyl	: 6519
4-Heptanol, 4-phenyl-	: 1867
1-Heptanol, 1-phenyl-	: 1881
3-Heptanol, 7-phenyl	: 2102
cis-2-Heptene	: 6535
trans-3-Heptene	: 6536
cis-3-Heptene	: 6537
1-Heptene, 1,1-diphenyl-	: 1744
2-Heptene, 3-methyl (*trans*)	: 6549
2-Heptene, 3-methyl (*cis*)	: 6550
2-Heptene, 4-methyl (*trans*)	: 6551
2-Heptene, 4-methyl (*cis*)	: 6552
1-Heptene-4-ol	: 6572
1-Heptene-4-ol, 4-methyl	: 6573
cis-2-Hepten-1-ol	: 6568
1-Hepten-5-one	: 6586
3-Hepten-2-one (*cis*)	: 6588
2-Hepten-6-one, 4-ethyl-2-methyl	: 6593
6-Hepten-4-yne-3-ol, 3-ethyl	: 6608
Heptyl acetate	: 193
Heptylacetylene	: 8196
Heptyl alcohol	: 6472
Heptylamine	: 6349
Heptyl bromide	: 6356
2-Heptyl bromide	: 6357
4-Heptyl bromide	: 6358
Heptyl chloride	: 6360
Heptylcyclopentane	: 4820
Heptyl formate	: 6008
Heptyl heptanoate	: 6464
Heptylidene acetone	: 5093
Heptyl mercaptan	: 6436
Heptyl methyl ketone	: 8172
1-Heptylnaphthalene	: 7895
2-Heptylnaphthalene	: 7896
Heptyl nitrite	: 8088
Heptyl propyl ketone	: 11960
Heptyl thiocyanate	: 11704
2-Heptynyl bromide	: 6613
4-Heptynyl chloride	: 6616
Herbicide 634	: 4900
Hesperetin	: 2800
Hesperetin 7-rutinoside	: 2794
Hesperidin	: 2794
Hexabromoethane	: 5594

Hexachlorane	: 4409
Hexachloroacetone	: 10440
Hexachlorobenzene	: 1749
Hexachloro-1,3-butadiene	: 3163
γ-Hexachlorocyclohexane	: 4409
Hexachloroethane	: 5595
1,2,3,4,5,6-Hexachloro-3-hexene	: 7035
Hexachlorophene	: 9409
1,1,1,2,2,3-Hexachloropropane	: 10051
1,1,1,3,3,3-Hexachloropropane	: 10052
1,1,2,2,3,3-Hexachloropropane	: 10053
Hexachloropropylene	: 10571
Hexacosane, 1-phenyl-	: 1750
Hexacose	: 6707
Hexacosylcyclohexane	: 4410
Hexacosylcyclopentane	: 4821
Hexadecamethylcyclooctasiloxane	: 4741
Hexadecamethylene dibromide	: 6635
Hexadecamethylheptasiloxane	: 6528
Hexadecanal dimethyl acetal	: 6636
Hexadecane, 1-cyclohexyl-	: 4411
Hexadecane, 1-cyclopentyl-	: 4822
Hexadecylcyclopentane	: 4822
Hexadecyl 3-hydroxy-2-naphthoate	: 7774
Hexadecyl mercaptan	: 6643
Hexadecyl pyridinium chloride	: 11030
1,5-Hexadiene, 2-methyl-3-ethyl	: 6694
1,3-Hexadiene, 1-phenyl	: 1753
1,5-Hexadiene, 1-phenyl-	: 1754
1,3-Hexadiene, 6-phenyl	: 1755
1,5-Hexadiene, 3-phenyl	: 1625
1,5-Hexadiene-3-yne	: 6713
1,3-Hexadien-5-ol	: 6708
1,5-Hexadienylbenzene	: 1754
Hexaethylbenzene	: 1756
1,1,1,3,3,3-Hexaethyldisiloxane	: 5235
Hexafluorenium bromide	: 6753
Hexafluoro-2,4-pentanedione	: 8802
Hexafluoropropylene oxide	: 8629
1,2,3,4,5,6-Hexahydroanthracene	: 422
Hexahydroaplotaxene	: 6322
Hexahydroazepine	: 492
Hexahydrobenzene	: 4305
Hexahydrobenzoic acid	: 4332
Hexahydrobenzyl alcohol	: 4419
Hexahydroequilenin	: 5402
Hexahydroindan	: 7281
Hexahydronicotinic acid	: 9641
Hexahydrophthalic anhydride	: 7345
Hexahydropyrazine	: 9602
3,3',4',5,5',7-Hexahydroxyflavylium chloride	: 2849
3,3',4',5,5',7-Hexahydroxy-2-phenylbenzopyrylium chloride	: 2849
Hexamethylbenzene	: 1758
Hexamethylcyclohexasiloxane	: 4625
1,1,3,3,5,5-Hexamethylcyclotrisilazane	: 5001
Hexamethylcyclotrisiloxane	: 5002
Hexamethyldisilane	: 5228
Hexamethyldisilazane	: 11363
Hexamethyldisiloxane	: 5236
Hexamethylenediamine	: 6749
Hexamethylene diiodide	: 6769
Hexamethylene diisocyanate	: 6770
Hexamethylene glycol	: 6796
Hexamethylenetetramine, allyl iodide	: 11825
Hexamethylenimine	: 492
Hexamethylolmelamine	: 7584
Hexane, 1-cyclohexyl-	: 4414
Hexane, 1-cyclopentyl-	: 4823
Hexane, 2,5-dibromo(*DL*)	: 6759
3,4-Dibromohexane	: 6760
Hexane, 2,5-dimethyl-2,5-diphenyl-	: 2281
Hexane, 2,2-dimethyl-3-ethyl	: 6816
Hexane, 2,2-dimethyl-4-ethyl	: 6821
Hexane, 2,3-dimethyl-3-ethyl	: 6817
Hexane, 2,3-dimethyl-4-ethyl	: 6822
Hexane, 2,4-dimethyl-3-ethyl	: 6818
Hexane, 2,4-dimethyl-4-ethyl	: 6823
Hexane, 2,5-dimethyl-3-ethyl	: 6819
Hexane, 3,3-dimethyl-4-ethyl	: 6824
Hexane, 3,4-dimethyl-3-ethyl	: 6820
Hexane, 2-iodo (*l*)	: 6832
Hexane, 3-isopropyl-2-methyl-	: 6839
Hexane, 2-methyl-3-isopropyl	: 6839

Hexane, 3-phenyl-	: 1669
Hexane, 1-phenyl-	: 1763
Hexane, 2-phenyl-	: 2009
5-Hexanoic acid	: 7051
Hexanoic acid, 2-amino(DL)	: 8215
Hexanoic acid, 6-cyclohexyl	: 4412
Hexanoic acid, 2-methyl(DL)	: 6893
Hexanoic acid, 2,2,3,4,4,5,6,6-octafluoro-3,5,6-trichloro	: 6910
3-Hexanol (l)	: 6915
3-Hexanol, 3,5-dimethyl (DL)	: 6928
1-Hexanol, 2-isopropyl-5-methyl-	: 6948
1-Hexanol, 2-methyl (DL)	: 6937
1-Hexanol, 4-methyl (DL)	: 6940
3-Hexanol, 5-methyl (DL)	: 6947
2-Hexanol-3-one, 2-methyl	: 6973
2-Hexanol-4-one, 5-methyl	: 6974
2-Hexanol, 2-phenyl-	: 1847
3-Hexanol, 3-phenyl	: 1875
2-Hexanol, 6-phenyl	: 2103
1-Hexanol, 3-phenyl	: 2159
Hexatriacontylcyclohexane	: 4413
Hexazole	: 11848
trans-2-Hexene	: 7000
cis-2-Hexene	: 7001
3-Hexene, 3-chloro (cis)	: 7009
1-Hexene, 1,2-dichloro (trans)	: 7011
1-Hexene 1,2-dichloro (cis)	: 7012
3-Hexene, 2,4-dimethyl (cis)	: 7031
3-Hexene, 3-(4-hydroxyphenyl)-4-(4-methoxyphenyl)	: 9358
5-Hexene-2-one,5-methyl	: 7099
2-Hexene, 1-phenyl	: 1760
2-Hexene, 2-phenyl	: 2008
2-Hexene-4-yne	: 7106
4-Hexenoic acid, 6-(4-hydroxy-6-methoxy-7-methyl-3-oxo-5-phthalanyl)-4-methyl-, (E)-	: 7052
2-Hexenoic acid, 3-phenyl-	: 4135
2-Hexen-4-ol, 5,5-dimethyl	: 7074
2-Hexen-4-ol, 2,4-dimethyl	: 7076
2-Hexen-5-one	: 7087
2-Hexen-5-one, 3,4-dimethyl	: 7092
3-Hexen-1-yl chloride	: 7008
Hex-5-en-1-yne	: 7105
Hexestrol	: 9298
Hexobarbital	: 11083
Hexocyclium methyl sulfate	: 9623
Hexylacetylene	: 8529
Hexylamine	: 6723
Hexylamine, 3-methyl-	: 6728
Hexylamine, 4-methyl-	: 6729
Hexyl bromide	: 6732
1-Hexyl-1,3-butadiene	: 5023
3-Hexylbutanol-1	: 8166
Hexyl chloride	: 6737
2-Hexyl chloride	: 6738
2-Hexyldecanoic acid	: 5065
7-Hexyldocosane	: 5274
Hexylene glycol	: 8793
Hexyl fluoride	: 6829
Hexyl formate	: 6009
9-Hexylheptadecane	: 6309
Hexyl heptanoate	: 6465
Hexyl iodide	: 6831
Hexyl mercaptan	: 6856
Hexyl methyl ether	: 6833
Hexyl methyl ketone	: 8471
1-Hexylnaphthalene	: 7900
Hexyl nitrite	: 8089
1-Hexylpiperidine	: 9672
Hexyl propyl ketone	: 5077
4-Hexylresorcinol	: 1519
Hexyltrichlorosilane	: 11464
Hexylvinylcarbinol	: 8189
Hexyl vinyl ether	: 6813
3-Hexyne, 2,5-dimethyl-2,5-dichloro	: 7117
3-Hexyne-2-one	: 7130
3-Hexynol	: 7126
Hippuric acid	: 6255
Hiragonic acid	: 6666
Histamine	: 7203
Histantin Richter	: 9502
Holocaine, hydrochloride	: 5670
Homatropine	: 911
Homochlorcyclizine	: 5140
Homochlorocyclizine	: 5140

Homogentisic acid	: 895
Homomesityl oxide	: 6600
o-Homosalicylic acid	: 2629
m-Homosalicylic acid	: 2630
p-Homosalicylic acid	: 2631
Homovanillic acid	: 913
Hordenine	: 9320
Humulene	: 5006
Humulon	: 4282
Hydantoin	: 7216
Hydantoin, 3-allyl-5-isobutyl-2-thio-	: 7229
Hydnocarpic acid	: 4940
Hydracrylic acid	: 10267
Hydracrylic acid, isopropyl ester	: 10277
Hydracrylonitrile	: 10106
Hydrangin	: 2826
Hydrastine	: 7357
Hydrastinine	: 5217
Hydrazinecarboxamide, N-amino-	: 4009
Hydrazinecarboxamide, 2-phenyl-N-(phenylimino)-	: 5122
Hydrazinecarboxylic acid, hydrazide	: 4009
Hydrazine, 1-isopentyl-1-phenyl	: 7162
Hydrazine, 1-methyl-2-(3-tolyl)	: 7163
Hydrazine, 1-phenyl-2-(3-tolyl)	: 7174
2-Hydrazinoethanol	: 5729
Hydrindantin	: 2999
Hydrocarbostyril	: 11327
Hydrochlorothiazide	: 2855
Hydrochol	: 4101
Hydrocinchonidine	: 4121
Hydrocinchonine	: 4122
Hydrocinnamaldehyde, p-isopropyl-α-methyl-	: 2117
Hydrocinnamaldehyde, o-methyl-	: 2115
Hydrocinnamic acid	: 2126
Hydrocinnamic acid, α-bromo-	: 2147
Hydrocinnamic acid, α,α-dimethyl-, methyl ester	: 2129
Hydrocinnamic acid, ester with 17β-hydroxyestr-4-en-3-one	: 5408
Hydrocinnamic acid, 4-(4-hydroxy-3-iodophenoxy)-3,5-diiodo-	: 2136
Hydrocinnamic aldehyde	: 2114
Hydrocinnamonitrile	: 2123
Hydrocinnamyl alcohol	: 2140
Hydrocodone	: 7678
Hydrocortisone	: 9753
Hydrocortisone acetate	: 9747
Hydrocotarnine	: 5214
Hydroflumethiazide	: 2858
Hydrofuramide	: 7488
Hydrogen carboxylic acid	: 6002
Hydrogen cyanate (HOCN)	: 4162
Hydrogen cyanide	: 7176
Hydrohydrastinine	: 5215
Hydroquinidine	: 4124
Hydroquinine	: 4123
p-Hydroquinone	: 1499
Hydroquinonecarboxylic acid	: 2570
Hydrosorbic acid	: 7050
Hydrosorbic acid, ethyl ester	: 7054
Hydrosorbic acid, 2-methyl-	: 7058
Hydrotrichlorothiazide	: 2854
2-Hydroxyacetaldehyde	: 15
Hydroxyacetic acid, propyl ester	: 199
4-Hydroxy-3-aminobutyric acid	: 3465
γ-Hydroxy-β-aminobutyric acid	: 3465
1-Hydroxy-3-aminomethyl-3,5,5-trimethylcyclohexane	: 4502
Hydroxyamphetamine	: 9215
2-Hydroxyaniline	: 9201
3-Hydroxyaniline	: 9202
4-Hydroxyaniline	: 9203
o-Hydroxyanisole	: 9379
p-Hydroxybenzaldehyde	: 566
Hydroxybenzene	: 9199
3-Hydroxybenzoic acid, 4-formyl	: 9598
3-Hydroxybenzoic acid, 6-iodo	: 2624
4-Hydroxybenzoic acid, 3-iodo	: 2623
2-Hydroxybenzoxazole	: 2881
o-Hydroxybenzyl alcohol	: 1882
p-Hydroxybenzyl alcohol	: 1884
2-Hydroxybiphenyl	: 3088
2-Hydroxy-3-butenoic acid	: 3842
Hydroxy-tert-butylamine	: 10341

N-(2-Hydroxybutyl)-N-(2-hydroxyethyl)amine	: 3623
3-Hydroxybutyraldehyde	: 3184
4'-Hydroxybutyranilide	: 3205
3-Hydroxycamphor	: 2957
3-Hydroxy-4-carboxyaniline	: 2498
3-Hydroxy-β-carotene	: 4061
2-Hydroxy-5-chloroaniline	: 9204
2-Hydroxy-3-chlorobiphenyl	: 3092
3α-Hydroxy-5α-cholestane	: 4108
3β-Hydroxy-5β-cholestanol	: 4109
3β-Hydroxycholest-5-ene	: 4110
3β-Hydroxycholest-4-ene	: 4111
DL-α-Hydroxycinnamic acid	: 2133
o-Hydroxycinnamic acid lactone	: 2792
Hydroxycodeinone	: 7675
7-Hydroxycoumarin 7-glucoside	: 2821
1-Hydroxycumene	: 1857
Hydroxycyclobutane	: 4196
Hydroxycyclopentane	: 4864
3-Hydroxy-2-p-cymenecarboxylic acid	: 2642
2-Hydroxy-1,4-dibromobutane	: 3614
12β-Hydroxydigitoxin	: 4041
2-Hydroxy-3,5-diiodobenzoic acid	: 2617
1-Hydroxy-2,11-dimethoxy-6aα-aporphine	: 5158
Hydroxydimethylphosphine oxide	: 9528
3-Hydroxy-1,1-dimethylpyrrolidinium bromide α-cyclopentylmandelate	: 11179
4-Hydroxydiphenyl	: 3090
3-Hydroxy-1,5-diphenylformazan	: 5122
16β-Hydroxyestradiol	: 5405
2-Hydroxyethanamine	: 5675
Hydroxyethanoic acid	: 196
2-Hydroxyethyl acetate	: 5577
2-Hydroxyethyl 2-hydroxybutyl amine	: 3623
1-Hydroxyethyl-4-methoxybenzene	: 1895
2-(2-Hydroxyethyl)piperidine	: 9663
4-(2-Hydroxyethyl)piperidine	: 9664
α-(2-Hydroxyethyl)styrene	: 2157
4-Hydroxy-1,6-heptadiene	: 6337
3-Hydroxy-1,5-hexadiene	: 6706
1-Hydroxy-2,4-hexadiene	: 6707
3-Hydroxy-2,5-hexanedione	: 6808
5-Hydroxy-2-hexanone	: 6969
5-Hydroxy-3-hexanone	: 6972
p-Hydroxyhydrocinnamic acid	: 2134
Hydroxyhydroquinone	: 2367
(Hydroxyimino)cyclohexane	: 4610
2-Hydroxyisobutyronitrile	: 10108
4-Hydroxyisophthalic acid	: 1352
β-Hydroxylactic acid	: 10252
Hydroxylamine, N,N-dipropyl-	: 9814
Hydroxylamine, N-ethyl-	: 5434
Hydroxylamine, N-methyl	: 7447
6α-Hydroxylithocholic acid	: 4096
Hydroxylupanine	: 7574
3-Hydroxy-5-methoxybenzaldehyde	: 573
1-Hydroxy-2-methoxybenzene	: 9379
4-Hydroxy-3-methoxybenzylacetone	: 3663
7-Hydroxy-4'-methoxyisoflavone	: 2830
3-Hydroxy-2-methoxy-5-methyl-p-benzoquinone	: 4265
(4-Hydroxy-3-methoxyphenyl)ethyl methyl ketone	: 3663
3-Hydroxy-4-methoxy-2,5-toluquinone	: 4265
p-Hydroxy-α-[(methylamino)-methyl]benzyl alcohol	: 1887
2-Hydroxymethylbenzofuran	: 2455
3-Hydroxy-3-methyl-1-butyne	: 3922
7-Hydroxy-4-methylcoumarin	: 2831
5-Hydroxymethylcytosine	: 11105
5-(Hydroxymethyl)-2-furaldehyde	: 6044
4-Hydroxy-4-methylheptane	: 6505
p-(Hydroxymethyl)nitrobenzene	: 1908
o-(Hydroxymethyl)phenol	: 1882
p-(Hydroxymethyl)phenol	: 1884
2'-Hydroxy-4'-methylpropiophenone	: 10445
3-Hydroxy-2-methylundecane	: 11956
1-Hydroxy-2-naphthoic acid	: 7771
3-Hydroxy-2-naphthoic acid	: 7772
2-Hydroxy-1-nitrobutane	: 3641
4-Hydroxynonanoic acid lactone	: 6149
2'-Hydroxypelargidenolon 1522	: 2816

4-Hydroxy-1-pentanol	: 8786
1-Hydroxy-2-pentanone	: 8971
2-Hydroxy-3-pentanone	: 8975
3-Hydroxy-1-pentyne	: 9140
1-Hydroxypent-2-yn-4-ene	: 9118
3-Hydroxy-9,10-phenanthraquinone	: 9176
2-(β-Hydroxyphenethyl-amino)pyridine	: 1913
Hydroxyphenylamine	: 9200
o-Hydroxyphenylamine	: 9201
m-Hydroxyphenylamine	: 9202
N-(4-Hydroxyphenyl)glycine	: 6267
1-Hydroxy-3-phenylpropane	: 2148
1-(m-Hydroxyphenyl)-1-propanol	: 1872
DL-2-Hydroxy-3-phenylpropionic acid	: 2133
4-Hydroxyphenylpropionic acid	: 2134
3-Hydroxy-3-phenylprop-1-yne	: 1876
4-Hydroxyphenylsulfonic acid	: 2226
17α-Hydroxyprogesterone	: 9750
4-Hydroxy-L-proline	: 9767
3-Hydroxy-1-propene	: 10697
2-Hydroxypropionitrile	: 10105
(3-Hydroxypropyl)benzene	: 2148
1-Hydroxy-1-propylcyclohexane	: 4557
3-Hydroxypropylene oxide	: 8618
N-(2-Hydroxypropyl)ethylenediamine	: 10340
3-Hydroxypropyl fluoride	: 10382
2-(2-Hydroxypropyl)pyridine	: 10951
2-(3-Hydroxypropyl)pyridine	: 11012
(2-Hydroxypropyl)trimethyl ammonium chloride	: 9831
4-Hydroxypyrazolo[3,4-d]pyrimidine	: 10852
5-Hydroxyquinoline	: 11305
7-Hydroxyquinoline	: 11307
8-Hydroxyquinoline	: 11308
8-Hydroxyquinoline sulfate	: 11324
8-Hydroxy-5-quinolinesulfonic acid	: 11294
p-Hydroxysalicylic acid	: 2569
5-Hydroxysalicylic acid	: 2570
2α-Hydroxytigogenin	: 11530
2-Hydroxy-m-toluic acid	: 2629
2-Hydroxy-p-toluic acid	: 2630
6-Hydroxy-m-toluic acid	: 2631
2-Hydroxytridecane	: 11884
11-Hydroxyundecanoic acid	: 11948
4-Hydroxyundecanoic acid lactone	: 6152
Hydroxyurea	: 12016
Hydroxyzine	: 5700
Hygrine	: 10471
Hymecromone	: 2831
Hyodeoxycholic acid	: 4096
Hyoscyamine	: 917
Hypothiazide	: 2855
Hypoxanthine	: 10770
Hypoxanthine riboside	: 7335
Ibogaine	: 7186
Ibrotamide	: 3199
Imecromone	: 2831
1H-Imidazole, 1-[2-(2,4-dichlorophenyl)-2-(2-propenyloxy)ethyl]-	: 7187
1H-Imidazole, 2-heptadecyl-4,5-dihydro-, monoacetate	: 6275
2-Imidazolidine, 2-benzyl	: 7198
1-Imidazolidinecarboxamide, 3-(3,5-dichlorophenyl)-N-(1-methylethyl)-2,4-dioxo-	: 7338
7H-Imidazo[4,5-d]pyrimidine	: 10752
2-Iminobenzothiazoline	: 2859
Iminodiacetic acid	: 6260
2,2'-Iminodibenzoic acid	: 2656
Iminourea	: 6278
Imperatorin	: 6189
Indaconitine	: 307
Indan	: 7250
Indan, 1-amino (DL)	: 7243
Indan, 1,2-dibromo-	: 7247
Indan, 1,2-dichloro	: 7249
Indan, 1,1-diemthyl	: 7251
Indan, 1,2-dimethyl-	: 7252
Indan, 4,7-dimethyl-	: 7253
Indan, 5,6-dimethyl-	: 7254
Indan, 1-ethyl-	: 7273
Indan, 5-hexyl-	: 7274
Indan, 1-methyl-	: 7255
Indan, 2-methyl-	: 7256

Indan, 4-methyl-	: 7257
Indan, 5-methyl-	: 7258
2-Indanone	: 7287
1-Indanone, 4-methyl-	: 7290
1-Indanone, 2-phenyl-	: 7291
Indan, 2-phenyl-	: 7259
Indan, 1-phenyl-1,3,3-trimethyl	: 7263
Indan, 1,1,4,7-tetramethyl-	: 7260
Indanthrene	: 434
Indan, 1,1,5-trimethyl-	: 7261
Indan, 1,4,7-trimethyl	: 7262
1-Indenecarboxylic acid	: 7246
Indene, 1,3-diphenyl-	: 7271
Indene, 2,3-diphenyl-	: 7272
Indene, 4-methyl-	: 7278
Indene, 6-methyl-	: 7279
Indene, 7-methyl-	: 7280
Indigo	: 7331
Indigo Yellow	: 2843
Indoleacetic acid	: 7295
Indolebutyric acid	: 7296
Indole, 7-methyl-	: 7317
Indole, 1,2,3-trimethyl	: 7322
Indole, 2,3,5-trimethyl-	: 7323
2-Indolinone, 1-ethyl-3-methyl-	: 7332
1H-Indolo[3,2,1-de]pyrido[3,2,1-ij][1,5]naphthyridine, eburnamenine-14-carboxylic acid deriv.	: 5349
Indolo[4,3-fg]quinoline, ergoline-8-carboxamide deriv.	: 5373
3-Indolylacetone	: 10442
Indonaphthene	: 7245
Indoxazene	: 2389
Inosine, 2-amino-	: 6289
Iodipamide	: 2594
5-Iodoacenaphthene	: 7
Iodoacetic acid	: 200
Iodoacetonitrile	: 283
Iodoalphionic acid	: 2135
p-Iodoaniline	: 761
o-Iodoanisole	: 1775
Iodobenzene	: 1766
p-Iodobenzenesulfonyl chloride	: 2247
o-Iodobenzoic acid	: 2657
4-Iodo-1,2-butadiene	: 3166
4-Iodocatechol	: 1523
Iodochlorhydroxyquin	: 11312
Iodocyanide (ICN)	: 7337
Iodocyclohexane	: 4415
Iodocyclopentane	: 4824
2-Iodo-p-cymene	: 1781
1-Iododecane	: 5053
Iododifluoromethane	: 7505
1-Iodo-3,5-dimethylbenzene	: 1771
4-Iododiphenyl	: 3079
1-Iododecane	: 5302
Iodoethane	: 5597
2-Iodoethanol	: 5731
Iodoethene	: 5907
Iodoethylene	: 5907
Iodoform	: 7560
2-Iodofuran	: 6110
3-Iodofuran	: 6111
1-Iodoheptane	: 6419
2-Iodoheptane	: 6420
1-Iodo-1-heptyne	: 6620
1-Iodohexadecane	: 6640
1-Iodohexane	: 6831
Iodomethane	: 7518
2-Iodo-4-methylaniline	: 762
4-Iodo-2-methylbutane	: 3384
(Iodomethyl) methyl ether	: 7519
1-Iodonaphthalene	: 7902
1-Iodooctadecane	: 8242
1-Iodooctane	: 8406
1-Iodopentane	: 8818
3-Iodopentane	: 8820
1-Iodo-1-pentyne	: 9133
o-Iodophenol	: 9371
p-Iodophenol	: 9373
1-Iodopropane	: 10054
2-Iodopropane	: 10055
3-Iodopropene	: 10573
3-Iodopropylene	: 10573
1-Iodopropyne	: 10734
3-Iodopropyne	: 10735
2-Iodopyridine	: 10970

Iodopyrrole	: 11140
Iodoquinol	: 11315
4-Iodoquinoline	: 11272
Iodostarin	: 8300
N-Iodosuccinimide	: 11159
Iodotrifluoroethylene	: 5925
Iodotrifluoromethane	: 7559
3-Iodotyrosine	: 11932
3-Iodoveratrole	: 1767
4-Iodo-o-xylene	: 1774
β-Ionol	: 3873
α-Ionol	: 3874
Irene	: 7999
Iron dicyclopentadienyl	: 5960
β-Irone	: 3883
α-Irone	: 3884
Iron pentacarbonyl	: 7339
Iron, tris(dimethylcarbamodithioato-S,S')-, (OC-6-11)-	: 5957
Isatin	: 7300
Isoamylaldehyde	: 3186
Isoamylamine	: 3220
Isoamyl benzoate	: 3632
Isoamyl bromide	: 3239
Isoamyl chloride	: 3252
Isoamyl cyanide	: 8831
Isoamyl ether	: 3418
Isoamyl iodide	: 3384
Isoamyl isovalerate	: 3551
Isoamyl 3-methylbutyrate	: 3551
Isoamyl nitrate	: 3637
Isoamyl nitrite	: 8090
Isoamyl salicylate	: 2633
Isoamyl sulfide	: 3436
Isobutamben	: 2505
Isobutanal	: 9785
Isobutane	: 10073
Isobutane, 1,1-dichloro	: 9899
Isobutane, 1,2-dichloro	: 9900
Isobutane, 1,1,1,2,3-pentachloro	: 10127
Isobutane, 1,1,1,2-tetrachloro	: 10147
Isobutane, 1,1,2-trichloro	: 10177
Isobutene	: 10578
2-Isobutoxyethanol	: 5746
1-Isobutoxy-2-nitrobenzene	: 2034
Isobutryic anhydride	: 10291
Isobutyl acetate	: 215
Isobutyl acrylate	: 10673
Isobutyl alcohol	: 10388
Isobutyl alcohol, 2-chloro	: 10352
Isobutyl alcohol, 3-chloro	: 10353
Isobutylamine	: 9817
Isobutyl p-aminobenzoate	: 2505
Isobutylbenzene	: 2039
Isobutyl bromide	: 9845
Isobutyl butanoate	: 3563
Isobutyl carbamate	: 3959
Isobutyl chloride	: 9862
Isobutyl chlorocarbonate	: 4024
Isobutyl chloroformate	: 4024
Isobutyl cyanide	: 3410
Isobutylcyclopentane	: 4838
Isobutyldimethylcarbinol	: 8928
Isobutylene chlorohydrin	: 10354
Isobutylene, 1,3-dibromo	: 10546
Isobutylene glycol, 3-chloro	: 9998
Isobutylene (trimer)	: 10582
Isobutylethene	: 9054
Isobutyl formate	: 6013
Isobutylhydrochlorothiazide	: 2856
Isobutyl iodide	: 10057
Isobutyl isobutanoate	: 10304
Isobutyl isocyanate	: 10061
Isobutyl isovalerate	: 3555
Isobutyl ketone	: 6514
Isobutyl mercaptan	: 10158
Isobutyl methacrylate	: 10666
Isobutyl methyl ether	: 10071
Isobutyl methyl ketone	: 8981
Isobutyl nitrate	: 8078
Isobutyl nitrite	: 8094
Isobutyloxycarbonyl chloride	: 4024
Isobutyl propionate	: 10310
Isobutyl propyl ketone	: 6524
4-Isobutylresorcinol	: 1539
Isobutyl salicylate	: 2643
Isobutyl stearate	: 8268

Isobutyl thiocyanate	: 11707
Isobutyl 2-tolyl ketone	: 3669
Isobutyl 4-tolyl ketone	: 3670
Isobutyl urethane	: 3961
Isobutyl vinyl ether	: 10035
Isobutyl vinyl ketone	: 7093
Isobutyramide, 2-bromo	: 9792
Isobutyric acid	: 10290
Isobutyric acid chloride	: 10504
Isobutyric acid, 2-chloroethyl ester	: 10292
Isobutyric acid, cyclohexyl ester	: 10293
Isobutyric acid, (2-nitropentyl) ester	: 10305
Isobutyric acid, 3-phenyl	: 2137
1-Isobutyronaphthone	: 10464
2-Isobutyronaphthone	: 10465
Isobutyronitrile	: 10111
Isobutyronitrile, 2-hydroxy-3-chloro	: 10098
Isobutyryl chloride, 3-chloro	: 10499
Isocaine	: 2505
Isocamphoric acid, (DL)	: 3931
Isochroman	: 2776
Isochrysene	: 11910
Isocinchomeronic acid	: 10929
Isocitrate lactone	: 6079
Isocorybulbine	: 5150
L-(+)-Isocorydine	: 5161
Isocorydine	: 5161
Isocoumarin	: 2791
Isodurene	: 2279
Isoestradiol	: 5401
8-Isoestrone	: 5407
Isoeugenol	: 9388
Isoeugenol, acetate (cis)	: 9394
Isoflavone	: 2841
Isoflavone, 4',7-dihydroxy-	: 2827
Isoflavone, 4',5-dihydroxy-7-methoxy-	: 2828
Isoflurophate	: 9580
Isogeronic acid	: 6461
Isoguanine	: 10767
Isohexane	: 8823
Isohexyl alcohol	: 8939
1H-Isoindole-1,3(2H)-dione, 2-[(trichloromethyl)thio]-	: 5984
Isolysergic acid	: 5375
β-Isomalic acid	: 9959
Isometheptene	: 6532
Isoniazid	: 10905
Isonicotinic acid	: 10895
Isonicotinic acid diethylamide	: 10892
Isonicotinic acid, (α-methylfurfurylidene)hydrazide	: 10904
Isonipecotic acid	: 9642
Isonipecotic acid, methyl ester	: 9644
Isonipecotic acid, 1-methyl-, methyl ester	: 9645
Isonitrosoacetone	: 9789
Isonitrosoacetophenone	: 876
β-Isonitrosopropane	: 10476
Isooctane	: 8848
Isopentadiene	: 3171
Isopentanal	: 3186
Isopentane	: 3397
Isopentanoic acid	: 3539
Isopentanoyl chloride	: 3693
Isopentyl acetate	: 3631
Isopentyl alcohol	: 3627
Isopentyl alcohol, orthoformate (3:1)	: 8552
Isopentyl alcohol, stearate	: 8265
Isopentylamine	: 3220
Isopentyl bromide	: 3239
Isopentyl 3-bromopropionate	: 10210
Isopentyl butanoate	: 3544
Isopentyl chloride	: 3252
N-Isopentylethanolamine	: 5739
Isopentyl ether	: 3418
Isopentyl formate	: 3636
Isopentyl iodide	: 3384
Isopentyl isovalerate	: 3551
Isopentyl mercaptan	: 3440
Isopentyl 3-methylbutanoate	: 3551
Isopentyl β-Naphthyl ether	: 7918
Isopentyl oleate	: 8306
Isopentyl propanoate	: 3639
Isopentyl salicylate	: 2633
2-Isopentylthiophene	: 11771
Isopentyl vinyl ether	: 3371

Isophorol	: 4702
Isophorone	: 4721
Isophthalamide, N,N,N',N'-tetraethyl	: 1326
Isophthalic acid	: 1328
Isophthalic acid, diethyl ester	: 1346
Isophytol	: 6671
Isopilosine	: 6144
Isoprene	: 3171
Isopropanesulfonic acid	: 10130
Isopropenyl acetate	: 10698
Isopropenylacetylene	: 3892
Isopropenyl carbinol	: 10702
Isopropenyl chloride	: 10535
Isopropenyl chloroformate	: 4022
4-Isopropenyl-1-cyclohexenecarboxaldehyde	: 4633
4-Isopropenyl-1-methylcyclohexene	: 4677
Isopropenyl methyl ketone	: 3879
9-Isopropenylphenanthrene	: 9182
2-Isopropoxyethanol	: 5740
1-Isopropoxy-2-propanol	: 10390
N-Isopropylacetanilide	: 74
Isopropyl acetate	: 209
Isopropylacetic acid	: 3539
Isopropyl acetoacetate	: 3574
Isopropyl alcohol	: 10335
Isopropylamine	: 9804
2-Isopropylaniline	: 788
N-Isopropylaniline	: 790
Isopropylbenzene hydroperoxide	: 7181
p-Isopropylbenzoic acid	: 2689
α-Isopropylbenzyl alcohol	: 1901
4-Isopropylbenzyl alcohol	: 1902
Isopropyl borate	: 3128
Isopropyl bromide	: 9837
α-Isopropyl-α-bromobutyramide	: 3199
Isopropyl-tert-butylcarbinol	: 8955
Isopropyl butyrate	: 3549
Isopropyl carbamate	: 3955
Isopropyl carbanilate	: 3966
Isopropyl chloride	: 9849
Isopropyl chloroacetate	: 139
Isopropyl 3-chlorocarbanilate	: 3939
Isopropyl chloroformate	: 4023
Isopropyl N-(3-chlorophenyl)carbamate	: 3030
Isopropyl 2-chloropropionate	: 10223
Isopropyl trans-cinnamate	: 10687
Isopropylcyclohexane	: 4431
cis-2-Isopropylcyclohexanol	: 4539
cis-4-Isopropylcyclohexanol	: 4540
trans-4-Isopropylcyclohexanol	: 4541
2-Isopropylcyclohexanone	: 4599
4-Isopropylcyclohexanone	: 4600
Isopropylcyclopentane	: 4828
2-Isopropylcyclopentanone	: 4887
7-Isopropyl-1,4-dimethylazulene	: 505
N-Isopropylethanolamine	: 5741
Isopropyl fluoride	: 10047
Isopropyl formate	: 6011
Isopropyl heptafluorobutyrate	: 3526
4-Isopropylheptane	: 6427
Isopropylidene glycerol	: 5207
Isopropyl iodide	: 10055
Isopropyl isobutyrate	: 10303
Isopropyl isocyanate	: 10060
Isopropyl isocyanide	: 10063
Isopropyl isonitrile	: 10063
Isopropyl lactate	: 10275
Isopropyl mercaptan	: 10157
Isopropylmethylamine	: 9819
4-Isopropyl-1-methylcyclopentene	: 4927
Isopropyl methyl ether	: 10070
2-Isopropyl-5-methyl-1-hexanol	: 6948
2-(4-Isopropyl-4-methyl-5-oxo-2-imidazolin-2-yl)nicotinic acid	: 7188
(RS)-2-(4-Isopropyl-4-methyl-5-oxo-2-imidazolin-2-yl)quinoline-3-carboxylic acid	: 7189
Isopropyl methyl sulfide	: 10089
Isopropyl myristate	: 11626
Isopropyl 1-naphthyl ketone	: 10464
Isopropyl nitrate	: 8076
Isopropyl nitrite	: 8092
2-Isopropyl-1-pentene	: 6836
Isopropyl phenylcarbamate	: 3966

Isopropylphenylcarbinol	: 1901
3-(p-Isopropylphenyl)-2-methylpropionaldehyde	: 2117
Isopropyl phenyl urethane	: 3966
Isopropyl propionate	: 10299
Isopropylpropylamine	: 9827
Isopropyl propyl sulfide	: 10077
3-Isopropylpyridine	: 10985
Isopropyl salicylate	: 2635
Isopropyl 2-thienyl ketone	: 10472
Isopropyl thiocyanate	: 11706
2-Isopropylthiophene	: 11772
3-Isopropylthiophene	: 11773
Isoquinaldine	: 7381
Isoquinoline, 5-amino-	: 7372
Isoquinoline, 3-benzyl	: 7387
Isoquinoline, 3-butyl-1-[2-(dimethylamino)ethoxy]-	: 5411
Isoquinoline, 6,7-diethoxy-1(3,4-diethoxybenzyl)	: 7377
Isoquinoline, 6,7-dimethoxy-1,2-dimethyl-1,2,3,4-tetrahydro	: 7391
Isoquinoline, 8-methoxy-2-methyl-6,7-methylenedioxy-1,2,3,4-tetrahydro	: 5214
Isorubijervine	: 11516
(+)-Isothebaine	: 5158
Isothebaine	: 5158
3(2H)-Isothiazolone, 2-octyl-	: 8528
Isothiocyanic acid, acetyl	: 303
Isothiocyanic acid, tert-butyl ester	: 10066
Isovaleraldehyde	: 3186
Isovaleramide	: 3206
Isovaleric acid	: 3539
Isovaleric acid chloride	: 3693
Isovaleric aldehyde	: 3186
Isovaleric amide	: 3206
Isovalerylamine	: 3220
Isovaleryl chloride	: 3693
Isovaleryl diethylamide	: 3201
Isoxylaldehyde	: 548
Itaconic acid	: 3328
Jasmone	: 4949
Javanicin	: 7855
Jervine	: 12046
Jilkon	: 6200
Juglone	: 7860
Julolidine	: 2737
Julolidine, 1,6-dioxo	: 2736
Kaempferol	: 2843
Kainic acid	: 11144
Kelthane	: 1851
Ketamine	: 4579
Ketene	: 5927
Ketene, diethyl-	: 3877
β-Ketobutyranilide	: 3207
α-Ketoglutaric acid	: 8781
3-Ketoglutaric acid	: 8782
Ketohydroxyestrin	: 5406
Ketone, 2-benzofuranyl methyl	: 5775
Ketone, 2-benzofuranyl phenyl	: 7588
Ketone, benzyl phenyl	: 5812
Ketone, 5-bromo-2-thienyl methyl	: 5783
Ketone, α-chlorobenzyl phenyl	: 5785
Ketone, 5-chloro-2-thienyl methyl	: 5791
Ketone, cyano phenyl	: 968
Ketone, cyclobutyl methyl	: 5792
Ketone, cyclobutyl phenyl	: 7605
Ketone, 1-cyclohexen-1-yl methyl	: 5793
Ketone, cyclohexyl methyl	: 5794
Ketone, cyclopentyl methyl	: 5795
Ketone, cyclopropyl methyl	: 5796
Ketone, dicyclohexyl	: 7606
Ketone, dicyclopropyl	: 7607
Ketone, 6,7-dimethoxy-1-isoquinolyl 3,4-dimethoxyphenyl	: 7611
Ketone, 10-[3-(dimethylamino)-propyl]phenothiazin-2-yl methyl	: 5807
Ketone, 2-furyl methyl	: 5817
Ketone, 2-furyl phenyl	: 7618
Ketone, 1-hydroxycyclohexyl methyl	: 5821
Ketone, 6-hydroxy-2-isopropenyl-5-benzofuranyl methyl	: 5825
Ketone, 3-hydroxy-2-thienyl methyl	: 5835
Ketone, indol-3-yl methyl	: 5819
Ketone, methyl 1-methylcyclohexyl	: 5843
Ketone, methyl 3-methylcyclohexyl	: 5844

Ketone, methyl 4-methylcyclohexyl	: 5845
Ketone, methyl 1-methylcyclopentyl	: 5846
Ketone, methyl 2-methylcyclopentyl	: 5847
Ketone, methyl 6-methyl-3-pyridyl	: 5855
Ketone, methyl 1-methylpyrrol-2-yl	: 5849
Ketone, methyl 5-methyl-2-thienyl	: 5856
Ketone, methyl 2-pyridyl	: 5867
Ketone, methyl 3-pyridyl	: 5868
Ketone, methyl 4-pyridyl	: 5869
Ketone, methyl pyrrol-2-yl	: 5820
Ketone, methyl styryl	: 3881
Ketone, methyl 2-thienyl	: 5870
Ketone, 1-naphthyl o-tolyl	: 7629
Ketone, phenyl 2-pyridyl	: 7638
Ketone, phenyl 4-pyridyl	: 7639
Ketone, phenyl 2-thienyl	: 7641
Ketone, phenyl 3-thienyl	: 7642
2-Ketosuccinic acid	: 3339
Khellin	: 6185
Kojic acid	: 10795
KSCN	: 11711
Kyanmethin	: 11045
Kynurenic acid	: 11232
Kynurenine	: 1102
Lactic acid (DL)	: 10266
Lactic acid, butyl ester (d)	: 10268
Lactic acid, isopropyl ester	: 10275
Lactic acid, methyl ester (DL)	: 10273
Lactic acid, 3,3,3-trichloro-	: 10333
Lactobionic acid	: 6210
Lactonic acid	: 6210
Lactonitrile	: 10105
Lactonitrile, 3-chloro-2-methyl-	: 10098
Lactonitrile, 3,3,3-trichloro-	: 10115
p-Lactophenetide	: 9796
Lactyl chloride, acetate (DL)	: 10494
Lamoryl	: 11520
Lanex	: 12007
Lanosterol	: 7411
Lanoxin	: 4041
L-Lanthionine	: 5016
Lapachol	: 7862
Laricic acid	: 8102
Laudanidine	: 9399
Laudanine	: 9398
Laudanosine	: 7379
Lauraldehyde	: 5288
Lauraldehyde, dimethyl acetal	: 5297
Laureline	: 2469
Lauric acid	: 5305
Lauric acid, phenyl ester	: 5317
Lauric acid, propyl ester	: 5320
Laurolenic acid	: 4912
Laurone	: 11857
Lauronitrile	: 5303
Lauroyl peroxide	: 9155
Lauryl alcohol	: 5321
Lauryl bromide	: 5293
Lawsone	: 7859
Leaf aldehyde violet	: 8108
Ledol	: 4989
Lemonol	: 8333
Lenacil	: 4900
Lenthionine	: 8998
Lepidine	: 11279
Lepidine, 2-amino-	: 11208
Lepidine, 2-ethyl-	: 11270
Lergitin	: 5506
Leucethane	: 3948
Leucoline	: 7373
Levodopa	: 11929
Levopimaric acid	: 9169
Levulinic acid	: 8903
Levulinic acid, benzyl ester	: 8911
Levulinic acid, isopropyl ester	: 8909
Levulinic acid, 3-methyl-, methyl ester	: 8898
Levulinic acid, propyl ester	: 8912
β-Levulose	: 6023
Lichenic acid	: 3764
Lidocaine	: 54
Lignoceric acid	: 11607
Limettin	: 2818
α-Limonene	: 4677
Limonene	: 4677
Linalool formate	: 8337

Linalool, tetrahydro (DL)	: 8454
Linalyl formate	: 8337
Linamarin	: 10104
Linarigenin	: 2809
β-Lindane	: 4406
δ-Lindane	: 4407
Lindane	: 4409
Linderic acid	: 5338
Linoleic acid	: 8223
Linolenic acid	: 8287
Lithocholic acid	: 4099
Lobelanidine	: 9646
Lobelanine	: 5854
Lobeline	: 5834
Loflucarban	: 11802
Longifolene	: 7568
Luciculine	: 5007
Lumazine	: 10748
Lumichrome	: 2472
Luminol	: 9600
Lumisterol	: 5382
Lunacrine	: 6182
Lupeol	: 7417
3,5-Lupetidine	: 9652
Lupinine	: 11331
Lupulon	: 4281
trans-Lutein	: 4058
Luteolin	: 2814
2,3-Lutidine	: 10936
2,4-Lutidine	: 10937
2,5-Lutidine	: 10938
2,6-Lutidine	: 10939
3,4-Lutidine	: 10940
3,5-Lutidine	: 10941
Lycomarasmine	: 472
trans-Lycopene	: 4053
Lycoremine	: 6200
Lycorine	: 6201
Lycoxanthin	: 4060
Lysergamide	: 5374
Lysergic acid	: 5376
D-Lysergic acid D-propanolamide	: 5373
Lysidine	: 7197
Lysilan	: 5140
Maclurin	: 7610
Magenta I	: 634
Magneson	: 1542
Malachite green	: 7455
Malathion	: 3316
Maleamic acid	: 3817
Maleanilic acid	: 3853
Maleic acid	: 3765
Maleic acid, chloro-, diethyl ester	: 3771
Maleic acid, diallyl ester	: 3782
Maleic acid, dihydroxy-	: 3777
Maleic anhydride	: 6091
Maleimide, N-phenyl-	: 11123
Malidone	: 8586
Malilum	: 11091
Malonic acid	: 9921
Malonic acid, amino-, diethyl ester	: 9924
Malonic acid, 2-cyclopentenyl, diethyl ester	: 9935
Malonic acid, cyclopentylidene, diethyl ester	: 9936
Malonic acid, dibenzyl-, diethyl ester	: 9926
Malonic acid, dibutyl-, diethyl ester	: 9937
Malonic acid, diethyl-, diethyl ester	: 9940
Malonic acid, dimethyl-, diethyl ester	: 9942
Malonic acid, dipropyl ester	: 9945
Malonic acid, di-2-propynyl-, diethyl ester	: 9946
Malonic acid, ethylisopropyl-, diethyl ester	: 9951
Malonic acid, (hydroxymethyl)-	: 9959
Malonic acid, isobutyl-, diethyl ester	: 9969
Malonic acid, isopropylidene-, diethyl ester	: 9966
Malonomononitrile, benzylidene,monoethyl ester	: 10645
Malononitrile	: 9911
Malononitrile, phenyl-	: 9916
Maltol	: 10796
Malvidin chloride	: 2847
Mandelic acid isoamyl ester	: 912
Mandelonitrile glucoside	: 955

Manganese, [[1,2-ethanediylbis-[carbamodithioato]](2-)]-	: 7420
Mannite	: 7422
Mannitol, hexanitrate	: 7424
Margaric acid	: 6314
Margaric aldehyde	: 6304
Matrine	: 7426
MCPA	: 141
Mechlorethamine	: 5413
Meclizine	: 9605
Meconic acid	: 10775
Meconin	: 7355
Mecrylate	: 10624
Medroxyprogesterone	: 9752
Melamine	: 11835
Melampyrin	: 6194
Melatonin	: 69
Meldrum's acid	: 5190
Melezitose	: 6221
Mellitene	: 1758
Mellitic acid	: 1748
Menadiol diacetate	: 7841
Menadione	: 7864
p-Mentha-1,8-dien-7-al	: 4633
p-Mentha-1(7),2-diene	: 4673
p-Menthane, 4-bromo	: 4321
trans-(-)-p-Menthan-3-one	: 4606
m-Menth-8-ene	: 4435
p-Menth-2-ene, 1,4-epidioxy-	: 5184
p-Menth-3-en-2-ol	: 4698
(±)-Menthol	: 4548
Menthol(DL)	: 4548
Menthone(d)	: 4605
l-Menthone	: 4606
(-)-Menthone	: 4606
Meparfynol	: 9146
Mephenesin	: 10019
Mephenoxalone	: 8588
Mephobarbital	: 11094
Mepivacaine	: 9639
Meprobamate	: 10021
β-Mercaptoalanine	: 5014
Mercaptobenzene	: 2289
Mercaptobenzimidazole	: 2386
2-Mercaptobenzoic acid	: 2663
2-Mercaptobenzothiazole	: 2866
1-Mercaptobutane	: 3437
Mercaptocyclopentane	: 4852
2-Mercaptoethanol	: 5732
3-Mercaptohexane	: 6858
Mercaptomethane	: 7546
6-Mercaptopurine	: 10763
Mercapturic acid	: 5015
Mercury, chloro(2-methoxyethyl)-	: 11822
Mesaconic acid	: 3785
Mesaconic acid, diethyl ester	: 3787
Mesalamine	: 2499
Mescaline	: 1586
Mesitylamine	: 859
Mesitylene	: 2359
Mesitylene, α-bromo-	: 1057
Mesitylene, 2,4-dibromo-	: 1313
Mesitylene, 2,4-dichloro-	: 1402
Mesitylenic acid	: 2583
Mesityl oxide	: 9108
Mesoxalylcarbamide	: 11073
Mesoxalylurea	: 11073
Mestanolone	: 350
Mestilbol	: 9358
Mesulphen	: 11661
Metacrolein	: 11906
Metaldehyde II (tetramer)	: 19
Metanilic acid	: 2216
Methacetin	: 71
Methacholine chloride	: 9830
Methacrolein	: 10515
Methacrolein diacetate	: 10564
Methacrylic acid	: 10648
Methacrylic acid anhydride	: 10649
Methacrylic acid, 3-bromo	: 10612
Methacrylic acid chloride	: 10715
Methacrylic acid, 2-ethylhexyl ester	: 10659
Methadone hydrochloride	: 6516
Methallatal	: 11064
Methallyl alcohol	: 10702
α-Methallyl bromide	: 3710
Methaminodiazepoxide	: 2409

4-Methylnaphthalene-1-carboxaldehyde	: 7756
Methyl nicotinate	: 10906
Methyl nitrate	: 8075
Methyl nitrite	: 8091
4-Methyl-2-nitrobenzophenone	: 7627
3-Methyl-1-nitrobutane	: 3400
trans-Methyl o-nitrocinnamate	: 10677
2-Methylnitrocyclohexane	: 4444
1-Methyl-1-nitrocyclohexane	: 4443
1-Methyl-4-nitrocyclohexane	: 4445
1-Methyl-1-nitrocyclopentane	: 4830
1-Methyl-2-nitrocyclopentane	: 4831
2-Methyl-2'-nitrodiphenyl ether	: 2005
4-Methyl-1-nitronaphthalene	: 7925
2-Methyl-1-nitronaphthalene	: 7926
N-Methyl-N'-nitro-N-nitrosoguanidine	: 6284
N-Methyl-N-nitroso-p-toluenesulfonamide	: 2212
Methylnitrosourea	: 12019
1-Methyl-1-nonadecanol	: 5359
2-Methylnonane	: 8140
3-Methylnonane	: 8141
4-Methylnonane	: 8142
5-Methylnonane	: 8143
3-Methyl-1-nonanol	: 8166
4-Methyl-4-nonanol	: 8170
5-Methyl-5-nonanol	: 8171
8-Methyl-1-nonanol	: 8168
2-Methyl-5-nonanone	: 8176
2-Methylnon-1-ene	: 8185
Methyl nonyl ketone	: 11958
8-Methyl-4-nonyne	: 8202
Methyl 9,12-octadecadienoate	: 8229
9-Methyloctadecanoic acid	: 8264
4-Methyl-3,5-octadiene	: 8329
2-Methyloctane	: 8408
3-Methyloctane	: 8409
4-Methyloctane	: 8410
3-Methyloctanoic acid	: 8433
3-Methyl-1-octanol	: 8461
4-Methyl-1-octanol	: 8462
2-Methyl-2-octanol	: 8464
3-Methyl-3-octanol	: 8466
6-Methyl-3-octanol	: 8468
4-Methyl-4-octanol	: 8469
3-Methyl-2-octanone	: 8475
4-Methyl-3-octanone	: 8476
7-Methyl-3-octanone	: 8477
7-Methyl-4-octanone	: 8479
2-Methyl-1-octene	: 8505
2-Methyl-5-octen-4-one	: 8523
Methyl octyl ketone	: 5075
7-Methyl-3-octyne	: 8537
Methyl (2-octyn-1-yl) ether	: 8536
Methyl oleate	: 8307
4-Methylolphenol	: 1884
Methyl orange	: 2222
4-Methyloxazole	: 8583
Methyloxirane	: 8621
Methyl β-oxoheptyl ketone	: 8137
2-Methyl-4-oxo-3-(2-propenyl)-2-cyclopenten-1-yl 2,2-dimethyl-3-(2-methyl-1-propenyl) cyclopropane carboxylate	: 336
Methylparaben	: 2637
Methyl pentadecanoate	: 8646
2-Methyl-1-pentadecanol	: 8649
3-Methyl-1,2-pentadiene	: 8666
cis-2-Methyl-1,3-pentadiene	: 8668
4-Methyl-1,3-pentadiene	: 8670
2-Methyl-1,4-pentadiene	: 8671
3-Methylpentane	: 8824
3-Methyl-2,4-pentanediol	: 8794
2-Methyl-1-pentanol	: 8937
2-Methyl-2-pentanol	: 8940
3-Methyl-2-pentanol	: 8941
4-Methyl-2-pentanol	: 8942
2-Methyl-3-pentanol	: 8943
2-Methyl-1-pentene	: 9052
3-Methyl-1-pentene	: 9053
4-Methyl-1-pentene	: 9054
2-Methyl-2-pentene	: 9055
(cis)-3-Methyl-2-pentene	: 9057
4-Methyl-trans-2-pentene	: 9058
cis-4-Methyl-2-pentene	: 9059

4-Methyl-2-pentenoic acid	: 9078
cis-3-Methyl-2-pentenoic acid methyl ester	: 9085
4-Methyl-3-pentenol	: 9090
2-Methyl-3-penten-2-ol	: 9097
3-Methyl-3-penten-2-ol	: 9098
Methyl 4-pentenyl ketone	: 6585
2-Methyl-1-penten-3-yne	: 9116
3-Methyl-3-penten-1-yne	: 9117
Methylpentylacetylene	: 8530
3-Methylpentylamine	: 8706
(3-Methylpentyl)benzene	: 2010
2-Methylpentyl bromide	: 8714
3-Methylpentyl bromide	: 8715
1-Methyl-2-pentylcyclohexane	: 4446
Methyl pentyl ether	: 8822
Methyl pentyl ketone	: 6510
Methyl pentyl sulfide	: 8826
2-Methylpentyl trifluoroacetate	: 268
3-Methyl-1-pentyne	: 9134
Methyl 2-pentynoate	: 9139
3-Methylphenanthrene	: 9180
4-Methylphenanthrene	: 9181
2-Methylphenazine	: 9195
Methyl phenoxyacetate	: 230
2-Methyl-4-phenyl-2-butanol	: 2150
1-(3-Methylphenyl)ethanol	: 1858
1-(4-Methylphenyl)ethanol	: 1859
1-(3-Methylphenyl)ethyl alcohol	: 1858
3-Methyl-5-phenylhydantoin	: 7221
1-Methyl-2-phenylhydrazine	: 7166
Methyl phenyl ketone	: 5864
Methyl β-phenylpropionate	: 2139
2-Methyl-3-phenylpropionic acid	: 2137
1-Methyl-3-phenylpropylamine	: 2119
3-Methyl-2-phenylpyridine	: 10991
2-Methyl-5-phenylthiophene	: 11774
Methyl 2-phenylvinyl ketone	: 3881
Methyl picrate	: 1946
3-Methylpicric acid	: 9434
2-Methylpiperazine	: 9619
N-Methylpiperidine	: 9679
1-Methyl-1,3-propanediol	: 3345
Methyl propionate	: 10296
2-Methyl-6-propionylphenol	: 10444
1-Methyl-2-propylacetylene	: 7115
Methylpropylcarbinol	: 8916
2-Methylpropyl chloroformate	: 4024
1-Methyl-1-propylcyclohexane	: 4449
1-Methyl-2-propylcyclohexane	: 4450
1-Methyl-3-propylcyclohexane	: 4451
1-Methyl-4-propylcyclohexane	: 4452
3-Methyl-1-propylcyclohexanol	: 4551
2-Methyl-1-propylcyclohexanol	: 4550
1-Methyl-1-propylcyclopentane	: 4833
cis-1-Methyl-2-propylcyclopentane	: 4834
trans-1-Methyl-2-propylcyclopentane	: 4835
cis-1-Methyl-3-propylcyclopentane	: 4836
trans-1-Methyl-3-propylcyclopentane	: 4837
Methyl propyl ether	: 10069
1-Methyl-1-propylhexanol	: 8170
Methyl propyl ketone	: 8957
Methyl propyl ketone oxime	: 8985
Methyl propyl sulfide	: 10088
2-Methyl-2-propyltrimethylene butylcarbamate carbamate	: 3933
2-Methylpyrazine	: 10810
1-Methylpyrazole	: 10834
4-Methylpyrazole	: 10836
3-Methylpyridazine	: 10856
Methyl 3-pyridyl ketone	: 5868
2-Methylpyrimidine	: 11070
4-Methylpyrimidine	: 11071
5-Methylpyrimidine	: 11072
N-(4-Methyl-2-pyrimidyl)sulfanilamide	: 2199
3-Methylpyrrole	: 11132
N-Methylpyrrolidine	: 11168
N-Methyl-2-pyrrolidone	: 11185
4-Methylquinaldine	: 11255
3-Methylquinoline	: 11278
5-Methylquinoline	: 11280
6-Methylquinoline	: 11281
7-Methylquinoline	: 11282
8-Methylquinoline	: 11283
6-Methyl-7-quinolinol	: 11320
2-Methylquinoxaline	: 11337

6-Methylquinoxaline	: 11338
Methyl red	: 2587
6-Methyl-β-resorcylic acid	: 2573
Methyl salicylate	: 2634
3-Methylsalicylic acid	: 2629
4-Methylsalicylic acid	: 2630
5-Methylsalicylic acid	: 2631
Methylsilane	: 11437
Methyl silicate	: 11508
Methyl silyl ether	: 11436
o-Methylstyrene	: 1640
m-Methylstyrene	: 1641
p-Methylstyrene	: 1642
α-Methylstyrene	: 1972
Methyl styryl ketone	: 3881
5-Methyl-3-sulfanilamidoisoxazole	: 2198
Methyl sulfate	: 11577
Methyl sulfocyanate	: 11705
3-Methylsulfolane	: 11789
Methylsulfonic acid	: 7530
N-Methyltaurine	: 5624
17-Methyltestosterone	: 358
2-Methyltetrahydrofuran	: 6171
3-Methyltetrahydrofuran	: 6172
2-Methylthiazole	: 11678
4-Methylthiazole	: 11679
4-Methyl-5-thiazolecarboxylic acid	: 11669
4-Methyl-5-thiazoleethanol	: 11677
N1-(4-Methyl-2-thiazolyl)sulfanilamide	: 2200
(Methylthio)acetonitrile	: 286
Methyl thiocyanate	: 11705
2-(Methylthio)ethanol	: 5749
2-Methylthiophene	: 11769
3-Methylthiophene	: 11770
3-(Methylthio)propionaldehyde	: 9786
Methylthiouracil	: 11108
8-Methyltocol	: 2785
Methyl trichloroacetate	: 253
Methyltrichlorosilane	: 11465
2-Methyltricosane	: 11856
7-Methyltridecane	: 11879
Methyl tridecanoate	: 11882
12-Methyl-1-tridecanol	: 11886
Methyltrimethylene glycol	: 3345
3-Methyl-2,4,6-trinitrophenol	: 0434
Methyltriphenoxysilane	: 11443
Methyltriphenylsilane	: 11444
N-Methyltryptophan	: 11923
β-Methylumbelliferone	: 2831
2-Methyl-1-undecanol	: 11955
Methyl undecyl ketone	: 11889
Methyl 10-undecynoate	: 11991
Methyl 9-undecynoate	: 11990
6-Methyluracil	: 11069
3-Methyluric acid	: 10765
7-Methyluric acid	: 10766
2-Methylvaleraldehyde	: 8693
Methyl valerate	: 8891
Methyl vincaminate	: 5349
Methylvinylacetylene	: 9111
Methyl vinyl ether	: 5908
Methyl vinyl ketone	: 3875
Methysticin	: 10790
Metiazinic acid	: 9500
Metoxon	: 3939
Metronidazole	: 7206
Metyrapone	: 10462
Metyridine	: 10979
Michler's Base	: 786
Michler's ketone	: 7593
Mimosine	: 11011
Mirex	: 7644
Molybdenum hexacarbonyl	: 7662
Monobenzone	: 9453
Monofluoroacetic acid	: 189
Monomethylarsine	: 446
Monophenylphosphine	: 9519
Montanic acid	: 8221
Montanyl alcohol	: 8222
Monuron	: 11999
Moquizone	: 11198
Morin	: 2816
Morphine	: 7666
Morphine, N-oxide	: 7668
Morpholine, 2,6-dimethyl-4-tridecyl-	: 11898
Morpholine, N-(2-ethoxyethyl)	: 7693

Phosphoric acid, 2-chloro-1-(2,4,5-trichlorophenyl)ethenyl dimethyl ester	: 6202
Phosphoric acid, 1,2-dibromo-2,2-dichloroethyl dimethyl ester	: 7702
Phosphoric acid dichloride,thiono,2,4,5-trichlorophenyl ester	: 9578
Phosphoric acid, 2,2-dichloroethenyl dimethyl ester	: 5170
Phosphoric acid, 3-(dimethylamino)-1-methyl-3-oxo-1-propenyl dimethyl ester, (E)-	: 5172
Phosphoric acid, triisobutyl ester	: 9572
Phosphorodithioic acid, O,O-bis(1-methylethyl) S-[2-[(phenylsulfonyl)amino]ethyl] ester	: 517
Phosphorodithioic acid, S-[[(4-chlorophenyl)thio]methyl] O,O-diethyl ester	: 4037
Phosphorodithioic acid, S-[(6-chloro-2-oxo-3(2H)-benzoxazolyl)methyl]O,O-diethyl ester	: 9514
Phosphorodithioic acid, S-[2-chloro-1-(1,3-dihydro-1,3-dioxo-2H-isoindol-2-yl)ethyl]	: 5105
Phosphorodithioic acid, O,O-diethyl S-[2-(ethylthio)ethyl] ester	: 5260
Phosphorodithioic acid, O,O-diethyl S-[(ethylthio)methyl] ester	: 9513
Phosphorodithioic acid, S-[(1,3-dihydro-1,3-dioxo-2H-isoindol-2-yl)methyl] O,O-dimethyl ester	: 9515
Phosphorodithioic acid, S-[[(1,1-dimethylethyl)thio]methyl] O,O-diethyl ester	: 11592
Phosphorodithioic acid, O,O-dimethyl S-[2-(methylamino)-2-oxoethyl] ester	: 5009
Phosphorodithioic acid, S,S'-1,4-dioxane-2,3-diyl O,O,O',O'-tetraethyl ester	: 5201
Phosphorodithioic acid, O-ethyl S,S-dipropyl ester	: 5934
Phosphorodithioic acid, S,S'-methylene O,O,O',O'-tetraethyl ester	: 5931
Phosphorodithioic acid, S-[(5-methoxy-2-oxo-1,3,4-thiadiazol-3(2H)-yl)methyl], O,O-dimethyl ester	: 7645
Phosphorothioic acid, O-(4-bromo-2-chlorophenyl) O-ethyl S-propyl ester	: 9762
Phosphorothioic acid, O-[2-chloro-l-(l-methylethyl)-lH-imidazol-4-yl] O,O-diethyl ester	: 7340
Phosphorothioic acid, O-[2,5-dichloro-4-(methylthio)phenyl] O,O-diethyl ester	: 4092
Phosphorothioic acid, O,O-diethyl O-(3,5,6-trichloro-2-pyridinyl) ester	: 4089
Phosphorothioic acid, O-[2-(diethylamino)-6-methyl-4-pyrimidinyl] O,O-dimethyl ester	: 9730
Phosphorothioic acid, O-[2-(diethylamino)-6-methyl-4-pyrimidinyl] O,O-diethyl ester	: 9729
Phosphorothioic acid, O,O-diethyl O-[4-(methylsulfinyl)phenyl] ester	: 5954
Phosphorothioic acid, O,O-diethyl O-pyrazinyl ester	: 11720
Phosphorothioic acid, O,O-dimethyl O-[3-methyl-4-(methylthio)phenyl] ester	: 5955
Phosphorothioic acid, O,O-dimethyl O-(3,5,6-trichloro-2-pyridinyl) ester	: 4090
Phosphorothioic acid, O-[4-[(dimethylamino)sulfonyl]phenyl] O,O-dimethyl ester	: 5946
Phosphorothioic acid, O,O-dimethyl O-(3-methyl-4-nitrophenyl) ester	: 5951
Phosphorothioic acid, O-(6-ethoxy-2-ethyl-4-pyrimidinyl) O,O-dimethyl ester	: 5945
Phosphorothioic acid, S-[2-(ethylsulfinyl)ethyl] O,O-dimethyl ester	: 8632

Phosphorothioic acid, S-[2-(ethylthio)ethyl] O,O-dimethyl ester	: 5103
Phosphorothioic acid, O,O'-(thiodi-4,1-phenylene) O,O,O',O'-tetramethyl ester	: 1
Phosphorotrithious acid, S,S,S-tributyl ester	: 7438
Phthalamide	: 1324
Phthalic acid	: 1327
Phthalic acid, 3,5-dichloro-	: 1340
Phthalic acid, dioctyl ester	: 1344
Phthalic acid, monoamide,N(1-naphthyl)	: 2702
Phthalic acid, tetrachloro-, ethyl ester	: 1353
Phthalic acid, tetrachloro-, monoethyl ester	: 1353
Phthalic anhydride	: 7342
Phthalic anhydride, 3,4-dichloro	: 7343
Phthalic anhydride, 4,5-dichloro-	: 7344
Phthalimide	: 7362
Phthaloyl chloride	: 1318
Phthaloylmalonic acid, diethyl ester	: 7248
2-(N4-Phthylaminobenzene-sulfonamide)thiazole	: 2726
Phthalylsulfonazole	: 2726
Phthalylsulphathiazole	: 2726
Phthiocol	: 7861
Phycitol	: 3431
Phylloquinone	: 7865
Physeteric acid	: 11638
Physostigmine	: 11191
Physoteric acid	: 11638
Phytanic acid	: 6658
Phytol	: 6672
Phytomenadione	: 7865
Piazthiole	: 11658
Picein	: 5818
Picloram	: 10896
2-Picoline	: 10980
3-Picoline	: 10981
4-Picoline	: 10982
γ-Picoline, 2-amino	: 10867
3-Picoline, 2-phenyl-	: 10991
Picolinic acid	: 10893
4-Picolyl alcohol	: 10977
Picramic acid	: 9207
Picric acid	: 9497
Picrolonic acid	: 10848
Picropodophyllin	: 6191
Picrotoxin	: 7579
Picryl chloride	: 1244
Picryl sulfide	: 2288
Pilocarpine	: 6151
Pimaric acid	: 9170
Pimeclone	: 4611
Pimelic acid	: 6393
Pimelic acid, diethyl ester	: 6394
Pimelic ketone	: 4571
Pinacol	: 3350
Pinacolone	: 3657
Pinacone	: 3350
Pinane (DL)	: 2934
(-)-2-Pinene	: 2971
(-)-α-Pinene	: 2971
(1S,5S)-(-)-α-Pinene	: 2971
(±)-α-Pinene	: 2972
β-Pinene	: 2925
2-Pinene	: 2973
2-Pinen-7-one	: 2977
2-Pinen-4-one, (1R,5R)-(+)-	: 2978
Pinic acid diethyl ester	: 4171
(±)-Pinol	: 8561
Pinosylvin	: 1545
Pipecolic acid	: 9640
N,N'-Piperazinedicarboxylic acid, diethyl ester	: 9606
Piperidic acid	: 3464
Δ 3-Piperidine	: 11020
Piperidine, 4-(aminomethyl)-	: 9673
Piperidine, 4-benzoyl-N-methyl	: 7409
Piperidine, 2-benzyl-	: 9699
Piperidine, 4-benzyl-	: 9700
Piperidine, N-tert-butyl-	: 9653
Piperidine, N-carboxyethyl, ethyl ester	: 9704

Piperidine, 4(4-chlorophenyl)-1(3(4-fluorobenzoyl)propyl)-4-hydroxy	: 3651
Piperidine, N,2-dimethyl (DL)	: 9648
Piperidine, 2-ethyl (DL)	: 9668
Piperidine, 3-ethyl (DL)	: 9669
Piperidine, 2-(2-hydroxyethyl)	: 9663
Piperidine, 4-(2-hydroxyethyl)	: 9664
Piperidine, 3-methyl (DL)	: 9681
Piperidine, N-(3-methyl-3-pentyl)	: 9671
1-Piperidinepropionic acid, ethyl ester	: 9704
Piperidine, 2(1-propionyl)	: 10484
Piperidine, 3-vinyl-	: 9666
Piperidinium, 1,1-dimethyl-, chloride	: 7430
Piperidione	: 9654
4-Piperidone, 1,2,3,6-tetramethyl	: 9725
Piperine	: 9632
Piperitenone	: 4718
(±)-Piperitone	: 4716
Piperonal	: 2417
Piperonaldehyde	: 2417
Piperonylic acid	: 2418
trans-Piperylene	: 8661
cis-Piperylene	: 8662
Pipradrol	: 9676
Pipsyl chloride	: 2247
Pivalamide, N,N-diethyl	: 9794
Pivaldehyde	: 9783
Pivalic acid	: 10258
Pivalic acid chloride	: 10503
Pivalil	: 6811
Pivalon	: 6975
Plasmocid	: 9877
Plumericin	: 11904
Podocarpa-7,13-dien-15-oic acid, 13-isopropyl-	: 9167
Podophyllotoxin	: 6192
Polythiazide	: 2857
Populin	: 6223
Pramoxine	: 7683
Prednisolone	: 9738
Prednisone	: 9739
Pregnanediol	: 9742
5β-Pregnane-3,20-dione	: 9743
Pregnan-3α-ol-20-one	: 9745
4-Pregnene-11β,17α,20β,21-tetrol-3-one	: 9757
Pregnenolone	: 9756
Prehnitenol	: 9473
Prenoxdiazine hydrochloride	: 9657
Primeverose	: 6237
Primidone	: 11066
Pristane	: 8644
Probarbital	: 11093
Probenecid	: 2595
Procainamide hydrochloride	: 602
Procaine	: 2493
Proflavine	: 316
Proformiphen	: 2149
Progallin P	: 2732
Promethazine	: 9504
Prontosil	: 2209
Propallylonal	: 11078
Propanal, 2-methyl-2-(methylsulfonyl)-, O-[(methylamino)carbonyl]oxime	: 335
Propanal, 2-methyl-2-(methylthio)-, O-[(methylamino)carbonyl]oxime	: 334
Propanamide, N-(3,4-dichlorophenyl)-	: 10200
Propanamide, N,N-diethyl-2-(1-naphthalenyloxy)-	: 8067
1-Propanearsonic acid	: 461
Propane, 1,3-bis(1-methyl-4-piperidyl)	: 9702
Propane, 3-bromo-1,2-epoxy (DL)	: 8600
1-Propanecarboxylic acid	: 3460
Propane, 2-chloromethyl-1,1,2,3-tetrachloro	: 10146
Propane, 2-chloro-1-phenyl (l)	: 1234
Propane, 1-cyclohexyl-3-cyclopentyl-	: 4341
Propane, 1-cyclopropyl-	: 4984
Propane, 1,1-dibenzyl	: 1698
Propane, 1,2-dichloro (DL)	: 9892
Propane, 1,2-dicyclohexyl-	: 4430
Propane, 2,2-dicyclohexyl-	: 4432
Propane, 1,3-dicyclohexyl-	: 4466

Propane, 1,1-dicyclohexyl-	: 4472
Propane, 1,3-dicyclohexyl-2-ethyl	: 4402
Propane, 1,3-dicyclohexyl-2-methyl-	: 4447
1,2-Propanediol, 3-amino(DL)	: 9984
1,3-Propanediol, 2-methyl-2-propyl,N-butyl carbamate,carbamate	: 3933
Propane, 1,2-diphenoxy	: 1970
Propane, 1,3-dipheny-1-ethyl	: 1697
Propane, 1,3-diphenyl-	: 2121
Propane, 1,1-diphenyl-	: 2171
Propanenitrile, 2-[[4-chloro-6-(cyclopropylamino)-1,3,5-triazin-2-yl]amino]-2-methyl-	: 9759
Propanenitrile, 2-[[4-chloro-6-(ethylamino)-1,3,5-triazin-2-yl]amino]-2-methyl-	: 4161
Propane, 1-nitro-3,3,3-trifluoro	: 10184
2-Propanephosphonic acid	: 9542
1,3-Propanesulfone	: 11694
Propane sultone	: 8570
Propane, 1,2,2,3-tetrachloro-tetrafluoro	: 10149
Propanoic acid, 3-amino-	: 323
Propanoic acid, 2-amino-3-mercapto-, (R)-	: 5014
Propanoic acid, chloride, 2-chloro(d)	: 10496
Propanoic acid, 2-(3-chlorophenoxy)-	: 4141
Propanoic acid, 2-[4-(2,4-dichlorophenoxy)phenoxy]-, methyl ester	: 5171
Propanoic acid, 2,2-dichloro-3,3,3-trifluoro, propyl ester	: 10721
Propanoic acid, 3,3'-dithiobis[2-amino-, [R-(R*,R*)]-	: 5020
Propanoic acid, methylene ester	: 7654
Propanoic acid, 2-(2,4,5-trichlorophenoxy)-	: 11512
Propanolamine	: 10337
Propanolamine, N-pantoyl-	: 3202
1-Propanol, 2-[2-(2-amino-2-methylpropyl)amino]-2-methyl-	: 10342
2-Propanol, 1-chloro-3-isopropoxy-	: 10355
2-Propanol, 1-diethylamino (DL)	: 10368
2-Propanol, 1,1',1'',1'''-(ethylenedinitrilo)tetra-	: 10379
2-Propanol, 1-isopropoxy-	: 10390
2-Propanol, 1-nitro-3,3,3-trichloro	: 10408
2-Propanol, 1,1,1-trifluoro(DL)	: 10409
2-Propanone, acetyl-	: 8798
2-Propanone, (methoxyphenyl)-	: 10459
Propantheline bromide	: 9832
Proparglacetic acid	: 9138
Propargyl alcohol	: 10743
Propargyl aldehyde	: 10723
Propargyl bromide	: 10728
Propargyl chloride	: 10729
Propargylic acid, (2-nitrophenyl)	: 10740
2-Propene-1,2-dicarboxylic acid	: 3328
Propene-3,3-diol, 2-methyl,diacetate	: 10564
Propene, 2-methyl(tetramer)	: 10580
2-Propene-1-ol, 1-phenyl	: 1868
Propenoic acid, 2-chloro,ethyl ester	: 10619
Propenoic acid, 2-methyl,butyl ester	: 10651
Propenoic acid, 3(1-naphthyl)(trans)	: 10675
2-Propenyl alcohol	: 10697
2-Propenyl bromide	: 10531
2-Propenyl chloride	: 10536
2-Propenyl isothiocyanate	: 10575
N-2-Propenylurea	: 12028
Propesin	: 2507
Propham	: 3966
Propiocine	: 5396
β-Propiolactone	: 8595
Propiolic acid	: 10738
Propiolic acid, propyl-	: 7123
Propionaldehyde	: 9774
Propionaldehyde, 3-(2-tolyl)	: 2115
Propionamide	: 9791
Propionamide, N-antipyrinyl-2-(dimethylamino)-	: 9795
Propionic acid	: 10201
Propionic acid bromide	: 10490
Propionic acid, 3-bromo, isopentyl ester	: 10210
Propionic acid, 2-chloro, isobutyl ester (d)	: 10224

Propionic acid, 2-diazo-, ethyl ester	: 10236
Propionic acid, 2-diazo-, methyl ester	: 10237
Propionic acid, 2,3-dibromo-, butyl ester	: 10239
Propionic acid, 2,3-dibromo-, propyl ester	: 10243
Propionic acid, 2,2-dichloro-3,3,3-trifluoro-, methyl ester	: 10251
Propionic acid, 3-(diethylamino), ethyl ester	: 328
Propionic acid, 2,3-dihydroxy-2-methyl-, methyl ester	: 10256
Propionic acid, 2,3-diphenyl (DL)	: 2143
Propionic acid, 3,3'-iminodi-N-methyl, diethyl ester	: 330
Propionic acid, 3,3'-(methylimino)-di-, diethyl ester	: 330
Propionic acid, 2,3,3,3-tetrafluoro-, ethyl ester	: 10328
Propionic acid, 3-(2-tetrahydrofuryl)	: 6164
Propionic acid, 3,3,3-trichloro-2-(2,2,2-trichloro-1-hydroxyethoxy)-, γ-lactone	: 5212
Propionic anhydride	: 10203
Propionitrile, 2-ethoxy-	: 10102
Propionitrile, 2-methoxy-	: 10109
Propionylacetone	: 6804
Propionyl bromide	: 10490
Propionyl chloride	: 10493
Propionyl chloride, 3,3-dichloro	: 10502
Propionyl fluoride	: 10506
Propiophenone	: 10480
Propiophenone, 2'-hydroxy-3'-methyl-	: 10444
Propiophenone, 2'-hydroxy-4'-methyl-	: 10445
Propiophenone, 3-methoxy	: 10456
Propiophenone, 2-methoxy-	: 10457
Propofol	: 9226
2-Propoxyethanol	: 5764
1-Propoxy-2-propanol	: 10401
N-Propylacetanilide	: 88
Propyl acetate	: 236
Propylacetylene	: 9122
Propyl alcohol	: 10334
Propylallene	: 6676
Propylamine	: 9803
Propylamine, N,N-dichloro	: 9805
Propyl p-aminobenzoate	: 2507
Propyl amyl ketone	: 8174
Propylbenzene	: 2169
Propyl benzenesulfonate	: 2235
Propyl benzoate	: 2724
Propyl bromide	: 9836
Propyl bromoacetate	: 120
Propyl butanoate	: 3588
Propyl-tert-butylcarbinol	: 6925
Propyl carbamate	: 3969
Propyl carbonate	: 4002
Propyl chloride	: 9848
Propyl chloroacetate	: 149
1-Propyl-3-(p-chlorobenzenesulfonyl)urea	: 2208
Propyl chlorocarbonate	: 4028
trans-Propyl cinnamate	: 10693
Propyl cyanide	: 3403
Propylcyclohexane	: 4471
Propyl cyclohexanecarboxylate	: 4337
trans-4-Propylcyclohexanol	: 4560
cis-2-Propylcyclohexanol	: 4558
trans-2-Propylcyclohexanol	: 4559
4-Propylcyclohexanone	: 4613
2-Propylcyclohexanone	: 4612
trans-2-Propylcyclopentanol	: 4874
Propylcyclopentane	: 4849
cis-2-Propylcyclopentanol	: 4873
2-Propylcyclopentanone	: 4894
3-Propylcyclopentanone	: 4895
1-Propyl-1-cyclopentene	: 4935
Propylcyclopropane	: 4984
Propyl 2,3-dibromopropionate	: 10243
Propyldimethylamine	: 9807
Propylene	: 10528
Propylene, 2-bromo-3-cyclohexyl	: 4322
Propylene, 3-bromo-3,3-difluoro	: 10532
Propylene chlorohydrin	: 10347

sec-Propylene chlorohydrin	: 10349
Propylenediamine	: 9872
Propylene dibromide	: 9881
Propylene, 1,3-dibromo (trans)	: 10543
Propylene, 1,3-dibromo (cis)	: 10544
Propylenedicarboxylic acid	: 3328
Propylene, 1,1-dichloro-2-fluoro	: 10554
Propylene, 1,2-dichloro-3,3,3-trifluoro	: 10558
1,2-Propylene glycol	: 9982
1,3-Propylene glycol	: 9983
Propylene glycol β-monoethyl ether	: 10380
1,2-Propylene glycol 1-propyl ether	: 10401
Propylene oxide	: 8621
1,2-Propylenimine	: 501
N-Propylethanolamine	: 5765
Propylethylene	: 9004
Propyl fluoride	: 10046
Propyl formate	: 6018
Propylformic acid	: 3460
2-Propylfuran	: 6167
Propyl gallate	: 2732
4-Propylheptane	: 6434
Propyl heptanoate	: 6471
4-Propyl-3-heptene	: 6559
1-Propylheptyl alcohol	: 5072
Propyl hexahydrobenzoate	: 4337
Propylhexedrine	: 4385
Propyl hexyl ketone	: 5077
Propyl p-hydroxybenzoate	: 2653
Propylidene chloride	: 9891
Propyl iodide	: 10054
Propyliodone	: 10873
Propyl isobutyrate	: 10308
Propyl isocyanate	: 10059
Propyl isopropyl sulfide	: 10077
Propyl isovalerate	: 3562
Propyl laurate	: 5320
Propyl levulinate	: 8912
Propyl mercaptan	: 10156
Propyl methacrylate	: 10672
Propyl nitrate	: 8080
Propyl nitrite	: 8098
5-Propyl-1,5-octadien-3-yne	: 8341
Propylparaben	: 2653
2-Propylpentanal	: 8697
2-Propyl-1-pentene	: 6426
3-Propyl-1-pentene	: 7033
Propyl phosphite	: 9594
4-Propylpiperidine	: 9709
Propyl propionate	: 10326
3-Propylpyrrole	: 11139
2-Propylquinoline	: 11292
Propyl sulfide, 3,3'-dichloro	: 10153
2-Propylthiophene	: 11778
3-Propylthiophene	: 11779
Propylthiouracil	: 11109
N-Propyl-p-toluidine	: 815
Propyltrichlorosilane	: 11476
Propyl trifluoroacetate	: 269
Propyl 3,4,5-trihydroxybenzoate	: 2732
Propyne, 1,3-dibromo-	: 10730
Protionamide	: 10886
Protoanemonin	: 6155
Protocatechualdehyde	: 542
Protocatechuic acid	: 2571
Protocatechuic acid methylene ether	: 2418
Protopine	: 3109
Protoverine	: 4070
Provitamin D3	: 4103
Prunetin	: 2828
Prunetol	: 2807
Pseudoaconitine	: 308
Pseudocinchonine	: 4122
Pseudocodeine	: 7672
Pseudoconhydrine	: 9718
Pseudocumene	: 2358
Pseudoionone	: 11965
Pseudojervine	: 12045
Pseudomorphine	: 3002
Pseudopelletierine	: 479
Pseudotropine	: 486
Pteroylglutamic acid	: 6243
Puleganic acid	: 4784
Pulegene	: 4931
Pulegone	: 4608

Salvarsan	: 9275
Sanguinarine	: 2432
α-Santalol	: 9093
β-Santalol	: 9103
Santene	: 2968
β-Santenol	: 2938
α-Santonin	: 8057
Sarcosine	: 6268
Sarmentogenin	: 4046
Sarpagine	: 11347
Sarsasapogenin	: 11531
Scarlet red	: 8033
Schradan	: 5220
Scilliroside	: 3142
Scopolamine	: 908
Scopoletin	: 2829
Scopoline	: 7575
Sebacic acid	: 5042
Sebacic acid di(2-ethylhexyl) ester	: 5043
Sebacic acid, mononitrile, methyl ester	: 8148
Sebacil	: 4203
Sebacoin	: 4206
7,13a-Secoberbin-13a-one, 7-methyl-2,3:9,10-bis(methylenedioxy)-	: 3109
Seedrin	: 5177
Selanthrene	: 11353
Semicarbazide hydrochloride	: 7143
Seminose	: 7425
Semioxamazide	: 105
Senecialdehyde	: 3700
Senfoel	: 10575
Septochol	: 4095
Serenase	: 3651
Serpentine	: 8572
Silane, allyldiethoxymethyl-	: 11408
Silane, methyl tri-(4-tolyl)	: 11445
Silane, (oxy-p-phenylene)bis[trimethyl-	: 11497
Silane, phenyl-tributyl	: 11455
Silane, propyl trimethyl	: 11496
Silane, tetravinyl-	: 11449
Silane, 4-trimethylsilylphenoxy-trimethyl	: 11497
Silane, tris(allyloxy)vinyl-	: 11427
Silanol, dimethyl ethyl	: 11501
3-Silapentane	: 11409
2-Silapropane	: 11414
Silicon tetraethoxide	: 11506
Silicon tetramethoxide	: 11508
Silin, 1,1-diphenyl perhydro	: 11360
Silmurin	: 3142
Siloxane, butylmethyl(cyclic tetramer)	: 4995
Sinomenine	: 7676
α1-Sitosterol	: 11556
β-Sitosterol	: 11558
γ-Sitosterol	: 11559
Skatole	: 7314
Skimmetin	: 2826
Skimmianine	: 6181
Skimmin	: 2821
Smilagenin	: 11533
Sodium cyanate	: 4164
Sodium N-lauroylsarcosinate	: 6269
Solanone	: 8113
Solasodine	: 11529
Solbrol A	: 2621
Solestro	: 5403
Sophoricoside	: 2824
Sorbic acid	: 6698
L-Sorbinose	: 11519
Sorbitol	: 6206
Spantol	: 2149
Sparteine	: 7571
Spasmocromona	: 2838
Spermine	: 3259
Spinulosin	: 4260
Spiro[3,5,8-ethanylylidene-1H-pyrano[3,4-c]pyridine-10,3'-[3H]indol]-2'(1'H)-one, 5-ethenyl	: 6203
Spiro[2.2]pentane	: 11527
Spirosal	: 2622
Spir[5.]undecane	: 11537
Squalane	: 11606

Squalene	: 11602
Stachydrine	: 11177
Stannane, hydroxytriphenyl-	: 11911
Stannane, tricyclohexylhydroxy-	: 5011
Stanolone	: 349
Staticin	: 2721
Stearic acid	: 8247
Stearic acid, 18-bromo	: 8250
Stearic acid, cyclohexyl ester	: 8253
Stearic acid, isopentyl ester	: 8265
Stearic acid, 14-methyl	: 8262
Stearic acid, 9-methyl	: 8264
Stearic acid, 12-oxo, ethyl ester	: 8269
Stearic acid, pentyl ester	: 8270
Stearic acid, phenyl ester	: 8271
Stearic acid, 9,10,12,13-tetrabromo, methyl ester	: 8274
Stearic acid, tetrahydrofurfuryl ester	: 8275
9-Stearolic acid	: 8316
Stearyl alcohol	: 8278
Stearyl mercaptan	: 8244
Stercorin	: 4109
Stigmastanol	: 11557
Stigmasterol	: 11555
trans-Stilbene	: 1619
cis-Stilbene	: 1620
Stilbene, 2-bromo (trans)	: 1080
Stilbene, 2-bromo (cis)	: 1081
4-Stilbenecarbonitrile (cis)	: 2768
Stilbene, 2,2'-dimethyl-	: 1622
Stilbene, 2,4-dinitro (trans)	: 1496
Stimulina	: 6249
Streptomycin, sulfate	: 11561
Strodival	: 4040
Strophanthidin	: 4049
Strychnine	: 11563
Strychnine nitrate	: 11565
Strychnine, sulfate	: 11566
Stylophorine	: 4071
Styphnic acid	: 1554
Styrallyl alcohol	: 1896
Styrene	: 1624
Styrene, α-chloro-2,4-dimethyl	: 1152
Styrene, β-chloro, α-4-dimethyl	: 1188
Styrene, 2,3-dichloro-	: 1367
Styrene, 2,6-dichloro-	: 1369
Styrene, 3,5-dichloro-	: 1370
Styrene, 2,5-dichloro-	: 1371
Styrene, 2,4-dichloro-	: 1373
Styrene epoxide	: 8624
Styrene glycol	: 5579
Styrylacetic acid	: 3855
Styryl methyl ketone	: 3881
Suberic acid	: 8386
Suberone	: 4234
Suberonitrile	: 8385
Succinamide	: 3257
Succinanil	: 11162
Succinanilic acid	: 3575
Succinanilic acid, 4'-(2-thiazolylsulfamoyl)-	: 3578
Succinic acid	: 3294
Succinic acid, amide,N-phenyl	: 3575
Succinic acid, sec-butyl-, diethyl ester	: 3332
Succinic acid, di-sec-butyl ester	: 3297
Succinic acid, dibutyl ester	: 3301
Succinic acid dichloride	: 3366
Succinic acid, 2,3-dichloro-, (±)-	: 3302
Succinic acid, diethyl ester	: 3303
Succinic acid N,N-dimethylhydrazide	: 3334
Succinic acid dinitrile	: 3291
Succinic acid, diphenyl ester	: 3320
Succinic acid, ethyl ester	: 3335
Succinic acid, mercapto-, diethyl ester, S-ester with O,O-dimethyl phosphorodithioate	: 3316
Succinic acid monoanilide	: 3575
Succinic acid, monopropyl ester	: 3338
Succinic acid, oxo,diethyl ester	: 3340
Succinic acid, pentyl ester	: 3337
Succinic acid, propyl ester	: 3338
Succinic anhydride	: 6093
Succinic anhydride, phenyl (DL)	: 6096
Succinimide	: 11152

Succinimide, N-benzyl-	: 11163
Succinonitrile	: 3291
Succinyl chloride	: 3366
Succinylsulphathiazole	: 3578
Sucrose	: 6218
Sucrose, octaacetate	: 6229
Sudan III	: 8038
Sulfabenzamide	: 603
Sulfabenzpyrazine	: 2205
Sulfacetamide	: 30
Sulfachlorpyridazine	: 2193
Sulfadiazine	: 2204
Sulfadigesin	: 3578
Sulfadimethoxine	: 2194
Sulfafurazole	: 2195
Sulfaguanidine	: 2191
Sulfamerazine	: 2199
Sulfamethazine	: 2196
Sulfamethoxazole	: 2198
Sulfamethoxypyridazine	: 2197
Sulfamethyldiazine	: 2199
Sulfamethylthiazole	: 2200
p-Sulfamidoaniline	: 2187
p-Sulfamoylbenzoic acid	: 2508
p-Sulfamylacetanilide	: 31
Sulfanilamide	: 2187
Sulfanilamide-N4-acetate	: 31
Sulfanilamide, N1-(5-methyl-3-isoxazolyl)-	: 2198
Sulfanilamide, N1-(thiocarbamoyl)-	: 2189
5-Sulfanilamido-1-phenylpyrazole	: 2201
2-Sulfanilamidopyrazine	: 2202
2-Sulfanilamidopyridine	: 2203
o-Sulfanilic acid	: 2215
m-Sulfanilic acid	: 2216
Sulfanilic acid	: 2217
2-(Sulfanilylamino)thiazole	: 2206
p-Sulfanilyl fluoride	: 2252
N4-Sulfanilylsulfanilamide	: 2188
Sulfanilylurea	: 2190
Sulfaphenazole	: 2201
Sulfapyrazine	: 2202
Sulfapyridine	: 2203
Sulfapyrimidine	: 2204
Sulfaquinoxaline	: 2205
Sulfasalazine	: 2654
Sulfasuccidine	: 3578
Sulfathalidine	: 2726
Sulfathiazole	: 2206
Sulfathiourea	: 2189
Sulfazole	: 2200
o-Sulfhydrylbenzoic acid	: 2663
Sulfide, 3-chloropropyl methyl	: 9864
Sulfide, 2-chloropropyl methyl	: 9865
Sulfide, ethyl isobutyl	: 10044
Sulfide, o-tolyl p-tolyl	: 1993
4,4'-Sulfinyldianiline	: 845
Sulfiram	: 11714
Sulfisoxazole	: 2195
Sulfoacetic acid	: 238
o-Sulfobenzoic acid imide	: 2388
Sulfodiamine	: 91
2-Sulfoethylamine	: 5623
Sulfolane	: 11786
Sulfolane, 2,4-dimethyl	: 11784
3-Sulfolene	: 11752
Sulfonethylmethane	: 3230
Sulfonmethane	: 9835
Sulfonyldiacetic acid	: 239
p-Sulfophenol	: 2226
p-Sulfophenylhydrazine	: 2224
Sulfosalicylic acid	: 2655
Sulfotep	: 11716
Sulfovinic acid	: 11576
Sulfurous acid, 2-[4-(1,1-dimethylethyl)phenoxy]cyclohexyl 2-propynyl ester	: 10507
Sulphamidine	: 2196
Sulphenone	: 1225
Surinamine	: 11933
Sylvic acid	: 9167
Symclosene	: 11840
Synephrine	: 1887
Syringaldehyde	: 569
Syringidin chloride	: 2847
Syringin	: 6225
2,4,5-T	: 256

2,2,4,6-Tetramethylpiperidine	: 9710
2,2,6,6-Tetramethylpiperidine	: 9711
1,2,5,6-Tetramethyl-4-piperidinone	: 9725
Tetramethylputrescine	: 3263
2,3,4,6-Tetramethylpyridine	: 11022
Tetramethylsilane	: 11451
Tetramethyl silicate	: 11508
Tetramethylstannane	: 11544
Tetramethylsuccinonitrile	: 3292
2,2,4,4-Tetramethyltetrahydrofuran	: 6174
N,N,N',N'-Tetramethyltetra-methylenediamine	: 3263
Tetramethylurea	: 12030
Tetranitrin	: 3432
Tetranitromethane	: 7544
Tetranitropentaerythritol	: 9990
2,3,4,6-Tetranitrophenol	: 9476
2,4,5,6-Tetraoxohexahydropyrimidine	: 11073
1,1,1,2-Tetraphenylethane	: 1618
1,1,2,2-Tetraphenyl-1,2-ethanediol	: 5580
1,1,2,2-Tetraphenylethylene glycol	: 5580
2,3,4,5-Tetraphenylfuran	: 6175
Tetraphenylfuran	: 6175
Tetraphenylmethane	: 1831
Tetraphenylsilane	: 11452
Tetraphenylurea	: 12031
Tetraphthene	: 9
Tetrapropylammonium iodide	: 9833
Tetratriacontylcyclohexane	: 4479
Tetravinylsilane	: 11449
sym-Tetrazine	: 11651
Tetrolaldehyde	: 3895
Tetrole	: 6025
Tetroquinone	: 4269
Thalidomide	: 7363
Thebaine	: 7679
Thebainone	: 7677
3-Thenoic acid	: 11738
Thenylpyramine	: 5507
Theobromine	: 10757
Theophylline	: 10756
THF glycol	: 7109
1-Thia-2-azaindene	: 2387
Thiabutazide	: 2856
Thiacetazone	: 34
Thiachromine-8-ethanol, 2,7-dimethyl	: 11111
Thiacyclohexane	: 11795
Thiacyclopentane	: 11783
1-Thia-2,4-diazacyclopentadiene	: 11657
1,2,4-Thiadiazole, 5-ethoxy-3-(trichloromethyl)-	: 11599
1,3,4-Thiadiazole-2-sulfonic acid,amide,5-acetamido	: 32
1-Thiaindene	: 2399
Thiamine hydrochloride	: 11686
Thiamorpholine	: 11719
Thianaphthene	: 2399
Thiazolsulfone	: 11663
2-Thiazolylamine	: 11662
4'-(2-Thiazolylsulfamoyl)succinanilic acid	: 3578
2-Thienonitrile	: 11732
2-Thienyl bromide	: 11726
2-Thienyl chloride	: 11740
Thioacetaldehyde trimer	: 11920
Thioacetamide	: 5643
Thioacetic acid	: 5646
Thiobenzyl alcohol	: 1832
Thiobutyl alcohol	: 3437
Thiocarbamide	: 11799
Thiochrome	: 11111
Thiocyanatobenzene	: 11709
Thiocyanic acid, tert-butyl ester	: 11701
Thiocyanic acid, isopropyl ester	: 11706
1-Thiocyanobutane	: 11699
Thiocyclohexanone	: 4481
2,2'-Thiodiethanol	: 5766
Thiodiglycolic acid	: 240
Thiodiphenylamine	: 9499
3,3'-Thiodipropionic acid	: 10329
Thiofuran	: 11724
Thioglycerol	: 10013
Thioglycolic acid	: 202
Thioguanine	: 10762

Thiolactic acid	: 10285
Thiomedon	: 7649
Thionalide	: 68
Thionaphthene-2-carboxylic acid	: 2400
Thiophene, 2-isobutyryl	: 10472
Thiophene, 2-isopentyl-	: 11771
Thiophene, 2-isopropyl-	: 11772
Thiophene, 3-isopropyl-	: 11773
Thiophenol	: 2289
Thiophenol, 3-ethyl	: 2297
Thiophosgene	: 4032
o-Thiosalicylic acid	: 2663
Thiosemicarbazide	: 7142
Thioserine	: 5014
Thiosinamine	: 11809
Thiotax	: 2866
3-Thiotolene	: 11770
2-Thiouracil	: 11110
Thioxanthone	: 11812
Thiram	: 11722
THP	: 10800
Threose (D-)	: 3193
4(10)-Thujene, (±)-	: 2981
4(10)-Thujene (d)	: 11815
4(10)-Thujene-3-ol	: 2983
Thujic acid	: 4241
Thujopsene	: 4954
Thymine	: 11068
Thymine 2-desoxyriboside	: 11816
Thymol	: 9418
Thymol, acetate	: 9420
Thymol blue	: 9369
2-Thymolcarboxylic acid	: 2642
Thymol, 2,6-dibromo	: 9281
Thymolphthalein	: 7352
o-Thymotic acid	: 2642
Thymyl iodide	: 1782
Thyroxine	: 11930
Tiglic acid	: 3844
Tigogenin	: 11532
Timonacic	: 11682
Tiocarlide	: 11800
δ-Tocopherol	: 2785
α-Tocopherol	: 2787
β-Tocopherol	: 2789
γ-Tocopherol	: 2790
α-Tocopheryl acetate	: 2788
Tokoferol acetate	: 2788
Tolazine	: 9195
Tolbutamide	: 2207
3,3'-Tolidine	: 3034
o-Tolidine	: 3034
o-Tolualdehyde	: 587
p-Tolualdehyde	: 589
α-Tolualdehyde	: 868
m-Tolualdehyde, 2-hydroxy	: 575
o-Toluamide	: 619
Toluene	: 1947
Toluene, 3-bromo-2-nitro-	: 1068
Toluene, 3-bromo-5-nitro-	: 1069
Toluene, 3-bromo-4-nitro-	: 1073
Toluene, 4-bromo-2-nitro-	: 1074
Toluene, 2-bromo-6-nitro-	: 1070
o-Toluenecarbonitrile	: 2763
Toluene, 4-chloro-3-iodo-	: 1173
Toluene, 5-chloro-2-iodo	: 1174
Toluene, 2-chloro-α-methylimino	: 7652
Toluene, 3-chloro-5-nitro-	: 1207
Toluene-3,4-diamine	: 1276
Toluene-2,6-diamine	: 1277
Toluene-2,4-diamine	: 1278
Toluene, 2,3-dibromo	: 1302
Toluene, 3,4-dibromo	: 1303
Toluene, 2,6-dibromo	: 1304
Toluene, 2,5-dibromo-	: 1305
Toluene, 2,3-dichloro-	: 1383
Toluene, 2,4-dichloro-	: 1387
Toluene, 2,4-dichloro-α-hydroxy	: 1853
Toluene, 2,5-diethoxy	: 1410
Toluene, α,α-dihydroxy	: 1884
Toluene 2,4-diisocyanate	: 1431
Toluene, 2,6-diisocyanate	: 1430
Toluene, 2,4-diisopropyl-	: 1948
Toluene-α,α-diol, m-nitro-, diacetate (ester)	: 7511
Toluene-3,4-dithiol	: 1564

Toluene, 3-fluoro-4-nitro-	: 1730
Toluene, 4-fluoro-3-nitro-	: 1729
Toluene, 3-(1-hydroxyethyl)	: 1858
Toluene, α-hydroxy-4-methoxy-α-methyl	: 1895
Toluene, α-hydroxy-4-nitro	: 1908
m-Toluenenitrile	: 2764
p-Toluenenitrile	: 2765
Toluene, p-(1-octyldecyl)-	: 8245
Toluene, α,α,2,3,4-pentachloro-	: 2326
Toluene, α,α,2,4,5-pentachloro-	: 2327
Toluene, 4-propoxy	: 2032
o-Toluenesulfinic acid	: 2181
p-Toluenesulfinic acid	: 2182
p-Toluenesulfonamide, N,N-dichloro-	: 2210
p-Toluenesulfonic acid	: 2229
p-Toluenesulfonic acid chloride	: 2250
p-Toluenesulfonic acid, ethyl ester	: 2231
o-Toluenesulfonyl chloride	: 2249
p-Toluenesulfonyl chloride	: 2250
Toluene, 2,3,4,6-tetrachloro-	: 2262
Toluene, 2,3,5,6-tetrachloro-	: 2263
α-Tolueneethiol, α-methyl-, (S)-(-)-	: 1834
Toluene, 2,4,6-tribromo-	: 2314
Toluene, 2,3,4-trichloro-	: 2334
Toluene, 3,4,5-trichloro-	: 2335
Toluene, 2,4,5-trichloro-	: 2336
Toluene, 2,3,5-trichloro-	: 2337
Toluene, 2,4,6-triiodo-	: 2352
Toluene, 2,3,4-trinitro-	: 2046
Toluene, 2,4,5-trinitro-	: 2047
o-Toluic acid	: 2676
m-Toluic acid	: 2677
p-Toluic acid	: 2678
o-Toluidine	: 778
m-Toluidine	: 779
p-Toluidine	: 780
m-Toluidine, N-benzyl	: 1824
p-Toluidine, N-benzyl	: 1825
m-Toluidine, 6-bromo-	: 648
o-Toluidine hydrochloride	: 791
p-Toluidine, 2-iodo-	: 762
o-Toluidine, 5-iodo	: 763
o-Toluidine, 4-(phenylsulfonyl)-N-[(trifluoromethyl)sultono]-	: 9149
p-Toluidine, N-propyl-	: 815
o-Tolunitrile	: 2763
m-Tolunitrile	: 2764
p-Tolunitrile	: 2765
m-Toluquinaldine	: 11258
m-Toluquinoline	: 11282
2-(p-Toluyl)benzoic acid	: 2684
o-Tolyl alcohol	: 1897
p-Tolyl alcohol	: 1899
2-o-Tolylaminoethanol	: 5744
2-p-Tolylaminoethanol	: 5745
p-Tolylcarbinol	: 1899
1-(4-Tolyl)ethanol	: 1859
m-Tolylmethanol	: 1898
p-Tolylsulfonylmethylnitrosamide	: 2212
o-Tolylthiourea	: 11806
Tomatidine	: 11528
Tomatine	: 6196
Topazone	: 8589
Torularhodin	: 4059
Tosic acid	: 2229
p-Tosyl chloride	: 2250
Tranexamic acid	: 4333
trans-1,3-Dimethylcyclohexanol	: 4511
Traumatic acid	: 5336
Trehalose	: 6220
Triacetamide	: 50
Triacetin	: 10195
N,N,N-Triacetylammonia	: 50
Triacontyl alcohol	: 11821
Triacontylcyclohexane	: 4482
Triacontylcyclopentane	: 4853
Triallylamine	: 10523
1,2,3-Triaminobenzene	: 2305
Triamterene	: 10750
1,2,3-Triaza-1H-indene	: 2871
1,2,4-Triazanaphthalene	: 2870
1,3,5-Triazine-2,4-diamine, N,N'-bis(1-methylethyl)-6-(methylthio)-	: 9770

Trimethylantimony	: 11550
Trimethylarsine	: 449
1,2,3-Trimethylbenzene	: 2357
Trimethylborane	: 3119
Trimethyl borate	: 3126
2,2,3-Trimethylbutane	: 3457
Trimethylbutylsilane	: 11371
Trimethylchlorosilane	: 11389
Trimethyl citrate	: 10170
1,1,2-Trimethylcycloheptane	: 4227
1,1,4-Trimethylcycloheptane	: 4228
3,3,7-Trimethylcycloheptanol	: 4233
2,6,6-Trimethylcycloheptanone	: 4238
1,4,4-Trimethyl-1-cycloheptene	: 4247
1,1,2-Trimethylcyclohexane	: 4485
1,1,3-Trimethylcyclohexane	: 4486
1,1,4-Trimethylcyclohexane	: 4487
1,2,3-Trimethylcyclohexane	: 4488
trans-1,3,5-Trimethylcyclohexane	: 4497
1,1,3-Trimethyl-2-cyclohexanol	: 4566
2,3,6-Trimethylcyclohexanol	: 4568
cis-3,3,5-Trimethylcyclohexanol	: 4569
trans-3,5,5-Trimethylcyclohexanol	: 4570
trans-3,3,5-Trimethylcyclohexanol	: 4570
1,2,2-Trimethylcyclohexanol	: 4562
1,2,6-Trimethylcyclohexanol	: 4563
1,3,5-Trimethylcyclohexanol	: 4564
2,2,5-Trimethylcyclohexanol	: 4565
2,2,3-Trimethylcyclohexanone	: 4614
2,2,4-Trimethylcyclohexanone	: 4615
2,3,6-Trimethylcyclohexanone	: 4618
2,4,4-Trimethylcyclohexanone	: 4619
2,4,5-Trimethylcyclohexanone	: 4620
3,3,5-Trimethylcyclohexanone	: 4622
3,4,4-Trimethylcyclohexanone	: 4623
2,2,5-Trimethylcyclohexanone	: 4616
2,4,6-Trimethylcyclohexanone	: 4621
1,4,4-Trimethylcyclohexene	: 4889
1,5,5-Trimethylcyclohexene	: 4691
1,2,3-Trimethylcyclohexene	: 4687
1,3,5-Trimethylcyclohexene	: 4688
1,4,5-Trimethylcyclohexene	: 4690
1,6,6-Trimethylcyclohexene	: 4692
2,4,4-Trimethyl-2-cyclohexen-1-ol	: 4701
3,6,6-Trimethyl-2-cyclohexenone	: 4722
4-(2,6,6-Trimethyl-1-cyclohexen-1-yl)-3-buten-2-ol	: 3873
1,1,2-Trimethylcyclopentane	: 4855
1,1,3-Trimethylcyclopentane	: 4856
1,2,2-Trimethylcyclopentanol	: 4875
1,2,4-Trimethylcyclopentanol	: 4876
1,2,5-Trimethylcyclopentanol	: 4877
2,3,5-Trimethylcyclopentanol	: 4898
2,2,4-Trimethylcyclopentanone	: 4896
2,2,5-Trimethylcyclopentanone	: 4897
2,4,4-Trimethylcyclopentanone	: 4899
1,5,5-Trimethylcyclopentene	: 4938
1,1,2-Trimethylcyclopropane	: 4985
trans-3,7,11-Trimethyl-1,3,6,10-dodecatetraene	: 5329
trans-2,6,10-Trimethyl-2,6,9,11-dodecatetraene	: 5329
Trimethylene	: 4955
Trimethylene chlorohydrin	: 10348
Trimethylene dichloride	: 9893
Trimethylene diiodide	: 9907
Trimethylene dimercaptan	: 10031
Trimethylene glycol	: 9983
Trimethylene oxide	: 8592
Trimethylene sulfide	: 11691
Trimethylene sulfone	: 11692
2,2,3-Trimethylheptane	: 6437
2,2,4-Trimethylheptane	: 6438
2,2,5-Trimethylheptane	: 6439
2,2,6-Trimethylheptane	: 6440
2,3,3-Trimethylheptane	: 6441
2,3,4-Trimethylheptane	: 6442
2,3,5-Trimethylheptane	: 6443
2,3,6-Trimethylheptane	: 6444
2,4,4-Trimethylheptane	: 6445
2,4,5-Trimethylheptane	: 6446
2,4,6-Trimethylheptane	: 6447
2,5,5-Trimethylheptane	: 6448
3,3,4-Trimethylheptane	: 6449
3,3,5-Trimethylheptane	: 6450
3,4,4-Trimethylheptane	: 6451
3,4,5-Trimethylheptane	: 6452

2,2,3-Trimethylhexane	: 6859
2,2,4-Trimethylhexane	: 6860
2,2,5-Trimethylhexane	: 6861
2,3,3-Trimethylhexane	: 6862
2,3,4-Trimethylhexane	: 6863
2,3,5-Trimethylhexane	: 6864
2,4,4-Trimethylhexane	: 6865
3,3,4-Trimethylhexane	: 6866
3,5,5-Trimethylhexan-3-ol	: 6958
1,1,5-Trimethylindan	: 7261
1,4,7-Trimethylindan	: 7262
1,2,3-Trimethylindole	: 7322
2,3,5-Trimethylindole	: 7323
Trimethyl methanetricarboxylate	: 7553
1,2,4-Trimethylnaphthalene	: 8009
1,2,5-Trimethylnaphthalene	: 8010
1,2,6-Trimethylnaphthalene	: 8011
2,4,6-Trimethylnaphthalene	: 8013
2,4,7-Trimethyloctane	: 8421
Trimethylolmelamine	: 7583
Trimethylolpropane	: 10006
Trimethylolpropane phosphite	: 11909
2,2,3-Trimethylpentane	: 8847
2,3,3-Trimethylpentane	: 8849
2,3,4-Trimethylpentane	: 8850
2,2,4-Trimethyl-1,3-pentanediol	: 8795
2,4,4-Trimethyl-2-pentanol	: 8954
2,2,4-Trimethyl-3-pentanol	: 8955
2,4,4-Trimethyl-1-pentene	: 9064
2,3,3-Trimethyl-1-pentene	: 9065
2,3,4-Trimethyl-1-pentene	: 9066
2,2,4-Trimethyl-4-pentene	: 9067
3,4,4-Trimethyl-1-pentene	: 9069
2,3,4-Trimethyl-2-pentene	: 9070
2,4,4-Trimethyl-2-pentene	: 9071
3,4,4-Trimethyl-2-pentene	: 9072
2,4,5-Trimethylphenyl acetate	: 9496
Trimethylphosphine	: 9522
Trimethyl phosphite	: 9592
1,2,4-Trimethylpiperazine	: 9622
2,3,6-Trimethylpiperidine	: 9713
2,4,6-Trimethylpiperidine	: 9714
1,1,2-Trimethylpropyl bromide	: 3236
Trimethylpropylsilane	: 11496
1,3,4-Trimethylpyrazole	: 10839
1,3,5-Trimethylpyrazole	: 10840
1,4,5-Trimethylpyrazole	: 10841
2,3,4-Trimethylpyridine	: 11024
2,3,5-Trimethylpyridine	: 11025
2,3,6-Trimethylpyridine	: 11026
2,4,5-Trimethylpyridine	: 11027
1,2,5-Trimethylpyrrole	: 11142
2,2,4-Trimethylpyrrolidine	: 11175
2,4,6-Trimethylquinoline	: 11302
2,4,7-Trimethylquinoline	: 11303
Trimethylsilane	: 11489
(Trimethylsilyl)methyl bromide	: 11366
Trimethylstibine	: 11550
2,3,4-Trimethylthiophene	: 11792
2,3,5-Trimethylthiophene	: 11793
5,7,8-Trimethyltocol	: 2787
1,1,1-Trimethyl-3,3,3-triphenyldisiloxane	: 5239
Trimethylurea	: 12032
Trimyristin	: 11628
Trinitroacetonitrile	: 290
2,3,5-Trinitroanisole	: 364
sym-Trinitrobenzene	: 2365
2,4,6-Trinitrobenzoic acid	: 2735
Trinitro-m-cresol	: 9434
2,4,6-Trinitro-3,5-dimethyl-tert-butylbenzene	: 1461
2,4,7-Trinitrofluorenone	: 5979
Trinitromethane	: 7562
2,3,4-Trinitrotoluene	: 2046
2,4,5-Trinitrotoluene	: 2047
2,4,6-Trinitrotoluene	: 2048
28,29,30-Trinorlanostane	: 4105
Triolein	: 8310
Trional	: 3230
Triopron	: 2136
s-Trioxane, 2,4,6-trivinyl-	: 11906
3,6,9-Trioxaundecane	: 5612
3,7,12-Trioxocholanic acid	: 4101
Tripalmitin	: 6656
Tripelennamine	: 5505
Triphenylamine	: 728

Triphenylarsane	: 450
Triphenylarsine	: 450
Triphenylbismuth	: 3111
Triphenylcarbinol	: 1866
Triphenylhydrazine	: 7173
Triphenylmethane	: 1982
Triphenylmethane, 3-methyl	: 7513
Triphenylmethanol	: 1866
Triphenylmethylsilane	: 11444
Triphenyl phosphate	: 9566
Triphenylphosphine	: 9523
Triphenyltetrazolium chloride	: 11653
Triphenyltin acetate	: 11539
Triphenyltin chloride	: 11541
Tripropylamine	: 9808
Tripropyl phosphate	: 9567
Tripropyl phosphite	: 9594
Tripropylsilane	: 11500
Triptycene	: 2375
2,4,6-Tripyridyl-s-triazine	: 11843
Tris(allyloxy)vinylsilane	: 11427
Tris-BP	: 10358
Tris-(2-chloroethyl) amine	: 5412
Tris(chloromethyl)phosphine	: 9524
Tris(2,3-dibromopropyl) phosphate	: 10358
Tris(dimethylamino)phosphine oxide	: 9573
Tris(hydroxymethyl)nitromethane	: 10012
Tristearin	: 8272
Trithioacetaldehyde	: 11920
Trithiocarbonic acid	: 4034
Tri-o-tolyl phosphate	: 9569
Trityl alcohol	: 1866
Tromasedan	: 2385
Tromethamine	: 9986
Tropacocaine	: 488
Tropane	: 484
Tropic acid	: 915
Tropilidene	: 4240
Tropine	: 489
Tryptamine	: 7302
Tryptophol	: 7305
Tsuduranine	: 5159
Tsuzuic acid	: 11637
Tuaminoheptane	: 6350
Tubatoxin	: 2783
Tuduic acid	: 11637
D-Turanose	: 6022
Tybamate	: 3933
Tylenol	: 67
Tyramine	: 9208
Tyrosine, 3,5-dibromo(L)	: 11927
Ugurol	: 4333
Umbelliferone	: 2826
Umbelliferone glucoside	: 2821
Undecafluorocyclohexane	: 4498
trans-9-Undecenoic acid	: 11975
(10-Undecenyl)acetic acid	: 11895
Undecylenic acid	: 11974
Undecyl ketone	: 11857
Undecyl mercaptan	: 11942
Uneicosane, 11-phenyl	: 6297
Unidigin	: 4043
Untriacontane	: 6299
Uracil	: 11059
Uracil mustard	: 11061
Uramil	: 11077
Urazole	: 11850
Urea, N'-(4-bromo-3-chlorophenyl)-N-methoxy-N-methyl-	: 4076
Urea, 1-(2-bromo-2-ethylbutanoyl)	: 3195
Urea, 1-(2-bromo-3-methylbutanoyl)	: 3196
Urea, 3-(p-bromophenyl)-1-methoxy-1-methyl-	: 7656
Urea, N'-[4-(4-chlorophenoxy)phenyl]-N,N-dimethyl-	: 4088
Urea, N,N'-diamino-	: 4009
Urea, N'-(3,4-dichlorophenyl)-N-methoxy-N-methyl-	: 7415
Urea, 1,1-diisobutyl	: 11996
Urea, N-[5-(1,1-dimethylethyl)-1,3,4-thiadiazol-2-yl]-N,N'-dimethyl-	: 11590
Urea, N,N-dimethyl-N'-(octahydro-4,7-methano-1H-indenN-5-yl)-, (3α,4α, 5α,7α,7aα)-	: 8212
Urea hydrochloride	: 12020
Urea nitrate	: 12021

Urea, N-phenyl-N'-1,2,3-thiadiazol-5-yl-	: 11687
Urea, 1-sulfanilyl-2-thio-	: 2189
Urea, 1,1,3,3-tetraethyl-	: 12029
5-Ureidohydantoin	: 12009
Urethan	: 3948
Urethane, sec-butyl	: 3960
Urethane, isopentyl	: 3952
Uric acid	: 10764
Uric acid, 3-methyl-	: 10765
Uric acid, 7-methyl-	: 10766
Uridine 5'-phosphoric acid	: 12035
Urinex	: 2852
Uriprim	: 10852
Urochloralic acid	: 6231
Urofort	: 11830
Ursodeoxycholic acid	: 4098
Ursodiol	: 4098
Ursolic acid	: 12037
Uteroverdine	: 3001
Uvic acid	: 3311
Uzarin	: 4045
Vaccenic acid	: 8295
Valeraldehyde	: 8690
Valeraldehyde, 2-propyl-	: 8697
Valeric acid	: 8855
Valeric acid, 4-acetyl, ethyl ester	: 6902
Valeric acid, 4-acetyl, methyl ester	: 6903
Valeric acid, sec-butyl ester (d)	: 8899
Valeric acid, 4-chloro-	: 8865
Valeric acid, 3-chloro	: 8864
Valeric acid, 4-chloro-, ethyl ester	: 8867
Valeric acid, 5-chloro-4-hydroxy-, γ-lactone	: 6141
Valeric acid, 2,3-dibromo-	: 8870
Valeric acid, 2,5-dibromo-	: 8871
Valeric acid, 2,3-dibromo, ethyl ester	: 8872
Valeric acid, 2,5-dibromo-, ethyl ester	: 8873
Valeric acid, 4,4-dimethyl-5-acetyl	: 6461
Valeric acid, 2,2-dimethyl-5-acetyl	: 6460
Valeric acid, 2-ethyl (DL)	: 8877
Valeric acid, 5-fluoro-	: 8879
Valeric acid, furfuryl ester	: 8880
Valeric acid, 4-hydroxy-, ethyl ester	: 8884
Valeric acid, 2-hydroxyethyl ester	: 8885
Valeric acid, 3-methyl (DL)	: 8888
Valeric acid, 2-methylene	: 8890
Valeric acid, 3-methyl-3-phenyl	: 7175
Valeric acid, octyl ester	: 8901
Valeric acid, 2-oxo-, ethyl ester	: 8905
Valeric acid, 2-phenyl-, (±)-	: 943
Valeronitrile	: 8827
Valeronitrile, 3-imino-2-methyl	: 8830
Valeronitrile, 4-methyl-2-phenyl-	: 964
Valeroyl chloride	: 8990
Valerylacetone	: 8393
Valerylacetyl	: 6402
Valeryl chloride, 2-methyl (DL)	: 8993
Valeryl chloride, 3-methyl (DL)	: 8994
Valimyl	: 3201
Valium	: 2410
Vanillic acid	: 2625
Vanillin	: 574
Vanillylacetone	: 3663
Vanilmandelic acid	: 896
Vasicine	: 11192

Vellosimine	: 11346
Veratraldehyde	: 545
Veratramine	: 12044
Veratric acid	: 2576
Veratrole	: 1436
Veratrole, 3-iodo	: 1767
(R)-(+)-Verbenone	: 2978
d-Verbenone	: 2978
Vernine	: 6289
Vesipyrin	: 2483
Vibeline	: 3540
Vicianose	: 6234
Vicine	: 11107
(+)-cis-Vincamine	: 5349
Vincamine	: 5349
Vinyl acetate	: 179
Vinylacetonitrile	: 3803
Vinylacetylene	: 3889
Vinyl amide	: 10519
α-Vinylbenzyl alcohol	: 1868
Vinyl bromide	: 5880
Vinyl butyl ketone	: 6583
Vinyl butyrate	: 3513
Vinylcarbinol	: 10697
Vinyl chloride	: 5884
Vinyl cyanide	: 10583
4-Vinyl-1-cyclohexene dioxide	: 8557
Vinylethylacetylene	: 7103
Vinylethylene	: 3146
Vinyl fluoride	: 5906
Vinyl formate	: 6006
Vinylformic acid	: 10609
Vinylglycolic acid	: 3842
1-Vinylheptanol	: 8189
Vinylidene chloride	: 5894
Vinylidene fluoride	: 5902
Vinyl isobutyl ketone	: 7093
2-Vinylnaphthalene	: 7882
5-Vinyl-2-picoline	: 10955
Vinyl propionate	: 10261
Vinyltrichlorosilane	: 11461
Violaxanthin	: 4056
Violaxanthin, all-trans-	: 4056
Violet-leaf aldehyde	: 8108
Violuric acid	: 11074
Viquidil	: 10434
Visnadine	: 3540
Visnagin	: 6186
Vitamin A	: 11341
Vitamin B1, hydrochloride	: 11686
Vitamin B4	: 10751
Vitamin Bc	: 6243
Vitamin BX	: 2486
Vitamin C	: 466
Vitamin D2	: 11350
Vitamin D3	: 11348
Vitamin E	: 2787
Vitamin E acetate	: 2788
Vitamin K1	: 7865
Vitamin K6	: 7800
Vitamin L1	: 2484
Vomicine	: 11351
Warfarin	: 2832
Wintergreen oil	: 2634
Woodward's Reagent K	: 7407
Xanthaurine	: 2817
9-Xanthenecarboxylic acid, 2(N,N-diethylamino)ethyl ester, methylbromide	: 5445

Xanthine	: 10755
Xanthone	: 12050
Xanthone, 2,7-dihydroxy	: 12052
Xanthophyll 3	: 4057
all-trans-(+)-Xanthophyll	: 4058
Xanthophyll	: 4058
Xanthopterin	: 10749
Xanthoxyletin	: 2398
Xanthurenic acid	: 11230
Xanthyletin	: 2397
o-Xenol	: 3088
p-Xenylamine	: 3011
o-Xylene	: 1452
m-Xylene	: 1453
p-Xylene	: 1454
m-Xylene, 4-sec-butyl-	: 1473
m-Xylene, 5-tert-butyl-	: 1460
m-Xylenediamine	: 1433
o-Xylene, 3,4-dibromo-	: 1297
o-Xylene, 3,4-dichloro-	: 1360
p-Xylene, α,α'-dimethyl-	: 1415
p-Xylene, 2,5-dimethyl-	: 2280
p-Xylene, α,α'-diphenyl-	: 995
o-Xylene, 3-ethyl-	: 1670
o-Xylene, 4-ethyl-	: 1675
m-Xylene, 2-ethyl-	: 1673
m-Xylene, 4-ethyl-	: 1671
m-Xylene, 5-ethyl-	: 1672
p-Xylene, 2-ethyl-	: 1674
o-Xylene, 4-iodo-	: 1774
m-Xylene, 5-iodo-	: 1771
o-Xylene, 3-isopropyl-	: 1469
m-Xylene, 2-isopropyl-	: 1471
o-Xylene, 3-propyl-	: 1485
m-Xylene, 2-propyl-	: 1482
m-Xylene, 4-propyl-	: 1487
m-Xylene, 5-propyl-	: 1483
p-Xylene, 2-propyl-	: 1486
p-Xylene, 2,3,5-trinitro-	: 1489
2,3-Xylenol	: 9307
2,4-Xylenol	: 9308
2,5-Xylenol	: 9309
2,6-Xylenol	: 9310
3,4-Xylenol	: 9311
3,5-Xylenol	: 9312
3,4-Xylenol, 6-tert-butyl-	: 9324
2,3-Xylidine	: 704
2,4-Xylidine	: 705
2,5-Xylidine	: 706
2,6-Xylidine	: 707
3,4-Xylidine	: 708
3,5-Xylidine	: 709
Xylite	: 12058
p-Xylotocopherol	: 2789
o-Xylotocopherol	: 2790
1-(2,5-Xylylazo)-2-naphthol	: 8029
Yellow AB	: 7729
Yohimbine	: 12062
Zeaxanthin	: 4057
Zinc, bis(dimethylcarbamodithioato-S,S')-, (T-4)-	: 12068
Zineb	: 12067
Zingerone	: 3663
Zoxazolamine	: 2876
Zymostanol	: 4107

CBrClF$_2$
 7464 - Methane, bromochlorodifluoro-
CBrClN$_2$O$_4$
 7465 - Methane, bromochlorodinitro-
CBrCl$_2$F
 7468 - Methane, bromodichlorofluoro-
CBrCl$_3$
 7474 - Methane, bromotrichloro-
CBrF$_3$
 7475 - Methane, bromotrifluoro-
CBrN
 4165 - Cyanogen bromide
CBrN$_3$O$_6$
 7476 - Methane, bromotrinitro-
CBr$_2$ClF
 7494 - Methane, dibromochlorofluoro-
CBr$_2$Cl$_2$
 7495 - Methane, dibromodichloro-
CBr$_2$F$_2$
 7496 - Methane, dibromodifluoro-
CBr$_2$N$_2$O$_4$
 7497 - Methane, dibromodinitro-
CBr$_2$O
 4006 - Carbonic dibromide
CBr$_3$Cl
 7549 - Methane, tribromochloro-
CBr$_3$F
 7550 - Methane, tribromofluoro-
CBr$_3$NO$_2$
 7551 - Methane, tribromonitro-
CBr$_4$
 7539 - Methane, tetrabromo-
CClFO
 4005 - Carbonic chloride fluoride
CClF$_3$
 7486 - Methane, chlorotrifluoro-
CClF$_3$O$_2$S
 7537 - Methanesulfonyl chloride, trifluoro-
CClN
 4166 - Cyanogen chloride
CClNO$_3$S
 11585 - Sulfuryl chloride isocyanate
CClN$_3$O$_6$
 7487 - Methane, chlorotrinitro-
CCl$_2$F$_2$
 7500 - Methane, dichlorodifluoro-
CCl$_2$N$_2$O$_4$
 7501 - Methane, dichlorodinitro-
CCl$_2$O
 4007 - Carbonic dichloride
CCl$_2$S
 4032 - Carbonothioic dichloride
CCl$_3$F
 7555 - Methane, trichlorofluoro-
CCl$_3$I
 7556 - Methane, trichloroiodo-
CCl$_3$NO$_2$
 7557 - Methane, trichloronitro-
CCl$_4$
 7540 - Methane, tetrachloro-
CCl$_4$O$_2$S
 7536 - Methanesulfonyl chloride, trichloro-
CCl$_4$O$_4$
 9148 - Perchloric acid, trichloromethyl ester
CCl$_4$S
 7527 - Methanesulfenyl chloride, trichloro-
CFN
 4167 - Cyanogen fluoride
CF$_2$O
 4008 - Carbonic difluoride
CF$_3$I
 7559 - Methane, trifluoroiodo-
CF$_4$
 7541 - Methane, tetrafluoro-
CF$_4$S
 7528 - Methanesulfenyl fluoride, trifluoro-

CF$_8$S
 11584 - Sulfur, pentafluoro(trifluoromethyl)-, (OC-6-21)-
CHBrClF
 7466 - Methane, bromochlorofluoro-
CHBrCl$_2$
 7467 - Methane, bromodichloro-
CHBrF$_2$
 7469 - Methane, bromodifluoro-
CHBr$_2$Cl
 7493 - Methane, dibromochloro-
CHBr$_2$F
 7498 - Methane, dibromofluoro-
CHBr$_3$
 7548 - Methane, tribromo-
CHClF$_2$
 7478 - Methane, chlorodifluoro-
CHClI$_2$
 7479 - Methane, chlorodiiodo-
CHClN$_2$O$_4$
 7480 - Methane, chlorodinitro-
CHCl$_2$F
 7502 - Methane, dichlorofluoro-
CHCl$_2$I
 7503 - Methane, dichloroiodo-
CHCl$_3$
 7554 - Methane, trichloro-
CHCl$_5$Si
 11459 - Silane, trichloro(dichloromethyl)-
CHFI$_2$
 7516 - Methane, fluorodiiodo-
CHFO
 6019 - Formyl fluoride
CHF$_2$I
 7505 - Methane, difluoroiodo-
CHF$_3$
 7558 - Methane, trifluoro-
CHI$_3$
 7560 - Methane, triiodo-
CHN
 7176 - Hydrocyanic acid
CHNO
 4162 - Cyanic acid
 6024 - Fulminic acid
 7360 - Isofulminic acid
CHNS
 7399 - Isothiocyanic acid
CHN$_3$O$_6$
 7562 - Methane, trinitro-
CHNaO$_3$
 4004 - Carbonic acid monosodium salt
CH$_2$BrCl
 7463 - Methane, bromochloro-
CH$_2$BrF
 7470 - Methane, bromofluoro-
CH$_2$BrI
 7471 - Methane, bromoiodo-
CH$_2$BrNO$_2$
 7473 - Methane, bromonitro-
CH$_2$Br$_2$
 7492 - Methane, dibromo-
CH$_2$ClF
 7481 - Methane, chlorofluoro-
CH$_2$ClI
 7482 - Methane, chloroiodo-
CH$_2$ClNO
 3971 - Carbamic chloride
CH$_2$ClNO$_2$
 7485 - Methane, chloronitro-
CH$_2$Cl$_2$
 7499 - Methane, dichloro-
CH$_2$Cl$_4$Si
 11456 - Silane, trichloro(chloromethyl)-
CH$_2$Cl$_6$Si$_2$
 11440 - Silane, methylenebis[trichloro-
CH$_2$Cu$_2$O$_5$
 511 - Basic copper carbonate

CH$_2$FI
 7517 - Methane, fluoroiodo-
CH$_2$F$_2$
 7504 - Methane, difluoro-
CH$_2$I$_2$
 7506 - Methane, diiodo-
CH$_2$N$_2$
 4155 - Cyanamide
 7491 - Methane, diazo-
CH$_2$N$_2$O$_4$
 7508 - Methane, dinitro-
CH$_2$N$_4$
 11652 - 1H-Tetrazole
CH$_2$O
 5987 - Formaldehyde
(CH$_2$O)$_x$
 8634 - Paraformaldehyde
CH$_2$O$_2$
 6002 - Formic acid
CH$_2$S$_3$
 4034 - Carbonotrithioic acid
CH$_3$AsF$_2$
 463 - Arsonous difluoride, methyl-
CH$_3$Br
 7462 - Methane, bromo-
CH$_3$Br$_3$Si
 11453 - Silane, tribromomethyl-
CH$_3$Cl
 7477 - Methane, chloro-
CH$_3$ClO$_2$S
 7535 - Methanesulfonyl chloride
CH$_3$ClO$_3$S
 4086 - Chlorosulfuric acid, methyl ester
CH$_3$Cl$_3$Si
 11465 - Silane, trichloromethyl-
CH$_3$F
 7515 - Methane, fluoro-
CH$_3$FO$_2$S
 7538 - Methanesulfonyl fluoride
CH$_3$I
 7518 - Methane, iodo-
CH$_3$NO
 5988 - Formaldehyde, oxime
 5989 - Formamide
CH$_3$NO$_2$
 7522 - Methane, nitro-
 8091 - Nitrous acid, methyl ester
CH$_3$NO$_3$
 8075 - Nitric acid, methyl ester
CH$_3$N$_3$
 7460 - Methane, azido-
CH$_3$N$_3$O$_3$
 12024 - Urea, nitro-
CH$_4$
 7459 - Methane
CH$_4$Cl$_2$Si
 11398 - Silane, dichloromethyl-
CH$_4$N$_2$
 7563 - Methanimidamide
CH$_4$N$_2$O
 7566 - Methanimidamide, N-hydroxy-
 11992 - Urea
CH$_4$N$_2$O$_2$
 7451 - Methanamine, N-nitro-
 12016 - Urea, hydroxy-
CH$_4$N$_2$S
 11799 - Thiourea
CH$_4$N$_2$Se
 11354 - Selenourea
CH$_4$N$_4$O$_2$
 6287 - Guanidine, nitro-
CH$_4$O
 7581 - Methanol
CH$_4$O$_2$
 7180 - Hydroperoxide, methyl
CH$_4$O$_3$S
 7530 - Methanesulfonic acid

CH_4O_4S
11577 - Sulfuric acid, monomethyl ester
$CH_4O_6S_2$
7514 - Methanedisulfonic acid
CH_4S
7546 - Methanethiol
CH_5As
446 - Arsine, methyl-
CH_5AsO_3
458 - Arsonic acid, methyl-
CH_5ClN_2O
12020 - Urea, monohydrochloride
CH_5ClSi
11378 - Silane, chloromethyl
CH_5N
7442 - Methanamine
CH_5NO
7447 - Methanamine, N-hydroxy-
CH_5NO_2
6003 - Formic acid, ammonium salt
CH_5N_3
6278 - Guanidine
$CH_5N_3O_4$
12021 - Urea, mononitrate
CH_5N_3S
7142 - Hydrazinecarbothioamide
CH_5O_3P
9538 - Phosphonic acid, methyl-
CH_6ClN
7446 - Methanamine, hydrochloride
CH_6ClN_3
6285 - Guanidine, monohydrochloride
CH_6ClN_3O
7143 - Hydrazinecarboxamide, monohydrochloride
CH_6N_2
7161 - Hydrazine, methyl-
CH_6N_4
7147 - Hydrazinecarboximidamide
CH_6N_4O
4009 - Carbonic dihydrazide
$CH_6N_4O_3$
6286 - Guanidine, mononitrate
CH_6N_4S
4033 - Carbonothioic dihydrazide
CH_6OSi
11436 - Silane, methoxy
CH_6Si
11437 - Silane, methyl-
CH_8B_2S
5168 - Diborane(6), mu-(methylthio)-
CH_8Si_2
11439 - Silane, methylenebis-
CIN
7337 - Iodine cyanide
CI_4
7542 - Methane, tetraiodo-
$CKNS$
11711 - Thiocyanic acid, potassium salt
$CNNaO$
4164 - Cyanic acid, sodium salt
CN_4O_8
7544 - Methane, tetranitro-
CO
4011 - Carbon monoxide
COS
4035 - Carbon oxide sulfide
CO_2
3992 - Carbon dioxide
CS_2
3993 - Carbon disulfide
CSe_2
4036 - Carbon selenide
C_2BrCl_3O
293 - Acetyl bromide, trichloro-
C_2Br_2
5939 - Ethyne, dibromo-

$C_2Br_2ClF_3$
5519 - Ethane, 1,2-dibromo-1-chloro-1,2,2-trifluoro-
$C_2Br_2Cl_4$
5524 - Ethane, 1,2-dibromo-1,1,2,2-tetrachloro-
$C_2Br_2F_4$
5525 - Ethane, 1,2-dibromo-1,1,2,2-tetrafluoro-
C_2Br_4
5914 - Ethene, tetrabromo-
C_2Br_6
5594 - Ethane, hexabromo-
C_2CaN_2
3930 - Calcium cyanide
C_2ClF_3
5889 - Ethene, chlorotrifluoro-
C_2ClF_5
5484 - Ethane, chloropentafluoro-
C_2Cl_2
5940 - Ethyne, dichloro-
C_2Cl_2FN
279 - Acetonitrile, dichlorofluoro-
$C_2Cl_2F_2$
5897 - Ethene, 1,1-dichloro-2,2-difluoro-
5898 - Ethene, 1,2-dichloro-1,2-difluoro-
$C_2Cl_2F_4$
5537 - Ethane, 1,1-dichloro-1,2,2,2-tetrafluoro-
5538 - Ethane, 1,2-dichloro-1,1,2,2-tetrafluoro-
$C_2Cl_2O_2$
5585 - Ethanedioyl dichloride
C_2Cl_3F
5923 - Ethene, trichlorofluoro-
$C_2Cl_3F_3$
5662 - Ethane, 1,1,1-trichloro-2,2,2-trifluoro-
5663 - Ethane, 1,1,2-trichloro-1,2,2-trifluoro-
C_2Cl_3N
288 - Acetonitrile, trichloro-
C_2Cl_4
5917 - Ethene, tetrachloro-
$C_2Cl_4F_2$
5636 - Ethane, 1,1,1,2-tetrachloro-2,2-difluoro-
5637 - Ethane, 1,1,2,2-tetrachloro-1,2-difluoro-
C_2Cl_4O
300 - Acetyl chloride, trichloro-
$C_2Cl_4O_2$
4029 - Carbonochloridic acid, trichloromethyl ester
C_2Cl_5F
5618 - Ethane, pentachlorofluoro-
C_2Cl_6
5595 - Ethane, hexachloro-
C_2F_3I
5925 - Ethene, trifluoroiodo-
C_2F_3N
289 - Acetonitrile, trifluoro-
C_2F_4
5918 - Ethene, tetrafluoro-
$C_2F_4N_2O_4$
5642 - Ethane, 1,1,2,2-tetrafluoro-1,2-dinitro-
C_2F_6
5596 - Ethane, hexafluoro-
$C_2F_6S_2$
5249 - Disulfide, bis(trifluoromethyl)
C_2HBr
5937 - Ethyne, bromo-
$C_2HBrClF_3$
5457 - Ethane, 1-bromo-2-chloro-1,1,2-trifluoro-
5458 - Ethane, 2-bromo-2-chloro-1,1,1-trifluoro-
C_2HBrCl_2
5882 - Ethene, 2-bromo-1,1-dichloro-
5935 - Ethylene, 1-bromo-1,2-dichloro-, trans-
C_2HBrF_2
5883 - Ethene, 1-bromo-1,2-difluoro-
$C_2HBr_2F_3$
5526 - Ethane, 1,2-dibromo-1,1,2-trifluoro-
C_2HBr_2N
277 - Acetonitrile, dibromo-
C_2HBr_3
5921 - Ethene, tribromo-

C_2HBr_3O
20 - Acetaldehyde, tribromo-
$C_2HBr_3O_2$
241 - Acetic acid, tribromo-
C_2HBr_5
5616 - Ethane, pentabromo-
C_2HCl
5938 - Ethyne, chloro-
C_2HClF_2
5885 - Ethene, 2-chloro-1,1-difluoro-
$C_2HClF_2O_2$
127 - Acetic acid, chlorodifluoro-
C_2HClF_3I
5490 - Ethane, 1-chloro-1,1,2-trifluoro-2-iodo-
C_2HClF_4
5485 - Ethane, 1-chloro-1,1,2,2-tetrafluoro-
5486 - Ethane, 1-chloro-1,2,2,2-tetrafluoro-
C_2HCl_2F
5900 - Ethene, 1,1-dichloro-2-fluoro-
C_2HCl_2FO
296 - Acetyl chloride, chlorofluoro-
$C_2HCl_2F_3$
5539 - Ethane, 1,2-dichloro-1,1,2-trifluoro-
5540 - Ethane, 2,2-dichloro-1,1,1-trifluoro-
C_2HCl_2N
278 - Acetonitrile, dichloro-
C_2HCl_3
5922 - Ethene, trichloro-
$C_2HCl_3F_2$
5658 - Ethane, 1,1,1-trichloro-2,2-difluoro
5659 - Ethane, 1,2,2-trichloro-1,1-difluoro-
5660 - Ethane, 1,2,2-trichloro-1,2-difluoro-
C_2HCl_3O
21 - Acetaldehyde, trichloro-
297 - Acetyl chloride, dichloro-
$C_2HCl_3O_2$
243 - Acetic acid, trichloro-
4016 - Carbonochloridic acid, dichloromethyl ester
C_2HCl_4F
5638 - Ethane, 1,1,2,2-tetrachloro-1-fluoro-
5639 - Ethane, 1,1,1,2-tetrachloro-2-fluoro-
C_2HCl_5
5617 - Ethane, pentachloro-
C_2HF
5943 - Ethyne, fluoro-
C_2HF_3
5924 - Ethene, trifluoro-
$C_2HF_3O_2$
263 - Acetic acid, trifluoro-
C_2H_2
5936 - Ethyne
C_2H_2BrCl
5881 - Ethene, 1-bromo-2-chloro-
$C_2H_2BrClO_2$
112 - Acetic acid, bromochloro-
$C_2H_2BrFO_2$
116 - Acetic acid, bromofluoro-
$C_2H_2BrF_3$
5466 - Ethane, 2-bromo-1,1,1-trifluoro-
$C_2H_2Br_2$
5890 - Ethene, 1,1-dibromo-
5891 - Ethene, 1,2-dibromo-, (E)-
5892 - Ethene, 1,2-dibromo-, (Z)-
$C_2H_2Br_2Cl_2$
5520 - Ethane, 1,2-dibromo-1,1-dichloro-
5521 - Ethane, 1,2-dibromo-1,2-dichloro-
$C_2H_2Br_2F_2$
5522 - Ethane, 1,2-dibromo-1,1-difluoro-
$C_2H_2Br_2O$
292 - Acetyl bromide, bromo-
$C_2H_2Br_2O_2$
159 - Acetic acid, dibromo-
$C_2H_2Br_3NO$
94 - Acetamide, 2,2,2-tribromo-
$C_2H_2Br_4$
5631 - Ethane, 1,1,1,2-tetrabromo-

5632 - Ethane, 1,1,2,2-tetrabromo-

$C_2H_2ClFO_2$
131 - Acetic acid, chlorofluoro-

$C_2H_2ClF_3$
5487 - Ethane, 1-chloro-1,1,2-trifluoro
5488 - Ethane, 1-chloro-1,2,2-trifluoro
5489 - Ethane, 2-chloro-1,1,1-trifluoro-

C_2H_2ClI
5888 - Ethene, 1-chloro-2-iodo-

C_2H_2ClN
274 - Acetonitrile, chloro-

C_2H_2ClNS
11700 - Thiocyanic acid, chloromethyl ester

$C_2H_2Cl_2$
5894 - Ethene, 1,1-dichloro-
5895 - Ethene, 1,2-dichloro-, (E)-
5896 - Ethene, 1,2-dichloro-, (Z)-

$C_2H_2Cl_2F_2$
5531 - Ethane, 1,2-dichloro-1,1-difluoro-
5532 - Ethane, 1,2-dichloro-1,2-difluoro-

$C_2H_2Cl_2O$
295 - Acetyl chloride, chloro-

$C_2H_2Cl_2O_2$
162 - Acetic acid, dichloro-
4014 - Carbonochloridic acid, chloromethyl ester

$C_2H_2Cl_3F$
5661 - Ethane, 1,1,2-trichloro-2-fluoro-

$C_2H_2Cl_3NO$
95 - Acetamide, 2,2,2-trichloro-

$C_2H_2Cl_4$
5634 - Ethane, 1,1,1,2-tetrachloro-
5635 - Ethane, 1,1,2,2-tetrachloro-

$C_2H_2F_2$
5902 - Ethene, 1,1-difluoro-
5903 - Ethene, 1,2-difluoro-, (Z)-

$C_2H_2F_2O_2$
175 - Acetic acid, difluoro-

$C_2H_2F_3I$
5667 - Ethane, 1,1,1-trifluoro-2-iodo-

$C_2H_2F_3NO_2$
5668 - Ethane, 1,1,1-trifluoro-2-nitro-

$C_2H_2F_4$
5640 - Ethane, 1,1,1,2-tetrafluoro-
5641 - Ethane, 1,1,2,2-tetrafluoro-

C_2H_2IN
283 - Acetonitrile, iodo-

$C_2H_2I_2$
5904 - Ethene, 1,2-diiodo-, (Z)-

$C_2H_2N_2O$
8563 - 1,3,4-Oxadiazole

$C_2H_2N_2S$
11657 - 1,2,4-Thiadiazole
11658 - 1,2,5-Thiadiazole

$C_2H_2N_4$
11651 - 1,2,4,5-Tetrazine

C_2H_2O
5927 - Ethenone

$C_2H_2O_2$
5491 - Ethanedial

$C_2H_2O_3$
223 - Acetic acid, oxo-

$C_2H_2O_4$
5553 - Ethanedioic acid

C_2H_3Br
5880 - Ethene, bromo-

C_2H_3BrO
291 - Acetyl bromide

$C_2H_3BrO_2$
111 - Acetic acid, bromo-

$C_2H_3Br_3$
5653 - Ethane, 1,1,2-tribromo-

$C_2H_3Br_3O$
5767 - Ethanol, 2,2,2-tribromo-

$C_2H_3Br_3O_2$
5581 - 1,1-Ethanediol, 2,2,2-tribromo-

C_2H_3Cl
5884 - Ethene, chloro-

$C_2H_3ClF_2$
5470 - Ethane, 1-chloro-1,1-difluoro
5471 - Ethane, 2-chloro-1,1-difluoro

C_2H_3ClO
12 - Acetaldehyde, chloro-
294 - Acetyl chloride

$C_2H_3ClO_2$
123 - Acetic acid, chloro-
4020 - Carbonochloridic acid, methyl ester

$C_2H_3Cl_2F$
5534 - Ethane, 1,1-dichloro-1-fluoro-
5535 - Ethane, 1,2-dichloro-1-fluoro-

$C_2H_3Cl_2NO$
51 - Acetamide, 2,2-dichloro-

$C_2H_3Cl_2NO_2$
5536 - Ethane, 1,1-dichloro-1-nitro-

$C_2H_3Cl_3$
5655 - Ethane, 1,1,1-trichloro-
5656 - Ethane, 1,1,2-trichloro-

$C_2H_3Cl_3O$
5768 - Ethanol, 2,2,2-trichloro-

$C_2H_3Cl_3O_2$
5582 - 1,1-Ethanediol, 2,2,2-trichloro-

$C_2H_3Cl_3Si$
11461 - Silane, trichloroethenyl-

C_2H_3F
5906 - Ethene, fluoro-

C_2H_3FO
301 - Acetyl fluoride

$C_2H_3FO_2$
189 - Acetic acid, fluoro-

$C_2H_3F_3$
5665 - Ethane, 1,1,1-trifluoro-
5666 - Ethane, 1,1,2-trifluoro-

$C_2H_3F_3O$
5770 - Ethanol, 2,2,2-trifluoro-

C_2H_3I
5907 - Ethene, iodo-

C_2H_3IO
302 - Acetyl iodide

$C_2H_3IO_2$
200 - Acetic acid, iodo-

$C_2H_3KO_2$
234 - Acetic acid, potassium salt

C_2H_3N
271 - Acetonitrile
7655 - Methylisonitrile

C_2H_3NO
282 - Acetonitrile, hydroxy-
7520 - Methane, isocyanato-

$C_2H_3NO_2$
5911 - Ethene, nitro-

$C_2H_3NO_3$
104 - Acetic acid, aminooxo-

$C_2H_3NO_4$
108 - Acetic acid, anhydride with nitric acid

C_2H_3NS
7521 - Methane, isothiocyanato-
11705 - Thiocyanic acid, methyl ester

$C_2H_3N_3$
11846 - 1H-1,2,3-Triazole

$C_2H_3N_3O_2$
109 - Acetic acid, azido-
11850 - 1,2,4-Triazolidine-3,5-dione

$C_2H_3NaO_2$
237 - Acetic acid, sodium salt

C_2H_4
5879 - Ethene

C_2H_4BrCl
5455 - Ethane, 1-bromo-1-chloro-
5456 - Ethane, 1-bromo-2-chloro-

$C_2H_4BrClO_2S$
5628 - Ethanesulfonyl chloride, 2-bromo-

C_2H_4BrF
5462 - Ethane, 1-bromo-2-fluoro-

C_2H_4BrI
5463 - Ethane, 1-bromo-2-iodo-

C_2H_4BrNO
36 - Acetamide, N-bromo-

$C_2H_4Br_2$
5517 - Ethane, 1,1-dibromo-
5518 - Ethane, 1,2-dibromo-

$C_2H_4Br_2O$
7524 - Methane, oxybis[bromo-

C_2H_4ClF
5475 - Ethane, 1-chloro-1-fluoro-
5476 - Ethane, 1-chloro-2-fluoro-

C_2H_4ClI
5477 - Ethane, 1-chloro-2-iodo-

C_2H_4ClNO
41 - Acetamide, 2-chloro-

$C_2H_4ClNO_2$
5482 - Ethane, 1-chloro-1-nitro-
5483 - Ethane, 1-chloro-2-nitro-

$C_2H_4Cl_2$
5528 - Ethane, 1,1-dichloro-
5529 - Ethane, 1,2-dichloro-

$C_2H_4Cl_2O$
5704 - Ethanol, 2,2-dichloro-
7525 - Methane, oxybis[chloro-

$C_2H_4Cl_2O_2S$
5629 - Ethanesulfonyl chloride, 2-chloro-

$C_2H_4Cl_3NO$
5680 - Ethanol, 1-amino-2,2,2-trichloro-

$C_2H_4Cl_4Si$
11391 - Silane, dichloro(dichloromethyl)methyl-

C_2H_4FNO
63 - Acetamide, 2-fluoro-

$C_2H_4F_2$
5544 - Ethane, 1,1-difluoro-
5545 - Ethane, 1,2-difluoro-

$C_2H_4F_2O$
5708 - Ethanol, 2,2-difluoro-

$C_2H_4I_2$
5546 - Ethane, 1,1-diiodo-
5547 - Ethane, 1,2-diiodo-

$C_2H_4N_2$
272 - Acetonitrile, amino-

$C_2H_4N_2O_2$
5492 - Ethanedial, dioxime
5493 - Ethanediamide

$C_2H_4N_2O_4$
5551 - Ethane, 1,1-dinitro-
5552 - Ethane, 1,2-dinitro-
8086 - Nitrous acid, 1,2-ethanediyl ester

$C_2H_4N_2O_6$
5574 - 1,2-Ethanediol, dinitrate

$C_2H_4N_2S_2$
5588 - Ethanedithioamide

$C_2H_4N_4$
6280 - Guanidine, cyano-
11845 - 1H-1,2,4-Triazol-3-amine

$C_2H_4N_4O_2$
5124 - Diazenedicarboxamide

C_2H_4O
8599 - Oxirane
11 - Acetaldehyde

$(C_2H_4O)_x$
7440 - Metaldehyde

C_2H_4OS
5646 - Ethanethioic acid

$C_2H_4O_2$
15 - Acetaldehyde, hydroxy-
100 - Acetic acid
6010 - Formic acid, methyl ester

$C_2H_4O_2S$
202 - Acetic acid, mercapto-

$C_2H_4O_3$
196 - Acetic acid, hydroxy-
5619 - Ethaneperoxoic acid

$C_2H_4O_3S$
5200 - 1,3,2-Dioxathiolane, 2-oxide

$C_2H_4O_4S$
5199 - 1,3,2-Dioxathiolane, 2,2-dioxide

$C_2H_4O_5S$
238 - Acetic acid, sulfo-
C_2H_4S
11695 - Thiirane
$C_2H_4S_2$
5589 - Ethane(dithioic) acid
$C_2H_4S_5$
8998 - 1,2,3,5,6-Pentathiepane
C_2H_4Si
11435 - Silane, ethynyl
$C_2H_5AsF_2$
445 - Arsine, ethyldifluoro-
C_2H_5Br
5454 - Ethane, bromo-
C_2H_5BrO
5685 - Ethanol, 2-bromo-
7472 - Methane, bromomethoxy-
C_2H_5Cl
5467 - Ethane, chloro-
C_2H_5ClO
5693 - Ethanol, 2-chloro-
7184 - Hypochlorous acid, ethyl ester
7483 - Methane, chloromethoxy-
$C_2H_5ClO_2S$
4087 - Chlorosulfurous acid, ethyl ester
5627 - Ethanesulfonyl chloride
$C_2H_5ClO_3S$
4085 - Chlorosulfuric acid, ethyl ester
C_2H_5ClS
5651 - Ethanethiol, 2-chloro-
7484 - Methane, chloro(methylthio)-
$C_2H_5Cl_2O_2P$
9577 - Phosphorodichloridic acid, ethyl ester
$C_2H_5Cl_3OSi$
11462 - Silane, trichloroethoxy-
$C_2H_5Cl_3Si$
11390 - Silane, dichloro(chloromethyl)methyl-
11463 - Silane, trichloroethyl-
C_2H_5F
5593 - Ethane, fluoro-
C_2H_5FO
5727 - Ethanol, 2-fluoro-
$C_2H_5F_3OSi$
11429 - Silane, ethoxytrifluoro-
$C_2H_5F_3Si$
11432 - Silane, ethyltrifluoro-
C_2H_5I
5597 - Ethane, iodo-
C_2H_5IO
5731 - Ethanol, 2-iodo-
7519 - Methane, iodomethoxy-
C_2H_5N
498 - Aziridine
C_2H_5NO
17 - Acetaldehyde, oxime
22 - Acetamide
5995 - Formamide, N-methyl-
$C_2H_5NO_2$
3953 - Carbamic acid, methyl ester
5607 - Ethane, nitro-
6250 - Glycine
8087 - Nitrous acid, ethyl ester
$C_2H_5NO_3$
5752 - Ethanol, 2-nitro-
8074 - Nitric acid, ethyl ester
C_2H_5NS
5643 - Ethanethioamide
$C_2H_5N_3O$
5681 - Ethanol, 2-azido-
$C_2H_5N_3O_2$
105 - Acetic acid, aminooxo-, hydrazide
7232 - Imidodicarbonic diamide
12019 - Urea, N-methyl-N-nitroso-
$C_2H_5N_3S_2$
11718 - Thioimidodicarbonic diamide
$C_2H_5N_5O_3$
6284 - Guanidine, N-methyl-N'-nitro-N-nitroso-

C_2H_6
5448 - Ethane
C_2H_6AsCl
453 - Arsinous chloride, dimethyl-
$C_2H_6BF_3O$
3138 - Boron, trifluoro[oxybis[methane]]-, (T-4)-
C_2H_6Cd
3929 - Cadmium, dimethyl-
$C_2H_6ClO_3P$
5929 - Ethephon
$C_2H_6Cl_2NPS$
9557 - Phosphoramidothioic dichloride, dimethyl-
$C_2H_6Cl_2Si$
11394 - Silane, dichlorodimethyl-
C_2H_6Hg
7435 - Mercury, dimethyl-
$C_2H_6N_2$
5125 - Diazene, dimethyl-
5126 - Diazene, dimethyl-, (E)-
$C_2H_6N_2O$
195 - Acetic acid, hydrazide
7450 - Methanamine, N-methyl-N-nitroso-
12017 - Urea, methyl-
$C_2H_6N_2O_2$
7149 - Hydrazinecarboxylic acid, methyl ester
7449 - Methanamine, N-methyl-N-nitro-
$C_2H_6N_4O$
11993 - Urea, (aminoiminomethyl)-
$C_2H_6N_4O_2$
5562 - Ethanedioic acid, dihydrazide
C_2H_6O
5672 - Ethanol
7523 - Methane, oxybis-
C_2H_6OS
5732 - Ethanol, 2-mercapto-
7529 - Methane, sulfinylbis-
$C_2H_6O_2$
5569 - 1,2-Ethanediol
7179 - Hydroperoxide, ethyl
9159 - Peroxide, dimethyl
$C_2H_6O_2S$
7534 - Methane, sulfonylbis-
$C_2H_6O_3S$
5622 - Ethanesulfonic acid
7532 - Methanesulfonic acid, methyl ester
11582 - Sulfurous acid, dimethyl ester
$C_2H_6O_4S$
11574 - Sulfuric acid, dimethyl ester
11576 - Sulfuric acid, monoethyl ester
$C_2H_6O_6$
5561 - Ethanedioic acid, dihydrate
$C_2H_6O_6S_2$
5587 - 1,2-Ethanedisulfonic acid
C_2H_6S
5648 - Ethanethiol
7545 - Methane, thiobis-
$C_2H_6S_2$
5254 - Disulfide, dimethyl
5590 - 1,2-Ethanedithiol
C_2H_6Se
7526 - Methane, selenobis-
C_2H_6Zn
12066 - Zinc, dimethyl
C_2H_7As
442 - Arsine, dimethyl-
444 - Arsine, ethyl-
$C_2H_7AsO_2$
451 - Arsinic acid, dimethyl-
$C_2H_7AsO_3$
457 - Arsonic acid, ethyl-
$C_2H_7BO_2$
3117 - Borane, dimethoxy-
3133 - Boronic acid, ethyl-
$C_2H_7ClN_2$
5671 - Ethanimidamide, monohydrochloride
C_2H_7N
5410 - Ethanamine

7448 - Methanamine, N-methyl-
C_2H_7NO
5434 - Ethanamine, N-hydroxy-
5674 - Ethanol, 1-amino-
5675 - Ethanol, 2-amino-
$C_2H_7NO_2$
106 - Acetic acid, ammonium salt
$C_2H_7NO_3S$
5623 - Ethanesulfonic acid, 2-amino-
C_2H_7NS
5649 - Ethanethiol, 2-amino-
$C_2H_7N_3O$
6266 - Glycine, hydrazide
$C_2H_7N_5$
7233 - Imidodicarbonimidic diamide
$C_2H_7O_2P$
9528 - Phosphinic acid, dimethyl-
$C_2H_7O_3P$
9588 - Phosphorous acid, dimethyl ester
$C_2H_7O_4P$
9559 - Phosphoric acid, dimethyl ester
9561 - Phosphoric acid, monoethyl ester
C_2H_8ClN
5433 - Ethanamine, hydrochloride
$C_2H_8NO_2PS$
7441 - Methamidophos
$C_2H_8N_2$
5496 - 1,2-Ethanediamine
7154 - Hydrazine, 1,1-dimethyl-
7155 - Hydrazine, 1,2-dimethyl-
$C_2H_8N_2O$
5729 - Ethanol, 2-hydrazino-
C_2H_8Si
11414 - Silane, dimethyl-
$C_2H_{10}N_2O$
5513 - 1,2-Ethanediamine, monohydrate
C_2I_2
5941 - Ethyne, diiodo-
C_2I_4
5919 - Ethene, tetraiodo-
C_2N_2
5550 - Ethanedinitrile
C_2N_2S
11572 - Sulfur cyanide
$C_2N_4O_6$
290 - Acetonitrile, trinitro-
$C_2Na_2O_4$
5566 - Ethanedioic acid, disodium salt
C_2O_4Pb
5568 - Ethanedioic acid, lead salt
C_3ClF_5
10541 - 1-Propene, 2-chloro-1,1,3,3,3-pentafluoro-
C_3ClF_5O
10421 - 2-Propanone, 1-chloro-1,1,3,3,3-pentafluoro-
$C_3Cl_2F_4$
10557 - 1-Propene, 1,2-dichloro-1,3,3,3-tetrafluoro-
$C_3Cl_2F_6$
9898 - Propane, 1,2-dichloro-1,1,2,3,3,3-hexafluoro-
$C_3Cl_2N_2$
9912 - Propanedinitrile, dichloro-
$C_3Cl_3F_3$
10607 - 1-Propene, 1,1,2-trichloro-3,3,3-trifluoro-
$C_3Cl_3F_5$
10179 - Propane, 1,1,1-trichloro-2,2,3,3,3-pentafluoro-
10180 - Propane, 1,2,2-trichloro-1,1,3,3,3-pentafluoro-
10181 - Propane, 1,2,3-trichloro-1,1,2,3,3-pentafluoro-
$C_3Cl_3N_3$
11837 - 1,3,5-Triazine, 2,4,6-trichloro-
$C_3Cl_3N_3O_3$
11840 - 1,3,5-Triazine-2,4,6(1H,3H,5H)-trione, 1,3,5-trichloro-
$C_3Cl_4F_4$
10149 - Propane, 1,2,2,3-tetrachloro-1,1,3,3-tetrafluoro-

$C_3H_4O_4$
9921 - Propanedioic acid
$C_3H_4O_5$
9956 - Propanedioic acid, hydroxy-
C_3H_5Br
10529 - 1-Propene, 1-bromo-, (Z)-
10530 - 1-Propene, 2-bromo-
10531 - 1-Propene, 3-bromo-
C_3H_5BrO
8600 - Oxirane, (bromomethyl)-, (±)-
9775 - Propanal, 2-bromo-
10416 - 2-Propanone, 1-bromo-
10490 - Propanoyl bromide
10699 - 2-Propen-1-ol, 2-bromo-
$C_3H_5BrO_2$
117 - Acetic acid, bromo-, methyl ester
10204 - Propanoic acid, 2-bromo-, (±)-
10205 - Propanoic acid, 3-bromo-
$C_3H_5Br_2Cl$
9885 - Propane, 1,2-dibromo-3-chloro-
$C_3H_5Br_3$
10160 - Propane, 1,1,1-tribromo-
10161 - Propane, 1,1,2-tribromo-
10162 - Propane, 1,2,2-tribromo-
10163 - Propane, 1,2,3-tribromo-
C_3H_5Cl
10533 - 1-Propene, 1-chloro-, (E)-
10534 - 1-Propene, 1-chloro-, (Z)-
10535 - 1-Propene, 2-chloro-
10536 - 1-Propene, 3-chloro-
$C_3H_5ClF_2$
9851 - Propane, 1-chloro-2,2-difluoro-
$C_3H_5ClN_2O_6$
9997 - 1,2-Propanediol, 3-chloro-, dinitrate
C_3H_5ClO
8605 - Oxirane, (chloromethyl)-, (±)-
9777 - Propanal, 2-chloro-
9778 - Propanal, 3-chloro-
10420 - 2-Propanone, 1-chloro-
10493 - Propanoyl chloride
10700 - 2-Propen-1-ol, 2-chloro-
10701 - 2-Propen-1-ol, 3-chloro-
$C_3H_5ClO_2$
138 - Acetic acid, chloro-, methyl ester
299 - Acetyl chloride, methoxy-
4018 - Carbonochloridic acid, ethyl ester
7582 - Methanol, chloro-, acetate
10215 - Propanoic acid, 2-chloro-
10216 - Propanoic acid, 3-chloro-
$C_3H_5Cl_2F$
9897 - Propane, 1,2-dichloro-2-fluoro-
$C_3H_5Cl_2NO_2$
3940 - Carbamic acid, dichloro-, ethyl ester
9902 - Propane, 1,1-dichloro-1-nitro-
$C_3H_5Cl_3$
10171 - Propane, 1,1,1-trichloro-
10172 - Propane, 1,1,2-trichloro-
10173 - Propane, 1,1,3-trichloro-
10174 - Propane, 1,2,2-trichloro-
10175 - Propane, 1,2,3-trichloro-
$C_3H_5Cl_3O$
10406 - 2-Propanol, 1,1,1-trichloro-
$C_3H_5Cl_3Si$
11475 - Silane, trichloro-2-propenyl-
C_3H_5F
10567 - 1-Propene, 1-fluoro- (Z)
10568 - 1-Propene, 1-fluoro- (E)
10569 - 1-Propene, 2-fluoro-
10570 - 1-Propene, 3-fluoro-
C_3H_5FO
8616 - Oxirane, (fluoromethyl)-
10437 - 2-Propanone, 1-fluoro-
10506 - Propanoyl fluoride
$C_3H_5FO_2$
192 - Acetic acid, fluoro-, methyl ester
$C_3H_5F_3O$
10409 - 2-Propanol, 1,1,1-trifluoro-, (±)-

$C_3H_5F_3O_3S$
7533 - Methanesulfonic acid, trifluoro-, ethyl ester
C_3H_5I
10573 - 1-Propene, 3-iodo-
C_3H_5IO
8617 - Oxirane, (iodomethyl)-
10454 - 2-Propanone, 1-iodo-
$C_3H_5IO_2$
10282 - Propanoic acid, 2-iodo-, (±)-
C_3H_5N
5599 - Ethane, isocyano-
10090 - Propanenitrile
10725 - 2-Propyn-1-amine
C_3H_5NO
284 - Acetonitrile, methoxy-
496 - 2-Azetidinone
4163 - Cyanic acid, ethyl ester
5598 - Ethane, isocyanato-
10105 - Propanenitrile, 2-hydroxy-
10106 - Propanenitrile, 3-hydroxy-
10519 - 2-Propenamide
$C_3H_5NO_2$
9789 - Propanal, 2-oxo-, 1-oxime
10589 - 1-Propene, 1-nitro-
10590 - 1-Propene, 2-nitro-
10591 - 1-Propene, 3-nitro-
$C_3H_5NO_3$
10475 - 2-Propanone, 1-nitro-
C_3H_5NS
286 - Acetonitrile, (methylthio)-
5600 - Ethane, isothiocyanato-
11703 - Thiocyanic acid, ethyl ester
$C_3H_5N_3$
10812 - 1H-Pyrazol-4-amine
$C_3H_5N_3O$
153 - Acetic acid, cyano-, hydrazide
$C_3H_5N_3O_9$
10197 - 1,2,3-Propanetriol, trinitrate
$C_3H_5NaO_2$
10327 - Propanoic acid, sodium salt
C_3H_6
4955 - Cyclopropane
10528 - 1-Propene
C_3H_6BrCl
9838 - Propane, 1-bromo-2-chloro-
9839 - Propane, 1-bromo-3-chloro-
9840 - Propane, 2-bromo-1-chloro-
9841 - Propane, 2-bromo-2-chloro-
C_3H_6BrF
9844 - Propane, 1-bromo-3-fluoro-
$C_3H_6BrNO_2$
9847 - Propane, 2-bromo-2-nitro-
$C_3H_6BrNO_4$
9991 - 1,3-Propanediol, 2-bromo-2-nitro-
$C_3H_6Br_2$
9880 - Propane, 1,1-dibromo-
9881 - Propane, 1,2-dibromo-
9882 - Propane, 1,3-dibromo-
9883 - Propane, 2,2-dibromo-
$C_3H_6Br_2O$
10357 - 2-Propanol, 1,3-dibromo-
C_3H_6ClF
9856 - Propane, 1-chloro-3-fluoro-
C_3H_6ClI
9857 - Propane, 1-chloro-3-iodo-
C_3H_6ClNO
3973 - Carbamic chloride, dimethyl-
$C_3H_6ClNO_2$
9866 - Propane, 1-chloro-1-nitro-
9867 - Propane, 1-chloro-2-nitro-
9868 - Propane, 1-chloro-3-nitro-
9869 - Propane, 2-chloro-1-nitro-
9870 - Propane, 2-chloro-2-nitro-
C_3H_6ClNS
3980 - Carbamothioic chloride, dimethyl-
$C_3H_6Cl_2$
9891 - Propane, 1,1-dichloro-

9892 - Propane, 1,2-dichloro-, (±)-
9893 - Propane, 1,3-dichloro-
9894 - Propane, 2,2-dichloro-
$C_3H_6Cl_2O$
10359 - 1-Propanol, 2,3-dichloro-
10360 - 1-Propanol, 3,3-dichloro-
10361 - 2-Propanol, 1,1-dichloro-
10362 - 2-Propanol, 1,3-dichloro-
$C_3H_6Cl_2Si$
11396 - Silane, dichloroethenylmethyl-
$C_3H_6Cl_3P$
9524 - Phosphine, tris(chloromethyl)-
$C_3H_6Cl_4Si$
11458 - Silane, trichloro(3-chloropropyl)-
$C_3H_6F_2$
9905 - Propane, 1,3-difluoro-
9906 - Propane, 2,2-difluoro-
$C_3H_6F_2O$
10369 - 2-Propanol, 1,3-difluoro-
$C_3H_6I_2$
9907 - Propane, 1,3-diiodo-
9908 - Propane, 2,2-diiodo-
$C_3H_6N_2$
4157 - Cyanamide, dimethyl-
10092 - Propanenitrile, 3-amino-
10822 - 1H-Pyrazole, 4,5-dihydro-
$C_3H_6N_2O$
7227 - 2-Imidazolidinone
$C_3H_6N_2OS$
33 - Acetamide, N-(aminothioxomethyl)-
$C_3H_6N_2O_2$
27 - Acetamide, N-(aminocarbonyl)-
7406 - 3-Isoxazolidinone, 4-amino-, (R)-
9784 - Propanal, 2-(hydroxyimino)-, oxime
$C_3H_6N_2O_4$
3963 - Carbamic acid, nitro-, ethyl ester
9918 - Propane, 1,1-dinitro-
9919 - Propane, 1,3-dinitro-
9920 - Propane, 2,2-dinitro-
$C_3H_6N_2O_7$
10191 - 1,2,3-Propanetriol, 1,3-dinitrate
$C_3H_6N_2S$
7225 - 2-Imidazolidinethione
11664 - 2-Thiazolamine, 4,5-dihydro-
$C_3H_6N_6$
11835 - 1,3,5-Triazine-2,4,6-triamine
$C_3H_6N_6O_6$
11833 - 1,3,5-Triazine, hexahydro-1,3,5-trinitro-
C_3H_6O
5908 - Ethene, methoxy-
8592 - Oxetane
8621 - Oxirane, methyl-
9774 - Propanal
10411 - 2-Propanone
10697 - 2-Propen-1-ol
$C_3H_6O_2$
16 - Acetaldehyde, methoxy-
208 - Acetic acid, methyl ester
5204 - 1,3-Dioxolane
6007 - Formic acid, ethyl ester
8618 - Oxiranemethanol
10201 - Propanoic acid
10443 - 2-Propanone, 1-hydroxy-
$C_3H_6O_2S$
204 - Acetic acid, mercapto-, methyl ester
216 - Acetic acid, (methylthio)-
5909 - Ethene, (methylsulfonyl)-
10285 - Propanoic acid, 2-mercapto-
10286 - Propanoic acid, 3-mercapto-
11692 - Thietane, 1,1-dioxide
$C_3H_6O_3$
198 - Acetic acid, hydroxy-, methyl ester
205 - Acetic acid, methoxy-
4000 - Carbonic acid, dimethyl ester
9782 - Propanal, 2,3-dihydroxy-, (±)-
$C_3H_6O_2$
10266 - Propanoic acid, 2-hydroxy-, (±)-

$C_3H_6O_3$
- 10267 - Propanoic acid, 3-hydroxy-
- 10429 - 2-Propanone, 1,3-dihydroxy-
- 11905 - 1,3,5-Trioxane

$C_3H_6O_3S$
- 8570 - 1,2-Oxathiolane, 2,2-dioxide
- 11694 - Thietane, 1-oxide

$C_3H_6O_4$
- 10252 - Propanoic acid, 2,3-dihydroxy-

C_3H_6S
- 5910 - Ethene, (methylthio)-
- 10600 - 2-Propene-1-thiol
- 11691 - Thietane
- 11696 - Thiirane, methyl-

$C_3H_6S_2$
- 5266 - 1,3-Dithiolane

$C_3H_6S_3$
- 11919 - 1,3,5-Trithiane

C_3H_7Br
- 9836 - Propane, 1-bromo-
- 9837 - Propane, 2-bromo-

C_3H_7BrO
- 5464 - Ethane, (bromomethoxy)-
- 5465 - Ethane, 1-bromo-2-methoxy-
- 10344 - 1-Propanol, 3-bromo-
- 10345 - 2-Propanol, 1-bromo-

C_3H_7Cl
- 9848 - Propane, 1-chloro-
- 9849 - Propane, 2-chloro-

C_3H_7ClHgO
- 11822 - Triadimenol

C_3H_7ClO
- 5478 - Ethane, (chloromethoxy)-
- 5479 - Ethane, 1-chloro-1-methoxy-
- 5480 - Ethane, 1-chloro-2-methoxy-
- 10347 - 1-Propanol, 2-chloro-
- 10348 - 1-Propanol, 3-chloro-
- 10349 - 2-Propanol, 1-chloro-

$C_3H_7ClO_2$
- 9994 - 1,2-Propanediol, 3-chloro-
- 0006 - 1,3 Propanediol, 2 chloro-

$C_3H_7ClO_2S$
- 10133 - 1-Propanesulfonyl chloride

$C_3H_7ClO_3S$
- 5699 - Ethanol, 2-chloro-, methanesulfonate

C_3H_7ClS
- 5481 - Ethane, 1-chloro-2-(methylthio)-

$C_3H_7Cl_2N$
- 9805 - 1-Propanamine, N,N-dichloro-

$C_3H_7Cl_3Si$
- 11373 - Silane, chloro(dichloromethyl)dimethyl-
- 11466 - Silane, trichloro(1-methylethyl)-
- 11476 - Silane, trichloropropyl-

C_3H_7F
- 10046 - Propane, 1-fluoro-
- 10047 - Propane, 2-fluoro-

C_3H_7FO
- 10382 - 1-Propanol, 3-fluoro-

C_3H_7I
- 10054 - Propane, 1-iodo-
- 10055 - Propane, 2-iodo-

C_3H_7IO
- 10385 - 1-Propanol, 3-iodo-

C_3H_7N
- 494 - Azetidine
- 500 - Aziridine, 1-methyl-
- 501 - Aziridine, 2-methyl-
- 4953 - Cyclopropanamine
- 10522 - 2-Propen-1-amine

$C_3H_7NNaS_2$
- 11514 - Sodium dimethyldithiocarbamate

C_3H_7NO
- 72 - Acetamide, N-methyl-
- 5992 - Formamide, N,N-dimethyl-
- 5994 - Formamide, N-ethyl-
- 9787 - Propanal, oxime
- 9791 - Propanamide

- 10476 - 2-Propanone, oxime

C_3H_7NOS
- 3978 - Carbamothioic acid, O-ethyl ester
- 3979 - Carbamothioic acid, S-ethyl ester

$C_3H_7NO_2$
- 323 - β-Alanine
- 324 - DL-Alanine
- 325 - L-Alanine
- 3948 - Carbamic acid, ethyl ester
- 6268 - Glycine, N-methyl-
- 8092 - Nitrous acid, 1-methylethyl ester
- 8098 - Nitrous acid, propyl ester
- 9797 - Propanamide, 2-hydroxy-, (±)-
- 10116 - Propane, 1-nitro-
- 10117 - Propane, 2-nitro-

$C_3H_7NO_2S$
- 5014 - L-Cysteine

$C_3H_7NO_3$
- 8076 - Nitric acid, 1-methylethyl ester
- 8080 - Nitric acid, propyl ester
- 10394 - 1-Propanol, 2-nitro-
- 10395 - 1-Propanol, 3-nitro-
- 10396 - 2-Propanol, 1-nitro-
- 11355 - D-Serine
- 11356 - DL-Serine
- 11357 - L-Serine

$C_3H_7NO_4$
- 10022 - 1,2-Propanediol, 1-nitrate

$C_3H_7NO_5$
- 10192 - 1,2,3-Propanetriol, 1-nitrate
- 10193 - 1,2,3-Propanetriol, 2-nitrate

C_3H_7NS
- 11681 - Thiazolidine

$C_3H_7N_3O_2$
- 6253 - Glycine, N-(aminoiminomethyl)-
- 12015 - Urea, N-ethyl-N-nitroso-

$C_3H_7N_3O_5$
- 11839 - 1,3,5-Triazine-2,4,6(1H,3H,5H)-trione, dihydrate

C_3H_8
- 9834 - Propane

$C_3H_8BrClSi$
- 11365 - Silane, (bromomethyl)chlorodimethyl-

$C_3H_8Cl_2Si$
- 11372 - Silane, chloro(chloromethyl)dimethyl-
- 11397 - Silane, dichloroethylmethyl-

$C_3H_8NO_5P$
- 6276 - Glyphosate

$C_3H_8N_2O$
- 12004 - Urea, N,N'-dimethyl-
- 12005 - Urea, N,N-dimethyl-
- 12014 - Urea, ethyl-

$C_3H_8N_2O_2$
- 326 - Alanine, 3-amino-
- 7148 - Hydrazinecarboxylic acid, ethyl ester
- 9824 - 1-Propanamine, N-nitro-

$C_3H_8N_2O_3$
- 11995 - Urea, N,N'-bis(hydroxymethyl)-

$C_3H_8N_2S$
- 11804 - Thiourea, N,N'-dimethyl-

C_3H_8O
- 5602 - Ethane, methoxy-
- 10334 - 1-Propanol
- 10335 - 2-Propanol

C_3H_8OS
- 5749 - Ethanol, 2-(methylthio)-

$C_3H_8OS_2$
- 10370 - 1-Propanol, 2,3-dimercapto-
- 10371 - 2-Propanol, 1,3-dimercapto-

$C_3H_8O_2$
- 5733 - Ethanol, 2-methoxy-
- 7507 - Methane, dimethoxy-
- 9982 - 1,2-Propanediol
- 9983 - 1,3-Propanediol

$C_3H_8O_2S$
- 10013 - 1,2-Propanediol, 3-mercapto-

$C_3H_8O_3$
- 10186 - 1,2,3-Propanetriol

$C_3H_8O_3S$
- 7531 - Methanesulfonic acid, ethyl ester
- 10130 - 2-Propanesulfonic acid

C_3H_8S
- 5606 - Ethane, (methylthio)-
- 10156 - 1-Propanethiol
- 10157 - 2-Propanethiol

$C_3H_8S_2$
- 7461 - Methane, bis(methylthio)-
- 10030 - 1,2-Propanedithiol
- 10031 - 1,3-Propanedithiol

C_3H_9Al
- 339 - Aluminum, trimethyl

C_3H_9As
- 449 - Arsine, trimethyl-

$C_3H_9AsO_3$
- 461 - Arsonic acid, propyl-

C_3H_9B
- 3119 - Borane, trimethyl-

$C_3H_9BO_2$
- 3136 - Boronic acid, propyl-

$C_3H_9BO_3$
- 3126 - Boric acid, trimethyl ester

C_3H_9BS
- 3129 - Borinic acid, dimethylthio-, methyl ester

$C_3H_9BS_3$
- 11698 - Thioboric acid, trimethyl ester

C_3H_9ClSi
- 11389 - Silane, chlorotrimethyl-

C_3H_9ClSn
- 11540 - Stannane, chlorotrimethyl-

C_3H_9N
- 7444 - Methanamine, N,N-dimethyl-
- 9803 - 1-Propanamine
- 9804 - 2-Propanamine

C_3H_9NO
- 5738 - Ethanol, 2-(methylamino)-
- 10336 - 1-Propanol, 2-amino-, (±)-
- 10337 - 1-Propanol, 3-amino-
- 10338 - 2-Propanol, 1-amino-, (±)-

$C_3H_9NO_2$
- 9984 - 1,2-Propanediol, 3-amino-, (±)-

$C_3H_9NO_3S$
- 5624 - Ethanesulfonic acid, 2-(methylamino)-

$C_3H_9O_2PS$
- 9550 - Phosphonothioic acid, methyl-, O,O-dimethyl ester

$C_3H_9O_3P$
- 9540 - Phosphonic acid, methyl-, dimethyl ester
- 9542 - Phosphonic acid, (1-methylethyl)-
- 9543 - Phosphonic acid, methyl-, monoethyl ester
- 9547 - Phosphonic acid, propyl-
- 9592 - Phosphorous acid, trimethyl ester

$C_3H_9O_4P$
- 9564 - Phosphoric acid, trimethyl ester

C_3H_9P
- 9522 - Phosphine, trimethyl-

C_3H_9Sb
- 11550 - Stibine, trimethyl-

$C_3H_{10}ClN$
- 5435 - Ethanamine, N-methyl-, hydrochloride
- 7445 - Methanamine, N,N-dimethyl-, hydrochloride

$C_3H_{10}N_2$
- 5512 - 1,2-Ethanediamine, N-methyl-
- 9872 - 1,2-Propanediamine
- 9873 - 1,3-Propanediamine

$C_3H_{10}Si$
- 11489 - Silane, trimethyl-

$C_3H_{12}BN$
- 3120 - Borane, trimethyl-, monoammoniate
- 3132 - Boron, (N,N-dimethylmethanamine)trihydro-, (T-4)-

C_3N_2O
- 9915 - Propanedinitrile, oxo-

C₃O₂
9773 - 1,2-Propadiene-1,3-dione
C₄Br₂F₈
3275 - Butane, 1,4-dibromo-1,1,2,2,3,3,4,4-octafluoro-
C₄Br₄S
11781 - Thiophene, tetrabromo-
C₄ClF₇
4177 - Cyclobutane, 1-chloro-1,2,2,3,3,4,4-heptafluoro-
C₄ClF₇O
3692 - Butanoyl chloride, heptafluoro-
C₄Cl₂F₄
4199 - Cyclobutene, 1,2-dichloro-3,3,4,4-tetrafluoro-
C₄Cl₂F₆
3754 - 2-Butene, 2,3-dichloro-1,1,1,4,4,4-hexafluoro-
C₄Cl₂F₈
3287 - Butane, 2,3-dichloro-1,1,1,2,3,4,4,4-octafluoro-
C₄Cl₃F₇
3452 - Butane, 2,2,3-trichloro-1,1,1,3,4,4,4-heptafluoro-
C₄Cl₄F₄
3808 - 1-Butene, 1,3,4,4-tetrachloro-1,2,3,4-tetrafluoro-
C₄Cl₄S
11782 - Thiophene, tetrachloro-
C₄Cl₆
3163 - 1,3-Butadiene, 1,1,2,3,4,4-hexachloro-
C₄Cl₆O₃
244 - Acetic acid, trichloro-, anhydride
C₄F₄O₃
6099 - 2,5-Furandione, 3,3,4,4-tetrafluorodihydro-
C₄F₆
3164 - 1,3-Butadiene, 1,1,2,3,4,4-hexafluoro-
3914 - 2-Butyne, 1,1,1,4,4,4-hexafluoro-
4200 - Cyclobutene, hexafluoro-
C₄F₆O₃
264 - Acetic acid, trifluoro-, anhydride
C₄F₈
3804 - 2-Butene, 1,1,1,2,3,4,4,4-octafluoro-
4194 - Cyclobutane, octafluoro-
C₄F₁₀
3255 - Butane, decafluoro-
10050 - Propane, 1,1,1,2,3,3,3-heptafluoro-2-(trifluoromethyl)-
C₄HBr₃S
11790 - Thiophene, 2,3,5-tribromo-
C₄HClN₂
3763 - 2-Butenedinitrile, 2-chloro-, (Z)-
C₄HClO₃
6092 - 2,5-Furandione, 3-chloro-
C₄HCl₂N₃O₂
11054 - Pyrimidine, 2,4-dichloro-5-nitro-
C₄HCl₃N₂
11075 - Pyrimidine, 2,4,6-trichloro-
C₄HCl₃O₂
3796 - 2-Butenedioyl dichloride, 2-chloro-, (Z)-
C₄HCl₃S
11791 - Thiophene, 2,3,5-trichloro-
C₄HCl₇O₂
5192 - 1,4-Dioxane, 2,2,3,3,5,5,6-heptachloro-
C₄HF₇O₂
3521 - Butanoic acid, heptafluoro-
C₄HI₄N
11140 - 1H-Pyrrole, 2,3,4,5-tetraiodo-
C₄H₂
3177 - 1,3-Butadiyne
C₄H₂BrClS
11728 - Thiophene, 2-bromo-5-chloro-
C₄H₂Br₂N₂O₃
11086 - 2,4,6(1H,3H,5H)-Pyrimidinetrione, 5,5-dibromo-
C₄H₂Br₂O
6077 - Furan, 2,3-dibromo
6078 - Furan, 2,5-dibromo-

C₄H₂Br₂S
11742 - Thiophene, 2,3-dibromo-
11743 - Thiophene, 2,4-dibromo-
11744 - Thiophene, 2,5-dibromo-
C₄H₂Cl₂O₂
3795 - 2-Butenedioyl dichloride, (E)-
C₄H₂Cl₂S
11746 - Thiophene, 2,3-dichloro-
11747 - Thiophene, 2,4-dichloro-
11748 - Thiophene, 2,5-dichloro-
11749 - Thiophene, 3,4-dichloro-
C₄H₂Cl₄
3173 - 1,3-Butadiene, 1,2,3,4-tetrachloro-
C₄H₂Cl₆
3798 - 2-Butene, 1,1,2,3,4,4-hexachloro-
C₄H₂Cl₈O
5615 - Ethane, 1,1'-oxybis[1,2,2,2-tetrachloro-
C₄H₂I₂S
11753 - Thiophene, 2,5-diiodo-
C₄H₂N₂
3762 - 2-Butenedinitrile, (E)-
C₄H₂N₂O₄
11073 - 2,4,5,6(1H,3H)-Pyrimidinetetrone
C₄H₂N₂O₄S
11758 - Thiophene, 2,5-dinitro-
C₄H₂O₃
6091 - 2,5-Furandione
C₄H₃BrClF₃
3715 - 1-Butene, 4-bromo-3-chloro-3,4,4-trifluoro-
C₄H₃BrN₂O₂
11062 - 2,4(1H,3H)-Pyrimidinedione, 5-bromo-
C₄H₃BrO
6033 - Furan, 2-bromo-
6034 - Furan, 3-bromo-
C₄H₃BrO₄
3769 - 2-Butenedioic acid, 2-bromo-, (E)-
3770 - 2-Butenedioic acid, 2-bromo-, (Z)-
C₄H₃BrS
11726 - Thiophene, 2-bromo-
11727 - Thiophene, 3-bromo-
C₄H₃Cl
3890 - 1-Buten-3-yne, 4-chloro-
C₄H₃ClO
6073 - Furan, 2-chloro-
6074 - Furan, 3-chloro-
C₄H₃ClO₂S₂
11780 - 2-Thiophenesulfonyl chloride
C₄H₃ClS
11740 - Thiophene, 2-chloro-
C₄H₃Cl₂F₃O₂
10251 - Propanoic acid, 2,2-dichloro-3,3,3-trifluoro-, methyl ester
C₄H₃Cl₃
3174 - 1,3-Butadiene, 1,2,3-trichloro-
C₄H₃Cl₅
3805 - 2-Butene, 1,1,1,4,4-pentachloro-
C₄H₃Cl₇
3380 - Butane, 1,1,2,2,3,4,4-heptachloro-
C₄H₃FN₂O₂
11067 - 2,4(1H,3H)-Pyrimidinedione, 5-fluoro-
C₄H₃F₅O₂
10317 - Propanoic acid, pentafluoro-, methyl ester
C₄H₃F₇O
3622 - 1-Butanol, 2,2,3,3,4,4,4-heptafluoro-
C₄H₃IO
6110 - Furan, 2-iodo-
6111 - Furan, 3-iodo-
C₄H₃IS
11767 - Thiophene, 2-iodo-
C₄H₃NO₂
11121 - 1H-Pyrrole-2,5-dione
C₄H₃NO₂S
11775 - Thiophene, 2-nitro-
11776 - Thiophene, 3-nitro-
C₄H₃NO₃
6134 - Furan, 2-nitro-

C₄H₃N₃O₄
11074 - 2,4,5,6(1H,3H)-Pyrimidinetetrone, 5-oxime
C₄H₃N₃O₅
11102 - 2,4,6(1H,3H,5H)-Pyrimidinetrione, 5-nitro-
C₄H₄
3889 - 1-Buten-3-yne
C₄H₄BrNO₂
11153 - 2,5-Pyrrolidinedione, 1-bromo-
C₄H₄Br₂
3904 - 2-Butyne, 1,4-dibromo-
C₄H₄Br₄
4195 - Cyclobutane, 1,2,3,4-tetrabromo-
C₄H₄ClNO₂
11154 - 2,5-Pyrrolidinedione, 1-chloro-
C₄H₄ClNS
10538 - 1-Propene, 2-chloro-3-isothiocyanato-
C₄H₄ClN₃O₂
4083 - 5-Chloro-3-methyl-4-nitro-1H-pyrazole
C₄H₄Cl₂
3156 - 1,3-Butadiene, 1,1-dichloro-
3157 - 1,3-Butadiene, 1,2-dichloro-
3158 - 1,3-Butadiene, 2,3-dichloro-
3905 - 2-Butyne, 1,4-dichloro-
C₄H₄Cl₂O₂
3366 - Butanedioyl dichloride
3687 - Butanoyl chloride, 4-chloro-3-oxo-
4152 - Crotonic acid, 4,4-dichloro-
C₄H₄Cl₂O₃
124 - Acetic acid, chloro-, anhydride
C₄H₄Cl₂O₄
3302 - Butanedioic acid, 2,3-dichloro-, (±)-
C₄H₄Cl₄
3806 - 1-Butene, 1,3,4,4-tetrachloro-
3807 - 1-Butene, 2,3,3,4-tetrachloro-
C₄H₄Cl₄O₂
246 - Acetic acid, trichloro-, 2-chloroethyl ester
C₄H₄INO₂
11159 - 2,5-Pyrrolidinedione, 1-iodo-
C₄H₄I₂
3906 - 2-Butyne, 1,4-diiodo-
C₄H₄N₂
3291 - Butanedinitrile
10801 - Pyrazine
10855 - Pyridazine
11049 - Pyrimidine
C₄H₄N₂OS
11110 - 4(1H)-Pyrimidinone, 2,3-dihydro-2-thioxo-
C₄H₄N₂O₂
3902 - 2-Butynediamide
11059 - 2,4(1H,3H)-Pyrimidinedione
C₄H₄N₂O₃
11076 - 2,4,6(1H,3H,5H)-Pyrimidinetrione
C₄H₄N₂S₂
11702 - Thiocyanic acid, 1,2-ethanediyl ester
C₄H₄N₆O
11851 - 7H-1,2,3-Triazolo[4,5-d]pyrimidin-7-one, 5-amino-1,4-dihydro-
C₄H₄Na₂O₆
3314 - Butanedioic acid, 2,3-dihydroxy- [R-(R*,R*)]-, disodium salt
C₄H₄O
3895 - 2-Butynal
3923 - 3-Butyn-2-one
6025 - Furan
C₄H₄OS
11777 - Thiophene-2-ol
C₄H₄O₂
3917 - 2-Butynoic acid
5202 - 1,4-Dioxin
8598 - 2-Oxetanone, 4-methylene-
C₄H₄O₃
6093 - 2,5-Furandione, dihydro-
C₄H₄O₄
3764 - 2-Butenedioic acid (E)-
3765 - 2-Butenedioic acid (Z)-
C₄H₄O₅
3339 - Butanedioic acid, oxo-

$C_4H_4O_6$
3777 - 2-Butenedioic acid, 2,3-dihydroxy-, (Z)-

C_4H_4S
11724 - Thiophene

C_4H_5Br
3147 - 1,2-Butadiene, 4-bromo-
3148 - 1,3-Butadiene, 1-bromo-
3149 - 1,3-Butadiene, 2-bromo-

$C_4H_5BrN_2$
10817 - 1H-Pyrazole, 4-bromo-3-methyl-

C_4H_5BrO
3696 - 2-Butenal, 2-bromo-

$C_4H_5BrO_2$
10612 - 2-Propenoic acid, 3-bromo-2-methyl-

$C_4H_5BrO_4$
3299 - Butanedioic acid, bromo-, (±)-
3300 - Butanedioic acid, bromo-

$C_4H_5Br_2FO_2$
161 - Acetic acid, dibromofluoro-, ethyl ester

$C_4H_5Br_3O_2$
242 - Acetic acid, tribromo-, ethyl ester

C_4H_5Cl
3150 - 1,2-Butadiene, 4-chloro-
3151 - 1,3-Butadiene, 1-chloro-
3152 - 1,3-Butadiene, 2-chloro-
3899 - 1-Butyne, 3-chloro-
3900 - 2-Butyne, 1-chloro-

$C_4H_5ClN_2$
10821 - 1H-Pyrazole, 3-chloro-5-methyl-

C_4H_5ClO
3697 - 2-Butenal, 2-chloro-
3887 - 2-Butenoyl chloride
10715 - 2-Propenoyl chloride, 2-methyl-

$C_4H_5ClO_2$
3820 - 2-Butenoic acid, 3-chloro-, (E)-
3821 - 2-Butenoic acid, 3-chloro-, (Z)-
3822 - 2-Butenoic acid, 4-chloro-
4022 - Carbonochloridic acid, 1-methylethenyl ester
4150 - Crotonic acid, 2-chloro-
4151 - Crotonic acid, 4-chloro-, (E)-
10621 - 2-Propenoic acid, 2-chloro-, methyl ester

$C_4H_5ClO_3$
144 - Acetic acid, chlorooxo-, ethyl ester

$C_4H_5Cl_3$
3809 - 1-Butene, 2,3,4-trichloro-
3810 - 2-Butene, 1,4-trichloro-
10605 - 1-Propene, 1,1,3-trichloro-2-methyl-
10606 - 1-Propene, 3,3,3-trichloro-2-methyl-

$C_4H_5Cl_3O$
3191 - Butanal, 2,2,3-trichloro-

$C_4H_5Cl_3O_2$
248 - Acetic acid, trichloro-, ethyl ester
3590 - Butanoic acid, 2,2,3-trichloro-

$C_4H_5Cl_3O_3$
249 - Acetic acid, trichloro-, 2-hydroxyethyl ester

$C_4H_5Cl_5$
3419 - Butane, 1,1,2,3,4-pentachloro-
3420 - Butane, 1,2,2,3,4-pentachloro-
10127 - Propane, 1,1,1,2,3-pentachloro-2-methyl-
10146 - Propane, 1,1,2,3-tetrachloro-2-(chloromethyl)-

C_4H_5F
3162 - 1,3-Butadiene, 2-fluoro-

$C_4H_5F_3O_2$
267 - Acetic acid, trifluoro-, ethyl ester

C_4H_5I
3165 - 1,3-Butadiene, 2-iodo-
3166 - 1,2-Butadiene, 4-iodo-

C_4H_5N
3802 - 2-Butenenitrile, (E)-
3803 - 3-Butenenitrile
4956 - Cyclopropanecarbonitrile
10574 - 1-Propene, 3-isocyano-
10586 - 2-Propenenitrile, 2-methyl-
11112 - 1H-Pyrrole

C_4H_5NO
7405 - Isoxazole, 5-methyl-

8583 - Oxazole, 4-methyl-

$C_4H_5NO_2$
154 - Acetic acid, cyano-, methyl ester
4030 - Carbonocyanidic acid, ethyl ester
10233 - Propanoic acid, 2-cyano-
11152 - 2,5-Pyrrolidinedione

$C_4H_5NO_3$
3817 - 2-Butenoic acid, 4-amino-4-oxo-, (Z)-

C_4H_5NS
10575 - 1-Propene, 3-isothiocyanato-
11678 - Thiazole, 2-methyl-
11679 - Thiazole, 4-methyl-
11712 - Thiocyanic acid, 2-propenyl ester

$C_4H_5NS_2$
11680 - 2(3H)-Thiazolethione, 4-methyl-

$C_4H_5N_3$
11042 - 2-Pyrimidinamine

$C_4H_5N_3O$
11104 - 2(1H)-Pyrimidinone, 4-amino-

$C_4H_5N_3O_2$
11831 - 1,2,4-Triazine-3,5(2H,4H)-dione, 6-methyl-

$C_4H_5N_3O_3$
11077 - 2,4,6(1H,3H,5H)-Pyrimidinetrione, 5-amino-

C_4H_6
3145 - 1,2-Butadiene
3146 - 1,3-Butadiene
3897 - 1-Butyne
3898 - 2-Butyne
4198 - Cyclobutene

$C_4H_6BaO_4$
110 - Acetic acid, barium salt

C_4H_6BrClO
3682 - Butanoyl chloride, 2-bromo-

$C_4H_6BrClO_2$
113 - Acetic acid, bromochloro-, ethyl ester

C_4H_6BrN
3404 - Butanenitrile, 4-bromo-
10096 - Propanenitrile, 2-bromo-2-methyl-

$C_4H_6Br_2$
3738 - 1-Butene, 1,1-dibromo-
3739 - 1-Butene, 1,2-dibromo-
3740 - 1-Butene, 3,4-dibromo-
3741 - 2-Butene, 1,3-dibromo-
3742 - 2-Butene, 1,4-dibromo-, (E)-
4178 - Cyclobutane, 1,1-dibromo-
4179 - Cyclobutane, 1,2-dibromo-
10546 - 1-Propene, 1,3-dibromo-2-methyl-

$C_4H_6Br_2O$
3653 - 2-Butanone, 3,4-dibromo-
5893 - Ethene, 1,1-dibromo-2-ethoxy-
10491 - Propanoyl bromide, 2-bromo-2-methyl-

$C_4H_6Br_2O_2$
160 - Acetic acid, dibromo-, ethyl ester
3497 - Butanoic acid, 2,3-dibromo-
10241 - Propanoic acid, 2,3-dibromo-, methyl ester

$C_4H_6Br_4$
3424 - Butane, 1,1,4,4-tetrabromo-
3425 - Butane, 1,2,2,3-tetrabromo-
3426 - Butane, 1,2,3,4-tetrabromo-, DL-

$C_4H_6CaO_4$
122 - Acetic acid, calcium salt

$C_4H_6ClFO_2$
133 - Acetic acid, chlorofluoro-, ethyl ester

C_4H_6ClN
3405 - Butanenitrile, 4-chloro-

C_4H_6ClNO
10098 - Propanenitrile, 3-chloro-2-hydroxy-2-methyl-

$C_4H_6Cl_2$
3743 - 1-Butene, 1,3-dichloro-
3744 - 1-Butene, 2,3-dichloro-
3745 - 1-Butene, 3,4-dichloro-
3746 - 2-Butene, 1,1-dichloro-
3747 - 2-Butene, 1,2-dichloro-
3748 - 2-Butene, 1,3-dichloro-, (E)-
3749 - 2-Butene, 1,3-dichloro-, (Z)-
3750 - 2-Butene, 1,4-dichloro-, (E)-
3751 - 2-Butene, 1,4-dichloro-, (Z)-

3752 - 2-Butene, 2,3-dichloro-, (E)-
3753 - 2-Butene, 2,3-dichloro-, (Z)-
10537 - 1-Propene, 3-chloro-2-(chloromethyl)-
10555 - 1-Propene, 1,1-dichloro-2-methyl-
10556 - 1-Propene, 3,3-dichloro-2-methyl-

$C_4H_6Cl_2O$
3182 - Butanal, 2,3-dichloro-
3654 - 2-Butanone, 1,3-dichloro-
3655 - 2-Butanone, 3,4-dichloro-
3683 - Butanoyl chloride, 2-chloro-
3684 - Butanoyl chloride, 3-chloro-
3685 - Butanoyl chloride, 4-chloro-
5899 - Ethene, 1,2-dichloro-1-ethoxy-
10498 - Propanoyl chloride, 2-chloro-2-methyl-
10499 - Propanoyl chloride, 3-chloro-2-methyl-

$C_4H_6Cl_2O_2$
126 - Acetic acid, chloro-, 2-chloroethyl ester
164 - Acetic acid, dichloro-, ethyl ester
3500 - Butanoic acid, 2,2-dichloro-
3926 - Butyric acid, 3,4-dichloro-
4015 - Carbonochloridic acid, 3-chloropropyl ester
5188 - 1,4-Dioxane, 2,3-dichloro-
10248 - Propanoic acid, 2,3-dichloro-, methyl ester

$C_4H_6Cl_2O_3$
165 - Acetic acid, dichloro-, 2-hydroxyethyl ester
10247 - Propanoic acid, 3,3-dichloro-2-hydroxy-2-methyl-

$C_4H_6Cl_3NO$
97 - Acetamide, 2,2,2-trichloro-N,N-dimethyl-

$C_4H_6Cl_4$
3427 - Butane, 1,1,1,2-tetrachloro-
3428 - Butane, 1,2,2,3-tetrachloro-
3429 - Butane, 1,2,3,3-tetrachloro-
10147 - Propane, 1,1,1,2-tetrachloro-2-methyl-
10148 - Propane, 1,1,2,3-tetrachloro-2-methyl-
10176 - Propane, 1,2,3-trichloro-2-(chloromethyl)-

$C_4H_6F_2O_2$
176 - Acetic acid, difluoro-, ethyl ester

$C_4H_6F_3NO$
99 - Acetamide, 2,2,2-trifluoro-N,N-dimethyl-

$C_4H_6MnN_2S_4$
7420 - Maneb

$C_4H_6N_2$
7208 - 1H-Imidazole, 1-methyl-
7209 - 1H-Imidazole, 2-methyl-
7210 - 1H-Imidazole, 4-methyl-
10834 - 1H-Pyrazole, 1-methyl-
10835 - 1H-Pyrazole, 3-methyl-
10836 - 1H-Pyrazole, 4-methyl-

$C_4H_6N_2O$
6176 - Furazan, dimethyl-

$C_4H_6N_2O_2$
158 - Acetic acid, diazo-, ethyl ester
9612 - 2,5-Piperazinedione
10237 - Propanoic acid, 2-diazo-, methyl ester

$C_4H_6N_2S$
7214 - 2H-Imidazole-2-thione, 1,3-dihydro-1-methyl-
11659 - 1,3,4-Thiadiazole, 2,5-dimethyl-
11665 - 2-Thiazolamine, 4-methyl-

$C_4H_6N_2S_4Zn$
12067 - Zinc, [[1,2-ethanediylbis [carbamodithioato]](2-)]-

$C_4H_6N_4O$
7191 - Imidazo[4,5-d]imidazole-2,5(1H,3H)-dione, tetrahydro-

$C_4H_6N_4O_3$
12009 - Urea, (2,5-dioxo-4-imidazolidinyl)-

$C_4H_6N_4O_3S_2$
32 - Acetamide, N-[5-(aminosulfonyl)-1,3,4-thiadiazol-2-yl]-

$C_4H_6N_4O_{12}$
3432 - 1,2,3,4-Butanetetrol, tetranitrate, (R*,S*)-

C_4H_6O
3175 - 2,3-Butadien-1-ol
3695 - 2-Butenal, (E)-
3875 - 3-Buten-2-one
3919 - 2-Butyn-1-ol

3920 - 3-Butyn-1-ol
3921 - 3-Butyn-2-ol
4197 - Cyclobutanone
5912 - Ethene, 1,1'-oxybis-
5942 - Ethyne, ethoxy-
6086 - Furan, 2,3-dihydro-
8613 - Oxirane, ethenyl-
10515 - 2-Propenal, 2-methyl-
10736 - 1-Propyne, 3-methoxy-

C$_4$H$_6$OS
11794 - 2(3H)-Thiophenone, dihydro-

C$_4$H$_6$O$_2$
179 - Acetic acid ethenyl ester
3007 - 2,2'-Bioxirane, (R*,S*)-
3256 - Butanedial
3359 - 2,3-Butanedione
3813 - 3-Butenoic acid
3814 - 2-Butenoic acid, (E)-
3815 - 2-Butenoic acid, (Z)-
3911 - 2-Butyne-1,4-diol
3912 - 3-Butyne-1,2-diol
4958 - Cyclopropanecarboxylic acid
5203 - 1,4-Dioxin, 2,3-dihydro-
6017 - Formic acid, 2-propenyl ester
6142 - 2(3H)-Furanone, dihydro-
10648 - 2-Propenoic acid, 2-methyl-
10655 - 2-Propenoic acid, methyl ester

C$_4$H$_6$O$_2$S
5913 - Ethene, 1,1'-sulfonylbis-
11752 - Thiophene, 2,5-dihydro-, 1,1-dioxide

C$_4$H$_6$O$_2$S$_2$
5250 - Disulfide, diacetyl

C$_4$H$_6$O$_3$
107 - Acetic acid, anhydride
3567 - Butanoic acid, 2-oxo-
3568 - Butanoic acid, 3-oxo-
3842 - 3-Butenoic acid, 2-hydroxy-
5213 - 1,3-Dioxolan-2-one, 4-methyl-
10316 - Propanoic acid, 2-oxo-, methyl ester

C$_4$H$_6$O$_4$
101 - Acetic acid, (acetyloxy)-
3294 - Butanedioic acid
5563 - Ethanedioic acid, dimethyl ester
5573 - 1,2-Ethanediol, diformate
9156 - Peroxide, diacetyl
9960 - Propanedioic acid, methyl-

C$_4$H$_6$O$_4$S
240 - Acetic acid, 2,2'-thiobis-

C$_4$H$_6$O$_5$
226 - Acetic acid, 2,2'-oxybis-
9959 - Propanedioic acid, (hydroxymethyl)-

C$_4$H$_6$O$_6$
3304 - Butanedioic acid, 2,3-dihydroxy-, (R*,S*)-
3309 - Butanedioic acid, 2,3-dihydroxy- [R-(R*,R*)]-
3311 - Butanedioic acid, 2,3-dihydroxy-, (R*,R*)-(±)-
3312 - Butanedioic acid, 2,3-dihydroxy-, [S-(R*,R*)]-

C$_4$H$_6$O$_6$S
239 - Acetic acid, 2,2'-sulfonylbis-

C$_4$H$_6$O$_8$
3342 - Butanedioic acid, tetrahydroxy-

C$_4$H$_6$S
5920 - Ethene, 1,1'-thiobis-
11750 - Thiophene, 2,3-dihydro-
11751 - Thiophene, 2,5-dihydro-

C$_4$H$_7$Br
3707 - 1-Butene, 1-bromo-, (E)-
3708 - 1-Butene, 1-bromo-, (Z)-
3709 - 1-Butene, 2-bromo-
3710 - 1-Butene, 3-bromo-
3711 - 1-Butene, 4-bromo-
3712 - 2-Butene, 1-bromo-
3713 - 2-Butene, 2-bromo-, (E)-
3714 - 2-Butene, 2-bromo-, (Z)-

C$_4$H$_7$BrO
3179 - Butanal, 2-bromo-
3680 - Butanoyl bromide
9776 - Propanal, 2-bromo-2-methyl-

10492 - Propanoyl bromide, 2-methyl-

C$_4$H$_7$BrO$_2$
115 - Acetic acid, bromo-, ethyl ester
3468 - Butanoic acid, 2-bromo-, (±)-
5686 - Ethanol, 2-bromo-, acetate
10209 - Propanoic acid, 2-bromo-2-methyl-
10211 - Propanoic acid, 2-bromo-, methyl ester, (±)-
10212 - Propanoic acid, 3-bromo-, methyl ester

C$_4$H$_7$Br$_2$Cl$_2$O$_4$P
7702 - Naled

C$_4$H$_7$Br$_3$
3442 - Butane, 1,1,2-tribromo-
3443 - Butane, 1,2,2-tribromo-
3444 - Butane, 1,2,3-tribromo-
3445 - Butane, 1,2,4-tribromo-
3446 - Butane, 1,3,3-tribromo-
3447 - Butane, 2,2,3-tribromo-
10164 - Propane, 1,2,3-tribromo-2-methyl-

C$_4$H$_7$Br$_3$O
10405 - 2-Propanol, 1,1,1-tribromo-2-methyl-

C$_4$H$_7$Cl
3720 - 1-Butene, 1-chloro-, (E)-
3721 - 1-Butene, 1-chloro-, (Z)-
3722 - 1-Butene, 2-chloro-
3723 - 1-Butene, 3-chloro-
3724 - 1-Butene, 4-chloro-
3725 - 2-Butene, 1-chloro-, (E)-
3726 - 2-Butene, 1-chloro-, (Z)-
3727 - 2-Butene, 2-chloro-, (E)-
3728 - 2-Butene, 2-chloro-, (Z)-
4966 - Cyclopropane, (chloromethyl)-
10539 - 1-Propene, 1-chloro-2-methyl-
10540 - 1-Propene, 3-chloro-2-methyl-

C$_4$H$_7$ClO
3180 - Butanal, 2-chloro-
3181 - Butanal, 4-chloro-
3646 - 2-Butanone, 1-chloro-
3647 - 2-Butanone, 3-chloro-
3648 - 2-Butanone, 4-chloro-
3681 - Butanoyl chloride
3863 - 2-Buten-1-ol, 2-chloro-
3864 - 3-Buten-1-ol, 3-chloro-
3865 - 2-Buten-1-ol, 4-chloro-
3866 - 3-Buten-2-ol, 1-chloro-
3867 - 3-Buten-1-ol, 2-chloro-
3868 - 3-Buten-2-ol, 3-chloro-
5886 - Ethene, 1-chloro-2-ethoxy-
5887 - Ethene, (2-chloroethoxy)-
8606 - Oxirane, 2-(chloromethyl)-2-methyl-
9779 - Propanal, 2-chloro-2-methyl-
10504 - Propanoyl chloride, 2-methyl-

C$_4$H$_7$ClO$_2$
130 - Acetic acid, chloro-, ethyl ester
298 - Acetyl chloride, ethoxy-
3480 - Butanoic acid, 2-chloro-
3481 - Butanoic acid, 3-chloro-, (±)-
3482 - Butanoic acid, 4-chloro-
4023 - Carbonochloridic acid, 1-methylethyl ester
4028 - Carbonochloridic acid, propyl ester
5694 - Ethanol, 1-chloro-, acetate
5695 - Ethanol, 2-chloro-, acetate
10222 - Propanoic acid, 2-chloro-, methyl ester
10225 - Propanoic acid, 3-chloro-, methyl ester
10228 - Propanoic acid, 2-chloro-2-methyl-
10230 - Propanoic acid, 3-chloro-2-methyl-

C$_4$H$_7$ClO$_3$
136 - Acetic acid, chloro-, 2-hydroxyethyl ester
4019 - Carbonochloridic acid, 2-methoxyethyl ester

C$_4$H$_7$Cl$_2$O$_4$P
5170 - Dichlorvos

C$_4$H$_7$Cl$_3$
3448 - Butane, 1,1,3-trichloro-
3449 - Butane, 1,2,3-trichloro-
3450 - Butane, 1,2,4-trichloro-
3451 - Butane, 2,2,3-trichloro-
10177 - Propane, 1,1,2-trichloro-2-methyl-
10178 - Propane, 1,2,3-trichloro-2-methyl-

C$_4$H$_7$Cl$_3$O
3643 - 2-Butanol, 1,1,1-trichloro-
10407 - 2-Propanol, 1,1,1-trichloro-2-methyl-

C$_4$H$_7$Cl$_3$O$_2$
5769 - Ethanol, 2,2,2-trichloro-1-ethoxy-

C$_4$H$_7$FO$_2$
191 - Acetic acid, fluoro-, ethyl ester
3927 - Butyric acid, 4-fluoro-

C$_4$H$_7$IO$_2$
201 - Acetic acid, iodo-, ethyl ester
10284 - Propanoic acid, 3-iodo-, methyl ester

C$_4$H$_7$N
3403 - Butanenitrile
10063 - Propane, 2-isocyano-
10111 - Propanenitrile, 2-methyl-
10726 - 2-Propyn-1-amine, N-methyl-
11118 - 1H-Pyrrole, 2,5-dihydro-

C$_4$H$_7$NO
3702 - 2-Butenamide, (E)-
8576 - Oxazole, 4,5-dihydro-2-methyl-
10059 - Propane, 1-isocyanato-
10060 - Propane, 2-isocyanato-
10108 - Propanenitrile, 2-hydroxy-2-methyl-
10109 - Propanenitrile, 2-methoxy-
10110 - Propanenitrile, 3-methoxy-
11180 - 2-Pyrrolidinone

C$_4$H$_7$NO$_2$
23 - Acetamide, N-acetyl-
495 - 2-Azetidinecarboxylic acid
3361 - 2,3-Butanedione, monooxime

C$_4$H$_7$NO$_2$S
11682 - 4-Thiazolidinecarboxylic acid

C$_4$H$_7$NO$_3$
6251 - Glycine, N-acetyl-

C$_4$H$_7$NO$_4$
218 - Acetic acid, nitro-, ethyl ester
470 - DL-Aspartic acid
471 - L-Aspartic acid
3466 - Butanoic acid, 4-amino-2-hydroxy-4-oxo-
6260 - Glycine, N-(carboxymethyl)-

C$_4$H$_7$NS
10064 - Propane, 1-isothiocyanato-
11672 - Thiazole, 4,5-dihydro-2-methyl-
11706 - Thiocyanic acid, 1-methylethyl ester

C$_4$H$_7$N$_3$
11849 - 1H-1,2,3-Triazole, 1,5-dimethyl-

C$_4$H$_7$N$_3$O
7230 - 4H-Imidazol-4-one, 2-amino-1,5-dihydro-1-methyl-

C$_4$H$_8$
3704 - 1-Butene
3705 - 2-Butene, (E)-
3706 - 2-Butene, (Z)-
4170 - Cyclobutane
4981 - Cyclopropane, methyl-
10578 - 1-Propene, 2-methyl-

C$_4$H$_8$BrCl
3233 - Butane, 1-bromo-4-chloro-
3234 - Butane, 2-bromo-1-chloro-
9842 - Propane, 1-bromo-3-chloro-2-methyl-

C$_4$H$_8$BrF
3237 - Butane, 1-bromo-4-fluoro-

C$_4$H$_8$BrNO
9792 - Propanamide, 2-bromo-2-methyl-

C$_4$H$_8$Br$_2$
3264 - Butane, 1,1-dibromo-
3265 - Butane, 1,2-dibromo-
3266 - Butane, 1,3-dibromo-
3267 - Butane, 1,4-dibromo-
3268 - Butane, 2,3-dibromo-
9889 - Propane, 1,2-dibromo-2-methyl-
9890 - Propane, 1,3-dibromo-2-methyl-

C$_4$H$_8$Br$_2$O
3612 - 1-Butanol, 2,3-dibromo-
3613 - 1-Butanol, 3,4-dibromo-
3614 - 2-Butanol, 1,4-dibromo-
5523 - Ethane, 1,2-dibromo-1-ethoxy-

5609 - Ethane, 1,1'-oxybis[2-bromo-
9888 - Propane, 1,2-dibromo-3-methoxy-

C₄H₈ClF
3250 - Butane, 1-chloro-4-fluoro-

C₄H₈Cl₂
3276 - Butane, 1,1-dichloro-
3277 - Butane, 1,2-dichloro-
3278 - Butane, 1,3-dichloro-
3279 - Butane, 1,4-dichloro-
3280 - Butane, 2,2-dichloro-
3281 - Butane, 2,3-dichloro-, (R*,R*)-(±)-
9899 - Propane, 1,1-dichloro-2-methyl-
9900 - Propane, 1,2-dichloro-2-methyl-
9901 - Propane, 1,3-dichloro-2-methyl-

C₄H₈Cl₂O
5468 - Ethane, 1-chloro-1-(2-chloroethoxy)-
5533 - Ethane, 1,2-dichloro-1-ethoxy-
5610 - Ethane, 1,1'-oxybis[1-chloro-
5611 - Ethane, 1,1'-oxybis[2-chloro-
10364 - 2-Propanol, 1,1-dichloro-2-methyl-
10365 - 2-Propanol, 1,3-dichloro-2-methyl-

C₄H₈Cl₂O₂
3349 - 2,3-Butanediol, 1,4-dichloro-

C₄H₈Cl₂S
5645 - Ethane, 1,1'-thiobis[2-chloro-

C₄H₈Cl₂Si
11402 - Silane, dichloromethyl-2-propenyl-

C₄H₈Cl₃O₄P
11853 - Trichlorfon

C₄H₈I₂
3288 - Butane, 1,4-diiodo-

C₄H₈N₂
281 - Acetonitrile, (dimethylamino)-
3903 - 2-Butyne-1,4-diamine
7197 - 1H-Imidazole, 4,5-dihydro-2-methyl-
10112 - Propanenitrile, 3-(methylamino)-

C₄H₈N₂O
11170 - Pyrrolidine, 1-nitroso-
12028 - Urea, 2-propenyl-

C₄H₈N₂O₂
73 - Acetamide, N-[(methylamino)carbonyl]-
3257 - Butanediamide
3360 - 2,3-Butanedione, dioxime
7699 - Morpholine, 4-nitroso-

C₄H₈N₂O₃
468 - L-Asparagine
3932 - Carbamic acid, (aminocarbonyl)-, ethyl ester
6265 - Glycine, N-glycyl-

C₄H₈N₂O₄
3293 - Butane, 1,4-dinitro-

C₄H₈N₂S
11809 - Thiourea, 2-propenyl-

C₄H₈O
3178 - Butanal
3645 - 2-Butanone
3858 - 3-Buten-2-ol, (±)-
3859 - 2-Buten-1-ol, (E)-
3860 - 2-Buten-1-ol, (Z)-
3861 - 3-Buten-1-ol
4196 - Cyclobutanol
4980 - Cyclopropane, methoxy-
5905 - Ethene, ethoxy-
6168 - Furan, tetrahydro-
8594 - Oxetane, 2-methyl-
8607 - Oxirane, 2,2-dimethyl-
8608 - Oxirane, 2,3-dimethyl-, cis-
8609 - Oxirane, 2,3-dimethyl-, trans-
9785 - Propanal, 2-methyl-
10576 - 1-Propene, 3-methoxy-
10702 - 2-Propen-1-ol, 2-methyl-

C₄H₈OS
5647 - Ethanethioic acid, S-ethyl ester
8569 - 1,4-Oxathiane
9786 - Propanal, 3-(methylthio)-

C₄H₈O₂
14 - Acetaldehyde, ethoxy-
186 - Acetic acid ethyl ester

3184 - Butanal, 3-hydroxy-
3460 - Butanoic acid
3660 - 2-Butanone, 1-hydroxy-
3661 - 2-Butanone, 3-hydroxy-
3662 - 2-Butanone, 4-hydroxy-
3792 - 2-Butene-1,4-diol, (E)-
3793 - 2-Butene-1,4-diol, (Z)-
3794 - 3-Butene-1,2-diol
5186 - 1,3-Dioxane
5187 - 1,4-Dioxane
5210 - 1,3-Dioxolane, 2-methyl-
5716 - Ethanol, 2-(ethenyloxy)-
6011 - Formic acid, 1-methylethyl ester
6018 - Formic acid, propyl ester
6137 - 3-Furanol, tetrahydro-
8615 - Oxirane, ethyl-
8620 - Oxirane, (methoxymethyl)-
10290 - Propanoic acid, 2-methyl-
10296 - Propanoic acid, methyl ester
10455 - 2-Propanone, 1-methoxy-

C₄H₈O₂S
188 - Acetic acid, (ethylthio)-
203 - Acetic acid, mercapto-, ethyl ester
3537 - Butanoic acid, 3-mercapto-
10287 - Propanoic acid, 3-mercapto-, methyl ester
10288 - Propanoic acid, 2-mercapto-2-methyl-
11786 - Thiophene, tetrahydro-, 1,1-dioxide

C₄H₈O₃
180 - Acetic acid, ethoxy-
197 - Acetic acid, hydroxy-, ethyl ester
207 - Acetic acid, methoxy-, methyl ester
3530 - Butanoic acid, 2-hydroxy-, (±)-
3531 - Butanoic acid, 3-hydroxy-, (±)-
4003 - Carbonic acid, ethyl methyl ester
5577 - 1,2-Ethanediol, monoacetate
10273 - Propanoic acid, 2-hydroxy-, methyl ester, (±)-
10276 - Propanoic acid, 3-hydroxy-, methyl ester
10278 - Propanoic acid, 2-hydroxy-2-methyl-

C₄H₈O₃S
5198 - 1,3,2-Dioxathiane, 4-methyl-, 2-oxide

C₄H₈O₄
3192 - Butanal, 2,3,4-trihydroxy-, [R-(R*,R*)]-
3193 - Butanal, 2,3,4-trihydroxy-, [S-(R*,S*)]-
3679 - 2-Butanone, 1,3,4-trihydroxy-, (S)-
10254 - Propanoic acid, 2,3-dihydroxy-, methyl ester, (±)-

C₄H₈S
10581 - 1-Propene, 3-(methylthio)-
11693 - Thietane, 2-methyl-
11783 - Thiophene, tetrahydro-

C₄H₈S₂
5261 - 1,2-Dithiane
5262 - 1,3-Dithiane
5263 - 1,4-Dithiane

C₄H₉Br
3231 - Butane, 1-bromo-
3232 - Butane, 2-bromo-, (±)-
9845 - Propane, 1-bromo-2-methyl-
9846 - Propane, 2-bromo-2-methyl-

C₄H₉BrO
3600 - 1-Butanol, 3-bromo-
3601 - 2-Butanol, 3-bromo-, (R*,S*)-(±)-
5461 - Ethane, 1-bromo-2-ethoxy-

C₄H₉BrO₂
5460 - Ethane, 2-bromo-1,1-dimethoxy-

C₄H₉Cl
3241 - Butane, 1-chloro-
3242 - Butane, 2-chloro-, (±)-
9862 - Propane, 1-chloro-2-methyl-
9863 - Propane, 2-chloro-2-methyl-

C₄H₉ClO
1605 - Benzeneethanol, α-ethyl-α-methyl-, (±)-
3602 - 1-Butanol, 2-chloro-
3604 - 1-Butanol, 4-chloro-
3605 - 2-Butanol, 1-chloro-
3606 - 2-Butanol, 3-chloro-
3607 - 2-Butanol, 4-chloro-

5473 - Ethane, 1-chloro-1-ethoxy-
5474 - Ethane, 1-chloro-2-ethoxy-
7183 - Hypochlorous acid, 1,1-dimethylethyl ester
9858 - Propane, 1-(chloromethoxy)-
9859 - Propane, 1-chloro-2-methoxy-
9860 - Propane, 1-chloro-3-methoxy-
9861 - Propane, 2-chloro-1-methoxy-
10352 - 1-Propanol, 2-chloro-2-methyl-
10353 - 1-Propanol, 3-chloro-2-methyl-
10354 - 2-Propanol, 1-chloro-2-methyl-

C₄H₉ClO₂
5472 - Ethane, 2-chloro-1,1-dimethoxy-
5698 - Ethanol, 2-(2-chloroethoxy)-
9998 - 1,2-Propanediol, 3-chloro-2-methyl-

C₄H₉ClO₂S
10134 - 1-Propanesulfonyl chloride, 2-methyl-

C₄H₉ClS
9864 - Propane, 1-chloro-3-(methylthio)-
9865 - Propane, 2-chloro-1-(methylthio)-

C₄H₉ClSi
11376 - Silane, chloroethenyldimethyl-

C₄H₉Cl₃Si
11370 - Silane, butyltrichloro-
11469 - Silane, trichloro(2-methylpropyl)-

C₄H₉F
3378 - Butane, 1-fluoro-
3379 - Butane, 2-fluoro-

C₄H₉FO
3621 - 1-Butanol, 4-fluoro-

C₄H₉I
3381 - Butane, 1-iodo-
3382 - Butane, 2-iodo-, (±)-
10057 - Propane, 1-iodo-2-methyl-
10058 - Propane, 2-iodo-2-methyl-

C₄H₉I₉N
4169 - Cyclobutanamine
10524 - 2-Propen-1-amine, N-methyl-
11143 - Pyrrolidine

C₄H₉NO
55 - Acetamide, N,N-dimethyl-
61 - Acetamide, N-ethyl-
499 - 1-Aziridineethanol
3188 - Butanal, oxime
3194 - Butanamide
3675 - 2-Butanone, oxime
7681 - Morpholine
9798 - Propanamide, 2-methyl-
9799 - Propanamide, N-methyl-

C₄H₉NO₂
65 - Acetamide, N-(2-hydroxyethyl)-
331 - Alanine, 2-methyl-
3412 - Butane, 1-nitro-
3413 - Butane, 2-nitro-, (±)-
3463 - Butanoic acid, 2-amino-, (±)-
3464 - Butanoic acid, 4-amino-
3954 - Carbamic acid, methyl-, ethyl ester
3955 - Carbamic acid, 1-methylethyl ester
3969 - Carbamic acid, propyl ester
6261 - Glycine, N,N-dimethyl-
6263 - Glycine, ethyl ester
8083 - Nitrous acid, butyl ester
8085 - Nitrous acid, 1,1-dimethylethyl ester
8093 - Nitrous acid, 1-methylpropyl ester
8094 - Nitrous acid, 2-methylpropyl ester
10084 - Propane, 2-methyl-1-nitro-
10085 - Propane, 2-methyl-2-nitro-

C₄H₉NO₃
3465 - Butanoic acid, 3-amino-4-hydroxy-
3640 - 1-Butanol, 2-nitro-
3641 - 2-Butanol, 1-nitro-
3642 - 2-Butanol, 3-nitro-
7134 - Homoserine
8072 - Nitric acid, butyl ester
8077 - Nitric acid, 1-methylpropyl ester
8078 - Nitric acid, 2-methylpropyl ester
10391 - 1-Propanol, 2-methyl-2-nitro-
11813 - L-Threonine

C_5F_{10}
9027 - 1-Pentene, 1,1,2,3,3,4,4,5,5,5-decafluoro-
C_5F_{12}
8807 - Pentane, dodecafluoro-
C_5FeO_5
7339 - Iron carbonyl
C_5HCl_4N
11019 - Pyridine, 2,3,5,6-tetrachloro-
$C_5H_2Cl_6O_3$
5212 - 1,3-Dioxolan-4-one, 2,5-bis(trichloromethyl)-
$C_5H_2F_6O_2$
8802 - 2,4-Pentanedione, 1,1,1,5,5,5-hexafluoro-
$C_5H_2O_5$
4939 - 4-Cyclopentene-1,2,3-trione, 4,5-dihydroxy-
$C_5H_3BrO_2$
6043 - 2-Furancarboxaldehyde, 5-bromo-
$C_5H_3Br_2N$
10926 - Pyridine, 3,5-dibromo-
$C_5H_3ClN_4$
10753 - 1H-Purine, 6-chloro-
$C_5H_3ClO_2$
6040 - 2-Furancarbonyl chloride
$C_5H_3F_7O_2$
3525 - Butanoic acid, heptafluoro-, methyl ester
C_5H_3NO
6039 - 2-Furancarbonitrile
$C_5H_3NO_4$
6046 - 2-Furancarboxaldehyde, 5-nitro-
$C_5H_3NO_5$
6066 - 2-Furancarboxylic acid, 5-nitro-
C_5H_3NS
11732 - 2-Thiophenecarbonitrile
C_5H_4
8684 - 1,3-Pentadiyne
C_5H_4BrN
10875 - Pyridine, 2-bromo-
10876 - Pyridine, 3-bromo-
10877 - Pyridine, 4-bromo-
C_5H_4ClN
10915 - Pyridine, 2-chloro-
10916 - Pyridine, 3-chloro-
10917 - Pyridine, 4-chloro-
C_5H_4ClOS
11735 - 2-Thiophenecarboxaldehyde, 5-chloro-
$C_5H_4Cl_2N_2$
10862 - 3-Pyridinamine, 2,6-dichloro-
11052 - Pyrimidine, 2,4-dichloro-5-methyl-
11053 - Pyrimidine, 2,4-dichloro-6-methyl-
$C_5H_4Cl_2O_2$
3367 - Butanedioyl dichloride, methylene-
$C_5H_4F_8O$
8953 - 1-Pentanol, 2,2,3,3,4,4,5,5-octafluoro-
C_5H_4IN
10970 - Pyridine, 2-iodo-
$C_5H_4N_2O_2$
10803 - Pyrazinecarboxylic acid
$C_5H_4N_2O_3$
10993 - Pyridine, 4-nitro-, 1-oxide
$C_5H_4N_2O_4$
7195 - 1H-Imidazole-4,5-dicarboxylic acid
11050 - 4-Pyrimidinecarboxylic acid, 1,2,3,6-tetrahydro-2,6-dioxo-
$C_5H_4N_4$
10752 - 1H-Purine
$C_5H_4N_4O$
10770 - 6H-Purin-6-one, 1,7-dihydro-
10852 - 4H-Pyrazolo[3,4-d]pyrimidin-4-one, 1,5-dihydro-
$C_5H_4N_4O_2$
10755 - 1H-Purine-2,6-dione, 3,7-dihydro-
$C_5H_4N_4O_3$
10764 - 1H-Purine-2,6,8(3H)-trione, 7,9-dihydro-
$C_5H_4N_4S$
10763 - 6H-Purine-6-thione, 1,7-dihydro-
C_5H_4OS
11733 - 2-Thiophenecarboxaldehyde
11734 - 3-Thiophenecarboxaldehyde

$C_5H_4O_2$
6041 - 2-Furancarboxaldehyde
6042 - 3-Furancarboxaldehyde
6155 - 2(5H)-Furanone, 5-methylene-
10788 - 2H-Pyran-2-one
10789 - 4H-Pyran-4-one
$C_5H_4O_2S$
11737 - 2-Thiophenecarboxylic acid
11738 - 3-Thiophenecarboxylic acid
$C_5H_4O_3$
6050 - 2-Furancarboxylic acid
6051 - 3-Furancarboxylic acid
6095 - 2,5-Furandione, dihydro-3-methylene-
6098 - 2,5-Furandione, 3-methyl-
10794 - 2H-Pyran-2-one, 3-hydroxy-
$C_5H_4O_4$
6054 - 3-Furancarboxylic acid, 4,5-dihydro-5-oxo-
C_5H_5BrO
6035 - Furan, 2-(bromomethyl)-
C_5H_5BrS
11729 - Thiophene, 2-bromo-3-methyl-
11730 - Thiophene, 2-bromo-5-methyl-
$C_5H_5ClN_2$
10861 - 2-Pyridinamine, 5-chloro-
C_5H_5ClO
6075 - Furan, 2-(chloromethyl)-
C_5H_5ClS
11741 - Thiophene, 2-chloro-5-methyl-
$C_5H_5Cl_3N_2OS$
11599 - Terrazole
C_5H_5N
8673 - 2,4-Pentadienenitrile
8674 - 2,4-Pentadienenitrile, (E)-
8675 - 2,4-Pentadienenitrile, (Z)-
10872 - Pyridine
C_5H_5NO
10994 - Pyridine, 1-oxide
11031 - 2-Pyridinol
11032 - 4-Pyridinol
11033 - 2(1H)-Pyridinone
11114 - 1H-Pyrrole-2-carboxaldehyde
$C_5H_5NO_2$
6047 - 2-Furancarboxaldehyde, oxime, (E)-
6048 - 2-Furancarboxaldehyde, oxime, (Z)-
10624 - 2-Propenoic acid, 2-cyano-, methyl ester
$C_5H_5NO_2S$
11669 - 5-Thiazolecarboxylic acid, 4-methyl-
$C_5H_5NO_3S$
11018 - 3-Pyridinesulfonic acid
$C_5H_5N_3O$
10802 - Pyrazinecarboxamide
$C_5H_5N_3O_3S$
85 - Acetamide, N-(5-nitro-2-thiazolyl)-
$C_5H_5N_5$
10751 - 1H-Purin-6-amine
$C_5H_5N_5O$
10767 - 2H-Purin-2-one, 6-amino-1,3-dihydro-
10768 - 6H-Purin-6-one, 2-amino-1,7-dihydro-
$C_5H_5N_5S$
10762 - 6H-Purine-6-thione, 2-amino-1,7-dihydro-
C_5H_6
3892 - 1-Buten-3-yne, 2-methyl-
4764 - 1,3-Cyclopentadiene
9111 - 1-Penten-3-yne
9112 - 1-Penten-4-yne
9113 - 3-Penten-1-yne, (E)-
9114 - 3-Penten-1-yne, (Z)-
$C_5H_6Br_2N_2O_2$
7217 - 2,4-Imidazolidinedione, 1,3-dibromo-5,5-dimethyl-
C_5H_6ClN
10969 - Pyridine, hydrochloride
$C_5H_6ClNO_2$
11155 - 2,5-Pyrrolidinedione, 1-(chloromethyl)-
$C_5H_6Cl_2O_2$
8806 - Pentanedioyl dichloride

$C_5H_6F_4O_2$
10328 - Propanoic acid, 2,3,3,3-tetrafluoro-, ethyl ester
$C_5H_6N_2$
8767 - Pentanedinitrile
9913 - Propanedinitrile, dimethyl-
10810 - Pyrazine, methyl-
10856 - Pyridazine, 3-methyl-
10857 - 2-Pyridinamine
10858 - 3-Pyridinamine
10859 - 4-Pyridinamine
11070 - Pyrimidine, 2-methyl-
11071 - Pyrimidine, 4-methyl-
11072 - Pyrimidine, 5-methyl-
$C_5H_6N_2OS$
11108 - 4(1H)-Pyrimidinone, 2,3-dihydro-6-methyl-2-thioxo-
$C_5H_6N_2O_2$
11068 - 2,4(1H,3H)-Pyrimidinedione, 5-methyl-
11069 - 2,4(1H,3H)-Pyrimidinedione, 6-methyl-
$C_5H_6N_6$
10754 - 1H-Purine-2,6-diamine
C_5H_6O
3891 - 1-Buten-3-yne, 1-methoxy-
4941 - 2-Cyclopenten-1-one
6130 - Furan, 2-methyl-
6131 - Furan, 3-methyl-
9118 - 4-Penten-2-yn-1-ol
10771 - Pyran
C_5H_6OS
6117 - 2-Furanmethanethiol
11768 - 2-Thiophenemethanol
$C_5H_6O_2$
6118 - 2-Furanmethanol
6127 - Furan, 2-methoxy-
6147 - 2(3H)-Furanone, dihydro-3-methylene-
6148 - 2(3H)-Furanone, dihydro-5-methylene-
6153 - 2(3H)-Furanone, 5-methyl-
6154 - 2(5H)-Furanone, 5-methyl-
8676 - 2,4-Pentadienoic acid
9137 - 2-Pentynoic acid
9138 - 4-Pentynoic acid
10739 - 2-Propynoic acid, ethyl ester
10744 - 2-Propyn-1-ol, acetate
$C_5H_6O_3$
4942 - 2-Cyclopenten-1-one, 2,3-dihydroxy-
6094 - 2,5-Furandione, dihydro-3-methyl-
10780 - 2H-Pyran-2,6(3H)-dione, dihydro-
$C_5H_6O_4$
3328 - Butanedioic acid, methylene-
3785 - 2-Butenedioic acid, 2-methyl-, (E)-
3786 - 2-Butenedioic acid, 2-methyl-, (Z)-
4970 - 1,2-Cyclopropanedicarboxylic acid, trans-(±)-
8875 - Pentanoic acid, 2,4-dioxo-
$C_5H_6O_5$
8781 - Pentanedioic acid, 2-oxo-
8782 - Pentanedioic acid, 3-oxo-
C_5H_6S
11769 - Thiophene, 2-methyl-
11770 - Thiophene, 3-methyl-
C_5H_7Br
9124 - 1-Pentyne, 1-bromo-
$C_5H_7BrN_2$
10816 - 1H-Pyrazole, 4-bromo-1,3-dimethyl-
$C_5H_7BrO_2$
10611 - 2-Propenoic acid, 2-bromoethyl ester
C_5H_7Cl
3153 - 1,3-Butadiene, 1-chloro-2-methyl-
3154 - 1,3-Butadiene, 1-chloro-3-methyl-
3155 - 1,3-Butadiene, 2-chloro-3-methyl-
3901 - 1-Butyne, 3-chloro-3-methyl-
4914 - Cyclopentene, 3-chloro-
8663 - 1,3-Pentadiene, 3-chloro-
$C_5H_7ClN_2$
10818 - 1H-Pyrazole, 3-chloro-1,5-dimethyl-
10819 - 1H-Pyrazole, 4-chloro-3,5-dimethyl-
10820 - 1H-Pyrazole, 5-chloro-1,3-dimethyl-

C_5H_7ClO
3888 - 2-Butenoyl chloride, 3-methyl-
4880 - Cyclopentanone, 2-chloro-

$C_5H_7ClO_2$
3830 - 2-Butenoic acid, 2-chloro-, methyl ester, (E)-
3831 - 2-Butenoic acid, 3-chloro-, methyl ester, (E)-
3832 - 2-Butenoic acid, 3-chloro-, methyl ester, (Z)-
3833 - 2-Butenoic acid, 4-chloro-, methyl ester
6141 - 2(3H)-Furanone, 5-(chloromethyl)dihydro-
10619 - 2-Propenoic acid, chloro-, ethyl ester
10620 - 2-Propenoic acid, 2-chloroethyl ester

$C_5H_7ClO_3$
10494 - Propanoyl chloride, 2-(acetyloxy)-, (±)-

$C_5H_7Cl_2NO$
43 - Acetamide, 2-chloro-N-3-chloroallyl-

$C_5H_7Cl_3O_2$
254 - Acetic acid, trichloro-, 1-methylethyl ester
261 - Acetic acid, trichloro-, propyl ester

$C_5H_7Cl_3O_3$
250 - Acetic acid, trichloro-, 2-methoxyethyl ester

$C_5H_7F_3O_2$
269 - Acetic acid, trifluoro-, propyl ester

C_5H_7I
9133 - 1-Pentyne, 1-iodo-

C_5H_7N
4173 - Cyclobutanecarbonitrile
9061 - 3-Pentenenitrile, (E)
9062 - 4-Pentenenitrile
11130 - 1H-Pyrrole, 1-methyl-
11131 - 1H-Pyrrole, 2-methyl-
11132 - 1H-Pyrrole, 3-methyl-

C_5H_7NO
6113 - 2-Furanmethanamine
7403 - Isoxazole, 3,5-dimethyl-
8577 - Oxazole, 2,4-dimethyl-
8578 - Oxazole, 2,5-dimethyl-
9063 - 3-Pentenenitrile, 2-hydroxy-, (Z)-

$C_5H_7NO_2$
152 - Acetic acid, cyano-, ethyl ester
7404 - 5-Isoxazolemethanol, 3-methyl-
10091 - Propanenitrile, 2-(acetyloxy)-
10234 - Propanoic acid, 3-cyano-, methyl ester
11160 - 2,5-Pyrrolidinedione, 1-methyl-

$C_5H_7NO_3$
8584 - 2,4-Oxazolidinedione, 5,5-dimethyl-
9768 - L-Proline, 5-oxo-

C_5H_7NS
11674 - Thiazole, 2,4-dimethyl-
11675 - Thiazole, 4,5-dimethyl-

$C_5H_7N_3$
10923 - 2,3-Pyridinediamine
10924 - 2,5-Pyridinediamine
10925 - 2,6-Pyridinediamine
11034 - 2(1H)-Pyridinone, hydrazone
11046 - 2-Pyrimidinamine, 4-methyl-
11047 - 2-Pyrimidinamine, 5-methyl-
11048 - 4-Pyrimidinamine, 6-methyl-

$C_5H_7N_3O$
11106 - 2(1H)-Pyrimidinone, 4-amino-5-methyl-

$C_5H_7N_3O_2$
7202 - 1H-Imidazole, 1,2-dimethyl-5-nitro-
11105 - 2(1H)-Pyrimidinone, 4-amino-5-
(hydroxymethyl)-

C_5H_8
3170 - 1,2-Butadiene, 3-methyl-
3171 - 1,3-Butadiene, 2-methyl-
3916 - 1-Butyne, 3-methyl-
4193 - Cyclobutane, methylene-
4201 - Cyclobutene, 1-methyl-
4903 - Cyclopentene
8658 - 1,2-Pentadiene
8659 - 1,4-Pentadiene
8660 - 2,3-Pentadiene
8661 - 1,3-Pentadiene, (E)-
8662 - 1,3-Pentadiene, (Z)-
9122 - 1-Pentyne
9123 - 2-Pentyne

11527 - Spiropentane

C_5H_8BrN
8828 - Pentanenitrile, 5-bromo-

$C_5H_8Br_2O_2$
8870 - Pentanoic acid, 2,3-dibromo-
8871 - Pentanoic acid, 2,5-dibromo-
10240 - Propanoic acid, 2,3-dibromo, ethyl ester
10242 - Propanoic acid, 2,3-dibromo-2-methyl-,
methyl ester

$C_5H_8Br_4$
9884 - Propane, 1,3-dibromo-2,2-bis(bromomethyl)-

$C_5H_8ClFO_2$
135 - Acetic acid, chlorofluoro-, propyl ester

$C_5H_8Cl_2$
3729 - 1-Butene, 3-chloro-2-(chloromethyl)-
3755 - 1-Butene, 3,3-dichloro-2-methyl-
3756 - 2-Butene, 1,3-dichloro-2-methyl-
3757 - 2-Butene, 1,4-dichloro-2-methyl-
9028 - 2-Pentene, 2,5-dichloro-

$C_5H_8Cl_2O$
3686 - Butanoyl chloride, 2-chloro-3-methyl-
8963 - 2-Pentanone, 3,5-dichloro-
8991 - Pentanoyl chloride, 2-chloro-
8992 - Pentanoyl chloride, 5-chloro-

$C_5H_8Cl_2O_2$
167 - Acetic acid, dichloro-, 1-methylethyl ester
3502 - Butanoic acid, 3,4-dichloro-, methyl ester
10246 - Propanoic acid, 2,3-dichloro-, ethyl ester
10363 - 2-Propanol, 1,3-dichloro-, acetate

$C_5H_8Cl_2O_3$
5697 - Ethanol, 2-chloro-, carbonate (2:1)

$C_5H_8Cl_4$
8838 - Pentane, 1,1,1,5-tetrachloro-
9895 - Propane, 1,3-dichloro-2,2-bis(chloromethyl)-

$C_5H_8NNaO_4$
6248 - L-Glutamic acid, monosodium salt

$C_5H_8N_2$
7200 - 1H-Imidazole, 1,2-dimethyl-
7201 - 1H-Imidazole, 1,4-dimethyl-
7207 - 1H-Imidazole, 1-ethyl-
10825 - 1H-Pyrazole, 1,3-dimethyl-
10826 - 1H-Pyrazole, 1,5-dimethyl-
10827 - 1H-Pyrazole, 3,4-dimethyl-
10828 - 1H-Pyrazole, 3,5-dimethyl-
10832 - 1H-Pyrazole, 1-ethyl-

$C_5H_8N_2O$
7205 - 1H-Imidazole-1-ethanol

$C_5H_8N_2O_2$
7218 - 2,4-Imidazolidinedione, 5,5-dimethyl-
10236 - Propanoic acid, 2-diazo-, ethyl ester

$C_5H_8N_2O_3$
40 - Acetamide, N,N'-carbonylbis-

$C_5H_8N_4O_{12}$
9990 - 1,3-Propanediol, 2,2-bis[(nitrooxy)methyl]-,
dinitrate (ester)

C_5H_8O
3167 - 1,2-Butadiene, 4-methoxy-
3168 - 1,3-Butadiene, 1-methoxy-
3169 - 1,3-Butadiene, 2-methoxy-
3698 - 2-Butenal, 2-methyl-
3699 - 2-Butenal, 2-methyl-, (E)-
3700 - 2-Butenal, 3-methyl-
3879 - 3-Buten-2-one, 3-methyl-
3915 - 2-Butyne, 1-methoxy-
3922 - 3-Butyn-2-ol, 2-methyl-
4878 - Cyclopentanone
5796 - Ethanone, 1-cyclopropyl-
8558 - 6-Oxabicyclo[3.1.0]hexane
8681 - 1,4-Pentadien-3-ol
9001 - 4-Pentenal
9104 - 1-Penten-3-one
9105 - 3-Penten-2-one, (E)-
9140 - 1-Pentyn-3-ol
9141 - 2-Pentyn-1-ol
9142 - 3-Pentyn-1-ol
9143 - 4-Pentyn-1-ol
10045 - Propane, 1-(ethynyloxy)-

10565 - 1-Propene, 3-(ethenyloxy)-
10732 - 1-Propyne, 1-ethoxy-
10733 - 1-Propyne, 3-ethoxy-
10777 - 2H-Pyran, 3,4-dihydro-

$C_5H_8O_2$
235 - Acetic acid, 2-propenyl ester
3844 - 2-Butenoic acid, 2-methyl-, (E)-
3845 - 2-Butenoic acid, 2-methyl-, (Z)-
3846 - 2-Butenoic acid, 3-methyl-
3847 - 2-Butenoic acid, methyl ester, (E)-
3848 - 2-Butenoic acid, methyl ester, (Z)-
4174 - Cyclobutanecarboxylic acid
4963 - Cyclopropanecarboxylic acid, methyl ester
4964 - Cyclopropanecarboxylic acid, 2-methyl-
6049 - 2-Furancarboxaldehyde, tetrahydro-
6145 - 2(3H)-Furanone, dihydro-3-methyl-
6146 - 2(3H)-Furanone, dihydro-5-methyl-, (±)-
8695 - Pentanal, 2-oxo-
8696 - Pentanal, 4-oxo-
8734 - Pentanedial
8797 - 2,3-Pentanedione
8798 - 2,4-Pentanedione
9073 - 4-Pentenoic acid
10261 - Propanoic acid, ethenyl ester
10635 - 2-Propenoic acid, ethyl ester
10663 - 2-Propenoic acid, 2-methyl-, methyl ester
10698 - 1-Propen-2-ol, acetate
10798 - 2H-Pyran-2-one, tetrahydro-
10799 - 4H-Pyran-4-one, tetrahydro-

$C_5H_8O_3$
3556 - Butanoic acid, 3-methyl-2-oxo-
3573 - Butanoic acid, 3-oxo-, methyl ester
6072 - 2-Furancarboxylic acid, tetrahydro-
8902 - Pentanoic acid, 2-oxo-
8903 - Pentanoic acid, 4-oxo-
10315 - Propanoic acid, 2-oxo-, ethyl ester
10412 - 2-Propanone, 1-(acetyloxy)-

$C_5H_8O_4$
3325 - Butanedioic acid, methyl-, (±)-
3336 - Butanedioic acid, monomethyl ester
5567 - Ethanedioic acid, ethyl methyl ester
8770 - Pentanedioic acid
9943 - Propanedioic acid, dimethyl ester
9948 - Propanedioic acid, ethyl-
10202 - Propanoic acid, 2-(acetyloxy)-, (±)-

C_5H_9Br
3716 - 1-Butene, 2-bromo-3-methyl-
3717 - 2-Butene, 1-bromo-3-methyl-
3718 - 2-Butene, 2-bromo-3-methyl-
4777 - Cyclopentane, bromo-
9007 - 1-Pentene, 1-bromo-
9008 - 1-Pentene, 2-bromo-
9009 - 1-Pentene, 5-bromo-
9010 - 1-Pentene, 5-bromo-
9011 - 2-Pentene, 1-bromo-
9012 - 2-Pentene, 2-bromo-
9013 - 2-Pentene, 4-bromo-
9014 - 2-Pentene, 4-bromo-
9015 - 2-Pentene, 5-bromo-

C_5H_9BrO
6036 - Furan, 2-(bromomethyl)tetrahydro-

$C_5H_9BrO_2$
120 - Acetic acid, bromo-, propyl ester
3471 - Butanoic acid, 2-bromo-3-methyl-, (±)-
3472 - Butanoic acid, 3-bromo-3-methyl-
3473 - Butanoic acid, 2-bromo-, methyl ester
3474 - Butanoic acid, 4-bromo-, methyl ester
8859 - Pentanoic acid, 2-bromo-, (±)-
10207 - Propanoic acid, 2-bromo-, ethyl ester
10208 - Propanoic acid, 3-bromo-, ethyl ester

C_5H_9Cl
3732 - 1-Butene, 1-chloro-2-methyl-
3733 - 1-Butene, 1-chloro-3-methyl-
3734 - 1-Butene, 3-chloro-2-methyl-
3735 - 2-Butene, 1-chloro-2-methyl-
3736 - 2-Butene, 1-chloro-3-methyl-
3737 - 2-Butene, 2-chloro-3-methyl-

4789 - Cyclopentane, chloro-
9016 - 1-Pentene, 2-chloro-
9017 - 1-Pentene, 3-chloro-
9018 - 1-Pentene, 4-chloro-
9019 - 1-Pentene, 5-chloro-
9020 - 2-Pentene, 1-chloro-
9021 - 2-Pentene, 2-chloro-
9022 - 2-Pentene, 3-chloro-
9023 - 2-Pentene, 4-chloro-
9024 - 2-Pentene, 5-chloro-

C_5H_9ClO
3603 - 1-Butanol, 3-chloro-
3649 - 2-Butanone, 3-chloro-3-methyl-
3693 - Butanoyl chloride, 3-methyl-
3694 - Butanoyl chloride, 2-methyl-, (±)-
8961 - 2-Pentanone, 5-chloro-
8962 - 3-Pentanone, 1-chloro-
8990 - Pentanoyl chloride
10503 - Propanoyl chloride, 2,2-dimethyl-

$C_5H_9ClO_2$
139 - Acetic acid, chloro-, 1-methylethyl ester
149 - Acetic acid, chloro-, propyl ester
3486 - Butanoic acid, 2-chloro-2-methyl-
3487 - Butanoic acid, 2-chloro-3-methyl-
3488 - Butanoic acid, 2-chloro-, methyl ester
3489 - Butanoic acid, 4-chloro-, methyl ester
4012 - Carbonochloridic acid, butyl ester
4024 - Carbonochloridic acid, 2-methylpropyl ester
8863 - Pentanoic acid, 2-chloro-, (±)-
8864 - Pentanoic acid, 3-chloro-
8865 - Pentanoic acid, 4-chloro-
8866 - Pentanoic acid, 5-chloro-
10219 - Propanoic acid, 3-chloro-2,2-dimethyl-
10220 - Propanoic acid, 2-chloro-, ethyl ester
10221 - Propanoic acid, 3-chloro-, ethyl ester
10229 - Propanoic acid, 2-chloro-2-methyl-, methyl ester
10350 - 1-Propanol, 3-chloro-, acetate
10351 - 2-Propanol, 1-chloro-, acetate

$C_5H_9ClO_3$
137 - Acetic acid, chloro-, 2-methoxyethyl ester
4017 - Carbonochloridic acid, 2-ethoxyethyl ester

$C_5H_9Cl_2N_3O_2$
11994 - Urea, N,N'-bis(2-chloroethyl)-N-nitroso-

$C_5H_9Cl_3$
3453 - Butane, 1,2,3-trichloro-2-methyl-
3454 - Butane, 2,2,3-trichloro-3-methyl-

$C_5H_9FO_2$
8879 - Pentanoic acid, 5-fluoro-

C_5H_9I
4824 - Cyclopentane, iodo-

$C_5H_9IO_2$
10283 - Propanoic acid, 3-iodo-, ethyl ester

C_5H_9N
3387 - Butane, 1-isocyano-
3409 - Butanenitrile, 2-methyl-
3410 - Butanenitrile, 3-methyl-
3896 - 3-Butyn-2-amine, 2-methyl-
8827 - Pentanenitrile
10100 - Propanenitrile, 2,2-dimethyl-
11020 - Pyridine, 1,2,3,6-tetrahydro-
11189 - 2-Pyrroline, 2-methyl-

C_5H_9NO
4892 - Cyclopentanone, oxime
8977 - 2-Pentanone, 4-imino-
9719 - 2-Piperidinone
10061 - Propane, 1-isocyanato-2-methyl-
10062 - Propane, 2-isocyanato-2-methyl-
10102 - Propanenitrile, 2-ethoxy-
10103 - Propanenitrile, 3-ethoxy-
11185 - 2-Pyrrolidinone, 1-methyl-
11186 - 2-Pyrrolidinone, 5-methyl-
11187 - 3-Pyrrolidinone, 1-methyl-

$C_5H_9NO_2$
25 - Acetamide, N-acetyl-N-methyl-
4839 - Cyclopentane, nitro-
7685 - 4-Morpholinecarboxaldehyde

8803 - 2,3-Pentanedione, 3-oxime
9764 - L-Proline
9765 - DL-Proline

$C_5H_9NO_2S_2$
10067 - Propane, 1-isothiocyanato-3-(methylsulfonyl)-

$C_5H_9NO_3$
9767 - L-Proline, 4-hydroxy-, trans-

$C_5H_9NO_3S$
5015 - L-Cysteine, N-acetyl-

$C_5H_9NO_4$
6240 - DL-Glutamic acid
6241 - L-Glutamic acid
10312 - Propanoic acid, 2-nitro-, ethyl ester

$C_5H_9NO_4S$
5017 - L-Cysteine, S-(carboxymethyl)-

$C_5H_9NO_5$
6246 - DL-Glutamic acid, 3-hydroxy-

C_5H_9NS
3388 - Butane, 1-isothiocyanato-
3389 - Butane, 2-isothiocyanato-, (±)-
10065 - Propane, 1-isothiocyanato-2-methyl-
10066 - Propane, 2-isothiocyanato-2-methyl-
11699 - Thiocyanic acid, butyl ester
11701 - Thiocyanic acid, 1,1-dimethylethyl ester
11707 - Thiocyanic acid, 2-methylpropyl ester

$C_5H_9N_3$
7203 - 1H-Imidazole-4-ethanamine

C_5H_{10}
3799 - 1-Butene, 2-methyl-
3800 - 1-Butene, 3-methyl-
3801 - 2-Butene, 2-methyl-
4192 - Cyclobutane, methyl-
4774 - Cyclopentane
4971 - Cyclopropane, 1,1-dimethyl-
4972 - Cyclopropane, 1,2-dimethyl-, cis-
4973 - Cyclopropane, 1,2-dimethyl-, trans-
4974 - Cyclopropane, ethyl-
9004 - 1-Pentene
9005 - 2-Pentene, (E)-
9006 - 2-Pentene, (Z)-

$C_5H_{10}BrF$
8713 - Pentane, 1-bromo-5-fluoro-

$C_5H_{10}Br_2$
3272 - Butane, 1,2-dibromo-2-methyl-
3273 - Butane, 1,4-dibromo-2-methyl-, (R)-
3274 - Butane, 2,3-dibromo-2-methyl-
8739 - Pentane, 1,3-dibromo-
8740 - Pentane, 1,4-dibromo-
8741 - Pentane, 1,5-dibromo-
8742 - Pentane, 2,3-dibromo-
8743 - Pentane, 2,4-dibromo-
9886 - Propane, 1,1-dibromo-2,2-dimethyl-
9887 - Propane, 1,3-dibromo-2,2-dimethyl-

$C_5H_{10}ClF$
8728 - Pentane, 1-chloro-5-fluoro-

$C_5H_{10}ClNO$
3972 - Carbamic chloride, diethyl-

$C_5H_{10}Cl_2$
3282 - Butane, 1,1-dichloro-3-methyl-
3283 - Butane, 1,2-dichloro-2-methyl-
3284 - Butane, 1,3-dichloro-3-methyl-
3285 - Butane, 1,4-dichloro-2-methyl-
3286 - Butane, 2,3-dichloro-2-methyl-
8744 - Pentane, 1,2-dichloro-
8745 - Pentane, 1,3-dichloro-
8746 - Pentane, 1,4-dichloro-
8747 - Pentane, 1,5-dichloro-
8748 - Pentane, 2,2-dichloro-
8749 - Pentane, 2,3-dichloro-
8750 - Pentane, 2,4-dichloro-
8751 - Pentane, 3,3-dichloro-

$C_5H_{10}I_2$
8758 - Pentane, 1,5-diiodo-
8759 - Pentane, 2,4-diiodo-

$C_5H_{10}N_2$
4156 - Cyanamide, diethyl-

10101 - Propanenitrile, 3-(dimethylamino)-

$C_5H_{10}N_2O$
9688 - Piperidine, 1-nitroso-

$C_5H_{10}N_2O_2$
8800 - 2,3-Pentanedione, dioxime
9687 - Piperidine, 1-nitro-

$C_5H_{10}N_2O_2S$
7650 - Methomyl

$C_5H_{10}N_2O_3$
6249 - L-Glutamine

$C_5H_{10}N_2O_4$
3950 - Carbamic acid, ethylnitro-, ethyl ester
8769 - Pentane, 1,5-dinitro-

$C_5H_{10}O$
3185 - Butanal, 2-methyl-, (±)-
3186 - Butanal, 3-methyl-
3668 - 2-Butanone, 3-methyl-
3869 - 2-Buten-1-ol, 3-methyl-
3870 - 3-Buten-2-ol, 2-methyl-
3871 - 3-Buten-2-ol, 3-methyl-
4864 - Cyclopentanol
4979 - Cyclopropanemethanol, α-methyl-
6171 - Furan, tetrahydro-2-methyl-
6172 - Furan, tetrahydro-3-methyl-
8593 - Oxetane, 3,3-dimethyl-
8690 - Pentanal
8957 - 2-Pentanone
8958 - 3-Pentanone
9087 - 1-Penten-3-ol
9088 - 4-Penten-1-ol
9089 - 4-Penten-2-ol
9090 - 2-Penten-1-ol, (E)-
9091 - 2-Penten-1-ol, (Z)-
9092 - 3-Penten-2-ol, (±)-
9783 - Propanal, 2,2-dimethyl-
10033 - Propane, 1-(ethenyloxy)-
10034 - Propane, 2-(ethenyloxy)-
10566 - 1-Propene, 3-ethoxy-
10577 - 1-Propene, 3-methoxy-2-methyl-
10800 - 2H-Pyran, tetrahydro-

$C_5H_{10}OS_2$
4031 - Carbonodithioic acid, S,S-diethyl ester

$C_5H_{10}O_2$
209 - Acetic acid, 1-methylethyl ester
236 - Acetic acid, propyl ester
3538 - Butanoic acid, 2-methyl-, (±)-
3539 - Butanoic acid, 3-methyl-
3547 - Butanoic acid, methyl ester
3664 - 2-Butanone, 4-hydroxy-3-methyl-
4806 - 1,3-Cyclopentanediol, cis-
4807 - 1,3-Cyclopentanediol, trans-
5193 - 1,3-Dioxane, 4-methyl-
5763 - Ethanol, 2-(2-propenyloxy)-
6004 - Formic acid, butyl ester
6012 - Formic acid, 1-methylpropyl ester
6013 - Formic acid, 2-methylpropyl ester
6124 - 2-Furanmethanol, tetrahydro-
8614 - Oxirane, (ethoxymethyl)-
8855 - Pentanoic acid
8971 - 2-Pentanone, 1-hydroxy-
8972 - 2-Pentanone, 3-hydroxy-
8973 - 2-Pentanone, 4-hydroxy-
8974 - 2-Pentanone, 5-hydroxy-
8975 - 3-Pentanone, 2-hydroxy-
10258 - Propanoic acid, 2,2-dimethyl-
10263 - Propanoic acid, ethyl ester
10301 - Propanoic acid, 2-methyl-, methyl ester
10562 - 1-Propene, 3,3-dimethoxy-
10787 - 2H-Pyran-4-ol, tetrahydro-

$C_5H_{10}O_2S$
10311 - Propanoic acid, 3-(methylthio)-, methyl ester
11789 - Thiophene, tetrahydro-3-methyl-, 1,1-dioxide

$C_5H_{10}O_3$
182 - Acetic acid, ethoxy-, methyl ester
199 - Acetic acid, hydroxy-, propyl ester
206 - Acetic acid, methoxy-, ethyl ester
3535 - Butanoic acid, 3-hydroxy-3-methyl-

3999 - Carbonic acid, diethyl ester
5197 - 1,3-Dioxan-5-ol, 2-methyl-
5208 - 1,3-Dioxolane-4-methanol, 2-methyl-
5734 - Ethanol, methoxy-, acetate
5735 - Ethanol, 2-methoxy-, acetate
8883 - Pentanoic acid, 2-hydroxy-
10269 - Propanoic acid, 2-hydroxy-, ethyl ester, (±)-
10270 - Propanoic acid, 3-hydroxy-, ethyl ester
10280 - Propanoic acid, 2-hydroxy-2-methyl-, methyl ester
10289 - Propanoic acid, 3-methoxy-, methyl ester

$C_5H_{10}O_4$
10187 - 1,2,3-Propanetriol, 1-acetate, (±)-
10253 - Propanoic acid, 2,3-dihydroxy-, ethyl ester, (±)-
10256 - Propanoic acid, 2,3-dihydroxy-2-methyl-, methyl ester
10272 - Propanoic acid, 3-hydroxy-2-(hydroxymethyl)-2-methyl-

$C_5H_{10}O_5$
436 - β-D-Arabinopyranose
11345 - D-Ribose
12059 - D-Xylose

$C_5H_{10}S$
4852 - Cyclopentanethiol
11787 - Thiophene, tetrahydro-2-methyl-
11788 - Thiophene, tetrahydro-3-methyl-
11795 - 2H-Thiopyran, tetrahydro-

$C_5H_{11}Br$
3238 - Butane, 1-bromo-2-methyl-, (S)-
3239 - Butane, 1-bromo-3-methyl-
3240 - Butane, 2-bromo-2-methyl-
8710 - Pentane, 1-bromo-
8711 - Pentane, 2-bromo-
8712 - Pentane, 3-bromo-
9843 - Propane, 1-bromo-2,2-dimethyl-

$C_5H_{11}Cl$
3251 - Butane, 1-chloro-2-methyl-, (±)-
3252 - Butane, 1-chloro-3-methyl-
3253 - Butane, 2-chloro-2-methyl-
8719 - Pentane, 1-chloro-
8720 - Pentane, 2-chloro-, (+)-
8721 - Pentane, 3-chloro-
9853 - Propane, 1-chloro-2,2-dimethyl-

$C_5H_{11}ClO$
3610 - 2-Butanol, 1-chloro-2-methyl-
3611 - 2-Butanol, 3-chloro-2-methyl-
8923 - 1-Pentanol, 5-chloro-
8924 - 2-Pentanol, 1-chloro-
8925 - 3-Pentanol, 1-chloro-
9854 - Propane, 1-(1-chloroethyl)-

$C_5H_{11}ClO_2$
9852 - Propane, 3-chloro-1,1-dimethoxy-

$C_5H_{11}ClSi$
11375 - Silane, chlorodimethyl-2-propenyl-

$C_5H_{11}Cl_2N$
5413 - Ethanamine, 2-chloro-N-(2-chloroethyl)-N-methyl-

$C_5H_{11}Cl_3Si$
11472 - Silane, trichloropentyl-

$C_5H_{11}F$
8817 - Pentane, 1-fluoro-

$C_5H_{11}FO$
8935 - 1-Pentanol, 5-fluoro-

$C_5H_{11}I$
3383 - Butane, 1-iodo-2-methyl-, (S)-
3384 - Butane, 1-iodo-3-methyl-
3385 - Butane, 2-iodo-2-methyl-
8818 - Pentane, 1-iodo-
8819 - Pentane, 2-iodo-, (±)-
8820 - Pentane, 3-iodo-
10056 - Propane, 1-iodo-2,2-dimethyl-

$C_5H_{11}N$
4771 - Cyclopentanamine
9629 - Piperidine
11168 - Pyrrolidine, 1-methyl-

$C_5H_{11}NO$
3187 - Butanal, 3-methyl-, oxime
3206 - Butanamide, 3-methyl-
5991 - Formamide, N,N-diethyl-
6116 - 2-Furanmethanamine, tetrahydro-
7696 - Morpholine, 4-methyl-
8698 - Pentanamide
8985 - 2-Pentanone, oxime

$C_5H_{11}NO_2$
3400 - Butane, 3-methyl-1-nitro-
3935 - Carbamic acid, butyl ester
3941 - Carbamic acid, diethyl-
3949 - Carbamic acid, ethyl-, ethyl ester
3959 - Carbamic acid, 2-methylpropyl ester
7400 - DL-Isovaline
7401 - L-Isovaline
7454 - Methanaminium, 1-carboxy-N,N,N-trimethyl-, hydroxide, inner salt
8090 - Nitrous acid, 3-methylbutyl ester
8096 - Nitrous acid, pentyl ester
8217 - DL-Norvaline
8218 - L-Norvaline
8832 - Pentane, 1-nitro-
8833 - Pentane, 2-nitro-
8834 - Pentane, 3-nitro-
8857 - Pentanoic acid, 5-amino-
12040 - DL-Valine
12041 - L-Valine

$C_5H_{11}NO_2S$
7647 - DL-Methionine
7648 - L-Methionine
12043 - D-Valine, 3-mercapto-

$C_5H_{11}NO_3$
3637 - 1-Butanol, 3-methyl-, nitrate
3638 - 1-Butanol, 3-methyl-2-nitro-
8950 - 1-Pentanol, 1-nitro-
8951 - 2-Pentanol, 1-nitro-
8952 - 2-Pentanol, 3-nitro-

$C_5H_{11}NO_4$
10008 - 1,3-Propanediol, 2-ethyl-2-nitro-

C_5H_{12}
3397 - Butane, 2-methyl-
8709 - Pentane
9910 - Propane, 2,2-dimethyl-

$C_5H_{12}NO_3PS_2$
5009 - Cygon

$C_5H_{12}N_2$
9618 - Piperazine, 1-methyl-
9619 - Piperazine, 2-methyl-
9624 - 1-Piperidinamine

$C_5H_{12}N_2O$
12001 - Urea, N,N'-diethyl-
12002 - Urea, N,N-diethyl-
12030 - Urea, tetramethyl-

$C_5H_{12}N_2O_2$
8550 - L-Ornithine

$C_5H_{12}N_2S$
11803 - Thiourea, N,N'-diethyl-
11810 - Thiourea, tetramethyl-

$C_5H_{12}N_4O_3$
7135 - L-Homoserine, O-[(aminoiminomethyl)amino]-

$C_5H_{12}O$
3391 - Butane, 1-methoxy-
3392 - Butane, 2-methoxy-, (S)-
3393 - Butane, 2-methoxy-, (±)-
3626 - 1-Butanol, 2-methyl-
3627 - 1-Butanol, 3-methyl-
3628 - 2-Butanol, 2-methyl-
3629 - 2-Butanol, 3-methyl-, (±)-
8915 - 1-Pentanol
8916 - 2-Pentanol
8917 - 3-Pentanol
10037 - Propane, 1-ethoxy-
10038 - Propane, 2-ethoxy-
10071 - Propane, 1-methoxy-2-methyl-
10072 - Propane, 2-methoxy-2-methyl-
10373 - 1-Propanol, 2,2-dimethyl-

$C_5H_{12}O_2$
3352 - 1,2-Butanediol, 3-methyl-
3353 - 1,3-Butanediol, 2-methyl-
3354 - 1,3-Butanediol, 3-methyl-
3355 - 2,3-Butanediol, 2-methyl-
3625 - 1-Butanol, 3-methoxy-
5591 - Ethane, 1-ethoxy-2-methoxy-
5603 - Ethane, 1,1'-[methylenebis(oxy)]bis-
5740 - Ethanol, 2-(1-methylethoxy)-
5764 - Ethanol, 2-propoxy-
8785 - 1,3-Pentanediol
8786 - 1,4-Pentanediol
8787 - 1,5-Pentanediol
8788 - 2,3-Pentanediol
8789 - 2,4-Pentanediol
8790 - 1,2-Pentanediol, (±)-
9909 - Propane, 2,2-dimethoxy-
10005 - 1,3-Propanediol, 2,2-dimethyl-
10380 - 1-Propanol, 2-ethoxy-
10381 - 2-Propanol, 1-ethoxy-

$C_5H_{12}O_3$
5669 - Ethane, 1,1,1-trimethoxy-
5737 - Ethanol, 2-(2-methoxyethoxy)-
8851 - 1,2,5-Pentanetriol
8852 - 1,3,5-Pentanetriol
9119 - Pentitol, 1,2-dideoxy-
10011 - 1,3-Propanediol, 2-(hydroxymethyl)-2-methyl-
10372 - 2-Propanol, 1,3-dimethoxy-

$C_5H_{12}O_4$
7543 - Methane, tetramethoxy-
9989 - 1,3-Propanediol, 2,2-bis(hydroxymethyl)-

$C_5H_{12}O_5$
11343 - Ribitol
12058 - Xylitol

$C_5H_{12}S$
3402 - Butane, 1-(methylthio)-
3439 - 1-Butanethiol, 2-methyl-, (S)-
3440 - 1-Butanethiol, 3-methyl-
3441 - 2-Butanethiol, 2-methyl-
8844 - 1-Pentanethiol
8845 - 2-Pentanethiol
8846 - 3-Pentanethiol
10042 - Propane, 1-(ethylthio)-
10043 - Propane, 2-(ethylthio)-
10082 - Propane, 2-methyl-1-(methylthio)-
10083 - Propane, 2-methyl-2-(methylthio)-

$C_5H_{12}Si$
11425 - Silane, ethenyltrimethyl-

$C_5H_{13}Cl_2N$
5442 - Ethanaminium, 2-chloro-N,N,N-trimethyl-, chloride

$C_5H_{13}N$
3220 - 1-Butanamine, 3-methyl-
3221 - 1-Butanamine, N-methyl-
3222 - 2-Butanamine, 2-methyl-
3223 - 2-Butanamine, 3-methyl-
5430 - Ethanamine, N-ethyl-N-methyl-
8700 - 1-Pentanamine
8701 - 2-Pentanamine
8702 - 3-Pentanamine
9806 - 1-Propanamine, 2,2-dimethyl-
9807 - 1-Propanamine, N,N-dimethyl-
9809 - 1-Propanamine, N-ethyl-
9810 - 2-Propanamine, N-ethyl-

$C_5H_{13}NO$
3599 - 1-Butanol, 2-amino-3-methyl-, (±)-
5426 - Ethanamine, 2-ethoxy-N-methyl-
5741 - Ethanol, 2-[(1-methylethyl)amino]-
5765 - Ethanol, 2-(propylamino)-
5878 - Ethanaminium, N,N,N-trimethyl-, hydroxide
8920 - 1-Pentanol, 5-amino-
8921 - 1-Pentanol, 2-amino-
10374 - 1-Propanol, 2-(dimethylamino)-
10375 - 1-Propanol, 3-(dimethylamino)-
10376 - 2-Propanol, 1-(dimethylamino)-

$C_5H_{13}NO_2$
5420 - Ethanamine, 2,2-dimethoxy-N-methyl-
5742 - Ethanol, 2,2'-(methylimino)bis-
9985 - 1,3-Propanediol, 2-amino-2-ethyl-
10383 - 2-Propanol, 1-[(2-hydroxyethyl)amino]-
$C_5H_{13}O_3P$
9539 - Phosphonic acid, methyl-, diethyl ester
$C_5H_{14}ClNO$
5446 - Ethanaminium, 2-hydroxy-N,N,N-trimethyl-, chloride
$C_5H_{14}N_2$
7490 - Methanediamine, N,N,N',N'-tetramethyl-
8735 - 1,5-Pentanediamine
$C_5H_{14}N_2O$
10340 - 2-Propanol, 1-[(2-aminoethyl)amino]-
$C_5H_{14}OSi$
11430 - Silane, ethoxytrimethyl-
$C_5H_{14}O_3Si$
11433 - Silane, ethyltrimethoxy-
$C_5H_{15}NSi$
11361 - Silanamine, pentamethyl-
$C_5H_{20}O_5Si_5$
4902 - Cyclopentasiloxane, 2,4,6,8,10-pentamethyl-
C_6BrF_5
1078 - Benzene, bromopentafluoro-
$C_6Br_2F_4$
1312 - Benzene, 1,2-dibromo-3,4,5,6-tetrafluoro-
C_6ClF_5
1220 - Benzene, chloropentafluoro-
$C_6Cl_2F_4$
1398 - Benzene, 1,2-dichloro-3,4,5,6-tetrafluoro
1399 - Benzene, 1,3-dichloro-2,4,5,6-tetrafluoro-
1400 - Benzene, 1,4-dichloro-2,3,5,6-tetrafluoro-
$C_6Cl_2F_8$
4647 - Cyclohexene, 1,2-dichloro-3,3,4,4,5,5,6,6-octafluoro-
$C_6Cl_3F_3$
2340 - Benzene, 1,3,5-trichloro-2,4,6-trifluoro-
$C_6Cl_4O_2$
4268 - 2,5-Cyclohexadiene-1,4-dione, 2,3,5,6-tetrachloro-
$C_6Cl_5NO_2$
2087 - Benzene, pentachloronitro-
C_6Cl_6
1749 - Benzene, hexachloro-
C_6Cl_{10}
6689 - 1,5-Hexadiene, 1,1,2,3,3,4,4,5,6,6-decachloro-
C_6F_5I
2092 - Benzene, pentafluoroiodo-
C_6F_6
1757 - Benzene, hexafluoro-
C_6F_8
4276 - 1,3-Cyclohexadiene, 1,2,3,4,5,5,6,6-octafluoro-
4277 - 1,4-Cyclohexadiene, 1,2,3,3,4,5,6,6-octafluoro-
C_6F_{10}
4643 - Cyclohexene, decafluoro-
C_6F_{12}
4381 - Cyclohexane, dodecafluoro-
7032 - 1-Hexene, 1,1,2,3,3,4,4,5,5,6,6,6-dodecafluoro-
C_6F_{14}
3414 - Butane, 1,1,1,2,3,4,4,4-octafluoro-2,3-bis(trifluoromethyl)-
6843 - Hexane, tetradecafluoro-
8853 - Pentane, 1,1,1,2,2,3,4,4,5,5,5-undecafluoro-3-(trifluoromethyl)-
8854 - Pentane, 1,1,1,2,2,3,3,4,5,5,5-undecafluoro-4-(trifluoromethyl)-
$C_6F_{15}N$
5436 - Ethanamine, 1,1,2,2,2-pentafluoro-N,N-bis(pentafluoroethyl)-
C_6HBr_5
2082 - Benzene, pentabromo-

C_6HBr_5O
9439 - Phenol, pentabromo-
$C_6HClF_8O_2$
6910 - Hexanoic acid, 3,5,6-trichloro-2,2,3,4,4,5,6,6-octafluoro-
$C_6HCl_4NO_2$
2264 - Benzene, 1,2,4,5-tetrachloro-3-nitro-
C_6HCl_5
2084 - Benzene, pentachloro-
C_6HCl_5O
9440 - Phenol, pentachloro-
C_6HF_5
2091 - Benzene, pentafluoro-
C_6HF_5O
9442 - Phenol, pentafluoro-
C_6HF_5S
2304 - Benzenethiol, pentafluoro-
C_6HF_{11}
4498 - Cyclohexane, undecafluoro-
C_6HI_5
2096 - Benzene, pentaiodo-
$C_6H_2Br_2ClNO$
4279 - 2,5-Cyclohexadien-1-one, 2,6-dibromo-4-(chloroimino)-
$C_6H_2Br_3NO_2$
2315 - Benzene, 1,2,3-tribromo-5-nitro-
2316 - Benzene, 1,3,5-tribromo-2-nitro-
$C_6H_2Br_4$
2253 - Benzene, 1,2,3,5-tetrabromo-
2254 - Benzene, 1,2,4,5-tetrabromo-
$C_6H_2Br_4O$
9467 - Phenol, 2,3,4,6-tetrabromo-
$C_6H_2Br_4O_2$
1551 - 1,4-Benzenediol, 2,3,5,6-tetrabromo-
$C_6H_2ClN_3O_6$
1244 - Benzene, 2-chloro-1,3,5-trinitro-
$C_6H_2Cl_2O_4$
4258 - 2,5-Cyclohexadiene-1,4-dione, 2,5-dichloro-3,6-dihydroxy-
$C_6H_2Cl_3NO$
4280 - 2,5-Cyclohexadien-1-one, 2,6-dichloro-4-(chloroimino)-
$C_6H_2Cl_3NO_2$
2338 - Benzene, 1,2,3-trichloro-5-nitro-
2339 - Benzene, 1,2,4-trichloro-5-nitro-
$C_6H_2Cl_3NO_3$
9490 - Phenol, 3,4,6-trichloro-2-nitro-
$C_6H_2Cl_4$
2259 - Benzene, 1,2,3,4-tetrachloro-
2260 - Benzene, 1,2,4,5-tetrachloro-
$C_6H_2Cl_4O$
9469 - Phenol, 2,3,4,5-tetrachloro-
9470 - Phenol, 2,3,4,6-tetrachloro-
9471 - Phenol, 2,3,5,6-tetrachloro-
$C_6H_2Cl_5OPS$
9578 - Phosphorodichloridothioic acid, O-(2,4,5-trichlorophenyl) ester
$C_6H_2F_4$
2270 - Benzene, 1,2,3,4-tetrafluoro-
2271 - Benzene, 1,2,3,5-tetrafluoro-
2272 - Benzene, 1,2,4,5-tetrafluoro-
$C_6H_2I_3NO_2$
2353 - Benzene, 1,2,3-triiodo-5-nitro-
$C_6H_2I_4$
2273 - Benzene, 1,2,3,4-tetraiodo-
2274 - Benzene, 1,2,3,5-tetraiodo-
2275 - Benzene, 1,2,4,5-tetraiodo-
$C_6H_2N_2O_8$
4259 - 2,5-Cyclohexadiene-1,4-dione, 2,5-dihydroxy-3,6-dinitro-
$C_6H_2N_4O_5$
2874 - 1,2,3-Benzoxadiazole, 5,7-dinitro-
$C_6H_2N_4O_9$
9476 - Phenol, 2,3,4,6-tetranitro-
$C_6H_2O_6$
4686 - 5-Cyclohexene-1,2,3,4-tetrone, 5,6-dihydroxy-

$C_6H_3BrCl_2$
1010 - Benzene, 1-bromo-2,3-dichloro-
1011 - Benzene, 1-bromo-2,4-dichloro-
1012 - Benzene, 1-bromo-3,5-dichloro-
1013 - Benzene, 2-bromo-1,3-dichloro-
1014 - Benzene, 2-bromo-1,4-dichloro-
1015 - Benzene, 4-bromo-1,2-dichloro-
$C_6H_3BrCl_2O$
9233 - Phenol, 2-bromo-4,6-dichloro-
$C_6H_3BrN_2O_4$
1024 - Benzene, 1-bromo-2,3-dinitro-
$C_6H_3BrN_2O_5$
9234 - Phenol, 2-bromo-4,6-dinitro-
9235 - Phenol, 4-bromo-2,6-dinitro-
$C_6H_3Br_2Cl$
1292 - Benzene, 1,2-dibromo-3-chloro-
1293 - Benzene, 1,2-dibromo-4-chloro-
1294 - Benzene, 1,3-dibromo-2-chloro-
1295 - Benzene, 1,4-dibromo-2-chloro-
1296 - Benzene, 2,4-dibromo-1-chloro-
$C_6H_3Br_2F$
1300 - Benzene, 2,4-dibromo-1-fluoro-
$C_6H_3Br_2NO_2$
1307 - Benzene, 1,2-dibromo-4-nitro-
1308 - Benzene, 1,3-dibromo-2-nitro-
1309 - Benzene, 1,3-dibromo-5-nitro-
1310 - Benzene, 1,4-dibromo-2-nitro-
1311 - Benzene, 2,4-dibromo-1-nitro-
$C_6H_3Br_3$
2307 - Benzene, 1,2,3-tribromo-
2308 - Benzene, 1,2,4-tribromo-
2309 - Benzene, 1,3,5-tribromo-
$C_6H_3Br_3O$
9479 - Phenol, 2,4,6-tribromo-
$C_6H_3ClNO_3$
9296 - Phenol, 2,6-dichloro-4-nitro-
$C_6H_3ClN_2O_4$
1144 - Benzene, 1-chloro-2,4-dinitro-
1145 - Benzene, 2-chloro-1,3-dinitro-
$C_6H_3ClN_2O_4S$
2178 - Benzenesulfenyl chloride, 2,4-dinitro-
$C_6H_3ClN_2O_5$
9258 - Phenol, 4-chloro-2,6-dinitro-
9259 - Phenol, 5-chloro-2,4-dinitro-
$C_6H_3Cl_2I$
1380 - Benzene, 1,4-dichloro-2-iodo-
$C_6H_3Cl_2NO_2$
1391 - Benzene, 1,2-dichloro-3-nitro-
1392 - Benzene, 1,2-dichloro-4-nitro-
1393 - Benzene, 1,3-dichloro-2-nitro-
1394 - Benzene, 1,3-dichloro-5-nitro-
1395 - Benzene, 1,4-dichloro-2-nitro-
1396 - Benzene, 2,4-dichloro-1-nitro-
4142 - Clopyralid
$C_6H_3Cl_3$
2321 - Benzene, 1,2,3-trichloro-
2322 - Benzene, 1,2,4-trichloro-
2323 - Benzene, 1,3,5-trichloro-
$C_6H_3Cl_3N_2O_2$
10896 - 2-Pyridinecarboxylic acid, 4-amino-3,5,6-trichloro-
$C_6H_3Cl_3O$
9481 - Phenol, 2,3,4-trichloro-
9482 - Phenol, 2,3,5-trichloro-
9483 - Phenol, 2,3,6-trichloro-
9484 - Phenol, 2,4,5-trichloro-
9485 - Phenol, 2,4,6-trichloro-
9486 - Phenol, 3,4,5-trichloro-
$C_6H_3Cl_3O_3$
2371 - 1,3,5-Benzenetriol, 2,4,6-trichloro-
$C_6H_3Cl_4N$
8071 - Nitrapyrin
$C_6H_3FN_2O_4$
1718 - Benzene, 1-fluoro-2,4-dinitro-
$C_6H_3F_2NO_2$
1426 - Benzene, 2,4-difluoro-1-nitro-

$C_6H_3F_3$
2346 - Benzene, 1,2,4-trifluoro-
2347 - Benzene, 1,3,5-trifluoro-
$C_6H_3F_7O_2$
3523 - Butanoic acid, heptafluoro-, ethenyl ester
$C_6H_3I_2NO_3$
9304 - Phenol, 2,6-diiodo-4-nitro-
$C_6H_3I_3$
2349 - Benzene, 1,2,3-triiodo-
2350 - Benzene, 1,2,4-triiodo-
2351 - Benzene, 1,3,5-triiodo-
$C_6H_3I_3O$
9492 - Phenol, 2,4,6-triiodo-
$C_6H_3N_3O_6$
2365 - Benzene, 1,3,5-trinitro-
$C_6H_3N_3O_7$
9497 - Phenol, 2,4,6-trinitro-
$C_6H_3N_3O_8$
1554 - 1,3-Benzenediol, 2,4,6-trinitro-
C_6H_4BrCl
1004 - Benzene, 1-bromo-2-chloro-
1005 - Benzene, 1-bromo-3-chloro-
1006 - Benzene, 1-bromo-4-chloro-
C_6H_4BrClO
9231 - Phenol, 3-bromo-5-chloro-
9232 - Phenol, 4-bromo-2-chloro-
$C_6H_4BrClO_2S$
2244 - Benzenesulfonyl chloride, 4-bromo-
$C_6H_4BrCl_2N$
641 - Benzenamine, 2-bromo-4,6-dichloro-
C_6H_4BrF
1043 - Benzene, 1-bromo-3-fluoro-
1044 - Benzene, 1-bromo-4-fluoro-
C_6H_4BrI
1046 - Benzene, 1-bromo-2-iodo-
1047 - Benzene, 1-bromo-3-iodo-
1048 - Benzene, 1-bromo-4-iodo-
$C_6H_4BrNO_2$
1075 - Benzene, 1-bromo-2-nitro-
1076 - Benzene, 1-bromo-3-nitro-
1077 - Benzene, 1-bromo-4-nitro-
$C_6H_4BrNO_3$
9243 - Phenol, 2-bromo-3-nitro-
9244 - Phenol, 4-bromo-2-nitro-
$C_6H_4BrN_3$
977 - Benzene, 1-azido-4-bromo-
$C_6H_4BrN_3O_4$
646 - Benzenamine, 2-bromo-4,6-dinitro-
$C_6H_4Br_2$
1288 - Benzene, 1,2-dibromo-
1289 - Benzene, 1,3-dibromo-
1290 - Benzene, 1,4-dibromo-
$C_6H_4Br_2O$
9276 - Phenol, 2,4-dibromo-
9277 - Phenol, 2,6-dibromo-
9278 - Phenol, 3,5-dibromo-
$C_6H_4Br_3N$
851 - Benzenamine, 2,4,6-tribromo-
C_6H_4ClF
1166 - Benzene, 1-chloro-2-fluoro-
1167 - Benzene, 1-chloro-3-fluoro-
1168 - Benzene, 1-chloro-4-fluoro-
$C_6H_4ClFO_2S$
2246 - Benzenesulfonyl chloride, 4-fluoro-
C_6H_4ClI
1171 - Benzene, 1-chloro-2-iodo-
1172 - Benzene, 1-chloro-4-iodo-
$C_6H_4ClIO_2S$
2247 - Benzenesulfonyl chloride, 4-iodo-
$C_6H_4ClNO_2$
1214 - Benzene, 1-chloro-2-nitro-
1215 - Benzene, 1-chloro-3-nitro-
1216 - Benzene, 1-chloro-4-nitro-
$C_6H_4ClNO_2S$
2179 - Benzenesulfenyl chloride, 4-nitro-
$C_6H_4ClNO_3$
9269 - Phenol, 5-chloro-2-nitro-

$C_6H_4ClN_3$
978 - Benzene, 1-azido-4-chloro-
$C_6H_4ClO_2P$
2413 - 1,3,2-Benzodioxaphosphole, 2-chloro-
$C_6H_4Cl_2$
1354 - Benzene, 1,2-dichloro-
1355 - Benzene, 1,3-dichloro-
1356 - Benzene, 1,4-dichloro-
$C_6H_4Cl_2N_2O_2$
692 - Benzenamine, 2,6-dichloro-4-nitro-
$C_6H_4Cl_2O$
9282 - Phenol, 2,3-dichloro-
9283 - Phenol, 2,4-dichloro-
9284 - Phenol, 2,5-dichloro-
9285 - Phenol, 2,6-dichloro-
9286 - Phenol, 3,4-dichloro-
9287 - Phenol, 3,5-dichloro-
$C_6H_4Cl_2O_2$
1510 - 1,3-Benzenediol, 4,6-dichloro-
$C_6H_4Cl_2O_2S$
2245 - Benzenesulfonyl chloride, 4-chloro-
$C_6H_4Cl_3N$
852 - Benzenamine, 2,3,4-trichloro-
853 - Benzenamine, 2,4,5-trichloro-
854 - Benzenamine, 2,4,6-trichloro-
11023 - Pyridine, 2-(trichloromethyl)-
$C_6H_4Cl_4Si$
11457 - Silane, trichloro(4-chlorophenyl)-
C_6H_4FI
1719 - Benzene, 1-fluoro-2-iodo-
1720 - Benzene, 1-fluoro-4-iodo-
$C_6H_4FNO_2$
1733 - Benzene, 1-fluoro-2-nitro-
1734 - Benzene, 1-fluoro-3-nitro-
1735 - Benzene, 1-fluoro-4-nitro-
$C_6H_4F_2$
1422 - Benzene, 1,2-difluoro-
1423 - Benzene, 1,3-difluoro-
1424 - Benzene, 1,4-difluoro-
$C_6H_4INO_2$
1783 - Benzene, 1-iodo-2-nitro-
1784 - Benzene, 1-iodo-3-nitro-
1785 - Benzene, 1-iodo-4-nitro-
$C_6H_4IN_3$
979 - Benzene, 1-azido-2-iodo-
$C_6H_4I_2$
1427 - Benzene, 1,2-diiodo-
1428 - Benzene, 1,3-diiodo-
1429 - Benzene, 1,4-diiodo-
$C_6H_4N_2$
10881 - 2-Pyridinecarbonitrile
10882 - 3-Pyridinecarbonitrile
$C_6H_4N_2O_4$
1491 - Benzene, 1,2-dinitro-
1492 - Benzene, 1,3-dinitro-
1493 - Benzene, 1,4-dinitro-
10804 - 2,3-Pyrazinedicarboxylic acid
$C_6H_4N_2O_5$
9341 - Phenol, 2,3-dinitro-
9342 - Phenol, 2,4-dinitro-
9343 - Phenol, 2,5-dinitro-
9344 - Phenol, 2,6-dinitro-
9345 - Phenol, 3,4-dinitro-
9346 - Phenol, 3,5-dinitro-
$C_6H_4N_2O_6$
1517 - 1,3-Benzenediol, 2,4-dinitro-
$C_6H_4N_4$
10747 - Pteridine
$C_6H_4N_4O_2$
2873 - 1H-Benzotriazole, 5-nitro-
10748 - 2,4(1H,3H)-Pteridinedione
$C_6H_4N_4O_6$
863 - Benzenamine, 2,4,6-trinitro-
C_6H_4O
6108 - Furan, 2-ethynyl-
$C_6H_4O_2$
4257 - 2,5-Cyclohexadiene-1,4-dione

$C_6H_4O_4$
10773 - 2H-Pyran-5-carboxylic acid, 2-oxo-
$C_6H_4O_4S$
11745 - 2,5-Thiophenedicarboxylic acid
$C_6H_4O_5$
6079 - 2,3-Furandicarboxylic acid
6080 - 2,4-Furandicarboxylic acid
6081 - 2,5-Furandicarboxylic acid
$C_6H_4O_6$
4269 - 2,5-Cyclohexadiene-1,4-dione, 2,3,5,6-
    tetrahydroxy-
$C_6H_4S_2$
11688 - Thieno[2,3-b]thiophene
$C_6H_5AsCl_2$
462 - Arsonous dichloride, phenyl-
$C_6H_5AsI_2$
464 - Arsonous diiodide, phenyl-
C_6H_5AsO
447 - Arsine, oxophenyl-
$C_6H_5BF_2$
3116 - Borane, difluorophenyl-
$C_6H_5BO_2$
2412 - 1,3,2-Benzodioxaborole
C_6H_5Br
998 - Benzene, bromo-
$C_6H_5BrN_2O_2$
653 - Benzenamine, 2-bromo-6-nitro-
654 - Benzenamine, 4-bromo-2-nitro-
C_6H_5BrO
9228 - Phenol, 2-bromo-
9229 - Phenol, 3-bromo-
9230 - Phenol, 4-bromo-
C_6H_5BrOS
5783 - Ethanone, 1-(5-bromo-2-thienyl)-
$C_6H_5BrO_2$
1502 - 1,3-Benzenediol, 4-bromo-
1503 - 1,4-Benzenediol, 2-bromo-
C_6H_5BrS
2292 - Benzenethiol, 4-bromo-
$C_6H_5Br_2N$
681 - Benzenamine, 2,4-dibromo-
682 - Benzenamine, 2,6-dibromo-
683 - Benzenamine, 3,4-dibromo-
C_6H_5Cl
1131 - Benzene, chloro-
C_6H_5ClFN
664 - Benzenamine, 2-chloro-6-fluoro-
$C_6H_5ClN_2O_2$
677 - Benzenamine, 5-chloro-2-nitro-
C_6H_5ClO
9251 - Phenol, 2-chloro-
9252 - Phenol, 3-chloro-
9253 - Phenol, 4-chloro-
C_6H_5ClOS
2185 - Benzenesulfinyl chloride
5791 - Ethanone, 1-(5-chloro-2-thienyl)-
$C_6H_5ClO_2$
1504 - 1,2-Benzenediol, 3-chloro-
1505 - 1,2-Benzenediol, 4-chloro-
1506 - 1,3-Benzenediol, 5-chloro-
1507 - 1,4-Benzenediol, 2-chloro-
$C_6H_5ClO_2S$
2242 - Benzenesulfonyl chloride
$C_6H_5ClO_3S$
2220 - Benzenesulfonic acid, 4-chloro-
C_6H_5ClS
2293 - Benzenethiol, 2-chloro-
2294 - Benzenethiol, 3-chloro-
2295 - Benzenethiol, 4-chloro-
$C_6H_5ClS_2$
1563 - 1,3-Benzenedithiol, 4-chloro-
$C_6H_5Cl_2N$
685 - Benzenamine, 2,3-dichloro-
686 - Benzenamine, 2,4-dichloro-
687 - Benzenamine, 2,5-dichloro-
688 - Benzenamine, 2,6-dichloro-
689 - Benzenamine, 3,4-dichloro-

690 - Benzenamine, 3,5-dichloro-

$C_6H_5Cl_2NO$
9205 - Phenol, 2-amino-4,6-dichloro-
9206 - Phenol, 4-amino-2,6-dichloro-

$C_6H_5Cl_2OP$
9548 - Phosphonic dichloride, phenyl-

$C_6H_5Cl_2P$
9556 - Phosphonous dichloride, phenyl-

$C_6H_5Cl_2PS$
9552 - Phosphonothioic dichloride, phenyl-

$C_6H_5Cl_2Sb$
11547 - Stibine, dichlorophenyl-

$C_6H_5Cl_3Si$
11473 - Silane, trichlorophenyl-

C_6H_5F
1717 - Benzene, fluoro-

C_6H_5FO
9360 - Phenol, 2-fluoro-
9361 - Phenol, 3-fluoro-
9362 - Phenol, 4-fluoro-

$C_6H_5FO_2S$
2251 - Benzenesulfonyl fluoride

$C_6H_5F_2N$
700 - Benzenamine, 2,4-difluoro-

$C_6H_5F_3Si$
11486 - Silane, trifluorophenyl-

$C_6H_5F_7O_2$
3524 - Butanoic acid, heptafluoro-, ethyl ester

C_6H_5I
1766 - Benzene, iodo-

C_6H_5IO
9371 - Phenol, 2-iodo-
9372 - Phenol, 3-iodo-
9373 - Phenol, 4-iodo-

$C_6H_5IO_2$
1523 - 1,2-Benzenediol, 4-iodo-
1524 - 1,3-Benzenediol, 5-iodo-

$C_6H_5I_2N$
701 - Benzenamine, 2,4-diiodo-

C_6H_5NO
2050 - Benzene, nitroso-
6031 - 2-Furanacetonitrile
10887 - 2-Pyridinecarboxaldehyde
10888 - 3-Pyridinecarboxaldehyde
10889 - 4-Pyridinecarboxaldehyde

C_6H_5NOS
844 - Benzenamine, N-sulfinyl-

$C_6H_5NO_2$
2049 - Benzene, nitro-
9438 - Phenol, 4-nitroso-
10893 - 2-Pyridinecarboxylic acid
10894 - 3-Pyridinecarboxylic acid
10895 - 4-Pyridinecarboxylic acid

$C_6H_5NO_3$
9435 - Phenol, 2-nitro-
9436 - Phenol, 3-nitro-
9437 - Phenol, 4-nitro-
10899 - 3-Pyridinecarboxylic acid, 1,6-dihydro-6-oxo-
10909 - 3-Pyridinecarboxylic acid, 1-oxide

$C_6H_5NO_4$
1541 - 1,3-Benzenediol, 4-nitro-
10900 - 4-Pyridinecarboxylic acid, 1,2-dihydro-6-
hydroxy-2-oxo-

C_6H_5NS
11725 - 2-Thiopheneacetonitrile

$C_6H_5N_3$
976 - Benzene, azido-
2871 - 1H-Benzotriazole

$C_6H_5N_3O_4$
724 - Benzenamine, 2,3-dinitro-
725 - Benzenamine, 2,4-dinitro-
726 - Benzenamine, 2,6-dinitro-
727 - Benzenamine, 3,5-dinitro-

$C_6H_5N_3O_5$
9207 - Phenol, 2-amino-4,6-dinitro-

$C_6H_5N_5O_2$
10749 - 4,6-Pteridinedione, 2-amino-1,5-dihydro-

C_6H_6
867 - Benzene
4767 - 1,3-Cyclopentadiene, 5-methylene-
6712 - 1,3-Hexadien-5-yne
6713 - 1,5-Hexadien-3-yne
6715 - 1,4-Hexadiyne
6716 - 1,5-Hexadiyne
6717 - 2,4-Hexadiyne

$C_6H_6AsNO_5$
459 - Arsonic acid, (4-nitrophenyl)-

C_6H_6BrN
638 - Benzenamine, 2-bromo-
639 - Benzenamine, 3-bromo-
640 - Benzenamine, 4-bromo-

$C_6H_6Br_2N_2$
8768 - Pentanedinitrile, 2-bromo-2-(bromomethyl)-

C_6H_6ClN
658 - Benzenamine, 2-chloro-
659 - Benzenamine, 3-chloro-
660 - Benzenamine, 4-chloro-
10919 - Pyridine, 2-chloro-4-methyl-
10920 - Pyridine, 2-chloro-6-methyl-

C_6H_6ClNO
9204 - Phenol, 2-amino-4-chloro-
10918 - Pyridine, 2-chloro-6-methoxy-

$C_6H_6ClNO_4$
1559 - 1,3-Benzenedisulfonamide, 4,5-dichloro-

$C_6H_6Cl_2Si$
11403 - Silane, dichlorophenyl-

$C_6H_6Cl_4$
6697 - 2,4-Hexadiene, 1,3,4,6-tetrachloro-

$C_6H_6Cl_6$
4406 - Cyclohexane, 1,2,3,4,5,6-hexachloro-,
(1α,2β,3α,4β,5α,6β)-
4407 - Cyclohexane, 1,2,3,4,5,6-hexachloro-,
(1α,2α,3α,4β,5α,6β)-
4408 - Cyclohexane, 1,2,3,4,5,6-hexachloro-,
(1α,2α,3β,4α,5β,6β)-(±)-
4409 - Cyclohexane, 1,2,3,4,5,6-hexachloro-,
(1α,2α,3β,4α,5α,6β)-
7035 - 3-Hexene, 1,2,3,4,5,6-hexachloro-

C_6H_6FN
752 - Benzenamine, 2-fluoro-
753 - Benzenamine, 3-fluoro-
754 - Benzenamine, 4-fluoro-

$C_6H_6FNO_2S$
2252 - Benzenesulfonyl fluoride, 4-amino-

C_6H_6IN
760 - Benzenamine, 3-iodo-
761 - Benzenamine, 4-iodo-

$C_6H_6N_2O$
828 - Benzenamine, 4-nitroso-
9914 - Propanedinitrile, (ethoxymethylene)-
10890 - 3-Pyridinecarboxamide

$C_6H_6N_2O_2$
822 - Benzenamine, 2-nitro-
823 - Benzenamine, 3-nitro-
824 - Benzenamine, 4-nitro-
825 - Benzenamine, N-nitro-
10897 - 3-Pyridinecarboxylic acid, 6-amino-

$C_6H_6N_2O_3$
9214 - Phenol, 2-amino-3-nitro-

$C_6H_6N_4O_3$
10765 - 1H-Purine-2,6,8(3H)-trione, 4,9-dihydro-3-
methyl-
10766 - 1H-Purine-2,6,8(3H)-trione, 7,9-dihydro-7-
methyl-

$C_6H_6N_4O_4$
7144 - Hydrazinecarboxamide, 2-[(5-nitro-2-
furanyl)methylene]-
7156 - Hydrazine, (2,4-dinitrophenyl)-

C_6H_6O
6102 - Furan, 2-ethenyl-
9199 - Phenol

C_6H_6OS
5870 - Ethanone, 1-(2-thienyl)-
9376 - Phenol, 2-mercapto-

9377 - Phenol, 3-mercapto-
9378 - Phenol, 4-mercapto-
11736 - 2-Thiophenecarboxaldehyde, 5-methyl-

$C_6H_6O_2$
1497 - 1,2-Benzenediol
1498 - 1,3-Benzenediol
1499 - 1,4-Benzenediol
5817 - Ethanone, 1-(2-furanyl)-
6045 - 2-Furancarboxaldehyde, 5-methyl-

$C_6H_6O_2S$
2180 - Benzenesulfinic acid
5835 - Ethanone, 1-(3-hydroxy-2-thienyl)-

$C_6H_6O_2Se$
2174 - Benzeneseleninic acid

$C_6H_6O_3$
2366 - 1,2,3-Benzenetriol
2367 - 1,2,4-Benzenetriol
2368 - 1,3,5-Benzenetriol
6026 - 2-Furanacetic acid
6044 - 2-Furancarboxaldehyde, 5-(hydroxymethyl)-
6059 - 2-Furancarboxylic acid, methyl ester
6060 - 2-Furancarboxylic acid, 5-methyl-
6063 - 3-Furancarboxylic acid, methyl ester
6065 - 3-Furancarboxylic acid, 5-methyl-
6097 - 2,5-Furandione, 3,4-dimethyl-
10796 - 4H-Pyran-4-one, 3-hydroxy-2-methyl-

$C_6H_6O_3S$
2214 - Benzenesulfonic acid

$C_6H_6O_4$
3910 - 2-Butynedioic acid, dimethyl ester
10795 - 4H-Pyran-4-one, 5-hydroxy-
2-(hydroxymethyl)-

$C_6H_6O_4S$
2225 - Benzenesulfonic acid, 2-hydroxy-
2226 - Benzenesulfonic acid, 4-hydroxy-

$C_6H_6O_7S_2$
1560 - 1,3-Benzenedisulfonic acid, 4-hydroxy

C_6H_6S
2289 - Benzenethiol

$C_6H_6S_2$
1561 - 1,2-Benzenedithiol
1562 - 1,3-Benzenedithiol

C_6H_6Se
2177 - Benzeneselenol

$C_6H_7AsO_3$
460 - Arsonic acid, phenyl-

$C_6H_7BO_2$
3135 - Boronic acid, phenyl-

$C_6H_7BrN_2$
7141 - Hydrazine, (4-bromophenyl)-

$C_6H_7ClN_2$
1264 - 1,2-Benzenediamine, 4-chloro-

C_6H_7ClO
6711 - 2,4-Hexadienoyl chloride, (E,E)-

$C_6H_7ClO_4$
3773 - 2-Butenedioic acid, 2-chloro-, dimethyl ester,
(E)-

C_6H_7ClSi
11384 - Silane, chlorophenyl-

$C_6H_7Cl_2F_3O_2$
10721 - Propionic acid, 2,2-dichloro-3,3,3-trifluoro-,
propyl ester

$C_6H_7Cl_2N$
665 - Benzenamine, 2-chloro-, hydrochloride

$C_6H_7F_3O_3$
3592 - Butanoic acid, 4,4,4-trifluoro-3-oxo-, ethyl
ester

$C_6H_7KO_2$
6704 - 2,4-Hexadienoic acid, potassium salt, (E,E)-

C_6H_7N
633 - Benzenamine
4907 - 1-Cyclopentenecarbonitrile
10980 - Pyridine, 2-methyl-
10981 - Pyridine, 3-methyl-
10982 - Pyridine, 4-methyl-

C_6H_7NO
757 - Benzenamine, N-hydroxy-

5820 - Ethanone, 1-(1H-pyrrol-2-yl)-
9200 - Phenol, amino-
9201 - Phenol, 2-amino-
9202 - Phenol, 3-amino-
9203 - Phenol, 4-amino-
10975 - 2-Pyridinemethanol
10976 - 3-Pyridinemethanol
10977 - 4-Pyridinemethanol
10988 - Pyridine, 3-methyl-, 1-oxide
11035 - 2(1H)-Pyridinone, 1-methyl-
11036 - 2(1H)-Pyridinone, 3-methyl-
11037 - 2(1H)-Pyridinone, 4-methyl-

$C_6H_7NO_2$
11122 - 1H-Pyrrole-2,5-dione, 1-ethyl-

$C_6H_7NO_3S$
2215 - Benzenesulfonic acid, 2-amino-
2216 - Benzenesulfonic acid, 3-amino-
2217 - Benzenesulfonic acid, 4-amino-

$C_6H_7NO_4$
6129 - Furan, 2-(methoxymethyl)-5-nitro-

C_6H_7NS
2290 - Benzenethiol, 2-amino-
2291 - Benzenethiol, 4-amino-

$C_6H_7N_3O$
10905 - 4-Pyridinecarboxylic acid, hydrazide

$C_6H_7N_3O_2$
1281 - 1,2-Benzenediamine, 4-nitro-
7168 - Hydrazine, (4-nitrophenyl)-

$C_6H_7N_5O$
10769 - 6H-Purin-6-one, 2-amino-1,7-dihydro-7-methyl

$C_6H_7O_3Sb$
11548 - Stibine oxide, dihydroxyphenyl-

C_6H_7P
9519 - Phosphine, phenyl-

C_6H_8
4172 - Cyclobutane, 1,2-bis(methylene)-
4248 - 1,3-Cyclohexadiene
4249 - 1,4-Cyclohexadiene
6992 - 1,3,5-Hexatriene, (E)-
6993 - 1,3,5-Hexatriene, (Z)-
7103 - 1-Hexen-3-yne
7104 - 1-Hexen-4-yne
7105 - 1-Hexen-5-yne
7106 - 2-Hexen-4-yne
9116 - 1-Penten-3-yne, 2-methyl-
9117 - 3-Penten-1-yne, 3-methyl-

$C_6H_8AsNO_3$
456 - Arsonic acid, (4-aminophenyl)-

C_6H_8ClN
756 - Benzenamine, hydrochloride

C_6H_8ClNS
11670 - Thiazole, 5-(2-chloroethyl)-4-methyl-

$C_6H_8ClN_3O_4S_2$
1558 - 1,3-Benzenedisulfonamide, 4-amino-6-chloro-

$C_6H_8Cl_2$
6690 - 2,4-Hexadiene, 1,3-dichloro-

$C_6H_8Cl_2O_3$
3503 - Butanoic acid, 2,2-dichloro-3-oxo-, ethyl ester

$C_6H_8Cl_2O_4$
129 - Acetic acid, chloro-, 1,2-ethanediyl ester
5554 - Ethanedioic acid, bis(2-chloroethyl) ester

$C_6H_8N_2$
1260 - 1,2-Benzenediamine
1261 - 1,3-Benzenediamine
1262 - 1,4-Benzenediamine
6777 - Hexanedinitrile
7169 - Hydrazine, phenyl-
10805 - Pyrazine, 2,3-dimethyl-
10806 - Pyrazine, 2,5-dimethyl-
10807 - Pyrazine, 2,6-dimethyl-
10866 - 2-Pyridinamine, 3-methyl-
10867 - 2-Pyridinamine, 4-methyl-
10868 - 2-Pyridinamine, 6-methyl-
10869 - 2-Pyridinamine, N-methyl-
10870 - 3-Pyridinamine, 4-methyl-
10971 - 2-Pyridinemethanamine

10972 - 3-Pyridinemethanamine
10973 - 4-Pyridinemethanamine
11055 - Pyrimidine, 2,4-dimethyl-
11056 - Pyrimidine, 2,5-dimethyl-
11057 - Pyrimidine, 4,5-dimethyl-
11058 - Pyrimidine, 4,6-dimethyl-

$C_6H_8N_2O$
9274 - Phenol, 2,4-diamino-
10114 - Propanenitrile, 3,3'-oxybis-
10808 - Pyrazineethanol
10865 - 3-Pyridinamine, 6-methoxy-

$C_6H_8N_2O_2S$
2187 - Benzenesulfonamide, 4-amino-

$C_6H_8N_2O_3$
821 - Benzenamine, nitrate

$C_6H_8N_2O_3S$
2224 - Benzenesulfonic acid, 4-hydrazino-

$C_6H_8N_6O_{18}$
7424 - D-Mannitol, hexanitrate

C_6H_8O
4703 - 2-Cyclohexen-1-one
4704 - 3-Cyclohexen-1-one
4908 - 1-Cyclopentene-1-carboxaldehyde
4944 - 2-Cyclopenten-1-one, 2-methyl-
4945 - 2-Cyclopenten-1-one, 3-methyl-
4946 - 2-Cyclopenten-1-one, 5-methyl-
6088 - Furan, 2,4-dimethyl-
6089 - Furan, 2,5-dimethyl-
6105 - Furan, 2-ethyl-
6675 - 2,4-Hexadienal, (E,E)-
7113 - 3-Hexynal
7130 - 3-Hexyn-2-one
7131 - 5-Hexyn-2-one

$C_6H_8O_2$
3176 - 1,3-Butadien-1-ol, acetate
3918 - 2-Butynoic acid, ethyl ester
4375 - 1,2-Cyclohexanedione
4376 - 1,3-Cyclohexanedione
4377 - 1,4-Cyclohexanedione
4909 - 1-Cyclopentene-1-carboxylic acid
6103 - Furan, 2-ethoxy-
6121 - 2-Furanmethanol, α-methyl-
6122 - 2-Furanmethanol, 5-methyl-
6128 - Furan, 2-(methoxymethyl)-
6698 - 2,4-Hexadienoic acid
7123 - 2-Hexynoic acid
9139 - 2-Pentynoic acid, methyl ester
10694 - 2-Propenoic acid, 2-propenyl ester
10791 - 2H-Pyran-2-one, 5,6-dihydro-6-methyl-, (S)-

$C_6H_8O_3$
6139 - 2(3H)-Furanone, 3-acetyldihydro-
10680 - 2-Propenoic acid, oxiranylmethyl ester

$C_6H_8O_4$
3778 - 2-Butenedioic acid (E)-, dimethyl ester
3779 - 2-Butenedioic acid (Z)-, dimethyl ester
3791 - 2-Butenedioic acid (E)-, monoethyl ester
4182 - 1,3-Cyclobutanedicarboxylic acid, cis-
4185 - 1,3-Cyclobutanedicarboxylic acid, trans-
5190 - 1,3-Dioxane-4,6-dione, 2,2-dimethyl-
9965 - Propanedioic acid, (1-methylethylidene)-
9979 - Propanedioic acid, 2-propenyl-

$C_6H_8O_5$
467 - L-Ascorbic acid, 6-deoxy-

$C_6H_8O_6$
465 - DL-Ascorbic acid
466 - L-Ascorbic acid
6239 - D-Glucuronic acid, γ-lactone
10165 - 1,2,3-Propanetricarboxylic acid

$C_6H_8O_7$
10167 - 1,2,3-Propanetricarboxylic acid, 2-hydroxy-

C_6H_8S
11754 - Thiophene, 2,3-dimethyl-
11755 - Thiophene, 2,4-dimethyl-
11756 - Thiophene, 2,5-dimethyl-
11757 - Thiophene, 3,4-dimethyl-
11759 - Thiophene, 2-ethyl-
11760 - Thiophene, 3-ethyl-

C_6H_8Si
11446 - Silane, phenyl-

C_6H_9Br
4627 - Cyclohexene, 1-bromo-
4628 - Cyclohexene, 3-bromo-

C_6H_9BrO
4574 - Cyclohexanone, 2-bromo-

$C_6H_9BrO_2$
3819 - 2-Butenoic acid, 4-bromo-, ethyl ester, (E)-

$C_6H_9BrO_3$
3477 - Butanoic acid, 2-bromo-3-oxo-, ethyl ester
3478 - Butanoic acid, 4-bromo-3-oxo-, ethyl ester

C_6H_9Cl
4642 - Cyclohexene, 1-chloro-
6688 - 1,3-Hexadiene, 3-chloro-
8664 - 1,2-Pentadiene, 1-chloro-3-methyl-
9126 - 1-Pentyne, 3-chloro-3-methyl-
9127 - 2-Pentyne, 4-chloro-4-methyl-

$C_6H_9ClN_2$
7172 - Hydrazine, phenyl-, monohydrochloride

C_6H_9ClO
4577 - Cyclohexanone, 2-chloro-
4578 - Cyclohexanone, 4-chloro-

$C_6H_9ClO_2$
3823 - 2-Butenoic acid, 2-chloro-, ethyl ester
3824 - 2-Butenoic acid, 2-chloro-, ethyl ester, (E)-
3825 - 2-Butenoic acid, 2-chloro-, ethyl ester, (Z)-
3826 - 2-Butenoic acid, 3-chloro-, ethyl ester
3827 - 2-Butenoic acid, 3-chloro-, ethyl ester, (E)-
3828 - 2-Butenoic acid, 3-chloro-, ethyl ester, (Z)-
3829 - 2-Butenoic acid, 4-chloro-, ethyl ester

$C_6H_9ClO_3$
3494 - Butanoic acid, 4-chloro-3-oxo-, ethyl ester
3570 - Butanoic acid, 3-oxo-, 2-chloroethyl ester

$C_6H_9Cl_3$
7048 - 1-Hexene, 1,1,2-trichloro-

$C_6H_9Cl_3O_2$
245 - Acetic acid, trichloro-, butyl ester
247 - Acetic acid, trichloro-, 1,1-dimethylethyl ester
251 - Acetic acid, trichloro-, 1-methylpropyl ester
252 - Acetic acid, trichloro-, 2-methylpropyl ester
3591 - Butanoic acid, 2,2,3-trichloro-, ethyl ester

$C_6H_9Cl_3O_3$
13 - Acetaldehyde, chloro-, trimer

$C_6H_9F_3O_2$
265 - Acetic acid, trifluoro-, butyl ester
266 - Acetic acid, trifluoro-, 1,1-dimethylethyl ester

C_6H_9N
4779 - Cyclopentanecarbonitrile
11119 - 1H-Pyrrole, 2,4-dimethyl-
11120 - 1H-Pyrrole, 2,5-dimethyl-
11124 - 1H-Pyrrole, 1-ethyl-
11125 - 1H-Pyrrole, 2-ethyl-

C_6H_9NO
6115 - 2-Furanmethanamine, N-methyl-
11183 - 2-Pyrrolidinone, 1-ethenyl-

C_6H_9NOS
11677 - 5-Thiazoleethanol, 4-methyl-

$C_6H_9NO_2$
10912 - 3-Pyridinecarboxylic acid, 1,2,5,6-tetrahydro-
11157 - 2,5-Pyrrolidinedione, 1-ethyl-

$C_6H_9NO_4$
50 - Acetamide, N,N-diacetyl-
8587 - 2,4-Oxazolidinedione, 3,5,5-trimethyl-

$C_6H_9NO_6$
6256 - Glycine, N,N-bis(carboxymethyl)-

$C_6H_9N_3$
2305 - 1,2,3-Benzenetriamine
2306 - 1,2,4-Benzenetriamine
11044 - 2-Pyrimidinamine, 4,5-dimethyl-
11045 - 4-Pyrimidinamine, 2,6-dimethyl-

$C_6H_9N_3O_2$
759 - Benzenamine, N-hydroxy-N-nitroso-, ammonium salt
7132 - L-Histidine

$C_6H_9N_3O_3$
7206 - 1H-Imidazole-1-ethanol, 2-methyl-5-nitro-

11841 - 1,3,5-Triazine-2,4,6(1H,3H,5H)-trione, 1,3,5-
trimethyl-

C_6H_{10}

3159 - 1,3-Butadiene, 2,3-dimethyl-
3907 - 1-Butyne, 3,3-dimethyl-
4626 - Cyclohexene
4827 - Cyclopentane, methylene-
4928 - Cyclopentene, 1-methyl-
4929 - Cyclopentene, 3-methyl-
4930 - Cyclopentene, 4-methyl-
4982 - Cyclopropane, (1-methylethenyl)-
6676 - 1,2-Hexadiene
6677 - 1,4-Hexadiene
6678 - 1,5-Hexadiene
6679 - 2,3-Hexadiene
6680 - 2,4-Hexadiene
6681 - 1,3-Hexadiene, (E)-
6682 - 1,3-Hexadiene, (Z)-
6683 - 1,4-Hexadiene, (E)-
6684 - 1,4-Hexadiene, (Z)-
6685 - 2,4-Hexadiene, (E,E)-
6686 - 2,4-Hexadiene, (E,Z)-
6687 - 2,4-Hexadiene, (Z,Z)-
7114 - 1-Hexyne
7115 - 2-Hexyne
7116 - 3-Hexyne
8666 - 1,2-Pentadiene, 3-methyl-
8667 - 1,2-Pentadiene, 4-methyl-
8668 - 1,3-Pentadiene, 2-methyl-, (Z)-
8669 - 1,3-Pentadiene, 3-methyl-
8670 - 1,3-Pentadiene, 4-methyl-
8671 - 1,4-Pentadiene, 2-methyl-
8672 - 2,3-Pentadiene, 2-methyl-
9060 - 1-Pentene, 3-methylene-
9134 - 1-Pentyne, 3-methyl-
9135 - 1-Pentyne, 4-methyl-
9136 - 2-Pentyne, 4-methyl-

$C_6H_{10}Br_2$

4345 - Cyclohexane, 1,1-dibromo-
4346 - Cyclohexane, 1,2-dibromo-, cis-
4347 - Cyclohexane, 1,2-dibromo-, trans-(±)-
4348 - Cyclohexane, 1,3-dibromo-, trans-
4349 - Cyclohexane, 1,4-dibromo-, cis-

$C_6H_{10}Br_2O_2$

3498 - Butanoic acid, 2,3-dibromo-, ethyl ester
3499 - Butanoic acid, 2,4-dibromo-, ethyl ester
10243 - Propanoic acid, 2,3-dibromo-, propyl ester

$C_6H_{10}ClFO_2$

132 - Acetic acid, chlorofluoro-, butyl ester

$C_6H_{10}Cl_2$

4358 - Cyclohexane, 1,1-dichloro-
4359 - Cyclohexane, 1,2-dichloro-, cis-
4360 - Cyclohexane, 1,2-dichloro-, trans-(±)-
4361 - Cyclohexane, 1,4-dichloro-, cis-
7011 - 1-Hexene, 1,2-dichloro-, (E)-
7012 - 1-Hexene, 1,2-dichloro-, (Z)-

$C_6H_{10}Cl_2O_2$

163 - Acetic acid, dichloro-, butyl ester
10249 - Propanoic acid, 2,3-dichloro-, 1-methylethyl
ester

$C_6H_{10}Cl_3NO$

96 - Acetamide, 2,2,2-trichloro-N,N-diethyl-

$C_6H_{10}N_2$

8830 - Pentanenitrile, 3-imino-2-methyl-
10833 - 1H-Pyrazole, 1-ethyl-3-methyl-
10839 - 1H-Pyrazole, 1,3,4-trimethyl-
10840 - 1H-Pyrazole, 1,3,5-trimethyl-
10841 - 1H-Pyrazole, 1,4,5-trimethyl-
10842 - 1H-Pyrazole, 3,4,5-trimethyl-

$C_6H_{10}N_2O_2$

4379 - 1,2-Cyclohexanedione, dioxime

$C_6H_{10}N_4$

11654 - 5H-Tetrazolo[1,5-a]azepine, 6,7,8,9-
tetrahydro-

$C_6H_{10}N_4O_3S_2$

3197 - Butanamide, N-[5-(aminosulfonyl)-1,3,4-
thiadiazol-2-yl]-

$C_6H_{10}O$

3160 - 1,3-Butadiene, 1-ethoxy-
3161 - 1,3-Butadiene, 2-ethoxy-
3377 - Butane, 1-(ethynyloxy)-
3877 - 1-Buten-1-one, 2-ethyl-
4571 - Cyclohexanone
4694 - 2-Cyclohexen-1-ol
4695 - 3-Cyclohexen-1-ol
4780 - Cyclopentanecarboxaldehyde
4885 - Cyclopentanone, 2-methyl-
4886 - Cyclopentanone, 3-methyl-, (±)-
5792 - Ethanone, 1-cyclobutyl-
6705 - 1,4-Hexadien-3-ol
6706 - 1,5-Hexadien-3-ol
6707 - 2,4-Hexadien-1-ol
6708 - 3,5-Hexadien-2-ol
6709 - 2,4-Hexadien-1-ol, (E,E)-
6996 - 2-Hexenal, (E)-
6997 - 3-Hexenal, (E)-
6998 - 3-Hexenal, (Z)-
7086 - 3-Hexen-2-one
7087 - 4-Hexen-2-one
7088 - 4-Hexen-3-one
7089 - 5-Hexen-2-one
7090 - 3-Hexen-2-one, (E)-
7124 - 1-Hexyn-3-ol
7125 - 2-Hexyn-1-ol
7126 - 3-Hexyn-1-ol
7127 - 5-Hexyn-2-ol
8554 - 7-Oxabicyclo[4.1.0]heptane
9002 - 2-Pentenal, 2-methyl-
9107 - 1-Penten-3-one, 2-methyl-
9108 - 3-Penten-2-one, 4-methyl-
9109 - 4-Penten-2-one, 4-methyl-
9146 - 1-Pentyn-3-ol, 3-methyl-
10592 - 1-Propene, 3,3'-oxybis-
10778 - 2H-Pyran, 3,6-dihydro-4-methyl-

$C_6H_{10}OS$

10596 - 1-Propene, 3,3'-sulfinylbis-

$C_6H_{10}OS_2$

10595 - 2-Propene-1-sulfinothioic acid, S-2-propenyl
ester

$C_6H_{10}O_2$

3513 - Butanoic acid, ethenyl ester
3835 - 3-Butenoic acid, 2,2-dimethyl-
3837 - 2-Butenoic acid, 2-ethyl-, (E)-
3838 - 2-Butenoic acid, ethyl ester
3839 - 2-Butenoic acid, ethyl ester, (E)-
3840 - 2-Butenoic acid, ethyl ester, (Z)-
3841 - 3-Butenoic acid, ethyl ester
3851 - 2-Butenoic acid, 2-methyl-, methyl ester, (E)-
3852 - 2-Butenoic acid, 3-methyl-, methyl ester
3862 - 3-Buten-1-ol, acetate
4176 - Cyclobutanecarboxylic acid, methyl ester
4781 - Cyclopentanecarboxylic acid
4961 - Cyclopropanecarboxylic acid, ethyl ester
6150 - 2(3H)-Furanone, 5-ethyldihydro-
6693 - 1,5-Hexadiene-3,4-diol, (R*,R*)-(±)-
6748 - Hexanedial
6804 - 2,4-Hexanedione
6805 - 2,5-Hexanedione
6806 - 3,4-Hexanedione
7049 - 2-Hexenoic acid
7050 - 3-Hexenoic acid
7051 - 5-Hexenoic acid
7118 - 3-Hexyne-2,5-diol
8591 - 2-Oxepanone
8626 - Oxirane, [(2-propenyloxy)methyl]-
8890 - Pentanoic acid, 2-methylene-
9076 - 2-Pentenoic acid, 2-methyl-, (E)-
9077 - 2-Pentenoic acid, 3-methyl-, (Z)-
9078 - 2-Pentenoic acid, 4-methyl-
9079 - 3-Pentenoic acid, 4-methyl-
10297 - Propanoic acid, 2-methyl-, ethenyl ester
10325 - Propanoic acid, 2-propenyl ester
10658 - 2-Propenoic acid, 2-methyl-, ethyl ester

$C_6H_{10}O_2S$

3589 - Butanoic acid, 3-thioxo-, ethyl ester
10597 - 1-Propene, 3,3'-sulfonylbis-

$C_6H_{10}O_2S_2$

5453 - Ethanebis(thioic) acid, S,S-diethyl ester

$C_6H_{10}O_3$

3510 - Butanoic acid, 3,3-dimethyl-2-oxo-
3558 - Butanoic acid, 2-methyl-3-oxo-, methyl ester
3571 - Butanoic acid, 3-oxo-, ethyl ester
3572 - Butanoic acid, 4-oxo-, ethyl ester
3843 - 3-Butenoic acid, 2-hydroxy-, ethyl ester
5206 - 1,3-Dioxolane-4-carboxaldehyde, 2,2-
dimethyl-
6087 - Furan, 2,5-dihydro-2,5-dimethoxy-
6143 - 2(3H)-Furanone, dihydro-3-hydroxy-4,4-
dimethyl-, (R)-
6808 - 2,5-Hexanedione, 3-hydroxy-
6905 - Hexanoic acid, 5-oxo-
8895 - Pentanoic acid, 2-methyl-4-oxo-
8896 - Pentanoic acid, 3-methyl-4-oxo-
8908 - Pentanoic acid, 4-oxo-, methyl ester
10203 - Propanoic acid, anhydride
10281 - Propanoic acid, 2-hydroxy-, 2-propenyl ester
10661 - 2-Propenoic acid, 2-methyl-, 2-hydroxyethyl
ester
10774 - 2H-Pyran-4-carboxylic acid, tetrahydro-

$C_6H_{10}O_4$

102 - Acetic acid, (acetyloxy)-, ethyl ester
3317 - Butanedioic acid, 2,3-dimethyl-
3318 - Butanedioic acid, dimethyl ester
3322 - Butanedioic acid, ethyl-, (R)-
3335 - Butanedioic acid, monoethyl ester
5560 - Ethanedioic acid, diethyl ester
5570 - 1,1-Ethanediol, diacetate
5571 - 1,2-Ethanediol, diacetate
6779 - Hexanedioic acid
7423 - D-Mannitol, 1,4:3,6-dianhydro-
8779 - Pentanedioic acid, 3-methyl-
8780 - Pentanedioic acid, monomethyl ester
9963 - Propanedioic acid, methyl-, dimethyl ester

$C_6H_{10}O_4S$

10329 - Propanoic acid, 3,3'-thiobis-

$C_6H_{10}O_4S_2$

5179 - Dimethipin

$C_6H_{10}O_5$

5169 - Dicarbonic acid, diethyl ester

$(C_6H_{10}O_5)_x$

6274 - Glycogen

$C_6H_{10}O_6$

3308 - Butanedioic acid, 2,3-dihydroxy-, dimethyl
ester, (R*,R*)-(±)-
3315 - Butanedioic acid, 2,3-dihydroxy- [R-(R*,R*)]-,
monoethyl ester
6211 - Gluconic acid, β-lactone
6212 - D-Gluconic acid, δ-lactone

$C_6H_{10}O_7$

6199 - D-Galacturonic acid
6238 - D-Glucuronic acid

$C_6H_{10}O_8$

6193 - Galactaric acid
6205 - D-Glucaric acid

$C_6H_{10}S$

4481 - Cyclohexanethione
10599 - 1-Propene, 3,3'-thiobis-

$C_6H_{10}S_3$

11917 - Trisulfide, di-2-propenyl

$C_6H_{11}Br$

4315 - Cyclohexane, bromo-

$C_6H_{11}BrN_2O_2$

3196 - Butanamide, N-(aminocarbonyl)-2-bromo-3-
methyl-

$C_6H_{11}BrO$

4503 - Cyclohexanol, 2-bromo-
6961 - 2-Hexanone, 6-bromo-

$C_6H_{11}BrO_2$

114 - Acetic acid, bromo-, 1,1-dimethylethyl ester
118 - Acetic acid, bromo-, 2-methylpropyl ester

3469 - Butanoic acid, 2-bromo-, ethyl ester
3470 - Butanoic acid, 4-bromo-, ethyl ester
3476 - Butanoic acid, 2-bromo-3-methyl-, methyl ester
6875 - Hexanoic acid, 2-bromo-, (±)-
6876 - Hexanoic acid, 6-bromo-
10213 - Propanoic acid, 2-bromo-2-methyl-, ethyl ester

$C_6H_{11}Cl$
3731 - 2-Butene, 1-chloro-2,3-dimethyl-
4339 - Cyclohexane, chloro-
7004 - 1-Hexene, 1-chloro-
7005 - 1-Hexene, 2-chloro-
7006 - 1-Hexene, 5-chloro-
7007 - 2-Hexene, 4-chloro-
7008 - 3-Hexene, 4-chloro-
7009 - 3-Hexene, 3-chloro-, (Z)-
9025 - 1-Pentene, 3-chloro-2-methyl-
9026 - 2-Pentene, 5-chloro-2-methyl-

$C_6H_{11}ClO$
3688 - Butanoyl chloride, 2,2-dimethyl-
3689 - Butanoyl chloride, 2,3-dimethyl-
3690 - Butanoyl chloride, 3,3-dimethyl-
3691 - Butanoyl chloride, 2-ethyl-
4505 - Cyclohexanol, 2-chloro-, trans-
4506 - Cyclohexanol, 2-chloro-, cis-(±)-
4507 - Cyclohexanol, 4-chloro-, trans-
6986 - Hexanoyl chloride
8993 - Pentanoyl chloride, 2-methyl-, (±)-
8994 - Pentanoyl chloride, 3-methyl-, (±)-

$C_6H_{11}ClO_2$
125 - Acetic acid, chloro-, butyl ester
128 - Acetic acid, chloro-, 1,1-dimethylethyl ester
142 - Acetic acid, chloro-, 1-methylpropyl ester
143 - Acetic acid, chloro-, 2-methylpropyl ester
3483 - Butanoic acid, 2-chloro-, ethyl ester
3484 - Butanoic acid, 3-chloro-, ethyl ester
3485 - Butanoic acid, 4-chloro-, ethyl ester
3730 - 2-Butene, 2-chloro-1,1-dimethoxy-
4021 - Carbonochloridic acid, 3-methylbutyl ester
4025 - Carbonochloridic acid, pentyl ester
10223 - Propanoic acid, 2-chloro-, 1-methylethyl ester
10232 - Propanoic acid, 3-chloro-, propyl ester
10292 - Propanoic acid, 2-methyl-, 2-chloroethyl ester

$C_6H_{11}Cl_3O_2$
5657 - Ethane, 1,1,1-trichloro-2,2-diethoxy-

$C_6H_{11}F$
4403 - Cyclohexane, fluoro-

$C_6H_{11}FO_2$
190 - Acetic acid, fluoro-, 1,1-dimethylethyl ester
6888 - Hexanoic acid, 6-fluoro-

$C_6H_{11}I$
4415 - Cyclohexane, iodo-

$C_6H_{11}IO_2$
3928 - Butyric acid, 2-iodo-, ethyl ester

$C_6H_{11}N$
478 - 7-Azabicyclo[4.1.0]heptane
3407 - Butanenitrile, 2-ethyl-
6840 - Hexanenitrile
8821 - Pentane, 1-isocyano-
8831 - Pentanenitrile, 4-methyl-
10526 - 2-Propen-1-amine, N-2-propenyl-

$C_6H_{11}NO$
493 - 2H-Azepin-2-one, hexahydro-
3386 - Butane, 1-isocyanato-3-methyl-
4610 - Cyclohexanone, oxime
9638 - 1-Piperidinecarboxaldehyde
9721 - 2-Piperidinone, 1-methyl-
9722 - 2-Piperidinone, 3-methyl-
9723 - 3-Piperidinone, 1-methyl-
9724 - 4-Piperidinone, 1-methyl-
11181 - 2-Pyrrolidinone, 1,5-dimethyl-
11182 - 2-Pyrrolidinone, 3,3-dimethyl-

$C_6H_{11}NO_2$
24 - Acetamide, N-acetyl-N-ethyl-

3816 - 2-Butenoic acid, 3-amino-, ethyl ester, (E)-
4454 - Cyclohexane, nitro-
4782 - Cyclopentanecarboxylic acid, 1-amino-
4830 - Cyclopentane, 1-methyl-1-nitro-
4831 - Cyclopentane, 1-methyl-2-nitro-
7682 - Morpholine, 4-acetyl-
9640 - 2-Piperidinecarboxylic acid
9641 - 3-Piperidinecarboxylic acid
9642 - 4-Piperidinecarboxylic acid
11184 - 2-Pyrrolidinone, 1-(2-hydroxyethyl)-

$C_6H_{11}NO_3S$
333 - L-Alanine, 3-(2-propenylsulfinyl)-, (S)-

$C_6H_{11}NO_4$
7231 - Imidodicarbonic acid, diethyl ester

$C_6H_{11}NS$
3390 - Butane, 1-isothiocyanato-3-methyl-
11673 - Thiazole, 4,5-dihydro-2,4,4-trimethyl-

$C_6H_{11}N_2O_4PS_3$
7645 - Methidathion

$C_6H_{11}N_3$
11847 - 1H-1,2,4-Triazole, 1-butyl

$C_6H_{11}NaO_7$
6213 - D-Gluconic acid, monosodium salt

$C_6H_{11}O_3P$
11909 - 2,6,7-Trioxa-1-phosphabicyclo[2.2.2]octane, 4-ethyl-

C_6H_{12}
3759 - 1-Butene, 2,3-dimethyl-
3760 - 1-Butene, 3,3-dimethyl-
3761 - 2-Butene, 2,3-dimethyl-
4188 - Cyclobutane, 1,2-dimethyl-, trans-
4189 - Cyclobutane, 1,3-dimethyl-, cis-
4190 - Cyclobutane, 1,3-dimethyl-, trans-
4191 - Cyclobutane, ethyl-
4305 - Cyclohexane
4826 - Cyclopentane, methyl-
4975 - Cyclopropane, 1-ethyl-1-methyl-
4976 - Cyclopropane, 1-ethyl-2-methyl-, cis-
4977 - Cyclopropane, 1-ethyl-2-methyl-, trans-
4983 - Cyclopropane, (1-methylethyl)-
4984 - Cyclopropane, propyl-
4985 - Cyclopropane, 1,1,2-trimethyl-
4986 - Cyclopropane, 1,2,3-trimethyl-, (1α,2α,3α)-
4987 - Cyclopropane, 1,2,3-trimethyl-, (1α,2α,3β)-
6999 - 1-Hexene
7000 - 2-Hexene, (E)-
7001 - 2-Hexene, (Z)-
7002 - 3-Hexene, (E)-
7003 - 3-Hexene, (Z)-
8825 - Pentane, 3-methylene-
9052 - 1-Pentene, 2-methyl-
9053 - 1-Pentene, 3-methyl-
9054 - 1-Pentene, 4-methyl-
9055 - 2-Pentene, 2-methyl-
9056 - 2-Pentene, 3-methyl-, (E)-
9057 - 2-Pentene, 3-methyl-, (Z)-
9058 - 2-Pentene, 4-methyl-, (E)-
9059 - 2-Pentene, 4-methyl-, (Z)-

$C_6H_{12}BrF$
6735 - Hexane, 1-bromo-6-fluoro-

$C_6H_{12}BrNO$
3198 - Butanamide, 2-bromo-2-ethyl-

$C_6H_{12}Br_2$
3269 - Butane, 1,2-dibromo-2,3-dimethyl-
3270 - Butane, 1,4-dibromo-2,3-dimethyl-
3271 - Butane, 2,3-dibromo-2,3-dimethyl-
6754 - Hexane, 1,2-dibromo-
6755 - Hexane, 1,4-dibromo-
6756 - Hexane, 1,5-dibromo-
6757 - Hexane, 1,6-dibromo-
6758 - Hexane, 2,3-dibromo-
6759 - Hexane, 2,5-dibromo-, (R*,R*)-(±)-
6760 - Hexane, 3,4-dibromo-

$C_6H_{12}ClF$
6742 - Hexane, 1-chloro-6-fluoro-

$C_6H_{12}ClNO$
47 - Acetamide, 2-chloro-N-(1-methylpropyl)-

$C_6H_{12}ClNO_2$
46 - Acetamide, 2-chloro-N-(3-methoxypropyl)-

$C_6H_{12}Cl_2$
6761 - Hexane, 1,2-dichloro-
6762 - Hexane, 1,6-dichloro-
6763 - Hexane, 1,2-dichloro-
6764 - Hexane, 2,3-dichloro-
6766 - Hexane, 2,5-dichloro-, (R*,R*)-(±)-

$C_6H_{12}Cl_2O$
10121 - Propane, 1,1'-oxybis[2-chloro-
10122 - Propane, 1,1'-oxybis[3-chloro-
10123 - Propane, 2,2'-oxybis[1-chloro-

$C_6H_{12}Cl_2O_2$
5449 - Ethane, 1,2-bis(2-chloroethoxy)-
5530 - Ethane, 1,1-dichloro-2,2-diethoxy-
5592 - Ethane, 1,1'-[ethylidenebis(oxy)]bis[2-chloro-
9896 - Propane, 1,3-dichloro-2-(ethoxymethoxy)-

$C_6H_{12}Cl_2S$
10153 - Propane, 1,1'-thiobis[3-chloro-

$C_6H_{12}Cl_3N$
5412 - Ethanamine, 2-chloro-N,N-bis(2-chloroethyl)-

$C_6H_{12}Cl_3O_3P$
5702 - Ethanol, 2-chloro-, phosphite (3:1)

$C_6H_{12}Cl_3O_4P$
5701 - Ethanol, 2-chloro-, phosphate (3:1)

$C_6H_{12}I_2$
6769 - Hexane, 1,6-diiodo-

$C_6H_{12}MnN_2S_4$
7421 - Manganous dimethyldithiocarbamate

$C_6H_{12}N_2$
273 - Acetonitrile, (butylamino)-
280 - Acetonitrile, (diethylamino)-
5111 - 1,4-Diazabicyclo[2.2.2]octane
10463 - 2-Propanone, (1-methylethylidene)hydrazone

$C_6H_{12}N_2O_2$
5494 - Ethanediamide, N,N'-diethyl-

$C_6H_{12}N_2O_3$
3334 - Butanedioic acid, mono (2,2-dimethylhydrazide)

$C_6H_{12}N_2O_4$
3957 - Carbamic acid, (1-methylethyl)nitro-, ethyl ester
3964 - Carbamic acid, nitropropyl-, ethyl ester
6778 - Hexane, 1,6-dinitro-
7150 - 1,1-Hydrazinedicarboxylic acid, diethyl ester
7151 - 1,2-Hydrazinedicarboxylic acid, diethyl ester

$C_6H_{12}N_2O_4S$
5016 - L-Cysteine, S-(2-amino-2-carboxyethyl)-, (R)-

$C_6H_{12}N_2O_4S_2$
5020 - L-Cystine

$C_6H_{12}N_2S_3$
11715 - Thiodicarbonic diamide ([[(H2N)C(S)]2S], tetramethyl-

$C_6H_{12}N_2S_4$
11722 - Thioperoxydicarbonic diamide ([[(H2N)C(S)]2S2), tetramethyl-

$C_6H_{12}N_2S_4Zn$
12068 - Ziram

$C_6H_{12}N_4$
11601 - 1,3,5,7-Tetraazatricyclo[3.3.1.13,7]decane

$C_6H_{12}N_6O_3$
7583 - Methanol, (1,3,5-triazine-2,4,6-triyltriimino)tris-

$C_6H_{12}O$
3183 - Butanal, 2-ethyl-
3369 - Butane, 1-(ethenyloxy)-
3370 - Butane, 2-(ethenyloxy)-
3657 - 2-Butanone, 3,3-dimethyl-
4500 - Cyclohexanol
4825 - Cyclopentanemethanol
4867 - Cyclopentanol, 1-methyl-
4868 - Cyclopentanol, 2-methyl-, cis-
4869 - Cyclopentanol, 2-methyl-, trans-
4870 - Cyclopentanol, 3-methyl-
6107 - Furan, 2-ethyltetrahydro-

6718 - Hexanal
6959 - 2-Hexanone
6960 - 3-Hexanone
7066 - 1-Hexen-3-ol
7067 - 4-Hexen-2-ol
7068 - 5-Hexen-2-ol
7069 - 2-Hexen-1-ol, (E)-
7070 - 2-Hexen-1-ol, (Z)-
7071 - 3-Hexen-1-ol, (Z)-
7072 - 4-Hexen-1-ol, (E)-
8590 - Oxepane
8627 - Oxirane, tetramethyl-
8693 - Pentanal, 2-methyl-
8694 - Pentanal, 3-methyl-
8980 - 2-Pentanone, 3-methyl-, (±)-
8981 - 2-Pentanone, 4-methyl-
8982 - 3-Pentanone, 2-methyl-
9094 - 2-Penten-1-ol, 2-methyl-
9095 - 2-Penten-1-ol, 4-methyl-
9096 - 3-Penten-1-ol, 4-methyl-
9097 - 3-Penten-2-ol, 2-methyl-
9098 - 3-Penten-2-ol, 3-methyl-
9099 - 3-Penten-2-ol, 4-methyl-
9100 - 4-Penten-2-ol, 2-methyl-
9101 - 4-Penten-2-ol, 3-methyl-
9102 - 4-Penten-2-ol, 4-methyl-
10035 - Propane, 1-(ethenyloxy)-2-methyl-
10036 - Propane, 2-(ethenyloxy)-2-methyl-
10579 - 1-Propene, 3-(1-methylethoxy)-
10594 - 1-Propene, 3-propoxy-

$C_6H_{12}O_2$
121 - Acetic acid, butyl ester
177 - Acetic acid, 1,1-dimethylethyl ester
214 - Acetic acid, 1-methylpropyl ester
215 - Acetic acid, 2-methylpropyl ester
3506 - Butanoic acid, 2,2-dimethyl-
3507 - Butanoic acid, 2,3-dimethyl-
3508 - Butanoic acid, 3,3-dimethyl-
3514 - Butanoic acid, 2-ethyl-
3515 - Butanoic acid, ethyl ester
3552 - Butanoic acid, 3-methyl-, methyl ester
3636 - 1-Butanol, 3-methyl-, formate
4072 - 1,2-Cyclohexanediol, cis-
4373 - 1,2-Cyclohexanediol, trans-(±)-
4374 - 1,4-Cyclohexanediol, trans-
5189 - 1,3-Dioxane, 2,4-dimethyl-
5901 - Ethene, 1,1-diethoxy-
6015 - Formic acid, pentyl ester
6170 - Furan, tetrahydro-2-(methoxymethyl)-
6870 - Hexanoic acid
6969 - 2-Hexanone, 5-hydroxy-
6970 - 2-Hexanone, 6-hydroxy-
6971 - 3-Hexanone, 4-hydroxy-
6972 - 3-Hexanone, 5-hydroxy-
7177 - Hydroperoxide, cyclohexyl
8691 - Pentanal, 3-hydroxy-2-methyl-
8887 - Pentanoic acid, 2-methyl-, (±)-
8888 - Pentanoic acid, 3-methyl-, (±)-
8889 - Pentanoic acid, 4-methyl-
8891 - Pentanoic acid, methyl ester
8976 - 2-Pentanone, 4-hydroxy-4-methyl-
10260 - Propanoic acid, 2,2-dimethyl-, methyl ester
10298 - Propanoic acid, 2-methyl-, ethyl ester
10299 - Propanoic acid, 1-methylethyl ester
10326 - Propanoic acid, propyl ester
10784 - 2H-Pyran-2-methanol, tetrahydro-

$C_6H_{12}O_2S$
11784 - Thiophene, tetrahydro-2,4-dimethyl-, 1,1-
dioxide

$C_6H_{12}O_3$
181 - Acetic acid, ethoxy-, ethyl ester
3532 - Butanoic acid, 2-hydroxy-, ethyl ester, (±)-
3533 - Butanoic acid, 3-hydroxy-, ethyl ester, (±)-
3534 - Butanoic acid, 3-hydroxy-, methyl ester
5207 - 1,3-Dioxolane-4-methanol, 2,2-dimethyl-
5718 - Ethanol, 2-ethoxy-, acetate
6169 - Furan, tetrahydro-2,5-dimethoxy-

6891 - Hexanoic acid, 2-hydroxy-, (±)-
7109 - Hexitol, 2,5-anhydro-3,4-dideoxy-
10275 - Propanoic acid, 2-hydroxy-, 1-methylethyl
ester
10277 - Propanoic acid, 3-hydroxy-, 1-methylethyl
ester
10279 - Propanoic acid, 2-hydroxy-2-methyl-, ethyl
ester
11908 - 1,3,5-Trioxane, 2,4,6-trimethyl-

$C_6H_{12}O_4$
5673 - Ethanol, 2-[2-(acetyloxy)ethoxy]-
7111 - D-ribo-Hexose, 2,6-dideoxy-
10255 - Propanoic acid, 2,3-dihydroxy-2-methyl-,
ethyl ester
10257 - Propanoic acid, 2,3-dihydroxy-, propyl ester

$C_6H_{12}O_4Si$
11419 - Silanediol, dimethyl-, diacetate

$C_6H_{12}O_5$
4074 - D-chiro-Inositol, 2-deoxy-
6235 - D-Glucose, 6-deoxy-
7110 - D-arabino-Hexose, 2-deoxy-

$C_6H_{12}O_5S$
184 - Acetic acid, (ethoxysulfonyl)-, ethyl ester

$C_6H_{12}O_6$
337 - D-Allose
338 - D-Altrose
4073 - DL-chiro-Inositol
6023 - β-D-Fructose
6197 - D-Galactose
6232 - β-D-Glucose
6290 - D-Gulose
7336 - scyllo-Inositol
7425 - D-Mannose
11518 - DL-Sorbose
11519 - L-Sorbose
11587 - D-Tagatose

$C_6H_{12}O_7$
6209 - D-Gluconic acid

$C_6H_{12}S$
4480 - Cyclohexanethiol
11690 - Thiepane
11765 - Thiophene, 2-ethyltetrahydro-
11766 - Thiophene, 3-ethyltetrahydro-
11785 - Thiophene, tetrahydro-2,5-dimethyl-, cis-
11796 - 2H-Thiopyran, tetrahydro-2-methyl-
11797 - 2H-Thiopyran, tetrahydro-3-methyl-
11798 - 2H-Thiopyran, tetrahydro-4-methyl-

$C_6H_{12}S_3$
11920 - 1,3,5-Trithiane, 2,4,6-trimethyl-
11921 - 1,3,5-Trithiane, 2,4,6-trimethyl-, (2α,4α,6β)-

$C_6H_{13}Br$
3235 - Butane, 1-bromo-3,3-dimethyl-
3236 - Butane, 2-bromo-2,3-dimethyl-
6732 - Hexane, 1-bromo-
6733 - Hexane, 2-bromo-
6734 - Hexane, 3-bromo-
8714 - Pentane, 1-bromo-2-methyl-
8715 - Pentane, 1-bromo-3-methyl-
8716 - Pentane, 1-bromo-4-methyl-
8717 - Pentane, 2-bromo-2-methyl-
8718 - Pentane, 3-bromo-3-methyl-

$C_6H_{13}BrO_2$
5459 - Ethane, 2-bromo-1,1-diethoxy-

$C_6H_{13}Cl$
3245 - Butane, 1-chloro-2,3-dimethyl-
3246 - Butane, 1-chloro-3,3-dimethyl-
3247 - Butane, 2-chloro-2,3-dimethyl-
3248 - Butane, 3-chloro-2,2-dimethyl-
6737 - Hexane, 1-chloro-
6738 - Hexane, 2-chloro-
6739 - Hexane, 3-chloro-
8729 - Pentane, 2-chloro-2-methyl-
8730 - Pentane, 2-chloro-4-methyl-
8731 - Pentane, 3-(chloromethyl)-
8732 - Pentane, 3-chloro-3-methyl-

$C_6H_{13}ClO$
3249 - Butane, 1-(2-chloroethoxy)-

3608 - 2-Butanol, 1-chloro-2,3-dimethyl-
3609 - 2-Butanol, 1-chloro-3,3-dimethyl-
6916 - 1-Hexanol, 6-chloro-
6917 - 2-Hexanol, 1-chloro-
6918 - 3-Hexanol, 1-chloro-
6919 - 3-Hexanol, 5-chloro-

$C_6H_{13}ClO_2$
3244 - Butane, 3-chloro-1,1-dimethoxy-
5469 - Ethane, 2-chloro-1,1-diethoxy-
10355 - 2-Propanol, 1-chloro-3-(1-methylethoxy)-
10356 - 2-Propanol, 1-chloro-3-propoxy-

$C_6H_{13}ClS$
9855 - Propane, 2-[(2-chloroethyl)thio]-2-methyl-

$C_6H_{13}Cl_3Si$
11464 - Silane, trichlorohexyl-

$C_6H_{13}F$
6829 - Hexane, 1-fluoro-
6830 - Hexane, 2-fluoro-

$C_6H_{13}FO$
6936 - 1-Hexanol, 6-fluoro-

$C_6H_{13}I$
6831 - Hexane, 1-iodo-
6832 - Hexane, 2-iodo-, (R)-

$C_6H_{13}IO_2$
5543 - Ethane, 1,1-diethoxy-2-iodo-

$C_6H_{13}N$
492 - 1H-Azepine, hexahydro-
4283 - Cyclohexanamine
4772 - Cyclopentanamine, 1-methyl-
4773 - Cyclopentanamine, 2-methyl-
9679 - Piperidine, 1-methyl-
9680 - Piperidine, 2-methyl-, (±)-
9681 - Piperidine, 3-methyl-, (±)-
9682 - Piperidine, 4-methyl-
10525 - 2-Propen-1-amine, N-(1-methylethyl)-
10527 - 2-Propen-1-amine, N-propyl-
11149 - Pyrrolidine, 2,2-dimethyl-
11150 - Pyrrolidine, 2,4-dimethyl-
11151 - Pyrrolidine, 2,5-dimethyl-, cis-
11166 - Pyrrolidine, 1-ethyl-

$C_6H_{13}NO$
38 - Acetamide, N-butyl-
53 - Acetamide, N,N-diethyl-
3203 - Butanamide, N,N-dimethyl-
4501 - Cyclohexanol, 2-amino-, trans-(±)-
6721 - Hexanamide
7687 - Morpholine, 2,6-dimethyl-
7694 - Morpholine, 4-ethyl-
8959 - 2-Pentanone, 4-amino-4-methyl-
9674 - 2-Piperidinemethanol
9675 - 3-Piperidinemethanol
9716 - 3-Piperidinol, 1-methyl-
9717 - 4-Piperidinol, 1-methyl-
9802 - Propanamide, N-propyl-
11165 - 1-Pyrrolidineethanol

$C_6H_{13}NO_2$
332 - β-Alanine, N-methyl-, ethyl ester
3633 - 1-Butanol, 3-methyl-, carbamate
3634 - 2-Butanol, 2-methyl-, carbamate
3937 - Carbamic acid, butyl-, methyl ester
3943 - Carbamic acid, (1,1-dimethylethyl)-, methyl
ester
3956 - Carbamic acid, (1-methylethyl)-, ethyl ester
3962 - Carbamic acid, (1-methylpropyl)-, methyl ester
3970 - Carbamic acid, propyl-, ethyl ester
6120 - 2-Furanmethanol, 5-(aminomethyl)tetrahydro-
6841 - Hexane, 1-nitro-
6873 - Hexanoic acid, 6-amino-
7370 - L-Isoleucine
7412 - DL-Leucine
7413 - L-Leucine
7691 - 4-Morpholineethanol
8089 - Nitrous acid, hexyl ester
8214 - Norleucine
8215 - DL-Norleucine

$C_6H_{13}NO_3S$
11568 - Sulfamic acid, cyclohexyl-

C$_6$H$_{16}$ClNO
9831 - 1-Propanaminium, 2-hydroxy-N,N,N-trimethyl-, chloride

C$_6$H$_{16}$Cl$_2$OSi$_2$
5230 - Disiloxane, 1,3-bis(chloromethyl)-1,1,3,3-tetramethyl-

C$_6$H$_{16}$N$_2$
5501 - 1,2-Ethanediamine, N,N'-diethyl-
5502 - 1,2-Ethanediamine, N,N-diethyl-
5516 - 1,2-Ethanediamine, N,N,N',N'-tetramethyl-
6749 - 1,6-Hexanediamine
7137 - Hydrazine, 1,2-bis(1-methylethyl)-

C$_6$H$_{16}$N$_2$O
5677 - Ethanol, 2-[(2-amino-2-methylpropyl)amino]-

C$_6$H$_{16}$N$_2$O$_2$
5715 - Ethanol, 2,2'-(1,2-ethanediyldiimino)bis-

C$_6$H$_{16}$OSi
10410 - 1-Propanol, 3-(trimethylsilyl)-
11503 - Silanol, triethyl-

C$_6$H$_{16}$O$_2$Si
11405 - Silane, diethoxydimethyl-

C$_6$H$_{16}$O$_3$Si
11477 - Silane, triethoxy-

C$_6$H$_{16}$Si
11483 - Silane, triethyl-
11496 - Silane, trimethylpropyl-

C$_6$H$_{17}$N$_3$
9874 - 1,3-Propanediamine, N-(3-aminopropyl)-

C$_6$H$_{18}$AlO$_9$P$_3$
6020 - Fosetyl-Al

C$_6$H$_{18}$Cl$_2$O$_2$Si$_3$
11913 - Trisiloxane, 1,5-dichloro-1,1,3,3,5,5-hexamethyl-

C$_6$H$_{18}$N$_3$OP
9573 - Phosphoric triamide, hexamethyl-

C$_6$H$_{18}$N$_4$
5499 - 1,2-Ethanediamine, N,N'-bis(2-aminoethyl)-

C$_6$H$_{18}$OSi$_2$
5236 - Disiloxane, hexamethyl-

C$_6$H$_{18}$O$_3$Si$_2$
5234 - Disiloxane, 1,3-dimethoxy-1,1,3,3-tetramethyl-

C$_6$H$_{18}$O$_3$Si$_3$
6002 - Cyclotrisiloxane, hexamethyl-

C$_6$H$_{18}$Si$_2$
5228 - Disilane, hexamethyl-

C$_6$H$_{19}$NSi$_2$
11363 - Silanamine, 1,1,1-trimethyl-N-(trimethylsilyl)-

C$_6$H$_{21}$N$_3$Si$_3$
5001 - Cyclotrisilazane, 2,2,4,4,6,6-hexamethyl-

C$_6$H$_{24}$O$_6$Si$_6$
4625 - Cyclohexasiloxane, 2,4,6,8,10,12-hexamethyl-

C$_6$H$_{26}$O$_8$
3351 - 2,3-Butanediol, 2,3-dimethyl-, hexahydrate

C$_6$MoO$_6$
7662 - Molybdenum carbonyl

C$_6$N$_4$
5915 - Ethenetetracarbonitrile

C$_6$N$_6$
11836 - 1,3,5-Triazine-2,4,6-tricarbonitrile

C$_7$F$_5$N
2767 - Benzonitrile, pentafluoro-

C$_7$F$_8$
2095 - Benzene, pentafluoro(trifluoromethyl)-

C$_7$F$_{14}$
4499 - Cyclohexane, undecafluoro(trifluoromethyl)-
6560 - 1-Heptene, 1,1,2,3,3,4,4,5,5,6,6,7,7,7-tetradecafluoro-

C$_7$F$_{16}$
6418 - Heptane, hexadecafluoro-

C$_7$HCl$_5$O$_2$
2712 - Benzoic acid, pentachloro-

C$_7$HCl$_7$
2085 - Benzene, pentachloro(dichloromethyl)-

C$_7$HF$_5$O
596 - Benzaldehyde, pentafluoro-

C$_7$HF$_{15}$
6432 - Heptane, 1,1,1,2,2,3,3,4,4,5,5,6,6,7,7-pentadecafluoro-

C$_7$H$_3$Br$_2$NO
2751 - Benzonitrile, 3,5-dibromo-4-hydroxy-

C$_7$H$_3$Br$_5$
2083 - Benzene, pentabromomethyl-

C$_7$H$_3$ClF$_3$NO$_2$
1217 - Benzene, 1-chloro-2-nitro-4-(trifluoromethyl)-
1218 - Benzene, 1-chloro-4-nitro-2-(trifluoromethyl)-
1219 - Benzene, 4-chloro-1-nitro-2-(trifluoromethyl)-

C$_7$H$_3$ClN$_2$O$_5$
2896 - Benzoyl chloride, 3,5-dinitro-

C$_7$H$_3$Cl$_2$N
2752 - Benzonitrile, 2,6-dichloro-

C$_7$H$_3$Cl$_3$O
2893 - Benzoyl chloride, 2,4-dichloro-
2894 - Benzoyl chloride, 3,4-dichloro-

C$_7$H$_3$Cl$_3$O$_2$
2727 - Benzoic acid, 2,3,6-trichloro-
2728 - Benzoic acid, 2,4,5-trichloro-

C$_7$H$_3$Cl$_5$
1401 - Benzene, 1,2-dichloro-4-(trichloromethyl)-
2086 - Benzene, pentachloromethyl-
2326 - Benzene, 1,2,3-trichloro-4-(dichloromethyl)-
2327 - Benzene, 1,2,4-trichloro-5-(dichloromethyl)-

C$_7$H$_3$F$_5$
2094 - Benzene, pentafluoromethyl-

C$_7$H$_3$F$_5$O
2093 - Benzene, pentafluoromethoxy-

C$_7$H$_3$N$_3$O$_8$
2735 - Benzoic acid, 2,4,6-trinitro-

C$_7$H$_4$BrClO
2887 - Benzoyl chloride, 2-bromo-
2888 - Benzoyl chloride, 4-bromo

C$_7$H$_4$BrF$_3$
1089 - Benzene, 1-bromo-2-(trifluoromethyl)-
1090 - Benzene, 1-bromo-3-(trifluoromethyl)-
1091 - Benzene, 1-bromo-4-(trifluoromethyl)-

C$_7$H$_4$BrN
2744 - Benzonitrile, 2-bromo-
2745 - Benzonitrile, 3-bromo-
2746 - Benzonitrile, 4-bromo-

C$_7$H$_4$BrNO
1049 - Benzene, 1-bromo-4-isocyanato-

C$_7$H$_4$BrNO$_4$
2527 - Benzoic acid, 2-bromo-4-nitro-
2528 - Benzoic acid, 2-bromo-5-nitro-
2529 - Benzoic acid, 4-bromo-3-nitro-
2530 - Benzoic acid, 5-bromo-2-nitro-

C$_7$H$_4$Br$_2$O$_2$
534 - Benzaldehyde, 3,5-dibromo-2-hydroxy-
2555 - Benzoic acid, 2,4-dibromo-
2556 - Benzoic acid, 2,5-dibromo-
2557 - Benzoic acid, 2,6-dibromo-

C$_7$H$_4$Br$_2$O$_3$
2558 - Benzoic acid, 3,5-dibromo-2-hydroxy-

C$_7$H$_4$Br$_2$O$_5$
2560 - Benzoic acid, 2,6-dibromo-3,4,5-trihydroxy-

C$_7$H$_4$Br$_4$O
2256 - Benzene, 1,2,3,5-tetrabromo-4-methoxy-
9468 - Phenol, 2,3,4,5-tetrabromo-6-methyl-

C$_7$H$_4$ClFO
2897 - Benzoyl chloride, 2-fluoro-
2898 - Benzoyl chloride, 3-fluoro-
2899 - Benzoyl chloride, 4-fluoro-

C$_7$H$_4$ClF$_3$
1240 - Benzene, 1-chloro-2-(trifluoromethyl)-
1241 - Benzene, 1-chloro-3-(trifluoromethyl)-
1242 - Benzene, 1-chloro-4-(trifluoromethyl)-

C$_7$H$_4$ClN
2747 - Benzonitrile, 2-chloro-
2748 - Benzonitrile, 3-chloro-
2749 - Benzonitrile, 4-chloro-

C$_7$H$_4$ClNO
1176 - Benzene, 1-chloro-2-isocyanato-
2878 - Benzoxazole, 2-chloro-

C$_7$H$_4$ClNO$_2$
2883 - 2(3H)-Benzoxazolone, 5-chloro-

C$_7$H$_4$ClNO$_3$
2905 - Benzoyl chloride, 3-nitro-
2906 - Benzoyl chloride, 4-nitro-

C$_7$H$_4$ClNO$_4$
2544 - Benzoic acid, 2-chloro-3-nitro-
2545 - Benzoic acid, 2-chloro-5-nitro-
2546 - Benzoic acid, 4-chloro-3-nitro-
2547 - Benzoic acid, 4-chloro-3-nitro-

C$_7$H$_4$ClNS
1177 - Benzene, 1-chloro-4-isothiocyanato-
2862 - Benzothiazole, 2-chloro-

C$_7$H$_4$Cl$_2$O
535 - Benzaldehyde, 2,5-dichloro-
536 - Benzaldehyde, 3,4-dichloro-
537 - Benzaldehyde, 3,5-dichloro-
2890 - Benzoyl chloride, 2-chloro-
2891 - Benzoyl chloride, 3-chloro-
2892 - Benzoyl chloride, 4-chloro-

C$_7$H$_4$Cl$_2$O$_2$
2561 - Benzoic acid, 2,4-dichloro-
2562 - Benzoic acid, 2,5-dichloro-
2563 - Benzoic acid, 2,6-dichloro-
2564 - Benzoic acid, 3,5-dichloro-

C$_7$H$_4$Cl$_2$O$_3$
2566 - Benzoic acid, 3,5-dichloro-2-hydroxy-

C$_7$H$_4$Cl$_3$F
1736 - Benzene, 1-fluoro-2-(trichloromethyl)-

C$_7$H$_4$Cl$_3$NO$_3$
11854 - Triclopyr

C$_7$H$_4$Cl$_4$
1237 - Benzene, 1-chloro-2-(trichloromethyl)-
1238 - Benzene, 1-chloro-3-(trichloromethyl)-
1239 - Benzene, 1-chloro-4-(trichloromethyl)-
1359 - Benzene, 1,2-dichloro-4-(dichloromethyl)-
2262 - Benzene, 1,2,3,5-tetrachloro-4-methyl-
2263 - Benzene, 1,2,4,5-tetrachloro-3-methyl-
2324 - Benzene, 1,2,4-trichloro-5-(chloromethyl)-

C$_7$H$_4$Cl$_4$O$_2$
9472 - Phenol, 2,3,5,6-tetrachloro-4-methoxy-

C$_7$H$_4$FN
2756 - Benzonitrile, 4-fluoro-

C$_7$H$_4$FNS
1721 - Benzene, 1-fluoro-3-isothiocyanato-

C$_7$H$_4$F$_3$NO$_2$
2060 - Benzene, 1-nitro-2-(trifluoromethyl)-
2061 - Benzene, 1-nitro-3-(trifluoromethyl)-

C$_7$H$_4$F$_4$
1737 - Benzene, 1-fluoro-2-(trifluoromethyl)-
1738 - Benzene, 1-fluoro-3-(trifluoromethyl)-
1739 - Benzene, 1-fluoro-4-(trifluoromethyl)-

C$_7$H$_4$I$_2$O$_3$
2617 - Benzoic acid, 2-hydroxy-3,5-diiodo-
2618 - Benzoic acid, 4-hydroxy-3,5-diiodo-

C$_7$H$_4$N$_2$O$_3$
1788 - Benzene, 1-isocyanato-2-nitro-
1789 - Benzene, 1-isocyanato-3-nitro-
1790 - Benzene, 1-isocyanato-4-nitro-

C$_7$H$_4$N$_2$O$_5$
551 - Benzaldehyde, 2,4-dinitro-

C$_7$H$_4$N$_2$O$_6$
2590 - Benzoic acid, 3,4-dinitro-
2591 - Benzoic acid, 3,5-dinitro-

C$_7$H$_4$O$_4$S
2875 - 3H-2,1-Benzoxathiol-3-one, 1,1-dioxide

C$_7$H$_4$O$_6$
10776 - 4H-Pyran-2,6-dicarboxylic acid, 4-oxo-

C$_7$H$_4$O$_7$
10775 - 4H-Pyran-2,6-dicarboxylic acid, 3-hydroxy-4-oxo-

C$_7$H$_5$BrO
526 - Benzaldehyde, 2-bromo-
527 - Benzaldehyde, 3-bromo-
528 - Benzaldehyde, 4-bromo-
2885 - Benzoyl bromide

$C_7H_5BrO_2$
2518 - Benzoic acid, 2-bromo-
2519 - Benzoic acid, 3-bromo-
2520 - Benzoic acid, 4-bromo-

$C_7H_5BrO_3$
2524 - Benzoic acid, 5-bromo-2-hydroxy-

$C_7H_5Br_3$
1291 - Benzene, 1,3-dibromo-5-(bromomethyl)-
2314 - Benzene, 1,3,5-tribromo-2-methyl-

$C_7H_5Br_3O$
2310 - Benzene, 1,2,3-tribromo-5-methoxy-
2311 - Benzene, 1,2,4-tribromo-5-methoxy-
2312 - Benzene, 1,2,5-tribromo-3-methoxy-
2313 - Benzene, 1,3,5-tribromo-2-methoxy-
9480 - Phenol, 2,4,6-tribromo-3-methyl-

$C_7H_5ClN_2$
7236 - 1H-Indazole, 3-chloro-

$C_7H_5ClN_2O$
2876 - 2-Benzoxazolamine, 5-chloro-

C_7H_5ClO
529 - Benzaldehyde, 2-chloro-
530 - Benzaldehyde, 3-chloro-
531 - Benzaldehyde, 4-chloro-
2886 - Benzoyl chloride

$C_7H_5ClO_2$
532 - Benzaldehyde, 3-chloro-4-hydroxy-
533 - Benzaldehyde, 5-chloro-2-hydroxy-
2534 - Benzoic acid, 2-chloro-
2535 - Benzoic acid, 3-chloro-
2536 - Benzoic acid, 4-chloro-
2900 - Benzoyl chloride, 2-hydroxy-
4026 - Carbonochloridic acid, phenyl ester

$C_7H_5ClO_3$
1127 - Benzenecarboperoxoic acid, 3-chloro-
2540 - Benzoic acid, 3-chloro-4-hydroxy-

$C_7H_5Cl_2F$
1379 - Benzene, (dichlorofluoromethyl)-

$C_7H_5Cl_2FN_2O_3$
5982 - Fluroxypyr

$C_7H_5Cl_2NO_2$
1135 - Benzene, 1-chloro-4-(chloromethyl)-2-nitro-
2491 - Benzoic acid, 2-amino-3,6-dichloro-
2492 - Benzoic acid, 3-amino-2,5-dichloro-

$C_7H_5Cl_2NO_4S$
2565 - Benzoic acid, 4-[(dichloroamino)sulfonyl]-

$C_7H_5Cl_3$
1357 - Benzene, 1,2-dichloro-4-(chloromethyl)-
1358 - Benzene, 1,3-dichloro-5-(chloromethyl)-
2333 - Benzene, (trichloromethyl)-
2334 - Benzene, 1,2,3-trichloro-4-methyl-
2335 - Benzene, 1,2,3-trichloro-5-methyl-
2336 - Benzene, 1,2,4-trichloro-5-methyl-
2337 - Benzene, 1,2,5-trichloro-3-methyl-

$C_7H_5Cl_3O$
2330 - Benzene, 1,2,4-trichloro-3-methoxy-
2331 - Benzene, 1,2,4-trichloro-5-methoxy-
2332 - Benzene, 1,3,5-trichloro-2-methoxy-
9487 - Phenol, 2,3,4-trichloro-6-methyl-
9488 - Phenol, 2,3,6-trichloro-4-methyl-
9489 - Phenol, 2,4,6-trichloro-3-methyl-

C_7H_5FO
559 - Benzaldehyde, 2-fluoro-
560 - Benzaldehyde, 3-fluoro-
561 - Benzaldehyde, 4-fluoro-
2909 - Benzoyl fluoride

$C_7H_5FO_2$
2603 - Benzoic acid, 2-fluoro-
2604 - Benzoic acid, 3-fluoro-
2605 - Benzoic acid, 4-fluoro-

$C_7H_5F_3$
2348 - Benzene, (trifluoromethyl)-

$C_7H_5F_3O$
9491 - Phenol, 3-(trifluoromethyl)-

$C_7H_5HgNO_3$
7437 - Mercury, [2-methyl-5-nitrophenolato(2-)-
C6,O1]-

C_7H_5IO
580 - Benzaldehyde, 2-iodo-
581 - Benzaldehyde, 3-iodo-
582 - Benzaldehyde, 4-iodo-
2910 - Benzoyl iodide

$C_7H_5IO_2$
2657 - Benzoic acid, 2-iodo-
2658 - Benzoic acid, 3-iodo-
2659 - Benzoic acid, 4-iodo-

$C_7H_5IO_3$
2623 - Benzoic acid, 4-hydroxy-3-iodo-
2624 - Benzoic acid, 5-hydroxy-2-iodo-

$C_7H_5I_3$
2352 - Benzene, 1,3,5-triiodo-2-methyl-

C_7H_5N
2740 - Benzonitrile

C_7H_5NO
1786 - Benzene, isocyanato-
2389 - 1,2-Benzisoxazole
2390 - 2,1-Benzisoxazole
2759 - Benzonitrile, 2-hydroxy-
2877 - Benzoxazole
10585 - 2-Propenenitrile, 3-(2-furanyl)-

C_7H_5NOS
2868 - 2(3H)-Benzothiazolone

$C_7H_5NO_2$
2881 - 2-Benzoxazolol
2882 - 2(3H)-Benzoxazolone

$C_7H_5NO_3$
591 - Benzaldehyde, 2-nitro-
592 - Benzaldehyde, 3-nitro-
593 - Benzaldehyde, 4-nitro-

$C_7H_5NO_3S$
2388 - 1,2-Benzisothiazol-3(2H)-one, 1,1-dioxide

$C_7H_5NO_4$
2703 - Benzoic acid, 2-nitro-
2704 - Benzoic acid, 3-nitro-
2705 - Benzoic acid, 4-nitro-
10927 - 2,3-Pyridinedicarboxylic acid
10928 - 2,4-Pyridinedicarboxylic acid
10929 - 2,5-Pyridinedicarboxylic acid
10930 - 3,4-Pyridinedicarboxylic acid
10931 - 3,5-Pyridinedicarboxylic acid

$C_7H_5NO_5$
2646 - Benzoic acid, 2-hydroxy-3-nitro-
2647 - Benzoic acid, 2-hydroxy-5-nitro-

C_7H_5NS
1792 - Benzene, isothiocyanato-
2387 - 1,2-Benzisothiazole
2861 - Benzothiazole
11709 - Thiocyanic acid, phenyl ester

$C_7H_5NS_2$
2866 - 2(3H)-Benzothiazolethione

$C_7H_5N_3$
2870 - 1,2,4-Benzotriazine

$C_7H_5N_3O$
2884 - Benzoyl azide

$C_7H_5N_3O_5$
614 - Benzamide, 3,5-dinitro-

$C_7H_5N_3O_6$
2046 - Benzene, 1-methyl-2,3,4-trinitro-
2047 - Benzene, 1-methyl-2,4,5-trinitro-
2048 - Benzene, 2-methyl-1,3,5-trinitro-

$C_7H_5N_3O_7$
364 - Anisole, 2,3,5-trinitro-
1946 - Benzene, 2-methoxy-1,3,5-trinitro-
9434 - Phenol, 3-methyl-2,4,6-trinitro-

$C_7H_5N_5O_8$
818 - Benzenamine, N-methyl-N,2,4,6-tetranitro-

$C_7H_5NaO_2$
2725 - Benzoic acid, sodium salt

C_7H_6BrCl
1007 - Benzene, 1-bromo-3-(chloromethyl)-
1008 - Benzene, 1-bromo-4-(chloromethyl)-

C_7H_6BrNO
604 - Benzamide, 2-bromo-
605 - Benzamide, 3-bromo-

$C_7H_6BrNO_2$
1066 - Benzene, 1-bromo-2-methyl-4-nitro-
1067 - Benzene, 1-(bromomethyl)-3-nitro-
1068 - Benzene, 1-bromo-3-methyl-2-nitro-
1069 - Benzene, 1-bromo-3-methyl-5-nitro-
1070 - Benzene, 1-bromo-4-methyl-2-nitro-
1071 - Benzene, 2-bromo-1-methyl-3-nitro-
1072 - Benzene, 2-bromo-1-methyl-4-nitro-
1073 - Benzene, 2-bromo-4-methyl-1-nitro-
1074 - Benzene, 4-bromo-1-methyl-2-nitro-
2487 - Benzoic acid, 2-amino-5-bromo-

$C_7H_6BrNO_3$
606 - Benzamide, 5-bromo-N,2-dihydroxy-

$C_7H_6Br_2$
999 - Benzene, 1-bromo-2-(bromomethyl)-
1000 - Benzene, 1-bromo-4-(bromomethyl)-
1301 - Benzene, (dibromomethyl)-
1302 - Benzene, 1,2-dibromo-3-methyl-
1303 - Benzene, 1,2-dibromo-4-methyl-
1304 - Benzene, 1,3-dibromo-2-methyl-
1305 - Benzene, 1,4-dibromo-2-methyl-
1306 - Benzene, 2,4-dibromo-1-methyl-

$C_7H_6Br_2O$
9279 - Phenol, 2,4-dibromo-6-methyl-
9280 - Phenol, 3,5-dibromo-2-methyl-

C_7H_6ClF
1169 - Benzene, 1-chloro-3-fluoro-2-methyl-
1170 - Benzene, 2-chloro-4-fluoro-1-methyl-
1194 - Benzene, 1-(chloromethyl)-2-fluoro-
1195 - Benzene, 1-(chloromethyl)-4-fluoro-

C_7H_6ClI
1173 - Benzene, 1-chloro-2-iodo-4-methyl-
1174 - Benzene, 4-chloro-1-iodo-2-methyl-
1175 - Benzene, 4-chloro-2-iodo-1-methyl-
11817 - Toluene, 3-chloro-2-iodo-

$C_7H_6ClNO_2$
1203 - Benzene, 1-(chloromethyl)-2-nitro-
1204 - Benzene, 1-(chloromethyl)-3-nitro-
1205 - Benzene, 1-chloro-2-methyl-4-nitro-
1206 - Benzene, 1-(chloromethyl)-3-nitro-
1207 - Benzene, 1-chloro-3-methyl-5-nitro-
1208 - Benzene, 1-(chloromethyl)-4-nitro-
1209 - Benzene, 1-chloro-4-methyl-2-nitro-
1210 - Benzene, 2-chloro-1-methyl-4-nitro-
1211 - Benzene, 2-chloro-4-methyl-1-nitro-
1212 - Benzene, 4-chloro-1-methyl-2-nitro-
1213 - Benzene, 4-chloro-2-methyl-1-nitro-
2490 - Benzoic acid, 5-amino-2-chloro-

$C_7H_6ClN_3O_4S_2$
2852 - 2H-1,2,4-Benzothiadiazine-7-sulfonamide, 6-
chloro-, 1,1-dioxide

$C_7H_6Cl_2$
1132 - Benzene, 1-chloro-2-(chloromethyl)-
1133 - Benzene, 1-chloro-3-(chloromethyl)-
1134 - Benzene, 1-chloro-4-(chloromethyl)-
1382 - Benzene, (dichloromethyl)-
1383 - Benzene, 1,2-dichloro-3-methyl-
1384 - Benzene, 1,2-dichloro-4-methyl-
1385 - Benzene, 1,3-dichloro-2-methyl-
1386 - Benzene, 1,3-dichloro-5-methyl-
1387 - Benzene, 1,4-dichloro-2-methyl-
1388 - Benzene, 2,4-dichloro-1-methyl-

$C_7H_6Cl_2O$
1381 - Benzene, 2,4-dichloro-1-methoxy-
1853 - Benzenemethanol, 2,4-dichloro-
9292 - Phenol, 2,4-dichloro-3-methyl-
9293 - Phenol, 2,4-dichloro-5-methyl-
9294 - Phenol, 2,6-dichloro-3-methyl-
9295 - Phenol, 2,6-dichloro-4-methyl-

$C_7H_6Cl_2S$
1136 - Benzene, 1-chloro-4-[(chloromethyl)thio]-

$C_7H_6Cl_4O_2$
4770 - 1,3-Cyclopentadiene, 1,2,3,4-tetrachloro-5,5-
dimethoxy-

$C_7H_6FNO_2$
1729 - Benzene, 1-fluoro-4-methyl-2-nitro-
1730 - Benzene, 2-fluoro-4-methyl-1-nitro-

1731 - Benzene, 4-fluoro-1-methyl-2-nitro-
1732 - Benzene, 4-fluoro-2-methyl-1-nitro-

$C_7H_6F_2$
1425 - Benzene, (difluoromethyl)-

$C_7H_6F_3N$
855 - Benzenamine, 2-(trifluoromethyl)-
856 - Benzenamine, 3-(trifluoromethyl)-
857 - Benzenamine, 4-(trifluoromethyl)-

$C_7H_6N_2$
2376 - 1H-Benzimidazole
2741 - Benzonitrile, 2-amino-
2742 - Benzonitrile, 3-amino-
7235 - 1H-Indazole

$C_7H_6N_2O$
10884 - 3-Pyridinecarbonitrile, 1,2-dihydro-1-methyl-
2-oxo-

$C_7H_6N_2O_3$
624 - Benzamide, 2-nitro-
625 - Benzamide, 3-nitro-

$C_7H_6N_2O_4$
1956 - Benzene, 1-methyl-2,4-dinitro-
1957 - Benzene, 1-methyl-3,5-dinitro-
1958 - Benzene, 2-methyl-1,3-dinitro-
1959 - Benzene, 2-methyl-1,4-dinitro-
1960 - Benzene, 4-methyl-1,2-dinitro-
2506 - Benzoic acid, 2-amino-3-nitro-

$C_7H_6N_2O_5$
1922 - Benzene, 1-methoxy-2,3-dinitro-
1923 - Benzene, 1-methoxy-2,4-dinitro-
1924 - Benzene, 1-methoxy-3,5-dinitro-
1925 - Benzene, 2-methoxy-1,3-dinitro-
1926 - Benzene, 2-methoxy-1,4-dinitro-
1927 - Benzene, 4-methoxy-1,2-dinitro-
9405 - Phenol, 2-methyl-4,6-dinitro-

$C_7H_6N_2S$
2386 - 2H-Benzimidazole-2-thione, 1,3-dihydro-
2859 - 2-Benzothiazolamine
2860 - 6-Benzothiazolamine

$C_7H_6N_4O$
7193 - 1H-Imidazole, 1,1'-carbonylbis-

C_7H_6O
525 - Benzaldehyde
4242 - 2,4,6-Cycloheptatrien-1-one

C_7H_6OS
1130 - Benzenecarbothioic acid

$C_7H_6OS_2$
1125 - Benzenecarbodithioic acid, 2-hydroxy-

$C_7H_6O_2$
564 - Benzaldehyde, 2-hydroxy-
565 - Benzaldehyde, 3-hydroxy-
566 - Benzaldehyde, 4-hydroxy-
2415 - 1,3-Benzodioxole
2475 - Benzoic acid
4243 - 2,4,6-Cycloheptatrien-1-one, 2-hydroxy-
4266 - 2,5-Cyclohexadiene-1,4-dione, 2-methyl-
10512 - 2-Propenal, 3-(2-furanyl)-
10513 - 2-Propenal, 3-(2-furanyl)-, (E)-

$C_7H_6O_2S$
2663 - Benzoic acid, 2-mercapto-

$C_7H_6O_3$
540 - Benzaldehyde, 2,3-dihydroxy-
541 - Benzaldehyde, 2,4-dihydroxy-
542 - Benzaldehyde, 3,4-dihydroxy-
1126 - Benzenecarboperoxoic acid
2612 - Benzoic acid, 2-hydroxy-
2613 - Benzoic acid, 4-hydroxy-
10638 - 2-Propenoic acid, 3-(2-furanyl)-

$C_7H_6O_4$
2568 - Benzoic acid, 2,3-dihydroxy-
2569 - Benzoic acid, 2,4-dihydroxy-
2570 - Benzoic acid, 2,5-dihydroxy-
2571 - Benzoic acid, 3,4-dihydroxy-
6183 - 4H-Furo[3,2-c]pyran-2(6H)-one, 4-hydroxy-

$C_7H_6O_5$
2730 - Benzoic acid, 2,3,4-trihydroxy-
2731 - Benzoic acid, 3,4,5-trihydroxy-

$C_7H_6O_6S$
2655 - Benzoic acid, 2-hydroxy-5-sulfo-

$C_7H_7BF_2$
3115 - Borane, difluoro(4-methylphenyl)-

C_7H_7Br
1053 - Benzene, (bromomethyl)-
1054 - Benzene, 1-bromo-2-methyl-
1055 - Benzene, 1-bromo-3-methyl-
1056 - Benzene, 1-bromo-4-methyl-

$C_7H_7BrN_2O$
11998 - Urea, (4-bromophenyl)-

C_7H_7BrO
1050 - Benzene, 1-bromo-2-methoxy-
1051 - Benzene, 1-bromo-3-methoxy-
1052 - Benzene, 1-bromo-4-methoxy-
9236 - Phenol, 2-bromo-4-methyl-
9237 - Phenol, 2-bromo-5-methyl-
9238 - Phenol, 3-bromo-2-methyl-
9239 - Phenol, 3-bromo-4-methyl-
9240 - Phenol, 3-bromo-5-methyl-
9241 - Phenol, 4-bromo-2-methyl-
9242 - Phenol, 4-bromo-3-methyl-

$C_7H_7BrO_2$
1843 - Benzenemethanol, 5-bromo-2-hydroxy-

$C_7H_7BrO_3$
6190 - 3-Furoic acid, 5-bromo-, ethyl ester

C_7H_7Cl
1182 - Benzene, (chloromethyl)-
1183 - Benzene, 1-chloro-2-methyl-
1184 - Benzene, 1-chloro-3-methyl-
1185 - Benzene, 1-chloro-4-methyl-

$C_7H_7ClN_2S$
11801 - Thiourea, (2-chlorophenyl)-

C_7H_7ClO
1178 - Benzene, 1-chloro-2-methoxy-
1179 - Benzene, 1-chloro-3-methoxy-
1180 - Benzene, 1-chloro-4-methoxy-
1848 - Benzenemethanol, 2-chloro-
1849 - Benzenemethanol, 4-chloro-
9260 - Phenol, 2-chloro-3-methyl-
9261 - Phenol, 2-chloro-4-methyl-
9262 - Phenol, 2-chloro-5-methyl-
9263 - Phenol, 2-chloro-6-methyl-
9264 - Phenol, 3-chloro-2-methyl-
9265 - Phenol, 3-chloro-4-methyl-
9266 - Phenol, 4-chloro-2-methyl-
9267 - Phenol, 4-chloro-3-methyl-

C_7H_7ClOS
2186 - Benzenesulfinyl chloride, 4-methyl-

$C_7H_7ClO_2S$
2249 - Benzenesulfonyl chloride, 2-methyl-
2250 - Benzenesulfonyl chloride, 4-methyl-

$C_7H_7ClO_3$
6053 - 2-Furancarboxylic acid, 5-chloro-, ethyl ester

$C_7H_7ClO_3S$
2248 - Benzenesulfonyl chloride, 2-methoxy-

C_7H_7ClS
1833 - Benzenemethanethiol, 4-chloro-

$C_7H_7Cl_2N$
1803 - Benzenemethanamine, 2,4-dichloro-

$C_7H_7Cl_2NO_2S$
2210 - Benzenesulfonamide, N,N-dichloro-4-methyl-

$C_7H_7Cl_3NO_3PS$
4090 - Chlorpyrifos-methyl

$C_7H_7Cl_3Si$
11467 - Silane, trichloro(2-methylphenyl)-
11468 - Silane, trichloro(3-methylphenyl)-

C_7H_7F
1725 - Benzene, (fluoromethyl)-
1726 - Benzene, 1-fluoro-2-methyl-
1727 - Benzene, 1-fluoro-3-methyl-
1728 - Benzene, 1-fluoro-4-methyl-

C_7H_7FO
1722 - Benzene, 1-fluoro-2-methoxy-
1723 - Benzene, 1-fluoro-3-methoxy-
1724 - Benzene, 1-fluoro-4-methoxy-
1879 - Benzenemethanol, 3-fluoro-

1880 - Benzenemethanol, 4-fluoro-

$C_7H_7F_7O_2$
3526 - Butanoic acid, heptafluoro-, 1-methylethyl
ester

C_7H_7I
1777 - Benzene, (iodomethyl)-
1778 - Benzene, 1-iodo-2-methyl-
1779 - Benzene, 1-iodo-3-methyl-
1780 - Benzene, 1-iodo-4-methyl-

C_7H_7IO
1775 - Benzene, 1-iodo-2-methoxy-
1776 - Benzene, 1-iodo-4-methoxy-
1888 - Benzenemethanol, 2-iodo-
9374 - Phenol, 2-iodo-4-methyl-
9375 - Phenol, 2-iodo-6-methyl-

C_7H_7N
10952 - Pyridine, 2-ethenyl-
10953 - Pyridine, 3-ethenyl-
10954 - Pyridine, 4-ethenyl-

C_7H_7NO
594 - Benzaldehyde, oxime, (E)-
595 - Benzaldehyde, oxime, (Z)-
600 - Benzamide
5867 - Ethanone, 1-(2-pyridinyl)-
5868 - Ethanone, 1-(3-pyridinyl)-
5869 - Ethanone, 1-(4-pyridinyl)-
5999 - Formamide, N-phenyl-

$C_7H_7NO_2$
579 - Benzaldehyde, 2-hydroxy-, oxime
616 - Benzamide, 2-hydroxy-
2000 - Benzene, 1-methyl-2-nitro-
2001 - Benzene, 1-methyl-3-nitro-
2002 - Benzene, 1-methyl-4-nitro-
2052 - Benzene, (nitromethyl)-
2484 - Benzoic acid, 2-amino-
2485 - Benzoic acid, 3-amino-
2486 - Benzoic acid, 4-amino-
8097 - Nitrous acid, phenylmethyl ester
10906 - 3-Pyridinecarboxylic acid, methyl ester
10908 - 4-Pyridinecarboxylic acid, methyl ester
11029 - Pyridinium, 3-carboxy-1-methyl-, hydroxide,
inner salt

$C_7H_7NO_2S$
2043 - Benzene, 1-(methylthio)-2-nitro-
2044 - Benzene, 1-(methylthio)-4-nitro-

$C_7H_7NO_3$
613 - Benzamide, N,2-dihydroxy-
1906 - Benzenemethanol, 2-nitro-
1907 - Benzenemethanol, 3-nitro-
1908 - Benzenemethanol, 4-nitro-
1935 - Benzene, 1-methoxy-2-nitro-
1936 - Benzene, 1-methoxy-3-nitro-
1937 - Benzene, 1-methoxy-4-nitro-
2498 - Benzoic acid, 4-amino-2-hydroxy-
2499 - Benzoic acid, 5-amino-2-hydroxy-
9422 - Phenol, 2-methyl-4-nitro-
9423 - Phenol, 2-methyl-6-nitro-
9424 - Phenol, 4-methyl-2-nitro-

$C_7H_7NO_4$
9386 - Phenol, 2-methoxy-3-nitro-

$C_7H_7NO_4S$
2508 - Benzoic acid, 4-(aminosulfonyl)-

$C_7H_7N_3$
980 - Benzene, (azidomethyl)-
981 - Benzene, 1-azido-2-methyl-
982 - Benzene, 1-azido-4-methyl-
2872 - 1H-Benzotriazole, 1-methyl-

$C_7H_7N_3O_3$
12025 - Urea, (4-nitrophenyl)-

C_7H_8
1947 - Benzene, methyl-
2913 - Bicyclo[2.2.1]hepta-2,5-diene
4240 - 1,3,5-Cycloheptatriene
4765 - 1,3-Cyclopentadiene, 5-ethylidene-
6341 - 1,6-Heptadien-3-yne
6342 - 1,5-Heptadiyne
6343 - 1,6-Heptadiyne

C7H8BrN
647 - Benzenamine, 2-bromo-4-methyl-
648 - Benzenamine, 2-bromo-5-methyl-
649 - Benzenamine, 3-bromo-4-methyl-
650 - Benzenamine, 4-bromo-2-methyl-
651 - Benzenamine, 4-bromo-3-methyl-
652 - Benzenamine, 5-bromo-2-methyl-
1799 - Benzenemethanamine, 4-bromo-

C7H8ClN
668 - Benzenamine, 2-chloro-4-methyl-
669 - Benzenamine, 2-chloro-5-methyl-
670 - Benzenamine, 2-chloro-N-methyl-
671 - Benzenamine, 3-chloro-2-methyl-
672 - Benzenamine, 3-chloro-4-methyl-
673 - Benzenamine, 4-chloro-2-methyl-
674 - Benzenamine, 4-chloro-3-methyl-
675 - Benzenamine, 4-chloro-N-methyl-
676 - Benzenamine, 5-chloro-2-methyl-

C7H8ClNO
666 - Benzenamine, 2-chloro-5-methoxy-
667 - Benzenamine, 4-chloro-2-methoxy-

C7H8ClN3O4S2
2855 - 2H-1,2,4-Benzothiadiazine-7-sulfonamide, 6-
chloro-3,4-dihydro-, 1,1-dioxide

C7H8Cl2Si
11401 - Silane, dichloromethylphenyl-

C7H8FN
755 - Benzenamine, 4-fluoro-2-methyl-

C7H8IN
762 - Benzenamine, 2-iodo-4-methyl-
763 - Benzenamine, 5-iodo-2-methyl-

C7H8N2
563 - Benzaldehyde, hydrazone
11137 - 1H-Pyrrole-1-propanenitrile

C7H8N2O
89 - Acetamide, N-3-pyridinyl-
601 - Benzamide, 2-amino-
806 - Benzenamine, N-methyl-N-nitroso-
2611 - Benzoic acid, hydrazide
12026 - Urea, phenyl-

C7H8N2O2
799 - Benzenamine, 2-methyl-3-nitro-
800 - Benzenamine, 2-methyl-4-nitro-
801 - Benzenamine, 2-methyl-5-nitro-
802 - Benzenamine, 2-methyl-6-nitro-
803 - Benzenamine, 4-methyl-2-nitro-
804 - Benzenamine, N-methyl-2-nitro-
805 - Benzenamine, N-methyl-4-nitro-
2552 - Benzoic acid, 2,3-diamino-
2553 - Benzoic acid, 2,4-diamino-
2554 - Benzoic acid, 3,5-diamino-

C7H8N2O3
775 - Benzenamine, 2-methoxy-5-nitro-
776 - Benzenamine, 3-methoxy-4-nitro-
777 - Benzenamine, 5-methoxy-2-nitro-

C7H8N2S
11808 - Thiourea, phenyl-

C7H8N4O2
10756 - 1H-Purine-2,6-dione, 3,7-dihydro-1,3-
dimethyl-
10757 - 1H-Purine-2,6-dione, 3,7-dihydro-3,7-
dimethyl-

C7H8O
1836 - Benzenemethanol
1915 - Benzene, methoxy-
9400 - Phenol, 2-methyl-
9401 - Phenol, 3-methyl-
9402 - Phenol, 4-methyl-

C7H8OS
2040 - Benzene, (methylsulfinyl)-
2299 - Benzenethiol, 3-methoxy-
2300 - Benzenethiol, 4-methoxy-
5856 - Ethanone, 1-(5-methyl-2-thienyl)-

C7H8O2
1527 - 1,2-Benzenediol, 3-methyl-
1528 - 1,2-Benzenediol, 4-methyl-
1529 - 1,3-Benzenediol, 2-methyl-

1530 - 1,3-Benzenediol, 4-methyl-
1531 - 1,3-Benzenediol, 5-methyl-
1532 - 1,4-Benzenediol, 2-methyl-
1882 - Benzenemethanol, 2-hydroxy-
1883 - Benzenemethanol, 3-hydroxy-
1884 - Benzenemethanol, 4-hydroxy-
9379 - Phenol, 2-methoxy-
9380 - Phenol, 3-methoxy-
9381 - Phenol, 4-methoxy-
10438 - 1-Propanone, 1-(2-furanyl)-
10439 - 2-Propanone, 1-(2-furanyl)-
10792 - 2H-Pyran-2-one, 4,6-dimethyl-
10793 - 4H-Pyran-4-one, 2,6-dimethyl-

C7H8O2S
2181 - Benzenesulfinic acid, 2-methyl-
2182 - Benzenesulfinic acid, 4-methyl-
11739 - 2-Thiophenecarboxylic acid, ethyl ester

C7H8O3
1520 - 1,4-Benzenediol, 2-(hydroxymethyl)-
1525 - 1,2-Benzenediol, 3-methoxy-
1526 - 1,3-Benzenediol, 5-methoxy-
2369 - 1,2,3-Benzenetriol, 5-methyl-
6029 - 2-Furanacetic acid, methyl ester
6056 - 2-Furancarboxylic acid, ethyl ester
6119 - 2-Furanmethanol, acetate
6161 - 2-Furanpropanoic acid

C7H8O3S
2227 - Benzenesulfonic acid, methyl ester
2228 - Benzenesulfonic acid, 2-methyl-
2229 - Benzenesulfonic acid, 4-methyl-

C7H8S
1832 - Benzenemethanethiol
2041 - Benzene, (methylthio)-
2301 - Benzenethiol, 2-methyl-
2302 - Benzenethiol, 3-methyl-
2303 - Benzenethiol, 4-methyl-

C7H8S2
1564 - 1,2-Benzenedithiol, 4-methyl-

C7H9AsN2O4
455 - Arsonic acid,
[4-[(aminocarbonyl)amino]phenyl]-

C7H9BO2
3134 - Boronic acid, (2-methylphenyl)-

C7H9Cl
7107 - 1-Hexen-3-yne, 5-chloro-5-methyl-

C7H9ClO
9147 - 2-Pentynoyl chloride, 4,4-dimethyl-

C7H9ClSi
11381 - Silane, chloromethylphenyl-

C7H9N
778 - Benzenamine, 2-methyl-
779 - Benzenamine, 3-methyl-
780 - Benzenamine, 4-methyl-
781 - Benzenamine, N-methyl-
1795 - Benzenemethanamine
4629 - 3-Cyclohexene-1-carbonitrile
4630 - 1-Cyclohexenecarbonitrile
4905 - 1-Cyclopentene-1-acetonitrile
10936 - Pyridine, 2,3-dimethyl-
10937 - Pyridine, 2,4-dimethyl-
10938 - Pyridine, 2,5-dimethyl-
10939 - Pyridine, 2,6-dimethyl-
10940 - Pyridine, 3,4-dimethyl-
10941 - Pyridine, 3,5-dimethyl-
10956 - Pyridine, 2-ethyl-
10957 - Pyridine, 3-ethyl-
10958 - Pyridine, 4-ethyl-

C7H9NO
758 - Benzenamine, N-hydroxy-4-methyl-
766 - Benzenamine, 2-methoxy-
767 - Benzenamine, 3-methoxy-
768 - Benzenamine, 4-methoxy-
1837 - Benzenemethanol, 2-amino-
5849 - Ethanone, 1-(1-methyl-1H-pyrrol-2-yl)-
9209 - Phenol, 2-amino-3-methyl-
9210 - Phenol, 2-amino-4-methyl-
9211 - Phenol, 3-amino-4-methyl-

9212 - Phenol, 3-amino-5-methyl-
9213 - Phenol, 4-amino-2-methyl-
10945 - Pyridine, 2,6-dimethyl-, 1-oxide
10950 - 2-Pyridineethanol

C7H9NO2
2509 - Benzoic acid, ammonium salt
10935 - Pyridine, 2,6-dimethoxy-

C7H9NO3
8586 - 2,4-Oxazolidinedione, 5-methyl-3-(2-
propenyl)-

C7H9NS
819 - Benzenamine, 2-(methylthio)-
820 - Benzenamine, 4-(methylthio)-

C7H9N3O
7145 - Hydrazinecarboxamide, 2-phenyl-
7146 - Hydrazinecarboxamide, N-phenyl-

C7H9N3O2
2500 - Benzoic acid, 4-amino-2-hydroxy-, hydrazide

C7H9N3O2S2
2189 - Benzenesulfonamide, 4-amino-N-
(aminothioxomethyl)-

C7H9N3O3S
2190 - Benzenesulfonamide, 4-amino-N-
(aminocarbonyl)-

C7H10
4219 - 1,3-Cycloheptadiene
4271 - 1,3-Cyclohexadiene, 2-methyl-
4272 - 1,3-Cyclohexadiene, 5-methyl-, (±)-
6529 - 1,3,5-Heptatriene
6606 - 1-Hepten-3-yne
9115 - 3-Penten-1-yne, 3-ethyl-

C7H10ClN
791 - Benzenamine, 2-methyl-, hydrochloride
792 - Benzenamine, N-methyl-, hydrochloride

C7H10ClN3
11043 - 4-Pyrimidinamine, 2-chloro-N,N,6-trimethyl-

C7H10N2
1275 - 1,2-Benzenediamine, 3-methyl-
1276 - 1,2-Benzenediamine, 4-methyl-
1277 - 1,3-Benzenediamine, 2-methyl-
1278 - 1,3-Benzenediamine, 4-methyl-
1279 - 1,4-Benzenediamine, 2-methyl-
1280 - 1,4-Benzenediamine, N-methyl-
1796 - Benzenemethanamine, 2-amino-
4159 - Cyanamide, di-2-propenyl-
6392 - Heptanedinitrile
7164 - Hydrazine, (3-methylphenyl)-
7165 - Hydrazine, 1-methyl-1-phenyl-
7166 - Hydrazine, 1-methyl-2-phenyl-
10863 - 2-Pyridinamine, N,N-dimethyl-
10864 - 4-Pyridinamine, 2,6-dimethyl-
10948 - 2-Pyridineethanamine
10949 - 4-Pyridineethanamine

C7H10N2O
1273 - 1,2-Benzenediamine, 4-methoxy-
1274 - 1,3-Benzenediamine, 4-methoxy-

C7H10N2OS
11109 - 4(1H)-Pyrimidinone, 2,3-dihydro-6-propyl-2-
thioxo-

C7H10N2O2S
7194 - 1H-Imidazole-1-carboxylic acid, 2,3-dihydro-3-
methyl-2-thioxo-, ethyl ester

C7H10N4O2S
2191 - Benzenesulfonamide, 4-amino-N-
(aminoiminomethyl)-

C7H10O
4631 - 1-Cyclohexene-1-carboxaldehyde
4632 - 3-Cyclohexene-1-carboxaldehyde
4710 - 2-Cyclohexen-1-one, 2-methyl-
4711 - 2-Cyclohexen-1-one, 3-methyl-
4712 - 2-Cyclohexen-1-one, 5-methyl-
4713 - 3-Cyclohexen-1-one, 3-methyl-
4714 - 3-Cyclohexen-1-one, 4-methyl-
4866 - Cyclopentanol, 1-ethynyl-
4943 - 2-Cyclopenten-1-one, 2,4-dimethyl-
6106 - Furan, 2-ethyl-5-methyl-
6167 - Furan, 2-propyl-

6325 - 2,4-Heptadienal, (*E,E*)-
6339 - 3,5-Heptadien-2-one
7607 - Methanone, dicyclopropyl-

$C_7H_{10}O_2$
3856 - 2-Butenoic acid, 2-propenyl ester
3857 - 2-Butenoic acid, 2-propenyl ester, (*E*)-
4225 - 1,2-Cycloheptanedione
4637 - 1-Cyclohexene-1-carboxylic acid
4638 - 3-Cyclohexene-1-carboxylic acid
4879 - Cyclopentanone, 2-acetyl-
4904 - 3-Cyclopentene-1-acetic acid
6104 - Furan, 2-(ethoxymethyl)-
6112 - Furan, 2-isopropoxy-
6702 - 2,4-Hexadienoic acid, methyl ester
6703 - 2,4-Hexadienoic acid, methyl ester, (*E,E*)-
10671 - 2-Propenoic acid, 2-methyl-, 2-propenyl ester

$C_7H_{10}O_3$
3577 - Butanoic acid, 3-oxo-, 2-propenyl ester
6454 - 2,4,6-Heptanetrione
10668 - 2-Propenoic acid, 2-methyl-, oxiranylmethyl
       ester

$C_7H_{10}O_4$
3330 - Butanedioic acid, methylene-, dimethyl ester
3331 - Butanedioic acid, (1-methylethylidene)-
3789 - 2-Butenedioic acid, 2-methyl-, dimethyl ester,
       (*E*)-
3790 - 2-Butenedioic acid, 2-methyl-, dimethyl ester,
       (*Z*)-
4793 - 1,3-Cyclopentanedicarboxylic acid, *cis*-
4969 - 1,2-Cyclopropanedicarboxylic acid, dimethyl
       ester
8876 - Pentanoic acid, 2,4-dioxo-, ethyl ester
10563 - 2-Propene-1,1-diol, diacetate

$C_7H_{10}O_5$
8784 - Pentanedioic acid, 3-oxo-, dimethyl ester
9973 - Propanedioic acid, oxo-, diethyl ester

$C_7H_{10}O_6$
7553 - Methanetricarboxylic acid, trimethyl ester

$C_7H_{10}S$
11761 - Thiophene, 2-ethyl-3-methyl-
11762 - Thiophene, 2-ethyl-5-methyl-
11763 - Thiophene, 3-ethyl-4-methyl-
11764 - Thiophene, 4-ethyl-2-methyl-
11772 - Thiophene, 2-(1-methylethyl)-
11773 - Thiophene, 3-(1-methylethyl)-
11778 - Thiophene, 2-propyl-
11779 - Thiophene, 3-propyl-
11792 - Thiophene, 2,3,4-trimethyl-
11793 - Thiophene, 2,3,5-trimethyl-

$C_7H_{10}Si$
11442 - Silane, methylphenyl-

$C_7H_{11}Br$
6612 - 1-Heptyne, 1-bromo-
6613 - 2-Heptyne, 1-bromo-

$C_7H_{11}BrO_4$
9928 - Propanedioic acid, bromo-, diethyl ester

$C_7H_{11}Cl$
2918 - Bicyclo[2.2.1]heptane, 2-chloro-
6614 - 1-Heptyne, 1-chloro-
6615 - 2-Heptyne, 7-chloro-
6616 - 3-Heptyne, 7-chloro-
9125 - 1-Pentyne, 3-chloro-3-ethyl-

$C_7H_{11}ClO$
4330 - Cyclohexanecarbonyl chloride

$C_7H_{11}ClO_3$
3492 - Butanoic acid, 2-chloro-2-methyl-3-oxo-, ethyl
       ester

$C_7H_{11}ClO_4$
9931 - Propanedioic acid, chloro-, diethyl ester
9996 - 1,2-Propanediol, 3-chloro-, diacetate

$C_7H_{11}Cl_3O_2$
255 - Acetic acid, trichloro-, 3-methylbutyl ester

$C_7H_{11}I$
6620 - 1-Heptyne, 1-iodo-

$C_7H_{11}N$
4328 - Cyclohexanecarbonitrile
11128 - 1H-Pyrrole, 3-ethyl-4-methyl-

11129 - Pyrrole, 2-isopropyl-
11138 - 1H-Pyrrole, 1-propyl-
11139 - 1H-Pyrrole, 3-propyl-
11142 - 1H-Pyrrole, 1,2,5-trimethyl-

$C_7H_{11}NO$
4329 - Cyclohexanecarbonitrile, 1-hydroxy-
4416 - Cyclohexane, isocyanato-

$C_7H_{11}NO_2$
151 - Acetic acid, cyano-, butyl ester
9720 - 4-Piperidinone, 1-acetyl-
10913 - 3-Pyridinecarboxylic acid, 1,2,5,6-tetrahydro-
        1-methyl-
11158 - 2,5-Pyrrolidinedione, 3-ethyl-3-methyl-

$C_7H_{11}NO_3$
11567 - Succinimide, *N*-(ethoxymethyl)-

$C_7H_{11}NO_4$
9766 - *L*-Proline, 1-acetyl-4-hydroxy-, *trans*-

$C_7H_{11}NO_5$
9958 - Propanedioic acid, (hydroxyimino)-, diethyl
       ester

$C_7H_{11}NS$
4417 - Cyclohexane, isothiocyanato-

C_7H_{12}
2914 - Bicyclo[2.2.1]heptane
2915 - Bicyclo[4.1.0]heptane
2980 - Bicyclo[3,1,0]hexane, 1-methyl-
4244 - Cycloheptene
4428 - Cyclohexane, methylene-
4670 - Cyclohexene, 1-methyl-
4671 - Cyclohexene, 3-methyl-, (±)-
4672 - Cyclohexene, 4-methyl-
4813 - Cyclopentane, ethylidene-
4918 - Cyclopentene, 1,2-dimethyl-
4919 - Cyclopentene, 1,3-dimethyl-
4920 - Cyclopentene, 1,4-dimethyl-
4921 - Cyclopentene, 1,5-dimethyl-
4922 - Cyclopentene, 4,4-dimethyl-
4924 - Cyclopentene, 1-ethyl-
4925 - Cyclopentene, 3-ethyl-
4926 - Cyclopentene, 4-ethyl-
6326 - 1,2-Heptadiene
6327 - 1,4-Heptadiene
6328 - 1,5-Heptadiene
6329 - 2,4-Heptadiene
6609 - 1-Heptyne
6610 - 2-Heptyne
6611 - 3-Heptyne
6695 - 1,5-Hexadiene, 2-methyl-
6696 - 2,4-Hexadiene, 2-methyl-
7120 - 1-Hexyne, 5-methyl-
7121 - 2-Hexyne, 5-methyl-
7122 - 3-Hexyne, 2-methyl-
8665 - 1,3-Pentadiene, 2,4-dimethyl-
9128 - 1-Pentyne, 3,3-dimethyl-
9129 - 1-Pentyne, 4,4-dimethyl-
9130 - 2-Pentyne, 4,4-dimethyl-
9131 - 1-Pentyne, 3-ethyl-

$C_7H_{12}Br_2O_2$
8872 - Pentanoic acid, 2,3-dibromo-, ethyl ester
8873 - Pentanoic acid, 2,5-dibromo-, ethyl ester
10239 - Propanoic acid, 2,3-dibromo-, butyl ester

$C_7H_{12}ClN_5$
11513 - Simazine

$C_7H_{12}N_2$
7196 - 1H-Imidazole, 1,2-diethyl-
11190 - Pyrrolo[1,2-a]pyrimidine, 2,3,4,6,7,8-
        hexahydro-

$C_7H_{12}O$
4234 - Cycloheptanone
4331 - Cyclohexanecarboxaldehyde
4537 - Cyclohexanol, 2-methylene-
4596 - Cyclohexanone, 2-methyl-, (±)-
4597 - Cyclohexanone, 3-methyl-, (±)-
4598 - Cyclohexanone, 4-methyl-
4882 - Cyclopentanone, 2,5-dimethyl-
5795 - Ethanone, 1-cyclopentyl-
6336 - 1,5-Heptadien-4-ol

6337 - 1,6-Heptadien-4-ol
6530 - 2-Heptenal
6531 - 3-Heptenal
6583 - 1-Hepten-3-one
6584 - 4-Hepten-2-one
6585 - 6-Hepten-2-one
6586 - 6-Hepten-3-one
6587 - 3-Hepten-2-one, (*E*)-
6588 - 3-Hepten-2-one, (*Z*)-
6589 - 5-Hepten-2-one, (*E*)-
7093 - 1-Hexen-3-one, 5-methyl-
7094 - 3-Hexen-2-one, 5-methyl-
7095 - 4-Hexen-2-one, 5-methyl-
7096 - 4-Hexen-3-one, 2-methyl-
7097 - 5-Hexen-2-one, 3-methyl-
7098 - 5-Hexen-2-one, 4-methyl-
7099 - 5-Hexyn-2-one, 5-methyl-
7128 - 1-Hexyn-3-ol, 3-methyl-
7129 - 3-Hexyn-2-ol, 2-methyl-
9144 - 1-Pentyn-3-ol, 3,4-dimethyl-
9145 - 1-Pentyn-3-ol, 3-ethyl-

$C_7H_{12}O_2$
3587 - Butanoic acid, 2-propenyl ester
3849 - 2-Butenoic acid, 2-methyl-, ethyl ester, (*E*)-
3850 - 2-Butenoic acid, 3-methyl-, ethyl ester
4175 - Cyclobutanecarboxylic acid, ethyl ester
4332 - Cyclohexanecarboxylic acid
4775 - Cyclopentaneacetic acid
4783 - Cyclopentanecarboxylic acid, 1-methyl-
4865 - Cyclopentanol, acetate
6005 - Formic acid, cyclohexyl ester
6402 - 2,3-Heptanedione
6403 - 2,6-Heptanedione
6561 - 2-Heptenoic acid
6562 - 4-Heptenoic acid
6563 - 5-Heptenoic acid
6564 - 6-Heptenoic acid
6809 - 2,3-Hexanedione, 5-methyl-
6810 - 2,5-Hexanedione, 3-methyl-
7055 - 2-Hexenoic acid, 2-methyl-, (*E*)-
7056 - 2-Hexenoic acid, 4-methyl-
7057 - 2-Hexenoic acid, 5-methyl-
7058 - 3-Hexenoic acid, 2-methyl-
7059 - 3-Hexenoic acid, 3-methyl-
7060 - 3-Hexenoic acid, 4-methyl-
7061 - 4-Hexenoic acid, 5-methyl-
7062 - 3-Hexenoic acid, methyl ester
8799 - 2,4-Pentanedione, 3,3-dimethyl-
8801 - 2,4-Pentanedione, 3-methyl-
9075 - 4-Pentenoic acid, 2,2-dimethyl-
9083 - 2-Pentenoic acid, 2-methyl-, methyl ester
9084 - 2-Pentenoic acid, 3-methyl-, methyl ester, (*E*)-
9085 - 2-Pentenoic acid, 3-methyl-, methyl ester, (*Z*)-
9086 - 3-Pentenoic acid, 3-methyl-, methyl ester
10616 - 2-Propenoic acid, butyl ester
10664 - 2-Propenoic acid, 2-methyl-, 1-methylethyl
        ester
10672 - 2-Propenoic acid, 2-methyl-, propyl ester
10673 - 2-Propenoic acid, 2-methylpropyl ester
10731 - 1-Propyne, 3,3-diethoxy-
10783 - 2H-Pyran, 2-ethoxy-3,4-dihydro-

$C_7H_{12}O_3$
3519 - Butanoic acid, 2-ethyl-3-oxo, methyl ester
3557 - Butanoic acid, 2-methyl-3-oxo-, ethyl ester
3574 - Butanoic acid, 3-oxo-, 1-methylethyl ester
6125 - 2-Furanmethanol, tetrahydro-, acetate
6164 - 2-Furanpropanoic acid, tetrahydro-
6469 - Heptanoic acid, 6-oxo-
6901 - Hexanoic acid, 4-methyl-5-oxo-
8874 - Pentanoic acid, 2,2-dimethyl-4-oxo-
8898 - Pentanoic acid, 3-methyl-4-oxo-, methyl ester
8905 - Pentanoic acid, 2-oxo-, ethyl ester
8906 - Pentanoic acid, 3-oxo-, ethyl ester
8907 - Pentanoic acid, 4-oxo-, ethyl ester
10634 - 2-Propenoic acid, 2-ethoxyethyl ester
10662 - 2-Propenoic acid, 2-methyl-, 2-hydroxypropyl
        ester

$C_7H_{12}O_4$
3323 - Butanedioic acid, ethyl methyl ester
3327 - Butanedioic acid, methyl-, dimethyl ester
3338 - Butanedioic acid, monopropyl ester
6393 - Heptanedioic acid
6788 - Hexanedioic acid, 2-methyl-
6789 - Hexanedioic acid, 3-methyl-, (±)-
6792 - Hexanedioic acid, monomethyl ester
7654 - Methylene bispropionate
8773 - Pentanedioic acid, 3,3-dimethyl-
8775 - Pentanedioic acid, dimethyl ester
9929 - Propanedioic acid, butyl-
9939 - Propanedioic acid, diethyl-
9941 - Propanedioic acid, diethyl ester
9968 - Propanedioic acid, (2-methylpropyl)-
10002 - 1,2-Propanediol, diacetate
10003 - 1,3-Propanediol, diacetate
11643 - 2,4,8,10-Tetraoxaspiro[5.5]undecane

$C_7H_{12}O_5$
9957 - Propanedioic acid, hydroxy-, diethyl ester
10189 - 1,2,3-Propanetriol, diacetate
10190 - 1,2,3-Propanetriol, 1,3-diacetate

$C_7H_{12}O_6$
4338 - Cyclohexanecarboxylic acid, 1,3,4,5-tetrahydroxy-, [1R-(1α,3α,4α,5β)]-

$C_7H_{12}O_6Si$
11499 - Silanetriol, methyl-, triacetate

$C_7H_{13}Br$
4222 - Cycloheptane, bromo-
4318 - Cyclohexane, (bromomethyl)-
4319 - Cyclohexane, 1-bromo-1-methyl-
4320 - Cyclohexane, 1-bromo-3-methyl-

$C_7H_{13}BrN_2O_2$
3195 - Butanamide, N-(aminocarbonyl)-2-bromo-2-ethyl-

$C_7H_{13}BrO$
4317 - Cyclohexane, 1-bromo-2-methoxy-

$C_7H_{13}BrO_2$
3475 - Butanoic acid, 2-bromo-3-methyl-, ethyl ester
6457 - Heptanoic acid, 2-bromo-
6458 - Heptanoic acid, 7-bromo-
8860 - Pentanoic acid, 2-bromo-, ethyl ester
8861 - Pentanoic acid, 5-bromo-, ethyl ester
10206 - Propanoic acid, 3-bromo-, butyl ester

$C_7H_{13}Cl$
6538 - 1-Heptene, 1-chloro-
6539 - 1-Heptene, 2-chloro-
6540 - 3-Heptene, 4-chloro-

$C_7H_{13}ClO$
6513 - 2-Heptanone, 1-chloro-
6527 - Heptanoyl chloride
6988 - Hexanoyl chloride, 3-methyl-
6989 - Hexanoyl chloride, 4-methyl-

$C_7H_{13}ClO_2$
3490 - Butanoic acid, 2-chloro-3-methyl-, ethyl ester
3491 - Butanoic acid, 2-chloro-2-methyl-, ethyl ester
8867 - Pentanoic acid, 4-chloro-, ethyl ester
8868 - Pentanoic acid, 5-chloro-, ethyl ester
8869 - Pentanoic acid, 2-chloroethyl ester
10217 - Propanoic acid, 2-chloro-, butyl ester
10218 - Propanoic acid, 3-chloro-, butyl ester
10224 - Propanoic acid, 2-chloro-, 2-methylpropyl ester, (R)-
10227 - Propanoic acid, 3-chloro-, 2-methylpropyl ester
12039 - Valeric acid, 3-chloro-, ethyl ester

$C_7H_{13}FO_2$
6463 - Heptanoic acid, 7-fluoro-

$C_7H_{13}N$
480 - 1-Azabicyclo[2.2.2]octane
6428 - Heptanenitrile
9121 - 1-Pentyn-3-amine, 3-ethyl-
9666 - Piperidine, 3-ethenyl-

$C_7H_{13}NO$
485 - 1-Azabicyclo[2.2.2]octan-3-ol
5990 - Formamide, N-cyclohexyl-
9630 - Piperidine, 1-acetyl-

$C_7H_{13}NO_2$
4443 - Cyclohexane, 1-methyl-1-nitro-
4444 - Cyclohexane, 1-methyl-2-nitro-
4445 - Cyclohexane, 1-methyl-4-nitro-
4455 - Cyclohexane, (nitromethyl)-
9644 - 4-Piperidinecarboxylic acid, methyl ester
10633 - 2-Propenoic acid, 2-(dimethylamino)ethyl ester
11177 - Pyrrolidinium, 2-carboxy-1,1-dimethyl-, hydroxide, inner salt, (S)-

$C_7H_{13}NO_3$
10107 - Propanenitrile, 3-[2-(2-hydroxyethoxy)ethoxy]-
11178 - Pyrrolidinium, 2-carboxy-4-hydroxy-1,1-dimethyl-, hydroxide, inner salt, (2S-trans)-

$C_7H_{13}NO_3S$
7649 - L-Methionine, N-acetyl-

$C_7H_{13}NO_4$
6245 - L-Glutamic acid, 5-ethyl ester
9924 - Propanedioic acid, amino-, diethyl ester

$C_7H_{13}N_3O_3S$
8568 - Oxamyl

$C_7H_{13}O_6P$
7659 - Mevinphos

C_7H_{14}
3797 - 1-Butene, 2-ethyl-3-methyl-
3812 - 1-Butene, 2,3,3-trimethyl-
4221 - Cycloheptane
4426 - Cyclohexane, methyl-
4800 - Cyclopentane, 1,1-dimethyl-
4801 - Cyclopentane, 1,2-dimethyl-, cis-
4802 - Cyclopentane, 1,2-dimethyl-, trans-
4803 - Cyclopentane, 1,3-dimethyl-, cis-
4804 - Cyclopentane, 1,3-dimethyl-, trans-
4812 - Cyclopentane, ethyl-
6533 - 1-Heptene
6534 - 2-Heptene, (E)-
6535 - 2-Heptene, (Z)-
6536 - 3-Heptene, (E)-
6537 - 3-Heptene, (Z)-
7036 - 1-Hexene, 2-methyl-
7037 - 1-Hexene, 3-methyl-
7038 - 1-Hexene, 4-methyl-
7039 - 1-Hexene, 5-methyl-
7040 - 2-Hexene, 2-methyl-
7041 - 2-Hexene, 3-methyl-, (Z)-
7042 - 2-Hexene, 4-methyl-, (E)-
7043 - 2-Hexene, 4-methyl-, (Z)-
7044 - 2-Hexene, 5-methyl-, (E)-
7045 - 2-Hexene, 5-methyl-, (Z)-
7046 - 3-Hexene, 2-methyl-, (E)-
7047 - 3-Hexene, 2-methyl-, (Z)-
9029 - 1-Pentene, 2,3-dimethyl-
9030 - 1-Pentene, 2,4-dimethyl-
9031 - 1-Pentene, 3,3-dimethyl-
9032 - 1-Pentene, 3,4-dimethyl-
9033 - 1-Pentene, 4,4-dimethyl-
9034 - 2-Pentene, 2,3-dimethyl-
9035 - 2-Pentene, 2,4-dimethyl-
9036 - 2-Pentene, 3,4-dimethyl-
9037 - 2-Pentene, 3,4-dimethyl-, (E)-
9038 - 2-Pentene, 3,4-dimethyl-, (Z)-
9039 - 2-Pentene, 4,4-dimethyl-, (E)-
9040 - 2-Pentene, 4,4-dimethyl-, (Z)-
9043 - 1-Pentene, 2-ethyl-
9044 - 1-Pentene, 3-ethyl-
9045 - 2-Pentene, 3-ethyl-

$C_7H_{14}BrF$
6359 - Heptane, 1-bromo-7-fluoro-

$C_7H_{14}BrNO$
3199 - Butanamide, 2-bromo-2-ethyl-3-methyl-

$C_7H_{14}Br_2$
6372 - Heptane, 1,2-dibromo-
6373 - Heptane, 1,5-dibromo-
6374 - Heptane, 1,7-dibromo-
6375 - Heptane, 2,3-dibromo-
6376 - Heptane, 3,4-dibromo-

$C_7H_{14}ClF$
6365 - Heptane, 1-chloro-7-fluoro-

$C_7H_{14}Cl_2$
6377 - Heptane, 1,1-dichloro-
6378 - Heptane, 1,2-dichloro-
6379 - Heptane, 1,7-dichloro-
6380 - Heptane, 2,2-dichloro-
6381 - Heptane, 4,4-dichloro-
8752 - Pentane, 1,2-dichloro-4,4-dimethyl-
8753 - Pentane, 1,5-dichloro-3,3-dimethyl-
8754 - Pentane, 2,4-dichloro-2,4-dimethyl-

$C_7H_{14}NO_5P$
7663 - Monocrotophos

$C_7H_{14}N_2$
9815 - 2-Propanamine, N,N'-methanetetraylbis-

$C_7H_{14}N_2O_2$
9604 - 1-Piperazinecarboxylic acid, ethyl ester

$C_7H_{14}N_2O_2S$
334 - Aldicarb

$C_7H_{14}N_2O_4$
3944 - Carbamic acid, (1,1-dimethylethyl)nitro-, ethyl ester
3951 - Carbamic acid, ethylnitro-, methyl ester

$C_7H_{14}N_2O_4S$
335 - Aldoxycarb

$C_7H_{14}N_2O_4S_2$
5018 - L-Cysteine, S,S'-methylenebis-

$C_7H_{14}O$
3371 - Butane, 1-(ethenyloxy)-3-methyl-
4229 - Cycloheptanol
4419 - Cyclohexanemethanol
4425 - Cyclohexane, methoxy-
4530 - Cyclohexanol, 1-methyl-
4531 - Cyclohexanol, 2-methyl-, cis-(±)-
4532 - Cyclohexanol, 2-methyl-, trans-(±)-
4533 - Cyclohexanol, 3-methyl-, cis-(±)-
4534 - Cyclohexanol, 3-methyl-, trans-(±)-
4535 - Cyclohexanol, 4-methyl-, cis-
4536 - Cyclohexanol, 4-methyl-, trans-
4811 - Cyclopentaneethanol
6173 - Furan, tetrahydro-2-propyl-
6344 - Heptanal
6510 - 2-Heptanone
6511 - 3-Heptanone
6512 - 4-Heptanone
6566 - 1-Hepten-4-ol
6567 - 2-Hepten-1-ol, (E)-
6568 - 2-Hepten-1-ol, (Z)-
6569 - 2-Hepten-1-ol, (±)-
6570 - 3-Hepten-2-ol
6571 - 6-Hepten-2-ol
6572 - 6-Hepten-3-ol
6720 - Hexanal, 3-methyl-
6976 - 2-Hexanone, 3-methyl-
6977 - 2-Hexanone, 4-methyl-
6978 - 2-Hexanone, 5-methyl-
6979 - 3-Hexanone, 2-methyl-
6980 - 3-Hexanone, 4-methyl-
6981 - 3-Hexanone, 5-methyl-
7077 - 1-Hexen-3-ol, 5-methyl-
7078 - 2-Hexen-1-ol, 5-methyl-
7079 - 3-Hexen-2-ol, 2-methyl-
7080 - 3-Hexen-2-ol, 3-methyl-
7081 - 4-Hexen-3-ol, 2-methyl-
7082 - 4-Hexen-3-ol, 3-methyl-
7083 - 5-Hexen-2-ol, 2-methyl-
7084 - 5-Hexen-3-ol, 3-methyl-
7085 - 5-Hexen-3-ol, 4-methyl-
8965 - 2-Pentanone, 4,4-dimethyl-
8966 - 3-Pentanone, 2,2-dimethyl-
8967 - 3-Pentanone, 2,4-dimethyl-

$C_7H_{14}O_2$
227 - Acetic acid, pentyl ester
3516 - Butanoic acid, 2-ethyl-2-methyl-
3517 - Butanoic acid, 2-ethyl-, methyl ester
3548 - Butanoic acid, 3-methyl-, ethyl ester
3549 - Butanoic acid, 1-methylethyl ester

3550 - Butanoic acid, 2-methyl-, ethyl ester, (S)-
3588 - Butanoic acid, propyl ester
3630 - 1-Butanol, 2-methyl-, acetate
3631 - 1-Butanol, 3-methyl-, acetate
6009 - Formic acid, hexyl ester
6455 - Heptanoic acid
6517 - 4-Heptanone, 2-hydroxy-
6893 - Hexanoic acid, 2-methyl-, (±)-
6894 - Hexanoic acid, 3-methyl-
6895 - Hexanoic acid, 4-methyl-, (±)-
6896 - Hexanoic acid, 5-methyl-
6898 - Hexanoic acid, methyl ester
6973 - 3-Hexanone, 2-hydroxy-2-methyl-
6974 - 3-Hexanone, 5-hydroxy-2-methyl-
8602 - Oxirane, (butoxymethyl)-
8610 - Oxirane, [(1,1-dimethylethoxy)methyl]-
8877 - Pentanoic acid, 2-ethyl-, (±)-
8878 - Pentanoic acid, ethyl ester
8918 - 2-Pentanol, acetate, (R)-
8919 - 3-Pentanol, acetate
8978 - 2-Pentanone, 4-methoxy-4-methyl-
10214 - Propanoic acid, butyl ester
10259 - Propanoic acid, 2,2-dimethyl-, ethyl ester
10303 - Propanoic acid, 2-methyl-, 1-methylethyl ester
10308 - Propanoic acid, 2-methyl-, propyl ester
10309 - Propanoic acid, 1-methylpropyl ester
10310 - Propanoic acid, 2-methylpropyl ester
10559 - 1-Propene, 1,1-diethoxy-
10560 - 1-Propene, 3,3-diethoxy-

$C_7H_{14}O_3$
3995 - Carbonic acid, bis(1-methylethyl) ester
4002 - Carbonic acid, dipropyl ester
8884 - Pentanoic acid, 4-hydroxy-, ethyl ester
8885 - Pentanoic acid, 2-hydroxyethyl ester
10262 - Propanoic acid, 3-ethoxy-, ethyl ester
10268 - Propanoic acid, 2-hydroxy-, butyl ester, (R)-

$C_7H_{14}O_4$
3505 - Butanoic acid, 2,3-dihydroxypropyl ester
7112 - ribo-Hexose, 2,6-dideoxy-3-O-methyl-

$C_7H_{14}O_5$
5736 - Ethanol, 2-methoxy-, carbonate (2:1)
6198 - Galactose, 6-deoxy-3-O-methyl-

$C_7H_{14}O_6$
6227 - α-D-Glucopyranoside, methyl

$C_7H_{15}Br$
6356 - Heptane, 1-bromo-
6357 - Heptane, 2-bromo-
6358 - Heptane, 4-bromo-

$C_7H_{15}BrO$
6736 - Hexane, 1-bromo-6-methoxy-

$C_7H_{15}Cl$
3254 - Butane, 2-chloro-2,3,3-trimethyl-
6360 - Heptane, 1-chloro-
6361 - Heptane, 2-chloro-
6362 - Heptane, 3-chloro-
6363 - Heptane, 4-chloro-
6743 - Hexane, 1-chloro-3-methyl-
6744 - Hexane, 2-chloro-2-methyl-
6745 - Hexane, 2-chloro-5-methyl-
6746 - Hexane, 3-chloro-3-methyl-
8722 - Pentane, 2-chloro-2,3-dimethyl-
8723 - Pentane, 2-chloro-2,4-dimethyl-
8724 - Pentane, 3-chloro-2,3-dimethyl-
8725 - Pentane, 4-chloro-2,2-dimethyl-
8726 - Pentane, 3-chloro-3-ethyl-

$C_7H_{15}ClO$
6477 - 1-Heptanol, 7-chloro-
6478 - 2-Heptanol, 1-chloro-

$C_7H_{15}ClO_2$
9850 - Propane, 3-chloro-1,1-diethoxy-

$C_7H_{15}F$
6417 - Heptane, 1-fluoro-

$C_7H_{15}FO$
6491 - 1-Heptanol, 7-fluoro-

$C_7H_{15}I$
6419 - Heptane, 1-iodo-

6420 - Heptane, 2-iodo-

$C_7H_{15}N$
502 - Azocine, octahydro-
4292 - Cyclohexanamine, 2-methyl-, cis-
4293 - Cyclohexanamine, 2-methyl-, trans-
4294 - Cyclohexanamine, 3-methyl-, cis-
4295 - Cyclohexanamine, 3-methyl-, trans-
4296 - Cyclohexanamine, 4-methyl-, cis-
4297 - Cyclohexanamine, 4-methyl-, trans-
4298 - Cyclohexanamine, N-methyl-
4418 - Cyclohexanemethanamine
9648 - Piperidine, 1,2-dimethyl-, (±)-
9649 - Piperidine, 2,5-dimethyl-
9650 - Piperidine, 2,6-dimethyl-
9651 - Piperidine, 3,3-dimethyl-
9652 - Piperidine, 3,5-dimethyl-
9667 - Piperidine, 1-ethyl-
9668 - Piperidine, 2-ethyl-, (±)-
9669 - Piperidine, 3-ethyl-, (±)-
9670 - Piperidine, 4-ethyl-
11173 - Pyrrolidine, 1-propyl-
11174 - Pyrrolidine, 3-propyl-
11175 - Pyrrolidine, 2,2,4-trimethyl-
11176 - Pyrrolidine, 2,2,5-trimethyl-

$C_7H_{15}NO$
6346 - Heptanal, oxime
6348 - Heptanamide
6982 - 2-Hexanone, 5-methyl-, oxime
8699 - Pentanamide, N,N-dimethyl-
8968 - 3-Pentanone, 2,4-dimethyl-, oxime
9662 - 1-Piperidineethanol
9663 - 2-Piperidineethanol
9664 - 4-Piperidineethanol
9678 - 3-Piperidinemethanol, 1-methyl-
9715 - 3-Piperidinol, 1-ethyl-
9793 - Propanamide, N,N-diethyl-
10427 - 2-Propanone, 1-(diethylamino)-

$C_7H_{15}NOS$
5930 - Ethiolate

$C_7H_{15}NO_2$
3541 - Butanoic acid, 3-(methylamino)-, ethyl ester
3936 - Carbamic acid, butyl-, ethyl ester
3942 - Carbamic acid, (1,1-dimethylethyl)-, ethyl ester
3947 - Carbamic acid, ethyl-, butyl ester
3960 - Carbamic acid, (1-methylpropyl)-, ethyl ester
3961 - Carbamic acid, (2-methylpropyl)-, ethyl ester
6429 - Heptane, 1-nitro-
7692 - 4-Morpholineethanol, α-methyl-
8088 - Nitrous acid, heptyl ester

$C_7H_{15}NO_5$
6215 - α-L-Glucopyranose, 2-(methylamino)-2-deoxy-

C_7H_{16}
3457 - Butane, 2,2,3-trimethyl-
6355 - Heptane
6834 - Hexane, 2-methyl-
6835 - Hexane, 3-methyl-, (S)-
6941 - 1-Hexanol, 5-methyl-
8760 - Pentane, 2,2-dimethyl-
8761 - Pentane, 2,3-dimethyl-
8762 - Pentane, 2,4-dimethyl-
8763 - Pentane, 3,3-dimethyl-
8809 - Pentane, 3-ethyl-

$C_7H_{16}BrNO_2$
5438 - Ethanaminium, 2-(acetyloxy)-N,N,N-trimethyl-, bromide

$C_7H_{16}NCl$
7430 - Mepiquat chloride

$C_7H_{16}ClNO_2$
5439 - Ethanaminium, 2-(acetyloxy)-N,N,N-trimethyl-, chloride

$C_7H_{16}N_2$
9622 - Piperazine, 1,2,4-trimethyl-
9625 - 1-Piperidinamine, 2,6-dimethyl-
9626 - 3-Piperidinamine, 1-ethyl-
11167 - 2-Pyrrolidinemethanamine, 1-ethyl-

$C_7H_{16}N_2O$
7701 - 4-Morpholinepropanamine
12012 - Urea, N,N'-dipropyl-

$C_7H_{16}O$
3374 - Butane, 1-ethoxy-3-methyl-
3375 - Butane, 2-ethoxy-2-methyl-
3399 - Butane, 1-(1-methylethoxy)-
3421 - Butane, 1-propoxy-
3644 - 2-Butanol, 2,3,3-trimethyl-
6472 - 1-Heptanol
6473 - 2-Heptanol, (±)-
6474 - 3-Heptanol, (S)-
6475 - 4-Heptanol
6833 - Hexane, 1-methoxy-
6937 - 1-Hexanol, 2-methyl-, (±)-
6938 - 1-Hexanol, 2-methyl-, (R)-
6939 - 1-Hexanol, 3-methyl-, (±)-
6940 - 1-Hexanol, 4-methyl-, (±)-
6942 - 2-Hexanol, 2-methyl-
6943 - 2-Hexanol, 3-methyl-
6944 - 2-Hexanol, 5-methyl-
6945 - 3-Hexanol, 2-methyl-, (±)-
6946 - 3-Hexanol, 3-methyl-
6947 - 3-Hexanol, 5-methyl-, (±)-
8808 - Pentane, 1-ethoxy-
8927 - 1-Pentanol, 2,4-dimethyl-, (±)-
8928 - 2-Pentanol, 2,4-dimethyl-
8929 - 3-Pentanol, 2,2-dimethyl-
8930 - 3-Pentanol, 2,3-dimethyl-
8931 - 3-Pentanol, 2,4-dimethyl-
8932 - 3-Pentanol, 3-ethyl-
10081 - Propane, 2-methyl-2-(1-methylethoxy)-
10086 - Propane, 2-methyl-1-propoxy-
10087 - Propane, 2-methyl-2-propoxy-

$C_7H_{16}O_2$
5757 - Ethanol, 2-(pentyloxy)-
6397 - 1,4-Heptanediol
6398 - 1,5-Heptanediol
6399 - 1,7-Heptanediol
6400 - 2,4-Heptanediol
6803 - 2,4-Hexanediol, 2-methyl-
8792 - 1,3-Pentanediol, 2,2-dimethyl-
9903 - Propane, 1,1-diethoxy-
9904 - Propane, 2,2-diethoxy-
10004 - 1,3-Propanediol, 2,2-diethyl-
10020 - 1,3-Propanediol, 2-methyl-2-propyl-
10074 - Propane, 1,1'-[methylenebis(oxy)]bis-
10346 - 2-Propanol, 1-butoxy-

$C_7H_{16}O_2Si$
11422 - Silane, ethenyldiethoxymethyl-

$C_7H_{16}O_3$
3455 - Butane, 1,1,3-trimethoxy-
3456 - Butane, 1,3,3-trimethoxy-
5604 - Ethane, 1,1',1''-[methylidynetris(oxy)]tris-
6453 - 1,4,7-Heptanetriol

$C_7H_{16}O_4$
10150 - Propane, 1,1,3,3-tetramethoxy-

$C_7H_{16}O_4S_2$
9835 - Propane, 2,2-bis(ethylsulfonyl)-

$C_7H_{16}S$
6436 - 1-Heptanethiol

$C_7H_{16}S_3$
5605 - Ethane, 1,1',1''-[methylidynetris(thio)]tris-

$C_7H_{17}N$
6349 - 1-Heptanamine
6350 - 2-Heptanamine
6351 - 4-Heptanamine
6728 - 1-Hexanamine, 3-methyl-
6729 - 1-Hexanamine, 4-methyl-
6730 - 2-Hexanamine, 4-methyl-
8703 - 1-Pentanamine, N,N-dimethyl-
8704 - 2-Pentanamine, 2,4-dimethyl-

$C_7H_{17}NO$
5739 - Ethanol, 2-[(3-methylbutyl)amino]-
10366 - 1-Propanol, 2-(diethylamino)-
10367 - 1-Propanol, 3-(diethylamino)-

10368 - 2-Propanol, 1-(diethylamino)-, (±)-

C₇H₁₇NO₂

7443 - Methanamine, 1,1-diethoxy-*N,N*-dimethyl-

C₇H₁₇NO₅

6208 - *D*-Glucitol, 1-deoxy-1-(methylamino)-

C₇H₁₇O₂PS

9549 - Phosphonothioic acid, methyl-, *O,O*-bis(1-methylethyl) ester

C₇H₁₇O₂PS₃

9513 - Phorate

C₇H₁₈N₂

6371 - 1,7-Heptanediamine
9876 - 1,3-Propanediamine, *N,N*-diethyl-
9879 - 1,3-Propanediamine, *N,N,N′,N′*-tetramethyl-

C₇H₁₈N₂O

10339 - 2-Propanol, 1-amino-3-(diethylamino)-
10343 - 2-Propanol, 1,3-bis(dimethylamino)-

C₇H₁₈N₂O₂

5679 - Ethanol, 2,2′-[(3-aminopropyl)imino]bis-

C₇H₁₈O₃Si

11479 - Silane, triethoxymethyl-

C₇H₁₈Si

11371 - Silane, butyltrimethyl-
11491 - Silane, trimethyl(2-methylpropyl)-

C₇H₁₉N₃

9875 - 1,3-Propanediamine, *N*-(3-aminopropyl)-*N*-methyl-

C₇H₂₂O₄Si₄

4992 - Cyclotetrasiloxane, heptamethyl-

C₈Cl₂N₂O₂

4251 - 1,4-Cyclohexadiene-1,2-dicarbonitrile, 4,5-dichloro-3,6-dioxo-

C₈Cl₄N₂

1317 - 1,3-Benzenedicarbonitrile, 2,4,5,6-tetrachloro-

C₈Cl₄O₃

7346 - 1,3-Isobenzofurandione, 4,5,6,7-tetrachloro-

C₈F₁₄O₃

3522 - Butanoic acid, heptafluoro-, anhydride

C₈F₁₆

4342 - Cyclohexane, 1,1,2,2,3,3,4,4,5,6-decafluoro-5,6-bis(trifluoromethyl)-
4343 - Cyclohexane, 1,1,2,2,3,4,4,5,5,6-decafluoro-3,6-bis(trifluoromethyl)-

C₈F₁₆O

6109 - Furan, 2,2,3,3,4,4,5-heptafluorotetrahydro-5-(nonafluorobutyl)-

C₈F₁₈

8415 - Octane, octadecafluoro-

C₈HCl₄F₁₁O₂

8442 - Octanoic acid, 3,5,7,8-tetrachloro-2,2,3,4,4,5,6,6,7,8,8-undecafluoro-

C₈H₂Cl₂O₃

7343 - 1,3-Isobenzofurandione, 4,5-dichloro-
7344 - 1,3-Isobenzofurandione, 5,6-dichloro-

C₈H₃Cl₂F₃N₂

2377 - 1H-Benzimidazole, 4,5-dichloro-2-(trifluoromethyl)-

C₈H₄Cl₂O₂

1318 - 1,2-Benzenedicarbonyl dichloride
1319 - 1,3-Benzenedicarbonyl dichloride
1320 - 1,4-Benzenedicarbonyl dichloride

C₈H₄Cl₂O₄

1340 - 1,2-Benzenedicarboxylic acid, 3,5-dichloro-
1349 - 1,4-Benzenedicarboxylic acid, 2,5-dichloro-

C₈H₄Cl₆

996 - Benzene, 1,4-bis(trichloromethyl)-

C₈H₄F₂O₂

1321 - 1,2-Benzenedicarbonyl difluoride

C₈H₄F₃N

2769 - Benzonitrile, 2-(trifluoromethyl)-
2770 - Benzonitrile, 3-(trifluoromethyl)-

C₈H₄F₆

997 - Benzene, 1,3-bis(trifluoromethyl)-

C₈H₄N₂

1315 - 1,3-Benzenedicarbonitrile
1316 - 1,4-Benzenedicarbonitrile

C₈H₄N₂S₂

1432 - Benzene, 1,4-diisothiocyanato-

C₈H₄O₂S

2401 - Benzo[b]thiophene-2,3-dione

C₈H₄O₃

7342 - 1,3-Isobenzofurandione

C₈H₅Br

1040 - Benzene, 1-bromo-2-ethynyl-
1041 - Benzene, 1-bromo-3-ethynyl-
1042 - Benzene, 1-bromo-4-ethynyl-

C₈H₅ClN₂

11334 - Quinoxaline, 6-chloro-

C₈H₅Cl₃O

5871 - Ethanone, 2,2,2-trichloro-1-phenyl-

C₈H₅Cl₃O₂

944 - Benzeneacetic acid, 2,3,6-trichloro-

C₈H₅Cl₃O₃

256 - Acetic acid, (2,4,5-trichlorophenoxy)-

C₈H₅F₃O

5872 - Ethanone, 2,2,2-trifluoro-1-phenyl-

C₈H₅F₆N

637 - Benzenamine, 3,5-bis(trifluoromethyl)-

C₈H₅NO

968 - Benzeneacetonitrile, α-oxo-
2757 - Benzonitrile, 3-formyl-
2758 - Benzonitrile, 4-formyl-

C₈H₅NO₂

2550 - Benzoic acid, 3-cyano-
7300 - 1H-Indole-2,3-dione
7362 - 1H-Isoindole-1,3(2H)-dione

C₈H₆

1715 - Benzene, ethynyl-

C₈H₆BrClO

5778 - Ethanone, 2-bromo-1-(4-chlorophenyl)-

C₈H₆BrN

947 - Benzeneacetonitrile, α-bromo-
948 - Benzeneacetonitrile, 2-bromo-

C₈H₆Br₂O

5777 - Ethanone, 2-bromo-1-(4-bromophenyl)-
5797 - Ethanone, 1-(3,5-dibromophenyl)-
5798 - Ethanone, 2,2-dibromo-1-phenyl-

C₈H₆Br₂O₃

2559 - Benzoic acid, 3,5-dibromo-2-hydroxy-, methyl ester

C₈H₆Br₄

2255 - Benzene, 1,2,3,4-tetrabromo-5,6-dimethyl-

C₈H₆ClN

949 - Benzeneacetonitrile, 2-chloro-
950 - Benzeneacetonitrile, 3-chloro-
2750 - Benzonitrile, 2-(chloromethyl)-

C₈H₆ClNO₄

2548 - Benzoic acid, 4-chloro-3-nitro-, methyl ester
2549 - Benzoic acid, 5-chloro-2-nitro-, methyl ester

C₈H₆Cl₂

1367 - Benzene, 1,2-dichloro-3-ethenyl-
1368 - Benzene, 1,2-dichloro-4-ethenyl-
1369 - Benzene, 1,3-dichloro-2-ethenyl-
1370 - Benzene, 1,3-dichloro-5-ethenyl-
1371 - Benzene, 1,4-dichloro-2-ethenyl-
1372 - Benzene, (2,2-dichloroethenyl)-
1373 - Benzene, 2,4-dichloro-1-ethenyl-

C₈H₆Cl₂O

973 - Benzeneacetyl chloride, α-chloro-
5784 - Ethanone, 2-chloro-1-(4-chlorophenyl)-
5799 - Ethanone, 1-(2,4-dichlorophenyl)-
5800 - Ethanone, 1-(2,5-dichlorophenyl)-
5801 - Ethanone, 1-(3,4-dichlorophenyl)-
5802 - Ethanone, 2,2-dichloro-1-phenyl-

C₈H₆Cl₂O₂

9288 - Phenol, 2,4-dichloro-, acetate

C₈H₆Cl₂O₃

169 - Acetic acid, (2,4-dichlorophenoxy)-
2567 - Benzoic acid, 3,6-dichloro-2-methoxy-

C₈H₆Cl₃NO

98 - Acetamide, 2,2,2-trichloro-*N*-phenyl-

C₈H₆Cl₄

989 - Benzene, 1,4-bis(dichloromethyl)-

2261 - Benzene, 1,2,3,5-tetrachloro-4,6-dimethyl-

C₈H₆FN

952 - Benzeneacetonitrile, 2-fluoro-
953 - Benzeneacetonitrile, 4-fluoro-

C₈H₆N₂

4136 - Cinnoline
8065 - 1,5-Naphthyridine
9599 - Phthalazine
11195 - Quinazoline
11333 - Quinoxaline

C₈H₆N₂O₂

965 - Benzeneacetonitrile, 2-nitro-
966 - Benzeneacetonitrile, 3-nitro-
967 - Benzeneacetonitrile, 4-nitro-
2766 - Benzonitrile, 2-methyl-5-nitro-

C₈H₆N₄O₅

7222 - 2,4-Imidazolidinedione, 1-[[(5-nitro-2-furanyl)methylene]amino]-

C₈H₆N₄O₈

3106 - [5,5′-Bipyrimidine]2,2′,4,4′,6,6′(1H,1′H,3H,3′H,5H,5′H)-hexone-5,5′-dihydroxy

C₈H₆O

1716 - Benzene, (ethynyloxy)-
2438 - Benzofuran

C₈H₆OS

2403 - Benzo[b]thiophene-4-ol

C₈H₆O₂

1322 - 1,3-Benzenedicarboxaldehyde
1323 - 1,4-Benzenedicarboxaldehyde
2461 - 2(3H)-Benzofuranone
2462 - 3(2H)-Benzofuranone
7349 - 1(3H)-Isobenzofuranone

C₈H₆O₃

934 - Benzeneacetic acid, α-oxo-
2417 - 1,3-Benzodioxole-5-carboxaldehyde
2607 - Benzoic acid, 2-formyl-

C₈H₆O₄

1327 - 1,2-Benzenedicarboxylic acid
1328 - 1,3-Benzenedicarboxylic acid
1329 - 1,4-Benzenedicarboxylic acid
2418 - 1,3-Benzodioxole-5-carboxylic acid
2609 - Benzoic acid, 3-formyl-4-hydroxy-
9598 - *p*-Phthalaldehydic acid, 3-hydroxy-

C₈H₆O₅

1352 - 1,3-Benzenedicarboxylic acid, 4-hydroxy-

C₈H₆S

2399 - Benzo[b]thiophene

C₈H₆S₂

2434 - 1,4-Benzodithiin
3112 - 2,2′-Bithiophene

C₈H₇Br

1025 - Benzene, (1-bromoethenyl)-
1026 - Benzene, 1-bromo-2-ethenyl-
1027 - Benzene, 1-bromo-3-ethenyl-
1028 - Benzene, 1-bromo-4-ethenyl-
1029 - Benzene, (2-bromoethenyl)-, (*E*)-
1031 - Benzene, (2-bromoethenyl)-, (*Z*)-
4743 - 1,3,5,7-Cyclooctatetraene, 1-bromo-

C₈H₇BrO

5780 - Ethanone, 1-(3-bromophenyl)-
5781 - Ethanone, 1-(4-bromophenyl)-
5782 - Ethanone, 2-bromo-1-phenyl-

C₈H₇BrO₂

119 - Acetic acid, bromo-, phenyl ester
888 - Benzeneacetic acid, 4-bromo-
2525 - Benzoic acid, 3-bromo-, methyl ester
2526 - Benzoic acid, 4-bromo-, methyl ester

C₈H₇BrO₃

889 - Benzeneacetic acid, 4-bromo-α-hydroxy-, (±)-

C₈H₇Cl

1148 - Benzene, (1-chloroethenyl)-
1149 - Benzene, 1-chloro-2-ethenyl-
1150 - Benzene, 1-chloro-3-ethenyl-
1151 - Benzene, 1-chloro-4-ethenyl-
1153 - Benzene, (2-chloroethenyl)-, (*E*)-
1155 - Benzene, (2-chloroethenyl)-, (*Z*)-
4745 - 1,3,5,7-Cyclooctatetraene, 1-chloro-

C_8H_7ClO
971 - Benzeneacetyl chloride
2902 - Benzoyl chloride, 2-methyl-
2903 - Benzoyl chloride, 3-methyl-
2904 - Benzoyl chloride, 4-methyl-
5788 - Ethanone, 1-(3-chlorophenyl)-
5789 - Ethanone, 1-(4-chlorophenyl)-
5790 - Ethanone, 2-chloro-1-phenyl-

$C_8H_7ClO_2$
147 - Acetic acid, chloro-, phenyl ester
2421 - 1,3-Benzodioxole, 5-(chloromethyl)-
2543 - Benzoic acid, 4-chloro-, methyl ester
2901 - Benzoyl chloride, 4-methoxy-
4027 - Carbonochloridic acid, phenylmethyl ester

$C_8H_7ClO_3$
146 - Acetic acid, (4-chlorophenoxy)-
2542 - Benzoic acid, 5-chloro-2-hydroxy-, methyl ester

C_8H_7F
1633 - Benzene, 1-ethenyl-2-fluoro-
1634 - Benzene, 1-ethenyl-3-fluoro-
1635 - Benzene, 1-ethenyl-4-fluoro-

C_8H_7FO
5815 - Ethanone, 1-(4-fluorophenyl)-
5816 - Ethanone, 2-fluoro-1-phenyl-

$C_8H_7FO_2$
901 - Benzeneacetic acid, 4-fluoro-

$C_8H_7F_3$
2045 - Benzene, 1-methyl-4-(trifluoromethyl)-

C_8H_7IO
5836 - Ethanone, 1-(3-iodophenyl)-
5837 - Ethanone, 1-(4-iodophenyl)-
5838 - Ethanone, 2-iodo-1-phenyl-

$C_8H_7IO_2$
2660 - Benzoic acid, 2-iodo-, methyl ester
2661 - Benzoic acid, 3-iodo-, methyl ester
2662 - Benzoic acid, 4-iodo-, methyl ester

C_8H_7N
945 - Benzeneacetonitrile
1791 - Benzene, (isocyanomethyl)-
2763 - Benzonitrile, 2-methyl-
2764 - Benzonitrile, 3-methyl-
2765 - Benzonitrile, 4-methyl-
7294 - 1H-Indole
7324 - Indolizine

C_8H_7NO
287 - Acetonitrile, phenoxy-
956 - Benzeneacetonitrile, α-hydroxy-, (±)-
1787 - Benzene, 1-isocyanato-2-methyl-
2760 - Benzonitrile, 2-methoxy-
2761 - Benzonitrile, 3-methoxy-
2762 - Benzonitrile, 4-methoxy-
2880 - Benzoxazole, 2-methyl-
7327 - 2H-Indol-2-one, 1,3-dihydro-
7369 - 1H-Isoindol-1-one, 2,3-dihydro-

C_8H_7NOS
1793 - Benzene, 1-isothiocyanato-2-methoxy-

$C_8H_7NO_2$
876 - Benzeneacetaldehyde, α-oxo-, aldoxime
1644 - Benzene, 1-ethenyl-3-nitro-
1645 - Benzene, 1-ethenyl-4-nitro-
2050 - Benzene, (2-nitroethenyl)-, (E)-
8414 - Octane, 3-nitro-

$C_8H_7NO_3$
875 - Benzeneacetaldehyde, 2-nitro-
2489 - Benzoic acid, 4-(aminocarbonyl)-
5860 - Ethanone, 1-(3-nitrophenyl)-
5861 - Ethanone, 1-(4-nitrophenyl)-
5862 - Ethanone, 2-nitro-1-phenyl-

$C_8H_7NO_4$
219 - Acetic acid, 2-nitrophenyl ester
933 - Benzeneacetic acid, 4-nitro-
1330 - 1,3-Benzenedicarboxylic acid, 2-amino-
1331 - 1,3-Benzenedicarboxylic acid, 5-amino-
2710 - Benzoic acid, 2-nitro-, methyl ester
2711 - Benzoic acid, 3-nitro-, methyl ester

C_8H_7NS
1794 - Benzene, (isothiocyanatomethyl)-
2863 - Benzothiazole, 2-methyl-
11710 - Thiocyanic acid, phenylmethyl ester

$C_8H_7NS_2$
2867 - 2(3H)-Benzothiazolethione, 3-methyl-

$C_8H_7N_3$
11332 - 6-Quinoxalinamine

$C_8H_7N_3O$
5863 - Ethanone, 1-(2-nitrophenyl)-

$C_8H_7N_3O_2$
9600 - 1,4-Phthalazinedione, 5-amino-2,3-dihydro-

$C_8H_7N_3O_5$
8589 - 2-Oxazolidinone, 3-[[(5-nitro-2-furanyl)methylene]amino]-

$C_8H_7N_3O_6$
1489 - Benzene, 1,4-dimethyl-2,3,5-trinitro-
1490 - Benzene, 2,4-dimethyl-1,3,5-trinitro-

C_8H_8
1624 - Benzene, ethenyl-
4742 - 1,3,5,7-Cyclooctatetraene
4769 - 1,3-Cyclopentadiene, 5-(2-propenylidene)-, (E)-

C_8H_8BrNO
37 - Acetamide, N-(4-bromophenyl)-

$C_8H_8Br_2$
983 - Benzene, 1,2-bis(bromomethyl)-
984 - Benzene, 1,3-bis(bromomethyl)-
985 - Benzene, 1,4-bis(bromomethyl)-
1297 - Benzene, 1,2-dibromo-4,5-dimethyl-
1298 - Benzene, 1,5-dibromo-2,4-dimethyl-
1299 - Benzene, (1,2-dibromoethyl)-

C_8H_8ClN
7652 - Methylamine, N-(o-chlorobenzylidene)-

C_8H_8ClNO
48 - Acetamide, N-(4-chlorophenyl)-
3974 - Carbamic chloride, methylphenyl-

$C_8H_8ClNO_3S$
2243 - Benzenesulfonyl chloride, 4-(acetylamino)-

$C_8H_8Cl_2$
986 - Benzene, 1,2-bis(chloromethyl)-
987 - Benzene, 1,3-bis(chloromethyl)-
988 - Benzene, 1,4-bis(chloromethyl)-
1360 - Benzene, 1,2-dichloro-3,4-dimethyl-
1361 - Benzene, 1,3-dichloro-2,5-dimethyl-
1362 - Benzene, 1,4-dichloro-2,5-dimethyl-
1363 - Benzene, 1,5-dichloro-2,4-dimethyl-
1364 - Benzene, 2,3-dichloro-1,4-dimethyl-
1376 - Benzene, (1,2-dichloroethyl)-
1377 - Benzene, 1,4-dichloro-2-ethyl-
1390 - Benzene, 1-(dichloromethyl)-4-methyl-

$C_8H_8Cl_2NaO_5S$
11358 - Sesone

$C_8H_8Cl_2O$
1375 - Benzene, 2,4-dichloro-1-ethoxy-
1854 - Benzenemethanol, 2,4-dichloro-α-methyl-
9290 - Phenol, 2,4-dichloro-3,5-dimethyl-
9291 - Phenol, 2,6-dichloro-3,5-dimethyl-

$C_8H_8Cl_2O_2$
4084 - Chloroneb

$C_8H_8Cl_3N_3O_4S_2$
2854 - 2H-1,2,4-Benzothiadiazine-7-sulfonamide, 6-chloro-3-(dichloromethyl)-3,4-dihydro-, 1,1-dioxide

$C_8H_8Cl_3O_3PS$
9584 - Phosphorothioic acid, O,O-dimethyl O-(2,4,5-trichlorophenyl) ester

$C_8H_8F_3N_3O_4S_2$
2858 - 2H-1,2,4-Benzothiadiazine-7-sulfonamide, 3,4-dihydro-6-(trifluoromethyl)-, 1,1-dioxide

$C_8H_8HgO_2$
7431 - Mercury, (acetato-O)phenyl-

$C_8H_8N_2$
946 - Benzeneacetonitrile, 4-amino-
2382 - 1H-Benzimidazole, 1-methyl-
2743 - Benzonitrile, 5-amino-2-methyl-
4160 - Cyanamide, methylphenyl-

7239 - 1H-Indazole, 1-methyl-
7240 - 1H-Indazole, 3-methyl-
7241 - 1H-Indazole, 5-methyl-
7242 - 2H-Indazole, 2-methyl-

$C_8H_8N_2O_2$
1324 - 1,2-Benzenedicarboxamide
10883 - 3-Pyridinecarbonitrile, 1,2-dihydro-4-methoxy-1-methyl-2-oxo-

$C_8H_8N_2O_3$
80 - Acetamide, 2-nitro-N-phenyl-
81 - Acetamide, N-(2-nitrophenyl)-
82 - Acetamide, N-(3-nitrophenyl)-
83 - Acetamide, N-(4-nitrophenyl)-

$C_8H_8N_2O_5$
1650 - Benzene, 2-ethoxy-1,3-dinitro-

C_8H_8O
587 - Benzaldehyde, 2-methyl-
588 - Benzaldehyde, 3-methyl-
589 - Benzaldehyde, 4-methyl-
868 - Benzeneacetaldehyde
1646 - Benzene, (ethenyloxy)-
2442 - Benzofuran, 2,3-dihydro-
5864 - Ethanone, 1-phenyl-
8624 - Oxirane, phenyl-
9348 - Phenol, 2-ethenyl-
9349 - Phenol, 3-ethenyl-

$C_8H_8O_2$
231 - Acetic acid, phenyl ester
575 - Benzaldehyde, 2-hydroxy-3-methyl-
576 - Benzaldehyde, 2-hydroxy-4-methyl-
577 - Benzaldehyde, 2-hydroxy-5-methyl-
578 - Benzaldehyde, 2-hydroxy-6-methyl-
583 - Benzaldehyde, 2-methoxy-
584 - Benzaldehyde, 3-methoxy-
585 - Benzaldehyde, 4-methoxy-
883 - Benzeneacetic acid
2414 - 1,4-Benzodioxin, 2,3-dihydro-
2676 - Benzoic acid, 2-methyl-
2677 - Benzoic acid, 3-methyl-
2678 - Benzoic acid, 4-methyl-
2687 - Benzoic acid, methyl ester
3878 - 3-Buten-2-one, 4-(2-furanyl)-
4263 - 2,5-Cyclohexadiene-1,4-dione, 2,3-dimethyl-
4264 - 2,5-Cyclohexadiene-1,4-dione, 2,6-dimethyl-
5830 - Ethanone, 1-(2-hydroxyphenyl)-
5831 - Ethanone, 1-(3-hydroxyphenyl)-
5832 - Ethanone, 1-(4-hydroxyphenyl)-
5833 - Ethanone, 2-hydroxy-1-phenyl-
6016 - Formic acid, phenylmethyl ester

$C_8H_8O_3$
228 - Acetic acid, phenoxy-
570 - Benzaldehyde, 2-hydroxy-3-methoxy-
571 - Benzaldehyde, 2-hydroxy-5-methoxy-
572 - Benzaldehyde, 3-hydroxy-4-methoxy-
573 - Benzaldehyde, 3-hydroxy-5-methoxy-
574 - Benzaldehyde, 4-hydroxy-3-methoxy-
903 - Benzeneacetic acid, α-hydroxy-, (S)-
904 - Benzeneacetic acid, 2-hydroxy-
905 - Benzeneacetic acid, 3-hydroxy-
906 - Benzeneacetic acid, 4-hydroxy-
1540 - 1,2-Benzenediol, monoacetate
2427 - 1,3-Benzodioxole-5-methanol
2629 - Benzoic acid, 2-hydroxy-3-methyl-
2630 - Benzoic acid, 2-hydroxy-4-methyl-
2631 - Benzoic acid, 2-hydroxy-5-methyl-
2632 - Benzoic acid, 3-hydroxy-5-methyl-
2634 - Benzoic acid, 2-hydroxy-, methyl ester
2636 - Benzoic acid, 3-hydroxy-, methyl ester
2637 - Benzoic acid, 4-hydroxy-, methyl ester
2664 - Benzoic acid, 2-methoxy-
2665 - Benzoic acid, 3-methoxy-
2666 - Benzoic acid, 4-methoxy-
5803 - Ethanone, 1-(2,4-dihydroxyphenyl)-
6070 - 2-Furancarboxylic acid, 2-propenyl ester
7347 - 1,3-Isobenzofurandione, 4,5,6,7-tetrahydro-
10640 - 2-Propenoic acid, 3-(2-furanyl)-, methyl ester

$C_8H_8O_4$

895 - Benzeneacetic acid, 2,5-dihydroxy-
2573 - Benzoic acid, 2,4-dihydroxy-6-methyl-
2625 - Benzoic acid, 4-hydroxy-3-methoxy-
4262 - 2,5-Cyclohexadiene-1,4-dione, 2,6-dimethoxy-
4265 - 2,5-Cyclohexadiene-1,4-dione, 3-hydroxy-2-methoxy-5-methyl-
5873 - Ethanone, 1-(2,3,4-trihydroxyphenyl)-
10779 - 2H-Pyran-2,4(3H)-dione, 3-acetyl-6-methyl-

$C_8H_8O_5$

4260 - 2,5-Cyclohexadiene-1,4-dione, 2,5-dihydroxy-3-methoxy-6-methyl-
6083 - 2,5-Furandicarboxylic acid, dimethyl ester

C_8H_9Br

1017 - Benzene, 1-bromo-2,3-dimethyl-
1018 - Benzene, 1-bromo-2,4-dimethyl-
1019 - Benzene, 1-bromo-3,5-dimethyl-
1020 - Benzene, 2-bromo-1,3-dimethyl-
1021 - Benzene, 2-bromo-1,4-dimethyl-
1022 - Benzene, 4-bromo-1,2-dimethyl-
1034 - Benzene, (1-bromoethyl)-, (±)-
1035 - Benzene, 1-bromo-2-ethyl-
1036 - Benzene, 1-bromo-3-ethyl-
1037 - Benzene, 1-bromo-4-ethyl-
1038 - Benzene, (2-bromoethyl)-
1039 - Benzene, (1-bromoethyl)-, (R)-
1062 - Benzene, 1-(bromomethyl)-2-methyl-
1063 - Benzene, 1-(bromomethyl)-3-methyl-
1064 - Benzene, 1-(bromomethyl)-4-methyl-

C_8H_9BrO

1032 - Benzene, 1-bromo-4-ethoxy-
1033 - Benzene, (2-bromoethoxy)-
1595 - Benzeneethanol, 4-bromo-
1844 - Benzenemethanol, α-(bromomethyl)-
1845 - Benzenemethanol, 2-bromo-α-methyl-

$C_8H_9BrO_2$

1016 - Benzene, 4-bromo-1,2-dimethoxy-

C_8H_9Cl

1137 - Benzene, 1-chloro-2,3-dimethyl-
1138 - Benzene, 1-chloro-2,4-dimethyl-
1139 - Benzene, 2-chloro-1,3-dimethyl-
1140 - Benzene, 2-chloro-1,4-dimethyl-
1141 - Benzene, 4-chloro-1,2-dimethyl-
1160 - Benzene, (1-chloroethyl)-, (±)-
1161 - Benzene, 1-chloro-2-ethyl-
1162 - Benzene, 1-chloro-3-ethyl-
1163 - Benzene, 1-chloro-4-ethyl-
1164 - Benzene, (2-chloroethyl)-
1198 - Benzene, 1-(chloromethyl)-2-methyl-
1199 - Benzene, 1-(chloromethyl)-3-methyl-
1200 - Benzene, 1-(chloromethyl)-4-methyl-

C_8H_9ClO

1156 - Benzene, 1-chloro-2-ethoxy-
1157 - Benzene, 1-chloro-3-ethoxy-
1158 - Benzene, 1-chloro-4-ethoxy-
1159 - Benzene, (2-chloroethoxy)-
1181 - Benzene, [(chloromethoxy)methyl]-
1197 - Benzene, 1-(chloromethyl)-4-methoxy-
1596 - Benzeneethanol, 3-chloro-
1597 - Benzeneethanol, 4-chloro-
1852 - Benzenemethanol, α-(chloromethyl)-
9256 - Phenol, 2-chloro-3,4-dimethyl-
9257 - Phenol, 4-chloro-3,5-dimethyl-

C_8H_9ClS

1165 - Benzene, [(2-chloroethyl)thio]-

$C_8H_9Cl_2NO$

691 - Benzenamine, 2,6-dichloro-4-ethoxy-

$C_8H_9Cl_2P$

9553 - Phosphonous dichloride, (2,5-dimethylphenyl)-
9554 - Phosphonous dichloride, (4-ethylphenyl)-

$C_8H_9Cl_3Si$

11474 - Silane, trichloro(2-phenylethyl)-

C_8H_9FO

1654 - Benzene, 1-ethoxy-2-fluoro-
1655 - Benzene, 1-ethoxy-3-fluoro-
1656 - Benzene, 1-ethoxy-4-fluoro-

C_8H_9I

1680 - Benzene, 1-ethyl-2-iodo-
1681 - Benzene, 1-ethyl-4-iodo-
1769 - Benzene, 1-iodo-2,3-dimethyl-
1770 - Benzene, 1-iodo-2,4-dimethyl-
1771 - Benzene, 1-iodo-3,5-dimethyl-
1772 - Benzene, 2-iodo-1,3-dimethyl-
1773 - Benzene, 2-iodo-1,4-dimethyl-
1774 - Benzene, 4-iodo-1,2-dimethyl-

C_8H_9IO

1657 - Benzene, 1-ethoxy-4-iodo-

$C_8H_9IO_2$

1767 - Benzene, 1-iodo-2,3-dimethoxy-
1768 - Benzene, 4-iodo-1,2-dimethoxy-

C_8H_9N

733 - Benzenamine, 2-ethenyl-
734 - Benzenamine, 3-ethenyl-
735 - Benzenamine, 4-ethenyl-
2966 - Bicyclo[2.2.1]hept-5-ene-2-carbonitrile
7297 - 1H-Indole, 2,3-dihydro-
7452 - Methanamine, N-(phenylmethylene)-
10955 - Pyridine, 5-ethenyl-2-methyl-
11014 - Pyridine, 2-(2-propenyl)-

C_8H_9NO

86 - Acetamide, N-phenyl-
619 - Benzamide, 2-methyl-
620 - Benzamide, N-methyl-
878 - Benzeneacetamide
5772 - Ethanone, 1-(3-aminophenyl)-
5773 - Ethanone, 1-(4-aminophenyl)-
5774 - Ethanone, 2-amino-1-phenyl-
5855 - Ethanone, 1-(6-methyl-3-pyridinyl)-
5866 - Ethanone, 1-phenyl-, oxime
5996 - Formamide, N-(2-methylphenyl)-
5997 - Formamide, N-(3-methylphenyl)-
5998 - Formamide, N-methyl-N-phenyl-

$C_8H_9NO_2$

67 - Acetamide, N-(4-hydroxyphenyl)-
618 - Benzamide, 4-methoxy-
885 - Benzeneacetic acid, α-amino-
886 - Benzeneacetic acid, 4-amino-
1474 - Benzene, 1,2-dimethyl-3-nitro-
1475 - Benzene, 1,2-dimethyl-4-nitro-
1476 - Benzene, 1,3-dimethyl-2-nitro-
1477 - Benzene, 1,3-dimethyl-5-nitro-
1478 - Benzene, 1,4-dimethyl-2-nitro-
1479 - Benzene, 2,4-dimethyl-1-nitro-
1692 - Benzene, 1-ethyl-2-nitro-
1693 - Benzene, 1-ethyl-3-nitro-
1694 - Benzene, 1-ethyl-4-nitro-
2004 - Benzene, 1-methyl-4-(nitromethyl)-
2051 - Benzene, (2-nitroethyl)-
2426 - 1,3-Benzodioxole-5-methanamine
2503 - Benzoic acid, 2-amino-, methyl ester
2504 - Benzoic acid, 3-amino-, methyl ester
2679 - Benzoic acid, 2-(methylamino)-
6270 - Glycine, N-phenyl-
10901 - 2-Pyridinecarboxylic acid, ethyl ester
10902 - 3-Pyridinecarboxylic acid, ethyl ester
10903 - 4-Pyridinecarboxylic acid, ethyl ester

$C_8H_9NO_3$

1612 - Benzeneethanol, 2-nitro-
1662 - Benzene, 1-ethoxy-2-nitro-
1663 - Benzene, 1-ethoxy-3-nitro-
1664 - Benzene, 1-ethoxy-4-nitro-
1934 - Benzene, 4-methoxy-1-methyl-2-nitro-
2501 - Benzoic acid, 3-amino-4-hydroxy-, methyl ester
6267 - Glycine, N-(4-hydroxyphenyl)-
9334 - Phenol, 3,6-dimethyl-2-nitro-

$C_8H_9NO_4$

1445 - Benzene, 1,2-dimethoxy-4-nitro-
1446 - Benzene, 1,3-dimethoxy-5-nitro-
1447 - Benzene, 1,4-dimethoxy-2-nitro-
1448 - Benzene, 2,4-dimethoxy-1-nitro-

C_8H_{10}

1452 - Benzene, 1,2-dimethyl-

1453 - Benzene, 1,3-dimethyl-
1454 - Benzene, 1,4-dimethyl-
1668 - Benzene, ethyl-
4749 - 1,3,5-Cyclooctatriene
4750 - 1,3,6-Cyclooctatriene
4768 - 1,3-Cyclopentadiene, 5-(1-methylethylidene)-
6714 - 1,5-Hexadien-3-yne, 2,5-dimethyl-
8342 - 1,7-Octadiyne
8343 - 2,6-Octadiyne
8344 - 3,5-Octadiyne
8483 - 1,3,5,7-Octatetraene

$C_8H_{10}BrN$

643 - Benzenamine, 2-bromo-N,N-dimethyl-
644 - Benzenamine, 3-bromo-N,N-dimethyl-
645 - Benzenamine, 4-bromo-N,N-dimethyl-
1570 - Benzeneethanamine, 4-bromo-

$C_8H_{10}ClN$

662 - Benzenamine, 2-chloro-N,N-dimethyl-
663 - Benzenamine, 4-chloro-N,N-dimethyl-
1802 - Benzenemethanamine, 3-chloro-N-methyl-

$C_8H_{10}Cl_2Si$

11400 - Silane, dichloromethyl(4-methylphenyl)-

$C_8H_{10}K_2O_{15}Sb_2$

11588 - Tartar emetic

$C_8H_{10}N_2$

18 - Acetaldehyde, phenylhydrazone
5134 - Diazene, ethylphenyl-
11339 - Quinoxaline, 1,2,3,4-tetrahydro-

$C_8H_{10}N_2O$

29 - Acetamide, N-(4-aminophenyl)-
720 - Benzenamine, N,N-dimethyl-4-nitroso-
750 - Benzenamine, N-ethyl-N-nitroso-
12027 - Urea, (phenylmethyl)-

$C_8H_{10}N_2O_2$

718 - Benzenamine, N,N-dimethyl-2-nitro-
719 - Benzenamine, N,N-dimethyl-3-nitro-

$C_8H_{10}N_2O_3$

739 - Benzenamine, 2-ethoxy-5-nitro-

$C_8H_{10}N_2O_3S$

30 - Acetamide, N-[(4-aminophenyl)sulfonyl]-
31 - Acetamide, N-[4-(aminosulfonyl)phenyl]-
2212 - Benzenesulfonamide, N,4-dimethyl-N-nitroso-

$C_8H_{10}N_2O_4$

11011 - 1(4H)-Pyridinepropanoic acid, α-amino-3-hydroxy-4-oxo-, (S)-

$C_8H_{10}N_2O_4S$

474 - Asulam

$C_8H_{10}N_2S$

10885 - 4-Pyridinecarbothioamide, 2-ethyl-
11806 - Thiourea, (2-methylphenyl)-

$C_8H_{10}N_4O_2$

10760 - 1H-Purine-2,6-dione, 3,7-dihydro-1,3,7-trimethyl-

$C_8H_{10}N_6O_7$

11097 - 2,4,6(1H,3H,5H)-Pyrimidinetrione, 5-[(hexahydro-2,4,6-trioxo-5-pyrimidinyl)imino]-, monoammonium salt

$C_8H_{10}Na_2O_5$

5370 - Endothall, disodium salt

$C_8H_{10}O$

1592 - Benzeneethanol
1649 - Benzene, ethoxy-
1896 - Benzenemethanol, α-methyl-
1897 - Benzenemethanol, 2-methyl-
1898 - Benzenemethanol, 3-methyl-
1899 - Benzenemethanol, 4-methyl-
1930 - Benzene, (methoxymethyl)-
1931 - Benzene, 1-methoxy-2-methyl-
1932 - Benzene, 1-methoxy-3-methyl-
1933 - Benzene, 1-methoxy-4-methyl-
2916 - Bicyclo[2.2.1]heptane-2-carboxaldehyde
2958 - Bicyclo[2.2.1]heptan-2-one, 3-methylene-
2967 - Bicyclo[2.2.1]hept-5-ene-2-carboxaldehyde
8484 - 2,4,6-Octatrienal
9307 - Phenol, 2,3-dimethyl-
9308 - Phenol, 2,4-dimethyl-
9309 - Phenol, 2,5-dimethyl-

9310 - Phenol, 2,6-dimethyl-
9311 - Phenol, 3,4-dimethyl-
9312 - Phenol, 3,5-dimethyl-
9353 - Phenol, 2-ethyl-
9354 - Phenol, 3-ethyl-
9355 - Phenol, 4-ethyl-

$C_8H_{10}OS$
2296 - Benzenethiol, 4-ethoxy-
10472 - 1-Propanone, 2-methyl-1-(2-thienyl)-

$C_8H_{10}O_2$
1434 - 1,3-Benzenedimethanol
1435 - 1,4-Benzenedimethanol
1436 - Benzene, 1,2-dimethoxy-
1437 - Benzene, 1,3-dimethoxy-
1438 - Benzene, 1,4-dimethoxy-
1511 - 1,2-Benzenediol, 4,5-dimethyl-
1512 - 1,3-Benzenediol, 2,5-dimethyl-
1513 - 1,3-Benzenediol, 4,5-dimethyl-
1514 - 1,3-Benzenediol, 4,6-dimethyl-
1518 - 1,3-Benzenediol, 4-ethyl-
1606 - Benzeneethanol, 3-hydroxy-
1889 - Benzenemethanol, 2-methoxy-
1890 - Benzenemethanol, 3-methoxy-
1891 - Benzenemethanol, 4-methoxy-
3659 - 2-Butanone, 4-(2-furanyl)-
5579 - 1,2-Ethanediol, 1-phenyl-
5758 - Ethanol, 2-phenoxy-
8685 - 1,5-Pentalenedione, hexahydro-
9350 - Phenol, 2-ethoxy-
9351 - Phenol, 3-ethoxy-
9352 - Phenol, 4-ethoxy-
9385 - Phenol, 2-methoxy-4-methyl-

$C_8H_{10}O_2S$
1704 - Benzene, (ethylsulfonyl)-

$C_8H_{10}O_3$
1885 - Benzenemethanol, 2-hydroxy-3-methoxy-
1886 - Benzenemethanol, 4-hydroxy-3-methoxy-
3818 - 2-Butenoic acid, anhydride
6028 - 2-Furanacetic acid, ethyl ester
6061 - 2-Furancarboxylic acid, 1-methylethyl ester
6064 - 3-Furancarboxylic acid, 2-methyl-, ethyl ester
6071 - 2-Furancarboxylic acid, propyl ester
6123 - 2-Furanmethanol, propanoate
6162 - 2-Furanpropanoic acid, methyl ester
7345 - 1,3-Isobenzofurandione, hexahydro-
9305 - Phenol, 2,6-dimethoxy-
9306 - Phenol, 3,5-dimethoxy-
10649 - 2-Propenoic acid, 2-methyl-, anhydride

$C_8H_{10}O_3S$
2223 - Benzenesulfonic acid, ethyl ester
2232 - Benzenesulfonic acid, 4-methyl-, methyl ester

$C_8H_{10}O_4$
3909 - 2-Butynedioic acid, diethyl ester
3913 - 2-Butyne-1,4-diol, diacetate
5564 - Ethanedioic acid, di-2-propenyl ester
6701 - 2,5-Hexadienoic acid, 3-methoxy-5-methyl-4-oxo-

$C_8H_{10}S$
1705 - Benzene, (ethylthio)-
1834 - Benzenemethanethiol, α-methyl-, (S)-
2042 - Benzene, [(methylthio)methyl]-
2297 - Benzenethiol, 3-ethyl-
2298 - Benzenethiol, 4-ethyl-

$C_8H_{11}ClO_4$
3771 - 2-Butenedioic acid, 2-chloro-, diethyl ester, (E)-
3772 - 2-Butenedioic acid, 2-chloro-, diethyl ester, (Z)-

$C_8H_{11}ClSi$
11374 - Silane, chlorodimethylphenyl-

$C_8H_{11}Cl_2N_3O_2$
11061 - 2,4(1H,3H)-Pyrimidinedione, 5-[bis(2-chloroethyl)amino]-

$C_8H_{11}Cl_3O_7$
6231 - β-D-Glucopyranosiduronic acid, 2,2,2-trichloroethyl

$C_8H_{11}N$
276 - Acetonitrile, cyclohexylidene-
704 - Benzenamine, 2,3-dimethyl-
705 - Benzenamine, 2,4-dimethyl-
706 - Benzenamine, 2,5-dimethyl-
707 - Benzenamine, 2,6-dimethyl-
708 - Benzenamine, 3,4-dimethyl-
709 - Benzenamine, 3,5-dimethyl-
710 - Benzenamine, N,2-dimethyl-
711 - Benzenamine, N,3-dimethyl-
712 - Benzenamine, N,4-dimethyl-
713 - Benzenamine, N,N-dimethyl-
740 - Benzenamine, 2-ethyl-
741 - Benzenamine, 3-ethyl-
742 - Benzenamine, 4-ethyl-
743 - Benzenamine, N-ethyl-
1569 - Benzeneethanamine
1814 - Benzenemethanamine, α-methyl-, (±)-
1815 - Benzenemethanamine, 2-methyl-
1816 - Benzenemethanamine, 3-methyl-
1817 - Benzenemethanamine, 4-methyl-
1818 - Benzenemethanamine, N-methyl-
10962 - Pyridine, 2-ethyl-4-methyl-
10963 - Pyridine, 2-ethyl-6-methyl-
10964 - Pyridine, 3-ethyl-4-methyl-
10965 - Pyridine, 4-ethyl-2-methyl-
10966 - Pyridine, 5-ethyl-2-methyl-
10984 - Pyridine, 2-(1-methylethyl)-
10985 - Pyridine, 3-(1-methylethyl)-
10986 - Pyridine, 4-(1-methylethyl)-
11015 - Pyridine, 2-propyl-
11016 - Pyridine, 4-propyl-
11024 - Pyridine, 2,3,4-trimethyl-
11025 - Pyridine, 2,3,5-trimethyl-
11026 - Pyridine, 2,3,6-trimethyl-
11027 - Pyridine, 2,4,5-trimethyl-
11028 - Pyridine, 2,4,6-trimethyl-

$C_8H_{11}NO$
736 - Benzenamine, 2-ethoxy-
737 - Benzenamine, 4-ethoxy-
770 - Benzenamine, 2-methoxy-5-methyl-
771 - Benzenamine, 3-methoxy-4-methyl-
772 - Benzenamine, 4-methoxy-2-methyl-
773 - Benzenamine, 5-methoxy-2-methyl-
774 - Benzenamine, 2-methoxy-6-methyl-
1593 - Benzeneethanol, β-amino-
1594 - Benzeneethanol, 2-amino-
1812 - Benzenemethanamine, 2-methoxy-
1813 - Benzenemethanamine, 4-methoxy-
1839 - Benzenemethanol, α-(aminomethyl)-
5759 - Ethanol, 2-(phenylamino)-
9208 - Phenol, 4-(2-aminoethyl)-
9317 - Phenol, 2-(dimethylamino)-
9318 - Phenol, 3-(dimethylamino)-
9319 - Phenol, 4-(dimethylamino)-
9356 - Phenol, 3-(ethylamino)-
10951 - 2-Pyridineethanol, α-methyl-
10979 - Pyridine, 2-(2-methoxyethyl)-
11012 - 2-Pyridinepropanol
11013 - 3-Pyridinepropanol

$C_8H_{11}NO_2$
702 - Benzenamine, 2,5-dimethoxy-
703 - Benzenamine, 2,6-dimethoxy-

$C_8H_{11}NO_2S$
2211 - Benzenesulfonamide, N,4-dimethyl-

$C_8H_{11}NO_3$
1500 - 1,2-Benzenediol, 4-(2-amino-1-hydroxyethyl)-, (R)-

$C_8H_{11}NO_4$
11145 - 1-Pyrrolidineacetic acid, 2,5-dioxo-, ethyl ester

$C_8H_{11}N_3O_6$
11832 - 1,2,4-Triazine-3,5(2H,4H)-dione, 2-β-D-ribofuranosyl-

$C_8H_{11}N_5$
7234 - Imidodicarbonimidic diamide, N-phenyl-

$C_8H_{11}O_4$
5191 - 1,3-Dioxane-2-ethanol, 4-hydroxy-α,6-dimethyl-

C_8H_{12}
2993 - Bicyclo[4.2.0]oct-2-ene
4186 - Cyclobutane, 1,2-diethenyl-, cis-
4187 - Cyclobutane, 1,2-diethenyl-, trans-
4252 - 1,3-Cyclohexadiene, 1,3-dimethyl-
4253 - 1,3-Cyclohexadiene, 1,4-dimethyl-
4254 - 1,3-Cyclohexadiene, 1,5-dimethyl-
4255 - 1,3-Cyclohexadiene, 2,5-dimethyl-
4256 - 1,3-Cyclohexadiene, 2,6-dimethyl-
4314 - Cyclohexane, 1,2-bis(methylene)-
4657 - Cyclohexene, 1-ethenyl-
4658 - Cyclohexene, 4-ethenyl-
4732 - 1,4-Cyclooctadiene
4733 - 1,5-Cyclooctadiene
4734 - 1,5-Cyclooctadiene, (Z,Z)-
4755 - Cyclooctyne
6994 - 1,3,5-Hexatriene, 2,5-dimethyl-
8489 - 2,4,6-Octatriene, (E,E,E)-
8526 - 1-Octen-3-yne

$C_8H_{12}ClN$
717 - Benzenamine, N,N-dimethyl-, hydrochloride

$C_8H_{12}ClNO$
44 - Acetamide, 2-chloro-N,N-di-2-propenyl-

$C_8H_{12}ClNO_3$
10934 - 3,4-Pyridinedimethanol, 5-hydroxy-6-methyl-, hydrochloride

$C_8H_{12}Cl_2$
7117 - 3-Hexyne, 2,5-dichloro-2,5-dimethyl-

$C_8H_{12}N_2$
1267 - 1,2-Benzenediamine, N,N-dimethyl-
1268 - 1,3-Benzenediamine, N,N-dimethyl-
1269 - 1,2-Benzenediamine, N,N'-dimethyl-
1270 - 1,4-Benzenediamine, N,N-dimethyl-
1433 - 1,3-Benzenedimethanamine
3292 - Butanedinitrile, tetramethyl-
7159 - Hydrazine, 1-ethyl-1-phenyl-
7160 - Hydrazine, 1-ethyl-2-phenyl-
7163 - Hydrazine, 1-methyl-2-(3-methylphenyl)-
7170 - Hydrazine, (1-phenylethyl)-
8385 - Octanedinitrile
10809 - Pyrazine, 3-ethyl-2,5-dimethyl-

$C_8H_{12}N_2O$
1272 - 1,2-Benzenediamine, 4-ethoxy-

$C_8H_{12}N_2O_2$
6770 - Hexane, 1,6-diisocyanato-

$C_8H_{12}N_2O_3$
11087 - 2,4,6(1H,3H,5H)-Pyrimidinetrione, 5,5-diethyl-

$C_8H_{12}N_4$
10093 - Propanenitrile, 2,2'-azobis[2-methyl-

$C_8H_{12}O$
4529 - Cyclohexanol, 1-ethynyl-
4594 - Cyclohexanone, 2-ethylidene-
4705 - 2-Cyclohexen-1-one, 2,3-dimethyl-
4706 - 2-Cyclohexen-1-one, 2,5-dimethyl-
4707 - 2-Cyclohexen-1-one, 3,5-dimethyl-
4708 - 2-Cyclohexen-1-one, 3,6-dimethyl-
4754 - 3-Cycloocten-1-one
4947 - 2-Cyclopenten-1-one, 3-(1-methylethyl)-
5793 - Ethanone, 1-(1-cyclohexen-1-yl)-
6090 - Furan, 2-(1,1-dimethylethyl)-
8509 - 1-Octen-4-ol
8555 - 7-Oxabicyclo[4.1.0]heptane, 3-ethenyl-
8688 - 2(1H)-Pentalenone, hexahydro-
8689 - 2(1H)-Pentalenone, hexahydro-, cis-

$C_8H_{12}O_2$
4378 - 1,3-Cyclohexanedione, 5,5-dimethyl-
4572 - Cyclohexanone, 2-acetyl-
4640 - 3-Cyclohexene-1-carboxylic acid, methyl ester
6699 - 2,4-Hexadienoic acid, ethyl ester
6700 - 2,4-Hexadienoic acid, ethyl ester, (E,E)-
8504 - 4-Octene-2,7-dione
8538 - 2-Octynoic acid
8539 - 3-Octynoic acid

8540 - 5-Octynoic acid
8541 - 7-Octynoic acid
8557 - 7-Oxabicyclo[4.1.0]heptane, 3-oxiranyl-
8597 - 2-Oxetanone, 3,3-dimethyl-4-
    (1-methylethylidene)-
8805 - 2,4-Pentanedione, 3-(2-propenyl)-

$C_8H_{12}O_3$
4336 - Cyclohexanecarboxylic acid, 3-methyl-4-oxo-
4785 - Cyclopentanecarboxylic acid, 2-oxo-, ethyl
    ester
10695 - 2-Propenoic acid, (tetrahydro-2-
    furanyl)methyl ester

$C_8H_{12}O_4$
3462 - Butanoic acid, 2-acetyl-3-oxo-, ethyl ester
3775 - 2-Butenedioic acid (E)-, diethyl ester
3776 - 2-Butenedioic acid (Z)-, diethyl ester
4181 - 1,2-Cyclobutanedicarboxylic acid, dimethyl
    ester, trans-
4183 - 1,3-Cyclobutanedicarboxylic acid, 2,2-
    dimethyl-, cis-
4354 - 1,4-Cyclohexanedicarboxylic acid, trans-
6030 - 3-Furanacetic acid, tetrahydro-2,2-dimethyl-5-
    oxo-
10564 - 2-Propene-1,1-diol, 2-methyl-, diacetate

$C_8H_{12}O_5$
3340 - Butanedioic acid, oxo-, diethyl ester

$C_8H_{12}O_6Si$
11498 - Silanetriol, ethenyl, triacetate

$C_8H_{12}S$
11731 - Thiophene, 2-butyl-

$C_8H_{12}Si$
11418 - Silane, dimethylphenyl-
11449 - Silane, tetraethenyl-

$C_8H_{12}Sn$
11543 - Stannane, tetraethenyl-

$C_8H_{13}Cl$
8533 - 1-Octyne, 1-chloro-

$C_8H_{13}F_3O_2$
268 - Acetic acid, trifluoro-, 2-methylpentyl ester

$C_8H_{13}N$
4290 - Cyclohexanamine, 1-ethynyl-
4313 - Cyclohexaneacetonitrile
11113 - 1H-Pyrrole, 1-butyl-
11126 - 1H-Pyrrole, 3-ethyl-2,4-dimethyl-
11127 - 1H-Pyrrole, 4-ethyl-2,3-dimethyl-
11141 - 1H-Pyrrole, 2,3,4,5-tetramethyl-

$C_8H_{13}NO$
490 - 8-Azabicyclo[3.2.1]octan-3-one, 8-methyl-

$C_8H_{13}NO_2$
7575 - 2,5-Methano-2H-furo[3,2-b]pyrrol-6-ol,
    hexahydro-4-methyl-, (2α,3aβ,5α,6β,6aβ
9656 - 2,6-Piperidinedione, 4-ethyl-4-methyl-
10914 - 3-Pyridinecarboxylic acid, 1,2,5,6-tetrahydro-
    1-methyl-, methyl ester

$C_8H_{13}NO_3$
8573 - 2H-1,3-Oxazine-2,4(3H)-dione, 5,5-
    diethyldihydro-

$C_8H_{13}NO_5$
9953 - Propanedioic acid, (formylamino)-, diethyl
    ester

$C_8H_{13}N_2O_3PS$
11720 - Thionazin

C_8H_{14}
2991 - Bicyclo[4.2.0]octane
4245 - Cycloheptene, 1-methyl-
4246 - Cycloheptene, 3-methyl-
4388 - Cyclohexane, ethenyl-
4391 - Cyclohexane, ethylidene-
4434 - Cyclohexane, 1-methyl-4-methylene-
4648 - Cyclohexene, 1,2-dimethyl-
4649 - Cyclohexene, 1,3-dimethyl-
4650 - Cyclohexene, 1,4-dimethyl-
4651 - Cyclohexene, 1,5-dimethyl-
4652 - Cyclohexene, 1,6-dimethyl-
4653 - Cyclohexene, 3,3-dimethyl-
4654 - Cyclohexene, 3,5-dimethyl-
4655 - Cyclohexene, 3,6-dimethyl-

4656 - Cyclohexene, 4,4-dimethyl-
4660 - Cyclohexene, 1-ethyl-
4661 - Cyclohexene, 3-ethyl-
4662 - Cyclohexene, 4-ethyl-
4751 - Cyclooctene, (E)-
4752 - Cyclooctene, (Z)-
4848 - Cyclopentane, 2-propenyl-
4935 - Cyclopentene, 1-propyl-
4937 - Cyclopentene, 1,2,3-trimethyl-
4938 - Cyclopentene, 1,5,5-trimethyl-
6335 - 1,5-Heptadiene, 3-methyl-
6691 - 1,5-Hexadiene, 2,5-dimethyl-
6692 - 2,4-Hexadiene, 2,5-dimethyl-
8319 - 1,7-Octadiene
8320 - 2,6-Octadiene
8529 - 1-Octyne
8530 - 2-Octyne
8531 - 3-Octyne
8532 - 4-Octyne
8686 - Pentalene, octahydro-, cis-
8687 - Pentalene, octahydro-, trans-
9132 - 1-Pentyne, 3-ethyl-3-methyl-

$C_8H_{14}ClNS_2$
4065 - CDEC

$C_8H_{14}ClN_5$
476 - Atrazine

$C_8H_{14}Cl_2N_2O_2$
10978 - 3-Pyridinemethanol, 4-(aminomethyl)-5-
    hydroxy-6-methyl-, dihydrochloride

$C_8H_{14}Cl_2O_2$
3501 - Butanoic acid, 3,4-dichloro-, butyl ester

$C_8H_{14}N_2$
275 - Acetonitrile, (cyclohexylamino)-
9703 - 1-Piperidinepropanenitrile

$C_8H_{14}N_4OS$
7658 - Metribuzin

$C_8H_{14}O$
4235 - Cycloheptanone, 2-methyl-
4580 - Cyclohexanone, 2,2-dimethyl-
4581 - Cyclohexanone, 2,3-dimethyl-, cis-
4582 - Cyclohexanone, 2,3-dimethyl-, trans-
4583 - Cyclohexanone, 2,4-dimethyl-, (2S-trans)-
4584 - Cyclohexanone, 2,4-dimethyl-, cis-
4585 - Cyclohexanone, 2,5-dimethyl-, trans-(±)-
4586 - Cyclohexanone, 2,6-dimethyl-
4587 - Cyclohexanone, 3,3-dimethyl-
4588 - Cyclohexanone, 3,4-dimethyl-
4589 - Cyclohexanone, 3,5-dimethyl-
4590 - Cyclohexanone, 4,4-dimethyl-
4593 - Cyclohexanone, 4-ethyl-
4740 - Cyclooctanone
4753 - 2-Cyclooocten-1-ol
4883 - Cyclopentanone, 2-ethyl-5-methyl-
4887 - Cyclopentanone, 2-(1-methylethyl)-
4894 - Cyclopentanone, 2-propyl-
4895 - Cyclopentanone, 3-propyl-
4896 - Cyclopentanone, 2,2,4-trimethyl-
4897 - Cyclopentanone, 2,2,5-trimethyl-
4898 - Cyclopentanone, 2,3,5-trimethyl-
4899 - Cyclopentanone, 2,4,4-trimethyl-
5794 - Ethanone, 1-cyclohexyl-
5846 - Ethanone, 1-(1-methylcyclopentyl)-
5847 - Ethanone, 1-(2-methylcyclopentyl)-
6338 - 1,6-Heptadien-4-ol, 4-methyl-
6595 - 1-Hepten-3-one, 5-methyl-
6596 - 2-Hepten-4-one, 5-methyl-
6597 - 2-Hepten-4-one, 6-methyl-
6598 - 3-Hepten-2-one, 6-methyl-
6599 - 4-Hepten-3-one, 6-methyl-
6600 - 5-Hepten-3-one, 5-methyl-
6601 - 5-Hepten-2-one, 3-methyl-
6602 - 5-Hepten-2-one, 5-methyl-
6603 - 5-Hepten-2-one, 6-methyl-
6604 - 5-Hepten-3-one, 2-methyl-
6605 - 5-Hepten-3-one, 5-methyl-
7091 - 3-Hexen-2-one, 3,4-dimethyl-
7092 - 4-Hexen-2-one, 3,4-dimethyl-

8543 - 2-Octyn-1-ol

$C_8H_{14}O_2$
155 - Acetic acid, cyclohexyl ester
3836 - 2-Butenoic acid, 2,3-dimethyl-, ethyl ester
4223 - Cycloheptanecarboxylic acid
4306 - Cyclohexaneacetic acid
4335 - Cyclohexanecarboxylic acid, methyl ester
4776 - Cyclopentaneacetic acid, 3-methyl-
4847 - Cyclopentanepropanoic acid
5821 - Ethanone, 1-(1-hydroxycyclohexyl)-
6140 - 2(3H)-Furanone, 5-butyldihydro-
6565 - 2-Heptenoic acid, 6-methyl-
6807 - 3,4-Hexanedione, 2,5-dimethyl-
6892 - Hexanoic acid, 3-hydroxy-2,2-dimethyl-, β-
    lactone
7053 - 2-Hexenoic acid, ethyl ester
7054 - 3-Hexenoic acid, ethyl ester
7119 - 3-Hexyne-2,5-diol, 2,5-dimethyl-
8364 - Octanedial
8393 - 2,4-Octanedione
8394 - 2,7-Octanedione
8395 - 3,6-Octanedione
8396 - 4,5-Octanedione
9080 - 2-Pentenoic acid, 4-methyl-, ethyl ester
9081 - 3-Pentenoic acid, 3-methyl-, ethyl ester
9082 - 3-Pentenoic acid, 4-methyl-, ethyl ester
10650 - 2-Propenoic acid, 2-methyl-, butyl ester
10651 - 2-Propenoic acid, methylbutyl ester
10652 - 2-Propenoic acid, 2-methylbutyl ester
10665 - 2-Propenoic acid, 2-methyl-, 1-methylpropyl
    ester
10666 - 2-Propenoic acid, 2-methyl-, 2-methylpropyl
    ester

$C_8H_{14}O_3$
3467 - Butanoic acid, anhydride
3511 - Butanoic acid, 2,2-dimethyl-3-oxo-, ethyl ester
3518 - Butanoic acid, 2-ethyl-3-oxo-, ethyl ester
3569 - Butanoic acid, 3-oxo-, butyl ester
6126 - 2-Furanmethanol, tetrahydro-, propanoate
6903 - Hexanoic acid, 4-methyl-5-oxo-, methyl ester
6906 - Hexanoic acid, 5-oxo-, ethyl ester
8897 - Pentanoic acid, 4-methyl-3-oxo-, ethyl ester
8909 - Pentanoic acid, 4-oxo-, 1-methylethyl ester
8912 - Pentanoic acid, 4-oxo-, propyl ester
10291 - Propanoic acid, 2-methyl-, anhydride

$C_8H_{14}O_4$
3303 - Butanedioic acid, diethyl ester
3333 - Butanedioic acid, monobutyl ester
3343 - Butanedioic acid, tetramethyl-
3348 - 1,4-Butanediol, diacetate
5556 - Ethanedioic acid, bis(1-methylethyl) ester
5565 - Ethanedioic acid, dipropyl ester
5576 - 1,2-Ethanediol, dipropanoate
6786 - Hexanedioic acid, dimethyl ester
6791 - Hexanedioic acid, monoethyl ester
8386 - Octanedioic acid
8777 - Pentanedioic acid, 3-ethyl-3-methyl-
9962 - Propanedioic acid, methyl-, diethyl ester

$C_8H_{14}O_4S$
10331 - Propanoic acid, 3,3'-thiobis-, dimethyl ester

$C_8H_{14}O_5$
5754 - Ethanol, 2,2'-oxybis-, diacetate

$C_8H_{14}O_6$
3307 - Butanedioic acid, 2,3-dihydroxy-, diethyl ester,
    (R*,R*)-(±)-

$C_8H_{15}Br$
4316 - Cyclohexane, (2-bromoethyl)-

$C_8H_{15}BrO_2$
4149 - Crotonaldehyde, 2-bromo-, diethyl acetal
6877 - Hexanoic acid, 2-bromo-, ethyl ester, (±)-
6878 - Hexanoic acid, 5-bromo-, ethyl ester
6879 - Hexanoic acid, 6-bromo-, ethyl ester
8425 - Octanoic acid, 2-bromo-
8426 - Octanoic acid, 8-bromo-
10210 - Propanoic acid, 3-bromo-, 3-methylbutyl
    ester

7076 - 4-Hexen-3-ol, 3,5-dimethyl-
8346 - Octanal
8471 - 2-Octanone
8472 - 3-Octanone
8473 - 4-Octanone
8510 - 2-Octen-1-ol
8511 - 2-Octen-4-ol
8512 - 4-Octen-1-ol
8513 - 5-Octen-1-ol
8612 - Oxirane, 2-(2,2-dimethylpropyl)-2-methyl-
8697 - Pentanal, 2-propyl-
8970 - 2-Pentanone, 3-ethyl-4-methyl-
8989 - 3-Pentanone, 2,2,4-trimethyl-

$C_8H_{16}O_2$
185 - Acetic acid, 2-ethylbutyl ester
194 - Acetic acid, hexyl ester
3479 - Butanoic acid, butyl ester
3509 - Butanoic acid, 1,1-dimethylethyl ester
3553 - Butanoic acid, 3-methyl-, 1-methylethyl ester
3562 - Butanoic acid, 3-methyl-, propyl ester
3563 - Butanoic acid, 2-methylpropyl ester
3564 - Butanoic acid, 1-methylpropyl ester, (S)-
3639 - 1-Butanol, 3-methyl-, propanoate
3758 - 2-Butene, 1,1-diethoxy-
4363 - 1,3-Cyclohexanedimethanol
6008 - Formic acid, heptyl ester
6466 - Heptanoic acid, methyl ester
6518 - 2-Heptanone, 4-hydroxy-3-methyl-
6519 - 3-Heptanone, 5-hydroxy-5-methyl-
6882 - Hexanoic acid, 2-ethyl-
6883 - Hexanoic acid, 3-ethyl-
6884 - Hexanoic acid, ethyl ester
8422 - Octanoic acid
8474 - 4-Octanone, 5-hydroxy-
8692 - Pentanal, 3-hydroxy-2,2,4-trimethyl-
8892 - Pentanoic acid, 4-methyl-, ethyl ester
8893 - Pentanoic acid, 1-methylethyl ester
8914 - Pentanoic acid, propyl ester
8945 - 1-Pentanol, 2-methyl-, acetate
8946 - 2-Pentanol, 2-methyl-, acetate
8947 - 2-Pentanol, 4-methyl-, acetate
8948 - 3-Pentanol, 2-methyl-, acetate
8949 - 3-Pentanol, 3-methyl-, acetate
10294 - Propanoic acid, 2-methyl-, 1,1-dimethylethyl ester
10304 - Propanoic acid, 2-methyl-, 2-methylpropyl ester
10318 - Propanoic acid, pentyl ester

$C_8H_{16}O_3$
3536 - Butanoic acid, 2-hydroxy-, 2-methylpropyl ester
8432 - Octanoic acid, 2-hydroxy-
10274 - Propanoic acid, 2-hydroxy-, 3-methylbutyl ester

$C_8H_{16}O_4$
174 - Acetic acid, diethoxy-, ethyl ester
5721 - Ethanol, 2-(2-ethoxyethoxy)-, acetate

$C_8H_{16}Si$
11417 - Silane, dimethyldi-2-propenyl-

$C_8H_{17}Br$
8356 - Octane, 1-bromo-
8357 - Octane, 2-bromo-, (±)-

$C_8H_{17}Cl$
6366 - Heptane, 2-chloro-2-methyl-
6367 - Heptane, 2-chloro-6-methyl-
6368 - Heptane, 3-(chloromethyl)-
6369 - Heptane, 3-chloro-3-methyl-
6370 - Heptane, 4-chloro-4-methyl-
6740 - Hexane, 2-chloro-2,5-dimethyl-
6741 - Hexane, 3-chloro-2,3-dimethyl-
8359 - Octane, 1-chloro-
8360 - Octane, 2-chloro-
8733 - Pentane, 2-chloro-2,4,4-trimethyl-

$C_8H_{17}ClO$
8448 - 1-Octanol, 8-chloro-

$C_8H_{17}ClO_2$
3243 - Butane, 3-chloro-1,1-diethoxy-

$C_8H_{17}Cl_3Si$
11471 - Silane, trichlorooctyl-

$C_8H_{17}F$
8405 - Octane, 1-fluoro-

$C_8H_{17}FO$
8460 - 1-Octanol, 8-fluoro-

$C_8H_{17}I$
8406 - Octane, 1-iodo-
8407 - Octane, 2-iodo-, (±)-

$C_8H_{17}N$
503 - 1H-Azonine, octahydro-
4286 - Cyclohexanamine, N,N-dimethyl-
4287 - Cyclohexanamine, 2-ethyl-
4288 - Cyclohexanamine, N-ethyl-
4723 - Cyclohexylamine, N,4-dimethyl-
4735 - Cyclooctanamine
9684 - Piperidine, 1-(1-methylethyl)-
9707 - Piperidine, 1-propyl-
9708 - Piperidine, 2-propyl-, (S)-
9709 - Piperidine, 4-propyl-
9712 - Piperidine, 2,2,4-trimethyl-
9713 - Piperidine, 2,3,6-trimethyl-
9714 - Piperidine, 2,4,6-trimethyl-
11146 - Pyrrolidine, 1-butyl-
11147 - Pyrrolidine, 2-butyl-

$C_8H_{17}NO$
58 - Acetamide, N,N-dipropyl-
3200 - Butanamide, N,N-diethyl-
3656 - 2-Butanone, 4-(diethylamino)-
4520 - Cyclohexanol, 3-(dimethylamino)-
7684 - Morpholine, 4-butyl-
8348 - Octanal, oxime
8349 - Octanamide
9677 - 2-Piperidinemethanol, α-ethyl-, (R*,S*)-(±)-
9718 - 3-Piperidinol, 6-propyl-, (3S-trans)-

$C_8H_{17}NO_2$
3952 - Carbamic acid, (3-methylbutyl)-, ethyl ester
7693 - Morpholine, 4-(2-ethoxyethyl)-
8095 - Nitrous acid, octyl ester
8412 - Octane, 1-nitro-
8413 - Octane, 2-nitro-, (+)-
8423 - Octanoic acid, 2-amino-, (±)-

$C_8H_{17}NO_3$
8079 - Nitric acid, octyl ester

C_8H_{18}
3430 - Butane, 2,2,3,3-tetramethyl-
6422 - Heptane, 2-methyl-
6423 - Heptane, 3-methyl-, (S)-
6424 - Heptane, 4-methyl-
6771 - Hexane, 2,2-dimethyl-
6772 - Hexane, 2,3-dimethyl-
6773 - Hexane, 2,4-dimethyl-
6774 - Hexane, 2,5-dimethyl-
6775 - Hexane, 3,3-dimethyl-
6776 - Hexane, 3,4-dimethyl-
6815 - Hexane, 3-ethyl-
8355 - Octane
8813 - Pentane, 3-ethyl-2-methyl-
8814 - Pentane, 3-ethyl-3-methyl-
8847 - Pentane, 2,2,3-trimethyl-
8848 - Pentane, 2,2,4-trimethyl-
8849 - Pentane, 2,3,3-trimethyl-
8850 - Pentane, 2,3,4-trimethyl-

$C_8H_{18}ClNO_2$
9830 - 1-Propanaminium, 2-(acetyloxy)-N,N,N-trimethyl-, chloride

$C_8H_{18}Hg$
7433 - Mercury, dibutyl-

$C_8H_{18}N_2$
4362 - 1,3-Cyclohexanedimethanamine
5123 - Diazene, dibutyl

$C_8H_{18}N_2O$
3212 - 1-Butanamine, N-butyl-N-nitroso-

$C_8H_{18}O$
3401 - Butane, 1-(2-methylpropoxy)-
3415 - Butane, 1,1'-oxybis-
3416 - Butane, 2,2'-oxybis-, (±)-

6421 - Heptane, 1-methoxy-
6492 - 1-Heptanol, 3-methyl-
6493 - 1-Heptanol, 4-methyl-
6494 - 1-Heptanol, 6-methyl-
6495 - 2-Heptanol, 2-methyl-
6496 - 2-Heptanol, 2-methyl-
6497 - 2-Heptanol, 5-methyl-
6498 - 2-Heptanol, 6-methyl-
6499 - 3-Heptanol, 2-methyl-, (±)-
6500 - 3-Heptanol, 3-methyl-
6501 - 3-Heptanol, 4-methyl-
6502 - 3-Heptanol, 5-methyl-
6503 - 4-Heptanol, 2-methyl-
6504 - 4-Heptanol, 3-methyl-, (R*,S*)-(±)-
6505 - 4-Heptanol, 4-methyl-
6814 - Hexane, 1-ethoxy-
6921 - 1-Hexanol, 2,5-dimethyl-
6922 - 2-Hexanol, 2,3-dimethyl-
6923 - 2-Hexanol, 2,5-dimethyl-
6924 - 2-Hexanol, 3,4-dimethyl-
6925 - 3-Hexanol, 2,2-dimethyl-
6926 - 3-Hexanol, 2,3-dimethyl-
6927 - 3-Hexanol, 2,5-dimethyl-
6928 - 3-Hexanol, 3,5-dimethyl-, (±)-
6929 - 3-Hexanol, 4,4-dimethyl-
6930 - 1-Hexanol, 2-ethyl-
6931 - 3-Hexanol, 3-ethyl-
8443 - 1-Octanol
8444 - 2-Octanol, (±)-
8445 - 3-Octanol
8446 - 4-Octanol, (±)-
8933 - 2-Pentanol, 3-ethyl-2-methyl-
8934 - 3-Pentanol, 3-ethyl-2-methyl-
8954 - 2-Pentanol, 2,4,4-trimethyl-
8955 - 3-Pentanol, 2,2,4-trimethyl-
8956 - 3-Pentanol, 2,3,4-trimethyl-
10124 - Propane, 2,2'-oxybis[2-methyl-

$C_8H_{18}OS$
3422 - Butane, 1,1'-sulfinylbis-

$C_8H_{18}OSi_2$
5233 - Disiloxane, 1,3-diethenyl-1,1,3,3-tetramethyl-

$C_8H_{18}O_2$
5586 - Ethane, 1,1-diproxy-
5728 - Ethanol, 2-(hexyloxy)-
6401 - 2,4-Heptanediol, 3-methyl-
6801 - 2,5-Hexanediol, 2,5-dimethyl-
6802 - 1,3-Hexanediol, 2-ethyl-
8389 - 1,8-Octanediol
8390 - 2,4-Octanediol
8391 - 4,5-Octanediol, (±)-
8795 - 1,3-Pentanediol, 2,2,4-trimethyl-
8796 - 1,4-Pentanediol, 2,2,4-trimethyl-
9153 - Peroxide, bis(1,1-dimethylethyl)
9993 - 1,3-Propanediol, 2-butyl-2-methyl-

$C_8H_{18}O_2S$
3423 - Butane, 1,1'-sulfonylbis-
10132 - Propane, 1,1'-sulfonylbis[2-methyl-

$C_8H_{18}O_2Si$
11408 - Silane, diethoxymethyl-2-propenyl-

$C_8H_{18}O_3$
5612 - Ethane, 1,1'-oxybis[2-ethoxy-
5664 - Ethane, 1,1,1-triethoxy-
5688 - Ethanol, 2-(2-butoxyethoxy)-
6920 - 1-Hexanol, 3,5-dimethoxy-

$C_8H_{18}O_3S$
11578 - Sulfurous acid, bis(2-methylpropyl) ester
11580 - Sulfurous acid, dibutyl ester

$C_8H_{18}O_3Si$
11424 - Silane, ethenyltriethoxy-

$C_8H_{18}O_4$
11642 - 2,5,8,11-Tetraoxadodecane

$C_8H_{18}O_4S_2$
3230 - Butane, 2,2-bis(ethylsulfonyl)-

$C_8H_{18}O_5$
5756 - Ethanol, 2,2'-[oxybis(2,1-ethanediyloxy)]bis-

$C_8H_{18}S$
3433 - Butane, 1,1'-thiobis-

3434 - Butane, 2,2'-thiobis-
8418 - 1-Octanethiol
8419 - 2-Octanethiol, (±)-
10154 - Propane, 1,1'-thiobis[2-methyl-
10155 - Propane, 2,2'-thiobis[2-methyl-

$C_8H_{18}S_2$
5242 - Disulfide, bis(1,1-dimethylethyl)
5252 - Disulfide, dibutyl

$C_8H_{18}S_4$
11649 - Tetrasulfide, bis(1,1-dimethylethyl)

$C_8H_{18}Si_2$
11434 - Silane, 1,2-ethynediylbis[trimethyl-

$C_8H_{19}N$
3211 - 1-Butanamine, N-butyl-
3227 - 2-Butanamine, N-(1-methylpropyl)-
6353 - 2-Heptanamine, 6-methyl-, (±)-
6726 - 1-Hexanamine, 2-ethyl-
8350 - 1-Octanamine
8351 - 2-Octanamine, (±)-
9811 - 2-Propanamine, N-ethyl-N-(1-methylethyl)-
9812 - 1-Propanamine, N-ethyl-N-propyl-
9823 - 1-Propanamine, 2-methyl-N-(2-methylpropyl)-

$C_8H_{19}NO$
5683 - Ethanol, 2-[bis(1-methylethyl)amino]-
5712 - Ethanol, 2-(dipropylamino)-

$C_8H_{19}NO_2$
3214 - 1-Butanamine, 4,4-diethoxy-
3624 - 2-Butanol, 3,3'-iminobis-
5416 - Ethanamine, 2,2-diethoxy-N,N-dimethyl-
5425 - Ethanamine, 2-ethoxy-N-(2-ethoxyethyl)-
5692 - Ethanol, 2,2'-(butylimino)bis-
5707 - Ethanol, 2-[2-(diethylamino)ethoxy]-

$C_8H_{19}O_2PS_2$
5934 - Ethoprop

$C_8H_{19}O_2PS_3$
5260 - Disulfoton

$C_8H_{19}O_3P$
9531 - Phosphonic acid, bis(1,1-dimethylethyl) ester
9532 - Phosphonic acid, dibutyl ester

$C_8H_{20}As_2$
5109 - Diarsine, tetraethyl-

$C_8H_{20}BrN$
5447 - Ethanaminium, N,N,N-triethyl-, bromide

$C_8H_{20}N_2$
3263 - 1,4-Butanediamine, N,N,N',N'-tetramethyl-
6751 - 2,5-Hexanediamine, 2,5-dimethyl-
7140 - Hydrazine, 1,2-bis(2-methylpropyl)-
8365 - 1,8-Octanediamine

$C_8H_{20}N_2O$
10342 - 1-Propanol, 2-[(2-amino-
2-methylpropyl)amino]-2-methyl-

$C_8H_{20}OSi$
11428 - Silane, ethoxytriethyl-

$C_8H_{20}O_3P_2$
9526 - Phosphinic acid, diethyl-, anhydride

$C_8H_{20}O_3Si$
11478 - Silane, triethoxyethyl-

$C_8H_{20}O_4Si$
11506 - Silicic acid (H4SiO4), tetraethyl ester

$C_8H_{20}O_5P_2S_2$
11716 - Thiodiphosphoric acid ([(HO)2P(S)]2O),
tetraethyl ester

$C_8H_{20}O_6P_2$
7185 - Hypophosphoric acid, tetraethyl ester

$C_8H_{20}O_7P_2$
5221 - Diphosphoric acid, tetraethyl ester

$C_8H_{20}Pb$
9731 - Plumbane, tetraethyl

$C_8H_{20}Si$
11450 - Silane, tetraethyl-

$C_8H_{20}Si_2$
11421 - Silane, 1,2-ethenediylbis[trimethyl-, (E)-

$C_8H_{23}N_5$
5498 - 1,2-Ethanediamine, N-(2-aminoethyl)-N'-[2-
[(2-aminoethyl)amino]ethyl]-

$C_8H_{24}N_4O_3P_2$
5220 - Diphosphoramide, octamethyl-

$C_8H_{24}O_2Si_3$
11914 - Trisiloxane, octamethyl-

$C_8H_{24}O_4Si_4$
4993 - Cyclotetrasiloxane, octamethyl-

$C_8H_{26}O_3Si_4$
11647 - Tetrasiloxane, 1,1,1,3,5,7,7,7-octamethyl-

$C_8I_4O_3$
7348 - 1,3-Isobenzofurandione, 4,5,6,7-tetraiodo-

C_9F_{20}
8139 - Nonane, icosofluoro

$C_9F_{21}N$
9813 - 1-Propanamine, 1,1,2,2,3,3,3-heptafluoro-
N,N-bis(heptafluoropropyl)-

$C_9H_3Cl_3O_3$
2317 - 1,3,5-Benzenetricarbonyl trichloride

$C_9H_3F_{15}O_2$
8438 - Octanoic acid, pentadecafluoro-, methyl ester

$C_9H_4Cl_3NO_2S$
5984 - Folpet

$C_9H_4O_5$
7341 - 5-Isobenzofurancarboxylic acid, 1,3-dihydro-
1,3-dioxo-

$C_9H_5BrN_2O_2$
11221 - Quinoline, 6-bromo-5-nitro-

$C_9H_5Br_2NO$
11313 - 8-Quinolinol, 5,7-dibromo-

C_9H_5ClINO
11312 - 8-Quinolinol, 5-chloro-7-iodo-

$C_9H_5ClO_6$
2320 - 1,3,5-Benzenetricarboxylic acid, 2-chloro-

$C_9H_5Cl_2N$
11248 - Quinoline, 2,4-dichloro-
11249 - Quinoline, 2,7-dichloro-
11250 - Quinoline, 4,5-dichloro-
11251 - Quinoline, 4,7-dichloro-
11252 - Quinoline, 5,8-dichloro-

$C_9H_5Cl_2NO$
11314 - 8-Quinolinol, 5,7-dichloro-

$C_9H_5Cl_3N_4$
360 - Anilazine

$C_9H_5I_2NO$
11315 - 8-Quinolinol, 5,7-diiodo-

$C_9H_5NO_4$
10740 - 2-Propynoic acid, 3-(2-nitrophenyl)-

C_9H_6BrN
7374 - Isoquinoline, 4-bromo-
11217 - Quinoline, 3-bromo-
11218 - Quinoline, 5-bromo-
11219 - Quinoline, 6-bromo-
11220 - Quinoline, 7-bromo-

C_9H_6BrNO
11310 - 8-Quinolinol, 5-bromo-
11326 - 2(1H)-Quinolinone, 4-bromo-

$C_9H_6Br_2O_2$
10629 - 2-Propenoic acid, 2,3-dibromo-3-phenyl-,
(Z)-

C_9H_6ClN
7376 - Isoquinoline, 1-chloro-
11237 - Quinoline, 2-chloro-
11238 - Quinoline, 4-chloro-
11239 - Quinoline, 5-chloro-
11240 - Quinoline, 6-chloro-
11241 - Quinoline, 7-chloro-
11242 - Quinoline, 8-chloro-

C_9H_6ClNO
11311 - 8-Quinolinol, 5-chloro-

$C_9H_6Cl_2N_2O_3$
7643 - Methazole

$C_9H_6Cl_6O_3S$
5369 - Endosulfan

C_9H_6IN
11272 - Quinoline, 4-iodo-
11273 - Quinoline, 6-iodo-

$C_9H_6INO_4S$
11295 - 5-Quinolinesulfonic acid, 8-hydroxy-7-iodo-

$C_9H_6N_2$
9916 - Propanedinitrile, phenyl-

$C_9H_6N_2O_2$
1430 - Benzene, 1,3-diisocyanato-2-methyl-
1431 - Benzene, 2,4-diisocyanato-1-methyl-
7386 - Isoquinoline, 5-nitro-
11284 - Quinoline, 5-nitro-
11285 - Quinoline, 6-nitro-
11286 - Quinoline, 7-nitro-

$C_9H_6N_2O_3$
11287 - Quinoline, 4-nitro-, 1-oxide
11323 - 8-Quinolinol, 5-nitro-

C_9H_6O
10724 - 2-Propynal, 3-phenyl-

$C_9H_6OS_2$
7617 - Methanone, di-2-thienyl-

$C_9H_6O_2$
2791 - 1H-2-Benzopyran-1-one
2792 - 2H-1-Benzopyran-2-one
2793 - 4H-1-Benzopyran-4-one
7264 - 1H-Indene-1,3(2H)-dione

$C_9H_6O_2S$
2400 - Benzo[b]thiophene-2-carboxylic acid

$C_9H_6O_3$
2439 - 2-Benzofurancarboxylic acid
2775 - 2H-1-Benzopyran-3-carboxylic acid, 2-oxo-
2825 - 2H-1-Benzopyran-2-one, 6-hydroxy-
2826 - 2H-1-Benzopyran-2-one, 7-hydroxy-

$C_9H_6O_4$
2804 - 2H-1-Benzopyran-2-one, 6,7-dihydroxy-
2805 - 2H-1-Benzopyran-2-one, 7,8-dihydroxy-
7267 - 1H-Indene-1,3(2H)-dione, 2,2-dihydroxy-

$C_9H_6O_6$
2318 - 1,2,3-Benzenetricarboxylic acid
2319 - 1,2,4-Benzenetricarboxylic acid

$C_9H_7BrN_2$
11207 - 4-Quinolinamine, 3-bromo-
11215 - Quinoline, 6-amino-3-bromo-

C_9H_7BrO
10510 - 2-Propenal, 3-bromo-3-phenyl-

$C_9H_7BrO_4$
2481 - Benzoic acid, 2-(acetyloxy)-5-bromo-

C_9H_7ClO
10716 - 2-Propenoyl chloride, 3-phenyl-
10717 - 2-Propenoyl chloride, 3-phenyl-, (E)-

$C_9H_7Cl_3O_3$
11512 - Silvex

$C_9H_7FO_2$
10637 - 2-Propenoic acid, 2-fluoro-3-phenyl-

C_9H_7N
7373 - Isoquinoline
10587 - 2-Propenenitrile, 3-phenyl-, (E)-
10588 - 2-Propenenitrile, 3-phenyl-, (Z)-
11214 - Quinoline

C_9H_7NO
2124 - Benzenepropanenitrile, β-oxo-
11305 - 5-Quinolinol
11306 - 6-Quinolinol
11307 - 7-Quinolinol
11308 - 8-Quinolinol
11325 - 2(1H)-Quinolinone

$C_9H_7NO_2$
7365 - 1H-Isoindole-1,3(2H)-dione, 2-methyl-

$C_9H_7NO_4$
10676 - 2-Propenoic acid, 3-(2-nitrophenyl)-, (E)-

$C_9H_7NO_4S$
11294 - 5-Quinolinesulfonic acid, 8-hydroxy-

$C_9H_7N_3S$
11858 - Tricyclazole

C_9H_8
2173 - Benzene, 1-propynyl-
7245 - 1H-Indene

$C_9H_8Br_2$
7247 - 1H-Indene, 1,2-dibromo-2,3-dihydro-

$C_9H_8Cl_2$
1397 - Benzene, (3,3-dichloro-2-propenyl)-
7249 - 1H-Indene, 1,2-dichloro-2,3-dihydro-

C₉H₈Cl₂O₃
172 - Acetic acid, (2,4-dichlorophenoxy)-, methyl ester
7653 - Methyl 3,6-dichloro-o-anisate
10250 - Propanoic acid, 2-(2,4-dichlorophenoxy)-

C₉H₈Cl₃NO₂S
7368 - 1H-Isoindole-1,3(2H)-dione, 3a,4,7,7a-tetrahydro-2-[(trichloromethyl)thio]-

C₉H₈NO₃S
11324 - 8-Quinolinol, sulfate (2:1) (salt)

C₉H₈N₂
7212 - 1H-Imidazole, 1-phenyl-
7213 - 1H-Imidazole, 2-phenyl-
7372 - 5-Isoquinolinamine
11202 - 2-Quinolinamine
11203 - 4-Quinolinamine
11204 - 5-Quinolinamine
11205 - 6-Quinolinamine
11206 - 8-Quinolinamine
11337 - Quinoxaline, 2-methyl-
11338 - Quinoxaline, 6-methyl-

C₉H₈N₂O
11336 - Quinoxaline, 6-methoxy-

C₉H₈N₂O₆
2593 - Benzoic acid, 3,5-dinitro-, ethyl ester

C₉H₈N₄OS
11687 - Thidiazuron

C₉H₈O
1876 - Benzenemethanol, α-ethynyl-
2456 - Benzofuran, 2-methyl-
2457 - Benzofuran, 3-methyl-
2458 - Benzofuran, 5-methyl-
2459 - Benzofuran, 7-methyl-
2772 - 2H-1-Benzopyran
7282 - 6H-Indeno[1,2-b]oxirene, 1a,6a-dihydro-
7286 - 1H-Inden-1-one, 2,3-dihydro-
7287 - 2H-Inden-2-one, 1,3-dihydro-
10518 - 2-Propenal, 3-phenyl-, (E)-
10745 - 2-Propyn-1-ol, 3-phenyl-

C₉H₈OS
2869 - 4H-1-Benzothiopyran-4-one, 2,3-dihydro-

C₉H₈O₂
927 - Benzeneacetic acid, α-methylene-
2425 - 1,3-Benzodioxole, 5-ethenyl-
2455 - 2-Benzofuranmethanol
2797 - 1H-2-Benzopyran-1-one, 3,4-dihydro-
2798 - 2H-1-Benzopyran-2-one, 3,4-dihydro-
2799 - 4H-1-Benzopyran-4-one, 2,3-dihydro-
6132 - Furan, 2,2'-methylenebis-
10028 - 1,2-Propanedione, 1-phenyl-
10682 - 2-Propenoic acid, 3-phenyl-, (E)-
10683 - 2-Propenoic acid, 3-phenyl-, (Z)-

C₉H₈O₃
2476 - Benzoic acid, 2-acetyl-
2477 - Benzoic acid, 3-acetyl-
2478 - Benzoic acid, 4-acetyl-
7580 - 4,7-Methanoisobenzofuran-1,3-dione, 3a,4,7,7a-tetrahydro-, (3aα,4α,7α,7aα)-
10644 - 2-Propenoic acid, 3-(4-hydroxyphenyl)-

C₉H₈O₄
2420 - 1,3-Benzodioxole-5-carboxylic acid, methyl ester
2480 - Benzoic acid, 2-(acetyloxy)-
9151 - Peroxide, acetyl benzoyl
10632 - 2-Propenoic acid, 3-(3,4-dihydroxyphenyl)-

C₉H₈S
2402 - Benzo[b]thiophene, 5-methyl-

C₉H₉Br
1082 - Benzene, 1-bromo-4-(1-propenyl)-
1083 - Benzene, 1-bromo-4-(2-propenyl)-
1084 - Benzene, (3-bromo-1-propenyl)-

C₉H₉BrO
5779 - Ethanone, 2-bromo-1-(4-methylphenyl)-
10418 - 1-Propanone, 1-(4-bromophenyl)-
10419 - 1-Propanone, 2-bromo-1-phenyl-

C₉H₉BrO₂
2147 - Benzenepropanoic-β-14C acid, α-bromo-

2521 - Benzoic acid, 2-bromo-, ethyl ester
2522 - Benzoic acid, 3-bromo-, ethyl ester
2523 - Benzoic acid, 4-bromo-, ethyl ester
5205 - 1,3-Dioxolane, 2-(3-bromophenyl)-

C₉H₉Br₂NO₃
11927 - L-Tyrosine, 3,5-dibromo-

C₉H₉Cl
1226 - Benzene, (1-chloro-1-propenyl)-
1227 - Benzene, (2-chloro-1-propenyl)-
1228 - Benzene, (3-chloro-2-propenyl)-
1229 - Benzene, 1-chloro-3-(1-propenyl)-, (E)-
1230 - Benzene, (3-chloro-1-propenyl)-, (E)-

C₉H₉ClO
1232 - Benzene, 1-chloro-4-(2-propenyloxy)-
2160 - Benzenepropanoyl chloride
2441 - Benzofuran, 2-(chloromethyl)-2,3-dihydro-
5787 - Ethanone, 2-chloro-1-(4-methylphenyl)-
7283 - 1H-Inden-4-ol, 7-chloro-2,3-dihydro-
9271 - Phenol, 4-chloro-2-(2-propenyl)-
10422 - 1-Propanone, 1-(4-chlorophenyl)-

C₉H₉ClO₂
140 - Acetic acid, chloro-, 4-methylphenyl ester
148 - Acetic acid, chloro-, phenylmethyl ester
2537 - Benzoic acid, 2-chloro-, ethyl ester
2538 - Benzoic acid, 3-chloro-, ethyl ester
2539 - Benzoic acid, 4-chloro-, ethyl ester
5696 - Ethanol, 2-chloro-, benzoate
10505 - Propanoyl chloride, 2-phenoxy-

C₉H₉ClO₃
141 - Acetic acid, (4-chloro-2-methylphenoxy)-
2541 - Benzoic acid, 3-chloro-2-hydroxy-, ethyl ester
4141 - Cloprop
5786 - Ethanone, 2-chloro-1-(4-hydroxy-3-methoxyphenyl)-

C₉H₉Cl₂NO
10200 - Propanil

C₉H₉FO₂
2606 - Benzoic acid, 4-fluoro-, ethyl ester

C₉H₉I₂NO₃
11928 - Tyrosine, 3,5-diiodo-

C₉H₉N
959 - Benzeneacetonitrile, α-methyl-
960 - Benzeneacetonitrile, 2-methyl-
961 - Benzeneacetonitrile, 3-methyl-
962 - Benzeneacetonitrile, 4-methyl-
2123 - Benzenepropanenitrile
7312 - 1H-Indole, 1-methyl-
7313 - 1H-Indole, 2-methyl-
7314 - 1H-Indole, 3-methyl-
7315 - 1H-Indole, 4-methyl-
7316 - 1H-Indole, 5-methyl-
7317 - 1H-Indole, 7-methyl-

C₉H₉NO
957 - Benzeneacetonitrile, 2-methoxy-
958 - Benzeneacetonitrile, 4-methoxy-
2754 - Benzonitrile, 2-ethoxy-
2755 - Benzonitrile, 4-ethoxy-
2879 - Benzoxazole, 2,5-dimethyl-
7311 - 1H-Indole, 7-methoxy-
11327 - 2(1H)-Quinolinone, 3,4-dihydro-

C₉H₉NO₂
2753 - Benzonitrile, 2,6-dimethoxy-
3363 - 1,3-Butanedione, 1-(2-pyridinyl)-
3364 - 1,3-Butanedione, 1-(3-pyridinyl)-
3365 - 1,3-Butanedione, 1-(4-pyridinyl)-

C₉H₉NO₃
6255 - Glycine, N-benzoyl-
7301 - 1H-Indole-5,6-dione, 2,3-dihydro-3-hydroxy-1-methyl-

C₉H₉NO₄
103 - Acetic acid, [2-(aminocarbonyl)phenoxy]-
524 - Benzadox
2707 - Benzoic acid, 2-nitro-, ethyl ester
2708 - Benzoic acid, 3-nitro-, ethyl ester
2709 - Benzoic acid, 4-nitro-, ethyl ester

C₉H₉N₃O₂
3990 - Carbendazim

C₉H₉N₃O₂S₂
2206 - Benzenesulfonamide, 4-amino-N-2-thiazolyl-
11663 - 2-Thiazolamine, 5-[(4-aminophenyl)sulfonyl]-

C₉H₉N₃O₄S₂
3975 - Carbamodithioic acid, dimethyl-, 2,4-dinitrophenyl ester

C₉H₉N₃O₆
2364 - Benzene, 1,3,5-trimethyl-2,4,6-trinitro-

C₉H₉N₃S
11671 - 2,4-Thiazolediamine, 5-phenyl-

C₉H₉N₅
11829 - 1,3,5-Triazine-2,4-diamine, 6-phenyl-
11830 - 1,3,5-Triazine-2,4-diamine, N-phenyl-

C₉H₁₀
1256 - Benzene, cyclopropyl-
1640 - Benzene, 1-ethenyl-2-methyl-
1641 - Benzene, 1-ethenyl-3-methyl-
1642 - Benzene, 1-ethenyl-4-methyl-
1972 - Benzene, (1-methylethenyl)-
2162 - Benzene, 1-propenyl-
2163 - Benzene, 1-propenyl-, (E)-
2164 - Benzene, 1-propenyl-, (Z)-
2165 - Benzene, 2-propenyl-
4747 - 1,3,5,7-Cyclooctatetraene, 1-methyl-
7250 - 1H-Indene, 2,3-dihydro-

C₉H₁₀BrClN₂O₂
4076 - Chlorbromuron

C₉H₁₀Br₂
1313 - Benzene, 2,4-dibromo-1,3,5-trimethyl-

C₉H₁₀Cl₂
1402 - Benzene, 2,4-dichloro-1,3,5-trimethyl-

C₉H₁₀Cl₂N₂O
12000 - Urea, N'-(3,4-dichlorophenyl)-N,N-dimethyl-

C₉H₁₀Cl₂N₂O₂
7415 - Linuron

C₉H₁₀INO₃
11932 - L-Tyrosine, 3-iodo-

C₉H₁₀N₂
2378 - 1H-Benzimidazole, 1,5-dimethyl-
2379 - 1H-Benzimidazole, 2,5-dimethyl-
2380 - 1H-Benzimidazole, 5,6-dimethyl-
7237 - 1H-Indazole, 1,5-dimethyl-
7238 - 2H-Indazole, 2,3-dimethyl-
10823 - 1H-Pyrazole, 4,5-dihydro-1-phenyl-
10824 - 1H-Pyrazole, 4,5-dihydro-3-phenyl-
11133 - 1H-Pyrrole, 2,2'-methylenebis-

C₉H₁₀N₂O
8574 - 2-Oxazolamine, 4,5-dihydro-5-phenyl-
10844 - 3-Pyrazolidinone, 1-phenyl-

C₉H₁₀N₂O₂
879 - Benzeneacetamide, N-(aminocarbonyl)-

C₉H₁₀N₂O₃
6252 - Glycine, N-(4-aminobenzoyl)-

C₉H₁₀N₂O₃S₂
2865 - 2-Benzothiazolesulfonamide, 6-ethoxy-

C₉H₁₀N₂O₄
2360 - Benzene, 1,3,5-trimethyl-2,4-dinitro-

C₉H₁₀O
547 - Benzaldehyde, 2,4-dimethyl-
548 - Benzaldehyde, 2,5-dimethyl-
549 - Benzaldehyde, 3,5-dimethyl-
558 - Benzaldehyde, 2-ethyl-
872 - Benzeneacetaldehyde, α-methyl-
873 - Benzeneacetaldehyde, 2-methyl-
874 - Benzeneacetaldehyde, 4-methyl-
1637 - Benzene, 1-ethenyl-2-methoxy-
1638 - Benzene, 1-ethenyl-3-methoxy-
1639 - Benzene, 1-ethenyl-4-methoxy-
1868 - Benzenemethanol, α-ethenyl-
1928 - Benzene, (2-methoxyethenyl)-
2114 - Benzenepropanal
2167 - Benzene, (2-propenyloxy)-
2443 - Benzofuran, 2,3-dihydro-2-methyl-
2444 - Benzofuran, 2,3-dihydro-5-methyl-
2776 - 1H-2-Benzopyran, 3,4-dihydro-
2777 - 2H-1-Benzopyran, 3,4-dihydro-
5851 - Ethanone, 1-(2-methylphenyl)-

$C_9H_{14}O_3$
6084 - Furan, 2-(diethoxymethyl)-
9074 - 4-Pentenoic acid, 2-acetyl-, ethyl ester
10674 - 2-Propenoic acid, 2-methyl-, (tetrahydro-2-furanyl)methyl ester
10703 - 2-Propen-1-ol, 2-methyl-, carbonate (2:1)

$C_9H_{14}O_3Si$
11488 - Silane, trimethoxyphenyl-

$C_9H_{14}O_4$
3329 - Butanedioic acid, methylene-, diethyl ester
3787 - 2-Butenedioic acid, 2-methyl-, diethyl ester, (E)-
3788 - 2-Butenedioic acid, 2-methyl-, diethyl ester, (Z)-
4792 - 1,2-Cyclopentanedicarboxylic acid, dimethyl ester, trans-(±)-
4967 - 1,1-Cyclopropanedicarboxylic acid, diethyl ester
4968 - 1,2-Cyclopropanedicarboxylic acid, diethyl ester, cis-
9041 - 2-Pentenedioic acid, diethyl ester
9042 - 2-Pentenedioic acid, diethyl ester, (E)-
9950 - Propanedioic acid, ethylidene-, diethyl ester

$C_9H_{14}O_5$
3324 - Butanedioic acid, formyl-, diethyl ester
8783 - Pentanedioic acid, 3-oxo-, diethyl ester
9923 - Propanedioic acid, acetyl-, diethyl ester

$C_9H_{14}O_6$
10195 - 1,2,3-Propanetriol, triacetate

$C_9H_{14}O_7$
10170 - 1,2,3-Propanetricarboxylic acid, 2-hydroxy-, trimethyl ester

$C_9H_{14}S$
11771 - Thiophene, 2-(3-methylbutyl)-

$C_9H_{14}Si$
11493 - Silane, trimethylphenyl-

$C_9H_{15}AlO_9$
340 - Aluminum, tris(2-hydroxypropanoato-O1,O2)-

$C_9H_{15}Br$
4322 - Cyclohexane, (2-bromo-2-propenyl)-

$C_9H_{15}Br_6O_4P$
10358 - 1-Propanol, 2,3-dibromo-, phosphate (3:1)

$C_9H_{15}Cl$
8200 - 1-Nonyne, 1-chloro-
8534 - 3-Octyne, 2-chloro-2-methyl-

$C_9H_{15}N$
10523 - 2-Propen-1-amine, N,N-di-2-propenyl-
11117 - 1H-Pyrrole, 3,4-diethyl-2-methyl-

$C_9H_{15}NO$
479 - 9-Azabicyclo[3.3.1]nonan-3-one, 9-methyl-

$C_9H_{15}NO_2$
3495 - Butanoic acid, 2-cyano-2-ethyl-, ethyl ester
6881 - Hexanoic acid, 2-cyano-, ethyl ester
9654 - 2,4-Piperidinedione, 3,3-diethyl-
10521 - 2-Propenamide, N-(1,1-dimethyl-3-oxobutyl)-

$C_9H_{15}NO_3$
483 - 8-Azabicyclo[3.2.1]octane-2-carboxylic acid, 3-hydroxy-8-methyl-, [1R-(exo,exo)]-
8585 - 2,4-Oxazolidinedione, 5,5-dipropyl-

$C_9H_{15}NO_5$
9922 - Propanedioic acid, (acetylamino)-, diethyl ester

$C_9H_{15}N_3O_2S$
7204 - 1H-Imidazole-4-ethanaminium, α-carboxy-2,3-dihydro-N,N,N-trimethyl-2-thioxo-, hydroxide, in

$C_9H_{15}N_3O_7$
472 - L-Aspartic acid, N-[2-[(2-amino-2-oxoethyl)amino]-2-carboxyethyl]-

C_9H_{16}
2990 - Bicyclo[3.3.1]nonane
2992 - Bicyclo[2.2.2]octane, 2-methyl-
4393 - Cyclohexane, 1-ethylidene-2-methyl-
4470 - Cyclohexane, 2-propenyl-
4663 - Cyclohexene, 1-ethyl-4-methyl-
4687 - Cyclohexene, 1,2,3-trimethyl-
4688 - Cyclohexene, 1,3,5-trimethyl-
4689 - Cyclohexene, 1,4,4-trimethyl-
4690 - Cyclohexene, 1,4,5-trimethyl-
4691 - Cyclohexene, 1,5,5-trimethyl-
4692 - Cyclohexene, 1,6,6-trimethyl-
4693 - Cyclohexene, 1,5,6-trimethyl-
4727 - Cyclononene, (E)-
4728 - Cyclononene, (Z)-
4906 - Cyclopentene, 1-butyl-
4927 - Cyclopentene, 4-isopropyl-1-methyl-
4931 - Cyclopentene, 3-methyl-1-(1-methylethyl)-
6330 - 1,3-Heptadiene, 2,6-dimethyl-
6331 - 1,5-Heptadiene, 2,6-dimethyl-
6332 - 2,4-Heptadiene, 2,6-dimethyl-
6333 - 2,4-Heptadiene, 3,5-dimethyl-
6617 - 3-Heptyne, 2,6-dimethyl-
6618 - 3-Heptyne, 5,5-dimethyl-
6694 - 1,5-Hexadiene, 3-ethyl-2-methyl-
7281 - 1H-Indene, octahydro-
8110 - 1,8-Nonadiene
8111 - 2,7-Nonadiene
8196 - 1-Nonyne
8197 - 2-Nonyne
8198 - 3-Nonyne
8199 - 4-Nonyne
8328 - 2,4-Octadiene, 7-methyl-
8329 - 3,5-Octadiene, 4-methyl-
8537 - 3-Octyne, 7-methyl-

$C_9H_{16}ClN_5$
10508 - Propazine
11593 - Terbuthylazine

$C_9H_{16}N_4OS$
11590 - Tebuthiuron

$C_9H_{16}O$
2938 - Bicyclo[2.2.1]heptan-2-ol, 1,7-dimethyl-, (exo,syn)-
4555 - Cyclohexanol, 1-(2-propenyl)-
4556 - Cyclohexanol, 2-(2-propenyl)-, trans-
4595 - Cyclohexanone, 2-ethyl-4-methyl-
4599 - Cyclohexanone, 2-(1-methylethyl)-
4600 - Cyclohexanone, 4-(1-methylethyl)-
4612 - Cyclohexanone, 2-propyl-
4613 - Cyclohexanone, 4-propyl-
4614 - Cyclohexanone, 2,2,3-trimethyl-
4615 - Cyclohexanone, 2,2,4-trimethyl-
4616 - Cyclohexanone, 2,2,5-trimethyl-
4617 - Cyclohexanone, 2,2,6-trimethyl-
4618 - Cyclohexanone, 2,3,6-trimethyl-
4619 - Cyclohexanone, 2,4,4-trimethyl-
4620 - Cyclohexanone, 2,4,5-trimethyl-
4621 - Cyclohexanone, 2,4,6-trimethyl-
4622 - Cyclohexanone, 3,3,5-trimethyl-
4623 - Cyclohexanone, 3,4,4-trimethyl-
4700 - 2-Cyclohexen-1-ol, 1,3,5-trimethyl-
4701 - 2-Cyclohexen-1-ol, 2,4,4-trimethyl-
4702 - 2-Cyclohexen-1-ol, 3,5,5-trimethyl-
4725 - Cyclononanone
4738 - Cyclooctanecarboxaldehyde
4884 - Cyclopentanone, 4-isopropyl-2-methyl-
4889 - Cyclopentanone, 2-methyl-2-(1-methylethyl)-
4890 - Cyclopentanone, 2-methyl-5-(1-methylethyl)-
4891 - Cyclopentanone, 4-methyl-2-(1-methylethyl)-
5843 - Ethanone, 1-(1-methylcyclohexyl)-
5844 - Ethanone, 1-(3-methylcyclohexyl)-
5845 - Ethanone, 1-(4-methylcyclohexyl)-
8112 - 2,6-Nonadien-1-ol
8179 - 2-Nonenal
8180 - 2-Nonenal, (E)-
8208 - 1-Nonyn-3-ol
8209 - 2-Nonyn-1-ol
8522 - 1-Octen-3-one, 5-methyl-
8523 - 5-Octen-4-one, 5-methyl-
8524 - 5-Octen-4-one, 7-methyl-
8525 - 7-Octen-4-one, 2-methyl-
8536 - 2-Octyne, 1-methoxy-
8544 - 1-Octyn-3-ol, 3-methyl-
10423 - 1-Propanone, 1-cyclohexyl-

$C_9H_{16}O_2$
4310 - Cyclohexaneacetic acid, methyl ester
4311 - Cyclohexaneacetic acid, 2-methyl-
4312 - Cyclohexaneacetic acid, 3-methyl-
4334 - Cyclohexanecarboxylic acid, ethyl ester
4468 - Cyclohexanepropanoic acid
6149 - 2(3H)-Furanone, dihydro-5-pentyl-
7063 - 2-Hexenoic acid, 2-methyl-, ethyl ester
7064 - 3-Hexenoic acid, 2-methyl-, ethyl ester
7065 - 3-Hexenoic acid, 3-methyl-, ethyl ester
8137 - 2,4-Nonanedione
8187 - 3-Nonenoic acid
8188 - 8-Nonenoic acid
10235 - Propanoic acid, cyclohexyl ester
10643 - 2-Propenoic acid, hexyl ester

$C_9H_{16}O_3$
3461 - Butanoic acid, 2-acetyl-3-methyl-, ethyl ester
6165 - 2-Furanpropanoic acid, tetrahydro-, ethyl ester
6460 - Heptanoic acid, 2,2-dimethyl-6-oxo-
6461 - Heptanoic acid, 4,4-dimethyl-6-oxo-
6902 - Hexanoic acid, 4-methyl-5-oxo-, ethyl ester
8856 - Pentanoic acid, 2-acetyl-, ethyl ester
8904 - Pentanoic acid, 4-oxo-, butyl ester
8910 - Pentanoic acid, 4-oxo-, 2-methylpropyl ester

$C_9H_{16}O_4$
3326 - Butanedioic acid, methyl-, diethyl ester
3337 - Butanedioic acid, monopentyl ester
6395 - Heptanedioic acid, dimethyl ester
6396 - Heptanedioic acid, monoethyl ester
8130 - Nonanedioic acid
8772 - Pentanedioic acid, diethyl ester
8774 - Pentanedioic acid, 3,3-dimethyl-, dimethyl ester
8791 - 1,5-Pentanediol, diacetate
9942 - Propanedioic acid, dimethyl-, diethyl ester
9945 - Propanedioic acid, dipropyl ester
9949 - Propanedioic acid, ethyl-, diethyl ester

$C_9H_{16}O_5$
8778 - Pentanedioic acid, 3-hydroxy-, diethyl ester

$C_9H_{17}ClN_3O_3PS$
7340 - Isazophos

$C_9H_{17}ClO$
8177 - Nonanoyl chloride

$C_9H_{17}FO_2$
8151 - Nonanoic acid, 9-fluoro-

$C_9H_{17}IN_4$
11825 - 3,5,7-Triaza-1-azoniatricyclo[3.3.1.13,7]decane, 1-(2-propenyl)-, iodide

$C_9H_{17}N$
4304 - Cyclohexanamine, N-2-propenyl-
8144 - Nonanenitrile
11246 - Quinoline, decahydro-, cis-
11247 - Quinoline, decahydro-, trans-(±)-

$C_9H_{17}NO$
4591 - Cyclohexanone, 2-[(dimethylamino)methyl]-
9003 - 4-Pentenamide, 2,2-diethyl-
9725 - 4-Piperidinone, 1,2,3,6-tetramethyl-
9726 - 4-Piperidinone, 2,2,6,6-tetramethyl-
10470 - 2-Propanone, 1-(1-methyl-2-piperidinyl)-, (±)-

$C_9H_{17}NOS$
7661 - Molinate

$C_9H_{17}NO_2$
10631 - 2-Propenoic acid, 2-(diethylamino)ethyl ester

$C_9H_{17}NO_4$
10305 - Propanoic acid, 2-methyl-, 2-nitropentyl ester

$C_9H_{17}NO_5$
329 - β-Alanine, N-(2,4-dihydroxy-3,3-dimethyl-1-oxobutyl)-, (R)-

$C_9H_{17}NS$
11708 - Thiocyanic acid, octyl ester

$C_9H_{17}N_5S$
341 - Ametryn

C_9H_{18}
4394 - Cyclohexane, 1-ethyl-1-methyl-
4395 - Cyclohexane, 1-ethyl-2-methyl-, cis-
4396 - Cyclohexane, 1-ethyl-2-methyl-, trans-

4397 - Cyclohexane, 1-ethyl-3-methyl-, *cis*-
4398 - Cyclohexane, 1-ethyl-3-methyl-, *trans*-
4399 - Cyclohexane, 1-ethyl-4-methyl-, *cis*-
4400 - Cyclohexane, 1-ethyl-4-methyl-, *trans*-
4431 - Cyclohexane, (1-methylethyl)-
4471 - Cyclohexane, propyl-
4485 - Cyclohexane, 1,1,2-trimethyl-
4486 - Cyclohexane, 1,1,3-trimethyl-
4487 - Cyclohexane, 1,1,4-trimethyl-
4488 - Cyclohexane, 1,2,3-trimethyl-
4489 - Cyclohexane, 1,2,3-trimethyl-, (1α,2β,3α)-
4490 - Cyclohexane, 1,2,3-trimethyl-, (1α,2α,3α)-
4491 - Cyclohexane, 1,2,3-trimethyl-, (1α,2α,3β)-
4492 - Cyclohexane, 1,2,4-trimethyl-, (1α,2α,4α)-
4493 - Cyclohexane, 1,2,4-trimethyl-, (1α,2β,4α)-
4494 - Cyclohexane, 1,2,4-trimethyl-, (1α,2α,4β)-
4495 - Cyclohexane, 1,2,4-trimethyl-, (1α,2β,4β)-
4496 - Cyclohexane, 1,3,5-trimethyl-, (1α,3α,5α)-
4497 - Cyclohexane, 1,3,5-trimethyl-, (1α,3α,5β)-
4724 - Cyclononane
4778 - Cyclopentane, butyl-
4796 - Cyclopentane, 1,1-diethyl-
4797 - Cyclopentane, 1,2-diethyl-, *cis*-
4798 - Cyclopentane, 1,2-diethyl-, *trans*-
4799 - Cyclopentane, 1,3-diethyl-
4805 - Cyclopentane, (1,1-dimethylethyl)-
4829 - Cyclopentane, 1-methyl-2-(1-methylethyl)-
4832 - Cyclopentane, (1-methylpropyl)-
4833 - Cyclopentane, 1-methyl-1-propyl-
4834 - Cyclopentane, 1-methyl-2-propyl-, *cis*-
4835 - Cyclopentane, 1-methyl-2-propyl-, *trans*-
4836 - Cyclopentane, 1-methyl-3-propyl-, *cis*-
4837 - Cyclopentane, 1-methyl-3-propyl-, *trans*-
4838 - Cyclopentane, (2-methylpropyl)-
4851 - Cyclopentane, 1,1,3,3-tetramethyl-
6542 - 2-Heptene, 2,3-dimethyl-
8182 - 1-Nonene
8183 - 3-Nonene, (*E*)-
8505 - 1-Octene, 2-methyl-
8506 - 2-Octene, 3-methyl-
8507 - 3-Octene, 7-methyl-
8508 - 4-Octene, 2-methyl-

C₉H₁₈Br₂
8127 - Nonane, 1,2-dibromo-

C₉H₁₈ClF
8126 - Nonane, 1-chloro-9-fluoro-

C₉H₁₈Cl₂
6382 - Heptane, 2,6-dichloro-2,6-dimethyl-
8128 - Nonane, 1,9-dichloro-

C₉H₁₈FeN₃S₆
5957 - Ferbam

C₉H₁₈NO₃PS₂
6021 - Fosthiazate

C₉H₁₈N₂O₂S
11717 - Thiofanox

C₉H₁₈N₂O₄
3938 - Carbamic acid, *N*-butyl-*N*-nitro,butyl ester
10021 - 1,3-Propanediol, 2-methyl-2-propyl-, dicarbamate

C₉H₁₈N₆O₆
7584 - Methanol, (1,3,5-triazine-2,4,6-triyltrinitrilo)hexakis-

C₉H₁₈O
4526 - Cyclohexanol, 1-ethyl-2-methyl-
4527 - Cyclohexanol, 1-ethyl-3-methyl-
4538 - Cyclohexanol, 1-(1-methylethyl)-
4539 - Cyclohexanol, 2-(1-methylethyl)-, *cis*-
4540 - Cyclohexanol, 4-(1-methylethyl)-, *cis*-
4541 - Cyclohexanol, 4-(1-methylethyl)-, *trans*-
4557 - Cyclohexanol, 1-propyl-
4558 - Cyclohexanol, 2-propyl-, *cis*-
4559 - Cyclohexanol, 2-propyl-, *trans*-
4560 - Cyclohexanol, 4-propyl-, *trans*-
4561 - Cyclohexanol, 4-propyl-, *cis*-
4562 - Cyclohexanol, 1,2,2-trimethyl-
4563 - Cyclohexanol, 1,2,6-trimethyl-
4564 - Cyclohexanol, 1,3,5-trimethyl-

4565 - Cyclohexanol, 2,2,5-trimethyl-
4566 - Cyclohexanol, 2,2,6-trimethyl-
4567 - Cyclohexanol, 2,3,3-trimethyl-
4568 - Cyclohexanol, 2,3,6-trimethyl-
4569 - Cyclohexanol, 3,3,5-trimethyl-, *cis*-
4570 - Cyclohexanol, 3,3,5-trimethyl-, *trans*-
6514 - 4-Heptanone, 2,6-dimethyl-
6515 - 4-Heptanone, 3,5-dimethyl-
8117 - Nonanal
8172 - 2-Nonanone
8173 - 3-Nonanone
8174 - 4-Nonanone
8175 - 5-Nonanone
8189 - 1-Nonen-3-ol
8190 - 8-Nonen-1-ol
8475 - 2-Octanone, 3-methyl-
8476 - 3-Octanone, 4-methyl-
8477 - 3-Octanone, 7-methyl-
8478 - 4-Octanone, 3-methyl-
8479 - 4-Octanone, 7-methyl-
8517 - 1-Octen-4-ol, 4-methyl-
8518 - 2-Octen-4-ol, 7-methyl-
8519 - 3-Octen-2-ol, 2-methyl-, (*Z*)-
8520 - 4-Octen-3-ol, 3-methyl-
8521 - 5-Octen-4-ol, 5-methyl-
8988 - 3-Pentanone, 2,2,4,4-tetramethyl-

C₉H₁₈O₂
193 - Acetic acid, heptyl ester
3543 - Butanoic acid, 3-methyl-, butyl ester
3544 - Butanoic acid, 3-methylbutyl ester
3545 - Butanoic acid, 2-methyl-, butyl ester
3555 - Butanoic acid, 3-methyl-, 2-methylpropyl ester
3579 - Butanoic acid, pentyl ester
6014 - Formic acid, octyl ester
6462 - Heptanoic acid, ethyl ester
6476 - 4-Heptanol, acetate
6886 - Hexanoic acid, 2-ethyl-3-methyl-
6887 - Hexanoic acid, 3-ethyl-3-methyl-
6899 - Hexanoic acid, 3-methyl-, ethyl ester
6900 - Hexanoic acid, 4-methyl-, ethyl ester
6909 - Hexanoic acid, propyl ester
8146 - Nonanoic acid
8433 - Octanoic acid, 3-methyl-
8434 - Octanoic acid, methyl ester
8862 - Pentanoic acid, butyl ester
8899 - Pentanoic acid, 1-methylpropyl ester, (+)-
8900 - Pentanoic acid, 2-methylpropyl ester
10265 - Propanoic acid, hexyl ester
10302 - Propanoic acid, 2-methyl-, 3-methylbutyl ester

C₉H₁₈O₃
3994 - Carbonic acid, bis(1,1-dimethylethyl) ester
3997 - Carbonic acid, bis(2-methylpropyl) ester
3998 - Carbonic acid, dibutyl ester

C₉H₁₈O₅
5719 - Ethanol, 2-ethoxy-, carbonate (2:1)

C₉H₁₉Br
8121 - Nonane, 1-bromo-

C₉H₁₉Cl
6364 - Heptane, 3-chloro-3-ethyl-
6747 - Hexane, 3-chloro-2,2,3-trimethyl-
8123 - Nonane, 1-chloro-
8124 - Nonane, 2-chloro-
8125 - Nonane, 5-chloro-
8362 - Octane, 3-chloro-3-methyl-
8363 - Octane, 4-chloro-4-methyl-
8727 - Pentane, 3-chloro-3-ethyl-2,2-dimethyl-.

C₉H₁₉ClO
8163 - 1-Nonanol, 9-chloro-

C₉H₁₉FO
8165 - 1-Nonanol, 9-fluoro-

C₉H₁₉N
4303 - Cyclohexanamine, *N*-(1-methylethyl)-
4810 - Cyclopentaneethanamine, *N*,α-dimethyl-
6532 - 5-Hepten-2-amine, *N*,6-dimethyl-
9635 - Piperidine, 1-butyl-
9636 - Piperidine, 3-butyl-

9637 - Piperidine, 4-butyl-
9647 - Piperidine, 2,5-diethyl-
9653 - Piperidine, 1-(1,1-dimethylethyl)-
9685 - Piperidine, 1-(2-methylpropyl)-
9686 - Piperidine, 1-methyl-2-propyl-, (*S*)-
9710 - Piperidine, 2,2,4,6-tetramethyl-
9711 - Piperidine, 2,2,6,6-tetramethyl-

C₉H₁₉NO
3201 - Butanamide, *N*,*N*-diethyl-3-methyl-
4144 - Conhydrine, *N*-methyl-
8118 - Nonanamide
8964 - 2-Pentanone, 5-(diethylamino)-
9794 - Propanamide, *N*,*N*-diethyl-2,2-dimethyl-

C₉H₁₉NOS
5372 - EPTC

C₉H₁₉NO₂
328 - β-Alanine, *N*,*N*-diethyl-, ethyl ester
3934 - Carbamic acid, butyl-, butyl ester

C₉H₁₉NO₄
3202 - Butanamide, 2,4-dihydroxy-*N*-(3-hydroxypropyl)-3,3-dimethyl-, (*R*)-

C₉H₂₀
6383 - Heptane, 2,2-dimethyl-
6384 - Heptane, 2,3-dimethyl-
6385 - Heptane, 2,4-dimethyl-
6386 - Heptane, 2,5-dimethyl-
6387 - Heptane, 2,6-dimethyl-
6388 - Heptane, 3,3-dimethyl-
6389 - Heptane, 3,4-dimethyl-
6390 - Heptane, 3,5-dimethyl-
6391 - Heptane, 4,4-dimethyl-
6407 - Heptane, 3-ethyl-
6408 - Heptane, 4-ethyl-
6825 - Hexane, 3-ethyl-2-methyl-
6826 - Hexane, 3-ethyl-3-methyl-
6827 - Hexane, 3-ethyl-4-methyl-
6828 - Hexane, 4-ethyl-2-methyl-
6859 - Hexane, 2,2,3-trimethyl-
6860 - Hexane, 2,2,4-trimethyl-
6861 - Hexane, 2,2,5-trimethyl-
6862 - Hexane, 2,3,3-trimethyl-
6863 - Hexane, 2,3,4-trimethyl-
6864 - Hexane, 2,3,5-trimethyl-
6865 - Hexane, 2,4,4-trimethyl-
6866 - Hexane, 3,3,4-trimethyl-
8120 - Nonane
8408 - Octane, 2-methyl-
8409 - Octane, 3-methyl-
8410 - Octane, 4-methyl-
8756 - Pentane, 3,3-diethyl-
8810 - Pentane, 3-ethyl-2,2-dimethyl-
8811 - Pentane, 3-ethyl-2,3-dimethyl-
8812 - Pentane, 3-ethyl-2,4-dimethyl-
8839 - Pentane, 2,2,3,3-tetramethyl-
8840 - Pentane, 2,2,3,4-tetramethyl-
8841 - Pentane, 2,2,4,4-tetramethyl-
8842 - Pentane, 2,3,3,4-tetramethyl-

C₉H₂₀N₂
9628 - 4-Piperidinamine, 2,2,6,6-tetramethyl-

C₉H₂₀N₂O
11996 - Urea, *N*,*N*-bis(2-methylpropyl)-
12029 - Urea, tetraethyl-

C₉H₂₀O
6406 - Heptane, 1-ethoxy-
6479 - 2-Heptanol, 2,6-dimethyl-
6480 - 3-Heptanol, 2,3-dimethyl-
6481 - 3-Heptanol, 2,6-dimethyl-
6482 - 3-Heptanol, 3,5-dimethyl-
6483 - 3-Heptanol, 3,6-dimethyl-
6484 - 4-Heptanol, 2,4-dimethyl-
6485 - 4-Heptanol, 2,6-dimethyl-
6486 - 4-Heptanol, 3,5-dimethyl-
6487 - 4-Heptanol, 4-ethyl-
6932 - 1-Hexanol, 2-ethyl-3-methyl-
6933 - 2-Hexanol, 2-ethyl-2-methyl-
6934 - 3-Hexanol, 3-ethyl-2-methyl-
6935 - 3-Hexanol, 3-ethyl-5-methyl-

6951 - 1-Hexanol, 3,5,5-trimethyl-
6952 - 2-Hexanol, 2,3,4-trimethyl-
6953 - 3-Hexanol, 2,2,3-trimethyl-
6954 - 3-Hexanol, 2,3,5-trimethyl-
6955 - 3-Hexanol, 2,4,4-trimethyl-
6956 - 3-Hexanol, 2,5,5-trimethyl-
6957 - 3-Hexanol, 3,4,4-trimethyl-
6958 - 3-Hexanol, 3,5,5-trimethyl-
8157 - 1-Nonanol
8158 - 2-Nonanol, (±)-
8159 - 3-Nonanol, (±)-
8160 - 4-Nonanol, (S)-
8161 - 5-Nonanol
8461 - 1-Octanol, 3-methyl-
8462 - 1-Octanol, 4-methyl-
8463 - 1-Octanol, 7-methyl-
8464 - 2-Octanol, 2-methyl-
8465 - 2-Octanol, 6-methyl-
8466 - 3-Octanol, 3-methyl-
8467 - 3-Octanol, 4-methyl-
8468 - 3-Octanol, 6-methyl-
8469 - 4-Octanol, 4-methyl-
8470 - 4-Octanol, 5-methyl-

$C_9H_{20}O_2$
3398 - Butane, 1,1'-[methylenebis(oxy)]bis-
8136 - 1,9-Nonanediol
8755 - Pentane, 1,1-diethoxy-
9992 - 1,3-Propanediol, 2-butyl-2-ethyl-
10075 - Propane, 1,1'-[methylenebis(oxy)]bis[2-methyl-

$C_9H_{20}O_3$
10182 - Propane, 1,1,1-triethoxy-
10183 - Propane, 1,1,3-triethoxy-

$C_9H_{20}O_3Si$
11482 - Silane, triethoxy-2-propenyl-

$C_9H_{20}O_4$
5601 - Ethane, 1,1',1'''-[methanetetrayl tetrakis(oxy)]tetrakis-

$C_9H_{20}S$
8145 - 1-Nonanethiol

$C_9H_{21}B$
3121 - Borane, tripropyl-

$C_9H_{21}BO_3$
3127 - Boric acid, tripropyl ester
3128 - Boric acid, tris(1-methylethyl) ester

$C_9H_{21}ClSn$
11542 - Stannane, chlorotripropyl-

$C_9H_{21}N$
8119 - 1-Nonanamine
9808 - 1-Propanamine, N,N-dipropyl-

$C_9H_{21}NO_3$
10393 - 2-Propanol, 1,1',1''-nitrilotris-

$C_9H_{21}NO_4$
5678 - Ethanol, 2-[2-[2-(3-aminopropoxy) ethoxy]ethoxy]-

$C_9H_{21}O_2PS_3$
11592 - Terbufos

$C_9H_{21}O_3P$
9594 - Phosphorous acid, tripropyl ester
9595 - Phosphorous acid, tris(1-methylethyl) ester

$C_9H_{21}O_4P$
9567 - Phosphoric acid, tripropyl ester
9568 - Phosphoric acid, tris(1-methylethyl) ester

$C_9H_{22}N_2$
7489 - Methanediamine, N,N,N',N'-tetraethyl-
8736 - 1,4-Pentanediamine, N1,N1-diethyl-

$C_9H_{22}O_4P_2S_4$
5931 - Ethion

$C_9H_{22}Si$
11500 - Silane, tripropyl-

$C_9H_{23}NO_3Si$
9828 - 1-Propanamine, 3-(triethoxysilyl)-

$C_9H_{24}O_3Si_3$
5003 - Cyclotrisiloxane, 2,4,6-triethyl-2,4,6-trimethyl-

$C_9H_{24}O_4Si_4$
4991 - Cyclotetrasiloxane, ethenylheptamethyl-

$C_{10}Cl_8$
7933 - Naphthalene, octachloro-

$C_{10}Cl_{10}O$
7408 - Kepone

$C_{10}Cl_{12}$
7644 - 1,3,4-Metheno-1H-cyclobuta[cd]pentalene, 1,1a,2,2,3,3a,4,5,5,5a,5b, 6-dodecachlorooctahydro-

$C_{10}F_8$
7935 - Naphthalene, octafluoro-

$C_{10}F_{18}$
7934 - Naphthalene, octadecafluorodecahydro-

$C_{10}F_{22}$
5050 - Decane, docosafluoro-

$C_{10}F_{23}N$
11912 - Tripropylamine, eicosafluoro-2-(trifluoromethyl)-

$C_{10}H_4Cl_2O_2$
7849 - 1,2-Naphthalenedione, 3,4-dichloro-
7850 - 1,4-Naphthalenedione, 2,3-dichloro-
7851 - 1,4-Naphthalenedione, 5,6-dichloro-

$C_{10}H_4N_4O_8$
8004 - Naphthalene, 1,3,6,8-tetranitro-

$C_{10}H_5BrO_2$
7844 - 1,2-Naphthalenedione, 3-bromo-

$C_{10}H_5BrO_3$
7845 - 1,4-Naphthalenedione, 2-bromo-3-hydroxy-

$C_{10}H_5ClO_2$
7847 - 1,4-Naphthalenedione, 5-chloro-

$C_{10}H_5ClO_3$
7848 - 1,4-Naphthalenedione, 2-chloro-3-hydroxy-

$C_{10}H_5Cl_2NO_2$
11201 - Quinclorac

$C_{10}H_5Cl_7$
7576 - 4,7-Methano-1H-indene, 1,4,5,6,7,8,8-heptachloro-3a,4,7,7a-tetrahydro-

$C_{10}H_5Cl_7O$
6301 - Heptachlor epoxide

$C_{10}H_5N_3O_6$
8016 - Naphthalene, 1,3,5-trinitro-

$C_{10}H_5NaO_5S$
7970 - 1-Naphthalenesulfonic acid, 3,4-dihydro-3,4-dioxo-, sodium salt

$C_{10}H_6BrCl$
7741 - Naphthalene, 1-bromo-4-chloro-

$C_{10}H_6BrNO_2$
7744 - Naphthalene, 1-bromo-3-nitro-
7745 - Naphthalene, 2-bromo-1-nitro-

$C_{10}H_6Br_2$
7801 - Naphthalene, 1,3-dibromo-
7802 - Naphthalene, 1,4-dibromo-
7803 - Naphthalene, 1,5-dibromo-

$C_{10}H_6ClNO_2$
7786 - Naphthalene, 1-chloro-5-nitro-
7787 - Naphthalene, 1-chloro-8-nitro-
7788 - Naphthalene, 2-chloro-1-nitro-

$C_{10}H_6Cl_2$
7807 - Naphthalene, 1,2-dichloro-
7808 - Naphthalene, 1,3-dichloro-
7809 - Naphthalene, 1,4-dichloro-
7810 - Naphthalene, 1,5-dichloro-
7811 - Naphthalene, 1,7-dichloro-
7812 - Naphthalene, 1,8-dichloro-
7813 - Naphthalene, 2,6-dichloro-

$C_{10}H_6Cl_2O$
8025 - 1-Naphthalenol, 2,3-dichloro-
8026 - 1-Naphthalenol, 2,4-dichloro-

$C_{10}H_6Cl_4O_4$
1353 - 1,2-Benzenedicarboxylic acid, 3,4,5,6-tetrachloro-, monoethyl ester
5180 - Dimethyl tetrachloroterephthalate

$C_{10}H_6Cl_8$
4077 - Chlordane

$C_{10}H_6N_2$
7375 - 5-Isoquinolinecarbonitrile
11223 - 2-Quinolinecarbonitrile
11224 - 4-Quinolinecarbonitrile

$C_{10}H_6N_2OS_2$
4072 - Chinomethionat

$C_{10}H_6N_2O_2$
5223 - 5H,10H-Dipyrrolo[1,2-a:1',2'-d]pyrazine-5,10-dione

$C_{10}H_6N_2O_4$
7831 - Naphthalene, 1,3-dinitro-
7832 - Naphthalene, 1,5-dinitro-
7833 - Naphthalene, 1,8-dinitro-

$C_{10}H_6O_2$
7842 - 1,2-Naphthalenedione
7843 - 1,4-Naphthalenedione

$C_{10}H_6O_3$
7859 - 1,4-Naphthalenedione, 2-hydroxy-
7860 - 1,4-Naphthalenedione, 5-hydroxy-

$C_{10}H_6O_3S$
7840 - 1,8-Naphthalenediol, cyclic sulfite

$C_{10}H_6O_4$
7852 - 1,4-Naphthalenedione, 2,3-dihydroxy-
7853 - 1,4-Naphthalenedione, 5,8-dihydroxy-

$C_{10}H_6O_8$
2257 - 1,2,4,5-Benzenetetracarboxylic acid

$C_{10}H_7Br$
7739 - Naphthalene, 1-bromo-
7740 - Naphthalene, 2-bromo-

$C_{10}H_7BrO$
8022 - 2-Naphthalenol, 1-bromo-

$C_{10}H_7Cl$
7780 - Naphthalene, 1-chloro-
7781 - Naphthalene, 2-chloro-

$C_{10}H_7ClO$
8023 - 2-Naphthalenol, 6-chloro-
8063 - 2-Naphthol, 8-chloro-

$C_{10}H_7ClO_2S$
7975 - 1-Naphthalenesulfonyl chloride
7976 - 2-Naphthalenesulfonyl chloride

$C_{10}H_7Cl_5O$
11899 - Tridiphane

$C_{10}H_7F$
7893 - Naphthalene, 1-fluoro-
7894 - Naphthalene, 2-fluoro-

$C_{10}H_7I$
7902 - Naphthalene, 1-iodo-
7903 - Naphthalene, 2-iodo-

$C_{10}H_7NO$
11225 - 4-Quinolinecarboxaldehyde

$C_{10}H_7NO_2$
7929 - Naphthalene, 1-nitro-
7930 - Naphthalene, 2-nitro-
8035 - Naphthalene, 1-nitroso-
11123 - 1H-Pyrrole-2,5-dione, 1-phenyl-
11227 - 2-Quinolinecarboxylic acid
11228 - 7-Quinolinecarboxylic acid
11229 - 8-Quinolinecarboxylic acid

$C_{10}H_7NO_3$
8034 - 2-Naphthalenol, 1-nitro-
11232 - 2-Quinolinecarboxylic acid, 4-hydroxy-

$C_{10}H_7NO_4$
11230 - 2-Quinolinecarboxylic acid, 4,8-dihydroxy-

$C_{10}H_7N_3$
7738 - Naphthalene, 1-azido-

$C_{10}H_7N_3S$
11656 - Thiabendazole

$C_{10}H_8$
504 - Azulene
7732 - Naphthalene

$C_{10}H_8BrN$
7711 - 1-Naphthalenamine, 5-bromo-

$C_{10}H_8ClN$
11243 - Quinoline, 2-chloro-4-methyl-
11244 - Quinoline, 2-chloro-8-methyl-
11245 - Quinoline, 4-chloro-2-methyl-

$C_{10}H_8ClN_3O$
4080 - Chloridazon

$C_{10}H_8N_2$
3099 - 2,2'-Bipyridine
3100 - 2,3'-Bipyridine

3101 - 2,4'-Bipyridine
3102 - 3,3'-Bipyridine
3103 - 3,4'-Bipyridine
3104 - 4,4'-Bipyridine
9917 - Propanedinitrile, (phenylmethyl)-

$C_{10}H_8N_2O_4$
5583 - Ethanedione, di-2-furanyl-, dioxime

$C_{10}H_8N_4O_5$
10848 - 3H-Pyrazol-3-one, 2,4-dihydro-5-methyl-4-nitro-2-(4-nitrophenyl)-

$C_{10}H_8O$
3924 - 3-Butyn-2-one, 4-phenyl-
6159 - Furan, 2-phenyl-
8017 - 1-Naphthalenol
8018 - 2-Naphthalenol

$C_{10}H_8OS$
6160 - Furan, 2-(phenylthio)-

$C_{10}H_8O_2$
2833 - 2H-1-Benzopyran-2-one, 3-methyl-
2834 - 2H-1-Benzopyran-2-one, 5-methyl-
2835 - 2H-1-Benzopyran-2-one, 6-methyl-
2836 - 2H-1-Benzopyran-2-one, 7-methyl-
2837 - 2H-1-Benzopyran-2-one, 8-methyl-
2838 - 4H-1-Benzopyran-4-one, 3-methyl-
5775 - Ethanone, 1-(2-benzofuranyl)-
6157 - Furan, 2-phenoxy-
7246 - 1H-Indene-1-carboxylic acid
7834 - 1,3-Naphthalenediol
7835 - 1,5-Naphthalenediol
7836 - 1,6-Naphthalenediol
7837 - 1,7-Naphthalenediol
7838 - 2,6-Naphthalenediol
7839 - 2,7-Naphthalenediol
10742 - 2-Propynoic acid, 3-phenyl-, methyl ester

$C_{10}H_8O_3$
2831 - 2H-1-Benzopyran-2-one, 7-hydroxy-4-methyl-
6096 - 2,5-Furandione, dihydro-3-phenyl-, (±)-

$C_{10}H_8O_3S$
7959 - 1-Naphthalenesulfonic acid
7960 - 2-Naphthalenesulfonic acid

$C_{10}H_8O_4$
2829 - 2H-1-Benzopyran-2-one, 7-hydroxy-6-methoxy-
5185 - 1,7-Dioxadispiro[4.0.4.2]dodeca-3,9-diene-2,8-dione, trans-
6057 - 2-Furancarboxylic acid, 2-furanylmethyl ester

$C_{10}H_8O_4S$
7971 - 1-Naphthalenesulfonic acid, 4-hydroxy-
7972 - 1-Naphthalenesulfonic acid, 7-hydroxy-
7973 - 2-Naphthalenesulfonic acid, 1-hydroxy-
7974 - 2-Naphthalenesulfonic acid, 6-hydroxy-

$C_{10}H_8O_5$
2808 - 2H-1-Benzopyran-2-one, 7,8-dihydroxy-6-methoxy-

$C_{10}H_8O_6S_2$
7867 - 1,5-Naphthalenedisulfonic acid
7868 - 1,6-Naphthalenedisulfonic acid
7869 - 2,7-Naphthalenedisulfonic acid

$C_{10}H_8O_7S_2$
7871 - 1,3-Naphthalenedisulfonic acid, 7-hydroxy-
7876 - 2,7-Naphthalenedisulfonic acid, 3-hydroxy-

$C_{10}H_8O_8S_2$
7875 - 2,7-Naphthalenedisulfonic acid, 4,5-dihydroxy-

$C_{10}H_8S$
8007 - 1-Naphthalenethiol
8008 - 2-Naphthalenethiol

$C_{10}H_9BrO_2$
10615 - 2-Propenoic acid, 2-bromo-3-phenyl-, methyl ester

$C_{10}H_9ClN_4O_2S$
2193 - Benzenesulfonamide, 4-amino-N-(6-chloro-3-pyridazinyl)-

$C_{10}H_9ClO$
4957 - Cyclopropanecarbonyl chloride, 2-phenyl-, trans-

$C_{10}H_9ClO_2$
10622 - 2-Propenoic acid, 2-chloro-3-phenyl-, methyl ester, (E)-
10623 - 2-Propenoic acid, 3-chloro-3-phenyl-, methyl ester

$C_{10}H_9ClO_3$
972 - Benzeneacetyl chloride, α-(acetyloxy)-

$C_{10}H_9Cl_2NO$
10520 - 2-Propenamide, N-(3,4-dichlorophenyl)-2-methyl-

$C_{10}H_9Cl_4O_4P$
6202 - Gardona

$C_{10}H_9N$
7381 - Isoquinoline, 1-methyl-
7382 - Isoquinoline, 3-methyl-
7383 - Isoquinoline, 4-methyl-
7384 - Isoquinoline, 6-methyl-
7385 - Isoquinoline, 8-methyl-
7708 - 1-Naphthalenamine
7709 - 2-Naphthalenamine
11134 - 1H-Pyrrole, 1-phenyl-
11135 - 1H-Pyrrole, 2-phenyl-
11277 - Quinoline, 2-methyl-
11278 - Quinoline, 3-methyl-
11279 - Quinoline, 4-methyl-
11280 - Quinoline, 5-methyl-
11281 - Quinoline, 6-methyl-
11282 - Quinoline, 7-methyl-
11283 - Quinoline, 8-methyl-

$C_{10}H_9NO$
5819 - Ethanone, 1-(1H-indol-3-yl)-
7380 - Isoquinoline, 7-methoxy-
8019 - 2-Naphthalenol, 6-amino-
8020 - 2-Naphthalenol, 8-amino-
11274 - Quinoline, 4-methoxy-
11275 - Quinoline, 6-methoxy-
11276 - Quinoline, 8-methoxy-
11319 - 6-Quinolinol, 2-methyl-
11320 - 7-Quinolinol, 6-methyl-
11321 - 8-Quinolinol, 2-methyl-
11322 - 8-Quinolinol, 6-methyl-
11328 - 2(1H)-Quinolinone, 1-methyl-
11329 - 2(1H)-Quinolinone, 6-methyl-
11330 - 4(1H)-Quinolinone, 1-methyl-

$C_{10}H_9NO_2$
1877 - Benzenemethanol, α-ethynyl-, carbamate
7295 - 1H-Indole-3-acetic acid
7364 - 1H-Isoindole-1,3(2H)-dione, N-ethyl-
8582 - Oxazole, 5-methoxy-2-phenyl-
11162 - 2,5-Pyrrolidinedione, 1-phenyl-

$C_{10}H_9NO_3$
3853 - 2-Butenoic acid, 4-oxo-4-(phenylamino)-, (Z)-

$C_{10}H_9NO_3S$
7961 - 1-Naphthalenesulfonic acid, 2-amino-
7962 - 1-Naphthalenesulfonic acid, 4-amino-
7963 - 1-Naphthalenesulfonic acid, 5-amino-
7964 - 1-Naphthalenesulfonic acid, 7-amino-
7965 - 1-Naphthalenesulfonic acid, 8-amino-
7967 - 2-Naphthalenesulfonic acid, 8-amino-

$C_{10}H_9NO_4$
10677 - 2-Propenoic acid, 3-(2-nitrophenyl)-, methyl ester, (E)-
10678 - 2-Propenoic acid, 3-(3-nitrophenyl)-, methyl ester, (E)-
10679 - 2-Propenoic acid, 3-(4-nitrophenyl)-, methyl ester, (E)-

$C_{10}H_9NO_4S$
7966 - 1-Naphthalenesulfonic acid, 4-amino-3-hydroxy-

$C_{10}H_9NO_6S_2$
7870 - 1,3-Naphthalenedisulfonic acid, 7-amino-
7872 - 1,6-Naphthalenedisulfonic acid, 4-amino-
7873 - 1,7-Naphthalenedisulfonic acid, 4-amino-

$C_{10}H_9NO_7S_2$
7874 - 2,7-Naphthalenedisulfonic acid, 4-amino-5-hydroxy-

$C_{10}H_9N_3$
10871 - 2-Pyridinamine, N-2-pyridinyl-

$C_{10}H_{10}$
1097 - Benzene, 1,3-butadienyl-, (E)-
1098 - Benzene, 1,3-butadienyl-, (Z)-
1124 - Benzene, 3-butynyl-
1403 - Benzene, 1,2-diethenyl-
1404 - Benzene, 1,3-diethenyl-
1405 - Benzene, 1,4-diethenyl-
1967 - Benzene, (1-methylene-2-propenyl)-
7275 - 1H-Indene, 1-methyl-
7276 - 1H-Indene, 2-methyl-
7277 - 1H-Indene, 3-methyl-
7278 - 1H-Indene, 4-methyl-
7279 - 1H-Indene, 6-methyl-
7280 - 1H-Indene, 7-methyl-
7814 - Naphthalene, 1,2-dihydro-
7815 - Naphthalene, 1,4-dihydro-

$C_{10}H_{10}Cl_2O_3$
3925 - Butyrac 118

$C_{10}H_{10}Fe$
5960 - Ferrocene

$C_{10}H_{10}N_2$
7167 - Hydrazine, 1-naphthalenyl-
7795 - 1,2-Naphthalenediamine
7796 - 1,4-Naphthalenediamine
7797 - 1,5-Naphthalenediamine
7798 - 1,8-Naphthalenediamine
7799 - 2,3-Naphthalenediamine
10837 - 1H-Pyrazole, 1-methyl-3-phenyl-
10838 - 1H-Pyrazole, 3-methyl-1-phenyl-
10987 - Pyridine, 3-(1-methyl-1H-pyrrol-2-yl)-
11208 - 2-Quinolinamine, 4-methyl-
11209 - 3-Quinolinamine, 2-methyl-
11210 - 4-Quinolinamine, 2-methyl-
11211 - 7-Quinolinamine, 2-methyl-
11212 - 8-Quinolinamine, 6-methyl-
11335 - Quinoxaline, 2,6-dimethyl-

$C_{10}H_{10}N_2O$
10849 - 3H-Pyrazol-3-one, 1,2-dihydro-5-methyl-2-phenyl-

$C_{10}H_{10}N_2O_2$
7221 - 2,4-Imidazolidinedione, 3-methyl-5-phenyl-

$C_{10}H_{10}N_4O_2S$
2202 - Benzenesulfonamide, 4-amino-N-pyrazinyl-
2204 - Benzenesulfonamide, 4-amino-N-2-pyrimidinyl-

$C_{10}H_{10}O$
1878 - Benzenemethanol, α-ethynyl-α-methyl-
2445 - Benzofuran, 2,5-dimethyl-
2446 - Benzofuran, 2,6-dimethyl-
2447 - Benzofuran, 2,7-dimethyl-
2448 - Benzofuran, 3,5-dimethyl-
2449 - Benzofuran, 3,6-dimethyl-
2450 - Benzofuran, 4,6-dimethyl-
2451 - Benzofuran, 4,7-dimethyl-
2452 - Benzofuran, 5,6-dimethyl-
2453 - Benzofuran, 5,7-dimethyl-
2454 - Benzofuran, 6,7-dimethyl-
3701 - 2-Butenal, 3-phenyl-
3880 - 2-Buten-1-one, 1-phenyl-
3881 - 3-Buten-2-one, 4-phenyl-
3882 - 3-Buten-2-one, 4-phenyl-, (E)-
7289 - 1H-Inden-1-one, 2,3-dihydro-2-methyl-
7290 - 1H-Inden-1-one, 2,3-dihydro-4-methyl-
8046 - 1(2H)-Naphthalenone, 3,4-dihydro-
8047 - 2(1H)-Naphthalenone, 3,4-dihydro-
10516 - 2-Propenal, 2-methyl-3-phenyl-
10517 - 2-Propenal, 3-(4-methylphenyl)-

$C_{10}H_{10}O_2$
2118 - Benzenepropanal, α-methyl-β-oxo-
2429 - 1,3-Benzodioxole, 5-(2-propenyl)-
2430 - 1,3-Benzodioxole, 5-(1-propenyl)-, (E)-
2722 - Benzoic acid, 2-propenyl ester
2803 - 4H-1-Benzopyran-4-one, 2,3-dihydro-6-methyl-
3362 - 1,3-Butanedione, 1-phenyl-

3854 - 2-Butenoic acid, 3-phenyl-, (Z)-
3855 - 3-Butenoic acid, 4-phenyl-
5865 - Ethanone, 1,1'-(1,3-phenylene)bis-
10670 - 2-Propenoic acid, 2-methyl-3-phenyl-, (Z)-
10686 - 2-Propenoic acid, 3-phenyl-, methyl ester,
    (E)-
10688 - 2-Propenoic acid, phenylmethyl ester
10707 - 2-Propen-1-ol, 3-phenyl-, formate

$C_{10}H_{10}O_2S_2$
6101 - Furan, 2,2'-[dithiobis(methylene)]bis-

$C_{10}H_{10}O_3$
935 - Benzeneacetic acid, α-oxo-, ethyl ester
2142 - Benzenepropanoic acid, β-oxo-, methyl ester
2479 - Benzoic acid, 4-acetyl-, methyl ester
5771 - Ethanone, 2-(acetyloxy)-1-phenyl-
6156 - Furan, 2,2'-[oxybis(methylene)]bis-
10514 - 2-Propenal, 3-(4-hydroxy-3-methoxyphenyl)-

$C_{10}H_{10}O_4$
1343 - 1,2-Benzenedicarboxylic acid, dimethyl ester
1347 - 1,3-Benzenedicarboxylic acid, dimethyl ester
1348 - 1,3-Benzenedicarboxylic acid, 4,6-dimethyl-
1351 - 1,4-Benzenedicarboxylic acid, dimethyl ester
1508 - 1,2-Benzenediol, diacetate
1509 - 1,4-Benzenediol, diacetate
2416 - 1,3-Benzodioxole-5-acetic acid, methyl ester
2419 - 1,3-Benzodioxole-5-carboxylic acid, ethyl
    ester
2482 - Benzoic acid, 2-(acetyloxy)-, methyl ester
3341 - Butanedioic acid, phenyl-, (±)-
7355 - 1(3H)-Isobenzofuranone, 6,7-dimethoxy-

$C_{10}H_{10}O_5$
2608 - Benzoic acid, 6-formyl-2,3-dimethoxy-

$C_{10}H_{11}Br$
1001 - Benzene, (3-bromo-3-butenyl)-
1002 - Benzene, (3-bromo-2-butenyl)-
3719 - 2-Butene, 2-bromo-3-phenyl-

$C_{10}H_{11}Cl$
1152 - Benzene, 1-(1-chloroethenyl)-2,4-dimethyl-
1188 - Benzene, 1-(2-chloro-1-methylethenyl)-4-
    methyl-

$C_{10}H_{11}ClO$
974 - Benzeneacetyl chloride, α-ethyl-
1231 - Benzene, 1-(3-chloro-2-propenyl)-4-methoxy-
3650 - 1-Butanone, 4-chloro-1-phenyl-

$C_{10}H_{11}ClO_2$
891 - Benzeneacetic acid, α-chloro-, ethyl ester, (S)-

$C_{10}H_{11}ClO_2S$
7978 - 2-Naphthalenesulfonyl chloride, 5,6,7,8-
    tetrahydro-

$C_{10}H_{11}F_3N_2O$
12007 - Urea, N,N-dimethyl-N'-
    [3-(trifluoromethyl)phenyl]-

$C_{10}H_{11}F_7O_2$
8398 - 3,5-Octanedione, 6,6,7,7,8,8,8-heptafluoro-
    2,2-dimethyl-

$C_{10}H_{11}I_2NO_3$
10873 - 1(4H)-Pyridineacetic acid, 3,5-diiodo-4-oxo-,
    propyl ester

$C_{10}H_{11}N$
951 - Benzeneacetonitrile, α-ethyl-
7298 - 1H-Indole, 1,3-dimethyl-
7299 - 1H-Indole, 2,5-dimethyl-
7307 - 1H-Indole, 1-ethyl-
7308 - 1H-Indole, 3-ethyl-
7714 - 1-Naphthalenamine, 5,8-dihydro-

$C_{10}H_{11}NO$
3411 - Butanenitrile, 4-phenoxy-
7305 - 1H-Indole-3-ethanol
7328 - 2H-Indol-2-one, 1,3-dihydro-1,3-dimethyl-
7329 - 2H-Indol-2-one, 1,3-dihydro-3,3-dimethyl-
7330 - 2H-Indol-2-one, 1,3-dihydro-4,7-dimethyl-

$C_{10}H_{11}NO_2$
26 - Acetamide, N-acetyl-N-phenyl-
3207 - Butanamide, 3-oxo-N-phenyl-
6114 - 2-Furanmethanamine, N-(2-furanylmethyl)-

$C_{10}H_{11}NO_3$
224 - Acetic acid, oxo(phenylamino)-, ethyl ester

327 - DL-Alanine, N-benzoyl-
3575 - Butanoic acid, 4-oxo-4-(phenylamino)-

$C_{10}H_{11}N_3$
10813 - 1H-Pyrazol-4-amine, 3-methyl-1-phenyl-
10814 - 1H-Pyrazol-3-amine, 3-methyl-1-phenyl-

$C_{10}H_{11}N_3O_2S_2$
2200 - Benzenesulfonamide, 4-amino-N-(4-methyl-2-
    thiazolyl)-

$C_{10}H_{11}N_3O_3S$
2198 - Benzenesulfonamide, 4-amino-N-(5-methyl-3-
    isoxazolyl)-

$C_{10}H_{12}$
1109 - Benzene, 1-butenyl-, (E)-
1110 - Benzene, 1-butenyl-, (Z)-
1111 - Benzene, 2-butenyl-
1112 - Benzene, 3-butenyl-
1245 - Benzene, cyclobutyl-
1626 - Benzene, 1-ethenyl-2,4-dimethyl-
1627 - Benzene, 1-ethenyl-3,5-dimethyl-
1628 - Benzene, 2-ethenyl-1,4-dimethyl-
1629 - Benzene, 4-ethenyl-1,2-dimethyl-
1630 - Benzene, 1-ethenyl-2-ethyl-
1631 - Benzene, 1-ethenyl-3-ethyl-
1632 - Benzene, 1-ethenyl-4-ethyl-
1955 - Benzene, (1-methylcyclopropyl)-
1968 - Benzene, (1-methylenepropyl)-
1984 - Benzene, 1-methyl-2-(1-methylethenyl)-
1985 - Benzene, 1-methyl-3-(1-methylethenyl)-
1986 - Benzene, 1-methyl-4-(1-methylethenyl)-
2023 - Benzene, 1-methyl-2-(2-propenyl)-
2024 - Benzene, (2-methyl-1-propenyl)-
2025 - Benzene, (1-methyl-1-propenyl)-, (E)-
2029 - Benzene, (1-methyl-1-propenyl)-, (Z)-
4746 - 1,3,5,7-Cyclooctatetraene, 1-ethyl-
7255 - 1H-Indene, 2,3-dihydro-1-methyl-
7256 - 1H-Indene, 2,3-dihydro-2-methyl-
7257 - 1H-Indene, 2,3-dihydro-4-methyl-
7258 - 1H-Indene, 2,3-dihydro-5-methyl-
7577 - 4,7-Methano-1H-indene, 3a,4,7,7a-tetrahydro-
    , (3aα,4α,7α,7aα)-
7979 - Naphthalene, 1,2,3,4-tetrahydro-

$C_{10}H_{12}Br_2O$
9281 - Phenol, 2,4-dibromo-3-methyl-6-(1-
    methylethyl)-

$C_{10}H_{12}ClNO_2$
3939 - Carbamic acid, (3-chlorophenyl)-, 1-
    methylethyl ester

$C_{10}H_{12}ClNO_4$
10001 - 1,2-Propanediol, 3-(4-chlorophenoxy)-, 1-
    carbamate

$C_{10}H_{12}ClN_3O_3S$
11197 - 6-Quinazolinesulfonamide, 7-chloro-2-ethyl-
    1,2,3,4-tetrahydro-4-oxo-

$C_{10}H_{12}N_2$
2381 - 1H-Benzimidazole, 1-ethyl-2-methyl-
3105 - 2,3'-Bipyridine, 1,2,3,6-tetrahydro-, (S)-
7198 - 1H-Imidazole, 4,5-dihydro-2-(phenylmethyl)-
7302 - 1H-Indole-3-ethanamine

$C_{10}H_{12}N_2O_3$
1102 - Benzenebutanoic acid, α,2-diamino-γ-oxo-
11091 - 2,4,6(1H,3H,5H)-Pyrimidinetrione, 5,5-di-2-
    propenyl-

$C_{10}H_{12}N_2O_3S$
518 - Bentazon

$C_{10}H_{12}N_2O_4$
2282 - Benzene, 1,2,4,5-tetramethyl-3,6-dinitro-

$C_{10}H_{12}N_2O_5$
5183 - Dinoseb
9328 - Phenol, 2-(1,1-dimethylethyl)-4,6-dinitro-

$C_{10}H_{12}N_3O_3PS_2$
497 - Azinphos-methyl

$C_{10}H_{12}N_4OS$
34 - Acetamide, N-[4-[[(aminothioxomethyl)
    hydrazono]methyl]phenyl]-

$C_{10}H_{12}N_4O_5$
7335 - Inosine

$C_{10}H_{12}O$
590 - Benzaldehyde, 4-(1-methylethyl)-
599 - Benzaldehyde, 2,4,6-trimethyl-
1651 - Benzene, (2-ethoxyethenyl)-
1869 - Benzenemethanol, α-ethenyl-α-methyl-
1911 - Benzenemethanol, α-2-propenyl-
1941 - Benzene, 1-methoxy-2-(1-propenyl)-
1942 - Benzene, 1-methoxy-4-(1-propenyl)-
1943 - Benzene, 1-methoxy-4-(2-propenyl)-
1944 - Benzene, 1-methoxy-4-(1-propenyl)-, (E)-
2026 - Benzene, 1-methyl-2-(2-propenyloxy)-
2027 - Benzene, 1-methyl-3-(2-propenyloxy)-
2028 - Benzene, 1-methyl-4-(2-propenyloxy)-
2115 - Benzenepropanal, 2-methyl-
2116 - Benzenepropanal, 4-methyl-
2157 - Benzenepropanol, γ-methylene-
2778 - 2H-1-Benzopyran, 3,4-dihydro-2-methyl-
2779 - 2H-1-Benzopyran, 3,4-dihydro-4-methyl-
2780 - 2H-1-Benzopyran, 3,4-dihydro-6-methyl-
2781 - 2H-1-Benzopyran, 3,4-dihydro-7-methyl-
3676 - 1-Butanone, 1-phenyl-
3677 - 2-Butanone, 1-phenyl-
3678 - 2-Butanone, 4-phenyl-
3872 - 2-Buten-1-ol, 3-phenyl-
5809 - Ethanone, 1-(2,4-dimethylphenyl)-
5810 - Ethanone, 1-(2,5-dimethylphenyl)-
5811 - Ethanone, 1-(3,4-dimethylphenyl)-
8041 - 1-Naphthalenol, 1,2,3,4-tetrahydro-
8042 - 1-Naphthalenol, 5,6,7,8-tetrahydro-
8043 - 2-Naphthalenol, 1,2,3,4-tetrahydro-
8044 - 2-Naphthalenol, 5,6,7,8-tetrahydro-
9427 - Phenol, 2-methyl-6-(2-propenyl)-
9428 - Phenol, 4-methyl-2-(1-propenyl)-
9429 - Phenol, 4-methyl-2-(2-propenyl)-
10466 - 1-Propanone, 1-(2-methylphenyl)-
10467 - 1-Propanone, 1-(3-methylphenyl)-
10468 - 1-Propanone, 1-(4-methylphenyl)-
10469 - 1-Propanone, 2-methyl-1-phenyl-

$C_{10}H_{12}O_2$
232 - Acetic acid, 2-phenylethyl ester
897 - Benzeneacetic acid, α-ethyl-
899 - Benzeneacetic acid, ethyl ester
1100 - Benzenebutanoic acid
2137 - Benzenepropanoic acid, α-methyl-
2138 - Benzenepropanoic acid, β-methyl-, (±)-
2139 - Benzenepropanoic acid, methyl ester
2431 - 1,3-Benzodioxole, 5-propyl-
2589 - Benzoic acid, 2,4-dimethyl-, methyl ester
2688 - Benzoic acid, 2-(1-methylethyl)-
2689 - Benzoic acid, 4-(1-methylethyl)-
2690 - Benzoic acid, 2-methyl-, ethyl ester
2691 - Benzoic acid, 3-methyl-, ethyl ester
2692 - Benzoic acid, 4-methyl-, ethyl ester
2693 - Benzoic acid, 1-methylethyl ester
2723 - Benzoic acid, o-propyl-
2724 - Benzoic acid, propyl ester
3584 - Butanoic acid, phenyl ester
3665 - 1-Butanone, 1-(2-hydroxyphenyl)-
3666 - 1-Butanone, 2-hydroxy-1-phenyl-
3667 - 1-Butanone, 4-hydroxy-1-phenyl-
4241 - 1,3,6-Cycloheptatriene-1-carboxylic acid, 5,5-
    dimethyl-
4267 - 2,5-Cyclohexadiene-1,4-dione, 2-methyl-5-(1-
    methylethyl)-
4270 - 2,5-Cyclohexadiene-1,4-dione, 2,3,5,6-
    tetramethyl-
5195 - 1,3-Dioxane, 2-phenyl-
5196 - 1,3-Dioxane, 4-phenyl-
5814 - Ethanone, 2-ethoxy-1-phenyl-
9313 - Phenol, 2,4-dimethyl-, acetate
9314 - Phenol, 2,5-dimethyl-, acetate
9315 - Phenol, 3,4-dimethyl-, acetate
9316 - Phenol, 3,5-dimethyl-, acetate
9388 - Phenol, 2-methoxy-4-(1-propenyl)-
9389 - Phenol, 2-methoxy-4-(2-propenyl)-
9390 - Phenol, 2-methoxy-5-(2-propenyl)-
9391 - Phenol, 2-methoxy-6-(1-propenyl)-

9322 - Phenol, 3-(1,1-dimethylethyl)-
9323 - Phenol, 4-(1,1-dimethylethyl)-
9414 - Phenol, 2-methyl-4-(1-methylethyl)-
9415 - Phenol, 2-methyl-5-(1-methylethyl)-
9416 - Phenol, 2-methyl-6-(1-methylethyl)-
9417 - Phenol, 4-methyl-2-(1-methylethyl)-
9418 - Phenol, 5-methyl-2-(1-methylethyl)-
9430 - Phenol, 2-(1-methylpropyl)-
9431 - Phenol, 4-(1-methylpropyl)-
9432 - Phenol, 4-(2-methylpropyl)-
9473 - Phenol, 2,3,4,5-tetramethyl-
9474 - Phenol, 2,3,4,6-tetramethyl-
9475 - Phenol, 2,3,5,6-tetramethyl-

$C_{10}H_{14}O_2$
1406 - Benzene, 1,2-diethoxy-
1407 - Benzene, 1,3-diethoxy-
1408 - Benzene, 1,4-diethoxy-
1516 - 1,2-Benzenediol, 4-(1,1-dimethylethyl)-
1536 - 1,2-Benzenediol, 3-methyl-6-(1-methylethyl)-
1538 - 1,4-Benzenediol, 2-methyl-5-(1-methylethyl)-
1539 - 1,3-Benzenediol, 4-(2-methylpropyl)-
1552 - 1,4-Benzenediol, 2,3,5,6-tetramethyl-
1652 - Benzene, (2-ethoxyethoxy)-
1873 - Benzenemethanol, α-ethyl-4-methoxy-
2154 - Benzenepropanol, 4-methoxy-
2926 - Bicyclo[2.2.1]heptane-2,3-dione, 1,7,7-
      trimethyl-, (1R)-
3356 - 1,2-Butanediol, 2-phenyl-
3357 - 1,3-Butanediol, 1-phenyl-
3358 - 1,3-Butanediol, 3-phenyl-
4758 - Cyclopenta[c]pyran-1(4aH)-one, 5,6,7,7a-
      tetrahydro-4,7-dimethyl-
6076 - Furan, 2-(cyclohexyloxy)-
9245 - Phenol, 2-butoxy-
10400 - 1-Propanol, 3-(phenylmethoxy)-
11862 - Tricyclo[2.2.1.02,6]heptane-1-carboxylic
      acid, 7,7-dimethyl-

$C_{10}H_{14}O_3$
1599 - Benzeneethanol, 2,5-dimethoxy-
2153 - Benzenepropanol, 4-hydroxy-3-methoxy-
6027 - 2-Furanacetic acid, butyl ester
6068 - 2-Furancarboxylic acid, pentyl ester
8559 - 3-Oxabicyclo[3.2.1]octane-2,4-dione, 1,8,8-
      trimethyl-, (±)-
8880 - Pentanoic acid, 2-furanylmethyl ester
10019 - 1,2-Propanediol, 3-(2-methylphenoxy)-

$C_{10}H_{14}O_3S$
2219 - Benzenesulfonic acid, sec-butyl ester
2233 - Benzenesulfonic acid, 4-methyl-, propyl ester

$C_{10}H_{14}O_4$
1914 - Benzenemethanol, 3,4,5-trimethoxy-
4646 - 1-Cyclohexene-1,4-dicarboxylic acid, dimethyl
      ester
10016 - 1,2-Propanediol, 3-(2-methoxyphenoxy)-
10656 - 2-Propenoic acid, 2-methyl-, 1,2-ethanediyl
      ester

$C_{10}H_{15}BrO$
2951 - Bicyclo[2.2.1]heptan-2-one, 1-(bromomethyl)-
      7,7-dimethyl-, (1S)-
2952 - Bicyclo[2.2.1]heptan-2-one, 3-bromo-1,7,7-
      trimethyl-, (1R-endo)-

$C_{10}H_{15}N$
655 - Benzenamine, 4-butyl-
656 - Benzenamine, N-butyl-
694 - Benzenamine, 2,6-diethyl-
695 - Benzenamine, N,N-diethyl-
714 - Benzenamine, 2-(1,1-dimethylethyl)-
715 - Benzenamine, 4-(1,1-dimethylethyl)-
716 - Benzenamine, N-(1,1-dimethylethyl)-
793 - Benzenamine, 2-methyl-5-(1-methylethyl)-
794 - Benzenamine, 4-methyl-N-(1-methylethyl)-
812 - Benzenamine, 2-(1-methylpropyl)-
813 - Benzenamine, 4-(1-methylpropyl)-
814 - Benzenamine, 4-(2-methylpropyl)-
815 - Benzenamine, 4-methyl-N-propyl-
816 - Benzenamine, N-(1-methylpropyl)-
817 - Benzenamine, N-(2-methylpropyl)-

847 - Benzenamine, 2,3,4,5-tetramethyl-
848 - Benzenamine, 2,3,4,6-tetramethyl-
849 - Benzenamine, N,N,2,6-tetramethyl-
1572 - Benzeneethanamine, α,α-dimethyl-
1573 - Benzeneethanamine, N,α-dimethyl-, (S)-
1574 - Benzeneethanamine, N,β-dimethyl-
1819 - Benzenemethanamine, 4-(1-methylethyl)-
1820 - Benzenemethanamine, N-(1-methylethyl)-
1830 - Benzenemethanamine, α-propyl-
2119 - Benzenepropanamine, α-methyl-
10967 - Pyridine, 2-(1-ethylpropyl)-
10968 - Pyridine, 4-(1-ethylpropyl)-

$C_{10}H_{15}NO$
1840 - Benzenemethanol, α-(1-aminopropyl)- ·
1900 - Benzenemethanol, α-[1-(methylamino)ethyl]-,
      (R*,S*)-(±)-
9297 - Phenol, 3-(diethylamino)-
9320 - Phenol, 4-[2-(dimethylamino)ethyl]-
9403 - Phenol, 4-[2-(methylamino)propyl]-
10983 - Pyridine, 2-[2-(1-methylethoxy)ethyl]-

$C_{10}H_{15}NO_2$
693 - Benzenamine, 3,4-diethoxy-
1571 - Benzeneethanamine, 3,4-dimethoxy-
5760 - Ethanol, 2,2′-(phenylimino)bis-

$C_{10}H_{15}NO_4$
9933 - Propanedioic acid, (2-cyanoethyl)-, diethyl
      ester
11144 - 3-Pyrrolidineacetic acid, 2-carboxy-4-(1-
      methylethenyl)-, [2S-(2α,3β,4β)]-

$C_{10}H_{15}N_3$
6281 - Guanidine, N,N′-dimethyl-N″-(phenylmethyl)-

$C_{10}H_{15}OPS_2$
5966 - Fonofos

$C_{10}H_{15}O_3PS_2$
5955 - Fenthion

$C_{10}H_{16}$
2921 - Bicyclo[2.2.1]heptane, 2,2-dimethyl-3-
      methylene-, (±)-
2922 - Bicyclo[2.2.1]heptane, 2,2-dimethyl-5-
      methylene-
2923 - Bicyclo[2.2.1]heptane, 7,7-dimethyl-2-
      methylene-, (±)-
2924 - Bicyclo[3.1.1]heptane, 6,6-dimethyl-2-
      methylene-, (1R)-
2925 - Bicyclo[3.1.1]heptane, 6,6-dimethyl-2-
      methylene-
2970 - Bicyclo[2.2.1]hept-2-ene, 1,7,7-trimethyl-
2971 - Bicyclo[3.1.1]hept-2-ene, 2,6,6-trimethyl-,
      (1S)-
2972 - Bicyclo[3.1.1]hept-2-ene, 2,6,6-trimethyl-, (±)-
2973 - Bicyclo[3.1.1]hept-2-ene, 2,6,6-trimethyl
2974 - Bicyclo[4.1.0]hept-3-ene, 3,7,7-trimethyl-,
      (1S)-
2981 - Bicyclo[3.1.0]hexane, 4-methylene-1-(1-
      methylethyl)-, (±)-
2986 - Bicyclo[3.1.0]hex-2-ene, 2-methyl-5-(1-
      methylethyl)-
4050 - 4-Carene, (1S,3R,6R)-(-)-
4209 - Cyclodecyne
4273 - 1,3-Cyclohexadiene, 1-methyl-4-(1-
      methylethyl)-
4274 - 1,3-Cyclohexadiene, 2-methyl-5-(1-
      methylethyl)-, (±)-
4275 - 1,3-Cyclohexadiene, 1-methyl-4-(1-
      methylethyl)-
4673 - Cyclohexene, 3-methylene-6-(1-methylethyl)-
4674 - Cyclohexene, 3-methylene-6-(1-methylethyl)-,
      (+)-
4675 - Cyclohexene, 4-methylene-1-(1-methylethyl)-
4676 - Cyclohexene, 1-methyl-3-(1-methylethenyl)-
4677 - Cyclohexene, 1-methyl-4-(1-methylethenyl)-
4678 - Cyclohexene, 1-methyl-5-(1-methylethenyl)-
4679 - Cyclohexene, 5-methyl-1-(1-methylethenyl)-,
      (±)-
4684 - Cyclohexene, 1-methyl-4-
      (1-methylethylidene)-
5094 - 1-Decen-3-yne

5095 - 1-Decen-4-yne
5096 - 2-Decen-4-yne
8330 - 1,6-Octadiene, 7-methyl-3-methylene-
8485 - 1,3,6-Octatriene, 3,7-dimethyl-
8486 - 1,3,7-Octatriene, 3,7-dimethyl-
8487 - 2,4,6-Octatriene, 2,6-dimethyl-, (E,E)-
8488 - 2,4,6-Octatriene, 2,6-dimethyl-, (E,Z)-
11815 - 4(10)-Thujene, (+)-
11860 - Tricyclo[3.3.1.13,7]decane
11866 - Tricyclo[2.2.1.02,6]heptane, 1,3,3-trimethyl-
11867 - Tricyclo[2.2.1.02,6]heptane, 1,7,7-trimethyl-

$C_{10}H_{16}ClNO$
865 - Benzenaminium, N-ethyl-3-hydroxy-N,N-
      dimethyl-, chloride

$C_{10}H_{16}Cl_2O_2$
5049 - Decanedioyl dichloride

$C_{10}H_{16}Cl_3NOS$
11823 - Triallate

$C_{10}H_{16}NO_5PS_2$
5946 - Famphur

$C_{10}H_{16}N_2$
1265 - 1,3-Benzenediamine, N,N-diethyl-
1266 - 1,4-Benzenediamine, N,N-diethyl-
1286 - 1,2-Benzenediamine, N,N,N′,N′-tetramethyl-
1287 - 1,4-Benzenediamine, N,N,N′,N′-tetramethyl-
5041 - Decanedinitrile

$C_{10}H_{16}N_2OS$
7229 - 4-Imidazolidinone, 5-(2-methylpropyl)-3-(2-
      propenyl)-2-thioxo-

$C_{10}H_{16}N_2O_2S$
11065 - 4,6(1H,5H)-Pyrimidinedione, 5-ethyldihydro-
      5-(1-methylpropyl)-2-thioxo-

$C_{10}H_{16}N_2O_3$
11082 - 2,4,6(1H,3H,5H)-Pyrimidinetrione, 5-butyl-5-
      ethyl-

$C_{10}H_{16}N_2O_3S$
11689 - 1H-Thieno[3,4-d]imidazole-4-pentanoic acid,
      hexahydro-2-oxo-, [3aS-(3aα,4β,6aα)]-

$C_{10}H_{16}N_2O_8$
6262 - Glycine, N,N′-1,2-ethanediylbis[N-
      (carboxymethyl)-

$C_{10}H_{16}N_4O_7$
11107 - 4(1H)-Pyrimidinone, 2,6-diamino-5-(β-D-
      glucopyranosyloxy)-

$C_{10}H_{16}O$
2960 - Bicyclo[2.2.1]heptan-2-one, 1,3,3-trimethyl-,
      (±)-
2961 - Bicyclo[2.2.1]heptan-2-one, 1,7,7-trimethyl-,
      (1R)-
2963 - Bicyclo[2.2.1]heptan-2-one, 4,7,7-trimethyl-,
      (1S)-
2964 - Bicyclo[3.1.1]heptan-2-one, 4,6,6-trimethyl-,
      [1R-(1α,4α,5α)]-
2965 - Bicyclo[3.1.1]heptan-3-one, 2,6,6-trimethyl-,
      [1S-(1α,2β,5α)]-
2975 - Bicyclo[3.1.1]hept-3-en-2-ol, 4,6,6-trimethyl-,
      (1α,2β,5α)-
2976 - Bicyclo[3.1.1]hept-3-en-2-ol, 4,6,6-trimethyl-,
      (1α,2α,5α)-
2983 - Bicyclo[3.1.0]hexan-3-ol, 4-methylene-1-(1-
      methylethyl)-
2984 - Bicyclo[3.1.0]hexan-3-ol, 4-methylene-1-(1-
      methylethyl)-, [1S-(1α,3β,5α)]-
2995 - [1,1′-Bicyclopentyl]-2-one
4576 - Cyclohexanone, 2-butylidene-
4601 - Cyclohexanone, 2-methyl-5-
      (1-methylethenyl)-, (2S-trans)-
4602 - Cyclohexanone, 5-methyl-2-(1-methylethenyl)-
4603 - Cyclohexanone, 5-methyl-2-
      (1-methylethenyl)-, trans-
4607 - Cyclohexanone, 5-methyl-2-
      (1-methylethylidene)-
4608 - Cyclohexanone, 5-methyl-2-
      (1-methylethylidene)-, (R)-
4635 - 1-Cyclohexene-1-carboxaldehyde, 4-(1-
      methylethyl)-

4636 - 1-Cyclohexene-1-carboxaldehyde, 2,6,6-
    trimethyl-
4664 - 1-Cyclohexene-1-methanol, 4-(1-
    methylethenyl)-
4696 - 2-Cyclohexen-1-ol, 2-methyl-5-(1-
    methylethenyl)-
4716 - 2-Cyclohexen-1-one, 3-methyl-6-(1-
    methylethyl)-, (±)-
4717 - 2-Cyclohexen-1-one, 6-methyl-3-(1-
    methylethyl)-, (±)-
8053 - 1(2H)-Naphthalenone, octahydro-, cis-
8054 - 1(2H)-Naphthalenone, octahydro-, trans-
8055 - 2(1H)-Naphthalenone, octahydro-
8317 - 2,6-Octadienal, 3,7-dimethyl-, (E)-
8318 - 2,6-Octadienal, 3,7-dimethyl-, (Z)-
8561 - 6-Oxabicyclo[3.2.1]oct-3-ene, 4,7,7-trimethyl-,
    (±)-

$C_{10}H_{16}O_2$
2957 - Bicyclo[2.2.1]heptan-2-one, 3-hydroxy-1,7,7-
    trimethyl-
3834 - 2-Butenoic acid, 3-cyclohexyl-
4203 - 1,2-Cyclodecanedione
4307 - Δ1,α-Cyclohexaneacetic acid, 3,3-dimethyl-
4709 - 2-Cyclohexen-1-one, 2-hydroxy-3-methyl-6-(1-
    methylethyl)-
4911 - 1-Cyclopentene-1-carboxylic acid, 2,3,3-
    trimethyl-, methyl ester
4959 - Cyclopropanecarboxylic acid, 2,2-dimethyl-3-
    (2-methyl-1-propenyl)-, (1R-cis)-
5025 - 2,4-Decadienoic acid, (E,E)-
5184 - 2,3-Dioxabicyclo[2.2.2]oct-5-ene, 1-methyl-4-
    (1-methylethyl)-
8331 - 2,4-Octadienoic acid, 3,7-dimethyl-
10653 - 2-Propenoic acid, 2-methyl-, cyclohexyl ester

$C_{10}H_{16}O_4$
3766 - 2-Butenedioic acid (E)-, bis(1-methylethyl)
    ester
3783 - 2-Butenedioic acid (E)-, dipropyl ester
3784 - 2-Butenedioic acid (Z)-, dipropyl ester
3931 - Camphoric acid, trans-(±)-
4180 - 1,1-Cyclobutanedicarboxylic acid, diethyl ester
4184 - 1,3-Cyclobutanedicarboxylic acid, 2,2-
    dimethyl-, dimethyl ester
4352 - 1,3-Cyclohexanedicarboxylic acid, dimethyl
    ester, cis-
4353 - 1,3-Cyclohexanedicarboxylic acid, dimethyl
    ester, trans-
4357 - 1,4-Cyclohexanedicarboxylic acid, dimethyl
    ester, cis-
4794 - 1,3-Cyclopentanedicarboxylic acid, 1,2,2-
    trimethyl-, (1R-cis)-
9966 - Propanedioic acid, (1-methylethylidene)-,
    diethyl ester
9980 - Propanedioic acid, 2-propenyl, diethyl ester

$C_{10}H_{16}O_4S$
2929 - Bicyclo[2.2.1]heptane-1-methanesulfonic acid,
    7,7-dimethyl-2-oxo-, (1S)-

$C_{10}H_{16}O_5$
3295 - Butanedioic acid, acetyl-, diethyl ester
9947 - Propanedioic acid, (ethoxymethylene)-, diethyl
    ester

$C_{10}H_{16}O_6$
7552 - Methanetricarboxylic acid, triethyl ester

$C_{10}H_{16}Si$
11490 - Silane, trimethyl(4-methylphenyl)-
11494 - Silane, trimethyl(phenylmethyl)-

$C_{10}H_{17}Br$
8321 - 2,6-Octadiene, 1-bromo-3,7-dimethyl-, (E)-

$C_{10}H_{17}Cl$
2919 - Bicyclo[2.2.1]heptane, 2-chloro-1,7,7-
    trimethyl-
2920 - Bicyclo[2.2.1]heptane, 2-chloro-1,7,7-
    trimethyl-, endo-
7782 - Naphthalene, 2-chlorodecahydro-

$C_{10}H_{17}ClO_4$
9932 - Propanedioic acid, (3-chloropropyl)-, diethyl
    ester

$C_{10}H_{17}Cl_2NOS$
5106 - Diallate

$C_{10}H_{17}N$
11859 - Tricyclo[3.3.1.13,7]decan-1-amine

$C_{10}H_{17}NO$
2962 - Bicyclo[2.2.1]heptan-2-one, 1,7,7-trimethyl-,
    oxime, (1S)-

$C_{10}H_{17}NO_6$
6258 - Glycine, N,N-bis(ethoxycarbonyl)-, ethyl ester
10104 - Propanenitrile, 2-(β-D-glucopyranosyloxy)-2-
    methyl-

$C_{10}H_{17}N_2O_4PS$
5945 - Etrimfos

$C_{10}H_{17}N_3$
11848 - 4H-1,2,4-Triazole, 4-cyclohexyl-3-ethyl-

$C_{10}H_{17}N_3O_6S$
6264 - Glycine, N-(N-L-γ-glutamyl-L-cysteinyl)-

$C_{10}H_{18}$
2931 - Bicyclo[2.2.1]heptane, 1,3,3-trimethyl-, (+)-
2932 - Bicyclo[2.2.1]heptane, 1,7,7-trimethyl-
2933 - Bicyclo[2.2.1]heptane, 2,2,3-trimethyl-
2934 - Bicyclo[3.1.1]heptane, 2,6,6-trimethyl-,
    (1α,2α,5α)-(±)-
2935 - Bicyclo[3.1.1]heptane, 2,6,6-trimethyl-, [1R-
    (1α,2β,5α)]-
2936 - Bicyclo[4.1.0]heptane, 3,7,7-trimethyl-, [1S-
    (1α,3β,6α)]-
2937 - Bicyclo[4.1.0]heptane, 3,7,7-trimethyl-, [1S-
    (1α,3α,6α)]-
2982 - Bicyclo[3.1.0]hexane, 4-methyl-1-(1-
    methylethyl)-
4207 - Cyclodecene, (E)-
4208 - Cyclodecene, (Z)-
4247 - Cycloheptene, 1,4,4-trimethyl-
4435 - Cyclohexane, 1-methyl-3-(1-methylethenyl)-
4441 - Cyclohexane, 1-methyl-3-
    (1-methylethylidene)-
4442 - Cyclohexane, 1-methyl-4-
    (1-methylethylidene)-
4680 - Cyclohexene, 1-methyl-4-(1-methylethyl)-
4681 - Cyclohexene, 1-methyl-4-(1-methylethyl)-,
    (R)-
4682 - Cyclohexene, 3-methyl-6-(1-methylethyl)-,
    (3R-trans)-
4683 - Cyclohexene, 4-methyl-1-(1-methylethyl)-,
    (R)-
4923 - Cyclopentene, 1,5-dimethyl-4-(1-methylethyl)-
4934 - Cyclopentene, 1-pentyl-
5023 - 1,3-Decadiene
5024 - 1,9-Decadiene
5097 - 1-Decyne
5098 - 3-Decyne
5099 - 4-Decyne
5100 - 5-Decyne
6619 - 3-Heptyne, 5-ethyl-5-methyl-
7429 - p-Menth-3-ene, (S)-(-)-
7789 - Naphthalene, decahydro-
7790 - Naphthalene, decahydro-, cis-
7791 - Naphthalene, decahydro-, trans-
8202 - 4-Nonyne, 8-methyl-
8322 - 2,4-Octadiene, 3,7-dimethyl-
8323 - 2,5-Octadiene, 2,6-dimethyl-
8324 - 2,6-Octadiene, 2,6-dimethyl-, (Z)-
8325 - 2,6-Octadiene, 2,7-dimethyl-
8326 - 2,6-Octadiene, 3,6-dimethyl-
8327 - 2,6-Octadiene, 4,5-dimethyl-
8535 - 3-Octyne, 2,2-dimethyl-
11814 - Thujane, (1S,4R,5S)-(+)-

$C_{10}H_{18}N_2O_2$
7223 - 4-Imidazolidinehexanoic acid, 5-methyl-2-
    oxo-, (4R-cis)-

$C_{10}H_{18}N_2O_4$
9606 - 1,4-Piperazinedicarboxylic acid, diethyl ester

$C_{10}H_{18}N_4O_4S_3$
11713 - Thiodicarb

$C_{10}H_{18}O$
2930 - Bicyclo[3.1.1]heptane-2-methanol, 6,6-
    dimethyl-, [1S-(1α,2β,5α)]-
2940 - Bicyclo[2.2.1]heptan-2-ol, 1,3,3-trimethyl-,
    endo-(±)-
2941 - Bicyclo[2.2.1]heptan-2-ol, 1,3,3-trimethyl-,
    (1S-endo)-
2942 - Bicyclo[2.2.1]heptan-2-ol, 1,7,7-trimethyl-,
    endo-(±)-
2943 - Bicyclo[2.2.1]heptan-2-ol, 1,7,7-trimethyl-,
    exo-(±)-
2944 - Bicyclo[3.1.1]heptan-2-ol, 4,6,6-trimethyl-,
    [1R-(1α,2α,4α,5α)]-
2985 - Bicyclo[3.1.0]hexan-3-ol, 4-methyl-1-(1-
    methylethyl)-
2994 - [1,1'-Bicyclopentyl]-2-ol
3652 - 1-Butanone, 1-cyclohexyl-
4205 - Cyclododecanone
4237 - Cycloheptanone, 2,2,6-trimethyl-
4238 - Cycloheptanone, 2,6,6-trimethyl-
4542 - Cyclohexanol, 1-methyl-4-(1-methylethenyl)-
4543 - Cyclohexanol, 2-methyl-5-(1-methylethenyl)-,
    [1S-(1α,2β,5α)]-
4544 - Cyclohexanol, 5-methyl-2-(1-methylethenyl)-,
    [1R-(1α,2β,5α)]-
4575 - Cyclohexanone, 2-butyl-
4604 - Cyclohexanone, 2-methyl-5-(1-methylethyl)-,
    (2S-trans)-
4605 - Cyclohexanone, 5-methyl-2-(1-methylethyl)-,
    (2R-cis)-
4606 - Cyclohexanone, 5-methyl-2-(1-methylethyl)-,
    (2S-trans)-
4609 - Cyclohexanone, 4-(2-methylpropyl)-
4665 - 1-Cyclohexene-1-methanol, 2,4,4-trimethyl-
4666 - 1-Cyclohexene-1-methanol, 2,6,6-trimethyl-
4667 - 2-Cyclohexene-1-methanol, 2,6,6-trimethyl-
4668 - 3-Cyclohexene-1-methanol, α,α,4-trimethyl-,
    (±)-
4697 - 2-Cyclohexen-1-ol, 3-methyl-6-(1-
    methylethyl)-
4698 - 2-Cyclohexen-1-ol, 6-methyl-3-(1-
    methylethyl)-
4699 - 3-Cyclohexen-1-ol, 4-methyl-1-(1-
    methylethyl)-
4978 - Cyclopropanemethanol, 2,2-dimethyl-3-(2-
    methyl-1-propenyl)-
5080 - 2-Decenal
5081 - 3-Decenal
5093 - 3-Decen-2-one
6590 - 3-Hepten-2-one, 3-ethyl-4-methyl-
6591 - 4-Hepten-2-one, 3-ethyl-4-methyl-
6592 - 4-Hepten-3-one, 5-ethyl-4-methyl-
6593 - 5-Hepten-2-one, 4-ethyl-6-methyl-
6594 - 5-Hepten-3-one, 5-ethyl-4-methyl-
8024 - 2-Naphthalenol, decahydro-
8191 - 5-Nonen-4-one, 6-methyl-
8192 - 6-Nonen-4-one, 6-methyl-
8332 - 1,6-Octadien-3-ol, 3,7-dimethyl-, (±)-
8333 - 2,6-Octadien-1-ol, 3,7-dimethyl-, (E)-
8334 - 2,6-Octadien-1-ol, 3,7-dimethyl-, (Z)-
8490 - 6-Octenal, 3,7-dimethyl-
8556 - 7-Oxabicyclo[2.2.1]heptane, 1-methyl-4-(1-
    methylethyl)-
8560 - 2-Oxabicyclo[2.2.2]octane, 1,3,3-trimethyl-

$C_{10}H_{18}O_2$
3496 - Butanoic acid, cyclohexyl ester
4206 - Cyclodecanone, 2-hydroxy-
4224 - Cycloheptanecarboxylic acid, ethyl ester
4308 - Cyclohexaneacetic acid, ethyl ester
4337 - Cyclohexanecarboxylic acid, propyl ester
4784 - Cyclopentanecarboxylic acid, 2-methyl-5-(1-
    methylethyl)-
4786 - Cyclopentanecarboxylic acid, 1,2,2,3-
    tetramethyl-
5088 - 2-Decenoic acid
5089 - 3-Decenoic acid

5090 - 4-Decenoic acid
5091 - 9-Decenoic acid
6811 - 3,4-Hexanedione, 2,2,5,5-tetramethyl-
10293 - Propanoic acid, 2-methyl-, cyclohexyl ester
10660 - 2-Propenoic acid, 2-methyl-, hexyl ester

$C_{10}H_{18}O_3$

3504 - Butanoic acid, 2,2-diethyl-3-oxo-, ethyl ester
3542 - Butanoic acid, 3-methyl-, anhydride
4309 - Cyclohexaneacetic acid, α-ethyl-1-hydroxy-
6872 - Hexanoic acid, 2-acetyl-, ethyl ester
8858 - Pentanoic acid, anhydride
11903 - 2,5,7-Trioxabicyclo[2.2.1]heptane, 1,3,3,4,6,6-hexamethyl-

$C_{10}H_{18}O_4$

3321 - Butanedioic acid, dipropyl ester
3512 - Butanoic acid, 1,2-ethanediyl ester
5042 - Decanedioic acid
5557 - Ethanedioic acid, bis(2-methylpropyl) ester
5558 - Ethanedioic acid, dibutyl ester
6785 - Hexanedioic acid, diethyl ester
8135 - Nonanedioic acid, 3-methyl-
8388 - Octanedioic acid, dimethyl ester
8601 - Oxirane, 2,2'-[1,4-butanediylbis (oxymethylene)]bis-
9964 - Propanedioic acid, (1-methylethyl)-, diethyl ester
9981 - Propanedioic acid, propyl-, diethyl ester

$C_{10}H_{18}O_4S$

10330 - Propanoic acid, 3,3'-thiobis-, diethyl ester

$C_{10}H_{18}O_6$

3305 - Butanedioic acid, 2,3-dihydroxy-, bis(1-methylethyl) ester, (R*,R*)-(±)-
3310 - Butanedioic acid, 2,3-dihydroxy- [R-(R*,R*)]-, dipropyl ester
5714 - Ethanol, 2,2'-[1,2-ethanediylbis(oxy)]bis-, diacetate

$C_{10}H_{19}Br$

4321 - Cyclohexane, 1-bromo-4-methyl-1-(1-methylethyl)-
5086 - 1-Decene, 2-bromo-
5087 - 2-Decene, 1-bromo-

$C_{10}H_{19}BrO_2$

5061 - Decanoic acid, 2-bromo-

$C_{10}H_{19}ClO$

5078 - Decanoyl chloride

$C_{10}H_{19}FO_2$

5064 - Decanoic acid, 10-fluoro-

$C_{10}H_{19}N$

2927 - Bicyclo[2.2.1]heptane-1-methanamine, 7,7-dimethyl-
2928 - Bicyclo[3.1.1]heptane-2-methanamine, 6,6-dimethyl-, [1S-(1α,2β,5α)]-
5057 - Decanenitrile
7712 - 1-Naphthalenamine, decahydro-

$C_{10}H_{19}NO$

11331 - 2H-Quinolizine-1-methanol, octahydro-, (1R-trans)-

$C_{10}H_{19}NO_2$

9704 - 1-Piperidinepropanoic acid, ethyl ester

$C_{10}H_{19}N_5O$

3143 - sec-Bumeton

$C_{10}H_{19}N_5S$

9770 - Prometryn
11594 - Terbutryn

$C_{10}H_{19}O_6PS_2$

3316 - Butanedioic acid, [(dimethoxyphosphinothioyl)thio]-, diethyl ester

$C_{10}H_{20}$

4202 - Cyclodecane
4227 - Cycloheptane, 1,1,2-trimethyl-
4228 - Cycloheptane, 1,1,4-trimethyl-
4325 - Cyclohexane, butyl-
4369 - Cyclohexane, (1,1-dimethylethyl)-
4436 - Cyclohexane, 1-methyl-2-(1-methylethyl)-
4437 - Cyclohexane, 1-methyl-3-(1-methylethyl)-
4438 - Cyclohexane, 1-methyl-4-(1-methylethyl)-

4439 - Cyclohexane, 1-methyl-4-(1-methylethyl)-, cis-
4440 - Cyclohexane, 1-methyl-4-(1-methylethyl)-, trans-
4448 - Cyclohexane, (1-methylpropyl)-
4449 - Cyclohexane, 1-methyl-1-propyl-
4450 - Cyclohexane, 1-methyl-2-propyl-
4451 - Cyclohexane, 1-methyl-3-propyl-
4452 - Cyclohexane, 1-methyl-4-propyl-
4453 - Cyclohexane, (2-methylpropyl)-
4477 - Cyclohexane, 1,1,3,4-tetramethyl-
4478 - Cyclohexane, 1,1,3,5-tetramethyl-, cis-
4846 - Cyclopentane, pentyl-
5082 - 1-Decene
5083 - 4-Decene
5084 - 5-Decene, (E)-
5085 - 5-Decene, (Z)-
6559 - 3-Heptene, 4-propyl-
8185 - 1-Nonene, 2-methyl-
8186 - 3-Nonene, 2-methyl-
8502 - 1-Octene, 3,7-dimethyl-
8503 - 2-Octene, 2,6-dimethyl-

$C_{10}H_{20}BrF$

5035 - Decane, 1-bromo-10-fluoro-

$C_{10}H_{20}Br_2$

5038 - Decane, 1,10-dibromo-
5039 - Decane, 5,6-dibromo-

$C_{10}H_{20}ClF$

5037 - Decane, 1-chloro-10-fluoro-

$C_{10}H_{20}Cl_2$

5040 - Decane, 1,10-dichloro-

$C_{10}H_{20}NO_4PS$

10719 - Propetamphos

$C_{10}H_{20}N_2$

3098 - 2,3'-Bipiperidine

$C_{10}H_{20}N_2O_2$

7690 - Morpholine, 4,4'-(1,2-ethanediyl)bis-

$C_{10}H_{20}N_2S_2$

9659 - Piperidine, 1,1'-dithiobis-

$C_{10}H_{20}N_2S_3$

11714 - Thiodicarbonic diamide ([(H2N)C(S)]2S), tetraethyl-

$C_{10}H_{20}N_2S_4$

11721 - Thioperoxydicarbonic diamide ([(H2N)C(S)]2S2), tetraethyl-

$C_{10}H_{20}O$

4204 - Cyclodecanol
4233 - Cycloheptanol, 2,6,6-trimethyl-
4324 - Cyclohexanebutanol
4469 - Cyclohexanepropanol, α-methyl-
4504 - Cyclohexanol, 2-butyl-, trans-
4521 - Cyclohexanol, 2-(1,1-dimethylethyl)-
4546 - Cyclohexanol, 1-methyl-4-(1-methylethyl)-
4547 - Cyclohexanol, 2-methyl-5-(1-methylethyl)-, [1R-(1α,2α,5β)]-
4548 - Cyclohexanol, 5-methyl-2-(1-methylethyl)-, (1α,2β,5α)-(±)-
4550 - Cyclohexanol, 2-methyl-1-propyl-
4551 - Cyclohexanol, 3-methyl-1-propyl-
5030 - Decanal
5075 - 2-Decanone
5076 - 3-Decanone
5077 - 4-Decanone
5092 - 9-Decen-1-ol
6405 - Heptane, 3-[(ethenyloxy)methyl]-
7428 - m-Menthan-6-ol
8176 - 5-Nonanone, 2-methyl-
8401 - Octane, 1-(ethenyloxy)-
8514 - 6-Octen-1-ol, 3,7-dimethyl-, (±)-
8515 - 6-Octen-3-ol, 3,7-dimethyl-
8516 - 7-Octen-1-ol, 3,7-dimethyl-, (S)-

$C_{10}H_{20}O_2$

187 - Acetic acid, 2-ethylhexyl ester
222 - Acetic acid, octyl ester
3528 - Butanoic acid, hexyl ester
3546 - Butanoic acid, 3-methyl-, 1,1-dimethylpropyl ester
3551 - Butanoic acid, 3-methyl-, 3-methylbutyl ester

3559 - Butanoic acid, 1-methylpentyl ester, (S)-
5059 - Decanoic acid
6471 - Heptanoic acid, propyl ester
6506 - 2-Heptanol, 6-methyl-, acetate
6507 - 3-Heptanol, 6-methyl-, acetate
6880 - Hexanoic acid, butyl ester
6885 - Hexanoic acid, 2-ethyl-, ethyl ester
6975 - 3-Hexanone, 4-hydroxy-2,2,5,5-tetramethyl-
8153 - Nonanoic acid, 3-methyl-
8154 - Nonanoic acid, methyl ester
8347 - Octanal, 7-hydroxy-3,7-dimethyl-
8428 - Octanoic acid, ethyl ester
8447 - 2-Octanol, acetate, (±)-
8894 - Pentanoic acid, 4-methyl-, 2-methylpropyl ester
8913 - Pentanoic acid, pentyl ester
10264 - Propanoic acid, heptyl ester

$C_{10}H_{20}O_3$

3811 - 2-Butene, 1,1,3-triethoxy-

$C_{10}H_{20}O_4$

5689 - Ethanol, 2-(2-butoxyethoxy)-, acetate

$C_{10}H_{21}Br$

5034 - Decane, 1-bromo-

$C_{10}H_{21}Cl$

5036 - Decane, 1-chloro-

$C_{10}H_{21}ClO$

5073 - 1-Decanol, 10-chloro-

$C_{10}H_{21}F$

5052 - Decane, 1-fluoro-

$C_{10}H_{21}FO$

5074 - 1-Decanol, 10-fluoro-

$C_{10}H_{21}I$

5053 - Decane, 1-iodo-

$C_{10}H_{21}N$

4285 - Cyclohexanamine, N,N-diethyl-
4385 - Cyclohexaneethanamine, N,α-dimethyl-
9691 - Piperidine, 1,2,2,6,6-pentamethyl-
9692 - Piperidine, 1-pentyl-

$C_{10}H_{21}NO$

4502 - Cyclohexanol, 3-(aminomethyl)-3,5,5-trimethyl-
5031 - Decanamide

$C_{10}H_{21}NOS$

3976 - Carbamothioic acid, butylethyl-, S-propyl ester
12047 - Vernolate

$C_{10}H_{21}NO_2$

8084 - Nitrous acid, decyl ester

$C_{10}H_{21}NO_3$

8073 - Nitric acid, decyl ester

$C_{10}H_{21}N_3O$

9603 - 1-Piperazinecarboxamide, N,N-diethyl-4-methyl-

$C_{10}H_{22}$

5033 - Decane
6409 - Heptane, 3-ethyl-2-methyl-
6410 - Heptane, 3-ethyl-3-methyl-
6411 - Heptane, 3-ethyl-4-methyl-
6412 - Heptane, 3-ethyl-5-methyl-
6413 - Heptane, 4-ethyl-2-methyl-
6414 - Heptane, 4-ethyl-3-methyl-
6415 - Heptane, 4-ethyl-4-methyl-
6416 - Heptane, 5-ethyl-2-methyl-
6427 - Heptane, 4-(1-methylethyl)-
6434 - Heptane, 4-propyl-
6437 - Heptane, 2,2,3-trimethyl-
6438 - Heptane, 2,2,4-trimethyl-
6439 - Heptane, 2,2,5-trimethyl-
6440 - Heptane, 2,2,6-trimethyl-
6441 - Heptane, 2,3,3-trimethyl-
6442 - Heptane, 2,3,4-trimethyl-
6443 - Heptane, 2,3,5-trimethyl-
6444 - Heptane, 2,3,6-trimethyl-
6445 - Heptane, 2,4,4-trimethyl-
6446 - Heptane, 2,4,5-trimethyl-
6447 - Heptane, 2,4,6-trimethyl-
6448 - Heptane, 2,5,5-trimethyl-
6449 - Heptane, 3,3,4-trimethyl-

6450 - Heptane, 3,3,5-trimethyl-
6451 - Heptane, 3,4,4-trimethyl-
6452 - Heptane, 3,4,5-trimethyl-
6767 - Hexane, 3,3-diethyl-
6768 - Hexane, 3,4-diethyl-
6816 - Hexane, 3-ethyl-2,2-dimethyl-
6817 - Hexane, 3-ethyl-2,3-dimethyl-
6818 - Hexane, 3-ethyl-2,4-dimethyl-
6819 - Hexane, 3-ethyl-2,5-dimethyl-
6820 - Hexane, 3-ethyl-3,4-dimethyl-
6821 - Hexane, 4-ethyl-2,2-dimethyl-
6822 - Hexane, 4-ethyl-2,3-dimethyl-
6823 - Hexane, 4-ethyl-2,4-dimethyl-
6824 - Hexane, 4-ethyl-3,3-dimethyl-
6839 - Hexane, 2-methyl-3-(1-methylethyl)-
6844 - Hexane, 2,2,3,3-tetramethyl-
6845 - Hexane, 2,2,3,4-tetramethyl-
6846 - Hexane, 2,2,3,5-tetramethyl-
6847 - Hexane, 2,2,4,4-tetramethyl-
6848 - Hexane, 2,2,4,5-tetramethyl-
6849 - Hexane, 2,2,5,5-tetramethyl-
6850 - Hexane, 2,3,3,4-tetramethyl-
6851 - Hexane, 2,3,3,5-tetramethyl-
6852 - Hexane, 2,3,4,4-tetramethyl-
6853 - Hexane, 2,3,4,5-tetramethyl-
6854 - Hexane, 3,3,4,4-tetramethyl-
8140 - Nonane, 2-methyl-
8141 - Nonane, 3-methyl-
8142 - Nonane, 4-methyl-
8143 - Nonane, 5-methyl-
8373 - Octane, 2,2-dimethyl-
8374 - Octane, 2,3-dimethyl-
8375 - Octane, 2,4-dimethyl-
8376 - Octane, 2,5-dimethyl-
8377 - Octane, 2,6-dimethyl-
8378 - Octane, 2,7-dimethyl-
8379 - Octane, 3,3-dimethyl-
8380 - Octane, 3,4-dimethyl-
8381 - Octane, 3,5-dimethyl-
8382 - Octane, 3,6-dimethyl-
8383 - Octane, 4,4-dimethyl-
8384 - Octane, 4,5-dimethyl-
8403 - Octane, 3-ethyl-
8404 - Octane, 4-ethyl-
8757 - Pentane, 3,3-diethyl-2-methyl-
8766 - Pentane, 2,4-dimethyl-3-(1-methylethyl)-
8815 - Pentane, 3-ethyl-2,2,3-trimethyl-
8816 - Pentane, 3-ethyl-2,3,4-trimethyl-
8836 - Pentane, 2,2,3,3,4-pentamethyl-
8837 - Pentane, 2,2,3,4,4-pentamethyl-

$C_{10}H_{22}O$
3418 - Butane, 1,1'-oxybis[3-methyl-
5070 - 1-Decanol
5071 - 2-Decanol, (±)-
5072 - 4-Decanol
6488 - 3-Heptanol, 3-ethyl-2-methyl-
6489 - 3-Heptanol, 3-ethyl-6-methyl-
6490 - 3-Heptanol, 5-ethyl-4-methyl-
6508 - 4-Heptanol, 4-propyl-
6509 - 4-Heptanol, 2,4,6-trimethyl-
6948 - 1-Hexanol, 5-methyl-2-(1-methylethyl)-
6950 - 3-Hexanol, 2,2,5,5-tetramethyl-
8166 - 1-Nonanol, 3-methyl-
8167 - 1-Nonanol, 4-methyl-
8168 - 1-Nonanol, 8-methyl
8169 - 3-Nonanol, 2-methyl-
8170 - 4-Nonanol, 4-methyl-
8171 - 5-Nonanol, 5-methyl-
8402 - Octane, 1-ethoxy-
8449 - 1-Octanol, 3,7-dimethyl-, (±)-
8450 - 2-Octanol, 2,6-dimethyl-
8451 - 2-Octanol, 2,3-dimethyl-
8452 - 3-Octanol, 2,7-dimethyl-
8453 - 3-Octanol, 3,6-dimethyl-
8454 - 3-Octanol, 3,6-dimethyl-(+-)-
8455 - 4-Octanol, 2,4-dimethyl-
8456 - 4-Octanol, 2,5-dimethyl-

8457 - 4-Octanol, 2,6-dimethyl-
8458 - 4-Octanol, 2,7-dimethyl-
8459 - 3-Octanol, 3-ethyl-
8835 - Pentane, 1,1'-oxybis-
$C_{10}H_{22}O_2$
5047 - 1,10-Decanediol
5048 - 2,9-Decanediol
5527 - Ethane, 1,2-dibutoxy
6800 - 3,4-Hexanediol, 3,4-diethyl-
8392 - 1,7-Octanediol, 3,7-dimethyl-
9154 - Peroxide, bis(1,1-dimethylpropyl)
10032 - Propane, 1,1'-[1,2-ethanediylbis(oxy)]bis[2-
    methyl-
10041 - Propane, 1,1'-[ethylidenebis(oxy)]bis[2-
    methyl-
$C_{10}H_{22}O_3$
10078 - Propane, 1,1',1''-[methylidynetris(oxy)]tris-
10079 - Propane, 2,2',2''-[methylidynetris(oxy)]tris-
$C_{10}H_{22}O_4$
5690 - Ethanol, 2-[2-(2-butoxyethoxy)ethoxy]-
$C_{10}H_{22}O_5$
8995 - 2,5,8,11,14-Pentaoxapentadecane
$C_{10}H_{22}S$
3435 - Butane, 1,1'-thiobis[2-methyl-
3436 - Butane, 1,1'-thiobis[3-methyl-
5058 - 1-Decanethiol
8843 - Pentane, 1,1'-thiobis-
$C_{10}H_{22}S_2$
5243 - Disulfide, bis(3-methylbutyl)
5257 - Disulfide, dipentyl
$C_{10}H_{23}N$
3226 - 1-Butanamine, 3-methyl-N-(3-methylbutyl)-
5032 - 1-Decanamine
8708 - 1-Pentanamine, N-pentyl-
$C_{10}H_{23}NO$
5684 - Ethanol, 2-[bis(2-methylpropyl)amino]-
5703 - Ethanol, 2-(dibutylamino)-
$C_{10}H_{23}NO_2$
5415 - Ethanamine, 2,2-diethoxy-N,N-diethyl-
$C_{10}H_{24}N_2$
5515 - 1,2-Ethanediamine, N,N,N',N'-tetraethyl-
6752 - 1,6-Hexanediamine, N,N,N',N'-tetramethyl-
$C_{10}H_{24}N_4$
9614 - 1,4-Piperazinedipropanamine
$C_{10}H_{26}N_4$
3259 - 1,4-Butanediamine, N,N'-bis(3-aminopropyl)-
$C_{10}H_{30}O_3Si_4$
11646 - Tetrasiloxane, decamethyl-
$C_{10}H_{30}O_5Si_5$
4901 - Cyclopentasiloxane, decamethyl-
$C_{11}H_7BrO_2$
7761 - 1-Naphthalenecarboxylic acid, 5-bromo-
7769 - 2-Naphthalenecarboxylic acid, 5-bromo-
7846 - 1,4-Naphthalenedione, 2-bromo-3-methyl-
$C_{11}H_7ClO$
7750 - 2-Naphthalenecarbonyl chloride
$C_{11}H_7ClO_2$
7762 - 1-Naphthalenecarboxylic acid, 5-chloro-
7763 - 1-Naphthalenecarboxylic acid, 8-chloro-
$C_{11}H_7N$
7748 - 1-Naphthalenecarbonitrile
7749 - 2-Naphthalenecarbonitrile
$C_{11}H_7NO$
7904 - Naphthalene, 1-isocyanato-
$C_{11}H_7NO_4$
7767 - 1-Naphthalenecarboxylic acid, 5-nitro-
$C_{11}H_7NS$
7906 - Naphthalene, 1-isothiocyanato-
$C_{11}H_8O$
7751 - 1-Naphthalenecarboxaldehyde
7752 - 2-Naphthalenecarboxaldehyde
$C_{11}H_8OS$
7641 - Methanone, phenyl-2-thienyl-
7642 - Methanone, phenyl-3-thienyl-
$C_{11}H_8O_2$
7618 - Methanone, 2-furanylphenyl-
7754 - 1-Naphthalenecarboxaldehyde, 2-hydroxy-

7758 - 1-Naphthalenecarboxylic acid
7759 - 2-Naphthalenecarboxylic acid
7864 - 1,4-Naphthalenedione, 2-methyl-
$C_{11}H_8O_3$
7771 - 2-Naphthalenecarboxylic acid, 1-hydroxy-
7772 - 2-Naphthalenecarboxylic acid, 3-hydroxy-
7861 - 1,4-Naphthalenedione, 2-hydroxy-3-methyl-
7863 - 1,4-Naphthalenedione, 5-hydroxy-2-methyl-
$C_{11}H_8O_4$
7856 - 1,4-Naphthalenedione, 3,5-dihydroxy-2-
    methyl-
$C_{11}H_8O_5$
2408 - 5H-Benzocyclohepten-5-one, 2,3,4,6-
    tetrahydroxy-
$C_{11}H_9Br$
7742 - Naphthalene, 1-(bromomethyl)-
7743 - Naphthalene, 2-(bromomethyl)-
$C_{11}H_9Cl$
7783 - Naphthalene, 1-(chloromethyl)-
7784 - Naphthalene, 1-chloro-2-methyl-
7785 - Naphthalene, 2-(chloromethyl)-
$C_{11}H_9Cl_2NO_2$
510 - Barban
$C_{11}H_9N$
10997 - Pyridine, 2-phenyl-
10998 - Pyridine, 3-phenyl-
10999 - Pyridine, 4-phenyl-
$C_{11}H_9NO$
7637 - Methanone, phenyl-1H-pyrrol-2-yl-
7757 - 1-Naphthalenecarboxamide
$C_{11}H_9NO_2$
7760 - 1-Naphthalenecarboxylic acid, 5-amino-
7768 - 2-Naphthalenecarboxylic acid, 3-amino-
7925 - Naphthalene, 1-methyl-4-nitro-
7926 - Naphthalene, 2-methyl-1-nitro-
$C_{11}H_9NO_3$
11234 - 4-Quinolinecarboxylic acid, 6-methoxy-
$C_{11}H_{10}$
7915 - Naphthalene, 1-methyl-
7916 - Naphthalene, 2-methyl-
$C_{11}H_{10}NO_3$
7695 - Morpholine, 4-(2-hydroxybenzoyl)-
$C_{11}H_{10}N_2O$
90 - Acetamide, N-4-quinolinyl-
12022 - Urea, 1-naphthalenyl-
12023 - Urea, 2-naphthalenyl-
$C_{11}H_{10}N_2O_3$
11100 - 2,4,6(1H,3H,5H)-Pyrimidinetrione, 5-methyl-
    5-phenyl-
$C_{11}H_{10}N_2S$
11807 - Thiourea, 1-naphthalenyl-
$C_{11}H_{10}O$
4951 - 2-Cyclopenten-1-one, 3-phenyl-
7908 - 1-Naphthalenemethanol
7909 - 2-Naphthalenemethanol
7913 - Naphthalene, 1-methoxy-
7914 - Naphthalene, 2-methoxy-
8030 - 1-Naphthalenol, 4-methyl-
8031 - 2-Naphthalenol, 1-methyl-
8657 - 2,4-Pentadienal, 5-phenyl-
$C_{11}H_{10}O_2$
2820 - 2H-1-Benzopyran-2-one, 3-ethyl-
6158 - Furan, 2-(phenoxymethyl)-
7764 - 1-Naphthalenecarboxylic acid, 3,4-dihydro-
10741 - 2-Propynoic acid, 3-phenyl-, ethyl ester
$C_{11}H_{10}O_3$
2440 - 2-Benzofurancarboxylic acid, ethyl ester
10781 - 2H-Pyran-2,6(3H)-dione, dihydro-3-phenyl-
10782 - 2H-Pyran-2,6(3H)-dione, dihydro-4-phenyl-
$C_{11}H_{10}O_4$
2818 - 2H-1-Benzopyran-2-one, 5,7-dimethoxy-
$C_{11}H_{10}S$
11774 - Thiophene, 2-methyl-5-phenyl-
$C_{11}H_{11}BrN_2O$
10846 - 3H-Pyrazol-3-one, 4-bromo-1,2-dihydro-1,5-
    dimethyl-2-phenyl-

C₁₁H₁₁BrO₂
10613 - 2-Propenoic acid, 2-bromo-3-phenyl-, ethyl ester
10614 - 2-Propenoic acid, 3-bromo-3-phenyl-, ethyl ester, (Z)-

C₁₁H₁₁Cl₃O₃
259 - Acetic acid, (2,4,5-trichlorophenoxy)-, 1-methylethyl ester

C₁₁H₁₁N
7720 - 1-Naphthalenamine, 4-methyl-
7721 - 1-Naphthalenamine, N-methyl-
7722 - 2-Naphthalenamine, N-methyl-
7907 - 2-Naphthalenemethanamine
11136 - 1H-Pyrrole, 1-(phenylmethyl)-
11254 - Quinoline, 2,3-dimethyl-
11255 - Quinoline, 2,4-dimethyl-
11256 - Quinoline, 2,5-dimethyl-
11257 - Quinoline, 2,6-dimethyl-
11258 - Quinoline, 2,7-dimethyl-
11259 - Quinoline, 2,8-dimethyl-
11260 - Quinoline, 3,4-dimethyl-
11261 - Quinoline, 3,8-dimethyl-
11262 - Quinoline, 4,6-dimethyl-
11263 - Quinoline, 5,8-dimethyl-
11264 - Quinoline, 6,8-dimethyl-
11266 - Quinoline, 2-ethyl-
11267 - Quinoline, 3-ethyl-
11268 - Quinoline, 6-ethyl-

C₁₁H₁₁NO
10442 - 2-Propanone, 1-(1H-indol-3-yl)-
11316 - 4-Quinolinol, 2,8-dimethyl-
11317 - 6-Quinolinol, 2,4-dimethyl-
11318 - 8-Quinolinol, 2,4-dimethyl-

C₁₁H₁₁NO₂
893 - Benzeneacetic acid, α-cyano-, ethyl ester
11161 - 2,5-Pyrrolidinedione, 1-methyl-3-phenyl-
11163 - 2,5-Pyrrolidinedione, 1-(phenylmethyl)-

C₁₁H₁₁NO₄S
7407 - Isoxazolium, 2-ethyl-5-(3-sulfophenyl)-, hydroxide, inner salt
11293 - 5-Quinolinesulfonic acid, 8-ethoxy-

C₁₁H₁₁NO₆
7511 - Methanediol, (3-nitrophenyl)-, diacetate (ester)

C₁₁H₁₁N₃O₂S
2203 - Benzenesulfonamide, 4-amino-N-2-pyridinyl-

C₁₁H₁₂
7816 - Naphthalene, 1,2-dihydro-3-methyl-
7817 - Naphthalene, 1,2-dihydro-4-methyl-

C₁₁H₁₂Cl₂N₂O₅
52 - Acetamide, 2,2-dichloro-N-[2-hydroxy-1-(hydroxymethyl)-2-(4-nitrophenyl)ethyl]-, [R-(R*,R*)]-

C₁₁H₁₂Cl₂O₃
173 - Acetic acid, (2,4-dichlorophenoxy)-, 1-methylethyl ester

C₁₁H₁₂Cl₃N
1583 - Benzeneethanamine, α-methyl-N-(2,2,2-trichloroethylidene)-

C₁₁H₁₂NO₄PS₂
9515 - Phosmet

C₁₁H₁₂N₂
7800 - 1,4-Naphthalenediamine, 2-methyl-
10829 - 1H-Pyrazole, 3,5-dimethyl-1-phenyl-

C₁₁H₁₂N₂O
10847 - 3H-Pyrazol-3-one, 1,2-dihydro-1,5-dimethyl-2-phenyl-
11192 - Pyrrolo[2,1-b]quinazolin-3-ol, 1,2,3,9-tetrahydro-, (R)-

C₁₁H₁₂N₂O₂
11922 - L-Tryptophan

C₁₁H₁₂N₂O₆
2592 - Benzoic acid, 3,5-dinitro-, butyl ester

C₁₁H₁₂N₄O₂S
2199 - Benzenesulfonamide, 4-amino-N-(4-methyl-2-pyrimidinyl)-

C₁₁H₁₂N₄O₃S
2197 - Benzenesulfonamide, 4-amino-N-(6-methoxy-3-pyridazinyl)-

C₁₁H₁₂O
2407 - 5H-Benzocyclohepten-5-one, 6,7,8,9-tetrahydro-
3190 - Butanal, 2-(phenylmethylene)-
7288 - 1H-Inden-1-one, 2,3-dihydro-3,3-dimethyl-
7605 - Methanone, cyclobutylphenyl-
8049 - 1(2H)-Naphthalenone, 3,4-dihydro-2-methyl-
8050 - 1(2H)-Naphthalenone, 3,4-dihydro-4-methyl-
9110 - 1-Penten-3-one, 1-phenyl-

C₁₁H₁₂O₂
7778 - 2-Naphthalenecarboxylic acid, 1,2,3,4-tetrahydro-
7779 - 2-Naphthalenecarboxylic acid, 5,6,7,8-tetrahydro-
8048 - 1(2H)-Naphthalenone, 3,4-dihydro-6-methoxy-
8804 - 1,4-Pentanedione, 1-phenyl-
10685 - 2-Propenoic acid, 3-phenyl-, ethyl ester, (E)-
10706 - 2-Propen-1-ol, 3-phenyl-, acetate, (E)-

C₁₁H₁₂O₃
2141 - Benzenepropanoic acid, β-oxo-, ethyl ester
2428 - 1,3-Benzodioxole, 4-methoxy-6-(2-propenyl)-
6138 - 2-Furanol, tetrahydro-, benzoate
10647 - 2-Propenoic acid, 3-(2-methoxyphenyl)-, methyl ester
10681 - 2-Propenoic acid, 2-phenoxyethyl ester

C₁₁H₁₂O₄
7512 - Methanediol, phenyl-, diacetate
9976 - Propanedioic acid, phenyl-, dimethyl ester

C₁₁H₁₃ClF₃N₃O₄S₃
2857 - 2H-1,2,4-Benzothiadiazine-7-sulfonamide, 6-chloro-3,4-dihydro-2-methyl-3-[[(2,2,2-trifluoroethyl)thio]methyl]-

C₁₁H₁₃ClO
2889 - Benzoyl chloride, 4-butyl-
2895 - Benzoyl chloride, 4-(1,1-dimethylethyl)-
9255 - Phenol, 4-chloro-2-cyclopentyl-

C₁₁H₁₃ClO₃
3493 - Butanoic acid, 4-(4-chloro-2-methylphenoxy)-

C₁₁H₁₃F₃N₂O₃S
7427 - Metluidide

C₁₁H₁₃F₃N₄O₄
5181 - Dinitramine

C₁₁H₁₃N
963 - Benzeneacetonitrile, α-(1-methylethyl)-
970 - Benzeneacetonitrile, α-propyl-
1823 - Benzenemethanamine, N-methyl-N-2-propynyl-
7309 - 1H-Indole, 3-ethyl-2-methyl-
7322 - 1H-Indole, 1,2,3-trimethyl-
7323 - 1H-Indole, 2,3,5-trimethyl-
10922 - Pyridine, 4-(3-cyclohexen-1-yl)-

C₁₁H₁₃NO
3703 - 3-Butenamide, 3-methyl-4-phenyl-
7332 - 2H-Indol-2-one, 1-ethyl-1,3-dihydro-3-methyl-

C₁₁H₁₃NO₂
5215 - 1,3-Dioxolo[4,5-g]isoquinoline, 5,6,7,8-tetrahydro-6-methyl-

C₁₁H₁₃NO₃
5217 - 1,3-Dioxolo[4,5-g]isoquinolin-5-ol, 5,6,7,8-tetrahydro-6-methyl-
7398 - 1(2H)-Isoquinolinone, 3,4-dihydro-6,7-dimethoxy-

C₁₁H₁₃NO₄
514 - Bendiocarb
2706 - Benzoic acid, 4-nitro-, butyl ester
8588 - 2-Oxazolidinone, 5-[(2-methoxyphenoxy)methyl]-

C₁₁H₁₃N₃O
10845 - 3H-Pyrazol-3-one, 4-amino-1,2-dihydro-1,5-dimethyl-2-phenyl-

C₁₁H₁₃N₃O₃S
2195 - Benzenesulfonamide, 4-amino-N-(3,4-dimethyl-5-isoxazolyl)-

C₁₁H₁₄
1254 - Benzene, cyclopentyl-
1643 - Benzene, 1-ethenyl-4-(1-methylethyl)-
1647 - Benzene, 2-ethenyl-1,3,5-trimethyl-
1699 - Benzene, (1-ethyl-1-propenyl)-
1700 - Benzene, (1-ethyl-2-propenyl)-
1950 - Benzene, (1-methyl-1-butenyl)-
2104 - Benzene, 1-pentenyl-
2105 - Benzene, 3-pentenyl-
2106 - Benzene, 4-pentenyl-
4748 - 1,3,5,7-Cyclooctatetraene, propyl-
7251 - 1H-Indene, 2,3-dihydro-1,1-dimethyl-
7252 - 1H-Indene, 2,3-dihydro-1,2-dimethyl-
7253 - 1H-Indene, 2,3-dihydro-4,7-dimethyl-
7254 - 1H-Indene, 2,3-dihydro-5,6-dimethyl-
7273 - 1H-Indene, 1-ethyl-2,3-dihydro-
7995 - Naphthalene, 1,2,3,4-tetrahydro-1-methyl-
7996 - Naphthalene, 1,2,3,4-tetrahydro-2-methyl-
7997 - Naphthalene, 1,2,3,4-tetrahydro-5-methyl-
7998 - Naphthalene, 1,2,3,4-tetrahydro-6-methyl-

C₁₁H₁₄ClNO
9771 - Propachlor

C₁₁H₁₄ClNO₂
607 - Benzamide, N-butyl-4-chloro-2-hydroxy-

C₁₁H₁₄N₂
7310 - 1H-Indole-3-methanamine, N,N-dimethyl-

C₁₁H₁₄N₂O
7304 - 1H-Indole-3-ethanamine, 5-methoxy-

C₁₁H₁₄N₂O₄
84 - Acetamide, N-(5-nitro-2-propoxyphenyl)-

C₁₁H₁₄O
598 - Benzaldehyde, 2,3,5,6-tetramethyl-
1665 - Benzene, 1-ethoxy-2-(1-propenyl)-
1666 - Benzene, (3-ethoxy-1-propenyl)-
3671 - 1-Butanone, 3-methyl-1-phenyl-
3672 - 1-Butanone, 1-(4-methylphenyl)-
4871 - Cyclopentanol, 1-phenyl-
5848 - Ethanone, 1-[4-(1-methylethyl)phenyl]-
5875 - Ethanone, 1-(2,4,5-trimethylphenyl)-
5876 - Ethanone, 1-(2,4,6-trimethylphenyl)-
5877 - Ethanone, 1-(3,4,5-trimethylphenyl)-
8986 - 1-Pentanone, 1-phenyl-
10431 - 1-Propanone, 2,2-dimethyl-1-phenyl-

C₁₁H₁₄O₂
900 - Benzeneacetic acid, α-ethyl-, methyl ester
930 - Benzeneacetic acid, α-(1-methylethyl)-
943 - Benzeneacetic acid, α-propyl-, (±)-
1104 - Benzenebutanoic acid, γ-methyl-
1449 - Benzene, 1,2-dimethoxy-4-(1-propenyl)-
1450 - Benzene, 1,2-dimethoxy-4-(2-propenyl)-
1451 - Benzene, 1,2-dimethoxy-4-(1-propenyl)-, (Z)-
2099 - Benzenepentanoic acid
2130 - Benzenepropanoic acid, β,β-dimethyl-
2131 - Benzenepropanoic acid, ethyl ester
2532 - Benzoic acid, butyl ester
2588 - Benzoic acid, 4-(1,1-dimethylethyl)-
2694 - Benzoic acid, 4-(1-methylethyl)-, methyl ester
2701 - Benzoic acid, 2-methylpropyl ester
3585 - Butanoic acid, phenylmethyl ester
5194 - 1,3-Dioxane, 4-methyl-4-phenyl-
9496 - Phenol, 2,4,5-trimethyl-, acetate
10307 - Propanoic acid, 2-methyl-, phenylmethyl ester
10324 - Propanoic acid, 2-phenylethyl ester
10436 - 1-Propanone, 3-ethoxy-1-phenyl-
10447 - 1-Propanone, 1-(2-hydroxy-5-methylphenyl)-2-methyl-
10448 - 1-Propanone, 1-(4-hydroxy-3-methylphenyl)-2-methyl-

C₁₁H₁₄O₃
538 - Benzaldehyde, 3,4-diethoxy-
923 - Benzeneacetic acid, 4-methoxy-, ethyl ester
1128 - Benzenecarboperoxoic acid, 1,1-dimethylethyl ester
2598 - Benzoic acid, 2-ethoxy-, ethyl ester
2599 - Benzoic acid, 4-ethoxy-, ethyl ester
2614 - Benzoic acid, 2-hydroxy-, butyl ester

2615 - Benzoic acid, 4-hydroxy-, butyl ester
2641 - Benzoic acid, 2-hydroxy-3-methyl-6-(1-methylethyl)-
2642 - Benzoic acid, 2-hydroxy-6-methyl-3-(1-methylethyl)-
2643 - Benzoic acid, 2-hydroxy-, 2-methylpropyl ester
3663 - 2-Butanone, 4-(4-hydroxy-3-methoxyphenyl)-
10321 - Propanoic acid, 2-phenoxy-, ethyl ester
10322 - Propanoic acid, 3-phenoxy-, ethyl ester

$C_{11}H_{14}O_4$
2578 - Benzoic acid, 3,4-dimethoxy-, ethyl ester
5874 - Ethanone, 1-(2,3,4-trimethoxyphenyl)-

$C_{11}H_{15}BrClO_3PS$
9762 - Profenofos

$C_{11}H_{15}BrN_2O_3$
11079 - 2,4,6(1H,3H,5H)-Pyrimidinetrione, 5-(2-bromo-2-propenyl)-1-methyl-5-(1-methylethyl)-
11080 - 2,4,6(1H,3H,5H)-Pyrimidinetrione, 5-(2-bromo-2-propenyl)-5-(1-methylpropyl)-

$C_{11}H_{15}Cl_2O_3PS_2$
4092 - Chlorthiophos

$C_{11}H_{15}N$
9693 - Piperidine, 1-phenyl-
9694 - Piperidine, 3-phenyl-
9695 - Piperidine, 4-phenyl-
11172 - Pyrrolidine, 1-(phenylmethyl)-

$C_{11}H_{15}NO$
74 - Acetamide, N-(1-methylethyl)-N-phenyl-
88 - Acetamide, N-phenyl-N-propyl-
539 - Benzaldehyde, 4-(diethylamino)-
882 - Benzeneacetamide, α-(1-methylethyl)-
7697 - Morpholine, 3-methyl-2-phenyl-
7698 - Morpholine, 4-(4-methylphenyl)-
7700 - Morpholine, 4-(phenylmethyl)-
8960 - 1-Pentanone, 1-(4-aminophenyl)-

$C_{11}H_{15}NO_2$
2488 - Benzoic acid, 4-amino-, butyl ester
2505 - Benzoic acid, 4-amino-, 2-methylpropyl ester
3968 - Carbamic acid, phenyl-, 2-methylpropyl ester
6272 - Glycine, N-(phenylmethyl)-, ethyl ester
7397 - 6-Isoquinolinol, 1,2,3,4-tetrahydro-7-methoxy-1-methyl-, (S)-
9512 - L-Phenylalanine, ethyl ester

$C_{11}H_{15}NO_2S$
7646 - Methiocarb

$C_{11}H_{15}NO_3$
7393 - 8-Isoquinolinol, 1,2,3,4-tetrahydro-6,7-dimethoxy-
9796 - Propanamide, N-(4-ethoxyphenyl)-2-hydroxy-
10722 - Propoxur

$C_{11}H_{15}N_2NaO_2S$
11063 - 4,6(1H,5H)-Pyrimidinedione, dihydro-5-(2-methylpropyl)-5-(2-propenyl)-2-thioxo-, monosodium salt

$C_{11}H_{16}$
1121 - Benzene, 1-butyl-2-methyl-
1122 - Benzene, 1-butyl-3-methyl-
1123 - Benzene, 1-butyl-4-methyl-
1416 - Benzene, 1,2-diethyl-3-methyl-
1417 - Benzene, 1,2-diethyl-4-methyl-
1418 - Benzene, 1,3-diethyl-2-methyl-
1419 - Benzene, 1,3-diethyl-5-methyl-
1420 - Benzene, 1,4-diethyl-2-methyl-
1421 - Benzene, 2,4-diethyl-1-methyl-
1464 - Benzene, 1-(1,1-dimethylethyl)-2-methyl-
1465 - Benzene, 1-(1,1-dimethylethyl)-3-methyl-
1466 - Benzene, 1-(1,1-dimethylethyl)-4-methyl-
1469 - Benzene, 1,2-dimethyl-3-(1-methylethyl)-
1470 - Benzene, 1,2-dimethyl-4-(1-methylethyl)-
1471 - Benzene, 1,3-dimethyl-2-(1-methylethyl)-
1472 - Benzene, 2,4-dimethyl-1-(1-methylethyl)-
1480 - Benzene, (1,1-dimethylpropyl)-
1481 - Benzene, (1,2-dimethylpropyl)-
1482 - Benzene, 1,3-dimethyl-2-propyl-
1483 - Benzene, 1,3-dimethyl-5-propyl-
1484 - Benzene, (2,2-dimethylpropyl)-
1485 - Benzene, 1,2-dimethyl-3-propyl-

1486 - Benzene, 1,4-dimethyl-2-propyl-
1487 - Benzene, 2,4-dimethyl-1-propyl-
1688 - Benzene, 1-ethyl-2-(1-methylethyl)-
1689 - Benzene, 1-ethyl-3-(1-methylethyl)-
1690 - Benzene, 1-ethyl-4-(1-methylethyl)-
1701 - Benzene, (1-ethylpropyl)-
1702 - Benzene, 1-ethyl-2-propyl-
1703 - Benzene, 1-ethyl-4-propyl-
1708 - Benzene, 1-ethyl-2,3,4-trimethyl-
1709 - Benzene, 1-ethyl-2,3,5-trimethyl-
1710 - Benzene, 1-ethyl-2,4,5-trimethyl-
1711 - Benzene, 2-ethyl-1,3,4-trimethyl-
1712 - Benzene, 2-ethyl-1,3,5-trimethyl-
1713 - Benzene, 5-ethyl-1,2,3-trimethyl-
1952 - Benzene, (1-methylbutyl)-
1953 - Benzene, (2-methylbutyl)-
1954 - Benzene, (3-methylbutyl)-
1994 - Benzene, 1-methyl-2-(1-methylpropyl)-
1995 - Benzene, 1-methyl-2-(2-methylpropyl)-
1996 - Benzene, 1-methyl-3-(1-methylpropyl)-
1997 - Benzene, 1-methyl-3-(2-methylpropyl)-
1998 - Benzene, 1-methyl-4-(1-methylpropyl)-
1999 - Benzene, 1-methyl-4-(2-methylpropyl)-
2097 - Benzene, pentamethyl-
2107 - Benzene, pentyl-
8341 - 1,5-Octadien-3-yne, 5-propyl-
11934 - 1,10-Undecadiyne
12057 - o-Xylene, 4-propyl-

$C_{11}H_{16}ClN_3O_2$
6001 - Formetanate hydrochloride

$C_{11}H_{16}ClN_3O_4S_2$
2856 - 2H-1,2,4-Benzothiadiazine-7-sulfonamide, 6-chloro-3,4-dihydro-3-(2-methylpropyl)-, 1,1-dioxide

$C_{11}H_{16}ClO_2PS_3$
4037 - Carbophenothion

$C_{11}H_{16}N_2$
9621 - Piperazine, 1-(phenylmethyl)-

$C_{11}H_{16}N_2O_2$
6151 - 2(3H)-Furanone, 3-ethyldihydro-4-[(1-methyl-1H-imidazol-5-yl)methyl]-, (3S-cis)-

$C_{11}H_{16}N_2O_3$
11099 - 2,4,6(1H,3H,5H)-Pyrimidinetrione, 1-methyl-5-(1-methylethyl)-5-(2-propenyl)-
11101 - 2,4,6(1H,3H,5H)-Pyrimidinetrione, 5-(2-methylpropyl)-5-(2-propenyl)-

$C_{11}H_{16}O$
363 - Anisole, 3-isopropyl-2-methyl-
1106 - Benzenebutanol, α-methyl-
1107 - Benzenebutanol, β-methyl-, (±)-
1115 - Benzene, (butoxymethyl)-
1116 - Benzene, 1-butoxy-3-methyl-
1117 - Benzene, 1-butoxy-4-methyl-
1462 - Benzene, 1-(1,1-dimethylethyl)-4-methoxy-
1611 - Benzeneethanol, β-(1-methylethyl)-
1616 - Benzeneethanol, α,α,β-trimethyl-
1617 - Benzeneethanol, α,β,β-trimethyl-
1667 - Benzene, 1-ethoxy-2-propyl-
1846 - Benzenemethanol, α-butyl-
1855 - Benzenemethanol, α,α-diethyl-
1864 - Benzenemethanol, α-(1,1-dimethylethyl)-
1865 - Benzenemethanol, 4-(1,1-dimethylethyl)-
1904 - Benzenemethanol, α-methyl-α-propyl-
1905 - Benzenemethanol, α-(2-methylpropyl)-
2033 - Benzene, [(2-methylpropoxy)methyl]-
2101 - Benzenepentanol
2109 - Benzene, (pentyloxy)-
2150 - Benzenepropanol, α,α-dimethyl-
2151 - Benzenepropanol, β,β-dimethyl-
2152 - Benzenepropanol, α-ethyl-, (S)-
4949 - 2-Cyclopenten-1-one, 3-methyl-2-(2-pentenyl)-, (Z)-
9250 - Phenol, 2-butyl-4-methyl-
9330 - Phenol, 2-(1,1-dimethylethyl)-4-methyl-
9331 - Phenol, 2-(1,1-dimethylethyl)-5-methyl-
9332 - Phenol, 2-(1,1-dimethylethyl)-6-methyl-
9333 - Phenol, 4-(1,1-dimethylethyl)-2-methyl-

9339 - Phenol, 4-(1,1-dimethylpropyl)-
9404 - Phenol, 4-(3-methylbutyl)-
9421 - Phenol, 5-methyl-2-(1-methylpropyl)-
9443 - Phenol, pentamethyl-
9444 - Phenol, 4-pentyl-

$C_{11}H_{16}O_2$
1105 - Benzenebutanol, 4-methoxy-
1410 - Benzene, 1,4-diethoxy-2-methyl-
1442 - Benzene, 1,4-dimethoxy-2-(1-methylethyl)-
1534 - 1,3-Benzenediol, 4-(3-methylbutyl)-
1544 - 1,3-Benzenediol, 4-pentyl-
9329 - Phenol, (1,1-dimethylethyl)-4-methoxy-
11864 - Tricyclo[2.2.1.0²,⁶]heptane-1-carboxylic acid, 7,7-dimethyl-, methyl ester

$C_{11}H_{16}O_3$
1409 - Benzene, (diethoxymethoxy)-
2917 - Bicyclo[2.2.1]heptane-2-carboxylic acid, 4,7,7-trimethyl-3-oxo-
4641 - 2-Cyclohexene-1-carboxylic acid, 2,6-dimethyl-4-oxo-, ethyl ester
6058 - 2-Furancarboxylic acid, hexyl ester

$C_{11}H_{16}O_3S$
2230 - Benzenesulfonic acid, 4-methyl-, butyl ester

$C_{11}H_{16}O_4$
4915 - 1-Cyclopentene-1,2-dicarboxylic acid, diethyl ester
4916 - 1-Cyclopentene-1,3-dicarboxylic acid, diethyl ester

$C_{11}H_{17}Cl_3N_2O_2S$
7220 - 2,4-Imidazolidinedione, 5-(1-ethylpentyl)-3-[(trichloromethyl)thio]-

$C_{11}H_{17}N$
362 - Aniline, p-isopentyl-
696 - Benzenamine, N,N-diethyl-2-methyl-
697 - Benzenamine, N,N-diethyl-4-methyl-
783 - Benzenamine, N-(3-methylbutyl)-
833 - Benzenamine, 2,3,4,5,6-pentamethyl-
1576 - Benzeneethanamine, N-ethyl-α-methyl-

$C_{11}H_{17}NO$
1863 - Benzenemethanol, α-[1-(dimethylamino)ethyl]-, [R-(R*,S*)]-
1871 - Benzenemethanol, α-[1-(ethylamino)ethyl]-, (R*,S*)-
5724 - Ethanol, 2-[ethyl(3-methylphenyl)amino]-
10392 - 1-Propanol, 3-[methyl(phenylmethyl)amino]-

$C_{11}H_{17}NO_3$
1522 - 1,2-Benzenediol, 4-[1-hydroxy-2-[(1-methylethyl)amino]ethyl]-
1586 - Benzeneethanamine, 3,4,5-trimethoxy-

$C_{11}H_{17}N_3O_3$
11096 - 2,4,6(1H,3H,5H)-Pyrimidinetrione, 5-ethyl-5-(1-piperidinyl)-

$C_{11}H_{17}N_3O_3S$
2192 - Benzenesulfonamide, 4-amino-N-[(butylamino)carbonyl]-

$C_{11}H_{17}O_3P$
9545 - Phosphonic acid, (phenylmethyl)-, diethyl ester

$C_{11}H_{17}O_4PS_2$
5954 - Fensulfothion

$C_{11}H_{18}$
7792 - Naphthalene, decahydro-2-methylene-, trans-
11983 - 1-Undecen-3-yne

$C_{11}H_{18}N_2$
5504 - 1,2-Ethanediamine, N,N-dimethyl-N'-(phenylmethyl)-
7162 - Hydrazine, 1-(3-methylbutyl)-1-phenyl-

$C_{11}H_{18}N_2O_3$
11092 - 2,4,6(1H,3H,5H)-Pyrimidinetrione, 5-ethyl-5-(3-methylbutyl)-

$C_{11}H_{18}N_4O_2$
9728 - Pirimicarb

$C_{11}H_{18}O$
2959 - Bicyclo[2.2.1]heptan-2-one, 1,4,7,7-tetramethyl-
2969 - Bicyclo[3.1.1]hept-2-ene-2-ethanol, 6,6-dimethyl-

4950 - 2-Cyclopenten-1-one, 3-methyl-2-pentyl-

$C_{11}H_{18}O_2$

2949 - Bicyclo[2.2.1]heptan-2-ol, 1,7,7-trimethyl-, formate, (1R-endo)-
3131 - 3-Bornanecarboxylic acid
4965 - Cyclopropanecarboxylic acid, 2,2-dimethyl-3-(2-methyl-1-propenyl)-, methyl ester, trans-(±)-
5026 - 2,4-Decadienoic acid, methyl ester, (E,E)-
5027 - 2,4-Decadienoic acid, methyl ester, (E,Z)-
5028 - 2,4-Decadienoic acid, methyl ester, (Z,E)-
5029 - 2,4-Decadienoic acid, methyl ester, (Z,Z)-
8337 - 1,6-Octadien-3-ol, 3,7-dimethyl-, formate
8338 - 2,6-Octadien-1-ol, 3,7-dimethyl-, formate, (E)-
11988 - 6-Undecynoic acid
11989 - 9-Undecynoic acid

$C_{11}H_{18}O_2Si$

11407 - Silane, diethoxymethylphenyl-

$C_{11}H_{18}O_3Si$

11427 - Silane, ethenyltris(2-propenyloxy)-

$C_{11}H_{18}O_4$

4791 - 1,1-Cyclopentanedicarboxylic acid, diethyl ester

$C_{11}H_{18}O_5$

8771 - Pentanedioic acid, 2-acetyl-, diethyl ester
9972 - Propanedioic acid, (1-oxobutyl)-, diethyl ester

$C_{11}H_{18}O_6$

5654 - 1,1,2-Ethanetricarboxylic acid, triethyl ester

$C_{11}H_{19}ClO$

11982 - 10-Undecenoyl chloride

$C_{11}H_{19}NOS$

8528 - Octhilinone

$C_{11}H_{19}NO_2$

8148 - Nonanoic acid, 9-cyano-, methyl ester

$C_{11}H_{19}N_3O$

5032 - Ethirimol

$C_{11}H_{20}$

4340 - Cyclohexane, cyclopentyl-
8201 - 4-Nonyne, 3,3-dimethyl-
11537 - Spiro[5.5]undecane
11984 - 1-Undecyne
11985 - 2-Undecyne
11986 - 4-Undecyne
11987 - 5-Undecyne

$C_{11}H_{20}ClN_6$

11828 - 1,3,5-Triazine-2,4-diamine, 6-chloro-N,N,N',N'-tetraethyl-

$C_{11}H_{20}N_2O_3$

7224 - 4-Imidazolidinehexanoic acid, 5-methyl-2-oxo-, methyl ester, (4R-cis)-

$C_{11}H_{20}N_3O_3PS$

9730 - Pirimiphos-methyl

$C_{11}H_{20}O$

4592 - Cyclohexanone, 4-(1,1-dimethylpropyl)-
4888 - Cyclopentanone, 3-methyl-2-(3-methylbutyl)-

$C_{11}H_{20}O_2$

6152 - 2(3H)-Furanone, 5-heptyldihydro-
6404 - 3,5-Heptanedione, 2,2,6,6-tetramethyl-
8149 - Nonanoic acid, ethenyl ester
10636 - 2-Propenoic acid, 2-ethylhexyl ester
11974 - 10-Undecenoic acid
11975 - 9-Undecenoic acid, (E)-
11976 - 9-Undecenoic acid, (Z)-

$C_{11}H_{20}O_3$

8436 - Octanoic acid, 2-methyl-3-oxo-, ethyl ester

$C_{11}H_{20}O_4$

6394 - Heptanedioic acid, diethyl ester
6790 - Hexanedioic acid, 3-methyl-, diethyl ester
8134 - Nonanedioic acid, dimethyl ester
9925 - Propanedioic acid, bis(1,1-dimethylethyl) ester
9930 - Propanedioic acid, butyl-, diethyl ester
9938 - Propanedioic acid, dibutyl ester
9940 - Propanedioic acid, diethyl-, diethyl ester
9967 - Propanedioic acid, (1-methylpropyl)-, diethyl ester
9969 - Propanedioic acid, (2-methylpropyl)-, diethyl ester
9971 - Propanedioic acid, octyl-

$C_{11}H_{20}O_{10}$

6234 - D-Glucose, 6-O-α-L-arabinopyranosyl-
6237 - D-Glucose, 6-O-β-D-xylopyranosyl-

$C_{11}H_{21}BrO_2$

11944 - Undecanoic acid, 11-bromo-
11945 - Undecanoic acid, 2-bromo-

$C_{11}H_{21}FO_2$

11947 - Undecanoic acid, 11-fluoro-

$C_{11}H_{21}N$

11941 - Undecanenitrile

$C_{11}H_{21}NOS$

4168 - Cycloate

$C_{11}H_{21}NO_4$

330 - β-Alanine, N-(3-ethoxy-3-oxopropyl)-N-methyl-, ethyl ester

$C_{11}H_{21}N_5S$

5222 - Dipropetryn

$C_{11}H_{22}$

4427 - Cyclohexane, (3-methylbutyl)-
4465 - Cyclohexane, pentyl-
4823 - Cyclopentane, hexyl-
11966 - 1-Undecene
11967 - 3-Undecene
11968 - 2-Undecene, (E)-
11969 - 2-Undecene, (Z)-
11970 - 4-Undecene, (E)-
11971 - 4-Undecene, (Z)-
11972 - 5-Undecene, (E)-
11973 - 5-Undecene, (Z)-

$C_{11}H_{22}BrF$

11939 - Undecane, 1-bromo-11-fluoro-

$C_{11}H_{22}Br_2$

11940 - Undecane, 1,11-dibromo-

$C_{11}H_{22}N_2$

9683 - Piperidine, 1,1'-methylenebis-

$C_{11}H_{22}O$

11935 - Undecanal
11958 - 2-Undecanone
11959 - 3-Undecanone
11960 - 4-Undecanone
11961 - 5-Undecanone
11962 - 6-Undecanone
11979 - 10-Undecen-1-ol
11980 - 2-Undecen-4-ol
11981 - 9-Undecen-1-ol

$C_{11}H_{22}O_2$

220 - Acetic acid, nonyl ester
3527 - Butanoic acid, heptyl ester
5066 - Decanoic acid, methyl ester
6459 - Heptanoic acid, butyl ester
6467 - Heptanoic acid, 2-methylpropyl ester
6897 - Hexanoic acid, 3-methylbutyl ester
6907 - Hexanoic acid, pentyl ester
8150 - Nonanoic acid, ethyl ester
8435 - Octanoic acid, 1-methylethyl ester
8441 - Octanoic acid, propyl ester
8882 - Pentanoic acid, hexyl ester
10313 - Propanoic acid, octyl ester
11943 - Undecanoic acid

$C_{11}H_{22}O_3$

3635 - 1-Butanol, 3-methyl-, carbonate (2:1)
11948 - Undecanoic acid, 11-hydroxy-

$C_{11}H_{22}O_6$

6228 - α-D-Glucopyranoside, methyl 2,3,4,6-tetra-O-methyl-

$C_{11}H_{23}N$

9671 - Piperidine, 1-(1-ethyl-1-methylpropyl)-
9672 - Piperidine, 1-hexyl-

$C_{11}H_{23}NOS$

11586 - Sutan

$C_{11}H_{24}$

5054 - Decane, 2-methyl-
5055 - Decane, 3-methyl-
5056 - Decane, 4-methyl-
8420 - Octane, 2,2,4-trimethyl-
8421 - Octane, 2,4,7-trimethyl-
11938 - Undecane

$C_{11}H_{24}O$

8164 - 5-Nonanol, 2,8-dimethyl-
11950 - 1-Undecanol
11951 - 2-Undecanol, (±)-
11952 - 3-Undecanol, (R)-
11953 - 5-Undecanol
11954 - 6-Undecanol

$C_{11}H_{24}O_3Si$

11426 - Silane, ethenyltris(1-methylethoxy)-

$C_{11}H_{24}S$

11942 - 1-Undecanethiol

$C_{11}H_{25}N$

11937 - 1-Undecanamine

$C_{11}H_{26}O_3Si$

11480 - Silane, triethoxypentyl-

$C_{11}H_{26}O_6P_2$

9546 - Phosphonic acid, 1,3-propanediylbis-, tetraethyl ester

$C_{11}H_{28}Br_2N_2$

8738 - 1,5-Pentanediaminium, N,N,N,N',N',N'-hexamethyl-, dibromide

$C_{12}F_{10}$

3027 - 1,1'-Biphenyl, 2,2',3,3',4,4',5,5',6,6'-decafluoro-

$C_{12}F_{27}N$

3228 - 1-Butanamine, 1,1,2,2,3,3,4,4,4-nonafluoro-N,N-bis(nonafluorobutyl)-

$C_{12}H_4N_6O_{12}S$

2288 - Benzene, 1,1'-thiobis[2,4,6-trinitro-

$C_{12}H_5N_7O_{12}$

864 - Benzenamine, 2,4,6-trinitro-N-(2,4,6-trinitrophenyl)-

$C_{12}H_6Cl_2O_2$

7804 - 1,8-Naphthalenedicarbonyl dichloride

$C_{12}H_6Cl_4O_2S$

2325 - Benzene, 1,2,4-trichloro-5-[(4-chlorophenyl)sulfonyl]-
9478 - Phenol, 2,2'-thiobis[4,6-dichloro-

$C_{12}H_6Cl_4S$

11648 - Tetrasul

$C_{12}H_6O_2$

8 - 1,2-Acenaphthylenedione

$C_{12}H_6O_{12}$

1748 - Benzenehexacarboxylic acid

$C_{12}H_7BrO$

5164 - Dibenzofuran, 2-bromo-
5165 - Dibenzofuran, 3-bromo-

$C_{12}H_7Cl_2NO_3$

8081 - Nitrofen

$C_{12}H_7NO_3$

5166 - Dibenzofuran, 3-nitro-

$C_{12}H_7NO_4$

9508 - 3H-Phenoxazin-3-one, 7-hydroxy-, 10-oxide

$C_{12}H_8$

3 - Acenaphthylene
7892 - Naphthalene, 1-ethynyl-

$C_{12}H_8Br_2$

3037 - 1,1'-Biphenyl, 4,4'-dibromo-

$C_{12}H_8Br_2O$

2072 - Benzene, 1,1'-oxybis[4-bromo-

$C_{12}H_8Br_2O_2S$

2238 - Benzene, 1,1'-sulfonylbis[4-bromo-

$C_{12}H_8Br_2S$

2284 - Benzene, 1,1'-thiobis[4-bromo-

$C_{12}H_8Cl_2$

3040 - 1,1'-Biphenyl, 3,3'-dichloro-
3041 - 1,1'-Biphenyl, 4,4'-dichloro-

$C_{12}H_8Cl_2O$

2073 - Benzene, 1,1'-oxybis[4-chloro-

$C_{12}H_8Cl_2O_2S$

2239 - Benzene, 1,1'-sulfonylbis[4-chloro-
9477 - Phenol, 2,2'-thiobis[4-chloro-

$C_{12}H_8Cl_2O_3S$

2221 - Benzenesulfonic acid, 4-chloro-, 4-chlorophenyl ester
9280 - Phenol, 2,4-dichloro-, benzenesulfonate

$C_{12}H_8Cl_6$
5177 - 1,4 5,8-Dimethanonaphthalene,1,2,3,4,10,10-
hexachloro-1,4,4a,5,8,8a-hexahydro-,

$C_{12}H_8Cl_6O$
5178 - 2,7 3,6-Dimethanonaphth[2,3-b]oxirene,
3,4,5,6,9,9-hexachloro-1a,2,2a,3,6,6a,7,7a-
octahydro-,(1aα,2β,2aα,3β,6β,6aα,7β,7aα)-

$C_{12}H_8F_2$
3042 - 1,1'-Biphenyl, 2,2'-difluoro-
3043 - 1,1'-Biphenyl, 3,3'-difluoro-
3044 - 1,1'-Biphenyl, 4,4'-difluoro-

$C_{12}H_8N_2$
9189 - 1,10-Phenanthroline
9190 - 1,7-Phenanthroline
9191 - 4,7-Phenanthroline
9193 - Phenazine

$C_{12}H_8N_2O$
9197 - 1-Phenazinol

$C_{12}H_8N_2O_4$
3062 - 1,1'-Biphenyl, 2,2'-dinitro-
3063 - 1,1'-Biphenyl, 2,4'-dinitro-

$C_{12}H_8N_2O_4S_2$
5247 - Disulfide, bis(3-nitrophenyl)

$C_{12}H_8N_2O_5$
1495 - Benzene, 2,4-dinitro-1-phenoxy-

$C_{12}H_8O$
5163 - Dibenzofuran

$C_{12}H_8O_3$
8060 - Naphtho[1,2-c]furan-1,3-dione, 4,5-dihydro-

$C_{12}H_8O_4$
6187 - 7H-Furo[3,2-g][1]benzopyran-7-one, 9-
methoxy-
7805 - 1,8-Naphthalenedicarboxylic acid

$C_{12}H_8S$
5167 - Dibenzothiophene

$C_{12}H_8S_2$
11660 - Thianthrene

$C_{12}H_8S_3$
11600 - 2,2':5',2''-Terthiophene

$C_{12}H_8Se_2$
11353 - Selenanthrene

$C_{12}H_9Br$
4 - Acenaphthylene, 5-bromo-1,2-dihydro-
3016 - 1,1'-Biphenyl, 2-bromo-
3017 - 1,1'-Biphenyl, 3-bromo-
3018 - 1,1'-Biphenyl, 4-bromo-

$C_{12}H_9BrN_2O$
5120 - Diazene, (4-bromophenyl)phenyl-, 1-oxide

$C_{12}H_9BrO$
1079 - Benzene, 1-bromo-4-phenoxy-

$C_{12}H_9Cl$
5 - Acenaphthylene, 5-chloro-1,2-dihydro-
3023 - 1,1'-Biphenyl, 2-chloro-
3024 - 1,1'-Biphenyl, 3-chloro-
3025 - 1,1'-Biphenyl, 4-chloro-

$C_{12}H_9ClF_3N_3O$
8213 - Norflurazon

$C_{12}H_9ClO$
1221 - Benzene, 1-chloro-4-phenoxy-
3092 - [1,1'-Biphenyl]-2-ol, 3-chloro-

$C_{12}H_9ClO_2S$
1225 - Benzene, 1-chloro-4-(phenylsulfonyl)-

$C_{12}H_9ClO_2S_2$
2236 - Benzenesulfonothioic acid, 4-chloro-, S-
phenyl ester

$C_{12}H_9Cl_2NO_3$
12048 - Vinclozolin

$C_{12}H_9Cl_3Si$
11364 - Silane, [1,1'-biphenyl]-4-yltrichloro-

$C_{12}H_9F$
3074 - 1,1'-Biphenyl, 2-fluoro-
3075 - 1,1'-Biphenyl, 4-fluoro-

$C_{12}H_9I$
7 - Acenaphthylene, 1,2-dihydro-5-iodo-
3077 - 1,1'-Biphenyl, 2-iodo-
3078 - 1,1'-Biphenyl, 3-iodo-
3079 - 1,1'-Biphenyl, 4-iodo-

$C_{12}H_9N$
3981 - 9H-Carbazole
7736 - 1-Naphthaleneacetonitrile
7737 - 2-Naphthaleneacetonitrile

$C_{12}H_9NO$
7638 - Methanone, phenyl-2-pyridinyl-
7639 - Methanone, phenyl-4-pyridinyl-
9507 - 10H-Phenoxazine

$C_{12}H_9NO_2$
3085 - 1,1'-Biphenyl, 2-nitro-
3086 - 1,1'-Biphenyl, 3-nitro-
3087 - 1,1'-Biphenyl, 4-nitro-
6180 - Furo[2,3-b]quinoline, 4-methoxy-

$C_{12}H_9NO_2S$
2057 - Benzene, 1-nitro-2-(phenylthio)-
2058 - Benzene, 1-nitro-4-(phenylthio)-

$C_{12}H_9NO_3$
2053 - Benzene, 1-nitro-2-phenoxy-
2054 - Benzene, 1-nitro-4-phenoxy-

$C_{12}H_9NS$
9499 - 10H-Phenothiazine

$C_{12}H_9N_3O_4$
1542 - 1,3-Benzenediol, 4-[(4-nitrophenyl)azo]-

$C_{12}H_{10}$
6 - Acenaphthylene, 1,2-dihydro-
3008 - 1,1'-Biphenyl
7881 - Naphthalene, 1-ethenyl-
7882 - Naphthalene, 2-ethenyl-

$C_{12}H_{10}AsCl$
454 - Arsinous chloride, diphenyl-

$C_{12}H_{10}ClN$
678 - Benzenamine, 3-chloro-N-phenyl-
679 - Benzenamine, 4-chloro-N-phenyl-
3012 - [1,1'-Biphenyl]-4-amine, 4'-chloro-
10921 - Pyridine, 2-[(4-chlorophenyl)methyl]-

$C_{12}H_{10}ClO_3P$
9575 - Phosphorochloridic acid, diphenyl ester

$C_{12}H_{10}Cl_2N_2$
3032 - [1,1'-Biphenyl]-4,4'-diamine, 2,2'-dichloro-

$C_{12}H_{10}Cl_2Si$
11395 - Silane, dichlorodiphenyl-

$C_{12}H_{10}F_2Si$
11411 - Silane, difluorodiphenyl-

$C_{12}H_{10}Hg$
7436 - Mercury, diphenyl-

$C_{12}H_{10}N_2$
5129 - Diazene, diphenyl-, (E)-
5130 - Diazene, diphenyl-, (Z)-
11040 - 9H-Pyrido[3,4-b]indole, 1-methyl-

$C_{12}H_{10}N_2O$
829 - Benzenamine, 4-nitroso-N-phenyl-
830 - Benzenamine, N-nitroso-N-phenyl-
5131 - Diazene, diphenyl-, 1-oxide, (Z)-
9448 - Phenol, 4-(phenylazo)-

$C_{12}H_{10}N_2O_2$
826 - Benzenamine, 4-nitro-N-phenyl-
9216 - Phenol, 2,2'-azobis-

$C_{12}H_{10}N_4$
9194 - 2,3-Phenazinediamine

$C_{12}H_{10}N_4O_2$
2472 - Benzo[g]pteridine-2,4(1H,3H)-dione, 7,8-
dimethyl-

$C_{12}H_{10}O$
2071 - Benzene, 1,1'-oxybis-
3088 - [1,1'-Biphenyl]-2-ol
3089 - [1,1'-Biphenyl]-3-ol
3090 - [1,1'-Biphenyl]-4-ol
5857 - Ethanone, 1-(1-naphthalenyl)-
5858 - Ethanone, 1-(2-naphthalenyl)-
7756 - 1-Naphthalenecarboxaldehyde, 4-methyl-

$C_{12}H_{10}OS$
2183 - Benzene, 1,1'-sulfinylbis-

$C_{12}H_{10}O_2$
3064 - [1,1'-Biphenyl]-2,2'-diol
3065 - [1,1'-Biphenyl]-2,4'-diol
3066 - [1,1'-Biphenyl]-3,3'-diol
5829 - Ethanone, 1-(1-hydroxy-2-naphthalenyl)-

7734 - 1-Naphthaleneacetic acid
7755 - 1-Naphthalenecarboxaldehyde, 4-methoxy-
7766 - 1-Naphthalenecarboxylic acid, methyl ester
7776 - 2-Naphthalenecarboxylic acid, methyl ester

$C_{12}H_{10}O_2S$
2237 - Benzene, 1,1'-sulfonylbis-

$C_{12}H_{10}O_3$
217 - Acetic acid, (2-naphthalenyloxy)-
6069 - 2-Furancarboxylic acid, phenylmethyl ester
7775 - 2-Naphthalenecarboxylic acid, 3-hydroxy-,
methyl ester

$C_{12}H_{10}O_4$
3097 - [1,1'-Biphenyl]-3,3',5,5'-tetrol
8677 - 2,4-Pentadienoic acid, 5-(1,3-benzodioxol-5-
yl)-
8678 - 2,4-Pentadienoic acid, 5-(1,3-benzodioxol-5-
yl)-, (E,E)-

$C_{12}H_{10}O_4S$
9466 - Phenol, 4,4'-sulfonylbis-

$C_{12}H_{10}O_6S_2$
3070 - [1,1'-Biphenyl]-4,4'-disulfonic acid

$C_{12}H_{10}O_7$
7857 - 1,4-Naphthalenedione, 2-ethyl-3,5,6,7,8-
pentahydroxy-
7858 - 1,4-Naphthalenedione, 6-ethyl-2,3,5,7,8-
pentahydroxy-

$C_{12}H_{10}S$
2283 - Benzene, 1,1'-thiobis-

$C_{12}H_{10}S_2$
5258 - Disulfide, diphenyl

$C_{12}H_{10}Se$
2175 - Benzene, 1,1'-selenobis-

$C_{12}H_{10}Se_2$
5225 - Diselenide, diphenyl

$C_{12}H_{11}As$
443 - Arsine, diphenyl-

$C_{12}H_{11}BO$
3130 - Borinic acid, diphenyl-

$C_{12}H_{11}Cl_2NO$
10746 - Propyzamide

$C_{12}H_{11}N$
2 - 1-Acenaphthylenamine, 1,2-dihydro-
836 - Benzenamine, N-phenyl-
3010 - [1,1'-Biphenyl]-2-amine
3011 - [1,1'-Biphenyl]-4-amine
10989 - Pyridine, 2-(4-methylphenyl)-
10990 - Pyridine, 2-methyl-5-phenyl-
10991 - Pyridine, 3-methyl-2-phenyl-
11004 - Pyridine, 2-(phenylmethyl)-
11005 - Pyridine, 3-(phenylmethyl)-
11006 - Pyridine, 4-(phenylmethyl)-

$C_{12}H_{11}NO$
834 - Benzenamine, 2-phenoxy-
835 - Benzenamine, 3-phenoxy-
7733 - 1-Naphthaleneacetamide
9445 - Phenol, 2-(phenylamino)-
9446 - Phenol, 3-(phenylamino)-
9447 - Phenol, 4-(phenylamino)-

$C_{12}H_{11}NOS$
68 - Acetamide, 2-mercapto-N-2-naphthalenyl-

$C_{12}H_{11}NO_2$
66 - Acetamide, N-(2-hydroxy-1-naphthalenyl)-
2736 - 1H,5H-Benzo[ij]quinolizine-1,6(7H)-dione, 2,3-
dihydro-
8032 - 1-Naphthalenol, methylcarbamate
10625 - 2-Propenoic acid, 2-cyano-3-phenyl-, ethyl
ester
10645 - 2-Propenoic acid, 2-isocyano-3-phenyl-,
ethyl ester

$C_{12}H_{11}NO_3S$
2234 - Benzenesulfonic acid, 4-(phenylamino)-

$C_{12}H_{11}NO_4$
5218 - 1,3-Dioxolo[4,5-h]quinolin-8(9H)-one, 6-
methoxy-9-methyl-
7361 - 2H-Isoindole-2-acetic acid, 1,3-dihydro-1,3-
dioxo-, ethyl ester

$C_{12}H_{11}N_3$
837 - Benzenamine, 4-(phenylazo)-
11826 - 1-Triazene, 1,3-diphenyl-

$C_{12}H_{11}N_3O_2$
10904 - 4-Pyridinecarboxylic acid, [1-(2-furanyl)ethylidene]hydrazide

$C_{12}H_{11}N_7$
10750 - 2,4,7-Pteridinetriamine, 6-phenyl-

$C_{12}H_{11}O_3P$
9534 - Phosphonic acid, diphenyl ester

$C_{12}H_{12}$
7818 - Naphthalene, 1,2-dimethyl-
7819 - Naphthalene, 1,3-dimethyl-
7820 - Naphthalene, 1,4-dimethyl-
7821 - Naphthalene, 1,5-dimethyl-
7822 - Naphthalene, 1,6-dimethyl-
7823 - Naphthalene, 1,7-dimethyl-
7824 - Naphthalene, 1,8-dimethyl-
7825 - Naphthalene, 2,3-dimethyl-
7826 - Naphthalene, 2,6-dimethyl-
7827 - Naphthalene, 2,7-dimethyl-
7885 - Naphthalene, 1-ethyl-
7886 - Naphthalene, 2-ethyl-

$C_{12}H_{12}N_2 \cdot 2Br$
5224 - Diquat dibromide

$C_{12}H_{12}ClNO_2S$
7977 - 1-Naphthalenesulfonyl chloride, 5-(dimethylamino)-

$C_{12}H_{12}ClN_5O_4S$
4091 - Chlorsulfuron

$C_{12}H_{12}N_2$
1282 - 1,2-Benzenediamine, N-phenyl-
1283 - 1,4-Benzenediamine, N-phenyl-
3028 - [1,1'-Biphenyl]-2,2'-diamine
3029 - [1,1'-Biphenyl]-2,4'-diamine
3030 - [1,1'-Biphenyl]-3,3'-diamine
3031 - [1,1'-Biphenyl]-4,4'-diamine
7157 - Hydrazine, 1,1-diphenyl-
7158 - Hydrazine, 1,2-diphenyl-

$C_{12}H_{12}N_2O$
832 - Benzenamine, 4,4'-oxybis-

$C_{12}H_{12}N_2OS$
845 - Benzenamine, 4,4'-sulfinylbis-

$C_{12}H_{12}N_2OS_2$
11684 - 4-Thiazolidinone, 5-[[4-(dimethylamino)phenyl]methylene]-2-thioxo-

$C_{12}H_{12}N_2O_2S$
846 - Benzenamine, 4,4'-sulfonylbis-

$C_{12}H_{12}N_2O_3$
8066 - 1,8-Naphthyridine-3-carboxylic acid, 1-ethyl-1,4-dihydro-7-methyl-4-oxo-
11095 - 2,4,6(1H,3H,5H)-Pyrimidinetrione, 5-ethyl-5-phenyl-

$C_{12}H_{12}N_2S$
850 - Benzenamine, 4,4'-thiobis-

$C_{12}H_{12}N_4$
635 - Benzenamine, 4,4'-azobis-

$C_{12}H_{12}O$
6710 - 3,5-Hexadien-2-one, 6-phenyl-
7883 - Naphthalene, 1-ethoxy-
7884 - Naphthalene, 2-ethoxy-
7911 - 1-Naphthalenemethanol, α-methyl-, (±)-
8027 - 2-Naphthalenol, 1,4-dimethyl-

$C_{12}H_{12}O_2$
8045 - 1(2H)-Naphthalenone, 2-acetyl-3,4-dihydro-
10692 - 2-Propenoic acid, 3-phenyl-, 2-propenyl ester

$C_{12}H_{12}O_5$
2845 - 2H-1-Benzopyran-2-one, 6,7,8-trimethoxy-
10785 - 4H,5H-Pyrano[4,3-b]pyran-4,5-dione, 2,3-dihydro-3-hydroxy-2-methyl-7-(1-propenyl)-, [2S-[2α,3β,7(E)]]-

$C_{12}H_{12}O_6$
2370 - 1,2,4-Benzenetriol, triacetate

$C_{12}H_{12}Si$
11420 - Silane, diphenyl-

$C_{12}H_{13}ClF_3N_3O_4$
5964 - Fluchloralin

$C_{12}H_{13}ClN_4$
1284 - 1,3-Benzenediamine, 4-(phenylazo)-, monohydrochloride
11051 - 2,4-Pyrimidinediamine, 5-(4-chlorophenyl)-6-ethyl-

$C_{12}H_{13}Cl_3O_3$
258 - Acetic acid, (2,4,5-trichlorophenoxy)-, butyl ester
260 - Acetic acid, (2,4,5-trichlorophenoxy)-, 1-methylpropyl ester

$C_{12}H_{13}N$
3989 - 1H-Carbazole, 2,3,4,9-tetrahydro-
7715 - 1-Naphthalenamine, N,N-dimethyl-
7716 - 2-Naphthalenamine, N,N-dimethyl-
7718 - 1-Naphthalenamine, N-ethyl-
7719 - 2-Naphthalenamine, N-ethyl-
8064 - 2-Naphthylamine, 1,4-dimethyl-
11269 - Quinoline, 2-ethyl-3-methyl-
11270 - Quinoline, 2-ethyl-4-methyl-
11271 - Quinoline, 2-ethyl-6-methyl-
11292 - Quinoline, 2-propyl-
11300 - Quinoline, 2,3,6-trimethyl-
11301 - Quinoline, 2,3,8-trimethyl-
11302 - Quinoline, 2,4,6-trimethyl-
11303 - Quinoline, 2,4,7-trimethyl-
11304 - Quinoline, 2,5,7-trimethyl-

$C_{12}H_{13}NO$
5750 - Ethanol, 2-(2-naphthalenylamino)-

$C_{12}H_{13}NO_2$
7296 - 1H-Indole-3-butanoic acid
11156 - 2,5-Pyrrolidinedione, 1,3-dimethyl-3-phenyl-

$C_{12}H_{13}NO_2S$
4039 - Carboxin

$C_{12}H_{13}N_3$
1263 - 1,4-Benzenediamine, N-(4-aminophenyl)-
10974 - 2-Pyridinemethanamine, N-(2-pyridinylmethyl)-

$C_{12}H_{13}N_3O_4S_2$
2100 - Benzenesulfonamide, 4-amino-N-[4-(aminosulfonyl)phenyl]-

$C_{12}H_{14}$
9 - Acenaphthylene, 1,2,2a,3,4,5-hexahydro-
1246 - Benzene, 1-cyclohexen-1-yl-
1247 - Benzene, 2-cyclohexen-1-yl-
1625 - Benzene, (1-ethenyl-3-butenyl)-
1753 - Benzene, 1,3-hexadienyl-
1754 - Benzene, 1,5-hexadienyl-
1755 - Benzene, 3,5-hexadienyl-

$C_{12}H_{14}ClNO_2$
4140 - Clomazone

$C_{12}H_{14}ClN_5O_2S$
2209 - Benzenesulfonamide, 4-[(2,4-diaminophenyl)azo]-

$C_{12}H_{14}Cl_2O_3$
171 - Acetic acid, (2,4-dichlorophenoxy)-, butyl ester

$C_{12}H_{14}Cl_3O_4P$
4078 - Chlorfenvinphos

$C_{12}H_{14}N_2$
5514 - 1,2-Ethanediamine, N-1-naphthalenyl-
11213 - 6-Quinolinamine, N,N,2-trimethyl-

$C_{12}H_{14}N_2O_2$
11066 - 4,6(1H,5H)-Pyrimidinedione, 5-ethyldihydro-5-phenyl-
11923 - L-Tryptophan, N-methyl-

$C_{12}H_{14}N_2O_3$
11085 - 2,4,6(1H,3H,5H)-Pyrimidinetrione, 5-(2-cyclopenten-1-yl)-5-(2-propenyl)-

$C_{12}H_{14}N_2O_5$
6242 - L-Glutamic acid, N-(4-aminobenzoyl)-
9273 - Phenol, 2-cyclohexyl-4,6-dinitro-

$C_{12}H_{14}N_4OS$
11111 - 5H-Pyrimido[4,5-d]thiazolo[3,2-a]pyrimidine-8-ethanol, 2,7-dimethyl-

$C_{12}H_{14}N_4O_2S$
2196 - Benzenesulfonamide, 4-amino-N-(4,6-dimethyl-2-pyrimidinyl)-

$C_{12}H_{14}N_4O_4S$
2194 - Benzenesulfonamide, 4-amino-N-(2,6-dimethoxy-4-pyrimidinyl)-

$C_{12}H_{14}N_4O_4S_2$
11723 - Thiophanate-methyl

$C_{12}H_{14}O$
2460 - Benzofuran, 4-methyl-7-(1-methylethyl)-
4893 - Cyclopentanone, 2-(phenylmethyl)-
7101 - 5-Hexen-2-one, 4-phenyl-
7102 - 5-Hexen-2-one, 6-phenyl-
8051 - 1(2H)-Naphthalenone, 7-ethyl-3,4-dihydro-
8987 - 2-Pentanone, 3-(phenylmethylene)-

$C_{12}H_{14}O_2$
4135 - Cinnamic acid, β-propyl-
6038 - Furan, 2,2'-sec-butylidenedi-
10687 - 2-Propenoic acid, 3-phenyl-, 1-methylethyl ester, (E)-
10693 - 2-Propenoic acid, 3-phenyl-, propyl ester, (E)-

$C_{12}H_{14}O_3$
884 - Benzeneacetic acid, α-acetyl-, ethyl ester
2100 - Benzenepentanoic acid, δ-oxo-, methyl ester
8604 - Oxiranecarboxylic acid, 3-methyl-3-phenyl-, ethyl ester
8911 - Pentanoic acid, 4-oxo-, phenylmethyl ester
9393 - Phenol, 2-methoxy-4-(1-propenyl)-, acetate, (E)-
9394 - Phenol, 2-methoxy-4-(1-propenyl)-, acetate, (Z)-
9395 - Phenol, 2-methoxy-4-(2-propenyl)-, acetate

$C_{12}H_{14}O_4$
1259 - 1,4-Benzenediacetic acid, dimethyl ester
1342 - 1,2-Benzenedicarboxylic acid, diethyl ester
1346 - 1,3-Benzenedicarboxylic acid, diethyl ester
1350 - 1,4-Benzenedicarboxylic acid, diethyl ester
2422 - 1,3-Benzodioxole, 4,7-dimethoxy-5-(1-propenyl)-, (E)-
2423 - 1,3-Benzodioxole, 4,7-dimethoxy-5-(2-propenyl)-
8625 - Oxirane, 2,2'-[1,3-phenylenebis(oxymethylene)]bis-

$C_{12}H_{15}Br$
1009 - Benzene, 1-bromo-4-cyclohexyl-

$C_{12}H_{15}ClNO_4PS_2$
9514 - Phosalone

$C_{12}H_{15}ClO$
2907 - Benzoyl chloride, 4-pentyl-
9254 - Phenol, 2-chloro-4-cyclohexyl-

$C_{12}H_{15}ClO_2$
2908 - Benzoyl chloride, 4-(pentyloxy)-

$C_{12}H_{15}ClO_3$
10231 - Propanoic acid, 2-(4-chlorophenoxy)-2-methyl-, ethyl ester

$C_{12}H_{15}N$
964 - Benzeneacetonitrile, α-(2-methylpropyl)-
1800 - Benzenemethanamine, N-3-butynyl-N-methyl-
2737 - 1H,5H-Benzo[ij]quinolizine, 2,3,6,7-tetrahydro-

$C_{12}H_{15}NO$
9634 - Piperidine, 1-benzoyl-

$C_{12}H_{15}NO_2$
1251 - Benzene, 1-cyclohexyl-2-nitro-
1252 - Benzene, 1-cyclohexyl-4-nitro-

$C_{12}H_{15}NO_3$
3991 - Carbofuran
5214 - 1,3-Dioxolo[4,5-g]isoquinoline, 5,6,7,8-tetrahydro-4-methoxy-6-methyl-

$C_{12}H_{15}NO_4$
5216 - 1,3-Dioxolo[4,5-g]isoquinolin-5-ol, 5,6,7,8-tetrahydro-4-methoxy-6-methyl-

$C_{12}H_{15}N_3O_3$
11842 - 1,3,5-Triazine-2,4,6(1H,3H,5H)-trione, 1,3,5-tri-2-propenyl-

$C_{12}H_{15}N_3O_6$
1461 - Benzene, 1-(1,1-dimethylethyl)-3,5-dimethyl-2,4,6-trinitro-

$C_{12}H_{16}$
1248 - Benzene, cyclohexyl-
1760 - Benzene, 2-hexenyl-
1761 - Benzene, 3-hexenyl-
1762 - Benzene, 5-hexenyl-
1973 - Benzene, 1-(1-methylethenyl)-4-(1-methylethyl)-
2008 - Benzene, (1-methyl-1-pentenyl)-
4744 - 1,3,5,7-Cyclooctatetraene, 1-butyl-
7261 - 1H-Indene, 2,3-dihydro-1,1,5-trimethyl-
7262 - 1H-Indene, 2,3-dihydro-1,4,7-trimethyl-
7888 - Naphthalene, 1-ethyl-1,2,3,4-tetrahydro-
7889 - Naphthalene, 2-ethyl-1,2,3,4-tetrahydro-
7890 - Naphthalene, 5-ethyl-1,2,3,4-tetrahydro-
7891 - Naphthalene, 6-ethyl-1,2,3,4-tetrahydro-
7980 - Naphthalene, 1,2,3,4-tetrahydro-1,1-dimethyl-
7981 - Naphthalene, 1,2,3,4-tetrahydro-1,2-dimethyl-
7982 - Naphthalene, 1,2,3,4-tetrahydro-1,3-dimethyl-
7983 - Naphthalene, 1,2,3,4-tetrahydro-1,4-dimethyl-
7984 - Naphthalene, 1,2,3,4-tetrahydro-1,5-dimethyl-
7985 - Naphthalene, 1,2,3,4-tetrahydro-2,2-dimethyl-
7986 - Naphthalene, 1,2,3,4-tetrahydro-2,3-dimethyl-
7987 - Naphthalene, 1,2,3,4-tetrahydro-2,5-dimethyl-
7988 - Naphthalene, 1,2,3,4-tetrahydro-2,6-dimethyl-
7989 - Naphthalene, 1,2,3,4-tetrahydro-2,7-dimethyl-
7990 - Naphthalene, 1,2,3,4-tetrahydro-2,8-dimethyl-
7991 - Naphthalene, 1,2,3,4-tetrahydro-5,6-dimethyl-
7992 - Naphthalene, 1,2,3,4-tetrahydro-5,7-dimethyl-
7993 - Naphthalene, 1,2,3,4-tetrahydro-5,8-dimethyl-
7994 - Naphthalene, 1,2,3,4-tetrahydro-6,7-dimethyl-

$C_{12}H_{16}ClNOS$
11697 - Thiobencarb

$C_{12}H_{16}N_2$
7303 - 1H-Indole-3-ethanamine, N,N-dimethyl-

$C_{12}H_{16}N_2O$
7326 - 1H-Indol-5-ol, 3-[2-(dimethylamino)ethyl]-
7578 - 1,5-Methano-8H-pyrido[1,2-a][1,5]diazocin-8-one, 1,2,3,4,5,6-hexahydro-3-methyl-, (1R)-

$C_{12}H_{16}N_2O_3$
11083 - 2,4,6(1H,3H,5H)-Pyrimidinetrione, 5-(1-cyclohexen-1-yl)-1,5-dimethyl-
11084 - 2,4,6(1H,3H,5H)-Pyrimidinetrione, 5-(1-cyclohexen-1-yl)-5-ethyl-

$C_{12}H_{16}N_2O_5$
1463 - Benzene, 1-(1,1-dimethylethyl)-2-methoxy-4-methyl-3,5-dinitro-

$C_{12}H_{16}O$
1253 - Benzene, (cyclohexyloxy)-
3669 - 1-Butanone, 3-methyl-1-(2-methylphenyl)-
3670 - 1-Butanone, 3-methyl-1-(4-methylphenyl)-
4552 - Cyclohexanol, 1-phenyl-
4553 - Cyclohexanol, 2-phenyl-, cis-(±)-
4554 - Cyclohexanol, 2-phenyl-, trans-(±)-
5808 - Ethanone, 1-[4-(1,1-dimethylethyl)phenyl]-
5850 - Ethanone, 1-[2-methyl-5-(1-methylethyl)phenyl]-
6983 - 1-Hexanone, 1-phenyl-
6984 - 2-Hexanone, 3-phenyl-
6985 - 3-Hexanone, 2-phenyl-
8983 - 1-Pentanone, 4-methyl-1-phenyl-
8984 - 2-Pentanone, 3-methyl-1-phenyl-
9272 - Phenol, 4-cyclohexyl-

$C_{12}H_{16}O_2$
890 - Benzeneacetic acid, α-butyl-
932 - Benzeneacetic acid, 2-methylpropyl ester
1759 - Benzenehexanoic acid
2129 - Benzenepropanoic acid, α,α-dimethyl-, methyl ester
2140 - Benzenepropanoic acid, 1-methylethyl ester
2146 - Benzenepropanoic acid, β-propyl-
2685 - Benzoic acid, 3-methyl-, butyl ester
2713 - Benzoic acid, pentamethyl-
3560 - Butanoic acid, 3-methyl-, phenylmethyl ester
3632 - 1-Butanol, 3-methyl-, benzoate

7175 - Hydrocinnamic acid, β-ethyl-β-methyl-
8979 - 1-Pentanone, 1-(3-methoxyphenyl)-
9419 - Phenol, 2-methyl-5-(1-methylethyl)-, acetate
9420 - Phenol, 5-methyl-2-(1-methylethyl)-, acetate
10306 - Propanoic acid, 2-methyl-, 2-phenylethyl ester

$C_{12}H_{16}O_3$
2356 - Benzene, 1,2,4-trimethoxy-5-(1-propenyl)-
2633 - Benzoic acid, 2-hydroxy-, 3-methylbutyl ester
2649 - Benzoic acid, 2-hydroxy-, pentyl ester
2668 - Benzoic acid, 4-methoxy-, butyl ester
3581 - Butanoic acid, 4-phenoxy-, ethyl ester
3582 - Butanoic acid, 2-phenoxyethyl ester
6962 - 1-Hexanone, 1-(2,4-dihydroxyphenyl)-
8886 - Pentanoic acid, 2-methoxyphenyl ester
10641 - 2-Propenoic acid, 3-(2-furanyl)-, pentyl ester

$C_{12}H_{16}O_4$
2628 - Benzoic acid, 4-hydroxy-3-methoxy-, 2-methylpropyl ester

$C_{12}H_{16}O_7$
6224 - β-D-Glucopyranoside, 4-hydroxyphenyl

$C_{12}H_{17}BrO$
1045 - Benzene, 1-bromo-4-(hexyloxy)-

$C_{12}H_{17}N$
680 - Benzenamine, N-cyclohexyl-
9698 - Piperidine, 1-(phenylmethyl)-
9699 - Piperidine, 2-(phenylmethyl)-
9700 - Piperidine, 4-(phenylmethyl)-

$C_{12}H_{17}NO$
39 - Acetamide, N-butyl-N-phenyl-
612 - Benzamide, N,N-diethyl-3-methyl-
6722 - Hexanamide, N-phenyl-

$C_{12}H_{17}NO_2$
9769 - Promecarb

$C_{12}H_{17}NO_3$
611 - Benzamide, N,N-diethyl-4-hydroxy-3-methoxy-
880 - Benzeneacetamide, 4-butoxy-N-hydroxy-
7395 - 8-Isoquinolinol, 1,2,3,4-tetrahydro-6,7-dimethoxy-1-methyl-, (S)-

$C_{12}H_{17}N_3O_4$
6247 - L-Glutamic acid, 5-[2-[4-(hydroxymethyl)phenyl]hydrazide]

$C_{12}H_{18}$
992 - Benzene, 1,2-bis(1-methylethyl)-
993 - Benzene, 1,3-bis(1-methylethyl)-
994 - Benzene, 1,4-bis(1-methylethyl)-
1455 - Benzene, 1-(1,1-dimethylbutyl)-
1460 - Benzene, 1-(1,1-dimethylethyl)-3,5-dimethyl-
1473 - Benzene, 2,4-dimethyl-1-(1-methylpropyl)-
1557 - Benzene, 1,4-dipropyl-
1669 - Benzene, (1-ethylbutyl)-
1691 - Benzene, (1-ethyl-1-methylpropyl)-
1758 - Benzene, hexamethyl-
1763 - Benzene, hexyl-
2009 - Benzene, (1-methylpentyl)-
2010 - Benzene, (3-methylpentyl)-
2011 - Benzene, (4-methylpentyl)-
2344 - Benzene, 1,2,4-triethyl-
2345 - Benzene, 1,3,5-triethyl-
2361 - Benzene, 1,2,4-trimethyl-5-(1-methylethyl)-
4212 - 1,5,9-Cyclododecatriene
4213 - 1,5,9-Cyclododecatriene, (Z,E,E)-
4214 - 1,5,9-Cyclododecatriene, (Z,Z,Z)-
5286 - 3,9-Dodecadiyne
5287 - 5,7-Dodecadiyne
8339 - 2,6-Octadien-4-yne, 3,6-diethyl-

$C_{12}H_{18}Cl_2N_4OS$
11686 - Thiazolium, 3-[(4-amino-2-methyl-5-pyrimidinyl)methyl]-5-(2-hydroxyethyl)-4-methyl- chloride, monohydrochlorid

$C_{12}H_{18}N_2O$
11188 - 2-Pyrrolidinone, 1-[4-(1-pyrrolidinyl)-2-butynyl]-

$C_{12}H_{18}N_2O_3$
11090 - 2,4,6(1H,3H,5H)-Pyrimidinetrione, 5-(2,2-dimethylpropyl)-5-(2-propenyl)-

$C_{12}H_{18}N_2O_3S$
2207 - Benzenesulfonamide, N-[(butylamino)carbonyl]-4-methyl-

$C_{12}H_{18}N_4O$
7171 - Hydrazine, phenyl-, hemihydrate

$C_{12}H_{18}N_4O_6S$
8553 - Oryzalin

$C_{12}H_{18}O$
1488 - Benzene, 1-(1,1-dimethylpropyl)-4-methoxy-
1765 - Benzene, (hexyloxy)-
1847 - Benzenemethanol, α-butyl-α-methyl-
1875 - Benzenemethanol, α-ethyl-α-propyl-
1909 - Benzenemethanol, α-pentyl-
1951 - Benzene, [(3-methylbutoxy)methyl]-
2103 - Benzenepentanol, α-methyl-
2159 - Benzenepropanol, γ-propyl-
6949 - 3-Hexanol, 1-phenyl-
9226 - Phenol, 2,6-bis(1-methylethyl)-
9324 - Phenol, 2-(1,1-dimethylethyl)-4,5-dimethyl-
9325 - Phenol, 2-(1,1-dimethylethyl)-4,6-dimethyl-
9326 - Phenol, 4-(1,1-dimethylethyl)-2,5-dimethyl-
9327 - Phenol, 4-(1,1-dimethylethyl)-2,6-dimethyl-
9340 - Phenol, 4-(1,1-dimethylpropyl)-2-methyl-
9347 - Phenol, 2,6-dipropyl-
9425 - Phenol, 5-methyl-2-pentyl-

$C_{12}H_{18}O_2$
1519 - 1,3-Benzenediol, 4-hexyl-
1555 - Benzene, 1,2-dipropoxy-
1556 - Benzene, 1,3-dipropoxy-
2950 - Bicyclo[2.2.1]heptan-2-one, 3-(acetyl-17O)-1,7,7-trimethyl-
11863 - Tricyclo[2.2.1.02,6]heptane-1-carboxylic acid, 7,7-dimethyl, ethyl ester

$C_{12}H_{18}O_3$
2343 - Benzene, 1,3,5-triethoxy-

$C_{12}H_{18}O_4$
4644 - 1-Cyclohexene-1,2-dicarboxylic acid, diethyl ester
4645 - 1-Cyclohexene-1,3-dicarboxylic acid, diethyl ester
9935 - Propanedioic acid, 2-cyclopenten-1-yl-, diethyl ester
9936 - Propanedioic acid, cyclopentylidene-, diethyl ester

$C_{12}H_{18}O_5$
10669 - 2-Propenoic acid, 2-methyl-, oxydi-2,1-ethanediyl ester

$C_{12}H_{18}O_6$
10601 - 1-Propene-1,2,3-tricarboxylic acid, triethyl ester, (E)-

$C_{12}H_{19}ClNO_3P$
4153 - Crufomate

$C_{12}H_{19}N$
636 - Benzenamine, 2,6-bis(1-methylethyl)-
730 - Benzenamine, N,N-dipropyl-

$C_{12}H_{19}NO$
738 - Benzenamine, 3-ethoxy-N,N-diethyl-

$C_{12}H_{20}$
5342 - 1-Dodecen-3-yne

$C_{12}H_{20}As_2Cl_2O_5$
9275 - Phenol, 4,4'-(1,2-diarsenediyl)bis[2-amino-, dihydrochloride

$C_{12}H_{20}N_2O_2$
10811 - 2(1H)-Pyrazinone, 1-hydroxy-6-(1-methylpropyl)-3-(2-methylpropyl)-, (+)-

$C_{12}H_{20}N_4O_2$
6995 - Hexazinone

$C_{12}H_{20}O$
2989 - [1,1'-Bicyclohexyl]-2-one
6032 - Furan, 2,5-bis(1,1-dimethylethyl)-

$C_{12}H_{20}O_2$
2945 - Bicyclo[2.2.1]heptan-2-ol, 1,7,7-trimethyl-, acetate, endo-(±)-
2946 - Bicyclo[2.2.1]heptan-2-ol, 1,7,7-trimethyl-, acetate, exo-(±)-
2947 - Bicyclo[2.2.1]heptan-2-ol, 4,7,7-trimethyl-, acetate, (1R-endo)-

4545 - Cyclohexanol, 5-methyl-2-(1-methylethenyl)-, acetate, [1R-(1α,2β,5α)]-
4669 - 3-Cyclohexene-1-methanol, α,α,4-trimethyl-, acetate
6135 - Furan, 2-(octyloxy)-
8335 - 1,6-Octadien-3-ol, 3,7-dimethyl-, acetate, (R)-
8336 - 2,6-Octadien-1-ol, 3,7-dimethyl-, acetate, (Z)-
11990 - 9-Undecynoic acid, methyl ester
11991 - 10-Undecynoic acid, methyl ester

$C_{12}H_{20}O_3Si$
11481 - Silane, triethoxyphenyl-

$C_{12}H_{20}O_4$
3767 - 2-Butenedioic acid (E)-, bis(2-methylpropyl) ester
3774 - 2-Butenedioic acid (E)-, dibutyl ester
4350 - 1,2-Cyclohexanedicarboxylic acid, diethyl ester, $trans$-(±)-
4351 - 1,3-Cyclohexanedicarboxylic acid, diethyl ester, cis-
4355 - 1,4-Cyclohexanedicarboxylic acid, diethyl ester, cis-
4356 - 1,4-Cyclohexanedicarboxylic acid, diethyl ester, $trans$-
4795 - 1,3-Cyclopentanedicarboxylic acid, 1,2,2-trimethyl-, dimethyl ester, (1R-cis)-
5336 - 2-Dodecenedioic acid, (E)-

$C_{12}H_{20}O_4Si$
11511 - Silicic acid (H_4SiO_4), triethyl phenyl ester

$C_{12}H_{20}O_6$
10198 - 1,2,3-Propanetriol, tripropanoate

$C_{12}H_{20}O_7$
10169 - 1,2,3-Propanetricarboxylic acid, 2-hydroxy-, triethyl ester

$C_{12}H_{20}Si$
11485 - Silane, triethylphenyl-

$C_{12}H_{21}NO$
4611 - Cyclohexanone, 2-(1-piperidinylmethyl)-

$C_{12}H_{21}N_2O_3PS$
9582 - Phosphorothioic acid, O,O-diethyl O-[6-methyl-2-(1-methylethyl)-4-pyrimidinyl] ester

$C_{12}H_{22}$
2988 - 1,1'-Bicyclohexyl
4215 - Cyclododecene, (E)-
4216 - Cyclododecene, (Z)-
5101 - 4-Decyne, 3,3-dimethyl-
5285 - 1,11-Dodecadiene
5343 - 1-Dodecyne
5344 - 2-Dodecyne
5345 - 3-Dodecyne
5346 - 6-Dodecyne

$C_{12}H_{22}O$
4211 - Cyclododecanone
4461 - Cyclohexane, 1,1'-oxybis-

$C_{12}H_{22}OSi_2$
11497 - Silane, trimethyl[4-[(trimethylsilyl)oxy]phenyl]-

$C_{12}H_{22}O_2$
2948 - Bicyclo[2.2.1]heptan-2-ol, 1,7,7-trimethyl-, acetate, $endo$-
4412 - Cyclohexanehexanoic acid
4549 - Cyclohexanol, 5-methyl-2-(1-methylethyl)-, acetate, [1R-(1α,2β,5α)]-
4787 - Cyclopentanecarboxylic acid, 3-isopropyl-1-methyl-
4788 - Cyclopentanecarboxylic acid, 2-methyl-5-(1-methylethyl)-, ethyl ester
5337 - 2-Dodecenoic acid
5338 - 4-Dodecenoic acid
5339 - 5-Dodecenoic acid
5340 - 11-Dodecenoic acid
8400 - 3,6-Octanedione, 2,2,7,7-tetramethyl-
10659 - 2-Propenoic acid, 2-methyl-, 2-ethylhexyl ester
11978 - 10-Undecenoic acid, methyl ester

$C_{12}H_{22}O_2Si_2$
11447 - Silane, [1,3 phenylenebis(oxy)]bis[trimethyl-

$C_{12}H_{22}O_3$
210 - Acetic acid, [[5-methyl-2-(1-methylethyl)cyclohexyl]oxy]-, [1R-(1α,2β,5α)]-
6874 - Hexanoic acid, anhydride

$C_{12}H_{22}O_4$
3296 - Butanedioic acid, bis(1,1-dimethylethyl) ester
3297 - Butanedioic acid, bis(1-methylpropyl) ester
3301 - Butanedioic acid, dibutyl ester
3332 - Butanedioic acid, (1-methylpropyl)-, diethyl ester
5046 - Decanedioic acid, dimethyl ester
5299 - Dodecanedioic acid
5555 - Ethanedioic acid, bis(3-methylbutyl) ester
6782 - Hexanedioic acid, bis(1-methylethyl) ester
6787 - Hexanedioic acid, dipropyl ester
8387 - Octanedioic acid, diethyl ester
9951 - Propanedioic acid, ethyl(1-methylethyl)-, diethyl ester
9961 - Propanedioic acid, (3-methylbutyl)-, diethyl ester
9974 - Propanedioic acid, pentyl-, diethyl ester

$C_{12}H_{22}O_6$
3313 - Butanedioic acid, 2,3-dihydroxy- [R-(R^*,R^*)]-, dibutyl ester

$C_{12}H_{22}O_{10}$
6236 - D-Glucose, 6-O-(6-deoxy-α-L-mannopyranosyl)-

$C_{12}H_{22}O_{11}$
6022 - D-Fructopyranose, 3-O-α-D-glucopyranosyl-
6214 - β-D-Glucopyranose, 4-O-β-D-galactopyranosyl-
6218 - α-D-Glucopyranoside, β-D-fructofuranosyl
6220 - α-D-Glucopyranoside, α-D-glucopyranosyl

$C_{12}H_{22}O_{12}$
6210 - D-Gluconic acid, 4-O-β-D-galactopyranosyl-

$C_{12}H_{23}BrO_2$
5307 - Dodecanoic acid, 2-bromo-

$C_{12}H_{23}ClO$
5328 - Dodecanoyl chloride

$C_{12}H_{23}N$
4284 - Cyclohexanamino, N-cyclohexyl-
5303 - Dodecanenitrile

$C_{12}H_{23}NSi_2$
11362 - Silanamine, 1,1,1-trimethyl-N-phenyl-N-(trimethylsilyl)-

$C_{12}H_{23}P$
9516 - Phosphine, dicyclohexyl-

$C_{12}H_{24}$
4210 - Cyclododecane
4414 - Cyclohexane, hexyl-
4446 - Cyclohexane, 1-methyl-2-pentyl-
4737 - Cyclooctane, butyl-
4820 - Cyclopentane, heptyl-
5335 - 1-Dodecene
10582 - 1-Propene, 2-methyl-, trimer

$C_{12}H_{24}BrF$
5294 - Dodecane, 1-bromo-12-fluoro-

$C_{12}H_{24}Br_2$
5296 - Dodecane, 1,12-dibromo-

$C_{12}H_{24}N_2$
9661 - Piperidine, 1,1'-(1,2-ethanediyl)bis-

$C_{12}H_{24}O$
5051 - Decane, 1-(ethenyloxy)-
5288 - Dodecanal
5325 - 2-Dodecanone
5326 - 6-Dodecanone
11936 - Undecanal, 2-methyl-

$C_{12}H_{24}O_2$
157 - Acetic acid, decyl ester
3566 - Butanoic acid, octyl ester
5063 - Decanoic acid, ethyl ester
5305 - Dodecanoic acid
6470 - Heptanoic acid, pentyl ester
6890 - Hexanoic acid, hexyl ester
8155 - Nonanoic acid, 3-methyl-, ethyl ester
8427 - Octanoic acid, butyl ester

8881 - Pentanoic acid, heptyl ester

$C_{12}H_{24}O_3$
11949 - Undecanoic acid, 11-hydroxy-, methyl ester

$C_{12}H_{24}O_4Si_4$
4996 - Cyclotetrasiloxane, 2,4,6,8-tetraethenyl-2,4,6,8-tetramethyl-

$C_{12}H_{25}Br$
5293 - Dodecane, 1-bromo-

$C_{12}H_{25}Cl$
5295 - Dodecane, 1-chloro-

$C_{12}H_{25}Cl_3Si$
11460 - Silane, trichlorododecyl-

$C_{12}H_{25}I$
5302 - Dodecane, 1-iodo-

$C_{12}H_{25}NO$
5289 - Dodecanamide

$C_{12}H_{26}$
5292 - Dodecane
6433 - Heptane, 2,2,4,6,6-pentamethyl-

$C_{12}H_{26}O$
5321 - 1-Dodecanol
5322 - 2-Dodecanol
5323 - 3-Dodecanol
5324 - 6-Dodecanol
6842 - Hexane, 1,1'-oxybis-
11955 - 1-Undecanol, 2-methyl-
11956 - 3-Undecanol, 2-methyl-
11957 - 5-Undecanol, 2-methyl-

$C_{12}H_{26}O_2$
5301 - 1,12-Dodecanediol

$C_{12}H_{26}O_3$
3417 - Butane, 1,1'-[oxybis(2,1-ethanediyloxy)]bis-

$C_{12}H_{26}O_6P_2S_4$
5201 - Dioxathion

$C_{12}H_{26}S$
5304 - 1-Dodecanethiol
6855 - Hexane, 1,1'-thiobis-

$C_{12}H_{27}B$
3123 - Borane, tris(2-methylpropyl)-

$C_{12}H_{27}BO_3$
3124 - Boric acid, tributyl ester

$C_{12}H_{27}N$
3213 - 1-Butanamine, N,N-dibutyl-
5291 - 1-Dodecanamine
6727 - 1-Hexanamine, N-hexyl-
9820 - 1-Propanamine, 2-methyl-N,N-bis(2-methylpropyl)-

$C_{12}H_{27}NO$
5682 - Ethanol, 2-[bis(3-methylbutyl)amino]-

$C_{12}H_{27}NO_2$
8936 - 2-Pentanol, 1,1'-iminobis[2-methyl-

$C_{12}H_{27}OPS_3$
11852 - S,S,S-Tributyl phosphorotrothioate

$C_{12}H_{27}O_3P$
9590 - Phosphorous acid, tributyl ester

$C_{12}H_{27}O_4P$
9562 - Phosphoric acid tributyl ester
9572 - Phosphoric acid, tris(2-methylpropyl) ester

$C_{12}H_{27}P$
9520 - Phosphine, tributyl-

$C_{12}H_{27}PS_3$
7438 - Merphos

$C_{12}H_{28}IN$
9833 - 1-Propanaminium, N,N,N-tripropyl-, iodide

$C_{12}H_{28}N_2$
3262 - 1,4-Butanediamine, N,N,N',N'-tetraethyl-

$C_{12}H_{28}O_4Si$
11510 - Silicic acid (H4SiO4), tetrapropyl ester

$C_{12}H_{28}Si$
11454 - Silane, tributyl-

$C_{12}H_{28}Sn$
11546 - Stannane, tetrapropyl-

$C_{12}H_{30}OSi_2$
5235 - Disiloxane, hexaethyl-

$C_{12}H_{30}O_{13}P_4$
11645 - Tetraphosphoric acid, hexaethyl ester

$C_{12}H_{32}O_4Si_4$
4997 - Cyclotetrasiloxane, 2,4,6,8-tetraethyl-2,4,6,8-tetramethyl-

$C_{12}H_{36}O_4Si_5$
8997 - Pentasiloxane, dodecamethyl-

$C_{12}H_{36}O_6Si_6$
4624 - Cyclohexasiloxane, dodecamethyl-

$C_{13}H_5N_3O_7$
5979 - 9H-Fluoren-9-one, 2,4,7-trinitro-

$C_{13}H_6Cl_6O_2$
9409 - Phenol, 2,2'-methylenebis[3,4,6-trichloro-

$C_{13}H_7ClF_3NO_3$
8082 - Nitrofluorfen

$C_{13}H_7ClO$
5976 - 9H-Fluoren-9-one, 2-chloro-

$C_{13}H_7Cl_4NO_2$
610 - Benzamide, 3,5-dichloro-N-(3,4-dichlorophenyl)-2-hydroxy-

$C_{13}H_7F_3N_2O_5$
5980 - Fluorodifen

$C_{13}H_7NO_3$
5977 - 9H-Fluoren-9-one, 2-nitro-

$C_{13}H_8Br_2O$
7590 - Methanone, bis(4-bromophenyl)-

$C_{13}H_8Br_3NO_2$
609 - Benzamide, 3,5-dibromo-N-(4-bromophenyl)-2-hydroxy-

$C_{13}H_8ClN$
315 - Acridine, 9-chloro-

$C_{13}H_8Cl_2$
5972 - 9H-Fluorene, 2,7-dichloro-

$C_{13}H_8Cl_2O$
7591 - Methanone, bis(3-chlorophenyl)-
7592 - Methanone, bis(4-chlorophenyl)-
7603 - Methanone, (2-chlorophenyl)(4-chlorophenyl)-

$C_{13}H_8Cl_3NO_2$
608 - Benzamide, 5-chloro-N-(3,4-dichlorophenyl)-2-hydroxy-

$C_{13}H_8I_2O$
7598 - Methanone, bis(4-iodophenyl)-

$C_{13}H_8O$
5975 - 9H-Fluoren-9-one

$C_{13}H_8OS$
11812 - 9H-Thioxanthen-9-one

$C_{13}H_8O_2$
12050 - 9H-Xanthen-9-one

$C_{13}H_8O_4$
12051 - 9H-Xanthen-9-one, 1,3-dihydroxy-
12052 - 9H-Xanthen-9-one, 2,7-dihydroxy-
12053 - 9H-Xanthen-9-one, 3,6-dihydroxy-

$C_{13}H_9Br$
5968 - 9H-Fluorene, 2-bromo-

$C_{13}H_9BrO$
7600 - Methanone, (2-bromophenyl)phenyl-
7601 - Methanone, (3-bromophenyl)phenyl-
7602 - Methanone, (4-bromophenyl)phenyl-

$C_{13}H_9ClO$
7604 - Methanone, (4-chlorophenyl)phenyl-

$C_{13}H_9Cl_2FN_2S$
11802 - Thiourea, N-(3,5-dichlorophenyl)-N'-(4-fluorophenyl)-

$C_{13}H_9N$
314 - Acridine
2437 - Benzo[f]quinoline
2473 - Benzo[g]quinoline
2474 - Benzo[h]quinoline
3020 - [1,1'-Biphenyl]-2-carbonitrile
9188 - Phenanthridine

$C_{13}H_9NOS$
9217 - Phenol, 2-(2-benzothiazolyl)-

$C_{13}H_9NO_2$
9218 - Phenol, 2-(2-benzoxazolyl)-

$C_{13}H_9NO_3$
7635 - Methanone, (3-nitrophenyl)phenyl-
7636 - Methanone, (4-nitrophenyl)phenyl-

$C_{13}H_9NS$
2864 - Benzothiazole, 2-phenyl-

$C_{13}H_9N_3O_5$
2648 - Benzoic acid, 2-hydroxy-5-[(4-nitrophenyl)azo]-

$C_{13}H_{10}$
5967 - 9H-Fluorene

$C_{13}H_{10}Br_2$
1962 - Benzene, 1,1'-methylenebis[2-bromo-

$C_{13}H_{10}Cl_2$
1389 - Benzene, 1,1'-(dichloromethylene)bis-
1963 - Benzene, 1,1'-methylenebis[4-chloro-

$C_{13}H_{10}Cl_2O_2$
1965 - Benzene, 1,1'-[methylenebis(oxy)]bis[4-chloro-
9408 - Phenol, 2,2'-methylenebis[4-chloro-

$C_{13}H_{10}Cl_2S$
4075 - Chlorbenside

$C_{13}H_{10}N_2$
312 - 4-Acridinamine
313 - 9-Acridinamine
764 - Benzenamine, N,N'-methanetetraylbis-
2383 - 1H-Benzimidazole, 1-phenyl-
2384 - 1H-Benzimidazole, 2-phenyl-
4158 - Cyanamide, diphenyl-
9195 - Phenazine, 2-methyl-

$C_{13}H_{10}N_2O$
9196 - Phenazinium, 1-hydroxy-5-methyl-, hydroxide, inner salt

$C_{13}H_{10}N_2O_4$
7363 - 1H-Isoindole-1,3(2H)-dione, 2-(2,6-dioxo-3-piperidinyl)-

$C_{13}H_{10}N_4O_5$
11997 - Urea, N,N'-bis(4-nitrophenyl)-

$C_{13}H_{10}O$
7616 - Methanone, diphenyl-
12049 - 9H-Xanthene

$C_{13}H_{10}O_2$
597 - Benzaldehyde, 3-phenoxy-
2716 - Benzoic acid, phenyl ester
3021 - [1,1'-Biphenyl]-2-carboxylic acid
3022 - [1,1'-Biphenyl]-4-carboxylic acid
7622 - Methanone, (2-hydroxyphenyl)phenyl-
10675 - 2-Propenoic acid, 3-(1-naphthalenyl)-, (E)-
10710 - 2-Propen-1-one, 3-(2-furanyl)-1-phenyl-

$C_{13}H_{10}O_3$
2650 - Benzoic acid, 2-hydroxy-, phenyl ester
2714 - Benzoic acid, 2-phenoxy-
4001 - Carbonic acid, diphenyl ester
7596 - Methanone, bis(2-hydroxyphenyl)-
7597 - Methanone, bis(4-hydroxyphenyl)-
7609 - Methanone, (2,4-dihydroxyphenyl)phenyl-
8682 - 1,4-Pentadien-3-one, 1,5-di-2-furanyl-

$C_{13}H_{10}O_4$
6186 - 5H-Furo[3,2-g][1]benzopyran-5-one, 4-methoxy-7-methyl-

$C_{13}H_{10}O_6$
7610 - Methanone, (3,4-dihydroxyphenyl)(2,4,6-trihydroxyphenyl)-

$C_{13}H_{10}S$
7547 - Methanethione, diphenyl-
11811 - 9H-Thioxanthene

$C_{13}H_{11}Br$
1058 - Benzene, 1,1'-(bromomethylene)bis-

$C_{13}H_{11}Cl$
1187 - Benzene, 1,1'-(chloromethylene)bis-
1224 - Benzene, 1-chloro-4-(phenylmethyl)-
3026 - Biphenyl, 2-chloro-2'-methyl-

$C_{13}H_{11}ClO$
9270 - Phenol, 4-chloro-2-(phenylmethyl)-

$C_{13}H_{11}Cl_2NO_2$
9760 - Procymidone

$C_{13}H_{11}N$
840 - Benzenamine, N-(phenylmethylene)-
1835 - Benzenemethanimine, α-phenyl-
3986 - 9H-Carbazole, 3-methyl-
3987 - 9H-Carbazole, 9-methyl-
11000 - Pyridine, 2-(2-phenylethenyl)-, (E)-
11001 - Pyridine, 4-(2-phenylethenyl)-, (E)-

11002 - Pyridine, 2-(2-phenylethenyl)-, (Z)-

$C_{13}H_{11}NO$
626 - Benzamide, N-phenyl-
5993 - Formamide, N,N-diphenyl-
9450 - Phenol, 2-[(phenylimino)methyl]-

$C_{13}H_{11}NO_2$
617 - Benzamide, 2-hydroxy-N-phenyl-
2715 - Benzoic acid, 2-(phenylamino)-
10911 - 3-Pyridinecarboxylic acid, phenylmethyl ester

$C_{13}H_{11}NO_3$
2005 - Benzene, 1-methyl-2-(2-nitrophenoxy)-
2006 - Benzene, 1-methyl-4-(4-nitrophenoxy)-
2502 - Benzoic acid, 4-amino-2-hydroxy-, phenyl ester
6179 - Furo[2,3-b]quinoline, 4,8-dimethoxy-

$C_{13}H_{11}NS$
1129 - Benzenecarbothioamide, N-phenyl-

$C_{13}H_{11}N_3$
316 - 3,6-Acridinediamine

$C_{13}H_{12}$
632 - 1H-Benz[e]indene, 2,3-dihydro-
1961 - Benzene, 1,1'-methylenebis-
3082 - 1,1'-Biphenyl, 2-methyl-
3083 - 1,1'-Biphenyl, 3-methyl-
3084 - 1,1'-Biphenyl, 4-methyl-
7954 - Naphthalene, 1-(2-propenyl)-

$C_{13}H_{12}Cl_2N_2$
785 - Benzenamine, 4,4'-methylenebis[2-chloro-

$C_{13}H_{12}Cl_2O_4$
168 - Acetic acid, [2,3-dichloro-4-(2-methylene-1-oxobutyl)phenoxy]-

$C_{13}H_{12}N_2$
5138 - Diazene, (3-methylphenyl)phenyl-
5139 - Diazene, (4-methylphenyl)phenyl-
5971 - 9H-Fluorene-2,7-diamine
7565 - Methanimidamide, N,N'-diphenyl-

$C_{13}H_{12}N_2O$
2717 - Benzoic acid, 2-phenylhydrazide
5135 - Diazene, (2-methoxyphenyl)phenyl-
5136 - Diazene, (3-methoxyphenyl)phenyl-
5137 - Diazene, (4-methoxyphenyl)phenyl-
7589 - Methanone, bis(3-aminophenyl)-
9426 - Phenol, 4-methyl-2-(phenylazo)-
11039 - 9H-Pyrido[3,4-b]indole, 7-methoxy-1-methyl-
12010 - Urea, N,N'-diphenyl-
12011 - Urea, N,N-diphenyl-

$C_{13}H_{12}N_2O_3$
11103 - 2,4,6(1H,3H,5H)-Pyrimidinetrione, 5-phenyl-5-(2-propenyl)-

$C_{13}H_{12}N_2O_3S$
603 - Benzamide, N-[(4-aminophenyl)sulfonyl]-

$C_{13}H_{12}N_2S$
11805 - Thiourea, N,N'-diphenyl-

$C_{13}H_{12}N_4O$
5122 - Diazenecarboxylic acid, phenyl-, 2-phenylhydrazide

$C_{13}H_{12}N_4S$
5121 - Diazenecarbothioic acid, phenyl-, 2-phenylhydrazide

$C_{13}H_{12}O$
1910 - Benzenemethanol, α-phenyl-
2012 - Benzene, 1-methyl-2-phenoxy-
2013 - Benzene, 1-methyl-3-phenoxy-
3080 - 1,1'-Biphenyl, 2-methoxy-
3081 - 1,1'-Biphenyl, 4-methoxy-
5978 - 9H-Fluoren-9-one, 1,2,3,4-tetrahydro-
9454 - Phenol, 2-(phenylmethyl)-
9455 - Phenol, 4-(phenylmethyl)-
10473 - 1-Propanone, 1-(1-naphthalenyl)-
10474 - 1-Propanone, 1-(2-naphthalenyl)-

$C_{13}H_{12}O_2$
1546 - 1,4-Benzenediol, 2-(phenylmethyl)-
1938 - Benzene, 1-methoxy-2-phenoxy-
7753 - 1-Naphthalenecarboxaldehyde, 2-ethoxy-
7765 - 1-Naphthalenecarboxylic acid, ethyl ester
7770 - 2-Naphthalenecarboxylic acid, ethyl ester
7952 - 1-Naphthalenepropanoic acid

9406 - Phenol, 3,3′-methylenebis-
9407 - Phenol, 4,4′-methylenebis-
9451 - Phenol, 2-(phenylmethoxy)-
9452 - Phenol, 3-(phenylmethoxy)-
9453 - Phenol, 4-(phenylmethoxy)-

$C_{13}H_{12}O_2S$
2112 - Benzene, [(phenylmethyl)sulfonyl]-

$C_{13}H_{12}O_2S_2$
11818 - o-Toluenesulfonic acid, thio-, S-phenyl ester

$C_{13}H_{12}O_3$
5825 - Ethanone, 1-[6-hydroxy-2-(1-methylethenyl)-5-
benzofuranyl]-
7773 - 2-Naphthalenecarboxylic acid, 3-hydroxy-,
ethyl ester
7953 - 2-Naphthalenepropanoic acid, 6-hydroxy-

$C_{13}H_{12}S$
2017 - Benzene, 1-methyl-2-(phenylthio)-
2018 - Benzene, 1-methyl-3-(phenylthio)-
2019 - Benzene, 1-methyl-4-(phenylthio)-
2113 - Benzene, [(phenylmethyl)thio]-

$C_{13}H_{13}ClSi$
11380 - Silane, chloromethyldiphenyl-

$C_{13}H_{13}Cl_2N_3O_3$
7338 - Iprodione

$C_{13}H_{13}N$
807 - Benzenamine, N-methyl-N-phenyl-
808 - Benzenamine, 3-methyl-N-phenyl-
839 - Benzenamine, 4-(phenylmethyl)-
1826 - Benzenemethanamine, α-phenyl-
1827 - Benzenemethanamine, N-phenyl-
3013 - [1,1′-Biphenyl]-2-amine, 6-methyl-
3014 - [1,1′-Biphenyl]-4-amine, 3-methyl-
3015 - [1,1′-Biphenyl]-4-amine, 4′-methyl-
11003 - Pyridine, 2-(2-phenylethyl)-

$C_{13}H_{13}N_3$
6282 - Guanidine, N,N′-diphenyl-

$C_{13}H_{13}N_3O_5S_2$
3578 - Butanoic acid, 4-oxo-4-[[4-
[(2-thiazolylamino)sulfonyl]phenyl]amino]-

$C_{13}H_{13}O_3P$
9541 - Phosphonic acid, methyl-, diphenyl ester

$C_{13}H_{14}$
7887 - Naphthalene, 1-ethyl-5-methyl-
7920 - Naphthalene, 1-(1-methylethyl)-
7921 - Naphthalene, 2-(1-methylethyl)-
7957 - Naphthalene, 1-propyl-
7958 - Naphthalene, 2-propyl-
8009 - Naphthalene, 1,2,4-trimethyl-
8010 - Naphthalene, 1,2,5-trimethyl-
8011 - Naphthalene, 1,2,6-trimethyl-
8012 - Naphthalene, 1,2,7-trimethyl-
8013 - Naphthalene, 1,3,7-trimethyl-

$C_{13}H_{14}F_3N_3O_4$
5409 - Ethalfluralin

$C_{13}H_{14}N_2$
784 - Benzenamine, 4,4′-methylenebis-
7174 - Hydrazobenzene, 3-methyl-

$C_{13}H_{14}N_2O$
1913 - Benzenemethanol, α-
[(2-pyridinylamino)methyl]-

$C_{13}H_{14}N_2O_2S$
2213 - Benzenesulfonamide,
4-[(phenylmethyl)amino]-

$C_{13}H_{14}N_2O_3$
11094 - 2,4,6(1H,3H,5H)-Pyrimidinetrione, 5-ethyl-1-
methyl-5-phenyl-

$C_{13}H_{14}N_2O_{13}$
11038 - 3H-Pyrido[3,4-b]indole, 4,9-dihydro-7-
methoxy-1-methyl-

$C_{13}H_{14}N_4O$
4010 - Carbonic dihydrazide, 2,2′-diphenyl-

$C_{13}H_{14}O$
7955 - Naphthalene, 1-propoxy-
7956 - Naphthalene, 2-propoxy-

$C_{13}H_{14}OSi$
11502 - Silanol, methyldiphenyl-

$C_{13}H_{14}O_2$
8680 - 2,4-Pentadienoic acid, 5-phenyl-, ethyl ester

$C_{13}H_{14}O_3$
3576 - Butanoic acid, 3-oxo-2-(phenylmethylene)-,
ethyl ester

$C_{13}H_{14}O_5$
2774 - 3H-2-Benzopyran-7-carboxylic acid, 4,6-
dihydro-8-hydroxy-3,4,5-trimethyl-6-oxo-, (3R-
trans)-

$C_{13}H_{14}Si$
11438 - Silane, methyldiphenyl-

$C_{13}H_{15}N$
11222 - Quinoline, 2-butyl-

$C_{13}H_{15}NO_4$
11148 - 1,2-Pyrrolidinedicarboxylic acid, 1-
(phenylmethyl) ester, (S)-

$C_{13}H_{15}N_3O_2$
3945 - Carbamic acid, dimethyl-, 3-methyl-1-phenyl-
1H-pyrazol-5-yl ester

$C_{13}H_{15}N_3O_3$
7188 - Imazapyr

$C_{13}H_{16}ClNO$
4579 - Cyclohexanone, 2-(2-chlorophenyl)-2-
(methylamino)-, (±)-

$C_{13}H_{16}F_3N_3O_4$
509 - Balan
11901 - Trifluralin

$C_{13}H_{16}N_2O_2$
69 - Acetamide, N-[2-(5-methoxy-1H-indol-3-yl)ethyl]-

$C_{13}H_{16}O$
4236 - Cycloheptanone, 2-phenyl-
7100 - 1-Hexen-3-one, 5-methyl-1-phenyl-
9106 - 1-Penten-3-one, 4,4-dimethyl-1-phenyl-

$C_{13}H_{16}O_2$
1412 - Benzene, (3,3-diethoxy-1-propynyl)-
2551 - Benzoic acid, cyclohexyl ester

$C_{13}H_{16}O_3$
2132 - Benzenepropanoic acid, α-ethyl-β-oxo-, ethyl
ester

$C_{13}H_{16}O_4$
9946 - Propanedioic acid, di-2-propynyl-, diethyl ester
9975 - Propanedioic acid, phenyl-, diethyl ester

$C_{13}H_{16}O_7$
562 - Benzaldehyde, 2-(β-D-glucopyranosyloxy)-

$C_{13}H_{17}F_3N_4O_4$
9761 - Prodiamine

$C_{13}H_{17}NO$
7409 - Ketone, 1-methyl-4-piperidyl phenyl

$C_{13}H_{17}NO_4$
10932 - 3,5-Pyridinedicarboxylic acid, 2,6-dimethyl-,
diethyl ester

$C_{13}H_{17}N_3O$
10851 - 3H-Pyrazol-3-one, 4-(dimethylamino)-1,2-
dihydro-1,5-dimethyl-2-phenyl-

$C_{13}H_{18}$
7260 - 1H-Indene, 2,3-dihydro-1,1,4,7-tetramethyl-
8000 - Naphthalene, 1,2,3,4-tetrahydro-1,1,6-
trimethyl-

$C_{13}H_{18}ClN_3O_4S_2$
2853 - 2H-1,2,4-Benzothiadiazine-7-sulfonamide, 6-
chloro-3-(cyclopentylmethyl)-3,4-dihydro-, 1,1-
dioxide

$C_{13}H_{18}N_2O_2$
4900 - 1H-Cyclopentapyrimidine-2,4(3H,5H)-dione,
3-cyclohexyl-6,7-dihydro-

$C_{13}H_{18}O$
1250 - Benzene, 1-cyclohexyl-2-methoxy-
2117 - Benzenepropanal, α-methyl-4-(1-methylethyl)-
6526 - 1-Heptanone, 1-phenyl-

$C_{13}H_{18}O_2$
2610 - Benzoic acid, hexyl ester
8611 - Oxirane, [[4-(1,1-dimethylethyl)
phenoxy]methyl]-

$C_{13}H_{18}O_3$
912 - Benzeneacetic acid, α-hydroxy-, 3-methylbutyl
ester

$C_{13}H_{18}O_4$
2277 - Benzene, 1,2,3,4-tetramethoxy-5-(2-
propenyl)-

$C_{13}H_{18}O_5S$
5933 - Ethofumesate

$C_{13}H_{18}O_7$
6222 - β-D-Glucopyranoside,
2-(hydroxymethyl)phenyl

$C_{13}H_{19}N$
9697 - Piperidine, 1-(1-phenylethyl)-

$C_{13}H_{19}NO$
10428 - 1-Propanone, 2-(diethylamino)-1-phenyl-

$C_{13}H_{19}NO_2$
7390 - Isoquinoline, 1,2,3,4-tetrahydro-6,7-
dimethoxy-1,2-dimethyl-, (±)-
7391 - Isoquinoline, 1,2,3,4-tetrahydro-6,7-
dimethoxy-1,2-dimethyl-

$C_{13}H_{19}NO_3$
7394 - 8-Isoquinolinol, 1,2,3,4-tetrahydro-6,7-
dimethoxy-1,2-dimethyl-, (S)-

$C_{13}H_{19}NO_4S$
2595 - Benzoic acid, 4-[(dipropylamino)sulfonyl]-

$C_{13}H_{19}N_3O_4$
8635 - Pendimethalin

$C_{13}H_{19}N_3O_6S$
8070 - Nitralin

$C_{13}H_{20}$
1745 - Benzene, heptyl-
1948 - Benzene, 1-methyl-2,4-bis(1-methylethyl)-
1949 - Benzene, 1-methyl-3,5-bis(1-methylethyl)-
1992 - Benzene, [2-methyl-1-(1-methylethyl)propyl]-
11868 - 1,12-Tridecadiyne

$C_{13}H_{20}N_2O_2$
2493 - Benzoic acid, 4-amino-, 2-(diethylamino)ethyl
ester
5748 - Ethanol, 2-[(2-methylpropyl)amino]-, 4-
aminobenzoate (ester)

$C_{13}H_{20}O$
1747 - Benzene, (heptyloxy)-
1842 - Benzenemethanol, α,α-bis(1-methylethyl)-
1867 - Benzenemethanol, α,α-dipropyl-
1881 - Benzenemethanol, α-hexyl-
2102 - Benzenepentanol, α-ethyl-
3885 - 3-Buten-2-one, 4-(2,6,6-trimethyl-1-
cyclohexen-1-yl)-, (E)-
3886 - 3-Buten-2-one, 4-(2,6,6-trimethyl-2-
cyclohexen-1-yl)-, (E)-(±)-
4148 - o-Cresol, 4-(3-methylpentyl)-
11965 - 3,5,9-Undecatrien-2-one, 6,10-dimethyl-

$C_{13}H_{20}O_3$
6067 - 2-Furancarboxylic acid, octyl ester

$C_{13}H_{20}O_4$
9944 - Propanedioic acid, di-2-propenyl-, diethyl
ester

$C_{13}H_{20}O_8$
9988 - 1,3-Propanediol, 2,2-bis[(acetyloxy)methyl]-,
diacetate

$C_{13}H_{21}N$
10874 - Pyridine, 2,6-bis(1,1-dimethylethyl)-

$C_{13}H_{22}$
5973 - 1H-Fluorene, dodecahydro-

$C_{13}H_{22}ClN_3O$
602 - Benzamide, 4-amino-N-[2-(diethylamino)ethyl]-,
monohydrochloride

$C_{13}H_{22}NO_3PS$
5947 - Fenamiphos

$C_{13}H_{22}N_2$
4291 - Cyclohexanamine, N,N′-methanetetraylbis-

$C_{13}H_{22}N_2O$
8212 - Norea

$C_{13}H_{22}O$
3873 - 3-Buten-2-ol, 4-(2,6,6-trimethyl-1-cyclohexen-
1-yl)-
3874 - 3-Buten-2-ol, 4-(2,6,6-trimethyl-2-cyclohexen-
1-yl)-
7606 - Methanone, dicyclohexyl-

8113 - 6,8-Nonadien-2-one, 8-methyl-5-(1-methylethyl)-, [S-(E)]-

$C_{13}H_{22}O_4$
4171 - Cyclobutaneacetic acid, 3-(ethoxycarbonyl)-2,2-dimethyl-, ethyl ester
9934 - Propanedioic acid, cyclohexyl-, diethyl ester

$C_{13}H_{24}$
4429 - Cyclohexane, 1,1'-methylenebis-
11897 - 1-Tridecyne

$C_{13}H_{24}N_2O$
10415 - 2-Propanone, 1,3-bis(1-methyl-2-pyrrolidinyl)-, (R*,S*)-

$C_{13}H_{24}N_3O_3PS$
9729 - Pirimiphos-ethyl

$C_{13}H_{24}O_2$
5341 - 11-Dodecenoic acid, methyl ester
10646 - 2-Propenoic acid, isodecyl ester
11894 - 2-Tridecenoic acid
11895 - 12-Tridecenoic acid
11977 - 10-Undecenoic acid, ethyl ester

$C_{13}H_{24}O_3$
6166 - 2-Furanpropanol, tetrahydro-α-[2-(tetrahydro-2-furanyl)ethyl]-

$C_{13}H_{24}O_4$
8133 - Nonanedioic acid, diethyl ester
9955 - Propanedioic acid, hexyl-, diethyl ester

$C_{13}H_{25}BrO_2$
5309 - Dodecanoic acid, 2-bromo-, methyl ester

$C_{13}H_{25}N$
11880 - Tridecanenitrile

$C_{13}H_{26}$
4327 - Cyclohexane, 2-butyl-1,1,3-trimethyl-
4405 - Cyclohexane, heptyl-
4843 - Cyclopentane, octyl-
8184 - 4-Nonene, 5-butyl-
11893 - 1-Tridecene

$C_{13}H_{26}Br_2$
11874 - Tridecane, 1,13-dibromo-

$C_{13}H_{26}N_2$
4299 - Cyclohexanamine, 4,4'-methylenebis-
4300 - Cyclohexanamine, 4,4'-methylenebis-, [cis(cis)]-
4301 - Cyclohexanamine, 4,4'-methylenebis-, [trans(cis)]-
4302 - Cyclohexanamine, 4,4'-methylenebis-, [trans(trans)]-
9701 - Piperidine, 4,4'-(1,3-propanediyl)bis-

$C_{13}H_{26}N_2O_4$
3933 - Carbamic acid, butyl-,2-[[(aminocarbonyl)oxy]methyl]-2-methylpentyl ester

$C_{13}H_{26}O$
11869 - Tridecanal
11889 - 2-Tridecanone
11890 - 3-Tridecanone
11891 - 7-Tridecanone

$C_{13}H_{26}O_2$
5067 - Decanoic acid, 1-methylethyl ester
5069 - Decanoic acid, propyl ester
5314 - Dodecanoic acid, 2-methyl-
5315 - Dodecanoic acid, methyl ester
6465 - Heptanoic acid, hexyl ester
6889 - Hexanoic acid, heptyl ester
8147 - Nonanoic acid, butyl ester
8439 - Octanoic acid, pentyl ester
8901 - Pentanoic acid, octyl ester
11881 - Tridecanoic acid
11946 - Undecanoic acid, ethyl ester

$C_{13}H_{27}Br$
11872 - Tridecane, 1-bromo-

$C_{13}H_{27}N$
9690 - Piperidine, 1-octyl-

$C_{13}H_{28}$
8122 - Nonane, 5-butyl-
11871 - Tridecane

$C_{13}H_{28}O$
8162 - 5-Nonanol, 5-butyl-
11883 - 1-Tridecanol

11884 - 2-Tridecanol
11885 - 3-Tridecanol

$C_{13}H_{28}O_2$
11875 - 1,12-Tridecanediol
11876 - 1,13-Tridecanediol

$C_{13}H_{28}O_3$
10080 - Propane, 1,1',1''-[methylidynetris(oxy)]tris[2-methyl-

$C_{13}H_{28}O_4$
10068 - Propane, 1,1',1'',1'''-[methanetetrayltetrakis(oxy)]tetrakis-

$C_{13}H_{29}N$
6354 - 2-Heptanamine, 6-methyl-N-(3-methylbutyl)-
11870 - 1-Tridecanamine

$C_{14}H_4N_2O_2S_2$
8056 - Naphtho[2,3-b]-1,4-dithiin-2,3-dicarbonitrile, 5,10-dihydro-5,10-dioxo-

$C_{14}H_4N_4O_{12}$
401 - 9,10-Anthracenedione, 1,8-dihydroxy-2,4,5,7-tetranitro-

$C_{14}H_4O_6$
2784 - [2]Benzopyrano[6,5,4-def][2]benzopyran-1,3,6,8-tetrone

$C_{14}H_6Br_2O_2$
390 - 9,10-Anthracenedione, 2,7-dibromo-

$C_{14}H_6Cl_4O_4$
9152 - Peroxide, bis(2,4-dichlorobenzoyl)

$C_{14}H_6N_2O_6$
405 - 9,10-Anthracenedione, 1,5-dinitro-

$C_{14}H_7BrO_2$
385 - 9,10-Anthracenedione, 1-bromo-
386 - 9,10-Anthracenedione, 2-bromo-

$C_{14}H_7ClF_3NO_5$
304 - Acifluorfen

$C_{14}H_7ClO_2$
387 - 9,10-Anthracenedione, 1-chloro-
388 - 9,10-Anthracenedione, 2-chloro-

$C_{14}H_7NO_4$
410 - 9,10-Anthracenedione, 1-nitro-
411 - 9,10-Anthracenedione, 2-nitro-

$C_{14}H_7NO_6$
399 - 9,10-Anthracenedione, 1,2-dihydroxy-3-nitro-
400 - 9,10-Anthracenedione, 1,2-dihydroxy-4-nitro-

$C_{14}H_8Br_2$
375 - Anthracene, 9,10-dibromo-

$C_{14}H_8Cl_2N_4$
4139 - Clofentezine

$C_{14}H_8Cl_4$
1374 - Benzene, 1,1'-(dichloroethenylidene)bis[4-chloro-

$C_{14}H_8O_2$
379 - 9,10-Anthracenedione
9174 - 9,10-Phenanthrenedione

$C_{14}H_8O_3$
407 - 9,10-Anthracenedione, 1-hydroxy-
408 - 9,10-Anthracenedione, 2-hydroxy-
5145 - Dibenz[c,e]oxepin-5,7-dione
5969 - 9H-Fluorene-2-carboxylic acid, 9-oxo-
9175 - 9,10-Phenanthrenedione, 2-hydroxy-
9176 - 9,10-Phenanthrenedione, 3-hydroxy-

$C_{14}H_8O_4$
391 - 9,10-Anthracenedione, 1,2-dihydroxy-
392 - 9,10-Anthracenedione, 1,4-dihydroxy-
393 - 9,10-Anthracenedione, 1,5-dihydroxy-
394 - 9,10-Anthracenedione, 1,7-dihydroxy-
395 - 9,10-Anthracenedione, 1,8-dihydroxy-
396 - 9,10-Anthracenedione, 2,7-dihydroxy-

$C_{14}H_8O_5$
415 - 9,10-Anthracenedione, 1,2,3-trihydroxy-
416 - 9,10-Anthracenedione, 1,2,4-trihydroxy-
417 - 9,10-Anthracenedione, 1,2,6-trihydroxy-
418 - 9,10-Anthracenedione, 1,2,7-trihydroxy-

$C_{14}H_8O_6$
412 - 9,10-Anthracenedione, 1,2,5,6-tetrahydroxy-
413 - 9,10-Anthracenedione, 1,2,5,8-tetrahydroxy-
414 - 9,10-Anthracenedione, 1,3,5,7-tetrahydroxy-

$C_{14}H_8O_8$
406 - 9,10-Anthracenedione, 1,2,3,5,6,7-hexahydroxy-

$C_{14}H_9Br$
9164 - Phenanthrene, 9-bromo-

$C_{14}H_9Cl$
374 - Anthracene, 1-chloro-

$C_{14}H_9ClF_2N_2O_2$
5176 - Diflubenzuron

$C_{14}H_9ClO_3$
4079 - Chlorflurecol

$C_{14}H_9Cl_2NO_5$
2997 - Bifenox

$C_{14}H_9Cl_5$
2328 - Benzene, 1,1'-(2,2,2-trichloroethylidene)bis[4-chloro-

$C_{14}H_9Cl_5O$
1851 - Benzenemethanol, 4-chloro-α-(4-chlorophenyl)-α-(trichloromethyl)-

$C_{14}H_9NO_2$
380 - 9,10-Anthracenedione, 1-amino-
381 - 9,10-Anthracenedione, 2-amino-
426 - Anthracene, 9-nitro-
7366 - 1H-Isoindole-1,3(2H)-dione, 2-phenyl-

$C_{14}H_9NO_3$
383 - 9,10-Anthracenedione, 2-amino-1-hydroxy-

$C_{14}H_9NO_4$
382 - 9,10-Anthracenedione, 3-amino-1,2-dihydroxy-

$C_{14}H_{10}$
366 - Anthracene
1714 - Benzene, 1,1'-(1,2-ethynediyl)bis-
9163 - Phenanthrene

$C_{14}H_{10}Cl_2$
1365 - Benzene, 1,1'-(1,2-dichloro-1,2-ethenediyl)bis-, (E)-
1366 - Benzene, 1,1'-(1,2-dichloro-1,2-ethenediyl)bis-, (Z)-

$C_{14}H_{10}Cl_4$
1378 - Benzene, 1,1'-(2,2-dichloroethylidene)bis[4-chloro-

$C_{14}H_{10}F_3NO_2$
2729 - Benzoic acid, 2-[[3-(trifluoromethyl)phenyl]amino]-

$C_{14}H_{10}N_2O$
8564 - 1,2,4-Oxadiazole, 3,5-diphenyl-

$C_{14}H_{10}N_2O_2$
389 - 9,10-Anthracenedione, 1,5-diamino-

$C_{14}H_{10}N_2O_4$
1494 - Benzene, 1,1'-(1,2-dinitro-1,2-ethenediyl)bis-, (Z)-
1496 - Benzene, 2,4-dinitro-1-(2-phenylethenyl)-, (E)-

$C_{14}H_{10}O$
431 - 1-Anthracenol
432 - 9-Anthracenol
433 - 9(10H)-Anthracenone
5928 - Ethenone, diphenyl-

$C_{14}H_{10}O_2$
5584 - Ethanedione, diphenyl-
9173 - 3,4-Phenanthrenediol

$C_{14}H_{10}O_2S_2$
5251 - Disulfide, dibenzoyl

$C_{14}H_{10}O_3$
429 - 1,2,10-Anthracenetriol
430 - 1,8,9-Anthracenetriol
2510 - Benzoic acid, anhydride
2511 - Benzoic acid, 3-benzoyl-
2512 - Benzoic acid, 4-benzoyl-

$C_{14}H_{10}O_4$
3038 - [1,1'-Biphenyl]-2,2'-dicarboxylic acid
9157 - Peroxide, dibenzoyl

$C_{14}H_{10}O_5$
2616 - Benzoic acid, 2-hydroxy-, 2-carboxyphenyl ester
12054 - 9H-Xanthen-9-one, 1,7-dihydroxy-3-methoxy-

$C_{14}H_{10}O_6$
2979 - [Bi-1,4-cyclohexadien-1-yl]-3,3',6,6'-tetrone,
2,2'-dihydroxy-4,4'-dimethyl-

$C_{14}H_{10}O_9$
2575 - Benzoic acid, 3,4-dihydroxy-5-[(3,4,5-
trihydroxybenzoyl)oxy]-

$C_{14}H_{11}Br$
1030 - Benzene, 1,1'-(bromoethenylidene)bis-
1080 - Benzene, 1-bromo-2-(2-phenylethenyl)-, (E)-
1081 - Benzene, 1-bromo-2-(2-phenylethenyl)-, (Z)-

$C_{14}H_{11}Cl$
1146 - Benzene, 1,1'-(1-chloro-1,2-ethenediyl)bis-,
(E)-
1147 - Benzene, 1,1'-(1-chloro-1,2-ethenediyl)bis-,
(Z)-
1222 - Benzene, 1-chloro-2-(2-phenylethenyl)-, (E)-
1223 - Benzene, 1-chloro-3-(2-phenylethenyl)-, (E)-

$C_{14}H_{11}ClO$
975 - Benzeneacetyl chloride, α-phenyl-
5785 - Ethanone, 2-chloro-1,2-diphenyl-

$C_{14}H_{11}N$
365 - 2-Anthracenamine
969 - Benzeneacetonitrile, α-phenyl-
7320 - 1H-Indole, 2-phenyl-
9162 - 9-Phenanthrenamine

$C_{14}H_{11}NO$
3983 - 9H-Carbazole, 9-acetyl-

$C_{14}H_{11}NO_2$
2055 - Benzene, 1-nitro-2-(2-phenylethenyl)-, (E)-
2056 - Benzene, 1-nitro-2-(2-phenylethenyl)-, (Z)-
3982 - 9H-Carbazole-9-acetic acid

$C_{14}H_{11}NO_3$
2513 - Benzoic acid, 3-(benzoylamino)-
7627 - Methanone, (4-methyl-2-nitrophenyl)phenyl-
7630 - Methanone, (4-methylphenyl)(4-nitrophenyl)-

$C_{14}H_{11}NO_4$
2514 - Benzoic acid, 4-(benzoylamino)-2-hydroxy-
2656 - Benzoic acid, 2,2'-iminobis-

$C_{14}H_{12}$
376 - Anthracene, 9,10-dihydro-
1619 - Benzene, 1,1'-(1,2-ethenediyl)bis-, (E)-
1620 - Benzene, 1,1'-(1,2-ethenediyl)bis-, (Z)-
1636 - Benzene, 1,1'-ethenylidenebis-
5974 - 9H-Fluorene, 9-methyl-
9171 - Phenanthrene, 9,10-dihydro-

$C_{14}H_{12}Cl_2$
1678 - Benzene, 1,1'-ethylidenebis[4-chloro-

$C_{14}H_{12}Cl_2O$
1850 - Benzenemethanol, 4-chloro-α-(4-
chlorophenyl)-α-methyl-

$C_{14}H_{12}F_3NO_4S_2$
9149 - Perfluidone

$C_{14}H_{12}N_2$
2385 - 1H-Benzimidazole, 2-(phenylmethyl)-
9192 - 1,10-Phenanthroline, 2,9-dimethyl-
11196 - Quinazoline, 3,4-dihydro-3-phenyl-

$C_{14}H_{12}N_2O_2$
5495 - Ethanediamide, N,N'-diphenyl-
9449 - Phenol, 4-(phenylazo)-, acetate (ester)

$C_{14}H_{12}N_2S$
782 - Benzenamine, 4-(6-methyl-2-benzothiazolyl)-

$C_{14}H_{12}N_4O_2S$
2205 - Benzenesulfonamide, 4-amino-N-2-
quinoxalinyl-

$C_{14}H_{12}O$
877 - Benzeneacetaldehyde, α-phenyl-
5776 - Ethanone, 1-[1,1'-biphenyl]-4-yl-
5812 - Ethanone, 1,2-diphenyl-
7631 - Methanone, (2-methylphenyl)phenyl-
7632 - Methanone, (3-methylphenyl)phenyl-
7633 - Methanone, (4-methylphenyl)phenyl-

$C_{14}H_{12}O_2$
936 - Benzeneacetic acid, α-phenyl-
1545 - 1,3-Benzenediol, 5-(2-phenylethenyl)-, (E)-
2699 - Benzoic acid, 2-methylphenyl ester
2700 - Benzoic acid, 4-methylphenyl ester
2718 - Benzoic acid, 2-(phenylmethyl)-

2719 - Benzoic acid, 4-(phenylmethyl)-
2720 - Benzoic acid, phenylmethyl ester
3009 - [1,1'-Biphenyl]-4-acetic acid
3091 - [1,1'-Biphenyl]-4-ol, acetate
5822 - Ethanone, 2-hydroxy-1,2-diphenyl-, (±)-
7623 - Methanone, (2-methoxyphenyl)phenyl-
7624 - Methanone, (3-methoxyphenyl)phenyl-
7625 - Methanone, (4-methoxyphenyl)phenyl-

$C_{14}H_{12}O_3$
918 - Benzeneacetic acid, α-hydroxy-α-phenyl-
2397 - 2H,8H-Benzo[1,2-b:5,4-b']dipyran-2-one, 8,8-
dimethyl-
2651 - Benzoic acid, 2-hydroxy-, phenylmethyl ester
7620 - Methanone, (2-hydroxy-
4-methoxyphenyl)phenyl-
9383 - Phenol, 2-methoxy-, benzoate

$C_{14}H_{12}O_4$
7608 - Methanone, (2,6-dihydroxy-
4-methoxyphenyl)phenyl-
7619 - Methanone, (2-hydroxy-4-methoxyphenyl)(2-
hydroxyphenyl)-

$C_{14}H_{12}O_5$
6185 - 5H-Furo[3,2-g][1]benzopyran-5-one, 4,9-
dimethoxy-7-methyl-

$C_{14}H_{12}S_2$
11661 - Thianthrene, 2,7-dimethyl-

$C_{14}H_{13}N$
809 - Benzenamine, 2-methyl-N-(phenylmethylene)-
810 - Benzenamine, 3-methyl-N-(phenylmethylene)-
811 - Benzenamine, 4-methyl-N-(phenylmethylene)-
3985 - 9H-Carbazole, 9-ethyl-

$C_{14}H_{13}NO$
35 - Acetamide, N-[1,1'-biphenyl]-2-yl-
57 - Acetamide, N,N-diphenyl-
621 - Benzamide, N-(2-methylphenyl)-
622 - Benzamide, N-(3-methylphenyl)-
623 - Benzamide, N-(4-methylphenyl)-
7626 - Methanone, [2-(methylamino)phenyl]phenyl-

$C_{14}H_{13}NO_4$
6181 - Furo[2,3-b]quinoline, 4,7,8-trimethoxy-

$C_{14}H_{13}NO_4S$
2721 - Benzoic acid, 4-[[(phenylmethyl)
sulfonyl]amino]-

$C_{14}H_{14}$
1587 - Benzene, 1,1'-(1,2-ethanediyl)bis-
1677 - Benzene, 1,1'-ethylidenebis-
2014 - Benzene, 1-methyl-2-(phenylmethyl)-
2015 - Benzene, 1-methyl-3-(phenylmethyl)-
2016 - Benzene, 1-methyl-4-(phenylmethyl)-
3050 - 1,1'-Biphenyl, 2,2'-dimethyl-
3051 - 1,1'-Biphenyl, 2,3'-dimethyl-
3052 - 1,1'-Biphenyl, 2,3-dimethyl-
3053 - 1,1'-Biphenyl, 2,4'-dimethyl-
3054 - 1,1'-Biphenyl, 2,4-dimethyl-
3055 - 1,1'-Biphenyl, 2,5-dimethyl-
3056 - 1,1'-Biphenyl, 2,6-dimethyl-
3057 - 1,1'-Biphenyl, 3,3'-dimethyl-
3058 - 1,1'-Biphenyl, 3,4'-dimethyl-
3059 - 1,1'-Biphenyl, 3,4-dimethyl-
3060 - 1,1'-Biphenyl, 3,5-dimethyl-
3061 - 1,1'-Biphenyl, 4,4'-dimethyl-
3072 - 1,1'-Biphenyl, 2-ethyl-
3073 - 1,1'-Biphenyl, 3-ethyl-
9187 - Phenanthrene, 1,2,3,4-tetrahydro-

$C_{14}H_{14}Cl_2N_2O$
7187 - Imazalil

$C_{14}H_{14}Hg$
7432 - Mercury, bis(4-methylphenyl)-

$C_{14}H_{14}NO_4PS$
9551 - Phosphonothioic acid, phenyl-, O-ethyl O-(4-
nitrophenyl) ester

$C_{14}H_{14}N_2$
732 - Benzenamine, 4,4'-(1,2-ethenediyl)bis-, (E)-
5114 - Diazene, bis(2-methylphenyl)-
5115 - Diazene, bis(3-methylphenyl)-, (E)-

$C_{14}H_{14}N_2O$
5116 - Diazene, bis(2-methylphenyl)-, 1-oxide, (Z)-

5117 - Diazene, bis(2-methylphenyl)-, 1-oxide, (E)-
5118 - Diazene, bis(3-methylphenyl)-, 1-oxide, (Z)-
5119 - Diazene, bis(3-methylphenyl)-, 1-oxide, (E)-
5133 - Diazene, (4-ethoxyphenyl)phenyl-
10462 - 1-Propanone, 2-methyl-1,2-di-3-pyridinyl-
12018 - Urea, N-methyl-N,N'-diphenyl-

$C_{14}H_{14}N_3NaO_3S$
2222 - Benzenesulfonic acid, 4-
[[4-(dimethylamino)phenyl]azo]-, sodium salt

$C_{14}H_{14}O$
1613 - Benzeneethanol, α-phenyl-
1614 - Benzeneethanol, α-phenyl-, (S)-
1615 - Benzeneethanol, β-phenyl-
2076 - Benzene, 1,1'-oxybis[2-methyl-
2077 - Benzene, 1,1'-oxybis[3-methyl-
2078 - Benzene, 1,1'-oxybis[4-methyl-
2079 - Benzene, 1,1'-[oxybis(methylene)]bis-
3071 - 1,1'-Biphenyl, 3-ethoxy-
3674 - 1-Butanone, 1-(1-naphthalenyl)-
10464 - 1-Propanone, 2-methyl-1-(1-naphthalenyl)-
10465 - 1-Propanone, 2-methyl-1-(2-naphthalenyl)-

$C_{14}H_{14}OS$
2184 - Benzene, 1,1'-[sulfinylbis(methylene)]bis-

$C_{14}H_{14}O_2$
1588 - Benzene, 1,1'-[1,2-ethanediylbis(oxy)]bis-
3045 - 1,1'-Biphenyl, 2,2'-dimethoxy-
3046 - 1,1'-Biphenyl, 3,3'-dimethoxy-
3047 - 1,1'-Biphenyl, 4,4'-dimethoxy-
3067 - [1,1'-Biphenyl]-2,2'-diol, 3,3'-dimethyl-
3068 - [1,1'-Biphenyl]-2,2'-diol, 5,5'-dimethyl-
5575 - 1,2-Ethanediol, 1,2-diphenyl-, (R*,R*)-(±)-
7735 - 1-Naphthaleneacetic acid, ethyl ester

$C_{14}H_{14}O_2S$
2240 - Benzene, 1,1'-sulfonylbis[4-methyl-
2241 - Benzene, 1,1'-[sulfonylbis(methylene)]bis-

$C_{14}H_{14}O_3$
1929 - Benzene, 1-methoxy-2-(3-methoxyphenoxy)-
2075 - Benzene, 1,1'-oxybis[2-methoxy-
2786 - 2H-1-Benzopyran-7-ol, 3,4-dihydro-3-(4-
hydroxyphenyl)-, (S)-

$C_{14}H_{14}O_8$
2258 - 1,2,4,5-Benzenetetracarboxylic acid,
tetramethyl ester

$C_{14}H_{14}S$
1993 - Benzene, 1-methyl-2-[(4-methylphenyl)thio]-
2285 - Benzene, 1,1'-thiobis[2-methyl-
2286 - Benzene, 1,1'-thiobis[4-methyl-
2287 - Benzene, 1,1'-[thiobis(methylene)]bis-
11570 - Sulfide, m-tolyl p-tolyl

$C_{14}H_{14}S_2$
5245 - Disulfide, bis(3-methylphenyl)
5246 - Disulfide, bis(4-methylphenyl)
5248 - Disulfide, bis(phenylmethyl)

$C_{14}H_{14}Se$
2176 - Benzene, 1,1'-selenobis[4-methyl-

$C_{14}H_{15}Cl_2N$
7710 - 2-Naphthalenamine, N,N-bis(2-chloroethyl)-

$C_{14}H_{15}N$
751 - Benzenamine, N-ethyl-N-phenyl-
795 - Benzenamine, 2-methyl-N-(2-methylphenyl)-
796 - Benzenamine, 3-methyl-N-(3-methylphenyl)-
797 - Benzenamine, 4-methyl-N-(4-methylphenyl)-
838 - Benzenamine, 2-(2-phenylethyl)-
1584 - Benzeneethanamine, α-phenyl-
1821 - Benzenemethanamine, N-(2-methylphenyl)-
1824 - Benzenemethanamine, N-(3-methylphenyl)-
1825 - Benzenemethanamine, N-(4-methylphenyl)-
1828 - Benzenemethanamine, N-(phenylmethyl)-
11007 - Pyridine, 4-(3-phenylpropyl)-

$C_{14}H_{15}N_3$
721 - Benzenamine, N,N-dimethyl-4-(phenylazo)-
798 - Benzenamine, 2-methyl-4-
[(2-methylphenyl)azo]-

$C_{14}H_{15}O_3P$
9586 - Phosphorous acid, bis(phenylmethyl) ester

$C_{14}H_{16}$
422 - Anthracene, 1,2,3,4,5,6-hexahydro-

7746 - Naphthalene, 1-butyl-
7828 - Naphthalene, 1-(1,1-dimethylethyl)-
7829 - Naphthalene, 2-(1,1-dimethylethyl)-
7922 - Naphthalene, 1-methyl-4-(1-methylethyl)-
7923 - Naphthalene, 1-methyl-7-(1-methylethyl)-
7924 - Naphthalene, 7-methyl-1-(1-methylethyl)-
7927 - Naphthalene, 1-(1-methylpropyl)-
8001 - Naphthalene, 1,2,4,7-tetramethyl-
8002 - Naphthalene, 1,2,5,6-tetramethyl-
8003 - Naphthalene, 1,3,6,8-tetramethyl-

$C_{14}H_{16}ClN_3O_2$
512 - Bayleton

$C_{14}H_{16}ClN_3O_4S_2$
2851 - 2H-1,2,4-Benzothiadiazine-7-sulfonamide, 3-bicyclo[2.2.1]hept-5-en-2-yl-6-chloro-3,4-dihydro-, 1,1-dioxide

$C_{14}H_{16}ClN_5O_5S$
11824 - Triasulfuron

$C_{14}H_{16}ClO_5PS$
4147 - Coumaphos

$C_{14}H_{16}F_3N_3O_4$
9763 - Profluralin

$C_{14}H_{16}N_2$
731 - Benzenamine, 4,4'-(1,2-ethanediyl)bis-
3034 - [1,1'-Biphenyl]-4,4'-diamine, 3,3'-dimethyl-
5508 - 1,2-Ethanediamine, N,N'-diphenyl-
7138 - Hydrazine, 1,2-bis(3-methylphenyl)-
7139 - Hydrazine, 1,2-bis(4-methylphenyl)-

$C_{14}H_{16}N_2O_2$
3033 - [1,1'-Biphenyl]-4,4'-diamine, 3,3'-dimethoxy-

$C_{14}H_{16}N_2O_3$
11089 - 2,4,6(1H,3H,5H)-Pyrimidinetrione, 5,5-diethyl-1-phenyl-

$C_{14}H_{16}O_2Si$
11413 - Silane, dimethoxydiphenyl-
11415 - Silane, dimethyldiphenoxy-

$C_{14}H_{16}O_4$
9978 - Propanedioic acid, (phenylmethylene)-, diethyl ester

$C_{14}H_{16}O_9$
10786 - Pyrano[3,2-c][2]benzopyran-6(2H)-one, 3,4,4a,10b-tetrahydro-3,4,8,10-tetrahydroxy-2-(hydroxymethyl)-9-methoxy-[2R-(2α,3β,4α,4aα,10bβ)]-

$C_{14}H_{16}Si$
11416 - Silane, dimethyldiphenyl-

$C_{14}H_{17}ClNO_4PS_2$
5105 - Dialifor

$C_{14}H_{17}Cl_3O_4$
257 - Acetic acid, (2,4,5-trichlorophenoxy)-, 2-butoxyethyl ester

$C_{14}H_{17}N$
7713 - 1-Naphthalenamine, N,N-diethyl-

$C_{14}H_{17}NO_3$
11231 - 1(2H)-Quinolinecarboxylic acid, 2-ethoxy-, ethyl ester

$C_{14}H_{17}NO_6$
955 - Benzeneacetonitrile, α-(β-D-glucopyranosyloxy)-

$C_{14}H_{18}$
427 - Anthracene, 1,2,3,4,5,6,7,8-octahydro-
7747 - Naphthalene, 2-butyl-
9184 - Phenanthrene, 1,2,3,4,4a,9,10,10a-octahydro-
9185 - Phenanthrene, 1,2,3,4,5,6,7,8-octahydro-

$C_{14}H_{18}Cl_2O_4$
170 - Acetic acid, (2,4-dichlorophenoxy)-, 2-butoxyethyl ester

$C_{14}H_{18}N_2O_4$
8566 - Oxadixyl

$C_{14}H_{18}N_2O_5$
9511 - L-Phenylalanine, N-L-α-aspartyl-, 1-methyl ester

$C_{14}H_{18}N_4O_3$
515 - Benomyl

$C_{14}H_{18}O$
6347 - Heptanal, 2-(phenylmethylene)-

$C_{14}H_{18}O_4$
9977 - Propanedioic acid, (phenylmethyl)-, diethyl ester

$C_{14}H_{18}O_7$
5818 - Ethanone, 1-[4-(β-D-glucopyranosyloxy) phenyl]-

$C_{14}H_{19}Cl_2NO_2$
1101 - Benzenebutanoic acid, 4-[bis (2-chloroethyl)amino]-

$C_{14}H_{19}NO$
11265 - Quinoline, 6-ethoxy-1,2-dihydro-2,2,4-trimethyl-

$C_{14}H_{19}N_3S$
5507 - 1,2-Ethanediamine, N,N-dimethyl-N'-2-pyridinyl-N'-(2-thienylmethyl)-

$C_{14}H_{19}O_6P$
4137 - Ciodrin

$C_{14}H_{20}$
1249 - Benzene, (1-cyclohexylethyl)-
1255 - Benzene, (3-cyclopentylpropyl)-
7999 - Naphthalene, 1,2,3,4-tetrahydro-1,1,2,6-tetramethyl-

$C_{14}H_{20}ClNO_2$
270 - Acetochlor
322 - Alachlor

$C_{14}H_{20}O$
8480 - 1-Octanone, 1-phenyl-

$C_{14}H_{20}O_2$
2695 - Benzoic acid, 4-(1-methylethyl)-, 2-methylpropyl ester

$C_{14}H_{20}O_8$
5916 - Ethenetetracarboxylic acid, tetraethyl ester

$C_{14}H_{21}N_3O_4$
3893 - Butralin

$C_{14}H_{22}$
990 - Benzene, 1,4-bis(1,1-dimethylethyl)-
1981 - Benzene, (1-methylheptyl)-
2067 - Benzene, octyl-
2268 - Benzene, 1,2,3,4-tetraethyl-
2269 - Benzene, 1,2,4,5-tetraethyl-
9178 - Phenanthrene, dodecahydro-
11610 - 6,8-Tetradecadiyne

$C_{14}H_{22}N_2O$
54 - Acetamide, 2-(diethylamino)-N-(2,6-dimethylphenyl)-

$C_{14}H_{22}N_2O_3$
11081 - 2,4,6(1H,3H,5H)-Pyrimidinetrione, 5-butyl-1-cyclohexyl-

$C_{14}H_{22}O$
2069 - Benzene, (octyloxy)-
3883 - 3-Buten-2-one, 4-(2,5,6,6-tetramethyl-1-cyclohexen-1-yl)-
3884 - 3-Buten-2-one, 4-(2,5,6,6-tetramethyl-2-cyclohexen-1-yl)-
9219 - Phenol, 2,4-bis(1,1-dimethylethyl)-
9220 - Phenol, 2,6-bis(1,1-dimethylethyl)-
9227 - Phenol, 2,6-bis(1-methylpropyl)-

$C_{14}H_{22}O_2$
1314 - Benzene, 1,4-dibutoxy-
1411 - Benzene, 1-(diethoxymethyl)-4-(1-methylethyl)-
4528 - Cyclohexanol, 1,1'-(1,2-ethynediyl)bis-

$C_{14}H_{22}O_4$
5559 - Ethanedioic acid, dicyclohexyl ester

$C_{14}H_{22}O_6$
10657 - 2-Propenoic acid, 2-methyl-, 1,2-ethanediylbis(oxy-2,1-ethanediyl) ester

$C_{14}H_{22}O_8$
5633 - 1,1,2,2-Ethanetetracarboxylic acid, tetraethyl ester
10166 - 1,2,3-Propanetricarboxylic acid, 2-(acetyloxy)-, triethyl ester

$C_{14}H_{23}N$
684 - Benzenamine, N,N-dibutyl-
831 - Benzenamine, 4-octyl-
10880 - Pyridine, 4-(1-butylpentyl)-

$C_{14}H_{23}N_3O_{10}$
6257 - Glycine, N,N-bis[2-[bis(carboxymethyl) amino]ethyl]-

$C_{14}H_{24}$
5240 - Dispiro[5.1.5.1]tetradecane
9186 - Phenanthrene, tetradecahydro-

$C_{14}H_{244}NO_4PS_3$
517 - Bensulide

$C_{14}H_{24}N_2$
1285 - 1,3-Benzenediamine, N,N,N',N'-tetraethyl-

$C_{14}H_{24}O_2$
10295 - Propanoic acid, 2-methyl-, 3,7-dimethyl-2,6-octadienyl ester, (E)-

$C_{14}H_{26}$
4341 - Cyclohexane, (3-cyclopentylpropyl)-
4386 - Cyclohexane, 1,1'-(1,2-ethanediyl)bis-
4392 - Cyclohexane, 1,1'-ethylidenebis-
11640 - 2-Tetradecyne
11641 - 7-Tetradecyne

$C_{14}H_{26}O_2$
11637 - 4-Tetradecenoic acid
11638 - 5-Tetradecenoic acid
11639 - 9-Tetradecenoic acid
11896 - 12-Tridecenoic acid, methyl ester

$C_{14}H_{26}O_3$
183 - Acetic acid, ethoxy-, 5-methyl-2-(1-methylethyl)cyclohexyl ester, (1α,2β,5α)-
6456 - Heptanoic acid, anhydride

$C_{14}H_{26}O_4$
3319 - Butanedioic acid, dipentyl ester
5045 - Decanedioic acid, diethyl ester
5300 - Dodecanedioic acid, dimethyl ester
6783 - Hexanedioic acid, bis(2-methylpropyl) ester
6784 - Hexanedioic acid, dibutyl ester

$C_{14}H_{27}BrO_2$
5308 - Dodecanoic acid, 2-bromoethyl ester

$C_{14}H_{27}ClO$
11635 - Tetradecanoyl chloride

$C_{14}H_{27}N$
11619 - Tetradecanenitrile

$C_{14}H_{28}$
4460 - Cyclohexane, octyl-
4841 - Cyclopentane, nonyl-
11636 - 1-Tetradecene

$C_{14}H_{28}Br_2$
11616 - Tetradecane, 1,14-dibromo-

$C_{14}H_{28}O$
11633 - 2-Tetradecanone
11634 - 3-Tetradecanone
11892 - 3-Tridecanone, 2-methyl-
11963 - 4-Undecanone, 7-ethyl-2-methyl-

$C_{14}H_{28}O_2$
178 - Acetic acid, dodecyl ester
5312 - Dodecanoic acid, ethyl ester
6464 - Heptanoic acid, heptyl ester
6904 - Hexanoic acid, octyl ester
8156 - Nonanoic acid, pentyl ester
8431 - Octanoic acid, hexyl ester
11622 - Tetradecanoic acid
11882 - Tridecanoic acid, methyl ester

$C_{14}H_{29}Br$
11614 - Tetradecane, 1-bromo-
11873 - Tridecane, 1-bromo-12-methyl-

$C_{14}H_{29}Cl$
11615 - Tetradecane, 1-chloro-

$C_{14}H_{29}N$
4289 - Cyclohexanamine, N-(2-ethylhexyl)-
9689 - Piperidine, 1-nonyl-

$C_{14}H_{29}NO$
11611 - Tetradecanamide

$C_{14}H_{30}$
11613 - Tetradecane
11879 - Tridecane, 7-methyl-

$C_{14}H_{30}O$
6430 - Heptane, 1,1'-oxybis-
11630 - 1-Tetradecanol
11631 - 2-Tetradecanol

11632 - 3-Tetradecanol
11886 - 1-Tridecanol, 12-methyl-
11887 - 3-Tridecanol, 2-methyl-
11888 - 4-Tridecanol, 2-methyl

$C_{14}H_{30}O_2$
5297 - Dodecane, 1,1-dimethoxy-
5298 - Dodecane, 1,12-dimethoxy-
11618 - 1,14-Tetradecanediol

$C_{14}H_{30}O_2Sn$
11538 - Stannane, (acetyloxy)tributyl-

$C_{14}H_{30}O_3S$
11620 - 1-Tetradecanesulfonic acid

$C_{14}H_{30}S$
6435 - Heptane, 1,1'-thiobis-
11621 - 1-Tetradecanethiol

$C_{14}H_{31}N$
6352 - 1-Heptanamine, N-heptyl-
11612 - 1-Tetradecanamine

$C_{14}H_{32}N_2$
6750 - 1,6-Hexanediamine, N,N'-dibutyl-

$C_{14}H_{32}N_2O_4$
10379 - 2-Propanol, 1,1',1'',1'''-
(1,2-ethanediyldinitrilo)tetrakis-

$C_{14}H_{42}O_5Si_6$
6990 - Hexasiloxane, tetradecamethyl-

$C_{14}H_{42}O_7Si_7$
4239 - Cycloheptasiloxane, tetradecamethyl-

$C_{15}H_8O_4$
372 - 2-Anthracenecarboxylic acid, 9,10-dihydro-
9,10-dioxo-

$C_{15}H_8O_5$
2466 - 6H-Benzofuro[3,2-c][1]benzopyran-6-one, 3,9-
dihydroxy-

$C_{15}H_8O_6$
371 - 2-Anthracenecarboxylic acid, 9,10-dihydro-4,5-
dihydroxy-9,10-dioxo-

$C_{15}H_9BrO_2$
7265 - 1H-Indene-1,3(2H)-dione, 2-(4-bromophenyl)-

$C_{15}H_9ClO_2$
7266 - 1H-Indene-1,3(2H)-dione, 2-(4-chlorophenyl)-

$C_{15}H_{10}ClF_3N_2O_6S$
5985 - Fomesafen

$C_{15}H_{10}O_2$
368 - 1-Anthracenecarboxylic acid
369 - 2-Anthracenecarboxylic acid
370 - 9-Anthracenecarboxylic acid
409 - 9,10-Anthracenedione, 2-methyl-
2840 - 4H-1-Benzopyran-4-one, 2-phenyl-
2841 - 4H-1-Benzopyran-4-one, 3-phenyl-
7270 - 1H-Indene-1,3(2H)-dione, 2-phenyl-
7588 - Methanone, 2-benzofuranylphenyl-
9165 - 9-Phenanthrenecarboxylic acid

$C_{15}H_{10}O_3$
10199 - Propanetrione, diphenyl-

$C_{15}H_{10}O_4$
398 - 9,10-Anthracenedione, 1,8-dihydroxy-3-methyl-
2811 - 4H-1-Benzopyran-4-one, 5,7-dihydroxy-2-
phenyl-
2827 - 4H-1-Benzopyran-4-one, 7-hydroxy-3-(4-
hydroxyphenyl)-

$C_{15}H_{10}O_5$
397 - 9,10-Anthracenedione, 1,8-dihydroxy-3-
(hydroxymethyl)-
419 - 9,10-Anthracenedione, 1,3,8-trihydroxy-6-
methyl-
1332 - 1,2-Benzenedicarboxylic acid, 3-benzoyl-
2533 - Benzoic acid, 4,4'-carbonylbis-
2806 - 4H-1-Benzopyran-4-one, 5,7-dihydroxy-2-(4-
hydroxyphenyl)-
2807 - 4H-1-Benzopyran-4-one, 5,7-dihydroxy-3-(4-
hydroxyphenyl)-
2844 - 4H-1-Benzopyran-4-one, 5,6,7-trihydroxy-2-
phenyl-

$C_{15}H_{10}O_6$
2813 - 4H-1-Benzopyran-4-one, 2-
(3,4-dihydroxyphenyl)-3,7-dihydroxy-

2814 - 4H-1-Benzopyran-4-one,
2-(3,4-dihydroxyphenyl)-5,7-dihydroxy-
2842 - 4H-1-Benzopyran-4-one, 3,5,7-trihydroxy-2-
(2-hydroxyphenyl)-
2843 - 4H-1-Benzopyran-4-one, 3,5,7-trihydroxy-2-
(4-hydroxyphenyl)-

$C_{15}H_{10}O_7$
2816 - 4H-1-Benzopyran-4-one,
2-(2,4-dihydroxyphenyl)-3,5,7-trihydroxy-
2817 - 4H-1-Benzopyran-4-one,
2-(3,4-dihydroxyphenyl)-3,5,7-trihydroxy-

$C_{15}H_{11}ClF_3NO_4$
8633 - Oxyfluorfen

$C_{15}H_{11}ClN_2O$
2411 - 2H-1,4-Benzodiazepin-2-one, 7-chloro-1,3-
dihydro-5-phenyl-

$C_{15}H_{11}ClO_5$
2848 - 1-Benzopyrylium, 3,5,7-trihydroxy-2-(4-
hydroxyphenyl)-, chloride

$C_{15}H_{11}ClO_7$
2849 - 1-Benzopyrylium, 3,5,7-trihydroxy-2-(3,4,5-
trihydroxyphenyl)-, chloride

$C_{15}H_{11}I_3O_4$
2136 - Benzenepropanoic acid, 4-(4-hydroxy-3-
iodophenoxy)-3,5-diiodo-

$C_{15}H_{11}I_4NO_4$
11930 - L-Tyrosine, O-(4-hydroxy-3,5-diiodophenyl)-
3,5-diiodo-

$C_{15}H_{11}N$
2768 - Benzonitrile, 4-(2-phenylethenyl)-, (Z)-
11288 - Quinoline, 2-phenyl-
11289 - Quinoline, 3-phenyl-
11290 - Quinoline, 4-phenyl-
11291 - Quinoline, 6-phenyl-

$C_{15}H_{11}NO$
8579 - Oxazole, 2,4-diphenyl-
8580 - Oxazole, 2,5-diphenyl-
8581 - Oxazole, 4,5-diphenyl-

$C_{15}H_{11}NO_2$
384 - 9,10-Anthracenedione, 1-amino-2-methyl-
7367 - 1H-Isoindole-1,3(2H)-dione, 2-(phenylmethyl)-

$C_{15}H_{11}NS$
11676 - Thiazole, 2,4-diphenyl-

$C_{15}H_{12}$
423 - Anthracene, 1-methyl-
424 - Anthracene, 2-methyl-
425 - Anthracene, 9-methyl-
9180 - Phenanthrene, 3-methyl-
9181 - Phenanthrene, 4-methyl-

$C_{15}H_{12}I_2O_3$
2135 - Benzenepropanoic acid, 4-hydroxy-3,5-diiodo-
α-phenyl-

$C_{15}H_{12}N_2$
10830 - 1H-Pyrazole, 1,3-diphenyl-
10831 - 1H-Pyrazole, 1,5-diphenyl-

$C_{15}H_{12}N_2O$
5144 - 5H-Dibenz[b,f]azepine-5-carboxamide

$C_{15}H_{12}N_2O_2$
7219 - 2,4-Imidazolidinedione, 5,5-diphenyl-
10843 - 3,5-Pyrazolidinedione, 1,4-diphenyl-

$C_{15}H_{12}N_2O_3$
7488 - Methanediamine, 1-(2-furanyl)-N,N'-bis(2-
furanylmethylene)-

$C_{15}H_{12}N_6O_4$
2494 - Benzoic acid, 4-[[(2-amino-1,4-dihydro-4-oxo-
6-pteridinyl)methyl]formylamino]-

$C_{15}H_{12}O$
5146 - 10H-Dibenzo[a,d]cyclohepten-10-one, 5,11-
dihydro-
5147 - 5H-Dibenzo[a,d]cyclohepten-5-one, 10,11-
dihydro-
7291 - 1H-Inden-1-one, 2,3-dihydro-2-phenyl-
10708 - 2-Propen-1-one, 1,3-diphenyl-
10709 - 2-Propen-1-one, 1,3-diphenyl-, (E)-

$C_{15}H_{12}O_2$
941 - Benzeneacetic acid, α-(phenylmethylene)-
10027 - 1,3-Propanedione, 1,3-diphenyl-

10689 - 2-Propenoic acid, 3-phenyl-, phenyl ester,
(E)-
10711 - 2-Propen-1-one, 3-hydroxy-1,3-diphenyl-

$C_{15}H_{12}O_3$
2516 - Benzoic acid, 2-benzoyl-, methyl ester
2684 - Benzoic acid, 2-(4-methylbenzoyl)-

$C_{15}H_{12}O_4$
2483 - Benzoic acid, 2-(acetyloxy)-, phenyl ester
2667 - Benzoic acid, 2-(4-methoxybenzoyl)-
7509 - Methanediol, dibenzoate

$C_{15}H_{12}O_5$
2802 - 4H-1-Benzopyran-4-one, 2,3-dihydro-5,7-
dihydroxy-2-(4-hydroxyphenyl)-, (S)-

$C_{15}H_{12}O_6$
2686 - Benzoic acid, 3,3'-methylenebis[6-hydroxy-
2812 - 4H-1-Benzopyran-4-one,
2-(3,4-dihydroxyphenyl)-2,3-dihydro-5,7-
dihydroxy-, (S)-

$C_{15}H_{13}N$
7318 - 1H-Indole, 3-methyl-2-phenyl-
7321 - 1H-Indole, 1-(phenylmethyl)-

$C_{15}H_{13}NO$
64 - Acetamide, N-9H-fluoren-2-yl-

$C_{15}H_{13}NO_2S$
9500 - 10H-Phenothiazine-2-acetic acid, 10-methyl-

$C_{15}H_{14}$
1966 - Benzene, 1,1'-(1-methylene-
1,2-ethanediyl)bis-
1971 - Benzene, 1,1'-(1-methyl-1,2-ethenediyl)bis-
2161 - Benzene, 1,1'-(1-propene-1,3-diyl)bis-
2166 - Benzene, 1,1'-(1-propenylidene)bis-
7259 - 1H-Indene, 2,3-dihydro-2-phenyl-

$C_{15}H_{14}F_3N_3$
723 - Benzenamine, N,N-dimethyl-4-[[3-
(trifluoromethyl)phenyl]azo]-

$C_{15}H_{14}N_2$
765 - Benzenamine, N,N'-methanetetraylbis[4-
methyl-

$C_{15}H_{14}N_2O$
7228 - 4-Imidazolidinone, 5,5-diphenyl-

$C_{15}H_{14}N_4O_2S$
2201 - Benzenesulfonamide, 4-amino-N-(1-phenyl-
1H-pyrazol-5-yl)-

$C_{15}H_{14}O$
1939 - Benzene, 1-methoxy-4-(2-phenylethenyl)-,
(E)-
2782 - 2H-1-Benzopyran, 3,4-dihydro-2-phenyl-
7599 - Methanone, bis(4-methylphenyl)-
7614 - Methanone, (2,4-dimethylphenyl)phenyl-
7615 - Methanone, (3,4-dimethylphenyl)phenyl-
7628 - Methanone, (2-methylphenyl)
(4-methylphenyl)-
10432 - 2-Propanone, 1,1-diphenyl-
10433 - 2-Propanone, 1,3-diphenyl-

$C_{15}H_{14}O_2$
931 - Benzeneacetic acid, α-methyl-α-phenyl-
2143 - Benzenepropanoic acid, α-phenyl-, (±)-
5839 - Ethanone, 2-methoxy-1,2-diphenyl-

$C_{15}H_{14}O_3$
920 - Benzeneacetic acid, α-hydroxy-α-phenyl-,
methyl ester
3996 - Carbonic acid, bis(2-methylphenyl) ester
7612 - Methanone, (2,4-dimethoxyphenyl)phenyl-
7862 - 1,4-Naphthalenedione, 2-hydroxy-3-(3-methyl-
2-butenyl)-

$C_{15}H_{14}O_4$
2398 - 2H,8H-Benzo[1,2-b:5,4-b']dipyran-2-one, 5-
methoxy-8,8-dimethyl-
6188 - 7H-Furo[3,2-g][1]benzopyran-7-one, 3-
methoxy-2-(1-methylethyl)-
7841 - 1,4-Naphthalenediol, 2-methyl-, diacetate

$C_{15}H_{14}O_5$
7595 - Methanone, bis(2-hydroxy-4-methoxyphenyl)-
9384 - Phenol, 2-methoxy-, carbonate (2:1)
10453 - 1-Propanone, 3-(4-hydroxyphenyl)-1-(2,4,6-
trihydroxyphenyl)-

10790 - 2H-Pyran-2-one, 6-[2-(1,3-benzodioxol-5-yl)ethenyl]-5,6-dihydro-4-methoxy-, [R-(E)]-

$C_{15}H_{14}O_6$

2846 - 2H-1-Benzopyran-3,5,7-triol, 2-(3,4-dihydroxyphenyl)-3,4-dihydro-, (2S-cis)-

7248 - 2H-Indene-2,2-dicarboxylic acid, 1,3-dihydro-1,3-dioxo-, diethyl ester

7855 - 1,4-Naphthalenedione, 5,8-dihydroxy-6-methoxy-2-methyl-3-(2-oxopropyl)-

11904 - 2H,4aH-1,4,5-Trioxadicyclopent[a,hi]indene-7-carboxylic acid, 3-ethylidene-3,3a,7a,9b-tetrahydro-2-oxo-, methyl ester, [3aS-(3E,3aα,4aβ,7aβ,9aR*,9bβ)]-

$C_{15}H_{15}ClF_3N_3O$

11900 - Triflumizole

$C_{15}H_{15}ClN_2O_2$

4088 - Chloroxuron

$C_{15}H_{15}ClN_4O_6S$

4081 - Chlorimuron

$C_{15}H_{15}NO$

2771 - Benzophenone, 3-(dimethylamino)-

7613 - Methanone, [4-(dimethylamino)phenyl]phenyl-

$C_{15}H_{15}NO_2$

3946 - Carbamic acid, diphenyl-, ethyl ester

$C_{15}H_{15}N_3O$

317 - 3,9-Acridinediamine, 7-ethoxy-

$C_{15}H_{15}N_3O_2$

2587 - Benzoic acid, 2-[[4-(dimethylamino)phenyl]azo]-

$C_{15}H_{16}$

1696 - Benzene, 1-ethyl-4-(phenylmethyl)-

1964 - Benzene, 1,1′-methylenebis[4-methyl-

1969 - Benzene, 1,1′-(1-methyl-1,2-ethanediyl)bis-

1976 - Benzene, 1,1′-(1-methylethylidene)bis-

2121 - Benzene, 1,1′-(1,3-propanediyl)bis-

2171 - Benzene, 1,1′-propylidenebis-

$C_{15}H_{16}F_5NO_2S_2$

5267 - Dithiopyr

$C_{15}H_{16}N_2O$

12006 - Urea, N,N′-dimethyl-N,N′-diphenyl-

$C_{15}H_{16}O$

1940 - Benzene, 1-methoxy-4-(1-phenylethyl)-

2158 - Benzenepropanol, β-phenyl-

$C_{15}H_{16}O_2$

1970 - Benzene, 1,1′-[(1-methyl-1,2-ethanediyl)bis(oxy)]bis-

2122 - Benzene, 1,1′-[1,3-propanediylbis(oxy)]bis-

9413 - Phenol, 4,4′-(1-methylethylidene)bis-

9464 - Phenol, 4,4′-propylidenebis-

$C_{15}H_{16}O_3$

10377 - 2-Propanol, 1,3-diphenoxy-

$C_{15}H_{16}O_8$

2821 - 2H-1-Benzopyran-2-one, 7-(β-D-glucopyranosyloxy)-

$C_{15}H_{16}O_9$

2822 - 2H-1-Benzopyran-2-one, 6-(β-D-glucopyranosyloxy)-7-hydroxy-

$C_{15}H_{17}Cl_2N_3O_2$

10720 - Propiconazole

$C_{15}H_{17}N$

1811 - Benzenemethanamine, N-ethyl-N-phenyl-

$C_{15}H_{17}N_3$

6279 - Guanidine, N,N′-bis(2-methylphenyl)-

$C_{15}H_{18}$

505 - Azulene, 1,4-dimethyl-7-(1-methylethyl)-

506 - Azulene, 4,8-dimethyl-2-(1-methylethyl)-

7830 - Naphthalene, 1,6-dimethyl-4-(1-methylethyl)-

7941 - Naphthalene, 1-pentyl-

7942 - Naphthalene, 2-pentyl-

$C_{15}H_{18}Cl_2N_2O_3$

8565 - Oxadiazon

$C_{15}H_{18}N_2O_6$

3003 - Binapacryl

$C_{15}H_{18}N_6O_6S$

8069 - Nicosulfuron

$C_{15}H_{18}O$

7917 - Naphthalene, 1-(3-methylbutoxy)-

7918 - Naphthalene, 2-(3-methylbutoxy)-

7943 - Naphthalene, 1-(pentyloxy)-

7944 - Naphthalene, 2-(pentyloxy)-

$C_{15}H_{18}O_2$

2939 - Bicyclo[2.2.1]heptan-2-ol, 1-phenyl-, acetate

$C_{15}H_{18}O_3$

8057 - Naphtho[1,2-b]furan-2,8(3H,4H)-dione, 3a,5,5a,9b-tetrahydro-3,5a,9-trimethyl-,

$C_{15}H_{18}O_6$

8058 - Naphtho[1,2-b]furan-2,8(3H,4H)-dione, 3a,5,5a,9b-tetrahydro-4,5-dihydroxy-9-methoxy-3,5a-dimethyl-

$C_{15}H_{19}NO_2$

487 - 8-Azabicyclo[3.2.1]octan-3-ol, 8-methyl-, benzoate (ester), endo-

488 - 8-Azabicyclo[3.2.1]octan-3-ol, 8-methyl-, benzoate (ester), exo-

$C_{15}H_{19}N_3O_3$

7190 - Imazethapyr

$C_{15}H_{20}N_2O$

7573 - 7,14-Methano-4H,6H-dipyrido[1,2-a1′,2′-e][1,5]diazocin-4-one, 7,7a,8,9,10,11,13,14-octahydro-,

$C_{15}H_{20}O_2$

6293 - Helenine

8059 - Naphtho[2,3-b]furan-2(3H)-one, 3a,5,6,7,8,8a,9,9a-octahydro-5,8a-dimethyl-3-methylene-;

$C_{15}H_{20}O_4$

8679 - 2,4-Pentadienoic acid, 5-(1-hydroxy-2,6,6-trimethyl-4-oxo-2-cyclohexen-1-yl)-3-methyl-, [S-(Z,E)]-

9952 - Propanedioic acid, ethylphenyl-, diethyl ester

$C_{15}H_{21}NO_4$

7439 - Metalaxyl

$C_{15}H_{21}N_3O_2$

11191 - Pyrrolo[2,3-b]indol-5-ol, 1,2,3,3a,8,8a-hexahydro-1,3a,8-trimethyl-, methylcarbamate (ester), (3aS-cis)-

$C_{15}H_{22}$

1468 - Benzene, 1-(1,5-dimethyl-4-hexenyl)-4-methyl-

7274 - 1H-Indene, 5-hexyl-2,3-dihydro-

$C_{15}H_{22}ClNO_2$

7657 - Metolachlor

$C_{15}H_{22}N_2O$

9639 - 2-Piperidinecarboxamide, N-(2,6-dimethylphenyl)-1-methyl-

$C_{15}H_{22}O$

8052 - 2(3H)-Naphthalenone, 4,4a,5,6,7,8-hexahydro-4,4a-dimethyl-6-(1-methylethylidene)-, (4R-cis)-

11521 - Spiro[4.5]dec-6-en-8-one, 6,10-dimethyl-2-(1-methylethylidene)-, (5R-cis)-

$C_{15}H_{23}ClO_4$

11579 - Sulfurous acid, 2-chloroethyl 2-[4-(1,1-dimethylethyl)phenoxy]-1-methylethyl ester

$C_{15}H_{23}NO_4$

9655 - 2,6-Piperidinedione, 4-[2-(3,5-dimethyl-2-oxocyclohexyl)-2-hydroxyethyl]-,

$C_{15}H_{23}N_3O_4$

7371 - Isopropalin

$C_{15}H_{24}$

2064 - Benzene, nonyl-

2372 - Benzene, 1,2,4-tris(1-methylethyl)-

2373 - Benzene, 1,3,5-tris(1-methylethyl)-

2996 - Bicyclo[7.2.0]undec-4-ene, 4,11,11-trimethyl-8-methylene-, [1R-(1R*,4E,9S*)]-

4066 - Cedrene

4389 - Cyclohexane, 1-ethenyl-1-methyl-2,4-bis(1-methylethenyl)-, (1α,2β,4β)-

4659 - Cyclohexene, 6-ethenyl-6-methyl-1-(1-methylethyl)-3-(1-methylethylidene)-, (S)-

4685 - Cyclohexene, 1-methyl-4-(5-methyl-1-methylene-4-hexenyl)-, (S)-

4954 - Cyclopropa[d]naphthalene, 1,1a,4,4a,5,6,7,8-octahydro-2,4a,8,8-tetramethyl-, [1aS-(1aα,4aβ,8aR*)]-

5006 - 1,4,8-Cycloundecatriene, 2,6,6,9-tetramethyl-, (E,E,E)-

5329 - 1,3,6,10-Dodecatetraene, 3,7,11-trimethyl-, (E,E)-

5331 - 1,6,10-Dodecatriene, 7,11-dimethyl-3-methylene-, (E)-

7568 - 1,4-Methanoazulene, decahydro-4,8,8-trimethyl-9-methylene-, [1S-(1α,3aβ,4α,8aβ)]-

7898 - Naphthalene, 1,2,4a,5,8,8a-hexahydro-4,7-dimethyl-1-(1-methylethyl)-, [1S-(1α,4aβ,8aα)]-

7899 - Naphthalene, 1,2,3,5,6,8a-hexahydro-4,7-dimethyl-1-(1-methylethyl)-, (1S-cis)-

7936 - Naphthalene, 1,2,3,4,4a,5,6,8a-octahydro-7-methyl-4-methylene-1-(1-methylethyl)-, (1α,4aβ,8aα)-

11861 - Tricyclo[4.4.0.02,7]dec-3-ene, 1,3-dimethyl-8-(1-methylethyl)-, stereoisomer

$C_{15}H_{24}NO_4PS$

7359 - Isofenphos

$C_{15}H_{24}N_2O$

7426 - Matridin-15-one

7572 - 7,14-Methano-2H,6H-dipyrido[1,2-a1′,2′-e][1,5]diazocin-6-one, dodecahydro-, [7R-(7α,7aβ,14α,14aalp

$C_{15}H_{24}N_2O_2$

7574 - 7,14-Methano-4H,6H-dipyrido[1,2-a1′,2′-e][1,5]diazocin-4-one, dodecahydro-9-hydroxy-,

$C_{15}H_{24}O$

5330 - 2,6,10-Dodecatrienal, 3,7,11-trimethyl-

9093 - 2-Penten-1-ol, 5-(2,3-dimethyltricyclo[2.2.1.02,6]hept-3-yl)-2-methyl, stereoisomer

9103 - 2-Penten-1-ol, 2-methyl-5-(2-methyl-3-methylenebicyclo[2.2.1]hept-2-yl)-, [1S-[1α,2α(Z),4α]

9222 - Phenol, 2,4-bis(1,1-dimethylethyl)-5-methyl-

9223 - Phenol, 2,4-bis(1,1-dimethylethyl)-6-methyl-

9224 - Phenol, 2,6-bis(1,1-dimethylethyl)-4-methyl-

$C_{15}H_{24}O_8$

10139 - 1,1,2,3-Propanetetracarboxylic acid, tetraethyl ester

10140 - 1,1,3,3-Propanetetracarboxylic acid, tetraethyl ester

$C_{15}H_{25}ClN_2O_2$

2531 - Benzoic acid, 4-(butylamino)-, 2-(dimethylamino)ethyl ester, monohydrochloride

$C_{15}H_{26}N_2$

7571 - 7,14-Methano-2H,6H-dipyrido[1,2-a1′,2′-e][1,5]diazocine, dodecahydro-,

$C_{15}H_{26}O$

507 - 5-Azulenemethanol, 1,2,3,4,5,6,7,8-octahydro-α,α,3,8-tetramethyl-,

4989 - 1H-Cycloprop[e]azulen-4-ol, decahydro-1,1,4,7tetramethyl-,(1aR,4R,4aS,7R,7aS,7bS)-

5332 - 1,6,10-Dodecatrien-3-ol, 3,7,11-trimethyl-, [S-(Z)-

5333 - 2,6,10-Dodecatrien-1-ol, 3,7,11-trimethyl-, (E,E)-

5334 - 2,6,10-Dodecatrien-1-ol, 3,7,11-trimethyl-, (Z,E)-

7569 - 1,4-Methanoazulen-9-ol, decahydro-1,5,5,8a-tetramethyl-, [1R-(1α,3aβ,4α,8aβ,9S*)]

7570 - 1H-3a,7-Methanoazulen-6-ol, octahydro-3,6,8,8-tetramethyl-,

7586 - 1,6-Methanonaphthalen-1(2H)-ol, octahydro-4,8a,9,9-tetramethyl-,

$C_{15}H_{26}O_2$

3565 - Butanoic acid, 3-methyl-, 1,7,7-trimethylbicyclo[2.2.1]hept-2-yl ester, (1R-endo)-

$C_{15}H_{26}O_6$

3586 - Butanoic acid, 1,2,3-propanetriyl ester

2144 - Benzenepropanoic acid, α-(phenylmethyl)-
2145 - Benzenepropanoic acid, phenylmethyl ester
3019 - [1,1'-Biphenyl]-4-butanoic acid
5813 - Ethanone, 2-ethoxy-1,2-diphenyl-

$C_{16}H_{16}O_3$
919 - Benzeneacetic acid, α-hydroxy-α-phenyl-, ethyl
ester
3583 - Butanoic acid, 2-phenoxy-, phenyl ester

$C_{16}H_{16}O_4$
7806 - 1,8-Naphthalenedicarboxylic acid, diethyl
ester

$C_{16}H_{16}O_5$
7854 - 1,4-Naphthalenedione, 5,8-dihydroxy-6-(1-
hydroxy-4-methyl-3-pentenyl)-, (S)-

$C_{16}H_{16}Si$
11404 - Silane, diethenyldiphenyl-

$C_{16}H_{17}N$
722 - Benzenamine, N,N-dimethyl-4-
(2-phenylethenyl)-, (E)-
3984 - 9H-Carbazole, 9-butyl-

$C_{16}H_{17}NO$
5219 - Diphenamid

$C_{16}H_{17}NO_3$
7665 - Morphinan-3,6-diol, 7,8-didehydro-4,5-epoxy-,
(5α,6α)-

$C_{16}H_{17}NO_4$
6201 - Galanthan-1,2-diol, 3,12-didehydro-9,10-
[methylenebis(oxy)]-, (1α,2β)-

$C_{16}H_{17}N_2NaO_4S$
11655 - 4-Thia-1-azabicyclo[3.2.0]heptane-2-
carboxylic acid, 3,3-dimethyl-7-oxo-6-
[(phenylacetyl)amino]-

$C_{16}H_{17}N_3O$
5374 - Ergoline-8-carboxamide, 9,10-didehydro-6-
methyl-, (8β)-

$C_{16}H_{18}$
1099 - Benzene, 1,1'-(1,4-butanediyl)bis-
1120 - Benzene, 1,1'-butylidenebis-
1589 - Benzene, 1,1'-(1,2-ethanediyl)bis[3-methyl-
1590 - Benzene, 1,1'-(1,2-ethanediyl)bis[4-methyl-
1676 - Benzene, 1,1'-(1-ethyl-1,2-ethanediyl)bis-
1979 - Benzene, 1-(1-methylethyl)-4-(phenylmethyl)-
2020 - Benzene, 1,1'-(1-methyl-1,3-propanediyl)bis-,
(±)-
2021 - Benzene, 1,1'-(2-methyl-1,3-propanediyl)bis-
2111 - Benzene, 1-(phenylmethyl)-4-propyl-
3095 - 1,1'-Biphenyl, 2,2',4,4'-tetramethyl-
3096 - 1,1'-Biphenyl, 2,2',5,5'-tetramethyl-

$C_{16}H_{18}ClN_3S$
9506 - Phenothiazin-5-ium, 3,7-bis(dimethylamino)-,
chloride

$C_{16}H_{18}N_2$
9613 - Piperazine, 1,4-diphenyl-

$C_{16}H_{18}N_2O_2$
5112 - Diazene, bis(2-ethoxyphenyl)-
5113 - Diazene, bis(4-ethoxyphenyl)-

$C_{16}H_{18}N_2O_3$
6144 - 2(3H)-Furanone, dihydro-
3-(hydroxyphenylmethyl)-4-[(1-methyl-1H-
imidazol-5-yl)methyl]-,

$C_{16}H_{18}N_4O_2$
10910 - 4-Pyridinecarboxylic acid, 2-[3-oxo-3-
[(phenylmethyl)amino]propyl]hydrazide

$C_{16}H_{18}N_4O_7S$
516 - Bensulfuron-methyl

$C_{16}H_{18}O$
1108 - Benzenebutanol, β-phenyl-, (R)-
2074 - Benzene, 1,1'-oxybis[2-ethyl-
2080 - Benzene, 1,1'-(oxydi-2,1-ethanediyl)bis-
2081 - Benzene, 1,1'-(oxydiethylidene)bis-, (R*,R*)-
(±)-

$C_{16}H_{18}O_2$
3048 - 1,1'-Biphenyl, 2,2'-dimethoxy-5,5'-dimethyl-
3049 - 1,1'-Biphenyl, 2,5'-dimethoxy-2',5-dimethyl-
3069 - [1,1'-Biphenyl]-4,4'-diol, 3,3',5,5'-tetramethyl-
9249 - Phenol, 4,4'-butylidenebis-
9433 - Phenol, 4,4'-(1-methylpropylidene)bis-

$C_{16}H_{18}O_{10}$
2823 - 2H-1-Benzopyran-2-one,
8-(β-D-glucopyranosyloxy)-7-hydroxy-6-
methoxy-

$C_{16}H_{19}Cl_2N$
1801 - Benzenemethanamine, N-(2-chloroethyl)-N-
(phenylmethyl)-, hydrochloride

$C_{16}H_{19}N$
744 - Benzenamine, 2-ethyl-N-(2-ethylphenyl)-
1585 - Benzeneethanamine, N-(2-phenylethyl)-
1822 - Benzeneethanamine, α-methyl-N-(1-
phenylethyl)-

$C_{16}H_{19}NO_3$
6182 - Furo[2,3-b]quinolin-4(2H)-one, 3,9-dihydro-8-
methoxy-9-methyl-2-(1-methylethyl)-, (R)-
8567 - 16(15H)-Oxaerythrinan-15-one, 1,2,6,7-
tetradehydro-14,17-dihydro-3-methoxy-, (3β)-

$C_{16}H_{19}NO_4$
481 - 8-Azabicyclo[3.2.1]octane-2-carboxylic acid, 3-
(benzoyloxy)-8-methyl-, [1R-(exo,exo)]-

$C_{16}H_{20}$
7900 - Naphthalene, 1-hexyl-
7901 - Naphthalene, 2-hexyl-
7905 - Naphthalene, 1-isopropyl-2,4,7-trimethyl-
8014 - Naphthalene, 1,2,5-trimethyl-8-(1-
methylethyl)-
8015 - Naphthalene, 1,3,8-trimethyl-5-(1-
methylethyl)-

$C_{16}H_{20}N_2$
3036 - [1,1'-Biphenyl]-4,4'-diamine, tetramethyl-
5500 - 1,2-Ethanediamine, N,N'-bis(phenylmethyl)-
11010 - 2-Pyridinepropanamine, N,N-dimethyl-γ-
phenyl-

$C_{16}H_{20}N_4$
361 - Aniline, 3,3'-azobis[N,N-dimethyl-

$C_{16}H_{20}O_2Si$
11406 - Silane, diethoxydiphenyl-

$C_{16}H_{20}O_6P_2S_3$
1 - Abate

$C_{16}H_{21}NO_3$
911 - Benzeneacetic acid, α-hydroxy-, 8-methyl-8-
azabicyclo[3.2.1]oct-3-yl ester, endo-(±)-
916 - Benzeneacetic acid, α-(hydroxymethyl)-, 8-
azabicyclo[3.2.1]oct-3-yl ester, [3(S)-endo]-

$C_{16}H_{21}N_3$
5505 - 1,2-Ethanediamine, N,N-dimethyl-N'-
(phenylmethyl)-N'-2-pyridinyl-

$C_{16}H_{22}ClNO_3$
5173 - Diethatyl-ethyl

$C_{16}H_{22}N_4O_2$
9795 - Propanamide, N-(2,3-dihydro-1,5-dimethyl-3-
oxo-2-phenyl-1H-pyrazol-4-yl)-
2-(dimethylamino)-

$C_{16}H_{22}OSi_2$
5238 - Disiloxane, 1,1,3,3-tetramethyl-1,3-diphenyl-

$C_{16}H_{22}O_4$
1336 - 1,2-Benzenedicarboxylic acid, bis(2-
methylpropyl) ester
1338 - 1,4-Benzenedicarboxylic acid, bis(2-
methylpropyl) ester
1339 - 1,2-Benzenedicarboxylic acid, dibutyl ester

$C_{16}H_{22}O_6$
1333 - 1,2-Benzenedicarboxylic acid, bis(2-
ethoxyethyl) ester

$C_{16}H_{22}O_8$
6226 - β-D-Glucopyranoside, 4-(3-hydroxy-1-
propenyl)-2-methoxyphenyl

$C_{16}H_{22}O_{11}$
6216 - α-D-Glucopyranose, pentaacetate
6217 - β-D-Glucopyranose, pentaacetate

$C_{16}H_{23}ClN_3O$
11589 - Tebuconazole

$C_{16}H_{24}N_2O_2$
1325 - 1,2-Benzenedicarboxamide, N,N,N',N'-
tetraethyl-
1326 - 1,3-Benzenedicarboxamide, N,N,N',N'-
tetraethyl-

$C_{16}H_{25}NO_2$
898 - Benzeneacetic acid, α-ethyl-,
2-(diethylamino)ethyl ester

$C_{16}H_{26}$
1258 - Benzene, decyl-
2090 - Benzene, pentaethyl-
6627 - 6,10-Hexadecadiyne
11609 - 5,9-Tetradecadien-7-yne, 6,9-dimethyl-

$C_{16}H_{26}O$
9221 - Phenol, 2,6-bis(1,1-dimethylethyl)-4-ethyl-

$C_{16}H_{26}O_2$
1501 - 1,4-Benzenediol, 2,5-bis(1,1-dimethylpropyl)-
6666 - 6,10,14-Hexadecatrienoic acid

$C_{16}H_{28}O_2$
4940 - 2-Cyclopentene-1-undecanoic acid, (R)-

$C_{16}H_{30}$
4323 - Cyclohexane, 1,1'-(1,4-butanediyl)bis-
4326 - Cyclohexane, 1,1'-butylidenebis-
4447 - Cyclohexane, 1,1'-(2-methyl-
1,3-propanediyl)bis-
6673 - 1-Hexadecyne
6674 - 2-Hexadecyne

$C_{16}H_{30}O$
4762 - Cyclopentadecanone, 2-methyl-
4763 - Cyclopentadecanone, 3-methyl-

$C_{16}H_{30}O_2$
6668 - 7-Hexadecenoic acid
6669 - 9-Hexadecenoic acid
6670 - 9-Hexadecenoic acid, (Z)-

$C_{16}H_{30}O_3$
8424 - Octanoic acid, anhydride

$C_{16}H_{31}ClO$
6665 - Hexadecanoyl chloride

$C_{16}H_{31}N$
6641 - Hexadecanenitrile

$C_{16}H_{32}$
4344 - Cyclohexane, decyl-
4863 - Cyclopentane, undecyl-
6667 - 1-Hexadecene
10580 - 1-Propene, 2-methyl-, tetramer

$C_{16}H_{32}Br_2$
6635 - Hexadecane, 1,16-dibromo-

$C_{16}H_{32}O$
6628 - Hexadecanal
6663 - 3-Hexadecanone

$C_{16}H_{32}O_2$
5065 - Decanoic acid, 2-hexyl-
6644 - Hexadecanoic acid
8152 - Nonanoic acid, heptyl ester
8437 - Octanoic acid, octyl ester
8646 - Pentadecanoic acid, methyl ester
11624 - Tetradecanoic acid, ethyl ester

$C_{16}H_{32}O_4$
5313 - Dodecanoic acid, 2-(2-hydroxyethoxy)ethyl
ester

$C_{16}H_{33}Br$
6633 - Hexadecane, 1-bromo-

$C_{16}H_{33}Cl$
6634 - Hexadecane, 1-chloro-

$C_{16}H_{33}F$
6639 - Hexadecane, 1-fluoro-

$C_{16}H_{33}I$
6640 - Hexadecane, 1-iodo-

$C_{16}H_{33}NO$
5290 - Dodecanamide, N,N-diethyl-
6629 - Hexadecanamide

$C_{16}H_{34}$
6632 - Hexadecane

$C_{16}H_{34}O$
6431 - Heptane, 3,3'-[oxybis(methylene)]bis-
6659 - 1-Hexadecanol
6660 - 2-Hexadecanol
6661 - 3-Hexadecanol
8416 - Octane, 1,1'-oxybis-
8649 - 1-Pentadecanol, 2-methyl-
8650 - 6-Pentadecanol, 6-methyl-

$C_{16}H_{34}O_2$
11617 - Tetradecane, 1,1-dimethoxy-
$C_{16}H_{34}O_3$
8552 - Orthoformic acid, triisopentyl ester
$C_{16}H_{34}S$
6643 - 1-Hexadecanethiol
8417 - Octane, 1,1'-thiobis-
$C_{16}H_{35}N$
6631 - 1-Hexadecanamine
8352 - 2-Octanamine, N-(1-methylheptyl)
8354 - 1-Octanamine, N-octyl-
$C_{16}H_{36}O_4Si$
11505 - Silicic acid (H_4SiO_4), tetrabutyl ester
$C_{16}H_{48}O_6Si_7$
6528 - Heptasiloxane, hexadecamethyl-
$C_{16}H_{48}O_8Si_8$
4741 - Cyclooctasiloxane, hexadecamethyl-
$C_{17}H_9NO_4$
8062 - Naphtho[2,3-f]quinoline-7,12-dione, 5,6-
dihydroxy-
$C_{17}H_{10}O$
6292 - 7H-Benz[de]anthracen-7-one
$C_{17}H_{10}O_2$
10854 - 1-Pyrenecarboxylic acid
$C_{17}H_{11}N$
630 - Benz[c]acridine
$C_{17}H_{12}$
2393 - 11H-Benzo[a]fluorene
$C_{17}H_{12}ClFN_2O$
8219 - Nuarimol
$C_{17}H_{12}Cl_2N_2O$
5948 - Fenarimol
$C_{17}H_{12}O_2$
8021 - 2-Naphthalenol, benzoate
$C_{17}H_{12}O_3$
2644 - Benzoic acid, 2-hydroxy-, 1-naphthalenyl
ester
2645 - Benzoic acid, 2-hydroxy-, 2-naphthalenyl
ester
$C_{17}H_{12}O_7$
6184 - 1H,12H-Furo[3',2':4,5]furo[2,3-h]pyrano[3,4-
c][1]benzopyran-1,12-dione, 3,4,7a,10a-
tetrahydro-5-meth
$C_{17}H_{13}N$
10947 - Pyridine, 2,6-diphenyl-
$C_{17}H_{13}N_3O_5S_2$
2726 - Benzoic acid, 2-[[[4-[(2-thiazolylamino)
sulfonyl]phenyl]amino]carbonyl]-
$C_{17}H_{14}$
4756 - 15H-Cyclopenta[a]phenanthrene, 16,17-
dihydro-
7949 - Naphthalene, 1-(phenylmethyl)-
7950 - Naphthalene, 2-(phenylmethyl)-
9182 - Phenanthrene, 9-(1-methylethenyl)-
$C_{17}H_{14}O$
7912 - 1-Naphthalenemethanol, α-phenyl-
8039 - 1-Naphthalenol, 2-(phenylmethyl)-
8040 - 2-Naphthalenol, 1-(phenylmethyl)-
8683 - 1,4-Pentadien-3-one, 1,5-diphenyl-
$C_{17}H_{14}O_6$
4757 - Cyclopenta[c]furo[3',2':4,5]furo[2,3-
h][1]benzopyran-1,11-dione, 2,3,6a,8,9,9a-
hexahydro-4-methoxy-, (6aR-
$C_{17}H_{15}ClF_3NO_4$
6291 - Haloxyfop-(2-ethoxyethyl)
$C_{17}H_{15}ClO_7$
2847 - 1-Benzopyrylium, 3,5,7-trihydroxy-2-(4-
hydroxy-3,5-dimethoxyphenyl)-, chloride
$C_{17}H_{15}N$
7723 - 1-Naphthalenamine, N-(2-methylphenyl)-
7724 - 1-Naphthalenamine, N-(4-methylphenyl)-
7725 - 2-Naphthalenamine, N-(2-methylphenyl)-
7730 - 2-Naphthalenamine, N-(phenylmethyl)-
$C_{17}H_{16}Br_2O_3$
3140 - Bromopropylate

$C_{17}H_{16}O_2$
942 - Benzeneacetic acid, α-(phenylmethylene)-,
ethyl ester
$C_{17}H_{16}O_4$
8776 - Pentanedioic acid, diphenyl ester
9927 - Propanedioic acid, bis(phenylmethyl) ester
$C_{17}H_{17}ClO_6$
11520 - Spiro[benzofuran-2(3H),1'-[2]cyclohexene]-
3,4'-dione, 7-chloro-2',4,6-trimethoxy-
$C_{17}H_{17}NO_2$
5155 - 4H-Dibenzo[de,g]quinoline-10,11-diol,
5,6,6a,7-tetrahydro-6-methyl-, (R)-
$C_{17}H_{17}N_3O_3$
7189 - Imazaquin
$C_{17}H_{18}O$
7594 - Methanone, bis(2,4-dimethylphenyl)-
8969 - 3-Pentanone, 1,5-diphenyl-
$C_{17}H_{18}O_4$
1103 - Benzenebutanoic acid, 4-hydroxy-γ-(4-
hydroxyphenyl)-γ-methyl-
9970 - Propanedioic acid, 1-naphthalenyl-, diethyl
ester
10378 - 2-Propanol, 1,3-diphenoxy-, acetate
$C_{17}H_{19}ClN_2S$
9505 - 10H-Phenothiazine-10-propanamine, 2-
chloro-N,N-dimethyl-
$C_{17}H_{19}NO_3$
7396 - 7-Isoquinolinol, 1,2,3,4-tetrahydro-1-[(4-
hydroxyphenyl)methyl]-6-methoxy-, (S)-
7666 - Morphinan-3,6-diol, 7,8-didehydro-4,5-epoxy-
17-methyl- (5α,6α)-
9632 - Piperidine, 1-[5-(1,3-benzodioxol-5-yl)-1-oxo-
2,4-pentadienyl]-, (E,E)-
9633 - Piperidine, 1-[5-(1,3-benzodioxol-5-yl)-1-oxo-
2,4-pentadienyl]-, (Z,Z)-
$C_{17}H_{19}NO_4$
5952 - Fenoxycarb
7668 - Morphinan-3,6-diol, 7,8-didehydro-4,5-epoxy-
17-methyl- (5α,6α)-, 17-oxide
$C_{17}H_{20}$
1697 - Benzene, 1,1'-(1-ethyl-1,3-propanediyl)bis-
1690 - Benzene, 1,1'-(2-ethyl-1,3-propanediyl)bis-
2098 - Benzene, 1,1'-(1,5-pentanediyl)bis-
2108 - Benzene, 1,1'-pentylidenebis-
$C_{17}H_{20}N_2O$
7593 - Methanone, bis[4-(dimethylamino)phenyl]-
12003 - Urea, N,N'-diethyl-N,N'-diphenyl-
$C_{17}H_{20}N_2S$
9504 - 10H-Phenothiazine-10-ethanamine, N,N,α-
trimethyl-
$C_{17}H_{20}N_4O_6$
11344 - Riboflavin
$C_{17}H_{20}O_6$
7052 - 4-Hexenoic acid, 6-(1,3-dihydro-4-hydroxy-6-
methoxy-7-methyl-3-oxo-5-isobenzofuranyl)-4-
methyl-, (E)-
$C_{17}H_{20}Si$
11360 - Silacyclohexane, 1,1-diphenyl-
$C_{17}H_{21}N$
1575 - Benzeneethanamine, N,α-dimethyl-N-
(phenylmethyl)-, (+)-
1804 - Benzenemethanamine, N,N-diethyl-α-phenyl-
$C_{17}H_{21}NO$
5422 - Ethanamine, 2-(diphenylmethoxy)-N,N-
dimethyl-
$C_{17}H_{21}NO_2$
928 - Benzeneacetic acid, α-methylene-, 8-methyl-8-
azabicyclo[3.2.1]oct-3-yl ester, endo-
8067 - Napropamide
$C_{17}H_{21}NO_3$
6200 - Galanthamine
$C_{17}H_{21}NO_4$
482 - 8-Azabicyclo[3.2.1]octane-2-carboxylic acid, 3-
(benzoyloxy)-8-methyl-, methyl ester, [1R-
(exo,exo)]-

908 - Benzeneacetic acid, α-(hydroxymethyl)-, 9-
methyl-3-oxa-9-azatricyclo[3.3.1.02,4]non-7-yl
ester, [(7S)-(1α,2β,4β,5α,7β)]-
$C_{17}H_{21}N_3$
657 - Benzenamine, 4,4'-carbonimidoylbis[N,N-
dimethyl-
$C_{17}H_{22}$
7895 - Naphthalene, 1-heptyl-
7896 - Naphthalene, 2-heptyl-
$C_{17}H_{22}N_2$
786 - Benzenamine, 4,4'-methylenebis[N,N-dimethyl-
5506 - 1,2-Ethanediamine, N,N-dimethyl-N'-phenyl-
N'-(phenylmethyl)-
$C_{17}H_{23}NO_3$
902 - Benzeneacetic acid, α-(hydroxymethyl)- 8-
methyl-8-azabicyclo[3.2.1]oct-3-yl ester endo-
(±)-
917 - Benzeneacetic acid, α-(hydroxymethyl)-, 8-
methyl-8-azabicyclo[3.2.1]oct-3-yl ester, [3(S)-
endo]-
$C_{17}H_{23}N_3O$
5511 - 1,2-Ethanediamine, N-[(4-methoxyphenyl)
methyl]-N',N'-dimethyl-N-2-pyridinyl-
$C_{17}H_{24}N_2O$
5411 - Ethanamine, 2-[(3-butyl-1-isoquinolinyl)oxy]-
N,N-dimethyl-
$C_{17}H_{24}O_9$
6225 - β-D-Glucopyranoside, 4-(3-hydroxy-1-
propenyl)-2,6-dimethoxyphenyl
$C_{17}H_{25}N$
9696 - Piperidine, 1-(1-phenylcyclohexyl)-
$C_{17}H_{25}NO_2$
7660 - MGK 264
$C_{17}H_{25}N_3O$
9877 - 1,3-Propanediamine, N,N-diethyl-N'-(6-
methoxy-8-quinolinyl)-
$C_{17}H_{26}ClNO_2$
3144 - Butachlor
$C_{17}H_{26}ClNO_3S$
4138 - Clethodim
$C_{17}H_{26}O_4$
4261 - 2,5-Cyclohexadiene-1,4-dione, 2,5-dihydroxy-
3-undecyl-
$C_{17}H_{27}NO$
11169 - Pyrrolidine, 1-[2-(3-methylbutoxy)-2-
phenylethyl]-
$C_{17}H_{27}NO_3$
7683 - Morpholine, 4-[3-(4-butoxyphenoxy)propyl]-
$C_{17}H_{28}$
2374 - Benzene, undecyl-
$C_{17}H_{28}O$
9225 - Phenol, 2,6-bis(1,1-dimethylpropyl)-4-methyl-
$C_{17}H_{29}NO_3S$
11359 - Sethoxydim
$C_{17}H_{30}O$
4217 - 9-Cycloheptadecen-1-one
4218 - 9-Cycloheptadecen-1-one, (Z)-
$C_{17}H_{30}O_3$
7182 - Hydroprene
$C_{17}H_{32}$
4402 - Cyclohexane, 1,1'-(2-ethyl-
1,3-propanediyl)bis-
4464 - Cyclohexane, 1,1'-(1,5-pentanediyl)bis-
$C_{17}H_{32}O_2$
6324 - 9-Heptadecenoic acid
$C_{17}H_{33}N$
6312 - Heptadecanenitrile
$C_{17}H_{34}$
4809 - Cyclopentane, dodecyl-
6322 - 1-Heptadecene
$C_{17}H_{34}Br_2$
6308 - Heptadecane, 1,17-dibromo-
$C_{17}H_{34}O$
6304 - Heptadecanal
6320 - 2-Heptadecanone
6321 - 9-Heptadecanone

$C_{17}H_{34}O_2$
6314 - Heptadecanoic acid
6653 - Hexadecanoic acid, methyl ester
11626 - Tetradecanoic acid, 1-methylethyl ester
11629 - Tetradecanoic acid, propyl ester

$C_{17}H_{35}Br$
6307 - Heptadecane, 1-bromo-

$C_{17}H_{35}N$
9660 - Piperidine, 1-dodecyl-

$C_{17}H_{36}$
6306 - Heptadecane

$C_{17}H_{36}O$
6317 - 1-Heptadecanol
6318 - 2-Heptadecanol
6319 - 9-Heptadecanol

$C_{17}H_{37}N$
6305 - 1-Heptadecanamine
8353 - 1-Octanamine, N-methyl-N-octyl-

$C_{18}H_{10}O_2$
4116 - 5,6-Chrysenedione
7707 - 1,2-Naphthacenedione

$C_{18}H_{10}O_6$
2999 - [2,2'-Bi-1H-indene]-1,1',3,3'(2H,2'H)-tetrone,
2,2'-dihydroxy-

$C_{18}H_{12}$
519 - Benz[a]anthracene
2406 - Benzo[c]phenanthrene
4114 - Chrysene
7703 - Naphthacene
11910 - Triphenylene

$C_{18}H_{12}N_2$
3107 - 2,3'-Biquinoline

$C_{18}H_{12}N_6$
11843 - 1,3,5-Triazine, 2,4,6-tri-2-pyridinyl-

$C_{18}H_{13}N$
4113 - 6-Chrysenamine

$C_{18}H_{13}NO_3$
2702 - Benzoic acid,
2-[(1-naphthalenylamino)carbonyl]-

$C_{18}H_{13}N_3O$
7334 - Indolo[2',3':3,4]pyrido[2,1-b]quinazolin-5(7H)-
one, 8,13-dihydro-

$C_{18}H_{14}$
11595 - 1,1':2',1''-Terphenyl
11596 - 1,1':3',1''-Terphenyl
11597 - 1,1':4',1''-Terphenyl

$C_{18}H_{14}N_4O_5S$
2654 - Benzoic acid, 2-hydroxy-5-[[4-[(2-
pyridinylamino)sulfonyl]phenyl]azo]-

$C_{18}H_{14}O$
5859 - Ethanone, 1-(1-naphthalenyl)-2-phenyl-
7629 - Methanone, (2-methylphenyl)-1-naphthalenyl-

$C_{18}H_{14}O_3$
10684 - 2-Propenoic acid, 3-phenyl-, anhydride

$C_{18}H_{15}As$
450 - Arsine, triphenyl-

$C_{18}H_{15}Bi$
3111 - Bismuthine, triphenyl-

$C_{18}H_{15}ClSn$
11541 - Stannane, chlorotriphenyl-

$C_{18}H_{15}N$
728 - Benzenamine, N,N-diphenyl-

$C_{18}H_{15}OP$
9518 - Phosphine oxide, triphenyl-

$C_{18}H_{15}O_3P$
9593 - Phosphorous acid, triphenyl ester

$C_{18}H_{15}O_4P$
9566 - Phosphoric acid, triphenyl ester

$C_{18}H_{15}P$
9523 - Phosphine, triphenyl-

$C_{18}H_{15}Sb$
11551 - Stibine, triphenyl-

$C_{18}H_{16}N_2$
1271 - 1,4-Benzenediamine, N,N'-diphenyl-
7173 - Hydrazine, triphenyl-

$C_{18}H_{16}N_2O$
8028 - 2-Naphthalenol, 1-[(2,4-dimethylphenyl)azo]-

8029 - 2-Naphthalenol, 1-[(2,5-dimethylphenyl)azo]-

$C_{18}H_{16}OSi$
11504 - Silanol, triphenyl-

$C_{18}H_{16}OSn$
11911 - Triphenyltin hydroxide

$C_{18}H_{16}O_2$
9177 - 9,10-Phenanthrenedione, 1-methyl-7-(1-
methylethyl)-
10691 - 2-Propenoic acid, 3-phenyl-, 3-phenyl-2-
propenyl ester

$C_{18}H_{16}O_4$
3768 - 2-Butenedioic acid (E)-, bis(phenylmethyl)
ester

$C_{18}H_{18}$
9183 - Phenanthrene, 1-methyl-7-(1-methylethyl)-

$C_{18}H_{18}O_2$
5397 - Estra-1,3,5,7,9-pentaen-17-one, 3-hydroxy-
9302 - Phenol, 4,4'-(1,2-diethylidene-1,2-
ethanediyl)bis-

$C_{18}H_{18}O_4$
3298 - Butanedioic acid, bis(phenylmethyl) ester

$C_{18}H_{18}O_5$
5755 - Ethanol, 2,2'-oxybis-, dibenzoate

$C_{18}H_{18}O_6$
3306 - Butanedioic acid, 2,3-dihydroxy- [R-(R*,R*)]-,
bis(phenylmethyl) ester

$C_{18}H_{19}NO_2$
5160 - 4H-Dibenzo[de,g]quinolin-11-ol, 5,6,6a,7-
tetrahydro-10-methoxy-6-methyl-, (R)- ·

$C_{18}H_{19}NO_3$
5159 - 4H-Dibenzo[de,g]quinolin-10-ol, 5,6,6a,7-
tetrahydro-1,2-dimethoxy-, (R)-

$C_{18}H_{19}NO_4$
7675 - Morphinan-6-one, 7,8-didehydro-4,5-epoxy-
14-hydroxy-3-methoxy-17-methyl-, (5α)-

$C_{18}H_{20}$
7263 - 1H-Indene, 2,3-dihydro-1,1,3-trimethyl-3-
phenyl-

$C_{18}H_{20}Cl_2$
9160 - Perthane

$C_{17}H_{17}N_2 \cdot CH_3O_4S$
5175 - Difenzoquat methyl sulfate

$C_{18}H_{20}O_2$
5398 - Estra-1,3,5(10),7-tetraen-17-one, 3-hydroxy-
9300 - Phenol, 4,4'-(1,2-diethyl-1,2-ethenediyl)bis-,
(E)-

$C_{18}H_{21}NO$
9676 - 2-Piperidinemethanol, α,α-diphenyl-

$C_{18}H_{21}NO_3$
7669 - Morphinan-6-ol, 7,8-didehydro-4,5-epoxy-3-
methoxy-17-methyl-, (5α,6α)-
7670 - Morphinan-6-ol, 7,8-didehydro-4,5-epoxy-3-
methoxy-17-methyl-, (5α,6β)-
7671 - Morphinan-6-ol, 8,14-didehydro-4,5-epoxy-3-
methoxy-17-methyl-, (5α,6α)-
7672 - Morphinan-8-ol, 6,7-didehydro-4,5-epoxy-3-
methoxy-17-methyl-, (5α,8β)-
7677 - Morphinan-6-one, 7,8-didehydro-4-hydroxy-3-
methoxy-17-methyl-
7678 - Morphinan-6-one, 4,5-epoxy-3-methoxy-17-
methyl-, (5α)-

$C_{18}H_{22}$
3076 - 1,1'-Biphenyl, 2,2',4,4',6,6'-hexamethyl-

$C_{18}H_{22}N_2S$
9503 - 10H-Phenothiazine-10-ethanamine, N,N-
diethyl-

$C_{18}H_{22}O_2$
5406 - Estra-1,3,5(10)-trien-17-one, 3-hydroxy-
5407 - Estra-1,3,5(10)-trien-17-one, 3-hydroxy-, (8α)-
9298 - Phenol, 4,4'-(1,2-diethyl-1,2-ethanediyl)bis-,
(R*,S*)-

$C_{18}H_{22}O_4$
1515 - 1,2-Benzenediol, 4,4'-(2,3-dimethyl-1,4-
butanediyl)bis-

$C_{18}H_{23}ClN_2O_2$
5670 - Ethanimidamide, N,N'-bis(4-ethoxyphenyl)-,
monohydrochloride

$C_{18}H_{23}NO_3$
7674 - Morphinan-6-ol, 4,5-epoxy-3-methoxy-17-
methyl-, (5α,6α)-

$C_{18}H_{24}$
7937 - Naphthalene, 1-octyl-
7938 - Naphthalene, 2-octyl-

$C_{18}H_{24}NO_7P$
7673 - Morphinan-6-ol, 7,8-didehydro-4,5-epoxy-3-
methoxy-17-methyl-, (5α,6α)-, phosphate (1
1) (salt)

$C_{18}H_{24}N_2O_6$
5182 - Dinocap

$C_{18}H_{24}O_2$
5399 - Estra-1,3,5(10)-triene-3,17-diol (17β)-
5400 - Estra-1,3,5(10)-triene-3,17-diol, (17α)-
5401 - Estra-1,3,5(10)-triene-3,17-diol, (8α,17β)-
5402 - Estra-5,7,9-triene-3,17-diol, (3β,17β)-

$C_{18}H_{24}O_3$
5404 - Estra-1,3,5(10)-triene-3,16,17-triol, (16α,17β)-
5405 - Estra-1,3,5(10)-triene-3,16,17-triol, (16β,17β)-

$C_{18}H_{26}O_4$
1335 - 1,2-Benzenedicarboxylic acid, bis(3-
methylbutyl) ester

$C_{18}H_{27}NO_3$
8181 - 6-Nonenamide, N-[(4-hydroxy-
3-methoxyphenyl)methyl]-8-methyl-, (E)-

$C_{18}H_{27}NO_5$
894 - Benzeneacetic acid, 4-[2-(diethylamino)-2-
oxoethoxy]-3-methoxy-, propyl ester

$C_{18}H_{27}N_3O$
8737 - 1,5-Pentanediamine, N-(6-methoxy-8-
quinolinyl)-N'-(1-methylethyl)-

$C_{18}H_{28}N_2O_4S$
1582 - Benzeneethanamine, α-methyl-, (S)-, sulfate
(2:1)

$C_{18}H_{28}O$
5327 - 1-Dodecanone, 1-phenyl-

$C_{18}H_{28}O_2$
5317 - Dodecanoic acid, phenyl ester

$C_{18}H_{30}$
1566 - Benzene, dodecyl-
1756 - Benzene, hexaethyl-
2276 - Benzene, 1,2,4,5-tetrakis(1-methylethyl)-
4118 - Chrysene, octadecahydro-
8230 - 7,11-Octadecadiyne

$C_{18}H_{30}O$
9498 - Phenol, 2,4,6-tris(1,1-dimethylethyl)-

$C_{18}H_{30}O_2$
8246 - 9,11,13-Octadecatrienoic acid (Z,Z,Z)
8282 - 9,11,13-Octadecatrienoic acid, (E,E,E)-
8284 - 9,11,13-Octadecatrienoic acid, (E,Z,E)-
8286 - 9,11,13-Octadecatrienoic acid, (Z,Z,E)-
8287 - 9,12,15-Octadecatrienoic acid, (Z,Z,Z)-

$C_{18}H_{30}O_4$
4933 - 1-Cyclopentene-1-pentanoic acid, δ-hydroxy-
3,5-bis(1-methylpropyl)-β-oxo-

$C_{18}H_{32}$
4731 - 1,3-Cyclooctadecadiene

$C_{18}H_{32}I_2$
8225 - 9,12-Octadecadienoic acid

$C_{18}H_{32}I_2O_2$
8300 - 6-Octadecenoic acid, 6,7-diiodo-

$C_{18}H_{32}N_2O_5S$
8926 - 1-Pentanol, 2-(diethylamino)-4-methyl-,
4-aminobenzoate (ester),
monomethanesulfonate (salt)

$C_{18}H_{32}O_2$
4936 - 2-Cyclopentene-1-tridecanoic acid, (S)-
8223 - 9,12-Octadecadienoic acid (Z,Z)-
8224 - 10,12-Octadecadienoic acid, (E,E)-
8316 - 9-Octadecynoic acid

$C_{18}H_{32}O_5$
9120 - Pentonic acid, 5-C-[3,5-bis(1-methylpropyl)-1-
cyclopenten-1-yl]-4-deoxy-

$C_{18}H_{32}O_6$
3561 - Butanoic acid, 3-methyl-, 1,2,3-propanetriyl
ester

$C_{18}H_{32}O_7$
10168 - 1,2,3-Propanetricarboxylic acid, 2-hydroxy-, tributyl ester

$C_{18}H_{32}O_{16}$
6219 - α-D-Glucopyranoside, β-D-fructofuranosyl O-α-D-galactopyranosyl-(1
6221 - α-D-Glucopyranoside, O-α-D-glucopyranosyl-(13)-β-D-fructofuranosyl

$C_{18}H_{32}Si$
11455 - Silane, tributylphenyl-

$C_{18}H_{33}N$
8293 - 9-Octadecenenitrile, (Z)-

$C_{18}H_{34}$
8313 - 1-Octadecyne
8314 - 2-Octadecyne
8315 - 9-Octadecyne

$C_{18}H_{34}O$
8288 - 9-Octadecenal
8289 - 9-Octadecenal, (Z)-

$C_{18}H_{34}OSn$
5011 - Cyhexatin

$C_{18}H_{34}O_2$
8294 - 9-Octadecenoic acid (Z)-
8295 - 11-Octadecenoic acid, (E)-
8296 - 2-Octadecenoic acid, (E)-
8297 - 9-Octadecenoic acid, (E)-

$C_{18}H_{34}O_3$
8303 - 9-Octadecenoic acid, 12-hydroxy-, [R-(Z)]-

$C_{18}H_{34}O_4$
5044 - Decanedioic acid, dibutyl ester
6780 - Hexanedioic acid, bis(2-ethylbutyl) ester

$C_{18}H_{34}O_6$
8239 - Octadecanedioic acid, 9,10-dihydroxy-, (R*,R*)-(±)-

$C_{18}H_{35}BrO_2$
8250 - Octadecanoic acid, 18-bromo-

$C_{18}H_{35}ClO$
8281 - Octadecanoyl chloride

$C_{18}H_{35}N$
8243 - Octadecanenitrile

$C_{18}H_{36}$
4382 - Cyclohexane, dodecyl-
4854 - Cyclopentane, tridecyl-
8291 - 1-Octadecene
8292 - 9-Octadecene

$C_{18}H_{36}O$
6638 - Hexadecane, 1-(ethenyloxy)-
8280 - 3-Octadecanone
8311 - 9-Octadecen-1-ol, (E)-
8312 - 9-Octadecen-1-ol, (Z)-

$C_{18}H_{36}O_2$
5068 - Decanoic acid, 2-octyl-
6316 - Heptadecanoic acid, methyl ester
6650 - Hexadecanoic acid, ethyl ester
6662 - 1-Hexadecanol, acetate
8247 - Octadecanoic acid

$C_{18}H_{36}O_3$
6652 - Hexadecanoic acid, 2-hydroxyethyl ester

$C_{18}H_{36}O_4$
8254 - Octadecanoic acid, 9,10-dihydroxy-

$C_{18}H_{36}O_5$
8276 - Octadecanoic acid, 9,10,18-trihydroxy-, (R*,R*)-

$C_{18}H_{37}Br$
8235 - Octadecane, 1-bromo-

$C_{18}H_{37}Cl$
8236 - Octadecane, 1-chloro-

$C_{18}H_{37}Cl_3Si$
11470 - Silane, trichlorooctadecyl-

$C_{18}H_{37}I$
8242 - Octadecane, 1-iodo-

$C_{18}H_{37}NO$
8231 - Octadecanamide

$C_{18}H_{38}$
6310 - Heptadecane, 2-methyl-
6311 - Heptadecane, 9-methyl-
8234 - Octadecane

$C_{18}H_{38}O$
8278 - 1-Octadecanol
8279 - 3-Octadecanol, (±)-

$C_{18}H_{38}O_2$
6636 - Hexadecane, 1,1-dimethoxy-

$C_{18}H_{38}S$
8244 - 1-Octadecanethiol

$C_{18}H_{39}N$
6725 - 1-Hexanamine, N,N-dihexyl-
8233 - 1-Octadecanamine

$C_{18}H_{54}O_7Si_8$
8482 - Octasiloxane, octadecamethyl-

$C_{18}H_{54}O_9Si_9$
4726 - Cyclononasiloxane, octadecamethyl-

$C_{19}H_{10}Br_4O_5S$
9366 - Phenol, 4,4'-(3H-2,1-benzoxathiol-3-ylidene)bis[2,6-dibromo-, S,S-dioxide

$C_{19}H_{12}O_6$
2839 - 2H-1-Benzopyran-2-one, 3,3'-methylenebis[4-hydroxy-

$C_{19}H_{13}N$
318 - Acridine, 9-phenyl-

$C_{19}H_{14}$
521 - Benz[a]anthracene, 11-methyl-
522 - Benz[a]anthracene, 3-methyl-
523 - Benz[a]anthracene, 8-methyl-
4117 - Chrysene, 1-methyl-

$C_{19}H_{14}F_3NO$
5981 - Fluridone

$C_{19}H_{14}O_3$
4278 - 2,5-Cyclohexadien-1-one, 4-[bis(4-hydroxyphenyl)methylene]-

$C_{19}H_{14}O_5S$
9363 - Phenol, 4,4'-(3H-2,1-benzoxathiol-3-ylidene)bis-, S,S-dioxide

$C_{19}H_{15}Br$
1061 - Benzene, 1,1',1''-(bromomethylidyne)tris-

$C_{19}H_{15}Cl$
1196 - Benzene, 1,1',1''-(chloromethylidyne)tris-

$C_{19}H_{15}ClF_3NO_7$
7410 - Lactofen

$C_{19}H_{15}ClN_4$
11653 - 2H-Tetrazolium, 2,3,5-triphenyl-, chloride

$C_{19}H_{15}N$
3988 - 9H-Carbazole, 9-(phenylmethyl)-

$C_{19}H_{15}NO_2$
11236 - 4-Quinolinecarboxylic acid, 2-phenyl-, 2-propenyl ester

$C_{19}H_{16}$
1982 - Benzene, 1,1',1''-methylidynetris-
3093 - 1,1'-Biphenyl, 2-(phenylmethyl)-
3094 - 1,1'-Biphenyl, 4-(phenylmethyl)-

$C_{19}H_{16}N_2$
11041 - 9H-Pyrido[3,4-b]indole, 1-[(2-methylphenyl)methyl]-

$C_{19}H_{16}O$
1866 - Benzenemethanol, α,α-diphenyl-

$C_{19}H_{16}O_3$
1983 - Benzene, 1,1',1''-[methylidynetris(oxy)]tris-

$C_{19}H_{16}O_4$
2832 - 2H-1-Benzopyran-2-one, 4-hydroxy-3-(3-oxo-1-phenylbutyl)-

$C_{19}H_{16}O_5$
225 - Acetic acid, [(4-oxo-2-phenyl-4H-1-benzopyran-7-yl)oxy]-, ethyl ester

$C_{19}H_{17}Cl_2N_3O_3$
5174 - Difenoconazole

$C_{19}H_{17}N$
729 - Benzenamine, 4-(diphenylmethyl)-

$C_{19}H_{17}NO_3$
11216 - Quinoline, 2-[2-(1,3-benzodioxol-5-yl)ethyl]-4-methoxy-

$C_{19}H_{17}N_3$
6288 - Guanidine, N,N',N''-triphenyl-

$C_{19}H_{17}N_3O$
7333 - Indolo[2',3':3,4]pyrido[2,1-b]quinazolin-5(7H)-one, 8,13,13b,14-tetrahydro-14-methyl-, (S)-

$C_{19}H_{18}O_3Si$
11443 - Silane, methyltriphenoxy-

$C_{19}H_{18}O_5$
2464 - 5-Benzofuranpropanol, 2-(1,3-benzodioxol-5-yl)-7-methoxy-

$C_{19}H_{18}Si$
11444 - Silane, methyltriphenyl-

$C_{19}H_{19}NO_3$
2469 - 5H-Benzo[g]-1,3-benzodioxolo[6,5,4-de]quinoline, 6,7,7a,8-tetrahydro-11-methoxy-7-methyl-, (R)-

$C_{19}H_{19}NO_4$
2470 - 5H-Benzo[g]-1,3-benzodioxolo[6,5,4-de]quinolin-12-ol, 6,7,7a,8-tetrahydro-11-methoxy-7-methyl-, (S)-

$C_{19}H_{19}N_7O_6$
6243 - L-Glutamic acid, N-[4-[[(2-amino-1,4-dihydro-4-oxo-6-pteridinyl)methyl]amino]benzoyl]-

$C_{19}H_{20}F_3NO_4$
5962 - Fluazipop-butyl
5963 - Fluazipop-P-butyl

$C_{19}H_{20}N_2O$
11346 - Sarpagan-17-al

$C_{19}H_{21}NO_3$
5158 - 4H-Dibenzo[de,g]quinolin-1-ol, 5,6,6a,7-tetrahydro-2,11-dimethoxy-6-methyl-, (S)-
7679 - Morphinan, 6,7,8,14-tetradehydro-4,5-epoxy-3,6-dimethoxy-17-methyl-, (5α)-

$C_{19}H_{21}NO_4$
5156 - 4H-Dibenzo[de,g]quinoline-2,9-diol, 5,6,6a,7-tetrahydro-1,10-dimethoxy-6-methyl-, (S)

$C_{19}H_{22}$
1744 - Benzene, 1,1'-(1-heptenylidene)bis-

$C_{19}H_{22}N_2O$
4120 - Cinchonan-9-ol, (9S)-
4133 - Cinchonan-9-ol, (8α,9R)-
10435 - 1-Propanone, 3-(3-ethenyl-4-piperidinyl)-1-(4-quinolinyl)-, (3R-cis)-

$C_{19}H_{22}N_2OS$
5807 - Ethanone, 1-[10-[3-(dimethylamino)propyl]-10H-phenothiazin-2-yl]-

$C_{19}H_{22}N_2O_2$
4119 - Cinchonan-6',9-diol, (8α,9R)-
11347 - Sarpagan-10,17-diol

$C_{19}H_{22}O_2$
9358 - Phenol, 4-[1-ethyl-2-(4-methoxyphenyl)-1-butenyl]-
9359 - Phenol, 4-[1-ethyl-2-(4-methoxyphenyl)-1-butenyl]-, (E)-

$C_{19}H_{22}O_6$
6204 - Gibb-3-ene-1,10-dicarboxylic acid, 2,4a,7-trihydroxy-1-methyl-8-methylene-, 1,4a-lactone,

$C_{19}H_{23}ClN_2$
5140 - 1H-1,4-Diazepine, 1-[(4chlorophenyl)phenylmethyl]hexahydro-4-methyl-

$C_{19}H_{23}NO$
1903 - Benzenemethanol, α-[1-[methyl(3-phenyl-2-propenyl)amino]ethyl]-

$C_{19}H_{23}NO_3$
7664 - Morphinan, 6,7-didehydro-4,5-epoxy-3,6-dimethoxy-17-methyl-, (5α)-

$C_{19}H_{23}NO_4$
7676 - Morphinan-6-one, 7,8-didehydro-4-hydroxy-3,7-dimethoxy-17-methyl-, (9α,13α,14α)-

$C_{19}H_{23}N_3$
342 - Amitraz

$C_{19}H_{23}N_3O_2$
5373 - Ergoline-8-carboxamide, 9,10-didehydro-N-(2-hydroxy-1-methylethyl)-6-methyl-, [8α(S)]-

$C_{19}H_{24}$
1746 - Benzene, 1,1'-heptylidenebis-

$C_{19}H_{24}N_2$
9627 - 4-Piperidinamine, 1-methyl-*N*-phenyl-*N*-(phenylmethyl)-
$C_{19}H_{24}N_2O$
4121 - Cinchonan-9-ol, 10,11-dihydro-, (8α,9*R*)-
4122 - Cinchonan-9-ol, 10,11-dihydro-, (9*S*)-
7306 - 1H-Indole-3-ethanol, 2-(5-ethenyl-1-azabicyclo[2.2.2]oct-2-yl)-
$C_{19}H_{24}N_2OS$
3977 - Carbamothioic acid, diphenyl-, *S*-[2-(diethylamino)ethyl] ester
$C_{19}H_{24}N_2O_2$
6177 - 3aH-Furo[2,3-b]indol-3a-ol, 8a-(5-ethenyl-1-azabicyclo[2.2.2]oct-2-yl)-2,3,8,8a-tetrahydro-
6178 - 3aH-Furo[2,3-b]indol-3a-ol, 8a-(5-ethenyl-1-azabicyclo[2.2.2]oct-2-yl)-2,3,8,8a-tetrahydro-
$C_{19}H_{24}O_3$
353 - Androst-4-ene-3,11,17-trione
$C_{19}H_{26}$
7931 - Naphthalene, 1-nonyl-
7932 - Naphthalene, 2-nonyl-
$C_{19}H_{26}N_2O$
4154 - Curan-17-ol, (16α)-
$C_{19}H_{26}O_2$
343 - Androsta-1,4-dien-3-one, 17-hydroxy-, (17β)-
352 - Androst-4-ene-3,17-dione
$C_{19}H_{26}O_3$
336 - Allethrin
$C_{19}H_{26}O_4S$
10507 - Propargite
$C_{19}H_{27}ClO_2$
354 - Androst-4-en-3-one, 4-chloro-17-hydroxy-, (17β)-
$C_{19}H_{27}NO_3$
2396 - 2H-Benzo[a]quinolizin-2-one, 1,3,4,6,7,11b-hexahydro-9,10-dimethoxy-3-(2-methylpropyl)-
$C_{19}H_{28}BrNO_3$
11179 - Pyrrolidinium, 3-[(cyclopentylhydroxyphenylacetyl)oxy]-1,1-dimethyl-, bromide
$C_{19}H_{28}O_2$
356 - Androst-4-en-3-one, 17-hydroxy-, (17β)-
$C_{19}H_{30}O_2$
347 - Androstan-17-one, 3-hydroxy-, (3β,5α)-
348 - Androstan-17-one, 3-hydroxy-, (3α,5α)-
349 - Androstan-3-one, 17-hydroxy-, (5α,17β)-
5318 - Dodecanoic acid, phenylmethyl ester
$C_{19}H_{30}O_5$
9727 - Piperonyl butoxide
$C_{19}H_{32}$
345 - Androstane
$C_{20}H_{27}O_4P$
9560 - Phosphoric acid, 2-ethylhexyl diphenyl ester
$C_{20}H_{28}$
7793 - Naphthalene, 1-decyl-
7794 - Naphthalene, 2-decyl-
$C_{20}H_{28}O_2$
344 - Androsta-1,4-dien-3-one, 17-hydroxy-17-methyl-, (17β)-
$C_{20}H_{28}O_3S$
7968 - 1-Naphthalenesulfonic acid, 3,6-bis(1,1-dimethylethyl)-, ethyl ester
$C_{20}H_{29}FO_3$
355 - Androst-4-en-3-one, 9-fluoro-11,17-dihydroxy-17-methyl-, (11β,17β)-
$C_{20}H_{30}ClN_3O_2$
11226 - 4-Quinolinecarboxamide, 2-butoxy-*N*-[2-(diethylamino)ethyl]-, monohydrochloride
$C_{20}H_{30}O$
11341 - Retinol
$C_{20}H_{30}O_2$
358 - Androst-4-en-3-one, 17-hydroxy-17-methyl-, (17β)-
9166 - 1-Phenanthrenecarboxylic acid, 1,2,3,4,4a,5,6,9,10,10a-decahydro-1,4a-dimethyl-7-(1-methylethyl)-,

$C_{20}H_{30}O_2$ (continued)
9167 - 1-Phenanthrenecarboxylic acid, 1,2,3,4,4a,4b,5,6,10,10a-decahydro-1,4a-dimethyl-7-(1-methylethyl)-,
9169 - 1-Phenanthrenecarboxylic acid, 1,2,3,4,4a,4b,5,9,10,10a-decahydro-1,4a-dimethyl-7-(1-methylethyl)-,
9170 - 1-Phenanthrenecarboxylic acid, 7-ethenyl-1,2,3,4,4a,4b,5,6,7,9,10,10a-dodecahydro-1,4a,7-trimethyl-, [1R-(1α,4aβ,4bα,7β,10aα)]-
$C_{20}H_{32}ClNO$
9705 - 1-Piperidinepropanol, α-cyclohexyl-α-phenyl-, hydrochloride
$C_{20}H_{32}N_2O_3S$
4038 - Carbosulfan
$C_{20}H_{32}O_2$
346 - Androstane-17-carboxylic acid, (5β,17β)-
350 - Androstan-3-one, 17-hydroxy-17-methyl-, (5α,17β)-
5360 - 5,8,11,14-Eicosatetraenoic acid
5361 - 5,8,11,14-Eicosatetraenoic acid, (all-Z)-
$C_{20}H_{34}$
2266 - Benzene, tetradecyl-
$C_{20}H_{34}O_2$
8283 - 9,12,15-Octadecatrienoic acid, ethyl ester, (Z,Z,Z)-
$C_{20}H_{36}O_2$
8226 - 9,12-Octadecadienoic acid (Z,Z)-, ethyl ester
8227 - 9,12-Octadecadienoic acid, ethyl ester
$C_{20}H_{38}$
5366 - 1-Eicosyne
$C_{20}H_{38}N_4O_4$
7686 - 4-Morpholinecarboxamide, N,N'-1,2-ethanediylbis[N-butyl-
$C_{20}H_{38}O_2$
4952 - Cycloprate
5363 - 11-Eicosenoic acid
5364 - 9-Eicosenoic acid
5365 - 9-Eicosenoic acid, (Z)-
8301 - 9-Octadecenoic acid (Z)-, ethyl ester
8302 - 9-Octadecenoic acid, ethyl ester, (E)-
$C_{20}H_{38}O_3$
5060 - Decanoic acid, anhydride
8269 - Octadecanoic acid, 12-oxo-, ethyl ester
8305 - 9-Octadecenoic acid, 12-hydroxy-, ethyl ester, [R-(Z)]-
$C_{20}H_{38}O_4$
5352 - Eicosanedioic acid
$C_{20}H_{40}$
4475 - Cyclohexane, tetradecyl-
4844 - Cyclopentane, pentadecyl-
5362 - 1-Eicosene
$C_{20}H_{40}O$
6671 - 1-Hexadecen-3-ol, 3,7,11,15-tetramethyl-
6672 - 2-Hexadecen-1-ol, 3,7,11,15-tetramethyl-, [R-[R*,R*-(E)]]-
8240 - Octadecane, 1-(ethenyloxy)-
$C_{20}H_{40}O_2$
221 - Acetic acid, octadecyl ester
5062 - Decanoic acid, decyl ester
5355 - Eicosanoic acid
6647 - Hexadecanoic acid, butyl ester
6658 - Hexadecanoic acid, 3,7,11,15-tetramethyl-
8257 - Octadecanoic acid, ethyl ester
$C_{20}H_{40}O_3$
8260 - Octadecanoic acid, 2-hydroxyethyl ester
$C_{20}H_{42}$
5350 - Eicosane
$C_{20}H_{42}O$
5358 - 1-Eicosanol
5359 - 2-Eicosanol
$C_{20}H_{42}O_2$
8237 - Octadecane, 1,1-dimethoxy-
$C_{20}H_{48}O_4Si_4$
4995 - Cyclotetrasiloxane, 2,4,6,8-tetrabutyl-2,4,6,8-tetramethyl-
$C_{20}H_{60}O_8Si_9$
8178 - Nonasiloxane, eicosamethyl-

$C_{21}H_{12}O$
5152 - 13H-Dibenzo[a,h]fluoren-13-one
$C_{21}H_{13}N$
5143 - Dibenz[a,j]acridine
$C_{21}H_{14}$
5148 - 13H-Dibenzo[a,g]fluorene
$C_{21}H_{14}Br_4O_5S$
9367 - Phenol, 4,4'-(3H-2,1-benzoxathiol-3-ylidene)bis[2,6-dibromo-3-methyl-, S,S-dioxide
$C_{21}H_{14}O$
7634 - Methanone, 1-naphthalenyl-2-naphthalenyl-
$C_{21}H_{16}$
2392 - Benz[j]aceanthrylene, 1,2-dihydro-3-methyl-
7271 - 1H-Indene, 1,3-diphenyl-
7272 - 1H-Indene, 2,3-diphenyl-
7919 - Naphthalene, 1,1'-methylenebis-
$C_{21}H_{16}Br_2O_5S$
9364 - Phenol, 4,4'-(3H-2,1-benzoxathiol-3-ylidene)bis[2-bromo-6-methyl-, S,S-dioxide
$C_{21}H_{16}N_2$
7215 - 1H-Imidazole, 2,4,5-triphenyl-
$C_{21}H_{16}N_2O$
12008 - Urea, N,N'-di-1-naphthalenyl-
$C_{21}H_{16}NaO_4$
11517 - Sonar
$C_{21}H_{18}N_2$
7199 - 1H-Imidazole, 4,5-dihydro-2,4,5-triphenyl-, cis-
$C_{21}H_{18}O_5S$
9368 - Phenol, 4,4'-(3H-2,1-benzoxathiol-3-ylidene)bis[2-methyl-, S,S-dioxide
$C_{21}H_{19}NO_5$
2433 - [1,3]Benzodioxolo[5,6-c]phenanthridinium, 1,2-dimethoxy-12-methyl-
$C_{21}H_{20}$
2110 - Benzene, 1,1'-(phenylmethylene)bis[4-methyl-
$C_{21}H_{20}Cl_2O_3$
9150 - Permethrin
$C_{21}H_{20}N_2O_3$
8572 - Oxayohimbanium, 3,4,5,6,16,17 hexadehydro-16-(methoxycarbonyl)-19-methyl-, (19α)-
$C_{21}H_{20}O_6$
6334 - 1,6-Heptadiene-3,5-dione, 1,7-bis(4-hydroxy-3-methoxyphenyl)-, (E,E)-
$C_{21}H_{20}O_{10}$
2824 - 4H-1-Benzopyran-4-one, 3-[4-(β-D-glucopyranosyloxy)phenyl]-5,7-dihydroxy-
$C_{21}H_{20}O_{11}$
2796 - 4H-1-Benzopyran-4-one, 3-[(6-deoxy-α-L-mannopyranosyl)oxy]-2-(3,4-dihydroxyphenyl)-5,7-dihydroxy-
$C_{21}H_{21}N$
1798 - Benzenemethanamine, N,N-bis(phenylmethyl)-
$C_{21}H_{21}NO_5$
3108 - Bis[1,3]benzodioxolo[4,5-c:5',6'-g]azecin-13(5H)-one, 4,6,7,14-tetrahydro-5,14-dimethyl-, (R)-
$C_{21}H_{21}NO_6$
7357 - 1(3H)-Isobenzofuranone, 6,7-dimethoxy-3-(5,6,7,8-tetrahydro-6-methyl-1,3-dioxolo[4,5-g]isoquinolin-5-yl)-, [S-
11342 - Rheadan, 8-methoxy-16-methyl-2,3 10,11-bis[methylenebis(oxy)]-, (8β)-
$C_{21}H_{21}N_3$
11834 - 1,3,5-Triazine, hexahydro-1,3,5-triphenyl-
$C_{21}H_{21}N_3O_7$
11352 - 9,10-Secostrychnidin-10-oic acid, 2,3-dihydro-4-nitro-2,3-dioxo-
$C_{21}H_{21}O_3P$
9596 - Phosphorous acid, tris(2-methylphenyl) ester
9597 - Phosphorous acid, tris(4-methylphenyl) ester
$C_{21}H_{21}O_4P$
9569 - Phosphoric acid, tris(2-methylphenyl) ester
9570 - Phosphoric acid, tris(3-methylphenyl) ester
9571 - Phosphoric acid, tris(4-methylphenyl) ester

$C_{21}H_{21}O_7P$
9387 - Phenol, 2-methoxy-, phosphate (3:1)

$C_{21}H_{21}Sb$
11552 - Stibine, tris(3-methylphenyl)-
11553 - Stibine, tris(4-methylphenyl)-

$C_{21}H_{22}N_2O_2$
11563 - Strychnidin-10-one

$C_{21}H_{23}ClFNO_2$
3651 - 1-Butanone, 4-[4-(4-chlorophenyl)-4-hydroxy-
1-piperidinyl]-1-(4-fluorophenyl)-

$C_{21}H_{23}NO_5$
2435 - Benzo[e]-1,3-dioxolo[4,5-l][2]benzazecin-
12(5H)-one, 4,6,7,13-tetrahydro-9,10-
dimethoxy-5-methyl-
7667 - Morphinan-3,6-diol, 7,8-didehydro-4,5-epoxy-
17-methyl- (5α,6α)-, diacetate (ester)

$C_{21}H_{23}NO_6$
92 - Acetamide, N-(5,6,7,9-tetrahydro-10-hydroxy-
1,2,3-trimethoxy-9-oxobenzo[a]heptalen-7-yl)-,
(S)-

$C_{21}H_{23}N_3O_5$
11565 - Strychnidin-10-one, mononitrate

$C_{21}H_{24}N_2O_3$
8571 - Oxayohimban-16-carboxylic acid, 16,17-
didehydro-19-methyl-, methyl ester, (19α)-

$C_{21}H_{24}OSi_2$
5239 - Disiloxane, 1,1,1-trimethyl-3,3,3-triphenyl-

$C_{21}H_{24}O_3Si_3$
5005 - Cyclotrisiloxane, 2,4,6-trimethyl-2,4,6-
triphenyl-

$C_{21}H_{24}O_4$
9926 - Propanedioic acid, bis(phenylmethyl)-, diethyl
ester

$C_{21}H_{24}O_7$
3540 - Butanoic acid, 2-methyl-, 10-(acetyloxy)-9,10-
dihydro-8,8-dimethyl-2-oxo-2H,8H-benzo[1,2-b
3,4-b′]dipyran-9-yl

$C_{21}H_{25}NO_4$
5150 - 6H-Dibenzo[a,g]quinolizin-2-ol, 5,8,13,13a-
tetrahydro-3,9,10-trimethoxy-13-methyl-, (13S-
trans)-
5151 - 6H-Dibenzo[a,g]quinolizin-3-ol, 5,8,13,13a-
tetrahydro-2,9,10-trimethoxy-13-methyl-, (13S-
trans)-
5157 - 4H-Dibenzo[de,g]quinoline, 5,6,6a,7-
tetrahydro-1,2,9,10-tetramethoxy-6-methyl-,
(S)-

$C_{21}H_{26}BrNO_3$
5445 - Ethanaminium, N,N-diethyl-N-methyl-2-[(9H-
xanthen-9-ylcarbonyl)oxy]-, bromide

$C_{21}H_{26}ClN_3OS$
9617 - 1-Piperazineethanol, 4-[3-(2-chloro-10H-
phenothiazin-10-yl)propyl]-

$C_{21}H_{26}N_2O_3$
5349 - Eburnamenine-14-carboxylic acid, 14,15-
dihydro-14-hydroxy-, methyl ester,
(3α,14β,16α)-
12062 - Yohimban-16-carboxylic acid, 17-hydroxy-,
methyl ester, (16α,17α)-

$C_{21}H_{26}N_2O_4$
4129 - Cinchonan-9-ol, 6′-methoxy-, (8α,9R)-,
monoformate (salt)

$C_{21}H_{26}O_2$
5154 - 6H-Dibenzo[b,d]pyran-1-ol, 6,6,9-trimethyl-3-
pentyl-

$C_{21}H_{26}O_3$
7621 - Methanone, [2-hydroxy-4-(octyloxy)
phenyl]phenyl-

$C_{21}H_{26}O_5$
9739 - Pregna-1,4-diene-3,11,20-trione, 17,21-
dihydroxy-

$C_{21}H_{27}ClN_2O_2$
5700 - Ethanol, 2-[2-[4-[(4-chlorophenyl)
phenylmethyl]-1-piperazinyl]ethoxy]-

$C_{21}H_{27}NO_4$
7379 - Isoquinoline, 1-[(3,4-dimethoxyphenyl)methyl]-
1,2,3,4-tetrahydro-6,7-dimethoxy-2-methyl-,
(S)-

$C_{21}H_{28}ClNO$
6516 - 3-Heptanone, 6-(dimethylamino)-4,4-
diphenyl-, hydrochloride

$C_{21}H_{28}O_2$
9758 - Pregn-4-en-20-yn-3-one, 17-hydroxy-, (17α)-

$C_{21}H_{28}O_3$
4960 - Cyclopropanecarboxylic acid, 2,2-dimethyl-3-
(2-methyl-1-propenyl)-, 2-methyl-4-oxo-3-(2,4-
pentadienyl)-2-cyclopenten-1-yl ester, [1R-
[1α[S*(Z)],3β]]-

$C_{21}H_{28}O_4$
9755 - Pregn-4-ene-3,11,20-trione, 21-hydroxy-

$C_{21}H_{28}O_5$
9738 - Pregna-1,4-diene-3,20-dione, 11,17,21-
trihydroxy-, (11β)-
9746 - Pregn-4-en-18-al, 11,21-dihydroxy-3,20-
dioxo-, (11β)-
9754 - Pregn-4-ene-3,11,20-trione, 17,21-dihydroxy-

$C_{21}H_{30}O_2$
1537 - 1,3-Benzenediol, 2-[3-methyl-6-(1-
methylethenyl)-2-cyclohexen-1-yl]-5-pentyl-,
(1R-trans)-

$C_{21}H_{30}O_3$
9750 - Pregn-4-ene-3,20-dione, 17-hydroxy-
9751 - Pregn-4-ene-3,20-dione, 21-hydroxy-

$C_{21}H_{30}O_4$
9748 - Pregn-4-ene-3,20-dione, 11,21-dihydroxy-,
(11β)-
9749 - Pregn-4-ene-3,20-dione, 17,21-dihydroxy-

$C_{21}H_{30}O_5$
4282 - 2,4-Cyclohexadien-1-one, 3,5,6-trihydroxy-
4,6-bis(3-methyl-2-butenyl)-2-(3-methyl-1-
oxobutyl)-, (R)-
9753 - Pregn-4-ene-3,20-dione, 11,17,21-trihydroxy-,
(11β)-

$C_{21}H_{32}O_2$
9108 - 1-Phenanthrenecarboxylic acid,
1,2,3,4,4a,4b,5,6,10,10a-decahydro-1,4a-
dimethyl-7-(1-methylethyl)-, methyl ester, [1R-
(1α,4aβ,4bα,10aα)]-
9743 - Pregnane-3,20-dione, (5β)-
9756 - Pregn-5-en-20-one, 3-hydroxy-, (3β)-

$C_{21}H_{32}O_5$
9744 - Pregnane-11,20-dione, 3,17,21-trihydroxy-,
(3α,5β)-
9757 - Pregn-4-en-3-one, 11,17,20,21-tetrahydroxy-,
(11β,20R)-

$C_{21}H_{34}BrNO_3$
5443 - Ethanaminium,
2-[(cyclohexylhydroxyphenylacetyl)
oxy]-N,N-diethyl-N-methyl-, bromide

$C_{21}H_{34}O_2$
9745 - Pregnan-20-one, 3-hydroxy-, (3α,5β)-
11627 - Tetradecanoic acid, phenylmethyl ester

$C_{21}H_{36}$
2089 - Benzene, pentadecyl-
9740 - Pregnane, (5α)-
9741 - Pregnane, (5β)-

$C_{21}H_{36}N_2O_5S$
9623 - Piperazinium, 4-(2-cyclohexyl-2-hydroxy-2-
phenylethyl)-1,1-dimethyl-, methyl sulfate
(salt)

$C_{21}H_{36}O$
9441 - Phenol, 3-pentadecyl-

$C_{21}H_{36}O_2$
1543 - 1,2-Benzenediol, 3-pentadecyl-
9742 - Pregnane-3,20-diol, (3α,5β,20S)-

$C_{21}H_{38}ClN$
11030 - Pyridinium, 1-hexadecyl-, chloride

$C_{21}H_{38}O_6$
6908 - Hexanoic acid, 1,2,3-propanetriyl ester

$C_{21}H_{39}N_7O_{12}$
11562 - Streptomycin

$C_{21}H_{40}O_4$
8131 - Nonanedioic acid, bis(2-ethylbutyl) ester
8299 - 9-Octadecenoic acid (Z)-, 2,3-dihydroxypropyl
ester

$C_{21}H_{41}O_2$
5357 - Eicosanoic acid, methyl ester

$C_{21}H_{42}$
4463 - Cyclohexane, pentadecyl-
4822 - Cyclopentane, hexadecyl-

$C_{21}H_{42}O_2$
8267 - Octadecanoic acid, 1-methylethyl ester
8273 - Octadecanoic acid, propyl ester

$C_{21}H_{42}O_4$
8255 - Octadecanoic acid, 2,3-dihydroxypropyl ester,
(±)-

$C_{21}H_{44}$
6294 - Heneicosane

$C_{21}H_{44}O_3$
10023 - 1,2-Propanediol, 3-(octadecyloxy)-

$C_{22}H_{12}$
2471 - Benzo[ghi]perylene

$C_{22}H_{14}$
5142 - Dibenz[a,h]anthracene
8636 - Pentacene
8996 - Pentaphene
9601 - Picene

$C_{22}H_{16}N_4O$
8038 - 2-Naphthalenol, 1-[[4-(phenylazo)phenyl]azo]-

$C_{22}H_{17}N$
7717 - 1-Naphthalenamine, N,N-diphenyl-

$C_{22}H_{18}Cl_2FNO_3$
5008 - Cyfluthrin

$C_{22}H_{18}O_4$
1337 - 1,2-Benzenedicarboxylic acid,
bis(phenylmethyl) ester
7353 - 1(3H)-Isobenzofuranone, 3,3-bis(4-hydroxy-3-
methylphenyl)-

$C_{22}H_{19}Br_2NO_3$
5102 - Deltamethrin

$C_{22}H_{19}Cl_2NO_3$
5012 - Cypermethrin

$C_{22}H_{19}NO_4$
9465 - Phenol, 4,4′-(2-pyridinylmethylene)bis-,
diacetate (ester)

$C_{22}H_{20}O_{13}$
373 - 2-Anthracenecarboxylic acid, 7-β-D-
glucopyranosyl-9,10-dihydro-3,5,6,8-
tetrahydroxy-1-methyl-9,10-dioxo-

$C_{22}H_{21}NO_2$
1797 - Benzenemethanamine, N-[bis
(4-methoxyphenyl)methylene]-

$C_{22}H_{22}O_8$
6191 - Furo[3′,4′:6,7]naphtho[2,3-d]-1,3-dioxol-
6(5aH)-one, 5,8,8a,9-tetrahydro-9-hydroxy-5-
(3,4,5-trimethoxyphenyl)-
6192 - Furo[3′,4′:6,7]naphtho[2,3-d]-1,3-dioxol-
6(5aH)-one, 5,8,8a,9-tetrahydro-9-hydroxy-5-
(3,4,5-trimetho

$C_{22}H_{23}ClN_2O_8$
7704 - 2-Naphthacenecarboxamide, 7-chloro-4-
(dimethylamino)-1,4,4a,5,5a,6,11,12a-
octahydro-3,6,10,12,12a-pentahydroxy-
6-methyl-1,11-dioxo, [4S
(4α,4aα,5aα,6β,12aα)]-

$C_{22}H_{23}NO_3$
5953 - Fenpropathrin

$C_{22}H_{23}NO_6$
10797 - 4H-Pyran-4-one, 2-methoxy-3,5-dimethyl-6-
[tetrahydro-4-[2-methyl-3-(4-nitrophenyl)-2-
propenylidene]-2-furanyl]

$C_{22}H_{23}NO_7$
7356 - 1(3H)-Isobenzofuranone, 6,7-dimethoxy-3-
(5,6,7,8-tetrahydro-4-methoxy-6-methyl-1,3-
dioxolo[4,5-g]isoquinolin-5

$C_{22}H_{24}N_2O_4$
3204 - Butanamide, N,N′-(3,3′-dimethyl[1,1′-
biphenyl]-4,4′-diyl)bis[3-oxo-

10772 - 2H-Pyran-5-carboxylic acid, 3,4-dihydro-3-
(hydroxymethyl)-2-methyl-4-[(9-methyl-9H-
pyrido[3,4-b]indol-1-yl)methyl]-, methyl ester,
[2S-(2α,3β,4β)]-

11351 - 16,19-Secostrychnidine-10,16-dione, 4-
hydroxy-19-methyl-

$C_{22}H_{24}N_2O_8$

7706 - 2-Naphthacenecarboxamide,
4-(dimethylamino)-1,4,4a,5,5a,6,11,12a-
octahydro-3,6,10,12,12a-pentahydroxy-6-
methyl-1,11-dioxo-,
[4S-(4α,4aα,5aα,6β,12aα)]-

$C_{22}H_{24}N_2O_9$

7705 - 2-Naphthacenecarboxamide,
4-(dimethylamino)-1,4,4a,5,5a,6,11,12a-
octahydro-3,5,6,10,12,12a-hexahydroxy-
6-methyl-1,11-dioxo-,[4S-
(4α,4aα,5aα,6β,12aα)]-

$C_{22}H_{24}Si$

11445 - Silane, methyltri-p-tolyl-

$C_{22}H_{25}NO_2$

5854 - Ethanone, 2,2'-(1-methyl-
2,6-piperidinediyl)bis[1-phenyl-, cis-

$C_{22}H_{25}NO_6$

93 - Acetamide, N-(5,6,7,9-tetrahydro-1,2,3,10-
tetramethoxy-9-oxobenzo[a]heptalen-7-yl)-,
(S)-

$C_{22}H_{25}N_3O$

10850 - 3H-Pyrazol-3-one, 1,2-dihydro-2-(1-methyl-4-
piperidinyl)-5-phenyl-4-(phenylmethyl)-

$C_{22}H_{26}N_2O_2$

4146 - Corynan-16-carboxylic acid, 16,17,18,19-
tetradehydro-17-methoxy-, methyl ester,
(16E)-

$C_{22}H_{26}O_3$

11340 - (-)-cis-Resmethrin

$C_{22}H_{27}NO_2$

5834 - Ethanone, 2-[6-(2-hydroxy-2-phenylethyl)-1-
methyl-2-piperidinyl]-1-phenyl-, [2R-
[2α,6α(S*)]]-

$C_{22}H_{27}NO_4$

5149 - 6H-Dibenzo[a,g]quinolizine, 5,8,13,13a-
tetrahydro-2,3,9,10-tetramethoxy-13-methyl-,
(13S-trans)-

$C_{22}H_{28}N_2O$

9801 - Propanamide, N-phenyl-N-[1-(2-phenylethyl)-
4-piperidinyl]-

$C_{22}H_{28}N_2O_5$

12063 - Yohimban-16-carboxylic acid, 18-hydroxy-
11,17-dimethoxy-, (3β,16β,17α,18β,20

$C_{22}H_{28}O_3$

9736 - Pregna-4,6-diene-21-carboxylic acid, 17-
hydroxy-3-oxo-, γ-lactone, (17α)-

$C_{22}H_{28}O_5$

4962 - Cyclopropanecarboxylic acid, 3-(3-methoxy-2-
methyl-3-oxo-1-propenyl)-2,2-dimethyl-, 2-
methyl-4-oxo-3-(2,4-pentadienyl)-2-
cyclopenten-1-yl ester [1R-[1α[S*(Z)],3β(E)]]-

$C_{22}H_{29}FO_5$

9737 - Pregna-1,4-diene-3,20-dione, 9-fluoro-
11,17,21-trihydroxy-16-methyl-, (11β,16α)-

$C_{22}H_{29}NO_2$

9646 - 2,6-Piperidinediethanol, 1-methyl-α,α'-
diphenyl-, [2α(R*),6α(S*)]-

$C_{22}H_{30}N_2O_2$

473 - Aspidospermidine, 1-acetyl-17-methoxy-

$C_{22}H_{30}N_2O_5$

4730 - 2,4(1H)-Cyclo-3,4-secoakuammilanium, 3,17-
dihydroxy-16-(methoxycarbonyl)-4-methyl-,
(3β,16R)-

$C_{22}H_{32}$

7879 - Naphthalene, 1-dodecyl-
7880 - Naphthalene, 2-dodecyl-

$C_{22}H_{32}O_3$

359 - Androst-4-en-3-one, 17-(1-oxopropoxy)-, (17β)-
9752 - Pregn-4-ene-3,20-dione, 17-hydroxy-6-
methyl-, (6α)-

$C_{22}H_{33}NO_2$

475 - Atisine

$C_{22}H_{35}NO_3$

5007 - 7,20-Cycloveatchane-1,12,15-triol, 21-ethyl-4-
methyl-16-methylene-, (1α,12α,15β)-

$C_{22}H_{36}O$

6664 - 1-Hexadecanone, 1-phenyl-

$C_{22}H_{37}NO$

6630 - Hexadecanamide, N-phenyl-

$C_{22}H_{38}$

1751 - Benzene, hexadecyl-

$C_{22}H_{38}O$

1752 - Benzene, (hexadecyloxy)-

$C_{22}H_{40}O_7$

8102 - 1,2,3-Nonadecanetricarboxylic acid, 2-
hydroxy-

$C_{22}H_{42}O_2$

5281 - 13-Docosenoic acid
5282 - 13-Docosenoic acid, (E)-
5283 - 13-Docosenoic acid, (Z)-
8298 - 9-Octadecenoic acid (Z)-, butyl ester
10667 - 2-Propenoic acid, 2-methyl-, octadecyl ester

$C_{22}H_{42}O_3$

8304 - 9-Octadecenoic acid, 12-hydroxy-, butyl ester,
[R-(Z)-

$C_{22}H_{42}O_4$

6781 - Hexanedioic acid, bis(2-ethylhexyl) ester
8238 - Octadecanedioic acid, diethyl ester

$C_{22}H_{44}$

4411 - Cyclohexane, hexadecyl-
5280 - 1-Docosene

$C_{20}H_{40}N_2 \cdot C_2H_4O_2$

6275 - Glyodin

$C_{22}H_{44}O$

5284 - 13-Docosen-1-ol, (Z)-

$C_{22}H_{44}O_2$

5276 - Docosanoic acid
5356 - Eicosanoic acid, ethyl ester
8251 - Octadecanoic acid, butyl ester
8268 - Octadecanoic acid, 2-methylpropyl ester

$C_{22}H_{46}$

5268 - Docosane

$C_{22}H_{46}O$

5279 - 1-Docosanol

$C_{22}H_{66}O_9Si_{10}$

5079 - Decasiloxane, docosamethyl-

$C_{23}H_{16}O_3$

7268 - 1H-Indene-1,3(2H)-dione, 2-(diphenylacetyl)-

$C_{23}H_{16}O_6$

7777 - 2-Naphthalenecarboxylic acid, 4,4'-
methylenebis[3-hydroxy-

$C_{23}H_{18}O$

7910 - 1-Naphthalenemethanol, α,α-diphenyl-

$C_{23}H_{19}ClF_3NO_3$

5010 - Cyhalothrin

$C_{23}H_{21}ClO_3$

1154 - Benzene, 1,1',1''-(1-chloro-1-ethenyl-2-
ylidene)tris[4-methoxy-

$C_{23}H_{22}ClF_3O_2$

2998 - Bifenthrin

$C_{23}H_{22}O_6$

2783 - [1]Benzopyrano[3,4-b]furo
[2,3-h][1]benzopyran-6(6aH)-one, 1,2,12,12a-
tetrahydro-8,9-dimethoxy-2-(1-methylethen

$C_{23}H_{22}O_7$

3110 - 3H-Bis[1]benzopyrano[3,4-b:6',5'-e]pyran-
7(7aH)-one, 13,13a-dihydro-7a-hydroxy-9,10-
dimethoxy-3,3-dimethyl

$C_{23}H_{23}O_6$

9747 - Pregn-4-ene-3,20-dione, 21-(acetyloxy)-
11,17-dihydroxy-, (11β)-

$C_{23}H_{25}ClN_2$

7455 - Methanaminium, N-[4-
[[4-(dimethylamino)phenyl]phenylmethylene]-
2,5-cyclohexadien-1-ylidene]-N-methyl-,
chlorid

$C_{23}H_{26}N_2O_4$

11564 - Strychnidin-10-one, 2,3-dimethoxy-

$C_{23}H_{27}NO_8$

2586 - Benzoic acid, 6-[[6-[2-(dimethylamino)ethyl]-4-
methoxy-1,3-benzodioxol-5-yl]acetyl]-2,3-
dimethoxy-

$C_{23}H_{28}ClN_3O$

9657 - Piperidine, 1-[2-[3-(2,2-diphenylethyl)-1,2,4-
oxadiazol-5-yl]ethyl]-, monohydrochloride

$C_{23}H_{30}BrNO_3$

9832 - 2-Propanaminium, N-methyl-N-(1-
methylethyl)-N-[2-[(9H-xanthen-9-
ylcarbonyl)oxy]ethyl]-, bromide

$C_{23}H_{30}O_2Si_3$

11915 - Trisiloxane, 1,1,3,5,5-pentamethyl-1,3,5-
triphenyl-

$C_{23}H_{32}N_2O_2S$

11800 - Thiourea, N,N'-bis[4-
(3-methylbutoxy)phenyl]-

$C_{23}H_{32}O_6$

4049 - Card-20(22)-enolide, 3,5,14-trihydroxy-19-
oxo-, (3β,5β)-

$C_{23}H_{33}IN_2O$

2120 - Benzenepropanaminium, γ-(aminocarbonyl)-
N-methyl-N,N-bis(1-methylethyl)-γ-phenyl-, iod

$C_{23}H_{34}O_4$

4044 - Card-20(22)-enolide, 3,14-dihydroxy-, (3β,5β)-

$C_{23}H_{34}O_5$

4046 - Card-20(22)-enolide, 3,11,14-trihydroxy-,
(3β,5β,11α)-
4047 - Card-20(22)-enolide, 3,12,14-trihydroxy-,
(3β,5β,12β)-
4048 - Card-20(22)-enolide, 3,14,16-trihydroxy-,
(3β,5β,16β)-

$C_{23}H_{36}O_3$

351 - Androstan-3-one, 2-methyl-17-(1-oxopropoxy)-,
(2α,5α,17β)-

$C_{23}H_{38}O_2$

6655 - Hexadecanoic acid, phenylmethyl ester
8210 - 24-Norcholan-23-oic acid, (5β)-

$C_{23}H_{40}$

1743 - Benzene, heptadecyl-

$C_{23}H_{44}O_2$

8306 - 9-Octadecenoic acid (Z)-, 3-methylbutyl ester

$C_{23}H_{44}O_3$

8275 - Octadecanoic acid, (tetrahydro-2-
furanyl)methyl ester

$C_{23}H_{44}O_4$

9954 - Propanedioic acid, hexadecyl-, diethyl ester

$C_{23}H_{46}$

4404 - Cyclohexane, heptadecyl-
4842 - Cyclopentane, octadecyl-

$C_{23}H_{46}O$

11857 - 12-Tricosanone

$C_{23}H_{46}O_2$

5278 - Docosanoic acid, methyl ester
8265 - Octadecanoic acid, 3-methylbutyl ester
8270 - Octadecanoic acid, pentyl ester

$C_{23}H_{48}$

6309 - Heptadecane, 9-hexyl-
11855 - Tricosane

$C_{24}H_{12}$

4145 - Coronene

$C_{24}H_{14}$

2850 - Benzo[rst]pentaphene
8061 - Naphtho[1,2,3,4-def]chrysene

$C_{24}H_{16}$

11644 - Tetraphenylene

$C_{24}H_{18}$

11193 - 1,1':2',1'':2'',1'''-Quaterphenyl
11194 - 1,1':4',1'':4'',1'''-Quaterphenyl
11598 - 1,1':3',1''-Terphenyl, 5'-phenyl-

$C_{24}H_{20}N_2$

3035 - [1,1'-Biphenyl]-4,4'-diamine, N,N'-diphenyl-

$C_{24}H_{20}N_4O$

8033 - 2-Naphthalenol, 1-[[2-methyl-4-[(2-
methylphenyl)azo]phenyl]azo]-

$C_{24}H_{20}O_4Si$
11509 - Silicic acid (H4SiO4), tetraphenyl ester
$C_{24}H_{20}O_6$
10196 - 1,2,3-Propanetriol, tribenzoate
$C_{24}H_{20}Pb$
9733 - Plumbane, tetraphenyl-
$C_{24}H_{20}Si$
11452 - Silane, tetraphenyl-
$C_{24}H_{20}Sn$
11545 - Stannane, tetraphenyl-
$C_{24}H_{27}O_4P$
9335 - Phenol, 2,4-dimethyl-, phosphate (3:1)
9336 - Phenol, 2,5-dimethyl-, phosphate (3:1)
9337 - Phenol, 2,6-dimethyl-, phosphate (3:1)
9338 - Phenol, 3,4-dimethyl-, phosphate (3:1)
$C_{24}H_{28}O_4$
9301 - Phenol, 4,4'-(1,2-diethyl-1,2-ethenediyl)bis-, dipropanoate, (E)-
$C_{24}H_{29}NO_4$
7377 - Isoquinoline, 1-[(3,4-diethoxyphenyl)methyl]-6,7-diethoxy-
$C_{24}H_{30}O_3Si_3$
5004 - Cyclotrisiloxane, 2,4,6-triethyl-2,4,6-triphenyl-
$C_{24}H_{34}O_5$
4101 - Cholan-24-oic acid, 3,7,12-trioxo-, (5β)-
$C_{24}H_{38}O_4$
1334 - 1,2-Benzenedicarboxylic acid, bis(2-ethylhexyl) ester
1344 - 1,2-Benzenedicarboxylic acid, dioctyl ester
$C_{24}H_{39}NO$
8290 - 9-Octadecenamide, N-phenyl-, (Z)-
$C_{24}H_{40}N_2$
4143 - Con-5-enin-3-amine, N,N-dimethyl-, (3β)-
$C_{24}H_{40}N_8O_4$
5711 - Ethanol, 2,2',2'',2'''-[(4,8-di-1-piperidinylpyrimido[5,4-d]pyrimidine-
$C_{24}H_{40}O_2$
1257 - Benzenedecanoic acid, ι-octyl-
2063 - Benzenenonanoic acid, θ-nonyl-
4094 - Cholan-24-oic acid
8271 - Octadecanoic acid, phenyl ester
$C_{24}H_{40}O_3$
4099 - Cholan-24-oic acid, 3-hydroxy-, (3α,5β)-
$C_{24}H_{40}O_4$
4095 - Cholan-24-oic acid, 3,12-dihydroxy-, (3α,5β,12α)-
4096 - Cholan-24-oic acid, 3,6-dihydroxy-, (3α,5β,6α)-
4097 - Cholan-24-oic acid, 3,7-dihydroxy-, (3α,5β,7α)-
4098 - Cholan-24-oic acid, 3,7-dihydroxy-, (3α,5β,7β)-
$C_{24}H_{40}O_5$
4100 - Cholan-24-oic acid, 3,7,12-trihydroxy-, (3α,5β,7α,12α)-
$C_{24}H_{41}NO_2$
8232 - Octadecanamide, N-(4-hydroxyphenyl)-
$C_{24}H_{42}$
2066 - Benzene, octadecyl-
4093 - Cholane
$C_{24}H_{46}O_2$
8253 - Octadecanoic acid, cyclohexyl ester
11608 - 15-Tetracosenoic acid
$C_{24}H_{46}O_3$
5306 - Dodecanoic acid, anhydride
$C_{24}H_{46}O_4$
5353 - Eicosanedioic acid, diethyl ester
9155 - Peroxide, bis(1-oxododecyl)
$C_{24}H_{48}$
4459 - Cyclohexane, octadecyl-
$C_{24}H_{48}O_2$
5277 - Docosanoic acid, ethyl ester
11607 - Tetracosanoic acid
$C_{24}H_{50}$
11603 - Tetracosane
11856 - Tricosane, 2-methyl-

$C_{24}H_{52}O_4Si$
11507 - Silicic acid (H4SiO4), tetrakis(2-ethylbutyl) ester
$C_{24}H_{54}OSn_2$
5241 - Distannoxane, hexabutyl-
$C_{24}H_{72}O_{10}Si_{11}$
11964 - Undecasiloxane, tetracosamethyl-
$C_{25}H_{20}$
1831 - Benzene, 1,1',1'',1'''-methanetetrayltetrakis-
$C_{25}H_{20}N_2O$
12031 - Urea, tetraphenyl-
$C_{25}H_{22}ClNO_3$
5956 - Fenvalerate
$C_{25}H_{22}OSi$
11441 - Silane, (2-methylphenoxy)triphenyl-
$C_{25}H_{24}F_6N_4$
7136 - Hydramethylnon
$C_{25}H_{27}ClN_2$
9605 - Piperazine, 1-[(4-chlorophenyl)phenylmethyl]-4-[(3-methylphenyl)methyl]-
$C_{25}H_{28}O_3$
5403 - Estra-1,3,5(10)-triene-3,17-diol (17β)-, 3-benzoate
$C_{25}H_{30}ClN_3$
7453 - Methanaminium, N-[4-[bis[4-(dimethylamino)phenyl]methylene]-2,5-cyclohexadien-1-ylidene]-N-methyl-, chloride
$C_{25}H_{30}O_4$
5108 - 6,6'-Diapo-ψ,ψ-carotenedioic acid, monomethyl ester, 9-cis-
$C_{25}H_{40}O_2$
8309 - 9-Octadecenoic acid (Z)-, phenylmethyl ester
$C_{25}H_{41}NO_9$
305 - Aconitane-3,8,13,14,15-pentol, 20-ethyl-1,6,16-trimethoxy-4-(methoxymethyl)-,
$C_{25}H_{44}$
2062 - Benzene, nonadecyl-
2070 - Benzene, (3-octylundecyl)-
8245 - Octadecane, 9-p-tolyl-
$C_{25}H_{48}O_4$
8132 - Nonanedioic acid, bis(2-ethylhexyl) ester
$C_{25}H_{50}$
4456 - Cyclohexane, nonadecyl-
5351 - Eicosane, 1-cyclopentyl-
6323 - 8-Heptadecene, 9-octyl-
$C_{25}H_{52}$
6313 - Heptadecane, 9-octyl-
8637 - Pentacosane
$C_{26}H_{16}$
6621 - Hexacene
$C_{26}H_{20}$
1623 - Benzene, 1,1',1'',1'''-(1,2-ethenediylidene)tetrakis-
$C_{26}H_{22}$
1591 - Benzene, 1,1',1'',1'''-(1,2-ethenediylidene)tetrakis-
1618 - Benzene, 1,1',1'',1'''-(1-ethanyl-2-ylidene)tetrakis-
$C_{26}H_{22}ClF_3N_2O_3$
5983 - Fluvalinate
$C_{26}H_{22}O_2$
5580 - 1,2-Ethanediol, 1,1,2,2-tetraphenyl-
$C_{26}H_{23}F_2NO_4$
5965 - Flucythrinate
$C_{26}H_{36}O_6$
3141 - Bufa-20,22-dienolide, 16-(acetyloxy)-3,14-dihydroxy-, (3β,5β,16β)-
$C_{26}H_{38}$
1567 - Benzene, 1,1'-dodecylidenebis[4-methyl-
2267 - Benzene, 1,1'-tetradecylidenebis-
$C_{26}H_{38}O_4$
4281 - 2,4-Cyclohexadien-1-one, 3,5-dihydroxy-2,6,6-tris(3-methyl-2-butenyl)-4-(3-methyl-1-oxobutyl)-
$C_{26}H_{43}NO_6$
6273 - Glycine, N-[(3α,5β,7α,12α)-3,7,12-trihydroxy-24-oxocholan-24-yl]-

$C_{26}H_{45}NO_7S$
5625 - Ethanesulfonic acid, 2-[[(3α,5β,7α,12α)-3,7,12-trihydroxy-24-oxocholan-24-yl]amino]-
$C_{26}H_{46}$
1119 - Benzene, (1-butylhexadecyl)-
1568 - Benzene, eicosyl-
1695 - Benzene, (1-ethyloctadecyl)-
2007 - Benzene, (1-methylnonadecyl)-
2068 - Benzene, (1-octyldodecyl)-
2170 - Benzene, (1-propylheptadecyl)-
$C_{26}H_{50}$
4476 - Cyclohexane, 1,1'-tetradecylidenebis-
$C_{26}H_{50}O_4$
5043 - Decanedioic acid, bis(2-ethylhexyl) ester
5311 - Dodecanoic acid, 1,2-ethanediyl ester
$C_{26}H_{52}$
4384 - Cyclohexane, eicosyl-
4819 - Cyclopentane, heneicosyl-
$C_{26}H_{52}O_2$
6624 - Hexacosanoic acid
$C_{26}H_{52}O_6$
8259 - Octadecanoic acid, 2-[2-[2-[2-(2-hydroxyethoxy)ethoxy]ethoxy]ethyl ester
$C_{26}H_{54}$
5269 - Docosane, 5-butyl-
5270 - Docosane, 7-butyl-
5271 - Docosane, 9-butyl-
5272 - Docosane, 11-butyl-
6296 - Heneicosane, 11-(2,2-dimethylpropyl)-
6622 - Hexacosane
6637 - Hexadecane, 6,11-dipentyl-
8241 - Octadecane, 3-ethyl-5-(2-ethylbutyl)-
11605 - Tetracosane, 3-ethyl-
$C_{26}H_{54}O$
6626 - 1-Hexacosanol
$C_{27}H_{18}O_3$
7587 - Methanone, 1,3,5-benzenetriyltris[phenyl-
$C_{27}H_{20}O_{12}$
2773 - 1H-2-Benzopyran-3-carboxylic acid, 6-[2-(4,9-dihydro-8-hydroxy-5,7-dimethoxy-4,9-dioxonaphtho[2,3-b]furan-2-yl
$C_{27}H_{28}Br_2O_5S$
9365 - Phenol, 4,4'-(3H-2,1-benzoxathiol-3-ylidene)bis[2-bromo-3-methyl-6-(1-methylethyl)-, S,S-dioxide
$C_{27}H_{30}O_5S$
9369 - Phenol, 4,4'-(3H-2,1-benzoxathiol-3-ylidene)bis[5-methyl-2-(1-methylethyl)-, S,S-dioxide
$C_{27}H_{32}N_2O_6$
4131 - Cinchonan-9-ol, 6'-methoxy-, (8α,9R)-, mono(2-hydroxybenzoate) (salt)
$C_{27}H_{32}O_{14}$
2795 - 4H-1-Benzopyran-4-one, 7-[[2-O-(6-deoxy-α-L-mannopyranosyl)-β-D-glucopyranosyl]oxy]-2,3-dihydro-5-hy
$C_{27}H_{34}N_2O_4S$
5444 - Ethanaminium, N-[4-[[4-(diethylamino)phenyl]phenylmethylene]-2,5-cyclohexadien-1-ylidene]-N-ethyl-, sulfate (1
$C_{27}H_{34}O_3$
5408 - Estr-4-en-3-one, 17-(1-oxo-3-phenylpropoxy)-, (17β)-
$C_{27}H_{39}NO_2$
12044 - Veratraman-3,23-diol, 14,15,16,17-tetradehydro-, (3β,23β)-
$C_{27}H_{39}NO_3$
12046 - Veratraman-11-one, 17,23-epoxy-3-hydroxy-, (3β,23β)-
$C_{27}H_{40}O_3$
7774 - 2-Naphthalenecarboxylic acid, 3-hydroxy-, hexadecyl ester

$C_{27}H_{42}Cl_2N_2O_6$
6648 - Hexadecanoic acid, 2-[(dichloroacetyl)amino]-3-hydroxy-3-(4-nitrophenyl)propyl ester, [R-(R*,R*)]-

$C_{27}H_{42}O_3$
11536 - Spirost-5-en-3-ol, (3β,25R)-

$C_{27}H_{42}O_4$
11534 - Spirostan-12-one, 3-hydroxy-, (3β,5α,25R)-

$C_{27}H_{43}NO_2$
11515 - Solanid-5-ene-3,12-diol, (3β,12α)-
11516 - Solanid-5-ene-3,18-diol, (3β)-
11529 - Spirosol-5-en-3-ol, (3β,22α,25R)-

$C_{27}H_{43}NO_8$
4068 - Cevane-3,4,7,14,15,16,20-heptol, 4,9-epoxy-, (3β,4α,7α,15α,16β)-

$C_{27}H_{43}NO_9$
4070 - Cevane-3,4,6,7,14,15,16,20-octol, 4,9-epoxy-, (3β,4α,6α,7α,15α,16β

$C_{27}H_{44}$
4102 - Cholesta-3,5-diene

$C_{27}H_{44}O$
4103 - Cholesta-5,7-dien-3-ol, (3β)-
4104 - Cholesta-8,24-dien-3-ol, (3β,5α)-
11348 - 9,10-Secocholesta-5,7,10(19)-trien-3-ol, (3β,5Z,7E)-

$C_{27}H_{44}O_3$
11531 - Spirostan-3-ol, (3β,5β,25S)-
11532 - Spirostan-3-ol, (3β,5α,25R)-
11533 - Spirostan-3-ol, (3β,5β,25R)-

$C_{27}H_{44}O_4$
11530 - Spirostan-2,3-diol, (2α,3β,5α,25R)-

$C_{27}H_{44}O_5$
11535 - Spirostan-2,3,15-triol, (2α,3β,5α,15β,25R)-

$C_{27}H_{45}NO_2$
11528 - Spirosolan-3-ol, (3β,5α,22β,25S)-

$C_{27}H_{46}$
6298 - 10-Heneicosene, 11-phenyl-

$C_{27}H_{46}O$
4110 - Cholest-5-en-3-ol (3β)-
4111 - Cholest-4-en-3-ol, (3β)-
4112 - Cholest-5-en-3-ol, (3α)-

$C_{27}H_{46}O_2$
2785 - 2H-1-Benzopyran-6-ol, 3,4-dihydro-2,8-dimethyl-2-(4,8,12-trimethyltridecyl)-, [2R-[2R*(4R*,8R*)]]-

$C_{27}H_{48}$
1741 - Benzene, heneicosyl-
4105 - Cholestane, (5α)-
4106 - Cholestane, (5β)-
6297 - Heneicosane, 11-phenyl-

$C_{27}H_{48}O$
4107 - Cholestanol
4108 - Cholestan-3-ol, (3α,5α)-
4109 - Cholestan-3-ol, (3β,5β)-

$C_{27}H_{50}O_6$
8440 - Octanoic acid, 1,2,3-propanetriyl ester

$C_{27}H_{54}$
4808 - Cyclopentane, docosyl-

$C_{27}H_{54}O$
6303 - 14-Heptacosanone

$C_{27}H_{54}O_2$
6625 - Hexacosanoic acid, methyl ester

$C_{27}H_{56}$
6302 - Heptacosane

$C_{28}H_{14}N_2O_4$
434 - 5,9,14,18-Anthrazinetetrone, 6,15-dihydro-

$C_{28}H_{20}O$
6175 - Furan, tetraphenyl-

$C_{28}H_{22}$
367 - Anthracene, 9,10-bis(phenylmethyl)-

$C_{28}H_{22}O_3$
937 - Benzeneacetic acid, α-phenyl-, anhydride

$C_{28}H_{24}N_2O_7$
8549 - Orcein

$C_{28}H_{30}O_4$
7352 - 1(3H)-Isobenzofuranone, 3,3-bis[4-hydroxy-2-methyl-5-(1-methylethyl)phenyl]-

$C_{28}H_{32}ClN_2O_3$
12056 - Xanthylium, 9-(2-carboxyphenyl)-3,6-bis(diethylamino)-, chloride

$C_{28}H_{32}O_4Si_4$
5000 - Cyclotetrasiloxane, 2,4,6,8-tetramethyl-2,4,6,8-tetraphenyl-

$C_{28}H_{34}O_{15}$
2794 - 4H-1-Benzopyran-4-one, 7-[[6-O-(6-deoxy-α-L-mannopyranosyl)-β-D-glucopyranosyl]oxy]-2,3-dih

$C_{28}H_{38}N_2O_4$
5367 - Emetan-6′-ol, 7′,10,11-trimethoxy-

$C_{28}H_{38}O_{19}$
6229 - α-D-Glucopyranoside, 1,3,4,6-tetra-O-acetyl-β-D-fructofuranosyl, tetraacetate

$C_{28}H_{42}O$
5380 - Ergosta-5,7,9(11),22-tetraen-3-ol, (3β,22E)-

$C_{28}H_{44}O$
5381 - Ergosta-5,7,22-trien-3-ol, (3β,10α,22E)-
5382 - Ergosta-5,7,22-trien-3-ol, (3β,9β,10α,22E)-
5383 - Ergosta-5,7,22-trien-3-ol, (3β,22E)-
11349 - 9,10-Secoergosta-5(10),6,8,22-tetraen-3-ol, (3β,6E,22E)-
11350 - 9,10-Secoergosta-5,7,10(19),22-tetraen-3-ol, (3β,5Z,7E,22E)-

$C_{28}H_{48}O$
357 - Androst-4-en-3-one, 17-hydroxy-7,17-dimethyl-, (7β,17β)-
5384 - Ergost-7-en-3-ol, (3β,5α)-
5385 - Ergost-8(14)-en-3-ol, (3β,5α)-
5386 - Ergost-5-en-3-ol, (3β,24R)-

$C_{28}H_{48}O_2$
2789 - 2H-1-Benzopyran-6-ol, 3,4-dihydro-2,5,8-trimethyl-2-(4,8,12-trimethyltridecyl)-
2790 - 2H-1-Benzopyran-6-ol, 3,4-dihydro-2,7,8-trimethyl-2-(4,8,12-trimethyltridecyl)-

$C_{28}H_{50}$
1565 - Benzene, docosyl-
5377 - Ergostane, (5α)-
5378 - Ergostane, (5β)-

$C_{28}H_{50}N_2O_4$
4064 - Carpaine

$C_{28}H_{50}O$
5379 - Ergostan-3-ol, (3β,5α)-

$C_{28}H_{54}O_3$
11623 - Tetradecanoic acid, anhydride

$C_{28}H_{56}$
4380 - Cyclohexane, docosyl-

$C_{28}H_{56}O_2$
8221 - Octacosanoic acid

$C_{28}H_{58}$
5274 - Docosane, 7-hexyl-
5354 - Eicosane, 9-octyl-
8220 - Octacosane
11877 - Tridecane, 2,2,4,10,12,12-hexamethyl-7-(3,5,5-trimethylhexyl)-

$C_{28}H_{58}O$
8222 - 1-Octacosanol

$C_{29}H_{40}N_2O_4$
5368 - Emetan, 6′,7′,10,11-tetramethoxy-

$C_{29}H_{44}O_{12}$
4040 - Card-20(22)-enolide, 3-[(6-deoxy-α-L-mannopyranosyl)oxy]-1,5,11,14,19-pentahydroxy-, (1β,3β

$C_{29}H_{47}NO_8$
4067 - Cevane-3,4,12,14,16,17,20-heptol, 3-acetate, (3β,4α,16β)-

$C_{29}H_{48}O$
11554 - Stigmasta-5,7-dien-3-ol, (3β)-
11555 - Stigmasta-5,22-dien-3-ol, (3β,22E)-

$C_{29}H_{50}O$
11558 - Stigmast-5-en-3-ol, (3β)-
11559 - Stigmast-5-en-3-ol, (3β,24S)-

$C_{29}H_{50}O_2$
2787 - 2H-1-Benzopyran-6-ol, 3,4-dihydro-2,5,7,8-tetramethyl-2-(4,8,12-trimethyltridecyl)-, [2R-[2R*(4R*,8R*)]]-

$C_{29}H_{52}$
2341 - Benzene, tricosyl-

$C_{29}H_{52}O$
11557 - Stigmastan-3-ol, (3β,5α)-

$C_{29}H_{58}$
4483 - Cyclohexane, tricosyl-

$C_{29}H_{60}$
8099 - Nonacosane

$C_{29}H_{60}O$
8100 - 1-Nonacosanol

$C_{30}H_{34}O_{13}$
7579 - 3,6-Methano-8H-1,5,7-trioxacyclopenta[ij]cycloprop[a]azulene-4,8(3H)-dione, hexahydro-2a-hydroxy

$C_{30}H_{37}N_5O_5$
5393 - Ergotaman-3′,6′,18-trione, 12′-hydroxy-2′-methyl-5′-(2-methylpropyl)-,

$C_{30}H_{40}O_2$
435 - 8′-Apo-β,ψ-carotenal, 3-hydroxy-, (3R)-

$C_{30}H_{42}FeO_2Si_4$
5961 - Ferrocene, 1,1′-bis(1,1,3,3-tetramethyl-3-phenyldisiloxanyl)-

$C_{30}H_{46}O_5$
8546 - Olean-12-en-28-oic acid, 3,16-dihydroxy-23-oxo-, (3β,4α,16α)-
12036 - Urs-12-ene-27,28-dioic acid, 3-hydroxy-, (3β)-

$C_{30}H_{48}O_3$
8547 - Olean-12-en-28-oic acid, 3-hydroxy-, (3β)-
12037 - Urs-12-en-28-oic acid, 3-hydroxy-, (3β)-

$C_{30}H_{48}O_4$
8545 - Olean-12-en-28-oic acid, 3,23-dihydroxy-, (3β,4α)-

$C_{30}H_{50}$
11602 - 2,6,10,14,18,22-Tetracosahexaene, 2,6,10,15,19,23-hexamethyl-, (all-E)-

$C_{30}H_{50}O$
7411 - Lanosta-8,24-dien-3-ol, (3β)-
7417 - Lup-20(29)-en-3-ol, (3β)-
8548 - Olean-12-en-3-ol, (3β)-
11556 - Stigmasta-7,24(28)-dien-3-ol, 4-methyl-, (3β,4α,5α,24Z)-
12038 - Urs-12-en-3-ol, (3β)-

$C_{30}H_{50}O_2$
7416 - Lup-20(29)-ene-3,28-diol, (3β)-

$C_{30}H_{54}$
2265 - Benzene, tetracosyl-

$C_{30}H_{60}$
4474 - Cyclohexane, tetracosyl-

$C_{30}H_{60}I_3N_3O_3$
5441 - Ethanaminium, 2,2′,2″-[1,2,3-benzenetriyltris(oxy)]tris[N,N,N-triethyl-, tr

$C_{30}H_{60}O_2$
11820 - Triacontanoic acid

$C_{30}H_{62}$
5275 - Docosane, 9-octyl-
11606 - Tetracosane, 2,6,10,15,19,23-hexamethyl-
11819 - Triacontane

$C_{30}H_{62}O$
11821 - 1-Triacontanol

$C_{31}H_{36}N_2O_8$
12064 - Yohimban-16-carboxylic acid, 17-hydroxy-18-[(3,4,5-trimethoxybenzoyl)oxy]-, methyl ester, (3β,16β

$C_{31}H_{39}N_5O_5$
5387 - Ergotaman-3′,6′,18-trione, 12′-hydroxy-2′,5′-bis(1-methylethyl)-,
5388 - Ergotaman-3′,6′,18-trione, 12′-hydroxy-2′,5′-bis(1-methylethyl)-,

$C_{31}H_{46}O_2$
7865 - 1,4-Naphthalenedione, 2-methyl-3-(3,7,11,15-tetramethyl-2-hexadecenyl)-, [R-[R*,R*-(E)]]-

$C_{31}H_{52}O_3$
2788 - 2H-1-Benzopyran-6-ol, 3,4-dihydro-2,5,7,8-tetramethyl-2-(4,8,12-trimethyltridecyl)-, acetate, [2R-[2R*(4R*,8R*

$C_{41}H_{64}O_{14}$

4041 - Card-20(22)-enolide, 3-[(O-2,6-dideoxy-β-D-ribo-hexopyranosyl-(1-4)-O-2,6-dideoxy-β-D-ribo-hexopyranosyl-(1-4)-2,6-dideoxy-β-D, 12,14-dihydroxy

4042 - Card-20(22)-enolide, 3-[(O-2,6-dideoxy-β-D-ribo-hexopyranosyl-(1-4)-O-2,6-dideoxy-β-D-ribo-hexopyranosyl-(1-4)-2,6-dideoxy-β-D, 14,16-dihydroxy

$C_{42}H_{46}N_4O_8S$

11566 - Strychnidin-10-one, sulfate (2:1)

$C_{42}H_{62}O_{16}$

6230 - α-D-Glucopyranosiduronic acid, (3β,20β)-20-carboxy-11-oxo-30-norolean-12-en-3-yl 2-O-

$C_{42}H_{84}$

4413 - Cyclohexane, hexatriacontyl-

$C_{42}H_{84}N_{14}O_{36}S_3$

11561 - D-Streptamine, O-2-deoxy-2-(methylamino)-α-L-glucopyranosyl-(12)-O-5-deoxy-3-C-formyl-α-L-lyxofuranosyl-(1-4)-N,N,N' -55-bis(aminoiminomethyl)-, sulfate (2:3)(salt)

$C_{43}H_{88}$

11918 - Tritetracontane

$C_{45}H_{96}O_6$

11628 - Tetradecanoic acid, 1,2,3-propanetriyl ester

$C_{48}H_{40}O_4Si_4$

4994 - Cyclotetrasiloxane, octaphenyl-

$C_{48}H_{72}O_{14}$

477 - Avermectin B1a

$C_{48}H_{93}AlO_6$

6645 - Hexadecanoic acid, aluminum salt

$C_{50}H_{83}NO_{21}$

6196 - β-D-Galactopyranoside, (3β,5α,22β,25S)-spirosolan-3-yl O-β-

$C_{51}H_{98}O_6$

6656 - Hexadecanoic acid, 1,2,3-propanetriyl ester

$C_{54}H_{105}AlO_6$

8248 - Octadecanoic acid, aluminum salt

$C_{56}H_{92}O_{29}$

6195 - β-D-Galactopyranoside, (2α,3β,5α,15β,25R)-2,15-dihydroxyspirostan-3-yl

$C_{57}H_{104}O_6$

8310 - 9-Octadecenoic acid (Z)-, 1,2,3-propanetriyl ester

$C_{57}H_{110}O_6$

8272 - Octadecanoic acid, 1,2,3-propanetriyl ester

$C_{60}H_{78}OSn_2$

5950 - Fenbutatin oxide

$C_{66}H_{103}N_{17}O_{16}S$

508 - Bacitracin

CAS	Page	CAS	Page	CAS	Page	CAS	Page	CAS	Page	CAS	Page
50-00-0	5987	54-95-5	11654	58-72-0	1648	63-74-1	2187	71-30-7	11104		
50-01-1	6285	55-10-7	896	58-73-1	5422	63-75-2	10914	71-36-3	3593		
50-02-2	9737	55-18-5	5432	58-74-2	7378	63-91-2	9510	71-41-0	8915		
50-03-3	9747	55-21-0	600	58-85-5	11689	63-98-9	879	71-43-2	867		
50-06-6	11095	55-22-1	10895	58-86-6	12059	64-04-0	1569	71-44-3	3259		
50-10-2	5443	55-38-9	5955	58-89-9	4409	64-10-8	12026	71-55-6	5655		
50-11-3	11088	55-43-6	1801	58-90-2	9470	64-17-5	5672	71-63-6	4043		
50-14-6	11350	55-63-0	10197	58-93-5	2855	64-18-6	6002	71-81-8	2120		
50-22-6	9748	55-73-2	6281	58-94-6	2852	64-19-7	100	71-91-0	5447		
50-23-7	9753	55-91-4	9580	58-95-7	2788	64-20-0	7456	72-14-0	2206		
50-24-8	9738	56-03-1	7233	58-96-8	12033	64-65-3	9656	72-18-4	12041		
50-27-1	5404	56-04-2	11108	58-97-9	12035	64-67-5	11573	72-19-5	11813		
50-28-2	5399	56-12-2	3464	59-00-7	11230	64-69-7	200	72-23-1	9755		
50-29-3	2328	56-18-8	9874	59-02-7	2787	64-77-7	2207	72-43-5	2329		
50-30-6	2563	56-23-5	7540	59-23-4	6197	64-85-7	9751	72-44-6	11199		
50-31-7	2727	56-25-7	5371	59-26-7	10891	64-86-8	93	72-48-0	391		
50-32-8	2395	56-29-1	11083	59-30-3	6243	65-29-2	5441	72-54-8	1378		
50-34-0	9832	56-33-7	5238	59-31-4	11325	65-45-2	616	72-55-9	1374		
50-35-1	7363	56-35-9	5241	59-40-5	2205	65-46-3	5021	72-56-0	9160		
50-36-2	482	56-36-0	11538	59-46-1	2493	65-49-6	2498	72-63-9	344		
50-42-0	938	56-38-2	9583	59-47-2	10019	65-64-5	7170	73-22-3	11922		
50-44-2	10763	56-40-6	6250	59-48-3	7327	65-71-4	11068	73-24-5	10751		
50-50-0	5403	56-41-7	325	59-49-4	2882	65-82-7	7649	73-31-4	69		
50-53-3	9505	56-45-1	11357	59-50-7	9267	65-85-0	2475	73-32-5	7370		
50-55-5	12061	56-49-5	2392	59-51-8	7647	65-86-1	11050	73-40-5	10768		
50-69-1	11345	56-53-1	9300	59-52-9	10370	66-02-4	11928	73-49-4	11197		
50-70-4	6206	56-54-2	4126	59-66-5	32	66-22-8	11059	74-11-3	2536		
50-71-5	11073	56-55-3	519	59-67-6	10894	66-23-9	5438	74-31-7	1271		
50-78-2	2480	56-57-5	11287	59-87-0	7144	66-25-1	6718	74-39-5	1542		
50-79-3	2562	56-72-4	4147	59-88-1	7172	66-27-3	7532	74-79-3	438		
50-81-7	466	56-75-7	52	59-89-2	7699	66-28-4	4049	74-82-8	7459		
50-82-8	2728	56-81-5	10186	59-92-7	11929	66-32-0	11565	74-83-9	7462		
50-84-0	2561	56-82-6	9782	59-98-3	7198	66-56-8	9341	74-84-0	5448		
50-85-1	2630	56-84-8	471	60-00-4	6262	66-71-7	9189	74-85-1	5879		
50-89-5	11816	56-85-9	6249	60-01-5	3586	66-75-1	11061	74-86-2	5936		
50-91-9	12034	56-86-0	6241	60-09-3	837	66-76-2	2839	74-87-3	7477		
51-03-6	9727	56-87-1	7419	60-10-6	5121	66-77-3	7751	74-88-4	7518		
51 12-7	10910	56-89-3	5020	60-11-9	721	66-81-9	9655	74-89-5	7442		
51-17-2	2376	57-00-1	6254	60-12-8	1592	66-99-9	7752	74-90-8	7176		
51 18 3	11844	57-06-7	10575	60-18-4	11926	67-03-8	11686	74-93-1	7546		
51-20-7	11062	57-08-9	6871	60-23-1	5649	67-20-9	7222	74-95-3	7492		
51-21-8	11067	57-10-3	6644	60-24-2	5732	67-43-6	6257	74-96-4	5454		
51-26-3	2136	57-11-4	8247	60-27-5	7230	67-45-8	8589	74-97-5	7463		
51-28-5	9342	57-13-6	11992	60-29-7	5608	67-47-0	6044	74-98-6	9834		
51-34-3	908	57-14-7	7154	60-31-1	5439	67-48-1	5446	74-99-7	10727		
51-35-4	9767	57-15-8	10407	60-32-2	6873	67-51-6	10828	75 00 3	5467		
51-38-5	2564	57-24-9	11563	60-33-3	8223	67-52-7	11076	75-01-4	5884		
51-41-2	1500	57-27-2	7666	60-34-4	7161	67-56-1	7581	75-02-5	5906		
51-43-4	1521	57-41-0	7219	60-35-5	22	67-63-0	10335	75-03-6	5597		
51-45-6	7203	57-43-2	11092	60-41-3	11566	67-64-1	10411	75-04-7	5410		
51-48-9	11930	57-44-3	11087	60-51-5	5009	67-66-3	7554	75-05-8	271		
51-52-5	11109	57-47-6	11191	60-54-8	7706	67-68-5	7529	75-07-0	11		
51-55-8	902	57-50-1	6218	60-56-0	7214	67-71-0	7534	75-08-1	5648		
51-63-8	1582	57-53-4	10021	60-57-1	5178	67-72-1	5595	75-09-2	7499		
51-66-1	71	57-55-6	9982	60-70-8	12044	67-97-0	11348	75-10-5	7504		
51-67-2	9208	57-57-8	8595	60-80-0	10847	68-11-1	202	75-11-6	7506		
51-75-2	5413	57-62-5	7704	60-82-2	10453	68-12-2	5992	75-12-7	5989		
51-79-6	3948	57-63-6	8216	60-87-7	9504	68-26-8	11341	75-15-0	3993		
51-80-9	7490	57-66-9	2595	60-91-3	9503	68-35-9	2204	75-17-2	5988		
51-83-2	5440	57-67-0	2191	61-00-7	5807	68-36-0	996	75-18-3	7545		
52-28-8	7673	57-68-1	2196	61-12-1	11226	68-41-7	7406	75-19-4	4955		
52-31-3	11084	57-71-6	3361	61-19-8	320	68-88-2	5700	75-21-8	8599		
52-39-1	9746	57-74-9	4077	61-50-7	7303	68-94-0	10770	75-22-9	3132		
52-43-7	11091	57-85-2	359	61-54-1	7302	68-96-2	9750	75-24-1	339		
52-49-3	9705	57-87-4	5383	61-73-4	9506	69-24-9	10435	75-25-2	7548		
52-51-7	9991	57-88-5	4110	61-78-9	6252	69-57-8	11655	75-26-3	9837		
52-52-8	4782	57-91-0	5400	61-80-3	2876	69-65-8	7422	75-27-4	7467		
52-67-5	12043	57-92-1	11562	61-82-5	11845	69-72-7	2612	75-28-5	10073		
52-68-6	11853	57-97-6	520	61-90-5	7413	69-89-6	10755	75-29-6	9849		
52-85-7	5946	58-00-4	5155	62-23-7	2705	69-91-0	885	75-30-9	10055		
52-86-8	3651	58-05-9	6244	62-38-4	7431	69-93-2	10764	75-31-0	9804		
52-90-4	5014	58-08-2	10760	62-44-2	60	70-07-5	8588	75-33-2	10157		
53-03-2	9739	58-14-0	11051	62-50-0	7531	70-11-1	5782	75-34-3	5528		
53-05-4	9744	58-15-1	10851	62-51-1	9830	70-18-8	6264	75-35-4	5894		
53-06-5	9754	58-18-4	358	62-53-3	633	70-22-4	11188	75-36-5	294		
53-16-7	5406	58-22-0	356	62-54-4	122	70-25-7	6284	75-37-6	5544		
53-41-8	348	58-25-3	2409	62-55-5	5643	70-26-8	8550	75-38-7	5902		
53-46-3	5445	58-27-5	7864	62-56-6	11799	70-29-1	5621	75-39-8	5674		
53-70-3	5142	58-32-2	5711	62-57-7	331	70-30-4	9409	75-43-4	7502		
53-89-4	10850	58-38-8	9501	62-59-9	4069	70-34-8	1718	75-44-5	4007		
53-96-3	64	58-39-9	9617	62-73-7	5170	70-47-3	468	75-45-6	7478		
54-04-6	1586	58-46-8	2396	62-75-9	7450	70-54-2	7418	75-46-7	7558		
54-06-8	7301	58-54-8	168	62-76-0	5566	70-69-9	10413	75-47-8	7560		
54-11-5	10992	58-55-9	10756	62-90-8	5408	70-70-2	10450	75-50-3	7444		
54-25-1	11832	58-56-0	10034	63-05-0	352	70-78-0	11932	75-52-5	7522		
54-36-4	10462	58-61-7	319	63-25-2	8032	71-00-1	7132	75-54-7	11398		
54-85-3	10905	58-63-9	7335	63-68-3	7648	71-23-8	10334	75-55-8	501		

CAS	Number	CAS	Number	CAS	Number	CAS	Number	CAS	Number
75-56-9	: 8621	77-09-8	: 7354	79-10-7	: 10609	81-92-5	: 1841	85-46-1	: 7975
75-57-0	: 7457	77-10-1	: 9696	79-11-8	: 123	82-02-0	: 6185	85-47-2	: 7959
75-58-1	: 7458	77-16-7	: 11904	79-14-1	: 196	82-05-3	: 6292	85-55-2	: 2684
75-60-5	: 451	77-25-8	: 9940	79-15-2	: 36	82-12-2	: 406	85-64-3	: 9398
75-61-6	: 7496	77-26-9	: 11101	79-16-3	: 72	82-28-0	: 384	85-66-5	: 9196
75-62-7	: 7474	77-28-1	: 11082	79-17-4	: 7147	82-29-1	: 417	85-73-4	: 2726
75-63-8	: 7475	77-40-7	: 9433	79-19-6	: 7142	82-34-8	: 410	85-74-5	: 7873
75-64-9	: 9818	77-41-8	: 11156	79-20-9	: 208	82-35-9	: 405	85-75-6	: 7872
75-65-0	: 10389	77-42-9	: 9103	79-21-0	: 5619	82-40-6	: 6182	85-82-5	: 8029
75-66-1	: 10159	77-46-3	: 91	79-22-1	: 4020	82-44-0	: 387	85-83-6	: 8033
75-68-3	: 5470	77-47-4	: 4766	79-24-3	: 5607	82-45-1	: 380	85-84-7	: 7729
75-69-4	: 7555	77-48-5	: 7217	79-27-6	: 5632	82-54-2	: 5216	85-86-9	: 8038
75-71-8	: 7500	77-49-6	: 10018	79-28-7	: 5914	82-57-5	: 6186	85-90-5	: 2838
75-72-9	: 7486	77-52-1	: 12037	79-29-8	: 3290	82-58-6	: 5376	85-91-6	: 2681
75-73-0	: 7541	77-53-2	: 7570	79-30-1	: 10504	82-62-2	: 9490	85-95-0	: 9303
75-74-1	: 9732	77-59-8	: 11528	79-31-2	: 10290	82-66-6	: 7268	85-97-2	: 3092
75-75-2	: 7530	77-60-1	: 11532	79-34-5	: 5635	82-68-8	: 2087	85-98-3	: 12003
75-76-3	: 11451	77-63-4	: 5000	79-35-6	: 5897	82-71-3	: 1554	86-00-0	: 3085
75-77-4	: 11389	77-65-6	: 3195	79-36-7	: 297	82-75-7	: 7965	86-21-5	: 11010
75-78-5	: 11394	77-67-8	: 11158	79-37-8	: 5585	82-86-0	: 8	86-26-0	: 3080
75-79-6	: 11465	77-71-4	: 7218	79-38-9	: 5889	83-07-8	: 10845	86-28-2	: 3985
75-80-9	: 5767	77-74-7	: 8944	79-40-3	: 5588	83-12-5	: 7270	86-29-3	: 969
75-81-0	: 5520	77-75-8	: 9146	79-41-4	: 10648	83-13-6	: 9975	86-30-6	: 830
75-82-1	: 5522	77-76-9	: 9909	79-42-5	: 10285	83-14-7	: 7394	86-34-0	: 11161
75-83-2	: 3289	77-77-0	: 5913	79-43-6	: 162	83-25-0	: 11162	86-48-6	: 7771
75-84-3	: 10373	77-78-1	: 11574	79-44-7	: 3973	83-27-2	: 9967	86-50-0	: 497
75-85-4	: 3628	77-79-2	: 11752	79-46-9	: 10117	83-32-9	: 6	86-52-2	: 7783
75-86-5	: 10108	77-83-8	: 8604	79-49-2	: 10477	83-33-0	: 7286	86-53-3	: 7748
75-87-6	: 21	77-84-9	: 10007	79-53-8	: 10421	83-34-1	: 7314	86-55-5	: 7758
75-88-7	: 5489	77-85-0	: 10011	79-54-9	: 9169	83-40-9	: 2629	86-56-6	: 7715
75-89-8	: 5770	77-86-1	: 9986	79-55-0	: 9691	83-41-0	: 1474	86-57-7	: 7929
75-91-2	: 7178	77-89-4	: 10166	79-57-2	: 7705	83-42-1	: 1204	86-59-9	: 11229
75-93-4	: 11577	77-92-9	: 10167	79-58-3	: 11515	83-44-3	: 4095	86-60-2	: 7964
75-94-5	: 11461	77-93-0	: 10169	79-63-0	: 7411	83-45-4	: 11557	86-65-7	: 7870
75-95-6	: 5616	77-94-1	: 10168	79-69-6	: 3884	83-46-5	: 11558	86-68-0	: 11234
75-96-7	: 241	77-95-2	: 4338	79-70-9	: 3883	83-47-6	: 11559	86-73-7	: 5967
75-97-8	: 3657	77-99-6	: 10006	79-74-3	: 1501	83-48-7	: 11555	86-74-8	: 3981
75-98-9	: 10258	78-00-2	: 9731	79-77-6	: 3885	83-49-8	: 4096	86-75-9	: 11309
75-99-0	: 10244	78-07-9	: 11478	79-83-4	: 329	83-53-4	: 7802	86-76-0	: 5164
76-00-6	: 10406	78-08-0	: 11424	80-00-2	: 1225	83-54-5	: 11330	86-78-2	: 8737
76-01-7	: 5617	78-09-1	: 5601	80-05-7	: 9413	83-56-7	: 7835	86-80-6	: 5411
76-02-8	: 300	78-10-4	: 11506	80-06-8	: 1850	83-60-3	: 12063	86-84-0	: 7904
76-03-9	: 243	78-11-5	: 9990	80-07-9	: 2239	83-66-9	: 1463	86-86-2	: 7733
76-04-0	: 127	78-13-7	: 11507	80-08-0	: 846	83-67-0	: 10757	86-87-3	: 7734
76-05-1	: 263	78-26-2	: 10020	80-09-1	: 9466	83-68-1	: 7800	86-88-4	: 11807
76-06-2	: 7557	78-27-3	: 4529	80-10-4	: 11395	83-72-7	: 7859	86-89-5	: 7941
76-08-4	: 10405	78-30-8	: 9569	80-11-5	: 2212	83-73-8	: 11315	86-98-6	: 11251
76-09-5	: 3350	78-32-0	: 9571	80-13-7	: 2565	83-74-9	: 7186	87-00-3	: 911
76-11-9	: 5636	78-34-2	: 5201	80-14-8	: 11915	83-79-4	: 2783	87-10-5	: 609
76-12-0	: 5637	78-38-6	: 9536	80-15-9	: 7181	83-81-8	: 1325	87-13-8	: 9947
76-13-1	: 5663	78-39-7	: 5664	80-18-2	: 2227	83-88-5	: 11344	87-17-2	: 617
76-14-2	: 5538	78-40-0	: 9563	80-26-2	: 4669	83-95-4	: 6181	87-19-4	: 2643
76-15-3	: 5484	78-48-8	: 11852	80-32-0	: 2193	84-11-7	: 9174	87-20-7	: 2633
76-16-4	: 5596	78-54-6	: 4528	80-33-1	: 2221	84-15-1	: 11595	87-24-1	: 2690
76-17-5	: 10181	78-59-1	: 4721	80-35-3	: 2197	84-16-2	: 9298	87-25-2	: 2495
76-19-7	: 10118	78-62-6	: 11405	80-40-0	: 2231	84-17-3	: 9302	87-28-5	: 2622
76-20-0	: 3230	78-67-1	: 10093	80-42-2	: 2235	84-26-4	: 7334	87-31-0	: 2874
76-24-4	: 3106	78-74-0	: 5653	80-46-6	: 9339	84-31-1	: 4134	87-39-8	: 11074
76-28-8	: 4046	78-75-1	: 9881	80-48-8	: 2232	84-52-6	: 5022	87-40-1	: 2332
76-30-2	: 3342	78-77-3	: 9845	80-55-7	: 10279	84-54-8	: 409	87-41-2	: 7349
76-36-8	: 3191	78-78-4	: 3397	80-56-8	: 2973	84-55-9	: 10434	87-42-3	: 10753
76-37-9	: 10404	78-79-5	: 3171	80-59-1	: 3844	84-58-2	: 4251	87-44-5	: 2996
76-39-1	: 10391	78-80-8	: 3892	80-62-6	: 10663	84-61-7	: 1341	87-47-8	: 3945
76-43-7	: 355	78-81-9	: 9817	80-63-7	: 10621	84-62-8	: 1345	87-51-4	: 7295
76-44-8	: 7576	78-82-0	: 10111	80-69-3	: 9956	84-65-1	: 379	87-52-5	: 7310
76-45-9	: 4070	78-83-1	: 10388	80-72-8	: 4942	84-66-2	: 1342	87-58-1	: 11140
76-49-3	: 2948	78-84-2	: 9785	80-92-2	: 9742	84-68-4	: 3032	87-59-2	: 704
76-57-3	: 7669	78-85-3	: 10515	80-97-7	: 4107	84-69-5	: 1336	87-60-5	: 671
76-59-5	: 9365	78-88-6	: 10552	81-04-9	: 7867	84-74-2	: 1339	87-61-6	: 2321
76-60-8	: 9367	78-89-7	: 10347	81-07-2	: 2388	84-79-7	: 7862	87-62-7	: 707
76-61-9	: 9369	78-90-0	: 9872	81-08-3	: 2875	84-80-0	: 7865	87-64-9	: 9263
76-62-0	: 7350	78-92-2	: 3594	81-13-0	: 3202	84-86-6	: 7962	87-65-0	: 9285
76-65-3	: 2463	78-93-3	: 3645	81-15-2	: 1461	84-87-7	: 7971	87-66-1	: 2366
76-67-5	: 9952	78-94-4	: 3875	81-16-3	: 7961	84-88-8	: 11294	87-68-3	: 3163
76-68-6	: 11085	78-95-5	: 10420	81-20-9	: 1476	84-89-9	: 7963	87-69-4	: 3309
76-76-6	: 11093	78-97-7	: 10105	81-23-2	: 4101	84-95-7	: 7713	87-73-0	: 6205
76-80-2	: 3110	78-98-8	: 9788	81-24-3	: 5625	84-99-1	: 2398	87-79-6	: 11519
76-83-5	: 1196	78-99-9	: 9891	81-25-4	: 4100	85-00-7	: 5224	87-81-0	: 11587
76-84-6	: 1866	79-00-5	: 5656	81-30-1	: 2784	85-01-8	: 9163	87-83-2	: 2083
76-87-9	: 11911	79-01-6	: 5922	81-38-9	: 2469	85-02-9	: 2437	87-85-4	: 1758
76-89-1	: 920	79-03-8	: 10493	81-54-9	: 416	85-23-4	: 4260	87-86-5	: 9440
76-93-7	: 918	79-04-9	: 295	81-61-8	: 413	85-29-0	: 7603	87-88-7	: 4258
76-94-8	: 11100	79-05-0	: 9791	81-64-1	: 392	85-34-7	: 944	87-90-1	: 11840
77-02-1	: 11098	79-06-1	: 10519	81-77-6	: 434	85-38-1	: 2646	87-92-3	: 3313
77-03-2	: 9654	79-07-2	: 41	81-81-2	: 2832	85-41-6	: 7362	87-99-0	: 12058
77-04-3	: 10946	79-08-3	: 111	81-88-9	: 12056	85-42-7	: 7345	88-04-0	: 9257
77-06-5	: 6204	79-09-4	: 10201	81-90-3	: 2517	85-44-9	: 7342	88-05-1	: 859

CAS	No.	CAS	No.	CAS	No.	CAS	No.	CAS	No.
88-06-2	9485	90-80-2	6212	93-15-2	1450	95-50-1	1354	97-30-3	6227
88-09-5	3514	90-81-3	1900	93-16-3	1449	95-51-2	658	97-39-2	6279
88-10-8	3972	90-84-6	10428	93-18-5	7884	95-52-3	1726	97-53-0	9389
88-12-0	11183	90-86-8	1903	93-22-1	7942	95 53 4	778	97-54-1	9388
88-13-1	11738	90-89-1	9603	93-25-4	921	95-54-5	1260	97-56-3	798
88-14-2	6050	90-90-4	7602	93-26-5	70	95-55-6	9201	97-59-0	12009
88-15-3	5870	90-94-8	7593	93-28-7	9395	95-56-7	9228	97-62-1	10298
88-16-4	1240	90-98-2	7592	93-35-6	2826	95-57-8	9251	97-63-2	10658
88-17-5	855	90-99-3	1187	93-37-8	11258	95-59-0	5188	97-65-4	3328
88-18-6	9321	91-00-9	1826	93-39-0	2821	95-63-6	2358	97-72-3	10291
88-20-0	2228	91-01-0	1910	93-42-5	68	95-64-7	708	97-74-5	11715
88-21-1	2215	91-02-1	7638	93-44-7	8021	95-65-8	9311	97-77-8	11721
88-60-8	9331	91-08-7	1430	93-50-5	667	95-66-9	1138	97-85-8	10304
88-65-3	2518	91-10-1	9305	93-51-6	9385	95-68-1	705	97-86-9	10666
88-67-5	2657	91-13-4	983	93-52-7	1299	95-69-2	673	97-88-1	10650
88-68-6	601	91-14-5	1403	93-53-8	872	95-70-5	1279	97-90-5	10656
88-69-7	9410	91-16-7	1436	93-54-9	1870	95-71-6	1532	97-94-9	3118
88-72-2	2000	91-17-8	7789	93-55-0	10480	95-72-7	1140	97-95-0	3620
88-73-3	1214	91-18-9	10747	93-56-1	5579	95-73-8	1388	97-96-1	3183
88-74-4	822	91-19-0	11333	93-58-3	2687	95-74-9	672	97-97-2	5472
88-75-5	9435	91-20-3	7732	93-59-4	1126	95-75-0	1384	97-99-4	6124
88-85-7	5183	91-21-4	7388	93-60-7	10906	95-76-1	689	98-00-0	6118
88-87-9	9258	91-22-5	11214	93-61-8	5998	95-77-2	9286	98-01-1	6041
88-88-0	1244	91-23-6	1935	93-71-0	44	95-78-3	706	98-02-2	6117
88-89-1	9497	91-40-7	2715	93-72-1	11512	95-79-4	676	98-03-3	11733
88-95-9	1318	91-49-6	39	93-76-5	256	95-80-7	1278	98-04-4	866
88-96-0	1324	91-53-2	11265	93-78-7	259	95-81-8	669	98-05-5	460
88-99-3	1327	91-56-5	7300	93-79-8	258	95-82-9	687	98-06-6	1459
89-00-9	10927	91-57-6	7916	93-88-9	1574	95-83-0	1264	98-07-7	2333
89-01-0	10804	91-58-7	7781	93-89-0	2602	95-84-1	9210	98-08-8	2348
89-05-4	2257	91-59-9	7709	93-90-3	5743	95-85-2	9204	98-09-9	2242
89-31-6	7397	91-60-1	8008	93-91-4	3362	95-86-3	9274	98-11-3	2214
89-37-2	3975	91-62-3	11281	93-97-0	2510	95-87-4	9309	98-13-5	11473
89-39-4	1447	91-63-4	11277	93-98-1	626	95-92-1	5560	98-15-7	1241
89-54-3	2490	91-64-5	2792	93-99-2	2716	95-93-2	2280	98-16-8	856
89-55-4	2524	91-65-6	4285	94-02-0	2141	95-94-3	2260	98-17-9	9491
89-56-5	2631	91-66-7	695	94-07-5	1887	95-95-4	9484	98-19-1	1460
89-57-6	2499	91-68-0	9297	94-00-0	2092	96-00-4	4206	98-27-1	9333
89-58-7	1478	91-73-6	1829	94-09-7	2497	96-01-5	4203	98-29-3	1516
89-59-8	1212	91-76-0	11829	94-11-1	173	96-02-6	6092	98-46-4	2061
89-60-1	1209	91-78-1	11834	94-12-2	2507	96-04-8	6402	98-50-0	456
89-61-2	1395	91-80-5	5507	94-13-3	2653	96-05-9	10671	98-51-1	1466
89-62-3	803	91-81-6	5505	94-14-4	2505	96-09-3	8624	98-54-4	9323
89-68-9	9268	91-84-9	5511	94-20-2	2208	96-11-7	10163	98-56-6	1242
89-69-0	2339	91-88-3	5724	94-25-7	2488	96-12-8	9085	98-58-8	2244
89-71-4	2696	91-96-3	3204	94-26-8	2615	96-14-0	8824	98-59-9	2250
89-72-5	9430	92-06-8	11596	94-30-4	2671	96-18-4	10175	98-60-2	2245
89-73-6	613	92-13-7	6151	94-33-7	5578	96-19-5	10603	98-61-3	2247
89-74-7	5809	92-24-0	7703	94-36-0	9157	96-21-9	10357	98-62-4	2252
89-75-8	2893	92-35-3	11111	94-41-7	10708	96-22-0	8958	98-66-8	2220
89-79-2	4544	92-36-4	782	94-44-0	10911	96-23-1	10362	98-67-9	2226
89-82-7	4608	92-41-1	7869	94-46-2	3632	96-24-2	9994	98-71-5	2224
89-83-8	9418	92-43-3	10844	94-49-5	5572	96-26-4	10429	98-72-6	459
89-84-9	5803	92-48-8	2835	94-53-1	2418	96-27-5	10013	98-73-7	2588
89-86-1	2569	92-51-3	2988	94-58-6	2431	96-29-7	3675	98-79-3	9768
89-87-2	1479	92-52-4	3008	94-59-7	2429	96-31-1	12004	98-80-6	3135
89-92-9	1062	92-54-6	9620	94-62-2	9632	96-32-2	117	98-81-7	1025
89-93-0	1815	92-59-1	1811	94-65-5	4612	96-33-3	10655	98-82-8	1975
89-95-2	1897	92-61-5	2829	94-67-7	579	96-34-4	138	98-83-9	1972
89-96-3	1161	92-62-6	316	94-68-8	746	96-35-5	198	98-85-1	1896
89-98-5	529	92-66-0	3018	94-69-9	5996	96-37-7	4826	98-86-2	5864
90-00-6	9353	92-67-1	3011	94-70-2	736	96-40-2	4914	98-87-3	1382
90-01-7	1882	92-69-3	3090	94-71-3	9350	96-41-3	4864	98-88-4	2886
90-02-8	564	92-70-6	7772	94-74-6	141	96-43-5	11740	98-89-5	4332
90-04-0	766	92-71-7	8580	94-75-7	169	96-45-7	7225	98-91-9	1130
90-05-1	9379	92-82-0	9193	94-80-4	171	96-47-9	6171	98-92-0	10890
90-11-9	7739	92-83-1	12049	94-81-5	3493	96-48-0	6142	98-94-2	4286
90-12-0	7915	92-84-2	9499	94-82-6	3925	96-49-1	5211	98-95-3	2049
90-13-1	7780	92-85-3	11660	94-96-2	6802	96-50-4	11662	98-96-4	10802
90-14-2	7902	92-86-4	3037	95-00-1	1803	96-54-8	11130	98-97-5	10803
90-15-3	8017	92-87-5	3031	95-01-2	541	96-76-4	9219	98-98-6	10893
90-19-7	2815	92-91-1	5776	95-05-6	11714	96-77-5	1560	99-02-5	5788
90-20-0	7874	92-92-2	3022	95-06-7	4065	96-80-0	5683	99-03-6	5772
90-26-6	881	92-93-3	3087	95-11-4	2966	96-82-2	6210	99-04-7	2677
90-27-7	897	92-94-4	11597	95-13-6	7245	96-88-8	9639	99-05-8	2485
90-33-5	2831	92-99-9	11213	95-14-3	2871	96-91-3	9207	99-07-0	9318
90-39-1	7571	93-01-6	7974	95-15-8	2399	96-97-9	2647	99-08-1	2001
90-41-5	3010	93-02-7	544	95-16-9	2861	96-99-1	2547	99-09-2	823
90-42-9	2989	93-03-8	1856	95-20-5	7313	97-00-7	1144	99-11-6	10900
90-43-7	3088	93-04-9	7914	95-21-6	2880	97-02-9	725	99-12-7	1477
90-44-8	433	93-05-0	1266	95-25-0	2883	97-05-2	2655	99-14-9	10165
90-45-9	313	93-07-2	2576	95-43-2	3193	97-16-5	9289	99-17-2	6223
90-47-1	12050	93-08-3	5858	95-45-4	3360	97-17-6	9581	99-20-7	6220
90-59-5	534	93-09-9	7759	95-46-5	1054	97-18-7	9478	99-30-9	692
90-65-3	6701	93-10-7	11227	95-47-6	1452	97-23-4	9408	99-31-0	1331
90-69-7	5834	93-11-8	7976	95-48-7	9400	97-24-5	9477	99-32-1	10776
90-74-4	6236	93-14-1	10016	95-49-8	1183	97-25-6	6231	99-33-2	2896

99-34-3	: 2591	100-71-0	: 10956	103-19-5	: 5246	105-07-7	: 2758	106-96-7	: 10728
99-35-4	: 2365	100-72-1	: 10784	103-23-1	: 6781	105-13-5	: 1891	106-97-8	: 3229
99-36-5	: 2697	100-74-3	: 7694	103-24-2	: 8132	105-14-6	: 8964	106-98-9	: 3704
99-45-6	: 5804	100-75-4	: 9688	103-25-3	: 2139	105-30-6	: 8937	106-99-0	: 3146
99-48-9	: 4696	100-76-5	: 480	103-28-6	: 10307	105-31-7	: 7124	107-00-6	: 3897
99-50-3	: 2571	100-79-8	: 5207	103-29-7	: 1587	105-34-0	: 154	107-02-8	: 10509
99-51-4	: 1475	100-80-1	: 1641	103-30-0	: 1619	105-36-2	: 115	107-03-9	: 10156
99-52-5	: 800	100-81-2	: 1816	103-32-2	: 1827	105-37-3	: 10263	107-04-0	: 5456
99-53-6	: 9422	100-83-4	: 565	103-34-4	: 7688	105-38-4	: 10261	107-05-1	: 10536
99-54-7	: 1392	100-84-5	: 1932	103-37-7	: 3585	105-39-5	: 130	107-06-2	: 5529
99-55-8	: 801	100-86-7	: 1600	103-38-8	: 3560	105-40-8	: 3954	107-07-3	: 5693
99-56-9	: 1281	100-88-9	: 11568	103-43-5	: 3298	105-41-9	: 6730	107-08-4	: 10054
99-59-2	: 775	100-97-0	: 11601	103-44-6	: 6405	105-42-0	: 6977	107-10-8	: 9803
99-61-6	: 592	101-01-9	: 6288	103-45-7	: 232	105-45-3	: 3573	107-11-9	: 10522
99-62-7	: 993	101-02-0	: 9593	103-48-0	: 10306	105-46-4	: 214	107-12-0	: 10090
99-63-8	: 1319	101-05-3	: 360	103-49-1	: 1828	105-48-6	: 139	107-13-1	: 10583
99-65-0	: 1492	101-10-0	: 4141	103-50-4	: 2079	105-50-0	: 8783	107-14-2	: 274
99-71-8	: 9431	101-14-4	: 785	103-55-9	: 5504	105-53-3	: 9941	107-15-3	: 5496
99-73-0	: 5777	101-17-7	: 678	103-63-9	: 1038	105-54-4	: 3515	107-16-4	: 282
99-75-2	: 2698	101-18-8	: 9446	103-65-1	: 2169	105-55-5	: 11803	107-18-6	: 10697
99-76-3	: 2637	101-21-3	: 3939	103-67-3	: 1818	105-56-6	: 152	107-19-7	: 10743
99-77-4	: 2709	101-27-9	: 510	103-69-5	: 743	105-57-7	: 5541	107-20-0	: 12
99-82-1	: 4438	101-31-5	: 917	103-70-8	: 5999	105-58-8	: 3999	107-21-1	: 5569
99-84-3	: 4675	101-38-2	: 4280	103-71-9	: 1786	105-59-9	: 5742	107-22-2	: 5491
99-85-4	: 4275	101-39-3	: 10516	103-72-0	: 1792	105-60-2	: 493	107-25-5	: 5908
99-86-5	: 4273	101-40-6	: 4385	103-73-1	: 1649	105-66-8	: 3588	107-29-9	: 17
99-87-6	: 1990	101-41-7	: 929	103-74-2	: 10950	105-67-9	: 9308	107-30-2	: 7483
99-88-7	: 789	101-43-9	: 10653	103-75-3	: 10783	105-68-0	: 3639	107-31-3	: 6010
99-89-8	: 9412	101-53-1	: 9455	103-76-4	: 9616	105-70-4	: 10190	107-34-6	: 461
99-90-1	: 5781	101-54-2	: 1283	103-79-7	: 10481	105-72-6	: 3512	107-35-7	: 5623
99-91-2	: 5789	101-55-3	: 1079	103-80-0	: 971	105-74-8	: 9155	107-37-9	: 11475
99-92-3	: 5773	101-57-5	: 2234	103-81-1	: 878	105-75-9	: 3774	107-39-1	: 9067
99-93-4	: 5832	101-60-0	: 9734	103-82-2	: 883	105-83-9	: 9875	107-40-4	: 9071
99-94-5	: 2678	101-61-1	: 786	103-83-3	: 1807	105-86-2	: 8338	107-41-5	: 8793
99-96-7	: 2613	101-76-8	: 1963	103-84-4	: 86	105-99-7	: 6784	107-43-7	: 7454
99-97-8	: 862	101-77-9	: 784	103-85-5	: 11808	106-02-5	: 8562	107-46-0	: 5236
99-98-9	: 1270	101-80-4	: 832	103-88-8	: 37	106-07-0	: 8259	107-47-1	: 10155
99-99-0	: 2002	101-81-5	: 1961	103-89-9	: 78	106-19-4	: 6787	107-48-2	: 8612
100-00-5	: 1216	101-82-6	: 11004	103-90-2	: 67	106-23-0	: 8490	107-49-3	: 5221
100-01-6	: 824	101-83-7	: 4284	103-95-7	: 2117	106-24-1	: 8333	107-50-6	: 4239
100-02-7	: 9437	101-84-8	: 2071	103-99-1	: 8232	106-25-2	: 8334	107-51-7	: 11914
100-06-1	: 5841	101-90-6	: 8625	104-01-8	: 922	106-26-3	: 8318	107-52-8	: 6990
100-07-2	: 2901	101-91-7	: 3205	104-03-0	: 933	106-27-4	: 3544	107-53-9	: 11964
100-09-4	: 2666	101-97-3	: 899	104-04-1	: 83	106-28-5	: 5333	107-59-5	: 128
100-10-7	: 550	101-99-5	: 3965	104-06-3	: 34	106-30-9	: 6462	107-68-6	: 5624
100-12-9	: 1694	102-01-2	: 3207	104-07-4	: 8253	106-31-0	: 3467	107-70-0	: 8978
100-13-0	: 1645	102-02-3	: 7234	104-09-6	: 874	106-32-1	: 8428	107-72-2	: 11472
100-14-1	: 1208	102-04-5	: 10433	104-13-2	: 655	106-33-2	: 5312	107-74-4	: 8392
100-15-2	: 805	102-06-7	: 6282	104-15-4	: 2229	106-35-4	: 6511	107-75-5	: 8347
100-16-3	: 7168	102-07-8	: 12010	104-21-2	: 1892	106-36-5	: 10326	107-80-2	: 3266
100-17-4	: 1937	102-08-9	: 11805	104-22-3	: 2213	106-37-6	: 1290	107-81-3	: 8711
100-18-5	: 994	102-09-0	: 4001	104-29-0	: 10000	106-38-7	: 1056	107-82-4	: 3239
100-19-6	: 5861	102-13-6	: 932	104-36-9	: 1314	106-39-8	: 1006	107-83-5	: 8823
100-20-9	: 1320	102-14-7	: 3575	104-45-0	: 1945	106-40-1	: 640	107-84-6	: 3252
100-21-0	: 1329	102-20-5	: 940	104-46-1	: 1942	106-41-2	: 9230	107-85-7	: 3220
100-22-1	: 1287	102-25-0	: 2345	104-47-2	: 958	106-42-3	: 1454	107-86-8	: 3700
100-25-4	: 1493	102-27-2	: 747	104-50-7	: 6140	106-43-4	: 1185	107-87-9	: 8957
100-26-5	: 10929	102-45-4	: 4810	104-51-8	: 1118	106-44-5	: 9402	107-88-0	: 3345
100-27-8	: 1790	102-47-6	: 1357	104-52-9	: 1235	106-45-6	: 2303	107-89-1	: 3184
100-29-8	: 1664	102-50-1	: 772	104-53-0	: 2114	106-46-7	: 1356	107-91-5	: 49
100-36-7	: 5502	102-51-2	: 1273	104-57-4	: 6016	106-47-8	: 660	107-92-6	: 3460
100-37-8	: 5706	102-52-3	: 10150	104-61-0	: 6149	106-48-9	: 9253	107-93-7	: 3814
100-39-0	: 1053	102-53-4	: 7489	104-63-2	: 5762	106-49-0	: 780	107-94-8	: 10216
100-40-3	: 4658	102-54-5	: 5960	104-65-4	: 10707	106-50-3	: 1262	107-95-9	: 323
100-41-4	: 1668	102-56-7	: 702	104-66-5	: 1588	106-51-4	: 4257	107-96-0	: 10286
100-42-5	: 1624	102-60-3	: 10379	104-67-6	: 6152	106-52-5	: 9717	107-97-1	: 6268
100-43-6	: 10954	102-69-2	: 9808	104-72-3	: 1258	106-53-6	: 2292	107-98-2	: 10387
100-44-7	: 1182	102-70-5	: 10523	104-75-6	: 6726	106-54-7	: 2295	108-01-0	: 5709
100-45-8	: 4629	102-71-6	: 5751	104-76-7	: 6930	106-57-0	: 9612	108-03-2	: 10116
100-46-9	: 1795	102-76-1	: 10195	104-78-9	: 9876	106-58-1	: 9608	108-05-4	: 179
100-47-0	: 2740	102-79-4	: 5692	104-80-3	: 7109	106-63-8	: 10673	108-08-7	: 8762
100-49-2	: 4419	102-81-8	: 5703	104-81-4	: 1064	106-65-0	: 3318	108-10-1	: 8981
100-50-5	: 4632	102-82-9	: 3213	104-82-5	: 1200	106-68-3	: 8472	108-11-2	: 8942
100-51-6	: 1836	102-85-2	: 9590	104-83-6	: 1134	106-69-4	: 6868	108-12-3	: 3693
100-52-7	: 525	102-86-3	: 6725	104-84-7	: 1817	106-70-7	: 6898	108-16-7	: 10376
100-53-8	: 1832	102-92-1	: 10716	104-85-8	: 2765	106-73-0	: 6466	108-18-9	: 9821
100-54-9	: 10882	102-94-3	: 10683	104-87-0	: 589	106-74-1	: 10634	108-19-0	: 7232
100-55-0	: 10976	102-97-6	: 1820	104-88-1	: 531	106-79-6	: 5046	108-20-3	: 10120
100-60-7	: 4298	103-01-5	: 6270	104-90-5	: 10966	106-86-5	: 8555	108-21-4	: 209
100-61-8	: 781	103-02-6	: 10482	104-91-6	: 9438	106-87-6	: 8557	108-22-5	: 10698
100-63-0	: 7169	103-03-7	: 7145	104-92-7	: 1052	106-88-7	: 8615	108-23-6	: 4023
100-64-1	: 4610	103-05-9	: 2150	104-93-8	: 1933	106-90-1	: 10680	108-24-7	: 107
100-65-2	: 757	103-09-3	: 187	104-94-9	: 768	106-91-2	: 10668	108-27-0	: 11186
100-66-3	: 1915	103-11-7	: 10636	104-96-1	: 820	106-92-3	: 8626	108-28-1	: 6155
100-68-5	: 2041	103-12-8	: 2209	105-03-3	: 8131	106-93-4	: 5518	108-30-5	: 6093
100-69-6	: 10952	103-16-2	: 9453	105-05-5	: 1415	106-94-5	: 9836	108-31-6	: 6091
100-70-9	: 10881	103-17-3	: 4075	105-06-6	: 1405	106-95-6	: 10531	108-32-7	: 5213

CAS	Page	CAS	Page	CAS	Page	CAS	Page	CAS	Page
108-36-1	1289	109-90-0	5598	111-17-1	10329	112-51-6	5257	117-37-3	7269
108-37-2	1005	109-92-2	5905	111-18-2	6752	112-52-7	5295	117-39-5	2817
108-38-3	1453	109-93-3	5912	111-19-3	5049	112-53-8	5321	117-73-7	12064
108-39-4	9401	109-94-4	6007	111-20-6	5042	112-54-9	5288	117-78-2	372
108-40-7	2302	109-95-5	8087	111-21-7	5714	112-55-0	5304	117-79-3	381
108-41-8	1184	109-96-6	11118	111-24-0	8741	112-57-2	5498	117-80-6	7850
108-42-9	659	109-97-7	11112	111-25-1	6732	112-58-3	6842	117-81-7	1334
108-43-0	9252	109-98-8	10822	111-26-2	6723	112-60-7	5756	117-84-0	1344
108-44-1	779	109-99-9	6168	111-27-3	6911	112-61-8	8266	117-93-1	66
108-45-2	1261	110-00-9	6025	111-28-4	6707	112-62-9	8307	117-99-7	7622
108-46-3	1498	110-01-0	11783	111-29-5	8787	112-63-0	8228	118-00-3	6289
108-47-4	10937	110-02-1	11724	111-30-8	8734	112-64-1	11635	118-08-1	7357
108-48-5	10939	110-03-2	6801	111-31-9	6856	112-66-3	178	118-10-5	4120
108-50-9	10807	110-04-3	5587	111-34-2	3369	112-67-4	6665	118-32-1	7871
108-52-1	11046	110-05-4	9153	111-40-0	5497	112-70-9	11883	118-34-3	6225
108-55-4	10780	110-06-5	5242	111-41-1	5676	112-71-0	11614	118-41-2	2734
108-56-5	3340	110-12-3	6978	111-42-2	5730	112-72-1	11630	118-44-5	7718
108-57-6	1404	110-13-4	6805	111-43-3	10119	112-73-2	3417	118-46-7	8020
108-59-8	9943	110-14-5	3257	111-44-4	5611	112-76-5	8281	118-55-8	2650
108-60-1	10123	110-15-6	3294	111-45-5	5763	112-79-8	8297	118-58-1	2651
108-64-5	3548	110-16-7	3765	111-46-6	5753	112-80-1	8294	118-61-6	2619
108-67-8	2359	110-17-8	3764	111-47-7	10151	112-82-3	6633	118-69-4	1385
108-68-9	9312	110-18-9	5516	111-48-8	5766	112-85-6	5276	118-71-8	10796
108-69-0	709	110-19-0	215	111-49-9	492	112-86-7	5283	118-74-1	1749
108-70-3	2323	110-22-5	9156	111-51-3	3263	112-88-9	8291	118-75-2	4268
108-73-6	2368	110-27-0	11626	111-55-7	5571	112-89-0	8235	118-76-3	4686
108-75-8	11028	110-38-3	5063	111-56-8	5307	112-91-4	8293	118-78-5	11077
108-77-0	11837	110-39-4	3566	111-60-4	8260	112-92-5	8278	118-79-6	9479
108-78-1	11835	110-40-7	5045	111-61-5	8257	112-95-8	5350	118-83-2	1219
108-82-7	6485	110-41-8	11936	111-62-6	8301	113-00-8	6278	118-90-1	2676
108-83-8	6514	110-42-9	5066	111-64-8	8481	113-15-5	5394	118-91-2	2534
108-84-9	8947	110-43-0	6510	111-65-9	8355	113-48-4	7660	118-92-3	2484
108-85-0	4315	110-44-1	6698	111-66-0	8491	114-07-8	5396	118-93-4	5830
108-86-1	998	110-45-2	3636	111-68-2	6349	114-25-0	3001	118-96-7	2048
108-87-2	4426	110-46-3	8090	111-69-3	6777	114-26-1	10722	119-13-1	2785
108-88-3	1947	110-49-6	5735	111-70-6	6472	114-91-0	10979	119-20-0	2494
108-89-4	10982	110-52-1	3267	111-71-7	6344	115-07-1	10528	119-26-6	7156
108-90-7	1131	110-53-2	8710	111-74-0	6501	115-10-6	7523	119-27-7	1923
108-91-8	4283	110-54-3	6731	111-75-1	5691	115-11-7	10578	119-28-8	7967
108-93-0	4500	110-56-5	3279	111-76-2	5687	115-17-3	20	119-33-5	9424
108-94-1	4571	110-57-6	3750	111-77-3	5737	115-18-4	3870	119-36-8	2634
108-95-2	9199	110-58-7	8700	111-78-4	4733	115-19-5	3922	119-41-5	225
108-97-4	10789	110-59-8	8827	111-81-9	11978	115-20-8	5768	119-44-8	10749
108-98-5	2289	110-60-1	3258	111-82-0	5315	115-21-9	11463	119-48-2	7686
108-99-6	10981	110-61-2	3291	111-83-1	8356	115-24-2	9835	119-59-5	845
109-01-3	9618	110-62-3	8690	111-84-2	8120	115-25-3	4194	119-60-8	7606
109-02-4	7696	110-63-4	3346	111-85-3	8359	115-29-7	5369	119-61-9	7616
109-04-6	10875	110-65-6	3911	111-86-4	8350	115-32-2	1851	119-64-2	7979
109-06-8	10980	110-66-7	8844	111-87-5	8443	115-37-7	7679	119-65-3	7373
109-07-9	9619	110-67-8	10110	111-88-6	8418	115-38-8	11094	119-67-5	2607
109-08-0	10810	110-68-9	3221	111-90-0	5720	115-39-9	9366	119-68-6	2679
109-09-1	10915	110-69-0	3188	111-92-2	3211	115-40-2	9364	119-84-6	2798
109-12-6	11042	110-70-3	5503	111-96-6	5613	115-43-5	11103	119-90-4	3033
109-16-0	10657	110-71-4	5549	112-04-9	11470	115-53-7	7676	119-93-7	3034
109-19-3	3543	110-72-5	5509	112-05-0	8146	115-56-0	11064	120-07-0	5760
109-21-7	3479	110-73-6	5722	112-06-1	193	115-61-7	11349	120-12-7	366
109-43-3	5044	110-74-7	6018	112-10-7	8267	115-63-9	9623	120-14-9	545
109-49-9	7089	110-75-8	5887	112-12-9	11958	115-69-5	9987	120-15-0	11298
109-52-4	8855	110-76-9	5423	112-13-0	5078	115-70-8	9985	120-18-3	7960
109-53-5	10035	110-77-0	5726	112-14-1	222	115-71-9	9093	120-20-7	1571
109-56-8	5741	110-78-1	10059	112-15-2	5721	115-76-4	10004	120-21-8	539
109-57-9	11809	110-80-5	5717	112-16-3	5328	115-77-5	9989	120-22-9	699
109-59-1	5740	110-81-6	5253	112-17-4	157	115-80-0	10182	120-23-0	217
109-60-4	236	110-82-7	4305	112-20-9	8119	115-84-4	9992	120-25-2	557
109-61-5	4028	110-83-8	4626	112-23-2	6008	115-86-6	9566	120-29-6	489
109-63-7	3137	110-85-0	9602	112-24-3	5499	115-90-2	5954	120-32-1	9270
109-64-8	9882	110-86-1	10872	112-25-4	5728	115-96-8	5701	120-33-2	2691
109-65-9	3231	110-87-2	10777	112-26-5	5449	115-99-1	8337	120-36-5	10250
109-66-0	8709	110-88-3	11905	112-27-6	5713	116-06-3	334	120-43-4	9604
109-67-1	9004	110-89-4	9629	112-29-8	5034	116-09-6	10443	120-46-7	10027
109-69-3	3241	110-91-8	7681	112-30-1	5070	116-14-3	5918	120-47-8	2621
109-70-6	9839	110-93-0	6603	112-31-2	5030	116-15-4	10572	120-48-9	2706
109-73-9	3209	110-94-3	8770	112-32-3	6014	116-16-5	10440	120-50-3	2701
109-74-0	3403	110-95-2	9879	112-34-5	5688	116-17-6	9595	120-51-4	2720
109-75-1	3803	110-96-3	9823	112-36-7	5612	116-26-7	4250	120-55-8	5755
109-76-2	9873	110-97-4	10384	112-37-8	11943	116-29-0	2325	120-57-0	2417
109-77-3	9911	110-99-6	226	112-38-9	11974	116-38-1	865	120-61-6	1351
109-78-4	10106	111-01-3	11606	112-39-0	6653	116-43-8	3578	120-66-1	76
109-79-5	3437	111-02-4	11602	112-40-3	5292	116-44-9	2202	120-71-8	770
109-80-8	10031	111-03-5	8299	112-41-4	5335	116-54-1	166	120-72-9	7294
109-81-9	5512	111-06-8	6647	112-42-5	11950	116-58-5	9757	120-73-0	10752
109-82-0	285	111-11-5	8434	112-43-6	11979	116-63-2	7966	120-75-2	2863
109-83-1	5738	111-12-6	8542	112-44-7	11935	117-08-8	7346	120-80-9	1497
109-84-2	5729	111-13-7	8471	112-46-9	11981	117-10-2	395	120-82-1	2322
109-86-4	5733	111-14-8	6455	112-47-0	5047	117-12-4	393	120-83-2	9203
109-87-5	7507	111-15-9	5718	112-48-1	5527	117-18-0	2264	120-85-4	9622
109-89-7	5427	111-16-0	6393	112-49-2	11642	117-34-0	936	120-87-6	8254

120-89-8	: 7226	123-11-5	: 585	126-68-1	: 9585	133-59-5	: 2249	140-80-7	: 8736
120-92-3	: 4878	123-15-9	: 8693	126-71-6	: 9572	133-67-5	: 2854	140-82-9	: 5707
120-93-4	: 7227	123-19-3	: 6512	126-72-7	: 10358	133-90-4	: 2492	140-87-4	: 153
120-94-5	: 11168	123-20-6	: 3513	126-73-8	: 9562	133-91-5	: 2617	140-88-5	: 10635
120-97-8	: 1559	123-22-8	: 9587	126-81-8	: 4378	134-04-3	: 2848	140-95-4	: 11995
121-06-2	: 9337	123-25-1	: 3303	126-84-1	: 9904	134-11-2	: 2596	141-03-7	: 3301
121-14-2	: 1956	123-29-5	: 8150	126-98-7	: 10586	134-20-3	: 2503	141-04-8	: 6783
121-17-5	: 1217	123-30-8	: 9203	126-99-8	: 3152	134-31-6	: 11324	141-05-9	: 3776
121-21-1	: 4960	123-31-9	: 1499	127-00-4	: 10349	134-32-7	: 7708	141-06-0	: 8914
121-29-9	: 4962	123-32-0	: 10806	127-06-0	: 10476	134-49-6	: 7697	141-10-6	: 11965
121-30-2	: 1558	123-35-3	: 8330	127-07-1	: 12016	134-55-4	: 2483	141-12-8	: 8336
121-32-4	: 555	123-38-6	: 9774	127-08-2	: 234	134-58-7	: 11851	141-20-8	: 5313
121-33-5	: 574	123-39-7	: 5995	127-09-3	: 237	134-62-3	: 612	141-22-0	: 8303
121-34-6	: 2625	123-42-2	: 8976	127-17-3	: 10314	134-81-6	: 5584	141-27-5	: 8317
121-43-7	: 3126	123-43-3	: 238	127-18-4	: 5917	134-84-9	: 7633	141-28-6	: 6785
121-44-8	: 5417	123-45-5	: 239	127-19-5	: 55	134-85-0	: 7604	141-32-2	: 10616
121-45-9	: 9592	123-51-3	: 3627	127-25-3	: 9168	134-96-3	: 569	141-43-5	: 5675
121-46-0	: 2913	123-54-6	: 8798	127-27-5	: 9170	135-00-2	: 7641	141-46-8	: 15
121-47-1	: 2216	123-56-8	: 11152	127-29-7	: 308	135-01-3	: 1413	141-57-1	: 11476
121-57-3	: 2217	123-62-6	: 10203	127-40-2	: 4058	135-02-4	: 583	141-62-8	: 11646
121-59-5	: 455	123-63-7	: 11908	127-48-0	: 8587	135-09-1	: 2858	141-63-9	: 8997
121-60-8	: 2243	123-66-0	: 6884	127-63-9	: 2237	135-19-3	: 8018	141-66-2	: 5172
121-61-9	: 31	123-72-8	: 3178	127-66-2	: 1878	135-20-6	: 759	141-75-3	: 3681
121-66-4	: 11666	123-73-9	: 3695	127-69-5	: 2195	135-44-4	: 8926	141-78-6	: 186
121-69-7	: 713	123-75-1	: 11143	127-71-9	: 603	135-48-8	: 8636	141-79-7	: 9108
121-71-1	: 5831	123-76-2	: 8903	127-79-7	: 2199	135-58-0	: 11661	141-82-2	: 9921
121-72-2	: 861	123-77-3	: 5124	127-91-3	: 2925	135-67-1	: 9507	141-83-3	: 11993
121-73-3	: 1215	123-80-8	: 5576	128-04-1	: 11514	135-68-2	: 3012	141-84-4	: 11685
121-75-5	: 3316	123-82-0	: 6350	128-08-5	: 11153	135-70-6	: 11194	141-86-6	: 10925
121-79-9	: 2732	123-83-1	: 5510	128-09-6	: 11154	135-88-6	: 7728	141-90-2	: 11110
121-81-3	: 614	123-84-2	: 10340	128-13-2	: 4098	135-97-7	: 486	141-91-3	: 7687
121-82-4	: 11833	123-86-4	: 121	128-20-1	: 9745	135-98-8	: 2035	141-93-5	: 1414
121-86-8	: 1210	123-88-6	: 11822	128-23-4	: 9743	136-35-6	: 11826	141-97-9	: 3571
121-89-1	: 5860	123-90-0	: 11719	128-27-8	: 5381	136-44-7	: 10188	142-04-1	: 756
121-90-4	: 2905	123-91-1	: 5187	128-33-6	: 4104	136-47-0	: 2531	142-08-5	: 11033
121-91-5	: 1328	123-92-2	: 3631	128-37-0	: 9224	136-60-7	: 2532	142-09-6	: 10660
121-92-6	: 2704	123-93-3	: 240	128-39-2	: 9220	136-72-1	: 8678	142-26-7	: 65
121-98-2	: 2675	123-95-5	: 8251	128-50-7	: 2969	136-77-6	: 1519	142-28-9	: 9893
122-00-9	: 5853	123-98-8	: 8138	128-53-0	: 11122	136-78-7	: 11358	142-29-0	: 4903
122-01-0	: 2892	123-99-9	: 8130	128-62-1	: 7356	136-79-9	: 739	142-30-3	: 7119
122-03-2	: 590	124-02-7	: 10526	128-68-7	: 2979	136-80-1	: 5744	142-47-2	: 6248
122-04-3	: 2906	124-04-9	: 6779	129-00-0	: 10853	136-95-8	: 2859	142-50-7	: 5332
122-07-6	: 5420	124-06-1	: 11624	129-43-1	: 407	137-00-8	: 11677	142-60-9	: 10313
122-09-8	: 1572	124-07-2	: 8422	129-44-2	: 389	137-04-2	: 665	142-61-0	: 6986
122-11-2	: 2194	124-09-4	: 6749	129-64-6	: 7580	137-05-3	: 10624	142-62-1	: 6870
122-14-5	: 5951	124-10-7	: 11625	129-66-8	: 2735	137-06-4	: 2301	142-64-3	: 9607
122-20-3	: 10393	124-11-8	: 8182	129-79-3	: 5979	137-07-5	: 2290	142-68-7	: 10800
122-25-8	: 2686	124-12-9	: 8411	130-15-4	: 7843	137-16-6	: 6269	142-73-4	: 6260
122-28-1	: 82	124-13-0	: 8346	130-16-5	: 11311	137-17-7	: 858	142-77-8	: 8298
122-32-7	: 8310	124-17-4	: 5689	130-26-7	: 11312	137-19-9	: 1510	142-82-5	: 6355
122-34-9	: 11513	124-18-5	: 5033	130-73-4	: 9299	137-26-8	: 11722	142-83-6	: 6675
122-35-0	: 10505	124-19-6	: 8117	130-80-3	: 9301	137-30-4	: 12068	142-84-7	: 9826
122-37-2	: 9447	124-21-0	: 5425	130-85-8	: 7777	137-32-6	: 3626	142-91-6	: 6654
122-39-4	: 836	124-22-1	: 5291	130-86-9	: 3109	137-40-6	: 10327	142-92-7	: 194
122-40-7	: 6347	124-26-5	: 8231	130-89-2	: 4130	137-43-9	: 4777	142-96-1	: 3415
122-42-9	: 3966	124-30-1	: 8233	130-90-5	: 4129	137-58-6	: 54	143-07-7	: 5305
122-46-3	: 212	124-38-9	: 3992	130-95-0	: 4127	138-41-0	: 2508	143-08-8	: 8157
122-48-5	: 3663	124-40-3	: 7448	131-01-1	: 12065	138-52-3	: 6222	143-10-2	: 5058
122-51-0	: 5604	124-42-5	: 5671	131-09-9	: 388	138-53-4	: 955	143-13-5	: 220
122-52-1	: 9591	124-47-0	: 12021	131-11-3	: 1343	138-86-3	: 4677	143-15-7	: 5293
122-57-6	: 3881	124-48-1	: 7493	131-28-2	: 2586	138-87-4	: 4542	143-16-8	: 6727
122-59-8	: 228	124-58-3	: 458	131-53-3	: 7619	138-89-6	: 720	143-22-6	: 5690
122-60-1	: 8623	124-63-0	: 7535	131-54-4	: 7595	139-13-9	: 6256	143-24-8	: 8995
122-62-3	: 5043	124-68-5	: 10341	131-56-6	: 7609	139-40-2	: 10508	143-27-1	: 6631
122-66-7	: 7158	124-70-9	: 11396	131-57-7	: 7620	139-45-7	: 10198	143-28-2	: 8312
122-69-0	: 10691	124-73-2	: 5525	131-58-8	: 7631	139-65-1	: 850	143-50-0	: 7408
122-70-3	: 10324	124-80-1	: 4067	131-73-7	: 864	139-66-2	: 2283	143-62-4	: 4044
122-73-6	: 1951	124-83-4	: 4794	131-89-5	: 9273	139-85-5	: 542	143-74-8	: 9363
122-78-1	: 868	124-87-8	: 7579	131-91-9	: 8035	139-87-7	: 5723	144-19-4	: 8795
122-79-2	: 231	125-15-5	: 11351	132-57-2	: 7972	139-93-5	: 9275	144-49-0	: 189
122-80-5	: 29	125-20-2	: 7352	132-60-5	: 11235	140-08-9	: 5702	144-55-8	: 4004
122-84-9	: 10461	125-24-6	: 3002	132-64-9	: 5163	140-10-3	: 10682	144-62-7	: 5553
122-87-2	: 6267	125-28-0	: 7674	132-65-0	: 5167	140-11-4	: 233	144-68-3	: 4057
122-88-3	: 146	125-29-1	: 7678	132-66-1	: 2702	140-18-1	: 148	144-79-6	: 11380
122-95-2	: 1408	125-33-7	: 11066	132-75-2	: 7736	140-22-7	: 4010	144-80-9	: 30
122-97-4	: 2148	125-55-3	: 11079	132-86-5	: 7834	140-25-0	: 5318	144-83-2	: 2203
122-98-5	: 5759	126-00-1	: 1103	133-06-2	: 7368	140-28-3	: 5500	145-13-1	: 9756
122-99-6	: 5758	126-07-8	: 11520	133-07-3	: 5984	140-29-4	: 945	145-73-3	: 5370
123-00-2	: 7701	126-11-4	: 10012	133-08-4	: 9930	140-31-8	: 9615	145-94-8	: 7283
123-01-3	: 1566	126-14-7	: 6229	133-11-9	: 2502	140-39-6	: 213	146-06-5	: 4128
123-02-4	: 2342	126-17-0	: 11529	133-13-1	: 9949	140-40-9	: 85	146-48-5	: 12062
123-03-5	: 11030	126-18-1	: 11533	133-14-2	: 9152	140-55-6	: 9718	147-71-7	: 3312
123-04-6	: 6368	126-19-2	: 11531	133-26-6	: 6188	140-57-8	: 11579	147-73-9	: 3304
123-05-7	: 6719	126-29-4	: 4056	133-32-4	: 7296	140-65-8	: 7683	147-82-0	: 851
123-06-8	: 9914	126-30-7	: 10005	133-37-9	: 3311	140-67-0	: 1943	147-85-3	: 9764
123-07-9	: 9355	126-33-0	: 11786	133-53-9	: 9290	140-76-1	: 10955	147-93-3	: 2663
123-08-0	: 566	126-54-5	: 11643	133-58-4	: 7437	140-77-2	: 4847	148-03-8	: 2789

CAS	No.	CAS	No.	CAS	No.	CAS	No.	CAS	No.
148-24-3	11308	258-31-1	6621	307-34-6	8415	353-50-4	4008	373-05-7	6888
148-25-4	7875	260-36-6	2473	307-45-9	5050	353-54-8	7550	373-14-8	6829
148-53-8	570	260-94-6	314	308-24-7	4498	353-55-9	7494	373-28-4	6735
148-75-4	7876	261-31-4	11811	309-00-2	5177	353-58-2	7468	373-32-0	6936
148-79-8	11656	262-30-6	11353	311-89-7	3228	353-59-3	7464	373-44-4	8365
148-86-7	3091	271-44-3	7235	312-40-3	11411	353-82-2	279	373-49-9	6670
149-29-1	6183	271-58-9	2390	312-84-5	11355	353-83-3	5667	373-52-4	7470
149-30-4	2866	271-89-6	2438	313-72-4	7935	353-85-5	289	373-53-5	7517
149-32-6	3431	271-95-4	2389	314-13-6	7877	353-89-9	11432	373-80-8	11584
149-57-5	6882	272-16-2	2387	314-04-9	3139	354-11-0	5639	374-07-2	5537
149-73-5	7561	273-53-0	2877	315-30-0	10852	354-12-1	5658	374-25-4	3715
149-74-6	11401	274-07-7	2412	317-52-2	6753	354-14-3	5638	374-77-6	4343
149-91-7	2731	274-09-9	2415	319-85-7	4406	354-15-4	5660	375-01-9	3622
150-13-0	2486	274-40-8	7324	319-86-8	4407	354-21-2	5659	375-16-6	3692
150-19-6	9380	275-51-4	504	319-88-0	2340	354-23-4	5539	375-22-4	3521
150-25-4	6259	278-30-8	2991	319-89-1	4269	354-25-6	5485	375-83-7	6432
150-30-1	9509	279-23-2	2914	320-72-9	2566	354-26-7	5490	375-96-2	8139
150-46-9	3125	280-57-9	5111	321-28-8	1722	354-51-8	5519	376-27-2	8438
150-50-5	7438	280-65-9	2990	321-38-0	7893	354-56-3	5618	376-87-4	9027
150-60-7	5248	281-23-2	11860	321-60-8	3074	354-58-5	5662	377-41-3	4177
150-61-8	5508	285-67-6	8558	323-09-1	7894	354-92-7	10050	377-70-8	4276
150-68-5	11999	286-08-8	2915	324-74-3	3075	354-96-1	3414	377-93-5	4199
150-69-6	12013	286-18-0	478	326-56-7	2420	355-02-2	4499	378-68-7	10251
150-76-5	9381	286-20-4	8554	326-59-0	2416	355-04-4	8854	378-75-6	10317
150-77-6	5515	287-23-0	4170	326-62-5	952	355-20-4	3287	381-61-3	3162
150-78-7	1438	287-27-4	11691	327-54-8	2272	355-25-9	3255	381-73-7	175
150-86-7	6672	287-92-3	4774	327-57-1	8214	355-42-0	6843	382-45-6	353
151-10-0	1437	288-13-1	10815	328-39-2	7412	355-63-5	6560	383-63-1	267
151-13-3	8304	288-14-2	7402	328-42-7	3339	355-68-0	4381	383-66-4	269
151-18-8	10092	288-32-4	7192	328-50-7	8781	355-75-9	4643	383-90-4	3114
151-19-9	8453	288-36-8	11846	328-74-5	637	355-80-6	8953	384-04-3	3172
151-56-4	498	288-39-1	11658	328-96-1	723	356-16-1	5642	384-22-5	2060
151-67-7	5458	288-42-6	8575	329-71-5	9343	356-24-1	3525	386-17-4	7351
152-16-9	5220	288-47-1	11667	330-54-1	12000	356-27-4	3524	388-82-9	3042
152-58-9	9749	288-92-6	11657	330-55-2	7415	356-28-5	3523	389-08-2	8066
152-93-2	11107	288-94-8	11652	331-39-5	10632	367-20-0	3000	389-40-2	2093
152-95-4	2824	288-99-3	8563	332-77-4	6087	357-49-3	10721	392-56-3	1757
154-17-6	7110	289-80-5	10855	333-20-0	11711	357-57-3	11564	392-89-8	1740
154-41-6	1838	289-95-2	11049	333-41-5	9582	357-67-5	11089	392-83-6	1089
154-42-7	10762	290-37-9	10801	333-93-7	3260	357-70-0	6200	392-85-8	1737
154-93-8	11994	290-67-5	5202	334-28-1	6463	358-06-5	11410	393-52-2	2897
156-08-1	1575	290-87-9	11827	334-42-9	6359	358-43-0	11484	394-46-7	1633
156-10-5	829	290-96-0	11651	334-43-0	6365	358-99-6	5883	396-01-0	10750
156-38-7	906	291-21-4	11919	334-48-5	5059	359-01-3	3379	396-64-5	3043
156-43-4	737	291-64-5	4221	334-56-5	5052	359-02-4	5900	398-23-2	3044
156-59-2	5896	292-46-6	8998	334-59-8	5064	359-10-4	5885	399-92-8	10328
156-60-5	5895	292-64-8	4736	334-61-2	5035	359-11-5	5924	400-52-2	266
156-87-6	10337	293-55-0	4724	334-62-3	5037	359-13-7	5708	401-56-9	133
157-40-4	11527	293-90-9	4202	334-64-5	5074	359-25-1	116	401-78-5	1090
180-43-8	11537	294-62-2	4210	334-88-3	7491	359-28-4	5661	401-80-9	1738
184-97-4	5240	295-48-7	4759	335-36-4	6109	359-29-5	5923	402-31-3	997
189-55-9	2850	297-97-2	11720	335-44-4	3452	359-32-0	296	402-43-7	1091
191-07-1	4145	298-02-2	9513	335-48-8	3275	359-35-3	5641	402-44-8	1739
191-24-2	2471	298-04-4	5260	335-57-9	6418	359-37-5	5925	402-67-5	1734
191-28-6	12055	298-12-4	223	336-19-6	4647	359-70-6	5436	403-42-9	5815
192-65-4	8061	298-45-3	2470	336-59-4	3522	360-68-9	4109	403-43-0	2899
192-97-2	2436	298-46-4	5144	338-65-8	5471	360-89-4	3804	404-72-8	1721
195-19-7	2406	298-81-7	6187	338-83-0	9813	363-51-9	664	404-86-4	8181
198-55-0	9161	298-96-4	11653	339-43-5	2192	363-72-4	2091	405-50-5	901
205-25-4	2405	299-84-3	9584	343-65-7	1102	366-18-7	3099	405-99-2	1635
205-82-3	2738	299-86-5	4153	344-04-7	1078	367-11-3	1422	406-74-6	190
205-99-2	631	300-38-9	11927	344-07-0	1220	367-12-4	9360	407-25-0	264
206-44-0	5966	300-57-2	2165	345-35-7	1194	367-23-7	2346	407-75-0	8879
207-08-9	2739	300-62-9	1578	346-18-9	2857	367-25-9	700	407-97-6	8713
207-83-0	5148	300-76-5	7702	348-51-6	1166	367-64-6	265	407-98-7	8728
208-96-8	3	300-87-8	7403	348-52-7	1719	368-34-3	132	408-16-2	6491
212-74-8	11644	301-00-8	8285	348-54-9	752	368-43-4	2251	408-27-5	8460
213-46-7	9601	301-12-2	8632	348-70-9	135	368-47-8	11486	408-38-8	6639
217-59-4	11910	301-21-3	9399	349-88-2	2246	368-77-4	2770	420-04-2	4155
218-01-9	4114	302-17-0	5582	350-03-8	5868	368-98-9	3116	420-05-3	4162
222-93-5	8996	302-27-2	306	350-46-9	1735	370-14-9	9403	420-12-2	11695
224-42-0	5143	302-72-7	324	350-50-5	1725	371-40-4	754	420-24-6	463
225-51-4	630	302-84-1	11356	350-51-6	1634	371-41-5	9362	420-26-8	10047
225-61-6	2394	303-04-8	3754	350-90-3	10637	371-62-0	5727	420-45-1	9906
229-87-8	9188	303-38-8	2568	352-11-4	1195	372-09-8	150	420-46-2	5665
230-07-9	9191	304-21-2	11038	352-32-9	1728	372-18-9	1423	420-90-6	10532
230-27-3	2474	304-84-7	611	352-33-0	1168	372-19-0	753	420-97-3	9897
230-46-6	9190	305-01-1	2804	352-34-1	1720	372-20-3	9361	420-99-5	9851
238-84-6	2393	305-03-3	1101	352-70-5	1727	372-31-6	3592	421-04-5	5487
250-84-0	11688	305-84-0	7133	352-91-0	9844	372-38-3	2347	421-06-7	5466
253-52-1	9599	305-85-1	9304	352-93-2	5644	372-46-3	4403	421-83-0	7537
253-66-7	4136	306-08-1	913	352-97-6	6253	372-54-3	6830	425-23-0	3526
253-82-7	11195	306-44-5	9789	353-25-3	8429	372-64-5	5249	425-75-2	7533
254-04-6	2772	306-83-2	5540	353-29-7	5294	372-75-8	8551	428-59-1	8629
254-79-5	8065	306-94-5	7934	353-36-6	5593	372-93-0	3621	430-40-0	445
254-87-5	2870	306-98-9	4342	353-42-4	3138			430-51-3	10437
255-50-5	2434	307-11-9	11912	353-49-1	4005			430-57-9	5535

CAS	No.	CAS	No.	CAS	No.	CAS	No.	CAS	No.
430-63-7	: 10561	461-98-3	: 11045	474-69-1	: 5382	486-25-9	: 5975	495-45-4	: 3876
430-66-0	: 5666	462-02-2	: 11907	474-77-1	: 4112	486-34-0	: 8012	495-61-4	: 4685
430-71-7	: 10506	462-06-6	: 1717	474-86-2	: 5398	486-35-1	: 2805	495-69-2	: 6255
430-95-5	: 10554	462-08-8	: 10858	475-03-6	: 8000	486-39-5	: 7396	495-76-1	: 2427
431-03-8	: 3359	462-18-0	: 11891	475-20-7	: 7568	486-47-5	: 7377	495-85-2	: 10790
431-06-1	: 5532	462-23-7	: 3927	475-25-2	: 628	486-66-8	: 2827	495-91-0	: 9633
431-07-2	: 5488	462-38-4	: 9856	475-31-0	: 6273	486-70-4	: 11331	496-10-6	: 7281
431-27-6	: 10558	462-39-5	: 9905	475-38-7	: 7853	486-84-0	: 11040	496-11-7	: 7250
431-52-7	: 10607	462-43-1	: 10382	475-67-2	: 5161	486-86-2	: 7578	496-15-1	: 7297
431-53-8	: 10557	462-72-6	: 3237	475-81-0	: 5157	486-89-5	: 7573	496-16-2	: 2442
432-25-7	: 4636	462-73-7	: 3250	476-28-8	: 6201	487-06-9	: 2818	496-41-3	: 2439
433-52-3	: 134	462-94-2	: 8735	476-32-4	: 4071	487-10-5	: 5127	496-46-8	: 7191
434-03-7	: 9758	462-95-3	: 5603	476-45-9	: 7855	487-19-4	: 10987	496-49-1	: 10471
434-13-9	: 4099	463-04-7	: 8096	476-69-7	: 5162	487-21-8	: 10748	496-64-0	: 10794
434-16-2	: 4103	463-11-6	: 8405	476-70-0	: 5156	487-27-4	: 7575	496-67-3	: 3196
434-45-7	: 5872	463-16-1	: 8151	477-27-0	: 92	487-68-3	: 599	496-72-0	: 1276
434-64-0	: 2095	463-17-2	: 11947	477-32-7	: 3540	487-79-6	: 11144	496-73-1	: 1530
434-90-2	: 3027	463-23-0	: 8126	477-47-4	: 6191	487-93-4	: 7326	496-74-2	: 1564
436-05-5	: 11925	463-24-1	: 8165	477-60-1	: 11924	488-10-8	: 4949	496-77-5	: 8474
437-38-7	: 9801	463-33-2	: 11939	477-75-8	: 2375	488-17-5	: 1527	496-78-6	: 9494
437-50-3	: 12054	463-40-1	: 8287	477-89-4	: 5218	488-23-3	: 2278	497-03-0	: 3699
438-08-4	: 346	463-49-0	: 9772	477-90-7	: 10786	488-43-7	: 6207	497-04-1	: 9995
439-14-5	: 2410	463-51-4	: 5927	478-40-0	: 7856	488-59-5	: 7336	497-06-3	: 3794
442-16-0	: 317	463-52-5	: 7563	478-43-3	: 371	488-70-0	: 9473	497-18-7	: 4009
442-51-3	: 11039	463-58-1	: 4035	478-61-5	: 2911	488-71-1	: 848	497-20-1	: 4767
443-48-1	: 7206	463-71-8	: 4032	478-71-7	: 9173	488-73-3	: 4074	497-26-7	: 5210
443-83-4	: 1169	463-72-9	: 3971	478-94-4	: 5374	488-81-3	: 11343	497-30-3	: 7204
444-27-9	: 11682	463-82-1	: 9910	478-95-5	: 5375	488-86-8	: 4939	497-32-5	: 2922
445-29-4	: 2603	463-88-7	: 5878	479-00-5	: 5373	488-87-9	: 1512	497-39-2	: 9222
445-69-2	: 1321	464-06-2	: 3457	479-13-0	: 2466	488-92-6	: 11128	497-59-6	: 10775
446-10-6	: 1731	464-10-8	: 7551	479-18-5	: 10761	488-93-7	: 6051	497-76-7	: 6224
446-11-7	: 1729	464-15-3	: 2932	479-21-0	: 7608	488-97-1	: 11866	498-00-0	: 1886
446-33-3	: 1732	464-17-5	: 2970	479-22-1	: 4259	488-98-2	: 1736	498-02-2	: 5824
446-34-4	: 1730	464-41-5	: 2920	479-23-2	: 2391	489-84-9	: 505	498-15-7	: 2974
446-35-5	: 1426	464-49-3	: 2961	479-27-6	: 7798	489-86-1	: 507	498-23-7	: 3786
446-52-6	: 559	464-72-2	: 5580	479-45-8	: 818	489-98-5	: 863	498-24-8	: 3785
446-72-0	: 2807	464-85-7	: 6178	479-59-4	: 2737	490-02-8	: 10811	498-59-9	: 5018
447-31-4	: 5785	464-86-8	: 6177	480-11-5	: 2810	490-03-9	: 4709	498-60-2	: 6042
447-53-0	: 7814	464-88-0	: 4786	480-15-9	: 2842	490-06-2	: 1536	498-62-4	: 11734
447-60-9	: 2769	465-24-7	: 7569	480-16-0	: 2816	490-10-8	: 4758	498-67-9	: 1379
450-95-3	: 5816	465-42-9	: 4063	480-40-0	: 2811	490-11-9	: 10930	498-94-2	: 9642
451-13-8	: 895	465-74-7	: 12036	480-41-1	: 2802	490-53-9	: 7390	498-95-3	: 9641
451-40-1	: 5812	465-99-6	: 8545	480-44-4	: 2809	490-55-1	: 11671	498-96-4	: 10912
451-46-7	: 2606	466-14-8	: 3199	480-64-8	: 2573	490-64-2	: 2733	499-04-7	: 10913
451-80-9	: 1654	466-43-3	: 475	480-68-2	: 11102	490-65-3	: 7923	499-06-9	: 2583
452-35-7	: 2865	466-49-9	: 473	480-72-8	: 9	490-79-9	: 2570	499-75-2	: 9415
452-58-4	: 10923	466-81-9	: 8567	480-91-1	: 7369	490-91-5	: 4267	499-80-9	: 10928
452-71-1	: 755	466-96-6	: 7672	481-06-1	: 8057	491-04-3	: 4697	499-81-0	: 10931
452-73-3	: 1170	466-97-7	: 7665	481-20-9	: 4106	491-09-8	: 4718	499-89-8	: 4241
452-86-8	: 1528	467-14-1	: 7671	481-21-0	: 4105	491-14-5	: 9120	500-05-0	: 10773
453-13-4	: 10369	467-55-0	: 11534	481-26-5	: 9741	491-18-9	: 11127	500-22-1	: 10888
453-18-9	: 192	467-60-7	: 9676	481-29-8	: 347	491-31-6	: 2791	500-38-9	: 1515
453-20-3	: 6137	467-85-6	: 6968	481-37-8	: 483	491-34-9	: 11299	500-44-7	: 11011
454-14-8	: 10415	467-98-1	: 7677	481-39-0	: 7860	491-35-0	: 11279	500-49-2	: 1550
454-31-9	: 176	468-28-0	: 4281	481-42-5	: 7863	491-37-2	: 2799	500-55-0	: 928
455-14-1	: 857	468-45-1	: 11516	481-49-2	: 8630	491-38-3	: 2793	500-99-2	: 9306
455-31-2	: 1425	469-59-0	: 12046	481-72-1	: 397	491-67-8	: 2844	501-15-5	: 1533
455-32-3	: 2909	470-40-6	: 4954	481-74-3	: 398	491-70-3	: 2814	501-19-9	: 9390
455-38-9	: 2604	470-49-5	: 8874	482-05-3	: 3038	491-88-3	: 6144	501-24-6	: 9441
456-22-4	: 2605	470-67-7	: 8556	482-28-0	: 7306	492-17-1	: 3029	501-30-4	: 10795
456-47-3	: 1879	470-82-6	: 8560	482-44-0	: 6189	492-22-8	: 11812	501-52-0	: 2126
456-48-4	: 560	470-90-6	: 4078	482-66-6	: 4756	492-27-3	: 11232	501-53-1	: 4027
456-49-5	: 1723	470-99-5	: 4702	482-68-8	: 11347	492-38-6	: 927	501-65-5	: 1714
457-87-4	: 1576	471-04-5	: 6461	482-74-6	: 2435	492-80-8	: 657	501-92-8	: 9458
458-03-7	: 1655	471-12-5	: 2982	482-89-3	: 7331	492-89-7	: 1543	501-97-3	: 2134
458-35-5	: 9370	471-16-9	: 2984	483-04-5	: 8571	492-97-7	: 3112	502-41-0	: 4229
458-36-6	: 10514	471-25-0	: 10738	483-17-0	: 5367	492-99-9	: 4379	502-42-1	: 4234
458-37-7	: 6334	471-35-2	: 5110	483-18-1	: 5368	493-01-6	: 7790	502-44-3	: 8591
458-88-8	: 9708	471-44-3	: 131	483-55-6	: 7861	493-02-7	: 7791	502-49-8	: 4740
459-22-3	: 953	471-46-5	: 5493	483-64-7	: 2672	493-05-0	: 2776	502-56-7	: 8175
459-26-7	: 1656	471-47-6	: 104	483-65-8	: 9183	493-08-3	: 2777	502-59-0	: 6354
459-56-3	: 1880	471-87-4	: 11177	483-76-1	: 7899	493-09-4	: 2414	502-61-4	: 5329
459-57-4	: 561	471-95-4	: 3141	483-78-3	: 7830	493-49-2	: 7398	502-65-8	: 4053
459-60-9	: 1724	472-20-8	: 4666	484-11-7	: 9192	493-52-7	: 2587	502-72-7	: 4761
459-67-6	: 4940	472-70-8	: 4061	484-29-7	: 6180	493-57-2	: 2401	502-73-8	: 6300
459-72-3	: 191	472-93-5	: 4052	484-47-9	: 7215	494-03-1	: 7710	502-99-8	: 8486
459-73-4	: 6263	473-02-9	: 11865	484-67-3	: 9472	494-12-2	: 2782	503-01-5	: 6532
460-00-4	: 1044	473-06-3	: 2977	484-73-1	: 5223	494-40-6	: 2356	503-09-3	: 8616
460-12-8	: 3177	473-19-8	: 2933	484-89-9	: 4265	494-47-3	: 7488	503-17-3	: 3898
460-13-9	: 10046	473-30-3	: 11663	484-93-5	: 491	494-52-0	: 11008	503-28-6	: 5125
460-19-5	: 5550	473-34-7	: 2210	485-31-4	: 3003	494-55-3	: 5215	503-29-7	: 494
460-35-5	: 9871	473-81-4	: 10252	485-47-2	: 7267	494-90-6	: 2465	503-30-0	: 8592
460-55-9	: 11429	473-91-6	: 4937	485-64-3	: 4121	494-97-3	: 11017	503-38-8	: 4029
461-11-0	: 6460	473-98-3	: 7416	485-65-4	: 4122	494-99-5	: 1440	503-40-2	: 7514
461-35-8	: 10184	474-25-9	: 4097	485-71-3	: 4133	495-08-9	: 1520	503-46-8	: 4691
461-58-5	: 6280	474-40-8	: 11556	485-72-3	: 2830	495-40-9	: 3676	503-60-6	: 3736
461-72-3	: 7216	474-62-4	: 5386	485-89-2	: 11233	495-41-0	: 3880	503-64-0	: 3815

503-66-2	: 10267	512-04-9	: 11536	521-04-0	: 11554	529-86-2	: 432	537-29-1	: 916
503-74-2	: 3539	512-12-9	: 8585	521-11-9	: 350	529-92-0	: 11216	537-45-1	: 4279
503-76-4	: 9869	512-13-0	: 2941	521-12-0	: 351	530-14-3	: 5818	537-46-2	: 1573
503-80-0	: 452	512-16-3	: 4309	521-18-6	: 349	530-22-3	: 2464	537-47-3	: 7146
503-93-5	: 4220	512-48-1	: 9003	521-24-4	: 7970	530-40-5	: 10892	537-49-5	: 11933
504-02-9	: 4376	512-56-1	: 9564	521-31-3	: 9600	530-43-8	: 6648	537-64-4	: 7432
504-03-0	: 9650	512-60-7	: 11862	521-35-7	: 5154	530-44-9	: 7613	537-65-5	: 1263
504-15-4	: 1531	512-69-6	: 6219	521-74-4	: 11313	530-48-3	: 1636	537-91-7	: 5247
504-20-1	: 6340	512-85-6	: 5184	521-85-7	: 3108	530-50-7	: 7157	537-92-8	: 77
504-24-5	: 10859	513-02-0	: 9568	522-12-3	: 2796	530-55-2	: 4262	538-08-9	: 4159
504-29-0	: 10857	513-08-6	: 9567	522-27-0	: 5583	530-62-1	: 7193	538-23-8	: 8440
504-31-4	: 10788	513-12-2	: 5725	522-57-6	: 7611	530-78-9	: 2729	538-24-9	: 5319
504-53-0	: 9000	513-23-5	: 2985	522-66-7	: 4123	530-85-8	: 10740	538-32-9	: 12027
504-57-4	: 8105	513-31-5	: 10545	522-75-8	: 2404	530-91-6	: 8043	538-39-6	: 1590
504-61-0	: 3859	513-35-9	: 3801	523-27-3	: 375	530-93-8	: 8047	538-41-0	: 635
504-63-2	: 9983	513-36-0	: 9862	523-31-9	: 1337	531-02-2	: 3097	538-43-2	: 10026
504-64-3	: 9773	513-37-1	: 10539	523-47-7	: 7898	531-18-0	: 7584	538-44-3	: 9106
504-78-9	: 11681	513-38-2	: 10057	523-80-8	: 2423	531-29-3	: 6226	538-49-8	: 11000
505-20-4	: 5261	513-42-8	: 10702	524-15-2	: 6179	531-37-3	: 9383	538-51-2	: 840
505-22-6	: 5186	513-44-0	: 10158	524-30-1	: 2823	531-39-5	: 8886	538-56-7	: 10684
505-23-7	: 5262	513-81-5	: 3159	524-34-5	: 11236	531-75-9	: 2822	538-58-9	: 8683
505-29-3	: 5263	513-86-0	: 3661	524-36-7	: 10978	531-81-7	: 2775	538-62-5	: 5122
505-32-8	: 6671	513-88-2	: 10425	524-40-3	: 10883	531-84-0	: 7774	538-64-7	: 3768
505-34-0	: 10067	513-92-8	: 5919	524-42-5	: 7842	531-91-9	: 3035	538-68-1	: 2107
505-48-6	: 8386	513-96-2	: 10115	524-63-0	: 4119	531-95-3	: 2786	538-74-9	: 2287
505-60-2	: 5645	514-10-3	: 9167	524-80-1	: 3982	532-18-3	: 7727	538-75-0	: 4291
505-84-0	: 10074	514-12-5	: 2953	524-96-9	: 1797	532-24-1	: 490	538-81-8	: 1095
505-92-0	: 5338	514-92-1	: 4059	525-06-4	: 5928	532-27-4	: 5790	538-86-3	: 1930
506-03-6	: 10010	515-12-8	: 4401	525-15-5	: 11041	532-32-1	: 2725	538-93-2	: 2039
506-12-7	: 6314	515-25-3	: 11178	525-37-1	: 7868	532-48-9	: 5825	539-03-7	: 48
506-23-0	: 8284	515-30-0	: 914	525-64-4	: 5971	532-54-7	: 876	539-08-2	: 9796
506-24-1	: 8316	515-40-2	: 1143	525-68-8	: 11253	532-82-1	: 1284	539-15-1	: 9320
506-30-9	: 5355	515-46-8	: 2223	525-82-6	: 2840	532-96-7	: 2717	539-23-1	: 10860
506-31-0	: 5364	515-49-1	: 2189	526-08-9	: 2201	533-18-6	: 211	539-30-0	: 1658
506-32-1	: 5361	515-59-3	: 2200	526-31-8	: 11923	533-24-4	: 1544	539-47-9	: 10638
506-33-2	: 5282	515-82-2	: 6000	526-35-2	: 8586	533-30-2	: 2860	539-52-6	: 6133
506-42-3	: 8311	515-83-3	: 5769	526-55-6	: 7305	533-32-4	: 3201	539-74-2	: 10208
506-46-7	: 6624	515-84-4	: 248	526-73-8	: 2357	533-45-9	: 11670	539-80-0	: 4242
506-48-9	: 8221	515-88-8	: 2944	526-75-0	: 9307	533-48-2	: 7223	539-82-2	: 8878
506-50-3	: 11820	515-90-2	: 2964	526-84-1	: 3777	533-50-6	: 3679	539-86-6	: 10595
506-52-5	: 6626	515-94-6	: 326	526-85-2	: 9493	533-58-4	: 9371	539-88-8	: 8907
506-68-3	: 4165	515-96-8	: 105	526-86-3	: 4263	533-68-6	: 3469	539-89-9	: 3961
506-77-4	: 4166	516-05-2	: 9960	526-95-4	: 6209	533-70-0	: 701	539-90-2	: 3563
506-78-5	: 7337	516-06-3	: 12040	526-99-8	: 6193	533-73-3	: 2367	539-92-4	: 3997
506-80-9	: 4036	516-12-1	: 11159	527-07-1	: 6213	533-75-5	: 4243	540-07-8	: 6907
506-82-1	: 3929	516-78-9	: 5384	527-17-3	: 4270	533-86-8	: 8300	540-08-9	: 6321
506-85-4	: 6024	516-85-8	: 5380	527-18-4	: 1552	533-98-2	: 3265	540-00-0	: 11857
506-89-8	: 12020	516-95-0	: 4108	527-35-5	: 9475	534-00-9	: 3238	540-10-3	: 6651
506-93-4	: 6286	517-04-4	: 5401	527-52-6	: 7111	534-07-0	: 10426	540-18-1	: 3579
506-96-7	: 291	517-06-6	: 5407	527-53-7	: 2279	534-13-4	: 11804	540-36-3	: 1424
507-02-8	: 302	517-07-7	: 5402	527-55-9	: 1513	534-15-6	: 5548	540-37-4	: 761
507-09-5	: 5646	517-09-9	: 5397	527-60-6	: 9495	534-22-5	: 6130	540-38-5	: 9373
507-19-7	: 9846	517-10-2	: 4111	527-61-7	: 4264	534-26-9	: 7197	540-42-1	: 10310
507-20-0	: 9863	517-22-6	: 11126	527-62-8	: 9205	534-52-1	: 9405	540-47-6	: 4980
507-25-5	: 7542	517-23-7	: 6139	527-69-5	: 6040	534-59-8	: 9929	540-51-2	: 5685
507-32-4	: 457	517-25-9	: 7562	527-72-0	: 11737	534-84-9	: 4611	540-54-5	: 9848
507-36-8	: 3240	517-28-2	: 629	527-73-1	: 7211	534-85-0	: 1282	540-61-4	: 272
507-40-4	: 7183	517-60-2	: 1748	527-84-4	: 1988	535-11-5	: 10207	540-63-6	: 5590
507-42-6	: 5581	517-66-8	: 2468	527-85-5	: 619	535-13-7	: 10220	540-67-0	: 5602
507-45-9	: 3286	517-82-8	: 7857	527-89-9	: 1125	535-15-9	: 164	540-69-2	: 6003
507-47-1	: 5680	517-92-0	: 401	527-93-5	: 5153	535-46-6	: 11548	540-73-8	: 7155
507-60-8	: 3142	517-97-5	: 5159	528-21-2	: 5873	535-75-1	: 9640	540-80-7	: 8085
508-02-1	: 8547	518-05-8	: 7805	528-23-4	: 4784	535-77-3	: 1989	540-82-9	: 11576
508-32-7	: 11867	518-17-2	: 7333	528-29-0	: 1491	535-80-8	: 2535	540-84-1	: 8848
508-44-1	: 5185	518-28-5	: 6192	528-44-9	: 2319	535-83-1	: 11029	540-88-5	: 177
508-54-3	: 7675	518-44-5	: 2572	528-45-0	: 2590	535-87-5	: 2554	540-97-6	: 4624
508-65-6	: 4068	518-61-6	: 9502	528-48-3	: 2813	535-89-7	: 11043	541-01-5	: 6528
509-12-6	: 2938	518-69-4	: 5149	528-53-0	: 2849	536-08-3	: 2575	541-02-6	: 4901
509-14-8	: 7544	518-75-2	: 2774	528-17-2	: 9197	536-17-4	: 11684	541-05-9	: 5002
509-15-9	: 6203	518-77-4	: 5151	528-75-6	: 551	536-25-4	: 2501	541-16-2	: 9925
509-20-6	: 305	518-82-1	: 419	528-76-7	: 2178	536-33-4	: 10885	541-20-8	: 8738
509-64-8	: 7670	519-02-8	: 7426	528-79-0	: 9420	536-38-9	: 5778	541-28-6	: 3384
509-87-5	: 11096	519-05-1	: 2608	528-81-4	: 467	536-50-5	: 1859	541-31-1	: 3440
510-15-6	: 4082	519-09-5	: 481	529-00-0	: 4602	536-57-2	: 2182	541-33-3	: 3276
510-20-3	: 9939	519-34-6	: 7610	529-08-8	: 506	536-59-4	: 4664	541-35-5	: 3194
510-90-7	: 11063	519-37-9	: 10758	529-16-8	: 2968	536-60-7	: 1902	541-41-3	: 4018
511-07-9	: 5392	519-44-8	: 1517	529-17-9	: 484	536-66-3	: 2689	541-42-4	: 8092
511-08-0	: 5391	519-72-2	: 1804	529-19-1	: 2763	536-69-6	: 10898	541-46-8	: 3206
511-09-1	: 5389	519-73-3	: 1982	529-20-4	: 587	536-74-3	: 1715	541-47-9	: 3846
511-10-4	: 5390	519-87-9	: 57	529-21-5	: 10964	536-75-4	: 10958	541-50-4	: 3568
511-18-2	: 8210	520-03-6	: 7366	529-28-2	: 1775	536-78-7	: 10957	541-53-7	: 11718
511-20-6	: 5377	520-18-3	: 2843	529-33-9	: 8041	536-88-9	: 10965	541-58-2	: 11674
511-21-7	: 5378	520-26-3	: 2794	529-34-0	: 8046	536-89-0	: 7164	541-59-3	: 11121
511-34-2	: 11535	520-33-2	: 2800	529-35-1	: 8042	536-90-3	: 767	541-73-1	: 1355
511-67-1	: 11086	520-36-5	: 2806	529-36-2	: 8007	536-95-8	: 2721	541-88-8	: 124
511-70-6	: 3190	520-45-8	: 10779	529-64-6	: 915	537-17-7	: 11830	541-91-3	: 4763
511-96-6	: 11530	520-85-4	: 9752	529-65-7	: 62	537-26-8	: 488	542-05-2	: 8782

CAS	No.	CAS	No.	CAS	No.	CAS	No.	CAS	No.
542-08-5	3574	551-09-7	5514	557-89-1	453	573-33-1	7199	583-53-9	1288
542-10-9	5570	551-62-2	2270	557-91-5	5517	573-56-8	9344	583-55-1	1046
542-15-4	821	551-76-8	9489	557-93-7	10530	573-58-0	7969	583-58-4	10940
542-18-7	4339	551-77-9	9488	557-98-2	10535	573-97-7	8022	583-61-9	10936
542-28-9	10798	551-78-0	9487	557-99-3	301	573-98-8	7818	583-68-6	647
542-37-0	3546	551-88-2	8834	558-13-4	7539	574-09-4	5813	583-69-7	1503
542-46-1	4218	551-92-8	7202	558-17-8	10058	574-12-9	2841	583-70-0	1018
542-50-7	6303	552-16-9	2703	558-25-8	7538	574-39-0	3983	583-71-1	1022
542-52-9	3998	552-30-7	7341	558-30-5	8607	574-84-5	2808	583-75-5	650
542-54-1	8831	552-32-9	81	558-37-2	3760	575-37-1	7823	583-78-8	9284
542-55-2	6013	552-41-0	5823	558-38-3	10352	575-38-2	7837	583-86-8	8276
542-56-3	8094	552-45-4	1198	558-42-9	10354	575-41-7	7819	584-02-1	8917
542-58-5	5695	552-58-9	2812	558-43-0	10017	575-43-9	7822	584-04-3	10371
542-59-6	5577	552-59-0	2828	559-70-6	8548	575-44-0	7836	584-12-3	6033
542-69-8	3381	552-70-5	479	560-08-7	3931	575-74-6	607	584-27-0	3331
542-76-7	10097	552-72-7	9646	560-21-4	8849	576-22-7	1020	584-70-3	621
542-81-4	5481	552-79-4	1863	560-22-5	9068	576-23-8	1017	584-79-2	336
542-85-8	5600	552-82-9	807	560-23-6	9065	576-24-9	9282	584-84-9	1431
542-88-1	7525	552-89-6	591	560-95-2	7476	576-26-1	9310	584-87-2	2609
542-90-5	11703	552-94-3	2616	561-07-9	311	576-55-6	9468	584-90-7	5114
542-91-6	11409	553-03-7	11327	561-20-6	11352	576-83-0	1094	584-92-9	8399
542-92-7	4764	553-17-3	9384	561-25-1	7664	577-16-2	5851	584-94-1	6772
543-20-4	3366	553-19-5	2397	561-27-3	7667	577-19-5	1075	584-98-5	3299
543-21-5	3902	553-20-8	84	561-83-1	11090	577-27-5	4989	584-99-6	3769
543-24-8	6251	553-26-4	3104	561-94-4	5393	577-33-3	429	585-32-0	1806
543-27-1	4024	553-39-9	7953	562-49-2	8763	577-37-7	7572	585-34-2	9322
543-28-2	3959	553-60-6	10907	562-74-3	4699	577-55-9	992	585-48-8	10874
543-29-3	8078	553-69-5	1913	563-03-1	9387	577-56-0	2476	585-68-2	6054
543-38-4	7135	553-79-7	827	563-04-2	9570	577-59-3	5863	585-74-0	5852
543-59-9	8719	553-82-2	1381	563-12-2	5931	577-71-9	9345	585-76-2	2519
543-67-9	8098	553-86-6	2461	563-16-6	6775	577-91-3	2135	585-79-5	1076
543-75-9	5203	553-90-2	5563	563-41-7	7143	578-07-4	312	585-81-9	2632
543-80-6	110	553-94-6	1021	563-45-1	3800	578-54-1	740	586-11-8	9346
543-83-9	6283	553-97-9	4266	563-46-2	3799	578-57-4	1050	586-27-6	4698
543-86-2	3633	554-00-7	686	563-47-3	10540	578-58-5	1931	586-37-8	5840
543-87-3	3637	554-01-8	11106	563-52-0	3723	578-66-5	11206	586-38-9	2665
544-00-3	3226	554-12-1	10296	563-57-5	10553	578-67-6	11305	586-39-0	1644
544-01-4	3418	554-14-3	11769	563-58-6	10547	578-68-7	11203	586-42-5	2477
544-02-5	3436	554-21-2	5212	563-70-2	7473	578-76-7	10769	586-61-8	1059
544-10-5	6737	554-35-8	10104	563-78-0	3759	579-04-4	7112	586-62-9	4684
544-13-8	8767	554-68-7	5418	563-79-1	3761	579-07-7	10028	586-75-4	2888
544-16-1	8083	554-70-1	9521	563-80-4	3668	579-10-2	79	586-76-5	2520
544-25-2	4240	554-84-7	9436	563-83-7	9798	579-18-0	2511	586-77-6	645
544-35-4	8226	555-03-3	1936	563-84-8	3606	579-21-5	5854	586-78-7	1077
544-40-1	3433	555-10-2	4673	564-00-1	3007	579-44-2	5822	586-84-5	6129
544-62-7	10023	555-16-8	593	564-02-3	8847	579-60-2	9392	586-89-0	2478
544-63-8	11622	555-21-5	967	564-03-4	9069	579-66-8	694	586-95-8	10977
544-65-0	11637	555-30-6	11931	564-04-5	8966	579-75-9	2664	586-96-9	2059
544-66-1	11638	555-35-1	6645	564-36-3	5387	579-92-0	2656	586-98-1	10975
544-72-9	8286	555-43-1	8272	564-37-4	5388	579-94-2	183	587-02-0	741
544-73-0	8282	555-44-2	6656	565-00-2	2921	580-02-9	2482	587-03-1	1898
544-76-3	6632	555-45-3	11628	565-53-7	161	580-13-2	7740	587-04-2	530
544-77-4	6640	555-57-7	1823	565-59-3	8761	580-15-4	11205	587-54-2	2513
544-85-4	5348	555-59-9	3853	565-60-6	8941	580-16-5	11306	587-61-1	10873
544-97-8	12066	555-77-1	5412	565-63-9	3845	580-20-1	11307	587-85-9	7436
545-06-2	288	555-89-5	1965	565-64-0	10245	580-22-3	11202	587-90-6	11997
545-26-6	4048	556-10-5	12025	565-67-3	8943	580-48-3	11828	588-01-2	11906
545-47-1	7417	556-22-9	6275	565-69-5	8982	580-51-8	3089	588-46-5	87
545-93-7	11078	556-24-1	3552	565-75-3	8850	581-08-8	59	588-47-6	817
546-06-5	4143	556-27-4	333	565-76-4	9066	581-40-8	7825	588-52-3	5113
546-33-8	5004	556-50-3	6265	565-77-5	9070	581-42-0	7826	588-63-6	1085
546-43-0	8059	556-52-5	8618	565-80-0	8967	581-43-1	7838	588-67-0	1115
546-45-2	5005	556-56-9	10573	567-18-0	7973	581-45-3	11009	588-72-7	1029
546-56-5	4994	556-61-6	7521	568-02-5	8062	581-46-4	3102	588-73-8	1031
546-97-4	5926	556-64-9	11705	568-21-8	5158	581-47-5	3101	588-96-5	1032
547-25-1	6022	556-67-2	4993	568-93-4	399	581-49-7	3105	589-08-2	1581
547-44-4	2190	556-68-3	4741	568-99-0	383	581-50-0	3100	589-09-3	841
547-52-4	2188	556-69-4	8482	569-08-4	394	581-55-5	7512	589-10-6	1033
547-58-0	2222	556-70-7	5079	569-31-3	7355	581-89-5	7930	589-14-0	6120
547-63-7	10301	556-71-8	4726	569-34-6	10759	582-08-1	5128	589-15-1	1000
547-65-9	6147	556-82-1	3869	569-41-5	7824	582-16-1	7827	589-16-2	742
547-81-9	5405	556-88-7	6287	569-51-7	2318	582-17-2	7839	589-17-3	1008
547-91-1	11295	556-89-8	12024	569-57-3	1154	582-22-9	1579	589-18-4	1899
548-40-3	8631	556-96-7	1019	569-64-2	7455	582-24-1	5833	589-21-9	7141
548-51-6	2642	557-00-6	3562	569-65-3	9605	582-33-2	2496	589-29-7	1435
548-62-9	7453	557-05-1	8277	569-77-7	2408	582-60-5	2380	589-33-3	11113
548-80-1	7878	557-11-9	12028	570-08-1	9923	582-61-6	2884	589-38-8	6960
548-98-1	4093	557-17-5	10069	570-22-9	7195	582-62-7	3671	589-40-2	6012
550-10-7	5214	557-24-4	3817	570-24-1	802	582-77-4	622	589-43-5	6773
550-24-3	4261	557-25-5	3505	571-58-4	7820	582-78-5	623	589-44-6	3465
550-44-7	7365	557-30-2	5492	571-61-9	7821	583-03-9	1846	589-53-7	6424
550-60-7	8034	557-31-3	10566	572-59-8	4125	583-04-0	2722	589-55-9	6475
550-74-1	10848	557-40-4	10592	572-83-8	386	583-05-1	8804	589-59-3	3555
550-82-3	9508	557-59-5	11607	572-93-0	396	583-33-5	6052	589-63-9	8473
550-97-0	2644	557-61-9	8222	573-17-1	9164	583-39-1	2386	589-75-3	8427
551-01-9	9877	557-66-4	5433	573-20-6	7841	583-48-2	6776	589-87-7	1048
551-06-4	7906	557-68-6	7471	573-26-2	75	583-50-6	3192	589-92-4	4598

589-93-5	: 10938	593-59-9	: 444	597-90-0	: 6487	603-39-4	: 2110	608-93-5	: 2084
589-98-0	: 8445	593-60-2	: 5880	597-93-3	: 8162	603-45-2	: 4278	608-96-8	: 2096
590-01-2	: 10214	593-61-3	: 5937	597-96-6	: 6946	603-50-9	: 9465	608-98-0	: 1553
590-02-3	: 125	593-63-5	: 5938	598-01-6	: 6505	603-52-1	: 3946	609-02-9	: 9963
590-11-4	: 5892	593-66-8	: 5907	598-02-7	: 9558	603-54-3	: 12011	609-08-5	: 9962
590-12-5	: 5891	593-70-4	: 7481	598-03-8	: 10131	603-69-0	: 3462	609-09-6	: 9973
590-13-6	: 10529	593-71-5	: 7482	598-04-9	: 3423	603-71-4	: 2363	609-11-0	: 3498
590-16-9	: 5463	593-74-8	: 7435	598-05-0	: 11575	603-76-9	: 7312	609-12-1	: 3475
590-18-1	: 3706	593-75-9	: 7655	598-09-4	: 8606	603-81-6	: 2552	609-13-2	: 3477
590-19-2	: 3145	593-77-1	: 7447	598-16-3	: 5921	603-83-8	: 799	609-14-3	: 3557
590-26-1	: 5904	593-79-3	: 7526	598-17-4	: 9880	603-85-0	: 9214	609-19-8	: 9486
590-35-2	: 8760	593-81-7	: 7445	598-19-6	: 10699	604-53-5	: 3004	609-22-3	: 9279
590-36-3	: 8940	593-88-4	: 449	598-21-0	: 292	604-61-5	: 2515	609-23-4	: 9492
590-42-1	: 10066	593-90-8	: 3119	598-22-1	: 10490	604-68-2	: 6216	609-25-6	: 2369
590-50-1	: 8965	593-92-0	: 5890	598-23-2	: 3916	604-69-3	: 6217	609-26-7	: 8813
590-60-3	: 3634	593-95-3	: 4006	598-25-4	: 3170	604-83-1	: 9172	609-29-0	: 8803
590-66-9	: 4364	593-96-4	: 5455	598-31-2	: 10416	604-88-6	: 1756	609-31-4	: 3640
590-67-0	: 4530	593-98-6	: 7466	598-38-9	: 5704	605-01-6	: 2090	609-36-9	: 9765
590-73-8	: 6771	594-02-5	: 5546	598-45-8	: 10063	605-02-7	: 7947	609-39-2	: 6134
590-86-3	: 3186	594-03-6	: 5589	598-50-5	: 12017	605-03-8	: 11290	609-40-5	: 11775
590-90-9	: 3662	594-04-7	: 7503	598-53-8	: 10070	605-27-6	: 411	609-46-1	: 2225
590-92-1	: 10205	594-08-1	: 4034	598-55-0	: 3953	605-32-3	: 408	609-65-4	: 2890
590-93-2	: 3917	594-09-2	: 9522	598-56-1	: 5421	605-37-8	: 7852	609-72-3	: 860
591-07-1	: 27	594-10-5	: 11550	598-57-2	: 7451	605-39-0	: 3050	609-73-4	: 1783
591-08-2	: 33	594-11-6	: 4981	598-58-3	: 8075	605-42-5	: 390	610-02-6	: 2730
591-09-3	: 108	594-15-0	: 7549	598-61-8	: 4192	605-50-5	: 1335	610-15-1	: 624
591-11-7	: 6154	594-16-1	: 9883	598-74-3	: 3223	605-54-9	: 1333	610-16-2	: 2584
591-12-8	: 6153	594-18-3	: 7495	598-76-5	: 9899	605-63-0	: 7786	610-17-3	: 718
591-17-3	: 1055	594-20-7	: 9894	598-77-6	: 10172	605-65-2	: 7977	610-25-3	: 2047
591-18-4	: 1047	594-22-9	: 7556	598-78-7	: 10215	605-71-0	: 7832	610-34-4	: 2707
591-19-5	: 639	594-27-4	: 11544	598-82-3	: 10266	605-79-8	: 7634	610-39-9	: 1960
591-20-8	: 9229	594-34-3	: 9889	598-88-9	: 5898	605-81-2	: 6228	610-46-8	: 377
591-22-0	: 10941	594-36-5	: 3253	598-92-5	: 5482	605-82-3	: 421	610-48-0	: 423
591-27-5	: 9202	594-37-6	: 9900	598-94-7	: 12005	605-83-4	: 420	610-50-4	: 431
591-31-1	: 584	594-38-7	: 3385	598-96-9	: 9072	605-85-6	: 5859	610-66-2	: 965
591-34-4	: 10309	594-39-8	: 3222	598-98-1	: 10260	605-99-2	: 10765	610-67-3	: 1662
591-35-5	: 9287	594-42-3	: 7527	598-99-2	: 253	606-17-7	: 2594	610-69-5	: 219
591-47-0	: 4672	594-44-5	: 5827	599-01-9	: 10333	606-18-8	: 2506	610-71-9	: 2556
591-49-1	: 4670	594-45-6	: 5622	599-04-2	: 6143	606-20-2	: 1958	610-72-0	: 2581
591-50-4	: 1766	594-47-8	: 94	599-66-6	: 2240	606-21-3	: 1145	610-89-9	: 9074
591-60-6	: 3569	594-51-4	: 3274	599-70-2	: 1704	606-22-4	: 726	610-91-3	: 2362
591-62-8	: 3936	594-52-5	: 3236	599-79-1	: 2654	606-23-5	: 7264	610-97-9	: 2660
591-68-4	: 8862	594-56-9	: 3812	599-91-7	: 2233	606-27-9	: 2710	610-99-1	: 10449
591-76-4	: 6834	594-57-0	: 3247	599-97-3	: 5657	606-28-0	: 2516	611-00-7	: 2555
591-78-6	: 6959	594-58-1	: 10228	600-00-5	: 242	606-35-9	: 1946	611-01-8	: 2580
591-80-0	: 9073	594-60-5	: 3618	600-00-0	: 10213	606-37-1	: 7831	611-03-0	: 2553
591-02-2	: 10085	594-61-6	: 10278	600-05-5	: 10238	606-43-9	: 11328	611-06-3	: 1396
591-84-4	: 11707	594-65-0	: 95	600-06-6	: 3245	606-45-1	: 2673	611-07-4	: 9269
591-87-7	: 235	594-70-7	: 10085	600-07-7	: 3538	606-46-2	: 696	611-10-9	: 4785
591-93-5	: 8659	594-71-8	: 9870	600-11-3	: 8749	606-81-5	: 3063	611-12-1	: 10366
591-95-7	: 8658	594-72-9	: 5536	600-13-5	: 4150	606-88-2	: 7173	611-13-2	: 6059
591-96-8	: 8660	594-73-0	: 5594	600-14-6	: 8797	606-97-3	: 3093	611-14-3	: 1685
592-01-8	: 3930	594-81-0	: 3271	600-15-7	: 3530	606-99-5	: 751	611-15-4	: 1640
592-13-2	: 6774	594-82-1	: 3430	600-17-9	: 3842	607-00-1	: 5993	611-19-8	: 1132
592-20-1	: 10412	594-83-2	: 3644	600-18-0	: 3567	607-31-8	: 11274	611-20-1	: 2759
592-22-3	: 5250	594-89-8	: 10048	600-22-6	: 10316	607-32-9	: 7386	611-21-2	: 710
592-27-8	: 6422	595-30-2	: 8559	600-25-9	: 9866	607-34-1	: 11284	611-32-5	: 11283
592-34-7	: 4012	595-37-9	: 3506	600-30-6	: 3497	607-42-1	: 368	611-33-6	: 11242
592-35-8	: 3935	595-38-0	: 8724	600-36-2	: 8931	607-50-1	: 7919	611-34-7	: 11204
592-41-6	: 6999	595-39-1	: 7400	600-40-8	: 5551	607-52-3	: 7939	611-35-8	: 11238
592-42-7	: 6678	595-40-4	: 7401	601-75-2	: 9948	607-53-4	: 8005	611-43-8	: 3051
592-44-9	: 6676	595-41-5	: 8930	601-76-3	: 9918	607-56-7	: 12008	611-45-0	: 7949
592-45-0	: 6677	595-44-8	: 9902	601-77-4	: 9822	607-81-8	: 9977	611-49-4	: 7928
592-46-1	: 6680	595-49-3	: 9920	601-84-3	: 2557	607-85-2	: 2635	611-61-0	: 3053
592-49-4	: 6679	595-89-1	: 9733	601-88-7	: 1393	607-86-3	: 2682	611-62-1	: 3065
592-50-7	: 8817	595-90-4	: 11545	602-03-9	: 724	607-90-9	: 2652	611-69-8	: 1901
592-51-8	: 9062	596-27-0	: 7353	602-21-1	: 1353	607-91-0	: 2428	611-70-1	: 10469
592-55-2	: 5461	596-29-2	: 7358	602-27-7	: 1489	607-97-6	: 3518	611-73-4	: 934
592-57-4	: 4248	596-43-0	: 1061	602-29-9	: 2046	607-99-8	: 2313	611-79-0	: 7589
592-65-4	: 10154	596-51-0	: 11179	602-37-9	: 7787	608-07-1	: 7304	611-92-7	: 12006
592-76-7	: 6533	596-75-8	: 9937	602-38-0	: 7833	608-21-9	: 2307	611-94-9	: 7625
592-80-3	: 8935	597-04-6	: 3511	602-55-1	: 428	608-23-1	: 1137	611-95-0	: 2512
592-82-5	: 3388	597-05-7	: 8934	602-56-2	: 318	608-25-3	: 1529	611-97-2	: 7599
592-84-7	: 6004	597-09-1	: 10008	602-60-8	: 426	608-26-4	: 9260	611-99-4	: 7597
592-88-1	: 10599	597-12-6	: 6221	602-64-2	: 415	608-27-5	: 685	612-00-0	: 1677
592-90-5	: 8590	597-32-0	: 10365	602-65-3	: 418	608-28-6	: 1772	612-01-1	: 12018
593-03-3	: 6661	597-33-1	: 9998	602-82-4	: 1332	608-29-7	: 2349	612-08-8	: 5109
593-08-8	: 11889	597-35-3	: 5626	602-92-6	: 2560	608-30-0	: 682	612-12-4	: 986
593-12-4	: 8358	597-49-9	: 8932	602-96-0	: 2364	608-31-1	: 688	612-13-5	: 2750
593-14-6	: 8361	597-50-2	: 9517	602-99-3	: 9434	608-32-2	: 2305	612-15-7	: 1637
593-45-3	: 8234	597-52-4	: 11503	603-26-9	: 7513	608-33-3	: 9277	612-16-8	: 1889
593-49-7	: 6302	597-55-7	: 9926	603-32-7	: 450	608-50-4	: 2360	612-17-9	: 7815
593-50-0	: 11821	597-67-1	: 11428	603-33-8	: 3111	608-66-2	: 6194	612-19-1	: 2600
593-51-1	: 7446	597-71-7	: 9988	603-34-9	: 728	608-69-5	: 3308	612-22-6	: 1692
593-52-2	: 446	597-72-8	: 10068	603 35 0	: 0623	608-71-9	: 9439	612-23-7	: 1203
593-53-3	: 7515	597-76-2	: 6931	603-36-1	: 11551	608-89-9	: 3315	612-25-9	: 1906
593-57-7	: 442	597-77-3	: 6935	603-38-3	: 729	608-90-2	: 2082	612-28-2	: 804

612-35-1	: 2718	615-57-6	: 681	618-85-9	: 1957	622-04-8	: 10377	623-96-1	: 4002
612-37-3	: 10766	615-58-7	: 9276	618-87-1	: 727	622-08-2	: 5761	623-97-2	: 5697
612-55-5	: 7903	615-59-8	: 1305	618-89-3	: 2525	622-09-3	: 7693	624-03-3	: 6649
612-57-7	: 11240	615-60-1	: 1141	618-91-7	: 2661	622-15-1	: 7565	624-04-4	: 5311
612-58-8	: 11278	615-62-3	: 9265	618-95-1	: 2711	622-16-2	: 764	624-08-8	: 6319
612-60-2	: 11282	615-65-6	: 668	618-98-4	: 2708	622-24-2	: 1164	624-09-9	: 6464
612-61-3	: 11241	615-67-8	: 1507	619-02-3	: 4601	622-26-4	: 9664	624-13-5	: 8441
612-62-4	: 11237	615-68-9	: 2350	619-12-5	: 9598	622-29-7	: 7452	624-16-8	: 5077
612-64-6	: 750	615-71-4	: 2306	619-15-8	: 1959	622-31-1	: 594	624-17-9	: 8133
612-71-5	: 11598	615-74-7	: 9262	619-20-5	: 2601	622-32-2	: 595	624-20-4	: 6754
612-75-9	: 3057	615-79-2	: 8876	619-23-8	: 1206	622-36-6	: 7166	624-24-8	: 8891
612-76-0	: 3066	615-81-6	: 5556	619-25-0	: 1907	622-37-7	: 976	624-29-3	: 4368
612-78-2	: 3005	615-83-8	: 8860	619-31-8	: 719	622-38-8	: 1705	624-31-7	: 1780
612-81-7	: 3107	615-84-9	: 5494	619-33-0	: 5530	622-39-9	: 11015	624-38-4	: 1429
612-94-2	: 7948	615-85-0	: 5453	619-39-6	: 5068	622-40-2	: 7691	624-41-9	: 3630
612-95-3	: 11291	615-87-2	: 1298	619-41-0	: 5779	622-42-4	: 2052	624-42-0	: 6523
612-96-4	: 11288	615-89-4	: 1514	619-42-1	: 2526	622-45-7	: 155	624-43-1	: 10192
612-97-5	: 11196	615-98-5	: 5565	619-44-3	: 2662	622-47-9	: 926	624-44-2	: 6982
613-03-6	: 2370	615-99-6	: 5564	619-52-3	: 4683	622-57-1	: 748	624-45-3	: 8908
613-08-1	: 369	616-02-4	: 6098	619-58-9	: 2659	622-60-6	: 1661	624-48-6	: 3779
613-12-7	: 424	616-06-8	: 8215	619-60-3	: 9319	622-61-7	: 1158	624-49-7	: 3778
613-13-8	: 365	616-12-6	: 9056	619-73-8	: 1908	622-62-8	: 9352	624-54-4	: 10318
613-21-8	: 11319	616-19-3	: 9901	619-82-9	: 4354	622-63-9	: 1707	624-57-7	: 8617
613-29-6	: 684	616-20-6	: 8721	619-84-1	: 2585	622-78-6	: 1794	624-60-2	: 5435
613-31-0	: 376	616-21-7	: 3277	619-99-8	: 6815	622-79-7	: 980	624-61-3	: 5939
613-33-2	: 3061	616-23-9	: 10359	620-02-0	: 6045	622-80-0	: 843	624-64-6	: 3705
613-37-6	: 3081	616-24-0	: 8702	620-03-1	: 6047	622-82-2	: 7160	624-65-7	: 10729
613-39-8	: 9613	616-25-1	: 9087	620-05-3	: 1777	622-85-5	: 2168	624-66-8	: 10734
613-42-3	: 3094	616-27-3	: 3646	620-11-1	: 8919	622-86-6	: 1159	624-67-9	: 10723
613-43-4	: 5112	616-28-4	: 3912	620-12-2	: 10193	622-93-5	: 10367	624-70-4	: 5477
613-45-6	: 543	616-31-9	: 8846	620-13-3	: 1063	622-96-8	: 1687	624-72-6	: 5545
613-46-7	: 7749	616-38-6	: 4000	620-14-4	: 1686	622-97-9	: 1642	624-73-7	: 5547
613-48-9	: 697	616-39-7	: 5430	620-16-6	: 1162	622-98-0	: 1163	624-74-8	: 5941
613-50-3	: 11285	616-40-0	: 7152	620-17-7	: 9354	623-00-7	: 2746	624-75-9	: 283
613-51-4	: 11286	616-42-2	: 11582	620-18-8	: 9349	623-03-0	: 2749	624-76-0	: 5731
613-59-2	: 7950	616-43-3	: 11132	620-19-9	: 1199	623-05-2	: 1884	624-79-3	: 5599
613-69-4	: 552	616-44-4	: 11770	620-20-2	: 1133	623-08-5	: 712	624-81-7	: 5434
613-70-7	: 9382	616-45-5	: 11180	620-22-4	: 2764	623-09-6	: 1280	624-82-8	: 7566
613-75-2	: 7510	616-47-7	: 7208	620-23-5	: 588	623-10-9	: 758	624-83-9	: 7520
613-77-4	: 11249	616-55-7	: 9223	620-24-6	: 1883	623-12-1	: 1180	624-85-1	: 7184
613-78-5	: 2645	616-91-1	: 5015	620-32-6	: 2241	623-15-4	: 3878	624-89-5	: 5606
613-80-9	: 7940	617-00-5	: 795	620-40-6	: 1798	623-17-6	: 6119	624-90-8	: 7460
613-81-0	: 8006	617-02-7	: 2699	620-42-8	: 9597	623-18-7	: 10640	624-91-9	: 8091
613-84-3	: 577	617-05-0	: 2626	620-43-9	: 3224	623-19-8	: 6123	624-92-0	: 5254
613-88-7	: 956	617-09-4	: 3996	620-47-3	: 2015	623-20-1	: 10639	624-95-3	: 3617
613-89-8	: 5774	617-30-1	: 6797	620-63-3	: 3561	623-21-2	: 3520	624-96-4	: 3284
613-90-1	: 968	617-31-2	: 8883	620-71-3	: 9800	623-22-3	: 10642	624-97-5	: 9100
613-91-2	: 5866	617-33-4	: 160	620-72-4	: 119	623-24-5	: 985	625-01-4	: 4085
613-93-4	: 620	617-35-6	: 10315	620-73-5	: 147	623-25-6	: 988	625-04-7	: 8959
613-94-5	: 2611	617-45-8	: 470	620-80-4	: 3576	623-26-7	: 1316	625-06-9	: 8928
613-97-8	: 749	617-50-5	: 10303	620-81-5	: 5495	623-27-8	: 1323	625-08-1	: 3535
614-00-6	: 806	617-51-6	: 10275	620-82-6	: 5559	623-30-3	: 10512	625-23-0	: 6942
614-16-4	: 2124	617-52-7	: 3330	620-83-7	: 2016	623-34-7	: 3285	625-25-2	: 6495
614-18-6	: 10902	617-53-8	: 3789	620-85-9	: 1696	623-36-9	: 9002	625-27-4	: 9055
614-21-1	: 5862	617-54-9	: 3790	620-86-0	: 2719	623-39-2	: 10014	625-28-5	: 3410
614-27-7	: 2142	617-60-7	: 5309	620-88-2	: 2054	623-40-5	: 8985	625-30-9	: 8701
614-29-9	: 1613	617-65-2	: 6240	620-92-8	: 9407	623-42-7	: 3547	625-31-0	: 9089
614-33-5	: 10196	617-73-2	: 8432	620-93-9	: 797	623-43-8	: 3847	625-36-5	: 10497
614-34-6	: 2700	617-75-4	: 448	620-94-0	: 2286	623-46-1	: 5533	625-37-6	: 3702
614-39-1	: 602	617-78-7	: 8809	620-95-1	: 11005	623-47-2	: 10739	625-38-7	: 3813
614-45-9	: 1128	617-80-1	: 3407	620-99-5	: 5670	623-48-3	: 201	625-44-5	: 10071
614-47-1	: 10709	617-83-4	: 4156	621-08-9	: 2184	623-49-4	: 4030	625-45-6	: 205
614-54-0	: 1881	617-84-5	: 5991	621-14-7	: 3320	623-50-7	: 197	625-46-7	: 10591
614-68-6	: 1787	617-85-6	: 11549	621-15-8	: 6722	623-51-8	: 203	625-47-8	: 5483
614-71-1	: 1659	617-86-7	: 11483	621-23-8	: 2355	623-53-0	: 4003	625-48-9	: 5752
614-72-2	: 1156	617-88-9	: 6075	621-25-0	: 7174	623-56-3	: 6981	625-50-3	: 61
614-75-5	: 904	617-89-0	: 6113	621-26-1	: 7138	623-59-6	: 73	625-52-5	: 12014
614-78-8	: 11806	617-90-3	: 6039	621-27-2	: 9462	623-60-9	: 10351	625-54-7	: 10038
614-96-0	: 7316	617-92-5	: 11124	621-31-8	: 9356	623-65-4	: 6646	625-55-8	: 6011
614-99-3	: 6056	617-94-7	: 1857	621-32-9	: 1660	623-66-5	: 8424	625-56-9	: 7582
615-05-4	: 1274	618-31-5	: 1301	621-34-1	: 9351	623-68-7	: 3818	625-57-0	: 3978
615-09-8	: 6163	618-32-6	: 2885	621-36-3	: 925	623-69-8	: 10372	625-58-1	: 8074
615-10-1	: 6071	618-34-8	: 1148	621-37-4	: 905	623-70-1	: 3839	625-59-2	: 11706
615-11-2	: 6057	618-36-0	: 1814	621-50-1	: 966	623-71-2	: 10221	625-60-5	: 5647
615-13-4	: 7287	618-38-2	: 2910	621-51-2	: 2597	623-72-3	: 10270	625-65-0	: 9035
615-18-9	: 2878	618-40-6	: 7165	621-52-3	: 1663	623-73-4	: 158	625-66-1	: 3282
615-20-3	: 2862	618-41-7	: 2180	621-59-0	: 572	623-76-7	: 12001	625-67-2	: 8750
615-30-5	: 8691	618-42-8	: 9630	621-62-5	: 5469	623-78-9	: 3949	625-68-3	: 3481
615-36-1	: 638	618-45-1	: 9411	621-63-6	: 5705	623-80-3	: 4031	625-69-4	: 8789
615-37-2	: 1778	618-46-2	: 2891	621-64-7	: 9825	623-81-4	: 11581	625-71-8	: 3531
615-38-3	: 4531	618-51-9	: 2658	621-70-5	: 6908	623-84-7	: 10002	625-74-1	: 10084
615-39-4	: 4532	618-62-2	: 1394	621-72-7	: 2385	623-85-8	: 3970	625-76-3	: 7508
615-41-8	: 1171	618-65-5	: 562	621-76-1	: 10078	623-86-9	: 102	625-77-4	: 23
615-42-9	: 1427	618-68-8	: 2144	621-77-2	: 8705	623-87-0	: 10191	625-80-9	: 10152
615-52-1	: 5567	618-71-3	: 2593	621-87-4	: 10479	623-91-6	: 3775	625-82-1	: 11119
615-54-3	: 2308	618-76-8	: 2618	621-95-4	: 731	623-93-8	: 8161	625-84-1	: 11120
615-55-4	: 683	618-80-4	: 9296	622-00-4	: 3306	623-95-0	: 12012	625-86-5	: 6089

625-88-7	: 11753	627-48-5	: 4163	629-39-0	: 8079	634-67-3	: 852	641-36-1	: 5160
625-92-3	: 10926	627-51-0	: 5920	629-40-3	: 8385	634-68-4	: 2273	641-74-7	: 7423
625-95-6	: 1779	627-52-1	: 11572	629-41-4	: 8389	634-89-9	: 2253	641-85-0	: 9740
625-96-7	: 4597	627-53-2	: 5620	629-43-6	: 8088	634-92-4	: 2274	641-91-8	: 8010
625-98-9	: 1167	627-54-3	: 5630	629-45-8	: 5252	634-93-5	: 854	641-96-3	: 11193
626-00-6	: 1428	627-58-7	: 6691	629-46-9	: 8095	634-95-7	: 12002	642-28-4	: 7912
626-01-7	: 760	627-59-8	: 6944	629-49-2	: 5344	635-03-0	: 10139	642-32-0	: 2268
626-02-8	: 9372	627-63-4	: 3795	629-50-5	: 11871	635-10-9	: 2258	642-44-4	: 11060
626-04-0	: 1562	627-67-8	: 3400	629-54-9	: 6629	635-11-0	: 2276	642-84-2	: 608
626-10-8	: 3771	627-70-3	: 10463	629-59-4	: 11613	635-27-8	: 11239	643-28-7	: 788
626-13-1	: 796	627-73-6	: 3323	629-60-7	: 11880	635-46-1	: 11296	643-58-3	: 3082
626-15-3	: 984	627-82-7	: 10025	629-62-9	: 8642	635-65-4	: 3000	643-60-7	: 7393
626-16-4	: 987	627-83-8	: 8256	629-63-0	: 11619	635-67-6	: 1508	643-65-2	: 7632
626-17-5	: 1315	627-88-3	: 8265	629-64-1	: 6430	635-81-4	: 2269	643-75-4	: 10199
626-18-6	: 1434	627-89-4	: 8306	629-65-2	: 6435	635-88-1	: 7918	643-84-5	: 2847
626-19-7	: 1322	627-90-7	: 11946	629-66-3	: 8106	635-90-5	: 11134	643-93-6	: 3083
626-20-0	: 1638	627-91-8	: 6792	629-70-9	: 6662	635-93-8	: 533	644-08-6	: 3084
626-21-1	: 10933	627-93-0	: 6786	629-72-1	: 8643	636-04-4	: 1129	644-15-5	: 7725
626-23-3	: 3227	627-96-3	: 6756	629-73-2	: 6667	636-09-9	: 1350	644-21-3	: 7159
626-26-6	: 3434	627-97-4	: 6548	629-74-3	: 6673	636-10-2	: 7162	644-30-4	: 1468
626-27-7	: 6456	627-98-5	: 6941	629-75-4	: 6674	636-21-5	: 791	644-31-5	: 9151
626-29-9	: 11623	628-02-4	: 6721	629-78-7	: 6306	636-28-2	: 2254	644-32-6	: 5251
626-31-3	: 3297	628-03-5	: 3390	629-79-8	: 6641	636-30-6	: 853	644-35-9	: 9461
626-33-5	: 6524	628-04-6	: 3374	629-80-1	: 6628	636-31-7	: 2275	644-49-5	: 10308
626-35-7	: 218	628-05-7	: 8832	629-82-3	: 8416	636-36-2	: 6891	644-71-3	: 9445
626-36-8	: 3932	628-08-0	: 3862	629-89-0	: 8313	636-41-9	: 11131	644-80-4	: 3770
626-37-9	: 3963	628-09-1	: 10350	629-90-3	: 6304	636-46-4	: 1352	644-90-6	: 8423
626-39-1	: 2309	628-11-5	: 4015	629-92-5	: 8101	636-53-3	: 1346	644-97-3	: 9556
626-41-5	: 9278	628-12-6	: 4019	629-93-6	: 8242	636-60-2	: 3325	644-98-4	: 10984
626-43-7	: 690	628-13-7	: 10969	629-94-7	: 6294	636-72-6	: 11768	645-00-1	: 1784
626-44-8	: 2351	628-16-0	: 6716	629-96-9	: 5358	636-73-7	: 11018	645-09-0	: 625
626-48-2	: 11069	628-17-1	: 8818	629-97-0	: 5268	636-82-8	: 4637	645-12-5	: 6066
626-51-7	: 8779	628-20-6	: 3405	629-98-1	: 5284	636-98-6	: 1785	645-13-6	: 5848
626-53-9	: 6454	628-21-7	: 3288	629-99-2	: 8637	637-03-6	: 447	645-36-3	: 5414
626-55-1	: 10876	628-28-4	: 3391	630-01-3	: 6622	637-07-0	: 10231	645-41-0	: 3225
626-58-4	: 9682	628-29-5	: 3402	630-02-4	: 8220	637-12-7	: 8248	645-45-4	: 2160
626-60-8	: 10916	628-30-8	: 10064	630-03-5	: 8099	637-27-4	: 10323	645-49-8	: 1620
626-61-9	: 10917	628-32-0	: 10037	630-04-6	: 6200	637-47-8	: 7139	645-55-6	: 825
626-62-0	: 4415	628-33-1	: 10733	630-06-8	: 6991	637-50-3	: 2162	645-56-7	: 9463
626-64-2	: 11032	628-34-2	: 5474	630-07-9	: 8999	637-55-8	: 8271	645-59-0	: 2123
626-67-5	: 9679	628-37-5	: 9158	630-08-0	: 4011	637-57-0	: 10679	645-66-9	: 5306
626-70-0	: 6788	628-41-1	: 4249	630-10-4	: 11354	637-59-2	: 1088	645-67-0	: 8912
626-77-7	: 6909	628-44-4	: 8464	630-13-7	: 9908	637-64-9	: 6125	645-69-2	: 3319
626-82-4	: 6880	628-46-6	: 6896	630-16-0	: 5631	637-65-0	: 6126	645-96-5	: 2177
626-84-6	: 5736	628-50-2	: 4021	630-17-1	: 9843	637-69-4	: 1639	646-04-8	: 9005
626-85-7	: 11580	628-61-5	: 8360	630-18-2	: 10100	637-78-7	: 10299	646-05-9	: 9111
626-86-8	: 6791	628-62-6	: 6348	630-19-3	: 9783	637-87-6	: 1172	646-06-0	: 5204
626-87-9	: 8740	628-63-7	: 227	630-20-6	: 5634	637-88-7	: 4377	646-07-1	: 8889
020-88-0	: 8716	628-64-8	: 4017	630-25-1	: 5524	637-89-8	: 9378	646-13-9	: 8268
626-89-1	: 8939	628-66-0	: 10003	630-51-3	: 3343	637-92-3	: 10040	646-14-0	: 6841
626-90-4	: 3187	628-67-1	: 3348	630-60-4	: 4040	637-98-9	: 3979	646-19-5	: 6371
626-91-5	: 3395	628-68-2	: 5754	630-72-8	: 290	638-00-6	: 11755	646-20-8	: 6392
626-92-6	: 8746	628-71-7	: 6609	630-76-2	: 1831	638-02-8	: 11756	646-30-0	: 8103
626-94-8	: 7068	628-72-8	: 6329	630-95-5	: 7910	638-04-0	: 4367	646-31-1	: 11603
626-95-9	: 8786	628-73-9	: 6840	631-01-6	: 8546	638-07-3	: 3494	650-69-1	: 435
626-96-0	: 8696	628-76-2	: 8747	631-28-7	: 10164	638-08-4	: 8249	653-37-2	: 596
626-97-1	: 8698	628-77-3	: 8758	631-36-7	: 11450	638-10-8	: 3850	655-48-1	: 5575
626-99-3	: 8676	628-80-8	: 8822	631-40-3	: 9833	638-11-9	: 3549	655-86-7	: 9194
627-00-9	: 3482	628-81-9	: 3372	631-57-2	: 10113	638-17-5	: 5265	659-04-1	: 10678
627-02-1	: 10039	628-83-1	: 11699	631-61-8	: 106	638-20-0	: 40	659-49-4	: 828
627-03-2	: 180	628-89-7	: 5698	631-64-1	: 159	638-21-1	: 9519	659-70-1	: 3551
627-04-3	: 188	628-92-2	: 4244	632-05-3	: 3444	638-23-3	: 5017	659-86-9	: 10735
627-05-4	: 3412	628-96-6	: 5574	632-07-5	: 10233	638-25-5	: 8439	660-68-4	: 5428
627-07-6	: 9824	628-97-7	: 6650	632-14-4	: 12032	638-28-8	: 6738	660-88-8	: 8857
627-08-7	: 10076	629-01-6	: 8349	632-15-5	: 11757	638-29-9	: 8990	661-11-0	: 6417
627-09-8	: 10744	629-03-8	: 6757	632-16-6	: 11754	638-37-9	: 3256	661-19-8	: 5279
627-11-2	: 4013	629-04-9	: 6356	632-21-3	: 10487	638-41-5	: 4025	661-68-7	: 5227
627-12-3	: 3969	629-05-0	: 8529	632-22-4	: 12030	638-45-9	: 6831	661-97-2	: 9898
627-13-4	: 8080	629-06-1	: 6360	632-32-6	: 5385	638-46-0	: 3376	666-99-9	: 8102
627-16-7	: 10730	629-08-3	: 6428	632-46-2	: 2582	638-49-3	: 6015	670-54-2	: 5915
627-18-9	: 10344	629-09-4	: 6769	632-50-8	: 1591	638-51-7	: 8089	670-96-2	: 7213
627-19-0	: 9122	629-11-8	: 6796	632-51-9	: 1623	638-53-9	: 11881	671-56-7	: 3866
627-20-3	: 9006	629-14-1	: 5542	632-56-4	: 5633	638-54-0	: 8364	672-13-9	: 571
627-21-4	: 9123	629-15-2	: 5573	632-77-9	: 412	638-58-4	: 11611	672-15-1	: 7134
627-22-5	: 3151	629-16-3	: 8086	632-80-4	: 7348	638-60-8	: 11640	673-31-4	: 2149
627-26-9	: 3802	629-17-4	: 11702	632-82-6	: 414	638-65-3	: 8243	673-32-5	: 2173
627-27-0	: 3861	629-19-6	: 5259	632-83-7	: 385	638-67-5	: 11855	673-79-0	: 10330
627-30-5	: 10348	629-20-9	: 4742	632-89-3	: 12031	638-68-6	: 11819	674-76-0	: 9058
627-31-6	: 9907	629-23-2	: 11634	632-92-8	: 1490	638-73-3	: 7479	674-82-8	: 8598
627-32-7	: 10385	629-27-6	: 8406	632-99-5	: 634	638-95-9	: 12038	675-09-2	: 10792
627-37-2	: 10524	629-30-1	: 6399	633-03-4	: 5444	639-46-3	: 7668	675-20-7	: 9719
627-39-4	: 9787	629-31-2	: 6346	634-19-5	: 5978	639-58-7	: 11541	677-21-4	: 10608
627-40-7	: 10576	629-32-3	: 6421	634-36-6	: 2354	639-81-6	: 5395	677-68-9	: 10149
627-41-8	: 10736	629-33-4	: 6009	634-41-3	: 7723	640-19-7	: 63	678-26-2	: 8807
627-42-9	: 5480	629-35-6	: 7433	634-43-5	: 7724	640-61-9	: 2211	679-37-8	: 7185
627-44-1	: 7434	629-36-7	: 10122	634-47-9	: 11243	641-06-5	: 50	680-31-9	: 9573
627-45-2	: 5994	629-37-8	: 8412	634-66-2	: 2259	641-16-7	: 9476	681-06-1	: 9550

681-84-5	: 11508	699-08-1	: 1657	764-96-5	: 11973	779-84-0	: 9450
682-01-9	: 11510	699-09-2	: 2296	764-97-6	: 11972	780-69-8	: 11481
682-30-4	: 9535	699-17-2	: 3659	765-03-7	: 5343	781-35-1	: 10432
683-08-9	: 9539	699-30-9	: 6099	765-05-9	: 5051	782-05-8	: 1457
683-50-1	: 9777	700-12-9	: 2097	765-09-3	: 11872	782-06-9	: 1456
683-51-2	: 10511	700-75-4	: 769	765-13-9	: 8656	784-50-9	: 5969
683-68-1	: 5521	700-88-9	: 1254	765-27-5	: 5366	786-19-6	: 4037
683-72-7	: 51	701-57-5	: 2044	765-30-0	: 4953	790-69-2	: 11802
684-16-2	: 10441	701-97-3	: 4468	765-34-4	: 8603	791-28-6	: 9518
684-84-4	: 3353	702-28-3	: 4969	765-42-4	: 4979	791-29-7	: 11444
684-93-5	: 12019	702-54-5	: 8573	765-43-5	: 5796	791-31-1	: 11504
685-63-2	: 3164	703-32-2	: 11252	765-47-9	: 4918	793-23-7	: 995
685-73-4	: 6199	703-61-7	: 11248	765-53-1	: 11730	799-51-1	: 5239
685-87-0	: 9928	703-80-0	: 5819	765-85-5	: 4176	804-63-7	: 4132
685-91-6	: 53	704-01-8	: 1286	765-87-7	: 4375	811-93-8	: 9878
688-14-2	: 7414	704-38-1	: 7617	766-00-7	: 4811	811-97-2	: 5640
688-52-8	: 7108	704-79-0	: 3854	766-05-2	: 4328	812-01-1	: 4086
688-71-1	: 3127	706-78-5	: 4932	766-07-4	: 7177	812-03-3	: 10141
688-74-4	: 3124	706-79-6	: 4917	766-08-5	: 11442	813-76-3	: 9525
688-84-6	: 10659	708-06-5	: 7754	766-09-6	: 9667	813-78-5	: 9559
689-89-4	: 6703	708-76-9	: 567	766-20-1	: 5189	814-49-3	: 9574
689-97-4	: 3889	708-82-7	: 10629	766-39-2	: 6097	814-67-5	: 10030
690-02-8	: 9159	709-09-1	: 1445	766-46-1	: 1040	814-68-6	: 10714
690-08-4	: 9039	709-98-8	: 10200	766-51-8	: 1178	814-78-8	: 3879
690-37-9	: 8954	710-43-0	: 4968	766-53-0	: 2992	815-17-8	: 3510
690-92-6	: 7029	711-79-5	: 5829	766-77-8	: 11418	815-24-7	: 8988
690-93-7	: 7028	712-48-1	: 454	766-79-0	: 9684	815-58-7	: 10718
691-37-2	: 9054	713-36-0	: 2014	766-81-4	: 1041	815-66-7	: 6975
691-38-3	: 9059	716-79-0	: 2384	766-84-7	: 2748	816-11-5	: 3517
691-83-8	: 3788	717-21-5	: 10710	766-90-5	: 2164	816-39-7	: 10424
692-24-0	: 7046	717-74-8	: 2373	766-92-7	: 2042	816-79-5	: 9045
692-35-3	: 9854	719-79-9	: 1120	766-93-8	: 5990	817-91-4	: 6493
692-42-2	: 441	723-46-6	: 2198	766-94-9	: 1646	817-95-8	: 181
692-45-5	: 6006	723-62-6	: 370	766-96-1	: 1042	818-23-5	: 8652
692-50-2	: 3914	726-42-1	: 765	767-03-3	: 4904	818-38-2	: 8772
692-86-4	: 11977	726-44-3	: 2122	767-10-2	: 11146	818-92-8	: 10570
693-02-7	: 7114	732-11-6	: 9515	767-12-4	: 4516	819-07-8	: 5668
693-05-0	: 10112	732-26-3	: 9498	767-54-4	: 4570	819-44-3	: 7107
693-13-0	: 9815	737-89-3	: 7726	767-58-8	: 7255	819-93-2	: 3502
693-23-2	: 5299	741-58-2	: 517	767-59-9	: 7275	821-06-7	: 3742
693-39-0	: 6429	742-20-1	: 2853	767-60-2	: 7277	821-07-8	: 6992
693-54-9	: 5075	747-90-0	: 4102	767-88-4	: 10884	821-08-9	: 6713
693-58-3	: 8121	750-90-3	: 4131	767-99-7	: 2029	821-09-0	: 9088
693-61-8	: 11968	753-89-9	: 9853	768-00-3	: 2025	821-10-3	: 3905
693-62-9	: 11970	754-05-2	: 11425	768-22-9	: 7282	821-11-4	: 3792
693-65-2	: 8835	755-25-9	: 7032	768-32-1	: 11493	821-25-0	: 6377
693-72-1	: 8295	756-79-6	: 9540	768-33-2	: 11374	821-55-6	: 8172
693-89-0	: 4928	757-58-4	: 11645	768-39-8	: 3115	821-67-0	: 5132
693-93-6	: 8583	759-05-7	: 3556	768-49-0	: 2024	821-76-1	: 6379
693-95-8	: 11679	759-36-4	: 9964	768-52-5	: 790	821-95-4	: 11966
693-98-1	: 7209	759-73-9	: 12015	768-56-9	: 1112	821-96-5	: 11969
694-05-3	: 11020	759-94-4	: 5372	768-66-1	: 9711	821-98-7	: 11971
694-28-0	: 4880	760-20-3	: 9053	768-94-5	: 11859	821-99-8	: 8128
694-31-5	: 10826	760-21-4	: 8825	769-06-2	: 849	822-06-0	: 6770
694-35-9	: 4925	760-23-6	: 3745	769-25-5	: 1647	822-23-1	: 221
694-48-4	: 10825	760-40-7	: 10408	769-68-6	: 951	822-28-6	: 6638
694-53-1	: 11446	760-55-4	: 9971	769-92-6	: 715	822-35-5	: 4198
694-59-7	: 10994	760-67-8	: 6987	770-09-2	: 11494	822-36-6	: 7210
694-80-4	: 1004	760-78-1	: 8217	770-35-4	: 10399	822-50-4	: 4802
694-81-5	: 11057	760-79-2	: 3203	771-56-2	: 2094	822-66-2	: 4695
694-85-9	: 11035	760-93-0	: 10649	771-61-9	: 9442	822-67-3	: 4694
695-06-7	: 6150	761-06-8	: 10015	771-62-0	: 2304	822-84-4	: 11776
695-12-5	: 4388	762-21-0	: 3909	771-97-1	: 7799	822-85-5	: 4574
695-28-3	: 4653	762-42-5	: 3910	771-98-2	: 1246	822-87-7	: 4577
695-34-1	: 10867	762-49-2	: 5462	771-99-3	: 9695	823-17-6	: 4654
695-53-4	: 8584	762-50-5	: 5476	772-00-9	: 5196	823-40-5	: 1277
695-84-1	: 9348	762-62-9	: 9033	772-01-0	: 5195	823-76-7	: 5794
695-98-7	: 11025	762-63-0	: 9040	772-17-8	: 2138	824-11-3	: 11909
696-28-6	: 462	762-72-1	: 11495	773-76-2	: 11314	824-22-6	: 7257
696-29-7	: 4431	763-29-1	: 9052	773-82-0	: 2767	824-42-0	: 575
696-30-0	: 10986	763-30-4	: 8671	774-40-3	: 909	824-47-5	: 1193
696-41-3	: 581	763-69-9	: 10262	775-12-2	: 11420	824-55-5	: 1186
696-44-6	: 711	763-89-3	: 9096	775-51-9	: 4277	824-63-5	: 7256
696-59-3	: 6169	763-93-9	: 7086	775-56-4	: 11407	824-72-6	: 9548
696-62-8	: 1776	764-01-2	: 3919	776-35-2	: 9171	824-94-2	: 1197
696-63-9	: 2300	764-13-6	: 6692	776-74-9	: 1058	825-25-2	: 4881
696-71-9	: 4739	764-33-0	: 7123	776-75-0	: 9634	825-51-4	: 8024
697-11-0	: 4200	764-35-2	: 7115	776-76-1	: 11438	825-94-5	: 11457
697-86-9	: 641	764-42-1	: 3762	777-22-0	: 1981	826-18-6	: 2104
698-00-0	: 643	764-47-6	: 10033	777-37-7	: 1218	826-36-8	: 9726
698-01-1	: 662	764-48-7	: 5716	778-22-3	: 1976	826-73-3	: 2407
698-27-1	: 576	764-49-8	: 11712	778-24-5	: 11416	826-74-4	: 7881
698-63-5	: 6046	764-56-7	: 6342	778-25-6	: 11502	826-77-7	: 11262
698-69-1	: 663	764-60-3	: 7125	778-28-9	: 2230	826-81-3	: 11321
698-87-3	: 1607	764-73-8	: 8343	778-66-5	: 2166	827-08-7	: 1312
698-88-4	: 1372	764-85-2	: 8177	779-02-2	: 425	827-15-6	: 2092
699-02-5	: 1610	764-93-2	: 5097	779-51-1	: 1971	827-16-7	: 11841

827-52-1	: 1248
827-54-3	: 7882
828-01-3	: 2133
829-83-4	: 443
829-84-5	: 9516
830-13-7	: 4211
830-89-7	: 7229
831-81-2	: 1224
831-91-4	: 2113
832-64-4	: 9181
832-71-3	: 9180
834-12-8	: 341
835-11-0	: 7596
835-64-3	: 9218
836-30-6	: 826
837-45-6	: 9165
838-41-5	: 8579
838-45-9	: 1697
838-95-9	: 722
841-73-6	: 11081
842-07-9	: 8037
846-48-0	: 343
848-53-3	: 5140
865-71-4	: 8853
868-18-8	: 3314
868-77-9	: 10661
868-85-9	: 9588
869-29-4	: 10563
869-50-1	: 9996
870-23-5	: 10600
870-63-3	: 3717
870-74-6	: 8977
871-28-3	: 9112
871-83-0	: 8140
871-84-1	: 8342
872-05-9	: 5082
872-10-6	: 8843
872-31-1	: 11727
872-50-4	: 11185
872-53-7	: 4780
872-55-9	: 11759
872-85-5	: 10889
872-93-5	: 11789
873-32-5	: 2747
873-49-4	: 1256
873-66-5	: 2163
873-76-7	: 1849
873-94-9	: 4622
874-23-7	: 4572
874-35-1	: 7258
874-41-9	: 1671
874-60-2	: 2904
874-63-5	: 1918
874-79-3	: 2172
874-90-8	: 2762
875-30-9	: 7298
875-40-1	: 2262
875-51-4	: 654
875-59-2	: 5828
876-05-1	: 4793
877-09-8	: 2261
877-11-2	: 2086
877-43-0	: 11257
877-44-1	: 2344
877-48-5	: 9314
877-53-2	: 9313
877-65-6	: 1865
877-82-7	: 9316
878-93-3	: 8064
879-97-0	: 9327
880-09-1	: 9683
880-93-3	: 7925
881-03-8	: 7926
881-67-4	: 1444
882-25-7	: 4148
882-33-7	: 5258
883-93-2	: 2864
883-99-8	: 7775
886-50-0	: 11594
886-58-8	: 1979
886-74-8	: 10001
886-77-1	: 8682
888-71-1	: 8564
897-78-9	: 4573
900-95-8	: 11539
910-86-1	: 11800
917-61-3	: 4164

917-92-0	: 3907	932-16-1	: 5849	982-43-4	: 9657	1070-87-7	: 8841	1119-49-9	: 38
917-93-1	: 9779	932-32-1	: 670	992-94-9	: 11437	1071-26-7	: 6383	1119-51-3	: 9010
918-00-3	: 10488	932-39-8	: 4797	993-00-0	: 11378	1071-73-4	: 8974	1119-60-4	: 6564
918-07-0	: 3254	932-40-1	: 4798	993-07-7	: 11489	1071-81-4	: 6849	1119-65-9	: 6610
918-79-6	: 8692	932-43-4	: 4834	993-13-5	: 9538	1071-83-6	: 6276	1120-07-6	: 8118
918-82-1	: 9128	932-44-5	: 4835	994-05-8	: 3396	1071-94-9	: 6589	1120-16-7	: 5289
918-84-3	: 8732	932-53-6	: 11831	994-25-2	: 8726	1071-98-3	: 3908	1120-21-4	: 11938
919-12-0	: 9132	932-56-9	: 4235	994-30-9	: 11388	1072-05-5	: 6387	1120-28-1	: 5357
919-19-7	: 9530	932-66-1	: 5793	994-49-0	: 5235	1072-16-8	: 8378	1120-29-2	: 5287
919-30-2	: 9828	932-77-4	: 1007	996-98-5	: 5562	1072-21-5	: 6748	1120-36-1	: 11636
919-86-8	: 5103	932-96-7	: 675	997-49-9	: 11698	1072-36-2	: 8224	1120-48-5	: 8354
919-94-8	: 3375	933-05-1	: 4865	998-29-8	: 11500	1072-39-5	: 5281	1120-56-5	: 4193
920-37-6	: 10584	933-21-1	: 4728	998-30-1	: 11477	1072-43-1	: 11696	1120-59-8	: 11750
920-46-7	: 10715	933-36-8	: 4616	998-40-3	: 9520	1072-44-2	: 500	1120-62-3	: 4929
921-08-4	: 3487	933-48-2	: 4569	998-41-4	: 11454	1072-52-2	: 499	1120-64-5	: 8576
921-13-1	: 7480	933-67-5	: 7317	998-93-6	: 6358	1072-53-3	: 5199	1120-71-4	: 8570
921-47-1	: 6863	933-75-5	: 9483	998-95-8	: 6363	1072-83-9	: 5820	1120-72-5	: 4885
922-28-1	: 6389	933-78-8	: 9482	999-21-3	: 3782	1072-91-9	: 10840	1120-73-6	: 4944
922-55-4	: 5016	933-88-0	: 2902	999-52-0	: 6362	1072-98-6	: 10861	1120-87-2	: 10877
922-59-8	: 9134	933-98-2	: 1670	999-55-3	: 10694	1073-06-9	: 1043	1120-97-4	: 5193
922-62-3	: 9057	934-00-9	: 1525	999-78-0	: 9130	1073-07-0	: 4732	1121-07-9	: 11160
922-65-6	: 8681	934-34-9	: 2868	999-79-1	: 9127	1073-23-0	: 10945	1121-18-2	: 4710
923-06-8	: 3300	934-53-2	: 1189	999-81-5	: 5442	1073-67-2	: 1151	1121-24-0	: 9376
923-26-2	: 10662	934-74-7	: 1672	999-97-3	: 11363	1074-11-9	: 1376	1121-37-5	: 7607
923-99-9	: 9594	934-80-5	: 1675	1000-49-3	: 11371	1074-17-5	: 2036	1121-55-7	: 10953
924-16-3	: 3212	935-05-7	: 8097	1000-50-6	: 11369	1074-43-7	: 2037	1121-60-4	: 10887
924-41-1	: 6706	935-07-9	: 18	1000-86-8	: 8665	1074-55-1	: 2038	1121-92-2	: 502
924-43-6	: 8093	935-08-0	: 5134	1001-53-2	: 28	1074-58-4	: 275	1121-95-5	: 4875
924-50-5	: 3852	935-12-6	: 6164	1001-89-4	: 6361	1074-92-6	: 1464	1122-20-9	: 4705
924-52-7	: 8077	935-13-7	: 6161	1002-28-4	: 7126	1075-22-5	: 7254	1122-25-4	: 4594
925-15-5	: 3321	935-31-9	: 4208	1002-69-3	: 5036	1075-38-3	: 1465	1122-39-0	: 11027
925-78-0	: 8173	935-95-5	: 9471	1002-84-2	: 8645	1075-74-7	: 2106	1122-42-5	: 1773
926-02-3	: 10036	936-48-1	: 10824	1003-03-8	: 4771	1076-26-2	: 8031	1122-54-9	: 5869
926-26-1	: 3296	936-53-8	: 10823	1003-09-4	: 11726	1076-55-7	: 9391	1122-60-7	: 4454
926-55-6	: 9116	936-58-3	: 1911	1003-10-7	: 11794	1076-61-5	: 7994	1122-62-9	: 5867
926-56-7	: 8670	936-72-1	: 2026	1003-29-8	: 11114	1076-66-0	: 1625	1122-69-6	: 10963
926-63-6	: 9807	937-14-4	: 1127	1003-30-1	: 6107	1077-16-3	: 1763	1122-81-2	: 11016
926-64-7	: 281	937-20-2	: 5784	1003-31-2	: 11732	1078-04-2	: 7260	1122-82-3	: 4417
926-65-8	: 10034	937-32-6	: 2179	1003-56-1	: 11036	1078-19-9	: 8048	1122-83-4	: 844
926-82-9	: 6390	937-33-7	: 716	1003-64-1	: 4391	1078-21-3	: 2128	1122-91-4	: 528
927-49-1	: 11962	938-16-9	: 10431	1003-73-2	: 10988	1078-30-4	: 11228	1123-00-8	: 4775
927-54-8	: 7006	938-25-0	: 7795	1003-78-7	: 11784	1078-71-3	: 1745	1123-09-7	: 4707
927-62-8	: 3215	938-33-0	: 11276	1003-90-3	: 11141	1079-71-6	: 427	1123-27-9	: 5821
927-68-4	: 5686	938-39-6	: 11326	1004-36-0	: 10793	1080-16-6	: 5130	1123-34-8	: 4555
927-73-1	: 3724	938-45-4	: 10446	1004-66-6	: 1919	1080-32-6	: 9545	1123-84-8	: 1371
927-74-2	: 3920	938-73-8	: 615	1004-77-9	: 4599	1081-34-1	: 11600	1123-85-9	: 1608
927-80-0	: 5942	939-23-1	: 10999	1005-64-7	: 1109	1081-75-0	: 2121	1123-86-0	: 10423
928-40-5	: 6795	939-26-4	: 7743	1005-67-0	: 7684	1081-77-2	: 2064	1123-95-1	: 11105
928-45-0	: 8072	939-27-5	: 7886	1006-06-0	: 1874	1083-50-3	: 1099	1123-96-2	: 10959
928-49-4	: 7116	939-48-0	: 2693	1006-31-1	: 2263	1086-80-2	: 2472	1124-05-6	: 1362
928-50-7	: 9019	939-52-6	: 3650	1007-26-7	: 1484	1088-11-5	: 2411	1124-07-8	: 9293
928-51-8	: 3604	939-55-9	: 5696	1007-32-5	: 3677	1093-58-9	: 354	1124-20-5	: 1985
928-56-3	: 5886	939-83-3	: 2766	1008-88-4	: 10998	1095-90-5	: 6516	1124-27-2	: 4442
928-65-4	: 11464	939-87-7	: 4957	1008-89-5	: 10997	1111-74-6	: 11414	1124-33-0	: 10993
928-68-7	: 6521	940-41-0	: 11474	1009-14-9	: 8986	1111-97-3	: 3901	1125-60-6	: 7372
928-80-3	: 5076	940-49-8	: 2033	1009-67-2	: 2137	1112-37-4	: 9529	1125-74-2	: 1547
928-92-7	: 7072	940-54-5	: 3189	1009-81-0	: 1754	1112-39-6	: 11412	1125-78-6	: 8044
928-94-9	: 7070	941-69-5	: 11123	1009-93-4	: 5001	1112-48-7	: 11368	1125-80-0	: 7382
928-95-0	: 7069	941-98-0	: 5857	1010-48-6	: 2130	1112-55-6	: 11449	1126-09-6	: 9643
928-96-1	: 7071	942-01-8	: 3989	1012-72-2	: 990	1112-56-7	: 11543	1126-18-7	: 4575
929-55-5	: 8348	942-06-3	: 7344	1012-84-6	: 2712	1113-12-8	: 11417	1126-46-1	: 2543
929-61-3	: 8402	942-92-7	: 6983	1013-08-7	: 9187	1113-30-0	: 3113	1126-75-6	: 2031
929-62-4	: 8401	943-14-6	: 2528	1013-83-8	: 7769	1113-68-4	: 25	1126-78-9	: 656
929-77-1	: 5278	943-15-7	: 1991	1013-88-3	: 1835	1113-74-2	: 8968	1126-79-0	: 1114
930-02-9	: 8240	943-27-1	: 5808	1013-96-3	: 10676	1114-09-6	: 3302	1127-76-0	: 7885
930-18-7	: 4972	944-22-9	: 5986	1016-97-9	: 7601	1114-51-8	: 9793	1128-08-1	: 4950
930-21-2	: 496	945-51-7	: 2183	1016-94-0	: 9195	1114-71-2	: 3976	1128-54-7	: 10838
930-22-3	: 8613	947-73-9	: 9162	1017-56-7	: 7583	1114-76-7	: 3200	1128-76-3	: 2538
930-27-8	: 6131	947-84-2	: 3021	1024-57-3	: 6301	1115-11-3	: 3698	1129-47-1	: 10293
930-28-9	: 4789	947-91-1	: 877	1025-15-6	: 11842	1115-12-4	: 9915	1129-50-6	: 3208
930-30-3	: 4941	948-32-3	: 2372	1048-08-4	: 11452	1115-30-6	: 3295	1129-89-1	: 4216
930-36-9	: 10834	948-65-2	: 7320	1056-77-5	: 6175	1116-24-1	: 58	1131-16-4	: 10829
930-37-0	: 8620	948-97-0	: 1966	1066-27-9	: 11435	1116-39-8	: 3123	1131-18-6	: 10814
930-55-2	: 11170	948-98-1	: 1146	1066-45-1	: 11540	1116-40-1	: 9820	1131-60-8	: 9272
930-66-5	: 4642	948-99-2	: 1147	1067-08-9	: 8814	1116-61-6	: 3121	1131-62-0	: 5805
930-68-7	: 4703	949-87-1	: 5139	1067-20-5	: 8756	1117-19-7	: 9945	1131-63-1	: 7779
930-87-0	: 11142	950-37-8	: 7645	1068-19-5	: 6391	1117-55-1	: 8431	1132-21-4	: 2577
930-89-2	: 4815	951-86-0	: 1365	1068-57-1	: 195	1117-59-5	: 8882	1132-39-4	: 2175
930-90-5	: 4816	952-47-6	: 9426	1068-87-7	: 8812	1118-09-8	: 11491	1132-66-7	: 1765
931-10-2	: 4293	952-97-6	: 2058	1068-90-2	: 9922	1118-68-9	: 6261	1133-80-8	: 5968
931-20-4	: 9721	955-83-9	: 6100	1069-53-0	: 6864	1118-71-4	: 6404	1134-23-2	: 4168
931-56-6	: 4425	957-51-7	: 5219	1070-14-0	: 6705	1118-84-9	: 3577	1134-62-9	: 7747
931-77-1	: 4319	959-33-1	: 10713	1070-32-2	: 6492	1119-06-8	: 6465	1135-12-2	: 839
931-87-3	: 4752	961-11-5	: 6202	1070-34-4	: 3335	1119-19-3	: 3895	1140-14-3	: 7614
931-89-5	: 4751	961-71-7	: 5506	1070-71-9	: 10737	1119-33-1	: 6245	1140-16-5	: 7628
931-97-5	: 4329	964-68-1	: 2533	1070-78-6	: 10142	1119-40-0	: 8775	1142-70-7	: 11080
932-01-4	: 4519	976-71-6	: 9736	1070-83-3	: 3508	1119-46-6	: 8866	1142-97-8	: 2236

CAS	No.
1731-79-9	: 5300
1731-84-6	: 8154
1731-88-0	: 11882
1731-92-6	: 6316
1732-08-7	: 8395
1732-09-8	: 8388
1732-10-1	: 8134
1733-12-6	: 9368
1738-25-6	: 10101
1738-36-9	: 284
1739-84-0	: 7200
1740-97-2	: 9414
1741-83-9	: 8826
1745-16-0	: 2105
1745-81-9	: 9457
1746-11-8	: 2443
1746-13-0	: 2167
1746-77-6	: 3955
1754-62-7	: 10686
1755-01-7	: 7577
1755-05-1	: 8686
1758-10-7	: 2027
1758-33-4	: 8608
1758-85-6	: 1421
1758-88-9	: 1674
1759-53-1	: 4958
1759-58-6	: 4804
1759-64-4	: 4652
1759-81-5	: 4930
1759-88-2	: 11439
1760-46-9	: 937
1761-71-3	: 4299
1768-31-6	: 10478
1772-10-7	: 1996
1776-37-0	: 7241
1777-82-8	: 1853
1779-81-3	: 11664
1780-31-0	: 11052
1781-78-8	: 4755
1790-22-3	: 3450
1792-41-2	: 2379
1792-81-0	: 4372
1794-84-9	: 7485
1795-01-3	: 11760
1795-04-6	: 11792
1795-05-7	: 11793
1795-09-1	: 11787
1795-15-9	: 4460
1795-16-0	: 4344
1795-17-1	: 4382
1795-18-2	: 4475
1795-20-6	: 4843
1795-21-7	: 4790
1795-22-8	: 4850
1795-26-2	: 4497
1795-27-3	: 4496
1795-48-8	: 10060
1796-05-0	: 1494
1804-15-5	: 9784
1805-61-4	: 9404
1806-29-7	: 3064
1807-36-9	: 2837
1809-02-5	: 3155
1809-05-8	: 8820
1809-10-5	: 8712
1809-19-4	: 9532
1809-26-3	: 9014
1812-51-7	: 3072
1814-88-6	: 10128
1817-47-6	: 1978
1817-57-8	: 3924
1817-73-8	: 646
1818-07-1	: 2069
1820-09-3	: 2976
1821-02-9	: 8902
1821-12-1	: 1100
1821-36-9	: 680
1821-39-2	: 842
1822-74-8	: 5910
1822-86-2	: 3149
1823-91-2	: 959
1824-81-3	: 10868
1825-30-5	: 7810
1825-31-6	: 7809
1825-62-3	: 11430
1825-82-7	: 11462
1826-14-8	: 11676

CAS	No.
1830-54-2	: 8784
1830-95-1	: 3120
1832-53-7	: 9543
1833-51-8	: 11379
1836-75-5	: 8081
1838-59-1	: 6017
1838-77-3	: 7085
1839-88-9	: 4490
1843-05-6	: 7621
1845-30-3	: 2975
1846-70-4	: 8203
1850-14-2	: 7543
1855-63-6	: 4630
1859-76-3	: 7626
1860-21-5	: 247
1860-27-1	: 3399
1861-21-8	: 11099
1861-32-1	: 5180
1861-40-1	: 509
1863-63-4	: 2509
1864-92-2	: 738
1866-31-5	: 10692
1867-37-4	: 9917
1868-53-7	: 7498
1871-52-9	: 4749
1871-57-4	: 10537
1871-58-5	: 10178
1871-76-7	: 975
1871-96-1	: 5041
1873-25-2	: 3605
1873-29-6	: 10061
1873-54-7	: 11267
1873-92-3	: 11402
1875-88-3	: 1597
1877-72-1	: 2550
1878-68-8	: 888
1879-06-7	: 5845
1879-09-0	: 9325
1883-32-5	: 1615
1885-14-9	: 4026
1885-29-6	: 2741
1885-38-7	: 10587
1888-57-9	: 6966
1888-71-7	: 10571
1888-85-3	: 3370
1896-62-4	: 3882
1897-45-6	: 1317
1899-24-7	: 6043
1904-98-9	: 10754
1912-21-6	: 10319
1912-24-9	: 476
1917-15-3	: 6060
1918-00-9	: 2567
1918-02-1	: 10896
1918-16-7	: 9771
1921-70-6	: 8644
1928-30-9	: 11856
1928-38-7	: 172
1929-73-3	: 170
1929-77-7	: 12047
1929-82-4	: 8071
1931-60-8	: 3367
1932-04-3	: 9661
1937-54-8	: 8113
1937-62-8	: 8308
1938-17-6	: 1257
1938-22-3	: 2063
1940-18-7	: 4522
1942-45-6	: 8532
1942-46-7	: 5100
1943-16-4	: 7487
1943-54-0	: 9291
1945-53-5	: 9166
1951-11-7	: 3684
1967-25-5	: 11998
1969-43-3	: 6406
1969-73-9	: 875
1971-46-6	: 7322
1973-22-4	: 1035
1974-04-5	: 6357
1975-44-6	: 7767
1975-78-6	: 5057
1982-47-4	: 4088
1985-57-5	: 1455
1985-59-7	: 7980
1985-97-3	: 1691
1990-29-0	: 338

CAS	No.
2004-67-3	: 9102
2004-69-5	: 9113
2004-70-8	: 8661
2006-14-6	: 10675
2008-41-5	: 11586
2009-74-7	: 6598
2009-83-8	: 6916
2016-42-4	: 11612
2016-57-1	: 5032
2018-90-8	: 7907
2021-28-5	: 2131
2025-40-3	: 10625
2027-17-0	: 7921
2027-19-2	: 7958
2028-63-9	: 3921
2029-94-9	: 538
2031-67-6	: 11479
2032-35-1	: 5459
2032-65-7	: 7646
2033-24-1	: 5190
2035-94-1	: 1602
2036-41-1	: 11072
2037-31-2	: 2294
2038-03-1	: 7689
2039-82-9	: 1028
2039-83-0	: 1368
2039-85-2	: 1150
2039-86-3	: 1027
2039-87-4	: 1149
2039-88-5	: 1026
2039-89-6	: 1628
2039-93-2	: 1968
2040-07-5	: 5875
2040-14-4	: 10466
2040-95-1	: 4778
2040-96-2	: 4849
2042-37-7	: 2744
2043-38-1	: 2856
2043-61-0	: 4331
2044-08-8	: 4627
2045-79-6	: 1577
2047-21-4	: 5877
2049-66-3	: 9951
2049-67-4	: 9041
2049-73-2	: 1407
2049-74-3	: 5719
2049-80-1	: 9980
2049-94-7	: 1954
2049-95-8	: 1480
2050-01-3	: 10302
2050-03-5	: 1488
2050-04-6	: 2109
2050-07-9	: 8983
2050-08-0	: 2649
2050-14-8	: 9216
2050-20-6	: 6394
2050-23-9	: 8387
2050-24-0	: 1419
2050-32-0	: 4256
2050-33-1	: 4255
2050-43-3	: 56
2050-46-6	: 1406
2050-47-7	: 2072
2050-48-8	: 2238
2050-60-4	: 5558
2050-61-5	: 5557
2050-67-1	: 3040
2050-68-2	: 3041
2050-69-3	: 7807
2050-73-9	: 7811
2050-74-0	: 7812
2050-76-2	: 8026
2050-77-3	: 5053
2050-78-4	: 8073
2050-87-5	: 11917
2050-89-7	: 3030
2050-92-2	: 8708
2050-95-5	: 3635
2051-00-5	: 5555
2051-04-9	: 5243
2051-10-7	: 4116
2051-25-4	: 5023
2051-30-1	: 8377
2051-31-2	: 5072
2051-32-3	: 8459
2051-33-4	: 6948
2051-49-2	: 6874

CAS	No.
2051-53-8	: 793
2051-60-7	: 3023
2051-61-8	: 3024
2051-62-9	: 3025
2051-78-7	: 3587
2051-84-5	: 783
2051-90-3	: 1389
2051-97-0	: 11136
2051-98-1	: 4
2052-01-9	: 10209
2052-06-4	: 642
2052-07-5	: 3016
2052-14-4	: 2614
2052-15-5	: 8904
2055-40-5	: 1643
2057-43-4	: 10996
2057-49-0	: 11007
2065-23-8	: 230
2065-70-5	: 7813
2077-13-6	: 1190
2078-54-8	: 9226
2079-95-0	: 11621
2081-44-9	: 10787
2082-59-9	: 8858
2082-76-0	: 5060
2083-91-2	: 11361
2084-19-7	: 8845
2084-45-9	: 1363
2085-88-3	: 8622
2086-83-1	: 2467
2090-14-4	: 4118
2090-89-3	: 5748
2091-29-4	: 6669
2093-20-1	: 5673
2095-57-0	: 11065
2100-17-6	: 9001
2101-86-2	: 982
2101-88-4	: 977
2103-64-2	: 11526
2104-64-5	: 9551
2106-54-9	: 6910
2108-92-1	: 4358
2109-66-2	: 7692
2110-78-3	: 10280
2111-75-3	: 4633
2113-47-5	: 35
2113-51-1	: 3077
2113-57-7	: 3017
2113-58-8	: 3086
2114-00-3	: 10419
2114-39-8	: 1087
2114-42-3	: 4470
2116-62-3	: 11003
2116-65-6	: 11006
2121-66-6	: 10140
2122-70-5	: 7735
2123-27-5	: 1373
2123-28-6	: 1367
2130-23-6	: 8504
2131-18-2	: 2089
2131-38-6	: 8013
2131-43-3	: 8002
2131-55-7	: 1177
2132-80-1	: 3047
2136-81-4	: 1238
2136-89-2	: 1237
2136-95-0	: 2085
2138-22-9	: 1505
2138-43-4	: 1535
2141-62-0	: 10103
2142-01-0	: 7367
2142-06-5	: 11163
2142-63-4	: 5780
2142-73-6	: 5810
2146-37-4	: 4813
2146-38-5	: 4924
2150-18-7	: 10962
2150-38-1	: 2579
2155-30-8	: 10273
2155-42-2	: 1370
2157-31-5	: 8792
2157-50-8	: 10483
2158-55-6	: 4228
2159-75-3	: 5123
2162-92-7	: 6761
2162-98-3	: 5040
2162-99-4	: 8371

CAS	No.
2163-00-0	: 6762
2163-44-2	: 9934
2163-48-6	: 9981
2164-08-1	: 4900
2164-09-2	: 10520
2164-17-2	: 12007
2164-19-4	: 4292
2167-39-7	: 8594
2168-93-6	: 3422
2170-03-8	: 6095
2171-96-2	: 11436
2173-56-0	: 8913
2174-64-3	: 1526
2175-91-9	: 4768
2176-62-7	: 10995
2176-98-9	: 11546
2177-32-4	: 7093
2177-47-1	: 7276
2180-43-0	: 6949
2180-68-9	: 8674
2180-69-0	: 8675
2182-66-3	: 11419
2189-60-8	: 2067
2196-23-8	: 6529
2197-37-7	: 8225
2198-20-1	: 4207
2198-61-0	: 6897
2198-72-3	: 6508
2198-75-6	: 7808
2199-51-1	: 11115
2199-52-2	: 11116
2203-34-1	: 3607
2203-35-2	: 3603
2203-80-7	: 7120
2206-38-4	: 1253
2206-48-6	: 1250
2206-89-5	: 10620
2207-01-4	: 4365
2207-03-6	: 4370
2207-04-7	: 4371
2207-27-4	: 4770
2207-50-3	: 8574
2210-28-8	: 10672
2211-67-8	: 3281
2211-68-9	: 3727
2211-69-0	: 3728
2211-70-3	: 3722
2212-67-1	: 7661
2213-08-3	: 332
2213-23-2	: 6385
2213-32-3	: 9030
2213-43-6	: 9624
2214-14-4	: 1955
2216-12-8	: 2053
2216-15-1	: 698
2216-30-0	: 6386
2216-32-2	: 6408
2216-33-3	: 8409
2216-34-4	: 8410
2216-36-6	: 8124
2216-67-3	: 7722
2216-68-4	: 7721
2216-69-5	: 7913
2216-70-8	: 8328
2216-81-1	: 10264
2216-87-7	: 11959
2216-92-4	: 6271
2216-94-6	: 10741
2217-06-3	: 2288
2217-07-4	: 730
2217-14-3	: 3310
2217-41-6	: 7731
2217-45-0	: 847
2217-60-9	: 1538
2219-66-1	: 3904
2219-82-1	: 9332
2224-37-5	: 8951
2227-13-6	: 11648
2230-70-8	: 4619
2231-57-4	: 4033
2233-00-3	: 10604
2233-29-6	: 11024
2234-13-1	: 7933
2234-16-4	: 5799
2234-20-0	: 1626
2237-30-1	: 2742
2239-78-3	: 6657

CAS	No.	CAS	No.	CAS	No.	CAS	No.	CAS	No.
2243-27-8	: 8144	2385-70-8	: 3468	2444-89-5	: 2073	2612-46-6	: 6993	2807-30-9	: 5764
2243-30-3	: 833	2385-85-5	: 7644	2445-76-3	: 10265	2613-61-8	: 6447	2807-54-7	: 3781
2243-32-5	: 2713	2388-14-9	: 1973	2445-82-1	: 2833	2613-65-2	: 4818	2808-71-1	: 4661
2243-35-8	: 5771	2393-23-9	: 1813	2445-83-2	: 2836	2613-66-3	: 4817	2808-76-6	: 4649
2243-42-7	: 2714	2396-53-4	: 2080	2447-54-3	: 2432	2613-69-6	: 4858	2808-77-7	: 4651
2243-43-8	: 3411	2396-60-3	: 5137	2450-71-7	: 10725	2613-72-1	: 4861	2808-79-9	: 4650
2243-53-0	: 3855	2396-63-6	: 6343	2455-24-5	: 10674	2614-88-2	: 6711	2808-80-2	: 4434
2243-55-2	: 7167	2396-65-8	: 8115	2459-05-4	: 3791	2620-50-0	: 2426	2809-64-5	: 7997
2243-61-0	: 7796	2396-78-3	: 7062	2459-09-8	: 10908	2621-46-7	: 1192	2809-67-8	: 8530
2243-62-1	: 7797	2396-83-0	: 7054	2459-24-7	: 7766	2621-78-5	: 3967	2809-69-0	: 6717
2243-71-2	: 400	2396-84-1	: 6700	2459-25-8	: 7776	2622-08-4	: 9596	2810-04-0	: 11739
2243-76-7	: 2648	2398-09-6	: 4189	2462-85-3	: 8229	2622-21-1	: 4657	2810-69-7	: 2255
2243-80-3	: 7635	2398-10-9	: 4190	2462-94-4	: 5363	2622-60-8	: 2383	2815-34-1	: 9610
2243-81-4	: 7757	2398-16-5	: 4182	2463-53-8	: 8179	2622-89-1	: 3130	2816-57-1	: 4586
2243-83-6	: 7750	2398-37-0	: 1051	2463-63-0	: 6530	2623-23-6	: 4549	2817-71-2	: 10832
2243-89-2	: 11302	2398-65-4	: 2068	2464-37-1	: 4079	2623-50-9	: 11263	2819-48-9	: 4314
2243-94-9	: 8016	2398-66-5	: 2007	2467-09-6	: 3098	2623-54-3	: 2923	2819-86-5	: 9443
2243-98-3	: 11984	2398-68-7	: 1568	2467-10-9	: 8838	2623-82-7	: 8425	2820-37-3	: 10827
2244-07-7	: 11941	2398-81-4	: 10909	2470-68-0	: 6352	2623-84-9	: 11945	2825-00-5	: 10797
2244-16-8	: 4715	2399-48-6	: 10695	2473-01-0	: 8123	2623-95-2	: 5061	2825-79-8	: 8296
2254-94-6	: 2867	2400-01-3	: 1764	2476-37-1	: 5800	2624-01-3	: 6457	2834-05-1	: 11944
2257-35-4	: 10696	2400-02-4	: 1695	2486-07-9	: 1495	2625-30-1	: 4455	2835-21-4	: 10574
2258-58-4	: 9086	2400-03-5	: 2170	2489-86-3	: 7954	2625-31-2	: 4840	2835-33-8	: 11014
2259-96-3	: 2851	2400-04-6	: 1119	2492-22-0	: 8324	2627-95-4	: 5233	2835-81-6	: 3463
2270-20-4	: 2099	2401-22-1	: 1173	2493-02-9	: 1049	2631-37-0	: 9769	2835-96-3	: 9213
2272-45-9	: 811	2401-24-3	: 666	2495-35-4	: 10688	2639-63-6	: 3528	2835-97-4	: 9209
2279-76-7	: 11542	2401-73-2	: 5232	2497-21-4	: 7088	2641-89-6	: 1551	2836-00-2	: 9211
2288-18-8	: 1967	2402-06-4	: 4973	2498-75-1	: 522	2642-63-9	: 5801	2836-03-5	: 1267
2291-80-7	: 3968	2402-79-1	: 11019	2499-95-8	: 10643	2642-98-0	: 4113	2836-04-6	: 1268
2294-43-1	: 1083	2404-35-5	: 4222	2505-06-8	: 4246	2648-61-5	: 5802	2837-89-0	: 5486
2294-71-5	: 900	2408-20-0	: 10325	2506-41-4	: 7785	2652-13-3	: 8178	2845-89-8	: 1179
2294-72-6	: 11987	2408-37-9	: 4617	2508-29-4	: 8920	2655-83-6	: 1157	2847-72-5	: 5056
2294-94-2	: 1618	2409-52-1	: 3329	2512-81-4	: 9648	2657-65-0	: 3426	2848-25-1	: 1540
2295-31-0	: 11683	2409-55-4	: 9330	2514-91-2	: 11815	2675-77-6	: 4084	2855-08-5	: 3246
2295-58-1	: 10489	2412-73-9	: 2551	2516-34-9	: 4169	2676-33-7	: 10269	2856-63-5	: 949
2303-16-4	: 5106	2415-72-7	: 4984	2516-96-3	: 2545	2678-54-8	: 5901	2858-66-4	: 10485
2303-17-5	: 11823	2417-04-1	: 3069	2517-04-6	: 495	2679-87-0	: 3373	2859-67-8	: 11013
2305-13-7	: 2153	2417-90-5	: 10095	2517-43-3	: 3625	2681-83-6	: 6875	2859-68-9	: 11012
2306-88-9	: 8437	2418-31-7	: 3787	2523-37-7	: 5974	2681-92-7	: 8859	2859-78-1	: 1016
2308-38-5	: 3509	2419-73-0	: 3349	2523-55-9	: 4297	2687-25-4	: 1275	2867-05-2	: 2986
2310-17-0	: 9514	2420-16-8	: 532	2523-56-0	: 4296	2688-77-9	: 7379	2867-47-2	: 10654
2310-98-7	: 9841	2423-01-0	: 4906	2524-04-1	: 9576	2688-84-8	: 834	2867-59-6	: 3596
2311-59-3	: 5067	2423-10-1	: 8289	2524-09-6	: 9579	2689-59-0	: 7618	2867-63-2	: 4893
2312-35-8	: 10507	2424-92-2	: 5352	2524-37-0	: 2574	2689-88-5	: 9946	2868-37-3	: 4963
2313-65-7	: 6943	2424-98-8	: 10297	2524-52-9	: 10901	2690-08-6	: 8417	2869-34-3	: 11870
2314-78-5	: 11157	2425-28-7	: 1844	2524-64-3	: 9575	2706-56-1	: 10948	2870-04-4	: 1673
2314-97-8	: 7559	2425-54-9	: 11615	2525-02-2	: 1548	2713-09-9	: 5943	2873-18-9	: 11728
2315-68-6	: 2724	2425-66-3	: 9867	2528-38-3	: 9565	2717-42-2	: 8009	2873-74-7	: 8806
2316-50-9	: 9234	2425-79-8	: 8601	2528-61-2	: 6527	2717-44-4	: 7816	2873-89-4	: 661
2316-64-5	: 1843	2426-02-0	: 7347	2531-80-8	: 10312	2718-25-4	: 11342	2873-97-4	: 10521
2319-29-1	: 5031	2426-08-6	: 8602	2532-58-3	: 4803	2719-27-9	: 4330	2874-74-0	: 5314
2319-61-1	: 4392	2426-54-2	: 10631	2536-38-1	: 8250	2719-52-0	: 1952	2876-35-9	: 7829
2320-30-1	: 4589	2430-00-4	: 96	2545-59-7	: 257	2720-41-4	: 4481	2876-44-0	: 7938
2321-07-5	: 11522	2430-22-0	: 8463	2547-61-7	: 7536	2721-38-2	: 4796	2876-45-1	: 7896
2323-81-1	: 8868	2431-24-5	: 6540	2550-04-1	: 11482	2722-36-3	: 2156	2876-46-2	: 7901
2328-24-7	: 924	2431-30-3	: 7117	2550-06-3	: 11458	2724-59-6	: 8263	2876-51-9	: 7937
2338-12-7	: 2873	2431-50-7	: 3809	2550-21-2	: 6976	2725-82-8	: 1036	2876-52-0	: 7895
2344-70-9	: 2155	2431-52-9	: 3420	2550-22-3	: 7097	2734-70-5	: 703	2876-53-1	: 7900
2344-71-0	: 1106	2431-54-1	: 3810	2550-26-7	: 3678	2736-37-0	: 10492	2876-62-2	: 3674
2344-80-1	: 11382	2432-12-4	: 9295	2550-28-9	: 7131	2736-40-5	: 3691	2876-63-3	: 10473
2344-83-4	: 11386	2432-63-5	: 3784	2550-36-9	: 4318	2738-19-4	: 7040	2882-98-6	: 4841
2345-26-8	: 10295	2436-84-2	: 6330	2553-04-0	: 7623	2739-12-0	: 792	2883-02-5	: 4457
2345-27-9	: 11633	2436-85-3	: 7716	2553-19-7	: 11406	2741-16-4	: 1974	2883-07-0	: 4341
2345-28-0	: 8651	2436-92-2	: 11260	2554-06-5	: 4996	2745-25-7	: 6031	2883-08-1	: 4447
2345-61-1	: 10617	2436-93-3	: 11264	2555-49-9	: 229	2745-26-8	: 6026	2883-12-7	: 1255
2346-00-1	: 11672	2436-96-6	: 3062	2562-38-1	: 4839	2747-53-7	: 9098	2883-45-6	: 6337
2346-81-8	: 6739	2437-03-8	: 7719	2563-97-5	: 98	2757-90-6	: 6247	2885-00-9	: 8244
2350-19-8	: 6367	2437-25-4	: 5303	2565-39-1	: 6586	2758-18-1	: 4945	2886-52-4	: 10445
2351-13-5	: 5229	2437-56-1	: 11893	2568-33-4	: 3354	2759-28-6	: 9621	2887-61-8	: 3665
2351-97-5	: 9080	2437-73-2	: 11300	2568-90-3	: 3398	2761-24-2	: 11480	2890-62-2	: 5843
2358-84-1	: 10669	2437-76-5	: 1065	2568-91-4	: 10075	2761-84-4	: 5339	2892-18-4	: 7100
2362-10-9	: 5230	2437-88-9	: 2343	2570-12-9	: 4747	2765-04-0	: 11920	2896-60-8	: 1518
2362-12-1	: 9241	2437-95-8	: 2972	2570-26-5	: 8641	2765-11-9	: 8640	2899-90-3	: 8950
2364-59-2	: 3581	2437-98-1	: 88	2571-39-3	: 7615	2765-18-6	: 7957	2902-69-4	: 5871
2365-48-2	: 204	2438-03-1	: 2723	2575-20-4	: 3658	2768-41-4	: 3915	2902-96-7	: 10394
2366-52-1	: 3378	2438-04-2	: 2688	2579-20-6	: 4362	2769-64-4	: 3387	2905-56-8	: 9698
2367-82-0	: 2271	2438-12-2	: 4668	2579-22-8	: 10724	2777-66-4	: 11991	2912-62-1	: 973
2370-88-9	: 4999	2438-19-9	: 10641	2583-25-7	: 9979	2778-68-9	: 2936	2917-26-2	: 6643
2371-19-9	: 6520	2438-72-4	: 880	2586-89-2	: 6611	2781-85-3	: 4990	2917-47-7	: 10410
2373-51-5	: 7484	2439-01-2	: 4072	2591-86-8	: 9638	2782-57-2	: 11838	2918-13-0	: 6583
2381-31-9	: 523	2439-10-3	: 5347	2593-15-9	: 11599	2782-91-4	: 11810	2919-05-3	: 9118
2382-43-6	: 9831	2439-35-2	: 10633	2594-20-9	: 3956	2785-74-2	: 1511	2919-23-5	: 4196
2384-73-8	: 6606	2439-99-8	: 6277	2594-21-0	: 3937	2785-89-9	: 9357	2921-88-2	: 4089
2384-75-0	: 6538	2443-03-0	: 4837	2595-97-3	: 337	2790-09-2	: 1348	2922-51-2	: 6320
2384-85-2	: 5098	2443-04-1	: 4836	2601-89-0	: 10099	2791-29-9	: 6636	2923-68-4	: 8442
2384-86-3	: 5099	2443-91-6	: 9886	2605-18-7	: 1940	2798-73-4	: 3891	2933-74-6	: 5745
2384-90-9	: 6326	2444-37-3	: 216	2612-33-1	: 9997	2804-50-4	: 10541	2935-44-6	: 6799

CAS	Page	CAS	Page	CAS	Page	CAS	Page	CAS	Page
2935-90-2	10287	3087-62-5	6211	3280-08-8	1616	3463-92-1	4064	3728-43-6	11490
2941-19-7	1830	3088-41-3	9703	3280-51-1	6708	3468-99-3	939	3730-60-7	6923
2941-20-0	1809	3096-47-7	5976	3282-30-2	10503	3476-89-9	11339	3731-51-9	10971
2941-55-1	5930	3096-52-4	5977	3283-12-3	9528	3478-88-4	7594	3731-52-0	10972
2943-70-6	5201	3101-00-8	8811	3288-04-8	1793	3497-00-5	9552	3731-53-1	10973
2944-49-2	1916	3102-33-8	9105	3289-28-9	4334	3508-00-7	6307	3735-90-8	3977
2947-60-6	961	3112-88-7	2112	3290-06-0	1358	3508-78-9	8805	3741-00-2	4846
2947-61-7	962	3113-98-2	4421	3296-05-7	978	3508-94-9	930	3741-38-6	5200
2955-65-9	8395	3118-97-6	8028	3299-38-5	3656	3510-70-1	11496	3742-38-9	4926
2955-88-6	11165	3121-83-3	9993	3319-15-1	1895	3512-80-9	10864	3742-42-5	4662
2959-96-8	10781	3128-06-1	6905	3320-83-0	1176	3521-91-3	6566	3743-22-4	9317
2961-47-9	10880	3128-07-2	6469	3320-86-3	1788	3522-94-9	6861	3744-02-3	9109
2969-81-5	3470	3129-39-3	7846	3320-87-4	1789	3524-62-7	5839	3757-32-2	8910
2971-79-1	9644	3129-90-6	7399	3321-50-4	4386	3524-73-0	7039	3761-94-2	4230
2973-00-4	7497	3132-99-8	527	3329-48-4	7073	3524-75-2	4848	3763-39-1	4991
2973-27-5	11224	3140-92-9	11743	3329-56-4	3377	3528-17-4	2869	3763-55-1	4062
2977-31-3	1617	3140-93-0	11742	3333-23-1	7764	3535-67-9	1925	3764-01-0	11075
2978-58-7	3896	3141-12-6	440	3333-44-6	6138	3536-29-6	2158	3768-43-2	9722
2979-19-3	4587	3141-24-0	11790	3333-52-6	3292	3538-65-6	3529	3768-60-3	11583
2980-71-4	8185	3141-27-3	11744	3337-71-1	474	3540-95-2	9658	3768-63-6	11569
2983-26-8	5523	3142-58-3	8799	3338-15-6	10444	3544-25-0	946	3769-23-1	7038
2983-37-1	6885	3142-66-3	8972	3344-70-5	5296	3547-04-4	1678	3769-41-3	9452
2984-40-9	3806	3142-84-5	6608	3347-22-6	8056	3552-01-0	4538	3773-93-1	5208
2987-53-3	819	3144-16-9	2929	3350-30-9	4725	3554-74-3	9716	3774-03-6	5197
2987-77-1	11485	3144-54-5	6962	3350-78-5	3888	3555-84-8	7612	3779-29-1	4180
2996-92-1	11488	3147-55-5	2558	3351-28-8	4117	3558-69-8	10947	3780-58-3	6894
2998-18-7	10665	3152-68-9	9110	3351-86-8	4055	3567-38-2	1877	3788-32-7	4838
3000-79-1	9680	3153-36-4	3485	3352-87-2	5290	3570-55-6	5652	3790-71-4	5334
3001-72-7	11190	3153-37-5	3489	3354-58-3	9427	3570-58-9	5699	3810-26-2	4951
3002-24-2	6804	3156-70-5	10589	3355-17-7	3900	3574-40-1	3157	3810-74-0	11561
3007-91-8	7770	3156-73-8	10396	3358-28-9	6174	3574-43-4	5682	3835-64-1	1864
3007-97-4	7765	3156-74-9	3641	3368-16-9	941	3581-87-1	11678	3837-38-5	404
3008-39-7	4225	3163-07-3	1541	3373-53-3	10767	3581-91-7	11675	3839-50-7	4765
3010-02-4	280	3163-27-7	7742	3375-23-3	8036	3582-71-6	11913	3840-31-1	1914
3010-04-6	273	3167-49-5	10897	3377-71-7	7321	3586-12-7	835	3848-12-2	126
3012-37-1	11710	3172-52-9	11748	3377-86-4	6733	3586-14-9	2013	3849-33-0	10049
3016-19-1	8487	3173-53-3	4416	3377-87-5	6734	3587-57-3	9858	3850-30-4	3217
3017-67-2	5086	3173-79-3	8597	3383-96-8	1	3587-60-8	1181	3852-09-3	10289
3017-68-3	3714	3174-67-2	8785	3384-04-1	1233	3588-30-5	3169	3853-80-3	10202
3017-70-7	3718	3174-83-2	2157	3386-33-2	8236	3593-00-8	6563	3856-25-5	11861
3017-71-8	3713	3175-23-3	10174	3393-64-4	3664	3594-37-4	3364	3857-25-8	6122
3017-95-6	9840	3176-62-3	7240	3393-78-0	2284	3598-14-9	287	3862-11-1	9338
3017-96-7	9838	3178-22-1	4369	3395-79-7	1443	3600-24-6	11916	3862-12-2	9335
3018-09-5	5881	3178-23-2	4429	3395-91-3	10212	3607-78-1	10052	3864-18-4	3052
3018-12-0	278	3178-24-3	4466	3399-21-1	4357	3609-53-8	2479	3867-18-3	733
3019-04-3	10454	3178-29-8	6434	3400-45-1	4781	3610-02-4	2403	3875-51-2	4828
3019-20-3	1980	3179-63-3	10375	3402-74-2	2006	3613-42-1	367	3875-52-3	4805
3019-25-8	5792	3180-09-4	9246	3402-76-4	7945	3615-21-2	2377	3875-68-1	12051
3022-41-1	4209	3187-04-8	6073	3404-57-7	6558	3616-57-7	5415	3876-50-0	10088
3024-72-4	2894	3188-13-4	5478	3404-58-8	7033	3634-92-2	8273	3877-15-4	10088
3027-13-2	10460	3189-22-8	7311	3404-61-3	7037	3637-01-2	5811	3877-19-8	7996
3031-05-8	8011	3194-15-8	10438	3404-63-5	9060	3637-61-4	4825	3878-55-5	3336
3031-66-1	7118	3195-24-2	9944	3404-67-9	6838	3638-35-5	4983	3884-88-6	8246
3031-73-0	7180	3196-15-4	3473	3404-71-5	9043	3638-64-0	5911	3910-35-8	7263
3031-74-1	7179	3199-61-9	2440	3404-72-6	9029	3643-64-9	4688	3913-71-1	5080
3032-32-4	2491	3202-84-4	7695	3404-73-7	9031	3657-07-6	245	3913-85-7	5088
3034-53-5	11668	3208-16-0	6105	3404-77-1	7016	3658-79-5	8755	3916-64-1	6339
3035-75-4	11877	3208-22-8	6173	3404-78-2	7024	3669-80-5	11948	3917-15-5	10565
3036-66-6	3168	3209-22-1	1391	3404-80-6	6837	3673-79-8	10242	3920-53-4	6453
3040-44-6	9662	3211-48-1	4183	3411-95-8	9217	3674-09-7	10248	3937-45-9	4422
3041-40-5	9916	3212-75-7	4753	3416-24-8	6233	3674-10-0	10363	3937-46-0	4423
3042-22-6	11135	3213-79-4	1269	3419-71-4	4689	3674-13-3	10240	3937-56-2	8136
3042-50-0	1486	3217-86-5	9802	3424-93-9	618	3674-75-7	9179	3938-95-2	10259
3043-33-2	8672	3218-02-8	4418	3426-01-5	10843	3675-68-1	3442	3938-96-3	206
3047-38-9	4907	3221-61-2	8408	3430-27-1	10870	3675-69-2	3443	3943-74-6	2627
3051-09-0	11097	3228-04-4	9416	3433-37-2	9674	3678-62-4	10919	3943-77-9	2578
3054-92-0	8956	3228-99-7	9895	3433-80-5	999	3680-02-2	5909	3944-36-3	10390
3054-95-3	10560	3229-00-3	9884	3435-28-7	11048	3681-78-5	5320	3955-26-8	2324
3055-14-9	1949	3230-23-7	9670	3437-89-6	8987	3682-35-7	11843	3958-03-0	11781
3058-47-7	2043	3232-84-6	11850	3437-95-4	11767	3683-19-0	7043	3958-38-1	4727
3060-89-7	7656	3234-64-8	9121	3438-46-8	11071	3683-22-5	7042	3958-57-4	1067
3062-81-5	3122	3238-38-8	9474	3439-97-2	11443	3689-24-5	11716	3959-07-7	1799
3066-71-5	10626	3238-40-2	6081	3440-02-6	11415	3690-04-8	9795	3962-77-4	1926
3068-00-6	3459	3238-62-8	9677	3445-11-2	11184	3697-27-6	4115	3963-78-8	382
3073-66-3	4486	3238-75-3	5712	3446-98-8	11147	3699-01-2	2019	3964-56-5	9232
3073-92-5	3421	3240-09-3	7099	3452-07-1	5362	3710-30-3	8319	3964-58-7	2540
3074-61-1	4935	3243-42-3	7952	3452-09-3	8196	3710-43-8	6088	3964-61-2	9254
3074-64-4	6542	3244-88-0	2218	3452-97-9	6951	3710-84-7	5429	3968-85-2	1953
3074-71-3	6384	3252-43-5	277	3454-07-7	1632	3711-34-0	7137	3970-35-2	2544
3074-75-7	6828	3254-93-1	7228	3457-58-7	2003	3711-37-3	7140	3970-62-5	8929
3074-76-8	6826	3256-88-0	10990	3458-28-4	7425	3721-95-7	4174	3971-28-6	4363
3074-77-9	6827	3260-87-5	9264	3460-18-2	1310	3722-73-4	2781	3972-56-3	1142
3075-84-1	3096	3261-62-9	1580	3460-20-6	2315	3722-74-5	2780	3972-65-4	1023
3077-16-5	7698	3268-49-3	9786	3460-24-0	1295	3725-05-1	6714	3973-62-4	9694
3081-24-1	9512	3268-79-9	11700	3460-67-1	10904	3725-07-3	8340	3974-99-0	254
3083-23-6	8628	3277-26-7	5237	3463-26-1	10837	3725-30-2	4750	3976-34-9	3056
3085-53-8	5997	3278-46-4	10332	3463-40-9	2316			3976-36-1	3095
								3978-81-2	10944

3982-64-7	: 1483	4200-95-7	: 6305	4388-88-9	: 6811	4562-36-1	: 4042	4806-58-0	: 4986
3982-66-9	: 12057	4205-23-6	: 6290	4389-53-1	: 2641	4573-05-1	: 4252	4806-59-1	: 4987
3982-67-0	: 1712	4206-75-1	: 11384	4390-75-4	: 3408	4573-09-5	: 4897	4806-61-5	: 4191
3988-03-2	: 7590	4208-49-5	: 6070	4392-24-9	: 1084	4588-18-5	: 8505	4810-09-7	: 6544
3991-61-5	: 2012	4208-64-4	: 6121	4392-30-7	: 1245	4593-16-2	: 9631	4815-57-0	: 1557
4001-73-4	: 604	4209-24-9	: 5787	4393-06-0	: 1868	4594-78-9	: 4467	4823-47-6	: 10611
4008-48-4	: 11323	4209-90-9	: 6925	4394-11-0	: 3103	4597-87-9	: 10869	4829-04-3	: 5266
4016-11-9	: 8614	4214-28-2	: 1770	4394-85-8	: 7685	4599-94-4	: 5152	4830-99-3	: 7273
4018-65-9	: 1504	4215-86-5	: 2011	4395-73-7	: 1819	4604-28-8	: 1155	4831-43-0	: 11182
4021-75-4	: 3612	4218-48-8	: 1690	4395-79-3	: 1201	4606-07-9	: 4961	4832-17-1	: 8055
4028-23-3	: 11375	4219-24-3	: 7050	4395-80-6	: 1202	4606-65-9	: 9675	4835-11-4	: 6750
4032-86-4	: 6388	4219-49-2	: 6652	4395-81-7	: 1782	4609-89-6	: 8833	4838-00-0	: 7242
4032-92-2	: 6445	4224-70-8	: 6876	4396-98-9	: 1847	4609-92-1	: 8414	4839-46-7	: 8773
4032-93-3	: 6444	4225-85-8	: 11244	4397-05-1	: 1842	4619-74-3	: 9480	4850-28-6	: 4862
4032-94-4	: 8375	4228-00-6	: 5317	4398-65-6	: 6744	4620-70-6	: 5710	4850-32-2	: 4832
4038-04-4	: 9044	4229-91-8	: 6167	4403-69-4	: 1796	4625-24-5	: 2980	4860-03-1	: 6634
4041-09-2	: 4882	4253-34-3	: 11499	4405-42-9	: 11385	4628-21-1	: 3726	4877-77-4	: 1081
4043-71-4	: 2430	4253-89-8	: 5244	4407-36-7	: 10704	4630-20-0	: 3986	4877-93-4	: 3045
4044-65-9	: 1432	4253-91-2	: 10129	4412-16-2	: 5337	4630-82-4	: 4335	4884-24-6	: 2995
4048-30-0	: 6578	4254-02-8	: 4779	4413-16-5	: 1249	4635-59-0	: 3685	4884-25-7	: 2994
4048-31-1	: 6574	4255-62-3	: 4590	4422-95-1	: 2317	4636-16-2	: 5838	4887-30-3	: 6904
4048-32-2	: 6581	4258-93-9	: 4449	4426-11-3	: 4173	4645-15-2	: 4461	4891-38-7	: 10742
4049-81-4	: 6695	4259-00-1	: 4855	4426-51-1	: 5198	4654-39-1	: 1595	4894-61-5	: 3725
4050-45-7	: 7000	4259-43-2	: 10179	4427-56-9	: 9417	4655-34-9	: 10664	4894-62-6	: 6335
4053-34-3	: 11329	4264-29-3	: 2055	4428-13-1	: 1601	4662-96-8	: 1589	4897-84-1	: 3474
4054-38-0	: 4219	4265-25-2	: 2456	4433-10-7	: 3054	4663-22-3	: 4982	4900-30-5	: 8110
4057-42-5	: 8503	4265-97-8	: 8430	4433-11-8	: 3059	4667-38-3	: 11392	4901-51-3	: 9469
4065-80-9	: 4537	4268-36-4	: 3933	4433-63-0	: 3133	4667-99-6	: 11387	4904-61-4	: 4212
4068-78-4	: 2542	4271-30-1	: 6242	4435-14-7	: 4313	4669-01-6	: 4844	4911-55-1	: 8684
4074-24-2	: 3322	4279-22-5	: 3280	4435-18-1	: 276	4670-56-8	: 2639	4911-65-3	: 1236
4074-43-5	: 9247	4279-70-3	: 1993	4435-50-1	: 3458	4671-03-8	: 11848	4912-92-9	: 7251
4079-52-1	: 5842	4279-76-9	: 1716	4436-59-3	: 4238	4672-38-2	: 9547	4914-91-4	: 9038
4088-60-2	: 3860	4280-28-8	: 1927	4436-96-8	: 1867	4675-18-7	: 8581	4914-92-5	: 9037
4088-81-7	: 9175	4282-24-0	: 6079	4437-18-7	: 6035	4676-51-1	: 3326	4915-21-3	: 6028
4091-39-8	: 3647	4282-28-4	: 6080	4437-20-1	: 6101	4685-47-6	: 1921	4915-22-4	: 6029
4095-09-4	: 11453	4282-31-9	: 11745	4437-22-3	: 6156	4694-12-6	: 4899	4915-23-5	: 6027
4095-22-1	: 11772	4282-32-0	: 6083	4437-23-4	: 6158	4695-31-2	: 10288	4916-63-6	: 6994
4096-20-2	: 9693	4282-40-0	: 6419	4437-50-7	: 6810	4701-36-4	: 1699	4920-84-7	: 1448
4096-21-3	: 11171	4283-80-1	: 8717	4437-51-8	: 6806	4702-34-5	: 2797	4920-99-4	: 1689
4096-34-8	: 4704	4285-42-1	: 3974	4439-20-7	: 5715	4705-34-4	: 1621	4923-77-7	: 4395
4100-80-5	: 6094	4286-47-9	: 6778	4439-24-1	: 5746	4705-94-6	: 8027	4923-78-8	: 4396
4101-68-2	: 5038	4286-49-1	: 3293	4441-57-0	: 4324	4706-81-4	: 11631	4923-84-6	: 9533
4104-01-2	: 9025	4288-84-0	: 10355	4441-90-1	: 9965	4706-89-2	: 1472	4926-76-5	: 4398
4107-62-4	: 10234	4291-79-6	: 4450	4442-79-9	: 4387	4712-55-4	: 9534	4926-78-7	: 4399
4110-44-5	: 8379	4291-80-9	: 4451	4443-55-4	: 4384	4721-34-0	: 9544	4926-90-3	: 4394
4110-50-3	: 10042	4291-81-0	: 4452	4444-12-6	: 6666	4721-37-3	: 9542	4930-98-7	: 11034
4110-77-4	: 1153	4291-98-9	: 4934	4445-06-1	: 4459	4726-14-1	: 8070	4944-94-9	: 632
4113-12-6	: 6642	4292-19-7	: 5302	4445-07-2	: 2066	4727-17-7	: 4760	4945-28-2	: 11301
4114-28-7	: 7151	4292-75-5	: 4414	4445-08-3	: 8245	4730-22-7	: 6498	4945-47-5	: 9627
4114-31-2	: 7148	4292-92-6	: 4465	4447-60-3	: 10079	4731-34-4	: 2076	4945-48-6	: 9635
4117-15-1	: 11934	4295-06-1	: 11245	4455-26-9	: 8353	4731-84-4	: 9249	4946-13-8	: 2298
4124-31-6	: 244	4301-39-7	: 6053	4457-00-5	: 4823	4734-90-1	: 4754	4949-44-4	: 8906
4124-88-3	: 8187	4303-44-0	: 9855	4463-33-6	: 1439	4736-48-5	: 4214	4957-14-6	: 1964
4126-78-7	: 4226	4313-03-5	: 6325	4467-06-5	: 10989	4737-41-1	: 8731	4964-71-0	: 11218
4127-45-1	: 4985	4318-37-0	: 5141	4467-88-3	: 7271	4740-00-5	: 11788	4965-36-0	: 11220
4128-31-8	: 8444	4318-76-7	: 10924	4468-42-2	: 1669	4744-08-5	: 9903	4972-29-6	: 2185
4130-08-9	: 11498	4319-49-7	: 7680	4471-05-0	: 1909	4747-05-1	: 3161	4974-21-4	: 246
4130-42-1	: 9221	4325-48-8	: 8729	4481-08-7	: 6198	4747-07-3	: 6833	4974-27-0	: 8320
4131-74-2	: 10331	4325-49-9	: 6366	4481-30-5	: 1481	4747-15-3	: 1928	4975-21-7	: 6176
4132-71-2	: 1442	4325-82-0	: 9099	4482-03-5	: 3076	4747-21-1	: 9819	4984-01-4	: 8502
4132-77-8	: 1470	4325-97-7	: 5651	4484-72-4	: 11460	4749-27-3	: 10606	4984-85-4	: 6971
4142-85-2	: 11440	4328-94-3	: 8852	4484-80-4	: 251	4749-28-4	: 10590	4985-70-0	: 374
4143-41-3	: 5126	4339-05-3	: 7128	4485-09-0	: 8174	4749-31-9	: 10127	4985-85-7	: 5679
4145-94-2	: 11673	4340-76-5	: 5359	4485-13-6	: 6559	4753-75-7	: 6115	4996-48-9	: 6068
4147-51-7	: 5222	4341-76-8	: 3918	4485-16-9	: 6556	4753-80-4	: 11690	5006-66-6	: 10899
4147-89-1	: 11362	4342-61-4	: 5226	4489-23-0	: 1082	4755-36-6	: 4247	5008-52-6	: 5007
4156-16-5	: 7407	4346-18-3	: 3584	4491-19-4	: 307	4755-77-5	: 144	5008-69-5	: 10082
4160-52-5	: 3672	4350-41-8	: 10921	4492-01-7	: 10830	4766-33-0	: 7711	5008-73-1	: 10077
4160-80-9	: 10782	4351-54-6	: 6005	4493-42-9	: 5027	4766-57-8	: 11505	5009-11-0	: 7104
4161-60-8	: 8973	4354-56-7	: 4412	4497-29-4	: 7524	4767-03-7	: 10272	5009-27-8	: 4988
4164-28-7	: 7449	4358-59-2	: 3848	4510-34-3	: 10705	4771-47-5	: 2546	5020-21-3	: 7246
4166-46-5	: 6926	4358-75-2	: 3216	4516-69-2	: 4856	4771-80-6	: 4638	5022-29-7	: 7364
4167-74-2	: 9432	4358-87-6	: 910	4518-10-9	: 2504	4773-53-9	: 10364	5026-76-6	: 6547
4167-77-5	: 4791	4358-88-7	: 907	4520-20-0	: 11447	4775-09-1	: 9527	5029-66-3	: 10284
4169-04-4	: 10398	4360-12-7	: 321	4523-45-9	: 7720	4775-86-4	: 8800	5029-67-4	: 10970
4170-24-5	: 3480	4360-96-7	: 9959	4524-77-0	: 9233	4780-79-4	: 7908	5035-52-9	: 11547
4170-46-1	: 11466	4361-06-2	: 9968	4525-36-4	: 6165	4784-77-4	: 3712	5053-43-0	: 11070
4171-83-9	: 2057	4362-01-0	: 4135	4525-44-4	: 11397	4789-40-6	: 6032	5058-19-5	: 10878
4173-44-8	: 6710	4363-04-6	: 8019	4534-74-1	: 4525	4795-29-3	: 6116	5069-26-1	: 11774
4175-54-6	: 7983	4363-93-3	: 11225	4535-66-4	: 5684	4795-86-2	: 2935	5075-33-2	: 7129
4179-19-5	: 1441	4373-13-1	: 7817	4537-00-2	: 7763	4798-44-1	: 7066	5075-92-3	: 11181
4180-23-8	: 1944	4375-14-8	: 7319	4537-05-7	: 2285	4798-46-3	: 7077	5076-20-0	: 8627
4185-62-0	: 7745	4376-23-2	: 7090	4542-61-4	: 3117	4798-60-1	: 7081	5077-67-8	: 3660
4185-63-1	: 7788	4377-37-1	: 11023	4549-31-9	: 6374	4798-61-2	: 8511	5090-41-5	: 8288
4187-86-4	: 9140	4383-05-5	: 1885	4549-32-0	: 8369	4798-62-3	: 6575	5097-93-8	: 11001
4187-87-5	: 1876	4383-18-0	: 1904	4549-74-0	: 8669	4798-64-5	: 8518	5100-98-1	: 11817
4192-77-2	: 10685	4388-87-8	: 6807	4553-07-5	: 893	4799-68-2	: 10400	5103-42-4	: 2999

CAS	No.	CAS	No.	CAS	No.	CAS	No.	CAS	No.
5112-21-0	: 9999	5362-55-0	: 9095	5535-49-9	: 1165	5842-00-2	: 5650	6094-02-6	: 7036
5113-93-9	: 4682	5363-64-4	: 6813	5556-07-0	: 5835	5846-43-5	: 4559	6102-15-4	: 4641
5131-66-8	: 10346	5367-28-2	: 1213	5557-31-3	: 8292	5856-77-9	: 3688	6108-61-8	: 6687
5132-75-2	: 6468	5368-81-0	: 2674	5558-29-2	: 963	5856-82-6	: 3680	6109-22-4	: 422
5137-45-1	: 5591	5370-25-2	: 5783	5558-31-6	: 964	5857-36-3	: 8989	6111-61-1	: 3643
5145-64-2	: 11138	5371-48-2	: 9119	5558-66-7	: 931	5857-86-3	: 4558	6111-78-0	: 521
5145-99-3	: 10043	5379-20-4	: 1627	5558-78-1	: 970	5859-45-0	: 7784	6111-88-2	: 8733
5150-80-1	: 6341	5380-40-5	: 6069	5560-69-0	: 7968	5867-45-8	: 89	6111-99-5	: 10236
5150-93-6	: 3333	5390-04-5	: 9143	5563-78-0	: 4547	5870-61-1	: 5882	6114-18-7	: 8302
5153-67-3	: 2050	5390-71-6	: 9861	5581-35-1	: 1583	5870-93-9	: 3527	6117-80-2	: 3793
5159-41-1	: 1888	5390-72-7	: 9859	5581-75-9	: 1759	5876-87-9	: 5285	6118-14-5	: 3858
5160-99-6	: 1997	5392-86-9	: 1375	5582-82-1	: 6500	5877-55-4	: 809	6119-92-2	: 5182
5161-04-6	: 1999	5395-43-7	: 11553	5586-15-2	: 5256	5877-58-7	: 810	6120-13-4	: 2919
5161-13-7	: 11785	5395-75-5	: 5451	5587-63-3	: 4910	5878-19-3	: 10455	6125-21-9	: 9919
5161-16-0	: 11796	5396-24-7	: 149	5587-79-1	: 4948	5881-17-4	: 8403	6125-24-2	: 2051
5161-17-1	: 11798	5396-38-3	: 1462	5588-20-5	: 7220	5882-44-0	: 717	6127-92-0	: 3827
5162-44-7	: 3711	5396-58-7	: 3355	5589-96-8	: 112	5891-21-4	: 8961	6127-93-1	: 3828
5162-48-1	: 8955	5396-91-8	: 8969	5597-27-3	: 2958	5897-76-7	: 1840	6130-75-2	: 2331
5166-35-8	: 3734	5398-08-3	: 9961	5597-81-9	: 7782	5902-51-2	: 11591	6131-24-4	: 6835
5166-53-0	: 7094	5398-10-7	: 9955	5597-89-7	: 8687	5906-73-0	: 11501	6131-25-5	: 6423
5171-84-6	: 6854	5398-25-4	: 8742	5598-13-0	: 4090	5908-87-2	: 5277	6134-66-3	: 3503
5182-44-5	: 1596	5398-75-4	: 9177	5609-09-6	: 6587	5910-87-2	: 8109	6135-29-1	: 262
5183-77-7	: 4347	5399-02-0	: 6312	5614-32-4	: 3160	5910-89-4	: 10805	6136-67-0	: 7624
5183-79-9	: 4360	5405-13-0	: 1821	5617-41-4	: 4405	5911-04-6	: 8141	6137-08-2	: 8475
5194-50-3	: 6686	5405-15-2	: 1825	5617-42-5	: 4820	5911-08-0	: 4966	6137-15-1	: 8476
5194-51-4	: 6685	5405-17-4	: 1824	5617-92-5	: 4978	5912-58-3	: 5694	6138-90-5	: 8321
5195-37-9	: 7982	5405-41-4	: 3534	5630-56-8	: 7598	5912-86-7	: 9397	6139-84-0	: 3181
5195-40-4	: 7981	5405-79-8	: 6965	5634-22-0	: 1565	5912-87-8	: 9393	6140-17-6	: 2045
5208-49-1	: 4050	5406-12-2	: 2116	5634-30-0	: 4809	5913-85-9	: 9706	6140-65-4	: 4908
5209-18-7	: 2161	5406-18-8	: 2154	5636-65-7	: 7061	5915-41-3	: 11593	6140-80-7	: 10579
5209-33-6	: 5	5407-98-7	: 7605	5637-96-7	: 2070	5917-47-5	: 9660	6142-95-6	: 1412
5216-25-1	: 1239	5408-57-1	: 8477	5638-26-6	: 6336	5921-54-0	: 3673	6149-41-3	: 10276
5216-32-0	: 1366	5408-86-6	: 3268	5648-29-3	: 5614	5921-73-3	: 8209	6149-46-8	: 8884
5217-05-0	: 4783	5410-97-9	: 5166	5650-40-8	: 10451	5921-84-6	: 6476	6152-31-4	: 7171
5221-17-0	: 9780	5411-50-7	: 1307	5650-41-9	: 10452	5930-28-9	: 9206	6153-16-8	: 4674
5223-59-6	: 1676	5411-56-3	: 1845	5650-75-9	: 8890	5932-68-3	: 9396	6153-56-6	: 5561
5234-68-4	: 4039	5412-92-0	: 8352	5661-71-2	: 503	5943-35-1	: 11649	6158-45-8	: 7920
5251-93-4	: 524	5413-05-8	: 884	5663-96-7	: 8538	5944-41-2	: 10942	6159-55-3	: 11192
5258-50-4	: 11797	5414-19-7	: 5609	5666-21-7	: 10527	5950-19-6	: 3649	6161-50-8	: 3046
5259-65-4	: 4714	5414-21-1	: 8828	5668-93-9	: 3073	5951-67-7	: 4659	6162-79-4	: 3951
5259-98-3	: 8923	5416-94-4	: 11704	5669-09-0	: 10041	5959-52-4	: 7768	6163-64-0	: 10083
5263-87-6	: 11275	5419-55-6	: 3128	5675-22-9	: 6327	5963-77-9	: 9137	6163-66-2	: 10124
5265-18-9	: 10414	5421-04-5	: 912	5675-51-4	: 5301	5965-66-2	: 6214	6163-75-3	: 9537
5271-38-5	: 5749	5424-21-5	: 11053	5676-58-4	: 2879	5966-51-8	: 10343	6164-98-3	: 7564
5271-39-6	: 1679	5426-65-3	: 10846	5683-33-0	: 10863	5976-47-6	: 10700	6165-44-2	: 4323
5272-02-6	: 6369	5429-85-6	: 8290	5683-44-3	: 8794	5984-58-7	: 6353	6166-86-5	: 4902
5274-50-0	: 9554	5432-61-1	: 4289	5685-06-3	: 11680	5985-23-9	: 6246	6166-87-6	: 4625
5281-18-5	: 563	5432-85-9	: 4600	5687-92-3	: 11692	5986-55-0	: 7586	6168-72-5	: 10336
5283-66-9	: 11471	5434-27-5	: 9887	5680-04-3	: 7292	6001-87-2	: 10225	6174-95-4	: 5910
5285-18-7	: 8677	5441-51-0	: 4593	5697-85-8	: 838	6004-60-0	: 5795	6175-23-1	: 8137
5285-87-0	: 11709	5441-52-1	: 4518	5699-58-1	: 8875	6006-06-0	: 11895	6175-49-1	: 5325
5292-21-7	: 4306	5445-24-9	: 4557	5699-74-1	: 8829	6006-33-3	: 4484	6195-92-2	: 4886
5292-43-3	: 114	5447-97-2	: 9847	5703-21-9	: 869	6006-34-4	: 4854	6196-58-3	: 5757
5292-53-5	: 9978	5447-99-4	: 8952	5703-26-4	: 871	6006-90-2	: 8638	6196-60-7	: 9047
5307-19-7	: 10951	5448-43-1	: 11334	5703-52-6	: 2146	6006-95-7	: 4463	6202-04-6	: 11839
5311-96-6	: 7150	5451-80-9	: 8881	5704-20-1	: 8975	6012-97-1	: 11782	6213-90-7	: 3821
5314-37-4	: 3070	5451-85-4	: 8901	5715-23-1	: 4517	6020-51-5	: 7224	6214-25-1	: 3832
5314-55-6	: 11433	5452-37-9	: 4735	5728-52-9	: 3009	6031-02-3	: 2009	6214-28-4	: 3820
5324-00-5	: 7272	5452-75-5	: 4308	5736-03-8	: 5206	6032-29-7	: 8916	6222-35-1	: 10235
5325-97-3	: 9185	5453-44-1	: 3857	5743-97-5	: 9186	6035-49-0	: 2845	6223-78-5	: 6765
5326-81-8	: 45	5453-80-5	: 2967	5744-03-6	: 5973	6040-62-6	: 8058	6225-06-5	: 8699
5327-44-6	: 1924	5454-28-4	: 6459	5750-00-5	: 3248	6044-68-4	: 10562	6231-18-1	: 10935
5328-01-8	: 7883	5454-79-5	: 4533	5756-43-4	: 6814	6050-13-1	: 5145	6236-88-0	: 4400
5332-06-9	: 3404	5455-24-3	: 8396	5763-61-1	: 1805	6051-43-0	: 2489	6258-66-8	: 1833
5332-24-1	: 11217	5458-48-0	: 1252	5765-44-6	: 7405	6051-52-1	: 1869	6261-22-9	: 9141
5332-25-2	: 11219	5458-59-3	: 8435	5773-56-8	: 1614	6057-60-9	: 3019	6265-26-5	: 6166
5332-52-5	: 11942	5459-58-5	: 151	5775-82-6	: 10816	6061-06-9	: 3156	6267-24-9	: 5605
5332-73-0	: 9816	5459-93-8	: 4288	5779-94-2	: 548	6061-10-5	: 8433	6268-37-7	: 1228
5335-75-1	: 10879	5460-32-2	: 1768	5779-95-3	: 549	6064-27-3	: 5326	6269-92-7	: 8366
5337-03-1	: 10774	5461-51-8	: 74	5781-53-3	: 145	6065-59-4	: 9974	6270-16-2	: 3642
5337-70-2	: 8552	5465-09-8	: 4588	5787-31-5	: 3232	6065-82-3	: 174	6270-34-4	: 6061
5337-72-4	: 4515	5465-13-4	: 2282	5792-36-9	: 10430	6065-90-3	: 8752	6270-55-9	: 10300
5337-93-9	: 10468	5468-97-3	: 1598	5794-88-7	: 2487	6065-93-6	: 10555	6270-56-0	: 6104
5339-85-5	: 1594	5469-19-2	: 1093	5798-75-4	: 2523	6068-69-5	: 816	6272-38-4	: 9451
5340-36-3	: 8466	5470-02-0	: 9707	5798-79-8	: 947	6069-71-2	: 2954	6272-40-8	: 7588
5340-41-0	: 6953	5470-28-0	: 9606	5798-88-9	: 3472	6069-97-2	: 2937	6274-16-4	: 3950
5342-31-4	: 7509	5470-47-3	: 882	5798-94-7	: 606	6069-98-3	: 4439	6279-86-3	: 7552
5344-20-7	: 8707	5470-91-7	: 10712	5802-82-4	: 6625	6078-26-8	: 310	6280-03-1	: 8149
5344-55-8	: 3590	5472-38-8	: 3324	5805-76-5	: 2381	6080-79-1	: 9125	6280-96-2	: 9460
5344-82-1	: 11801	5500-21-0	: 4956	5807-64-7	: 3039	6086-22-2	: 11847	6280-98-4	: 1555
5344-90-1	: 1837	5502-88-5	: 4680	5807-76-1	: 1096	6088-94-4	: 8114	6284-40-8	: 6208
5345-01-7	: 8777	5503-12-8	: 4634	5809-07-4	: 6103	6089-09-4	: 9138	6284-84-0	: 9609
5349-18-8	: 2032	5507-44-8	: 11422	5813-64-9	: 9806	6089-12-9	: 3600	6285-05-8	: 10422
5349-55-3	: 10281	5510-99-6	: 9227	5814-42-6	: 10392	6091-52-7	: 4716	6285-06-9	: 9145
5349-60-0	: 1873	5515-83-3	: 328	5814-85-7	: 1969	6091-58-3	: 3351	6287-38-3	: 536
5350-47-0	: 7630	5519-42-0	: 10842	5836-68-8	: 9888	6091-64-1	: 2521	6290-24-0	: 2598
5356-83-2	: 11423	5519-50-6	: 9723	5837-78-5	: 3849	6093-68-1	: 2825	6290-49-9	: 207

CAS	Ref	CAS	Ref	CAS	Ref	CAS	Ref	CAS	Ref
6292-82-6	9710	6630-33-7	526	6943-58-4	10356	7225-64-1	6313	7567-22-8	4743
6294-31-1	6855	6639-30-1	2336	6946-29-8	2500	7225-66-3	11878	7568-92-5	1593
6294-89-9	7149	6639-82-3	11336	6946-35-6	2668	7235-40-7	4051	7568-93-6	1839
6298-37-9	11332	6640-27-3	9261	6946-88-9	3338	7236-47-7	3703	7570-26-5	5552
6302-94-9	5264	6640-77-3	2685	6950-43-2	2530	7239-24-9	9829	7570-98-1	7951
6304-18-3	10985	6641-83-4	8895	6950-84-1	12022	7242-17-3	3780	7572-29-4	5940
6306-30-5	6984	6669-13-2	1458	6951-08-2	2419	7250-95-5	7712	7583-53-1	9678
6308-98-1	1585	6682-71-9	7253	6952-59-6	2745	7251-61-8	11337	7611-43-0	472
6309-50-8	5308	6688-11-5	4738	6959-06-4	3624	7252-83-7	5460	7611-86-1	3721
6310-09-4	5791	6693-29-4	4302	6963-44-6	8791	7283-69-4	3767	7611-87-2	3720
6310-21-0	714	6693-30-7	4301	6964-04-1	6401	7283-70-7	3766	7614-93-9	2022
6311-60-0	1309	6693-31-8	4300	6967-43-7	3623	7283-96-7	11735	7615-57-8	11836
6314-28-9	2400	6697-12-7	6664	6969-16-0	11894	7287-19-6	9770	7616-22-0	2790
6315-03-3	2695	6703-78-2	6298	6971-51-3	1890	7287-81-2	1858	7619-08-1	8227
6315-60-2	330	6703-80-6	6297	6973-79-1	10206	7291-33-0	97	7623-01-0	4997
6315-96-4	10474	6703-82-8	4819	6974-10-3	7361	7297-25-8	3432	7623-10-1	10499
6319-21-7	5135	6704-31-0	8596	6974-77-2	9842	7300-03-0	6555	7623-11-2	3683
6319-40-0	2529	6705-31-3	10808	6975-60-6	10439	7307-55-3	11937	7623-13-4	10501
6320-03-2	2293	6709-22-4	4913	6975-92-4	7015	7311-34-4	546	7642-04-8	8493
6320-40-7	2314	6709-39-3	6331	6975-98-0	5054	7314-06-9	732	7642-09-3	7003
6322-01-6	10339	6728-26-3	6996	6975-99-1	5346	7314-37-6	11620	7642-10-6	6537
6322-49-2	3648	6734-98-1	7007	6976-00-7	11963	7315-32-4	2425	7642-15-1	8497
6326-44-9	9953	6738-23-4	1917	6976-72-3	6889	7318-67-4	6684	7658-08-4	6235
6338-45-0	7201	6739-34-0	4337	6982-25-8	3347	7319-00-8	6683	7661-39-4	11222
6341-60-2	9970	6740-88-1	4579	6983-79-5	5108	7321-48-4	8500	7661-55-4	11280
6341-97-5	9288	6742-54-7	2374	6986-48-7	5610	7321-55-3	9913	7665-72-7	8610
6344-28-1	5786	6748-95-4	436	6995-30-8	7905	7328-33-8	5026	7667-55-2	4491
6344-72-5	11338	6753-98-6	5006	6995-79-5	4374	7332-93-6	8695	7667-58-5	4494
6346-09-4	3214	6765-39-5	6322	6996-92-5	2174	7335-01-5	9672	7667-59-6	4493
6351-10-6	7284	6776-19-8	3840	6998-82-9	4352	7335-02-6	9690	7667-60-9	4495
6361-23-5	535	6780-13-8	5513	7005-72-3	1221	7335-06-0	11166	7681-84-7	6049
6372-01-6	3831	6780-49-0	7567	7012-16-0	5972	7335-07-1	11173	7683-59-2	1522
6374-70-5	5146	6781-42-6	5865	7013-11-8	3744	7335-25-3	2537	7688-21-3	7001
6378-11-6	4087	6781-98-2	1139	7021-04-7	889	7335-26-4	2669	7693-52-9	9244
6378-65-0	6890	6784-18-5	4890	7035-03-2	957	7335-27-5	2539	7696-51-7	11129
6380-21-8	9456	6785-23-5	4863	7036-98-8	8191	7344-34-5	7278	7697-46-3	7637
6380-24-1	1451	6789-80-6	6998	7040-43-9	6090	7351-74-8	7803	7700-17-6	4137
6380-28-5	9419	6789-88-4	2610	7044-96-4	7654	7357-93-9	3797	7712-60-9	9026
6380-34-3	464	6789-94-2	9626	7045-89-8	11903	7359-72-0	2334	7712-66-5	11814
6382-13-4	8270	6790-27-8	5345	7058-01-7	4448	7367-38-6	8184	7731-28-4	4535
6402-36-4	5336	6790-47-2	4307	7065-46-5	3690	7369-50-8	1693	7731-29-5	4536
6414-69-3	10283	6795-75-1	5935	7069-38-7	10548	7371-67-7	4181	7735-33-3	8869
6423-02-5	6758	6802-75-1	9966	7081-78-9	5473	7372-85-2	3055	7740-33-2	3589
6423-29-6	7804	6812-38-0	4411	7086-07-9	10147	7372-92-1	7280	7745-91-7	649
6436-90-4	6272	6812-39-1	4822	7087-68-5	9811	7379-12-6	6979	7745-93-9	1072
6443-91-0	5944	6812-78-8	8516	7094-26-0	4485	7383-20-2	8208	7756-94-7	10582
6443-92-1	6535	6814-58-0	4083	7094-27-1	4487	7383-90-6	3058	7766-23-6	9238
6453-99-2	7642	6818-07-1	6901	7094-34-0	7591	7385-78-6	9032	7771-44-0	5360
6482-24-2	5465	6829-40-9	9924	7098-07-9	7207	7385-82-2	7044	7773-60-6	9358
6482-34-4	3995	6829-41-0	9958	7098-21-7	11918	7391-39-1	6881	7778-87-2	6471
6493-77-2	4640	6831-89-6	10831	7101-96-4	11215	7398-82-5	989	7779-80-8	6467
6493-83-0	10457	6832-98-0	6630	7116-86-1	7021	7399-49-7	1984	7781-98-8	2620
6496-48-6	11176	6836-38-0	5324	7119-92-8	5431	7399-50-0	10967	7782-21-0	943
6498-47-1	3868	6843-66-9	11413	7119-94-0	9687	7400-08-0	10644	7785-26-4	2971
6513-13-9	12039	6846-11-3	7221	7132-64-1	8646	7403-66-9	42	7786-34-7	7659
6526-72-3	1977	6848-84-6	8769	7133-68-8	3358	7415-31-8	3748	7786-44-9	8112
6531-35-7	403	6849-18-9	9082	7137-56-6	1251	7417-19-8	1599	7787-72-6	7792
6538-02-9	5379	6850-36-8	4287	7144-05-0	9673	7417-48-3	8666	7789-89-1	10171
6553-48-6	4187	6850-57-3	1812	7145-20-2	7022	7423-69-0	7018	7789-90-4	9790
6556-12-3	6238	6852-54-6	5437	7146-60-3	8374	7425-45-8	3483	7789-92-6	10183
6568-58-7	4731	6861-64-9	7	7148-74-5	10495	7425-47-0	3928	7789-99-3	8945
6573-52-0	4729	6871-44-9	4730	7149-70-4	1066	7425-48-1	3484	9005-79-2	6274
6579-55-1	10383	6873-15-0	11200	7149-75-9	674	7428-48-0	8261	10008-73-8	6148
6592-85-4	5217	6874-98-2	11346	7152-15-0	8897	7433-56-9	5084	10021-92-8	4353
6597-78-0	7653	6876-23-9	4366	7152-88-7	2628	7433-78-5	5085	10022-60-3	6780
6600-40-4	8218	6886-16-4	6921	7153-14-2	4576	7439-33-0	4185	10024-74-5	1822
6607-66-5	3456	6890-03-5	4520	7154-66-7	2887	7443-55-2	4534	10031-82-0	554
6609-56-9	2760	6891-45-8	9250	7154-73-6	11164	7446-43-7	9814	10031-87-5	185
6609-57-0	2754	6894-69-5	4912	7154-75-8	9135	7459-46-3	5654	10042-29-2	268
6617-04-5	4311	6897-76-3	8014	7154-79-2	8839	7462-74-0	9792	10048-32-5	10791
6622-76-0	3851	6897-88-7	8015	7154-80-5	6450	7466-38-8	5133	10061-01-5	10551
6624-76-6	8100	6912-05-6	4923	7163-25-9	7773	7474-10-4	4720	10061-02-6	10550
6626-15-9	1502	6917-76-6	11377	7164-98-9	7212	7474-83-1	7844	10071-60-0	9020
6626-32-0	9347	6921-35-3	8593	7168-55-0	3048	7488-99-5	4054	10075-38-4	3749
6627-55-0	9236	6921-40-0	7738	7169-34-8	2462	7493-82-5	6470	10088-95-6	10785
6627-72-1	2942	6921-64-8	5826	7170-38-9	10320	7495-97-8	9526	10124-73-9	7005
6627-74-3	4667	6923-20-2	10549	7187-01-1	10585	7498-57-9	7737	10130-87-7	2248
6628-00-8	4304	6923-22-4	7663	7193-78-4	5844	7499-65-2	7744	10130-91-3	4566
6628-04-2	11210	6931-71-1	6800	7200-25-1	437	7521-80-4	11370	10136-52-4	6324
6628-06-4	9429	6933-10-4	651	7205-90-5	1136	7523-44-6	3865	10138-10-0	3572
6628-18-8	5452	6938-94-9	6782	7208-05-1	8577	7524-63-2	7988	10138-89-3	3455
6628-21-3	10246	6939-75-9	8911	7208-92-6	5554	7525-62-4	1631	10140-89-3	9781
6628-77-9	10865	6940-09-6	8275	7209-38-3	9614	7526-26-3	9541	10143-54-1	10107
6628-79-1	8896	6940-76-7	9857	7214-50-8	4712	7529-63-7	9697	10143-60-9	6431
6628-80-4	4505	6940-78-9	3233	7216-56-0	8488	7539-25-5	364	10147-36-1	10133
6628-97-3	7866	6941-69-1	129	7220-26-0	6818	7554-65-6	10836	10151-95-8	80
6629-96-5	5192	6942-99-0	1313	7220-81-7	4757	7564-63-8	1630	10152-76-8	10581

10160-87-9	: 10731	10574-36-4	: 7041	13402-36-3	: 4748	14077-58-8	: 298	15068-90-3	: 2950
10160-99-3	: 6627	10574-37-5	: 9034	13402-37-4	: 4744	14090-87-0	: 8393	15104-61-7	: 10126
10166-08-2	: 873	10575-87-8	: 6378	13403-37-7	: 8605	14092-20-7	: 7106	15111-56-5	: 4639
10193-50-7	: 9406	10577-44-3	: 1941	13417-43-1	: 3735	14108-88-4	: 7993	15116-17-3	: 1534
10203-08-4	: 537	10588-10-0	: 8900	13423-15-9	: 6172	14116-69-9	: 6234	15121-84-3	: 1612
10203-28-8	: 5322	10602-34-3	: 3758	13436-48-1	: 7239	14122-00-0	: 9280	15150-39-7	: 140
10203-30-2	: 5323	10604-59-8	: 7307	13444-24-1	: 9715	14159-48-9	: 10130	15164-44-0	: 582
10203-58-4	: 9969	10605-21-7	: 3990	13463-39-3	: 8068	14167-59-0	: 11650	15192-80-0	: 8489
10210-64-7	: 2912	10606-73-2	: 891	13463-40-6	: 7339	14179-94-3	: 9126	15220-85-6	: 10580
10212-74-5	: 3960	11024-24-1	: 6195	13466-40-5	: 8657	14180-63-3	: 10596	15232-76-5	: 8531
10220-20-9	: 9659	11028-42-5	: 4066	13466-41-6	: 11037	14181-05-6	: 11145	15232-96-9	: 1247
10222-95-4	: 2361	12069-69-1	: 511	13471-42-6	: 6915	14188-79-5	: 7794	15240-91-2	: 1800
10226-29-6	: 6961	12122-67-7	: 12067	13475-76-8	: 5272	14214-31-4	: 10538	15250-29-0	: 10670
10229-10-4	: 9142	12321-04-9	: 5961	13475-77-9	: 5354	14222-60-7	: 10886	15268-49-2	: 10086
10230-68-9	: 10475	12427-38-2	: 7420	13475-78-0	: 6416	14248-22-7	: 2129	15299-99-7	: 8067
10233-13-3	: 5316	13012-59-4	: 9864	13475-79-1	: 8766	14252-42-7	: 5450	15301-40-3	: 11293
10236-47-2	: 2795	13014-24-9	: 1401	13475-81-5	: 6844	14255-23-3	: 7023	15320-72-6	: 3833
10240-08-1	: 8030	13019-20-0	: 6522	13475-82-6	: 6433	14272-54-9	: 9115	15329-10-9	: 4708
10250-47-2	: 8949	13019-22-2	: 5092	13481-70-4	: 9294	14273-06-4	: 10732	15333-22-9	: 3829
10257-92-8	: 7318	13023-00-2	: 3500	13482-62-7	: 9524	14282-76-9	: 11729	15339-36-3	: 7421
10259-22-0	: 2670	13024-80-1	: 1750	13491-79-7	: 4521	14287-61-7	: 3507	15356-70-4	: 4548
10265-92-6	: 7441	13027-43-5	: 1650	13505-34-5	: 6403	14296-80-1	: 4172	15358-48-2	: 7574
10273-90-2	: 10991	13030-26-7	: 2778	13511-38-1	: 10219	14303-70-9	: 11629	15361-99-6	: 2277
10276-09-2	: 3835	13038-21-6	: 8190	13513-82-1	: 1893	14304-30-4	: 5087	15409-60-6	: 4591
10283-15-5	: 9256	13045-99-3	: 3798	13529-27-6	: 6084	14315-14-1	: 2402	15411-43-5	: 734
10289-68-6	: 11885	13047-06-8	: 7600	13532-18-8	: 10311	14321-27-8	: 1810	15430-98-5	: 8654
10292-98-5	: 2963	13054-87-0	: 3595	13545-04-5	: 3317	14331-54-5	: 11055	15457-05-3	: 5980
10293-06-8	: 2952	13057-17-5	: 7472	13552-31-3	: 9984	14352-61-5	: 4310	15466-93-0	: 4511
10297-09-3	: 8541	13057-19-7	: 7519	13556-55-3	: 7985	14371-10-9	: 10518	15466-94-1	: 4510
10297-38-8	: 11571	13063-43-9	: 9912	13556-58-6	: 7888	14379-80-7	: 6266	15469-77-9	: 5089
10299-30-6	: 6867	13065-07-1	: 7989	13602-13-6	: 3747	14400-94-3	: 9467	15489-90-4	: 5959
10302-94-0	: 3772	13071-79-9	: 11592	13603-04-8	: 11150	14401-73-1	: 5797	15501-33-4	: 10056
10307-61-6	: 3550	13073-29-5	: 9423	13603-21-9	: 9636	14411-75-7	: 1471	15519-73-0	: 3068
10309-50-9	: 2959	13073-35-3	: 991	13618-93-4	: 7325	14436-32-9	: 5091	15521-18-3	: 10374
10311-74-7	: 1622	13086-84-5	: 9531	13632-93-4	: 1416	14444-77-0	: 1409	15543-70-1	: 4965
10311-84-9	: 5105	13099-26-8	: 1227	13632-94-5	: 1420	14446-69-6	: 9653	15570-12-4	: 2299
10312-83-1	: 16	13102-31-3	: 9449	13632-95-6	: 1418	14452-30-3	: 5836	15600-01-8	: 10053
10315-89-6	: 9685	13105-52-7	: 3934	13643-05-5	: 8667	14459-29-1	: 9735	15606-05-8	: 439
10316-00-4	: 7700	13114-62-0	: 12023	13665-04-8	: 5798	14472-14-1	: 9242	15644-80-1	: 11316
10317-10-9	: 10353	13116-53-5	: 10145	13672-18-9	: 7730	14476-30-3	: 10162	15647-08-2	: 9589
10321-71-8	: 9078	13116-60-4	: 10143	13678-54-1	: 11764	14476-37-0	: 11960	15647-11-7	: 4502
10323-40-7	: 3471	13121-70-5	: 5011	13679-46-4	: 6128	14484-64-1	: 5957	15657-96-2	: 10091
10324-58-0	: 9692	13129-23-2	: 6063	13679-70-4	: 11736	14490-79-0	: 11608	15679-02-4	: 4188
10327-08-9	: 10204	13138-51-7	: 3429	13679-74-8	: 5856	14548-31-3	: 7105	15706-73-7	: 3545
10334-26-6	: 2926	13147-06-3	: 11639	13684-56-5	: 5104	14548-46-0	: 7639	15721-05-8	: 4992
10339-67-0	: 6602	13151-04-7	: 6546	13684-63-4	: 9198	14558-14-6	: 8003	15726-15-5	: 6525
10340-91-7	: 1791	13151-05-8	: 6545	13688-75-0	: 11468	14575-93-0	: 2905	15727-05-8	: 7892
10342-83-3	: 10418	13151-17-2	: 7045	13698-16-3	: 3940	14595-35-8	: 3783	15764-04-2	: 8052
10343-99-4	: 11246	13151-34-3	: 5055	13698-87-8	: 1326	14596-92-0	: 6682	15764-16-6	: 547
10373-81-6	: 2957	13153-11-2	: 11694	13706-86-0	: 6809	14602-62-1	: 10161	15797-21-4	: 4795
10394-35-1	: 2378	13163-73-0	: 4604	13721-54-5	: 7103	14620-52-1	: 5297	15802-97-8	: 10841
10403-60-8	: 3451	13165-77-0	: 2181	13732-80-4	: 1417	14620-53-2	: 11617	15802-99-0	: 10839
10410-35-2	: 2448	13172-91-3	: 6557	13794-14-4	: 3580	14620-55-4	: 8237	15825-70-4	: 7424
10420-90-3	: 6712	13176-46-0	: 3478	13799-90-1	: 1349	14630-40-1	: 11434	15826-80-9	: 2981
10420-91-4	: 6715	13194-48-4	: 5934	13808-64-5	: 10817	14631-45-9	: 10102	15840-60-5	: 7047
10424-29-0	: 3341	13195-80-7	: 10542	13811-01-3	: 4274	14660-52-7	: 8861	15843-48-8	: 5568
10428-64-5	: 3272	13206-46-7	: 9776	13828-34-7	: 4441	14676-29-0	: 6409	15854-58-7	: 10647
10435-81-1	: 8419	13222-26-9	: 10498	13831-30-6	: 101	14685-29-1	: 4995	15869-80-4	: 6407
10436-75-6	: 794	13249-58-6	: 1030	13835-81-9	: 11467	14686-13-6	: 6534	15869-85-9	: 8143
10439-23-3	: 2186	13250-46-9	: 303	13855-77-1	: 3964	14686-14-7	: 6536	15869-86-0	: 8404
10443-65-9	: 10610	13254-34-7	: 6479	13862-07-2	: 9665	14689-97-5	: 5592	15869-87-1	: 8373
10453-86-8	: 11340	13258-63-4	: 10949	13865-48-0	: 2018	14697-46-2	: 8851	15869-89-3	: 8376
10463-14-0	: 3740	13269-52-8	: 7002	13877-91-3	: 8485	14716-89-3	: 7404	15869-92-8	: 8380
10473-14-0	: 3871	13279-87-3	: 3448	13898-58-3	: 2514	14719-83-6	: 2548	15869-93-9	: 8381
10476-95-6	: 10564	13290-74-9	: 1205	13898-73-2	: 4678	14720-74-2	: 6438	15869-94-0	: 8382
10478-02-1	: 2592	13313-91-2	: 261	13905-48-1	: 4320	14721-66-5	: 6658	15869-95-1	: 8383
10478-12-3	: 9083	13325-10-5	: 3597	13909-73-4	: 5874	14737-94-1	: 10622	15869-96-2	: 8384
10486-19-8	: 11869	13327-31-6	: 11273	13937-08-1	: 9957	14752-60-4	: 9124	15870-10-7	: 6543
10487-71-5	: 3887	13329-40-3	: 5837	13939-06-5	: 7662	14752-61-5	: 9133	15874-03-0	: 6637
10487-96-4	: 4871	13347-42-7	: 9255	13956-29-1	: 1537	14752-75-1	: 1743	15877-57-3	: 8694
10488-25-2	: 4539	13351-61-6	: 2151	13963-35-4	: 2017	14804-32-1	: 1682	15890-36-5	: 4541
10488-94-5	: 8436	13351-73-0	: 2872	13979-26-9	: 3836	14804-59-2	: 10510	15890-40-1	: 4859
10489-97-1	: 4345	13356-08-6	: 5950	13984-57-1	: 6906	14813-68-4	: 8345	15901-49-2	: 5003
10495-45-1	: 10132	13360-45-7	: 4076	13993-65-2	: 9500	14845-35-3	: 4706	15909-76-9	: 10254
10498-35-8	: 4359	13360-61-7	: 8653	13997-70-1	: 1232	14845-55-7	: 4887	15918-08-8	: 6426
10500-57-9	: 11297	13360-63-9	: 3218	13997-73-4	: 9271	14847-51-9	: 9237	15922-53-9	: 11849
10503-96-5	: 3249	13360-65-1	: 10809	14007-64-8	: 898	14850-22-7	: 8495	15932-80-6	: 4607
10508-09-5	: 9154	13361-35-8	: 143	14010-23-2	: 6315	14850-23-8	: 8496	15950-66-0	: 9481
10519-06-9	: 1117	13361-36-9	: 137	14021-92-2	: 5048	14852-31-4	: 6660	15953-45-4	: 10821
10519-33-2	: 5093	13361-63-2	: 9129	14045-26-2	: 9671	14869-39-7	: 5814	15953-73-8	: 10819
10521-91-2	: 2101	13362-52-2	: 11876	14062-18-1	: 923	14919-01-8	: 8494	15971-29-6	: 7755
10522-26-6	: 11955	13367-29-8	: 2993	14062-78-3	: 627	14924-53-9	: 4175	15972-60-8	: 322
10524-08-0	: 9018	13388-93-7	: 5846	14064-10-9	: 9931	14938-35-3	: 9444	15980-15-1	: 8569
10528-67-3	: 4469	13389-42-9	: 8492	14064-43-8	: 1223	14963-40-7	: 4946	16003-14-8	: 6160
10544-63-5	: 3838	13398-94-2	: 1606	14064-68-7	: 2768	14979-39-6	: 6501	16009-13-5	: 5958
10568-38-4	: 1683	13402-32-9	: 1308	14072-86-7	: 4656	14996-78-2	: 4236	16020-17-0	: 8001
10574-17-1	: 2030	13402-35-2	: 4746	14073-97-3	: 4606	15014-25-2	: 9927	16021-20-8	: 1702

16066-09-4	: 11647	16789-46-1	: 6825	17745-45-8	: 3136	18913-31-0	: 3175	19780-66-6	: 9046
16096-32-5	: 7315	16789-51-8	: 7034	17773-65-8	: 3737	18917-91-4	: 340	19780-67-7	: 9049
16106-59-5	: 7020	16790-49-1	: 3197	17788-00-0	: 9292	18932-14-4	: 8420	19780-90-6	: 6798
16109-68-5	: 5120	16805-78-0	: 9578	17789-14-9	: 5205	18936-17-9	: 3409	19780-96-2	: 8164
16133-49-6	: 777	16813-18-6	: 6318	17804-35-2	: 515	18937-76-3	: 11990	19781-10-3	: 8451
16136-84-8	: 10534	16837-43-7	: 2118	17837-41-1	: 11693	18963-00-3	: 10176	19781-11-4	: 8458
16136-85-9	: 10533	16841-48-8	: 10597	17849-38-6	: 1848	18963-01-4	: 10148	19781-68-1	: 4976
16147-07-2	: 1446	16874-33-2	: 6072	17851-27-3	: 1710	18970-44-0	: 1688	19781-69-2	: 4977
16177-46-1	: 4186	16883-48-0	: 4860	17878-44-3	: 11367	18971-59-0	: 8821	19781-73-8	: 4404
16179-36-5	: 8885	16889-72-8	: 10294	17880-86-3	: 10502	18979-60-7	: 1549	19781-77-2	: 6572
16183-46-3	: 3666	16898-52-5	: 9701	17912-17-3	: 8368	18979-62-9	: 1539	19812-64-7	: 11618
16188-57-1	: 9374	16932-49-3	: 2753	17937-68-7	: 11404	18992-39-7	: 10402	19814-71-2	: 2077
16197-90-3	: 3822	16939-57-4	: 1097	17952-11-3	: 8808	18992-80-8	: 5806	19819-98-8	: 1609
16245-79-7	: 831	16957-70-3	: 9076	17972-08-6	: 7652	18999-28-5	: 6561	19821-84-2	: 6612
16246-07-4	: 3947	16980-61-3	: 4512	17988-31-7	: 11427	19037-72-0	: 4922	19832-98-5	: 8050
16292-88-9	: 776	16995-35-0	: 10486	18002-79-4	: 11360	19044-88-3	: 8553	19860-56-1	: 10256
16302-35-5	: 10778	17005-59-3	: 11383	18023-33-1	: 11426	19074-59-0	: 9336	19866-51-4	: 9077
16306-39-1	: 9184	17015-11-1	: 6914	18023-36-4	: 11511	19089-47-5	: 10380	19879-11-9	: 4508
16308-92-2	: 1860	17021-26-0	: 357	18030-61-0	: 11364	19113-78-1	: 6928	19879-12-0	: 4509
16315-07-4	: 1922	17042-16-9	: 6980	18036-81-2	: 11497	19145-60-9	: 487	19889-37-3	: 3516
16326-97-9	: 4806	17057-82-8	: 7252	18169-57-8	: 11469	19145-96-1	: 4356	19890-00-7	: 2956
16326-98-0	: 4807	17057-88-4	: 3060	18171-59-0	: 11373	19155-24-9	: 7329	19891-74-8	: 4060
16357-59-8	: 11231	17057-92-0	: 7887	18178-59-1	: 11421	19163-05-4	: 3129	19902-08-0	: 2924
16369-05-4	: 3599	17059-44-8	: 1485	18181-80-1	: 3140	19177-04-9	: 1987	19915-11-8	: 8689
16369-17-8	: 8922	17059-45-9	: 1482	18187-24-1	: 5234	19184-10-2	: 10567	19942-78-0	: 11708
16369-21-4	: 5765	17059-52-8	: 2459	18231-53-3	: 8764	19184-67-9	: 8774	19947-22-9	: 1700
16386-93-9	: 9075	17071-47-5	: 3401	18236-57-2	: 11400	19221-28-4	: 3165	19961-27-4	: 9810
16387-70-5	: 8344	17082-09-6	: 10717	18236-89-0	: 11399	19230-27-4	: 1294	19967-57-8	: 9775
16387-72-7	: 11610	17082-12-1	: 5129	18240-50-1	: 6114	19246-38-9	: 4346	19987-13-4	: 9865
16409-46-4	: 3554	17085-91-5	: 7828	18243-41-9	: 11366	19246-39-0	: 3601	20009-26-1	: 7955
16419-60-6	: 3134	17091-40-6	: 5747	18261-92-2	: 8280	19269-28-4	: 6720	20024-90-2	: 1703
16420-13-6	: 3980	17094-21-2	: 3558	18262-85-6	: 1709	19317-11-4	: 5330	20027-77-4	: 7991
16423-68-0	: 11524	17096-37-6	: 3796	18281-05-5	: 5356	19329-89-6	: 10274	20030-32-4	: 4690
16426-64-5	: 2527	17102-64-6	: 6709	18282-40-1	: 1680	19354-27-9	: 6170	20063-92-7	: 8183
16432-33-4	: 6794	17113-33-6	: 6159	18282-59-2	: 1015	19374-46-0	: 4857	20089-07-0	: 3439
16433-43-5	: 1104	17176-77-1	: 9586	18295-59-5	: 8697	19393-45-4	: 3773	20098-48-0	: 2338
16435-50-0	: 9024	17199-29-0	: 903	18309-32-5	: 2978	19393-92-1	: 1013	20177-02-0	: 9028
16452-01-0	: 771	17206-52-9	: 4595	18338-40-4	: 3449	19395-58-5	: 11198	20184-89-8	: 8198
16484-17-6	: 7293	17216-08-9	: 8045	18362-36-2	: 578	19396-83-9	: 2916	20184-91-2	: 8199
16491-15-9	: 4921	17216-62-5	: 9933	18362-97-5	: 8893	19398-47-1	: 3614	20185-55-1	: 2694
16503-53-0	: 2147	17226-28-7	: 3416	18368-63-3	: 10920	19398-53-9	: 8743	20193-20-8	: 9809
16509-46-9	: 8335	17226-34-5	: 7009	18381-45-8	: 3406	19398-61-9	: 1387	20213-30-3	: 7917
16518-62-0	: 644	17228-64-7	: 10918	18388-45-9	: 11408	19398-75-5	: 4799	20231-81-6	: 4045
16520-62-0	: 1124	17236-58-7	: 11777	18397-07-4	: 4154	19398-77-7	: 6768	20232-11-5	: 7279
16532-02-8	: 11365	17249-75-1	: 11747	18398-36-2	: 7849	19402-87-0	: 3988	20237-34-7	: 6681
16538-93-5	: 4737	17249-76-2	: 11749	18423-69-3	: 6136	19420-29-2	: 7946	20244-71-7	: 5888
16545-78-1	: 9061	17249-77-3	: 11791	18441-43-5	: 2458	19434-65-2	: 9778	20247-89-6	: 2008
16560-43-3	: 11272	17249-79-5	: 11746	18444-79-6	: 11521	19438-10-9	: 2636	20266-65-3	: 10022
16580-23-7	: 4436	17249-82-0	: 11741	18476-57-8	: 8327	19447-29-1	: 8197	20278-84-6	: 6446
16580-24-8	: 4437	17283-45-3	: 2946	18479-51-1	: 8515	19472-74-3	: 948	20278-85-7	: 6443
16582-38-0	: 1207	17297-04-0	: 1108	18479-57-7	: 8450	19481-82-4	: 10094	20278-87-9	: 6449
16587-71-6	: 4592	17299-41-1	: 4719	18492-37-0	: 2960	19482-57-6	: 8535	20278-88-0	: 6451
16605-36-0	: 4435	17301-94-9	: 8142	18495-23-3	: 8536	19489-10-2	: 4397	20278-89-1	: 6452
16619-12-8	: 7291	17302-01-1	: 6410	18495-26-6	: 6613	19493-31-3	: 3026	20279-53-2	: 6062
16629-19-9	: 11780	17302-02-2	: 6767	18495-30-2	: 10144	19493-44-8	: 7376	20281-83-8	: 8938
16631-63-3	: 4833	17302-04-4	: 6415	18510-29-7	: 11455	19523-57-0	: 7753	20281-85-0	: 3616
16650-52-5	: 7762	17312-63-9	: 8122	18521-07-8	: 8519	19549-71-4	: 6480	20281-86-1	: 6912
16661-99-7	: 4349	17348-59-3	: 10081	18523-48-3	: 109	19549-73-6	: 6481	20281-91-8	: 3619
16672-87-0	: 5929	17356-19-3	: 4866	18530-30-8	: 2917	19549-74-7	: 6482	20291-91-2	: 6816
16674-04-7	: 10230	17372-87-1	: 11525	18530-56-8	: 8212	19549-77-0	: 6484	20291-95-6	: 6439
16694-52-3	: 9868	17406-45-0	: 6196	18589-29-2	: 6420	19549-79-2	: 6486	20327-65-5	: 10568
16696-65-4	: 11940	17432-37-0	: 598	18598-63-5	: 5019	19549-84-9	: 6515	20333-41-9	: 5245
16714-77-5	: 3654	17447-00-6	: 9085	18649-64-4	: 6108	19549-86-1	: 6617	20354-26-1	: 7643
16726-67-3	: 7761	17447-01-7	: 9084	18668-68-3	: 3147	19549-97-4	: 6617	20357-79-3	: 7247
16736-42-8	: 8325	17478-66-9	: 5138	18685-70-6	: 4073	19550-03-9	: 6922	20359-55-1	: 8979
16737-44-3	: 1983	17484-36-5	: 1934	18699-48-4	: 1338	19550-05-1	: 6924	20395-25-9	: 10173
16744-89-1	: 7083	17496-14-9	: 7289	18719-43-2	: 9932	19550-07-3	: 6927	20442-79-9	: 3078
16745-94-1	: 7017	17534-15-5	: 1561	18720-65-5	: 6502	19550-09-5	: 6929	20474-93-5	: 3856
16746-02-4	: 6836	17556-48-8	: 10409	18729-48-1	: 4870	19550-10-8	: 6964	20485-41-0	: 11669
16746-86-4	: 7013	17577-93-4	: 8523	18747-42-7	: 10470	19550-14-2	: 6967	20589-85-9	: 10598
16746-87-5	: 7014	17587-22-3	: 8398	18748-27-1	: 11578	19550-40-4	: 4655	20599-21-7	: 8534
16747-25-4	: 6859	17609-31-3	: 8484	18752-92-6	: 11445	19550-48-2	: 4920	20599-27-3	: 9011
16747-26-5	: 6860	17618-76-7	: 6554	18758-34-4	: 4998	19564-40-0	: 2115	20634-92-8	: 9812
16747-27-6	: 6862	17627-77-9	: 7395	18773-77-8	: 4160	19594-02-6	: 1652	20637-49-4	: 10185
16747-30-1	: 6865	17639-93-9	: 10222	18786-24-8	: 8572	19613-76-4	: 1024	20668-29-5	: 11261
16747-31-2	: 6866	17640-26-5	: 182	18787-64-9	: 6663	19617-44-8	: 7231	20699-48-3	: 3261
16747-32-3	: 8810	17648-05-4	: 2281	18791-49-6	: 10282	19653-33-9	: 9704	20731-93-5	: 1875
16747-33-4	: 8811	17655-74-2	: 1651	18794-84-8	: 5331	19655-60-8	: 11268	20734-71-8	: 9386
16747-38-9	: 8842	17658-32-1	: 8924	18812-62-9	: 7079	19666-30-9	: 8565	20739-58-6	: 8543
16747-42-5	: 6848	17672-88-7	: 2422	18829-56-6	: 8180	19686-73-8	: 10345	20748-86-1	: 6400
16747-44-7	: 8836	17679-92-4	: 8526	18839-90-2	: 9359	19689-18-0	: 5083	20754-04-5	: 8478
16747-45-8	: 8837	17696-11-6	: 8426	18840-45-4	: 12042	19694-02-1	: 10854	20769-85-1	: 10491
16747-50-5	: 4814	17696-37-6	: 9326	18854-56-3	: 5586	19700-97-1	: 6693	20809-46-5	: 8479
16749-11-4	: 4361	17696-64-9	: 142	18858-65-6	: 11441	19718-45-7	: 7956	20820-82-0	: 11022
16751-59-0	: 6351	17704-30-2	: 165	18869-72-2	: 6311	19735-89-8	: 10849	20830-75-5	: 4041
16752-77-5	: 7650	17715-00-3	: 4473	18904-54-6	: 4146	19752-55-7	: 1012	20850-43-5	: 2421
16754-49-7	: 10080	17742-04-0	: 7528	18905-91-4	: 6973	19780-63-3	: 8933	20893-30-5	: 11725

CAS	Page	CAS	Page	CAS	Page	CAS	Page	CAS	Page
20972-43-4	5131	22401-25-8	9546	24342-04-9	9139	25871-69-6	7587	29091-05-2	5181
20984-33-2	11322	22409-91-2	10322	24346-56-3	2132	26073-26-7	10500	29091-21-2	9761
21020-24-6	3899	22418-49-1	10577	24356-00-1	9017	26106-95-6	3602	29106-32-9	4936
21020-26-8	9131	22421-97-2	10229	24393-70-2	2943	26116-12-1	11167	29110-74-5	7236
21020-27-9	9136	22445-41-6	1771	24395-10-6	6571	26118-38-7	6963	29113-63-1	6055
21040-45-9	10706	22483-09-6	5419	24398-88-7	2522	26120-52-5	4253	29136-19-4	2062
21040-59-5	7392	22520-40-7	8391	24410-50-2	2449	26153-88-8	8703	29172-20-1	6111
21087-29-6	1230	22524-32-9	113	24410-51-3	2446	26171-83-5	3344	29204-02-2	5365
21087-64-9	7658	22531-06-2	8051	24410-52-4	2452	26186-01-6	8107	29232-93-7	9730
21102-13-6	6952	22531-20-0	7891	24425-40-9	7244	26186-02-7	11897	29276-40-2	7707
21129-27-1	4546	22539-65-7	1469	24425-97-6	10051	26190-82-9	11256	29289-13-2	762
21179-48-6	10239	22545-12-6	1603	24438-17-3	7328	26198-63-0	9892	29342-44-7	6740
21211-65-4	11133	22545-13-7	1604	24446-63-7	942	26225-79-6	5933	29342-53-8	2918
21232-53-1	361	22564-83-6	7429	24544-04-5	636	26259-45-0	3143	29394-58-9	3383
21266-88-6	6375	22564-99-4	8332	24545-81-1	2987	26260-02-6	580	29400-42-8	8211
21266-90-0	6376	22587-21-1	4543	24549-06-2	745	26303-90-2	5090	29472-95-5	9421
21293-29-8	8679	22610-63-5	8255	24557-10-6	10627	26311-17-1	5424	29488-27-5	11773
21296-92-4	7323	22618-23-1	9315	24579-70-2	3941	26311-33-1	7063	29538-77-0	4507
21299-26-3	4578	22621-74-5	2947	24580-48-1	4550	26330-51-8	3476	29554-49-2	4745
21326-62-5	3611	22646-79-3	4646	24580-52-7	4551	26340-58-9	4151	29555-02-0	4964
21352-22-7	11209	22663-64-5	8166	24602-86-6	11898	26370-28-5	8108	29559-27-1	2004
21370-71-8	8054	22672-74-8	5150	24612-75-7	4477	26399-36-0	9763	29559-52-2	10177
21391-98-0	4635	22692-70-2	11560	24618-31-3	4317	26438-26-6	7931	29560-84-7	10701
21400-25-9	10602	22717-57-3	2640	24622-72-8	11169	26438-27-7	7793	29604-75-9	1296
21436-44-2	2683	22726-00-7	605	24634-61-5	6704	26438-28-8	7879	29606-79-9	4603
21472-86-6	2335	22734-04-9	4905	24644-78-8	7290	26464-32-4	3488	29624-17-7	4348
21490-63-1	8609	22767-95-9	2140	24677-78-9	540	26465-81-6	7288	29653-38-1	3926
21504-43-8	10559	22767-96-0	2145	24716-09-4	890	26472-41-3	4282	29682-41-5	1380
21531-91-9	6793	22768-19-0	6550	24724-07-0	11949	26473-49-4	3537	29780-00-5	11896
21535-97-7	2457	22768-20-3	6549	24734-68-7	2125	26489-01-0	8514	29823-16-3	8873
21564-91-0	7984	22781-23-3	514	24764-97-4	3179	26513-20-2	1265	29843-58-1	3757
21564-92-1	7986	22839-47-0	9511	24796-87-0	4503	26530-20-1	8528	29865-90-5	568
21570-35-4	6503	22868-76-4	11056	24815-24-5	12060	26531-85-1	6237	29882-57-3	8720
21571-91-5	8751	22900-08-9	4540	24840-05-9	10588	26533-33-5	8169	29883-15-6	954
21617-18-5	11250	22922-67-4	7004	24844-28-8	4556	26549-24-6	6913	29897-82-3	11172
21633-79-4	9972	22924-15-8	553	24885-45-8	9375	26549-25-7	6474	29916-45-8	3269
21655-48-1	9611	22927-13-5	558	24887-75-0	345	26608-06-0	5165	29921-50-4	8063
21693-54-9	7992	22936-86-3	5013	24910-63-2	9036	26644-46-2	11902	29943-42-8	10700
21700-31-2	10125	23010-00-6	3733	24932-49-8	1297	26730-14-3	11879	29949-19-7	7511
21702-79-4	2559	23010-04-0	3283	24948-66-1	5095	26754-48-3	6030	29972-79-0	5341
21725-46-2	4161	23012-11-5	8578	24964-64-5	2757	26771-11-9	4959	30090-17-6	814
21777-84-4	1992	23048-13-7	4943	24965-84-2	4596	26963-33-7	11771	30122-12-4	8745
21849-97-8	7840	23063-36-7	1390	24973-49-7	3020	27063-27-0	11208	30168-50-4	4830
21856-89-3	6970	23074-36-4	3709	25006-60-4	167	27129-86-8	1057	30273-11-1	813
21884-26-4	8909	23079-65-4	1116	25013-16-5	9329	27129-87-9	1861	30361-33-2	5025
21889-88-3	6585	23097-98-5	6618	25044-01-3	9107	27267-69-2	2773	30389-18-5	4290
21890-35-7	3148	23103-98-2	9728	25057-89-0	518	27314-13-2	8213	30433-57-9	10833
21915-53-7	8619	23132-52-7	1467	25109-28-8	1009	27356-52-1	11269	30515-28-7	6458
21923-23-9	4092	23135-22-0	8568	25109-57-3	3853	27387-79-7	10218	30525-89-4	8634
21948-47-0	8372	23144-52-7	8448	25117-74-2	2755	27464-82-0	11659	30538-80-8	9689
21964-23-8	9013	23184-66-9	3144	25118-28-9	6755	27532-76-9	378	30544-34-4	6077
21964-44-3	8189	23204-80-0	1229	25144-04-1	4869	27598-85-2	9200	30560-19-1	10
21968-17-2	9827	23287-26-5	2638	25144-05-2	4868	27610-88-4	8400	30600-19-2	10619
21974-48-1	9714	23308-82-9	1894	25148-87-2	3756	27644-49-1	8465	30614-77-8	6082
21984-93-0	6065	23378-11-2	3732	25167-70-8	9064	27655-41-0	7375	30673-60-0	5069
21987-21-3	11886	23399-70-4	1174	25172-42-3	4873	27701-01-5	3943	30685-95-1	3886
21994-77-4	7065	23418-37-3	8469	25172-43-4	4874	27779-01-7	4158	30689-41-9	4916
21994-78-5	7064	23418-38-4	8170	25182-84-7	10395	27831-13-6	1629	30752-19-3	1045
22029-76-1	3873	23422-53-9	6001	25186-47-4	1386	27876-94-4	5107	30926-60-4	1188
22037-28-1	6034	23431-48-3	2028	25201-40-5	6338	27993-42-6	8984	30964-01-3	8207
22037-73-6	3710	23438-77-9	4722	25229-42-9	3834	28056-54-4	4896	31053-99-3	1092
22038-57-9	3830	23444-65-7	7854	25234-83-7	8898	28123-70-8	8125	31059-19-5	7944
22056-82-2	14	23470-12-4	7127	25245-33-4	1767	28148-04-1	9890	31116-84-4	10459
22077-55-0	2176	23505-41-1	9729	25265-71-8	10397	28249-77-6	11697	31161-71-4	11627
22104-78-5	8510	23511-70-8	3582	25309-64-2	1681	28261-97-4	5273	31202-12-7	363
22118-12-3	3682	23564-05-8	11723	25311-71-1	7359	28300-74-5	11588	31218-83-4	10719
22128-62-7	4014	23578-35-0	6751	25312-34-9	3874	28320-88-9	8362	31252-42-3	9700
22128-63-8	4016	23580-51-0	8544	25312-65-6	4094	28320-89-0	6364	31283-43-9	8498
22139-77-1	1545	23580-52-1	10960	25346-31-0	8718	28332-44-7	7095	31340-44-0	10247
22144-60-1	1912	23617-71-2	2589	25346-32-1	8730	28466-26-4	10812	31366-00-4	1713
22160-12-9	6893	23619-59-2	9459	25346-33-2	8714	28467-92-7	3190	31367-46-1	6496
22160-39-0	8887	23676-09-7	2599	25354-97-6	5065	28469-92-3	1369	31508-08-4	6607
22160-40-3	8888	23708-56-7	11954	25354-98-7	8649	28664-03-1	4171	31508-12-0	8194
22160-41-4	6895	23733-68-8	4717	25394-75-6	10772	28685-60-1	3958	31543-75-6	1306
22202-65-9	11989	23769-39-3	11921	25395-22-6	103	28715-26-6	2451	31599-60-7	1769
22224-92-6	5947	23843-52-9	8239	25395-31-7	10189	28788-62-7	2889	31599-61-8	1774
22227-75-4	10556	23947-60-6	5932	25414-22-6	6127	28823-41-8	6696	31642-67-8	8188
22232-54-8	7194	23950-58-5	10746	25415-67-2	8892	28832-55-5	3180	31656-92-5	981
22253-11-8	7259	24127-62-6	10615	25415-70-7	8894	28905-12-6	6232	31699-35-1	8116
22287-02-1	8176	24152-39-4	9637	25419-36-7	7990	28921-35-9	6064	31702-33-7	10605
22319-28-1	6592	24247-53-8	3696	25419-37-8	7987	28994-41-4	9454	31720-69-1	4562
22331-38-0	5351	24264-08-2	3877	25542-62-5	6879	28995-89-3	8004	31766-07-1	2771
22349-03-7	4456	24280-93-1	7052	25611-78-3	1584	29003-73-4	163	31772-05-1	11874
22364-68-7	960	24306-18-1	6323	25641-98-4	1340	29022-31-9	8201	31805-48-8	11761
22374-89-6	2119	24307-26-4	7430	25659-22-7	7087	29030-84-0	6689	31844-95-8	9008
22381-86-8	11980	24331-72-4	9794	25695-77-6	10689	29040-46-8	2445	31844-96-9	3716
22398-09-0	9709	24332-22-7	6584	25790-55-0	3150	29072-93-3	10087	31849-78-2	3708

31883-98-4	: 4713	34840-79-4	: 1364	37493-31-5	: 6162	40575-42-6	: 8509	50710-40-2	: 3244
31915-94-3	: 1098	34859-98-8	: 8946	37493-70-2	: 11953	40589-38-6	: 3890	50840-23-8	: 11047
32018-88-5	: 7760	34874-30-1	: 11117	37549-83-0	: 7056	40596-69-8	: 7651	50868-72-9	: 773
32064-78-1	: 8524	34885-03-5	: 4424	37549-89-6	: 7031	40604-49-7	: 8023	50868-73-0	: 774
32116-65-7	: 4889	34887-14-4	: 8748	37674-63-8	: 9079	40605-42-3	: 3864	50876-32-9	: 4478
32121-06-5	: 10544	34893-50-0	: 9649	37674-67-2	: 7074	40614-52-6	: 3363	50876-33-0	: 4851
32121-07-6	: 10543	34973-41-6	: 3380	37688-96-3	: 11616	40646-07-9	: 6397	50902-82-4	: 6887
32161-06-1	: 9720	35000-22-7	: 10525	37710-49-9	: 8663	40649-36-3	: 4613	51034-46-9	: 3154
32166-40-8	: 8053	35008-86-7	: 1761	37746-78-4	: 3819	40650-41-7	: 7261	51060-05-0	: 7360
32315-10-9	: 7585	35018-15-6	: 11149	37806-29-4	: 1808	40734-75-6	: 1002	51115-02-7	: 4931
32322-73-9	: 47	35018-28-1	: 11175	37845-14-0	: 8060	40738-26-9	: 5310	51116-73-5	: 8715
32328-03-3	: 8778	35021-70-6	: 1752	37866-06-1	: 3731	40745-44-6	: 2	51152-12-6	: 2930
32337-74-9	: 8664	35099-89-9	: 4227	37924-13-3	: 9149	40775-09-5	: 1741	51218-45-2	: 7657
32360-05-7	: 10667	35120-10-6	: 286	37935-39-0	: 3492	40960-69-8	: 4554	51225-20-8	: 691
32367-54-7	: 7889	35161-71-8	: 10726	37951-49-8	: 10456	40960-73-4	: 4553	51235-04-2	: 6995
32395-96-3	: 1747	35182-51-5	: 10968	37973-84-5	: 11975	41055-92-9	: 6513	51282-49-6	: 2549
32433-28-6	: 90	35194-37-7	: 6562	37985-18-5	: 11701	41065-91-2	: 6883	51284-68-5	: 5116
32449-92-6	: 6239	35205-71-1	: 7059	38031-78-6	: 10943	41085-43-2	: 1071	51299-55-9	: 4620
32460-00-7	: 6078	35205-79-9	: 8153	38204-89-6	: 1361	41096-46-2	: 7182	51308-99-7	: 8163
32460-20-1	: 6190	35216-11-6	: 11641	38237-74-0	: 8960	41198-08-7	: 9762	51309-10-5	: 5073
32541-13-6	: 8925	35223-80-4	: 120	38256-94-9	: 5426	41223-14-7	: 4773	51333-70-1	: 8111
32607-54-8	: 8865	35242-05-8	: 4504	38260-54-7	: 5945	41284-12-2	: 5615	51338-27-3	: 5171
32620-08-9	: 3707	35246-18-5	: 7309	38300-67-3	: 3445	41295-64-1	: 3687	51437-67-3	: 5115
32665-23-9	: 3553	35274-05-6	: 10271	38460-95-6	: 11982	41433-81-2	: 9954	51500-32-4	: 10818
32666-56-1	: 7714	35280-53-6	: 136	38487-94-4	: 2103	41589-42-8	: 9936	51525-97-4	: 10137
32718-56-2	: 5734	35305-13-6	: 9686	38511-07-8	: 4645	41635-77-2	: 1753	51556-10-6	: 6614
32723-67-4	: 586	35323-91-2	: 2846	38514-02-2	: 8461	41653-96-7	: 7057	51575-85-0	: 6616
32750-14-4	: 3067	35355-36-3	: 2454	38514-03-3	: 8462	41658-69-9	: 10096	51630-58-1	: 5956
32764-34-4	: 6507	35364-90-0	: 1411	38552-72-6	: 6601	41692-47-1	: 6899	51637-47-9	: 8864
32768-54-0	: 1383	35365-59-4	: 8315	38577-97-8	: 11523	41755-60-6	: 6655	51686-60-3	: 4883
32777-26-7	: 4224	35367-38-5	: 5176	38624-62-3	: 3826	41761-12-0	: 6766	51686-78-3	: 1311
32809-16-8	: 9760	35383-51-8	: 3153	38628-39-6	: 11868	41814-78-2	: 11858	51707-55-2	: 11687
32830-97-0	: 8962	35383-59-6	: 3686	38651-65-9	: 2955	41851-34-7	: 4430	51750-65-3	: 6847
32838-55-4	: 9699	35413-38-8	: 4615	38661-81-3	: 1034	41867-20-3	: 3816	51756-08-2	: 3519
32854-75-4	: 309	35432-36-1	: 10134	38661-82-4	: 1160	41901-72-8	: 1410	51760-90-8	: 3689
32889-48-8	: 9759	35449-34-4	: 250	38727-55-4	: 5173	42027-23-6	: 8788	51772-30-6	: 10467
33018-91-6	: 6396	35472-56-1	: 2680	38738-60-2	: 4676	42061-72-3	: 4609	51786-70-0	: 3243
33083-83-9	: 11961	35502-06-8	: 9852	38827-66-6	: 11517	42067-48-1	: 9051	51806-20-3	: 5543
33089-61-1	: 342	35554-44-0	: 7187	38862-78-1	: 1397	42067-49-2	: 9050	51807-53-5	: 7196
33092-82-9	: 4501	35569-54-1	: 11552	38870-89-2	: 299	42101-38-2	: 3755	51952-42-2	: 9015
33184-48-4	: 1175	35573-93-4	: 9850	38939-88-7	: 1211	42131-85-1	: 9016	52033-97-3	: 4888
33245-39-5	: 5964	35608-64-1	: 3533	39029-41-9	: 7936	42131-89-5	: 6763	52152-71-3	: 3382
33357-44-7	: 11270	35673-03-1	: 1226	39052-12-5	: 693	42185-47-7	: 4618	52152-72-4	: 8819
33401-05-7	: 10098	35691-65-7	: 8768	39052-57-8	: 2736	42235-39-2	: 5353	52182-15-7	: 919
33429-70-8	: 2327	35794-11-7	: 9652	39075-90-6	: 9245	42245-37-4	: 8706	52208-62-5	: 2056
33429-72-0	: 8725	35897-16-6	: 8948	39076-02-3	: 3962	42286-84-0	: 2834	52244-70-9	: 1105
33467-76-4	: 6567	35911-17-2	: 10546	39118-35-9	: 8527	42345-82-4	: 5899	52262-38-1	: 9281
33515-82-1	: 4233	35951-33-8	: 8723	39121-37-4	: 6519	42398-73-2	: 7384	52315-07-8	: 5012
33553-93-4	: 8754	35951-36-1	: 6382	39135-39-2	: 9625	42412-84-0	: 10321	52324-03-5	: 9147
33560-15-5	: 252	36045-56-4	: 6903	39148-24-8	: 6020	42474-20-4	: 8739	52358-73-3	: 7801
33560-17-7	: 249	36049-78-2	: 8407	39178-11-5	: 5255	42474-21-5	: 6372	52387-50-5	: 7067
33567-59-8	: 870	36065-15-3	: 2962	39191-07-6	: 1802	42551-55-3	: 3598	52390-72-4	: 6473
33622-26-3	: 5094	36069-05-3	: 12045	39196-18-4	: 11717	42569-16-4	: 9092	52414-97-8	: 1068
33629-47-9	: 3893	36076-25-2	: 1259	39228-29-0	: 10677	42576-02-3	: 2997	52418-81-2	: 6918
33662-96-3	: 10292	36190-77-9	: 5750	39251-86-0	: 6058	42597-26-2	: 2127	52460-86-3	: 892
33683-44-2	: 6972	36215-07-3	: 9860	39251-88-2	: 6067	42775-75-7	: 7890	52488-28-5	: 1069
33695-59-9	: 10109	36301-29-8	: 1995	39257-08-4	: 4614	42842-08-0	: 10045	52497-07-1	: 3743
33738-48-6	: 7756	36357-38-7	: 5855	39273-81-9	: 6599	42852-95-9	: 6215	52562-19-3	: 787
33742-81-3	: 4178	36386-49-9	: 2940	39478-78-9	: 652	42874-01-1	: 8082	52645-53-1	: 9150
33820-53-0	: 7371	36386-52-4	: 2945	39511-08-5	: 10513	42874-03-3	: 8633	52708-03-9	: 8160
33829-48-0	: 7465	36402-31-0	: 3609	39513-75-2	: 2803	43036-06-2	: 11137	52744-85-1	: 10645
33840-38-9	: 4877	36441-31-3	: 8040	39515-51-0	: 597	43121-43-3	: 512	52780-23-1	: 1506
33843-55-9	: 4700	36441-32-4	: 8039	39622-79-2	: 1330	43197-78-0	: 6746	52802-07-0	: 6917
33877-11-1	: 1834	36448-60-9	: 10472	39669-95-9	: 1755	43222-48-6	: 5175	52896-87-4	: 6427
33880-83-0	: 4389	36556-06-6	: 7389	39713-71-8	: 11151	44829-76-7	: 6595	52896-88-5	: 6413
33933-78-7	: 8171	36566-80-0	: 7122	39755-03-8	: 3667	44855-57-4	: 8351	52896-89-6	: 6414
33933-79-8	: 8455	36653-82-4	: 6659	39761-61-0	: 7027	44914-03-6	: 10652	52896-90-9	: 6412
33941-07-0	: 10771	36678-06-5	: 3167	39782-38-2	: 3371	45223-18-5	: 6635	52896-91-0	: 6411
33966-50-6	: 3210	36691-77-7	: 11888	39782-43-9	: 7026	47172-89-4	: 8776	52896-92-1	: 6437
33978-71-1	: 11957	36701-01-6	: 8880	39928-72-8	: 6141	48145-04-6	: 10681	52896-93-2	: 6441
34008-71-4	: 10436	36734-19-7	: 7338	39966-95-5	: 3427	49542-66-7	: 5678	52896-95-4	: 6442
34014-18-1	: 11590	36768-62-4	: 9628	39989-39-4	: 7380	49763-65-7	: 2907	52896-99-8	: 6821
34069-94-8	: 293	36823-84-4	: 2908	40058-87-5	: 10223	49845-33-2	: 11054	52897-00-4	: 6817
34103-97-4	: 4876	36825-36-2	: 11207	40203-74-5	: 156	49852-35-9	: 6597	52897-01-5	: 6822
34238-52-3	: 9022	36847-51-5	: 3499	40225-75-0	: 8468	50375-10-5	: 2330	52897-03-7	: 6823
34238-81-8	: 5136	36854-57-6	: 974	40239-35-8	: 11892	50402-72-7	: 9713	52897-04-8	: 6819
34240-76-1	: 5739	36895-62-2	: 11825	40242-15-7	: 7847	50407-18-6	: 7238	52897-05-9	: 6824
34256-82-1	: 270	36903-89-6	: 8363	40248-63-3	: 210	50461-74-0	: 8905	52897-06-0	: 6820
34263-68-8	: 6729	36903-92-1	: 8521	40248-84-8	: 9377	50468-22-9	: 3352	52897-08-2	: 6845
34314-06-2	: 3036	36903-95-4	: 8329	40323-88-4	: 11762	50471-44-8	: 12048	52897-09-3	: 6846
34316-15-9	: 2433	36917-36-9	: 10961	40385-54-4	: 1073	50552-30-2	: 7101	52897-10-6	: 6850
34451-18-8	: 10268	36994-79-3	: 2352	40441-35-8	: 4623	50592-87-5	: 6736	52897-11-7	: 6851
34570-59-7	: 10136	37025-57-3	: 1871	40466-95-3	: 9235	50594-66-6	: 304	52897-12-8	: 6852
34619-03-9	: 3994	37050-06-9	: 8537	40487-42-1	: 8635	50599-73-0	: 6989	52897-15-1	: 6853
34757-14-7	: 10237	37273-91-9	: 7440	40514-70-3	: 4393	50623-57-9	: 8147	52897-16-2	: 8757
34786-24-8	: 1706	37434-59-6	: 9976	40527-16-0	: 8582	50670-64-9	: 2743	52897-17-3	: 8815
		37470-46-5	: 2623	40571-45-7	: 4772	50689-17-3	: 6074	52897-19-5	: 8816

52914-66-6	: 4762	55130-16-0	: 8309	57968-71-5	: 3307	61828-19-1	: 4479	66235-62-9	: 3608
52918-63-5	: 5102	55156-16-6	: 8980	58138-08-2	: 11899	61828-21-5	: 4413	66295-45-2	: 9549
53022-14-3	: 3565	55182-83-7	: 11988	58138-81-1	: 3669	61828-25-9	: 1742	66395-23-1	: 4585
53045-71-9	: 9009	55268-62-7	: 1567	58167-01-4	: 3305	61828-26-0	: 2065	66398-52-5	: 3466
53068-88-5	: 6902	55268-63-8	: 2267	58190-94-6	: 3491	61838-78-6	: 10464	66577-03-5	: 11763
53078-85-6	: 648	55282-10-5	: 6296	58210-03-0	: 4568	61847-78-7	: 7030	66577-17-1	: 7924
53109-18-5	: 4933	55282-12-7	: 8241	58371-98-5	: 8963	61847-80-1	: 9048	66610-39-7	: 3392
53152-98-0	: 9681	55282-14-9	: 5271	58474-80-9	: 5081	61847-97-0	: 8314	66719-29-7	: 8759
53172-84-2	: 1950	55282-15-0	: 5270	58588-28-6	: 6919	61886-67-7	: 7932	66719-37-7	: 6488
53175-28-3	: 3697	55282-16-1	: 5269	58593-78-5	: 1563	61892-85-1	: 6808	66719-40-2	: 6489
53178-20-4	: 3242	55282-17-2	: 11605	58616-95-8	: 4970	61949-83-5	: 7243	66719-53-7	: 8456
53188-23-1	: 6023	55283-68-6	: 5409	58618-39-6	: 7806	61971-91-3	: 3670	66719-54-8	: 8457
53220-82-9	: 7741	55285-14-8	: 4038	58625-77-7	: 3655	62016-13-1	: 6839	66719-55-9	: 8452
53252-08-7	: 6591	55289-35-5	: 1070	58766-17-9	: 6745	62016-38-0	: 8421	66731-94-0	: 6490
53252-19-0	: 7096	55290-64-7	: 5179	58873-44-2	: 9182	62016-48-2	: 3394	66793-72-4	: 6956
53252-20-3	: 8192	55319-83-0	: 5275	59020-74-5	: 2447	62016-53-9	: 4842	66793-74-6	: 6957
53252-21-4	: 7092	55320-06-4	: 6295	59022-71-8	: 7330	62016-55-1	: 4808	66793-80-4	: 8467
53346-71-7	: 8648	55334-08-2	: 4476	59113-36-9	: 10024	62016-58-4	: 4821	66793-92-8	: 6955
53381-90-1	: 9725	55335-06-3	: 11854	59204-02-3	: 8449	62059-56-7	: 4351	66794-02-3	: 6933
53440-12-3	: 7778	55373-86-9	: 5274	59255-95-7	: 653	62067-44-1	: 4527	66794-03-4	: 6934
53491-32-0	: 2441	55429-84-0	: 11604	59357-07-2	: 6974	62080-86-8	: 8274	66794-04-5	: 6932
53535-68-5	: 10417	55449-46-2	: 1234	59368-15-9	: 4443	62088-36-2	: 4663	66810-87-5	: 6958
53566-37-3	: 7121	55454-22-3	: 6568	59376-54-4	: 184	62108-06-9	: 8517	66898-39-3	: 2820
53584-24-5	: 10009	55505-26-5	: 8168	59433-08-8	: 9669	62108-66-1	: 10232	66997-45-3	: 10146
53588-92-4	: 5464	55517-73-2	: 6623	59434-05-8	: 11758	62108-68-3	: 10227	67191-37-1	: 1665
53608-85-8	: 4322	55517-89-0	: 8639	59654-13-6	: 6832	62108-70-7	: 10226	67329-04-8	: 7340
53608-93-8	: 9935	55563-72-9	: 10458	59669-26-0	: 11713	62127-46-2	: 3446	67451-76-7	: 4312
53660-21-2	: 6478	55665-79-7	: 7060	59697-51-7	: 7011	62127-47-3	: 3447	67485-29-4	: 7136
53663-23-3	: 2353	55696-02-1	: 3944	59697-55-1	: 7012	62127-49-5	: 10135	67535-07-3	: 10255
53663-26-6	: 8148	55702-49-3	: 3894	59734-23-5	: 8470	62127-61-1	: 10160	67590-77-6	: 1760
53776-69-5	: 2081	55751-54-7	: 812	59756-60-4	: 5981	62154-77-2	: 2297	67632-66-0	: 9148
53778-43-1	: 4975	55775-41-2	: 3182	59777-92-3	: 4231	62155-41-3	: 2111	67700-22-5	: 2159
53780-34-0	: 7427	55789-02-7	: 1872	59889-45-1	: 8722	62168-25-6	: 3264	67700-26-9	: 8512
53859-78-2	: 4213	55877-01-5	: 6790	59905-55-4	: 2983	62184-67-2	: 11766	67747-70-0	: 9021
53878-96-9	: 3903	55882-21-8	: 10153	59956-48-8	: 118	62184-82-1	: 4919	67885-08-9	: 3166
53889-94-4	: 9428	55944-70-2	: 6477	60012-63-7	: 2102	62184-83-2	: 4938	67952-57-2	: 6506
53894-19-2	: 10361	56072-27-6	: 6518	60096-09-5	: 6398	62261-05-6	: 3957	68057-83-0	: 3532
53966-53-3	: 8186	56103-67-4	: 9225	60111-14-0	: 7082	62353-69-9	: 8685	68077-28-1	: 10601
53977-99-4	: 7048	56106-37-7	: 1556	60168-88-9	: 5948	62476-56-6	: 10862	68107-05-1	: 6724
54105-76-9	: 4427	56180-61-1	: 8688	60207-90-1	: 10720	62496-53-1	: 8753	68165-06-0	: 11187
54106-40-0	: 2005	56216-12-7	: 3823	60212-29-5	: 11985	62521-69-1	: 3453	68266-67-1	: 1360
54131-44-1	: 2927	56245-02-4	: 2034	60212-31-9	: 11986	62624-30-0	: 465	68359-37-5	: 5008
54244-90-5	: 6590	56456-42-9	: 1496	60251-57-2	: 8357	62690-62-4	: 7897	68384-70-3	: 3541
54268-02-9	: 10138	56523-26-3	: 8322	60291-32-9	: 4408	62692-40-4	: 1001	68420-93-9	: 11212
54298-97-4	: 8525	56630-31-0	: 8206	60468-23-7	: 9007	62706-16-5	: 7008	68684-11-1	: 11900
54305-87-2	: 6764	56630-33-2	: 8204	60512-85-8	: 10687	62882-00-2	: 7385	68723-25-1	: 7629
54340-87-3	: 7262	56630-34-3	: 8205	60669-40-1	: 11967	63019-98-7	: 3014	68750-16-3	: 5677
54383-22-1	: 4373	56688-75-6	: 4671	60671-36-5	: 11956	63157-81-3	: 10194	69112-21-6	: 6997
54410-69-4	: 2010	56694-97-4	: 11604	60690-31-9	: 6157	63204-71-9	: 6219	69190-05-4	: 3011
54411-01-7	: 4446	56694-98-5	: 11863	60703-31-3	: 7075	63370-69-4	: 4776	69248-35-7	: 3332
54429-56-0	: 5628	56739-96-8	: 1781	60710-39-6	: 9239	63468-05-3	: 9097	69295-21-2	: 3729
54454-10-3	: 10820	56745-61-0	: 6969	60761-04-8	: 8561	63790-14-7	: 1086	69305-42-6	: 9496
54460-46-7	: 4952	56800-09-0	: 3746	60814-29-1	: 11335	63927-44-6	: 6877	69321-60-4	: 1304
54460-96-7	: 10121	56908-88-4	: 1291	60836-07-9	: 6509	64057-79-0	: 10703	69371-59-1	: 7102
54462-68-9	: 6759	56911-77-4	: 2074	60899-39-0	: 7880	64161-50-8	: 2951	69377-81-7	: 5982
54462-70-3	: 3270	56920-82-2	: 6037	60956-23-2	: 1303	64165-34-0	: 11317	69498-28-8	: 3273
54467-95-7	: 979	56961-23-0	: 2371	60956-24-3	: 1293	64168-11-2	: 9702	69542-91-2	: 4514
54484-63-8	: 1377	56961-24-1	: 2320	60956-26-5	: 1074	64207-03-0	: 10447	69564-68-7	: 2881
54527-68-3	: 3570	56961-32-1	: 2541	61050-68-8	: 8397	64257-84-7	: 5953	69668-88-8	: 6588
54553-14-9	: 11155	56961-77-4	: 1010	61228-10-2	: 6619	64287-26-9	: 1285	69704-44-5	: 3262
54573-13-6	: 6620	56961-82-1	: 2326	61231-45-6	: 7717	64334-96-9	: 11211	69745-49-9	: 4692
54630-50-1	: 6497	56961-84-3	: 1359	61531-45-1	: 8156	64339-43-1	: 1524	69770-98-5	: 4891
54638-10-7	: 8918	56961-86-5	: 2337	61539-75-1	: 8370	64346-45-8	: 3807	69806-50-4	: 5962
54644-23-4	: 1231	56961-93-4	: 7851	61551-49-3	: 7978	64379-30-2	: 8704	69813-63-4	: 1970
54676-39-0	: 4327	56962-04-0	: 9231	61563-25-5	: 1302	64501-77-5	: 8508	69833-18-7	: 469
54715-57-0	: 9259	56962-07-3	: 7343	61589-68-2	: 8991	64502-89-2	: 8971	69947-01-9	: 9805
54737-45-0	: 1080	57074-96-1	: 8539	61592-89-0	: 1962	64531-26-6	: 8533	70110-24-6	: 10496
54774-99-1	: 10249	57129-69-8	: 6146	61675-14-7	: 7098	64632-72-0	: 12052	70116-68-6	: 3629
54819-86-2	: 10217	57150-86-0	: 9553	61764-94-1	: 6370	64653-59-4	: 744	70124-77-5	: 5965
54829-48-0	: 6110	57179-35-8	: 573	61792-07-2	: 260	64902-72-3	: 4091	70127-93-4	: 7237
54833-31-7	: 4464	57223-18-4	: 8193	61827-85-8	: 1487	64930-16-1	: 3741	70202-92-5	: 4915
54833-34-0	: 4402	57296-00-1	: 10613	61827-86-9	: 1708	64965-91-9	: 2453	70396-13-3	: 6615
54837-90-0	: 815	57323-93-0	: 6988	61827-87-0	: 1711	65144-02-7	: 9797	70401-56-8	: 2779
54852-73-2	: 3071	57386-83-1	: 1402	61827-89-2	: 5286	65248-43-3	: 6985	70644-46-1	: 10922
54862-92-9	: 6517	57392-55-9	: 255	61827-93-8	: 4853	65423-25-8	: 5340	70690-24-3	: 8367
54885-00-6	: 7640	57403-35-7	: 1135	61827-98-3	: 4845	65438-35-9	: 13	70715-19-4	: 10983
54889-55-3	: 7274	57404-54-3	: 9176	61828-04-4	: 2341	65469-88-7	: 3357	70732-43-3	: 8202
54890-00-5	: 4326	57456-98-1	: 3185	61828-05-5	: 2265	65530-93-0	: 6728	70732-44-4	: 8341
54922-86-0	: 10032	57502-49-5	: 9666	61828-06-6	: 2088	65732-90-3	: 11518	70732-45-5	: 5101
54934-90-6	: 4432	57526-28-0	: 3694	61828-07-7	: 4380	65786-09-6	: 6380	70786-82-2	: 8867
54934-91-7	: 4472	57576-09-7	: 4545	61828-08-8	: 4483	65786-11-0	: 6539	71159-31-4	: 10378
55030-56-3	: 3739	57605-95-5	: 7911	61828-09-9	: 4474	65927-60-8	: 6954	71172-57-1	: 8970
55038-01-2	: 2455	57706-88-4	: 8454	61828-10-2	: 4462	66050-98-4	: 6938	71173-15-4	: 2939
55057-45-9	: 10494	57772-57-3	: 2624	61828-11-3	: 4410	66056-06-2	: 19	71200-79-8	: 3763
55066-53-0	: 8305	57837-19-1	: 7439	61828-13-5	: 4458	66225-16-9	: 6552	71284-96-3	: 8025
55073-86-4	: 6950	57885-43-5	: 10211	61828-15-7	: 4482	66225-17-0	: 6551	71297-97-7	: 5118
55124-79-3	: 6309	57933-83-2	: 4022	61828-17-9	: 4383	66225-35-2	: 6803	71605-85-1	: 8152

71608-10-1 : 9334	77753-21-0 : 3805	89896-73-1 : 6531	99129-21-2 : 4138	112456-46-9 : 2934	
71626-11-4 : 513	77753-24-3 : 3419	90112-26-8 : 4898	99172-57-3 : 6076	112489-85-7 : 6096	
71633-43-7 : 11303	77825-53-7 : 3824	90112-75-7 : 6565	99365-26-1 : 2801	113426-22-5 : 8506	
71745-63-6 : 9340	77825-54-8 : 3825	90112-82-6 : 6892	99706-68-0 : 11875	113666-64-1 : 11951	
71747-37-0 : 2152	77866-58-1 : 4560	90162-24-6 : 8390	99814-65-0 : 8655	114180-21-1 : 3251	
71751-41-2 : 477	77928-86-0 : 5039	90200-61-6 : 4232	99980-00-4 : 4152	114369-43-6 : 5949	
71783-56-7 : 7391	77958-21-5 : 6604	90200-83-2 : 8513	100121-73-1 : 6038	114489-96-2 : 10224	
72178-02-0 : 5985	78277-23-3 : 10690	90202-08-7 : 6878	100130-25-4 : 43	115113-98-9 : 6569	
72312-48-2 : 4687	78738-36-0 : 10368	90202-21-4 : 8499	100224-75-7 : 7627	115310-98-0 : 11318	
72622-74-3 : 10342	78738-37-1 : 9668	90210-56-3 : 7387	100295-82-7 : 6945	116435-30-4 : 2931	
72762-00-6 : 11031	78938-11-1 : 11448	90226-23-6 : 4723	100295-83-8 : 6947	116435-95-1 : 10257	
72935-12-7 : 3049	79127-80-3 : 5952	90357-58-7 : 199	100296-26-2 : 6499	116529-73-8 : 3906	
73120-52-2 : 5298	79241-46-6 : 5963	90499-41-5 : 1611	100314-90-7 : 6135	116530-63-3 : 10305	
73178-43-5 : 9042	79504-01-1 : 3234	90638-48-5 : 10651	100319-48-0 : 8339	116530-99-5 : 4788	
73206-57-2 : 10448	79630-70-9 : 3428	90644-60-3 : 8195	100367-45-1 : 6697	116632-47-4 : 11174	
73210-25-0 : 4565	79762-76-8 : 10243	90645-54-8 : 4621	100791-95-5 : 11189	116668-34-9 : 2450	
73383-24-1 : 3738	79912-55-3 : 8872	90645-61-7 : 4884	100983-12-8 : 10210	116668-40-7 : 5096	
73398-15-9 : 6694	79912-57-5 : 8870	90676-55-4 : 8520	101144-90-5 : 1667	116668-44-1 : 6577	
73454-83-8 : 6690	80204-19-9 : 9012	90722-14-8 : 8200	101257-63-0 : 6743	116668-45-2 : 6576	
73513-50-5 : 7058	80325-37-7 : 6541	90760-75-1 : 4564	101257-71-0 : 9712	116668-48-5 : 8323	
73522-42-6 : 2928	80656-12-8 : 4792	90769-70-3 : 4927	101257-79-8 : 4195	116668-49-6 : 8326	
73548-71-7 : 7025	80832-54-8 : 556	90841-14-8 : 3719	101427-54-7 : 11609	116668-50-9 : 8501	
73548-72-8 : 6553	80864-10-4 : 8796	90925-48-7 : 3356	101567-36-6 : 4336	116697-35-9 : 4524	
73557-60-5 : 2310	81177-02-8 : 6789	90952-10-6 : 6920	101654-30-2 : 6741	116723-89-8 : 5117	
73642-91-8 : 8127	81280-12-8 : 6579	90982-32-4 : 4081	101715-60-0 : 8331	116723-90-1 : 5119	
73741-61-4 : 4701	81334-34-1 : 7188	91049-43-3 : 8790	101870-06-8 : 6688	116723-93-4 : 3863	
73758-54-0 : 3486	81335-37-7 : 7189	91057-79-3 : 4184	101935-28-8 : 6333	116723-94-5 : 3501	
73918-56-6 : 1570	81335-77-5 : 7190	91465-08-6 : 5010	101935-40-4 : 9243	116723-95-6 : 3536	
73931-44-9 : 2312	81726-82-1 : 6308	91572-15-5 : 5168	102153-88-8 : 4831	116723-96-7 : 3559	
74051-80-2 : 11359	81777-89-1 : 4140	91840-99-2 : 3438	102370-18-3 : 4526	116723-97-8 : 3591	
74112-36-0 : 8447	81925-81-7 : 6596	91890-87-8 : 3843	102449-95-6 : 6747	116724-10-8 : 3219	
74115-24-5 : 4139	82097-50-5 : 11824	91913-99-4 : 3490	102851-06-9 : 5983	116724-11-9 : 3389	
74204-00-5 : 9240	82322-93-8 : 4895	91965-23-0 : 3131	102871-67-0 : 11304	116724-15-3 : 4355	
74219-20-8 : 2949	82507-04-8 : 7010	92040-00-1 : 7409	103095-51-8 : 10813	116724-17-5 : 4567	
74221-86-6 : 4444	82584-73-4 : 8600	92687-41-7 : 4644	103275-78-1 : 2219	116724-18-6 : 4693	
74244-64-7 : 4217	82657-04-3 : 2998	92737-91-2 : 6504	103697-14-9 : 8230	116779-77-2 : 3424	
74283-48-0 : 3610	82906-78-3 : 9555	93713-40-7 : 10187	103985-40-6 : 4665	116779-78-3 : 3425	
74346-30-8 : 1152	83055-99-6 : 516	93921-42-7 : 4583	104035-66-7 : 6593	116781-85-2 : 3413	
74421-05-9 : 6332	83682-72-8 : 10360	93972-93-1 : 6869	104068-42-0 : 4787	116781-86-3 : 4272	
74513-58-9 : 10693	83863-33-6 : 763	94347-45-2 : 8863	104177-72-2 : 362	116783-11-0 : 1107	
74581-94-5 : 6886	84087-01-4 : 11201	94434-64-7 : 8262	104514-49-0 : 1292	116783-12-1 : 1605	
74683-66-2 : 8158	84098-45-3 : 10277	94826-08-1 : 4144	105065-24-5 : 4679	116783-21-2 : 2020	
74742-08-8 : 8159	84258-49-1 : 7332	94942-89-9 : 2143	105401-50-9 : 7175	116783-23-4 : 3393	
74742-10-2 : 5071	84583-53-9 : 11320	95642-49-2 : 8830	105688-74-2 : 11271	116783-28-9 : 4506	
74744-28-8 : 11983	85733-97-7 : 8936	95835-77-1 : 2460	105728-23-2 : 11247	116783-29-0 : 4584	
74744-36-8 : 5342	86051-37-8 : 8155	95970-07-3 : 2256	107494-37-1 : 11952	116836-06-7 : 7163	
74764-56-0 : 6594	86073-38-3 : 8264	95970-10-8 : 2311	107534-96-3 : 11589	116836-10-3 : 4321	
74778-22-6 : 8446	86623-80-5 : 9081	96034-00-3 : 3435	107574-57-2 : 10465	116836-12-5 : 8413	
74925-48-7 : 7249	86661-53-2 : 8727	96097-69-7 : 11818	107770-92-3 : 11570	116836-13-6 : 8522	
75055-73-1 : 3365	86668-33-9 : 8507	96244-13-2 : 4563	108365-83-9 : 3730	116836-16-9 : 8921	
75455-41-3 : 3867	87018-30-2 : 3613	96836-97-4 : 4350	108438-40-0 : 7943	116836-19-2 : 3583	
75579-88-3 : 6345	87237-48-7 : 6291	97073-23-9 : 10484	108836-86-8 : 8650	116836-20-5 : 7248	
76078-85-8 : 8135	87541-87-5 : 10623	97412-23-2 : 9394	108965-84-0 : 5028	116836-21-6 : 9647	
76149-14-9 : 1523	88472-61-1 : 8269	97479-78-2 : 3337	108965-86-2 : 5029	116836-30-7 : 11976	
76429-68-0 : 2444	88947-16-4 : 10403	97886-45-8 : 5267	110318-09-7 : 6699	116836-32-9 : 8899	
76469-08-4 : 8540	89033-70-5 : 4179	97961-66-5 : 7055	110383-31-8 : 7080	116836-55-6 : 3564	
76472-83-8 : 3013	89123-63-7 : 10612	98070-91-8 : 3454	110795-27-2 : 3938	116862-65-8 : 4769	
76619-89-1 : 9212	89223-57-4 : 4829	98203-04-4 : 11221	111768-02-6 : 8927	116908-78-2 : 9063	
77053-92-0 : 7078	89300-69-6 : 10253	98272-34-5 : 6112	111768-04-8 : 6937	116908-83-9 : 8877	
77295-79-5 : 5893	89533-67-5 : 7113	98431-97-1 : 11567	111768-05-9 : 6940	116908-84-0 : 8993	
77392-54-2 : 3638	89583-12-0 : 6760	98551-14-5 : 4149	111768-08-2 : 6939	116908-85-1 : 8994	
77437-98-0 : 6582	89583-61-9 : 9896	98790-22-8 : 4561	111772-90-8 : 11873	119446-68-3 : 5174	
77464-06-3 : 11996	89796-76-9 : 6381	98886-44-3 : 6021	111823-35-9 : 8765	127665-32-1 : 6258	
77501-63-4 : 7410	89886-25-9 : 4523	98930-89-3 : 11887	111897-18-8 : 8279		
77732-09-3 : 8566	89895-45-4 : 4445	98991-54-9 : 6570	111991-09-4 : 8069		

Section 4
Properties of the Elements and Inorganic Compounds

THE ELEMENTS
C. R. Hammond

One of the most striking facts about the elements is their unequal distribution and occurrence in nature. Present knowledge of the chemical composition of the universe, obtained from the study of the spectra of stars and nebulae, indicates that hydrogen is by far the most abundant element and may account for more than 90% of the atoms or about 75% of the mass of the universe. Helium atoms make up most of the remainder. All of the other elements together contribute only slightly to the total mass.

The chemical composition of the universe is undergoing continuous change. Hydrogen is being converted into helium, and helium is being changed into heavier elements. As time goes on, the ratio of heavier elements increases relative to hydrogen. Presumably, the process is not reversible.

Burbidge, Burbidge, Fowler, and Hoyle, and more recently, Peebles, Penzias, and others have studied the synthesis of elements in stars. To explain all of the features of the nuclear abundance curve — obtained by studies of the composition of the earth, meteorites, stars, etc. — it is necessary to postulate that the elements were originally formed by at least eight different processes: (1) hydrogen burning, (2) helium burning, (3) χ process, (4) e process, (5) s process, (6) r process, (7) p process, and (8) the X process. The X process is thought to account for the existence of light nuclei such as D, Li, Be, and B. Common metals such as Fe, Cr, Ni, Cu, Ti, Zn, etc. were likely produced early in the history of our galaxy. It is also probable that most of the heavy elements on earth and elsewhere in the universe were originally formed in supernovae, or in the hot interior of stars.

Studies of the solar spectrum have led to the identification of 67 elements in the sun's atmosphere; however, all elements cannot be identified with the same degree of certainty. Other elements may be present in the sun, although they have not yet been detected spectroscopically. The element helium was discovered on the sun before it was found on earth. Some elements such as scandium are relatively more plentiful in the sun and stars than here on earth.

Minerals in lunar rocks brought back from the moon on the Apollo missions consist predominantly of *plagioclase* $\{(Ca,Na)(Al,Si)O_4O_8\}$ and *pyroxene* $\{(Ca,Mg,Fe)_2Si_2O_6\}$ — two minerals common in terrestrial volcanic rock. No new elements have been found on the moon that cannot be accounted for on earth; however, three minerals, *armalcolite* $\{(Fe,Mg)Ti_2O_5\}$, *pyroxferroite* $\{CaFe_6(SiO_3)_7\}$, and *tranquillityite* $\{Fe_8(Zr,Y)Ti_3Si_3O_2\}$, are new. The oldest known terrestrial rocks are about 4 billion years old. One rock, known as the "Genesis Rock," brought back from the Apollo 15 Mission, is about 4.15 billion years old. This is only about one-half billion years younger than the supposed age of the moon and solar system. Lunar rocks appear to be relatively enriched in refractory elements such as chromium, titanium, zirconium, and the rare earths, and impoverished in volatile elements such as the alkali metals, in chlorine, and in noble metals such as nickel, platinum, and gold.

Even older than the "Genesis Rock" are *carbonaceous chondrites,* a type of meteorite that has fallen to earth and has been studied. These are some of the most primitive objects of the solar system yet found. The grains making up these objects probably condensed directly out the gaseous nebula from which the sun and planets were born. Most of the condensation of the grains probably was completed within 50,000 years of the time the disk of the nebula was first formed about 4.6 billion years ago. It is now thought that this type of meteorite may contain a small percentage of presolar dust grains. The relative abundances of the elements of these meteorites are about the same as the abundances found in the solar chromosphere.

The X-ray fluorescent spectrometer sent with the Viking I spacecraft to Mars shows that the Martian soil contains about 12 to 16% iron, 14 to 15% silicon, 3 to 8% calcium, 2 to 7% aluminum, and one-half to 2% titanium. The gas chromatograph — mass spectrometer on Viking II found no trace of organic compounds that should be present if life ever existed there.

F. W. Clarke and others have carefully studied the composition of rocks making up the crust of the earth. Oxygen accounts for about 47% of the crust, by weight, while silicon comprises about 28% and aluminum about 8%. These elements, plus iron, calcium, sodium, potassium, and magnesium, account for about 99% of the composition of the crust.

Many elements such as tin, copper, zinc, lead, mercury, silver, platinum, antimony, arsenic, and gold, which are so essential to our needs and civilization, are among some of the rarest elements in the earth's crust. These are made available to us only by the processes of concentration in ore bodies. Some of the so-called *rare-earth* elements have been found to be much more plentiful than originally thought and are about as abundant as uranium, mercury, lead, or bismuth. The least abundant rare-earth or *lanthanide* element, thulium, is now believed to be more plentiful on earth than silver, cadmium, gold, or iodine, for example. Rubidium, the 16th most abundant element, is more plentiful than chlorine while its compounds are little known in chemistry and commerce.

It is now thought that at least 24 elements are essential to living matter. The four most abundant in the human body are hydrogen, oxygen, carbon, and nitrogen. The seven next most common, in order of abundance, are calcium, phosphorus, chlorine, potassium, sulfur, sodium, and magnesium. Iron, copper, zinc, silicon, iodine, cobalt, manganese, molybdenum, fluorine, tin, chromium, selenium, and vanadium are needed and play a role in living matter. Boron is also thought essential for some plants, and it is possible that aluminum, nickel, and germanium may turn out to be necessary.

Ninety-one elements occur naturally on earth. Minute traces of plutonium-244 have been discovered in rocks mined in Southern California. This discovery supports the theory that heavy elements were produced during creation of the solar system. While technetium and promethium have not yet been found naturally on earth, they have been found to be present in stars. Technetium has been identified in the spectra of certain "late" type stars, and promethium lines have been identified in the spectra of a faintly visible star HR465 in Andromeda. Promethium must have been made very recently near the star's surface for no known isotope of this element has a half-life longer than 17.7 years.

It has been suggested that californium is present in certain stellar explosions known as supernovae; however, this has not been proved. At present no elements are found elsewhere in the universe that cannot be accounted for here on earth.

All atomic mass numbers from 1 to 238 are found naturally on earth except for masses 5 and 8. About 285 relatively stable and 67 naturally radioactive isotopes occur on earth totaling 352. In addition, the neutron, technetium, promethium, and the transuranic elements (lying beyond uranium) up to Element 112 have been produced artificially. Scientists at G.S.I., Darmstadt, Germany, in February, 1996, reported they had found evidence of an isotope of Element 112. Laboratory processes have now extended the radioactive element mass numbers beyond 238 to 277. Each element from atomic numbers 1 to 108 is known to have at least one radioactive isotope. As of January 1996, 3018 isotopes and isomers were thought to be known and recognized. Many stable and radioactive isotopes are now produced and distributed by the Oak Ridge National Laboratory, Oak Ridge, Tenn., U.S.A., to customers licensed by the U.S. Department of Energy.

The nucleus of an atom is characterized by the number of protons it contains, denoted by Z, and by the number of neutrons, N. Isotopes of an element have the same value of Z, but different values of N. The *mass number A,* is the sum of Z and N. For example, Uranium-238 has a mass number of 238, and contains 92 protons and 146 neutrons.

THE ELEMENTS (continued)

There is evidence that the definition of chemical elements must be broadened to include the electron. Several compounds known as *electrides,* have recently been made of alkaline metal elements and electrons. A relatively stable combination of a positron and electron, known as *positronium,* has also been studied.

In addition to the proton, neutron, and electron, there are considerably more than 100 other fundamental particles which have been discovered or hypothesized. The majority of these fall into one of two classes, *leptons* or *hadrons.* The leptons comprise just four known particles, the *electron,* the *muon* (μ meson), and two kinds of *neutrinos.* The muon is essentially similar to the electron and has a charge of –1, but it is 200 times heavier. The neutrino is either of two stable particles of small (probably zero) rest mass, carrying no charge. Also there are four *antileptons,* identical to the corresponding leptons in some respects, such as mass, but they have properties exactly opposite those of the leptons. The *positron,* for example, is an antilepton, with a charge of +1. Leptons cannot be broken into smaller units and are considered to be elementary. On the other hand, hadrons are complex and thought to have internal structure. Protons and neutrons, which make up atomic nuclei, are hadrons.

Elementary particle physics is not yet clearly understood, but groupings and arrangements of these particles have been made resembling the periodic table of chemical elements. This has led to the speculation that hadrons are composed of simpler components called *quarks.* Quarks are presumed to be elementary particles. Quarks come in a variety of "flavors". Two kinds of quarks — up and down — combine to form the protons and neutrons of ordinary matter. Three other quarks, known as {strange", "charm", and "bottom", are found in particle accelerators, and one of these may exist in the center of very dense, massive stars. Theorists have thought that a sixth "top" quark, much more massive than the others, should exist to complete the quark family of sub-atomic particles. In 1994, scientists using Fermilab's Tevatron accelerator may have "glimpsed" to top quark. There is presently no evidence that quarks exist in isolation. Many physicists now hold that all the matter and energy in the universe is controlled by four fundamental natural forces: the electromagnetic force, gravity, a weak nuclear force, and a strong nuclear force. Each of these natural forces is passed back and forth among the basic particles of matter by unique force-carrying particles. The electromagnetic force is carried by the *photon,* the weak nuclear force by the intermediate vector *boson,* and gravity by the *gravitron.* There is now evidence of the existence of a particle, known as the *gluon,* that binds quarks together by carrying the strong nuclear force. *Hadrons* are thought to be particles composed of quarks and gluons.

The available evidence leads to the conclusion that elements 89 (actinium) through 103 (lawrencium) are chemically similar to the rare-earth or lanthanide elements (elements 57 to 71, inclusive). These elements therefore have been named *actinides* after the first member of this series. Those elements beyond uranium that have been produced artificially have the following names and symbols: neptunium, 93 (Np); plutonium, 94 (Pu); americium, 95 (Am); curium, 96 (Cm); berkelium, 97 (Bk); californium, 98 (Cf); einsteinium, 99 (Es); fermium, 100 (Fm); mendelevium, 101 (Md); nobelium, 102 (No); and lawrencium, 103 (Lr). It is now claimed that Elements 104 through 112 have been produced and identified. In late 1994 the International Union of Pure and Applied Chemistry (IUPAC) announced their recommendations for naming Elements 101 to 109 and their symbols as follows: Element 101, Mendelevium (Md); 102, Nobelium (No); 103, Lawrencium (Lr); 104, Dubnium (Db); 105, Joliotium (Jl); 106, Rutherfordium (Rf); 107, Bohrium (Bh); 108, Hahnium (Hn); 109, Meitnerium (Mt). However, the U.S. Berkeley team that discovered 105 suggested the name Hahnium (Ha) for that element and proposed the name Seaborgium (Sg) for 106, named after Glenn T. Seaborg, codiscoverer of many of the transuranic elements. This team also proposed the name Rutherfordium for 104. The Joint Institute for Nuclear Research in Dubna had competing claims for the discovery of 104 and 105. They proposed the names Kurchatovium (Ku) for 104 and Joliotium for 105. Recently the G.S.I. in Darmstadt, proposed the name Hassium (after Hassia, the Latin name of the German State where the G.S.I. is located) for Element 108. Earlier, the IUPAC suggested the temporary names of unnilquadium for Element 104; unnilpentium for 105; unnilhexium for 106; unnilseptium for 107; unniloctium for 108; unnilennium for 109; ununnilium for 110; unununium for 111; and ununbium for 112. The matter of naming 103 to 112 with their symbols is yet to be settled.

Element 104 is expected to have chemical properties similar to those of hafnium and would not be a member of the actinide series. Element 105 probably would have chemical properties similar to those of tantalum, Element 106 similar to tungsten, Element 107 similar to rhenium, and Element 109 similar to iridium, etc.

There is still thought to be a possibility of producing elements beyond Element 112 either by bombardment of heavy isotopic targets with heavy ions, or by the irradiation of uranium or transuranic elements with the instantaneous high flux of neutrons produced by underground nuclear explosions. The limit will be set by the yields of the nuclear reactions and by the half-lives of radioactive decay. It has been suggested that Elements 102 and 103 have abnormally short lives only because they are in a pocket of instability, and that this region of instability might begin to "heal" around Element 105. If so, it may be possible to produce heavier isotopes with longer half-lives. It has also been suggested that Element 114, with a mass number of 298, and Element 126, with a mass number of 310, may be sufficiently stable to make discovery and identification possible. Searches have been made by workers in a number of laboratories for Element 110 and its neighboring elements in naturally occurring platinum. Recent studies of the xenon component (^{131}Xe to ^{136}Xe) of certain carbonaceous chronditic meteorites suggest that Elements 113, 114, or 115 may have been its progenitor.

There are many claims in the literature of the existence of various allotropic modifications of the elements, some of which are based on doubtful or incomplete evidence. Also, the physical properties of an element may change drastically by the presence of small amounts of impurities. With new methods of purification, which are now able to produce elements with 99.9999% purity, it has been necessary to restudy the properties of the elements. For example, the melting point of thorium changes by several hundred degrees by the presence of a small percentage of ThO_2 as an impurity. Ordinary commercial tungsten is brittle and can be worked only with difficulty. Pure tungsten, however, can be cut with a hacksaw, forged, spun, drawn, or extruded. In general, the value of a physical property given here applies to the pure element, when it is known.

Many of the chemical elements and their compounds are toxic and should be handled with due respect and care. In recent years there has been a greatly increased knowledge and awareness of the health hazards associated with chemicals, radioactive materials, and other agents. Anyone working with the elements and certain of their compounds should become thoroughly familiar with the proper safeguards to be taken. Information on specific hazards and recommended exposure limits may be found in Section 16. Reference should be made to publications such as the following:

1. *Code of Federal Regulations,* Title 29, Labor, chapter XVII, section 1910.93 of subpart G, redesignated as 1910.1000 at 40 FR (Federal Register) 23072. May 28, 1975; amended at 41 FR 11505, March 19, 1976; 41 FR 35184, August 20, 1976; FR 46784, October 22, 1976; 42 FR 3304, January 18, 1977 (corrections and additional amendments and corrections as issued, U.S. Government Printing Office, Supt. of Documents, Washington, D.C.

2. *Code of Federal Regulations,* Title 10, Energy, Chapter 1, Nuclear Regulatory Commission, section 20.103 — 108; 20.201 — 207; 20.301 — 305; 20.401 — 409; 20.501 — 502; 20.601; appendices, corrections, and amendments.

3. *Occupational Safety and Health Reporter* (latest edition with amendments and corrections), Bureau of National Affairs, Washington, D.C.

4. *Atomic Energy Law Reporter,* Commerce Clearing House, Chicago, IL.

5. *Nuclear Regulation Reporter,* Commerce Clearing House, Chicago, IL.

6. *TLVs® Threshold Limit Values for Chemical Substances and Physical Agents* is issued annually be the American Conference of Governmental Industrial Hygienists, Cincinnati, Ohio.

7. *The Sigma-Aldrich Library of Regulatory and Safety Data*, Vol. 3 by Robert E. Lenga and Kristine L. Volonpal, published by the Sigma Chemical Co. and The Aldrich Chemical Co., Inc., 1993, (pp. 3211—3254).

Actinium — (Gr. *aktis, aktinos,* beam or ray), Ac; at. wt. (227); at. no. 89; m.p. 1051°C, b.p. 3200 ± 300°C (est.); sp. gr. 10.07 (calc.). Discovered by Andre Debierne in 1899 and independently by F. Giesel in 1902. Occurs naturally in association with uranium minerals. Thirty two isotopes and isomers are now recognized. All are radioactive. Actinium-227, a decay product of uranium-235, is an alpha and beta emitter with a 21.77-year half-life. Its principal decay products are thorium-227 (18.72-day half-life), radium-223 (11.4-day half-life), and a number of short-lived products including radon, bismuth, polonium, and lead isotopes. In equilibrium with its decay products, it is a powerful source of alpha rays. Actinium metal has been prepared by the reduction of actinium fluoride with lithium vapor at about 1100 to 1300°C. The chemical behavior of actinium is similar to that of the rare earths, particularly lanthanum. Purified actinium comes into equilibrium with its decay products at the end of 185 days, and then decays according to its 21.77-year half-life. It is about 150 times as active as radium, making it of value in the production of neutrons.

Aluminum — (L. *alumen, alum*), Al; at. wt. 26.981539(5); at. no. 13; f.p. 660.323°C; b.p. 2519°C; sp. gr. 2.6989 (20°C); valence 3. The ancient Greeks and Romans used *alum* in medicine as an astringent, and as a mordant in dyeing. In 1761 de Morveau proposed the name *alumine* for the base in alum, and Lavoisier, in 1787, thought this to be the oxide of a still undiscovered metal. Wohler is generally credited with having isolated the metal in 1827, although an impure form was prepared by Oersted two years earlier. In 1807, Davy proposed the name *alumium* for the metal, undiscovered at that time, and later agreed to change it to *aluminum*. Shortly thereafter, the name *aluminium* was adopted to conform with the "ium" ending of most elements, and this spelling is now in use elsewhere in the world. *Aluminium* was also the accepted spelling in the U.S. until 1925, at which time the American Chemical Society officially decided to use the name *aluminum* thereafter in their publications. The method of obtaining aluminum metal by the electrolysis of alumina dissolved in *cryolite* was discovered in 1886 by Hall in the U.S. and at about the same time by Heroult in France. Cryolite, a natural ore found in Greenland, is no longer widely used in commercial production, but has been replaced by an artificial mixture of sodium, aluminum, and calcium fluorides. *Bauxite,* an impure hydrated oxide ore, is found in large deposits in Jamaica, Australia, Surinam, Guyana, Arkansas, and elsewhere. The Bayer process is most commonly used today to refine bauxite so it can be accommodated in the Hall-Heroult refining process, used to make most aluminum. Aluminum can now be produced from clay, but the process is not economically feasible at present. Aluminum is the most abundant metal to be found in the earth's crust (8.1%), but is never found free in nature. In addition to the minerals mentioned above, it is found in feldspars, granite, and in many other common minerals. Seventeen isotopes and isomers are known. Natural aluminum is made of one isotope, ^{27}Al. Pure aluminum, a silvery-white metal, possesses many desirable characteristics. It is light, nontoxic, has a pleasing appearance, can easily be formed, machined, or cast, has a high thermal conductivity, and has excellent corrosion resistance. It is nonmagnetic and nonsparking, stands second among metals in the scale of malleability, and sixth in ductility. It is extensively used for kitchen utensils, outside building decoration, and in thousands of industrial applications where a strong, light, easily constructed material is needed. Although its electrical conductivity is only about 60% that of copper, it is used in electrical transmission lines because of its light weight. Pure aluminum is soft and lacks strength, but it can be alloyed with small amounts of copper, magnesium, silicon, manganese, and other elements to impart a variety of useful properties. These alloys are of vital importance in the construction of modern aircraft and rockets. Aluminum, evaporated in a vacuum, forms a highly reflective coating for both visible light and radiant heat. These coatings soon form a thin layer of the protective oxide and do not deteriorate as do silver coatings. They have found application in coatings for telescope mirrors, in making decorative paper, packages, toys, and in many other uses. The compounds of greatest importance are aluminum oxide, the sulfate, and the soluble sulfate with potassium (alum). The oxide, alumina, occurs naturally as ruby, sapphire, corundum, and emery, and is used in glassmaking and refractories. Synthetic ruby and sapphire have found application in the construction of lasers for producing coherent light. In 1852, the price of aluminum was about $1200/kg, and just before Hall's discovery in 1886, about $25/kg. The price rapidly dropped to 60¢ and has been as low as 33¢/kg. The price in December 1995 was about $1.70/kg.

Americium — (the Americas), Am; at. wt. 243; at. no. 95; m.p. 1176°C; b.p. 2011°C; sp. gr. 13.67 (20°C); valence 2, 3, 4, 5, or 6. Americium was the fourth transuranium element to be discovered; the isotope ^{241}Am was identified by Seaborg, James, Morgan, and Ghiorso late in 1944 at the wartime Metallurgical Laboratory of the University of Chicago as the result of successive neutron capture reactions by plutonium isotopes in a nuclear reactor:

$$^{239}\text{Pu}(n,\gamma) \rightarrow {}^{240}\text{Pu}(n,\gamma) \rightarrow {}^{241}\text{Pu} \xrightarrow{\beta} {}^{241}\text{Am}$$

Since the isotope ^{241}Am can be prepared in relatively pure form by extraction as a decay product over a period of years from strongly neutron-bombarded plutonium, ^{241}Pu, this isotope is used for much of the chemical investigation of this element. Better suited is the isotope ^{243}Am due to its longer half-life (7.37 × 10³ years as compared to 432.2 years for ^{241}Am). A mixture of the isotopes ^{241}Am, ^{242}Am, and ^{243}Am can be prepared by intense neutron irradiation of ^{241}Am according to the reactions ^{241}Am (n, γ) → ^{242}Am (n, γ) → ^{243}Am. Nearly isotopically pure ^{243}Am can be prepared by a sequence of neutron bombardments and chemical separations as follows: neutron bombardment of ^{241}Am yields ^{242}Pu by the reactions ^{241}Am (n, γ) → ^{242}Am → ^{242}Pu, after chemical separation the ^{242}Pu can be transformed to ^{243}Am via the reactions ^{242}Pu (n, γ) → ^{243}Pu → ^{243}Am, and the ^{243}Am can be chemically separated. Fairly pure ^{242}Pu can be prepared more simply by very intense neutron irradiation of ^{239}Pu as the result of successive neutron-capture reactions. Sixteen radioactive isotopes and isomers are now recognized. Americium metal has been prepared by reducing the trifluoride with barium vapor at 1000 to 1200°C or the dioxide by lanthanum metal. The luster of freshly prepared americium metal is white and more silvery than plutonium or neptunium prepared in the same manner. It appears to be more malleable than uranium or neptunium and tarnishes slowly in dry air at room temperature. Americium is thought to exist in two forms: an alpha form which has a double hexagonal close-packed structure and a loose-packed cubic beta form. Americium must be handled with great care to avoid personal contamination. As little as $0.03 \mu Ci$ of ^{241}Am is the maximum permissible total body burden. The alpha activity from ^{241}Am is about three times that of radium. When gram quantities of ^{241}Am are handled, the intense gamma activity makes exposure a serious problem. Americium dioxide, AmO_2, is the most important oxide. AmF_3, AmF_4, $AmCl_3$, $AmBr_3$, AmI_3, and other compounds have been prepared. The isotope ^{241}Am has been used as a portable source for gamma radiography. It has also been used as a radioactive glass thickness gage for the flat glass industry, and as a source of ionization for smoke detectors. Americium-243 is available from the Oak Ridge National Laboratory at a cost of $160/mg plus packing charges.

Antimony — (Gr. *anti* plus *monos* — a metal not found alone), Sb; at. wt. 121.760(1); at. no. 51; m.p. 630.63°C; b.p. 1587°C; sp. gr. 6.691 (20°C); valence 0, −3, +3, or +5. Antimony was recognized in compounds by the ancients and was known as a metal at the beginning of the 17th century and possibly much earlier. It is not abundant, but is found in over 100 mineral species. It is sometimes found native, but more frequently as the sulfide, *stibnite* (Sb_2S_3); it is also found as antimonides of the heavy metals, and as oxides. It is extracted from the sulfide by roasting to the oxide, which is reduced by salt and scrap iron; from its oxides it is also prepared by reduction with carbon. Two allotropic forms of antimony exist: the normal stable, metallic form, and the amorphous gray form. The so-called explosive antimony is an ill-defined material always containing an appreciable amount of halogen; therefore, it no longer warrants consideration as a separate allotrope. The yellow form, obtained by oxidation of *stibine*, SbH_3, is probably impure, and is not a distinct form. Natural antimony is made of two stable isotopes, ^{121}Sb and ^{123}Sb. Forty four other radioactive isotopes and isomers are now recognized. Metallic antimony is an extremely brittle metal of a flaky, crystalline texture. It is bluish white and has a metallic luster. It is not acted on by air at room temperature, but burns brilliantly when heated with the formation of white fumes of Sb_2O_3. It is a poor conductor of heat and electricity, and has a hardness of 3 to 3.5. Antimony, available commercially with a purity of 99.999 + %, is finding use in semiconductor technology for making infrared detectors, diodes, and Hall-effect devices. Commercial-grade antimony is widely used in alloys with percentages ranging from 1 to 20. It greatly increases the hardness and mechanical strength of lead. Batteries, antifriction alloys, type metal, small arms and tracer bullets, cable sheathing, and minor products use about half the metal produced. Compounds taking up the other half are oxides, sulfides, sodium antimonate, and antimony trichloride. These are used in manufacturing flame-proofing compounds, paints, ceramic enamels, glass, and pottery. Tartar emetic (hydrated potassium antimonyl tartate) has been used in medicine. Antimony and many of its compounds are toxic. Antimony costs about $80/kg (99.5%), and about 75¢/g (99.999%).

Argon — (Gr. *argos*, inactive), Ar; at. wt. 39.948(1); at. no. 18; triple pt. −189.3442; b.p. −185.89°C; density 1.7837 g/l. Its presence in air was suspected by Cavendish in 1785, discovered by Lord Rayleigh and Sir William Ramsay in 1894. The gas is prepared by fractionation of liquid air, the atmosphere containing 0.94% argon. The atmosphere of Mars contains 1.6% of ^{40}Ar and 5 p.p.m. of ^{36}Ar. Argon is two and one half times as soluble in water as nitrogen, having about the same solubility as oxygen. It is recognized by the characteristic lines in the red end of the spectrum. It is used in electric light bulbs and in fluorescent tubes at a pressure of about 400 Pa, and in filling photo tubes, glow tubes, etc. Argon is also used as an inert gas shield for arc welding and cutting, as a blanket for the production of titanium and other reactive elements, and as a protective atmosphere for growing silicon and germanium crystals. Argon is colorless and odorless, both as a gas and liquid. It is available in high-purity form. Commercial argon is available at a cost of about 3¢ per cubic foot. Argon is considered to be a very inert gas and is not known to form true chemical compounds, as do krypton, xenon, and radon. However, it does form a hydrate having a dissociation pressure of 105 atm at 0°C. Ion molecules such as $(ArKr)^+$, $(ArXe)^+$, $(NeAr)^+$ have been observed spectroscopically. Argon also forms a clathrate with β-hydroquinone. This clathrate is stable and can be stored for a considerable time, but a true chemical bond does not exist. Van der Waals' forces act to hold the argon. Naturally occurring argon is a mixture of three isotopes. Seventeen other radioactive isotopes are now known to exist. Commercial argon is priced at about $70/300 cu. ft. or 8.5 cu. meters.

Arsenic — (L. *arsenicum*, Gr. *arsenikon*, yellow orpiment, identified with *arsenikos*, male, from the belief that metals were different sexes; Arabic, *Az-zernikh*, the orpiment from Persian *zerni-zar*, gold), As; at. wt. 74.92159(2); at. no. 33; valence −3, 0, +3 or +5. Elemental arsenic occurs in two solid modifications: yellow, and gray or metallic, with specific gravities of 1.97, and 5.73, respectively. Gray arsenic, the ordinary stable form, has a m.p. of 817°C (28 atm) and sublimes at 614°C. Several other allotropic forms of arsenic are reported in the literature. It is believed that Albertus Magnus obtained the element in 1250 A.D. In 1649 Schroeder published two methods of preparing the element. It is found native, in the sulfides *realgar* and *orpiment*, as arsenides and sulfarsenides of heavy metals, as the oxide, and as arsenates. *Mispickel*, arsenopyrite, (FeSAs) is the most common mineral, from which on heating the arsenic sublimes leaving ferrous sulfide. The element is a steel gray, very brittle, crystalline, semimetallic solid; it tarnishes in air, and when heated is rapidly oxidized to arsenous oxide (As_2O_3) with the odor of garlic. Arsenic and its compounds are poisonous. These values, however, are being studied, and may be lowered. Arsenic is also used in bronzing, pyrotechny, and for hardening and improving the sphericity of shot. The most important compounds are white arsenic (As_2O_3), the sulfide, Paris green $3Cu(AsO_2)_2 \cdot Cu(C_2H_3O_2)_2$, calcium arsenate, and lead arsenate; the last three have been used as agricultural insecticides and poisons. Marsh's test makes use of the formation and ready decomposition of arsine (AsH_3). Arsenic is available in high-purity form. It is finding increasing uses as a doping agent in solid-state devices such as transistors. Gallium arsenide is used as a laser material to convert electricity directly into coherent light. Natural arsenic is made of one isotope ^{75}As. Twenty-five other radioactive isotopes and isomers are known. Arsenic (99%) costs about $175/kg. Purified arsenic (99.9995%) costs about $2/gm.

Astatine — (Gr. *astatos*, unstable), At; at. wt. (210); at. no. 85; m.p. 302°C; b.p. 337°C (est.); valence probably 1, 3, 5, or 7. Synthesized in 1940 by D. R. Corson, K. R. MacKenzie, and E. Segre at the University of California by bombarding bismuth with alpha particles. The longest-lived isotopes, ^{210}At, have a half-life of only 8.1 hours. Thirty-four isotopes and isomers are now known. Minute quantities of ^{215}At, ^{218}At, and ^{219}At exist in equilibrium in nature with naturally occurring uranium and thorium isotopes, and traces of ^{217}At are equilibrium with ^{233}U and ^{239}Np resulting from interaction of thorium and uranium with naturally produced neutrons. The total amount of astatine present in the earth's crust, however, is less than 1 oz. Astatine can be produced by bombarding bismuth with energetic alpha particles to obtain the relatively long-lived $^{209-211}At$, which can be distilled from the target by heating it in air. Only about 0.05 μg of astatine has been prepared to date. The "time of flight" mass spectrometer has been used to confirm that this highly radioactive halogen behaves chemically very much like other halogens, particularly iodine. The interhalogen compounds AtI, AtBr, and AtCl are known to form, but it is not yet known if astatine forms diatomic astatine molecules. HAt and CH_3At (methyl astatide) have been detected. Astatine is said to be more metallic that iodine, and, like iodine, it probably accumulates in the thyroid gland. Workers at the Brookhaven National Laboratory have recently used reactive scattering in crossed molecular beams to identify and measure elementary reactions involving astatine.

Barium — (Gr. *barys*, heavy), Ba; at. wt. 137.327(7), at. no. 56; m.p. 727°C; b.p. 1897°C; sp. gr. 3.5 (20°C); valence 2. Baryta was distinguished from lime by Scheele in 1774; the element was discovered by Sir Humphrey Davy in 1808. It is found only in combination with other elements, chiefly in *barite* or *heavy spar* (sulfate) and *witherite* (carbonate) and is prepared by electrolysis of the chloride. Barium is a metallic element, soft, and when pure is silvery white like lead; it belongs to the alkaline earth group, resembling calcium chemically. The metal oxidizes very easily and should be kept under petroleum or other suitable oxygen-free liquids to exclude air. It is decomposed by water or alcohol. The metal is used as a "getter" in vacuum tubes. The most important compounds are the peroxide (BaO_2), chloride, sulfate, carbonate, nitrate, and chlorate. Lithopone, a pigment containing barium sulfate and zinc sulfide, has good covering power, and does not darken in the presence of sulfides. The sulfate, as permanent white or *blanc fixe*, is also used in paint, in X-ray diagnostic work, and in glassmaking. *Barite* is extensively used as a weighting agent in oilwell drilling fluids, and also in making rubber. The carbonate has been used as a rat poison, while the nitrate and chlorate give colors in pyrotechny. The impure sulfide phosphoresces after exposure to the light. The compounds and the metal are not expensive. Barium metal (99.7 + % pure) costs about 40¢/gm. All barium compounds that are water or acid soluble are poisonous. Naturally occurring barium is a mixture of seven stable isotopes. Thirty nine other radioactive isotopes and isomers are known to exist.

Berkelium — (*Berkeley*, home of the University of California), Bk; at. wt. (247); at. no. 97; m.p. 1050°C; valence 3 or 4; sp. gr. 14 (est.). Berkelium, the eighth member of the actinide transition series, was discovered in December 1949 by Thompson, Ghiorso, and Seaborg, and was the fifth transuranium element synthesized. It was produced by cyclotron bombardment of milligram amounts of ^{241}Am with helium ions at Berkeley, California. The first isotope produced had a mass number of 243 and decayed with a half-life of 4.5 hours. Eleven isotopes are now known and have been synthesized. The existence of ^{249}Bk, with a half-life of 320 days, makes it feasible to isolate berkelium in weighable amounts so that its properties can be investigated with macroscopic quantities. One of the first visible amounts of a pure berkelium compound, berkelium chloride, was produced in 1962. It weighed 3 billionth of a gram. Berkelium probably has not yet been prepared in elemental form, but it is expected to be a silvery metal, easily soluble in dilute mineral acids, and readily oxidized by air or oxygen at elevated temperatures to form the oxide. X-ray diffraction methods have been used to identify the following compounds: BkO_2, BkO_3, BkF_3, $BkCl$, and $BkOCl$. As with other actinide elements, berkelium tends to accumulate in the skeletal system. The maximum permissible body burden of ^{249}Bk in the human skeleton is about 0.0004 μg. Because of its rarity, berkelium presently has no commercial or technological use. Berkelium-249 is available from O.R.N.L. at a cost of $160/μg plus packing charges.

Beryllium — (Gr. *beryllos*, beryl; also called Glucinium or Glucinum, Gr. *glykys*, sweet), Be; at. wt. 9.012182(3); at no. 4; m.p. 1287°C; b.p. 2471°C; sp. gr. 1.848 (20°C); valence 2. Discovered as the oxide by Vauquelin in beryl and in emeralds in 1798. The metal was isolated in 1828 by Wohler and by Bussy independently by the action of potassium on beryllium chloride. Beryllium is found in some 30 mineral species, the most important of which are *bertrandite, beryl, chrysoberyl*, and *phenacite. Aquamarine* and *emerald* are precious forms of *beryl. Beryl* (3BeO · Al_2O_3 · $6SiO_2$) and *bertrandite* (4BeO · $2SiO_2$ · H_2O) are the most important commercial sources of the element and its compounds. Most of the metal is now prepared by reducing beryllium fluoride with magnesium metal. Beryllium metal did not become readily available to industry until 1957. The metal, steel gray in color, has many desirable properties. It is one of the lightest of all metals, and has one of the highest melting points of the light metals. Its modulus of elasticity is about one third greater than that of steel. It resists attack by concentrated nitric acid, has excellent thermal conductivity, and is nonmagnetic. It has a high permeability to X-rays, and when bombarded by alpha particles, as from radium or polonium, neutrons are produced in the ratio of about 30 neutrons/million alpha particles. At ordinary temperatures beryllium resists oxidation in air, although its ability to scratch glass is probably due to the formation of a thin layer of the oxide. Beryllium is used as an alloying agent in producing beryllium copper which is extensively used for springs, electrical contacts, spot-welding electrodes, and nonsparking tools. It has found application as a structural material for high-speed aircraft, missiles, spacecraft, and communication satellites. It is being used in the windshield frame, brake discs, support beams, and other structural components of the space shuttle. Because beryllium is relatively transparent to X-rays, ultra-thin Be-foil is finding use in X-ray lithography for reproduction of microminiature integrated circuits. Natural beryllium is made of ^9Be and is stable. Eight other radioactive isotopes are known.

Beryllium is used in nuclear reactors as a reflector or moderator for it has a low thermal neutron absorption cross section. It is used in gyroscopes, computer parts, and instruments where lightness, stiffness, and dimensional stability are required. The oxide has a very high melting point and is also used in nuclear work and ceramic applications. Beryllium and its salts are toxic and should be handled with the greatest of care. Beryllium and its compounds should not be tasted to verify the sweetish nature of beryllium (as did early experimenters). The metal, its alloys, and its salts can be handled safely if certain work codes are observed, but no attempt should be made to work with beryllium before becoming familiar with proper safeguards. Beryllium metal is available at a cost of about $2.50/g (99.5% pure).

Bismuth — (Ger. *Weisse Masse*, white mass; later *Wisuth* and *Bisemutum*), Bi; at. wt. 208.98037(3); at. no. 83; m.p. 271.4°C; b.p. 1564 ± 5°C; sp. gr. 9.747 (20°C); valence 3 or 5. In early times bismuth was confused with tin and lead. Claude Geoffroy the Younger showed it to be distinct from lead in 1753. It is a white crystalline, brittle metal with a pinkish tinge. It occurs native. The most important ores are *bismuthinite* or bismuth glance (Bi_2S_3) and *bismite* (Bi_2O_3). Peru, Japan, Mexico, Bolivia, and Canada are major bismuth producers. Much of the bismuth produced in the U.S. is obtained as a by-product in refining lead, copper, tin, silver, and gold ores. Bismuth is the most diamagnetic of all metals, and the thermal conductivity is lower than any metal, except mercury. It has a high electrical resistance, and has the highest Hall effect of any metal (i.e., greatest increase in electrical resistance when placed in a magnetic field). "Bismanol" is a permanent magnet of high coercive force, made of MnBi, by the U.S. Naval Surface Weapons Center. Bismuth expands 3.32% on solidification. This property makes bismuth alloys particularly suited to the making of sharp castings of objects subject to damage by high temperatures. With other metals such as tin, cadmium, etc., bismuth forms low-melting alloys which are extensively used for safety devices in fire detection and extinguishing systems. Bismuth is used in producing malleable irons and is finding use as a catalyst for making acrylic fibers. When bismuth is heated in air it burns with a blue flame, forming yellow fumes of the oxide. The metal is also used as a thermocouple material, and has found application as a carrier for U^{235} or U^{233} fuel in atomic reactors. Its soluble salts are characterized by forming insoluble basic salts on the addition of water, a property sometimes used in detection work. Bismuth oxychloride is used extensively in cosmetics. Bismuth subnitrate and subcarbonate are used in medicine. Natural bismuth contains only one isotope ^{209}Bi. Forty one isotopes and isomers of bismuth are known. Bismuth metal costs about $90/kg (99.999%).

Bohrium — See Element 107.

Boron — (Ar. *Buraq*, Pers. *Burah*), B; at. wt. 10.81(5); at. no. 5; m.p. 2075°C; b.p. 4000°C; sp. gr. of crystals 2.34, of amorphous variety 2.37; valence 3. Boron compounds have been known for thousands of years, but the element was not discovered until 1808 by Sir Humphry Davy and by Gay-Lussac and Thenard. The element is not found free in nature, but occurs as orthoboric acid usually in certain volcanic spring waters and as borates in *borax* and *colemanite. Ulexite*, another boron mineral, is interesting as it is nature's own version of "fiber optics." Important sources of boron are the ores *rasorite (kernite)* and *tincal (borax ore)*. Both of these ores are found in the Mojave Desert. *Tincal* is the most important source of boron from the Mojave. Extensive *borax* deposits are also found in Turkey. Boron exists naturally as 19.9% ^{10}B isotope and 80.1% ^{11}B isotope. Eleven isotopes of boron are known. High-purity crystalline boron may be prepared by the vapor phase reduction of boron trichloride or tribromide with hydrogen on electrically heated filaments. The impure, or amorphous, boron, a brownish-black powder, can be obtained by heating the trioxide with magnesium powder. Boron of 99.9999% purity has been produced and is available commercially. Elemental boron has an energy band gap of 1.50 to 1.56 eV, which is higher than that of either silicon or germanium. It has interesting optical characteristics, transmitting portions of the infrared, and is a poor conductor of electricity at room temperature, but a good conductor at high temperature. Amorphous boron is used in pyrotechnic flares to provide a distinctive green color, and in rockets as an igniter. By far the most commercially important boron compound in terms of dollar sales is $Na_2B_4O_7 \cdot 5H_2O$. This pentahydrate is used in very large quantities in the manufacture of insulation fiberglass and sodium perborate bleach. Boric acid is also an important boron compound with major markets in textile fiberglass and in cellulose insulation as a flame retardant. Next in order of importance is borax ($Na_2B_4O_7 \cdot 10H_2O$) which is used principally in laundry products. Use of borax as a mild antiseptic is minor in terms of dollars and tons. Boron compounds are also extensively used in the manufacture of borosilicate glasses. Other boron compounds show promise in treating arthritis. The isotope boron-10 is used as a control for nuclear reactors, as a shield for nuclear radiation, and in instruments used for detecting neutrons. Boron nitride has remarkable properties and can be used to make a material as hard as diamond. The nitride also behaves like an electrical insulator but conducts heat like a metal. It also has lubricating properties similar to graphite. The hydrides are easily oxidized with considerable energy liberation, and have been studied for use as rocket fuels. Demand is increasing for boron filaments, a high-strength, lightweight material chiefly employed for advanced aerospace structures. Boron is similar to carbon in that it has a capacity to form stable covalently bonded molecular networks. Carboranes, metalloboranes, phosphacarboranes, and other families comprise thousands of compounds. Crystalline boron (99%) costs about $8/g. Amorphous boron costs about $4/g. Elemental boron and the borates are not considered to be toxic, and they do not require special care in handling. However, some of the more exotic boron hydrogen compounds are definitely toxic and do require care.

Bromine — (Gr. *bromos,* stench), Br; at. wt. 79.904(1); at. no. 35; m.p. –7.2°C; b.p. 58.78°C; density of gas 7.59 g/l, liquid 3.12 (20°C); valence 1, 3, 5, or 7. Discovered by Balard in 1826, but not prepared in quantity until 1860. A member of the halogen group of elements, it is obtained from natural brines from wells in Michigan and Arkansas. Little bromine is extracted today from seawater, which contains only about 85 ppm. Bromine is the only liquid nonmetallic element. It is a heavy, mobile, reddish-brown liquid, volatilizing readily at room temperature to a red vapor with a strong disagreeable odor, resembling chlorine, and having a very irritating effect on the eyes and throat; it is readily soluble in water or carbon disulfide, forming a red solution, is less active than chlorine but more so than iodine; it unites readily with many elements and has a bleaching action; when spilled on the skin it produces painful sores. It presents a serious health hazard, and maximum safety precautions should be taken when handling it. Much of the bromine output in the U.S. was used in the production of ethylene dibromide, a lead scavenger used in making gasoline antiknock compounds. Lead in gasoline, however, has been drastically reduced, due to environmental considerations. This will greatly affect future production of bromine. Bromine is also used in making fumigants, flameproofing agents, water purification compounds, dyes, medicinals, sanitizers, inorganic bromides for photography, etc. Organic bromides are also important. Natural bromine is made of two isotopes, [79]Br and [81]Br. Thirty three isotopes and isomers are known. Bromine costs about $100/kg.

Cadmium — (L. *cadmia;* Gr. *kadmeia* — ancient name for calamine, zinc carbonate), Cd; at. wt. 112.411(8); at. no. 48; m.p. 321.07°C; b.p. 767°C; sp. gr. 8.65 (20°C); valence 2. Discovered by Stromeyer in 1817 from an impurity in zinc carbonate. Cadmium most often occurs in small quantities associated with zinc ores, such as *sphalerite* (ZnS). *Greenockite* (CdS) is the only mineral of any consequence bearing cadmium. Almost all cadmium is obtained as a by-product in the treatment of zinc, copper, and lead ores. It is a soft, bluish-white metal which is easily cut with a knife. It is similar in many respects to zinc. It is a component of some of the lowest melting alloys; it is used in bearing alloys with low coefficients of friction and great resistance to fatigue; it is used extensively in electroplating, which accounts for about 60% of its use. It is also used in many types of solder, for standard E.M.F. cells, for Ni-Cd batteries, and as a barrier to control atomic fission. Cadmium compounds are used in black and white television phosphors and in blue and green phosphors for color TV tubes. It forms a number of salts, of which the sulfate is most common; the sulfide is used as a yellow pigment. Cadmium and solutions of its compounds are toxic. Failure to appreciate the toxic properties of cadmium may cause workers to be unwittingly exposed to dangerous fumes. Some silver solders, for example, contain cadmium and should be handled with care. Serious toxicity problems have been found from long-term exposure and work with cadmium plating baths. In 1927 the International Conference on Weights and Measures redefined the meter in terms of the wavelength of the red cadmium spectral line (i.e. 1 m = 1,553,164.13 wavelengths). This definition has been changed (see under Krypton). The current price of cadmium is about $100/kg (99.5%). It is available in high purity form for about $300/kg. Natural cadmium is made of eight isotopes. Thirty four other isotopes and isomers are now known and recognized.

Calcium — (L. *calx,* lime), Ca; at. wt. 40.078(4); at. no. 20; m.p. 842 ± 2°C; b.p. 1484°C; sp. gr. 1.55 (20°C); valence 2. Though lime was prepared by the Romans in the first century under the name calx, the metal was not discovered until 1808. After learning that Berzelius and Pontin prepared calcium amalgam by electrolyzing lime in mercury, Davy was able to isolate the impure metal. Calcium is a metallic element, fifth in abundance in the earth's crust, of which it forms more than 3%. It is an essential constituent of leaves, bones, teeth, and shells. Never found in nature uncombined, it occurs abundantly as *limestone* ($CaCO_3$), *gypsum* ($CaSO_4 \cdot 2H_2O$), and *fluorite* (CaF_2); *apatite* is the fluorophosphate or chlorophosphate of calcium. The metal has a silvery color, is rather hard, and is prepared by electrolysis of the fused chloride to which calcium fluoride is added to lower the melting point. Chemically it is one of the alkaline earth elements; it readily forms a white coating of nitride in air, reacts with water, burns with a yellow-red flame, forming largely the nitride. The metal is used as a reducing agent in preparing other metals such as thorium, uranium, zirconium, etc., and is used as a deoxidizer, desulfurizer, or decarburizer for various ferrous and nonferrous alloys. It is also used as an alloying agent for aluminum, beryllium, copper, lead, and magnesium alloys, and serves as a "getter" for residual gases in vacuum tubes, etc. Its natural and prepared compounds are widely used. Quicklime (CaO), made by heating limestone and changed into slaked lime by the careful addition of water, is the great cheap base of chemical industry with countless uses. Mixed with sand it hardens as mortar and plaster by taking up carbon dioxide from the air. Calcium from limestone is an important element in Portland cement. The solubility of the carbonate in water containing carbon dioxide causes the formation of caves with stalactites and stalagmites and is responsible for hardness in water. Other important compounds are the carbide (CaC_2), chloride ($CaCl_2$), cyanamide ($CaCN_2$), hypochlorite ($Ca(OCl)_2$), nitrate ($Ca(NO_3)_2$), and sulfide (CaS). Natural calcium contains six isotopes. Thirteen other radioactive isotopes are known. Metallic calcium (99.5%) costs about $200/kg.

Californium — (State and University of California), Cf; at. wt. (251); m.p. (est) 900°C; at. no. 98. Californium, the sixth transuranium element to be discovered, was produced by Thompson, Street, Ghiorso, and Seaborg in 1950 by bombarding microgram quantities of [242]Cm with 35 MeV helium ions in the Berkeley 60-inch cyclotron. Californium (III) is the only ion stable in aqueous solutions, all attempts to reduce or oxidize californium

(III) having failed. The isotope ^{249}Cf results from the beta decay of ^{249}Bk while the heavier isotopes are produced by intense neutron irradiation by the reactions:

$$^{249}\text{Bk}(n,\gamma)\rightarrow\,^{250}\text{Bk}\xrightarrow{\;\beta\;}\,^{250}\text{Cf} \text{ and } ^{249}\text{Cf}(n,\gamma)\rightarrow\,^{250}\text{Cf}$$

followed by

$$^{250}\text{Cf}(n,\gamma)\rightarrow\,^{251}\text{Cf}(n,\gamma)\rightarrow\,^{252}\text{Cf}$$

The existence of the isotopes ^{249}Cf, ^{250}Cf, ^{251}Cf, and ^{252}Cf makes it feasible to isolate californium in weighable amounts so that its properties can be investigated with macroscopic quantities. Californium-252 is a very strong neutron emitter. One microgram releases 170 million neutrons per minute, which presents biological hazards. Proper safeguards should be used in handling californium. Eighteen isotopes of californium are now recognized. ^{249}Cf and ^{252}Cf have half-lives of 351 years and 900 years, respectively. In 1960 a few tenths of a microgram of californium trichloride, $CfCl_3$, californium oxychloride, CfOCl, and californium oxide, Cf_2O_3, were first prepared. Reduction of californium to its metallic state has not yet been accomplished. Because californium is a very efficient source of neutrons, many new uses are expected for it. It has already found use in neutron moisture gages and in well-logging (the determination of water and oil-bearing layers). It is also being used as a portable neutron source for discovery of metals such as gold or silver by on-the-spot activation analysis. ^{252}Cf is now being offered for sale by the Oak Ridge National Laboratory (O.R.N.L.) at a cost of \$50/μg and ^{249}Cf at a cost of \$160/μg plus packing charges. It has been suggested that californium may be produced in certain stellar explosions, called *supernovae,* for the radioactive decay of ^{254}Cf (55-day half-life) agrees with the characteristics of the light curves of such explosions observed through telescopes. This suggestion, however, is questioned.

Carbon — (L. *carbo*, charcoal), C; at. wt. 12.011(1); at. no. 6; m.p. ~3550°C, graphite sublimes at 3825°C; triple point; (graphite-liquid-gas), 4492°C at a pressure of 10.3 MPa and (graphite-diamond-liquid), 3830—3930° at a pressure of 12—13 GPa; sp. gr. amorphous 1.8 to 2.1, graphite 1.9 to 2.3, diamond 3.15 to 3.53 (depending on variety); gem diamond 3.513 (25°C); valence 2, 3, or 4. Carbon, an element of prehistoric discovery, is very widely distributed in nature. It is found in abundance in the sun, stars, comets, and atmospheres of most planets. Carbon in the form of microscopic diamonds is found in some meteorites. Natural diamonds are found in *kimberlite* of ancient volcanic "pipes," such as found in South Africa, Arkansas, and elsewhere. Diamonds are now also being recovered from the ocean floor off the Cape of Good Hope. About 30% of all industrial diamonds used in the U.S. are now made synthetically. The energy of the sun and stars can be attributed at least in part to the well-known carbon-nitrogen cycle. Carbon is found free in nature in three allotropic forms: amorphous, graphite, and diamond. A fourth form, known as "white" carbon, is now thought to exist. Graphite is one of the softest known materials while diamond is one of the hardest. Graphite exists in two forms: alpha and beta. These have identical physical properties, except for their crystal structure. Naturally occurring graphites are reported to contain as much as 30% of the rhombohedral (beta) form, whereas synthetic materials contain only the alpha form. The hexagonal alpha type can be converted to the beta by mechanical treatment, and the beta form reverts to the alpha on heating it above 1000°C. In 1969 a new allotropic form of carbon was produced during the sublimation of pyrolytic graphite at low pressures. Under free-vaporization conditions above ~2550 K, "white" carbon forms as small transparent crystals on the edges of the basal planes of graphite. The interplanar spacings of "white" carbon are identical to those of carbon form noted in the graphitic gneiss from the Ries (meteoritic) Crater of Germany. "White" carbon is a transparent birefringent material. Little information is presently available about this allotrope. Of recent interest is the discovery of all-carbon molecules, known as "buckyballs" or fullerenes, which have a number of unusual properties. These interesting molecules, consisting of 60 or 70 carbon atoms linked together, seem capable of withstanding great pressure and trapping foreign atoms inside their network of carbon. They are said to be capable of magnetism and superconductivity and have potential as a nonlinear optical material. Buckyball films are reported to remain superconductive at temperatures as high as 45 K. In combination, carbon is found as carbon dioxide in the atmosphere of the earth and dissolved in all natural waters. It is a component of great rock masses in the form of carbonates of calcium (limestone), magnesium, and iron. Coal, petroleum, and natural gas are chiefly hydrocarbons. Carbon is unique among the elements in the vast number and variety of compounds it can form. With hydrogen, oxygen, nitrogen, and other elements, it forms a very large number of compounds, carbon atom often being linked to carbon atom. There are close to ten million known carbon compounds, many thousands of which are vital to organic and life processes. Without carbon, the basis for life would be impossible. While it has been thought that silicon might take the place of carbon in forming a host of similar compounds, it is now not possible to form stable compounds with very long chains of silicon atoms. The atmosphere of Mars contains 96.2% CO_2. Some of the most important compounds of carbon are carbon dioxide (CO_2), carbon monoxide (CO), carbon disulfide (CS_2), chloroform ($CHCl_3$), carbon tetrachloride (CCl_4), methane (CH_4), ethylene (C_2H_4), acetylene (C_2H_2), benzene (C_6H_6), ethyl alcohol (C_2H_5OH), acetic acid (CH_3COOH), and their derivatives. Carbon has thirteen isotopes. Natural carbon consists of 98.89% ^{12}C and 1.11% ^{13}C. In 1961 the International Union of Pure and Applied Chemistry adopted the isotope carbon-12 as the basis for atomic weights. Carbon-14, an isotope with a half-life of 5715 years, has been widely used to date such materials as wood, archeological specimens, etc.

Cerium — (named for the asteroid *Ceres*, which was discovered in 1801 only 2 years before the element), Ce; at. wt. 140.115(4); at. no. 58; m.p. 798°C; b.p. 3424°C; sp. gr. 6.770 (25°C); valence 3 or 4. Discovered in 1803 by Klaproth and by Berzelius and Hisinger; metal prepared by Hillebrand and Norton in 1875. Cerium is the most abundant of the metals of the so-called rare earths. It is found in a number of minerals including *allanite* (also known as *orthite*), *monazite, bastnasite, cerite*, and *samarskite*. Monazite and bastnasite are presently the two most important sources of cerium. Large deposits of monazite found on the beaches of Travancore, India, in river sands in Brazil, and deposits of *allanite* in the western United States, and *bastnasite* in Southern California will supply cerium, thorium, and the other rare-earth metals for many years to come. Metallic cerium is prepared by metallothermic reduction techniques, such as by reducing cerous fluoride with calcium, or by electrolysis of molten cerous chloride or other cerous halides. The metallothermic technique is used to produce high-purity cerium. Cerium is especially interesting because of its variable electronic structure. The energy of the inner 4f level is nearly the same as that of the outer or valence electrons, and only small amounts of energy are required to change the relative occupancy of these electronic levels. This gives rise to dual valency states. For example, a volume change of about 10% occurs when cerium is subjected to high pressures or low temperatures. It appears that the valence changes from about 3 to 4 when it is cooled or compressed. The low temperature behavior of cerium is complex. Four allotropic modifications are thought to exist: cerium at room temperature and at atmospheric pressure is known as γ cerium. Upon cooling to −16°C, γ cerium changes to β cerium. The remaining γ cerium starts to change to α cerium when cooled to −172°C, and the transformation is complete at −269°C. α Cerium has a density of 8.16; δ cerium exists above 726°C. At atmospheric pressure, liquid cerium is more dense than its solid form at the melting point. Cerium is an iron-gray lustrous metal. It is malleable, and oxidizes very readily at room temperature, especially in moist air. Except for europium, cerium is the most reactive of the "rare-earth" metals. It slowly decomposes in cold water, and rapidly in hot water. Alkali solutions and dilute and concentrated acids attack the metal rapidly. The pure metal is likely to ignite if scratched with a knife. Ceric salts are orange red or yellowish; cerous salts are usually white. Cerium is a component of misch metal, which is extensively used in the manufacture of pyrophoric alloys for cigarette lighters, etc. Natural cerium is stable and contains four isotopes. Thirty five other radioactive isotopes and isomers are known. While cerium is not radioactive, the impure commercial grade may contain traces of thorium, which is radioactive. The oxide is an important constituent of incandescent gas mantles and it is emerging as a hydrocarbon catalyst in "self-cleaning" ovens. In this application it can be incorporated into oven walls to prevent the collection of cooking residues. As ceric sulfate it finds extensive use as a volumetric oxidizing agent in quantitative analysis. Cerium compounds are used in the manufacture of glass, both as a component and as a decolorizer. The oxide is finding increased use as a glass polishing agent instead of rouge, for it is much faster than rouge in polishing glass surfaces. Cerium, with other rare earths, is used in carbon-arc lighting, especially in the motion picture industry. It is also finding use as an important catalyst in petroleum refining and in metallurgical and nuclear applications. In small lots, 99.9% cerium costs about \$4/g (99.9%).

Cesium — (L. *caesius,* sky blue), Cs; at. wt. 132.90543(5); at. no. 55; m.p. 28.44 ± 0.01°C; b.p. 671°C; sp. gr. 1.873 (20°C); valence 1. Cesium was discovered spectroscopically by Bunsen and Kirchhoff in 1860 in mineral water from Durkheim. Cesium, an alkali metal, occurs in *lepidolite, pollucite* (a hydrated silicate of aluminum and cesium), and in other sources. One of the world's richest sources of cesium is located at Bernic Lake, Manitoba. The deposits are estimated to contain 300,000 tons of pollucite, averaging 20% cesium. It can be isolated by electrolysis of the fused cyanide and by a number of other methods. Very pure, gas-free cesium can be prepared by thermal decomposition of cesium azide. The metal is characterized by a spectrum containing two bright lines in the blue along with several others in the red, yellow, and green. It is silvery white, soft, and ductile. It is the most electropositive and most alkaline element. Cesium, gallium, and mercury are the only three metals that are liquid at room temperature. Cesium reacts explosively with cold water, and reacts with ice at temperatures above –116°C. Cesium hydroxide, the strongest base known, attacks glass. Because of its great affinity for oxygen the metal is used as a "getter" in electron tubes. It is also used in photoelectric cells, as well as a catalyst in the hydrogenation of certain organic compounds. The metal has recently found application in ion propulsion systems. Cesium is used in atomic clocks, which are accurate to 5 s in 300 years. Its chief compounds are the chloride and the nitrate. Cesium has 52 isotopes and isomers with masses ranging from 112 to 148. The present price of cesium is about \$40/g (99.98%).

Chlorine — (Gr. *chloros,* greenish yellow), Cl; at. wt. 35.4527(9); at. no. 17; m.p. –101.5°C; b.p. –34.04°C; density 3.214 g/l; sp. gr. 1.56 (–33.6°C); valence 1, 3, 5, or 7. Discovered in 1774 by Scheele, who thought it contained oxygen; named in 1810 by Davy, who insisted it was an element. In nature it is found in the combined state only, chiefly with sodium as common salt (NaCl), *carnallite* ($KMgCl_3 \cdot 6H_2O$), and *sylvite* (KCl). It is a member of the halogen (salt-forming) group of elements and is obtained from chlorides by the action of oxidizing agents and more often by electrolysis; it is a greenish-yellow gas, combining directly with nearly all elements. At 10°C one volume of water dissolves 3.10 volumes of chlorine, at 30°C only 1.77 volumes. Chlorine is widely used in making many everyday products. It is used for producing safe drinking water the world over. Even the smallest water supplies are now usually chlorinated. It is also extensively used in the production of paper products, dyestuffs, textiles, petroleum products, medicines, antiseptics, insecticides, foodstuffs, solvents, paints, plastics, and many other consumer products. Most of the chlorine produced is used in the manufacture of chlorinated compounds for sanitation, pulp bleaching, disinfectants, and textile processing. Further use is in the manufacture of chlorates, chloroform, carbon tetrachloride, and in the extraction of bromine. Organic chemistry demands much from chlorine, both as an oxidizing agent and in substitution, since it often brings desired properties in an organic compound when substituted for hydrogen, as in one form of synthetic rubber. Chlorine is a respiratory irritant. The gas irritates the mucous membranes and the liquid burns the skin. As little as 3.5 ppm can be detected as an odor, and 1000 ppm is likely to be fatal after a few deep breaths. It was used as a war gas in 1915. Natural chlorine contains two isotopes. Sixteen other isotopes and isomers are known.

Chromium — (Gr. *chroma,* color), Cr; at. wt. 51.9961(6); at. no. 24; m.p. 1907°C; b.p. 2671°C; sp. gr. 7.18 to 7.20 (20°C); valence chiefly 2, 3, or 6. Discovered in 1797 by Vauquelin, who prepared the metal the next year, chromium is a steel-gray, lustrous, hard metal that takes a high polish. The principal ore is *chromite* ($FeCr_2O_4$), which is found in Zimbabwe, Russia, Transvaal, Turkey, Iran, Albania, Finland, Democratic Republic of Madagascar, and the Philippines. The metal is usually produced by reducing the oxide with aluminum. Chromium is used to harden steel, to manufacture stainless steel, and to form many useful alloys. Much is used in plating to produce a hard, beautiful surface and to prevent corrosion. Chromium is used to give glass an emerald green color. It finds wide use as a catalyst. All compounds of chromium are colored; the most important are the chromates of sodium and potassium (K_2CrO_4) and the dichromates ($K_2Cr_2O_7$) and the potassium and ammonium chrome alums, as $KCr(SO_4)_2$ $\cdot$ $12H_2O$. The dichromates are used as oxidizing agents in quantitative analysis, also in tanning leather. Other compounds are of industrial value; lead chromate is chrome yellow, a valued pigment. Chromium compounds are used in the textile industry as mordants, and by the aircraft and other industries for anodizing aluminum. The refractory industry has found chromite useful for forming bricks and shapes, as it has a high melting point, moderate thermal expansion, and stability of crystalline structure. Many chromium compounds are toxic and should be handled with proper safeguards. Natural chromium contains four isotopes. Sixteen other isotopes are known. Chromium metal (99.99%) costs about \$200/kg. Commercial grade chromium (99%) costs about \$75/kg.

Cobalt — (*Kobald,* from the German, goblin or evil spirit, *cobalos,* Greek, mine), Co; at. wt. 58.93320(1); at. no. 27; m.p. 1495°C; b.p. 2927°C; sp. gr. 8.9 (20°C); valence 2 or 3. Discovered by Brandt about 1735. Cobalt occurs in the mineral *cobaltite, smaltite,* and *erythrite,* and is often associated with nickel, silver, lead, copper, and iron ores, from which it is most frequently obtained as a by-product. It is also present in meteorites. Important ore deposits are found in Zaire, Morocco, and Canada. The U.S. Geological Survey has announced that the bottom of the north central Pacific Ocean may have cobalt-rich deposits at relatively shallow depths in waters close to the Hawaiian Islands and other U.S. Pacific territories. Cobalt is a brittle, hard metal, closely resembling iron and nickel in appearance. It has a magnetic permeability of about two thirds that of iron. Cobalt tends to exist as a mixture of two allotropes over a wide temperature range; the β-form predominates below 400°C, and the α above that temperature. The transformation is sluggish and accounts in part for the wide variation in reported data on physical properties of cobalt. It is alloyed with iron, nickel and other metals to make Alnico, an alloy of unusual magnetic strength with many important uses. Stellite alloys, containing cobalt, chromium, and tungsten, are used for high-speed, heavy-duty, high temperature cutting tools, and for dies. Cobalt is also used in other magnet steels and stainless steels, and in alloys used in jet turbines and gas turbine generators. The metal is used in electroplating because of its appearance, hardness, and resistance to oxidation. The salts have been used for centuries for the production of brilliant and permanent blue colors in porcelain, glass, pottery, tiles, and enamels. It is the principal ingredient in Sevre's and Thenard's blue. A solution of the chloride ($CoCl_2 \cdot 6H_2O$) is used as sympathetic ink. The cobalt ammines are of interest; the oxide and the nitrate are important. Cobalt carefully used in the form of the chloride, sulfate, acetate, or nitrate has been found effective in correcting a certain mineral deficiency disease in animals. Soils should contain 0.13 to 0.30 ppm of cobalt for proper animal nutrition. Cobalt is found in Vitamin B-12, which is essential for human nutrition. Cobalt-60, an artificial isotope, is an important gamma ray source, and is extensively used as a tracer and a radiotherapeutic agent. Single compact sources of Cobalt-60 vary from about \$1 to \$10/curie, depending on quantity and specific activity. Twenty six isotopes and isomers of cobalt are known.

Columbium — See Niobium.

Copper — (L. *cuprum,* from the island of Cyprus), Cu; at. wt. 63.546(3); at. no. 29; f.p. 1084.62 °C; b.p. 2562°C; sp. gr. 8.96 (20°C); valence 1 or 2. The discovery of copper dates from prehistoric times. It is said to have been mined for more than 5000 years. It is one of man's most important metals. Copper is reddish colored, takes on a bright metallic luster, and is malleable, ductile, and a good conductor of heat and electricity (second only to silver in electrical conductivity). The electrical industry is one of the greatest users of copper. Copper occasionally occurs native, and is found in many minerals such as *cuprite, malachite, azurite, chalcopyrite,* and *bornite.* Large copper ore deposits are found in the U.S., Chile, Zambia, Zaire, Peru, and Canada. The most important copper ores are the sulfides, oxides, and carbonates. From these, copper is obtained by smelting, leaching, and by electrolysis. Its alloys, brass and bronze, long used, are still very important; all American coins are now copper alloys; monel and gun metals also contain copper. The most important compounds are the oxide and the sulfate, blue vitriol; the latter has wide use as an agricultural poison and as an algicide in water purification. Copper compounds such as Fehling's solution are widely used in analytical chemistry in tests for sugar. High-purity copper (99.999 + %) is available commercially. Natural copper contains two isotopes. Twenty five other radioactive isotopes and isomers are known.

Curium — (Pierre and Marie Curie), Cm; at. wt. (247); at. no. 96; m.p. 1345±40°C; sp. gr. 13.51 (calc.); valence 3 and 4. Although curium follows americium in the periodic system, it was actually known before americium and was the third transuranium element to be discovered. It was identified by Seaborg, James, and Ghiorso in 1944 at the wartime Metallurgical Laboratory in Chicago as a result of helium-ion bombardment of ^{239}Pu in the Berkeley, California, 60-inch cyclotron. Visible amounts (30 μg) of ^{242}Cm, in the form of the hydroxide, were first isolated by Werner and Perlman of the University of California in 1947. In 1950, Crane, Wallmann, and Cunningham found that the magnetic susceptibility of microgram samples of CmF_3 was of the same magnitude as that of GdF_3. This provided direct experimental evidence for assigning an electronic configuration to Cm^{+3}. In 1951, the same workers prepared curium in its elemental form for the first time. Sixteen isotopes of curium are now known. The most stable, ^{247}Cm, with a half-life of 16 million years, is so short compared to the earth's age that any primordial curium must have disappeared long ago from the natural scene. Minute amounts of curium probably exist in natural deposits of uranium, as a result of a sequence of neutron captures and β decays sustained by the very low flux of neutrons naturally present in uranium ores. The presence of natural curium, however, has never been detected. ^{242}Cm and ^{244}Cm are available in multigram quantities. ^{248}Cm has been produced only in milligram amounts. Curium is similar in some regards to gadolinium, its rare-earth homolog, but it has a more complex crystal structure. Curium is silver in color, is chemically reactive, and is more electropositive than aluminum. CmO_2, Cm_2O_3, CmF_3, CmF_4, $CmCl_3$, $CmBr_3$, and CmI_3 have been prepared. Most compounds of trivalent curium are faintly yellow in color. ^{242}Cm generates about three watts of thermal energy per gram. This compares to one-half watt per gram of ^{238}Pu. This suggests use for curium as a power source. ^{244}Cm is now offered for sale at \$160/mg plus packing charges. ^{248}Cm is available at a cost of \$160/μg, plus packing charges, from the O.R.N.L. Curium absorbed into the body accumulates in the bones, and is therefore very toxic as its radiation destroys the red-cell forming mechanism. The maximum permissible total body burden of ^{244}Cm (soluble) in a human being is 0.3 μCi (microcurie).

Deuterium, an isotope of hydrogen — see Hydrogen.

Dubnium — see Element 104.

Dysprosium — (Gr. *dysprositos,* hard to get at), Dy; at. wt. 162.50(3); at. no. 66; m.p. 1412°C; b.p. 2567°C; sp. gr. 8.551 (25°C); valence 3. Dysprosium was discovered in 1886 by Lecoq de Boisbaudran, but not isolated. Neither the oxide nor the metal was available in relatively pure form until the development of ion-exchange separation and metallographic reduction techniques by Spedding and associates about 1950. Dysprosium occurs along with other so-called rare-earth or lanthanide elements in a variety of minerals such as *xenotime, fergusonite, gadolinite, euxenite, polycrase,* and *blomstrandine.* The most important sources, however, are from *monazite* and *bastnasite.* Dysprosium can be prepared by reduction of the trifluoride with calcium. The element has a metallic, bright silver luster. It is relatively stable in air at room temperature, and is readily attacked and dissolved, with the evolution of hydrogen, by dilute and concentrated mineral acids. The metal is soft enough to be cut with a knife and can be machined without sparking if overheating is avoided. Small amounts of impurities can greatly affect its physical properties. While dysprosium has not yet found many applications, its thermal neutron absorption cross-section and high melting point suggest metallurgical uses in nuclear control applications and for alloying with special stainless steels. A dysprosium oxide-nickel cermet has found use in cooling nuclear reactor rods. This cermet absorbs neutrons readily without swelling or contracting under prolonged neutron bombardment. In combination with vanadium and other rare earths, dysprosium has been used in making laser materials. Dysprosium-cadmium chalcogenides, as sources of infrared radiation, have been used for studying chemical reactions. The cost of dysprosium metal has dropped in recent years since the development of ion-exchange and solvent extraction techniques, and the discovery of large ore bodies. Thirty two isotopes and isomers are now known. The metal costs about \$4/g (99.9% purity).

Einsteinium — (Albert Einstein), Es; at. wt. (252); m.p. 860°C (est.); at. no. 99. Einsteinium, the seventh transuranic element of the actinide series to be discovered, was identified by Ghiorso and co-workers at Berkeley in December 1952 in debris from the first large thermonuclear explosion, which took place in the Pacific in November 1952. The isotope produced was the 20-day ^{253}Es isotope. In 1961, a sufficient amount of einsteinium was produced to permit separation of a macroscopic amount of ^{253}Es. This sample weighed about 0.01 μg. A special magnetic-type balance was used in making this determination. ^{253}Es so produced was used to produce mendelevium (Element 101). About 3 μg of einsteinium has been produced at Oak Ridge National Laboratories by irradiating for several years kilogram quantities of ^{239}Pu in a reactor to produce ^{242}Pu. This was then fabricated into pellets of plutonium oxide and aluminum powder, and loaded into target rods for an initial 1-year irradiation at the Savannah River Plant, followed by irradiation in a HFIR (High Flux Isotopic Reactor). After 4 months in the HFIR the targets were removed for chemical separation of the einsteinium from californium. Seventeen isotopes and isomers of einsteinium are now recognized. ^{254}Es has the longest half-life (276 days). Tracer studies using ^{253}Es show that einsteinium has chemical properties typical of a heavy trivalent, actinide element.

Element 104 — (at. wt. 261). In 1964, workers of the Joint Nuclear Research Institute at Dubna (U.S.S.R.) bombarded plutonium with accelerated 113 to 115 MeV neon ions. By measuring fission tracks in a special glass with a microscope, they detected an isotope that decays by spontaneous fission. They suggested that this isotope, which had a half-life of 0.3 ± 0.1 s might be ^{260}104, produced by the following reaction:

$$^{242}_{94}Pu + ^{22}_{10}Ne \rightarrow ^{260}104 + 4n$$

Element 104, the first *transactinide* element, is expected to have chemical properties similar to those of hafnium. It would, for example, form a relatively volatile compound with chlorine (a tetrachloride). The Soviet scientists have performed experiments aimed at chemical identification, and have attempted to show that the 0.3-s activity is more volatile than that of the relatively nonvolatile actinide trichlorides. This experiment does not fulfill the test of chemically separating the new element from all others, but it provides important evidence for evaluation. New data, reportedly issued by Soviet scientists, have reduced the half-life of the isotope they worked with from 0.3 to 0.15 s. The Dubna scientists suggest the name *kurchatovium* and symbol *Ku* for Element 104, in honor of Igor Vasilevich Kurchatov (1903—1960), late Head of Soviet Nuclear Research. The Dubna Group also has proposed the name *dubnium* for Element 104. In 1969, Ghiorso, Nurmia, Harris, K. A. Y. Eskola, and P. L. Eskola of the University of California at Berkeley reported they had positively identified two, and possibly three, isotopes of Element 104. The group also indicated that after repeated attempts so far they have been unable to produce isotope $^{260}104$ reported by the Dubna groups in 1964. The discoveries at Berkeley were made by bombarding a target of ^{249}Cf with ^{12}C nuclei of 71 MeV, and ^{13}C nuclei of 69 MeV. The combination of ^{12}C with ^{249}Cf followed by instant emission of four neutrons produced Element $^{257}104$. This isotope has a half-life of 4 to 5 s, decaying by emitting an alpha particle into ^{253}No, with a half-life of 105 s. The same reaction, except with the emission of three neutrons, was thought to have produced $^{258}104$ with a half-life of about 1/100 s. Element $^{259}104$ is formed by the merging of a ^{13}C nuclei with ^{249}Cf, followed by emission of three neutrons. This isotope has a half-life of 3 to 4 s, and decays by emitting an alpha particle into ^{255}No, which has a half-life of 185 s. Thousands of atoms of $^{257}104$ and $^{259}104$ have been detected. The Berkeley group believe their identification of $^{258}104$ was correct. As of January 1995 it was thought that eleven isotopes of Element 104 have been identified. The Berkeley group proposed for the new element the name *rutherfordium* (symbol Rf), in honor of Ernest Rutherford, New Zealand physicist. The claims for discovery and the naming of Element 104 are still in question. The International Union of Pure and Applied Physics proposed using the neutral temporary name, "unnilquadium" and more recently has recommended the name *dubnium*, with symbol Db, be used, even though many scientists have used the name *rutherfordium* for many years. The naming of this element is not yet settled.

Element 105 — In 1967 G. N. Flerov reported that a Soviet team working at the Joint Institute for Nuclear Research at Dubna may have produced a few atoms of $^{260}105$ and $^{261}105$ by bombarding ^{243}Am with ^{22}Ne. Their evidence was based on time-coincidence measurements of alpha energies. More recently, it was reported that early in 1970 Dubna scientists synthesized Element 105 and that by the end of April 1970 "had investigated all the types of decay of the new element and had determined its chemical properties." The Soviet group proposed the name *joliotium* for Element 105. In late April 1970, it was announced that Ghiorso, Nurmia, Harris, K. A. Y. Eskola, and P. L. Eskola, working at the University of California at Berkeley, had positively identified Element 105. The discovery was made by bombarding a target of ^{249}Cf with a beam of 84 MeV nitrogen nuclei in the Heavy Ion Linear Accelerator (HILAC). When a ^{15}N nuclear is absorbed by a ^{249}Cf nucleus, four neutrons are emitted and a new atom of $^{260}105$ with a half-life of 1.6 s is formed. While the first atoms of Element 105 are said to have been detected conclusively on March 5, 1970, there is evidence that Element 105 had been formed in Berkeley experiments a year earlier by the method described. Ghiorso and his associates have attempted to confirm Soviet findings by more sophisticated methods without success. The Berkeley Group proposed the name *hahnium,* after the late German scientist Otto Hahn (1879—1968), and *Ha* for the chemical symbol.

In October 1971, it was announced that two new isotopes of Element 105 were synthesized with the heavy ion linear accelerator by A. Ghiorso and co-workers at Berkeley. Element $^{261}105$ was produced both by bombarding ^{250}Cf with ^{15}N and by bombarding ^{249}Bk with ^{16}O. The isotope emits 8.93-MeV α particles and decays to ^{257}Lr with a half-life of about 1.8 s. Element $^{262}105$ was produced by bombarding ^{249}Bk with ^{18}O. It emits 8.45 MeV α particles and decays to ^{258}Lr with a half-life of about 40 s. Eight isotopes of Element 105 (unnilpentium) are now recognized. In 1994 the International Union of Pure and Applied Chemistry recommended the name *joliotium* for Element 105; however, the matter has not been settled.

Element 106 — The discovery of Element 106 took place in 1974 almost simultaneously at the Lawrence-Berkeley Laboratory and at the Joint Institute for Nuclear Research at Dubna (near Moscow). The Berkeley Group, under direction of Ghiorso, used the Super-Heavy Ion Linear Accelerator (Super HILAC) as a source of heavy ^{18}O ions to bombard a 259-μg target of ^{249}Cf. This resulted in the production and positive identification of $^{263}106$, which decayed with a half-life of 0.9 ± 0.2 s by the emission of alpha particles as follows:

$$^{263}106 \xrightarrow{\alpha} ^{259}104 \xrightarrow{\alpha} ^{255}No \xrightarrow{\alpha}.$$

The Dubna Team, directed by Flerov and Organessian, produced heavy ions of ^{54}Cr with their 310-cm heavy-ion cyclotron to bombard ^{207}Pb and ^{208}Pb and found a product that decayed with a half-life of 7 ms. They assigned $^{259}106$ to this isotope. Earlier, the IUPAC proposed a temporary name of *unnilhexium* for this element. The Berkeley Group, who first claimed positive identification, proposed the name *seaborgium* to Element 106. In 1994 the IUPAC recommended the name *rutherfordium* with the symbol Rf to this element; however, this name is being challenged by the Berkeley Group. The name *rutherfordium* has been used for many years for Element 104, which would lead to confusion if the name for Element 104 is changed. The naming of this element is now unsettled. It is now thought six isotopes of Element 106 have been identified. Two of the isotopes are believed to have half-lives of about 30 s.

Element 107 — This element, expected to have chemical properties similar to rhenium, was synthesized and unambiguously identified in 1981, using the Universal Linear Accelerator (UNILAC) at the Gesellschaft für Schwerionenforschung (G.S.I.) in Darmstadt, Germany. The discovery team was led by Armbruster and Münzenberg. The reaction producing the element was proposed and applied earlier by a Dubna Group led by Oganessian in 1976. A target of ^{209}Bi was bombarded by a beam of ^{54}Cr ions. In 1983 experiments at Dubna using the 157 -inch cyclotron, produced $^{262}107$ by the reaction $^{209}Bi + ^{54}Cr$. The alpha decay of ^{246}Cf, the sixth member in the decay chain of $^{262}107$, served to establish a 1-neutron reaction channel. The IUPAC has recently recommended the name *bohrium* with symbol Bh for Element 107. The naming of this element is still unsettled. Three isotopes of 107 are now recognized.

Element 108 — This element was first synthesized and identified in 1984 by the same G.S.I. Darmstadt group who first identified Elements 107 and 109. Presumably this element will have chemical properties similar to osmium. Isotope $^{265}108$ was produced using a beam of ^{58}Fe projectiles, produced by the Universal Linear Accelerator (UNILAC) to bombard a ^{208}Pb target. Discovery of Elements 107 and 109 was made using detection of isotopes with odd proton and neutron numbers. Elements having even atomic numbers are thought to be less stable against spontaneous fusion than odd elements. The production of $^{265}108$ in the same reaction as was used at G.S.I. was confirmed at Dubna with detection of the 7th member of the decay chain ^{253}Es. Isotopes of Element 108 were believed to decay by spontaneous fission, explaining why 109 was produced before 108. Isotope $^{265}108$

and $^{266}108$ are thought to decay to $^{261}106$, which in turn decays to $^{257}104$ and ^{253}No. The IUPAC recommends the name *hahnium* (Hn), honoring Otto Hahn, for this element; however, Element 105 has already been referred to as *hahnium* for many years, which would lead to confusion. The naming of Element 108 has not yet been settled.

Element 109 — On August 29, 1992, Element 109 was made and identified by physicists at the Heavy Ion Research Laboratory (G.S.I.), Darmstadt, Germany, by bombarding a target of ^{209}Bi with accelerated nuclei of ^{58}Fe. The production of Element 109 has been extremely small. It took a week of target bombardment (10^{11} nuclear encounters) to produce a single atom of 109. Oganessian and his team at Dubna in 1994 repeated the Darmstadt experiment using a 10-fold irradiation dose. One fission event from seven alpha decays of 109 was observed, thus indirectly confirming the existence of isotope $^{266}109$. The IUPAC has recommended adoption of the name *meitnerium* (Mt) for this element, honoring L. Meitner.

Element 110 — In 1987 Oganessian, et al., at Dubna, claimed discovery of this element. Their experiments indicated the spontaneous fissioning nuclide $^{272}110$ with a half-life of 10 ms. More recently a group led by Armbruster at G.S.I. in Darmstadt, Germany, reported evidence of $^{269}110$, which was produced by bombarding lead for many days with more than 10^{18} nickel atoms. A detector searched each collision for Element 110's distinct decay sequence. On November 9, 1994, evidence of 110 was detected. Berkeley scientists, in 1991, performed similar experiments and reported evidence of 110, but this was not confirmed. Workers at Dubna have experiments underway to produce $^{273}110$ by bombarding plutonium with sulfur atoms. Other experiments at G.S.I. are now searching for heavier isotopes. Several years ago the IUPAC suggested the use of the temporary name *ununnilium* for Element 110 when it was found.

Element 111 — On December 20, 1994, scientists at GSI Darmstadt, Germany announced they had detected three atoms of a new element with 111 protons and 161 neutrons. This element was made by bombarding ^{83}Bi with ^{28}Ni. Signals of Element 111 appeared for less than 0.002 sec, then decayed into lighter elements including Element 109_{268} and Element 107_{264}. These isotopes had not previously been observed. A name for Element 111 has not been suggested although IUPAC has suggested a temporary name of Unununium.

Element 112 — In late February 1996, Siguard Hofmann and his collaborators at GSI Darmstadt announced their discovery of Element 112, having 112 protons and 165 neutrons, with an atomic mass of 277. This element was made by bombarding a lead target with high-energy zinc ions. A single nucleus of Element 112 was detected, which decayed after less than 0.001 sec by emitting an α particle, consisting of two protons and two neutrons. This created Element 110_{273}, which in turn decayed by emitting an α particle to form a new isotope of Element 108 and so on. Evidence indicates that nuclei with 162 neutrons are held together more strongly than nuclei with a smaller or larger number of neutrons. This suggests a narrow "peninsula" of relatively stable isotopes around Element 114. GSI scientists are experimenting to bombard targets with ions heavier than zinc to produce Elements 113 and 114. A name has not yet been suggested for Element 112, although the IUPAC suggested the temproary name of ununbium when the element was discovered.

Erbium — (*Ytterby*, a town in Sweden), Er; at. wt. 167.26(3); at. no. 68; m.p. 1529°C; b.p. 2868°C; sp. gr. 9.066 (25°C); valence 3, Erbium, one of the so-called rare-earth elements of the lanthanide series, is found in the minerals mentioned under dysprosium above. In 1842 Mosander separated "yttria," found in the mineral *gadolinite*, into three fractions which he called *yttria, erbia,* and *terbia*. The names *erbia* and *terbia* became confused in this early period. After 1860, Mosander's *terbia* was known as *erbia*, and after 1877, the earlier known *erbia* became *terbia*. The *erbia* of this period was later shown to consist of five oxides, now known as *erbia, scandia, holmia, thulia* and *ytterbia*. By 1905 Urbain and James independently succeeded in isolating fairly pure Er_2O_3. Klemm and Bommer first produced reasonably pure erbium metal in 1934 by reducing the anhydrous chloride with potassium vapor. The pure metal is soft and malleable and has a bright, silvery, metallic luster. As with other rare-earth metals, its properties depend to a certain extent on the impurities present. The metal is fairly stable in air and does not oxidize as rapidly as some of the other rare-earth metals. Naturally occurring erbium is a mixture of six isotopes, all of which are stable. Twenty four radioactive isotopes of erbium are also recognized. Recent production techniques, using ion-exchange reactions, have resulted in much lower prices of the rare-earth metals and their compounds in recent years. The cost of 99.9% erbium metal is about $4/gm. Erbium is finding nuclear and metallurgical uses. Added to vanadium, for example, erbium lowers the hardness and improves workability. Most of the rare-earth oxides have sharp absorption bands in the visible, ultraviolet, and near infrared. This property, associated with the electronic structure, gives beautiful pastel colors to many of the rare-earth salts. Erbium oxide gives a pink color and has been used as a colorant in glasses and porcelain enamel glazes.

Europium — (Europe), Eu; at. wt. 151.965(9); at. no. 63; m.p. 822°C; b.p. 1529°C; sp. gr. 5.244 (25°C); valence 2 or 3. In 1890 Boisbaudran obtained basic fractions from samarium-gadolinium concentrates which had spark spectral lines not accounted for by samarium or gadolinium. These lines subsequently have been shown to belong to europium. The discovery of europium is generally credited to Demarcay, who separated the rare earth in reasonably pure form in 1901. The pure metal was not isolated until recent years. Europium is now prepared by mixing Eu_2O_3 with a 10%-excess of lanthanum metal and heating the mixture in a tantalum crucible under high vacuum. The element is collected as a silvery-white metallic deposit on the walls of the crucible. As with other rare-earth metals, except for lanthanum, europium ignites in air at about 150 to 180°C. Europium is about as hard as lead and is quite ductile. It is the most reactive of the rare-earth metals, quickly oxidizing in air. It resembles calcium in its reaction with water. *Bastnasite* and *monazite* are the principal ores containing europium. Europium has been identified spectroscopically in the sun and certain stars. Europium isotopes are good neutron absorbers and are being studied for use in nuclear control applications. Europium oxide is now widely used as a phosphor activator and europium-activated yttrium vanadate is in commercial use as the red phosphor in color TV tubes. Europium-doped plastic has been used as a laser material. With the development of ion-exchange techniques and special processes, the cost of the metal has been greatly reduced in recent years. Natural europium contains two stable isotopes. Thirty five other radioactive isotopes and isomers are known. Europium is one of the rarest and most costly of the rare-earth metals. It is priced at about $50/g (99.9% pure).

Fermium — (Enrico Fermi), Fm; at. wt. (257); at. no. 100; m.p. 1527°C. Fermium, the eighth transuranium element of the actinide series to be discovered, was identified by Ghiorso and co-workers in 1952 in the debris from a thermonuclear explosion in the Pacific in work involving the University of California Radiation Laboratory, the Argonne National Laboratory, and the Los Alamos Scientific Laboratory. The isotope produced was the 20-hour ^{255}Fm. During 1953 and early 1954, while discovery of elements 99 and 100 was withheld from publication for security reasons, a group from the Nobel Institute of Physics in Stockholm bombarded ^{238}U with ^{16}O ions, and isolated a 30-min α-emitter, which they ascribed to $^{250}100$, without claiming discovery of the element. This isotope has since been identified positively, and the 30-min half-life confirmed. The chemical properties of fermium have been studied solely with tracer amounts, and in normal aqueous media only the (III) oxidation state appears to exist. The isotope ^{254}Fm and heavier isotopes can be produced by intense neutron irradiation of lower elements such as plutonium by a process of successive neutron capture interspersed with beta decays until these mass numbers and atomic numbers are reached. Twenty isotopes and isomers of fermium are known to exist. ^{257}Fm, with a half-life of about 100.5 days, is the longest lived. ^{250}Fm, with a half-life of 30 min, has been shown to be a product of decay of Element ^{254}102. It was by chemical identification of ^{250}Fm that production of Element 102 (nobelium) was confirmed.

Fluorine — (L. and F. *fluere*, flow, or flux), F; at. wt. 18.9984032(9); at. no. 9; m.p. –219.62°C (1 atm); b.p. –188.12°C (1 atm); density 1.696 g/L (0°C, 1 atm); liq. den. at b.p. 1.50 g/cm^3; valence 1. In 1529, Georigius Agricola described the use of fluorspar as a flux, and as early as 1670 Schwandhard found that glass was etched when exposed to fluorspar treated with acid. Scheele and many later investigators, including Davy, Gay-Lussac, Lavoisier, and Thenard, experimented with hydrofluoric acid, some experiments ending in tragedy. The element was finally isolated in 1886 by Moisson after nearly 74 years of continuous effort. Fluorine occurs chiefly in *fluorspar* (CaF$_2$) and *cryolite* (Na$_2$AlF$_6$), but is rather widely distributed in other minerals. It is a member of the halogen family of elements, and is obtained by electrolyzing a solution of potassium hydrogen fluoride in anhydrous hydrogen fluoride in a vessel of metal or transparent fluorspar. Modern commercial production methods are essentially variations on the procedures first used by Moisson. Fluorine is the most electronegative and reactive of all elements. It is a pale yellow, corrosive gas, which reacts with practically all organic and inorganic substances. Finely divided metals, glass, ceramics, carbon, and even water burn in fluorine with a bright flame. Until World War II, there was no commercial production of elemental fluorine. The atom bomb project and nuclear energy applications, however, made it necessary to produce large quantities. Safe handling techniques have now been developed and it is possible at present to transport liquid fluorine by the ton. Fluorine and its compounds are used in producing uranium (from the hexafluoride) and more than 100 commercial fluorochemicals, including many well-known high-temperature plastics. Hydrofluoric acid is extensively used for etching the glass of light bulbs, etc. Fluorochloro hydrocarbons are extensively used in air conditioning and refrigeration. It has been suggested that fluorine can be substituted for hydrogen wherever it occurs in organic compounds, which could lead to an astronomical number of new fluorine compounds. The presence of fluorine as a soluble fluoride in drinking water to the extent of 2 ppm may cause mottled enamel in teeth, when used by children acquiring permanent teeth; in smaller amounts, however, fluorides are said to be beneficial and used in water supplies to prevent dental cavities. Elemental fluorine has been studied as a rocket propellant as it has an exceptionally high specific impulse value. Compounds of fluorine with rare gases have now been confirmed. Fluorides of xenon, radon, and krypton are among those known. Elemental fluorine and the fluoride ion are highly toxic. The free element has a characteristic pungent odor, detectable in concentrations as low as 20 ppb, which is below the safe working level. The recommended maximum allowable concentration for a daily 8-hour time-weighted exposure is 1 ppm. Fluorine is known to have thirteen isotopes.

Francium — (France), Fr; at. no. 87; at. wt. (223); m.p. 27°C; b.p. 677°C; valence 1. Discovered in 1939 by Mlle. Marguerite Perey of the Curie Institute, Paris. Francium, the heaviest known member of the alkali metal series, occurs as a result of an alpha disintegration of actinium. It can also be made artificially by bombarding thorium with protons. While it occurs naturally in uranium minerals, there is probably less than an ounce of francium at any time in the total crust of the earth. It has the highest equivalent weight of any element, and is the most unstable of the first 101 elements of the periodic system. Thirty-five isotopes and isomers of francium are recognized. The longest lived ^{223}Fr(Ac, K), a daughter of ^{227}Ac, has a half-life of 21.8 min. This is the only isotope of francium occurring in nature. Because all known isotopes of francium are highly unstable, knowledge of the chemical properties of this element comes from radiochemical techniques. No weighable quantity of the element has been prepared or isolated. The chemical properties of francium most closely resemble cesium.

Gadolinium — (*gadolinite*, a mineral named for Gadolin, a Finnish chemist), Gd; at. wt. 157.25(3); at. no. 64; m.p. 1313°C; b.p. 3273°C; sp. gr. 7.901 (25°C); valence 3. Gadolinia, the oxide of gadolinium, was separated by Marignac in 1880 and Lecoq de Boisbaudran independently isolated the element from Mosander's "yttria" in 1886. The element was named for the mineral *gadolinite* from which this rare earth was originally obtained. Gadolinium is found in several other minerals, including *monazite* and *bastnasite,* which are of commercial importance. The element has been isolated only in recent years. With the development of ion-exchange and solvent extraction techniques, the availability and price of gadolinium and the other rare-earth metals have greatly improved. Thirty isotopes and isomers of gadolinium are now recognized; seven are stable and occur naturally. The metal can be prepared by the reduction of the anhydrous fluoride with metallic calcium. As with other related rare-earth metals, it is silvery white, has a metallic luster, and is malleable and ductile. At room temperature, gadolinium crystallizes in the hexagonal, close-packed α form. Upon heating to 1235°C, α gadolinium transforms into the β form, which has a body-centered cubic structure. The metal is relatively stable in dry air, but in moist air it tarnishes with the formation of a loosely adhering oxide film which spalls off and exposes more surface to oxidation. The metal reacts slowly with water and is soluble in dilute acid. Gadolinium has the highest thermal neutron capture cross-section of any known element (49,000 barns). Natural gadolinium is a mixture of seven isotopes. Two of these, ^{155}Gd and ^{157}Gd, have excellent capture characteristics, but they are present naturally in low concentrations. As a result, gadolinium has a very fast burnout rate and has limited use as a nuclear control rod material. It has been used in making gadolinium yttrium garnets, which have microwave applications. Compounds of gadolinium are used in making phosphors for color TV tubes. The metal has unusual superconductive properties. As little as 1% gadolinium has been found to improve the workability and resistance of iron, chromium, and related alloys to high temperatures and oxidation. Gadolinium ethyl sulfate has extremely low noise characteristics and may find use in duplicating the performance of amplifiers, such as the maser. The metal is ferromagnetic. Gadolinium is unique for its high magnetic moment and for its special Curie temperature (above which ferromagnetism vanishes) lying just at room temperature. This suggests uses as a magnetic component that senses hot and cold. The price of the metal is $2/g (99.9% purity).

Gallium — (L. *Gallia*, France; also from Latin, *gallus,* a translation of Lecoq, a cock), Ga; at. wt. 69.723(1); at. no. 31; m.p. 29.7646°C; b.p. 2204°C; sp. gr. 5.904 (29.6°C) solid; sp. gr. 6.095 (29.6°C) liquid; valence 2 or 3. Predicted and described by Mendeleev as ekaaluminum, and discovered spectroscopically by Lecoq de Boisbaudran in 1875, who in the same year obtained the free metal by electrolysis of a solution of the hydroxide in KOH. Gallium is often found as a trace element in *diaspore, sphalerite, germanite, bauxite,* and *coal.* Some flue dusts from burning coal have been shown to contain as much as 1.5% gallium. It is the only metal, except for mercury, cesium, and rubidium, which can be liquid near room temperatures; this makes possible its use in high-temperature thermometers. It has one of the longest liquid ranges of any metal and has a low vapor pressure even at high temperatures. There is a strong tendency for gallium to supercool below its freezing point. Therefore, seeding may be necessary to initiate solidification. Ultra-pure gallium has a beautiful, silvery appearance, and the solid metal exhibits a conchoidal fracture similar to glass. The metal expands 3.1% on solidifying; therefore, it should not be stored in glass or metal containers, as they may break as the metal solidifies. Gallium wets glass or porcelain, and forms a brilliant mirror when it is painted on glass. It is widely used in doping semiconductors and producing solid-state devices such as transistors. High-purity gallium is attacked only slowly by mineral acids. Magnesium gallate containing divalent impurities such as Mn^{+2} is finding use in commercial ultraviolet activated powder phosphors. Gallium arsenide is capable of converting electricity directly into coherent light. Gallium readily alloys with most metals, and has been used as a component in low-melting alloys. Its toxicity appears to be of a low order, but should be handled with care until more data are forthcoming. Natural gallium contains two stable isotopes. Twenty three other isotopes, one of which is an isomer, are known. The metal can be supplied in ultrapure form (99.99999+%). The cost is about $4/g.

Germanium — (L. *Germania,* Germany), Ge; at. wt. 72.61(2); at. no. 32; m.p. 938.25°C; b.p. 2833°C; sp. gr. 5.323 (25°C); valence 2 and 4. Predicted by Mendeleev in 1871 as ekasilicon, and discovered by Winkler in 1886. The metal is found in *argyrodite,* a sulfide of germanium and silver; in *germanite,* which contains 8% of the element; in zinc ores; in coal; and in other minerals. The element is frequently obtained commercially from flue dusts of smelters processing zinc ores, and has been recovered from the by-products of combustion of certain coals. Its presence in coal insures a large reserve of the element in the years to come. Germanium can be separated from other metals by fractional distillation of its volatile tetrachloride. The tetrachloride may then be hydrolyzed to give GeO_2; the dioxide can be reduced with hydrogen to give the metal. Recently developed zone-refining techniques permit the production of germanium of ultra-high purity. The element is a gray-white metalloid, and in its pure state is crystalline and brittle, retaining its luster in air at room temperature. It is a very important semiconductor material. Zone-refining techniques have led to production of crystalline germanium for semiconductor use with an impurity of only one part in 10^{10}. Doped with arsenic, gallium, or other elements, it is used as a transistor element in thousands of electronic applications. Its application as a semiconductor element now provides the largest use for germanium. Germanium is also finding many other applications including use as an alloying agent, as a phosphor in fluorescent lamps, and as a catalyst. Germanium and germanium oxide are transparent to the infrared and are used in infrared spectroscopes and other optical equipment, including extremely sensitive infrared detectors. Germanium oxide's high index of refraction and dispersion has made it useful as a component of glasses used in wide-angle camera lenses and microscope objectives. The field of organogermanium chemistry is becoming increasingly important. Certain germanium compounds have a low mammalian toxicity, but a marked activity against certain bacteria, which makes them of interest as chemotherapeutic agents. The cost of germanium is about $3/g (99.999% purity). Twenty nine isotopes and isomers are known, five of which occur naturally.

Gold — (Sanskrit *Jval;* Anglo-Saxon *gold*), Au (L. *aurum,* gold); at. wt. 196.96654(3); at. no. 79; m.p. 1064.18°C; b.p. 2856°C; sp. gr. ~19.3 (20°C); valence 1 or 3. Known and highly valued from earliest times, gold is found in nature as the free metal and in tellurides; it is very widely distributed and is almost always associated with quartz or pyrite. It occurs in veins and alluvial deposits, and is often separated from rocks and other minerals by sluicing and panning operations. About two thirds of the world's gold output comes from South Africa, and about two thirds of the total U.S. production comes from South Dakota and Nevada. The metal is recovered from its ores by cyaniding, amalgamating, and smelting processes. Refining is also frequently done by electrolysis. Gold occurs in sea water to the extent of 0.1 to 2 mg/ton, depending on the location where the sample is taken. As yet, no method has been found for recovering gold from sea water profitably. It is estimated that all the gold in the world, so far refined, could be placed in a single cube 60 ft on a side. Of all the elements, gold in its pure state is undoubtedly the most beautiful. It is metallic, having a yellow color when in a mass, but when finely divided it may be black, ruby, or purple. The Purple of Cassius is a delicate test for auric gold. It is the most malleable and ductile metal; 1 oz. of gold can be beaten out to 300 ft^2. It is a soft metal and is usually alloyed to give it more strength. It is a good conductor of heat and electricity, and is unaffected by air and most reagents. It is used in coinage and is a standard for monetary systems in many countries. It is also extensively used for jewelry, decoration, dental work, and for plating. It is used for coating certain space satellites, as it is a good reflector of infrared and is inert. Gold, like other precious metals, is measured in troy weight; when alloyed with other metals, the term *carat* is used to express the amount of gold present, 24 carats being pure gold. For many years the value of gold was set by the U.S. at $20.67/troy ounce; in 1934 this value was fixed by law at $35.00/troy ounce, 9/10th fine. On March 17, 1968, because of a gold crisis, a two-tiered pricing system was established whereby gold was still used to settle international accounts at the old $35.00/troy ounce price while the price of gold on the private market would be allowed to fluctuate. Since this time, the price of gold on the free market has fluctuated widely. The price of gold on the free market reached a price of $620/troy oz. in January 1980. The most common gold compounds are auric chloride ($AuCl_3$) and chlorauric acid ($HAuCl_4$), the latter being used in photography for toning the silver image. Gold has forty six recognized isotopes and isomers; ^{198}Au, with a half-life of 2.7 days, is used for treating cancer and other diseases. Disodium aurothiomalate is administered intramuscularly as a treatment for arthritis. A mixture of one part nitric acid with three of hydrochloric acid is called *aqua regia* (because it dissolved gold, the King of Metals). Gold is available commercially with a purity of 99.999+%. For many years the temperature assigned to the freezing point of gold has been 1063.0°C; this has served as a calibration point for the International Temperature Scales (ITS-27 and ITS-48) and the International Practical Temperature Scale (IPTS-48). In 1968, a new International Practical Temperature Scale (IPTS-68) was adopted, which demanded that the freezing point of gold be changed to 1064.43°C. In 1990 a new International Temperature Scale (ITS-90) was adopted bringing the t.p. (triple point) of H_2O (t$_{90}$ (°C)) to 0.01°C and the freezing point of gold to 1064.18°C. The specific gravity of gold has been found to vary considerably depending on temperature, how the metal is precipitated, and cold-worked. As of January 1996, gold was priced at about $390/troy oz. ($12.50/g).

Hafnium — (*Hafnia,* Latin name for Copenhagen), Hf; at. wt. 178.49(2); at. no. 72; m.p. 2233 ± 20°C; b.p. 4603°C; sp. gr. 13.31 (20°C); valence 4. Hafnium was thought to be present in various minerals and concentrations many years prior to its discovery, in 1923, credited to D. Coster and G. von Hevesey. On the basis of the Bohr theory, the new element was expected to be associated with zirconium. It was finally identified in *zircon* from Norway, by means of X-ray spectroscopic analysis. It was named in honor of the city in which the discovery was made. Most zirconium minerals contain 1 to 5% hafnium. It was originally separated from zirconium by repeated recrystallization of the double ammonium or potassium fluorides by von Hevesey and Jantzen. Metallic hafnium was first prepared by van Arkel and deBoer by passing the vapor of the tetraiodide over a heated tungsten filament. Almost all hafnium metal now produced is made by reducing the tetrachloride with magnesium or with sodium (Kroll Process). Hafnium is a ductile metal with a brilliant silver luster. Its properties are considerably influenced by the impurities of zirconium present. Of all the elements, zirconium and hafnium are two of the most difficult to separate. Their chemistry is almost identical, however, the density of zirconium is about half that of hafnium. Very pure hafnium has been produced, with zirconium being the major impurity. Natural hafnium contains six isotopes, one of which is slightly radioactive. Hafnium has a total of 40 recognized isotopes and isomers. Because hafnium has a good absorption cross section for thermal neutrons (almost 600 times that of zirconium), has excellent mechanical properties, and is extremely corrosion resistant, it is used for reactor control rods. Such rods are used in nuclear submarines. Hafnium has been successfully alloyed with iron, titanium, niobium, tantalum, and other metals. Hafnium carbide is the most refractory binary composition known, and the nitride is the most refractory of all known metal nitrides (m.p. 3310°C). Hafnium is used in gas-filled and incandescent lamps, and is an efficient "getter" for scavenging oxygen and nitrogen. Finely divided hafnium is pyrophoric and can ignite spontaneously in air. Care should be taken when machining the metal or when handling hot sponge hafnium. At 700°C hafnium rapidly absorbs hydrogen to form the composition $HfH_{1.86}$. Hafnium is resistant to concentrated alkalis, but at elevated temperatures reacts with oxygen, nitrogen, carbon, boron, sulfur, and silicon. Halogens react directly to form tetrahalides. The price of the metal is in the broad range of $1/g to $3/g, depending on purity and quantity. The yearly demand for hafnium in the U.S. is now in excess of 50,000 kg.

Hahnium — see Element 105 (unnilpentium) and also under Element 108 (unniloctium).

Helium — (Gr. *helios*, the sun), He; at. wt. 4.002602(2); at. no. 2; m.p. below — 272.2°C (26 atm); b.p. — 268.93°C; density 0.1785 g/l (0°C, 1 atm); liquid density 7.62 lb/ft^3 at. b.p.; valence usually 0. Evidence of the existence of helium was first obtained by Janssen during the solar eclipse of 1868 when he detected a new line in the solar spectrum; Lockyer and Frankland suggested the name *helium* for the new element; in 1895, Ramsay discovered helium in the uranium mineral *clevite*, and it was independently discovered in cleveite by the Swedish chemists Cleve and Langlet about the same time. Rutherford and Royds in 1907 demonstrated that α particles are helium nuclei. Except for hydrogen, helium is the most abundant element found throughout the universe. Helium is extracted from natural gas; all natural gas contains at least trace quantities of helium. It has been detected spectroscopically in great abundance, especially in the hotter stars, and it is an important component in both the proton-proton reaction and the carbon cycle, which account for the energy of the sun and stars. The fusion of hydrogen into helium provides the energy of the hydrogen bomb. The helium content of the atmosphere is about 1 part in 200,000. While it is present in various radioactive minerals as a decay product, the bulk of the Free World's supply is obtained from wells in Texas, Oklahoma, and Kansas. The only known helium extraction plants, outside the United States, in 1996 were in Eastern Europe (Poland), the Russia, China, and India. The cost of helium fell from $2500/ft^3 in 1915 to 1.5¢/ft^3 in 1940. Commercial helium in 1996 was priced at about $70/300 cu. ft. (8.5 cu. meters). Helium has the lowest melting point of any element and has found wide use in cryogenic research, as its boiling point is close to absolute zero. Its use in the study of superconductivity is vital. Using liquid helium, Kurti and co-workers, and others, have succeeded in obtaining temperatures of a few microkelvins by the adiabatic demagnetization of copper nuclei, starting from about 0.01 K. Seven isotopes of helium are known. Liquid helium (He^4) exists in two forms: He^{4}I and He^{4}II, with a sharp transition point at 2.174 K (3.83 cm Hg). He^{4}I (above this temperature) is a normal liquid, but He^{4}II (below it) is unlike any other known substance. It expands on cooling; its conductivity for heat is enormous; and neither its heat conduction nor viscosity obeys normal rules. It has other peculiar properties. Helium is the only liquid that cannot be solidified by lowering the temperature. It remains liquid down to absolute zero at ordinary pressures, but it can readily be solidified by increasing the pressure. Solid ^{3}He and ^{4}He are unusual in that both can readily be changed in volume by more than 30% by application of pressure. The specific heat of helium gas is unusually high. The density of helium vapor at the normal boiling point is also very high, with the vapor expanding greatly when heated to room temperature. Containers filled with helium gas at 5 to 10 K should be treated as though they contained liquid helium due to the large increase in pressure resulting from warming the gas to room temperature. While helium normally has a 0 valence, it seems to have a weak tendency to combine with certain other elements. Means of preparing helium diflouride have been studied, and species such as HeNe and the molecular ions He$^+$ and He^{++} have been investigated. Helium is widely used as an inert gas shield for arc welding; as a protective gas in growing silicon and germanium crystals, and in titanium and zirconium production; as a cooling medium for nuclear reactors, and as a gas for supersonic wind tunnels. A mixture of helium and oxygen is used as an artificial atmosphere for divers and others working under pressure. Different ratios of He/O$_2$ are used for different depths at which the diver is operating. Helium is extensively used for filling balloons as it is a much safer gas than hydrogen. One of the recent largest uses for helium has been for pressuring liquid fuel rockets. A Saturn booster such as used on the Apollo lunar missions required about 13 million ft^3 of helium for a firing, plus more for checkouts. Liquid helium's use in magnetic resonance imaging (MRI) continues to increase as the medical profession accepts and develops new uses for the equipment. This equipment is providing accurate diagnoses of problems where exploratory surgery has previously been required to determine problems. Another medical application that is being developed uses MRI to determine by blood analysis whether a patient has any form of cancer. Lifting gas applications are increasing. Various companies in addition to Goodyear, are now using "blimps" for advertising. The Navy and the Air Force are investigating the use of airships to provide early warning systems to detect low-flying cruise missiles. The Drug Enforcement Agency has used radar-equipped blimps to detect drug smugglers along the southern border of the U.S. In addition, NASA is currently using helium-filled balloons to sample the atmosphere in Antarctica to determine what is depleting the ozone layer that protects Earth from harmful U.V. radiation. Research on and development of materials which become superconductive at temperatures well above the boiling point of helium could have a major impact on the demand for helium. Less costly refrigerants having boiling points considerably higher could replace the present need to cool such superconductive materials to the boiling point of helium.

Holmium — (L. *Holmia* , for Stockholm), Ho; at. wt. 164.93032(3); at. no 67; m.p. 1474°C; b.p. 2700°C; sp. gr. 8.795 (25°C); valence + 3. The spectral absorption bands of holmium were noticed in 1878 by the Swiss chemists Delafontaine and Soret, who announced the existence of an "Element X". Cleve, of Sweden, later independently discovered the element while working on erbia earth. The element is named after Cleve's native city. Pure holmia, the yellow oxide, was prepared by Homberg in 1911. Holmium occurs in *gadolinite, monazite,* and in other rare-earth minerals. It is commercially obtained from monazite, occurring in that mineral to the extent of about 0.05%. It has been isolated by the reduction of its anhydrous chloride or fluoride with calcium metal. Pure holmium has a metallic to bright silver luster. It is relatively soft and malleable, and is stable in dry air at room temperature, but rapidly oxidizes in moist air and at elevated temperatures. The metal has unusual magnetic properties. Few uses have yet been found for the element. The element, as with other rare earths, seems to have a low acute toxic rating. Natural holmium is not radioactive. Holmium has 46 other isotopes known, all of which are radioactive. The price of 99.9% holmium metal is about $15/g.

Hydrogen — (Gr. *hydro*, water, and *genes*, forming), H; at. wt. 1.00794(7); at. no. 1; m.p. –259.34°C; b.p. –252.87°C; density 0.08988 g/l; density (liquid) 70.8 g/l (–253°C); density (solid) 70.6 g/l (–262°C); valence 1. Hydrogen was prepared many years before it was recognized as a distinct substance by Cavendish in 1766. It was named by Lavoisier. Hydrogen is the most abundant of all elements in the universe, and it is thought that the heavier elements were, and still are, being built from hydrogen and helium. It has been estimated that hydrogen makes up more than 90% of all the atoms or three quarters of the mass of the universe. It is found in the sun and most stars, and plays an important part in the proton-proton reaction and carbon-nitrogen cycle, which accounts for the energy of the sun and stars. It is thought that hydrogen is a major component of the planet Jupiter and that at some depth in the planet's interior the pressure is so great that solid molecular hydrogen is converted into solid metallic hydrogen. In 1973, it was reported that a group of Russian experimenters may have produced metallic hydrogen at a pressure of 2.8 Mbar. At the transition the density changed from 1.08 to 1.3 g/cm^3. Earlier, in 1972, a Livermore (California) group also reported on a similar experiment in which they observed a pressure-volume point centered at 2 Mbar. It has been predicted that metallic hydrogen may be metastable; others have predicted it would be a superconductor at room temperature. On earth, hydrogen occurs chiefly in combination with oxygen in water, but it is also present in organic matter such as living plants, petroleum, coal, etc. It is present as the free element in the atmosphere, but only to the extent of less than 1 ppm by volume. It is the lightest of all gases, and combines with other elements, sometimes explosively, to form compounds. Great quantities of hydrogen are required commercially for the fixation of nitrogen from the air in the Haber ammonia process and for the hydrogenation of fats and oils. It is also used in large quantities in methanol production, in hydrodealkylation, hydrocracking, and hydrodesulfurization. It is also used as a rocket fuel, for welding, for production of hydrochloric acid, for the reduction of metallic ores, and for filling balloons. The lifting power of 1 ft^3 of hydrogen gas is about 0.076 lb at 0°C, 760 mm pressure. Production of hydrogen in the U.S. alone now amounts to about 3 billion cubic feet per year. It is prepared by the action of steam on heated carbon, by decomposition of certain hydrocarbons with heat, by the electrolysis of water, or by the displacement from acids by certain metals. It is also produced by the action of sodium or potassium hydroxide on aluminum. Liquid hydrogen is important in cryogenics and in the study of superconductivity, as its melting point is only a 20 degrees above absolute zero. The ordinary isotope of hydrogen, H, is known as *protium*. In 1932, Urey announced the discovery of a stable isotope, deuterium (^{2}H or D) with an atomic weight of 2. Deuterium is present in natural hydrogen to the extent of 0.015%. Two years later an unstable isotope, tritium (H), with an atomic weight of 3 was discovered. Tritium has a half-life of about 12.5 years. Tritium atoms are also present in hydrogen but in much smaller proportion. Tritium is readily produced in nuclear reactors and is used in the production of the hydrogen bomb. It is also used as a radioactive agent in making luminous paints, and as a tracer. Deuterium gas is readily available, without permit, at about $1/l. Heavy water, deuterium oxide (D_2O), which is used as a moderator to slow down neutrons, is available without permit at a cost of 6c to $1/g, depending on quantity and purity. Quite apart from isotopes, it has been shown that hydrogen gas under ordinary conditions is a mixture of two kinds of molecules, known as *ortho-* and *para*-hydrogen, which differ from one another by the spins of their electrons and nuclei. Normal hydrogen at room temperature contains 25% of the *para* form and 75% of the *ortho* form. The *ortho* form cannot be prepared in the pure state. Since the two forms differ in energy, the physical properties also differ. The melting and boiling points of *para*hydrogen are about 0.1°C lower than those of normal hydrogen. Consideration is being given to an entire economy based on solar- and nuclear-generated hydrogen. Located in remote regions, power plants would electrolyze sea water; the hydrogen produced would travel to distant cities by pipelines. Pollution-free hydrogen could replace natural gas, gasoline, etc., and could serve as a reducing agent in metallurgy, chemical processing, refining, etc. It could also be used to convert trash into methane and ethylene. Public acceptance, high capital investment, and the high present cost of hydrogen with respect to present fuels are but a few of the problems facing establishment of such an economy.

Indium — (from the brilliant indigo line in its spectrum),In; at. wt. 114.818(3); at. no. 49; f.p. 156.5985°C; b.p. 2072°C; sp. gr. 7.31 (20°C); valence 1, 2, or 3. Discovered by Reich and Richter, who later isolated the metal. Indium is most frequently associated with zinc materials, and it is from these that most commercial indium is now obtained; however, it is also found in iron, lead, and copper ores. Until 1924, a gram or so constituted the world's supply of this element in isolated form. It is probably about as abundant as silver. About 4 million troy ounces of indium are now produced annually in the Free World. Canada is presently producing more than 1,000,000 troy ounces annually. The present cost of indium is about $2 to $10/g, depending on quantity and purity. It is available in ultrapure form. Indium is a very soft, silvery-white metal with a brilliant luster. The pure metal gives a high-pitched "cry" when bent. It wets glass, as does gallium. It has found application in making low-melting alloys; an alloy of 24% indium-76% gallium is liquid at room temperature. It is used in making bearing alloys, germanium transistors, rectifiers, thermistors, and photoconductors. It can be plated onto metal and evaporated onto glass, forming a mirror as good as that made with silver but with more resistance to atmospheric corrosion. There is evidence that indium has a low order of toxicity; however, care should be taken until further information is available. Sixty seven isotopes and isomers are now recognized (more than any other element). Natural indium contains two isotopes. One is stable. The other, ^{115}In, comprising 95.71% of natural indium is slightly radioactive with a very long half-life.

Iodine — (Gr. *iodes*, violet), I; at. wt. 126.90447(3); at. no. 53; m.p. 113.7°C; b.p. 184.4°C; density of the gas 11.27 g/l; sp. gr. solid 4.93 (20°C); valence 1, 3, 5, or 7. Discovered by Courtois in 1811. Iodine, a halogen, occurs sparingly in the form of iodides in sea water from which it is assimilated by seaweeds, in Chilean saltpeter and nitrate-bearing earth, known as *caliche* in brines from old sea deposits, and in brackish waters from oil and salt wells. Ultrapure iodine can be obtained from the reaction of potassium iodide with copper sulfate. Several other methods of isolating the element are known. Iodine is a bluish-black, lustrous solid, volatilizing at ordinary temperatures into a blue-violet gas with an irritating odor; it forms compounds with many elements, but is less active than the other halogens, which displace it from iodides. Iodine exhibits some metallic-like properties. It dissolves readily in chloroform, carbon tetrachloride, or carbon disulfide to form beautiful purple solutions. It is only slightly soluble in water. Iodine compounds are important in organic chemistry and very useful in medicine. Forty two isotopes and isomers are recognized. Only one stable isotope, ^{127}I is found in nature. The artificial radioisotope ^{131}I, with a half-life of 8 days, has been used in treating the thyroid gland. The most common compounds are the iodides of sodium and potassium (KI) and the iodates (KIO_3). Lack of iodine is the cause of goiter. Iodides, and thyroxin which contains iodine, are used internally in medicine, and a solution of KI and iodine in alcohol is used for external wounds. Potassium iodide finds use in photography. The deep blue color with starch solution is characteristic of the free element. Care should be taken in handling and using iodine, as contact with the skin can cause lesions; iodine vapor is intensely irritating to the eyes and mucous membranes. Elemental iodine costs about 25 to 75¢/g depending on purity and quantity.

Iridium — (L. *iris*, rainbow), Ir; at. wt. 192.217(3); at. no. 77; m.p. 2446°C; b.p. 4428°C; sp. gr. 22.42 (17°C); valence 3 or 4. Discovered in 1803 by Tennant in the residue left when crude platinum is dissolved by aqua regia. The name iridium is appropriate, for its salts are highly colored. Iridium, a metal of the platinum family, is white, similar to platinum, but with a slight yellowish cast. It is very hard and brittle, making it very hard to machine, form, or work. It is the most corrosion-resistant metal known, and was used in making the standard meter bar of Paris, which is a 90% platinum-10% iridium alloy. This meter bar was replaced in 1960 as a fundamental unit of length (see under Krypton). Iridium is not attacked by any of the acids nor by aqua regia, but is attacked by molten salts, such as NaCl and NaCN. Iridium occurs uncombined in nature with platinum and other metals of this family in alluvial deposits. It is recovered as a by-product from the nickel mining industry. The presence of iridium has recently been used in examining the Cretaceous-Tertiary (K-T) boundary. Meteorites contain small amounts of iridium. Because iridium is found widely distributed at the K-T boundary, it has been suggested that a large meteorite or asteroid collided with the earth, killing the dinosaurs, and creating a large dust cloud and crater. Searches for such a crater point to one in the Yucatan, known as Chicxulub. Iridium has found use in making crucibles and apparatus for use at high temperatures. It is also used for electrical contacts. Its principal use is as a hardening agent for platinum. With osmium, it forms an alloy which is used for tipping pens and compass bearings. The specific gravity of iridium is only very slightly lower than that of osmium, which has been generally credited as being the heaviest known element. Calculations of the densities of iridium and osmium from the space lattices gives values of 22.65 and 22.61 g/cm^3, respectively. These values may be more reliable than actual physical measurements. Natural iridium contains two isotopes. Forty two other isotopes, all radioactive, are now recognized. At present, therefore, we know that either iridium or osmium is the densest known element, but the data do not yet allow selection between the two. Iridium costs about $50/g.

Iron — (Anglo-Saxon, *iron*), Fe (L. *ferrum*); at. wt. 55.845(2); at. no. 26; m.p. 1538°C; b.p. 2861°C; sp. gr. 7.874 (20°C); valence 2, 3, 4, or 6. The use of iron is prehistoric. Genesis mentions that Tubal-Cain, seven generations from Adam, was "an instructor of every artificer in brass and iron." A remarkable iron pillar, dating to about A.D. 400, remains standing today in Delhi, India. This solid shaft of wrought iron is about $7^1/_4$ m high by 40 cm in diameter. Corrosion to the pillar has been minimal although it has been exposed to the weather since its erection. Iron is a relatively abundant element in the universe. It is found in the sun and many types of stars in considerable quantity. Its nuclei are very stable. Iron is found native as a principal component of a class of meteorites known as "siderites, and is a minor constituent of the other two classes. The core of the earth, 2150 miles in radius, is thought to be largely composed of iron with about 10% occluded hydrogen. The metal is the fourth most abundant element, by weight, making up the crust of the earth. The most common ore is *hematite* (Fe_2O_3), which is frequently seen as *black sands* along beaches and banks of streams. *Taconite* is becoming increasingly important as a commercial ore. Common iron is a mixture of four isotopes. Ten other isotopes are known to exist. Iron is a vital constituent of plant and animal life, and appears in hemoglobin. The pure metal is not often encountered in commerce, but is usually alloyed with carbon or other metals. The pure metal is very reactive chemically, and rapidly corrodes, especially in moist air or at elevated temperatures. It has four allotropic forms, or ferrites, known as α, β, γ, and δ, with transition points at 700, 928, and 1530°C. The α form is magnetic, but when transformed into the β form, the magnetism disappears although the lattice remains unchanged. The relations of these forms are peculiar. Pig iron is an alloy containing about 3% carbon with varying amounts of S, Si, Mn, and P. It is hard, brittle, fairly fusible, and is used to produce other alloys, including steel. Wrought iron contains only a few tenths of a percent of carbon, is tough, malleable, less fusible, and has usually a "fibrous" structure. Carbon steel is an alloy of iron with carbon, with small amounts of Mn, S, P, and Si. Alloy steels are carbon steels with other additives such as nickel, chromium, vanadium, etc. Iron is the cheapest and most abundant, useful, and important of all metals. Natural iron contains four isotopes and isomers. Twenty one other isotopes and isomers, all radioactive, are now recognized.

Joliotium — See Element 105.

Krypton — (Gr. *kryptos*, hidden), Kr; at. wt. 83.80(1); at. no. 36; m.p. –157.36°C; b.p. –153.22 ± 0.10°C; density 3.733 g/l (0°C); valence usually 0. Discovered in 1898 by Ramsay and Travers in the residue left after liquid air had nearly boiled away. Krypton is present in the air to the extent of about 1 ppm. The atmosphere of Mars has been found to contain 0.3 ppm of krypton. It is one of the "noble" gases. It is characterized by its brilliant green and orange spectral lines. Naturally occurring krypton contains six stable isotopes. Twenty four other unstable isotopes and isomers are now recognized. The spectral lines of krypton are easily produced and some are very sharp. In 1960 it was internationally agreed that the fundamental unit of length, the meter, should be defined in terms of the orange-red spectral line of ^{86}Kr. This replaced the standard meter of Paris, which was defined in terms of a bar made of a platinum-iridium alloy. In October 1983 the meter, which originally was defined as being one ten millionth of a quadrant of the earth's polar circumference, was again redefined by the International Bureau of Weights and Measures as being the length of path traveled by light in a vacuum during a time interval of 1/299,792,458 of a second. Solid krypton is a white crystalline substance with a face-centered cubic structure which is common to all the "rare gases". While krypton is generally thought of as a rare gas that normally does not combine with other elements to form compounds, it now appears that the existence of some krypton compounds is established. Krypton difluoride has been prepared in gram quantities and can be made by several methods. A higher fluoride of krypton and a salt of an oxyacid of krypton also have been reported. Molecule-ions of $ArKr^+$ and KrH^+ have been identified and investigated, and evidence is provided for the formation of KrXe or $KrXe^+$. Krypton clathrates have been prepared with hydroquinone and phenol. ^{85}Kr has found recent application in chemical analysis. By imbedding the isotope in various solids, *kryptonates* are formed. The activity of these kryptonates is sensitive to chemical reactions at the surface. Estimates of the concentration of reactants are therefore made possible. Krypton is used in certain photographic flash lamps for high-speed photography. Uses thus far have been limited because of its high cost. Krypton gas presently costs about $690/100 l.

Kurchatovium — see Element 104.

Lanthanum — (Gr. *lanthanein*, to lie hidden), La; at. wt. 138.9055(2); at. no. 57; m.p. 918°C; b.p. 3464°C; sp. gr. 6.145 (25°C); valence 3. Mosander in 1839 extracted a new earth *lanthana*, from impure cerium nitrate, and recognized the new element. Lanthanum is found in rare-earth minerals such as *cerite, monazite, allanite,* and *bastnasite.* Monazite and bastnasite are principal ores in which lanthanum occurs in percentages up to 25 and 38%, respectively. Misch metal, used in making lighter flints, contains about 25% lanthanum. Lanthanum was isolated in relatively pure form in 1923. Iron-exchange and solvent extraction techniques have led to much easier isolation of the so-called "rare-earth" elements. The availability of lanthanum and other rare earths has improved greatly in recent years. The metal can be produced by reducing the anhydrous fluoride with calcium. Lanthanum is silvery white, malleable, ductile, and soft enough to be cut with a knife. It is one of the most reactive of the rare-earth metals. It oxidizes rapidly when exposed to air. Cold water attacks lanthanum slowly, and hot water attacks it much more rapidly. The metal reacts directly with elemental carbon, nitrogen, boron, selenium, silicon, phosphorus, sulfur, and with halogens. At 310°C, lanthanum changes from a hexagonal to a face-centered cubic structure, and at 865°C it again transforms into a body-centered cubic structure. Natural lanthanum is mixture of two isotopes, one of which is stable and one of which is radioactive with a very long half-life. Twenty nine other radioactive isotopes are recognized. Rare-earth compounds containing lanthanum are extensively used in carbon lighting applications, especially by the motion picture industry for studio lighting and projection. This application consumes about 25% of the rare-earth compounds produced. La_2O_3 improves the alkali resistance of glass, and is used in making special optical glasses. Small amounts of lanthanum, as an additive, can be used to produce nodular cast iron. There is current interest in hydrogen sponge alloys containing lanthanum. These alloys take up to 400 times their own volume of hydrogen gas, and the process is reversible. Heat energy is released every time they do so; therefore these alloys have possibilities in energy conservation systems. Lanthanum and its compounds have a low to moderate acute toxicity rating; therefore, care should be taken in handling them. The metal costs about $2/g (99.9%).

Lawrencium — (Ernest O. Lawrence, inventor of the cyclotron), Lr; at. no. 103; at. mass no. (262); valence + 3(?). This member of the 5f transition elements (actinide series) was discovered in March 1961 by A. Ghiorso, T. Sikkeland, A. E. Larsh, and R. M. Latimer. A 3-μg californium target, consisting of a mixture of isotopes of mass number 249, 250, 251, and 252, was bombarded with either ^{10}B or ^{11}B. The electrically charged transmutation nuclei recoiled with an atmosphere of helium and were collected on a thin copper conveyor tape which was then moved to place collected atoms in front of a series of solid-state detectors. The isotope of element 103 produced in this way decayed by emitting an 8.6-MeV alpha particle with a half-life of 8 s. In 1967, Flerov and associates of the Dubna Laboratory reported their inability to detect an alpha emitter with a half-life of 8 s which was assigned by the Berkeley group to $^{257}103$. This assignment has been changed to ^{258}Lr or ^{259}Lr. In 1965, the Dubna workers found a longer-lived lawrencium isotope, ^{256}Lr, with a half-life of 35 s. In 1968, Ghiorso and associates at Berkeley were able to use a few atoms of this isotope to study the oxidation behavior of lawrencium. Using solvent extraction techniques and working very rapidly, they extracted lawrencium ions from a buffered aqueous solution into an organic solvent, completing each extraction in about 30 s. It was found that lawrencium behaves differently from dipositive nobelium and more like the tripositive elements earlier in the actinide series. Ten isotopes of lawrencium are now recognized.

Lead — (Anglo-Saxon *lead*), Pb (L. *plumbum*); at. wt. 207.2(1); at. no. 82; m.p. 327.46°C; b.p. 1749°C; sp. gr. 11.35 (20°C); valence 2 or 4. Long known, mentioned in Exodus. The alchemists believed lead to be the oldest metal and associated it with the planet Saturn. Native lead occurs in nature, but it is rare. Lead is obtained chiefly from *galena* (PbS) by a roasting process. *Anglesite* ($PbSO_4$), *cerussite* ($PbCO_3$), and *minim* (Pb_3O_4) are other common lead minerals. Lead is a bluish-white metal of bright luster, is very soft, highly malleable, ductile, and a poor conductor of electricity. It is very resistant to corrosion; lead pipes bearing the insignia of Roman emperors, used as drains from the baths, are still in service. It is used in containers for corrosive liquids (such as sulfuric acid) and may be toughened by the addition of a small percentage of antimony or other metals. Natural lead is a mixture of four stable isotopes: ^{204}Pb (1.4%), ^{206}Pb (24.1%), ^{207}Pb (22.1%), and ^{208}Pb (52.4%). Lead isotopes are the end products of each of the three series of naturally occurring radioactive elements: ^{206}Pb for the uranium series, ^{207}Pb for the actinium series, and ^{208}Pb for the thorium series. Forty other isotopes of lead, all of which are radioactive, are recognized. Its alloys include solder, type metal, and various antifriction metals. Great quantities of lead, both as the metal and as the dioxide, are used in storage batteries. Much metal also goes into cable covering, plumbing, ammunition, and in the manufacture of lead tetraethyl. The metal is very effective as a sound absorber, is used as a radiation shield around X-ray equipment and nuclear reactors, and is used to absorb vibration. White lead, the basic carbonate, sublimed white lead ($PbSO_4$) chrome yellow ($PbCrO_4$), red lead (Pb_3O_4), and other lead compounds are used extensively in paints, although in recent years the use of lead in paints has been drastically curtailed to eliminate or reduce health hazards. Lead oxide is used in producing fine "crystal glass" and "flint glass" of a high index of refraction for achromatic lenses. The nitrate and the acetate are soluble salts. Lead salts such as lead arsenate have been used as insecticides, but their use in recent years has been practically eliminated in favor of less harmful organic compounds. Care must be used in handling lead as it is a cumulative poison. Environmental concern with lead poisoning has resulted in a national program to eliminate the lead in gasoline. Lead is priced at about $1/kg (99.9%).

Lithium — (Gr. *lithos*, stone), Li; at. wt. 6.941(2); at. no. 3; m.p. 180.5°C; b.p. 1342°C; sp. gr. 0.534 (20°C); valence 1. Discovered by Arfvedson in 1817. Lithium is the lightest of all metals, with a density only about half that of water. It does not occur free in nature; combined it is found in small amounts in nearly all igneous rocks and in the waters of many mineral springs. *Lepidolite, spodumene, petalite,* and *amblygonite* are the more important minerals containing it. Lithium is presently being recovered from brines of Searles Lake, in California, and from Nevada, Chile, and Argentina. Large deposits of spodumene are found in North Carolina. The metal is produced electrolytically from the fused chloride. Lithium is silvery in appearance, much like Na and K, other members of the alkali metal series. It reacts with water, but not as vigorously as sodium. Lithium imparts a beautiful crimson color to a flame, but when the metal burns strongly the flame is a dazzling white. Since World War II, the production of lithium metal and its compounds has increased greatly. Because the metal has the highest specific heat of any solid element, it has found use in heat transfer applications; however, it is corrosive and requires special handling. The metal has been used as an alloying agent, is of interest in synthesis of organic compounds, and has nuclear applications. It ranks as a leading contender as a battery anode material as it has a high electrochemical potential. Lithium is used in special glasses and ceramics. The glass for the 200-inch telescope at Mt. Palomar contains lithium as a minor ingredient. Lithium chloride is one of the most hygroscopic materials known, and it, as well as lithium bromide, is used in air conditioning and industrial drying systems. Lithium stearate is used as an all-purpose and high-temperature lubricant. Other lithium compounds are used in dry cells and storage batteries. Seven isotopes of lithium are recognized. Natural lithium contains two isotopes. The metal is priced at about $500/kg (99.9%).

Lutetium — (Lutetia, ancient name for Paris, sometimes called *cassiopeium* by the Germans), Lu; at. wt. 174.967(1); at. no. 71; m.p. 1663°C; b.p. 3402°C; sp. gr. 9.841 (25°C); valence 3. In 1907, Urbain described a process by which Marignac's ytterbium (1879) could be separated into the two elements, ytterbium (neoytterbium)and lutetium. These elements were identical with "aldebaranium" and "cassiopeium," independently discovered by von Welsbach about the same time. Charles James of the University of New Hampshire also independently prepared the very pure oxide, *lutecia*, at this time. The spelling of the element was changed from *lutecium* to *lutetium* in 1949. Lutetium occurs in very small amounts in nearly all minerals containing yttrium, and is present in *monazite* to the extent of about 0.003%, which is a commercial source. The pure metal has been isolated only in recent years and is one of the most difficult to prepare. It can be prepared by the reduction of anhydrous $LuCl_3$ or LuF_3 by an alkali or alkaline earth metal. The metal is silvery white and relatively stable in air. While new techniques, including ion-exchange reactions, have been developed to separate the various rare-earth elements, lutetium is still the most costly of all rare earths. It is priced at about $100/g. ^{176}Lu occurs naturally (97.41%) with ^{175}Lu (2.59%), which is radioactive with a very long half-life. It is radioactive with a half-life of about 4×10^{10} years. Lutetium has 49 isotopes and isomers that are now recognized. Stable lutetium nuclides, which emit pure beta radiation after thermal neutron activation, can be used as catalysts in cracking, alkylation, hydrogenation, and polymerization. Virtually no other commercial uses have been found yet for lutetium. While lutetium, like other rare-earth metals, is thought to have a low toxicity rating, it should be handled with care until more information is available.

Magnesium — (*Magnesia,* district in Thessaly) Mg; at. wt. 24.3050(6); at. no. 12; m.p. 650°C; b.p. 1090°C; sp. gr. 1.738 (20°C); valence 2. Compounds of magnesium have long been known. Black recognized magnesium as an element in 1755. It was isolated by Davy in 1808, and prepared in coherent form by Bussy in 1831. Magnesium is the eighth most abundant element in the earth's crust. It does not occur uncombined, but is found in large deposits in the form of *magnesite, dolomite,* and other minerals. The metal is now principally obtained in the U.S. by electrolysis of fused magnesium chloride derived from brines, wells, and sea water. Magnesium is a light, silvery-white, and fairly tough metal. It tarnishes slightly in air, and finely divided magnesium readily ignites upon heating in air and burns with a dazzling white flame. It is used in flashlight photography, flares, and pyrotechnics, including incendiary bombs. It is one third lighter than aluminium, and in alloys is essential for airplane and missile contruction. The metal improves the mechanical, fabrication, and welding characteristics of aluminum when used as an alloying agent. Magnesium is used in producing nodular graphite in cast iron,and is used as an additive to conventional propellants. It is also used as a reducing agent in the production of pure uranium and other metals from their salts. The hydroxide (*milk of magnesia*), chloride, sulfate (*Epsom salts*), and citrate are used in medicine. Dead-burned magnesite is employed for refractory purposes such as brick and liners in furnaces and converters. Organic magnesium compounds (Grignard's reagents) are important. Magnesium is an important element in both plant and animal life. Chlorophylls are magnesium-centered porphyrins. The adult daily requirement of magnesium is about 300 mg/day, but this is affected by various factors. Great care should be taken in handling magnesium metal, especially in the finely divided state, as serious fires can occur. Water should not be used on burning magnesium or on magnesium fires. Natural magnesium contains three isotopes. Twelve other isotopes are recognized. Magnesium metal costs about $170/kg (99.99%).

Manganese — (L. *magnes*, magnet, from magnetic properties of pyrolusite; It. *manganese*, corrupt form of *magnesia*), Mn; at. wt. 54.93805(1); at. no. 25; m.p. 1246±3°C; b.p. 2061°C; sp. gr. 7.21 to 7.44, depending on allotropic form; valence 1, 2, 3, 4, 6, or 7. Recognized by Scheele, Bergman, and others as an element and isolated by Gahn in 1774 by reduction of the dioxide with carbon. Manganese minerals are widely distributed; oxides, silicates, and carbonates are the most common. The discovery of large quantities of manganese nodules on the floor of the oceans holds promise as a source of manganese. These nodules contain about 24% manganese together with many other elements in lesser abundance. Most manganese today is obtained from ores found in the Ukraine, Brazil, Australia, Republic of So. Africa, Gabon, China, and India. *Pyrolusite* (MnO_2) and *rhodochrosite* ($MnCO_3$) are among the most common manganese minerals. The metal is obtained by reduction of the oxide with sodium, magnesium, aluminum, or by electrolysis. It is gray-white, resembling iron, but is harder and very brittle. The metal is reactive chemically, and decomposes cold water slowly. Manganese is used to form many important alloys. In steel, manganese improves the rolling and forging qualities, strength, toughness, stiffness, wear resistance, hardness, and hardenability. With aluminum and antimony, especially with small amounts of copper, it forms highly ferromagnetic alloys. Manganese metal is ferromagnetic only after special treatment. The pure metal exists in four allotropic forms. The alpha form is stable at ordinary temperature; gamma manganese, which changes to alpha at ordinary temperatures, is said to be flexible, soft, easily cut, and capable of being bent. The dioxide (pyrolusite) is used as a depolarizer in dry cells, and is used to "decolorize" glass that is colored green by impurities of iron. Manganese by itself colors glass an amethyst color, and is responsible for the color of true amethyst. The dioxide is also used in the preparation of oxygen and chlorine, and in drying black paints. The permanganate is a powerful oxidizing agent and is used in quantitative analysis and in medicine. Manganese is widely distributed throughout the animal kingdom. It is an important trace element and may be essential for utilization of vitamin B_1. Twenty three isotopes and isomers are known. Manganese metal (99.95%) is priced at about \$400/kg. Metal of 99.6% purity is priced at about \$60/kg.

Meitnerium — See Element 109.

Mendelevium — (Dmitri Mendeleev), Md; at. wt. (258); at. no. 101; m.p. 827°C; valence +2, +3. Mendelevium, the ninth transuranium element of the actinide series to be discovered, was first identified by Ghiorso, Harvey, Choppin, Thompson, and Seaborg early in 1955 as a result of the bombardment of the isotope ^{253}Es with helium ions in the Berkeley 60-inch cyclotron. The isotope produced was ^{256}Md, which has a half-life of 78 min. This first identification was notable in that ^{256}Md was synthesized on a one-atom-at-a-time basis. Sixteen isotopes and isomers are now recognized. ^{258}Md has a half-life of 51.5 days. This isotope has been produced by the bombardment of an isotope of einsteinium with ions of helium. It now appears possible that eventually enough ^{258}Md can be made so that some of its physical properties can be determined. ^{256}Md has been used to elucidate some of the chemical properties of mendelevium in aqueous solution. Experiments seem to show that the element possesses a moderately stable dipositive (II) oxidation state in addition to the tripositive (III) oxidation state, which is characteristic of actinide elements.

Mercury — (Planet *Mercury*), Hg (*hydrargyrum*, liquid silver); at. wt. 200.59(2); at. no. 80; t.p. –38.8344°C; b.p. 356.73°C; sp. gr. 13.546 (20°C); valence 1 or 2. Known to ancient Chinese and Hindus; found in Egyptian tombs of 1500 B.C. Mercury is the only common metal liquid at ordinary temperatures. It only rarely occurs free in nature. The chief ore is *cinnabar* (HgS). Spain and Italy produce about 50% of the world's supply of the metal. The commercial unit for handling mercury is the "flask," which weighs 76 lb (34.46 kg) and is priced at about \$250. The metal is obtained by heating cinnabar in a current of air and by condensing the vapor. It is a heavy, silvery-white metal; a rather poor conductor of heat, as compared with other metals, and a fair conductor of electricity. It easily forms alloys with many metals, such as gold, silver, and tin, which are called *amalgams*. Its ease in amalgamating with gold is made use of in the recovery of gold from its ores. The metal is widely used in laboratory work for making thermometers, barometers, diffusion pumps, and many other instruments. It is used in making mercury-vapor lamps and advertising signs, etc. and is used in mercury switches and other electrical apparatus. Other uses are in making pesticides, mercury cells for caustic soda and chlorine production, dental preparations, antifouling paint, batteries, and catalysts. The most important salts are mercuric chloride $HgCl_2$ (corrosive sublimate — a violent poison), mercurous chloride Hg_2Cl_2 (calomel, occasionally still used in medicine), mercury fulminate ($Hg(ONC)_2$), a detonator widely used in explosives, and mercuric sulfide (HgS, vermillion, a high-grade paint pigment). Organic mercury compounds are important. It has been found that an electrical discharge causes mercury vapor to combine with neon, argon, krypton, and xenon. These products, held together with van der Waals' forces, correspond to HgNe, HgAr, HgKr, and HgXe. Mercury is a virulent poison and is readily absorbed through the respiratory tract, the gastrointestinal tract, or through unbroken skin. It acts as a cumulative poison and dangerous levels are readily attained in air. Air saturated with mercury vapor at 20°C contains a concentration that exceeds the toxic limit many times. The danger increases at higher temperatures. *It is therefore important that mercury be handled with care.* Containers of mercury should be securely covered and spillage should be avoided. If it is necessary to heat mercury or mercury compounds, it should be done in a well-ventilated hood. Methyl mercury is a dangerous pollutant and is now widely found in water and streams. The triple point of mercury, –38.8344°C, is a fixed point on the International Temperature Scale (ITS-90). Native mercury contains seven isotopes. Thirty five other isotopes and isomers are known.

Meitnerium — See Element 109.

Molybdenum — (Gr. *molybdos*, lead), Mo; at. wt. 95.94(1); at. no. 42; m.p. 2623°C; b.p. 4639°C; sp. gr. 10.22 (20°C); valence 2, 3, 4?, 5?, or 6. Before Scheele recognized molybdenite as a distinct ore of a new element in 1778, it was confused with graphite and lead ore. The metal was prepared in an impure form in 1782 by Hjelm. Molybdenum does not occur native, but is obtained principally from *molybdenite* (MoS_2). *Wulfenite* ($PbMoO_4$) and *Powellite* ($Ca(MoW)O_4$) are also minor commercial ores. Molybdenum is also recovered as a by-product of copper and tungsten mining operations. The metal is prepared from the powder made by the hydrogen reduction of purified molybdic trioxide or ammonium molybdate. The metal is silvery white, very hard, but is softer and more ductile than tungsten. It has a high elastic modulus, and only tungsten and tantalum, of the more readily available metals, have higher melting points. It is a valuable alloying agent, as it contributes to the hardenability and toughness of quenched and tempered steels. It also improves the strength of steel at high temperatures. It is used in certain nickel-based alloys, such as the "Hastelloys®" which are heat-resistant and corrosion-resistant to chemical solutions. Molybdenum oxidizes at elevated temperatures. The metal has found recent application as electrodes for electrically heated glass furnaces and forehearths. The metal is also used in nuclear energy applications and for missile and aircraft parts. Molybdenum is valuable as a catalyst in the refining of petroleum. It has found application as a filament material in electronic and electrical applications. Molybdenum is an essential trace element in plant nutrition. Some lands are barren for lack of this element in the soil. Molybdenum sulfide is useful as a lubricant, especially at high temperatures where oils would decompose. Almost all ultra-high strength steels with minimum yield points up to 300,000 psi(lb/in.^{2}) contain molybdenum in amounts from 0.25 to 8%. Natural molybdenum contains seven isotopes. Twenty three other isotopes and isomers are known, all of which are radioactive. Molybdenum metal costs about \$1/g (99.999% purity). Commercial molybdenum metal (99.9%) costs about \$200/kg.

Neodymium — (Gr. *neos*, new, and *didymos*, twin), Nd; at. wt. 144.24(3); at. no. 60; m.p. 1021°C; b.p. 3074°C; sp. gr. 7.008 (25°C); valence 3. In 1841, Mosander, extracted from *cerite* a new rose-colored oxide, which he believed contained a new element. He named the element *didymium*, as it was *an inseparable twin brother of lanthanum*. In 1885 von Welsbach separated didymium into two new elemental components, *neodymia* and *praseodymia*, by repeated fractionation of ammonium didymium nitrate. While the free metal is in *misch metal*, long known and used as a pyrophoric alloy for light flints, the element was not isolated in relatively pure form until 1925. Neodymium is present in misch metal to the extent of about 18%. It is present in the minerals *monazite* and *bastnasite*, which are principal sources of rare-earth metals. The element may be obtained by separating neodymium salts from other rare earths by ion-exchange or solvent extraction techniques, and by reducing anhydrous halides such as NdF_3 with calcium metal. Other separation techniques are possible. The metal has a bright silvery metallic luster. Neodymium is one of the more reactive rare-earth metals and quickly tarnishes in air, forming an oxide that spalls off and exposes metal to oxidation. The metal, therefore, should be kept under light mineral oil or sealed in a plastic material. Neodymium exists in two allotropic forms, with a transformation from a double hexagonal to a body-centered cubic structure taking place at 863°C. Natural neodymium is a mixture of seven isotopes, one of which has a very long half-life. Twenty seven other radioactive isotopes and isomers are recognized. Didymium, of which neodymium is a component, is used for coloring glass to make welder's goggles. By itself, neodymium colors glass delicate shades ranging from pure violet through wine-red and warm gray. Light transmitted through such glass shows unusually sharp absorption bands. The glass has been used in astronomical work to produce sharp bands by which spectral lines may be calibrated. Glass containing neodymium can be used as a laser material to produce coherent light. Neodymium salts are also used as a colorant for enamels. The element is also being used with iron and boron to produce extremely strong magnets having energy densities as high as 27 to 35 million gauss oersteds. These are the most compact magnets commercially available. The price of the metal is about $2/g. Neodymium has a low-to-moderate acute toxic rating. As with other rare earths, neodymium should be handled with care.

Neon — (Gr. *neos*, new), Ne; at. wt. 20.1797(6); at. no. 10; t.p. –248.5938°C; b.p. –246.08°C (1 atm); density of gas 0.89990 g/l (1 atm, 0°C); density of liquid at b.p. 1.207 g/cm³; valence 0. Discovered by Ramsay and Travers in 1898. Neon is a rare gaseous element present in the atmosphere to the extent of 1 part in 65,000 of air. It is obtained by liquefaction of air and separated from the other gases by fractional distillation. Natural neon is a mixture of three isotopes. Six other unstable isotopes are known. It is very inert element; however, it is said to form a compound with fluorine. It is still questionable if true compounds of neon exist, but evidence is mounting in favor of their existence. The following ions are known from optical and mass spectrometric studies: Ne^+, $(NeAr)^+$, $(NeH)^+$, and $(HeNe^+)$. Neon also forms an unstable hydrate. In a vacuum discharge tube, neon glows reddish orange. Of all the rare gases, the discharge of neon is the most intense at ordinary voltages and currents. Neon is used in making the common neon advertising signs, which accounts for its largest use. It is also used to make high-voltage indicators, lightning arrestors, wave meter tubes, and TV tubes. Neon and helium are used in making gas lasers. Liquid neon is now commercially available and is finding important application as an economical cryogenic refrigerant. It has over 40 times more refrigerating capacity per unit volume than liquid helium and more than three times that of liquid hydrogen. It is compact, inert, and is less expensive than helium when it meets refrigeration requirements. Neon costs about $800/80 cu. ft. (2265 l).

Neptunium — (Planet *Neptune*), Np; at. wt. (237); at. no. 93; m.p. 644°C; b.p. 3902°C (est.); sp. gr. 20.25 (20°C); valence 3, 4, 5, and 6. Neptunium was the first synthetic transuranium element of the actinide series discovered; the isotope ^{239}Np was produced by McMillan and Abelson in 1940 at Berkeley, California, as the result of bombarding uranium with cyclotron-produced neutrons. The isotope ^{237}Np (half-life of 2.14×10^6 years) is currently obtained in gram quantities as a by-product from nuclear reactors in the production of plutonium. Twenty isotopes and isomers of neptunium are now recognized. Trace quantities of the element are actually found in nature due to transmutation reactions in uranium ores produced by the neutrons which are present. Neptunium is prepared by the reduction of NpF_3 with barium or lithium vapor at about 1200°C. Neptunium metal has a silvery appearance, is chemically reactive, and exists in at least three structural modifications: α-neptunium, orthorhombic, density 20.25 g/cm³; β-neptunium (above 280°C), tetragonal, density (313°C) 19.36 g/cm³; γ-neptunium (above 577°C), cubic, density (600°C) 18.0 g/cm³. Neptunium has four ionic oxidation states in solution: Np^{+3} (pale purple), analogous to the rare earth ion Pm^{+3}, Np^{+4} (yellow green); NpO^+ (green blue); and NpO^{++} (pale pink). These latter oxygenated species are in contrast to the rare earths which exhibit only simple ions of the (II), (III), and (IV) oxidation states in aqueous solution. The element forms tri- and tetrahalides such as NpF_3, NpF_4, $NpCl_4$, $NpBr_3$, NpI_3, and oxides of various compositions such as are found in the uranium-oxygen system, including Np_3O_8 and NpO_2. Fifteen isotopes of neptunium are now recognized. The O.R.N.L. has ^{237}Np available for sale to its licensees and for export. This isotope can be used as a component in neutron detection instruments. It is offered at a price of $660/g plus packing costs.

Nickel — (Ger. *Nickel*, Satan or Old Nick's and from *kupfernickel*, Old Nick's copper), Ni; at. wt. 58.6934(2); at. no. 28; m.p. 1455°C; b.p. 2913°C; sp. gr. 8.902 (25°C); valence 0, 1, 2, 3. Discovered by Cronstedt in 1751 in kupfernickel (*niccolite*). Nickel is found as a constituent in most meteorites and often serves as one of the criteria for distinguishing a meteorite from other minerals. Iron meteorites, or *siderites*, may contain iron alloyed with from 5 to nearly 20% nickel. Nickel is obtained commercially from *pentlandite* and *pyrrhotite* of the Sudbury region of Ontario, a district that produces much of the world's nickel. It is now thought that the Sudbury deposit is the result of an ancient meteorite impact. Other deposits of nickel are found in Russia, New Caledonia, Australia, Cuba, Indonesia, and elsewhere. Nickel is silvery white and takes on a high polish. It is hard, malleable, ductile, somewhat ferromagnetic, and a fair conductor of heat and electricity. It belongs to the iron-cobalt group of metals and is chiefly valuable for the alloys it forms. It is extensively used for making stainless steel and other corrosion-resistant alloys such as Invar®, Monel®, Inconel®, and the Hastelloys®. Tubing made of a copper-nickel alloy is extensively used in making desalination plants for converting sea water into fresh water. Nickel is also now used extensively in coinage and in making nickel steel for armor plate and burglar-proof vaults, and is a component in Nichrome®, Permalloy®, and constantan. Nickel added to glass gives a green color. Nickel plating is often used to provide a protective coating for other metals, and finely divided nickel is a catalyst for hydrogenating vegetable oils. It is also used in ceramics, in the manufacture of Alnico magnets, and in the Edison® storage battery. The sulfate and the oxides are important compounds. Natural nickel is a mixture of five stable isotopes; nineteen other unstable isotopes are known. Nickel sulfide fume and dust is recognized as having carcinogenic potential. Nickel metal (99.9%) is priced at about $100/kg or less in larger quantities.

Niobium — (*Niobe*, daughter of Tantalus), Nb; or Columbium (*Columbia*, name for America); at. wt. 92.90638(2); at. no. 41; m.p. 2477 ± 10°C; b.p. 4744°C, sp. gr. 8.57 (20°C); valence 2, 3, 4?, 5. Discovered in 1801 by Hatchett in an ore sent to England more that a century before by John Winthrop the Younger, first governor of Connecticut. The metal was first prepared in 1864 by Blomstrand, who reduced the chloride by heating it in a hydrogen atmosphere. The name *niobium* was adopted by the International Union of Pure and Applied Chemistry in 1950 after 100 years of controversy. Many leading chemical societies and government organizations refer to it by this name. Most metallurgists, leading metal societies, and all but one of the leading U.S. commercial producers, however, still refer to the metal as "columbium". The element is found in *niobite* (or *columbite*), *niobite-tantalite, pyrochlore,* and *euxenite*. Large deposits of niobium have been found associated with *carbonatites* (carbon-silicate rocks), as a constituent of *pyrochlore*. Extensive ore reserves are found in Canada, Brazil, Nigeria, Zaire, and in Russia. The metal can be isolated from tantalum, and prepared in several ways. It is a shiny, white, soft, and ductile metal, and takes on a bluish cast when exposed to air at room temperatures for a long time. The metal starts to oxidize in air at 200°C, and when processed at even moderate·temperatures must be placed in a protective atmosphere. It is used in arc-welding rods for stabilized grades of stainless steel. Thousands of pounds of niobium have been used in advance air frame systems such as were used in the Gemini space program. The element has superconductive properties; superconductive magnets have been made with Nb-Zr wire, which retains its superconductivity in strong magnetic fields. This type of application offers hope of direct large-scale generation of electric power. Natural niobium is composed of only one isotope, ^{93}Nb. Forty one other isotopes and isomers of niobium are now recognized. Niobium metal (99.9% pure) is priced at about $400/kg.

Nitrogen — (L. *nitrum*, Gr. *nitron*, native soda; genes, *forming*, N; at. wt. 14.00674(7); at. no. 7; m.p. –210.00°C; b.p. –195.8°C; density 1.2506 g/l; sp. gr. liquid 0.808 (–195.8°C), solid 1.026 (–252°C); valence 3 or 5. Discovered by Daniel Rutherford in 1772, but Scheele, Cavendish,Priestley, and others about the same time studied "burnt or dephlogisticated air," as air without oxygen was then called. Nitrogen makes up 78% of the air, by volume. The atmosphere of Mars, by comparison, is 2.6% nitrogen. The estimated amount of this element in our atmosphere is more than 4000 trillion tons. From this incxhaustible source it can be obtained by liquefaction and fractional distillation. Nitrogen molecules give the orange-red, blue-green, blue-violet, and deep violet shades to the aurora.The element is so inert that Lavoisier named it *azote*, meaning without life, yet its compounds are so active as to be most important in foods, poisons, fertilizers, and explosives. Nitrogen can be also easily prepared by heating a water solution of ammonium nitrite. Nitrogen, as a gas, is colorless, odorless, and a generally inert element. As a liquid it is also colorless and odorless, and is similar in appearance to water. Two allotropic forms of solid nitrogen exist, with the transition from the α to the β form taking place at –237°C. When nitrogen is heated, it combines directly with magnesium, lithium, or calcium; when mixed with oxygen and subjected to electric sparks, it forms first nitric oxide (NO) and then the dioxide (NO_2); when heated under pressure with a catalyst with hydrogen, ammonia is formed (Haber process). The ammonia thus formed is of the utmost importance as it is used in fertilizers, and it can be oxidized to nitric acid (Ostwald process). The ammonia industry is the largest consumer of nitrogen. Large amounts of gas are also used by the electronics industry, which uses the gas as a blanketing medium during production of such components as transistors, diodes, etc. Large quantities of nitrogen are used in annealing stainless steel and other steel mill products. The drug industry also uses large quantities. Nitrogen is used as a refrigerant both for the immersion freezing of food products and for transportation of foods. Liquid nitrogen is also used in missile work as a purge for components, insulators for space chambers, etc., and by the oil industry to build up great pressures in wells to force crude oil upward. Sodium and potassium nitrates are formed by the decomposition of organic matter with compounds of the metals present. In certain dry areas of the world these saltpeters are found in quantity. Ammonia, nitric acid, the nitrates, the five oxides (N_2O, NO, N_2O_3, NO_2, and N_2O_5), TNT, the cyanides, etc. are but a few of the important compounds. Nitrogen gas prices vary from 2¢ to $2.75 per 100 ft^3 (2.83 cu. meters), depending on purity, etc. Production of elemental nitrogen in the U.S. is more than 9 million short tons per year. Natural nitrogen contains two isotopes, ^{14}N and ^{15}N. Ten other isotopes are known.

Nobelium — (Alfred Nobel, discoverer of dynamite), No; at. wt. (259); at. no. 102; valence +2, +3. Nobelium was unambiguously discovered and identified in April 1958 at Berkeley by A. Ghiorso, T. Sikkeland, J. R. Walton, and G. T. Seaborg, who used a new double-recoil technique. A heavy-ion linear accelerator (HILAC) was used to bombard a thin target of curium (95% ^{244}Cm and 4.5% ^{246}Cm) with ^{12}C ions to produce 102^{254} according to the ^{246}Cm (^{12}C, 4n) reaction. Earlier in 1957 workers of the U.S., Britain, and Sweden announced the discovery of an isotope of Element 102 with a 10-min half-life at 8.5 MeV, as a result of bombarding ^{244}Cm with ^{13}C nuclei. On the basis of this experiment the name *nobelium* was assigned and accepted by the Commission on Atomic Weights of the International Union of Pure and Applied Chemistry. The acceptance of the name was premature, for both Russian and American efforts now completely rule out the possibility of any isotope of Element 102 having a half-life of 10 min in the vicinity of 8.5 MeV. Early work in 1957 on the search for this element, in Russia at the Kurchatov Institute, was marred by the assignment of 8.9 ± 0.4 MeV alpha radiation with a half-life of 2 to 40 sec, which was too indefinite to support claim to discovery. Confirmatory experiments at Berkeley in 1966 have shown the existence of ^{254}102 with a 55-s half-life, ^{252}102 with a 2.3-s half-life, and ^{257}102 with a 25-s half-life. Twelve isotopes are now recognized, one of which — ^{255}102 has a half-life of 3.1 min. In view of the discover's traditional right to name an element, the Berkeley group, in 1967, suggested that the hastily given name *nobelium,* along with the symbol No, be retained.

Osmium — (Gr. *osme*, a smell), Os; at. wt. 190.23(3); at. no. 76; m.p. 3033 ± 30°C; b.p. 5012 ± 100°C; sp. gr. 22.57; valence 0 to +8, more usually +3, +4, +6, and +8. Discovered in 1803 by Tennant in the residue left when crude platinum is dissolved by *aqua regia*. Osmium occurs in *iridosmine* and in platinum-bearing river sands of the Urals, North America, and South America. It is also found in the nickel-bearing ores of Sudbury, Ontario, region along with other platinum metals. While the quantity of platinum metals in these ores is very small, the large tonnages of nickel ores processed make commercial recovery possible. The metal is lustrous, bluish white, extremely hard, and brittle even at high temperatures. It has the highest melting point and the lowest vapor pressure of the platinum group. The metal is very difficult to fabricate, but the powder can be sintered in a hydrogen atmosphere at a temperature of 2000°C. The solid metal is not affected by air at room temperature, but the powdered or spongy metal slowly gives off osmium tetroxide, which is a powerful oxidizing agent and has a strong smell. The tetroxide is highly toxic, and boils at 130°C (760 mm). Concentrations in air as low as 10^{-7} g/m^3 can cause lung congestion, skin damage, or eye damage. The tetroxide has been used to detect fingerprints and to stain fatty tissue for microscope slides. The metal is almost entirely used to produce very hard alloys, with other metals of the platinum group, for fountain pen tips, instrument pivots, phonograph needles, and electrical contacts. The price of 99.9% pure osmium powder — the form usually supplied commercially — is about $100/g, depending on quantity and supplier. Natural osmium contains seven isotopes, one of which, ^{186}Os, is radioactive with a very long half-life. Thirty four other isotopes and isomers are known, all of which are radioactive.The measured densities of iridium and osmium seem to indicate that osmium is slightly more dense than iridium, so osmium has generally been credited with being the heaviest known element. Calculations of the density from the space lattice, which may be more reliable for these elements than actual measurements, however, give a density of 22.65 for iridium compared to 226.61 for osmium. At present, therefore, we know either iridium or osmium is the heaviest element, but the data do not allow selection between the two.

Oxygen — (Gr. *oxys*, sharp, acid, and *genes*, forming; acid former), O; at. wt. 15.9994(3); at. no. 8; t.p. –218.7916°C; b.p. –182.95°C; valence 2. For many centuries, workers occasionally realized air was composed of more than one component. The behavior of oxygen and nitrogen as components of air led to the advancement of the phlogiston theory of combustion, which captured the minds of chemists for a century. Oxygen was prepared by several workers, including Bayen and Borch, but they did not know how to collect it, did not study its properties, and did not recognize it as an elementary substance. Priestley is generally credited with its discovery, although Scheele also discovered it independently. Oxygen is the third most abundant element found in the sun, and it plays a part in the carbon-nitrogen cycle, one process thought to give the sun and stars their energy. Oxygen under excited conditions is responsible for the bright red and yellow-green colors of the aurora. Oxygen, as a gaseous element, forms 21% of the atmosphere by volume from which it can be obtained by liquefaction and fractional distillation. The atmosphere of Mars contains about 0.15% oxygen. The element and its compounds make up 49.2%, by weight, of the earth's crust. About two thirds of the human body and nine tenths of water is oxygen. In the laboratory it can be prepared by the electrolysis of water or by heating potassium chlorate with manganese dioxide as a catalyst. The gas is colorless, odorless, and tasteless. The liquid and solid forms are a pale blue color and are strongly paramagnetic. Ozone (O_3), a highly active compound, is formed by the action of an electrical discharge or ultraviolet light on oxygen. Ozone's presence in the atmosphere (amounting to the equivalent of a layer 3 mm thick at ordinary pressures and temperatures) is of vital importance in preventing harmful ultraviolet rays of the sun from reaching the earth's surface. There has been recent concern that pollutants in the atmosphere may have a detrimental effect on this ozone layer. Ozone is toxic and exposure should not exceed 0.2 mg/m^3 (8-hour time-weighted average — 40-hour work week). Undiluted ozone has a bluish color. Liquid ozone is bluish black, and solid ozone is violet-black. Oxygen is very reactive and capable of combining with most elements. It is a component of hundreds of thousands of organic compounds. It is essential for respiration of all plants and animals and for practically all combustion. In hospitals it is frequently used to aid respiration of patients. Its atomic weight was used as a standard of comparison for each of the other elements until 1961 when the International Union of Pure and Applied Chemistry adopted carbon 12 as the new basis. Oxygen has thirteen recognized isotopes. Natural oxygen is a mixture of three isotopes. Oxygen 18 occurs naturally, is stable, and is available commercially. Water (H_2O with 1.5% ^{18}O) is also available. Commercial oxygen consumption in the U.S. is estimated to be 20 million short tons per year and the demand is expected to increase substantially in the next few years. Oxygen enrichment of steel blast furnaces accounts for the greatest use of the gas. Large quantities are also used in making synthesis gas for ammonia and methanol, ethylene oxide, and for oxy-acetylene welding. Air separation plants produce about 99% of the gas, electrolysis plants about 1%. The gas costs 5¢/ft^3 ($1.75/cu. meters) in small quantities.

Palladium — (named after the asteroid *Pallas*, discovered about the same time; Gr. *Pallas*, goddess of wisdom), Pd. at. wt. 106.42(1) at. no. 46; m.p. 1554.9°C; b.p. 2963°C; sp. gr. 1202 (20°C); valence 2, 3, or 4. Discovered in 1803 by Wollaston. Palladium is found along with platinum and other metals of the platinum group in placer deposits of Russia, South and North America, Ethiopia, and Australia. Natural palladium contains six stable isotopes. Twenty five other isotopes are recognized, all of which are radioactive. It is also found associated with the nickel-copper deposits of South Africa and Ontario. Its separation from the platinum metals depends upon the type of ore in which it is found. It is a steel-white metal, does not tarnish in air, and is the least dense and lowest melting of the platinum group of metals. When annealed, it is soft and ductile; cold working greatly increases its strength and hardness. Palladium is attacked by nitric and sulfuric acid. At room temperatures the metal has the unusual property of absorbing up to 900 times its own volume of hydrogen, possibly forming Pd$_2$H. It is not yet clear if this a true compound. Hydrogen readily diffuses through heated palladium and this provides a means of purifying the gas. Finely divided palladium is a good catalyst and is used for hydrogenation and dehydrogenation reactions. It is alloyed and used in jewelry trades. White gold is an alloy of gold decolorized by the addition of palladium. Like gold, palladium can be beaten into leaf as thin as 1/250,000 in. The metal is used in dentistry, watchmaking, and in making surgical instruments and electrical contacts. The metal sells for about $140/troy oz. ($4.50/g).

Phosphorus — (Gr. *phosphoros*, light bearing; ancient name for the planet Venus when appearing before sunrise), P; at. wt. 30.973762(4); at. no. 15; m.p. (white) 44.15°C; b.p. 280.5°C; sp. gr. (white) 1.82 (red) 2.20, (black) 2.25 to 2.69; valence 3 or 5. Discovered in 1669 by Brand, who prepared it from urine. Phosphorus exists in four or more allotropic forms: white (or yellow), red, and black (or violet). White phosphorus has two modifications: α and β with a transition temperature at –3.8°C. Never found free in nature, it is widely distributed in combination with minerals. Seventeen isotopes of phosphorus are recognized. *Phosphate* rock, which contains the mineral *apatite*, an impure tri-calcium phosphate, is an important source of the element. Large deposits are found in the U.S.S.R., in Morocco, and in Florida, Tennessee, Utah, Idaho, and elsewhere. Phosphorus in an essential ingredient of all cell protoplasm, nervous tissue, and bones. Ordinary phosphorus is a waxy white solid; when pure it is colorless and transparent. It is insoluble in water, but soluble in carbon disulfide. It takes fire spontaneously in air, burning to the pentoxide. It is very poisonous, 50 mg constituting an approximate fatal dose. Exposure to white phosphorus should not exceed 0.1 mg/m^3 (8-hour time-weighted average — 40-hour work week). White phosphorus should be kept under water, as it is dangerously reactive in air, and it should be handled with forceps, as contact with the skin may cause severe burns. When exposed to sunlight or when heated in its own vapor to 250°C, it is converted to the red variety, which does not phosphoresce in air as does the white variety. This form does not ignite spontaneously and it is not as dangerous as white phosphorus. It should, however, be handled with care as it does convert to the white form at some temperatures and it emits highly toxic fumes of the oxides of phosphorus when heated. The red modification is fairly stable, sublimes with a vapor pressure of 1 atm at 417°C, and is used in the manufacture of safety matches, pyrotechnics, pesticides, incendiary shells, smoke bombs, tracer bullets, etc. White phosphorus may be made by several methods. By one process, tri-calcium phosphate, the essential ingredient of phosphate rock, is heated in the presence of carbon and silica in an electric furnace or fuel-fired furnace. Elementary phosphorus is liberated as vapor and may be collected under water. If desired, the phosphorus vapor and carbon monoxide produced by the reaction can be oxidized at once in the presence of moisture to produce phosphoric acid, an important compound in making super-phosphate fertilizers. In recent years, concentrated phosphoric acids, which may contain as much as 70 to 75% P_2O_5 content, have become of great importance to agriculture and farm production. World-wide demand for fertilizers has caused record phosphate production. Phosphates are used in the production of special glasses, such as those used for sodium lamps. Bone-ash, calcium phosphate, is also used to produce fine chinaware and to produce mono-calcium phosphate used in baking powder. Phosphorus is also important in the production of steels, phosphor bronze, and many other products. Trisodium phosphate is important as a cleaning agent, as a water softener, and for preventing boiler scale and corrosion of pipes and boiler tubes. Organic compounds of phosphorus are important. Amorphous (red) phosphorus costs about $60/kg (99.5%).

Platinum — (Sp. *platina*, silver), Pt; at. wt. 195.08(3); at. no. 78; m.p. 1768.4°C; b.p. 3825 ± 100°C; sp. gr. 21.45 (20°C); valence 1?, 2, 3, or 4. Discovered in South America by Ulloa in 1735 and by Wood in 1741. The metal was used by pre-Columbian Indians. Platinum occurs native, accompanied by small quantities of iridium, osmium, palladium, ruthenium, and rhodium, all belonging to the same group of metals. These are found in the alluvial deposits of the Ural mountains, of Columbia, and of certain western American states. *Sperrylite* (PtAs$_2$), occurring with the nickel-bearing deposits of Sudbury, Ontario, is the source of a considerable amount of metal. The large production of nickel offsets there being only one part of the platinum metals in two million parts of ore. Platinum is a beautiful silvery-white metal, when pure, and is malleable and ductile. It has a coefficient of expansion almost equal to that of soda-lime-silica glass, and is therefore used to make sealed electrodes in glass systems. The metal does not oxidize in air at any temperature, but is corroded by halogens, cyanides, sulfur, and caustic alkalis. It is insoluble in hydrochloric and nitric acid, but dissolves when they are mixed as *aqua regia*, forming chloroplatinic acid (H_2PtCl_6), an important compound. Natural platinum contains six isotopes, one of which, ^{190}Pt, is radioactive with a long half-life. Thirty five other radioactive isotopes and isomers are recognized. The metal is extensively used in jewelry, wire, and vessels for laboratory use, and in many valuable instruments including thermocouple elements. It is also used for electrical contacts, corrosion-resistant apparatus, and in dentistry. Platinum-cobalt alloys have magnetic properties. One such alloy made of 76.7% Pt and 23.3% Co, by weight, is an extremely powerful magnet that offers a B-H (max) almost twice that of Alnico V. Platinum resistance wires are used for constructing high-temperature electric furnaces. The metal is used for coating missile nose cones, jet engine fuel nozzles, etc., which must perform reliably for long periods of time at high temperatures. The metal, like palladium, absorbs large volumes, of hydrogen, retaining it at ordinary temperatures but giving it up at red heat. In the finely divided state platinum is an excellent catalyst, having long been used in the contact process for producing sulfuric acid. It is also used as a catalyst in cracking petroleum products. There is also much current interest in the use of platinum as a catalyst in fuel cells and in its use as antipollution devices for automobiles. Platinum anodes are extensively used in cathodic protection systems for large ships and ocean-going vessels, pipelines, steel piers, etc. Pure platinum wire will glow red hot when placed in the vapor of methyl alcohol. It acts here as a catalyst, converting the alcohol to formaldehyde. This phenomenon has been used commercially to produce cigarette lighters and hand warmers. Hydrogen and oxygen explode in the presence of platinum. The price of platinum has varied widely; more than a century ago it was used to adulterate gold. It was nearly eight times as valuable as gold in 1920. The price in January 1996 was about $400/troy oz. ($13/g), about the same price as gold.

Plutonium — (Planet *pluto*), Pu; at. wt. (244); at. no. 94; sp. gr. (α modification) 19.84 (25°C); m.p. 640°C; b.p. 3228°C; valence 3, 4, 5, or 6. Plutonium was the second transuranium element of the actinide series to be discovered. The isotope ^{238}Pu was produced in 1940 by Seaborg, McMillan, Kennedy, and Wahl by deuteron bombardment of uranium in the 60-inch cyclotron at Berkeley, California. Plutonium also exists in trace quantities in naturally occurring uranium ores. It is formed in much the same manner as neptunium, by irradiation of natural uranium with the neutrons which are present. By far of greatest importance is the isotope Pu^{239}, with a half-life of 24,100 years, produced in extensive quantities in nuclear reactors from natural uranium:

$$^{238}U(n,\gamma) \rightarrow \,^{239}U \xrightarrow{\beta} \,^{239}Np \xrightarrow{\beta} \,^{239}Pu$$

Eighteen isotopes of plutonium are now known. Plutonium has assumed the position of dominant importance among the transuranium elements because of its successful use as an explosive ingredient in nuclear weapons and the place which it holds as a key material in the development of industrial use of nuclear power. One kilogram is equivalent to about 22 million kilowatt hours of heat energy. The complete detonation of a kilogram of plutonium produces an explosion equal to about 20,000 tons of chemical explosive. Its importance depends on the nuclear property of being readily fissionable with neutrons and its availability in quantity. The world's nuclear-power reactors are now producing about 20,000 kg of plutonium/yr. By 1982 it was estimated that about 300,000 kg had accumulated. The various nuclear applications of plutonium are well known. ^{238}Pu has been used in the Apollo lunar missions to power seismic and other equipment on the lunar surface. As with neptunium and uranium, plutonium metal can be prepared by reduction of the trifluoride with alkaline-earth metals. The metal has a silvery appearance and takes on a yellow tarnish when slightly oxidized. It is chemically reactive. A relatively large piece of plutonium is warm to the touch because of the energy given off in alpha decay. Larger pieces will produce enough heat to boil water. The metal readily dissolves in concentrated hydrochloric acid, hydroiodic acid, or perchloric acid with formation of the Pu^{+3} ion. The metal exhibits six allotropic modifications having various crystalline structures. The densities of these vary from 16.00 to 19.86 g/cm^3. Plutonium also exhibits four ionic valence states in aqueous solutions: Pu^{+3}(blue lavender), Pu^{+4} (yellow brown), PuO^+ (pink?), and PuO^{+2} (pink orange). The ion PuO^+ is unstable in aqueous solutions, disproportionating into Pu^{+4} and PuO^{+2}. The Pu^{+4} thus formed, however, oxidizes the PuO^+ into PuO^{+2}, itself being reduced to Pu^{+3}, giving finally Pu^{+3} and PuO^{+2}. Plutonium forms binary compounds with oxygen: PuO, PuO_2, and intermediate oxides of variable composition; with the halides: PuF_3, PuF_4, $PuCl_3$, $PuBr_3$, PuI_3; with carbon, nitrogen, and silicon: PuC, PuN, $PuSi_2$. Oxyhalides are also well known: $PuOCl$, $PuOBr$, $PuOI$. Because of the high rate of emission of alpha particles and the element being specifically absorbed by bone marrow, plutonium, as well as all of the other transuranium elements except neptunium, are radiological poisons and must be handled with very special equipment and precautions. Plutonium is a very dangerous radiological hazard. Precautions must also be taken to prevent the unintentional formation of a critical mass. Plutonium in liquid solution is more likely to become critical than solid plutonium. The shape of the mass must also be considered where criticality is concerned. Plutonium-238 is available to authorized users from the O.R.N.L. at a cost of about $7.50/mg (97%) plus packing costs of $1250 per package.

Polonium — (Poland, native country of Mme. Curie), Po; at. wt. (209); at. no. 84; m.p. 254°C; b.p. 962°C; sp. gr. (alpha modification) 9.32; valence –2, 0, +2, +3(?), +4, and +6. Polonium was the first element discovered by Mme. Curie in 1898, while seeking the cause of radioactivity of pitchblende from Joachimsthal, Bohemia. The electroscope showed it separating with bismuth. Polonium is also called Radium F. Polonium is a very rare natural element. Uranium ores contain only about 100 μg of the element per ton. Its abundance is only about 0.2% of that of radium. In 1934, it was found that when natural bismuth (^{209}Bi) was bombarded by neutrons, ^{210}Bi, the parent of polonium, was obtained. Milligram amounts of polonium may now be prepared this way, by using the high neutron fluxes of nuclear reactors. Polonium-210 is a low-melting, fairly volatile metal, 50% of which is vaporized in air in 45 hours at 55°C. It is an alpha emitter with a half-life of 138.39 days. A milligram emits as many alpha particles as 5 g of radium. The energy released by its decay is so large (140 W/g) that a capsule containing about half a gram reaches a temperature above 500°C. The capsule also presents a contact gamma-ray dose rate of 0.012 Gy/h. A few curies (1 curie = 3.7×10^{10} Bq) of polonium exhibit a blue glow, caused by excitation of the surrounding gas. Because almost all alpha radiation is stopped within the solid source and its container, giving up its energy, polonium has attracted attention for uses as a lightweight heat source for thermoelectric power in space satellites. Thirty six isotopes and isomers of polonium are known, with atomic masses ranging from 192 to 218. All are radioactive. Polonium-210 is the most readily available. Isotopes of mass 209 (half-life 102 years) and mass 208 (half-life 2.9 years) can be prepared by alpha, proton, or deuteron bombardment of lead or bismuth in a cyclotron, but these are expensive to produce. Metallic polonium has been prepared from polonium hydroxide and some other polonium compounds in the presence of concentrated aqueous or anhydrous liquid ammonia. Two allotropic modifications are known to exist. Polonium is readily dissolved in dilute acids, but is only slightly soluble in alkalis. Polonium salts of organic acids char rapidly; halide amines are reduced to the metal. Polonium can be mixed or alloyed with beryllium to provide a source of neutrons. It has been used in devices for eliminating static charges in textile mills, etc.; however, beta sources are more commonly used and are less dangerous. It is also used on brushes for removing dust from photographic films. The polonium for these is carefully sealed and controlled, minimizing hazards to the user. Polonium-210 is very dangerous to handle in even milligram or microgram amounts, and special equipment and strict control is necessary. Damage arises from the complete absorption of the energy of the alpha particle into tissue. The maximum permissible body burden for ingested polonium is only 0.03 μCi, which represents a particle weighing only 6.8×10^{-12} g. Weight for weight it is about 2.5×10^{11} times as toxic as hydrocyanic acid. The maximum allowable concentration for soluble polonium compounds in air is about 2×10^{11} μCi/cm^3. Polonium-209 is available commercially on special order from the Oak Ridge National Laboratory at a cost of $3195/$\mu$Ci.

Potassium — (English, *potash* — pot ashes; L. *kalium*, Arab. *qali*, alkali), K; at. wt. 39.0983(1); at. no. 19; m.p. 63.28°C; b.p. 759°C; sp. gr. 0.862 (20°C); valence 1. Discovered in 1807 by Davy, who obtained it from caustic potash (KOH); this was the first metal isolated by electrolysis. The metal is the seventh most abundant and makes up about 2.4% by weight of the earth's crust. Most potassium minerals are insoluble and the metal is obtained from them only with great difficulty. Certain minerals, however, such as *sylvite, carnallite, langbeinite,* and *polyhalite* are found in ancient lake and sea beds and form rather extensive deposits from which potassium and its salts can readily be obtained. Potash is mined in Germany, New Mexico, California, Utah, and elsewhere. Large deposits of potash, found at a depth of some 1000 m in Saskatchewan, promise to be important in coming years. Potassium is also found in the ocean, but is present only in relatively small amounts, compared to sodium. The greatest demand for potash has been in its use for fertilizers. Potassium is an essential constituent for plant growth and it is found in most soils. Potassium is never found free in nature, but is obtained by electrolysis of the hydroxide, much in the same manner as prepared by Davy. Thermal methods also are commonly used to produce potassium (such as by reduction of potassium compounds with CaC_2, C, Si, or Na). It is one of the most reactive and electropositive of metals. Except for lithium, it is the lightest known metal. It is soft, easily cut with a knife, and is silvery in appearance immediately after a fresh surface is exposed. It rapidly oxidizes in air and should be preserved in a mineral oil. As with other metals of the alkali group, it decomposes in water with the evolution of hydrogen. It catches fire spontaneously on water. Potassium and its salts impart a violet color to flames. Twenty one isotopes, one of which is an isomer, of potassium are known. Ordinary potassium is composed of three isotopes, one of which is ^{40}K (0.0117%), a radioactive isotope with a half-life of 1.28×10^9 years. The radioactivity presents no appreciable hazard. An alloy of sodium and potassium (NaK) is used as a heat-transfer medium. Many potassium salts are of utmost importance, including the hydroxide, nitrate, carbonate, chloride, chlorate, bromide, iodide, cyanide, sulfate, chromate, and dichromate. Metallic potassium is available commercially for about $650/kg (98% purity) or $10/g (99.95% purity).

Praseodymium — (Gr. *prasios,* green, and *didymos,* twin), Pr; at. wt. 140.90765(3); at. no. 59; m.p. 931°C; b.p. 3520°C; sp. gr. 6.773; valence 3. In 1841 Mosander extracted the rare earth *didymia* from *lanthana;* in 1879, Lecoq de Boisbaudran isolated a new earth, *samaria,* from didymia obtained from the mineral *samarskite.* Six years later, in 1885, von Welsbach separated didymia into two others, *praseodymia* and *neodymia,* which gave salts of different colors. As with other rare earths, compounds of these elements in solution have distinctive sharp spectral absorption bands or lines, some of which are only a few Angstroms wide. The element occurs along with other rare-earth elements in a variety of minerals. *Monazite* and *bastnasite* are the two principal commercial sources of the rare-earth metals. Ion-exchange and solvent extraction techniques have led to much easier isolation of the rare earths and the cost has dropped greatly in the past few years. Thirty six isotopes and isomers are now recognized. Praseodymium can be prepared by several methods, such as by calcium reduction of the anhydrous chloride of fluoride. Misch metal, used in making cigarette lighters, contains about 5% praseodymium metal. Praseodymium is soft, silvery, malleable, and ductile. It was prepared in relatively pure form in 1931. It is somewhat more resistant to corrosion in air than europium, lanthanum, cerium, or neodymium, but it does develop a green oxide coating that spalls off when exposed to air. As with other rare-earth metals it should be kept under a light mineral oil or sealed in plastic. The rare-earth oxides, including Pr_2O_3, are among the most refractory substances known. Along with other rare earths, it is widely used as a core material for carbon arcs used by the motion picture industry for studio lighting and projection. Salts of praseodymium are used to color glasses and enamels; when mixed with certain other materials, praseodymium produces an intense and unusually clean yellow color in glass. Didymium glass, of which praseodymium is a component, is a colorant for welder's goggles. The metal (99.9% pure) is priced at about $2/g.

Promethium — (*Prometheus,* who, according to mythology, stole fire from heaven), Pm; at. no. 61; at. wt. (145); m.p. 1042°C; b.p. 3000°C (est.); sp. gr. 7.264 (25°C); valence 3. In 1902 Branner predicted the existence of an element between neodymium and samarium, and this was confirmed by Moseley in 1914. Unsuccessful searches were made for this predicted element over two decades, and various investigators proposed the names "illinium", "florentium", and "cyclonium" for this element. In 1941, workers at Ohio State University irradiated neodymium and praseodymium with neutrons, deuterons, and alpha particles, resp., and produced several new radioactivities, which most likely were those of element 61. Wu and Segre, and Bethe, in 1942, confirmed the formation; however, chemical proof of the production of element 61 was lacking because of the difficulty in separating the rare earths from each other at that time. In 1945, Marinsky, Glendenin, and Coryell made the first chemical identification by use of ion-exchange chromatography. Their work was done by fission of uranium and by neutron bombardment of neodymium. These investigators named the newly discovered element. Searches for the element on earth have been fruitless, and it now appears that promethium is completely missing from the earth's crust. Promethium, however, has been reported to be in the spectrum of the star HR^{465} in Andromeda. This element is being formed recently near the star's surface, for no known isotope of promethium has a half-life longer than 17.7 years. Thirty five isotopes and isomers of promethium, with atomic masses from 130 to 158 are now known. Promethium-145, with a half-life of 17.7 years, is the most useful. Promethium-145 has a specific activity of 940 Ci/g. It is a soft beta emitter; although no gamma rays are emitted, X-radiation can be generated when beta particles impinge on elements of a high atomic number, and great care must be taken in handling it. Promethium salts luminesce in the dark with a pale blue or greenish glow, due to their high radioactivity. Ion-exchange methods led to the preparation of about 10 g of promethium from atomic reactor fuel processing wastes in early 1963. Little is yet generally known about the properties of metallic promethium. Two allotropic modifications exist. The element has applications as a beta source for thickness gages, and it can be absorbed by a phosphor to produce light. Light produced in this manner can be used for signs or signals that require dependable operation; it can be used as a nuclear-powered battery by capturing light in photocells which convert it into electric current. Such a battery, using ^{147}Pm, would have a useful life of about 5 years. It is being used for fluorescent lighting starter sand coatings for self-luminous watch dials. Promethium shows promise as a portable X-ray source, and it may become useful as a heat source to provide auxiliary power for space probes and satellites. More than 30 promethium compounds have been prepared. Most are colored. Promethium-147 is available upon special order from the Idaho National Engineering Laboratory, Idaho Falls, ID, or from the Westinghouse Hanford Co., Richland, WA.

Protactinium — (Gr. *protos,* first), Pa; at. wt. 231.03588(2); at. no. 91; m.p. 1572°C; sp. gr. 15.37 (calc.); valence 4 or 5. The first isotope of element 91 to be discovered was ^{234}Pa, also known as UX$_2$, a short-lived member of the naturally occurring ^{238}U decay series. It was identified by K. Fajans and O. H. Gohring in 1913 and they named the new element *brevium.* When the longer-lived isotope ^{231}Pa was identified by Hahn and Meitner in 1918, the name protoactinium was adopted as being more consistent with the characteristics of the most abundant isotope. Soddy, Cranson, and Fleck were also active in this work. The name *protoactinium* was shortened to *protactinium* in 1949. In 1927, Grosse prepared 2 mg of a white powder, which was shown to be Pa_2O_5. Later, in 1934, from 0.1 g of pure Pa_2O_5 he isolated the element by two methods, one of which was by converting the oxide to an iodide and "cracking" it in a high vacuum by an electrically heated filament by the reaction

$$2PaI_5 \rightarrow 2Pa + 5I_2$$

Protactinium has a bright metallic luster which it retains for some time in air. The element occurs in *pitchblende* to the extent of about 1 part ^{231}Pa to 10 million of ore. Ores from Zaire have about 3 ppm. Protactinium has twenty four isotopes and isomers, the most common of which is ^{231}Pr with a half-life of 32,500 years. A number of protactinium compounds are known, some of which are colored. The element is superconductive below 1.4 K. The element is a dangerous toxic material and requires precautions similar to those used when handling plutonium. In 1959 and 1961, it was announced that the Great Britain Atomic Energy Authority extracted by a 12-stage process 125 g of 99.9% protactinium, the world's only stock of the metal for many years to come. The extraction was made from 60 tons of waste material at a cost of about $500,000. Protactinium is one of the rarest and most expensive naturally occurring elements.

Radium — (L. *radius*, ray), Ra; at. wt. (226); at. no. 88; m.p. 700°C; b.p. 1140°C; sp. gr. 5; valence 2. Radium was discovered in 1898 by M. and Mme. Curie in the *pitchblende* or *uraninite* of North Bohemia, where it occurs. There is about 1 g of radium in 7 tons of pitchblende. The element was isolated in 1911 by Mme. Curie and Debierne by the electrolysis of a solution of pure radium chloride, employing a mercury cathode; on distillation in an atmosphere of hydrogen this amalgam yielded the pure metal. Originally, radium was obtained from the rich pitchblende ore found at Joachimsthal, Bohemia. The *carnotite* sands of Colorado furnish some radium, but richer ores are found in the Republic of Zaire and the Great Bear Lake region of Canada. Radium is present in all uranium minerals, and could be extracted, if desired, from the extensive wastes of uranium processing. Large uranium deposits are located in Ontario, New Mexico, Utah, Australia, and elsewhere. Radium is obtained commercially as the bromide or chloride; it is doubtful if any appreciable stock of the isolated element now exists. The pure metal is brilliant white when freshly prepared, but blackens on exposure to air, probably due to formation of the nitride. It exhibits luminescence, as do its salts; it decomposes in water and is somewhat more volatile than barium. It is a member of the alkaline-earth group of metals. Radium imparts a carmine red color to a flame. Radium emits alpha, beta, and gamma rays and when mixed with beryllium produce neutrons. One gram of ^{226}Ra undergoes 3.7×10^{10} disintegrations per s. The *curie (Ci)* is defined as that amount of radioactivity which has the same disintegration rate as 1 g of ^{226}Ra. Twenty nine isotopes and one isomer are now known; radium 226, the common isotope, has a half-life of 1599 years. One gram of radium produces about 0.0001 ml (stp) of emanation, or radon gas, per day. This is pumped from the radium and sealed in minute tubes, which are used in the treatment of cancer and other diseases. One gram of radium yields about 4186 kJ per year. Radium is used in producing self-luminous paints, neutron sources, and in medicine for the treatment of disease. Some of the more recently discovered radioisotopes, such as ^{60}Co, are now being used in place of radium. Some of these sources are much more powerful, and others are safer to use. Radium loses about 1% of its activity in 25 years, being transformed into elements of lower atomic weight. Lead is a final product of disintegration. Stored radium should be ventilated to prevent build-up of radon. Inhalation, injection, or body exposure to radium can cause cancer and other body disorders. The maximum permissible burden in the total body for ^{226}Ra is 7400 becquerel.

Radon — (from *radium*; called *niton* at first, L. *nitens*, shining), Rn; at. wt. (222); at. no. 86; m.p. –71°C; b.p. –61.7°C; density of gas 9.73 g/l; sp. gr. liquid 4.4 at –62°C, solid 4; valence usually 0. The element was discovered in 1900 by Dorn, who called it *radium emanation*. In 1908 Ramsay and Gray, who named it *niton*, isolated the element and determined its density, finding it to be the heaviest known gas. It is essentially inert and occupies the last place in the zero group of gases in the Periodic Table. Since 1923, it has been called radon. Thirty nine isotopes and isomers are known. Radon-222, coming from radium, has a half-life of 3.823 days and is an alpha emitter; Radon-220, emanating naturally from thorium and called *thoron*, has a half-life of 55.6 s and is also an alpha emitter. Radon-219 emanates from actinium and is called *actinon*. It has a half-life of 3.96 s and is also on alpha emitter. It is estimated that every square mile of soil to a depth of 6 inch contains about 1 g of radium, which releases radon in tiny amounts to the atmosphere. Radon is present in some spring waters, such as those at Hot Springs, Arkansas. On the average, one part of radon is present to 1×10^{21} part of air. At ordinary temperatures radon is a colorless gas; when cooled below the freezing point, radon exhibits a brilliant phosphorescence which becomes yellow as the temperature is lowered and orange-red at the temperature of liquid air. It has been reported that fluorine reacts with radon, forming radon fluoride. Radon clathrates have also been reported. Radon is still produced for therapeutic use by a few hospitals by pumping it from a radium source and sealing it in minute tubes, called seeds or needles, for application to patients. This practice has now been largely discontinued as hospitals can order the seeds directly from suppliers, who make up the seeds with the desired activity for the day of use. Radon is available at a cost of about $4/m Ci. Care must be taken in handling radon, as with other radioactive materials. The main hazard is from inhalation of the element and its solid daughters, which are collected on dust in the air. Good ventilation should be provided where radium, thorium, or actinium is stored to prevent build-up of this element. Radon build-up is a health consideration in uranium mines. Recently radon build-up in homes has been a concern. Many deaths from lung cancer are caused by radon exposure. In the U.S. it is recommended that remedial action be taken if the air in homes exceeds 4 pCi/l.

Rhenium — (L. *Rhenus*, Rhine), Re; at. wt. 186.207(1); at. no. 75; m.p. 3186°C; b.p. 5596°C; sp. gr. 21.02 (20°C); valence –1, +1, 2, 3, 4, 5, 6, 7. Discovery of rhenium is generally attributed to Noddack, Tacke, and Berg, who announced in 1925 they had detected the element in platinum ores and *columbite*. They also found the element in *gadolinite* and *molybdenite*. By working up 660 kg of molybdenite they were able in 1928 to extract 1 g of rhenium. The price in 1928 was $10,000/g. Rhenium does not occur free in nature or as a compound in a distinct mineral species. It is, however, widely spread throughout the earth's crust to the extent of about 0.001 ppm. Commercial rhenium in the U.S. today is obtained from molybdenite roaster-flue dusts obtained from copper-sulfide ores mined in the vicinity of Miami, Arizona, and elsewhere in Arizona and Utah. Some molybdenites contain from 0.002 to 0.2% rhenium. More than 150,000 troy ounces of rhenium are now being produced yearly in the United States. The total estimated Free World reserve of rhenium metal is 3500 tons. Natural rhenium is a mixture of two isotopes, one of which has a very long half-life. Thirty nine other unstable isotopes are recognized. Rhenium metal is prepared by reducing ammonium perrhenate with hydrogen at elevated temperatures. The element is silvery white with a metallic luster; its density is exceeded only by that of platinum, iridium, and osmium, and its melting point is exceeded only by that of tungsten and carbon. It has other useful properties. The usual commercial form of the element is a powder, but it can be consolidated by pressing and resistance-sintering in a vacuum or hydrogen atmosphere. This produces a compact shape in excess of 90% of the density of the metal. Annealed rhenium is very ductile, and can be bent, coiled, or rolled. Rhenium is used as an additive to tungsten and molybdenum-based alloys to impart useful properties. It is widely used for filaments for mass spectrographs and ion gages. Rhenium-molybdenum alloys are superconductive at 10 K.

Rhenium is also used as an electrical contact material as it has good wear resistance and withstands arc corrosion. Thermocouples made of Re-W are used for measuring temperatures up to 2200°C, and rhenium wire has been used in photoflash lamps for photography. Rhenium catalysts are exceptionally resistant to poisoning from nitrogen, sulfur, and phosphorus, and are used for hydrogenation of fine chemicals, hydrocracking, reforming, and disproportionation of olefins. Rhenium costs about $12/g (99.99% pure). Little is known of its toxicity; therefore, it should be handled with care until more data are available.

Rhodium — (Gr. *rhodon*, rose), Rh; at. wt. 102.90550(3); at. no. 45; m.p. 1964 ± 3°C; b.p. 3695 ± 100°C; sp. gr. 12.41 (20°C); valence 2, 3, 4, 5, and 6. Wollaston discovered rhodium in 1803-4 in crude platinum ore he presumably obtained from South America. Rhodium occurs native with other platinum metals in river sands of the Urals and in North and South America. It is also found with other platinum metals in the copper-nickel sulfide ores of the Sudbury, Ontario region. Although the quantity occurring here is very small, the large tonnages of nickel processed make the recovery commercially feasible. The annual world production of rhodium is only 7 or 8 tons. The metal is silvery white and at red heat slowly changes in air to the sesquioxide. At higher temperatures it converts back to the element. Rhodium has a higher melting point and lower density than platinum. Its major use is as an alloying agent to harden platinum and palladium. Such alloys are used for furnace windings, thermocouple elements, bushings for glass fiber production, electrodes for aircraft spark plugs, and laboratory crucibles. It is useful as an electrical contact material as it has a low electrical resistance, a low and stable contact resistance, and is highly resistant to corrosion. Plated rhodium, produced by electroplating or evaporation, is exceptionally hard and is used for optical instruments. It has a high reflectance and is hard and durable. Rhodium is also used for jewelry, for decoration, and as a catalyst. Forty four isotopes and isomers are now known. Soluble salts should not exceed 0.01 mg/m^3. Rhodium metal (powder) costs about $300/g (99.9% pure).

Rubidium — (L. *rubidus*, deepest red), Rb; at. wt. 85.4678(3); at. no. 37; m.p. 39.31°C; b.p. 688°C; sp. gr. (solid) 1.532 (20°C), (liquid) 1.475 (39°C); valence 1, 2, 3, 4. Discovered in 1861 by Bunsen and Kirchoff in the mineral *lepidolite* by use of the spectroscope. The element is much more abundant than was thought several years ago. It is now considered to be the 16th most abundant element in the earth's crust. Rubidium occurs in *pollucite, carnallite, leucite*, and *zinnwaldite*, which contains traces up to 1%, in the form of the oxide. It is found in lepidolite to the extent of about 1.5%, and is recovered commercially from this source. Potassium minerals, such as those found at Searles Lake, California, and potassium chloride recovered from brines in Michigan also contain the element and are commercial sources. It is also found along with cesium in the extensive deposits of *pollucite* at Bernic Lake, Manitoba. Rubidium can be liquid at room temperature. It is a soft, silvery-white metallic element of the alkali group and is the second most electropositive and alkaline element. It ignites spontaneously in air and reacts violently in water, setting fire to the liberated hydrogen. As with other alkali metals, it forms amalgams with mercury and it alloys with gold, cesium, sodium, and potassium. It colors a flame yellowish violet. Rubidium metal can be prepared by reducing rubidium chloride with calcium, and by a number of other methods. It must be kept under a dry mineral oil or in a vacuum or inert atmosphere. Thirty five isotopes and isomers of rubidium are known. Naturally occurring rubidium is made of two isotopes, ^{85}Rb and ^{87}Rb. Rubidium-87 is present to the extent of 27.83% in natural rubidium and is a beta emitter with a half-life of 4.9×10^{10} years. Ordinary rubidium is sufficiently radioactive to expose a photographic film in about 30 to 60 days. Rubidium forms four oxides: Rb_2O, Rb_2O_2, Rb_2O_3, Rb_2O_4. Because rubidium can be easily ionized, it has been considered for use in "ion engines" for space vehicles; however, cesium is somewhat more efficient for this purpose. It is also proposed for use as a working fluid for vapor turbines and for use in a thermoelectric generator using the magnetohydro-dynamic principle where rubidium ions are formed by heat at high temperature and passed through a magnetic field. These conduct electricity and act like an armature of a generator thereby generating an electric current. Rubidium is used as a getter in vacuum tubes and as a photocell component. It has been used in making special glasses. $RbAg_4I_5$ is important, as it has the highest room conductivity of any known ionic crystal. At 20°C its conductivity is about the same as dilute sulfuric acid. This suggests use in thin film batteries and other applications. The present cost in small quantities is about $20/g (99.8% pure).

Ruthenium — (L. *Ruthenia*, Russia), Ru; at. wt. 101.07(2); at. no. 44, m.p. 2334°C; b.p. 4150°C; sp. gr. 12.41 (20°C); valence 0, 1, 2, 3, 4, 5, 6, 7, 8. Berzelius and Osann in 1827 examined the residues left after dissolving crude platinum from the Ural mountains in *aqua regia*. While Berzelius found no unusual metals, Osann thought he found three new metals, one of which he named ruthenium. In 1844 Klaus, generally recognized as the discoverer, showed that Osann's ruthenium oxide was very impure and that it contained a new metal. Klaus obtained 6 g of ruthenium from the portion of crude platinum that is insoluble in *aqua regia*. A member of the platinum group, ruthenium occurs native with other members of the group in ores found in the Ural mountains and in North and South America. It is also found along with other platinum metals in small but commercial quantities in *pentlandite* of the Sudbury, Ontario, nickel-mining region, and in *pyroxinite* deposits of South Africa. Natural ruthenium contains seven isotopes. Twenty one other isotopes and isomers are known, all of which are radioactive. The metal is isolated commercially by a complex chemical process, the final stage of which is the hydrogen reduction of ammonium ruthenium chloride, which yields a powder. The powder is consolidated by powder metallurgy techniques or by argon-arc welding. Ruthenium is a hard, white metal and has four crystal modifications. It does not tarnish at room temperatures, but oxidizes in air at about 800°C. The metal is not attacked by hot or cold acids or *aqua regia*, but when potassium chlorate is added to the solution, it oxidizes explosively. It is attacked by halogens, hydroxides, etc. Ruthenium can be plated by electrodeposition or by thermal decomposition methods. The metal is one of the most effective hardeners for platinum and palladium, and is alloyed with these metals to make electrical contacts for severe wear resistance. A ruthenium-molybdenum alloy is said to be superconductive at 10.6 K. The corrosion resistance of titanium is improved a hundredfold by addition of 0.1% ruthenium. It is a versatile catalyst. Hydrogen sulfide can be split catalytically by light using an aqueous suspension of CdS particles loaded with ruthenium dioxide. It is thought this may have application to removal of H_2S from oil refining and other industrial processes. Compounds in at least eight oxidation states have been found, but of these, the +2. +3. and +4 states are the most common. Ruthenium tetroxide, like osmium tetroxide, is highly toxic. In addition, it may explode. Ruthenium compounds show a marked resemblance to those of osmium. The metal is priced at about $30/g (99.95% pure).

Rutherfordium — See Element 104 (unnilquadium) and also Element 106 (unnilhexium).

Samarium — (*Samarskite* a mineral), Sm; at. wt. 150.36(3); at. no. 62; m.p. 1074°C; b.p. 1794°C; sp. gr (α) 7.520 (25°C); valence 2 or 3. Discovered spectroscopically by its sharp absorption lines in 1879 by Lecoq de Boisbaudran in the mineral *samarskite*, named in honor of a Russian mine official, Col. Samarski. Samarium is found along with other members of the rare-earth elements in many minerals, including *monazite* and *bastnasite*, which are commercial sources. It occurs in monazite to the extent of 2.8%. While *misch metal* containing about 1% of samarium metal, has long been used, samarium has not been isolated in relatively pure form until recent years. Ion-exchange and solvent extraction techniques have recently simplified separation of the rare earths from one another; more recently, electrochemical deposition, using an electrolytic solution of lithium citrate and a mercury electrode, is said to be a simple, fast, and highly specific way to separate the rare earths. Samarium metal can be produced by reducing the oxide with barium or lanthanum. Samarium has a bright silver luster and is reasonably stable in air. Three crystal modifications of the metal exist, with transformations at 734 and 922°C. The metal ignites in air at about 150°C. Thirty two isotopes and isomers of samarium are now recognized. Natural samarium is a mixture of seven isotopes, three of which are unstable but have long half-lives. Samarium, along with other rare earths, is used for carbon-arc lighting for the motion picture industry. The sulfide has excellent high-temperature stability and good thermoelectric efficiencies up to 1100°C. $SmCo_5$ has been used in making a new permanent magnet material with the highest resistance to demagnetization of any known material. It is said to have an intrinsic coercive force as high as 2200 kA/m. Samarium oxide has been used in optical glass to absorb the infrared. Samarium is used to dope calcium fluoride crystals for use in optical masers or lasers. Compounds of the metal act as sensitizers for phosphors excited in the infrared; the oxide exhibits catalytic properties in the dehydration and dehydrogenation of ethyl alcohol. It is used in infrared absorbing glass and as a neutron absorber in nuclear reactors. The metal is priced at about $2/g (99.9%). Little is known of the toxicity of samarium; therefore, it should be handled carefully.

Scandium — (L. *Scandia*, Scandinavia, Sc; at. wt. 44.955910(9); at. no. 21; m.p. 1541°C; b.p. 2836°C; sp. gr. 2.989 (25°C); valence 3. On the basis of the Periodic System, Mendeleev predicted the existence of *ekaboron*, which would have an atomic weight between 40 of calcium and 48 of titanium. The element was discovered by Nilson in 1878 in the minerals *euxenite* and *gadolinite*, which had not yet been found anywhere except in Scandinavia. By processing 10 kg of euxenite and other residues of rare-earth minerals, Nilson was able to prepare about 2 g of scandium oxide of high purity. Cleve later pointed out that Nilson's scandium was identical with Mendeleev's ekaboron. Scandium is apparently a much more abundant element in the sun and certain stars than here on earth. It is about the 23rd most abundant element in the sun, compared to the 50th most abundant on earth. It is widely distributed on earth, occurring in very minute quantities in over 800 mineral species. The blue color of beryl (aquamarine variety) is said to be due to scandium. It occurs as a principal component in the rare mineral *thortveitite*, found in Scandinavia and Malagasy. It is also found in the residues remaining after the extraction of tungsten from Zinnwald *wolframite*, and in *wiikite* and *bazzite*. Most scandium is presently being recovered from *thortveitite* or is extracted as a by-product from uranium mill tailings. Metallic scandium was first prepared in 1937 by Fischer, Brunger, and Grieneisen, who electrolyzed a eutectic melt of potassium, lithium, and scandium chlorides at 700 to 800°C. Tungsten wire and a pool of molten zinc served as the electrodes in a graphite crucible. Pure scandium is now produced by reducing scandium fluoride with calcium metal. The production of the first pound of 99% pure scandium metal was announced in 1960. Scandium is a silver-white metal which develops a slightly yellowish or pinkish cast upon exposure to air. It is relatively soft, and resembles yttrium and the rare-earth metals more than it resembles aluminum or titanium. It is a very light metal and has a much higher melting point than aluminum, making it of interest to designers of spacecraft. Scandium is not attacked by a 1:1 mixture of conc. HNO_3 and 48% HF. Scandium reacts rapidly with many acids. Nineteen isotopes and isomers of scandium are recognized. The metal is expensive, costing about $120/g with a purity of about 99.9%. Scandium oxide costs about $40/g. About 20 kg of scandium (as Sc_2O_3) are now being used yearly in the U.S. to produce high-intensity lights, and the radioactive isotope ^{46}Sc is used as a tracing agent in refinery crackers for crude oil, etc. Scandium iodide added to mercury vapor lamps produces a highly efficient light source resembling sunlight, which is important for indoor or night-time color TV. Little is yet known about the toxicity of scandium; therefore, it should be handled with care.

Selenium — (Gr. *Selene*, moon), Se; at. wt. 78.96(3); at. no. 34; m.p. (gray) 221°C; b.p. (gray) 685°C; sp. gr. (gray) 4.79, (vitreous) 4.28; valence –2, +4, or +6. Discovered by Berzelius in 1817, who found it associated with tellurium, named for the earth. Selenium is found in a few rare minerals, such as *crooksite* and *clausthalite*. In years past it has been obtained from flue dusts remaining from processing copper sulfide ores, but the anode muds from electrolytic copper refineries now provide the source of most of the world's selenium. Selenium is recovered by roasting the muds with soda or sulfuric acid, or by smelting them with soda and niter. Selenium exists in several allotropic forms. Three are generally recognized, but as many as six have been claimed. Selenium can be prepared with either an amorphous or crystalline structure. The color of amorphous selenium is either red, in powder form, or black, in vitreous form. Crystalline monoclinic selenium is a deep red; crystalline hexagonal selenium, the most stable variety, is a metallic gray. Natural selenium contains six stable isotopes. Twenty four other isotopes and isomers have been characterized. The element is a member of the sulfur family and resembles sulfur both in its various forms and in its compounds. Selenium exhibits both photovoltaic action, where light is converted directly into electricity, and photoconductive action, where the electrical resistance decreases with increased illumination. These properties make selenium useful in the production of photocells and exposure meters for photographic use, as well as solar cells. Selenium is also able to convert a.c. electricity to d.c., and is extensively used in rectifiers. Below its melting point selenium is a p-type semiconductor and is finding many uses in electronic and solid-state applications. It is used in Xerography for reproducing and copying documents, letters, etc. It is used by the glass industry to decolorize glass and to make ruby-colored glasses and enamels. It is also used as a photographic toner, and as an additive to stainless steel. Elemental selenium has been said to be practically nontoxic and is considered to be an essential trace element; however, hydrogen selenide and other selenium compounds are extremely toxic, and resemble arsenic in their physiological reactions. Hydrogen selenide in a concentration of 1.5 ppm is intolerable to man. Selenium occurs in some soils in amounts sufficient to produce serious effects on animals feeding on plants, such as locoweed, grown in such soils. Selenium is priced at about $150/kg. It is also available in high-purity form at a cost of about $250/kg (99.999%).

Silicon — (L. *silex, silicis*, flint), Si; at. wt. 28.0855(3); at. no. 14; m.p. 1414°C; b.p. 3265°C; sp. gr. 2.33 (25°C); valence 4. Davy in 1800 thought silica to be a compound and not an element; later in 1811, Gay Lussac and Thenard probably prepared impure amorphous silicon by heating potassium with silicon tetrafluoride. Berzelius, generally credited with the discovery, in 1824 succeeded in preparing amorphous silicon by the same general method as used earlier, but he purified the product by removing the fluosilicates by repeated washings. Deville in 1854 first prepared crystalline silicon, the second allotropic form of the element. Silicon is present in the sun and stars and is a principal component of a class of meteorites known as "aerolites". It is also a component of *tektites*, a natural glass of uncertain origin. Natural silicon contains three isotopes. Fourteen other radioactive isotopes are recognized. Silicon makes up 25.7% of the earth's crust, by weight, and is the second most abundant element, being exceeded only by oxygen. Silicon is not found free in nature, but occurs chiefly as the oxide and as silicates. *Sand, quartz, rock crystal, amethyst, agate, flint, jasper,* and *opal* are some of the forms in which the oxide appears. *Granite, hornblende, asbestos, feldspar, clay mica,* etc. are but a few of the numerous silicate minerals. Silicon is prepared commercially by heating silica and carbon in an electric furnace, using carbon electrodes. Several other methods can be used for preparing the element. Amorphous silicon can be prepared as a brown powder, which can be easily melted or vaporized. Crystalline silicon has a metallic luster and grayish color. The Czochralski process is commonly used to produce single crystals of silicon used for solid-state or semiconductor devices. Hyperpure silicon can be prepared by the thermal decomposition of ultra-pure trichlorosilane in a hydrogen atmosphere, and by a vacuum float zone process. This product can be doped with boron, gallium, phosphorus, or arsenic to produce silicon for use in transistors, solar cells, rectifiers, and other solid-state devices which are used extensively in the electronics and space-age industries. Hydrogenated amorphous silicon has shown promise in producing economical cells for converting solar energy into electricity. Silicon is a relatively inert element, but it is attacked by halogens and dilute alkali. Most acids except hydrofluoric, do not affect it. Silicones are important products of silicon. They may be prepared by hydrolyzing a silicon organic chloride, such as dimethyl silicon chloride. Hydrolysis and condensation of various substituted chlorosilanes can be used to produce a very great number of polymeric products, or silicones, ranging from liquids to hard, glasslike solids with many useful properties. Elemental silicon transmits more than 95% of all wavelengths of infrared, from 1.3 to 6.7 μm. Silicon is one of man's most useful elements. In the form of sand and clay it is used to make concrete and brick; it is a useful refractory material for high-temperature work, and in the form of silicates it is used in making enamels, pottery, etc. Silica, as sand, is a principal ingredient of glass, one of the most inexpensive of materials with excellent mechanical, optical, thermal, and electrical properties. Glass can be made in a very great variety of shapes, and is used as containers, window glass, insulators, and thousands of other uses. Silicon tetrachloride can be used to iridize glass. Silicon is important in plant and animal life. Diatoms in both fresh and salt water extract silica from the water to build up their cell walls. Silica is present in ashes of plants and in the human skeleton. Silicon is an important ingredient in steel; silicon carbide is one of the most important abrasives and has been used in lasers to produce coherent light of 4560 Å. Regular grade silicon (99.5%) costs about $140/kg. Silicon 99.96% pure costs about $250/kg; hyperpure silicon may cost as much as $400/kg. Miners, stonecutters, and other engaged in work where siliceous dust is breathed in large quantities often develop a serious lung disease known as *silicosis*.

Silver — (Anglo-Saxon, *Seolfor siolfur*), Ag (L. argentum), at. wt. 107.8682(2); at. no. 47; f.p. 961.78°C; b.p. 2162°C; sp. gr. 10.50 (20°C); valence 1, 2. Silver has been known since ancient times. It is mentioned in Genesis. Slag dumps in Asia Minor and on islands in the Aegean Sea indicate that man learned to separate silver from lead as early as 3000 B.C. Silver occurs native and in ores such as *argentite* (Ag_2S) and *horn silver* (AgCl); lead, lead-zinc, copper, gold, and copper-nickel ores are principal sources. Mexico, Canada, Peru, and the U.S. are the principal silver producers in the western hemisphere. Silver is also recovered during electrolytic refining of copper. Commercial fine silver contains at least 99.9% silver. Purities of 99.999+% are available commercially. Pure silver has a brilliant white metallic luster. It is a little harder than gold and is very ductile and malleable, being exceeded only by gold and perhaps palladium. Pure silver has the highest electrical and thermal conductivity of all metals, and possesses the lowest contact resistance. It is stable in pure air and water, but tarnishes when exposed to ozone, hydrogen sulfide, or air containing sulfur. The alloys of silver are important. Sterling silver is used for jewelry, silverware, etc. where appearance is paramount. This alloy contains 92.5% silver, the remainder being copper or some other metal. Silver is of utmost importance in photography, about 30% of the U.S. industrial consumption going into this application. It is used for dental alloys. Silver is used in making solder and brazing alloys, electrical contacts, and high capacity silver-zinc and silver-cadmium batteries. Silver paints are used for making printed circuits. It is used in mirror production and may be deposited on glass or metals by chemical deposition, electrodeposition, or by evaporation. When freshly deposited, it is the best reflector of visible light known, but is rapidly tarnishes and loses much of its reflectance. It is a poor reflector of ultraviolet. Silver fulminate ($Ag_2C_2N_2O_2$), a powerful explosive, is sometimes formed during the silvering process. Silver iodide is used in seeding clouds to produce rain. Silver chloride has interesting optical properties as it can be made transparent; it also is a cement for glass. Silver nitrate, or *lunar caustic*, the most important silver compound, is used extensively in photography. While silver itself is not considered to be toxic, most of its salts are poisonous. Natural silver contains two stable isotopes. Forty nine other radioactive isotopes and isomers are known. Silver compounds can be absorbed in the circulatory system and reduced silver deposited in the various tissues of the body. A condition, known as *argyria*, results, with a greyish pigmentation of the skin and mucous membranes. Silver has germicidal effects and kills many lower organisms effectively without harm to higher animals. Silver for centuries has been used traditionally for coinage by many countries of the world. In recent times, however, consumption of silver has at times greatly exceeded the output. In 1939, the price of silver was fixed by the U.S. Treasury at 71¢/troy oz., and at 90.5¢/troy oz. in 1946. In November 1961 the U.S. Treasury suspended sales of nonmonetized silver, and the price stabilized for a time at about $1.29, the melt-down value of silver U.S. coins. The Coinage Act of 1965 authorized a change in the metallic composition of the three U.S. subsidiary denominations to clad or composite type coins. This was the first change in U.S. coinage since the monetary system was established in 1792. Clad dimes and quarters are made of an outer layer of 75% Cu and 25% Ni bonded to a central core of pure Cu. The composition of the one- and five-cent pieces remains unchanged. One-cent coins are 95% Cu and 5% Zn. Five-cent coins are 75% Cu and 25% Ni. Old silver dollars are 90% Ag and 10% Cu. Earlier subsidiary coins of 90% Ag and 10% Cu officially were to circulate alongside the clad coins; however, in practice they have largely disappeared (Gresham's Law), as the value of the silver is now greater than their exchange value. Silver coins of other countries have largely been replaced with coins made of other metals. On June 24, 1968, the U.S. Government ceased to redeem U.S. Silver Certificates with silver. Since that time, the price of silver has fluctuated widely. As of January 1996, the price of silver was about $5.30/troy oz. (17¢/g); however the price has fluctuated considerably due to market instability.

Sodium — (English, *soda*; Medieval Latin, *sodanum*, headache remedy), Na (L. *natrium*); at. wt. 22.989768(6); at. no. 11; m.p. 97.72 ± 0.03°C; b.p. 883°C; sp. gr. 0.971 (20°C); valence 1. Long recognized in compounds, sodium was first isolated by Davy in 1807 by electrolysis of caustic soda. Sodium is present in fair abundance in the sun and stars. The D lines of sodium are among the most prominent in the solar spectrum. Sodium is the sixth most abundant element on earth, comprising about 2.6% of the earth's crust; it is the most abundant of the alkali group of metals of which it is a member. The most common compound is sodium chloride, but it occurs in many other minerals, such as *soda niter, cryolite, amphibole, zeolite, sodalite,* etc. It is a very reactive element and is never found free in nature. It is now obtained commercially by the electrolysis of absolutely dry fused sodium chloride. This method is much cheaper than that of electrolyzing sodium hydroxide, as was used several years ago. Sodium is a soft, bright, silvery metal which floats on water, decomposing it with the evolution of hydrogen and the formation of the hydroxide. It may or may not ignite spontaneously on water, depending on the amount of oxide and metal exposed to the water. It normally does not ignite in air at temperatures below 115°C. Sodium should be handled with respect, as it can be dangerous when improperly handled. Metallic sodium is vital in the manufacture of sodamide and esters, and in the preparation of organic compounds. The metal may be used to improve the structure of certain alloys, to descale metal, to purify molten metals, and as a heat transfer agent. An alloy of sodium with potassium, NaK, is also an important heat transfer agent. Sodium compounds are important to the paper, glass, soap, textile, petroleum, chemical, and metal industries. Soap is generally a sodium salt of certain fatty acids. The importance of common salt to animal nutrition has been recognized since prehistoric times. Among the many compounds that are of the greatest industrial importance are common salt (NaCl), soda ash (Na_2CO_3), baking soda ($NaHCO_3$), caustic soda (NaOH), Chile saltpeter ($NaNO_3$), di- and tri-sodium phosphates, sodium thiosulfate (hypo, $Na_2S_2O_3 \cdot 5H_2O$), and borax ($Na_2B_4O_7 \cdot 10H_2O$). Seventeen isotopes of sodium are recognized. Metallic sodium is priced at about $250/kg (99.95%). On a volume basis, it is the cheapest of all metals. Sodium metal should be handled with great care. It should be kept in an inert atmosphere and contact with water and other substances with which sodium reacts should be avoided.

THE ELEMENTS (continued)

Strontium — (*Strontian*, town in Scotland), Sr; at. wt. 87.62(1); at. no. 38; m.p. 777°C; b.p. 1382°C; sp. gr. 2.54; valence 2. Isolated by Davey by electrolysis in 1808; however, Adair Crawford in 1790 recognized a new mineral (strontianite) as differing from other barium minerals (baryta). Strontium is found chiefly as *celestite* ($SrSO_4$) and *strontianite* ($SrCO_3$). The metal can be prepared by electrolysis of the fused chloride mixed with potassium chloride, or is made by reducing strontium oxide with aluminum in a vacuum at a temperature at which strontium distills off. Three allotropic forms of the metal exist, with transition points at 235 and 540°C. Strontium is softer than calcium and decomposes water more vigorously. It does not absorb nitrogen below 380°C. It should be kept under mineral oil to prevent oxidation. Freshly cut strontium has a silvery appearance, but rapidly turns a yellowish color with the formation of the oxide. The finely divided metal ignites spontaneously in air. Volatile strontium salts impart a beautiful crimson color to flames, and these salts are used in pyrotechnics and in the production of flares. Natural strontium is a mixture of four stable isotopes. Twenty six other unstable isotopes and isomers are known to exist. Of greatest importance is ^{90}Sr with a half-life of 29 years. It is a product of nuclear fallout and presents a health problem. This isotope is one of the best long-lived high-energy beta emitters known, and is used in SNAP (Systems for Nuclear Auxiliary Power) devices. These devices hold promise for use in space vehicles, remote weather stations, navigational buoys, etc., where a lightweight, long-lived, nuclear-electric power source is needed. The major use for strontium at present is in producing glass for color television picture tubes. It has also found use in producing ferrite magnets and in refining zinc. Strontium titanate is an interesting optical material as it has an extremely high refractive index and an optical dispersion greater than that of diamond. It has been used as a gemstone, but it is very soft. It does not occur naturally. Strontium metal (99% pure) costs about $1/g while the metal, 99.95% pure, sells for about $10/g.

Sulfur — (Sanskrit, *sulvere*; L. *sulphurium*), S; at. wt. 32.066(6); at. no. 16; m.p. 115.21°C; b.p. 444.60°C; sp. gr. (rhombic) 2.07, (monoclinic) 1.957 (20°C); valence 2, 4, or 6. Known to the ancients; referred to in Genesis as *brimstone*. Sulfur is found in meteorites. A dark area near the crater Aristarchus on the moon has been studied by R. W. Wood with ultraviolet light. This study suggests strongly that it is a sulfur deposit. Sulfur occurs native in the vicinity of volcanoes and hot springs. It is widely distributed in nature as *iron pyrites, galena, sphalerite, cinnabar, stibnite, gypsum, Epsom salts, celestite, barite,*etc. Sulfur is commercially recovered from wells sunk into the salt domes along the Gulf Coast of the U.S. It is obtained from these wells by the Frasch process, which forces heated water into the wells to melt the sulfur, which is then brought to the surface. Sulfur also occurs in natural gas and petroleum crudes and must be removed from these products. Formerly this was done chemically, which wasted the sulfur. New processes now permit recovery, and these sources promise to be very important. Large amounts of sulfur are being recovered from Alberta gas fields. Sulfur is a pale yellow, odorless, brittle solid, which is insoluble in water but soluble in carbon disulfide. In every state, whether gas, liquid or solid, elemental sulfur occurs in more than one allotropic form or modification; these present a confusing multitude of forms whose relations are not yet fully understood. Amorphous or "plastic" sulfur is obtained by fast cooling of the crystalline form. X-ray studies indicate that amorphous sulfur may have a helical structure with eight atoms per spiral. Crystalline sulfur seems to be made of rings, each containing eight sulfur atoms, which fit together to give a normal X-ray pattern. Seventeen isotopes of sulfur are now recognized. Four occur in natural sulfur, none of which is radioactive. A finely divided form of sulfur, known as *flowers of sulfur*, is obtained by sublimation. Sulfur readily forms sulfides with many elements. Sulfur is a component of black gunpowder, and is used in the vulcanization of natural rubber and a fungicide. It is also used extensively is making phosphatic fertilizers. A tremendous tonnage is used to produce sulfuric acid, the most important manufactured chemical. It is used in making sulfite paper and other papers, as a fumigant, and in the bleaching of dried fruits. The element is a good electrical insulator. Organic compounds containing sulfur are very important. Calcium sulfate, ammonium sulfate, carbon disulfide, sulfur dioxide, and hydrogen sulfide are but a few of the many other important compounds of sulfur. Sulfur is essential to life. It is a minor constituent of fats, body fluids, and skeletal minerals. Carbon disulfide, hydrogen sulfide, and sulfur dioxide should be handled carefully. Hydrogen sulfide in small concentrations can be metabolized, but in higher concentrations it quickly can cause death by respiratory paralysis. It is insidious in that it quickly deadens the sense of smell. Sulfur dioxide is a dangerous component in atmospheric air pollution. In 1975, University of Pennsylvania scientists reported synthesis of polymeric sulfur nitride, which has the properties of a metal, although it contains no metal atoms. The material has unusual optical and electrical properties. High-purity sulfur is commercially available in purities of 99.999+%, at a cost of about $50/100 g.

Tantalum — (Gr. *Tantalos*, mythological character, father of *Niobe*), Ta; at. wt. 180.9479(1); at. no. 73; m.p. 3017°C; b.p. 5458 ± 100°C; sp. gr. 16.654; valence 2?, 3, 4?, or 5. Discovered in 1802 by Ekeberg, but many chemists thought niobium and tantalum were identical elements until Rose, in 1844, and Marignac, in1866, showed that niobic and tantalic acids were two different acids. The early investigators only isolated the impure metal. The first relatively pure ductile tantalum was produced by von Bolton in 1903. Tantalum occurs principally in the mineral *columbite-tantalite* (Fe, Mn)(Nb, Ta)$_2O_6$. Tantalum ores are found in Australia, Brazil, Mozambique, Thailand, Portugal, Nigeria, Zaire, and Canada. Separation of tantalum from niobium requires several complicated steps. Several methods are used to commercially produce the element, including electrolysis of molten potassium fluorotantalate, reduction of potassium fluorotantalate with sodium, or reacting tantalum carbide with tantalum oxide. Thirty four isotopes and isomers of tantalum are known to exist. Natural tantalum contains two isotopes, one of which is radioactive with a very long half-life. Tantalum is a gray, heavy, and very hard metal. When pure, it is ductile and can be drawn into fine wire, which is used as a filament for evaporating metals such as aluminum. Tantalum is almost completely immune to chemical attack at temperatures below 150°C, and is attacked only by hydrofluoric acid, acidic solutions containing the fluoride ion, and free sulfur trioxide. Alkalis attack it only slowly. At high temperatures, tantalum becomes much more reactive. The element has a melting point exceeded only by tungsten and rhenium. Tantalum is used to make a variety of alloys with desirable properties such as high melting point, high strength, good ductility, etc. Scientists at Los Alamos have produced a tantalum carbide graphite composite material, which is said to be one of the hardest materials ever made. The compound has a melting point of 3738°C. Tantalum has good "gettering" ability at high temperatures, and tantalum oxide films are stable and have good rectifying and dielectric properties. Tantalum is used to make electrolytic capacitors and vacuum furnace parts, which account for about 60% of its use. The metal is also widely used to fabricate chemical process equipment, nuclear reactors, and aircraft and missile parts. Tantalum is completely immune to body liquids and is a nonirritating metal. It has, therefore, found wide use in making surgical appliances. Tantalum oxide is used to make special glass with high index of refraction for camera lenses. The metal has many other uses. The price of (99.9%) tantalum is about $900/kg. The metal of 99.995% purity sells for about $2/g.

Technetium — (Gr. *technetos*, artificial), Tc; at. wt. (98); at. no. 43; m.p. 2157°C; b.p. 4265°C; sp. gr. 11.50 (calc.); valence 0, +2, +4, +5, +6, and +7. Element 43 was predicted on the basis of the periodic table, and was erroneously reported as having been discovered in 1925, at which time it was named *masurium*. The element was actually discovered by Perrier and Segre in Italy in 1937. It was found in a sample of molybdenum, which was bombarded by deuterons in the Berkeley cyclotron, and which E. Lawrence sent to these investigators. Technetium was the first element to be produced artificially. Since its discovery, searches for the element in terrestrial materials have been made without success. If it does exist, the concentration must be very small. Technetium has been found in the spectrum of S-, M-, and N-type stars, and its presence in stellar matter is leading to new theories of the production of heavy elements in the stars. Thirty one isotopes and isomers of technetium, with atomic masses ranging from 86 to 113, are known. ^{97}Tc has a half-life of 2.6×10^6 years. ^{98}Tc has a half-life of 4.2×10^6 years. The isomeric isotope ^{95m}Tc, with a half-life of 61 days, is useful for tracer work, as it produces energetic gamma rays. Technetium metal has been produced in kilogram quantities. The metal was first prepared by passing hydrogen gas at 1100°C over Tc_2S_7. It is now conveniently prepared by the reduction of ammonium pertechnetate with hydrogen. Technetium is a silvery-gray metal that tarnishes slowly in moist air. Until 1960, technetium was available only in small amounts and the price was as high as $2800/g. It is now commercially available to holders of O.R.N.L. permits at a price of $74/g plus packing charges. The chemistry of technetium is said to be similar to that of rhenium. Technetium dissolves in nitric acid, aqua regia, and conc. sulfuric acid, but is not soluble in hydrochloric acid of any strength. The element is a remarkable corrosion inhibitor for steel. It is reported that mild carbon steels may be effectively protected by as little as 55 ppm of $KTcO_4$ in aerated distilled water at temperatures up to 250°C. This corrosion protection is limited to closed systems, since technetium is radioactive and must be confined. ^{99}Tc has a specific activity of 6.2×10^8 Bq/g. Activity of this level must not be allowed to spread. ^{99}Tc is a contamination hazard and should be handled in a glove box. The metal is an excellent superconductor at 11 K and below.

Tellurium — (L. *tellus*, earth), Te; at. wt. 127.60(3); at. no. 52; m.p. 449.51 ± 0.3°C; b.p. 988°C; sp. gr. 6.24 (20°C); valence 2, 4, or 6. Discovered by Muller von Reichenstein in 1782; named by Klaproth, who isolated it in 1798. Tellurium is occasionally found native, but is more often found as the telluride of gold (*calaverite*), and combined with other metals. It is recovered commercially from the anode muds produced during the electrolytic refining of blister copper. The U.S., Canada, Peru, and Japan are the largest Free World producers of the element. Crystalline tellurium has a silvery-white appearance, and when pure exhibits a metallic luster. It is brittle and easily pulverized. Amorphous tellurium is formed by precipitating tellurium from a solution of telluric or tellurous acid. Whether this form is truly amorphous, or made of minute crystals, is open to question. Tellurium is a p-type semiconductor, and shows greater conductivity in certain directions, depending on alignment of the atoms. Its conductivity increases slightly with exposure to light. It can be doped with silver, copper, gold, tin, or other elements. In air, tellurium burns with a greenish-blue flame, forming the dioxide. Molten tellurium corrodes iron, copper, and stainless steel. Tellurium and its compounds are probably toxic and should be handled with care. Workmen exposed to as little as 0.01 mg/m^3 of air, or less, develop "tellurium breath," which has a garlic-like odor. Forty two isotopes and isomers of tellurium are known, with atomic masses ranging from 106 to 138. Natural tellurium consists of eight isotopes, two of which are radioactive with very long half-lives. Tellurium improves the machinability of copper and stainless steel, and its addition to lead decreases the corrosive action of sulfuric acid on lead and improves its strength and hardness. Tellurium is used as a basic ingredient in blasting caps, and is added to cast iron for chill control. Tellurium is used in ceramics. Bismuth telluride has been used in thermoelectric devices. Tellurium costs about 20¢/g, with a purity of about 99.5%. The metal with a purity of 99.9999% costs about $2/g.

Terbium — (*Ytterby*, village in Sweden), Tb; at. wt. 158.92534(3); at. no. 65; m.p. 1356°C; b.p. 3230°C; sp. gr. 8.230; valence 3, 4. Discovered by Mosander in 1843. Terbium is a member of the lanthanide or "rare earth" group of elements. It is found in *cerite, gadolinite*, and other minerals along with other rare earths. It is recovered commercially from *monazite* in which it is present to the extent of 0.03%, from *xenotime*, and from *euxenite*, a complex oxide containing 1% of more of terbia. Terbium has been isolated only in recent years with the development of ion-exchange techniques for separating the rare-earth elements. As with other rare earths, it can be produced by reducing the anhydrous chloride or fluoride with calcium metal in a tantalum crucible. Calcium and tantalum impurities can be removed by vacuum remelting. Other methods of isolation are possible. Terbium is reasonably stable in air. It is a silver-gray metal, and is malleable, ductile, and soft enough to be cut with a knife. Two crystal modifications exist, with a transformation temperature of 1289°C. Forty one isotopes and isomers are recognized. The oxide is a chocolate or dark maroon color. Sodium terbium borate is used as a laser material and emits coherent light at 0.546 μm. Terbium is used to dope calcium fluoride, calcium tungstate, and strontium molybdate, used in solid-state devices. The oxide has potential application as an activator for green phosphors used in color TV tubes. It can be used with ZrO_2 as a crystal stabilizer of fuel cells which operate at elevated temperature. Few other uses have been found. The element is priced at about $30/g (99.9%). Little is known of the toxicity of terbium. It should be handled with care as with other lanthanide elements.

Thallium — (Gr. *thallos*, a green shoot or twig), Tl; at. wt. 204.3833(2); at. no. 81; m.p. 304°C; b.p. 1473 ± 10°C; sp. gr. 11.85 (20°C); valence 1, or 3. Thallium was discovered spectroscopically in 1861 by Crookes. The element was named after the beautiful green spectral line, which identified the element. The metal was isolated both by Crookes and Lamy in 1862 about the same time. Thallium occurs in *crooksite, lorandite*, and *hutchinsonite*. It is also present in *pyrites* and is recovered from the roasting of this ore in connection with the production of sulfuric acid. It is also obtained from the smelting of lead and zinc ores. Extraction is somewhat complex and depends on the source of the thallium. Manganese nodules, found on the ocean floor, contain thallium. When freshly exposed to air, thallium exhibits a metallic luster, but soon develops a bluish-gray tinge, resembling lead in appearance. A heavy oxide builds up on thallium if left in air, and in the presence of water the hydroxide is formed. The metal is very soft and malleable. It can be cut with a knife. forty seven isotopic forms of thallium, with atomic masses ranging from 179 to 210 are recognized. Natural thallium is a mixture of two isotopes. The element and its compounds are toxic and should be handled carefully. Contact of the metal with skin is dangerous, and when melting the metal adequate ventilation should be provided. Thallium is suspected of carcinogenic potential for man. Thallium sulfate has been widely employed as a rodenticide and ant killer. It is odorless and tasteless, giving no warning of its presence. Its use, however, has been prohibited in the U.S. since 1975 as a household insecticide and rodenticide. The electrical conductivity of thallium sulfide changes with exposure to infrared light, and this compound is used in photocells. Thallium bromide-iodide crystals have been used as infrared optical materials. Thallium has been used, with sulfur or selenium and arsenic, to produce low melting glasses which become fluid between 125 and 150°C. These glasses have properties at room temperatures similar to ordinary glasses and are said to be durable and insoluble in water. Thallium oxide has been used to produce glasses with a high index of refraction. Thallium has been used in treating ringworm and other skin infections; however, its use has been limited because of the narrow margin between toxicity and therapeutic benefits. A mercury-thallium alloy, which forms a eutectic at 8.5% thallium, is reported to freeze at –60°C, some 20° below the freezing point of mercury. Thallium metal (99.999%) costs about $1/g.

Thorium — (*Thor*, Scandinavian god of war), Th; at. wt. 232.0381(1); at. no. 90; m.p. 1750°C; b.p. 4788°C; sp. gr. 11.72; valence +2(?), +3(?), +4. Discovered by Berzelius in 1828. Thorium occurs in *thorite* ($ThSiO_4$) and in *thorianite* ($ThO_2 + UO_2$). Large deposits of thorium minerals have been reported in New England and elsewhere, but these have not yet been exploited. Thorium is now thought to be about three times as abundant as uranium and about as abundant as lead or molybdenum. The metal is a source of nuclear power. There is probably more energy available for use from thorium in the minerals of the earth's crust than from both uranium and fossil fuels. Any sizable demand for thorium as a nuclear fuel is still several years in the future. Work has been done in developing thorium cycle converter-reactor systems. Several prototypes, including the HTGR (high-temperature gas-cooled reactor) and MSRE (molten salt converter reactor experiment), have operated. While the HTGR reactors are efficient, they are not expected to become important commercially for many years because of certain operating difficulties. Thorium is recovered commercially from the mineral *monazite*, which contains from 3 to 9% ThO_2 along with rare-earth minerals. Much of the internal heat the earth produces has been attributed to thorium and uranium. Several methods are available for producing thorium metal: it can be obtained by reducing thorium oxide with calcium, by electrolysis of anhydrous thorium chloride in a fused mixture of sodium and potassium chlorides, by calcium reduction of thorium tetrachloride mixed with anhydrous zinc chloride, and by reduction of thorium tetrachloride with an alkali metal. Thorium was originally assigned a position in Group IV of the periodic table. Because of its atomic weight, valence, etc., it is now considered to be the second member of the *actinide* series of elements. When pure, thorium is a silvery-white metal which is air-stable and retains its luster for several months. When contaminated with the oxide, thorium slowly tarnishes in air, becoming gray and finally black. The physical properties of thorium are greatly influenced by the degree of contamination with the oxide. The purest specimens often contain several tenths of a percent of the oxide. High-purity thorium has been made. Pure thorium is soft, very ductile, and can be cold-rolled, swaged, and drawn. Thorium is dimorphic, changing at 1400°C from a cubic to a body-centered cubic structure. Thorium oxide has a melting point of 3300°C, which is the highest of all oxides. Only a few elements, such as tungsten, and a few compounds, such as tantalum carbide, have higher melting points. Thorium is slowly attacked by water, but does not dissolve readily in most common acids, except hydrochloric. Powdered thorium metal is often pyrophoric and should be carefully handled. When heated in air, thorium turnings ignite and burn brilliantly with a white light. The principal use of thorium has been in the preparation of the Welsbach mantle, used for portable gas lights. These mantles, consisting of thorium oxide with about 1% cerium oxide and other ingredients, glow with a dazzling light when heated in a gas flame. Thorium is an important alloying element in magnesium, imparting high strength and creep resistance at elevated temperatures. Because thorium has a low work-function and high electron emission, it is used to coat tungsten wire used in electronic equipment. The oxide is also used to control the grain size of tungsten used for electric lamps; it is also used for high-temperature laboratory crucibles. Glasses containing thorium oxide have a high refractive index and low dispersion. Consequently, they find application in high quality lenses for cameras and scientific instruments. Thorium oxide has also found use as a catalyst in the conversion of ammonia to nitric acid, in petroleum cracking, and in producing sulfuric acid. Twenty seven isotopes of thorium are known with atomic masses ranging from 212 to 237. All are unstable. ^{232}Th occurs naturally and has a half-life of 1.4×10^{10} years. It is an alpha emitter. ^{232}Th goes through six alpha and four beta decay steps before becoming the stable isotope ^{208}Pb. ^{232}Th is sufficiently radioactive to expose a photographic plate in a few hours. Thorium disintegrates with the production of "thoron" (^{220}Rn), which is an alpha emitter and presents a radiation hazard. Good ventilation of areas where thorium is stored or handled is therefore essential. Thorium metal (99.8%) costs about $15/g.

Thulium — (*Thule*, the earliest name for Scandinavia), Tm; at. wt. 168.93421(3); at. no. 69; m.p. 1545°C; b.p. 1950°C; sp. gr. 9.321 (25°C); valence 3. Discovered in 1879 by Cleve. Thulium occurs in small quantities along with other rare earths in a number of minerals. It is obtained commercially from *monazite*, which contains about 0.007% of the element. Thulium is the least abundant of the rare earth elements, but with new sources recently discovered, it is now considered to be about as rare as silver, gold, or cadmium. Ion-exchange and solvent extraction techniques have recently permitted much easier separation of the rare earths, with much lower costs. Only a few years ago, thulium metal was not obtainable at any cost; in 1996 the oxide cost $20/g. Thulium metal powder now costs $60/g (99.9%). Thulium can be isolated by reduction of the oxide with lanthanum metal or by calcium reduction of the anhydrous fluoride. The pure metal has a bright, silvery luster. It is reasonably stable in air, but the metal should be protected from moisture in a closed container. The element is silver-gray, soft, malleable, and ductile, and can be cut with a knife. Thirty eight isotopes and isomers are known, with atomic masses ranging from 146 to 176. Natural thulium, which is 100% ^{169}Tm, is stable. Because of the relatively high price of the metal, thulium has not yet found many practical applications. ^{169}Tm bombarded in a nuclear reactor can be used as a radiation source in portable X-ray equipment. ^{171}Tm is potentially useful as an energy source. Natural thulium also has possible use in *ferrites* (ceramic magnetic materials) used in microwave equipment. As with other lanthanides, thulium has a low-to-moderate acute toxic rating. It should be handled with care.

Tin — (anglo-Saxon, *tin*), Sn (L. *stannum*); at. wt. 118.710(7); at. no. 50; f.p. 231.928°C; b.p. 2602°C; sp. gr. (gray) 5.75, (white) 7.31; valence 2, 4. Known to the ancients. Tin is found chiefly in *cassiterite* (SnO_2). Most of the world's supply comes from Malaysia, Bolivia, China, Indonesia, Russia, Zaire, Thailand, and Nigeria. The U.S. produces almost none, although occurrences have been found in Alaska and Colorado. Tin is obtained by reducing the ore with coal in a reverberatory furnace. Ordinary tin is composed of ten stable isotopes; thirty five unstable isotopes and isomers are also known. Ordinary tin is a silver-white metal, is malleable, somewhat ductile, and has a highly crystalline structure. Due to the breaking of these crystals, a "tin cry" is heard when a bar is bent. The element has two allotropic forms at normal pressure. On warming, gray, or α tin, with a cubic structure, changes at 13.2°C into white, or β tin, the ordinary form of the metal. White tin has a tetragonal structure. When tin is cooled below 13.2°C, it changes slowly from white to gray. This change is affected by impurities such as aluminum and zinc, and can be prevented by small additions of antimony or bismuth. This change from the α to β form is called the tin pest. There are few if any uses for gray tin. Tin takes a high polish and is used to coat other metals to prevent corrosion or other chemical action. Such tin plate over steel is used in the so-called tin can for preserving food. Alloys of tin are very important. Soft solder, type metal, fusible metal, pewter, bronze, bell metal, Babbitt metal, White metal, die casting alloy, and phosphor bronze are some of the important alloys using tin. Tin resists distilled sea and soft tap water, but is attacked by strong acids, alkalis, and acid salts. Oxygen in solution accelerates the attack. When heated in air, tin forms SnO_2, which is feebly acid, forming stannate salts with basic oxides. The most important salt is the chloride ($SnCl_2 \cdot H_2O$), which is used as a reducing agent and as a mordant in calico printing. Tin salts sprayed onto glass are used to produce electrically conductive coatings. These have been used for panel lighting and for frost-free windshields. Most window glass is now made by floating molten glass on molten tin (float glass) to produce a flat surface (Pilkington process). Of recent interest is a crystalline tin-niobium alloy that is superconductive at very low temperatures. This promises to be important in the construction of superconductive magnets that generate enormous field strengths but use practically no power. Such magnets, made of tin-niobium wire, weigh but a few pounds and produce magnetic fields that, when started with a small battery, are comparable to that of a 100 ton electromagnet operated continuously with a large power supply. The small amount of tin found in canned foods is quite harmless. The agreed limit of tin content in U.S. foods is 300 mg/kg. The trialkyl and triaryl tin compounds are used as biocides and must be handled carefully. Over the past 25 years the price of commercial tin has varied from 50¢/lb ($1.10/kg) to its present price of about $4.30/lb ($9.50/kg) as of January 1996. Tin with a purity of 99.9999% is available at a price of about $4/g.

Titanium — (L. *Titans*, the first sons of the Earth, myth.), Ti; at. wt. 47.867(1); at. no. 22; m.p. 1668 ± 10°C; b.p. 3287°C; sp. gr. 4.54; valence 2, 3, or 4. Discovered by Gregor in 1791; named by Klaproth in 1795. Impure titanium was prepared by Nilson and Pettersson in 1887; however, the pure metal (99.9%) was not made until 1910 by Hunter by heating $TiCl_4$ with sodium in a steel bomb. Titanium is present in meteorites and in the sun. Rocks obtained during the Apollo 17 lunar mission showed presence of 12.1% TiO_2. Analyses of rocks obtained during earlier Apollo missions show lower percentages. Titanium oxide bands are prominent in the spectra of M-type stars. The element is the ninth most abundant in the crust of the earth. Titanium is almost always present in igneous rocks and in the sediments derived from them. It occurs in the minerals *rutile, ilmenite,* and *sphene*, and is present in titanates and in many iron ores. Deposits of ilmenite and rutile are found in Florida, California, Tennessee, New York, and elsewhere. Titanium is present in the ash of coal, in plants, and in the human body. The metal was a laboratory curiosity until Kroll, in 1946, showed that titanium could be produced commercially by reducing titanium tetrachloride with magnesium. This method is largely used for producing the metal today. The metal can be purified by decomposing the iodide. Titanium, when pure, is a lustrous, white metal. It has a low density, good strength, is easily fabricated, and has excellent corrosion resistance. It is ductile only when it is free of oxygen. The metal burns in air and is the only element that burns in nitrogen. Titanium is resistant to dilute sulfuric and hydrochloric acid, most organic acids, moist chlorine gas, and chloride solutions. Natural titanium consists of five isotopes with atomic masses from 46 to 50. All are stable. Fifteen other unstable isotopes are known. The metal is dimorphic. The hexagonal α form changes to the cubic β form very slowly at about 880°C. The metal combines with oxygen at red heat, and with chlorine at 550°C. Titanium is important as an alloying agent with aluminum, molybdenum, manganese, iron, and other metals. Alloys of titanium are principally used for aircraft and missiles where lightweight strength and ability to withstand extremes of temperature are important. Titanium is as strong as steel, but 45% lighter. It is 60% heavier than aluminum, but twice as strong. Titanium has potential use in desalination plants for converting sea water into fresh water. The metal has excellent resistance to sea water and is used for propeller shafts, rigging, and other parts of ships exposed to salt water. A titanium anode coated with platinum has been used to provide cathodic protection from corrosion by salt water. Titanium metal is considered to be physiologically inert; however, titanium powder may be a carcinogenic hazard. When pure, titanium dioxide is relatively clear and has an extremely high index of refraction with an optical dispersion higher than diamond. It is produced artificially for use as a gemstone, but it is relatively soft. Star sapphires and rubies exhibit their asterism as a result of the presence of TiO_2. Titanium dioxide is extensively used for both house paint and artist's paint, as it is permanent and has good covering power. Titanium oxide pigment accounts for the largest use of the element. Titanium paint is an excellent reflector of infrared, and is extensively used in solar observatories where heat causes poor seeing conditions. Titanium tetrachloride is used to iridize glass. This compound fumes strongly in air and has been used to produce smoke screens. The price of titanium metal (99.8%) is about $550/kg.

Tungsten — (Swedish, *tung sten*, heavy stone); also known as *wolfram* (from *wolframite*, said to be named from *wolf rahm* or *spumi lupi*, because the ore interfered with the smelting of tin and was supposed to devour the tin), W; at. wt. 183.84(1); at. no. 74; m.p. 3422 ± 20°C; b.p. 5555°C; sp. gr. 19.3 (20°C); valence 2, 3, 4, 5, or 6. In 1779 Peter Woulfe examined the mineral now known as *wolframite* and concluded it must contain a new substance. Scheele, in 1781, found that a new acid could be made from *tung sten* (a name first applied about 1758 to a mineral now known as *scheelite*). Scheele and Berman suggested the possibility of obtaining a new metal by reducing this acid. The de Elhuyar brothers found an acid in *wolframite* in 1783 that was identical to the acid of *tungsten*(tungstic acid) of Scheele, and in that year they succeeded in obtaining the element by reduction of this acid with charcoal. Tungsten occurs in *wolframite*, (Fe, Mn)WO_4; *scheelite*, $CaWO_4$; *huebnerite*, $MnWO_4$; and *ferberite*, $FeWO_4$. Important deposits of tungsten occur in California, Colorado, South Korea, Bolivia, Russia, and Portugal. China is reported to have about 75% of the world's tungsten resources. Natural tungsten contains five stable isotopes. Thirty two other unstable isotopes and isomers are recognized. The metal is obtained commercially by reducing tungsten oxide with hydrogen or carbon. Pure tungsten is a steel-gray to tin-white metal. Very pure tungsten can be cut with a hacksaw, and can be forged, spun, drawn, and extruded. The impure metal is brittle and can be worked only with difficulty. Tungsten has the highest melting point of all metals, and at temperatures over 1650°C has the highest tensile strength. The metal oxidizes in air and must be protected at elevated temperatures. It has excellent corrosion resistance and is attacked only slightly by most mineral acids. The thermal expansion is about the same as borosilicate glass, which makes the metal useful for glass-to-metal seals. Tungsten and its alloys are used extensively for filaments for electric lamps, electron and television tubes, and for metal evaporation work; for electrical contact points for automobile distributors, X-ray targets, windings and heating elements for electrical furnaces; and for numerous spacecraft and high-temperature applications. High-speed tool steels, Hastelloy®, Stellite®, and many other alloys contain tungsten. Tungsten carbide is of great importance to the metal-working, mining, and petroleum industries. Calcium and magnesium tungstates are widely used in fluorescent lighting; other salts of tungsten are used in the chemical and tanning industries. Tungsten disulfide is a dry, high-temperature lubricant, stable to 500°C. Tungsten bronzes and other tungsten compounds are used in paints. Tungsten powder (99.95%) costs about $325/kg.

Uranium — (Planet *Uranus*), U; at. wt. 238.0289(1); at. no. 92; m.p. 1135°C; b.p. 4131°C; sp. gr. ~18.95; valence 2, 3, 4, 5, or 6. Yellow-colored glass, containing more than 1% uranium oxide and dating back to 79 A.D., has been found near Naples, Italy. Klaproth recognized an unknown element in *pitchblende* and attempted to isolate the metal in 1789. The metal apparently was first isolated in 1841 by Peligot, who reduced the anhydrous chloride with potassium. Uranium is not as rare as it was once thought. It is now considered to be more plentiful than mercury, antimony, silver, or cadmium, and is about as abundant as molybdenum or arsenic. It occurs in numerous minerals such as *pitchblende, uraninite, carnotite, autunite, uranophane, davidite,* and *tobernite*. It is also found in *phosphate rock, lignite, monazite sands,* and can be recovered commercially from these sources. Large deposits of uranium ore occur in Utah, Colorado, New Mexico, Canada, and elsewhere. The U.S.D.O.E. purchases uranium in the form of acceptable U_3O_8 concentrates. This incentive program has greatly increased the known uranium reserves. Uranium can be made by reducing uranium halides with alkali or alkaline earth metals or by reducing uranium oxides by calcium, aluminum, or carbon at high temperatures. The metal can also be produced by electrolysis of KUF_5 or UF_4, dissolved in a molten mixture of $CaCl_2$ and NaCl. High-purity uranium can be prepared by the thermal decomposition of uranium halides on a hot filament. Uranium exhibits three crystallographic modifications as follows:

$$\alpha \xrightarrow{688°C} \beta \xrightarrow{776°C} \gamma$$

Uranium is a heavy, silvery-white metal which is pyrophoric when finely divided. It is a little softer than steel, and is attacked by cold water in a finely divided state. It is malleable, ductile, and slightly paramagnetic. In air, the metal becomes coated with a layer of oxide. Acids dissolve the metal, but it is unaffected by alkalis. Uranium has twenty three isotopes, one of which is an isomer and all of which are radioactive. Naturally occurring uranium contains 99.2745% by weight ^{238}U, 0.720% ^{235}U, and 0.0055% ^{234}U. Studies show that the percentage weight of ^{235}U in natural uranium varies by as much as 0.1%, depending on the source. The U.S.D.O.E. has adopted the value of 0.711 as being their "official" percentage of ^{235}U in natural uranium. Natural uranium is sufficiently radioactive to expose a photographic plate in an hour or so. Much of the internal heat of the earth is thought to be attributable to the presence of uranium and thorium. ^{238}U with a half-life of 4.46×10^9 years, has been used to estimate the age of igneous rocks. The origin of uranium, the highest member of the naturally occurring elements — except perhaps for traces of neptunium or plutonium — is not clearly understood, although it may be presumed that uranium is a decay product of elements of higher atomic weight, which may have once been present on earth or elsewhere in the universe. These original elements may have been formed as a result of a primordial "creation," known as "the big bang," in a supernova, or in some other stellar processes. Uranium is of great importance as a nuclear fuel. ^{238}U can be converted into fissionable plutonium by the following reactions:

$$^{238}U(n, \gamma) \rightarrow {}^{239}U \xrightarrow{\beta} {}^{239}Np \xrightarrow{\beta} {}^{239}Pu$$

This nuclear conversion can be brought about in "breeder" reactors where it is possible to produce more new fissionable material than the fissionable material used in maintaining the chain reaction. ^{235}U is of even greater importance, for it is the key to the utilization of uranium. ^{235}U, while occurring in natural uranium to the extent of only 0.71%, is so fissionable with slow neutrons that a self-sustaining fission chain reaction can be made to occur in a reactor constructed from natural uranium and a suitable moderator, such as heavy water or graphite, alone. ^{235}U can be concentrated by gaseous diffusion and other physical processes, if desired, and used directly as a nuclear fuel, instead of natural uranium, or used as an explosive. Natural uranium, slightly enriched with ^{235}U by a small percentage, is used to fuel nuclear power reactors for the generation of electricity. Natural thorium can be irradiated with neutrons as follows to produce the important isotope ^{233}U.

$$^{232}Th(n, \gamma) \rightarrow {}^{233}Th \xrightarrow{\beta} {}^{233}Pa \xrightarrow{\beta} {}^{233}U$$

While thorium itself is not fissionable, ^{233}U is, and in this way may be used as a nuclear fuel. One pound of completely fissioned uranium has the fuel value of over 1500 tons of coal. The uses of nuclear fuels to generate electrical power, to make isotopes for peaceful purposes, and to make explosives are well known. The estimated world-wide production of the 430 nuclear power reactors in operation in January 1994 amounted to about 338,000 megawatts. Uranium in the U.S.A. is controlled by the U.S. Nuclear Regulatory Commission. New uses are being found for "depleted" uranium, i.e., uranium with the percentage of ^{235}U lowered to about 0.2%. It has found use in inertial guidance devices, gyro compasses, counterweights for aircraft control surfaces, as ballast for missile reentry vehicles, and as a shielding material. Uranium metal is used for X-ray targets for production of high-energy X-rays; the nitrate has been used as photographic toner, and the acetate is used in analytical chemistry. Crystals of uranium nitrate are triboluminescent. Uranium salts have also been used for producing yellow "vaseline" glass and glazes. Uranium and its compounds are highly toxic, both from a chemical and radiological standpoint. Finely divided uranium metal, being pyrophoric, presents a fire hazard. The maximum permissible total body burden of natural uranium (based on radiotoxicity) is 0.2 μCi for soluble compounds. Recently, the natural presence of uranium in many soils has become of concern to homeowners because of the generation of radon and its daughters (see under Radon). Uranium metal is available commercially at a cost of about $200/kg (99.7%) in air-tight glass under argon.

Unnilquadium, etc. — See under Elements 104 to 112.

Vanadium — (Scandinavian goddess, *Vanadis*), V; at. wt. 50.9415(1); at. no. 23; m.p. 1910 ± 10°C; b.p. 3407°C; sp. gr. 6.11 (18.7°C); valence 2, 3, 4, or 5. Vanadium was first discovered by del Rio in 1801. Unfortunately, a French chemist incorrectly declared del Rio's new element was only impure chromium; del Rio thought himself to be mistaken and accepted the French chemist's statement. The element was rediscovered in 1830 by Sefstrom, who named the element in honor of the Scandinavian goddess *Vanadis* because of its beautiful multicolored compounds. It was isolated in nearly pure form by Roscoe, in 1867, who reduced the chloride with hydrogen. Vanadium of 99.3 to 99.8% purity was not produced until 1927. Vanadium is found in about 65 different minerals among which are *carnotite, roscoelite, vanadinite,* and *patronite* important sources of the metal. Vanadium is also found in phosphate rock and certain iron ores, and is present in some crude oils in the form of organic complexes. It is also found in small percentages in meteorites. Commercial production from petroleum ash holds promise as an important source of the element. High-purity ductile vanadium can be obtained by reduction of vanadium trichloride with magnesium or with magnesium-sodium mixtures. Much of the vanadium metal being produced is now made by calcium reduction of V_2O_5 in a pressure vessel, an adaption of a process developed by McKechnie and Seybolt. Natural vanadium is a mixture of two isotopes, ^{50}V (0.25%) and ^{51}V (99.75%). ^{50}V is slightly radioactive, having a long half-life. Seventeen other unstable isotopes are recognized. Pure vanadium is a bright white metal, and is soft and ductile. It has good corrosion resistance to alkalis, sulfuric and hydrochloric acid, and salt water, but the metal oxidizes readily above 660°C. The metal has good structural strength and a low fission neutron cross section, making it useful in nuclear applications. Vanadium is used in producing rust resistant, spring, and highspeed tool steels. It is an important carbide stabilizer in making steels. About 80% of the vanadium now produced is used as ferrovanadium or as a steel additive. Vanadium foil is used as a bonding agent in cladding titanium to steel. Vanadium pentoxide is used in ceramics and as a catalyst. It is also used in producing a superconductive magnet with a field of 175,000 gauss. Vanadium and its compounds are toxic and should be handled with care. Ductile vanadium is commercially available. Commercial vanadium metal, of about 95% purity, costs about $50kg. Vanadium metal (99.7%) costs about $1.50/g or $700/kg.

Wolfram — see Tungsten.

Xenon — (Gr. *xenon*, stranger), Xe; at. wt. 131.29(2); at. no. 54; m.p. –111.75°C; b.p. –108.0°C; density (gas) 5.887 ± 0.009 g/l, sp. gr (liquid) 3.52 (–109°C); valence usually 0. Discovered by Ramsay and Travers in 1898 in the residue left after evaporating liquid air components. Xenon is a member of the so-called noble or "inert" gases. It is present in the atmosphere to the extent of about one part in twenty million. Xenon is present in the Martian atmosphere to the extent of 0.08 ppm. The element is found in the gases evolved from certain mineral springs, and is commercially obtained by extraction from liquid air. Natural xenon is composed of nine stable isotopes. In addition to these, thirty five unstable isotopes and isomers have been characterized. Before 1962, it had generally been assumed that xenon and other noble gases were unable to form compounds. Evidence has been mounting in the past few years that xenon, as well as other members of the zero valence elements, do form compounds. Among the "compounds" of xenon now reported are xenon hydrate, sodium perxenate, xenon deuterate, difluoride, tetrafluoride, hexafluoride, and $XePtF_6$ and $XeRhF_6$. Xenon trioxide, which is highly explosive, has been prepared. More than 80 xenon compounds have been made with xenon chemically bonded to fluorine and oxygen. Some xenon compounds are colored. Metallic xenon has been produced, using several hundred kilobars of pressure. Xenon in a vacuum tube produces a beautiful blue glow when excited by an electrical discharge. The gas is used in making electron tubes, stroboscopic lamps, bactericidal lamps, and lamps used to excite ruby lasers for generating coherent light. Xenon is used in the atomic energy field in bubble chambers, probes, and other applications where its high molecular weight is of value. The perxenates are used in analytical chemistry as oxidizing agents. ^{133}Xe and ^{135}Xe are produced by neutron irradiation in air cooled nuclear reactors. ^{133}Xe has useful applications as a radioisotope. The element is available in sealed glass containers for about $20/l of gas at standard pressure. Xenon is not toxic, but its compounds are highly toxic because of their strong oxidizing characteristics.

Ytterbium — (Ytterby, village in Sweden), Yb; at. wt. 173.04(3); at. no. 70; m.p. 819°C; b.p. 1196°C; sp. gr (α) 6.903 (β) 6.966; valence 2, 3. Marignac in 1878 discovered a new component, which he called *ytterbia*, in the earth then known as *erbia*. In 1907, Urbain separated ytterbia into two components, which he called *neoytterbia* and *lutecia*. The elements in these earths are now known as *ytterbium* and *lutetium*, respectively. These elements are identical with *aldebaranium* and *cassiopeium*, discovered independently and at about the same time by von Welsbach. Ytterbium occurs along with other rare earths in a number of rare minerals. It is commercially recovered principally from *monazite sand*, which contains about 0.03%. Ion-exchange and solvent extraction techniques developed in recent years have greatly simplified the separation of the rare earths from one another. The element was first prepared by Klemm and Bonner in 1937 by reducing ytterbium trichloride with potassium. Their metal was mixed, however, with KCl. Daane, Dennison, and Spedding prepared a much purer form in 1953 from which the chemical and physical properties of the element could be determined. Ytterbium has a bright silvery luster, is soft, malleable, and quite ductile. While the element is fairly stable, it should be kept in closed containers to protect it from air and moisture. Ytterbium is readily attacked and dissolved by dilute and concentrated mineral acids and reacts slowly with water. Ytterbium has three allotropic forms with transformation points at −13° and 795°C. The beta form is a room-temperature, face-centered, cubic modification, while the high-temperature gamma form is a body-centered cubic form. Another body-centered cubic phase has recently been found to be stable at high pressures at room temperatures. The beta form ordinarily has metallic-type conductivity, but becomes a semiconductor when the pressure is increased above 16,000 atm. The electrical resistance increases tenfold as the pressure is increased to 39,000 atm and drops to about 80% of its standard temperature-pressure resistivity at a pressure of 40,000 atm. Natural ytterbium is a mixture of seven stable isotopes. Twenty six other unstable isotopes and isomers are known. Ytterbium metal has possible use in improving the grain refinement, strength, and other mechanical properties of stainless steel. One isotope is reported to have been used as a radiation source as a substitute for a portable X-ray machine where electricity is unavailable. Few other uses have been found. Ytterbium metal is commercially available with a purity of about 99.9% for about $5/g. Ytterbium has a low acute toxic rating, but may present a carcinogenic hazard.

Yttrium — (*Ytterby*, village in Sweden near Vauxholm), Y; at. wt. 88.90585(2); at. no. 39; m.p. 1522°C; b.p. 3345°C; sp. gr. 4.469 (25°C); valence 3. *Yttria*, which is an earth containing yttrium, was discovered by Gadolin in 1794. *Ytterby* is the site of a quarry which yielded many unusually minerals containing rare earths and other elements. This small town, near Stockholm, bears the honor of giving names to *erbium, terbium*, and *ytterbium* as well as *yttrium*. In 1843 Mosander showed that yttria could be resolved into the oxides (or earths) of three elements. The name yttria was reserved for the most basic one; the others were named *erbia* and *terbia*. Yttrium occurs in nearly all of the rare-earth minerals. Analysis of lunar rock samples obtained during the Apollo missions show a relatively high yttrium content. It is recovered commercially from *monazite sand*, which contains about 3%, and from *bastnasite*, which contains about 0.2%. Wohler obtained the impure element in 1828 by reduction of the anhydrous chloride with potassium. The metal is now produced commercially by reduction of the fluoride with calcium metal. It can also be prepared by other techniques. Yttrium has a silver-metallic luster and is relatively stable in air. Turnings of the metal, however, ignite in air if their temperature exceeds 400°C, and finely divided yttrium is very unstable in air. Yttrium oxide is one of the most important compounds of yttrium and accounts for the largest use. It is widely used in making YVO_4 europium, and Y_2O_3 europium phosphors to give the red color in color television tubes. Many hundreds of thousands of pounds are now used in this application. Yttrium oxide also is used to produce yttrium-iron-garnets, which are very effective microwave filters. Yttrium iron, aluminum, and gadolinium garnets, with formulas such as $Y_3Fe_5O_{12}$ and $Y_3Al_5O_{12}$, have interesting magnetic properties. Yttrium iron garnet is also exceptionally efficient as both a transmitter and transducer of acoustic energy. Yttrium aluminum garnet, with a hardness of 8.5, is also finding use as a gemstone (simulated diamond). Small amounts of yttrium (0.1 to 0.2%) can be used to reduce the grain size in chromium, molybdenum, zirconium, and titanium, and to increase strength of aluminum and magnesium alloys. Alloys with other useful properties can be obtained by using yttrium as an additive. The metal can be used as a deoxidizer for vanadium and other nonferrous metals. The metal has a low cross section for nuclear capture. ^{90}Y, one of the isotopes of yttrium, exists in equilibrium with its parent ^{90}Sr, a product of atomic explosions. Yttrium has been considered for use as a nodulizer for producing nodular cast iron, in which the graphite forms compact nodules instead of the usual flakes. Such iron has increased ductility. Yttrium is also finding application in laser systems and as a catalyst for ethylene polymerization.It has also potential use in ceramic and glass formulas, as the oxide has a high melting point and imparts shock resistance and low expansion characteristics to glass. Natural yttrium contains but one isotope, ^{89}Y. Thirty seven other unstable isotopes and isomers have been characterized. Yttrium metal of 99.9% purity is commercially available at a cost of about $4/g.

Zinc — (Ger. *Zink*, of obscure origin), Zn; at. wt. 65.39(2); at. no. 30; f.p. 419.527°C; b.p. 907°C; sp. gr. 7.133 (25°C); valence 2. Centuries before zinc was recognized as a distinct element, zinc ores were used for making brass. Tubal-Cain, seven generations from Adam, is mentioned as being an "instructor in every artificer in brass and iron." An alloy containing 87% zinc has been found in prehistoric ruins in Transylvania. Metallic zinc was produced in the 13th century A.D. in India by reducing calamine with organic substances such as wool. The metal was rediscovered in Europe by Marggraf in 1746, who showed that it could be obtained by reducing *calamine* with charcoal. The principal ores of zinc are *sphalerite* or *blende* (sulfide), *smithsonite* (carbonate), *calamine* (silicate), and *franklinite* (zinc, manganese, iron oxide). Zinc can be obtained by roasting its ores to form the oxide and by reduction of the oxide with coal or carbon, with subsequent distillation of the metal. Other methods of extraction are possible. Naturally occurring zinc contains five stable isotopes. Twenty three other unstable isotopes and isomers are recognized. Zinc is a bluish-white, lustrous metal. It is brittle at ordinary temperatures but malleable at 100 to 150°C. It is a fair conductor of electricity, and burns in air at high red heat with evolution of white clouds of the oxide. The metal is employed to form numerous alloys with other metals. Brass, nickel silver, typewriter metal, commercial bronze, spring brass, German silver, soft solder, and aluminum solder are some of the more important alloys. Large quantities of zinc are used to produce die castings, used extensively by the automotive, electrical, and hardware industries. An alloy called *Prestal®*, consisting of 78% zinc and 22% aluminum is reported to be almost as strong as steel but as easy to mold as plastic. It is said to be so plastic that it can be molded into form by relatively inexpensive die casts made of ceramics and cement. It exhibits superplasticity. Zinc is also extensively used to galvanize other metals such as iron to prevent corrosion. Neither zinc nor zirconium is ferromagnetic; but $ZrZn_2$ exhibits ferromagnetism at temperatures below 35 K. Zinc oxide is a unique and very useful material to modern civilization. It is widely used in the manufacture of paints, rubber products, cosmetics, pharmaceuticals, floor coverings, plastics, printing inks, soap, storage batteries, textiles, electrical equipment, and other products. It has unusual electrical, thermal, optical, and solid-state properties that have not yet been fully investigated. Lithopone, a mixture of zinc sulfide and barium sulfate, is an important pigment. Zinc sulfide is used in making luminous dials, X-ray and TV screens, and fluorescent lights. The chloride and chromate are also important compounds. Zinc is an essential element in the growth of human beings and animals. Tests show that zinc-deficient animals require 50% more food to gain the same weight as an animal supplied with sufficient zinc. Zinc is not considered to be toxic, but when freshly formed ZnO is inhaled a disorder known as the *oxide shakes* or *zinc chills* sometimes occurs. It is recommended that where zinc oxide is encountered good ventilation be provided. The commercial price of zinc was roughly 50¢/lb ($1.10 kg) in January 1996. Zinc metal with a purity of 99.9999% is priced at about 50¢/g.

Zirconium — (Persian *zargun*, gold like), Zr; at. wt. 91.224(2); at. no. 40; m.p. $1855 \pm 2°C$; b.p. 4409°C; sp. gr. 6.506 (20°C); valence +2, +3, and +4. The name *zircon* probably originated from the Persian word *zargun*, which describes the color of the gemstone now known as *zircon, jargon, hyacinth, jacinth,* or *ligure.* This mineral, or its variations, is mentioned in biblical writings. The mineral was not known to contain a new element until Klaproth, in 1789, analyzed a jargon from Ceylon and found a new earth, which Werner named zircon (*silex circonius*), and Klaproth called *Zirkonerde (zirconia).* The impure metal was first isolated by Berzelius in 1824 by heating a mixture of potassium and potassium zirconium fluoride in a small iron tube. Pure zirconium was first prepared in 1914. Very pure zirconium was first produced in 1925 by van Arkel and de Boer by an iodide decomposition process they developed. Zirconium is found in abundance in S-type stars, and has been identified in the sun and meteorites. Analyses of lunar rock samples obtained during the various Apollo missions to the moon show a surprisingly high zirconium oxide content, compared with terrestial rocks. Naturally occurring zirconium contains five isotopes. Twenty six other radioactive isotopes and isomers are known to exist. *Zircon,* $ZrSiO_4$, the principal ore, is found in deposits in Florida, South Carolina, Australia, and Brazil. *Baddeleyite,* found in Brazil, is an important zirconium mineral. It is principally pure ZrO_2 in crystalline form having a hafnium content of about 1%. Zirconium also occurs in some 30 other recognized mineral species. Zirconium is produced commercially by reduction of the chloride with magnesium (the Kroll Process), and by other methods. It is a grayish-white lustrous metal. When finely divided, the metal may ignite spontaneously in air, especially at elevated temperatures. The solid metal is much more difficult to ignite. The inherent toxicity of zirconium compounds is low. Hafnium is invariably found in zirconium ores, and the separation is difficult. Commercial-grade zirconium contains from 1 to 3% hafnium. Zirconium has a low absorption cross section for neutrons, and is therefore used for nuclear energy applications, such as for cladding fuel elements. Commercial nuclear power generation now takes more than 90% of zirconium metal production. Reactors of the size now being made may use as much as a half-million lineal feet of zirconium alloy tubing. Reactor-grade zirconium is essentially free of hafnium. *Zircaloy*® is an important alloy developed specifically for nuclear applications. Zirconium is exceptionally resistant to corrosion by many common acids and alkalis, by sea water, and by other agents. It is used extensively by the chemical industry where corrosive agents are employed. Zirconium is used as a getter in vacuum tubes, as an alloying agent in steel, in surgical appliances, photoflash bulbs, explosive primers, rayon spinnerets, lamp filaments, etc. It is used in poison ivy lotions in the form of the carbonate as it combines with *urushiol.* With niobium, zirconium is superconductive at low temperatures and is used to make superconductive magnets, which offer hope of direct large-scale generation of electric power. Alloyed with zinc, zirconium becomes magnetic at temperatures below 35 K. Zirconium oxide (zircon) has a high index of refraction and is used as a gem material. The impure oxide, zirconia, is used for laboratory crucibles that will withstand heat shock, for linings of metallurgical furnaces, and by the glass and ceramic industries as a refractory material. Its use as a refractory material accounts for a large share of all zirconium consumed. Zirconium of about 99.8% purity is available at a cost of about $170/kg.

PHYSICAL CONSTANTS OF INORGANIC COMPOUNDS

The compounds in this table were selected on the basis of their laboratory and industrial importance, as well as their value in illustrating trends in the variation of physical properties with position in the periodic table. An effort has been made to include the most frequently encountered inorganic substances; organometallics are not covered, with the exception of metal salts of organic acids. Many, if not most, of the compounds that are solids at ambient temperature can exist in more than one crystalline modification. The information given here applies to the most stable or common crystalline form. In cases where two or more forms are of practical importance, separate entries will be found in the table.

Compounds are arranged primarily in alphabetical order by the most commonly used name. However, adjustments are made in many instances in order to bring closely related compounds together. For example, hydrides of elements such as boron, silicon, and germanium are grouped together immediately following the entry for the parent element, since they would otherwise be scattered throughout the table. Likewise, the oxoacids of an element are given in one group when a strict alphabetical order would separate them (e.g., sulfuric acid and fluorosulfuric acid). Whenever deviations from alphabetical order occur, the names are included in the Synonym Index to provide an easy way to locate these substances.

The data fields in the table are described below:

- **No.:** An identification number used in the indexes.
- **Name:** Common name for the substance. The valence state of a metallic element is indicated by a Roman numeral , e.g., copper in the +2 state is written as copper(II) rather than cupric (but names beginning with cupric, ferrous, etc. are included in the synonym index).
- **Formula:** The simplest descriptive formula is given, but this does not necessarily specify the actual structure of the compound. For example, aluminum chloride is designated as $AlCl_3$, even though a more accurate representation of the structure in the solid phase (and, under some conditions, in the gas phase) is Al_2Cl_6. A few exceptions are made, such as the use of Hg_2^{+2} for the mercury(I) ion.
- **CAS RN:** Chemical Abstracts Service Registry Number. An asterisk * following the CAS RN for a hydrate indicates that the number refers to the anhydrous compound. In most cases the generic CAS RN for the compound is given rather than the number for a specific crystalline form or mineral.
- **Mol. Wt.:** Molecular weight (relative molar mass) as calculated with the 1993 IUPAC Recommended Atomic Weights. The number of decimal places corresponds to the number of places in the atomic weight of the least accurately known element (e.g., one place for lead compounds, two places for compounds of selenium, germanium, etc.); a maximum of three places is given. For compounds of radioactive elements for which IUPAC makes no recommendation, the mass number of the isotope with the longest half-life is used, and the result is rounded to the nearest integer.
- **Physical form:** The crystal system is given ,when available, for compounds that are solid at room temperature, together with color and other descriptive features. Abbreviations are listed below.
- **mp:** Normal melting point in °C. A "tp" following the number indicates a triple point, where solid, liquid, and gas are in equilibrium at a pressure greater than one atmosphere (i.e., the normal melting point does not exist). When available, the triple point pressure is listed.
- **bp:** Normal boiling point in °C (referred to 101.325 kPa or 760 mmHg pressure). An "sp" following the number indicates a sublimation point, where the pressure of the vapor in equilibrium with the solid reaches 101.325 kPa. See Reference 8, p. 23, for further discussion of sublimation points and triple points. Note: The abbreviation "subl" indicates that there is a perceptible sublimation pressure above the solid, but does not imply any quantitative information on the pressure-temperature behavior.
- **den:** Density. Values for solids and liquids are always in grams per cubic centimeter and can be assumed to refer to temperatures near room temperature unless otherwise stated. Values for gases are the calculated ideal gas densities in grams per liter at 25°C and 101.325 kPa; the unit is always specified with a gas value.
- **Solubility:** Qualitative information on the solubility in various solvents is given. The abbreviations are listed below.
- **Other Data:** The codes refer to other tables in this handbook which include an entry for the compound in question. These codes are:

    a. Data on enthalpy of formation, entropy, and heat capacity in the table "Standard Thermodynamic Properties of Chemical Substances" (Section 5).
    b. Critical temperature, pressure, and volume in the table "Critical Constants, Boiling Points, and Melting Points of Selected Compounds" (Section 6).
    c. Table of "Enthalpy of Fusion" (Section 6).
    d. Table of "Enthalpy of Vaporization" (Section 6).
    e. Vapor pressure data in one or more of the vapor pressure tables in Section 6.
    f. Optical constants and hardness in the table "Physical and Optical Properties of Minerals" (Section 4).

A Synonym Index and a CAS Registry Number Index follow the table.

LIST OF ABBREVIATIONS

Ac	acetyl	gl	glass, glassy	rhom	rhombohedral
ace	acetone	grn	green	s	soluble in
acid	acid solutions	hc	hydrocarbon solvents	silv	silvery
alk	alkaline solutions	hex	hexagonal	sl	slightly soluble in
amorp	amorphous	hp	heptane	soln	solution
anh	anhydrous	hx	hexane	sp	sublimation point
aq	aqueous	hyd	hydrate	stab	stable
blk	black	hyg	hygroscopic	subl	sublimes
brn	brown	i	insoluble in	temp	temperature
bz	benzene	liq	liquid	tetr	tetragonal
chl	chloroform	MeOH	methanol	thf	tetrahydrofuran
col	colorless	mono	monoclinic	tol	toluene
conc	concentrated	octahed	octahedral	tp	triple point
cry	crystals, crystalline	oran	orange	trans	transition, transformation
cub	cubic	orth	orthorhombic	tricl	triclinic
cyhex	cyclohexane	os	organic solvents	trig	trigonal
dec	decomposes	peth	petroleum ether	unstab	unstable
dil	dilute	pow	powder	viol	violet
diox	dioxane	prec	precipitate	visc	viscous
eth	ethyl ether	pur	purple	vs	very soluble in
EtOH	ethanol	py	pyridine	wh	white
exp	explodes, explosive	reac	reacts with	xyl	xylene
flam	flammable	refrac	refractory	yel	yellow

REFERENCES

1. Phillips, S.L., and Perry, D.L., *Handbook of Inorganic Compounds*, CRC Press, Boca Raton, FL, 1995.
2. Trotman-Dickenson, A.F., Executive Editor, *Comprehensive Inorganic Chemistry*, Vol. 1-5, Pergamon Press, Oxford, 1973.
3. Greenwood, N.N., and Earnshaw, A., *Chemistry of the Elements*, Pergamon Press, Oxford, 1984.
4. Budavari, S., Editor, *The Merck Index, Eleventh Edition*, Merck & Co., Rahway, NJ, 1989.
5. *GMELIN Handbook of Inorganic and Organometallic Chemistry*, Springer-Verlag, Heidelberg.
6. Chase, M.W., Davies, C.A., Downey, J.R., Frurip, D.J., McDonald, R.A., and Syverud, A.N.; *JANAF Thermochemical Tables, Third Edition*, *J. Phys. Chem. Ref. Data*, Vol. 14, Suppl. 1, 1985.
7. Donnay, J.D.H., and Ondik, H.M., *Crystal Data Determinative Tables, Third Edition, Volumes 2 and 4, Inorganic Compounds*, Joint Committee on Powder Diffraction Standards, Swarthmore, PA, 1973.
8. Lide, D.R., and Kehiaian, H.V., *CRC Handbook of Thermophysical and Thermochemical Data*, CRC Press, Boca Raton, FL, 1994.
9. *Kirk-Othmer Concise Encyclopedia of Chemical Technology*, Wiley-Interscience, New York, 1985.
10. *Dictionary of Inorganic Compounds*, Chapman & Hall, New York, 1992.
11. Massalski, T.B., Editor, *Binary Alloy Phase Diagrams*, American Society for Metals, Metals Park, Ohio, 1986.
12. *Landolt-Börnstein, Numerical Data and Functional Relationships in Science and Technology, Sixth Edition, II/4, Caloric Quantities of State*, Springer-Verlag, Heidelberg, 1961.
13. Deer, W.A., Howie, R.A., and Zussman, J., *An Introduction to the Rock-Forming Minerals, 2nd Edition*, Longman Scientific & Technical, Harlow, Essex, 1992.
14. Carmichael, R.S., *Practical Handbook of Physical Properties of Rocks and Minerals*, CRC Press, Boca Raton, FL, 1989.
15. Dinsdale, A.T., "SGTE Data for Pure Elements", *CALPHAD*, 15, 317-425, 1991.
16. Madelung, O., *Semiconductors: Group IV Elements and III-IV Compounds*, Springer-Verlag, Heidelberg, 1991.
17. Daubert, T.E., Danner, R.P., Sibul, H.M., and Stebbins, C.C., *Physical and Thermodynamic Properties of Pure Compounds: Data Compilation*, extant 1994 (core with 4 supplements), Taylor & Francis, Bristol, PA.

No.	Name / Formula	CAS RN / Mol. Wt.	Physical Form	mp/°C / den/g cm^{-3}	bp/°C / Other Data	Solubility
1	Actinium Ac	7440-34-8 227	silv metal; cub	1051 10	3198 a	
2	Actinium bromide AcBr$_3$	33689-81-5 467	wh hex cry	 5.85		s H$_2$O
3	Actinium chloride AcCl$_3$	22986-54-5 333	wh hex cry	 4.81		
4	Actinium fluoride AcF$_3$	33689-80-4 284	wh hex cry	 7.88		i H$_2$O
5	Actinium iodide AcI$_3$	33689-82-6 608	wh cry			s H$_2$O
6	Actinium oxide Ac$_2$O$_3$	12002-61-8 502	wh hex cry	1977 9.19		i H$_2$O
7	Aluminum Al	7429-90-5 26.982	silv-wh metal; cub cry	660.32 2.70	2519 a,c,d,e	i H$_2$O; s acid, alk
8	Aluminum ammonium sulfate	7784-25-0 237.148	wh powder			sl H$_2$O; i EtOH
9	Aluminum ammonium sulfate dodecahydrate AlNH$_4$(SO$_4$)$_2$•12H$_2$O	7784-26-1 453.331	col cry or powder	94.5 1.65	dec >280	s H$_2$O; i EtOH
10	Aluminum antimonide AlSb	25152-52-7 148.742	cub cry	1065 4.26		
11	Aluminum arsenide AlAs	22831-42-1 101.903	oran cub cry; hyg	1740 3.76		
12	Aluminum borate 2Al$_2$O$_3$•B$_2$O$_3$	11121-16-7 273.543	needles	≈1050		i H$_2$O
13	Aluminum boride AlB$_2$	12041-50-8 48.604	powder	dec >920 3.19		s dil HCl
14	Aluminum borohydride Al(BH$_4$)$_3$	16962-07-5 71.510	flam liq	-64.5	44.5 a,d	reac H$_2$O
15	Aluminum bromate nonahydrate Al(BrO$_3$)$_3$•9H$_2$O	11126-81-1* 572.826	wh hyg cry	62	dec >100	s H$_2$O
16	Aluminum bromide AlBr$_3$	7727-15-3 266.694	wh-yel monocl cry; hyg	97.5 3.2	255 a,b,c,d,e	reac H$_2$O; s bz, tol
17	Aluminum bromide hexahydrate AlBr$_3$•6H$_2$O	7784-11-4 374.785	col-yel hyg cry	93 2.54		s H$_2$O, EtOH, CS$_2$
18	Aluminum carbide Al$_4$C$_3$	1299-86-1 143.959	yel hex cry	2100 2.36	dec >2200	reac H$_2$O
19	Aluminum chlorate nonahydrate Al(ClO$_3$)$_3$•9H$_2$O	15477-33-5 439.472	hyg cry			vs H$_2$O; s EtOH
20	Aluminum chloride AlCl$_3$	7446-70-0 133.340	wh hex cry or powder; hyg	192.6 2.48	 a,b,c,e	reac H$_2$O; s bz, ctc, chl
21	Aluminum chloride hexahydrate AlCl$_3$•6H$_2$O	7784-13-6 241.431	col hyg cry	dec 100 2.398		vs H$_2$O; s EtOH, eth
22	Aluminum diacetate Al(OH)(C$_2$H$_3$O$_2$)$_2$	142-03-0 162.079	wh amorp powder			i H$_2$O
23	Aluminum ethoxide Al(C$_2$H$_5$O)$_3$	555-75-9 162.165	liq, condenses to wh solid	140		reac H$_2$O; sl xyl
24	Aluminum fluoride AlF$_3$	7784-18-1 83.977	wh hex cry	≈2250 tp (220 MPa) 3.10	1276 sp a,c,e	sl H$_2$O
25	Aluminum fluoride monohydrate AlF$_3$•H$_2$O	32287-65-3 101.992	orth cry	 2.17		sl H$_2$O
26	Aluminum fluoride trihydrate AlF$_3$•3H$_2$O	15098-87-0 138.023	wh hyg cry	 1.914		sl H$_2$O
27	Aluminum hexafluorosilicate nonahydrate Al$_2$(SiF$_6$)$_3$•9H$_2$O	17099-70-6 642.329	hex prisms	dec >500		s H$_2$O
28	Aluminum hydride AlH$_3$	7784-21-6 30.006	col hex cry	dec >150	 a	reac H$_2$O
29	Aluminum hydroxide Al(OH)$_3$	21645-51-2 78.004	wh amorp powder	 2.42	 f	i H$_2$O; s alk, acid
30	Aluminum hydroxychloride Al$_2$(OH)$_5$Cl•2H$_2$O	1327-41-9 210.483	gl solid			s H$_2$O
31	Aluminum hypophosphite Al(H$_2$PO$_2$)$_3$	7784-22-7 221.948	cry powder	dec 220		i H$_2$O; s alk, acid
32	Aluminum iodide AlI$_3$	7784-23-8 407.695	wh leaflets	188.32 3.98	382 a,b,c,d,e	reac H$_2$O
33	Aluminum iodide hexahydrate AlI$_3$•6H$_2$O	10090-53-6 515.786	yel hyg cry powder			vs H$_2$O; s EtOH, eth
34	Aluminum lactate Al(C$_3$H$_5$O$_3$)$_3$	18917-91-4 294.195	powder			vs H$_2$O
35	Aluminum nitrate nonahydrate Al(NO$_3$)$_3$•9H$_2$O	7784-27-2 375.134	wh hyg cry; monocl	73 1.72	dec 135	vs H$_2$O, EtOH; i pyr
36	Aluminum nitride AlN	24304-00-5 40.989	blue-wh hex cry	3000 3.255	 a	reac H$_2$O
37	Aluminum oleate Al(C$_{18}$H$_{33}$O$_2$)$_3$	688-37-9 871.358	yel solid			i H$_2$O; s EtOH, bz
38	Aluminum phosphate AlPO$_4$	7784-30-7 121.953	wh rhomb plates	>1460 2.56	 a	i H$_2$O; sl acid

No.	Name Formula	CAS RN Mol. Wt.	Physical Form	mp/°C den/g cm^{-3}	bp/°C Other Data	Solubility
39	Aluminum metaphosphate $Al(PO_3)_3$	32823-06-6 263.898	col powder; tetr	≈1525 2.78		i H_2O
40	Aluminum oxide Al_2O_3	1344-28-1 101.961	wh powder; hex	2054 3.97	≈3000 a,c,e,f	i H_2O, os; sl alk
41	Aluminum oxyhydroxide $AlO(OH)$	14457-84-2 59.989	ortho cry	 3.44	f	i H_2O; s acid, alk
42	Aluminum palmitate $Al(C_{15}H_{31}COO)_3$	555-35-1 793.244	wh-yel powder			i H_2O, EtOH; s peth
43	Aluminum perchlorate nonahydrate $Al(ClO_4)_3 \cdot 9H_2O$	14452-39-2 487.470	wh hyg cry	82 dec 2.0		
44	Aluminum phosphide AlP	20859-73-8 57.956	grn or yel cub cry	2550 2.40	a	reac H_2O
45	Aluminum selenide Al_2Se_3	1302-82-5 290.84	yel-brown powder	960 3.437		reac H_2O
46	Aluminum silicate Al_2SiO_5	12183-80-1 162.046	gray-grn cry	 3.145	f	
47	Aluminum silicate dihydrate $Al_2O_3 \cdot 2SiO_2 \cdot 2H_2O$	1332-58-7 258.161	wh-yel powder; tricl	 2.59		i H_2O, acid, alk
48	Aluminum stearate $Al(C_{18}H_{35}O_2)_3$	637-12-7 877.406	wh powder	115 1.070		i H_2O, EtOH, eth; s alk
49	Aluminum sulfate $Al_2(SO_4)_3$	10043-01-3 342.154	wh cry	dec 1040		s H_2O; i EtOH
50	Aluminum sulfate octadecahydrate $Al_2(SO_4)_3 \cdot 18H_2O$	7784-31-8 666.429	col monocl cry	dec 86 1.69	f	s H_2O
51	Aluminum sulfide Al_2S_3	1302-81-4 150.161	yel-gray powder	1100 2.02	a	
52	Aluminum telluride Al_2Te_3	12043-29-7 436.76	gray-blk hex cry	≈895 4.5		
53	Aluminum thiocyanate $Al(CNS)_3$	538-17-0 201.233	yel powder			s H_2O; i EtOH, eth
54	Americium Am	7440-35-9 243	silv metal; hex or cub	1176 12	2011 a,c	s acid
55	Americium(III) oxide Am_2O_3	12254-64-7 534	tan hex cry	 11.77		s acid
56	Americium(III) bromide $AmBr_3$	14933-38-1 483	wh orth cry	 6.85		s H_2O
57	Americium(III) chloride $AmCl_3$	13464-46-5 349	pink hex cry	500 5.87		
58	Americium(III) fluoride AmF_3	13708-80-0 300	pink hex cry	1393 9.53		
59	Americium(III) iodide AmI_3	13813-47-3 624	yel ortho cry	≈950 6.9		
60	Americium(IV) fluoride AmF_4	15947-41-8 319	tan monocl cry	 7.23		
61	Americium(IV) oxide AmO_2	12005-67-3 275	blk cub cry	dec >1000 11.68		s acid
62	Ammonia NH_3	7664-41-7 17.031	col gas	-77.74 0.747 g/L	-33.33 a,b,c,d,e	vs H_2O; s EtOH, eth
63	Ammonium acetate $NH_4C_2H_3O_2$	631-61-8 77.084	wh hyg cry	114 1.073		vs H_2O
64	Ammonium azide NH_4N_3	12164-94-2 60.059	ortho cry; flam	160 1.346	exp a,e	s H_2O
65	Ammonium benzoate $NH_4C_7H_5O_2$	1863-63-4 139.154	wh cry or powder	198 1.26		s H_2O; sl EtOH
66	Ammonium bimalate $NH_4OOCCH_2CH(OH)COOH$	5972-71-4 151.119	ortho cry	160 1.15		s H_2O; sl EtOH
67	Ammonium borate tetrahydrate $(NH_4)_2B_4O_7 \cdot 4H_2O$	12228-87-4 263.377	tetr cry			s H_2O; i EtOH
68	Ammonium bromide NH_4Br	12124-97-9 97.943	wh hyg tetr cry	542 dec 2.429	396 sp a,e	s H_2O, EtOH, ace; sl eth
69	Ammonium caprylate $NH_4C_8H_{15}O_2$	5972-76-9 161.245	hyg monocl cry	≈75		reac H_2O; s EtOH; i chl, bz
70	Ammonium carbamate NH_2COONH_4	1111-78-0 78.071	cry powder			vs H_2O; s EtOH
71	Ammonium carbonate $(NH_4)_2CO_3$	506-87-6 96.086	col cry powder	dec 58		
72	Ammonium cerium(III) sulfate tetrahydrate $NH_4Ce(SO_4)_2 \cdot 4H_2O$	21995-38-0* 422.342	monocl cry			s H_2O
73	Ammonium cerium(IV) nitrate $(NH_4)_2Ce(NO_3)_6$	16774-21-3 548.222	red-oran cry			vs H_2O
74	Ammonium chlorate NH_4ClO_3	10192-29-7 101.490	wh cry	102 exp 1.80		s H_2O
75	Ammonium chloride NH_4Cl	12125-02-9 53.492	col cub cry	520 tp (dec) 1.519	338 sp a,b,e	s H_2O
76	Ammonium chromate $(NH_4)_2CrO_4$	7788-98-9 152.071	yel cry	dec 185 1.90		s H_2O; sl ace, MeOH; i EtOH

No.	Name Formula	CAS RN Mol. Wt.	Physical Form	mp/°C den/g cm^{-3}	bp/°C Other Data	Solubility
77	Ammonium chromic sulfate dodecahydrate $NH_4Cr(SO_4)_2 \cdot 12H_2O$	10022-47-6 478.345	blue-viol cry	94 dec 1.72		s H_2O; sl EtOH
78	Ammonium cobalt(II) phosphate $CoNH_4PO_4$	14590-13-7 171.943	red-viol powder; hyg			i H_2O; s acid
79	Ammonium cobalt(II) sulfate hexahydrate $(NH_4)_2Co(SO_4)_2 \cdot 6H_2O$	13586-38-4 395.229	red monocl prisms	1.90		s H_2O; i EtOH
80	Ammonium copper(II) chloride dihydrate $2NH_4Cl \cdot CuCl_2 \cdot 2H_2O$	10060-13-6 277.464	blue-grn tetr cry	≈115 dec 1.99		s H_2O, EtOH
81	Ammonium cyanide NH_4CN	12211-52-8 44.056	col tetr cry	dec 1.10	a,e	vs H_2O
82	Ammonium dichromate $(NH_4)_2Cr_2O_7$	7789-09-5 252.065	oran-red monocl cry; hyg	dec 180 2.155		vs H_2O
83	Ammonium dihydrogen arsenate $NH_4H_2AsO_4$	13462-93-6 158.975	tetr cry	300 dec 2.311		vs H_2O
84	Ammonium dihydrogen phosphate $NH_4H_2PO_4$	7722-76-1 115.026	wh tetr cry	190 1.80		vs H_2O; sl EtOH; i ace
85	Ammonium dithiocarbamate NH_4NH_2CSS	513-74-6 110.204	yel ortho cry	99 dec 1.45		s H_2O
86	Ammonium ferric chromate $NH_4Fe(CrO_4)_2$	7789-08-4 305.871	red powder			i H_2O
87	Ammonium ferric oxalate trihydrate $(NH_4)_3Fe(C_2O_4)_3 \cdot 3H_2O$	13268-42-3 428.065	grn monocl cry; hyg	≈160 dec 1.780		vs H_2O; i EtOH
88	Ammonium ferric sulfate dodecahydrate $NH_4Fe(SO_4)_2 \cdot 12H_2O$	10138-04-2 482.194	col to viol cry	≈37 1.71		vs H_2O; i EtOH
89	Ammonium ferricyanide trihydrate $(NH_4)_3Fe(CN)_6 \cdot 3H_2O$	14221-48-8* 320.113	red cry			s H_2O; i EtOH
90	Ammonium ferrocyanide trihydrate $(NH_4)_4Fe(CN)_6 \cdot 3H_2O$	14481-29-9* 338.151	yel cry	dec		s H_2O; i EtOH
91	Ammonium ferrous sulfate hexahydrate $(NH_4)_2Fe(SO_4)_2 \cdot 6H_2O$	10045-89-3 392.141	blue-grn monocl cry	dec ≈100 1.86		s H_2O; i EtOH
92	Ammonium fluoride NH_4F	12125-01-8 37.037	wh hex cry; hyg	dec 1.015	a	vs H_2O; sl EtOH
93	Ammonium fluoroborate NH_4BF_4	13826-83-0 104.844	wh powder; orth	487 dec 1.871		sl H_2O
94	Ammonium fluorosulfonate NH_4SO_3F	13446-08-7 117.101	col needles	245		s H_2O, EtOH, MeOH
95	Ammonium formate NH_4CHO_2	540-69-2 63.057	hyg cry	116 1.27		vs H_2O; s EtOH
96	Ammonium hexachloroiridate(IV) $(NH_4)_2IrCl_6$	16940-92-4 441.010	blk cry powder	dec 2.856		sl H_2O
97	Ammonium hexachloroosmiate(IV) $(NH_4)_2OsCl_6$	12125-08-5 439.02	red cry or powder	2.93	subl	s H_2O, EtOH
98	Ammonium hexachloropalladate(IV) $(NH_4)_2PdCl_6$	19168-23-1 355.21	red-brn hyg cry	dec 2.418		
99	Ammonium hexachloroplatinate(IV) $(NH_4)_2PtCl_6$	16919-58-7 443.87	red-oran cub cry	dec 380 3.065		sl H_2O; i EtOH
100	Ammonium hexafluoroaluminate $(NH_4)_3AlF_6$	7784-19-2 195.087	cub cry	1.78		s H_2O
101	Ammonium hexafluorogallate $(NH_4)_3GaF_6$	14639-94-2 237.828	col cub cry	dec >200 2.10		
102	Ammonium hexafluorogermanate $(NH_4)_2GeF_6$	16962-47-3 222.68	wh cry	380 2.564	subl	s H_2O; i EtOH
103	Ammonium hexafluorophosphate NH_4PF_6	16941-11-0 163.003	wh cub cry	dec 58 2.180		vs H_2O; s ace, EtOH, MeOH
104	Ammonium hexafluorosilicate $(NH_4)_2SiF_6$	16919-19-0 178.153	wh cub or trig cry	dec 2.011	a	s H_2O; i EtOH, ace
105	Ammonium hydrogen arsenate $(NH_4)_2HAsO_4$	7784-44-3 176.004	wh powder	1.99		s H_2O
106	Ammonium hydrogen borate trihydrate $NH_4HB_4O_7 \cdot 3H_2O$	10135-84-9 228.332	col cry	≈2.5		s H_2O
107	Ammonium hydrogen carbonate NH_4HCO_3	1066-33-7 79.056	col or wh prisms	107 dec 1.586	a	s H_2O; i EtOH, bz
108	Ammonium hydrogen citrate $(NH_4)_2HC_6H_5O_7$	3012-65-5 226.186	col cry	1.48		vs H_2O; sl EtOH
109	Ammonium hydrogen fluoride NH_4HF_2	1341-49-7 57.044	wh orth cry	125 1.50	240 dec	s H_2O
110	Ammonium hydrogen oxalate monohydrate $NH_4HC_2O_4 \cdot H_2O$	5972-72-5* 125.082	col rhomb cry	dec 1.56		sl H_2O, EtOH
111	Ammonium hydrogen phosphate $(NH_4)_2HPO_4$	7783-28-0 132.055	wh cry	155 dec 1.619	a	s H_2O; i EtOH, ace
112	Ammonium hydrogen sulfate NH_4HSO_4	7803-63-6 115.111	wh hyg cry	147 1.78	a	vs H_2O; i EtOH, ace, py
113	Ammonium hydrogen sulfide NH_4HS	12124-99-1 51.113	wh tetr or orth cry	dec 1.17	e	vs H_2O; sl ace; i bz, eth
114	Ammonium hydrogen sulfite NH_4HSO_3	10192-30-0 99.111	col cry	dec 2.03	a	vs H_2O

No.	Name Formula	CAS RN Mol. Wt.	Physical Form	mp/°C den/g cm^{-3}	bp/°C Other Data	Solubility
115	Ammonium hydrogen tartrate $NH_4OOCCH(OH)CH(OH)COOH$	3095-65-6 167.118	wh cry	 1.68		sl H_2O; s alk; i EtOH
116	Ammonium hydroxide NH_4OH	1336-21-6 35.046	exists only in soln		a	
117	Ammonium hypophosphite $NH_4H_2PO_2$	7803-65-8 83.028	wh hyg cry	dec		vs H_2O; sl EtOH; i ace
118	Ammonium iodate NH_4IO_3	13446-09-8 192.941	wh powder	150 3.3		
119	Ammonium iodide NH_4I	12027-06-4 144.943	wh tetr cry; hyg	551 dec 2.514	405 sp a,c,e	vs H_2O; sl EtOH, MeOH
120	Ammonium lactate $NH_4C_3H_5O_3$	52003-58-4 107.109	col cry	92		s H_2O, EtOH; sl MeOH; i ace, eth
121	Ammonium metatungstate hexahydrate $(NH_4)_6W_7O_{24}\cdot 6H_2O$	12028-48-7 1887.19	wh cry			s H_2O; i EtOH
122	Ammonium metavanadate $NH_4VO_3\cdot$	7803-55-6 116.979	wh-yel cry	dec 2.326		sl H_2O
123	Ammonium molybdate(IV) tetrahydrate $(NH_4)_6Mo_7O_{24}\cdot 4H_2O$	12054-85-2 1235.86	col or grn-yel cry	dec 2.498		s H_2O; i EtOH
124	Ammonium nickel chloride hexahydrate $NH_4NiCl_3\cdot 6H_2O$	16122-03-5* 291.181	grn hyg cry	 1.65		s H_2O
125	Ammonium nickel sulfate hexahydrate $(NH_4)_2Ni(SO_4)_2\cdot 6H_2O$	7785-20-8 394.989	blue-grn cry	dec 1.923		sl H_2O; i EtOH
126	Ammonium nitrate NH_4NO_3	6484-52-2 80.043	wh hyg cry; orth	dec 210; exp 1.72	a,c	vs H_2O; sl MeOH
127	Ammonium nitroferricyanide $(NH_4)_2Fe(CN)_5NO$	14402-70-1 252.017	red-brn cry			s H_2O, EtOH
128	Ammonium oleate $NH_4C_{18}H_{33}O_2$	544-60-5 299.498	yel-brn paste	21		s H_2O; sl ace
129	Ammonium oxalate monohydrate $(NH_4)_2C_2O_4\cdot H_2O$	6009-70-7 142.111	wh orth cry	dec 1.50		sl H_2O, EtOH
130	Ammonium palmitate $NH_4C_{15}H_{31}COO$	593-26-0 273.460	yel-wh powder	22		s H_2O; sl bz, xyl; i ace, EtOH, ctc
131	Ammonium pentachlorozincate $(NH_4)_3ZnCl_5$	14639-98-6 296.77	hyg orth cry	 1.81		vs H_2O
132	Ammonium perchlorate NH_4ClO_4	7790-98-9 117.490	wh orth cry	dec, exp 1.95	a	vs H_2O; s MeOH; sl EtOH, ace; i eth
133	Ammonium permanganate NH_4MnO_4	13446-10-1 136.975	purp rhomb cry	dec 70 2.22		sl H_2O
134	Ammonium peroxydisulfate $(NH_4)_2S_2O_8$	7727-54-0 228.204	monocl cry or wh powder	dec 1.982		vs H_2O
135	Ammonium perrhenate NH_4ReO_4	13598-65-7 268.244	col powder	 3.97		sl H_2O
136	Ammonium phosphite, dibasic, monohydrate $(NH_4)_2HPO_3\cdot H_2O$	51503-61-8 134.071	hyg cry			s H_2O
137	Ammonium phosphomolybdate monohydrate $(NH_4)_3PO_4\cdot 12MoO_3\cdot H_2O$	54723-94-3 1894.36	yel cry or powder	dec		sl H_2O
138	Ammonium phosphotungstate dihydrate $(NH_4)_3PO_4\cdot 12WO_3\cdot 2H_2O$	1311-90-6 2967.18	cry powder			sl H_2O
139	Ammonium picrate $NH_4C_6H_2N_3O_7$	131-74-8 246.137	yel orth cry	exp 1.72		sl H_2O
140	Ammonium salicylate $NH_4C_7H_5O_3$	528-94-9 155.153	wh cry powder			vs H_2O; s EtOH
141	Ammonium selenate $(NH_4)_2SeO_4$	7783-21-3 179.04	wh monocl cry	dec 2.194		vs H_2O; i EtOH, ace
142	Ammonium selenite $(NH_4)_2SeO_3$	7783-19-9 163.04	wh or red hyg cry	dec		s H_2O
143	Ammonium stearate $NH_4C_{18}H_{35}O_2$	1002-89-7 301.514	yel-wh powder	22 0.89		sl H_2O, bz; s EtOH, MeOH; i ace
144	Ammonium sulfamate $NH_4NH_2SO_3$	7773-06-0 114.125	wh hyg cry	131	dec 160	vs H_2O; sl EtOH
145	Ammonium sulfate $(NH_4)_2SO_4$	7783-20-2 132.141	wh or brn orth cry	dec 280 1.77	a,f	vs H_2O; i EtOH, ace
146	Ammonium sulfide $(NH_4)_2S$	12135-76-1 68.143	yel-oran cry	dec ≈0		s H_2O, EtOH, alk
147	Ammonium sulfite monohydrate $(NH_4)_2SO_3\cdot H_2O$	7783-11-1 134.156	col cry	dec 1.41		s H_2O; i EtOH, ace
148	Ammonium tartrate $(NH_4)_2C_4H_4O_6$	3164-29-2 184.148	wh cry	dec 1.601		s H_2O
149	Ammonium tellurate $(NH_4)_2TeO_4$	13453-06-0 227.68	wh powder	dec 3.024		
150	Ammonium tetrachloroaluminate NH_4AlCl_4	7784-14-7 186.832	wh hyg solid	304		s H_2O, eth
151	Ammonium tetrachloroplatinate(II) $(NH_4)_2PtCl_4$	13820-41-2 372.97	red cry	dec 2.936		s H_2O; i EtOH
152	Ammonium tetrachlorozincate $(NH_4)_2ZnCl_4$	14639-97-5 243.28	wh orth plates; hyg	150 dec 1.879		vs H_2O

No.	Name Formula	CAS RN Mol. Wt.	Physical Form	mp/°C den/g cm^{-3}	bp/°C Other Data	Solubility
153	Ammonium tetrathiotungstate $(NH_4)_2WS_4$	13862-78-7 348.18	oran cry	dec 2.71		s H_2O
154	Ammonium thiocyanate NH_4SCN	1762-95-4 76.122	col hyg cry	≈149 1.30	dec	vs H_2O, EtOH; s ace; i chl
155	Ammonium thiosulfate $(NH_4)_2S_2O_3$	7783-18-8 148.207	wh cry	dec 150 1.678		vs H_2O; i EtOH, eth
156	Ammonium titanium oxalate monohydrate $(NH_4)_2TiO(C_2O_4)_2 \cdot H_2O$	10580-03-7 293.997	hyg cry			vs H_2O
157	Ammonium uranate(VI) $(NH_4)_2U_2O_7$	7783-22-4 624.131	red-yel amorp powder			i H_2O, akl; s acid
158	Ammonium uranium fluoride $UO_2(NH_4)_3F_5$	18433-40-4 419.135	grn-yel monocl cry			s H_2O; i EtOH
159	Ammonium valerate $NH_4C_4H_9COO$	42739-38-8 119.164	hyg cry	108		vs H_2O, EtOH; s eth
160	Ammonium zirconyl carbonate dihydrate $(NH_4)_3ZrOH(CO_3)_3 \cdot 2H_2O$	12616-24-9* 362.405	prisms; unstable			s H_2O
161	Antimony Sb	7440-36-0 121.760	silv metal; hex	630.63 6.68	1587 a,c,e,f	i dil acid
162	Stibine SbH_3	7803-52-3 124.784	col gas; flam	-88 5.475 g/L	-17 a,d	sl H_2O; s EtOH
163	Antimony arsenide SbAs	12322-34-8 196.682	hex cry	≈680 6.0		
164	Antimony(III) bromide $SbBr_3$	7789-61-9 361.472	yel orth cry; hyg	96.6 4.35	280 a,b,d,e	reac H_2O; s ace, bz, chl
165	Antimony(III) chloride $SbCl_3$	10025-91-9 228.118	col orth cry; hyg	73.4 3.14	220.3 a,b,c,d,e	vs H_2O; s acid, EtOH, bz, ace
166	Antimony(III) fluoride SbF_3	7783-56-4 178.755	wh orth cry; hyg	292 4.38	≈345 a	vs H_2O
167	Antimony(III) iodide SbI_3	7790-44-5 502.473	red rhomb cry	168 4.92	401 a,b,d,e	reac H_2O; s EtOH, ace; i ctc
168	Antimony(III) oxide (senarmontite) Sb_2O_3	1309-64-4 291.518	col cub cry	570 trans 5.58	1425 e,f	sl H_2O; i os
169	Antimony(III) oxide (valentinite) Sb_2O_3	1309-64-4 291.518	wh orth cry	655 5.7	1425 e,f	sl H_2O; i os
170	Antimony(III) oxychloride SbOCl	7791-08-4 173.212	wh momo cry	170 dec		reac H_2O; i EtOH, eth
171	Antimony(III) selenide Sb_2Se_3	1315-05-5 480.40	grn orth cry	611 5.81		sl H_2O
172	Antimony(III) sulfate $Sb_2(SO_4)_3$	7446-32-4 531.711	wh cry powder; hyg	dec 3.62		sl H_2O
173	Antimony(III) sulfide Sb_2S_3	1345-04-6 339.718	gray-blk orth cry	550 4.562	f	i H_2O; s conc HCl
174	Antimony(III) telluride Sb_2Te_3	1327-50-0 626.32	gray cry	620 6.5		
175	Antimony(III,V) oxide Sb_2O_4	1332-81-6 307.518	yel orth cry	6.64	f	
176	Antimony(V) chloride $SbCl_5$	7647-18-9 299.024	col or yel liq	4 2.34	140 dec e	reac H_2O; s chl, ctc
177	Antimony(V) fluoride SbF_5	7783-70-2 216.752	hyg visc liq	8.3 3.10	141	reac H_2O
178	Antimony(V) dichlorotrifluoride $SbCl_2F_3$	7791-16-4 249.660	visc liq			reac H_2O
179	Antimony(V) oxide Sb_2O_5	1314-60-9 323.517	yel powder; cub	dec 3.78	a	sl H_2O
180	Antimony(V) sulfide Sb_2S_5	1315-04-4 403.850	oran-yel powder	75 dec 4.120		i H_2O; s acid, alk
181	Argon Ar	7440-37-1 39.948	col gas	-189.35 1.753 g/L	-185.85 a,b,c,d,e	sl H_2O
182	Arsenic As	7440-38-2 74.922	gray metal; rhomb	817 tp 5.75	614 sp a,b,c,e,f	i H_2O
183	Arsine AsH_3	7784-42-1 77.946	col gas	-116 3.420 g/L	-62.5 a,b,d,e	sl H_2O
184	Diarsine As_2H_4	15942-63-9 153.875	unstable liq		≈100	
185	Arsenic acid H_3AsO_4	7778-39-4 141.944	exists only in soln		a	
186	Arsenic acid hemihydrate $H_3AsO_4 \cdot 0.5H_2O$	7778-39-4* 150.951	wh hyg cry	35.5 ≈2		vs H_2O, EtOH
187	Arsenious acid H_3AsO_3	13464-58-9 125.944	exists only in soln			
188	Arsenic diiodide As_2I_4	13770-56-4 657.461	red cry	137		reac H_2O; s os
189	Arsenic hemiselenide As_2Se	1303-35-1 228.80	blk cry			i H_2O, os; dec acid, alk
190	Arsenic sulfide As_4S_4	12279-90-2 427.950	red monocl cry	320 3.5	565 f	i H_2O; sl bz; s alk

No.	Name / Formula	CAS RN / Mol. Wt.	Physical Form	mp/°C / den/g cm^{-3}	bp/°C / Other Data	Solubility
191	Arsenic(III) bromide / $AsBr_3$	7784-33-0 / 314.634	yel orth cry; hyg	31.1 / 3.40	221 / a,c,d,e	reac H_2O; s hc, ctc; vs eth, bz
192	Arsenic(III) chloride / $AsCl_3$	7784-34-1 / 181.280	col liq	-16 / 2.150	130 / a,b,c,d,e	reac H_2O; vs chl, ctc, eth
193	Arsenic(III) fluoride / AsF_3	7784-35-2 / 131.917	col liq	-5.9 / 2.7	57.8 / a,c,d,e	reac H_2O; s EtOH, eth, bz
194	Arsenic(III) iodide / AsI_3	7784-45-4 / 455.635	red hex cry	140.9 / 4.73	424 / a,d	sl H_2O, EtOH, eth; s bz. tol
195	Arsenic(III) oxide (arsenolite) / As_2O_3	1327-53-3 / 197.841	wh cub cry	274 / 3.86	460 / e,f	sl H_2O
196	Arsenic(III) oxide (claudetite) / As_2O_3	1327-53-3 / 197.841	wh monocl cry	313 / 3.74	460 / e,f	sl H_2O; s dil acid, alk; i EtOH
197	Arsenic(III) selenide / As_2Se_3	1303-36-2 / 386.72	brn-blk solid	260 / 4.75		i H_2O; s alk
198	Arsenic(III) sulfide / As_2S_3	1303-33-9 / 246.041	yel-oran monocl cry	310 / 3.46	707 / a,f	i H_2O; s alk
199	Arsenic(III) telluride / As_2Te_3	12044-54-1 / 532.64	blk monocl cry	621 / 6.50		
200	Arsenic(V) chloride / $AsCl_5$	22441-45-8 / 252.186	stable at low temp	≈-50 dec		
201	Arsenic(V) fluoride / AsF_5	7784-36-3 / 169.914	col gas	-79.8 / 7.456 g/L	-52.8 / d,e	reac H_2O; s EtOH, bz, eth
202	Arsenic(V) oxide / As_2O_5	1303-28-2 / 229.840	wh amorp powder	315 / 4.32	a	vs H_2O, EtOH
203	Arsenic(V) selenide / As_2Se_5	1303-37-3 / 544.64	blk solid	dec		i H_2O, EtOH, eth; s alk
204	Arsenic(V) sulfide / As_2S_5	1303-34-0 / 310.173	brn-yel amorp solid	dec		i H_2O; s alk
205	Astatine / At	7440-68-8 / 210	cry	302	a	s HNO_3, os
206	Barium / Ba	7440-39-3 / 137.327	silv-yel metal; cub	727 / 3.62	1897 / a,c,d,e	reac H_2O; sl EtOH
207	Barium acetate / $Ba(C_2H_3O_2)_2$	543-80-6 / 255.417	wh powder	/ 2.47		vs H_2O
208	Barium acetate monohydrate / $Ba(C_2H_3O_2)_2 \cdot H_2O$	5908-64-5 / 273.432	wh cry	dec 110 / 2.19		vs H_2O; sl EtOH
209	Barium aluminate / $BaAl_2O_4$	12004-04-5 / 255.288	hex cry	1827		
210	Barium azide / $Ba(N_3)_2$	18810-58-7 / 221.367	monocl cry; exp	dec ≈120 / 2.936		s H_2O; sl EtOH; i eth
211	Barium bromate monohydrate / $Ba(BrO_3)_2 \cdot H_2O$	10326-26-8 / 411.147	wh monocl cry	dec 260 / 3.99		sl H_2O; i EtOH
212	Barium bromide / $BaBr_2$	10553-31-8 / 297.135	wh orth cry	857 / 4.781	1835 / a,c	vs H_2O
213	Barium bromide dihydrate / $BaBr_2 \cdot 2H_2O$	7791-28-8 / 333.166	wh cry	dec 75 / 3.7		vs H_2O; s MeOH; i EtOH, ace, diox
214	Barium carbide / BaC_2	50813-65-5 / 161.349	gray tetr cry	dec / 3.74		reac H_2O
215	Barium carbonate / $BaCO_3$	513-77-9 / 197.336	wh orth cry	1555 / 4.2865	a,f	i H_2O; s acid
216	Barium chlorate monohydrate / $Ba(ClO_3)_2 \cdot H_2O$	10294-38-9 / 322.244	wh monocl cry	414 / 3.179		s H_2O, acid; sl EtOH, ace
217	Barium chloride / $BaCl_2$	10361-37-2 / 208.232	wh orth cry; hyg	962 / 3.9	1560 / a,c	s H_2O
218	Barium chloride dihydrate / $BaCl_2 \cdot 2H_2O$	10326-27-9 / 244.263	wh monocl cry	dec ≈120 / 3.097		s H_2O; i EtOH
219	Barium chromate(V) / $Ba_3(CrO_4)_2$	12345-14-1 / 643.968	grn-blk hex cry	/ 5.25		s H_2O
220	Barium chromate(VI) / $BaCrO_4$	10294-40-3 / 253.321	yel orth cry	dec / 4.50		i H_2O; reac acid
221	Barium citrate monohydrate / $Ba_3(C_6H_5O_7)_2 \cdot H_2O$	512-25-4* / 808.199	gray-wh cry			s H_2O, acid
222	Barium cyanide / $Ba(CN)_2$	542-62-1 / 189.362	wh cry powder			vs H_2O; s EtOH
223	Barium dichromate dihydrate / $BaCr_2O_7 \cdot 2H_2O$	10031-16-0 / 389.346	brn-red needles	dec		reac H_2O
224	Barium dithionate dihydrate / $Ba(SO_3)_2 \cdot 2H_2O$	13845-17-5 / 333.486	cry	/ 4.54		s H_2O; sl EtOH
225	Barium ferrocyanide hexahydrate / $Ba_2Fe(CN)_6 \cdot 6H_2O$	13821-06-2* / 594.696	yel monocl cry	dec 80		i H_2O, EtOH
226	Barium fluoride / BaF_2	7787-32-8 / 175.324	wh cub cry	1368 / 4.893	2260 / a,c	sl H_2O
227	Barium formate / $Ba(CHO_2)_2$	541-43-5 / 227.363	cry	/ 3.21		s H_2O; i EtOH
228	Barium hexaboride / BaB_6	12046-08-1 / 202.193	blk cub cry	2070 / 4.36		i H_2O; s acid; i EtOH
229	Barium hexafluorosilicate / $BaSiF_6$	17125-80-3 / 279.403	wh orth needles	dec 300 / 4.29		i H_2O, EtOH; sl acid

No.	Name Formula	CAS RN Mol. Wt.	Physical Form	mp/°C den/g cm⁻³	bp/°C Other Data	Solubility
230	Barium hydride BaH_2	13477-09-3 139.343	gray orth cry	1200 4.16	a	reac H_2O
231	Barium hydrogen phosphate $BaHPO_4$	10048-98-3 233.306	wh cry powder	dec 400 4.16		i H_2O; s dil acid
232	Barium hydrosulfide $Ba(HS)_2$	25417-81-6 203.475	yel hyg cry			s H_2O
233	Barium hydrosulfide tetrahydrate $Ba(HS)_2 \cdot 4H_2O$	12230-74-9 275.536	yel rhomb cry	dec 50		s H_2O
234	Barium hydroxide $Ba(OH)_2$	17194-00-2 171.342	wh powder	408 	a,c	sl H_2O
235	Barium hydroxide monohydrate $Ba(OH)_2 \cdot H_2O$	22326-55-2 189.357	wh powder	3.743		sl H_2O; s acid
236	Barium hydroxide octahydrate $Ba(OH)_2 \cdot 8H_2O$	12230-71-6 315.464	wh monocl cry	dec 78 2.18		sl H_2O
237	Barium hypophosphite monohydrate $Ba(H_2PO_2)_2 \cdot H_2O$	14871-79-5* 285.320	monocl plates	2.90		s H_2O; i EtOH
238	Barium iodate $Ba(IO_3)_2$	10567-69-8 487.132	wh cry powder	dec 476 5.23		sl H_2O
239	Barium iodate monohydrate $Ba(IO_3)_2 \cdot H_2O$	7787-34-0 505.148	cry	dec 130 5.00		sl H_2O; s acid; i EtOH
240	Barium iodide BaI_2	13718-50-8 391.136	wh orth cry	711 5.15	a,c	vs H_2O
241	Barium iodide dihydrate $BaI_2 \cdot 2H_2O$	7787-33-9 427.167	col cry	740 dec 5.0		vs H_2O; s EtOH, ace
242	Barium manganate(VI) $BaMnO_4$	7787-35-1 256.263	grn-gray hyg cry	4.85		reac H_2O
243	Barium metaborate monohydrate $Ba(BO_2)_2 \cdot H_2O$	26124-86-7 240.962	wh powder	>900 3.3		sl H_2O
244	Barium molybdate $BaMoO_4$	7787-37-3 297.27	wh powder	1450 4.975		i H_2O
245	Barium niobate $Ba(NbO_3)_2$	12009-14-2 419.136	yel orth cry	1455 5.44		i H_2O
246	Barium nitrate $Ba(NO_3)_2$	10022-31-8 261.336	wh cub cry	590 3.24	a	s H_2O; sl EtOH, ace
247	Barium nitride Ba_3N_2	12047-79-9 439.994	yel-brn cry	4.78		reac H_2O
248	Barium nitrite $Ba(NO_2)_2$	13465-94-6 229.338	col hex cry	267 3.234	a	
249	Barium nitrite monohydrate $Ba(NO_2)_2 \cdot H_2O$	7787-38-4 247.353	yel-wh hex cry	217 dec 3.18		vs H_2O; i EtOH
250	Barium oxalate BaC_2O_4	516-02-9 225.347	wh powder	400 dec 2.658		i H_2O
251	Barium oxalate monohydrate $BaC_2O_4 \cdot H_2O$	13463-22-4 243.362	wh cry powder	2.66		i H_2O; s acid
252	Barium oxide BaO	1304-28-5 153.326	wh-yel powder; cub and hex	1973 5.72(cub)	a,c	sl H_2O; s dil acid, EtOH; i ace
253	Barium perchlorate $Ba(ClO_4)_2$	13465-95-7 336.227	col hex cry	505 3.20		vs H_2O, EtOH
254	Barium perchlorate trihydrate $Ba(ClO_4)_2 \cdot 3H_2O$	10294-39-0 390.273	col cry	2.74		s H_2O, MeOH; sl EtOH, ace; i eth
255	Barium permanganate $Ba(MnO_4)_2$	7787-36-2 375.198	brn-viol cry	dec 200 3.77		sl H_2O; reac EtOH
256	Barium peroxide BaO_2	1304-29-6 169.326	gray-wh tetr cry	450 dec 4.96		i H_2O; reac dil acid
257	Barium metaphosphate $Ba(PO_3)_2$	13466-20-1 295.271	wh powder	1560		i H_2O; sl acid
258	Barium potassium chromate $BaK_2(CrO_4)_2$	27133-66-0 447.511	yel hex cry	3.63		vs H_2O
259	Barium pyrophosphate $Ba_2P_2O_7$	13466-21-2 448.597	wh powder	3.9		sl H_2O; s acid
260	Barium selenate $BaSeO_4$	7787-41-9 280.29	wh rhomb cry	dec 4.75		i H_2O
261	Barium selenide $BaSe$	1304-39-8 216.29	cub cry powder	1780 5.02		reac H_2O
262	Barium selenite $BaSeO_3$	13718-59-7 264.29	solid			i H_2O
263	Barium disilicate $BaSi_2O_5$	12650-28-1 273.495	wh orth cry	1420 3.70		
264	Barium metasilicate $BaSiO_3$	13255-26-0 213.411	col rhomb powder	1605 4.40		i H_2O; s acid
265	Barium silicide $BaSi_2$	1304-40-1 193.498	gray lumps	1180		reac H_2O
266	Barium sodium niobate $Ba_2Na(NbO_3)_5$	12323-03-4 1002.167	wh orth cry	1437 5.40		i H_2O
267	Barium stannate $BaSnO_3$	12009-18-6 304.035	cub cry	7.24		sl H_2O
268	Barium stannate trihydrate $BaSnO_3 \cdot 3H_2O$	12009-18-6* 358.081	wh cry powder			sl H_2O; s acid

No.	Name Formula	CAS RN Mol. Wt.	Physical Form	mp/°C den/g cm^{-3}	bp/°C Other Data	Solubility
269	Barium stearate $Ba(C_{18}H_{35}O_2)_2$	6865-35-6 704.277	wh powder	160 1.145		i H_2O, alc
270	Barium sulfate $BaSO_4$	7727-43-7 233.391	wh orth cry	1580 4.49	a,c,f	i H_2O, EtOH
271	Barium sulfide BaS	21109-95-5 169.393	col cub cry or gray powder	2230 4.3	a	sl H_2O
272	Barium sulfite $BaSO_3$	7787-39-5 217.391	wh monocl cry	dec 4.44		i H_2O, EtOH
273	Barium tartrate $BaC_4H_4O_6$	5908-81-6 285.399	wh cry	2.98		s H_2O; i EtOH
274	Barium tetracyanoplatinate(II) tetrahydrate $BaPt(CN)_4\cdot4H_2O$	13755-32-3 508.54	yel powder or cry	2.076		sl H_2O; i EtOH
275	Barium tetraiodomercurate(II) $BaHgI_4$	10048-99-4 845.54	yel-red hyg cry			vs H_2O, EtOH
276	Barium thiocyanate $Ba(SCN)_2$	2092-17-3 253.494	hyg cry			vs H_2O; s ace, MeOH, EtOH
277	Barium thiocyanate dihydrate $Ba(SCN)_2\cdot2H_2O$	2092-17-3* 289.525	hyg wh cry			s H_2O, EtOH
278	Barium thiocyanate trihydrate $Ba(SCN)_2\cdot3H_2O$	68016-36-4 307.540	wh needles; hyg	2.286		s H_2O, EtOH
279	Barium thiosulfate BaS_2O_3	35112-53-9 249.457	wh cry powder	dec 220		sl H_2O; i EtOH
280	Barium thiosulfate monohydrate $BaS_2O_3\cdot H_2O$	7787-40-8 267.473	wh cry powder	dec 3.5		sl H_2O; i EtOH
281	Barium titanate $BaTiO_3$	12047-27-7 233.192	wh tetr cry	1625 6.02		i H_2O
282	Barium tungstate $BaWO_4$	7787-42-0 385.17	wh tetr cry	5.04		i H_2O
283	Barium uranium oxide BaU_2O_7	10380-31-1 725.381	oran-yel powder			i H_2O; s acid
284	Barium orthovanadate $Ba_3(VO_4)_2$	39416-30-3 641.859	hex cry	707 5.14		
285	Barium zirconate $BaZrO_3$	12009-21-1 276.549	gray-wh cub cry	2500 5.52		i H_2O, alk; sl acid
286	Berkelium (α form) Bk	7440-40-6 247	hex	1050 14.78	a	
287	Berkelium (β form) Bk	7440-40-6 247	cub cry	986 13.25	a	
288	Beryllium Be	7440-41-7 9.012	hex	1287 1.85	2471 a,c,e	s acid, alk
289	Beryllium acetate $Be(C_2H_3O_2)_2$	543-81-7 127.102	wh cry	dec 60		i H_2O, EtOH
290	Beryllium acetylacetonate $Be(C_5H_7O_2)_2$	10210-64-7 207.231	monocl cry powder	108 1.168	270	i H_2O, vs EtOH, eth
291	Beryllium aluminate $BeAl_2O_4$	12004-06-7 126.973	orth cry	3.65	f	
292	Beryllium aluminum metasilicate $Be_3Al_2(SiO_3)_6$	1302-52-9 537.502	col or grn-yel cry; hex	2.64	f	
293	Beryllium basic acetate $Be_4O(C_2H_3O_2)_6$	1332-52-1 406.316	wh cry	285 1.25	330	i H_2O; s eth, os
294	Beryllium boride BeB_2	12228-40-9 30.634	refrac solid	>1970		
295	Beryllium borohydride $Be(BH_4)_2$	17440-85-6 36.682	solid	dec 125; flam	subl e	reac H_2O
296	Beryllium bromide $BeBr_2$	7787-46-4 168.820	orth cry; hyg	508 3.465	520 a,c,e	vs H_2O; s EtOH, pyr
297	Beryllium carbide Be_2C	506-66-1 30.035	red cub cry	dec >2100 1.90		reac H_2O
298	Beryllium carbonate tetrahydrate $BeCO_3\cdot4H_2O$	60883-64-9 93.085	unstable solid			
299	Beryllium chloride $BeCl_2$	7787-47-5 79.917	wh-yel orth cry; hyg	415 1.90	482 a,c,d,e	vs H_2O; s EtOH, eth, py; i bz, tol
300	Beryllium fluoride BeF_2	7787-49-7 47.009	tetr cry or gl; hyg	552 2.1	1169 a,c	vs H_2O; sl EtOH
301	Beryllium formate $Be(CHO_2)_2$	1111-71-3 99.048	powder	dec >250		reac H_2O; i os
302	Beryllium hydride BeH_2	7787-52-2 11.028	wh amorp solid	dec 250 0.65		reac H_2O; i eth, tol
303	Beryllium hydroxide $Be(OH)_2$	13327-32-7 43.027	amorp powder or cry	1.92	a	sl H_2O, alk; s acid
304	Beryllium iodide BeI_2	7787-53-3 262.821	hyg needles	470 4.32	487 a,c,d,e	reac H_2O; s EtOH
305	Beryllium nitrate trihydrate $Be(NO_3)_2\cdot3H_2O$	13597-99-4 187.068	yel-wh hyg cry	≈30	dec	vs H_2O; s EtOH
306	Beryllium nitride Be_3N_2	1304-54-7 55.050	gray refrac cry; cub	2200 2.71		reac acid, alk
307	Beryllium oxide BeO	1304-56-9 25.011	wh amorp powder	2578 3.01	a,c	i H_2O; sl acid, alk

No.	Name / Formula	CAS RN / Mol. Wt.	Physical Form	mp/°C / den/g cm^{-3}	bp/°C / Other Data	Solubility
308	Beryllium perchlorate tetrahydrate $Be(ClO_4)_2 \cdot 4H_2O$	7787-48-6 279.974	hyg cry			vs H_2O
309	Beryllium selenate tetrahydrate $BeSeO_4 \cdot 4H_2O$	10039-31-3 224.03	orth cry	dec 100 2.03		vs H_2O
310	Beryllium sulfate $BeSO_4$.	13510-49-1 105.076	col tetr cry; hyg	1127 2.5	a	s H_2O
311	Beryllium sulfate tetrahydrate $BeSO_4 \cdot 4H_2O$	7787-56-6 177.137	col tetr cry	dec ≈100 1.71		vs H_2O; i EtOH
312	Beryllium sulfide BeS	13598-22-6 41.078	col cub cry	dec 2.36	a	reac hot H_2O
313	Bismuth Bi	7440-69-9 208.980	gray-wh soft metal	271.40 9.79	1564 a,c,d,e,f	s acid
314	Bismuth basic carbonate $(BiO)_2CO_3$	5892-10-4 509.969	wh powder	6.86		i H_2O; s acid
315	Bismuth bromide $BiBr_3$	7787-58-8 448.692	yel cub cry	218 5.72	453 b,d,e	reac H_2O; s dil acid, ace; i EtOH
316	Bismuth chloride $BiCl_3$	7787-60-2 315.338	yel-wh cub cry; hyg	230 4.75	447 a,b,c,d,e	reac H_2O; s acid, EtOH, ace
317	Bismuth citrate $BiC_6H_5O_7$	813-93-4 398.082	wh powder	3.458		i H_2O; sl EtOH
318	Bismuth fluoride BiF_3	7787-61-3 265.975	wh-gray cub cry	725 8.3	900	i H_2O
319	Bismuth pentafluoride BiF_5	7787-62-4 303.972	wh tetr needles; hyg	154 5.55	230	reac H_2O
320	Bismuth hydride BiH_3	18288-22-7 212.004	col gas; unstable	-67 9.303 g/L	≈17	
321	Bismuth hydroxide $Bi(OH)_3$	10361-43-0 260.002	wh-yel amorp powder	4.962	a	i H_2O; s acid
322	Bismuth iodide BiI_3	7787-64-6 589.693	blk hex cry	408.6 5.778	542 a	i H_2O; s EtOH
323	Bismuth molybdate $Bi_2(MoO_4)_3$	51898-99-8 897.77	monocl cry	5.95		
324	Bismuth nitrate pentahydrate $Bi(NO_3)_3 \cdot 5H_2O$	10035-06-0 485.071	col tricl cry; hyg	≈75 dec 2.83		reac H_2O; s ace; i EtOH
325	Bismuth oleate $Bi(C_{18}H_{33}O_2)_3$	52951-38-9 1053.356	soft yel-brn solid			i H_2O; s eth; sl bz
326	Bismuth orthovanadate $BiVO_4$	14059-33-7 323.920	oran-yel orth cry	trans 500 6.25		
327	Bismuth oxalate $Bi_2(C_2O_4)_3$	6591-55-5 682.020	wh powder			i H_2O, EtOH; s dil acid
328	Bismuth oxide Bi_2O_3	1304-76-3 465.959	yel monocl cry or powder	817 8.9	1890 a	i H_2O, s acid
329	Bismuth oxybromide $BiOBr$	7787-57-7 304.883	col tetr cry	8.08		i H_2O, EtOH; s acid
330	Bismuth oxychloride $BiOCl$	7787-59-9 260.432	wh tetr cry	7.72	a	i H_2O
331	Bismuth oxyiodide $BiOI$	7787-63-5 351.883	red tetr cry	>300 dec 7.92		i H_2O, EtOH, chl; s HCl
332	Bismuth oxynitrate $BiONO_3$	10361-46-3 286.985	wh powder	260 dec 4.93		i H_2O, EtOH; s acid
333	Bismuth phosphate $BiPO_4$	10049-01-1 303.951	monocl cry	6.32		sl H_2O, dil acid; i EtOH
334	Bismuth selenide Bi_2Se_3	12068-69-8 654.84	blk hex cry	710 dec 7.5		i H_2O
335	Bismuth stannate pentahydrate $Bi_2(SnO_3)_3 \cdot 5H_2O$	12777-45-6 1008.162	wh cry			i H_2O
336	Bismuth sulfate $Bi_2(SO_4)_3$	7787-68-0 706.152	wh needles or powder	405 dec 5.08	a	reac H_2O, EtOH
337	Bismuth sulfide Bi_2S_3	1345-07-9 514.159	blk-brn orth cry	850 6.78	a,f	i H_2O; s acid
338	Bismuth telluride Bi_2Te_3	1304-82-1 800.76	gray hex plates	580 7.74	f	i H_2O; s EtOH
339	Bismuth tetroxide Bi_2O_4	12048-50-9 481.959	red-oran powder	305 5.6		reac H_2O
340	Bismuth titanate $Bi_4(TiO_4)_3$	12048-51-0 1171.515	wh orth cry	7.85		
341	Boron B	7440-42-8 10.811	blk rhomb cry	2075 2.34	4000 a,c,d	i H_2O
342	Diborane(6) B_2H_6	19287-45-7 27.670	col gas; flam	-165.5 1.214 g/L	-92.4 a,b,d,e	reac H_2O
343	Tetraborane(10) B_4H_{10}	18283-93-7 53.323	col gas	-121 2.340 g/L	18 a,d,e	reac H_2O
344	Pentaborane(9) B_5H_9	19624-22-7 63.126	flam liq	-46.6 0.60	60 a,e	reac hot H_2O
345	Pentaborane(11) B_5H_{11}	18433-84-6 65.142	col liq; unstable	-123	63 a,d,e	reac H_2O
346	Hexaborane(10) B_6H_{10}	23777-80-2 74.945	col liq	-62.3 0.67	108 dec a	reac hot H_2O

No.	Name / Formula	CAS RN / Mol. Wt.	Physical Form	mp/°C / den/g cm⁻³	bp/°C / Other Data	Solubility
347	Hexaborane(12) B_6H_{12}	12008-19-4 76.961	col liq; unstable	-82	≈80	reac H_2O
348	Nonaborane(15) B_9H_{15}	19465-30-6 112.418	col liq	2.6		
349	Decaborane(14) $B_{10}H_{14}$	17702-41-9 122.221	wh orth cry	99.6 0.94	≈213 e	sl H_2O; s EtOH, bz, CS_2, ctc
350	Borane carbonyl BH_3CO	13205-44-2 41.845	col gas	-137	-64 e	reac H_2O
351	Borazine $B_3N_3H_6$	6569-51-3 80.501	col liq	-58 0.824	53 a	reac H_2O
352	Boric acid (orthoboric acid) $B(OH)_3$	10043-35-3 61.833	col tricl cry	170.9 1.5	a	sl H_2O, EtOH
353	Metaboric acid (α form) HBO_2	13460-50-9 43.818	col orth cry; hyg	176 1.784	a	s H_2O
354	Metaboric acid (β form) HBO_2	13460-50-9 43.818	col monocl cry; hyg	201 2.045	a	s H_2O
355	Metaboric acid (γ form) HBO_2	13460-50-9 43.818	col cub cry	236 2.487	a	s H_2O
356	Fluoroboric acid HBF_4	16872-11-0 87.813	col liq	≈1.8	130 dec	vs H_2O, EtOH
357	Boron arsenide BAs	12005-69-5 85.733	cub cry	dec 920 5.22		
358	Boron bromide BBr_3	10294-33-4 250.523	col liq; hyg	-45 2.6	91 a,b,d,e	reac H_2O, EtOH
359	Boron carbide B_4C	12069-32-8 55.255	hard blk cry	2350 2.50	>3500	i H_2O, acid
360	Boron chloride BCl_3	10294-34-5 117.169	col liq or gas	-107 5.141 g/L	12.65 a,b,c,d,e	reac H_2O, EtOH
361	Boron chloride B_2Cl_4	13701-67-2 163.433	col liq; flam	-92.6	65 a	reac H_2O
362	Boron fluoride BF_3	7637-07-2 67.806	col gas	-126.8 2.975 g/L	-101 a,b,c,d,e	s H_2O
363	Boron fluoride B_2F_4	13965-73-6 97.616	col gas; flam	-56 4.283 g/L	-34 a,d	reac H_2O
364	Boron iodide BI_3	13517-10-7 391.524	needles	43 3.96	210 a,b,d	
365	Boron nitride BN	10043-11-5 24.818	wh powder; hex or cub cry	2967 2.18	a	i H_2O, acid
366	Boron oxide B_2O_3	1303-86-2 69.620	col gl or hex cry; hyg	450 2.55	a,c	sl H_2O; s EtOH
367	Boron phosphide BP	20205-91-8 41.785	red cub cry or powder	dec 1125		reac H_2O, acid
368	Boron sulfide B_2S_3	12007-33-9 117.820	yel amorp solid	softens ≈320 ≈1.7	a	
369	Bromine Br_2	7726-95-6 159.808	red liq	-7.2 3.1028	58.8 a,b,c,d,e	sl H_2O
370	Bromic acid $HBrO_3$	7789-31-3 128.910	stable only in aq soln			s H_2O
371	Bromine oxide Br_2O	21308-80-5 175.807	brn solid	-17.5 dec		
372	Bromine dioxide BrO_2	21255-83-4 111.903	unstable yel cry	dec ≈0	a	
373	Bromine azide BrN_3	13973-87-0 121.924	red cry; exp	≈45	exp	
374	Bromine chloride BrCl	13863-41-7 115.357	unstable red-brn gas	≈-66 5.062 g/L	≈5 dec a	reac H_2O; s eth, CS_2
375	Bromine fluoride BrF	13863-59-7 98.902	unstable red-brn gas	≈-33 4.340 g/L	≈20 dec a,d	
376	Bromine trifluoride BrF_3	7787-71-5 136.899	col hyg liq	8.77 2.803	125.8 a,d	reac H_2O
377	Bromine·pentafluoride BrF_5	7789-30-2 174.896	col liq	-60.5 2.460	40.76 a,c,d,e	reac H_2O (exp)
378	Cadmium Cd	7440-43-9 112.411	silv-wh metal	321.07 8.69	767 a,c,d,e	i H_2O; reac acid
379	Cadmium acetate $Cd(C_2H_3O_2)_2$	543-90-8 230.501	col cry	255 2.34		s H_2O, EtOH
380	Cadmium acetate dihydrate $Cd(C_2H_3O_2)_2 \cdot 2H_2O$	5743-04-4 266.530	wh cry	dec 130 2.01		vs H_2O; s EtOH
381	Cadmium antimonide CdSb	12014-29-8 234.171	orth cry	456 6.92		
382	Cadmium arsenide Cd_3As_2	12006-15-4 487.076	gray tetr cry	721 6.25		
383	Cadmium azide $Cd(N_3)_2$	14215-29-3 196.451	yel-wh orth cry; exp	exp 3.24		
384	Cadmium bromide $CdBr_2$	7789-42-6 272.219	wh hex powder or flakes; hyg	568 5.19	844 a,c,d	s H_2O; sl ace, eth
385	Cadmium bromide tetrahydrate $CdBr_2 \cdot 4H_2O$	13464-92-1 344.281	wh-yel cry			s H_2O, ace, EtOH

No.	Name Formula	CAS RN Mol. Wt.	Physical Form	mp/°C den/g cm⁻³	bp/°C Other Data	Solubility
386	Cadmium carbonate $CdCO_3$	513-78-0 172.420	wh hex cry	dec 500 4.258	a	i H_2O; s acid
387	Cadmium chlorate dihydrate $Cd(ClO_3)_2 \cdot 2H_2O$	22750-54-5* 315.343	col hyg cry	80 dec 2.28		vs H_2O
388	Cadmium chloride $CdCl_2$	10108-64-2 183.316	rhomb cry; hyg	564 4.08	960 a,c,d,e	vs H_2O; s ace; sl EtOH; i eth
389	Cadmium chloride hemipentahydrate $CdCl_2 \cdot 2.5H_2O$	7790-78-5 228.354	wh rhomb leaflets	3.327		vs H_2O; s ace
390	Cadmium chloride monohydrate $CdCl_2 \cdot H_2O$	34330-64-8 201.331	wh cry			vs H_2O
391	Cadmium chromate $CdCrO_4$	14312-00-6 228.405	yel orth cry	4.5		i H_2O
392	Cadmium cyanide $Cd(CN)_2$	542-83-6 164.446	wh cub cry	2.23		sl H_2O
393	Cadmium fluoride CdF_2	7790-79-6 150.408	cub cry	1110 6.33	1748 a,c,d,e	sl H_2O; s acid; i EtOH
394	Cadmium hydroxide $Cd(OH)_2$	21041-95-2 146.426	wh trig or hex cry	dec 130 4.79	a	i H_2O, s dil acid
395	Cadmium iodate $Cd(IO_3)_2$	7790-81-0 462.216	wh powder	6.48		i H_2O; s HNO_3
396	Cadmium iodide CdI_2	7790-80-9 366.220	hex flakes	387 5.64	742 a,c,d,e	s H_2O, EtOH, eth, ace
397	Cadmium metasilicate $CdSiO_3$	13477-19-5 188.495	grn monocl cry	1252 5.10		
398	Cadmium molybdate $CdMoO_4$	13972-68-4 272.35	col tetr cry	≈900 dec 5.4		i H_2O; s acid
399	Cadmium niobate $Cd_2Nb_2O_7$	12187-14-3 522.631	cub cry	≈1410 6.28		i H_2O
400	Cadmium nitrate $Cd(NO_3)_2$	10325-94-7 236.420	wh cub cry; hyg	350 3.6		vs H_2O; s EtOH
401	Cadmium nitrate tetrahydrate $Cd(NO_3)_2 \cdot 4H_2O$	10022-68-1 308.482	col orth cry; hyg	59.5 2.45		vs H_2O; s EtOH, ace
402	Cadmium oxalate CdC_2O_4	814-88-0 200.431	wh solid	3.32		i H_2O
403	Cadmium oxalate trihydrate $CdC_2O_4 \cdot 3H_2O$	20712-42-9 254.477	wh amorp powder	340 dec		i H_2O, EtOH; s dil acid
404	Cadmium oxide CdO	1306-19-0 128.410	brn cub cry	8.15	1559 sp a,e	i H_2O; s dil acid
405	Cadmium perchlorate hexahydrate $Cd(ClO_4)_2 \cdot 6H_2O$	10326-28-0 419.403	wh hex cry	2.37		vs H_2O
406	Cadmium phosphide Cd_3P_2	12014-28-7 399.181	gr tetr needles	700 5.96		s dil HCl
407	Cadmium selenate dihydrate $CdSeO_4 \cdot 2H_2O$	10060-09-0 291.40	orth cry	dec 100 3.62		vs H_2O
408	Cadmium selenide $CdSe$	1306-24-7 191.37	wh cub cry	1240 5.81		i H_2O
409	Cadmium sulfate $CdSO_4$	10124-36-4 208.475	col orth cry	1000 4.69	a	vs H_2O; i EtOH
410	Cadmium sulfate monohydrate $CdSO_4 \cdot H_2O$	7790-84-3 226.490	monocl cry	105 3.79		
411	Cadmium sulfate octahydrate $CdSO_4 \cdot 8H_2O$	15244-35-6 352.597	col monocl cry	dec 40 3.08		vs H_2O
412	Cadmium sulfide CdS	1306-23-6 144.477	yel-oran cub cry	1750 4.83	a,f	i H_2O; s acid
413	Cadmium telluride $CdTe$	1306-25-8 240.01	brn-blk cub cry	1042 6.2	a	i H_2O, dil acid
414	Cadmium titanate $CdTiO_3$	12014-14-1 208.276	orth cry	6.5		
415	Cadmium tungstate $CdWO_4$	7790-85-4 360.25	wh monocl cry	8.0		i H_2O, acid; s NH_4OH
416	Calcium Ca	7440-70-2 40.078	silv-wh metal	842 1.54	1484 a,c,e	reac H_2O; i bz
417	Calcium acetate $Ca(C_2H_3O_2)_2$	62-54-4 158.168	wh hyg cry	dec 160 1.50		s H_2O; sl EtOH
418	Calcium acetate monohydrate $Ca(C_2H_3O_2)_2 \cdot H_2O$	5743-26-0 176.183	wh needles or powder	dec ≈150		s H_2O; sl EtOH
419	Calcium aluminate $CaAl_2O_4$	12042-68-1 158.039	wh monocl cry	1605 2.98		reac H_2O
420	Calcium aluminate (β form) $Ca_3Al_2O_6$	12042-78-3 270.193	wh cub cry; refr	1535 3.04		i H_2O
421	Calcium arsenate $Ca_3(AsO_4)_2$	7778-44-1 398.072	wh powder	dec 3.6		sl H_2O; s dil acid
422	Calcium boride CaB_6	12007-99-7 104.944	refrac solid	2235 2.49		
423	Calcium bromide $CaBr_2$	7789-41-5 199.886	rhomb cry; hyg	742 3.38	1815 a,c	vs H_2O; s EtOH, ace
424	Calcium bromide hexahydrate $CaBr_2 \cdot 6H_2O$	13477-20-6 307.977	wh hyg powder	38 dec 2.29		vs H_2O

No.	Name Formula	CAS RN Mol. Wt.	Physical Form	mp/°C den/g cm^{-3}	bp/°C Other Data	Solubility
425	Calcium carbide CaC_2	75-20-7 64.100	gray-blk orth cry	2300 2.22	a	reac H_2O
426	Calcium carbonate (aragonite) $CaCO_3$	471-34-1 100.087	wh orth cry or powder	825 dec 2.83	a,c,f	i H_2O; s dil acid
427	Calcium carbonate (calcite) $CaCO_3$	471-34-1 100.087	wh hex cry or powder	1330 2.71	a,c,f	i H_2O; s dil acid
428	Calcium chlorate $Ca(ClO_3)_2$	10137-74-3 206.979	wh cry	340		vs H_2O
429	Calcium chlorate dihydrate $Ca(ClO_3)_2 \cdot 2H_2O$	10035-05-9 243.010	wh monocl cry; hyg	100 dec 2.711		vs H_2O; s EtOH
430	Calcium chloride $CaCl_2$	10043-52-4 110.983	wh cub cry or powder; hyg	775 2.15	1935.5 a,c	vs H_2O, EtOH
431	Calcium chloride dihydrate $CaCl_2 \cdot 2H_2O$	10035-04-8 147.014	hyg flakes or powder	175 dec 1.85		vs H_2O, EtOH
432	Calcium chloride hexahydrate $CaCl_2 \cdot 6H_2O$	7774-34-7 219.074	wh hex cry; hyg	30 dec 1.71		vs H_2O
433	Calcium chloride monohydrate $CaCl_2 \cdot H_2O$	13477-29-7 128.998	wh hyg cry	260 dec 2.24		s H_2O, EtOH
434	Calcium chromate dihydrate $CaCrO_4 \cdot 2H_2O$	13765-19-0 192.102	yel orth cry	2.50		s H_2O
435	Calcium cyanamide $CaCN_2$	156-62-7 80.102	col hex cry	≈1340 2.29	subl	reac H_2O
436	Calcium cyanide $Ca(CN)_2$	592-01-8 92.113	wh rhomb cry; hyg		a	s H_2O, EtOH
437	Calcium dichromate trihydrate $CaCr_2O_7 \cdot 3H_2O$	14307-33-6* 310.112	red-oran cry	dec 100 2.37		vs H_2O; reac EtOH; i eth, ctc
438	Calcium dihydrogen phosphate monohydrate $Ca(H_2PO_4)_2 \cdot H_2O$	10031-30-8 252.068	col tricl plates	dec ≈100 2.22		sl H_2O; s dil acid
439	Calcium fluoride CaF_2	7789-75-5 78.075	wh cub cry or powder	1418 3.18	2533.4 a,c,f	i H_2O; sl acid
440	Calcium formate $Ca(CHO_2)_2$	544-17-2 130.114	orth cry	300 dec 2.02		s H_2O; i EtOH
441	Calcium hexafluorosilicate dihydrate $CaSiF_6 \cdot 2H_2O$	16925-39-6 218.185	col tetr cry	2.25		i H_2O, ace; reac hot H_2O
442	Calcium hydride CaH_2	7789-78-8 42.094	gray orth cry or powder	1000 1.7	a	reac H_2O, EtOH
443	Calcium hydrogen phosphate $CaHPO_4$	7757-93-9 136.057	wh tricl cry	dec 2.92	f	i H_2O, EtOH
444	Calcium hydrogen phosphate dihydrate $CaHPO_4 \cdot 2H_2O$	7789-77-7 172.088	monocl cry	dec ≈100 2.31		i H_2O, EtOH; s dil acid
445	Calcium hydroxide $Ca(OH)_2$	1305-62-0 74.093	soft hex cry	≈2.2	a	sl H_2O; s acid
446	Calcium hypophosphite $Ca(H_2PO_2)_2$	7789-79-9 170.055	wh monocl cry	dec 300		s H_2O; i EtOH
447	Calcium iodate $Ca(IO_3)_2$	7789-80-2 389.883	wh monocl cry	4.52		sl H_2O; s HNO_3; i EtOH
448	Calcium iodide CaI_2	10102-68-8 293.887	hyg hex cry	783 3.96	a,c	vs H_2O; s MeOH, EtOH, ace; i eth
449	Calcium iodide hexahydrate $CaI_2 \cdot 6H_2O$	71626-98-7 401.978	wh hex needles or powder	42 dec 2.55		vs H_2O, EtOH
450	Calcium molybdate $CaMoO_4$	7789-82-4 200.02	wh tetr cry	965 dec 4.35	f	i H_2O, EtOH; s conc acid
451	Calcium nitrate $Ca(NO_3)_2$	10124-37-5 164.087	wh cub cry; hyg	561 2.5	a	vs H_2O; s EtOH, MeOH, ace
452	Calcium nitrate tetrahydrate $Ca(NO_3)_2 \cdot 4H_2O$	13477-34-4 236.149	wh cry	dec ≈40 1.82		vs H_2O; s EtOH, ace
453	Calcium nitride Ca_3N_2	12013-82-0 148.247	red-brn cub cry	1195 2.67		s H_2O, acid; i EtOH
454	Calcium nitrite $Ca(NO_2)_2$	13780-06-8 132.089	wh-yel hex cry; hyg	2.23		vs H_2O; sl EtOH
455	Calcium oxalate CaC_2O_4	563-72-4 128.098	wh cry powder	2.2	a	i H_2O
456	Calcium oxalate monohydrate $CaC_2O_4 \cdot H_2O$	5794-28-5 146.113	cub cry	dec 200 2.2	f	i H_2O; s dil acid
457	Calcium oxide CaO	1305-78-8 56.077	gray-wh cub cry	2899 3.34	a,c	reac H_2O; s acid
458	Calcium perchlorate $Ca(ClO_4)_2$	13477-36-6 238.978	wh cry	dec 270 2.65		vs H_2O; s EtOH
459	Calcium permanganate $Ca(MnO_4)_2$	10118-76-0 277.949	purp hyg cry	2.4		vs H_2O; reac EtOH
460	Calcium peroxide CaO_2	1305-79-9 72.077	wh-yel tetr cry; hyg	dec ≈200 2.9		sl H_2O; s acid
461	Calcium phosphate $Ca_3(PO_4)_2$	7758-87-4 310.177	wh amorp powder	1670 3.14	a	i H_2O, EtOH; s dil acid
462	Calcium dihydrogen phosphate monohydrate $Ca(H_2PO_4)_2 \cdot H_2O$	7758-23-8* 252.068	col hyg cry	dec ≈100 2.22		s H_2O, acid

No.	Name Formula	CAS RN Mol. Wt.	Physical Form	mp/°C den/g cm^{-3}	bp/°C Other Data	Solubility
463	Calcium phosphide Ca_3P_2	1305-99-3 182.182	red-brn hyg cry	≈1600 2.51		reac H_2O; i EtOH, eth
464	Calcium pyrophosphate $Ca_2P_2O_7$	7790-76-3 254.099	wh powder	1353 3.09		i H_2O; s dil acid
465	Calcium selenate dihydrate $CaSeO_4 \cdot 2H_2O$	7790-74-1 219.07	wh monocl cry	 2.75		s H_2O
466	Calcium selenide CaSe	1305-84-6 119.04	wh-brn cub cry	 3.8		reac H_2O
467	Calcium metasilicate $CaSiO_3$	1344-95-2 116.162	wh monocl cry	 2.92	f	i H_2O
468	Calcium silicide $CaSi_2$	12013-56-8 96.249	gray hex cry	1040 2.50		i cold H_2O; reac hot H_2O; s acid
469	Calcium silicide CaSi	12013-55-7 68.164	orth cry	1324 2.39		
470	Calcium sulfate $CaSO_4$	7778-18-9 136.142	orth cry	1460 2.96	a,c,f	sl H_2O
471	Calcium sulfate dihydrate $CaSO_4 \cdot 2H_2O$	10101-41-4 172.172	monocl cry or powder	dec 150 2.32	f	s H_2O; i os
472	Calcium sulfate hemihydrate $CaSO_4 \cdot 0.5H_2O$	10034-76-1 145.149	wh powder			s H_2O
473	Calcium sulfide CaS	20548-54-3 72.144	wh-yel cub cry; hyg	2525 2.59	a,f	sl H_2O; i EtOH
474	Calcium sulfite dihydrate $CaSO_3 \cdot 2H_2O$	10257-55-3 156.173	wh powder			sl H_2O, EtOH; s acid
475	Calcium telluride CaTe	12013-57-9 167.68	cub cry	 4.87		
476	Calcium tetrahydroaluminate $Ca(AlH_4)_2$	16941-10-9 102.105	gray powder; flam			reac H_2O; s thf; i eth, bz
477	Calcium thiocyanate tetrahydrate $Ca(SCN)_2 \cdot 4H_2O$	2092-16-2 228.307	hygr cry	dec 160		vs H_2O; s EtOH, ace
478	Calcium thiosulfate hexahydrate $CaS_2O_3 \cdot 6H_2O$	10124-41-1 260.300	tricl cry	dec 45 1.87		s H_2O; i EtOH
479	Calcium titanate $CaTiO_3$	12049-50-2 135.943	cub cry	1980 3.98	f	
480	Calcium tungstate $CaWO_4$	7790-75-2 287.92	wh tetr cry	1620 6.06	f	i H_2O; s hot acid
481	Californium Cf	7440-71-3 251	hex or cub metal	900 15.1	a	
482	Carbon (diamond) C	7782-40-3 12.011	col cub cry	4440 (12.4 GPa) 3.513	e,f	i H_2O
483	Carbon (graphite) C	7440-44-0 12.011	soft blk hex cry	4492 lp (10.3 MPa) 2.2	3825 sp a,c,f	i H_2O
484	Carbon (fullerene-C_{60}) C_{60}	99685-96-8 720.660	yel needles or plates	>280		s os
485	Carbon (fullerene-C_{70}) C_{70}	115383-22-7 840.770	red-brn solid	>280		s bz, tol
486	Fullerene fluoride $C_{60}F_{60}$	134929-59-2 1860.564	col plates	287		vs ace; s thf; i chl
487	Carbon monoxide CO	630-08-0 28.010	col gas	-205 1.229 g/L	-191.5 a,b,c,d,e	sl H_2O; s chl, EtOH
488	Carbon dioxide CO_2	124-38-9 44.010	col gas	-56.57 tp 1.931 g/L	-78.4 sp a,b,c,e	s H_2O
489	Carbon diselenide CSe_2	506-80-9 169.93	yel liq	-45.5 2.6626	125.5 a,e	i H_2O; vs ctc, tol
490	Carbon disulfide CS_2	75-15-0 76.143	col or yel liq	-111.5 1.2555	46 a,b,c,d,e	i H_2O; vs EtOH, bz, os
491	Carbon oxyselenide COSe	1603-84-5 106.97	col gas; unstable	-122 4.694 g/L	-21.5 e	reac H_2O
492	Carbon oxysulfide COS	463-58-1 60.076	col gas	-138.8 2.636 g/L	-50 a,b,e	s H_2O, EtOH
493	Carbon suboxide C_3O_2	504-64-3 68.032	col gas	-111.3 2.985 g/L	6.8 e	reac H_2O
494	Carbon sulfide telluride CSTe	10340-06-4 171.68	red-yel liq; unstable			reac H_2O
495	Carbonyl bromide $COBr_2$	593-95-3 187.818	col liq	 2.5	64.5 a	reac H_2O
496	Carbonyl chloride $COCl_2$	75-44-5 98.915	col gas	-127.9 4.340 g/L	8 a,b,c,e	sl H_2O; s bz, tol
497	Carbonyl fluoride COF_2	353-50-4 66.007	col gas	-111.26 2.896 g/L	-84.57 a	reac H_2O
498	Cyanogen C_2N_2	460-19-5 52.035	col gas	-27.9 2.283 g/L	-21.1 a,b,d,e	sl H_2O, eth; s EtOH
499	Cyanogen bromide BrCN	506-68-3 105.922	wh hyg needles	52 2.005	61.5 a	s H_2O, EtOH, eth
500	Cyanogen chloride ClCN	506-77-4 61.471	col gas	-6.55 2.697 g/L	13 a	s H_2O, EtOH, eth

No.	Name Formula	CAS RN Mol. Wt.	Physical Form	mp/°C den/g cm^{-3}	bp/°C Other Data	Solubility
501	Cyanogen fluoride FCN	1495-50-7 45.016	col gas	-82 1.975 g/L	-46 a	
502	Cyanogen iodide ICN	506-78-5 152.922	col needles	146.7 1.84	a	s H_2O, EtOH, eth
503	Cerium Ce	7440-45-1 140.115	silv metal; cub or hex	798 6.770	3443 a,c	s dil acid
504	Cerium boride CeB_6	12008-02-5 204.981	blue refrac solid; hex	2550 4.87		i H_2O, HCl
505	Cerium carbide CeC_2	12012-32-7 164.137	red hex cry	2250 5.47		reac H_2O
506	Cerium nitride CeN	25764-08-3 154.122	refrac cub cry	2557 7.89		
507	Cerium silicide $CeSi_2$	12014-85-6 196.286	tetr cry	1620 5.31		i H_2O
508	Cerium(II) hydride CeH_2	13569-50-1 142.131	cub cry	 5.45		reac H_2O
509	Cerium(II) iodide CeI_2	19139-47-0 393.924	bronze cry	808		
510	Cerium(II) sulfide CeS	12014-82-3 172.181	yel cub cry	2445 5.9	a	
511	Cerium(III) bromide $CeBr_3$	14457-87-5 379.827	wh hex cry; hyg	733	1457	s H_2O
512	Cerium(III) bromide heptahydrate $CeBr_3 \cdot 7H_2O$	14457-87-5* 505.934	col hyg needles	732		s H_2O, EtOH
513	Cerium(III) carbide Ce_2C_3	12115-63-8 316.263	yel-brn cub cry	1505 6.9		
514	Cerium(III) carbonate hydrate $Ce_2(CO_3)_3 \cdot 5H_2O$	72520-94-6 550.334	wh powder			i H_2O; s dil acid
515	Cerium(III) chloride $CeCl_3$	7790-86-5 246.473	wh hex cry	817 3.97	a	s H_2O, EtOH
516	Cerium(III) chloride heptahydrate $CeCl_3 \cdot 7H_2O$	18618-55-8 372.580	yel orth cry; hyg	dec 90		vs H_2O, EtOH
517	Cerium(III) fluoride CeF_3	7758-88-5 197.110	wh hex cry; hyg	1430 6.157		i H_2O
518	Cerium(III) iodide CeI_3	7790-87-6 520.828	yel orth cry; hyg	766		s H_2O
519	Cerium(III) iodide nonahydrate $CeI_3 \cdot 9H_2O$	7790-87-6* 682.966	wh-red cry			vs H_2O; s EtOH
520	Cerium(III) oxide Ce_2O_3	1345-13-7 328.228	yel-grn cub cry	2230 6.2	a	i H_2O; s acid
521	Cerium(III) sulfate octahydrate $Ce_2(SO_4)_3 \cdot 8H_2O$	13454-94-9 712.543	wh orth cry	dec ≈250 2.87		s H_2O
522	Cerium(III) sulfide Ce_2S_3	12014-93-6 376.428	red cub cry	2450 5.02		i H_2O
523	Cerium(IV) fluoride CeF_4	10060-10-3 216.109	wh hyg powder	dec ≈600 4.77		i H_2O
524	Cerium(IV) oxide CeO_2	1306-38-3 172.114	wh-yel powder; cub	2400 7.65	a	i H_2O, dil acid
525	Cerium(IV) sulfate tetrahydrate $Ce(SO_4)_2 \cdot 4H_2O$	10294-42-5 404.304	yel-oran orth cry	dec 180 3.91		reac H_2O
526	Cesium Cs	7440-46-2 132.905	silv-wh metal	28.44 1.93	671 a,c,e	reac H_2O
527	Cesium amide $CsNH_2$	22205-57-8 148.928	wh tetr cry	 3.70	a	
528	Cesium azide CsN_3	22750-57-8 174.925	hyg tetr cry; exp	326 ≈3.5		
529	Cesium bromate $CsBrO_3$	13454-75-6 260.807	col hex cry	 4.11		sl H_2O
530	Cesium bromide CsBr	7787-69-1 212.809	wh cub cry; hyg	636 4.43	≈1300 a,e	vs H_2O; s EtOH; i ace
531	Cesium carbonate Cs_2CO_3	534-17-8 325.820	wh monocl cry; hyg	792 4.24	a	s H_2O, EtOH, eth
532	Cesium chlorate $CsClO_3$	13763-67-2 216.356	col hex cry	 3.57		sl H_2O
533	Cesium chloride CsCl	7647-17-8 168.358	wh cub cry; hyg	645 3.988	1297 a,c,e	vs H_2O; s EtOH
534	Cesium cyanide CsCN	21159-32-0 158.923	wh cub cry; hyg	350 3.34		vs H_2O
535	Cesium fluoride CsF	13400-13-0 151.903	wh cub cry; hyg	703 4.64	a,c,e	vs H_2O; s MeOH; i diox, py
536	Cesium hydride CsH	58724-12-2 133.913	wh cub cry; flam	dec ≈170 3.42	a	reac H_2O
537	Cesium hydrogen fluoride $CsHF_2$	12280-52-3 171.910	tetr cry	170 3.86	a	
538	Cesium hydroxide CsOH	21351-79-1 149.912	wh-yel hyg cry	272 3.68	a	vs H_2O; s EtOH
539	Cesium iodide CsI	7789-17-5 259.809	col cub cry; hyg	621 4.51	≈1280 a,e	vs H_2O; s EtOH, MeOH, ace

No.	Name Formula	CAS RN Mol. Wt.	Physical Form	mp/°C den/g cm^{-3}	bp/°C Other Data	Solubility
540	Cesium metaborate $CsBO_2$	92141-86-1 175.715	cub cry	732 ≈3.7	a	
541	Cesium nitrate $CsNO_3$	7789-18-6 194.910	wh hex or cub cry	414 3.66	a	s H_2O, ace; sl EtOH
542	Cesium oxide Cs_2O	20281-00-9 281.810	yel-oran hex cry	490 4.65	a	vs H_2O
543	Cesium perchlorate $CsClO_4$	13454-84-7 232.356	wh orth cry; hyg	250 3.327	a	sl H_2O
544	Cesium superoxide CsO_2	12018-61-0 164.904	yel tetr cry	432 3.77	a	reac H_2O
545	Cesium sulfate Cs_2SO_4	10294-54-9 361.875	wh orth cry or hex prisms; hyg	1005 4.24	a,c	vs H_2O; i EtOH, ace, py
546	Chlorine Cl_2	7782-50-5 70.905	grn-yel gas	-101.5 3.111 g/L	-34.04 a,b,c,d,e	sl H_2O
547	Hypochlorous acid HOCl	7790-92-3 52.460	grn-yel; stable only in aq soln		a	s H_2O
548	Perchloric acid $HClO_4$	7601-90-3 100.459	col hyg liq	-112 1.77	dec ≈90 a	s H_2O
549	Chlorine oxide Cl_2O	7791-21-1 86.904	yel-brn gas	-120.6 3.813 g/L	2.2 a,d,e	vs H_2O
550	Chlorine dioxide ClO_2	10049-04-4 67.452	oran-grn gas	-59 2.960 g/L	11 a,d,e	sl H_2O
551	Chlorine trioxide Cl_2O_3	17496-59-2 118.903	dark brn solid	exp <25		
552	Chlorine hexoxide Cl_2O_6	12442-63-6 166.901	red liq	3.5	≈200 e	reac H_2O
553	Chlorine heptoxide Cl_2O_7	10294-48-1 182.901	col oily liq; exp	-91.5 1.9	82 e	reac H_2O
554	Chlorine fluoride ClF	7790-89-8 54.451	col gas	-155.6 2.389 g/L	-101.1 a,d,e	reac H_2O
555	Chlorine trifluoride ClF_3	7790-91-2 92.448	gas	-76.34 4.057 g/L	11.75 a,d,e	reac H_2O
556	Chlorine pentafluoride ClF_5	13637-63-3 130.445	col gas	-103 5.724 g/L	-13.1 b	
557	Chlorine trioxide fluoride (perchloryl fluoride) ClO_3F	7616-94-6 102.449	col gas	-147 4.495 g/L	-46.75 a,b,d	
558	Chlorine perchlorate $ClOClO_3$	27218-16-2 134.903	unstable liq	-117 1.81(0°C)	dec ≈25	
559	Chromium Cr	7440-47-3 51.996	blue-wh metal; cub	1907 7.15	2671 a,c,e	reac dil acid
560	Chromium acetylacetonate $Cr(CH_3COCHCOCH_3)_3$	21679-31-2 349.324	red monocl cry	208 1.34	345	i H_2O; s bz
561	Chromium antimonide CrSb	12053-12-2 173.756	hex cry	1110 7.11		
562	Chromium arsenide Cr_2As	12254-85-2 178.914	tetr cry	7.04		
563	Chromium boride CrB	12006-79-0 62.807	refrac orth cry	2100 6.1		
564	Chromium boride CrB_2	12007-16-8 73.618	refrac solid; hex	2200 5.22		
565	Chromium boride Cr_5B_3	12007-38-4 292.414	tetr cry	1900 6.10		
566	Chromium carbide Cr_3C_2	12012-35-0 180.010	gray orth cry	1895 6.68		
567	Chromium carbonyl $Cr(CO)_6$	13007-92-6 220.058	col orth cry	dec 130 1.77	subl e	i H_2O, EtOH; s eth, chl
568	Chromium nitride Cr_2N	12053-27-9 117.999	hex cry	1650 6.8		
569	Chromium nitride CrN	24094-93-7 66.003	gray cub cry	dec 1080 5.9		
570	Chromium phosphide CrP	26342-61-0 82.970	orth cry	5.25		
571	Chromium selenide CrSe	12053-13-3 130.96	hex cry	≈1500 6.1		
572	Chromium silicide Cr_3Si	12018-36-9 184.074	cub cry	1770 6.4		
573	Chromium silicide $CrSi_2$	12018-09-6 108.167	gray hex cry	1490 4.91		
574	Chromium(II) acetate monohydrate $Cr(C_2H_3O_2)_2 \cdot H_2O$	628-52-4* 188.101	red monocl cry	1.79		sl H_2O
575	Chromium(II) bromide $CrBr_2$	10049-25-9 211.804	wh monocl cry; aq soln blue	842 4.236	a	s H_2O, EtOH
576	Chromium(II) chloride $CrCl_2$	10049-05-5 122.901	hyg needles; aq soln blue	814 2.88	1300 a,c,d	s H_2O
577	Chromium(II) chloride tetrahydrate $Cr(H_2O)_4Cl_2 \cdot 4H_2O$	13931-94-7 267.023	blue hyg cry	dec 51		s H_2O

No.	Name Formula	CAS RN Mol. Wt.	Physical Form	mp/°C den/g cm^{-3}	bp/°C Other Data	Solubility
578	Chromium(II) fluoride CrF_2	10049-10-2 89.993	blue-grn monocl cry	894 3.79	a	sl H_2O; i EtOH
579	Chromium(II) iodide CrI_2	13478-28-9 305.805	red-brn cry; hyg	868 5.1	a	
580	Chromium(II) oxalate monohydrate $CrC_2O_4 \cdot H_2O$	814-90-4* 158.031	yel-grn powder	2.468		sl H_2O
581	Chromium(II) sulfate pentahydrate $CrSO_4 \cdot 5H_2O$	13825-86-0 238.136	blue cry			s H_2O, dil acid; sl EtOH; i ace
582	Chromium(II,III) oxide Cr_3O_4	12018-34-7 219.986	cub cry	6.1	a	
583	Chromium(III) acetate hexahydrate $Cr(C_2H_3O_2)_3 \cdot 6H_2O$	1066-30-4* 285.226	blue needles			s H_2O
584	Chromium(III) bromide $CrBr_3$	10031-25-1 291.708	dark grn hex cry	1130 4.68		s hot H_2O
585	Chromium(III) bromide hexahydrate $Cr(H_2O)_6Br_3$	10031-25-1* 399.799	viol hyg cry			s H_2O; i EtOH, eth
586	Chromium(III) bromide hexahydrate $CrBr_3(H_2O)_4 \cdot 2H_2O$	18721-05-6 399.799	grn hyg cry			s H_2O, EtOH
587	Chromium(III) chloride $CrCl_3$	10025-73-7 158.354	purp hex plates	1152 2.87	dec 1300 a	sl H_2O
588	Chromium(III) chloride hexahydrate $[CrCl_2(H_2O)_4]Cl \cdot 2H_2O$	10060-12-5 266.445	grn monocl cry; hyg			s H_2O, EtOH; sl ace; i eth
589	Chromium(III) fluoride CrF_3	7788-97-8 108.991	grn needles	1400 3.8	a	i H_2O; EtOH
590	Chromium(III) fluoride trihydrate $CrF_3 \cdot 3H_2O$	16671-27-5 163.037	grn hex cry	2.2		sl H_2O
591	Chromium(III) hydroxide trihydrate $Cr(OH)_3 \cdot 3H_2O$	1308-14-1 157.063	blue-grn powder			i H_2O; s acid
592	Chromium(III) iodide CrI_3	13569-75-0 432.709	dark grn hex cry	dec 500 5.32	a	sl H_2O
593	Chromium(III) nitrate $Cr(NO_3)_3$	13548-38-4 238.011	grn hyg powder	dec >60		vs H_2O
594	Chromium(III) nitrate nonahydrate $Cr(NO_3)_3 \cdot 9H_2O$	7789-02-8 400.148	grn-blk monocl cry	66.3 1.80	dec >100	vs H_2O
595	Chromium(III) oxide Cr_2O_3	1308-38-9 151.990	grn hex cry	2330 5.22	≈3000 a,c	i H_2O, EtOH; sl acid, alk
596	Chromium(III) phosphate $CrPO_4$	7789-04-0 146.967	blue orth cry	>1800 4.6		i H_2O, acid, aqua regia
597	Chromium(III) phosphate hemiheptahydrate $CrPO_4 \cdot 3.5H_2O$	84359-31-9 210.021	blue-grn powder	2.15		i H_2O; s acid
598	Chromium(III) phosphate hexahydrate $CrPO_4 \cdot 6H_2O$	84359-31-9 255.059	viol cry	dec >500 2.121		i H_2O; s acid, alk
599	Chromium(III) potassium sulfate dodecahydrate $CrK(SO_4)_2 \cdot 12H_2O$	7788-99-0 499.405	viol-blk cub cry	89 dec 1.83		s H_2O; i EtOH
600	Chromium(III) sulfate $Cr_2(SO_4)_3$	10101-53-8 392.183	red-brn hex cry	3.1		i H_2O, acid
601	Chromium(III) sulfide Cr_2S_3	12018-22-3 200.190	brn-blk hex cry	3.8		
602	Chromium(III) telluride Cr_2Te_3	12053-39-3 486.79	hex cry	≈1300 7.0		
603	Chromium(IV) chloride $CrCl_4$	15597-88-3 193.807	gas, stable at high temp	8.504 g/L	dec >600	
604	Chromium(IV) fluoride CrF_4	10049-11-3 127.990	grn cry	277		
605	Chromium(IV) oxide CrO_2	12018-01-8 83.995	brn-blk tetr powder	dec ~400 4.89	a	i H_2O; s acid
606	Chromium(V) fluoride CrF_5	14884-42-5 146.988	red orth cry	34	117	
607	Chromium(VI) fluoride CrF_6	13843-28-2 165.986	yel solid; stable at low temp	dec -100		
608	Chromium(VI) oxide CrO_3	1333-82-0 99.994	red orth cry	197 2.7	dec ≈250	vs H_2O
609	Chromyl chloride CrO_2Cl_2	14977-61-8 154.900	red liq	-96.5 1.91	117 a,d,e	reac H_2O; s ctc, chl, bz
610	Cobalt Co	7440-48-4 58.933	gray metal; hex or cub	1495 8.86	2927 a,c	s dil acid
611	Cobalt antimonide CoSb	12052-42-5 180.69	hex cry	1202 8.8		
612	Cobalt arsenic sulfide CoAsS	12254-82-9 165.92	silv-wh solid	≈6.1	f	
613	Cobalt arsenide CoAs	27016-73-5 133.855	orth cry	1180 8.22		
614	Cobalt arsenide $CoAs_2$	12044-42-7 208.776	monocl cry	7.2		
615	Cobalt arsenide $CoAs_3$	12256-04-1 283.698	cub cry	942 6.84		

No.	Name / Formula	CAS RN / Mol. Wt.	Physical Form	mp/°C / den/g cm⁻³	bp/°C / Other Data	Solubility
616	Cobalt boride Co_2B	12045-01-1 128.677	refrac solid	1280 8.1		
617	Cobalt boride CoB	12006-77-8 69.744	refrac solid	1460 7.25		reac H_2O, HNO_3
618	Cobalt carbonyl $Co_2(CO)_8$	10210-68-1 341.949	oran cry	51 dec 1.78		i H_2O; s EtOH, eth, CS_2
619	Cobalt phosphide Co_2P	12134-02-0 148.840	gray needles	1386 6.4		i H_2O; s HNO_3
620	Cobalt silicide $CoSi_2$	12017-12-8 115.10	gray cub cry	1326 4.9		s hot HCl
621	Cobalt disulfide CoS_2	12013-10-4 123.07	cub cry	4.3		
622	Cobalt dodecacarbonyl $Co_4(CO)_{12}$	17786-31-1 571.858	blk cry	dec 60 2.09		
623	Cobalt(II) acetate $Co(C_2H_3O_2)_2$	71-48-7 177.023	pink cry			vs H_2O; s EtOH
624	Cobalt(II) acetate tetrahydrate $Co(C_2H_3O_2)_2 \cdot 4H_2O$	6147-53-1 249.083	red monocl cry	1.705		s H_2O, EtOH, dil acid
625	Cobalt(II) aluminate $CoAl_2O_4$	13820-62-7 176.894	blue cub cry	4.37		i H_2O
626	Cobalt(II) arsenate octahydrate $Co_3(AsO_4)_2 \cdot 8H_2O$	24719-19-5 598.760	red monocl needles	dec 1000 3.0		i H_2O; s dil acid
627	Cobalt(II) bromate hexahydrate $Co(BrO_3)_2 \cdot 6H_2O$	13476-01-2 422.829	viol cry	≈2.5		vs H_2O
628	Cobalt(II) bromide $CoBr_2$	7789-43-7 218.741	grn hex cry; hyg	678 4.91	a	vs H_2O; s MeOH, EtOH, ace
629	Cobalt(II) bromide hexahydrate $CoBr_2 \cdot 6H_2O$	13762-12-4 326.832	red hyg cry	47 dec 2.46	dec 100	s H_2O
630	Cobalt(II) carbonate $CoCO_3$	513-79-1 118.942	pink rhomb cry	4.2		i H_2O, EtOH
631	Cobalt(II) chloride $CoCl_2$	7646-79-9 129.838	blue hyg leaflets	740 3.36	1049 a,c,e	s H_2O, EtOH, eth, ace, py
632	Cobalt(II) chloride dihydrate $CoCl_2 \cdot 2H_2O$	16544-92-6 165.869	viol-blue cry	2.477		
633	Cobalt(II) chloride hexahydrate $CoCl_2 \cdot 6H_2O$	7791-13-1 237.929	pink-red monocl cry	87 dec 1.924		vs H_2O; s EtOH, ace, eth
634	Cobalt(II) chromate $CoCrO_4$	24613-38-5 174.927	yel-brn orth cry	≈4.0		i H_2O; s acid
635	Cobalt(II) chromite $CoCr_2O_4$	13455-25-9 226.923	blue-grn cub cry	5.14		i H_2O, conc acid
636	Cobalt(II) cyanide $Co(CN)_2$	542-84-7 110.960	blue hyg cry	1.872		i H_2O
637	Cobalt(II) cyanide dihydrate $Co(CN)_2 \cdot 2H_2O$	20427-11-6 146.999	pink-brn needles			i H_2O, acid
638	Cobalt(II) ferricyanide $Co_3[Fe(CN)_6]_2$	14049-81-1 600.703	red needles			i H_2O, HCl; s NH_4OH
639	Cobalt(II) fluoride CoF_2	10026-17-2 96.930	red tetr cry	1127 4.46	≈1400 a,c	sl H_2O; s acid
640	Cobalt(II) fluoride tetrahydrate $CoF_2 \cdot 4H_2O$	13817-37-3 168.992	red orth cry	dec 2.22		s H_2O
641	Cobalt(II) formate dihydrate $Co(CHO_2)_2 \cdot 2H_2O$	6424-20-0 184.999	red cry powder	dec 140 2.13		s H_2O; i EtOH
642	Cobalt(II) hexafluorosilicate hexahydrate $CoSiF_6 \cdot 6H_2O$	12021-68-0 309.100	pale red cry	2.087		vs H_2O
643	Cobalt(II) hydroxide $Co(OH)_2$	21041-93-0 92.948	blue-grn cry	dec ≈160 3.60	a	sl H_2O; s acid
644	Cobalt(II) iodate $Co(IO_3)_2$	13455-28-2 408.738	blk-viol needles	dec 200 5.09		sl H_2O
645	Cobalt(II) iodide CoI_2	15238-00-3 312.742	blk hex cry; hyg	520 5.60	a	vs H_2O
646	Cobalt(II) iodide hexahydrate $CoI_2 \cdot 6H_2O$	15238-00-3* 420.833	red hex prisms	dec 130 2.90		s H_2O, EtOH, eth, ace
647	Cobalt(II) titanate $CoTiO_3$	12017-01-5 154.798	grn rhomb cry	5.0		
648	Cobalt(II) molybdate $CoMoO_4$	13762-14-6 218.87	blk monocl cry	1040 4.7		
649	Cobalt(II) nitrate $Co(NO_3)_2$	10141-05-6 182.942	pale red powder	dec 100 2.49	a	vs H_2O
650	Cobalt(II) nitrate hexahydrate $Co(NO_3)_2 \cdot 6H_2O$	10026-22-9 291.034	red monocl cry; hyg	≈55 1.88		vs H_2O; s EtOH
651	Cobalt(II) oxalate CoC_2O_4	814-89-1 146.953	pink powder	250 dec 3.02		i H_2O; s acid, NH_4OH
652	Cobalt(II) oxalate dihydrate $CoC_2O_4 \cdot 2H_2O$	814-89-1* 166.984	pink needles	dec		i H_2O; sl acid; s NH_4OH
653	Cobalt(II) oxide CoO	1307-96-6 74.932	gray cub cry	1830 6.44	a	i H_2O; s acid
654	Cobalt(II) perchlorate $Co(ClO_4)_2$	13455-31-7 257.833	red needles	3.33		vs H_2O; i EtOH, ace

No.	Name / Formula	CAS RN / Mol. Wt.	Physical Form	mp/°C / den/g cm^{-3}	bp/°C / Other Data	Solubility
655	Cobalt(II) phosphate octahydrate / Co$_3$(PO$_4$)$_2$·8H$_2$O	10294-50-5 / 510.865	pink amorp powder	/ 2.77		i H$_2$O; s acid
656	Cobalt(II) potassium sulfate hexahydrate / CoK$_2$(SO$_4$)$_2$·6H$_2$O	10026-20-7 / 437.349	red monocl cry	dec 75 / 2.22		vs H$_2$O
657	Cobalt(II) selenate pentahydrate / CoSeO$_4$·5H$_2$O	14590-19-3 / 291.97	red tricl cry	dec / 2.51		vs H$_2$O
658	Cobalt(II) selenide / CoSe	1307-99-9 / 137.89	yel hex cry	1055 / 7.65		i H$_2$O, alk; s aqua regia
659	Cobalt(II) selenite dihydrate / CoSeO$_3$·2H$_2$O	19034-13-0 / 221.92	blue-red powder			i H$_2$O
660	Cobalt(II) orthosilicate / Co$_2$SiO$_4$	12017-08-2 / 209.950	red-viol orth cry	1345 / 4.63		i H$_2$O; s dil HCl
661	Cobalt(II) stannate / Co$_2$SnO$_4$	12139-93-4 / 300.574	grn-blue cub cry	/ 6.30		i H$_2$O; s alk
662	Cobalt(II) sulfate / CoSO$_4$	10124-43-3 / 154.997	red orth cry	>700 / 3.71	a	s H$_2$O
663	Cobalt(II) sulfate heptahydrate / CoSO$_4$·7H$_2$O	10026-24-1 / 281.103	pink monocl cry	41 dec / 2.03		vs H$_2$O; sl EtOH, MeOH
664	Cobalt(II) sulfate monohydrate / CoSO$_4$·H$_2$O	13455-34-0 / 173.012	red monocl cry	/ 3.08		sl H$_2$O
665	Cobalt(II) sulfide / CoS	1317-42-6 / 91.00	blk amorp powder	1182 / 5.45	a	i H$_2$O; s acid
666	Cobalt(II) telluride / CoTe	12017-13-9 / 186.53	hex cry	≈8.8		
667	Cobalt(II) thiocyanate trihydrate / Co(SCN)$_2$·3H$_2$O	97126-35-7 / 229.146	viol rhomb cry			s H$_2$O, EtOH, eth, ace
668	Cobalt(II) tungstate / CoWO$_4$	12640-47-0 / 306.77	blue monocl cry	≈7.8		i H$_2$O; s hot conc acid
669	Cobalt(II,III) oxide / Co$_3$O$_4$	1308-06-1 / 240.798	blk cub cry	dec 900 / 6.11	a	i H$_2$O; s acid, alk
670	Cobalt(III) acetate / Co(C$_2$H$_3$O$_2$)$_3$	917-69-1 / 236.066	grn hyg cry	dec 100		s H$_2$O, EtOH
671	Cobalt(III) ammonium tetranitrodiammine / NH$_4$[Co(NH$_3$)$_2$(NO$_2$)$_4$]	13600-89-0 / 295.054	red-brn orth cry	/ 1.97		s H$_2$O
672	Cobalt(III) fluoride / CoF$_3$	10026-18-3 / 115.928	brn hex cry	927 / 3.88		reac H$_2$O
673	Cobalt(III) hexammine chloride / Co(NH$_3$)$_6$Cl$_3$	10534-89-1 / 267.474	red monocl cry	/ 1.71		s H$_2$O; i EtOH
674	Cobalt(III) hydroxide / Co(OH)$_3$	1307-86-4 / 109.955	brn powder	dec / ≈4		i H$_2$O; s acid
675	Cobalt(III) nitrate / Co(NO$_3$)$_3$	15520-84-0 / 244.948	grn cub cry; hyg	≈3.0		s H$_2$O; reac os
676	Cobalt(III) oxide / Co$_2$O$_3$	1308-04-9 / 165.864	gray-blk powder	dec 895 / 5.18		i H$_2$O; s conc acid
677	Cobalt(III) oxide monohydrate / Co$_2$O$_3$·H$_2$O	12016-80-7 / 183.880	brn-blk hex cry	dec 150		i H$_2$O; s acid
678	Cobalt(III) potassium nitrite sesquihydrate / CoK$_3$(NO$_2$)$_6$·1.5H$_2$O	13782-01-9* / 479.284	yel cub cry	/ 2.6		sl H$_2$O; reac acid; i EtOH
679	Cobalt(III) sulfide / Co$_2$S$_3$	1332-71-4 / 214.064	blk cub cry	/ 4.8	a	reac acid
680	Cobalt(III) titanate / Co$_2$TiO$_4$	12017-38-8 / 229.731	grn-blk cub cry	/ 5.1		s conc HCl
681	Copper / Cu	7440-50-8 / 63.546	red metal; cub	1084.62 / 8.96	2562 / a,c,e,f	sl dil acid
682	Copper acetylacetonate / Cu(CH$_3$COCHCOCH$_3$)$_2$	13395-16-9 / 261.765	blue powder	284 dec	subl	sl H$_2$O; s chl
683	Copper nitride / Cu$_3$N	1308-80-1 / 204.645	cub cry	300 dec / 5.84		
684	Copper phosphide / CuP$_2$	12019-11-3 / 125.494	monocl cry	≈900 / 4.20		
685	Copper(I) acetate / CuC$_2$H$_3$O$_2$	598-54-9 / 122.591	col cry	dec	subl	reac H$_2$O
686	Copper(I) acetylide / Cu$_2$C$_2$	1117-94-8 / 151.114	red amorp powder; exp			
687	Copper(I) azide / CuN$_3$	14336-80-2 / 105.566	tetr cry; exp			
688	Copper(I) bromide / CuBr	7787-70-4 / 143.450	wh cub cry; hyg	497 / 4.98	1345 / a,e	sl H$_2$O; i ace
689	Copper(I) chloride / CuCl	7758-89-6 / 98.999	wh cub cry	430 / 4.14	≈1400 / a,c,e,f	sl H$_2$O; i EtOH, ace
690	Copper(I) cyanide / CuCN	544-92-3 / 89.564	wh powder or grn orth cry	474 / 2.9	dec / a	i H$_2$O, EtOH; s KCN soln
691	Copper(I) fluoride / CuF	13478-41-6 / 82.544	cub cry	/ 7.1		
692	Copper(I) hydride / CuH	13517-00-5 / 64.554	red-brn solid	dec 60		

No.	Name Formula	CAS RN Mol. Wt.	Physical Form	mp/°C den/g cm^{-3}	bp/°C Other Data	Solubility
693	Copper(I) iodide CuI	7681-65-4 190.450	wh cub cry	606 5.67	≈1290 a,f	i H$_2$O, dil acid
694	Copper(I) mercury iodide Cu$_2$HgI$_4$	13876-85-2 835.30	red cry powder	trans ≈60 (brn)		i H$_2$O, EtOH
695	Copper(I) oxide Cu$_2$O	1317-39-1 143.091	red-brn cub cry	1235 6.0	dec 1800 a,f	i H$_2$O
696	Copper(I) selenide Cu$_2$Se	20405-64-5 206.05	blue-blk tetr cry	1113 6.84		i H$_2$O; s acid
697	Copper(I) sulfide Cu$_2$S	22205-45-4 159.158	blue-blk orth cry	≈1100 5.6	a,f	i H$_2$O; sl acid
698	Copper(I) sulfite hemihydrate Cu$_2$SO$_3$·0.5H$_2$O	13982-53-1* 216.164	wh-yel hex cry			sl H$_2$O; s acid, alk; i EtOH, eth
699	Copper(I) telluride Cu$_2$Te	12019-52-2 254.69	blue hex cry	1127 4.6		
700	Copper(I) thiocyanate CuSCN	1111-67-7 121.630	wh-yel amorp powder	1084 2.85		i H$_2$O, dil acid, EtOH, ace; s eth
701	Copper(I,II) sulfite dihydrate Cu$_2$SO$_3$·CuSO$_3$·2H$_2$O	13814-81-8 386.797	red prisms or powder			i H$_2$O, EtOH; s HCl
702	Copper(II) acetate Cu(C$_2$H$_3$O$_2$)$_2$	142-71-2 181.636	blue-grn hyg powder			
703	Copper(II) acetate metaarsenite Cu(C$_2$H$_3$O$_2$)$_2$·3Cu(AsO$_2$)$_2$	12002-03-8 1013.796	grn cry powder			i H$_2$O; reac acid
704	Copper(II) acetate monohydrate Cu(C$_2$H$_3$O$_2$)$_2$·H$_2$O	6046-93-1 199.651	grn monocl cry	115 1.88	dec 240	s H$_2$O, EtOH; sl eth
705	Copper(II) acetylide CuC$_2$	12540-13-5 87.568	brn-blk solid; exp	exp 100		
706	Copper(II) arsenate Cu$_3$(AsO$_4$)$_2$	10103-61-4 468.476	blue-grn cry			i H$_2$O, EtOH; s dil acid
707	Copper(II) arsenite CuHAsO$_3$	10290-12-7 187.474	yel-grn powder			i H$_2$O, EtOH; s acid
708	Copper(II) azide Cu(N$_3$)$_2$	14215-30-6 147.586	brn orth cry; exp	≈2.6		
709	Copper(II) basic acetate Cu(C$_2$H$_3$O$_2$)$_2$·CuO·6H$_2$O	52503-64-7 369.272	blue-grn cry or powder			sl H$_2$O, EtOH; s dil acid, NH$_4$OH
710	Copper(II) borate Cu(BO$_2$)$_2$	39290-85-2 149.166	blue-grn powder	3.859		i H$_2$O; s acid
711	Copper(II) bromide CuBr$_2$	7789-45-9 223.354	blk monocl cry; hyg	498 4.710	900 a	vs H$_2$O; s EtOH, ace; i bz, eth
712	Copper(II) butanoate monohydrate Cu(C$_4$H$_7$O$_2$)$_2$·H$_2$O	540-16-9 255.758	grn monocl plates			s H$_2$O, diox, bz; sl EtOH
713	Copper(II) carbonate hydroxide CuCO$_3$·Cu(OH)$_2$	12069-69-1 221.116	grn monocl cry	200 dec 4.0		i H$_2$O, EtOH; s dil acid
714	Copper(II) chlorate hexahydrate Cu(ClO$_3$)$_2$·6H$_2$O	14721-21-2 338.539	blue-grn hyg cry	65	dec 100	vs H$_2$O, EtOH
715	Copper(II) chloride CuCl$_2$	7447-39-4 134.451	yel-brn monocl cry; hyg	630 dec 3.4	a,c	s H$_2$O, EtOH, ace
716	Copper(II) chloride dihydrate CuCl$_2$·2H$_2$O	10125-13-0 170.482	grn-blue orth cry; hyg	100 dec 2.51		vs H$_2$O, EtOH, MeOH; s ace; i eth
717	Copper(II) chromate CuCrO$_4$	13548-42-0 179.540	red-brn cry			i H$_2$O; s EtOH
718	Copper(II) chromite CuCr$_2$O$_4$	12018-10-9 231.536	gray-blk tetr cry	5.4		i H$_2$O, dil acid
719	Copper(II) citrate hemipentahydrate Cu$_2$C$_6$H$_4$O$_7$·2.5H$_2$O	10402-15-0 360.223	blue-grn cry	dec 100		sl H$_2$O; s dil acid
720	Copper(II) cyanide Cu(CN)$_2$	14763-77-0 115.581	grn powder			i H$_2$O; s acid, alk
721	Copper(II) dichromate dihydrate CuCr$_2$O$_7$·2H$_2$O	13675-47-3 315.565	red-brn tricl cry	2.286		vs H$_2$O
722	Copper(II) ferrocyanide Cu$_2$Fe(CN)$_6$	13601-13-3 339.043	red-br cub cry or powder	2.2		i H$_2$O, acid, os
723	Copper(II) ferrous sulfide CuFeS$_2$	1308-56-1 183.523	yel tetr cry	950 4.2	f	i H$_2$O, HCl; s HNO$_3$
724	Copper(II) fluoride CuF$_2$	7789-19-7 101.543	wh monocl cry	836 4.23	1676 a,c	sl H$_2$O
725	Copper(II) fluoride dihydrate CuF$_2$·2H$_2$O	13454-88-1 137.574	blue monocl cry	dec 130 2.934		sl H$_2$O
726	Copper(II) formate Cu(CHO$_2$)$_2$	544-19-4 153.582	blue cry			s H$_2$O; i os
727	Copper(II) formate tetrahydrate Cu(CHO$_2$)$_2$·4H$_2$O	5893-61-8 225.642	blue monocl cry			s H$_2$O; sl EtOH; i os
728	Copper(II) hexafluorosilicate tetrahydrate CuSiF$_6$·4H$_2$O	12062-24-7 277.684	blue monocl cry	dec 2.56		vs H$_2$O; sl EtOH
729	Copper(II) hydroxide Cu(OH)$_2$	20427-59-2 97.561	blue-grn powder	3.37	a	i H$_2$O; s acid, conc alk
730	Copper(II) molybdate CuMoO$_4$	13767-34-5 223.48	grn cry	820 dec 3.4		i H$_2$O
731	Copper(II) nitrate Cu(NO$_3$)$_2$	3251-23-8 187.555	blue-grn orth cry; hyg	255	subl a	s H$_2$O, diox; reac eth

No.	Name / Formula	CAS RN / Mol. Wt.	Physical Form	mp/°C / den/g cm⁻³	bp/°C / Other Data	Solubility
732	Copper(II) nitrate hexahydrate $Cu(NO_3)_2 \cdot 6H_2O$	13478-38-1 295.647	blue rhomb cry; hyg	2.07		vs H_2O; s EtOH
733	Copper(II) nitrate trihydrate $Cu(NO_3)_2 \cdot 3H_2O$	10031-43-3 241.602	blue rhomb cry	114 2.32	170 dec	vs H_2O, EtOH
734	Copper(II) oleate $Cu(C_{18}H_{33}O_2)_2$	1120-44-1 626.464	blue-grn solid			i H_2O; sl EtOH; s eth
735	Copper(II) oxalate CuC_2O_4	814-91-5 151.566	blue-wh powder	dec 310		i H_2O, EtOH, eth; s NH_4OH
736	Copper(II) oxide CuO	1317-38-0 79.545	blk powder or monocl cry	1446 6.31	a,c	i H_2O, EtOH; s dil acid
737	Copper(II) perchlorate $Cu(ClO_4)_2$	13770-18-8 262.446	grn hyg cry	dec 130		s H_2O, eth, diox, i bz, ctc
738	Copper(II) perchlorate hexahydrate $Cu(ClO_4)_2 \cdot 6H_2O$	10294-46-9 370.538	blue monocl cry; hyg	82 2.22	dec 120	vs H_2O, EtOH, HOAc, ace; sl eth
739	Copper(II) phosphate trihydrate $Cu_3(PO_4)_2 \cdot 3H_2O$	10031-48-8 434.627	blue-grn orth cry			i H_2O; s acid, NH_4OH
740	Copper(II) selenate pentahydrate $CuSeO_4 \cdot 5H_2O$	10031-45-5 296.58	blue tricl cry	dec 80 2.56		s H_2O, acid, NH_4OH; sl ace; i EtOH
741	Copper(II) selenide $CuSe$	1317-41-5 142.51	blue-blk needles or plates	550 dec 5.99	a	reac acid
742	Copper(II) selenite dihydrate $CuSeO_3 \cdot 2H_2O$	15168-20-4 226.54	blue orth cry	3.31		i H_2O; s acid, NH_4OH
743	Copper(II) stearate $Cu(C_{18}H_{35}O_2)_2$	660-60-6 630.496	blue-grn amorp powder	≈250		i H_2O, EtOH, eth; s py
744	Copper(II) sulfate $CuSO_4$	7758-98-7 159.610	wh-grn amorp powder or rhomb cry	560 dec 3.60	a	s H_2O, i EtOH
745	Copper(II) sulfate pentahydrate $CuSO_4 \cdot 5H_2O$	7758-99-8 249.686	blue tricl cry	dec 110 2.286	f	vs H_2O; s MeOH; sl EtOH
746	Copper(II) sulfate, basic $Cu_3(OH)_4SO_4$	1332-14-5 200.126	grn rhomb cry	3.88		i H_2O
747	Copper(II) sulfide CuS	1317-40-4 95.612	blk hex cry	trans 507 4.76	a,f	i H_2O, EtOH, dil acid, alk
748	Copper(II) tartrate trihydrate $CuC_4H_4O_6 \cdot 3H_2O$	815-82-7 265.664	blue-grn powder			sl H_2O; s acid, alk
749	Copper(II) telluride $CuTe$	12019-23-7 191.15	yel orth cry	trans ≈400 7.09		
750	Copper(II) tungstate $CuWO_4$	13587-35-4 311.38	yel-brn powder	7.5	a	
751	Copper(II) tungstate dihydrate $CuWO_4 \cdot 2H_2O$	13587-35-4* 347.41	grn powder			i H_2O; sl HOAc; reac conc acid
752	Curium Cm	7440-51-9 247	silv metal; hex or cub	1345 13.51	≈3100 a	
753	Dysprosium Dy	7429-91-6 162.50	silv metal; hex	1412 8.55	2567 a,c	s dil acid
754	Dysprosium boride DyB_4	12310-43-9 205.74	tetr cry	2500 6.98		
755	Dysprosium nitride DyN	12019-88-4 176.51	cub cry	9.93		
756	Dysprosium silicide $DySi_2$	12133-07-2 218.67	orth cry	5.2		
757	Dysprosium(II) chloride $DyCl_2$	13767-31-2 233.41	blk cry	721 dec		reac H_2O
758	Dysprosium(II) iodide DyI_2	36377-94-3 416.31	purp cry	659		reac H_2O
759	Dysprosium(III) bromide $DyBr_3$	14456-48-5 402.21	wh hyg cry	879		s H_2O
760	Dysprosium(III) chloride $DyCl_3$	10025-74-8 268.86	yel cry	680 3.67	a	s H_2O
761	Dysprosium(III) fluoride DyF_3	13569-80-7 219.50	grn cry	1154		
762	Dysprosium(III) hydride DyH_3	13537-09-2 165.52	hex cry	7.1		
763	Dysprosium(III) iodide DyI_3	15474-63-2 543.21	grn cry	978		
764	Dysprosium(III) oxide Dy_2O_3	1308-87-8 373.00	wh cub cry	2408 7.81	a	s acid
765	Dysprosium(III) sulfide Dy_2S_3	12133-10-7 421.20	red-brn monocl cry	6.08		
766	Einsteinium Es	7429-92-7 252	metal; cub	860	a	
767	Erbium Er	7440-52-0 167.26	silv metal; hex	1529 9.07	2868 a,c	i H_2O; s acid
768	Erbium boride ErB_4	12310-44-0 210.50	tetr cry	2450 7.0		
769	Erbium bromide $ErBr_3$	13536-73-7 406.97	viol hyg cry	923		s H_2O
770	Erbium chloride $ErCl_3$	10138-41-7 273.62	viol monocl cry; hyg	776 4.1	a	s H_2O

No.	Name / Formula	CAS RN / Mol. Wt.	Physical Form	mp/°C / den/g cm^{-3}	bp/°C / Other Data	Solubility
771	Erbium chloride hexahydrate / $ErCl_3 \cdot 6H_2O$	10025-75-9 / 381.71	pink hyg cry	dec		s H_2O; sl EtOH
772	Erbium fluoride / ErF_3	13760-83-3 / 224.26	pink orth cry	1147 / 7.8	a	i H_2O
773	Erbium hydride / ErH_3	13550-53-3 / 170.28	hex cry	≈7.6		
774	Erbium iodide / ErI_3	13813-42-8 / 547.97	viol hex cry; hyg	1014 / ≈5.5		s H_2O
775	Erbium nitride / ErN	12020-21-2 / 181.27	cub cry	10.6		
776	Erbium oxide / Er_2O_3	12061-16-4 / 382.52	pink powder	2418 / 8.64	a	i H_2O; s acid
777	Erbium silicide / $ErSi_2$	12020-28-9 / 223.43	orth cry	7.26		
778	Erbium sulfate / $Er_2(SO_4)_3$	13478-49-4 / 622.71	hyg powder	dec / 3.68		s H_2O
779	Erbium sulfate octahydrate / $Er_2(SO_4)_3 \cdot 8H_2O$	10031-52-4 / 766.83	pink monocl cry	dec / 3.20		s H_2O
780	Erbium sulfide / Er_2S_3	12159-66-9 / 430.72	red-brn monocl cry	1730 / 6.07		
781	Erbium telluride / Er_2Te_3	12020-39-2 / 717.32	orth cry	1213 / 7.11		
782	Europium / Eu	7440-53-1 / 151.965	soft silv metal; cub	822 / 5.24	1529 / a,c	reac H_2O
783	Europium boride / EuB_6	12008-05-8 / 216.831	cub cry	≈2600 / 4.91		
784	Europium nitride / EuN	12020-58-5 / 165.972	cub cry	8.7		
785	Europium silicide / $EuSi_2$	12434-24-1 / 208.136	tetr cry	1500 / 5.46		
786	Europium(II) bromide / $EuBr_2$	13780-48-8 / 311.773	wh cry	683		s H_2O
787	Europium(II) chloride / $EuCl_2$	13769-20-5 / 222.870	wh orth cry	731 / 4.9		s H_2O
788	Europium(II) fluoride / EuF_2	14077-39-5 / 189.962	grn-yel cub cry	≈1380 / 6.5		
789	Europium(II) iodide / EuI_2	22015-35-6 / 405.774	grn cry	580		s H_2O
790	Europium(II) selenide / FuSe	12020-66-5 / 230.93	brn cub cry	6.45		
791	Europium(II) sulfide / EuS	12020-65-4 / 184.031	cub cry	5.7		
792	Europium(II) telluride / EuTe	12020-69-8 / 279.57	blk cub cry	1526 / 6.48		
793	Europium(III) bromide / $EuBr_3$	13759-88-1 / 391.677	gray cry	dec		s H_2O
794	Europium(III) chloride / $EuCl_3$	10025-76-0 / 258.323	grn-yel needles	623 / 4.89	a	
795	Europium(III) chloride hexahydrate / $EuCl_3 \cdot 6H_2O$	13759-92-7 / 366.414	wh-yel hyg cry	850 / 4.89		s H_2O
796	Europium(III) fluoride / EuF_3	13765-25-8 / 208.960	wh hyg cry	1276		i H_2O
797	Europium(III) nitrate hexahydrate / $Eu(NO_3)_3 \cdot 6H_2O$	10031-53-5 / 446.071	wh-pink hyg cry	dec 85		s H_2O
798	Europium(III) oxide / Eu_2O_3	1308-96-9 / 351.928	pink powder	2350 / 7.42	a	i H_2O; s acid
799	Europium(III) sulfate / $Eu_2(SO_4)_3$	13537-15-0 / 592.121	pale pink cry	4.99		sl H_2O
800	Europium(III) sulfate octahydrate / $Eu_2(SO_4)_3 \cdot 8H_2O$	10031-52-4 / 736.243	pink cry	dec 375		sl H_2O
801	Fermium / Fm	7440-72-4 / 257	metal	1527	a	
802	Fluorine / F_2	7782-41-4 / 37.997	pale yel gas	-219.66 / 1.667 g/L	-188.12 / a,b,c,d,e	reac H_2O
803	Fluorine monoxide / F_2O	7783-41-7 / 53.996	col gas	-223.8 / 2.369 g/L	-144.75 / a,b,d,e	sl H_2O
804	Fluorine dioxide / F_2O_2	7783-44-0 / 69.996	gas, stable only at low temp	-154 / 3.071 g/L	-57 / a,d	
805	Fluorine nitrate / FNO_3	7789-26-6 / 81.003	col gas	-175 / 3.554 g/L	-46	reac H_2O, EtOH, eth; s ace
806	Fluorine perchlorate / $FOClO_3$	10049-03-3 / 118.449	col gas; exp	-167.3 / 5.197 g/L	-16	reac H_2O
807	Francium / Fr	7440-73-5 / 223	short-lived alkali metal	27	a	
808	Gadolinium / Gd	7440-54-2 / 157.25	silv metal; hex	1313 / 7.90	3273 / a,c	s dil acid
809	Gadolinium boride / GdB_6	12008-06-0 / 222.12	blk-brn cub cry	2510 / 5.31		

No.	Name Formula	CAS RN Mol. Wt.	Physical Form	mp/°C den/g cm^{-3}	bp/°C Other Data	Solubility
810	Gadolinium nitride GdN	25764-15-2 171.26	cub cry	9.10		
811	Gadolinium silicide GdSi$_2$	12134-75-7 213.42	orth cry	5.9		
812	Gadolinium(II) iodide GdI$_2$	13814-72-7 411.06	bronze cry	831		
813	Gadolinium(II) selenide GdSe	12024-81-6 236.21	cub cry	2170 8.1		
814	Gadolinium(III) bromide GdBr$_3$	13818-75-2 396.96	wh monocl cry; hyg	770 4.56		
815	Gadolinium(III) chloride GdCl$_3$	10138-52-0 263.61	wh monocl cry; hyg	609 4.52	a	s H$_2$O
816	Gadolinium(III) chloride hexahydrate GdCl$_3$•6H$_2$O	19423-81-5 371.70	col hyg cry	2.424		s H$_2$O
817	Gadolinium(III) fluoride GdF$_3$	13765-26-9 214.25	wh cry	1231	a	
818	Gadolinium(III) iodide GdI$_3$	13572-98-0 537.96	yel cry	925		
819	Gadolinium(III) nitrate hexahydrate Gd(NO$_3$)$_3$•6H$_2$O	19598-90-4 451.36	hyg tricl cry	91 dec 2.33		s H$_2$O, EtOH
820	Gadolinium(III) nitrate pentahydrate Gd(NO$_3$)$_3$•5H$_2$O	52788-53-1 433.34	wh cry	92 dec 2.41		i H$_2$O
821	Gadolinium(III) oxide Gd$_2$O$_3$	12064-62-9 362.50	wh hyg powder	2420 7.07	a	i H$_2$O; s acid
822	Gadolinium(III) sulfate octahydrate Gd$_2$(SO$_4$)$_3$•8H$_2$O	13450-87-8 746.81	col monocl cry	dec 400 4.14		s H$_2$O
823	Gadolinium(III) sulfide Gd$_2$S$_3$	12134-77-9 410.70	yel cub cry	6.1		
824	Gadolinium(III) telluride Gd$_2$Te$_3$	12160-99-5 697.30	orth cry	1255 7.7		
825	Gallium Ga	7440-55-3 69.723	silv liq or gray orth cry	29.76 5.91	2204 a,c,d,e	reac alk
826	Gallium antimonide GaSb	12064-03-8 191.483	cub cry	712 5.6137	a,c	
827	Gallium arsenide GaAs	1303-00-0 144.645	gray cub cry	1238 5.3176	a	
828	Gallium nitride GaN	25617-97-4 83.730	gray hex cry	600 6.1	dec >600 a	
829	Gallium phosphide GaP	12063-98-8 100.697	yel cub cry	1457 4.138	a	
830	Gallium suboxide Ga$_2$O	12024-20-3 155.445	brn powder	>660 4.77	dec >800 a	
831	Gallium(II) chloride GaCl$_2$	24597-12-4 140.628	wh orth cry	172.4 2.74	535	
832	Gallium(II) selenide GaSe	12024-11-2 148.68	hex cry	960 5.03		
833	Gallium(II) sulfide GaS	12024-10-1 101.789	hex cry	965 3.86		
834	Gallium(II) telluride GaTe	12024-14-5 197.32	monocl cry	824 5.44		
835	Gallium(III) acetylacetonate Ga(CH$_3$COCHCOCH$_3$)$_3$	14405-43-7 367.051	wh powder	193 1.42	subl	
836	Gallium(III) bromide GaBr$_3$	13450-88-9 309.435	wh orth cry	121.5 3.69	279 a,b,c,d	
837	Gallium(III) chloride GaCl$_3$	13450-90-3 176.081	col needles or gl solid	77.9 2.47	201 a,b,c,d,e	
838	Gallium(III) fluoride GaF$_3$	7783-51-9 126.718	wh powder or col needles	>1000 4.47	a	i H$_2$O
839	Gallium(III) fluoride trihydrate GaF$_3$•3H$_2$O	22886-66-4 180.764	wh cry	dec >140		sl H$_2$O
840	Gallium(III) hydride GaH$_3$	13572-93-5 72.747	visc liq	-15	dec ≈0	
841	Gallium(III) hydroxide Ga(OH)$_3$	12023-99-3 120.745	unstable prec		a	
842	Gallium(III) iodide GaI$_3$	13450-91-4 450.436	monocl cry	212 4.5	340 a,b,c,d	
843	Gallium(III) nitrate Ga(NO$_3$)$_3$	13494-90-1 255.738	wh cry powder			s H$_2$O, EtOH, eth
844	Gallium(III) oxide Ga$_2$O$_3$	12024-21-4 187.444	wh cry	1807 ≈6.0	a	s hot acid
845	Gallium(III) oxide hydroxide GaOOH	20665-52-5 102.730	orth cry	5.23		
846	Gallium(III) selenide Ga$_2$Se$_3$	12024-24-7 376.33	cub cry	937 4.92		
847	Gallium(III) sulfate Ga$_2$(SO$_4$)$_3$	13494-91-2 427.637	hex cry			
848	Gallium(III) sulfate octadecahydrate Ga$_2$(SO$_4$)$_3$•18H$_2$O	13780-42-2 751.912	octahed cry			s H$_2$O, EtOH

No.	Name Formula	CAS RN Mol. Wt.	Physical Form	mp/°C den/g cm^{-3}	bp/°C Other Data	Solubility
849	Gallium(III) sulfide Ga_2S_3	12024-22-5 235.644	monocl cry	1090 3.7		
850	Gallium(III) telluride Ga_2Te_3	12024-27-0 522.25	cub cry	790 5.57		
851	Germanium Ge	7440-56-4 72.61	gray-wh cub cry	938.25 5.3234	2833 a,c,d	i H_2O, dil acid, alk
852	Germanium nitride Ge_3N_4	12065-36-0 273.86	orth cry	dec 900		i H_2O, acid, aqua regia
853	Germane GeH_4	7782-65-2 76.64	col gas; flam	-165 3.363 g/L	-88.1 a,b,d,e	i H_2O
854	Bromogermane GeH_3Br	13569-43-2 155.54	col liq	-32 2.34	52 a	reac H_2O
855	Chlorogermane GeH_3Cl	13637-65-5 111.09	col liq	-52 1.75	28 a	reac H_2O
856	Chlorotrifluorogermane GeF_3Cl	14188-40-0 165.06	gas	-66.2 7.243 g/L	-20.3	
857	Dibromogermane GeH_2Br_2	13769-36-3 234.43	col liq	-15 2.80	89	reac H_2O
858	Dichlorogermane GeH_2Cl_2	15230-48-5 145.53	col liq	-68 1.90	69.5	reac H_2O
859	Dichlorodifluorogermane GeF_2Cl_2	24422-21-7 181.51	col gas	-51.8 7.965 g/L	-2.8	
860	Fluorogermane GeH_3F	13537-30-9 94.63	col gas	 4.152 g/L	 a	reac H_2O
861	Iodogermane GeH_3I	13573-02-9 202.54	liq	-15	≈90 a	reac H_2O
862	Tribromogermane $GeHBr_3$	14779-70-5 313.33	col liq	-25	dec	reac H_2O
863	Trichlorogermane $GeHCl_3$	1184-65-2 179.98	liq	-71 1.93	75.3 e	reac H_2O
864	Trichlorofluorogermane $GeCl_3F$	24422-20-6 197.97	liq	-49.8	37.5	
865	Methylgermane GeH_3CH_3	1449-65-6 90.67	col gas	-158 3.979 g/L	-23	
866	Digermane Ge_2H_6	13818-89-8 151.27	col liq; flam	-109	30.8 a,d,e	
867	Trigermane Ge_3H_8	14691-44-2 225.89	col liq	-105.6 2.20(-105°C)	110.5 a,d,e	i H_2O
868	Germanium(II) bromide $GeBr_2$	24415-00-7 232.42	yel monocl cry	122	dec 150	reac H_2O
869	Germanium(II) chloride $GeCl_2$	10060-11-4 143.51	wh-yel hyg powder	dec		reac H_2O; s eth, bz
870	Germanium(II) fluoride GeF_2	13940-63-1 110.61	wh orth cry; hyg	110 3.64	dec 130	reac H_2O
871	Germanium(II) iodide GeI_2	13573-08-5 326.42	oran-yel hex cry	dec 550 5.4		reac H_2O
872	Germanium(II) oxide GeO	20619-16-3 88.61	blk solid	dec 700	 a	
873	Germanium(II) selenide GeSe	12065-10-0 151.57	gray orth cry or brn powder	667 5.6		
874	Germanium(II) sulfide GeS	12025-32-0 104.68	gray orth cry	615 4.1	 a	
875	Germanium(II) telluride GeTe	12025-39-7 200.21	cub cry	725 6.16		i H_2O; s conc HNO_3
876	Germanium(IV) bromide $GeBr_4$	13450-92-5 392.23	wh cry	26.1 3.132	186.35 a,b,d,e	reac H_2O
877	Germanium(IV) chloride $GeCl_4$	10038-98-9 214.42	col liq	-49.5 1.88	86.55 a,b,d,e	reac H_2O; s bz, eth, EtOH, ctc
878	Germanium(IV) fluoride GeF_4	7783-58-6 148.60	col gas	-15 tp 6.521 g/L	-36.5 sp a	reac H_2O
879	Germanium(IV) iodide GeI_4	13450-95-8 580.23	red-oran cub cry	144 4.4	377 a,b	reac H_2O
880	Germanium(IV) oxide GeO_2	1310-53-8 104.61	wh hex cry	1115 4.25	 a	i H_2O
881	Germanium(IV) selenide $GeSe_2$	12065-11-1 230.53	yel-oran orth ccry	707 dec 4.56		
882	Germanium(IV) sulfide GeS_2	12025-34-2 136.74	blk orth cry	530 3.01		
883	Gold Au	7440-57-5 196.967	soft yel metal	1064.18 19.3	2856 a,c,d,e,f	s aqua regia
884	Bromoauric acid pentahydrate $HAuBr_4 \cdot 5H_2O$	17083-68-0 607.667	red-brn hyg cry	27		s H_2O, EtOH
885	Chloroauric acid tetrahydrate $HAuCl_4 \cdot 4H_2O$	16903-35-8 411.847	yel monocl cry; hyg	 ≈3.9		vs H_2O, EtOH; s eth
886	Gold(I) bromide AuBr	10294-27-6 276.871	yel-gray tetr cry	dec 165 8.20	 a	i H_2O
887	Gold(I) chloride AuCl	10294-29-8 232.420	yel orth cry	dec 289 7.6	 a	i H_2O

No.	Name Formula	CAS RN Mol. Wt.	Physical Form	mp/°C den/g cm^{-3}	bp/°C Other Data	Solubility
888	Gold(I) cyanide AuCN	506-65-0 222.985	yel hex cry	dec 7.2		i H_2O, EtOH, eth, dil acid
889	Gold(I) iodide AuI	10294-31-2 323.871	yel-grn powder; tetr	dec 120 8.25	a	i H_2O; s CN soln
890	Gold(I) sulfide Au_2S	1303-60-2 425.999	brn-blk cub cry; unstable	240 dec ≈11		i H_2O, acid; s aqua regia
891	Gold(III) bromide $AuBr_3$	10294-28-7 436.679	red-br monocl cry	dec ≈160 	a	s H_2O, EtOH
892	Gold(III) chloride $AuCl_3$	13453-07-1 303.325	red monocl cry	dec >160 4.7	a	
893	Gold(III) cyanide trihydrate $Au(CN)_3$·$3H_2O$	535-37-5* 329.066	wh hyg cry	dec 50		vs H_2O; sl EtOH
894	Gold(III) fluoride AuF_3	14720-21-9 253.962	oran-yel hex cry	>300 6.75	subl a	
895	Gold(III) hydroxide $Au(OH)_3$	1303-52-2 247.989	brn powder	dec ≈100		i H_2O; s acid
896	Gold(III) iodide AuI_3	31032-13-0 577.680	unstable grn powder			
897	Gold(III) oxide Au_2O_3	1303-58-8 441.931	brn powder	dec ≈150		i H_2O; s acid
898	Gold(III) selenate $Au_2(SeO_4)_3$	10294-32-3 822.81	yel cry			i H_2O; s acid
899	Gold(III) selenide Au_2Se_3	1303-62-4 630.81	blk amorp solid	dec 4.65		s aqua regia
900	Gold(III) sulfide Au_2S_3	1303-61-3 490.131	unstable blk powder	dec 200		
901	Hafnium Hf	7440-58-6 178.49	gray metal; hex	2233 13.3	4603 a,c	s HF
902	Hafnium boride HfB_2	12007-23-7 200.11	gray hex cry	3100 10.5		
903	Hafnium bromide $HfBr_4$	13777-22-5 498.11	wh cub cry	424 tp 4.90	323 sp b	
904	Hafnium carbide HfC	12069-85-1 190.50	refrac cub cry	≈3000 12.2		
905	Hafnium chloride $HfCl_4$	13499-05-3 320.30	wh monocl cry	432 tp a,b	317 sp 	reac H_2O
906	Hafnium fluoride HfF_4	13709-52-9 254.48	wh monocl cry	>970 7.1	970 sp a	
907	Hafnium hydride HfH_2	12770-26-2 180.51	refrac tetr cry	 11.4		
908	Hafnium iodide HfI_4	13777-23-6 686.11	yel-oran cub cry	449 tp 5.6	394 sp b	
909	Hafnium nitride HfN	25817-87-2 192.50	yel-brn cub cry	3305 13.8		
910	Hafnium oxide HfO_2	12055-23-1 210.49	wh cub cry	2774 9.68	a	i H_2O
911	Hafnium oxychloride octahydrate $HfOCl_2$·$8H_2O$	14456-34-9 409.52	wh tetr cry	dec		s H_2O
912	Hafnium phosphide HfP	12325-59-6 209.46	hex cry	 9.78		
913	Hafnium selenide $HfSe_2$	12162-21-9 336.41	brn hex cry	 7.46		
914	Hafnium orthosilicate $HfSiO_4$	13870-13-8 270.57	tetr cry	 7.0		
915	Hafnium silicide $HfSi_2$	12401-56-8 234.66	gray orth cry	≈1700 7.6		
916	Hafnium sulfate $Hf(SO_4)_2$	15823-43-5 370.62	wh cry	dec >500		
917	Hafnium sulfide HfS_2	18855-94-2 242.62	purp-brn hex cry	 6.03		
918	Helium He	7440-59-7 4.003	col gas	 0.176 g/L	-268.93 a,b,d,e	sl H_2O; i EtOH
919	Holmium Ho	7440-60-0 164.930	silv metal; hex	1474 8.80	2700 a,c	s dil acid
920	Holmium bromide $HoBr_3$	13825-76-8 404.642	yel hyg cry	919	1470	
921	Holmium chloride $HoCl_3$	10138-62-2 271.288	yel monocl cry; hyg	718 3.7	1500 a	s H_2O
922	Holmium fluoride HoF_3	13760-78-6 221.925	pink-yel orth cry; hyg	1143 7.664	>2200 a	s H_2O
923	Holmium iodide HoI_3	13813-41-7 545.643	yel hex cry	994 5.4		
924	Holmium nitride HoN	12029-81-1 178.937	cub cry	 10.6		
925	Holmium oxide Ho_2O_3	12055-62-8 377.859	yel cub cry	2415 8.41	a	s acid
926	Holmium silicide $HoSi_2$	12136-24-2 221.10	hex cry	 7.1		

No.	Name Formula	CAS RN Mol. Wt.	Physical Form	mp/°C den/g cm^{-3}	bp/°C Other Data	Solubility
927	Holmium sulfide Ho_2S_3	12162-59-3 426.059	yel-oran monocl cry	 5.92		
928	Hydrazine N_2H_4	302-01-2 32.045	col oily liq	1.4 1.0036	113.55 a,b,c,d	vs H_2O, EtOH, MeOH
929	Hydrazine hydrate $N_2H_4 \cdot H_2O$	7803-57-8 50.060	fuming liq	-51.7 1.030	119	vs H_2O, EtOH; i chl, eth
930	Hydrazine hydrobromide $N_2H_4 \cdot HBr$	13775-80-9 112.957	wh monocl cry flakes	84 2.3	dec ≈190	s H_2O, EtOH
931	Hydrazine hydrochloride $N_2H_4 \cdot HCl$	2644-70-4 68.506	wh orth cry	89 1.5	dec 240	s H_2O; i os
932	Hydrazine dihydrochloride $N_2H_4 \cdot 2HCl$	5341-61-7 104.966	wh orth cry	198 dec 1.42		s H_2O; sl EtOH
933	Hydrazine hydroiodide $N_2H_4 \cdot HI$	10039-55-1 159.957	hyg cry	125		s H_2O
934	Hydrazine nitrate $N_2H_4 \cdot HNO_3$	13464-97-6 95.058	monocl cry; exp	70		vs H_2O
935	Hydrazine sulfate $N_2H_4 \cdot H_2SO_4$	10034-93-2 130.125	col orth cry	254 1.378		sl H_2O; i EtOH
936	Hydrazoic acid HN_3	7782-79-8 43.028	col liq; exp	-80	35.7 a,d	s H_2O
937	Hydroxylamine H_2NOH	7803-49-8 33.030	wh orth flakes or needles	33.1 1.21	58 a	vs H_2O, MeOH
938	Hydroxylamine sulfate $(H_2NOH)_2 \cdot H_2SO_4$	10039-54-0 164.139	cry	170		vs H_2O
939	Hydrogen H_2	1333-74-0 2.016	col gas; flam	-259.34 0.088 g/L	-252.87 a,b,c,d,e	sl H_2O
940	Hydrogen bromide HBr	10035-10-6 80.912	col gas	-86.81 3.550 g/L	-66.38 a,b,c,d,e	vs H_2O; s EtOH
941	Hydrogen chloride HCl	7647-01-0 36.461	col gas	-114.18 1.600 g/L	-85 a,b,c,d,e	vs H_2O
942	Hydrogen cyanide HCN	74-90-8 27.026	col liq	-13.4 0.684	26 a,b,c,e	vs H_2O, EtOH; sl eth
943	Hydrogen fluoride HF	7664-39-3 20.006	col gas	-83.36 0.878 g/L	20 a,b,c,e	vs H_2O, EtOH; sl eth
944	Hydrogen iodide HI	10034-85-2 127.912	col or yel gas	-50.77 5.613 g/L	-35.55 a,b,c,d,e	vs H_2O; s os
945	Hydrogen peroxide H_2O_2	7722-84-1 34.015	col liq	-0.43 1.44	150.2 a,c,d,e	vs H_2O
946	Hydrogen selenide H_2Se	7783-07-5 80.98	col gas; flam	-65.73 3.553 g/L	-41.25 a,b,d,e	s H_2O
947	Hydrogen sulfide H_2S	7783-06-4 34.082	col gas; flam	-85.5 1.496 g/L	-59.55 a,b,c,d,e	s H_2O
948	Hydrogen disulfide H_2S_2	13465-07-1 66.148	col liq	 1.334	70.7 a,d,e	
949	Hydrogen telluride H_2Te	7783-09-7 129.62	col gas	-49 5.687 g/L	-2 a,d,e	s H_2O, EtOH, alk
950	Indium In	7440-74-6 114.818	soft wh metal	156.60 7.31	2072 a,c	s acid
951	Indium antimonide InSb	1312-41-0 236.578	blk cub cry	525 5.7747	 a,c	
952	Indium arsenide InAs	1303-11-3 189.740	gray cub cry	942 5.67	 a	i acids
953	Indium nitride InN	25617-98-5 128.825	hex cry	1100 6.88		
954	Indium phosphide InP	22398-80-7 145.792	blk cub cry	1062 4.81	 a	sl acid
955	Indium(I) bromide InBr	14280-53-6 194.722	oran-red orth cry	290 4.96	656 a,d	reac H_2O
956	Indium(I) chloride InCl	13465-10-6 150.271	yel cub cry	211 4.19	608 a,c	reac H_2O
957	Indium(I) iodide InI	13966-94-4 241.722	orth cry	364.4 5.32	712 a,d	
958	Indium(II) bromide $InBr_2$	21264-43-7 274.626	orth cry			reac H_2O
959	Indium(II) chloride $InCl_2$	13465-11-7 185.723	col orth cry	235 3.64		reac H_2O
960	Indium(II) sulfide InS	12030-14-7 146.884	red-brn orth cry	692 5.2	 a	
961	Indium(III) bromide $InBr_3$	13465-09-3 354.530	hyg yel-wh monocl cry	420 4.74	 a	
962	Indium(III) chloride $InCl_3$	10025-82-8 221.176	yel monocl cry; hyg	583 4.0	 a	vs H_2O; s EtOH
963	Indium(III) fluoride InF_3	7783-52-0 171.813	wh hex cry; hyg	1170 4.39	>1200	sl H_2O; s dil acid
964	Indium(III) fluoride trihydrate $InF_3 \cdot 3H_2O$	14166-78-0 225.859	wh cry	dec 100		s H_2O
965	Indium(III) hydroxide $In(OH)_3$	20661-21-6 165.840	cub cry	 4.4		

No.	Name / Formula	CAS RN / Mol. Wt.	Physical Form	mp/°C / den/g cm⁻³	bp/°C / Other Data	Solubility
966	Indium(III) iodide InI_3	13510-35-5 / 495.531	yel-red monocl cry; hyg	207 / 4.69	a	
967	Indium(III) oxide In_2O_3	1312-43-2 / 277.634	yel cub cry	1913 / 7.18	a	i H_2O; s hot acid
968	Indium(III) perchlorate octahydrate $In(ClO_4)_3 \cdot 8H_2O$	13465-15-1 / 557.291	wh cry	≈80	dec 200	
969	Indium(III) phosphate $InPO_4$	14693-82-4 / 209.789	wh orth cry	4.9		i H_2O
970	Indium(III) selenide In_2Se_3	1312-42-1 / 466.52	blk hex cry	660 / 5.8		
971	Indium(III) sulfate $In_2(SO_4)_3$	13464-82-9 / 517.827	hyg wh powder	3.44		s H_2O
972	Indium(III) sulfide In_2S_3	12030-24-9 / 325.834	oran cub cry	1050 / 4.45	a	
973	Indium(III) telluride In_2Te_3	1312-45-4 / 612.44	blk cub cry	667 / 5.75		
974	Iodine I_2	7553-56-2 / 253.809	blue-blk plates	113.7 / 4.933	184.4 / a,b,c,d,e	sl H_2Os; s bz, EtOH, eth, ctc, chl
975	Iodic acid HIO_3	7782-68-5 / 175.910	col orth cry	110 dec / 4.63	a	vs H_2O; i EtOH, eth
976	Periodic acid dihydrate $HIO_4 \cdot 2H_2O$	10450-60-9 / 227.940	monocl hyg cry	122 dec		s H_2O, EtOH; sl eth
977	Iodine tetroxide I_2O_4	12399-08-5 / 317.807	yel cry	dec 85 / 4.2		sl H_2O
978	Iodine pentoxide I_2O_5	12029-98-0 / 333.806	hyg wh cry	≈300 dec / 4.98		s H_2O; i EtOH, eth, CS_2
979	Iodine nonaoxide I_4O_9	73560-00-6 / 651.613	hyg yel powder	dec 75		
980	Iodine bromide IBr	7789-33-5 / 206.808	blk orth cry	40 / 4.3	116 dec / a,b	s H_2O, EtOH, eth
981	Iodine chloride ICl	7790-99-0 / 162.357	red cry or oily liq	27.38 / 3.24	100 dec / a,c	reac H_2O; s EtOH
982	Iodine trichloride ICl_3	865-44-1 / 233.262	yel tricl cry; hyg	101 tp (16 atm) / 3.2	64 sp (dec)	reac H_2O; s EtOH, bz
983	Iodine fluoride IF	13873-84-2 / 145.902	disproportionates at room temp		a	
984	Iodine trifluoride IF_3	22520-96-3 / 183.899	yel solid, stable at low temp	dec -28		
985	Iodine pentafluoride IF_5	7783-66-6 / 221.896	yel liq	9.43 / 3.19	100.5 / a,d,e	reac H_2O
986	Iodine heptafluoride IF_7	16921-96-3 / 259.893	col liq	6.5 tp / 2.67	4.8 sp / e	s H_2O
987	Iridium Ir	7439-88-5 / 192.217	silv-wh metal; cub	2446 / 22.5	4428 / a,c	s aqua regia
988	Iridium(III) sulfide Ir_2S_3	12136-42-4 / 480.632	orth cry	10.2	a	
989	Iridium(III) bromide $IrBr_3$	10049-24-8 / 431.929	red-brn monocl cry	6.82		i H_2O, acid, alk
990	Iridium(III) bromide tetrahydrate $IrBr_3 \cdot 4H_2O$	10049-24-8* / 503.991	grn-brn cry			s H_2O; i EtOH
991	Iridium(III) chloride $IrCl_3$	10025-83-9 / 298.575	brn monocl cry	dec 763 / 5.30	a	i H_2O, acid, alk
992	Iridium(III) fluoride IrF_3	23370-59-4 / 249.212	blk hex cry	dec 250 / ≈8.0		i H_2O, dil acid
993	Iridium(III) iodide IrI_3	7790-41-2 / 572.930	dark brn monocl cry	≈7.4		i H_2O, acid, bz, chl; s alk
994	Iridium(III) oxide Ir_2O_3	1312-46-5 / 432.432	blue-blk cry	dec 1000		i H_2O; sl hot HCl
995	Iridium(IV) oxide IrO_2	12030-49-8 / 224.216	brn tetr cry	1100 dec / 11.7	a	
996	Iridium(IV) sulfide IrS_2	12030-51-2 / 256.349	orth cry	9.3	a	
997	Iridium(VI) fluoride IrF_6	7783-75-7 / 306.207	yel cub cry; hyg	44 / 4.8	53 / a,c,d	reac H_2O
998	Iron Fe	7439-89-6 / 55.845	silv-wh or gray met	1538 / 7.87	2861 / a,c,e	s dil acid
999	Iron arsenide $FeAs$	12044-16-5 / 130.767	gray orth cry	1030 / 7.85		
1000	Iron boride Fe_2B	12006-86-9 / 122.501	refr solid; tetr	1389 / 7.3		
1001	Iron boride FeB	12006-84-7 / 66.656	refr solid; orth	1650 / ≈7		
1002	Iron pentacarbonyl $Fe(CO)_5$	13463-40-6 / 195.897	yel oily liq; flam	-20 / 1.490	103 / a,e	i H_2O; s eth, bz, ace
1003	Iron nonacarbonyl $Fe_2(CO)_9$	15321-51-4 / 363.784	oran-yel cry	100 dec / 2.85		
1004	Iron dodecacarbonyl $Fe_3(CO)_{12}$	12088-65-2 / 503.660	blk cry	140 / 2.00		

No.	Name / Formula	CAS RN / Mol. Wt.	Physical Form	mp/°C den/g cm⁻³	bp/°C Other Data	Solubility
1005	Iron hydrocarbonyl / $H_2Fe(CO)_4$	17440-90-3 / 169.903	col liq; unstable	-70	dec	s alk
1006	Iron phosphide / Fe_2P	1310-43-6 / 142.664	gray hex needles	1370 / 6.8		i H_2O, dil acid, alk
1007	Iron disulfide / FeS_2	1317-66-4 / 119.977	blk cub cry	dec >600 / 5.02	a,f	i H_2O
1008	Iron silicide / $FeSi$	12022-95-6 / 83.931	gray cub cry	1410 / 6.1		
1009	Iron silicide / $FeSi_2$	12022-99-0 / 112.016	gray tetr cry	1220 / 4.74		
1010	Iron(II) aluminate / $Fe(AlO_2)_2$	12068-49-4 / 173.806	blk cub cry	4.3		
1011	Iron(II) bromide / $FeBr_2$	7789-46-0 / 215.653	yel-brn hex cry; hyg	691 / 4.636	dec / a,c	vs H_2O, EtOH
1012	Iron(II) bromide hexahydrate / $FeBr_2 \cdot 6H_2O$	13463-12-2 / 323.744	grn hyg cry	27 dec / 4.64		s H_2O, EtOH
1013	Iron(II) carbonate / $FeCO_3$	563-71-3 / 115.854	gray-brn hex cry	3.9	f	
1014	Iron(II) chloride / $FeCl_2$	7758-94-3 / 126.750	wh hex cry; hyg	677 / 3.16	1023 / a,c,e	vs H_2O, EtOH, ace; sl bz
1015	Iron(II) chloride dihydrate / $FeCl_2 \cdot 2H_2O$	16399-77-2 / 162.781	wh-grn monocl cry	dec 120 / 2.39		s H_2O
1016	Iron(II) chloride tetrahydrate / $FeCl_2 \cdot 4H_2O$	13478-10-9 / 198.812	grn monocl cry	dec 105 / 1.93		s H_2O, EtOH
1017	Iron(II) chromite / $FeCr_2O_4$	1308-31-2 / 223.835	blk cub cry	5.0	f	
1018	Iron(II) fluoride / FeF_2	7789-28-8 / 93.842	wh tetr cry	1100 / 4.09	a,c	sl H_2O; s dil HF; i EtOH, eth
1019	Iron(II) fluoride tetrahydrate / $FeF_2 \cdot 4H_2O$	13940-89-1 / 165.904	col hex cry	2.20		
1020	Iron(II) hydroxide / $Fe(OH)_2$	18624-44-7 / 89.860	wh-grn hex cry	3.4		i H_2O
1021	Iron(II) iodide / FeI_2	7783-86-0 / 309.654	red-viol hex cry; hyg	587 / 5.3	a,c	s H_2O, EtOH, eth
1022	Iron(II) iodide tetrahydrate / $FeI_2 \cdot 4H_2O$	7783-86-0* / 381.716	blk hyg leaflets	dec 90 / 2.87		s H_2O, EtOH
1023	Iron(II) molybdate / $FeMoO_4$	13718-70-2 / 215.78	brn-yel monocl cry	1115 / 5.6	a	i H_2O
1024	Iron(II) nitrate / $Fe(NO_3)_2$	14013-86-6 / 179.854	grn solid			s H_2O
1025	Iron(II) nitrate hexahydrate / $Fe(NO_3)_2 \cdot 6H_2O$	13476-08-9 / 287.946	grn solid	60 dec		vs H_2O
1026	Iron(II) oxalate dihydrate / $FeC_2O_4 \cdot 2H_2O$	6047-25-2 / 179.895	yel cry	dec 150 / 2.28		sl H_2O; s acid
1027	Iron(II) oxide / FeO	1345-25-1 / 71.844	blk cub cry	1377 / 6.0	a,c	i H_2O, alk; s acid
1028	Iron(II) phosphate octahydrate / $Fe_3(PO_4)_2 \cdot 8H_2O$	14940-41-1 / 501.600	gray-blue monocl cry; hyg	2.58	f	i H_2O; s acid
1029	Iron(II) selenide / $FeSe$	1310-32-3 / 134.81	blk hex cry	6.7		i H_2O
1030	Iron(II) orthosilicate / Fe_2SiO_4	10179-73-4 / 203.774	brn orth cry	4.30	a,f	
1031	Iron(II) sulfate / $FeSO_4$	7720-78-7 / 151.909	wh orth cry; hyg	3.65	a	s H_2O
1032	Iron(II) sulfate monohydrate / $FeSO_4 \cdot H_2O$	17375-41-6 / 169.924	wh-yel monocl cry	dec 300 / 3.0		s H_2O
1033	Iron(II) sulfate heptahydrate / $FeSO_4 \cdot 7H_2O$	7782-63-0 / 278.015	blue-grn monocl cry	dec ≈60 / 1.895	f	vs H_2O; i EtOH
1034	Iron(II) sulfide / FeS	1317-37-9 / 87.911	col hex or tetr cry; hyg	1188 / 4.7	dec / a,c	i H_2O; reac acid
1035	Iron(II) telluride / $FeTe$	12125-63-2 / 183.45	tetr cry	914 / 6.8		
1036	Iron(II) thiocyanate trihydrate / $Fe(SCN)_2 \cdot 3H_2O$	6010-09-9 / 226.058	grn monocl cry			s H_2O, EtOH, eth
1037	Iron(II) titanate / $FeTiO_3$	12168-52-4 / 151.710	blk rhomb cry	≈1470 / 4.72	f	
1038	Iron(II) tungstate / $FeWO_4$	13870-24-1 / 303.68	monocl cry	7.51	a,f	
1039	Iron(II,III) oxide / Fe_3O_4	1317-61-9 / 231.533	blk cub cry or amorp powder	1597 / 5.17	a,c,f	i H_2O; s acid
1040	Iron(III) acetate, basic / $FeOH(C_2H_3O_2)_2$	10450-55-2 / 190.942	brn-red amorp powder			i H_2O; s EtOH, acid
1041	Iron(III) acetylacetonate / $Fe(OC(CH_3)=CHC(O)CH_3)_3$	14024-18-1 / 353.173	red-oran cry	179 / 5.24		sl H_2O; s os
1042	Iron(III) arsenate dihydrate / $FeAsO_4 \cdot 2H_2O$	10102-49-5 / 230.795	grn-brn powder	dec / 3.18		i H_2O; s dil acid
1043	Iron(III) bromide / $FeBr_3$	10031-26-2 / 295.557	dark red hex cry; hyg	dec / 4.5	a	s H_2O, EtOH, eth

No.	Name Formula	CAS RN Mol. Wt.	Physical Form	mp/°C den/g cm^{-3}	bp/°C Other Data	Solubility
1044	Iron(III) chloride $FeCl_3$	7705-08-0 162.203	grn hex cry; hyg	304 2.90	≈316 a,c,e	s H_2O, EtOH, eth, ace
1045	Iron(III) chloride hexahydrate $FeCl_3 \cdot 6H_2O$	10025-77-1 270.294	yel-oran monocl cry; hyg	dec 37 1.82		s H_2O, EtOH, eth, ace
1046	Iron(III) chromate $Fe_2(CrO_4)_3$	10294-52-7 459.671	yel powder			i H_2O, EtOH; s acid
1047	Iron(III) citrate pentahydrate $FeC_6H_5O_7 \cdot 5H_2O$	3522-50-7 335.023	red-brn cry			s H_2O; i EtOH
1048	Iron(III) dichromate $Fe_2(Cr_2O_7)_3$	10294-53-8 759.654	red-brn solid			s H_2O, acid
1049	Iron(III) ferrocyanide $Fe_4[Fe(CN)_6]_3$	14038-43-8 859.234	dark blue powder	1.80		i H_2O, dil acid, os
1050	Iron(III) fluoride FeF_3	7783-50-8 112.840	grn hex cry	>1000 3.87		sl H_2O; i EtOH, eth, bz
1051	Iron(III) fluoride trihydrate $FeF_3 \cdot 3H_2O$	15469-38-2 166.886	yel-brn tetr cry	2.3		
1052	Iron(III) hydroxide $Fe(OH)_3$	1309-33-7 106.867	yel monocl cry	3.12		
1053	Iron(III) hydroxide oxide $FeO(OH)$	20344-49-4 88.852	red-brn orth cry	4.26	f	i H_2O; s acid
1054	Iron(III) nitrate nonahydrate $Fe(NO_3)_3 \cdot 9H_2O$	7782-61-8 403.997	viol-gray hyg cry	47 dec 1.68		vs H_2O, EtOH, ace
1055	Iron(III) oxalate $Fe_2(C_2O_4)_3$	19469-07-9 375.749	yel amorp powder	dec 100		s H_2O, acid; i alk
1056	Iron(III) oxide Fe_2O_3	1309-37-1 159.688	red-brn hex cry	1565 5.25	a,f	i H_2O; s acid
1057	Iron(III) phosphate dihydrate $FePO_4 \cdot 2H_2O$	10045-86-0 186.847	gray-wh orth cry	2.87	f	i H_2O; s HCl
1058	Iron(III) pyrophosphate nonahydrate $Fe_4(P_2O_7)_3 \cdot 9H_2O$	10058-44-3 907.348	yel powder			i H_2O; s acid
1059	Iron(III) hypophosphite $Fe(H_2PO_2)_3$	7783-84-8 250.811	wh-gray powder			i H_2O
1060	Iron(III) sulfate $Fe_2(SO_4)_3$	10028-22-5 399.881	gray-wh rhomb cry; hyg	3.10		sl H_2O
1061	Iron(III) sulfate nonahydrate $Fe_2(SO_4)_3 \cdot 9H_2O$	13520-56-4 562.018	yel hex cry	dec 400 2.1	f	vs H_2O
1062	Iron(III) thiocyanate monohydrate $Fe(SCN)_3 \cdot H_2O$	4119-52-2 248.111	red hyg cry	dec		s H_2O, EtOH, ace; i tol, chl
1063	Iron(III) metavanadate $Fe(VO_3)_3$	65842-03-7 352.665	gray-brn powder			i H_2O, EtOH; s acid
1064	Krypton Kr	7439-90-9 83.80	col gas	-157.36 3.677 g/L	-153.22 a,b,c,d,e	sl H_2O
1065	Krypton difluoride KrF_2	13773-81-4 121.80	col tetr cry	dec ≈25 3.24		reac H_2O
1066	Lanthanum La	7439-91-0 138.906	silv metal; hex	918 6.15	3464 a,c	s dil acid
1067	Lanthanum boride LaB_6	12008-21-8 203.772	blk cub cry; refrac	2715 4.76		
1068	Lanthanum bromide $LaBr_3$	13536-79-3 378.618	wh hex cry; hyg	788 5.1		s H_2O
1069	Lanthanum carbide LaC_2	12071-15-7 162.928	tetr cry	2360 5.29		
1070	Lanthanum carbonate octahydrate $La_2(CO_3)_3 \cdot 8H_2O$	6487-39-4 601.961	wh cry powder	2.6		i H_2O; s dil acid
1071	Lanthanum chloride $LaCl_3$	10099-58-8 245.264	wh hex cry; hyg	859 3.84	a	s H_2O
1072	Lanthanum chloride heptahydrate $LaCl_3 \cdot 7H_2O$	20211-76-1 371.371	wh tricl cry; hyg	dec 91		s H_2O, EtOH
1073	Lanthanum fluoride LaF_3	13709-38-1 195.901	wh hex cry; hyg	1493 5.9		i H_2O, acid
1074	Lanthanum hydride LaH_3	13864-01-2 141.930	blk cub cry	5.36		
1075	Lanthanum hydroxide $La(OH)_3$	14507-19-8 189.928	wh amorp solid	dec		i H_2O
1076	Lanthanum iodide LaI_3	13813-22-4 519.619	wh orth cry; hyg	778 5.6		s H_2O
1077	Lanthanum nitrate hexahydrate $La(NO_3)_3 \cdot 6H_2O$	10277-43-7 433.012	wh hyg cry	dec ≈40		vs H_2O, EtOH
1078	Lanthanum nitride LaN	25764-10-7 152.913	cub cry	6.73		
1079	Lanthanum oxide La_2O_3	1312-81-8 325.809	wh amorp powder	2305 6.51	4200 a	i H_2O; s dil acid
1080	Lanthanum silicide $LaSi_2$	12056-90-5 195.077	gray tetr cry	5.0		
1081	Lanthanum sulfate nonahydrate $La_2(SO_4)_3 \cdot 9H_2O$	10294-62-9 728.139	hex cry	2.82		sl H_2O; i EtOH
1082	Lanthanum sulfide La_2S_3	12031-49-1 374.009	red cub cry	2110 4.9		

No.	Name Formula	CAS RN Mol. Wt.	Physical Form	mp/°C den/g cm^{-3}	bp/°C Other Data	Solubility
1083	Lanthanum sulfide LaS	12031-30-0 170.972	yel cub cry	2300 5.61	a	
1084	Lawrencium Lr	22537-19-5 262	metal	1627	a	
1085	Lead Pb	7439-92-1 207.2	soft silv-gray metal; cub	327.46 11.3	1749 a,c,d,e	s conc acid
1086	Lead(II) acetate Pb(C$_2$H$_3$O$_2$)$_2$	301-04-2 325.3	wh cry	280 3.25	dec	vs H$_2$O
1087	Lead(II) acetate trihydrate Pb(C$_2$H$_3$O$_2$)$_2$·3H$_2$O	6080-56-4 427.3	col cry	75 dec 2.55		vs H$_2$O; sl EtOH
1088	Lead(II) acetate, basic Pb(C$_2$H$_3$O$_2$)$_2$·2PbO·2H$_2$O	1335-32-6 807.7	wh powder	dec		
1089	Lead(II) antimonate Pb$_3$(SbO$_4$)$_2$	13510-89-9 993.1	oran-yel powder	 6.58		i H$_2$O, dil acid
1090	Lead(II) arsenate Pb$_3$(AsO$_4$)$_2$	3687-31-8 899.4	wh cry	1042 dec 5.8		i H$_2$O; s HNO$_3$
1091	Lead(II) arsenite Pb(AsO$_2$)$_2$	10031-13-7 421.0	wh powder	 5.85		i H$_2$O; s dil HNO$_3$
1092	Lead(II) azide Pb(N$_3$)$_2$	13424-46-9 291.2	col orth needles; exp	exp ≈350 4.7		i H$_2$O; vs HOAc
1093	Lead(II) borate monohydrate Pb(BO$_2$)$_2$·H$_2$O	10214-39-8 310.8	wh powder	500 dec 5.6		i H$_2$O; s dil HNO$_3$
1094	Lead(II) bromate monohydrate Pb(BrO$_3$)$_2$·H$_2$O	10031-21-7 481.0	col cry	≈180 dec 5.53		sl H$_2$O
1095	Lead(II) bromide PbBr$_2$	10031-22-8 367.0	wh orth cry	371 6.69	892 a,c,d,e	sl H$_2$O; i EtOH
1096	Lead(II) butanoate Pb(C$_4$H$_7$O$_2$)$_2$	819-73-8 381.4	col solid	≈90		i H$_2$O; s dil HNO$_3$
1097	Lead(II) carbonate PbCO$_3$	598-63-0 267.2	col orth cry	dec ≈315 6.6	a,f	i H$_2$O
1098	Lead(II) carbonate, basic Pb(OH)$_2$·2PbCO$_3$	1319-46-6 775.6	wh hex cry	400 dec ≈6.5		i H$_2$O, EtOH; s acid
1099	Lead(II) chlorate Pb(ClO$_3$)$_2$	10294-47-0 374.1	col hyg cry	dec 230 3.9		vs H$_2$O, EtOH
1100	Lead(II) chloride PbCl$_2$	7758-95-4 278.1	wh orth needles or powder	501 5.98	951 a,c,d,e,f	sl H$_2$O; s alk
1101	Lead(II) chloride fluoride PbClF	13847-57-9 261.7	tetr cry	 7.05		i H$_2$O
1102	Lead(II) chromate PbCrO$_4$	7758-97-6 323.2	yel-oran monocl cry	844 6.12	a,f	i H$_2$O; s akl, dil acid
1103	Lead(II) chromate(VI) oxide PbCrO$_4$·PbO	18454-12-1 546.4	red powder			i H$_2$O
1104	Lead(II) citrate trihydrate Pb$_3$(C$_6$H$_5$O$_7$)$_2$·3H$_2$O	512-26-5 1053.8	wh cry powder			s H$_2$O; sl EtOH
1105	Lead(II) cyanide Pb(CN)$_2$	592-05-2 259.2	wh-yel powder			sl H$_2$O; reac acid
1106	Lead(II) fluoride PbF$_2$	7783-46-2 245.2	wh orth cry	830 8.44	1293 a,c,d,e	sl H$_2$O
1107	Lead(II) fluoroborate Pb(BF$_4$)$_2$	13814-96-5 380.8	stable only in aq soln			s H$_2$O
1108	Lead(II) formate Pb(CHO$_2$)$_2$	811-54-1 297.2	wh prisms or needles	dec 190 4.63		sl H$_2$O; i EtOH
1109	Lead(II) hydrogen arsenate PbHAsO$_4$	7784-40-9 347.1	wh monocl cry	dec 280 5.943		i H$_2$O; s HNO$_3$, alk
1110	Lead(II) hydrogen phosphate PbHPO$_4$	15845-52-0 303.2	wh monocl cry	dec 5.66		
1111	Lead(II) hydroxide Pb(OH)$_2$	19783-14-3 241.2	wh powder	145 dec 7.59		i H$_2$O; s acid
1112	Lead(II) iodate Pb(IO$_3$)$_2$	25659-31-8 557.0	wh orth cry	 6.50		i H$_2$O
1113	Lead(II) iodide PbI$_2$	10101-63-0 461.0	yel hex cry or powder	410 6.16	872 dec a,c,d,e	sl H$_2$O; i EtOH
1114	Lead(II) lactate Pb(C$_3$H$_5$O$_3$)$_2$	18917-82-3 385.3	wh cry powder			s H$_2$O, hot EtOH
1115	Lead(II) molybdate PbMoO$_4$	10190-55-3 367.1	yel tetr cry	≈1060 6.7	a,f	i H$_2$O; HNO$_3$, NaOH
1116	Lead(II) niobate Pb(NbO$_3$)$_2$	12034-88-7 489.0	rhomb or tetr cry	1343 6.6		i H$_2$O
1117	Lead(II) nitrate Pb(NO$_3$)$_2$	10099-74-8 331.2	col cub cry	470 4.53	a	vs H$_2$O; sl EtOH
1118	Lead(II) oleate Pb(C$_{18}$H$_{33}$O$_2$)$_2$	1120-46-3 770.1	wax-like solid			i H$_2$O; s EtOH, bz, eth
1119	Lead(II) oxalate PbC$_2$O$_4$	814-93-7 295.2	wh powder	dec 300 5.28	a	i H$_2$O; s dil HNO$_3$
1120	Lead(II) oxide (litharge) PbO	1317-36-8 223.2	red tetr cry	888 9.35	a,e,f	i H$_2$O, EtOH; s dil HNO$_3$
1121	Lead(II) oxide (massicot) PbO	1317-36-8 223.2	yel orth cry	stab >489 9.64	a,e,f	i H$_2$O, EtOH; s dil HNO$_3$

No.	Name Formula	CAS RN Mol. Wt.	Physical Form	mp/°C den/g cm^{-3}	bp/°C Other Data	Solubility
1122	Lead(II) oxide hydrate $3PbO \cdot H_2O$	1311-11-1 687.6	wh powder	 7.41		i H_2O; s dil acid
1123	Lead(II) perchlorate $Pb(ClO_4)_2$	13453-62-8 406.1	wh cry			vs H_2O
1124	Lead(II) perchlorate trihydrate $Pb(ClO_4)_2 \cdot 3H_2O$	13637-76-8 460.1	wh cry	100 dec 2.6		vs cold H_2O; s EtOH
1125	Lead(II) phosphate $Pb_3(PO_4)_2$	7446-27-7 811.5	wh hex cry	1014 7.01		i H_2O, EtOH
1126	Lead(II) hypophosphite $Pb(H_2PO_2)_2$	10294-58-3 337.2	hyg cry powder	dec		sl H_2O; i EtOH
1127	Lead(II) metasilicate $PbSiO_3$	10099-76-0 283.3	wh monocl cry powder	764 6.49	a	i H_2O, os
1128	Lead(II) orthosilicate Pb_2SiO_4	13566-17-1 506.5	monocl cry	743 7.60	a	
1129	Lead(II) hexafluorosilicate dihydrate $PbSiF_6 \cdot 2H_2O$	1310-03-8 385.3	col cry	dec		vs H_2O
1130	Lead(II) selenate $PbSeO_4$	7446-15-3 350.2	orth cry	 6.37	a	i H_2O; s conc acid
1131	Lead(II) selenide $PbSe$	12069-00-0 286.2	gray cub cry	1078 8.1	a	i H_2O; s HNO_3
1132	Lead(II) selenite $PbSeO_3$	7488-51-9 334.2	wh monocl cry	≈500 7.0		i H_2O
1133	Lead(II) sodium thiosulfate $Na_4Pb(S_2O_3)_3$	10101-94-7 635.6	wh cry			sl H_2O
1134	Lead(II) stearate $Pb(C_{18}H_{35}O_2)_2$	1072-35-1 774.2	wh powder	≈100 1.4		i H_2O, s hot EtOH
1135	Lead(II) sulfate $PbSO_4$	7446-14-2 303.3	orth cry	1087 6.29	a,f	i H_2O, acid; sl alk
1136	Lead(II) sulfide PbS	1314-87-0 239.3	blk powder or silv cub cry	1118 7.60	a,c,e,f	i H_2O; s acid
1137	Lead(II) sulfite $PbSO_3$	7446-10-8 287.3	wh powder	dec	a	i H_2O; s HNO_3
1138	Lead(II) tantalate $PbTa_2O_6$	12065-68-8 665.1	orth cry	 7.9		i H_2O
1139	Lead(II) telluride $PbTe$	1314-91-6 334.8	gray cub cry	924 8.164	a,f	i H_2O, acid
1140	Lead(II) thiocyanate $Pb(SCN)_2$	592-87-0 323.4	wh-yel powder	 3.82		sl H_2O
1141	Lead(II) thiosulfate PbS_2O_3	13478-50-7 319.3	wh cry	dec 5.18		i H_2O; s acid
1142	Lead(II) titanate $PbTiO_3$	12060-00-3 303.1	yel tetr cry	 7.9		i H_2O; reac HCl
1143	Lead(II) tungstate (stolzite) $PbWO_4$	7759-01-5 455.0	yel tetr cry	1130 8.24	f	i H_2O; s alk
1144	Lead(II) tungstate (raspite) $PbWO_4$	7759-01-5 455.0	monocl cry	trans 400 8.46	f	i H_2O; s alk
1145	Lead(II) metavanadate $Pb(VO_3)_2$	10099-79-3 405.1	yel powder			i H_2O; reac HNO_3
1146	Lead(II) zirconate $PbZrO_3$	12060-01-4 346.4	col orth cry	≈8		i H_2O, alk; s acid
1147	Lead(II,IV) oxide Pb_2O_3	1314-27-8 462.4	blk monocl cry or red amorp powder	dec 530 10.05		i H_2O; s alk; reac conc HCl
1148	Lead(II,II,IV) oxide Pb_3O_4	1314-41-6 685.6	red tetr cry	830 8.92	a,f	i H_2O, EtOH; s hot HCl
1149	Lead(IV) acetate $Pb(C_2H_3O_2)_4$	546-67-8 443.4	col monocl cry	≈175 2.23		reac H_2O, EtOH; s bz, chl
1150	Lead(IV) bromide $PbBr_4$	13701-91-2 526.8	unstable liq			
1151	Lead(IV) chloride $PbCl_4$	13463-30-4 349.0	yel oily liq	-15	dec ≈50 a	
1152	Lead(IV) fluoride PbF_4	7783-59-7 283.2	wh tetr cry; hyg	≈600 6.7		
1153	Lead(IV) oxide PbO_2	1309-60-0 239.2	red tetr cry or brn powder	290 dec 9.64	a	
1154	Lithium Li	7439-93-2 6.941	soft silv-wh metal	180.5 0.534	1342 a,c,e	reac H_2O
1155	Lithium acetate dihydrate $LiC_2H_3O_2 \cdot 2H_2O$	6108-17-4 102.017	wh rhomb cry	58 dec 1.3		vs H_2O; s EtOH
1156	Lithium aluminum hydride $LiAlH_4$	16853-85-3 37.955	gray-wh monocl cry	dec >125 0.917	a	reac H_2O, EtOH; s eth, thf
1157	Lithium amide $LiNH_2$	7782-89-0 22.964	tetr cry	380 1.18	a	reac H_2O
1158	Lithium arsenate Li_3AsO_4	13478-14-3 159.743	col orth cry	 3.07		sl H_2O; s HOAc
1159	Lithium azide LiN_3	19597-69-4 48.961	hyg monocl cry; exp	 1.83		vs H_2O
1160	Lithium borate $LiBO_2$	13453-69-5 49.751	wh monocl cry; hyg	849 2.18	a	vs H_2O; s EtOH

No.	Name Formula	CAS RN Mol. Wt.	Physical Form	mp/°C den/g cm^{-3}	bp/°C Other Data	Solubility
1161	Lithium borohydride $LiBH_4$	16949-15-8 21.784	wh-gray orth cry or powder	268 0.66	dec 380 a	s alk, eth, thf
1162	Lithium bromide $LiBr$	7550-35-8 86.845	wh cub cry; hyg	552 3.464	≈1300 a,c,e	s H_2O, EtOH, eth
1163	Lithium carbonate Li_2CO_3	554-13-2 73.891	wh monocl cry	723 2.11	dec 1300 a,c	sl H_2O; s acid; i EtOH
1164	Lithium chloride $LiCl$	7447-41-8 42.394	wh cub cry or powder; hyg	610 2.07	1383 a,c,e	s H_2O, EtOH, ace, py
1165	Lithium chromate dihydrate $Li_2CrO_4 \cdot 2H_2O$	7789-01-7 165.906	yel orth cry; hyg	dec 75 2.15		vs H_2O; s EtOH
1166	Lithium dichromate dihydrate $Li_2Cr_2O_7 \cdot 2H_2O$	10022-48-7 265.901	yel-red hyg cry	130 dec 2.34		vs H_2O
1167	Lithium fluoride LiF	7789-24-4 25.939	wh cub cry or powder	848.2 2.640	1673 a,c,d,e	sl H_2O; s acid
1168	Lithium formate monohydrate $Li(CHO_2) \cdot H_2O$	6108-23-2 69.974	col-wh cry	1.46		s H_2O
1169	Lithium hydride LiH	7580-67-8 7.949	gray cub cry or powder; hyg	688.7 0.78	a,c	reac H_2O, EtOH
1170	Lithium hydroxide $LiOH$	1310-65-2 23.948	col tetr cry	471.2 1.45	1626 a,c,d	s H_2O; sl EtOH
1171	Lithium hydroxide monohydrate $LiOH \cdot H_2O$	1310-66-3 41.964	wh monocl cry or powder	1.51		s H_2O; sl EtOH
1172	Lithium iodide LiI	10377-51-2 133.845	wh cub cry; hyg	469 4.06	1171 a,c,e	vs H_2O
1173	Lithium iodide trihydrate $LiI \cdot 3H_2O$	7790-22-9 187.891	wh hyg cry	73 3.48		vs H_2O, EtOH, ace
1174	Lithium niobate $LiNbO_3$	12031-63-9 147.845	wh hex cry	≈1240 4.30		
1175	Lithium nitrate $LiNO_3$	7790-69-4 68.946	col hex cry; hyg	253 2.38	a,c	vs H_2O; s EtOH
1176	Lithium nitride Li_3N	26134-62-3 34.830	red hex cry	813 1.27		reac H_2O
1177	Lithium phosphate Li_3PO_4	10377-52-3 115.794	wh orth cry	1205 2.46	a	i H_2O
1178	Lithium oxide Li_2O	12057-24-8 29.881	wh cub cry	1570 2.013	a	
1179	Lithium perchlorate $LiClO_4$	7791-03-9 106.392	wh orth cry or powder	236 2.428	dec 430 a,c	s H_2O, EtOH, ace, eth
1180	Lithium peroxide Li_2O_2	12031-80-0 45.881	wh hex cry	2.31	a	s H_2O; i EtOH
1181	Lithium selenate monohydrate $Li_2SeO_4 \cdot H_2O$	7790-71-8 174.86	monocl cry	2.56		vs H_2O
1182	Lithium metasilicate Li_2SiO_3	10102-24-6 89.966	wh orth needles	1201 2.52	a,c	i cold H_2O; reac dil acid
1183	Lithium sulfate Li_2SO_4	10377-48-7 109.946	wh monocl cry; hyg	859 2.21	a,c	s H_2O
1184	Lithium sulfate monohydrate $Li_2SO_4 \cdot H_2O$	10102-25-7 127.961	col cry	dec 130 2.06		vs H_2O; sl EtOH
1185	Lithium sulfide Li_2S	12136-58-2 45.948	wh cub cry; hyg	1372 1.64	a	
1186	Lutetium Lu	7439-94-3 174.967	silv metal; hex	1663 9.84	3402 a,c	s dil acid
1187	Lutetium boride LuB_4	12688-52-7 218.211	tetr cry	2600 ≈7.0		
1188	Lutetium bromide $LuBr_3$	14456-53-2 414.679	wh hyg cry	1025		vs H_2O
1189	Lutetium chloride $LuCl_3$	10099-66-8 281.325	wh monocl cry; hyg	925 3.98	a	s H_2O
1190	Lutetium fluoride LuF_3	13760-81-1 231.962	orth cry	1182 8.3	2200	i H_2O
1191	Lutetium iodide LuI_3	13813-45-1 555.680	brn hex cry; hyg	1050 ≈5.6	a	vs H_2O
1192	Lutetium nitride LuN	12125-25-6 188.974	cub cry	11.6		
1193	Lutetium oxide Lu_2O_3	12032-20-1 397.932	wh cub cry or powder	2490 9.41	a	
1194	Lutetium sulfate octahydrate $Lu_2(SO_4)_3 \cdot 8H_2O$	13473-77-3 782.247	wh cry			vs H_2O
1195	Lutetium sulfide Lu_2S_3	12163-20-1 446.132	gray rhomb cry	1750 dec 6.26		
1196	Lutetium telluride Lu_2Te_3	12163-22-3 732.73	orth cry	7.8		
1197	Magnesium Mg	7439-95-4 24.305	silv-wh metal	650 1.74	1090 a,c,e	s dil acid
1198	Magnesium acetate tetrahydrate $Mg(C_2H_3O_2)_2 \cdot 4H_2O$	16674-78-5 214.455	col monocl cry; hyg	80 dec 1.45		vs H_2O, EtOH
1199	Magnesium amide $Mg(NH_2)_2$	7803-54-5 56.350	wh powder; flam	dec 1.39		reac H_2O

No.	Name Formula	CAS RN Mol. Wt.	Physical Form	mp/°C den/g cm^{-3}	bp/°C Other Data	Solubility
1200	Magnesium antimonide Mg_3Sb_2	12057-75-9 316.435	hex cry	1245 3.99		
1201	Magnesium boride MgB_2	12007-25-9 45.927	hex cry	dec 800 2.57		
1202	Magnesium bromate hexahydrate $Mg(BrO_3)_2 \cdot 6H_2O$	7789-36-8 388.201	col cub cry	dec 200 2.29		vs H_2O
1203	Magnesium bromide $MgBr_2$	7789-48-2 184.113	wh hex cry; hyg	711 3.72	a,c	vs H_2O
1204	Magnesium bromide hexahydrate $MgBr_2 \cdot 6H_2O$	13446-53-2 292.204	col monocl cry	dec 165 2.0		vs H_2O; s EtOH
1205	Magnesium carbonate $MgCO_3$	546-93-0 84.314	wh hex cry	990 3.05	f	i H_2O, EtOH; s acid
1206	Magnesium chlorate hexahydrate $Mg(ClO_3)_2 \cdot 6H_2O$	10326-21-3* 299.298	wh hyg cry	dec ≈35 1.80		vs H_2O; sl EtOH
1207	Magnesium chloride $MgCl_2$	7786-30-3 95.210	wh hex leaflets; hyg	714 2.325	1412 a,c,e	vs H_2O
1208	Magnesium chloride hexahydrate $MgCl_2 \cdot 6H_2O$	7791-18-6 203.301	wh hyg cry	dec ≈100 1.56		vs H_2O; s EtOH
1209	Magnesium fluoride MgF_2	7783-40-6 62.302	wh tetr cry	1263 3.148	2227 a,c,f	sl H_2O
1210	Magnesium formate dihydrate $Mg(CHO_2)_2 \cdot 2H_2O$	6150-82-9 150.371	wh cry	dec		s H_2O; i EtOH
1211	Magnesium germanide Mg_2Ge	1310-52-7 121.22	cub cry	1117 3.09		
1212	Magnesium hydride MgH_2	60616-74-2 26.321	wh tetr cry	327 1.45		reac H_2O
1213	Magnesium hydrogen phosphate trihydrate $MgHPO_4 \cdot 3H_2O$	7757-86-0 174.331	wh powder	dec 550 2.13	f	sl H_2O; s dil acid
1214	Magnesium hydroxide $Mg(OH)_2$	1309-42-8 58.320	wh hex cry	350 2.37	a,f	i H_2O; s dil acid
1215	Magnesium iodide MgI_2	10377-58-9 278.114	wh hex cry; hyg	634 4.43	a,c	vs H_2O
1216	Magnesium iodide octahydrate $MgI_2 \cdot 8H_2O$	7790-31-0 422.236	wh orth cry; hyg	41 dec 2.10		vs H_2O; s EtOH
1217	Magnesium nitrate $Mg(NO_3)_2$	10377-60-3 148.314	wh cub cry	≈2.3	a	vs H_2O
1218	Magnesium nitrate dihydrate $Mg(NO_3)_2 \cdot 2H_2O$	15750-45-5 184.345	wh cry	≈100 dec 1.45		s H_2O, EtOH
1219	Magnesium nitrate hexahydrate $Mg(NO_3)_2 \cdot 6H_2O$	13446-18-9 256.406	col monocl cry; hyg	dec ≈95 1.46		vs H_2O; s EtOH
1220	Magnesium nitride Mg_3N_2	12057-71-5 100.928	yel cub cry	dec 270 2.71		
1221	Magnesium oxalate dihydrate $MgC_2O_4 \cdot 2H_2O$	6150-88-5 148.355	wh powder			i H_2O, EtOH; s dil acid
1222	Magnesium oxide MgO	1309-48-4 40.304	wh cub cry	2826 3.6	3600 a,c,f	sl H_2O; i EtOH
1223	Magnesium perchlorate $Mg(ClO_4)_2$	10034-81-8 223.205	wh hyg powder	dec 250 2.2		vs H_2O
1224	Magnesium perchlorate hexahydrate $Mg(ClO_4)_2 \cdot 6H_2O$	13446-19-0 331.297	wh hyg cry	dec 190 1.98		vs H_2O; s EtOH
1225	Magnesium permanganate hexahydrate $Mg(MnO_4)_2 \cdot 6H_2O$	10377-62-5 370.268	blue-blk cry	dec 2.18		s H_2O
1226	Magnesium peroxide MgO_2	1335-26-8 56.304	wh cub cry	dec 100 ≈3.0		i H_2O; s dil acid
1227	Magnesium phosphate octahydrate $Mg_3(PO_4)_2 \cdot 8H_2O$	13446-23-6 406.980	wh monocl cry	2.17		i H_2O; s acid
1228	Magnesium pyrophosphate trihydrate $Mg_2P_2O_7 \cdot 3H_2O$	10102-34-8 276.600	wh powder	dec 100 2.56		i H_2O; s acid
1229	Magnesium phosphide Mg_3P_2	12057-74-8 134.863	yel cub cry	2.06		reac H_2O
1230	Magnesium selenate hexahydrate $MgSeO_4 \cdot 6H_2O$	13446-28-1 275.35	wh monocl cry	1.928		s H_2O
1231	Magnesium selenide $MgSe$	1313-04-8 103.27	brn cub cry	4.2		reac H_2O
1232	Magnesium selenite hexahydrate $MgSeO_3 \cdot 6H_2O$	15593-61-0 259.36	col hex cry	2.09		i H_2O; s dil acid
1233	Magnesium metasilicate $MgSiO_3$	13776-74-4 100.389	wh monocl cry	dec ≈1550 3.19	f	i H_2O; sl HF
1234	Magnesium orthosilicate Mg_2SiO_4	26686-77-1 140.694	wh orth cry	1898 3.21	a,c,f	i H_2O
1235	Magnesium hexafluorosilicate hexahydrate $MgSiF_6 \cdot 6H_2O$	60950-56-3 274.472	wh cry	120 dec 1.79		s H_2O; i EtOH
1236	Magnesium silicide Mg_2Si	22831-39-6 76.696	gray cub cry	1102 1.99		reac H_2O
1237	Magnesium stannide Mg_2Sn	1313-08-2 167.320	blue cub cry	771 3.60		s H_2O, dil HCl

No.	Name / Formula	CAS RN / Mol. Wt.	Physical Form	mp/°C / den/g cm⁻³	bp/°C / Other Data	Solubility
1238	Magnesium sulfate / $MgSO_4$	7487-88-9 / 120.369	col orth cry	1127 / 2.66	a,c	s H_2O
1239	Magnesium sulfate heptahydrate / $MgSO_4 \cdot 7H_2O$	10034-99-8 / 246.475	col orth cry	dec 150 / 1.67	f	vs H_2O; sl EtOH
1240	Magnesium sulfate monohydrate / $MgSO_4 \cdot H_2O$	14168-73-1 / 138.384	col monocl cry	dec 150 / 2.57	f	vs H_2O
1241	Magnesium sulfide / MgS	12032-36-9 / 56.371	red-brn cub cry	2227 / 2.68	a	reac H_2O
1242	Magnesium sulfite hexahydrate / $MgSO_3 \cdot 6H_2O$	13446-29-2 / 212.461	wh hex cry	dec 200 / 1.72		sl H_2O; i EtOH
1243	Magnesium sulfite trihydrate / $MgSO_3 \cdot 3H_2O$	19086-20-5 / 158.415	col orth cry	/ 2.12		sl H_2O
1244	Magnesium thiosulfate hexahydrate / $MgS_2O_3 \cdot 6H_2O$	13446-30-5 / 244.527	col cry	dec 170 / 1.82		vs H_2O; i EtOH
1245	Magnesium titanate / $MgTiO_3$	1312-99-8 / 120.170	col hex cry	1565 / 3.85	f	
1246	Magnesium tungstate / $MgWO_4$	13573-11-0 / 272.14	wh monocl cry	/ 6.89		i H_2O, EtOH
1247	Manganese / Mn	7439-96-5 / 54.938	hard gray metal	1246 / 7.3	2061 / a,c,e	s dil acids
1248	Manganese antimonide / $MnSb$	12032-82-5 / 176.698	hex cry	840 / 6.9		
1249	Manganese antimonide / Mn_2Sb	12032-97-2 / 231.636	tetr cry	948 / 7.0		
1250	Manganese boride / MnB	12045-15-7 / 65.749	orth cry	1890 / 6.45		
1251	Manganese boride / MnB_2	12228-50-1 / 76.560	hex cry	1827 / 5.3		
1252	Manganese boride / Mn_2B	12045-16-8 / 120.687	red-brn tetr cry	1580 / 7.20		
1253	Manganese carbide / Mn_3C	12266-65-8 / 176.825	refrac solid	1520 / 6.89		
1254	Manganese carbonyl / $Mn_2(CO)_{10}$	10170-69-1 / 389.980	yel monocl cry	154 / 1.75		i H_2O; s os
1255	Manganese phosphide / Mn_2P	12333-54-9 / 140.850	hex cry	1327 / 6.0		
1256	Manganese phosphide / MnP	12032-78-9 / 85.912	orth cry	1147 / 5.49		
1257	Manganese(II) acetate tetrahydrate / $Mn(C_2H_3O_2)_2 \cdot 4H_2O$	15243-27-3 / 245.088	red monocl cry	80 / 1.59		s H_2O, EtOH
1258	Manganese(II) tetraborate octahydrate / $MnB_4O_7 \cdot 8H_2O$	12228-91-0 / 354.300	red solid			i H_2O, EtOH; s dil acid
1259	Manganese(II) bromide / $MnBr_2$	13446-03-2 / 214.746	pink hex cry	698 / 4.385	a	vs H_2O
1260	Manganese(II) bromide tetrahydrate / $MnBr_2 \cdot 4H_2O$	10031-20-6 / 286.808	red hyg cry	64 dec		vs H_2O
1261	Manganese(II) carbonate / $MnCO_3$	598-62-9 / 114.947	pink hex cry	dec >200 / 3.70	f	sl H_2O; s dil acid
1262	Manganese(II) chloride / $MnCl_2$	7773-01-5 / 125.843	pink trig cry; hyg	650 / 2.977	1190 / a,c,e	s H_2O, py, EtOH; i eth
1263	Manganese(II) chloride tetrahydrate / $MnCl_2 \cdot 4H_2O$	13446-34-9 / 197.905	red monocl cry; hyg	87.5 / 1.913		vs H_2O; s EtOH; i eth
1264	Manganese(II) dihydrogen phosphate dihydrate / $Mn(H_2PO_4)_2 \cdot 2H_2O$	18718-07-5 / 284.944	col hyg cry			s H_2O; i EtOH
1265	Manganese(II) fluoride / MnF_2	7782-64-1 / 92.935	red tetr cry	930 / 3.98		sl H_2O; i EtOH
1266	Manganese(II) hydroxide / $Mn(OH)_2$	18933-05-6 / 88.953	pink hex cry	dec / 3.26	f	i H_2O
1267	Manganese(II) iodide / MnI_2	7790-33-2 / 308.747	wh hex cry; hyg	638 / 5.04		s H_2O, EtOH
1268	Manganese(II) iodide tetrahydrate / $MnI_2 \cdot 4H_2O$	7790-33-2* / 380.809	red cry			vs H_2O; s EtOH
1269	Manganese(II) molybdate / $MnMoO_4$	14013-15-1 / 214.88	yel monocl cry	/ 4.05		
1270	Manganese(II) nitrate / $Mn(NO_3)_2$	10377-93-2 / 178.948	col orth cry; hyg	/ 2.2		vs H_2O; s diox, thf
1271	Manganese(II) nitrate hexahydrate / $Mn(NO_3)_2 \cdot 6H_2O$	10377-66-9 / 287.040	rose monocl cry	28 dec / 1.8		vs H_2O, EtOH
1272	Manganese(II) nitrate tetrahydrate / $Mn(NO_3)_2 \cdot 4H_2O$	20694-39-7 / 251.010	pink hyg cry	37.1 dec / 2.13		vs H_2O; s EtOH
1273	Manganese(II) oxalate dihydrate / $MnC_2O_4 \cdot 2H_2O$	6556-16-7 / 178.988	wh cry powder	dec 150 / 2.45		sl H_2O; s acid
1274	Manganese(II) oxide / MnO	1344-43-0 / 70.937	gr cub cry or powder	1840 / 5.37	a,c,f	i H_2O; s acid
1275	Manganese(II) perchlorate hexahydrate / $Mn(ClO_4)_2 \cdot 6H_2O$	15364-94-0 / 361.930	pink hex cry	/ 2.10		
1276	Manganese(II) pyrophosphate / $Mn_2P_2O_7$	53731-35-4 / 283.819	wh monocl cry	1196 / 3.71		i H_2O

No.	Name Formula	CAS RN Mol. Wt.	Physical Form	mp/°C den/g cm^{-3}	bp/°C Other Data	Solubility
1277	Manganese(II) metasilicate $MnSiO_3$	7759-00-4 131.022	red orth cry	1291 3.48	a,f	i H_2O
1278	Manganese(II) orthosilicate Mn_2SiO_4	13568-32-6 201.960	orth cry	4.11	a	i H_2O
1279	Manganese(II) selenide MnSe	1313-22-0 133.90	gray cub cry	1460 5.45	a	i H_2O
1280	Manganese(II) sulfate $MnSO_4$	7785-87-7 151.002	wh orth cry	700 3.25	dec 850	vs H_2O
1281	Manganese(II) sulfate monohydrate $MnSO_4 \cdot H_2O$	10034-96-5 169.017	red monocl cry	2.95		vs H_2O; i EtOH
1282	Manganese(II) sulfate tetrahydrate $MnSO_4 \cdot 4H_2O$	10101-68-5 223.063	red monocl cry	38 dec 2.26		s H_2O; i EtOH
1283	Manganese(II) sulfide (α) MnS	18820-29-6 87.004	grn cub cry	1610 4.0	a,f	i H_2O; s dil acid
1284	Manganese(II) sulfide (β) MnS	18820-29-6 87.004	red cub cry	3.3	a,f	i H_2O; s dil acid
1285	Manganese(II) sulfide (γ) MnS	18820-29-6 87.004	red hex cry	≈3.3	a,f	i H_2O; s dil acid
1286	Manganese(II) telluride MnTe	12032-88-1 182.54	hex cry	≈1150 6.0		
1287	Manganese(II) titanate $MnTiO_3$	12032-74-5 150.803	red hex cry	1360 4.55		
1288	Manganese(II) tungstate $MnWO_4$	13918-22-4 302.78	wh monocl cry	7.2	f	
1289	Manganese(II,III) oxide Mn_3O_4	1317-35-7 228.812	brn tetr cry	1567 4.84	a,f	i H_2O; s HCl
1290	Manganese(III) fluoride MnF_3	7783-53-1 111.933	red monocl cry; hyg	dec >600 3.54		reac H_2O
1291	Manganese(III) hydroxide MnO(OH)	1332-63-4 87.945	blk monocl cry	dec 250 ≈4.3	f	i H_2O
1292	Manganese(III) oxide Mn_2O_3	1317-34-6 157.874	blk cub cry	dec 1080 ≈5.0	a	i H_2O
1293	Manganese(IV) oxide MnO_2	1313-13-9 86.937	blk tetr cry	535 dec 5.08	a,f	i H_2O, HNO_3
1294	Manganese(VII) oxide Mn_2O_7	12057-92-0 221.872	grn oil; exp	5.9 2.40	exp 95	vs H_2O
1295	Mendelevium Md	7440-11-1 258	Metal	827	a	
1296	Mercury Hg	7439-97-6 200.59	heavy silv liq	-38.83 13.5336	356.73 a,b,c,d,e	i H_2O
1297	Mercury(I) acetate $Hg_2(C_2H_3O_2)_2$	631-60-7 519.27	col scales	dec		sl H_2O; i EtOH, eth
1298	Mercury(I) bromide Hg_2Br_2	15385-58-7 560.99	wh tetr cry or powder	407 7.307	a	i H_2O, EtOH, eth
1299	Mercury(I) chlorate $Hg_2(ClO_3)_2$	10294-44-7 568.08	wh cry	dec ≈250 6.41		sl H_2O; s EtOH
1300	Mercury(I) chloride Hg_2Cl_2	10112-91-1 472.09	wh tetr cry	525 tp 7.16	383 sp a,f	i H_2O, EtOH, eth
1301	Mercury(I) fluoride Hg_2F_2	13967-25-4 439.18	yel cub cry	570 dec 8.73	subl	reac H_2O
1302	Mercury(I) iodide Hg_2I_2	15385-57-6 654.99	yel amorp powder	290 7.70	a,c	i H_2O, EtOH, eth
1303	Mercury(I) nitrate dihydrate $Hg_2(NO_3)_2 \cdot 2H_2O$	7782-86-7 561.22	col cry	70 dec 4.8		reac H_2O
1304	Mercury(I) oxide Hg_2O	15829-53-5 417.18	prob mixture of HgO+Hg	100 dec 9.8		i H_2O; s HNO_3
1305	Mercury(I) sulfate Hg_2SO_4	7783-36-0 497.24	wh-yel cry powder	7.56	a	i H_2O; s dil HNO_3
1306	Mercury(II) acetate $Hg(C_2H_3O_2)_2$	1600-27-7 318.68	wh-yel cry or powder	179 dec 3.28		s H_2O, EtOH
1307	Mercury(II) amide chloride $Hg(NH_2)Cl$	10124-48-8 252.07	wh solid	5.38	subl	i H_2O, EtOH; s warm acid
1308	Mercury(II) bromide $HgBr_2$	7789-47-1 360.40	wh rhomb cry or powder	236 6.05	322 a,b,c,d,e	sl H_2O, chl; s EtOH, MeOH
1309	Mercury(II) chloride $HgCl_2$	7487-94-7 271.50	wh orth cry	276 5.6	304 a,b,c,d,e	sl H_2O, bz; s EtOH, MeOH, ace, eth
1310	Mercury(II) chromate $HgCrO_4$	13444-75-2* 316.58	red monocl cry	6.06		sl H_2O
1311	Mercury(II) cyanide $Hg(CN)_2$	592-04-1 252.63	col tetr cry	dec 320 4.00		s H_2O, EtOH; sl eth
1312	Mercury(II) dichromate $HgCr_2O_7$	7789-10-8 416.58	red cry powder			i H_2O; s acid
1313	Mercury(II) fluoride HgF_2	7783-39-3 238.59	wh cub cry; hyg	dec 645 8.95		reac H_2O
1314	Mercury(II) fulminate $Hg(CNO)_2$	628-86-4 284.62	gray cry	exp 4.42		sl H_2O; s EtOH, NH_4OH
1315	Mercury(II) hydrogen arsenate $HgHAsO_4$	7784-37-4 340.52	yel powder			i H_2O; s acid

No.	Name Formula	CAS RN Mol. Wt.	Physical Form	mp/°C den/g cm^{-3}	bp/°C Other Data	Solubility
1316	Mercury(II) iodate $Hg(IO_3)_2$	7783-32-6 550.40	wh powder	dec 175		i H_2O
1317	Mercury(II) iodide HgI_2	7774-29-0 454.40	red tetr cry or powder	259 6.28	354 a,b,c,d,e	i H_2O; sl EtOH, ace, eth
1318	Mercury(II) nitrate $Hg(NO_3)_2$	10045-94-0 324.60	col hyg cry	79 4.3		s H_2O; i EtOH
1319	Mercury(II) nitrate dihydrate $Hg(NO_3)_2 \cdot 2H_2O$	10045-94-0* 360.63	monocl cry	4.78		s H_2O
1320	Mercury(II) nitrate monohydrate $Hg(NO_3)_2 \cdot H_2O$	7783-34-8 342.62	wh-yel hyg cry	4.3		s H_2O, dil acid
1321	Mercury(II) oxide HgO	21908-53-2 216.59	red or yel orth cry	dec 500 11.14	a,f	i H_2O, EtOH; s dil acid
1322	Mercury(II) oxycyanide $Hg(CN)_2 \cdot HgO$	1335-31-5 469.21	wh orth cry	exp 4.44		s H_2O
1323	Mercury(II) phosphate $Hg_3(PO_4)_2$	7782-66-3 791.71	wh-yel powder			i H_2O, EtOH; s acid
1324	Mercury(II) selenide HgSe	20601-83-6 279.55	gray cub cry	8.21	subl	i H_2O
1325	Mercury(II) sulfate $HgSO_4$	7783-35-9 296.65	wh monocl cry	6.47	a	reac H_2O
1326	Mercury(II) sulfide (black) HgS	1344-48-5 232.66	blk cub cry or powder	583 7.70	a,f	i H_2O; s acid, EtOH
1327	Mercury(II) sulfide (red) HgS	1344-48-5 232.66	red hex cry	trans to blk HgO 386 8.17	a,f	i H_2O, acid; s aqua regia
1328	Mercury(II) telluride HgTe	12068-90-5 328.19	gray cub cry	673 8.63	a	
1329	Mercury(II) thiocyanate $Hg(SCN)_2$	592-85-8 316.76	monocl cry	dec ≈165 3.71		i H_2O; s dil HCl
1330	Molybdenum Mo	7439-98-7 95.94	gray-blk metal; cub	2623 10.2	4639 a,c,e	i H_2O, dil acid, alk
1331	Molybdenum boride Mo_2B	12006-99-4 202.69	refrac tetr cry	2000 9.2		
1332	Molybdenum boride Mo_2B_5	12007-97-5 245.94	refrac hex cry	1600 ≈7.2		
1333	Molybdenum carbide MoC	12011-97-1 107.95	refrac solid, cub	2577		
1334	Molybdenum carbide Mo_2C	12069-89-5 203.89	gray orth cry	2687 9.18		
1335	Molybdenum carbonyl $Mo(CO)_6$	13939-06-5 264.00	wh orth cry	dec 150 1.96	subl a,d	i H_2O; s bz; sl eth
1336	Molybdenum nitride MoN	12033-19-1 109.95	hex cry	1750 9.20		
1337	Molybdenum nitride Mo_2N	12033-31-7 205.89	gray cub cry	790 dec 9.46		
1338	Molybdenum phosphide MoP	12163-69-8 126.91	blk hex cry	7.34		
1339	Molybdenum silicide $MoSi_2$	12136-78-6 152.11	gray tetr cry	≈1900 6.2		i H_2O; s HF
1340	Molybdenum(II) bromide $MoBr_2$	13446-56-5 255.75	yel-red cry	dec 900		
1341	Molybdenum(II) chloride $MoCl_2$	13478-17-6 166.85	yel cry	dec 530		
1342	Molybdenum(II) iodide MoI_2	14055-74-4 349.75	blk hyg cry	5.278		
1343	Molybdenum(III) bromide $MoBr_3$	13446-57-6 335.65	grn hex cry	977 4.89		i H_2O
1344	Molybdenum(III) chloride $MoCl_3$	13478-18-7 202.30	dark red monocl cry	1027 3.74		i H_2O
1345	Molybdenum(III) fluoride MoF_3	20193-58-2 152.94	brn hex cry	>600 4.64		i H_2O
1346	Molybdenum(III) iodide MoI_3	14055-75-5 476.65	blk solid	927		i H_2O
1347	Molybdenum(III) oxide Mo_2O_3	1313-29-7 239.88	gray-blk powder			i H_2O; sl acid
1348	Molybdenum(IV) bromide $MoBr_4$	13520-59-7 415.56	blk cry	dec		reac H_2O
1349	Molybdenum(IV) chloride $MoCl_4$	13320-71-3 237.75	blk cry	dec >170		reac H_2O
1350	Molybdenum(IV) fluoride MoF_4	23412-45-5 171.93	grn cry	dec		reac H_2O
1351	Molybdenum(IV) oxide MoO_2	18868-43-4 127.94	brn-viol tetr cry	dec ≈1100 6.47	a	sl H_2O
1352	Molybdenum(IV) selenide $MoSe_2$	12058-18-3 253.86	gray hex cry	>1200 6.90		
1353	Molybdenum(IV) sulfide MoS_2	1317-33-5 160.07	blk powder or hex cry	2375 5.06	a,f	i H_2O; s conc acid

No.	Name Formula	CAS RN Mol. Wt.	Physical Form	mp/°C den/g cm^{-3}	bp/°C Other Data	Solubility
1354	Molybdenum(IV) telluride MoTe$_2$	12058-20-7 351.14	gray hex cry	 7.7		
1355	Molybdenum(V) chloride MoCl$_5$	10241-05-1 273.20	gr-blk monocl cry; hyg	194 2.93	268 b,c,d	s EtOH, eth
1356	Molybdenum(V) fluoride MoF$_5$	13819-84-6 190.93	yel monocl cry	67 3.5	213	
1357	Molybdenum(V) oxytrichloride MoOCl$_3$	13814-74-9 218.30	blk monocl cry	297	subl	reac H$_2$O
1358	Molybdenum(VI) acid monohydrate H$_2$MoO$_4$·H$_2$O	7782-91-4 179.97	wh powder	 3.1		sl H$_2$O; s alk
1359	Molybdenum(VI) fluoride MoF$_6$	7783-77-9 209.93	wh cub cry or col liq; hyg	17.5 2.54	34 a,b,c,d,e	reac H$_2$O
1360	Molybdenum(VI) oxytetrafluoride MoOF$_4$	14459-59-7 187.93	volatile solid	98		
1361	Molybdenum(VI) oxytetrachloride MoOCl$_4$	13814-75-0 253.75	grn hyg powder	101		
1362	Molybdenum(VI) dioxydichloride MoO$_2$Cl$_2$	13637-68-8 198.84	yel-oran solid	≈175 3.31		reac H$_2$O
1363	Molybdenum(VI) oxide MoO$_3$	1313-27-5 143.94	wh-yel rhomb cry	801 4.70	1155 a,c,d,e	sl H$_2$O; s conc acid
1364	Molybdenum(VI) metaphosphate Mo(PO$_3$)$_6$	133863-98-6 569.77	yel powder	 3.28		i H$_2$O, acid
1365	Neodymium Nd	7440-00-8 144.24	silv metal; hex	1021 7.01	3074 a,c	
1366	Neodymium boride NdB$_6$	12008-23-0 209.11	blk cub cry	2610 4.93		
1367	Neodymium bromide NdBr$_3$	13536-80-6 383.95	viol orth cry; hyg	682 5.3	1540	s H$_2$O
1368	Neodymium chloride NdCl$_3$	10024-93-8 250.60	viol hex cry	758 4.13	1600 a	vs H$_2$O, EtOH; i eth, chl
1369	Neodymium chloride hexahydrate NdCl$_3$·6H$_2$O	13477-89-9 358.69	purp cry	dec 124 2.3		vs H$_2$O; s EtOH
1370	Neodymium fluoride NdF$_3$	13709-42-7 201.24	viol hex cry; hyg	1377 6.51	2300 a	i H$_2$O
1371	Neodymium iodide NdI$_3$	13813-24-6 524.95	grn orth cry; hyg	784 5.85		s H$_2$O
1372	Neodymium nitrate hexahydrate Nd(NO$_3$)$_3$·6H$_2$O	14517-29-4 438.35	purp hyg cry			vs H$_2$O; s EtOH, ace
1373	Neodymium nitride NdN	25764-11-8 158.25	blk cub cry	 7.69		
1374	Neodymium oxide Nd$_2$O$_3$	1313-97-9 336.48	blue hex cry; hyg	2320 7.24	 a	i H$_2$O; s dil acid
1375	Neodymium sulfate Nd$_2$(SO$_4$)$_3$	13477-91-3 576.67	pink needles	dec ≈700		s H$_2$O
1376	Neodymium sulfide Nd$_2$S$_3$	12035-32-4 384.68	orth cry	2207 5.46		
1377	Neodymium telluride Nd$_2$Te$_3$	12035-35-7 671.28	gray orth cry	1377 7.0		
1378	Neon Ne	7440-01-9 20.180	col gas	-248.59 0.885 g/L	-246.08 a,b,c,d,e	sl H$_2$O
1379	Neptunium Np	7439-99-8 237	silv metal	644 20.2	 c	s HCl
1380	Neptunium(IV) oxide NpO$_2$	12035-79-9 269	grn cub cry	2547 11.1		
1381	Nickel Ni	7440-02-0 58.693	wh metal; cub	1455 8.90	2913 a,c,e	i H$_2$O; sl dil acid
1382	Nickel antimonide NiSb	12035-52-8 180.453	hex cry	1147 8.74	 f	
1383	Nickel arsenide NiAs	27016-75-7 133.615	hex cry	967 7.77	 f	
1384	Nickel boride Ni$_3$B	12007-02-2 186.891	refrac solid	1156 8.17		
1385	Nickel boride NiB	12007-00-0 69.504	grn refrac solid	1035 7.13		
1386	Nickel boride Ni$_2$B	12007-01-1 128.198	refrac solid	1125 7.90		
1387	Nickel carbonyl Ni(CO)$_4$	13463-39-3 170.735	col liq	-19.3 1.31	43 (exp ≈60) a,e	i H$_2$O; s EtOH, bz, ace, ctc
1388	Nickel phosphide Ni$_2$P	12035-64-2 148.361	hex cry	1100 7.33		
1389	Nickel silicide Ni$_2$Si	12059-14-2 145.473	orth cry	1255 7.40		
1390	Nickel silicide NiSi$_2$	12201-89-7 114.864	cub cry	993 4.83		
1391	Nickel(II) arsenate octahydrate Ni$_3$(AsO$_4$)$_2$·8H$_2$O	7784-48-7 598.040	yel-grn powder	dec 4.98		i H$_2$O; s acid
1392	Nickel(II) bromide NiBr$_2$	13462-88-9 218.501	yel hex cry; hyg	963 5.10	subl a	vs H$_2$O

No.	Name Formula	CAS RN Mol. Wt.	Physical Form	mp/°C den/g cm^{-3}	bp/°C Other Data	Solubility
1393	Nickel(II) bromide trihydrate NiBr$_2$·3H$_2$O	13462-88-9* 272.547	yel-grn hyg cry	dec 200		vs H$_2$O; s EtOH, eth
1394	Nickel(II) carbonate NiCO$_3$	3333-67-3 118.702	grn rhomb cry	4.39		i H$_2$O; s dil acid
1395	Nickel(II) chloride NiCl$_2$	7718-54-9 129.598	yel hex cry; hyg	1009 3.51	subl a,c,e	s H$_2$O, EtOH
1396	Nickel(II) chloride hexahydrate NiCl$_2$·6H$_2$O	7791-20-0 237.689	grn monocl cry			vs H$_2$O; s EtOH
1397	Nickel(II) cyanide tetrahydrate Ni(CN)$_2$·4H$_2$O	13477-95-7 182.790	grn plates	dec 200		i H$_2$O; sl dil acid; s NH$_4$OH
1398	Nickel(II) fluoride NiF$_2$	10028-18-9 96.690	yel tetr cry	1474 4.7	a	sl H$_2$O; i EtOH, eth
1399	Nickel(II) hydroxide Ni(OH)$_2$	12054-48-7 92.708	grn hex cry	dec 230 4.1	a	i H$_2$O
1400	Nickel(II) hydroxide monohydrate Ni(OH)$_2$·H$_2$O	36897-37-7 110.723	grn powder			i H$_2$O; s dil acid
1401	Nickel(II) iodide NiI$_2$	13462-90-3 312.502	blk hex cry; hyg	780 5.22	subl a	vs H$_2$O
1402	Nickel(II) iodide hexahydrate NiI$_2$·6H$_2$O	7790-34-3 420.593	grn monocl cry; hyg			vs H$_2$O, EtOH
1403	Nickel(II) nitrate hexahydrate Ni(NO$_3$)$_2$·6H$_2$O	13478-00-7 290.794	grn monocl cry; hyg	dec 56 2.05		vs H$_2$O; s EtOH
1404	Nickel(II) oxide NiO	1313-99-1 74.692	grn cub cry	1955 6.72	f	i H$_2$O; s acid
1405	Nickel(II) selenide NiSe	1314-05-2 137.65	yel-grn hex cry	980 7.2		
1406	Nickel(II) sulfate NiSO$_4$	7786-81-4 154.757	grn-yel orth cry	dec 840 4.01	a	vs H$_2$O
1407	Nickel(II) sulfate heptahydrate NiSO$_4$·7H$_2$O	10101-98-1 280.863	grn orth cry	1.98		s H$_2$O, EtOH
1408	Nickel(II) sulfate hexahydrate NiSO$_4$·6H$_2$O	10101-97-0 262.848	blue-grn tetr cry	dec ≈100 2.07		vs H$_2$O; sl EtOH
1409	Nickel(II) sulfide NiS	16812-54-7 90.759	yel hex cry	976 5.5	a,c,f	i H$_2$O
1410	Nickel(II) titanate NiTiO$_3$	12035-39-1 154.558	brn hex cry	5.0		
1411	Nickel(II,III) sulfide Ni$_3$S$_4$	12137-12-1 304.344	cub cry	995 4.77		
1412	Nickel(III) oxide Ni$_2$O$_3$	1314-06-3 165.385	gray-blk cub cry	dec ≈600	a	i H$_2$O; s hot acid
1413	Nickel(III) sulfide Ni$_3$S$_2$	12035-72-2 240.212	hex cry	787 5.87		
1414	Niobium Nb	7440-03-1 92.906	gray metal; cub	2477 8.57	4744 a,c	i acid
1415	Niobium boride NbB	12045-19-1 103.717	gray orth cry	2270 7.5		
1416	Niobium boride NbB$_2$	12007-29-3 114.528	gray hex cry	2900 6.8		
1417	Niobium carbide NbC	12069-94-2 104.917	gray cub cry	3608 7.82	4300	i H$_2$O, acid
1418	Niobium carbide Nb$_2$C	12011-99-3 197.824	refrac hex cry	3080 7.8		i H$_2$O
1419	Niobium nitride NbN	24621-21-4 106.913	gray cry; cub	2300 8.47		i HCl, acid
1420	Niobium phosphide NbP	12034-66-1 123.880	tetr cry	6.5		
1421	Niobium silicide NbSi$_2$	12034-80-9 149.077	gray hex cry	1950 5.7		
1422	Niobium(II) oxide NbO	12034-57-0 108.905	gray cub cry	1937 7.30	a,c	
1423	Niobium(III) bromide NbBr$_3$	15752-41-7 332.618	dark brn solid		subl	
1424	Niobium(III) chloride NbCl$_3$	13569-59-0 199.264	blk solid			
1425	Niobium(III) fluoride NbF$_3$	15195-53-6 149.901	blue cub cry	4.2		
1426	Niobium(IV) chloride NbCl$_4$	13569-70-5 234.717	viol-blk monocl cry	3.2		
1427	Niobium(IV) fluoride NbF$_4$	13842-88-1 168.900	blk tetr cry; hyg	dec >350 4.01		
1428	Niobium(IV) iodide NbI$_4$	13870-21-8 600.524	gray orth cry	503 5.6		
1429	Niobium(IV) oxide NbO$_2$	12034-59-2 124.905	wh tetr cry or powder	1902 5.9	a,c	
1430	Niobium(IV) selenide NbSe$_2$	12034-77-4 250.83	gray hex cry	>1300 6.3		
1431	Niobium(IV) sulfide NbS$_2$	12136-97-9 157.038	blk rhomb cry	4.4		

No.	Name / Formula	CAS RN / Mol. Wt.	Physical Form	mp/°C / den/g cm⁻³	bp/°C / Other Data	Solubility
1432	Niobium(IV) telluride / NbTe₂	12034-83-2 / 348.11	hex cry	7.6		
1433	Niobium(V) bromide / NbBr₅	13478-45-0 / 492.426	oran orth cry	254 / 4.36	360	s H₂O, EtOH
1434	Niobium(V) chloride / NbCl₅	10026-12-7 / 270.170	yel monocl cry; hyg	204.7 / 2.78	254.0 / a,b,c,d	reac H₂O; s HCl, ctc
1435	Niobium(V) fluoride / NbF₅	7783-68-8 / 187.898	col monocl cry; hyg	80 / 2.70	229 / a,b,c,d,e	reac H₂O; sl CS₂, chl
1436	Niobium(V) iodide / NbI₅	13779-92-5 / 727.428	yel-blk monocl cry	dec ≈200 / 5.32		
1437	Niobium(V) oxide / Nb₂O₅	1313-96-8 / 265.810	wh orth cry	1512 / 4.6	a,c	i H₂O; s HF
1438	Niobium(V) oxybromide / NbOBr₃	14459-75-7 / 348.617	yel-brn cry	dec ≈320	subl	
1439	Niobium(V) oxychloride / NbOCl₃	13597-20-1 / 215.263	wh tetr cry	3.72	subl	
1440	Niobium(V) dioxyfluoride / NbO₂F	15195-33-2 / 143.903	wh cub cry	4.0		
1441	Nitrogen / N₂	7727-37-9 / 28.013	col gas	-210.00 / 1.229 g/L	-195.79 / a,b,c,d,e	sl H₂O; i EtOH
1442	Nitramide / NO₂NH₂	7782-94-7 / 62.028	unstable wh cry	72 dec	a	s H₂O, EtOH, ace, eth; i chl
1443	Nitric acid / HNO₃	7697-37-2 / 63.013	col liq; hyg	-41.6 / 1.55	83 / a,c,d	vs H₂O
1444	Nitrous acid / HNO₂	7782-77-6 / 47.014	stable only in soln		a	
1445	Nitrous oxide / N₂O	10024-97-2 / 44.012	col gas	-90.8 / 1.931 g/L	-88.48 / a,b,c,d,e	sl H₂O; s EtOH, eth
1446	Nitric oxide / NO	10102-43-9 / 30.006	col gas	-163.6 / 1.317 g/L	-151.74 / a,b,c,d,e	sl H₂O
1447	Nitrogen dioxide / NO₂	10102-44-0 / 46.006	brn gas; equil with N₂O₄	2.019 g/L	see N₂O₄ / a	reac H₂O
1448	Nitrogen trioxide / N₂O₃	10544-73-7 / 76.011	blue solid or liq (low temp)	-101.1 / 1.4(2°C)	≈3 dec / a	reac H₂O
1449	Nitrogen tetroxide / N₂O₄	10544-72-6 / 92.011	col liq; equil with NO₂	-9.3 / 1.45(20°C)	21.15 / a,b,c,d,e	reac H₂O
1450	Nitrogen pentoxide / N₂O₅	10102-03-1 / 108.010	col hex cry	30 / 2.0	47 / a,e	s chl; sl ctc
1451	Nitrogen tribromide / NBr₃	15162-90-0 / 253.719	unstable solid	exp -100		
1452	Nitrogen trichloride / NCl₃	10025-85-1 / 120.365	yel oily liq; exp	-40 / 1.653	71 / a	i H₂O; s CS₂, bz, ctc
1453	Nitrogen trifluoride / NF₃	7783-54-2 / 71.002	col gas	-206.79 / 3.116 g/L	-128.75 / a,b,d,e	i H₂O
1454	Nitrogen triiodide / NI₃	13444-85-4 / 394.720	unstable blk cry; exp			
1455	Difluoramine / NHF₂	10405-27-3 / 53.012	col gas	2.326 g/L	-23 / a,b	
1456	cis-Difluorodiazine / N₂F₂	13812-43-6 / 66.010	col gas	<-195 / 2.896 g/L	-105.75 / a,b	
1457	trans-Difluorodiazine / N₂F₂	13776-62-0 / 66.010	col gas	-172 / 2.896 g/L	-111.45 / a,b	
1458	Tetrafluorohydrazine / N₂F₄	10036-47-2 / 104.007	col gas	-164.5 / 4.564 g/L	-74 / a,b,d	
1459	Nitrosyl bromide / NOBr	13444-87-6 / 109.910	red gas	-56 / 4.823 g/L	≈0 / a	reac H₂O
1460	Nitrosyl chloride / NOCl	2696-92-6 / 65.459	yel gas	-59.6 / 2.872 g/L	-5.5 / a,b,d,e	reac H₂O
1461	Nitrosyl fluoride / NOF	7789-25-5 / 49.004	col gas	-132.5 / 2.150 g/L	-59.9 / a,d,e	
1462	Trifluoramine oxide / NOF₃	13847-65-9 / 87.001	col gas	-161 / 3.818 g/L	-87.5 / b	
1463	Nitryl chloride / NO₂Cl	13444-90-1 / 81.459	col gas	-145 / 3.574 g/L	-15 / a,d	
1464	Nitryl fluoride / NO₂F	10022-50-1 / 65.004	col gas	-166 / 2.852 g/L	-72.4 / a,b,d,e	reac H₂O
1465	Nitrogen selenide / N₄Se₄	12033-88-4 / 371.87	red monocl cry; hyg	exp / 4.2		i H₂O, eth, EtOH; sl bz, CS₂
1466	Nobelium / No	10028-14-5 / 259	metal	827	a	
1467	Osmium / Os	7440-04-2 / 190.23	blue-wh metal; hex	3033 / 22.59	5012 / a,c	s aqua regia
1468	Osmium carbonyl / Os₃(CO)₁₂	15696-40-9 / 906.82	yel cry	3.48		
1469	Osmium(III) bromide / OsBr₃	59201-51-3 / 429.94	dark gray cry	dec 340		
1470	Osmium(III) chloride / OsCl₃	13444-93-4 / 296.59	gray cub cry	dec >450	a	i H₂O; s HNO₃

PHYSICAL CONSTANTS OF INORGANIC COMPOUNDS (continued)

No.	Name Formula	CAS RN Mol. Wt.	Physical Form	mp/°C den/g cm⁻³	bp/°C Other Data	Solubility
1471	Osmium(IV) chloride OsCl₄	10026-01-4 332.04	red-blk orth cry	4.38	450 sp	reac H₂O
1472	Osmium(IV) fluoride OsF₄	54120-05-7 266.22	yel cry	230		
1473	Osmium(IV) oxide OsO₂	12036-02-1 222.23	yel-brn tetr cry	11.4		i H₂O, acid
1474	Osmium(V) fluoride OsF₅	31576-40-6 285.22	blue cry	70		reac H₂O
1475	Osmium(VI) fluoride OsF₆	13768-38-2 304.22	yel cub cry	33.2 4.1	46 a	reac H₂O
1476	Osmium(VIII) oxide OsO₄	20816-12-0 254.23	yel monocl cry	41 5.1	135 a,b,c,e	sl H₂O
1477	Oxygen O₂	7782-44-7 31.999	col gas	-218.79 1.404 g/L	-182.95 a,b,c,d,e	sl H₂O, EtOH, os
1478	Ozone O₃	10028-15-6 47.998	blue gas	-193 2.106 g/L	-111.35 a,b,e	sl H₂O
1479	Palladium Pd	7440-05-3 106.42	silv-wh metal; cub	1554.9 12.0	2963 a,c	s aqua regia
1480	Palladium(II) sulfide PdS	12125-22-3 138.49	gray tetr cry	6.7	a	
1481	Palladium(II) bromide PdBr₂	13444-94-5 266.23	red-blk monocl cry; hyg	250 dec ≈5.2		i H₂O
1482	Palladium(II) chloride PdCl₂	7647-10-1 177.33	red rhomb cry; hyg	679 4.0		s H₂O, EtOH, ace
1483	Palladium(II) fluoride PdF₂	13444-96-7 144.42	viol tetr cry; hyg	952 5.76		reac H₂O
1484	Palladium(II) iodide PdI₂	7790-38-7 360.23	blk cry	360 dec 6.0		i H₂O, EtOH, eth
1485	Palladium(II) nitrate Pd(NO₃)₂	10102-05-3 230.43	brn hyg cry	dec		sl H₂O; s dil HNO₃
1486	Palladium(II) oxide PdO	1314-08-5 122.42	grn-blk tetr cry	750 dec 8.3	a	i H₂O, acid; sl aqua regia
1487	Phosphorus (white) P	7723-14-0 30.974	col waxlike cub cry	44.15 1.823	280.5 a,b,c,d,e	i H₂O; sl bz, EtOH, chl; s CS₂
1488	Phosphorus (red) P	7723-14-0 30.974	red-viol amorp powder	590tp 2.16	431sp a,b,c,d,e	i H₂O, os
1489	Phosphorus (black) P	7723-14-0 30.974	blk orth cry or amorp solid	610 2.69	a,b,c,d,e	i os
1490	Phosphine PH₃	7803-51-2 33.998	col gas; flam	-133 1.492 g/L	-87.75 a,b,d,e	i H₂O; sl EtOH, eth
1491	Diphosphine P₂H₄	13445-50-6 65.980	col liq	-99	63.5 dec a,d	reac H₂O
1492	Phosphoric acid (orthophosphoric acid) H₃PO₄	7664-38-2 97.995	col visc liq	42.4	407 a,c	s H₂O, EtOH
1493	Metaphosphoric acid HPO₃	37267-86-0 79.980	gl solid; hyg			sl H₂O; s EtOH
1494	Pyrophosphoric acid H₄P₂O₇	2466-09-3 177.975	wh cry	71.5	a	s H₂O
1495	Hypophosphoric acid H₄P₂O₆	7803-60-3 161.976	col orth cry	73 dec		vs H₂O
1496	Difluorophosphoric acid HPO₂F₂	13779-41-4 101.978	col liq	≈-94 1.583	dec 110	reac H₂O
1497	Phosphorous acid H₃PO₃	13598-36-2 81.996	wh hyg cry	74.4 1.65	200 a,c	vs H₂O, EtOH
1498	Hypophosphorous acid HPH₂O₂	6303-21-5 65.997	hyg cry or col oily liq	26.5 1.49	130 a,c	vs H₂O, EtOH, eth
1499	Phosphonium iodide PH₄I	12125-09-6 161.910	col tetr cry	18.5 2.86	62.5 e	reac H₂O, EtOH
1500	Phosphorus heptasulfide P₄S₇	12037-82-0 348.357	pale yel monocl cry	308 2.19	523	sl CS₂
1501	Phosphorus nitride P₃N₅	12136-91-3 162.955	wh amorp solid	dec 800		i H₂O; s os
1502	Phosphonitrilic chloride trimer (PNCl₂)₃	940-71-6 347.657	wh hyg cry	128.8 1.98		reac H₂O
1503	Phosphorus(III) bromide PBr₃	7789-60-8 270.686	col liq	-40 2.8	172.95 a,b,d,e	reac H₂O, EtOH; s ace, CS₂
1504	Phosphorus(III) chloride PCl₃	7719-12-2 137.332	col liq	-112 1.574	75.95 a,b,c,d,e	reac H₂O, EtOH; s bz, chl, eth
1505	Phosphorus(III) chloride difluoride PClF₂	14335-40-1 104.424	col gas	4.582 g/L	-47.25 b,d	
1506	Phosphorus(III) dichloride fluoride PCl₂F	15597-63-4 120.877	col gas	5.304 g/L	14 b,d	
1507	Phosphorus(III) fluoride PF₃	7783-55-3 87.969	col gas	-151.5 3.860 g/L	-101.5 a,b,d,e	reac H₂O
1508	Phosphorus(III) iodide PI₃	13455-01-1 411.687	red-oran hex cry; hyg	61.5 4.18	227 dec a,d	reac H₂O; s EtOH
1509	Phosphorus(III) oxide P₂O₃	1314-24-5 109.946	col monocl cry or liq	23.8 2.13	173 e	reac H₂O

No.	Name / Formula	CAS RN / Mol. Wt.	Physical Form	mp/°C den/g cm⁻³	bp/°C Other Data	Solubility
1510	Phosphorus(III) selenide P_2Se_3	1314-86-9 298.83	oran-red cry	245 1.31	≈380	reac H_2O; s bz, ctc, CS_2, ace
1511	Phosphorus(III) sulfide P_2S_3	12165-69-4 158.146	yel solid	290	490	reac H_2O; s EtOH, eth, CS_2
1512	Phosphorus(V) bromide PBr_5	7789-69-7 430.494	yel orth cry, Donnay	dec >100 3.61	a	reac H_2O, EtOH; s CS_2, ctc
1513	Phosphorus(V) chloride PCl_5	10026-13-8 208.238	wh-yel tetr cry; hyg	167 tp 2.1	160 sp a,b,e	reac H_2O; s CS_2, ctc
1514	Phosphorus(V) fluoride PF_5	7647-19-0 125.966	col gas	-93.8 5.527 g/L	-84.6 a,d	reac H_2O
1515	Phosphorus(V) oxide P_2O_5	1314-56-3 141.945	wh orth cry; hyg	562 2.30	605 c,e	reac H_2O, EtOH
1516	Phosphorus(V) oxybromide $POBr_3$	7789-59-5 286.685	faint oran plates	56 2.822	191.7 a,d	reac H_2O; s bz, eth, chl
1517	Phosphorus(V) oxychloride $POCl_3$	10025-87-3 153.331	col liq	1 1.645	105.5 a,c,d,e	reac H_2O, EtOH
1518	Phosphorus(V) oxyfluoride POF_3	13478-20-1 103.968	col gas	-39.1 tp 4.562 g/L	-39.7 sp a	reac H_2O
1519	Phosphorus(V) selenide P_2Se_5	1314-82-5 456.75	blk-purp amorp solid			reac hot H_2O, ctc; i CS_2
1520	Phosphorus(V) sulfide P_2S_5	1314-80-3 222.278	grn-yel hyg cry	285 2.03	515	reac H_2O; s CS_2
1521	Phosphorus(V) sulfide trifluoride PSF_3	2404-52-6 120.035	gas	-148.8 5.267 g/L	-52.25 b,d	
1522	Phosphorus(V) sulfide trichloride $PSCl_3$	3982-91-0 169.398	fuming liq	-36.2 1.635	125 e	reac H_2O, s bz, ctc, chl, CS_2
1523	Phosphorus(V) sulfide chloride difluoride $PSClF_2$	2524-02-9 136.490	col gas	-155 5.989 g/L	6.3 b	
1524	Platinum Pt	7440-06-4 195.08	silv-gray metal; cub	1768.4 21.5	3825 a,c,e,f	i acid; s aqua regia
1525	Hexachloroplatinic acid hexahydrate $H_2PtCl_6 \cdot 6H_2O$	16941-12-1 517.90	brn-yel hyg cry	60 2.43		vs H_2O, EtOH
1526	Platinum silicide $PtSi$	12137-83-6 223.17	orth cry	1229 12.4		
1527	Platinum(II) bromide $PtBr_2$	13455-12-4 354.89	red-brn powder	dec 250 6.65	a	i H_2O
1528	Platinum(II) chloride $PtCl_2$	10025-65-7 265.99	grn hex cry	dec 581 6.0	a	i H_2O, EtOH, eth; s HCl
1529	Platinum(II) iodide PtI_2	7790-39-8 448.89	blk powder	dec 325 6.4		i H_2O
1530	Platinum(II) oxide PtO	12035-82-4 211.08	blk tetr cry	dec 325 14.1		i H_2O, EtOH; s aqua regia
1531	Platinum(II) sulfide PtS	12038-20-9 227.15	tetr cry	10.25	a,f	
1532	Platinum(III) bromide $PtBr_3$	25985-07-3 434.79	grn-blk cry	dec 200	a	
1533	Platinum(III) chloride $PtCl_3$	25909-39-1 301.44	grn-blk cry	435 dec 5.26	a	
1534	Platinum(IV) bromide $PtBr_4$	68938-92-1 514.70	brn-blk cry	dec 180	a	i H_2O; sl EtOH, eth
1535	Platinum(IV) chloride $PtCl_4$	37773-49-2 336.89	red-brn cub cry	327 dec 4.30		sl H_2O
1536	Platinum(IV) chloride pentahydrate $PtCl_4 \cdot 5H_2O$	13454-96-1 426.97	red cry	2.43		s H_2O, EtOH
1537	Platinum(IV) fluoride PtF_4	13455-15-7 271.07	red cry	600		
1538	Platinum(IV) iodide PtI_4	7790-46-7 702.70	brn-blk powder	dec 130	a	s H_2O
1539	Platinum(IV) oxide PtO_2	1314-15-4 227.08	blk hex cry	450 11.8		i H_2O; s conc acid, dil alk
1540	Platinum(IV) sulfide PtS_2	12038-21-0 259.21	hex cry	7.85	a	
1541	Platinum(VI) fluoride PtF_6	13693-05-5 309.07	red cub cry	61.3 ≈4.0	69.1 a	
1542	Plutonium Pu	7440-07-5 244	silv-wh metal; monocl	640 19.7	3228 a,c	
1543	Plutonium nitride PuN	12033-54-4 258	gray cub cry	2550 14.4		
1544	Plutonium(II) oxide PuO	12035-83-5 260	cub cry	14.0		
1545	Plutonium(III) bromide $PuBr_3$	15752-46-2 484	grn orth cry	681 6.75		s H_2O
1546	Plutonium(III) chloride $PuCl_3$	13569-62-5 350	grn hex cry	760 5.71		s H_2O
1547	Plutonium(III) fluoride PuF_3	13842-83-6 301	purp hex cry	1425 9.33		i H_2O; sl acid
1548	Plutonium(III) iodide PuI_3	13813-46-2 625	grn orth cry; hyg	777 6.92		s H_2O

No.	Name Formula	CAS RN Mol. Wt.	Physical Form	mp/°C den/g cm^{-3}	bp/°C Other Data	Solubility
1549	Plutonium(III) oxide Pu_2O_3	12036-34-9 536	blk cub cry	10.5		
1550	Plutonium(IV) fluoride PuF_4	13709-56-3 320	red-brn monocl cry	1027 7.1		
1551	Plutonium(IV) oxide PuO_2	12059-95-9 276	yel-brn cub cry	2400 11.5		
1552	Plutonium(VI) fluoride PuF_6	13693-06-6 358	red-brn orth cry	52 5.08		
1553	Polonium Po	7440-08-6 209	silv metal; cub	254 9.20	962 a	
1554	Polonium(IV) chloride $PoCl_4$	10026-02-5 351	yel hyg cry	≈300	390	s H_2O, EtOH, ace
1555	Polonium(IV) oxide PoO_2	7446-06-2 241	yel cub cry	dec 500 8.9		
1556	Potassium K	7440-09-7 39.098	soft silv-wh metal; cub	63.38 0.89	759 a,c,e	reac H_2O
1557	Potassium acetate $KC_2H_3O_2$	127-08-2 98.143	wh hyg cry	292 1.57	a	vs H_2O; s EtOH; i eth
1558	Potassium aluminate trihydrate $K_2Al_2O_4 \cdot 3H_2O$	12003-63-3* 250.204	wh orth cry	2.13		vs H_2O; i EtOH
1559	Potassium aluminum silicate $KAlSi_3O_8$	1327-44-2 278.332	col monocl cry	2.56		i H_2O
1560	Potassium aluminum sulfate $KAl(SO_4)_2$	10043-67-1 258.207	wh hyg powder			sl H_2O
1561	Potassium aluminum sulfate dodecahydrate $KAl(SO_4)_2 \cdot 12H_2O$	7784-24-9 474.391	col cry	dec ≈100 1.72		sl H_2O
1562	Potassium arsenate K_3AsO_4	7784-41-0 256.215	col cry	2.8		s H_2O
1563	Potassium azide KN_3	20762-60-1 81.118	tetr cry; exp	2.04		vs H_2O
1564	Potassium borohydride KBH_4	13762-51-1 53.941	wh cub cry	dec ≈500 1.11	a	s H_2O
1565	Potassium bromate $KBrO_3$	7758-01-2 167.000	wh hex cry	434 dec 3.27	a	sl H_2O; i EtOH
1566	Potassium bromide KBr	7758-02-3 119.002	col cub cry; hyg	734 2.74	1435 a,c,e	vs H_2O; sl EtOH
1567	Potassium carbonate K_2CO_3	584-08-7 138.206	wh monocl cry; hyg	898 2.29	dec a,c	s H_2O; i EtOH
1568	Potassium carbonate sesquihydrate $K_2CO_3 \cdot 1.5H_2O$	6381-79-9 165.229	granular cry			vs H_2O
1569	Potassium chlorate $KClO_3$	3811-04-9 122.549	wh monocl cry	368 2.32	dec a	s H_2O
1570	Potassium chloride KCl	7447-40-7 74.551	wh cub cry	771 1.988	a,c,e,f	s H_2O; i eth, ace
1571	Potassium chromate K_2CrO_4	7789-00-6 194.191	yel orth cry	975 2.73		vs H_2O
1572	Potassium cyanate KCNO	590-28-3 81.115	wh tetr cry	dec ≈700 2.05		s H_2O; sl EtOH
1573	Potassium cyanide KCN	151-50-8 65.116	wh cub cry; hyg	634 1.55	a	vs H_2O; sl EtOH
1574	Potassium dichromate $K_2Cr_2O_7$	7778-50-9 294.185	oran-red tricl cry	398 2.68	dec ≈500	s H_2O
1575	Potassium dihydrogen arsenate KH_2AsO_4	7784-41-0 180.034	col cry	288 2.87		s H_2O; i EtOH
1576	Potassium dihydrogen phosphate KH_2PO_4	7778-77-0 136.085	wh tetr cry	253 2.34	a	vs H_2O; sl EtOH
1577	Potassium ferricyanide $K_3Fe(CN)_6$	13746-66-2 329.246	red cry	dec 1.89		vs H_2O
1578	Potassium ferrocyanide trihydrate $K_4Fe(CN)_6 \cdot 3H_2O$	14459-95-1 422.390	yel monocl cry	dec 60 1.85		s H_2O; i EtOH, eth
1579	Potassium fluoride KF	7789-23-3 58.096	wh cub cry	858 2.48	1502 a,c,e	s H_2O
1580	Potassium fluoride dihydrate $KF \cdot 2H_2O$	13455-21-5 94.127	monocl cry	41 dec 2.5		vs H_2O
1581	Potassium fluoroborate KBF_4	14075-53-7 125.903	col orth cry	530 2.505		sl H_2O, EtOH
1582	Potassium formate $KCHO_2$	590-29-4 84.116	col hyg cry	167 1.91	a	vs H_2O
1583	Potassium hexachloroosmate(IV) K_2OsCl_6	16871-60-6 481.14	red cub cry			vs H_2O; sl EtOH
1584	Potassium hexachloroplatinate K_2PtCl_6	16921-30-5 485.99	yel-oran cub cry	dec 250 3.50		sl H_2O; i EtOH
1585	Potassium hexacyanocobalt $K_3Co(CN)_6$	13963-58-1 332.334	yel monocl cry	dec 1.91		vs H_2O; i EtOH
1586	Potassium hexafluoromanganate(IV) K_2MnF_6	16962-31-5 247.125	yel hex cry			reac H_2O
1587	Potassium hexafluorosilicate K_2SiF_6	16871-90-2 220.273	wh cry	dec 2.27	a	sl H_2O; i EtOH

No.	Name / Formula	CAS RN / Mol. Wt.	Physical Form	mp/°C / den/g cm^{-3}	bp/°C / Other Data	Solubility
1588	Potassium hydride / KH	7693-26-7 / 40.106	cub cry	/ 1.43	/ a	reac H_2O
1589	Potassium hydrogen arsenite / $KAsO_2 \cdot HAsO_2$	10124-50-2 / 253.947	wh hyg powder			s H_2O
1590	Potassium hydrogen carbonate / $KHCO_3$	298-14-6 / 100.115	col monocl cry	dec ≈100 / 2.17	/ a	vs H_2O; i EtOH
1591	Potassium hydrogen fluoride / KHF_2	7789-29-9 / 78.103	col tetr cry	238.9 / 2.37	/ a,c	s H_2O; i EtOH
1592	Potassium hydrogen phosphate / K_2HPO_4	7758-11-4 / 174.176	wh hyg cry	dec		vs H_2O; sl EtOH
1593	Potassium hydrogen phosphite / K_2HPO_3	13492-26-7 / 158.177	wh hyg powder	dec		vs H_2O; i EtOH
1594	Potassium hydrogen selenite / $KHSeO_3$	7782-70-9 / 167.06	hyg orth cry	dec >100		s H_2O; sl EtOH
1595	Potassium hydrogen sulfate / $KHSO_4$	7646-93-7 / 136.170	wh monocl cry; hyg	≈200 / 2.32	/ a	vs H_2O
1596	Potassium hydrogen sulfide / KHS	1310-61-8 / 72.172	wh hex cry; hyg	≈450 / 1.69		s H_2O, EtOH
1597	Potassium hydrogen sulfide hemihydrate / $KHS \cdot 0.5H_2O$	1310-61-8* / 81.179	wh-yel hyg cry	≈175 / 1.7		vs H_2O, EtOH
1598	Potassium hydrogen sulfite / $KHSO_3$	7773-03-7 / 120.170	wh cry powder	dec 190		s H_2O; i EtOH
1599	Potassium hydrogen tartrate / $KHC_4H_4O_6$	868-14-4 / 188.178	wh cry	/ 1.98		sl H_2O; s acid, alk; i EtOH
1600	Potassium hydroxide / KOH	1310-58-3 / 56.105	wh rhomb cry; hyg	406 / 2.044	1327 / a,c,e	vs H_2O; s EtOH
1601	Potassium hypophosphite / KH_2PO_2	7782-87-8 / 104.087	wh hyg cry	dec		vs H_2O; s EtOH
1602	Potassium iodate / KIO_3	7758-05-6 / 214.001	wh monocl cry	560 dec / 3.89	/ a	sl H_2O
1603	Potassium iodide / KI	7681-11-0 / 166.003	col cub cry	681 / 3.12	1323 / a,c,e	vs H_2O; sl EtOH
1604	Potassium manganate / K_2MnO_4	10294-64-1 / 197.133	grn cry	190 dec		s H_2O; reac HCl
1605	Potassium metabisulfite / $K_2S_2O_5$	16731-55-8 / 222.326	wh powder	dec ≈150 / 2.3		vs H_2O; reac acid; i EtOH
1606	Potassium metaborate / KBO_2	13709-94-9 / 81.908	wh hex cry	≈2.3	/ a	
1607	Potassium molybdate / K_2MoO_4	13446-49-6 / 238.14	wh hyg cry	919 / 2.3		vs H_2O; i EtOH
1608	Potassium niobate / $KNbO_3$	12030-85-2 / 180.002	wh rhomb cry	≈1100 / 4.64		i H_2O
1609	Potassium nitrate / KNO_3	7757-79-1 / 101.103	col rhomb cry or powder	337 / 2.11	dec 400 / a,c	s H_2O; i EtOH
1610	Potassium nitrite / KNO_2	7758-09-0 / 85.104	wh hyg cry	441 / 1.915	exp 537 / a	vs H_2O; sl EtOH
1611	Potassium oxalate monohydrate / $K_2C_2O_4 \cdot H_2O$	6487-48-5 / 184.232	col cry	dec 160 / 2.13		vs H_2O
1612	Potassium oxide / K_2O	12136-45-7 / 94.196	gray cub cry	350 dec / 2.35	/ a	s H_2O, EtOH, eth
1613	Potassium perchlorate / $KClO_4$	7778-74-7 / 138.549	col orth cry; hyg	525 / 2.52	/ a	sl H_2O
1614	Potassium periodate / KIO_4	7790-21-8 / 230.001	col tetr cry	582 / 3.618	exp / a	sl H_2O
1615	Potassium permanganate / $KMnO_4$	7722-64-7 / 158.034	purp orth cry	dec / 2.7	/ a	s H_2O; reac EtOH
1616	Potassium peroxide / K_2O_2	17014-71-0 / 110.196	yel amorp solid	490	/ a	reac H_2O
1617	Potassium persulfate / $K_2S_2O_8$	7727-21-1 / 270.324	col cry	dec ≈100 / 2.48		sl H_2O
1618	Potassium phosphate / K_3PO_4	7778-53-2 / 212.266	wh orth cry; hyg	1340 / 2.564	/ a	vs H_2O; i EtOH
1619	Potassium pyrophosphate trihydrate / $K_4P_2O_7 \cdot 3H_2O$	7320-34-5* / 384.383	col hyg cry	1090 / 2.33		vs H_2O; i EtOH
1620	Potassium pyrosulfate / $K_2S_2O_7$	7790-62-7 / 254.325	col needles	≈325 / 2.28		s H_2O
1621	Potassium selenate / K_2SeO_4	7790-59-2 / 221.16	wh powder	/ 3.07		vs H_2O
1622	Potassium selenide / K_2Se	1312-74-9 / 157.16	red cub cry; hyg	800 / 2.29		s H_2O
1623	Potassium stannate trihydrate / $K_2SnO_3 \cdot 3H_2O$	12142-33-5* / 298.951	col cry	/ 3.20		vs H_2O; i EtOH
1624	Potassium sulfate / K_2SO_4	7778-80-5 / 174.261	wh orth cry	1069 / 2.66	/ a,c,f	s H_2O; i EtOH
1625	Potassium sulfide / K_2S	1312-73-8 / 110.263	red-yel cub cry; hyg	948 / 1.74	/ a,c	s H_2O, EtOH; i eth
1626	Potassium sulfide pentahydrate / $K_2S \cdot 5H_2O$	1312-73-8* / 200.339	col rhomb cry	60		vs H_2O, EtOH; i eth

No.	Name Formula	CAS RN Mol. Wt.	Physical Form	mp/°C den/g cm^{-3}	bp/°C Other Data	Solubility
1627	Potassium sulfite dihydrate $K_2SO_3 \cdot 2H_2O$	7790-56-9 194.292	wh monocl cry	dec		s H_2O; sl EtOH; dec dil acid
1628	Potassium superoxide KO_2	12030-88-5 71.097	yel tetr cry; hyg	380 2.16	a	reac H_2O
1629	Potassium tellurate(VI) trihydrate $K_2TeO_4 \cdot 3H_2O$	15571-91-2* 323.84	wh cry powder			s H_2O
1630	Potassium tellurite K_2TeO_3	7790-58-1 253.80	wh hyg cry	dec ≈460		vs H_2O
1631	Potassium tetraborate pentahydrate $K_2B_4O_7 \cdot 5H_2O$	1332-77-0 323.513	wh cry powder			s H_2O; sl EtOH
1632	Potassium tetrachloroaurate dihydrate $KAuCl_4 \cdot 2H_2O$	13682-61-6 413.907	yel monocl cry			s H_2O, EtOH, eth
1633	Potassium tetrachloroplatinate K_2PtCl_4	10025-99-7 415.09	pink-red tetr cry	dec 500 3.38		s H_2O; i EtOH
1634	Potassium tetracyanoplatinate(II) trihydrate $K_2Pt(CN)_4 \cdot 3H_2O$	562-76-5* 431.39	col rhomb prisms			s H_2O
1635	Potassium tetraiodomercurate(II) K_2HgI_4	7783-33-7 786.40	yel hyg cry	4.29		vs H_2O; s EtOH, eth, ace
1636	Potassium thiocyanate KSCN	333-20-0 97.182	col tetr cry; hyg	173 1.88	dec 500 a	vs H_2O; s EtOH
1637	Potassium thiosulfate $K_2S_2O_3$	10294-66-3 190.327	col hyg cry			vs H_2O; i EtOH
1638	Potassium titanate K_2TiO_3	12030-97-6 174.062	wh orth cry	1515 3.1		reac H_2O
1639	Potassium triiodide monohydrate $KI_3 \cdot H_2O$	7790-42-3 437.827	brn monocl cry; hyg	dec 225 3.5		s H_2O; reac EtOH, eth
1640	Potassium thiocarbonate K_2CS_3	26750-66-3 186.406	yel-red hyg cry			vs H_2O
1641	Potassium tungstate K_2WO_4	7790-60-5 326.04	hyg cry	921 3.12		vs H_2O; i EtOH
1642	Potassium uranate $K_2U_2O_7$	7790-63-8 666.251	oran cub cry	6.12		i H_2O; s acid
1643	Praseodymium Pr	7440-10-0 140.908	silv metal; hex	931 6.77	3520 a,c	
1644	Praseodymium boride PrB_6	12008-27-4 205.774	blk cub cry	2610 4.84		
1645	Praseodymium bromide $PrBr_3$	13536-53-3 380.620	grn hex cry; hyg	693 5.28		s H_2O
1646	Praseodymium chloride $PrCl_3$	10361-79-2 247.266	grn hex needles; hyg	786 4.0	a	s H_2O, EtOH
1647	Praseodymium chloride heptahydrate $PrCl_3 \cdot 7H_2O$	10025-90-8 373.373	grn cry	dec 110		s H_2O, EtOH
1648	Praseodymium fluoride PrF_3	13709-46-1 197.903	grn hex cry	1395 6.3		
1649	Praseodymium iodide PrI_3	13813-23-5 521.621	orth hyg cry	737 ≈5.8		s H_2O
1650	Praseodymium nitrate hexahydrate $Pr(NO_3)_3 \cdot 6H_2O$	14483-17-1 435.014	ligtt grn cry			vs H_2O
1651	Praseodymium nitride PrN	25764-09-4 154.915	cub cry	7.46		
1652	Praseodymium oxide Pr_2O_3	12036-32-7 329.813	wh hex cry	2300 6.9	a	
1653	Praseodymium silicide $PrSi_2$	12066-83-0 197.079	tetr cry	1712 5.46		
1654	Praseodymium sulfide Pr_2S_3	12038-13-0 378.013	cub cry	1765 5.1		
1655	Praseodymium telluride Pr_2Te_3	12038-12-9 664.62	cub cry	1500 ≈7.0		
1656	Promethium Pm	7440-12-2 145	silv metal; hex	1042 7.26	3000 a	
1657	Protactinium Pa	7440-13-3 231.036	shiny metal; tetr or cub	1572 15.4	a,c	
1658	Protactinium(V) chloride $PaCl_5$	13760-41-3 408.300	yel monocl cry	306 3.74	a	
1659	Radium Ra	7440-14-4 226	wh metal; cub	700 5	a	
1660	Radium bromide $RaBr_2$	10031-23-9 386	wh orth cry	728 5.79		s H_2O, EtOH
1661	Radium chloride $RaCl_2$	10025-66-8 297	wh orth cry	1000 4.9		s H_2O, EtOH
1662	Radium fluoride RaF_2	20610-49-5 264	wh cub cry	6.7		
1663	Radium sulfate $RaSO_4$	7446-16-4 322	wh cry		a	i H_2O, acid
1664	Radon Rn	10043-92-2 222	col gas	-71 9.741 g/L	-61.7 a,b,e	sl H_2O
1665	Rhenium Re	7440-15-5 186.207	sllv-gray metal	3186 20.8	5596 a,c	i HCl

No.	Name Formula	CAS RN Mol. Wt.	Physical Form	mp/°C den/g cm^{-3}	bp/°C Other Data	Solubility
1666	Perrhenic acid $HReO_4$	13768-11-1 251.213	exists only in soln		a	vs H_2O, os
1667	Rhenium carbonyl $Re_2(CO)_{10}$	14285-68-8 652.518	yel-wh cry	170 dec 2.87		s os
1668	Rhenium(III) bromide $ReBr_3$	13569-49-8 425.919	red-brn monocl cry	6.10	subl 500 a	s ace, MeOH, EtOH
1669	Rhenium(III) chloride $ReCl_3$	13569-63-6 292.565	red-blk hyg cry	dec 500 4.81	a	s H_2O
1670	Rhenium(III) iodide ReI_3	15622-42-1 566.920	blk solid	dec		
1671	Rhenium(IV) chloride $ReCl_4$	13569-71-6 328.018	purp-blk cry; hyg	dec 300 4.9		
1672	Rhenium(IV) fluoride ReF_4	15192-42-4 262.201	blue tetr cry	7.49	subl >300	
1673	Rhenium(IV) oxide ReO_2	12036-09-8 218.206	gray orth cry	dec 900 11.4		
1674	Rhenium(IV) sulfide ReS_2	12038-63-0 250.339	tricl cry	7.6		
1675	Rhenium(IV) telluride $ReTe_2$	12067-00-4 441.41	orth cry	8.50		
1676	Rhenium(V) bromide $ReBr_5$	30937-53-2 585.727	brn solid	110 dec		
1677	Rhenium(V) chloride $ReCl_5$	39368-69-9 363.471	brn-blk solid	220 4.9		reac H_2O
1678	Rhenium(V) fluoride ReF_5	30937-52-1 281.199	yel-grn solid	48	≈220	
1679	Rhenium(V) oxide Re_2O_5	12165-05-8 452.411	blue-blk tetr cry	≈7		
1680	Rhenium(VI) chloride $ReCl_6$	31234-26-1 398.923	red-grn solid	29		
1681	Rhenium(VI) fluoride ReF_6	10049-17-9 300.197	yel liq or cub cry	18.5 4.06(cry)	33.7	s HNO_3
1682	Rhenium(VI) oxide ReO_3	1314-28-9 234.205	redcub cry	400 dec 6.9		i H_2O, acid, alk
1683	Rhenium(VI) oxytetrachloride $ReOCl_4$	13814-76-1 344.017	brn cry	29.3	223 b	reac H_2O
1684	Rhenium(VI) oxytetrafluoride $ReOF_4$	17026-29-8 278.200	blue solid	108	171	
1685	Rhenium(VII) fluoride ReF_7	17029-21-9 319.196	yel cub cry	48.3 4.32	73.7	
1686	Rhenium(VII) oxide Re_2O_7	1314-68-7 484.410	yel hyg cry	297 6.10	360 a,b,c,e	s H_2O, EtOH, eth, diox, py
1687	Rhenium(VII) trioxychloride ReO_3Cl	7791-09-5 269.658	col liq	4.5 3.87	128	reac H_2O
1688	Rhenium(VII) trioxyfluoride ReO_3F	42246-24-2 253.203	yel solid	147	164	
1689	Rhenium(VII) dioxytrifluoride ReO_2F_3	57246-89-6 275.201	yel solid	90	185	reac H_2O
1690	Rhenium(VII) oxypentafluoride $ReOF_5$	23377-53-9 297.198	cream solid	43.8	73.0	
1691	Rhenium(VII) sulfide Re_2S_7	12038-67-4 596.876	brn-blk tetr cry	4.87		i H_2O
1692	Rhodium Rh	7440-16-6 102.906	silv-wh metal; cub	1964 12.4	3695 a,c	i acid, sl aqua regia
1693	Rhodium carbonyl chloride $[Rh(CO)_2Cl]_2$	14523-22-9 388.758	red-oran cry	124		s os
1694	Rhodium dodecacarbonyl $Rh_4(CO)_{12}$	19584-30-6 747.747	red hyg cry	2.52		reac H_2O
1695	Rhodium(III) chloride $RhCl_3$	10049-07-7 209.264	red monocl cry	5.38	717 a	i H_2O; s alk
1696	Rhodium(III) fluoride RhF_3	60804-25-3 159.901	red hex cry	5.4		
1697	Rhodium(III) iodide RhI_3	15492-38-3 483.619	blk monocl cry; hyg	6.4		
1698	Rhodium(III) oxide Rh_2O_3	12036-35-0 253.809	gray hex cry	1100 dec 8.2	a	
1699	Rhodium(III) sulfate $Rh_2(SO_4)_3$	10489-46-0 494.002	red-yel solid	dec >500		
1700	Rhodium(IV) oxide RhO_2	12137-27-8 134.905	blk tetr cry	7.2		
1701	Rhodium(VI) fluoride RhF_6	13693-07-7 216.896	blk cub cry	≈70 3.1		
1702	Rubidium Rb	7440-17-7 85.468	soft silv metal; cub	39.31 1.53	688 a,c,e	reac H_2O
1703	Rubidium aluminum sulfate $RbAl(SO_4)_2$	13530-57-9 304.577	hex cry	≈3.1		sl H_2O; i EtOH

No.	Name Formula	CAS RN Mol. Wt.	Physical Form	mp/°C den/g cm^{-3}	bp/°C Other Data	Solubility
1704	Rubidium aluminum sulfate dodecahydrate RbAl(SO$_4$)$_2$·12H$_2$O	7784-29-4 520.761	col cub cry	≈100 dec ≈1.9		s H$_2$O; i EtOH
1705	Rubidium azide RbN$_3$	22756-36-1 127.488	tetr cry; exp	317 2.79		
1706	Rubidium bromide RbBr	7789-39-1 165.372	wh cub cry; hyg	682 3.35	1340 a,c,e	vs H$_2$O
1707	Rubidium carbonate Rb$_2$CO$_3$	584-09-8 230.945	col monocl cry; hyg	837	a	vs H$_2$O
1708	Rubidium chloride RbCl	7791-11-9 120.921	wh cub cry; hyg	715 2.76	1390 a,c,e	s H$_2$O; sl EtOH
1709	Rubidium cyanide RbCN	19073-56-4 111.486	wh cub cry	 2.3		s H$_2$O; i EtOH, eth
1710	Rubidium fluoride RbF	13446-74-7 104.466	wh cub cry; hyg	833 3.2	1410 a,c,e	vs H$_2$O; i EtOH
1711	Rubidium fluoride RbHF$_2$	12280-64-7 124.473	tetr cry	188 3.3	a	
1712	Rubidium hydride RbH	13446-75-8 86.476	wh cub cry; flam	dec ≈170 2.60	a	reac H$_2$O
1713	Rubidium hydrogen carbonate RbHCO$_3$	19088-74-5 146.485	wh rhomb cry	dec 175		vs H$_2$O
1714	Rubidium hydrogen sulfate RbHSO$_4$	15587-72-1 182.540	col monocl cry	208 2.9	a	s H$_2$O
1715	Rubidium hydroxide RbOH	1310-82-3 102.475	gray-wh orth cry; hyg	300 3.2	a	vs H$_2$O; s EtOH
1716	Rubidium iodide RbI	7790-29-6 212.372	wh cub cry	642 3.55	1300 a,c,e	s H$_2$O, EtOH
1717	Rubidium nitrate RbNO$_3$	13126-12-0 147.473	wh hex cry; hyg	305 3.11	a,c	vs H$_2$O
1718	Rubidium oxide Rb$_2$O	18088-11-4 186.935	yel-brn cub cry; hyg	dec 400 4.0	a	reac H$_2$O
1719	Rubidium perchlorate RbClO$_4$	13510-42-4 184.919	wh hyg cry	281 2.8	dec 600 a	sl H$_2$O
1720	Rubidium peroxide Rb$_2$O$_2$	23611-30-5 202.935	wh orth cry	 3.8	a	reac H$_2$O
1721	Rubidium selenide Rb$_2$Se	31052-43-4 249.90	wh cub cry	733 3.22		reac H$_2$O
1722	Rubidium sulfate Rb$_2$SO$_4$	7488-54-2 267.000	wh orth cry	1050 3.6	a	vs H$_2$O
1723	Rubidium sulfide Rb$_2$S	31083-74-6 203.002	wh cub cry	425 2.91		s H$_2$O
1724	Rubidium superoxide RbO$_2$	12137-25-6 117.467	tetr cry	412 ≈3.0	a	
1725	Ruthenium Ru	7440-18-8 101.07	silv-wh metal; hex	2334 12.1	4150 a,c	i acid, aqua regia
1726	Ruthenium dodecacarbonyl Ru$_3$(CO)$_{12}$	15243-33-1 639.34	oran cry	dec 150		
1727	Ruthenium(III) acetylacetonate Ru(CH$_3$COCHCOCH$_3$)$_3$	14284-93-6 398.40	red-brn cry	230		
1728	Ruthenium(III) bromide RuBr$_3$	14014-88-1 340.78	brn hex cry	dec >400 5.3	a	
1729	Ruthenium(III) chloride RuCl$_3$	10049-08-8 207.43	brn hex cry	dec >500 3.1	a	i H$_2$O; s EtOH
1730	Ruthenium(III) fluoride RuF$_3$	51621-05-7 158.07	brn rhomb cry	dec >600 5.36		
1731	Ruthenium(III) iodide RuI$_3$	13896-65-6 481.78	blk hex cry	 6.0	a	
1732	Ruthenium(IV) fluoride RuF$_4$	71500-16-8 177.06	yel cry			reac H$_2$O
1733	Ruthenium(IV) oxide RuO$_2$	12036-10-1 133.07	gray-blk tetr cry	 7.05	a	i H$_2$O, acid
1734	Ruthenium(V) fluoride RuF$_5$	14521-18-7 196.06	grn monocl cry	86.5 3.90	227	
1735	Ruthenium(VI) fluoride RuF$_6$	13693-08-8 215.06	dark brn orth cry	54 3.54		reac H$_2$O
1736	Ruthenium(VIII) oxide RuO$_4$	20427-56-9 165.07	yel monocl prisms	25.4 3.29	40 a	sl H$_2$O; vs ctc; reac EtOH
1737	Samarium Sm	7440-19-9 150.36	silv metal; rhomb	1074 7.52	1794 a,c	
1738	Samarium boride SmB$_6$	12008-29-6 215.23	refrac solid	2580 5.07		
1739	Samarium silicide SmSi$_2$	12300-22-0 206.53	orth cry	 5.14		
1740	Samarium(II) bromide SmBr$_2$	50801-97-3 310.17	brn cry	669		reac H$_2$O
1741	Samarium(II) chloride SmCl$_2$	13874-75-4 221.27	brn cry	855 3.69		reac H$_2$O
1742	Samarium(II) fluoride SmF$_2$	15192-17-3 188.36	purp cry			reac H$_2$O

No.	Name Formula	CAS RN Mol. Wt.	Physical Form	mp/°C den/g cm^{-3}	bp/°C Other Data	Solubility
1743	Samarium(II) iodide SmI_2	32248-43-4 404.17	grn cry	520		reac H_2O
1744	Samarium(III) bromide $SmBr_3$	13759-87-0 390.07	yel cry	640		reac H_2O
1745	Samarium(III) chloride $SmCl_3$	10361-82-7 256.72	yel cry	682 4.46	a	reac H_2O
1746	Samarium(III) chloride hexahydrate $SmCl_3 \cdot 6H_2O$	13465-55-9 364.81	yel cry	dec 2.38		reac H_2O
1747	Samarium(III) fluoride SmF_3	13765-24-7 207.36	wh cry	1306	a	reac H_2O
1748	Samarium(III) iodide SmI_3	13813-25-7 531.07	oran cry	850		reac H_2O
1749	Samarium(III) oxide Sm_2O_3	12060-58-1 348.72	yel-wh cub cry	2335 7.6	a	
1750	Samarium(III) sulfate octahydrate $Sm_2(SO_4)_3 \cdot 8H_2O$	13465-58-2 733.03	yel cry	2.93		sl H_2O
1751	Samarium(III) sulfide Sm_2S_3	12067-22-0 396.92	gray-brn cub cry	1720 5.87		
1752	Samarium(III) telluride Sm_2Te_3	12040-00-5 683.52	orth cry	7.31		
1753	Scandium Sc	7440-20-2 44.956	silv metal; hex	1541 2.99	2836 a,c	
1754	Scandium boride ScB_2	12007-34-0 66.578	refrac solid	2250 3.17		
1755	Scandium bromide $ScBr_3$	13465-59-3 284.668	wh hyg cry	969 9.33	a	s H_2O
1756	Scandium chloride $ScCl_3$	10361-84-9 151.314	wh hyg cry	967 2.4	a	s H_2O; i EtOH
1757	Scandium fluoride ScF_3	13709-47-2 101.951	wh powder	1515	a	sl H_2O
1758	Scandium oxide Sc_2O_3	12060-08-1 137.910	wh cub cry	2485 3.864	a	s conc acid
1759	Scandium sulfide Sc_2S_3	12166-29-9 186.110	yel orth cry	1775 2.91		
1760	Scandium telluride Sc_2Te_3	12166-44-8 472.71	blk hex cry	5.29		
1761	Selenium (α form) Se	7782-49-2 78.96	red monocl cry	221 4.39	685 a,b,c,d,e	i H_2O, EtOH; sl eth
1762	Selenium (gray) Se	7782-49-2 78.96	gray metallic cry; hex	221 4.81	685 a,b,c,d,e	i H_2O, CS_2
1763	Selenium (vitreous) Se	7782-49-2 78.96	blk amorp solid	trans to gray Se 180 4.28	685 a,b,c,d,e	i H_2O; sl CS_2
1764	Selenic acid H_2SeO_4	7783-08-6 144.97	wh hyg solid	58 2.95	260 dec a	vs H_2O; reac EtOH
1765	Selenous acid H_2SeO_3	7783-00-8 128.97	wh hyg cry	dec 70 3.0		vs H_2O; s EtOH
1766	Selenium dioxide SeO_2	7446-08-4 110.96	wh tetr needles or powder	340 tp 3.95	315 sp a,e	s H_2O, EtOH, MeOH; sl ace
1767	Selenium trioxide SeO_3	13768-86-0 126.96	wh tetr cry; hyg	118 3.44	subl	s H_2O, os
1768	Selenium bromide Se_2Br_2	7789-52-8 317.73	red liq	3.60	225 dec	reac H_2O; s CS_2, chl
1769	Selenium chloride Se_2Cl_2	10025-68-0 228.83	yel-brn oily liq	-85 2.774	130 dec	reac H_2O; s CS_2, bz, ctc, chl
1770	Selenium tetrabromide $SeBr_4$	7789-65-3 398.58	oran-red cry	123		reac H_2O; s CS_2, chl
1771	Selenium tetrachloride $SeCl_4$	10026-03-6 220.77	wh-yel cry	305 2.6	subl e	reac H_2O
1772	Selenium tetrafluoride SeF_4	13465-66-2 154.95	col liq	-10 2.75	106 d	reac H_2O, vs EtOH, eth
1773	Selenium hexafluoride SeF_6	7783-79-1 192.95	col gas	-34.6 tp 8.467 g/L	-46.6 sp a,b,e	i H_2O
1774	Selenium oxybromide $SeOBr_2$	7789-51-7 254.77	red-yel solid	41.6 3.38	220 dec	reac H_2O; s CS_2, bz, ctc
1775	Selenium oxychloride $SeOCl_2$	7791-23-3 165.86	col or yel liq	8.5 2.44	177 b,e	reac H_2O; s ctc, chl, bz, tol
1776	Selenium oxyfluoride $SeOF_2$	7783-43-9 132.96	col liq	15 2.8	125	reac H_2O
1777	Selenium dioxydifluoride SeO_2F_2	14984-81-7 148.96	col gas	-99.5 6.536 g/L	-8.4	reac H_2O
1778	Selenium sulfide Se_2S_6	75926-26-0 350.32	oran needles	121.5 2.44		s CS_2; sl bz
1779	Selenium sulfide Se_4S_4	75926-28-2 444.10	red cry	113 dec 3.29		s bz; sl CS_2
1780	Silicon Si	7440-21-3 28.086	gray cry or brn amorp solid	1414 2.3290	3265 a,c,e	i H_2O, acid; s alk
1781	Silane SiH_4	7803-62-5 32.118	col gas; flam	-185 1.409 g/L	-111.9 a,d,e	reac H_2O; i EtOH, bz

No.	Name Formula	CAS RN Mol. Wt.	Physical Form	mp/°C den/g cm^{-3}	bp/°C Other Data	Solubility
1782	Bromosilane SiH$_3$Br	13465-73-1 111.014	col gas	-94 4.871 g/L	1.9 a,d,e	
1783	Bromotrichlorosilane SiCl$_3$Br	13465-74-2 214.348	col liq	-62 1.826	80.3 a	reac H$_2$O
1784	Chlorosilane SiH$_3$Cl	13465-78-6 66.563	col gas	-118 2.921 g/L	-30.4 a,d,e	
1785	Chlorotrifluorosilane SiF$_3$Cl	14049-36-6 120.534	col gas	-138 5.289 g/L	-70.0 b,d	reac H$_2$O
1786	Dibromosilane SiH$_2$Br$_2$	13768-94-0 189.910	liq	-70.1 	66 a,d,e	
1787	Dichlorosilane SiH$_2$Cl$_2$	4109-96-0 101.007	col gas; flam	-122 4.432 g/L	8.3 a,d	reac H$_2$O
1788	Dichlorodifluorosilane SiCl$_2$F$_2$	18356-71-3 136.988	col gas	-44 6.011 g/L	-32 b,d,e	reac H$_2$O
1789	Difluorosilane SiH$_2$F$_2$	13824-36-7 68.099	col gas	-122 2.988 g/L	-77.8 d,e	
1790	Diiodosilane SiH$_2$I$_2$	13760-02-6 283.911	col liq	-1 	150 e	
1791	Fluorosilane SiH$_3$F	13537-33-2 50.108	col gas	 2.199 g/L	-98.6 a,d,e	
1792	Iodosilane SiH$_3$I	13598-42-0 158.014	col liq	-57 	45.6 a,e	
1793	Tetrabromosilane SiBr$_4$	7789-66-4 347.702	col fuming liq	5.2 2.8	154 a,b,d	reac H$_2$O
1794	Tetrachlorosilane SiCl$_4$	10026-04-7 169.897	col fuming liq	-68.85 1.5	57.65 a,b,c,d,e	reac H$_2$O
1795	Tetrafluorosilane SiF$_4$	7783-61-1 104.080	col gas	-90.2 4.567 g/L	-86 a,b,e	reac H$_2$O
1796	Tetraiodosilane SiI$_4$	13465-84-4 535.704	wh powder	120.5 4.1	287.35 a,b,c,d	
1797	Tribromosilane SiHBr$_3$	7789-57-3 268.806	flam liq	-73 2.7	109 a,b,d,e	reac H$_2$O
1798	Tribromochlorosilane SiBr$_3$Cl	13465-76-4 303.251	col liq	-20.8 2.497	127 a	reac H$_2$O
1799	Trichlorosilane SiHCl$_3$	10025-78-2 135.452	fuming liq	-128.2 1.331	33 a,b,d,e	reac H$_2$O
1800	Trichlorofluorosilane SiCl$_3$F	14965-52-7 153.442	col gas	 6.733 g/L	12.25 b,e	
1801	Trifluorosilane SiHF$_3$	13465-71-9 86.089	col gas	-131 3.778 g/L	-95 a,d,e	
1802	Triiodosilane SiHI$_3$	13465-72-0 409.807	liq	8 	220 dec 	
1803	Disilane Si$_2$H$_6$	1590-87-0 62.219	col gas; flam	-132.5 2.730 g/L	-14.3 a,d,e	reac H$_2$O, etc, chl, s EtOH, bz
1804	Trisilane Si$_3$H$_8$	7783-26-8 92.321	flam liq	-117.4 0.725(-117°C)	52.9 a,d,e	reac H$_2$O
1805	Tetrasilane Si$_4$H$_{10}$	7783-29-1 122.421	col liq	-84.3 0.79(-84°C)	107.4 e	reac H$_2$O
1806	Disiloxane (SiH$_3$)$_2$O	13597-73-4 78.218	gas	-144 3.432 g/L	-15.2 e	
1807	Metasilicic acid H$_2$SiO$_3$	7699-41-4 78.100	wh amorp powder	 	 a,f	i H$_2$O; s HF
1808	Orthosilicic acid H$_4$SiO$_4$	10193-36-9 96.116	exists only in soln	 	 a	
1809	Fluorosilicic acid H$_2$SiF$_6$	16961-83-4 144.092	stable only in aq soln	 	 	s H$_2$O
1810	Silicon carbide SiC	409-21-2 40.097	hard grn-black hex cry	2830 3.16	 a,f	i H$_2$O, EtOH
1811	Silicon nitride Si$_3$N$_4$	12033-89-5 140.284	gray refrac solid; hex	1900 3.17	 a	
1812	Silicon monoxide SiO	10097-28-6 44.085	blk cub cry, stable >1200	 2.18	 a	
1813	Silicon dioxide (α quartz) SiO$_2$	14808-60-7 60.085	col hex cry	573 trans to quartz 2.648	2950 a,c,f	i H$_2$O, acid; s HF
1814	Silicon dioxide (β quartz) SiO$_2$	14808-60-7 60.085	col hex cry	867 trans to tridymite 2.533(600°C)	2950 a,c,f	i H$_2$O, acid; s HF
1815	Silicon dioxide (tridymite) SiO$_2$	15468-32-3 60.085	col hex cry	1470 trans to cristobalite 2.265	2950 	i H$_2$O, acid; s HF
1816	Silicon dioxide (cristobalite) SiO$_2$	14464-46-1 60.085	col hex cry	1713 2.334	2950 e	i H$_2$O, acid; s HF
1817	Silicon dioxide (vitreous) SiO$_2$	60676-86-0 60.085	col amorp solid	1713 2.196	2950 	i H$_2$O, acid; s HF
1818	Silicon monosulfide SiS	50927-81-6 60.152	yel-red hyg powder	≈900 1.85	940 	reac H$_2$O
1819	Silicon disulfide SiS$_2$	13759-10-9 92.218	wh rhomb cry	1090 2.04	subl 	reac H$_2$O, EtOH; i bz

No.	Name Formula	CAS RN Mol. Wt.	Physical Form	mp/°C den/g cm^{-3}	bp/°C Other Data	Solubility
1820	Silver Ag	7440-22-4 107.868	silv metal; cub	961.78 10.5	2162 a,c,e,f	
1821	Silver azide AgN$_3$	13863-88-2 149.888	orth cry; exp	exp ≈250 4.9		
1822	Silver subfluoride Ag$_2$F	1302-01-8 234.734	yel hex cry	dec 100 8.6		reac H$_2$O
1823	Silver(I) acetate AgC$_2$H$_3$O$_2$	563-63-3 166.913	wh needles or powder	dec 3.26		s H$_2$O
1824	Silver(I) acetylide Ag$_2$C$_2$	7659-31-6 239.758	wh powder; exp			
1825	Silver(I) acetylide AgC$_2$H	13092-75-6 132.898	wh powder; exp			
1826	Silver(I) bromate AgBrO$_3$	7783-89-3 235.770	wh tetr cry	dec 5.2	a	sl H$_2$O
1827	Silver(I) bromide AgBr	7785-23-1 187.772	yel cub cry	432 6.47	1502 a,c,d,f	i H$_2$O, acid
1828	Silver(I) carbonate Ag$_2$CO$_3$	534-16-7 275.745	yel monocl cry	218 6.077		sl H$_2$O; s acid
1829	Silver(I) chlorate AgClO$_3$	7783-92-8 191.319	wh tetr cry	230 4.430	dec 270 a	s H$_2$O; sl EtOH
1830	Silver(I) chloride AgCl	7783-90-6 143.321	wh cub cry	455 5.56	1547 a,c,d,e,f	i H$_2$O
1831	Silver(I) chromate Ag$_2$CrO$_4$	7784-01-2 331.730	brn-red monocl cry	5.625	a	i H$_2$O
1832	Silver(I) citrate Ag$_3$C$_6$H$_5$O$_7$	126-45-4 512.707	wh cry powder			i H$_2$O; s HNO$_3$
1833	Silver(I) cyanide AgCN	506-64-9 133.886	wh-gray hex cry	320 dec 3.95	a	i H$_2$O, EtOH, dil acid
1834	Silver(I) dichromate Ag$_2$Cr$_2$O$_7$	7784-02-3 431.724	red cry	4.770		sl H$_2$O
1835	Silver(I) fluoride AgF	7775-41-9 126.866	yel-brn cub cry; hyg	435 5.852	1159 a	s H$_2$O
1836	Silver(I) hydrogen fluoride AgHF$_2$	12249-52-4 146.873	hyg cry	dec		
1837	Silver(I) iodate AgIO$_3$	7783-97-3 282.770	wh orth cry	>200 5.53	a	i H$_2$O
1838	Silver(I) iodide AgI	7783-96-2 234.772	yel powder; hex	558 5.68	1506 a,c,d,e,f	i H$_2$O, acid
1839	Silver(I) lactate monohydrate AgC$_3$H$_5$O$_3$·H$_2$O	128-00-7 214.955	gray cry powder			sl H$_2$O, EtOH
1840	Silver(I) molybdate Ag$_2$MoO$_4$	13765-74-7 375.67	yel cub cry	483 6.18		sl H$_2$O
1841	Silver(I) nitrate AgNO$_3$	7761-88-8 169.873	col rhomb cry	212 4.35	dec 440 a,c	vs H$_2$O; sl EtOH, ace
1842	Silver(I) nitrite AgNO$_2$	7783-99-5 153.874	yel needles	dec 140 4.453		sl H$_2$O; i EtOH; reac acid
1843	Silver(I) oxalate Ag$_2$C$_2$O$_4$	533-51-7 303.756	wh cry powder	exp 140 5.03		i H$_2$O
1844	Silver(I) oxide Ag$_2$O	20667-12-3 231.735	brn-blk cub cry	dec ≈200 7.2	a	i H$_2$O, EtOH; s acid, alk
1845	Silver(I) perchlorate AgClO$_4$	7783-93-9 207.319	col cub cry; hyg	dec 486 2.806	a	vs H$_2$O; s bz, py, os
1846	Silver(I) perchlorate monohydrate AgClO$_4$·H$_2$O	14242-05-8 225.334	hyg wh cry	dec 43		
1847	Silver(I) permanganate AgMnO$_4$	7783-98-4 226.804	viol monocl cry	dec 4.49		sl H$_2$O; reac EtOH
1848	Silver(I) phosphate Ag$_3$PO$_4$	7784-09-0 418.576	yel powder	849 6.37		i H$_2$O; sl dil acid
1849	Silver(I) picrate monohydrate AgC$_6$H$_2$N$_3$O$_7$·H$_2$O	146-84-9 353.981	yel cry			sl H$_2$O, EtOH; i chl, eth
1850	Silver(I) selenate Ag$_2$SeO$_4$	7784-07-8 358.69	orth cry	5.72		sl H$_2$O
1851	Silver(I) selenide Ag$_2$Se	1302-09-6 294.70	gray hex needles	880 8.216		i H$_2$O
1852	Silver(I) selenite Ag$_2$SeO$_3$	7784-05-6 342.69	needles	530 5.930	dec >550	sl H$_2$O; s acid
1853	Silver(I) sulfate Ag$_2$SO$_4$	10294-26-5 311.800	col cry or powder	652 5.45	a	sl H$_2$O
1854	Silver(I) sulfide Ag$_2$S	21548-73-2 247.802	gray-blk orth powder	825 7.23	a,c,f	i H$_2$O; s acid
1855	Silver(I) telluride Ag$_2$Te	12002-99-2 343.34	blk orth cry	955 8.4	f	
1856	Silver(I) tetraiodomercurate(II) Ag$_2$HgI$_4$	7784-03-4 923.94	yel tetr cry	trans to red cub ≈40 6.1		i H$_2$O, dil acid
1857	Silver(I) thiocyanate AgSCN	14104-20-2 165.952	wh powder	dec		i H$_2$O

No.	Name Formula	CAS RN Mol. Wt.	Physical Form	mp/°C den/g cm⁻³	bp/°C Other Data	Solubility
1858	Silver(II) oxide AgO	1301-96-8 123.867	gray powder; monocl or cub	dec >100 7.5		i H_2O; s alk; reac acid
1859	Silver(II) fluoride AgF_2	7783-95-1 145.865	wh or gray hyg cry	690 4.58	a	reac H_2O
1860	Sodium Na	7440-23-5 22.990	soft silv met; cub	97.80 0.97	883 a,c,e	reac H_2O
1861	Sodium acetate $NaC_2H_3O_2$	127-09-3 82.035	col cry	324 1.528	a	vs H_2O
1862	Sodium acetate trihydrate $NaC_2H_3O_2 \cdot 3H_2O$	6131-90-4 136.080	col cry	58 dec 1.45		vs H_2O; sl EtOH
1863	Sodium aluminate $NaAlO_2$	1302-42-7 81.971	wh orth cry; hyg	1650 4.63		vs H_2O; i EtOH
1864	Sodium aluminum sulfate dodecahydrate $NaAl(SO_4)_2 \cdot 12H_2O$	10102-71-3 458.283	col cry	≈60 1.61		vs H_2O; i EtOH
1865	Sodium amide $NaNH_2$	7782-92-5 39.013	wh-grn orth cry	210 1.39	dec 500 a	reac H_2O
1866	Sodium ammonium phosphate tetrahydrate $NaNH_4HPO_4 \cdot 4H_2O$	13011-54-6 209.069	monocl cry	≈80 dec 1.54		s H_2O; i EtOH
1867	Sodium arsenite $NaAsO_2$	7784-46-5 129.911	wh-gray hyg powder	 1.87		vs H_2O; i EtOH
1868	Sodium azide NaN_3	26628-22-8 65.010	col hex cry	dec 300; exp 1.846	a	vs H_2O; sl EtOH; i eth
1869	Sodium borohydride $NaBH_4$	16940-66-2 37.833	wh cub cry; hyg	dec ≈400 1.07	a	vs H_2O; reac EtOH
1870	Sodium bromate $NaBrO_3$	7789-38-0 150.892	col cub cry	381 3.34	a,c	s H_2O; i EtOH
1871	Sodium bromide NaBr	7647-15-6 102.894	wh cub cry	747 3.200	1390 a,c,e	vs H_2O; s EtOH
1872	Sodium bromide dihydrate $NaBr \cdot 2H_2O$	13466-08-5 138.925	wh cry	36 dec 2.18		s H_2O; sl EtOH
1873	Sodium carbonate Na_2CO_3	497-19-8 105.989	wh hyg powder	858.1 2.54	a,c	s H_2O; i EtOH
1874	Sodium carbonate decahydrate $Na_2CO_3 \cdot 10H_2O$	6132-02-1 286.142	col cry	34 dec 1.46		vs H_2O; i EtOH
1875	Sodium carbonate monohydrate $Na_2CO_3 \cdot H_2O$	5968-11-6 124.005	col orth cry	dec 100 2.25	f	vs H_2O; i EtOH
1876	Sodium chlorate $NaClO_3$	7775-09-9 106.441	col cub cry	248 2.5	dec >300 a,c	vs H_2O; sl EtOH
1877	Sodium chloride NaCl	7647-14-5 58.443	col cub cry	800.7 2.17	1465 a,c,e,f	s H_2O; sl EtOH
1878	Sodium chlorite $NaClO_2$	7758-19-2 90.442	wh hyg cry	dec ≈180	a	vs H_2O
1879	Sodium chromate Na_2CrO_4	7775-11-3 161.974	yel orth cry	792 2.72		vs H_2O; sl EtOH
1880	Sodium chromate tetrahydrate $Na_2CrO_4 \cdot 4H_2O$	10034-82-9 234.035	yel hyg cry	dec		vs H_2O; sl EtOH
1881	Sodium citrate dihydrate $Na_3C_6H_5O_7 \cdot 2H_2O$	6132-04-3 294.101	wh cry	dec 150		vs H_2O; i EtOH, eth
1882	Sodium cyanate NaOCN	917-61-3 65.007	col needles	550 1.89	a	s H_2O; sl EtOH; i eth
1883	Sodium cyanide NaCN	143-33-9 49.008	wh cub cry; hyg	563 1.6	a,e	vs H_2O; sl EtOH
1884	Sodium cyanoborohydride $NaBH_3(CN)$	25895-60-7 62.843	wh hyg powder	240 dec 1.12		vs H_2O; s thf; sl EtOH; i bz, eth
1885	Sodium dichromate $Na_2Cr_2O_7$	10588-01-9 261.968	red hyg cry	357	dec 400	vs H_2O
1886	Sodium dihydrogen phosphate monohydrate $NaH_2PO_4 \cdot H_2O$	10049-21-5 137.993	wh hyg cry	dec 100		vs H_2O; i EtOH
1887	Sodium dihydrogen phosphate dihydrate $NaH_2PO_4 \cdot 2H_2O$	13472-35-0 156.008	col orth cry	60 dec 1.91		vs H_2O; i EtOH
1888	Sodium dihydrogen pyrophosphate $Na_2H_2P_2O_7$	7758-16-9 221.939	wh powder	dec 220 ≈1.9		s H_2O
1889	Sodium dithionate dihydrate $Na_2S_2O_6 \cdot 2H_2O$	7631-94-9* 242.139	col orth cry	dec 110 2.19		s H_2O; i EtOH
1890	Sodium ferricyanide monohydrate $Na_3Fe(CN)_6 \cdot H_2O$	14217-21-1* 298.935	red hyg cry			s H_2O; i EtOH
1891	Sodium ferrocyanide decahydrate $Na_4Fe(CN)_6 \cdot 10H_2O$	13601-19-9 484.063	yel monocl cry	dec ≈50 1.46		s H_2O; i os
1892	Sodium fluoride NaF	7681-49-4 41.988	col cub or tetr cry	996 2.78	1704 a,c,e,f	sl H_2O; i EtOH
1893	Sodium fluoroborate $NaBF_4$	13755-29-8 109.795	wh orth prisms	384 2.47	a	vs H_2O; sl EtOH
1894	Sodium formate $NaCHO_2$	141-53-7 68.008	wh hyg cry	253 1.92	dec a	vs H_2O; sl EtOH
1895	Sodium hexafluoroaluminate Na_3AlF_6	13775-53-6 209.941	col monocl cry; trans cub	1009 2.97	560 f	i H_2O

No.	Name Formula	CAS RN Mol. Wt.	Physical Form	mp/°C den/g cm^{-3}	bp/°C Other Data	Solubility
1896	Sodium hexafluorosilicate Na_2SiF_6	16893-85-9 188.056	wh hex cry	dec 2.7	a	sl H_2O; i EtOH
1897	Sodium hydride NaH	7646-69-7 23.998	silv cub cry; flam	425 dec 1.39	a	reac H_2O, EtOH
1898	Sodium hydrogen arsenate heptahydrate $Na_2HAsO_4 \cdot 7H_2O$	10048-95-0 312.014	wh monocl cry	dec ≈50 1.87		vs H_2O; sl EtOH
1899	Sodium hydrogen carbonate $NaHCO_3$	144-55-8 84.007	wh monocl cry	dec ≈50 2.20	a	s H_2O; i EtOH
1900	Sodium hydrogen fluoride $NaHF_2$	1333-83-1 61.995	wh hex cry	dec >160 2.08	a	s H_2O
1901	Sodium hydrogen phosphate Na_2HPO_4	7558-79-4 141.959	wh hyg powder	1.7	a	s H_2O
1902	Sodium hydrogen phosphate dodecahydrate $Na_2HPO_4 \cdot 12H_2O$	10039-32-4 358.143	col cry	dec ≈35 ≈1.5		s H_2O; i EtOH
1903	Sodium hydrogen phosphate heptahydrate $Na_2HPO_4 \cdot 7H_2O$	7782-85-6 268.066	col cry	≈1.7		s H_2O; i EtOH
1904	Sodium hydrogen sulfate $NaHSO_4$	7681-38-1 120.062	wh hyg cry	≈315 2.43	a	vs H_2O
1905	Sodium hydrogen sulfate monohydrate $NaHSO_4 \cdot H_2O$	10034-88-5 138.077	wh monocl cry	2.10		vs H_2O; reac EtOH
1906	Sodium hydrogen sulfide NaHS	16721-80-5 56.064	col rhomb cry	350 1.79		s H_2O, EtOh, eth
1907	Sodium hydrogen sulfide dihydrate $NaHS \cdot 2H_2O$	16721-80-5 92.095	yel hyg needles	dec 55		vs H_2O, EtOH, eth
1908	Sodium hydrogen sulfite $NaHSO_3$	7631-90-5 104.062	wh cry	1.48		s H_2O; sl EtOH
1909	Sodium hydroxide NaOH	1310-73-2 39.997	wh orth cry; hyg	323 2.13	1388 a,c,d,e	vs H_2O; s EtOH, MeOH
1910	Sodium hypochlorite NaClO	7681-52-9 74.442	stable in aq soln	anh form exp		vs H_2O
1911	Sodium hypochlorite pentahydrate $NaOCl \cdot 5H_2O$	7681-52-9* 164.518	pale grn orth cry	18 1.6		s H_2O
1912	Sodium iodate $NaIO_3$	7681-55-2 197.892	wh orth cry	dec 4.28	a	s H_2O; i EtOH
1913	Sodium iodide NaI	7681-82-5 149.894	wh cub cry; hyg	660 3.67	1304 a,c,e	vs H_2O; s EtOH, ace
1914	Sodium bismuthate $NaBiO_3$	12232-99-4 279.968	yel-brn hyg cry			i cold H_2O, reac acid
1915	Sodium metabisulfite $Na_2S_2O_5$	7681-57-4 190.109	wh cry			vs H_2O; sl EtOH
1916	Sodium metaborate $NaBO_2$	7775-19-1 65.800	wh hex cry	966 2.46	1434 a,c	s H_2O
1917	Sodium metasilicate Na_2SiO_3	6834-92-0 122.064	wh amorp solid; hyg	1089 2.61	a,c	s cold H_2O; reac hot H_2O
1918	Sodium molybdate Na_2MoO_4	7631-95-0 205.92	col cub cry	687 ≈3.5	a	vs H_2O
1919	Sodium molybdate dihydrate $Na_2MoO_4 \cdot 2H_2O$	10102-40-6 241.95	cry powder	dec 100 ≈3.5		vs H_2O
1920	Sodium niobate $NaNbO_3$	12034-09-2 163.894	rhomb cry	1422 4.55		i H_2O
1921	Sodium nitrate $NaNO_3$	7631-99-4 84.995	col hex cry; hyg	307 2.26	a,c	vs H_2O; sl EtOH, MeOH
1922	Sodium nitrite $NaNO_2$	7632-00-0 68.996	wh orth cry; hyg	271 2.17	dec >320 a	vs H_2O; sl EtOH; reac acid
1923	Sodium oxalate $Na_2C_2O_4$	62-76-0 134.000	wh powder	dec ≈250 2.34	a	sl H_2O; i EtOH
1924	Sodium oxide Na_2O	1313-59-3 61.979	wh amorp powder	1132 dec 2.27	a,c	reac H_2O
1925	Sodium perchlorate $NaClO_4$	7601-89-0 122.441	wh orth cry; hyg	480 dec 2.52	a	vs H_2O
1926	Sodium perchlorate monohydrate $NaClO_4 \cdot H_2O$	7791-07-3 140.456	wh hyg cry	dec ≈130 2.02		vs H_2O
1927	Sodium periodate $NaIO_4$	7790-28-5 213.892	wh tetr cry	dec ≈300 3.86	a	s H_2O, acid
1928	Sodium periodate trihydrate $NaIO_4 \cdot 3H_2O$	13472-31-6 267.938	wh hex cry	dec 175 3.22		s H_2O
1929	Sodium permanganate trihydrate $NaMnO_4 \cdot 3H_2O$	10101-50-5* 195.972	red-blk hyg cry	dec 170 2.47		vs H_2O; reac EtOH
1930	Sodium peroxide Na_2O_2	1313-60-6 77.979	yel hyg powder	675 2.805	a	reac H_2O
1931	Sodium persulfate $Na_2S_2O_8$	7775-27-1 238.107	wh hyg cry			vs H_2O; reac EtOH
1932	Sodium phosphate dodecahydrate $Na_3PO_4 \cdot 12H_2O$	7601-54-9 380.124	col hex cry	≈75 1.62		s H_2O; i EtOH
1933	Sodium potassium tartrate tetrahydrate $NaK(C_4H_4O_6) \cdot 4H_2O$	304-59-6 282.221	wh cry	dec ≈70 1.79	anh at 130	vs H_2O; i EtOH

No.	Name Formula	CAS RN Mol. Wt.	Physical Form	mp/°C den/g cm⁻³	bp/°C Other Data	Solubility
1934	Sodium pyrophosphate $Na_4P_2O_7$	7722-88-5 265.902	col cry	988 2.53		sl H_2O
1935	Sodium selenate decahydrate $Na_2SeO_4 \cdot 10H_2O$	10102-23-5 369.09	wh cry	1.61		vs H_2O
1936	Sodium selenide Na_2Se	1313-85-5 124.94	amorp solid	>875 2.62		reac H_2O
1937	Sodium selenite Na_2SeO_3	10102-18-8 172.94	wh tetr cry			vs H_2O; i EtOH
1938	Sodium sulfate Na_2SO_4	7757-82-6 142.044	wh orth cry or powder	884 2.7	a,c,f	s H_2O; i EtOH
1939	Sodium sulfate decahydrate $Na_2SO_4 \cdot 10H_2O$	7727-73-3 322.197	col monocl cry	32 dec 1.46	f	s H_2O; i EtOH
1940	Sodium sulfide Na_2S	1313-82-2 78.046	wh cub cry; hyg	1172 1.856	a,c	s H_2O; sl EtOH; i eth
1941	Sodium sulfide nonahydrate $Na_2S \cdot 9H_2O$	1313-84-4 240.184	wh-yel hyg cry	dec ≈50 1.43		vs H_2O; sl EtOH; i eth
1942	Sodium sulfide pentahydrate $Na_2S \cdot 5H_2O$	1313-83-3 168.122	col orth cry	120 dec 1.58		vs H_2O; s EtOH; i eth
1943	Sodium sulfite Na_2SO_3	7757-83-7 126.044	wh hex cry	dec 2.63	a	s H_2O; i EtOH
1944	Sodium sulfite heptahydrate $Na_2SO_3 \cdot 7H_2O$	10102-15-5 252.151	wh monocl cry; unstable	1.56		vs H_2O; sl EtOH
1945	Sodium superoxide NaO_2	12034-12-7 54.989	yel cub cry	552 2.2	a	reac H_2O
1946	Sodium tellurate Na_2TeO_4	10102-20-2 237.58	wh powder			sl H_2O
1947	Sodium tetraborate $Na_2B_4O_7$	1330-43-4 201.220	col gl solid; hyg	743 2.4	1575 a	sl H_2O, MeOH
1948	Sodium tetraborate decahydrate $Na_2B_4O_7 \cdot 10H_2O$	1303-96-4 381.373	wh monocl cry	75 dec 1.73	f	sl H_2O; i EtOH
1949	Sodium tetraborate pentahydrate $Na_2B_4O_7 \cdot 5H_2O$	12045-88-4 291.296	hex cry	dec 1.88		sl H_2O
1950	Sodium tetraborate tetrahydrate $Na_2B_4O_7 \cdot 4H_2O$	12045-87-3 273.281	wh monocl cry	1.95	f	sl H_2O
1951	Sodium tetrachloroaluminate $NaAlCl_4$	7784-16-9 191.783	orth cry	2.01		s H_2O
1952	Sodium tetrafluoroberyllate Na_2BeF_4	13871-27-7 130.986	orth cry	575 2.47		sl H_2O
1953	Sodium thiocyanate NaSCN	540-72-7 81.074	col hyg cry	287		vs H_2O
1954	Sodium thiosulfate pentahydrate $Na_2S_2O_3 \cdot 5H_2O$	10102-17-7 240.186	col cry	dec ≈50 1.69		vs H_2O; i EtOH
1955	Sodium tungstate dihydrate $Na_2WO_4 \cdot 2H_2O$	10213-10-2 329.85	wh orth cry	dec 100 3.25		vs H_2O; i EtOH
1956	Strontium Sr	7440-24-6 87.62	silv-wh metal; cub	777 2.64	1382 a,c,e	reac H_2O; s EtOH
1957	Strontium bromate monohydrate $Sr(BrO_3)_2 \cdot H_2O$	14519-18-7 361.44	wh-yel hyg cry	dec 240 3.77		s H_2O
1958	Strontium bromide $SrBr_2$	10476-81-0 247.43	wh tetr cry	657 4.216	a,c	vs H_2O
1959	Strontium bromide hexahydrate $SrBr_2 \cdot 6H_2O$	7789-53-9 355.52	col hyg cry	88 dec		vs H_2O; s EtOH; i eth
1960	Strontium carbide SrC_2	12071-29-3 111.64	tetr cry	3.19		
1961	Strontium carbonate $SrCO_3$	1633-05-2 147.63	wh orth cry; hyg	1494 3.5	a,f	i H_2O; s dil acid
1962	Strontium chlorate $Sr(ClO_3)_2$	7791-10-8 254.52	col cry	120 dec 3.15		vs H_2O; sl EtOH
1963	Strontium chloride $SrCl_2$	10476-85-4 158.53	wh cub cry; hyg	874 3.052	1250 a,c	vs H_2O
1964	Strontium chloride hexahydrate $SrCl_2 \cdot 6H_2O$	10025-70-4 266.62	col hyg cry	dec 100 1.96		vs H_2O; s EtOH
1965	Strontium chromate $SrCrO_4$	7789-06-2 203.61	yel monocl cry	dec 3.9		sl H_2O; s dil acid
1966	Strontium fluoride SrF_2	7783-48-4 125.62	wh cub cry or powder	1477 4.24	2460 a,c	s H_2O, dil acid
1967	Strontium hydride SrH_2	13598-33-9 89.64	orth cry	1050 3.26	a	reac H_2O
1968	Strontium hydroxide $Sr(OH)_2$	18480-07-4 121.64	col orth cry; hyg	535 3.625	dec 710 a,c	s H_2O
1969	Strontium iodide SrI_2	10476-86-5 341.43	wh hyg cry	538 4.55	1773 dec a,c	vs H_2O
1970	Strontium iodide hexahydrate $SrI_2 \cdot 6H_2O$	73796-25-5 449.52	wh-yel hex cry; hyg	dec 120 4.4		vs H_2O; s EtOH
1971	Strontium niobate $SrNb_2O_6$	12034-89-8 369.43	monocl cry	1225 5.11		i H_2O
1972	Strontium nitrate $Sr(NO_3)_2$	10042-76-9 211.63	wh cub cry	570 2.99	645 a	vs H_2O; sl EtOH, ace

No.	Name / Formula	CAS RN / Mol. Wt.	Physical Form	mp/°C / den/g cm^{-3}	bp/°C / Other Data	Solubility
1973	Strontium nitrite / $Sr(NO_2)_2$	13470-06-9 / 179.63	wh-yel hyg needles	dec 240 / 2.8	/ a	vs H_2O
1974	Strontium oxide / SrO	1314-11-0 / 103.62	col cub cry	2532 / 5.1	/ a,c,e	reac H_2O
1975	Strontium peroxide / SrO_2	1314-18-7 / 119.62	wh tetr cry; unstable	215 dec / 4.78		reac H_2O
1976	Strontium phosphate / $Sr_3(PO_4)_2$	7446-28-8 / 452.80	wh powder			i H_2O; s acid
1977	Strontium selenate / $SrSeO_4$	7446-21-1 / 230.58	orth cry	/ 4.25		i H_2O; s hot HCl
1978	Strontium selenide / $SrSe$	1315-07-7 / 166.58	wh cub cry	1600 / 4.54	/ a	
1979	Strontium orthosilicate / Sr_2SiO_4	13597-55-2 / 267.32	orth cry	/ 4.5	/ a	
1980	Strontium silicide / $SrSi_2$	12138-28-2 / 143.79	silv-gray cub cry	1100 / 3.35		
1981	Strontium sulfate / $SrSO_4$	7759-02-6 / 183.68	wh orth cry	1607 / 3.96	/ a,f	i H_2O, EtOH; sl acid
1982	Strontium sulfide / SrS	1314-96-1 / 119.69	gray cub cry	2227 / 3.70	/ a	sl H_2O; s acid
1983	Strontium titanate / $SrTiO_3$	12060-59-2 / 183.49	wh cub cry	2080 / 5.1		i H_2O
1984	Sulfur (α form) / S	7704-34-9 / 32.066	yel orth cry	trans to β form 94.5 / 2.07	444.60 / a,b,c,d,e,f	i H_2O; sl EtOH, bz, eth; s CS_2
1985	Sulfur (β form) / S	7704-34-9 / 32.066	yel monocl needles, stable 94.5-115	115.21 / 2.07	444.60 / a,b,c,d,e,f	i H_2O; sl EtOH, bz, eth; s CS_2
1986	Sulfuric acid / H_2SO_4	7664-93-9 / 98.080	col oily liq	10.31 / 1.8	337 / a,c,e	vs H_2O
1987	Peroxysulfuric acid / H_2SO_5	7722-86-3 / 114.079	wh cry; unstable	dec 45		vs H_2O
1988	Chlorosulfonic acid / $SO_2(OH)Cl$	7790-94-5 / 116.525	col-yel liq	-80 / 1.75	152	reac H_2O; s py
1989	Fluorosulfonic acid / $SO_2(OH)F$	7789-21-1 / 100.070	col liq	-89 / 1.726	163	reac H_2O
1990	Nitrosylsulfuric acid / $HNOSO_4$	7782-78-7 / 127.078	prisms	dec 73		reac H_2O; s H_2SO_4
1991	Sulfurous acid / H_2SO_3	7782-99-2 / 82.080	exists only in soln			soln of SO_2 in H_2O
1992	Sulfur dioxide / SO_2	7446-09-5 / 64.065	col gas	-75.5 / 2.811 g/L	-10.05 / a,b,d,e	s H_2O, EtOh, eth, chl
1993	Sulfur trioxide / SO_3	7446-11-9 / 80.064	col liq	16.8 / 1.92	45 / a,b,d,e	reac H_2O
1994	Sulfur bromide / $SSBr_2$	13172-31-1 / 223.940	red oily liq	-46 / 2.63	dec >25 / a	reac H_2O
1995	Sulfur chloride / $SSCl_2$	10025-67-9 / 135.037	yel-red oily liq	-77 / 1.69	137 / a,e	reac H_2O; s EtOH, bz, eth, ctc
1996	Sulfur fluoride / SSF_2	16860-99-4 / 102.129	col gas	-164.6 / 4.481 g/L	-10.6	reac H_2O
1997	Sulfur fluoride / FSSF	13709-35-8 / 102.129	col gas	-133 / 4.481 g/L	15	reac H_2O
1998	Sulfur dichloride / SCl_2	10545-99-0 / 102.971	red visc liq	-122 / 1.62	59.6 / a	reac H_2O
1999	Sulfur tetrafluoride / SF_4	7783-60-0 / 108.060	col gas	-125 / 4.742 g/L	-40.45 / a,b,d	reac H_2O
2000	Sulfur hexafluoride / SF_6	2551-62-4 / 146.056	col gas	-50.7 tp / 6.409 g/L	-63.8 sp / a,b,c,d,e	sl H_2O; s EtOH
2001	Sulfur bromide pentafluoride / SF_5Br	15607-89-3 / 206.962	col gas	-79 / 9.081 g/L	3.1	
2002	Sulfur chloride pentafluoride / SF_5Cl	13780-57-9 / 162.511	col gas	-64 / 7.131 g/L	-19.05 / a,b	
2003	Sulfur decafluoride / S_2F_{10}	5714-22-7 / 254.116	liq	-52.7 / 2.08	30; dec 150	i H_2O
2004	Sulfuryl amide / $SO_2(NH_2)_2$	7803-58-9 / 96.110	orth plates	93	dec 250	s H_2O, hot EtOH, ace
2005	Sulfuryl chloride / SO_2Cl_2	7791-25-5 / 134.970	col liq	-51 / 1.680	69.4 / a,d,e	reac H_2O; s bz, tol, eth
2006	Sulfuryl fluoride / SO_2F_2	2699-79-8 / 102.062	col gas	-135.8 / 4.478 g/L	-55.4 / a	sl H_2O, EtOH; s tol, ctc
2007	Thionyl bromide / $SOBr_2$	507-16-4 / 207.873	yel liq	-50	140 / e	reac H_2O
2008	Thionyl chloride / $SOCl_2$	7719-09-7 / 118.970	yel fuming liq	-101 / 1.631	75.6 / a,d,e	reac H_2O; s bz, ctc, chl
2009	Thionyl fluoride / SOF_2	7783-42-8 / 86.062	col gas	-129.5 / 3.776 g/L	-43.8 / a,d	reac H_2O; s bz, eth
2010	Sulfur fluoride hypofluorite / F_5SOF	15179-32-5 / 162.055	col gas	-86 / 7.111 g/L	-35.1	
2011	Tantalum / Ta	7440-25-7 / 180.948	gray metal; cub	3017 / 16.4	5458 / a,c	reac HF

No.	Name Formula	CAS RN Mol. Wt.	Physical Form	mp/°C den/g cm^{-3}	bp/°C Other Data	Solubility
2012	Tantalum aluminide TaAl$_3$	12004-76-1 261.893	gray refrac powder	≈1400 7.02		i H$_2$O, acid, alk
2013	Tantalum boride TaB	12007-07-7 191.759	refrac orth cry	2040 14.2		
2014	Tantalum boride TaB$_2$	12007-35-1 202.570	blk hex cry	3140 11.2		i H$_2$O, acid, alk
2015	Tantalum carbide TaC	12070-06-3 192.959	gold-brown powder; cub	3880 14.3	4780	s HF-HNO$_3$ mixture
2016	Tantalum carbide Ta$_2$C	12070-07-4 373.907	refrac hex cry	3327 15.1		
2017	Tantalum nitride TaN	12033-62-4 194.955	blk hex cry	3090 13.7		i H$_2$O; sl aqua regia; reac alk
2018	Tantalum silicide TaSi$_2$	12039-79-1 237.119	gray powder	2200 9.14		
2019	Tantalum(IV) oxide TaO$_2$	12036-14-5 212.947	tetr cry	 10.0		
2020	Tantalum(IV) selenide TaSe$_2$	12039-55-3 338.87	hex cry	 6.7		
2021	Tantalum(IV) sulfide TaS$_2$	12143-72-5 245.080	blk hex cry	>3000 6.86		i H$_2$O
2022	Tantalum(IV) telluride TaTe$_2$	12067-66-2 436.15	monocl cry	 9.4		
2023	Tantalum(V) bromide TaBr$_5$	13451-11-1 580.468	yel cry powder	265 4.99	349 a,b,c,d	
2024	Tantalum(V) chloride TaCl$_5$	7721-01-9 358.212	yel monocl cry; hyg	216 3.68	239.35 a,b,c,d	reac H$_2$O; s EtOH
2025	Tantalum(V) fluoride TaF$_5$	7783-71-3 275.940	wh monocl cry; hyg	95.1 5.0	229.2 a,d,e	s H$_2$O, eth; sl CS$_2$, ctc
2026	Tantalum(V) iodide TaI$_5$	14693-81-3 815.470	blk hex cry; hyg	496 5.80	543	
2027	Tantalum(V) oxide Ta$_2$O$_5$	1314-61-0 441.893	wh rhomb cry or powder	1785 8.2	 a,c	i H$_2$O, EtOH, acid; s HF
2028	Technetium Tc	7440-26-8 98	hex cry	2157 11	4265 a,c	
2029	Technetium(V) fluoride TcF$_5$	31052-14-9 193	yel solid	50	dec	
2030	Technetium(VI) fluoride TcF$_6$	13842-93-8 212	yel cub cry	37.4 3.0	55.3	
2031	Tellurium Te	13494-80-9 127.60	gray-wh rhomb cry	449.51 6.24	988 a,c,d,e	i H$_2$O, bz, CS$_2$
2032	Telluric(VI) acid H$_6$TeO$_6$	7803-68-1 229.64	wh monocl cry	136 3.07		s H$_2$O
2033	Tellurous acid H$_2$TeO$_3$	10049-23-7 177.61	wh cry	dec 40 3.0		sl H$_2$O; s dil acid, alk
2034	Tellurium dioxide TeO$_2$	7446-07-3 159.60	wh orth cry	733 5.9	1245 a	i H$_2$O; s alk, acid
2035	Tellurium trioxide TeO$_3$	13451-18-8 175.60	yel-oran cry	430 5.07		i H$_2$O
2036	Tellurium dibromide TeBr$_2$	7789-54-0 287.41	grn-brn hyg cry	210	339	reac H$_2$O; s eth; sl chl
2037	Tellurium dichloride TeCl$_2$	10025-71-5 198.51	blk amorp solid; hyg	208 6.9	328	reac H$_2$O; i ctc
2038	Tellurium tetrabromide TeBr$_4$	10031-27-3 447.22	yel-oran monocl cry	388 4.3	≈420 dec a	reac H$_2$O; s eth
2039	Tellurium tetrachloride TeCl$_4$	10026-07-0 269.41	wh monocl cry; hyg	224 3.0	387 a,b,d,e	reac H$_2$O; s EtOH, tol
2040	Tellurium tetrafluoride TeF$_4$	15192-26-4 203.59	col cry	129	dec 195	reac H$_2$O
2041	Tellurium tetraiodide TeI$_4$	7790-48-9 635.22	blk orth cry	280 5.05		reac H$_2$O; sl ace
2042	Tellurium hexafluoride TeF$_6$	7783-80-4 241.59	col gas	-37.6 tp 10.601 g/L	-38.9 sp a,b,e	reac H$_2$O
2043	Terbium Tb	7440-27-9 158.925	silv metal; hex	1356 8.23	3230 a,c	
2044	Terbium chloride TbCl$_3$	10042-88-3 265.283	wh orth cry; hyg	588 4.35	 a	s H$_2$O
2045	Terbium chloride hexahydrate TbCl$_3$·6H$_2$O	13798-24-8 373.374	hyg cry	 4.35		vs H$_2$O
2046	Terbium iodide TbI$_3$	13813-40-6 539.638	hex cry; hyg	957 ≈5.2		s H$_2$O
2047	Terbium nitride TbN	12033-64-6 172.932	cub cry	 9.55		
2048	Terbium oxide Tb$_2$O$_3$	12036-41-8 365.849	wh cub cry	2410 7.91	 a	
2049	Terbium silicide TbSi$_2$	12039-80-4 215.096	orth cry	 6.66		
2050	Terbium sulfide Tb$_2$S$_3$	12138-11-3 414.049	cub cry	 6.35		

No.	Name Formula	CAS RN Mol. Wt.	Physical Form	mp/°C den/g cm^{-3}	bp/°C Other Data	Solubility
2051	Thallium Tl	7440-28-0 204.383	soft blue-wh metal	304 11.8	1473 a,c,e	i H_2O; reac acid
2052	Thallium(I) acetate $TlC_2H_3O_2$	563-68-8 263.428	hyg wh cry	131 3.68		s H_2O, EtOH
2053	Thallium(I) bromide TlBr	7789-40-4 284.287	yel cub cry	460 7.5	819 a,c,d,e	i H_2O
2054	Thallium(I) carbonate Tl_2CO_3	6533-73-9 468.776	wh monocl cry	272 7.11	a,c	sl H_2O; i EtOH
2055	Thallium(I) chlorate $TlClO_3$	13453-30-0 287.834	col hex cry	5.5		sl H_2O
2056	Thallium(I) chloride TlCl	7791-12-0 239.836	wh cub cry	430 7.0	720 a,c,d,e	sl H_2O; i EtOH
2057	Thallium(I) cyanide TlCN	13453-34-4 230.401	wh hex plates	6.523		s H_2O, acid, EtOH
2058	Thallium(I) ethoxide TlC_2H_5O	20398-06-5 249.444	cloudy liq	-3 3.49	dec 130	reac H_2O
2059	Thallium(I) fluoride TlF	7789-27-7 223.381	wh orth cry	326 8.36	826 a,c	s H_2O
2060	Thallium(I) formate $TlCHO_2$	992-98-3 249.401	hyg col needles	101 4.97		vs H_2O; s MeOH
2061	Thallium(I) hexafluorophosphate $TlPF_6$	60969-19-9 349.347	wh cub cry	4.6		
2062	Thallium(I) hydroxide TlOH	12026-06-1 221.390	yel needles	139 dec 7.44	a	vs H_2O
2063	Thallium(I) iodide TlI	7790-30-9 331.287	yel cry powder	441.8 7.1	824 a,c,d,e	i H_2O, EtOH
2064	Thallium(I) molybdate Tl_2MoO_4	34128-09-1 568.71	yel-wh cub cry			i H_2O
2065	Thallium(I) nitrate $TlNO_3$	10102-45-1 266.388	wh cry	206 5.55	450 dec a,c	s H_2O; i EtOH
2066	Thallium(I) nitrite $TlNO_2$	13826-63-6 250.389	cub cry	5.7		vs H_2O
2067	Thallium(I) oxalate $Tl_2C_2O_4$	30737-24-7 496.787	wh powder	6.31		
2068	Thallium(I) oxide Tl_2O	1314-12-1 424.766	blk rhomb cry; hyg	579 9.52	≈1080 a	s H_2O, EtOH
2069	Thallium(I) perchlorate $TlClO_4$	13453-40-2 303.834	col orth cry	4.8		s H_2O
2070	Thallium(I) selenate Tl_2SeO_4	7446-22-2 551.73	orth cry	>400 6.875		sl H_2O; i EtOH, eth
2071	Thallium(I) selenide Tl_2Se	15572-25-5 487.73	gray plates	340	a	i H_2O, acid
2072	Thallium(I) sulfate Tl_2SO_4	7446-18-6 504.831	wh rhomb prisms	632 6.77	a,c	sl H_2O
2073	Thallium(I) sulfide Tl_2S	1314-97-2 440.833	blue-blk cry	448 8.39	1367 a,c,d	sl H_2O, alk; s acid
2074	Thallium(III) bromide tetrahydrate $TlBr_3•4H_2O$	13701-90-1 516.157	yel orth cry	3.65		s H_2O, EtOH
2075	Thallium(III) chloride $TlCl_3$	13453-32-2 310.741	monocl cry	155 4.7	a	vs H_2O, EtOH, eth
2076	Thallium(III) chloride tetrahydrate $TlCl_3•4H_2O$	13453-32-2* 382.803	orth cry	3.00		s H_2O
2077	Thallium(III) fluoride TlF_3	7783-57-5 261.378	wh orth cry; hyg	550 dec 8.65		reac H_2O
2078	Thallium(III) nitrate $Tl(NO_3)_3$	13746-98-0 390.398	col cry			reac H_2O
2079	Thallium(III) oxide Tl_2O_3	1314-32-5 456.765	brn cub cry	834 10.2		i H_2O; reac acid
2080	Thorium Th	7440-29-1 232.038	soft gray-wh metal; cub	1750 11.7	4788 a,c	s acids
2081	Thorium hydride ThH_2	16689-88-6 234.054	tetr cry	9.5	a	
2082	Thorium boride ThB_6	12229-63-9 296.904	refrac solid	2450 6.99		
2083	Thorium bromide $ThBr_4$	13453-49-1 551.654	wh hyg cry	679		reac H_2O
2084	Thorium carbide ThC	12012-16-7 244.049	cub cry	2500 10.6		reac H_2O
2085	Thorium dicarbide ThC_2	12071-31-7 256.060	yel monocl cry	≈2650 9.0		reac H_2O
2086	Thorium chloride $ThCl_4$	10026-08-1 373.849	gray-wh tetr needles; hyg	770 4.59	921 a,c,d	s H_2O, EtOH
2087	Thorium fluoride ThF_4	13709-59-6 308.032	wh monocl cry; hyg	1110 6.1	1680 a,d	
2088	Thorium iodide ThI_4	7790-49-0 739.656	wh-yel monocl cry	570	837	
2089	Thorium nitrate tetrahydrate $Th(NO_3)_4•4H_2O$	33088-16-3 552.119	wh hyg cry	dec 500		vs H_2O; s EtOH

No.	Name Formula	CAS RN Mol. Wt.	Physical Form	mp/°C den/g cm^{-3}	bp/°C Other Data	Solubility
2090	Thorium nitride ThN	12033-65-7 246.045	refrac cub cry	2820 11.6		reac H_2O
2091	Thorium oxide ThO_2	1314-20-1 264.037	wh cub cry	3390 10.0	4400 a,f	i H_2O, alk; sl acid
2092	Thorium selenide $ThSe_2$	60763-24-8 389.96	orth cry	 8.5		
2093	Thorium orthosilicate $ThSiO_4$	14553-44-7 324.122	brn tetr cry	 6.7	f	
2094	Thorium silicide $ThSi_2$	12067-54-8 288.209	tetr cry	1850 7.9		
2095	Thorium sulfate nonahydrate $Th(SO_4)_2 \cdot 9H_2O$	10381-37-0 586.303	wh monocl cry	dec 2.8		s H_2O
2096	Thorium sulfide ThS_2	12138-07-7 296.170	dark brn cry	1905 7.30		i H_2O; s acid
2097	Thulium Tm	7440-30-4 168.934	silv metal; hex	1545 9.32	1950 a,c	s dil acid
2098	Thulium bromide $TmBr_3$	14456-51-0 408.646	wh hyg cry	954		s H_2O
2099	Thulium chloride $TmCl_3$	13537-18-3 275.292	yel hyg cry	824	a	s H_2O
2100	Thulium chloride heptahydrate $TmCl_3 \cdot 7H_2O$	13778-39-7 401.399	hyg cry			s H_2O, EtOH
2101	Thulium fluoride TmF_3	13760-79-7 225.929	wh cry	1158		s H_2O
2102	Thulium iodide TmI_3	13813-43-9 549.647	yel hyg cry	1021		
2103	Thulium oxide Tm_2O_3	12036-44-1 385.866	grn-wh cub cry	2425 8.6	a	sl acid
2104	Tin (gray) Sn	7440-31-5 118.710	cub cry	trans to wh Sn 13.2 5.769	2602 a,c,e	
2105	Tin (white) Sn	7440-31-5 118.710	silv tetr cry	231.93 7.265	2602 a,c,e	
2106	Stannane SnH_4	2406-52-2 122.742	unstable col gas	-146 5.386 g/L	-51.8 a,d,e	
2107	Methylstannane SnH_3CH_3	1631-78-3 136.769	col gas	 6.001 g/L	0	reac H_2O
2108	Tin(II) acetate $Sn(C_2H_3O_2)_2$	638-39-1 236.800	wh orth cry	183 2.31	subl	s dil HCl
2109	Tin(II) bromide $SnBr_2$	10031-24-0 278.518	yel powder	215 5.12	639 a,d	s H_2O, EtOH, eth, ace
2110	Tin(II) chloride $SnCl_2$	7772-99-8 189.615	wh orth cry	247 3.90	623 a,o,d,o	s H_2O, EtOH, ace, eth; i xyl
2111	Tin(II) chloride dihydrate $SnCl_2 \cdot 2H_2O$	10025-69-1 225.646	wh monocl cry	37 dec 2.71		s H_2O, EtOH, NaOH; vs HCl
2112	Tin(II) fluoride SnF_2	7783-47-3 156.707	wh monocl cry; hyg	213 4.57	850	s H_2O; i EtOH, eth, chl
2113	Tin(II) hexafluorozirconate $SnZrF_6$	12419-43-1 323.924	cry	 4.21		s H_2O
2114	Tin(II) hydroxide $Sn(OH)_2$	12026-24-3 152.725	wh amorp solid		a	
2115	Tin(II) iodide SnI_2	10294-70-9 372.519	red-oran powder	320 5.28	714 a,d	sl H_2O; s bz, chl, CS_2
2116	Tin(II) oxalate SnC_2O_4	814-94-8 206.730	wh powder	280 dec 3.56		i H_2O; s dil HCl
2117	Tin(II) oxide SnO	21651-19-4 134.709	blue-blk tetr cry	1080 dec 6.45	a	i H_2O, EtOH; s acid
2118	Tin(II) pyrophosphate $Sn_2P_2O_7$	15578-26-4 411.363	wh amorp powder	dec 400 4.009		i H_2O; s conc acid
2119	Tin(II) selenide SnSe	1315-06-6 197.67	gray orth cry	861 6.18		i H_2O; s aqua regia
2120	Tin(II) sulfate $SnSO_4$	7488-55-3 214.774	wh orth cry	dec 378 4.15		reac H_2O
2121	Tin(II) sulfide SnS	1314-95-0 150.776	gray orth cry	880 5.08	1210 a	i H_2O; s conc acid
2122	Tin(II) tartrate $SnC_4H_4O_6$	815-85-0 266.782	wh cry powder			s H_2), dil HCl
2123	Tin(II) telluride SnTe	12040-02-7 246.31	gray cub cry	790 6.5		
2124	Tin(IV) bromide $SnBr_4$	7789-67-5 438.326	wh cry	31 3.34	205 a,b,c,d,e	vs H_2O; s EtOH
2125	Tin(IV) chloride $SnCl_4$	7646-78-8 260.521	col fuming liq	-33 2.234	114.15 a,b,c,d,e	reac H_2O; s EtOH, ctc, bz, ace
2126	Tin(IV) chloride pentahydrate $SnCl_4 \cdot 5H_2O$	10026-06-9 350.597	wh-yel cry	dec 56 2.04		vs H_2O; s EtOH
2127	Tin(IV) chromate $Sn(CrO_4)_2$	38455-77-5 350.697	brn-yel cry powder	dec		s H_2O
2128	Tin(IV) fluoride SnF_4	7783-62-2 194.704	wh tetr cry	 4.78	subl 705	reac H_2O

No.	Name Formula	CAS RN Mol. Wt.	Physical Form	mp/°C den/g cm⁻³	bp/°C Other Data	Solubility
2129	Tin(IV) iodide SnI_4	7790-47-8 626.328	yel-brn cub cry	143 4.46	364.35 a,b,d,e	reac H_2O; s EtOH, bz, chl, eth
2130	Tin(IV) oxide SnO_2	18282-10-5 150.709	gray tetr cry	1630 6.85	a,f	i H_2O, EtOH; s hot conc alk
2131	Tin(IV) selenide $SnSe_2$	20770-09-6 276.63	red-brn cry	650 ≈5.0		i H_2O; s alk, conc acid
2132	Tin(IV) selenite $Sn(SeO_3)_2$	7446-25-5 372.63	cry powder			i H_2O; s hot HCl
2133	Tin(IV) sulfide SnS_2	1315-01-1 182.842	gold-yel hex cry	dec 600 4.5		i H_2O; s alk, aqua regia
2134	Titanium Ti	7440-32-6 47.867	gray metal; hex	1668 4.506	3287 a,c	
2135	Titanium hydride TiH_2	7704-98-5 49.883	gray-blk powder	dec ≈450 3.75		i H_2O
2136	Titanium boride TiB_2	12045-63-5 69.489	gray refrac solid; hex	3225 4.38		
2137	Titanium carbide TiC	12070-08-5 59.878	cub cry	3067 4.93		i H_2O; s HNO_3
2138	Titanium nitride TiN	25583-20-4 61.874	yel-brn cub cry	3290 5.43		i H_2O; s aqua regia
2139	Titanium phosphide TiP	12037-65-9 78.841	gray hex cry	1990 4.08		
2140	Titanium silicide $TiSi_2$	12039-83-7 104.038	blk orth cry	1500 4.0		i H_2O, acid, alk; s HF
2141	Titanium(II) bromide $TiBr_2$	13783-04-5 207.675	blk powder	4.0	a	reac H_2O
2142	Titanium(II) chloride $TiCl_2$	10049-06-6 118.772	blk hex cryc	1035 3.13	1500 a,d	reac H_2O; s EtOH; i chl, eth
2143	Titanium(II) iodide TiI_2	13783-07-8 301.676	blk hex cry	5.02		reac H_2O
2144	Titanium(II) oxide TiO	12137-20-1 63.866	cub cry	1750 4.95	a	
2145	Titanium(II) sulfide TiS	12039-07-5 79.933	brn hex cry	1780 3.85		s conc acid
2146	Titanium(III) bromide $TiBr_3$	13135-31-4 287.579	blue-blk hex cry		a	s H_2O
2147	Titanium(III) chloride $TiCl_3$	7705-07-9 154.225	red-viol hex cry; hyg	425 dec 2.64	a,d	reac H_2O
2148	Titanium(III) fluoride TiF_3	13470-08-1 104.862	viol hex cry	1200 2.98	1400	i H_2O, dil acid, alk
2149	Titanium(III) oxide Ti_2O_3	1344-54-3 143.732	viol hex cry	1842 4.486	a	s hot HF
2150	Titanium(III) sulfate $Ti_2(SO_4)_3$	10343-61-0 383.925	grn cry			i H_2O, EtOH; s dil HCl
2151	Titanium(III) sulfide Ti_2S_3	12039-16-6 191.932	blk hex cry	3.56		
2152	Titanium(III,IV) oxide Ti_3O_5	12065-65-5 223.598	blk monocl cry	1777 4.24	a	
2153	Titanium(IV) bromide $TiBr_4$	7789-68-6 367.483	yel-oran cub cry; hyg	39 3.37	230 a,b,c,d	reac H_2O
2154	Titanium(IV) chloride $TiCl_4$	7550-45-0 189.678	col or yel liq	-25 1.73	136.45 a,b,c,d,e	reac H_2O; s EtOH
2155	Titanium(IV) fluoride TiF_4	7783-63-3 123.861	wh hyg powder	284 2.798	subl	reac H_2O; s EtOH, py
2156	Titanium(IV) iodide TiI_4	7720-83-4 555.485	red hyg powder	150 4.3	377 a,b,c,d	reac H_2O
2157	Titanium(IV) oxide TiO_2	13463-67-7 79.866	wh tetr cry	1843 4.23	a,f	i H_2O, dil acid; s conc acid
2158	Titanium(IV) oxysulfate monohydrate $TiOSO_4 \cdot H_2O$	13825-74-6* 177.945	col orth cry	2.71		reac H_2O
2159	Titanium(IV) sulfide TiS_2	12039-13-3 111.999	yel-brn hex cry; hyg	3.37		s H_2SO_4
2160	Tungsten W	7440-33-7 183.84	gray-wh metal; cub	3422 19.3	5555 a,c,e	
2161	Tungstic acid H_2WO_4	7783-03-1 249.85	yel amorp powder	dec 100 5.5		i H_2O, acid; s alk
2162	Tungsten boride W_2B	12007-10-2 378.49	refrac blk powder	2670 16.0		i H_2O
2163	Tungsten boride WB	12007-09-9 194.65	blk refrac powder	2665 15.2		i H_2O
2164	Tungsten boride W_2B_5	12007-98-6 421.74	refrac solid	2365 11.0		i H_2O
2165	Tungsten carbide W_2C	12070-13-2 379.69	refrac hex cry	≈2800 14.8		i H_2O
2166	Tungsten carbide WC	12070-12-1 195.85	gray hex cry	2785 15.6		i H_2O; s HNO_3/HF
2167	Tungsten carbonyl $W(CO)_6$	14040-11-0 351.90	wh cry	dec 170 2.65	subl	i H_2O; s os

No.	Name Formula	CAS RN Mol. Wt.	Physical Form	mp/°C den/g cm^{-3}	bp/°C Other Data	Solubility
2168	Tungsten nitride WN_2	60922-26-1 211.85	hex cry	dec 600 7.7		
2169	Tungsten nitride W_2N	12033-72-6 381.69	gray cub cry	dec 17.8		
2170	Tungsten silicide WSi_2	12039-88-2 240.01	blue-gray tetr cry	2160 9.3		i H_2O
2171	Tungsten silicide W_5Si_3	12039-95-1 1003.46	blue-gray refrac solid	2320 14.4		
2172	Tungsten(II) bromide WBr_2	13470-10-5 343.65	yel powder	dec 400		
2173	Tungsten(II) chloride WCl_2	13470-12-7 254.75	yel solid	dec >500		s H_2O
2174	Tungsten(II) iodide WI_2	13470-17-2 437.65	oran cry	6.79		
2175	Tungsten(III) bromide WBr_3	15163-24-3 423.55	blk hex cry	dec >80		i H_2O
2176	Tungsten(III) chloride WCl_3	20193-56-0 290.20	red solid	550 dec	subl	reac H_2O
2177	Tungsten(IV) bromide WBr_4	14055-81-3 503.46	blk orth cry			reac H_2O
2178	Tungsten(IV) chloride WCl_4	13470-13-8 325.65	blk hyg powder	4.62		reac H_2O
2179	Tungsten(IV) fluoride WF_4	13766-47-7 259.83	red-brn cry	dec >800		
2180	Tungsten(IV) iodide WI_4	14055-84-6 691.46	blk powder	dec		reac H_2O; s EtOH; i eth chl
2181	Tungsten(IV) oxide WO_2	12036-22-5 215.84	blue monocl cry	dec 1500-1700 10.8	a	i H_2O, os
2182	Tungsten(IV) selenide WSe_2	12067-46-8 341.76	gray hex cry	9.2		
2183	Tungsten(IV) sulfide WS_2	12138-09-9 247.97	gray hex cry	1250 dec 7.6		i H_2O, HCl, alk
2184	Tungsten(IV) telluride WTe_2	12067-76-4 439.04	gray orth cry	1020 9.43		
2185	Tungsten(V) bromide WBr_5	13470-11-6 583.36	brn-blk hyg solid	286	333	
2186	Tungsten(V) chloride WCl_5	13470-14-9 361.10	blk hyg cry	242	286	reac H_2O
2187	Tungsten(V) fluoride WF_5	19357-83-6 278.83	yel solid	dec >80		
2188	Tungsten(V) oxytribromide $WOBr_3$	20213-56-3 439.55	dark brn tetr cry	≈5.9		
2189	Tungsten(V) oxytrichloride $WOCl_3$	14249-98-0 306.20	grn tetr cry	≈4.6		
2190	Tungsten(VI) bromide WBr_6	13701-86-5 663.26	blue-blk cry	309	a	
2191	Tungsten(VI) chloride WCl_6	13283-01-7 396.56	purp hex cry; hyg	275 3.52	346.75 a,b,c,d	s EtOH, os
2192	Tungsten(VI) dioxydibromide WO_2Br_2	13520-75-7 375.65	red cry			
2193	Tungsten(VI) dioxydichloride WO_2Cl_2	13520-76-8 286.74	yel orth cry	265 4.67		i H_2O
2194	Tungsten(VI) dioxydiiodide WO_2I_2	14447-89-3 469.65	monocl cry	6.39		
2195	Tungsten(VI) fluoride WF_6	7783-82-6 297.83	col gas	2.3 13.069 g/L	17 a,b,c,d,e	reac H_2O
2196	Tungsten(VI) oxide WO_3	1314-35-8 231.84	yel powder	1472 7.2	a,c	i H_2O; sl acid; s alk
2197	Tungsten(VI) oxytetrabromide $WOBr_4$	13520-77-9 519.46	red tetr cry	277 ≈5.5	327	reac H_2O
2198	Tungsten(VI) oxytetrachloride $WOCl_4$	13520-78-0 341.65	red hyg cry	211 11.92	227.55 b,c,d	reac H_2O; s bz, CS_2
2199	Tungsten(VI) oxytetrafluoride WOF_4	13520-79-1 275.83	wh monocl cry	106 5.07	186	reac H_2O
2200	Tungsten(VI) sulfide WS_3	12125-19-8 280.04	brn powder			sl H_2O; s alk
2201	Uranium U	7440-61-1 238.029	silv-wh orth cry	1135 19.1	4131 a,c	
2202	Uranium boride UB_2	12007-36-2 259.651	refrac solid	2430 12.7		
2203	Uranium boride UB_4	12007-84-0 281.273	refrac solid	2530 9.32		i H_2O
2204	Uranium carbide UC	12070-09-6 250.040	gray cub cry	2790		
2205	Uranium carbide UC_2	12071-33-9 262.051	gray tetr cry	2350 11.3	4370	reac H_2O; sl EtOH
2206	Uranium carbide U_2C_3	12076-62-9 512.091	gray cub cry	dec ≈1700 12.7		

No.	Name Formula	CAS RN Mol. Wt.	Physical Form	mp/°C den/g cm^{-3}	bp/°C Other Data	Solubility
2207	Uranium nitride UN	25658-43-9 252.036	gray cub cry	2805 14.3		i H$_2$O
2208	Uranium nitride U$_2$N$_3$	12033-83-9 518.078	cub cry	dec 11.3		
2209	Uranium(III) bromide UBr$_3$	13470-19-4 477.741	red hyg cry	727		s H$_2$O
2210	Uranium(III) chloride UCl$_3$	10025-93-1 344.387	grn hyg cry	837 5.51	a	vs H$_2$O; i bz, ctc
2211	Uranium(III) fluoride UF$_3$	13775-06-9 295.024	blk hex cry	dec 8.9	a	i H$_2$O; s acid
2212	Uranium(III) hydride UH$_3$	13598-56-6 241.053	gray-blk cub cry	11.1	a	
2213	Uranium(III) iodide UI$_3$	13775-18-3 618.742	blk hyg cry	766		s H$_2$O
2214	Uranium(IV) bromide UBr$_4$	13470-20-7 557.645	brn hyg cry	519		s H$_2$O, EtOH
2215	Uranium(IV) chloride UCl$_4$	10026-10-5 379.840	grn octahed cry	590 4.72	791 a,c	reac H$_2$O; s EtOH
2216	Uranium(IV) fluoride UF$_4$	10049-14-6 314.023	grn monocl cry	1036 6.7	1417 a	i H$_2$O; s conc acid, alk
2217	Uranium(IV) iodide UI$_4$	13470-22-9 745.647	blk hyg cry	506		s H$_2$O, EtOH
2218	Uranium(IV) oxide UO$_2$	1344-57-6 270.028	brn cub cry	2827 10.97	a,f	i H$_2$O, dil acid; s conc acid
2219	Uranium(IV,V) oxide U$_4$O$_9$	12037-15-9 1096.111	cub cry	11.2	a	
2220	Uranium(V) bromide UBr$_5$	13775-16-1 637.549	brn hyg cry			reac H$_2$O
2221	Uranium(V) chloride UCl$_5$	13470-21-8 415.293	brn hyg cry	287		reac H$_2$O
2222	Uranium(V) fluoride UF$_5$	13775-07-0 333.021	pale blue tetr cry; hyg	348 5.81		s H$_2$O
2223	Uranium(V,VI) oxide U$_3$O$_8$	1344-59-8 842.082	grn-blk orth cry	1300 dec 8.38	a	
2224	Uranium(VI) chloride UCl$_6$	13763-23-0 450.745	green hex cry	177 3.6	a	
2225	Uranium(VI) fluoride UF$_6$	7783-81-5 352.019	wh monocl solid	64.0 tp 5.09	56.5 sp a,b,c,e	reac H$_2$O; s ctc, chl
2226	Uranium(VI) oxide UO$_3$	1344-58-7 286.027	oran-yel cry	≈7.3	a	i H$_2$O; s acid
2227	Uranium(VI) oxide monohydrate UO$_3$·H$_2$O	12326-21-5 304.043	yel orth cry	dec 570 7.05		
2228	Uranyl chloride UO$_2$Cl$_2$	7791-26-6 340.933	yel orth cry; hyg	577	a	vs H$_2$O; s EtOH, ace; i bz
2229	Uranyl nitrate hexahydrate UO$_2$(NO$_3$)$_2$·6H$_2$O	13520-83-7 502.129	yel orth cry; hyg	60 2.81	dec 118	vs H$_2$O; s EtOH, eth
2230	Uranyl sulfate trihydrate UO$_2$SO$_4$·3H$_2$O	20910-28-5 420.138	yel cry	3.28		s H$_2$O, sl EtOH
2231	Vanadium V	7440-62-2 50.942	gray-wh metal; cub	1910 6.0	3407 a,c	i H$_2$O; s acid
2232	Vanadium boride VB	12045-27-1 61.753	refrac solid	2250		i H$_2$O
2233	Vanadium boride VB$_2$	12007-37-3 72.564	refrac solid	2450		
2234	Vanadium carbide VC	12070-10-9 62.953	refrac blk cry; cub	2810 5.77		i H$_2$O
2235	Vanadium carbide V$_2$C	12012-17-8 113.894	hex cry	2167		
2236	Vanadium carbonyl V(CO)$_6$	20644-87-5 219.004	blue-grn cry; flam	dec 60	subl	
2237	Vanadium nitride VN	24646-85-3 64.949	blk powder; cub	2050 6.13		i H$_2$O; s aqua regia
2238	Vanadium silicide VSi$_2$	12039-87-1 107.113	metallic prisms	4.42		s HF
2239	Vanadium silicide V$_3$Si	12039-76-8 180.911	cub cry	1935 5.70		
2240	Vanadium(II) bromide VBr$_2$	14890-41-6 210.750	oran-brn hex cry	4.58	subl 800	reac H$_2$O
2241	Vanadium(II) chloride VCl$_2$	10580-52-6 121.847	grn hex plates	3.23	subl 910	reac H$_2$O; s EtOH, eth
2242	Vanadium(II) fluoride VF$_2$	13842-80-3 88.939	blue hyg cry			reac H$_2$O
2243	Vanadium(II) iodide VI$_2$	15513-84-5 304.751	red-viol hex cry	5.44		reac H$_2$O
2244	Vanadium(II) oxide VO	12035-98-2 66.941	grn cry	1790 5.758	a,c	s acid
2245	Vanadium(II) sulfate heptahydrate VSO$_4$·7H$_2$O	36907-42-3 273.112	viol cry			

No.	Name Formula	CAS RN Mol. Wt.	Physical Form	mp/°C den/g cm^{-3}	bp/°C Other Data	Solubility
2246	Vanadium(III) acetylacetonate $V(CHCOCHCOCH_3)_3$	13476-99-8 348.270	brn cry	≈185 ≈1.0	subl	s MeOH, ace, bz chl
2247	Vanadium(III) bromide VBr_3	13470-26-3 290.654	gray-brn hyg cry	4.00		reac H_2O
2248	Vanadium(III) chloride VCl_3	7718-98-1 157.300	red-viol hex cry; hyg	3.00	a	reac H_2O; s EtOH, eth
2249	Vanadium(III) fluoride VF_3	10049-12-4 107.937	yel-grn hex cry	≈1400 3.363	subl	i H_2O, EtOH
2250	Vanadium(III) fluoride trihydrate $VF_3 \cdot 3H_2O$	10049-12-4* 161.983	grn rhomb cry	dec ≈100		sl H_2O
2251	Vanadium(III) iodide VI_3	15513-94-7 431.655	brn-blk rhomb cry; hyg	5.21		reac H_2O
2252	Vanadium(III) oxide V_2O_3	1314-34-7 149.881	blk powder	2067 4.87	a	i H_2O
2253	Vanadium(III) sulfate $V_2(SO_4)_3$	13701-70-7 390.074	yel powder	dec ≈400		sl H_2O
2254	Vanadium(III) sulfide V_2S_3	1315-03-3 198.081	grn-blk powder	dec 4.7		i H_2O, s hot HCl
2255	Vanadium(IV) bromide VBr_4	13595-30-7 370.558	unstable magenta cry	dec -23	a	
2256	Vanadium(IV) chloride VCl_4	7632-51-1 192.753	unstable red liq	-25.7 1.816	148 a,c,d	reac H_2O; s EtOH, eth
2257	Vanadium(IV) fluoride VF_4	10049-16-8 126.936	grn hyg powder	dec 325 3.15	subl a	vs H_2O
2258	Vanadium(IV) oxide VO_2	12036-21-4 82.941	blue-blk powder	1967 4.339		i H_2O; s acid, alk
2259	Vanadium(V) fluoride VF_5	7783-72-4 145.934	col liq	19.5 2.50	48.3 a,c,d	reac H_2O
2260	Vanadium(V) oxide V_2O_5	1314-62-1 181.880	yel-brn orth cry	670 3.35	1800 dec a,c	sl H_2O; s conc acid, alk; i EtOH
2261	Vanadyl bromide $VOBr$	13520-88-2 146.845	viol cry	dec 480		
2262	Vanadyl chloride $VOCl$	13520-87-1 102.394	brn orth cry	dec >700 1.72	a,b	
2263	Vanadyl dibromide $VOBr_2$	13520-89-3 226.749	yel-brn cry	dec 180		
2264	Vanadyl dichloride $VOCl_2$	10213-09-9 137.846	grn hyg cry	dec 380 2.88		reac H_2O; s EtOH
2265	Vanadyl difluoride VOF_2	13814-83-0 104.938	yel cry			
2266	Vanadyl selenite hydrate $VOSeO_3 \cdot H_2O$	133578-89-9 211.92	grn tricl plates	3.506		
2267	Vanadyl sulfate dihydrate $VOSO_4 \cdot 2H_2O$	27774-13-6 199.036	blue cry powder			s H_2O
2268	Vanadyl tribromide $VOBr_3$	13520-90-6 306.653	deep red liq		dec 180	reac H_2O
2269	Vanadyl trichloride $VOCl_3$	7727-18-6 173.299	fuming red liq	-79 1.829	127 a,d,e	reac H_2O; s MeOH, eth, ace
2270	Vanadyl trifluoride VOF_3	13709-31-4 123.936	yel hyg powder	300 2.459	480	reac H_2O
2271	Water H_2O	7732-18-5 18.015	col liq	0.00 0.9970	100.0 a,b,c,d,e	s EtOH, MeOH, ace
2272	Xenon Xe	7440-63-3 131.29	col gas	-111.75 5.761 g/L	-108.04 a,b,c,d,e	sl H_2O
2273	Xenon trioxide XeO_3	13776-58-4 179.29	col orth cry	exp ≈25 4.55		s H_2O
2274	Xenon tetroxide XeO_4	12340-14-6 195.29	yel solid; exp	-35.9	dec ≈0	
2275	Xenon difluoride XeF_2	13709-36-9 169.29	col tetr cry	129.03 tp 4.32	114.35 sp b	sl H_2O
2276	Xenon tetrafluoride XeF_4	13709-61-0 207.28	col monocl cry	117.10 tp 4.04	115.75 sp a,b	reac H_2O
2277	Xenon hexafluoride XeF_6	13693-09-9 245.28	col monocl cry	49.5 3.56	75.6	reac H_2O
2278	Xenon dioxydifluoride XeO_2F_2	13875-06-4 201.29	col orth cry	30.8 exp 4.10		
2279	Xenon oxytetrafluoride $XeOF_4$	13774-85-1 223.28	col liq	-46.2 3.17(0°C)		reac H_2O
2280	Xenon fluoride hexafluororuthenate $XeFRuF_6$	22527-13-5 365.35	yel-grn monocl cry	110 3.78		
2281	Xenon fluoride undecafluoroantimonate $XeFSb_2F_{11}$	15364-10-0 602.79	yel monocl cry	63 3.69		
2282	Xenon fluoride hexafluoroarsenate $Xe_2F_3AsF_6$	50432-32-1 508.49	yel-grn monocl cry	99 3.62		reac H_2O
2283	Xenon fluoride hexafluoroantimonate XeF_3SbF_6	39797-63-2 424.04	yel-grn monocl cry	≈110 3.92		
2284	Xenon trifluoride undecafluoroantimonate $XeF_3Sb_2F_{11}$	35718-37-7 640.79	yel-grn tricl cry	82 3.98		

No.	Name Formula	CAS RN Mol. Wt.	Physical Form	mp/°C den/g cm^{-3}	bp/°C Other Data	Solubility
2285	Xenon pentafluoride hexafluoroarsenate XeF_5AsF_6	20328-94-3 415.19	wh monocl cry	130.5 3.51		
2286	Xenon pentafluoride hexafluororuthenate XeF_5RuF_6	39796-98-0 441.34	grn orth cry	152 3.79		
2287	Ytterbium Yb	7440-64-4 173.04	silv metal; cub	819 6.90	1196 a,c	s dil acid
2288	Ytterbium silicide $YbSi_2$	12039-89-3 229.21	hex cry	7.54		
2289	Ytterbium(II) bromide $YbBr_2$	25502-05-0 332.85	yel cry	673		reac H_2O
2290	Ytterbium(II) chloride $YbCl_2$	13874-77-6 243.95	grn cry	721 5.27		reac H_2O
2291	Ytterbium(II) iodide YbI_2	19357-86-9 426.85	blk cry	772		reac H_2O
2292	Ytterbium(III) chloride $YbCl_3$	10361-91-8 279.40	wh hyg powder	875	a	s H_2O
2293	Ytterbium(III) chloride hexahydrate $YbCl_3 \cdot 6H_2O$	19423-87-1 387.49	grn hyg cry	dec 150 2.57		vs H_2O
2294	Ytterbium(III) fluoride YbF_3	13760-80-0 230.04	wh cry	1157 8.2		i H_2O
2295	Ytterbium(III) oxide Yb_2O_3	1314-37-0 394.08	col cub cry	2435 9.2	a	s dil acid
2296	Ytterbium(III) sulfate octahydrate $Yb_2(SO_4)_3 \cdot 8H_2O$	10034-98-7 778.39	col cry	3.3		s H_2O
2297	Yttrium Y	7440-65-5 88.906	silv metal; hex	1522 4.47	3345 a,c	reac H_2O; s dil acid
2298	Yttrium aluminum oxide $Y_3Al_5O_{12}$	12005-21-9 593.619	grn cub cry	≈4.5		
2299	Yttrium antimonide YSb	12186-97-9 210.666	cub cry	2310 5.97		
2300	Yttrium arsenide YAs	12255-48-0 163.828	cub cry	5.59		
2301	Yttrium boride YB_6	12008-32-1 153.772	refrac solid	2600 3.72		
2302	Yttrium bromide YBr_3	13469-98-2 328.618	col hyg cry	904		vs H_2O
2303	Yttrium carbide YC_2	12071-35-1 112.928	refrac solid	≈2400 4.13		
2304	Yttrium carbonate trihydrate $Y_2(CO_3)_3 \cdot 3H_2O$	5970-44-5 411.886	red-brn powder			i H_2O; s dil acid
2305	Yttrium chloride YCl_3	10361-92-9 195.264	wh monocl cry; hyg	721 2.61	a	vs H_2O
2306	Yttrium fluoride YF_3	13709-49-4 145.901	wh hyg powder	≈1150 4.0	a	i H_2O
2307	Yttrium nitrate hexahydrate $Y(NO_3)_3 \cdot 6H_2O$	13494-98-9 383.012	hyg cry			vs H_2O
2308	Yttrium oxide Y_2O_3	1314-36-9 225.810	wh cry; cub	2439 5.03	a,c	s dil acid
2309	Yttrium phosphide YP	12294-01-8 119.880	cub cry	≈4.4		
2310	Yttrium sulfate octahydrate $Y_2(SO_4)_3 \cdot 8H_2O$	7446-33-5 610.125	red monocl cry	2.6		s H_2O
2311	Yttrium sulfide Y_2S_3	12039-19-9 274.010	yel cub cry	1925 3.87		
2312	Zinc Zn	7440-66-6 65.39	blue-wh metal; hex	419.53 7.14	907 a,c,e	s acid, alk
2313	Zinc acetate dihydrate $Zn(C_2H_3O_2)_2 \cdot 2H_2O$	5970-45-6 219.51	wh powder	237 dec 1.735		vs H_2O; s EtOH
2314	Zinc antimonide ZnSb	12039-35-9 187.15	silv-wh orth cry	565 6.33		reac H_2O
2315	Zinc arsenate $Zn_3(AsO_4)_2$	13464-44-3 474.01	wh powder			i H_2O; s acid, alk
2316	Zinc arsenate octahydrate $Zn_3(AsO_4)_2 \cdot 8H_2O$	13464-45-4 618.13	wh monocl cry	3.33		i H_2O; s acid, alk
2317	Zinc arsenite $Zn(AsO_2)_2$	10326-24-6 279.23	col powder			i H_2O; s acid
2318	Zinc borate $3ZnO \cdot 2B_2O_3$	27043-84-1 383.41	wh amorp powder	3.64		sl H_2O; s dil acid
2319	Zinc borate hemiheptahydrate $2ZnO \cdot 3B_2O_3 \cdot 3.5H_2O$	12513-27-8 434.69	wh cry	980 4.22		i H_2O
2320	Zinc bromate hexahydrate $Zn(BrO_3)_2 \cdot 6H_2O$	13517-27-6 429.29	wh hyg solid	100 2.57		vs H_2O
2321	Zinc bromide $ZnBr_2$	7699-45-8 225.20	wh hex cry; hyg	394 4.5	697 a,c,d	vs H_2O, EtOH; s eth
2322	Zinc caprylate $Zn(C_8H_{15}O_2)_2$	557-09-5 351.80	wh hyg cry	136		sl H_2O
2323	Zinc carbonate $ZnCO_3$	3486-35-9 125.40	wh rhomb cry	dec 4.4	a,f	i H_2O; s dil acid, alk

No.	Name Formula	CAS RN Mol. Wt.	Physical Form	mp/°C den/g cm^{-3}	bp/°C Other Data	Solubility
2324	Zinc chlorate $Zn(ClO_3)_2$	10361-95-2 232.29	yel hyg cry	dec 60 2.15		vs H_2O
2325	Zinc chloride $ZnCl_2$	7646-85-7 136.29	wh hyg cry	290 2.907	732 a,d,c	vs H_2O; s EtOH, ace
2326	Zinc chromite $ZnCr_2O_4$	12018-19-8 233.38	grn cub cry	5.29		
2327	Zinc citrate dihydrate $Zn_3(C_6H_5O_7)_2 \cdot 2H_2O$	546-46-3 610.40	col powder			sl H_2O; s dil acid, alk
2328	Zinc cyanide $Zn(CN)_2 \cdot$	557-21-1 117.43	wh powder	1.852		i H_2O; reac acid
2329	Zinc fluoride ZnF_2	7783-49-5 103.39	wh tetr needles; hyg	872 4.9	1500 a,d,e	sl H_2O
2330	Zinc fluoride tetrahydrate $ZnF_2 \cdot 4H_2O$	13986-18-0 175.45	wh orth cry	2.30		sl H_2O
2331	Zinc fluoroborate hexahydrate $Zn(BF_4)_2 \cdot 6H_2O$	27860-83-9 347.09	hex cry	2.12		vs H_2O; s EtOH
2332	Zinc formate dihydrate $Zn(CHO_2)_2 \cdot 2H_2O$	5970-62-7 191.46	wh cry	2.207		s H_2O, i EtOH
2333	Zinc hexafluorosilicate hexahydrate $ZnSiF_6 \cdot 6H_2O$	16871-71-9 315.56	wh cry			s H_2O
2334	Zinc hydroxide $Zn(OH)_2$	20427-58-1 99.41	col orth cry	125 dec 3.05	a	sl H_2O
2335	Zinc iodate $Zn(IO_3)_2$	7790-37-6 415.20	wh cry powder			sl H_2O
2336	Zinc iodide ZnI_2	10139-47-6 319.20	wh hyg cry	446 4.74	625 a	vs H_2O; s EtOH, eth
2337	Zinc laurate $Zn(C_{12}H_{23}O_2)_2$	2452-01-9 464.02	wh powder	128		sl H_2O
2338	Zinc molybdate $ZnMoO_4$	13767-32-3 225.33	wh tetr cry	>700 4.3		i H_2O
2339	Zinc nitrate $Zn(NO_3)_2$	7779-88-6 189.40	wh powder		a	vs H_2O
2340	Zinc nitrate hexahydrate $Zn(NO_3)_2 \cdot 6H_2O$	10196-18-6 297.49	col orth cry	36 dec 2.067		vs H_2O, EtOH
2341	Zinc nitride Zn_3N_2	1313-49-1 224.18	blue-gray cub cry			
2342	Zinc nitrite $Zn(NO_2)_2$	10102-02-0 157.40	hyg solid			reac H_2O
2343	Zinc oleate $Zn(C_{18}H_{33}O_2)_2$	557-07-3 628.31	wh powder	70 dec		i H_2O; s EtOH, eth, bz
2344	Zinc oxalate dihydrate $ZnC_2O_4 \cdot 2H_2O$	547-68-2 189.44	wh powder	100 dec 2.56		i H_2O; s dil acid
2345	Zinc oxide ZnO	1314-13-2 81.39	wh powder; hex	1075 5.6	a,c,f	i H_2O; s dil acid
2346	Zinc perchlorate hexahydrate $Zn(ClO_4)_2 \cdot 6H_2O$	10025-64-6 372.38	wh cub cry; hyg	106 dec 2.2		vs H_2O; s EtOH
2347	Zinc permanganate hexahydrate $Zn(MnO_4)_2 \cdot 6H_2O$	23414-72-4 411.35	blk orth cry; hyg	2.45		s H_2O; reac EtOH
2348	Zinc peroxide ZnO_2	1314-22-3 97.39	yel-wh powder	dec >150 1.57	exp 212	i H_2O; reac acid, EtOH, ace
2349	Zinc phosphate $Zn_3(PO_4)_2$	7779-90-0 386.11	wh monocl cry	900 4.0		i H_2O
2350	Zinc phosphate tetrahydrate $Zn_3(PO_4)_2 \cdot 4H_2O$	7543-51-3 458.17	col orth cry	3.04	f	i H_2O, EtOH; s dil acid, alk
2351	Zinc phosphide Zn_3P_2	1314-84-7 258.12	gray tetr cry	1160 4.55		i H_2O, EtOH; reac acid; s bz
2352	Zinc pyrophosphate $Zn_2P_2O_7$	7446-26-6 304.72	wh cry powder	3.75		i H_2O; s dil acid
2353	Zinc selenate pentahydrate $ZnSeO_4 \cdot 5H_2O$	13597-54-1 298.42	tricl cry	dec 50 2.59		s H_2O
2354	Zinc selenide $ZnSe$	1315-09-9 144.35	yel-red cub cry	>1100 5.65	subl a	i H_2O; s dil acid
2355	Zinc orthosilicate Zn_2SiO_4	13597-65-4 222.86	wh hex cry	1509 4.1	a,f	i H_2O, dil acid
2356	Zinc stearate $Zn(C_{18}H_{35}O_2)_2$	557-05-1 632.34	wh powder	130 1.095		i H_2O, EtOH, eth; s bz
2357	Zinc sulfate $ZnSO_4$	7733-02-0 161.45	col orth cry	680 dec 3.8	a	vs H_2O
2358	Zinc sulfate monohydrate $ZnSO_4 \cdot H_2O$	7446-19-7 179.47	wh monocl cry	238 dec 3.20		vs H_2O; i EtOH
2359	Zinc sulfate heptahydrate $ZnSO_4 \cdot 7H_2O$	7446-20-0 287.56	col orth cry	100 dec 1.97	f	vs H_2O; i EtOH
2360	Zinc sulfide (sphalerite) ZnS	1314-98-3 97.46	gray-wh cub cry	1700 4.04	a,f	i H_2O, EtOH; s dil acid
2361	Zinc sulfide (wurtzite) ZnS	1314-98-3 97.46	wh hex cry	1700 4.09	a,f	i H_2O; s dil acid
2362	Zinc sulfite dihydrate $ZnSO_3 \cdot 2H_2O$	123714-57-8 181.49	wh powder	200 dec		i H_2O, EtOH; reac hot H_2O

No.	Name Formula	CAS RN Mol. Wt.	Physical Form	mp/°C den/g cm^{-3}	bp/°C Other Data	Solubility
2363	Zinc telluride ZnTe	1315-11-3 192.99	red cub cry	1239 5.9		i H_2O
2364	Zinc thiocyanate Zn(SCN)$_2$	557-42-6 181.56	wh hyg cry			sl H_2O; s EtOH
2365	Zirconium Zr	7440-67-7 91.224	gray-wh metal; hex	1855 6.52	4409 a,c	s hot conc acid
2366	Zirconium boride ZrB$_2$	12045-64-6 112.846	gray refrac solid; hex	3245 6.17		
2367	Zirconium bromide ZrBr$_4$	13777-25-8 410.840	wh cub cry	450 tp 3.98	360 sp a,b,e	
2368	Zirconium carbide ZrC	12020-14-3 103.235	gray refrac solid; cub	3532 6.73		s HF
2369	Zirconium chloride ZrCl$_4$	10026-11-6 233.035	wh monocl cry; hyg	437 tp 2.80	331 sp a,b,c,e	reac H_2O; s EtOH, eth
2370	Zirconium fluoride ZrF$_4$	7783-64-4 167.218	wh monocl cry	932 tp 4.43	912 sp a,c	sl H_2O
2371	Zirconium hydride ZrH$_2$	7704-99-6 93.240	gray tetr cry	 5.6	 a	i H_2O
2372	Zirconium hydroxide Zr(OH)$_4$	14475-63-9 159.254	wh amorp powder	dec 3.25		i H_2O; s acid
2373	Zirconium iodide ZrI$_4$	13986-26-0 598.842	oran cub cry; hyg	499 tp 4.85	431 sp a,b,e	vs H_2O
2374	Zirconium nitrate pentahydrate Zr(NO$_3$)$_4$•5H$_2$O	13746-89-9 429.320	wh hyg cry	dec 100		vs H_2O; s EtOH
2375	Zirconium nitride ZrN	25658-42-8 105.231	yel cub cry	2960 7.09		s conc HF; sl dil acid
2376	Zirconium oxide ZrO$_2$	1314-23-4 123.223	wh amorp powder	2710 5.68	 a,c,f	i H_2O; sl acid
2377	Zirconium phosphide ZrP$_2$	12037-80-8 153.172	orth cry	 ≈5.1		
2378	Zirconium orthosilicate ZrSiO$_4$	10101-52-7 183.308	wh tetr cry	dec 1540 4.6	 a,f	i H_2O, acid
2379	Zirconium silicide ZrSi$_2$	12039-90-6 147.395	gray powder	1620 4.88		i H_2O, aqua regia; s HF
2380	Zirconium sulfate tetrahydrate Zr(SO$_4$)$_2$•4H$_2$O	7446-31-3 355.413	wh tetr cry	dec 100 2.80		vs H_2O
2381	Zirconium sulfide ZrS$_2$	12039-15-5 155.356	red-brn hex cry	1480 3.82		i H_2O
2382	Zirconyl chloride octahydrate ZrOCl$_2$•8H$_2$O	13520-92-8 322.251	tetr cry	dec 400 1.91		vs H_2O, EtOH

Caustic baryta	234	Chrysoberyl	291	Cupric acetoarsenite	703
Caustic soda	1909	Cianurina	1311	Cupric acetylide	705
Celestite	1981	Cinnabar	1327	Cupric ammonium chloride	80
Celphos	44	Claudetite	196	Cupric arsenate	706
Cerargyrite	1830	Clausthalite	1131	Cupric azide	708
Ceria	524	Clinoenstatite	1233	Cupric borate	710
Ceric ammonium nitrate	73	Cobalt black	676	Cupric bromide	711
Ceric fluoride	523	Cobalt hydroxide oxide	677	Cupric butyrate monohydrate	712
Ceric oxide	524	Cobalt yellow	678	Cupric chlorate hexahydrate	714
Ceric sulfate tetrahydrate	525	Cobaltic acetate	670	Cupric chloride	715
Cerium hexaboride	504	Cobaltic ammonium phosphate	78	Cupric chloride dihydrate	716
Cerous ammonium sulfate tetrahydrate	72	Cobaltic ammonium sulfate hexahydrate	79	Cupric chromate	717
Cerous bromide	511	Cobaltic fluoride	672	Cupric chromite	718
Cerous bromide heptahydrate	512	Cobaltic hexafluorosilicate hexahydrate	642	Cupric cyanide	720
Cerous carbide	513	Cobaltic hydroxide	674	Cupric dichromate dihydrate	721
Cerous carbonate hydrate	514	Cobaltic nitrate	675	Cupric ferrocyanide	722
Cerous chloride	515	Cobaltic orthotitanate	680	Cupric fluoride	724
Cerous chloride heptahydrate	516	Cobaltic oxide	676	Cupric fluoride dihydrate	725
Cerous fluoride	517	Cobaltic sulfide	679	Cupric formate	726
Cerous iodide	518	Cobaltic-cobaltous oxide	669	Cupric formate tetrahydrate	727
Cerous iodide nonahydrate	519	Cobaltite	612	Cupric hexafluorosilicate tetrahydrate	728
Cerous nitride	506	Cobaltous acetate	623	Cupric hydroxide	729
Cerous oxide	520	Cobaltous acetate tetrahydrate	624	Cupric molybdate	730
Cerous sulfate octahydrate	521	Cobaltous arsenate octahydrate	626	Cupric nitrate	731
Cerous sulfide	522	Cobaltous bromate hexahydrate	627	Cupric nitrate hexahydrate	732
Cerussite	1097	Cobaltous bromide	628	Cupric nitrate trihydrate	733
Cervantite	175	Cobaltous bromide hexahydrate	629	Cupric oleate	734
Chalcanthite	745	Cobaltous chloride	631	Cupric oxalate	735
Chalcocite	697	Cobaltous chloride dihydrate	632	Cupric oxide	736
Chalcocyanite	744	Cobaltous chloride hexahydrate	633	Cupric perchlorate	737
Chalcopyrite	723	Cobaltous chromate	634	Cupric perchlorate hexahydrate	738
Chereul's salt	701	Cobaltous chromate(III)	635	Cupric phosphate trihydrate	739
Chile saltpeter	1921	Cobaltous cyanide	636	Cupric selenate pentahydrate	740
China clay	47	Cobaltous cyanide dihydrate	637	Cupric selenide	741
Chlorine monofluoride	554	Cobaltous dipotassium sulfate hexahydrate	656	Cupric selenite dihydrate	742
Chlorine tetroxyfluoride	806	Cobaltous ferricyanide	638	Cupric stearate	743
Chlorocyanide	500	Cobaltous fluoride	639	Cupric sulfate	744
Chlorogermane	855	Cobaltous fluoride tetrahydrate	640	Cupric sulfate pentahydrate	745
Chlorosilane	1784	Cobaltous formate dihydrate	641	Cupric sulfide	747
Chlorosulfuric acid	1988	Cobaltous hydroxide	643	Cupric tartrate trihydrate	748
Chlorotrifluorogermane	856	Cobaltous iodate	644	Cupric tungstate	750
Chrome alum	599	Cobaltous iodide	645	Cupric tungstate dihydrate	751
Chrome alum ammonium	77	Cobaltous iodide hexahydrate	646	Cuprite	695
Chrome yellow	1102	Cobaltous metatitanate	647	Cuprous acetate	685
Chromic acetate hexahydrate	583	Cobaltous molybdate	648	Cuprous acetylide	686
Chromic acid	608	Cobaltous nitrate	649	Cuprous azide	687
Chromic anhydride	608	Cobaltous nitrate hexahydrate	650	Cuprous bromide	688
Chromic bromide	584	Cobaltous orthosilicate	660	Cuprous chloride	689
Chromic bromide hexahydrate	585,586	Cobaltous oxalate	651	Cuprous cyanide	690
Chromic chloride	587	Cobaltous oxalate dihydrate	652	Cuprous fluoride	691
Chromic chloride hexahydrate	588	Cobaltous oxide	653	Cuprous hydride	692
Chromic fluoride	589	Cobaltous perchlorate	654	Cuprous iodide	693
Chromic fluoride trihydrate	590	Cobaltous phosphate octahydrate	655	Cuprous oxide	695
Chromic hydroxide trihydrate	591	Cobaltous selenate pentahydrate	657	Cuprous selenide	696
Chromic iodide	592	Cobaltous selenide	658	Cuprous sulfide	697
Chromic nitrate	593	Cobaltous selenite dihydrate	659	Cuprous sulfite hemihydrate	698
Chromic nitrate nonahydrate	594	Cobaltous stannate	661	Cuprous telluride	699
Chromic oxide	595	Cobaltous sulfate	662	Cuprous tetraiodomercurate	694
Chromic phosphate	596	Cobaltous sulfate monohydrate	664	Cuprous thiocyanate	700
Chromic phosphate hemiheptahydrate	597	Cobaltous sulfide	665	Cyanogas	436
Chromic phosphate hexahydrate	598	Cobaltous telluride	666	Cyanogran	1883
Chromic sulfate	600	Cobaltous thiocyanate trihydrate	667	Decaborane	349
Chromic sulfide	601	Cobaltous tungstate	668	Dehydrite	1223
Chromic telluride	602	Coccinite	1317	Detia	44
Chromite	1017	Coloradoite	1328	Diammonium citrate	108
Chromium diboride	601	Copper hydroxide sulfate	746	Diammonium hydrogen phosphite	
Chromium dioxide	605	Coquimbite	1061	monohydrate	136
Chromium trioxide	608	Corundum	40	Diamond	482
Chromium(VI) oxychloride	609	Cotunnite	1100	Diarsane	184
Chromous acetate monohydrate	574	Covellite	747	Diaspore	41
Chromous bromide	575	Cristobalite	1816	Diborane	342
Chromous chloride	576	Crocoite	1102	Diboron tetrachloride	361
Chromous chloride tetrahydrate	577	Cryolite	1895	Dibromogermane	857
Chromous fluoride	578	Cryptohalite	104	Dibromosilane	1786
Chromous oxalate monohydrate	580	Cupric acetate	702	Dichlorine hexoxide	552
Chromous sulfate pentahydrate	581	Cupric acetate monohydrate	704	Dichlorine trioxide	551

Lead diacetate	1086
Lead dioxide	1153
Lead hydroxide	1122
Lead metasilicate	1127
Lead metatantalate	1138
Lead metatitanate	1142
Lead monetite	1110
Lead orthosilicate	1128
Lead sesquioxide	1147
Lead tetraacetate	1149
Lead tetrabromide	1150
Lead tetrachloride	1151
Lead tetrafluoride	1152
Lead tetroxide	1148
Lead vanadate	1145
Lepidocrocite	1053
Lime	457
Litharge	1120
Lithia	1178
Lithiophosphate	1177
Lithium metaborate	1160
Lithium metasilicate	1182
Lutetium triiodide	1191
Macquer's salt	1575
Maekinenite	1405
Maghemite	1056
Magnesia	1222
Magnesite	1205
Magnesium dioxide	1226
Magnesium fluorosilicate hexahydrate	1235
Magnesium hyposulfite hexahydrate	1244
Magnesium metasilicate	1233
Magnesium orthosilicate	1234
Magnetite	1039
Magnogene	1207
Malacite	713
Malladrite	1896
Mallebrin	19
Manganese dibromide	1259
Manganese dichloride	1261
Manganese difluoride	1265
Manganese diiodide	1267
Manganese dioxide	1293
Manganese green	242
Manganese heptoxide	1294
Manganese trioxide	1292
Manganite	1291
Manganjustite	1277
Manganosite	1274
Marshite	693
Mascagnite	145
Massicot	1121
Matlockite	1101
Melanterite	1033
Mercallite	1595
Mercuric acetate	1306
Mercuric arsenate	1315
Mercuric bromide	1308
Mercuric chloride	1309
Mercuric chromate	1310
Mercuric cyanide	1311
Mercuric dichromate	1312
Mercuric fluoride	1313
Mercuric fulminate	1314
Mercuric iodate	1316
Mercuric iodide	1317
Mercuric nitrate	1318
Mercuric nitrate dihydrate	1319
Mercuric nitrate monohydrate	1320
Mercuric oxide	1321
Mercuric oxycyanide	1322
Mercuric phosphate	1323
Mercuric potassium iodide	1635
Mercuric silver iodide	1856
Mercuric sulfate	1325
Mercuric sulfide	1327

Mercuric thiocyanate	1329
Mercurous acetate	1297
Mercurous bromide	1298
Mercurous chlorate	1299
Mercurous chloride	1300
Mercurous fluoride	1301
Mercurous iodide	1302
Mercurous nitrate dihydrate	1303
Mercurous oxide	1304
Mercurous sulfate	1305
Metaboric acid	353,354, 355
Metacinnabar	1326
Metaphosphoric acid	1493
Metasilicic acid	1807
Microcline	1559
Microcosmic salt	1866
Millerite	1409
Minium	1148
Mirabilite	1939
Misenite	1595
Mohr's salt	91
Moissanite	1810
Molybdenite	1353
Molybdenum dibromide	1340
Molybdenum dichloride	1341
Molybdenum diiodide	1342
Molybdenum dioxide	1351
Molybdenum diselenide	1352
Molybdenum disilicide	1339
Molybdenum disulfide	1353
Molybdenum ditelluride	1354
Molybdenum hexacarbonyl	1335
Molybdenum hexafluoride	1359
Molybdenum metaphosphate	1364
Molybdenum oxychloride	1362
Molybdenum pentaboride	1332
Molybdenum pentachloride	1355
Molybdenum pentafluoride	1356
Molybdenum sesquioxide	1347
Molybdenum tetrabromide	1348
Molybdenum tetrachloride	1349
Molybdenum tetrafluoride	1350
Molybdenum tribromide	1343
Molybdenum trichloride	1344
Molybdenum trichloride oxide	1357
Molybdenum trifluoride	1345
Molybdenum triiodide	1346
Molybdenum trioxide	1363
Molybdic acid monohydrate	1358
Molybdic anhydride	1363
Molysite	1044
Monetite	443
Monosilver acetylide	1825
Montroydite	1321
Nantokite	689
Naples yellow	1089
Natrium	1860
Neodymia	1374
Neptunium dioxide	1380
Newberyite	1213
Niccolite	1383
Nickel ammonium chloride	124
Nickel ammonium sulfate	125
Nickel sesquioxide	1412
Niobium dioxide	1429
Niobium fluorodioxide	1440
Niobium pentabromide	1433
Niobium pentachloride	1434
Niobium pentafluoride	1435
Niobium pentaiodide	1436
Niobium pentoxide	1437
Niobium tetrachloride	1426
Niobium tetrafluoride	1427
Niobium tetraiodide	1428
Niobium tribromide	1423

Niobium trichloride	1424
Niobium trifluoride	1425
Niter	1921
Nitrobarite	246
Nitromagnesite	1219
Nitrosylsulfuric acid	1990
Nitroxy fluoride	805
Norbide	359
Oil of vitriol	1986
Oldhamite	473
Opal	1807
Orpiment	198
Orthoarsenic acid	185
Orthoboric acid	352
Orthoclase	1559
Orthophosphoric acid	1492
Orthophosphorous acid	1497
Orthosilicic acid	1808
Orthotelluric acid	2032
Osmic acid	1476
Osmium dioxide	1473
Osmium hexafluoride	1475
Osmium pentafluoride	1474
Osmium tetrachloride	1471
Osmium tetrafluoride	1472
Osmium tetroxide	1476
Osmium tribromide	1469
Osmium trichloride	1470
Otavite	386
Oxygen difluoride	803
Palladium monoxide	1486
Palladous bromide	1481
Palladous chloride	1482
Palladous fluoride	1483
Palladous iodide	1484
Palladous nitrate	1485
Palladous sulfide	1480
Parawollastonite	467
Paris green	703
Pearl ash	1567
Pentaborane(11)	345
Pentaborane(9)	344
Perchloric acid	548
Perchloric anhydride	553
Perchloryl fluoride	557
Periclase	1222
Perovskite	479
Perrhenic acid	1666
Phosgene	496
Phosphorus hydride	1490
Phosphorus pentabromide	1512
Phosphorus pentachloride	1513
Phosphorus pentafluoride	1514
Phosphorus pentaselenide	1519
Phosphorus pentasulfide	1520
Phosphorus pentoxide	1515
Phosphorus sulfochloride	1522
Phosphorus tribromide	1503
Phosphorus trichloride	1504
Phosphorus trifluoride	1507
Phosphorus triiodide	1508
Phosphorus trioxide	1509
Phosphorus trisulfide	1511
Phosphoryl bromide	1516
Phosphoryl chloride	1517
Phosphoryl fluoride	1518
Phostoxin	44
Photophor	463
Pigment E	258
Plaster of Paris	472
Platinic acid hexahydrate	1525
Platinic oxide	1539
Platinum dibromide	1527
Platinum dichloride	1528
Platinum diiodide	1529
Platinum hexafluoride	1541

Platinum monoxide	1530
Platinum tetrabromide	1534
Platinum tetrachloride	1535
Platinum tetrafluoride	1537
Platinum tetraiodide	1538
Platinum tribromide	1532
Platinum trichloride	1533
Plattnerite	1153
Plessy's green	597
Plutonium dioxide	1551
Plutonium hexafluoride	1552
Plutonium monoxide	1544
Plutonium sesquioxide	1549
Plutonium tetrafluoride	1550
Plutonium tribromide	1545
Plutonium trichloride	1546
Plutonium trifluoride	1547
Plutonium triiodide	1548
Polonium dioxide	1555
Polonium tetrachloride	1554
Polydymite	1411
Portlandite	445
Potassium bicarbonate	1590
Potassium bichromate	1574
Potassium bifluoride	1591
Potassium biselenite	1594
Potassium bisulfate	1595
Potassium bisulfide	1596
Potassium bisulfite	1598
Potassium bitartrate	1599
Potassium chloroplatinate	1633
Potassium cobalticyanide	1585
Potassium dioxide	1628
Potassium hexacyanoferrate(II) trihydrate	1578
Potassium hexacyanoferrate(III)	1577
Potassium hyposulfite	1637
Potassium monoxide	1612
Potassium orthodiuranate	1642
Potassium peroxydisulfate	1617
Potassium phosphate, dibasic	1592
Potassium pyrosulfite	1605
Potassium sodium tartrate	1933
Potassium trithiocarbonate	1640
Potcrate	1569
Powellite	450
Prussic acid	942
Pucherite	326
Pyrite	1007
Pyrochroite	1266
Pyrolusite	1293
Pyrophenite	1287
Pyrophosphoric acid	1494
Quartz	1813-1816
Quartz	1815
Quicklime	457
Quicksilver	1296
Quilonorm	1155
Raspite	1144
Realgar	190
Red lead	1148
Red lead chromate	1103
Red phosphorus	1488
Retgersite	1408
Rhenium dioxide	1673
Rhenium disulfide	1674
Rhenium ditelluride	1675
Rhenium heptafluoride	1685
Rhenium heptasulfide	1691
Rhenium heptoxide	1686
Rhenium hexachloride	1680
Rhenium hexafluoride	1681
Rhenium pentabromide	1676
Rhenium pentachloride	1677
Rhenium pentafluoride	1678
Rhenium pentoxide	1679

Rhenium tetrachloride	1671
Rhenium tetrafluoride	1672
Rhenium tribromide	1668
Rhenium trichloride	1669
Rhenium triiodide	1670
Rhenium trioxide	1682
Rhodium dioxide	1700
Rhodium hexafluoride	1701
Rhodium trichloride	1695
Rhodium trifluoride	1696
Rhodium triiodide	1697
Rhodium trioxide	1698
Rhodochrosite	1261
Rhodonite	1277
Rochelle salt	1933
Rock salt	1877
Rohrbach's solution	275
Rubidium alum	1704
Rubidium bicarbonate	1713
Rubidium bifluoride	1711
Rubidium bisulfate	1714
Rubidium dioxide	1724
Ruthenium dioxide	1733
Ruthenium hexafluoride	1735
Ruthenium monoxide	1718
Ruthenium pentafluoride	1734
Ruthenium tetrafluoride	1732
Ruthenium tetroxide	1736
Ruthenium tribromide	1728
Ruthenium trichloride	1729
Ruthenium trifluoride	1730
Ruthenium triiodide	1731
Rutile	2157
Sacchite	1262
Sal ammoniac	75
Sal soda	1874
Salt of tartar	1567
Saltpeter	1609
Samaria	1749
Sanbornite	263
Sassolite	352
Scandia	1758
Scheele's green	707
Scheelite	480
Schultenite	1109
Selenium hexasulfide	1778
Selenium tetrasulfide	1779
Selenium(IV) bromide	1770
Selenium(IV) oxide	1766
Sellaite	1209
Senarmontite	168
Siderite	1013
Silane	1781
Silica	1813
Silicon chloride trifluoride	1785
Silicon decahydride	1805
Silicon octahydride	1804
Silicon tetrabromide	1793
Silicon tetrachloride	1794
Silicon tetrafluoride	1795
Silicon tetraiodide	1796
Silicon(II) oxide	1812
Silver difluoride	1859
Sinjarite	431
Skutterudite	615
Smaltite	614
Smite	1868
Smithsonite	2323
Soda ash	1873
Soda lye	1909
Sodamide	1865
Sodium acid pyrophosphate	1888
Sodium alum	1864
Sodium bicarbonate	1899
Sodium bifluoride	1900
Sodium bisulfate	1904

Sodium bisulfate monohydrate	1905
Sodium bisulfide	1906
Sodium bisulfide dihydrate	1907
Sodium bisulfite	1908
Sodium borate	1947
Sodium borofluoride	1893
Sodium dioxide	1945
Sodium metaarsenite	1867
Sodium metabismuthate	1914
Sodium metaperiodate	1928
Sodium metasilicate	1917
Sodium monobasic phosphate	1886
Sodium monoxide	1924
Sodium peroxydisulfate	1931
Sodium phosphate, dibasic	1901
Sodium prussiate yellow	1891
Sodium pyrosulfite	1915
Sodium tetrafluoroborate	1893
Sphalerite	2360
Spherocobaltite	630
Stannane	2106
Stannic bromide	2124
Stannic chloride	2125
Stannic chloride pentahydrate	2126
Stannic chromate	2127
Stannic fluoride	2128
Stannic iodide	2129
Stannic oxide	2130
Stannic selenide	2131
Stannic selenite	2132
Stannic sulfide	2133
Stannous acetate	2108
Stannous bromide	2109
Stannous chloride	2110
Stannous chloride dihydrate	2111
Stannous fluoride	2112
Stannous hexafluorozirconate	2113
Stannous hydroxide	2114
Stannous iodide	2115
Stannous monoxide	2117
Stannous oxalate	2116
Stannous pyrophosphate	2118
Stannous selenide	2119
Stannous sulfate	2120
Stannous sulfide	2121
Stannous tartrate	2122
Stannous telluride	2123
Stibine	162
Stibium	161
Stibnite	173
Stilleite	2354
Stolzite	1143
Strengite	1057
Strontia	1974
Strontianite	1961
Strontium orthosilicate	1979
Sulfamide	2004
Sulfinyl dibromide	2007
Sulfinyl dichloride	2008
Sulfinyl difluoride	2009
Sulfur chloride	1998
Sulfur decafluoride	2003
Sulfur monobromide	1994
Sulfur monochloride	1995
Sulfur monofluoride	1996, 1997
Sycoporite	665
Sylvite	1570
Szmikite	1281
Szomolnokite	1032
Tantalum boride	2014
Tantalum dioxide	2019
Tantalum diselenide	2020
Tantalum disulfide	2021
Tantalum ditelluride	2022
Tantalum pentabromide	2023

CAS No.	Page	CAS No.	Page	CAS No.	Page	CAS No.	Page	CAS No.	Page
62-54-4	417	555-75-9	23	1303-35-1	189	1313-29-7	1347	1332-52-1	293
62-76-0	1923	557-05-1	2356	1303-36-2	197	1313-49-1	2341	1332-58-7	47
71-48-7	623	557-07-3	2343	1303-37-3	203	1313-59-3	1924	1332-63-4	1291
74-90-8	942	557-09-5	2322	1303-52-2	895	1313-60-6	1930	1332-71-4	679
75-15-0	490	557-21-1	2328	1303-58-8	897	1313-82-2	1940	1332-77-0	1631
75-20-7	425	557-42-6	2364	1303-60-2	890	1313-83-3	1942	1332-81-6	175
75-44-5	496	562-76-5*	1634	1303-61-3	900	1313-84-4	1941	1333-74-0	939
124-38-9	488	563-63-3	1823	1303-62-4	899	1313-85-5	1936	1333-82-0	608
126-45-4	1832	563-68-8	2052	1303-86-2	366	1313-96-8	1437	1333-83-1	1900
127-08-2	1557	563-71-3	1013	1303-96-4	1948	1313-97-9	1374	1335-26-8	1226
127-09-3	1861	563-72-4	455	1304-28-5	252	1313-99-1	1404	1335-31-5	1322
128-00-7	1839	584-08-7	1567	1304-29-6	256	1314-05-2	1405	1335-32-6	1088
131-74-8	139	584-09-8	1707	1304-39-8	261	1314-06-3	1412	1336-21-6	116
141-53-7	1894	590-28-3	1572	1304-40-1	265	1314-08-5	1486	1341-49-7	109
142-03-0	22	590-29-4	1582	1304-54-7	306	1314-11-0	1974	1344-28-1	40
142-71-2	702	592-01-8	436	1304-56-9	307	1314-12-1	2068	1344-43-0	1274
143-33-9	1883	592-04-1	1311	1304-76-3	328	1314-13-2	2345	1344-48-5	1327
144-55-8	1899	592-05-2	1105	1304-82-1	338	1314-15-4	1539	1344-48-5	1326
146-84-9	1849	592-85-8	1329	1305-62-0	445	1314-18-7	1975	1344-54-3	2149
151-50-8	1573	592-87-0	1140	1305-78-8	457	1314-20-1	2091	1344-57-6	2218
156-62-7	435	593-60-0	130	1305-79-9	460	1314-22-3	2348	1344-58-7	2226
298-14-6	1590	593-95-3	495	1305-84-6	466	1314-23-4	2376	1344-59-8	2223
301-04-2	1086	598-54-9	685	1305-99-3	463	1314-24-5	1509	1344-95-2	467
302-01-2	928	598-62-9	1261	1306-19-0	404	1314-27-8	1147	1345-04-6	173
304-59-6	1933	598-63-0	1097	1306-23-6	412	1314-28-9	1682	1345-07-9	337
333-20-0	1636	628-52-4*	574	1306-24-7	408	1314-32-5	2079	1345-13-7	520
353-50-4	497	628-86-4	1314	1306-25-8	413	1314-34-7	2252	1345-25-1	1027
409-21-2	1810	630-08-0	487	1306-38-3	524	1314-35-8	2196	1449-65-6	865
460-19-5	498	631-60-7	1297	1307-86-4	674	1314-36-9	2308	1495-50-7	501
463-58-1	492	631-61-8	63	1307-96-6	653	1314-37-0	2295	1590-87-0	1803
471-34-1	426	637-12-7	48	1307-99-9	658	1314-41-6	1148	1600-27-7	1306
471-34-1	427	638-39-1	2108	1308-04-9	676	1314-56-3	1515	1603-84-5	491
497-19-8	1873	660-60-6	743	1308-06-1	669	1314-60-9	179	1631-78-3	2107
504-64-3	493	688-37-9	37	1308-14-1	591	1314-61-0	2027	1633-05-2	1961
506-64-9	1833	811-54-1	1108	1308-31-2	1017	1314-62-1	2260	1762-95-4	154
506-65-0	888	813-93-4	317	1308-38-9	595	1314-68-7	1686	1863-63-4	65
506-66-1	297	814-88-0	402	1308-56-1	723	1314-80-3	1520	2092-16-2	477
506-68-3	499	814-89-1	651	1308-80-1	683	1314-82-5	1519	2092-17-3	276
506-77-4	500	814-89-1*	652	1308-87-8	764	1314-84-7	2351	2092-17-3*	277
506-78-5	502	814-90-4*	580	1308-96-9	798	1314-86-9	1510	2404-52-6	1521
506-80-9	489	814-91-5	735	1309-33-7	1052	1314-87-0	1136	2406-52-2	2106
506-87-6	71	814-93-7	1119	1309-37-1	1056	1314-91-6	1139	2452-01-9	2337
507-16-4	2007	814-94-8	2116	1309-42-8	1214	1314-95-0	2121	2466-09-3	1404
512-25-4*	221	815-82-7	748	1309-40-4	1222	1314-96-1	1982	2524-02-9	1523
512-26-5	1104	815-85-0	2122	1309-60-0	1153	1314-97-2	2073	2551-62-4	2000
513-74-6	85	819-73-8	1096	1309-64-4	169	1314-98-3	2361	2644-70-4	931
513-77-9	215	865-44-1	982	1309-64-4	168	1314-98-3	2360	2696-92-6	1460
513-78-0	386	868-14-4	1599	1310-03-8	1129	1315-01-1	2133	2699-79-8	2006
513-79-1	630	917-61-3	1882	1310-32-3	1029	1315-03-3	2254	3012-65-5	108
516-02-9	250	917-69-1	670	1310-43-6	1006	1315-04-4	180	3095-65-6	115
528-94-9	140	940-71-6	1502	1310-52-7	1211	1315-05-5	171	3164-29-2	148
533-51-7	1843	992-98-3	2060	1310-53-8	880	1315-06-6	2119	3251-23-8	731
534-16-7	1828	1002-89-7	143	1310-58-3	1600	1315-07-7	1978	3333-67-3	1394
534-17-8	531	1066-30-4*	583	1310-61-8	1596	1315-09-9	2354	3486-35-9	2323
535-37-5*	893	1066-33-7	107	1310-61-8*	1597	1315-11-3	2363	3522-50-7	1047
538-17-0	53	1072-35-1	1134	1310-65-2	1170	1317-33-5	1353	3687-31-8	1090
540-16-9	712	1111-67-7	700	1310-66-3	1171	1317-34-6	1292	3811-04-9	1569
540-69-2	95	1111-71-3	301	1310-73-2	1909	1317-35-7	1289	3982-91-0	1522
540-72-7	1953	1111-78-0	70	1310-82-3	1715	1317-36-8	1120	4109-96-0	1787
541-43-5	227	1117-94-8	686	1311-11-1	1122	1317-36-8	1121	4119-52-2	1062
542-62-1	222	1120-44-1	734	1311-90-6	138	1317-37-9	1034	5341-61-7	932
542-83-6	392	1120-46-3	1118	1312-41-0	951	1317-38-0	736	5714-22-7	2003
542-84-7	636	1184-65-2	863	1312-42-1	970	1317-39-1	695	5743-04-4	380
543-80-6	207	1299-86-1	18	1312-43-2	967	1317-40-4	747	5743-26-0	418
543-81-7	289	1301-96-8	1858	1312-45-4	973	1317-41-5	741	5794-28-5	456
543-90-8	379	1302-01-8	1822	1312-46-5	994	1317-42-6	665	5892-10-4	314
544-17-2	440	1302-09-6	1851	1312-73-8	1625	1317-61-9	1039	5893-61-8	727
544-19-4	726	1302-42-7	1863	1312-73-8*	1626	1317-66-4	1007	5908-64-5	208
544-60-5	128	1302-52-9	292	1312-74-9	1622	1319-46-6	1098	5908-81-6	273
544-92-3	690	1302-81-4	51	1312-81-8	1079	1327-41-9	30	5968-11-6	1875
546-46-3	2327	1302-82-5	45	1312-99-8	1245	1327-44-2	1559	5970-44-5	2304
546-67-8	1149	1303-00-0	827	1313-04-8	1231	1327-50-0	174	5970-45-6	2313
546-93-0	1205	1303-11-3	952	1313-08-2	1237	1327-53-3	196	5970-62-7	2332
547-68-2	2344	1303-28-2	202	1313-13-9	1293	1327-53-3	195	5972-71-4	66
554-13-2	1163	1303-33-9	198	1313-22-0	1279	1330-43-4	1947	5972-72-5*	110
555-35-1	42	1303-34-0	204	1313-27-5	1363	1332-14-5	746	5972-76-9	69

CAS	No.	CAS	No.	CAS	No.	CAS	No.	CAS	No.
6009-70-7	129	7440-34-8	1	7580-67-8	1169	7757-86-0	1213	7783-11-1	147
6010-09-9	1036	7440-35-9	54	7601-54-9	1932	7757-93-9	443	7783-18-8	155
6046-93-1	704	7440-36-0	161	7601-89-0	1925	7758-01-2	1565	7783-19-9	142
6047-25-2	1026	7440-37-1	181	7601-90-3	548	7758-02-3	1566	7783-20-2	145
6080-56-4	1087	7440-38-2	182	7616-94-6	557	7758-05-6	1602	7783-21-3	141
6108-17-4	1155	7440-39-3	206	7631-90-5	1908	7758-09-0	1610	7783-22-4	157
6108-23-2	1168	7440-40-6	286	7631-94-9*	1889	7758-11-4	1592	7783-26-8	1804
6131-90-4	1862	7440-40-6	287	7631-95-0	1918	7758-16-9	1888	7783-28-0	111
6132-02-1	1874	7440-41-7	288	7631-99-4	1921	7758-19-2	1878	7783-29-1	1805
6132-04-3	1881	7440-42-8	341	7632-00-0	1922	7758-23-8*	462	7783-32-6	1316
6147-53-1	624	7440-43-9	378	7632-51-1	2256	7758-87-4	461	7783-33-7	1635
6150-82-9	1210	7440-44-0	483	7637-07-2	362	7758-88-5	517	7783-34-8	1320
6150-88-5	1221	7440-45-1	503	7646-69-7	1897	7758-89-6	689	7783-35-9	1325
6303-21-5	1498	7440-46-2	526	7646-78-8	2125	7758-94-3	1014	7783-36-0	1305
6381-79-9	1568	7440-47-3	559	7646-79-9	631	7758-95-4	1100	7783-39-3	1313
6424-20-0	641	7440-48-4	610	7646-85-7	2325	7758-97-6	1102	7783-40-6	1209
6484-52-2	126	7440-50-8	681	7646-93-7	1595	7758-98-7	744	7783-41-7	803
6487-39-4	1070	7440-51-9	752	7647-01-0	941	7758-99-8	745	7783-42-8	2009
6487-48-5	1611	7440-52-0	767	7647-10-1	1482	7759-00-4	1277	7783-43-9	1776
6533-73-9	2054	7440-53-1	782	7647-14-5	1877	7759-01-5	1143	7783-44-0	804
6556-16-7	1273	7440-54-2	808	7647-15-6	1871	7759-01-5	1144	7783-46-2	1106
6569-51-3	351	7440-55-3	825	7647-17-8	533	7759-02-6	1981	7783-47-3	2112
6591-55-5	327	7440-56-4	851	7647-18-9	176	7761-88-8	1841	7783-48-4	1966
6834-92-0	1917	7440-57-5	883	7647-19-0	1514	7772-99-8	2110	7783-49-5	2329
6865-35-6	269	7440-58-6	901	7659-31-6	1824	7773-01-5	1262	7783-50-8	1050
7320-34-5*	1619	7440-59-7	918	7664-38-2	1492	7773-03-7	1598	7783-51-9	838
7429-90-5	7	7440-60-0	919	7664-39-3	943	7773-06-0	144	7783-52-0	963
7429-91-6	753	7440-61-1	2201	7664-41-7	62	7774-29-0	1317	7783-53-1	1290
7429-92-7	766	7440-62-2	2231	7664-93-9	1986	7774-34-7	432	7783-54-2	1453
7469-88-5	987	7440-63-3	2272	7681-11-0	1603	7775-09-9	1876	7783-55-3	1507
7439-89-6	998	7440-64-4	2287	7681-38-1	1904	7775-11-3	1879	7783-56-4	166
7439-90-9	1064	7440-65-5	2297	7681-49-4	1892	7775-19-1	1916	7783-57-5	2077
7439-91-0	1066	7440-66-6	2312	7681-52-9	1910	7775-27-1	1931	7783-58-6	878
7439-92-1	1085	7440-67-7	2365	7681-52-9*	1911	7775-41-9	1835	7783-59-7	1152
7439-93-2	1154	7440-68-8	205	7681-55-2	1912	7778-18-9	470	7783-60-0	1999
7439-94-3	1186	7440-69-9	313	7681-57-4	1915	7778-39-4*	186	7783-61-1	1795
7439-95-4	1197	7440-70-2	416	7681-65-4	693	7778-39-4	185	7783-62-2	2128
7439-96-5	1247	7440-71-3	481	7681-82-5	1913	7778-44-1	421	7783-63-3	2155
7439-97-6	1296	7440-72-4	801	7693-26-7	1588	7778-50-9	1574	7783-64-4	2370
7439-98-7	1330	7440-73-5	807	7697-37-2	1443	7778-53-2	1618	7783-66-6	985
7439-99-8	1379	7440-74-6	950	7699-41-4	1807	7778-74-7	1613	7783-68-8	1435
7440-00-8	1365	7446-06-2	1555	7699-45-8	2321	7778-77-0	1576	7783-70-2	177
7440-01-9	1378	7446-07-3	2034	7704-34-9	1985	7778-80-5	1624	7783-71-3	2025
7440-02-0	1381	7446-08-4	1766	7704-34-9	1984	7779-88-6	2339	7783-72-4	2259
7440-03-1	1414	7446-09-5	1992	7704-98-5	2135	7779-90-0	2349	7783-75-7	997
7440-04-2	1467	7446-10-8	1137	7704-99-6	2371	7782-40-3	482	7783-77-9	1359
7440-05-3	1479	7446-11-9	1993	7705-07-9	2147	7782-41-4	802	7783-79-1	1773
7440-06-4	1524	7446-14-2	1135	7705-08-0	1044	7782-44-7	1477	7783-80-4	2042
7440-07-5	1542	7446-15-3	1130	7718-54-9	1395	7782-49-2	1761	7783-81-5	2225
7440-08-6	1553	7446-16-4	1663	7718-98-1	2248	7782-49-2	1762	7783-82-6	2195
7440-09-7	1556	7446-18-6	2072	7719-09-7	2008	7782-49-2	1763	7783-84-8	1059
7440-10-0	1643	7446-19-7	2358	7719-12-2	1504	7782-50-5	546	7783-86-0	1021
7440-11-1	1295	7446-20-0	2359	7720-78-7	1031	7782-61-8	1054	7783-86-0*	1022
7440-12-2	1656	7446-21-1	1977	7720-83-4	2156	7782-63-0	1033	7783-89-3	1826
7440-13-3	1657	7446-22-2	2070	7721-01-9	2024	7782-64-1	1265	7783-90-6	1830
7440-14-4	1659	7446-25-5	2132	7722-64-7	1615	7782-65-2	853	7783-92-8	1829
7440-15-5	1665	7446-26-6	2352	7722-76-1	84	7782-66-3	1323	7783-93-9	1845
7440-16-6	1692	7446-27-7	1125	7722-84-1	945	7782-68-5	975	7783-95-1	1859
7440-17-7	1702	7446-28-8	1976	7722-86-3	1987	7782-70-9	1594	7783-96-2	1838
7440-18-8	1725	7446-31-3	2380	7722-88-5	1934	7782-77-6	1444	7783-97-3	1837
7440-19-9	1737	7446-32-4	172	7723-14-0	1489	7782-78-7	1990	7783-98-4	1847
7440-20-2	1753	7446-33-5	2310	7723-14-0	1487	7782-79-8	936	7783-99-5	1842
7440-21-3	1780	7446-70-0	20	7723-14-0	1488	7782-85-6	1903	7784-01-2	1831
7440-22-4	1820	7447-39-4	715	7726-95-6	369	7782-86-7	1303	7784-02-3	1834
7440-23-5	1860	7447-40-7	1570	7727-15-3	16	7782-87-8	1601	7784-03-4	1856
7440-24-6	1956	7447-41-8	1164	7727-18-6	2269	7782-89-0	1157	7784-05-6	1852
7440-25-7	2011	7487-88-9	1238	7727-21-1	1617	7782-91-4	1358	7784-07-8	1850
7440-26-8	2028	7487-94-7	1309	7727-37-9	1441	7782-92-5	1865	7784-09-0	1848
7440-27-9	2043	7488-51-9	1132	7727-43-7	270	7782-94-7	1442	7784-11-4	17
7440-28-0	2051	7488-54-2	1722	7727-54-0	134	7782-99-2	1991	7784-13-6	21
7440-29-1	2080	7488-55-3	2120	7727-73-3	1939	7783-00-8	1765	7784-14-7	150
7440-30-4	2097	7543-51-3	2350	7732-18-5	2271	7783-03-1	2161	7784-16-9	1951
7440-31-5	2104	7550-35-8	1162	7733-02-0	2357	7783-06-4	947	7784-18-1	24
7440-31-5	2105	7550-45-0	2154	7757-79-1	1609	7783-07-5	946	7784-19-2	100
7440-32-6	2134	7553-56-2	974	7757-82-6	1938	7783-08-6	1764	7784-21-6	28
7440-33-7	2160	7558-79-4	1901	7757-83-7	1943	7783-09-7	949	7784-22-7	31

CAS	No.	CAS	No.	CAS	No.	CAS	No.	CAS	No.
7784-23-8	32	7789-28-8	1018	7790-92-3	547	10026-24-1	663	10058-44-3	1058
7784-24-9	1561	7789-29-9	1591	7790-94-5	1988	10028-14-5	1466	10060-09-0	407
7784-25-0	8	7789-30-2	377	7790-98-9	132	10028-15-6	1478	10060-10-3	523
7784-26-1	9	7789-31-3	370	7790-99-0	981	10028-18-9	1398	10060-11-4	869
7784-27-2	35	7789-33-5	980	7791-03-9	1179	10028-22-5	1060	10060-12-5	588
7784-29-4	1704	7789-36-8	1202	7791-07-3	1926	10031-13-7	1091	10060-13-6	80
7784-30-7	38	7789-38-0	1870	7791-08-4	170	10031-16-0	223	10090-53-6	33
7784-31-8	50	7789-39-1	1706	7791-09-5	1687	10031-20-6	1260	10097-28-6	1812
7784-33-0	191	7789-40-4	2053	7791-10-8	1962	10031-21-7	1094	10099-58-8	1071
7784-34-1	192	7789-41-5	423	7791-11-9	1708	10031-22-8	1095	10099-66-8	1189
7784-35-2	193	7789-42-6	384	7791-12-0	2056	10031-23-9	1660	10099-74-8	1117
7784-36-3	201	7789-43-7	628	7791-13-1	633	10031-24-0	2109	10099-76-0	1127
7784-37-4	1315	7789-45-9	711	7791-16-4	178	10031-25-1	584	10099-79-3	1145
7784-40-9	1109	7789-46-0	1011	7791-18-6	1208	10031-25-1*	585	10101-41-4	471
7784-41-0	1562	7789-47-1	1308	7791-20-0	1396	10031-26-2	1043	10101-50-5*	1929
7784-41-0	1575	7789-48-2	1203	7791-21-1	549	10031-27-3	2038	10101-52-7	2378
7784-42-1	183	7789-51-7	1774	7791-23-3	1775	10031-30-8	438	10101-53-8	600
7784-44-3	105	7789-52-8	1768	7791-25-5	2005	10031-43-3	733	10101-63-0	1113
7784-45-4	194	7789-53-9	1959	7791-26-6	2228	10031-45-5	740	10101-68-5	1282
7784-46-5	1867	7789-54-0	2036	7791-28-8	213	10031-48-8	739	10101-94-7	1133
7784-48-7	1391	7789-57-3	1797	7803-49-8	937	10031-52-4	779	10101-97-0	1408
7785-20-8	125	7789-59-5	1516	7803-51-2	1490	10031-52-4	800	10101-98-1	1407
7785-23-1	1827	7789-60-8	1503	7803-52-3	162	10031-53-5	797	10102-02-0	2342
7785-87-7	1280	7789-61-9	164	7803-54-5	1199	10034-76-1	472	10102-03-1	1450
7786-30-3	1207	7789-65-3	1770	7803-55-6	122	10034-81-8	1223	10102-05-3	1485
7786-81-4	1406	7789-66-4	1793	7803-57-8	929	10034-82-9	1880	10102-15-5	1944
7787-32-8	226	7789-67-5	2124	7803-58-9	2004	10034-85-2	944	10102-17-7	1954
7787-33-9	241	7789-68-6	2153	7803-60-3	1495	10034-88-5	1905	10102-18-8	1937
7787-34-0	239	7789-69-7	1512	7803-62-5	1781	10034-93-2	935	10102-20-2	1946
7787-35-1	242	7789-75-5	439	7803-63-6	112	10034-96-5	1281	10102-23-5	1935
7787-36-2	255	7789-77-7	444	7803-65-8	117	10034-98-7	2296	10102-24-6	1182
7787-37-3	244	7789-78-8	442	7803-68-1	2032	10034-99-8	1239	10102-25-7	1184
7787-38-4	249	7789-79-9	446	10022-31-8	246	10035-04-8	431	10102-34-8	1228
7787-39-5	272	7789-80-2	447	10022-47-6	77	10035-05-9	429	10102-40-0	1919
7787-40-8	280	7789-82-4	450	10022-48-7	1166	10035-06-0	324	10102-43-9	1446
7787-41-9	260	7790-21-8	1614	10022-50-1	1464	10035-10-6	940	10102-44-0	1447
7787-42-0	282	7790-22-9	1173	10022-68-1	401	10036-47-2	1458	10102-45-1	2065
7787-46-4	296	7790-28-5	1927	10024-93-8	1368	10038-98-9	877	10102-49-5	1042
7787-47-5	299	7790-29-6	1716	10024-97-2	1445	10039-31-3	309	10102-68-8	448
7787-48-6	308	7790-30-9	2063	10025-64-6	2346	10039-32-4	1902	10102-71-3	1864
7787-49-7	300	7790-31-0	1216	10025-65-7	1528	10039-54-0	938	10103-61-4	706
7787-52-2	302	7790-33-2	1267	10025-66-8	1661	10042-76-9	1972	10108-64-2	388
7787-53-3	304	7790-33-2*	1268	10025-67-9	1995	10042-88-3	2044	10112-91-1	1300
7787-56-6	311	7790-34-3	1402	10025-68-0	1769	10043-01-3	49	10118-76-0	459
7787-57-7	329	7790-37-6	2335	10025-69-1	2111	10043-11-5	365	10124-36-4	409
7787-58-8	315	7790-38-7	1484	10025-70-4	1964	10043-35-3	352	10124-37-5	451
7787-59-9	330	7790-39-8	1529	10025-71-5	2037	10043-52-4	430	10124-41-1	478
7787-60-2	316	7790-41-2	993	10025-73-7	587	10043-67-1	1560	10124-43-3	662
7787-61-3	318	7790-42-3	1639	10025-74-8	760	10043-92-2	1664	10124-48-8	1307
7787-62-4	319	7790-44-5	167	10025-75-9	771	10045-86-0	1057	10124-50-2	1589
7787-63-5	331	7790-46-7	1538	10025-76-0	794	10045-89-3	91	10125-13-0	716
7787-64-6	322	7790-47-8	2129	10025-77-1	1045	10045-94-0	1318	10135-84-9	106
7787-68-0	336	7790-48-9	2041	10025-78-2	1799	10045-94-0*	1319	10137-74-3	428
7787-69-1	530	7790-49-0	2088	10025-82-8	962	10048-95-0	1898	10138-04-2	88
7787-70-4	688	7790-56-9	1627	10025-83-9	991	10048-98-3	231	10138-41-7	770
7787-71-5	376	7790-58-1	1630	10025-87-3	1517	10048-99-4	275	10138-52-0	815
7788-97-8	589	7790-59-2	1621	10025-90-8	1647	10049-01-1	333	10138-62-2	921
7788-98-9	76	7790-60-5	1641	10025-91-9	165	10049-03-3	806	10139-47-6	2336
7788-99-0	599	7790-62-7	1620	10025-93-1	2210	10049-04-4	550	10141-05-6	649
7789-00-6	1571	7790-63-8	1642	10025-99-7	1633	10049-05-5	576	10170-69-1	1254
7789-01-7	1165	7790-69-4	1175	10026-01-4	1471	10049-06-6	2142	10179-73-4	1030
7789-02-8	594	7790-71-8	1181	10026-02-5	1554	10049-07-7	1695	10190-55-3	1115
7789-04-0	596	7790-74-1	465	10026-03-6	1771	10049-08-8	1729	10192-29-7	74
7789-06-2	1965	7790-75-2	480	10026-04-7	1794	10049-10-2	578	10192-30-0	1
7789-08-4	86	7790-76-3	464	10026-06-9	2126	10049-11-3	604	10193-36-9	1808
7789-09-5	82	7790-78-5	389	10026-07-0	2039	10049-12-4	2249	10196-18-6	2340
7789-10-8	1312	7790-79-6	393	10026-08-1	2086	10049-12-4*	2250	10210-64-7	290
7789-17-5	539	7790-80-9	396	10026-10-5	2215	10049-14-6	2216	10210-68-1	618
7789-18-6	541	7790-81-0	395	10026-11-6	2369	10049-16-8	2257	10213-09-9	2264
7789-19-7	724	7790-84-3	410	10026-12-7	1434	10049-17-9	1681	10213-10-2	1955
7789-21-1	1989	7790-85-4	415	10026-13-8	1513	10049-21-5	1886	10214-39-8	1093
7789-23-3	1579	7790-86-5	515	10026-17-2	639	10049-23-7	2033	10241-05-1	1355
7789-24-4	1167	7790-87-6	518	10026-18-3	672	10049-24-8*	990	10257-55-3	474
7789-25-5	1461	7790-87-6*	519	10026-20-7	656	10049-24-8	989	10277-43-7	1077
7789-26-6	805	7790-89-8	554	10026-22-9	650	10049-25-9	575	10290-12-7	707
7789-27-7	2059	7790-91-2	555					10294-26-5	1853

CAS	No.	CAS	No.	CAS	No.	CAS	No.	CAS	No.	CAS	No.
10294-27-6	886	12004-76-1	2012	12019-52-2	699	12035-79-9	1380	12057-71-5	1220
10294-28-7	891	12005-21-9	2298	12019-88-4	755	12035-82-4	1530	12057-74-8	1229
10294-29-8	887	12005-67-3	61	12020-14-3	2368	12035-83-5	1544	12057-75-9	1200
10294-31-2	889	12005-69-5	357	12020-21-2	775	12035-98-2	2244	12057-92-0	1294
10294-32-3	898	12006-15-4	382	12020-28-9	777	12036-02-1	1473	12058-18-3	1352
10294-33-4	358	12006-77-8	617	12020-39-2	781	12036-09-8	1673	12058-20-7	1354
10294-34-5	360	12006-79-0	563	12020-58-5	784	12036-10-1	1733	12059-14-2	1389
10294-38-9	216	12006-84-7	1001	12020-65-4	791	12036-14-5	2019	12059-95-9	1551
10294-39-0	254	12006-86-9	1000	12020-66-5	790	12036-21-4	2258	12060-00-3	1142
10294-40-3	220	12006-99-4	1331	12020-69-8	792	12036-22-5	2181	12060-01-4	1146
10294-42-5	525	12007-00-0	1385	12021-68-0	642	12036-32-7	1652	12060-08-1	1758
10294-44-7	1299	12007-01-1	1386	12022-95-6	1008	12036-34-9	1549	12060-58-1	1749
10294-46-9	738	12007-02-2	1384	12022-99-0	1009	12036-35-0	1698	12060-59-2	1983
10294-47-0	1099	12007-07-7	2013	12023-99-3	841	12036-41-8	2048	12061-16-4	776
10294-48-1	553	12007-09-9	2163	12024-10-1	833	12036-44-1	2103	12062-24-7	728
10294-50-5	655	12007-10-2	2162	12024-11-2	832	12037-15-9	2219	12063-98-8	829
10294-52-7	1046	12007-16-8	564	12024-14-5	834	12037-65-9	2139	12064-03-8	826
10294-53-8	1048	12007-23-7	902	12024-20-3	830	12037-80-8	2377	12064-62-9	821
10294-54-9	545	12007-25-9	1201	12024-21-4	844	12037-82-0	1500	12065-10-0	873
10294-58-3	1126	12007-29-3	1416	12024-22-5	849	12038-12-9	1655	12065-11-1	881
10294-62-9	1081	12007-33-9	368	12024-24-7	846	12038-13-0	1654	12065-36-0	852
10294-64-1	1604	12007-34-0	1754	12024-27-0	850	12038-20-9	1531	12065-65-5	2152
10294-66-3	1637	12007-35-1	2014	12024-81-6	813	12038-21-0	1540	12065-68-8	1138
10294-70-9	2115	12007-36-2	2202	12025-32-0	874	12038-63-0	1674	12066-83-0	1653
10325-94-7	400	12007-37-3	2233	12025-34-2	882	12038-67-4	1691	12067-00-4	1675
10326-21-3*	1206	12007-38-4	565	12025-39-7	875	12039-07-5	2145	12067-22-0	1751
10326-24-6	2317	12007-84-0	2203	12026-06-1	2062	12039-13-3	2159	12067-46-8	2182
10326-26-8	211	12007-97-5	1332	12026-24-3	2114	12039-15-5	2381	12067-54-8	2094
10326-27-9	218	12007-98-6	2164	12027-06-4	119	12039-16-6	2151	12067-66-2	2022
10326-28-0	405	12007-99-7	422	12028-48-7	121	12039-19-9	2311	12067-76-4	2184
10340-06-4	494	12008-02-5	504	12029-81-1	924	12039-35-9	2314	12068-49-4	1010
10343-61-0	2150	12008-05-8	783	12029-98-0	978	12039-55-3	2020	12068-69-8	334
10361-37-2	217	12008-06-9	809	12030-14-7	960	12039-76-8	2239	12068-90-5	1328
10361-43-0	321	12008-19-4	347	12030-24-9	972	12039-79-1	2018	12069-00-0	1131
10361-46-3	332	12008-21-8	1067	12030-49-8	995	12039-80-4	2049	12069-32-8	359
10361-79-2	1646	12008-23-0	1366	12030-51-2	996	12039-83-7	2140	12069-69-1	713
10361-82-7	1745	12008-27-4	1644	12030-85-2	1608	12039-87-1	2238	12069-85-1	904
10361-84-9	1756	12008-29-6	1738	12030-88-5	1628	12039-88-2	2170	12069-89-5	1334
10361-91-8	2292	12008-32-1	2301	12030-97-6	1638	12039-89-3	2288	12069-94-2	1417
10361-92-9	2305	12009-14-2	245	12031-30-0	1083	12039-90-6	2379	12070-06-3	2015
10361-95-2	2324	12009-18-6*	268	12031-49-1	1082	12039-95-1	2171	12070-07-4	2016
10377-48-7	1183	12009-18-6	267	12031-63-9	1174	12040-00-5	1752	12070-08-5	2137
10377-51-2	1172	12009-21-1	285	12031-80-0	1180	12040-02-7	2123	12070-09-6	2204
10377-52-3	1177	12011-97-1	1333	12032-20-1	1193	12041-50-8	13	12070-10-9	2234
10377-58-9	1215	12011-99-3	1418	12032-36-9	1241	12042-68-1	419	12070-12-1	2166
10377-60-3	1217	12012-16-7	2084	12032-74-5	1287	12042-78-3	420	12070-13-2	2165
10377-62-5	1225	12012-17-8	2235	12032-78-9	1256	12043-29-7	52	12071-15-7	1069
10377-66-9	1271	12012-32-7	505	12032-82-5	1248	12044-16-5	999	12071-29-3	1960
10377-93-2	1270	12012-35-0	566	12032-88-1	1286	12044-42-7	614	12071-31-7	2085
10380-31-1	283	12013-10-4	621	12032-97-2	1249	12044-54-1	199	12071-33-9	2205
10381-37-0	2095	12013-55-7	469	12033-19-1	1336	12045-01-1	616	12071-35-1	2303
10402-15-0	719	12013-56-8	468	12033-31-7	1337	12045-15-7	1250	12076-62-9	2206
10405-27-3	1455	12013-57-9	475	12033-54-4	1543	12045-16-8	1252	12088-65-2	1004
10450-55-2	1040	12013-82-0	453	12033-62-4	2017	12045-19-1	1415	12115-63-8	513
10450-60-9	976	12014-14-1	414	12033-64-6	2047	12045-27-1	2232	12124-97-9	68
10476-81-0	1958	12014-28-7	406	12033-65-7	2090	12045-63-5	2136	12124-99-1	113
10476-85-4	1963	12014-29-8	381	12033-72-6	2169	12045-64-6	2366	12125-01-8	92
10476-86-5	1969	12014-82-3	510	12033-83-9	2208	12045-87-3	1950	12125-02-9	75
10489-46-0	1699	12014-85-6	507	12033-88-4	1465	12045-88-4	1949	12125-08-5	97
10534-89-1	673	12014-93-6	522	12033-89-5	1811	12046-08-1	228	12125-09-6	1499
10544-72-6	1449	12016-80-7	677	12034-09-2	1920	12047-27-7	281	12125-19-8	2200
10544-73-7	1448	12017-01-5	647	12034-12-7	1945	12047-79-9	247	12125-22-3	1480
10545-99-0	1998	12017-08-2	660	12034-57-0	1422	12048-50-9	339	12125-25-6	1192
10553-31-8	212	12017-12-8	620	12034-59-2	1429	12048-51-0	340	12125-63-2	1035
10567-69-8	238	12017-13-9	666	12034-66-1	1420	12049-50-2	479	12133-07-2	756
10580-03-7	156	12017-38-8	680	12034-77-4	1430	12052-42-5	611	12133-10-7	765
10580-52-6	2241	12018-01-8	605	12034-80-9	1421	12053-12-2	561	12134-02-0	619
10588-01-9	1885	12018-09-6	573	12034-83-2	1432	12053-13-3	571	12134-75-7	811
11121-16-7	12	12018-10-9	718	12034-88-7	1116	12053-27-9	568	12134-77-9	823
11126-81-1*	15	12018-19-8	2326	12034-89-8	1971	12053-39-3	602	12135-76-1	146
12002-03-8	703	12018-22-3	601	12035-32-4	1376	12054-48-7	1399	12136-24-2	926
12002-61-8	6	12018-34-7	582	12035-35-7	1377	12054-85-2	123	12136-42-4	988
12002-99-2	1855	12018-36-9	572	12035-39-1	1410	12055-23-1	910	12136-45-7	1612
12003-63-3*	1558	12018-61-0	544	12035-52-8	1382	12055-62-8	925	12136-58-2	1185
12004-04-5	209	12019-11-3	684	12035-64-2	1388	12056-90-5	1080	12136-78-6	1339
12004-06-7	291	12019-23-7	749	12035-72-2	1413	12057-24-8	1178	12136-91-3	1501

12136-97-9	1431	13126-12-0	1717	13463-22-4	251	13494-90-1	843	13637-76-8	1124
12137-12-1	1411	13135-31-4	2146	13463-30-4	1151	13494-91-2	847	13675-47-3	721
12137-20-1	2144	13172-31-1	1994	13463-39-3	1387	13494-98-9	2307	13682-61-6	1632
12137-25-6	1724	13205-44-2	350	13463-40-6	1002	13499-05-3	905	13693-05-5	1541
12137-27-8	1700	13255-26-0	264	13463-67-7	2157	13510-35-5	966	13693-06-6	1552
12137-83-6	1526	13268-42-3	87	13464-44-3	2315	13510-42-4	1719	13693-07-7	1701
12138-07-7	2096	13283-01-7	2191	13464-45-4	2316	13510-49-1	310	13693-08-8	1735
12138-09-9	2183	13320-71-3	1349	13464-46-5	57	13510-89-9	1089	13693-09-9	2277
12138-11-3	2050	13327-32-7	303	13464-58-9	187	13517-00-5	692	13701-67-2	361
12138-28-2	1980	13395-16-9	682	13464-82-9	971	13517-10-7	364	13701-70-7	2253
12139-93-4	661	13400-13-0	535	13464-92-1	385	13517-27-6	2320	13701-86-5	2190
12142-33-5*	1623	13424-46-9	1092	13464-97-6	934	13520-56-4	1061	13701-90-1	2074
12143-72-5	2021	13444-75-2*	1310	13465-07-1	948	13520-59-7	1348	13701-91-2	1150
12159-66-9	780	13444-85-4	1454	13465-09-3	961	13520-75-7	2192	13708-80-0	58
12160-99-5	824	13444-87-6	1459	13465-10-6	956	13520-76-8	2193	13709-31-4	2270
12162-21-9	913	13444-90-1	1463	13465-11-7	959	13520-77-9	2197	13709-35-8	1997
12162-59-3	927	13444-93-4	1470	13465-15-1	968	13520-78-0	2198	13709-36-9	2275
12163-20-1	1195	13444-94-5	1481	13465-55-9	1746	13520-79-1	2199	13709-38-1	1073
12163-22-3	1196	13444-96-7	1483	13465-58-2	1750	13520-83-7	2229	13709-42-7	1370
12163-69-8	1338	13445-50-6	1491	13465-59-3	1755	13520-87-1	2262	13709-46-1	1648
12164-94-2	64	13446-03-2	1259	13465-66-2	1772	13520-88-2	2261	13709-47-2	1757
12165-05-8	1679	13446-08-7	94	13465-71-9	1801	13520-89-3	2263	13709-49-4	2306
12165-69-4	1511	13446-09-8	118	13465-72-0	1802	13520-90-6	2268	13709-52-9	906
12166-29-9	1759	13446-10-1	133	13465-73-1	1782	13520-92-8	2382	13709-56-3	1550
12166-44-8	1760	13446-18-9	1219	13465-74-2	1783	13530-57-9	1703	13709-59-6	2087
12168-52-4	1037	13446-19-0	1224	13465-76-4	1798	13536-53-3	1645	13709-61-0	2276
12183-80-1	46	13446-23-6	1227	13465-78-6	1784	13536-73-7	769	13709-94-9	1606
12186-97-9	2299	13446-28-1	1230	13465-84-4	1796	13536-80-6	1367	13718-50-8	240
12187-14-3	399	13446-29-2	1242	13465-94-6	248	13537-09-2	762	13718-59-7	262
12201-89-7	1390	13446-30-5	1244	13465-95-7	253	13537-15-0	799	13718-70-2	1023
12211-52-8	81	13446-34-9	1263	13466-08-5	1872	13537-18-3	2099	13746-66-2	1577
12228-40-9	294	13446-49-6	1607	13466-20-1	257	13537-30-9	860	13746-89-9	2374
12228-50-1	1251	13446-53-2	1204	13466-21-2	259	13537-33-2	1791	13746-98-0	2078
12228-87-4	67	13446-56-5	1340	13469-98-2	2302	13548-38-4	593	13755-29-8	1893
12228-91-0	1258	13446-57-6	1343	13470-06-9	1973	13548-42-0	717	13755-32-3	274
12229-63-9	2082	13446-74-7	1710	13470-08-1	2148	13550-53-3	773	13759-10-9	1819
12230-71-6	236	13446-75-8	1712	13470-10-5	2172	13566-17-1	1128	13759-87-0	1744
12230-74-9	233	13450-87-8	822	13470-11-6	2185	13568-32-6	1278	13759-88-1	793
12232-99-4	1914	13450-88-9	836	13470-12-7	2173	13569-43-2	854	13759-92-7	795
12249-52-4	1836	13450-90-3	837	13470-13-8	2178	13569-49-8	1668	13760-02-6	1790
12254-64-7	55	13450-91-4	842	13470-14-9	2186	13569-50-1	508	13760-41-3	1658
12254-82-9	612	13450-92-5	876	13470-17-2	2174	13569-59-0	1424	13760-78-6	922
12254-85-2	562	13450-95-8	879	13470-19-4	2209	13569-62-5	1546	13760-79-7	2101
12255-48-0	2300	13451-11-1	2023	13470-20-7	2214	13569-63-6	1669	13760-80-0	2294
12256-04-1	615	13451-18-8	2035	13470-21-8	2221	13569-70-5	1426	13760-81-1	1190
12266-65-8	1253	13453-06-0	149	13470-22-9	2217	13569-71-6	1671	13760-83-3	772
12279-90-2	190	13453-07-1	892	13470-26-3	2247	13569-75-0	592	13762-12-4	629
12280-52-3	537	13453-30-0	2055	13472-31-6	1928	13569-80-7	761	13762-14-6	648
12280-64-7	1711	13453-32-2	2075	13472-35-0	1887	13572-93-5	840	13762-51-1	1564
12294-01-8	2309	13453-32-2*	2076	13473-77-3	1194	13572-98-0	818	13763-23-0	2224
12300-22-0	1739	13453-34-4	2057	13476-01-2	627	13573-02-9	861	13763-67-2	532
12310-43-9	754	13453-40-2	2069	13476-02-8	1025	13573-08-5	871	13765-19-0	434
12310-44-0	768	13453-49-1	2083	13476-99-8	2246	13573-11-0	1246	13765-24-7	1747
12322-34-8	163	13453-62-8	1123	13477-09-3	230	13586-38-4	79	13765-25-8	796
12323-03-4	266	13453-69-5	1160	13477-19-5	397	13587-35-4	750	13765-26-9	817
12325-59-6	912	13454-75-6	529	13477-28-6	424	13587-35-4*	751	13765-74-7	1840
12326-21-5	2227	13454-84-7	543	13477-29-7	433	13595-30-7	2255	13766-47-7	2179
12333-54-9	1255	13454-88-1	725	13477-34-4	452	13597-20-1	1439	13767-31-2	757
12340-14-6	2274	13454-94-9	521	13477-36-6	458	13597-54-1	2353	13767-32-3	2338
12345-14-1	219	13454-96-1	1536	13477-89-9	1369	13597-55-2	1979	13767-34-5	730
12399-08-5	977	13455-01-1	1508	13477-91-3	1375	13597-65-4	2355	13768-11-1	1666
12401-56-8	915	13455-12-4	1527	13477-95-7	1397	13597-73-4	1806	13768-38-2	1475
12419-43-1	2113	13455-15-7	1537	13478-00-7	1403	13597-99-4	305	13768-86-0	1767
12434-24-1	785	13455-21-5	1580	13478-10-9	1016	13598-22-6	312	13768-94-0	1786
12442-63-6	552	13455-25-9	635	13478-14-3	1158	13598-33-9	1967	13769-20-5	787
12513-27-8	2319	13455-28-2	644	13478-17-6	1341	13598-36-2	1497	13769-36-3	857
12540-13-5	705	13455-31-7	654	13478-18-7	1344	13598-42-0	1792	13770-18-8	737
12616-24-9*	160	13455-34-0	664	13478-20-1	1518	13598-56-6	2212	13770-56-4	188
12640-47-0	668	13460-50-9	353	13478-28-9	579	13598-65-7	135	13773-81-4	1065
12650-28-1	263	13460-50-9	354	13478-38-1	732	13600-89-0	671	13774-85-1	2279
12688-52-7	1187	13460-50-9	355	13478-41-6	691	13601-13-3	722	13775-06-9	2211
12770-26-2	907	13462-88-9	1392	13478-45-0	1433	13601-19-9	1891	13775-07-0	2222
12777-45-6	335	13462-88-9*	1393	13478-49-2	778	13637-63-3	556	13775-16-1	2220
13007-92-6	567	13462-90-3	1401	13478-50-7	1141	13637-65-5	855	13775-18-3	2213
13011-54-6	1866	13462-93-6	83	13492-26-7	1593	13637-68-8	1362	13775-53-6	1895
13092-75-6	1025	13463-12-2	1012	13494-80-9	2031			13775-80-9	930

13776-58-4	2273	13940-89-1	1019	14763-77-0	720	16893-85-9	1896	20328-94-3	2285
13776-62-0	1457	13963-58-1	1585	14779-70-5	862	16903-35-8	885	20344-49-4	1053
13776-74-4	1233	13965-73-6	363	14808-60-7	1813	16919-19-0	104	20398-06-5	2058
13777-22-5	903	13966-94-4	957	14808-60-7	1814	16919-58-7	99	20405-64-5	696
13777-23-6	908	13967-25-4	1301	14871-79-5*	237	16921-30-5	1584	20427-11-6	637
13777-25-8	2367	13972-68-4	398	14884-42-5	606	16921-96-3	986	20427-56-9	1736
13778-39-7	2100	13973-87-0	373	14890-41-6	2240	16925-39-6	441	20427-58-1	2334
13779-41-4	1496	13982-53-1*	698	14933-38-1	56	16940-66-2	1869	20427-59-2	729
13779-92-5	1436	13986-18-0	2330	14940-41-1	1028	16940-92-4	96	20548-54-3	473
13780-06-8	454	13986-26-0	2373	14965-52-7	1800	16941-10-9	476	20601-83-6	1324
13780-42-2	848	14013-15-1	1269	14977-61-8	609	16941-11-0	103	20610-49-5	1662
13780-48-8	786	14013-86-6	1024	14984-81-7	1777	16941-12-1	1525	20619-16-3	872
13780-57-9	2002	14014-88-1	1728	15098-87-0	26	16949-15-8	1161	20644-87-5	2236
13782-01-9*	678	14024-18-1	1041	15162-90-0	1451	16961-83-4	1809	20661-21-6	965
13783-04-5	2141	14038-43-8	1049	15163-24-3	2175	16962-07-5	14	20665-52-5	845
13783-07-8	2143	14040-11-0	2167	15168-20-4	742	16962-31-5	1586	20667-12-3	1844
13798-24-8	2045	14049-36-6	1785	15179-32-5	2010	16962-47-3	102	20694-39-7	1272
13812-43-6	1456	14049-81-1	638	15192-17-3	1742	17014-71-0	1616	20712-42-9	403
13813-22-4	1076	14055-74-4	1342	15192-26-4	2040	17026-29-8	1684	20762-60-1	1563
13813-23-5	1649	14055-75-5	1346	15192-42-4	1672	17029-21-9	1685	20770-09-6	2131
13813-24-6	1371	14055-81-3	2177	15195-33-2	1440	17083-68-0	884	20816-12-0	1476
13813-25-7	1748	14055-84-6	2180	15195-53-6	1425	17099-70-6	27	20859-73-8	44
13813-40-6	2046	14059-33-7	326	15230-48-5	858	17125-80-3	229	20910-28-5	2230
13813-41-7	923	14075-53-7	1581	15238-00-3	645	17194-00-2	234	21041-93-0	643
13813-42-8	774	14077-39-5	788	15238-00-3*	646	17375-41-6	1032	21041-95-2	394
13813-43-9	2102	14104-20-2	1857	15243-27-3	1257	17440-85-6	295	21109-95-5	271
13813-45-1	1191	14166-78-0	964	15243-33-1	1726	17440-90-3	1005	21159-32-0	534
13813-46-2	1548	14168-73-1	1240	15244-35-6	411	17496-59-2	551	21255-83-4	372
13813-47-3	59	14188-40-0	856	15321-51-4	1003	17702-41-9	349	21264-43-7	958
13814-72-7	812	14215-29-3	383	15364-10-0	2281	17786-31-1	622	21308-80-5	371
13814-74-9	1357	14215-30-6	708	15364-94-0	1275	18088-11-4	1718	21351-79-1	538
13814-75-0	1361	14217-21-1*	1890	15385-57-6	1302	18282-10-5	2130	21548-73-2	1854
13814-76-1	1683	14221-48-8*	89	15385-58-7	1298	18283-93-7	343	21645-51-2	29
13814-81-8	701	14242-05-8	1846	15468-32-3	1815	18288-22-7	320	21651-19-4	2117
13814-83-0	2265	14249-98-0	2189	15469-38-2	1051	18356-71-3	1788	21679-31-2	560
13814-96-5	1107	14280-53-6	955	15474-63-2	763	18433-40-4	158	21908-53-2	1321
13817-37-3	640	14284-93-6	1727	15477-33-5	19	18433-84-6	345	21995-38-0*	72
13818-75-2	814	14285-68-8	1667	15492-38-3	1697	18454-12-1	1103	22015-35-6	789
13818-89-8	866	14307-33-6*	437	15513-84-5	2243	18480-07-4	1968	22205-45-4	697
13819-84-6	1356	14312-00-6	391	15513-94-7	2251	18618-55-8	516	22205-57-8	527
13820-41-2	151	14335-40-1	1505	15520-84-0	675	18624-44-7	1020	22326-55-2	235
13820-62-7	625	14336-80-2	687	15571-91-2*	1629	18718-07-5	1264	22398-80-7	954
13821-06-2*	225	14402-70-1	127	15572-25-5	2071	18721-05-6	586	22441-45-8	200
13824-36-7	1789	14405-43-7	835	15578-26-4	2118	18810-58-7	210	22520-96-3	984
13825-74-6*	2158	14447-89-3	2194	15587-72-1	1714	18820-29-6	1285	22527-13-5	2280
13825-76-8	920	14452-39-2	43	15593-61-0	1232	18820-29-6	1283	22537-19-5	1084
13825-86-0	581	14456-34-9	911	15597-63-4	1506	18820-29-6	1284	22750-54-5*	387
13826-63-6	2066	14456-48-5	759	15597-88-3	603	18855-94-2	917	22750-57-8	528
13826-83-0	93	14456-51-0	2098	15607-89-3	2001	18868-43-4	1351	22756-36-1	1705
13842-80-3	2242	14456-53-2	1188	15622-42-1	1670	18917-82-3	1114	22831-39-6	1236
13842-83-6	1547	14457-84-2	41	15696-40-9	1468	18917-91-4	34	22831-42-1	11
13842-88-1	1427	14457-87-5	511	15750-45-5	1218	18933-05-6	1266	22886-66-4	839
13842-93-8	2030	14457-87-5*	512	15752-41-7	1423	19034-13-0	659	22986-54-5	3
13843-28-2	607	14459-59-7	1360	15752-46-2	1545	19073-56-4	1709	23370-59-4	992
13845-17-5	224	14459-75-7	1438	15823-43-5	916	19086-20-5	1243	23377-53-9	1690
13847-57-9	1101	14459-95-1	1578	15829-53-5	1304	19088-74-5	1713	23412-45-5	1350
13847-65-9	1462	14464-46-1	1816	15845-52-0	1110	19139-47-0	509	23414-72-4	2347
13862-78-7	153	14475-63-9	2372	15942-63-9	184	19168-23-1	98	23611-30-5	1720
13863-41-7	374	14481-29-9*	90	15947-41-8	60	19287-45-7	342	23777-80-2	346
13863-59-7	375	14483-17-1	1650	16122-03-5*	124	19357-83-6	2187	24094-93-7	569
13863-88-2	1821	14507-19-8	1075	16399-77-2	1015	19357-86-9	2291	24304-00-5	36
13864-01-2	1074	14517-29-4	1372	16544-92-6	632	19423-81-5	816	24415-00-7	868
13870-13-8	914	14519-18-7	1957	16671-27-5	590	19423-87-1	2293	24422-20-6	864
13870-21-8	1428	14521-18-7	1734	16674-78-5	1198	19465-30-6	348	24422-21-7	859
13870-24-1	1038	14523-22-9	1693	16689-88-6	2081	19469-07-9	1055	24597-12-4	831
13871-27-7	1952	14553-44-7	2093	16721-80-5	1906	19584-30-6	1694	24613-38-5	634
13873-84-2	983	14590-13-7	78	16721-80-5	1907	19597-69-4	1159	24621-21-4	1419
13874-75-4	1741	14590-19-3	657	16731-55-8	1605	19598-90-4	819	24646-85-3	2237
13874-77-6	2290	14639-94-2	101	16774-21-3	73	19624-22-7	344	24719-19-5	626
13875-06-4	2278	14639-97-5	152	16812-54-7	1409	19783-14-3	1111	25152-52-7	10
13876-85-2	694	14639-98-6	131	16853-85-3	1156	20193-56-0	2176	25417-81-6	232
13896-65-6	1731	14691-44-2	867	16860-99-4	1996	20193-58-2	1345	25502-05-0	2289
13918-22-4	1288	14693-81-3	2026	16871-60-6	1583	20205-91-8	367	25583-20-4	2138
13931-94-7	577	14693-82-4	969	16871-71-9	2333	20211-76-1	1072	25617-97-4	828
13939-06-5	1335	14720-21-9	894	16871-90-2	1587	20213-56-3	2188	25617-98-5	953
13940-63-1	870	14721-21-2	714	16872-11-0	356	20281-00-9	542	25658-42-8	2375

25658-43-9	2207	27218-16-2	558	35112-53-9	279	51621-05-7	1730	65842-03-7	1063
25659-31-8	1112	27774-13-6	2267	35718-37-7	2284	51898-99-8	323	68016-36-4	278
25764-08-3	506	27860-83-9	2331	36377-94-3	758	52003-58-4	120	68938-92-1	1534
25764-09-4	1651	30737-24-7	2067	36897-37-7	1400	52503-64-7	709	71500-16-8	1732
25764-10-7	1078	30937-52-1	1678	36907-42-3	2245	52788-53-1	820	71626-98-7	449
25764-11-8	1373	30937-53-2	1676	37267-86-0	1493	52951-38-9	325	72520-94-6	514
25764-15-2	810	31032-13-0	896	37773-49-2	1535	53731-35-4	1276	73560-00-6	979
25817-87-2	909	31052-14-9	2029	38455-77-5	2127	54120-05-7	1472	73796-25-5	1970
25895-60-7	1884	31052-43-4	1721	39290-85-2	710	54723-94-3	137	75926-26-0	1778
25909-39-1	1533	31083-74-6	1723	39368-69-9	1677	57246-89-6	1689	75926-28-2	1779
25985-07-3	1532	31234-26-1	1680	39416-30-3	284	58724-12-2	536	84359-31-9	597
26124-86-7	243	31576-40-6	1474	39796-98-0	2286	59201-51-3	1469	84359-31-9	598
26134-62-3	1176	32248-43-4	1743	39797-63-2	2283	60616-74-2	1212	92141-86-1	540
26342-61-0	570	32287-65-3	25	42246-24-2	1688	60676-86-0	1817	97126-35-7	667
26628-22-8	1868	32823-06-6	39	42739-38-8	159	60763-24-8	2092	99685-96-8	484
26686-77-1	1234	33088-16-3	2089	50432-32-1	2282	60804-25-3	1696	115383-22-7	485
26750-66-3	1640	33689-80-4	4	50801-97-3	1740	60883-64-9	298	123714-57-8	2362
27016-73-5	613	33689-81-5	2	50813-65-5	214	60922-26-1	2168	133578-89-9	2266
27016-75-7	1383	33689-82-6	5	50927-81-6	1818	60950-56-3	1235	133863-98-6	1364
27043-84-1	2318	34128-09-1	2064	51503-61-8	136	60969-19-9	2061	134929-59-2	486
27133-66-0	258	34330-64-8	390						

* Registry number refers to the anhydrous compound.

PHYSICAL PROPERTIES OF THE RARE EARTH METALS

K.A. Gschneidner, Jr.

Table 1
Data for the Trivalent Ions of the Rare Earth Elements

Rare earth	Symbol	Atomic no.	Atomic wt.[a]	Electronic configuration for R^{3+} No. 4f electrons	S	L	J	Spectroscopic ground state symbol
Scandium	Sc	21	44.955910	0	—	—	—	—
Yttrium	Y	39	88.90585	0	—	—	—	—
Lanthanum	La	57	138.9055	0	—	—	—	—
Cerium	Ce	58	140.115	1	1/2	3	5/2	$^2F_{5/2}$
Praseodymium	Pr	59	140.90765	2	1	5	4	$^3H_4$
Neodymium	Nd	60	144.24	3	3/2	6	9/2	$^4I_{9/2}$
Promethium	Pm	61	(145)	4	2	6	4	$^5I_4$
Samarium	Sm	62	150.36	5	5/2	5	5/2	$^6H_{5/2}$
Europium	Eu	63	151.965	6	3	3	0	$^7F_0$
Gadolinium	Gd	64	157.25	7	7/2	0	7/2	$^8S_{7/2}$
Terbium	Tb	65	158.92534	8	3	3	6	$^7F_6$
Dysprosium	Dy	66	162.50	9	5/2	5	15/2	$^6H_{15/2}$
Holmium	Ho	67	164.93032	10	2	6	8	$^5I_8$
Erbium	Er	68	167.26	11	3/2	6	15/2	$^4I_{15/2}$
Thulium	Tm	69	168.93421	12	1	5	6	$^3H_6$
Ytterbium	Yb	70	173.04	13	1/2	3	7/2	$^2F_{7/2}$
Lutetium	Lu	71	174.967	14	—	—	—	—

Note: For additional information, see Goldschmidt, Z.B., in *Handbook on the Physics and Chemistry of Rare Earths,* Vol. 1, Gschneidner, K.A., Jr. and Eyring, L., Eds., North-Holland Physics, Amsterdam, 1978; DeLaeter, J.R., and Heumann, K.G., *J. Phys. Chem. Ref. Data,* 20, 1313, 1991; *Pure Appl. Chem.,* 66, 2423, 1994.

[a] 1993 standard atomic weights.

Table 2
Crystallographic Data for the Rare Earth Metals at 24°C (297 K) or Below

Rare earth metal	Crystal structure[a]	Lattice constants(Å) a_o	b_o	c_o	Metallic radius CN = 12 (Å)	Atomic volume (cm^3/mol)	Density (g/cm^3)
αSc	hcp	3.3088	—	5.2680	1.6406	15.039	2.989
αY	hcp	3.6482	—	5.7318	1.8012	19.893	4.469
αLa	dhcp	3.7740	—	12.171	1.8791	22.602	6.146
αCe[b]	fcc	4.85[b]	—	—	1.72[b]	17.2[b]	8.16[b]
βCe	dhcp	3.6810	—	11.857	1.8321	20.947	6.689
γCe[c]	fcc	5.1610	—	—	1.8247	20.696	6.770
αPr	dhcp	3.6721	—	11.8326	1.8279	20.803	6.773
αNd	dhcp	3.6582	—	11.7966	1.8214	20.583	7.008
αPm	dhcp	3.65	—	11.65	1.811	20.24	7.264
αSm	rhomb[d]	3.6290[d]	—	26.207	1.8041	20.000	7.520
Eu	bcc	4.5827	—	—	2.0418	28.979	5.244
αGd	hcp	3.6336	—	5.7810	1.8013	19.903	7.901
α'Tb[e]	ortho	3.605[e]	6.244[e]	5.706[e]	1.784[e]	19.34[e]	8.219[e]
αTb	hcp	3.6055	—	5.6966	1.7833	19.310	8.230
α'Dy[f]	ortho	3.595[f]	6.184[f]	5.678[f]	1.774[f]	19.00[f]	8.551[f]
αDy	hcp	3.5915	—	5.6501	1.7740	19.004	8.551
Ho	hcp	3.5778	—	5.6178	1.7661	18.752	8.795

Table 2
Crystallographic Data for the Rare Earth Metals at 24°C (297 K) or Below (continued)

Rare earth metal	Crystal structure[a]	Lattice constants(Å)			Metallic radius CN = 12 (Å)	Atomic volume (cm³/mol)	Density (g/cm³)
		a_0	b_0	c_0			
Er	hcp	3.5592	—	5.5850	1.7566	18.449	9.066
Tm	hcp	3.5375	—	5.5540	1.7462	18.124	9.321
αYb[g]	hcp	3.8799[g]	—	6.3859[g]	1.9451[g]	25.067[g]	6.903[g]
βYb	fcc	5.4848	—	—	1.9392	24.841	6.966
Lu	hcp	3.5052	—	5.5494	1.7349	17.779	9.841

Note: For additional information, see Gschneidner, K.A., Jr. and Calderwood, F.W., in *Handbook on the Physics and Chemistry of Rare Earths,* Vol. 8, Gschneidner, K.A., Jr. and Eyring, L., Eds., North-Holland Physics, Amsterdam, 1986; Gschneidner, K.A., Jr., Pecharsky, V.K., Cho, Jaephil and Martin, S.W., *Scripta Mater.,* 1996, to be published.

[a] hcp = hexagonal close-packed; P6₃/mmc, hP2, A3, Mg-type; dhcp = double-c hexagonal close-packed; P6₃/mmc, hP4, A3′, αLa-type; fcc = face-centered cubic; Fm3̄m, cF4, A1, Cu-type; rhomb = rhombohedral; R3̄m, hR3, αSm-type; bcc = body-centered cubic; Im3̄m, cI2, A2, W-type; ortho = orthorhombic; Cmcm, oC4, α′ Dy-type.
[b] At 77 K (–196°C).
[c] Equilibrium room temperature (standard state) phase.
[d] Rhombohedral is the primitive cell. Lattice parameters given are for the nonprimitive hexagonal cell.
[d] At 220 K (–53°C).
[f] At 86 K (–187°C).
[g] At 23°C.

Table 3
Crystallographic Data for Rare Earth Metals at High Temperature

Rare earth metal	Structure	Lattice parameter (Å)	Temp. (°C)	Metallic radius CN = 8 (Å)	Metallic radius CN = 12 (Å)	Atomic volume (cm³/mol)	Density (g/cm³)
βSc	bcc	3.73 (est.)	1337	1.62	1.66	15.6	2.88
βY	bcc	4.10[a]	1478	1.78	1.83	20.8	4.28
βLa	fcc	5.303	325	—	1.875	22.45	6.187
γLa	bcc	4.26	887	1.84	1.90	23.3	5.97
δCe	bcc	4.12	757	1.78	1.84	21.1	6.65
βPr	bcc	4.13	821	1.79	1.84	21.2	6.64
βNd	bcc	4.13	883	1.79	1.84	21.2	6.80
βPm	bcc	4.10 (est.)	890	1.78	1.83	20.8	6.99
βSm	hcp	$a = 3.6630$ $c = 5.8448$	450[b]	—	1.8176	20.450	7.353
γSm	bcc	4.10 (est.)	922	1.77	1.82	20.8	7.25
βGd	bcc	4.06	1265	1.76	1.81	20.2	7.80
βTb	bcc	4.07[a]	1289	1.76	1.81	20.3	7.82
βDy	bcc	4.03[a]	1381	1.75	1.80	19.7	8.23
γYb	bcc	4.44	763[c]	1.92	1.98	26.4	6.57

Note: The rare earths Eu, Ho, Er, Tm, and Lu are monomorphic. For additional information, see Gschneidner, K.A., Jr. and Calderwood, F.W., in *Handbook on the Physics and Chemistry of Rare Earths,* Vol. 8, Gschneidner, K.A., Jr. and Eyring, L., Eds., North-Holland Physics, Amsterdam, 1986, 1.

[a] Determined by extrapolation to 0% solute of *a* vs. composition data for R-Mg alloys at 24°C and corrected for thermal expansion to temperature given.
[b] The hcp phase was stabilized by impurities and the temperature of measurement was below the equilibrium transition temperature (see Table 4).
[c] The bcc phase was stabilized by impurities and the temperature of measurement was below the equilibrium transition temperature (see Table 4).

Table 4
High Temperature Transition Temperatures and Melting Point of Rare Earth Metals

Rare earth metal	Transition I ($\alpha - \beta$)[a]		Transition II ($\beta - \gamma$)[a]		Melting point (C°)
	Temp. (°C)	Phases	Temp. (C°)	Phases	
Sc	1337	hcp ⇌ bcc	—	—	1541
Y	1478	hcp ⇌ bcc	—	—	1522
La[b]	310	dhcp → fcc	865	fcc ⇌ bcc	918
Ce[c,d]	139	dhcp → fcc ($\beta - \gamma$)	726	fcc ⇌ bcc ($\gamma - \delta$)	798
Pr	795	dhcp ⇌ bcc	—	—	931
Nd	863	dhcp ⇌ bcc	—	—	1021
Pm	890	dhcp ⇌ bcc	—	—	1042
Sm[e]	734	rhom → hcp	922	hcp ⇌ bcc	1074
Eu	—	—	—	—	822
Gd	1235	hcp ⇌ bcc	—	—	1313
Tb	1289	hcp ⇌ bcc	—	—	1356
Dy	1381	hcp ⇌ bcc	—	—	1412
Ho	—	—	—	—	1474
Er	—	—	—	—	1529
Tm	—	—	—	—	1545
Yb	795	fcc ⇌ bcc ($\beta - \gamma$)	—	—	819
Lu	—	—	—	—	1663

Note: For additional information, see Gschneidner, K.A., Jr. and Calderwood, F.W., in *Handbook on the Physics and Chemistry of Rare Earths,* Vol. 8, Gschneidner, K.A., Jr. and Eyring, L., Eds., North-Holland Physics, Amsterdam, 1986; Gschneidner, K.A., Jr., Pecharsky, V.K., Cho, Jaephil and Martin, S.W., *Scripta Mater.,* 34, 1717, 1996.

[a] For all the transformations listed, unless otherwise noted.
[b] On cooling, fcc → dhcp ($\beta \to \alpha$), 260°C.
[c] The $\beta \rightleftharpoons \gamma$ equilibrium transition temperature is 10 ± 5°C.
[d] On cooling, fcc → dhcp ($\gamma \to \beta$), –16°C.
[e] On cooling, hcp → rhomb ($\beta \to \alpha$), 727°C.

Table 5
Low Temperature Transition Temperatures of the Rare Earth Metals

Rare earth metal	Cooling			Rare earth metal	Heating		
	Transformation	°C	K		Transformation	°C	K
Ce	$\gamma \to \beta$[a]	–16	257	Ce	$\alpha \to \beta$	–148	125
	$\gamma \to \alpha$	–172	101		$\alpha \to \beta + \gamma$	–104	169
	$\beta \to \alpha$	–228	45		$\beta \to \gamma$[a]	139	412
Tb	$\alpha \to \alpha'$	–53	220	Yb	$\alpha \to \beta$	7	280
Dy	$\alpha \to \alpha'$	–187	86				
Yb	$\beta \to \alpha$	–13	260				

Note: For additional information, see Beaudry, B.J. and Gschneidner, K.A., Jr., in *Handbook on the Physics and Chemistry of Rare Earths,* Vol. 1, Gschneidner, K.A., Jr. and Eyring, L., Eds., North-Holland Physics, Amsterdam, 1978, 173. Koskenmaki, D.C. and Gschneidner, K.A., Jr., 1978, in *Handbook on the Physics and Chemistry of Rare Earths,* Vol. 1, Gschneidner, K.A., Jr. and Eyring, L., Eds., North-Holland Physics, Amsterdam, 1978, 337. Gschneidner, K.A., Jr., Pecharsky, V.K., Cho, Jaephil and Martin, S.W., *Scripta Mater.,* 34, 1717, 1996.

[a] The $\beta \rightleftharpoons \gamma$ equilibrium transition temperature is 10 ± 5°C (283 ± 5K).

Table 6
Heat Capacity, Standard Entropy, Heats of Transformation, and Fusion of the Rare Earth Metals

Rare earth metal	Heat capacity at 298 K (J/mol K)	Standard entropy $S°_{298}$ (J/mol K)	Heat of transformation (kJ/mol)				Heat of fusion (kJ/mol)
			trans. 1	$\Delta H_{tr}{}^1$	trans. 2	$\Delta H_{tr}{}^2$	
Sc	25.5	34.6	$\alpha \rightleftharpoons \beta$	4.00	—	—	14.1
Y	26.5	44.4	$\alpha \rightleftharpoons \beta$	4.99	—	—	11.4
La	27.1	56.9	$\alpha \rightleftharpoons \beta$	0.36	$\beta \rightleftharpoons \gamma$	3.12	6.20
Ce	26.9	72.0	$\beta \rightleftharpoons \gamma$	0.05	$\gamma \rightleftharpoons \delta$	2.99	5.46
Pr	27.2	73.2	$\alpha \rightleftharpoons \beta$	3.17	—	—	6.89
Nd	27.5	71.5	$\alpha \rightleftharpoons \beta$	3.03	—	—	7.14
Pm	27.3[a]	71.6[a]	$\alpha \rightleftharpoons \beta$	3.0[a]	—	—	7.7[a]
Sm	29.5	69.6	$\alpha \rightleftharpoons \beta$	0.2[a]	$\beta \rightleftharpoons \gamma$	3.11	8.62
Eu	27.7	77.8	—	—	—	—	9.21
Gd	37.0	68.1	$\alpha \rightleftharpoons \beta$	3.91	—	—	10.0
Tb	28.9	73.2	$\alpha \rightleftharpoons \beta$	5.02	—	—	10.79
Dy	27.7	75.6	$\alpha \rightleftharpoons \beta$	4.16	—	—	11.06
Ho	27.2	75.3	—	—	—	—	17.0[a]
Er	28.1	73.2	—	—	—	—	19.9
Tm	27.0	74.0	—	—	—	—	16.8
Yb	26.7	59.9	$\beta \rightleftharpoons \gamma$	1.75	—	—	7.66
Lu	26.9	51.0	—	—	—	—	22[a]

Note: For additional information, see Hultgren, R., Desai, P.D., Hawkins, D.T., Gleiser, M., Kelley, K.K., and Wagman, D.D., *Selected Values of the Thermodynamic Properties of the Elements,* ASM International, Metals Park, Ohio, 1973; Wagman, D.D., Evans, W.H., Parker, V.B., Schumm, R.H., Halow, I., Bailey, S.M., Churney, K.L., and Nuttall, R.L., *The NBS Tables of Chemical Thermodynamic Properties, J. Phys. Chem. Ref. Data,* Vol. 11, Suppl 2, 1982; Amitin, E.B., Bessergenev, W.G., Kovalevskaya, Yu. A., and Paukov, I.E., *J. Chem. Thermodyn.,* 15, 181, 1983; Amitin, E.B., Bessergenev, W.G., Kovalevskaya, Yu. A., and Paukov, I.E., *J. Chem. Thermodyn.,* 15, 181, 1983.

[a] Estimated.

Table 7
Vapor Pressures, Boiling Points, and Heats of Sublimation of Rare Earth Metals

Rare earth metal	Temperature in °C[a] for a vapor pressure of				Boiling point[a] (°C)	Heat of sublimation at 25°C (kJ/mol)
	10^{-8} atm (0.001 Pa)	10^{-6} atm (0.101 Pa)	10^{-4} atm (10.1Pa)	10^{-2} atm (1013 Pa)		
Sc	1036	1243	1533	1999	2836	377.8
Y	1222	1460	1812	2360	3345	424.7
La	1301	1566	1938	2506	3464	431.0
Ce	1290	1554	1926	2487	3443	422.6
Pr	1083	1333	1701	2305	3520	355.6
Nd	955	1175	1500	2029	3074	327.6
Pm	—	—	—	—	3000[b]	348[b]
Sm	508	642	835	1150	1794	206.7
Eu	399	515	685	964	1529	175.3
Gd	1167	1408	1760	2306	3273	397.5
Tb	1124	1354	1698	2237	3230	388.7
Dy	804	988	1252	1685	2567	290.4
Ho	845	1036	1313	1771	2700	300.8
Er	908	1113	1405	1896	2868	317.1
Tm	599	748	964	1300	1950	232.2
Yb	301	400	541	776	1196	152.1
Lu	1241	1483	1832	2387	3402	427.6

Note: For additional information, see Hultgren, R., Desai, P.D., Hawkins, D.T., Gleiser, M., Kelley, K.K., and Wagman, D.D., *Selected Values of the Thermodynamic Properties of the Elements,* ASM International, Metals Park, Ohio, 1973 and Beaudry, B.J. and Gschneidner, K.A., Jr., in *Handbook on the Physics and Chemistry of Rare Earths,* Vol. 1, Gschneidner, K.A., Jr. and Eyring, L., Eds., North-Holland Physics, Amsterdam, 1978, 173.

[a] International Temperature Scale of 1990 (ITS-90) values.
[b] Estimated.

Table 8
Magnetic Properties of the Rare Earth Metals

Rare earth metal	$\chi_A \times 10^6$ at 298 K (emu/mol)	Effective magnetic moment Paramagnetic at ~298 K Theory[a]	Obs.	Ferromagnetic at ~0 K Theory[b]	Obs.	Easy axis	Néel temp. T_N (K) Hex sites	Cubic sites	Curie temp. T_C (K)	θ_p (K) ‖c	⊥c	Polycryst. or avg.
αSc	295.2	—	—	—	—	—	—	—	—	—	—	—
αY	187.7	—	—	—	—	—	—	—	—	—	—	—
αLa	95.9	—	—	—	—	—	—	—	—	—	—	—
βLa	105	—	—	—	—	—	—	—	—	—	—	—
γCe	2,270	2.54	2.52	2.14	—	—	13.7	14.4	—	—	—	-50
βCe	2,500	2.54	2.61	2.14	—	—	0.03	12.5	—	—	—	-41
αPr	5,530	3.58	3.56	3.20	2.7[c]	a	—	—	—	—	—	0
αNd	5,930	3.62	3.45	3.27	2.2[c]	b	19.9	7.5	—	0	5	3.3
αPm	—	2.68	—	2.40	—	—	—	14.0	—	—	—	—
αSm	1,278[d]	0.85	1.74	0.71	0.5[c]	a	109	90.4	—	—	—	—
Eu	30,900	7.94	8.48	7.0	5.9	<110>	—	—	—	—	—	100
αGd	185,000[e]	7.94	7.98	7.0	7.63	30° to c	230.0	—	293.4	317	317	317
αTb	170,000	9.72	9.77	—	—	b	230.0	—	219.5	195	239	224
α'Tb	—	—	—	9.0	9.34	b	180.2	—	219.5	—	—	—
αDy	98,000	10.64	10.83	10.0	10.33	a	—	—	90.5[g]	121	169	153
α'Dy	—	—	—	10.0	—	a	—	—	—	—	—	—
Ho	72,900	10.60	11.2	10.0	10.34	b	132	—	19.5	73.0	88.0	83.0
Er	48,000	9.58	9.9	9.0	9.1	30° to c	85	—	18.7	61.7	32.5	42.2
Tm	24,700	7.56	7.61	7.0	7.14	c	58	—	32.0	41.0	-17.0	2.3
βYb	67[d]	—	—	—	—	—	—	—	—	—	—	—
Lu	182.9	—	—	—	—	—	—	—	—	—	—	—

Note: For additional information, see McEwen, K.A., in *Handbook on the Physics and Chemistry of Rare Earths*, Vol. 1, Gschneider, K.A., Jr. and Eyring, L., Eds., North-Holland Physics, Amsterdam, 1978, 411 and Legvold, S., in *Ferromagnetic Materials*, Vol. 1, Wohlfarth, E.P., Ed., North-Holland Physics, Amsterdam, 1980, 183; Pecharsky, V.K., Gschneider, K.A., Jr. and Fort, D., *Phys. Rev. B*, 47, 5063, 1993; Pecharsky, V.K., Gschneider, K.A., Jr. and Fort, D., 1996, to be published; Steward, A.M. and Collocott, S.J., *J. Phys.: Condens. Matter*, 1, 677, 1988.

[a] $g[J(J + 1)]^{1/2}$.
[b] gJ.
[c] At 38 T and 4.2 K.
[d] At 290 K.
[e] At 350 K.
[g] On cooling $T_C = 89.6$ K and on warming $T_C = 91.5$ K.

Table 9

Room Temperature Coefficient of Thermal Expansion, Thermal Conductivity, Electrical Resistance, and Hall Coefficient

Rare earth metal	Expansion $\alpha_i \times 10^6$ ($°C^{-1}$)			Thermal conductivity (W/cm-K)	Electrical resistance ($\mu\Omega\cdot cm$)			Hall coefficient ($R_i \times 10^{12}$) (V·cm/A·Oe)		
	α_a	α_c	α_{poly}		ρ_a	ρ_c	ρ_{poly}	R_a	R_c	R_{poly}
αSc	7.6	15.3	10.2	0.158	70.9	26.9	56.2[a]	—	—	-0.13
αY	6.0	19.7	10.6	0.172	72.5	35.5	59.6	-0.27	-1.6	—
aLa	4.5	27.2	12.1	0.134	—	—	61.5	—	—	-0.35
bCe	—	—	—	—	—	—	82.8	—	—	—
γCe	6.3	—	6.3	0.113	—	—	74.4	—	—	+1.81
αPr	4.5	11.2	6.7	0.125	—	—	70.0	—	—	+0.709
αNd	7.6	13.5	9.6	0.165	—	—	64.3	—	—	+0.971
αPm	9[b]	16[b]	11[b]	0.15[b]	—	—	75[b]	—	—	—
αSm	9.6	19.0	12.7	0.133	—	—	94.0	—	—	-0.21
Eu	35.0	—	35.0	0.139[b]	—	—	90.0	—	—	+24.4
αGd	9.1[c]	10.0[c]	9.4[c]	0.105	135.1	121.7	131.0	-10	-54	-4.48[d]
αTb	9.3	12.4	10.3	0.111	123.5	101.5	115.0	-1.0	-3.7	—
αDy	7.1	15.6	9.9	0.107	111.0	76.6	92.6	-0.3	-3.7	—
Ho	7.0	19.5	11.2	0.162	101.5	60.5	81.4	+0.2	-3.2	—
Er	7.9	20.9	12.2	0.145	94.5	60.3	86.0	+0.3	-3.6	—
Tm	8.8	22.2	13.3	0.169	88.0	47.2	67.6	—	—	-1.8
βYb	26.3	—	26.3	0.385	—	—	25.0	—	—	+3.77
Lu	4.8	20.0	9.9	0.164	76.6	34.7	58.2	+0.45	-2.6	-0.535

Note: For additional information, see Beaudry, B. J. and Gschneidner, K.A., Jr., in *Handbook on the Physics and Chemistry of Rare Earths*, Vol. 1, Gschneidner, K.A., Jr. and Eyring, L., Eds., North-Holland Physics, Amsterdam, 1978, 173 and McEwen, K.A., in *Handbook on the Physics and Chemistry of Rare Earths*, Vol. 1, Gschneidner, K.A., Jr. and Eyring, L., Eds., North-Holland Physics, Amsterdam, 1978, 411.

[a] Calculated from single crystal values.
[b] Estimated.
[c] At 100°C.
[d] At 77°C.

Table 10
Electronic Specific Heat Constant (γ), Electron-Electron (Coulomb) Coupling Constant (μ^*), Electron-Phonon Coupling Constant (λ), Debye Temperature at 0 K(θ_D), and Superconducting Transition Temperature

Rare earth metal	γ (mJ/mol·K^2)	μ^*	λ	θ_D (K) from Heat capacity	θ_D (K) from Elastic constants	Superconducting temperature (K)
αSc	10.334	0.16	0.30	345.3	—	0.050[a]
αY	7.878	0.15	0.30	244.4	258	1.3[b]
αLa	9.45	0.08	0.76	150	154	5.10
βLa	11.5	—	—	140	—	6.00
αCe	12.8	—	—	179	—	0.022[c]
αPr	20	—	1.07[d]	155[e]	153	—
αNd	f	—	0.86[d]	157[e]	163	—
αPm	—	—	—	159[e]	—	—
αSm	8.1 ± 1.5[g]	—	0.81[d]	162[e,f]	169	—
Eu	f	—	—	f	118	—
αGd	4.48	—	0.30	169	182	—
α'Tb	3.71	—	0.34[d]	169.6	177	—
α'Dy	4.9	—	0.32[d]	192	183	—
Ho	2.1	—	0.30[d]	175[e]	190	—
Er	8.7	—	0.33[d]	176.9	188	—
Tm	f	—	0.36[d]	179[e]	200	—
αYb	3.30	—	—	117.6	118	—
βYb	8.36	—	—	109	—	—
Lu	8.194	0.14	0.31	183.2	185	0.022[h]

Note: For additional information, see Sundström, L.J., in *Handbook on the Physics and Chemistry of Rare Earths,* Vol. 1, Gschneidner, K.A., Jr., and Eyring, L., Eds., North-Holland Physics, Amsterdam, 1978, 379, Scott, T., in *Handbook on the Physics and Chemistry of Rare Earths,* Vol. 1, Gschneidner, K.A., Jr. and Eyring, L., Eds., North-Holland Physics, Amsterdam, 1978, 591, Probst, C. and Wittig, J., in *Handbook on the Physics and Chemistry of Rare Earths,* Vol. 1, Gschneidner, K.A., Jr. and Eyring, L., Eds., North-Holland Physics, Amsterdam, 1978, 749, and Tsang, T.-W.E., Gschneidner, K.A., Jr., Schmidt, F.A., and Thome, D.K., *Phys. Rev.,* B, 31, 235, 1985. Collocott, S.J., Hill, R.W. and Stewart, A.M., *J. Phys. F,* 18, L223, 1988. Hill, R.W. and Gschneidner, K.A., Jr., *J. Phys. F,* 18, 2545, 1988. Skriver, H.L. and Mertig, I., *Phys. Rev. B,* 41, 6553, 1990. Collocott, S.J. and Stewart, A.M., *J. Phys.: Condens. Matter,* 4, 6743, 1992. Pecharsky, V.K., Gschneidner, K.A., Jr. and Fort, D., *Phys. Rev. B,* 47, 5063, 1993.

[a] At 18.6 GPa.
[b] At 11 GPa.
[c] At 2.2 GPa.
[d] Calculated value.
[e] Estimated.
[f] Heat capacity results have been reported, but the resultant γ and θ_D values are unreliable because of the presence of impurities and/or there was no reliable procedure or model to correct for the magnetic contribution to the heat capacity.
[g] Based on the values reported for the purer Sm sample (IV).
[h] At 4.5 GPa.

Table 11
Room Temperature Elastic Moduli and Mechanical Properties

Rare earth metal	Elastic moduli (GPa)				Mechanical properties (MPa)				Recryst. temp. (°C)
	Young's (elastic) modulus	Shear modulus	Bulk modulus	Poisson's ratio	Yield strength 0.2% offset	Ultimate tensile strength	Uniform elongation (%)	Reduction in area (%)	
Sc	74.4	29.1	56.6	0.279	173[a]	255[a]	5.0[a]	8.0[a]	550
Y	63.5	25.6	41.2	0.243	42	129	34.0	—	550
αLa	36.6	14.3	27.9	0.280	126[a]	130	7.9[a]	—	300
βCe	—	—	—	—	86	138	—	24.0	—
γCe	33.6	13.5	21.5	0.24	28	117	22.0	30.0	325
αPr	37.3	14.8	28.8	0.281	73	147	15.4	67.0	400
αNd	41.4	16.3	31.8	0.281	71	164	25.0	72.0	400
αPm	46[b]	18[b]	33[b]	0.28[b]	—	—	—	—	400[b]
αSm	49.7	19.5	37.8	0.274	68	156	17.0	29.5	440
Eu	18.2	7.9	8.3	0.152	—	—	—	—	300
αGd	54.8	21.8	37.9	0.259	15	118	37.0	56.0	500
αTb	55.7	22.1	38.7	0.261	—	—	—	—	500
αDy	61.4	24.7	40.5	0.247	43	139	30.0	30.0	550
Ho	64.8	26.3	40.2	0.231	—	—	—	—	520
Er	69.9	28.3	44.4	0.237	60	136	11.5	11.9	520
Tm	74.0	30.5	44.5	0.213	—	—	—	—	600
βYb	23.9	9.9	30.5	0.207	7	58	43.0	92.0	300
Lu	68.6	27.2	47.6	0.261	—	—	—	—	600

Note: For additional information, see Scott, T., in *Handbook on the Physics and Chemistry of Rare Earths,* Vol. 1, Gschneidner, K.A., Jr. and Eyring, L., Eds., North-Holland Physics, Amsterdam, 1978, 591.

[a] Value is questionable.
[b] Estimated.

Table 12
Liquid Metal Properties Near the Melting Point

Rare earth metal	Density (g/cm³)	Surface tension (N/m)	Viscosity (centipoise)	Heat capacity (J/mol K)	Thermal conductivity (W/cm K)	Magnetic susceptibility $\chi \times 10^4$ (emu/mol)	Electrical resistivity (μΩ·cm)	$\Delta V_{L \to s}$[a] (%)	Spectral emittance at λ = 645 nm ε (%)	Temp. range (°C)
Sc	2.80	0.954	—	44.2[b]	—	—	—	—	36.8	1522–1647
Y	4.24	0.871	—	43.1	—	—	—	—	25.4	920–1287
La	5.96	0.718	2.65	34.3	0.238	1.20	133	-0.6	32.2	877–1547
Ce	6.68	0.706	3.20	37.7	0.210	9.37	130	+1.1	28.4	931–1537
Pr	6.59	0.707	2.85	43.0	0.251	17.3	139	-0.02	39.4	1021–1567
Nd	6.72	0.687	—	48.8	0.195	18.7	151	-0.9	—	—
Pm	6.9[b]	0.680[b]	—	50[b]	—	—	160[b]	—	—	—
Sm	7.16	0.431	—	50.2[b]	—	18.3	182	-3.6	43.7	1075
Eu	4.87	0.264	—	38.1	—	97	242	-4.8	—	—
Gd	7.4	0.664	—	37.2	0.149	67	195	-2.0	34.2	1313–1600
Tb	7.65	0.669	—	46.5	—	82	193	-3.1	—	—
Dy	8.2	0.648	—	49.9	0.187	95	210	-4.5	29.7	1412–1437
Ho	8.34	0.650	—	43.9	—	88	221	-7.4	—	—
Er	8.6	0.637	—	38.7	—	69	226	-9.0	37.2	1529–1587
Tm	9.0[b]	—	—	41.4	—	41	235[b]	-6.9	—	—
Yb	6.21	0.320	2.67	36.8	—	—	113	-5.1	—	—
Lu	9.3	0.940	—	47.9[b]	—	—	224	-3.6	—	—

Note: For additional information, see Van Zytveld, J., in *Handbook on the Physics and Chemistry of Rare Earths*, Vol. 12, Gschneidner, K.A., Jr. and Eyring, L., Eds., North-Holland Physics, Amsterdam, 1989, 357. Stretz, L.A. and Bautista, R.G., in *Temperature, Its Measurement and Control in Science and Industry*, Vol. 4, part I, H.H. Plumb, Ed., Instrument Society of America, Pittsburgh, 1972, 489. King, T.S., Baria, D.N., and Bautista, R.G., *Met. Trans. B*, 7, 411, 1976. Baria, D.N., King, T.S., and Bautista; R.G., *Met. Trans. B*, 7, 577, 1976.

a Volume change on freezing.
b Estimated.

Table 13
Ionization Potentials (Electronvolts)

Rare earth	I Neutral atom	II Singly ionized	III Doubly ionized	IV Triply ionized	V Quadruply ionized
Sc	6.56144	12.79967	24.75666	73.4894	91.65
Y	6.217	12.24	20.52	60.597	77.0
La	5.5770	11.060	19.1773	49.95	61.6
Ce	5.5387	10.85	20.198	36.758	65.55
Pr	5.464	10.55	21.624	38.98	57.53
Nd	5.5250	10.73	22.1	40.41	—
Pm	5.554	10.90	22.3	41.1	—
Sm	5.6437	11.07	23.4	41.4	—
Eu	5.6704	11.241	24.92	42.7	—
Gd	6.1500	12.09	20.63	44.0	—
Tb	5.8639	11.52	21.91	39.79	—
Dy	5.9389	11.67	22.8	41.47	—
Ho	6.0216	11.80	22.84	42.5	—
Er	6.1078	11.93	22.74	42.7	—
Tm	6.18431	12.05	23.68	42.7	—
Yb	6.25416	12.1761	25.05	43.56	—
Lu	5.42585	13.9	20.9594	45.25	66.8

Note: For references, see the table "Ionization Potentials of Atoms and Atomic Ions" in Section 10.

Table 14
Effective Ionic Radii (Å)[a]

Rare earth ion	R^{2+}		R^{3+}			R^{4+}	
	CN = 6	CN = 8	CN = 6	CN = 8	CN = 12	CN = 6	CN = 8
Sc	—	—	0.745	0.87	1.116	—	—
Y	—	—	0.900	1.015	1.220	—	—
La	—	—	1.045	1.18	1.320	—	—
Ce	—	—	1.010	1.14	1.290	0.80	0.97
Pr	—	—	0.997	1.14	1.286	0.78	0.96
Nd	—	—	0.983	1.12	1.276	—	—
Pm	—	—	0.97	1.10	1.267	—	—
Sm	1.19	1.27	0.958	1.09	1.260	—	—
Eu	1.17	1.25	0.947	1.07	1.252	—	—
Gd	—	—	0.938	1.06	1.246	—	—
Tb	—	—	0.923	1.04	1.236	0.76	0.88
Dy	—	—	0.912	1.03	1.228	—	—
Ho	—	—	0.901	1.02	1.221	—	—
Er	—	—	0.890	1.00	1.214	—	—
Tm	—	—	0.880	0.99	1.207	—	—
Yb	1.00	1.07	0.868	0.98	1.199	—	—
Lu	—	—	0.861	0.97	1.194	—	—

Note: For additional information, see Shannon, R.D. and Prewitt, C.T., *Acta Cryst.*, 25, 925, 1969 and Shannon, R.D. and Prewitt, C.T., *Acta Cryst.*, 26, 1046, 1970.

[a] Radius of O^{2-} is 1.40 Å for a coordination number (CN) of 6.

MELTING, BOILING, AND CRITICAL TEMPERATURES OF THE ELEMENTS

This table gives the melting point (T_m), normal boiling point (T_b) at a pressure of 101.325 kPa (1 atmosphere), and critical temperature (T_c) for all the elements for which measurements or reliable estimates are available. Values are given in °C on the ITS-90 scale. A "t" after a value indicates a triple point, and "s" indicates sublimation temperature (i.e., vapor pressure of the solid phase reaches 101.325 kPa, or 1 atmosphere).

Name	T_m/°C	T_b/°C	T_c/°C	Name	T_m/°C	T_b/°C	T_c/°C
Actinium	1051	3198		Mercury	-38.83	356.73	1477
Aluminum	660.32	2519		Molybdenum	2623	4639	
Americium	1176	2011		Neodymium	1021	3074	
Antimony	630.63	1587		Neon	-248.59	-246.08	-228.7
Argon	-189.35	-185.85	-122.28	Neptunium	644		
Arsenic	817 t	614 s	1400	Nickel	1455	2913	
Astatine	302			Niobium	2477	4744	
Barium	727	1897		Nitrogen	-210.00	-195.79	-146.94
Berkelium	1050			Nobelium	827		
Beryllium	1287	2471		Osmium	3033	5012	
Bismuth	271.40	1564		Oxygen	-218.79	-182.95	-118.56
Boron	2075	4000		Palladium	1554.9	2963	
Bromine	-7.2	58.8	315	Phosphorus	44.15	280.5	721
Cadmium	321.07	767		Platinum	1768.4	3825	
Calcium	842	1484		Plutonium	640	3228	
Californium	900			Polonium	254	962	
Carbon	4492 t	3642 s		Potassium	63.38	759	
Cerium	798	3443		Praseodymium	931	3520	
Cesium	28.44	671		Promethium	1042	3000	
Chlorine	-101.5	-34.04	143.8	Protactinium	1572		
Chromium	1907	2671		Radium	700		
Cobalt	1495	2927		Radon	-71	-61.7	104
Copper	1084.62	2562		Rhenium	3186	5596	
Curium	1345			Rhodium	1964	3695	
Dysprosium	1412	2567		Rubidium	39.31	688	
Einsteinium	860			Ruthenium	2334	4150	
Erbium	1529	2868		Samarium	1074	1794	
Europium	822	1596		Scandium	1541	2836	
Fermium	1527			Selenium	221	685	1493
Fluorine	-219.62	-188.12	-129.02	Silicon	1414	3265	
Francium	27			Silver	961.78	2162	
Gadolinium	1313	3273		Sodium	97.80	883	
Gallium	29.76	2204		Strontium	777	1382	
Germanium	938.25	2833		Sulfur	115.21	444.60	1041
Gold	1064.18	2856		Tantalum	3017	5458	
Hafnium	2233	4603		Technetium	2157	4265	
Helium		-268.93	-267.96	Tellurium	449.51	988	
Holmium	1474	2700		Terbium	1356	3230	
Hydrogen	-259.34	-252.87	-240.18	Thallium	304	1473	
Indium	156.60	2072		Thorium	1750	4788	
Iodine	113.7	184.4	546	Thulium	1545	1950	
Iridium	2446	4428		Tin	231.93	2602	
Iron	1538	2861		Titanium	1668	3287	
Krypton	-157.36	-153.22	-63.74	Tungsten	3422	5555	
Lanthanum	918	3464		Uranium	1135	4131	
Lawrencium	1627			Vanadium	1910	3407	
Lead	327.46	1749		Xenon	-111.75	-108.04	16.58
Lithium	180.5	1342		Ytterbium	819	1196	
Lutetium	1663	3402		Yttrium	1522	3345	
Magnesium	650	1090		Zinc	419.53	907	
Manganese	1246	2061		Zirconium	1855	4409	
Mendelevium	827						

HEAT CAPACITY OF THE ELEMENTS AT 25°C

This table gives the specific heat capacity (c_p) in J/g K and the molar heat capacity (C_p) in J/mol K at a temperature of 25°C and a pressure of 100 kPa (1 bar or 0.987 standard atmospheres) for all the elements for which reliable data are available.

Name	c_p J/g K	C_p J/mol K	Name	c_p J/g K	C_p J/mol K
Actinium	0.120	27.2	Molybdenum	0.251	24.06
Aluminum	0.897	24.200	Neodymium	0.190	27.45
Antimony	0.207	25.23	Neon	1.030	20.786
Argon	0.520	20.786	Nickel	0.444	26.07
Arsenic	0.329	24.64	Niobium	0.265	24.60
Barium	0.204	28.07	Nitrogen (N_2)	1.040	29.124
Beryllium	1.825	16.443	Osmium	0.130	24.7
Bismuth	0.122	25.52	Oxygen (O_2)	0.918	29.378
Boron	1.026	11.087	Palladium	0.246	25.98
Bromine (Br_2)	0.226	36.057	Phosphorous (white)	0.769	23.824
Cadmium	0.232	26.020	Platinum	0.133	25.86
Calcium	0.647	25.929	Potassium	0.757	29.600
Carbon (graphite)	0.709	8.517	Praseodymium	0.193	27.20
Cerium	0.192	26.94	Radon	0.094	20.786
Cesium	0.242	32.210	Rhenium	0.137	25.48
Chlorine (Cl_2)	0.479	33.949	Rhodium	0.243	24.98
Chromium	0.449	23.35	Rubidium	0.363	31.060
Cobalt	0.421	24.81	Ruthenium	0.238	24.06
Copper	0.385	24.440	Samarium	0.197	29.54
Dysprosium	0.170	27.7	Scandium	0.568	25.52
Erbium	0.168	28.12	Selenium	0.321	25.363
Europium	0.182	27.66	Silicon	0.705	19.789
Fluorine (F_2)	0.824	31.304	Silver	0.235	25.350
Gadolinium	0.236	37.03	Sodium	1.228	28.230
Gallium	0.371	25.86	Strontium	0.301	26.4
Germanium	0.320	23.222	Sulfur (rhombic)	0.710	22.75
Gold	0.129	25.418	Tantalum	0.140	25.36
Hafnium	0.144	25.73	Tellurium	0.202	25.73
Helium	5.193	20.786	Terbium	0.182	28.91
Holmium	0.165	27.15	Thallium	0.129	26.32
Hydrogen (H_2)	14.304	28.836	Thorium	0.113	26.230
Indium	0.233	26.74	Thulium	0.160	27.03
Iodine (I_2)	0.145	36.888	Tin (white)	0.228	27.112
Iridium	0.131	25.10	Titanium	0.523	25.060
Iron	0.449	25.10	Tungsten	0.132	24.27
Krypton	0.248	20.786	Uranium	0.116	27.665
Lanthanum	0.195	27.11	Vanadium	0.489	24.89
Lead	0.129	26.650	Xenon	0.158	20.786
Lithium	3.582	24.860	Ytterbium	0.155	26.74
Lutetium	0.154	26.86	Yttrium	0.298	26.53
Magnesium	1.023	24.869	Zinc	0.388	25.390
Manganese	0.479	26.32	Zirconium	0.278	25.36
Mercury	0.140	27.983			

VAPOR PRESSURE OF THE METALLIC ELEMENTS

C. B. Alcock

This table gives coefficients in an equation for the vapor pressure of 65 metallic elements in both the solid and liquid state. Vapor pressures in the range 10^{-10} to 10^2 Pa (10^{-15} to 10^{-3} atm) are covered. The equation is:

for p in pascals: $\log(p/\text{Pa}) = 5.006 + A + BT^{-1} + C\log T + DT^{-3}$

for p in atmospheres: $\log(p/\text{atm}) = A + BT^{-1} + C\log T + DT^{-3}$, where T is the temperature in K

This equation reproduces the observed vapor pressures to an accuracy of $\pm5\%$ or better. Further details are given in:

C. B. Alcock, V. P. Itkin, and M. K. Horrigan, *Canadian Metallurgical Quarterly*, 23, 309, 1984.

Reprinted with permission of the publisher, Pergamon Press.

Element, state	A	B	C	D	Temperature range
Li sol	5.667	−8310			298–m.p.
Li liq	5.055	−8023			m.p.–1000
Na sol	5.298	−5603			298–m.p.
Na liq	4.704	−5377			m.p.–700
K sol	4.961	−4646			298–m.p.
K liq	4.402	−4453			m.p.–600
Rb sol	4.857	−4215			298–m.p.
Rb liq	4.312	−4040			m.p.–550
Cs sol	4.711	−3999			298–m.p.
Cs liq	4.165	−3830			m.p.–550
Be sol	8.042	−17020	−0.4440		298–m.p.
Be liq	5.786	−15731			m.p.–1800
Mg sol	8.489	−7813	−0.8253		298–m.p.
Ca sol	10.127	−9517	−1.4030		298–m.p.
Sr sol	9.226	−8572	−1.1926		298–m.p.
Ba sol	12.405	−9690	−2.2890		298–m.p.
Ba liq	4.007	−8163			m.p.–1200
Al sol	9.459	−17342	−0.7927		298–m.p.
Al liq	5.911	−16211			m.p.–1800
Ga sol	6.657	−14208			298–m.p.
Ga liq	6.754	−13984	−0.3413		m.p.–1600
In sol	5.991	−12548			298–m.p.
In liq	5.374	−12276			m.p.–1500
Tl sol	5.971	−9447			298–m.p.
Tl liq	5.259	−9037			m.p.–1100
Sn sol	6.036	−15710			298–m.p.
Sn liq	5.262	−15332			m.p.–1850
Pb sol	5.643	−10143			298–m.p.
Pb liq	4.911	−9701			m.p.–1200
Sc sol	6.650	−19721	0.2885	−0.3663	298–m.p.
Sc liq	5.795	−17681			m.p.–2000
Y sol	9.735	−22306	−0.8705		298–m.p.
Y liq	5.795	−20341			m.p.–2300
La sol	7.463	−22551	−0.3142		298–m.p.
La liq	5.911	−21855			m.p.–2450
Ti sol	11.925	−24991	−1.3376		298–m.p.
Ti liq	6.358	−22747			m.p.–2400
Zr sol	10.008	−31512	−0.7890		298–m.p.
Zr liq	6.806	−30295			m.p.–2500
Hf sol	9.445	−32482	−0.6735		298–m.p.
V sol	9.744	−27132	−0.5501		298–m.p.
V liq	6.929	−25011			m.p.–2500
Nb sol	8.822	−37818	−0.2575		298–2500
Ta sol	16.807	−41346	−3.2152	0.7437	298–2500
Cr sol	6.800	−20733	0.4391	−0.4094	298–2000
Mo sol	11.529	−34626	−1.1331		298–2500

Element, state	A	B	C	D	Temperature range
W sol	2.945	−44 094	1.3677		298–2350
W sol	−54.527	−57 687	−12.2231		2200–2500
Mn sol	12.805	−15 097	−1.7896		298–m.p.
Re sol	11.543	−40 726	−1.1629		298–2500
Fe sol	7.100	−21 723	0.4536	−0.5846	298–m.p.
Fe liq	6.347	−19 574			m.p.–2100
Ru sol	9.755	−34 154	−0.4723		298–m.p.
Os sol	9.419	−41 198	−0.3896		298–2500
Co sol	10.976	−22 576	−1.0280		298–m.p.
Co liq	6.488	−20 578			m.p.–2150
Rh sol	10.168	−29 010	−0.7068		298–m.p.
Rh liq	6.802	−26 792			m.p.–2500
Ir sol	10.506	−35 099	−0.7500		298–2500
Ni sol	10.557	−22 606	−0.8717		298–m.p.
Ni liq	6.666	−20 765			m.p.–2150
Pd sol	9.502	−19 813	−0.9258		298–m.p.
Pd liq	5.426	−17 899			m.p.–2100
Pt sol	4.882	−29 387	1.1039	−0.4527	298–m.p.
Pt liq	6.386	−26 856			m.p.–2500
Cu sol	9.123	−17 748	−0.7317		298–m.p.
Cu liq	5.849	−16 415			m.p.–1850
Ag sol	9.127	−14 999	−0.7845		298–m.p.
Ag liq	5.752	−13 827			m.p.–1600
Au sol	9.152	−19 343	−0.7479		298–m.p.
Au liq	5.832	−18 024			m.p.–2050
Zn sol	6.102	−6776			298–m.p.
Zn liq	5.378	−6286			m.p.–750
Cd sol	5.939	−5799			298–m.p.
Cd liq	5.242	−5392			m.p.–650
Hg liq	5.116	−3190			298–400
Ce sol	6.139	−21 752			298–m.p.
Ce liq	5.611	−21 200			m.p.–2450
Pr sol	8.859	−18 720	0.9512		298–m.p.
Pr liq	4.772	−17 315			m.p.–2200
Nd sol	8.996	−17 264	−0.9519		298–m.p.
Nd liq	4.912	−15 824			m.p.–2000
Sm sol	9.988	−11 034	−1.3287		298–m.p.
Eu sol	9.240	−9459	−1.1661		298–m.p.
Gd sol	8.344	−20 861	−0.5775		298–m.p.
Gd liq	5.557	−19 389			m.p.–2250
Tb sol	9.510	−20 457	−0.9247		298–m.p.
Tb liq	5.411	−18 639			m.p.–2200
Dy sol	9.579	−15 336	−1.1114		298–m.p.
Ho sol	9.785	−15 899	−1.1753		298–m.p.
Er sol	9.916	−16 642	−1.2154		298–m.p.
Er liq	4.688	−14 380			m.p.–1900
Tm sol	8.882	−12 270	−0.9564		298–1400
Yb sol	9.111	−8111	−1.0849		298–900
Lu sol	8.793	−22 423	−0.6200		298–m.p.
Lu liq	5.648	−20 302			m.p.–2350
Th sol	8.668	−31 483	−0.5288		298–m.p.
Th liq	−18.453	−24 569	6.6473		m.p.–2500
Pa sol	10.552	−34 869	−1.0075		298–m.p.
Pa liq	6.177	−32 874			m.p.–2500
U sol	0.770	−27 729	2.6982	−1.5471	298–m.p.
U liq	20.735	−28 776	−4.0962		m.p.–2500
Np sol	19.643	−24 886	−3.9991		298–m.p.
Np liq	10.076	−23 378	−1.3250		m.p.–2500
Pu sol	26.160	−19 162	−6.6675		298–600
Pu sol	18.858	−18 460	−4.4720		500–m.p.
Pu liq	3.666	−16 658			m.p.–2450
Am sol	11.311	−15 059	−1.3449		298–m.p.
Cm sol	8.369	−20 364	−0.5770		298–m.p.
Cm liq	5.223	−18 292			m.p.–2200

DENSITY OF MOLTEN ELEMENTS AND REPRESENTATIVE SALTS

This table lists the liquid density at the melting point, ρ_m, for elements that are solid at room temperature, as well as for some representative salts of these elements. Densities at higher temperatures (up to the t_{max} given in the last column) may be estimated from the equation

$$\rho(t) = \rho_m - k(t - t_m)$$

where t_m is the melting point and k is given in the fifth column of the table. If a value of t_{max} is not given, the equation should not be used to extrapolate more than about 20°C beyond the melting point.

Data for the elements were selected from the primary literature; the assistance of Gernot Lang in compiling these data is gratefully acknowledged. The molten salt data were derived from Reference 1.

REFERENCE

1. Janz, G. J., Thermodynamic and Transport Properties of Molten Salts: Correlation Equations for Critically Evaluated Density, Surface Tension, Electrical Conductance, and Viscosity Data, *J. Phys. Chem. Ref. Data*, 17, Suppl. 2, 1988.
2. Nasch, P. M., and Steinemann, S. G., *Phys. Chem. Liq.*, 29, 43, 1995.

Formula	Name	t_m/°C	ρ_m/g cm^{-3}	k/g cm^{-3} °C^{-1}	t_{max}
Ag	Silver	961.78	9.320	0.0009	1500
AgBr	Silver(I) bromide	432	5.577	0.001035	667
AgCl	Silver(I) chloride	455	4.83	0.00094	627
AgI	Silver(I) iodide	558	5.58	0.00101	802
AgNO$_3$	Silver(I) nitrate	212	3.970	0.001098	360
Ag$_2$SO$_4$	Silver(I) sulfate	652	4.84	0.001089	770
Al	Aluminum	660.32	2.375	0.000233	1340
AlBr$_3$	Aluminum bromide	97.5	2.647	0.002435	267
AlCl$_3$	Aluminum chloride	192.6	1.302	0.002711	296
AlI$_3$	Aluminum iodide	188.32	3.223	0.0025	240
As	Arsenic	817	5.22	0.000544	
Au	Gold	1064.18	17.31	0.001343	1200
B	Boron	2075	2.08		
Ba	Barium	727	3.338	0.000299	1550
BaBr$_2$	Barium bromide	857	3.991	0.000924	900
BaCl$_2$	Barium chloride	962	3.174	0.000681	1081
BaF$_2$	Barium fluoride	1368	4.14	0.000999	1727
BaI$_2$	Barium iodide	711	4.26	0.000977	975
Be	Beryllium	1287	1.690	0.00011	
BeCl$_2$	Beryllium chloride	415	1.54	0.0011	473
BeF$_2$	Beryllium fluoride	552	1.96	0.000015	850
Bi	Bismuth	271.40	10.05	0.00135	800
BiBr$_3$	Bismuth bromide	218	4.76	0.002637	927
BiCl$_3$	Bismuth chloride	230	3.916	0.0023	350
Ca	Calcium	842	1.378	0.000230	1570
CaBr$_2$	Calcium bromide	742	3.111	0.0005	791
CaCl$_2$	Calcium chloride	775	2.085	0.000422	950
CaF$_2$	Calcium fluoride	1418	2.52	0.000391	2027
CaI$_2$	Calcium iodide	783	3.443	0.000751	1028
Cd	Cadmium	321.07	7.996	0.001218	500
CdBr$_2$	Cadmium bromide	568	4.075	0.00108	720
CdCl$_2$	Cadmium chloride	564	3.392	0.00082	807
CdI$_2$	Cadmium iodide	387	4.396	0.001117	700
Ce	Cerium	799	6.55	0.000710	1460
CeCl$_3$	Cerium(III) chloride	817	3.25	0.00092	950
CeF$_3$	Cerium(III) fluoride	1430	4.659	0.000936	1927
Co	Cobalt	1495	7.75	0.00165	1580
Cr	Chromium	1907	6.3	0.0011	2100
Cs	Cesium	28.44	1.843	0.000556	510
CsBr	Cesium bromide	636	3.133	0.001223	860
CsCl	Cesium chloride	645	2.79	0.001065	906
CsF	Cesium fluoride	703	3.649	0.001282	912
CsI	Cesium iodide	621	3.197	0.001183	907

DENSITY OF MOLTEN ELEMENTS AND REPRESENTATIVE SALTS (continued)

Formula	Name	$t_m/°C$	$\rho_m/\text{g cm}^{-3}$	$k/\text{g cm}^{-3}\,°C^{-1}$	t_{max}
$CsNO_3$	Cesium nitrate	414	2.820	0.001166	491
Cs_2SO_4	Cesium sulfate	1005	3.1	0.00095	1530
Cu	Copper	1084.62	8.02	0.000609	1630
CuCl	Copper(I) chloride	430	3.692	0.00076	585
Dy	Dysprosium	1411	8.37	0.00143	1540
$DyCl_3$	Dysprosium(III) chloride	680	3.62	0.00068	987
Er	Erbium	1529	8.86	0.00157	1700
Eu	Europium	822	5.13	0.0028	980
Fe	Iron	1538	6.98	0.000572	1680
$FeCl_2$	Iron(II) chloride	677	2.348	0.000555	877
Ga	Gallium	29.76	6.08	0.00062	400
$GaBr_3$	Gallium(III) bromide	121.5	3.116	0.00246	135
$GaCl_3$	Gallium(III) chloride	77.9	2.053	0.002083	141
GaI_3	Gallium(III) iodide	212	3.630	0.002377	252
Gd	Gadolinium	1314	7.4		
$GdCl_3$	Gadolinium(III) chloride	609	3.56	0.000671	1007
GdI_3	Gadolinium(III) iodide	925	4.12	0.000908	1032
Ge	Germanium	938.25	5.60	0.00055	1600
Hf	Hafnium	2233	12		
$HgBr_2$	Mercury(II) bromide	236	5.126	0.003233	319
$HgCl_2$	Mercury(II) chloride	276	4.368	0.002862	304
HgI_2	Mercury(II) iodide	259	5.222	0.003235	354
Ho	Holmium	1472	8.34		
In	Indium	156.60	7.02	0.000836	500
$InBr_3$	Indium(III) bromide	420	3.121	0.0015	528
$InCl_3$	Indium(III) chloride	583	2.140	0.0021	666
InI_3	Indium(III) iodide	207	3.820	0.0015	360
Ir	Iridium	2446	19		
K	Potassium	63.38	0.828	0.000232	500
KBr	Potassium bromide	734	2.127	0.000825	930
KCl	Potassium chloride	771	1.527	0.000583	939
KF	Potassium fluoride	858	1.910	0.000651	1037
KI	Potassium iodide	681	2.448	0.000956	904
KNO_3	Potassium nitrate	337	1.865	0.000723	457
La	Lanthanum	920	5.94	0.00061	1600
$LaBr_3$	Lanthanum bromide	788	4.933	0.000096	912
$LaCl_3$	Lanthanum chloride	859	3.209	0.000777	973
LaF_3	Lanthanum fluoride	1493	4.589	0.000682	2177
LaI_3	Lanthanum iodide	778	4.29	0.001110	907
Li	Lithium	180.5	0.512	0.00052	285
LiBr	Lithium bromide	552	2.528	0.000652	739
LiCl	Lithium chloride	610	1.502	0.000432	781
LiF	Lithium fluoride	848.2	1.81	0.000490	1047
LiI	Lithium iodide	469	3.109	0.000917	667
$LiNO_3$	Lithium nitrate	253	1.781	0.000546	441
Li_2SO_4	Lithium sulfate	859	2.003	0.000407	1214
Lu	Lutetium	1663	9.3		
Mg	Magnesium	650	1.584	0.000234	900
$MgBr_2$	Magnesium bromide	711	2.62	0.000478	935
$MgCl_2$	Magnesium chloride	714	1.68	0.000271	826
MgI_2	Magnesium iodide	634	3.05	0.000651	888
Mn	Manganese	1246	5.95	0.00105	1590
$MnCl_2$	Manganese(II) chloride	650	2.353	0.000437	850
Mo	Molybdenum	2623	9.33		
Na	Sodium	97.80	0.927	0.00023	600
NaBr	Sodium bromide	747	2.342	0.000816	945
Na_2CO_3	Sodium carbonate	858.1	1.972	0.000448	1004
NaCl	Sodium chloride	800.7	1.556	0.000543	1027
NaF	Sodium fluoride	996	1.948	0.000636	1097
NaI	Sodium iodide	660	2.742	0.000949	912

DENSITY OF MOLTEN ELEMENTS AND REPRESENTATIVE SALTS (continued)

Formula	Name	$t_m/°C$	$\rho_m/g\ cm^{-3}$	$k/g\ cm^{-3}\ °C^{-1}$	t_{max}
$NaNO_3$	Sodium nitrate	307	1.90	0.000715	370
Na_2SO_4	Sodium sulfate	884	2.069	0.000483	1077
Nd	Neodymium	1016	6.89	0.00076	1350
Ni	Nickel	1455	7.81	0.000726	1700
$NiCl_2$	Nickel(II) chloride	1009	2.653	0.00066	1057
Os	Osmium	3033	20		
Pb	Lead	327.46	10.66	0.00122	700
$PbBr_2$	Lead(II) bromide	371	5.73	0.00165	600
$PbCl_2$	Lead(II) chloride	501	4.951	0.0015	710
PbI_2	Lead(II) iodide	410	5.691	0.001594	697
Pd	Palladium	1554.9	10.38	0.001169	1700
Pr	Praseodymium	931	6.50	0.00093	1460
$PrCl_3$	Praseodymium chloride	786	3.23	0.00074	977
Pt	Platinum	1768.4	19.77	0.0024	2200
Pu	Plutonium	640	16.63	0.001419	950
Rb	Rubidium	39.31	1.46	0.000451	800
RbBr	Rubidium bromide	682	2.715	0.001072	907
Rb_2CO_3	Rubidium carbonate	837	2.84	0.000640	1007
RbCl	Rubidium chloride	715	2.248	0.000883	923
RbF	Rubidium fluoride	833	2.87	0.00102	1067
RbI	Rubidium iodide	642	2.904	0.001143	902
$RbNO_3$	Rubidium nitrate	305	2.519	0.001068	417
Rb_2SO_4	Rubidium sulfate	1050	2.56	0.000665	1545
Re	Rhenium	3186	18.9		
Rh	Rhodium	1964	10.7	0.000895	2200
Ru	Ruthenium	2334	10.65		
S	Sulfur	115.21	1.819	0.00080	160
Sb	Antimony	630.63	6.53	0.00067	745
$SbCl_3$	Antimony(III) chloride	73.4	2.681	0.002293	77
$SbCl_5$	Antimony(V) chloride	4	2.37	0.001869	77
SbI_3	Antimony(III) iodide	168	4.171	0.002483	322
Sc	Scandium	1541	2.80		
Se	Selenium	221	3.99		
Si	Silicon	1414	2.57	0.000936	1500
Sm	Samarium	1072	7.16		
Sn	Tin	231.93	6.99	0.000601	1200
$SnCl_2$	Tin(II) chloride	247	3.36	0.001253	480
$SnCl_4$	Tin(IV) chloride	-33	2.37	0.002687	138
Sr	Strontium	777	6.980		
$SrBr_2$	Strontium bromide	657	3.70	0.000745	1004
$SrCl_2$	Strontium chloride	874	2.727	0.000578	1037
SrF_2	Strontium fluoride	1477	3.470	0.000751	1927
SrI_2	Strontium iodide	538	4.085	0.000885	1026
Ta	Tantalum	3017	15		
$TaCl_5$	Tantalum(V) chloride	216	2.700	0.004316	457
Tb	Terbium	1359	7.65		
Te	Tellurium	449.51	5.70	0.00035	600
$ThCl_4$	Thorium chloride	770	3.363	0.0014	847
ThF_4	Thorium fluoride	1110	6.058	0.000759	1378
Ti	Titanium	1668	4.11		
$TiCl_4$	Titanium(IV) chloride	-25	1.807	0.001735	137
Tl	Thallium	304	11.22	0.00144	600
TlBr	Thallium(I) bromide	460	5.98	0.001755	647
TlCl	Thallium(I) chloride	430	5.628	0.0018	642
TlI	Thallium(I) iodide	441.8	6.15	0.001761	737
$TlNO_3$	Thallium(I) nitrate	206	4.91	0.001873	279
Tl_2SO_4	Thallium(I) sulfate	632	5.62	0.00130	927
Tm	Thulium	1545	8.56	0.00050	1675
U	Uranium	1135	17.3		
UCl_3	Uranium(III) chloride	837	4.84	0.007943	1057

Formula	Name	$t_m/°C$	$\rho_m/g\ cm^{-3}$	$k/g\ cm^{-3}\ °C^{-1}$	t_{max}
UCl_4	Uranium(IV) chloride	590	3.572	0.001945	667
UF_4	Uranium(IV) fluoride	1036	6.485	0.000992	1341
V	Vanadium	1910	5.5		
W	Tungsten	3422	17.6		
Y	Yttrium	1526	4.24		
YCl_3	Yttrium chloride	721	2.510	0.0005	845
Yb	Ytterbium	824	6.21		
Zn	Zinc	419.53	6.57	0.0011	700
$ZnBr_2$	Zinc bromide	394	3.47	0.000959	602
$ZnCl_2$	Zinc chloride	290	2.54	0.00053	557
ZnI_2	Zinc iodide	446	3.878	0.00136	588
$ZnSO_4$	Zinc sulfate	680	3.14	0.00047	987
Zr	Zirconium	1855	5.8		
$ZrCl_4$	Zirconium chloride	437	1.643	0.007464	492

INDEX OF REFRACTION OF INORGANIC LIQUIDS

This table gives the index of refraction *n* of several inorganic substances in the liquid state at specified temperatures. The measurements refer to ambient atmospheric pressure except for substances whose normal boiling points are greater than the indicated temperature; in this case the pressure is the saturated vapor pressure of the substance. All values refer to a wavelength of 589 nm unless otherwise indicated. Entries are arranged in alphabetical order by chemical formula as normally written.

Data on the index of refraction at other temperatures and wavelengths may be found in Reference 1.

REFERENCES

1. Wohlfarth, C., and Wohlfarth, B., *Landolt-Börnstein, Numerical Data and Functional Relationships in Science and Technology, New Series,* III/38A, Martienssen, W., Editor, Springer-Verlag, Heidelberg, 1996.
2. Francis, A.W., *J. Chem. Eng. Data*, 5, 534, 1960.

Formula	Name	t/°C	n
Ar	Argon	-188	1.2312
$AsCl_3$	Arsenic(III) chloride	16	1.604
BBr_3	Boron tribromide	16	1.312
BrF_3	Bromine trifluoride	25	1.4536
BrF_5	Bromine pentafluoride	25	1.3529
Br_2	Bromine	15	1.659
COS	Carbon oxysulfide	25	1.3506
CO_2	Carbon dioxide	24	1.6630
CS_2	Carbon disulfide	20	1.62774
C_3O_2	Carbon suboxide	0	1.453
Cl_2	Chlorine	20	1.3834
CrO_2Cl_2	Chromyl chloride	23	1.524
$Fe(CO)_5$	Iron pentacarbonyl	14	1.523
$GeBr_4$	Germanium(IV) bromide	26	1.6269
$GeCl_4$	Germanium(IV) chloride	25	1.4614
HBr	Hydrogen bromide	10	1.325
HCN	Hydrogen cyanide	20	1.26136
HCl	Hydrogen chloride	18	1.3287 [a]
$HClO_4$	Perchloric acid	50	1.3819
HF	Hydrogen fluoride	25	1.1574
HI	Hydrogen iodide	16	1.466
HNO_3	Nitric acid	25	1.393
H_2	Hydrogen	-253	1.1096
H_2O	Water	20	1.33336
H_2O_2	Hydrogen peroxide	28	1.4061
H_2S	Hydrogen sulfide	-80	1.460
		20	1.3682
H_2SO_4	Sulfuric acid	20	1.4183
H_2S_2	Hydrogen disulfide	20	1.630
He	Helium	-269	1.02451 [c]
Kr	Krypton	-157	1.3032 [c]
NH_3	Ammonia	-77	1.3944 [b]
		20	1.3327
NO	Nitric oxide	-90	1.330
N_2	Nitrogen	-196	1.19876 [b]
N_2H_4	Hydrazine	22	1.470
N_2O	Nitrous oxide	25	1.238
O_2	Oxygen	-183	1.2243 [c]
PBr_3	Phosphorus(III) bromide	25	1.687
PCl_3	Phosphorus(III) chloride	21	1.5122
PH_3	Phosphine	17	1.317
P_2O_3	Phosphorus(III) oxide	27	1.540
S	Sulfur	125	1.9170
SCl_2	Sulfur dichloride	14	1.557
SF_6	Sulfur hexafluoride	25	1.167
$SOCl_2$	Thionyl chloride	10	1.527
SO_2	Sulfur dioxide	25	1.3396
SO_2Cl_2	Sulfuryl chloride	12	1.444

Formula	Name	$t/°C$	n
SO_3	Sulfur trioxide	20	1.40965
$SSCl_2$	Sulfur chloride	20	1.671
$SbCl_5$	Antimony(V) chloride	22	1.5925
$SiBr_4$	Tetrabromosilane	31	1.5685
$SiCl_4$	Tetrachlorosilane	25	1.41156
$SnBr_4$	Tin(IV) bromide	31	1.6628
$SnCl_4$	Tin(IV) chloride	25	1.5086
$TiCl_4$	Titanium(IV) chloride	18	1.6076
Xe	Xenon	-112	1.3918 [c]

[a] At 581 nm
[b] At 578 nm
[c] At 546 nm

PHYSICAL AND OPTICAL PROPERTIES OF MINERALS

The chemical formula, crystal system, density, hardness, and index of refraction of some common minerals are given in this table. Entries are arranged alphabetically by mineral name. The columns are:

- **Formula:** Chemical formula for a typical sample of the mineral. Composition often varies considerably with the origin of the sample.
- **Crystal system:** tricl = triclinic; monocl = monoclinic; orth = orthorhombic; tetr = tetragonal; hex = hexagonal; rhomb = rhombohedral; cub = cubic.
- **Density:** Typical density in g/cm^3. Individual samples may vary by a few percent.
- **Hardness:** On the Mohs' scale (range of 1 to 10, with talc = 1 and diamond = 10).
- **Index of refraction:** Values are given for the three coordinate axes in the order of least, intermediate, and greatest index. For cubic crystals there is only a single value. See Reference 1 for details on the axis systems. Variations of several percent, depending on the origin and exact composition of the sample, are common.

REFERENCES

1. Deer, W.A., Howie, R.A., and Zussman, J., *An Introduction to the Rock-Forming Minerals*, 2nd Edition, Longman Scientific & Technical, Harlow, Essex, 1992.
2. Carmichael, R.S., *Practical Handbook of Physical Properties of Rocks and Minerals*, CRC Press, Boca Raton, FL, 1989.
3. Donnay, J.D.H., and Ondik, H.M., *Crystal Data Determinative Tables, Third Edition, Volume 2, Inorganic Compounds*, Joint Committee on Powder Diffraction Standards, Swarthmore, PA, 1973.

Name	Formula	Crystal system	Density g/cm³	Hard-ness	Index of refraction n_α	n_β	n_γ
Acanthite	Ag_2S	orth	7.2	2.3			
Actinolite	$Ca_2(Mg,Fe)_5Si_8O_{22}(OH,F)_2$	monocl	3.23	5.5	1.624	1.655	1.664
Aegirine	$NaFe(SiO_3)_2$	monocl	3.58	6	1.763	1.800	1.815
Akermanite	$Ca_2MgSi_2O_7$	tetr	2.94	5.5	1.632	1.640	
Alabandite	MnS	cub	4.0	3.8			
Albite	$NaAlSi_3O_8$	tricl	2.63	6.3	1.527	1.531	1.538
Allanite	$(Ca,Mn,Ce,La,Y,Th)_2(Fe,Ti)(Al,Fe)$ $O\cdot OH(Si_2O_7)(SiO_4)$	monocl	3.8	5.8	1.75	1.78	1.80
Allemontite	$SbAs$	hex	6.0	3.5			
Almandine	$Fe_3Al_2Si_3O_{12}$	cub	4.32	6.8	1.830		
Altaite	$PbTe$	cub	8.16	3			
Aluminite	$Al_2(SO_4)(OH)_4\cdot 7H_2O$	monocl	1.74	1.5	1.459	1.464	1.470
Alunite	$(K,Na)Al_3(SO_4)_2(OH)_6$	rhomb	2.8	3.8	1.572	1.592	
Alunogen	$Al_2(SO_4)_3\cdot 18H_2O$	monocl	1.69	1.8	1.467	1.47	1.478
Amblygonite	$(Li,Na)Al(PO_4)(F,OH)$	tricl	3.1	5.8	1.591	1.604	1.613
Analcite	$NaAlSi_2O_6\cdot H_2O$	cub	2.27	5.5	1.486		
Anatase	TiO_2	tetr	4.23	5.8	2.488	2.561	
Andalusite	Al_2OSiO_4	orth	3.15	7.5	1.635	1.639	1.644
Andesine	$([NaSi]_{0.7-0.5}[CaAl]_{0.3-0.5}AlSi_2O_8$	tricl	2.67	6.3	1.550	1.553	1.557
Andorite	$PbAgSb_3S_6$	rhomb	5.35	3.3			
Andradite	$Ca_3(Fe,Ti)_2Si_3O_{12}$	cub	3.86	6.8	1.887		
Anglesite	$PbSO_4$	orth	6.29	2.8	1.877	1.883	1.894
Anhydrite	$CaSO_4$	orth	2.96	3.5	1.570	1.575	1.614
Ankerite	$Ca(Fe,Mg,Mn)(CO_3)_2$	rhomb	3.0	3.8	1.529	1.720	
Anorthite	$CaAl_2Si_2O_8$	tricl	2.76	6.3	1.577	1.585	1.590
Anorthoclase	$(Na,K)AlSi_3O_8$	tricl	2.58	6	1.523	1.528	1.529
Anthophyllite	$(Mg,Fe)_7Si_8O_{22}(OH,F)_2$	rhomb	3.21	5.8	1.645	1.658	1.668
Apatite	$Ca_5(PO_4)_3(OH,F,Cl)$	hex	3.2	5	1.645	1.648	
Apophyllite	$KFCa_4Si_8O_{20}\cdot 8H_2O$	tetr	2.35	4.8	1.535	1.536	
Aragonite	$CaCO_3$	orth	2.83	3.5	1.531	1.680	1.686
Arcanite	K_2SO_4	orth	2.66		1.494	1.494	1.497
Argentite	Ag_2S	orth	7.2	2.3			
Arsenolite	As_2O_3	cub	3.86	1.5	1.755		
Arsenopyrite	$FeAsS$	monocl	6.1	5.8			
Atacamite	$Cu_2(OH)_3Cl$	rhomb	3.76	3.3	1.831	1.861	1.880
Augelite	$Al_2(PO_4)(OH)_3$	monocl	2.70	4.8	1.574	1.576	1.588
Augite	$(Ca,Mg,Fe,Ti,Al)_2(Si,Al)_2O_6$	monocl	3.38	6	1.703	1.707	1.738
Autunite	$Ca(UO_{22})(PO_4)_2\cdot 10H_2O$	tetr	3.2	2.3	1.553	1.577	
Axinite	$(Ca,Mn,Fe)_3Al_2BO_3Si_4O_{12}(OH)$	tricl	3.31	6.8	1.684	1.691	1.694

Name	Formula	Crystal system	Density g/cm³	Hardness	Index of refraction		
					n_α	n_β	n_γ
Azurite	$Cu_3(OH)_2(CO_3)_2$	monocl	3.77	3.8	1.730	1.758	1.838
Baddeleyite	ZrO_2	monocl	5.7	6.5	2.13	2.19	2.20
Barite	$BaSO_4$	orth	4.49	3.3	1.636	1.637	1.648
Benitoite	$BaTi(SiO_3)_3$	rhomb	3.65	6.3	1.757	1.804	
Bertrandite	$Be_4Si_2O_7(OH)_2$	rhomb	2.6	6	1.589	1.602	1.613
Beryl	$Be_3Al_2(SiO_3)_6$	hex	2.64	7.8	1.582	1.589	
Beryllonite	$NaBe(PO)_4$	monocl	2.81	5.8	1.552	1.558	1.561
Biotite	$K(Mg,Fe)_3AlSi_3O_{10}(OH,F)_2$	monocl	3.0	2.8	1.595	1.651	1.651
Bismuthinite	Bi_2S_3	orth	6.78	2			
Bixbyite	$(Mn,Fe)_2O_3$	cub	4.95	6.3			
Bloedite	$Na_2Mg(SO_4)_2 \cdot 4H_2O$	monocl	2.25	2.8	1.483	1.486	1.487
Boehmite	$AlO(OH)$	orth	3.44	3.8	1.64	1.65	1.66
Boracite	$Mg_3B_7O_{13}Cl$	rhomb	2.94	7.3	1.66	1.66	1.67
Borax	$Na_2B_4O_7 \cdot 10H_2O$	monocl	1.73	2.3	1.447	1.469	1.472
Bornite	Cu_5FeS_4	cub	5.07	3			
Boulangerite	$Pb_5Sb_4S_{11}$	monocl	6.1	2.8			
Bournonite	$PbCuSbS_3$	rhomb	5.83	2.8			
Braggite	PtS	tetr	10.2				
Braunite	$(Mn,Si)_2O_3$	tetr	4.78	6.3			
Bravoite	$(Ni,Fe)S_2$	cub	4.62	5.8			
Breithauptite	$NiSb$	hex	≈8.7	5.5			
Brochantite	$Cu_4(SO_4)(OH)_6$	monocl	3.79	3.8	1.728	1.771	1.800
Bromyrite	$AgBr$	cub	6.47	2.5	2.253		
Brookite	TiO_2	tetr	4.23	5.8	2.583	2.584	2.700
Brucite	$Mg(OH)_2$	hex	2.37	2.5	1.575	1.59	
Bunsenite	NiO	cub	6.72	5.5			
Cacoxenite	$Fe_4(PO_4)_3(OH)_3 \cdot 12H_2O$	hex	2.3	3.5	1.580	1.646	
Calcite	$CaCO_3$	hex	2.71	3	1.486	1.658	
Caledonite	$Cu_2Pb_5(SO_4)_3(CO_3)(OH)_6$	rhomb	5.76	2.8	1.818	1.866	1.909
Calomel	Hg_2Cl_2	tetr	7.16	1.5	1.973	2.656	
Cancrinite	$(Na,Ca,K)_7[Al_6Si_6O_{24}]$ $(CO_3,SO_4,Cl,OH)_2 \cdot H_2O$	hex	2.42	5.5	1.495	1.509	
Carnalite	$KMgCl_3 \cdot 6H_2O$	rhomb	1.60	2.5	1.466	1.475	1.494
Carnotite	$K_2(UO_2)_2(VO_4)_2 \cdot 3H_2O$	rhomb		1.5	1.75	1.92	1.95
Cassiterite	SnO_2	tetr	6.85	6.5	2.006	2.097	
Celestite	$SrSO_4$	orth	3.96	3.3	1.622	1.624	1.631
Celsian	$BaAl_2Si_2O_8$	monocl	3.25	6.3	1.583	1.588	1.594
Cerargyrite	$AgCl$	cub	5.56	2.5	2.071		
Cerussite	$PbCO_3$	orth	6.6	3.3	1.804	2.076	2.079
Cervantite	Sb_2O_4	orth	6.64	4.5			
Chabazite	$Ca[Al_2Si_4O_{12}] \cdot 6H_2O$	trig	2.08	4.5	1.482		
Chalcanthite	$CuSO_4 \cdot 5H_2O$	tricl	2.29	2.5	1.514	1.537	1.543
Chalcocite	Cu_2S	orth	5.6	2.8			
Chalcopyrite	$CuFeS_2$	tetr	4.2	3.8			
Chiolite	$Na_5Al_3F_{14}$	tetr	3.00	3.8	1.342	1.349	
Chlorite	$(Mg,Al,Fe)_{12}(Si,Al)_8O_{20}(OH)_{16}$	monocl	3.0	2.5	1.61	1.62	1.62
Chloritoid	$FeAl_4O_2(SiO_4)_2(OH)_4$	monocl	3.66	6.5	1.717	1.721	1.726
Chondrodite	$Mg(OH,F)_2 \cdot 2Mg_2SiO_4$	monocl	3.21	6.5	1.604	1.615	1.634
Chromite	$FeCr_2O_4$	cub	5.0	5.5	2.16		
Chrysoberyl	$BeAl_2O_4$	orth	3.65	8.5	1.746	1.748	1.756
Chrysocolla	$CuSiO_3 \cdot 2H_2O$	rhomb	2.4	2	1.575	1.597	1.598
Cinnabar	HgS	hex	8.17	2.3	2.814	3.143	
Claudetite	As_2O_3	monocl	3.74	2.5	1.87	1.92	2.01
Clinohumite	$Mg(OH,F)_2 \cdot 4Mg_2SiO_4$	monocl	3.21	6	1.633	1.647	1.668
Clinozoisite	$Ca_2Al_3Si_3O_{12}(OH)$	monocl	3.30	6.5	1.693	1.700	1.712
Cobaltite	$CoAsS$	cub	≈6.1	5.5			
Colemanite	$Ca_2B_6O_{11} \cdot 5H_2O$	monocl	2.42	4.5	1.586	1.592	1.614
Columbite	$(Fe,Mn)(Nb,Ta)_2O_6$	rhomb	5.20	6			
Connellite	$Cu_{19}(SO_4)Cl_4(OH)_{32} \cdot 3H_2O$	hex	3.36	3	1.731	1.752	

Name	Formula	Crystal system	Density g/cm³	Hard-ness	Index of refraction n_α	n_β	n_γ
Copiapite	$(Fe,Mg)Fe_4(SO_4)_6(OH)_2 \cdot 20H_2O$	tricl	2.13	2.8	1.52	1.54	1.59
Coquimbite	$Fe_2(SO_4)_3 \cdot 9H_2O$	hex	2.1	2.5	1.54	1.56	
Cordierite	$Al_3(Mg,Fe)_2Si_5AlO_{18}$	rhomb	2.66	7	1.540	1.549	1.553
Corundum	Al_2O_3	hex	3.97	9	1.761	1.769	
Cotunnite	$PbCl_2$	orth	5.98	2.5	2.199	2.217	2.260
Covellite	CuS	hex	4.8	1.8			
Cristobalite	SiO_2	hex	2.33	6.5	1.484	1.487	
Crocoite	$PbCrO_4$	monocl	6.12	2.8	2.29	2.36	2.66
Cryolite	Na_3AlF_6	monocl	2.97	2.5	1.338	1.338	1.339
Cryolithionite	$Na_3Li_3Al_2F_{12}$	cub	2.77	2.8	1.340		
Cubanite	$CuFe_2S_3$	rhomb	4.11	3.5			
Cummingtonite	$(Mg,Fe)_7Si_8O_{22}(OH)_2$	monocl	3.4	5.5	1.650	1.660	1.676
Cuprite	Cu_2O	cub	6.0	3.8			
Danburite	$CaSi_2B_2O_8$	rhomb	3.0	7	1.63	1.63	1.63
Datolite	$CaBSiO_4(OH)$	monocl	2.98	5.3	1.624	1.652	1.668
Daubreelite	Cr_2FeS_4	cub	3.81				
Derbylite	$Fe_6Ti_6Sb_2O_{23}$	rhomb	4.53	5	2.45	2.45	2.51
Diamond	C	cub	3.51	10	2.418		
Diaspore	$AlO(OH)$	orth	3.4	6.8	1.694	1.715	1.741
Digenite	$Cu_{2-x}S$	cub	5.55	2.8			
Diopside	$CaMgSi_2O_6$	monocl	3.30	6	1.680	1.687	1.708
Dioptase	$CuSiO_2(OH)_2$	rhomb	3.5	5	1.65	1.70	
Dolomite	$CaMg(CO_3)_2$	rhomb	2.86	3.5	1.500	1.679	
Douglasite	$K_2FeCl_4 \cdot 2H_2O$	orth	2.16		1.488	1.500	
Dyscrasite	Ag_3Sb	rhomb	9.74	3.8			
Eddingtonite	$BaAl_2Si_3O_{10} \cdot 4H_2O$	rhomb	2.8		1.541	1.553	1.557
Eglestonite	Hg_4OCl_2	cub	8.4	2.5	2.49		
Emplectite	$CuBiS_2$	rhomb	6.38	2			
Enargite	Cu_3AsS_4	rhomb	4.5	3			
Enstatite	$MgSiO_3$	monocl	3.19	5.5	1.656	1.662	1.669
Epidote	$Ca_2Al_2(Al,Fe)OH(SiO_4)_3$	monocl	3.44	6	1.733	1.755	1.765
Epsomite	$MgSO_4 \cdot 7H_2O$	orth	1.67	2.3	1.433	1.455	1.461
Erythrite	$(Co,Ni)_3(AsO_4)_2 \cdot 8H_2O$	monocl	3.06	2	1.626	1.661	1.699
Eucairite	$CuAgSe$	orth	7.7	2.5			
Euclasite	$BeAlSiO_4(OH)$	monocl	3.1	7.5	1.651	1.655	1.671
Eudialite	$(Na,Ca,Ce)_5(Fe,Mn)(Zr,Ti)(Si_3O_9)_2$ (OH,Cl)	hex	3.0	5.5	1.623	1.600	1.615
Eulytite	$Bi_4Si_3O_{12}$	cub	6.6	4.5	2.05		
Euxenite	$(Y,Ca,Ce,U,Th)(Nb,Ta,Ti)_2O_6$	rhomb	5.5	6	2.2		
Fayalite	Fe_2SiO_4	orth	4.30	6.5	1.827	1.869	1.879
Ferberite	$FeWO_4$	monocl	7.51	4.3			
Fergussonite	$(Y,Er,Ce,Fe)(Nb,Ta,Ti)O_4$	tetr	5.7	6	2.1		
Fluorite	CaF_2	cub	3.18	4	1.434		
Forsterite	Mg_2SiO_4	orth	3.21	7	1.635	1.651	1.670
Franklinite	$ZnFe_2O_4$	cub	5.21	6	2.36		
Gahnite	$ZnAl_2O_4$	cub	4.62	7.8	1.805		
Galaxite	$MnAl_2O_4$	cub	4.04	7.8	1.92		
Galena	PbS	cub	7.60	2.5	3.91		
Galenabismuthite	$PbBi_2S_4$	rhomb	7.04	3			
Ganomalite	$(Ca,Pb)_{10}(OH,Cl)_2(Si_2O_7)_3$	hex	5.6	3.5	1.910	1.945	
Gaylussite	$Na_2Ca(CO_3)_2 \cdot 5H_2O$	monocl	1.99	2.8	1.444	1.516	1.523
Gehlenite	$Ca_2Al_2SiO_7$	tetr	3.04	5.5	1.658	1.669	
Geikielite	$MgTiO_3$	hex	3.85	5.5	1.95	2.31	
Gibbsite	$Al(OH)_3$	monocl	2.42	3	1.57	1.57	1.59
Glauberite	$Na_2Ca(SO_4)_2$	monocl	2.80	2.8	1.515	1.535	1.536
Glauconite	$(K,Na,Ca)_{1.6}(Fe,Al,Mg)_{4.0}Si_{7.3}Al_{0.7}$ $O_{20}(OH)_4$	monocl	2.7	2	1.60	1.63	1.63
Glaucophane	$Na_2Mg_3Al_2Si_8O_{22}(OH)_2$	monocl	3.19	6	1.634	1.645	1.648
Gmelinite	$(Ca,Na_2)[Al_2Si_4O_{12}] \cdot 6H_2O$	hex	2.10	4.5	1.477	1.485	

Name	Formula	Crystal system	Density g/cm³	Hard-ness	Index of refraction n_α	n_β	n_γ
Goethite	$FeO(OH)$	orth	4.3	5.3	2.268	2.401	2.457
Goslarite	$ZnSO_4 \cdot 7H_2O$	orth	1.97	2.3	1.457	1.480	1.484
Greenockite	CdS	cub	4.8	3.3	2.506	2.529	
Grossularite	$Ca_3Al_2Si_3O_{12}$	cub	3.59	6.8	1.734		
Gummite	$UO_3 \cdot H_2O$	orth	7.05	3.8			
Gypsum	$CaSO_4 \cdot 2H_2O$	monocl	2.32	2	1.520	1.525	1.530
Halite	$NaCl$	cub	2.17	2	1.544		
Hambergite	$Be_2(OH)(BO_3)$	rhomb	2.36	7.5	1.56	1.59	1.63
Hanksite	$Na_{22}K(SO_4)_9(CO_3)_2Cl$	hex	2.56	3.3	1.461	1.481	
Harmotome	$Ba[Al_2Si_6O_{16}] \cdot 6H_2O$	monocl	2.44	4.5	1.506	1.507	1.511
Hausmannite	Mn_3O_4	tetr	4.84	5.5	2.15	2.46	
Haüyne	$(Na,Ca)_{4-8}Al_6Si_6O_{24}(SO_4,S)_{1-2}$	cub	2.47	5.8	1.502		
Hedenbergite	$CaFeSi_2O_6$	monocl	3.53	6	1.721	1.727	1.746
Helvite	$Mn_4Be_3Si_3O_{12}S$	cub	3.32	6	1.739		
Hematite	Fe_2O_3	hex	5.25	6	2.91	3.19	
Hemimorphite	$Zn_4Si_2O_7(OH)_2 \cdot H_2O$	rhomb	3.45	5	1.614	1.617	1.636
Hercynite	$Fe(AlO_2)_2$	cub	4.3	7.8	1.835		
Herderite	$CaBe(PO_4)(Fe,OH)$	monocl	2.98	5.3	1.592	1.612	1.621
Hessite	Ag_2Te	orth	8.4	2.5			
Heulandite	$(Ca,Na_2,K_2)[Al_2Si_7O_{18}] \cdot 6H_2O$	monocl	2.2	3.8	1.498	1.498	1.506
Hopeite	$Zn_3(PO_4)_2 \cdot 4H_2O$	orth	3.0	3.2	1.58	1.59	1.59
Hornblende	$Ca_2(Mg,Fe)_4Al(Si_7AlO_{22})(OH)_2$	monocl	3.24	5.5	1.67	1.67	1.69
Huebnerite	$MnWO_4$	monocl	7.2	4.3	2.17	2.22	2.32
Humite	$Mg(OH,F)_2 \cdot 3Mg_2SiO_4$	orth	3.3	6	1.625	1.636	1.657
Huntite	$Mg_3Ca(CO_3)_4$	trig	2.70				
Hydrogrossularite	$Ca_3Al_2Si_2O_8(SiO_4)_{1-m}(OH)_{4m}$	cub	3.4	6.8	1.70		
Hydromagnesite	$3MgCO_3 \cdot Mg(OH)_2 \cdot 3H_2O$	monocl	2.24	3.5	1.523	1.527	1.545
Illite	$KAl_4[Si_7AlO_{20}](OH)_4$	monocl	2.8	1.5	1.56	1.59	1.59
Ilmenite	$FeTiO_3$	rhomb	4.72	5.5			
Iodyrite	AgI	hex	5.68	1.5	2.21	2.22	
Jacobsite	$MnFe_2O_4$	cub	4.87	7.8	2.3		
Jadeite	$NaAlSi_2O_6$	monocl	3.34	6	1.649	1.654	1.663
Jamesonite	$Pb_4FeSb_6S_{14}$	monocl	5.63	2.5			
Jarosite	$KFe_3(SO_4)_2(OH)_6$	rhomb	3.09	3	1.715	1.820	
Kainite	$KMg(SO_4)Cl \cdot 3H_2O$	monocl	2.15	2.8	1.494	1.505	1.516
Kaliophylite	$KAlSiO_4$	hex	2.61	6	1.532	1.537	
Kaolinite	$Al_4Si_4O_{10}(OH)_8$	tricl	2.65	2.3	1.549	1.564	1.565
Kernite	$Na_2B_4O_7 \cdot 4H_2O$	monocl	1.95	2.5	1.454	1.472	1.488
Kieserite	$MgSO_4 \cdot H_2O$	monocl	2.57	3.5	1.520	1.533	1.584
Kyanite	Al_2OSiO_4	tricl	3.59	6.3	1.715	1.722	1.731
Lanarkite	$Pb_2(SO_4)O$	monocl	6.92	2.3	1.928	2.007	2.036
Lanthanite	$(La,Ce)_2(CO_3)_3 \cdot 8H_2O$	rhomb	2.72	2.8	1.52	1.587	1.613
Laumontite	$Ca_4[Al_8Si_{16}O_{48}] \cdot 16H_2O$	monocl	2.3	3.3	1.508	1.517	1.519
Laurionite	$Pb(OH)Cl$	rhomb	6.24	3.3	2.08	2.12	2.16
Lawsonite	$CaAl_2(OH)_2Si_2O_7 \cdot H_2O$	rhomb	3.08	6	1.655	1.675	1.685
Lazulite	$(Mg,Fe)Al_2(PO_4)_2(OH)_2$	monocl	3.23	5.8	1.615	1.64	1.650
Lazurite	$Na_4SSi_3Al_3O_{12}$	cub	2.42	5.3	1.500		
Leadhillite	$Pb_4(SO_4)(CO_3)_2(OH)_2$	monocl	6.55	2.8	1.87	2.00	2.01
Lepidocrocite	$FeO(OH)$	orth	4.26	5	1.94	2.20	2.51
Lepidolite	$K_2(Li,Al)_{5-6}[Si_{6-7}Al_{2-1}O_{20}](OH,F)_4$	monocl	2.85	3.3	1.536	1.565	1.566
Leucite	$KAlSi_2O_6$	tetr	2.49	5.8	1.510		
Levyne	$(Ca,Na_2)Al_2Si_4O_{12} \cdot 6H_2O$	rhomb	2.10	4.5	1.496	1.501	
Litharge	PbO	tetr	9.35	2	2.535	2.665	
Loellingite	$FeAs_2$	rhomb	7.40	5.3			
Maghemite	Fe_2O_3	cub	4.88	7.8	2.63		
Magnesite	$MgCO_3$	hex	3.05	4	1.536	1.741	
Magnetite	Fe_3O_4	cub	5.17	6	2.42		
Malachite	$Cu_2(OH)_2(CO_3)$	monocl	4.05	3.8	1.655	1.875	1.909
Manganite	$MnO(OH)$	monocl	≈ 4.3	4	2.25	2.25	2.53

Name	Formula	Crystal system	Density g/cm^3	Hardness	Index of refraction n_α	n_β	n_γ
Manganosite	MnO	cub	5.37	5.5			
Marcasite	FeS_2	cub	5.02	6.3			
Marialite	$Na_4Al_3Si_9O_{24}Cl$	tetr	2.56	5.5	1.541	1.548	
Marshite	CuI	cub	5.67	2.5	2.346		
Mascagnite	$(NH_4)_2SO_4$	orth	1.77	2.3	1.520	1.523	1.533
Matlockite	$PbClF$	tetr	7.05	2.8	2.006	2.145	
Meionite	$Ca_4Al_6Si_6O_{24}CO_3$	tetr	2.78	5.5	1.559	1.595	
Melanterite	$FeSO_4 \cdot 7H_2O$	monocl	1.89	2	1.47	1.48	1.49
Melilite	$(Ca,Na)_2(Mg,Fe,Al,Si)_3O_7$	tetr	3.00	5.5	1.639	1.645	
Mellite	$Al_2C_{12}O_{12} \cdot 18H_2O$	tetr	1.64	2.3	1.511	1.539	
Mendipite	$Pb_3O_2Cl_2$	rhomb	7.24	2.5	2.24	2.27	2.31
Mesolite	$Na_2Ca_2(Al_2Si_3O_{10})_3 \cdot 8H_2O$	orth	2.26	5	1.506		
Metacinnabar	HgS	cub	7.70	3			
Microcline	$KAlSi_3O_8$	monocl	2.56	6.3	1.522	1.526	1.530
Miersite	AgI	hex	5.68	2.5	2.20		
Millerite	NiS	hex	5.5	3.3			
Mimetite	$Pb_5(AsO_4,PO_4)_3Cl$	hex	7.24	3.8	2.128	2.147	
Minium	Pb_3O_4	tetr	8.9	2.5			
Mirabilite	$Na_2SO_4 \cdot 10H_2O$	monocl	1.46	1.8	1.394	1.396	1.398
Moissanite	SiC	hex	3.16	9.5	2.648	2.691	
Molybdenite	MoS_2	hex	5.06	1.3			
Monazite	$(Ce,La,Th)PO_4$	monocl	5.2	5	1.787	1.789	1.840
Monetite	$CaHPO_4$	tricl	2.92	3.5	1.587	1.61	1.640
Monticellite	$Ca(Mg,Fe)SiO_4$	orth	3.18	5.5	1.647	1.655	1.664
Montmorillonite	$(0.5Ca,Na)_{0.7}(Al,Mg,Fe)_4$ $[(Si,Al)_8O_{20}](OH)_4 \cdot nH_2O$	monocl	2.5	1.5	1.55	1.57	1.57
Montroydite	HgO	orth	11.14	2.5	2.37	2.50	2.65
Mordenite	$(Na,K,Ca)[Al_2Si_{10}O_{24}] \cdot 7H_2O$	orth	2.13	3.5	1.478	1.480	1.482
Muscovite	$KAl_2Si_3AlO_{10}(OH,F)_2$	monocl	2.83	2.8	1.563	1.596	1.602
Nantokite	$CuCl$	cub	4.14	2.5	1.930		
Natrolite	$Na_2Al_2Si_3O_{10} \cdot 2H_2O$	orth	2.23	5	1.478	1.481	1.491
Nepheline	$Na_3KAl_4Si_4O_{16}$	hex	2.61	5.8	1.534	1.538	
Newberyite	$MgHPO_4 \cdot 3H_2O$	orth	2.13	3.3	1.514	1.517	1.533
Niccolite	$NiAs$	hex	7.77	5.3			
Norbergite	$Mg(OH,F)_2 \cdot Mg_2SiO_4$	orth	3.21	6.5	1.565	1.573	1.592
Nosean	$Na_8Al_6Si_6O_{24}SO_4$	cub	2.35	5.5	1.495		
Oldhamite	CaS	cub	2.59	4	2.137		
Oligoclase	$([NaSi]_{0.9-0.7}[CaAl]_{0.1-0.3})AlSi_2O_8$	tricl	2.64	6.3	1.539	1.543	1.547
Olivenite	$Cu_2(AsO_4)(OH)$	rhomb	4.2	3	1.77	1.80	1.85
Olivine	$(Mg,Fe)SiO_4$	rhomb	3.81	6.8	1.73	1.76	1.78
Opal	$SiO_2 \cdot nH_2O$	amorp	1.9	5	1.44		
Orpiment	As_2S_3	monocl	3.46	1.8	2.40	2.81	3.02
Orthoclase	$KAlSi_3O_8$	monocl	2.56	6	1.523	1.527	1.531
Orthopyroxene	$(Mg,Fe)SiO_3$	rhomb	3.6	5.5	1.709	1.712	1.723
Paragonite	$NaAl_2Si_3AlO_{10}(OH)_2$	monocl	2.85	2.5	1.572	1.602	1.605
Parisite	$(Ce,La,Na)FCO_3 \cdot CaCO_3$	hex	4.42	4.5	1.672	1.771	
Pectolite	$Ca_2NaH(SiO_3)_3$	tricl	2.88	4.8	1.603	1.610	1.639
Penfieldite	$Pb_4Cl_6(OH)_2$	hex	6.6		2.13	2.21	
Pentlandite	$(Fe,Ni)_9S_8$	cub	4.8	3.8			
Percylite	$PbCuCl_2(OH)_2$	cub		2.5	2.05		
Periclase	MgO	cub	3.6	5.5	1.735		
Perovskite	$CaTiO_3$	cub	3.98	5.5	2.34		
Petalite	$LiAlSi_4O_{10}$	monocl	2.42	6.5	1.506	1.511	1.519
Pharmacosiderite	$Fe_3(AsO_4)_2(OH)_3 \cdot 5H_2O$	cub	2.80	2.5	1.690		
Phenakite	Be_2SiO_4	rhomb	2.98	7.5	1.654	1.670	
Phillipsite	$K(Ca_{0.5},Na)_2[Al_3Si_5O_{16}] \cdot 6H_2O$	monocl	2.2	4.3	1.494	1.497	1.505
Phlogopite	$KMg_3AlSi_3O_{10}(OH,F)_2$	monocl	2.83	2.3	1.560	1.597	1.598
Phosgenite	$Pb_2(CO_3)Cl_2$	tetr	6.13	2.5	2.118	2.145	
Piemontite	$Ca_2(Mn,Fe,Al)_3O(Si_2O_7)(SiO_4)(OH)$	monocl	3.49	6	1.762	1.773	1.796

Name	Formula	Crystal system	Density g/cm^3	Hard-ness	Index of refraction n_α	n_β	n_γ
Pigeonite	$(Mg,Fe,Ca)(Mg,Fe)Si_2O_6$	monocl	3.38	6	1.702	1.703	1.728
Pollucite	$CsAlSi_2O_6$	tetr	2.9	6.5	1.517		
Polybasite	$(Ag,Cu)_{16}Sb_2S_{11}$	monocl	6.1	2.5			
Powellite	$Ca(Mo,W)O_4$	tetr	4.35	3.8	1.971	1.980	
Prehnite	$Ca_2Al_2Si_3O_{10}(OH)_2$	rhomb	2.93	6.3	1.622	1.628	1.648
Proustite	Ag_3AsS_3	rhomb	5.57	2.3	2.792	3.088	
Pseudobrookite	Fe_2TiO_5	rhomb	4.36	6	2.38	2.39	2.42
Psilomelane	$BaMn_9O_{16}(OH)_4$	rhomb	4.71	5.5			
Pumpellyite	$Ca_2Al_2(Al,Fe,Mg)[Si_2(O,OH)_7]$ $(SiO_4)(OH,O)_3$	monocl	3.21	5.5	1.688	1.695	1.705
Pyrargyrite	Ag_3SbS_3	rhomb	5.85	2.5	2.88	3.08	
Pyrite	FeS_2	cub	5.02	6.3			
Pyrochlore	$NaCaNb_2O_6F$	cub	5.3	5.3			
Pyrochroite	$Mn(OH)_2$	hex	3.26	2.5	1.68	1.72	
Pyrolusite	MnO_2	tetr	5.08	6.3			
Pyromorphite	$Pb_5(PO_4,AsO_4)_3Cl$	hex	7.04	3.8	2.048	2.058	
Pyrope	$Mg_3Al_2Si_3O_{12}$	cub	3.58	6.8	1.714		
Pyrophyllite	$Al_2Si_4O_{10}(OH)_2$	monocl	2.78	1.5	1.545	1.579	1.599
Pyrrhotite	Fe_7S_8	hex	4.62	4			
Quartz	SiO_2	hex	2.65	7	1.544	1.553	
Rammelsbergite	$NiAs_2$	orth	7.1	5.8			
Raspite	$PbWO_4$	monocl	8.46	2.8	1.27	1.27	1.30
Realgar	As_4S_4	monocl	3.5	1.8	2.538	2.684	2.704
Rhodochrosite	$MnCO_3$	hex	3.70	3.8	1.597	1.816	
Rhodonite	$(Mn,Fe,Ca)SiO_3$	orth	3.48	6	1.725	1.729	1.737
Riebeckite	$Na_2Fe_5(Si_8O_{22})(OH)_2$	monocl	3.3	5	1.675	1.683	1.694
Rutile	TiO_2	tetr	4.23	6.2	2.609	2.900	
Safflorite	$(Co,Fe)As_2$	rhomb	7.3	4.8			
Samarskite	(Y,Er,Ce,U,Ca,Fe,Pb,Th) $(Nb,Ta,Ti,Sn)_2O_6$	rhomb	5.69	5.5	2.200		
Sapphirine	$(Mg,Fe)_2Al_4O_6SiO_4$	monocl	3.49	7.5	1.709	1.712	1.715
Scapolite	$(Na,Ca)_4Al_3(Al,Si)_3Si_6O_{24}$ (Cl,F,OH,CO_3,SO_4)	tetr	2.64	5.5	1.551	1.573	
Scheelite	$CaWO_4$	tetr	6.06	4.8	1.920	1.936	
Scolecite	$CaAl_2Si_3O_{10}\cdot3H_2O$	monocl	2.27	5	1.510	1.518	1.519
Scorodite	$Fe(AsO_4)\cdot2H_2O$	rhomb	3.28	3.8	1.784	1.795	1.814
Sellaite	MgF_2	tetr	3.15	5	1.378	1.390	
Senarmontite	Sb_2O_3	cub	5.58	2.3	2.087		
Serpentine	$Mg_3Si_2O_5(OH)_4$	monocl	2.55	3	1.55	1.56	1.56
Siderite	$FeCO_3$	hex	3.9	4.3	1.635	1.875	
Sillimanite	Al_2OSiO_4	rhomb	3.25	7	1.658	1.660	1.660
Skutterudite	$(Co,Ni)As_3$	cub	6.8	5.8			
Smithsonite	$ZnCO_3$	rhomb	4.4	4.3	1.621	1.848	
Sodalite	$Na_8Al_6Si_6O_{24}Cl_2$	cub	2.30	5.8	1.485		
Sperrylite	$PtAs_2$	cub	10.58	6.5			
Spessartite	$Mn_3Al_2Si_3O_{12}$	cub	4.19	6.8	1.800		
Sphalerite	ZnS	cub	4.0	3.8	2.369		
Sphene	$CaTiSiO_4(O,OH,F)$	monocl	3.50	5	1.90	1.95	2.03
Spinel	$MgAl_2O_4$	cub	3.55	7.8	1.719		
Spodumene	$LiAlSi_2O_6$	monocl	3.13	6.8	1.656	1.662	1.671
Stannite	Cu_2FeSn_4	tetr	4.4	4			
Staurolite	$(Fe,Mg,Zn)_2(Al,Fe,Ti)_9O_6$ $[(Si,Al)O]_4(O,OH)_2$	monocl	3.79	7.5	1.743	1.747	1.755
Stercorite	$Na(NH_4)H(PO_4)\cdot4H_2O$	tricl	1.62	2	1.439	1.442	1.469
Stibiotantalite	$Sb(Ta,Nb)O_4$	rhomb	6.6	5.5	2.38	2.41	2.46
Stibnite	Sb_2S_3	orth	4.56	2			
Stilbite	$NaCa_2[Al_5Si_{13}O_{36}]\cdot14H_2O$	monocl	2.2	3.8	1.492	1.499	1.503
Stilpnomelane	$(K,Na,Ca)_{0.6}(Fe,Mg)_6Si_8Al$						

Name	Formula	Crystal system	Density g/cm³	Hard-ness	Index of refraction n_α	n_β	n_γ
	$(O,OH)_{27} \cdot 2H_2O$	monocl	2.8	3.5	1.585	1.665	1.665
Stolzite	$PbWO_4$	tetr	8.2	2.8	2.19	2.27	
Strengite	$FePO_4 \cdot 2H_2O$	orth	2.87	4	1.707	1.719	1.741
Strontianite	$SrCO_3$	orth	3.5	3.5	1.518	1.666	1.668
Struvite	$Mg(NH_4)(PO_4) \cdot 6H_2O$	rhomb	1.71	2	1.495	1.496	1.504
Sulfur	S	orth	2.07	2	1.958	2.038	2.245
Sylvanite	$(Ag,Au)Te_2$	monocl	8.16	1.8			
Sylvite	KCl	cub	1.99	2	1.490		
Talc	$Mg_3Si_4O_{10}(OH)_2$	monocl	2.71	1	1.545	1.592	1.595
Tantalite	$(Fe,Mn)(Ta,Nb)_2O_6$	rhomb	7.95	6.5	2.26	2.32	2.43
Tapiolite	$FeTa_2O_6$	tetr	7.9	6.3	2.27	2.42	
Tellurobismuthite	Bi_2Te_3	hex	7.74	1.8			
Terlinguaite	Hg_2OCl	monocl	8.73	2.5	2.35	2.64	2.66
Tetrahedrite	$(Cu,Fe)_{12}Sb_4S_{13}$	cub	4.9	3.8			
Thenardite	Na_2SO_4	orth	2.7	2.8	1.468	1.475	1.483
Thermonatrite	$Na_2CO_3 \cdot H_2O$	orth	2.25	1.3	1.420	1.506	1.524
Thomsenolite	$NaCaAlF_6 \cdot H_2O$	monocl	2.98	2	1.407	1.414	1.415
Thorianite	ThO_2	cub	10.0	6.5	2.200		
Thorite	$ThSiO_4$	tetr	6.7	4.8	1.8		
Topaz	$Al_2SiO_3(OH,F)_2$	rhomb	3.53	8	1.618	1.620	1.627
Torbernite	$Cu(UO_2)_2(PO_4)_2 \cdot 8H_2O$	tetr	3.22	2.3	1.582	1.592	
Tourmaline	$Na(Mg,Fe,Mn,Li,Al)_3Al_6Si_6O_{18}$ $(BO_3)_3$	rhomb	3.14	7	1.62	1.65	
Tremolite	$Ca_2Mg_5Si_8O_{22}(OH,F)_2$	monocl	3.0	5.5	1.599	1.612	1.622
Trevorite	$NiFe_2O_4$	cub	5.33	7.8	2.3		
Tridymite	SiO_2	hex	2.27	7	1.475	1.476	1.479
Triphyllite-Lithiophyllite	$Li(Fe,Mn)PO_4$	rhomb	3.46	4.5	1.68	1.68	1.69
Troegerite	$(UO_2)_3(AsO_4)_2 \cdot 12H_2O$	tetr		2.5	1.59	1.630	
Troilite	FeS	hex	4.7	4			
Trona	$Na_3H(CO_3)_2 \cdot 2H_2O$	monocl	2.14	2.8	1.412	1.492	1.540
Turquois	$Cu(Al,Fe)_6(PO_4)_4(OH)_8 \cdot 4H_2O$	tricl	2.9	5.3	1.70	1.73	1.75
Ullmannite	$NiSbS$	cub	6.65	5.3			
Uraninite	UO_2	cub	11.0	5.5			
Uvarovite	$Ca_3Cr_2Si_3O_{12}$	cub	3.83	6.8	1.865		
Valentinite	Sb_2O_3	orth	5.7	2.8	2.18	2.35	2.35
Vanadinite	$Pb_5(VO_4)_3Cl$	hex	6.8	2.9	2.350	2.416	
Variseite-Strengite	$(Al,Fe)(PO_4) \cdot 2H_2O$	rhomb	2.72	4	1.635	1.654	1.668
Vaterite	$CaCO_3$	hex	2.71		1.550	1.645	
Vermiculite	$(Mg,Ca)_{0.7}(Mg,Fe,Al)_6[(Al,Si)_8O_{20}]$ $(OH)_4 \cdot 8H_2O$	monocl	2.3	1.5	1.542	1.556	1.556
Vesuvianite	$Ca_{10}(Mg,Fe)_2Al_4(Si_2O_7)_2(SiO_4)_5$ $(OH,F)_4$	tetr	3.33	6.5	1.72	1.73	
Villiaumite	NaF	cub	2.78	2.3	1.327		
Vivianite	$Fe_3(PO_4)_2 \cdot 8H_2O$	monocl	2.58	1.8	1.598	1.629	1.652
Wagnerite	$Mg_2(PO_4)F$	monocl	3.15	5.3	1.568	1.572	1.582
Wavellite	$Al_3(OH)_3(PO_4)_2 \cdot 5H_2O$	rhomb	2.36	3.6	1.527	1.535	1.553
Whewellite	$CaC_2O_4 \cdot H_2O$	cub	2.2	2.8	1.491	1.554	1.650
Willemite	Zn_2SiO_4	hex	4.1	5.5	1.691	1.719	
Witherite	$BaCO_3$	orth	4.29	3.5	1.529	1.676	1.677
Wolframite	$(Fe,Mn)WO_4$	monocl	7.3	4.3	2.26	2.32	2.42
Wollastonite	$CaSiO_3$	monocl	2.92	4.8	1.628	1.639	1.642
Wulfenite	$PbMoO_4$	tetr	6.7	2.9	2.283	2.403	
Wurtzite	ZnS	hex	4.09	3.8	2.356	2.378	
Xenotime	YPO_4	tetr	4.8	4.5	1.721	1.816	
Zeunerite	$Cu(UO_2)_2(AsO_4)_2 \cdot 10H_2O$	tetr			1.606		
Zincite	ZnO	hex	5.6	4	2.013	2.029	
Zircon	$ZrSiO_4$	tetr	4.6	7.5	1.94	1.99	
Zoisite	$Ca_2Al_3Si_3O_{12}(OH)$	rhomb	3.26	6	1.695	1.699	1.711

CRYSTALLOGRAPHIC DATA ON MINERALS

This table contains x-ray crystallographic data on about 400 common minerals, as well as selected crystalline elements. Entries are arranged alphabetically by mineral name. The columns are:

Name: Common name of the mineral.
Formula: Chemical formula for a typical sample of the mineral. Composition often varies considerably with the origin of the sample.
Crystal system: tricl = triclinic; monocl = monoclinic; orth = orthorhombic; tetr = tetragonal; hex = hexagonal; rhomb = rhombohedral; cubic = cubic.
Structure type: Prototype for the structural arrangement of the crystallographic cell.
Z: Number of formula units per the unit cell.
a, b, c: Lengths of the cell edges in Å (1Å = 10^{-8} cm).
α, β, γ : Angles between cell axes.

REFERENCES

1. Robie, R.A., Bethke, P.M., and Beardsley, K.M., *U. S. Geological Survey Bulletin 1248*, U. S. Government Printing Office, Washington, D.C.
2. Donnay, J.D.H., and Ondik, H.M., *Crystal Data Determinative Tables, Third Edition, Volume 2, Inorganic Compounds*, Joint Committee on Powder Diffraction Standards, Swarthmore, PA, 1973.
3. Deer, W.A., Howie, R.A., and Zussman, J., *An Introduction to the Rock-Forming Minerals*, *2nd Edition*, Longman Scientific & Technical, Harlow, Essex, 1992.

Name	Formula	Crystal system	Structure type	Z	a/Å	b/Å	c/Å	α	β	γ
Acanthite	Ag_2S	monocl		4	4.228	6.928	7.862		99.58°	
Acmite (Aegirine)	$NaFe(SiO_3)_2$	monocl	diopside	4	9.658	8.795	5.294		107.42°	
Akermanite	$Ca_2MgSi_2O_7$	tetr	melilite	2	7.8435		5.010			
Alabandite	MnS	cubic	rock salt	4	5.223					
Almandine (Almandite)	$Fe_3Al_2Si_3O_{12}$	cubic	garnet	8	11.526					
Altaite	$PbTe$	cubic	rock salt	4	6.4606					
Aluminum	Al	cubic	copper	4	4.049					
Alunite	$KAl_3(SO_4)_2(OH)_6$	rhomb		3	6.982		17.32			
Analcite	$NaAlSi_2O_6 \cdot H_2O$	cubic		16	13.733					
Anatase	TiO_2	tetr		4	3.785		9.514			
Andalusite	Al_2OSiO_4	orth		4	7.7959	7.8983	5.5583			
Andradite	$Ca_3Fe_2Si_3O_{12}$	cubic	garnet	8	12.048					
Anglesite	$PbSO_4$	orth	barite	4	8.480	5.398	6.958			
Anhydrite	$CaSO_4$	orth	anhydrite	4	6.991	6.996	6.238			
Annite	$KFe_3[AlSi_3O_{10}](OH)_2$	monocl	1M mica	2	10.29	9.33	5.39		105.1°	
Anorthite	$CaAl_2Si_2O_8$	tricl	primitive cell	8	8.177	12.877	14.169	93.17°	115.85°	91.22°
Anthophyllite	$Mg_7Si_8O_{22}(OH)_2$	orth		4	18.61	18.01	5.24			
Antimony	Sb	rhomb	arsenic	6	4.2996		11.2516			
Aragonite	$CaCO_3$	orth	aragonite	4	5.741	7.968	4.959			
Arcanite	K_2SO_4	orth	arcanite	4	5.772	10.072	7.483			
Argentite	Ag_2S	cubic		2	4.870					
Argentopyrite	$AgFe_2S_3$	orth		4	6.64	11.47	6.45			
Arsenic	As	rhomb	arsenic	6	3.760		10.555			
Arsenolite	As_2O_3	cubic	diamond	16	11.074					
Arsenopyrite	$FeAsS$	tricl		4	5.760	5.690	5.785	90.00°	112.23°	90.00°
Azurite	$Cu_3(OH)_2(CO_3)_2$	monocl		2	5.008	5.844	10.336		92.45°	
Baddeleyite	ZrO_2	monocl	baddeleyite	4	5.1454	5.2075	5.3107		99.23°	
Banalsite	$BaNa_2Al_4Si_4O_{16}$	orth		4	8.50	9.97	16.72			
Barite	$BaSO_4$	orth	barite	4	8.878	5.450	7.152			
Berlinite	$AlPO_4$	hex	α-quartz	3	4.942		10.97			
Beryl	$Be_3Al_2(SiO_3)_6$	hex	beryl	2	9.215		9.192			
Berzelianite	Cu_2Se	cubic		4	5.85					
Bismite	Bi_2O_3	monocl	pseudo-orth	4	7.48	8.14	5.83		112.9°	
Bismuth	Bi	rhomb	arsenic	6	4.5367		11.8383			
Bismuthinite	Bi_2S_3	orth	stibnite	4	11.150	11.300	3.981			
Bixbyite	Mn_2O_3	cubic	thallium trioxide	16	9.411					
Boehmite	$AlO(OH)$	orth	lepidocrocite	4	2.868	12.227	3.700			
Borax	$Na_2B_4O_7 \cdot 10H_2O$	monocl		4	11.858	10.674	12.197		106.68°	
Bornite (metastable)	Cu_5FeS_4	cubic		8	10.94					
Breithauptite	$NiSb$	hex	niccolite	2	3.942		5.155			
Brochantite	$Cu_4SO_4(OH)_6$	monocl		4	13.066	9.85	6.022		103.27°	
Bromargyrite	$AgBr$	cubic	rock salt	4	5.7745					
Bromellite	BeO	hex	zincite	2	2.6979		4.3772			
Brookite	TiO_2	orth		8	5.456	9.182	5.143			
Brucite	$Mg(OH)_2$	hex	cadmium iodide	1	3.147		4.769			

Name	Formula	Crystal system	Structure type	Z	a/Å	b/Å	c/Å	α	β	γ
Bunsenite	NiO	cubic	rock salt	4	4.177					
Bustamite	$CaMn(SiO_3)_2$	tricl		6	7.736	7.157	13.824	90.52°	94.58°	103.87°
Cadmium telluride	$CdTe$	cubic	sphalerite	4	6.4805					
Cadmoselite	$CdSe$	hex	zincite	2	4.2977		7.0021			
Calcite	$CaCO_3$	rhomb	calcite	6	4.9899		17.064			
Calomel	Hg_2Cl_2	tetr		4	4.478		10.910			
Carbonate-apatite	$Ca_{10}(PO_4)_6CO_3 \cdot H_2O$	hex	apatite	1	9.436		6.883			
Cassiterite	SnO_2	tetr	rutile	2	4.738		3.188			
Cattierite	CoS_2	cubic	pyrite	4	5.5345					
Celestite	$SrSO_4$	orth	barite	4	8.359	5.352	6.866			
Celsian	$BaAl_2Si_2O_8$	monocl		8	8.627	13.045	14.408		115.20°	
Cerianite	CeO_2	cubic	fluorite	4	5.4110					
Cerussite	$PbCO_3$	orth	aragonite	4	6.152	8.436	5.195			
Cervantite	Sb_2O_4	orth		4	5.424	11.76	4.804			
Chalcanthite	$CuSO_4 \cdot 5H_2O$	tricl		2	6.1045	10.72	5.949	97.57°	107.28°	77.43°
Chalcocite	Cu_2S	orth		96	11.881	27.323	13.491			
Chalcopyrite	$CuFeS_2$	tetr		2	5.2988		10.434			
Chlorapatite	$Ca_5(PO_4)_3Cl$	hex	apatite	2	9.629		6.777			
Chlorargyrite	$AgCl$	cubic	rock salt	4	5.5491					
Chloritoid	$FeAl_4O_2(SiO_4)_2(OH)_4$	monocl		8	9.48	5.48	18.18		101.77°	
Chloromagnesite	$MgCl_2$	rhomb		3	3.632		17.795			
Chondrodite	$2Mg_2SiO_4 \cdot MgF_2$	monocl		2	7.89	4.743	10.29		109.03°	
Chrysoberyl	$BeAl_2O_4$	orth	olivine	4	5.4756	9.4041	4.4267			
Cinnabar	HgS	hex	cinnabar	3	4.149		9.495			
Claudetite	As_2O_3	monocl		4	5.339	12.984	4.5405		94.27°	
Clausthalite	$PbSe$	cubic	rock salt	4	6.1255					
Clinoenstatite	$MgSiO_3$	monocl		8	9.620	8.825	5.188		108.33°	
Clinoferrosilite	$FeSiO_3$	monocl		8	9.7085	9.0872	5.2284		108.43°	
Clinohumite	$4Mg_2SiO_4 \cdot MgF_2$	monocl		2	13.68	4.75	10.27		100.83°	
Clinozoisite	$Ca_2Al_3(SiO_4)_3OH$	monocl		2	8.887	5.581	10.14		115.93°	
Cobalt olivine	Co_2SiO_4	orth	olivine	4	4.782	10.301	6.003			
Cobalt oxide	CoO	cubic	rock salt	4	4.260					
Cobalt sulfide	CoS	cubic	sphalerite	4	5.339					
Cobalt titanate	$CoTiO_3$	rhomb	ilmenite	6	5.066		13.918			
Cobaltcalcite	$CoCO_3$	rhomb	calcite	6	4.6581		14.958			
Cobaltite	$CoAsS$	cubic	NiSbS	4	5.60					
Coesite	SiO_2	monocl		16	7.152	12.379	7.152		120.00°	
Coffinite	$USiO_4$	tetr	zircon	4	6.995		6.263			
Colemanite	$Ca_2B_6O_{11} \cdot 5H_2O$	monocl		4	8.743	11.264	6.102		110.12°	
Coloradoite	$HgTe$	cubic	sphalerite	4	5.4600					
Cooperite	PtS	tetr		2	3.4699		6.1098			
Copper	Cu	cubic	face-centered cubic	4	3.6150					
Corundum	Al_2O_3	rhomb	corundum	6	4.7591		12.9894			
Cotunnite	$PbCl_2$	orth		4	4.535	7.62	9.05			
Covellite	CuS	hex		6	3.792		16.34			

Name	Formula	Crystal system	Structure type	Z	a/Å	b/Å	c/Å	α	β	γ
Cristobalite (α)	SiO_2	tetr		4	4.971		6.918			
Cristobalite (β)	SiO_2	cubic		8	7.1382					
Cryolite	Na_3AlF_6	monocl		2	5.40	5.60	7.776		90.18°	
Cubanite	$CuFe_2S_3$	orth		4	6.46	11.12	6.23			
Cummingtonite	$(Mg,Fe,Mn)_7(Si_4O_{11})_2(OH)_2$	monocl	tremolite	2	9.522	18.223	5.332		101.92°	
Cuprite	Cu_2O	cubic		2	4.2696					
Danburite	$CaB_2Si_2O_8$	orth		4	8.04	8.77	7.74			
Datolite	$CaBSiO_4(OH)$	monocl		4	9.62	7.60	4.84		90.15°	
Daubreeite	$FeCr_2S_4$	cubic	spinel	8	9.966					
Diamond	C	cubic	diamond	8	3.5670					
Diaspore	$AlO(OH)$	orth		4	4.401	9.421	2.845			
Dickite	$Al_2Si_2O_5(OH)_4$	monocl		4	5.150	8.940	14.736		103.58°	
Digenite	$Cu_{1.79}S$	cubic	deformed fluorite	4	5.5695					
Diopside	$CaMg(SiO_3)_2$	monocl	diopside	4	9.743	8.923	5.251		105.93°	
Dioptase	$CuSiO_2(OH)_2$	rhomb	phenacite	18	14.61		7.80			
Dolerophanite	$Cu_2O(SO_4)$	monocl		4	8.334	6.312	7.628		108.4°	
Dolomite	$CaMg(CO_3)_2$	rhomb	calcite	3	4.8079		16.010			
Dravite	$NaMg_3Al_6B_3Si_6O_{27}(OH)_4$	rhomb	tourmaline	3	15.942		7.224			
Elbaite	$NaLiAl_{1.67}B_3Si_6O_{27}(OH)_4$	rhomb	tourmaline	3	15.842		7.009			
Enargite	Cu_3AsS_4	orth		4	6.426	7.422	6.144			
Enstatite	$MgSiO_3$	orth		16	8.829	18.22	5.192			
Epidote	$Ca_2Al_2(Al,Fe)OH(SiO_4)_3$	monocl		2	8.89	5.63	10.19		115.40°	
Epsomite	$MgSO_4 \cdot 7H_2O$	orth		4	11.86	11.99	6.858			
Eskolaite	Cr_2O_3	rhomb	corundum	6	4.9607		13.599			
Eucairite	$AgCuSe$	orth		10	4.105	20.35	6.31			
Euclase	$AlBeSiO_4(OH)$	monocl		4	4.763	14.29	4.618		100.25°	
Famatinite	Cu_3SbS_4	tetr		2	5.384		10.770			
Fayalite	Fe_2SiO_4	orth	olivine	4	4.817	10.477	6.105			
Fe-Cordierite	$Fe_2Al_3(AlSi_5O_{18})$	orth	cordierite	4	9.726	17.065	9.287			
Fe-Gehlenite	$Ca_2Fe_2SiO_7$	tetr.	melilite	2	7.54		4.855			
Fe-Indialite	$Fe_2Al_3(AlSi_5O_{18})$	hex	beryl	2	9.860		9.285			
Fe-Leucite	$KFeSi_2O_6$	tetr		16	13.205		13.970			
Fe-Microcline	$KFeSi_3O_8$	tricl		4	8.68	13.10	7.340	90.75°	116.05°	86.23°
Fe-Sanidine	$KFeSi_3O_8$	monocl		4	8.689	13.12	7.319		116.10°	
Fe-Skutterudite	$FeAs_{2.95}$	cubic		8	8.1814					
Ferberite	$FeWO_4$	monocl	wolframite	2	4.732	5.708	4.965		90.00°	
Ferriannite	$KFe_3[FeSi_3O_{10}](OH)_2$	monocl		2	5.430	9.404	10.341		100.07°	
Ferroselite	$FeSe_2$	orth	marcasite	2	4.801	5.778	3.587			
Ferrotremolite	$Ca_2Fe_5[Si_8O_{22}](OH)_2$	monocl	tremolite	2	9.97	18.34	5.30		104.50°	
Fluor-edenite	$NaCa_2Mg_5[AlSi_7O_{22}]F_2$	monocl	tremolite	2	9.847	18.00	5.282		104.83°	
Fluor-humite	$3Mg_2SiO_4 \cdot MgF_2$	orth		4	10.243	20.72	4.735			
Fluor-norbergite	$Mg_2SiO_4 \cdot MgF_2$	orth		4	8.727	10.271	4.709			
Fluor-phlogopite	$KMg_3[AlSi_3O_{10}]F_2$	monocl	1M mica	2	5.299	9.188	10.135		99.92°	
Fluor-richterite	$Na_2CaMg_5[Si_8O_{22}]F_2$	monocl	tremolite	2	9.823	17.96	5.268		104.33°	
Fluor-tremolite	$Ca_2Mg_5[Si_8O_{22}]F_2$	monocl	tremolite	2	9.781	18.01	5.267		104.52°	

Name	Formula	Crystal system	Structure type	Z	a/Å	b/Å	c/Å	α	β	γ
Fluorapatite	$Ca_5(PO_4)_3F$	hex	apatite	2	9.3684		6.8841			
Fluorite	CaF_2	cubic	fluorite	4	5.4638					
Forsterite	Mg_2SiO_4	orth	olivine	4	4.758	10.214	5.984			
Frohbergite	$FeTe_2$	orth	marcasite	2	5.265	6.265	3.869			
Gahnite	$ZnAl_2O_4$	cubic	spinel	8	8.0848					
Galaxite	$MnAl_2O_4$	cubic	spinel	8	8.258					
Galena	PbS	cubic	rock salt	4	5.9360					
Gallium oxide	Ga_2O_3	rhomb	corundum	6	4.9793		13.429			
Gehlenite	$Ca_2Al_2SiO_7$	tetr	melilite	2	7.690		5.0675			
Geikielite	$MgTiO_3$	rhomb	ilmenite	6	5.054		13.898			
Gerhardite	$Cu_2(NO_3)(OH)_3$	orth		4	6.075	13.812	5.592			
Gersdorffite	$NiAsS$	cubic		4	5.693					
Gibbsite	$Al(OH)_3$	monocl		8	9.719	5.0705	8.6412		94.57°	
Glauchroite	$CaMnSiO_4$	orth	olivine	4	4.944	11.19	6.529			
Glaucodot	$(Co,Fe)AsS$	orth		24	6.64	28.39	5.64			
Glaucophane I	$Na_2Mg_3Al_2[Si_8O_{22}](OH)_2$	monocl	tremolite	2	9.748	17.915	5.273		102.78°	
Glaucophane II	$Na_2Mg_3Al_2[Si_8O_{22}](OH)_2$	monocl	tremolite	2	9.663	17.696	5.277		103.67°	
Goethite	$FeO(OH)$	orth		4	4.596	9.957	3.021			
Gold	Au	cubic	face-centered cubic	4	4.0786					
Goldmanite	$Ca_3V_2Si_3O_{12}$	cubic	garnet	8	12.070					
Goslarite	$ZnSO_4 \cdot 7H_2O$	orth	epsomite	4	11.779	12.050	6.822			
Graphite	C	hex	graphite	4	2.4612		6.7079			
Greenockite	CdS	hex	zincite	2	4.1354		6.7120			
Greigite	Fe_3S_4	cubic	spinel	8	9.876					
Grossularite	$Ca_3Al_2Si_3O_{12}$	cubic	garnet	8	11.851					
Grunerite	$Fe_7[Si_8O_{22}](OH)_2$	monocl	tremolite	2	9.572	18.44	5.342		101.77°	
Gudmundite	$FeSbS$	monocl		8	10.00	5.93	6.73		90.00°	
Gypsum	$CaSO_4 \cdot 2H_2O$	monocl		8	5.68	15.18	6.29		113.83°	
Hafnia	HfO_2	monocl	baddeleyite	4	5.1156	5.1722	5.2948		99.18°	
Halite	$NaCl$	cubic	rock salt	4	5.6402					
Hambergite	$Be_2(OH,F)BO_3$	orth		8	9.755	12.201	4.426			
Hardystonite	$Ca_2ZnSi_2O_7$	tetr	melilite	2	7.87		5.01			
Hauerite	MnS_2	cubic	pyrite	4	6.1014					
Hausmannite	Mn_3O_4	tetr		8	8.136		9.422			
Hawleyite	CdS	cubic	sphalerite	4	5.833					
Heazelwoodite	Ni_3S_2	rhomb		3	5.746		7.134			
Hedenbergite	$CaFe(SiO_3)_2$	monocl	diopside	4	9.854	9.024	5.263		104.23°	
Hematite	Fe_2O_3	rhomb	corundum	6	5.0329		13.749			
Hemimorphite	$Zn_4(OH)_2Si_2O_7 \cdot H_2O$	orth		2	8.370	10.719	5.120			
Hercynite	$Fe(AlO_2)_2$	cubic	spinel	8	8.150					
Herzenbergite	SnS	orth	germanium sulfide	4	4.328	11.190	3.978			
Hessite	Ag_2Te	monocl		4	8.13	4.48	8.09		111.9°	
Hexahydrite	$MgSO_4 \cdot 6H_2O$	monocl		8	10.110	7.212	24.41		98.30°	
High albite (Analbite)	$NaAlSi_3O_8$	tricl		4	8.160	12.870	7.106	93.54°	116.36°	90.19°
High argentite	Ag_2S	cubic		4	6.269					

Name	Formula	Crystal system	Structure type	Z	a/Å	b/Å	c/Å	α	β	γ
High bornite	Cu_5FeS_4	cubic		1	5.50					
High carnegeite	$NaAlSiO_4$	cubic		4	7.325					
High chalcocite	Cu_2S	hex		2	3.961		6.722			
High clinoenstatite	$MgSiO_3$	tricl		8	10.000	8.934	5.170	88.27°	70.03°	91.01°
High digenite	Cu_2S	cubic		4	5.725					
High germania	GeO_2	hex	α–quartz	3	4.987		5.652			
High leucite	$KAlSi_2O_6$	cubic		16	13.43					
High naumanite	Ag_2Se	cubic		2	4.993					
High sanidine	$KAlSi_3O_8$	monocl		4	8.615	13.031	7.177		115.98°	
Huebnerite	$MnWO_4$	monocl	wolframite	2	4.834	5.758	4.999		91.18°	
Huntite	$Mg_3Ca(CO_3)_4$	rhomb	calcite	3	9.498		7.816			
Hydroxylapatite	$Ca_5(PO_4)_3OH$	hex	apatite	2	9.418		6.883			
Ice	H_2O	hex		4	4.5212		7.3666			
Ilmenite	$FeTiO_3$	rhomb	ilmenite	6	5.093		14.055			
Indialite (Cordierite)	$Mg_2Al_3(AlSi_5O_{18})$	hex	beryl	2	9.7698		9.3517			
Iodargyrite	AgI	hex	zincite	2	4.5955		7.5005			
Iron (α)	Fe	cubic	body-centered cubic	2	2.8664					
Jacobsite	$MnFe_2O_4$	cubic	spinel	8	8.499					
Jadeite	$NaAl(SiO_3)_2$	monocl	diopside	4	9.409	8.564	5.220		107.50°	
Jalpaite	$Ag_{1.55}Cu_{0.45}S$	tetr		16	8.673		11.756			
Johannsenite	$CaMn(SiO_3)_2$	monocl	diopside	4	9.83	9.04	5.27		105.00°	
Kaliophilite	$KAlSiO_4$	hex		54	26.930		8.522			
Kalsilite	$KAlSiO_4$	hex		2	5.1597		8.7032			
Kaolinite	$Al_2Si_2O_5(OH)_4$	tricl		2	5.155	8.959	7.407	91.68°	104.87°	89.93°
Karelianite	V_2O_3	rhomb	corundum	6	4.952		14.002			
Keatite	SiO_2	tetr		12	7.456		8.604			
Kernite	$Na_2B_4O_7 \cdot 4H_2O$	monocl		4	7.022	9.151	15.676		108.83°	
Kerschsteinite	$CaFeSiO_4$	orth	olivine	4	4.886	11.146	6.434			
Klockmannite	$CuSe$	hex	deformed covellite	78	14.206		17.25			
Knebelite	$MnFeSiO_4$	orth	olivine	4	4.854	10.602	6.162			
Kyanite	Al_2OSiO_4	tricl		4	7.123	7.848	5.564	89.92°	101.25°	105.97°
Larnite	Ca_2SiO_4	monocl		4	5.48	6.76	9.28		94.55°	
Laurite	RuS_2	cubic	pyrite	4	5.60					
Lawrencite	$FeCl_2$	rhomb		3	3.593		17.58			
Lawsonite	$CaAl_2Si_2O_7(OH)_2 \cdot H_2O$	orth		4	8.787	5.836	13.123			
Lead	Pb	cubic	face-centered cubic	4	4.9505					
Leonhardtite	$MgSO_4 \cdot 4H_2O$	monocl		4	5.922	13.604	7.905		90.85°	
Lepidocrocite	$FeO(OH)$	orth		4	3.868	12.525	3.066			
Lepidolite	$K_2Al_3Li_2AlSi_7O_{20}(OH)_4$	monocl	2M2 mica	2	9.2	5.3	20.0		98.00°	
Leucite	$KAlSi_2O_6$	tetr		16	13.074		13.738			
Lime	CaO	cubic	rock salt	4	4.8108					
Lime olivine	Ca_2SiO_4	orth	olivine	4	5.091	11.371	6.782			
Linnaeite	Co_3S_4	cubic	spinel	8	9.401					
Litharge	PbO	tetr		2	3.9759		5.023			
Loellingite	$FeAs_2$	orth	marcasite	2	5.300	5.981	2.882			

CRYSTALLOGRAPHIC DATA ON MINERALS (continued)

Name	Formula	Crystal system	Structure type	Z	a/Å	b/Å	c/Å	α	β	γ
Low albite	$NaAlSi_3O_8$	tricl		4	8.139	12.788	7.160	94.27°	116.57°	87.68°
Low bornite	Cu_5FeS_4	tetr		16	10.94		21.88			
Low cordierite	$Mg_2Al_3(AlSi_5O_{18})$	orth		4	9.721	17.062	9.339			
Low germania	GeO_2	tetr	rutile	2	4.3963		2.8626			
Low nepheline	$NaAlSiO_4$	hex		8	9.986		8.330			
Luzonite	Cu_3AsS_4	tetr		2	5.289		10.440			
Mackinawite	FeS	tetr		2	3.675		5.030			
Magnesioriebeckite	$Na_2Mg_3Fe_2[Si_8O_{22}](OH)_2$	monocl	tremolite	2	9.733	17.946	5.299		103.30°	
Magnesite	$MgCO_3$	rhomb	calcite	6	4.6330		15.016			
Magnetite	Fe_3O_4	cubic	spinel	8	8.3940					
Malachite	$Cu_2(OH)_2CO_3$	monocl		4	9.502	11.974	3.240		98.75°	
Maldonite	Au_2Bi	cubic		8	7.958					
Manganese sulfide (γ)	MnS	hex	zincite	2	3.976		6.432			
Manganese sulfide (β)	MnS	cubic	sphalerite	4	5.611					
Manganosite	MnO	cubic	rock salt	4	4.4448					
Marcasite	FeS_2	orth	marcasite	2	4.443	5.423	3.3876			
Margarite	$CaAl_2[AlSi_2O_{10}](OH)_2$	monocl	2M mica	4	5.13	8.92	19.50		95.00°	
Marialite	$Na_4Al_3Si_9O_{24}Cl$	tetr		2	12.064		7.514			
Marshite	CuI	cubic	sphalerite	4	6.0507					
Mascagnite	$(NH_4)_2SO_4$	orth	arcanite	4	7.782	5.993	10.636			
Massicot	PbO	orth		4	5.489	4.755	5.891			
Matlockite	$PbClF$	tetr		2	4.106		7.23			
Maucherite	$Ni_{11}As_8$	tetr		4	6.870		21.81			
Meionite	$Ca_4Al_6Si_6O_{24}CO_3$	tetr		2	12.174		7.652			
Melanophlogite	SiO_2	cubic	clathrate type	46	13.402					
Melanterite	$FeSO_4 \cdot 7H_2O$	monocl		4	14.072	6.503	11.041		105.57°	
Melonite	$NiTe_2$	hex	cadmium iodide	1	3.869		5.308			
Metacinnabar	HgS	cubic	sphalerite	4	5.8517					
Miargyrite	$AgSbS_2$	monocl		8	12.862	4.111	13.220		98.63°	
Microcline	$KAlSi_3O_8$	tricl		4	8.582	12.964	7.222	90.62°	115.92°	87.68°
Miersite	AgI	cubic	sphalerite	4	6.4963					
Millerite	NiS	rhomb		9	9.616		3.152			
Minium	Pb_3O_4	tetr		4	8.815		6.565			
Minnesotaite	$Fe_3Si_4O_{10}(OH)_2$	monocl		4	5.4	9.42	19.4		100.00°	
Mirabilite	$Na_2SO_4 \cdot 10H_2O$	monocl		4	11.51	10.38	12.83		107.75°	
Mn-Indialite	$Mn_2Al_3(AlSi_5O_{18})$	hex	beryl	2	9.925		9.297			
Molybdenite	MoS_2	hex	molybdenite	2	3.1604		12.295			
Molybdenum	Mo	cubic		2	3.1653					
Molybdite	MoO_3	orth		4	3.962	13.858	3.697			
Monteponite	CdO	cubic	rock salt	4	4.6953					
Monticellite	$CaMgSiO_4$	orth	olivine	4	4.827	11.084	6.376			
Montroydite	HgO	orth		4	6.608	5.518	3.519			
Mullite (2:1)	$2Al_2O_3 \cdot SiO_2$	orth		6	7.5788	7.6909	2.8883			
Mullite (3:2)	$3Al_2O_3 \cdot 2SiO_2$	orth		3	7.557	7.6876	2.8842			
Muscovite	$KAl_2AlSi_3O_{10}(OH)_2$	monocl	2M2 mica	4	5.203	8.995	20.030		94.47°	

Name	Formula	Crystal system	Structure type	Z	a/Å	b/Å	c/Å	α	β	γ
Nacrite	$Al_2Si_2O_5(OH)_4$	monocl		4	8.909	5.146	15.697		113.70°	
Nantokite	$CuCl$	cubic	sphalerite	4	5.416					
Natroalunite	$NaAl_3(SO_4)_2(OH)_6$	rhomb		3	6.974		16.69			
Natrolite	$Na_2Al_2Si_3O_{10}\cdot2H_2O$	orth		8	18.30	18.63	6.60			
Neighborite	$NaMgF_3$	orth	perovskite	4	5.363	7.676	5.503			
Ni-Skutterudite	$NiAs_{2.95}$	cubic		8	8.3300					
Niccolite	$NiAs$	hex	niccolite	2	3.618		5.034			
Nickel	Ni	cubic	face-centered cubic	4	3.5238					
Nickel carbonate	$NiCO_3$	rhomb	calcite	6	4.5975		14.723			
Nickel olivine	Ni_2SiO_4	orth	olivine	4	4.727	10.121	5.915			
Nickel selenide	$NiSe_2$	cubic	pyrite	4	5.9604					
Niter	KNO_3	orth	aragonite	4	6.431	9.164	5.414			
Norsethite	$BaMg(CO_3)_2$	rhomb	calcite	3	5.020		16.75			
Oldhamite	CaS	cubic	rock salt	4	5.689					
Orpiment	As_2S_3	monocl		4	11.49	9.59	4.25		90.45°	
Orthoclase	$KAlSi_3O_8$	monocl		4	8.562	12.996	7.193		116.02°	
Orthoferrosilite	$FeSiO_3$	orth	enstatite	16	9.080	18.431	5.238			
Otavite	$CdCO_3$	rhomb	calcite	6	4.9204		16.298			
Paracelsian	$BaAl_2Si_2O_8$	monocl		4	8.58	9.583	9.08		90.00°	
Paragonite	$NaAl_2AlSi_3O_{10}(OH)_2$	monocl	2M1 mica	4	5.13	8.89	19.32		95.17°	
Pararammelsbergite	$NiAs_2$	orth		8	5.75	5.82	11.428			
Paratellurite	TeO_2	tetr		4	4.810		7.613			
Parawollastonite	$CaSiO_3$	monocl		12	15.417	7.321	7.066		95.40°	
Pectolite	$Ca_2NaH(SiO_3)_3$	tricl		2	7.99	7.04	7.02	90.05°	95.27°	102.47°
Pentlandite	$Fe_{5.25}Ni_{3.75}S_8$	cubic		4	10.196					
Pentlandite	$Fe_{4.75}Ni_{5.25}S_8$	cubic		4	10.095					
Periclase	MgO	cubic	rock salt	4	4.2117	7.6438	5.4439			
Perovskite	$CaTiO_3$	orth	perovskite	4	5.3670	5.14	7.62		105.90°	
Petalite	$LiAlSi_4O_{10}$	monocl		2	11.32					
Petzite	Ag_3AuTe_2	cubic		8	10.38					
Phenacite	Be_2SiO_4	rhomb	phenacite	18	12.472	9.210	8.252			
Phlogopite	$KMg_3AlSi_3O_{10}(OH)_2$	monocl	1M mica	2	5.326		10.311		100.17°	
Picrochromite	$MgCr_2O_4$	cubic	spinel	8	8.333	5.70	9.41		115.70°	
Piemontite	$Ca_2Al_{1.5}Mn_{1.5}(SiO_4)_3OH$	monocl		2	8.95					
Platinum	Pt	cubic	face-centered cubic	4	3.9231					
Polymidite	Ni_3S_4	cubic	spinel	8	9.480					
Portlandite	$Ca(OH)_2$	hex	cadmium iodide	1	3.5933		4.9086			
Powellite	$CaMoO_4$	tetr	scheelite	4	5.226	8.74	11.43			
Protoenstatite	$MgSiO_3$	orth		8	9.25		5.32			
Proustite	Ag_3AsS_3	rhomb		6	10.816	11.78	8.6948			
Pseudowollastonite	$CaSiO_3$	tricl		24	6.90		19.65	90.00°	90.80°	90.00°
Pyrargyrite	Ag_3SbS_3	rhomb		6	11.052		8.7177			
Pyrite	FeS_2	cubic	pyrite	4	5.4175					
Pyrolusite	MnO_2	tetr	rutile	2	4.388		2.865			
Pyrope	$Mg_3Al_2Si_3O_{12}$	cubic	garnet	8	11.459					

Name	Formula	Crystal system	Structure type	Z	a/Å	b/Å	c/Å	α	β	γ
Pyrophanite	$MnTiO_3$	rhomb	ilmenite	6	5.155		14.18			
Pyrophyllite	$Al_2Si_4O_{10}(OH)_2$	monocl	2M1 mica	4	5.14	8.90	18.55		99.92°	
Pyroxmangite	$MnFe(SiO_3)_2$	tricl		7	7.56	17.45	6.67	84.00°	94.30°	113.70°
Pyrrhotite	$Fe_{0.980}S$	hex	defect niccolite	2	3.446		5.848			
Pyrrhotite	$Fe_{0.885}S$	hex	defect niccolite	2	3.440		5.709			
Quartz (α)	SiO_2	hex		3	4.9136		5.4051			
Quartz (β)	SiO_2	hex		3	4.999		5.4592			
Rammelsbergite	$NiAs_2$	orth	marcasite	2	4.757	5.797	3.542			
Realgar	AsS	monocl		16	9.29	13.53	6.57		106.55°	
Retgersite	$NiSO_4 \cdot 4H_2O$	tetr		4	6.782		18.28			
Rhodochrosite	$MnCO_3$	rhomb	calcite	6	4.7771		15.664			
Rhodonite	$MnSiO_3$	tricl		10	7.682	11.818	6.707	92.36°	93.95°	105.66°
Riebeckite	$Na_2Fe_5FSi_8O_{22}(OH)_2$	monocl	tremolite	2	9.729	18.065	5.334		103.31°	
Rutile	TiO_2	tetr		2	4.5937		2.9618			
Safflorite	$Co_{0.5}Fe_{0.5}As_2$	orth	marcasite	2	5.231	5.953	2.962			
Sanmartinite	$ZnWO_4$	monocl	wolframite	2	4.691	5.720	4.925		89.36°	
Sapphirine	$Mg_2Al_2O_6SiO_4$	monocl		8	9.96	28.60	9.85		110.5°	
Scacchite	$MnCl_2$	rhomb		3	3.711		17.59			
Scheelite	$CaWO_4$	tetr	scheelite	4	5.242		11.372			
Schorl	$NaFe_3Al_6B_3Si_6O_{27}(OH)_4$	rhomb	tourmaline	3	16.032		7.149			
Selenium	Se	hex		3	4.3642		4.9588			
Selenclite	SeO_2	tetr		8	8.35		5.05			
Sellaite	MgF_2	tetr	rutile	2	4.621		3.050			
Senarmontite	Sb_2O_3	cubic	arsenic trioxide	16	11.152					
Shandite	$Ni_3Pb_2S_2$	rhomb		3	5.576		13.658			
Shortite	$Na_2Ca_2(CO_3)_3$	orth		2	4.961	11.03	7.12			
Siderite	$FeCO_3$	rhomb	calcite	6	4.6887		15.373			
Silicon	Si	cubic	diamond	8	5.4305					
Sillimanite	Al_2OSiO_4	orth		4	7.4843	7.6730	5.7711			
Silver	Ag	cubic	face-centered cubic	4	4.0862					
Silver telluride I	Ag_2Te	cubic		2	5.29					
Silver telluride II	Ag_2Te	cubic		4	6.585					
Smithsonite	$ZnCO_3$	rhomb	calcite	6	4.6528		15.025			
Soda niter	$NaNO_3$	rhomb	calcite	6	5.0696		16.829			
Sodium melilite	$NaCaAlSi_2O_7$	tetr	melilite	2	8.511		4.809			
Sperrylite	$PtAs_2$	cubic	pyrite	4	5.968					
Spessartite	$Mn_3Al_2Si_3O_{12}$	cubic	garnet	8	11.621					
Sphalerite	ZnS	cubic	sphalerite	4	5.4093					
Sphene	$CaTiSiO_5$	monocl		4	7.07	8.72	6.56		113.95°	
Spinel	$MgAl_2O_4$	cubic	spinel	8	8.080					
Spodumene	$LiAl(SiO_3)_2$	monocl	diopside	4	9.451	8.387	5.208		110.07°	
Spodumene (β)	$LiAl(SiO_3)_2$	tetr		4	7.5332		9.1540			
Staurolite	$Fe_2Al_9Si_4O_{22}(OH)_2$	monocl		2	7.90	16.65	5.63		90.00°	
Sternbergite	$AgFe_2S_3$	orth		8	11.60	12.675	6.63			
Stibnite	Sb_2S_3	orth	stibnite	4	11.229	11.310	3.8389			

Name	Formula	Crystal system	Structure type	Z	a/Å	b/Å	c/Å	α	β	γ
Stilleite	$ZnSe$	cubic	sphalerite	4	5.6685					
Stishovite	SiO_2	tetr	rutile	2	4.1790		2.6649			
Stolzite	$PbWO_4$	tetr	scheelite	4	5.4616		12.046			
Stromeyerite	$Ag_{0.93}Cu_{1.07}S$	orth		4	4.066	6.628	7.972			
Strontianite	$SrCO_3$	orth	aragonite	4	6.029	8.414	5.107			
Sulfur (monoclinic)	S	monocl	S8 ring molecules	48	11.04	10.98	10.92		96.73°	
Sulfur (orthorhombic)	S	orth	S8 ring molecules	128	10.4646	12.8660	24.4860			
Sulfur (rhombohedral)	S	rhomb	S6 ring molecules	18	10.818		4.280			
Sylvite	KCl	cubic	rock salt	4	6.2931					
Syngenite	$K_2Ca(SO_4)_2 \cdot H_2O$	monocl		2	9.775	7.156	6.251		104.00°	
Synthetic anorthite	$CaAl_2Si_2O_8$	hex		2	5.10		14.72			
Synthetic anorthite	$CaAl_2Si_2O_8$	orth		2	8.22	8.60	4.83			
Talc	$Mg_3Si_4O_{10}(OH)_2$	monocl	2M1 mica	4	5.287	9.158	18.95		99.50°	
Tantalum	Ta	cubic	tungsten	2	3.3058					
Teallite	$PbSnS_2$	orth	germanium sulfide	8	4.266	11.419	4.090			
Tellurite	TeO_2	orth	tellurite	8	5.607	12.034	5.463			
Tellurium	Te	hex	selenium	3	4.4570		5.9290			
Tellurobismuthite	Bi_2Te_3	rhomb		3	4.3835		30.487			
Tennantite	$Cu_{12}As_4S_{13}$	cubic	tetrahedrite	2	10.190					
Tenorite	CuO	monocl		4	4.684	3.425	5.129		99.47°	
Tephroite	Mn_2SiO_4	orth	olivine	4	4.871	10.636	6.232			
Tetrahedrite	$Cu_{12}Sb_4S_{13}$	cubic	tetrahedrite	2	10.327					
Thenardite	Na_2SO_4	orth	thenardite	8	5.863	12.304	9.821			
Thorianite	ThO_2	cubic	fluorite	4	5.5952					
Thorite	$ThSiO_4$	tetr	zircon	4	7.143		6.327			
Tiemannite	$HgSe$	cubic	sphalerite	4	6.0853					
Tin	Sn	tetr		4	5.8315		3.1813			
Titanium	Ti	hex		2	2.953		4.729			
Titanium(III) oxide	Ti_2O_3	rhomb	corundum	6	5.149		13.642			
Topaz	$Al_2(SiO_4)(OH)$	orth		4	8.394	8.792	4.649			
Tremolite	$Ca_2Mg_5Si_8O_{22}(OH)_2$	monocl	tremolite	2	9.840	18.052	5.275		104.70°	
Trevorite	$NiFe_2O_4$	cubic	spinel	8	8.339					
Tridymite (β)	SiO_2	hex		4	5.0463		8.2563			
Trogtalite	$CoSe_2$	cubic	pyrite	4	5.8588					
Troilite	FeS	hex	niccolite	2	3.446		5.877			
Tschermakite	$CaAl_2SiO_6$	monocl	diopside	4	9.615	8.661	5.272		106.12°	
Tungsten	W	cubic		2	3.1653					
Tungstenite	WS_2	hex	molybdenite	2	3.154		12.362			
Turquois	$CuAl_6(PO_4)_4(OH)_8 \cdot 4H_2O$	tricl		1	7.424	7.629	9.910	68.61°	69.71°	65.08°
Umangite	Cu_3Se_2	tetr		2	6.402		4.276			
Uraninite	UO_2	cubic	fluorite	4	5.4682					
Ureyite	$NaCr(SiO_3)_2$	monocl	diopside	4	9.550	8.712	5.273		107.44°	
Uvarovite	$Ca_3Cr_2Si_3O_{12}$	cubic	garnet	8	11.999					
Uvite	$CaMg_4Al_5B_3Si_6O_{27}(OH)_4$	rhomb	tourmaline	3	15.86		7.19			
Vaesite	NiS_2	cubic	pyrite	4	5.6873					

Name	Formula	Crystal system	Structure type	Z	a/Å	b/Å	c/Å	α	β	γ
Valentinite	Sb_2O_3	orth	antimony trioxide	4	4.914	12.468	5.421			
Vanthoffite	$MgSO_4 \cdot 3Na_2SO_4$	monocl		2	9.797	9.217	8.199		113.50°	
Vaterite	$CaCO_3$	hex		6	7.135		8.524			
Villiaumite	NaF	cubic	rock salt	4	4.6342					
Violarite	$FeNi_2S_4$	cubic	spinel	8	9.464					
Willemite	Zn_2SiO_4	rhomb	phenacite	18	13.94		9.309			
Witherite	$BaCO_3$	orth	aragonite	4	6.430	8.904	5.314			
Wolframite	$Fe_{0.5}Mn_{0.5}WO_4$	monocl	wolframite	2	4.782	5.731	4.982		90.57°	
Wollastonite	$CaSiO_3$	tricl		6	7.94	7.32	7.07	90.03°	95.37°	103.43°
Wulfenite	$PbMoO_4$	tetr	scheelite	4	5.435		12.110			
Wurtzite	ZnS	hex	zincite	2	3.8230		6.2565			
Wustite	$Fe_{0.953}O$	cubic	defect rock salt	4	4.3088					
Xenotime	YPO_4	tetr	zircon	4	6.885		5.982			
Zinc	Zn	hex	hexagonal close pack	2	2.665		4.947			
Zinc telluride	ZnTe	cubic	sphalerite	4	5.1020					
Zincite	ZnO	hex	zincite	2	3.2495		5.2069			
Zinkosite	$ZnSO_4$	orth	barite	4	3.588	6.740	4.770			
Zircon	$ZrSiO_4$	tetr	zircon	4	6.604		5.979			
Zoisite	$Ca_2Al_3(SiO_4)_3OH$	orth		4	16.15	5.581	10.06			

Section 5
Thermochemistry, Electrochemistry, and Kinetics

CODATA KEY VALUES FOR THERMODYNAMICS

The Committee on Data for Science and Technology (CODATA) has conducted a project to establish internationally agreed values for the thermodynamic properties of key chemical substances. This table presents the final results of the project. Use of these recommended, internally consistent values is encouraged in the analysis of thermodynamic measurements, data reduction, and preparation of other thermodynamic tables.

The table includes the standard enthalpy of formation at 298.15 K, the entropy at 298.15 K, and the quantity $H°$ (298.15 K)–$H°$ (0). A value of 0 in the $\Delta_f H°$ column for an element indicates the reference state for that element. The standard state pressure is 100000 Pa (1 bar). See the reference for information on the dependence of gas-phase entropy on the choice of standard state pressure.

Substances are listed in alphabetical order of their chemical formulas when written in the most common form.

The table is reprinted with permission of CODATA.

REFERENCE

Cox, J. D., Wagman, D. D., and Medvedev, V. A., *CODATA Key Values for Thermodynamics,* Hemisphere Publishing Corp., New York, 1989.

Substance	State	$\Delta_f H°$ (298.15 K) kJ·mol^{-1}	$S°$ (298.15 K) J·K^{-1}·mol^{-1}	$H°$ (298.15 K)–$H°$ (0) kJ·mol^{-1}
Ag	cr	0	42.55 ± 0.20	5.745 ± 0.020
Ag	g	284.9 ± 0.8	172.997 ± 0.004	6.197 ± 0.001
Ag$^+$	aq	105.79 ± 0.08	73.45 ± 0.40	
AgCl	cr	-127.01 ± 0.05	96.25 ± 0.20	12.033 ± 0.020
Al	cr	0	28.30 ± 0.10	4.540 ± 0.020
Al	g	330.0 ± 4.0	164.554 ± 0.004	6.919 ± 0.001
Al^{+3}	aq	-538.4 ± 1.5	-325 ± 10	
AlF$_3$	cr	-1510.4 ± 1.3	66.5 ± 0.5	11.62 ± 0.04
Al$_2$O$_3$	cr, corundum	-1675.7 ± 1.3	50.92 ± 0.10	10.016 ± 0.020
Ar	g	0	154.846 ± 0.003	6.197 ± 0.001
B	cr, rhombic	0	5.90 ± 0.08	1.222 ± 0.008
B	g	565 ± 5	153.436 ± 0.015	6.316 ± 0.002
BF$_3$	g	-1136.0 ± 0.8	254.42 ± 0.20	11.650 ± 0.020
B$_2$O$_3$	cr	-1273.5 ± 1.4	53.97 ± 0.30	9.301 ± 0.040
Be	cr	0	9.50 ± 0.08	1.950 ± 0.020
Be	g	324 ± 5	136.275 + 0.003	6.197 ± 0.001
BeO	cr	-609.4 ± 2.5	13.77 ± 0.04	2.837 ± 0.008
Br	g	111.87 ± 0.12	175.018 ± 0.004	6.197 ± 0.001
Br$^-$	aq	-121.41 ± 0.15	82.55 ± 0.20	
Br$_2$	l	0	152.21 ± 0.30	24.52 ± 0.01
Br$_2$	g	30.91 ± 0.11	245.468 ± 0.005	9.725 ± 0.001
C	cr, graphite	0	5.74 ± 0.10	1.050 ± 0.020
C	g	716.68 ± 0.45	158.100 ± 0.003	6.536 ± 0.001
CO	g	-110.53 ± 0.17	197.660 ± 0.004	8.671 ± 0.001
CO$_2$	g	-393.51 ± 0.13	213.785 ± 0.010	9.365 ± 0.003
CO$_2$	aq, undissoc.	-413.26 ± 0.20	119.36 ± 0.60	
CO$_3^{-2}$	aq	-675.23 ± 0.25	-50.0 ± 1.0	
Ca	cr	0	41.59 ± 0.40	5.736 ± 0.040
Ca	g	177.8 ± 0.8	154.887 ± 0.004	6.197 ± 0.001
Ca^{+2}	aq	-543.0 ± 1.0	-56.2 ± 1.0	
CaO	cr	-634.92 ± 0.90	38.1 ± 0.4	6.75 ± 0.06
Cd	cr	0	51.80 ± 0.15	6.247 ± 0.015
Cd	g	111.80 ± 0.20	167.749 ± 0.004	6.197 ± 0.001
Cd^{+2}	aq	-75.92 ± 0.60	-72.8 ± 1.5	
CdO	cr	-258.35 ± 0.40	54.8 ± 1.5	8.41 ± 0.08
CdSO$_4$·8/3H$_2$O	cr	-1729.30 ± 0.80	229.65 ± 0.40	35.56 ± 0.04
Cl	g	121.301 ± 0.008	165.190 ± 0.004	6.272 ± 0.001
Cl$^-$	aq	-167.080 ± 0.10	56.60 ± 0.20	
ClO$_4^-$	aq	-128.10 ± 0.40	184.0 ± 1.5	
Cl$_2$	g	0	223.081 ± 0.010	9.181 ± 0.001
Cs	cr	0	85.23 ± 0.40	7.711 ± 0.020
Cs	g	76.5 ± 1.0	175.601 ± 0.003	6.197 ± 0.001
Cs$^+$	aq	-258.00 ± 0.50	132.1 + 0.5	

Substance	State	$\Delta_f H°$ (298.15 K) kJ·mol⁻¹	$S°$ (298.15 K) J·K⁻¹·mol⁻¹	$H°$ (298.15 K)–$H°$ (0) kJ·mol⁻¹
Cu	cr	0	33.15 ± 0.08	5.004 ± 0.008
Cu	g	337.4 ± 1.2	166.398 ± 0.004	6.197 ± 0.001
Cu^{+2}	aq	64.9 ± 1.0	-98 ± 4	
CuSO$_4$	cr	-771.4 ± 1.2	109.2 ± 0.4	16.86 ± 0.08
F	g	79.38 ± 0.30	158.751 ± 0.004	6.518 ± 0.001
F$^-$	aq	-335.35 ± 0.65	-13.8 ± 0.8	
F$_2$	g	0	202.791 ± 0.005	8.825 ± 0.001
Ge	cr	0	31.09 ± 0.15	4.636 ± 0.020
Ge	g	372 ± 3	167.904 ± 0.005	7.398 ± 0.001
GeF$_4$	g	-1190.20 ± 0.50	301.9 ± 1.0	17.29 ± 0.10
GeO$_2$	cr, tetragonal	-580.0 ± 1.0	39.71 ± 0.15	7.230 ± 0.020
H	g	217.998 ± 0.006	114.717 ± 0.002	6.197 ± 0.001
H$^+$	aq	0	0	
HBr	g	-36.29 ± 0.16	198.700 ± 0.004	8.648 ± 0.001
HCO$_3^-$	aq	-689.93 ± 0.20	98.4 ± 0.5	
HCl	g	-92.31 ± 0.10	186.902 ± 0.005	8.640 ± 0.001
HF	g	-273.30 ± 0.70	173.779 ± 0.003	8.599 ± 0.001
HI	g	26.50 ± 0.10	206.590 ± 0.004	8.657 ± 0.001
HPO$_4^{-2}$	aq	-1299.0 ± 1.5	-33.5 ± 1.5	
HS$^-$	aq	-16.3 ± 1.5	67 ± 5	
HSO$_4^-$	aq	-886.9 ± 1.0	131.7 ± 3.0	
H$_2$	g	0	130.680 ± 0.003	8.468 ± 0.001
H$_2$O	l	-285.830 ± 0.040	69.95 ± 0.03	13.273 ± 0.020
H$_2$O	g	-241.826 ± 0.040	188.835 ± 0.010	9.905 ± 0.005
H$_2$PO$_4^-$	aq	-1302.6 ± 1.5	92.5 ± 1.5	
H$_2$S	g	-20.6 ± 0.5	205.81 ± 0.05	9.957 ± 0.010
H$_2$S	aq, undissoc.	-38.6 ± 1.5	126 ± 5	
H$_3$BO$_3$	cr	-1094.8 ± 0.8	89.95 ± 0.60	13.52 ± 0.04
H$_3$BO$_3$	aq, undissoc.	-1072.8 ± 0.8	162.4 ± 0.6	
He	g	0	126.153 ± 0.002	6.197 ± 0.001
Hg	l	0	75.90 ± 0.12	9.342 ± 0.008
Hg	g	61.38 ± 0.04	174.971 ± 0.005	6.197 ± 0.001
Hg^{+2}	aq	170.21 ± 0.20	-36.19 ± 0.80	
HgO	cr, red	-90.79 ± 0.12	70.25 ± 0.30	9.117 ± 0.025
Hg$_2^{+2}$	aq	166.87 ± 0.50	65.74 ± 0.80	
Hg$_2$Cl$_2$	cr	-265.37 ± 0.40	191.6 ± 0.8	23.35 ± 0.20
Hg$_2$SO$_4$	cr	-743.09 ± 0.40	200.70 ± 0.20	26.070 ± 0.030
I	g	106.76 ± 0.04	180.787 ± 0.004	6.197 ± 0.001
I$^-$	aq	-56.78 ± 0.05	106.45 ± 0.30	
I$_2$	cr	0	116.14 ± 0.30	13.196 ± 0.040
I$_2$	g	62.42 ± 0.08	260.687 ± 0.005	10.116 ± 0.001
K	cr	0	64.68 ± 0.20	7.088 ± 0.020
K	g	89.0 ± 0.8	160.341 ± 0.003	6.197 ± 0.001
K$^+$	aq	-252.14 ± 0.08	101.20 ± 0.20	
Kr	g	0	164.085 ± 0.003	6.197 ± 0.001
Li	cr	0	29.12 ± 0.20	4.632 ± 0.040
Li	g	159.3 ± 1.0	138.782 ± 0.010	6.197 ± 0.001
Li$^+$	aq	-278.47 ± 0.08	12.24 ± 0.15	
Mg	cr	0	32.67 ± 0.10	4.998 ± 0.030
Mg	g	147.1 ± 0.8	148.648 ± 0.003	6.197 ± 0.001
Mg^{+2}	aq	-467.0 ± 0.6	-137 ± 4	
MgF$_2$	cr	-1124.2 ± 1.2	57.2 ± 0.5	9.91 ± 0.06
MgO	cr	-601.60 ± 0.30	26.95 ± 0.15	5.160 ± 0.020
N	g	472.68 ± 0.40	153.301 ± 0.003	6.197 ± 0.001
NH$_3$	g	-45.94 ± 0.35	192.77 ± 0.05	10.043 ± 0.010
NH$_4^+$	aq	-133.26 ± 0.25	111.17 ± 0.40	
NO$_3^-$	aq	-206.85 ± 0.40	146.70 ± 0.40	

Substance	State	$\Delta_f H°$ (298.15 K) kJ·mol^{-1}	$S°$ (298.15 K) J·K^{-1}·mol^{-1}	$H°$ (298.15 K)–$H°$ (0) kJ·mol^{-1}
N_2	g	0	191.609 ± 0.004	8.670 ± 0.001
Na	cr	0	51.30 ± 0.20	6.460 ± 0.020
Na	g	107.5 ± 0.7	153.718 ± 0.003	6.197 ± 0.001
Na$^+$	aq	-240.34 ± 0.06	58.45 ± 0.15	
Ne	g	0	146.328 ± 0.003	6.197 ± 0.001
O	g	249.18 ± 0.10	161.059 ± 0.003	6.725 ± 0.001
OH$^-$	aq	-230.015 ± 0.040	-10.90 ± 0.20	
O_2	g	0	205.152 ± 0.005	8.680 ± 0.002
P	cr, white	0	41.09 ± 0.25	5.360 ± 0.015
P	g	316.5 ± 1.0	163.199 ± 0.003	6.197 ± 0.001
P_2	g	144.0 ± 2.0	218.123 ± 0.004	8.904 ± 0.001
P_4	g	58.9 ± 0.3	280.01 ± 0.50	14.10 ± 0.20
Pb	cr	0	64.80 ± 0.30	6.870 ± 0.030
Pb	g	195.2 ± 0.8	175.375 ± 0.005	6.197 ± 0.001
Pb^{+2}	aq	0.92 ± 0.25	18.5 ± 1.0	
$PbSO_4$	cr	-919.97 ± 0.40	148.50 ± 0.60	20.050 ± 0.040
Rb	cr	0	76.78 ± 0.30	7.489 ± 0.020
Rb	g	80.9 ± 0.8	170.094 ± 0.003	6.197 ± 0.001
Rb$^+$	aq	-251.12 ± 0.10	121.75 ± 0.25	
S	cr, rhombic	0	32.054 ± 0.050	4.412 ± 0.006
S	g	277.17 ± 0.15	167.829 ± 0.006	6.657 ± 0.001
SO_2	g	-296.81 ± 0.20	248.223 ± 0.050	10.549 ± 0.010
SO_4^{-2}	aq	-909.34 ± 0.40	18.50 ± 0.40	
S_2	g	128.60 ± 0.30	228.167 ± 0.010	9.132 ± 0.002
Si	cr	0	18.81 ± 0.08	3.217 ± 0.008
Si	g	450 ± 8	167.981 ± 0.004	7.550 ± 0.001
SiF_4	g	-1615.0 ± 0.8	282.76 ± 0.50	15.36 ± 0.05
SiO_2	cr, alpha quartz	-910.7 ± 1.0	41.46 ± 0.20	6.916 ± 0.020
Sn	cr, white	0	51.18 ± 0.08	6.323 ± 0.008
Sn	g	301.2 ± 1.5	168.492 ± 0.004	6.215 ± 0.001
Sn^{+2}	aq	-8.9 ± 1.0	-16.7 ± 4.0	
SnO	cr, tetragonal	-280.71 ± 0.20	57.17 ± 0.30	8.736 ± 0.020
SnO_2	cr, tetragonal	-577.63 ± 0.20	49.04 ± 0.10	8.384 ± 0.020
Th	cr	0	51.8 ± 0.5	6.35 ± 0.05
Th	g	602 ± 6	190.17 ± 0.05	6.197 ± 0.003
ThO_2	cr	-1226.4 ± 3.5	65.23 ± 0.20	10.560 ± 0.020
Ti	cr	0	30.72 ± 0.10	4.824 ± 0.015
Ti	g	473 ± 3	180.298 ± 0.010	7.539 ± 0.002
$TiCl_4$	g	-763.2 ± 3.0	353.2 ± 4.0	21.5 ± 0.5
TiO_2	cr, rutile	-944.0 ± 0.8	50.62 ± 0.30	8.68 ± 0.05
U	cr	0	50.20 ± 0.20	6.364 ± 0.020
U	g	533 ± 8	199.79 ± 0.10	6.499 ± 0.020
UO_2	cr	-1085.0 ± 1.0	77.03 ± 0.20	11.280 ± 0.020
UO_2^{+2}	aq	-1019.0 ± 1.5	-98.2 ± 3.0	
UO_3	cr, gamma	-1223.8 ± 1.2	96.11 ± 0.40	14.585 ± 0.050
U_3O_8	cr	-3574.8 ± 2.5	282.55 ± 0.50	42.74 ± 0.10
Xe	g	0	169.685 ± 0.003	6.197 ± 0.001
Zn	cr	0	41.63 ± 0.15	5.657 ± 0.020
Zn	g	130.40 ± 0.40	160.990 ± 0.004	6.197 ± 0.001
Zn^{+2}	aq	-153.39 ± 0.20	-109.8 ± 0.5	
ZnO	cr	-350.46 ± 0.27	43.65 ± 0.40	6.933 ± 0.040

STANDARD THERMODYNAMIC PROPERTIES OF CHEMICAL SUBSTANCES

This table gives the standard state chemical thermodynamic properties of about 2400 individual substances in the crystalline, liquid, and gaseous states. Substances are listed by molecular formula in a modified Hill order; all compounds not containing carbon appear first, followed by those that contain carbon. The properties tabulated are:

$\Delta_f H°$ Standard molar enthalpy (heat) of formation at 298.15 K in kJ/mol
$\Delta_f G°$ Standard molar Gibbs energy of formation at 298.15 K in kJ/mol
$S°$ Standard molar entropy at 298.15 K in J/mol K
C_p Molar heat capacity at constant pressure at 298.15 K in J/mol K

The standard state pressure is 100 kPa (1 bar). The standard states are defined for different phases by:

- The standard state of a pure gaseous substance is that of the substance as a (hypothetical) ideal gas at the standard state pressure.
- The standard state of a pure liquid substance is that of the liquid under the standard state pressure.
- The standard state of a pure crystalline substance is that of the crystalline substance under the standard state pressure.

An entry of 0.0 for $\Delta_f H°$ for an element indicates the reference state of that element. See References 1 and 2 for further information on reference states. A blank means no value is available.

We are indebted to M.V. Korobov for providing data on fullerene compounds.

REFERENCES

1. Cox, J.D., Wagman, D.D., and Medvedev, V.A., *CODATA Key Values for Thermodynamics*, Hemisphere Publishing Corp., New York, 1989.
2. Wagman, D.D., Evans, W.H., Parker, V.B., Schumm, R.H., Halow, I., Bailey, S.M., Churney, K.L., and Nuttall, R.L., *The NBS Tables of Chemical of Chemical Thermodynamic Properties, J. Phys. Chem. Ref. Data*, Vol. 11, Suppl. 2, 1982.
3. Chase M.W., Davies, C.A., Downey, J.R., Frurip, D.J., McDonald, R.A., and Syverud, A.N., *JANAF Thermochemical Tables, Third Edition, J. Phys. Chem. Ref. Data*, Vol. 14, Suppl. 1, 1985.
4. Daubert, T.E., Danner, R.P., Sibul, H.M., and Stebbins, C.C., *Physical and Thermodynamic Properties of Pure Compounds: Data Compilation*, extant 1994 (core with 4 supplements), Taylor & Francis, Bristol, PA.
5. Pedley, J.B., Naylor, R.D., and Kirby, S.P., *Thermochemical Data of Organic Compounds, Second Edition*, Chapman & Hall, London, 1986.
6. Pedley, J.B., *Thermochemical Data and Structures of Organic Compounds*, Thermodynamic Research Center, Texas A & M University, College Station, TX, 1994.
7. Dolmalski, E.S., Evans, W.H., Hearing E.D., *Heat Capacities and Entropies of Organic Compounds in the Condensed Phase, J. Phys. Chem. Ref. Data*, Vol. 13, Suppl. 1, 1984; Vol. 19, No. 4, 881—1047, 1990.
8. Gurvich, L.V., Veyts, I.V., and Alcock, C.B., *Thermodynamic Properties of Individual Substances, Fourth Edition*, Vol. 1, Hermisphere Publishing Corp., New York, 1989.
9. Gurvich, L.V., Veyts, I.V., and Alcock, C.B., *Thermodynamic Properties of Individual Substances, Fourth Edition*, Vol. 3, CRC Press, Boca Raton, FL, 1994.

STANDARD THERMODYNAMIC PROPERTIES OF CHEMICAL SUBSTANCES (continued)

Molecular formula	Name	ΔfH°/kJ mol⁻¹ Crys.	Liq.	Gas	ΔfG°/kJ mol⁻¹ Crys.	Liq.	Gas	S°/J mol⁻¹ K⁻¹ Crys.	Liq.	Gas	Cp/J mol⁻¹ K⁻¹ Crys.	Liq.	Gas
Ac	Actinium	0.0		406.0			366.0	56.5		188.1	27.2		20.8
Ag	Silver	0.0		284.9			246.0	42.6		173.0	25.4		20.8
AgBr	Silver(I) bromide	-100.4			-96.9			107.1			52.4		
AgBrO₃	Silver(I) bromate	-10.5			71.3			151.9					
AgCl	Silver(I) chloride	-127.0			-109.8			96.3			50.8		
AgClO₃	Silver(I) chlorate	-30.3			64.5			142.0					
AgClO₄	Silver(I) perchlorate	-31.1											
AgF	Silver(I) fluoride	-204.6											
AgF₂	Silver(II) fluoride	-360.0											
AgI	Silver(I) iodide	-61.8			-66.2			115.5			56.8		
AgIO₃	Silver(I) iodate	-171.1			-93.7			149.4			102.9		
AgNO₃	Silver(I) nitrate	-124.4			-33.4			140.9			93.1		
Ag₂	Disilver			410.0			358.8			257.1			37.0
Ag₂CrO₄	Silver(I) chromate	-731.7			-641.8			217.6			142.3		
Ag₂O	Silver(I) oxide	-31.1			-11.2			121.3			65.9		
Ag₂O₂	Silver(II) oxide	-24.3			27.6			117.0			88.0		
Ag₂O₃	Silver(III) oxide	33.9			121.4			100.0					
Ag₂O₄S	Silver(I) sulfate	-715.9			-618.4			200.4			131.4		
Ag₂S	Silver(I) sulfide (argentite)	-32.6			-40.7			144.0			76.5		
Al	Aluminum	0.0		330.0			289.4	28.3		164.6	24.4		21.4
AlB₃H₁₂	Aluminum borohydride		-16.3	13.0		145.0	147.0		289.1	379.2		194.6	
AlBr	Aluminum monobromide			-4.0			-42.0			239.5			35.6
AlBr₃	Aluminum tribromide	-527.2		-425.1				180.2			100.6		
AlCl	Aluminum monochloride			-47.7			-74.1			228.1			35.0
AlCl₂	Aluminum dichloride			-331.0									
AlCl₃	Aluminum trichloride	-704.2		-583.2	-623.8		-283.7	109.3		215.0	91.1		31.9
AlF	Aluminum monofluoride			-258.2									
AlF₃	Aluminum trifluoride	-1510.4		-1204.6	-1431.1		-1188.2	66.5		277.1	75.1		62.6
AlF₄Na	Sodium tetrafluoroaluminate			-1869.0			-1827.5			345.7			105.9
AlH	Aluminum hydride			259.2			231.2	30.0		187.9	40.2		29.4
AlH₃	Aluminum hydride	-46.0											
AlH₄K	Potassium aluminum hydride	-183.7											
AlH₄Li	Lithium aluminum hydride	-116.3			-44.7			78.7			83.2		
AlH₄Na	Sodium aluminum hydride	-15.5											
AlI	Aluminum monoiodide			65.5			65.3			218.4			30.9
AlI₃	Aluminum triiodide	-313.8		-207.5	-300.8			159.0			98.7		36.0
AlN	Aluminum nitride	-318.0			-287.0			20.2			30.1		
AlO	Aluminum monoxide			91.2									
AlO₄P	Aluminum phosphate	-1733.8			-1617.9			90.8			93.2		
AlP	Aluminum phosphide	-166.5											
AlS	Aluminum monosulfide			200.9			150.1			230.6			33.4
Al₂	Dialuminum			485.9			433.3			233.2			36.4
Al₂Br₆	Aluminum hexabromide			-970.7									
Al₂Cl₆	Aluminum hexachloride			-1290.8			-1220.4			490.0			

Molecular formula	Name	$\Delta_f H°$/kJ mol⁻¹			$\Delta_f G°$/kJ mol⁻¹			$S°$/J mol⁻¹ K⁻¹			C_p/J mol⁻¹ K⁻¹		
		Crys.	Liq.	Gas	Crys.	Liq.	Gas	Crys.	Liq.	Gas	Crys.	Liq.	Gas
Al_2F_6	Aluminum hexafluoride	-2628.0											
Al_2I_6	Aluminum hexaiodide	-516.7											
Al_2O	Aluminum oxide (Al_2O)			-130.0			-159.0			259.4			45.7
Al_2O_3	Aluminum oxide (corundum)	-1675.7			-1582.3			50.9			79.0		
Al_2S_3	Aluminum sulfide	-724.0						116.9			105.1		
Am	Americium	0.0											
Ar	Argon			0.0						154.8			20.8
As	Arsenic (gray)	0.0		302.5			261.0	35.1		174.2	24.6		20.8
As	Arsenic (yellow)	14.6											
$AsBr_3$	Arsenic(III) bromide	-197.5		-130.0			-159.0			363.9			79.2
$AsCl_3$	Arsenic(III) chloride		-305.0	-261.5		-259.4	-248.9		216.3	327.2			75.7
AsF_3	Arsenic(III) fluoride		-821.3	-785.8		-774.2	-770.8		181.2	289.1		126.6	65.6
AsGa	Gallium arsenide	-71.0			-67.8			64.2			46.2		
AsH_3	Arsine			66.4			68.9			222.8			38.1
AsH_3O_4	Arsenic acid	-906.3											
AsI_3	Arsenic(III) iodide	-58.2			-59.4			213.1		388.3	105.8		80.6
AsIn	Indium arsenide	-58.6			-53.6			75.7			47.8		
AsO	Arsenic monoxide			70.0			171.9			239.4			35.0
As_2	Diarsenic			222.2									
As_2O_5	Arsenic(V) oxide	-924.9			-782.3			105.4			116.5		
As_2S_3	Arsenic(III) sulfide	-169.0			-168.6			163.6			116.3		
At	Astatine	0.0											
Au	Gold	0.0		366.1			326.3	47.4		180.5	25.4		20.8
AuBr	Gold(I) bromide	-14.0											
$AuBr_3$	Gold(III) bromide	-53.3											
AuCl	Gold(I) chloride	-34.7											
$AuCl_3$	Gold(III) chloride	-117.6											
AuF_3	Gold(III) fluoride	-363.6											
AuH	Gold hydride			295.0			265.7			211.2			29.2
AuI	Gold(I) iodide	0.0											
Au_2	Digold			515.1						225.0			36.9
B	Boron (rhombic)	0.0		565.0			521.0	5.9		153.4	11.1		20.8
BBr	Bromoborane			238.1			195.4						32.9
BBr_3	Boron tribromide		-239.7	-205.6		-238.5	-232.5		229.7	324.2			67.8
BCl	Chloroborane			149.5			120.9			213.2			31.7
BClO	Chloroxyborane			-314.0									
BCl_3	Boron trichloride		-427.2	-403.8		-387.4	-388.7		206.3	290.1		106.7	62.7
$BCsO_2$	Cesium metaborate	-972.0			-915.0			104.4			80.6		
BF	Fluoroborane			-122.2			-149.8			200.5			29.6
BFO	Fluorooxyborane			-607.0									
BF_3	Boron trifluoride			-1136.0			-1119.4			254.4			
BF_3H_3N	Aminetrifluoroboron	-1353.9											
BF_3H_3P	Trihydro(phosphorus trifluoride)boron			-854.0									
BF_4Na	Sodium tetrafluoroborate	-1844.7			-1750.1			145.3			120.3		

STANDARD THERMODYNAMIC PROPERTIES OF CHEMICAL SUBSTANCES (continued)

Molecular formula	Name	$\Delta_f H°$/kJ mol⁻¹ Crys.	Liq.	Gas	$\Delta_f G°$/kJ mol⁻¹ Crys.	Liq.	Gas	$S°$/J mol⁻¹ K⁻¹ Crys.	Liq.	Gas	C_p/J mol⁻¹ K⁻¹ Crys.	Liq.	Gas
BH	Boron monohydride			449.6			419.6			171.9			29.2
BHO_2	Metaboric acid (monoclinic)	-794.3		-561.9	-723.4		-551.0	38.0		240.1			42.2
BH_3	Borane			100.0									
BH_3O_3	Boric acid (orthoboric acid)	-1094.3		-994.1	-968.9			90.0			86.1		
BH_4K	Potassium borohydride	-227.4			-160.3			106.3			96.1		
BH_4Li	Lithium borohydride	-190.8			-125.0			75.9			82.6		
BH_4Na	Sodium borohydride	-188.6			-123.9			101.3			86.8		
BI_3	Boron triiodide			71.1			20.7			349.2			70.8
BKO_2	Potassium metaborate	-981.6			-923.4			80.0			66.7		
$BLiO_2$	Lithium metaborate	-1032.2			-976.1			51.5			59.8		
BN	Boron nitride	-254.4		647.5	-228.4		614.5	14.8		212.3	19.7		29.5
$BNaO_2$	Sodium metaborate	-977.0			-920.7			73.5			65.9		
BO	Boron monoxide			25.0			-4.0			203.5			29.2
BO_2	Boron dioxide			-300.4			-305.9			229.6			43.0
BO_2Rb	Rubidium metaborate	-971.0			-913.0			94.3			74.1		
BS	Boron monosulfide			342.0			288.8			216.2			30.0
B_2	Diboron			830.5			774.0			201.9			30.5
B_2Cl_4	Tetrachlorodiborane		-523.0	-490.4		-464.8	-460.6		262.3	357.4		137.7	95.4
B_2F_4	Tetrafluorodiborane			-1440.1			-1410.4			317.3			79.1
B_2H_6	Diborane			35.6			86.7			232.1			56.9
B_2O_2	Diboron dioxide			-454.8			-462.3			242.5			57.3
B_2O_3	Boron oxide	-1273.5		-843.8	-1194.3		-832.0	54.0		279.8	62.8		66.9
B_2S_3	Boron sulfide	-240.6		67.0				100.0			111.7		
$B_3H_6N_3$	Borazine		-541.0	66.1		-392.7			199.6				
B_4H_{10}	Tetraborane			66.1									
$B_4Na_2O_7$	Sodium tetraborate	-3291.1			-3095.0			189.5	184.2		186.8	151.1	
B_5H_9	Pentaborane(9)		42.7	73.2		171.8	175.0			275.9			96.8
B_5H_{11}	Pentaborane(11)		73.2	103.3									
B_6H_{10}	Hexaborane		56.3	94.6									
Ba	Barium	0.0		180.0			146.0	62.5		170.2	28.1		20.8
$BaBr_2$	Barium bromide	-757.3			-736.8			146.0					
$BaCl_2$	Barium chloride	-855.0			-806.7			123.7			75.1		
$BaCl_2H_4O_2$	Barium chloride dihydrate	-1456.9			-1293.2			203.0					
BaF_2	Barium fluoride	-1207.1			-1156.8			96.4			71.2		
BaH_2	Barium hydride	-177.0			-138.2			63.0			46.0		
BaH_2O_2	Barium hydroxide	-944.7											
BaI_2	Barium iodide	-602.1											
BaN_2O_4	Barium nitrite	-768.2											
BaN_2O_6	Barium nitrate	-988.0			-792.6			214.0			151.4		
BaO	Barium oxide	-548.0			-520.3			72.1			47.3		
BaO_4S	Barium sulfate	-1473.2			-1362.2			132.2			101.8		
BaS	Barium sulfide	-460.0			-456.0			78.2			49.4		
Be	Beryllium	0.0		324.0			286.6	9.5		136.3	16.4		20.8
$BeBr_2$	Beryllium bromide	-353.5						108.0			69.4		
$BeCl_2$	Beryllium chloride	-490.4			-445.6			75.8			62.4		

STANDARD THERMODYNAMIC PROPERTIES OF CHEMICAL SUBSTANCES (continued)

Molecular formula	Name	ΔfH°/kJ mol⁻¹ Crys.	ΔfH°/kJ mol⁻¹ Liq.	ΔfH°/kJ mol⁻¹ Gas	ΔfG°/kJ mol⁻¹ Crys.	ΔfG°/kJ mol⁻¹ Liq.	ΔfG°/kJ mol⁻¹ Gas	S°/J mol⁻¹ K⁻¹ Crys.	S°/J mol⁻¹ K⁻¹ Liq.	S°/J mol⁻¹ K⁻¹ Gas	Cp/J mol⁻¹ K⁻¹ Crys.	Cp/J mol⁻¹ K⁻¹ Liq.	Cp/J mol⁻¹ K⁻¹ Gas
BeF_2	Beryllium fluoride	-1026.8			-979.4			53.4			51.8		
BeH_2O_2	Beryllium hydroxide	-902.5			-815.0			45.5			62.1		
BeI_2	Beryllium iodide	-192.5						121.0			71.1		
BeO	Beryllium oxide	-609.4			-580.1			13.8			25.6		
BeO_4S	Beryllium sulfate	-1205.2			-1093.8			77.9			85.7		
BeS	Beryllium sulfide	-234.3						34.0			34.0		
Bi	Bismuth	0.0		207.1			168.2	56.7		187.0	25.5		20.8
$BiClO$	Bismuth oxychloride	-366.9			-322.1			120.5					
$BiCl_3$	Bismuth trichloride	-379.1		-265.7	-315.0		-256.0	177.0		358.9	105.0		79.7
BiH_3O_3	Bismuth hydroxide	-711.3											
BiI_3	Bismuth triiodide				-175.3								
Bi_2	Dibismuth			219.7									36.9
Bi_2O_3	Bismuth oxide	-573.9			-493.7			151.5			113.5		
$Bi_2O_{12}S_3$	Bismuth sulfate	-2544.3											
Bi_2S_3	Bismuth sulfide	-143.1			-140.6			200.4			122.2		
Bk	Berkelium	0.0											
Br	Bromine (atomic)			111.9			82.4			175.0			20.8
$BrCl$	Bromine chloride			14.6			-1.0			240.1			35.0
$BrCl_3Si$	Bromotrichlorosilane									350.1			90.9
$BrCs$	Cesium bromide	-405.8			-391.4			113.1			52.9		
$BrCu$	Copper(I) bromide	-104.6			-100.8			96.1			54.7		
BrF	Bromine fluoride			-93.8			-109.2			229.0			33.0
BrF_3	Bromine trifluoride		-300.8	-255.6		-240.5	-229.4		178.2	292.5		124.6	66.6
BrF_5	Bromine pentafluoride		-458.6	-428.9		-351.8	-350.6		225.1	320.2			99.6
$BrGe$	Germanium monobromide			235.6									37.1
$BrGeH_3$	Bromogermane									274.8			56.4
BrH	Hydrogen bromide			-36.3			-53.4			198.7			29.1
$BrHSi$	Bromosilylene			-464.4									
BrH_3Si	Bromosilane									262.4			52.8
BrH_4N	Ammonium bromide	-270.8			-175.2			113.0			96.0		
BrI	Iodine bromide			40.8			3.7			258.8			36.4
$BrIn$	Indium(I) bromide	-175.3		-56.9	-169.0		-94.3			259.5			36.7
BrK	Potassium bromide	-393.8			-380.7			95.9			52.3		
$BrKO_3$	Potassium bromate	-360.2			-271.2			149.2			105.2		
$BrKO_4$	Potassium perbromate	-287.9			-174.4			170.1			120.2		
$BrLi$	Lithium bromide	-351.2			-342.0			74.3					
$BrNO$	Nitrosyl bromide			82.2			82.4			273.7			45.5
$BrNa$	Sodium bromide	-361.1		-143.1	-349.0		-177.1	86.8		241.2	51.4		36.3
$BrNaO_3$	Sodium bromate	-334.1			-242.6			128.9					
BrO	Bromine monoxide			125.8			108.2			237.6			32.1
BrO_2	Bromine dioxide	48.5											
$BrRb$	Rubidium bromide	-394.6			-381.8			110.0			52.8		
$BrSi$	Bromosilyldyne			209.0									38.6
$BrTl$	Thallium(I) bromide	-173.2		-37.7	-167.4			120.5					
Br_2	Bromine		0.0	30.9			3.1		152.2	245.5		75.7	36.0

Molecular formula	Name	$\Delta_f H°$/kJ mol⁻¹ Crys.	Liq.	Gas	$\Delta_f G°$/kJ mol⁻¹ Crys.	Liq.	Gas	$S°$/J mol⁻¹ K⁻¹ Crys.	Liq.	Gas	C_p/J mol⁻¹ K⁻¹ Crys.	Liq.	Gas
Br₂Ca	Calcium bromide	-682.8			-663.6			130.0					
Br₂Cd	Cadmium bromide	-316.2			-296.3			137.2			76.7		
Br₂Co	Cobalt(II) bromide	-220.9									79.5		
Br₂Cr	Chromium(II) bromide	-302.1											
Br₂Cu	Copper(II) bromide	-141.8											
Br₂Fe	Iron(II) bromide	-249.8			-238.1			140.6					
Br₂H₂Si	Dibromosilane									309.7			65.5
Br₂Hg	Mercury(II) bromide	-170.7			-153.1			172.0					
Br₂Hg₂	Mercury(I) bromide	-206.9			-181.1			218.0					
Br₂Mg	Magnesium bromide	-524.3			-503.8			117.2					
Br₂Mn	Manganese(II) bromide	-384.9											
Br₂Ni	Nickel(II) bromide	-212.1											
Br₂Pb	Lead(II) bromide	-278.7			-261.9			161.5			80.1		
Br₂Pt	Platinum(II) bromide	-82.0											
Br₂S₂	Sulfur bromide		-13.0										
Br₂Se	Selenium bromide			-21.0									
Br₂Sn	Tin(II) bromide	-243.5											
Br₂Sr	Strontium bromide	-717.6			-697.1			135.1			75.3		
Br₂Ti	Titanium(II) bromide	-402.0											
Br₂Zn	Zinc bromide	-328.7			-312.1			138.5					
Br₃ClSi	Tribromochlorosilane									377.1			95.3
Br₃Fe	Iron(III) bromide	-268.2											
Br₃Ga	Gallium(III) bromide	-386.6		-317.6	-359.8		-328.5	180.0					
Br₃HSi	Tribromosilane		-355.6			-336.4							
Br₃In	Indium(III) bromide	-428.9		-282.0					248.1	348.6			80.8
Br₃OP	Phosphorus(V) oxybromide	-458.6								359.8			89.9
Br₃P	Phosphorus(III) bromide		-184.5	-139.3		-175.7	-162.8		240.2	348.1			76.0
Br₃Pt	Platinum(III) bromide	-120.9											
Br₃Re	Rhenium(III) bromide	-167.0											
Br₃Ru	Ruthenium(III) bromide	-138.0											
Br₃Sb	Antimony(III) bromide	-259.4		-194.6	-239.3		-223.9	207.1		372.9			80.2
Br₃Sc	Scandium bromide	-743.1											
Br₃Ti	Titanium(III) bromide	-548.5			-523.8			176.6			101.7		
Br₄Ge	Germanium(IV) bromide		-347.7	-300.0		-331.4	-318.0		280.7	396.2			101.8
Br₄Pa	Protactinium(IV) bromide	-824.0			-787.8			234.0					
Br₄Pt	Platinum(IV) bromide	-156.5											
Br₄Si	Tetrabromosilane		-457.3	-415.5		-443.9	-431.8		277.8	377.9			97.1
Br₄Sn	Tin(IV) bromide	-377.4		-314.6	-350.2		-331.4	264.4		411.9			103.4
Br₄Te	Tellurium tetrabromide	-190.4											
Br₄Ti	Titanium(IV) bromide	-616.7		-549.4	-589.5		-568.2	243.5		398.4	131.5		100.8
Br₄V	Vanadium(IV) bromide			-336.8									
Br₄Zr	Zirconium(IV) bromide	-760.7											
Br₅P	Phosphorus(V) bromide	-269.9											
Br₅Ta	Tantalum(V) bromide	-598.3											
Br₆W	Tungsten(VI) bromide	-348.5											

Molecular formula	Name	$\Delta_f H°$/kJ mol⁻¹ Crys.	Liq.	Gas	$\Delta_f G°$/kJ mol⁻¹ Crys.	Liq.	Gas	$S°$/J mol⁻¹ K⁻¹ Crys.	Liq.	Gas	C_p/J mol⁻¹ K⁻¹ Crys.	Liq.	Gas
Ca	Calcium	0.0		177.8			144.0	41.6		154.9	25.9		20.8
CaCl₂	Calcium chloride	-795.4			-748.8			108.4			72.9		
CaF₂	Calcium fluoride	-1228.0			-1175.6			68.5			67.0		
CaH₂	Calcium hydride	-181.5			-142.5			41.4			41.0		
CaH₂O₂	Calcium hydroxide	-985.2			-897.5			83.4			87.5		
CaI₂	Calcium iodide	-533.5			-528.9			142.0					
CaN₂O₆	Calcium nitrate	-938.2			-742.8			193.2			149.4		
CaO	Calcium oxide	-634.9			-603.3			38.1			42.0		
CaO₄S	Calcium sulfate	-1434.5			-1322.0			106.5			99.7		
CaS	Calcium sulfide	-482.4			-477.4			56.5			47.4		
Ca₃O₈P₂	Calcium phosphate	-4120.8			-3884.7			236.0			227.8		
Cd	Cadmium	0.0		111.8				51.8		167.7	26.0		20.8
CdCl₂	Cadmium chloride	-391.5			-343.9			115.3			74.7		
CdF₂	Cadmium fluoride	-700.4			-647.7			77.4					
CdH₂O₂	Cadmium hydroxide	-560.7			-473.6			96.0					
CdI₂	Cadmium iodide	-203.3			-201.4			161.1			80.0		
CdO	Cadmium oxide	-258.4			-228.7			54.8			43.4		
CdO₄S	Cadmium sulfate	-933.3			-822.7			123.0			99.6		
CdS	Cadmium sulfide	-161.9			-156.5			64.9					
CdTe	Cadmium telluride	-92.5			-92.0			100.0					
Ce	Cerium (γ, fcc)	0.0		423.0			385.0	72.0		191.8	26.9		23.1
CeCl₃	Cerium(III) chloride	-1060.5			-984.8			151.0			87.4		
CeI₃	Cerium(III) iodide	-669.3											
CeO₂	Cerium(IV) oxide	-1088.7			-1024.6			62.3			61.6		
CeS	Cerium(II) sulfide	-459.4			-451.5			78.2			50.0		
Ce₂O₃	Cerium(III) oxide	-1796.2			-1706.2			150.6			114.6		
Cf	Californium	0.0											
Cl	Chlorine (atomic)			121.3			105.3			165.2			21.8
ClCs	Cesium chloride	-443.0			-414.5			101.2			52.5		
ClCsO₄	Cesium perchlorate	-443.1			-314.3			175.1			108.3		
ClCu	Copper(I) chloride	-137.2			-119.9			86.2			48.5		
ClF	Chlorine fluoride			-50.3			-51.8			217.9			32.1
ClFO₃	Perchloryl fluoride			-23.8			48.2			279.0			64.9
ClF₃	Chlorine trifluoride			-163.2			-123.0			281.6			63.9
ClF₅S	Sulfur chloride pentafluoride		-189.5										
ClGe	Germanium monochloride			155.2			124.2			247.0			36.9
ClGeH₃	Chlorogermane		-1065.7								263.7			54.7
ClH	Hydrogen chloride			-92.3			-95.3			186.9			29.1
ClHO	Hypochlorous acid			-78.7			-66.1			236.7			37.2
ClHO₄	Perchloric acid		-40.6										
ClH₃Si	Chlorosilane									250.7			51.0
ClH₄N	Ammonium chloride	-314.4			-202.9			94.6			84.1		
ClH₄NO₄	Ammonium perchlorate	-295.3			-88.8			186.2					
ClH₄P	Phosphonium chloride	-145.2											
ClI	Iodine chloride		-23.9	17.8		-13.6	-5.5		135.1	247.6			35.6

Molecular formula	Name	$\Delta_f H°$/kJ mol⁻¹ Crys.	Liq.	Gas	$\Delta_f G°$/kJ mol⁻¹ Crys.	Liq.	Gas	$S°$/J mol⁻¹ K⁻¹ Crys.	Liq.	Gas	C_p/J mol⁻¹ K⁻¹ Crys.	Liq.	Gas
ClIn	Indium(I) chloride	-186.2		-75.0									
ClK	Potassium chloride	-436.5		-214.6	-408.5		-233.3	82.6		239.1	51.3		36.5
ClKO₃	Potassium chlorate	-397.7			-296.3			143.1			100.3		
ClKO₄	Potassium perchlorate	-432.8			-303.1			151.0			112.4		
ClLi	Lithium chloride	-408.6			-384.4			59.3			48.0		
ClLiO₄	Lithium perchlorate	-381.0											
ClNO	Nitrosyl chloride			51.7			66.1			261.7			44.7
ClNO₂	Nitryl chloride			12.6			54.4			272.2			53.2
ClNa	Sodium chloride	-411.2			-384.1			72.1			50.5		
ClNaO₂	Sodium chlorite	-307.0											
ClNaO₃	Sodium chlorate	-365.8			-262.3			123.4					
ClNaO₄	Sodium perchlorate	-383.3			-254.9			142.3					
ClO	Chlorine monoxide			101.8			98.1			226.6			31.5
ClOV	Vanadyl chloride	-607.0			-556.0			75.0					
ClO₂	Chlorine dioxide			102.5			120.5			256.8			42.0
ClO₂	Chlorine superoxide (ClOO)			89.1			105.0			263.7			46.0
ClO₄Rb	Rubidium perchlorate	-437.2			-306.9			161.1					
ClRb	Rubidium chloride	-435.4			-407.8			95.9					
ClSi	Chlorosilylidyne			189.9							52.4		36.9
ClTl	Thallium(I) chloride	-204.1		-67.8	-184.9			111.3			50.9		
Cl₂	Chlorine			0.0						223.1			33.9
Cl₂Co	Cobalt(II) chloride	-312.5			-269.8			109.2			78.5		
Cl₂Cr	Chromium(II) chloride	-395.4			-356.0			115.3			71.2		
Cl₂CrO₂	Chromyl chloride		-579.5	-538.1		-510.8	-501.6		221.8	329.8			84.5
Cl₂Cu	Copper(II) chloride	-220.1			-175.7			108.1			71.9		
Cl₂Fe	Iron(II) chloride	-341.8			-302.3			118.0			76.7		
Cl₂H₂Si	Dichlorosilane									285.7			60.5
Cl₂Hg	Mercury(II) chloride	-224.3			-178.6			146.0					
Cl₂Hg₂	Mercury(I) chloride	-265.4			-210.7			191.6					
Cl₂Mg	Magnesium chloride	-641.3			-591.8			89.6			71.4		
Cl₂Mn	Manganese(II) chloride	-481.3			-440.5			118.2			72.9		
Cl₂Ni	Nickel(II) chloride	-305.3			-259.0			97.7			71.7		
Cl₂O	Chlorine oxide			80.3			97.9			266.2			45.4
Cl₂OS	Thionyl chloride		-245.6	-212.5			-198.3			309.8		121.0	66.5
Cl₂O₂S	Sulfuryl chloride		-394.1	-364.0			-320.0			311.9		134.0	77.0
Cl₂O₂U	Uranyl chloride	-1243.9			-1146.4			150.5			107.9		
Cl₂Pb	Lead(II) chloride	-359.4			-314.1			136.0					
Cl₂Pt	Platinum(II) chloride	-123.4											
Cl₂S	Sulfur dichloride		-50.0										
Cl₂S₂	Sulfur chloride		-59.4										
Cl₂Sn	Tin(II) chloride	-325.1											
Cl₂Sr	Strontium chloride	-828.9			-781.1			114.9			75.6		
Cl₂Ti	Titanium(II) chloride	-513.8			-464.4			87.4			69.8		
Cl₂Zn	Zinc chloride	-415.1			-369.4			111.5			71.3		
Cl₂Zr	Zirconium(II) chloride	-502.0		-266.1									

STANDARD THERMODYNAMIC PROPERTIES OF CHEMICAL SUBSTANCES (continued)

Molecular formula	Name	$\Delta_fH°$/kJ mol⁻¹			$\Delta_fG°$/kJ mol⁻¹			$S°$/J mol⁻¹ K⁻¹			C_p/J mol⁻¹ K⁻¹		
		Crys.	Liq.	Gas	Crys.	Liq.	Gas	Crys.	Liq.	Gas	Crys.	Liq.	Gas
Cl_3Cr	Chromium(III) chloride	-556.5			-486.1			123.0			91.8		
Cl_3Dy	Dysprosium(III) chloride	-1000.0									100.0		
Cl_3Er	Erbium chloride	-998.7											
Cl_3Eu	Europium(II) chloride	-936.0									96.7		
Cl_3Fe	Iron(III) chloride	-399.5			-334.0			142.3					
Cl_3Ga	Gallium(III) chloride	-524.7			-454.8			142.0					
Cl_3Gd	Gadolinium(III) chloride	-1008.0									88.0		
Cl_3HSi	Trichlorosilane		-539.3	-513.0		-482.5	-482.0		227.6	313.9			75.8
Cl_3Ho	Holmium chloride	-1005.4									88.0		
Cl_3In	Indium(III) chloride	-537.2		-374.0									
Cl_3Ir	Iridium(III) chloride	-245.6											
Cl_3La	Lanthanum chloride	-1072.2									108.8		
Cl_3Lu	Lutetium chloride	-945.6		-649.0									
Cl_3N	Nitrogen trichloride		230.0										
Cl_3Nd	Neodymium chloride	-1041.0									113.0		
Cl_3OP	Phosphorus(V) oxychloride		-597.1	-558.5		-520.8	-512.9		222.5	325.5		138.8	84.9
Cl_3OV	Vanadyl trichloride		-734.7	-695.6		-668.5	-659.3		244.3	344.3			89.9
Cl_3Os	Osmium(III) chloride	-190.4											
Cl_3P	Phosphorus(III) chloride		-319.7	-287.0		-272.3	-267.8		217.1	311.8	100.0		71.8
Cl_3Pr	Praseodymium(III) chloride	-1056.9											
Cl_3Pt	Platinum(III) chloride	-182.0			-188.0			123.8			92.4		
Cl_3Re	Rhenium(III) chloride	-264.0											
Cl_3Rh	Rhodium(III) chloride	-299.2											
Cl_3Ru	Ruthenium(III) chloride	-205.0											
Cl_3Sb	Antimony(III) chloride	-382.2			-323.7			184.1			107.9		
Cl_3Sc	Scandium chloride	-925.1											
Cl_3Sm	Samarium(III) chloride	-1025.9											
Cl_3Tb	Terbium chloride	-997.0											
Cl_3Ti	Titanium(III) chloride	-720.9			-653.5			139.7			97.2		
Cl_3Tl	Thallium(III) chloride	-315.1											
Cl_3Tm	Thulium chloride	-986.6											
Cl_3U	Uranium(III) chloride	-866.5			-799.1			159.0			102.5		
Cl_3V	Vanadium(III) chloride	-580.7			-511.2			131.0			93.2		
Cl_3Y	Yttrium chloride	-1000.0		-750.2									75.0
Cl_3Yb	Ytterbium(III) chloride	-959.8											
Cl_4Ge	Germanium(IV) chloride		-531.8	-495.8		-462.7	-457.3		245.6	347.7	120.5		96.1
Cl_4Hf	Hafnium(IV) chloride	-990.4		-884.5	-901.3			190.8					
Cl_4Pa	Protactinium(IV) chloride	-1043.0			-953.0			192.0					
Cl_4Pb	Lead(IV) chloride		-329.3										
Cl_4Pt	Platinum(IV) chloride	-231.8											
Cl_4Si	Tetrachlorosilane		-687.0	-657.0		-619.8	-617.0		239.7	330.7		145.3	90.3
Cl_4Sn	Tin(IV) chloride		-511.3	-471.5		-440.1	-432.2		258.6	365.8		165.3	98.3
Cl_4Te	Tellurium tetrachloride	-326.4									138.5		
Cl_4Th	Thorium(IV) chloride	-1186.2		-964.4	-1094.1		-932.0	190.4		390.7	120.3		107.5
Cl_4Ti	Titanium(IV) chloride		-804.2	-763.2		-737.2	-726.3		252.3	353.2		145.2	95.4

STANDARD THERMODYNAMIC PROPERTIES OF CHEMICAL SUBSTANCES (continued)

Molecular formula	Name	$\Delta_f H°$/kJ mol⁻¹			$\Delta_f G°$/kJ mol⁻¹			S°/J mol⁻¹ K⁻¹			C_p/J mol⁻¹ K⁻¹		
		Crys.	Liq.	Gas	Crys.	Liq.	Gas	Crys.	Liq.	Gas	Crys.	Liq.	Gas
Cl₄U	Uranium(IV) chloride	-1019.2		-809.6	-930.0		-786.6	197.1		419.0	122.0		
Cl₄V	Vanadium(IV) chloride		-569.4	-525.5		-503.7	-492.0		255.0	362.4			96.2
Cl₄Zr	Zirconium(IV) chloride	-980.5			-889.9			181.6			119.8		
Cl₅Nb	Niobium(V) chloride	-797.5		-703.7	-683.2		-646.0	210.5		400.6	148.1		120.8
Cl₅P	Phosphorus(V) chloride	-443.5		-374.9			-305.0			364.6			112.8
Cl₅Pa	Protactinium(V) chloride	-1145.0			-1034.0			238.0					
Cl₅Ta	Tantalum(V) chloride	-859.0											
Cl₆U	Uranium(VI) chloride	-1092.0		-1013.0	-962.0		-928.0	285.8		431.0	175.7		
Cl₆W	Tungsten(VI) chloride	-602.5		-513.8									
Cm	Curium	0.0											
Co	Cobalt	0.0		424.7			380.3	30.0		179.5	24.8		23.0
CoF₂	Cobalt(II) fluoride	-692.0			-647.2			82.0			68.8		
CoH₂O₂	Cobalt(II) hydroxide	-539.7			-454.3			79.0					
CoI₂	Cobalt(II) iodide	-88.7											
CoN₂O₆	Cobalt(II) nitrate	-420.5											
CoO	Cobalt(II) oxide	-237.9			-214.2			53.0			55.2		
CoO₄S	Cobalt(II) sulfate	-888.3			-782.3			118.0					
CoS	Cobalt(II) sulfide	-82.8											
Co₂S₃	Cobalt(III) sulfide	-147.3											
Co₃O₄	Cobalt(II,III) oxide	-891.0			-774.0			102.5			123.4		
Cr	Chromium	0.0		396.6			351.8	23.8		174.5	23.4		20.8
CrF₂	Chromium(II) fluoride	-778.0											
CrF₃	Chromium(III) fluoride	-1159.0			-1083.0			93.9			78.7		
CrI₂	Chromium(II) iodide	-156.9											
CrI₃	Chromium(III) iodide	-205.0											
CrO₂	Chromium(IV) oxide	-598.0											
CrO₄Pb	Lead(II) chromate	-930.9											
Cr₂FeO₄	Chromium iron oxide	-1444.7			-1343.8			146.0			133.6		
Cr₂O₃	Chromium(III) oxide	-1139.7			-1058.1			81.2			118.7		
Cr₃O₄	Chromium(II,III) oxide	-1531.0											
Cs	Cesium	0.0		76.5			49.6	85.2		175.6	32.2		20.8
CsF	Cesium fluoride	-553.5			-525.5			92.8			51.1		
CsF₂H	Cesium hydrogen fluoride	-923.8			-858.9			135.2			87.3		
CsH	Cesium hydride	-54.2											
CsHO	Cesium hydroxide	-417.2											
CsHO₄S	Cesium hydrogen sulfate	-1158.1											
CsH₂N	Cesium amide	-118.4											
CsI	Cesium iodide	-346.6			-340.6			123.1			52.8		
CsNO₃	Cesium nitrate	-506.0			-406.5			155.2					
CsO₂	Cesium superoxide	-286.2											
Cs₂O	Cesium oxide	-345.8			-308.1			146.9			76.0		
Cs₂O₃S	Cesium sulfite	-1134.7											
Cs₂O₄S	Cesium sulfate	-1443.0			-1323.6			211.9			134.9		
Cs₂S	Cesium sulfide	-359.8											
Cu	Copper	0.0		337.4			297.7	33.2		166.4	24.4		20.8

Molecular formula	Name	$\Delta_f H°$/kJ mol⁻¹			$\Delta_f G°$/kJ mol⁻¹			$S°$/J mol⁻¹ K⁻¹			C_p/J mol⁻¹ K⁻¹		
		Crys.	Liq.	Gas	Crys.	Liq.	Gas	Crys.	Liq.	Gas	Crys.	Liq.	Gas
CuF_2	Copper(II) fluoride	-542.7											
CuH_2O_2	Copper(II) hydroxide	-449.8											
CuI	Copper(I) iodide	-67.8			-69.5			96.7			54.1		
CuN_2O_6	Copper(II) nitrate	-302.9											
CuO	Copper(II) oxide	-157.3			-129.7			42.6			42.3		
CuO_4S	Copper(II) sulfate	-771.4			-662.2			109.2					
CuO_4W	Copper(II) tungstate	-1105.0											
CuS	Copper(II) sulfide	-53.1			-53.6			66.5			47.8		
$CuSe$	Copper(II) selenide	-39.5											
Cu_2	Dicopper			484.2			431.9			241.6			36.6
Cu_2O	Copper(I) oxide	-168.6			-146.0			93.1			63.6		
Cu_2S	Copper(I) sulfide	-79.5			-86.2			120.9			76.3		
Dy	Dysprosium	0.0		290.4			254.4	75.6		196.6	27.7		20.8
Dy_2O_3	Dysprosium(III) oxide	-1863.1			-1771.5			149.8			116.3		
Er	Erbium	0.0		317.1			280.7	73.2		195.6	28.1		20.8
ErF_3	Erbium fluoride	-1711.0											
Er_2O_3	Erbium oxide	-1897.9			-1808.7			155.6			108.5		
Es	Einsteinium	0.0											
Eu	Europium	0.0		175.3			142.2	77.8		188.8	27.7		20.8
Eu_2O_3	Europium(III) oxide	-1651.4			-1556.8			146.0			122.2		
Eu_3O_4	Europium(II,III) oxide	-2272.0			-2142.0			205.0					
F	Fluorine (atomic)			79.4			62.3			158.8			22.7
FGa	Gallium monofluoride			-251.9									33.3
FGe	Germanium monofluoride			-33.4									34.7
$FGeH_3$	Fluorogermane									252.8			51.6
FH	Hydrogen fluoride		-299.8	-273.3			-275.4			173.8			
FH_3Si	Fluorosilane									238.4			47.4
FH_4N	Ammonium fluoride	-464.0			-348.7			72.0			65.3		
FI	Iodine monofluoride			-95.7			-118.5			236.2			33.4
FIn	Indium monofluoride			-203.4									
FK	Potassium fluoride	-567.3			-537.8			66.6			49.0		
FLi	Lithium fluoride	-616.0			-587.7			35.7			41.6		
FNO	Nitrosyl fluoride			-66.5			-51.0			248.1			41.3
FNO_2	Nitryl fluoride									260.4			49.8
FNS	Thionitrosyl fluoride (NSF)									259.8			44.1
FNa	Sodium fluoride	-576.6			-546.3			51.1			46.9		
FO	Fluorine monoxide			109.0			105.0			216.8			30.5
FRb	Rubidium fluoride	-557.7											
FSi	Fluorosilylidyne			7.1			-24.3			225.8			32.6
FTl	Thallium(I) fluoride	-324.7		-182.4									
F_2	Fluorine			0.0						202.8			31.3
F_2Fe	Iron(II) fluoride	-711.3			-668.6			87.0			68.1		
F_2HK	Potassium hydrogen fluoride	-927.7			-859.7			104.3			76.9		
F_2HN	Difluoramine									252.8			43.4
F_2HNa	Sodium hydrogen fluoride	-920.3			-852.2			90.9			75.0		

Molecular formula	Name	ΔfH°/kJ mol⁻¹ Crys.	Liq.	Gas	ΔfG°/kJ mol⁻¹ Crys.	Liq.	Gas	S°/J mol⁻¹ K⁻¹ Crys.	Liq.	Gas	Cp/J mol⁻¹ K⁻¹ Crys.	Liq.	Gas
F2HRb	Rubidium hydrogen fluoride	-922.6			-855.6			120.1			79.4		
F2Mg	Magnesium fluoride	-1124.2			-1071.1			57.2			61.6		
F2N	Difluoroamidogen			43.1			57.8			249.9			41.0
F2N2	cis-Difluorodiazine			69.5									
F2N2	trans-Difluorodiazine			82.0									
F2Ni	Nickel(II) fluoride	-651.4			-604.1			73.6			64.1		
F2O	Fluorine monoxide			24.7			41.9			247.4			43.3
F2OS	Thionyl fluoride									278.7			56.8
F2O2	Fluorine dioxide			18.0									
F2O2S	Sulfuryl fluoride									284.0			66.0
F2O2U	Uranyl fluoride	-1653.5			-1557.4			135.6			103.2		
F2Pb	Lead(II) fluoride	-664.0			-617.1			110.5					
F2Si	Difluorosilylene			-619.0			-628.0			252.7			43.9
F2Sr	Strontium fluoride	-1216.3			-1164.8			82.1			70.0		
F2Zn	Zinc fluoride	-764.4			-713.3			73.7			65.7		
F3Ga	Gallium(III) fluoride	-1163.0			-1085.3			84.0					
F3Gd	Gadolinium(III) fluoride			-1297.0									
F3HSi	Trifluorosilane									271.9			60.5
F3Ho	Holmium fluoride	-1707.0											
F3N	Nitrogen trifluoride			-132.1			-90.6			260.8			53.4
F3Nd	Neodymium fluoride	-1657.0											
F3OP	Phosphorus(V) oxyfluoride			-1254.3			-1205.8			285.4			68.8
F3P	Phosphorus(III) fluoride			-958.4			-936.9			273.1			58.7
F3Sb	Antimony(III) fluoride	-915.5		-1247.0			-1234.0			300.5			67.8
F3Sc	Scandium fluoride	-1629.2						92.0					
F3Sm	Samarium(III) fluoride	-1778.0											
F3Th	Thorium(III) fluoride	-1502.1		-1166.1	-1433.4		-1160.6	123.4		339.2	95.1		73.3
F3U	Uranium(III) fluoride	-1718.8		-1058.5	-1644.7		-1051.9	100.0		331.9			74.3
F3Y	Yttrium fluoride			-1288.7			-1277.8			311.8			
F4Ge	Germanium(IV) fluoride			-1190.2			-1150.0			301.9			70.3
F4Hf	Hafnium fluoride	-1930.5		-1669.8	-1830.4			113.0		301.2			79.2
F4N2	Tetrafluorohydrazine			-8.4			79.9						
F4Pb	Lead(IV) fluoride	-941.8											
F4S	Sulfur tetrafluoride			-763.2			-722.0			299.6			77.6
F4Si	Tetrafluorosilane			-1615.0			-1572.8			282.8			73.6
F4Th	Thorium(IV) fluoride	-2097.8		-1759.0	-2003.4		-1724.0	142.0		341.7	110.7		93.0
F4U	Uranium(IV) fluoride	-1914.2		-1598.7	-1823.3		-1572.7	151.7		368.0	116.0		91.2
F4V	Vanadium(IV) fluoride	-1403.3											
F4Xe	Xenon tetrafluoride	-261.5											
F4Zr	Zirconium(IV) fluoride	-1911.3			-1809.9			104.6			103.7		
F5I	Iodine pentafluoride		-864.8	-822.5			-751.7			327.7			99.2
F5Nb	Niobium(V) fluoride	-1813.8		-1739.7	-1699.0		-1673.6	160.2		321.9	134.7		97.1
F5P	Phosphorus(V) fluoride			-1594.4			-1520.7			300.8			84.8
F5Ta	Tantalum(V) fluoride	-1903.6											
F5V	Vanadium(V) fluoride		-1480.3	-1433.9		-1373.1	-1369.8		175.7	320.9			98.6

Molecular formula	Name	$\Delta_f H°$/kJ mol⁻¹ Crys.	Liq.	Gas	$\Delta_f G°$/kJ mol⁻¹ Crys.	Liq.	Gas	$S°$/J mol⁻¹ K⁻¹ Crys.	Liq.	Gas	C_p/J mol⁻¹ K⁻¹ Crys.	Liq.	Gas
F₆H₈N₂Si	Ammonium hexafluorosilicate	-2681.7			-2365.3			280.2			228.1		
F₆Ir	Iridium(VI) fluoride	-579.7		-544.0	-461.6		-460.0	247.7		357.8			121.1
F₆K₂Si	Potassium hexafluorosilicate	-2956.0			-2798.6			226.0					
F₆Mo	Molybdenum(VI) fluoride		-1585.5	-1557.7		-1473.0	-1472.2		259.7	350.5		169.8	120.6
F₆Na₂Si	Sodium hexafluorosilicate	-2909.6			-2754.2			207.1			187.1		
F₆Os	Osmium(VI) fluoride							246.0		358.1			120.8
F₆Pt	Platinum(VI) fluoride							235.6		348.3			122.8
F₆S	Sulfur hexafluoride			-1220.5			-1116.5			291.5			97.0
F₆Se	Selenium hexafluoride			-1117.0			-1017.0			313.9			110.5
F₆Te	Tellurium hexafluoride			-1318.0									
F₆U	Uranium(VI) fluoride	-2197.0		-2147.4	-2068.5		-2063.7	227.6		377.9	166.8		129.6
F₆W	Tungsten(VI) fluoride		-1747.7	-1721.7		-1631.4	-1632.1		251.5	341.1			119.0
Fe	Iron	0.0		416.3			370.7	27.3		180.5	25.1		25.7
FeI₂	Iron(II) iodide	-113.0											
FeI₃	Iron(III) iodide			71.0									
FeMoO₄	Iron(II) molybdate	-1075.0			-975.0			129.3			118.5		
FeO	Iron(II) oxide	-272.0											
FeO₄S	Iron(II) sulfate	-928.4			-820.8			107.5			100.6		
FeO₄W	Iron(II) tungstate	-1155.0			-1054.0			131.8			114.6		
FeS	Iron(II) sulfide	-100.0			-100.4			60.3			50.5		
FeS₂	Iron disulfide	-178.2			-166.9			52.9			62.2		
Fe₂O₃	Iron(III) oxide	-824.2			-742.2			87.4			103.9		
Fe₂O₄Si	Iron(II) orthosilicate	-1479.9			-1379.0			145.2			132.9		
Fe₃O₄	Iron(II,III) oxide	-1118.4			-1015.4			146.4			143.4		
Fm	Fermium	0.0											
Fr	Francium	0.0						95.4					
Ga	Gallium	0.0	5.6	272.0	0.0		233.7	40.8		169.0	26.1		25.3
GaH₃O₃	Gallium(III) hydroxide	-964.4			-831.3			100.0			100.0		
GaI₃	Gallium(III) iodide	-238.9						205.0					
GaN	Gallium nitride	-110.5											
GaO	Gallium monoxide			279.5			253.5			231.1			32.1
GaP	Gallium phosphide	-88.0											
GaSb	Gallium antimonide	-41.8			-38.9			76.1			48.5		
Ga₂	Digallium			438.5									
Ga₂O	Gallium suboxide	-356.0											
Ga₂O₃	Gallium(III) oxide	-1089.1			-998.3			85.0			92.1		
Gd	Gadolinium	0.0		397.5			359.8	68.1		194.3	37.0		27.5
Gd₂O₃	Gadolinium(III) oxide	-1819.6									106.7		
Ge	Germanium	0.0		372.0			331.2	31.1		167.9	23.3		30.7
GeH₃I	Iodogermane									283.2			57.5
GeH₄	Germane			90.8			113.4			217.1			45.0
GeI₄	Germanium(IV) iodide	-141.8		-56.9	-144.3		-106.3	271.1		428.9			104.1
GeO	Germanium(II) oxide	-261.9		-46.2	-237.2		-73.2	50.0		224.3			30.9
GeO₂	Germanium(IV) oxide	-580.0			-521.4			39.7			52.1		

Molecular formula	Name	$\Delta_f H°$/kJ mol⁻¹			$\Delta_f G°$/kJ mol⁻¹			$S°$/J mol⁻¹ K⁻¹			C_p/J mol⁻¹ K⁻¹		
		Crys.	Liq.	Gas	Crys.	Liq.	Gas	Crys.	Liq.	Gas	Crys.	Liq.	Gas
GeP	Germanium phosphide	-21.0			-17.0			63.0					
GeS	Germanium(II) sulfide	-69.0		92.0	-71.5		42.0	71.0		234.0			33.7
GeTe	Germanium(II) telluride	20.0											
Ge₂	Digermanium			473.1			416.3			252.8			35.6
Ge₂H₆	Digermane		137.3	162.3									
Ge₃H₈	Trigermane		193.7	226.8									
H	Hydrogen (atomic)			218.0			203.3			114.7			20.8
HI	Hydrogen iodide			26.5			1.7			206.6			29.2
HIO₃	Iodic acid	-230.1											
HK	Potassium hydride	-57.7											
HKO	Potassium hydroxide	-424.6		-228.0	-378.7		-229.7	78.9		238.3	64.9		49.2
HKO₄S	Potassium hydrogen sulfate	-1160.6			-1031.3			138.1					
HLi	Lithium hydride	-90.5			-68.3			20.0			27.9		
HLiO	Lithium hydroxide	-484.9			-439.0			42.8			49.7		
HN	Imidogen			351.5			345.6			181.2			29.2
HNO₂	Nitrous acid			-79.5			-46.0			254.1			45.6
HNO₃	Nitric acid		-174.1	-133.9		-80.7	-73.5		155.6	266.9		109.9	54.1
HN₃	Hydrazoic acid		264.0	294.1		327.3	328.1		140.6	239.0			43.7
HNa	Sodium hydride	-56.3			-33.5			40.0			36.4		
HNaO	Sodium hydroxide	-425.6			-379.5			64.5			59.5		
HNaO₄S	Sodium hydrogen sulfate	-1125.5			-992.8			113.0					
HNa₂O₄P	Sodium hydrogen phosphate	-1748.1			-1608.2			150.5			135.3		
HO	Hydroxyl			39.0			34.2			183.7			29.9
HORb	Rubidium hydroxide	-418.2			-195.8			88.0					
HOTl	Thallium(I) hydroxide	-238.9											
HO₂	Hydroperoxy			10.5			22.6			229.0			34.9
HO₃P	Metaphosphoric acid	-948.5			-656.4								
HO₄RbS	Rubidium hydrogen sulfate	-1159.0											
HO₄Re	Perrhenic acid	-762.3						158.2					
HRb	Rubidium hydride	-52.3											
HS	Mercapto			142.7			113.3			195.7			32.3
HSi	Silylidyne			361.0									
HTa₂	Tantalum hydride	-32.6			-69.0			79.1			90.8		
H₂	Hydrogen			0.0						130.7			28.8
H₂KN	Potassium amide	-128.9											
H₂KO₄P	Potassium dihydrogen phosphate	-1568.3			-1415.9			134.9			116.6		
H₂LiN	Lithium amide	-179.5											
H₂Mg	Magnesium hydride	-75.3			-35.9			31.1			35.4		
H₂MgO₂	Magnesium hydroxide	-924.5			-833.5			63.2			77.0		
H₂N	Amidogen			184.9			194.6			195.0			33.9
H₂NNa	Sodium amide	-123.8			-64.0			76.9			66.2		
H₂NRb	Rubidium amide	-113.0											
H₂N₂O₂	Nitramide	-89.5											
H₂NiO₂	Nickel(II) hydroxide	-529.7			-447.2			88.0					

Molecular formula	Name	ΔfH°/kJ mol⁻¹			ΔfG°/kJ mol⁻¹			S°/J mol⁻¹ K⁻¹			C_p/J mol⁻¹ K⁻¹		
		Crys.	Liq.	Gas	Crys.	Liq.	Gas	Crys.	Liq.	Gas	Crys.	Liq.	Gas
H_2O	Water		-285.8	-241.8		-237.1	-228.6		70.0	188.8		75.3	33.6
H_2O_2	Hydrogen peroxide		-187.8	-136.3		-120.4	-105.6		109.6	232.7		89.1	43.1
H_2O_2Sn	Tin(II) hydroxide	-561.1			-491.6			155.0					
H_2O_2Sr	Strontium hydroxide	-959.0											
H_2O_2Zn	Zinc hydroxide	-641.9			-553.5			81.2					
H_2O_3Si	Metasilicic acid	-1188.7			-1092.4			134.0					
H_2O_4S	Sulfuric acid		-814.0			-690.0			156.9			138.9	
H_2O_4Se	Selenic acid	-530.1											
H_2S	Hydrogen sulfide			-20.6			-33.4			205.8			34.2
H_2S_2	Hydrogen disulfide		-18.1	15.5								84.1	51.5
H_2Se	Hydrogen selenide			29.7			15.9			219.0			34.7
H_2Sr	Strontium hydride	-180.3											
H_2Te	Hydrogen telluride			99.6									
H_2Th	Thorium hydride	-139.7			-100.0			50.7			36.7		
H_2Zr	Zirconium(II) hydride	-169.0			-128.8			35.0			31.0		
H_3ISi	Iodosilane									270.9			54.4
H_3N	Ammonia			-45.9			-16.4			192.8·			35.1
H_3NO	Hydroxylamine	-114.2											
H_3O_2P	Hypophosphorous acid	-604.6	-595.4										
H_3O_3P	Phosphorous acid	-964.4											
H_3O_4P	Phosphoric acid (orthophosphoric acid)	-1284.4	-1271.7		-1124.3	-1123.6		110.5	150.8		106.1	145.0	
H_3P	Phosphine			5.4			13.4			210.2			37.1
H_3Sb	Stibine			145.1			147.8			232.8			41.1
H_3U	Uranium(III) hydride	-127.2			-72.8			63.7			49.3		
H_4IN	Ammonium iodide	-201.4			-112.5			117.0					
H_4N_2	Hydrazine		50.6	95.4		149.3	159.4		121.2	238.5		98.9	48.4
$H_4N_2O_2$	Ammonium nitrite	-256.5			-183.9			151.1			139.3		
$H_4N_2O_3$	Ammonium nitrate	-365.6						112.5					
H_4N_4	Ammonium azide	115.5			274.2			192.0					
H_4O_4Si	Orthosilicic acid	-1481.1			-1332.9								
$H_4O_7P_2$	Pyrophosphoric acid	-2241.0	-2231.7										
H_4P_2	Diphosphine		-5.0	20.9			56.9						
H_4Si	Silane			34.3			188.3			204.6			42.8
H_4Sn	Stannane			162.8					165.6	227.7		154.9	49.0
H_5NO	Ammonium hydroxide		-361.2			-254.0							
H_5NO_3S	Ammonium hydrogen sulfite	-768.6											
H_5NO_4S	Ammonium hydrogen sulfate	-1027.0											
H_6Si_2	Disilane			80.3			127.3			272.7			80.8
$H_8N_2O_4S$	Ammonium sulfate	-1180.9			-901.7			220.1			187.5		
H_8Si_3	Trisilane		92.5	120.9									
$H_9N_2O_4P$	Diammonium hydrogen phosphate	-1566.9									188.0		
$H_{12}N_3O_4P$	Ammonium phosphate	-1671.9								126.2			
He	Helium			0.0									20.8

STANDARD THERMODYNAMIC PROPERTIES OF CHEMICAL SUBSTANCES (continued)

Molecular formula	Name	$\Delta_f H°$/kJ mol^{-1}			$\Delta_f G°$/kJ mol^{-1}			$S°$/J mol^{-1} K^{-1}			C_p/J mol^{-1} K^{-1}		
		Crys.	Liq.	Gas	Crys.	Liq.	Gas	Crys.	Liq.	Gas	Crys.	Liq.	Gas
Hf	Hafnium	0.0		619.2			576.5	43.6		186.9	25.7		20.8
HfO_2	Hafnium oxide	-1144.7			-1088.2			59.3			60.3		
Hg	Mercury		0.0	61.4			31.8		75.9	175.0		28.0	20.8
HgI_2	Mercury(II) iodide	-105.4			-101.7			180.0					
HgO	Mercury(II) oxide	-90.8			-58.5			70.3			44.1		
HgO_4S	Mercury(II) sulfate	-707.5											
HgS	Mercury(II) sulfide	-58.2			-50.6			82.4			48.4		
HgTe	Mercury(II) telluride	-42.0											
Hg_2	Dimercury			108.8			68.2			288.1			37.4
Hg_2I_2	Mercury(I) iodide	-121.3			-111.0			233.5					
Hg_2O_4S	Mercury(I) sulfate	-743.1			-625.8			200.7			132.0		
Ho	Holmium	0.0		300.8			264.8	75.3		195.6	27.2		20.8
Ho_2O_3	Holmium oxide	-1880.7			-1791.1			158.2			115.0		
I	Iodine (atomic)			106.8			70.2			180.8			20.8
IIn	Indium(I) iodide	-116.3		7.5	-120.5		-37.7	130.0		267.3			36.8
IK	Potassium iodide	-327.9			-324.9			106.3			52.9		
IKO_3	Potassium iodate	-501.4			-418.4			151.5			106.5		
IKO_4	Potassium periodate	-467.2			-361.4			175.7					
ILi	Lithium iodide	-270.4			-270.3			86.8			51.0		
INa	Sodium iodide	-287.8			-286.1			98.5			52.1		
$INaO_3$	Sodium iodate	-481.8									92.0		
$INaO_4$	Sodium periodate	-429.3			-323.0			163.0					
IO	Iodine monoxide			175.1			149.8			245.5			32.9
IRb	Rubidium iodide	-333.8			-328.9			118.4			53.2		
ITl	Thallium(I) iodide	-123.8		7.1	-125.4			127.6					
I_2	Iodine (rhombic)	0.0		62.4			19.3	116.1		260.7	54.4		36.9
I_2Mg	Magnesium iodide	-364.0			-358.2			129.7					
I_2Ni	Nickel(II) iodide	-78.2											
I_2Pb	Lead(II) iodide	-175.5			-173.6			174.9			77.4		
I_2Sn	Tin(II) iodide	-143.5											
I_2Sr	Strontium iodide	-558.1									81.6		
I_2Zn	Zinc iodide	-208.0		-120.5	-209.0			161.1					
I_3In	Indium(III) iodide	-238.0											
I_3La	Lanthanum iodide	-668.9											
I_3Lu	Lutetium iodide	-548.0											
I_3P	Phosphorus(III) iodide	-45.6								374.4			78.4
I_3Ru	Ruthenium(III) iodide	-65.7											
I_3Sb	Antimony(III) iodide	-100.4											
I_4Pt	Platinum(IV) iodide	-72.8											
I_4Si	Tetraiodosilane	-189.5											
I_4Sn	Tin(IV) iodide										84.9		
I_4Ti	Titanium(IV) iodide	-375.7		-277.8	-371.5			249.4		446.1	125.7		105.4
I_4V	Vanadium(IV) iodide			-122.6									
I_4Zr	Zirconium(IV) iodide	-481.6											
In	Indium	0.0		243.3			208.7	57.8		173.8	26.7		20.8

STANDARD THERMODYNAMIC PROPERTIES OF CHEMICAL SUBSTANCES (continued)

Molecular formula	Name	$\Delta_f H°$/kJ mol⁻¹ Crys.	Liq.	Gas	$\Delta_f G°$/kJ mol⁻¹ Crys.	Liq.	Gas	$S°$/J mol⁻¹ K⁻¹ Crys.	Liq.	Gas	C_p/J mol⁻¹ K⁻¹ Crys.	Liq.	Gas
InO	Indium monoxide			387.0			364.4			236.5			32.6
InP	Indium phosphide	-88.7			-77.0			59.8			45.4		
InS	Indium(II) sulfide	-138.1		238.0	-131.8			67.0			49.5		
InSb	Indium antimonide	-30.5		344.3	-25.5			86.2					
In₂	Diindium		6.3	380.9									
In₂O₃	Indium(III) oxide	-925.8			-830.7			104.2			92.0		
In₂S₃	Indium(III) sulfide	-427.0			-412.5			163.6			118.0		
Ir	Iridium	0.0		665.3	0.0		617.9	35.5		193.6	25.1		20.8
IrO₂	Iridium(IV) oxide	-274.1									57.3		
IrS₂	Iridium(IV) sulfide	-138.0											
Ir₂S₃	Iridium(III) sulfide	-234.0											
K	Potassium	0.0		89.0	0.0		60.5	64.7		160.3	29.6		20.8
KMnO₄	Potassium permanganate	-837.2			-737.6			171.7			117.6		
KNO₂	Potassium nitrite	-369.8			-306.6			152.1			107.4		
KNO₃	Potassium nitrate	-494.6			-394.9			133.1			96.4		
KNa	Potassium sodium												
KO₂	Potassium superoxide	-284.9			-239.4			116.7			77.5		
K₂	Dipotassium			123.7			87.5			249.7			37.9
K₂O	Potassium oxide	-361.5						102.1					
K₂O₂	Potassium peroxide	-494.1			-425.1								
K₂O₄S	Potassium sulfate	-1437.8			-1321.4			175.6			131.5		
K₂S	Potassium sulfide	-380.7			-364.0			105.0					
K₃O₄P	Potassium phosphate	-1950.2											
Kr	Krypton			0.0						164.1			20.8
La	Lanthanum	0.0		431.0	0.0		393.6	56.9		182.4	27.1		22.8
LaS	Lanthanum sulfide	-456.0			-451.5			73.2			59.0		
La₂O₃	Lanthanum oxide	-1793.7			-1705.8			127.3			108.8		
Li	Lithium	0.0		159.3	0.0		126.6	29.1		138.8	24.8		20.8
LiNO₂	Lithium nitrite	-372.4			-302.0			96.0					
LiNO₃	Lithium nitrate	-483.1			-381.1			90.0					
Li₂	Dilithium			215.9			174.4			197.0			36.1
Li₂O	Lithium oxide	-597.9			-561.2			37.6			54.1		
Li₂O₂	Lithium peroxide	-634.3											
Li₂O₃Si	Lithium metasilicate	-1648.1			-1557.2			79.8			99.1		
Li₂O₄S	Lithium sulfate	-1436.5			-1321.7			115.1			117.6		
Li₂S	Lithium sulfide	-441.4											
Li₃O₄P	Lithium phosphate	-2095.8											
Lr	Lawrencium	0.0		427.6	0.0		387.8			184.8			20.9
Lu	Lutetium	0.0			0.0			51.0			26.9		
Lu₂O₃	Lutetium oxide	-1878.2			-1789.0			110.0			101.8		
Md	Mendelevium	0.0			0.0								
Mg	Magnesium	0.0		147.1	0.0		112.5	32.7		148.6	24.9		20.8
MgN₂O₆	Magnesium nitrate	-790.7			-589.4			164.0			141.9		
MgO	Magnesium oxide	-601.6			-569.3			27.0			37.2		
MgO₄S	Magnesium sulfate	-1284.9			-1170.6			91.6			96.5		

Molecular formula	Name	$\Delta_f H°$/kJ mol⁻¹			$\Delta_f G°$/kJ mol⁻¹			$S°$/J mol⁻¹ K⁻¹			C_p/J mol⁻¹ K⁻¹		
		Crys.	Liq.	Gas	Crys.	Liq.	Gas	Crys.	Liq.	Gas	Crys.	Liq.	Gas
MgO₄Se	Magnesium selenate	-968.5											
MgS	Magnesium sulfide	-346.0			-341.3			50.3			45.6		
Mg₂	Dimagnesium			287.7									
Mg₂O₄Si	Magnesium orthosilicate	-2174.0			-2055.1			95.1			118.5		
Mn	Manganese	0.0		280.7			238.5	32.0		173.7	26.3		20.8
MnN₂O₆	Manganese(II) nitrate	-576.3											
MnNa₂O₄	Sodium permanganate	-1156.0											
MnO	Manganese(II) oxide	-385.2			-352.9			59.7			45.4		
MnO₂	Manganese(IV) oxide	-520.0			-455.1			53.1			54.1		
MnO₃Si	Manganese(II) metasilicate	-1320.9			-1240.5			89.1			86.4		
MnS	Manganese(II) sulfide	-214.2			-218.4			78.2			50.0		
MnSe	Manganese(II) selenide	-106.7			-111.7			90.8			51.0		
Mn₂O₃	Manganese(III) oxide	-959.0			-881.1			110.5			107.7		
Mn₂O₄Si	Manganese(II) orthosilicate	-1730.5			-1632.1			163.2			129.9		
Mn₃O₄	Manganese(II,III) oxide	-1387.8			-1283.2			155.6			139.7		
Mo	Molybdenum	0.0		658.1			612.5	28.7		182.0	24.1		20.8
MoNa₂O₄	Sodium molybdate	-1468.1			-1354.3			159.7			141.7		
MoO₂	Molybdenum(IV) oxide	-588.9			-533.0			46.3			56.0		
MoO₃	Molybdenum(VI) oxide	-745.1			-668.0			77.7			75.0		
MoO₄Pb	Lead(II) molybdate	-1051.9			-951.4			166.1			119.7		
MoS₂	Molybdenum(IV) sulfide	-235.1			-225.9			62.6			63.6		
N	Nitrogen (atomic)			472.7			455.5			153.3			20.8
NNaO₂	Sodium nitrite	-358.7			-284.6			103.8					
NNaO₃	Sodium nitrate	-467.9			-367.0			116.5			92.9		
NO	Nitric oxide			91.3			87.6			210.8			29.9
NO₂	Nitrogen dioxide			33.2			51.3			240.1			37.2
NO₂Rb	Rubidium nitrite	-367.4			-306.2			172.0					
NO₃Rb	Rubidium nitrate	-495.1			-395.8			147.3			102.1		
NO₃Tl	Thallium(I) nitrate	-243.9			-152.4			160.7			99.5		
NP	Phosphorus nitride	-63.0											
N₂	Nitrogen			0.0						191.6			29.1
N₂O	Nitrous oxide			81.6			103.7			220.0			38.6
N₂O₃	Nitrogen trioxide		50.3	86.6			142.4			314.7			72.7
N₂O₄	Nitrogen tetroxide		-19.5	11.1		97.5	99.8		209.2	304.4		142.7	79.2
N₂O₄Sr	Strontium nitrite	-762.3											
N₂O₅	Nitrogen pentoxide	-43.1		13.3	113.9		117.1	178.2		355.7	143.1		95.3
N₂O₆Pb	Lead(II) nitrate	-451.9											
N₂O₆Ra	Radium nitrate	-992.0			-796.1			222.0					
N₂O₆Sr	Strontium nitrate	-978.2			-780.0			194.6			149.9		
N₂O₆Zn	Zinc nitrate	-483.7											
N₃Na	Sodium azide	21.7			93.8			96.9			76.6		
N₄Si₃	Silicon nitride	-743.5			-642.6			101.3					
Na	Sodium	0.0		107.5			77.0	51.3		153.7	28.2		20.8
NaO₂	Sodium superoxide	-260.2			-218.4			115.9			72.1		
Na₂	Disodium			142.1			103.9			230.2			37.6

Molecular formula	Name	$\Delta_f H°$/kJ mol⁻¹ Crys.	Liq.	Gas	$\Delta_f G°$/kJ mol⁻¹ Crys.	Liq.	Gas	$S°$/J mol⁻¹ K⁻¹ Crys.	Liq.	Gas	C_p/J mol⁻¹ K⁻¹ Crys.	Liq.	Gas
Na_2O	Sodium oxide	-414.2			-375.5			75.1			69.1		
Na_2O_2	Sodium peroxide	-510.9			-447.7			95.0			89.2		
Na_2O_3S	Sodium sulfite	-1100.8			-1012.5			145.9			120.3		
Na_2O_3Si	Sodium metasilicate	-1554.9			-1462.8			113.9			128.2		
Na_2O_4S	Sodium sulfate	-1387.1			-1270.2			149.6					
Na_2S	Sodium sulfide	-364.8			-349.8			83.7					
Nb	Niobium	0.0		725.9			681.1	36.4		186.3	24.6		30.2
NbO	Niobium(II) oxide	-405.8			-378.6			48.1			41.3		
NbO_2	Niobium(IV) oxide	-796.2			-740.5			54.5			57.5		
Nb_2O_5	Niobium(V) oxide	-1899.5			-1766.0			137.2			132.1		
Nd	Neodymium	0.0		327.6			292.4	71.5		189.4	27.5		22.1
Nd_2O_3	Neodymium oxide	-1807.9			-1720.8			158.6			111.3		
Ne	Neon			0.0						146.3			20.8
Ni	Nickel	0.0		429.7			384.5	29.9		182.2	26.1		23.4
NiO_4S	Nickel(II) sulfate	-872.9			-759.7			92.0			138.0		
NiS	Nickel(II) sulfide	-82.0			-79.5			53.0			47.1		
Ni_2O_3	Nickel(III) oxide	-489.5											
No	Nobelium	0.0											
O	Oxygen (atomic)			249.2			231.7			161.1			21.9
OP	Phosphorus monoxide			-28.5			-51.9			222.8			31.8
OPb	Lead(II) oxide (massicot)	-217.3			-187.9			68.7			45.8		
OPb	Lead(II) oxide (litharge)	-219.0			-188.9			66.5			45.8		
OPd	Palladium(II) oxide	-85.4		348.9			325.9			218.0	31.4		
ORa	Radium oxide	-523.0											
ORb_2	Rubidium oxide	-339.0											
ORh	Rhodium monoxide			385.0									
OS	Sulfur monoxide			6.3			-19.9			222.0			30.2
OSe	Selenium monoxide			53.4			26.8			234.0			31.3
OSi	Silicon monoxide			-99.6			-126.4			211.6			29.9
OSn	Tin(II) oxide	-280.7		15.1	-251.9		-8.4	57.2		232.1	44.3		31.6
OSr	Strontium oxide	-592.0			-561.9			54.4			45.0		
OTi	Titanium(II) oxide	-519.7			-495.0			50.0			40.0		
OTl_2	Thallium(I) oxide	-178.7			-147.3			126.0					
OU	Uranium(II) oxide			21.0									
OV	Vanadium(II) oxide	-431.8			-404.2			38.9			45.4		
OZn	Zinc oxide	-350.5			-320.5			43.7			40.3		
O_2	Oxygen			0.0						205.2			29.4
O_2P	Phosphorus dioxide			-279.9			-281.6			252.1			39.5
O_2Pb	Lead(IV) oxide	-277.4			-217.3			68.6			64.6		
O_2Rb	Rubidium superoxide	-278.7											
O_2Rb_2	Rubidium peroxide	-472.0											
O_2Ru	Ruthenium(IV) oxide	-305.0											
O_2S	Sulfur dioxide		-320.5	-296.8			-300.1			248.2			39.9
O_2Se	Selenium dioxide	-225.4											
O_2Si	Silicon dioxide (α-quartz)	-910.7		-322.0	-856.3			41.5			44.4		

Molecular formula	Name	$\Delta_f H°$/kJ mol⁻¹ Crys.	$\Delta_f H°$/kJ mol⁻¹ Liq.	$\Delta_f H°$/kJ mol⁻¹ Gas	$\Delta_f G°$/kJ mol⁻¹ Crys.	$\Delta_f G°$/kJ mol⁻¹ Liq.	$\Delta_f G°$/kJ mol⁻¹ Gas	$S°$/J mol⁻¹ K⁻¹ Crys.	$S°$/J mol⁻¹ K⁻¹ Liq.	$S°$/J mol⁻¹ K⁻¹ Gas	C_p/J mol⁻¹ K⁻¹ Crys.	C_p/J mol⁻¹ K⁻¹ Liq.	C_p/J mol⁻¹ K⁻¹ Gas
O₂Sn	Tin(IV) oxide	-577.6			-515.8			49.0			52.6		
O₂Te	Tellurium dioxide	-322.6			-270.3			79.5					
O₂Th	Thorium(IV) oxide	-1226.4			-1169.2			65.2			61.8		
O₂Ti	Titanium(IV) oxide	-944.0			-888.8			50.6			55.0		
O₂U	Uranium(IV) oxide	-1085.0		-465.7	-1031.8		-471.5	77.0		274.6	63.6		51.4
O₂W	Tungsten(IV) oxide	-589.7			-533.9			50.5			56.1		
O₂Zr	Zirconium(IV) oxide	-1100.6			-1042.8			50.4			56.2		
O₃	Ozone			142.7			163.2			238.9			39.2
O₃PbS	Lead(II) sulfite	-669.9											
O₃PbSi	Lead(II) metasilicate	-1145.7			-1062.1			109.6			90.0		
O₃Pr₂	Praseodymium oxide	-1809.6									117.4		
O₃Rh₂	Rhodium(III) oxide	-343.0									103.8		
O₃S	Sulfur trioxide	-454.5	-441.0	-395.7	-374.2	-373.8	-371.1	70.7	113.8	256.8			50.7
O₃Sc₂	Scandium oxide	-1908.8			-1819.4			77.0			94.2		
O₃SiSr	Strontium metasilicate	-1633.9			-1549.7			96.7			88.5		
O₃Sm₂	Samarium(III) oxide	-1823.0			-1734.6			151.0			114.5		
O₃Tb₂	Terbium oxide	-1865.2									115.9		
O₃Ti₂	Titanium(III) oxide	-1520.9			-1434.2			78.8			97.4		
O₃Tm₂	Thulium oxide	-1888.7			-1794.5			139.7			116.7		
O₃U	Uranium(VI) oxide	-1223.8			-1145.7			96.1			81.7		
O₃V₂	Vanadium(III) oxide	-1218.8			-1139.3			98.3			103.2		
O₃W	Tungsten(VI) oxide	-842.9			-764.0			75.9			73.8		
O₃Y₂	Yttrium oxide	-1905.3			-1816.6			99.1			102.5		
O₃Yb₂	Ytterbium(III) oxide	-1814.6			-1726.7			133.1			115.4		
O₄Os	Osmium(VIII) oxide	-394.1		-337.2	-304.9		-292.8	143.9		293.8			74.1
O₄PbS	Lead(II) sulfate	-920.0			-813.0			148.5			103.2		
O₄PbSe	Lead(II) selenate	-609.2			-504.9			167.8			137.2		
O₄Pb₂Si	Lead(II) orthosilicate	-1363.1			-1252.6			186.6			146.9		
O₄Pb₃	Lead(II,II,IV) oxide	-718.4			-601.2			211.3					
O₄RaS	Radium sulfate	-1471.1			-1365.6			138.0					
O₄Rb₂S	Rubidium sulfate	-1435.6			-1315.9			197.4			134.1		
O₄Ru	Ruthenium(VIII) oxide	-239.3			-152.2			146.4					
O₄SSr	Strontium sulfate	-1453.1			-1340.9			117.0					
O₄STl₂	Thallium(I) sulfate	-931.8			-830.4			230.5					
O₄SZn	Zinc sulfate	-982.8			-871.5			110.5			99.2		
O₄SiSr₂	Strontium orthosilicate	-2304.5			-2191.0			153.1			134.3		
O₄SiZn₂	Zinc orthosilicate	-1636.7			-1523.2			131.4			123.3		
O₄SiZr	Zirconium(IV) orthosilicate	-2033.4			-1919.1			84.1			98.7		
O₅Sb₂	Antimony(V) oxide	-971.9			-829.2			125.1					
O₅Ta₂	Tantalum(V) oxide	-2046.0			-1911.2			143.1			135.1		
O₅Ti₃	Titanium(III,IV) oxide	-2459.4			-2317.4			129.3			154.8		
O₅V₂	Vanadium(V) oxide	-1550.6			-1419.5			131.0			127.7		
O₅V₃	Vanadium(III,IV) oxide	-1933.0			-1803.0			163.0					
O₇Re₂	Rhenium(VII) oxide	-1240.1		-1100.0	-1066.0		-994.0	207.1		452.0	166.1		
O₇U₃	Uranium(IV,VI) oxide	-3427.1			-3242.9			250.5			215.5		

Molecular formula	Name	$\Delta_f H°$/kJ mol⁻¹ Crys.	$\Delta_f H°$/kJ mol⁻¹ Liq.	$\Delta_f H°$/kJ mol⁻¹ Gas	$\Delta_f G°$/kJ mol⁻¹ Crys.	$\Delta_f G°$/kJ mol⁻¹ Liq.	$\Delta_f G°$/kJ mol⁻¹ Gas	S°/J mol⁻¹ K⁻¹ Crys.	S°/J mol⁻¹ K⁻¹ Liq.	S°/J mol⁻¹ K⁻¹ Gas	C_p/J mol⁻¹ K⁻¹ Crys.	C_p/J mol⁻¹ K⁻¹ Liq.	C_p/J mol⁻¹ K⁻¹ Gas
O₈S₂Zr	Zirconium(IV) sulfate	-2217.1									172.0		
O₈U₃	Uranium(V,VI) oxide	-3574.8			-3369.5			282.6			238.4		
O₉U₄	Uranium(IV,V) oxide	-4510.4			-4275.1			334.1			293.3		
Os	Osmium	0.0		791.0			745.0	32.6		192.6	24.7		20.8
P	Phosphorus (white)	0.0		316.5			280.1	41.1		163.2	23.8		20.8
P	Phosphorus (red)	-17.6									21.2		
P	Phosphorus (black)	-39.3						22.8					
P₂	Diphosphorus			144.0			103.5			218.1			32.1
P₄	Tetraphosphorus			58.9			24.4			280.0			67.2
Pa	Protactinium	0.0		607.0			563.0	51.9		198.1			22.9
Pb	Lead	0.0		195.2			162.2	64.8		175.4	26.4		20.8
PbS	Lead(II) sulfide	-100.4			-98.7			91.2			49.5		
PbSe	Lead(II) selenide	-102.9			-101.7			102.5			50.2		
PbTe	Lead(II) telluride	-70.7			-69.5			110.0			50.5		
Pd	Palladium	0.0		378.2			339.7	37.6		167.1	26.0		20.8
PdS	Palladium(II) sulfide	-75.0			-67.0			46.0					
Pm	Promethium	0.0											
Po	Polonium	0.0								187.1			24.3
Pr	Praseodymium	0.0		355.6			320.9	73.2		189.8	27.2		21.4
Pt	Platinum	0.0		565.3			520.5	41.6		192.4	25.9		25.5
PtS	Platinum(II) sulfide	-81.6			-76.1			55.1			43.4		
PtS₂	Platinum(IV) sulfide	-108.8			-99.6			74.7			65.9		
Pu	Plutonium	0.0											
Ra	Radium	0.0		159.0			130.0	71.0		176.5			20.8
Rb	Rubidium	0.0		80.9			53.1	76.8		170.1	31.1		20.8
Re	Rhenium	0.0		769.9			724.6	36.9		188.9	25.5		20.8
Rh	Rhodium	0.0		556.9			510.8	31.5		185.8	25.0		21.0
Rn	Radon			0.0						176.2			20.8
Ru	Ruthenium	0.0		642.7			595.8	28.5		186.5	24.1		21.5
S	Sulfur (rhombic)	0.0		277.2			236.7	32.1		167.8	22.6		23.7
S	Sulfur (monoclinic)	0.3											
SSi	Silicon monosulfide			112.5			60.9			223.7			32.3
SSn	Tin(II) sulfide	-100.0			-98.3			77.0			49.3		
SSr	Strontium sulfide	-472.4			-467.8			68.2			48.7		
STl₂	Thallium(I) sulfide	-97.1			-93.7			151.0					
SZn	Zinc sulfide (wurtzite)	-192.6											
SZn	Zinc sulfide (sphalerite)	-206.0			-201.3			57.7			46.0		
S₂	Disulfur			128.6			79.7			228.2			32.5
Sb	Antimony	0.0		262.3			222.1	45.7		180.3	25.2		20.8
Sb₂	Diantimony			235.6			187.0			254.9			36.4
Sc	Scandium	0.0		377.8			336.0	34.6		174.8	25.5		22.1
Se	Selenium	0.0		227.1			187.0	42.4		176.7	25.4		20.8
SeSr	Strontium selenide	-385.8						172.0					
SeTl₂	Thallium(I) selenide	-59.0			-59.0								
SeZn	Zinc selenide	-163.0			-163.0			84.0					

STANDARD THERMODYNAMIC PROPERTIES OF CHEMICAL SUBSTANCES (continued)

Molecular formula	Name	$\Delta_f H°$/kJ mol⁻¹ Crys.	Liq.	Gas	$\Delta_f G°$/kJ mol⁻¹ Crys.	Liq.	Gas	$S°$/J mol⁻¹ K⁻¹ Crys.	Liq.	Gas	C_p/J mol⁻¹ K⁻¹ Crys.	Liq.	Gas
Se₂	Diselenium			146.0			96.2			252.0			35.4
Si	Silicon	0.0		450.0			405.5	18.8		168.0	20.0		22.3
Si₂	Disilicon			594.0			536.0			229.9			34.4
Sm	Samarium	0.0		206.7			172.8	69.6		183.0	29.5		30.4
Sn	Tin (white)	0.0		301.2			266.2	51.2		168.5	27.0		21.3
Sn	Tin (gray)	-2.1			0.1			44.1			25.8		
Sr	Strontium	0.0		164.4			130.9	55.0		164.6	26.8		20.8
Ta	Tantalum	0.0		782.0			739.3	41.5		185.2	25.4		20.9
Tb	Terbium	0.0		388.7			349.7	73.2		203.6	28.9		24.6
Tc	Technetium	0.0		678.0						181.1			20.8
Te	Tellurium	0.0		196.7			157.1	49.7		182.7	25.7		20.8
Te₂	Ditellurium			168.2			118.0			268.1			36.7
Th	Thorium	0.0		602.0			560.7	51.8		190.2	27.3		20.8
Ti	Titanium	0.0		473.0			428.4	30.7		180.3	25.0		24.4
Tl	Thallium	0.0		182.2			147.4	64.2		181.0	26.3		20.8
Tm	Thulium	0.0		232.2			197.5	74.0		190.1	27.0		20.8
U	Uranium	0.0		533.0			488.4	50.2		199.8	27.7		23.7
V	Vanadium	0.0		514.2			470.7	28.9		182.3	24.9		26.0
W	Tungsten	0.0		849.4			807.1	32.6		174.0	24.3		21.3
Xe	Xenon			0.0			0.0			169.7			20.8
Y	Yttrium	0.0		421.3			381.1	44.4		179.5	26.5		25.9
Yb	Ytterbium	0.0		152.3			118.4	59.9		173.1	26.7		20.8
Zn	Zinc	0.0		130.4			94.8	41.6		161.0	25.4		20.8
Zr	Zirconium	0.0		608.8			566.5	39.0		181.4	25.4		26.7
C	Carbon (graphite)	0.0		716.7			671.3	5.7		158.1	8.5		20.8
C	Carbon (diamond)	1.9			2.9			2.4			6.1		
CAgN	Silver(I) cyanide	146.0			156.9			107.2			66.7		
CAg₂O₃	Silver(I) carbonate	-505.8			-436.8			167.4			112.3		
CBaO₃	Barium carbonate	-1213.0			-1134.4			112.1			86.0		
CBeO₃	Beryllium carbonate	-1025.0						52.0			65.0		
CBrClF₂	Bromochlorodifluoro-methane									318.5			74.6
CBrCl₂F	Bromodichlorofluoro-methane									330.6			80.0
CBrCl₃	Bromotrichloromethane			-41.1									85.3
CBrF₃	Bromotrifluoromethane			-648.3									69.3
CBrN	Cyanogen bromide	140.5		186.2			165.3			248.3			46.9
CBrN₃O₆	Bromotrinitromethane		32.5	80.3									
CBr₂ClF	Dibromochlorofluoro-methane									342.8			82.4
CBr₂Cl₂	Dibromodichloromethane									347.8			87.1
CBr₂F₂	Dibromodifluoromethane									325.3			77.0
CBr₂O	Carbonyl bromide		-127.2	-96.2			-110.9			309.1			61.8
CBr₃Cl	Tribromochloromethane									357.8			89.4
CBr₃F	Tribromofluoromethane									345.9			84.4

Molecular formula	Name	$\Delta_f H°/\text{kJ mol}^{-1}$			$\Delta_f G°/\text{kJ mol}^{-1}$			$S°/\text{J mol}^{-1}\text{K}^{-1}$			$C_p/\text{J mol}^{-1}\text{K}^{-1}$		
		Crys.	Liq.	Gas	Crys.	Liq.	Gas	Crys.	Liq.	Gas	Crys.	Liq.	Gas
CBr_4	Tetrabromomethane	29.4		83.9	47.7		67.0	212.5		358.1	144.3		91.2
$CCaO_3$	Calcium carbonate (calcite)	-1207.6			-1129.1			91.7			83.5		
$CCaO_3$	Calcium carbonate (aragonite)	-1207.8			-1128.2			88.0			82.3		
$CCdO_3$	Cadmium carbonate	-750.6			-669.4			92.5					
$CClFO$	Carbonyl chloride fluoride			-706.3						276.7			52.4
$CClF_3$	Chlorotrifluoromethane		112.1	138.0						236.2			66.9
$CClN$	Cyanogen chloride		-27.1	18.4			131.0						45.0
$CClN_3O_6$	Chlorotrinitromethane												
CCl_2F_2	Dichlorodifluoromethane			-477.4			-439.4			300.8			72.3
CCl_2O	Carbonyl chloride			-219.1			-204.9			283.5			57.7
CCl_3	Trichloromethyl			59.0									
CCl_3F	Trichlorofluoromethane		-301.3	-268.3		-236.8			225.4			121.6	78.1
CCl_4	Tetrachloromethane		-128.2	-95.7								130.7	83.3
$CCoO_3$	Cobalt(II) carbonate	-713.0			-1054.3			204.5			123.9		
CCs_2O_3	Cesium carbonate	-1139.7						84.5					
$CCuN$	Copper(I) cyanide	96.2			111.3					224.7			41.8
CFN	Cyanogen fluoride												
CF_2O	Carbonyl fluoride			-639.8									46.8
CF_3	Trifluoromethyl			-477.0			-464.0			264.5			49.6
CF_3I	Trifluoroiodomethane			-587.8						307.4			70.9
CF_4	Tetrafluoromethane			-933.6						261.6			61.1
$CFeO_3$	Iron(II) carbonate	-740.6			-666.7			92.9			82.1		
CFe_3	Iron carbide	25.1			20.1			104.6			105.9		
CH	Methylidyne			595.8									
$CHBrClF$	Bromochlorofluoromethane									304.3			63.2
$CHBrCl_2$	Bromodichloromethane									316.4			67.4
$CHBrF_2$	Bromodifluoromethane			-424.9						295.1			58.7
$CHBr_2Cl$	Chlorodibromomethane									327.7			69.2
$CHBr_2F$	Dibromofluoromethane									316.8			65.1
$CHBr_3$	Tribromomethane		-22.3	23.8		-5.0	8.0		220.9	330.9		130.7	71.2
$CHClF_2$	Chlorodifluoromethane			-482.6						280.9			55.9
$CHCl_2F$	Dichlorofluoromethane									293.1			60.9
$CHCl_3$	Trichloromethane		-134.1	-102.7		-73.7	6.0		201.7	295.7		114.2	65.7
$CHCsO_3$	Cesium hydrogen carbonate	-966.1											
$CHFO$	Formyl fluoride									246.6			39.9
CHF_3	Trifluoromethane	-181.1		-695.4						259.7			51.0
CHI_3	Triiodomethane			251.0						356.2			75.0
$CHKO_2$	Potassium formate	-679.7			-863.5			115.5					
$CHKO_3$	Potassium hydrogen carbonate	-963.2											
CHN	Hydrogen cyanide		108.9	135.1		125.0	124.7		112.8	201.8		70.6	35.9
$CHNO$	Isocyanic acid (HNCO)									238.0			44.9
$CHNS$	Isothiocyanic acid		-32.8	127.6			113.0			247.8			46.9
CHN_3O_6	Trinitromethane			-0.2									
$CHNaO_2$	Sodium formate	-666.5			-599.9			103.8			82.7		

Molecular formula	Name	$\Delta_f H°$/kJ mol⁻¹			$\Delta_f G°$/kJ mol⁻¹			$S°$/J mol⁻¹ K⁻¹			C_p/J mol⁻¹ K⁻¹		
		Crys.	Liq.	Gas	Crys.	Liq.	Gas	Crys.	Liq.	Gas	Crys.	Liq.	Gas
$CHNaO_3$	Sodium hydrogen carbonate	-950.8			-851.0			101.7			87.6		
CHO	Oxomethyl (HCO)			43.1			28.0			224.7			34.6
CH_2	Methylene			390.4			372.9			194.9			33.8
CH_2BrCl	Bromochloromethane									287.6			52.7
CH_2BrF	Bromofluoromethane									276.3			49.2
CH_2Br_2	Dibromomethane									293.2			54.7
CH_2ClF	Chlorofluoromethane									264.4			47.0
CH_2Cl_2	Dichloromethane		-124.2	-95.4					177.8	270.2		101.2	51.0
CH_2F_2	Difluoromethane			-452.3						246.7			42.9
CH_2I_2	Diiodomethane		68.5	119.5		90.4	95.8		174.1	309.7		134.0	57.7
CH_2N_2	Diazomethane									242.9			52.5
CH_2N_2	Cyanamide	58.8											
$CH_2N_2O_4$	Dinitromethane		-104.9	-58.9									
CH_2O	Formaldehyde			-108.6			-102.5			218.8			35.4
$(CH_2O)_x$	Paraformaldehyde	-177.6											
CH_2O_2	Formic acid		-425.0	-378.7		-361.4			129.0			99.0	
CH_2S_3	Trithiocarbonic acid		24.0										
CH_3	Methyl			145.7			147.9			194.2			38.7
CH_3BO	Borane carbonyl			-111.2			-92.9			249.4			59.5
CH_3Br	Bromomethane		-59.8	-35.4			-26.3			246.4			42.4
CH_3Cl	Chloromethane			-81.9						234.6			40.8
CH_3F	Fluoromethane									222.9			37.5
CH_3I	Iodomethane		-13.6	14.4					163.2	254.1		126.0	44.1
CH_3NO	Formamide		-254.0	-193.9									
CH_3NO_2	Nitromethane		-112.6	-74.3		-14.4	-6.8		171.8	275.0		106.6	57.3
CH_3NO_2	Methyl nitrite			-66.1									
CH_3NO_3	Methyl nitrate		-156.3	-122.4		-43.4	-39.2		217.1	318.5		157.3	
CH_4	Methane			-74.6			-50.5			186.3	134.0		35.7
CH_4N_2	Ammonium cyanide	0.4											
CH_4N_2O	Urea	-333.1		-245.8									
CH_4N_2S	Thiourea	-89.1		22.9									
$CH_4N_4O_2$	Nitroguanidine	-92.4											
CH_4O	Methanol		-239.2	-201.0		-166.6	-162.3		126.8	239.9		81.1	44.1
CH_4S	Methanethiol		-46.7	-22.9		-7.7	-9.3		169.2	255.2		90.5	50.3
CH_5N	Methylamine		-47.3	-22.5		35.7	32.7		150.2	242.9		102.1	50.1
CH_5NO_3	Ammonium hydrogen carbonate	-849.4			-665.9			120.9					
CH_5N_3	Guanidine	-56.0											
CH_5N_3S	Hydrazinecarbothioamide	24.7											
$CH_5N_5O_2$	3-Amino-1-nitroguanidine	22.1											
CH_6ClN	Methylamine hydrochloride	-298.1											
CH_6N_2	Methylhydrazine		54.2	94.7		180.0	187.0		165.9	278.8		134.9	71.1
CH_6Si	Methylsilane									256.5			65.9
CHg_2O_3	Mercury(I) carbonate	-553.5			-468.1			180.0					
CIN	Cyanogen iodide	166.2		225.5	185.0		196.6	96.2		256.8			48.3

STANDARD THERMODYNAMIC PROPERTIES OF CHEMICAL SUBSTANCES (continued)

Molecular formula	Name	$\Delta_f H°$/kJ mol⁻¹			$\Delta_f G°$/kJ mol⁻¹			S°/J mol⁻¹ K⁻¹			C_p/J mol⁻¹ K⁻¹		
		Crys.	Liq.	Gas	Crys.	Liq.	Gas	Crys.	Liq.	Gas	Crys.	Liq.	Gas
CI₄	Tetraiodomethane	-392.9		474.0						391.9			95.9
CKN	Potassium cyanide	-113.0						128.5			66.3		
CKNS	Potassium thiocyanate	-200.2						124.3			88.5		
CK₂O₃	Potassium carbonate	-1151.0						155.5			114.4		
CLi₂O₃	Lithium carbonate	-1215.9						90.4			99.1		
CMgO₃	Magnesium carbonate	-1095.8						65.7			75.5		
CMnO₃	Manganese(II) carbonate	-894.1						85.8			81.5		
CN	Cyanide			437.6			407.5			202.6			29.2
CNNa	Sodium cyanide	-87.5			-76.4			115.6			70.4		
CNNaO	Sodium cyanate	-405.4			-358.1			96.7			86.6		
CN₄O₈	Tetranitromethane		38.4	88.3									
CNa₂O₃	Sodium carbonate	-1130.7			-1044.4			135.0			112.3		
CO	Carbon monoxide			-110.5			-137.2			197.7			29.1
COS	Carbon oxysulfide			-142.0			-169.2			231.6			41.5
CO₂	Carbon dioxide			-393.5			-394.4			213.8			37.1
CO₃Pb	Lead(II) carbonate	-699.1			-625.5			131.0			87.4		
CO₃Rb₂	Rubidium carbonate	-1136.0			-1051.0			181.3			117.6		
CO₃Sr	Strontium carbonate	-1220.1			-1140.1			97.1			81.4		
CO₃Tl₂	Thallium(I) carbonate	-700.0			-614.6			155.2					
CO₃Zn	Zinc carbonate	-812.8			-731.5			82.4			79.7		
CS	Carbon sulfide			234.0			184.0			210.6			29.8
CS₂	Carbon disulfide		89.0	116.7		64.6	67.1		151.3	237.8		76.4	45.4
CSe₂	Carbon diselenide		164.8										
CSi	Silicon carbide (cubic)	-65.3			-62.8			16.6			26.9		
CSi	Silicon carbide (hexagonal)	-62.8			-60.2			16.5			26.7		
C₂	Dicarbon			831.9			775.9			199.4			43.2
C₂BrF₅	Bromopentafluoroethane			-1064.4									
C₂Br₂ClF₃	1,2-Dibromo-1-chloro-1,2,2-trifluoroethane		-691.7	-656.6									
C₂Br₂F₄	1,2-Dibromotetrafluoro-ethane		-817.7	-789.1									
C₂Br₄	Tetrabromoethylene									387.1			102.7
C₂Br₆	Hexabromoethane									441.9			139.3
C₂Ca	Calcium carbide	-59.8			-64.9			70.0			62.7		
C₂CaN₂	Calcium cyanide	-184.5											
C₂CaO₄	Calcium oxalate	-1360.6											
C₂ClF₃	Chlorotrifluoroethylene		-522.7	-505.5			-523.8			322.1			83.9
C₂ClF₅	Chloropentafluoroethane			-1118.8									184.2
C₂Cl₂F₄	1,2-Dichlorotetrafluoro-ethane		-960.2	-937.0								111.7	
C₂Cl₂O₂	Ethanedioyl dichloride		-367.6	-335.8									
C₂Cl₃F₃	1,1,2-Trichlorotrifluoro-ethane		-745.0	-716.8								170.1	
C₂Cl₃N	Trichloroacetonitrile					3.0			266.9	336.6			96.1
C₂Cl₄	Tetrachloroethylene		-50.6	-10.9								143.4	

STANDARD THERMODYNAMIC PROPERTIES OF CHEMICAL SUBSTANCES (continued)

Molecular formula	Name	$\Delta_f H°$/kJ mol⁻¹ Crys.	Liq.	Gas	$\Delta_f G°$/kJ mol⁻¹ Crys.	Liq.	Gas	$S°$/J mol⁻¹ K⁻¹ Crys.	Liq.	Gas	C_p/J mol⁻¹ K⁻¹ Crys.	Liq.	Gas
$C_2Cl_4F_2$	1,1,1,2-Tetrachloro-2,2-difluoroethane			-489.9			-407.0			382.9			123.4
$C_2Cl_4F_2$	1,1,2,2-Tetrachloro-1,2-difluoroethane											173.6	
C_2Cl_4O	Trichloroacetyl chloride		-280.8	-239.8									
C_2Cl_6	Hexachloroethane	-202.8		-143.6				237.3			198.2		
C_2F_3N	Trifluoroacetonitrile			-497.9						298.1			77.9
C_2F_4	Tetrafluoroethylene			-658.9						300.1			80.5
C_2F_6	Hexafluoroethane	-820.5		-1344.2						332.3			106.7
C_2HBr	Bromoacetylene									253.7			55.7
$C_2HBrClF_3$	1-Bromo-2-chloro-1,1,2-trifluoroethane		-675.3	-644.8									
$C_2HBrClF_3$	2-Bromo-2-chloro-1,1,1-trifluoroethane		-720.0	-690.4									
C_2HCl	Chloroacetylene									242.0			54.3
C_2HClF_2	1-Chloro-2,2-difluoroethylene			-315.5			-289.1			303.0			72.1
C_2HCl_2F	1,1-Dichloro-2-fluoroethylene									313.9			76.5
$C_2HCl_2F_3$	2,2-Dichloro-1,1,1-trifluoroethane									352.8			102.5
C_2HCl_3	Trichloroethylene		-43.6	-9.0						324.8			80.3
C_2HCl_3O	Trichloroacetaldehyde		-234.5	-196.6					228.4			124.4	
C_2HCl_3O	Dichloroacetyl chloride		-280.4	-241.0								151.0	
$C_2HCl_3O_2$	Trichloroacetic acid	-503.3											
C_2HCl_5	Pentachloroethane		-187.6	-142.0									
C_2HF	Fluoroacetylene									231.7			52.4
C_2HF_3	Trifluoroethylene			-490.5									
$C_2HF_3O_2$	Trifluoroacetic acid		-1069.9	-1031.4								173.8	
C_2HF_5	Pentafluoroethane			-1100.4									
C_2H_2	Acetylene			227.4			209.9			200.9			44.0
$C_2H_2BrF_3$	2-Bromo-1,1,1-trifluoroethane			-694.5									
$C_2H_2Br_2$	cis-1,2-Dibromoethylene									311.3			68.8
$C_2H_2Br_2$	trans-1,2-Dibromoethylene									313.5			70.3
$C_2H_2Br_2Cl_2$	1,2-Dibromo-1,2-dichloroethane			-36.9									
$C_2H_2Br_4$	1,1,2,2-Tetrabromoethane											165.7	
$C_2H_2ClF_3$	2-Chloro-1,1,1-trifluoroethane									326.5			89.1
$C_2H_2Cl_2$	1,1-Dichloroethylene		-23.9	2.8		24.1	25.4		201.5	289.0		111.3	67.1
$C_2H_2Cl_2$	cis-1,2-Dichloroethylene		-26.4	4.6					198.4	289.6		116.4	67.1
$C_2H_2Cl_2$	trans-1,2-Dichloroethylene		-24.3	5.0		27.3	28.6		195.9	290.0		116.8	65.1
$C_2H_2Cl_2O$	Chloroacetyl chloride		-283.7										66.7
$C_2H_2Cl_2O_2$	Dichloroacetic acid		-496.3	-244.8									

Molecular formula	Name	$\Delta_f H°/\text{kJ mol}^{-1}$ Crys.	Liq.	Gas	$\Delta_f G°/\text{kJ mol}^{-1}$ Crys.	Liq.	Gas	$S°/\text{J mol}^{-1}\text{K}^{-1}$ Crys.	Liq.	Gas	$C_p/\text{J mol}^{-1}\text{K}^{-1}$ Crys.	Liq.	Gas
C₂H₂Cl₃NO	2,2,2-Trichloroacetamide	-358.0											
C₂H₂Cl₄	1,1',2-Tetrachloroethane		-195.0	-149.2					246.9	356.0		162.3	102.7
C₂H₂Cl₄	1,1,2,2-Tetrachloroethane			-335.0						362.8			100.8
C₂H₂F₂	1,1-Difluoroethylene									266.2			60.1
C₂H₂F₂	cis-1,2-Difluoroethylene									268.3			58.2
C₂H₂F₃I	1,1,1-Trifluoro-2-iodoethane			-644.5									
C₂H₂I₂	cis-1,2-Diiodoethylene			-207.4									
C₂H₂O	Ketene		-67.9	-47.5			-48.3			247.6			51.8
C₂H₂O₂	Glyoxal			-212.0									
C₂H₂O₄	Oxalic acid	-821.7		-723.7				109.8			91.0		
C₂H₂O₄Sr	Strontium formate	-1393.3											
C₂H₂S	Thiirene			300.0			275.8			255.3			54.7
C₂H₃Br	Bromoethylene			79.2			81.8			275.8			55.5
C₂H₃BrO	Acetyl bromide		-223.5	-190.4									
C₂H₃Cl	Chloroethylene		14.6	37.2			53.6			264.0	59.4		53.7
C₂H₃ClF₂	1-Chloro-1,1-difluoroethane	-94.1								307.2			82.5
C₂H₃ClO	Acetyl chloride		-272.9	-242.8		-208.0	-205.8		200.8	295.1		117.0	67.8
C₂H₃ClO₂	Chloroacetic acid	-510.5											
C₂H₃Cl₂F	1,1-Dichloro-1-fluoroethane			-435.2						320.2			88.7
C₂H₃Cl₃	1,1,1-Trichloroethane		-177.4	-144.4					227.4	323.1		144.3	93.3
C₂H₃Cl₃	1,1,2-Trichloroethane		-190.8	-151.3					232.6	337.2		150.9	89.0
C₂H₃F	Fluoroethylene			-138.8									
C₂H₃FO	Acetyl fluoride		-467.2	-442.1									
C₂H₃F₃	1,1,1-Trifluoroethane			-744.6						279.9			78.2
C₂H₃F₃	1,1,2-Trifluoroethane			-730.7									
C₂H₃F₃O	2,2,2-Trifluoroethanol		-932.4	-888.4						285.0			57.9
C₂H₃I	Iodoethylene		-163.5	-126.4									
C₂H₃IO	Acetyl iodide												
C₂H₃KO₂	Potassium acetate	-723.0											
C₂H₃N	Acetonitrile		40.6	74.0		86.5	91.9		149.6	243.4		91.5	52.2
C₂H₃N	Isocyanomethane		130.8	163.5		159.5	165.7		159.0	246.9			52.9
C₂H₃NO	Methylisocyanate		-92.0										
C₂H₃NO₃	Oxamic acid	-661.2		-552.3									
C₂H₃NS	Methyl isothiocyanate	79.4											
C₂H₃NaO₂	Sodium acetate	-708.8			-607.2			123.0			79.9		
C₂H₄	Ethylene			52.4			68.4			219.3			42.9
C₂H₄BrCl	1-Bromo-2-chloroethane		-66.2									130.1	
C₂H₄Br₂	1,1-Dibromoethane												
C₂H₄Br₂	1,2-Dibromoethane		-79.2	-37.5					223.3	327.7		136.0	80.8
C₂H₄ClF	1-Chloro-1-fluoroethane			-313.4									
C₂H₄Cl₂	1,1-Dichloroethane		-158.4	-127.7					211.8	305.1		126.3	76.2
C₂H₄Cl₂	1,2-Dichloroethane		-166.8	-126.4						308.4		128.4	78.7

Molecular formula	Name	$\Delta_f H°/\mathrm{kJ\ mol^{-1}}$			$\Delta_f G°/\mathrm{kJ\ mol^{-1}}$			$S°/\mathrm{J\ mol^{-1}\ K^{-1}}$			$C_p/\mathrm{J\ mol^{-1}\ K^{-1}}$		
		Crys.	Liq.	Gas	Crys.	Liq.	Gas	Crys.	Liq.	Gas	Crys.	Liq.	Gas
$C_2H_4F_2$	1,1-Difluoroethane			-497.0						282.5			67.8
$C_2H_4I_2$	1,2-Diiodoethane	9.3		75.0									
$C_2H_4N_2O_2$	Oxamide	-504.4		-387.1									
$C_2H_4N_2O_2$	Ethanedial, dioxime	-90.5											
$C_2H_4N_2O_4$	1,1-Dinitroethane		-148.2										
$C_2H_4N_2O_4$	1,2-Dinitroethane		-165.2										
$C_2H_4N_2S_2$	Ethanedithioamide	-20.8		83.0									
$C_2H_4N_4$	1H-1,2,4-Triazol-3-amine	76.8											
C_2H_4O	Acetaldehyde		-192.2	-166.2		-127.6	-133.0		160.2	263.8		89.0	55.3
C_2H_4O	Ethylene oxide		-78.0	-52.6		-11.8	-13.0		153.9	242.5		88.0	47.9
C_2H_4OS	Thioacetic acid		-216.9	-175.1									
$C_2H_4O_2$	Acetic acid		-484.3	-432.2		-389.9	-374.2		159.8	283.5		123.3	63.4
$C_2H_4O_2$	Methyl formate		-386.1	-357.4						285.3		119.1	64.4
C_2H_4S	Thiirane		51.6	82.0			96.8			255.2			53.3
C_2H_4Si	Ethynylsilane									269.4			72.6
C_2H_5Br	Bromoethane		-90.5	-61.9		-25.8	-23.9		198.7	286.7		100.8	64.5
C_2H_5Cl	Chloroethane		-136.8	-112.1		-59.3	-60.4		190.8	276.0		104.3	62.8
C_2H_5ClO	Ethylene chlorohydrin		-295.4										
C_2H_5F	Fluoroethane									264.5			58.6
C_2H_5I	Iodoethane		-40.0	-8.1		14.7	19.2		211.7	306.0		115.1	66.9
C_2H_5N	Ethyleneimine		91.9	126.5									
C_2H_5NO	Acetamide	-317.0		-238.3				115.0			91.3		
C_2H_5NO	N-Methylformamide		-143.9									123.8	
$C_2H_5NO_2$	Nitroethane			-102.3								134.4	
$C_2H_5NO_2$	Glycine	-528.5		-392.1									
$C_2H_5NO_3$	2-Nitroethanol		-350.7										
$C_2H_5NO_3$	Ethyl nitrate		-190.4	-154.1									
C_2H_5NS	Thioacetamide	-71.7		11.4									
C_2H_6	Ethane			-84.0			-32.0			229.2			52.5
C_2H_6Cd	Dimethyl cadmium		63.6	101.6		139.0	146.9		201.9	303.0		132.0	
C_2H_6Hg	Dimethyl mercury		59.8	94.4		140.3	146.1		209.0	306.0			83.3
$C_2H_6N_2O$	N-Methylurea	-332.8											
$C_2H_6N_4O_2$	1,2-Hydrazinedicarboxamide	-498.7											
$C_2H_6N_4O_2$	Ethanedioic acid, dihydrazide	-295.2											
C_2H_6O	Ethanol		-277.6	-234.8		-174.8	-167.9		160.7	281.6		112.3	65.6
C_2H_6O	Dimethyl ether		-203.3	-184.1			-112.6			266.4			64.4
C_2H_6OS	Dimethyl sulfoxide		-204.2	-151.3		-99.9			188.3	303.8		153.0	82.7
$C_2H_6O_2$	Ethylene glycol		-460.0	-392.2					163.2	310.6		148.6	100.0
$C_2H_6O_2S$	Dimethyl sulfone	-450.1		-373.1	-302.4		-272.7	142.0					
$C_2H_6O_3S$	Dimethyl sulfite		-523.6	-483.4									
$C_2H_6O_4S$	Dimethyl sulfate		-735.5	-687.0									
C_2H_6S	Ethanethiol		-73.6	-46.1		-5.5	-4.8		207.0	296.2		117.9	72.7
C_2H_6S	Dimethyl sulfide		-65.3	-37.4					196.4	286.0		118.1	74.1
$C_2H_6S_2$	1,2-Ethanedithiol		-54.3	-9.7									
$C_2H_6S_2$	Dimethyl disulfide		-62.6	-24.7					235.4			146.1	

Molecular formula	Name	$\Delta_f H°$/kJ mol⁻¹			$\Delta_f G°$/kJ mol⁻¹			$S°$/J mol⁻¹ K⁻¹			C_p/J mol⁻¹ K⁻¹		
		Crys.	Liq.	Gas	Crys.	Liq.	Gas	Crys.	Liq.	Gas	Crys.	Liq.	Gas
C₂H₆Zn	Dimethyl zinc		23.4	53.0					201.6			129.2	
C₂H₇N	Ethylamine		-74.1	-47.5			36.3			283.8		130.0	71.5
C₂H₇N	Dimethylamine		-43.9	-18.8		70.0	68.5		182.3	273.1		137.7	70.7
C₂H₇NO	Ethanolamine											195.5	
C₂H₈ClN	Dimethylamine hydrochloride	-289.3											
C₂H₈N₂	1,2-Ethanediamine		-63.0	-18.0								172.6	
C₂H₈N₂	1,1-Dimethylhydrazine		48.9	84.1		206.4			198.0			164.1	
C₂H₈N₂	1,2-Dimethylhydrazine		52.7	92.2									
C₂H₈N₂O₄	Ammonium oxalate	-1123.0									226.0		
C₂HgO₄	Mercury(II) oxalate	-678.2											
C₂I₂	Diiodoacetylene									313.1			70.3
C₂I₄	Tetraiodoethylene	305.0											
C₂K₂O₄	Potassium oxalate	-1346.0											
C₂MgO₄	Magnesium oxalate	-1269.0											
C₂N₂	Cyanogen		285.9	306.7						241.9			56.8
C₂N₄O₆	Trinitroacetonitrile		183.7										
C₂Na₂O₄	Sodium oxalate	-1318.0											
C₂O₄Pb	Lead(II) oxalate	-851.4			-750.1			146.0			105.4		
C₃F₈	Perfluoropropane			-1783.2									
C₃H₂N₂	Malononitrile	186.4		265.5									
C₃H₂O₂	2-Propynoic acid		-193.2										
C₃H₂O₃	1,3-Dioxol-2-one		-459.9	-418.6									
C₃H₃Cl₃	1,2,3-Trichloropropene		-101.8										
C₃H₃F₃	3,3,3-Trifluoropropene			-614.2									
C₃H₃N	2-Propenenitrile		147.1	180.6									
C₃H₃NO	Oxazole		-48.0	-15.5									
C₃H₃NO	Isoxazole		42.1	78.6									
C₃H₄	Allene			190.5									
C₃H₄	Propyne			184.9									
C₃H₄	Cyclopropene			277.1									
C₃H₄Cl₂	2,3-Dichloropropene		-73.3										
C₃H₄Cl₄	1,1,1,3-Tetrachloropropane		-208.7										
C₃H₄Cl₄	1,2,2,3-Tetrachloropropane		-251.8										
C₃H₄F₄O	2,2,3,3-Tetrafluoro-1-propanol		-1114.9	-1061.3									
C₃H₄N₂	1H-Pyrazole		105.4	179.4									
C₃H₄N₂	Imidazole	49.8		132.9									
C₃H₄O₂	1,2-Propanedione		-309.1	-271.0									
C₃H₄O₂	Propenoic acid		-383.8										
C₃H₄O₂	2-Oxetanone		-329.9	-282.9					175.3				
C₃H₄O₃	Ethylene carbonate		-571.5	-508.4									
C₃H₅Br	cis-1-Bromopropene		7.9	40.8								145.7	
C₃H₅Br	3-Bromopropene		12.2	45.2								122.1	
C₃H₅BrO	Bromoacetone			-181.0								133.9	

STANDARD THERMODYNAMIC PROPERTIES OF CHEMICAL SUBSTANCES (continued)

Molecular formula	Name	$\Delta_f H°$/kJ mol⁻¹ Crys.	Liq.	Gas	$\Delta_f G°$/kJ mol⁻¹ Crys.	Liq.	Gas	$S°$/J mol⁻¹ K⁻¹ Crys.	Liq.	Gas	C_p/J mol⁻¹ K⁻¹ Crys.	Liq.	Gas
C₃H₅Cl	2-Chloropropene			-21.0								125.1	
C₃H₅Cl	3-Chloropropene		-148.4	-107.8								131.6	
C₃H₅ClO	Epichlorohydrin		-522.5	-475.8								183.6	
C₃H₅ClO₂	2-Chloropropanoic acid	-549.3											
C₃H₅ClO₂	3-Chloropropanoic acid		-505.3	-462.9									
C₃H₅Cl₃	1,2,3-Trichloropropane		-230.6	-182.9									
C₃H₅I	3-Iodopropene		53.7	91.5								119.3	
C₃H₅IO	Iodoacetone			-130.5									
C₃H₅IO₂	3-Iodopropanoic acid	-460.0											
C₃H₅N	Propanenitrile		15.5	51.7									
C₃H₅N	2-Propyn-1-amine		205.7										
C₃H₅N	Ethyl isocyanide		108.6	141.7								149.1	
C₃H₅NO₃	Nitroacetone		-278.6										
C₃H₅NO₄	Methyl nitroacetate		-464.0										
C₃H₅N₃O₉	Trinitroglycerol		-370.9	-270.9									
C₃H₆	Propene		4.0	20.0									
C₃H₆	Cyclopropane		35.2	53.3			104.5			237.5			55.6
C₃H₆Br₂	1,2-Dibromopropane		-113.6	-71.6									
C₃H₆Cl₂	1,2-Dichloropropane		-198.8	-162.8									
C₃H₆Cl₂	1,3-Dichloropropane		-199.9	-159.2									
C₃H₆Cl₂	2,2-Dichloropropane		-205.8	-173.2									
C₃H₆Cl₂O	2,3-Dichloro-1-propanol		-381.5	-316.3									
C₃H₆Cl₂O	1,3-Dichloro-2-propanol		-385.3	-318.4									
C₃H₆I₂	1,2-Diiodopropane		-9.0	35.6									
C₃H₆I₂	1,3-Diiodopropane												
C₃H₆N₂O₂	Propanediamide	-546.1											
C₃H₆N₂O₂	N-(Aminocarbonyl)acetamide	-544.2		-441.2									
C₃H₆N₂O₄	1,1-Dinitropropane		-163.2	-100.7									
C₃H₆N₂O₄	1,3-Dinitropropane		-207.1										
C₃H₆N₂O₄	2,2-Dinitropropane		-181.2										
C₃H₆O	Allyl alcohol		-171.8	-124.5								138.9	
C₃H₆O	Propanal		-215.6	-185.6						304.5			80.7
C₃H₆O	Acetone		-248.4	-217.1			-152.7		199.8	295.3		126.3	74.5
C₃H₆O	Methyloxirane		-123.0	-94.7					196.5	286.9		120.4	72.6
C₃H₆O	Oxetane		-110.8	-80.5									
C₃H₆O₂	Propanoic acid		-510.7	-455.7					191.0			152.8	
C₃H₆O₂	Ethyl formate											149.3	
C₃H₆O₂	Methyl acetate		-445.9	-413.3						324.4		141.9	86.0
C₃H₆O₂	1,3-Dioxolane		-333.5	-298.0								118.0	
C₃H₆O₂S	Thiolactic acid		-468.4										
C₃H₆O₃	Trioxane	-522.5		-465.9				133.0			111.4		
C₃H₆S	Thiacyclobutane		24.7	60.6			107.1		184.9	285.0			
C₃H₆S	Methylthiirane		11.3	45.8									68.3

Molecular formula	Name	$\Delta_f H°$/kJ mol⁻¹ Crys.	Liq.	Gas	$\Delta_f G°$/kJ mol⁻¹ Crys.	Liq.	Gas	$S°$/J mol⁻¹ K⁻¹ Crys.	Liq.	Gas	C_p/J mol⁻¹ K⁻¹ Crys.	Liq.	Gas
C₃H₆S₂	1,2-Dithiolane			0.0			47.7			313.5			86.5
C₃H₆S₂	1,3-Dithiolane			10.0			54.7			323.3			84.7
C₃H₇Br	1-Bromopropane		-121.9	-87.0									
C₃H₇Br	2-Bromopropane		-130.5	-99.4									
C₃H₇Cl	1-Chloropropane		-160.5	-131.9									
C₃H₇Cl	2-Chloropropane		-172.3	-144.9									
C₃H₇ClO₂	3-Chloro-1,2-propanediol		-525.3										
C₃H₇ClO₂	2-Chloro-1,3-propanediol		-517.5										
C₃H₇F	1-Fluoropropane			-285.9									
C₃H₇F	2-Fluoropropane			-293.5									
C₃H₇I	1-Iodopropane		-66.0	-30.0									
C₃H₇I	2-Iodopropane		-74.8	-40.3									
C₃H₇N	Allylamine		-10.0										
C₃H₇N	Cyclopropylamine		45.8	77.0					187.7			147.1	
C₃H₇NO	N,N-Dimethylformamide		-239.3	-192.4								150.6	
C₃H₇NO	Propanamide	-338.2		-259.0									
C₃H₇NO₂	1-Nitropropane		-167.2	-123.8·									
C₃H₇NO₂	2-Nitropropane		-180.3	-138.9								170.3	
C₃H₇NO₂	DL-Alanine	-563.6											
C₃H₇NO₂	D-Alanine	-561.2											
C₃H₇NO₂	L-Alanine	-604.0		-465.9									
C₃H₇NO₂	β-Alanine	-558.0		-424.0									
C₃H₇NO₂	N-Methylglycine	-513.3		-367.3									
C₃H₇NO₂S	L-Cysteine	-534.1											
C₃H₇NO₃	Propyl nitrate		-214.5	-173.9									
C₃H₇NO₃	Isopropyl nitrate		-229.7	-191.0									
C₃H₇NO₃	DL-Serine	-739.0											
C₃H₇NO₃	L-Serine	-732.7											
C₃H₈	Propane		-120.9	-103.8			-23.4			270.3			73.6
C₃H₈N₂O	N-Ethylurea	-357.8											
C₃H₈N₂O	N,N-Dimethylurea	-319.1											
C₃H₈N₂O	N,N′-Dimethylurea	-312.1											
C₃H₈N₂O₃	Oxymethurea	-717.0											
C₃H₈O	1-Propanol		-302.6	-255.1					193.6	322.6		143.9	85.6
C₃H₈O	2-Propanol		-318.1	-272.6					181.1	309.2		156.5	89.3
C₃H₈O	Ethyl methyl ether			-216.4						309.2			93.3
C₃H₈O₂	1,2-Propylene glycol		-501.0	-429.8								190.8	
C₃H₈O₂	1,3-Propylene glycol		-480.8	-408.0									
C₃H₈O₂	Ethylene glycol monomethyl ether		-377.8	-348.5								171.1	
C₃H₈O₂	Dimethoxymethane								244.0			162.0	
C₃H₈O₃	Glycerol		-669.6	-577.9					206.3			218.9	
C₃H₈S	1-Propanethiol		-99.9	-67.8					242.5			144.6	
C₃H₈S	2-Propanethiol		-105.9	-76.2					233.5			145.3	
C₃H₈S	Ethyl methyl sulfide		-91.6	-59.6					239.1			144.6	

Molecular formula	Name	$\Delta_f H°$/kJ mol⁻¹ Crys.	Liq.	Gas	$\Delta_f G°$/kJ mol⁻¹ Crys.	Liq.	Gas	$S°$/J mol⁻¹ K⁻¹ Crys.	Liq.	Gas	C_p/J mol⁻¹ K⁻¹ Crys.	Liq.	Gas
C₃H₈S₂	1,3-Propanedithiol		-79.4	-29.8									
C₃H₉Al	Trimethyl aluminum		-136.4	-74.1		-9.9			209.4	314.7		155.6	
C₃H₉B	Trimethylborane		-143.1	-124.3		-32.1	-35.9		238.9	369.1			88.5
C₃H₉ClSi	Trimethylchlorosilane		-382.8	-352.8		-246.4	-243.5		278.2				
C₃H₉N	Propylamine		-101.5	-70.1			39.9			325.4		164.1	91.2
C₃H₉N	Isopropylamine		-112.3	-83.7			32.2		218.3	312.2		163.8	97.5
C₃H₉N	Trimethylamine		-45.7	-23.6					208.5	287.1		137.9	91.8
C₃H₁₀ClN	Propylamine hydrochloride	-354.7											
C₃H₁₀ClN	Trimethylamine hydrochloride	-282.9											
C₃H₁₀N₂	1,2-Propanediamine, (±)-		-97.8	-53.6									
C₃H₁₀Si	Trimethylsilane									331.0			117.9
C₃H₁₂BN	Trimethylamine borane	-142.5			70.7			187.0					
C₃H₁₂BN	Aminetrimethylboron	-284.1			-79.3			218.0					
C₄F₈	Perfluorocyclobutane											127.2	
C₄F₁₀	Perfluorobutane			-1542.6									
C₄H₂N₂	trans-2-Butenedinitrile	-268.2		340.2									
C₄H₂O₃	Maleic anhydride	-469.8		-398.3									
C₄H₂O₄	2-Butynedioic acid	-577.3											
C₄H₃NO₃	2-Nitrofuran	-104.1		-28.8									
C₄H₂BrNO₂	N-Bromosuccinimide	-335.9											
C₄H₂ClNO₂	N-Chlorosuccinimide	-357.9											
C₄H₄N₂	Succinonitrile		139.7	209.7					191.6			145.6	
C₄H₄N₂	Pyrazine	139.8		196.1									
C₄H₄N₂	Pyrimidine		145.9	195.7									
C₄H₄N₂	Pyridazine		224.9	278.3									
C₄H₄N₂O₂	Uracil	-429.4									120.5		
C₄H₄N₂O₃	2,4,6-Trihydroxypyrimidine	-634.7		-302.9									
C₄H₄O	Furan		-62.3	-34.8					177.0	267.2		114.8	65.4
C₄H₄O₂	Diketene		-233.1	-190.3									
C₄H₄O₃	Succinic anhydride	-608.6		-527.9									
C₄H₄O₄	Maleic acid	-789.4		-679.4				160.8			137.0		
C₄H₄O₄	Fumaric acid	-811.7		-675.8				168.0			142.0		
C₄H₄S	Thiophene		80.2	114.9			126.1		181.2	278.8		123.8	72.8
C₄H₅N	trans-2-Butenenitrile		95.1	134.3									
C₄H₅N	3-Butenenitrile		117.8	159.7									
C₄H₅N	Pyrrole		63.1	108.2					156.4			127.7	
C₄H₅N	Cyclopropanecarbonitrile		140.8	182.8									
C₄H₅NO₂	Succinimide	-459.0		-375.4									
C₄H₅NS	4-Methylthiazole		67.9	111.8									
C₄H₅N₃O	Cytosine	-221.3									132.6		
C₄H₆	1,2-Butadiene		138.6	162.3									
C₄H₆	1,3-Butadiene		88.5	110.0					199.0			123.6	
C₄H₆	1-Butyne		141.4	165.2									
C₄H₆	2-Butyne		119.1	145.7									

Molecular formula	Name	$\Delta_f H°$/kJ mol⁻¹ Crys.	Liq.	Gas	$\Delta_f G°$/kJ mol⁻¹ Crys.	Liq.	Gas	$S°$/J mol⁻¹ K⁻¹ Crys.	Liq.	Gas	C_p/J mol⁻¹ K⁻¹ Crys.	Liq.	Gas
C₄H₆	Cyclobutene			156.7									
C₄H₆N₂O₂	2,5-Piperazinedione	-446.5											
C₄H₆O	Divinyl ether		-39.8	-13.6									
C₄H₆O	trans-2-Butenal		-138.7	-100.6									
C₄H₆O₂	trans-Crotonic acid												
C₄H₆O₂	Methacrylic acid		-349.2	-314.4								161.1	
C₄H₆O₂	Vinyl acetate		-362.2	-333.0									
C₄H₆O₂	Methyl acrylate		-420.9	-366.5					239.5			158.8	
C₄H₆O₂	γ-Butyrolactone											141.4	
C₄H₆O₃	Acetic anhydride		-624.4	-572.5									
C₄H₆O₃	Propylene carbonate		-613.2	-582.5								218.6	
C₄H₆O₄	Succinic acid	-940.5		-823.0				167.3			153.1		
C₄H₆O₄	Dimethyl oxalate	-756.3		-708.9									
C₄H₆S	2,3-Dihydrothiophene		52.9	90.7			133.5			303.5			79.8
C₄H₆S	2,5-Dihydrothiophene		47.0	86.9			131.6			297.1			83.3
C₄H₇ClO	2-Chloroethyl vinyl ether		-208.1	-170.1									
C₄H₇ClO₂	2-Chlorobutanoic acid		-575.5										
C₄H₇ClO₂	3-Chlorobutanoic acid		-556.3										
C₄H₇ClO₂	4-Chlorobutanoic acid		-566.3										
C₄H₇ClO₂	Propyl chlorocarbonate		-533.4	-492.7									
C₄H₇N	Butanenitrile		-5.8	33.6									
C₄H₇N	2-Methylpropanenitrile		-13.8	23.4									
C₄H₇NO	2-Pyrrolidone		-286.2										
C₄H₇NO	2-Methyl-2-oxazoline		-169.5	-130.5									
C₄H₇NO₄	Iminodiacetic acid	-932.6											
C₄H₇NO₄	Ethyl nitroacetate		-487.1										
C₄H₇NO₄	L-Aspartic acid	-973.3											
C₄H₇N₃O	4H-Imidazol-4-one, 2-amino-1,5-dihydro-1-methyl-	-238.5											
C₄H₈	1-Butene		-20.8	0.1					227.0			118.0	
C₄H₈	cis-2-Butene		-29.8	-7.1					219.9			127.0	
C₄H₈	trans-2-Butene		-33.3	-11.4									
C₄H₈	Isobutene		-37.5	-16.9									
C₄H₈	Cyclobutane		3.7	27.7									
C₄H₈	Methylcyclopropane		1.7										
C₄H₈Br₂	1,2-Dibromobutane		-142.1	-91.6									
C₄H₈Br₂	1,3-Dibromobutane		-148.0										
C₄H₈Br₂	1,4-Dibromobutane		-140.3	-87.8									
C₄H₈Br₂	2,3-Dibromobutane		-139.6	-102.0									
C₄H₈Br₂	1,2-Dibromo-2-methylpropane		-156.6	-113.3									
C₄H₈Cl₂	1,3-Dichlorobutane		-237.3	-195.0									
C₄H₈Cl₂	1,4-Dichlorobutane		-229.8	-183.4									
C₄H₈Cl₂O	Bis(2-chloroethyl) ether											220.9	

Molecular formula	Name	$\Delta_f H°$/kJ mol⁻¹ Crys.	Liq.	Gas	$\Delta_f G°$/kJ mol⁻¹ Crys.	Liq.	Gas	S°/J mol⁻¹ K⁻¹ Crys.	Liq.	Gas	C_p/J mol⁻¹ K⁻¹ Crys.	Liq.	Gas
$C_4H_8I_2$	1,4-Diiodobenzene		-30.0										
$C_4H_8N_2O_2$	Succinamide	-581.2											
$C_4H_8N_2O_2$	Dimethylglyoxime	-199.7											
$C_4H_8N_2O_3$	L-Asparagine	-789.4											
$C_4H_8N_2O_3$	N-Glycylglycine	-747.7											
$C_4H_8N_2O_4$	1,4-Dinitrobenzene		-237.5										
C_4H_8O	Ethyl vinyl ether		-167.4	-140.8					230.9			147.0	
C_4H_8O	1,2-Epoxybutane		-168.9						246.6	343.7		163.7	103.4
C_4H_8O	Butanal		-239.2	-204.8									
C_4H_8O	Isobutanal		-247.3	-215.7									
C_4H_8O	2-Butanone		-273.3	-238.5					239.1	339.9		158.7	101.7
C_4H_8O	Tetrahydrofuran		-216.2	-184.1					204.3	302.4		124.0	76.3
C_4H_8OS	S-Ethyl thioacetate		-268.2	-228.1									
$C_4H_8O_2$	Butanoic acid		-533.8	-475.9					222.2			178.6	
$C_4H_8O_2$	2-Methylpropanoic acid											173.0	
$C_4H_8O_2$	Propyl formate		-500.3	-462.7									
$C_4H_8O_2$	Ethyl acetate		-479.3	-443.6					257.7			170.7	
$C_4H_8O_2$	1,3-Dioxane		-379.7	-340.6								143.9	
$C_4H_8O_2$	1,4-Dioxane		-353.9	-315.3					270.2			152.1	
$C_4H_8O_2$	2-Methyl-1,3-dioxolane		-386.9	-352.0									
$C_4H_8O_2S$	Sulfolane								180.0			180.0	
C_4H_8S	Tetrahydrothiophene		-72.9	-34.1			45.8			309.6			92.5
$C_4H_8S_2$	1,3-Dithiane			-10.0			72.4			333.5			110.4
C_4H_9Br	1-Bromobutane		-143.8	-107.1									
C_4H_9Br	2-Bromobutane		-154.9	-120.3									
C_4H_9Br	2-Bromo-2-methylpropane		-164.4	-132.4									
C_4H_9Cl	1-Chlorobutane		-188.1	-154.4									
C_4H_9Cl	2-Chlorobutane		-192.8	-161.1									
C_4H_9Cl	1-Chloro-2-methylpropane		-191.1	-159.3									
C_4H_9Cl	2-Chloro-2-methylpropane		-211.3	-182.2									
C_4H_9ClO	2-Chloroethyl ethyl ether		-335.6	-301.3									
C_4H_9I	1-Iodo-2-methylpropane		-107.5	-72.1								162.3	
C_4H_9I	2-Iodo-2-methylpropane		5.6	41.2									
C_4H_9N	Cyclobutanamine		-41.1	-3.6									
C_4H_9N	Pyrrolidine								204.1			156.6	
C_4H_9NO	Butanamide	-368.6		-282.0									
C_4H_9NO	N-Methylpropanamide		-346.9	-282.6								179.0	
C_4H_9NO	2-Methylpropanamide			-228.0									
C_4H_9NO	N,N-Dimethylacetamide		-278.3									175.6	
C_4H_9NO	Morpholine											164.8	
$C_4H_9NO_2$	1-Nitrobutane		-192.5	-143.9									
$C_4H_9NO_2$	2-Nitroisobutane		-217.2	-177.1									
$C_4H_9NO_2$	Propyl carbamate	-552.6		-471.4									
$C_4H_9NO_2$	4-Aminobutanoic acid	-581.0		-441.0									
$C_4H_9NO_3$	3-Nitro-2-butanol		-390.0										

Molecular formula	Name	$\Delta_f H°$/kJ mol⁻¹ Crys.	$\Delta_f H°$/kJ mol⁻¹ Liq.	$\Delta_f H°$/kJ mol⁻¹ Gas	$\Delta_f G°$/kJ mol⁻¹ Crys.	$\Delta_f G°$/kJ mol⁻¹ Liq.	$\Delta_f G°$/kJ mol⁻¹ Gas	$S°$/J mol⁻¹ K⁻¹ Crys.	$S°$/J mol⁻¹ K⁻¹ Liq.	$S°$/J mol⁻¹ K⁻¹ Gas	C_p/J mol⁻¹ K⁻¹ Crys.	C_p/J mol⁻¹ K⁻¹ Liq.	C_p/J mol⁻¹ K⁻¹ Gas
C₄H₉NO₃	2-Methyl-2-nitro-1-propanol	-410.1											
C₄H₉NO₃	DL-Threonine	-758.8											
C₄H₉NO₃	L-Threonine	-807.2											
C₄H₉N₃O₂	Creatine	-537.2											
C₄H₁₀	Butane		-147.3	-125.7								140.9	
C₄H₁₀	Isobutane		-154.2	-134.2									
C₄H₁₀Hg	Diethyl mercury		30.1	75.3								182.8	
C₄H₁₀N₂	Piperazine	-45.6											
C₄H₁₀N₂O	Trimethylurea	-330.5											
C₄H₁₀N₂O₂	N-Nitrodiethylamine		-106.2	-53.0									
C₄H₁₀N₂O₄	L-Asparagine, monohydrate	-1086.6											
C₄H₁₀O	1-Butanol		-327.3	-274.9					225.8			177.2	
C₄H₁₀O	2-Butanol		-342.6	-292.8					214.9	359.5		196.9	112.7
C₄H₁₀O	2-Methyl-1-propanol		-334.7	-283.8					214.7			181.5	
C₄H₁₀O	2-Methyl-2-propanol		-359.2	-312.5					193.3	326.7		218.6	113.6
C₄H₁₀O	Diethyl ether		-279.5	-252.1					172.4	342.7		175.6	119.5
C₄H₁₀O	Methyl propyl ether		-266.0	-238.1					262.9			165.4	
C₄H₁₀O	Isopropyl methyl ether		-278.8	-252.0					253.8			161.9	
C₄H₁₀OS	Diethyl sulfoxide		-268.0	-205.6									
C₄H₁₀O₂	1,2-Butanediol, (±)-		-523.6										
C₄H₁₀O₂	1,3-Butanediol		-501.0	-433.2					223.4				
C₄H₁₀O₂	1,4-Butanediol		-505.3	-428.7								200.1	
C₄H₁₀O₂	2,3-Butanediol		-541.5	-482.3								213.0	
C₄H₁₀O₂	2-Methyl-1,2-propanediol		-539.7										
C₄H₁₀O₂	Ethylene glycol monoethyl ether		-376.6	-389.7								210.8	
C₄H₁₀O₂	Ethylene glycol dimethyl ether		-420.6									193.3	
C₄H₁₀O₂	Dimethylacetal		-293.6	-245.9									
C₄H₁₀O₂	tert-Butyl hydroperoxide												
C₄H₁₀O₃	Diethylene glycol		-628.5	-571.2								244.8	
C₄H₁₀O₃S	Diethyl sulfite		-600.7	-552.2									
C₄H₁₀O₄S	Sulfuric acid, diethyl ester		-813.2	-756.3									
C₄H₁₀S	1-Butanethiol		-124.7	-88.0								171.2	
C₄H₁₀S	2-Butanethiol		-131.0	-96.9									
C₄H₁₀S	2-Methyl-1-propanethiol		-132.0	-97.3									
C₄H₁₀S	2-Methyl-2-propanethiol		-140.5	-109.6									
C₄H₁₀S	Diethyl sulfide		-119.4	-83.5					269.3	368.1		171.4	117.0
C₄H₁₀S	Methyl propyl sulfide		-118.5	-82.2					272.5			171.6	
C₄H₁₀S	Isopropyl methyl sulfide		-124.7	-90.5					263.1			172.4	
C₄H₁₀S₂	1,4-Butanedithiol		-105.7	-50.6								171.4	
C₄H₁₀S₂	Diethyl disulfide		-120.1	-79.4					269.3			179.2	
C₄H₁₁N	Butylamine		-127.6	-91.9								192.1	
C₄H₁₁N	sec-Butylamine		-137.5	-104.6									
C₄H₁₁N	tert-Butylamine		-150.6	-121.0									
C₄H₁₁N	Isobutylamine		-132.6	-98.7								183.2	
C₄H₁₁N	Diethylamine		-103.7	-72.2								169.2	

Molecular formula	Name	$\Delta_f H°$/kJ mol⁻¹			$\Delta_f G°$/kJ mol⁻¹			$S°$/J mol⁻¹ K⁻¹			C_p/J mol⁻¹ K⁻¹		
		Crys.	Liq.	Gas	Crys.	Liq.	Gas	Crys.	Liq.	Gas	Crys.	Liq.	Gas
$C_4H_{11}NO$	N,N-Dimethylethanolamine		-253.7	-203.6							233.5		
$C_4H_{11}NO_2$	Diethanolamine	-493.8		-397.1									
$C_4H_{11}NO_3$	2-Amino-2-(hydroxymethyl)-1,3-propanediol	-717.8											
$C_4H_{12}BrN$	Tetramethylammonium bromide	-251.0											
$C_4H_{12}ClN$	Diethylamine hydrochloride	-358.6											
$C_4H_{12}ClN$	Tetramethylammonium chloride	-276.4											
$C_4H_{12}IN$	Tetramethylammonium iodide	-203.9											
$C_4H_{12}N_2$	2-Methyl-1,2-propanediamine		-133.9	-90.3									
$C_4H_{12}Pb$	Tetramethyl lead		97.9	135.9									
$C_4H_{12}Si$	Tetramethylsilane		-264.0	-239.1		-100.0	-99.9		277.3	359.0		204.1	143.9
$C_4H_{12}Sn$	Tetramethylstannane		-52.3	-18.8								254.0	
$C_4H_{13}N_3$	Diethylenetriamine												
C_4N_2	2-Butynedinitrile		500.4	529.2									
C_4NiO_4	Nickel carbonyl		-633.0	-602.9		-588.2	-587.2		313.4	410.6		204.6	145.2
C_4FeO_5	Iron pentacarbonyl		-774.0			-705.3			338.1			240.6	
$C_5H_2F_6O_2$	Hexafluoroacetylacetone	-2286.7											
$C_5H_3NO_5$	5-Nitro-2-furancarboxylic acid	-516.8											
$C_5H_4N_4$	1H-Purine	169.4						145.6			134.5		
$C_5H_4N_4O$	Hypoxanthine	-110.8						161.1			151.3		
$C_5H_4N_4O_2$	Xanthine	-379.6						173.2			166.1		
$C_5H_4N_4O_3$	Uric acid	-618.8											
$C_5H_4O_2$	Furfural		-201.6	-151.0								163.2	
$C_5H_4O_3$	2-Furancarboxylic acid	-498.4		-390.0									
$C_5H_4O_3$	3-Methyl-2,5-furandione		-504.5	-447.2									
$C_5H_5F_3O_2$	1,1,1-Trifluoro-2,4-pentanedione		-1040.2	-993.3									
C_5H_5N	Pyridine		100.2	140.4								132.7	
C_5H_5NO	1H-Pyrrole-2-carboxaldehyde	-106.4											
$C_5H_5N_5$	Adenine	96.9		205.7							147.0		
$C_5H_5N_5O$	Guanine	-183.9											
C_5H_6	cis-3-Penten-1-yne		226.5										
C_5H_6	trans-3-Penten-1-yne		228.2										
C_5H_6	1,3-Cyclopentadiene		105.9	134.3									
$C_5H_6N_2O_2$	Thymine	-462.8		-328.7							150.8		
$C_5H_6O_2$	Furfuryl alcohol		-276.2	-211.8								204.0	
$C_5H_6O_4$	trans-1-Propene-1,2-dicarboxylic acid	-824.4											
C_5H_6S	2-Methylthiophene		44.6	83.5					218.5			149.8	
C_5H_6S	3-Methylthiophene		43.1	82.5									
C_5H_7N	trans-3-Pentenenitrile		80.9	125.7									
C_5H_7N	Cyclobutanecarbonitrile		103.0	147.4									

Molecular formula	Name	$\Delta_f H°$/kJ mol⁻¹ Crys.	Liq.	Gas	$\Delta_f G°$/kJ mol⁻¹ Crys.	Liq.	Gas	$S°$/J mol⁻¹ K⁻¹ Crys.	Liq.	Gas	C_p/J mol⁻¹ K⁻¹ Crys.	Liq.	Gas
C₅H₇N	1-Methylpyrrole		62.4	103.1									
C₅H₇N	2-Methylpyrrole		23.3	74.0									
C₅H₇N	3-Methylpyrrole		20.5	70.2									
C₅H₇NO₂	Ethyl cyanoacetate											220.2	
C₅H₈	1,2-Pentadiene			140.7									
C₅H₈	cis-1,3-Pentadiene			81.4									
C₅H₈	trans-1,3-Pentadiene			76.1									
C₅H₈	1,4-Pentadiene			105.7									
C₅H₈	2,3-Pentadiene			133.1									
C₅H₈	3-Methyl-1,2-butadiene		101.2										
C₅H₈	2-Methyl-1,3-butadiene		48.2	75.5					229.3			152.6	
C₅H₈	Cyclopentene		4.3	34.0					201.2			122.4	
C₅H₈	Methylenecyclobutane		93.8	121.6									
C₅H₈	Spiropentane		157.5	185.2					193.7			134.5	
C₅H₈N₄O₁₂	Pentaerythritol tetranitrate	-538.6											
C₅H₈O	Cyclopentanone		-235.9	-192.1									
C₅H₈O₂	4-Pentenoic acid	-430.6											
C₅H₈O₂	Methyl trans-2-butenoate		-382.9	-341.9									
C₅H₈O₂	Methyl methacrylate											191.2	
C₅H₈O₂	Allyl acetate											184.1	
C₅H₈O₂	2,4-Pentanedione		-423.8	-382.0									
C₅H₈O₂	4-Methyl-gamma-butyrolactone		-461.3	-406.5									
C₅H₈O₂	Tetrahydro-2H-pyran-2-one		-436.7	-379.6									
C₅H₈O₄	Glutaric acid	-960.0											
C₅H₉ClO₂	Propyl chloroacetate		-515.5	-467.0									
C₅H₉N	Pentanenitrile		-33.1	10.5									
C₅H₉N	2,2-Dimethylpropanenitrile		-39.8	-2.3									
C₅H₉N	1,2,5,6-Tetrahydropyridine		33.5										
C₅H₉NO	2-Piperidinone	-306.6											
C₅H₉NO	N-Methyl-2-pyrrolidone		-262.2	-366.2								307.8	
C₅H₉NO₂	L-Proline	-515.2											
C₅H₉NO₄	D-Glutamic acid	-1005.3											
C₅H₉NO₄	L-Glutamic acid	-1009.7											
C₅H₁₀	1-Pentene		-46.9	-21.1					262.6			154.0	
C₅H₁₀	cis-2-Pentene		-53.7	-27.6					258.6			151.7	
C₅H₁₀	trans-2-Pentene		-58.2	-31.9					256.5			157.0	
C₅H₁₀	2-Methyl-1-butene		-61.1	-35.2					254.0			157.2	
C₅H₁₀	3-Methyl-1-butene		-51.5	-27.5					253.3			156.1	
C₅H₁₀	2-Methyl-2-butene		-68.6	-41.7					251.0			152.8	
C₅H₁₀	Cyclopentane		-105.1	-76.4					204.5			128.8	
C₅H₁₀	Methylcyclobutane		-44.5										
C₅H₁₀	Ethylcyclopropane		-24.8										
C₅H₁₀	1,1-Dimethylcyclopropane		-33.3	-8.2									
C₅H₁₀	cis-1,2-Dimethylcyclopropane		-26.3										

Molecular formula	Name	$\Delta_f H°$/kJ mol⁻¹ Crys.	Liq.	Gas	$\Delta_f G°$/kJ mol⁻¹ Crys.	Liq.	Gas	$S°$/J mol⁻¹ K⁻¹ Crys.	Liq.	Gas	C_p/J mol⁻¹ K⁻¹ Crys.	Liq.	Gas
C₅H₁₀	*trans*-1,2-Dimethyl-cyclopropane		-30.7										
C₅H₁₀Br₂	1,3-Dibromo-2-methylpropane			-137.6									
C₅H₁₀N₂O	*N*-Nitrosopiperidine		-31.1	16.6									
C₅H₁₀N₂O₂	*N*-Nitropiperidine		-93.0	-44.5									
C₅H₁₀N₂O₃	L-Glutamine	-826.4											
C₅H₁₀O	Cyclopentanol		-300.1	-242.5					206.3			184.1	
C₅H₁₀O	Pentanal		-267.2	-228.4									
C₅H₁₀O	2-Pentanone		-297.3	-258.8					266.0			184.1	
C₅H₁₀O	3-Pentanone		-296.5	-257.9					268.5			190.9	
C₅H₁₀O	3-Methyl-2-butanone		-299.5	-262.6								179.9	
C₅H₁₀O	3,3-Dimethyloxetane		-182.2	-148.2									
C₅H₁₀O	Tetrahydropyran		-258.3	-223.4									
C₅H₁₀OS	S-Propyl thioacetate		-294.5	-250.4									
C₅H₁₀O₂	Pentanoic acid		-559.4	-491.9					259.8			210.3	
C₅H₁₀O₂	2-Methylbutanoic acid		-554.5										
C₅H₁₀O₂	3-Methylbutanoic acid		-561.6	-510.0									
C₅H₁₀O₂	2,2-Dimethylpropanoic acid	-564.5		-491.3									
C₅H₁₀O₂	Butyl formate											200.2	
C₅H₁₀O₂	Propyl acetate											196.2	
C₅H₁₀O₂	Isopropyl acetate		-518.9	-481.6								199.4	
C₅H₁₀O₂	Ethyl propanoate		-502.7	-463.4									
C₅H₁₀O₂	*cis*-1,2-Cyclopentanediol	-485.0											
C₅H₁₀O₂	*trans*-1,2-Cyclopentanediol	-490.1											
C₅H₁₀O₂	4-Methyl-1,3-dioxane	-416.1		-376.9									
C₅H₁₀O₂	(Ethoxymethyl)oxirane		-296.5										
C₅H₁₀O₂	Tetrahydrofurfuryl alcohol		-435.7	-369.1									
C₅H₁₀O₂	Diethyl carbonate		-681.5	-637.9									
C₅H₁₀O₃	Ethylene glycol monomethyl ether acetate											310.0	
C₅H₁₀O₃	Ethyl lactate											254.0	
C₅H₁₀O₄	1,2,3-Propanetriol, 1-acetate, (±)-		-909.2										
C₅H₁₀O₅	D-Ribose	-1047.2											
C₅H₁₀O₅	D-Xylose	-1057.8											
C₅H₁₀O₅	α-D-Arabinopyranose	-1057.9					53.1						
C₅H₁₀S	Thiacyclohexane		-106.3	-63.5					218.2	323.0		163.3	109.7
C₅H₁₀S	Cyclopentanethiol		-89.5	-48.1					256.9			165.2	
C₅H₁₁Br	1-Bromopentane		-170.2	-128.9									
C₅H₁₁Cl	1-Chloropentane		-213.2	-174.9									
C₅H₁₁Cl	2-Chloro-2-methylbutane		-235.7	-202.2									
C₅H₁₁Cl	1-Chloro-3-methylbutane		-216.0	-179.7									
C₅H₁₁Cl	2-Chloro-3-methylbutane		-226.6	-185.1									
C₅H₁₁N	Cyclopentylamine		-95.1	-54.9					241.0			181.2	

Molecular formula	Name	$\Delta_f H°$/kJ mol⁻¹ Crys.	Liq.	Gas	$\Delta_f G°$/kJ mol⁻¹ Crys.	Liq.	Gas	$S°$/J mol⁻¹ K⁻¹ Crys.	Liq.	Gas	C_p/J mol⁻¹ K⁻¹ Crys.	Liq.	Gas
C₅H₁₁N	Piperidine		-86.4	-47.1					210.0			179.9	
C₅H₁₁NO	Pentanamide	-379.5		-290.2									
C₅H₁₁NO	2,2-Dimethylpropanamide	-399.7		-313.1									
C₅H₁₁NO₂	1-Nitropentane		-215.4										
C₅H₁₁NO₂	DL-Valine	-628.9		-455.1									
C₅H₁₁NO₂	L-Valine	-617.9											
C₅H₁₁NO₂	5-Aminopentanoic acid	-604.1		-460.0									
C₅H₁₁NO₂S	L-Methionine	-577.5		-413.5									
C₅H₁₁NO₄	2-Ethyl-2-nitro-1,3-propanediol	-606.4											
C₅H₁₂	Pentane		-173.5	-146.9								167.2	
C₅H₁₂	Isopentane		-178.4	-153.6					260.4			164.8	
C₅H₁₂	Neopentane		-190.2	-168.0									
C₅H₁₂N₂O	Butylurea	-419.5											
C₅H₁₂N₂O	tert-Butylurea	-417.4											
C₅H₁₂N₂O	N,N-Diethylurea	-372.2											
C₅H₁₂N₂O	Tetramethylurea	-262.2											
C₅H₁₂N₂S	Tetramethylthiourea	-38.1		44.9									
C₅H₁₂O	1-Pentanol		-351.6	-294.6								208.1	
C₅H₁₂O	2-Pentanol		-365.2	-311.0									
C₅H₁₂O	3-Pentanol		-368.9	-314.9								239.7	
C₅H₁₂O	2-Methyl-1-butanol		-356.6	-301.4									
C₅H₁₂O	3-Methyl-1-butanol		-356.4	-300.7									
C₅H₁₂O	2-Methyl-2-butanol		-379.5	-329.3								247.1	
C₅H₁₂O	3-Methyl-2-butanol		-366.6	-313.5									
C₅H₁₂O	2,2-Dimethyl-1-propanol		-399.4										
C₅H₁₂O	Butyl methyl ether		-290.6	-258.1					295.3			192.7	
C₅H₁₂O	tert-Butyl methyl ether		-313.6	-283.7					265.3			187.5	
C₅H₁₂O	Ethyl propyl ether		-303.6	-272.0					295.0			197.2	
C₅H₁₂O₂	1,5-Pentanediol		-528.8	-450.8									
C₅H₁₂O₂	2,2-Dimethyl-1,3-propanediol	-551.2											
C₅H₁₂O₂	Diethoxymethane		-450.5	-414.7									
C₅H₁₂O₂	1,1-Dimethoxypropane		-443.6										
C₅H₁₂O₂	2,2-Dimethoxypropane		-459.4	-429.9									
C₅H₁₂O₃	Diethylene glycol monomethyl ether											271.1	
C₅H₁₂O₃	2-(Hydroxymethyl)-2-methyl-1,3-propanediol	-744.6											
C₅H₁₂O₄	Pentaerythritol	-920.6		-776.7									
C₅H₁₂O₅	Xylitol	-1118.5											
C₅H₁₂S	1-Pentanethiol		-151.3	-110.0									
C₅H₁₂S	2-Methyl-1-butanethiol		-154.4	-114.9									
C₅H₁₂S	3-Methyl-1-butanethiol		-154.4	-114.9									
C₅H₁₂S	2-Methyl-2-butanethiol		-162.8	-127.1					290.1			198.1	
C₅H₁₂S	3-Methyl-2-butanethiol		-158.8	-121.3									

Molecular formula	Name	$\Delta_f H°$/kJ mol^{-1}			$\Delta_f G°$/kJ mol^{-1}			$S°$/J mol^{-1} K^{-1}			C_p/J mol^{-1} K^{-1}		
		Crys.	Liq.	Gas	Crys.	Liq.	Gas	Crys.	Liq.	Gas	Crys.	Liq.	Gas
$C_5H_{12}S$	2,2-Dimethyl-1-propanethiol		-165.4	-129.0								200.9	
$C_5H_{12}S$	Butyl methyl sulfide		-142.9	-102.4					307.5			199.9	
$C_5H_{12}S$	tert-Butyl methyl sulfide		-157.1	-121.3					276.1			198.4	
$C_5H_{12}S$	Ethyl propyl sulfide		-144.8	-104.8					309.5				
$C_5H_{12}S$	Ethyl isopropyl sulfide		-156.1	-118.3								218.0	
$C_5H_{13}N$	Pentylamine		-51.1	-18.2									
$C_5H_{14}N_2$	N,N,N',N'-Tetramethyl-methanediamine												
C_6ClF_5	Chloropentafluorobenzene	-858.4		-809.3									
C_6Cl_6	Hexachlorobenzene	-127.6		-35.5				260.2	280.8		201.2	221.6	
C_6F_6	Hexafluorobenzene		-991.3	-955.4									
C_6F_{10}	Perfluorocyclohexene		-1963.5	-1932.7									
C_6F_{12}	Perfluorocyclohexane		-2406.3	-2370.4									
C_6HF_5	Pentafluorobenzene	-852.7	-841.8	-806.5									
C_6HF_5O	Pentafluorophenol	-1024.1	-1007.7										
$C_6H_2F_4$	1,2,4,5-Tetrafluorobenzene	-683.8		3.8									
$C_6H_3Cl_3$	1,2,3-Trichlorobenzene	-70.8		-8.1									
$C_6H_3Cl_3$	1,2,4-Trichlorobenzene		-63.1										
$C_6H_3Cl_3$	1,3,5-Trichlorobenzene	-78.4		-13.4									
$C_6H_3N_3O_8$	2,4,6-Trinitro-1,3-benzenediol	-467.5											
$C_6H_4Cl_2$	o-Dichlorobenzene		-17.5	30.2								162.4	
$C_6H_4Cl_2$	m-Dichlorobenzene		-20.7	25.7				175.4			147.8		
$C_6H_4Cl_2$	p-Dichlorobenzene	-42.3		22.5									
$C_6H_4F_2$	o-Difluorobenzene		-330.0	-293.8					222.6			159.0	
$C_6H_4F_2$	m-Difluorobenzene		-343.9	-309.2					223.8			159.1	
$C_6H_4F_2$	p-Difluorobenzene		-342.3	-306.7								157.5	
$C_6H_4N_2O_5$	2,4-Dinitrophenol	-232.7		-128.1									
$C_6H_4O_2$	p-Benzoquinone	-185.7		-122.9					219.2		129.0		
C_6H_5Br	Bromobenzene		60.9									154.3	
C_6H_5Cl	Chlorobenzene		11.1	52.0								150.1	
C_6H_5ClO	m-Chlorophenol	-206.4	-189.3										
C_6H_5ClO	p-Chlorophenol	-197.7	-181.3										
C_6H_5F	Fluorobenzene		-150.6	-115.9					205.9			146.4	
C_6H_5I	Iodobenzene		117.2	164.9					205.4			158.7	
$C_6H_5NO_2$	Nitrobenzene		12.5	67.5								185.8	
$C_6H_5NO_2$	3-Pyridinecarboxylic acid	-344.9		-221.5									
$C_6H_5NO_3$	2-Nitrophenol	-202.4											
$C_6H_5N_3$	1H-Benzotriazole	236.5		335.5									
$C_6H_5N_3O_4$	2,3-Dinitroaniline	-11.7											
$C_6H_5N_3O_4$	2,4-Dinitroaniline	-67.8											
$C_6H_5N_3O_4$	2,5-Dinitroaniline	-44.3											
$C_6H_5N_3O_4$	2,6-Dinitroaniline	-50.6											
$C_6H_5N_3O_4$	3,5-Dinitroaniline	-38.9											
C_6H_6	1,5-Hexadiyne		384.2										
C_6H_6	Benzene		49.1	82.9		124.5	129.7		173.4	269.2		136.0	82.4

Molecular formula	Name	$\Delta_f H°$/kJ mol⁻¹			$\Delta_f G°$/kJ mol⁻¹			$S°$/J mol⁻¹ K⁻¹			C_p/J mol⁻¹ K⁻¹		
		Crys.	Liq.	Gas	Crys.	Liq.	Gas	Crys.	Liq.	Gas	Crys.	Liq.	Gas
$C_6H_6N_2O_2$	o-Nitroaniline	-26.1	-9.4	63.8							166.0		
$C_6H_6N_2O_2$	m-Nitroaniline	-38.3	-14.4	58.4							158.8		
$C_6H_6N_2O_2$	p-Nitroaniline	-42.0	-20.7	58.8							167.0		
C_6H_6O	Phenol	-165.1		-96.4				144.0			127.4		
C_6H_6O	2-Vinylfuran		-10.3	27.8									
$C_6H_6O_2$	p-Hydroquinone	-364.5		-265.3							136.0		
$C_6H_6O_2$	Pyrocatechol	-354.1		-267.5									
$C_6H_6O_2$	Resorcinol	-368.0		-274.7									
$C_6H_6O_3$	1,2,3-Benzenetriol	-551.1		-434.2									
$C_6H_6O_3$	1,2,4-Benzenetriol	-563.8		-444.0									
$C_6H_6O_3$	1,3,5-Benzenetriol	-584.6		-452.9									
$C_6H_6O_3$	3,4-Dimethyl-2,5-furandione	-581.4											
$C_6H_6O_6$	cis-1-Propene-1,2,3-tricarboxylic acid	-1224.4											
$C_6H_6O_6$	trans-1-Propene-1,2,3-tricarboxylic acid	-1232.7											
C_6H_6S	Benzenethiol		63.7	111.3					222.8			173.2	
C_6H_7N	Aniline		31.6	87.5			-7.0			317.9		191.9	107.9
C_6H_7N	2-Methylpyridine		56.7	99.2								158.6	
C_6H_7N	3-Methylpyridine		61.9	106.4					216.3			158.7	
C_6H_7N	4-Methylpyridine		59.2	103.8					209.1			159.0	
C_6H_7N	1-Cyclopentenecarbonitrile		111.5	156.5									
$C_6H_8N_2$	Adiponitrile		85.1	149.5								128.7	
$C_6H_8N_2$	o-Phenylenediamine	-0.3											
$C_6H_8N_2$	m-Phenylenediamine	-7.8						154.5			159.6		
$C_6H_8N_2$	p-Phenylenediamine	3.0											
$C_6H_8N_2$	Phenylhydrazine		141.0	202.9									
$C_6H_8N_2S$	Bis(2-cyanoethyl) sulfide		96.3									217.0	
$C_6H_8O_4$	Dimethyl maleate											263.2	
$C_6H_8O_6$	L-Ascorbic acid	-1164.6											
$C_6H_8O_7$	2-Hydroxy-1,2,3-propane-tricarboxylic acid	-1543.8											
$C_6H_9Cl_3O_2$	Butyl trichloroacetate		-545.8	-492.3									
$C_6H_9Cl_3O_2$	Isobutyl trichloroacetate		-553.4	-500.2									
C_6H_9N	Cyclopentanecarbonitrile		0.7	44.1									
C_6H_9N	2,4-Dimethylpyrrole	-422.3											
C_6H_9N	2,5-Dimethylpyrrole		-16.7	39.8									
$C_6H_9NO_3$	Triacetamide	-610.5		-550.1									
$C_6H_9NO_6$	Nitrilotriacetic acid	-1311.9											
$C_6H_9N_3O_2$	L-Histidine	-466.7											
C_6H_{10}	1,5-Hexadiene		54.1	84.2									
C_6H_{10}	3,3-Dimethyl-1-butyne		78.4										
C_6H_{10}	Cyclohexene		-38.5	-5.0					214.6			148.3	
C_6H_{10}	1-Methylcyclopentene		-36.4	-3.8									
C_6H_{10}	3-Methylcyclopentene		-23.7	7.4									

Molecular formula	Name	ΔfH°/kJ mol⁻¹			ΔfG°/kJ mol⁻¹			S°/J mol⁻¹ K⁻¹			Cp/J mol⁻¹ K⁻¹		
		Crys.	Liq.	Gas	Crys.	Liq.	Gas	Crys.	Liq.	Gas	Crys.	Liq.	Gas
C_6H_{10}	4-Methylcyclopentene		-17.6	14.6									
$C_6H_{10}Cl_2O_2$	Butyl dichloroacetate		-550.1	-497.8									
$C_6H_{10}O$	Cyclohexanone		-271.2	-226.1								182.2	
$C_6H_{10}O$	2-Methylcyclopentanone		-265.2										
$C_6H_{10}O$	Mesityl oxide											212.5	
$C_6H_{10}O_2$	Ethyl *trans*-2-butenoate		-420.0	-375.6									
$C_6H_{10}O_2$	Methyl cyclobutanecarboxylate		-395.0	-350.2									
$C_6H_{10}O_3$	Ethyl acetoacetate											248.0	
$C_6H_{10}O_3$	Propanoic anhydride		-679.1	-626.5									
$C_6H_{10}O_4$	Adipic acid	-994.3											
$C_6H_{10}O_4$	Diethyl oxalate		-805.5	-742.0									
$C_6H_{10}O_4$	Ethylene glycol diacetate											310.0	
$C_6H_{11}Cl$	Chlorocyclohexane		-207.2	-163.7									
$C_6H_{11}ClO_2$	Ethyl 4-chlorobutanoate		-566.5	-513.8									
$C_6H_{11}ClO_2$	Propyl 3-chloropropanoate		-537.6	-485.7									
$C_6H_{11}ClO_2$	Butyl chloroacetate		-538.4	-487.4									
$C_6H_{11}NO$	Caprolactam	-329.4	-293.0	-239.6							156.8		
$C_6H_{11}NO$	1-Methyl-2-piperidinone												
C_6H_{12}	1-Hexene		-74.2	-43.5					295.2			183.3	
C_6H_{12}	*cis*-2-Hexene		-83.9	-52.3									
C_6H_{12}	*trans*-2-Hexene		-85.5	-53.9									
C_6H_{12}	*cis*-3-Hexene		-78.9	-47.6									
C_6H_{12}	*trans*-3-Hexene		-86.1	-54.4									
C_6H_{12}	2-Methyl-1-pentene		-90.0	-59.4									
C_6H_{12}	3-Methyl-1-pentene		-78.2	-49.5									
C_6H_{12}	4-Methyl-1-pentene		-80.0	-51.3									
C_6H_{12}	3-Methyl-*cis*-2-pentene		-94.5	-62.3									
C_6H_{12}	4-Methyl-*cis*-2-pentene		-87.0	-57.5									
C_6H_{12}	3-Methyl-*trans*-2-pentene		-94.6	-63.1									
C_6H_{12}	4-Methyl-*trans*-2-pentene		-91.6	-61.5									
C_6H_{12}	2-Ethyl-1-butene		-87.1	-56.0									
C_6H_{12}	2,3-Dimethyl-1-butene		-93.2	-62.4									
C_6H_{12}	3,3-Dimethyl-1-butene		-87.5	-60.3									
C_6H_{12}	2,3-Dimethyl-2-butene		-101.4	-68.1									
C_6H_{12}	Cyclohexane		-156.4	-123.4					270.2			174.7	
C_6H_{12}	Methylcyclopentane		-137.9	-106.2								154.9	
C_6H_{12}	Ethylcyclobutane		-59.0	-27.5									
C_6H_{12}	1,1,2-Trimethylcyclopropane		-96.2										
C_6H_{12}	2-Methyl-2-pentene		-98.5	-66.9									
$C_6H_{12}N_2O_4S_2$	L-Cystine	-1032.7											
$C_6H_{12}O$	Cyclohexanol		-348.2	-286.2								208.2	
$C_6H_{12}O$	*cis*-2-Methylcyclopentanol		-345.5										
$C_6H_{12}O$	Butyl vinyl ether		-218.8	-182.6								232.0	

Molecular formula	Name	$\Delta_f H°$/kJ mol⁻¹			$\Delta_f G°$/kJ mol⁻¹			S°/J mol⁻¹ K⁻¹			C_p/J mol⁻¹ K⁻¹		
		Crys.	Liq.	Gas	Crys.	Liq.	Gas	Crys.	Liq.	Gas	Crys.	Liq.	Gas
C₆H₁₂O	2-Hexanone		-322.0	-278.9								213.3	
C₆H₁₂O	3-Hexanone		-320.2	-277.6					305.3			216.9	
C₆H₁₂O	Methyl isobutyl ketone											213.3	
C₆H₁₂O	2-Methyl-3-pentanone		-325.9	-286.0									
C₆H₁₂O	3,3-Dimethyl-2-butanone		-328.6	-290.6									
C₆H₁₂O₂	Hexanoic acid		-583.8	-511.9									
C₆H₁₂O₂	Butyl acetate		-529.2	-485.3								227.8	
C₆H₁₂O₂	Isobutyl acetate											233.8	
C₆H₁₂O₂	Ethyl butanoate											228.0	
C₆H₁₂O₂	Methyl pentanoate		-514.2	-471.1								229.3	
C₆H₁₂O₂	Methyl 2,2-dimethyl-propanoate		-530.0	-491.2									
C₆H₁₂O₂	Diacetone alcohol											257.9	
C₆H₁₂O₃	Ethylene glycol ethyl ether acetate											221.3	
C₆H₁₂O₃	Paraldehyde		-673.1	-631.7								376.0	
C₆H₁₂O₆	β-D-Fructose	-1265.6											
C₆H₁₂O₆	D-Galactose	-1286.3											
C₆H₁₂O₆	α-D-Glucose	-1273.3											
C₆H₁₂O₆	D-Mannose	-1263.0											
C₆H₁₂O₆	L-Sorbose	-1271.5											
C₆H₁₂S	Cyclohexanethiol		-140.7	-96.2					255.6			192.6	
C₆H₁₂S	Cyclopentyl methyl sulfide		-109.8	-64.7									
C₆H₁₃Br	1-Bromohexane		-194.2	-148.3					453.0			203.5	
C₆H₁₃Cl	2-Chlorohexane		-246.1	-204.3									
C₆H₁₃N	Cyclohexylamine		-147.6	-104.0									
C₆H₁₃N	(±)-2-Methylpiperidine		-124.9	-84.4									
C₆H₁₃NO	Hexanamide		-397.9	-324.2									
C₆H₁₃NO	N-Butylacetamide		-380.9	-305.9									
C₆H₁₃NO₂	DL-Leucine	-640.6											
C₆H₁₃NO₂	D-Leucine	-637.3											
C₆H₁₃NO₂	L-Leucine	-637.4		-486.8							200.1		
C₆H₁₃NO₂	DL-Isoleucine	-635.3											
C₆H₁₃NO₂	L-Isoleucine	-637.8											
C₆H₁₃NO₂	Norleucine	-639.1											
C₆H₁₃NO₂	6-Aminohexanoic acid	-637.3											
C₆H₁₄	Hexane		-198.7	-166.9					290.6			195.6	
C₆H₁₄	2-Methylpentane		-204.6	-174.6					292.5			193.7	
C₆H₁₄	3-Methylpentane		-202.4	-171.9					272.5			190.7	
C₆H₁₄	2,2-Dimethylbutane		-213.8	-185.9					287.8			191.9	
C₆H₁₄	2,3-Dimethylbutane		-207.4	-178.1								189.7	
C₆H₁₄N₂	Azopropane		11.5	51.3									
C₆H₁₄N₂O₂	Lysine	-678.7						250.6					
C₆H₁₄N₄O₂	D-Arginine	-623.5									232.0		
C₆H₁₄O	1-Hexanol		-377.5	-315.9					287.4			240.4	

Molecular formula	Name	$\Delta_f H°$/kJ mol⁻¹			$\Delta_f G°$/kJ mol⁻¹			$S°$/J mol⁻¹ K⁻¹			C_p/J mol⁻¹ K⁻¹		
		Crys.	Liq.	Gas	Crys.	Liq.	Gas	Crys.	Liq.	Gas	Crys.	Liq.	Gas
C₆H₁₄O	2-Hexanol		-392.0	-333.5								286.2	
C₆H₁₄O	3-Hexanol		-392.4									248.0	
C₆H₁₄O	2-Methyl-1-pentanol											275.9	
C₆H₁₄O	3-Methyl-2-pentanol		-394.7									273.0	
C₆H₁₄O	4-Methyl-2-pentanol		-396.4										
C₆H₁₄O	2-Methyl-3-pentanol											293.4	
C₆H₁₄O	3-Methyl-3-pentanol											221.6	
C₆H₁₄O	Dipropyl ether		-328.8	-293.0					323.9			216.8	
C₆H₁₄O	Diisopropyl ether		-351.5	-319.2								159.0	
C₆H₁₄O	Butyl ethyl ether												
C₆H₁₄O	Ethyl *tert*-butyl ether			-313.9									
C₆H₁₄OS	Dipropyl sulfoxide		-329.4	-254.9									
C₆H₁₄O₂	1,2-Hexanediol		-577.1	-490.1									
C₆H₁₄O₂	1,6-Hexanediol	-569.9	-548.6	-461.2									
C₆H₁₄O₂	Hexylene glycol											336.0	
C₆H₁₄O₂	Ethylene glycol monobutyl ether											281.0	
C₆H₁₄O₂	1,1-Diethoxyethane		-491.4	-453.5									
C₆H₁₄O₂	1,2-Diethoxyethane		-451.4	-408.1								259.4	
C₆H₁₄O₃	Diethylene glycol monoethyl ether											301.0	
C₆H₁₄O₃	Diethylene glycol dimethyl ether											274.1	
C₆H₁₄O₃	Trimethylolpropane	-750.9											
C₆H₁₄O₄	Triethylene glycol		-804.3	-725.0									
C₆H₁₄O₄S	Dipropyl sulfate		-859.0	-792.0									
C₆H₁₄O₆	Galactitol		-1317.0										
C₆H₁₄O₆	D-Mannitol		-1314.5										
C₆H₁₄S	1-Hexanethiol		-175.7	-129.9									
C₆H₁₄S	2-Methyl-2-pentanethiol		-188.3	-148.3									
C₆H₁₄S	2,3-Dimethyl-2-butanethiol		-187.1	-147.9									
C₆H₁₄S	Diisopropyl sulfide		-181.6	-142.0					313.0			232.0	
C₆H₁₄S	Butyl ethyl sulfide		-172.3	-127.8									
C₆H₁₄S	Methyl pentyl sulfide		-167.1	-121.8									
C₆H₁₄S₂	Dipropyl disulfide		-171.5	-118.3									
C₆H₁₅B	Triethylborane		-194.6	-157.7		9.4	16.1		336.7	437.8		241.2	
C₆H₁₅N	Dipropylamine		-156.1	-116.0									
C₆H₁₅N	Diisopropylamine		-178.5	-143.8									
C₆H₁₅N	Triethylamine		-127.7	-92.7								219.9	
C₆H₁₅NO	2-Diethylaminoethanol		-305.9										
C₆H₁₅NO₃	Triethanolamine	-664.2		-558.3				389.0			389.0	321.0	
C₆H₁₃N₃OP	Hexamethylphosphoric triamide					-541.5	-534.5						
C₆H₁₈OSi₂	Hexamethyldisiloxane		-815.0	-777.7					433.8	535.0		311.4	238.5
C₆MoO₆	Molybdenum hexacarbonyl	-982.8	-912.1		-877.7		-856.0	325.9		490.0	242.3		205.0

Molecular formula	Name	$\Delta_f H°$/kJ mol⁻¹			$\Delta_f G°$/kJ mol⁻¹			$S°$/J mol⁻¹ K⁻¹			C_p/J mol⁻¹ K⁻¹		
		Crys.	Liq.	Gas	Crys.	Liq.	Gas	Crys.	Liq.	Gas	Crys.	Liq.	Gas
C_6N_4	Tetracyanoethylene	623.8		705.0									
C_7F_8	Perfluorotoluene		-1311.1						355.5			262.3	
C_7F_{14}	Perfluoromethylcyclohexane		-2931.1	-2897.2								353.1	
C_7F_{16}	Perfluoroheptane		-3420.0	-3383.6					561.8			419.0	
$C_7H_3F_5$	2,3,4,5,6-Pentafluorotoluene		-883.8	-842.7					306.4			225.8	
$C_7H_4Cl_2O$	m-Chlorobenzoyl chloride		-189.7										
$C_7H_4N_2O_6$	3,5-Dinitrobenzoic acid	-409.8											
C_7H_5ClO	Benzoyl chloride		-158.0	-103.2									
$C_7H_5ClO_2$	o-Chlorobenzoic acid	-404.5		-325.0									
$C_7H_5ClO_2$	o-Chlorobenzoic acid	-424.3		-342.3									
$C_7H_5ClO_2$	p-Chlorobenzoic acid	-428.9		-341.0									
$C_7H_5F_3$	(Trifluoromethyl)benzene										163.2		
C_7H_5N	Benzonitrile		163.2	215.7					209.1			188.4	
C_7H_5NO	Benzoxazole	-24.2		44.8								165.2	
$C_7H_5NO_4$	2-Nitrobenzoic acid	-378.8											
$C_7H_5NO_4$	3-Nitrobenzoic acid	-394.7											
$C_7H_5NO_4$	4-Nitrobenzoic acid	-392.2											
$C_7H_6N_2$	1H-Benzimidazole	79.5		181.7									
$C_7H_6N_2$	1H-Indazole	151.9		243.0					221.2			172.0	
C_7H_6O	Benzaldehyde		-87.0	-36.7									
$C_7H_6O_2$	Benzoic acid	-385.2		-294.0				167.6			146.8		
$C_7H_6O_2$	3-(2-Furanyl)-2-propenal	-182.0		-105.9									
$C_7H_6O_2$	Salicylaldehyde											222.0	
$C_7H_6O_3$	Salicylic acid	-589.9		-494.8									
C_7H_7Cl	o-Chlorotoluene		-32.5	18.9								166.8	
C_7H_7Cl	(Chloromethyl)benzene												
C_7H_7F	p-Fluorotoluene		-186.9	-147.4								171.2	
C_7H_7NO	Benzamide	-202.6		-100.9									
$C_7H_7NO_2$	2-Aminobenzoic acid		-380.4	-296.0									
$C_7H_7NO_2$	Aniline-3-carboxylic acid		-389.8	-283.6									
$C_7H_7NO_2$	Aniline-4-carboxylic acid		-391.9	-296.7									
$C_7H_7NO_2$	o-Nitrotoluene		-9.7										
$C_7H_7NO_2$	m-Nitrotoluene		-31.5	31.0									
$C_7H_7NO_2$	p-Nitrotoluene	-48.1		30.7							172.3		
$C_7H_7NO_2$	(Nitromethyl)benzene		-22.8										
$C_7H_7NO_2$	Salicylaldoxime	-183.7											
C_7H_8	Toluene		12.4	50.5								157.3	
$C_7H_8N_2O$	Phenylurea	-218.6											
C_7H_8O	o-Cresol	-204.6		-128.6				165.4			154.6		
C_7H_8O	m-Cresol		-194.0	-132.3					212.6			224.9	
C_7H_8O	p-Cresol	-199.3		-125.4				167.3			150.2		
C_7H_8O	Benzyl alcohol		-160.7	-100.4									
C_7H_8O	Anisole		-114.8	-67.9					216.7			217.9	
C_7H_9N	Benzylamine		34.2	94.4									
C_7H_9N	o-Methylaniline		-6.3	56.4			167.6			351.0			130.2

Molecular formula	Name	$\Delta_f H°$/kJ mol⁻¹ Crys.	Liq.	Gas	$\Delta_f G°$/kJ mol⁻¹ Crys.	Liq.	Gas	$S°$/J mol⁻¹ K⁻¹ Crys.	Liq.	Gas	C_p/J mol⁻¹ K⁻¹ Crys.	Liq.	Gas
C₇H₉N	m-Methylaniline		-8.1	54.6			165.4			352.5			125.5
C₇H₉N	p-Methylaniline	-23.5		55.3			167.7			347.0			126.2
C₇H₉N	N-Methylaniline											207.1	
C₇H₉N	1-Cyclohexenecarbonitrile		48.1	101.6									
C₇H₉N	2,3-Dimethylpyridine		19.4	67.1					243.7			189.5	
C₇H₉N	2,4-Dimethylpyridine		16.1	63.6					248.5			184.8	
C₇H₉N	2,5-Dimethylpyridine		18.7	66.5					248.8			184.7	
C₇H₉N	2,6-Dimethylpyridine		12.7	58.1					244.2			185.2	
C₇H₉N	3,4-Dimethylpyridine		18.3	68.8					240.7			191.8	
C₇H₉N	3,5-Dimethylpyridine		22.5	72.0					241.7			184.5	
C₇H₁₀O₂	Ethyl 2-pentynoate		-301.8	-250.3									
C₇H₁₀O₂	Methyl 2-hexynoate		-242.7										
C₇H₁₁Cl₃O₂	3-Methylbutyl trichloroacetate		-580.9	-523.1									
C₇H₁₁N	Cyclohexanecarbonitrile		-47.2	4.8									
C₇H₁₁N	Methylenecyclohexane		-61.3	-25.2									
C₇H₁₂	Vinylcyclopentane		-34.8	-19.8									
C₇H₁₂	1-Ethylcyclopentene		-53.3	-54.8									
C₇H₁₂	Bicyclo[2.2.1]heptane	-95.1											
C₇H₁₂	1-Methylbicyclo(3,1,0) hexane		-33.2	1.7									
C₇H₁₂O	2-Methylenecyclohexanol	-277.6											
C₇H₁₂O₄	Diethyl malonate		-571.7	-517.3								285.0	
C₇H₁₃ClO₂	Butyl 2-chloropropanoate		-603.1	-549.6									
C₇H₁₃ClO₂	Isobutyl 2-chloropropanoate		-557.9	-502.3									
C₇H₁₃ClO₂	Butyl 3-chloropropanoate		-572.6	-517.3									
C₇H₁₃ClO₂	Isobutyl 3-chloropropanoate		-630.7	-578.4									
C₇H₁₃ClO₂	Propyl 2-chlorobutanoate												
C₇H₁₃N	Heptanenitrile		-82.8	-31.0									
C₇H₁₄	1-Heptene		-97.9	-62.3					327.6			211.8	
C₇H₁₄	cis-2-Heptene		-105.1										
C₇H₁₄	trans-2-Heptene		-109.5										
C₇H₁₄	cis-3-Heptene		-104.3										
C₇H₁₄	trans-3-Heptene		-109.3										
C₇H₁₄	5-Methyl-1-hexene		-100.0	-65.7									
C₇H₁₄	3-Methyl-cis-3-hexene		-115.9	-79.4									
C₇H₁₄	3-Methyl-trans-3-hexene		-112.7	-76.8									
C₇H₁₄	2,4-Dimethyl-1-pentene		-117.0	-83.8									
C₇H₁₄	4,4-Dimethyl-1-pentene		-110.6	-81.6									
C₇H₁₄	2,4-Dimethyl-2-pentene		-123.1	-88.7									
C₇H₁₄	cis-4,4-Dimethyl-2-pentene		-105.3	-72.6									
C₇H₁₄	trans-4,4-Dimethyl-2-pentene		-121.7	-88.8									
C₇H₁₄	2-Ethyl-3-methyl-1-butene		-114.1	-79.5									
C₇H₁₄	2,3,3-Trimethyl-1-butene		-117.7	-85.5									
C₇H₁₄	Cycloheptane		-156.6	-118.1									
C₇H₁₄	Methylcyclohexane		-190.1	-154.7								184.8	

Molecular formula	Name	$\Delta_f H°$/kJ mol⁻¹			$\Delta_f G°$/kJ mol⁻¹			$S°$/J mol⁻¹ K⁻¹			C_p/J mol⁻¹ K⁻¹		
		Crys.	Liq.	Gas	Crys.	Liq.	Gas	Crys.	Liq.	Gas	Crys.	Liq.	Gas
C_7H_{14}	Ethylcyclopentane		-163.4	-126.9					279.9				
C_7H_{14}	1,1-Dimethylcyclopentane		-172.1	-138.2									
C_7H_{14}	cis-1,2-Dimethylcyclo-pentane		-165.3	-129.5					269.2				
C_7H_{14}	trans-1,2-Dimethylcyclo-pentane		-171.2	-136.6									
C_7H_{14}	cis-1,3-Dimethylcyclo-pentane		-170.1	-135.8									
C_7H_{14}	trans-1,3-Dimethylcyclo-pentane		-168.1	-133.6									
C_7H_{14}	1,1,2,2-Tetramethylcyclo-propane		-119.8										
$C_7H_{14}Br_2$	1,2-Dibromoheptane		-212.3	-157.9								230.1	
$C_7H_{14}O$	1-Heptanal		-311.5	-263.8					335.4			232.6	
$C_7H_{14}O$	2-Heptanone		-356.1	-313.6									
$C_7H_{14}O$	2,2-Dimethyl-3-pentanone		-352.9	-311.3									
$C_7H_{14}O$	2,4-Dimethyl-3-pentanone		-390.2	-327.0					318.0			233.7	
$C_7H_{14}O$	cis-2-Methylcyclohexanol		-415.7	-352.5									
$C_7H_{14}O$	trans-2-Methylcyclohexanol		-416.1	-350.9									
$C_7H_{14}O$	cis-3-Methylcyclohexanol		-394.4	-329.1									
$C_7H_{14}O$	trans-3-Methylcyclohexanol		-413.2	-347.5									
$C_7H_{14}O$	cis-4-Methylcyclohexanol		-433.3	-367.2									
$C_7H_{14}O$	trans-4-Methylcyclohexanol		-610.2	-536.2									
$C_7H_{14}O_2$	Heptanoic acid											265.4	
$C_7H_{14}O_2$	Pentyl acetate											261.0	
$C_7H_{14}O_2$	Isopentyl acetate											248.5	
$C_7H_{14}O_2$	Ethyl pentanoate		-553.0	-505.9									
$C_7H_{14}O_2$	Ethyl 3-methylbutanoate		-571.0	-527.0									
$C_7H_{14}O_2$	Ethyl 2,2-dimethyl-propanoate		-577.2	-536.0									
$C_7H_{14}O_2$	Methyl hexanoate		-540.2	-492.2									
$C_7H_{14}O_6$	α-Methylglucoside	-1233.3											
$C_7H_{15}Br$	1-Bromoheptane		-218.4	-167.8									
C_7H_{16}	Heptane		-224.2	-187.6					323.3			224.7	
C_7H_{16}	2-Methylhexane		-229.5	-194.5								222.9	
C_7H_{16}	3-Methylhexane		-226.4	-191.3									
C_7H_{16}	3-Ethylpentane		-224.9	-189.5					314.5			219.6	
C_7H_{16}	2,2-Dimethylpentane		-238.3	-205.7					300.3			221.1	
C_7H_{16}	2,3-Dimethylpentane		-233.1	-198.7									
C_7H_{16}	2,4-Dimethylpentane		-234.6	-201.6					303.2			224.2	
C_7H_{16}	3,3-Dimethylpentane		-234.2	-201.0									
C_7H_{16}	2,2,3-Trimethylbutane		-236.5	-204.4					292.2			213.5	
$C_7H_{16}O$	1-Heptanol		-403.3	-336.5								272.1	
$C_7H_{16}O$	tert-Butyl isopropyl ether		-392.8	-358.1									
$C_7H_{16}O_2$	1,7-Heptanediol		-574.2										

Molecular formula	Name	$\Delta_f H°$/kJ mol⁻¹			$\Delta_f G°$/kJ mol⁻¹			$S°$/J mol⁻¹ K⁻¹			C_p/J mol⁻¹ K⁻¹		
		Crys.	Liq.	Gas	Crys.	Liq.	Gas	Crys.	Liq.	Gas	Crys.	Liq.	Gas
C₇H₁₆O₂	2,2-Diethoxypropane		-538.9	-506.9									
C₇H₁₆S	1-Heptanethiol		-200.5	-149.9									
C₈H₄O₃	Phthalic anhydride	-460.1		-371.4				180.0			160.0		
C₈H₅NO₂	1H-Indole-2,3-dione	-268.2											
C₈H₆O₄	Phthalic acid	-782.0						207.9			188.1		
C₈H₆O₄	Isophthalic acid	-803.0		-696.3									
C₈H₆O₄	Terephthalic acid	-816.1		-717.9									
C₈H₆S	Benzo[b]thiophene	100.6		166.3									
C₈H₇N	1H-Indole	86.6		156.5									
C₈H₈	Styrene		103.8	147.9								182.0	
C₈H₈O	Phenyl vinyl ether		-26.2	22.7									
C₈H₈O	Acetophenone		-142.5	-86.7									
C₈H₈O₂	o-Toluic acid	-416.5									174.9		
C₈H₈O₂	m-Toluic acid	-426.1									163.6		
C₈H₈O₂	p-Toluic acid	-429.2									169.0		
C₈H₈O₂	Methyl benzoate		-343.5	-287.9								221.3	
C₈H₈O₃	Methyl salicylate											249.0	
C₈H₁₀	1,7-Octadiyne		334.4										
C₈H₁₀	Ethylbenzene		-12.3	29.9								183.2	
C₈H₁₀	o-Xylene		-24.4	19.1								186.1	
C₈H₁₀	m-Xylene		-25.4	17.3								183.0	
C₈H₁₀	p-Xylene		-24.4	18.0								181.5	
C₈H₁₀O	o-Ethylphenol		-208.8	-145.2									
C₈H₁₀O	m-Ethylphenol		-214.3	-146.1									
C₈H₁₀O	p-Ethylphenol	-224.4		-144.1							206.9		
C₈H₁₀O	2,3-Xylenol	-241.1		-157.2									
C₈H₁₀O	2,4-Xylenol		-228.7	-163.8									
C₈H₁₀O	2,5-Xylenol	-246.6		-161.6									
C₈H₁₀O	2,6-Xylenol	-237.4		-162.1									
C₈H₁₀O	3,4-Xylenol	-242.3		-157.3									
C₈H₁₀O	3,5-Xylenol	-244.4		-162.4									
C₈H₁₀O	Phenetole		-152.6	-101.6									
C₈H₁₀O₂	Veratrole		-290.3	-223.3									
C₈H₁₁N	N-Ethylaniline		4.0	56.3									
C₈H₁₁N	N,N-Dimethylaniline		47.7	100.5									
C₈H₁₁N	2,6-Xylidine												
C₈H₁₂	1-Octen-3-yne		140.7										
C₈H₁₂	cis-1,2-Divinylcyclobutane		124.3	166.5								228.5	
C₈H₁₂	trans-1,2-Divinylcyclobutane		101.3	143.5									
C₈H₁₂N₄	2,2'-Azobis[isobutyronitrile]	228.9											
C₈H₁₂O₂	2,2,4-Tetramethyl-1,3-cyclobutanedione	-379.9		-307.6								238.9	
C₈H₁₄	Ethylidenecyclohexane		-103.5	-59.5									

Molecular formula	Name	$\Delta_f H°$/kJ mol⁻¹ Crys.	Liq.	Gas	$\Delta_f G°$/kJ mol⁻¹ Crys.	Liq.	Gas	$S°$/J mol⁻¹ K⁻¹ Crys.	Liq.	Gas	C_p/J mol⁻¹ K⁻¹ Crys.	Liq.	Gas
C_8H_{14}	Allylcyclopentane		-64.5	-24.1									283.7
$C_8H_{14}O_3$	Butanoic anhydride		-627.3										
$C_8H_{15}ClO_2$	3-Methylbutyl 2-chloropropanoate			-575.0									
$C_8H_{15}ClO_2$	3-Methylbutyl 3-chloropropanoate		-593.4	-539.4									
$C_8H_{15}N$	Octanenitrile		-107.3	-50.5									
C_8H_{16}	1-Octene		-121.8	-81.3								241.0	
C_8H_{16}	cis-2-Octene		-135.7									239.0	
C_8H_{16}	trans-2-Octene		-135.7									239.0	
C_8H_{16}	2,2-Dimethyl-cis-3-hexene		-126.4	-89.3									
C_8H_{16}	2,2-Dimethyl-trans-3-hexene		-144.9	-107.7									
C_8H_{16}	3-Ethyl-2-methyl-1-pentene		-137.9	-100.3									
C_8H_{16}	2,4,4-Trimethyl-1-pentene		-145.9	-110.5									
C_8H_{16}	2,4,4-Trimethyl-2-pentene		-142.4	-104.9									
C_8H_{16}	Cyclooctane		-167.7	-124.4									
C_8H_{16}	Ethylcyclohexane		-212.1	-171.5					280.9			211.8	
C_8H_{16}	1,1-Dimethylcyclohexane		-218.7	-180.9					267.2			209.2	
C_8H_{16}	cis-1,2-Dimethylcyclohexane		-211.8	-172.1					274.1			210.2	
C_8H_{16}	trans-1,2-Dimethylcyclohexane		-218.2	-179.9					273.2			209.4	
C_8H_{16}	cis-1,3-Dimethylcyclohexane		-222.9	-184.6					272.6			209.4	
C_8H_{16}	trans-1,3-Dimethylcyclohexane		-215.7	-176.5					276.3			212.8	
C_8H_{16}	cis-1,4-Dimethylcyclohexane		-215.6	-176.6					271.1			212.1	
C_8H_{16}	trans-1,4-Dimethylcyclohexane		-222.4	-184.5					268.0			210.2	
C_8H_{16}	Propylcyclopentane		-188.8	-147.7					310.8			216.3	
C_8H_{16}	1-Ethyl-1-methylcyclopentane		-193.8										
C_8H_{16}	cis-1-Ethyl-2-methylcyclopentane		-190.8										
C_8H_{16}	trans-1-Ethyl-2-methylcyclopentane		-195.1	-156.2									
C_8H_{16}	cis-1-Ethyl-3-methylcyclopentane		-194.4										
C_8H_{16}	trans-1-Ethyl-3-methylcyclopentane		-196.0										
$C_8H_{16}O$	2-Ethylhexanal		-348.5	-299.6									
$C_8H_{16}O$	2-Octanone		-381.6	-338.3								273.3	
$C_8H_{16}O$	2,2,4-Trimethyl-3-pentanone												
$C_8H_{16}O_2$	Octanoic acid		-636.0	-554.3								297.9	
$C_8H_{16}O_2$	2-Ethylhexanoic acid		-635.1	-559.5									
$C_8H_{16}O_2$	Hexyl acetate											282.8	
$C_8H_{16}O_2$	Propyl pentanoate		-583.0	-533.6									
$C_8H_{16}O_2$	Isopropyl pentanoate		-592.2	-544.9									

Molecular formula	Name	$\Delta_f H°$/kJ mol^{-1}			$\Delta_f G°$/kJ mol^{-1}			$S°$/J mol^{-1} K^{-1}			C_p/J mol^{-1} K^{-1}		
		Crys.	Liq.	Gas	Crys.	Liq.	Gas	Crys.	Liq.	Gas	Crys.	Liq.	Gas
C$_8$H$_{16}$O$_2$	Methyl heptanoate		-567.1	-515.5								285.1	
C$_8$H$_{17}$Br	1-Bromooctane		-245.1	-189.3									
C$_8$H$_{17}$Cl	1-Chlorooctane		-291.3	-238.9									
C$_8$H$_{17}$NO	Octanamide	-473.2		-362.7									
C$_8$H$_{18}$	Octane		-250.1	-208.5								254.6	
C$_8$H$_{18}$	2-Methylheptane		-255.0	-215.3					356.4			252.0	
C$_8$H$_{18}$	3-Methylheptane		-252.3	-212.5					362.6			250.2	
C$_8$H$_{18}$	4-Methylheptane		-251.6	-211.9								251.1	
C$_8$H$_{18}$	3-Ethylhexane		-250.4	-210.7									
C$_8$H$_{18}$	2,2-Dimethylhexane		-261.9	-224.5									
C$_8$H$_{18}$	2,3-Dimethylhexane		-252.6	-213.8									
C$_8$H$_{18}$	2,4-Dimethylhexane		-257.0	-219.2									
C$_8$H$_{18}$	2,5-Dimethylhexane		-260.4	-222.5								249.2	
C$_8$H$_{18}$	3,3-Dimethylhexane		-257.5	-219.9								246.6	
C$_8$H$_{18}$	3,4-Dimethylhexane		-251.8	-212.8									
C$_8$H$_{18}$	3-Ethyl-2-methylpentane		-249.6	-211.0									
C$_8$H$_{18}$	3-Ethyl-3-methylpentane		-252.8	-214.8									
C$_8$H$_{18}$	2,2,3-Trimethylpentane		-256.9	-220.0								239.1	
C$_8$H$_{18}$	2,2,4-Trimethylpentane		-259.2	-224.0								245.6	
C$_8$H$_{18}$	2,3,3-Trimethylpentane		-253.5	-216.3								247.3	
C$_8$H$_{18}$	2,3,4-Trimethylpentane		-255.0	-217.3					329.3				
C$_8$H$_{18}$	2,2,3,3-Tetramethylbutane	-269.0		-226.0				273.7			239.2		
C$_8$H$_{18}$N$_2$	Azobutane		-40.1	9.2									
C$_8$H$_{18}$O	1-Octanol		-426.5	-355.6								305.2	
C$_8$H$_{18}$O	2-Octanol											330.1	
C$_8$H$_{18}$O	2-Ethyl-1-hexanol		-432.8	-365.3								317.5	
C$_8$H$_{18}$O	Dibutyl ether		-377.9	-332.8					347.0			278.2	
C$_8$H$_{18}$O	Di-sec-butyl ether		-401.5	-360.6									
C$_8$H$_{18}$O	Di-tert-butyl ether		-399.6	-362.0									
C$_8$H$_{18}$O	tert-Butyl isobutyl ether		-409.1	-369.0								276.1	
C$_8$H$_{18}$O$_2$	1,8-Octanediol	-626.6											
C$_8$H$_{18}$O$_2$	2,5-Dimethyl-2,5-hexanediol	-681.7											
C$_8$H$_{18}$O$_3$	Diethylene glycol diethyl ether											341.4	
C$_8$H$_{18}$O$_3$S	Dibutyl sulfite		-693.1	-625.3									
C$_8$H$_{18}$O$_5$	Tetraethylene glycol		-981.7	-883.0					405.1			428.8	
C$_8$H$_{18}$S	Dibutyl sulfide		-220.7	-167.7								284.3	
C$_8$H$_{18}$S	Di-sec-butyl sulfide		-220.7	-167.7									
C$_8$H$_{18}$S	Di-tert-butyl sulfide		-232.6	-188.8									
C$_8$H$_{18}$S	Diisobutyl sulfide		-229.2	-180.5									
C$_8$H$_{18}$S$_2$	Dibutyl disulfide		-222.9	-160.6									
C$_8$H$_{18}$S$_2$	Di-tert-butyl disulfide		-255.2	-201.0									
C$_8$H$_{19}$N	Dibutylamine		-206.0	-156.6								292.9	
C$_8$H$_{19}$N	Diisobutylamine		-218.5	-179.2									

STANDARD THERMODYNAMIC PROPERTIES OF CHEMICAL SUBSTANCES (continued)

Molecular formula	Name	$\Delta_f H°$/kJ mol⁻¹ Crys.	Liq.	Gas	$\Delta_f G°$/kJ mol⁻¹ Crys.	Liq.	Gas	S°/J mol⁻¹ K⁻¹ Crys.	Liq.	Gas	C_p/J mol⁻¹ K⁻¹ Crys.	Liq.	Gas
C₈H₂₀BrN	Tetraethylammonium bromide	-342.7											
C₈H₂₀Pb	Tetraethyl lead		52.7	109.6					464.6			307.4	
C₈H₂₀Si	Tetraethylsilane											298.1	
C₉H₇N	Quinoline		141.2	200.5					216.0			196.2	
C₉H₇N	Isoquinoline		144.3	204.6					215.3				
C₉H₇NO	2-Hydroxyquinoline	-144.9		-25.5									
C₉H₈	Indene		110.6	163.4								186.9	
C₉H₈O₄	2-(Acetyloxy)benzoic acid	-815.6											
C₉H₁₀	Cyclopropylbenzene		100.3	150.5									
C₉H₁₀	Indan		11.5	60.3					56.0			190.2	
C₉H₁₀N₂	2,2'-Dipyrrolylmethane	126.2											
C₉H₁₀O₂	Benzyl acetate											148.5	
C₉H₁₀O₂	Ethyl benzoate	-466.9		-312.9					287.8			246.0	
C₉H₁₁NO₂	L-Phenylalanine	-685.1						213.6			203.0		
C₉H₁₁NO₃	L-Tyrosine							214.0			216.4		
C₉H₁₂	Propylbenzene		-38.3	7.9								214.7	
C₉H₁₂	Cumene		-41.1	4.0								210.7	
C₉H₁₂	o-Ethyltoluene		-46.4	1.3									
C₉H₁₂	m-Ethyltoluene		-48.7	-1.8									
C₉H₁₂	p-Ethyltoluene		-49.8	-3.2					267.9				
C₉H₁₂	1,2,3-Trimethylbenzene		-58.5	-9.5								216.4	
C₉H₁₂	1,2,4-Trimethylbenzene		-61.8	-13.8								215.0	
C₉H₁₂	Mesitylene		-63.4	-15.9								209.3	
C₉H₁₂O	2-Isopropylphenol		-233.7	-182.2									
C₉H₁₂O	3-Isopropylphenol		-252.5	-196.0									
C₉H₁₂O	4-Isopropylphenol		-265.9	-209.4									
C₉H₁₂O₂	Hydroperoxide, 1-methyl-1-phenylethyl		-148.3	-78.4									
C₉H₁₃NO₂	Ethyl 3,5-dimethylpyrrole-2-carboxylate	-474.5											
C₉H₁₃NO₂	Ethyl 2,4-dimethylpyrrole-3-carboxylate	-463.2											
C₉H₁₃NO₂	Ethyl 2,5-dimethylpyrrole-3-carboxylate	-478.7											
C₉H₁₃NO₂	Ethyl 4,5-dimethylpyrrole-3-carboxylate	-470.3											
C₉H₁₄O	Isophorone											253.5	
C₉H₁₄O₆	Triacetin		-1330.8	-1245.0					458.3			384.7	
C₉H₁₅N	3-Ethyl-2,4,5-trimethylpyrrole	-89.2											
C₉H₁₆O₄	Nonanedioic acid	-1054.3											
C₉H₁₇NO	2,2,6,6-Tetramethyl-4-piperidinone	-334.2		-273.4									
C₉H₁₈	Propylcyclohexane		-237.4	-192.3					311.9			242.0	

Molecular formula	Name	$\Delta_f H°$/kJ mol⁻¹			$\Delta_f G°$/kJ mol⁻¹			$S°$/J mol⁻¹ K⁻¹			C_p/J mol⁻¹ K⁻¹		
		Crys.	Liq.	Gas	Crys.	Liq.	Gas	Crys.	Liq.	Gas	Crys.	Liq.	Gas
C₉H₁₈	1,3,5-Trimethylcyclohexane (1′,3′,5′)			-212.1									
C₉H₁₈O	2-Nonanone		-397.2	-340.7								303.6	
C₉H₁₈O	5-Nonanone		-398.2	-344.9					401.4			297.3	
C₉H₁₈O	2,6-Dimethyl-4-heptanone		-408.5	-357.6								362.4	
C₉H₁₈O₂	Nonanoic acid		-659.7	-577.3									
C₉H₁₈O₂	Butyl pentanoate	-613.3		-560.2									
C₉H₁₈O₂	sec-Butyl pentanoate		-624.2	-573.2									
C₉H₁₈O₂	Isobutyl pentanoate		-620.0	-568.6									
C₉H₁₈O₂	Methyl octanoate		-590.3	-533.9									
C₉H₁₉N	N-Butylpiperidine		-171.8										
C₉H₁₉N	2,2,6,6-Tetramethyl-piperidine		-206.9	-159.9									
C₉H₂₀	Nonane		-274.7	-228.2								284.4	
C₉H₂₀	2,2-Dimethylheptane		-288.1										
C₉H₂₀	2,2,3-Trimethylhexane		-282.7										
C₉H₂₀	2,2,4-Trimethylhexane		-282.8										
C₉H₂₀	2,2,5-Trimethylhexane		-293.3										
C₉H₂₀	2,3,3-Trimethylhexane		-281.1										
C₉H₂₀	2,3,5-Trimethylhexane		-284.0	-242.6									
C₉H₂₀	2,4,4-Trimethylhexane		-280.2										
C₉H₂₀	3,3,4-Trimethylhexane		-277.5										
C₉H₂₀	3,3-Diethylpentane		-275.4	-233.3								278.2	
C₉H₂₀	3-Ethyl-2,2-dimethylpentane		-272.7										
C₉H₂₀	3-Ethyl-2,4-dimethylpentane		-269.7										
C₉H₂₀	2,2,3,3-Tetramethylpentane		-278.3	-237.1								271.5	
C₉H₂₀	2,2,3,4-Tetramethylpentane		-277.7	-236.9									
C₉H₂₀	2,2,4,4-Tetramethylpentane		-280.0	-241.6								266.3	
C₉H₂₀	2,3,3,4-Tetramethylpentane		-277.9	-236.1									
C₉H₂₀N₂O	Tetraethylurea	-403.0	-453.4	-376.5									
C₉H₂₀O	1-Nonanol	-657.6											
C₉H₂₀O₂	1,9-Nonanediol												
C₉H₂₁N	Tripropylamine		-207.1	-161.0									
C₁₀H₆N₂O₄	1,5-Dinitronaphthalene	29.8		119.8									
C₁₀H₆N₂O₄	1,8-Dinitronaphthalene	39.7		137.4									
C₁₀H₇Cl	1-Chloronaphthalene		54.6										212.6	
C₁₀H₇Cl	2-Chloronaphthalene	55.4											
C₁₀H₇I	1-Iodonaphthalene		161.5	233.8									
C₁₀H₇I	2-Iodonaphthalene	144.3		235.1									
C₁₀H₇NO₂	1-Nitronaphthalene	42.6		111.2									
C₁₀H₈	Naphthalene	78.5		150.6	201.6		224.1	167.4		333.1	165.7		131.9
C₁₀H₈	Azulene	212.3		289.1							166.9		
C₁₀H₈O	1-Naphthol	-121.5		-30.4									
C₁₀H₈O	2-Naphthol		-124.1	-29.9									
C₁₀H₉N	1-Naphthalenamine	67.8		132.8									

Molecular formula	Name	$\Delta_f H°$/kJ mol^{-1} Crys.	Liq.	Gas	$\Delta_f G°$/kJ mol^{-1} Crys.	Liq.	Gas	$S°$/J mol^{-1} K^{-1} Crys.	Liq.	Gas	C_p/J mol^{-1} K^{-1} Crys.	Liq.	Gas
C$_{10}$H$_9$N	2-Naphthalenamine	60.2		134.3									
C$_{10}$H$_{10}$	1,2-Dihydronaphthalene		71.6										
C$_{10}$H$_{10}$	1,4-Dihydronaphthalene		84.2										
C$_{10}$H$_{10}$O	1-Tetralone	-209.6										303.1	
C$_{10}$H$_{10}$O$_4$	Dimethyl phthalate												
C$_{10}$H$_{10}$O$_4$	Dimethyl isophthalate	-730.9									261.1		
C$_{10}$H$_{10}$O$_4$	Dimethyl terephthalate	-732.6											
C$_{10}$H$_{12}$	1,2,3,4-Tetrahydronaphthalene		-29.2	26.0								217.5	
C$_{10}$H$_{14}$	Butylbenzene		-63.2	-11.8					321.2			243.4	
C$_{10}$H$_{14}$	sec-Butylbenzene		-66.4	-18.4									
C$_{10}$H$_{14}$	tert-Butylbenzene		-70.7	-23.0									
C$_{10}$H$_{14}$	Isobutylbenzene		-69.8	-21.9									
C$_{10}$H$_{14}$	o-Cymene		-73.3										
C$_{10}$H$_{14}$	m-Cymene		-78.6										
C$_{10}$H$_{14}$	p-Cymene		-78.0									236.4	
C$_{10}$H$_{14}$	o-Diethylbenzene		-68.5										
C$_{10}$H$_{14}$	m-Diethylbenzene		-73.5										
C$_{10}$H$_{14}$	p-Diethylbenzene		-72.8										
C$_{10}$H$_{14}$	3-Ethyl-o-xylene		-80.5										
C$_{10}$H$_{14}$	4-Ethyl-1,2-dimethylbenzene		-86.0										
C$_{10}$H$_{14}$	2-Ethyl-1,3-dimethylbenzene		-80.1										
C$_{10}$H$_{14}$	2-Ethyl-1,4-dimethylbenzene		-84.8										
C$_{10}$H$_{14}$	1-Ethyl-2,4-dimethylbenzene		-84.1										
C$_{10}$H$_{14}$	1-Ethyl-3,5-dimethylbenzene		-87.8										
C$_{10}$H$_{14}$	1,2,4,5-Tetramethylbenzene	-119.9		-218.5				245.6			215.1		
C$_{10}$H$_{14}$O	Thymol	-309.7											
C$_{10}$H$_{16}$	cis,cis-2,6-Dimethyl-2,4,6-octatriene		-24.0	-2.6									
C$_{10}$H$_{16}$	Dipentene		-50.8									249.4	
C$_{10}$H$_{16}$	D-Limonene		-54.5									249.0	
C$_{10}$H$_{16}$	β-Myrcene		14.5										
C$_{10}$H$_{16}$	α-Pinene		-16.4	28.3									
C$_{10}$H$_{16}$	β-Pinene		-7.7	38.7									
C$_{10}$H$_{16}$	α-Terpinene			-20.6									
C$_{10}$H$_{16}$N$_2$O$_8$	Glycine, N,N'-1,2-ethanediylbis[N-(carboxymethyl)]-	-1759.5											
C$_{10}$H$_{16}$O	Camphor	-319.4		-267.5							271.2		
C$_{10}$H$_{18}$	1,1'-Bicyclopentyl		-178.9										
C$_{10}$H$_{18}$	cis-Decahydronaphthalene		-219.4	-169.2					265.0			232.0	
C$_{10}$H$_{18}$	trans-Decahydronaphthalene		-230.6	-182.1					264.9			228.5	
C$_{10}$H$_{18}$O$_4$	Sebacic acid	-1082.6	-158.4	-921.9									
C$_{10}$H$_{19}$N	Decanenitrile		-173.8	-91.5									
C$_{10}$H$_{20}$	1-Decene		-163.6	-123.3					425.0			300.8	
C$_{10}$H$_{20}$	cis-1,2-Di-tert-butylethylene												

STANDARD THERMODYNAMIC PROPERTIES OF CHEMICAL SUBSTANCES (continued)

Molecular formula	Name	$\Delta_f H°$/kJ mol⁻¹			$\Delta_f G°$/kJ mol⁻¹			$S°$/J mol⁻¹ K⁻¹			C_p/J mol⁻¹ K⁻¹		
		Crys.	Liq.	Gas	Crys.	Liq.	Gas	Crys.	Liq.	Gas	Crys.	Liq.	Gas
$C_{10}H_{20}$	Butylcyclohexane		-263.1	-213.7					345.0			271.0	
$C_{10}H_{20}O_2$	Decanoic acid	-713.7	-684.3	-594.9									
$C_{10}H_{20}O_2$	Methyl nonanoate		-616.2	-554.2									
$C_{10}H_{21}NO_2$	1-Nitrodecane		-351.5										
$C_{10}H_{22}$	Decane		-300.9	-249.5								314.4	
$C_{10}H_{22}$	2-Methylnonane		-309.8	-260.2					420.1			313.3	
$C_{10}H_{22}$	5-Methylnonane		-307.9	-258.6					423.8			314.4	
$C_{10}H_{22}O$	1-Decanol		-478.1	-396.6								370.6	
$C_{10}H_{22}O$	Dipentyl ether											250.0	
$C_{10}H_{22}O$	Diisopentyl ether											379.0	
$C_{10}H_{22}O_2$	1,10-Decanediol	-678.9											
$C_{10}H_{22}O_2$	Ethylene glycol dibutyl ether								476.1			350.0	
$C_{10}H_{22}S$	1-Decanethiol	-309.9	-276.5	-211.5								350.4	
$C_{10}H_{22}S$	Dipentyl sulfide		-266.4	-204.9									
$C_{10}H_{22}S$	Diisopentylsulfide		-281.8	-221.5									
$C_{11}H_8O_2$	1-Naphthalenecarboxylic acid	-333.5		-223.1									
$C_{11}H_8O_2$	2-Naphthoic acid	-346.1		-232.5									
$C_{11}H_{10}$	1-Methylnaphthalene	44.9	56.3	106.7				220.0	254.8		196.0	224.4	
$C_{11}H_{10}$	2-Methylnaphthalene							251.0			238.1		
$C_{11}H_{12}N_2O_2$	L-Tryptophan	-415.3											
$C_{11}H_{14}$	1,1-Dimethylindan		-53.6	-1.6									
$C_{11}H_{16}$	Pentamethylenzene	-144.6		-67.2									
$C_{11}H_{20}$	Spiro[5.5]undecane		-244.5	-188.3									
$C_{11}H_{22}$	1-Undecene											344.9	
$C_{11}H_{22}O_2$	Methyl decanoate		-640.5	-573.8									
$C_{11}H_{24}$	Undecane		-327.2	-270.8								344.9	
$C_{11}H_{24}O$	1-Undecanol		-504.8										
$C_{12}F_{27}N$	Trinonafluorobutylamine											418.4	
$C_{12}H_8$	Acenaphthylene	186.7		259.7							166.4		
$C_{12}H_8N_2$	Phenazine	237.0		328.8									
$C_{12}H_8O$	Dibenzofuran	-5.3		83.4									
$C_{12}H_8S$	Dibenzothiophene	120.0		205.1									
$C_{12}H_8S_2$	Thianthrene	182.0		286.0									
$C_{12}H_9N$	Carbazole	101.7		200.7									
$C_{12}H_{10}$	Acenaphthene	70.3		156.0				188.9			190.4		
$C_{12}H_{10}$	Biphenyl	99.4		181.4				209.4			198.4		
$C_{12}H_{10}N_2O$	trans-Azoxybenzene	243.4		342.0									
$C_{12}H_{10}N_2O$	N-Nitrosodiphenylamine	227.2											
$C_{12}H_{10}O$	Diphenyl ether	-32.1	-14.9	52.0				233.9			216.6		
$C_{12}H_{10}O_2$	1-Naphthaleneacetic acid	-359.2											
$C_{12}H_{10}O_2$	2-Naphthaleneacetic acid	-371.9											
$C_{12}H_{11}N$	2-Aminobiphenyl	93.8		184.4									
$C_{12}H_{11}N$	p-Biphenylamine	81.0											
$C_{12}H_{11}N$	Diphenylamine	130.2		219.3									
$C_{12}H_{12}N_2$	p-Benzidine	70.7											

Molecular formula	Name	$\Delta_f H°$/kJ mol^{-1}			$\Delta_f G°$/kJ mol^{-1}			$S°$/J mol^{-1} K^{-1}			C_p/J mol^{-1} K^{-1}		
		Crys.	Liq.	Gas	Crys.	Liq.	Gas	Crys.	Liq.	Gas	Crys.	Liq.	Gas
$C_{12}H_{14}O_4$	Diethyl phthalate		-776.6	-688.4					425.1				366.1
$C_{12}H_{16}$	Cyclohexylbenzene		-76.6	-16.7									
$C_{12}H_{17}NO_4$	Diethyl 3,5-dimethyl-pyrrole-2,4-dicarboxylate	-916.7											
$C_{12}H_{18}$	3,9-Dodecadiyne		197.8										
$C_{12}H_{18}$	5,7-Dodecadiyne		181.5										
$C_{12}H_{18}$	Hexamethylbenzene	-162.4		-77.4				306.3			245.6		
$C_{12}H_{22}$	Cyclohexylcyclohexane		-273.7	-215.7									
$C_{12}H_{22}O_4$	Dodecanedioic acid	-1130.0		-976.9									
$C_{12}H_{22}O_{11}$	Sucrose	-2226.1											
$C_{12}H_{22}O_{11}$	β-Lactose	-2236.7											
$C_{12}H_{24}$	1-Dodecene		-226.2	-165.4					484.8			360.7	
$C_{12}H_{24}O_2$	Dodecanoic acid	-774.6	-737.9	-642.0							404.3		
$C_{12}H_{24}O_2$	Methyl undecanoate		-665.2	-593.8									
$C_{12}H_{24}O_{12}$	α-Lactose monohydrate	-2484.1											
$C_{12}H_{25}Br$	1-Bromododecane		-344.7	-269.9									
$C_{12}H_{25}Cl$	1-Chlorododecane		-392.3	-321.1									
$C_{12}H_{26}$	Dodecane		-350.9	-289.4								375.8	
$C_{12}H_{26}O$	1-Dodecanol		-528.5	-436.6								438.1	
$C_{12}H_{26}O_3$	Diethylene glycol dibutyl ether											452.0	
$C_{12}H_{27}N$	Tributylamine		-281.6										
$C_{13}H_8O_2$	Xanthone	-191.5											
$C_{13}H_9N$	Acridine	179.4		273.9									
$C_{13}H_9N$	Phenanthridine	141.9		240.5									
$C_{13}H_9N$	Benzo[f]quinoline	150.6		233.7									
$C_{13}H_{10}N_2$	9-Acridinamine	159.2											
$C_{13}H_{10}O$	Benzophenone	-34.5		54.9									
$C_{13}H_{11}N$	9-Methylcarbazole	105.5		201.0				239.3			224.8		
$C_{13}H_{12}$	Diphenylmethane	71.5	89.7	139.0									
$C_{13}H_{24}O_4$	Tridecanedioic acid	-1148.3											
$C_{13}H_{26}$	1-Tridecene												
$C_{13}H_{26}O_2$	Methyl dodecanoate		-693.0	-614.9								391.8	
$C_{13}H_{28}$	Tridecane											406.7	
$C_{13}H_{28}O$	1-Tridecanol	-599.4											
$C_{14}H_8O_2$	9,10-Anthracenedione	-188.5		-75.7									
$C_{14}H_8O_2$	9,10-Phenanthrenedione	-154.7		-46.6									
$C_{14}H_8O_4$	1,4-Dihydroxy-9,10-anthracenedione	-595.8		-471.7									
$C_{14}H_{10}$	Anthracene	129.2		230.9				207.5			210.5		
$C_{14}H_{10}$	Phenanthrene	116.2		207.5				215.1			220.6		
$C_{14}H_{10}$	Diphenylacetylene	312.4									225.9		
$C_{14}H_{10}O_2$	Benzil	-153.9		-55.5									
$C_{14}H_{12}$	cis-Stilbene		183.3	252.3									
$C_{14}H_{12}$	trans-Stilbene	136.9		236.1									

Molecular formula	Name	$\Delta_f H°$/kJ mol⁻¹			$\Delta_f G°$/kJ mol⁻¹			$S°$/J mol⁻¹ K⁻¹			C_p/J mol⁻¹ K⁻¹		
		Crys.	Liq.	Gas	Crys.	Liq.	Gas	Crys.	Liq.	Gas	Crys.	Liq.	Gas
C₁₄H₁₄	1,1-Diphenylethane		48.7										
C₁₄H₁₄	1,2-Diphenylethane	51.5		142.9									
C₁₄H₂₃N₃O₁₀	Glycine, N,N-bis[2-[bis(carboxymethyl)amino]ethyl]	-2225.2											
C₁₄H₂₇N	Tetradecanenitrile		-260.2	-174.9									
C₁₄H₂₈O₂	Tetradecanoic acid	-833.5	-788.8	-693.7							432.0		
C₁₄H₂₈O₂	Methyl tridecanoate		-717.9	-635.3									
C₁₄H₃₀O	1-Tetradecanol	-629.6	-580.6								388.0		
C₁₅H₁₆O₂	2,2-Bis(4-hydroxyphenyl)propane	-368.6											
C₁₅H₃₀	Decylcyclopentane		-367.3										
C₁₅H₃₀O₂	Pentadecanoic acid	-861.7	-811.7	-699.0							443.3		
C₁₅H₃₀O₂	Methyl tetradecanoate		-743.9	-656.9									
C₁₅H₃₂O	1-Pentadecanol	-658.2											
C₁₆H₁₀	Fluoranthene	189.9		289.0				230.6			230.2		
C₁₆H₁₀	Pyrene	125.5		225.7				224.9			229.7		
C₁₆H₂₂O₄	Dibutyl phthalate		-842.6	-750.9									
C₁₆H₂₂O₁₁	α-D-Glucose pentaacetate	-2249.4											
C₁₆H₂₂O₁₁	β-D-Glucose pentaacetate	-2232.6											
C₁₆H₂₆	Decylbenzene		-218.3	-138.5									
C₁₆H₃₂	1-Hexadecene		-328.7	-248.4					587.9			488.9	
C₁₆H₃₂O₂	Hexadecanoic acid	-891.5	-838.1	-737.1				452.4			460.7		
C₁₆H₃₂O₂	Methyl pentadecanoate		-771.0	-680.0									
C₁₆H₃₃Br	1-Bromohexadecane		-444.5	-350.2									
C₁₆H₃₄	Hexadecane		-456.1	-374.8								501.6	
C₁₆H₃₄O	1-Hexadecanol	-686.5		-517.0							422.0		
C₁₆H₃₆IN	Tetrabutyl ammonium iodide	-498.6											
C₁₇H₃₄O₂	Margaric acid	-924.4	-865.6								475.7		
C₁₈H₁₂	Benz[a]anthracene	170.8		293.0									
C₁₈H₁₂	Chrysene	145.3		269.8									
C₁₈H₁₅N	Triphenylamine	234.7		326.8									
C₁₈H₃₄O₂	Oleic acid											577.0	
C₁₈H₃₄O₄	Dibutyl sebacate											619.0	
C₁₈H₃₆O₂	Stearic acid	-947.7	-884.7	-781.2							501.5		
C₁₈H₃₇Cl	1-Chlorooctadecane		-544.1	-446.0									
C₁₈H₃₈	Octadecane	-567.4		-414.6				480.2			485.6		
C₁₈H₃₉N	Trihexylamine		-433.0										
C₁₉H₁₆O	Triphenylmethanol	-2.5											
C₁₉H₃₆O₂	Methyl oleate		-734.5	-649.9									
C₁₉H₃₆O₂	Methyl trans-9-octadecenoate		-737.0										
C₂₀H₁₂	Perylene	182.8						264.6			274.9		
C₂₀H₁₄O₄	1,2-Benzenedicarboxylic acid, diphenyl ester	-489.2											
C₂₀H₃₈O₂	Ethyl cis-9-octadecenoate		-775.8										

Molecular formula	Name	$\Delta_f H°$/kJ mol⁻¹			$\Delta_f G°$/kJ mol⁻¹			$S°$/J mol⁻¹ K⁻¹			C_p/J mol⁻¹ K⁻¹		
		Crys.	Liq.	Gas	Crys.	Liq.	Gas	Crys.	Liq.	Gas	Crys.	Liq.	Gas
$C_{20}H_{38}O_2$	Ethyl *trans*-9-octa-decenoate		-773.3										
$C_{20}H_{40}O_2$	Arachidic acid	-1011.9	-940.0	-812.4							545.1		
$C_{22}H_{42}O_2$	Brassidic acid	-960.7											
$C_{22}H_{42}O_2$	Butyl oleate		-816.9										
$C_{22}H_{44}O_2$	Butyl stearate												
$C_{24}H_{38}O_4$	Bis(2-ethylhexyl) phthalate											704.7	
$C_{24}H_{51}N$	Trioctylamine		-585.0										
$C_{26}H_{18}$	9,10-Diphenylanthracene	308.7		465.6									
$C_{26}H_{54}$	5-Butyldocosane		-713.5	-587.6									
$C_{26}H_{54}$	11-Butyldocosane		-716.0	-593.4									
$C_{28}H_{18}$	9,9'-Bianthracene	326.2		454.3									
$C_{31}H_{64}$	11-Decylheneicosane		-848.0	-705.8									
$C_{32}H_{66}$	Dotriacontane		-968.3	-697.2									
C_{60}	Fullerene-C₆₀	2327.0		2502.0	2302.0		2442.0	426.0		544.0	520.0		512.0
C_{70}	Fullerene-C₇₀	2555.0		2755.0	2537.0		2692.0	464.0		614.0	650.0		585.0

THERMODYNAMIC PROPERTIES AS A FUNCTION OF TEMPERATURE

L. V. Gurvich, V. S. Iorish, V. S. Yungman, and O. V. Dorofeeva

The thermodynamic properties $C_p^\circ(T)$, $S^\circ(T)$, $H^\circ(T)-H^\circ(T_r)$, $-[G^\circ(T)-H^\circ(T_r)]/T$ and formation properties $\Delta_f H^\circ(T)$, $\Delta_f G^\circ(T)$, $\log K_f^\circ(T)$ are tabulated as functions of temperature in the range 298.15 to 1500 K for 80 substances in the standard state. The reference temperature, T_r, is equal to 298.15 K. The standard state pressure is taken as 1 bar (100,000 Pa). The tables are presented in the JANAF Thermochemical Tables format (Reference 2). The numerical data are extracted from IVTANTHERMO databases except for C_2H_4O, C_3H_6O, C_6H_6, C_6H_6O, $C_{10}H_8$, and CH_5N, which are based upon TRC Tables. See the references for information on standard states and other details.

REFERENCES

1. Gurvich, L. V., Veyts, I. V., and Alcock, C. B., Eds., *Thermodynamic Properties of Individual Substances, 4th ed.*, Hemisphere Publishing Corp., New York, 1989.
2. Chase, M. W., et al., *JANAF Thermochemical Tables, 3rd ed., J. Phys. Chem. Ref. Data*, 14, Suppl. 1, 1985.

Order of Listing of Tables

No.	Formula	Name	State	No.	Formula	Name	State
1	Ar	Argon	g	41	$CuCl_2$	Copper dichloride	cr, l
2	Br	Bromine	g	42	$CuCl_2$	Copper dichloride	g
3	Br_2	Dibromine	g	43	F	Fluorine	g
4	BrH	Hydrogen bromide	g	44	F_2	Difluorine	g
5	C	Carbon (graphite)	cr	45	FH	Hydrogen fluoride	g
6	C	Carbon (diamond)	cr	46	Ge	Germanium	cr, l
7	C_2	Dicarbon	g	47	Ge	Germanium	g
8	C_3	Tricarbon	g	48	GeO_2	Germanium dioxide	cr, l
9	CO	Carbon oxide	g	49	$GeCl_4$	Germanium tetrachloride	g
10	CO_2	Carbon dioxide	g	50	H	Hydrogen	g
11	CH_4	Methane	g	51	H_2	Dihydrogen	g
12	C_2H_2	Acetylene	g	52	HO	Hydroxyl	g
13	C_2H_4	Ethylene	g	53	H_2O	Water	l
14	C_2H_6	Ethane	g	54	H_2O	Water	g
15	C_3H_6	Cyclopropane	g	55	I	Iodine	g
16	C_3H_8	Propane	g	56	I_2	Diiodine	cr, l
17	C_6H_6	Benzene	l	57	I_2	Diiodine	g
18	C_6H_6	Benzene	g	58	IH	Hydrogen iodide	g
19	$C_{10}H_8$	Naphthalene	cr, l	59	K	Potassium	cr, l
20	$C_{10}H_8$	Naphthalene	g	60	K	Potassium	g
21	CH_2O	Formaldehyde	g	61	K_2O	Dipotassium oxide	cr, l
22	CH_4O	Methanol	g	62	KOH	Potassium hydroxide	cr, l
23	C_2H_4O	Acetaldehyde	g	63	KOH	Potassium hydroxide	g
24	C_2H_6O	Ethanol	g	64	KCl	Potassium chloride	cr, l
25	$C_2H_4O_2$	Acetic acid	g	65	KCl	Potassium chloride	g
26	C_3H_6O	Acetone	g	66	N_2	Dinitrogen	g
27	C_6H_6O	Phenol	g	67	NO	Nitric oxide	g
28	CF_4	Carbon tetrafluoride	g	68	NO_2	Nitrogen dioxide	g
29	CHF_3	Trifluoromethane	g	69	NH_3	Ammonia	g
30	$CClF_3$	Chlorotrifluoromethane	g	70	O	Oxygen	g
31	CCl_2F_2	Dichlorodifluoromethane	g	71	O_2	Dioxygen	g
32	$CHClF_2$	Chlorodifluoromethane	g	72	S	Sulfur	cr, l
33	CH_5N	Methylamine	g	73	S	Sulfur	g
34	Cl	Chlorine	g	74	S_2	Disulfur	g
35	Cl_2	Dichlorine	g	75	S_8	Octasulfur	g
36	ClH	Hydrogen chloride	g	76	SO_2	Sulfur dioxide	g
37	Cu	Copper	cr, l	77	Si	Silicon	cr
38	Cu	Copper	g	78	Si	Silicon	g
39	CuO	Copper oxide	cr	79	SiO_2	Silicon dioxide	cr
40	Cu_2O	Dicopper oxide	cr	80	$SiCl_4$	Silicon tetrachloride	g

T/K	C_p°	S°	$-(G^\circ-H^\circ(T_r))/T$	$H^\circ-H^\circ(T_r)$	$\Delta_f H^\circ$	$\Delta_f G^\circ$	Log K_f
	J/K·mol			kJ/mol			

1. ARGON   Ar (g)

T/K	C_p°	S°	$-(G^\circ-H^\circ(T_r))/T$	$H^\circ-H^\circ(T_r)$	$\Delta_f H^\circ$	$\Delta_f G^\circ$	Log K_f
298.15	20.786	154.845	154.845	0.000	0.000	0.000	0.000
300	20.786	154.973	154.845	0.038	0.000	0.000	0.000
400	20.786	160.953	155.660	2.117	0.000	0.000	0.000
500	20.786	165.591	157.200	4.196	0.000	0.000	0.000
600	20.786	169.381	158.924	6.274	0.000	0.000	0.000
700	20.786	172.585	160.653	8.353	0.000	0.000	0.000
800	20.786	175.361	162.322	10.431	0.000	0.000	0.000
900	20.786	177.809	163.909	12.510	0.000	0.000	0.000
1000	20.786	179.999	165.410	14.589	0.000	0.000	0.000
1100	20.786	181.980	166.828	16.667	0.000	0.000	0.000
1200	20.786	183.789	168.167	18.746	0.000	0.000	0.000
1300	20.786	185.453	169.434	20.824	0.000	0.000	0.000
1400	20.786	186.993	170.634	22.903	0.000	0.000	0.000
1500	20.786	188.427	171.773	24.982	0.000	0.000	0.000

2. BROMINE   Br (g)

T/K	C_p°	S°	$-(G^\circ-H^\circ(T_r))/T$	$H^\circ-H^\circ(T_r)$	$\Delta_f H^\circ$	$\Delta_f G^\circ$	Log K_f
298.15	20.786	175.017	175.017	0.000	111.870	82.379	−14.432
300	20.786	175.146	175.018	0.038	111.838	82.196	−14.311
400	20.787	181.126	175.833	2.117	96.677	75.460	−9.854
500	20.798	185.765	177.373	4.196	96.910	70.129	−7.326
600	20.833	189.559	179.097	6.277	97.131	64.752	−5.637
700	20.908	192.776	180.827	8.364	97.348	59.338	−4.428
800	21.027	195.575	182.499	10.461	97.568	53.893	−3.519
900	21.184	198.061	184.093	12.571	97.796	48.420	−2.810
1000	21.365	200.302	185.604	14.698	98.036	42.921	−2.242
1100	21.559	202.347	187.034	16.844	98.291	37.397	−1.776
1200	21.752	204.231	188.390	19.010	98.560	31.850	−1.386
1300	21.937	205.980	189.676	21.195	98.844	26.279	−1.056
1400	22.107	207.612	190.900	23.397	99.141	20.686	−0.772
1500	22.258	209.142	192.065	25.615	99.449	15.072	−0.525

3. DIBROMINE   Br$_2$ (g)

T/K	C_p°	S°	$-(G^\circ-H^\circ(T_r))/T$	$H^\circ-H^\circ(T_r)$	$\Delta_f H^\circ$	$\Delta_f G^\circ$	Log K_f
298.15	36.057	245.467	245.467	0.000	30.910	3.105	−0.544
300	36.074	245.690	245.468	0.067	30.836	2.933	−0.511
332.25	36.340	249.387	245.671	1.235		pressure = 1 bar	
400	36.729	256.169	246.892	3.711	0.000	0.000	0.000
500	37.082	264.406	249.600	7.403	0.000	0.000	0.000
600	37.305	271.188	252.650	11.123	0.000	0.000	0.000
700	37.464	276.951	255.720	14.862	0.000	0.000	0.000
800	37.590	281.962	258.694	18.615	0.000	0.000	0.000
900	37.697	286.396	261.530	22.379	0.000	0.000	0.000
1000	37.793	290.373	264.219	26.154	0.000	0.000	0.000
1100	37.883	293.979	266.763	29.938	0.000	0.000	0.000
1200	37.970	297.279	269.170	33.730	0.000	0.000	0.000
1300	38.060	300.322	271.451	37.532	0.000	0.000	0.000
1400	38.158	303.146	273.615	41.343	0.000	0.000	0.000
1500	38.264	305.782	275.673	45.164	0.000	0.000	0.000

4. HYDROGEN BROMIDE   HBr (g)

T/K	C_p°	S°	$-(G^\circ-H^\circ(T_r))/T$	$H^\circ-H^\circ(T_r)$	$\Delta_f H^\circ$	$\Delta_f G^\circ$	Log K_f
298.15	29.141	198.697	198.697	0.000	−36.290	−53.360	9.348
300	29.141	198.878	198.698	0.054	−36.333	−53.466	9.309
400	29.220	207.269	199.842	2.971	−52.109	−55.940	7.305
500	29.454	213.811	202.005	5.903	−52.484	−56.854	5.939
600	29.872	219.216	204.436	8.868	−52.844	−57.694	5.023
700	30.431	223.861	206.886	11.882	−53.168	−58.476	4.363
800	31.063	227.965	209.269	14.957	−53.446	−59.214	3.866

	J/K·mol			kJ/mol			
T/K	C_p°	S°	$-(G^\circ - H^\circ(T_r))/T$	$H^\circ - H^\circ(T_r)$	$\Delta_f H^\circ$	$\Delta_f G^\circ$	Log K_f

4. HYDROGEN BROMIDE   HBr (g) (continued)

900	31.709	231.661	211.555	18.095	−53.677	−59.921	3.478
1000	32.335	235.035	213.737	21.298	−53.864	−60.604	3.166
1100	32.919	238.145	215.816	24.561	−54.012	−61.271	2.909
1200	33.454	241.032	217.799	27.880	−54.129	−61.925	2.696
1300	33.938	243.729	219.691	31.250	−54.220	−62.571	2.514
1400	34.374	246.261	221.499	34.666	−54.291	−63.211	2.358
1500	34.766	248.646	223.230	38.123	−54.348	−63.846	2.223

5. CARBON (GRAPHITE)   C (cr; graphite)

298.15	8.536	5.740	5.740	0.000	0.000	0.000	0.000
300	8.610	5.793	5.740	0.016	0.000	0.000	0.000
400	11.974	8.757	6.122	1.054	0.000	0.000	0.000
500	14.537	11.715	6.946	2.385	0.000	0.000	0.000
600	16.607	14.555	7.979	3.945	0.000	0.000	0.000
700	18.306	17.247	9.113	5.694	0.000	0.000	0.000
800	19.699	19.785	10.290	7.596	0.000	0.000	0.000
900	20.832	22.173	11.479	9.625	0.000	0.000	0.000
1000	21.739	24.417	12.662	11.755	0.000	0.000	0.000
1100	22.452	26.524	13.827	13.966	0.000	0.000	0.000
1200	23.000	28.502	14.968	16.240	0.000	0.000	0.000
1300	23.409	30.360	16.082	18.562	0.000	0.000	0.000
1400	23.707	32.106	17.164	20.918	0.000	0.000	0.000
1500	23.919	33.749	18.216	23.300	0.000	0.000	0.000

6. CARBON (DIAMOND)   C (cr; diamond)

298.15	6.109	2.362	2.362	0.000	1.850	2.857	−0.501
300	6.201	2.400	2.362	0.011	1.846	2.863	0.499
400	10.321	4.783	2.659	0.850	1.645	3.235	−0.422
500	13.404	7.431	3.347	2.042	1.507	3.649	−0.381
600	15.885	10.102	4.251	3.511	1.415	4.087	−0.356
700	17.930	12.709	5.274	5.205	1.361	4.537	−0.339
800	19.619	15.217	6.361	7.085	1.338	4.993	−0.326
900	21.006	17.611	7.479	9.118	1.343	5.450	−0.316
1000	22.129	19.884	8.607	11.277	1.372	5.905	−0.308
1100	23.020	22.037	9.731	13.536	1.420	6.356	−0.302
1200	23.709	24.071	10.842	15.874	1.484	6.802	−0.296
1300	24.222	25.990	11.934	18.272	1.561	7.242	−0.291
1400	24.585	27.799	13.003	20.714	1.646	7.675	−0.286
1500	24.824	29.504	14.047	23.185	1.735	8.103	−0.282

7. DICARBON   C$_2$ (g)

298.15	43.548	197.095	197.095	0.000	830.457	775.116	−135.795
300	43.575	197.365	197.096	0.081	830.506	774.772	−134.898
400	42.169	209.809	198.802	4.403	832.751	755.833	−98.700
500	39.529	218.924	201.959	8.483	834.170	736.423	−76.933
600	37.837	225.966	205.395	12.342	834.909	716.795	−62.402
700	36.984	231.726	208.758	16.078	835.148	697.085	−52.016
800	36.621	236.637	211.943	19.755	835.020	677.366	−44.227
900	36.524	240.943	214.931	23.411	834.618	657.681	−38.170
1000	36.569	244.793	217.728	27.065	834.012	638.052	−33.328
1100	36.696	248.284	220.349	30.728	833.252	618.492	−29.369
1200	36.874	251.484	222.812	34.406	832.383	599.006	−26.074
1300	37.089	254.444	225.133	38.104	831.437	579.596	−23.288
1400	37.329	257.201	227.326	41.824	830.445	560.261	−20.903
1500	37.589	259.785	229.405	45.570	829.427	540.997	−18.839

	J/K·mol			kJ/mol			
T/K	C_p°	S°	$-(G^\circ\text{-}H^\circ(T_r))/T$	$H^\circ\text{-}H^\circ(T_r)$	$\Delta_f H^\circ$	$\Delta_f G^\circ$	Log K_f
8. TRICARBON	**C$_3$ (g)**						
298.15	42.202	237.611	237.611	0.000	839.958	774.249	−135.643
300	42.218	237.872	237.611	0.078	839.989	773.841	−134.736
400	43.383	250.164	239.280	4.354	841.149	751.592	−98.147
500	44.883	260.003	242.471	8.766	841.570	729.141	−76.172
600	46.406	268.322	246.104	13.331	841.453	706.659	−61.519
700	47.796	275.582	249.807	18.042	840.919	684.230	−51.057
800	48.997	282.045	253.440	22.884	840.053	661.901	−43.217
900	50.006	287.876	256.948	27.835	838.919	639.698	−37.127
1000	50.844	293.189	260.310	32.879	837.572	617.633	−32.261
1100	51.535	298.069	263.524	37.999	836.059	595.711	−28.288
1200	52.106	302.578	266.593	43.182	834.420	573.933	−24.982
1300	52.579	306.768	269.524	48.417	832.690	552.295	−22.191
1400	52.974	310.679	272.326	53.695	830.899	530.793	−19.804
1500	53.307	314.346	275.006	59.010	829.068	509.421	−17.739
9. CARBON OXIDE	**CO (g)**						
298.15	29.141	197.658	197.658	0.000	−110.530	−137.168	24.031
300	29.142	197.838	197.659	0.054	−110.519	−137.333	23.912
400	29.340	206.243	198.803	2.976	−110.121	−146.341	19.110
500	29.792	212.834	200.973	5.930	−110.027	−155.412	16.236
600	30.440	218.321	203.419	8.941	−110.157	−164.480	14.319
700	31.170	223.067	205.895	12.021	−110.453	−173.513	12.948
800	31.898	227.277	208.309	15.175	−110.870	−182.494	11.915
900	32.573	231.074	210.631	18.399	−111.378	−191.417	11.109
1000	33.178	234.538	212.851	21.687	−111.952	−200.281	10.461
1100	33.709	237.726	214.969	25.032	−112.573	−209.084	9.928
1200	34.169	240.679	216.990	28.426	−113.228	−217.829	9.482
1300	34.568	243.430	218.920	31.864	−113.904	−226.518	9.101
1400	34.914	246.005	220.763	35.338	−114.594	−235.155	8.774
1500	35.213	248.424	222.527	38.845	−115.291	−243.742	8.488
10. CARBON DIOXIDE	**CO$_2$ (g)**						
298.15	37.135	213.783	213.783	0.000	−393.510	−394.373	69.092
300	37.220	214.013	213.784	0.069	−393.511	−394.379	68.667
400	41.328	225.305	215.296	4.004	−393.586	−394.656	51.536
500	44.627	234.895	218.280	8.307	−393.672	−394.914	41.256
600	47.327	243.278	221.762	12.909	−393.791	−395.152	34.401
700	49.569	250.747	225.379	17.758	−393.946	−395.367	29.502
800	51.442	257.492	228.978	22.811	−394.133	−395.558	25.827
900	53.008	263.644	232.493	28.036	−394.343	−395.724	22.967
1000	54.320	269.299	235.895	33.404	−394.568	−395.865	20.678
1100	55.423	274.529	239.172	38.893	−394.801	−395.984	18.803
1200	56.354	279.393	242.324	44.483	−395.035	−396.081	17.241
1300	57.144	283.936	245.352	50.159	−395.265	−396.159	15.918
1400	57.818	288.196	248.261	55.908	−395.488	−396.219	14.783
1500	58.397	292.205	251.059	61.719	−395.702	−396.264	13.799
11. METHANE	**CH$_4$ (g)**						
298.15	35.695	186.369	186.369	0.000	−74.600	−50.530	8.853
300	35.765	186.590	186.370	0.066	−74.656	−50.381	8.772
400	40.631	197.501	187.825	3.871	−77.703	−41.827	5.462
500	46.627	207.202	190.744	8.229	−80.520	−32.525	3.398
600	52.742	216.246	194.248	13.199	−82.969	−22.690	1.975
700	58.603	224.821	198.008	18.769	−85.023	−12.476	0.931
800	64.084	233.008	201.875	24.907	−86.693	−1.993	0.130

T/K	J/K·mol			kJ/mol			
	C_p°	S°	$-(G^\circ - H^\circ(T_r))/T$	$H^\circ - H^\circ(T_r)$	$\Delta_f H^\circ$	$\Delta_f G^\circ$	Log K_f

11. METHANE   CH₄ (g) (continued)

T/K	C_p°	S°	$-(G^\circ-H^\circ(T_r))/T$	$H^\circ-H^\circ(T_r)$	$\Delta_f H^\circ$	$\Delta_f G^\circ$	Log K_f
900	69.137	240.852	205.773	31.571	−88.006	8.677	−0.504
1000	73.746	248.379	209.660	38.719	−88.996	19.475	−1.017
1100	77.919	255.607	213.511	46.306	−89.698	30.358	−1.442
1200	81.682	262.551	217.310	54.289	−90.145	41.294	−1.797
1300	85.067	269.225	221.048	62.630	−90.367	52.258	−2.100
1400	88.112	275.643	224.720	71.291	−90.390	63.231	−2.359
1500	90.856	281.817	228.322	80.242	−90.237	74.200	−2.584

12. ACETYLENE   C₂H₂ (g)

T/K	C_p°	S°	$-(G^\circ-H^\circ(T_r))/T$	$H^\circ-H^\circ(T_r)$	$\Delta_f H^\circ$	$\Delta_f G^\circ$	Log K_f
298.15	44.036	200.927	200.927	0.000	227.400	209.879	−36.769
300	44.174	201.199	200.927	0.082	227.397	209.770	−36.524
400	50.388	214.814	202.741	4.829	227.161	203.928	−26.630
500	54.751	226.552	206.357	10.097	226.846	198.154	−20.701
600	58.121	236.842	210.598	15.747	226.445	192.452	−16.754
700	60.970	246.021	215.014	21.704	225.968	186.823	−13.941
800	63.511	254.331	219.418	27.931	225.436	181.267	−11.835
900	65.831	261.947	223.726	34.399	224.873	175.779	−10.202
1000	67.960	268.995	227.905	41.090	224.300	170.355	−8.898
1100	69.909	275.565	231.942	47.985	223.734	164.988	−7.835
1200	71.686	281.725	235.837	55.067	223.189	159.672	−6.950
1300	73.299	287.528	239.592	62.317	222.676	154.400	−6.204
1400	74.758	293.014	243.214	69.721	222.203	149.166	−5.565
1500	76.077	298.218	246.709	77.264	221.774	143.964	−5.013

13. ETHYLENE   C₂H₄ (g)

T/K	C_p°	S°	$-(G^\circ-H^\circ(T_r))/T$	$H^\circ-H^\circ(T_r)$	$\Delta_f H^\circ$	$\Delta_f G^\circ$	Log K_f
298.15	42.883	219.316	219.316	0.000	52.400	68.358	−11.976
300	43.059	219.582	219.317	0.079	52.341	68.457	−11.919
400	53.045	233.327	221.124	4.881	49.254	74.302	−9.703
500	62.479	246.198	224.864	10.667	46.533	80.887	−8.450
600	70.673	258.332	229.441	17.335	44.221	87.982	−7.659
700	77.733	269.770	234.393	24.764	42.278	95.434	−7.121
800	83.868	280.559	239.496	32.851	40.655	103.142	−6.734
900	89.234	290.754	244.630	41.512	39.310	111.036	−6.444
1000	93.939	300.405	249.730	50.675	38.205	119.067	−6.219
1100	98.061	309.556	254.756	60.280	37.310	127.198	−6.040
1200	101.670	318.247	259.688	70.271	36.596	135.402	−5.894
1300	104.829	326.512	264.513	80.599	36.041	143.660	−5.772
1400	107.594	334.384	269.225	91.223	35.623	151.955	−5.669
1500	110.018	341.892	273.821	102.107	35.327	160.275	−5.581

14. ETHANE   C₂H₆ (g)

T/K	C_p°	S°	$-(G^\circ-H^\circ(T_r))/T$	$H^\circ-H^\circ(T_r)$	$\Delta_f H^\circ$	$\Delta_f G^\circ$	Log K_f
298.15	52.487	229.161	229.161	0.000	−84.000	−32.015	5.609
300	52.711	229.487	229.162	0.097	−84.094	−31.692	5.518
400	65.459	246.378	231.379	5.999	−88.988	−13.473	1.759
500	77.941	262.344	235.989	13.177	−93.238	5.912	−0.618
600	89.188	277.568	241.660	21.545	−96.779	26.086	−2.271
700	99.136	292.080	247.835	30.972	−99.663	46.800	−3.492
800	107.936	305.904	254.236	41.334	−101.963	67.887	−4.433
900	115.709	319.075	260.715	52.525	−103.754	89.231	−5.179
1000	122.552	331.628	267.183	64.445	−105.105	110.750	−5.785
1100	128.553	343.597	273.590	77.007	−106.082	132.385	−6.286
1200	133.804	355.012	279.904	90.131	−106.741	154.096	−6.708
1300	138.391	365.908	286.103	103.746	−107.131	175.850	−7.066
1400	142.399	376.314	292.178	117.790	−107.292	197.625	−7.373
1500	145.905	386.260	298.121	132.209	−107.260	219.404	−7.640

T/K	C_p° J/K·mol	S°	$-(G^\circ-H^\circ(T_r))/T$	$H^\circ-H^\circ(T_r)$ kJ/mol	$\Delta_f H^\circ$	$\Delta_f G^\circ$	Log K_f

15. CYCLOPROPANE C_3H_6 (g)

T/K	C_p°	S°	$-(G^\circ-H^\circ(T_r))/T$	$H^\circ-H^\circ(T_r)$	$\Delta_f H^\circ$	$\Delta_f G^\circ$	Log K_f
298.15	55.571	237.488	237.488	0.000	53.300	104.514	−18.310
300	55.941	237.832	237.489	0.103	53.195	104.832	−18.253
400	76.052	256.695	239.924	6.708	47.967	122.857	−16.043
500	93.859	275.637	245.177	15.230	43.730	142.091	−14.844
600	108.542	294.092	251.801	25.374	40.405	162.089	−14.111
700	120.682	311.763	259.115	36.854	37.825	182.583	−13.624
800	130.910	328.564	266.755	49.447	35.854	203.404	−13.281
900	139.658	344.501	274.516	62.987	34.384	224.441	−13.026
1000	147.207	359.616	282.277	77.339	33.334	245.618	−12.830
1100	153.749	373.961	289.965	92.395	32.640	266.883	−12.673
1200	159.432	387.588	297.538	108.060	32.249	288.197	−12.545
1300	164.378	400.549	304.967	124.257	32.119	309.533	−12.437
1400	168.689	412.892	312.239	140.915	32.215	330.870	−12.345
1500	172.453	424.662	319.344	157.976	32.507	352.193	−12.264

16. PROPANE C_3H_8 (g)

T/K	C_p°	S°	$-(G^\circ-H^\circ(T_r))/T$	$H^\circ-H^\circ(T_r)$	$\Delta_f H^\circ$	$\Delta_f G^\circ$	Log K_f
298.15	73.597	270.313	270.313	0.000	−103.847	−23.458	4.110
300	73.931	270.769	270.314	0.136	−103.972	−22.959	3.997
400	94.014	294.739	273.447	8.517	−110.33	15.029	−0.657
500	112.591	317.768	280.025	18.872	−115.658	34.507	−3.605
600	128.700	339.753	288.162	30.955	−119.973	64.961	−5.655
700	142.674	360.668	297.039	44.540	−123.384	96.065	−7.168
800	154.766	380.528	306.245	59.427	−126.016	127.603	−8.331
900	165.352	399.381	315.555	75.444	−127.982	159.430	−9.253
1000	174.598	417.293	324.841	92.452	−129.380	191.444	−10.000
1100	182.673	434.321	334.026	110.325	−130.296	223.574	−10.617
1200	189.745	450.526	343.064	128.954	−130.802	255.770	−11.133
1300	195.853	465.961	351.929	148.241	−130.961	287.993	−11.572
1400	201.209	480.675	360.604	168.100	−130.829	320.217	−11.947
1500	205.895	494.721	369.080	188.460	−130.445	352.422	−12.272

17. BENZENE C_6H_6 (l)

T/K	C_p°	S°	$-(G^\circ-H^\circ(T_r))/T$	$H^\circ-H^\circ(T_r)$	$\Delta_f H^\circ$	$\Delta_f G^\circ$	Log K_f
298.15	135.950	173.450	173.450	0.000	49.080	124.521	−21.815
300	136.312	174.292	173.453	.252	49.077	124.989	−21.762
400	161.793	216.837	179.082	15.102	48.978	150.320	−19.630
500	207.599	257.048	190.639	33.204	50.330	175.559	−18.340

18. BENZENE C_6H_6 (g)

T/K	C_p°	S°	$-(G^\circ-H^\circ(T_r))/T$	$H^\circ-H^\circ(T_r)$	$\Delta_f H^\circ$	$\Delta_f G^\circ$	Log K_f
298.15	82.430	269.190	269.190	0.000	82.880	129.750	−22.731
300	83.020	269.700	269.190	0.153	82.780	130.040	−22.641
400	113.510	297.840	272.823	10.007	77.780	146.570	−19.140
500	139.340	326.050	280.658	22.696	73.740	164.260	−17.160
600	160.090	353.360	290.517	37.706	70.490	182.680	−15.903
700	176.790	379.330	301.360	54.579	67.910	201.590	−15.042
800	190.460	403.860	312.658	72.962	65.910	220.820	−14.418
900	201.840	426.970	324.084	92.597	64.410	240.280	−13.945
1000	211.430	448.740	335.473	113.267	63.340	259.890	−13.575
1100	219.580	469.280	346.710	134.827	62.620	277.640	−13.184
1200	226.540	488.690	357.743	157.137	62.200	299.320	−13.029
1300	232.520	507.070	368.534	180.097	62.000	319.090	−12.821
1400	237.680	524.490	379.056	203.607	61.990	338.870	−12.643
1500	242.140	541.040	389.302	227.607	62.110	358.640	−12.489

T/K	C_p°	S°	-(G°-H°(T_r))/T	H°-H°(T_r)	Δ_fH°	Δ_fG°	Log K_f
		J/K·mol			kJ/mol		

19. NAPHTHALENE $C_{10}H_8$ (cr, l)

T/K	C_p°	S°	$-(G^\circ-H^\circ(T_r))/T$	$H^\circ-H^\circ(T_r)$	Δ_fH°	Δ_fG°	Log K_f
298.15	165.720	167.390	167.390	0.000	78.530	201.585	−35.316
300	167.001	168.419	167.393	0.308	78.466	202.349	−35.232
353.43	208.722	198.948	169.833	10.290	96.099	224.543	−33.186
			PHASE TRANSITION: $\Delta_{trs} H$ = 18.980 kJ/mol, $\Delta_{trs} S$ = 53.702 J/K·mol, cr–l				
353.43	217.200	252.650	169.833	29.270	96.099	224.543	−33.186
400	241.577	280.916	181.124	39.917	96.067	241.475	−31.533
470	276.409	322.712	199.114	58.091	97.012	266.859	−29.658

20. NAPHTHALENE $C_{10}H_8$ (g)

T/K	C_p°	S°	$-(G^\circ-H^\circ(T_r))/T$	$H^\circ-H^\circ(T_r)$	Δ_fH°	Δ_fG°	Log K_f
298.15	131.920	333.150	333.150	0.000	150.580	224.100	−39.260
300	132.840	333.970	333.157	0.244	150.450	224.560	−39.098
400	180.070	378.800	338.950	15.940	144.190	250.270	−32.681
500	219.740	423.400	351.400	36.000	139.220	277.340	−28.973
600	251.530	466.380	367.007	59.624	135.350	305.330	−26.581
700	277.010	507.140	384.146	86.096	132.330	333.950	−24.919
800	297.730	545.520	401.935	114.868	130.050	362.920	−23.696
900	314.850	581.610	419.918	145.523	128.430	392.150	−22.759
1000	329.170	615.550	437.806	177.744	127.510	421.700	−22.027
1100	341.240	647.500	455.426	211.281	127.100	450.630	−21.398
1200	351.500	677.650	472.707	245.932	126.960	480.450	−20.913
1300	360.260	706.130	489.568	281.531	127.060	509.770	−20.482
1400	367.780	733.110	506.009	317.941	127.390	539.740	−20.137
1500	374.270	758.720	522.019	355.051	127.920	568.940	−19.812

21. FORMALDEHYDE H_2CO (g)

T/K	C_p°	S°	$-(G^\circ-H^\circ(T_r))/T$	$H^\circ-H^\circ(T_r)$	Δ_fH°	Δ_fG°	Log K_f
298.15	35.387	218.760	218.760	0.000	−108.700	−102.667	17.987
300	35.443	218.979	218.761	0.066	−108.731	−102.630	17.869
400	39.240	229.665	220.192	3.789	−110.438	−100.340	13.103
500	43.736	238.900	223.028	7.936	−112.073	−97.623	10.198
600	48.181	247.270	226.381	12.534	−113.545	−94.592	8.235
700	52.280	255.011	229.924	17.560	−114.833	−91.328	6.815
800	55.941	262.236	233.517	22.975	−115.942	−87.893	5.739
900	59.156	269.014	237.088	28.734	−116.889	−84.328	4.894
1000	61.951	275.395	240.603	34.792	−117.696	−80.666	4.213
1100	64.368	281.416	244.042	41.111	−118.382	−76.929	3.653
1200	66.453	287.108	247.396	47.655	−118.966	−73.134	3.183
1300	68.251	292.500	250.660	54.392	−119.463	−69.294	2.784
1400	69.803	297.616	253.833	61.297	−119.887	−65.418	2.441
1500	71.146	302.479	256.915	68.346	−120.249	−61.514	2.142

22. METHANOL CH_3OH (g)

T/K	C_p°	S°	$-(G^\circ-H^\circ(T_r))/T$	$H^\circ-H^\circ(T_r)$	Δ_fH°	Δ_fG°	Log K_f
298.15	44.101	239.865	239.865	0.000	−201.000	−162.298	28.434
300	44.219	240.139	239.866	0.082	−201.068	−162.057	28.216
400	51.713	253.845	241.685	4.864	−204.622	−148.509	19.393
500	59.800	266.257	245.374	10.442	−207.750	−134.109	14.010
600	67.294	277.835	249.830	16.803	−210.387	−119.125	10.371
700	73.958	288.719	254.616	23.873	−212.570	−103.737	7.741
800	79.838	298.987	259.526	31.569	−214.350	−88.063	5.750
900	85.025	308.696	264.455	39.817	−215.782	−72.188	4.190
1000	89.597	317.896	269.343	48.553	−216.916	−56.170	2.934
1100	93.624	326.629	274.158	57.718	−217.794	−40.050	1.902
1200	97.165	334.930	278.879	67.262	−218.457	−23.861	1.039
1300	100.277	342.833	283.497	77.137	−218.936	−7.624	0.306
1400	103.014	350.367	288.007	87.304	−219.261	8.644	−0.322
1500	105.422	357.558	292.405	97.729	−219.456	24.930	−0.868

	J/K·mol			kJ/mol			
T/K	C_p°	S°	$-(G^\circ-H^\circ(T_r))/T$	$H^\circ-H^\circ(T_r)$	$\Delta_f H^\circ$	$\Delta_f G^\circ$	Log K_f

23. ACETALDEHYDE   C$_2$H$_4$O (g)

T/K	C_p°	S°	$-(G^\circ-H^\circ(T_r))/T$	$H^\circ-H^\circ(T_r)$	$\Delta_f H^\circ$	$\Delta_f G^\circ$	Log K_f
298.15	55.318	263.840	263.840	0.000	−166.190	−133.010	23.302
300	55.510	264.180	263.837	0.103	−166.250	−132.800	23.122
400	66.282	281.620	266.147	6.189	−169.530	−121.130	15.818
500	76.675	297.540	270.850	13.345	−172.420	−108.700	11.356
600	85.942	312.360	276.550	21.486	−174.870	−95.720	8.334
700	94.035	326.230	282.667	30.494	−176.910	−82.350	6.145
800	101.070	339.260	288.938	40.258	−178.570	−68.730	4.487
900	107.190	351.520	295.189	50.698	−179.880	−54.920	3.187
1000	112.490	363.100	301.431	61.669	−180.850	−40.930	2.138
1100	117.080	374.040	307.537	73.153	−181.560	−27.010	1.283
1200	121.060	384.400	313.512	85.065	−182.070	−12.860	0.560
1300	124.500	394.230	319.350	97.344	−182.420	1.240	−0.050
1400	127.490	403.570	325.031	109.954	−182.640	15.470	−0.577
1500	130.090	412.460	330.571	122.834	−182.750	29.580	−1.030

24. ETHANOL   C$_2$H$_5$OH (g)

T/K	C_p°	S°	$-(G^\circ-H^\circ(T_r))/T$	$H^\circ-H^\circ(T_r)$	$\Delta_f H^\circ$	$\Delta_f G^\circ$	Log K_f
298.15	65.652	281.622	281.622	0.000	−234.800	−167.874	29.410
300	65.926	282.029	281.623	0.122	−234.897	−167.458	29.157
400	81.169	303.076	284.390	7.474	−239.826	−144.216	18.832
500	95.400	322.750	290.115	16.318	−243.940	−119.820	12.517
600	107.656	341.257	297.112	26.487	−247.260	−94.672	8.242
700	118.129	358.659	304.674	37.790	−249.895	−69.023	5.151
800	127.171	375.038	312.456	50.065	−251.951	−43.038	2.810
900	135.049	390.482	320.276	63.185	−253.515	−16.825	0.976
1000	141.934	405.075	328.033	77.042	−254.662	9.539	−0.498
1100	147.958	418.892	335.670	91.543	−255.454	36.000	−1.709
1200	153.232	431.997	343.156	106.609	−255.947	62.520	−2.721
1300	157.849	444.448	350.473	122.168	−256.184	89.070	−3.579
1400	161.896	456.298	357.612	138.160	−256.206	115.630	−4.314
1500	165.447	467.591	364.571	154.531	−256.044	142.185	−4.951

25. ACETIC ACID   C$_2$H$_4$O$_2$ (g)

T/K	C_p°	S°	$-(G^\circ-H^\circ(T_r))/T$	$H^\circ-H^\circ(T_r)$	$\Delta_f H^\circ$	$\Delta_f G^\circ$	Log K_f
298.15	63.438	283.470	283.470	0.000	−432.249	−374.254	65.567
300	63.739	283.863	283.471	0.118	−432.324	−373.893	65.100
400	79.665	304.404	286.164	7.296	−436.006	−353.840	46.206
500	93.926	323.751	291.765	15.993	−438.875	−332.950	34.783
600	106.181	341.988	298.631	26.014	−440.993	−311.554	27.123
700	116.627	359.162	306.064	37.169	−442.466	−289.856	21.629
800	125.501	375.331	313.722	49.287	−443.395	−267.985	17.497
900	132.989	390.558	321.422	62.223	−443.873	−246.026	14.279
1000	139.257	404.904	329.060	75.844	−443.982	−224.034	11.702
1100	144.462	418.429	336.576	90.039	−443.798	−202.046	9.594
1200	148.760	431.189	343.933	104.707	−443.385	−180.086	7.839
1300	152.302	443.240	351.113	119.765	−442.795	−158.167	6.355
1400	155.220	454.637	358.105	135.146	−442.071	−136.299	5.085
1500	157.631	465.432	364.903	150.793	−441.247	−114.486	3.987

26. ACETONE   C$_3$H$_6$O (g)

T/K	C_p°	S°	$-(G^\circ-H^\circ(T_r))/T$	$H^\circ-H^\circ(T_r)$	$\Delta_f H^\circ$	$\Delta_f G^\circ$	Log K_f
298.15	74.517	295.349	295.349	0.000	−217.150	−152.716	26.757
300	74.810	295.809	295.349	0.138	−217.233	−152.339	26.521
400	91.755	319.658	298.498	8.464	−222.212	−129.913	16.962
500	107.864	341.916	304.988	18.464	−226.522	−106.315	11.107
600	122.047	362.836	312.873	29.978	−230.120	−81.923	7.133
700	134.306	382.627	321.470	42.810	−233.049	−56.986	4.252
800	144.934	401.246	330.265	56.785	−235.350	−31.673	2.068

	J/K·mol			kJ/mol			
T/K	C_p°	S°	$-(G^\circ-H^\circ(T_r))/T$	$H^\circ-H^\circ(T_r)$	$\Delta_f H^\circ$	$\Delta_f G^\circ$	Log K_f

26. ACETONE C_3H_6O (g) (continued)

900	154.097	418.860	339.141	71.747	−237.149	−6.109	0.353
1000	162.046	435.513	347.950	87.563	−238.404	19.707	−1.030
1100	168.908	451.286	356.617	104.136	−239.283	45.396	−2.157
1200	174.891	466.265	365.155	121.332	−239.827	71.463	−3.110
1300	180.079	480.491	373.513	139.072	−240.120	97.362	−3.912
1400	184.556	493.963	381.596	157.314	−240.203	123.470	−4.607
1500	188.447	506.850	389.533	175.975	−240.120	149.369	−5.202

27. PHENOL C_6H_6O (g)

298.15	103.220	314.810	314.810	0.000	−96.400	−32.630	5.720
300	103.860	315.450	314.810	0.192	−96.490	−32.230	5.610
400	135.790	349.820	319.278	12.217	−100.870	−10.180	1.330
500	161.910	383.040	328.736	27.152	−104.240	12.970	−1.360
600	182.480	414.450	340.430	44.412	−106.810	36.650	−3.190
700	198.840	443.860	353.134	63.508	−108.800	60.750	−4.530
800	212.140	471.310	366.211	84.079	−110.300	85.020	−5.550
900	223.190	496.950	379.327	105.861	−111.370	109.590	−6.360
1000	232.490	520.960	392.302	128.658	−111.990	134.280	−7.010
1100	240.410	543.500	405.033	152.314	−112.280	158.620	−7.530
1200	247.200	564.720	417.468	176.703	−112.390	183.350	−7.980
1300	253.060	584.740	429.568	201.723	−112.330	208.070	−8.360
1400	258.120	603.680	441.331	227.288	−112.120	233.050	−8.700
1500	262.520	621.650	452.767	253.325	−111.780	257.540	−8.970

28. CARBON TETRAFLUORIDE CF_4 (g)

298.15	61.050	261.455	261.455	0.000	−933.200	−888.518	155.663
300	61.284	261.833	261.456	0.113	−933.219	−888.240	154.654
400	72.399	281.057	264.001	6.822	−933.986	−873.120	114.016
500	80.713	298.153	269.155	14.499	−934.372	−857.852	89.618
600	86.783	313.434	275.284	22.890	−934.490	−842.533	73.348
700	91.212	327.162	281.732	31.801	−934.431	−827.210	61.726
800	94.479	339.566	288.199	41.094	−934.261	−811.903	53.011
900	96.929	350.842	294.542	50.670	−934.024	−796.622	46.234
1000	98.798	361.156	300.695	60.460	−933.745	−781.369	40.814
1100	100.250	370.643	306.629	70.416	−933.442	−766.146	36.381
1200	101.396	379.417	312.334	80.500	−933.125	−750.952	32.688
1300	102.314	387.571	317.811	90.687	−932.800	−735.784	29.564
1400	103.059	395.181	323.069	100.957	−932.470	−720.641	26.887
1500	103.671	402.313	328.116	111.295	−932.137	−705.522	24.568

29. TRIFLUOROMETHANE CHF_3 (g)

298.15	51.069	259.675	259.675	0.000	−696.700	−662.237	116.020
300	51.258	259.991	259.676	0.095	−696.735	−662.023	115.267
400	61.148	276.113	261.807	5.722	−698.427	−650.186	84.905
500	69.631	290.700	266.149	12.275	−699.715	−637.969	66.647
600	76.453	304.022	271.368	19.593	−700.634	−625.528	54.456
700	81.868	316.230	276.917	27.519	−701.253	−612.957	45.739
800	86.201	327.455	282.542	35.930	−701.636	−600.315	39.196
900	89.719	337.818	288.116	44.732	−701.832	−587.636	34.105
1000	92.617	347.426	293.572	53.854	−701.879	−574.944	30.032
1100	95.038	356.370	298.879	63.240	−701.805	−562.253	26.699
1200	97.084	364.730	304.022	72.849	−701.629	−549.574	23.922
1300	98.833	372.571	308.997	82.647	−701.368	−536.913	21.573
1400	100.344	379.952	313.804	92.607	−701.033	−524.274	19.561
1500	101.660	386.921	318.449	102.709	−700.635	−511.662	17.817

	J/K·mol			kJ/mol			
T/K	C_p°	S°	$-(G^\circ\text{-}H^\circ(T_r))/T$	$H^\circ\text{-}H^\circ(T_r)$	$\Delta_f H^\circ$	$\Delta_f G^\circ$	Log K_f

30. CHLOROTRIFLUOROMETHANE   CClF$_3$ (g)

298.15	66.886	285.419	285.419	0.000	−707.800	−667.238	116.896
300	67.111	285.834	285.421	0.124	−707.810	−666.986	116.131
400	77.528	306.646	288.187	7.383	−708.153	−653.316	85.313
500	85.013	324.797	293.734	15.532	−708.170	−639.599	66.818
600	90.329	340.794	300.271	24.314	−707.975	−625.901	54.489
700	94.132	355.020	307.096	33.547	−707.654	−612.246	45.686
800	96.899	367.780	313.897	43.106	−707.264	−598.642	39.087
900	98.951	379.317	320.536	52.903	−706.837	−585.090	33.957
1000	100.507	389.827	326.947	62.880	−706.396	−571.586	29.856
1100	101.708	399.465	333.108	72.993	−705.950	−558.126	26.503
1200	102.651	408.357	339.013	83.213	−705.505	−544.707	23.710
1300	103.404	416.604	344.668	93.517	−705.064	−531.326	21.349
1400	104.012	424.290	350.084	103.889	−704.628	−517.977	19.326
1500	104.512	431.484	355.273	114.316	−704.196	−504.660	17.574

31. DICHLORODIFLUOROMETHANE   CCl$_2$F$_2$ (g)

298.15	72.476	300.903	300.903	0.000	−486.000	−447.030	78.317
300	72.691	301.352	300.905	0.134	−486.002	−446.788	77.792
400	82.408	323.682	303.883	7.919	−485.945	−433.716	56.637
500	89.063	342.833	309.804	16.514	−485.618	−420.692	43.949
600	93.635	359.500	316.729	25.663	−485.136	−407.751	35.497
700	96.832	374.189	323.909	35.196	−484.576	−394.897	29.467
800	99.121	387.276	331.027	44.999	−483.984	−382.126	24.950
900	100.801	399.053	337.942	55.000	−483.388	−369.429	21.441
1000	102.062	409.742	344.596	65.146	−482.800	−356.799	18.637
1100	103.030	419.517	350.969	75.402	−482.226	−344.227	16.346
1200	103.786	428.515	357.061	85.745	−481.667	−331.706	14.439
1300	104.388	436.847	362.882	96.154	−481.121	−319.232	12.827
1400	104.874	444.602	368.445	106.618	−480.588	−306.799	11.447
1500	105.270	451.851	373.767	117.126	−480.065	−294.404	10.252

32. CHLORODIFLUOROMETHANE   CHClF$_2$ (g)

298.15	55.853	280.915	280.915	0.000	−475.000	−443.845	77.759
300	56.039	281.261	280.916	0.104	−475.028	−443.652	77.246
400	65.395	298.701	283.231	6.188	−476.390	−432.978	56.540
500	73.008	314.145	287.898	13.123	−477.398	−422.001	44.086
600	78.940	328.003	293.448	20.733	−478.103	−410.851	35.767
700	83.551	340.533	299.294	28.867	−478.574	−399.603	29.818
800	87.185	351.936	305.172	37.411	−478.870	−388.299	25.353
900	90.100	362.379	310.956	46.280	−479.031	−376.967	21.878
1000	92.475	371.999	316.586	55.413	−479.090	−365.622	19.098
1100	94.433	380.908	322.033	64.761	−479.068	−354.276	16.823
1200	96.066	389.196	327.289	74.289	−478.982	−342.935	14.927
1300	97.438	396.941	332.352	83.966	−478.843	−331.603	13.324
1400	98.601	404.206	337.228	93.769	−478.661	−320.283	11.950
1500	99.593	411.044	341.923	103.681	−478.443	−308.978	10.759

33. METHYLAMINE   CH$_5$N (g)

298.15	50.053	242.881	242.881	0.000	−22.529	32.734	−5.735
300	50.227	243.196	242.893	0.091	−22.614	33.077	−5.759
400	60.171	258.986	244.975	5.604	−26.846	52.294	−6.829
500	70.057	273.486	249.244	12.121	−30.431	72.510	−7.575
600	78.929	287.063	254.431	19.579	−33.364	93.382	−8.129
700	86.711	299.826	260.008	27.873	−35.712	114.702	−8.559
800	93.545	311.865	265.749	36.893	−37.548	136.316	−8.900

	J/K·mol			kJ/mol			
T/K	C_p°	S°	$-(G^\circ-H^\circ(T_r))/T$	$H^\circ-H^\circ(T_r)$	$\Delta_f H^\circ$	$\Delta_f G^\circ$	Log K_f

33. METHYLAMINE   CH$_5$N (g) (continued)

900	99.573	323.239	271.511	46.555	−38.949	158.138	−9.178
1000	104.886	334.006	277.220	56.786	−39.967	180.098	−9.407
1100	109.576	344.233	282.861	67.509	−40.681	201.822	−9.584
1200	113.708	353.944	288.374	78.685	−41.136	224.240	−9.761
1300	117.341	363.190	293.775	90.239	−41.376	246.364	−9.899
1400	120.542	372.012	299.061	102.131	−41.451	268.504	−10.018
1500	123.353	380.426	304.209	114.326	−41.381	290.639	−10.121

34. CHLORINE   Cl (g)

298.15	21.838	165.190	165.190	0.000	121.302	105.306	−18.449
300	21.852	165.325	165.190	0.040	121.311	105.207	−18.318
400	22.467	171.703	166.055	2.259	121.795	99.766	−13.028
500	22.744	176.752	167.708	4.522	122.272	94.203	−9.841
600	22.781	180.905	169.571	6.800	122.734	88.546	−7.709
700	22.692	184.411	171.448	9.074	123.172	82.813	−6.179
800	22.549	187.432	173.261	11.337	123.585	77.019	−5.029
900	22.389	190.079	174.986	13.584	123.971	71.175	−4.131
1000	22.233	192.430	176.615	15.815	124.334	65.289	−3.410
1100	22.089	194.542	178.150	18.031	124.675	59.368	−2.819
1200	21.959	196.458	179.597	20.233	124.996	53.416	−2.325
1300	21.843	198.211	180.963	22.423	125.299	47.439	−1.906
1400	21.742	199.826	182.253	24.602	125.587	41.439	−1.546
1500	21.652	201.323	183.475	26.772	125.861	35.418	−1.233

35. DICHLORINE   Cl$_2$ (g)

298.15	33.949	223.079	223.079	0.000	0.000	0.000	0.000
300	33.981	223.290	223.080	0.063	0.000	0.000	0.000
400	35.296	233.263	224.431	3.533	0.000	0.000	0.000
500	36.064	241.229	227.021	7.104	0.000	0.000	0.000
600	36.547	247.850	229.956	10.736	0.000	0.000	0.000
700	36.874	253.510	232.926	14.408	0.000	0.000	0.000
800	37.111	258.450	235.815	18.108	0.000	0.000	0.000
900	37.294	262.832	238.578	21.829	0.000	0.000	0.000
1000	37.442	266.769	241.203	25.566	0.000	0.000	0.000
1100	37.567	270.343	243.692	29.316	0.000	0.000	0.000
1200	37.678	273.617	246.052	33.079	0.000	0.000	0.000
1300	37.778	276.637	248.290	36.851	0.000	0.000	0.000
1400	37.872	279.440	250.416	40.634	0.000	0.000	0.000
1500	37.961	282.056	252.439	44.426	0.000	0.000	0.000

36. HYDROGEN CHLORIDE   HCl (g)

298.15	29.136	186.902	186.902	0.000	−92.310	−95.298	16.696
300	29.137	187.082	186.902	0.054	−92.314	−95.317	16.596
400	29.175	195.468	188.045	2.969	−92.587	−96.278	12.573
500	29.304	201.990	190.206	5.892	−92.911	−97.164	10.151
600	29.576	207.354	192.630	8.835	−93.249	−97.983	8.530
700	29.988	211.943	195.069	11.812	−93.577	−98.746	7.368
800	30.500	215.980	197.435	14.836	−93.879	−99.464	6.494
900	31.063	219.604	199.700	17.913	−94.149	−100.145	5.812
1000	31.639	222.907	201.858	21.049	−94.384	−100.798	5.265
1100	32.201	225.949	203.912	24.241	−94.587	−101.430	4.816
1200	32.734	228.774	205.867	27.488	−94.760	−102.044	4.442
1300	33.229	231.414	207.732	30.786	−94.908	−102.645	4.124
1400	33.684	233.893	209.513	34.132	−95.035	−103.235	3.852
1500	34.100	236.232	211.217	37.522	−95.146	103.817	3.615

	J/K·mol			kJ/mol			
T/K	C_p°	S°	$-(G^\circ-H^\circ(T_r))/T$	$H^\circ-H^\circ(T_r)$	$\Delta_f H^\circ$	$\Delta_f G^\circ$	Log K_f

37. COPPER   Cu (cr, l)

298.15	24.440	33.150	33.150	0.000	0.000	0.000	0.000
300	24.460	33.301	33.150	0.045	0.000	0.000	0.000
400	25.339	40.467	34.122	2.538	0.000	0.000	0.000
500	25.966	46.192	35.982	5.105	0.000	0.000	0.000
600	26.479	50.973	38.093	7.728	0.000	0.000	0.000
700	26.953	55.090	40.234	10.399	0.000	0.000	0.000
800	27.448	58.721	42.322	13.119	0.000	0.000	0.000
900	28.014	61.986	44.328	15.891	0.000	0.000	0.000
1000	28.700	64.971	46.245	18.726	0.000	0.000	0.000
1100	29.553	67.745	48.075	21.637	0.000	0.000	0.000
1200	30.617	70.361	49.824	24.644	0.000	0.000	0.000
1300	31.940	72.862	51.501	27.769	0.000	0.000	0.000
1358	32.844	74.275	52.443	29.647	0.000	0.000	0.000

PHASE TRANSITION: $\Delta_{trs} H$ = 13.141 kJ/mol, $\Delta_{trs} S$ = 9.676 J/K·mol, cr–l

1358	32.800	83.951	52.443	42.788	0.000	0.000	0.000
1400	32.800	84.950	53.403	44.166	0.000	0.000	0.000
1500	32.800	87.213	55.583	47.446	0.000	0.000	0.000

38. COPPER   Cu (g)

298.15	20.786	166.397	166.397	0.000	337.600	297.873	−52.185
300	20.786	166.525	166.397	0.038	337.594	297.626	−51.821
400	20.786	172.505	167.213	2.117	337.179	284.364	−37.134
500	20.786	177.143	168.752	4.196	336.691	271.215	−28.333
600	20.786	180.933	170.476	6.274	336.147	258.170	−22.475
700	20.786	184.137	172.205	8.353	335.554	245.221	−18.298
800	20.786	186.913	173.874	10.431	334.913	232.359	−15.171
900	20.786	189.361	175.461	12.510	334.219	219.581	−12.744
1000	20.786	191.551	176.963	14.589	333.463	206.883	−10.806
1100	20.788	193.532	178.380	16.667	332.631	194.265	−9.225
1200	20.793	195.341	179.719	18.746	331.703	181.726	−7.910
1300	20.803	197.006	180.986	20.826	330.657	169.270	−6.801
1400	20.823	198.548	182.186	22.907	316.342	157.305	−5.869
1500	20.856	199.986	183.325	24.991	315.146	145.987	−5.084

39. COPPER OXIDE   CuO (cr)

298.15	42.300	42.740	42.740	0.000	−162.000	−134.277	23.524
300	42.417	43.002	42.741	0.078	−161.994	−134.105	23.349
400	46.783	55.878	44.467	4.564	−161.487	−124.876	16.307
500	49.190	66.596	47.852	9.372	−160.775	−115.803	12.098
600	50.827	75.717	51.755	14.377	−159.973	−106.883	9.305
700	52.099	83.651	55.757	19.526	−159.124	−98.102	7.320
800	53.178	90.680	59.691	24.791	−158.247	−89.444	5.840
900	54.144	97.000	63.491	30.158	−157.356	−80.897	4.695
1000	55.040	102.751	67.134	35.617	−156.462	−72.450	3.784
1100	55.890	108.037	70.615	41.164	−155.582	−64.091	3.043
1200	56.709	112.936	73.941	46.794	−154.733	−55.812	2.429
1300	57.507	117.507	77.118	52.505	−153.940	−47.601	1.913
1400	58.288	121.797	80.158	58.295	−166.354	−39.043	1.457
1500	59.057	125.845	83.070	64.163	−165.589	−29.975	1.044

40. DICOPPER OXIDE   Cu₂O (cr)

298.15	62.600	92.550	92.550	0.000	−173.100	−150.344	26.339
300	62.721	92.938	92.551	0.116	−173.102	−150.203	26.152
400	67.587	111.712	95.078	6.654	−173.036	−142.572	18.618
500	70.784	127.155	99.995	13.580	−172.772	−134.984	14.101

T/K	J/K·mol			kJ/mol			
	C_p°	S°	$-(G^\circ-H^\circ(T_r))/T$	$H^\circ-H^\circ(T_r)$	$\Delta_f H^\circ$	$\Delta_f G^\circ$	Log K_f

40. DICOPPER OXIDE Cu_2O (cr) (continued)

T/K	C_p°	S°	$-(G^\circ-H^\circ(T_r))/T$	$H^\circ-H^\circ(T_r)$	$\Delta_f H^\circ$	$\Delta_f G^\circ$	Log K_f
600	73.323	140.291	105.643	20.789	−172.389	−127.460	11.096
700	75.552	151.764	111.429	28.235	−171.914	−120.009	8.955
800	77.616	161.989	117.121	35.894	−171.363	−112.631	7.354
900	79.584	171.245	122.629	43.755	−170.750	−105.325	6.113
1000	81.492	179.729	127.920	51.809	−170.097	−98.091	5.124
1100	83.360	187.584	132.992	60.052	−169.431	−90.922	4.317
1200	85.202	194.917	137.850	68.480	−168.791	−83.814	3.648
1300	87.026	201.808	142.507	77.092	−168.223	−76.756	3.084
1400	88.836	208.324	146.978	85.885	−194.030	−68.926	2.572
1500	90.636	214.515	151.276	94.858	−193.438	−60.010	2.090

41. COPPER DICHLORIDE $CuCl_2$ (cr, l)

T/K	C_p°	S°	$-(G^\circ-H^\circ(T_r))/T$	$H^\circ-H^\circ(T_r)$	$\Delta_f H^\circ$	$\Delta_f G^\circ$	Log K_f
298.15	71.880	108.070	108.070	0.000	−218.000	−173.826	30.453
300	71.998	108.515	108.071	0.133	−217.975	−173.552	30.218
400	76.338	129.899	110.957	7.577	−216.494	−158.962	20.758
500	78.654	147.204	116.532	15.336	−214.873	−144.765	15.123
600	80.175	161.687	122.884	23.282	−213.182	−130.901	11.396
675	81.056	171.183	127.732	29.329	−211.185	−120.693	9.340
PHASE TRANSITION: $\Delta_{trs} H = 0.700$ kJ/mol, $\Delta_{trs} S = 1.037$ J/K·mol, crII–crI							
675	82.400	172.220	127.732	30.029	−211.185	−120.693	9.340
700	82.400	175.216	129.375	32.089	−210.719	−117.350	8.757
800	82.400	186.219	135.808	40.329	−208.898	−104.137	6.799
871	82.400	193.226	140.207	46.179	−192.649	−94.893	5.691
PHASE TRANSITION: $\Delta_{trs} H = 15.001$ kJ/mol, $\Delta_{trs} S = 17.221$ J/K·mol, crI–l							
871	100.000	210.447	140.207	61.180	−192.649	−94.893	5.691
900	100.000	213.723	142.523	64.080	−191.640	−91.655	5.319
1000	100.000	224.259	150.179	74.080	−188.212	−80.730	4.217
1100	100.000	233.790	157.353	84.080	−184.873	−70.144	3.331
1130.75	100.000	236.547	159.470	87.155	−183.867	−66.951	3.093

42. COPPER DICHLORIDE $CuCl_2$ (g)

T/K	C_p°	S°	$-(G^\circ-H^\circ(T_r))/T$	$H^\circ-H^\circ(T_r)$	$\Delta_f H^\circ$	$\Delta_f G^\circ$	Log K_f
298.15	56.814	278.418	278.418	0.000	−43.268	−49.883	8.739
300	56.869	278.769	278.419	0.105	−43.271	−49.924	8.692
400	58.992	295.456	280.679	5.911	−43.428	−52.119	6.806
500	60.111	308.752	285.010	11.871	−43.606	−54.271	5.670
600	60.761	319.774	289.911	17.918	−43.814	−56.385	4.909
700	61.168	329.173	294.865	24.015	−44.060	−58.462	4.362
800	61.439	337.360	299.677	30.147	−44.349	−60.500	3.950
900	61.630	344.608	304.274	36.301	−44.688	−62.499	3.627
1000	61.776	351.109	308.638	42.471	−45.088	−64.457	3.367
1100	61.900	357.003	312.771	48.655	−45.566	−66.372	3.152
1200	62.022	362.394	316.685	54.851	−46.139	−68.239	2.970
1300	62.159	367.364	320.395	61.060	−46.829	−70.053	2.815
1400	62.325	371.976	323.916	67.284	−60.784	−71.404	2.664
1500	62.531	376.283	327.265	73.526	−61.613	−72.133	2.512

43. FLUORINE   F (g)

T/K	C_p°	S°	$-(G^\circ-H^\circ(T_r))/T$	$H^\circ-H^\circ(T_r)$	$\Delta_f H^\circ$	$\Delta_f G^\circ$	Log K_f
298.15	22.746	158.750	158.750	0.000	79.380	62.280	−10.911
300	22.742	158.891	158.750	0.042	79.393	62.173	−10.825
400	22.432	165.394	159.639	2.302	80.043	56.332	−7.356
500	22.100	170.363	161.307	4.528	80.587	50.340	−5.259
600	21.832	174.368	163.161	6.724	81.046	44.246	−3.852
700	21.629	177.717	165.008	8.897	81.442	38.081	−2.842
800	21.475	180.595	166.780	11.052	81.792	31.862	−2.080
900	21.357	183.117	168.458	13.193	82.106	25.601	−1.486

		J/K·mol			kJ/mol		
T/K	C_p°	S°	$-(G^\circ-H^\circ(T_r))/T$	$H^\circ-H^\circ(T_r)$	$\Delta_f H^\circ$	$\Delta_f G^\circ$	Log K_f

43. FLUORINE   F (g) (continued)

1000	21.266	185.362	170.039	15.324	82.391	19.308	−1.009
1100	21.194	187.386	171.525	17.447	82.654	12.986	−0.617
1200	21.137	189.227	172.925	19.563	82.897	6.642	−0.289
1300	21.091	190.917	174.245	21.675	83.123	0.278	−0.011
1400	21.054	192.479	175.492	23.782	83.335	−6.103	0.228
1500	21.022	193.930	176.673	25.886	83.533	−12.498	0.435

44. DIFLUORINE F_2 (g)

298.15	31.304	202.790	202.790	0.000	0.000	0.000	0.000
300	31.337	202.984	202.790	0.058	0.000	0.000	0.000
400	32.995	212.233	204.040	3.277	0.000	0.000	0.000
500	34.258	219.739	206.453	6.643	0.000	0.000	0.000
600	35.171	226.070	209.208	10.117	0.000	0.000	0.000
700	35.839	231.545	212.017	13.669	0.000	0.000	0.000
800	36.343	236.365	214.765	17.279	0.000	0.000	0.000
900	36.740	240.669	217.409	20.934	0.000	0.000	0.000
1000	37.065	244.557	219.932	24.625	0.000	0.000	0.000
1100	37.342	248.103	222.334	28.346	0.000	0.000	0.000
1200	37.588	251.363	224.619	32.093	0.000	0.000	0.000
1300	37.811	254.381	226.794	35.863	0.000	0.000	0.000
1400	38.019	257.191	228.866	39.654	0.000	0.000	0.000
1500	38.214	259.820	230.843	43.466	0.000	0.000	0.000

45. HYDROGEN FLUORIDE   HF (g)

298.15	29.137	173.776	173.776	0.000	−273.300	−275.399	48.248
300	29.137	173.956	173.776	0.054	−273.302	−275.412	47.953
400	29.149	182.340	174.919	2.968	−273.450	−276.096	36.054
500	29.172	188.846	177.078	5.884	−273.679	−276.733	28.910
600	29.230	194.169	179.496	8.804	−273.961	−277.318	24.142
700	29.350	198.683	181.923	11.732	−274.277	−277.852	20.733
800	29.549	202.614	184.269	14.676	−274.614	−278.340	18.174
900	29.827	206.110	186.505	17.645	−274.961	−278.785	16.180
1000	30.169	209.270	188.626	20.644	−275.309	−279.191	14.583
1100	30.558	212.163	190.636	23.680	−275.652	−279.563	13.275
1200	30.974	214.840	192.543	26.756	−275.988	−279.904	12.184
1300	31.403	217.336	194.355	29.875	−276.315	−280.217	11.259
1400	31.831	219.679	196.081	33.037	−276.631	−280.505	10.466
1500	32.250	221.889	197.729	36.241	−276.937	−280.771	9.777

46. GERMANIUM   Ge (cr, l)

298.15	23.222	31.090	31.090	0.000	0.000	0.000	0.000
300	23.249	31.234	31.090	0.043	0.000	0.000	0.000
400	24.310	38.083	32.017	2.426	0.000	0.000	0.000
500	24.962	43.582	33.798	4.892	0.000	0.000	0.000
600	25.452	48.178	35.822	7.414	0.000	0.000	0.000
700	25.867	52.133	37.876	9.980	0.000	0.000	0.000
800	26.240	55.612	39.880	12.586	0.000	0.000	0.000
900	26.591	58.723	41.804	15.227	0.000	0.000	0.000
1000	26.926	61.542	43.639	17.903	0.000	0.000	0.000
1100	27.252	64.124	45.386	20.612	0.000	0.000	0.000
1200	27.571	66.509	47.048	23.353	0.000	0.000	0.000
1211.4	27.608	66.770	47.232	23.668	0.000	0.000	0.000
PHASE TRANSITION: $\Delta_{trs} H = 37.030$ kJ/mol, $\Delta_{trs} S = 30.568$ J/K·mol, cr–l							
1211.4	27.600	97.338	47.232	60.698	0.000	0.000	0.000
1300	27.600	99.286	50.714	63.143	0.000	0.000	0.000

T/K	C_p°	S°	$-(G^\circ-H^\circ(T_r))/T$	$H^\circ-H^\circ(T_r)$	$\Delta_f H^\circ$	$\Delta_f G^\circ$	Log K_f
		J/K·mol			kJ/mol		

46. GERMANIUM   Ge (cr, l) (continued)

T/K	C_p°	S°	$-(G^\circ-H^\circ(T_r))/T$	$H^\circ-H^\circ(T_r)$	$\Delta_f H^\circ$	$\Delta_f G^\circ$	Log K_f
1400	27.600	101.331	54.258	65.903	0.000	0.000	0.000
1500	27.600	103.236	57.460	68.663	0.000	0.000	0.000

47. GERMANIUM   Ge (g)

T/K	C_p°	S°	$-(G^\circ-H^\circ(T_r))/T$	$H^\circ-H^\circ(T_r)$	$\Delta_f H^\circ$	$\Delta_f G^\circ$	Log K_f
298.15	30.733	167.903	167.903	0.000	367.800	327.009	−57.290
300	30.757	168.094	167.904	0.057	367.814	326.756	−56.893
400	31.071	177.025	169.119	3.162	368.536	312.959	−40.868
500	30.360	183.893	171.415	6.239	369.147	298.991	−31.235
600	29.265	189.334	173.965	9.222	369.608	284.914	−24.804
700	28.102	193.758	176.487	12.090	369.910	270.773	−20.205
800	27.029	197.439	178.882	14.845	370.060	256.598	−16.754
900	26.108	200.567	181.122	17.501	370.073	242.414	−14.069
1000	25.349	203.277	183.205	20.072	369.969	228.234	−11.922
1100	24.741	205.664	185.141	22.575	369.763	214.069	−10.165
1200	24.264	207.795	186.941	25.025	369.471	199.928	−8.703
1300	23.898	209.722	188.621	27.432	332.088	188.521	−7.575
1400	23.624	211.483	190.192	29.807	331.704	177.492	−6.622
1500	23.426	213.105	191.666	32.159	331.296	166.491	−5.798

48. GERMANIUM DIOXIDE   GeO$_2$ (cr, l)

T/K	C_p°	S°	$-(G^\circ-H^\circ(T_r))/T$	$H^\circ-H^\circ(T_r)$	$\Delta_f H^\circ$	$\Delta_f G^\circ$	Log K_f
298.15	50.166	39.710	39.710	0.000	−580.200	−521.605	91.382
300	50.475	40.021	39.711	0.093	−580.204	−521.242	90.755
400	61.281	56.248	41.850	5.759	−579.893	−501.610	65.503
500	66.273	70.519	46.191	12.164	579.013	−482.134	50.368
600	69.089	82.872	51.299	18.943	−577.915	−462.859	40.295
700	70.974	93.671	56.597	25.952	−576.729	−443.776	33.115
800	72.449	103.247	61.841	33.125	−575.498	−424.866	27.741
900	73.764	111.857	66.928	40.436	−574.235	−406.113	23.570
1000	75.049	119.696	71.819	47.877	−572.934	−387.502	20.241
1100	76.378	126.910	76.504	55.447	−571.582	−369.024	17.523
1200	77.796	133.616	80.987	63.155	−570.166	−350.671	15.264
1300	79.332	139.903	85.279	71.010	−605.685	−329.732	13.249
1308	79.460	140.390	85.615	71.646	−584.059	−328.034	13.100

PHASE TRANSITION: $\Delta_{trs} H$ = 21.500 kJ/mol, $\Delta_{trs} S$ = 16.437 J/K·mol, crII–crI

T/K	C_p°	S°	$-(G^\circ-H^\circ(T_r))/T$	$H^\circ-H^\circ(T_r)$	$\Delta_f H^\circ$	$\Delta_f G^\circ$	Log K_f
1308	80.075	156.827	85.615	93.146	−584.059	−328.034	13.100
1388	81.297	161.617	89.858	99.601	−565.504	−312.415	11.757

PHASE TRANSITION: $\Delta_{trs} H$ = 17.200 kJ/mol, $\Delta_{trs} S$ = 12.392 J/K·mol, crI–l

T/K	C_p°	S°	$-(G^\circ-H^\circ(T_r))/T$	$H^\circ-H^\circ(T_r)$	$\Delta_f H^\circ$	$\Delta_f G^\circ$	Log K_f
1388	78.500	174.009	89.858	116.801	−565.504	−312.415	11.757
1400	78.500	174.685	90.582	117.743	−565.328	−310.228	11.575
1500	78.500	180.100	96.372	125.593	−563.882	−292.057	10.170

49. GERMANIUM TETRACHLORIDE   GeCl$_4$ (g)

T/K	C_p°	S°	$-(G^\circ-H^\circ(T_r))/T$	$H^\circ-H^\circ(T_r)$	$\Delta_f H^\circ$	$\Delta_f G^\circ$	Log K_f
298.15	95.918	348.393	348.393	0.000	−500.000	−461.582	80.866
300	96.041	348.987	348.395	0.178	−499.991	−461.343	80.326
400	100.750	377.342	352.229	10.045	−499.447	−448.540	58.573
500	103.206	400.114	359.604	20.255	−498.845	−435.882	45.536
600	104.624	419.067	367.980	30.652	−498.234	−423.347	36.855
700	105.509	435.266	376.463	41.162	−497.634	−410.914	30.662
800	106.096	449.396	384.715	51.744	−497.057	−398.565	26.023
900	106.504	461.917	392.611	62.375	−496.509	−386.287	22.419
1000	106.799	473.155	400.113	73.041	−495.993	−374.068	19.539
1100	107.020	483.344	407.224	83.733	−495.512	−361.899	17.185
1200	107.189	492.664	413.961	94.444	−495.067	−349.772	15.225
1300	107.320	501.249	420.349	105.169	−531.677	−334.973	13.459
1400	107.425	509.206	426.416	115.907	−531.265	−319.857	11.934
1500	107.509	516.621	432.185	126.654	−530.861	−304.771	10.613

	J/K·mol			kJ/mol			
T/K	C_p°	S°	$-(G^\circ-H^\circ(T_r))/T$	$H^\circ-H^\circ(T_r)$	$\Delta_f H^\circ$	$\Delta_f G^\circ$	Log K_f

50. HYDROGEN   H (g)

298.15	20.786	114.716	114.716	0.000	217.998	203.276	−35.613
300	20.786	114.845	114.716	0.038	218.010	203.185	−35.377
400	20.786	120.824	115.532	2.117	218.635	198.149	−25.875
500	20.786	125.463	117.071	4.196	219.253	192.956	−20.158
600	20.786	129.252	118.795	6.274	219.867	187.639	−16.335
700	20.786	132.457	120.524	8.353	220.476	182.219	−13.597
800	20.786	135.232	122.193	10.431	221.079	176.712	−11.538
900	20.786	137.680	123.780	12.510	221.670	171.131	−9.932
1000	20.786	139.870	125.282	14.589	222.247	165.485	−8.644
1100	20.786	141.852	126.700	16.667	222.806	159.781	−7.587
1200	20.786	143.660	128.039	18.746	223.345	154.028	−6.705
1300	20.786	145.324	129.305	20.824	223.864	148.230	−5.956
1400	20.786	146.864	130.505	22.903	224.360	142.393	−5.313
1500	20.786	148.298	131.644	24.982	224.835	136.522	−4.754

51. DIHYDROGEN   H₂ (g)

298.15	28.836	130.680	130.680	0.000	0.000	0.000	0.000
300	28.849	130.858	130.680	0.053	0.000	0.000	0.000
400	29.181	139.217	131.818	2.960	0.000	0.000	0.000
500	29.260	145.738	133.974	5.882	0.000	0.000	0.000
600	29.327	151.078	136.393	8.811	0.000	0.000	0.000
700	29.440	155.607	138.822	11.749	0.000	0.000	0.000
800	29.623	159.549	141.172	14.702	0.000	0.000	0.000
900	29.880	163.052	143.412	17.676	0.000	0.000	0.000
1000	30.204	166.217	145.537	20.680	0.000	0.000	0.000
1100	30.580	169.113	147.550	23.719	0.000	0.000	0.000
1200	30.991	171.791	149.460	26.797	0.000	0.000	0.000
1300	31.422	174.288	151.275	29.918	0.000	0.000	0.000
1400	31.860	176.633	153.003	33.082	0.000	0.000	0.000
1500	32.296	178.846	154.653	36.290	0.000	0.000	0.000

52. HYDROXYL   OH (g)

298.15	29.886	183.737	183.737	0.000	39.349	34.631	−6.067
300	29.879	183.922	183.738	0.055	39.350	34.602	−6.025
400	29.604	192.476	184.906	3.028	39.384	33.012	−4.311
500	29.495	199.067	187.104	5.982	39.347	31.422	−3.283
600	29.513	204.445	189.560	8.931	39.252	29.845	−2.598
700	29.655	209.003	192.020	11.888	39.113	28.287	−2.111
800	29.914	212.979	194.396	14.866	38.945	26.752	−1.747
900	30.265	216.522	196.661	17.874	38.763	25.239	−1.465
1000	30.682	219.731	198.810	20.921	38.577	23.746	−1.240
1100	31.135	222.677	200.848	24.012	38.393	22.272	−1.058
1200	31.603	225.406	202.782	27.149	38.215	20.814	−0.906
1300	32.069	227.954	204.621	30.332	38.046	19.371	−0.778
1400	32.522	230.347	206.374	33.562	37.886	17.941	−0.669
1500	32.956	232.606	208.048	36.836	37.735	16.521	−0.575

53. WATER   H₂O (l)

298.15	75.300	69.950	69.950	0.000	−285.830	−237.141	41.546
300	75.281	70.416	69.951	0.139	−285.771	−236.839	41.237
373.21	76.079	86.896	71.715	5.666	−283.454	−225.160	31.513

54. WATER   H₂O (g)

298.15	33.598	188.832	188.832	0.000	−241.826	−228.582	40.046

| T/K | J/K·mol | | | kJ/mol | | | |
	C_p°	S°	$-(G^\circ - H^\circ(T_r))/T$	$H^\circ - H^\circ(T_r)$	$\Delta_f H^\circ$	$\Delta_f G^\circ$	Log K_f
54. WATER	**H_2O (g) (continued)**						
300	33.606	189.040	188.833	0.062	−241.844	−228.500	39.785
400	34.283	198.791	190.158	3.453	−242.845	−223.900	29.238
500	35.259	206.542	192.685	6.929	−243.822	−219.050	22.884
600	36.371	213.067	195.552	10.509	−244.751	−214.008	18.631
700	37.557	218.762	198.469	14.205	−245.620	−208.814	15.582
800	38.800	223.858	201.329	18.023	−246.424	−203.501	13.287
900	40.084	228.501	204.094	21.966	−247.158	−198.091	11.497
1000	41.385	232.792	206.752	26.040	−247.820	−192.603	10.060
1100	42.675	236.797	209.303	30.243	−248.410	−187.052	8.882
1200	43.932	240.565	211.753	34.574	−248.933	−181.450	7.898
1300	45.138	244.129	214.108	39.028	−249.392	−175.807	7.064
1400	46.281	247.516	216.374	43.599	−249.792	−170.132	6.348
1500	47.356	250.746	218.559	48.282	−250.139	−164.429	5.726
55. IODINE	**I (g)**						
298.15	20.786	180.787	180.787	0.000	106.760	70.172	−12.294
300	20.786	180.915	180.787	0.038	106.748	69.945	−12.178
400	20.786	186.895	181.602	2.117	97.974	58.060	−7.582
500	20.786	191.533	183.142	4.196	75.988	50.202	−5.244
600	20.786	195.323	184.866	6.274	76.190	45.025	−3.920
700	20.786	198.527	186.594	8.353	76.385	39.816	−2.971
800	20.787	201.303	188.263	10.432	76.574	34.579	−2.258
900	20.789	203.751	189.851	12.510	76.757	29.319	−1.702
1000	20.795	205.942	191.352	14.589	76.936	24.038	−1.256
1100	20.806	207.924	192.770	16.669	77.109	18.740	−0.890
1200	20.824	209.735	194.110	18.751	77.277	13.426	−0.584
1300	20.851	211.403	195.377	20.835	77.440	8.098	−0.325
1400	20.889	212.950	196.577	22.921	77.596	2.758	−0.103
1500	20.936	214.392	197.717	25.013	77.745	−2.592	0.090
56. DIIODINE	**I_2 (cr, l)**						
298.15	54.440	116.139	116.139	0.000	0.000	0.000	0.000
300	54.518	116.476	116.140	0.101	0.000	0.000	0.000
386.75	61.531	131.039	117.884	5.088	0.000	0.000	0.000
PHASE TRANSITION: $\Delta_{trs} H$ = 15.665 kJ/mol, $\Delta_{trs} S$ = 40.504 J/K·mol, cr–l							
386.75	79.555	171.543	117.884	20.753	0.000	0.000	0.000
400	79.555	174.223	119.706	21.807	0.000	0.000	0.000
457.67	79.555	184.938	127.266	26.395	0.000	0.000	0.000
57. DIIODINE	**I_2 (g)**						
298.15	36.887	260.685	260.685	0.000	62.420	19.324	−3.385
300	36.897	260.913	260.685	0.068	62.387	19.056	−3.318
400	37.256	271.584	262.138	3.778	44.391	5.447	−0.711
457.67	37.385	276.610	263.652	5.931	pressure = 1 bar		
500	37.464	279.921	264.891	7.515	0.000	0.000	0.000
600	37.613	286.765	267.983	11.269	0.000	0.000	0.000
700	37.735	292.573	271.092	15.037	0.000	0.000	0.000
800	37.847	297.619	274.099	18.816	0.000	0.000	0.000
900	37.956	302.083	276.965	22.606	0.000	0.000	0.000
1000	38.070	306.088	279.681	26.407	0.000	0.000	0.000
1100	38.196	309.722	282.249	30.220	0.000	0.000	0.000
1200	38.341	313.052	284.679	34.047	0.000	0.000	0.000
1300	38.514	316.127	286.981	37.890	0.000	0.000	0.000
1400	38.719	318.989	289.166	41.751	0.000	0.000	0.000
1500	38.959	321.668	291.245	45.635	0.000	0.000	0.000

	J/K·mol			kJ/mol			
T/K	C_p°	S°	$-(G^\circ-H^\circ(T_r))/T$	$H^\circ-H^\circ(T_r)$	$\Delta_f H^\circ$	$\Delta_f G^\circ$	Log K_f
58. HYDROGEN IODIDE		**HI (g)**					
298.15	29.157	206.589	206.589	0.000	26.500	1.700	−0.298
300	29.158	206.769	206.589	0.054	26.477	1.546	−0.269
400	29.329	215.176	207.734	2.977	17.093	−6.289	0.821
500	29.738	221.760	209.904	5.928	−5.481	−9.946	1.039
600	30.351	227.233	212.348	8.931	−5.819	−10.806	0.941
700	31.070	231.965	214.820	12.002	−6.101	−11.614	0.867
800	31.807	236.162	217.230	15.145	−6.323	−12.386	0.809
900	32.511	239.950	219.548	18.362	−6.489	−13.133	0.762
1000	33.156	243.409	221.763	21.646	−6.608	−13.865	0.724
1100	33.735	246.597	223.878	24.991	−6.689	−14.586	0.693
1200	34.249	249.555	225.896	28.391	−6.741	−15.302	0.666
1300	34.703	252.314	227.823	31.839	−6.775	−16.014	0.643
1400	35.106	254.901	229.666	35.330	−6.797	−16.723	0.624
1500	35.463	257.336	231.430	38.858	−6.814	−17.432	0.607
59. POTASSIUM		**K (cr, l)**					
298.15	29.600	64.680	64.680	0.000	0.000	0.000	0.000
300	29.671	64.863	64.681	0.055	0.000	0.000	0.000
336.86	32.130	68.422	64.896	1.188	0.000	0.000	0.000
PHASE TRANSITION: $\Delta_{trs} H$ = 2.321 kJ/mol, $\Delta_{trs} S$ = 6.891 J/K·mol, cr–l							
336.86	32.129	75.313	64.896	3.509	0.000	0.000	0.000
400	31.552	80.784	66.986	5.519	0.000	0.000	0.000
500	30.741	87.734	70.469	8.632	0.000	0.000	0.000
600	30.158	93.283	73.824	11.675	0.000	0.000	0.000
700	29.851	97.905	76.943	14.673	0.000	0.000	0.000
800	29.838	101.887	79.818	17.655	0.000	0.000	0.000
900	30.130	105.415	82.470	20.651	0.000	0.000	0.000
1000	30.730	108.618	84.927	23.691	0.000	0.000	0.000
1039.4	31.053	109.812	85.847	24.908	0.000	0.000	0.000
60. POTASSIUM		**K (g)**					
298.15	20.786	160.340	160.340	0.000	89.000	60.479	−10.596
300	20.786	160.468	160.340	0.038	88.984	60.302	−10.499
400	20.786	166.448	161.155	2.117	85.598	51.332	−6.703
500	20.786	171.086	162.695	4.196	84.563	42.887	−4.480
600	20.786	174.876	164.419	6.274	83.599	34.643	−3.016
700	20.786	178.080	166.148	8.353	82.680	26.557	−1.982
800	20.786	180.856	167.817	10.431	81.776	18.601	−1.215
900	20.786	183.304	169.404	12.510	80.859	10.759	−0.624
1000	20.786	185.494	170.905	14.589	79.897	3.021	−0.158
1039.4	20.786	186.297	171.474	15.408		pressure = 1 bar	
1100	20.786	187.475	172.323	16.667	0.000	0.000	0.000
1200	20.786	189.284	173.662	18.746	0.000	0.000	0.000
1300	20.789	190.948	174.929	20.825	0.000	0.000	0.000
1400	20.793	192.489	176.129	22.904	0.000	0.000	0.000
1500	20.801	193.923	177.268	24.983	0.000	0.000	0.000
61. DIPOTASSIUM OXIDE		**K$_2$O (cr, l)**					
298.15	72.000	96.000	96.000	0.000	−361.700	−321.171	56.267
300	72.130	96.446	96.001	0.133	−361.704	−320.920	55.876
400	79.154	118.158	98.914	7.698	−366.554	−306.416	40.013
500	86.178	136.575	104.647	15.964	−366.043	−291.423	30.444
590	92.500	151.348	110.662	24.005	−364.204	−278.079	24.619
PHASE TRANSITION: $\Delta_{trs} H$ = 0.700 kJ/mol, $\Delta_{trs} S$ = 1.186 J/K·mol, crIII–crII							

	J/K·mol			kJ/mol			
T/K	C_p°	S°	$-(G^\circ - H^\circ(T_r))/T$	$H^\circ - H^\circ(T_r)$	$\Delta_f H^\circ$	$\Delta_f G^\circ$	Log K_f

61. DIPOTASSIUM OXIDE   K₂O (cr, l) (continued)

590	100.000	152.534	110.662	24.705	−364.204	−278.079	24.619
600	100.000	154.215	111.374	25.705	−363.968	−276.621	24.082
645	100.000	161.447	114.618	30.205	−358.901	−270.109	21.874
PHASE TRANSITION: $\Delta_{trs} H$ = 4.000 kJ/mol, $\Delta_{trs} S$ = 6.202 J/K·mol, crII–crI							
645	100.000	167.649	114.618	34.205	−358.901	−270.109	21.874
700	100.000	175.832	119.111	39.705	−357.592	−262.592	19.595
800	100.000	189.185	127.054	49.705	−355.224	−249.183	16.270
900	100.000	200.963	134.625	59.705	−352.919	−236.067	13.701
1000	100.000	211.499	141.794	69.705	−350.732	−223.202	11.659
1013	100.000	212.791	142.697	71.005	−323.459	−221.546	11.424
PHASE TRANSITION: $\Delta_{trs} H$ = 27.000 kJ/mol, $\Delta_{trs} S$ = 26.654 J/K·mol, crI–l							
1013	100.000	239.444	142.697	98.005	−323.459	−221.546	11.424
1100	100.000	247.684	150.679	106.705	−479.439	−203.633	9.670
1200	100.000	256.385	159.131	116.705	−475.371	−178.740	7.780
1300	100.000	264.389	166.924	126.705	−471.321	−154.185	6.195
1400	100.000	271.800	174.154	136.705	−467.287	−129.941	4.848
1500	100.000	278.699	180.896	146.705	−463.268	−105.986	3.691

62. POTASSIUM HYDROXIDE   KOH (cr, l)

298.15	64.900	78.870	78.870	0.000	−424.580	−378.747	66.354
300	65.038	79.272	78.871	0.120	−424.569	−378.463	65.895
400	72.519	99.007	81.512	6.998	−426.094	−362.765	47.372
500	80.000	115.993	86.745	14.624	−424.572	−347.093	36.260
520	81.496	119.159	87.931	16.239	−417.725	−344.002	34.555
PHASE TRANSITION: $\Delta_{trs} H$ = 6.450 kJ/mol, $\Delta_{trs} S$ = 12.404 J/K·mol, crII–crI							
520	79.000	131.563	87.931	22.689	−417.725	−344.002	34.555
600	79.000	142.868	94.520	29.009	−416.274	−332.766	28.969
678	79.000	152.523	100.649	35.171	−405.464	−321.998	24.807
PHASE TRANSITION: $\Delta_{trs} H$ = 9.400 kJ/mol, $\Delta_{trs} S$ = 13.865 J/K·mol, crI–l							
678	83.000	166.388	100.649	44.571	−405.464	−321.998	24.807
700	83.000	169.038	102.757	46.397	−404.981	−319.297	23.826
800	83.000	180.121	111.750	54.697	−402.808	−307.206	20.058
900	83.000	189.897	119.901	62.997	−400.694	−295.383	17.143
1000	83.000	198.642	127.345	71.297	−398.668	−283.791	14.824
1100	83.000	206.553	134.192	79.597	−475.618	−267.780	12.716
1200	83.000	213.775	140.527	87.897	−472.711	−249.014	10.839
1300	83.000	220.418	146.421	96.197	−469.843	−230.490	9.261
1400	83.000	226.569	151.929	104.497	−467.011	−212.184	7.917
1500	83.000	232.296	157.098	112.797	−464.217	−194.080	6.758

63. POTASSIUM HYDROXIDE   KOH (g)

298.15	49.184	238.283	238.283	0.000	−227.989	−229.685	40.239
300	49.236	238.588	238.284	0.091	−228.007	−229.696	39.993
400	51.178	253.053	240.243	5.124	−231.377	−229.667	29.991
500	52.178	264.591	243.998	10.296	−232.309	−229.129	23.937
600	52.804	274.163	248.251	15.547	−233.145	−228.413	19.885
700	53.296	282.340	252.551	20.853	−233.934	−227.562	16.981
800	53.758	289.487	256.730	26.206	−234.708	−226.599	14.795
900	54.229	295.846	260.730	31.605	−235.495	−225.538	13.090
1000	54.713	301.585	264.533	37.052	−236.322	−224.388	11.721
1100	55.203	306.823	268.143	42.548	−316.077	−218.535	10.377
1200	55.686	311.647	271.570	48.092	−315.925	−209.674	9.127
1300	56.153	316.122	274.827	53.684	−315.764	−200.826	8.069
1400	56.598	320.300	277.927	59.322	−315.595	−191.991	7.163
1500	57.016	324.220	280.884	65.003	−315.420	−183.169	6.378

	J/K·mol			kJ/mol			
T/K	$C_p°$	$S°$	$-(G°-H°(T_r))/T$	$H°-H°(T_r)$	$\Delta_f H°$	$\Delta_f G°$	Log K_f

64. POTASSIUM CHLORIDE   KCl (cr, l)

298.15	51.300	82.570	82.570	0.000	−436.490	−408.568	71.579
300	51.333	82.887	82.571	0.095	−436.481	−408.395	71.107
400	52.977	97.886	84.605	5.312	−438.463	−398.651	52.058
500	54.448	109.867	88.498	10.685	−437.990	−388.749	40.612
600	55.885	119.921	92.919	16.201	−437.332	−378.960	32.991
700	57.425	128.649	97.413	21.865	−436.502	−369.295	27.557
800	59.205	136.430	101.812	27.694	−435.505	−359.760	23.490
900	61.361	143.523	106.058	33.719	−434.337	−350.360	20.334
1000	64.032	150.121	110.138	39.983	−432.981	−341.100	17.817
1044	65.405	152.908	111.882	42.830	−485.450	−336.720	16.847
PHASE TRANSITION: $\Delta_{trs} H = 26.320$ kJ/mol, $\Delta_{trs} S = 25.210$ J/K·mol, cr–l							
1044	72.000	178.118	111.882	69.150	−485.450	−336.720	16.847
1100	72.000	181.880	115.351	73.182	−483.633	−328.790	15.613
1200	72.000	188.145	121.160	80.382	−480.393	−314.856	13.705
1300	72.000	193.908	126.537	87.582	−477.158	−301.192	12.102
1400	72.000	199.244	131.542	94.782	−473.928	−287.778	10.737
1500	72.000	204.211	136.223	101.982	−470.704	−274.594	9.562

65. POTASSIUM CHLORIDE   KCl (g)

298.15	36.505	239.091	239.091	0.000	−214.575	−233.320	40.876
300	36.518	239.317	239.092	0.068	−214.594	−233.436	40.644
400	37.066	249.904	240.532	3.749	−218.112	−239.107	31.224
500	37.384	258.212	243.267	7.473	−219.287	−244.219	25.513
600	37.597	265.048	246.344	11.222	−220.396	−249.100	21.686
700	37.769	270.857	249.441	14.991	−221.461	−253.799	18.938
800	37.907	275.910	252.441	18.775	−222.509	−258.347	16.868
900	38.041	280.382	255.302	22.572	−223.568	−262.764	15.250
1000	38.162	284.397	258.014	26.383	−224.667	−267.061	13.950
1100	38.279	288.039	260.581	30.205	−304.696	−266.627	12.661
1200	38.401	291.375	263.010	34.039	−304.821	−263.161	11.455
1300	38.518	294.454	265.312	37.885	−304.941	−259.684	10.434
1400	38.639	297.313	267.496	41.743	−305.053	−256.199	9.559
1500	38.761	299.983	269.574	45.613	−305.159	−252.706	8.800

66. DINITROGEN   N₂ (g)

298.15	29.124	191.608	191.608	0.000	0.000	0.000	0.000
300	29.125	191.788	191.608	0.054	0.000	0.000	0.000
400	29.249	200.180	192.752	2.971	0.000	0.000	0.000
500	29.580	206.738	194.916	5.911	0.000	0.000	0.000
600	30.109	212.175	197.352	8.894	0.000	0.000	0.000
700	30.754	216.864	199.812	11.936	0.000	0.000	0.000
800	31.433	221.015	202.208	15.046	0.000	0.000	0.000
900	32.090	224.756	204.509	18.222	0.000	0.000	0.000
1000	32.696	228.169	206.706	21.462	0.000	0.000	0.000
1100	33.241	231.311	208.802	24.759	0.000	0.000	0.000
1200	33.723	234.224	210.801	28.108	0.000	0.000	0.000
1300	34.147	236.941	212.708	31.502	0.000	0.000	0.000
1400	34.517	239.485	214.531	34.936	0.000	0.000	0.000
1500	34.842	241.878	216.275	38.404	0.000	0.000	0.000

67. NITRIC OXIDE   NO (g)

298.15	29.862	210.745	210.745	0.000	91.277	87.590	−15.345
300	29.858	210.930	210.746	0.055	91.278	87.567	−15.247
400	29.954	219.519	211.916	3.041	91.320	86.323	−11.272
500	30.493	226.255	214.133	6.061	91.340	85.071	−8.887

T/K	C_p°	S°	$-(G^\circ-H^\circ(T_r))/T$	$H^\circ-H^\circ(T_r)$	$\Delta_f H^\circ$	$\Delta_f G^\circ$	Log K_f
	J/K·mol			kJ/mol			

67. NITRIC OXIDE NO (g) (continued)

T/K	C_p°	S°	$-(G^\circ-H^\circ(T_r))/T$	$H^\circ-H^\circ(T_r)$	$\Delta_f H^\circ$	$\Delta_f G^\circ$	Log K_f
600	31.243	231.879	216.635	9.147	91.354	83.816	−7.297
700	32.031	236.754	219.168	12.310	91.369	82.558	−6.160
800	32.770	241.081	221.642	15.551	91.386	81.298	−5.308
900	33.425	244.979	224.022	18.862	91.405	80.036	−4.645
1000	33.990	248.531	226.298	22.233	91.426	78.772	−4.115
1100	34.473	251.794	228.469	25.657	91.445	77.505	−3.680
1200	34.883	254.811	230.540	29.125	91.464	76.237	−3.318
1300	35.234	257.618	232.516	32.632	91.481	74.967	−3.012
1400	35.533	260.240	234.404	36.170	91.495	73.697	−2.750
1500	35.792	262.700	236.209	39.737	91.506	72.425	−2.522

68. NITROGEN DIOXIDE NO$_2$ (g)

T/K	C_p°	S°	$-(G^\circ-H^\circ(T_r))/T$	$H^\circ-H^\circ(T_r)$	$\Delta_f H^\circ$	$\Delta_f G^\circ$	Log K_f
298.15	37.178	240.166	240.166	0.000	34.193	52.316	−9.165
300	37.236	240.397	240.167	0.069	34.181	52.429	−9.129
400	40.513	251.554	241.666	3.955	33.637	58.600	−7.652
500	43.664	260.939	244.605	8.167	33.319	64.882	−6.778
600	46.383	269.147	248.026	12.673	33.174	71.211	−6.199
700	48.612	276.471	251.575	17.427	33.151	77.553	−5.787
800	50.405	283.083	255.107	22.381	33.213	83.893	−5.478
900	51.844	289.106	258.555	27.496	33.334	90.221	−5.236
1000	53.007	294.631	261.891	32.741	33.495	96.534	−5.042
1100	53.956	299.729	265.102	38.090	33.686	102.828	−4.883
1200	54.741	304.459	268.187	43.526	33.898	109.105	−4.749
1300	55.399	308.867	271.148	49.034	34.124	115.363	−4.635
1400	55.960	312.994	273.992	54.603	34.360	121.603	−4.537
1500	56.446	316.871	276.722	60.224	34.604	127.827	−4.451

69. AMMONIA NH$_3$ (g)

T/K	C_p°	S°	$-(G^\circ-H^\circ(T_r))/T$	$H^\circ-H^\circ(T_r)$	$\Delta_f H^\circ$	$\Delta_f G^\circ$	Log K_f
298.15	35.630	192.768	192.768	0.000	−45.940	−16.407	2.874
300	35.678	192.989	192.769	0.066	−45.981	−16.223	2.825
400	38.674	203.647	194.202	3.778	−48.087	−5.980	0.781
500	41.994	212.633	197.011	7.811	−49.908	4.764	−0.498
600	45.229	220.578	200.289	12.174	−51.430	15.846	−1.379
700	48.269	227.781	203.709	16.850	−52.682	27.161	−2.027
800	51.112	234.414	207.138	21.821	−53.695	38.639	−2.523
900	53.769	240.589	210.516	27.066	−54.499	50.231	−2.915
1000	56.244	246.384	213.816	32.569	−55.122	61.903	−3.233
1100	58.535	251.854	217.027	38.309	−55.589	73.629	−3.496
1200	60.644	257.039	220.147	44.270	−55.920	85.392	−3.717
1300	62.576	261.970	223.176	50.432	−56.136	97.177	−3.905
1400	64.339	266.673	226.117	56.779	−56.251	108.975	−4.066
1500	65.945	271.168	228.971	63.295	−56.282	120.779	−4.206

70. OXYGEN O (g)

T/K	C_p°	S°	$-(G^\circ-H^\circ(T_r))/T$	$H^\circ-H^\circ(T_r)$	$\Delta_f H^\circ$	$\Delta_f G^\circ$	Log K_f
298.15	21.911	161.058	161.058	0.000	249.180	231.743	−40.600
300	21.901	161.194	161.059	0.041	249.193	231.635	−40.331
400	21.482	167.430	161.912	2.207	249.874	225.677	−29.470
500	21.257	172.197	163.511	4.343	250.481	219.556	−22.937
600	21.124	176.060	165.290	6.462	251.019	213.319	−18.571
700	21.040	179.310	167.067	8.570	251.500	206.997	−15.446
800	20.984	182.115	168.777	10.671	251.932	200.610	−13.098
900	20.944	184.584	170.399	12.767	252.325	194.171	−11.269
1000	20.915	186.789	171.930	14.860	252.686	187.689	−9.804
1100	20.893	188.782	173.372	16.950	253.022	181.173	−8.603
1200	20.877	190.599	174.733	19.039	253.335	174.628	−7.601
1300	20.864	192.270	176.019	21.126	253.630	168.057	−6.753

T/K	J/K·mol			kJ/mol			
	C_p°	S°	$-(G^\circ - H^\circ(T_r))/T$	$H^\circ - H^\circ(T_r)$	$\Delta_f H^\circ$	$\Delta_f G^\circ$	Log K_f

70. OXYGEN   O (g) (continued)

1400	20.853	193.815	177.236	23.212	253.908	161.463	–6.024
1500	20.845	195.254	178.389	25.296	254.171	154.851	–5.392

71. DIOXYGEN   O$_2$ (g)

298.15	29.378	205.148	205.148	0.000	0.000	0.000	0.000
300	29.387	205.330	205.148	0.054	0.000	0.000	0.000
400	30.109	213.873	206.308	3.026	0.000	0.000	0.000
500	31.094	220.695	208.525	6.085	0.000	0.000	0.000
600	32.095	226.454	211.045	9.245	0.000	0.000	0.000
700	32.987	231.470	213.612	12.500	0.000	0.000	0.000
800	33.741	235.925	216.128	15.838	0.000	0.000	0.000
900	34.365	239.937	218.554	19.244	0.000	0.000	0.000
1000	34.881	243.585	220.878	22.707	0.000	0.000	0.000
1100	35.314	246.930	223.096	26.217	0.000	0.000	0.000
1200	35.683	250.019	225.213	29.768	0.000	0.000	0.000
1300	36.006	252.888	227.233	33.352	0.000	0.000	0.000
1400	36.297	255.568	229.162	36.968	0.000	0.000	0.000
1500	36.567	258.081	231.007	40.611	0.000	0.000	0.000

72. SULFUR   S (cr, l)

298.15	22.690	32.070	32.070	0.000	0.000	0.000	0.000
300	22.737	32.210	32.070	0.042	0.000	0.000	0.000
368.3	24.237	37.030	32.554	1.649	0.000	0.000	0.000
PHASE TRANSITION: $\Delta_{trs} H = 0.401$ kJ/mol, $\Delta_{trs} S = 1.089$ J/K·mol, crII–crI							
368.3	24.773	38.119	32.553	2.050	0.000	0.000	0.000
388.36	25.180	39.444	32.875	2.551	0.000	0.000	0.000
PHASE TRANSITION: $\Delta_{trs} H = 1.722$ kJ/mol, $\Delta_{trs} S = 4.431$ J/K·mol, crI–l							
388.36	31.710	43.875	32.872	4.273	0.000	0.000	0.000
400	32.369	44.824	33.206	4.647	0.000	0.000	0.000
500	38.026	53.578	36.411	8.584	0.000	0.000	0.000
600	34.371	60.116	39.842	12.164	0.000	0.000	0.000
700	32.451	65.278	43.120	15.511	0.000	0.000	0.000
800	32.000	69.557	46.163	18.715	0.000	0.000	0.000
882.38	32.000	72.693	48.496	21.351	0.000	0.000	0.000

73. SULFUR   S (g)

298.15	23.673	167.828	167.828	0.000	277.180	236.704	–41.469
300	23.669	167.974	167.828	0.044	277.182	236.453	–41.170
400	23.233	174.730	168.752	2.391	274.924	222.962	–29.115
500	22.741	179.860	170.482	4.689	273.286	210.145	–21.953
600	22.338	183.969	172.398	6.942	271.958	197.646	–17.206
700	22.031	187.388	174.302	9.160	270.829	185.352	–13.831
800	21.800	190.314	176.125	11.351	269.816	173.210	–11.309
900	21.624	192.871	177.847	13.522	215.723	162.258	–9.417
1000	21.489	195.142	179.465	15.677	216.018	156.301	–8.164
1100	21.386	197.185	180.985	17.821	216.284	150.317	–7.138
1200	21.307	199.043	182.413	19.955	216.525	144.309	–6.282
1300	21.249	200.746	183.759	22.083	216.743	138.282	–5.556
1400	21.209	202.319	185.029	24.206	216.940	132.239	–4.934
1500	21.186	203.781	186.231	26.325	217.119	126.182	–4.394

74. DISULFUR   S$_2$ (g)

298.15	32.505	228.165	228.165	0.000	128.600	79.696	–13.962
300	32.540	228.366	228.165	0.060	128.576	79.393	–13.823
400	34.108	237.956	229.462	3.398	122.703	63.380	–8.276
500	35.133	245.686	231.959	6.863	118.296	49.031	–5.122

T/K	J/K·mol			kJ/mol			
	C_p°	S°	$-(G^\circ - H^\circ(T_r))/T$	$H^\circ - H^\circ(T_r)$	$\Delta_f H^\circ$	$\Delta_f G^\circ$	Log K_f

74. DISULFUR S₂ (g) (continued)

T/K	C_p°	S°	$-(G^\circ - H^\circ(T_r))/T$	$H^\circ - H^\circ(T_r)$	$\Delta_f H^\circ$	$\Delta_f G^\circ$	Log K_f
600	35.815	252.156	234.800	10.413	114.685	35.530	−3.093
700	36.305	257.715	237.686	14.020	111.599	22.588	−1.685
800	36.697	262.589	240.501	17.671	108.841	10.060	−0.657
882.38	36.985	266.200	242.734	20.706		pressure = 1 bar	
900	37.045	266.932	243.201	21.358	0.000	0.000	0.000
1000	37.377	270.852	245.773	25.079	0.000	0.000	0.000
1100	37.704	274.430	248.218	28.833	0.000	0.000	0.000
1200	38.030	277.725	250.541	32.620	0.000	0.000	0.000
1300	38.353	280.781	252.751	36.439	0.000	0.000	0.000
1400	38.669	283.635	254.856	40.290	0.000	0.000	0.000
1500	38.976	286.314	256.865	44.173	0.000	0.000	0.000

75. OCTASULFUR S₈ (g)

T/K	C_p°	S°	$-(G^\circ - H^\circ(T_r))/T$	$H^\circ - H^\circ(T_r)$	$\Delta_f H^\circ$	$\Delta_f G^\circ$	Log K_f
298.15	156.500	432.536	432.536	0.000	101.277	48.810	−8.551
300	156.768	433.505	432.539	0.290	101.231	48.484	−8.442
400	167.125	480.190	438.834	16.542	80.642	32.003	−4.179
500	173.181	518.176	451.022	33.577	66.185	21.409	−2.237
600	177.936	550.180	464.951	51.137	55.101	13.549	−1.180
700	182.441	577.948	479.152	69.157	46.349	7.343	−0.548
800	186.764	602.596	493.071	87.620	39.177	2.263	−0.148
900	190.595	624.821	506.495	106.494	−392.062	6.554	−0.380
1000	193.618	645.067	519.355	125.712	−387.728	50.614	−2.644
1100	195.684	663.625	531.639	145.185	−383.272	94.233	−4.475
1200	196.825	680.707	543.359	164.817	−378.786	137.444	−5.983
1300	197.195	696.480	554.539	184.524	−374.356	180.283	−7.244
1400	196.988	711.089	565.206	204.237	−370.048	222.785	−8.312
1500	196.396	724.662	575.389	223.909	−365.905	264.984	−9.227

76. SULFUR DIOXIDE SO₂ (g)

T/K	C_p°	S°	$-(G^\circ - H^\circ(T_r))/T$	$H^\circ - H^\circ(T_r)$	$\Delta_f H^\circ$	$\Delta_f G^\circ$	Log K_f
298.15	39.842	248.219	248.219	0.000	−296.810	−300.090	52.574
300	39.909	248.466	248.220	0.074	−296.833	−300.110	52.253
400	43.427	260.435	249.828	4.243	−300.240	−300.935	39.298
500	46.490	270.465	252.978	8.744	−302.735	−300.831	31.427
600	48.938	279.167	256.634	13.520	−304.699	−300.258	26.139
700	50.829	286.859	260.413	18.513	−306.308	−299.386	22.340
800	52.282	293.746	264.157	23.671	−307.691	−298.302	19.477
900	53.407	299.971	267.796	28.958	−362.075	−295.987	17.178
1000	54.290	305.646	271.301	34.345	−362.012	−288.647	15.077
1100	54.993	310.855	274.664	39.810	−361.934	−281.314	13.358
1200	55.564	315.665	277.882	45.339	−361.849	−273.989	11.926
1300	56.033	320.131	280.963	50.920	−361.763	−266.671	10.715
1400	56.426	324.299	283.911	56.543	−361.680	−259.359	9.677
1500	56.759	328.203	286.735	62.203	−361.605	−252.053	8.777

77. SILICON Si (cr)

T/K	C_p°	S°	$-(G^\circ - H^\circ(T_r))/T$	$H^\circ - H^\circ(T_r)$	$\Delta_f H^\circ$	$\Delta_f G^\circ$	Log K_f
298.15	19.789	18.810	18.810	0.000	0.000	0.000	0.000
300	19.855	18.933	18.810	0.037	0.000	0.000	0.000
400	22.301	25.023	19.624	2.160	0.000	0.000	0.000
500	23.610	30.152	21.231	4.461	0.000	0.000	0.000
600	24.472	34.537	23.092	6.867	0.000	0.000	0.000
700	25.124	38.361	25.006	9.348	0.000	0.000	0.000
800	25.662	41.752	26.891	11.888	0.000	0.000	0.000
900	26.135	44.802	28.715	14.478	0.000	0.000	0.000
1000	26.568	47.578	30.464	17.114	0.000	0.000	0.000
1100	26.974	50.130	32.138	19.791	0.000	0.000	0.000
1200	27.362	52.493	33.737	22.508	0.000	0.000	0.000

	J/K·mol			kJ/mol			
T/K	C_p°	S°	$-(G^\circ-H^\circ(T_r))/T$	$H^\circ-H^\circ(T_r)$	$\Delta_f H^\circ$	$\Delta_f G^\circ$	Log K_f

77. SILICON   Si (cr) (continued)

1300	27.737	54.698	35.265	25.263	0.000	0.000	0.000
1400	28.103	56.767	36.728	28.055	0.000	0.000	0.000
1500	28.462	58.719	38.130	30.883	0.000	0.000	0.000

78. SILICON   Si (g)

298.15	22.251	167.980	167.980	0.000	450.000	405.525	−71.045
300	22.234	168.117	167.980	0.041	450.004	405.249	−70.559
400	21.613	174.416	168.843	2.229	450.070	390.312	−50.969
500	21.316	179.204	170.456	4.374	449.913	375.388	−39.216
600	21.153	183.074	172.246	6.497	449.630	360.508	−31.385
700	21.057	186.327	174.032	8.607	449.259	345.682	−25.795
800	21.000	189.135	175.748	10.709	448.821	330.915	−21.606
900	20.971	191.606	177.375	12.808	448.329	316.205	−18.352
1000	20.968	193.815	178.911	14.904	447.791	301.553	−15.751
1100	20.989	195.815	180.358	17.002	447.211	286.957	−13.626
1200	21.033	197.643	181.723	19.103	446.595	272.416	−11.858
1300	21.099	199.329	183.014	21.209	445.946	257.927	−10.364
1400	21.183	200.895	184.236	23.323	445.268	243.489	−9.085
1500	21.282	202.360	185.396	25.446	444.563	229.101	−7.978

79. SILICON DIOXIDE   SiO$_2$ (cr)

298.15	44.602	41.460	41.460	0.000	−910.700	−856.288	150.016
300	44.712	41.736	41.461	0.083	−910.708	−855.951	149.032
400	53.477	55.744	43.311	4.973	−910.912	−837.651	109.385
500	60.533	68.505	47.094	10.705	−910.540	−819.369	85.598
600	64.452	79.919	51.633	16.971	−909.841	−801.197	69.749
700	68.234	90.114	56.414	23.590	−908.958	−783.157	58.439
800	76.224	99.674	61.226	30.758	−907.668	−765.265	49.966
848	82.967	104.298	63.533	34.569	−906.310	−756.747	46.613
PHASE TRANSITION: $\Delta_{trs} H$ = 0.411 kJ/mol, $\Delta_{trs} S$ = 0.484 J/K·mol, crII–crII′							
848	67.446	104.782	63.532	34.980	−906.310	−756.747	46.613
900	67.953	108.811	66.033	38.500	−905.922	−747.587	43.388
1000	68.941	116.021	70.676	45.345	−905.176	−730.034	38.133
1100	69.940	122.639	75.104	52.289	−904.420	−712.557	33.836
1200	70.947	128.768	79.323	59.333	−901.382	−695.148	30.259
PHASE TRANSITION: $\Delta_{trs} H$ = 2.261 kJ/mol, $\Delta_{trs} S$ = 1.883 J/K·mol, crII′–crI							
1200	71.199	130.651	79.323	61.594	−901.382	−695.148	30.259
1300	71.743	136.372	83.494	68.742	−900.574	−677.994	27.242
1400	72.249	141.707	87.463	75.941	−899.782	−660.903	24.658
1500	72.739	146.709	91.248	83.191	−899.004	−643.867	22.421

80. SILICON TETRACHLORIDE   SiCl$_4$ (g)

298.15	90.404	331.446	331.446	0.000	−662.200	−622.390	109.039
300	90.562	332.006	331.448	0.167	−662.195	−622.143	108.323
400	96.893	359.019	335.088	9.572	−661.853	−608.841	79.505
500	100.449	381.058	342.147	19.456	−661.413	−595.637	62.225
600	102.587	399.576	350.216	29.616	−660.924	−582.527	50.713
700	103.954	415.500	358.432	39.948	−660.417	−569.501	42.496
800	104.875	429.445	366.455	50.392	−659.912	−556.548	36.338
900	105.523	441.837	374.155	60.914	−659.422	−543.657	31.553
1000	105.995	452.981	381.490	71.491	−658.954	−530.819	27.727
1100	106.349	463.101	388.456	82.109	−658.515	−518.027	24.599
1200	106.620	472.366	395.068	92.758	−658.107	−505.274	21.994
1300	106.834	480.909	401.347	103.431	−657.735	−492.553	19.791
1400	107.003	488.833	407.316	Î14.123	−657.400	−479.860	17.904
1500	107.141	496.220	413.000	124.830	−657.104	−467.189	16.269

THERMODYNAMIC PROPERTIES OF AQUEOUS SYSTEMS

This table contains standard state thermodynamic properties of ions and neutral species in aqueous solution. It includes enthalpy and Gibbs energy of formation, entropy, and heat capacity, and thus serves as a companion to the preceding table, "Standard Thermodynamic Properties of Chemical Substances". The standard state is the hypothetical ideal solution with molality $m = 1$ mol/kg (mean ionic molality $m_\pm$ in the case of a species which is assumed to dissociate at infinite dilution). Further details on conventions may be found in Reference 1.

Cations are listed by formula in the first part of the table, followed by anions and finally neutral species. All values refer to standard conditions of 25°C and 100 kPa pressure.

REFERENCES

1. Wagman, D. D., Evans, W. H., Parker, V. B., Schumm, R. H., Halow, I., Bailey, S. M., Churney, K. L., and Nuttall, R. L., *The NBS Tables of Chemical Thermodynamic Properties, J. Phys. Chem. Ref. Data*, Vol. 11, Suppl. 2, 1982.
2. Zemaitis, J. F., Clark, D. M., Rafal, M., and Scrivner, N. C., *Handbook of Aqueous Electrolyte Thermodynamics*, American Institute of Chemical Engineers, New York, 1986.

Species	$\Delta_f H°$/ kJ mol⁻¹	$\Delta_f G°$/ kJ mol⁻¹	$S°$/ J mol⁻¹K⁻¹	C_p/ J mol⁻¹K⁻¹
Cations				
Ag⁺	105.6	77.1	72.7	21.8
Al⁺³	-531.0	-485.0	-321.7	
AlOH⁺²		-694.1		
Ba⁺²	-537.6	-560.8	9.6	
BaOH⁺		-730.5		
Be⁺²	-382.8	-379.7	-129.7	
Bi⁺³		82.8		
BiOH⁺²		-146.4		
Ca⁺²	-542.8	-553.6	-53.1	
CaOH⁺		-718.4		
Cd⁺²	-75.9	-77.6	-73.2	
CdOH⁺		-261.1		
Ce⁺³	-696.2	-672.0	-205.0	
Ce⁺⁴	-537.2	-503.8	-301.0	
Co⁺²	-58.2	-54.4	-113.0	
Co⁺³	92.0	134.0	-305.0	
Cr⁺²	-143.5			
Cs⁺	-258.3	-292.0	133.1	-10.5
Cu⁺	71.7	50.0	40.6	
Cu⁺²	64.8	65.5	-99.6	
Dy⁺³	-699.0	-665.0	-231.0	21.0
Er⁺³	-705.4	-669.1	-244.3	21.0
Eu⁺²	-527.0	-540.2	-8.0	
Eu⁺³	-605.0	-574.1	-222.0	8.0
Fe⁺²	-89.1	-78.9	-137.7	
Fe⁺³	-48.5	-4.7	-315.9	
FeOH⁺	-324.7	-277.4	-29.0	
FeOH⁺²	-290.8	-229.4	-142.0	
Fe(OH)₂⁺		-438.0		
Ga⁺²		-88.0		
Ga⁺³	-211.7	-159.0	-331.0	
GaOH⁺²		-380.3		
Ga(OH)₂⁺		-597.4		
Gd⁺³	-686.0	-661.0	-205.9	
H⁺	0	0	0	0
Hg⁺²	171.1	164.4	-32.2	
Hg₂⁺²	172.4	153.5	84.5	
HgOH⁺	-84.5	-52.3	71.0	
Ho⁺³	-705.0	-673.7	-226.8	17.0
In⁺		-12.1		
In⁺²		-50.7		
In⁺³	-105.0	-98.0	-151.0	
InOH⁺²	-370.3	-313.0	-88.0	
In(OH)₂⁺	-619.0	-525.0	25.0	
K⁺	-252.4	-283.3	102.5	21.8
La⁺³	-707.1	-683.7	-217.6	-13.0
Li⁺	-278.5	-293.3	13.4	68.6
Lu⁺³	-665.0	-628.0	-264.0	25.0
LuF⁺²		-931.4		
Mg⁺²	-466.9	-454.8	-138.1	
MgOH⁺		-626.7		
Mn⁺²	-220.8	-228.1	-73.6	50.0
MnOH⁺	-450.6	-405.0	-17.0	
NH₄⁺	-132.5	-79.3	113.4	79.9
N₂H₅⁺	-7.5	82.5	151.0	70.3
Na⁺	-240.1	-261.9	59.0	46.4
Nd⁺³	-696.2	671.6	-206.7	-21.0
Ni⁺²	-54.0	-45.6	-128.9	
NiOH⁺	-287.9	-227.6	-71.0	
PH₄⁺		92.1		
Pa⁺⁴	-619.0			
Pb⁺²	-1.7	-24.4	10.5	
PbOH⁺		-226.3		
Pd⁺²	149.0	176.5	-184.0	
Po⁺²		71.0		
Po⁺⁴		293.0		
Pr⁺³	-704.6	-679.1	-209.0	-29.0
Pt⁺²		254.8		
Ra⁺²	-527.6	-561.5	54.0	
Rb⁺	-251.2	-284.0	121.5	
Re⁺		-33.0		
Sc⁺³	-614.2	-586.6	-255.0	
ScOH⁺²	-861.5	-801.2	-134.0	
Sm⁺²		-497.5		
Sm⁺³	-691.6	-666.6	-211.7	-21.0
Sn⁺²	-8.8	-27.2	-17.0	
SnOH⁺	-286.2	-254.8	50.0	
Sr⁺²	-545.8	-559.5	-32.6	
SrOH⁺		-721.3		
Tb⁺³	-682.8	-651.9	-226.0	17.0
Te(OH)₃⁺	-608.4	-496.1	111.7	
Th⁺⁴	-769.0	-705.1	-422.6	
Th(OH)⁺³	-1030.1	-920.5	-343.0	
Th(OH)₂⁺²	-1282.4	-1140.9	-218.0	
Tl⁺	5.4	-32.4	125.5	
Tl⁺³	196.6	214.6	-192.0	

Species	$\Delta_f H°/$ kJ mol^{-1}	$\Delta_f G°/$ kJ mol^{-1}	$S°/$ J mol^{-1}K^{-1}	$C_p/$ J mol^{-1}K^{-1}	Species	$\Delta_f H°/$ kJ mol^{-1}	$\Delta_f G°/$ kJ mol^{-1}	$S°/$ J mol^{-1}K^{-1}	$C_p/$ J mol^{-1}K^{-1}
TlOH^{+2}		-15.9			H$_2$PO$_4^-$	-1296.3	-1130.2	90.4	
Tl(OH)$_2^+$		-244.7			H$_2$P$_2$O$_7^{-2}$	-2278.6	-2010.2	163.0	
Tm^{+3}	-697.9	-662.0	-243.0	25.0	I$^-$	-55.2	-51.6	111.3	-142.3
U^{+3}	-489.1	-476.2	-188.0		IO$^-$	-107.5	-38.5	-5.4	
U^{+4}	-591.2	-531.9	-410.0		IO$_3^-$	-221.3	-128.0	118.4	
Y^{+3}	-723.4	-693.8	-251.0		IO$_4^-$	-151.5	-58.5	222.0	
Y$_2$(OH)$_2^{+4}$		-1780.3			MnO$_4^-$	-541.4	-447.2	191.2	-82.0
Yb^{+2}		-527.0			MnO$_4^{-2}$	-653.0	-500.7	59.0	
Yb^{+3}	-674.5	-644.0	-238.0	25.0	MoO$_4^{-2}$	-997.9	-836.3	27.2	
Y(OH)$^{+2}$		-879.1			NO$_2^-$	-104.6	-32.2	123.0	-97.5
Zn^{+2}	-153.9	-147.1	-112.1	46.0	NO$_3^-$	-207.4	-111.3	146.4	-86.6
ZnOH$^+$		-330.1			N$_3^-$	275.1	348.2	107.9	
					OCN$^-$	-146.0	-97.4	106.7	
Anions					OH$^-$	-230.0	-157.2	-10.8	-148.5
					PO$_4^{-3}$	-1277.4	-1018.7	-220.5	
AlO$_2^-$	-930.9	-830.9	-36.8		P$_2$O$_7^{-4}$	-2271.1	-1919.0	-117.0	
Al(OH)$_4^-$	-1502.5	-1305.3	102.9		Re$^-$	46.0	10.1	230.0	
AsO$_2^-$	-429.0	-350.0	40.6		S^{-2}	33.1	85.8	-14.6	
AsO$_4^{-3}$	-888.1	-648.4	-162.8		SCN$^-$	76.4	92.7	144.3	-40.2
BF$_4^-$	-1574.9	-1486.9	180.0		SO$_3^{-2}$	-635.5	-486.5	-29.0	
BH$_4^-$	48.2	114.4	110.5		SO$_4^{-2}$	-909.3	-744.5	20.1	-293.0
BO$_2^-$	-772.4	-678.9	-37.2		S$_2^{-2}$	30.1	79.5	28.5	
B$_4$O$_7^{-2}$		-2604.8			S$_2$O$_3^{-2}$	-652.3	-522.5	67.0	
BeO$_2^{-2}$	-790.8	-640.1	-159.0		S$_2$O$_4^{-2}$	-753.5	-600.3	92.0	
Br$^-$	-121.6	-104.0	82.4	-141.8	S$_2$O$_8^{-2}$	-1344.7	-1114.9	244.3	
BrO$^-$	-94.1	-33.4	42.0		Se^{-2}		129.3		
BrO$_3^-$	-67.1	18.6	161.7		SeO$_3^{-2}$	-509.2	-369.8	13.0	
BrO$_4^-$	13.0	118.1	199.6		SeO$_4^{-2}$	-599.1	-441.3	54.0	
CHOO$^-$	-425.6	-351.0	92.0	-87.9	VO$_3^-$	-888.3	-783.6	50.0	
CH$_3$COO$^-$	-486.0	-369.3	86.6	-6.3	VO$_4^{-3}$		-899.0		
C$_2$O$_4^{-2}$	-825.1	-673.9	45.6		WO$_4^{-2}$	-1075.7			
C$_2$O$_4$H$^-$	-818.4	-698.3	149.4						
Cl$^-$	-167.2	-131.2	56.5	-136.4	**Neutral species**				
ClO$^-$	-107.1	-36.8	42.0						
ClO$_2^-$	-66.5	17.2	101.3		AgBr	-16.0	-26.9	155.2	-120.1
ClO$_3^-$	-104.0	-8.0	162.3		AgCl	-61.6	-54.1	129.3	-114.6
ClO$_4^-$	-129.3	-8.5	182.0		AgF	-227.1	-201.7	59.0	-84.9
CN$^-$	150.6	172.4	94.1		AgI	50.4	25.5	184.1	-120.5
CO$_3^{-2}$	-677.1	-527.8	-56.9		AgNO$_3$	-101.8	-34.2	219.2	-64.9
CrO$_4^{-2}$	-881.2	-727.8	50.2		Ag$_2$SO$_4$	-698.1	-590.3	165.7	-251.0
Cr$_2$O$_7^{-2}$	-1490.3	-1301.1	261.9		AlBr$_3$	-895.0	-799.0	-74.5	
F$^-$	-332.6	-278.8	-13.8	-106.7	AlCl$_3$	-1033.0	-879.0	-152.3	
Fe(CN)$_6^{-3}$	561.9	729.4	270.3		AlF$_3$	-1531.0	-1322.0	-363.2	
Fe(CN)$_6^{-4}$	455.6	695.1	95.0		AlI$_3$	-699.0	-640.0	12.1	
HB$_4$O$_7^-$		-2685.1			Al$_2$(SO$_4$)$_3$	-3791.0	-3205.0	-583.2	
HCO$_3^-$	-692.0	-586.8	91.2		BaBr$_2$	-780.7	-768.7	174.5	
HF$_2^-$	-649.9	-578.1	92.5		BaCO$_3$	-1214.8	-1088.6	-47.3	
HPO$_3$F$^-$		-1198.2			BaCl$_2$	-872.0	-823.2	122.6	
HPO$_4^{-2}$	-1292.1	-1089.2	-33.5		BaF$_2$	-1202.9	-1118.4	-18.0	
HP$_2$O$_7^{-3}$	-2274.8	-1972.2	46.0		Ba(HCO$_3$)$_2$	-1921.6	-1734.3	192.0	
HS$^-$	-17.6	12.1	62.8		BaI$_2$	-648.0	-663.9	232.2	
HSO$_3^-$	-626.2	-527.7	139.7		Ba(NO$_3$)$_2$	-952.4	-783.3	302.5	
HSO$_4^-$	-887.3	-755.9	131.8	-84.0	BaSO$_4$	-1446.9	-1305.3	29.7	
HS$_2$O$_4^-$		-614.5			BeSO$_4$	-1292.0	-1124.3	-109.6	
HSe$^-$	15.9	44.0	79.0		CCl$_3$COOH	-516.3			
HSeO$_3^-$	-514.6	-411.5	135.1		CHCl$_2$COOH	-512.1			
HSeO$_4^-$	-581.6	-452.2	149.4		CHOOCs	-683.8	-643.0	226.0	
H$_2$AsO$_3^-$	-714.8	-587.1	110.5		CHOOH	-425.6	-351.0	92.0	-87.9
H$_2$AsO$_4^-$	-909.6	-753.2	117.0		CHOOK	-677.9	-634.2	192.0	-66.1

Species	$\Delta_f H°/$ kJ mol^{-1}	$\Delta_f G°/$ kJ mol^{-1}	$S°/$ J mol^{-1}K^{-1}	$C_p/$ J mol^{-1}K^{-1}	Species	$\Delta_f H°/$ kJ mol^{-1}	$\Delta_f G°/$ kJ mol^{-1}	$S°/$ J mol^{-1}K^{-1}	$C_p/$ J mol^{-1}K^{-1}
CHOONH$_4$	-558.1	-430.4	205.0	-7.9	GdCl$_3$	-1188.0	-1059.0	-36.8	-410.0
CHOONa	-665.7	-612.9	151.0	-41.4	HBr	-121.6	-104.0	82.4	-141.8
CHOORb	-676.7	-635.1	213.0		HCN	150.6	172.4	94.1	
CH$_2$ClCOOH	-501.3				HCl	-167.2	-131.2	56.5	-136.4
CH$_3$COOCs	-744.3	-661.3	219.7		HF	-332.6	-278.8	-13.8	-106.7
CH$_3$COOH	-486.0	-369.3	86.6	-6.3	HI	-55.2	-51.6	111.3	-142.3
CH$_3$COOK	-738.4	-652.6	189.1	15.5	HNO$_3$	-207.4	-111.3	146.4	-86.6
CH$_3$COONH$_4$	-618.5	-448.6	200.0	73.6	HSCN	76.4	92.7	144.3	-40.2
CH$_3$COONa	-726.1	-631.2	145.6	40.2	H$_2$SO$_4$	-909.3	-744.5	20.1	-293.0
CH$_3$COORb	-737.2	-653.3	207.9		HoCl$_3$	-1206.7	-1067.3	-57.7	-393.0
(COOH)$_2$	-825.1	-673.9	45.6		KBr	-373.9	-387.2	184.9	-120.1
(CH$_3$)$_3$N	-76.0	93.1	133.5		KCl	-419.5	-414.5	159.0	-114.6
CaBr$_2$	-785.9	-761.5	111.7		KF	-585.0	-562.1	88.7	-84.9
CaCO$_3$	-1220.0	-1081.4	-110.0		KHCO$_3$	-944.4	-870.0	193.7	
CaCl$_2$	-877.1	-816.0	59.8		KHSO$_4$	-1139.7	-1039.2	234.3	-63.0
CaF$_2$	-1208.1	-1111.2	-80.8		KI	-307.6	-334.9	213.8	-120.5
CaI$_2$	-653.2	-656.7	169.5		KNO$_3$	-459.7	-394.5	248.9	-64.9
Ca(NO$_3$)$_2$	-957.6	-776.1	239.7		K$_2$CO$_3$	-1181.9	-1094.4	148.1	
CaSO$_4$	-1452.1	-1298.1	-33.1		K$_2$S	-471.5	-480.7	190.4	
CdBr$_2$	-319.0	-285.5	91.6		K$_2$SO$_4$	-1414.0	-1311.1	225.1	-251.0
CdCl$_2$	-410.2	-340.1	39.7		K$_2$Se		-437.2		
CdF$_2$	-741.2	-635.2	-100.8		LaCl$_3$	-1208.8	-1077.3	-50.0	-423.0
CdI$_2$	-186.3	-180.8	149.4		LiBr	-400.0	-397.3	95.8	-73.2
Cd(NO$_3$)$_2$	-490.6	-300.1	219.7		LiCl	-445.6	-424.6	69.9	-67.8
CdSO$_4$	-985.2	-822.1	-53.1		LiF	-611.1	-571.9	-0.4	-38.1
CeCl$_3$	-1197.5	-1065.6	-38.0		LiI	-333.7	-344.8	124.7	-73.6
CoBr$_2$	-301.2	-262.3	50.0		LiNO$_3$	-485.9	-404.5	160.2	-18.0
CoCl$_2$	-392.5	-316.7			Li$_2$CO$_3$	-1234.1	-1114.6	-29.7	
CoI$_2$	-168.6	-157.7	109.0		Li$_2$SO$_4$	-1466.2	-1331.2	47.3	-155.6
Co(NO$_3$)$_2$	-472.8	-276.9	180.0		LuCl$_3$	-1167.0	-1021.0	-96.0	-385.0
CoSO$_4$	-967.3	-799.1	-92.0		MgBr$_2$	-709.9	-662.7	26.8	
CsBr	-379.8	-396.0	215.5		MgCl$_2$	-801.2	-717.1	-25.1	
CsCl	-425.4	-423.2	189.5	-146.9	MgI$_2$	-577.2	-558.1	84.5	
CsF	-590.9	-570.8	119.2		Mg(NO$_3$)$_2$	-881.6	-677.3	154.8	
CsHCO$_3$	-950.3	-878.8	224.3		MgSO$_4$	-1376.1	-1199.5	-118.0	
CsHSO$_4$	-1145.6	-1047.9	264.8		MnBr$_2$	-464.0			
CsI	-313.5	-343.6	244.3	-152.7	MnCl$_2$	-555.1	-490.8	38.9	-222.0
CsNO$_3$	-465.6	-403.3	279.5	-99.0	MnI$_2$	-331.0			
Cs$_2$CO$_3$	-1193.7	-1111.9	209.2		Mn(NO$_3$)$_2$	-635.5	-450.9	218.0	-121.0
Cs$_2$S	-483.7	-498.3	251.0		MnSO$_4$	-1130.1	-972.7	-53.6	-243.0
Cs$_2$SO$_4$	-1425.8	-1328.6	286.2		NH$_4$Br	-254.1	-183.3	195.8	-61.9
Cs$_2$Se		-454.8			NH$_4$BrO$_3$	-199.6	-60.7	275.1	
Cu(NO$_3$)$_2$	-350.0	-157.0	193.3		NH$_4$CN	18.0	93.0	207.5	
CuSO$_4$	-844.5	-679.0	-79.5		NH$_4$Cl	-299.7	-210.5	169.9	-56.5
DyCl$_3$	-1197.0	-1059.0	-61.9	-389.0	NH$_4$ClO$_3$	-236.5	-87.3	275.7	
ErCl$_3$	-1207.1	-1062.7	-75.3	-389.0	NH$_4$ClO$_4$	-261.8	-87.8	295.4	
EuCl$_2$	-862.0				NH$_4$F	-465.1	-358.1	99.6	-26.8
EuCl$_3$	-1106.2	-967.7	-54.0	-402.0	NH$_4$HCO$_3$	-824.5	-666.1	204.6	
FeBr$_2$	-332.2	-286.8	27.2		NH$_4$HS	-150.2	-67.2	176.1	
FeBr$_3$	-413.4	-316.7	-68.6		NH$_4$HSO$_3$	-758.7	-607.0	253.1	
FeCl$_2$	-423.4	-341.3	-24.7		NH$_4$HSO$_4$	-1019.9	-835.2	245.2	-3.8
FeCl$_3$	-550.2	-398.3	-146.4		NH$_4$HSeO$_4$	-714.2	-531.6	262.8	
FeF$_2$	-754.4	-636.5	-165.3		NH$_4$H$_2$AsO$_3$	-847.3	-666.4	223.8	
FeF$_3$	-1046.4	-840.9	-357.3		NH$_4$H$_2$AsO$_4$	-1042.1	-832.5	230.5	
FeI$_2$	-199.6	-182.1	84.9		NH$_4$H$_2$PO$_4$	-1428.8	-1209.6	203.8	
FeI$_3$	-214.2	-159.4	18.0		NH$_4$H$_3$P$_2$O$_7$	-2409.1	-2102.6	326.0	
Fe(NO$_3$)$_3$	-670.7	-338.3	123.4		NH$_4$I	-187.7	-130.9	224.7	-62.3
FeSO$_4$	-998.3	-823.4	-117.6		NH$_4$IO$_3$	-354.0	-207.4	231.8	
Fe$_2$(SO$_4$)$_3$	-2825.0	-2242.8	-571.5		NH$_4$NO$_2$	-237.2	-111.6	236.4	-17.6

Species	$\Delta_f H°/$ kJ mol^{-1}	$\Delta_f G°/$ kJ mol^{-1}	$S°/$ J mol^{-1}K^{-1}	$C_p/$ J mol^{-1}K^{-1}	Species	$\Delta_f H°/$ kJ mol^{-1}	$\Delta_f G°/$ kJ mol^{-1}	$S°/$ J mol^{-1}K^{-1}	$C_p/$ J mol^{-1}K^{-1}
NH_4NO_3	-339.9	-190.6	259.8	-6.7	$RaSO_4$	-1436.8	-1306.2	75.0	
NH_4OH	-362.5	-236.5	102.5	-68.6	$RbBr$	-372.7	-387.9	203.9	
NH_4SCN	-56.1	13.4	257.7	39.7	$RbCl$	-418.3	-415.2	178.0	
$(NH_4)_2CO_3$	-942.2	-686.4	169.9		RbF	-583.8	-562.8	107.5	
$(NH_4)_2CrO_4$	-1146.2	-886.4	277.0		$RbHCO_3$	-943.2	-870.8	212.7	
$(NH_4)_2Cr_2O_7$	-1755.2	-1459.5	488.7		$RbHSO_4$	-1138.5	-1039.9	253.1	
$(NH_4)_2HAsO_4$	-1171.4	-873.2	225.1		RbI	-306.4	-335.6	232.6	
$(NH_4)_2HPO_4$	-1557.2	-1247.8	193.3		$RbNO_3$	-458.5	-395.2	267.8	
$(NH_4)_2S$	-231.8	-72.6	212.1		Rb_2CO_3	-1179.5	-1095.8	186.2	
$(NH_4)_2SO_3$	-900.4	-645.0	197.5		Rb_2S	-469.4	-482.0	228.4	
$(NH_4)_2SO_4$	-1174.3	-903.1	246.9	-133.1	Rb_2SO_4	-1411.6	-1312.5	263.2	
$(NH_4)_2SeO_4$	-864.0	-599.8	280.7		$SmCl_3$	-1193.3	-1060.2	-42.7	-431.0
$(NH_4)_3PO_4$	-1674.9	-1256.6	117.0		$SrBr_2$	-788.9	-767.4	132.2	
$NaBr$	-361.7	-365.8	141.4	-95.4	$SrCO_3$	-1222.9	-1087.3	-89.5	
$NaCl$	-407.3	-393.1	115.5	-90.0	$SrCl_2$	-880.1	-821.9	80.3	
NaF	-572.8	-540.7	45.2	-60.2	SrI_2	-656.2	-662.6	190.0	
$NaHCO_3$	-932.1	-848.7	150.2		$Sr(NO_3)_2$	-960.5	-782.0	260.2	
$NaHSO_4$	-1127.5	-1017.8	190.8	-38.0	$SrSO_4$	-1455.1	-1304.0	-12.6	
NaI	-295.3	-313.5	170.3	-95.8	$TbCl_3$	-1184.1	-1045.5	-59.0	-393.0
$NaNO_3$	-447.5	-373.2	205.4	-40.2	$TlBr$	-116.2	-136.4	207.9	
Na_2CO_3	-1157.4	-1051.6	61.1		$TlBr_3$	-168.2	-97.1	54.0	
Na_2S	-447.3	-438.1	103.3		$TlCl$	-161.8	-163.6	182.0	
Na_2SO_4	-1389.5	-1268.4	138.1	-201.0	$TlCl_3$	-305.0	-179.0	-23.0	
Na_2Se		-394.6			TlF	-327.3	-311.2	111.7	
$NdCl_3$	-1197.9	-1065.6	-37.7	-431.0	TlI	-49.8	-84.0	236.8	
$NiBr_2$	-297.1	-253.6	36.0		$TlNO_3$	-202.0	-143.7	272.0	
$NiCl_2$	-388.3	-307.9	-15.1		Tl_2SO_4	-898.6	-809.3	271.1	
NiF_2	-719.2	-603.3	-156.5		$TmCl_3$	-1199.1	-1055.6	-75.0	-385.0
NiI_2	-164.4	-149.0	93.7		UCl_4	-1259.8	-1056.8	-184.0	
$Ni(NO_3)_2$	-468.6	-268.5	164.0		UO_2CO_3	-1696.6	-1481.5	-154.4	
$NiSO_4$	-963.2	-790.3	-108.8		$UO_2(NO_3)_2$	-1434.3	-1176.0	195.4	
$PbBr_2$	-244.8	-232.3	175.3		UO_2SO_4	-1928.8	-1698.2	-77.4	
$PbCl_2$	-336.0	-286.9	123.4		$YbCl_3$	-1176.1	-1037.6	-71.0	-385.0
PbF_2	-666.9	-582.0	-17.2		$ZnBr_2$	-397.0	-355.0	52.7	-238.0
PbI_2	-112.1	-127.6	233.0		$ZnCl_2$	-488.2	-409.5	0.8	-226.0
$Pb(NO_3)_2$	-416.3	-246.9	303.3		ZnF_2	-819.1	-704.6	-139.7	-167.0
$PrCl_3$	-1206.2	-1072.7	-42.0	-439.0	ZnI_2	-264.3	-250.2	110.5	-238.0
$RaCl_2$	-861.9	-823.8	167.0		$Zn(NO_3)_2$	-568.6	-369.6	180.7	-126.0
$Ra(NO_3)_2$	-942.2	-784.0	347.0		$ZnSO_4$	-1063.2	-891.6	-92.0	-247.0

HEAT OF COMBUSTION

The heat of combustion of a substance at 25°C can be calculated from the enthalpy of formation ($\Delta_f H°$) data in the table "Standard Thermodynamic Properties of Chemical Substances" in this Section. We can write the general combustion reaction as

$$X + O_2 \rightarrow CO_2(g) + H_2O(l) + \text{other products}$$

For a compound containing only carbon, hydrogen, and oxygen, the reaction is simply

$$C_a H_b O_c + \left(a + \frac{1}{4}b - \frac{1}{2}c\right)O_2 \rightarrow a\ CO_2(g) + \frac{1}{2}b\ H_2O(l)$$

and the standard heat of combustion $\Delta_c H°$, which is defined as the negative of the enthalpy change for the reaction (i.e., the heat released in the combustion process), is given by

$$\Delta_c H° = -a\ \Delta_f H°(CO_2, g) - \frac{1}{2}b\ \Delta_f H°(H_2O, l) + \Delta_f H°(C_a H_b O_c)$$

$$393.51\ a + 142.915\ b + \Delta_f H°(C_a H_b O_c)$$

This equation applies if the reactants start in their standard states (25°C and one atmosphere pressure) and the products return to the same conditions. The same equation applies to a compound containing another element if that element ends in its standard reference state (e.g., nitrogen, if the product is N_2); in general, however, the exact products containing the other elements must be known in order to calculate the heat of combustion.

The following table gives the standard heat of combustion calculated in this manner for a few representative substances.

Molecular formula	Name	$\Delta_c H°/\text{kJ mol}^{-1}$
Inorganic substances		
C	Carbon (graphite)	393.5
CO	Carbon monoxide (g)	283.0
H_2	Hydrogen (g)	285.8
H_3N	Ammonia (g)	382.8
H_4N_2	Hydrazine (g)	667.1
N_2O	Nitrous oxide (g)	82.1
Hydrocarbons		
CH_4	Methane (g)	890.8
C_2H_2	Acetylene (g)	1301.1
C_2H_4	Ethylene (g)	1411.2
C_2H_6	Ethane (g)	1560.7
C_3H_6	Propylene (g)	2058.0
C_3H_6	Cyclopropane (g)	2091.3
C_3H_8	Propane (g)	2219.2
C_4H_6	1,3-Butadiene (g)	2541.5
C_4H_{10}	Butane (g)	2877.6
C_5H_{12}	Pentane (l)	3509.0
C_6H_6	Benzene (l)	3267.6
C_6H_{12}	Cyclohexane (l)	3919.6
C_6H_{14}	Hexane (l)	4163.2
C_7H_8	Toluene (l)	3910.3
C_7H_{16}	Heptane (l)	4817.0
$C_{10}H_8$	Naphthalene (s)	5156.3
Alcohols and ethers		
CH_4O	Methanol (l)	726.1
C_2H_6O	Ethanol (l)	1366.8
C_2H_6O	Dimethyl ether (g)	1460.4
$C_2H_6O_2$	Ethylene glycol (l)	1189.2

Molecular formula	Name	$\Delta_c H°/\text{kJ mol}^{-1}$
C_3H_8O	1-Propanol (l)	2021.3
$C_3H_8O_3$	Glycerol (l)	1655.4
$C_4H_{10}O$	Diethyl ether (l)	2723.9
$C_5H_{12}O$	1-Pentanol (l)	3330.9
C_6H_6O	Phenol (s)	3053.5
Carbonyl compounds		
CH_2O	Formaldehyde (g)	570.7
C_2H_2O	Ketene (g)	1025.4
C_2H_4O	Acetaldehyde (l)	1166.9
C_3H_6O	Acetone (l)	1789.9
C_3H_6O	Propanal (l)	1822.7
C_4H_8O	2-Butanone (l)	2444.1
Acids and esters		
CH_2O_2	Formic acid (l)	254.6
$C_2H_4O_2$	Acetic acid (l)	874.2
$C_2H_4O_2$	Methyl formate (l)	972.6
$C_3H_6O_2$	Methyl acetate (l)	1592.2
$C_4H_8O_2$	Ethyl acetate (l)	2238.1
$C_7H_6O_2$	Benzoic acid (s)	3226.9
Nitrogen compounds		
CHN	Hydrogen cyanide (g)	671.5
CH_3NO_2	Nitromethane (l)	709.2
CH_5N	Methylamine (g)	1085.6
C_2H_3N	Acetonitrile (l)	1247.2
C_2H_5NO	Acetamide (s)	1184.6
C_3H_9N	Trimethylamine (g)	2443.1
C_5H_5N	Pyridine (l)	2782.3
C_6H_7N	Aniline (l)	3392.8

MOLAR CONDUCTIVITY OF AQUEOUS HF, HCl, HBr, AND HI

The molar conductivity Λ of an electrolyte solution is defined as the conductivity divided by amount-of-substance concentration. The customary unit is S cm^{2}mol^{-1} (i.e., Ω^{-1} cm^{2}mol^{-1}). The first part of this table gives the molar conductivity of the hydrohalogen acids at 25°C as a function of the concentration in mol/L. The second part gives the temperature dependence of Λ for HCl and HBr. More extensive tables and mathematical representations may be found in the reference.

REFERENCE

Hamer, W.J., and DeWane, H.J., *Electrolytic Conductance and the Conductances of the Hydrohalogen Acids in Water*, Natl. Stand. Ref. Data Sys.- Natl. Bur. Standards (U.S.), No. 33, 1970.

c/mol L^{-1}	HF	HCl	HBr	HI	c/mol L^{-1}	HF	HCl	HBr	HI
Inf. dil.	405.1	426.1	427.7	426.4	3.5		218.3	217.5	215.4
0.0001		424.5	425.9	424.6	4.0		200.0	199.4	195.1
0.0005		422.6	424.3	423.0	4.5		183.1	182.4	176.8
0.001		421.2	422.9	421.7	5.0		167.4	166.5	160.4
0.005	128.1	415.7	417.6	416.4	5.5		152.9	151.8	145.5
0.01	96.1	411.9	413.7	412.8	6.0		139.7	138.2	131.7
0.05	50.1	398.9	400.4	400.8	6.5		127.7	125.7	118.6
0.10	39.1	391.1	391.9	394.0	7.0		116.9	114.2	105.7
0.5	26.3	360.7	361.9	369.8	7.5		107.0	103.8	
1.0	24.3	332.2	334.5	343.9	8.0		98.2	94.4	
1.5		305.8	307.6	316.4	8.5		90.3	85.8	
2.0		281.4	281.7	288.9	9.0		83.1		
2.5		258.9	257.8	262.5	9.5		76.6		
3.0		237.6	236.8	237.9	10.0		70.7		

c/mol L^{-1}	–20°C	–10°C	0°C	10°C	20°C	30°C	40°C	50°C
				HCl				
0.5			228.7	283.0	336.4	386.8	436.9	482.4
1.0			211.7	261.6	312.2	359.0	402.9	445.3
1.5			196.2	241.5	287.5	331.1	371.6	410.8
2.0			182.0	222.7	262.9	303.3	342.4	378.2
2.5		131.7	168.5	205.1	239.8	277.0	315.2	347.6
3.0		120.8	154.6	188.5	219.3	253.3	289.3	319.0
3.5	85.5	111.3	139.6	172.2	201.6	232.9	263.9	292.1
4.0	79.3	102.7	129.2	158.1	185.6	214.2	242.2	268.2
4.5	73.7	94.9	119.5	145.4	170.6	196.6	222.5	246.7
5.0	68.5	87.8	110.3	133.5	156.6	180.2	204.1	226.5
5.5	63.6	81.1	101.7	122.5	143.6	165.0	187.1	207.7
6.0	58.9	74.9	93.7	112.3	131.5	151.0	171.3	190.3
6.5	54.4	69.1	86.2	103.0	120.4	138.2	156.9	174.3
7.0	50.2	63.7	79.3	94.4	110.2	126.4	143.3	159.7
7.5	46.3	58.6	73.0	86.5	100.9	115.7	131.6	146.2
8.0	42.7	54.0	67.1	79.4	92.4	106.1	120.6	134.0
8.5	39.4	49.8	61.7	72.9	84.7	97.3	110.7	123.0
9.0	36.4	45.9	56.8	67.1	77.8	89.4	101.7	112.9
9.5	33.6	42.3	52.3	61.8	71.5	82.3	93.6	103.9
10.0	31.2	39.1	48.2	57.0	65.8	75.9	86.3	95.7
10.5	28.9	36.1	44.5	52.7	60.7	70.1	79.6	88.4
11.0	26.8	33.4	41.1	48.8	56.1	64.9	73.6	81.7
11.5	24.9	31.0	38.0	45.3	51.9	60.1	68.0	75.6
12.0	23.1	28.7	35.3	42.0	48.0	55.6	62.8	70.0
12.5	21.4	26.7	32.7	39.0	44.4	51.4	57.9	64.8

MOLAR CONDUCTIVITY OF AQUEOUS HF, HCl, HBr, AND HI (continued)

c/mol L^{-1}	–20°C	–10°C	0°C	10°C	20°C	30°C	40°C	50°C
				HBr				
0.5			240.9	295.9	347.0	398.9	453.6	496.8
1.0			229.6	276.0	329.0	380.4	418.6	465.2
1.5			209.5	254.9	298.9	340.6	381.8	421.4
2.0		150.8	188.6	231.3	271.8	314.1	350.5	387.4
2.5		136.8	171.7	208.3	244.8	281.7	316.0	349.1
3.0		125.7	157.2	189.5	222.2	255.0	287.8	318.6
3.5		116.1	144.1	174.6	203.2	234.4	263.7	291.9
4.0	84.0	107.5	132.3	160.2	186.8	214.2	239.7	266.9
4.5	78.0	99.0	123.0	146.4	171.2	195.1	218.8	242.6
5.0	72.3	91.4	112.6	134.0	155.7	178.2	199.6	221.3
5.5	67.0	84.2	103.1	122.7	142.1	162.8	181.4	201.8
6.0	61.8	77.2	94.3	112.0	129.6	148.0	165.4	183.4
6.5	56.8	70.7	86.0	102.0	118.0	134.1	150.5	166.3
7.0	51.9	64.6	78.4	92.6	107.1	121.4	136.3	150.8

STANDARD SOLUTIONS FOR CALIBRATING CONDUCTIVITY CELLS

Grams of KCl per kg of H$_2$O	k/(S/m)			
	0°C	18°C	25°C	25°C (Ref. 2)
76.5829	6.514	9.781	11.131	
7.47458	0.7134	1.1163	1.2852	1.2854
0.745819	0.07733	0.12201	0.14083	0.14086
7.45510 (0.1 m)				1.28217
0.745510 (0.01 m)				0.14079

Data are from Reference 1 except last column.

REFERENCES

1. K. N. Marsh, Ed., *Recommended Reference Materials for the Realization of Physicochemical Properties*, Blackwell, Oxford, 1987.
2. Y. C. Wu, K. W. Pratt, and W. F. Koch, *J. Solution Chem.*, 18, 515, 1989.

EQUIVALENT CONDUCTIVITY OF ELECTROLYTES IN AQUEOUS SOLUTION
Petr Vanýsek

This table gives the equivalent (molar) conductivity Λ at 25°C for some common electrolytes in aqueous solution at concentrations up to 0.1 mol/L. The units of Λ are 10^{-4} m^2 S mol^{-1}.

For very dilute solutions, the equivalent conductivity for any electrolyte of concentration c can be approximately calculated using the Debye-Hückel-Onsager equation, which can be written for a symmetrical (equal charge on cation and anion) electrolyte as

$$\Lambda = \Lambda° - (A + B\Lambda°)c^{1/2}$$

For a solution at 25°C and both cation and anion with charge $|1|$, the constants are $A = 60.20$ and $B = 0.229$. $\Lambda°$ can be found from the next table, "Ionic Conductivity and Diffusion at Infinite Dilution". The equation is reliable for $c < 0.001$ mol/L; with higher concentration the error increases.

Compound	Infinite dilution $\Lambda°$	Concentration (mol/L)						
		0.0005	0.001	0.005	0.01	0.02	0.05	0.1
					Λ			
AgNO$_3$	133.29	131.29	130.45	127.14	124.70	121.35	115.18	109.09
1/2BaCl$_2$	139.91	135.89	134.27	127.96	123.88	119.03	111.42	105.14
1/2CaCl$_2$	135.77	131.86	130.30	124.19	120.30	115.59	108.42	102.41
1/2Ca(OH)$_2$	258	—	—	233	226	214	—	—
1/2CuSO$_4$	133.6	121.6	115.20	94.02	83.08	72.16	59.02	50.55
HCl	425.95	422.53	421.15	415.59	411.80	407.04	398.89	391.13
KBr	151.9	149.8	148.9	146.02	143.36	140.41	135.61	131.32
KCl	149.79	147.74	146.88	143.48	141.20	138.27	133.30	128.90
KClO$_4$	139.97	138.69	137.80	134.09	131.39	127.86	121.56	115.14
1/3K$_3$Fe(CN)$_6$	174.5	166.4	163.1	150.7	—	—	—	—
1/4K$_4$Fe(CN)$_6$	184	—	167.16	146.02	134.76	122.76	107.65	97.82
KHCO$_3$	117.94	116.04	115.28	112.18	110.03	107.17	—	—
KI	150.31	148.2	143.32	144.30	142.11	139.38	134.90	131.05
KIO$_4$	127.86	125.74	124.88	121.18	118.45	114.08	106.67	98.2
KNO$_3$	144.89	142.70	141.77	138.41	132.75	132.34	126.25	120.34
KMnO$_4$	134.8	132.7	131.9	—	126.5	—	—	113
KOH	271.5	—	234	230	228	—	219	213
KReO$_4$	128.20	126.03	125.12	121.31	118.49	114.49	106.40	97.40
1/3LaCl$_3$	145.9	139.6	137.0	127.5	121.8	115.3	106.2	99.1
LiCl	114.97	113.09	112.34	109.35	107.27	104.60	100.06	95.81
LiClO$_4$	105.93	104.13	103.39	100.52	98.56	96.13	92.15	88.52
1/2MgCl$_2$	129.34	125.55	124.15	118.25	114.49	109.99	103.03	97.05
NH$_4$Cl	149.6	147.5	146.7	134.4	141.21	138.25	133.22	128.69
NaCl	126.39	124.44	123.68	120.59	118.45	115.70	111.01	106.69
NaClO$_4$	117.42	115.58	114.82	111.70	109.54	106.91	102.35	98.38
NaI	126.88	125.30	124.19	121.19	119.18	116.64	112.73	108.73
NaOOCCH$_3$	91.0	89.2	88.5	85.68	83.72	81.20	76.88	72.76
NaOH	247.7	245.5	244.6	240.7	237.9	—	—	—
Na picrate	80.45	78.7	78.6	75.7	73.7	—	66.3	61.8
1/2Na$_2$SO$_4$	129.8	125.68	124.09	117.09	112.38	106.73	97.70	89.94
1/2SrCl$_2$	135.73	131.84	130.27	124.18	120.23	115.48	108.20	102.14
1/2ZnSO$_4$	132.7	121.3	114.47	95.44	84.87	74.20	61.17	52.61

IONIC CONDUCTIVITY AND DIFFUSION AT INFINITE DILUTION
Petr Vanýsek

This table gives the molar (equivalent) conductivity λ for common ions at infinite dilution. All values refer to aqueous solutions at 25°C. It also lists the diffusion coefficient D of the ion in dilute aqueous solution, which is related to λ through the equation

$$D = \left(RT / F^2\right)\left(\lambda / |z|\right)$$

where R is the molar gas constant, T the temperature, F the Faraday constant, and z the charge on the ion. The variation with temperature is fairly sharp; for typical ions, λ and D increase by 2 to 3% per degree as the temperature increases from 25°C.

The diffusion coefficient for a salt, D_{salt}, may be calculated from the D_+ and D_- values of the constituent ions by the relation

$$D_{salt} = \frac{\left(z_+ + |z_-|\right)D_+ D_-}{z_+ D_+ + |z_-| D_-}$$

For solutions of simple, pure electrolytes (one positive and one negative ionic species), such as NaCl or $CaCl_2$, equivalent ionic conductivity $\Lambda°$, which is the conductivity per unit concentration of charge, is defined as

$$\Lambda° = \lambda_+ + \lambda_-$$

where λ_+ and λ_- are equivalent ionic conductivities of the cation and anion.

The ions are arranged in four groups, Inorganic Cations, Inorganic Anions, Organic Cations, and Organic Anions. Within each group, entries are listed in alphabetical order either by chemical formula (for inorganic ions) or name (for organic ions).

REFERENCES

1. Gray, D. E., Ed., *American Institute of Physics Handbook,* McGraw-Hill, New York, 1972, 2—226.
2. Robinson, R. A., and Stokes, R. H., *Electrolyte Solutions*, Butterworths, London, 1959.
3. Lobo, V. M. M., and Quaresma, J. L., *Handbook of Electrolyte Solutions*, Physical Science Data Series 41, Elsevier, Amsterdam, 1989.

Ion	λ 10^{-4} m^2 S mol^{-1}	D 10^{-5} cm^2 s^{-1}	Ion	λ 10^{-4} m^2 S mol^{-1}	D 10^{-5} cm^2 s^{-1}
Inorganic Cations			$1/3La^{3+}$	69.7	0.619
			Li^+	38.66	1.029
Ag^+	61.9	1.648	$1/2Mg^{2+}$	53.0	0.706
$1/3Al^{3+}$	61	0.541	$1/2Mn^{2+}$	53.5	0.712
$1/2Ba^{2+}$	63.6	0.847	NH_4^+	73.5	1.957
$1/2Be^{2+}$	45	0.599	$N_2H_5^+$	59	1.571
$1/2Ca^{2+}$	59.47	0.792	Na^+	50.08	1.334
$1/2Cd^{2+}$	54	0.719	$1/3Nd^{3+}$	69.4	0.616
$1/3Ce^{3+}$	69.8	0.620	$1/2Ni^{2+}$	49.6	0.661
$1/2Co^{2+}$	55	0.732	$1/4[Ni_2(trien)_3]^{4+}$	52	0.346
$1/3[Co(NH_3)_6]^{3+}$	101.9	0.904	$1/2Pb^{2+}$	71	0.945
$1/3[Co(en)_3]^{3+}$	74.7	0.663	$1/3Pr^{3+}$	69.5	0.617
$1/6[Co_2(trien)_3]^{6+}$	69	0.306	$1/2Ra^{2+}$	66.8	0.889
$1/3Cr^{3+}$	67	0.595	Rb^+	77.8	2.072
Cs^+	77.2	2.056	$1/3Sc^{3+}$	64.7	0.574
$1/2Cu^{2+}$	53.6	0.714	$1/3Sm^{3+}$	68.5	0.608
D^+	249.9	6.655	$1/2Sr^{2+}$	59.4	0.791
$1/3Dy^{3+}$	65.6	0.582	Tl^+	74.7	1.989
$1/3Er^{3+}$	65.9	0.585	$1/3Tm^{3+}$	65.4	0.581
$1/3Eu^{3+}$	67.8	0.602	$1/2UO_2^{2+}$	32	0.426
$1/2Fe^{2+}$	54	0.719	$1/3Y^{3+}$	62	0.550
$1/3Fe^{3+}$	68	0.604	$1/3Yb^{3+}$	65.6	0.582
$1/3Gd^{3+}$	67.3	0.597	$1/2Zn^{2+}$	52.8	0.703
H^+	349.65	9.311			
$1/2Hg^{2+}$	68.6	0.913			
$1/2Hg^{2+}$	63.6	0.847	**Inorganic Anions**		
$1/3Ho^{3+}$	66.3	0.589	$Au(CN)_2^-$	50	1.331
K^+	73.48	1.957	$Au(CN)_4^-$	36	0.959

Ion	λ 10^{-4} m^2 S mol^{-1}	D 10^{-5} cm^2 s^{-1}	Ion	λ 10^{-4} m^2 S mol^{-1}	D 10^{-5} cm^2 s^{-1}
B(C$_6$H$_5$)$_4^-$	21	0.559	**Organic Cations**		
Br$^-$	78.1	2.080			
Br$_3^-$	43	1.145	Benzyltrimethylammonium$^+$	34.6	0.921
BrO$_3^-$	55.7	1.483	Isobutylammonium$^+$	38	1.012
CN$^-$	78	2.077	Butyltrimethylammonium$^+$	33.6	0.895
CNO$^-$	64.6	1.720	Decylpyridinium$^+$	29.5	0.786
1/2CO$_3^{2-}$	69.3	0.923	Decyltrimethylammonium$^+$	24.4	0.650
Cl$^-$	76.31	2.032	Diethylammonium$^+$	42.0	1.118
ClO$_2^-$	52	1.385	Dimethylammonium$^+$	51.8	1.379
ClO$_3^-$	64.6	1.720	Dipropylammonium$^+$	30.1	0.802
ClO$_4^-$	67.3	1.792	Dodecylammonium$^+$	23.8	0.634
1/3[Co(CN)$_6$]$^{3-}$	98.9	0.878	Dodecyltrimethylammonium$^+$	22.6	0.602
1/2CrO$_4^{2-}$	85	1.132	Ethanolammonium$^+$	42.2	1.124
F$^-$	55.4	1.475	Ethylammonium$^+$	47.2	1.257
1/4[Fe(CN)$_6$]$^{4-}$	110.4	0.735	Ethyltrimethylammonium$^+$	40.5	1.078
1/3[Fe(CN)$_6$]$^{3-}$	100.9	0.896	Hexadecyltrimethylammonium$^+$	20.9	0.557
H$_2$AsO$_4^-$	34	0.905	Hexyltrimethylammonium$^+$	29.6	0.788
HCO$_3^-$	44.5	1.185	Histidyl$^+$	23.0	0.612
HF$_2^-$	75	1.997	Hydroxyethyltrimethylarsonium$^+$	39.4	1.049
1/2HPO$_4^{2-}$	33	0.439	Methylammonium$^+$	58.7	1.563
H$_2$PO$_4^-$	33	0.879	Octadecylpyridinium$^+$	20	0.533
H$_2$PO$_2^-$	46	1.225	Octadecyltributylammonium$^+$	16.6	0.442
HS$^-$	65	1.731	Octadecyltriethylammonium$^+$	17.9	0.477
HSO$_3^-$	50	1.331	Octadecyltrimethylammonium$^+$	19.9	0.530
HSO$_4^-$	50	1.331	Octadecyltripropylammonium$^+$	17.2	0.458
H$_2$SbO$_4^-$	31	0.825	Octyltrimethylammonium$^+$	26.5	0.706
I$^-$	76.8	2.045	Pentylammonium$^+$	37	0.985
IO$_3^-$	40.5	1.078	Piperidinium$^+$	37.2	0.991
IO$_4^-$	54.5	1.451	Propylammonium$^+$	40.8	1.086
MnO$_4^-$	61.3	1.632	Pyrilammonium$^+$	24.3	0.647
MoO$_4^-$	74.5	1.984	Tetrabutylammonium$^+$	19.5	0.519
N(CN)$_2^-$	54.5	1.451	Tetradecyltrimethylammonium$^+$	21.5	0.573
NO$_2^-$	71.8	1.912	Tetraethylammonium$^+$	32.6	0.868
NO$_3^-$	71.42	1.902	Tetramethylammonium$^+$	44.9	1.196
NH$_2$SO$_3^-$	48.3	1.286	Tetraisopentylammonium$^+$	17.9	0.477
N$_3^-$	69	1.837	Tetrapentylammmonium$^+$	17.5	0.466
OCN$^-$	64.6	1.720	Tetrapropylammonium$^+$	23.4	0.623
OH$^-$	198	5.273	Triethylammonium$^+$	34.3	0.913
PF$_6^-$	56.9	1.515	Triethylsulfonium$^+$	36.1	0.961
1/2PO$_3$F^{2-}	63.3	0.843	Trimethylammonium$^+$	47.23	1.258
1/3PO$_4^{3-}$	69.0	0.612	Trimethylhexylammonium$^+$	34.6	0.921
1/4P$_2$O$_7^{4-}$	96	0.639	Trimethylsulfonium$^+$	51.4	1.369
1/3P$_3$O$_9^{3-}$	83.6	0.742	Tripropylammonium$^+$	26.1	0.695
1/5P$_3$O$_{10}^{5-}$	109	0.581			
ReO$_4^-$	54.9	1.462	**Organic Anions**		
SCN$^-$	66	1.758			
1/2SO$_3^{2-}$	79.9	1.064	Acetate$^-$	40.9	1.089
1/2SO$_4^{2-}$	80.0	1.065	*p*-Anisate$^-$	29.0	0.772
1/2S$_2$O$_3^{2-}$	85.0	1.132	1/2Azelate^{2-}	40.6	0.541
1/2S$_4$O$_4^{2-}$	66.5	0.885	Benzoate$^-$	32.4	0.863
1/2S$_2$O$_6^{2-}$	93	1.238	Bromoacetate$^-$	39.2	1.044
1/2S$_2$O$_8^{2-}$	86	1.145	Bromobenzoate$^-$	30	0.799
Sb(OH)$_6^-$	31.9	0.849	Butyrate$^-$	32.6	0.868
SeCN$^-$	64.7	1.723	Chloroacetate$^-$	42.2	1.124
1/2SeO$_4^{2-}$	75.7	1.008	*m*-Chlorobenzoate$^-$	31	0.825
1/2WO$_4^{2-}$	69	0.919	*o*-Chlorobenzoate$^-$	30.2	0.804

Ion	λ 10^{-4} m^2 S mol^{-1}	D 10^{-5} cm^2 s^{-1}	Ion	λ 10^{-4} m^2 S mol^{-1}	D 10^{-5} cm^2 s^{-1}
1/3Citrate^{3-}	70.2	0.623	Iodoacetate$^-$	40.6	1.081
Crotonate$^-$	33.2	0.884	Lactate$^-$	38.8	1.033
Cyanoacetate$^-$	43.4	1.156	1/2Malate^{2-}	58.8	0.783
Cyclohexane carboxylate$^-$	28.7	0.764	1/2Maleate^{2-}	61.9	0.824
1/2 1,1-Cyclopropanedicarboxylate^{2-}	53.4	0.711	1/2Malonate^{2-}	63.5	0.845
Decylsulfate$^-$	26	0.692	Methylsulfate$^-$	48.8	1.299
Dichloroacetate$^-$	38.3	1.020	Naphthylacetate$^-$	28.4	0.756
1/2Diethylbarbiturate^{2-}	26.3	0.350	1/2Oxalate^{2-}	74.11	0.987
Dihydrogencitrate$^-$	30	0.799	Octylsulfate$^-$	29	0.772
1/2Dimethylmalonate^{2-}	49.4	0.658	Phenylacetate$^-$	30.6	0.815
3,5-Dinitrobenzoate$^-$	28.3	0.754	1/2o-Phthalate^{2-}	52.3	0.696
Dodecylsulfate$^-$	24	0.639	1/2m-Phthalate^{2-}	54.7	0.728
Ethylmalonate$^-$	49.3	1.313	Picrate$^-$	30.37	0.809
Ethylsulfate$^-$	39.6	1.055	Pivalate$^-$	31.9	0.849
Fluoroacetate$^-$	44.4	1.182	Propionate$^-$	35.8	0.953
Fluorobenzoate$^-$	33	0.879	Propylsulfate$^-$	37.1	0.988
Formate$^-$	54.6	1.454	Salicylate$^-$	36	0.959
1/2Fumarate^{2-}	61.8	0.823	1/2Suberate^{2-}	36	0.479
1/2Glutarate^{2-}	52.6	0.700	1/2Succinate^{2-}	58.8	0.783
Hydrogenoxalate$^-$	40.2	1.070	1/2Tartarate^{2-}	59.6	0.794
Isovalerate$^-$	32.7	0.871	Trichloroacetate$^-$	36.6	0.975

ACTIVITY COEFFICIENTS OF ACIDS, BASES, AND SALTS
Petr Vanýsek

This table gives mean activity coefficients at 25°C for molalities in the range 0.1 to 1.0. See the following table for definitions, references, and data over a wider concentration range.

	0.1	0.2	0.3	0.4	0.5	0.6	0.7	0.8	0.9	1.0
AgNO$_3$	0.734	0.657	0.606	0.567	0.536	0.509	0.485	0.464	0.446	0.429
AlCl$_3$	0.337	0.305	0.302	0.313	0.331	0.356	0.388	0.429	0.479	0.539
Al$_2$(SO$_4$)$_3$	0.035	0.0225	0.0176	0.0153	0.0143	0.014	0.0142	0.0149	0.0159	0.0175
BaCl$_2$	0.500	0.444	0.419	0.405	0.397	0.391	0.391	0.391	0.392	0.395
BeSO$_4$	0.150	0.109	0.0885	0.0769	0.0692	0.0639	0.0600	0.0570	0.0546	0.0530
CaCl$_2$	0.518	0.472	0.455	0.448	0.448	0.453	0.460	0.470	0.484	0.500
CdCl$_2$	0.2280	0.1638	0.1329	0.1139	0.1006	0.0905	0.0827	0.0765	0.0713	0.0669
Cd(NO$_3$)$_2$	0.513	0.464	0.442	0.430	0.425	0.423	0.423	0.425	0.428	0.433
CdSO$_4$	0.150	0.103	0.0822	0.0699	0.0615	0.0553	0.0505	0.0468	0.0438	0.0415
CoCl$_2$	0.522	0.479	0.463	0.459	0.462	0.470	0.479	0.492	0.511	0.531
CrCl$_3$	0.331	0.298	0.294	0.300	0.314	0.335	0.362	0.397	0.436	0.481
Cr(NO$_3$)$_3$	0.319	0.285	0.279	0.281	0.291	0.304	0.322	0.344	0.371	0.401
Cr$_2$(SO$_4$)$_3$	0.0458	0.0300	0.0238	0.0207	0.0190	0.0182	0.0181	0.0185	0.0194	0.0208
CsBr	0.754	0.694	0.654	0.626	0.603	0.586	0.571	0.558	0.547	0.538
CsCl	0.756	0.694	0.656	0.628	0.606	0.589	0.575	0.563	0.553	0.544
CsI	0.754	0.692	0.651	0.621	0.599	0.581	0.567	0.554	0.543	0.533
CsNO$_3$	0.733	0.655	0.602	0.561	0.528	0.501	0.478	0.458	0.439	0.422
CsOH	0.795	0.761	0.744	0.739	0.739	0.742	0.748	0.754	0.762	0.771
CsOAc	0.799	0.771	0.761	0.759	0.762	0.768	0.776	0.783	0.792	0.802
Cs$_2$SO$_4$	0.456	0.382	0.338	0.311	0.291	0.274	0.262	0.251	0.242	0.235
CuCl$_2$	0.508	0.455	0.429	0.417	0.411	0.409	0.409	0.410	0.413	0.417
Cu(NO$_3$)$_2$	0.511	0.460	0.439	0.429	0.426	0.427	0.431	0.437	0.445	0.455
CuSO$_4$	0.150	0.104	0.0829	0.0704	0.0620	0.0559	0.0512	0.0475	0.0446	0.0423
FeCl$_2$	0.5185	0.473	0.454	0.448	0.450	0.454	0.463	0.473	0.488	0.506
HBr	0.805	0.782	0.777	0.781	0.789	0.801	0.815	0.832	0.850	0.871
HCl	0.796	0.767	0.756	0.755	0.757	0.763	0.772	0.783	0.795	0.809
HClO$_4$	0.803	0.778	0.768	0.766	0.769	0.776	0.785	0.795	0.808	0.823
HI	0.818	0.807	0.811	0.823	0.839	0.860	0.883	0.908	0.935	0.963

	0.1	0.2	0.3	0.4	0.5	0.6	0.7	0.8	0.9	1.0
HNO_3	0.791	0.754	0.735	0.725	0.720	0.717	0.717	0.718	0.721	0.724
H_2SO_4	0.2655	0.2090	0.1826	—	0.1557	—	0.1417	—	—	0.1316
KBr	0.772	0.722	0.693	0.673	0.657	0.646	0.636	0.629	0.622	0.617
KCl	0.770	0.718	0.688	0.666	0.649	0.637	0.626	0.618	0.610	0.604
$KClO_3$	0.749	0.681	0.635	0.599	0.568	0.541	0.518	—	—	—
K_2CrO_4	0.456	0.382	0.340	0.313	0.292	0.276	0.263	0.253	0.243	0.235
KF	0.775	0.727	0.700	0.682	0.670	0.661	0.654	0.650	0.646	0.645
$K_3Fe(CN)_6$	0.268	0.212	0.184	0.167	0.155	0.146	0.140	0.135	0.131	0.128
$K_4Fe(CN)_6$	0.139	0.0993	0.0808	0.0693	0.0614	0.0556	0.0512	0.0479	0.0454	—
KH_2PO_4	0.731	0.653	0.602	0.561	0.529	0.501	0.477	0.456	0.438	0.421
KI	0.778	0.733	0.707	0.689	0.676	0.667	0.660	0.654	0.649	0.645
KNO_3	0.739	0.663	0.614	0.576	0.545	0.519	0.496	0.476	0.459	0.443
$KOAc$	0.796	0.766	0.754	0.750	0.751	0.754	0.759	0.766	0.774	0.783
KOH	0.798	0.760	0.742	0.734	0.732	0.733	0.736	0.742	0.749	0.756
$KSCN$	0.769	0.716	0.685	0.663	0.646	0.633	0.623	0.614	0.606	0.599
K_2SO_4	0.441	0.360	0.316	0.286	0.264	0.246	0.232	—	—	—
$LiBr$	0.796	0.766	0.756	0.752	0.753	0.758	0.767	0.777	0.789	0.803
$LiCl$	0.790	0.757	0.744	0.740	0.739	0.743	0.748	0.755	0.764	0.774
$LiClO_4$	0.812	0.794	0.792	0.798	0.808	0.820	0.834	0.852	0.869	0.887
LiI	0.815	0.802	0.804	0.813	0.824	0.838	0.852	0.870	0.888	0.910
$LiNO_3$	0.788	0.752	0.736	0.728	0.726	0.727	0.729	0.733	0.737	0.743
$LiOH$	0.760	0.702	0.665	0.638	0.617	0.599	0.585	0.573	0.563	0.554
$LiOAc$	0.784	0.742	0.721	0.709	0.700	0.691	0.689	0.688	0.688	0.689
Li_2SO_4	0.468	0.398	0.361	0.337	0.319	0.307	0.297	0.289	0.282	0.277
$MgCl_2$	0.529	0.489	0.477	0.475	0.481	0.491	0.506	0.522	0.544	0.570
$MgSO_4$	0.150	0.107	0.0874	0.0756	0.0675	0.0616	0.0571	0.0536	0.0508	0.0485
$MnCl_2$	0.516	0.469	0.450	0.442	0.440	0.443	0.448	0.455	0.466	0.479
$MnSO_4$	0.150	0.105	0.0848	0.0725	0.0640	0.0578	0.0530	0.0493	0.0463	0.0439
NH_4Cl	0.770	0.718	0.687	0.665	0.649	0.636	0.625	0.617	0.609	0.603
NH_4NO_3	0.740	0.677	0.636	0.606	0.582	0.562	0.545	0.530	0.516	0.504
$(NH_4)_2SO_4$	0.439	0.356	0.311	0.280	0.257	0.240	0.226	0.214	0.205	0.196
$NaBr$	0.782	0.741	0.719	0.704	0.697	0.692	0.689	0.687	0.687	0.687
$NaCl$	0.778	0.735	0.710	0.693	0.681	0.673	0.667	0.662	0.659	0.657
$NaClO_3$	0.772	0.720	0.688	0.664	0.645	0.630	0.617	0.606	0.597	0.589
$NaClO_4$	0.775	0.729	0.701	0.683	0.668	0.656	0.648	0.641	0.635	0.629
Na_2CrO_4	0.464	0.394	0.353	0.327	0.307	0.292	0.280	0.269	0.261	0.253
NaF	0.765	0.710	0.676	0.651	0.632	0.616	0.603	0.592	0.582	0.573
NaH_2PO_4	0.744	0.675	0.629	0.593	0.563	0.539	0.517	0.499	0.483	0.468
NaI	0.787	0.751	0.735	0.727	0.723	0.723	0.724	0.727	0.731	0.736
$NaNO_3$	0.762	0.703	0.666	0.638	0.617	0.599	0.583	0.570	0.558	0.548
$NaOAc$	0.791	0.757	0.744	0.737	0.735	0.736	0.740	0.745	0.752	0.757
$NaOH$	0.766	0.727	0.708	0.697	0.690	0.685	0.681	0.679	0.678	0.678
$NaSCN$	0.787	0.750	—	0.720	0.715	0.712	0.710	0.710	0.711	0.712
Na_2SO_4	0.445	0.365	0.320	0.289	0.266	0.248	0.233	0.221	0.210	0.201
$NiCl_2$	0.522	0.479	0.463	0.460	0.464	0.471	0.482	0.496	0.515	0.563
$NiSO_4$	0.150	0.105	0.0841	0.0713	0.0627	0.0562	0.0515	0.0478	0.0448	0.0425
$Pb(NO_3)_2$	0.395	0.308	0.260	0.228	0.205	0.187	0.172	0.160	0.150	0.141
$RbBr$	0.763	0.706	0.673	0.650	0.632	0.617	0.605	0.595	0.586	0.578
$RbCl$	0.764	0.709	0.675	0.652	0.634	0.620	0.608	0.599	0.590	0.583
RbI	0.762	0.705	0.671	0.647	0.629	0.614	0.602	0.591	0.583	0.575
$RbNO_3$	0.734	0.658	0.606	0.565	0.534	0.508	0.485	0.465	0.446	0.430
$RbOAc$	0.796	0.767	0.756	0.753	0.755	0.759	0.766	0.773	0.782	0.792
Rb_2SO_4	0.451	0.374	0.331	0.301	0.279	0.263	0.249	0.238	0.228	0.219
$SrCl_2$	0.511	0.462	0.442	0.433	0.430	0.431	0.434	0.441	0.449	0.461
$TlClO_4$	0.730	0.652	0.599	0.559	0.527	—	—	—	—	—
$TlNO_3$	0.702	0.606	0.545	0.500	—	—	—	—	—	—
UO_2Cl_2	0.544	0.510	0.520	0.505	0.517	0.532	0.549	0.571	0.595	0.620
UO_2SO_4	0.150	0.102	0.0807	0.0689	0.0611	0.0566	0.0515	0.0483	0.0458	0.0439
$ZnCl_2$	0.515	0.462	0.432	0.411	0.394	0.380	0.369	0.357	0.348	0.339
$Zn(NO_3)_2$	0.531	0.489	0.474	0.469	0.473	0.480	0.489	0.501	0.518	0.535
$ZnSO_4$	0.150	0.140	0.0835	0.0714	0.0630	0.0569	0.0523	0.0487	0.0458	0.0435

MEAN ACTIVITY COEFFICIENTS OF ELECTROLYTES AS A FUNCTION OF CONCENTRATION

The mean activity coefficient γ of an electrolyte X_aY_b is defined as

$$\gamma = \left(\gamma_+^a \gamma_-^b\right)^{1/(a+b)}$$

where γ_+ and γ_- are activity coefficients of the individual ions (which cannot be directly measured). This table gives the mean activity coefficients of about 100 electrolytes in aqueous solution as a function of concentration, expressed in molality terms. All values refer to a temperature of 25°C. Substances are arranged in alphabetical order by formula.

REFERENCES

1. Hamer, W. J., and Wu, Y. C., *J. Phys. Chem. Ref. Data*, 1, 1047, 1972.
2. Staples, B. R., *J. Phys. Chem. Ref. Data*, 6, 385, 1977; 10, 767, 1981; 10, 779, 1981.
3. Goldberg, R. N. et al., *J. Phys. Chem. Ref. Data*, 7, 263, 1978; 8, 923, 1979; 8, 1005, 1979; 10, 1, 1981; 10, 671, 1981.

Mean Activity Coefficient at 25°C

m/mol kg^{-1}	$AgNO_3$	$BaBr_2$	$BaCl_2$	BaI_2	$CaBr_2$	$CaCl_2$	CaI_2
0.001	0.964	0.881	0.887	0.890	0.890	0.888	0.890
0.002	0.950	0.850	0.849	0.853	0.853	0.851	0.853
0.005	0.924	0.785	0.782	0.792	0.791	0.787	0.791
0.010	0.896	0.727	0.721	0.737	0.735	0.727	0.736
0.020	0.859	0.661	0.653	0.678	0.674	0.664	0.677
0.050	0.794	0.573	0.559	0.600	0.594	0.577	0.600
0.100	0.732	0.517	0.492	0.551	0.540	0.517	0.552
0.200	0.656	0.463	0.436	0.520	0.502	0.469	0.524
0.500	0.536	0.435	0.391	0.536	0.500	0.444	0.554
1.000	0.430	0.470	0.393	0.664	0.604	0.495	0.729
2.000	0.316	0.654		1.242	1.125	0.784	
5.000	0.181				18.7	5.907	
10.000	0.108					43.1	
15.000	0.085						

m/mol kg^{-1}	$Cd(NO_2)_2$	$Cd(NO_3)_2$	$CoBr_2$	$CoCl_2$	CoI_2	$Co(NO_3)_2$	$CsBr$
0.001	0.881	0.888	0.890	0.889	0.887	0.888	0.965
0.002	0.837	0.851	0.854	0.852	0.849	0.850	0.951
0.005	0.759	0.787	0.794	0.789	0.783	0.786	0.925
0.010	0.681	0.728	0.740	0.732	0.724	0.728	0.898
0.020	0.589	0.664	0.681	0.670	0.661	0.663	0.864
0.050	0.451	0.576	0.605	0.586	0.582	0.576	0.806
0.100	0.344	0.515	0.556	0.528	0.540	0.516	0.752
0.200	0.247	0.465	0.523	0.483	0.527	0.469	0.691
0.500	0.148	0.428	0.538	0.465	0.596	0.446	0.605
1.000	0.098	0.437	0.685	0.532	0.845	0.492	0.540
2.000	0.069	0.517	1.421	0.864	2.287	0.722	0.485
5.000	0.054		13.9		55.3	3.338	0.454
10.000					196		

m/mol kg^{-1}	$CsCl$	CsF	CsI	$CsNO_3$	$CsOH$	Cs_2SO_4	$CuBr_2$
0.001	0.965	0.965	0.965	0.964	0.966	0.885	0.889
0.002	0.951	0.952	0.951	0.951	0.953	0.845	0.853
0.005	0.925	0.929	0.925	0.924	0.930	0.775	0.791
0.010	0.898	0.905	0.898	0.897	0.906	0.709	0.735
0.020	0.864	0.876	0.863	0.860	0.878	0.634	0.674
0.050	0.805	0.830	0.804	0.796	0.836	0.526	0.594
0.100	0.751	0.792	0.749	0.733	0.802	0.444	0.541

MEAN ACTIVITY COEFFICIENTS OF ELECTROLYTES AS A FUNCTION
OF CONCENTRATION (continued)

m/mol kg^{-1}	CsCl	CsF	CsI	CsNO$_3$	CsOH	Cs$_2$SO$_4$	CuBr$_2$
0.200	0.691	0.755	0.688	0.655	0.772	0.369	0.504
0.500	0.607	0.721	0.601	0.529	0.755	0.285	0.503
1.000	0.546	0.726	0.534	0.421	0.782	0.233	0.591
2.000	0.496	0.803	0.470				0.859
5.000	0.474						
10.000	0.508						

m/mol kg^{-1}	CuCl$_2$	Cu(ClO$_4$)$_2$	Cu(NO$_3$)$_2$	FeCl$_2$	HBr	HCl	HClO$_4$
0.001	0.887	0.890	0.888	0.888	0.966	0.965	0.966
0.002	0.849	0.854	0.851	0.850	0.953	0.952	0.953
0.005	0.783	0.795	0.787	0.785	0.930	0.929	0.929
0.010	0.722	0.741	0.729	0.725	0.907	0.905	0.906
0.020	0.654	0.685	0.664	0.659	0.879	0.876	0.878
0.050	0.561	0.613	0.577	0.570	0.837	0.832	0.836
0.100	0.495	0.572	0.516	0.509	0.806	0.797	0.803
0.200	0.441	0.553	0.466	0.462	0.783	0.768	0.776
0.500	0.401	0.617	0.431	0.443	0.790	0.759	0.769
1.000	0.405	0.892	0.456	0.500	0.872	0.811	0.826
2.000	0.453	2.445	0.615	0.782	1.167	1.009	1.055
5.000	0.601		2.083		3.800	2.380	3.100
10.000					33.4	10.4	30.8
15.000							323

m/mol kg^{-1}	HF	HI	HNO$_3$	H$_2$SO$_4$	KBr	KCNS	KCl
0.001	0.551	0.966	0.965	0.804	0.965	0.965	0.965
0.002	0.429	0.953	0.952	0.740	0.952	0.951	0.951
0.005	0.302	0.931	0.929	0.634	0.927	0.927	0.927
0.010	0.225	0.909	0.905	0.542	0.902	0.901	0.901
0.020	0.163	0.884	0.875	0.445	0.870	0.869	0.869
0.050	0.106	0.847	0.829	0.325	0.817	0.815	0.816
0.100	0.0766	0.823	0.792	0.251	0.771	0.768	0.768
0.200	0.0550	0.811	0.756	0.195	0.772	0.716	0.717
0.500	0.0352	0.845	0.725	0.146	0.658	0.647	0.649
1.000	0.0249	0.969	0.730	0.125	0.617	0.598	0.604
2.000	0.0175	1.363	0.788	0.119	0.593	0.556	0.573
5.000	0.0110	4.760	1.063	0.197	0.626	0.525	0.593
10.000	0.0085	49.100	1.644	0.527			
15.000	0.0077		2.212	1.077			
20.000	0.0075		2.607	1.701			

m/mol kg^{-1}	KClO$_3$	K$_2$CrO$_4$	KF	KH$_2$PO$_4$*	K$_2$HPO$_4$**	KI	KNO$_3$
0.001	0.965	0.886	0.965	0.964	0.886	0.965	0.964
0.002	0.951	0.847	0.952	0.950	0.847	0.952	0.950
0.005	0.926	0.779	0.927	0.924	0.779	0.927	0.924
0.010	0.899	0.715	0.902	0.896	0.715	0.902	0.896
0.020	0.865	0.643	0.870	0.859	0.643	0.871	0.860
0.050	0.805	0.539	0.818	0.793	0.538	0.820	0.797
0.100	0.749	0.460	0.773	0.730	0.457	0.776	0.735
0.200	0.681	0.385	0.726	0.652	0.379	0.731	0.662
0.500	0.569	0.296	0.670	0.529	0.283	0.676	0.546
1.000		0.239	0.645	0.422		0.646	0.444

m/mol kg^{-1}	KClO$_3$	K$_2$CrO$_4$	KF	KH$_2$PO$_4$*	K$_2$HPO$_4$**	KI	KNO$_3$
2.000		0.199	0.658			0.638	0.332
5.000			0.871				
10.000			1.715				
15.000			3.120				

m/mol kg^{-1}	KOH	K$_2$SO$_4$	LiBr	LiCl	LiClO$_4$	LiI	LiNO$_3$
0.001	0.965	0.885	0.965	0.965	0.966	0.966	0.965
0.002	0.952	0.844	0.952	0.952	0.953	0.953	0.952
0.005	0.927	0.772	0.929	0.928	0.931	0.930	0.928
0.010	0.902	0.704	0.905	0.904	0.908	0.908	0.904
0.020	0.871	0.625	0.877	0.874	0.882	0.882	0.874
0.050	0.821	0.511	0.832	0.827	0.843	0.843	0.827
0.100	0.779	0.424	0.797	0.789	0.815	0.817	0.788
0.200	0.740	0.343	0.767	0.756	0.795	0.802	0.753
0.500	0.710	0.251	0.754	0.739	0.806	0.824	0.726
1.000	0.733		0.803	0.775	0.887	0.912	0.743
2.000	0.860		1.012	0.924	1.161	1.197	0.837
5.000	1.697		2.696	2.000			1.298
10.000	6.110		20.0	9.600			2.500
15.000	19.9		147	30.9			3.960
20.000	46.4		486				4.970

m/mol kg^{-1}	LiOH	Li$_2$SO$_4$	MgBr$_2$	MgCl$_2$	MgI$_2$	MnBr$_2$	MnCl$_2$
0.001	0.964	0.887	0.889	0.889	0.889	0.889	0.888
0.002	0.950	0.847	0.852	0.852	0.853	0.853	0.850
0.005	0.923	0.780	0.790	0.790	0.791	0.791	0.786
0.010	0.895	0.716	0.733	0.734	0.736	0.735	0.727
0.020	0.858	0.645	0.672	0.672	0.677	0.674	0.662
0.050	0.794	0.544	0.593	0.590	0.602	0.595	0.574
0.100	0.735	0.469	0.543	0.535	0.556	0.543	0.513
0.200	0.668	0.400	0.512	0.493	0.535	0.508	0.464
0.500	0.579	0.325	0.540	0.485	0.594	0.519	0.437
1.000	0.522	0.284	0.715	0.577	0.858	0.650	0.477
2.000	0.484	0.270	1.590	1.065	2.326	1.224	0.661
5.000	0.493		36.1	14.40	109.8	6.697	1.539

m/mol kg^{-1}	Mn(ClO$_4$)$_2$	NH$_4$Cl	NH$_4$ClO$_4$	(NH$_4$)$_2$HPO$_4$**	NH$_4$NO$_3$	NaBr	NaBrO$_3$
0.001	0.892	0.965	0.964	0.882	0.964	0.965	0.965
0.002	0.858	0.952	0.950	0.839	0.951	0.952	0.951
0.005	0.801	0.927	0.924	0.763	0.925	0.928	0.926
0.010	0.752	0.901	0.895	0.688	0.897	0.903	0.900
0.020	0.700	0.869	0.859	0.600	0.862	0.873	0.867
0.050	0.637	0.816	0.794	0.469	0.801	0.824	0.811
0.100	0.604	0.769	0.734	0.367	0.744	0.783	0.759
0.200	0.596	0.718	0.663	0.273	0.678	0.742	0.698
0.500	0.686	0.649	0.560	0.171	0.582	0.697	0.605
1.000	1.030	0.603	0.479	0.114	0.502	0.687	0.528
2.000	3.072	0.569	0.399	0.074	0.419	0.730	0.449
5.000		0.563			0.303	1.083	
10.000					0.220		
15.000					0.179		
20.000					0.154		

m/mol kg^{-1}	Na_2CO_3	NaCl	$NaClO_3$	$NaClO_4$	Na_2CrO_4	NaF	Na_2HPO_4*
0.001	0.887	0.965	0.965	0.965	0.887	0.965	0.887
0.002	0.847	0.952	0.952	0.952	0.849	0.951	0.848
0.005	0.780	0.928	0.927	0.928	0.783	0.926	0.780
0.010	0.716	0.903	0.902	0.903	0.722	0.901	0.717
0.020	0.644	0.872	0.870	0.872	0.653	0.868	0.644
0.050	0.541	0.822	0.818	0.821	0.554	0.813	0.539
0.100	0.462	0.779	0.771	0.777	0.479	0.764	0.456
0.200	0.385	0.734	0.719	0.729	0.406	0.710	0.373
0.500	0.292	0.681	0.646	0.668	0.318	0.633	0.266
1.000	0.229	0.657	0.590	0.630	0.261	0.573	0.191
2.000	0.182	0.668	0.537	0.608	0.231		0.133
5.000		0.874		0.648			

m/mol kg^{-1}	NaI	$NaNO_3$	NaOH	Na_2SO_3	Na_2SO_4	Na_2WO_4	$NiBr_2$
0.001	0.965	0.965	0.965	0.887	0.886	0.886	0.889
0.002	0.952	0.951	0.952	0.847	0.846	0.846	0.853
0.005	0.928	0.926	0.927	0.779	0.777	0.777	0.791
0.010	0.904	0.900	0.902	0.716	0.712	0.712	0.735
0.020	0.874	0.866	0.870	0.644	0.637	0.638	0.675
0.050	0.827	0.810	0.819	0.540	0.529	0.534	0.596
0.100	0.789	0.759	0.775	0.462	0.446	0.457	0.546
0.200	0.753	0.701	0.731	0.386	0.366	0.388	0.514
0.500	0.722	0.617	0.685	0.296	0.268	0.320	0.535
1.000	0.734	0.550	0.674	0.237	0.204	0.291	0.692
2.000	0.823	0.480	0.714	0.196	0.155	0.291	1.476
5.000	1.402	0.388	1.076				
10.000	4.011	0.329	3.258				
15.000			9.796				
20.000			19.410				

m/mol kg^{-1}	$NiCl_2$	$Ni(ClO_4)_2$	$Ni(NO_3)_2$	$Pb(ClO_4)_2$	$Pb(NO_3)_2$	RbBr	RbCl
0.001	0.889	0.891	0.889	0.889	0.882	0.965	0.965
0.002	0.852	0.855	0.851	0.851	0.840	0.951	0.951
0.005	0.789	0.797	0.787	0.787	0.764	0.926	0.926
0.010	0.732	0.745	0.730	0.729	0.690	0.900	0.900
0.020	0.669	0.690	0.666	0.666	0.604	0.866	0.867
0.050	0.584	0.621	0.581	0.580	0.476	0.811	0.811
0.100	0.527	0.582	0.524	0.522	0.379	0.760	0.761
0.200	0.482	0.567	0.481	0.476	0.291	0.705	0.707
0.500	0.465	0.639	0.467	0.458	0.195	0.630	0.633
1.000	0.538	0.946	0.528	0.516	0.136	0.578	0.583
2.000	0.915	2.812	0.797	0.799		0.535	0.546
5.000	4.785			4.043		0.514	0.544
10.000				33.8			

m/mol kg^{-1}	RbF	RbI	$RbNO_3$	Rb_2SO_4	$SrBr_2$	$SrCl_2$	SrI_2
0.001	0.965	0.965	0.964	0.886	0.889	0.888	0.890
0.002	0.952	0.951	0.950	0.845	0.852	0.850	0.854
0.005	0.927	0.926	0.924	0.776	0.790	0.785	0.793
0.010	0.902	0.900	0.896	0.710	0.734	0.725	0.740
0.020	0.871	0.866	0.859	0.635	0.673	0.659	0.681

MEAN ACTIVITY COEFFICIENTS OF ELECTROLYTES AS A FUNCTION
OF CONCENTRATION (continued)

m/mol kg^{-1}	RbF	RbI	RbNO$_3$	Rb$_2$SO$_4$	SrBr$_2$	SrCl$_2$	SrI$_2$
0.050	0.821	0.810	0.795	0.526	0.591	0.569	0.606
0.100	0.780	0.759	0.733	0.443	0.535	0.506	0.557
0.200	0.739	0.703	0.657	0.365	0.492	0.455	0.526
0.500	0.701	0.627	0.536	0.274	0.476	0.421	0.542
1.000	0.697	0.574	0.430	0.217	0.545	0.451	0.686
2.000	0.724	0.532	0.320		0.921	0.650	
5.000		0.517					

m/mol kg^{-1}	UO$_2$Cl$_2$	UO$_2$(NO$_3$)$_2$	ZnBr$_2$	ZnCl$_2$	ZnI$_2$
0.001	0.888	0.888	0.890	0.887	0.893
0.002	0.851	0.849	0.854	0.847	0.859
0.005	0.787	0.784	0.794	0.781	0.804
0.010	0.729	0.726	0.741	0.719	0.757
0.020	0.666	0.663	0.683	0.652	0.708
0.050	0.583	0.583	0.606	0.561	0.644
0.100	0.529	0.535	0.553	0.499	0.601
0.200	0.493	0.509	0.515	0.447	0.574
0.500	0.501	0.532	0.516	0.384	0.635
1.000	0.601	0.673	0.558	0.330	0.836
2.000	0.948	1.223	0.578	0.283	1.062
5.000		3.020	0.788	0.342	1.546
10.000			2.317	0.876	4.698
15.000			5.381	1.914	
20.000			7.965	2.968	

*   The anion is H$_2$PO$_4^-$.
**   The anion is HPO$_4^{-2}$.

ENTHALPY OF DILUTION OF ACIDS

The quantity given in this table is $-\Delta_{dil}H$, the negative of the enthalpy (heat) of dilution to infinite dilution for aqueous solutions of several common acids; i.e., the negative of the enthalphy change when a solution of molality m at a temperature of 25°C is diluted with an infinite amount of water. The tabulated numbers thus represent the heat produced (or, if the value is negative, the heat absorbed) when the acid is diluted. The initial molality m is given in the first column. The second column gives the dilution ratio, which is the number of moles of water that must be added to one mole of the acid to produce a solution of the molality in the first column.

REFERENCE

Parker, V. B., *Thermal Properties of Aqueous Uni-Univalent Electrolytes*, Natl. Stand. Ref. Data Ser. - Natl. Bur. Stand. (U.S.) 2, U.S. Government Printing Office, 1965.

$-\Delta_{dil}H$ in kJ/mol at 25°C

m	Dil. ratio	HF	HCl	HClO$_4$	HBr	HI	HNO$_3$	CH$_2$O$_2$	C$_2$H$_4$O$_2$
55.506	1.0		45.61		48.83		19.73	0.046	2.167
20	2.775	14.88	19.87	13.81	19.92	21.71	9.498	0.038	2.075
15	3.700	14.34	15.40	7.920	14.29	14.02	6.883	0.109	1.962
10	5.551	13.87	10.24	2.013	8.694	7.615	3.933	0.205	1.824
9	6.167	13.81	9.213	1.280	7.719	6.569	3.368	0.230	1.782
8	6.938	13.77	8.201	0.611	6.786	5.607	2.791	0.255	1.724
7	7.929	13.73	7.217	0.046	5.925	4.728	2.251	0.272	1.648
6	9.251	13.69	6.268	-0.351	5.004	3.975	1.749	0.280	1.540
5.5506	10	13.66	5.841	-0.490	4.590	3.577	1.540	0.285	1.477
5	11.10	13.62	5.318	-0.628	4.113	3.197	1.310	0.289	1.393
4.5	12.33	13.58	4.899	-0.732	3.711	2.828	1.109	0.289	1.310
4	13.88	13.53	4.402	-0.787	3.330	2.460	0.958	0.289	1.218
3.5	15.86	13.47	3.958	-0.820	2.966	2.105	0.791	0.289	1.121
3	18.50	13.45	3.506	-0.782	2.611	1.787	0.665	0.289	1.025
2.5	22.20	13.43	3.063	-0.724	2.301	1.527	0.582	0.285	0.912
2	27.75	13.40	2.623	-0.623	1.996	1.318	0.527	0.276	0.803
1.5	37.00	13.36	2.167	-0.431	1.665	1.125	0.506	0.259	0.678
1	55.51	13.30	1.695	-0.201	1.314	0.933	0.506	0.226	0.544
0.5551	100	13.22	1.234	0.050	0.983	0.736	0.502	0.184	0.423
0.5	111.0	13.20	1.172	0.075	0.941	0.711	0.498	0.176	0.406
0.2	277.5	13.09	0.761	0.247	0.649	0.536	0.439	0.146	0.331
0.1	555.1	12.80	0.556	0.272	0.498	0.439	0.372	0.134	0.289
0.0925	600	12.79	0.540	0.272	0.481	0.427	0.368	0.134	0.285
0.0793	700	12.70	0.502	0.272	0.452	0.402	0.351	0.134	0.285
0.0694	800	12.61	0.473	0.268	0.427	0.385	0.339	0.130	0.280
0.0617	900	12.50	0.448	0.264	0.406	0.368	0.326	0.126	0.276
0.05551	1000	12.42	0.427	0.259	0.385	0.351	0.318	0.121	0.272
0.05	1110	12.24	0.406	0.259	0.372	0.339	0.305	0.121	0.272
0.02775	2000	11.29	0.310	0.226	0.285	0.264	0.247	0.117	0.264
0.01850	3000	10.66	0.251	0.197	0.234	0.218	0.213	0.117	0.259
0.01388	4000	10.25	0.226	0.180	0.205	0.192	0.192	0.113	0.259
0.01110	5000	9.874	0.197	0.167	0.184	0.172	0.176	0.109	0.255
0.00555	10000	8.912	0.142	0.126	0.130	0.121	0.130	0.105	0.243
0.00278	20000	7.531	0.105	0.092	0.092	0.084	0.096	0.096	0.230
0.00111	50000	5.439	0.067	0.059	0.054	0.050	0.063	0.084	0.222
0.000555	100000	3.766	0.042	0.042	0.038	0.038	0.046	0.054	0.209
0.000111	500000	1.255	0.021	0.021	0.021	0.021	0.021	0.038	0.167
0	∞	0	0	0	0	0	0	0	0

ENTHALPY OF SOLUTION OF ELECTROLYTES

This table gives the molar enthalpy (heat) of solution at infinite dilution for some common uni-univalent electrolytes. This is the enthalpy change when 1 mol of solute in its standard state is dissolved in an infinite amount of water. Values are given in kilojoules per mole at 25°C.

REFERENCE

Parker, V. B., *Thermal Properties of Uni-Univalent Electrolytes*, Natl. Stand. Ref. Data Series — Natl. Bur. Stand.(U.S.), No.2, 1965.

Solute	State	$\Delta_{sol} H°$ kJ/mol	Solute	State	$\Delta_{sol} H°$ kJ/mol	Solute	State	$\Delta_{sol} H°$ kJ/mol
HF	g	−61.50	$LiBr \cdot 2H_2O$	c	−9.41	KCl	c	17.22
HCl	g	−74.84	$LiBrO_3$	c	1.42	$KClO_3$	c	41.38
$HClO_4$	l	−88.76	LiI	c	−63.30	$KClO_4$	c	51.04
$HClO_4 \cdot H_2O$	c	−32.95	$LiI \cdot H_2O$	c	−29.66	KBr	c	19.87
HBr	g	−85.14	$LiI \cdot 2H_2O$	c	−14.77	$KBrO_3$	c	41.13
HI	g	−81.67	$LiI \cdot 3H_2O$	c	0.59	KI	c	20.33
HIO_3	c	8.79	$LiNO_2$	c	−11.00	KIO_3	c	27.74
HNO_3	l	−33.28	$LiNO_2 \cdot H_2O$	c	7.03	KNO_2	c	13.35
HCOOH	l	−0.86	$LiNO_3$	c	−2.51	KNO_3	c	34.89
CH_3COOH	l	−1.51				$KC_2H_3O_2$	c	−15.33
			NaOH	c	−44.51	KCN	c	11.72
NH_3	g	−30.50	$NaOH \cdot H_2O$	c	−21.41	KCNO	c	20.25
NH_4Cl	c	14.78	NaF	c	0.91	KCNS	c	24.23
NH_4ClO_4	c	33.47	NaCl	c	3.88	$KMnO_4$	c	43.56
NH_4Br	c	16.78	$NaClO_2$	c	0.33			
NH_4I	c	13.72	$NaClO_2 \cdot 3H_2O$	c	28.58	RbOH	c	−62.34
NH_4IO_3	c	31.80	$NaClO_3$	c	21.72	$RbOH \cdot H_2O$	c	−17.99
NH_4NO_2	c	19.25	$NaClO_4$	c	13.88	$RbOH \cdot 2H_2O$	c	0.88
NH_4NO_3	c	25.69	$NaClO_4 \cdot H_2O$	c	22.51	RbF	c	−26.11
$NH_4C_2H_3O_2$	c	−2.38	NaBr	c	−0.60	$RbF \cdot H_2O$	c	−0.42
NH_4CN	c	17.57	$NaBr \cdot 2H_2O$	c	18.64	$RbF \cdot 1.5H_2O$	c	1.34
NH_4CNS	c	22.59	$NaBrO_3$	c	26.90	RbCl	c	17.28
CH_3NH_3Cl	c	5.77	NaI	c	7.53	$RbClO_3$	c	47.74
$(CH_3)_3NHCl$	c	1.46	$NaI \cdot 2H_2O$	c	16.13	$RbClO_4$	c	56.74
$N(CH_3)_4Cl$	c	4.08	$NaIO_3$	c	20.29	RbBr	c	21.88
$N(CH_3)_4Br$	c	24.27	$NaNO_2$	c	13.89	$RbBrO_3$	c	48.95
$N(CH_3)_4I$	c	42.07	$NaNO_3$	c	20.50	RbI	c	25.10
			$NaC_2H_3O_2$	c	−17.32	$RbNO_3$	c	36.48
$AgClO_4$	c	7.36	$NaC_2H_3O_2 \cdot 3H_2O$	c	19.66			
$AgNO_2$	c	36.94	NaCN	c	1.21	CsOH	c	−71.55
$AgNO_3$	c	22.59	$NaCN \cdot 0.5H_2O$	c	3.31	$CsOH \cdot H_2O$	c	−20.50
			$NaCN \cdot 2H_2O$	c	18.58	CsF	c	−36.86
LiOH	c	−23.56	NaCNO	c	19.20	$CsF \cdot H_2O$	c	−10.46
$LiOH \cdot H_2O$	c	−6.69	NaCNS	c	6.83	$CsF \cdot 1.5H_2O$	c	−5.44
LiF	c	4.73				CsCl	c	17.78
LiCl	c	−37.03	KOH	c	−57.61	$CsClO_4$	c	55.44
$LiCl \cdot H_2O$	c	−19.08	$KOH \cdot H_2O$	c	−14.64	CsBr	c	25.98
$LiClO_4$	c	−26.55	$KOH \cdot 1.5H_2O$	c	−10.46	$CsBrO_3$	c	50.46
$LiClO_4 \cdot 3H_2O$	c	32.61	KF	c	−17.73	CsI	c	33.35
LiBr	c	−48.83	$KF \cdot 2H_2O$	c	6.97	$CsNO_3$	c	40.00
$LiBr \cdot H_2O$	c	−23.26						

KINETIC AND PHOTOCHEMICAL DATA FOR ATMOSPHERIC CHEMISTRY

R. Atkinson, D. L. Baulch, R. A. Cox, R. F. Hampson, Jr., J. A. Kerr, and J. Troe
IUPAC Subcommittee on Gas Kinetic Data Evaluation for Atmospheric Chemistry

The International Union of Pure and Applied Chemistry (IUPAC) maintains a Subcommittee on Gas Kinetic Data evaluation for Atmospheric Chemistry which issues periodic recommendations on rate constants and related data for chemical reactions that are important in the chemistry of the atmosphere. These recommended values are based on the latest (1992) evaluation of all relevant data reported in the literature. The table below provides a summary of the recommended values and their assigned uncertainties. Full details of the data and evaluation process may be found in the references.

This table is reprinted with permission of IUPAC and of the American Institute of Physics and American Chemical Society, who are the publishers of the *Journal of Physical and Chemical Reference Data*.

REFERENCE

1. Atkinson, R., et al., *Journal of Physical and Chemical Reference Data*, 21, 1125-1568, 1992; 18, 881-1097, 1989; 13, 1259-1380, 1984; 11, 327-496, 1982; 9, 295-471, 1980.

Reaction	k_{298} cm^3 molecule^{-1} s^{-1}		$\Delta\log k_{298}$	Temp. dependence of k/cm^3 molecule^{-1} s^{-1}	Temp. range/K	$\Delta(E/R)$/ K
O_x Reactions						
$O + O_2 + M \rightarrow O_3 + M$	6.0×10^{-34} [O$_2$]	(k_0)	±0.05	$6.0 \times 10^{-34}(T/300)^{-2.8}$[O$_2$]	100–300	$\Delta n = ±0.5$
	5.6×10^{-34} [N$_2$]	(k_0)	±0.05	$5.6 \times 10^{-34}(T/300)^{-2.8}$[N$_2$]	100–300	$\Delta n = ±0.5$
$O + O_2 \rightarrow O_3^*$	See previous evaluation					
$O_3^* + M \rightarrow O_3 + M$	See previous evaluation					
$O + O_3 \rightarrow 2 O_2$	8.0×10^{-15}		±0.08	$8.0 \times 10^{-12}\exp(-2060/T)$	200–400	±200
$O + O_3^* \rightarrow$ products	See previous evaluation					
$O(^1D) + O_2 \rightarrow O(^3P) + O_2$	4.0×10^{-11}		±0.05	$3.2 \times 10^{-11}\exp(67/T)$	200–350	±100
$O(^1D) + O_3 \rightarrow O_2 + 2 O(^3P)$	1.2×10^{-10}		±0.1⎫		100–400	
$\rightarrow 2 O_2(^3\Sigma_g^-)$	1.2×10^{-10}		±0.1⎭	2.4×10^{-10}		$\Delta\log k = ±0.05$
$O_2^* + O_3 \rightarrow O + 2 O_2$	See data sheet					
$O_2(^1\Delta_g) + M \rightarrow O_2(^3\Sigma_g^-) + M$	1.6×10^{-18}	(M = O$_2$)	±0.2	$3.0 \times 10^{-18}\exp(-200/T)$	100–450	±200
	$\leq 1.4 \times 10^{-19}$	(M = N$_2$)				
	5×10^{-18}	(M = H$_2$O)	±0.3			
	$\leq 2 \times 10^{-20}$	(M = CO$_2$)				
$O_2(^1\Sigma_g^+) + M \rightarrow O_2(^3\Sigma_g^-) + M$	See previous evaluation					
$O_2(^1\Sigma_g^+) + O_3 \rightarrow$ products	See previous evaluation					
$O_2(^1\Sigma_g^+)^* + O_2 \rightarrow O_2(^1\Sigma_g^+) + O_2$	See previous evaluation					
$O_2 + h\nu \rightarrow$ products	See data sheets					
$O_3 + h\nu \rightarrow$ products	See data sheets					
HO_x Reactions						
$H + HO_2 \rightarrow H_2 + O_2$	5.6×10^{-12}		±0.5	5.6×10^{-12}	245–300	$\Delta\log k = ±0.5$
$\rightarrow 2 HO$	7.2×10^{-11}		±0.1	7.2×10^{-11}	245–300	$\Delta\log k = ±0.1$
$\rightarrow H_2O + O$	2.4×10^{-12}		±0.5	2.4×10^{-12}	245–300	$\Delta\log k = ±0.5$
$H + O_2 + M \rightarrow HO_2 + M$	6.2×10^{-32}[N$_2$]	(k_0)	±0.05	$6.2 \times 10^{-32}(T/300)^{-1.6}$[N$_2$]	200–600	$\Delta n = ±0.6$
	7.5×10^{-11}	(k_∞)	±0.3	7.5×10^{-11}	200–300	$\Delta n = ±0.6$
	$F_c = 0.55$		$\Delta F_c = ±0.15$	$F_c = \exp(-T/498)$	200–300	
$H + O_3 \rightarrow HO + O_2$	See previous evaluation					
$H + O_3 \rightarrow HO^* + O_2$	See previous evaluation					
$O + H_2 \rightarrow HO + H$	See previous evaluation					
$O + HO \rightarrow O_2 + H$	3.3×10^{-11}		±0.1	$2.3 \times 10^{-11}\exp(110/T)$	220–500	±100
$O + HO_2 \rightarrow HO + O_2$	5.8×10^{-11}		±0.08	$2.7 \times 10^{-11}\exp(224/T)$	200–400	±100
$O + H_2O_2 \rightarrow HO + HO_2$	1.7×10^{-15}		±0.3	$1.4 \times 10^{-12}\exp(-2000/T)$	250–390	±1000
$O(^1D) + H_2 \rightarrow HO + H$	1.1×10^{-10}		±0.1	1.1×10^{-10}	200–350	±100
$O(^1D) + H_2O \rightarrow 2 HO$	2.2×10^{-10}		±0.1	2.2×10^{-10}	200–350	±100
$HO + H_2 \rightarrow H_2O + H$	6.7×10^{-15}		±0.1	$7.7 \times 10^{-12}\exp(-2100/T)$	200–450	±200
$HO + H_2(v=1) \rightarrow H_2O + H$	See previous evaluation					
$HO + HO \rightarrow H_2O + O$	1.9×10^{-12}		±0.15	$4.2 \times 10^{-12}\exp(-240/T)$	250–500	±240
$HO + HO + M \rightarrow H_2O_2 + M$	8×10^{-31}[N$_2$]	(k_0)	±0.3	$8 \times 10^{-31}(T/300)^{-0.8}$[N$_2$]	200–300	$\Delta n = ±0.5$
	3×10^{-11}	(k_∞)	±0.3	3×10^{-11}	200–300	$\Delta n = ±0.5$
	$F_c = 0.5$			$F_c = 0.5$	200–300	
$HO + HO_2 \rightarrow H_2O + O_2$	1.1×10^{-10}		±0.1	$4.8 \times 10^{-11}\exp(250/T)$	250–400	±200
$HO + H_2O_2 \rightarrow H_2O + HO_2$	1.7×10^{-12}		±0.1	$2.9 \times 10^{-12}\exp(-160/T)$	240–460	±100
$HO + O_3 \rightarrow HO_2 + O_2$	6.7×10^{-14}		±0.15	$1.9 \times 10^{-12}\exp(-1000/T)$	220–450	±300
$HO^* + M \rightarrow HO + M$	See previous evaluation					
$HO^* + O_3 \rightarrow$ products	See previous evaluation					
$HO_2 + HO_2 \rightarrow H_2O_2 + O_2$	1.6×10^{-12}		±0.15	$2.2 \times 10^{-13}\exp(600/T)$	230–420	±200
$HO_2 + HO_2 + M \rightarrow H_2O_2 + O_2 + M$	5.2×10^{-32} [N$_2$]		±0.15	$1.9 \times 10^{-33}\exp(980/T)$[N$_2$]	230–420	±300
	4.5×10^{-32} [O$_2$]		±0.15			
	See data sheets for effect of H$_2$O					
$HO_2 + O_3 \rightarrow HO + 2 O_2$	2.0×10^{-15}		±0.2	$1.4 \times 10^{-14}\exp(-600/T)$	250–350	+500 −100
$H_2O + h\nu \rightarrow HO + H$	See data sheets					
$H_2O_2 + h\nu \rightarrow 2 HO$	See data sheets					

Reaction	k_{298} cm^3 molecule^{-1} s^{-1}	$\Delta\log k_{298}$	Temp. dependence of k/cm^3 molecule^{-1} s^{-1}	Temp. range/K	$\Delta(E/R)/$ K
NO_x Reactions					
$O + NO + M \rightarrow NO_2 + M$	1.0×10^{-31} [N$_2$] $\quad(k_0)$	±0.1	$1.0 \times 10^{-31}(T/300)^{-1.6}$[N$_2$]	200–300	$\Delta n = \pm0.3$
	$3.0 \times 10^{-11}\quad(k_\infty)$	±0.3	$3.0 \times 10^{-11}(T/300)^{0.3}$	200–1500	$\Delta n = \pm0.3$
	$F_c = 0.85$	$\Delta F_c = \pm0.1$	$F_c = \exp(-T/1850)$	200–300	
$O + NO_2 \rightarrow O_2 + NO$	9.7×10^{-12}	±0.06	$6.5 \times 10^{-12}\exp(120/T)$	230–350	±120
$O + NO_2 + M \rightarrow NO_3 + M$	9.0×10^{-32} [N$_2$] $\quad(k_0)$	±0.1	$9.0 \times 10^{-32}(T/300)^{-2.0}$[N$_2$]	200–400	$\Delta n = \pm1$
	$2.2 \times 10^{-11}\quad(k_\infty)$	±0.2	2.2×10^{-11}	200–400	$\Delta n = \pm0.5$
	$F_c = 0.8$	$\Delta F_c = \pm0.1$	$F_c = \exp(-T/1300)$	200–400	
$O + NO_3 \rightarrow O_2 + NO_2$	1.7×10^{-11}	±0.3			
$O + N_2O_5 \rightarrow$ products	See previous evaluation				
$O(^1D) + N_2 \rightarrow O(^3P) + N_2$	2.6×10^{-11}	±0.1	$1.8 \times 10^{-11}\exp(107/T)$	200–350	±100
$O(^1D) + N_2O \rightarrow N_2 + O_2$	4.4×10^{-11}	±0.15	4.4×10^{-11}	200–350	±100
$\quad\rightarrow 2\,NO$	7.2×10^{-11}	±0.15	7.2×10^{-11}	200–350	±100
$N + HO \rightarrow NO + H$	See previous evaluation				
$N + O_2 \rightarrow NO + O$	See previous evaluation				
$N + O_2(^1\Delta_g) \rightarrow NO + O$	See previous evaluation				
$N + O_3 \rightarrow NO + O_2$	See previous evaluation				
$N + NO \rightarrow N_2 + O$	See previous evaluation				
$N + NO_2 \rightarrow N_2O + O$	See previous evaluation				
$HO + NH_3 \rightarrow H_2O + NH_2$	1.6×10^{-13}	±0.1	$3.5 \times 10^{-12}\exp(-925/T)$	230–450	±200
$HO + HONO \rightarrow H_2O + NO_2$	4.9×10^{-12}	±0.3	$1.8 \times 10^{-11}\exp(-390/T)$	280–340	±400
$HO + HONO_2 \rightarrow H_2O + NO_3$	1.5×10^{-13} (1 bar)	±0.1	See data sheets		
$HO + HO_2NO_2 \rightarrow$ products	5.0×10^{-12}	±0.2	$1.5 \times 10^{-12}\exp(360/T)$	240–340	$^{+300}_{-600}$
$HO + NO + M \rightarrow HONO + M$	7.4×10^{-31} [N$_2$] $\quad(k_0)$	±0.1	$7.4 \times 10^{-31}(T/300)^{-2.4}$[N$_2$]	200–300	$\Delta n = \pm0.5$
	$3.2 \times 10^{-11}\quad(k_\infty)$	±0.3	3.2×10^{-11}	200–400	$\Delta\log k = \pm0.3$
	$F_c = 0.8$				
$HO + NO_2 + M \rightarrow HONO_2 + M$	2.6×10^{-30} [N$_2$] $\quad(k_0)$	±0.1	$2.6 \times 10^{-30}(T/300)^{-2.9}$[N$_2$]	200–300	$\Delta n = \pm0.5$
	$6.0 \times 10^{-11}\quad(k_\infty)$	±0.1	6.0×10^{-11}	200–300	$\Delta n = \pm0.5$
	$F_c = 0.43$				
$HO + NO_3 \rightarrow HO_2 + NO_2$	2.3×10^{-11}	±0.2			
$HO_2 + NO \rightarrow HO + NO_2$	8.3×10^{-12}	±0.1	$3.7 \times 10^{-12}\exp(240/T)$	230–500	±100
$HO_2 + NO_2 + M \rightarrow HO_2NO_2 + M$	1.8×10^{-31} [N$_2$] $\quad(k_0)$	±0.1	$1.8 \times 10^{-31}(T/300)^{-3.2}$[N$_2$]	200–300	$\Delta n = \pm1$
	$4.7 \times 10^{-12}\quad(k_\infty)$	±0.2	4.7×10^{-12}	200–300	$\Delta n = \pm1$
	$F_c = 0.6$		$F_c = 0.6$		
$HO_2NO_2 + M \rightarrow HO_2 + NO_2 + M$	1.3×10^{-20} [N$_2$] $\quad(k_0/\mathrm{s}^{-1})$	±0.3	$5 \times 10^{-6}\exp(-10000/T)$[N$_2$]	260–300	±500
	$0.34\quad(k_\infty/\mathrm{s}^{-1})$	±0.5	$2.6 \times 10^{15}\exp(-10900/T)$	260–300	±500
	$F_c = 0.6$				
$HO_2 + NO_3 \rightarrow O_2 + HONO_2$ $\}$ $\quad\rightarrow HO + NO_2 + O_2$	4.3×10^{-12}	±0.2			
$NH_2 + HO \rightarrow$ products	See previous evaluation				
$NH_2 + HO_2 \rightarrow$ products	See previous evaluation				
$NH_2 + O_2 \rightarrow$ products	$<3 \times 10^{-18}$				
$NH_2 + O_3 \rightarrow$ products	1.7×10^{-13}	±0.5	$4.9 \times 10^{-12}\exp(-1000/T)$	250–380	±500
$NH_2 + NO \rightarrow$ products	1.6×10^{-11}	±0.2	$1.6 \times 10^{-11}(T/298)^{-1.5}$	210–500	$\Delta n = \pm0.5$
$NH_2 + NO_2 \rightarrow$ products	2.0×10^{-11}	±0.2	$2.0 \times 10^{-11}(T/298)^{-2.0}$	250–500	$\Delta n = \pm0.7$
$2\,NO + O_2 \rightarrow 2\,NO_2$	2.0×10^{-38} (cm^6 molecule^{-2} s^{-1})	±0.1	$3.3 \times 10^{-39}\exp(530/T)$	273–600	±400
$NO + O_3 \rightarrow NO_2 + O_2$	1.8×10^{-14}	±0.08	$1.8 \times 10^{-12}\exp(-1370/T)$	195–304	±200
$NO + NO_3 \rightarrow 2\,NO_2$	2.6×10^{-11}	±0.1	$1.8 \times 10^{-11}\exp(110/T)$	220–400	±100
$NO_2 + O_3 \rightarrow NO_3 + O_2$	3.2×10^{-17}	±0.06	$1.2 \times 10^{-13}\exp(-2450/T)$	230–360	±150
$NO_2 + NO_3 + M \rightarrow N_2O_5 + M$	2.7×10^{-30} [N$_2$] $\quad(k_0)$	±0.1	$2.7 \times 10^{-30}(T/300)^{-3.4}$[N$_2$]	200–400	$\Delta n = \pm0.5$
	$2.0 \times 10^{-12}\quad(k_\infty)$	±0.2	$2.0 \times 10^{-12}(T/300)^{0.2}$	200–500	$\Delta n = \pm0.6$
	$F_c = 0.33$		$F_c = [\exp(-T/250) + \exp(-1050/T)]$	200–500	
$N_2O_5 + M \rightarrow NO_2 + NO_3 + M$	1.6×10^{-19} [N$_2$] $\quad(k_0/\mathrm{s}^{-1})$	±0.2	$2.2 \times 10^{-3}(T/300)^{-4.4}$ $\exp(-11080/T)$[N$_2$]	220–300	$\Delta n = \pm0.5$
	$6.9 \times 10^{-2}\quad(k_\infty/\mathrm{s}^{-1})$	±0.3	$9.7 \times 10^{14}(T/300)^{0.1}$ $\exp(-11080/T)$	200–300	$\Delta n = \pm0.2$
	$F_c = 0.33$		$F_c = [\exp(-T/250) + \exp(-1050/T)]$	200–300	
$N_2O_5 + H_2O \rightarrow 2\,HONO_2$	$<2 \times 10^{-21}$				
$HONO + h\nu \rightarrow$ products	See data sheets				
$HONO_2 + h\nu \rightarrow$ products	See data sheets				
$HO_2NO_2 + h\nu \rightarrow$ products	See data sheets				
$NO + h\nu \rightarrow$ products	See previous evaluation				
$NO_2 + h\nu \rightarrow$ products	See data sheets				
$NO_3 + h\nu \rightarrow$ products	See data sheets				
$N_2O + h\nu \rightarrow$ products	See data sheets				
$N_2O_5 + h\nu \rightarrow$ products	See data sheets				
Organic Reactions					
$O + CH_3 \rightarrow HCHO + H$	1.4×10^{-10}	±0.1	1.4×10^{-10}	200–900	±100
$O + CN \rightarrow CO + N(^2D)$	See previous evaluation				
$\quad\rightarrow CO + N(^4S)$	See previous evaluation				
$O(^1D) + CH_4 \rightarrow HO + CH_3$	1.35×10^{-10}	±0.1	1.35×10^{-10}	200–300	±100
$\quad\rightarrow HCHO + H_2$	1.5×10^{-11}	±0.1	1.5×10^{-11}	200–300	±100
$HO + CH_4 \rightarrow H_2O + CH_3$	7.0×10^{-15}	±0.10	$3.9 \times 10^{-12}\exp(-1885/T)$	240–300	±100
$HO + C_2H_2 + M \rightarrow C_2H_2OH + M$	5×10^{-30} [N$_2$] $\quad(k_0)$	±0.1	$5 \times 10^{-30}(T/300)^{-1.5}$[N$_2$]	200–300	$\Delta n = \pm1.5$
	$9.0 \times 10^{-13}\quad(k_\infty)$	±0.1	$9.0 \times 10^{-13}(T/300)^2$	200–300	$\Delta n = \pm1$
	$F_c = 0.62$		$F_c = \exp(-T/623)$	200–300	

KINETIC AND PHOTOCHEMICAL DATA FOR ATMOSPHERIC CHEMISTRY (continued)

Reaction	k_{298} cm^3 molecule^{-1} s^{-1}		$\Delta \log k_{298}$	Temp. dependence of k/cm^3 molecule^{-1} s^{-1}	Temp. range/K	$\Delta(E/R)/$ K
HO + C$_2$H$_4$ + M → C$_2$H$_4$OH + M	7×10^{-29}[N$_2$]	(k_0)	±0.3	$7 \times 10^{-29}(T/300)^{-3.1}$[N$_2$]	200–300	$\Delta n = \pm 2$
	9×10^{-12}	(k_∞)	±0.3	9×10^{-12}	200–300	$\Delta n = \pm 0.5$
	$F_c = 0.7$					
HO + C$_2$H$_6$ → H$_2$O + C$_2$H$_5$	2.5×10^{-13}		±0.10	$7.8 \times 10^{-12}\exp(-1020/T)$	240–300	±100
HO + C$_3$H$_6$ + M → C$_3$H$_6$OH + M	8×10^{-27}[N$_2$]	(k_0)	±1	$8 \times 10^{-27}(T/300)^{-3.5}$[N$_2$]	200–300	$\Delta n = \pm 1$
	3.0×10^{-11}	(k_∞)	±0.1	3.0×10^{-11}	200–300	$\Delta n = \pm 1$
	$F_c = 0.5$					
HO + C$_3$H$_8$ → H$_2$O + C$_3$H$_7$	1.14×10^{-12}		±0.10	$9.8 \times 10^{-12}\exp(-640/T)$	~300	±150
HO + CO → H + CO$_2$	1.5×10^{-13} (1 + 0.6 P/bar)		±0.1	1.5×10^{-13}(1 + 0.6 P/bar)	200–300	±300
HO + HCHO → H$_2$O + HCO	9.6×10^{-12}		±0.10	$8.8 \times 10^{-12}\exp(25/T)$	240–300	±150
HO + CH$_3$CHO → H$_2$O + CH$_3$CO	1.6×10^{-11}		±0.10	$5.6 \times 10^{-12}\exp(310/T)$	240–530	±200
HO + C$_2$H$_5$CHO → products	2.0×10^{-11}		±0.15			
HO + (CHO)$_2$ → products	1.1×10^{-11}		±0.3			
HO + HOCH$_2$CHO → H$_2$O + HOCH$_2$CO	8.0×10^{-12}		±0.3			
→ H$_2$O + HOCHCHO	2.0×10^{-12}		±0.3			
HO + CH$_3$COCHO → H$_2$O + CH$_3$COCO	1.7×10^{-11}		±0.3			
HO + CH$_3$COCH$_3$ → H$_2$O + CH$_2$COCH$_3$	2.3×10^{-13}		±0.2	$1.7 \times 10^{-12}\exp(-600/T)$	240–440	±300
HO + CH$_3$OH → H$_2$O + CH$_2$OH	7.8×10^{-13}		±0.15	$3.3 \times 10^{-12}\exp(-380/T)$	240–300	±200
→ H$_2$O + CH$_3$O	1.4×10^{-13}		±0.15			
HO + C$_2$H$_5$OH → H$_2$O + CH$_2$CH$_2$OH	1.6×10^{-13}		±0.15			
→ H$_2$O + CH$_3$CHOH	2.9×10^{-12}		±0.15	$4.1 \times 10^{-12}\exp(-70/T)$	270–340	±200
→ H$_2$O + CH$_3$CH$_2$O	1.6×10^{-13}		±0.15			
HO + n-C$_3$H$_7$OH → products	5.5×10^{-12}		±0.2			
HO + i-C$_3$H$_7$OH → products	5.7×10^{-12}		±0.2	5.7×10^{-12}	240–440	±200
HO + CH$_3$COCH$_2$OH → products	3.0×10^{-12}		±0.3			
HO + CH$_3$OOH → H$_2$O + CH$_2$OOH	1.9×10^{-12}		±0.2	$1.0 \times 10^{-12}\exp(190/T)$	220–430	±150
→ H$_2$O + CH$_3$OO	3.6×10^{-12}		±0.2	$1.9 \times 10^{-12}\exp(190/T)$	220–430	±150
HO + HCOOH → products	4.5×10^{-13}		±0.15	4.5×10^{-13}	290–450	±250
HO + CH$_3$COOH → products	8×10^{-13}		±0.3			
HO + CH$_3$ONO$_2$ → products	3.5×10^{-13}	(1 bar)	±0.10	$1.0 \times 10^{-14}\exp(1060/T)$ (1 bar)	290–400	±500
HO + C$_2$H$_5$ONO$_2$ → products	4.9×10^{-13}	(1 bar)	±0.15	$4.4 \times 10^{-14}\exp(720/T)$ (1 bar)	290–380	±500
HO + n-C$_3$H$_7$ONO$_2$ → products	7.3×10^{-13}	(1 bar)	±0.15	7.3×10^{-13} (1 bar)	290–370	±500
HO + i-C$_3$H$_7$ONO$_2$ → products	4.9×10^{-13}	(1 bar)	±0.25			
HO + CH$_3$CO$_3$NO$_2$ → products	1.1×10^{-13}		±0.2	$9.5 \times 10^{-13}\exp(-650/T)$	270–300	±400
HO + HCN → products	3×10^{-14}	(1 bar)	±0.5	$1.2 \times 10^{-13}(-400/T)$ (1 bar)	290–440	±300
HO + CH$_3$CN → products	2.2×10^{-14}	(1 bar)	±0.15	$8.1 \times 10^{-13}\exp(-1080/T)$ (1 bar)	250–390	±200
HO$_2$ + CH$_3$O$_2$ → O$_2$ + CH$_3$O$_2$H	5.2×10^{-12}		±0.3	$3.8 \times 10^{-13}\exp(780/T)$	225–580	±500
HO$_2$ + HOCH$_2$O$_2$ → O$_2$ + HOCH$_2$O$_2$H	1.2×10^{-11}		±0.3	$5.6 \times 10^{-15}\exp(2300/T)$	275–335	±1500
→ O$_2$ + HCO$_2$H + H$_2$O						
HO$_2$ + C$_2$H$_5$O$_2$ → O$_2$ + C$_2$H$_5$O$_2$H	5.8×10^{-12}		±0.2	$6.5 \times 10^{-13}\exp(650/T)$	240–380	±200
HO$_2$ + CH$_3$CO$_3$ → O$_2$ + CH$_3$CO$_3$H	4.2×10^{-12}		±0.3	$1.3 \times 10^{-13}\exp(1040/T)$	250–370	±500
→ O$_3$ + CH$_3$CO$_2$H	1.0×10^{-12}		±0.3	$3.0 \times 10^{-13}\exp(1040/T)$	250–370	±500
HO$_2$ + HOCH$_2$CH$_2$O$_2$ → products	1.0×10^{-11}		±0.3			
HO$_2$ + HCHO → HOCH$_2$O$_2$	7.9×10^{-14}		±0.3	$9.7 \times 10^{-15}\exp(625/T)$	275–333	±600
HOCH$_2$OO → HO$_2$ + HCHO	1.5×10^{2}	(k/s^{-1})	±0.3	$2.4 \times 10^{12}\exp(-7000/T)$	275–333	±2000
NO$_3$ + C$_2$H$_2$ → products	$<1 \times 10^{-16}$					
NO$_3$ + C$_2$H$_4$ → products	2.1×10^{-16}		±0.2	$3.3 \times 10^{-12}\exp(-2880/T)$	270–330	±500
NO$_3$ + C$_3$H$_6$ → products	9.4×10^{-15}		±0.2			
NO$_3$ + HCHO → HNO$_3$ + HCO	5.8×10^{-16}		±0.3			
NO$_3$ + CH$_3$CHO → HNO$_3$ + CH$_3$CO	2.7×10^{-15}		±0.2	$1.4 \times 10^{-12}\exp(-1860/T)$	260–370	±500
NO$_3$ + CH$_3$OH → products	2.4×10^{-16}		±0.5	$1.3 \times 10^{-12}\exp(-2560/T)$	290–480	±700
NO$_3$ + C$_2$H$_5$OH → products	$<2 \times 10^{-15}$					
NO$_3$ + i-C$_3$H$_7$OH → products	$<5 \times 10^{-15}$					
CH$_3$ + O$_2$ + M → CH$_3$O$_2$ + M	1.0×10^{-30} [N$_2$]	(k_0)	±0.2	$1.0 \times 10^{-30}(T/300)^{-3.3}$[N$_2$]	200–300	$\Delta n = \pm 1$
	2.2×10^{-12}	(k_∞)	±0.3	$2.2 \times 10^{-12}(T/300)^{1.0}$	200–300	$\Delta n = \pm 1$
	$F_c = 0.27$					
C$_2$H$_5$ + O$_2$ → C$_2$H$_4$ + HO$_2$	3.8×10^{-15}	(1 bar air)	±0.5			
	1.9×10^{-14}	(0.133 bar air)	±0.5			
C$_2$H$_5$ + O$_2$ + M → C$_2$H$_5$O$_2$ + M	5.9×10^{-29} [N$_2$]	(k_0)	±0.3	$5.9 \times 10^{-29}(T/300)^{-3.8}$[N$_2$]	200–300	$\Delta n = \pm 1$
	7.8×10^{-12}	(k_∞)	±0.2	7.8×10^{-12}	200–300	$\Delta \log k = \pm 0.2$
	$F_c = 0.54$			$F_c = \{0.58 \exp(-T/1250) + 0.42 \exp(-T/183)\}$	200–300	
n-C$_3$H$_7$ + O$_2$ + M → n-C$_3$H$_7$O$_2$ + M	8×10^{-12}	(k_∞)	±0.2	8×10^{-12}	200–300	$\Delta \log k = \pm 0.2$
i-C$_3$H$_7$ + O$_2$ + M → i-C$_3$H$_7$O$_2$ + M	1.1×10^{-11}	(k_∞)	±0.3	1.1×10^{-11}	200–300	$\Delta \log k = \pm 0.3$
CH$_3$COCH$_2$+O$_2$+M → CH$_3$COCH$_2$O$_2$ + M	1.5×10^{-12}	(k_∞)	±0.5			
HCO + O$_2$ → CO + HO$_2$	5.5×10^{-12}		±0.15	5.5×10^{-12}	200–400	±150
CH$_3$CO + O$_2$ + M → CH$_3$CO$_3$ + M	5.0×10^{-12}	(k_∞)	±0.5	5.0×10^{-12}	200–300	$\Delta \log k = \pm 0.5$
CH$_2$OH + O$_2$ → HCHO + HO$_2$	9.4×10^{-12}		±0.12			
CH$_3$CHOH + O$_2$ → CH$_3$CHO + HO$_2$	1.9×10^{-11}		±0.3			
CH$_2$CH$_2$OH + O$_2$ → products	3.0×10^{-12}		±0.3			
CH$_3$O + O$_2$ → HCHO + HO$_2$	1.9×10^{-15}		±0.2	$7.2 \times 10^{-14}\exp(-1080/T)$	298–610	±300
C$_2$H$_5$O + O$_2$ → CH$_3$CHO + HO$_2$	9.5×10^{-15}		±0.2	$6.0 \times 10^{-14}\exp(-550/T)$	295–425	±300
n-C$_3$H$_7$O + O$_2$ → C$_2$H$_5$CHO + HO$_2$	8×10^{-15}		±0.5			
i-C$_3$H$_7$O + O$_2$ → CH$_3$COCH$_3$ + HO$_2$	8×10^{-15}		±0.3	$1.5 \times 10^{-14}\exp(-200/T)$	290–390	±200
CH$_3$ + O$_3$ → products	2.5×10^{-12}		±0.3	$5.1 \times 10^{-12}\exp(-210/T)$	240–400	±200
CH$_3$O + NO + M → CH$_3$ONO + M	1.6×10^{-29}[N$_2$]	(k_0)	±0.1	$1.6 \times 10^{-29}(T/300)^{-3.5}$[N$_2$]	200–400	$\Delta n = \pm 0.5$
	3.6×10^{-11}	(k_∞)	±0.5	$3.6 \times 10^{-11}(T/300)^{-0.6}$	200–400	$\Delta n = \pm 0.5$
	$F_c = 0.6$					
CH$_3$O + NO → HCHO + HNO	4×10^{-12}			$4 \times 10^{-12}(T/300)^{-0.7}$	200–400	$\Delta n = \pm 0.5$
C$_2$H$_5$O + NO + M → C$_2$H$_5$ONO + M	4.4×10^{-11}	(k_∞)	±0.3	4.4×10^{-11}	200–300	$\Delta n = \pm 0.5$
C$_2$H$_5$O + NO → CH$_3$CHO + HNO	1.3×10^{-11}					
i-C$_3$H$_7$O + NO + M → i-C$_3$H$_7$ONO + M	3.4×10^{-11}	(k_∞)	±0.3	3.4×10^{-11}	200–300	$\Delta n = \pm 0.5$

Reaction	k_{298} cm^3 molecule^{-1} s^{-1}		$\Delta\log k_{298}$	Temp. dependence of k/cm^3 molecule^{-1} s^{-1}	Temp. range/K	$\Delta(E/R)$/ K
$i\text{-}C_3H_7O + NO \rightarrow CH_3COCH_3 + HNO$	6.5×10^{-12}		± 0.5			
$CH_3O + NO_2 + M \rightarrow CH_3ONO_2 + M$	$2.8 \times 10^{-29}[N_2]$	(k_0)	± 0.5	$2.8 \times 10^{-29}(T/300)^{-4.5}[N_2]$	200–400	$\Delta n = \pm 1$
	2×10^{-11}	(k_∞)	± 0.3	2×10^{-11}	200–400	$\Delta n = \pm 0.5$
	$F_c = 0.44$					
$CH_3O + NO_2 \rightarrow HCHO + HONO$	See data sheets					
$C_2H_5O + NO_2 + M \rightarrow C_2H_5ONO_2 + M$	2.8×10^{-11}	(k_∞)	± 0.3	2.8×10^{-11}	200–300	$\Delta n = \pm 0.5$
$C_2H_5O + NO_2 \rightarrow CH_3CHO + HONO$	See data sheets					
$i\text{-}C_3H_7O + NO_2 + M \rightarrow i\text{-}C_3H_7ONO_2 + M$	3.5×10^{-11}	(k_∞)	± 0.3	3.5×10^{-11}	200–300	$\Delta n = \pm 0.5$
$i\text{-}C_3H_7O + NO_2 \rightarrow CH_3COCH_3 + HONO$	See data sheets					
$CH_3O_2 + NO \rightarrow CH_3O + NO_2$	7.6×10^{-12}		± 0.1	$4.2 \times 10^{-12}\exp(180/T)$	240–360	± 180
$C_2H_5O_2 + NO \rightarrow C_2H_5O + NO_2$	8.9×10^{-12}		± 0.3			
$C_2H_5O_2 + NO\ (+M) \rightarrow C_2H_5ONO_2\ (+M)$	$\leqslant 1.3 \times 10^{-13}$	(1 bar)				
$n\text{-}C_3H_7O_2 + NO \rightarrow n\text{-}C_3H_7O + NO_2$	8.7×10^{-12}		± 0.3			
$n\text{-}C_3H_7O_2 + NO\ (+M) \rightarrow n\text{-}C_3H_7ONO_2\ (+M)$	1.8×10^{-13}	(1 bar)	± 0.5			
$i\text{-}C_3H_7O_2 + NO \rightarrow i\text{-}C_3H_7O + NO_2$	8.5×10^{-12}		± 0.5			
$i\text{-}C_3H_7O_2 + NO\ (+M) \rightarrow i\text{-}C_3H_7ONO_2\ (+M)$	3.7×10^{-13}	(1 bar)	± 0.5			
$CH_3CO_3 + NO \rightarrow CH_3 + CO_2 + NO_2$	2.0×10^{-11}		± 0.2	2.0×10^{-11}	280–325	± 600
$CH_3O_2 + NO_2 + M \rightarrow CH_3O_2NO_2 + M$	$2.5 \times 10^{-30}[N_2]$	(k_0)	± 0.3	$2.5 \times 10^{-30}(T/300)^{-5.5}[N_2]$	250–350	$\Delta n = \pm 1$
	7.5×10^{-12}	(k_∞)	± 0.3	7.5×10^{-12}	250–350	$\Delta n = \pm 0.5$
	$F_c = 0.4$					
$CH_3O_2NO_2 + M \rightarrow CH_3O_2 + NO_2 + M$	$6.8 \times 10^{-19}[N_2]$	(k_0/s^{-1})	± 0.3	$9 \times 10^{-5}\exp(-9690/T)[N_2]$	250–300	± 500
	4.5	(k_∞/s^{-1})	± 0.3	$1.1 \times 10^{16}\exp(-10560/T)$	250–300	± 500
	$F_c = 0.4$					
$C_2H_5O_2 + NO_2 + M \rightarrow C_2H_5O_2NO_2 + M$	$1.3 \times 10^{-29}[N_2]$	(k_0)	± 0.3	$1.3 \times 10^{-29}(T/300)^{-6.2}[N_2]$	200–300	$\Delta n = \pm 1$
	8.8×10^{-12}	(k_∞)	± 0.3	8.8×10^{-12}	200–300	$\Delta\log k = \pm 0.3$
	$F_c = 0.31$			$F_c = 0.31$	250–300	
$C_2H_5O_2NO_2 + M \rightarrow C_2H_5O_2 + NO_2 + M$	$1.4 \times 10^{-17}[N_2]$	(k_0/s^{-1})	± 0.5	$4.8 \times 10^{-4}\exp(-9285/T)[N_2]$	250–300	± 1000
	5.4	(k_∞/s^{-1})	± 0.5	$8.8 \times 10^{15}\exp(-10440/T)$	250–300	± 1000
	$F_c = 0.31$			$F_c = 0.31$	250–300	
$CH_3CO_3 + NO_2 + M \rightarrow CH_3CO_3NO_2 + M$	$2.7 \times 10^{-28}[N_2]$	(k_0)	± 0.4	$2.7 \times 10^{-28}(T/300)^{-7.1}[N_2]$	250–300	$\Delta n = \pm 2$
	1.2×10^{-11}	(k_∞)	± 0.2	$1.2 \times 10^{-11}(T/300)^{-0.9}$	250–300	$\Delta n = \pm 1$
	$F_c = 0.3$					
$CH_3CO_3NO_2 + M \rightarrow CH_3CO_3 + NO_2 + M$	$1.1 \times 10^{-20}[N_2]$	(k_0/s^{-1})	± 0.4	$4.9 \times 10^{-3}\exp(-12100/T)[N_2]$	300–330	± 1000
	6.1×10^{-4}	(k_∞/s^{-1})	± 0.2	$4.0 \times 10^{16}\exp(-13600/T)$	280–330	± 200
	$F_c = 0.3$					
$CH_3O_2 + NO_3 \rightarrow products$	No recommendation (see data sheets)					
$CH_3O_2 + CH_3O_2 \rightarrow CH_3OH + HCHO + O_2$						
$ \rightarrow 2\ CH_3O + O_2$	3.7×10^{-13}		± 0.12	$1.1 \times 10^{-13}\exp(365/T)$	200–400	± 200
$ \rightarrow CH_3OOCH_3 + O_2$						
$CH_3O_2 + CH_3CO_3 \rightarrow CH_3O + CH_3CO_2 + O_2$	5.5×10^{-12}		± 0.5			
$ \rightarrow CH_3CO_2H + HCHO + O_2$	5.5×10^{-12}		± 0.5			
$HOCH_2O_2 + HOCH_2O_2 \rightarrow HCOOH + CH_2(OH)_2 + O_2$	7.0×10^{-13}		± 0.3	$5.7 \times 10^{-14}\exp(750/T)$	275–325	± 750
$ \rightarrow 2\ HOCH_2O + O_2$	5.5×10^{-12}		± 0.3			
$C_2H_5O_2 + C_2H_5O_2 \rightarrow C_2H_5OH + CH_3CHO + O_2$						
$ \rightarrow 2\ C_2H_5O + O_2$	6.8×10^{-14}		± 0.12	$9.8 \times 10^{-14}\exp(-110/T)$	250–450	$^{+300}_{-100}$
$ \rightarrow C_2H_5OOC_2H_5 + O_2$						
$CH_3CO_3 + CH_3CO_3 \rightarrow 2\ CH_3CO_2 + O_2$	1.6×10^{-11}		± 0.5	$2.8 \times 10^{-12}\exp(530/T)$	250–370	± 500
$HOCH_2CH_2O_2 + HOCH_2CH_2O_2$						
$ \rightarrow HOCH_2CH_2OH + HOCH_2CHO + O_2$	1.5×10^{-12}		± 0.3			
$ \rightarrow 2\ HOCH_2CH_2O + O_2$	8.3×10^{-13}		± 0.3			
$n\text{-}C_3H_7O_2 + n\text{-}C_3H_7O_2 \rightarrow n\text{-}C_3H_7OH + C_2H_5CHO + O_2$						
$ \rightarrow 2\ n\text{-}C_3H_7O + O_2$	3×10^{-13}		± 0.5			
$i\text{-}C_3H_7O_2 + i\text{-}C_3H_7O_2 \rightarrow i\text{-}C_3H_7OH + CH_3COCH_3 + O_2$	4.4×10^{-16}		± 0.3	$1.6 \times 10^{-12}\exp(-2200/T)$	300–400	± 300
$ \rightarrow 2\ i\text{-}C_3H_7O + O_2$	5.6×10^{-16}		± 0.3			
$CH_3COCH_2O_2 + CH_3COCH_2O_2$						
$ \rightarrow CH_3COCH_2OH + CH_3COCHO + O_2$	$\leqslant 1 \times 10^{-11}$					
$ \rightarrow 2\ CH_3COCH_2O + O_2$						
$RCHOO + H_2O \rightarrow RCOOH + H_2O$						
$RCHOO + NO_2 \rightarrow RCHO + NO_3$	No recommendations (see data sheets)					
$RCHOO + SO_2 \rightarrow products$						
$RCHOO + HCHO \rightarrow products$						
$CN + O_2 \rightarrow products$	See previous evaluation					
$O_3 + C_2H_2 \rightarrow products$	1×10^{-20}		± 1.0			
$O_3 + C_2H_4 \rightarrow products$	1.7×10^{-18}		± 0.10	$1.2 \times 10^{-14}\exp(-2630/T)$	180–360	± 100
$O_3 + C_3H_6 \rightarrow products$	1.2×10^{-17}		± 0.15	$6.5 \times 10^{-15}\exp(-1880/T)$	230–370	± 400
$HCHO + h\nu \rightarrow products$	See data sheets					
$CH_3CHO + h\nu \rightarrow products$	See data sheets					
$C_2H_5CHO + h\nu \rightarrow products$	See data sheets					
$(CHO)_2 + h\nu \rightarrow products$	See data sheets					
$CH_3COCHO + h\nu \rightarrow products$	See data sheets					
$CH_3COCH_3 + h\nu \rightarrow products$	See data sheets					
$CH_3OOH + h\nu \rightarrow products$	See data sheets					
$CH_3ONO_2 + h\nu \rightarrow products$	See data sheets					
$C_2H_5ONO_2 + h\nu \rightarrow products$	See data sheets					
$n\text{-}C_3H_7ONO_2 + h\nu \rightarrow products$	See data sheets					
$i\text{-}C_3H_7ONO_2 + h\nu \rightarrow products$	See data sheets					
$CH_3O_2NO_2 + h\nu \rightarrow products$	See data sheets					
$CH_3CO_3NO_2 + h\nu \rightarrow products$	See data sheets					

KINETIC AND PHOTOCHEMICAL DATA FOR ATMOSPHERIC CHEMISTRY (continued)

Reaction	k_{298} cm^3 molecule^{-1} s^{-1}		$\Delta \log k_{298}$	Temp. dependence of $k/$cm^3 molecule^{-1} s^{-1}	Temp. range/K	$\Delta(E/R)/$ K
SO$_x$ Reactions						
O + H$_2$S → HO + HS	See previous evaluation					
O + CS → CO + S	2.1×10^{-11}		±0.1	$2.7 \times 10^{-10}\exp(-760/T)$	150–300	±250
O + CH$_3$SCH$_3$ → CH$_3$SO + CH$_3$	5.0×10^{-11}		±0.1	$1.3 \times 10^{-11}\exp(409/T)$	270–560	±100
O + CS$_2$ → SO + CS	3.6×10^{-12}		±0.2	$3.2 \times 10^{-11}\exp(-650/T)$	200–500	±100
O + CH$_3$SSCH$_3$ → CH$_3$SO + CH$_3$S	1.3×10^{-10}		±0.3	$5.5 \times 10^{-11}\exp(250/T)$	290–570	±100
O + OCS → SO + CO	1.2×10^{-14}		±0.2	$1.6 \times 10^{-11}\exp(-2150/T)$	220–500	±150
O + SO$_2$ + M → SO$_3$ + M	1.4×10^{-33}[N$_2$]	(k_0)	±0.3	$4.0 \times 10^{-32}\exp(-1000/T)$[N$_2$]	200–400	±200
S + O$_2$ → SO + O	2.1×10^{-12}		±0.2	2.1×10^{-12}	230–400	±200
S + O$_3$ → SO + O$_2$	1.2×10^{-11}		±0.3			
Cl + H$_2$S → HCl + HS	5.7×10^{-11}		±0.3	5.7×10^{-11}	210–350	±100
HO + H$_2$S → H$_2$O + HS	4.8×10^{-12}		±0.08	$6.3 \times 10^{-12}\exp(-80/T)$	200–300	±80
HO + SO$_2$ + M → HOSO$_2$ + M	4.0×10^{-31}[N$_2$] $F_c = 0.8$	(k_0)	±0.3	$4.0 \times 10^{-31}(T/300)^{-3.3}$[N$_2$]	300–400	$\Delta n = \pm 1$
	2×10^{-12} $F_c = 0.45$	(k_∞)	±0.3	2×10^{-12}	200–300	$\Delta \log k = \pm 0.3$
HOSO$_2$ + O$_2$ → HO$_2$ + SO$_3$	4.0×10^{-13}		±0.1	$1.3 \times 10^{-12}\exp(-330/T)$	290–420	±200
HO + OCS → products	2.0×10^{-15}		±0.3	$1.1 \times 10^{-13}\exp(-1200/T)$	250–500	±500
HO + CS$_2$ + M → HOCS$_2$ + M	8×10^{-31}[N$_2$]	(k_0)	±0.5	8×10^{-31}[N$_2$]	270–300	$\Delta \log k = \pm 0.5$
	8×10^{-12} $F_c = 0.8$	(k_∞)	±0.5	8×10^{-12}	250–300	$\Delta \log k = \pm 0.5$
HO + CS$_2$ → HS + OCS	$< 2 \times 10^{-15}$					
HOCS$_2$ + M → HO + CS$_2$ + M	4.8×10^{-14}[N$_2$]	($k_0/$s^{-1})	±0.5	$1.6 \times 10^{-6}\exp(-5160/T)$[N$_2$]	250–300	±500
	4.8×10^{5} $F_c = 0.8$	($k_\infty/$s^{-1})	±0.5	$1.6 \times 10^{13}\exp(-5160/T)$	250–300	±500
HOCS$_2$ + O$_2$ → products	2.8×10^{-14}		±0.3	2.8×10^{-14}	240–300	$\Delta \log k = \pm 0.3$
HO + CH$_3$SH → products	3.3×10^{-11}		±0.10	$9.9 \times 10^{-12}\exp(356/T)$	240–430	±100
HO + CH$_3$SCH$_3$ → H$_2$O + CH$_2$SCH$_3$	4.4×10^{-12}		±0.10	$9.6 \times 10^{-12}\exp(-234/T)$	250–400	±300
→ CH$_3$S(OH)CH$_3$	See data sheets					
HO + CH$_3$SSCH$_3$ → products	2.0×10^{-10}		±0.10	$6.0 \times 10^{-11}\exp(380/T)$	250–370	±300
HO$_2$ + SO$_2$ → products	$\le 1 \times 10^{-18}$					
NO$_3$ + H$_2$S → products	$< 1 \times 10^{-15}$					
NO$_3$ + CS$_2$ → products	$< 1 \times 10^{-15}$					
NO$_3$ + OCS → products	$< 1 \times 10^{-16}$					
NO$_3$ + SO$_2$ → products	$< 1 \times 10^{-19}$					
NO$_3$ + CH$_3$SH → products	9.2×10^{-13}		±0.15	9.2×10^{-13}	250–370	±400
NO$_3$ + CH$_3$SCH$_3$ → products	1.1×10^{-12}		±0.15	$1.9 \times 10^{-13}\exp(520/T)$	250–380	±200
NO$_3$ + CH$_3$SSCH$_3$ → products	7×10^{-13}		±0.3	7×10^{-13}	300–380	±500
CH$_3$O$_2$ + SO$_2$ → CH$_3$O + SO$_3$	See previous evaluation					
→ CH$_3$O$_2$SO$_2$	See previous evaluation					
HS + O$_2$ → products	$\le 4 \times 10^{-19}$					
HS + O$_3$ → HSO + O$_2$	3.7×10^{-12}		±0.2	$9.5 \times 10^{-12}\exp(-280/T)$	290–450	±250
HS + NO + M → HSNO + M	2.4×10^{-31}[N$_2$]	(k_0)	±0.3	$2.4 \times 10^{-31}(T/300)^{-2.5}$[N$_2$]	200–300	$\Delta n = \pm 1$
	2.7×10^{-11} $F_c = 0.6$	(k_∞)	±0.5	2.7×10^{-11}	200–300	$\Delta \log k = \pm 0.5$
HS + NO$_2$ → HSO + NO	5.8×10^{-11}		±0.3	$2.6 \times 10^{-11}\exp(240/T)$	220–450	±200
HSO + O$_2$ → products	$\le 2.0 \times 10^{-17}$					
HSO + O$_3$ → products	1.1×10^{-13}		±0.3			
HSO + NO → products	$\le 1.0 \times 10^{-15}$					
HSO + NO$_2$ → products	9.6×10^{-12}		±0.3			
HSO$_2$ + O$_2$ → products	3.0×10^{-13}		±0.8			
SO + O$_2$ → SO$_2$ + O	6.7×10^{-17}		±0.15	$1.4 \times 10^{-13}\exp(-2280/T)$	230–420	±500
SO + O$_3$ → SO$_2$ + O$_2$	8.9×10^{-14}		±0.10	$4.5 \times 10^{-12}\exp(-1170/T)$	230–420	±150
SO + NO$_2$ → SO$_2$ + NO	1.4×10^{-11}		±0.1	1.4×10^{-11}	210–360	±100
SO$_3$ + H$_2$O → products	$< 6 \times 10^{-15}$					
CS + O$_2$ → products	2.9×10^{-19}		±0.6			
CS + O$_3$ → OCS + O$_2$	3.0×10^{-16}		±0.5			
CS + NO$_2$ → OCS + NO	7.6×10^{-17}		±0.5			
CH$_3$S + O$_2$ → products	$< 2.5 \times 10^{-18}$					
CH$_3$S + O$_3$ → products	4.1×10^{-12}		±0.5			
CH$_3$S + NO + M → CH$_3$SNO + M	3.2×10^{-29}[N$_2$]	(k_0)	±0.3	$3.2 \times 10^{-29}(T/298)^{-4}$[N$_2$]	250–450	$\Delta n = \pm 2$
	4×10^{-11} $F_c = 0.60$	(k_∞)	±0.5	4×10^{-11} $F_c = \exp(-T/580)$	250–450	$\Delta \log k = \pm 0.5$
CH$_3$S + NO$_2$ → CH$_3$SO + NO	5.6×10^{-11}		±0.2			
CH$_3$SO + O$_3$ → products	1×10^{-12}		±0.7			
CH$_3$SO + NO$_2$ → products	1.2×10^{-11}		±0.5			
O$_3$ + CH$_3$SCH$_3$ → products	$< 1 \times 10^{-18}$					
OCS + $h\nu$ → products	See data sheets					
CS$_2$ + $h\nu$ → products	See data sheets					
CH$_3$SSCH$_3$ + $h\nu$ → products	See data sheets					
CH$_3$SNO + $h\nu$ → products	See data sheets					
FO$_x$ Reactions						
O + FO → O$_2$ + F	5×10^{-11}		±0.5			
O + FO$_2$ → O$_2$ + FO	5×10^{-11}		±0.7			
O(^{1}D) + HF → HO + F	1×10^{-10}		±0.5			
→ O(^{3}P) + HF						
O(^{1}D) + COF$_2$ → CO$_2$ + F$_2$	2.2×10^{-11}		±0.2			
→ O(^{3}P) + COF$_2$	5.2×10^{-11}		±0.2			
O(^{1}D) + CH$_3$F → products	1.4×10^{-10}		±0.5			
O(^{1}D) + CH$_2$F$_2$ → products	9×10^{-11}		±0.5			
O(^{1}D) + CHF$_3$ → products	8.4×10^{-12}		±0.5			
O(^{1}D) + CH$_3$CHF$_2$ → products	1×10^{-10}		±0.7			

Reaction	k_{298} cm^3 molecule^{-1} s^{-1}	$\Delta\log k_{298}$	Temp. dependence of k/cm^3 molecule^{-1} s^{-1}	Temp. range/K	$\Delta(E/R)/$ K
$O(^1D) + CH_3CF_3 \rightarrow$ products	1×10^{-10}	± 0.5			
$O(^1D) + CH_2FCF_3 \rightarrow$ products	1×10^{-10}	± 0.7			
$O(^1D) + CHF_2CF_3 \rightarrow$ products	5×10^{-11}	± 0.5			
$F + H_2 \rightarrow HF + H$	2.6×10^{-11}	± 0.1	$1.4 \times 10^{-10}\exp(-500/T)$	200–375	± 200
$F + H_2O \rightarrow HF + HO$	1.4×10^{-11}	± 0.1	1.4×10^{-11}	240–370	± 200
$F + O_2 + M \rightarrow FO_2 + M$	$3.7 \times 10^{-33}[N_2]$ (k_0)	± 0.3	$3.7 \times 10^{-33}(T/300)^{-1}[N_2]$	300–400	$\Delta n = \pm 0.5$
$FO_2 + M \rightarrow F + O_2 + M$	No recommendation (see data sheets)				
$F + O_3 \rightarrow FO + O_2$	1.3×10^{-11}	± 0.3	$2.8 \times 10^{-11}\exp(-230/T)$	250–365	± 200
$F + HONO_2 \rightarrow HF + NO_3$	2.3×10^{-11}	± 0.1	$6.0 \times 10^{-12}\exp(400/T)$	260–320	± 200
$F + NO_2 + M \rightarrow FONO + M$	See previous evaluation				
$F + CH_4 \rightarrow HF + CH_3$	8×10^{-11}	± 0.2	$3.0 \times 10^{-10}\exp(-400/T)$	250–450	± 200
$HO + CH_3F \rightarrow H_2O + CH_2F$	1.7×10^{-14}	± 0.10	$3.7 \times 10^{-12}\exp(-1600/T)$	270–340	± 300
$HO + CH_2F_2 \rightarrow H_2O + CHF_2$	1.1×10^{-14}	± 0.10	$2.0 \times 10^{-12}\exp(-1545/T)$	240–300	± 200
$HO + CHF_3 \rightarrow H_2O + CF_3$	2.4×10^{-16}	± 0.5	$1.0 \times 10^{-12}\exp(-2490/T)$	270–340	± 500
$HO + CH_3CH_2F \rightarrow$ products	2.3×10^{-13}	± 0.3			
$HO + CH_3CHF_2 \rightarrow$ products	3.6×10^{-14}	± 0.10	$1.0 \times 10^{-12}\exp(-990/T)$	240–300	± 200
$HO + CH_3CF_3 \rightarrow H_2O + CH_2CF_3$	1.3×10^{-15}	± 0.15	$1.05 \times 10^{-12}\exp(-1990/T)$	240–300	± 300
$HO + CH_2FCHF_2 \rightarrow$ products	1.8×10^{-14}	± 0.3			
$HO + CH_2FCF_3 \rightarrow H_2O + CHFCF_3$	4.9×10^{-15}	± 0.2	$8.4 \times 10^{-13}\exp(-1535/T)$	240–300	± 300
$HO + CHF_2CHF_2 \rightarrow H_2O + CF_2CHF_2$	5.7×10^{-15}	± 0.3			
$HO + CHF_2CF_3 \rightarrow H_2O + CF_2CF_3$	1.9×10^{-15}	± 0.2	$4.9 \times 10^{-13}\exp(-1655/T)$	240–300	± 300
$HO + CF_3CHO \rightarrow H_2O + CF_3CO$	1.1×10^{-12}	± 0.3			
$FO + O_3 \rightarrow$ products	No recommendation (see data sheets)				
$FO + NO \rightarrow F + NO_2$	2.6×10^{-11}	± 0.3			
$FO + NO_2 + M \rightarrow FONO_2 + M$	See previous evaluation				
$FO + FO \rightarrow$ products	1.5×10^{-11}	± 0.3			
$COF_2 + h\nu \rightarrow$ products	See data sheets				
$HCOF + h\nu \rightarrow$ products	See data sheets				
$CF_3COF + h\nu \rightarrow$ products	See data sheets				

ClO$_x$ Reactions

Reaction	k_{298} cm^3 molecule^{-1} s^{-1}	$\Delta\log k_{298}$	Temp. dependence of k/cm^3 molecule^{-1} s^{-1}	Temp. range/K	$\Delta(E/R)/$ K
$O + HCl \rightarrow HO + Cl$	See previous evaluation				
$O + HOCl \rightarrow HO + ClO$	No recommendation (see data sheets)				
$O + ClO \rightarrow Cl + O_2$	3.8×10^{-11}	± 0.1	3.8×10^{-11}	200–300	± 250
$O + OClO \rightarrow O_2 + ClO$	1.0×10^{-13}	± 0.3	$2.5 \times 10^{-12}\exp(-950/T)$	240–400	± 300
$O + OClO + M \rightarrow ClO_3 + M$	$1.8 \times 10^{-31}[N_2]$ (k_0)	± 0.3	$1.8 \times 10^{-31}(T/298)^{-1}[N_2]$	250–300	$\Delta n = \pm 0.5$
	3.1×10^{-11} (k_∞)	± 0.3	$3.1 \times 10^{-11}(T/298)^1$	250–300	$\Delta n = \pm 1$
	$F_c = 0.48$				
$O + Cl_2O \rightarrow ClO + ClO$	3.5×10^{-12}	± 0.15	$2.9 \times 10^{-11}\exp(-630/T)$	235–300	± 200
$O + ClONO_2 \rightarrow$ products	2.0×10^{-13}	± 0.1	$3.0 \times 10^{-12}\exp(-800/T)$	213–295	± 200
$O(^1D) + CHF_2Cl \rightarrow$ products	9.5×10^{-11}	± 0.2	9.5×10^{-11}	175–340	$\Delta\log k = \pm 0.2$
$O(^1D) + CHFCl_2 \rightarrow$ products	1.9×10^{-10}	± 0.2	1.9×10^{-10}	175–340	$\Delta\log k = \pm 0.2$
$O(^1D) + CH_3CF_2Cl \rightarrow$ products	1.4×10^{-10}	± 0.3			
$O(^1D) + CH_3CFCl_2 \rightarrow$ products	1.5×10^{-10}	± 0.5			
$O(^1D) + CH_2ClCF_3 \rightarrow$ products	1.5×10^{-10}	± 0.3			
$O(^1D) + CH_2ClCF_2Cl \rightarrow$ products	1.6×10^{-10}	± 0.3			
$O(^1D) + CHFClCF_3 \rightarrow$ products	1.0×10^{-10}	± 0.5			
$O(^1D) + CHCl_2CF_3 \rightarrow$ products	2.2×10^{-10}	± 0.3			
$O(^1D) + CF_2Cl_2 \rightarrow$ products	1.4×10^{-10}	± 0.1			
$O(^1D) + CFCl_3 \rightarrow$ products	2.3×10^{-10}	± 0.1			
$O(^1D) + CCl_4 \rightarrow$ products	3.3×10^{-10}	± 0.1			
$O(^1D) + COFCl \rightarrow$ products	1.9×10^{-10}	± 0.3			
$O(^1D) + COCl_2 \rightarrow$ products	3.6×10^{-10}	± 0.3			
$Cl + H_2 \rightarrow HCl + H$	1.6×10^{-14}	± 0.1	$3.7 \times 10^{-11}\exp(-2300/T)$	200–300	± 200
$Cl + HO_2 \rightarrow HCl + O_2$	3.2×10^{-11}	± 0.2	$1.8 \times 10^{-11}\exp(170/T)$	250–420	± 250
$\rightarrow ClO + HO$	9.1×10^{-12}	± 0.3	$4.1 \times 10^{-11}\exp(-450/T)$	250–420	± 250
$Cl + H_2O_2 \rightarrow HCl + HO_2$	4.1×10^{-13}	± 0.2	$1.1 \times 10^{-11}\exp(-980/T)$	265–424	± 500
$Cl + O_2 + M \rightarrow ClOO + M$	$1.4 \times 10^{-33}[N_2]$ (k_0)	± 0.2	$1.4 \times 10^{-33}(T/300)^{-3.9}[N_2]$	160–300	$\Delta n = \pm 1$
	$1.6 \times 10^{-33}[O_2]$ (k_0)	± 0.2	$1.6 \times 10^{-33}(T/300)^{-2.9}[O_2]$	160–300	$\Delta n = \pm 1$
$ClOO + M \rightarrow Cl + O_2 + M$	$6.2 \times 10^{-13}[N_2]$ (k_0/s^{-1})	± 0.3	$2.8 \times 10^{-10}\exp(-1820/T)[N_2]$	200–300	± 200
$Cl + O_3 \rightarrow ClO + O_2$	1.2×10^{-11}	± 0.06	$2.9 \times 10^{-11}\exp(-260/T)$	205–298	± 100
$Cl + HONO_2 \rightarrow HCl + NO_3$	$<2.0 \times 10^{-16}$				
$Cl + NO_3 \rightarrow ClO + NO_2$	2.6×10^{-11}	± 0.3	2.6×10^{-11}	200–300	± 400
$Cl + OClO \rightarrow ClO + ClO$	5.8×10^{-11}	± 0.1	$3.4 \times 10^{-11}\exp(160/T)$	298–450	± 200
$Cl + Cl_2O \rightarrow Cl_2 + ClO$	9.8×10^{-11}	± 0.1	9.8×10^{-11}	200–300	± 250
$Cl + Cl_2O_2 \rightarrow Cl_2 + ClOO$	1.0×10^{-10}	± 0.3	1.0×10^{-10}	230–298	± 300
$Cl + ClONO_2 \rightarrow Cl_2 + NO_3$	1.2×10^{-11}	± 0.12	$6.8 \times 10^{-12}\exp(160/T)$	219–298	± 200
$Cl + CH_4 \rightarrow HCl + CH_3$	1.0×10^{-13}	± 0.08	$9.6 \times 10^{-12}\exp(-1350/T)$	200–300	± 250
$Cl + C_2H_2 + M \rightarrow C_2H_2Cl + M$	$6 \times 10^{-30}[N_2]$ (k_0)	± 0.3	$6 \times 10^{-30}(T/300)^{-3.5}[N_2]$	200–300	$\Delta n = \pm 1$
	2.3×10^{-10} (k_∞)	± 0.5	2.3×10^{-10}	200–300	$\Delta n = \pm 1$
	$F_c = 0.6$				
$Cl + C_2H_4 + M \rightarrow C_2H_4Cl + M$	$1.6 \times 10^{-29}[air]$ (k_0)	± 0.5	$1.6 \times 10^{-29}(T/298)^{-3.5}[air]$	250–300	$\Delta n = \pm 1$
	3×10^{-10} (k_∞)	± 0.3	3×10^{-10}	250–300	$\Delta n = \pm 1$
	$F_c = 0.6$				
$Cl + C_2H_6 \rightarrow HCl + C_2H_5$	5.9×10^{-11}	± 0.06	$8.2 \times 10^{-11}\exp(-100/T)$	220–600	± 100
$Cl + C_3H_8 \rightarrow HCl + C_3H_7$	1.4×10^{-10}	± 0.12	$1.2 \times 10^{-10}\exp(40/T)$	220–600	± 200
$Cl + HCHO \rightarrow HCl + HCO$	7.3×10^{-11}	± 0.06	$8.2 \times 10^{-11}\exp(-34/T)$	200–500	± 100
$Cl + CH_3CHO \rightarrow HCl + CH_3CO$	7.2×10^{-11}	± 0.15	7.2×10^{-11}	210–340	± 300
$Cl + C_2H_5CHO \rightarrow$ products	1.2×10^{-10}	± 0.3			
$Cl + CH_3COCH_3 \rightarrow HCl + CH_3COCH_2$	3.5×10^{-12}	± 0.3			
$Cl + CH_3OH \rightarrow HCl + CH_2OH$	5.3×10^{-11}	± 0.15	5.3×10^{-11}	200–500	± 200
$Cl + C_2H_5OH \rightarrow$ products	9.4×10^{-11}	± 0.2			
$Cl + n\text{-}C_3H_7OH \rightarrow$ products	1.5×10^{-10}	± 0.2			
$Cl + i\text{-}C_3H_7OH \rightarrow$ products	8.4×10^{-11}	± 0.3			
$Cl + CH_3OOH \rightarrow$ products	5.9×10^{-11}	± 0.5			

Reaction	k_{298} cm^3 molecule^{-1} s^{-1}		$\Delta\log k_{298}$	Temp. dependence of k/cm^3 molecule^{-1} s^{-1}	Temp. range/K	$\Delta(E/R)$/ K
Cl + HCOOH → products	2.0×10^{-13}		±0.2			
Cl + CH$_3$COOH → products	2.8×10^{-14}		±0.3			
Cl + CH$_3$ONO$_2$ → products	2.4×10^{-13}		±0.3			
Cl + C$_2$H$_5$ONO$_2$ → products	4.7×10^{-12}		±0.2			
Cl + n-C$_3$H$_7$ONO$_2$ → products	2.7×10^{-11}		±0.2			
Cl + i-C$_3$H$_7$ONO$_2$ → products	5.8×10^{-12}		±0.3			
Cl + CH$_3$CO$_3$NO$_2$ → products	$<2 \times 10^{-14}$					
Cl + CH$_3$CN → products	$\leqslant 2 \times 10^{-15}$					
Cl + HCOCl → HCl + ClCO	7.8×10^{-13}		±0.15	$1.2 \times 10^{-11}\exp(-815/T)$	265–325	±300
Cl + CH$_3$Cl → HCl + CH$_2$Cl	4.9×10^{-13}		±0.15	$3.3 \times 10^{-11}\exp(-1250/T)$	233–322	±300
Cl + CH$_2$Cl$_2$ → HCl + CHCl$_2$	4.1×10^{-13}		±0.25	$8.7 \times 10^{-12}\exp(-910/T)$	270–330	±400
Cl + CHCl$_3$ → HCl + CCl$_3$	7.6×10^{-14}		±0.3	$4.9 \times 10^{-12}\exp(-1240/T)$	240–330	±400
Cl + CH$_3$CCl$_3$ → HCl + CH$_2$CCl$_3$	$<4 \times 10^{-14}$					
HO + HCl → H$_2$O + Cl	8.1×10^{-13}		±0.1	$2.4 \times 10^{-12}\exp(-330/T)$	200–300	±150
HO + HOCl → ClO + H$_2$O	5.0×10^{-13}		±0.5	$3.0 \times 10^{-12}\exp(-500/T)$	200–300	±500
HO + ClO → HO$_2$ + Cl → HCl + O$_2$	1.7×10^{-11}		±0.2	$1.1 \times 10^{-11}\exp(120/T)$	200–373	±150
HO + OClO → HOCl + O$_2$	7.0×10^{-12}		±0.3	$4.5 \times 10^{-13}\exp(800/T)$	290–480	±200
HO + ClNO$_2$ → HOCl + NO$_2$	3.5×10^{-14}		±0.3			
HO + ClONO$_2$ → products	3.9×10^{-13}		±0.2	$1.2 \times 10^{-12}\exp(-330/T)$	246–387	±200
HO + CH$_3$Cl → H$_2$O + CH$_2$Cl	4.3×10^{-14}		±0.10	$1.8 \times 10^{-12}\exp(-1115/T)$	240–300	±200
HO + CH$_2$FCl → H$_2$O + CHFCl	4.4×10^{-14}		±0.10	$2.0 \times 10^{-12}\exp(-1135/T)$	240–300	±200
HO + CHF$_2$Cl → H$_2$O + CF$_2$Cl	4.6×10^{-15}		±0.10	$7.8 \times 10^{-13}\exp(-1530/T)$	240–300	±200
HO + CHFCl$_2$ → H$_2$O + CFCl$_2$	3.0×10^{-14}		±0.10	$8.8 \times 10^{-13}\exp(-1010/T)$	240–300	±200
HO + CH$_2$Cl$_2$ → H$_2$O + CHCl$_2$	1.4×10^{-13}		±0.10	$4.4 \times 10^{-12}\exp(-1030/T)$	240–300	±250
HO + CHCl$_3$ → H$_2$O + CCl$_3$	1.0×10^{-13}		±0.10	$3.3 \times 10^{-12}\exp(-1030/T)$	240–300	±100
HO + CFCl$_3$ → HOCl + CFCl$_2$	$<5 \times 10^{-18}$			$<1 \times 10^{-12}\exp(-3650/T)$	250–480	
HO + CF$_2$Cl$_2$ → HOCl + CF$_2$Cl	$<7 \times 10^{-18}$			$<1 \times 10^{-12}\exp(-3540/T)$	250–478	
HO + CCl$_4$ → HOCl + CCl$_3$	$<5 \times 10^{-16}$			$<1 \times 10^{-12}\exp(-2260/T)$	250–300	
HO + C$_2$HCl$_3$ → products	2.2×10^{-12}		±0.10	$5.0 \times 10^{-13}\exp(445/T)$	230–420	±200
HO + C$_2$Cl$_4$ → products	1.7×10^{-13}		±0.10	$9.4 \times 10^{-12}\exp(-1200/T)$	300–420	±200
HO + CH$_3$CF$_2$Cl → H$_2$O + CH$_2$CF$_2$Cl	3.0×10^{-15}		±0.10	$9.2 \times 10^{-13}\exp(-1705/T)$	240–300	±200
HO + CH$_3$CFCl$_2$ → H$_2$O + CH$_2$CFCl$_2$	5.9×10^{-15}		±0.2	$7.0 \times 10^{-13}\exp(-1425/T)$	240–300	±300
HO + CH$_3$CCl$_3$ → H$_2$O + CH$_2$CCl$_3$	9.5×10^{-15}		±0.10	$1.2 \times 10^{-12}\exp(-1440/T)$	240–300	±200
HO + CH$_2$ClCF$_3$ → H$_2$O + CHClCF$_3$	1.3×10^{-14}		±0.2	$5.2 \times 10^{-13}\exp(-1100/T)$	260–380	±250
HO + CH$_2$ClCF$_2$Cl → H$_2$O + CHClCF$_2$Cl	1.6×10^{-14}		±0.3	$3.2 \times 10^{-13}\exp(-1580/T)$	250–350	±500
HO + CHFClCF$_3$ → H$_2$O + CFClCF$_3$	9.5×10^{-15}		±0.10	$5.4 \times 10^{-13}\exp(-1205/T)$	240–300	±200
HO + CHCl$_2$CF$_3$ → H$_2$O + CCl$_2$CF$_3$	3.6×10^{-14}		±0.15	$5.5 \times 10^{-13}\exp(-815/T)$	240–300	±200
HO + CHCl$_2$CF$_2$CF$_3$ → H$_2$O + CCl$_2$CF$_2$CF$_3$	2.5×10^{-14}		±0.15	$1.1 \times 10^{-12}\exp(-1130/T)$	270–400	±300
HO + CHFClCF$_2$CF$_2$Cl → H$_2$O + CFClCF$_2$CF$_2$Cl	8.9×10^{-15}		±0.10	$5.5 \times 10^{-13}\exp(-1230/T)$	290–400	±300
HO + CH$_3$CF$_2$CFCl$_2$ → H$_2$O + CH$_2$CF$_2$CFCl$_2$	2.4×10^{-15}		±0.3	$7.0 \times 10^{-13}\exp(-1690/T)$	290–370	±300
HO + HCOCl → H$_2$O + ClCO	$<5 \times 10^{-13}$					
HO + COCl$_2$ → products	$<5 \times 10^{-15}$					
HO + CH$_2$ClCHO → products	3.0×10^{-12}		±0.3			
HO + CHCl$_2$CHO → products	2.4×10^{-12}		±0.3			
HO + CCl$_3$CHO → H$_2$O + CCl$_3$CO	1.4×10^{-12}		±0.3			
HO + CH$_3$COCl → H$_2$O + CH$_2$COCl	9×10^{-15}		±1.0			
HO + CHF$_2$OCHClCF$_3$ → products	2.1×10^{-14}		±0.5			
HO + CHF$_2$OCF$_2$CHFCl → products	1.6×10^{-14}		±0.5	$6.1 \times 10^{-13}\exp(-1080/T)$	300–430	±500
NO$_3$ + C$_2$HCl$_3$ → products	2.9×10^{-16}		±0.3			
NO$_3$ + C$_2$Cl$_4$ → products	$<1 \times 10^{-16}$					
ClO + HO$_2$ → HOCl + O$_2$ → HCl + O$_3$	5.0×10^{-12}		±0.15	$4.6 \times 10^{-13}\exp(710/T)$	200–300	±300
ClO + O$_2(^1\Delta_g)$ → sym-ClO$_3$	See previous evaluation					
ClO + O$_3$ → ClOO + O$_2$	$<1.5 \times 10^{-17}$					
→ OClO + O$_2$	$<1 \times 10^{-18}$					
ClO + NO → Cl + NO$_2$	1.7×10^{-11}		±0.1	$6.2 \times 10^{-12}\exp(294/T)$	202–415	±100
ClO + NO$_2$ + M → ClONO$_2$ + M	1.6×10^{-31}[N$_2$]	(k_0)	±0.1	$1.6 \times 10^{-31}(T/300)^{-3.4}$[N$_2$]	200–300	$\Delta n = \pm 1$
	2×10^{-11}	(k_∞)	±0.3	2×10^{-11}	200–300	$\Delta\log k = \pm 0.3$
	$F_c = 0.5$			$F_c = \exp(-T/430)$	200–300	
ClO + NO$_3$ → ClOO + NO$_2$ → OClO + NO$_2$	4.0×10^{-13}		±0.3			
ClO + HCHO → products	See previous evaluation					
ClO + ClO → Cl + ClOO	3.4×10^{-15}		±0.3			
→ Cl + OClO	1.7×10^{-15}		±0.3			
→ Cl$_2$ + O$_2$	4.9×10^{-15}		±0.3			
ClO + ClO + M → Cl$_2$O$_2$ + M	1.7×10^{-32}[N$_2$]	(k_0)	±0.1	$1.7 \times 10^{-32}(T/300)^{-4}$[N$_2$]	200–260	$\Delta n = \pm 1.5$
	5.4×10^{-12}	(k_∞)	±0.3	5.4×10^{-12}	200–300	$\Delta\log k = \pm 0.3$
	$F_c = 0.6$					
Cl$_2$O$_2$ + M → ClO + ClO + M	2.7×10^{-18}[N$_2$]	(k_0/s^{-1})	±0.3	$1.35 \times 10^{-5}(T/300)^{-5}$ $\exp(-8720/T)$[N$_2$]	200–300	±900
	8.7×10^{2}	(k_∞/s^{-1})	±0.3	$1.8 \times 10^{15}\exp(-8450/T)$	200–300	±900
	$F_c = 0.6$					
ClO + OClO + M → Cl$_2$O$_3$ + M	2.8×10^{-31}[N$_2$]	$(k_0; 226\,\text{K})$	±0.5 (226 K)			
Cl$_2$O$_3$ + M → ClO + OClO + M	2.8×10^{-18}[N$_2$]	$(k_0/\text{s}^{-1}; 226\,\text{K})$	±0.5 (226 K)			
ClO + CH$_3$O$_2$ → ClOO + CH$_3$O	$<4 \times 10^{-12}$	$(200\,\text{K})$				
→ OClO + CH$_3$O	$<1 \times 10^{-15}$	$(200\,\text{K})$				
OClO + O$_3$ → ClO$_3$ + O$_2$	3.0×10^{-19}		±0.4	$2.1 \times 10^{-12}\exp(-4700/T)$	262–298	±1000
OClO + NO → NO$_2$ + ClO	3.4×10^{-13}		±0.3			
Cl$_2$O$_2$ + O$_3$ → ClO + ClOO + O$_2$	$<1 \times 10^{-19}$	$(200\,\text{K})$				

Reaction	k_{298} cm^3 molecule^{-1} s^{-1}		Δlog k_{298}	Temp. dependence of k/cm^3 molecule^{-1} s^{-1}	Temp. range/K	$\Delta(E/R)/$ K
CF$_3$ + O$_2$ + M → CF$_3$O$_2$ + M	1.9×10^{-29}[N$_2$]	(k_0)	± 0.2	$1.9 \times 10^{-29}(T/300)^{-4.7}$[N$_2$]	200–300	$\Delta n = \pm 1$
	1.0×10^{-11}	(k_∞)	± 0.3	1.0×10^{-11}	200–400	Δlog $k = \pm 0.3$
	$F_c = 0.6$					
CF$_2$Cl + O$_2$ + M → CF$_2$ClO$_2$ + M	1.4×10^{-29}[N$_2$]	(k_0)	± 0.5	$1.4 \times 10^{-29}(T/300)^{-5}$[N$_2$]	200–300	$\Delta n = \pm 2$
	9×10^{-12}	(k_∞)	± 0.5	9×10^{-12}	200–300	Δlog $k = \pm 0.5$
	$F_c = 0.6$					
CFCl$_2$ + O$_2$ + M → CFCl$_2$O$_2$ + M	5.5×10^{-30}[N$_2$]	(k_0)	± 0.3	$5.5 \times 10^{-30}(T/300)^{-6}$[N$_2$]	200–300	$\Delta n = \pm 2$
	9×10^{-12}	(k_∞)	± 0.5	9×10^{-12}	200–300	$\Delta n = \pm 1$
	$F_c = 0.6$					
CCl$_3$ + O$_2$ + M → CCl$_3$O$_2$ + M	1.6×10^{-30}[N$_2$]	(k_0)	± 0.3	$1.6 \times 10^{-30}(T/300)^{-6}$[N$_2$]	200–300	$\Delta n = \pm 2$
	3.6×10^{-12}	(k_∞)	± 0.5	3.6×10^{-12}	200–300	Δlog $k = \pm 0.5$
	$F_c = 0.6$					
CF$_3$O → COF$_2$ + F	$< 10^{-5}$	(k_∞/s^{-1})				
CF$_2$ClO → COF$_2$ + Cl	7×10^{5}	(k/s^{-1})	± 1.0	$3 \times 10^{13}\exp(-5250/T)$	220–300	± 1000
CFCl$_2$O → COFCl + Cl	7×10^{5}	(k/s^{-1})	± 1.0	$3 \times 10^{13}\exp(-5250/T)$	220–300	± 1000
CCl$_3$O → COCl$_2$ + Cl	8×10^{6}	(k/s^{-1})	± 1.0	$4 \times 10^{13}\exp(-4600/T)$	220–300	± 1000
CF$_3$O$_2$ + NO → CF$_3$O + NO$_2$	1.6×10^{-11}		± 0.2	$1.6 \times 10^{-11}(T/300)^{-1.2}$	230–430	Δlog $k = \pm 0.2$
CF$_2$ClO$_2$ + NO → CF$_2$ClO + NO$_2$	1.6×10^{-11}		± 0.3	$1.6 \times 10^{-11}(T/300)^{-1.5}$	230–430	Δlog $k = \pm 0.3$
CFCl$_2$O$_2$ + NO → CFCl$_2$O + NO$_2$	1.5×10^{-11}		± 0.2	$1.5 \times 10^{-11}(T/300)^{-1.3}$	230–430	Δlog $k = \pm 0.2$
CCl$_3$O$_2$ + NO → CCl$_3$O + NO$_2$	1.8×10^{-11}		± 0.2	$1.8 \times 10^{-11}(T/300)^{-1.0}$	230–430	Δlog $k = \pm 0.2$
CF$_3$O$_2$ + NO$_2$ + M → CF$_3$O$_2$NO$_2$ + M	4.5×10^{-29}[N$_2$]	(k_0)	± 0.3	$4.5 \times 10^{-29}(T/300)^{-6.4}$[N$_2$]	220–300	$\Delta n = \pm 1$
	7.5×10^{-12}	(k_∞)	± 0.5	7.5×10^{-12}	200–300	Δlog $k = \pm 0.5$
	$F_c = 0.28$				220–300	
CF$_3$O$_2$NO$_2$ + M → CF$_3$O$_2$ + NO$_2$ + M	3.6×10^{-19}[N$_2$]	(k_0/s^{-1})	± 0.4	$5 \times 10^{-1}(T/300)^{-6}$ $\exp(-12460/T)$[N$_2$]	233–373	± 500
	5.6×10^{-2}	(k_∞/s^{-1})	± 0.5	$1.2 \times 10^{17}\exp(-12580/T)$	233–373	± 500
	$F_c = 0.28$			$F_c = 0.28$	220–300	
CF$_2$ClO$_2$ + NO$_2$ + M → CF$_2$ClO$_2$NO$_2$ + M	1.4×10^{-28}[N$_2$]	(k_0)	± 0.5	$1.4 \times 10^{-28}(T/300)^{-6.4}$[N$_2$]	200–300	$\Delta n = \pm 2$
	7.5×10^{-12}	(k_∞)	± 0.3	7.5×10^{-12}	200–300	Δlog $k = \pm 0.3$
	$F_c = 0.26$			$F_c = 0.26$	220–300	
CF$_2$ClO$_2$NO$_2$ + M → CF$_2$ClO$_2$ + NO$_2$ + M	9.0×10^{-19}[N$_2$]	(k_0/s^{-1})	± 0.3	$1.8 \times 10^{-3}\exp(-10500/T)$[N$_2$]	260–300	± 500
	5.4×10^{-2}	(k_∞/s^{-1})	± 0.3	$1.6 \times 10^{16}\exp(-11990/T)$	260–300	± 500
	$F_c = 0.26$			$F_c = 0.26$	250–300	
CFCl$_2$O$_2$ + NO$_2$ + M → CFCl$_2$O$_2$NO$_2$ + M	1.7×10^{-28}[N$_2$]	(k_0)	± 0.3	$1.7 \times 10^{-28}(T/300)^{-6.7}$[N$_2$]	230–300	$\Delta n = \pm 2$
	7.5×10^{-12}	(k_∞)	± 0.3	7.5×10^{-12}	250–300	Δlog $k = \pm 0.3$
	$F_c = 0.23$			$F_c = 0.23$	230–300	
CFCl$_2$O$_2$NO$_2$ + M → CFCl$_2$O$_2$ + NO$_2$ + M	1.5×10^{-18}[N$_2$]	(k_0/s^{-1})	± 0.3	$1.0 \times 10^{-2}\exp(-10860/T)$	250–300	$+500$
	9.6×10^{-2}	(k_∞/s^{-1})	± 0.3	$6.6 \times 10^{16}\exp(-12240/T)$	250–300	± 500
	$F_c = 0.23$			$F_c = 0.23$	250–300	
CCl$_3$O$_2$ + NO$_2$ + M → CCl$_3$O$_2$NO$_2$ + M	3.2×10^{-28}[N$_2$]	(k_0)	± 0.5	$3.2 \times 10^{-28}(T/300)^{-7.7}$[N$_2$]	230–300	$\Delta n = \pm 3$
	7.5×10^{-12}	(k_∞)	± 0.3	7.5×10^{-12}	250–300	Δlog $k = \pm 0.3$
	$F_c = 0.21$			$F_c = 0.21$	250–300	
CCl$_3$O$_2$NO$_2$ + M → CCl$_3$O$_2$ + NO$_2$ + M	7.6×10^{-18}[N$_2$]	(k_0/s^{-1})	± 0.3	$6.3 \times 10^{-3}\exp(-10235/T)$[N$_2$]	250–300	± 500
	0.29	(k_∞/s^{-1})	± 0.3	$4.8 \times 10^{16}\exp(-11820/T)$	250–300	± 500
	$F_c = 0.20$			$F_c = 0.20$	250–300	
O$_3$ + C$_2$HCl$_3$ → products	$< 5 \times 10^{-20}$					
O$_3$ + C$_2$Cl$_4$ → products	$< 10^{-21}$					
HCl + $h\nu$ → products	See data sheets					
HOCl + $h\nu$ → products	See data sheets					
OClO + $h\nu$ → products	See data sheets					
Cl$_2$O + $h\nu$ → products	See data sheets					
Cl$_2$O$_2$ + $h\nu$ → products	See data sheets					
Cl$_2$O$_3$ + $h\nu$ → products	See data sheets					
ClNO + $h\nu$ → products	See data sheets					
ClONO + $h\nu$ → products	See data sheets					
ClNO$_2$ + $h\nu$ → products	See data sheets					
ClONO$_2$ + $h\nu$ → products	See data sheets					
Cl$_2$ + $h\nu$ → products	See data sheets					
CH$_3$Cl + $h\nu$ → products	See data sheets					
CHF$_2$Cl + $h\nu$ → products	See data sheets					
CF$_2$Cl$_2$ + $h\nu$ → products	See data sheets					
CFCl$_3$ + $h\nu$ → products	See data sheets					
CCl$_4$ + $h\nu$ → products	See data sheets					
CH$_3$CF$_2$Cl + $h\nu$ → products	See data sheets					
CH$_3$CFCl$_2$ + $h\nu$ → products	See data sheets					
CH$_3$CCl$_3$ + $h\nu$ → products	See data sheets					
CF$_3$CHFCl + $h\nu$ → products	See data sheets					
CF$_3$CHCl$_2$ + $h\nu$ → products	See data sheets					
CF$_2$ClCFCl$_2$ + $h\nu$ → products	See data sheets					
CF$_2$ClCF$_2$Cl + $h\nu$ → products	See data sheets					
CF$_3$CF$_2$Cl + $h\nu$ → products	See data sheets					
CF$_3$CF$_2$CHCl$_2$ + $h\nu$ → products	See data sheets					
CF$_2$ClCF$_2$CHFCl + $h\nu$ → products	See data sheets					
HCOCl + $h\nu$ → products	See data sheets					
COFCl + $h\nu$ → products	See data sheets					
COCl$_2$ + $h\nu$ → products	See data sheets					
CCl$_3$CHO + $h\nu$ → products	See data sheets					
CF$_3$COCl + $h\nu$ → products	See data sheets					

BrO$_x$ Reactions

Reaction	k_{298} cm^3 molecule^{-1} s^{-1}		Δlog k_{298}			
O + HBr → HO + Br	See previous evaluation					
O + Br$_2$ → BrO + Br	See previous evaluation					
O + BrO → O$_2$ + Br	3×10^{-11}		± 0.5			

Reaction	k_{298} cm^3 molecule^{-1} s^{-1}		$\Delta\log k_{298}$	Temp. dependence of k/cm^3 molecule^{-1} s^{-1}	Temp. range/K	$\Delta(E/R)$/K
Br + HO$_2$ → HBr + O$_2$	2.0×10^{-12}		±0.3	$1.4 \times 10^{-11}\exp(-590/T)$	260–390	±200
Br + H$_2$O$_2$ → HBr + HO$_2$ → HOBr + HO}	$<5 \times 10^{-16}$					
Br + O$_3$ → BrO + O$_2$	1.2×10^{-12}		±0.08	$1.7 \times 10^{-11}\exp(-800/T)$	195–392	±200
Br + NO$_2$ + M → BrNO$_2$ + M	4.2×10^{-31} [N$_2$]	(k_0)	±0.3	$4.2 \times 10^{-31}(T/300)^{-2.4}$[N$_2$]	200–300	$\Delta n = \pm 1$
	2.7×10^{-11}	(k_∞)	±0.4	2.7×10^{-11}	200–300	$\Delta\log k = \pm 0.4$
	$F_c = 0.55$					
Br + OClO → BrO + ClO	3.4×10^{-13}		±0.3	$2.6 \times 10^{-11}\exp(-1300/T)$	200–450	±300
Br + Cl$_2$O → BrCl + ClO	3.8×10^{-12}		±0.3	$2.1 \times 10^{-11}\exp(-520/T)$	220–298	±300
Br + Cl$_2$O$_2$ → BrCl + ClOO	3.0×10^{-12}		±0.3			
Br + HCHO → HBr + HCO	1.0×10^{-12}		±0.15	$1.7 \times 10^{-11}\exp(-800/T)$	223–480	±250
Br + CH$_3$CHO → HBr + CH$_3$CO	3.9×10^{-12}		±0.2	$1.3 \times 10^{-11}\exp(-360/T)$	250–400	±200
HO + HBr → H$_2$O + Br	1.1×10^{-11}		±0.1	1.1×10^{-11}	249–416	±250
HO + Br$_2$ → HOBr + Br	4.5×10^{-11}		±0.15	$1.2 \times 10^{-11}\exp(400/T)$	260–360	±400
HO + CH$_3$Br → H$_2$O + CH$_2$Br	3.0×10^{-14}		±0.10	$1.9 \times 10^{-12}\exp(-1240/T)$	240–300	±200
HO + CHF$_2$Br → H$_2$O + CF$_2$Br	9.5×10^{-15}		±0.2	$7.7 \times 10^{-13}\exp(-1310/T)$	240–300	±200
HO + CF$_3$Br → products	$<1 \times 10^{-16}$					
HO + CF$_2$ClBr → products	$<1 \times 10^{-16}$					
HO + CF$_2$Br$_2$ → products	$<5 \times 10^{-16}$					
HO + CF$_3$CHFBr → H$_2$O + CF$_3$CFBr	1.7×10^{-14}		±0.3	$1.1 \times 10^{-12}\exp(-1250/T)$	270–430	±500
HO + CF$_3$CHClBr → H$_2$O + CF$_3$CClBr	5.8×10^{-14}		±0.3			
HO + CF$_2$BrCF$_2$Br → products	$<1.3 \times 10^{-16}$					
BrO + HO$_2$ → HOBr + O$_2$ → HBr + O$_3$}	3.3×10^{-11}		±0.5	$6.2 \times 10^{-12}\exp(500/T)$	200–300	±500
BrO + O$_3$ → Br + 2O$_2$	$<5 \times 10^{-15}$					
BrO + NO → Br + NO$_2$	2.1×10^{-11}		±0.1	$8.7 \times 10^{-12}\exp(260/T)$	224–425	±100
BrO + NO$_2$ + M → BrONO$_2$ + M	4.7×10^{-31} [N$_2$]	(k_0)	±0.1	$4.7 \times 10^{-31}(T/300)^{-3.1}$[N$_2$]	200–300	$\Delta n = \pm 1$
	1.7×10^{-11}	(k_∞)	±0.1	$1.7 \times 10^{-11}(T/298)^{-0.6}$	200–300	$\Delta n = \pm 1$
	$F_c = 0.40$			$F_c = \exp(-T/327)$	200–300	
BrO + ClO → Br + OClO	6.8×10^{-12}		±0.1	$1.6 \times 10^{-12}\exp(430/T)$	220–400	±200
→ Br + ClOO	6.1×10^{-12}		±0.1	$2.9 \times 10^{-12}\exp(220/T)$	220–400	±200
→ BrCl + O$_2$	1.0×10^{-12}		±0.1	$5.8 \times 10^{-13}\exp(170/T)$	220–400	±200
BrO + BrO → 2Br + O$_2$	2.1×10^{-12}		±0.1 }	$1.1 \times 10^{-12}\exp(250/T)$	223–400	±200
→ Br$_2$ + O$_2$	4.1×10^{-13}		±0.2 }			
HOBr + $h\nu$ → products	See data sheets					
BrO + $h\nu$ → products	See data sheets					
BrONO$_2$ + $h\nu$ → products	See data sheets					
CH$_3$Br + $h\nu$ → products	See data sheets					
CF$_3$Br + $h\nu$ → products	See data sheets					
CF$_2$ClBr + $h\nu$ → products	See data sheets					
CF$_2$Br$_2$ + $h\nu$ → products	See data sheets					
CHBr$_3$ + $h\nu$ → products	See data sheets					
CF$_2$BrCF$_2$Br + $h\nu$ → products	See data sheets					

IO$_x$ Reactions

Reaction	k_{298} cm^3 molecule^{-1} s^{-1}		$\Delta\log k_{298}$	Temp. dependence of k/cm^3 molecule^{-1} s^{-1}	Temp. range/K	$\Delta(E/R)$/K
O + I$_2$ → IO + I	1.4×10^{-10}		±0.3	1.4×10^{-10}	200–400	±250
O + IO → O$_2$ + I	3×10^{-11}		±0.5			
I + HO$_2$ → HI + O$_2$	3.8×10^{-13}		±0.3	$1.5 \times 10^{-11}\exp(-1090/T)$	250–350	±500
I + O$_3$ → IO + O$_2$	1.0×10^{-12}		±0.2	$2.0 \times 10^{-11}\exp(-890/T)$	200–350	±300
I + NO + M → INO + M	1.8×10^{-32} [N$_2$]	(k_0)	±0.1	$1.8 \times 10^{-32}(T/300)^{-1.0}$[N$_2$]	200–400	$\Delta n = \pm 0.5$
	1.7×10^{-11}	(k_∞)	±0.3	1.7×10^{-11}	200–400	$\Delta n = \pm 0.5$
	$F_c = 0.75$			$F_c = [\exp(-T/1040) + \exp(-4160/T)]$	200–400	
I + NO$_2$ + M → INO$_2$ + M	3.0×10^{-31} [N$_2$]	(k_0)	±0.2	$3.0 \times 10^{-31}(T/300)^{-1}$[N$_2$]	200–400	$\Delta n = \pm 1$
	6.6×10^{-11}	(k_∞)	±0.3	6.6×10^{-11}	200–400	$\Delta\log k = \pm 0.3$
	$F_c = 0.63$			$F_c = [\exp(-T/650) + \exp(-2600/T)]$	200–400	
HO + HI → H$_2$O + I	3.0×10^{-11}		±0.3			
HO + I$_2$ → HOI + I	1.8×10^{-10}		±0.3			
HO + CH$_3$I → H$_2$O + CH$_2$I	7.2×10^{-14}		±0.5	$3.1 \times 10^{-12}\exp(-1120/T)$	270–430	±500
NO$_3$ + HI → HNO$_3$ + I	No recommendation (see data sheets)					
IO + HO$_2$ → HOI + O$_2$	6.4×10^{-11}		±0.3			
IO + IO → products	5.2×10^{-11}		±0.3	$1.7 \times 10^{-12}\exp(1020/T)$	250–373	±500
IO + NO → I + NO$_2$	2.2×10^{-11}		±0.3	$7.3 \times 10^{-12}\exp(330/T)$	200–400	±150
IO + NO$_2$ + M → IONO$_2$ + M	7.7×10^{-31} [N$_2$]	(k_0)	±0.3	$7.7 \times 10^{-31}(T/300)^{-5}$[N$_2$]	250–350	$\Delta n = \pm 2$
	1.6×10^{-11}	(k_∞)	±0.3	1.6×10^{-11}	250–350	$\Delta\log k = \pm 0.3$
	$F_c = 0.4$					
IO + CH$_3$SCH$_3$ → products	1.2×10^{-14}		±0.3			
INO + INO → I$_2$ + 2 NO	1.3×10^{-14}		±0.4	$8.4 \times 10^{-11}\exp(-2620/T)$	300–450	±600
INO$_2$ + INO$_2$ → I$_2$ + 2 NO$_2$	4.7×10^{-15}		±0.5	$2.9 \times 10^{-11}\exp(-2600/T)$	298–400	±1000
HOI + $h\nu$ → products	See data sheets					
IO + $h\nu$ → products	See data sheets					
INO + $h\nu$ → products	See data sheets					
INO$_2$ + $h\nu$ → products	See data sheets					
IONO$_2$ + $h\nu$ → products	See data sheets					

*No data sheet or recommendation presented in this article. See our earlier evaluation; J. Phys. Chem. Ref. Data **18**, 881 (1989) for our most recent recommendation.

KINETIC DATA FOR COMBUSTION MODELLING

D. L. Baulch, C. J. Cobos, R. A. Cox, C. Esser, P. Frank, Th. Just, J. A. Kerr, M. J. Pilling, J. Troe, R. W. Walker, and J. Warnatz

The following tables present evaluated rate constants and other chemical kinetic data required for modelling the combustion of hydrocarbons. The compilation was prepared as part of the project "Kinetics and Mechanisms of Chemical Processes in Combustion", which is one of the projects in the third European Community Energy Research and Development Program. The tables are reprinted from the *Journal of Physical and Chemical Reference Data* by permission of the authors and the American Institute of Physics.

Table 1 lists all the reactions studied and gives the recommended rate constant k for every bimolecular reaction, as well as the applicable temperature range and the associated error limits. Where more than one set of products is possible, rate constants or branching ratios are given for all channels considered feasible. The data for decomposition reactions and combination reactions are given in Tables 2 and 3, respectively. The reference includes a detailed data sheet for each reaction listed here, covering the thermodynamic data, kinetic measurements, and reliability assessments.

REFERENCE

Baulch, D. L., et al., *J. Phys. Chem. Ref. Data*, 21, 411-734, 1992.

Table 1
BIMOLECULAR REACTIONS

Reaction	k/cm³ molecule^{-1}s^{-1}	Temp/K	Error limits ($\Delta \log k$)
O Atom Reactions			
$O + H_2 \rightarrow OH + H$	$8.5 \times 10^{-20} T^{2.67} \exp(-3160/T)$	300–2500	± 0.5 at 300 K falling to ± 0.2 for $T > 500$ K
$O + OH \rightarrow O_2 + H$	$2.0 \times 10^{-11} \exp(112/T)$ $2.4 \times 10^{-11} \exp(-353/T)$	220–500 1000–2000	± 0.2 ± 0.1
$O + HO_2 \rightarrow OH + O_2$	5.3×10^{-11}	300–1000	± 0.3 at 300 K rising to ± 0.5 at 1000 K.
$O + H_2O_2 \rightarrow OH + HO_2$	$1.1 \times 10^{-12} \exp(-2000/T)$	300–500	± 0.3
$O + NH_3 \rightarrow OH + NH_2$	$1.6 \times 10^{-11} \exp(-3670/T)$	500–2500	± 0.5
$O + CH \rightarrow CO + H$ $\rightarrow CHO^+ + e$	6.6×10^{-11} $4.2 \times 10^{-13} \exp(-850/T)$	300–2000 300–2500	± 0.5 ± 0.5
$O + {}^3CH_2 \rightarrow CO + 2H$ $\rightarrow CO + H_2$	2×10^{-10} $k_1/k = 0.6 \pm 0.3$ over whole range	300–2500	± 0.2 at 300 K rising to ± 0.7 at 2500 K.
$O + CH_3 \rightarrow HCHO + H$	1.4×10^{-10}	300–2500	± 0.2
$O + CH_4 \rightarrow OH + CH_3$	$1.5 \times 10^{-15} T^{1.56} \exp(-4270/T)$	300–2500	± 0.3 at 300 K falling to ± 0.15 at 2500 K.
$O + CHO \rightarrow OH + CO$ $\rightarrow CO_2 + H$	5.0×10^{-11} 5.0×10^{-11}	300–2500 300–2500	± 0.3 ± 0.3
$O + HCHO \rightarrow OH + CHO$	$6.9 \times 10^{-13} T^{0.57} \exp(-1390/T)$	250–2200	± 0.1 at 250 K rising to ± 0.3 at 2200 K.
$O + CH_3O \rightarrow O_2 + CH_3$ $\rightarrow OH + HCHO$	2.5×10^{-11} $k_2/k = (0.12 \pm 0.1)$ at 300 K	300–1000	± 0.3 at 300 K rising to ± 0.7 at 1000 K.
$O + CN \rightarrow CO + N(^4S)$ $\rightarrow CO + N(^2D)$	1.7×10^{-11}	300–5000	± 0.2 at 300 K rising to $\pm 0.6 \times 5000$ K.
$O + NCO \rightarrow NO + CO$ $\rightarrow O_2 + CN$	7.0×10^{-11}	1450–2600	± 0.8

Reaction	k/cm^3 molecule^{-1}s^{-1}	Temp/K	Error limits ($\Delta \log k$)
$O + HCN \rightarrow NCO + H$ $\rightarrow CO + NH$ $\rightarrow OH + CN$	$2.3 \times 10^{-18} T^{2.1} \exp(-3075/T)$	450–2500	± 0.2 at 450 K rising to ± 0.3 at 2500 K.
$O + CH_3OOH \rightarrow OH + CH_2COOH$ $\rightarrow OH + CH_3O_2$	$6.9 \times 10^{-13} T^{0.57} \exp(-1390/T)$ [estimate]	250–2200	± 0.1 at 250 K rising to ± 0.3 at 2200 K.
$O + C_2H \rightarrow CO + CH$	1.7×10^{-11}	300–2500	± 1.0
$O + C_2H_2 \rightarrow CO + {}^3CH_2$ $\rightarrow CHCO + H$	$3.6 \times 10^{-20} T^{2.8} \exp(-250/T)$ $k_1/k = 0.5 \pm 0.3$ over whole range.	300–2500	± 0.2
$O + C_2H_3 \rightarrow OH + C_2H_2$ $\rightarrow CO + CH_3$ $\rightarrow HCO + CH_2$	5×10^{-11}	300–2000	± 0.5
$O + C_2H_4 \rightarrow CH_2CHO + H$ $\rightarrow HCO + CH_3$ $\rightarrow HCHO + CH_2$ $\rightarrow CH_2CO + H_2$	$5.75 \times 10^{-18} T^{2.08}$ $k_1/k = 0.35 \pm 0.05$ at p > 3 Torr $k_2/k = 0.6 \pm 0.10$	300–2000 over whole temperature range	± 0.1 for T < 1000 K rising to ± 0.3 at 2000 K.
$O + C_2H_5 \rightarrow CH_3CHO + H$ $\rightarrow HCHO + CH_3$	1.1×10^{-10} $k_2/k = 0.17 \pm 0.2$ at 300 K	300–2500	± 0.3 from 300 to 1000 K ± 0.5 from 1000 to 2500 K
$O + C_2H_6 \rightarrow OH + C_2H_5$	$1.66 \times 10^{-15} T^{1.5} \exp(-2920/T)$	300–1200	± 0.3 at 300 K falling to ± 0.15 at 1200 K.
$O + CHCO \rightarrow 2CO + H$	1.6×10^{-10}	300–2500	± 0.3
$O + CH_2CO \rightarrow CH_2O + CO$ $\rightarrow HCO + H + CO$ $\rightarrow HCO + HCO$	$3.8 \times 10^{-12} \exp(-680/T)$	230–500	± 0.3
$O + CH_3CHO \rightarrow OH + CH_3CO$ $\rightarrow OH + CH_2CHO$	$9.7 \times 10^{-12} \exp(-910/T)$	300–1500	± 0.05 at 300 K rising to ± 0.5 at 1500 K.
$O + C_2H_5OOH \rightarrow OH + C_2H_4OOH$ $\rightarrow OH + C_2H_5OO$	$6.9 \times 10^{-13} T^{0.57} \exp(-1390/T)$ [estimate]	250–2200	± 0.1 at 150 K rising to ± 0.3 at 2200 K.
$O + C_6H_6 \rightarrow OH + C_6H_5$ $\rightarrow C_6H_5OH$	$1.2 \times 10^{-22} T^{3.7} \exp(-570/T)$	300–1000	± 0.5
$O + C_6H_5CH_2 \rightarrow HCO + C_6H_6$ $\rightarrow C_6H_5CH + H$ $\rightarrow CH_2O + C_6H_5$	5.5×10^{-10} No recommendation	300	± 0.3
$O + C_6H_5CH_3 \rightarrow$ products	$5.3 \times 10^{-15} T^{1.21} \exp(-1260/T)$	300–2800	± 0.1 at 300 K rising to ± 0.4 at 2800 K
$O + p\text{-}C_6H_4(CH_3)_2 \rightarrow$ products	$2.6 \times 10^{-11} \exp(-1409/T)$	300–600	± 0.3
$O + C_6H_5C_2H_5 \rightarrow$ products	1.0×10^{-13}	298	± 0.3

O₂ Reactions

Reaction	k/cm^3 molecule^{-1}s^{-1}	Temp/K	Error limits ($\Delta \log k$)
$O_2 + CH_4 \rightarrow HO_2 + CH_3$	$6.6 \times 10^{-11} \exp(-28630/T)$	500–2000	± 0.5 at 500 K rising to ± 1.0 at 2000 K.

Reaction	k/cm³ molecule⁻¹s⁻¹	Temp/K	Error limits ($\Delta \log k$)
$O_2 + C_2H_6 \rightarrow HO_2 + C_2H_5$	$1.0 \times 10^{-10} \exp(-26100/T)$	500–2000	$\pm$ 0.5 at 500 K rising to $\pm$ 1.0 at 2000 K
$O_2 + HCHO \rightarrow HO_2 + HCO$	$1.0 \times 10^{-10} \exp(-20460/T)$	700–1000	$\pm$ 0.5
$O_2 + CH_3CHO \rightarrow HO_2 + CH_3CO$	$5.0 \times 10^{-11} \exp(-19700/T)$	600–1100	$\pm$ 0.5 at 600 K rising to $\pm$ 1.0 at 1100 K.
H Atom Reactions			
$H + O_2 \rightarrow OH + O$	$3.3 \times 10^{-10} \exp(-8460/T)$	300–2500	$\pm$ 0.1 at 300 K rising to $\pm$ 0.2 at 2500 K.
$H + O_2 + Ar \rightarrow HO_2 + Ar$	See Table 3		
$H + O_2 + H_2 \rightarrow HO_2 + H_2$	See Table 3		
$H + O_2 + N_2 \rightarrow HO_2 + N_2$	See Table 3		
$H + O_2 + H_2O \rightarrow HO_2 + H_2O$	See Table 3		
$H + H + Ar \rightarrow H_2 + Ar$	See Table 3		
$H + H + H_2 \rightarrow H_2 + H_2$	See Table 3		
$H + OH + H_2O \rightarrow H_2O + H_2O$	See Table 3		
$H + OH + Ar \rightarrow H_2O + Ar$	See Table 3		
$H + HO + N_2 \rightarrow H_2O + N_2$	See Table 3		
$H + HO_2 \rightarrow H_2 + O_2$	$7.1 \times 10^{-11} \exp(-710/T)$	300–1000	$\pm$ 0.3
$\rightarrow 2 OH$	$2.8 \times 10^{-10} \exp(-440/T)$	300–1000	$\pm$ 0.3
$\rightarrow H_2O + O$	$5.0 \times 10^{-11} \exp(-866/T)$	300–1000	$\pm$ 0.3
$H + H_2O \rightarrow OH + H_2$	$7.5 \times 10^{-16} T^{1.6} \exp(-9270/T)$	300–2500	$\pm$0.2
$H + H_2O_2 \rightarrow H_2 + HO_2$	$2.8 \times 10^{-12} \exp(-1890/T)$	300–1000	$\pm$ 0.3
$\rightarrow OH + H_2O$	$1.7 \times 10^{-11} \exp(-1800/T)$	300–1000	$\pm$ 0.3
$H + NH \rightarrow H_2 + N$	1.7×10^{-11}	1500–2500	$\pm$ 1.0
$H + NH_2 \rightarrow H_2 + NH$	1.0×10^{-11}	2000–3000	$\pm$ 1.0
$H + {}^3CH_2 \rightarrow H_2 + CH$	$1.0 \times 10^{-11} \exp(900/T)$	300–3000	$\pm$ 0.7
$H + CH_3 \rightarrow H_2 + {}^1CH_2$	$1.0 \times 10^{-10} \exp(-7600/T)$	300–2500	$\pm$ 1.0
$\rightarrow CH_4$	See Table 3		
$H + CH_4 \rightarrow H_2 + CH_3$	$2.2 \times 10^{-20} T^{3.0} \exp(-4045/T)$	300–2500	$\pm$ 0.2
$H + CHO \rightarrow H_2 + CO$	1.5×10^{-10}	300–2500	$\pm$ 0.3
$H + HCHO \rightarrow H_2 + HCO$	$3.8 \times 10^{-14} T^{1.05} \exp(-1650/T)$	300–2200	$\pm$ 0.1 at 300 K rising to $\pm$ 0.5 at 2200 K
$H + CH_3O \rightarrow H_2 + HCHO$	3.0×10^{-11}	300–1000	$\pm$ 0.5
$H + HNCO \rightarrow NH_2 + CO$	No recommendation		
$\rightarrow H_2 + NCO$	$3.4 \times 10^{-10} T^{-0.27} \exp(-10190/T)$	500–1000	$\pm$ 1.0
$H + NCO \rightarrow NH + CO$ $\rightarrow HCN + O$	8.7×10^{-11}	1400–1500	$\pm$ 0.5

Reaction	k/cm^3 molecule^{-1}s^{-1}	Temp/K	Error limits ($\Delta \log k$)
H + C$_2$H$_2$ → H$_2$ + C$_2$H → C$_2$H$_3$	1.0×10^{-10} exp($-14000/T$) See Table 3	1000–3000	± 1.0
H + C$_2$H$_3$ → H$_2$ + C$_2$H$_2$ → C$_2$H$_4$	2.0×10^{-11} See Table 3	300–2500	± 0.5
H + C$_2$H$_4$ → C$_2$H$_3$ + H$_2$ → C$_2$H$_5$	9.0×10^{-10} exp($-7500/T$) See Table 3	700–2000	± 0.5
H + C$_2$H$_5$ → 2CH$_3$ → C$_2$H$_6$	6.0×10^{-11} See Table 3	300–2000	± 0.3
H + C$_2$H$_6$ → H$_2$ + C$_2$H$_5$	2.4×10^{-15} $T^{1.5}$ exp($-3730/T$)	300–2000	± 0.15 at 300 K rising to ± 0.3 at 2000 K
H + CHCO → CH$_2$ + CO → H$_2$ + C$_2$O → HCCOH	2.5×10^{-10}	300–2500	± 0.4
H + CH$_2$CO → CH$_3$ + CO → CH$_2$CHO	3.0×10^{-11} exp($-1700/T$) k_2/k very small	200–2000	± 0.5 at 200 K rising to ± 1.0 at 2000 K.
H + CH$_3$CHO → H$_2$ + CH$_3$CO → H$_2$ + CH$_2$CHO	6.8×10^{-15} $T^{1.16}$ exp($-1210/T$)	300–2000	± 0.1 at 300 rising to ± 0.4 at 2000 K.
H + C$_6$H$_5$ + M → C$_6$H$_6$ + M	See Table 3		
H + C$_6$H$_6$ → H$_2$ + C$_6$H$_5$ → C$_6$H$_7$	No recommendation See Table 3		
H + C$_6$H$_5$O + M → C$_6$H$_5$OH + M	See Table 3		
H + C$_6$H$_5$OH → C$_6$H$_5$O + H$_2$ → C$_6$H$_6$ + OH	1.9×10^{-10} exp($-6240/T$) 3.7×10^{-11} exp($-3990/T$)	1000–1150 1000–1150	± 0.3 ± 0.3
H + C$_6$H$_5$CH$_2$ + M → C$_6$H$_5$CH$_3$ + M	See Table 3		
H + C$_6$H$_5$CH$_3$ → H$_2$ + C$_6$H$_5$CH$_2$ → H$_2$ + C$_6$H$_4$CH$_3$ → C$_6$H$_6$ + CH$_3$ → C$_6$H$_6$CH$_3$	6.6×10^{-22} $T^{3.44}$ exp($-1570/T$) No recommendation No recommendation See Table 3	600–2800	± 0.3 at 600 K rising to ± 0.5 at 2800 K.
H + p-C$_6$H$_4$(CH$_3$)$_2$ → products	5.8×10^{-13}	298	± 0.1
H + C$_6$H$_5$C$_2$H$_5$ → H$_2$ + C$_6$H$_5$C$_2$H$_4$ → C$_6$H$_6$C$_2$H$_5$	2.4×10^{-12} See Table 3	773	± 0.1

H$_2$ Reactions

Reaction	k/cm^3 molecule^{-1}s^{-1}	Temp/K	Error limits ($\Delta \log k$)
H$_2$ + Ar → 2H + Ar	See Table 2		
H$_2$ + H$_2$ → 2H + H$_2$	See Table 2		

OH Radical Reactions

Reaction	k/cm^3 molecule^{-1}s^{-1}	Temp/K	Error limits ($\Delta \log k$)
OH + H$_2$ → H$_2$O + H	1.7×10^{-16} $T^{1.6}$ exp($-1660/T$)	300–2500	± 0.1 at 300 K rising to ± 0.3 at 2500 K
OH + OH → H$_2$O + O	2.5×10^{-15} $T^{1.14}$ exp($-50/T$)	250–2500	± 0.2
OH + OH + M → H$_2$O$_2$ + M	See Table 3		

Reaction	k/cm^3 molecule^{-1}s^{-1}	Temp/K	Error limits ($\Delta \log k$)
$OH + HO_2 \rightarrow H_2O + O_2$	4.8×10^{-11} exp(250/T)	300–2000	$\pm$ 0.2 at 300 K rising to $\pm$ 0.5 at 2000 K.
$OH + H_2O_2 \rightarrow H_2O + HO_2$	1.3×10^{-11} exp(−670/T)	300–1000	$\pm$ 0.2
$OH + NH \rightarrow NO + H_2$ $\rightarrow H_2O + N$	8.0×10^{-11}	300–1000	$\pm$ 0.5
$OH + NH_2 \rightarrow O + NH_3$ $\rightarrow H_2O + NH$	3.3×10^{-14} $T^{0.405}$ exp(−250/T) No recommendation	500–2500	$\pm$ 0.5
$OH + CO \rightarrow H + CO_2$	1.05×10^{-17} $T^{1.5}$ exp(250/T)	300–2000	$\pm$ 0.2 at 300 K rising to $\pm$ 0.5 at 2000 K.
$OH + CH_3 \rightarrow H + CH_2OH$ $\rightarrow H + CH_3O$ $\rightarrow H_2O + {}^1CH_2$ $\rightarrow CH_3OH$	6.0×10^{-11} See Table 3	300–2000	$\pm$ 0.7
$OH + CH_4 \rightarrow H_2O + CH_3$	2.6×10^{-17} $T^{1.83}$ exp(−1400/T)	250–2500	$\pm$ 0.07 at 250 K rising to $\pm$ 0.15 at 1200 K.
$OH + CHO \rightarrow H_2O + CO$	1.7×10^{-10}	300–2500	$\pm$ 0.3
$OH + HCHO \rightarrow H_2O + CHO$	5.7×10^{-15} $T^{1.18}$ exp(225/T)	300–3000	$\pm$ 0.1 at 300 K rising to $\pm$ 0.7 at 3000 K.
$OH + CN \rightarrow O + HCN$ $\rightarrow NCO + H$	1.0×10^{-10}	1500–3000	$\pm$ 0.5
$OH + HCN \rightarrow H_2O + CN$ $\rightarrow HOCN + H$ $\rightarrow HNCO + H$	1.5×10^{-11} exp(−5400/T) No recommendation	1500–2500	$\pm$ 0.5
$OH + CH_3OOH \rightarrow H_2O + CH_3OO$ $\rightarrow H_2O + CH_2OOH$	1.2×10^{-12} exp(130/T) 1.8×10^{-12} exp(220/T)	300–1000 300–1000	$\pm$ 0.2 at 300 K rising to $\pm$ 0.4 at 1000 K $\pm$ 0.1 at 300 K rising to $\pm$0.3 at 1000 K.
$OH + C_2H_2 \rightarrow H_2O + C_2H$ $\rightarrow H + CH_2CO$ $\rightarrow C_2H_2OH$	1.0×10^{-10} exp(−6500/T) See Table 3	1000–2000	$\pm$ 1.0
$OH + C_2H_4 \rightarrow H_2O + C_2H_3$	3.4×10^{-11} exp(−2990/T)	650–1500	$\pm$ 0.5
$OH + C_2H_6 \rightarrow H_2O + C_2H_5$	1.2×10^{-17} $T^{2.0}$ exp(−435/T)	250–2000	$\pm$ 0.07 at 250 K rising to $\pm$ 0.15 at 2000 K.
$OH + CH_2CO \rightarrow CH_2OH + CO$ $\rightarrow H_2CO + HCO$	1.7×10^{-11}	300–2000	$\pm$ 1.0
$OH + CH_3CHO \rightarrow H_2O + CH_3CO$ $\rightarrow H_2O + CH_2CHO$	3.9×10^{-14} $T^{0.73}$ exp(560/T)	250–1200	$\pm$ 0.1 at 250 K rising to $\pm$ 0.3 at 1200 K.
$OH + C_2H_5OOH \rightarrow H_2O + C_2H_5OO$ $\rightarrow H_2O + C_2H_4OOH$	3.0×10^{-12} exp(190/T) [estimate]	250–1000	$\pm$ 0.3 at 250 K rising to $\pm$ 0.7 at 1000 K

Reaction	k/cm^3 molecule^{-1}s^{-1}	Temp/K	Error limits ($\Delta \log k$)
OH + C$_6$H$_6$ → H$_2$O + C$_6$H$_5$	$2.7 \times 10^{-16}\, T^{1.42} \exp(-730/T)$	400–1500	± 0.3
→ H + C$_6$H$_5$OH	$2.2 \times 10^{-11} \exp(-5330/T)$	1000–1150	± 0.3
→ C$_6$H$_6$OH	See Table 3		
OH + C$_6$H$_5$OH → C$_6$H$_5$(OH)$_2$	See Table 3		
→ H$_2$O + C$_6$H$_5$O			
→ H$_2$O + C$_6$H$_4$OH $\rbrack$	1.0×10^{-11}	1000–1150	± 0.5
OH + C$_6$H$_5$CH$_3$ → H$_2$O + C$_6$H$_5$CH$_2$	$8.6 \cdot 10^{-15}\, T\exp(-1440/T)$ See Table 3	400–1200	± 0.5 at 400 K reducing to ± 0.3 at 1200 K.
OH + p-C$_6$H$_4$(CH$_3$)$_2$ → C$_6$H$_4$CH$_2$CH$_3$ + H$_2$O	$6.4 \times 10^{-11} \exp(-1440/T)$	500–960	±0.1
→ p-C$_6$H$_4$(CH$_3$)$_2$OH	See Table 3		
OH + C$_6$H$_5$C$_2$H$_5$ → HOC$_6$H$_5$C$_2$H$_5$	See Table 3		
→ H$_2$O + C$_6$H$_5$C$_2$H$_4$ $\rbrack$	8.7×10^{-12}	773	± 0.1
→ H$_2$O + C$_6$H$_4$C$_2$H$_5$			

H$_2$O Reactions

H$_2$O + M → H + OH + M	See Table 2		

HO$_2$ Radical Reactions

HO$_2$ + HO$_2$ → H$_2$O$_2$ + O$_2$	$3.1 \times 10^{-12} \exp(-775/T)$	550–1250	± 0.15 at 550 K rising to ± 0.3 at 1250 K.
HO$_2$ + NH$_2$ → NH$_3$ + O$_2$			
→ HNO + H$_2$O $\rbrack$	2.6×10^{-11}	300–400	± 0.4
HO$_2$ + CH$_3$ → OH + CH$_3$O	3×10^{-11}	300–2500	± 0.7
→O$_2$ + CH$_4$	No recommendation		
HO$_2$ + CH$_4$ → H$_2$O$_2$ + CH$_3$	$1.5 \times 10^{-11} \exp(-12400/T)$	600–1000	± 0.2 at 600 K rising to ± 0.3 at 1000 K.
HO$_2$ + HCHO → H$_2$O$_2$ + CHO	$5.0 \times 10^{-12} \exp(-6580/T)$	600–1000	± 0.5
HO$_2$ + C$_2$H$_4$ → OH + C$_2$H$_4$O	$3.7 \times 10^{-12} \exp(-8650/T)$	600–900	± 0.15 at 600 K rising to ± 0.25 at 900 K.
HO$_2$ + C$_2$H$_6$ → H$_2$O$_2$ + C$_2$H$_5$	$2.2 \times 10^{-11} \exp(-10300/T)$	500–1000	± 0.2 at 500 K rising to ± 0.3 at 1000 K.
HO$_2$ + CH$_3$CHO → H$_2$O$_2$ + CH$_3$CO	$5.0 \times 10^{-12} \exp(-6000/T)$	900–1200	± 0.7

H$_2$O$_2$ Reactions

H$_2$O$_2$ + M → 2OH + M	See Table 2		

N Atom Reactions

N + CN → N$_2$ + C	3×10^{-10}	300–2500	± 1.0
N + NCO → NO + CN	No recommendation		
→ N$_2$ + CO	3.3×10^{-11}	1700	± 0.5

Reaction	k/cm^3 molecule^{-1}s^{-1}	Temp/K	Error limits ($\Delta \log k$)
NH Radical Reactions			

NH + O$_2$ → NO + OH ⎤
     → NO$_2$ + H
     → HNO + O ⎦
	$1.26 \times 10^{-13} \exp(-770/T)$	270–550	± 0.2 at 270 K rising to ± 0.5 at 550 K.

NH + NO → N$_2$O + H ⎤
     → HN$_2$ + O
     → N$_2$ + OH ⎦
	5.0×10^{-11}	270–380	± 0.2

NH$_2$ Radical Reactions			
NH$_2$ + O$_2$ → products	$< 3 \times 10^{-18}$	298	

NH$_2$ + NO → N$_2$ + H$_2$O ⎤
     → N$_2$ + H + OH
     → N$_2$H + OH
     → N$_2$O + H$_2$ ⎦
	$1.8 \times 10^{-12} \exp(650/T)$
	$(k_2 + k_3)/k \approx 0.12$ at 298 K.
		220–2000	± 0.5

^{1}C$_2$ and ^{3}C$_2$ Radical Reactions	See data sheets.		
CH Radical Reactions			
CH + O$_2$ → CHO + O ⎤ → CO + OH ⎦	5.5×10^{-11}	300–2000	± 0.3 at 300 K rising to ± 0.5 at 2000 K.
CH + H$_2$ → CH$_2$ + H ⎤ → CH$_3$ ⎦	$2.4 \times 10^{-10} \exp(-1760/T)$	300–1000	± 0.3
CH + H$_2$O → products	$9.5 \times 10^{-12} \exp(380/T)$	300–1000	± 1.0
CH + CO → products	$4.6 \times 10^{-13} \exp(860/T)$	300–1000	± 1.0
CH + CO$_2$ → products	$5.7 \times 10^{-12} \exp(-345/T)$	300–1000	± 1.0
CH + CH$_4$ → products	$5.0 \times 10^{-11} \exp(200/T)$	200–700	± 1.0
CH + C$_2$H$_2$ → products	$3.5 \times 10^{-10} \exp(61/T)$	200–700	± 1.0
CH + C$_2$H$_4$ → products	$2.2 \times 10^{-10} \exp(173/T)$	200–700	± 1.0
CH + C$_2$H$_6$ → products	$1.8 \times 10^{-10} \exp(132/T)$	200–700	± 1.0
CH + C$_3$H$_8$ → products	$1.9 \times 10^{-10} \exp(240/T)$	300–700	± 1.0
CH + n-C$_4$H$_{10}$ → products	$4.4 \times 10^{-10} \exp(28/T)$	250–700	± 1.0
CH + i-C$_4$H$_{10}$ → products	$2.0 \times 10^{-10} \exp(240/T)$	300–700	± 1.0
CH + neo-C$_5$H$_{12}$ → products	$1.6 \times 10^{-10} \exp(340/T)$	300–700	± 1.0
CH + CH$_3$C$_2$H → products	No recommendation		
CH + CH$_2$O → products	$1.6 \times 10^{-10} \exp(260/T)$	300–700	± 1.0

Reaction	k/cm³ molecule⁻¹s⁻¹	Temp/K	Error limits ($\Delta \log k$)

³CH₂ Radical Reactions

Reaction	k/cm³ molecule^{-1}s^{-1}	Temp/K	Error limits ($\Delta \log k$)
$^3CH_2 + O_2 \rightarrow CO + H + OH$ $\rightarrow CO_2 + H + H$ $\rightarrow CO + H_2O$ $\rightarrow CO_2 + H_2$ $\rightarrow HCHO + O$	$4.1 \times 10^{-11} \exp(-750/T)$	300–1000	± 0.3 at 300 K rising to ± 0.5 at 1000 K.
$^3CH_2 + {}^3CH_2 \rightarrow C_2H_2 + H_2$ $\rightarrow C_2H_2 + 2H$	$2.0 \times 10^{-10} \exp(-400/T)$ $k_2/k = 0.9 \pm 0.1$ over range 300–3000 K.	300–3000	± 0.5
$^3CH_2 + CH_3 \rightarrow C_2H_4 + H$	7.0×10^{-11}	300–3000	± 0.3 at 300 K rising to ± 0.5 at 3000 K.
$^3CH_2 + C_2H_2 \rightarrow C_3H_4$	See Table 3		
$^3CH_2 + C_2H_4 \rightarrow C_3H_6$ $\rightarrow c\text{-}C_3H_6$ $\rightarrow CH_2CHCH_2 + H$	See Table 3		

¹CH₂ Radical Reactions

Reaction	k/cm³ molecule^{-1}s^{-1}	Temp/K	Error limits ($\Delta \log k$)
$^1CH_2 + Ar \rightarrow {}^3CH_2 + Ar$	6.0×10^{-12}	300–2000	± 0.3
$^1CH_2 + N_2 \rightarrow {}^3CH_2 + N_2$	1.0×10^{-11}	300–2000	± 0.3
$^1CH_2 + CH_4 \rightarrow {}^3CH_2 + CH_4$	1.2×10^{-11}	300–2000	± 0.4
$^1CH_2 + C_2H_2 \rightarrow {}^3CH_2 + C_2H_2$	8.0×10^{-11}	300–2000	± 0.4
$^1CH_2 + C_2H_4 \rightarrow {}^3CH_2 + C_2H_4$	2.3×10^{-11}	300–2000	± 0.4
$^1CH_2 + C_2H_6 \rightarrow {}^3CH_2 + C_2H_6$	3.6×10^{-11}	300–2000	± 0.4
$^1CH_2 + O_2 \rightarrow CO + H + OH$ $\rightarrow CO_2 + H_2$ $\rightarrow CO + H_2O$ $\rightarrow {}^3CH_2 + O_2$	5.2×10^{-11}	300–1000	± 0.3 at 300 K rising to ± 0.5 at 1000 K
$^1CH_2 + H_2 \rightarrow CH_3 + H$	1.2×10^{-10}	300–1000	± 0.1 at 300 K rising to ± 0.3 at 1000 K
$^1CH_2 + C_2H_2 \rightarrow CH_2CCH_2$ $\rightarrow CH_3CCH$ $\rightarrow CH_2CCH + H$ $\rightarrow {}^3CH_2 + C_2H_2$	See Table 3 See earlier entry		
$^1CH_2 + C_2H_4 \rightarrow C_3H_6$ $\rightarrow {}^3CH_2 + C_2H_4$	See Table 3 See earlier entry		

CH₃ Radical Reactions

Reaction	k/cm³ molecule^{-1}s^{-1}	Temp/K	Error limits ($\Delta \log k$)
$CH_3 + M \rightarrow CH_2 + H + M$	See Table 2		
$CH_3 + O_2 \rightarrow CH_3O + O$ $\rightarrow HCHO + OH$ $\rightarrow CH_3O_2$	$2.2 \times 10^{-10} \exp(-15800/T)$ $5.5 \times 10^{-13} \exp(-4500/T)$ See Table 3	300–2500 1000–2500	± 0.5 ± 0.5

Reaction	k/cm^3 molecule^{-1}s^{-1}	Temp/K	Error limits ($\Delta \log k$)
$CH_3 + H_2 \rightarrow CH_4 + H$	$1.14 \times 10^{-20} T^{2.74} \exp(-4740/T)$	300–2500	$\pm$ 0.15 in the range 300–700 K. $\pm$ 0.3 in the range 700–2500 K.
$CH_3 + CH_3 \rightarrow C_2H_5 + H$	$5 \times 10^{-11} \exp(-6800/T)$	1300–2500	$\pm$ 0.6
$\rightarrow C_2H_4 + H_2$	No recommendation (see data sheets)		
$\rightarrow C_2H_6$	See Table 3		
$CH_3 + HCHO \rightarrow CH_4 + HCO$	$6.8 \times 10^{-12} \exp(-4450/T)$	300–1000	$\pm$ 0.3
$CH_3 + C_2H_2 + M \rightarrow C_3H_5 + M$	See Table 3		
$\rightarrow CH_4 + C_2H$	No recommendation		
$CH_3 + C_2H_4 \rightarrow CH_4 + C_2H_3$	$6.9 \times 10^{-12} \exp(-5600/T)$	400–3000	$\pm$ 0.5
$\rightarrow n\text{-}C_3H_7$	See Table 3		
$CH_3 + C_2H_5 \rightarrow CH_4 + C_2H_4$	1.9×10^{-12}	300–800	$\pm$ 0.4
$\rightarrow C_3H_8$	See Table 3		
$CH_3 + C_2H_6 \rightarrow CH_4 + C_2H_5$	$2.5 \times 10^{-31} T^{6.0} \exp(-3043/T)$	300–1500	$\pm$ 0.1 at 300 K rising to $\pm$ 0.2 at 1500 K.
$CH_3 + CH_3CHO \rightarrow CH_4 + CH_3CO$	$3.3 \times 10^{-30} T^{5.64} \exp(-1240/T)$	300–1250	$\pm$ 0.3
$\rightarrow CH_4 + CH_2CHO$	No recommendation (see data sheets)		

CH$_4$ Reactions

Reaction	k/cm^3 molecule^{-1}s^{-1}	Temp/K	Error limits ($\Delta \log k$)
$CH_4 + M \rightarrow CH_3 + H + M$	See Table 2		

CHO Radical Reactions

Reaction	k/cm^3 molecule^{-1}s^{-1}	Temp/K	Error limits ($\Delta \log k$)
$CHO + O_2 \rightarrow CO + HO_2$ $\rightarrow OH + CO_2$ $\rightarrow HCO_3$	5.0×10^{-12}	300–2500	$\pm$ 0.3
$CHO + CHO \rightarrow HCHO + CO$	5.0×10^{-11}	300	$\pm$ 0.3

HCHO Reactions

Reaction	k/cm^3 molecule^{-1}s^{-1}	Temp/K	Error limits ($\Delta \log k$)
$HCHO + M \rightarrow H + CHO + M$ $\rightarrow H_2 + CO + M$	See Table 2		

CH$_2$OH Reactions

Reaction	k/cm^3 molecule^{-1}s^{-1}	Temp/K	Error limits ($\Delta \log k$)
$CH_2OH + O_2 \rightarrow CH_2O + HO_2$	$2.6 \times 10^{-9} T^{-1.0} +$ $1.2 \times 10^{-10} \exp(-1800/T)$	300–1200	$\pm$ 0.1 at 300 K rising to $\pm$ 0.3 at 1200 K.

CH$_3$O Radical Reactions

Reaction	k/cm^3 molecule^{-1}s^{-1}	Temp/K	Error limits ($\Delta \log k$)
$CH_3O + M \rightarrow HCHO + H + M$	See Table 2		
$CH_3O + O_2 \rightarrow HCHO + HO_2$	$6.7 \times 10^{-14} \exp(-1070/T)$	300–1000	$\pm$ 0.2 at 500 K rising to $\pm$ 0.3 at 300 K and 1000 K.

Reaction	$k/\mathrm{cm^3\ molecule^{-1}s^{-1}}$	Temp/K	Error limits ($\Delta \log k$)
CH₃OOH Reactions			
$CH_3OOH + M \rightarrow CH_3O + OH + M$	See Table 2		
CN Radical Reactions			
$CN + O_2 \rightarrow NCO + O$	$1.1 \times 10^{-11} \exp(205/T)$	300–2500	± 0.25 at 300 K rising to ± 0.5 at 2500 K.
$CN + H_2O \rightarrow HCN + OH$ $\rightarrow HOCN + H$	$1.3 \times 10^{-11} \exp(-3750/T)$	500–3000	± 0.3 at 500 K rising to ± 0.5 at 3000 K.
$CN + CH_4 \rightarrow HCN + CH_3$	$1.5 \times 10^{-11} \exp(-940/T)$	260–400	± 0.3
NCO Radical Reactions			
$NCO + M \rightarrow N + CO + M$	See Table 2		
$NCO + NO \rightarrow N_2O + CO$ $\rightarrow N_2 + CO_2$ $\rightarrow N_2 + CO + O$	$1.7 \times 10^{-11} \exp(200/T)$	300–600	± 0.5
C₂H Radical Reactions			
$C_2H + O_2 \rightarrow CO_2 + CH$ $\rightarrow 2CO + H$ $\rightarrow C_2HO + O$ $\rightarrow CO + HCO$	3.0×10^{-11}	300	± 0.5
$C_2H + H_2 \rightarrow C_2H_2 + H$	$2.5 \times 10^{-11} \exp(-1560/T)$	300–2500	± 0.3 at 300 K rising to ± 0.7 at 2500 K
$C_2H + C_2H_2 \rightarrow C_4H_2 + H$	5.0×10^{-11}	300–2700	± 0.3
$C_2H + CH_4 \rightarrow$ products	2.0×10^{-12}	298	± 1
$C_2H + C_2H_6 \rightarrow$ products	No recommendation		
C₂H₃ Radical Reactions			
$C_2H_3 + M \rightarrow C_2H_2 + H + M$	See Table 2		
$C_2H_3 + O_2 \rightarrow HCHO + CHO$	9.0×10^{-12}	300–2000	± 0.3 at 300 K rising to ± 0.5 at 2000 K
C₂H₅ Radical Reactions			
$C_2H_5 + O_2 \rightarrow C_2H_4 + HO_2$	$1.7 \times 10^{-14} \exp(1100/T)$	600–1200	± 0.3
$C_2H_5 + C_2H_5 \rightarrow C_2H_6 + C_2H_4$ $\rightarrow n\text{-}C_4H_{10}$	2.4×10^{-12} See Table 3	300–1200	± 0.4
C₂H₆ Reactions			
$C_2H_6 + M \rightarrow CH_3 + CH_3 + M$	See Table 2		

Reaction	k/cm³ molecule⁻¹s⁻¹	Temp/K	Error limits ($\Delta \log k$)

CHCO Reactions

$$CHCO + O_2 \rightarrow CO_2 + HCO$$
$$\rightarrow 2CO + OH$$
$$\rightarrow C_2O + HO_2$$
$$\rightarrow CHO_2CO$$

$2.7 \times 10^{-12} \exp(430/T)$
M = He, 2 Torr
300–550 ± 0.7

CH₂CHO Radical Reactions

$$CH_2CHO + O_2 \rightarrow HO_2 + CH_2CHO$$
$$\rightarrow HCHO + CO + OH$$
$$\rightarrow O_2CH_2CHO$$

$k_\infty = 2.6 \times 10^{-13}$ — 250–1000 — ± 0.2
$k_2 = 3.0 \times 10^{-14}$ — 300 — ± 0.3

CH₃CO Radical Reactions

$$CH_3CO + O_2 + M \rightarrow CH_3CO_3 + M$$ See Table 3

CH₃CHO Reactions

$$CH_3CHO + M \rightarrow CH_3 + HCO + M$$ See Table 2

C₂H₅O Reactions

$$C_2H_5O + M \rightarrow HCHO + CH_3 + M$$
$$\rightarrow CH_3CHO + H + M$$

See Table 2

$$C_2H_5O + O_2 \rightarrow CH_3CHO + HO_2$$

$1.0 \times 10^{-13} \exp(-830/T)$ 300–1000 ± 0.3 at 300 K rising to ± 0.5 at 1000 K

C₂H₅OOH Reactions

$$C_2H_5OOH + M \rightarrow C_2H_5O + OH + M$$ See Table 2

C₆H₅ Radical Reactions

$$C_6H_5 + M \rightarrow C_2H_2 + C_4H_3 + M$$
$$\rightarrow C_2H_3 + C_4H_2 + M$$
$$\rightarrow linear\text{-}C_6H_5 + M$$

See Table 2

C₆H₆ Reactions

$$C_6H_6 + M \rightarrow C_6H_5 + H + M$$
$$\rightarrow C_4H_4 + C_2H_2 + M$$

See Table 2

C₆H₅O Radical Reactions

$$C_6H_5O + M \rightarrow C_5H_5 + CO + M$$ See Table 2

C₆H₅CH₂ Radical Reactions

$$C_6H_5CH_2 + M \rightarrow C_3H_3 + 2C_2H_2 + M$$
$$\rightarrow C_4H_4 + C_3H_3 + M$$
$$\rightarrow C_5H_5 + C_2H_2 + M$$
$$\rightarrow C_7H_7(BCH) + M$$

See Table 2

Table 1
BIMOLECULAR REACTIONS (continued)

Reaction	k/cm^3 molecule^{-1}s^{-1}	Temp/K	Error limits ($\Delta \log k$)

$C_6H_5CH_3$ Reactions

$C_6H_5CH_3 + M \rightarrow C_6H_5CH_2 + H + M$
$ \rightarrow C_6H_5 + CH_3 + M$ See Table 2

p-$C_6H_4(CH_3)_2$ Reactions

p-$C_6H_4(CH_3)_2 + M \rightarrow C_6H_4CH_2CH_3 + H + M$ See Table 2

$C_6H_5C_2H_5$ Reactions

$C_6H_5C_2H_5 + M \rightarrow C_6H_5CH_2 + CH + M$
$ \rightarrow C_6H_6 + C_2H_4 + M$
$ \rightarrow C_6H_5CHCH_2 + H_2 + M$ See Table 2
$ \rightarrow C_6H_5 + C_2H_5 + M$
$ \rightarrow C_6H_5CHCH_3 + H + M$

Table 2
DECOMPOSITION REACTIONS

Reaction	k_∞/s^{-1} k_0/cm^3 molecule^{-1} s^{-1} F_c k/s$^{-1} = \dfrac{k_0 k_\infty [M]}{k_0[M] + k_\infty} F$	Temp/K	Error limits ($\Delta \log k$)
$H_2 + Ar \rightarrow 2H + Ar$	$k_0 = 3.7 \times 10^{-10} \exp(-48350/T)$	2500–8000	± 0.3
$H_2 + H_2 \rightarrow 2H + H_2$	$k_0 = 1.5 \times 10^{-9} \exp(-48350/T)$	2500–8000	± 0.5
$H_2O + N_2 \rightarrow H + OH + N_2$	$k_0 = 5.8 \times 10^{-9} \exp(-52920/T)$	2000–6000	± 0.5
$H_2O_2 + M \rightarrow 2\,OH + M$	$k_0(Ar) = 3 \times 10^{-8} \exp(-21600/T)$ $k_0(N_2) = 2 \times 10^{-7} \exp(-22900/T)$ $k_\infty = 3 \times 10^{14} \exp(-24400/T)$ $F_c(Ar) = 0.5$	1000–1500 700–1500 1000–1500 700–1500	± 0.2 ± 0.2 ± 0.5 $\Delta F_c = $ ± 0.1
$CH_3 + M \rightarrow CH_2 + H + M$	$k_0 = 1.7 \times 10^{-8} \exp(-45600/T)$	1500–3000	± 0.5
$CH_4 + M \rightarrow CH_3 + H + M$	$k_0(Ar) = 1.2 \times 10^{-6} \exp(-47000/T)$ $k_0(CH_4) = 1.4 \times 10^{-5} \exp(-48100/T)$ $k_\infty = 2.4 \times 10^{16} \exp(-52800/T)$ $F_c(Ar) = \exp(-0.45 - T/3231)$ $F_c(CH_4) = \exp(-0.37 - T/2210)$	1000–3000 1000–2000 1000–3000 1000–3000 1000–2000	± 0.3 ± 0.3 ± 0.5 $\Delta F_c = $ ± 0.1 $\Delta F_c = $ ± 0.1
$HCHO + M \rightarrow H + CHO + M$ $ \rightarrow H_2 + CO + M$	$k_0(1) = 2.1 \times 10^{-8} \exp(-39200/T)$ $k_1/k_2 = 0.5$ at 2200 K	1500–2500	± 0.3
$CH_3O + M \rightarrow HCHO + H + M$	$k_0 = 3.16 \times 10^2 \, T^{-2.7} \exp(-15400/T)$ [estimate]	300–1000	± 1.0
$CH_3OOH + M \rightarrow CH_3O + OH + M$	$k_\infty = 4 \times 10^{15} \exp(-21600/T)$	400–1000	± 0.5 at 600 K rising to ± 1.0 at 400 and 1000 K
$NCO + Ar \rightarrow N + CO + Ar$	$k_0 = 1.7 \times 10^{-9} \exp(-23500/T)$	1450–2600	± 0.4

Table 2
DECOMPOSITION REACTIONS (continued)

Reaction	k_∞/s^{-1} k_0/cm^3 molecule^{-1} s^{-1} F_c $k/s^{-1} = \dfrac{k_0\,k_\infty\,[M]}{k_0[M] + k_\infty}\,F$	Temp/K	Error limits ($\Delta \log k$)
$C_2H_3 + M \rightarrow C_2H_2 + H + M$	$k_0 = 6.9 \times 10^{17}\,T^{-7.5}\exp(-22900/T)$ $k_\infty = 2 \times 10^{14}\exp(-20000/T)$ $F_c = 0.35$	500–2500 500–2500 500–2500	± 0.5 ± 0.5 $\Delta F_c = \pm 0.1$
$C_2H_6 + M \rightarrow 2CH_3 + M$	$k_0(Ar) = 1.1 \times 10^{25}\,T^{-8.24}\exp(-47090/T)$ $k_0(C_2H_6) = 4.5 \times 10^{-2}\exp(-41930/T)$ $k_\infty = 1.8 \times 10^{21}\,T^{-1.24}\exp(-45700/T)$ $F_c(Ar) = 0.38\exp(-T/73) + 0.62\exp(-T/1180)$ $F_c(C_2H_6) = 0.54\exp(-T/1250)$	300–2000 800–1000 300–2000 300–2000 800–1000	± 0.5 ± 0.5 ± 0.3 $\Delta F_c = \pm 0.1$ $\Delta F_c = \pm 0.1$
$CH_3CHO + M \rightarrow CH_3 + CHO + M$	$k(1\text{ atm.}) = 7 \times 10^{15}\exp(-41100/T)$ (pressure dependent region)	750–1200	± 0.4
$C_2H_5O + M \rightarrow HCHO + CH_3 + M$	$k_\infty = 8 \times 10^{13}\exp(-10830/T)$ [estimate]	300–600	± 1.0
$C_2H_5OOH + M \rightarrow C_2H_5O + OH + M$	$k_\infty\ 4 \times 10^{15}\exp(-21600/T)$	400–1000	± 1.0
$C_6H_5 + M \rightarrow C_2H_2 + C_4H_3 + M$ $\rightarrow C_2H_3 + C_4H_2 + M$ $\rightarrow$ linear-$CH_5 + M$	No recommendation $4.0 \times 10^{13}\exp(-36700/T)$	 1450–1900	 ± 0.4
$C_6H_6 + M \rightarrow C_6H_5 + H + M$ $\rightarrow C_4H_4 + H_2 + M$	$9.0 \times 10^{15}\exp(-54060/T)$	1200–2500	± 0.4 at 1200 K reducing to ± 0.3 at 2500 K
$C_6H_5O + M \rightarrow C_5H_5 + CO + M$	$2.5 \times 10^{11}\exp(-22100/T)$	1000–1580	± 0.2
$C_6H_5CH_2 + M \rightarrow C_3H_3 + 2C_2H_2 + M$ $\rightarrow C_4H_4 + C_3H_3 + M$ $\rightarrow C_5H_5 + C_2H_2 + M$ $\rightarrow C_7H_7\ (BCH) + M$	$5.1 \times 10^{13}\exp(-36370/T)$	1350–1900	± 0.3 at 1350 K rising to ± 0.5 1900 K
$C_6H_5CH_3 + M \rightarrow C_6H_5CH_2 + H + M$ $\rightarrow C_6H_5 + CH_3 + M$	$3.1 \times 10^{15}\exp(-44890/T)$ No recommendation	920–2200	± 0.3 at 900 K rising to ± 0.5 at 2200 K
$p\text{-}C_6H_4(CH_3)_2 + M \rightarrow p\text{-}C_6H_4CH_2CH_3 + H + M$	$4.0 \times 10^{15}\exp(-42600/T)$	1400–1800	± 0.5
$C_6H_5C_2H_5 + M \rightarrow C_6H_5CH_2 + CH + M$ $\rightarrow C_6H_6 + C_2H_4 + M$ $\rightarrow C_6H_5CHCH_2 + H_2 + M$ $\rightarrow C_6H_5 + C_2H_5 + M$ $\rightarrow C_6H_5CHCH_3 + H + M$	$6.1 \times 10^{15}\exp(-37800/T)$ No recommendations	770–1800	± 0.1 at 770 K rising to ± 0.4 at

Table 3
COMBINATION REACTIONS

Reaction	k_∞/cm^3 molecule^{-1}s^{-1} k_0/cm^6 molecule^{-2} s^{-1} F_c k/cm^3 molecule^{-1}s$^{-1} = \dfrac{k_0\,k_\infty\,[M]}{k_0[M] + k_\infty}\,F$	Temp/K	Error limits ($\Delta \log k$)
$H + O_2 + Ar \rightarrow HO_2 + Ar$	$k_0 = 1.7 \times 10^{-30}\,T^{-0.8}$	300–2000	± 0.5

Reaction	k_∞/cm³ molecule⁻¹s⁻¹ k_0/cm⁶ molecule⁻² s⁻¹ F_c k/cm³ molecule⁻¹s⁻¹ $= \dfrac{k_0\, k_\infty\, [M]}{k_0[M] + k_\infty} F$	Temp/K	Error limits ($\Delta \log k$)
$H + O_2 + H_2 \rightarrow HO_2 + H_2$	$k_0 = 5.8 \times 10^{-30} T^{-0.8}$	300–2000	± 0.5
$H + O_2 + N_2 \rightarrow HO_2 + N_2$	$k_0 = 3.9 \times 10^{-30} T^{-0.8}$	300–2000	± 0.5
$H + O_2 + H_2O \rightarrow HO_2 + H_2O$	$k_0 = 4.3 \times 10^{-30} T^{-0.8}$	300–2000	± 0.5
$H + H + Ar \rightarrow H_2 + Ar$	$k_0 = 1.8 \times 10^{-30} T^{-1.0}$	300–2500	± 0.5
$H + H + H_2 \rightarrow H_2 + H_2$	$k_0 = 2.7 \times 10^{-31} T^{-0.6}$	100–5000	± 0.5
$H + OH + H_2O \rightarrow H_2O + H_2O$	$k_0 = 3.9 \times 10^{-25} T^{-2.0}$	300–3000	± 0.3
$H + OH + Ar \rightarrow H_2O + Ar$	$k_0 = 2.3 \times 10^{-26} T^{-2.0}$	300–3000	± 0.3
$H + OH + N_2 \rightarrow H_2O + N_2$	$k_0 = 6.1 \times 10^{-26} T^{-2.0}$	300–3000	± 0.3
$H + CH_3 + M \rightarrow CH_4 + M$	$k_0(He) = 6.2 \times 10^{-29} (T/3000)^{-1.8}$ $k_0(Ar) = 6 \times 10^{-29} (T/300)^{-1.8}$ $k_0(C_2H_6) = 3 \times 10^{-28} (T/300)^{-1.8}$ $k_\infty = 3.5 \times 10^{-10}$ $F_c(He,Ar) = \exp(-0.45 - T/3231)$ $F_c(C_2H_6) = \exp(-0.34 - T/3053)$	300–1000 300–1000 300–1000 300–1000 300–1000 300–1000	± 0.3 ± 0.5 ± 0.5 ± 0.3 $\Delta F_c = \pm 0.1$ $\Delta F_c = \pm 0.1$
$H + C_2H_2 + He \rightarrow C_2H_3 + He$	$k_\infty = 1.4 \times 10^{-11} \exp(-1300/T)$ $k_0 = 3.3 \times 10^{-30} \exp(-740/T)$ $F_c = 0.44$	200–400 200–400 200–400	± 0.3 ± 0.5 $\Delta F_c = \pm 0.1$
$H + C_2H_3 + M \rightarrow C_2H_4 + M$	No recommendation		
$H + C_2H_4 + M \rightarrow C_2H_5 + M$	No recommendation		
$H + C_2H_5 + M \rightarrow C_2H_6 + M$	No recommendation		
$H + C_6H_5 + M \rightarrow C_6H_6 + M$	$k_\infty = 1.3 \times 10^{-10}$	1400–1700	± 0.5
$H + C_6H_6 + M \rightarrow C_6H_7 + M$	$k_\infty = 6.7 \times 10^{-11} \exp(-2170/T)$	300–1000	± 0.2
$H + C_6H_5O + M \rightarrow C_6H_5OH + M$	$k_\infty = 4.2 \times 10^{-10}$	1000	± 0.3
$H + C_6H_5CH_2 + M \rightarrow C_6H_5CH_3 + M$	$k_\infty = 5.5 \times 10^{-10}$	300–2000	± 0.2 at 300 K rising to ± 0.7 at 2000 K.
$H + C_6H_5CH_3 + M \rightarrow C_6H_6CH_3 + M$	$k_\infty = 1.2 \times 10^{-13}$	298	± 0.2
$H + C_6H_5C_2H_5 + M \rightarrow C_6H_6C_2H_5 + M$	$k_\infty = 3.3 \times 10^{-13}$	298	± 0.1
$OH + OH + M \rightarrow H_2O_2 + M$	$k_0(N_2) = 8 \times 10^{-31} (T/300)^{-0.76}$ $k_0(H_2O) = 4 \times 10^{-30}$ $k_\infty = 1.5 \times 10^{-11} (T/300)^{-0.37}$ $F_c(N_2) = 0.5$	250–1400 300–400 200–1500 200–1500	± 0.4 ± 0.5 $\Delta F_c = \pm 0.2$
$OH + CH_3 + M \rightarrow CH_3OH + M$	No data available for this channel (See Table 1)		

Reaction	k_∞/cm³ molecule⁻¹s⁻¹ k_0/cm⁶ molecule⁻² s⁻¹ F_c k/cm³ molecule⁻¹s⁻¹ $= \dfrac{k_0 k_\infty [M]}{k_0[M] + k_\infty} F$	Temp/K	Error limits ($\Delta \log k$)
$OH + C_2H_2 + M \to C_2H_2OH + M$	See data sheet		
$OH + C_6H_6 + M \to C_6H_6OH + M$	$k_\infty = 3.8 \times 10^{-12} \exp(-340/T)$	240–340	± 0.2
$OH + C_6H_5OH + M \to C_6H_5(OH)_2 + M$	$k_\infty = 2.8 \times 10^{-11}$	298	± 0.1
$OH + C_6H_5CH_3 + M \to HOC_6H_5CH_3 + M$	$k_\infty = 3.8 \times 10^{-12} \exp(180/T)$	200–300	± 0.4
$OH + C_6H_4(CH_3)_2 + M \to C_6H_4(CH_3)_2OH + M$	$k_\infty = 1.4 \times 10^{-11}$	300–320	± 0.1
$OH + C_6H_5C_2H_5 + M \to HOC_6H_5C_2H_5 + M$	7.5×10^{-12} at p ≤ 1 atm.	298	± 0.1
$^3CH_2 + C_2H_2 + M \to C_3H_4 + M$	$2.0 \times 10^{-11} \exp(-3330/T)$ at $p = $ ≤ 10 Torr.	300–1000	± 0.3
$^3CH_2 + C_2H_4 + M \to C_3H_6 + M$ $\to c\text{-}C_3H_6 + M$ $\to C_3H_5 + H + M$	$5.3 \times 10^{-12} \exp(-2660/T)$	300–1000	± 0.2 at 300 K rising to ± 0.3 at 1000 K
$^1CH_2 + C_2H_2 + M \to CH_2CCH_2 + M$ $\to CH_3CCH + M$ $\to CH_2CCH + H + M$	3.7×10^{-10} independent of p	300–1000	± 0.3 at 300 K rising to ± 0.7 at 1000 K.
$^1CH_2 + C_2H_4 + M \to C_3H_6$	1.1×10^{-10} independent of p	300–1000	± 0.2 at 300 K rising to ± 0.5 at 1000 K.
$CH_3 + O_2 + M \to CH_3O_2 + M$	$k_0(Ar) = 1.5 \times 10^{-22} T^{-3.3}$ $k_0(N_2) = 1.6 \times 10^{-22} T^{-3.3}$ $k_\infty = 1.3 \times 10^{-15} T^{-1.2}$ $F_c = 0.466 - 1.30 \times 10^{-4} T$	300–800 300–800 300–800 300–800	± 0.3 ± 0.3 ± 0.3
$CH_3 + CH_3 + Ar \to C_2H_6 + Ar$	$k_\infty = 6 \times 10^{-11}$	300–2000	± 0.05 at 300 K rising to ± 0.3 at 2000 K
	$k_0 = 3.5 \times 10^{-7} T^{-7.0} \exp(-1390/T)$ $F_c = 0.38 \exp(-T/73)$ $+ 0.62\exp(-T/1180)$	300–2000 300–2000	± 0.3 $\Delta F_c = $ ± 0.1
$CH_3 + C_2H_2 + M \to C_3H_5 + M$	$k_\infty = 1 \times 10^{-12} \exp(-3900/T)$	300–600	± 0.5
$CH_3 + C_2H_4 + M \to n\text{-}C_3H_7 + M$	$3.5 \times 10^{-13} \exp(-3700/T)$	300–600	± 0.3
$CH_3 + C_2H_5 + M \to C_3H_8 + M$	$k_\infty = 4.7 \times 10^{-11}$	300–800	± 0.3
$C_2H_5 + C_2H_5 + M \to n\text{-}C_4H_{10} + M$	$k_\infty = 1.9 \times 10^{-11}$	300–1200	± 0.3
$CH_3CO + O_2 + M \to CH_3CO_3 + M$	2×10^{-12} for $p = $ 1–4 Torr.	300	± 0.3

Section 6
Fluid Properties

THERMODYNAMIC PROPERTIES OF AIR

These tables summarize the thermodynamic properties of air in the liquid and gaseous states, as well as along the saturation line. In the table for the saturation state, P(boil) is the bubble point temperature (i.e., the pressure at which boiling begins as the temperature of the liquid is raised), and P(con) is the dew point temperature (pressure at which condensation begins as the temperature of the gas is lowered). The other properties tabulated are density (ρ), enthalpy (H), entropy (S), and isobaric heat capacity (C_p). More detailed tables may be found in the references.

REFERENCES

1. Vasserman, A.A., and Rabinovich, V.A., *Thermophysical Properties of Liquid Air and its Components*, Izdatel'stvo Komiteta, Standartov, Moscow, 1968.
2. Vasserman, A.A., et al., *Thermophysical Properties of Air and Air Components,* Izdatel'stvo Nauka, Moscow, 1966.

Properties in the saturation state:

T K	P(boil) bar	P(con) bar	ρ (liq) g/cm^3	ρ (gas) g/L
65	0.1468	0.0861	0.939	0.464
70	0.3234	0.2052	0.917	1.033
75	0.6366	0.4321	0.894	2.048
80	1.146	0.8245	0.871	3.709
85	1.921	1.453	0.845	6.258
90	3.036	2.397	0.819	9.980
95	4.574	3.748	0.792	15.21
100	6.621	5.599	0.763	22.39
110	12.59	11.22	0.699	45.15
120	21.61	20.14	0.622	87.34
130	34.16	33.32	0.487	184.33
132.55	37.69	37.69	0.313	312.89

Properties of liquid air:

P bar	T K	ρ g/cm^3	H J/g	S J/g K	C_p J/g K
1	75	0.8935	−131.7	2.918	1.843
5	75	0.8942	−131.4	2.916	1.840
5	80	0.8718	−122.3	3.031	1.868
5	85	0.8482	−112.9	3.143	1.901
5	90	0.8230	−103.3	3.250	1.941
5	95	0.7962	−93.5	3.356	1.991
10	75	0.8952	−131.1	2.913	1.836
10	80	0.8729	−122.0	3.028	1.863
10	90	0.8245	−103.1	3.246	1.932
10	100	0.7695	−83.2	3.452	2.041
50	75	0.9025	−128.2	2.892	1.806
50	100	0.7859	−81.8	3.415	1.939
50	125	0.6222	−28.3	3.889	2.614
50	150	0.1879	91.9	4.764	2.721
100	75	0.9111	−124.5	2.867	1.774
100	100	0.8033	−79.4	3.376	1.852
100	125	0.6746	−31.4	3.805	2.062
100	150	0.4871	32.8	4.271	2.832

Properties of air in the gaseous state:

P bar	T K	ρ g/L	H J/g	S J/g K	C_p J/g K
1	100	3.556	98.3	5.759	1.032
1	200	1.746	199.7	6.463	1.007
1	300	1.161	300.3	6.871	1.007

P bar	T K	ρ g/L	H J/g	S J/g K	C_p J/g K
1	500	0.696	503.4	7.389	1.030
1	1000	0.348	1046.6	8.138	1.141
10	200	17.835	195.2	5.766	1.049
10	300	11.643	298.3	6.204	1.021
10	500	6.944	502.9	6.727	1.034
10	1000	3.471	1047.2	7.477	1.142
100	200	213.950	148.8	4.949	1.650
100	300	116.945	279.9	5.486	1.158
100	500	66.934	499.0	6.048	1.073
100	1000	33.613	1052.4	6.812	1.151

PROPERTIES OF WATER IN THE RANGE 0 — 100 °C

This table summarizes the best available values of the density, specific heat capacity at constant pressure (C_p), vapor pressure, viscosity, thermal conductivity, dielectric constant, and surface tension for liquid water in the range 0 — 100 °C. All values (except vapor pressure) refer to a pressure of 100 kPa (1 bar). The temperature scale is IPTS-68.

t °C	Density g/cm³	C_p J/g K	Vap. pres. kPa	Visc. µPa s	Ther. cond. mW/K m	Diel. const.	Surf. ten. mN/m
0	0.99984	4.2176	0.6113	1793	561.0	87.90	75.64
10	0.99970	4.1921	1.2281	1307	580.0	83.96	74.23
20	0.99821	4.1818	2.3388	1002	598.4	80.20	72.75
30	0.99565	4.1784	4.2455	797.7	615.4	76.60	71.20
40	0.99222	4.1785	7.3814	653.2	630.5	73.17	69.60
50	0.98803	4.1806	12.344	547.0	643.5	69.88	67.94
60	0.98320	4.1843	19.932	466.5	654.3	66.73	66.24
70	0.97778	4.1895	31.176	404.0	663.1	63.73	64.47
80	0.97182	4.1963	47.373	354.4	670.0	60.86	62.67
90	0.96535	4.2050	70.117	314.5	675.3	58.12	60.82
100	0.95840	4.2159	101.325	281.8	679.1	55.51	58.91
Ref.	1—3	2	1, 3	3	3	4	5

REFERENCES

1. L. Harr, J. S. Gallagher, and G. S. Kell, *NBS/NRC Steam Tables*, Hemisphere Publishing Corp., 1984.
2. K. N. Marsh, Ed., *Recommended Reference Materials for the Realization of Physicochemical Properties*, Blackwell Scientific Publications, Oxford, 1987.
3. J. V. Sengers and J. T. R. Watson, Improved international formulations for the viscosity and thermal conductivity of water substance, *J. Phys. Chem. Ref. Data*, 15, 1291, 1986.
4. D. G. Archer and P. Wang, The dielectric constant of water and Debye-Hückel limiting law slopes, *J. Phys. Chem. Ref. Data*, 19, 371, 1990.
5. N. B. Vargaftik, et al., International tables of the surface tension of water, *J. Phys. Chem. Ref. Data*, 12, 817, 1983.

ENTHALPY OF VAPORIZATION OF WATER

The enthalpy (heat) of vaporization of water is tabulated as a function of temperature on the IPTS-68 scale.

REFERENCE

Marsh, K. N., Ed., *Recommended Reference Materials for the Realization of Physicochemical Properties,* Blackwell, Oxford, 1987.

t °C	$\Delta_{vap}H$ kJ/mol	t °C	$\Delta_{vap}H$ kJ/mol
0	45.054	200	34.962
25	43.990	220	33.468
40	43.350	240	31.809
60	42.482	260	29.930
80	41.585	280	27.795
100	40.657	300	25.300
120	39.684	320	22.297
140	38.643	340	18.502
160	37.518	360	12.966
180	36.304	374	2.066

FIXED POINT PROPERTIES OF H_2O AND D_2O

Temperatures are given on the IPTS-68 scale.

REFERENCES

1. Haar, L., Gallagher, J.S., and Kell, G.S., *NBS/NRC Steam Tables*, Hemisphere Publishing Corp., 1984.
2. Levelt Sengers, J.M.H., Straub, J., Watanabe, K., and Hill, P.G., Assessment of critical parameter values for H_2O and D_2O, *J. Phys. Chem. Ref. Data,* 14, 193, 1985.
3. Kestin, J. et. al., Thermophysical properties of fluid D_2O, *J. Phys. Chem. Ref. Data,* 13, 601, 1984.
4. Kestin, J. et. al., Thermophysical properties of fluid H_2O, *J. Phys. Chem. Ref. Data*, 13, 175, 1984.
5. Hill, P.G., MacMillan, R.D.C., and Lee, V., A fundamental equation of state for heavy water, *J. Phys. Chem. Ref. Data*, 11, 1, 1982.

	Unit	H_2O	D_2O
Molar mass	g/mol	18.01528	20.02748
Melting point(101.325 kPa)	°C	0.00	3.82
Boiling point(101.325 kPa)	°C	100.00	101.42
Triple point temperature	°C	0.01	3.82
Triple point pressure	Pa	611.73	661
Triple point density(l)	g/cm³	0.99978	1.1055
Triple point density(g)	mg/L	4.885	5.75
Critical temperature	°C	373.99	370.74
Critical pressure	MPa	22.064	21.671
Critical density	g/cm³	0.322	0.356
Critical specific volume	cm³/g	3.11	2.81
Maximum density(saturated liquid)	g/cm³	0.99995	1.1053
Temperature of maximum density	°C	4.0	11.2

THERMAL CONDUCTIVITY OF SATURATED H_2O AND D_2O

This table gives the thermal conductivity λ for water (H_2O or D_2O) in equilibrium with its vapor. Values for the liquid (λ_l) and vapor (λ_v) are listed, as well as the vapor pressure.

REFERENCES

1. Sengers, J.V. and Watson, J.T.R., Improved international formulations for the viscosity and thermal conductivity of water substance, *J. Phys. Chem. Ref. Data*, 15, 1291, 1986.
2. Matsunaga, N. and Nagashima, A., Transport properties of liquid and gaseous D_2O over a wide range of temperature and pressure, *J. Phys. Chem. Ref. Data*, 12, 933, 1983.

	H_2O			D_2O		
$t/°C$	P/kPa	$\lambda_l/(mW/K\,m)$	$\lambda_v/(mW/K\,m)$	P/kPa	$\lambda_l/(mW/K\,m)$	$\lambda_v/(mW/K\,m)$
0	0.6	561.0	16.49			
10	1.2	580.0	17.21	1.0	575	17.0
20	2.3	598.4	17.95	2.0	589	17.8
30	4.2	615.4	18.70	3.7	600	18.5
40	7.4	630.5	19.48	6.5	610	19.3
50	12.3	643.5	20.28	11.1	618	20.2
60	19.9	654.3	21.10	18.2	625	21.0
70	31.2	663.1	21.96	28.8	629	21.9
80	47.4	670.0	22.86	44.2	633	22.8
90	70.1	675.3	23.80	66.1	635	23.8
100	101.3	679.1	24.79	96.2	636	24.8
150	476	682.1	30.77	465	625	30.8
200	1555	663.4	39.10	1546	592	39.0
250	3978	621.4	51.18	3995	541	52.0
300	8593	547.7	71.78	8688	473	75.2
350	16530	447.6	134.59	16820	391	143.0

STANDARD DENSITY OF WATER

This table gives the density ρ of standard mean ocean water (SMOW), free from dissolved salts and gases, at a pressure of 101325 Pa. SMOW is a standard water sample of high purity and known isotopic composition. Methods of correcting for different isotopic compositions are discussed in the reference. The table below is reprinted with the permission of IUPAC. Note that the temperature scale is IPTS-68.

REFERENCE

Marsh, K. N., Ed., *Recommended Reference Materials for the Realization of Physicochemical Properties,* Blackwell Scientific Publications, Oxford, 1987.

$\rho/\text{kg m}^{-3}$

$t_{68}/°C$	0.0	0.1	0.2	0.3	0.4	0.5	0.6	0.7	0.8	0.9
0	999.8426	8493	8558	8622	8683	8743	8801	8857	8912	8964
1	999.9015	9065	9112	9158	9202	9244	9284	9323	9360	9395
2	999.9429	9461	9491	9519	9546	9571	9595	9616	9636	9655
3	999.9672	9687	9700	9712	9722	9731	9738	9743	9747	9749
4	999.9750	9748	9746	9742	9736	9728	9719	9709	9696	9683
5	999.9668	9651	9632	9612	9591	9568	9544	9518	9490	9461
6	999.9430	9398	9365	9330	9293	9255	9216	9175	9132	9088
7	999.9043	8996	8948	8898	8847	8794	8740	8684	8627	8569
8	999.8509	8448	8385	8321	8256	8189	8121	8051	7980	7908
9	999.7834	7759	7682	7604	7525	7444	7362	7279	7194	7108
10	999.7021	6932	6842	6751	6658	6564	6468	6372	6274	6174
11	999.6074	5972	5869	5764	5658	5551	5443	5333	5222	5110
12	999.4996	4882	4766	4648	4530	4410	4289	4167	4043	3918
13	999.3792	3665	3536	3407	3276	3143	3010	2875	2740	2602
14	999.2464	2325	2184	2042	1899	1755	1609	1463	1315	1166
15	999.1016	0864	0712	0558	0403	0247	0090	9932*	9772*	9612*
16	998.9450	9287	9123	8957	8791	8623	8455	8285	8114	7942
17	998.7769	7595	7419	7243	7065	6886	6706	6525	6343	6160
18	998.5976	5790	5604	5416	5228	5038	4847	4655	4462	4268
19	998.4073	3877	3680	3481	3282	3081	2880	2677	2474	2269
20	998.2063	1856	1649	1440	1230	1019	0807	0594	0380	0164
21	997.9948	9731	9513	9294	9073	8852	8630	8406	8182	7957
22	997.7730	7503	7275	7045	6815	6584	6351	6118	5883	5648
23	997.5412	5174	4936	4697	4456	4215	3973	3730	3485	3240
24	997.2994	2747	2499	2250	2000	1749	1497	1244	0990	0735
25	997.0480	0223	9965*	9707*	9447*	9186*	8925*	8663*	8399*	8135*
26	996.7870	7604	7337	7069	6800	6530	6259	5987	5714	5441
27	996.5166	4891	4615	4337	4059	3780	3500	3219	2938	2655
28	996.2371	2087	1801	1515	1228	0940	0651	0361	0070	9778*
29	995.9486	9192	8898	8603	8306	8009	7712	7413	7113	6813
30	995.6511	6209	5906	5602	5297	4991	4685	4377	4069	3760
31	995.3450	3139	2827	2514	2201	1887	1572	1255	0939	0621
32	995.0302	9983*	9663*	9342*	9020*	8697*	8373*	8049*	7724*	7397*
33	994.7071	6743	6414	6085	5755	5423	5092	4759	4425	4091
34	994.3756	3420	3083	2745	2407	2068	1728	1387	1045	0703
35	994.0359	0015	9671*	9325*	8978*	8631*	8283*	7934*	7585*	7234*
36	993.6883	6531	6178	5825	5470	5115	4759	4403	4045	3687
37	993.3328	2968	2607	2246	1884	1521	1157	0793	0428	0062
38	992.9695	9328	8960	8591	8221	7850	7479	7107	6735	6361
39	992.5987	5612	5236	4860	4483	4105	3726	3347	2966	2586
40	992.2204									

* The leading figure decreases by 1.

VOLUMETRIC PROPERTIES OF AQUEOUS SODIUM CHLORIDE SOLUTIONS

This table gives the following properties of aqueous solutions of NaCl as a function of temperature and concentration:

Specific volume v (reciprocal of density) in cm^3/g
Isothermal compressibility $\kappa_T = -(1/v)(\partial v/\partial P)_T$ in GPa^{-1}
Cubic expansion coefficient $\alpha_v = (1/v)(\partial v/\partial T)_P$ in kK^{-1}

All data refer to a pressure of 100 kPa (1 bar). The reference gives properties over a wider range of temperature and pressure.

REFERENCE

Rogers, P. S. Z., and Pitzer, K. S., *J. Phys. Chem. Ref. Data,* 11, 15, 1982.

Molality in mol/kg

$T/°C$	0.100	0.250	0.500	0.750	1.000	2.000	3.000	4.000	5.000
				Specific volume v in cm^3/g					
0	0.995732	0.989259	0.978889	0.968991	0.959525	0.925426	0.896292	0.870996	0.848646
10	0.995998	0.989781	0.979804	0.970256	0.961101	0.927905	0.899262	0.874201	0.851958
20	0.997620	0.991564	0.981833	0.972505	0.963544	0.930909	0.902565	0.877643	0.855469
25	0.998834	0.992832	0.983185	0.973932	0.965038	0.932590	0.904339	0.879457	0.857301
30	1.000279	0.994319	0.984735	0.975539	0.966694	0.934382	0.906194	0.881334	0.859185
40	1.003796	0.997883	0.988374	0.979243	0.970455	0.938287	0.910145	0.885276	0.863108
50	1.008064	1.002161	0.992668	0.983551	0.974772	0.942603	0.914411	0.889473	0.867241
60	1.0130	1.0071	0.9976	0.9885	0.9797	0.9474	0.9191	0.8940	0.8716
70	1.0186	1.0127	1.0031	0.9939	0.9851	0.9526	0.9240	0.8987	0.8762
80	1.0249	1.0188	1.0092	0.9999	0.9909	0.9581	0.9293	0.9037	0.8809
90	1.0317	1.0256	1.0157	1.0063	0.9972	0.9640	0.9348	0.9089	0.8858
100	1.0391	1.0329	1.0228	1.0133	1.0040	0.9703	0.9406	0.9144	0.8910
				Compressibility κ_T in GPa^{-1}					
0	0.503	0.492	0.475	0.459	0.443	0.389	0.346	0.315	0.294
10	0.472	0.463	0.449	0.436	0.423	0.377	0.341	0.313	0.294
20	0.453	0.446	0.433	0.422	0.411	0.371	0.338	0.313	0.294
25	0.447	0.440	0.428	0.417	0.407	0.369	0.337	0.313	0.294
30	0.443	0.436	0.425	0.414	0.404	0.367	0.337	0.313	0.294
40	0.438	0.432	0.421	0.411	0.401	0.367	0.338	0.315	0.296
50	0.438	0.431	0.421	0.411	0.402	0.369	0.340	0.317	0.299
60	0.44	0.44	0.43	0.42	0.41	0.38	0.35	0.32	0.30
70	0.45	0.44	0.43	0.42	0.42	0.38	0.36	0.33	0.31
80	0.46	0.45	0.44	0.43	0.43	0.39	0.37	0.34	0.32
90	0.47	0.47	0.46	0.45	0.44	0.41	0.38	0.35	0.33
100	0.49	0.48	0.47	0.46	0.45	0.42	0.39	0.37	0.34
				Cubic expansion coefficient α_v in kK^{-1}					
0	-0.058	-0.026	0.024	0.069	0.110	0.237	0.313	0.355	
10	0.102	0.123	0.156	0.186	0.213	0.297	0.349	0.380	
20	0.218	0.232	0.254	0.274	0.292	0.349	0.384	0.406	
25	0.267	0.278	0.296	0.312	0.327	0.373	0.401	0.420	
30	0.311	0.320	0.334	0.347	0.359	0.395	0.418	0.433	
40	0.389	0.394	0.402	0.410	0.417	0.438	0.451	0.460	
50	0.458	0.460	0.464	0.467	0.470	0.479	0.484	0.486	
60	0.52	0.52	0.52	0.52	0.52	0.52	0.52	0.52	
70	0.58	0.58	0.58	0.57	0.57	0.56	0.55	0.54	
80	0.64	0.63	0.63	0.62	0.61	0.60	0.58	0.56	
90	0.69	0.68	0.67	0.67	0.66	0.63	0.61	0.59	
100	0.74	0.73	0.72	0.71	0.70	0.66	0.64	0.61	

DENSITY OF D₂O

Density of liquid D_2O in g/cm^3 at a pressure of 100 kPa (1 bar).

REFERENCE

Kirillin, V.A., Ed., *Heavy Water: Thermophysical Properties*, Gosudarstvennoe Energeticheskoe Izdatel'stvo, Moscow, 1963.

t/°C	3.8	5	10	15	20	25	30
Density	1.1053	1.1055	1.1057	1.1056	1.105	1.1044	1.1034

t/°C	35	40	45	50	55	60	65
Density	1.1019	1.1001	1.0979	1.0957	1.0931	1.0905	1.0875

t/°C	70	75	80	85	90	95	100
Density	1.0847	1.0815	1.0783	1.0748	1.0712	1.0673	1.0635

VAPOR PRESSURE OF ICE

The values of the vapor (sublimation) pressure of ice in this table were calculated from the equation recommended by the International Association for the Properties of Steam (IAPS) in 1993. Temperature values correspond to the ITS-90 temperature scale. The uncertainty in the pressure is estimated to be 0.1% for $t > -25°C$ and 0.5% for $t < -25°C$. The first entry in the table is the triple point of water.

REFERENCE

Wagner, W., Saul, A., and Pruss, A., *J. Phys. Chem. Ref. Data,* 23, 515, 1994.

t/°C	p/Pa	t/°C	p/Pa	t/°C	p/Pa
0.01	611.657	−16	150.68	−33	27.71
0	611.15	−17	137.25	−34	24.90
−1	562.67	−18	124.92	−35	22.35
−2	517.72	−19	113.62	−36	20.04
−3	476.06	−20	103.26	−37	17.96
−4	437.47	−21	93.77	−38	16.07
−5	401.76	−22	85.10	−39	14.37
−6	368.73	−23	77.16	−40	12.84
−7	338.19	−24	69.91	−45	7.202
−8	309.98	−25	63.29	−50	3.936
−9	283.94	−26	57.25	−55	2.093
−10	259.90	−27	51.74	−60	1.080
−11	237.74	−28	46.73	−65	0.540
−12	217.32	−29	42.16	−70	0.261
−13	198.52	−30	38.01	−75	0.122
−14	181.22	−31	34.24	−80	0.055
−15	165.30	−32	30.82		

VAPOR PRESSURE OF WATER FROM 0 TO 370° C

This table gives the vapor pressure of water at intervals of 1° C from the melting point to the critical point.

REFERENCE

Haar, L., Gallagher, J.S., and Kell, G.S., *NBS/NRC Steam Tables*, Hemisphere Publishing Corp., New York, 1984.

t/°C	P/kPa	t/°C	P/kPa	t/°C	P/kPa	t/°C	P/kPa
0	0.61129	53	14.303	106	125.03	159	602.11
1	0.65716	54	15.012	107	129.39	160	617.66
2	0.70605	55	15.752	108	133.88	161	633.53
3	0.75813	56	16.522	109	138.50	162	649.73
4	0.81359	57	17.324	110	143.24	163	666.25
5	0.87260	58	18.159	111	148.12	164	683.10
6	0.93537	59	19.028	112	153.13	165	700.29
7	1.0021	60	19.932	113	158.29	166	717.83
8	1.0730	61	20.873	114	163.58	167	735.70
9	1.1482	62	21.851	115	169.02	168	753.94
10	1.2281	63	22.868	116	174.61	169	772.52
11	1.3129	64	23.925	117	180.34	170	791.47
12	1.4027	65	25.022	118	186.23	171	810.78
13	1.4979	66	26.163	119	192.28	172	830.47
14	1.5988	67	27.347	120	198.48	173	850.53
15	1.7056	68	28.576	121	204.85	174	870.98
16	1.8185	69	29.852	122	211.38	175	891.80
17	1.9380	70	31.176	123	218.09	176	913.03
18	2.0644	71	32.549	124	224.96	177	934.64
19	2.1978	72	33.972	125	232.01	178	956.66
20	2.3388	73	35.448	126	239.24	179	979.09
21	2.4877	74	36.978	127	246.66	180	1001.9
22	2.6447	75	38.563	128	254.25	181	1025.2
23	2.8104	76	40.205	129	262.04	182	1048.9
24	2.9850	77	41.905	130	270.02	183	1073.0
25	3.1690	78	43.665	131	278.20	184	1097.5
26	3.3629	79	45.487	132	286.57	185	1122.5
27	3.5670	80	47.373	133	295.15	186	1147.9
28	3.7818	81	49.324	134	303.93	187	1173.8
29	4.0078	82	51.342	135	312.93	188	1200.1
30	4.2455	83	53.428	136	322.14	189	1226.9
31	4.4953	84	55.585	137	331.57	190	1254.2
32	4.7578	85	57.815	138	341.22	191	1281.9
33	5.0335	86	60.119	139	351.09	192	1310.1
34	5.3229	87	62.499	140	361.19	193	1338.8
35	5.6267	88	64.958	141	371.53	194	1368.0
36	5.9453	89	67.496	142	382.11	195	1397.6
37	6.2795	90	70.117	143	392.92	196	1427.8
38	6.6298	91	72.823	144	403.98	197	1458.5
39	6.9969	92	75.614	145	415.29	198	1489.7
40	7.3814	93	78.494	146	426.85	199	1521.4
41	7.7840	94	81.465	147	438.67	200	1553.6
42	8.2054	95	84.529	148	450.75	201	1586.4
43	8.6463	96	87.688	149	463.10	202	1619.7
44	9.1075	97	90.945	150	475.72	203	1653.6
45	9.5898	98	94.301	151	488.61	204	1688.0
46	10.094	99	97.759	152	501.78	205	1722.9
47	10.620	100	101.32	153	515.23	206	1758.4
48	11.171	101	104.99	154	528.96	207	1794.5
49	11.745	102	108.77	155	542.99	208	1831.1
50	12.344	103	112.66	156	557.32	209	1868.4
51	12.970	104	116.67	157	571.94	210	1906.2
52	13.623	105	120.79	158	586.87	211	1944.6

t/°C	P/kPa	t/°C	P/kPa	t/°C	P/kPa	t/°C	P/kPa
212	1983.6	253	4178.9	294	7881.3	335	13701
213	2023.2	254	4249.1	295	7995.2	336	13876
214	2063.4	255	4320.2	296	8110.3	337	14053
215	2104.2	256	4392.2	297	8226.8	338	14232
216	2145.7	257	4465.1	298	8344.5	339	14412
217	2187.8	258	4539.0	299	8463.5	340	14594
218	2230.5	259	4613.7	300	8583.8	341	14778
219	2273.8	260	4689.4	301	8705.4	342	14964
220	2317.8	261	4766.1	302	8828.3	343	15152
221	2362.5	262	4843.7	303	8952.6	344	15342
222	2407.8	263	4922.3	304	9078.2	345	15533
223	2453.8	264	5001.8	305	9205.1	346	15727
224	2500.5	265	5082.3	306	9333.4	347	15922
225	2547.9	266	5163.8	307	9463.1	348	16120
226	2595.9	267	5246.3	308	9594.2	349	16320
227	2644.6	268	5329.8	309	9726.7	350	16521
228	2694.1	269	5414.3	310	9860.5	351	16725
229	2744.2	270	5499.9	311	9995.8	352	16931
230	2795.1	271	5586.4	312	10133	353	17138
231	2846.7	272	5674.0	313	10271	354	17348
232	2899.0	273	5762.7	314	10410	355	17561
233	2952.1	274	5852.4	315	10551	356	17775
234	3005.9	275	5943.1	316	10694	357	17992
235	3060.4	276	6035.0	317	10838	358	18211
236	3115.7	277	6127.9	318	10984	359	18432
237	3171.8	278	6221.9	319	11131	360	18655
238	3228.6	279	6317.0	320	11279	361	18881
239	3286.3	280	6413.2	321	11429	362	19110
240	3344.7	281	6510.5	322	11581	363	19340
241	3403.9	282	6608.9	323	11734	364	19574
242	3463.9	283	6708.5	324	11889	365	19809
243	3524.7	284	6809.2	325	12046	366	20048
244	3586.3	285	6911.1	326	12204	367	20289
245	3648.8	286	7014.1	327	12364	368	20533
246	3712.1	287	7118.3	328	12525	369	20780
247	3776.2	288	7223.7	329	12688	370	21030
248	3841.2	289	7330.2	330	12852	371	21283
249	3907.0	290	7438.0	331	13019	372	21539
250	3973.6	291	7547.0	332	13187	373	21799
251	4041.2	292	7657.2	333	13357	373.98	22055
252	4109.6	293	7768.6	334	13528		

BOILING POINT OF WATER AT VARIOUS PRESSURES

Data are based on the equation of state recommended by the International Association for the Properties of Steam in 1984, as presented in Haar, Gallagher, and Kell, *NBS-NRC Steam Tables* (Hemisphere Publishing Corp., New York, 1984). The temperature scale is IPTS-68. Note that: 1 mbar = 100 Pa = 0.000986923 atmos = 0.750062 mmHg.

P/mbar	T/°C	P/mbar	T/°C	P/mbar	T/°C	P/mbar	T/°C
50	32.88	915	97.17	1013.25	100.00	1200	104.81
100	45.82	920	97.32	1015	100.05	1250	105.99
150	53.98	925	97.47	1020	100.19	1300	107.14
200	60.07	930	97.62	1025	100.32	1350	108.25
250	64.98	935	97.76	1030	100.46	1400	109.32
300	69.11	940	97.91	1035	100.60	1450	110.36
350	72.70	945	98.06	1040	100.73	1500	111.38
400	75.88	950	98.21	1045	100.87	1550	112.37
450	78.74	955	98.35	1050	101.00	1600	113.33
500	81.34	960	98.50	1055	101.14	1650	114.26
550	83.73	965	98.64	1060	101.27	1700	115.18
600	85.95	970	98.78	1065	101.40	1750	116.07
650	88.02	975	98.93	1070	101.54	1800	116.94
700	89.96	980	99.07	1075	101.67	1850	117.79
750	91.78	985	99.21	1080	101.80	1900	118.63
800	93.51	990	99.35	1085	101.93	1950	119.44
850	95.15	995	99.49	1090	102.06	2000	120.24
900	96.71	1000	99.63	1095	102.19	2050	121.02
905	96.87	1005	99.77	1100	102.32	2100	121.79
910	97.02	1010	99.91	1150	103.59	2150	122.54

MELTING POINT OF ICE AS A FUNCTION OF PRESSURE

This table gives values of the melting temperature of ice at various pressures, as calculated from the equation for the ice I - liquid water phase boundary recommended by the International Association for the Properties of Steam (IAPS). Temperatures are on the ITS-90 scale. See the Reference for information on forms of ice that exist at higher pressures. The transition points for transformations of the various forms of ice (in each case in equilibrium with liquid water) are:

ice I - ice III	209.9 MPa	−21.985°C
ice III - ice V	350.1	−16.986
ice V - ice VI	632.4	0.16
ice VI - ice VII	2216	82

REFERENCE

Wagner, W., Saul, A., and Pruss, A., *J. Phys. Chem. Ref. Data*, 23, 515, 1994.

p/MPa	t/°C	p/MPa	t/°C	p/MPa	t/°C
0.1	0.00	40	-3.15	130	-12.07
1	-0.06	50	-4.02	140	-13.22
2	-0.14	60	-4.91	150	-14.40
3	-0.21	70	-5.83	160	-15.62
4	-0.29	80	-6.79	170	-16.85
5	-0.36	90	-7.78	180	-18.11
10	-0.74	100	-8.80	190	-19.39
20	-1.52	110	-9.86	200	-20.69
30	-2.32	120	-10.95	210	-22.00

STEAM TABLES

This table gives properties of compressed water and superheated steam at selected pressures and temperatures. It was generated from the formulation approved by the International Association for the Properties of Steam in 1984. The properties tabulated are density (ρ), enthalpy (H), entropy (S), and isobaric specific heat capacity (C_p).

The reference state for these tables is the liquid at the triple point, at which the internal energy and entropy are taken as zero. A duplicate entry in the temperature column indicates a phase transition (liquid-vapor) at that temperature; property values are then given for both phases.

REFERENCE

Haar, L., Gallagher, J.S., and Kell, G.S., *NBS/NRC Steam Tables*, Hemisphere Press, New York, 1984

P MPa	T K	ρ kg/m³	H J/g	S J/g K	C_p J/g K
0.1	273.15	999.83	0.06	−0.00015	4.2282
0.1	300	996.57	112.58	0.3928	4.1831
0.1	372.78	958.66	417.51	1.3027	4.2166
0.1	372.78	0.59021	2675.1	7.3589	2.0427
0.1	373.15	0.58958	2675.9	7.3609	2.0418
0.1	400	0.54765	2730.0	7.5010	1.9973
0.1	500	0.43517	2927.9	7.9427	1.9816
0.1	600	0.36186	3128.2	8.3076	2.0272
0.1	700	0.30988	3333.8	8.6245	2.0869
0.1	800	0.27102	3545.8	8.9074	2.1526
0.1	900	0.24085	3764.4	9.1649	2.2216
0.1	1000	0.21673	3990.1	9.4026	2.2921
0.1	1100	0.19701	4222.8	9.6244	2.3621
0.1	1200	0.18058	4462.5	9.8329	2.4302
0.1	1300	0.16668	4708.8	10.030	2.4954
0.1	1400	0.15477	4961.4	10.217	2.5568
0.1	1500	0.14445	5220.0	10.396	2.6143
0.1	1600	0.13542	5484.2	10.566	2.6676
0.1	1700	0.12745	5753.4	10.729	2.7166
0.1	1800	0.12037	6027.3	10.886	2.7617
0.1	1900	0.11403	6305.6	11.036	2.8031
0.1	2000	0.10833	6587.9	11.181	2.8412
1	273.15	1000.30	0.98	−0.0001	4.2233
1	300	996.97	113.41	0.3926	4.1806
1	373.15	958.81	419.74	1.3062	4.2150
1	400	937.92	533.47	1.6005	4.2593
1	453.07	887.15	762.88	2.1388	4.4030
1	453.07	5.1445	2777.7	6.5859	2.5569
1	500	4.5348	2890.0	6.8220	2.2760
1	600	3.6877	3108.0	7.2199	2.1332
1	700	3.1307	3321.1	7.5483	2.1386
1	800	2.7266	3537.0	7.8365	2.1821
1	900	2.4175	3758.0	8.0968	2.2402
1	1000	2.1724	3985.2	8.3361	2.3046
1	1100	1.9730	4219.0	8.5588	2.3710
1	1200	1.8074	4459.4	8.7680	2.4368
1	1300	1.6677	4706.3	8.9656	2.5004
1	1400	1.5481	4959.3	9.1531	2.5608
1	1500	1.4445	5218.3	9.3317	2.6175
1	1600	1.3540	5482.7	9.5024	2.6702
1	1700	1.2742	5752.2	9.6657	2.7188
1	1800	1.2034	6026.3	9.8224	2.7636
1	1900	1.1399	6304.8	9.9729	2.8047
1	2000	1.0829	6587.2	10.1180	2.8426

P MPa	T K	ρ kg/m³	H J/g	S J/g K	C_p J/g K
10	273.15	1004.80	10.10	0.0004	4.1765
10	300	1001.00	121.66	0.3900	4.1564
10	373.15	962.98	426.52	1.2992	4.1948
10	400	942.47	539.67	1.5921	4.2360
10	500	838.17	977.04	2.5666	4.5903
10	584.18	688.63	1407.3	3.3591	6.1244
10	584.18	55.477	2724.5	5.6139	6.8973
10	600	49.830	2817.8	5.7716	5.1447
10	700	35.392	3176.1	6.3279	2.8769
10	800	29.128	3442.6	6.6844	2.5324
10	900	25.132	3690.6	6.9764	2.4463
10	1000	22.243	3934.5	7.2334	2.4395
10	1100	20.015	4179.5	7.4670	2.4654
10	1200	18.227	4428.0	7.6832	2.5059
10	1300	16.751	4680.9	7.8856	2.5529
10	1400	15.508	4938.6	8.0766	2.6018
10	1500	14.443	5201.3	8.2577	2.6503
10	1600	13.520	5468.6	8.4303	2.6969
10	1700	12.710	5740.6	8.5951	2.7410
10	1800	11.994	6016.8	8.7530	2.7823
10	1900	11.356	6296.9	8.9044	2.8206
10	2000	10.784	6580.8	9.0500	2.8563
100	273.15	1045.30	95.40	−0.0085	3.9092
100	300	1037.20	201.35	0.3614	3.9846
100	373.15	999.70	495.00	1.2371	4.0392
100	400	981.74	603.77	1.5186	4.0633
100	500	899.23	1015.3	2.4365	4.1854
100	600	791.46	1447.6	3.2237	4.5025
100	700	651.37	1925.2	3.9588	5.0954
100	800	482.17	2466.3	4.6806	5.6041
100	900	343.48	2997.0	5.3065	4.8280
100	1000	265.84	3431.3	5.7649	3.9368
100	1100	220.96	3798.6	6.1154	3.4632
100	1200	191.59	4131.4	6.4051	3.2174
100	1300	170.56	4445.8	6.6568	3.0848
100	1400	154.56	4750.3	6.8825	3.0129
100	1500	141.85	5049.5	7.0890	2.9759
100	1600	131.43	5346.2	7.2804	2.9600
100	1700	122.68	5641.9	7.4597	2.9570
100	1800	115.18	5937.8	7.6289	2.9620
100	1900	108.67	6234.5	7.7893	2.9720
100	2000	102.94	6532.4	7.9420	2.9852

PERMITTIVITY (DIELECTRIC CONSTANT) OF WATER AT VARIOUS FREQUENCIES

The permittivity of liquid water in the radiofrequency and microwave regions can be represented by the Debye equation (References 1 and 2):

$$\varepsilon' = \varepsilon_\infty + \frac{\varepsilon_s - \varepsilon_\infty}{1 + \omega^2 \tau^2}$$

$$\varepsilon'' = \frac{(\varepsilon_s - \varepsilon_\infty)\omega\tau}{1 + \omega^2 \tau^2}$$

where $\varepsilon = \varepsilon' + i\,\varepsilon''$ is the (complex) relative permittivity (i.e., the absolute permittivity divided by the permittivity of free space $\varepsilon_0 = 8.854 \cdot 10^{-12}$ F m^{-1}). Here ε_s is the static permittivity (see Reference 3 and the table "Properties of Water in the Range 0—100°C" in this Section); ε_∞ is a parameter describing the permittivity in the high frequency limit; τ is the relaxation time for molecular orientation; and $\omega = 2\pi f$ is the angular frequency. The values in this table have been calculated from parameters given in Reference 2:

	0°C	25°C	50°C
ε_∞	5.7	5.2	4.0
τ/ps	17.67	8.27	4.75

Other useful quantities that can be calculated from the values in the table are the loss tangent:

$$\tan\delta = \varepsilon'' / \varepsilon'$$

and the absorption coefficient α which describes the power attentuation per unit length ($P = P_0\,e^{-\alpha l}$):

$$\alpha = \frac{\pi f\,\varepsilon''}{c\sqrt{\varepsilon'}}$$

and c is the speed of light. The last equation is valid when $\varepsilon''/\varepsilon' \ll 1$.

REFERENCES

1. Fernendez, D.P., Mulev, Y., Goodwin, A.R.H., and Levelt Sengers, J.M.H., *J. Phys. Chem. Ref. Data*, 24, 33, 1995.
2. Kaatze, U., *J. Chem. Eng. Data*, 34, 371, 1989.
3. Archer, D.G., and Wang, P., *J. Phys. Chem. Ref. Data*, 12, 817, 1983.

Frequency	0°C ε'	0°C ε''	25°C ε'	25°C ε''	50°C ε'	50°C ε''
0	87.90	0.00	78.36	0.00	69.88	0.00
1 kHz	87.90	0.00	78.36	0.00	69.88	0.00
1 MHz	87.90	0.01	78.36	0.00	69.88	0.00
10 MHz	87.90	0.09	78.36	0.04	69.88	0.02
100 MHz	87.89	0.91	78.36	0.38	69.88	0.20
200 MHz	87.86	1.82	78.35	0.76	69.88	0.39
500 MHz	87.65	4.55	78.31	1.90	69.87	0.98
1 GHz	86.90	9.01	78.16	3.79	69.82	1.96
2 GHz	84.04	17.39	77.58	7.52	69.65	3.92
3 GHz	79.69	24.64	76.62	11.13	69.36	5.85
4 GHz	74.36	30.49	75.33	14.58	68.95	7.75
5 GHz	68.54	34.88	73.73	17.81	68.45	9.62
10 GHz	42.52	40.88	62.81	29.93	64.49	18.05
20 GHz	19.56	30.78	40.37	36.55	52.57	28.99
30 GHz	12.50	22.64	26.53	33.25	40.57	32.74
40 GHz	9.67	17.62	18.95	28.58	31.17	32.43
50 GHz	8.28	14.34	14.64	24.53	24.42	30.47

THERMOPHYSICAL PROPERTIES OF FLUIDS

These tables give thermodynamic and transport properties of some important fluids, as generated from the equations of state presented in the references below. The properties tabulated are density (ρ), energy (E), enthalpy (H), entropy (S), isochoric heat capacity (C_v), isobaric heat capacity (C_p), speed of sound (v_s), viscosity (η), thermal conductivity (λ), and dielectric constant (D). All extensive properties are given on a molar basis. Not all properties are included for every substance. The references should be consulted for information on the uncertainties and the reference states for E, H, and S.

Values are given as a function of temperature for several isobars. The phase can be determined by noting the sharp decrease in density between two successive temperature entries; all lines above this point refer to the liquid phase, and all lines below refer to the gas phase. If there is no sharp discontinuity in density, all data in the table refer to the supercritical region (i.e., the isobar is above the critical pressure).

REFERENCES

1. Younglove, B.A., *Thermophysical Properties of Fluids. Part I, J. Phys. Chem. Ref. Data,* 11, Suppl. 1, 1982.
2. Younglove, B.A., and Ely, J.F., *Thermophysical Properties of Fluids. Part II, J. Phys. Chem. Ref. Data,* 16, 577, 1987.
3. McCarty, R.D., *Thermodynamic Properties of Helium, J. Phys. Chem. Ref. Data,* 2, 923, 1973.

Nitrogen (N$_2$)

T K	ρ mol/L	E J/mol	H J/mol	S J/mol K	C_v J/mol K	C_p J/mol K	η μPa s	λ mW/m K	D
P = 0.1 MPa (1 bar)									
70	30.017	−3828	−3824	73.8	28.5	57.2	203.9	143.5	1.45269
77.25	28.881	−3411	−3407	79.5	27.8	57.8	152.2	133.8	1.43386
77.25	0.163	1546	2161	151.6	21.6	31.4	5.3	7.6	1.00215
100	0.123	2041	2856	159.5	21.1	30.0	6.8	9.6	1.00162
200	0.060	4140	5800	179.9	20.8	29.2	12.9	18.4	1.00079
300	0.040	6223	8717	191.8	20.8	29.2	18.0	25.8	1.00053
400	0.030	8308	11635	200.2	20.9	29.2	22.2	32.3	1.00040
500	0.024	10414	14573	206.7	21.2	29.6	26.1	38.5	1.00032
600	0.020	12563	17554	212.2	21.8	30.1	29.5	44.5	1.00026
700	0.017	14770	20593	216.8	22.4	30.7	32.8	50.5	1.00023
800	0.015	17044	23698	221.0	23.1	31.4	35.8	56.3	1.00020
900	0.013	19383	26869	224.7	23.7	32.0	38.7	62.0	1.00017
1000	0.012	21786	30103	228.1	24.3	32.6	41.5	67.7	1.00016
1500	0.008	34530	47004	241.8	26.4	34.7	54.0	93.3	1.00010
P = 1 MPa									
70	30.070	−3838	−3805	73.6	28.9	56.9	205.9	144.1	1.45355
80	28.504	−3267	−3232	81.3	27.8	57.7	139.5	130.7	1.42760
90	26.721	−2685	−2648	88.2	26.7	59.4	100.1	115.3	1.39824
100	24.634	−2073	−2032	94.6	26.2	64.4	73.1	98.5	1.36417
103.75	23.727	−1828	−1786	97.1	26.2	67.8	64.8	91.8	1.34947
103.75	1.472	1788	2467	138.1	24.1	45.0	7.6	12.5	1.01954
200	0.614	4048	5675	160.3	21.0	30.4	13.2	19.3	1.00812
300	0.402	6171	8661	172.5	20.9	29.6	18.1	26.3	1.00529
400	0.300	8273	11609	180.9	20.9	29.5	22.4	32.7	1.00395
500	0.240	10389	14563	187.5	21.3	29.7	26.1	38.8	1.00315
600	0.200	12544	17554	193.0	21.8	30.2	29.6	44.8	1.00262
700	0.171	14756	20600	197.7	22.4	30.8	32.8	50.7	1.00224
800	0.150	17032	23709	201.8	23.1	31.4	35.9	56.5	1.00196
900	0.133	19374	26884	205.6	23.7	32.1	38.8	62.2	1.00174
1000	0.120	21778	30121	209.0	24.3	32.7	41.5	67.8	1.00157
1500	0.080	34527	47029	222.7	26.4	34.8	54.0	93.4	1.00104
P = 10 MPa									
65.32	31.120	−4176	−3855	68.6	31.8	53.8	275.7	153.8	1.47067
100	26.201	−2328	−1946	92.0	27.4	56.3	90.2	112.3	1.38942
200	7.117	3037	4442	136.4	22.7	45.5	17.6	30.4	1.09698
300	3.989	5667	8174	151.7	21.4	33.4	20.1	31.9	1.05347
400	2.898	7941	11392	161.0	21.3	31.3	23.7	36.7	1.03860
500	2.302	10148	14492	167.9	21.5	30.8	27.1	42.0	1.03055

T K	ρ mol/L	E J/mol	H J/mol	S J/mol K	C_v J/mol K	C_p J/mol K	η μPa s	λ mW/m K	D
600	1.918	12361	17575	173.5	21.9	30.9	30.4	47.4	1.02538
700	1.647	14613	20683	178.3	22.5	31.3	33.5	53.0	1.02175
800	1.445	16919	23837	182.5	23.2	31.8	36.4	58.6	1.01904
900	1.288	19283	27046	186.3	23.8	32.4	39.3	64.1	1.01694
1000	1.162	21705	30308	189.8	24.4	32.9	42.0	69.6	1.01526
1500	0.783	34504	47283	203.5	26.5	34.8	54.3	94.7	1.01020

Oxygen (O_2)

T K	ρ mol/L	E J/mol	H J/mol	S J/mol K	C_v J/mol K	C_p J/mol K	η μPa s	λ mW/m K	D
P = 0.1 MPa (1 bar)									
60	40.049	−5883	−5880	72.4	34.9	53.4	425.2	188.2	1.55619
80	37.204	−4814	−4812	87.7	31.0	53.6	251.7	166.1	1.51114
100	0.123	2029	2840	172.9	21.4	30.5	7.5	9.3	1.00146
120	0.102	2458	3442	178.4	21.0	29.8	9.0	11.2	1.00121
140	0.087	2881	4035	182.9	20.9	29.5	10.5	13.1	1.00103
160	0.076	3301	4624	186.9	20.9	29.4	11.9	15.0	1.00090
180	0.067	3720	5210	190.3	20.8	29.3	13.3	16.7	1.00080
200	0.060	4138	5796	193.4	20.8	29.3	14.6	18.4	1.00072
220	0.055	4556	6381	196.2	20.8	29.3	15.9	20.1	1.00065
240	0.050	4974	6966	198.8	20.9	29.3	17.2	21.7	1.00060
260	0.046	5393	7552	201.1	20.9	29.3	18.4	23.2	1.00055
280	0.043	5812	8138	203.3	21.0	29.4	19.5	24.8	1.00051
300	0.040	6234	8726	205.3	21.1	29.4	20.6	26.3	1.00048
320	0.038	6657	9316	207.2	21.2	29.5	21.7	27.8	1.00045
340	0.035	7082	9908	209.0	21.3	29.7	22.8	29.3	1.00042
360	0.033	7510	10503	210.7	21.5	29.8	23.8	30.8	1.00040
380	0.032	7941	11100	212.3	21.6	30.0	24.8	32.2	1.00038
P = 1 MPa									
60	40.084	−5887	−5863	72.3	34.9	53.3	428.5	188.4	1.55674
80	37.254	−4822	−4795	87.6	31.0	53.5	253.8	166.4	1.51192
100	34.153	−3741	−3712	99.7	28.5	55.2	155.6	137.9	1.46381
120	1.198	2163	2997	156.7	24.0	40.6	9.4	13.9	1.01429
140	0.950	2683	3735	162.4	22.2	34.4	10.8	14.9	1.01133
160	0.802	3151	4398	166.8	21.5	32.2	12.2	16.3	1.00955
180	0.698	3598	5030	170.5	21.2	31.2	13.5	17.7	1.00831
200	0.620	4035	5647	173.8	21.1	30.6	14.8	19.3	1.00738
220	0.559	4466	6255	176.7	21.0	30.3	16.1	20.8	1.00665
240	0.509	4894	6858	179.3	21.0	30.1	17.3	22.3	1.00606
260	0.468	5321	7458	181.7	21.0	29.9	18.5	23.8	1.00556
280	0.433	5748	8056	183.9	21.1	29.9	19.6	25.2	1.00515
300	0.403	6174	8654	186.0	21.1	29.9	20.7	26.7	1.00479
320	0.377	6602	9252	187.9	21.2	29.9	21.8	28.2	1.00448
340	0.355	7032	9851	189.7	21.4	30.0	22.8	29.6	1.00421
360	0.335	7463	10452	191.4	21.5	30.1	23.9	31.1	1.00397
380	0.317	7898	11056	193.1	21.7	30.2	24.9	32.6	1.00376
P = 10 MPa									
60	40.419	−5931	−5684	71.5	35.1	53.0	461.8	189.9	1.56210
80	37.727	−4893	−4628	86.7	31.6	52.7	274.4	168.6	1.51936
100	34.881	−3856	−3570	98.5	29.1	53.4	171.0	141.2	1.47500
120	31.721	−2796	−2481	108.4	27.3	55.9	113.0	115.1	1.42677
140	27.890	−1662	−1304	117.5	26.2	62.9	76.3	91.8	1.36972
160	22.379	-322	125	127.0	26.1	84.8	48.6	71.2	1.29037
180	13.232	1489	2245	139.5	26.6	105.9	26.2	46.8	1.16560
200	8.666	2681	3835	147.9	24.0	60.6	21.2	34.0	1.10650

T K	ρ mol/L	E J/mol	H J/mol	S J/mol K	C_v J/mol K	C_p J/mol K	η μPa s	λ mW/m K	D
220	6.868	3424	4880	152.9	22.6	46.4	20.5	30.8	1.08380
240	5.836	4029	5742	156.6	22.0	40.6	20.8	30.1	1.07090
260	5.134	4573	6521	159.7	21.8	37.6	21.4	30.2	1.06219
280	4.613	5086	7254	162.5	21.6	35.8	22.1	30.8	1.05575
300	4.205	5581	7959	164.9	21.6	34.7	22.9	31.6	1.05073
320	3.874	6063	8645	167.1	21.7	33.9	23.7	32.6	1.04667
340	3.598	6538	9318	169.1	21.8	33.4	24.6	33.7	1.04329
360	3.363	7009	9982	171.0	21.9	33.0	25.4	34.9	1.04043
380	3.161	7477	10641	172.8	22.0	32.8	26.3	36.1	1.03796

Hydrogen (H_2)

T K	ρ mol/L	E J/mol	H J/mol	S J/mol K	C_v J/mol K	C_p J/mol K	v_s m/s	D
$P = 0.1$ MPa (1 bar)								
15	37.738	−605	−603	11.2	9.7	14.4	1319	1.24827
20	35.278	−524	−521	15.8	11.3	19.1	1111	1.23093
40	0.305	491	818	75.6	12.5	21.3	521	1.00186
60	0.201	748	1244	84.3	13.1	21.6	636	1.00122
80	0.151	1030	1694	90.7	15.3	23.7	714	1.00091
100	0.120	1370	2202	96.4	18.7	27.1	773	1.00073
120	0.100	1777	2776	101.6	21.8	30.2	827	1.00061
140	0.086	2237	3401	106.4	23.8	32.2	883	1.00052
160	0.075	2723	4054	110.8	24.6	33.0	940	1.00046
180	0.067	3216	4714	114.7	24.6	32.9	998	1.00041
200	0.060	3703	5367	118.1	24.1	32.4	1054	1.00037
220	0.055	4179	6009	121.2	23.4	31.8	1110	1.00033
240	0.050	4641	6638	123.9	22.8	31.2	1163	1.00030
260	0.046	5093	7256	126.4	22.3	30.6	1214	1.00028
280	0.043	5535	7865	128.6	21.9	30.2	1263	1.00026
300	0.040	5970	8466	130.7	21.6	29.9	1310	1.00024
400	0.030	8093	11421	139.2	21.0	29.3	1518	1.00018
$P = 1$ MPa								
15	38.109	−609	−583	10.9	10.1	14.1	1315	1.25089
20	35.852	−532	−504	15.5	11.4	18.4	1155	1.23496
40	3.608	399	676	54.1	12.9	28.4	498	1.02209
60	2.098	697	1173	64.3	13.2	23.5	635	1.01280
80	1.523	994	1651	71.1	15.4	24.7	719	1.00928
100	1.204	1343	2174	77.0	18.8	27.7	779	1.00733
120	0.999	1756	2758	82.3	21.9	30.6	835	1.00608
140	0.854	2219	3390	87.1	23.9	32.5	891	1.00520
160	0.747	2709	4048	91.5	24.7	33.2	949	1.00454
180	0.663	3204	4712	95.4	24.6	33.1	1006	1.00404
200	0.597	3693	5368	98.9	24.1	32.5	1063	1.00363
220	0.543	4170	6012	102.0	23.5	31.9	1118	1.00330
240	0.498	4634	6643	104.7	22.9	31.2	1171	1.00303
260	0.460	5087	7263	107.2	22.3	30.7	1222	1.00279
280	0.427	5530	7873	109.5	21.9	30.3	1271	1.00259
300	0.399	5966	8475	111.5	21.6	30.0	1317	1.00242
400	0.299	8091	11433	120.1	21.0	29.4	1525	1.00182
$P = 10$ MPa								
20	39.669	−568	−316	13.0	10.9	15.0	1458	1.26198
40	31.344	−209	110	27.3	13.2	27.0	1171	1.20354
60	21.273	255	725	39.7	13.8	32.5	931	1.13527
80	14.830	686	1360	48.8	15.9	31.1	886	1.09303
100	11.417	1110	1986	55.8	19.3	31.9	904	1.07109

T K	ρ mol/L	E J/mol	H J/mol	S J/mol K	C_v J/mol K	C_p J/mol K	v_s m/s	D
120	9.357	1571	2640	61.8	22.4	33.5	941	1.05801
140	7.969	2068	3323	67.0	24.3	34.6	989	1.04925
160	6.963	2583	4020	71.7	25.0	34.9	1042	1.04294
180	6.195	3099	4713	75.7	24.9	34.4	1096	1.03814
200	5.588	3604	5393	79.3	24.4	33.6	1150	1.03436
220	5.094	4094	6057	82.5	23.7	32.8	1203	1.03129
240	4.683	4569	6704	85.3	23.1	32.0	1254	1.02874
260	4.336	5030	7336	87.8	22.6	31.3	1302	1.02659
280	4.038	5481	7958	90.1	22.1	30.8	1349	1.02475
300	3.780	5924	8570	92.3	21.8	30.4	1394	1.02315
400	2.869	8073	11559	100.9	21.2	29.6	1592	1.01753

Helium (He-4)

T K	ρ mol/L	E J/mol	H J/mol	S J/mol K	C_v J/mol K	C_p J/mol K	v_s m/s	η μPa s	D
P = 0.1 MPa (1 bar)									
3	35.794	−39	−36	9.8	7.6	9.4	222	3.85	1.05646
4	32.477	−27	−24	13.3	9.1	16.3	185	3.33	1.05114
5	2.935	52	86	39.1	12.7	27.1	120	1.39	1.00456
10	1.238	120	201	55.2	12.5	21.7	185	2.26	1.00192
20	0.602	247	413	69.9	12.5	21.0	264	3.58	1.00093
50	0.240	623	1039	89.0	12.5	20.8	417	6.36	1.00037
100	0.120	1247	2079	103.4	12.5	20.8	589	9.78	1.00019
200	0.060	2494	4158	117.8	12.5	20.8	833	15.14	1.00009
300	0.040	3741	6237	126.3	12.5	20.8	1020	19.93	1.00006
400	0.030	4988	8315	132.3	12.5	20.8	1177	24.29	1.00005
500	0.024	6236	10394	136.9	12.5	20.8	1316	28.36	1.00004
600	0.020	7483	12472	140.7	12.5	20.8	1441	32.22	1.00003
700	0.017	8730	14551	143.9	12.5	20.8	1557	35.89	1.00003
800	0.015	9977	16630	146.7	12.5	20.8	1664	39.43	1.00002
900	0.013	11224	18708	149.1	12.5	20.8	1765	42.85	1.00002
1000	0.012	12471	20787	151.3	12.5	20.8	1861	46.16	1.00002
1500	0.008	18707	31179	159.7	12.5	20.8	2279	61.55	1.00001
P = 1 MPa									
3	39.703	−42	−16	8.6	7.1	7.8	300	5.63	1.06274
4	38.210	−34	−7	11.2	8.3	10.9	290	5.01	1.06034
5	35.818	−22	6	14.0	9.7	15.1	269	4.38	1.05650
10	15.378	78	143	32.2	12.3	30.5	198	3.07	1.02402
20	6.067	228	393	49.8	12.6	22.9	274	3.94	1.00943
50	2.353	617	1042	69.8	12.5	21.1	428	6.53	1.00365
100	1.186	1245	2089	84.3	12.5	20.9	597	9.89	1.00184
200	0.597	2495	4170	98.7	12.5	20.8	838	15.21	1.00093
300	0.399	3742	6249	107.1	12.5	20.8	1024	19.96	1.00062
400	0.300	4990	8327	113.1	12.5	20.8	1180	24.32	1.00046
500	0.240	6237	10406	117.8	12.5	20.8	1319	28.38	1.00037
600	0.200	7485	12484	121.5	12.5	20.8	1444	32.23	1.00031
700	0.172	8732	14562	124.7	12.5	20.8	1559	35.91	1.00027
800	0.150	9979	16641	127.5	12.5	20.8	1666	39.44	1.00023
900	0.133	11227	18719	130.0	12.5	20.8	1767	42.86	1.00021
1000	0.120	12474	20798	132.2	12.5	20.8	1862	46.17	1.00019
1500	0.080	18710	31190	140.6	12.5	20.8	2280	61.55	1.00012
P = 10 MPa									
4	51.978	−24	169	6.7	6.0	7.3	586	24.27	1.08262
5	51.118	−18	177	8.5	7.9	9.3	576	18.16	1.08122
10	46.872	23	236	16.6	11.0	14.5	546	9.31	1.07432

T K	ρ mol/L	E J/mol	H J/mol	S J/mol K	C_v J/mol K	C_p J/mol K	v_s m/s	η μPa s	D
20	37.092	154	423	29.5	12.6	20.7	498	6.99	1.05854
50	19.192	572	1093	49.9	12.9	22.4	541	8.07	1.03003
100	10.525	1231	2181	65.0	12.8	21.3	674	10.93	1.01640
200	5.605	2500	4284	79.6	12.6	20.9	889	15.82	1.00871
300	3.829	3755	6367	88.0	12.6	20.8	1063	20.25	1.00595
400	2.908	5006	8445	94.0	12.6	20.8	1212	24.54	1.00452
500	2.344	6256	10522	98.6	12.5	20.8	1346	28.56	1.00364
600	1.963	7505	12599	102.4	12.5	20.8	1467	32.38	1.00305
700	1.689	8754	14676	105.6	12.5	20.8	1580	36.04	1.00262
800	1.481	10003	16753	108.4	12.5	20.8	1685	39.56	1.00230
900	1.320	11252	18830	110.9	12.5	20.8	1784	42.96	1.00205
1000	1.189	12500	20907	113.0	12.5	20.8	1877	46.26	1.00185
1500	0.797	18742	31294	121.5	12.5	20.8	2289	61.62	1.00124

Argon (Ar)

T K	ρ mol/L	E J/mol	H J/mol	S J/mol K	C_v J/mol K	C_p J/mol K	v_s m/s	η μPa s	λ mW/m K
$P = 0.1$ MPa (1 bar)									
85	35.243	−4811	−4808	53.6	23.1	44.7	820	278.8	132.4
90	0.138	1077	1802	129.4	13.1	22.5	174	7.5	6.0
100	0.123	1211	2024	131.8	12.9	21.9	184	8.2	6.6
120	0.102	1471	2456	135.7	12.6	21.4	203	9.8	7.8
140	0.087	1727	2881	139.0	12.6	21.1	220	11.4	9.0
160	0.076	1980	3302	141.8	12.5	21.0	235	13.0	10.2
180	0.067	2232	3722	144.3	12.5	21.0	250	14.5	11.4
200	0.060	2483	4141	146.5	12.5	20.9	263	16.0	12.5
220	0.055	2734	4559	148.5	12.5	20.9	276	17.5	13.7
240	0.050	2984	4976	150.3	12.5	20.9	289	18.9	14.8
260	0.046	3234	5394	152.0	12.5	20.9	300	20.3	15.8
280	0.043	3484	5811	153.5	12.5	20.8	312	21.6	16.9
300	0.040	3734	6227	155.0	12.5	20.8	323	22.9	17.9
320	0.038	3984	6644	156.3	12.5	20.8	333	24.2	18.9
340	0.035	4234	7060	157.6	12.5	20.8	344	25.4	19.9
360	0.033	4484	7477	158.7	12.5	20.8	354	26.6	20.8
380	0.032	4734	7893	159.9	12.5	20.8	363	27.8	21.7
$P = 1$ MPa									
85	35.307	−4820	−4792	53.5	23.1	44.6	823	281.3	133.0
90	34.542	−4598	−4569	56.1	21.6	44.7	808	242.7	124.2
100	32.909	−4145	−4115	60.9	19.9	46.2	753	185.0	109.2
120	1.181	1210	2057	114.3	14.7	30.1	189	10.3	9.3
140	0.945	1544	2603	118.5	13.5	25.4	212	11.8	10.1
160	0.799	1838	3089	121.8	13.0	23.6	231	13.3	11.1
180	0.697	2116	3551	124.5	12.8	22.7	247	14.8	12.1
200	0.619	2384	3999	126.9	12.7	22.2	262	16.3	13.2
220	0.559	2648	4438	128.9	12.6	21.8	275	17.7	14.2
240	0.509	2908	4873	130.8	12.6	21.6	288	19.1	15.3
260	0.468	3167	5304	132.6	12.6	21.5	301	20.4	16.3
280	0.433	3423	5732	134.2	12.6	21.4	312	21.8	17.3
300	0.403	3679	6159	135.6	12.5	21.3	324	23.1	18.3
320	0.377	3934	6583	137.0	12.5	21.2	334	24.3	19.2
340	0.355	4188	7007	138.3	12.5	21.2	345	25.5	20.2
360	0.335	4441	7429	139.5	12.5	21.1	355	26.7	21.1
380	0.317	4694	7851	140.6	12.5	21.1	365	27.9	22.0
$P = 10$ MPa									
90	35.208	−4694	−4410	55.0	21.9	43.2	846	265.2	129.5

T K	ρ mol/L	E J/mol	H J/mol	S J/mol K	C_v J/mol K	C_p J/mol K	v_s m/s	η µPa s	λ mW/m K
100	33.744	−4271	−3974	59.6	20.4	44.0	800	205.0	115.1
120	30.525	−3396	−3069	67.8	18.8	46.9	672	131.2	92.1
140	26.609	−2447	−2072	75.5	17.6	54.1	526	85.9	71.7
160	20.816	−1279	−799	83.9	17.4	78.6	357	51.3	52.8
180	12.296	228	1042	94.8	17.3	83.6	257	27.8	32.0
200	8.442	1118	2302	101.4	15.3	48.6	268	23.3	23.6
220	6.776	1661	3137	105.4	14.2	36.8	284	22.8	21.6
240	5.787	2087	3815	108.4	13.7	31.6	300	23.2	21.3
260	5.105	2458	4416	110.8	13.4	28.8	314	23.9	21.4
280	4.596	2798	4974	112.9	13.2	27.1	327	24.8	21.8
300	4.195	3119	5503	114.7	13.1	25.9	339	25.7	22.3
320	3.869	3427	6012	116.3	13.0	25.0	350	26.7	22.9
340	3.596	3726	6506	117.8	13.0	24.4	361	27.7	23.5
360	3.364	4017	6989	119.2	12.9	23.9	372	28.7	24.2
380	3.164	4303	7464	120.5	12.9	23.5	381	29.7	24.9

Methane (CH$_4$)

T K	ρ mol/L	E J/mol	H J/mol	S J/mol K	C_v J/mol K	C_p J/mol K	η µPa s	λ mW/m K	D
$P = 0.1$ MPa (1 bar)									
100	27.370	−5258	−5254	73.0	33.4	54.1	156.3	208.1	1.65504
125	0.099	3026	4039	156.5	25.4	34.6	5.0	13.4	1.00193
150	0.081	3667	4896	162.7	25.2	34.0	5.9	16.2	1.00159
175	0.069	4301	5743	168.0	25.2	33.8	6.9	19.1	1.00136
200	0.061	4935	6587	172.5	25.3	33.8	7.8	21.9	1.00119
225	0.054	5571	7434	176.5	25.5	34.0	8.7	24.8	1.00105
250	0.048	6216	8288	180.1	26.0	34.4	9.6	27.8	1.00095
275	0.044	6875	9156	183.4	26.6	35.0	10.4	30.9	1.00086
300	0.040	7552	10042	186.4	27.5	35.9	11.2	34.1	1.00079
325	0.037	8252	10951	189.4	28.5	36.9	12.0	37.6	1.00073
350	0.034	8979	11887	192.1	29.7	38.0	12.8	41.2	1.00068
375	0.032	9737	12853	194.8	30.9	39.3	13.5	45.1	1.00063
400	0.030	10528	13852	197.4	32.3	40.7	14.3	49.1	1.00059
425	0.028	11354	14886	199.9	33.7	42.1	15.0	53.3	1.00056
450	0.027	12215	15956	202.3	35.2	43.5	15.7	57.6	1.00053
500	0.024	14047	18204	207.1	38.0	46.4	17.0	66.5	1.00047
600	0.020	18111	23101	216.0	42.9	51.3	19.4	84.1	1.00039
$P = 1$ MPa									
100	27.413	−5268	−5231	72.9	33.4	54.0	158.1	208.9	1.65617
125	25.137	−3882	−3842	85.3	32.4	57.4	89.2	168.2	1.59261
150	0.969	3282	4315	140.9	27.9	45.2	6.2	18.4	1.01911
175	0.765	4041	5348	147.3	26.4	38.9	7.1	20.6	1.01507
200	0.644	4736	6289	152.3	25.9	36.8	8.0	23.1	1.01268
225	0.560	5410	7197	156.6	25.9	36.0	8.9	25.8	1.01102
250	0.497	6081	8093	160.4	26.2	35.8	9.7	28.7	1.00979
275	0.448	6758	8991	163.8	26.8	36.1	10.6	31.7	1.00882
300	0.408	7449	9901	167.0	27.6	36.7	11.4	34.9	1.00803
325	0.375	8160	10829	169.9	28.6	37.6	12.1	38.3	1.00738
350	0.347	8897	11781	172.8	29.7	38.6	12.9	41.9	1.00683
375	0.323	9662	12760	175.5	31.0	39.8	13.6	45.7	1.00636
400	0.302	10460	13770	178.1	32.4	41.1	14.4	49.6	1.00595
425	0.284	11291	14814	180.6	33.8	42.4	15.1	53.8	1.00559
450	0.268	12157	15892	183.1	35.2	43.8	15.7	58.1	1.00527
500	0.241	13997	18153	187.8	38.1	46.6	17.0	66.9	1.00474
600	0.200	18073	23070	196.8	43.0	51.4	19.5	84.5	1.00394

T K	ρ mol/L	E J/mol	H J/mol	S J/mol K	C_v J/mol K	C_p J/mol K	η μPa s	λ mW/m K	D
P = 10 MPa									
100	27.815	−5362	−5003	72.0	33.8	53.2	175.4	217	1.66668
125	25.754	−4036	−3648	84.1	32.7	55.3	100.4	178.8	1.60895
150	23.441	−2655	−2229	94.4	31.4	58.6	65.7	144.6	1.54553
175	20.613	−1175	−689	103.9	30.3	65.5	44.9	113.4	1.47021
200	16.602	542	1144	113.6	30.1	84.7	29.4	85.8	1.36789
225	10.547	2680	3628	125.3	30.8	102.2	17.6	61.0	1.22352
250	7.013	4289	5714	134.1	29.3	67.4	14.3	47.6	1.14481
275	5.530	5387	7195	139.8	28.7	53.4	13.8	44.1	1.11297
300	4.685	6320	8454	144.2	28.9	48.0	13.9	44.6	1.09513
325	4.115	7192	9622	147.9	29.6	45.8	14.3	46.6	1.08322
350	3.695	8047	10753	151.3	30.5	44.9	14.7	49.2	1.07450
375	3.366	8903	11874	154.4	31.7	44.8	15.2	52.3	1.06773
400	3.101	9774	12999	157.3	32.9	45.2	15.8	55.7	1.06227
425	2.880	10666	14138	160.0	34.3	46.0	16.3	59.4	1.05775
450	2.692	11584	15298	162.7	35.7	46.9	16.9	63.3	1.05392
500	2.389	13507	17692	167.7	38.5	48.9	18.0	71.6	1.04775
600	1.963	17700	22795	177.0	43.3	52.9	20.2	88.3	1.03911

Ethane (C$_2$H$_6$)

T K	ρ mol/L	E J/mol	H J/mol	S J/mol K	C_v J/mol K	C_p J/mol K	v_s m/s	D
P = 0.1 MPa (1 bar)								
95	21.50	−14555	−14550	80.2	47.2	68.7	1970	1.93480
100	21.32	−14210	−14205	83.8	47.1	69.3	1943	1.92500
125	20.41	−12468	−12463	99.3	45.0	69.8	1775	1.87634
150	19.47	−10717	−10712	112.1	43.4	70.4	1587	1.82726
175	18.49	−8938	−8933	123.1	42.7	72.1	1396	1.77671
200	0.062	5503	7123	210.1	34.5	43.8	258	1.00208
225	0.054	6401	8238	215.4	36.5	45.5	273	1.00183
250	0.049	7349	9401	220.3	38.9	47.7	287	1.00164
275	0.044	8360	10624	224.9	41.6	50.2	300	1.00148
300	0.040	9439	11914	229.4	44.5	53.1	312	1.00136
325	0.037	10592	13278	233.8	47.6	56.1	324	1.00125
350	0.035	11823	14719	238.1	50.7	59.2	335	1.00116
375	0.032	13133	16240	242.3	54.0	62.4	345	1.00108
400	0.030	14525	17841	246.4	57.2	65.6	355	1.00101
450	0.027	17548	21282	254.5	63.6	72.0	375	1.00090
500	0.024	20883	25035	262.4	69.7	78.1	393	1.00081
600	0.020	28429	33415	277.6	80.9	89.3	428	1.00067
P = 1 MPa								
95	21.514	−14562	−14515	80.2	47.3	68.7	1972	1.93537
100	21.334	−14217	−14170	83.7	47.2	69.3	1946	1.92560
125	20.427	−12478	−12429	99.2	45.0	69.8	1778	1.87709
150	19.494	−10731	−10679	112.0	43.4	70.3	1592	1.82823
175	18.515	−8957	−8903	123.0	42.7	72.0	1402	1.77800
200	17.464	−7127	−7070	132.7	42.9	74.9	1209	1.72513
225	16.288	−5199	−5137	141.8	43.8	80.2	1008	1.66733
250	0.564	6762	8534	198.7	41.6	57.5	260	1.01909
275	0.489	7902	9949	204.1	43.2	56.2	280	1.01650
300	0.435	9063	11363	209.0	45.5	57.2	297	1.01467
325	0.393	10273	12815	213.7	48.3	59.1	311	1.01327
350	0.360	11546	14321	218.1	51.3	61.5	325	1.01214
375	0.333	12889	15893	222.5	54.4	64.2	337	1.01121
400	0.310	14306	17534	226.7	57.5	67.1	349	1.01043

T K	ρ mol/L	E J/mol	H J/mol	S J/mol K	C_v J/mol K	C_p J/mol K	v_s m/s	D
450	0.272	17367	21038	234.9	63.8	73.0	370	1.00917
500	0.244	20730	24836	242.9	69.9	78.9	390	1.00819
600	0.201	28313	33278	258.3	81.0	89.8	427	1.00677
P = 10 MPa								
95	21.624	−14626	−14163	79.5	47.4	68.5	2000	1.94104
100	21.448	−14286	−13819	83.0	47.4	69.1	1974	1.93146
125	20.570	−12572	−12086	98.5	45.5	69.3	1814	1.88436
150	19.678	−10858	−10350	111.1	43.9	69.6	1637	1.83753
175	18.758	−9130	−8596	121.9	43.3	70.8	1459	1.79010
200	17.793	−7363	−6801	131.5	43.5	73.0	1284	1.74134
225	16.760	−5535	−4938	140.3	44.3	76.4	1110	1.69017
250	15.620	−3609	−2969	148.6	45.8	81.5	935	1.63488
275	14.301	−1539	−839	156.7	47.9	89.4	758	1.57249
300	12.666	757	1547	165.0	50.8	102.7	577	1.49740
325	10.398	3443	4404	174.1	54.7	129.1	399	1.39745
350	7.292	6643	8015	184.8	58.8	150.1	290	1.26832
375	5.182	9419	11349	194.1	60.0	115.7	289	1.18570
400	4.182	11577	13968	200.8	61.4	96.9	310	1.14797
450	3.204	15379	18500	211.5	65.8	87.5	347	1.11193
500	2.677	19135	22870	220.7	71.2	88.0	378	1.09288
600	2.076	27160	31978	237.3	81.8	94.7	427	1.07142

Propane (C_3H_8)

T K	ρ mol/L	E J/mol	H J/mol	S J/mol K	C_v J/mol K	C_p J/mol K	v_s m/s	D
P = 0.1 MPa (1 bar)								
90	16.526	−21486	−21426	87.3	59.2	84.5	2126	2.07988
100	16.295	−20639	−20577	96.2	59.6	85.2	2041	2.05806
125	15.726	−18495	−18432	115.4	59.2	86.5	1856	2.00674
150	15.156	−16319	−16253	131.3	58.9	88.0	1685	1.95796
175	14.577	−14096	−14028	145.0	59.5	90.3	1521	1.91036
200	13.982	−11806	−11735	157.3	61.0	93.5	1359	1.86300
225	13.339	−9395	−9387	168.5	63.4	97.9	1197	1.81487
250	0.050	9194	11213	257.6	57.2	66.8	228	1.00238
275	0.045	10691	12930	264.1	61.6	70.7	239	1.00215
300	0.041	12297	14752	270.5	66.2	75.1	249	1.00195
325	0.037	14019	16689	276.7	71.1	79.8	259	1.00179
350	0.035	15862	18744	282.8	76.0	84.6	269	1.00166
375	0.032	17827	20921	288.8	80.9	89.5	278	1.00154
400	0.030	19912	23217	294.7	85.7	94.3	286	1.00144
450	0.027	24441	28166	306.4	95.2	103.6	303	1.00128
500	0.024	29428	33573	317.7	104.1	112.6	318	1.00115
600	0.020	40677	45658	339.7	120.4	128.8	347	1.00095
P = 1 MPa								
90	16.526	−21486	−21426	87.2	59.3	84.5	2128	2.08034
100	16.295	−20639	−20577	96.2	59.7	85.2	2043	2.05856
125	15.726	−18495	−18432	115.3	59.2	86.4	1859	2.00736
150	15.156	−16319	−16253	131.2	59.0	88.0	1690	1.95873
175	14.577	−14096	−14028	144.9	59.6	90.2	1526	1.91132
200	13.982	−11806	−11735	157.2	61.1	93.4	1365	1.86421
225	13.361	−9424	−9349	168.4	63.4	97.7	1205	1.81642
250	12.696	−6919	−6840	179.0	66.4	103.3	1045	1.76672
275	11.962	−4252	−4169	189.1	70.0	110.8	881	1.71316
300	11.102	−1360	−1270	199.2	74.1	121.9	708	1.65216
325	0.428	13278	15614	255.2	74.1	89.6	233	1.02067

T K	ρ mol/L	E J/mol	H J/mol	S J/mol K	C_v J/mol K	C_p J/mol K	v_s m/s	D
350	0.383	15259	17869	261.9	78.0	91.2	248	1.01846
375	0.349	17318	20183	268.3	82.2	94.2	261	1.01678
400	0.322	19472	22582	274.4	86.7	97.8	272	1.01544
450	0.279	24092	27672	286.4	95.7	105.9	293	1.01337
500	0.248	29137	33172	298.0	104.4	114.1	312	1.01184
600	0.203	40455	45374	320.2	120.5	129.7	344	1.00968
$P = 10$ MPa								
90	16.590	−21553	−20951	86.5	59.9	84.4	2146	2.08489
100	16.364	−20714	−20103	95.4	60.1	85.1	2068	2.06350
125	15.810	−18595	−17962	114.5	59.6	86.1	1895	2.01342
150	15.259	−16448	−15793	130.3	59.3	87.5	1733	1.96617
175	14.705	−14261	−13581	144.0	59.9	89.5	1577	1.92048
200	14.141	−12016	−11309	156.1	61.4	92.4	1425	1.87557
225	13.562	−9692	−8955	167.2	63.7	96.1	1277	1.83076
250	12.960	−7268	−6496	177.5	66.7	100.7	1133	1.78529
275	12.322	−4721	−3909	187.4	70.2	106.4	991	1.73826
300	11.631	−2027	−1167	196.9	74.1	113.2	851	1.68849
325	10.860	843	1764	206.3	78.4	121.5	715	1.63437
350	9.973	3924	4927	215.7	82.9	132.0	582	1.57361
375	8.905	7270	8393	225.2	87.7	146.1	455	1.50271
400	7.561	10957	12279	235.3	93.0	165.7	339	1.41671
450	4.614	18845	21013	255.8	101.8	167.8	249	1.24060
500	3.241	25567	28652	272.0	107.8	142.7	276	1.16439
600	2.242	38131	42591	297.4	121.7	140.5	332	1.11122

VIRIAL COEFFICIENTS OF SELECTED GASES

Henry V. Kehiaian

This table gives second virial coefficients of about 110 inorganic and organic gases as a function of temperature. Selected data from the literature have been fitted by least squares to the equation

$$B/\text{cm}^3\,\text{mol}^{-1} = \sum_{i=1}^{n} a(i)\big[(T_o/T)-1\big]^{i-1}$$

where $T_o = 298.15$ K. The table gives the coefficients $a(i)$ and values of B at fixed temperature increments, as calculated from this smoothing equation.

The equation may be used with the tabulated coefficients for interpolation within the indicated temperature range. It should not be used for extrapolation beyond this range.

Compounds are listed in the modified Hill order (see Introduction), with carbon-containing compounds following those compounds not containing carbon.

A useful compilation of virial coefficient data from the literature may be found in:

J. H. Dymond and E. B. Smith, *The Virial Coefficients of Pure Gases and Mixtures, A Critical Compilation*, Oxford University Press, Oxford, 1980.

Compounds Not Containing Carbon

Mol. form.	Name		T/K	B/cm³ mol⁻¹
Ar	Argon		100	−184
			120	−131
			140	−98
		a(1) = −16	160	−76
		a(2) = −60	80	−60
		a(3) = −10	200	−48
			300	−16
			400	−1
			500	7
			600	12
			700	15
			800	18
			900	20
			1000	22
BF₃	Boron trifluoride		200	−338
			240	−202
			280	−129
		a(1) = −106	320	−85
		a(2) = −330	360	−56
		a(3) = −251	400	−37
		a(4) = −80	440	−23
ClH	Hydrogen chloride		190	−451
			230	−269
			270	−181
		a(1) = −144	310	−132
		a(2) = −325	350	−102
		a(3) = −277	390	−81
		a(4) = −170	430	−66
			470	−54
Cl₂	Chlorine		210	−508
			220	−483

Mol. form.	Name	T/K	$B/\mathrm{cm}^3\ \mathrm{mol}^{-1}$
		230	−457
	$a(1) = -303$	240	−432
	$a(2) = -555$	250	−407
	$a(3) = 9$	260	−383
	$a(4) = 329$	270	−360
	$a(5) = 68$	280	−339
		290	−318
		300	−299
		350	−221
		400	−166
		450	−126
		500	−97
		600	−59
		700	−36
		800	−22
		900	−12
F_2	Fluorine	80	−386
		110	−171
		140	−113
	$a(1) = -25$	170	−73
	$a(2) = 21$	200	−47
	$a(3) = -185$	230	−32
	$a(4) = 113$	260	−25
F_4Si	Silicon tetrafluoride	210	−268
		240	−213
		270	−170
	$a(1) = -138$	300	−136
	$a(2) = -312$	330	−108
		360	−84
		390	−64
		420	−47
		450	−32
F_5I	Iodine pentafluoride	320	−2540
		330	−2344
		340	−2172
	$a(1) = -3077$	350	−2021
	$a(2) = -8474$	360	−1890
	$a(3) = -9116$	370	−1775
		380	−1674
		390	−1587
		400	−1510
		410	−1443
F_5P	Phosphorus pentafluoride	320	−162
		340	−143
		360	−127
	$a(1) = -186$	380	−112
	$a(2) = -345$	400	−98
		420	−86
		440	−75
		460	−64
F_6Mo	Molybdenum hexafluoride	300	−896
		310	−810
		320	−737

Mol. form.	Name	T/K	$B/cm^3\ mol^{-1}$
	a(1) = –914	330	–677
	a(2) = –2922	340	–627
	a(3) = –4778	350	–586
		360	–553
		370	–527
		380	–506
		390	–491
F_6S	Sulfur hexafluoride	200	–685
		250	–416
		300	–275
	a(1) = –279	350	–190
	a(2) = –647	400	–135
	a(3) = –335	450	–96
	a(4) = –72	500	–68
F_6U	Uranium hexafluoride	320	–1030
		340	–905
		360	–805
	a(1) = –1204	380	–724
	a(2) = –2690	400	–658
	a(3) = –2144	420	–604
		440	–560
F_6W	Tungsten hexafluoride	320	–641
		340	–578
		360	–523
	a(1) = –719	380	–473
	a(2) = –1143	400	–428
		420	–387
		440	–350
		460	–317
H_2	Hydrogen	15	–230
		20	–151
		25	–108
	a(1) = 15.4	30	–82
	a(2) = –9.0	35	–64
	a(3) = –0.2	40	–52
		45	–42
		50	–35
		60	–24
		70	–16
		80	–11
		90	–7
		100	–3
		200	11
		300	15
		400	18
H_2O	Water	300	–1126
		320	–850
		340	–660
	a(1) = –1158	360	–526
	a(2) = –5157	380	–428
	a(3) = –10301	400	–356
	a(4) =–10597	420	–301
	a(5) = –4415	440	–258

Mol. form.	Name	T/K	$B/\text{cm}^3 \text{ mol}^{-1}$
		460	−224
		480	−197
		500	−175
		600	−104
		700	−67
		800	−44
		900	−30
		1000	−20
		1100	−14
		1200	−11
H_3N	Ammonia	290	−302
		300	−265
		310	−236
a(1) = −271		320	−213
a(2) = −1022		330	−194
a(3) = −2715		340	−179
a(4) = −4189		350	−166
		360	−154
		370	−144
		380	−135
		400	−118
		420	−101
H_3P	Phosphine	190	−457
		200	−404
		210	−364
a(1) = −146		220	−332
a(2) = −733		230	−305
a(3) = 1022		240	−281
a(4) = −1220		250	−258
		260	−235
		270	−213
		280	−190
		290	−166
He	Helium	2	−172
		6	−48
		10	−24
a(1) = 12		14	−13
a(2) = −1		18	−7
		22	−3
		26	−1
		30	1
		50	6
		70	8
		90	10
		110	10
		150	11
		250	12
		650	13
		700	13
Kr	Krypton	110	−363
		120	−307
		130	−263
a(1) = −51		140	−229
a(2) = −118		150	−201

Mol. form.	Name	T/K	$B/cm^3\ mol^{-1}$
	a(3) = −29	160	−178
	a(4) = −5	170	−159
		180	−143
		190	−129
		200	−117
		250	−75
		300	−51
		400	−23
		500	−8
		600	2
		700	8
NO	Nitric oxide	120	−232
		130	−176
		140	−138
	a(1) = −12	150	−113
	a(2) = −119	160	−96
	a(3) = 89	170	−83
	a(4) = −73	180	−73
		190	−65
		200	−58
		210	−52
		230	−42
		250	−32
		270	−24
N_2	Nitrogen	75	−274
		100	−161
		125	−104
	a(1) = −4	150	−71
	a(2) = −56	175	49
	a(3) = −12	200	−34
		225	−24
		250	−15
		300	−4
		400	9
		500	16
		600	21
		700	24
N_2O	Nitrous oxide	240	−219
		260	−181
		280	−151
	a(1) = −130	300	−128
	a(2) = −307	320	−110
	a(3) = −248	340	−96
		360	−85
		380	−76
		400	−68
Ne	Neon	60	−25
		80	−13
		100	−6
	a(1) = 10.8	120	−1
	a(2) = −7.5	140	2
	a(3) = 0.4	160	4
		180	6
		200	7

Mol. form.	Name	T/K	$B/cm^3\ mol^{-1}$
		300	11
		400	13
		500	14
		600	15
O_2	Oxygen	90	−241
		110	−161
		130	−117
a(1) = −16		150	−88
a(2) = −62		170	−69
a(3) = −8		190	−55
a(4) = −3		210	−44
		230	−36
		250	−29
		270	−23
		290	−18
		310	−14
		330	−10
		350	−7
		400	−1
O_2S	Sulfur dioxide	290	−465
		320	−354
		350	−276
a(1) = −430		380	−221
a(2) = −1193		410	−181
a(3) = −1029		440	−153
		470	−132
Xe	Xenon	160	−421
		170	−377
		180	−340
a(1) = −130		190	−307
a(2) = −262		200	−280
a(3) = −87		210	−255
		220	−234
		230	−215
		240	−199
		250	−184
		300	−129
		350	−93
		400	−69
		500	−39
		600	−21
		650	−14

Compounds Containing Carbon

Mol. form.	Name	T/K	$B/cm^3\ mol^{-1}$
$CClF_3$	Chlorotrifluoromethane	240	−369
		290	−237
		340	−165
a(1) = −223		390	−119
a(2) = −504		440	−86
a(3) = −340		490	−60
a(4) = −291		540	−39

Mol. form.	Name	T/K	$B/cm^3\ mol^{-1}$
CCl_2F_2	Dichlorodifluoromethane	250	−769
		280	−570
		310	−441
	a(1) = −486	340	−353
	a(2) = −1217	370	−289
	a(3) = −1188	400	−241
	a(4) = −698	430	−204
		460	−174
CCl_3F	Trichlorofluoromethane	240	−1140
		280	−879
		320	−689
	a(1) = −786	360	−545
	a(2) = −1428	400	−431
	a(3) = −142	440	−340
		480	−265
CCl_4	Tetrachloromethane	320	−1345
		340	−1171
		360	−1040
	a(1) = −1600	380	−942
	a(2) = −4059	400	−868
	a(3) = −4653	420	−814
CF_4	Tetrafluoromethane	250	−137
		300	−87
		350	−55
	a(1) = −88	400	−32
	a(2) = −238	450	−16
	a(3) = −70	500	−4
		600	14
		700	25
		800	33
$CHClF_2$	Chlorodifluoromethane	300	−343
		325	−298
		350	−257
	a(1) = −347	375	−221
	a(2) = −575	400	−188
	a(3) = 187	425	−158
$CHCl_2F$	Dichlorofluoromethane	250	−728
		275	−634
		300	−557
	a(1) = −562	325	−491
	a(2) = −862	350	−434
		375	−385
		400	−343
		425	−305
		450	−271
$CHCl_3$	Trichloromethane	320	−1001
		330	−926
		340	−858
	a(1) = −1193	350	−797
	a(2) = −2936	360	−740
	a(3) = −1751	370	−689
		380	−642

Mol. form.	Name		T/K	$B/cm^3 mol^{-1}$
			390	−599
			400	−559
CHF$_3$	Trifluoromethane		200	−433
			220	−350
			240	−288
		a(1) = −177	260	−241
		a(2) = −399	280	−204
		a(3) = −250	300	−174
			320	−151
			340	−132
			360	−116
			380	−103
			400	−91
CH$_2$Cl$_2$	Dichloromethane		320	−706
			330	−634
			340	−574
		a(1) = −913	350	−524
		a(2) = −3371	360	−482
		a(3) = −5013	370	−447
			380	−420
			400	−380
			420	−357
CH$_2$F$_2$	Difluoromethane		280	−375
			290	−343
			300	−316
		a(1) = −321	310	−294
		a(2) = −754	320	−275
		a(3) = −1300	330	−260
			340	−248
			350	−238
CH$_3$Br	Bromomethane		280	−645
			290	−596
			300	−551
		a(1) = −559	310	−509
		a(2) = −1324	320	−469
			340	−396
			360	−332
			380	−274
CH$_3$Cl	Chloromethane		280	−466
			300	−402
			320	−348
		a(1) = −407	340	−304
		a(2) = −887	360	−266
		a(3) = −385	380	−234
			400	−206
			420	−182
			440	−161
			460	−142
			480	−126
			500	−112
			600	−58
CH$_3$F	Fluoromethane		280	−244

Mol. form.	Name		T/K	B/cm^3 mol^{-1}
			300	−205
			320	−174
		$a(1) = -209$	340	−150
		$a(2) = -525$	360	−129
		$a(3) = -365$	380	−112
			400	−99
			420	−87
CH_3I	Iodomethane		310	−725
			320	−646
			330	−582
		$a(1) = -844$	340	−531
		$a(2) = -3353$	350	−492
		$a(3) = -6590$	360	−462
			370	−441
			380	−427
CH_4	Methane		110	−328
			120	−276
			130	−237
		$a(1) = -43$	140	−206
		$a(2) = -114$	150	−181
		$a(3) = -19$	160	−160
		$a(4) = -7$	170	−143
			180	−128
			190	−116
			200	−105
			250	−66
			300	−43
			350	−27
			400	−16
			500	0
			600	10
CH_4O	Methanol		320	−1431
			330	−1299
			340	−1174
		$a(1) = -1752$	350	−1056
		$a(2) = -4694$	360	−945
			370	−840
			380	−741
			390	−646
			400	−557
CH_5N	Methylamine		300	−451
			325	−367
			350	−304
		$a(1) = -459$	375	−257
		$a(2) = -1191$	400	−220
		$a(3) = -995$	425	−192
			450	−170
			500	−140
			550	−122
CO	Carbon monoxide		210	−36
			240	−24
			270	−15
		$a(1) = -9$	300	−8

Mol. form.	Name		T/K	$B/cm^3 \; mol^{-1}$
		a(2) = −58	330	−3
		a(3) = −18	360	1
			420	7
			480	11
CO_2	Carbon dioxide		220	−244
			240	−204
			260	−172
		a(1) = −127	280	−146
		a(2) = −288	300	−126
		a(3) = −118	320	−108
			340	−94
			360	−81
			380	−71
			400	−62
			500	−30
			600	−13
			700	−1
			800	7
			900	12
			1000	16
			1100	19
CS_2	Carbon disulfide		280	−932
			310	−740
			340	−603
		a(1) = −807	370	−504
		a(2) = −1829	400	−431
		a(3) = −1371	430	−375
$C_2Cl_2F_4$	1,2-Dichloro-1,1,2,2-tetrafluoroethane		300	−801
			320	−695
			340	−608
		a(1) = −812	360	−536
		a(2) = −1773	380	−475
		a(3) = −963	400	−423
			420	−379
			440	−341
			460	−307
			480	−279
			500	−253
$C_2Cl_3F_3$	1,1,2-Trichloro-1,2,2-trifluoroethane		290	−1041
			310	−943
			330	−856
		a(1) = −999	350	−780
		a(2) = −1479	370	−712
			390	−651
			410	−596
			430	−546
			450	−500
C_2H_2	Ethyne		200	−573
			210	−500
			220	−440
		a(1) = −216	230	−390
		a(2) = −375	240	−349
		a(3) = −716	250	−315

Mol. form.	Name		T/K	B/cm^3 mol^{-1}
			260	−287
			270	−263
C_2H_3N	Ethanenitrile		330	−3468
			340	−2971
			350	−2563
		$a(1) = -5840$	360	−2233
		$a(2) = -29175$	370	−1970
		$a(3) = -47611$	380	−1765
			390	−1610
			400	−1499
			410	−1425
C_2H_4	Ethene		240	−218
			270	−172
			300	−139
		$a(1) = -140$	330	−113
		$a(2) = -296$	360	−92
		$a(3) = -101$	390	−76
			420	−63
			450	−52
$C_2H_4Cl_2$	1,2-Dichloroethane		370	−812
			390	−716
			410	−635
		$a(1) = -1362$	430	−566
		$a(2) = -3240$	450	−508
		$a(3) = -2100$	470	−458
			490	−416
			510	−379
			530	−347
			550	−319
			570	−295
C_2H_4O	Ethanal		290	−1352
			320	−927
			350	−654
		$a(1) = -1217$	380	−482
		$a(2) = -4647$	410	−375
		$a(3) = -5725$	440	−314
			470	−283
$C_2H_4O_2$	Methyl methanoate		320	−821
			330	−744
			340	−677
		$a(1) = -1035$	350	−620
		$a(2) = -3425$	360	−571
		$a(3) = -4203$	370	−528
			380	−492
			390	−461
			400	−435
C_2H_5Cl	Chloroethane		320	−634
			360	−450
			400	−330
		$a(1) = -777$	440	−249
		$a(2) = -2205$	480	−195
		$a(3) = -1764$	520	−157

Mol. form.	Name		T/K	$B/cm^3\ mol^{-1}$
			560	−131
			600	−114
C_2H_6	Ethane		200	−409
			220	−337
			240	−284
		$a(1) = -184$	260	−242
		$a(2) = -376$	280	−209
		$a(3) = -143$	300	−181
		$a(4) = -54$	320	−159
			340	−140
			360	−123
			380	−109
			400	−96
			500	−52
			600	−24
C_2H_6O	Ethanol		320	−2710
			330	−2135
			340	−1676
		$a(1) = -4475$	350	−1317
		$a(2) = -29719$	360	−1043
		$a(3) = -56716$	370	−843
			380	−705
			390	−622
C_2H_6O	Dimethyl ether		275	−536
			280	−517
			285	−499
		$a(1) = -455$	290	−482
		$a(2) = -965$	295	−465
			300	−449
			305	−433
			310	−418
C_2H_7N	Dimethylamine		310	−606
			320	−563
			330	−523
		$a(1) = -662$	340	−487
		$a(2) = -1504$	350	−454
		$a(3) = -667$	360	−423
			370	−395
			380	−369
			390	−345
			400	−322
C_2H_7N	Ethylamine		300	−773
			310	−710
			320	−654
		$a(1) = -785$	330	−604
		$a(2) = -2012$	340	−558
		$a(3) = -1397$	350	−517
			360	−480
			370	−447
			380	−416
			390	−389
			400	−363
C_3H_6	Cyclopropane		300	−383

Mol. form.	Name		T/K	$B/cm^3\ mol^{-1}$
			310	−356
			320	−332
	a(1) = −388		330	−310
	a(2) = −861		340	−290
	a(3) = −538		350	−272
			360	−256
			370	−241
			380	−227
			390	−215
			400	−204
C_3H_6	Propene		280	−395
			300	−342
			320	−299
	a(1) = −347		340	−262
	a(2) = −727		360	−232
	a(3) = −325		380	−205
			400	−183
			420	−163
			440	−146
			460	−131
			480	−118
			500	−106
C_3H_6O	2-Propanone		300	−1996
			320	−1522
			340	−1198
	a(1) = −2051		360	−971
	a(2) = −8903		380	−806
	a(3) = −18056		400	−683
	a(4) = −16448		420	−586
			440	−506
			460	−437
			480	−375
C_3H_6O	Ethyl methanoate		330	−1003
			340	−916
			350	−839
	a(1) = −1371		360	−771
	a(2) = −4231		370	−712
	a(3) = −4312		380	−660
			390	−614
C_3H_6O	Methyl ethanoate		320	−1320
			330	−1186
			340	−1074
	a(1) = −1709		350	−980
	a(2) = −6348		360	−903
	a(3) = −9650		370	−840
			380	−789
			390	−749
C_3H_7Cl	1-Chloropropane		310	−1001
			340	−772
			370	−614
	a(1) = −1121		400	−501
	a(2) = −3271		430	−417
	a(3) = −3786		460	−352
	a(4) = −1974		490	−302

Mol. form.	Name		T/K	$B/cm^3\ mol^{-1}$
			520	−261
			550	−227
			580	−198
C_3H_8	Propane		240	−641
			260	−527
			280	−444
	$a(1) = -386$		300	−381
	$a(2) = -844$		320	−331
	$a(3) = -720$		340	−292
	$a(4) = -574$		360	−259
			380	−232
			400	−208
			440	−169
			480	−138
			520	−112
			560	−90
C_3H_8O	1-Propanol		380	−873
			385	−826
			390	−783
	$a(1) = -2690$		395	−744
	$a(2) = -12040$		400	−709
	$a(3) = -16738$		405	−679
			410	−651
			415	−627
			420	−606
C_3H_8O	2-Propanol		380	−821
			385	−766
			390	−717
	$a(1) = -3165$		395	−674
	$a(2) = -16092$		400	−636
	$a(3) = -24197$		405	−604
			410	−576
			415	−552
			420	−533
C_3H_9N	Trimethylamine		310	−675
			320	−628
			330	−585
	$a(1) = -737$		340	−547
	$a(2) = -1669$		350	−512
	$a(3) = -986$		360	−480
			370	−450
C_4H_8	1-Butene		300	−624
			320	−539
			340	−470
	$a(1) = -633$		360	−413
	$a(2) = -1442$		380	−366
	$a(3) = -932$		400	−327
			420	−294
C_4H_8O	2-Butanone		310	−2056
			320	−1878
			330	−1712
	$a(1) = -2282$		340	−1555

Mol. form.	Name		T/K	B/cm^3 mol^{-1}
	a(2) = −5907		350	−1407
			360	−1267
			370	−1135
C$_4$H$_8$O$_2$	Propyl methanoate		330	−1496
			340	−1354
			350	−1231
	a(1) = −2118		360	−1126
	a(2) = −7299		370	−1035
	a(3) = −8851		380	−957
			390	−890
			400	−834
C$_4$H$_8$O$_2$	Ethyl ethanoate		330	−1543
			340	−1385
			350	−1254
	a(1) = −2272		360	−1144
	a(2) = −8818		370	−1055
	a(3) = −13130		380	−982
			390	−923
			400	−878
C$_4$H$_8$O$_2$	Methyl propanoate		330	−1588
			340	−1444
			350	−1319
	a(1) = −2216		360	−1211
	a(2) = −7339		370	−1117
	a(3) = −8658		380	−1037
			390	−968
			400	−908
C$_4$H$_9$Cl	1-Chlorobutane		330	1224
			370	−898
			410	−691
	a(1) = −1643		450	−551
	a(2) = −4897		490	−449
	a(3) = −6178		530	−371
	a(4) = −3718		570	−309
C$_4$H$_{10}$	Butane		250	−1170
			280	−863
			310	−668
	a(1) = −735		340	−536
	a(2) = −1835		370	−442
	a(3) = −1922		400	−371
	a(4) = −1330		430	−315
			460	−270
			490	−232
			520	−199
			550	−171
C$_4$H$_{10}$	2-Methylpropane		270	−900
			300	−697
			330	−553
	a(1) = −707		360	−450
	a(2) = −1719		390	−374
	a(3) = −1282		420	−317
			450	−273

Mol. form.	Name	T/K	$B/cm^3\ mol^{-1}$
		480	−240
		510	−215
$C_4H_{10}O$	1-Butanol	350	−1693
		360	−1544
		370	−1402
	a(1) = −2629	380	−1268
	a(2) = −6315	390	−1141
		400	−1021
		420	−796
		440	−593
$C_4H_{10}O$	2-Methyl-1-propanol	390	−1076
		400	−979
		410	−887
	a(1) = −2269	420	−800
	a(2) = −5065	430	−716
		440	−636
$C_4H_{10}O$	2-Butanol	380	−1110
		390	−1005
		400	−906
	a(1) = −2232	410	−811
	a(2) = −5209	420	−721
$C_4H_{10}O$	2-Methyl-2-propanol	380	−924
		390	−827
		400	−736
	a(1) = −1952	410	−649
	a(2) = −4775	420	−567
$C_4H_{10}O$	Diethyl ether	280	−1550
		300	−1199
		320	−954
	a(1) = −1226	340	−776
	a(2) = −4458	360	−638
	a(3) = −7746	380	−525
	a(4) = −10005	400	−428
		420	−340
$C_4H_{11}N$	Diethylamine	320	−1228
		330	−1134
		340	−1056
	a(1) = −1522	350	−988
	a(2) = −5204	360	−926
	a(3) = −15047	370	−868
	a(4) = −28835	380	−812
		390	−755
		400	−697
C_5H_5N	Pyridine	350	−1257
		360	−1176
		370	−1099
	a(1) = −1765	380	−1026
	a(2) = −3431	390	−957
		400	−892
		420	−770
		440	−659

Mol. form.	Name		T/K	B/cm^3 mol^{-1}
C_5H_{10}	Cyclopentane		300	−1049
			305	−1015
			310	−981
		$a(1) = -1062$	315	−949
		$a(2) = -2116$	320	−918
C_5H_{10}	1-Pentene		310	−966
			320	−898
			330	−836
		$a(1) = -1055$	340	−780
		$a(2) = -2377$	350	−729
		$a(3) = -1189$	360	−681
			370	−638
			380	−598
			390	−561
			400	−527
			410	−495
$C_5H_{10}O$	2-Pentanone		330	−2850
			340	−2420
			350	−2076
		$a(1) = -4962$	360	−1804
		$a(2) = -26372$	370	−1595
		$a(3) = -46537$	380	−1440
			390	−1332
C_5H_{12}	Pentane		300	−1234
			310	−1130
			320	−1038
		$a(1) = -1254$	330	−957
		$a(2) = -3345$	340	−884
		$a(3) = -2726$	350	−818
			400	−579
			450	−436
			500	−348
			550	−294
C_5H_{12}	2-Methylbutane		280	−1263
			290	−1166
			300	−1079
		$a(1) = -1095$	310	−1001
		$a(2) = -2503$	320	−931
		$a(3) = -1534$	330	−867
			340	−810
			350	−757
			400	−557
			450	−424
C_5H_{12}	2,2-Dimethylpropane		300	−916
			310	−843
			320	−780
		$a(1) = -931$	330	−724
		$a(2) = -2387$	340	−674
		$a(3) = -2641$	350	−629
		$a(4) = -1810$	360	−590
			370	−554
			380	−521
			390	−492

Mol. form.	Name		T/K	B/cm^3 mol^{-1}
			400	−464
			450	−357
			500	−279
			550	−218
C_6H_6	Benzene		290	−1588
			300	−1454
			310	−1335
		$a(1) = -1477$	320	−1231
		$a(2) = -3851$	330	−1139
		$a(3) = -3683$	340	−1056
		$a(4) = -1423$	350	−983
			400	−712
			450	−542
			500	−429
			550	−349
			600	−291
C_6H_7N	2-Methylpyridine		360	−1656
			370	−1523
			380	−1404
		$a(1) = -2940$	390	−1297
		$a(2) = -8813$	400	−1202
		$a(3) = -7809$	410	−1117
			420	−1040
			430	−972
C_6H_7N	3-Methylpyridine		380	−1819
			390	−1612
			400	−1448
		$a(1) = -6304$	410	−1322
		$a(2) = -30415$	420	−1230
		$a(3) = -44549$	430	−1166
C_6H_7N	4-Methylpyridine		380	−1787
			390	−1578
			400	−1417
		$a(1) = -6553$	410	−1297
		$a(2) = -32873$	420	−1214
		$a(3) = -49874$	430	−1163
C_6H_{12}	Cyclohexane		300	−1698
			320	−1391
			340	−1170
		$a(1) = -1733$	360	−1007
		$a(2) = -5618$	380	−883
		$a(3) = -9486$	400	−786
		$a(4) = -7936$	420	−707
			440	−641
			460	−584
			480	−534
			500	−488
			520	−446
			540	−406
			560	−368
C_6H_{12}	Methylcyclopentane		305	−1447
			315	−1357

Mol. form.	Name		T/K	$B/\text{cm}^3\ \text{mol}^{-1}$
	$a(1) = -1512$		325	−1272
	$a(2) = -2910$		335	−1192
			345	−1117
C_6H_{14}	Hexane		300	−1920
			310	−1724
			320	−1561
	$a(1) = -1961$		330	−1424
	$a(2) = -6691$		340	−1309
	$a(3) = -13167$		350	−1209
	$a(4) = -15273$		360	−1123
			370	−1046
			380	−978
			390	−916
			400	−859
			410	−806
			430	−707
			450	−616
$C_6H_{15}N$	Triethylamine		330	−1562
			340	−1444
			350	−1340
	$a(1) = -2061$		360	−1249
	$a(2) = -5735$		370	−1169
	$a(3) = -5899$		380	−1099
			390	−1037
			400	−983
C_7H_8	Toluene		350	−1641
			360	−1511
			370	−1394
	$a(1) = -2620$		380	−1289
	$a(2) = -7548$		390	−1195
	$a(3) = -6349$		400	−1110
			410	−1034
			420	−965
			430	−903
C_7H_{14}	1-Heptene		340	−1781
			350	−1651
			360	−1532
	$a(1) = -2491$		370	−1424
	$a(2) = -6230$		380	−1324
	$a(3) = -3780$		390	−1233
			400	−1150
			410	−1073
C_7H_{16}	Heptane		300	−2782
			320	−2297
			340	−1928
	$a(1) = -2834$		360	−1641
	$a(2) = -8523$		380	−1415
	$a(3) = -10068$		400	−1233
	$a(4) = -5051$		420	−1085
			440	−963
			460	−862
			480	−775
			500	−702

Mol. form.	Name		T/K	B/cm^3 mol^{-1}
			540	−583
			580	−490
			620	−416
			660	−355
			700	−304
C_8H_{10}	1,2-Dimethylbenzene		380	−2046
			390	−1848
			400	−1681
	a(1) = −5632		410	−1543
	a(2) = −22873		420	−1428
	a(3) = −28900		430	−1335
			440	−1261
C_8H_{10}	1,3-Dimethylbenzene		380	−2082
			390	−1865
			400	−1679
	a(1) = −5808		410	−1521
	a(2) = −23244		420	−1388
	a(3) = −27607		430	−1276
			440	−1184
C_8H_{10}	1,4-Dimethylbenzene		380	−2043
			390	−1851
			400	−1680
	a(1) = −4921		410	−1529
	a(2) = −16843		420	−1395
	a(3) = −16159		430	−1276
			440	−1171
C_8H_{16}	1-Octene		360	−2147
			370	−2000
			380	−1861
	a(1) = −3273		390	−1729
	a(2) = −6557		400	−1604
			410	−1485
C_8H_{18}	Octane		300	−4042
			350	−2511
			400	−1704
	a(1) = −4123		450	−1234
	a(2) = −13120		500	−936
	a(3) = −16408		550	−732
	a(4) = −8580		600	−583
			650	−468
			700	−375

VAN DER WAALS' CONSTANTS FOR GASES

The van der Waals' equation of state for a real gas is

$$(P + n^2a/V^2)(V - nb) = nRT$$

where P is the pressure, V the volume, T the temperature, n the amount of substance (in moles), and R the gas constant. The van der Waals constants a and b are characteristic of the substance and are independent of temperature. They are related to the critical temperature and pressure, T_c and P_c, by

$$a = 27R^2T_c^2/64P_c \qquad b = RT_c/8P_c$$

This table gives values of a and b for over 200 common gases. Most of the values have been calculated from the critical constants.

To convert the van der Waals constants to SI units, note that

$$1 \text{ bar L}^2/\text{mol}^2 = 0.1 \text{ Pa m}^6/\text{mol}^2$$
$$1 \text{ L/mol} = 0.001 \text{ m}^3/\text{mol}$$

Substances are listed by molecular formula in the modified Hill order, with substances not containing carbon preceding those that do contain carbon.

REFERENCE

Reid, R.C, Prausnitz, J. M., and Poling, B. E., *The Properties of Gases and Liquids*, Fourth Edition, McGraw-Hill, New York, 1987.

Molecular formula	Name	a bar L^2/mol^2	b L/mol
$AlCl_3$	Aluminum trichloride	42.63	0.2450
Ar	Argon	1.355	0.03201
BCl_3	Boron trichloride	15.60	0.1222
BF_3	Boron trifluoride	3.98	0.05443
B_2H_6	Diborane	6.048	0.07437
BrH	Hydrogen bromide	4.500	0.04415
Br_2	Bromine	9.75	0.0591
$ClFO_3$	Perchloryl fluoride	7.371	0.07130
ClF_5	Chlorine pentafluoride	9.58	0.08204
ClH	Hydrogen chloride	3.700	0.04061
ClH_4N	Ammonium chloride	2.380	0.00734
ClH_4P	Phosphonium chloride	4.111	0.04545
Cl_2	Chlorine	6.343	0.05422
Cl_3FSi	Trichlorofluorosilane	15.67	0.1273
Cl_4Ge	Germanium tetrachloride	23.12	0.1489
Cl_4Si	Silicon tetrachloride	20.96	0.1470
Cl_4Sn	Tin chloride ($SnCl_4$)	27.25	0.1641
Cl_4Ti	Titanium chloride ($TiCl_4$)	25.47	0.1423
FH	Hydrogen fluoride	9.565	0.0739
F_2	Fluorine	1.171	0.02896
F_2Xe	Xenon difluoride	12.46	0.07037
F_3N	Nitrogen trifluoride	3.58	0.05453
F_3P	Phosphorus trifluoride	4.954	0.06510
F_4N_2	Tetrafluorohydrazine	7.426	0.08564
F_4Si	Silicon tetrafluoride	5.259	0.07236
F_4Xe	Xenon tetrafluoride	15.52	0.09035
F_6S	Sulfur fluoride (SF_6)	7.857	0.08786
F_6U	Uranium fluoride (UF_6)	16.01	0.1128
F_6W	Tungsten fluoride (WF_6)	13.25	0.1063
GeH_4	Germane	5.743	0.06555
HI	Hydrogen iodide	6.309	0.05303
H_2	Hydrogen	0.2453	0.02651
H_2O	Water	5.537	0.03049
H_2S	Hydrogen sulfide	4.544	0.04339
H_2Se	Hydrogen selenide	5.523	0.0479

Molecular formula	Name	a bar L^2/mol^2	b L/mol
H_3N	Ammonia	4.225	0.03713
H_3P	Phosphine	4.696	0.05157
H_4N_2	Hydrazine	8.46	0.0462
H_4Si	Silane	4.38	0.0579
He	Helium	0.0346	0.0238
Hg	Mercury	5.193	0.01057
Kr	Krypton	2.325	0.0396
NO	Nitric oxide	1.46	0.0289
NO_2	Nitrogen dioxide	5.36	0.0443
N_2	Nitrogen	1.370	0.0387
N_2O	Nitrous oxide	3.852	0.04435
Ne	Neon	0.208	0.01672
O_2	Oxygen	1.382	0.03186
O_2S	Sulfur dioxide	6.865	0.05679
O_3	Ozone	3.570	0.0487
P	Phosphorus	53.6	0.157
Rn	Radon	6.601	0.06239
S	Sulfur	24.3	0.0660
Se	Selenium	33.4	0.0675
Xe	Xenon	4.192	0.05156
$CClF_3$	Chlorotrifluoromethane	6.873	0.08110
CCl_3F	Trichlorofluoromethane	14.68	0.1111
CCl_4	Tetrachloromethane	20.01	0.1281
CF_4	Tetrafluoromethane	4.040	0.06325
CO	Carbon monoxide	1.472	0.03948
COS	Carbon oxysulfide	6.975	0.06628
CO_2	Carbon dioxide	3.658	0.04286
CS_2	Carbon disulfide	11.25	0.07262
$CHCl_3$	Trichloromethane	15.34	0.1019
CHF_3	Trifluoromethane	5.378	0.06403
CHN	Hydrogen cyanide	11.29	0.08806
CH_2Cl_2	Dichloromethane	12.44	0.08689
CH_2F_2	Difluoromethane	6.184	0.06268
CH_3Cl	Chloromethane	7.566	0.06477
CH_3F	Fluoromethane	5.009	0.05617
CH_3NO_2	Nitromethane	17.18	0.1041
CH_4	Methane	2.300	0.04301
CH_4O	Methanol	9.472	0.06584
CH_4S	Methanethiol	8.911	0.06756
CH_5N	Methylamine	7.106	0.05879
$C_2Cl_3F_3$	1,1,2-Trichlorotrifluoroethane	20.25	0.1481
C_2F_4	Tetrafluoroethylene	6.954	0.08085
C_2N_2	Cyanogen	7.803	0.06952
C_2H_2	Acetylene	4.516	0.05220
$C_2H_2F_2$	1,1-Difluoroethylene	6.000	0.07058
$C_2H_3Cl_3$	1,1,1-Trichloroethane	20.15	0.1317
C_2H_3F	Fluoroethylene	5.984	0.06504
$C_2H_3F_3$	1,1,1-Trifluoroethane	9.302	0.09572
C_2H_3N	Acetonitrile	17.89	0.1169
C_2H_4	Ethylene	4.612	0.05821
$C_2H_4Cl_2$	1,1-Dichloroethane	15.73	0.1072
$C_2H_4Cl_2$	1,2-Dichloroethane	17.0	0.108
C_2H_4O	Ethylene oxide	8.922	0.06779
$C_2H_4O_2$	Acetic acid	17.71	0.1065
$C_2H_4O_2$	Methyl formate	11.54	0.08442
C_2H_5Br	Bromoethane	11.89	0.08406
C_2H_5Cl	Chloroethane	11.7	0.090

Molecular formula	Name	a bar L^2/mol^2	b L/mol
C_2H_5F	Fluoroethane	8.170	0.07758
C_2H_6	Ethane	5.570	0.06499
C_2H_6O	Dimethyl ether	8.690	0.07742
C_2H_6O	Ethanol	12.56	0.08710
C_2H_6S	Dimethyl sulfide	13.34	0.09453
C_2H_6S	Ethanethiol	13.23	0.09447
C_2H_7N	Dimethylamine	10.44	0.08510
C_2H_7N	Ethylamine	10.79	0.08433
C_3F_8	Perfluoropropane	12.96	0.1338
C_3H_5N	Propanenitrile	21.57	0.1369
C_3H_6	Propene	8.438	0.08242
C_3H_6	Cyclopropane	8.293	0.07420
C_3H_6O	Acetone	16.02	0.1124
C_3H_6O	Propanal	14.08	0.09947
$C_3H_6O_2$	Ethyl formate	15.91	0.1115
$C_3H_6O_2$	Methyl acetate	15.75	0.1108
$C_3H_6O_2$	Propanoic acid	23.49	0.1386
C_3H_7Cl	1-Chloropropane	16.11	0.1141
C_3H_8	Propane	9.385	0.09044
C_3H_8O	1-Propanol	16.26	0.1080
C_3H_8O	2-Propanol	15.82	0.1109
C_3H_8O	Ethyl methyl ether	12.70	0.1034
C_3H_8S	Ethyl methyl sulfide	19.45	0.1300
C_3H_9N	Propylamine	15.26	0.1094
C_3H_9N	Trimethylamine	13.37	0.1101
C_4H_4O	Furan	12.74	0.0926
C_4H_4S	Thiophene	17.21	0.1058
C_4H_5N	Pyrrole	18.82	0.1049
C_4H_6	1,3-Butadiene	12.17	0.1020
$C_4H_6O_3$	Acetic anhydride	26.8	0.157
C_4H_7N	Butanenitrile	25.76	0.1568
C_4H_8	1-Butene	12.76	0.1084
C_4H_8	Cyclobutane	12.39	0.0960
C_4H_8O	2-Butanone	19.97	0.1326
C_4H_8O	Tetrahydrofuran	16.39	0.1082
$C_4H_8O_2$	1,4-Dioxane	19.29	0.1171
$C_4H_8O_2$	Ethyl acetate	20.57	0.1401
$C_4H_8O_2$	Methyl propanoate	20.51	0.1377
$C_4H_8O_2$	Propyl formate	20.79	0.1377
$C_4H_8O_2$	Butanoic acid	28.18	0.1609
C_4H_9N	Pyrrolidine	16.84	0.1056
C_4H_{10}	Butane	13.93	0.1168
C_4H_{10}	Isobutane	13.36	0.1168
$C_4H_{10}O$	1-Butanol	20.90	0.1323
$C_4H_{10}O$	2-Methyl-2-propanol	18.81	0.1324
$C_4H_{10}O$	2-Methyl-1-propanol	20.35	0.1324
$C_4H_{10}O$	Diethyl ether	17.46	0.1333
$C_4H_{10}S$	Diethyl sulfide	22.85	0.1462
$C_4H_{11}N$	Butylamine	19.41	0.1301
$C_4H_{11}N$	Diethylamine	19.40	0.1383
$C_4H_{12}Si$	Tetramethylsilane	20.81	0.1653
$C_5H_4O_2$	Furfural	22.23	0.1182
C_5H_5N	Pyridine	19.77	0.1137
C_5H_8	Cyclopentene	15.61	0.1097
C_5H_{10}	1-Pentene	17.86	0.1370
C_5H_{10}	2-Methyl-1-butene	16.9	0.129
C_5H_{10}	2-Methyl-2-butene	17.26	0.1279

Molecular formula	Name	a bar L^2/mol^2	b L/mol
C_5H_{10}	Cyclopentane	16.94	0.1180
$C_5H_{10}O$	Tetrahydropyran	20.02	0.1247
$C_5H_{10}O_2$	Isobutyl formate	22.82	0.1476
$C_5H_{10}O_2$	Propyl acetate	26.23	0.1700
$C_5H_{10}O_2$	Ethyl propanoate	25.86	0.1688
$C_5H_{10}O_2$	Methyl butanoate	25.83	0.1661
$C_5H_{10}O_2$	Methyl isobutanoate	24.87	0.1639
$C_5H_{11}N$	Piperidine	20.84	0.1250
C_5H_{12}	Pentane	19.13	0.1451
C_5H_{12}	Isopentane	18.29	0.1415
C_5H_{12}	Neopentane	17.17	0.1410
$C_5H_{12}O$	1-Pentanol	25.81	0.1564
C_6H_5Br	Bromobenzene	28.96	0.1541
C_6H_5Cl	Chlorobenzene	25.80	0.1454
C_6H_5F	Fluorobenzene	20.10	0.1279
C_6H_5I	Iodobenzene	33.54	0.1658
C_6H_6	Benzene	18.82	0.1193
C_6H_6O	Phenol	22.93	0.1177
C_6H_7N	Aniline	29.14	0.1486
$C_6H_{10}O$	Cyclohexanone	31.1	0.170
$C_6H_{11}N$	Hexanenitrile	35.50	0.1996
C_6H_{12}	Cyclohexane	21.95	0.1413
$C_6H_{12}O$	Cyclohexanol	28.93	0.1586
$C_6H_{12}O_2$	Pentyl formate	27.97	0.1730
$C_6H_{12}O_2$	Isobutyl acetate	29.05	0.1845
$C_6H_{12}O_2$	Ethyl butanoate	30.53	0.1922
$C_6H_{12}O_2$	Ethyl 2-methylpropanoate	29.05	0.1872
$C_6H_{12}O_2$	Methyl pentanoate	29.39	0.1847
C_6H_{14}	Hexane	24.97	0.1753
C_6H_{14}	2,3-Dimethylbutane	23.29	0.1660
$C_6H_{14}O$	1-Hexanol	31.35	0.1829
$C_6H_{15}N$	Triethylamine	27.59	0.1836
$C_6H_{15}N$	Dipropylamine	24.82	0.1591
C_7H_5N	Benzonitrile	33.89	0.1727
C_7H_6O	Benzaldehyde	30.30	0.1553
C_7H_8	Toluene	24.89	0.1499
C_7H_8O	o-Cresol	28.33	0.1447
C_7H_8O	m-Cresol	31.86	0.1609
C_7H_8O	p-Cresol	28.11	0.1422
C_7H_8O	Benzyl alcohol	34.7	0.173
C_7H_8O	Anisole	28.60	0.1579
C_7H_{16}	Heptane	30.89	0.2038
$C_7H_{16}O$	1-Heptanol	37.22	0.2097
C_8H_{10}	Ethylbenzene	30.86	0.1782
C_8H_{10}	o-Xylene	31.06	0.1756
C_8H_{10}	m-Xylene	31.41	0.1814
C_8H_{10}	p-Xylene	31.54	0.1824
$C_8H_{10}O$	Phenetole	35.70	0.1966
$C_8H_{11}N$	N,N-Dimethylaniline	37.92	0.1967
C_8H_{18}	Octane	37.86	0.2372
C_8H_{18}	2,5-Dimethylhexane	35.49	0.2299
$C_8H_{18}O$	1-Octanol	43.42	0.2371
C_9H_7N	Quinoline	36.70	0.1672
C_9H_{12}	Cumene	36.20	0.2044
C_9H_{12}	Propylbenzene	37.14	0.2073
C_9H_{12}	1,2,4-Trimethylbenzene	38.03	0.2088
C_9H_{12}	Mesitylene	37.87	0.2118

Molecular formula	Name	a bar L^2/mol^2	b L/mol
C_9H_{20}	Nonane	45.11	0.2702
$C_9H_{20}O$	1-Nonanol	50.00	0.2654
$C_{10}H_8$	Naphthalene	40.32	0.1920
$C_{10}H_{14}$	Butylbenzene	44.07	0.2378
$C_{10}H_{14}$	Isobutylbenzene	40.40	0.2215
$C_{10}H_{14}$	o-Cymene	42.7	0.234
$C_{10}H_{14}$	p-Cymene	45.27	0.2478
$C_{10}H_{14}$	p-Diethylbenzene	45.03	0.2439
$C_{10}H_{14}$	1,2,4,5-Tetramethylbenzene	45.8	0.24
$C_{10}H_{22}$	Decane	52.88	0.3051
$C_{10}H_{22}O$	1-Decanol	57.45	0.2971
$C_{11}H_{24}$	Undecane	60.88	0.3396
$C_{12}H_{10}$	Biphenyl	47.16	0.2130
$C_{12}H_{26}$	Dodecane	69.14	0.3741
$C_{12}H_{26}O$	1-Dodecanol	72.69	0.3598
$C_{13}H_{12}$	Diphenylmethane	60.46	0.2798
$C_{13}H_{28}$	Tridecane	77.94	0.4109
$C_{13}H_{28}O$	1-Tridecanol	81.20	0.3942
$C_{14}H_{30}O$	1-Tetradecanol	89.91	0.4289
$C_{15}H_{32}$	Pentadecane	96.50	0.4857

MEAN FREE PATH AND RELATED PROPERTIES OF GASES

In the simplest version of the kinetic theory of gases, molecules are treated as hard spheres of diameter d which make binary collisions only. In this approximation the mean distance traveled by a molecule between successive collisions, the mean free path l, is related to the collision diameter by:

$$l = \frac{kT}{\pi\sqrt{2}Pd^2}$$

where P is the pressure, T the absolute temperature, and k the Boltzmann constant. At standard conditions ($P = 100\,000$ Pa and $T = 298.15$ K) this relation becomes:

$$l = \frac{9.27 \cdot 10^{-27}}{d^2}$$

where l and d are in meters.

Using the same model and the same standard pressure, the collision diameter can be calculated from the viscosity η by the kinetic theory relation:

$$\eta = \frac{2.67 \cdot 10^{-20}(MT)^{1/2}}{d^2}$$

where η is in units of μPa s and M is the molar mass in g/mol. Kinetic theory also gives a relation for the mean velocity $\bar{v}$ of molecules of mass m:

$$\bar{v} = \left(\frac{8kT}{\pi m}\right)^{1/2} = 145.5(T/M)^{1/2} \text{ m/s}$$

Finally, the mean time τ between collisions can be calculated from the relation $\tau = l/\bar{v}$.

The table below gives values of l, $\bar{v}$, and τ for some common gases at 25°C and atmospheric pressure, as well as the value of d, all calculated from measured gas viscosities (see References 2 and 3 and the table "Viscosity of Gases" in this section). It is seen from the above equations that the mean free path varies directly with T and inversely with P, while the mean velocity varies as the square root of T and, in this approximation, is independent of P.

A more accurate model, in which molecular interactions are described by a Lennard-Jones potential, gives mean free path values about 5% lower than this table (see Reference 4).

REFERENCES

1. Reid, R.C., Prausnitz, J.M., and Poling, B.E., *The Properties of Gases and Liquids, Fourth Edition*, McGraw-Hill, New York, 1987.
2. Lide, D.R., and Kehiaian, H.V., *CRC Handbook of Thermophysical and Thermochemical Data*, CRC Press, Boca Raton, FL, 1994.
3. Vargaftik, N.B., *Tables of Thermophysical Properties of Liquids and Gases, Second Edition*, John Wiley, New York, 1975.
4. Kaye, G.W.C., and Laby, T.H., *Tables of Physical and Chemical Constants, 15th Edition*, Longman, London, 1986.

Gas	d	l	$\bar{v}$	τ
Air	$3.66 \cdot 10^{-10}$ m	$6.91 \cdot 10^{-8}$ m	467 m/s	148 ps
Ar	3.58	7.22	397	182
CO_2	4.53	4.51	379	119
H_2	2.71	12.6	1769	71
He	2.15	20.0	1256	159
Kr	4.08	5.58	274	203
N_2	3.70	6.76	475	142
NH_3	4.32	4.97	609	82
Ne	2.54	14.3	559	256
O_2	3.55	7.36	444	166
Xe	4.78	4.05	219	185

INFLUENCE OF PRESSURE ON FREEZING POINTS

This table illustrates the variation of the freezing point of representative types of liquids with pressure. Substances are listed in alphabetical order. Note that 1 MPa = 0.01 kbar = 9.87 atm.

REFERENCES

1. Isaacs, N.S., *Liquid Phase High Pressure Chemistry*, John Wiley, New York, 1981.
2. Merrill, L., *J. Phys. Chem. Ref. Data*, 6, 1205, 1977; 11, 1005, 1982.

Substance	Molecular formula	Freezing point in °C at:		
		0.1 MPa	100 MPa	1000 MPa
Acetic acid	$C_2H_4O_2$	16.6	37	
Acetophenone	C_8H_8O	20.0	41.2	
Aniline	C_6H_7N	−6.0	13.5	140
Benzene	C_6H_6	5.5	33.4	
Benzonitrile	C_7H_5N	−12.8	7.6	
Benzyl alcohol	C_7H_8O	−15.2	0.2	
Bromobenzene	C_6H_5Br	−30.6	−12	108
Bromoethane	C_2H_5Br	−118.6	−108	
1-Bromonaphthalene	$C_{10}H_7Br$	−1.8	6.1	
1-Bromopropane	C_3H_7Br	−110	−98	
p-Bromotoluene	C_7H_7Br	28.0	56.7	
Butanoic acid	$C_4H_8O_2$	−5.7	13.8	
1-Butanol	$C_4H_{10}O$	−89.8	−77.2	
Carbon disulfide	CS_2	−111.5	−98	
Chlorobenzene	C_6H_5Cl	−45.2	−28	84
p-Chlorotoluene	C_7H_7Cl	6.9	33.1	
o-Cresol	C_7H_8O	29.8	47.7	
m-Cresol	C_7H_8O	11.8	25.6	
p-Cresol	C_7H_8O	35.8	56.2	
Cyclohexane	C_6H_{12}	6.6	32.5	
Cyclohexanol	$C_6H_{12}O$	25.5	62.3	
1,2-Dibromoethane	$C_2H_4Br_2$	9.9	34.0	
p-Dichlorobenzene	$C_6H_4Cl_2$	52.7	79.1	
Dichloromethane	CH_2Cl_2	−95.1	−83	
N,N-Dimethylaniline	$C_8H_{11}N$	2.5	26.3	
1,4-Dioxane	$C_4H_8O_2$	11	23	
Ethanol	C_2H_6O	−114.1	−108	
Formamide	CH_3NO	−15.5	10.8	
Formic acid	CH_2O_2	8.3	20.6	
Furan	C_4H_4O	−85.6	−73	
Hexamethyldisiloxane	$C_6H_{18}OSi_2$	−66	−37	
Menthol	$C_{10}H_{20}O$	42	60	
Methyl benzoate	$C_8H_8O_2$	−15	31.8	
2-Methyl-2-butanol	$C_5H_{12}O$	−8.8	13.4	
2-Methyl-2-propanol	$C_4H_{10}O$	25.4	58.1	
Naphthalene	$C_{10}H_8$	78.2	115.7	
Nitrobenzene	$C_6H_5NO_2$	5.7	13.5	
m-Nitrotoluene	$C_7H_7NO_2$	15.5	40.6	
Pentachloroethane	C_2HCl_5	−29.0	−6.3	
Potassium	K	63.7	78	170
Potassium chloride	ClK	771		945
Propanoic acid	$C_3H_6O_2$	−20.7	−1.2	
Silver chloride	AgCl	455		545
Sodium	Na	97.8	106	167
Sodium chloride	ClNa	800.7		997
Sodium fluoride	FNa	996		1115
Tetrachloromethane	CCl_4	−23.0	14.2	
Tribromomethane	$CHBr_3$	8.1	31.5	
Trichloromethane	$CHCl_3$	−63.6	−45.2	
Water	H_2O	0.0	−9.0	
o-Xylene	C_8H_{10}	−25.2	−3.5	
m-Xylene	C_8H_{10}	−47.8	−25.2	
p-Xylene	C_8H_{10}	13.2	46.0	

CRITICAL CONSTANTS

Douglas Ambrose

This table gives the liquid-gas critical constants and normal boiling points of about 650 inorganic and organic substances. The properties tabulated are:

T_b: Normal boiling point in K at a pressure of 101.325 kPa (1 atmosphere); an "s" following the value indicates a sublimation point (temperature at which the solid is in equilibrium with the gas at 101.325 kPa)
T_c: Critical temperature in K
P_c: Critical pressure in MPa
V_c: Critical molar volume in cm^3/mol

The number of digits given for T_b, T_c, and P_c indicates the estimated accuracy of these quantities; however, values of T_c greater than 750 K may be in error by 10 K or more. Although V_c values are given to three figures, they cannot be assumed accurate to better than a few percent.

The values of the critical constants in this table are taken from References 1 to 5 and from the current literature. Compounds are listed by molecular formula in the modified Hill order, with compounds not containing carbon preceding those that do contain carbon.

* An asterisk following an entry indicates the value was obtained by extrapolation.

REFERENCES

1. Stevenson, R.M., and Malanowski, S., *Handbook of the Thermodynamics of Organic Compounds*, Elsevier, New York, 1987, pp. 527—552.
2. Ambrose, D., and Tsonopoulos, C., *J. Chem. Eng. Data*, 40, 531, 1995.
3. Tsonopoulos, C., and Ambrose, D., *J. Chem. Eng. Data*, 40, 547, 1995.
4. Gude, M., and Teja, A.S., *J. Chem. Eng. Data*, 40, 1025, 1995.
5. Das, A., Frenkel, M., Gadalla, N.A.M., Kudchadker, S., Marsh, K.N., Rodgers, A.S., and Wilhoit, R.C., *J. Phys. Chem. Ref. Data*, 22, 659, 1993.
6. Vargaftik, N.B., *Tables of Thermophysical Properties of Liquids and Gases, Second Edition*, John Wiley, New York, 1975.
7. Daubert, T.E., Danner, R.P., Sibul, H.M., and Stebbins, C.C., *Physical and Thermodynamic Properties of Pure Compounds: Data Compilation*, extant 1994 (core with 4 supplements), Taylor & Francis, Bristol, PA.
8. Reid, R.C., Prausnitz, J.M., and Poling, B.E., *The Properties of Gases and Liquids, Fourth Edition*, McGraw-Hill, New York, 1987.

Molecular formula	Name	T_b/K	T_c/K	P_c/MPa	V_c/cm^3 mol^{-1}
AlBr$_3$	Aluminum tribromide	528	763	2.89	310
AlCl$_3$	Aluminum trichloride	453 s	620	2.63	257
AlI$_3$	Aluminum triiodide	655	983		408
Ar	Argon	87.30	150.87	4.898	75
As	Arsenic	887 s	1673		35
AsCl$_3$	Arsenic(III) chloride	403	654		252
AsH$_3$	Arsine	210.6	373.1		
BBr$_3$	Boron tribromide	364	581		272
BCl$_3$	Boron trichloride	285.8	455	3.87	239
BF$_3$	Boron trifluoride	172	260.8	4.98	115
BI$_3$	Boron triiodide	483	773		356
B$_2$H$_6$	Diborane	180.7	289.8	4.05	
BiBr$_3$	Bismuth tribromide	726	1220		301
BiCl$_3$	Bismuth trichloride	720	1179	12.0	261
BrH	Hydrogen bromide	206.77	363.2	8.55	
BrI	Iodine bromide	389	719		139
Br$_2$	Bromine	331.9	588	10.34	127
Br$_2$Hg	Mercury(II) bromide	595	1012		
Br$_3$Ga	Gallium(III) bromide	552	806.7		303
Br$_3$HSi	Tribromosilane	382	610.0		305
Br$_3$P	Phosphorus(III) bromide	446.1	711		300
Br$_3$Sb	Antimony(III) bromide	553	904		300
Br$_4$Ge	Germanium(IV) bromide	459.5	718		392
Br$_4$Hf	Hafnium bromide	596 s	746		415
Br$_4$Si	Tetrabromosilane	427	663		382
Br$_4$Sn	Tin(IV) bromide	478	744		417

CRITICAL CONSTANTS (continued)

Molecular formula	Name	T_b/K	T_c/K	P_c/MPa	V_c/cm^3 mol^{-1}
Br$_4$Ti	Titanium(IV) bromide	503	795.7		391
Br$_4$Zr	Zirconium(IV) bromide	633 s	805		424
Br$_5$Ta	Tantalum(V) bromide	622	974		461
ClFO$_3$	Perchloryl fluoride	226.4	368.4	5.37	161
ClF$_2$N	Nitrogen chloride difluoride	207	337.5	5.15	
ClF$_2$P	Phosphorus(III) chloride difluoride	225.9	362.4	4.52	
ClF$_2$PS	Phosphorus(V) sulfide chloride difluoride	279.4	439.2	4.14	
ClF$_3$Si	Chlorotrifluorosilane	203.1	307.7	3.46	
ClF$_5$	Chlorine pentafluoride	260.0	416	5.27	233
ClF$_5$S	Sulfur chloride pentafluoride	254.1	390.9		
ClH	Hydrogen chloride	188	324.7	8.31	81
ClH$_4$N	Ammonium chloride	611 s	1155	163.5	
ClH$_4$P	Phosphonium chloride	246 s	322.3	7.37	
ClNO	Nitrosyl chloride	267.6	440		
ClOV	Vanadyl chloride	400	636		171
Cl$_2$	Chlorine	239.11	416.9	7.991	123
Cl$_2$FP	Phosphorus(III) dichloride fluoride	287	463.0	4.96	
Cl$_2$F$_2$Si	Dichlorodifluorosilane	241	369.0	3.5	
Cl$_2$Hg	Mercury(II) chloride	577	973		174
Cl$_2$OSe	Selenium oxychloride	450	730	7.09	235
Cl$_3$FSi	Trichlorofluorosilane	285.4	438.6	3.58	
Cl$_3$Ga	Gallium(III) chloride	474	694		263
Cl$_3$HSi	Trichlorosilane	306	479		268
Cl$_3$P	Phosphorus(III) chloride	349.1	563		264
Cl$_3$Sb	Antimony(III) chloride	493.4	794		272
Cl$_4$Ge	Germanium(IV) chloride	359.7	553.2	3.861	330
Cl$_4$Hf	Hafnium(IV) chloride	590 s	725.7	5.42	314
Cl$_4$ORe	Rhenium(VI) oxytetrachloride	496	781		362
Cl$_4$OW	Tungsten(VI) oxytetrachloride	500.7	782		338
Cl$_4$Si	Tetrachlorosilane	330.8	508.1	3.593	326
Cl$_4$Sn	Tin(IV) chloride	387.3	591.9	3.75	351
Cl$_4$Te	Tellurium tetrachloride	660	1002	8.56	310
Cl$_4$Ti	Titanium(IV) chloride	409.6	638	4.66	339
Cl$_4$Zr	Zirconium(IV) chloride	604 s	778	5.77	319
Cl$_5$Mo	Molybdenum(V) chloride	541	850		369
Cl$_5$Nb	Niobium(V) chloride	527.1	803.5	4.88	397
Cl$_5$P	Phosphorus(V) chloride	433 s	646		
Cl$_5$Ta	Tantalum(V) chloride	512.5	767		402
Cl$_6$W	Tungsten(VI) chloride	619.9	923		422
Cs	Cesium	944	1938	9.4	341
FH	Hydrogen fluoride	293	461	6.48	69
FNO$_2$	Nitryl fluoride	200.7	349.5		
F$_2$	Fluorine	85.03	144.13	5.172	66
F$_2$HN	Difluoramine	250	403		
F$_2$N$_2$	cis-Difluorodiazine	167.4	272	7.09	
F$_2$N$_2$	trans-Difluorodiazine	161.7	260	5.57	
F$_2$O	Fluorine monoxide	128.4	215		
F$_2$Xe	Xenon difluoride	387.5	631	9.32	148
F$_3$N	Nitrogen trifluoride	144.40	234.0	4.46	126
F$_3$NO	Trifluoramine oxide	185.6	303	6.43	147
F$_3$P	Phosphorus(III) fluoride	171.6	271.2	4.33	
F$_3$PS	Phosphorus(V) sulfide trifluoride	220.9	346.0	3.82	
F$_4$N$_2$	Tetrafluorohydrazine	199	309	3.75	
F$_4$S	Sulfur tetrafluoride	232.7	364		
F$_4$Si	Tetrafluorosilane	187	259.0	3.72	
F$_4$Xe	Xenon tetrafluoride	388.9	612	7.04	188
F$_5$Nb	Niobium(V) fluoride	502	737	6.28	155
F$_6$Mo	Molybdenum(VI) fluoride	307	473	4.75	226
F$_6$S	Sulfur hexafluoride	209.3	318.69	3.77	199

Molecular formula	Name	T_b/K	T_c/K	P_c/MPa	V_c/cm^3 mol^{-1}
F_6Se	Selenium hexafluoride	226.5	345.5		
F_6Te	Tellurium hexafluoride	234.2	356		
F_6U	Uranium(VI) fluoride	329.6	505.8	4.66	250
F_6W	Tungsten(VI) fluoride	290	444	4.34	233
GaI_3	Gallium(III) iodide	613	951		395
GeH_4	Germane	185.0	312.2	4.95	147
GeI_4	Germanium(IV) iodide	650	973		500
HI	Hydrogen iodide	237.60	424.0	8.31	
H_2	Hydrogen	20.28	32.97	1.293	65
H_2O	Water	373.15	647.14	22.06	56
H_2O_2	Hydrogen peroxide	423.3	728*	22*	
H_2S	Hydrogen sulfide	213.60	373.2	8.94	99
H_2Se	Hydrogen selenide	231.90	411	8.92	
H_3N	Ammonia	239.82	405.5	11.35	72
H_3P	Phosphine	185.40	324.5	6.54	
H_4N_2	Hydrazine	386.70	653	14.7	
He	Helium	4.22	5.19	0.227	57
HfI_4	Hafnium iodide	667 s	916		528
Hg	Mercury	629.88	1750	172.00	43
HgI_2	Mercury(II) iodide	627	1072		
I_2	Iodine	457.5	819		155
I_3Sb	Antimony(III) iodide	674	1102		
I_4Si	Tetraiodosilane	560.5	944		558
I_4Sn	Tin(IV) iodide	637.5	968		531
I_4Ti	Titanium(IV) iodide	650	1040		505
I_4Zr	Zirconium(IV) iodide	704 s	960		530
K	Potassium	1032	2223*	16*	209*
Kr	Krypton	119.93	209.41	5.50	91
Li	Lithium	1615	3223*	67*	66*
NO	Nitric oxide	121.41	180	6.48	58
N_2	Nitrogen	77.36	126.21	3.39	90
N_2O	Nitrous oxide	184.67	309.57	7.255	97
N_2O_4	Nitrogen tetroxide	294.30	431	10.1	167
Na	Sodium	1156	2573*	35*	116*
Ne	Neon	27.07	44.4	2.76	42
O_2	Oxygen	90.20	154.59	5.043	73
O_2S	Sulfur dioxide	263.10	430.8	7.884	122
O_3	Ozone	161.80	261.1	5.57	89
O_3S	Sulfur trioxide	318	491.0	8.2	127
O_4Os	Osmium(VIII) oxide	408	678		
O_7Re_2	Rhenium(VII) oxide	633	942		334
P	Phosphorus	553.6	994		
Rb	Rubidium	961	2093*	16*	247*
Rn	Radon	211.4	377	6.28	
S	Sulfur	717.75	1314	20.7	
Se	Selenium	958	1766	27.2	
Xe	Xenon	165.11	289.73	5.84	118
$CBrClF_2$	Bromochlorodifluoromethane	269.4	426.88	4.254	246
$CBrF_3$	Bromotrifluoromethane	215.3	340.2	3.97	196
CBr_2F_2	Dibromodifluoromethane	298	471.3		
$CClF_3$	Chlorotrifluoromethane	191.7	302	3.870	180
CCl_2F_2	Dichlorodifluoromethane	243.3	384.95	4.136	217
CCl_2O	Carbonyl chloride	281	455	5.67	190
CCl_3F	Trichlorofluoromethane	296.8	471.1	4.47	247
CCl_4	Tetrachloromethane	349.9	556.6	4.516	276
CF_4	Tetrafluoromethane	145.1	227.6	3.74	140
$CHClF_2$	Chlorodifluoromethane	232.4	369.5	5.035	164
$CHCl_2F$	Dichlorofluoromethane	282.0	451.58	5.18	196
$CHCl_3$	Trichloromethane	334.2	536.4	5.47	239

Molecular formula	Name	T_b/K	T_c/K	P_c/MPa	V_c/cm^3 mol^{-1}
CHF_3	Trifluoromethane	191.0	299.3	4.858	133
CHN	Hydrogen cyanide	299	456.7	5.39	139
CH_2Cl_2	Dichloromethane	313	510	6.10	
CH_2F_2	Difluoromethane	221.5	351.6	5.830	121
CH_2O_2	Formic acid	374	588		
CH_3Cl	Chloromethane	249.1	416.25	6.679	139
CH_3Cl_3Si	Methyltrichlorosilane	338.7	517	3.28	348
CH_3F	Fluoromethane	194.7	317.8	5.88	113
CH_3I	Iodomethane	315.6	528		
CH_3NO_2	Nitromethane	374.2	588	5.87	173
CH_4	Methane	111.6	190.56	4.599	98.60
CH_4O	Methanol	337.7	512.5	8.084	117
CH_4S	Methanethiol	279.0	470.0	7.23	145
CH_5ClSi	Chloromethylsilane	280	517.8		
CH_5N	Methylamine	266.8	430.7	7.614	
CH_6N_2	Methylhydrazine	360.6	567	8.24	271
CH_6Si	Methylsilane	215.6	352.5		
CO	Carbon monoxide	81.6	132.91	3.499	93
COS	Carbon oxysulfide	223	375	5.88	137
CO_2	Carbon dioxide	194.7	304.14	7.375	94
CS_2	Carbon disulfide	319	552	7.90	173
$C_2Br_2ClF_3$	1,2-Dibromo-1-chloro-1,2,2-trifluoroethane	366	560.7	3.61	368
$C_2Br_2F_4$	1,2-Dibromotetrafluoroethane	320.4	487.8	3.393	341
C_2ClF_3	Chlorotrifluoroethylene	245.3	379	4.05	212
C_2ClF_5	Chloropentafluoroethane	235.2	353.2	3.229	252
$C_2Cl_2F_4$	1,1-Dichlorotetrafluoroethane	277	418.6	3.30	294
$C_2Cl_2F_4$	1,2-Dichlorotetrafluoroethane	276.9	418.78	3.252	297
$C_2Cl_3F_3$	1,1,2-Trichlorotrifluoroethane	320.8	487.3	3.42	325
C_2Cl_4	Tetrachloroethylene	394.4	620.2		
$C_2Cl_4F_2$	1,1,2,2-Tetrachloro-1,2-difluoroethane	366	551		
C_2F_3N	Trifluoroacetonitrile	204.3	311.11	3.618	202
C_2F_4	Tetrafluoroethylene	197.2	306.5	3.94	172
C_2F_6	Hexafluoroethane	195.0	293		222
C_2HClF_2	1-Chloro-2,2-difluoroethylene	254.6	400.6	4.46	197
C_2HClF_4	1-Chloro-1,1,2,2-tetrafluoroethane	263	399.9	3.72	244
$C_2HCl_2F_3$	1,2-Dichloro-1,1,2-trifluoroethane	301	461.6		278
$C_2HCl_2F_3$	2,2-Dichloro-1,1,1-trifluoroethane	300.97	456.83	3.661	278
C_2HCl_3	Trichloroethylene	360.3	544.2	5.02	
$C_2HF_3O_2$	Trifluoroacetic acid	346	491.3	3.258	204
C_2HF_5O	Trifluoromethyl difluoromethyl ether		353.8	3.326	
C_2H_2	Acetylene	188.4	308.33	6.139	113
$C_2H_2Cl_2$	cis-1,2-Dichloroethylene	333.2	544.2		
$C_2H_2Cl_2$	trans-1,2-Dichloroethylene	321.8	516.5	5.51	
$C_2H_2Cl_4$	1,1,2,2-Tetrachloroethane	419.6	661.15		
$C_2H_2F_2$	1,1-Difluoroethylene	187.4	302.9	4.46	154
$C_2H_2F_4$	1,1,1,2-Tetrafluoroethane	247.07	374.18	4.065	198
$C_2H_2F_4$	1,1,2,2-Tetrafluoroethane	253.2	391.8		191
$C_2H_3ClF_2$	1-Chloro-1,1-difluoroethane	263.4	410.3	4.041	225
$C_2H_3Cl_2F$	1,1-Dichloro-1-fluoroethane	305	477.3		254
$C_2H_3Cl_3$	1,1,1-Trichloroethane	347.1	545	4.30	
C_2H_3F	Fluoroethylene	201	327.9	5.24	144
$C_2H_3F_3$	1,1,1-Trifluoroethane	225.6	346.3	3.76	194
$C_2H_3F_3O$	Methyl trifluoromethyl ether		378.0	3.680	228
C_2H_3N	Acetonitrile	354.7	545.5	4.85	173
C_2H_4	Ethylene	169.4	282.34	5.041	131
$C_2H_4Br_2$	1,2-Dibromoethane	404.7	583.0	7.2	
$C_2H_4Cl_2$	1,1-Dichloroethane	330.5	523	5.07	236
$C_2H_4Cl_2$	1,2-Dichloroethane	356.6	561	5.4	225
$C_2H_4F_2$	1,1-Difluoroethane	248.2	386.7	4.50	181

Molecular formula	Name	T_b/K	T_c/K	P_c/MPa	$V_c/cm^3 mol^{-1}$
C_2H_4O	Acetaldehyde	293.2	466		154
C_2H_4O	Ethylene oxide	283.7	469	7.19	140
$C_2H_4O_2$	Acetic acid	391.0	592.71	5.786	171
$C_2H_4O_2$	Methyl formate	304.8	487.2	5.998	172
C_2H_5Br	Bromoethane	311.6	503.9	6.23	215
C_2H_5Cl	Chloroethane	285.4	460.4	5.3	
C_2H_5F	Fluoroethane	235.5	375.31	5.028	
C_2H_6	Ethane	184.5	305.32	4.872	145.5
$C_2H_6Cl_2Si$	Dichlorodimethylsilane	343.4	520.4	3.49	350
C_2H_6O	Ethanol	351.3	514.0	6.137	168
C_2H_6O	Dimethyl ether	248.3	400.0	5.37	190
$C_2H_6O_2$	Ethylene glycol	470.4	718		
C_2H_6S	Ethanethiol	308.2	499	5.49	207
C_2H_6S	Dimethyl sulfide	310.4	503.0	5.53	201
C_2H_7N	Ethylamine	289.6	456	5.62	182
C_2H_7N	Dimethylamine	279.9	437.22	5.340	
C_2N_2	Cyanogen	252.0	400	5.98	
C_3ClF_5O	Chloropentafluoroacetone	281	410.6	2.878	
C_3F_6O	Perfluoroacetone	245.7	357.14	2.84	329
C_3F_6O	Perfluorooxetane		361.8	3.094	272
C_3F_8	Perfluoropropane	236.5	345.1	2.680	299
$C_3F_8O_2$	Perfluorodimethoxymethane		372.3	2.333	363
C_3HF_7O	Trifluoromethyl 1,1,2,2-tetrafluoroethyl ether		387.7	2.640	337
$C_3H_3F_3$	3,3,3-Trifluoropropene	256	376.2	3.80	211
$C_3H_3F_5$	1,1,1,2,2-Pentafluoropropane	255.7	380.11	3.137	273
C_3H_3NO	Isoxazole	368	552.0		
C_3H_4	Allene	238.7	393		
C_3H_4	Propyne	249.9	402.38	5.628	164
C_3H_5Cl	3-Chloropropene	318.2	514		
C_3H_5N	Propanenitrile	370.2	561.3	4.26	229
C_3H_6	Propene	225.5	364.85	4.601	181
C_3H_6	Cyclopropane	240.3	398.3	5.579	162
C_3H_6O	Allyl alcohol	370.1	545.1		
C_3H_6O	Propanal	321	504.4	5.27	204
C_3H_6O	Acetone	329.1	508.1	4.700	209
C_3H_6O	Methyloxirane	308	482.2	4.92	186
$C_3H_6O_2$	Propanoic acid	414.2	604	4.53	222
$C_3H_6O_2$	Ethyl formate	327.5	508.5	4.74	229
$C_3H_6O_2$	Methyl acetate	329.9	506.55	4.75	228
C_3H_7Cl	1-Chloropropane	319.6	503	4.58	
C_3H_7NO	N,N-Dimethylformamide	426	649.6		262
C_3H_8	Propane	231.0	369.83	4.248	200
C_3H_8O	1-Propanol	370.3	536.8	5.169	218
C_3H_8O	2-Propanol	355.4	508.3	4.764	222
C_3H_8O	Ethyl methyl ether	280.5	437.8	4.40	221
C_3H_8S	1-Propanethiol	340.9	536.6		286
C_3H_8S	Ethyl methyl sulfide	339.8	533	4.26	
$C_3H_9BO_3$	Trimethyl borate	340.6	501.7	3.59	
C_3H_9ClSi	Trimethylchlorosilane	333	497.8	3.20	366
C_3H_9N	Propylamine	320.3	497.0	4.72	
C_3H_9N	Isopropylamine	304.8	471.8	4.54	221
C_3H_9N	Trimethylamine	275.9	432.79	4.087	254
$C_4Br_2F_8$	1,4-Dibromooctafluorobutane	370	532.5	2.39	
C_4F_8	Perfluorocyclobutane	267.2	388.46	2.784	324
C_4F_{10}	Perfluorobutane	271.2	386.4	2.323	378
C_4F_{10}	Perfluoroisobutane		395.4		
C_4H_4O	Furan	304.6	490.2	5.50	218
C_4H_4S	Thiophene	357.1	579.4	5.69	219
C_4H_5N	Pyrrole	402.94	639.7	6.34	200

Molecular formula	Name	T_b/K	T_c/K	P_c/MPa	V_c/cm^3 mol^{-1}
C_4H_6	1,3-Butadiene	268.7	425	4.33	221
C_4H_6	1-Butyne	281.1	463.7		
C_4H_6	2-Butyne	300.0	488.7		
$C_4H_6O_3$	Acetic anhydride	412.6	606	4.0	
C_4H_7N	Butanenitrile	390.7	585.4	3.88	
C_4H_8	1-Butene	266.9	419.57	4.023	240
C_4H_8	cis-2-Butene	276.8	435.58	4.197	234
C_4H_8	trans-2-Butene	273.9	428.63	3.985	238
C_4H_8	Isobutene	266.2	417.9	4.000	239
C_4H_8	Cyclobutane	285.7	460.0	4.98	210
C_4H_8O	Ethyl vinyl ether	308.6	475	4.07	
C_4H_8O	Butanal	347.9	537.2	4.32	258
C_4H_8O	2-Butanone	352.6	536.78	4.207	267
C_4H_8O	Tetrahydrofuran	338	540.1	5.19	224
$C_4H_8O_2$	Butanoic acid	436.8	624	4.03	290
$C_4H_8O_2$	2-Methylpropanoic acid	427.5	605	3.7	292
$C_4H_8O_2$	Propyl formate	354.0	538.0	4.06	285
$C_4H_8O_2$	Ethyl acetate	350.2	523.3	3.882	286
$C_4H_8O_2$	Methyl propanoate	352.9	530.6	4.004	282
$C_4H_8O_2$	1,4-Dioxane	374.6	587	5.21	238
C_4H_8S	Tetrahydrothiophene	394.1	632.0		
C_4H_9N	Pyrrolidine	359.71	568	6.00	238
C_4H_{10}	Butane	272.6	425.12	3.796	255
C_4H_{10}	Isobutane	261.4	407.85	3.630	257
$C_4H_{10}O$	1-Butanol	390.8	563.0	4.414	274
$C_4H_{10}O$	2-Butanol	372.6	536.2	4.202	269
$C_4H_{10}O$	2-Methyl-1-propanol	380.9	547.8	4.295	274
$C_4H_{10}O$	2-Methyl-2-propanol	355.5	506.2	3.972	275
$C_4H_{10}O$	Diethyl ether	307.6	466.74	3.638	280
$C_4H_{10}O$	Methyl propyl ether	312.2	476.25	3.801	
$C_4H_{10}O$	Isopropyl methyl ether	303.8	464.48	3.762	
$C_4H_{10}O_2$	Ethylene glycol dimethyl ether	358	536	3.87	271
$C_4H_{10}S$	1-Butanethiol	371.6	570.1		324
$C_4H_{10}S$	Diethyl sulfide	365.2	557	3.96	318
$C_4H_{11}N$	Butylamine	350.15	531.9	4.25	277
$C_4H_{11}N$	sec-Butylamine	335.88	514.3	4.20	278
$C_4H_{11}N$	tert-Butylamine	317.19	483.9	3.84	292
$C_4H_{11}N$	Isobutylamine	340.90	519	4.07	278
$C_4H_{11}N$	Diethylamine	328.6	499.99	3.758	
$C_4H_{12}Si$	Tetramethylsilane	299.7	448.64	2.821	362
$C_4H_{12}Sn$	Tetramethylstannane	351	521.8	2.981	
C_5F_{12}	Perfluoropentane	302.3	420.59	2.045	473
$C_5H_2F_6O_2$	Hexafluoroacetylacetone	327.2	485.1	2.767	
$C_5H_4O_2$	Furfural	434.8	670	5.89	
C_5H_5N	Pyridine	388.38	620.0	5.67	243
$C_5H_6N_2$	2-Methylpyrazine	410	634.3	5.01	283
C_5H_6O	2-Methylfuran	338	527	4.72	247
C_5H_7N	1-Methylpyrrole	385.96	596.0	4.86	271
C_5H_7N	2-Methylpyrrole	420.7	654	5.08	266
C_5H_7N	3-Methylpyrrole	416.0	647	5.08	266
C_5H_8	1-Pentyne	313.2	493.5		
C_5H_8	Cyclopentene	317.3	507.0	4.802	245
C_5H_8O	Cyclopentanone	403.6	624.5	4.60	
C_5H_8O	3,4-Dihydro-2H-pyran	359	561.7	4.56	268
C_5H_9N	Pentanenitrile	414.4	610.3	3.58	
C_5H_9NO	N-Methyl-2-pyrrolidone	475	721.8		311
C_5H_{10}	1-Pentene	303.0	464.78	3.527	293
C_5H_{10}	cis-2-Pentene	310.0	475	3.69	
C_5H_{10}	trans-2-Pentene	309.4	471	3.52	

CRITICAL CONSTANTS (continued)

Molecular formula	Name	T_b/K	T_c/K	P_c/MPa	V_c/cm³ mol⁻¹
C_5H_{10}	2-Methyl-1-butene	304.3	470	3.8	
C_5H_{10}	3-Methyl-1-butene	293.2	450	3.52	
C_5H_{10}	2-Methyl-2-butene	311.6	481	3.91	
C_5H_{10}	Cyclopentane	322.4	511.7	4.508	260
$C_5H_{10}O$	Cyclopentanol	413.5	619.5	4.9	
$C_5H_{10}O$	Allyl ethyl ether	340.7	518		
$C_5H_{10}O$	Pentanal	376	566.1	3.97	313
$C_5H_{10}O$	2-Pentanone	375.3	561.08	3.694	301
$C_5H_{10}O$	3-Pentanone	375.0	561.46	3.729	336
$C_5H_{10}O$	3-Methyl-2-butanone	367.4	553.4	3.85	310
$C_5H_{10}O$	Tetrahydropyran	361	572.2	4.77	263
$C_5H_{10}O$	2-Methyltetrahydrofuran	351	537	3.76	267
$C_5H_{10}O_2$	Pentanoic acid	459.2	643	3.58	340
$C_5H_{10}O_2$	3-Methylbutanoic acid	449.6	629	3.40	
$C_5H_{10}O_2$	Isobutyl formate	371.3	551	3.88	352
$C_5H_{10}O_2$	Propyl acetate	374.6	549.7	3.36	345
$C_5H_{10}O_2$	Isopropyl acetate	361.7	531		
$C_5H_{10}O_2$	Ethyl propanoate	372.2	546.0	3.362	345
$C_5H_{10}O_2$	Methyl butanoate	375.9	554.4	3.47	340
$C_5H_{10}O_2$	Methyl isobutanoate	365.6	540.8	3.43	339
$C_5H_{11}N$	Piperidine	379.37	594	4.94	288
C_5H_{12}	Pentane	309.1	469.7	3.370	311
C_5H_{12}	Isopentane	300.9	460.43	3.381	306
C_5H_{12}	Neopentane	282.5	433.8	3.197	307
$C_5H_{12}O$	1-Pentanol	411.0	588.1	3.897	326
$C_5H_{12}O$	2-Pentanol	392.4	560.3	3.675	329
$C_5H_{12}O$	3-Pentanol	389.3	559.6		325
$C_5H_{12}O$	2-Methyl-1-butanol	400.6	575.4	3.94	
$C_5H_{12}O$	3-Methyl-1-butanol	404.2	577.2	3.93	
$C_5H_{12}O$	2-Methyl-2-butanol	375.5	543.7	3.71	
$C_5H_{12}O$	3-Methyl-2-butanol	386.0	556.1	3.87	
$C_5H_{12}O$	Butyl methyl ether	343.2	512.78	3.371	329
$C_5H_{12}O$	tert-Butyl methyl ether	328.3	497.1	3.430	
$C_5H_{12}O$	Ethyl propyl ether	336.3	500.23	3.370	339
$C_5H_{12}O_2$	2-Propoxyethanol	422.9	614.6		364
$C_5H_{12}S$	3-Methyl-1-butanethiol	393	604		
C_6BrF_5	Bromopentafluorobenzene	410	601	3.0	
C_6ClF_5	Chloropentafluorobenzene	391.0	570.81	3.238	376
$C_6Cl_2F_4$	1,2-Dichlorotetrafluorobenzene	430.9	626	5.32	
$C_6Cl_3F_3$	1,3,5-Trichloro-2,4,6-trifluorobenzene	471.5	684.8	3.27	448
C_6F_6	Hexafluorobenzene	353.3	516.73	3.273	335
C_6F_{10}	Perfluorocyclohexene	325.1	461.8		
C_6F_{12}	Perfluoro-1-hexene	330.1	454.4		
C_6F_{12}	Perfluorocyclohexane	325.9	457.2	2.43	
C_6F_{14}	Perfluorohexane	329.7	448.77	1.868	606
C_6F_{14}	Perfluoro-2-methylpentane	330.7	455.3	1.923	532
C_6F_{14}	Perfluoro-3-methylpentane	331.5	450	1.69	
C_6F_{14}	Perfluoro-2,3-dimethylbutane	332.9	463	1.87	525
C_6HF_5	Pentafluorobenzene	358.8	530.97	3.531	324
C_6HF_5O	Pentafluorophenol	418.7	609	4.0	348
C_6HF_{11}	Undecafluorocyclohexane	335.1	477.7		
$C_6H_2F_4$	1,2,3,4-Tetrafluorobenzene	367.4	550.83	3.791	313
$C_6H_2F_4$	1,2,3,5-Tetrafluorobenzene	357.5	535.25	3.747	
$C_6H_2F_4$	1,2,4,5-Tetrafluorobenzene	363.3	543.35	3.801	
$C_6H_4F_2$	p-Difluorobenzene	362	556	4.40	
C_6H_5Br	Bromobenzene	429.1	670	4.52	324
C_6H_5Cl	Chlorobenzene	404.8	632.4	4.52	308
C_6H_5F	Fluorobenzene	357.8	560.09	4.551	269
C_6H_5I	Iodobenzene	461.5	721	4.52	351

Molecular formula	Name	T_b/K	T_c/K	P_c/MPa	V_c/cm^3 mol^{-1}
C_6H_6	Benzene	353.1	562.05	4.895	256
C_6H_6O	Phenol	454.9	694.2	6.13	
C_6H_7N	Aniline	457.2	699	4.89	287
C_6H_7N	2-Methylpyridine	402.53	621.0	4.60	292
C_6H_7N	3-Methylpyridine	417.29	645.0	4.65	288
C_6H_7N	4-Methylpyridine	418.51	645.7	4.70	292
C_6H_{10}	1,5-Hexadiene	332.5	507		
C_6H_{10}	Cyclohexene	356.0	560.48		
$C_6H_{10}O$	Cyclohexanone	428.5	653.0	4.0	
$C_6H_{10}O_2$	Ethyl trans-2-butenoate	411	599		
$C_6H_{10}S$	Diallylsulfide	411.7	653		
$C_6H_{11}N$	Hexanenitrile	436.7	633.8	3.30	
C_6H_{12}	1-Hexene	336.5	504.1	3.206	348
C_6H_{12}	Cyclohexane	353.8	553.5	4.07	308
C_6H_{12}	Methylcyclopentane	344.9	532.73	3.784	319
$C_6H_{12}O$	Cyclohexanol	433.9	650.1	4.26	
$C_6H_{12}O$	Hexanal	404	591	3.46	
$C_6H_{12}O$	2-Hexanone	400.7	587.0	3.32	
$C_6H_{12}O$	3-Hexanone	396.6	582.82	3.320	
$C_6H_{12}O$	4-Methyl-2-pentanone	389.6	571	3.27	
$C_6H_{12}O_2$	Hexanoic acid	478.3	662	3.20	
$C_6H_{12}O_2$	Pentyl formate	403.5	576	3.46	
$C_6H_{12}O_2$	Isopentyl formate	396.6	578		
$C_6H_{12}O_2$	Butyl acetate	399.2	579		
$C_6H_{12}O_2$	Isobutyl acetate	389.6	561	3.16	
$C_6H_{12}O_2$	Propyl propanoate	395.6	578		
$C_6H_{12}O_2$	Ethyl butanoate	394.6	566	3.06	421
$C_6H_{12}O_2$	Ethyl 2-methylpropanoate	383.2	553	3.07	421
$C_6H_{12}O_2$	Methyl pentanoate	400.5	567	3.19	
$C_6H_{12}O_3$	Ethylene glycol ethyl ether acetate	429.5	607.3	3.166	443
$C_6H_{12}O_3$	Paraldehyde	397.4	563		
C_6H_{14}	Hexane	341.8	507.6	3.025	368
C_6H_{14}	2-Methylpentane	333.3	497.7	3.031	367
C_6H_{14}	3-Methylpentane	336.3	504.5	3.126	367
C_6H_{14}	2,2-Dimethylbutane	322.8	488.8	3.090	359
C_6H_{14}	2,3-Dimethylbutane	331.0	500.0	3.131	358
$C_6H_{14}O$	1-Hexanol	430.7	610.3	3.417	387
$C_6H_{14}O$	2-Hexanol	413	585.9	3.31	384
$C_6H_{14}O$	3-Hexanol	408	582.4	3.36	383
$C_6H_{14}O$	2-Methyl-1-pentanol	422	604.4	3.45	
$C_6H_{14}O$	4-Methyl-1-pentanol	425.0	603.5		
$C_6H_{14}O$	2-Methyl-2-pentanol	394.2	559.5		
$C_6H_{14}O$	4-Methyl-2-pentanol	404.7	574.4		
$C_6H_{14}O$	2-Methyl-3-pentanol	399.6	576.0	3.46	
$C_6H_{14}O$	3-Methyl-3-pentanol	395.5	575.6	3.52	
$C_6H_{14}O$	Dipropyl ether	363.1	530.6	3.028	
$C_6H_{14}O$	Diisopropyl ether	341.6	500.32	2.832	386
$C_6H_{14}O$	Methyl pentyl ether	372	546.53	3.042	391
$C_6H_{14}O_2$	Ethylene glycol monobutyl ether	441.5	633.9		424
$C_6H_{14}O_2$	1,1-Diethoxyethane	375.3	527		
$C_6H_{15}N$	Dipropylamine	382.4	555.8	3.63	
$C_6H_{15}N$	Diisopropylamine	357.0	523.1	3.02	
$C_6H_{15}N$	Triethylamine	362	535.6	3.032	389
C_7F_8	Perfluorotoluene	377.6	534.47	2.705	428
C_7F_{14}	Perfluoro-1-heptene	354.1	478.2		
C_7F_{14}	Perfluoromethylcyclohexane	349.4	485.91	2.019	570
C_7F_{16}	Perfluoroheptane	355.6	474.8	1.62	664
C_7HF_{15}	1H-Pentadecafluoroheptane	369.1	495.8		
$C_7H_3F_5$	2,3,4,5,6-Pentafluorotoluene	390.6	566.52	3.126	384

Molecular formula	Name	T_b/K	T_c/K	P_c/MPa	V_c/cm³ mol⁻¹
C_7H_5N	Benzonitrile	464.2	699.4	4.21	
C_7H_6O	Benzaldehyde	452.1	695	4.65	
C_7H_8	Toluene	383.7	591.75	4.108	316
C_7H_8O	o-Cresol	464.1	697.6	5.01	
C_7H_8O	m-Cresol	475.3	705.8	4.56	309
C_7H_8O	p-Cresol	475.0	704.6	5.15	
C_7H_8O	Benzyl alcohol	478.4	715	4.3	
C_7H_8O	Anisole	426.8	645.6	4.25	
C_7H_9N	o-Methylaniline	473.4	707	4.37	
C_7H_9N	m-Methylaniline	476.4	707	4.28	
C_7H_9N	p-Methylaniline	473.5	706	4.58	
C_7H_9N	N-Methylaniline	469.3	701	5.20	
C_7H_9N	2,3-Dimethylpyridine	434.3	655.4	4.10	356
C_7H_9N	2,4-Dimethylpyridine	431.6	647	3.95	361
C_7H_9N	2,5-Dimethylpyridine	430.1	645	3.85	361
C_7H_9N	2,6-Dimethylpyridine	417.2	624	3.85	361
C_7H_9N	3,4-Dimethylpyridine	452.2	684	4.20	355
C_7H_9N	3,5-Dimethylpyridine	445	668	4.05	361
C_7H_{14}	1-Heptene	366.7	537.3	2.921	402
C_7H_{14}	Cycloheptane	391.5	604.2	3.81	353
C_7H_{14}	Methylcyclohexane	374.0	572.2	3.471	368
C_7H_{14}	Ethylcyclopentane	376.6	569.5	3.397	375
C_7H_{14}	1,1-Dimethylcyclopentane	360.6	547	3.45	
C_7H_{14}	cis-1,2-Dimethylcyclopentane	372.6	565	3.45	
C_7H_{14}	trans-1,2-Dimethylcyclopentane	365.0	553	3.45	
C_7H_{14}	cis-1,3-Dimethylcyclopentane	363.9	551	3.45	
C_7H_{14}	trans-1,3-Dimethylcyclopentane	364.8	553	3.45	
$C_7H_{14}O$	2-Heptanone	424.1	611.5	3.436	
$C_7H_{14}O_2$	Heptanoic acid	495.3	679	2.90	
$C_7H_{14}O_2$	Isopentyl acetate	415.6	599		
$C_7H_{14}O_2$	Isobutyl propanoate	410	592		
$C_7H_{14}O_2$	Propyl butanoate	416.1	600		
$C_7H_{14}O_2$	Propyl 2-methylpropanoate	408.5	589		
$C_7H_{14}O_2$	Ethyl pentanoate	419.2	570		
$C_7H_{14}O_2$	Ethyl 3-methylbutanoate	408.1	588		
C_7H_{16}	Heptane	371.6	540.2	2.74	428
C_7H_{16}	2-Methylhexane	363.1	530.4	2.734	421
C_7H_{16}	3-Methylhexane	365	535.3	2.814	404
C_7H_{16}	3-Ethylpentane	366.6	540.7	2.891	416
C_7H_{16}	2,2-Dimethylpentane	352.3	520.5	2.773	416
C_7H_{16}	2,3-Dimethylpentane	362.8	537.4	2.908	393
C_7H_{16}	2,4-Dimethylpentane	353.5	519.8	2.737	418
C_7H_{16}	3,3-Dimethylpentane	359.1	536.4	2.946	414
C_7H_{16}	2,2,3-Trimethylbutane	353.9	531.2	2.954	398
$C_7H_{16}O$	1-Heptanol	449.5	632.6	3.058	435
$C_7H_{16}O$	2-Heptanol	432	608.3	3.021	442
$C_7H_{16}O$	3-Heptanol	430	605.4		434
$C_7H_{16}O$	4-Heptanol	429	602.6		432
$C_8F_{16}O$	Perfluoro-2-butyltetrahydrofuran	375.7	500.2	1.607	588
C_8F_{18}	Perfluorooctane	379.0	502	1.66	
C_8H_7N	4-Methylbenzonitrile	490.1	723		
C_8H_7N	1H-Indole	526.7	794	4.8	356
C_8H_8O	Acetophenone	475	709.5		386
$C_8H_8O_3$	Methyl salicylate	496.0	709		
C_8H_{10}	Ethylbenzene	409.2	617.15	3.609	374
C_8H_{10}	o-Xylene	417.6	630.3	3.732	370
C_8H_{10}	m-Xylene	412.2	617.0	3.541	375
C_8H_{10}	p-Xylene	411.4	616.2	3.511	378
$C_8H_{10}O$	o-Ethylphenol	477.6	703.0		

Molecular formula	Name	T_b/K	T_c/K	P_c/MPa	V_c/cm^3 mol^{-1}
$C_8H_{10}O$	m-Ethylphenol	491.5	718.8		
$C_8H_{10}O$	p-Ethylphenol	491.0	716.4		
$C_8H_{10}O$	2,3-Xylenol	490.0	722.8		
$C_8H_{10}O$	2,4-Xylenol	484.0	707.6		
$C_8H_{10}O$	2,5-Xylenol	484.2	706.9		
$C_8H_{10}O$	2,6-Xylenol	474.1	701.0		
$C_8H_{10}O$	3,4-Xylenol	500	729.8		
$C_8H_{10}O$	3,5-Xylenol	494.8	715.6		
$C_8H_{10}O$	Phenetole	442.9	647	3.42	
$C_8H_{11}N$	N-Ethylaniline	476.1	698		
$C_8H_{11}N$	N,N-Dimethylaniline	467.2	687	3.63	
$C_8H_{14}O_4$	Ethyl succinate	490.8	663		
$C_8H_{15}N$	Octanenitrile	478.3	674.4	2.85	
C_8H_{16}	1-Octene	394.3	566.7	2.675	464
C_8H_{16}	Cyclooctane	422	647.2	3.56	410
C_8H_{16}	Ethylcyclohexane	405.0	609	3.04	
C_8H_{16}	cis-1,2-Dimethylcyclohexane	402.9	606	2.95	
C_8H_{16}	trans-1,2-Dimethylcyclohexane	396.6	596	2.94	
C_8H_{16}	cis-1,3-Dimethylcyclohexane	393.2	591	2.94	
C_8H_{16}	trans-1,3-Dimethylcyclohexane	397.6	598	2.94	
C_8H_{16}	trans-1,4-Dimethylcyclohexane	392.5	587.7		
$C_8H_{16}O_2$	Octanoic acid	512	695	2.64	
$C_8H_{16}O_2$	2-Ethylhexanoic acid	501	674.6	2.778	528
$C_8H_{16}O_2$	Isopentyl propanoate	433.3	611		
$C_8H_{16}O_2$	Isobutyl butanoate	430.0	611		
$C_8H_{16}O_2$	Isobutyl 2-methylpropanoate	421.7	602		
$C_8H_{16}O_2$	Propyl 3-methylbutanoate	429.0	609		
C_8H_{18}	Octane	398.7	568.7	2.49	492
C_8H_{18}	2-Methylheptane	390.7	559.7	2.484	488
C_8H_{18}	3-Methylheptane	389.6	563.7	2.546	464
C_8H_{18}	4-Methylheptane	390.8	561.8	2.542	476
C_8H_{18}	3-Ethylhexane	391.7	565.5	2.608	455
C_8H_{18}	2,2-Dimethylhexane	379.9	549.9	2.529	478
C_8H_{18}	2,3-Dimethylhexane	388.7	563.5	2.628	468
C_8H_{18}	2,4-Dimethylhexane	382.6	553.6	2.556	472
C_8H_{18}	2,5-Dimethylhexane	382.2	550.1	2.487	482
C_8H_{18}	3,3-Dimethylhexane	385.0	562.1	2.654	443
C_8H_{18}	3,4-Dimethylhexane	390.8	568.9	2.692	466
C_8H_{18}	3-Ethyl-2-methylpentane	388.7	567.1	2.700	443
C_8H_{18}	3-Ethyl-3-methylpentane	391.3	576.6	2.808	455
C_8H_{18}	2,2,3-Trimethylpentane	383	563.5	2.730	436
C_8H_{18}	2,2,4-Trimethylpentane	372.3	544.0	2.568	468
C_8H_{18}	2,3,3-Trimethylpentane	387.9	573.6	2.820	455
C_8H_{18}	2,3,4-Trimethylpentane	386.6	566.5	2.730	461
C_8H_{18}	2,2,3,3-Tetramethylbutane	379.5	567.8	2.87	461
$C_8H_{18}O$	1-Octanol	468.2	652.5	2.777	497
$C_8H_{18}O$	2-Octanol	453	629.6	2.754	519
$C_8H_{18}O$	3-Octanol	444	628.5		515
$C_8H_{18}O$	4-Octanol	449.4	625.1		515
$C_8H_{18}O$	4-Methyl-3-heptanol	443	623.5		
$C_8H_{18}O$	5-Methyl-3-heptanol	445	621.2		
$C_8H_{18}O$	2-Ethyl-1-hexanol	457.7	640.6	2.8	
$C_8H_{18}O$	Dibutyl ether	413.3	584.1	3.01	
$C_8H_{18}O$	Di-tert-butyl ether	380.3	550		
$C_8H_{19}N$	Dibutylamine	432.7	607.5	3.11	
$C_8H_{19}N$	Diisobutylamine	412.7	584.4	3.20	
$C_8H_{20}Si$	Tetraethylsilane	427.8	603.7	2.602	
C_9F_{20}	Perfluorononane	398.4	524	1.56	
C_9H_7N	Quinoline	510.31	782	4.86	371

Molecular formula	Name	T_b/K	T_c/K	P_c/MPa	V_c/cm^3 mol^{-1}
C_9H_7N	Isoquinoline	516.37	803	5.10	374
C_9H_{10}	Indan	451.0	684.9	3.95	
C_9H_{12}	Propylbenzene	432.3	638.35	3.200	440
C_9H_{12}	Isopropyl benzene [Cumene]	425.5	631.0	3.209	
C_9H_{12}	o-Ethyltoluene	438.3	651	3.38	
C_9H_{12}	m-Ethyltoluene	434.4	637	3.25	
C_9H_{12}	p-Ethyltoluene	435	640.2	3.23	
C_9H_{12}	1,2,3-Trimethylbenzene	449.2	664.5	3.454	
C_9H_{12}	1,2,4-Trimethylbenzene	442.4	649.1	3.232	
C_9H_{12}	1,3,5-Trimethylbenzene [Mesitylene]	437.8	637.3	3.127	
$C_9H_{13}N$	N,N-Dimethyl-o-toluidine	467.2	668	3.12	
C_9H_{18}	Cyclononane	451.5	682	3.34	
C_9H_{18}	1,3,5-Trimethylcyclohexane ($1\alpha,3\alpha,5\alpha$)	413.6	602.2		
$C_9H_{18}O$	5-Nonanone	461.5	640		
$C_9H_{18}O_2$	Nonanoic acid	527.6	711	2.40	
$C_9H_{18}O_2$	Isopentyl butanoate	452	619		
$C_9H_{18}O_2$	Isobutyl 3-methylbutanoate	441.6	621		
C_9H_{20}	Nonane	423.9	594.6	2.29	555
C_9H_{20}	2-Methyloctane	416.3	587.0	2.310	
C_9H_{20}	2,2-Dimethylheptane	405.8	576.8	2.350	
C_9H_{20}	2,2,5-Trimethylhexane	397.1	568		
C_9H_{20}	2,2,3,3-Tetramethylpentane	413.3	607.7	2.741	
C_9H_{20}	2,2,3,4-Tetramethylpentane	406.1	592.7	2.602	
C_9H_{20}	2,2,4,4-Tetramethylpentane	395.3	574.7	2.485	
C_9H_{20}	2,3,3,4-Tetramethylpentane	414.6	607.7	2.716	
$C_9H_{20}O$	1-Nonanol	486.4	670.7	2.528	572
$C_9H_{20}O$	2-Nonanol	466.6	649.6	2.53	575
$C_9H_{20}O$	3-Nonanol	468	648.0		577
$C_9H_{20}O$	4-Nonanol	465.6	645.1		575
$C_{10}F_8$	Perfluoronaphthalene	482	673.1		
$C_{10}F_{18}$	Perfluorodecalin	415	566	1.52	
$C_{10}F_{22}$	Perfluorodecane	417.3	542	1.45	
$C_{10}H_8$	Naphthalene	491.0	748.4	4.05	407
$C_{10}H_9N$	1-Naphthalenamine	573.8	850	5.0	438
$C_{10}H_9N$	2-Naphthalenamine	579.3	850	4.9	438
$C_{10}H_{12}$	1,2,3,4-Tetrahydronaphthalene	480.7	720	3.65	408
$C_{10}H_{14}$	Butylbenzene	456.4	660.5	2.89	497
$C_{10}H_{14}$	Isobutylbenzene	445.8	650	3.05	
$C_{10}H_{14}$	1-Isopropyl-4 methylbenzene [p-Cymene]	450.2	652	2.8	
$C_{10}H_{14}$	p-Diethylbenzene	456.8	657.9	2.803	
$C_{10}H_{14}$	1,2,4,5-Tetramethylbenzene	469.9	676	2.9	
$C_{10}H_{14}O$	Thymol	505.6	698		
$C_{10}H_{18}$	1,3-Decadiene	442	615		
$C_{10}H_{18}$	cis-Decahydronaphthalene	468.9	702.3	3.20	
$C_{10}H_{18}$	trans-Decahydronaphthalene	460.4	687.1		
$C_{10}H_{20}$	1-Decene	443.6	616.4	2.218	584
$C_{10}H_{20}O$	Decanal	481.6	674.2		
$C_{10}H_{20}O$	Menthol	489	694		
$C_{10}H_{20}O_2$	Decanoic acid	541.8	726	2.23	
$C_{10}H_{20}O_2$	Ethyl octanoate	481.6	659		
$C_{10}H_{22}$	Decane	447.2	617.7	2.11	624
$C_{10}H_{22}$	3,3,5-Trimethylheptane	428.8	609.7	2.317	
$C_{10}H_{22}$	2,2,3,3-Tetramethylhexane	433.4	623.2	2.510	
$C_{10}H_{22}$	2,2,5,5-Tetramethylhexane	410.5	581.6	2.186	
$C_{10}H_{22}O$	1-Decanol	504.2	687.3	2.315	649
$C_{10}H_{22}O$	2-Decanol	484	668.6		646
$C_{10}H_{22}O$	3-Decanol		666.1		643
$C_{10}H_{22}O$	4-Decanol	483.6	663.7		643
$C_{10}H_{22}O$	5-Decanol		663.2		646

Molecular formula	Name	T_b/K	T_c/K	P_c/MPa	V_c/cm^3 mol^{-1}
$C_{10}H_{22}S$	Diisopentylsulfide	484	664		
$C_{11}H_{10}$	1-Methylnaphthalene	517.8	772	3.60	
$C_{11}H_{10}$	2-Methylnaphthalene	514.2	761		
$C_{11}H_{22}O_2$	Ethyl nonanoate	500.1	674		
$C_{11}H_{24}$	Undecane	469.0	639	1.98	689
$C_{11}H_{24}O$	1-Undecanol	516	703.6	2.147	718
$C_{12}H_8$	Acenaphthylene	553	792	3.20	
$C_{12}H_9N$	Carbazole	627.84	901.8	3.13	454
$C_{12}H_{10}$	Biphenyl	529.2	773	3.38	497
$C_{12}H_{10}O$	Diphenyl ether	531.1	766.8		
$C_{12}H_{12}$	2,7-Dimethylnaphthalene	538	775	3.23	601
$C_{12}H_{18}$	Hexamethylbenzene	536.5	758		
$C_{12}H_{24}$	1-Dodecene	486.9	657.6	1.930	
$C_{12}H_{26}$	Dodecane	489.4	658	1.82	754
$C_{12}H_{26}O$	1-Dodecanol	532	719.4	1.994	
$C_{13}H_9N$	Acridine	618.01	891.1	3.21	548
$C_{13}H_9N$	Phenanthridine	622.0	895	3.6	548
$C_{13}H_{11}N$	9-Methylcarbazole	616.79	890	3.38	572
$C_{13}H_{12}$	Diphenylmethane	538.1	760	2.71	563
$C_{13}H_{26}O_2$	Methyl dodecanoate	540	712		
$C_{13}H_{28}$	Tridecane	508.5	675	1.68	823
$C_{13}H_{28}O$	1-Tridecanol	546.8	734	1.935	
$C_{14}H_{10}$	Anthracene	613.0	869.3		554
$C_{14}H_{10}$	Phenanthrene	613	869		
$C_{14}H_{30}$	Tetradecane	526.6	693	1.57	894
$C_{14}H_{30}O$	1-Tetradecanol	562	747	1.81	
$C_{15}H_{32}$	Pentadecane	543.7	708	1.48	966
$C_{16}H_{34}$	Hexadecane	559.9	723	1.40	1034
$C_{16}H_{34}O$	1-Hexadecanol	607	770	1.61	
$C_{17}H_{36}$	Heptadecane	575.1	736	1.34	1103
$C_{17}H_{36}O$	1-Heptadecanol	606	780	1.50	
$C_{18}H_{14}$	o-Terphenyl	605	857	2.99	731
$C_{18}H_{14}$	m-Terphenyl	636	883	2.48	724
$C_{18}H_{14}$	p-Terphenyl	649	908	2.99	729
$C_{18}H_{38}$	Octadecane	589.4	747	1.29	1189
$C_{18}H_{38}O$	1-Octadecanol	616.7	790	1.44	
$C_{19}H_{40}$	Nonadecane	603.0	755	1.16	
$C_{20}H_{42}$	Eicosane	616	768	1.07	
$C_{20}H_{42}O$	1-Eicosanol	582	809	1.30	
$C_{21}H_{44}$	Heneicosane	629.6	778	1.03	
$C_{22}H_{46}$	Docosane	641.7	786	0.98	
$C_{23}H_{48}$	Tricosane	653	790	0.92	
$C_{24}H_{50}$	Tetracosane	664.4	800	0.87	

SUBLIMATION PRESSURE OF SOLIDS

This table gives the sublimation pressure of some representative solids as a function of temperature. The last entry for each substance is the solid-liquid-gas triple point. Substances are listed by molecular formula in the Hill order. Note that the pressure is given in kilopascals (kPa), except for water and naphthalene, for which the pressure unit is pascals. For conversion, 1 kPa = 7.506 Torr = 0.0098692 atm.

Substance									
Ar Argon	T/K P/kPa	55 0.2	60 0.8	65 2.8	70 7.7	75 18.7	80 40.7	83.81 68.8	
BrH Hydrogen bromide	T/K P/kPa	135 0.1	140 0.3	150 1.1	160 3.3	170 8.7	180 20.1	185.1 27.4	
ClH Hydrogen chloride	T/K P/kPa	120 0.1	130 0.5	140 1.9	150 5.8	155 9.5	159.0 13.5		
F_4Si Silicon tetrafluoride	T/K P/kPa	130 0.2	140 0.9	150 3.9	160 14.0	170 43.8	175 74.2	180 122.4	186.3 220.8
F_6S Sulfur hexafluoride	T/K P/kPa	150 0.4	165 2.6	180 11.3	190 25.9	200 54.5	210 106.1	220 195.1	223.1 232.7
HI Hydrogen iodide	T/K P/kPa	160 0.2	170 0.8	180 2.2	190 5.3	200 11.7	210 23.6	220 44.1	222.4 49.3
H_2O Water	T/K P/Pa	190 0.032	210 0.702	225 4.942	240 27.28	250 76.04	260 195.8	270 470.1	273.16 611.66
H_2S Hydrogen sulfide	T/K P/kPa	140 0.2	150 0.6	160 1.9	165 3.2	170 5.2	175 8.3	180 12.7	187.6 22.7
H_3N Ammonia	T/K P/kPa	160 0.1	170 0.4	180 1.2	190 3.5	195 5.8	195.4 6.12		
Kr Krypton	T/K P/kPa	80 0.4	90 2.7	95 6.0	100 12.1	105 22.8	110 40.4	115.8 73.1	
NO Nitric oxide	T/K P/kPa	85 0.1	90 0.4	95 1.3	100 3.8	105 10.0	109.5 21.9		
Xe Xenon	T/K P/kPa	110 0.3	120 1.5	130 4.9	140 14.0	150 34.2	155 51.1	160 74.2	161.4 81.7
CO Carbon monoxide	T/K P/kPa	50 0.1	55 0.6	60 2.6	65 8.2	68.13 15.4			
CO_2 Carbon dioxide	T/K P/kPa	130 0.032	140 0.187	155 1.674	170 9.987	185 44.02	194.7 101.3	205 227.1	216.58 518.0
CHN Hydrogen cyanide	T/K P/kPa	200 0.2	210 0.4	220 1.0	230 2.2	240 4.8	250 9.7	255 13.6	259.83 18.62
CH_4 Methane	T/K P/kPa	65 0.1	70 0.3	75 0.8	80 2.1	85 4.9	90.69 11.70		
C_2H_2 Acetylene	T/K P/kPa	130 0.2	140 0.7	150 2.6	160 7.8	170 20.6	180 49.0	190 106.3	192.4 126.0
$C_2H_4O_2$ Acetic acid	T/K P/kPa	250 0.092	260 0.199	270 0.406	280 0.79	289.7 1.29			
C_5H_{12} Neopentane	T/K P/kPa	200 0.7	210 1.6	220 3.6	230 7.3	240 13.9	250 24.8	255 32.4	256.58 35.8
$C_{10}H_8$ Naphthalene	T/K P/Pa	250 0.036	270 0.514	280 1.662	290 4.918	300 13.43	310 34.15	330 182.9	353.43 999.6

VAPOR PRESSURE

This table gives vapor pressure data for about 1800 inorganic and organic substances. In order to accommodate elements and compounds ranging from refractory to highly volatile in a single table, the temperature at which the vapor pressure reaches specified pressure values is listed. The pressure values run in decade steps from 1 Pa (about 7.5 μm Hg) to 100 kPa (about 750 mm Hg). All temperatures are given in °C.

The data used in preparing the table came from a large number of sources; the main references used for each substance are indicated in the last column. Since the data were refit in most cases, values appearing in this table may not be identical with values in the source cited. The temperature entry in the 100 kPa column is close to, but not identical with, the normal boiling point (which is defined as the temperature at which the vapor pressure reaches 101.325 kPa). Although some temperatures are quoted to 0.1°C, uncertainties of several degrees should generally be assumed. Values followed by an "e" were obtained by extrapolating (usually with an Antoine equation) beyond the region for which experimental measurements were available and are thus subject to even greater uncertainty.

Compounds are listed by molecular formula following the Hill convention. Substances not containing carbon are listed first, followed by those that contain carbon. To locate an organic compound by name or CAS Registry Number when the molecular formula is not known, use the table *Physical Constants of Organic Compounds* in Section 3 and its indexes to determine the molecular formula. The indexes to *Physical Constants of Inorganic Compounds* in Section 4 can be used in a similar way.

More extensive and detailed vapor pressure data on selected important substances appear in other tables in this section of the *Handbook*. These substances are flagged by a symbol following the name as follows:

   *   See *Vapor Pressure of Fluids below 300 K*
  **   See *IUPAC Recommended Data for Vapor Pressure Calibration*
***   See *Vapor Pressure of Ice* and *Vapor Pressure of Water from 0 to 370°C*

The following notations appear after individual temperature entries:

  s — Indicates the substance is a solid at this temperature.
  e — Indicates an extrapolation beyond the region where experimental measurements exist.
  i — Indicates the value was calculated from ideal gas thermodynamic functions, such as those in the *JANAF Thermochemical Tables* (see Reference 8).

REFERENCES

1. Lide, D.R., and Kehiaian, H.V., *CRC Handbook of Thermophysical and Thermochemical Data*, CRC Press, Boca Raton, FL, 1994.
2. Stull, D., in *American Institute of Physics Handbook, Third Edition*, Gray, D.E., Ed., McGraw Hill, New York, 1972.
3. Hultgren, R., Desai, P.D., Hawkins, D.T., Gleiser, M., Kelley, K.K., and Wagman, D.D., *Selected Values of Thermodynamic Properties of the Elements*, American Society for Metals, Metals Park, OH, 1973.
4. Stull, D., *Ind. Eng. Chem.*, 39, 517, 1947.
5. *TRCVP, Vapor Pressure Database, Version 2.2P*, Thermodynamic Research Center, Texas A&M University, College Station, TX.
6. *TRC Thermodynamic Tables*, Thermodynamic Research Center, Texas A&M University, College Station, TX.
7. Ohe, S., *Computer Aided Data Book of Vapor Pressure*, Data Book Publishing Co., Tokyo, 1976.
8. Chase, M.W., Davies, C.A., Downey, J.R., Frurip, D.J., McDonald, R.A., and Syverud, A.N., *JANAF Thermochemical Tables, Third Edition*, *J. Phys. Chem. Ref. Data*, Vol. 14, Suppl. 1, 1985.
9. Barin, I., *Thermochemical Data of Pure Substances*, VCH Publishers, New York, 1993.
10. Jacobsen, R.T., et. al, *International Thermodynamic Tables of the Fluid State, No. 10. Ethylene*, Blackwell Scientific Publications, Oxford, 1988.
11. Wakeham, W.A., *International Thermodynamic Tables of the Fluid State, No. 12. Methanol*, Blackwell Scientific Publications, Oxford, 1993.
12. Janz, G.J., *Molten Salts Handbook*, Academic Press, New York, 1967.
13. Ohse, R.W. *Handbook of Thermodynamic and Transport Properties of Alkali Metals*, Blackwell Scientific Publications, Oxford, 1994.
14. Gschneidner, K.A., in *CRC Handbook of Chemistry and Physics, 77th Edition*, p. 4-112, CRC Press, Boca Raton, FL, 1996.
15. Leider, H.R., Krikorian, O.H., and Young, D.A., *Carbon*, 11, 555, 1973.
16. Ruzicka, K., and Majer, V., *J. Phys. Chem. Ref. Data*, 23, 1, 1994.
17. Tillner-Roth, R., and Baehr, H.D., *J. Phys. Chem. Ref. Data*, 23, 657, 1994.
18. Younglove, B.A., and McLinden, M.O., *J. Phys. Chem. Ref. Data*, 23, 731, 1994.
19. Outcalt, S.L., and McLinden, M.O., *J. Phys. Chem. Ref. Data*, 25, 605, 1996.
20. Weber, L.A., and Defibaugh, D.R., *J. Chem. Eng. Data*, 41, 382, 1996.
21. Rodrigues, M.F., and Bernardo-Gil, M.G., *J. Chem. Eng. Data*, 41, 581, 1996.
22. Piacente, V., Gigli, G., Scardala, P., and Giustini, A., *J. Phys. Chem.*, 100, 9815, 1996.
23. Barton, J.L., and Bloom, H., *J. Phys. Chem.*, 60, 1413, 1956.
24. Sense, K.A., Alexander, C.A., Bowman, R.E., and Filbert, R.B., *J. Phys. Chem.*, 61, 337, 1957.
25. Ewing, C.T., and Stern, K.H., *J. Phys. Chem.* 78, 1998, 1974.
26. Cady, G.H., and Hargreaves, G.B., *J. Chem. Soc.*, 1961, 1563; 1961, 1568.
27. Skudlarski, K., Dudek, J., and Kapala, J., *J. Chem. Thermodynamics*, 19, 857, 1987.
28. Wagner, W., and de Reuck, K.M., *International Thermodynamic Tables of the Fluid State, No. 9. Oxygen*, Blackwell Scientific Publications, Oxford, 1987.

VAPOR PRESSURE (continued)

29. Marsh, K.N., Editor, *Recommended Reference Materials for the Realization of Physicochemical Properties*, Blackwell Scientific Publications, Oxford, 1987.
30. Alcock, C.B., Itkin, V.P., and Horrigan, M.K., *Canadian Metallurgical Quarterly*, 23, 309, 1984.
31. Stewart, R.B., and Jacobsen, R.T., *J. Phys. Chem. Ref. Data*, 18, 639, 1989.
32. Sifner, O., and Klomfar, J., *J. Phys. Chem. Ref. Data*, 23, 63, 1994.
33. Bah, A., and Dupont-Pavlovsky, N., *J. Chem. Eng. Data,* 40, 869, 1995.
34. Behrens, R.G., and Rosenblatt, G., *J. Chem. Thermodynamics*, 4, 175, 1972.
35. Behrens, R.G., and Rosenblatt, G., *J. Chem. Thermodynamics*, 5, 173, 1973.
36. Haar, L., Gallagher, J.S., and Kell, G.S., *NBS/NRC Steam Tables*, Hemisphere Publishing Corp., New York, 1984.
37. Wagner, W., Saul, A., and Pruss, A., *J. Phys. Chem. Ref. Data*, 23, 515. 1994.
38. Behrens, R.G., Lemons, R.S., and Rosenblatt, G., *J. Chem. Thermodynamics*, 6, 457, 1974.
39. Boublik, T., Fried, V., and Hala, E., *The Vapor Pressure of Pure Substances, Second Edition,* Elsevier, Amsterdam, 1984.
40. Goodwin, R.D., *J. Phys. Chem. Ref. Data*, 14, 849, 1985.
41. Younglove, B.A., and Ely, J.F., *J. Phys. Chem. Ref. Data*, 16, 577, 1987.

Mol. Form.	Name	Temperature in °C for the indicated pressure						Ref.
		1 Pa	10 Pa	100 Pa	1 kPa	10 kPa	100 kPa	
Substances not containing carbon:								
Ag	Silver	1010	1140	1302	1509	1782	2160	2
AgBr	Silver(I) bromide	569 i	656 i	765 i	905 i	1093 i	1359 i	9
AgCl	Silver(I) chloride	670	769	873	1052	1264	1561	4
AgI	Silver(I) iodide	594	686	803	959	1177	1503	4
Al	Aluminum	1209	1359	1544	1781	2091	2517	2
AlB_3H_{12}	Aluminum borohydride				-46.8	-9.4	45.5	4
$AlCl_3$	Aluminum trichloride	58.4 s	76.5 s	97.1 s	120.7 s	148.2 s	180.5 s	4
AlF_3	Aluminum trifluoride	744 s	819 s	906 s	1008 s	1130 s	1276 s	8
AlI_3	Aluminum triiodide				218	285	385	4
Al_2O_3	Aluminum oxide			2122	2351	2629	2975	4
Ar	Argon*		-226.4 s	-220.3 s	-212.4 s	-201.7 s	-186.0	1,5,31
As	Arsenic	280 s	323 s	373 s	433 s	508 s	601 s	3
$AsCl_3$	Arsenic(III) chloride			-8 e	21.3	63.1	129.4	1
AsF_3	Arsenic(III) fluoride					8.1	56.0	4
AsI_3	Arsenic(III) iodide				187	261	367 e	7
As_2O_3	Arsenic(III) oxide (arsenolite)	133.7 s	163.0 s	196.8 s	236.2 s	283.0		34
At	Astatine	88 s	119 s	156 s	202 s	258 s	334	2
Au	Gold	1373	1541	1748	2008	2347	2805	2
B	Boron	2075	2289	2549	2868	3272	3799	2
BBr_3	Boron tribromide			-45 e	-15 e	27.5	90.4	1
BCl_3	Boron trichloride*			-94.0	-70.5	-37.4	12.3	4
BF_3	Boron trifluoride*	-173.9 s	-166.0 s	-156.0 s	-143.0 s	-125.9	-101.1	4
B_2F_4	Tetrafluorodiborane						-34	1
B_2H_6	Diborane			-162 e	-147.0	-125.8	-92.6	1
B_5H_9	Pentaborane(9)				-34.8	3.8	57.6	4
Ba	Barium	638 s	765	912	1115	1413	1897	9
Be	Beryllium	1189 s	1335	1518	1750	2054	2469	2
$BeBr_2$	Beryllium bromide	203 s	240 s	283 s	335 s	397 s	473 s	4
$BeCl_2$	Beryllium chloride	196 s	237 s	284 s	339 s	402 s	487	4
BeF_2	Beryllium fluoride		686 e	767 e	869	999	1172 e	7
BeI_2	Beryllium iodide	188 s	229 s	276 s	333 s	402 s	487	4
Bi	Bismuth	668	768	892	1052	1265	1562	2
$BiBr_3$	Bismuth tribromide			217 s	273 i	348 i	455 i	4,9
$BiCl_3$	Bismuth trichloride				248.9	328.6	438.7	1,4
BrCs	Cesium bromide	531 s	601 s	701 i	834 i	1019 i	1293 e	9
BrH	Hydrogen bromide*		-153.3 s	-140.4 s	-123.8 s	-101.5 s	-67.0	5
BrH_3Si	Bromosilane				-81.0	-47.3	2.2	4
BrH_4N	Ammonium bromide	121 s	154 s	195 s	246 s	310.4 s	395.1 s	5
BrK	Potassium bromide	597 s	674 s	773				25
BrLi	Lithium bromide		630	733	868	1049	1308	4
BrNa	Sodium bromide			791	931	1120	1389	4
BrRb	Rubidium bromide			766	903	1087	1350	4
BrTl	Thallium(I) bromide				509	635	817	4
Br_2	Bromine*	-87.7 s	-71.8 s	-52.7 s	-29.3 s	2.5	58.4	1
Br_2Cd	Cadmium bromide	373 s	435 s	509 s				27
Br_2Hg	Mercury(II) bromide	71 s	98 s	132 s	174 s	227 s	318	4
Br_2OS	Thionyl bromide	-49 e	-29 e	-5 e	27.8	72.9	139.6	5

Mol. Form.	Name	Temperature in °C for the indicated pressure						Ref.
		1 Pa	10 Pa	100 Pa	1 kPa	10 kPa	100 kPa	
Br_2Pb	Lead(II) bromide	374	431	502	597	726	914	4
Br_2S_2	Sulfur bromide	-7 e	15 e	42 e	78.4	128.1	200.9	5
Br_3In	Indium(III) bromide			304.6 s	328.7 s	364.8 s		1
Br_3OP	Phosphorus(V) oxybromide				64 e	115.5	191.4	5
Br_3P	Phosphorus(III) bromide		-23 e	5 e	42.3	94.6	172.6	5
Br_3Sb	Antimony(III) bromide				136.5	196.9	286.5	1
Br_4Ge	Germanium(IV) bromide				51	105	188	4
Br_4Sn	Tin(IV) bromide				67	122	204	4
Br_4Zr	Zirconium(IV) bromide	136 s	167 s	203 s	245 s	295 s	356 s	4
Br_5P	Phosphorus(V) bromide		-19 s	4 s	31 s	65.5 s	110.1	5
Ca	Calcium	591 s	683 s	798 s	954	1170	1482	2
Cd	Cadmium	257 s	310 s	381	472	594	767	2
$CdCl_2$	Cadmium chloride	412 s	471 s	541 s	634	768	959	23, 27
CdF_2	Cadmium fluoride				1257	1461	1742	4
CdI_2	Cadmium iodide	296 s	344 s	406	498	622	795	4,27
CdO	Cadmium oxide	770 s	866 s	983 s	1128 s	1314 s	1558 s	4
Ce	Cerium	1719	1921	2169	2481	2886	3432	14
ClCs	Cesium chloride			730	864	1043	1297	4
ClCu	Copper(I) chloride		459	543	675	914	1477	4
ClF	Chlorine fluoride*				-144.4	-122.6	-90.2	5
ClF_2P	Phosphorus(III) chloride difluoride				-119.5	-91.1	-47.6	5
ClF_3	Chlorine trifluoride				-63.7	-33.0	11.4	5
ClF_5	Chlorine pentafluoride				-88 e	-59	-14	7
ClH	Hydrogen chloride*				-138.2 s	-118.0	-85.2	1,5
$ClHO_3S$	Chlorosulfonic acid	-40 e	-20 e	5 e	38.7	85.0	153.6	5
ClH_4N	Ammonium chloride	91 s	121 s	159 s	204.7 s	263.1 s	339.5 s	5
ClK	Potassium chloride	625 s	704 s	804	945	1137	1411	23,25
ClLi	Lithium chloride		649 i	761 i	905 i	1101 i	1381 i	8
ClNO	Nitrosyl chloride		-116 s	-100 s	-78.7 s	-50.2	-5.7	5
$ClNO_2$	Nitryl chloride	-121 e	-113 e	-102 e	-86.1	-60.9	-15.7	5
ClNa	Sodium chloride	653 s	733 s	835	987	1182	1461	23,25
ClO_2	Chlorine dioxide*					-34.3	10.5	5
ClRb	Rubidium chloride			777	916	1105	1379	4
ClTl	Thallium(I) chloride				504	626	806	4
Cl_2	Chlorine*	-145 s	-133.7 s	-120.2 s	-103.6 s	-76.1	-34.2	1
Cl_2Co	Cobalt(II) chloride					818	1048	4
Cl_2FP	Phosphorus(III) dichloride fluoride				-71.1	-37.4	13.5	5
Cl_2F_3P	Phosphorus(V) dichloride trifluoride		-120 e	-101 e	-77.1	-44.3	3 e	7
Cl_2Fe	Iron(II) chloride				685	821	1025	4
Cl_2Hg	Mercury(II) chloride	64.4 s	94.7 s	130.8 s	174.5 s	228.5 s	304.0	4
Cl_2Mg	Magnesium chloride			762	908	1111	1414	4
Cl_2Mn	Manganese(II) chloride				760	933	1189	4
Cl_2Ni	Nickel(II) chloride	534 s	592 s	662 s	747 s	852 s	985 s	4
Cl_2OS	Thionyl chloride	-99 e	-81 e	-58 e	-27.1	14.6	75.2	5
Cl_2O_2S	Sulfuryl chloride				-27 e	11.8	69.0	5
Cl_2Pb	Lead(II) chloride			541 e	637	765	949	23
Cl_2S	Sulfur dichloride	-76 e	-61 e	-41 e	-16.7	15.3	58.7	5
Cl_2S_2	Sulfur chloride	-55 e	-36 e	-12 e	21.0	67.2	137.1	5
Cl_2Sn	Tin(II) chloride		253	308	381	479	622	4
Cl_2Zn	Zinc chloride	305 i	356 i	419 i	497 i	596 i	726 i	4,9,12
Cl_3Fe	Iron(III) chloride	118 s	153 s	190 s	229 s	268 s	319	4
Cl_3HSi	Trichlorosilane			-81 e	-56 e	-21 e	31.6	7
Cl_3N	Nitrogen trichloride				-25 e	13.2	70.6	5
Cl_3OP	Phosphorus(V) oxychloride					39.9	105.0	5
Cl_3P	Phosphorus(III) chloride	-93 e	-77 e	-55 e	-26.0	14.5	75.7	5
Cl_4Po	Polonium(IV) chloride					300.6	389.4	5
Cl_4Se	Selenium tetrachloride	23 s	45 s	71 s	102 s	141.4 s	191.1 s	5
Cl_4Si	Tetrachlorosilane*				-39 e	0 e	57.3	1
Cl_4Te	Tellurium tetrachloride				237 e	299.4	387.8	5
Cl_4Zr	Zirconium(IV) chloride	117 s	146 s	181 s	222 s	272 s	336 s	9
Cl_5P	Phosphorus(V) chloride	-2 s	19 s	44 s	74 s	111.4 s	158.9 s	5
Co	Cobalt	1517	1687	1892	2150	2482	2925	2
Cr	Chromium	1383 s	1534 s	1718 s	1950	2257	2669	2
Cs	Cesium	144.5	195.6	260.9	350.0	477.1	667.0	13,30

Mol. Form.	Name	Temperature in °C for the indicated pressure						Ref.
		1 Pa	10 Pa	100 Pa	1 kPa	10 kPa	100 kPa	
CsF	Cesium fluoride				825	999	1249	4
CsI	Cesium iodide	523 s	595 s	692	854	1029	1278	4,25
Cu	Copper	1236	1388	1577	1816	2131	2563	2
CuI	Copper(I) iodide				636	864	1331	4
Dy	Dysprosium	1105 s	1250 s	1431 i	1681 i	2031 i	2558 i	3
Er	Erbium	1231 s	1390 s	1612 i	1890 i	2279 i	2859 i	3
Eu	Europium	590 s	684 s	799 s	961	1179	1523	14
FH	Hydrogen fluoride*				-71.1	-33.7	19.2	1,5
FHO$_3$S	Fluorosulfonic acid	-14 e	4 e	28 e	59.1	101.3	162.2	5
FK	Potassium fluoride			869	1017	1216	1499	4
FLi	Lithium fluoride	801 s	896	1024	1188	1395	1672	4,12,25
FNO	Nitrosyl fluoride			-131 e	-116.1	-94.3	-60.1	5
FNO$_2$	Nitryl fluoride		-156 e	-144 e	-128.1	-106.0	-72.6	5
FNO$_3$	Fluorine nitrate	-160 e	-149 e	-135 e	-115.1	-87.4	-45.0	5
FNa	Sodium fluoride		920 s	1058	1218	1426	1702	4,12,24
FRb	Rubidium fluoride			910	1001	1145	1409	4,12
F$_2$	Fluorine*	-235 s	-229.5 s	-222.9 s	-214.8	-204.3	-188.3	1,5
F$_2$O	Fluorine monoxide*	-211.7	-204.7	-195.9	-184.2	-168.2	-144.9	5
F$_2$OS	Thionyl fluoride			-124 e	-106.5	-81.5	-44.1	5
F$_2$O$_2$Re	Rhenium(VI) dioxydifluoride				89.2	131.9	185 e	26
F$_2$Pb	Lead(II) fluoride				865	1054	1292	4
F$_2$Xe	Xenon difluoride			2.9 s	31.8 s	67.9 s	114 s	1,5
F$_2$Zn	Zinc fluoride	731 s	813 s	911 i	1048 i	1237 i	1503 i	9
F$_3$N	Nitrogen trifluoride*	-201 e	-194 e	-185 e	-172.8	-155.5	-129.2	5
F$_3$OP	Phosphorus(V) oxyfluoride	-124 s	-113 s	-100 s	-83.7 s	-64.1 s	-39.7 s	5
F$_3$P	Phosphorus(III) fluoride*				-152 e	-132.6	-101.4	5
F$_4$MoO	Molybdenum(VI) oxytetrafluoride	-21 s	3 s	33 s	69.3 s	117.3	184.1	26
F$_4$ORe	Rhenium(VI) oxytetrafluoride	5 s	26 s	50.7 s	80.1 s	117.1	171.2	26
F$_4$OW	Tungsten(VI) oxytetrafluoride	2 s	25 s	52.1 s	84.3 s	126.7	185.4	26
F$_4$S	Sulfur tetrafluoride				-110.0	-82.1	-40.3	5
F$_4$Se	Selenium tetrafluoride				13.6	51.6	104.7	5
F$_4$Si	Tetrafluorosilane*	-166 s	-157 s	-145.6 s	-132.3 s	-115.7 s	-94.9 s	4,7
F$_5$Mo	Molybdenum(V) fluoride				86.6	140.3	213 e	26
F$_5$Nb	Niobium(V) fluoride				80	140	224	4
F$_5$ORe	Rhenium(VII) oxypentafluoride	-103 s	-84 s	-59 s	-28 s	13.7 s	72.8	26
F$_5$Os	Osmium(V) fluoride			74.1	113.2	162.3	226 e	26
F$_5$P	Phosphorus(V) fluoride	-157 s	-148 s	-137 s	-124.5 s	-108.6 s	-84.8	5
F$_5$Re	Rhenium(V) fluoride			58.8	99.5	152 e	221 e	26
F$_5$Ta	Tantalum(V) fluoride				119		229	4
F$_6$Ir	Iridium(VI) fluoride	-88 s	-71 s	-51 s	-27 s	3.8 s	53.1	26
F$_6$Mo	Molybdenum(VI) fluoride	-98 s	-82 s	-64 s	-41.2 s	-13.4 s	33.5	26
F$_6$Os	Osmium(VI) fluoride	-89 s	-73 s	-54 s	-30.6 s	-1.7 s	47.4	26
F$_6$Re	Rhenium(VI) fluoride	-97 s	-82 s	-63 s	-40.2 s	-11.9 s	33.4	26
F$_6$S	Sulfur hexafluoride*	-158 s	-147 s	-133.6 s	-116.6 s	-94.4 s	-64.1 s	5
F$_6$Se	Selenium hexafluoride	-143 s	-132 s	-118 s	-100.7 s	-77.8 s	-46.5 s	5
F$_6$Te	Tellurium hexafluoride	-142 s	-130 s	-115 s	-96 s	-71.8 s	-39.1 s	5
F$_6$W	Tungsten(VI) fluoride	-107 s	-92 s	-74 s	-52.1 s	-24.8 s	16.9	26
F$_{10}$S$_2$	Sulfur decafluoride					-22.0	28.5	5
Fe	Iron	1455 s	1617	1818	2073	2406	2859	2
Fr	Francium	131 e	181 e	246 e	335 e	465 e	673 e	2
Ga	Gallium	1037	1175	1347	1565	1852	2245	2
Gd	Gadolinium	1563 i	1755 i	1994 i	2300 i	2703 i	3262 i	3
Ge	Germanium	1371	1541	1750	2014	2360	2831	2
HI	Hydrogen iodide*	-146 s	-135.2 s	-120.8 s	-101.9 s	-75.9 s	-35.9	5
HKO	Potassium hydroxide	520 e	601 e	704	842	1035	1325	4
HNO$_3$	Nitric acid				-37 e	-9 e	28.4	82.2	5
HN$_3$	Hydrazoic acid			-79 e	-54 e	-18.0	35.7	5
HNaO	Sodium hydroxide	513	605	722	874	1080	1377	4
H$_2$	Hydrogen*					-258.6	-252.8	1
H$_2$I$_2$Si	Diiodosilane				11.8	70.5	149.4	4
H$_2$O	Water***	-60.7 s	-42.2 s	-20.3 s	7.0	45.8	99.6	36,37
H$_2$O$_2$	Hydrogen peroxide				13 e	45 e	89.0	149.8	5
H$_2$O$_4$S	Sulfuric acid	72	103	140	187	248	330	4
H$_2$S	Hydrogen sulfide*		-149 s	-136 s	-118.9 s	-95.9 s	-60.5	1,5
H$_2$S$_2$	Hydrogen disulfide				-27 e	12.2	70.7	5
H$_2$Se	Hydrogen selenide	-145 s	-134 s	-120 s	-102.8 s	-78.9 s	-41.5	5

Mol. Form.	Name	Temperature in °C for the indicated pressure						Ref.
		1 Pa	10 Pa	100 Pa	1 kPa	10 kPa	100 kPa	
H_2Te	Hydrogen telluride					-46.6	-2.3	5
H_3ISi	Iodosilane				-47.7	-10.1	45.2	4
H_3N	Ammonia*	-139 s	-127 s	-112 s	-94.5 s	-71.3	-33.6	1,5,6
H_3NO	Hydroxylamine				43.7	73.3	109.8	4
H_3P	Phosphine*	-182 s	-173 s	-161 s	-145 s	-122.7	-88.0	5
H_4IN	Ammonium iodide	125 s	159 s	201 s	253 s	318.4 s	405.2 s	5
H_4N_2	Hydrazine				14.7	55.6	113 e	5
H_4Si	Silane*			-181	-165.4	-143.7	-111.8	4
He	Helium*					-270.6	-268.9	2
Hf	Hafnium	2416	2681	3004	3406	3921	4603	9
Hg	Mercury**	42.0	76.6	120.0	175.6	250.3	355.9	29,30
HgI_2	Mercury(II) iodide	85.1 s	115.6 s	152.4 s	197.8 s	255.1 s	353.6	4
Ho	Holmium	1159 s	1311 s	1502 i	1767 i	2137 i	2691 i	3
IK	Potassium iodide			731	866	1052	1322	4
ILi	Lithium iodide	545	619	710	824	972	1170	4
INa	Sodium iodide			753	883	1058	1301	4
IRb	Rubidium iodide			733	866	1045	1302	4
ITl	Thallium(I) iodide				520	644	821	4
I_2	Iodine (rhombic)	-12.8 s	9.3 s	35.9 s	68.7 s	108 s	184.0	1,2
I_2Pb	Lead(II) iodide			470	558	682	869	4
I_2Zn	Zinc iodide	301 s	351 s	409 s	488 i	598 i	750 i	9
I_3Sb	Antimony(III) iodide				214.9	292.0	401.2	4
I_4Sn	Tin(IV) iodide				167.1	242.7	347.7	4
I_4Zr	Zirconium(IV) iodide	187 s	220 s	259 s	305 s	361 s	430 s	4
In	Indium	923	1052	1212	1417	1689	2067	2
Ir	Iridium	2440 s	2684	2979	3341	3796	4386	2
K	Potassium	200.2	256.5	328	424	559	756.2	13,30
Kr	Krypton*	-214.0 s	-208.0 s	-199.4 s	-188.9 s	-174.6 s	-153.6	5
La	Lanthanum	1732 i	1935 i	2185 i	2499 i	2905 i	3453 i	3
Li	Lithium	524.3	612.3	722.1	871.2	1064.3	1337.1	13,30
Lu	Lutetium	1633 s	1829.8	2072.8	2380 i	2799 i	3390 i	3
Mg	Magnesium	428 s	500 s	588 s	698	859	1088	2
Mn	Manganese	955 s	1074 s	1220 s	1418	1682	2060	2
Mo	Molybdenum	2469 s	2721	3039	3434	3939	4606	2
MoO_3	Molybdenum(VI) oxide				801	935	1151	4
NO	Nitric oxide*	-201 s	-195 s	-188 s	-179.3 s	-168.1 s	-151.9	5
N_2	Nitrogen*	-236 s	-232 s	-226.8 s	-220.2 s	-211.1 s	-195.9	1,5
N_2O	Nitrous oxide*	-167 s	-157 s	-145.4 s	-131.1 s	-112.9 s	-88.7	5
N_2O_4	Nitrogen tetroxide	-92 s	-78 s	-61 s	-41.1 s	-16.6 s	28.7	5
N_2O_5	Nitrogen pentoxide	-71 s	-56 s	-40 s	-19.9 s	3.9 s	33.2	5
Na	Sodium	280.6	344.2	424.3	529	673	880.2	13,30
Nb	Niobium	2669	2934	3251	3637	4120	4740	2
Nd	Neodymium	1322.3	1501.2	1725.3	2023 i	2442 i	3063 i	3
Ne	Neon*	-261 s	-260 s	-258 s	-255 s	-252 s	-246.1	2
Ni	Nickel	1510	1677	1881	2137	2468	2911	2
OPb	Lead(II) oxide	724	816	928	1065	1241	1471	4
OSr	Strontium oxide	1789 s	1903 s	2047 s	2235 s	2488 s		4
O_2	Oxygen*				-211.9	-200.5	-183.1	1,28
O_2S	Sulfur dioxide*			-98 s	-80 s	-52.2	-10.3	1,5
O_2Se	Selenium dioxide	124.5 s	153.9 s	188 s	228 s	275 s	315 s	38
O_2Si	Silicon dioxide	1966 i	2149 i	2368 i				8
O_3	Ozone*	-189 e	-182 e	-172 e	-158 e	-139.7	-111.5	5
O_3P_2	Phosphorus(III) oxide				47.3	100.3	172.8	4
O_3S	Sulfur trioxide				-20 s	6.6 s	44.5	5
O_3Sb_2	Antimony(III) oxide (valentinite)	426.1 s	478 s	539 s	610 s	907	1420	4,35
O_5P_2	Phosphorus(V) oxide	285 s	328 s	377.5 s	434.4 s	500.5 s	591	4
O_7Re_2	Rhenium(VII) oxide	147 s	176 s	208 s	244 s	284 s	362	4
Os	Osmium	2887 s	3150	3478	3875	4365	4983	2
P	Phosphorus (white)	6 s	34 s	69	115	180	276	3,9
P	Phosphorus (red)	182 s	216 s	256 s	303 s	362 s	431 s	2,3
Pb	Lead	705	815	956	1139	1387	1754	2
PbS	Lead(II) sulfide	656 s	741 s	838 s	953 s	1088 s	1280	4
Pd	Palladium	1448 s	1624	1844	2122	2480	2961	2
Po	Polonium				573 e	730.2	963.3	5
Pr	Praseodymium	1497.7	1699.4	1954 i	2298 i	2781 i	3506 i	3
Pt	Platinum	2057	2277 e	2542	2870	3283	3821	2

Mol. Form.	Name	Temperature in °C for the indicated pressure						Ref.
		1 Pa	10 Pa	100 Pa	1 kPa	10 kPa	100 kPa	
Pu	Plutonium	1483	1680	1925	2238	2653	3226	2
Ra	Radium	546 s	633 s	764	936	1173	1526	2
Rb	Rubidium	160.4	212.5	278.9	368	496.1	685.3	13,30
Re	Rhenium	3030 s	3341	3736	4227	4854	5681	2
Rh	Rhodium	2015	2223	2476	2790	3132	3724	2
Rn	Radon*	-163 s	-152 s	-139 s	-121.4 s	-97.6 s	-62.3	5
Ru	Ruthenium	2315 s	2538	2814	3151	3572	4115	2
S	Sulfur	102 s	135	176	235	318	444	3
Sb	Antimony	534 s	603 s	738	946	1218	1585	2,3
Sc	Scandium	1372 s	1531 s	1733 i	1993 i	2340 i	2828 i	3
Se	Selenium	227	279	344	431	540	685	3
Si	Silicon	1635	1829	2066	2363	2748	3264	2
Sm	Samarium	728 s	833 s	967 s	1148 i	1402 i	1788 i	3
Sn	Tin	1224	1384	1582	1834	2165	2620	2
Sr	Strontium	523 s	609 s	717 s	866	1072	1373	2
Ta	Tantalum	3024	3324	3684	4122	4666	5361	2
Tb	Terbium	1516.1	1706.1	1928 i	2232 i	2640 i	3218 i	3
Tc	Technetium	2454 e	2725 e	3051 e	3453 e	3961 e	4621 e	2
Te	Tellurium			502 e	615 e	768.8	992.4	5
Th	Thorium	2360	2634	2975	3410	3986	4782	2
Ti	Titanium	1709	1898	2130 e	2419	2791	3285	2
Tl	Thallium	609	704	824	979	1188	1485	2
Tm	Thulium	844 s	962 s	1108 s	1297 s	1548 i	1944 i	3
U	Uranium	2052	2291	2586	2961	3454	4129	2
V	Vanadium	1828 s	2016	2250	2541	2914	3406	2
W	Tungsten	3204 s	3500	3864	4306	4854	5550	2
Xe	Xenon*	-190 s	-181 s	-170 s	-155.8 s	-136.6 s	-108.4	5,32
Y	Yttrium	1610.1	1802.3	2047 i	2354 i	2763 i	3334 i	3
Yb	Ytterbium	463 s	540 s	637 s	774 s	993 i	1192 i	3
Zn	Zinc	337 s	397 s	477	579	717	912 e	2
Zr	Zirconium	2366	2618	2924	3302	3780	4405	2

Substances containing carbon:

Mol. Form.	Name	1 Pa	10 Pa	100 Pa	1 kPa	10 kPa	100 kPa	Ref.
C	Carbon (graphite)		2566 s	2775 s	3016 s	3299 s	3635 s	15
CBrClF$_2$	Bromochloro-difluoromethane	-136 e	-123 e	-106 e	-83.4	-51.8	-4.3	1
CBrCl$_3$	Bromotrichloromethane				-6 e	38.9	104.4	5
CBrF$_3$	Bromotrifluoromethane*	-168 e	-156 e	-142 e	-122.8	-96.6	-58.1	5
CBrN	Cyanogen bromide				-13 s	17.7 s	61.0	1
CBr$_2$F$_2$	Dibromodifluoromethane		-110 e	-91 e	-66 e	-30 e	22.5	1
CBr$_4$	Tetrabromomethane				25.6 s	65.8 s	111.6	188.9	5
CClF$_3$	Chlorotrifluoromethane	-176 e	-167 e	-155 e	-139 e	-116 e	-81.7	5
CClN	Cyanogen chloride		-94.6 s	-78.1 s	-57 s	-29 s	13.0	5
CCl$_2$F$_2$	Dichlorodifluoromethane*	-150 e	-138 e	-122 e	-101.8	-73.1	-30.0	5
CCl$_2$O	Carbonyl chloride	-127 e	-113 e	-96 e	-73 e	-40.6	7.2	5
CCl$_3$F	Trichlorofluoromethane*		-107 e	-89 e	-63 e	-28.5	23.3	1,5
CCl$_3$NO$_2$	Trichloronitromethane		-59 e	-30 e	4.4	47.8	112.0	5
CCl$_4$	Tetrachloromethane*	-79.4 s	-70.8 s	-53.5 s	-24.4 s	15.8	76.2	1,5
CFN	Cyanogen fluoride		-135 s	-121.2 s	-104.1 s	-82.8 s	-46.2	1,5
CF$_4$	Tetrafluoromethane*	-199.9 s	-193 s	-183.9 s	-171.6	-153.9	-128.3	1,5
CHBrF$_2$	Bromodifluoromethane		-128 s	-111.4 s	-89.7 s	-59.7 s	-16 s	5
CHBr$_3$	Tribromomethane				30.5	78.3	148.8	1
CHClF$_2$	Chlorodifluoromethane*	-152	-141 e	-126 e	-107.1	-80.5	-41.1	5
CHCl$_2$F	Dichlorofluoromethane	-76 e	-70 e	-61 e	-49 e	-28.7	8.6	1
CHCl$_3$	Trichloromethane*			-61 e	-34 e	4.3	60.8	1
CHF$_3$	Trifluoromethane*			-152 e	-136 e	-114.4	-82.3	1
CHI$_3$	Triiodomethane	51.1 s	82.7 s	121 e			218.0	5
CHN	Hydrogen cyanide*			-77 s	-52.6 s	-22.7 s	25.4	1,5
CHNO	Cyanic acid			-81.1	-56.8	-23.9	23 e	5
CH$_2$BrCl	Bromochloromethane	-83 e	-69 e	-50 e	-25 e	11.4	67.7	1
CH$_2$Br$_2$	Dibromomethane			-37 e	-7 e	35.2	96.5	5
CH$_2$ClF	Chlorofluoromethane		-124 e	-108 e	-86.2	-55.7	-9.4	5
CH$_2$Cl$_2$	Dichloromethane*		-92 e	-73 e	-48 e	-12.5	39.3	1
CH$_2$F$_2$	Difluoromethane*	-156.7	-145.8	-131.9	-113.6	-88.6	-51.9	1
CH$_2$I$_2$	Diiodomethane			17 e	55 e	106.1	181.6	5
CH$_2$O	Formaldehyde*				-91 e	-61.7	-19.3	1

Mol. Form.	Name	Temperature in °C for the indicated pressure						Ref.
		1 Pa	10 Pa	100 Pa	1 kPa	10 kPa	100 kPa	
CH$_2$O$_2$	Formic acid	-56 s	-40.4 s	-22.3 s	-0.8 s	37.0	100.2	1,5
CH$_3$AsF$_2$	Methyldifluoroarsine				-15 e	22.1	76.1	5
CH$_3$BO	Borane carbonyl				-124	-99	-64	4
CH$_3$Br	Bromomethane				-77 e	-44.3	3.3	1
CH$_3$Cl	Chloromethane*	-140.2 s	-128.6 s	-114.7 s	-96 e	-67.1	-24.4	1,33
CH$_3$Cl$_3$Si	Methyltrichlorosilane		-83 e	-61 e	-33 e	7 e	65.7	1
CH$_3$F	Fluoromethane*				-130 e	-111 e	-78.6	1
CH$_3$I	Iodomethane				-49 e	-12.4	42.1	1
CH$_3$NO	Formamide		22 e	53 e	93 e	145.0	218 e	5
CH$_3$NO$_2$	Nitromethane				-2 e	40 e	100.8	1
CH$_3$NO$_3$	Methyl nitrate		-75 e	-55 e	-27 e	9.8	63 e	5
CH$_4$	Methane*	-220 s	-214.2 s	-206.8 s	-197 s	-183.6 s	-161.7	5,41
CH$_4$Cl$_2$Si	Dichloromethylsilane			-77 e	-51 e	-14 e	40.5	1
CH$_4$O	Methanol*	-87 e	-69 e	-47.5	-20.4	15.2	64.2	11
CH$_4$S	Methanethiol		-115 e	-97 e	-74 e	-41.7	5.7	1
CH$_5$ClSi	Chloromethylsilane	-129 e	-115 e	-97.9	-74.4	-41.5	8.3	5
CH$_5$N	Methylamine				-76.7	-48.1	-6.6	1
CH$_6$N$_2$	Methylhydrazine			-31 e	-4.7	32.9	91 e	1
CH$_6$OSi	Methyl silyl ether				-90.2	-61.8	-18 e	1
CH$_6$Si	Methylsilane			-144 e	-124.6	-97.5	-57.5	5
CIN	Cyanogen iodide						153.8	5
CNNa	Sodium cyanide		672 e	798	961	1182	1497	4
CN$_4$O$_8$	Tetranitromethane				18.0	61.8	124 e	5
CO	Carbon monoxide*			-223 s	-216.5 s	-207.2 s	-191.7	40
COS	Carbon oxysulfide*			-136 e	-117 e	-90.0	-50.4	1
COSe	Carbon oxyselenide			-120	-98	-67	-22	4
CO$_2$	Carbon dioxide*	-159.1 s	-148.9 s	-136.7 s	-121.6 s	-103.1 s	-78.6 s	5
CS$_2$	Carbon disulfide		-96 e	-76 e	-49 e	-10.9	45.9	1
CSe$_2$	Carbon diselenide			-24 e	9.4	56.2	127 e	1
C$_2$Br$_2$ClF$_3$	1,2-Dibromo-1-chloro-1,2,2-trifluoroethane						92.3	5
C$_2$Br$_2$F$_4$	1,2-Dibromotetrafluoroethane		-97 e	-75 e	-46 e	-7.2	47.1	5
C$_2$Br$_4$	Tetrabromoethylene		-54.5 s	-31.7 s	-3.5 s	32.2 s	226.0	5
C$_2$ClF$_3$	Chlorotrifluoroethylene	-146 e	-134 e	-119 e	-99 e	-71 e	-28.4	1
C$_2$ClF$_5$	Chloropentafluoroethane					-80.3	-39.4	1
C$_2$Cl$_2$F$_4$	1,1-Dichlorotetrafluoroethane					-45.4	2.7	5
C$_2$Cl$_2$F$_4$	1,2-Dichlorotetrafluoroethane				-76.8	-44.9	3.2	5
C$_2$Cl$_3$F$_3$	1,1,1-Trichlorotrifluoroethane						45.6	1,5
C$_2$Cl$_3$F$_3$	1,1,2-Trichlorotrifluoroethane					-8.2	47.3	1,5
C$_2$Cl$_3$N	Trichloroacetonitrile				-16 e	25.3	85.1	1
C$_2$Cl$_4$	Tetrachloroethylene			-22 e	10 e	54.4	120.7	1
C$_2$Cl$_4$F$_2$	1,1,1,2-Tetrachloro-2,2-difluoroethane				-7 e	31.0	91.1	5
C$_2$Cl$_4$F$_2$	1,1,2,2-Tetrachloro-1,2-difluoroethane					32.3	92.5	1
C$_2$Cl$_4$O	Trichloroacetyl chloride			-25 e	7 e	51.7	117.8	1,5
C$_2$Cl$_6$	Hexachloroethane	-7.6 s	9.9 s	33.6 s	67.7 s	116.9 s	184.2 s	5
C$_2$F$_3$N	Trifluoroacetonitrile				-126.1	-102.5	-67.8	1
C$_2$F$_4$	Tetrafluoroethylene				-132.3	-109.7	-75.8	1
C$_2$F$_4$N$_2$O$_4$	1,1,2,2-Tetrafluoro-1,2-dinitroethane				-30 e	6.4	59.5	5
C$_2$F$_6$	Hexafluoroethane**				-155.2 s	-137.5 s	-113.4 s	-78.4 s	1,5
C$_2$HBrClF$_3$	2-Bromo-2-chloro-1,1,1-trifluoroethane				-41.4	-4.8	49.8	1
C$_2$HBr$_3$O	Tribromoacetaldehyde			15.0	52.7	103.0	173.5	5
C$_2$HClF$_4$	1-Chloro-1,1,2,2-tetrafluoroethane				-110 e	-87.6	-57.0	-12.1	5
C$_2$HCl$_2$F$_3$	2,2-Dichloro-1,1,1-trifluoroethane		-101.0	-82.2	-57.4	-23.3	26.7	18
C$_2$HCl$_3$	Trichloroethylene	-74 e	-59 e	-39 e	-12 e	26.7	86.8	1
C$_2$HCl$_3$O	Trichloroacetaldehyde			-41.6	-9.8	33.8	97.4	5
C$_2$HCl$_3$O$_2$	Trichloroacetic acid				83.8	130.0	197.2	1,5
C$_2$HCl$_5$	Pentachloroethane		-23 e	3 e	37.4	86.0	159.4	1
C$_2$HF$_3$O$_2$	Trifluoroacetic acid					16.8	71.4	1,5
C$_2$HF$_5$O	Trifluoromethyl difluoromethyl ether	-147 e	-136 e	-121 e	-102 e	-75.0	-35.4	20
C$_2$H$_2$	Acetylene*			-146.6 s	-130.7 s	-110.6 s	-84.8 s	5

Mol. Form.	Name	Temperature in °C for the indicated pressure						Ref.
		1 Pa	10 Pa	100 Pa	1 kPa	10 kPa	100 kPa	
$C_2H_2Br_2$	cis-1,2-Dibromoethylene		-45 e	-21 e	10 e	52.2	114.8	1
$C_2H_2Br_2$	trans-1,2-Dibromoethylene				-4 e	42.2	107.4	5
$C_2H_2Br_2Cl_2$	1,2-Dibromo-1,1-dichloroethane					103.6	177.8	5
$C_2H_2Br_2Cl_2$	1,2-Dibromo-1,2-dichloroethane		-11 e	22 e	64.1	119 e	193 e	5
$C_2H_2Br_4$	1,1,2,2-Tetrabromoethane	14 e	38 e	69 e	109 e	163.7	242.9	5
$C_2H_2Cl_2$	1,1-Dichloroethylene	-116 e	-101 e	-82 e	-57 e	-21.4	31.2	1
$C_2H_2Cl_2$	cis-1,2-Dichloroethylene			-62 e	-34 e	3.8	60.3	1
$C_2H_2Cl_2$	trans-1,2-Dichloroethylene				-44 e	-7.5	47.3	1
$C_2H_2Cl_2F_2$	1,2-Dichloro-1,1-difluoroethane	-101 e	-87 e	-68 e	-42.2	-6.8	46.3	5
$C_2H_2Cl_2O$	Chloroacetyl chloride			-23.7	5.6	46.1	105.6	5
$C_2H_2Cl_4$	1,1,1,2-Tetrachloroethane	-58 e	-40 e	-15 e	17 e	62.2	129.7	1
$C_2H_2Cl_4$	1,1,2,2-Tetrachloroethane		-22 e	1 e	32.4	76.9	144.7	1
$C_2H_2F_4$	1,1,1,2-Tetrafluoroethane				-94.3	-66.8	-26.4	17
$C_2H_2F_4$	1,1,2,2-Tetrafluoroethane				-96.0	-66.9	-23.3	5
C_2H_2O	Ketene		-151 e	-135 e	-115 e	-88.2	-50.0	1
C_2H_3Br	Bromoethylene	-124 e	-110 e	-92 e	-68 e	-34.5	15.4	5
C_2H_3BrO	Acetyl bromide	-78 e	-65 e	-49 e	-25 e	13.9	84 e	5
$C_2H_3Br_3$	1,1,2-Tribromoethane	-18 e	4 e	32 e	68 e	117.1	188.4	5
C_2H_3Cl	Chloroethylene	-139 e	-127 e	-110 e	-89 e	-59.0	-14.1	1
$C_2H_3ClF_2$	1-Chloro-1,1-difluoroethane		-123 e	-107 e	-85.3	-55.4	-10.5	5
C_2H_3ClO	Acetyl chloride	-100 e	-85 e	-66 e	-40 e	-3.6	50.4	1
$C_2H_3ClO_2$	Chloroacetic acid				78.4	123.9	188.9	1
$C_2H_3Cl_2F$	1,1-Dichloro-1-fluoroethane		-101 e	-83 e	-57.9	-22.7	31.4	5
$C_2H_3Cl_2F$	1,2-Dichloro-1-fluoroethane			-50 e	-23.8	14.1	73.4	5
$C_2H_3Cl_3$	1,1,1-Trichloroethane				-25.3	14.2	73.7	5
$C_2H_3Cl_3$	1,1,2-Trichloroethane			-23 e	7 e	49.9	113.4	1
C_2H_3F	Fluoroethylene				-153.3	-135.2	-109.9	-72.2	5
C_2H_3FO	Acetyl fluoride					-64.1	17.0	5
$C_2H_3F_3$	1,1,1-Trifluoroethane				-113 e	-86.6	-47.8	1
$C_2H_3F_3O$	2,2,2-Trifluoroethanol			-33 e	-8 e	26.0	74 e	5
C_2H_3I	Iodoethylene				-41 e	-3 e	55.6	5
C_2H_3IO	Acetyl iodide				-0.6	47 e	107.0	5
C_2H_3N	Acetonitrile				-20 e	21.4	81.2	1
C_2H_3NO	Methylisocyanate				-43.5	-10.2	38.8	1
C_2H_3NS	Methyl thiocyanate			-18.4	16.2	63.5	132.5	5
C_2H_4	Ethylene*				-155.6	-135.1	-104.0	1,10
C_2H_4BrCl	1-Bromo-2-chloroethane				-0.4	41.7	105.7	6
$C_2H_4Br_2$	1,1-Dibromoethane		-49 e	-26 e	5 e	46.4	107.6	5
$C_2H_4Br_2$	1,2-Dibromoethane				18 e	62.2	130.9	1
C_2H_4ClF	1-Chloro-1-fluoroethane				-69.9	-36.1	15.8	5
$C_2H_4Cl_2$	1,1-Dichloroethane		-84 e	-64 e	-36.7	1.0	56.9	1
$C_2H_4Cl_2$	1,2-Dichloroethane				-16.4	23.7	83.1	1
$C_2H_4F_2$	1,1-Difluoroethane				-115.2	-94.6	-66.1	-24.3	19
$C_2H_4N_2O_6$	Ethylene glycol dinitrate	4 e	25.6	51.0	81 e	117 e	162 e	5
C_2H_4O	Acetaldehyde		-105 e	-87 e	-62.8	-29.4	20.0	5
C_2H_4O	Ethylene oxide		-111 e	-93 e	-70 e	-37.0	10.2	1
$C_2H_4O_2$	Acetic acid	-42.8 s	-26.7 s	-8 s	14.2 s	55.9	117.5	1,5
$C_2H_4O_2$	Methyl formate		-95 e	-76 e	-51.8	-18.1	31.4	5
$C_2H_4O_3$	Peroxyacetic acid				14.4	55.3	109.7	5
$C_2H_4O_3$	Glycolic acid						99.9	5
$C_2H_5AsF_2$	Ethyldifluoroarsine			-36 e	-6.0	35.0	93.1	5
C_2H_5Br	Bromoethane	-111 e	-96 e	-77 e	-51.3	-15.5	38.0	5
C_2H_5Cl	Chloroethane	-126 e	-112 e	-94 e	-70 e	-37.0	12.0	1
C_2H_5ClO	2-Chloroethanol	-61 e	-39 e	-12 e	23 e	67.1	127.3	5
C_2H_5ClO	Chloromethyl methyl ether	-96 e	-80 e	-59 e	-32 e	6 e	61 e	5
$C_2H_5Cl_3OSi$	Trichloroethoxysilane	-78 e	-60 e	-36.0	-4.6	38.7	102.0	5
$C_2H_5Cl_3Si$	Trichloroethylsilane	-79 e	-61 e	-38 e	-8 e	34.9	98.7	5
C_2H_5F	Fluoroethane			-142 e	-127 e	-106.3	-78.7	-37.9	1
C_2H_5FO	2-Fluoroethanol				-22 e	8.3	47.5	99 e	5
C_2H_5I	Iodoethane	-94 e	-78 e	-56 e	-27.9	11.9	71.9	5
C_2H_5N	Ethyleneimine		-74 e	-55 e	-30 e	4.1	55 e	5
C_2H_5NO	Acetamide	16.7 s	39.1 s	65.2 s	102.8	150.8	218.2	5
C_2H_5NO	N-Methylformamide		13 e	41 e	78 e	127.9	199.1	1
$C_2H_5NO_2$	Nitroethane	-61 e	-44 e	-21 e	8.3	50.1	113.5	5
$C_2H_5NO_3$	Ethyl nitrate	-81 e	-63 e	-41 e	-12 e	28.2	87 e	1
C_2H_6	Ethane*	-183.3 s	-173.2	-161.3	-145.3	-122.8	-88.8	41
$C_2H_6Cl_2Si$	Dichlorodimethylsilane					11.1	70.1	5

Mol. Form.	Name	Temperature in °C for the indicated pressure						Ref.
		1 Pa	10 Pa	100 Pa	1 kPa	10 kPa	100 kPa	
C_2H_6Hg	Dimethyl mercury				-13.5	29.0	92.1	5
$C_2H_6N_2O$	N-Nitrosodimethylamine				30.7	80.5	149.8	5
C_2H_6O	Ethanol	-73 e	-56 e	-34 e	-7 e	29.2	78.0	1,5
C_2H_6O	Dimethyl ether*		-135 e	-118 e	-96.8	-67.6	-25.1	1,5
C_2H_6OS	Dimethyl sulfoxide			27.4	65.0	115.9	188.6	1
$C_2H_6O_2$	Ethylene glycol	2 e	24 e	51.1	86.1	132.5	196.9	1
$C_2H_6O_2$	Ethyl hydroperoxide	-70 e	-49 e	-25 e	6.8	47.0	101 e	5
$C_2H_6O_2S$	Dimethyl sulfone				109 e	166.8	248.9	5
C_2H_6S	Ethanethiol	-112 e	-97 e	-78 e	-53 e	-18 e	34.7	1
C_2H_6S	Dimethyl sulfide			-96 e	-77 e	-51.2	-16.0	37.0	1,5
$C_2H_6S_2$	Dimethyl disulfide	-71 e	-53 e	-29 e	1.7	45.0	109.3	5
$C_2H_7BO_2$	Dimethoxyborane	-116 e	-101.9	-83.5	-59.2	-25.4	25 e	5
C_2H_7N	Ethylamine			-71 e	-53 e	-27 e	16.4	1
C_2H_7N	Dimethylamine			-88 e	-66.9	-37.2	6.6	1
C_2H_7NO	Ethanolamine		11 e	35 e	66.2	109.0	170.6	1
$C_2H_8N_2$	1,2-Ethanediamine				17.0	57.5	116.6	1,5
$C_2H_8N_2$	1,1-Dimethylhydrazine			-52 e	-25.6	10.5	63 e	5
$C_2H_8N_2$	1,2-Dimethylhydrazine		-49 e	-33 e	-9 e	26.4	88 e	1
C_2N_2	Cyanogen	-127 s	-114.1 s	-98.5 s	-79.2 s	-54.9 s	-21.4	5
C_3ClF_5O	Chloropentafluoroacetone	-122 e	-109 e	-93 e	-71 e	-39.4	7.4	5
C_3Cl_6	Hexachloropropene	-12 e	11 e	40 e	79 e	132.8	213.6	5
C_3F_6	Perfluoropropene	-150 e	-138 e	-122 e	-101 e	-72 e	-30.6	5
C_3F_6O	Perfluoroacetone			-113 e	-94 e	-67.8	-27.6	5
C_3F_8	Perfluoropropane		-139 e	-124 e	-105 e	-77.5	-37.0	1
C_3HN	Cyanoacetylene			-58.7 s	-35.6 s	-7 s	42.0	5
$C_3H_2F_6O$	1,1,1,3,3,3-Hexafluoro-2-propanol					12.7	57.1	5
$C_3H_3F_5$	1,1,1,2,2-Pentafluoropropane					-60 e	-17.9	5
C_3H_3N	2-Propenenitrile		-72 e	-50 e	-22 e	17.7	77.0	1
C_3H_3NS	Thiazole					54.4	117.8	5
C_3H_4	Allene*		-129 e	-118 e	-101.4	-76.7	-34.7	5
C_3H_4	Propyne				-94 e	-65.3	-23.2	1
$C_3H_4ClF_3$	3-Chloro-1,1,1-trifluoropropane	-102 e	-87 e	-68 e	-43 e	-8 e	45.3	5
$C_3H_4Cl_2O$	1,1-Dichloroacetone				1 e	47.8	118.0	5
$C_3H_4Cl_2O_2$	Methyl dichloroacetate	-44 e	-25 e	0 e	33 e	77.7	142.3	5
$C_3H_4Cl_4$	1,1,1,2-Tetrachloropropane	-48 e	-28 e	-2 e	32 e	79.1	149.5	5
$C_3H_4F_4O$	2,2,3,3-Tetrafluoro-1-propanol			-10 e	17 e	53.9	107.2	5
C_3H_4O	Acrolein		-87 e	-67 e	-40 e	-3.0	52.8	1
$C_3H_4O_2$	Propenoic acid				35 e	78.0	140.7	1
$C_3H_4O_2$	Vinyl formate			-58 e	-34 e	-1.6	46.2	1
$C_3H_4O_2$	2-Oxetanone		-21 e	8 e	45.5	93.8	159.3	5
$C_3H_4O_3$	Ethylene carbonate	12.7 s	37 e				247	5
C_3H_5Br	cis-1-Bromopropene	-100 e	-84 e	-64 e	-37 e	1.0	57.4	5
C_3H_5Br	2-Bromopropene	-112 e	-95 e	-75 e	-47 e	-9 e	48.0	5
C_3H_5Br	3-Bromopropene	-98 e	-80 e	-58 e	-28 e	12 e	69.6	5
C_3H_5Cl	cis-1-Chloropropene	-114 e	-100 e	-81 e	-55 e	-20.1	32.4	5
C_3H_5Cl	trans-1-Chloropropene		-97 e	-77 e	-52 e	-16.2	37.0	5
C_3H_5Cl	2-Chloropropene	-120 e	-106 e	-87 e	-63 e	-28.7	22.3	5
C_3H_5Cl	3-Chloropropene	-107 e	-92 e	-72.4	-46.3	-9.8	44.6	5
C_3H_5ClO	Epichlorohydrin			-21 e	11 e	53.8	115.5	5
$C_3H_5ClO_2$	Methyl chloroacetate		-28 e	-5 e	25 e	66.9	129.1	5
$C_3H_5Cl_3$	1,1,3-Trichloropropane	-51 e	-31 e	-5 e	28 e	75.3	145.1	5
$C_3H_5Cl_3$	1,2,3-Trichloropropane			2 e	37 e	84.9	156.3	5
$C_3H_5Cl_3Si$	Trichloro-2-propenylsilane					53.0	116.5	5
C_3H_5I	3-Iodopropene	-80 e	-62 e	-39 e	-8 e	36 e	101.5	5
C_3H_5N	Propanenitrile	-69.4	-55.3	-36.0	-7.9	35.2	97.4	1,5
C_3H_5NO	Acrylamide			109.6	161 e			5
C_3H_5NO	3-Hydroxypropanenitrile	-11 e	18 e	53 e	96.1	150.3	220.8	5
C_3H_5NS	Ethyl thiocyanate	-39 e	-20 e	4 e	35 e	79.1	143.4	5
C_3H_5NS	Ethyl isothiocyanate				17.4	66 e	136 e	5
$C_3H_5N_3O_9$	Trinitroglycerol	48.6	75.7	118 e	191 e	353 e	1007 e	5
C_3H_6	Propene*	-160.6	-149.0	-134.3	-114.9	-88.2	-47.9	1,5
C_3H_6	Cyclopropane			-124 e	-104 e	-75.7	-33.1	1
C_3H_6BrCl	1-Bromo-3-chloropropane	-51 e	-31 e	-6 e	28 e	74.1	142.9	5
$C_3H_6Br_2$	1,2-Dibromopropane	-46 e	-26 e	-2 e	31 e	75.3	139.5	5

Mol. Form.	Name	Temperature in °C for the indicated pressure						Ref.
		1 Pa	10 Pa	100 Pa	1 kPa	10 kPa	100 kPa	
$C_3H_6Br_2$	1,3-Dibromopropane	-30 e	-9 e	17 e	52 e	98.7	166.8	5
$C_3H_6Cl_2$	1,1-Dichloropropane				-14 e	27.0	87.7	5
$C_3H_6Cl_2$	1,2-Dichloropropane	-78 e	-61 e	-38.1	-8.1	33.7	95.9	5
$C_3H_6Cl_2$	1,3-Dichloropropane	-65 e	-46 e	-22 e	10 e	54.0	119.9	5
$C_3H_6Cl_2$	2,2-Dichloropropane				-28 e	10.8	68.9	5
$C_3H_6Cl_2O$	1,3-Dichloro-2-propanol			21.8	59.0	107.6	173.9	5
$C_3H_6N_2O_4$	1,1-Dinitropropane	-9 e	12 e	39 e	73.2	120 e	187 e	5
C_3H_6O	Allyl alcohol	-63 e	-48 e	-21.9	6.8	44.5	96.2	5
C_3H_6O	Methyl vinyl ether			-114 e	-89 e	-52.7	4.6	1
C_3H_6O	Propanal			-69 e	-42 e	-6 e	47.7	1
C_3H_6O	Acetone	-95	-81.8	-62.8	-35.6	1.3	55.7	1,5
C_3H_6O	Methyloxirane	-109 e	-95 e	-76 e	-51.5	-17.2	33.9	5
$C_3H_6O_2$	Propanoic acid			0 e	35.1	79.9	140.8	1,5
$C_3H_6O_2$	Ethyl formate		-80 e	-61 e	-35 e	1 e	54.0	1
$C_3H_6O_2$	Methyl acetate	-95 e	-79 e	-59 e	-33 e	3.3	56.6	1
$C_3H_6O_2$	1,3-Dioxolane		-72 e	-50 e	-22 e	17.0	75.3	1
$C_3H_6O_3$	1,3,5-Trioxane				53 e	113.7		1
C_3H_6S	Thiacyclobutane		-62 e	-40 e	-9 e	32.5	94.5	5
C_3H_7Br	1-Bromopropane	-95 e	-78 e	-57 e	-28 e	11.6	70.6	1
C_3H_7Br	2-Bromopropane		-84 e	-65 e	-39.6	-1.7	59.1	1,5
C_3H_7Cl	1-Chloropropane	-106 e	-90 e	-71 e	-44.5	-8.1	46.2	1
C_3H_7Cl	2-Chloropropane		-91 e	-74 e	-51.1	-17.8	35.4	1,5
C_3H_7ClO	2-Chloro-1-propanol				23 e	63.8	125.7	5
C_3H_7F	1-Fluoropropane	-133 e	-120 e	-103 e	-80.7	-49.4	-2.8	5
C_3H_7I	1-Iodopropane	-78 e	-60 e	-37 e	-6 e	36.9	102.0	5
C_3H_7I	2-Iodopropane	-89 e	-71 e	-47 e	-16.3	26.5	89.2	5
C_3H_7N	Allylamine		-88 e	-65 e	-37 e	0.4	52 e	5
C_3H_7NO	N,N-Dimethylformamide	-39 e	-20 e	5 e	38.0	83.9	152.6	1
C_3H_7NO	N-Methylacetamide	-13.3 s	13 s	43 e	83.8	136.1	206.3	5
$C_3H_7NO_2$	1-Nitropropane	-56 e	-37 e	-13 e	20 e	64.8	130.8	1
$C_3H_7NO_2$	2-Nitropropane		-48 e	-22 e	10.7	55.6	119.8	1
$C_3H_7NO_3$	Propyl nitrate			-23.9	6.1	48.1	111 e	5
C_3H_8	Propane*	-156.9	-145.6	-130.9	-111.4	-83.8	-42.3	1,41
C_3H_8O	1-Propanol	-54 e	-38 e	-16 e	10 e	47 e	96.9	1,5
C_3H_8O	2-Propanol	-65 e	-49 e	-28 e	-1.3	33.6	82.0	1,5
C_3H_8O	Ethyl methyl ether	-98 e	-89 e	-77 e	-60 e	-34.8	7.0	5
$C_3H_8O_2$	1,2-Propylene glycol	-11 e	13 e	42 e	78 e	125.0	187.2	5
$C_3H_8O_2$	1,3-Propylene glycol	4 e	30 e	62 e	101 e	149.9	214.0	5
$C_3H_8O_2$	Ethylene glycol monomethyl ether	-57 e	-37 e	-12 e	21 e	63.8	124.3	1
$C_3H_8O_2$	Dimethoxymethane	-93 e	-81 e	-64 e	-42 e	-9.3	41.7	5
$C_3H_8O_3$	Glycerol	96 e	113 e	136 e	168 e	213.4	287 e	1
C_3H_8S	1-Propanethiol	-94 e	-78 e	-57 e	-29.1	9.6	67.4	1,5
C_3H_8S	2-Propanethiol	-102 e	-87 e	-67 e	-41 e	-3 e	52.2	1
C_3H_8S	Ethyl methyl sulfide	-94 e	-78 e	-57 e	-29.7	8.8	66.3	1
$C_3H_8S_2$	1,3-Propanedithiol	-53 e	-28 e	3 e	43 e	97 e	172.4	5
C_3H_9As	Trimethylarsine			-74 e	-45 e	-5.4	52.0	5
$C_3H_9BO_3$	Trimethyl borate			-14 e	15.6	67.9		5
C_3H_9BS	Methyl dimethylthioborane			-62 e	-30.4	11.4	70.7	5
C_3H_9ClSi	Trimethylchlorosilane			-37.8	0.4	57.3		5
C_3H_9N	Propylamine		-81 e	-63 e	-38.3	-4.1	46.9	1,5
C_3H_9N	Isopropylamine		-91 e	-74 e	-50.4	-17.6	31.5	1,5
C_3H_9N	Trimethylamine		-114 e	-97 e	-75.0	-43.8	2.6	1,5
C_3H_9NO	1-Amino-2-propanol			18 e	53.2	98.2	157.9	5
$C_3H_9O_4P$	Trimethyl phosphate	-31 e	-7 e	23.6	62.8	116.0	192.0	5
C_3H_9P	Trimethylphosphine		-81 e	-53 e	-15.0	37.1		5
C_3H_9Sb	Trimethylstibine		-56 e	-23.8	19 e	80 e		5
$C_3H_{10}N_2$	1,2-Propanediamine		-35.4	-12.0	18.8	61 e	119 e	5
C_3N_2O	Carbonyl dicyanide			-21.7	15.3	65.2		5
C_4Cl_6	Hexachloro-1,3-butadiene	-1 e	22 e	50 e	86.7	137.0	209.7	5
$C_4F_6O_3$	Trifluoroacetic acid anhydride			-63 e	-39 e	-7.1	38.8	5
C_4F_8	Perfluorocyclobutane						-6.2	1
C_4F_{10}	Perfluorobutane		-122 e	-105 e	-82 e	-49.8	-2.5	1,5
$C_4H_2Cl_2O_2$	trans-2-Butenedioyl dichloride			8.0	45.6	94.3	159.8	5
$C_4H_2Cl_2S$	2,5-Dichlorothiophene			-20 e	22 e	81.4	171 e	5

Mol. Form.	Name	1 Pa	10 Pa	100 Pa	1 kPa	10 kPa	100 kPa	Ref.
					Temperature in °C for the indicated pressure			
$C_4H_2O_3$	Maleic anhydride				73.7	127.9	201.7	5
C_4H_3ClS	2-Chlorothiophene		-62 e	-35 e	2 e	51.8	123 e	5
C_4H_3IS	2-Iodothiophene			-25 e	23 e	94.9	181.0	5
C_4H_4	1-Buten-3-yne			-96.1	-73.4	-41.8	4.9	5
$C_4H_4N_2$	Succinonitrile	24.8 s					266.0	5
C_4H_4O	Furan			-78 e	-54 e	-20 e	31.0	1
$C_4H_4O_2$	Diketene				19.3	63.3	126 e	5
$C_4H_4O_3$	Succinic anhydride				121 e	180.8	260.8	5
$C_4H_4O_4$	Fumaric acid	123.9 s	150 s	180 s				5
C_4H_4S	Thiophene				-17 e	23.7	83.7	5
C_4H_5Cl	2-Chloro-1,3-butadiene	-113 e	-95 e	-71 e	-41 e	0.3	59.0	5
C_4H_5ClO	2-Methyl-2-propenoyl chloride		-57 e	-35 e	-5 e	36.4	98.2	5
$C_4H_5Cl_3O_2$	Ethyl trichloroacetate			15.3	51.9	100.1	166.6	5
C_4H_5N	3-Butenenitrile	-67 e	-48 e	-23.1	9.3	53.7	118.4	5
C_4H_5N	Methylacrylonitrile				-12 e	29.0	89.8	5
C_4H_5N	Pyrrole			-8 e	24 e	66.7	129.4	1
$C_4H_5NO_2$	Methyl cyanoacetate	-3 e	19 e	48 e	84 e	134.0	204.6	5
C_4H_5NS	Allyl isothiocyanate	-45 e	-27 e	-3 e	32.1	89 e	198 e	5
C_4H_5NS	4-Methylthiazole						67.0	5
C_4H_6	1,2-Butadiene	-132 e	-117 e	-98 e	-72.8	-38.9	10.5	5
C_4H_6	1,3-Butadiene*			-106 e	-83 e	-51.9	-4.7	1
C_4H_6	1-Butyne	-125 e	-111 e	-94 e	-71.2	-39.4	7.8	1
C_4H_6	2-Butyne		-89.2 s	-73.8 s	-53.5 s	-23.9	26.6	5
$C_4H_6Cl_2O_2$	Ethyl dichloroacetate			2.6	40.1	89.1	156.3	5
C_4H_6O	Divinyl ether		-99 e	-80 e	-56 e	-22.1	28.0	5
C_4H_6O	trans-2-Butenal	-74 e	-56 e	-33 e	-3 e	39.7	102.4	5
C_4H_6O	3-Buten-2-one					21 e	81.0	5
C_4H_6O	Cyclobutanone			-34 e	-4 e	37.1	97 e	5
$C_4H_6O_2$	cis-Crotonic acid			30 e	63 e	106.7	168.9	5
$C_4H_6O_2$	trans-Crotonic acid				74 e	120.8	184.9	5
$C_4H_6O_2$	3-Butenoic acid	-19 e	2 e	27 e	61 e	105.6	168.6	5
$C_4H_6O_2$	Methacrylic acid			22 e	56 e	99.9	161.5	5
$C_4H_6O_2$	Vinyl acetate	-88 e	-71 e	-50 e	-22 e	16.2	72.2	1
$C_4H_6O_2$	Methyl acrylate		-71 e	-48 e	-18 e	22 e	79.9	5
$C_4H_6O_2$	2,3-Butanedione					30.7	84.8	5
$C_4H_6O_2$	gamma-Butyrolactone		-17 e	24 e	72 e	130.2	203 e	5
$C_4H_6O_3$	Acetic anhydride	-44 e	-25 e	-1 e	31 e	75.1	139.7	1
$C_4H_6O_3$	Propylene carbonate	-40 e	-5 e	43 e	112 e	220 e	410 e	5
$C_4H_6O_4$	Dimethyl oxalate			50.5		98.1	163.0	5
C_4H_7Br	trans-1-Bromo-1-butene	-87 e	-68 e	-43.3	-11.4	31.9	94.4	5
C_4H_7Br	2-Bromo-1-butene	-87 e	-70 e	-48 e	-20 e	20.7	80.6	5
C_4H_7Br	cis-2-Bromo-2-butene	-90 e	-72 e	-49.0	-18.5	23.5	85.2	5
C_4H_7Br	trans-2-Bromo-2-butene	-86 e	-67 e	-43.4	-12.0	31.0	93.5	5
$C_4H_7Br_3$	1,2,3-Tribromobutane	0 e	23 e	53 e	91 e	143.7	219.5	5
$C_4H_7Br_3$	1,2,4-Tribromobutane	-3 e	20 e	49 e	87 e	139.4	214.5	5
C_4H_7Cl	3-Chloro-1-butene			-64 e	-36 e	4 e	63.6	5
C_4H_7Cl	cis-2-Chloro-2-butene	-100 e	-83 e	-62 e	-34 e	6 e	66.4	5
C_4H_7Cl	trans-2-Chloro-2-butene	-102 e	-86 e	-65 e	-37 e	3 e	62.2	5
C_4H_7Cl	3-Chloro-2-methylpropene		-75 e	-54 e	-25 e	13.8	71.5	5
$C_4H_7ClO_2$	Ethyl chloroacetate			-2.6	32.6	79.1	143.8	5
C_4H_7N	Butanenitrile	-67 e	-48 e	-24 e	8 e	52.3	117.2	1
C_4H_8	1-Butene	-139.0	-125.2	-107.8	-85.3	-53.7	-6.6	1,5
C_4H_8	cis-2-Butene	-131.2	-117.4	-99.8	-76.7	-44.8	3.4	1,5
C_4H_8	trans-2-Butene			-102 e	-80 e	-47.6	0.6	1
C_4H_8	Isobutene	-139.1	-125.5	-108.2	-85.5	-54.5	-7.3	1,5
C_4H_8	Cyclobutane				-71.8	-38.1	12.1	5
C_4H_8	Methylcyclopropane	-130 e	-116 e	-99.3	-76.3	-44.2	4.2	5
$C_4H_8Br_2$	1,2-Dibromobutane	-54 e	-30 e	0.4	39.6	92.1	166.1	5
$C_4H_8Br_2$	1,4-Dibromobutane	-13 e	9 e	37 e	74 e	124.0	196.5	5
$C_4H_8Cl_2$	1,1-Dichlorobutane			-25 e	6 e	49.3	113.4	5
$C_4H_8Cl_2$	1,2-Dichlorobutane			-28.4	5.8	53.1	123.1	5
$C_4H_8Cl_2$	1,4-Dichlorobutane		-26 e	0 e	35 e	82.4	153.4	5
$C_4H_8Cl_2$	2,2-Dichlorobutane		-58 e	-35 e	-5 e	37.8	102.1	5
$C_4H_8Cl_2O$	Bis(2-chloroethyl) ether	-32 e	-9 e	19.8	56.9	106.9	177.9	5
C_4H_8O	Ethyl vinyl ether		-102 e	-81 e	-53.1	-16.5	34.7	5
C_4H_8O	1,2-Epoxybutane	-135 e	-114 e	-87 e	-53 e	-5.5	62.1	5

Mol. Form.	Name	Temperature in °C for the indicated pressure						Ref.
		1 Pa	10 Pa	100 Pa	1 kPa	10 kPa	100 kPa	
C_4H_8O	Butanal	-88 e	-72 e	-50 e	-22 e	16.6	74.5	1,5
C_4H_8O	Isobutanal		-56 e	-29 e	8 e	63.8		1
C_4H_8O	2-Butanone	-85 e	-68 e	-46 e	-18.1	21.2	79.2	1
C_4H_8O	Tetrahydrofuran	-94 e	-78 e	-57.3	-29.8	9 e	65.6	1
$C_4H_8O_2$	Butanoic acid			12.9	52.2	101.4	163.3	1,5
$C_4H_8O_2$	2-Methylpropanoic acid	-30.1	-8.2	18.1	50.5	92.9	154.0	5
$C_4H_8O_2$	Propyl formate	-78 e	-62 e	-42 e	-15.1	23.0	80.4	1,5
$C_4H_8O_2$	Isopropyl formate	-80 e	-65 e	-47 e	-22.2	13.2	67.7	5
$C_4H_8O_2$	Ethyl acetate	-83 e	-66 e	-45 e	-18 e	20.4	76.8	1
$C_4H_8O_2$	Methyl propanoate	-80 e	-64 e	-43 e	-15.8	22.2	79.0	1
$C_4H_8O_2$	cis-2-Butene-1,4-diol	17 e	44 e	77 e	117.4	168.5	234.9	5
$C_4H_8O_2$	1,3-Dioxane			-37 e	-3 e	43.4	106.0	5
$C_4H_8O_2$	1,4-Dioxane					39.6	101.0	1
$C_4H_8O_2S$	Sulfolane		49 e	87 e	135 e	198.0	283.5	5
C_4H_8S	Tetrahydrothiophene	-66 e	-47 e	-23 e	9.4	54.1	120.5	1
C_4H_9Br	1-Bromobutane	-68.4	-53.9	-34.1	-5.4	37.6	101.1	1,5
C_4H_9Br	2-Bromobutane	-86 e	-68 e	-46 e	-16 e	26.6	90.7	5
C_4H_9Br	1-Bromo-2-methylpropane	-85 e	-68 e	-46 e	-16 e	26.8	91.1	5
C_4H_9Br	2-Bromo-2-methylpropane					11.7	72.4	1,5
C_4H_9Cl	1-Chlorobutane	-87 e	-71 e	-49 e	-21 e	18.4	78.1	1
C_4H_9Cl	2-Chlorobutane	-96 e	-80 e	-59 e	-31.0	8.5	67.9	1
C_4H_9Cl	1-Chloro-2-methylpropane	-94 e	-78 e	-56.6	-28.7	10.2	68.5	5
C_4H_9Cl	2-Chloro-2-methylpropane					-4.2	50.3	5
$C_4H_9Cl_3Si$	Butyltrichlorosilane					77.2	148.4	5
C_4H_9F	1-Fluorobutane	-114 e	-99 e	-80 e	-55 e	-20.0	32.1	5
C_4H_9F	2-Fluorobutane	-117 e	-103 e	-85 e	-60.7	-26.7	24.7	5
C_4H_9I	1-Iodobutane	-62 e	-43 e	-19 e	14 e	60.5	130.0	5
C_4H_9I	2-Iodobutane	-70 e	-51 e	-27 e	5 e	50 e	119.5	5
C_4H_9I	1-Iodo-2-methylpropane		-47 e	-21.4	12.0	56.8	120.0	5
C_4H_9I	2-Iodo-2-methylpropane	-75.1 s	-58.8 s	-39.5 s	-5.2	41 e	100.0	5
C_4H_9N	Pyrrolidine		-59 e	-38 e	-10 e	28.5	86.2	1
C_4H_9NO	N-Methylpropanamide				81.1	105 e		5
C_4H_9NO	N,N-Dimethylacetamide	-8 e	8 e	28.0	56.4	98.2	165.7	1
C_4H_9NO	2-Butanone oxime		-18 e	7 e	38.9	81.9	142.9	5
C_4H_9NO	Morpholine				21 e	64.5	128.5	1
$C_4H_9NO_3$	Isobutyl nitrate				-18 e	15.1	59.2	123.0	5
C_4H_{10}	Butane*	-134.3	-121.0	-103.9	-81.1	-49.1	-0.8	1,41
C_4H_{10}	Isobutane*		-129.0	-113.0	-90.9	-59.4	-12.0	1,41
$C_4H_{10}O$	1-Butanol	-37 e	-20 e	0 e	28 e	64 e	117.4	1
$C_4H_{10}O$	2-Butanol	-50 e	-34 e	-14 e	12.6	48.2	99.2	1,5
$C_4H_{10}O$	2-Methyl-1-propanol	-39 e	-24 e	-5 e	20.9	56.0	107.6	1,5
$C_4H_{10}O$	2-Methyl-2-propanol					34.4	82.1	1,5
$C_4H_{10}O$	Diethyl ether	-111 e	-96 e	-77 e	-52.6	-17.8	34.1	1
$C_4H_{10}O$	Methyl propyl ether				-40 e	-11.3	38.7	5
$C_4H_{10}O$	Isopropyl methyl ether				-56 e	-21.2	30.4	5
$C_4H_{10}O_2$	1,3-Butanediol	-4 e	23 e	55 e	94 e	142.9	206.1	5
$C_4H_{10}O_2$	1,4-Butanediol		45 e	77 e	116 e	164.7	227.6	5
$C_4H_{10}O_2$	2,3-Butanediol		15 e	43 e	77 e	121.2	180.3	5
$C_4H_{10}O_2$	Ethylene glycol monoethyl ether	-49 e	-29 e	-3 e	30 e	73.6	135.3	1
$C_4H_{10}O_2$	Ethylene glycol dimethyl ether			-44 e	-15 e	25.2	85.2	1
$C_4H_{10}O_2$	Dimethylacetal	-89 e	-74 e	-55 e	-29 e	7.7	64.1	5
$C_4H_{10}O_2$	Diethylperoxide			-39 e	3.6	65.0		5
$C_4H_{10}O_2S$	Bis(2-hydroxyethyl) sulfide			31 e	114.2		282.0	5
$C_4H_{10}O_3$	Diethylene glycol	35 e	58 e	86 e	123 e	173.6	245.2	1
$C_4H_{10}O_4S$	Diethyl sulfate	3 e	36 e	79 e	134 e	208.3		5
$C_4H_{10}S$	1-Butanethiol	-77 e	-59 e	-37 e	-6 e	35.4	98.0	5
$C_4H_{10}S$	2-Butanethiol	-86 e	-69 e	-47 e	-17 e	23.4	84.5	5
$C_4H_{10}S$	2-Methyl-1-propanethiol		-66 e	-44 e	-15 e	26.5	88.1	5
$C_4H_{10}S$	2-Methyl-2-propanethiol					5.8	63.8	5
$C_4H_{10}S$	Diethyl sulfide	-80 e	-62 e	-40 e	-10.8	30.3	91.7	1
$C_4H_{10}S$	Methyl propyl sulfide	-78 e	-61 e	-38 e	-8 e	33.1	95.1	5
$C_4H_{10}S$	Isopropyl methyl sulfide	-85 e	-68 e	-46 e	-17 e	23.4	84.3	5
$C_4H_{10}S_2$	1,4-Butanedithiol	-17 e	5 e	32 e	69.1	119.9	195.1	5
$C_4H_{10}S_2$	Diethyl disulfide	-46 e	-26 e	0 e	35 e	82.4	153.5	5

Mol. Form.	Name	Temperature in °C for the indicated pressure						Ref.
		1 Pa	10 Pa	100 Pa	1 kPa	10 kPa	100 kPa	
$C_4H_{11}N$	Butylamine			-46 e	-18.1	20.0	75.9	5
$C_4H_{11}N$	*sec*-Butylamine			-55 e	-29.1	7.5	62.3	5
$C_4H_{11}N$	*tert*-Butylamine			-67 e	-42.4	-8.1	43.7	5
$C_4H_{11}N$	Isobutylamine	-85 e	-70 e	-50 e	-24.5	12.0	67.3	5
$C_4H_{11}N$	Diethylamine			-46 e	-26 e	5 e	55.2	1
$C_4H_{11}NO$	*N,N*-Dimethylethanolamine	-52 e	-31 e	-6 e	27 e	70.9	133 e	5
$C_4H_{11}NO_2$	Diethanolamine	53 e	77 e	107 e	146 e	197.3	268 e	5
$C_4H_{12}BN$	(Dimethylamino)dimethyl-borane		-81 e	-60.1	-31.9	7.0	64.2	5
$C_4H_{12}Cl_2OSi_2$	1,3-Dichloro-1,1,3,3-tetramethyldisiloxane		-33 e	-9 e	23.8	69.1	136.5	5
$C_4H_{12}O_4Si$	Tetramethyl silicate				14.4	59.3	119.7	5
$C_4H_{12}Si$	Tetramethylsilane			-83 e	-59 e	-25 e	26.7	5
$C_4H_{12}Sn$	Tetramethylstannane			-55.0	-25.6	16.6	77.7	5
$C_4H_{13}N_3$	Diethylenetriamine	-10 e	13 e	43 e	80 e	129.6	198 e	5
C_4NiO_4	Nickel carbonyl					-12	42	4
C_5F_{12}	Perfluoropentane				-54.7	-20.9	28.6	5
C_5FeO_5	Iron pentacarbonyl				0	44	105	4
C_5H_4ClN	2-Chloropyridine			7.4	45.8	97.3	169.9	5
$C_5H_4O_2$	Furfural	-26 e	-8 e	16 e	47 e	92.4	161.4	1
C_5H_5N	Pyridine			-23 e	8 e	51.0	114.9	1
C_5H_6	1,3-Cyclopentadiene			-77 e	-51 e	-14 e	39.8	5
$C_5H_6N_2$	Pentanedinitrile	24.1	52 e	85 e	126 e	178 e	245 e	5
C_5H_6O	2-Methylfuran			-66 e	-35 e	6 e	64.5	1
$C_5H_6O_2$	Furfuryl alcohol	-30 e	-5 e	25 e	62.6	109.3	169.7	5
C_5H_6S	2-Methylthiophene		-58 e	-32 e	2 e	47.9	112.2	1
C_5H_6S	3-Methylthiophene		-53 e	-28 e	6 e	50.6	115.1	1
C_5H_7N	1-Methylpyrrole				8 e	49.9	112.3	5
$C_5H_7NO_2$	Ethyl cyanoacetate	16 e	39 e	67.0	102.1	146.7	205.6	5
C_5H_8	1,2-Pentadiene	-109 e	-93 e	-73 e	-46.1	-9.7	44.5	5
C_5H_8	*cis*-1,3-Pentadiene	-109 e	-93 e	-73 e	-47.0	-10.5	43.7	1,5
C_5H_8	*trans*-1,3-Pentadiene			-75 e	-49.0	-13 e	42 e	1
C_5H_8	1,4-Pentadiene	-120 e	-105 e	-86 e	-60.9	-26.2	25.6	5
C_5H_8	2,3-Pentadiene	-106 e	-90 e	-70 e	-42.9	-6.3	47.9	5
C_5H_8	3-Methyl-1,2-butadiene	-111 e	-95 e	-75 c	-49.2	-13.1	40.4	5
C_5H_8	2-Methyl-1,3-butadiene	-115 e	-100 e	-81 e	-55.4	-19.7	33.7	1,5
C_5H_8	1-Pentyne			-75 e	-49.1	-13.5	39.9	5
C_5H_8	2-Pentyne	-100 e	-85 e	-65 e	-37.9	-0.5	55.7	5
C_5H_8	3-Methyl-1-butyne			-82 e	-57.5	-23.1	28.6	5
C_5H_8	Cyclopentene	-109 e	-94 e	-74 e	-48 e	-11.1	43.8	5
C_5H_8	Spiropentane	-110 e	-95 e	-76 e	-51 e	-15 e	38.6	5
C_5H_8O	3-Methyl-3-buten-2-one			-35 e	-5 e	36.0	97.3	5
C_5H_8O	Cyclopropyl methyl ketone		-57 e	-31 e	3 e	49 e	112 e	5
C_5H_8O	Cyclopentanone		-39 e	-14 e	19 e	64 e	130.3	1
C_5H_8O	3,4-Dihydro-2H-pyran				-22 e	22.0	84.9	5
$C_5H_8O_2$	4-Pentenoic acid	0 e	19 e	44 e	77 e	122.0	187.5	5
$C_5H_8O_2$	Vinyl propanoate					31.2	94 e	5
$C_5H_8O_2$	Ethyl acrylate		-55 e	-32.7	-2.8	38.5	99.2	5
$C_5H_8O_2$	Methyl methacrylate			-31 e	-1 e	39.7	100.0	1
$C_5H_8O_2$	2,4-Pentanedione			-5 e	24.7	67.8	137.4	1
$C_5H_8O_2$	Tetrahydro-2H-pyran-2-one			5 e	35.1	74.4	128.3	207.0	5
$C_5H_8O_3$	Methyl acetoacetate				50.1	101.1	171.3	5
$C_5H_8O_4$	Glutaric acid		121 e	153.2	191.9	240.3	302.5	5
$C_5H_8O_4$	Dimethyl malonate	-22 e	1 e	30.0	66.7	114.7	180.2	5
$C_5H_9ClO_2$	Ethyl 2-chloropropanoate			1.4	36.4	82.5	146.0	5
$C_5H_9ClO_2$	Isopropyl chloroacetate			-2 e	35.0	83.3	148.1	5
C_5H_9N	Pentanenitrile	-54 e	-34 e	-8 e	26 e	72.2	140.9	1
C_5H_9N	2,2-Dimethylpropanenitrile					41.1	104.8	5
C_5H_9NO	*N*-Methyl-2-pyrrolidone	1 e	24 e	53.1	92.3	147.2	229 e	5
C_5H_{10}	1-Pentene	-118.9	-103.4	-84.0	-58.8	-23.3	29.6	1,5
C_5H_{10}	*cis*-2-Pentene	-113.8	-98.1	-78.4	-52.7	-16.8	36.6	1,5
C_5H_{10}	*trans*-2-Pentene	-114.5	-98.9	-79.1	-53.3	-17.5	36.0	1,5
C_5H_{10}	2-Methyl-1-butene	-117.7	-102.2	-82.7	-57.2	-21.9	30.8	1,5
C_5H_{10}	3-Methyl-1-butene	-125.0	-110.1	-91.2	-66.7	-32.1	19.7	1,5
C_5H_{10}	2-Methyl-2-butene	-113.4	-97.6	-77.7	-51.6	-15.8	38.2	1,5

Mol. Form.	Name	1 Pa	10 Pa	100 Pa	1 kPa	10 kPa	100 kPa	Ref.
				Temperature in °C for the indicated pressure				
C_5H_{10}	Cyclopentane			-77.0	-45.4	-7.1	48.8	5
C_5H_{10}	Ethylcyclopropane	-118 e	-102 e	-83 e	-57 e	-20 e	35.5	5
C_5H_{10}	cis-1,2-Dimethylcyclo-propane	-118 e	-103 e	-83 e	-57 e	-20 e	36.6	5
C_5H_{10}	trans-1,2-Dimethylcyclo-propane	-122 e	-108 e	-89 e	-63 e	-27 e	27.8	5
$C_5H_{10}Br_2$	1,5-Dibromopentane	1 e	25 e	54 e	93 e	145.6	221.8	5
$C_5H_{10}Cl_2$	1,2-Dichloropentane				30 e	77.4	147.8	5
$C_5H_{10}Cl_2$	1,5-Dichloropentane	-31 e	-10 e	17 e	54 e	104.1	178.9	5
$C_5H_{10}N_2$	3-(Dimethylamino)-propanenitrile				51.1	101.8	171.4	5
$C_5H_{10}O$	Cyclopentanol		-13 e	11.5	42.2	82.5	140.0	5
$C_5H_{10}O$	Allyl ethyl ether			-56 e	-28.7	9.8	67.2	5
$C_5H_{10}O$	Pentanal	-71 e	-53 e	-31 e	-1 e	40.8	102.6	5
$C_5H_{10}O$	2-Pentanone				-1 e	40.3	101.9	1,5
$C_5H_{10}O$	3-Pentanone			-31 e	-1 e	40 e	101.6	1
$C_5H_{10}O$	3-Methyl-2-butanone	-69 e	-54 e	-34 e	-6.9	32.2	94.0	1,5
$C_5H_{10}O$	Tetrahydropyran				-15 e	26.0	88 e	5
$C_5H_{10}O$	2-Methyltetrahydrofuran				-20 e	19.7	79.8	5
$C_5H_{10}O_2$	Pentanoic acid	-7.4	15.3	42.7	76.3	122.1	185.7	5
$C_5H_{10}O_2$	2-Methylbutanoic acid	-10 e	10 e	36 e	69 e	112.8	175.2	5
$C_5H_{10}O_2$	3-Methylbutanoic acid	-15.8	4 e	30.0	64.7	110.6	176.1	5
$C_5H_{10}O_2$	Butyl formate			-29 e	2 e	44.4	105.7	5
$C_5H_{10}O_2$	Isobutyl formate	-69 e	-53 e	-31 e	-3 e	37.4	97.6	5
$C_5H_{10}O_2$	Propyl acetate	-69 e	-51 e	-29 e	0 e	40.9	101.2	1
$C_5H_{10}O_2$	Isopropyl acetate		-61 e	-40 e	-11 e	29.8	88.2	5
$C_5H_{10}O_2$	Ethyl propanoate	-69 e	-52 e	-30 e	-1 e	38.9	98.7	1
$C_5H_{10}O_2$	Methyl butanoate	-68 e	-50 e	-28 e	0.9	41.7	102.3	5
$C_5H_{10}O_2$	Methyl isobutanoate	-83 e	-65 e	-41 e	-11 e	31 e	92.1	5
$C_5H_{10}O_2$	Tetrahydrofurfuryl alcohol	-40 e	-16 e	15 e	55 e	106 e	176.8	5
$C_5H_{10}O_3$	Diethyl carbonate	-42 e	-17 e	17 e	61.6	125.9		5
$C_5H_{10}O_3$	Ethylene glycol monomethyl ether acetate	-47 e	-26 e	0 e	34 e	79.4	144.1	5
$C_5H_{10}S$	Thiacyclohexane				24 e	71.1	141.2	5
$C_5H_{10}S$	Cyclopentanethiol				18 e	64 e	131.7	5
$C_5H_{11}Br$	1-Bromopentane	-60 e	-41 e	-16 e	16 e	61.5	129.1	5
$C_5H_{11}Br$	2-Bromopentane	-69 e	-51 e	-27 e	5 e	49.7	116.9	5
$C_5H_{11}Br$	3-Bromopentane	-68 e	-50 e	-26 e	6 e	50.8	118.1	5
$C_5H_{11}Br$	1-Bromo-3-methylbutane	-67 e	-49 e	-25 e	8 e	52.4	119.9	5
$C_5H_{11}Cl$	1-Chloropentane	-73 e	-55 e	-32 e	-1 e	42.5	107.9	5
$C_5H_{11}Cl$	2-Chloropentane	-80 e	-62 e	-39 e	-9 e	33.2	96.1	5
$C_5H_{11}Cl$	3-Chloropentane	-77 e	-60 e	-37 e	-7 e	34.9	97.3	5
$C_5H_{11}Cl$	2-Chloro-2-methylbutane			-52 e	-21 e	21.8	85.2	5
$C_5H_{11}Cl$	1-Chloro-2,2-dimethyl-propane				-17 e	23.5	83.9	5
$C_5H_{11}F$	1-Fluoropentane	-97 e	-80 e	-60 e	-32 e	5.7	62.4	5
$C_5H_{11}I$	1-Iodopentane	-47 e	-27 e	-1 e	34 e	83.0	156.5	5
$C_5H_{11}I$	1-Iodo-3-methylbutane		-34 e	-6.6	28.8	77.3	147.8	5
$C_5H_{11}N$	Cyclopentylamine	-66 e	-48 e	-26 e	4 e	45.8	108 e	5
$C_5H_{11}N$	Piperidine				2 e	43.3	105.8	5
$C_5H_{11}N$	N-Methylpyrrolidine				-23 e	18.5	78 e	5
$C_5H_{11}NO_3$	3-Methylbutyl nitrate		-26 e	1.0	35.5	81.7	147.0	5
C_5H_{12}	Pentane**	-115.5	-99.8	-80.0	-54.0	-18.1	35.7	16
C_5H_{12}	Isopentane	-119 e	-105 e	-86 e	-61 e	-26 e	27.5	1
C_5H_{12}	Neopentane*		-107.5 s	-90.8 s	-68.8 s	-38.5 s	9.2	1,5
$C_5H_{12}N_2O$	Tetramethylurea			20.7	58.0	106.7	179.5	5
$C_5H_{12}O$	1-Pentanol	-27 e	-10 e	12 e	41 e	79.8	137.4	5
$C_5H_{12}O$	2-Pentanol	-35 e	-19 e	1 e	28.0	64.9	118.7	1
$C_5H_{12}O$	3-Pentanol	-41 e	-25 e	-4 e	24 e	61.1	114.9	5
$C_5H_{12}O$	2-Methyl-1-butanol	-27 e	-11 e	9 e	36.2	73.4	128.3	1
$C_5H_{12}O$	3-Methyl-1-butanol	-22 e	-7 e	13 e	39.1	75.7	130.1	5
$C_5H_{12}O$	2-Methyl-2-butanol			-5 e	17.7	50.6	101.7	1,5
$C_5H_{12}O$	3-Methyl-2-butanol			-3 e	22.7	58.2	111.1	5
$C_5H_{12}O$	2,2-Dimethyl-1-propanol				59.2	112.7		5
$C_5H_{12}O$	Butyl methyl ether			-54 e	-27 e	12 e	69.8	1
$C_5H_{12}O$	Methyl tert-butyl ether			-66 e	-39 e	-2 e	54.8	1
$C_5H_{12}O$	Ethyl propyl ether	-92 e	-77 e	-57 e	-30.5	6.7	63.4	1,5

Mol. Form.	Name	Temperature in °C for the indicated pressure						Ref.
		1 Pa	10 Pa	100 Pa	1 kPa	10 kPa	100 kPa	
$C_5H_{12}O_2$	1,5-Pentanediol	25 e	52 e	85 e	125 e	175.1	238.9	5
$C_5H_{12}O_2$	Ethylene glycol monopropyl ether				40 e	85.6	149.3	5
$C_5H_{12}O_2$	Diethoxymethane		-65 e	-43 e	-14 e	27.3	87.7	5
$C_5H_{12}O_3$	Diethylene glycol monomethyl ether		12 e	40 e	76 e	124.2	193.7	1
$C_5H_{12}S$	1-Pentanethiol	-60 e	-41 e	-17 e	15 e	60 e	126.2	1
$C_5H_{12}S$	2-Pentanethiol	-70 e	-52 e	-28 e	3 e	46.6	111.9	5
$C_5H_{12}S$	3-Pentanethiol	-70 e	-51 e	-28 e	4 e	47.7	113.4	5
$C_5H_{12}S$	2-Methyl-1-butanethiol				8.0	52.3	118.5	5
$C_5H_{12}S$	3-Methyl-1-butanethiol				7.8	51.9	117.9	5
$C_5H_{12}S$	2-Methyl-2-butanethiol				-8.0	34.6	98.7	5
$C_5H_{12}S$	Butyl methyl sulfide		-43 e	-19 e	13 e	57 e	123.0	1
$C_5H_{12}S$	tert-Butyl methyl sulfide				-7.8	34.7	98.4	5
$C_5H_{12}S$	Ethyl propyl sulfide	-64 e	-46 e	-23 e	9 e	52.7	118.0	5
$C_5H_{12}S$	Ethyl isopropyl sulfide	-72 e	-54 e	-31 e	0 e	42.7	106.9	5
$C_5H_{13}N$	Pentylamine		-52 e	-29 e	1 e	42.8	104.0	5
C_6BrF_5	Bromopentafluorobenzene			-10 e	23 e	68 e	136.0	5
C_6ClF_5	Chloropentafluorobenzene		-44 e	-21 e	11 e	53.8	117.6	1
$C_6Cl_3F_3$	1,3,5-Trichloro-2,4,6-trifluorobenzene	-19 e	4 e	32 e	70 e	121.7	197.9	1
C_6F_6	Hexafluorobenzene		-56.9 s	-36 s	-11.5 s	22.6	79.9	1,5
C_6F_{12}	Perfluorocyclohexane				-46.2 s	-7.6 s	48.9 s	5
C_6F_{14}	Perfluorohexane		-75	-57 e	-32 e	2.8	56.8	5
C_6F_{14}	Perfluoro-2-methylpentane				-33 e	2.9	57.1	5
C_6F_{14}	Perfluoro-3-methylpentane	-95 e	-80 e	-60 e	-34 e	2.8	57.9	5
C_6F_{14}	Perfluoro-2,3-dimethylbutane					4.3	59.3	5
C_6HF_5	Pentafluorobenzene			-41 e	-13 e	27 e	85.3	5
C_6HF_5O	Pentafluorophenol				39 e	82 e	145.2	5
$C_6H_2F_4$	1,2,3,4-Tetrafluorobenzene			-36 e	-7 e	33.8	94.0	1
$C_6H_2F_4$	1,2,3,5-Tetrafluorobenzene			-43 e	-14 e	25.5	84.1	1
$C_6H_2F_4$	1,2,4,5-Tetrafluorobenzene					30.7	89.9	1
$C_6H_3Cl_3O$	2,4,6-Trichlorophenol			71.8	114.0	169.5	245.7	5
$C_6H_3F_3$	1,3,5-Trifluorobenzene					18.2	75.0	5
$C_6H_4Br_2$	m-Dibromobenzene	-7 e	16 e	44 e	83 e	137.0	218.2	5
$C_6H_4ClNO_2$	1-Chloro-4-nitrobenzene	15.4 s	35.8 s		97 e	156.0	238 e	5
$C_6H_4Cl_2$	o-Dichlorobenzene		-13 e	16.3	53.9	104.6	180.0	1,5
$C_6H_4Cl_2$	m-Dichlorobenzene		-22 e	8.0	46.7	97.8	172.5	1,5
$C_6H_4Cl_2$	p-Dichlorobenzene	-45.5 s	-21.8 s	8 s	46.7 s	99.0	173.6	1,5
$C_6H_4O_2$	p-Benzoquinone	-4.1 s	17.8 s	43.5 s	74.3 s	111.6 s		5
$C_6H_5AsCl_2$	Dichlorophenylarsine	6.9	35.2	70 e	113 e	170 e	245 e	5
C_6H_5Br	Bromobenzene		-25 e	1 e	34.9	83.1	155.4	1
C_6H_5Cl	Chlorobenzene		-43 e	-17 e	16.8	62.9	131.3	1,5
C_6H_5ClO	o-Chlorophenol				45.8	97.9	173.9	5
C_6H_5ClO	m-Chlorophenol			39.7	80.2	135.1	213.4	5
C_6H_5ClO	p-Chlorophenol			45.0	86.5	142.0	219.9	5
$C_6H_5Cl_3Si$	Trichlorophenylsilane			33 e	70.2	122.6	201 e	5
C_6H_5F	Fluorobenzene				-16.9	24.2	84.4	1
C_6H_5I	Iodobenzene	-30 e	-7 e	20.9	58.5	110.6	187.8	1
$C_6H_5NO_2$	Nitrobenzene		10 e	40 e	78 e	132 e	210.3	1
$C_6H_5NO_3$	p-Nitrophenol	72.6 s	97.4 s					5
C_6H_6	1,5-Hexadien-3-yne	-82 e	-66 e	-44.3	-16.0	23.7	83.6	5
C_6H_6	Benzene**			-40 e	-15.1 s	20.0	79.7	1,5
C_6H_6ClN	o-Chloroaniline		10 e	39.0	75.2	131.4	208.3	5
C_6H_6ClN	m-Chloroaniline	-5 e	19.7	49.4	94.2	162 e	1069 e	5
$C_6H_6N_2O_2$	p-Nitroaniline	87.8 s			192.0	252.6	331.2	5
C_6H_6O	Phenol	-9.7 s	9.6 s	34.1 s	68.9	113.7	181.4	1,5
$C_6H_6O_3$	1,2,3-Benzenetriol				162.0	222.8	308.3	5
C_6H_6S	Benzenethiol		-15 e	12 e	47 e	96.0	168.6	5
C_6H_7N	Aniline		-2.5	26.7	63.5	112.5	183.5	1,5
C_6H_7N	2-Methylpyridine	-56.5	-37.8	-13.9	18.3	62.9	129.0	1,5
C_6H_7N	3-Methylpyridine			-5 e	28.8	75.2	143.7	1
C_6H_7N	4-Methylpyridine	-58.2 s	-43.1 s	-3.9 s	29.6	76.1	144.9	1,5
C_6H_8	cis-1,3,5-Hexatriene					21 e	78 e	5
C_6H_8	1,3-Cyclohexadiene	-88 e	-71 e	-50 e	-21 e	19 e	79.9	5
C_6H_8	1,4-Cyclohexadiene				-15 e	27.3	85.0	5
$C_6H_8N_2$	Adiponitrile	30 e	61 e	100 e	148.6	211.8	297 e	5

Mol. Form.	Name	1 Pa	10 Pa	100 Pa	1 kPa	10 kPa	100 kPa	Ref.
				Temperature in °C for the indicated pressure				
$C_6H_8N_2$	m-Phenylenediamine			94.5	140.2	200.8	285.0	5
$C_6H_8N_2$	Phenylhydrazine		38 e	69 e	109 e	163.9	242.5	5
$C_6H_8O_4$	Dimethyl maleate		5 e	36 e	76 e	127.3	197 e	5
C_6H_8S	2,5-Dimethylthiophene		-43 e	-16 e	20 e	67.5	134.8	5
C_6H_{10}	trans-1,3-Hexadiene	-86 e	-70 e	-51 e	-24 e	14 e	72 e	5
C_6H_{10}	trans-1,4-Hexadiene	-98 e	-81 e	-60 e	-33 e	7 e	65 e	5
C_6H_{10}	1,5-Hexadiene	-99 e	-84 e	-64 e	-37 e	0.9	59.2	5
C_6H_{10}	cis,cis-2,4-Hexadiene					18 e	79.6	5
C_6H_{10}	trans,cis-2,4-Hexadiene	-89 e	-73 e	-52 e	-23 e	18 e	79.6	5
C_6H_{10}	trans,trans-2,4-Hexadiene				-23 e	18 e	79.6	5
C_6H_{10}	trans-2-Methyl-1,3-pentadiene	-92 e	-75 e	-54 e	-26 e	14 e	75.6	5
C_6H_{10}	2,3-Dimethyl-1,3-butadiene			-59 e	-30 e	9.7	68.1	5
C_6H_{10}	1-Hexyne	-91 e	-75 e	-54 e	-26 e	12.8	71.0	5
C_6H_{10}	2-Hexyne	-84 e	-67 e	-46 e	-17 e	23.6	84.1	5
C_6H_{10}	3-Hexyne	-86 e	-69 e	-48 e	-19.1	21.0	81.0	1,5
C_6H_{10}	4-Methyl-1-pentyne	-97 e	-81 e	-61 e	-34 e	4.1	60.7	5
C_6H_{10}	4-Methyl-2-pentyne	-91 e	-74 e	-54 e	-26 e	13.8	72.7	5
C_6H_{10}	Cyclohexene	-87 e	-70 e	-49 e	-19 e	21 e	82.6	1
$C_6H_{10}Cl_2$	1,1-Dichlorocyclohexane	-39 e	-19 e	8 e	43 e	93.5	170.5	5
$C_6H_{10}Cl_2$	cis-1,2-Dichlorocyclohexane			27 e	69 e	125.7	206.2	5
$C_6H_{10}O$	4-Methyl-4-penten-2-one	-59 e	-41 e	-17 e	14 e	57.0	121.0	5
$C_6H_{10}O$	Cyclohexanone		-25 e	1 e	36 e	84 e	155.2	1
$C_6H_{10}O$	Mesityl oxide	-56 e	-37 e	-13 e	19 e	63.5	129.3	5
$C_6H_{10}O_2$	Vinyl butanoate					53 e	114.5	5
$C_6H_{10}O_2$	Ethyl methacrylate				8 e	53.2	116.8	5
$C_6H_{10}O_2$	Allyl glycidyl ether				40.1	85.7	152.8	5
$C_6H_{10}O_3$	Ethyl acetoacetate	-25 e	-3 e	25.7	62.3	111.3	180.2	5
$C_6H_{10}O_3$	Propanoic anhydride	-32 e	-15 e	6 e	36 e	77.6	142.9	5
$C_6H_{10}O_4$	Diethyl oxalate	-5 e	18 e	44.9	79.4	124.3	185.2	5
$C_6H_{10}O_4$	Dimethyl succinate			30 e	70.4	123.3	195.4	5
$C_6H_{10}O_4$	Ethylene glycol diacetate	-17 e	6 e	35.0	71.9	121.1	190.0	5
$C_6H_{10}S$	Diallylsulfide	-58 e	-38 e	-12.4	21.7	68.8	138.1	5
$C_6H_{11}Cl$	Chlorocyclohexane		-35 e	-9 e	25 e	71.6	142.1	5
$C_6H_{11}N$	Hexanenitrile	-40 e	-19 e	8 e	43 e	91.5	163.2	1,5
$C_6H_{11}N$	4-Methylpentanenitrile		-50 e	-20 e	20 e	75.2	155.2	5
$C_6H_{11}NO$	Caprolactam	36.8 s	58.9 s	86.6 s			270	5
C_6H_{12}	1-Hexene	-99.8	-82.8	-61.4	-33.7	5.2	63.1	1,5
C_6H_{12}	cis-2-Hexene	-97 e	-80 e	-58 e	-30 e	9.9	68.5	5
C_6H_{12}	trans-2-Hexene	-94 e	-78 e	-57 e	-30 e	9.3	67.5	5
C_6H_{12}	cis-3-Hexene	-96 e	-79 e	-59 e	-30.8	7.9	66.0	5
C_6H_{12}	trans-3-Hexene	-95 e	-79 e	-58 e	-30.0	8.8	66.7	5
C_6H_{12}	2-Methyl-1-pentene	-98 e	-82 e	-62 e	-34.2	4.1	61.7	5
C_6H_{12}	3-Methyl-1-pentene	-104 e	-88 e	-68 e	-41.5	-3.6	53.8	5
C_6H_{12}	4-Methyl-1-pentene	-105 e	-89 e	-69 e	-41.6	-3.6	53.5	5
C_6H_{12}	2-Methyl-2-pentene	-95 e	-78 e	-58 e	-30 e	9.0	66.9	5
C_6H_{12}	3-Methyl-cis-2-pentene	-95 e	-79 e	-58 e	-30 e	8.9	67.3	5
C_6H_{12}	3-Methyl-trans-2-pentene	-93 e	-77 e	-55 e	-27.4	11.7	70.0	5
C_6H_{12}	4-Methyl-cis-2-pentene	-102 e	-86 e	-66 e	-38.7	-0.9	56.0	5
C_6H_{12}	4-Methyl-trans-2-pentene	-100 e	-84 e	-64 e	-36.8	1.2	58.2	5
C_6H_{12}	2-Ethyl-1-butene	-98 e	-81 e	-60 e	-32 e	6.6	64.3	5
C_6H_{12}	2,3-Dimethyl-1-butene	-103 e	-87 e	-67 e	-39.9	-1.9	55.2	5
C_6H_{12}	3,3-Dimethyl-1-butene	-110 e	-95 e	-76 e	-50.8	-14.5	40.8	5
C_6H_{12}	2,3-Dimethyl-2-butene		-75 e	-54 e	-25 e	14 e	72.9	1
C_6H_{12}	Cyclohexane	-85.6 s	-68.9 s	-47.6 s	-19.8 s	19.3	80.4	1,5
C_6H_{12}	Methylcyclopentane	-97 e	-80 e	-58 e	-28.8	11.6	71.4	1,5
C_6H_{12}	Ethylcyclobutane	-99 e	-82 e	-61 e	-32 e	9 e	70.2	5
C_6H_{12}	Isopropylcyclopropane	-104 e	-88 e	-68 e	-40 e	-1 e	57.9	5
C_6H_{12}	1-Ethyl-1-methylcyclopropane	-105 e	-89 e	-69 e	-41 e	-3 e	56.3	5
C_6H_{12}	1,1,2-Trimethylcyclopropane	-109 e	-94 e	-73 e	-46 e	-7 e	52.0	5
$C_6H_{12}Cl_2$	1,2-Dichlorohexane				49 e	98.1	171.7	5
$C_6H_{12}Cl_2O$	2,2′-Dichlorodiisopropyl ether		-1 e	27.3	63.4	112.3	182.1	5
$C_6H_{12}O$	Butyl vinyl ether	-87 e	-67 e	-42 e	-9.3	33.6	93.2	5
$C_6H_{12}O$	Isobutyl vinyl ether	-87 e	-68 e	-44 e	-13 e	26.5	80.7	5
$C_6H_{12}O$	Hexanal	-56 e	-37 e	-13 e	19 e	62.6	127.8	5
$C_6H_{12}O$	2-Hexanone	-43 e	-21 e	4.2	34.5	61.9	127.2	1,5
$C_6H_{12}O$	3-Hexanone		-40 e	-16 e	15 e	58.5	123.1	1
$C_6H_{12}O$	3-Methyl-2-pentanone				8.5	52.7	117.0	5

Mol. Form.	Name	Temperature in °C for the indicated pressure						Ref.
		1 Pa	10 Pa	100 Pa	1 kPa	10 kPa	100 kPa	
$C_6H_{12}O$	4-Methyl-2-pentanone	-61 e	-43 e	-21 e	9 e	51.5	116.1	5
$C_6H_{12}O$	2-Methyl-3-pentanone					50.2	113.0	5
$C_6H_{12}O$	3,3-Dimethyl-2-butanone			-30 e	0 e	42.5	105.7	1
$C_6H_{12}O$	Cyclohexanol			34 e	61 e	99.2	160.7	1
$C_6H_{12}O_2$	Hexanoic acid		33 e	59 e	93 e	139.3	204.5	1
$C_6H_{12}O_2$	4-Methylpentanoic acid	36 e	49 e	67.1	92.9	133.6	206.8	5
$C_6H_{12}O_2$	Diethylacetic acid	-9 e	16 e	46 e	83 e	130.7	192.5	5
$C_6H_{12}O_2$	Isopentyl formate	-60 e	-41 e	-17 e	15 e	59.1	124 e	5
$C_6H_{12}O_2$	Butyl acetate	-63 e	-43 e	-19 e	14 e	61.0	125.6	1,5
$C_6H_{12}O_2$	Isobutyl acetate	-63 e	-45 e	-21 e	10 e	53.4	116 e	5
$C_6H_{12}O_2$	Propyl propanoate	-62 e	-42 e	-18 e	14 e	58.3	122.0	5
$C_6H_{12}O_2$	Ethyl butanoate	-49 e	-34 e	-14 e	14.3	55.2	121.1	5
$C_6H_{12}O_2$	Ethyl 2-methylpropanoate	-65 e	-47 e	-24.6	5.4	47.3	109.8	5
$C_6H_{12}O_2$	Methyl pentanoate				19.2	63.7	127.4	5
$C_6H_{12}O_2$	Methyl isopentanoate					53.3	116.3	5
$C_6H_{12}O_2$	Diacetone alcohol	-41 e	-17 e	13 e	50.1	98.5	164 e	5
$C_6H_{12}O_3$	Ethylene glycol monoethyl ether acetate	-25 e	-8 e	14 e	44.6	88.0	155.6	5
$C_6H_{12}O_3$	Paraldehyde				17 e	62.2	124 e	5
$C_6H_{12}S$	Cyclohexanethiol					84.8	158.3	5
$C_6H_{12}S$	cis-Tetrahydro-2,5-dimethylthiophene	-53 e	-34 e	-8 e	25 e	72.0	142.1	5
$C_6H_{12}S$	Tetrahydro-3-methyl-2H-thiopyran	-48 e	-27 e	0 e	35 e	84.1	157.5	5
$C_6H_{13}Br$	1-Bromohexane	-45 e	-25 e	2 e	36 e	83.7	154.8	5
$C_6H_{13}Cl$	1-Chlorohexane	-55 e	-36 e	-11 e	21 e	66.7	134.6	5
$C_6H_{13}F$	1-Fluorohexane	-80 e	-62 e	-40 e	-11 e	30.4	91.1	5
$C_6H_{13}I$	1-Iodohexane	-33 e	-11 e	16 e	53 e	104.0	180.8	5
$C_6H_{13}N$	Cyclohexylamine			-9 e	22 e	66.6	133.5	1
C_6H_{14}	Hexane	-96.4 s	-79.2	-57.6	-29.3	9.8	68.3	16
C_6H_{14}	2-Methylpentane	-100 e	-84 e	-64 e	-36 e	2 e	59.9	1
C_6H_{14}	3-Methylpentane	-99 e	-83 e	-62 e	-34.3	4.6	62.9	1
C_6H_{14}	2,2-Dimethylbutane		-90 e	-71.5	-45.5	-7.7	49.4	1
C_6H_{14}	2,3-Dimethylbutane	-103 e	-87 e	-66 e	-39.0	-0.4	57.6	1
$C_6H_{14}O$	1-Hexanol		5 e	28 e	56.8	97.3	157.1	1
$C_6H_{14}O$	2-Hexanol	-28 e	-10 e	12 e	41.4	81.5	139.6	1
$C_6H_{14}O$	3-Hexanol	-43 e	-23 e	1 e	33 e	75.4	135.1	1
$C_6H_{14}O$	2-Methyl-1-pentanol			14 e	45.9	88.3	147.6	5
$C_6H_{14}O$	4-Methyl-1-pentanol			24 e	53 e	92.4	151.4	5
$C_6H_{14}O$	2-Methyl-2-pentanol	-29 e	-15 e	3 e	27.1	63.0	120.9	5
$C_6H_{14}O$	3-Methyl-2-pentanol				36.5	76.1	133.8	5
$C_6H_{14}O$	4-Methyl-2-pentanol	-43 e	-24 e	0 e	30 e	71.9	131.3	5
$C_6H_{14}O$	2-Methyl-3-pentanol					29.8	68.8	126.0	5
$C_6H_{14}O$	3-Methyl-3-pentanol		-23 e	-4 e	22.9	61.1	121.1	5
$C_6H_{14}O$	2-Ethyl-1-butanol		-5 e	17 e	46 e	85.7	146.1	5
$C_6H_{14}O$	3,3-Dimethyl-1-butanol	-37 e	-16 e	9 e	42 e	84.3	142.5	5
$C_6H_{14}O$	2,3-Dimethyl-2-butanol			-5 e	23 e	61.3	118.2	5
$C_6H_{14}O$	Dipropyl ether	-80 e	-63 e	-41 e	-12 e	28.8	89.7	1
$C_6H_{14}O$	Diisopropyl ether		-76 e	-55 e	-28 e	11 e	68.1	1
$C_6H_{14}O$	Butyl ethyl ether	-78 e	-61 e	-39 e	-10 e	31.0	91.9	1
$C_6H_{14}O$	tert-Butyl ethyl ether	-90 e	-74 e	-53 e	-24.6	14.4	72.6	5
$C_6H_{14}O_2$	2-Methyl-2,4-pentanediol	-8 e	17 e	48 e	86 e	134.4	197.5	5
$C_6H_{14}O_2$	Ethylene glycol monobutyl ether	-31 e	-8 e	20 e	55 e	103.2	170.2	5
$C_6H_{14}O_2$	1,1-Diethoxyethane	-68 e	-49 e	-26 e	3.7	44.2	101.9	5
$C_6H_{14}O_2$	Ethylene glycol diethyl ether		-59 e	-35.3	-2.8	44.4	118.8	5
$C_6H_{14}O_3$	1,2,6-Hexanetriol	92 e	114.8	146.0	191 e			5
$C_6H_{14}O_3$	Dipropylene glycol				110 e	162.6	231.4	5
$C_6H_{14}O_3$	Diethylene glycol monoethyl ether			40 e	80.3	132.4	201.4	5
$C_6H_{14}O_3$	Diethylene glycol dimethyl ether	-42 e	-20 e	8.3	44.3	92.3	159.4	5
$C_6H_{14}O_3$	Trimethylolpropane	73 e	98 e	128 e	167.8	220.5	295 e	5
$C_6H_{14}O_4$	Triethylene glycol	44 e	74 e	109.0	152.6	207.2	277.9	5
$C_6H_{14}S$	1-Hexanethiol	-45 e	-25 e	1 e	35 e	81.7	152.2	5
$C_6H_{14}S$	2-Hexanethiol	-50 e	-32 e	-8 e	25 e	69.9	138.4	5
$C_6H_{14}S$	Dipropyl sulfide	-50 e	-30 e	-6 e	28 e	73.6	142.4	5

VAPOR PRESSURE (continued)

Mol. Form.	Name	Temperature in °C for the indicated pressure						Ref.
		1 Pa	10 Pa	100 Pa	1 kPa	10 kPa	100 kPa	
$C_6H_{14}S$	Diisopropyl sulfide	-65 e	-47 e	-23 e	9 e	53.1	119.6	5
$C_6H_{14}S$	Isopropyl propyl sulfide				18.5	63.8	131.6	5
$C_6H_{14}S$	Butyl ethyl sulfide	-49 e	-30 e	-5 e	29 e	74.8	143.8	5
$C_6H_{15}N$	Hexylamine			-10 e	22 e	66.0	130.6	5
$C_6H_{15}N$	Butylethylamine				6.1	47.7	107.0	5
$C_6H_{15}N$	Dipropylamine		-48 e	-25 e	6 e	47.5	108.8	5
$C_6H_{15}N$	Diisopropylamine			-47 e	-17.5	23.5	84.0	5
$C_6H_{15}N$	Triethylamine	-58 e	-45 e	-29 e	-5 e	29.9	88.5	1
$C_6H_{15}NO$	2-Diethylaminoethanol				97 e		160.6	5
$C_6H_{15}NO_3$	Triethanolamine	75 e	108 e	148 e	196 e	256.7	334 e	5
$C_6H_{15}O_4P$	Triethyl phosphate			34	76	132	211	4
$C_6H_{16}N_2$	Hexamethylenediamine				76.0	128.2	199.0	5
$C_6H_{16}O_2Si$	Diethoxydimethylsilane	-62 e	-44 e	-21.2	9.1	51.0	113.0	5
$C_6H_{18}Cl_2O_2Si_3$	1,5-Dichloro-1,1,3,3,5,5-hexamethyltrisiloxane	-29 e	-7 e	22.2	59.7	110.5	183.4	5
$C_6H_{18}OSi_2$	Hexamethyldisiloxane		-56 e	-34 e	-5 e	37.1	100.1	5
C_6MoO_6	Molybdenum hexacarbonyl		17.4 s	42.8 s	73.1 s	109.9 s	155.4 s	5
C_7F_{14}	Perfluoromethylcyclohexane				-21 e	18 e	75.9	1
C_7F_{16}	Perfluoroheptane		-62 e	-41 e	-14 e	24.7	82.1	1
C_7HF_{15}	1H-Pentadecafluoroheptane				-7 e	35.9	96.0	5
$C_7H_3ClF_3NO_2$	1-Chloro-2-nitro-4-(trifluoromethyl)benzene	3 e	26 e	55 e	92.8	145.2	222.0	5
$C_7H_3F_5$	2,3,4,5,6-Pentafluorotoluene			-20 e	11 e	53.6	117.0	5
$C_7H_4ClF_3$	1-Chloro-2-(trifluoromethyl)benzene			1 e	34.5	81.8	151.8	5
$C_7H_4ClF_3$	1-Chloro-3-(trifluoromethyl)benzene	-53 e	-34 e	-9 e	24.2	69.8	137.2	5
$C_7H_4ClF_3$	1-Chloro-4-(trifluoromethyl)benzene			-9 e	24.2	70.4	138.1	5
$C_7H_4Cl_2O$	o-Chlorobenzoyl chloride				93 e	149 e	237.0	5
$C_7H_4Cl_2O$	m-Chlorobenzoyl chloride				87.8	147 e	225.0	5
$C_7H_4F_3NO_2$	1-Nitro-3-(trifluoromethyl)benzene		11 e	39 e	76.2	127.3	202.2	5
$C_7H_4F_4$	1-Fluoro-4-(trifluoromethyl)benzene			-38 e	-6 e	38.6	102.3	5
C_7H_5BrO	Benzoyl bromide	-15 e	11 e	42.6	83.9	139.5	218.0	5
C_7H_5ClO	Benzoyl chloride			27.5	67.0	120.4	196.7	5
$C_7H_5Cl_3$	(Trichloromethyl)benzene		9 e	40.6	81.5	136.2	213.0	5
$C_7H_5F_3$	(Trifluoromethyl)benzene			-3 e	39 e	101.6		5
C_7H_5N	Benzonitrile		-6 e	23.9	63.1	115.7	190.0	5
C_7H_5NS	Phenyl isothiocyanate				79.4	105 e	117 e	5
$C_7H_6Cl_2$	2,4-Dichlorotoluene		6 e	33 e	68.3	119.5	199.1	5
$C_7H_6Cl_2$	3,4-Dichlorotoluene	-13 e	9 e	38 e	76 e	129.3	208.4	5
$C_7H_6Cl_2$	(Dichloromethyl)benzene			31	72	130	213	4
C_7H_6O	Benzaldehyde		-9 e	19 e	54.6	104.6	178.3	1
$C_7H_6O_2$	Salicylaldehyde		-1 e	29 e	68 e	120.7	196.2	5
C_7H_7Br	o-Bromotoluene		-10 e	17 e	54 e	104.8	181.1	5
C_7H_7Br	m-Bromotoluene	-34 e	-11 e	19.4	58.1	109.9	183.1	5
C_7H_7Br	p-Bromotoluene				57 e	107.8	183.8	5
C_7H_7Br	(Bromomethyl)benzene			25.4	66.8	121.7	198.3	5
C_7H_7Cl	o-Chlorotoluene		-24 e	3 e	38 e	86.3	158.7	1,5
C_7H_7Cl	m-Chlorotoluene	-41 e	-21 e	6 e	41 e	89 e	161.8	5
C_7H_7Cl	p-Chlorotoluene				40 e	88.9	161.5	1,5
C_7H_7Cl	(Chloromethyl)benzene	-34 e	-11 e	17.7	55.4	106.3	178.9	5
C_7H_7ClO	1-Chloro-2-methoxy-benzene	-22 e	2 e	33 e	72 e	125.2	201 e	5
C_7H_7F	o-Fluorotoluene		-50 e	-26 e	5 e	49.0	113.9	5
C_7H_7F	m-Fluorotoluene	-67 e	-48 e	-25 e	7 e	51.0	116.1	5
C_7H_7F	p-Fluorotoluene		-48 e	-24 e	7 e	51 e	116.2	5
$C_7H_7NO_2$	o-Nitrotoluene	23 e	40 e	62 e	94 e	141.9	221.9	5
$C_7H_7NO_2$	m-Nitrotoluene			45 e	89.7	148.7	231.3	5
$C_7H_7NO_3$	2-Nitroanisole	15 e	45 e	82 e	129 e	189.4	271.8	5
C_7H_8	Toluene	-78.1	-57.1	-31.3	1.5	45.2	110.1	5
C_7H_8	Bicyclo[2.2.1]hepta-2,5-diene				-15 e	27.4	91 e	5
$C_7H_8Cl_2Si$	Dichloromethylphenylsilane			32.4	71.8	126.0	205.0	5
C_7H_8O	o-Cresol	-6.4 s	12.8 s	40.2	72.3	120.3	190.5	1,5
C_7H_8O	m-Cresol	20.8	33.6	52.4	82.6	130.6	201.8	1,5

Mol. Form.	Name	Temperature in °C for the indicated pressure						Ref.
		1 Pa	10 Pa	100 Pa	1 kPa	10 kPa	100 kPa	
C$_7$H$_8$O	p-Cresol	-0.2 s	20.7 s	52.7	83.1	130.7	201.5	1,5
C$_7$H$_8$O	Benzyl alcohol	8 e	28 e	54 e	88 e	134.7	204.9	1
C$_7$H$_8$O	Anisole		-21 e	4 e	38 e	84 e	153.2	1,5
C$_7$H$_8$S	3-Methylbenzenethiol		0 e	29 e	66 e	117.9	194.6	5
C$_7$H$_9$N	Benzylamine			25.6	62.6	112.7	183.9	5
C$_7$H$_9$N	o-Methylaniline	1.0	18.8	42.6	76.1	125.6	199.9	1,5
C$_7$H$_9$N	m-Methylaniline	3.8	22.0	46.2	80.1	128.8	202.9	1,5
C$_7$H$_9$N	p-Methylaniline				77.1	126.2	199.9	5
C$_7$H$_9$N	N-Methylaniline	-16 e	6 e	34 e	70.3	121.1	195.8	1
C$_7$H$_9$N	2-Ethylpyridine	-46 e	-26 e	-1 e	33 e	79.3	149.0	5
C$_7$H$_9$N	3-Ethylpyridine	-38 e	-17 e	9 e	44 e	92.7	166.5	5
C$_7$H$_9$N	4-Ethylpyridine	-35 e	-15 e	11 e	46 e	94.4	168.6	5
C$_7$H$_9$N	2,3-Dimethylpyridine				42 e	89.9	160.6	5
C$_7$H$_9$N	2,4-Dimethylpyridine		-25 e		3.7	40.0	87.5	157.9	1,5
C$_7$H$_9$N	2,5-Dimethylpyridine			4 e	39 e	86.2	156.6	1
C$_7$H$_9$N	2,6-Dimethylpyridine			-3 e	29.9	75.8	143.6	1
C$_7$H$_9$N	3,4-Dimethylpyridine		-9 e	19 e	55 e	104.8	178.6	5
C$_7$H$_9$N	3,5-Dimethylpyridine			11 e	48 e	98 e	171.5	1
C$_7$H$_{10}$N$_2$	Toluene-2,4-diamine			100.4	145.3	202.9	279.5	5
C$_7$H$_{12}$	1-Heptyne	-75 e	-57 e	-35 e	-5 e	37.1	99.5	5
C$_7$H$_{12}$	2-Heptyne		-51 e	-27 e	4 e	46.9	111.5	5
C$_7$H$_{12}$	3-Heptyne	-71 e	-53 e	-31 e	0 e	42.7	106.4	5
C$_7$H$_{12}$	5-Methyl-1-hexyne	-80 e	-62 e	-40 e	-11 e	30.1	91.4	5
C$_7$H$_{12}$	5-Methyl-2-hexyne	-75 e	-57 e	-34 e	-4 e	38.6	102.0	5
C$_7$H$_{12}$	2-Methyl-3-hexyne	-78 e	-61 e	-39 e	-9 e	32.6	94.8	5
C$_7$H$_{12}$	4,4-Dimethyl-1-pentyne		-73 e	-52 e	-24 e	15.9	75.6	5
C$_7$H$_{12}$	4,4-Dimethyl-2-pentyne		-70 e	-48 e	-19 e	21.4	82.6	5
C$_7$H$_{12}$	Bicyclo[4.1.0]heptane					49.9	116.3	5
C$_7$H$_{12}$	Cycloheptene			-30.0	3.4	47.5	108 e	5
C$_7$H$_{12}$	1-Methylbicyclo(3,1,0)hexane					29.8	92.6	5
C$_7$H$_{12}$	Methylenecyclohexane	-76 e	-58 e	-35 e	-5 e	38 e	103.0	5
C$_7$H$_{12}$	1-Methylcyclohexene	-72 e	-53 e	-30 e	1 e	45 e	109.8	5
C$_7$H$_{12}$	4-Methylcyclohexene	-76 e	-59 e	-36 e	-5 e	37.9	102.3	5
C$_7$H$_{12}$	1-Ethylcyclopentene	-75 e	-57 e	-34 e	-3 e	40.7	105.8	5
C$_7$H$_{12}$	1,2-Dimethylcyclopentene	-75 e	-57 e	-34 e	-3 e	40.2	105.3	5
C$_7$H$_{12}$	1,5-Dimethylcyclopentene	-77 e	-59 e	-36 e	-5.5	37.3	101.5	5
C$_7$H$_{12}$O	Cycloheptanone			18 e	53.7	104.0	178.7	5
C$_7$H$_{12}$O$_2$	Butyl acrylate	-52 e	-31 e	-4.5	30.4	78.0	146.9	5
C$_7$H$_{12}$O$_2$	Propyl methacrylate			26 e	73.8	139.7	5
C$_7$H$_{12}$O$_3$	Ethyl levulinate		17 e	45.3	82.6	133.2	205.7	5
C$_7$H$_{12}$O$_4$	Diethyl malonate	-23 e	4 e	36.0	76.4	128.5	198.3	5
C$_7$H$_{12}$O$_4$	Dimethyl glutarate	-11 e	15 e	47 e	87.7	139.8	209.5	5
C$_7$H$_{13}$ClO	Heptanoyl chloride	-17 e	4 e	29.4	59.7	96.9	144.0	5
C$_7$H$_{14}$	1-Heptene	-82.1	-63.8	-40.6	-10.7	31.1	93.2	1,5
C$_7$H$_{14}$	cis-2-Heptene	-79 e	-61 e	-38 e	-8 e	34.3	98.0	5
C$_7$H$_{14}$	trans-2-Heptene	-79 e	-61 e	-39 e	-8 e	34.0	97.5	5
C$_7$H$_{14}$	cis-3-Heptene	-80 e	-62 e	-40 e	-10 e	32.3	95.3	5
C$_7$H$_{14}$	trans-3-Heptene	-80 e	-62 e	-40 e	-10 e	32.2	95.2	5
C$_7$H$_{14}$	2-Methyl-1-hexene	-81 e	-64 e	-42 e	-12 e	29.3	91.6	5
C$_7$H$_{14}$	4-Methyl-1-hexene	-84 e	-67 e	-45 e	-16 e	25.3	86.3	5
C$_7$H$_{14}$	2-Methyl-2-hexene	-80 e	-63 e	-40 e	-10 e	32.0	95.0	5
C$_7$H$_{14}$	cis-3-Methyl-2-hexene	-79 e	-62 e	-39 e	-9 e	33.4	96.8	5
C$_7$H$_{14}$	trans-4-Methyl-2-hexene	-83 e	-66 e	-44 e	-15 e	25.9	87.1	5
C$_7$H$_{14}$	trans-5-Methyl-2-hexene	-83 e	-66 e	-44 e	-15 e	26.3	87.7	5
C$_7$H$_{14}$	trans-2-Methyl-3-hexene	-84 e	-67 e	-45 e	-16 e	24.6	85.5	5
C$_7$H$_{14}$	3-Ethyl-1-pentene	-85 e	-68 e	-46 e	-17 e	23.2	83.7	5
C$_7$H$_{14}$	2,3-Dimethyl-1-pentene	-85 e	-68 e	-46 e	-17 e	23.4	83.8	5
C$_7$H$_{14}$	2,4-Dimethyl-1-pentene	-88 e	-71 e	-50 e	-21 e	20.0	81.2	5
C$_7$H$_{14}$	3,3-Dimethyl-1-pentene	-87 e	-71 e	-50 e	-21 e	18.1	77.1	5
C$_7$H$_{14}$	4,4-Dimethyl-1-pentene	-94 e	-78 e	-57 e	-28 e	11.5	72.1	5
C$_7$H$_{14}$	2,3-Dimethyl-2-pentene	-79 e	-62 e	-39 e	-9 e	33.5	96.9	5
C$_7$H$_{14}$	2,4-Dimethyl-2-pentene	-84 e	-68 e	-46 e	-18 e	22.6	82.9	5
C$_7$H$_{14}$	cis-3,4-Dimethyl-2-pentene	-83 e	-65 e	-43 e	-14 e	27.2	88.8	5
C$_7$H$_{14}$	trans-3,4-Dimethyl-2-pentene	-82 e	-64 e	-42 e	-13 e	29.0	91.1	5
C$_7$H$_{14}$	cis-4,4-Dimethyl-2-pentene	-90 e	-73 e	-51 e	-22 e	18.6	80.0	5
C$_7$H$_{14}$	trans-4,4-Dimethyl-2-pentene	-90 e	-73 e	-52 e	-23 e	16.6	76.3	5
C$_7$H$_{14}$	2,3,3-Trimethyl-1-butene	-91 e	-75 e	-53 e	-24.2	16.3	77.5	5
C$_7$H$_{14}$	Cycloheptane				6 e	51.1	118.4	1

Temperature in °C for the indicated pressure

Mol. Form.	Name	1 Pa	10 Pa	100 Pa	1 kPa	10 kPa	100 kPa	Ref.
C_7H_{14}	Methylcyclohexane	-79 e	-62 e	-39 e	-7.9	35.5	100.5	1
C_7H_{14}	Ethylcyclopentane	-76 e	-59 e	-35 e	-5 e	38.4	103.0	5
C_7H_{14}	1,1-Dimethylcyclopentane		-69 e	-47 e	-17 e	24.8	87.4	5
C_7H_{14}	cis-1,2-Dimethylcyclopentane			-38 e	-8 e	34.9	99.0	5
C_7H_{14}	trans-1,2-Dimethylcyclopentane	-83 e	-66 e	-43 e	-13 e	28.4	91.4	5
C_7H_{14}	cis-1,3-Dimethylcyclopentane	-84 e	-66 e	-44 e	-14 e	28.2	91.1	5
C_7H_{14}	trans-1,3-Dimethylcyclopentane	-84 e	-67 e	-44 e	-14 e	27.4	90.3	5
$C_7H_{14}O$	1-Heptanal	-41 e	-21 e	4 e	37 e	83.7	152.3	5
$C_7H_{14}O$	2-Heptanone		-22 e	3 e	36 e	82.2	150.6	1
$C_7H_{14}O$	3-Heptanone		-28 e	0 e	36 e	83.2	147.0	5
$C_7H_{14}O$	4-Heptanone	-27 e	-6 e	18.8	50.2	90.3	143.4	5
$C_7H_{14}O$	5-Methyl-2-hexanone		-27 e	-2 e	31.0	76.6	144.4	5
$C_7H_{14}O$	2,4-Dimethyl-3-pentanone	-61 e	-42 e	-18 e	14 e	58.5	124.8	1
$C_7H_{14}O_2$	Heptanoic acid	24 e	46 e	72 e	107 e	154.6	222.6	5
$C_7H_{14}O_2$	Pentyl acetate	-58 e	-39 e	-14 e	20 e	70.1	149 e	5
$C_7H_{14}O_2$	Isopentyl acetate	-51 e	-30 e	-4 e	30.3	76.2	141.4	5
$C_7H_{14}O_2$	Isobutyl propanoate	-35 e	-19 e	2 e	31 e	72.0	136.1	5
$C_7H_{14}O_2$	Propyl butanoate	-35 e	-19 e	3 e	32.0	74.9	142.8	5
$C_7H_{14}O_2$	Propyl isobutanoate		-28 e	-5.7	24.5	67.5	133.3	5
$C_7H_{14}O_2$	Isopropyl isobutanoate		-44 e	-19.7	12.2	56.0	120.1	5
$C_7H_{14}O_2$	Ethyl 3-methylbutanoate	-57 e	-36 e	-10 e	23.9	69.5	134.4	5
$C_7H_{14}O_2$	Methyl hexanoate	-47 e	-26 e	2 e	36.6	83.3	149 e	5
$C_7H_{14}O_2$	4-Methoxy-4-methyl-2-pentanone				43 e	89.8	160 e	5
$C_7H_{15}Br$	1-Bromoheptane	-30 e	-9 e	18 e	54 e	104.4	178.4	5
$C_7H_{15}Cl$	1-Chloroheptane	-39 e	-19 e	7 e	41 e	88.6	159.9	5
$C_7H_{15}F$	1-Fluoroheptane	-64 e	-45 e	-22 e	10 e	53.3	117.4	5
$C_7H_{15}I$	1-Iodoheptane	-19 e	3 e	32 e	71 e	123.8	203.4	5
C_7H_{16}	Heptane	-78.6	-60.2	-37.0	-6.6	35.4	98.0	16
C_7H_{16}	2-Methylhexane	-82 e	-65 e	-43 e	-13 e	27.8	89.7	1
C_7H_{16}	3-Methylhexane	-81 e	-64 e	-42 e	-12 e	29.2	91.5	1
C_7H_{16}	3-Ethylpentane	-81 e	-63 e	-41 e	-11 e	30.5	93.1	1
C_7H_{16}	2,2-Dimethylpentane	-90 e	-73 e	-52 e	-22.9	17.6	78.8	1
C_7H_{16}	2,3-Dimethylpentane	-87 e	-68.4	-45.3	-14.9	26.8	89.3	5
C_7H_{16}	2,4-Dimethylpentane	-89 e	-72 e	-50 e	-21.3	19.2	80.1	1
C_7H_{16}	3,3-Dimethylpentane	-88 e	-71 e	-49 e	-18.8	22.9	85.6	1
C_7H_{16}	2,2,3-Trimethylbutane				-23.2	18.1	80.4	5
$C_7H_{16}O$	1-Heptanol		17 e	40 e	70.1	112.5	176 e	1
$C_7H_{16}O$	2-Heptanol	-9 e	7 e	27 e	55.0	95.2	158.7	5
$C_7H_{16}O$	3-Heptanol	-8 e	7 e	27 e	54.5	93.9	156.3	5
$C_7H_{16}O$	4-Heptanol	-16 e	1 e	22 e	51 e	91.9	154.6	5
$C_7H_{16}O$	2,2-Dimethyl-3-pentanol				9 e	35 e	73.1	135.5	5
$C_7H_{16}S$	1-Heptanethiol	-30 e	-9 e	18 e	53 e	102.7	176.4	5
$C_7H_{17}N$	Heptylamine			5 e	39 e	86.7	156.4	5
$C_7H_{18}N_2$	N,N-Diethyl-1,3-propanediamine				50.1	99.9	167.7	5
C_8F_{18}	Perfluoróoctane				5 e	45.0	105.6	5
$C_8H_4O_3$	Phthalic anhydride	48.2 s	72.4 s			192.7	284.2	5
C_8H_6O	Benzofuran		-16 e	12 e	47.9	97.7	170.7	5
C_8H_7Cl	o-Chlorostyrene	-33 e	-10 e	20 e	58 e	110.8	188 e	5
C_8H_7N	2-Methylbenzonitrile		1 e	32.1	72.2	126.6	204.7	5
C_8H_7N	4-Methylbenzonitrile			40.1	78.7	134.3	221.3	5
C_8H_7N	Benzeneacetonitrile	-3 e	23 e	55.3	97.4	153.7	233.1	5
C_8H_7N	Indole	20.6 s	44.5 s				254.0	5
$C_8H_7NO_4$	Methyl 2-nitrobenzoate	17 e	49 e	89 e	140 e	208 e	302 e	5
C_8H_8	Styrene		-31 e	-5 e	28.6	75.4	144.7	1
C_8H_8	1,3,5,7-Cyclooctatetraene				24.3	71.0	140.1	5
C_8H_8O	Acetophenone			36 e	73 e	125.3	201.5	5
$C_8H_8O_2$	Phenyl acetate		3 e	33.1	72.2	123.9	195.5	5
$C_8H_8O_2$	Methyl benzoate		-1 e	29 e	68 e	121.2	198.9	5
$C_8H_8O_2$	4-Methoxybenzaldehyde	9 e	35 e	68.1	110.8	167.9	248.5	5
$C_8H_8O_3$	Methyl salicylate	-1 e	22 e	51 e	88.8	141.8	219.9	5
C_8H_9Cl	1-Chloro-2-ethylbenzene	-30 e	-9 e	18 e	54 e	103.7	177.9	5
C_8H_9Cl	1-Chloro-4-ethylbenzene	-27 e	-6 e	22 e	58 e	108.7	183.9	5
$C_8H_9NO_2$	1-Ethyl-4-nitrobenzene	10 e	36 e	69 e	111.6	168 e	245 e	5
C_8H_{10}	Ethylbenzene	-56.2	-36.8	-12.0	21.1	67.1	135.7	1
C_8H_{10}	o-Xylene			-7 e	27 e	74.2	143.9	1
C_8H_{10}	m-Xylene		-35 e	-10 e	23.4	69.8	138.7	1
C_8H_{10}	p-Xylene				22.4	68.9	137.9	1

Mol. Form.	Name	Temperature in °C for the indicated pressure						Ref.
		1 Pa	10 Pa	100 Pa	1 kPa	10 kPa	100 kPa	
$C_8H_{10}O$	o-Ethylphenol		16.9	44.5	81.1	130.9	204.0	5
$C_8H_{10}O$	m-Ethylphenol	5.6	29.2	57.5	91.9	144.8	217.9	5
$C_8H_{10}O$	p-Ethylphenol			60 e	95.5	144.6	217.5	5
$C_8H_{10}O$	2,3-Xylenol	14.3 s	34.3 s	57.2 s	91.4	141.7	216.4	1,5
$C_8H_{10}O$	2,4-Xylenol			50.2	85.5	137.2	210.5	1,5
$C_8H_{10}O$	2,5-Xylenol	13.4 s	33.2 s	55.9 s	87.4	137.0	210.6	5
$C_8H_{10}O$	2,6-Xylenol	-3.1 s	16.7 s	39.6 s	75.3	125.9	200.6	1,5
$C_8H_{10}O$	3,4-Xylenol	19.7 s	40.2 s	63.7 s	102.1	152.3	226.4	1,5
$C_8H_{10}O$	3,5-Xylenol	16.5 s	37.2 s	61.1 s	98.0	147.9	221.3	1,5
$C_8H_{10}O$	Benzeneethanol	2 e	25 e	54 e	92 e	143.6	217.7	5
$C_8H_{10}O$	Phenetole		-9 e	17 e	51 e	99 e	169.3	5
$C_8H_{10}O_2$	2-Phenoxyethanol	21 e	46 e	75.9	115.4	168.7	244.8	5
$C_8H_{10}O_2$	1,3-Dimethoxybenzene	18 e	34 e	56 e	86.7	135.5	223 e	5
$C_8H_{11}N$	p-Ethylaniline	-2 e	21 e	49 e	87 e	139.4	216.7	5
$C_8H_{11}N$	N-Ethylaniline	-15 e	8 e	38 e	76.4	128.8	204.2	5
$C_8H_{11}N$	N,N-Dimethylaniline			28 e	66 e	118.1	193.6	1
$C_8H_{11}N$	2,4-Xylidine	-2 e	21 e	51 e	88 e	139.1	210.9	5
$C_8H_{11}N$	2,6-Xylidine			37 e	80 e	137.7	217.7	5
$C_8H_{11}N$	5-Ethyl-2-picoline	-33 e	-9.3	20 e			178.0	5
$C_8H_{11}NO$	o-Phenetidine	0 e	27 e	60 e	102.2	156.0	228.1	5
C_8H_{12}	1,5-Cyclooctadiene		-37 e	-8 e	30 e	80.2	150 e	5
C_8H_{12}	4-Vinylcyclohexene	-62 e	-43 e	-19 e	14.1	59.9	129 e	5
$C_8H_{12}O_4$	Diethyl maleate	-6 e	20 e	52.2	93.5	148.4	224.8	5
C_8H_{14}	2,5-Dimethyl-1,5-hexadiene	-38 e	-26 e	-10 e	14 e	50.8	115.1	5
C_8H_{14}	1-Octyne	-59 e	-40 e	-16 e	16 e	60.3	125.8	1
C_8H_{14}	2-Octyne	-52 e	-33 e	-8 e	25 e	70.6	137.8	1
C_8H_{14}	3-Octyne	-55 e	-35 e	-11 e	22 e	66.8	132.8	1
C_8H_{14}	4-Octyne	-56 e	-36 e	-12 e	21 e	65.6	131.4	1
C_8H_{14}	1-Ethylcyclohexene	-55 e	-35 e	-11 e	22 e	68 e	136.5	5
$C_8H_{14}O_2$	Cyclohexyl acetate					103.1	172.9	5
$C_8H_{14}O_2$	Butyl methacrylate				47 e	93.3	159.0	5
$C_8H_{14}O_3$	Butanoic anhydride	-28 e	-2 e	30 e	71 e	123.8	196.5	5
$C_8H_{14}O_4$	Ethyl succinate	-6 e	20 e	51.0	91.1	143.7	216.1	5
$C_8H_{14}O_4$	Dipropyl oxalate	-4 e	20 e	49.9	88.6	140.4	213.0	5
$C_8H_{14}O_4$	Dimethyl adipate		28 e	61 c	103 e	156.1	227.3	5
$C_8H_{15}Br$	(2-Bromoethyl)cyclohexane	-14 e	8 e	36.9	75.3	129.7	212.5	5
$C_8H_{15}ClO$	Octanoyl chloride	1 e	22 e	46 e	74.7	109 e	150 e	5
$C_8H_{15}N$	Octanenitrile	-15 e	8 e	37 e	75 e	127.7	204.4	5
C_8H_{16}	1-Octene	-65.7	-46.1	-21.4	10.5	54.9	120.9	1,5
C_8H_{16}	cis-2-Octene	-59 e	-41 e	-17 e	15 e	59 e	125.2	5
C_8H_{16}	trans-2-Octene	-59 e	-41 e	-17 e	14 e	59 e	124.5	5
C_8H_{16}	cis-3-Octene	-65 e	-46 e	-22 e	10 e	55.1	122.4	5
C_8H_{16}	trans-3-Octene	-61 e	-43 e	-19 e	13 e	57 e	122.8	5
C_8H_{16}	cis-4-Octene	-63 e	-44 e	-20 e	11 e	56 e	122.1	5
C_8H_{16}	trans-4-Octene	-65 e	-46 e	-22 e	10 e	54.6	121.8	5
C_8H_{16}	2-Methyl-1-heptene	-66 e	-48 e	-24 e	8 e	52.3	118.7	5
C_8H_{16}	2,2-Dimethyl-cis-3-hexene	-74 e	-56 e	-33 e	-3 e	40.1	105.0	5
C_8H_{16}	2,3-Dimethyl-2-hexene	-65 e	-47 e	-23 e	10 e	54.3	121.3	5
C_8H_{16}	2,3,3-Trimethyl-1-pentene		-53 e	-30 e	1 e	43.8	107.9	5
C_8H_{16}	2,4,4-Trimethyl-1-pentene	-79 e	-61 e	-38 e	-7 e	36.2	101.0	5
C_8H_{16}	2,3,4-Trimethyl-2-pentene	-68 e	-49 e	-26 e	6 e	50.0	115.8	5
C_8H_{16}	2,4,4-Trimethyl-2-pentene	-73 e	-56 e	-33 e	-2 e	40.4	104.5	5
C_8H_{16}	Cyclooctane				30 e	78 e	150.7	1
C_8H_{16}	Ethylcyclohexane	-61 e	-42 e	-17 e	15.8	61.9	131.3	5
C_8H_{16}	1,1-Dimethylcyclohexane			-27 e	5 e	50.6	119.1	5
C_8H_{16}	cis-1,2-Dimethylcyclohexane		-44 e	-20 e	14 e	59.7	129.2	5
C_8H_{16}	trans-1,2-Dimethylcyclohexane	-68 e	-49 e	-25 e	8 e	53.9	122.9	5
C_8H_{16}	cis-1,3-Dimethylcyclohexane	-68 e	-48 e	-23 e	10 e	55.6	123.1	5
C_8H_{16}	trans-1,3-Dimethylcyclohexane	-62 e	-45 e	-23 e	8 e	51.5	120.9	5
C_8H_{16}	cis-1,4-Dimethylcyclohexane	-66 e	-47 e	-23 e	10 e	55.3	123.8	5
C_8H_{16}	trans-1,4-Dimethylcyclohexane			-27 e	5 e	50.6	118.9	5
C_8H_{16}	Propylcyclopentane	-60 e	-41 e	-16 e	16.5	62.1	130.5	5
C_8H_{16}	Isopropylcyclopentane	-65 e	-46 e	-21 e	12 e	57.3	125.9	5
C_8H_{16}	1-Ethyl-1-methylcyclopentane	-67 e	-49 e	-24 e	8 e	53.2	121.0	5
C_8H_{16}	cis-1-Ethyl-2-methylcyclopentane	-63 e	-44 e	-19 e	13.3	59.1	127.6	5
C_8H_{16}	1,1,2-Trimethylcyclopentane				2 e	46.2	113.2	5
C_8H_{16}	1,1,3-Trimethylcyclopentane	-77 e	-59 e	-36 e	-5 e	38.7	104.4	5

Mol. Form.	Name	Temperature in °C for the indicated pressure						Ref.
		1 Pa	10 Pa	100 Pa	1 kPa	10 kPa	100 kPa	
C$_8$H$_{16}$	1′,2′,4a-1,2,4-Trimethylcyclo- pentane	-70 e	-52 e	-28 e	4 e	48.9	116.2	5
C$_8$H$_{16}$	1′,2a,4′-1,2,4-Trimethylcyclo- pentane	-74 e	-56 e	-33 e	-1 e	42.8	108.8	5
C$_8$H$_{16}$O	1-Propylcyclopentanol	9 e	24 e	43 e	69.0	108.4	173.5	5
C$_8$H$_{16}$O	Octanal			6 e	45.7	97.8	170.2	5
C$_8$H$_{16}$O	2-Octanone		-3 e	23 e	57 e	103.8	172.1	5
C$_8$H$_{16}$O	3-Octanone			8 e	47.7	97 e	161 e	5
C$_8$H$_{16}$O	2,2,4-Trimethyl-3-pentanone			11.3	42.1	81.7	134.6	5
C$_8$H$_{16}$O$_2$	Octanoic acid	37 e	58 e	85 e	120 e	165.5	238.4	1,5
C$_8$H$_{16}$O$_2$	2-Ethylhexanoic acid				108 e	159.6	226.6	5
C$_8$H$_{16}$O$_2$	Hexyl acetate	-37 e	-13 e	16 e	52.8	100.4	164 e	5
C$_8$H$_{16}$O$_2$	Isopentyl propanoate			3.1	40.7	90.6	159.8	5
C$_8$H$_{16}$O$_2$	Isobutyl isobutanoate	-47 e	-26 e	0.4	34.8	81.1	147.0	5
C$_8$H$_{16}$O$_2$	Propyl 3-methylbutanoate			1.8	38.9	87.9	155.6	5
C$_8$H$_{16}$O$_2$	Ethyl hexanoate	-31 e	-9 e	18.7	53.9	100.7	166.2	5
C$_8$H$_{16}$O$_2$	Methyl heptanoate	-30 e	-9 e	19 e	54.2	102.4	172 e	5
C$_8$H$_{16}$O$_4$	Diethylene glycol monoethyl ether acetate	-16 e	10.6	43.9	86.2	141.3	216.6	5
C$_8$H$_{17}$Br	1-Bromooctane	-17 e	6 e	34 e	72 e	123.8	200.3	5
C$_8$H$_{17}$Cl	1-Chlorooctane	-25 e	-4 e	23 e	59 e	108.8	182.9	5
C$_8$H$_{17}$Cl	3-(Chloromethyl)heptane					100.3	172.4	5
C$_8$H$_{17}$F	1-Fluorooctane				29 e	74.6	141.8	5
C$_8$H$_{17}$I	1-Iodooctane	-6 e	18 e	48 e	87 e	142.5	224.5	5
C$_8$H$_{18}$	Octane		-42.6	-17.9	14.4	58.9	125.3	16
C$_8$H$_{18}$	2-Methylheptane	-69 e	-49.1	-24.5	7.6	51.6	117.2	1,5
C$_8$H$_{18}$	3-Methylheptane	-67 e	-48.1	-23.6	8.5	52.7	118.5	1,5
C$_8$H$_{18}$	4-Methylheptane	-65 e	-47 e	-24 e	7.8	51.6	117.2	5
C$_8$H$_{18}$	3-Ethylhexane				8 e	52.1	118.1	5
C$_8$H$_{18}$	2,2-Dimethylhexane	-73 e	-55 e	-32 e	-1.5	41.6	106.4	5
C$_8$H$_{18}$	2,3-Dimethylhexane				5 e	49.2	115.1	5
C$_8$H$_{18}$	2,4-Dimethylhexane				0.6	43.9	109.0	5
C$_8$H$_{18}$	2,5-Dimethylhexane	-71 e	-53 e	-30 e	0.7	43.8	108.6	5
C$_8$H$_{18}$	3,3-Dimethylhexane	-72 e	-54 e	-30 e	1.4	45.4	111.5	5
C$_8$H$_{18}$	3,4-Dimethylhexane				7 e	50.9	117.3	5
C$_8$H$_{18}$	3-Ethyl-2-methylpentane	-69 e	-50 e	-27 e	5 e	48.9	115.2	5
C$_8$H$_{18}$	3-Ethyl-3-methylpentane	-70 e	-51 e	-27 e	5 e	50.2	117.8	5
C$_8$H$_{18}$	2,2,3-Trimethylpentane	-74 e	-56 e	-32 e	-0.8	43.1	109.4	5
C$_8$H$_{18}$	2,2,4-Trimethylpentane	-81.9	-63.4	-39.8	-8.9	34.0	98.8	5
C$_8$H$_{18}$	2,3,3-Trimethylpentane	-72 e	-54 e	-30 e	2.1	46.9	114.3	5
C$_8$H$_{18}$	2,3,4-Trimethylpentane	-74 e	-54.5	-30.0	2.2	46.7	113.1	1,5
C$_8$H$_{18}$	2,2,3,3-Tetramethylbutane	-62.5 s	-44 s	-20.9 s	8.9 s	48.8 s	105.8	5
C$_8$H$_{18}$O	1-Octanol	12 e	30 e	53 e	84 e	128.2	194.8	1,39
C$_8$H$_{18}$O	2-Octanol			40 e	69.9	112.5	179.4	1,39
C$_8$H$_{18}$O	3-Octanol	12 e	24 e	40 e	64 e	102.8	174.1	1
C$_8$H$_{18}$O	4-Octanol			40 e	66.9	107.3	176.0	1,39
C$_8$H$_{18}$O	4-Methyl-3-heptanol	-52 e	-28 e	1 e	39 e	87.6	155.0	5
C$_8$H$_{18}$O	5-Methyl-3-heptanol	-35 e	-16 e	8 e	40 e	84.8	153.0	5
C$_8$H$_{18}$O	4-Methyl-4-heptanol	-17 e	1 e	24 e	55 e	97.2	160.7	5
C$_8$H$_{18}$O	2-Ethyl-1-hexanol				45 e	75 e	118.3	184.2	1
C$_8$H$_{18}$O	2-Ethyl-2-hexanol	-13 e	4 e	26 e	55 e	96.3	160.3	5
C$_8$H$_{18}$O	2,4,4-Trimethyl-2-pentanol		-7 e	13 e	40 e	79.8	146.1	5
C$_8$H$_{18}$O	2,2,4-Trimethyl-3-pentanol	-2 e	9 e	24 e	47 e	82.6	150.4	5
C$_8$H$_{18}$O	Dibutyl ether	-55 e	-35 e	-8 e	26 e	73.0	141.2	5
C$_8$H$_{18}$O	Di-sec-butyl ether			-19 e	12.1	55.4	120.6	5
C$_8$H$_{18}$O	Di-tert-butyl ether			-33 e	-2 e	41.7	106.8	1
C$_8$H$_{18}$O$_2$	Ethylene glycol monohexyl ether	-13 e	14 e	46 e	86 e	137.7	206.9	5
C$_8$H$_{18}$O$_2$	1,2-Dipropoxyethane			-44.2	-2.0	63.6	179.2	5
C$_8$H$_{18}$O$_2$	Di-tert-butyl peroxide			-26 e	4.3	46.6	110.5	5
C$_8$H$_{18}$O$_3$	Diethylene glycol monobutyl ether	14 e	37 e	66.8	104.9	153 e	230.4	5
C$_8$H$_{18}$O$_3$	Diethylene glycol diethyl ether	-32 e	-7 e	25 e	64.9	117.1	189 e	5
C$_8$H$_{18}$O$_5$	Tetraethylene glycol	89 e	117 e	151.1	192.2	242.9	307.3	5
C$_8$H$_{18}$S	1-Octanethiol	-15 e	6 e	34 e	71 e	122.1	198.5	5
C$_8$H$_{18}$S	Dibutyl sulfide	-22 e	0 e	27 e	63 e	113.5	188.4	5

Mol. Form.	Name	Temperature in °C for the indicated pressure						Ref.
		1 Pa	10 Pa	100 Pa	1 kPa	10 kPa	100 kPa	
$C_8H_{19}N$	Dibutylamine	-37 e	-16 e	10 e	44 e	90.8	159.1	5
$C_8H_{19}N$	Diisobutylamine	-57 e	-36 e	-9.0	25.5	72.2	139.0	5
$C_8H_{20}O_4Si$	Ethyl silicate	-77 e	-52 e	-21 e	21.6	80.5	164.1	5
$C_8H_{20}Si$	Tetraethylsilane			-6.5	30.5	80.6	152.6	5
C_9F_{20}	Perfluorononane					40 e	114.7	5
$C_9H_6N_2O_2$	Toluene-2,4-diisocyanate		39 e	72 e	113.9	169.7	247 e	5
C_9H_7N	Quinoline	-1.3	23.7	55.4	96.8	153.4	236.5	1,5
C_9H_7N	Isoquinoline		30.2	60.7	101.3	157.9	242.7	1,5
C_9H_8	Indene			12 e	53.0	106.8	181.0	5
C_9H_{10}	cis-1-Propenylbenzene	-38 e	-15.4	13.3	51.4	103.7	178.4	5
C_9H_{10}	trans-1-Propenylbenzene		-16 e	13.3	51.6	103.7	178.4	5
C_9H_{10}	Isopropenylbenzene			3.2	41.5	92.8	164.9	5
C_9H_{10}	Indan	-33 e	-12 e	16 e	52 e	102.3	177.5	1
$C_9H_{10}O$	2,4-Dimethylbenzaldehyde	-3 e	23 e	54 e	93.2	144.6	214.5	5
$C_9H_{10}O_2$	Ethyl benzoate	-18 e	8 e	39 e	80.1	135.1	212.8	5
$C_9H_{10}O_2$	Benzyl acetate	-11 e	15 e	46.6	86.9	139.5	211 e	5
$C_9H_{11}Br$	1-Bromo-4-isopropylbenzene	-8 e	15 e	45 e	84 e	138.1	218.5	5
$C_9H_{11}Cl$	1-Chloro-2-isopropylbenzene	-23 e	-1 e	27 e	64 e	114.6	190.5	5
$C_9H_{11}Cl$	1-Chloro-4-isopropylbenzene		3 e	31 e	69 e	120.5	197.8	5
C_9H_{12}	Propylbenzene	-43 e	-23 e	4 e	38 e	86.7	158.8	1
C_9H_{12}	Isopropylbenzene	-46 e	-26 e	-1 e	33 e	80.9	152.0	1
C_9H_{12}	o-Ethyltoluene	-40 e	-19 e	8 e	43 e	92.1	164.7	5
C_9H_{12}	m-Ethyltoluene	-42 e	-21 e	5 e	40.4	88.9	160.8	5
C_9H_{12}	p-Ethyltoluene	-41 e	-21 e	6 e	41 e	89.2	161.5	5
C_9H_{12}	1,2,3-Trimethylbenzene		-12 e	15 e	52 e	101.5	175.6	1
C_9H_{12}	1,2,4-Trimethylbenzene	-37 e	-16 e	11 e	47 e	95.9	168.9	1
C_9H_{12}	1,3,5-Trimethylbenzene	-39 e	-18 e	9 e	43.7	92.4	164.3	1
$C_9H_{12}O$	Benzyl ethyl ether		-10 e	20.4	59.3	111.3	184.5	5
$C_9H_{12}O$	Phenyl propyl ether		-10 e	21 e	61 e	113.9	189.3	5
$C_9H_{12}O$	Phenyl isopropyl ether	-20 e	-1 e	23 e	56 e	103.7	176.9	5
$C_9H_{13}N$	2,4,6-Trimethylaniline	12 e	36 e	66 c	104.1	154.9	226 e	5
$C_9H_{13}N$	N,N-Dimethyl-o-toluidine	-25 e	-3 e	24.4	60.6	110.7	184.5	5
$C_9H_{13}N$	Amphetamine			33 e	70.1	118 e	202.0	5
$C_9H_{14}O$	Isophorone		1 e	33.1	75.1	132.4	215.1	5
$C_9H_{14}O_6$	Triacetin	37.6	62 e	90 c	124 e	165 e	214 e	5
$C_9H_{16}O_4$	Diethyl glutarate	-1 e	26 e	60.2	103.3	159.6	236.5	5
$C_9H_{17}N$	Nonanenitrile	-3 e	21 e	50.9	90.7	145.4	225.1	5
C_9H_{18}	1-Nonene	-50.1	-29.4	-3.3	30.4	77.1	146.4	1,5
C_9H_{18}	2-Methyl-1-octene	-53 e	-34 e	-9 e	25 e	72 e	144.1	5
C_9H_{18}	Butylcyclopentane	-45 e	-24 e	1 e	36 e	84 e	156.1	5
C_9H_{18}	Propylcyclohexane	-46 e	-26 e	0 e	35.1	83.6	156.2	5
C_9H_{18}	Isopropylcyclohexane	-48 e	-28 e	-2 e	33 e	81.3	154.0	5
C_9H_{18}	trans-1-Ethyl-4-methylcyclo-hexane	-53 e	-33 e	8 e	25 e	71.8	141.5	5
C_9H_{18}	1,1,2-Trimethylcyclohexane			-.2 e	23 e	71.5	145.5	5
C_9H_{18}	1,1,3-Trimethylcyclohexane	-60 e	-41 e	-16 e	18 e	65.2	136.1	5
C_9H_{18}	1′,2a,4a-1,2,4-Trimethylcyclo-hexane	-71 e	-50 e	-22 e	15 e	65.7	140.7	5
C_9H_{18}	1′,3′,5′-1,3,5-Trimethylcyclo-hexane	-72 e	-50 e	-22 e	14 e	65.1	140.0	5
C_9H_{18}	Isobutylcyclopentane	-105 e	-88 e	-64 e	-28 e	31 e	147.0	5
C_9H_{18}	cis-1-Methyl-2-propylcyclo-pentane	-52 e	-33 e	-7 e	28 e	77 e	152.0	5
C_9H_{18}	trans-1-Methyl-2-propylcyclo-pentane	-56 e	-36 e	-11 e	23 e	72 e	145.8	5
C_9H_{18}	1,1,3,3-Tetramethylcyclo-pentane	-72 e	-54 e	-30 e	2 e	47 e	117.4	5
$C_9H_{18}O$	Nonanal		-3 e	27.4	65.5	115.6	184.6	5
$C_9H_{18}O$	2-Nonanone		8 e	35 e	71 e	121.0	194.0	5
$C_9H_{18}O$	5-Nonanone			-1 e	39.1	94 e	188 e	5
$C_9H_{18}O$	2,6-Dimethyl-4-heptanone	-32 e	-12 e	14 e	48 e	96.2	167.7	5
$C_9H_{18}O_2$	Nonanoic acid	48 e	69 e	97 e	133 e	182.7	255.1	5
$C_9H_{18}O_2$	Heptyl acetate	-16 e	6 e	34 e	70 e	119.9	191.9	5
$C_9H_{18}O_2$	Isopentyl butanoate				55 e	105.6	178.4	5
$C_9H_{18}O_2$	Isobutyl 3-methylbutanoate			11.3	48.3	97.9	168.3	5
$C_9H_{18}O_2$	Propyl hexanoate	-26 e	-2 e	28 e	65.1	113.4	178 e	5
$C_9H_{18}O_2$	Methyl octanoate	-26 e	-9 e	13 e	40 e	76 e	127.9	5

VAPOR PRESSURE (continued)

Mol. Form.	Name	1 Pa	10 Pa	100 Pa	1 kPa	10 kPa	100 kPa	Ref.
				Temperature in °C for the indicated pressure				
$C_9H_{19}Cl$	1-Chlorononane	-11 e	11 e	39 e	76 e	127.8	204.7	5
C_9H_{20}	Nonane	-46.8	-26.0	0.0	34.0	80.8	150.3	16
C_9H_{20}	2-Methyloctane	-49 e	-30 e	-5 e	28 e	73.9	142.8	5
C_9H_{20}	3-Methyloctane	-49 e	-29 e	-5 e	29 e	74.7	143.7	5
C_9H_{20}	4-Methyloctane	-50 e	-30 e	-6 e	27 e	73.2	141.9	5
C_9H_{20}	2,2-Dimethylheptane	-58 e	-39 e	-15 e	18 e	63.6	132.3	5
C_9H_{20}	2,3-Dimethylheptane	-53 e	-33 e	-9 e	25 e	70.8	140.0	5
C_9H_{20}	2,6-Dimethylheptane	-55 e	-36 e	-12 e	21 e	66.4	134.7	5
C_9H_{20}	3-Ethyl-4-methylhexane			-9 e	24 e	70.6	139.9	5
C_9H_{20}	2,2,4-Trimethylhexane	-66.1	-46.4	-21.3	11.8	57.7	126.0	5
C_9H_{20}	2,2,5-Trimethylhexane	-65.1	-45.8	-21.2	11.2	56.2	123.7	1,5
C_9H_{20}	2,3,3-Trimethylhexane	-58 e	-38 e	-13 e	20 e	66.7	137.2	5
C_9H_{20}	2,3,5-Trimethylhexane	-60 e	-41 e	-16 e	17 e	62.3	130.9	5
C_9H_{20}	2,4,4-Trimethylhexane	-62 e	-43 e	-18 e	15 e	61.0	130.2	5
C_9H_{20}	3,3,4-Trimethylhexane	-53 e	-33 e	-7 e	28 e	76.3	148.9	1
C_9H_{20}	3,3-Diethylpentane			-9 e	26 e	73.7	145.7	1
C_9H_{20}	3-Ethyl-2,4-dimethylpentane	-58 e	-38 e	-13 e	20 e	66.7	136.2	5
C_9H_{20}	2,2,3,3-Tetramethylpentane				21 e	68.5	139.8	1
C_9H_{20}	2,2,3,4-Tetramethylpentane	-61 e	-42 e	-17 e	16 e	62.5	132.6	1
C_9H_{20}	2,2,4,4-Tetramethylpentane		-49 e	-25 e	8 e	53.2	121.8	1
C_9H_{20}	2,3,3,4-Tetramethylpentane	-57 e	-37 e	-12 e	22 e	69.7	141.1	1
$C_9H_{20}O$	1-Nonanol		40 e	64 e	96.9	141.0	213.0	5,39
$C_9H_{20}O$	3-Nonanol		24 e	47 e	78 e	123.0	194.2	5
$C_9H_{20}O$	4-Nonanol			45 e	76.4	121.3	192.0	5
$C_9H_{20}O$	5-Nonanol	13 e	31 e	54 e	84.5	128.1	194.7	5
$C_9H_{20}O$	2,2,4,4-Tetramethyl-3-pentanol				58	100	167	5
$C_9H_{20}S$	1-Nonanethiol	-2 e	21 e	49 e	87 e	140.4	219.2	5
$C_9H_{21}BO_3$	Triisopropyl borate					73.1	139.0	5
$C_9H_{21}N$	Nonylamine		9 e	37 e	75 e	126.2	202.1	5
$C_9H_{21}N$	Tripropylamine	-39 e	-18 e	8 e	42 e	88.2	156.0	5
$C_{10}F_8$	Perfluoronaphthalene	5.2 s	25.1 s	48.1 s				5
$C_{10}F_{22}$	Perfluorodecane					52 e	132.9	5
$C_{10}H_7Br$	1-Bromonaphthalene	17 e	45 e	80.3	126.7	189.8	280.5	5
$C_{10}H_7Cl$	1-Chloronaphthalene	14 e	39 e	70.5	112.8	171.6	258.6	5
$C_{10}H_8$	Naphthalene**	3.2 s	24.1 s	49.3 s	80.7	135.6	217.5	1,5
$C_{10}H_8$	Azulene	24.1 s	46 s	71.5 s	103.3	162.6	244.0	5
$C_{10}H_8O$	1-Naphthol				137.2	196.7	281.8	5
$C_{10}H_8O$	2-Naphthol				140.7	200.5	286.8	5
$C_{10}H_9N$	1-Naphthalenamine		62 e	99.0	146.9	210.7	300.1	5
$C_{10}H_9N$	2-Naphthalenamine	36.3 s	65.9 s	103 s	150.9	215.1	305.5	5
$C_{10}H_9N$	2-Methylquinoline	5.3	31.9	63.8	102.9	165.8	247.2	5
$C_{10}H_9N$	4-Methylquinoline	29 e	54 e	85 e	127 e	183.0	265.1	5
$C_{10}H_9N$	6-Methylquinoline	27 e	51 e	81 e	122 e	179.2	264.5	5
$C_{10}H_9N$	8-Methylquinoline	15 e	40 e	70 e	111 e	166.1	247.3	5
$C_{10}H_{10}$	m-Divinylbenzene	-29 e	-4 e	27.1	67.6	122.1	199 e	5
$C_{10}H_{10}O_4$	Dimethyl phthalate	27 e	56 e	92.7	137.8	195.8	272.7	5
$C_{10}H_{10}O_4$	Dimethyl isophthalate			85 e	129.5	189.2	273 e	5
$C_{10}H_{10}O_4$	Dimethyl terephthalate	56.6 s	79.4 s	106.1 s	137.9 s	197.9	282 e	5
$C_{10}H_{12}$	1,2,3,4-Tetrahydronaphthalene	-21 e	3 e	33.2	74.1	127.4	207.8	5
$C_{10}H_{12}$	2-Ethylstyrene	-31 e	-8 e	21 e	60 e	111.7	187 e	5
$C_{10}H_{12}$	3-Ethylstyrene	-28 e	-5.3	24.1	62.6	116 e	193 e	5
$C_{10}H_{12}$	4-Ethylstyrene	-31 e	-8.2	21.3	60.5	115 e	196 e	5
$C_{10}H_{12}O$	Estragole			48.5	88.0	140.7	214.6	5
$C_{10}H_{12}O$	4-Isopropylbenzaldehyde			54.1	96.0	152.2	231.5	5
$C_{10}H_{12}O_2$	4-Allyl-2-methoxyphenol	9 e	37 e	72 e	115.9	173.8	252.9	5
$C_{10}H_{12}O_2$	2-Phenylethyl acetate	-4 e	22 e	54 e	96 e	152.3	232.0	5
$C_{10}H_{12}O_2$	Propyl benzoate	-8 e	18 e	50.2	92.3	149.2	230.5	5
$C_{10}H_{12}O_2$	Ethyl phenylacetate	-9 e	19 e	52 e	95 e	150.2	225 e	5
$C_{10}H_{12}O_2$	Isoeugenol				125 e	185.3	267.1	5
$C_{10}H_{14}$	Butylbenzene	-28 e	-7 e	21 e	56.9	107.6	182.8	1,5
$C_{10}H_{14}$	sec-Butylbenzene	-35 e	-14 e	13 e	48 e	98.3	172.8	5
$C_{10}H_{14}$	tert-Butylbenzene	-37 e	-16 e	10 e	46 e	94.9	168.6	5
$C_{10}H_{14}$	Isobutylbenzene	-36 e	-15 e	12 e	47.9	97.8	172.3	5
$C_{10}H_{14}$	o-Cymene	-39 e	-16 e	13 e	51 e	103.1	177.8	5
$C_{10}H_{14}$	m-Cymene	-34 e	-13 e	14 e	50 e	99.9	174.6	5
$C_{10}H_{14}$	p-Cymene	-33 e	-12 e	16 e	52 e	102.2	176.6	5
$C_{10}H_{14}$	o-Diethylbenzene	-28 e	-6 e	21 e	58 e	107.9	182.9	5

VAPOR PRESSURE (continued)

Mol. Form.	Name	Temperature in °C for the indicated pressure						Ref.
		1 Pa	10 Pa	100 Pa	1 kPa	10 kPa	100 kPa	
$C_{10}H_{14}$	m-Diethylbenzene	-28 e	-7 e	20 e	56 e	106.2	180.6	5
$C_{10}H_{14}$	p-Diethylbenzene	-28 e	-6 e	21 e	57 e	108.1	183.3	5
$C_{10}H_{14}$	3-Ethyl-1,2-dimethylbenzene	-22 e	0 e	28 e	66 e	117.2	193.4	5
$C_{10}H_{14}$	4-Ethyl-1,2-dimethylbenzene	-24 e	-2 e	26 e	63 e	113.6	189.2	5
$C_{10}H_{14}$	2-Ethyl-1,3-dimethylbenzene		-2 e	26 e	63 e	113.7	189.5	5
$C_{10}H_{14}$	2-Ethyl-1,4-dimethylbenzene	-27 e	-5 e	23 e	60 e	110.6	186.4	5
$C_{10}H_{14}$	1-Ethyl-2,4-dimethylbenzene	-25 e	-4 e	24 e	61 e	112.2	187.9	5
$C_{10}H_{14}$	1-Ethyl-3,5-dimethylbenzene	-28 e	-6 e	21 e	58 e	108.3	183.2	5
$C_{10}H_{14}$	1-Methyl-2-propylbenzene	-27 e	-6 e	22 e	58.2	108.9	184.3	5
$C_{10}H_{14}$	1-Methyl-3-propylbenzene	-29 e	-8 e	20 e	56.1	106.5	181.3	5
$C_{10}H_{14}$	1-Methyl-4-propylbenzene	-29 e	-7 e	20 e	56.6	107.4	182.8	5
$C_{10}H_{14}$	1,2,3,4-Tetramethylbenzene		7 e	36 e	74 e	126.6	204.5	5
$C_{10}H_{14}$	1,2,3,5-Tetramethylbenzene	-19 e	3 e	32 e	69 e	120.9	197.5	5
$C_{10}H_{14}$	1,2,4,5-Tetramethylbenzene					119.9	196.3	5
$C_{10}H_{14}O$	2-Butylphenol	7 e	31 e	61 e	101 e	155.2	234.4	5
$C_{10}H_{14}O$	Butyl phenyl ether	-16 e	8 e	38 e	77 e	131.3	209.7	5
$C_{10}H_{14}O$	Thymol	18.9 s	37.9 s	59.5	101.2	155.0	230.4	5
$C_{10}H_{15}N$	2-Methyl-5-isopropylaniline	19 e	43 e	72 e	107.4	150 e	204 e	5
$C_{10}H_{15}N$	N-Butylaniline	11 e	35 e	66 e	106 e	160.9	241.0	5
$C_{10}H_{15}N$	N,N-Diethylaniline	-11 e	14 e	44.3	84.2	138.4	216.3	5
$C_{10}H_{16}$	Dipentene	-42 e	-19 e	10.6	48.7	100.2	173.9	5
$C_{10}H_{16}$	d-Limonene	-45 e	-21 e	9.1	48.0	100.4	174.5	5
$C_{10}H_{16}$	l-Limonene	-33 e	-12 e	16 e	52.0	102.3	177.0	21
$C_{10}H_{16}$	β-Myrcene			9.4	47.3	98.3	171.0	5
$C_{10}H_{16}$	α-Pinene	-48 e	-27 e	-1 e	33.6	82.2	155.1	21
$C_{10}H_{16}$	β-Pinene	-43 e	-22 e	5.0	40.6	90.5	165.5	21
$C_{10}H_{16}$	Camphene					90.7	160.1	4
$C_{10}H_{16}$	Terpinolene			26.5	64.9	115.4	184.6	5
$C_{10}H_{16}$	β-Phellandrene			16 e	53.2	104 e	171.0	5
$C_{10}H_{16}O$	(+)-Camphor	-15.8 s	10 s	41.5 s	80.8 s	131.4 s	207.6	5
$C_{10}H_{16}O$	Pulegone	37 e	49.1	66.4	92.2	135.1	220.2	5
$C_{10}H_{18}$	1-Decyne	-34 e	-13 e	14 e	51 e	100.3	173.5	5
$C_{10}H_{18}$	cis-Decahydronaphthalene	-26 e	-4 e	24 e	62.4	115.5	195.3	1
$C_{10}H_{18}$	trans-Decahydronaphthalene		-10 e	18 e	55.3	107.9	186.8	1
$C_{10}H_{18}O$	α-Terpineol			48	89	142	217	4
$C_{10}H_{18}O$	Eucalyptol			10.6	48.5	100.3	175.4	5
$C_{10}H_{18}O$	trans-Geraniol	4 e	31 e	63.2	104.3	157.7	229.6	5
$C_{10}H_{18}O_4$	Sebacic acid	125.9 s						5
$C_{10}H_{18}O_4$	Dipropyl succinate	11 e	38 e	72.1	115.4	172.3	250.4	5
$C_{10}H_{18}O_4$	Diethyl adipate	4 e	35 e	72 e	116.6	171.2	239.5	5
$C_{10}H_{19}N$	Decanenitrile	13 e	36 e	66 e	105.8	160.6	241.6	5
$C_{10}H_{20}$	1-Decene	-35.5	-13.7	13.7	49.0	97.9	170.1	1,5
$C_{10}H_{20}$	Cyclodecane			29 e	68 e	121.3	201.8	1
$C_{10}H_{20}$	Butylcyclohexane	-31 e	-9 e	18 e	54 e	104.7	180.4	5
$C_{10}H_{20}$	Isobutylcyclohexane	-37 e	-16 e	10 e	46 e	95.9	170.8	5
$C_{10}H_{20}$	tert-Butylcyclohexane	-39 e	-18 e	9 e	45 e	95.3	171.1	5
$C_{10}H_{20}O$	Decanal		16 e	47.2	86.3	137.7	208.0	5
$C_{10}H_{20}O_2$	Decanoic acid	58 e	80 e	108 e	145 e	195.2	269.5	5
$C_{10}H_{20}O_2$	Octyl acetate	-26 e	-3 e	27 e	66.3	120.0	198.2	5
$C_{10}H_{20}O_2$	2-Ethylhexyl acetate	-11 e	5 e	26 e	57.6	107.1	197.2	5
$C_{10}H_{20}O_2$	Isopentyl isopentanoate			22 e	62.8	116.9	193.6	5
$C_{10}H_{20}O_2$	Ethyl octanoate	-17 e	9 e	41 e	81.4	133.2	203 e	5
$C_{10}H_{20}O_4$	Diethylene glycol monobutyl ether acetate	6 e	34 e	69 e	112.6	169.2	245.4	5
$C_{10}H_{21}Br$	1-Bromodecane	9 e	33 e	63 e	104 e	159.2	240.0	5
$C_{10}H_{21}Cl$	1-Chlorodecane	2 e	25 e	54 e	92 e	145.7	225.3	5
$C_{10}H_{21}F$	1-Fluorodecane	-22 e	0 e	27 e	64 e	113.3	185.7	5
$C_{10}H_{22}$	Decane		-10.6	16.7	52.3	101.1	173.7	16
$C_{10}H_{22}$	2-Methylnonane	-34 e	-14 e	12 e	47 e	94.8	166.5	5
$C_{10}H_{22}$	3-Methylnonane	-34 e	-14 e	12 e	47 e	95.1	167.3	5
$C_{10}H_{22}$	4-Methylnonane	-36 e	-16 e	10 e	45 e	93.1	165.2	5
$C_{10}H_{22}$	5-Methylnonane	-36 e	-16 e	10 e	45 e	92.6	164.6	5
$C_{10}H_{22}$	2,4-Dimethyloctane				38 e	84.9	155.4	5
$C_{10}H_{22}$	2,7-Dimethyloctane	-39 e	-19 e	7 e	41 e	88.4	159.4	5
$C_{10}H_{22}$	2,2,6-Trimethylheptane	-46 e	-27 e	-2 e	32 e	78.5	148.4	5
$C_{10}H_{22}$	3,3,5-Trimethylheptane			0 e	35 e	82.7	155.2	5
$C_{10}H_{22}$	2,2,3,3-Tetramethylhexane	-46 e	-25 e	1 e	36 e	85.6	159.8	5

		Temperature in °C for the indicated pressure						
Mol. Form.	Name	1 Pa	10 Pa	100 Pa	1 kPa	10 kPa	100 kPa	Ref.
$C_{10}H_{22}$	2,2,5,5-Tetramethylhexane			-10 e	22 e	68.3	137.0	5
$C_{10}H_{22}$	2,4-Dimethyl-3-isopropylpentane	-46 e	-26 e	0 e	35 e	83.2	156.5	5
$C_{10}H_{22}$	2,2,3,3,4-Pentamethylpentane		-24 e	3 e	39 e	89.1	165.5	5
$C_{10}H_{22}$	2,2,3,4,4-Pentamethylpentane		-29 e	-3 e	33 e	82.8	158.7	5
$C_{10}H_{22}O$	1-Decanol	30 e	50 e	75 e	109 e	157.3	230.6	1,39
$C_{10}H_{22}O$	4-Decanol	18 e	37 e	61 e	93 e	139 e	210 e	5
$C_{10}H_{22}O$	Dipentyl ether	-31 e	-8 e	22 e	60 e	111.6	186.2	5
$C_{10}H_{22}O$	Diisopentyl ether			14.0	51.5	101.8	172.8	5
$C_{10}H_{22}O_2$	Ethylene glycol dibutyl ether	0 e	20 e	44 e	78.4	127.1	202.9	5
$C_{10}H_{22}O_5$	Tetraethylene glycol dimethyl ether				138 e	200.9	275.3	5
$C_{10}H_{22}S$	1-Decanethiol	11 e	34 e	64 e	103 e	157.5	238.6	5
$C_{10}H_{22}S$	Diisopentylsulfide			7 e	82 e	118 e	139 e	5
$C_{10}H_{23}N$	Dipentylamine				77 e	127.7	202.0	5
$C_{10}H_{30}O_3Si_4$	Decamethyltetrasiloxane	-31 e	-6 e	26 e	66.8	118.8	193.9	5
$C_{10}H_{30}O_5Si_5$	Decamethylcyclopentasiloxane	-2 e	19 e	46 e	82 e	132.9	210.4	5
$C_{11}H_8O_2$	1-Naphthalenecarboxylic acid				191.9	239.3	299.6	5
$C_{11}H_8O_2$	2-Naphthalenecarboxylic acid				197.9	246.0	308.1	5
$C_{11}H_{10}$	1-Methylnaphthalene	5 e	29 e	60 e	102 e	159.1	244.1	1
$C_{11}H_{10}$	2-Methylnaphthalene			57 e	99 e	156.0	240.5	1
$C_{11}H_{12}O_2$	Ethyl trans-cinnamate			79	125	187	271	4
$C_{11}H_{12}O_3$	Myristicin	23 e	53 e	88.9	135.2	196.0	279.4	5
$C_{11}H_{14}$	4-Isopropylstyrene	-25 e	-1 e	30.2	70.3	124.5	202.1	5
$C_{11}H_{14}$	1,2,3,4-Tetrahydro-5-methylnaphthalene	9 e	31 e	60 e	99 e	153.1	233.8	5
$C_{11}H_{14}$	1,2,3,4-Tetrahydro-6-methylnaphthalene	17 e	36 e	62 e	97 e	147.8	228.5	5
$C_{11}H_{14}O_2$	Butyl benzoate	6 e	34 e	67.9	110.3	165 e	237 e	5
$C_{11}H_{16}$	Pentylbenzene	-14 e	8 e	37 e	74 e	126.7	204.9	5
$C_{11}H_{16}$	p-tert-Butyltoluene	-24 e	-2 e	27 e	64.1	115.5	190.8	5
$C_{11}H_{16}$	1,3-Diethyl-5-methylbenzene	-26 e	-1 e	29.5	69.5	123.5	200.2	5
$C_{11}H_{16}$	2-Ethyl-1,3,5-trimethylbenzene		6 e	36 e	75.7	129.6	207.6	5
$C_{11}H_{16}$	1-Ethyl-2,4,5-trimethylbenzene	-13 e	11 e	40 e	79.4	132.1	207.7	5
$C_{11}H_{20}$	1-Undecyne	-22 e	0 e	29 e	67 e	118.5	194.5	5
$C_{11}H_{20}$	2-Undecyne	-17 e	6 e	35 e	74 e	127.4	205.4	5
$C_{11}H_{20}O_2$	10-Undecenoic acid	35 e	67 e	105 e	150.0	205.4	274.5	5
$C_{11}H_{20}O_4$	Ethyl diethylmalonate			74 e	105 e	149.4	219 e	5
$C_{11}H_{21}N$	Undecanenitrile			78.6	120.3	177.3	259.9	5
$C_{11}H_{22}$	1-Undecene	-21.6	1.2	29.7	66.4	117.1	192.2	5
$C_{11}H_{22}$	cis-2-Undecene	-14 e	7 e	34 e	70.2	120.6	196 e	5
$C_{11}H_{22}$	trans-2-Undecene	-14 e	7 e	33 e	69.3	119.6	195 e	5
$C_{11}H_{22}$	cis-4-Undecene	-19 e	3 e	30 e	66.6	117.1	192 e	5
$C_{11}H_{22}$	trans-4-Undecene	-17 e	4 e	31 e	67.1	117.4	193 e	5
$C_{11}H_{22}$	cis-5-Undecene	-19 e	2 e	30 e	66.2	116.7	191 e	5
$C_{11}H_{22}$	trans-5-Undecene	-18 e	3 e	31 e	67.0	117.4	192 e	5
$C_{11}H_{22}$	Pentylcyclohexane	-17 e	6 e	34 e	72 e	124.2	202.7	5
$C_{11}H_{22}$	Hexylcyclopentane	-15 e	7 e	36 e	73 e	125.0	202.5	5
$C_{11}H_{22}O$	2-Undecanone	17 e	37 e	64.3	103.0	153.6	232.6	1,5
$C_{11}H_{22}O$	6-Undecanone		28 e	57 e	95 e	148.4	226.9	1
$C_{11}H_{22}O_2$	Undecanoic acid	68 e	90 e	118 e	156 e	207.2	283.6	5
$C_{11}H_{22}O_2$	Heptyl butanoate	2 e	29 e	62 e	102.6	155.1	224.7	5
$C_{11}H_{22}O_2$	Propyl octanoate	-2 e	23 e	55 e	94.0	145.2	215 e	5
$C_{11}H_{22}O_2$	Methyl decanoate	10 e	33 e	62 e	100.9	154.0	232 e	5
$C_{11}H_{24}$	Undecane	-18.4	4.3	32.6	69.5	120.2	195.4	16
$C_{11}H_{24}$	2-Methyldecane	-20 e	1 e	28 e	64 e	114.0	188.7	5
$C_{11}H_{24}$	3-Methyldecane	-35 e	-10 e	22 e	61.9	115.6	190.4	5
$C_{11}H_{24}$	4-Methyldecane	-38 e	-12 e	20 e	60.8	113.9	186.4	5
$C_{11}H_{24}$	2,4,7-Trimethyloctane				43 e	94 e	170.4	5
$C_{11}H_{24}O$	1-Undecanol	52.2	80.0	82 e	118 e	167.6	244.1	5
$C_{11}H_{24}S$	1-Undecanethiol	23 e	47 e	77 e	118 e	173.6	256.8	5
$C_{12}F_{27}N$	Trinonafluorobutylamine		3 e	29.0	63.3	109.9	176.8	5
$C_{12}H_8$	Acenaphthylene	24 s	49.8 s	80.6 s				5
$C_{12}H_9N$	Carbazole					254.7	354.0	5
$C_{12}H_{10}$	Acenaphthene				126.2	187 e	276 e	1
$C_{12}H_{10}$	Biphenyl			69.0	111.1	169.5	254.7	1
$C_{12}H_{10}N_2$	Azobenzene			98.1	144.8	206.7	292.7	4
$C_{12}H_{10}O$	Diphenyl ether		44 e	75 e	116 e	173 e	257.4	5

VAPOR PRESSURE (continued)

Mol. Form.	Name	Temperature in °C for the indicated pressure						Ref.
		1 Pa	10 Pa	100 Pa	1 kPa	10 kPa	100 kPa	
$C_{12}H_{10}O$	1-Acetonaphthone	37 e	69 e	107.0	154.6	215.2	294.9	5
$C_{12}H_{10}O$	2-Acetonaphthone	48.3 s		118.7	163.0	221.1	300.3	5
$C_{12}H_{10}S$	Diphenyl sulfide	20 e	51 e	88.7	137.5	202.2	291.8	5
$C_{12}H_{11}N$	Diphenylamine	48 s		102.8	150.5	213.7	301.4	5
$C_{12}H_{12}$	1-Ethylnaphthalene	16 e	41 e	72 e	114 e	171.8	257.7	5
$C_{12}H_{12}$	2-Ethylnaphthalene	14 e	39 e	71 e	113 e	171.2	257.3	5
$C_{12}H_{12}$	1,2-Dimethylnaphthalene	26 e	51 e	82 e	123 e	180.5	265.7	5
$C_{12}H_{12}$	2,7-Dimethylnaphthalene	31.5 s	53.1 s	78.8 s	115.9	175 e	260 e	5
$C_{12}H_{14}O_4$	Diethyl phthalate	12 e	51 e	96 e	150.5	215.9	296.2	5
$C_{12}H_{16}$	p-Isopropenylisopropylbenzene	-11 e	15 e	46 e	87 e	142.4	221 e	5
$C_{12}H_{16}$	Cyclohexylbenzene		28 e	58 e	98 e	154.7	239.5	5
$C_{12}H_{16}O_2$	3-Methylbutyl benzoate			66 e	115.0	177.7	261.4	5
$C_{12}H_{18}$	Hexylbenzene	-2 e	22 e	51 e	90 e	144.5	225.5	5
$C_{12}H_{18}$	1,2-Diisopropylbenzene	-14 e	9 e	37 e	74 e	125.9	203.2	5
$C_{12}H_{18}$	1,3-Diisopropylbenzene	-14 e	8 e	36 e	74 e	125.5	202.6	5
$C_{12}H_{18}$	1,4-Diisopropylbenzene	-6 e	18 e	49 e	90 e	148.8	238 e	5
$C_{12}H_{18}$	Hexamethylbenzene	46.3 s	72.5 s	81.7 s	121.8 s	178.3	263.7	5
$C_{12}H_{18}$	1,5,9-Cyclododecatriene	-14 e	11 e	44 e	87 e	145.0	229.8	5
$C_{12}H_{20}O_2$	Geranyl acetate			67.7	110.8	166.9	242.9	5
$C_{12}H_{20}O_4$	Dibutyl maleate	12.3	50.4	94.0	144.2	203 e	272 e	5
$C_{12}H_{22}$	1-Dodecyne	-11 e	13 e	43 e	82 e	135.8	214.4	5
$C_{12}H_{22}$	Cyclohexylcyclohexane		20 e	53.1	96.0	154.1	237.2	5
$C_{12}H_{22}O_2$	Methyl 10-undecenoate	10 e	38 e	73 e	116 e	172.2	247.1	5
$C_{12}H_{22}O_4$	Dimethyl sebacate		53 e	97	150	214	293	4
$C_{12}H_{23}N$	Dodecanenitrile	36 e	60 e	92 e	133 e	190.5	275.5	5
$C_{12}H_{24}$	1-Dodecene	-8.3	15.2	44.8	82.9	135.4	212.8	5
$C_{12}H_{24}$	Hexylcyclohexane	-3 e	20 e	50 e	89 e	143.1	224.2	5
$C_{12}H_{24}$	Heptylcyclopentane	-1 e	22 e	51 e	90 e	143.5	223.5	5
$C_{12}H_{24}O$	Dodecanal			70 e	116.2	175.9	256.6	5
$C_{12}H_{24}O_2$	Dodecanoic acid	78 e	100 e	128 e	166 e	219.1	298.1	5
$C_{12}H_{24}O_2$	Decyl acetate	12 e	40 e	74 e	115.1	168.1	238 e	5
$C_{12}H_{24}O_2$	Ethyl decanoate	8 e	35 e	69 e	111.8	166.1	238 e	5
$C_{12}H_{25}Br$	1-Bromododecane	31 e	57 e	90 e	132 e	190.8	275.3	5
$C_{12}H_{25}Cl$	1-Chlorododecane	27 e	51 e	81 e	122 e	178.7	262.6	5
$C_{12}H_{26}$	Dodecane	-5.4	18.2	47.6	85.8	138.2	215.8	16
$C_{12}H_{26}O$	1 Dodecanol				133 c	185.0	264.1	1
$C_{12}H_{26}O_3$	Diethylene glycol dibutyl ether	5 e	34.4	70.2	115.3	174.1	253.8	5
$C_{12}H_{27}N$	Tributylamine	-26 e	1 e	35 e	77.7	134.5	213.4	5
$C_{12}H_{27}N$	Triisobutylamine		1 e	28.9	64.9	112.5	178.5	5
$C_{12}H_{27}O_4P$	Tributyl phosphate					205 e	288.3	5
$C_{12}H_{36}O_6Si_6$	Dodecamethylcyclohexasiloxane	18 e	41 e	69 e	108 e	162.2	244.7	5
$C_{13}H_9N$	Acridine			124.4	176.2	246.0	345.4	5
$C_{13}H_9N$	Phenanthridine	79 s						5
$C_{13}H_{10}$	Fluorene	48.4 s			137.4	205.4	295 e	5
$C_{13}H_{10}O_2$	Phenyl benzoate			102.3	151.4	217.9	313.3	5
$C_{13}H_{10}O_3$	Phenyl salicylate				166.0	224.8	312.4	5
$C_{13}H_{12}$	Diphenylmethane		45 e	77 e	119.3	177.7	263.6	1,5
$C_{13}H_{13}N$	Methyldiphenylamine	35 e	63 e	98.4	143.1	201.6	281.6	5
$C_{13}H_{14}$	1-Isopropylnaphthalene	27 e	51 e	82 e	123.2	180.8	267.3	5
$C_{13}H_{20}$	Heptylbenzene	12 e	36 e	66 e	107 e	162.7	246.2	5
$C_{13}H_{24}O_2$	Ethyl 10-undecenoate	32 e	55 e	86 e	125.2	179.5	258.4	5
$C_{13}H_{26}$	1-Tridecene	4.1	28.5	59.0	98.3	152.5	232.3	5
$C_{13}H_{26}$	Heptylcyclohexane	11 e	34 e	65 e	105 e	160.9	244.3	5
$C_{13}H_{26}$	Octylcyclopentane	13 e	36 e	66 e	106 e	160.9	243.1	5
$C_{13}H_{26}O_2$	Tridecanoic acid	87 e	109 e	138 e	176 e	230.3	311.5	5
$C_{13}H_{26}O_2$	Methyl dodecanoate	38 e	61 e	90 e	130 e	184.9	269 e	5
$C_{13}H_{28}$	Tridecane	7.2	31.5	61.8	101.1	155.1	234.9	16
$C_{13}H_{28}O$	1-Tridecanol	71.6	101.0	103 e	140 e	192.3	273.1	5
$C_{14}H_{10}$	Anthracene	89.2 s	125.9 s	151.5 s	165 s	238.8	340.2	1,5
$C_{14}H_{10}$	Phenanthrene	53 s	83 s	120.8	170.4	238.4	337.7	5
$C_{14}H_{10}O_2$	Benzil			123	175	246	346	4
$C_{14}H_{12}$	cis-Stilbene	26 e	54 e	88 e	130.4	183 e	253 e	5
$C_{14}H_{12}$	trans-Stilbene				155.6	218.1	305.8	5
$C_{14}H_{12}O_2$	Benzoin				181	248	342	4
$C_{14}H_{14}$	1,1-Diphenylethane	19 e	47 e	82.0	125.3	181 e	254 e	5
$C_{14}H_{15}N$	Dibenzylamine	48 e	77 e	113.1	158.9	218.5	299.4	5
$C_{14}H_{16}$	1-Butylnaphthalene	67 e	82 e	103 e	135 e	186.7	288.6	5

Mol. Form.	Name	Temperature in °C for the indicated pressure						Ref.
		1 Pa	10 Pa	100 Pa	1 kPa	10 kPa	100 kPa	
$C_{14}H_{16}$	2-Butylnaphthalene	44 e	67 e	98 e	139 e	197.5	287.4	5
$C_{14}H_{22}$	Octylbenzene	20.1	46.2	79.1	121.9	178.1	263.8	5
$C_{14}H_{26}O_4$	Diethyl sebacate		83 e	120	166	225	305	4
$C_{14}H_{27}N$	Tetradecanenitrile	52 e	79 e	114.0	159.0	219.7	306.3	5
$C_{14}H_{28}$	1-Tetradecene	16.1	41.3	72.7	113.2	168.7	250.6	5
$C_{14}H_{28}$	Octylcyclohexane	16.9	44.3	77.8	120.0	177.6	263.2	5
$C_{14}H_{28}$	Nonylcyclopentane	25 e	49 e	80 e	120 e	177.2	261.5	5
$C_{14}H_{28}O_2$	Tetradecanoic acid	96 e	118 e	147 e	186 e	241.3	325.6	5
$C_{14}H_{30}$	Tetradecane	19.1	44.1	75.3	115.7	171.1	253.0	16
$C_{14}H_{30}O$	1-Tetradecanol	80.0	110.5	149.6	152 e	205.3	286.7	5
$C_{14}H_{31}N$	Tetradecylamine			104 e	147 e	206.1	290.9	5
$C_{14}H_{42}O_5Si_6$	Tctradccamethylhexasiloxane	6 e	36 e	72 e	117 e	176.0	259.1	5
$C_{15}H_{18}$	1-Pentylnaphthalene	34 e	62 e	96 e	141.3	202.2	289 e	5
$C_{15}H_{24}$	Nonylbenzene	33.0	58.9	92.0	135.4	193.7	281.4	5
$C_{15}H_{30}$	Nonylcyclohexane	35 e	60 e	92 e	134 e	193.4	280.9	5
$C_{15}H_{30}$	Decylcyclopentane	37 e	61 e	93 e	134 e	192.5	278.8	5
$C_{15}H_{30}O_2$	Methyl tetradecanoate		75 e	110	155	214	295	4
$C_{15}H_{32}$	Pentadecane	30.5	56.1	88.1	129.6	186.3	270.1	16
$C_{16}H_{22}O_4$	Dibutyl phthalate		104.0	142.7	191.5	254.5	339.4	4
$C_{16}H_{32}$	1-Hexadecene	38.4	65.0	98.1	140.5	198.8	284.3	5
$C_{16}H_{32}O_2$	Hexadecanoic acid		136 e	165 e	205 e	261.9	350.2	5
$C_{16}H_{34}$	Hexadecane	41.1	67.4	100.3	142.7	200.7	286.3	16
$C_{16}H_{34}O$	1-Hexadecanol	99.5	130.6	171.9	175 e	229.0	311.7	5
$C_{16}H_{35}N$	Hexadecylamine	63 e	91 e	126 e	171 e	232.6	320.5	5
$C_{17}H_{10}O$	Benzanthrone		184 e	229.3	290.3	377.2	511 e	5
$C_{17}H_{34}O_2$	Methyl hexadecanoate	65 e	93	129	177			4
$C_{17}H_{36}$	Heptadecane	51.5	78.5	112.0	155.3	214.5	302 e	16
$C_{17}H_{36}O$	1-Heptadecanol	94 e	117 e	146 e	185 e	240.1	323.3	5
$C_{18}H_{14}$	o-Terphenyl	66 e	94 e	129 e	176 e	241.3	336.3	5
$C_{18}H_{14}$	m-Terphenyl	87 e	118 e	156 e	206.6	275.3	374.6	5
$C_{18}H_{14}$	p-Terphenyl	127.1 s	154.7 s		217.2	284.0	383.0	5
$C_{18}H_{30}$	Hexaethylbenzene				144.1	206.8	297.5	5
$C_{18}H_{34}O_2$	Oleic acid	94 e	126 e	165.5	214.5	277.0	359.7	5
$C_{18}H_{34}O_2$	Elaidic acid		124 e	166	216	280	361	4
$C_{18}H_{36}O$	Stearaldehyde			142 e	186 e	246.9	336.7	5
$C_{18}H_{36}O_2$	Stearic acid		153 e	183 e	223 e	281.6	374.5	5
$C_{18}H_{38}$	Octadecane	61.5	89.0	123.1	167.3	227.6	316 e	16
$C_{18}H_{38}O$	1-Octadecanol	106 e	130 e	160 e	200.5	257.3	343.0	5
$C_{19}H_{16}$	Triphenylmethane	81 s		112 e	175 e	254.6	360.0	5
$C_{19}H_{36}O_2$	Methyl oleate	85 e	114 e	149.7	195.6	256 e	340 e	5
$C_{19}H_{40}$	Nonadecane	71.1	99.1	133.8	178.8	240.1	330 e	16
$C_{20}H_{42}$	Eicosane	80.4	108.9	144.2	189.8	252.1	344 e	16
$C_{20}H_{42}O$	1-Eicosanol	119 e	143 e	173 e	213 e	270.0	355.1	5
$C_{20}H_{60}O_8Si_9$	Eicosamethylnonasiloxane			141 e	183.1	236.7	307.1	5
$C_{21}H_{21}O_4P$	Tri-o-cresyl phosphate	119.0	156.1	201.0	256.3	326.3	418 e	5
$C_{21}H_{21}O_4P$	Tri-m-cresyl phosphate	147.8	177.3	211.4	251.3	298 e	355 e	5
$C_{21}H_{21}O_4P$	Tri-p-cresyl phosphate	140.6	174 e	214 e	262 e	320 e	392 e	5
$C_{21}H_{44}$	Heneicosane	82.3	113.5	152.2	201.6	263.8	355.9	5
$C_{22}H_{42}O_2$	Brassidic acid	134 e	166 e	203.6	249.8	307.6	382.0	5
$C_{22}H_{42}O_2$	Erucic acid	126 e	160 e	199.4	247.4	306.5	381.1	5
$C_{22}H_{42}O_2$	Butyl oleate	95.5	124.2	158 e	198 e	245 e	304 e	5
$C_{22}H_{44}O_2$	Behenic acid	145.4	176.5	213.7	259.3	316.2	390 e	5
$C_{22}H_{44}O_2$	Butyl stearate	99.6	128 e	162 e	201 e	249 e	307 e	5
$C_{22}H_{46}$	Docosane	83.5	115.0	154.0	203.6	274.8	368.0	5
$C_{23}H_{48}$	Tricosane	102.9	135.1	174.8	221 e	285.3	379.5	5
$C_{24}H_{38}O_4$	Dioctyl phthalate	130 e	163.7	203.8	252 e	311 e	385 e	5
$C_{24}H_{38}O_4$	Bis(2-ethylhexyl) phthalate	122.0	153.2	189.2	231.3	281.1	341.1	5
$C_{24}H_{50}$	Tetracosane	115.0	148.1	188.5	239.1	295.4	390.6	5
$C_{25}H_{52}$	Pentacosane	119.7	152.7	193.2	244.4	305.0	401.1	5
$C_{26}H_{54}$	Hexacosane	125.1	158.8	200.1	252.1	314.3	411.3	5
$C_{27}H_{56}$	Heptacosane	136.7	168.8	206.5	255.8	323.3	421.2	5
$C_{28}H_{58}$	Octacosane	136.5	169.8	210.9	263.1	332.0	430.6	5
$C_{29}H_{60}$	Nonacosane	148.2	182.8	221.2	271.5	340.2	439.7	5
$C_{30}H_{62}$	Squalane	66 e	84 e	105.8	131.9	163.7	203.2	5
C_{70}	Carbon (fullerene-C_{70})	598 s	662 s					22

VAPOR PRESSURE OF FLUIDS AT TEMPERATURES BELOW 300 K

This table gives vapor pressures of 67 important fluids in the temperature range 2 to 300 K. Helium (^{4}He), hydrogen (H_2), and neon (Ne) are covered on this page. The remaining fluids are listed on subsequent pages by molecular formula in the Hill order (see Introduction). The data have been taken from evaluated sources; references are listed at the end of the table.

Pressures are given in kilopascals (kPa). Note that:

1 kPa = 7.50062 Torr
100 kPa = 1 bar
101.325 kPa = 1 atmos

s following an entry indicates that the compound is solid at that temperature.

Helium		Hydrogen		Neon	
T/K	P/kPa	T/K	P/kPa	T/K	P/kPa
2.2	5.3	14.0	7.90	25.0	51.3
2.3	6.7	14.5	10.38	26.0	71.8
2.4	8.3	15.0	13.43	27.0	98.5
2.5	10.2	15.5	17.12	28.0	132.1
2.6	12.4	16.0	21.53	29.0	173.5
2.7	14.8	16.5	26.74	30.0	223.8
2.8	17.5	17.0	32.84	31.0	284.0
2.9	20.6	17.5	39.92	32.0	355.2
3.0	24.0	18.0	48.08	33.0	438.6
3.1	27.8	18.5	57.39	34.0	535.2
3.2	32.0	19.0	67.96	35.0	646.2
3.3	36.5	19.5	79.89	36.0	772.8
3.4	41.5	20.0	93.26	37.0	916.4
3.5	47.0	20.5	108.2	38.0	1078
3.6	52.9	21.0	124.7	39.0	1260
3.7	59.3	21.5	143.1	40.0	1462
3.8	66.1	22.0	163.2	41.0	1688
3.9	73.5	22.5	185.3	42.0	1939
4.0	81.5	23.0	209.4	43.0	2216
4.1	90.0	23.5	235.7	44.0	2522
4.2	99.0	24.0	264.2		
4.3	108.7	24.5	295.1		
4.4	119.0	25.0	328.5		
4.5	129.9	25.5	364.3		
4.6	141.6	26.0	402.9		
4.7	153.9	26.5	444.3		
4.8	167.0	27.0	488.5		
4.9	180.8	27.5	535.7		
5.0	195.4	28.0	586.1		
5.1	210.9	28.5	639.7		
		29.0	696.7		
		29.5	757.3		
		30.0	821.4		
		30.5	889.5		
		31.0	961.5		
		31.5	1038.0		
		32.0	1119.0		
		32.5	1204.0		

| Ref. | 17,18 | | 1 | | 13 |

T/K	Ar Argon		BCl_3 Boron trichloride	BF_3 Boron trifluoride	BrH Hydrogen bromide		Br_2 Bromine		ClF Chlorine fluoride	ClH Hydrogen chloride	
50	0.1	s									
55	0.2	s									
60	0.8	s									
65	2.8	s									
70	7.7	s									
75	18.7	s									
80	40.7	s									
85	79.0										
90	134										
95	213										
100	324										
105	473										
110	666										
115	910									0.1	
120	1214								0.3	0.1	s
125	1584								0.6	0.3	s
130	2027								1.2	0.5	s
135	2553				0.1	s			2.1	1.0	s
140	3170				0.3	s			3.6	1.9	s
145	3892			7.7	0.6	s			6.0	3.4	s
150	4736			13.4	1.1	s			9.5	5.8	s
155				22.3	1.9	s			14.6	9.5	s
160				35.2	3.3	s			21.8	14.7	
165				53.7	5.4	s			31.7	22.0	
170				79.1	8.7	s			44.8	31.9	
175				113	13.4	s			62.0	45.1	
180			0.1	157	20.1	s			84.2	62.5	
185			0.2	214	29.5	s			112	84.7	
190			0.3	285	37.9				147	113	
195			0.5	372	51.8				190	148	
200			0.8	479	69.5				242	190	
205			1.2	608	91.8				304	242	
210			1.8	762	119				378	304	
215			2.6	944	153				464	377	
220			3.8	1160	194		0.1	s	564	463	
225			5.2	1413	242		0.2	s	680	563	
230			7.2	1709	299		0.3	s	812	678	
235			9.7	2056	366		0.4	s	961	811	
240			12.9	2460	443		0.7	s	1130	961	
245			17.0	2913	532		1.1	s	1319	1132	
250			22.0	3481	633		1.7	s	1529	1325	
255			28.1	4123	748		2.6	s	1762	1542	
260			35.6	4874	878		3.8	s	2019	1784	
265			44.5		1023		5.5	s	2301	2054	
270			55.1		1185		7.3		2608	2354	
275			67.6		1364		9.5		2941	2686	
280			82.2		1562		12.3		3303	3053	
285			99.1		1780		15.6		3693	3457	
290			119		2018		19.7		4111	3901	
295			141		2278		24.6		4560	4388	
300			166		2561		30.5		5039	4921	
Ref.	15		12	12	12		12		12	12	

T/K	ClO_2 Chlorine dioxide	Cl_2 Chlorine	Cl_4Si Silicon tetrachloride	FH Hydrogen fluoride	F_2 Fluorine	F_2O Difluorine oxide	F_3N Nitrogen trifluoride
50							
55					0.4		
60					1.5		
65					4.8		
70					12.3		
75					27.6	0.1	
80					55.3	0.2	
85					101	0.5	0.1
90					172	1.2	0.2
95					276	2.6	0.4
100					420	5.3	0.9
105					615	10.1	2.0
110					870	18.0	4.0
115					1196	30.5	7.3
120					1605	49.3	12.8
125					2108	76.7	21.1
130					2721	115	33.5
135					3458	168	51.1
140					4339	237	75.4
145						328	108
150						444	150
155						588	205
160						766	273
165						981	357
170						1238	459
175		1.8				1541	581
180		2.8				1895	726
185		4.2				2303	896
190		6.1		0.3		2771	1092
195	0.1	8.7		0.5		3302	1319
200	0.3	12.3		0.8		3899	1578
205	0.5	16.9		1.2		4567	1871
210	0.9	22.9	0.1	1.7		5308	2203
215	1.4	30.5	0.2	2.3			2577
220	2.3	40.1	0.3	3.2			2995
225	3.5	51.9	0.5	4.4			3464
230	5.3	66.4	0.7	5.9			3991
235	7.6	84.0	1.0	7.9			
240	10.8	105	1.5	10.3			
245	14.9	130	2.0	13.4			
250	20.1	160	2.8	17.2			
255	26.6	194	3.8	21.8			
260	34.6	234	5.0	27.4			
265	44.4	280	6.6	34.2			
270	56.1	332	8.6	42.2			
275	69.9	392	11.1	51.8			
280	86.2	459	14.2	63.1			
285	105	535	17.9	76.3			
290	127	619	22.3	91.7			
295	151	714	27.7	110			
300	179	818	34.0	130			
Ref.	12	5	12	12	12	12	1

T/K	F₃P Phosphorous trifluoride	F₄Si Silicon tetrafluoride		F₆S Sulfur hexafluoride		HI Hydrogen iodide		H₂S Hydrogen sulfide		H₃N Ammonia		H₃P Phosphine
50												
55												
60												
65												
70												
75												
80												
85												
90												
95												
100												
105	0.1											
110	0.2											0.1
115	0.5											0.2
120	1.0											0.4
125	1.9	0.1	s									0.7
130	3.5	0.2	s									1.3
135	5.9	0.4	s					0.1	s			2.3
140	9.5	0.9	s	0.1	s			0.2	s			3.9
145	14.9	1.9	s	0.2	s			0.3	s			6.2
150	22.5	3.8	s	0.4	s			0.6	s			9.6
155	33.1	7.5	s	0.8	s	0.1	s	1.1	s			14.5
160	47.3	14.0	s	1.5	s	0.2	s	1.9	s	0.1	s	21.1
165	66.0	25.2	s	2.6	s	0.4	s	3.2	s	0.2	s	30.0
170	90.1	43.8	s	4.4	s	0.8	s	5.2	s	0.3	s	41.6
175	121	74.2	s	7.1	s	1.3	s	8.3	s	0.6	s	56.6
180	159	122	s	11.3	s	2.2	s	12.7	s	1.2	s	75.6
185	206	197	s	17.3	s	3.4	s	18.9	s	2.1	s	99.2
190	262	280		25.9	s	5.3	s	26.6	s	3.5	s	128
195	330	376		38.0	s	8.0	s	36.7	s	5.8	s	163
200	410	488		54.4	s	11.7	s	49.8		8.7		205
205	503	618		76.6	s	16.8	s	66.4		12.6		254
210	611	766		106	s	23.6	s	87.1		17.9		312
215	736	932		145	s	32.5	s	113		24.9		379
220	877	1117		195	s	44.0	s	144		34.1		456
225	1037	1324		249		56.2		182		45.9		544
230	1217	1555		305		71.4		227		60.8		644
235	1418	1816		371		89.7		281		79.6		756
240	1640	2111		448		112		344		103		881
245	1885	2449		536		137		416		131		1019
250	2154	2841		636		168		500		165		1172
255	2448	3301		750		203		597		207		1341
260	2767			878		244		706		256		1525
265	3112			1021		290		830		313		1725
270				1181		343		969		381		1942
275				1358		404		1124		460		2176
280				1554		472		1297		552		2428
285				1768		548		1488		655		2699
290				2003		633		1698		774		2987
295				2258		727		1929		909		3295
300				2534		831		2181		1062		3621
Ref.	12	12		12,15		12		12,15		11		12

T/K	H_4Si Silane	Kr Krypton		NO Nitric oxide		N_2 Nitrogen		N_2O Nitrous oxide	O_2 Oxygen	O_2S Sulfur dioxide
50						0.4	s			
55						1.8	s		0.2	
60						6.3	s		0.7	
65						17.4			2.3	
70						38.6			6.3	
75		0.1	s			76.1			14.5	
80		0.4	s			137			30.1	
85		1.1	s	0.1	s	229			56.8	
90		2.7	s	0.4	s	361			99.3	
95	0.1	6.0	s	1.3	s	541			163	
100	0.2	12.1	s	3.8	s	779			254	
105	0.4	22.8	s	10.0	s	1084			379	
110	1.0	40.4	s	23.5		1467			543	
115	1.9	68.0	s	46.8		1939		0.1	756	
120	3.5	103		86.5		2513		0.1	1022	
125	6.1	150		151		3209		0.3	1351	
130	10.0	211		248				0.7	1749	
135	15.8	290		391				1.3	2225	
140	24.1	388		592				2.5	2788	
145	35.3	509		867				4.3	3448	
150	50.3	655		1231				7.1	4219	
155	69.8	830		1703				11.4		
160	94.6	1037		2302				17.6		
165	126	1278		3050				26.4		
170	164	1557		3971				38.5		0.1
175	210	1877		5089				54.7		0.2
180	265	2241		6433				75.9		0.3
185	331	2655						103		0.5
190	408	3120						138		0.8
195	498	3641						181		1.3
200	602	4223						234		2.0
205	722	4870						298		3.0
210	859							374		4.4
215	1017							465		6.3
220	1196							571		9.0
225	1398							694		12.6
230	1628							835		17.3
235	1888							996		23.3
240	2180							1179		31.1
245	2509							1385		40.9
250	2880							1615		53.2
255	3296							1870		68.3
260	3763							2152		86.7
265	4288							2462		109
270								2802		136
275								3172		168
280								3573		205
285								4006		249
290								4473		300
295								4973		359
300								5508		426
Ref.	12	13, 15		12, 15		1		12	3	12

T/K	O_3 Ozone	Rn Radon	Xe Xenon		$CBrF_3$ Bromotri- fluoromethane	$CClF_3$ Chlorotri- fluoromethane	CCl_2F_2 Dichlorodi- fluoromethane	CCl_3F Trichloro- fluoromethane
50								
55								
60								
65								
70								
75								
80								
85								
90								
95								
100	0.1		0.1	s				
105	0.2		0.1	s				
110	0.4		0.3	s				
115	1.0		0.7	s		0.1		
120	2.0		1.5	s		0.2		
125	3.8		2.7	s		0.3		
130	6.8	0.1	4.9	s		0.6		
135	11.5	0.3	8.5	s	0.1	1.1		
140	18.7	0.5	14.0	s	0.3	2.0		
145	29.1	0.9	22.2	s	0.5	3.3		
150	43.7	1.5	34.2	s	0.9	5.3		
155	63.6	2.4	51.1	s	1.5	8.3	0.1	
160	89.9	3.8	74.2	s	2.5	12.6	0.3	
165	124	5.8	101		3.9	18.6	0.5	
170	168	8.6	134		5.9	26.8	0.8	
175	222	12.5	173		8.8	37.6	1.3	
180	289	17.7	222		12.8	51.7	2.1	
185	367	24.5	280		18.1	69.7	3.2	
190	468	33.2	348		25.1	92.3	4.8	0.2
195	584	44.4	428		34.1	120	6.9	0.3
200	721	58.2	521		45.6	155	9.9	0.4
205	881	75.3	628		60.0	196	13.7	0.6
210	1068	96	750		77.8	246	18.8	1.0
215	1285	121	889		99.5	304	25.2	1.4
220	1536	151	1045		126	372	33.3	2.0
225	1824	185	1220		157	451	43.3	2.9
230	2155		1416		194	542	55.5	4.1
235	2534		1633		237	646	70.4	5.6
240	2968		1872		287	763	88.1	7.6
245	3464		2136		344	896	109	10.1
250	4031		2425		410	1044	134	13.3
255	4678		2742		485	1210	163	17.2
260	5417		3087		570	1394	196	22.1
265			3462		665	1598	234	28.0
270			3869		771	1823	278	35.1
275			4310		889	2071	327	43.7
280			4786		1021	2343	383	53.8
285			5299		1166	2641	445	65.7
290					1325	2968	515	79.6
295					1501	3325	593	95.6
300					1692	3716	679	114.1
Ref.	12	15	12,13		12	12	12	12

T/K	CCl_4 Tetrachloro-methane	CF_4 Tetrafluoro-methane	CO Carbon monoxide	COS Carbon oxysulfide	CO_2 Carbon dioxide	$CHClF_2$ Chlorodifluo-methane	$CHCl_3$ Trichloro-methane
50			0.1 s				
55			0.6 s				
60			2.6 s				
65			8.2 s				
70			21.0				
75			44.4				
80			83.7				
85			147				
90		0.1	239				
95		0.3	371				
100		0.8	545				
105		1.7	771				
110		3.4	1067				
115		6.5	1428				
120		11.5	1877				
125		19.3	2400				
130		30.8	3064				
135		47.4			0.1		
140		70.2		0.1	0.2		
145		101		0.2	0.4		
150		141		0.4	0.8	0.1	
155		191		0.8	1.7	0.3	
160		254		1.3	3.1	0.5	
165		332		2.2	5.7	0.8	
170		425		3.4	9.9	1.4	
175		537		5.2	16.8	2.3	
180		669		7.8	27.6	3.6	
185		824		11.3	44.0	5.5	
190		1005		15.9	68.4	8.1	
195		1216		22.1	104	11.8	
200		1460		30.0	155	16.7	
205		1743		40.1	227	23.1	
210		2073		52.7	327	31.5	
215		2457		68.2	465	42.1	0.1
220		2907		87.2	600	55.3	0.2
225		3438		110	735	71.7	0.3
230				137	894	91.6	0.4
235				169	1075	116	0.7
240				207	1283	144	1.0
245				250	1519	178	1.4
250				301	1786	218	2.0
255	1.5			358	2085	264	2.7
260	2.1			423	2419	317	3.7
265	2.8			497	2790	377	5.0
270	3.7			580	3203	446	6.6
275	4.9			673	3658	525	8.7
280	6.4			777	4160	613	11.3
285	8.2			892	4712	711	14.4
290	10.5			1019	5315	821	18.3
295	13.2			1159	5984	944	22.9
300	16.5			1313	6710	1080	28.5
Ref.	12	12	9	12	6	12	12

T/K	CHF₃ Trifluoro-methane	CHN Hydrogen cyanide		CH₂Cl₂ Dichloro-methane	CH₂F₂ Difluoro methane	CH₂O Formaldehyde	CH₃Cl Chloromethane	CH₃F Fluoromethane
50								
55								
60								
65								
70								
75								
80								
85								
90								
95								
100								
105								
110								
115								
120	0.1							
125	0.2							
130	0.4							
135	0.7							0.6
140	1.4				0.1			1.2
145	2.5				0.2			2.1
150	4.3				0.3			3.6
155	7.1				0.6			5.9
160	11.1				1.0			9.3
165	17.0				1.7			14.1
170	25.3				2.8			20.9
175	36.5				4.4			29.9
180	51.4				6.8			42.0
185	70.9				10.2	1.3	2.1	57.6
190	95.8				14.8	2.0	3.1	77.4
195	127				21.2	3.0	4.6	102
200	166	0.1	s	0.1	29.5	4.4	6.7	133
205	214	0.2	s	0.2	40.5	6.4	9.5	171
210	271	0.4	s	0.3	54.5	9.1	13.1	216
215	340	0.6	s	0.4	72.1	12.7	17.9	270
220	421	1	s	0.6	94.1	17.4	24.0	333
225	516	1.5	s	0.9	121	23.4	31.8	408
230	626	2.2	s	1.4	154	31.0	41.4	495
235	754	3.3	s	2.0	193	40.6	53.3	595
240	900	4.7	s	2.8	240	52.5	67.7	711
245	1067	6.8	s	3.8	295	67.0	85.1	843
250	1257	9.7	s	5.3	360	84.6	106	993
255	1472	13.6	s	7.1	434	106	131	1163
260	1713	18.8		9.5	521	131	159	1355
265	1984	24.1		12.4	620	161	193	1571
270	2287	30.5		16.1	732	196	232	1813
275	2624	38.3		20.7	860	236	277	2084
280	3000	47.7		26.3	1004	283	327	2387
285	3418	58.8		33.0	1165	337	385	2724
290	3881	72.1		41.1	1346	399	450	3099
295	4393	87.6		50.8	1547	470	524	3516
300		105.9		62.1	1770	549	606	3978
Ref.	12	12,16		12	12	12	12	12

T/K	CH_4 Methane	CH_4O Methanol	C_2H_2 Acetylene	C_2H_4 Ethylene	C_2H_6 Ethane	C_2H_6O Dimethyl ether	C_3H_4 Propadiene
50							
55							
60							
65	0.1						
70	0.3						
75	0.8						
80	2.1						
85	4.9						
90	10.6						
95	20.0						
100	34.5						
105	57.0						
110	88.4			0.3			
115	133			0.8	0.1		
120	192			1.4	0.4		
125	269			2.7	0.7		
130	368		0.1 s	4.5	1.3		
135	491		0.3 s	7.7	2.2		
140	642		0.7 s	11.9	3.8		
145	824		1.3 s	18.3	6.0		
150	1041		2.6 s	27.5	9.7		0.1
155	1297		4.6 s	39.9	15.0	0.1	0.2
160	1594		7.8 s	56.4	21.5	0.2	0.3
165	1937		12.8 s	77.9	31.0	0.3	0.6
170	2331		20.6 s	105	42.9	0.5	1.0
175	2779		32.2 s	140	59.0	0.9	1.7
180	3288		49.0 s	182	78.7	1.4	2.7
185	3865		72.9 s	234	104	2.1	4.1
190	4520		106 s	296	135	3.2	6.1
195			146	369	172	4.7	8.9
200			190	456	217	6.8	12.5
205			244	557	271	9.6	17.4
210			309	673	334	13.3	23.7
215			385	806	407	18.1	31.6
220			475	958	492	24.3	41.4
225			579	1128	590	32.1	53.5
230		0.1	699	1321	700	41.9	68.2
235		0.2	837	1535	826	53.9	85.8
240		0.4	993	1774	967	68.6	107
245		0.5	1170	2039	1125	86.3	131
250		0.8	1370	2331	1301	108	160
255		1.2	1593	2652	1496	133	193
260		1.7	1843	3005	1712	162	230
265		2.4	2121	3391	1949	197	273
270		3.3	2429	3813	2210	237	322
275		4.5	2771	4275	2495	283	376
280		6.2	3150		2806	335	438
285		8.3	3567		3146	395	506
290		11	4028		3515	463	582
295		14.4	4535		3917	538	666
300		18.7	5093		4355	623	759
Ref.	2,16	12	12,16	4	2	12	12

T/K	C_3H_6 Propylene	C_3H_8 Propane	C_4H_6 Buta-1,3-diene	C_4H_{10} n-Butane	C_4H_{10} Isobutane	C_5H_{12} n-Pentane	C_5H_{12} Neopentane	
50								
55								
60								
65								
70								
75								
80								
85								
90								
95								
100								
105								
110								
115								
120								
125								
130								
135								
140	0.1							
145	0.2							
150	0.4							
155	0.7							
160	1.2	0.8			0.1			
165	2.0	1.4			0.1			
170	3.1	2.2	0.1	0.1	0.3			
175	4.7	3.3	0.2	0.2	0.4			
180	7.0	5.0	0.4	0.3	0.7			
185	10.1	7.3	0.6	0.5	1.1		0.1	s
190	14.2	10.5	1.0	0.8	1.7		0.2	s
195	19.7	15.0	1.5	1.3	2.5		0.4	s
200	26.9	20.1	2.3	1.9	3.7		0.7	s
205	35.9	27.0	3.4	2.8	5.3		1.1	s
210	47.3	36.0	4.8	4.0	7.4		1.6	s
215	61.3	47.0	6.7	5.7	10.2		2.4	s
220	78.5	60.0	9.2	7.8	13.8	1.0	3.6	s
225	99.2	77.0	12.5	10.6	18.3	1.5	5.2	s
230	124	97.0	16.7	14.1	24.0	2.1	7.3	s
235	153	120	21.9	18.5	31.1	3.0	10.2	s
240	188	148	28.4	24.1	39.8	4.2	13.9	s
245	228	180	36.3	30.9	50.3	5.7	18.7	s
250	274	218	46.0	39.1	62.9	7.6	24.8	s
255	327	261	57.6	49.1	77.8	10.0	32.4	s
260	387	311	71.3	61.0	95.4	13.0	41.6	
265	456	367	87.6	75.0	116	16.6	51.4	
270	533	431	107	91.5	140	21.1	63.0	
275	619	502	129	111	167	26.6	76.6	
280	715	582	154	133	198	33.1	92.3	
285	822	671	184	159	234	40.8	111	
290	940	769	217	188	274	50.0	131	
295	1069	878	255	221	319	60.7	155	
300	1212	998	297	258	370	73.2	182	
Ref.	7	2	12	2	2	14	12,16	

REFERENCES

1. B. A. Younglove, Thermophysical properties of fluids. I. Ethylene, parahydrogen, nitrogen trifluoride, and oxygen, *J. Phys. Chem. Ref. Data*, 11, Supp. 1, 1982.
2. B. A. Younglove and J. F. Ely, Thermophysical properties of fluids. II. Methane, ethane, propane, isobutane, and normal butane, *J. Phys. Chem. Ref. Data*, 16, 577, 1987.
3. W. Wagner, et al., *International Tables for the Fluid State: Oxygen*, Blackwell Scientific Publications, Oxford, 1987.
4. R. T. Jacobsen, et al., *International Tables for the Fluid State: Ethylene*, Blackwell Scientific Publications, Oxford, 1988.
5. S. Angus, et al., *International Tables for the Fluid State: Chlorine*, Pergamon Press, Oxford, 1985.
6. S. Angus, et al., *International Tables for the Fluid State: Carbon Dioxide*, Pergamon Press, Oxford, 1976.
7. S. Angus, et al., *International Tables for the Fluid State: Propylene*, Pergamon Press, Oxford, 1980.
8. R. B. Stewart and R. T. Jacobsen, Thermophysical properties of argon, *J. Phys. Chem. Ref. Data*, 18, 639, 1989.
9. R. D. Goodwin, Carbon monoxide thermophysical properties, *J. Phys. Chem. Ref. Data*, 14, 849, 1985.
10. R. D. Goodwin, Methanol thermophysical properties, *J. Phys. Chem. Ref. Data*, 16, 799, 1987.
11. L. Haar, Thermodynamic properties of ammonia, *J. Phys. Chem. Ref. Data*, 7, 635, 1978.
12. DIPPR Data Compilation of Pure Compound Properties, Design Institute for Physical Properties Data, American Institute of Chemical Engineers, 1987.
13. V. A. Rabinovich, et al., *Thermophysical Properties of Neon, Argon, Krypton, and Xenon*, Hemisphere Publishing Corp., New York, 1987.
14. K. N. Marsh, *Recommended Reference Methods for the Realization of Physicochemical Properties*, Blackwell Scientific Publications, Oxford, 1987.
15. TRC Thermodynamic Tables: Non-Hydrocarbons, Thermodynamic Research Center, Texas A & M University, College Station, Texas, 1985.
16. R. M. Stevenson and S. Malanowski, *Handbook of the Thermodynamics of Organic Compounds*, Elsevier, New York, 1987.
17. S. Angus and K. M. de Reuck, *International Tables of the Fluid State: Helium-4*, Pergamon Press, Oxford, 1977.
18. R. D. McCarty, *J. Phys. Chem. Ref. Data*, 2, 923, 1973.

IUPAC RECOMMENDED DATA FOR VAPOR PRESSURE CALIBRATION

Vapor pressures are given in kPa (1 kPa = 0.0098692 atmos = 7.5006 Torr). Reprinted with permission if IUPAC.

REFERENCE

Marsh, K.N., Ed., *Recommended Reference Materials for the Realization of Physicochemical Properties*, Blackwell Scientific Publications, Oxford (1987).

T/K	$CO_2(s)$	$H_2O(s)$	$C_{10}H_8(s)$	$n\text{-}C_5H_{12}$	C_6H_6	C_6F_6	H_2O	Hg
180	27.62							
190	68.44							
200	155.11	0.0002						
210	327.17	0.0007						
220		0.0026						
230		0.0089						
240		0.0273						
250		0.0760		7.60				
260		0.1958	0.0001	12.98				
270		0.4701	0.0005	21.15			0.485	
280			0.0017	33.11	5.148	4.322	0.991	
290			0.0049	50.01	8.606	7.463	1.919	
300			0.0134	73.17	13.816	12.328	3.535	
310			0.0341	104.07	21.389	19.576	6.228	
320			0.0814	144.3	32.054	30.009	10.540	
330			0.1829	195.7	46.656	44.578	17.202	
340			0.3899	260.1	66.152	64.380	27.167	
350			0.7920	339.4	91.609	90.664	41.647	
360				435.9	124.192	124.816	62.139	
370				551.5	165.2	168.4	90.453	
380				688.8	215.9	223.0	128.74	
390				850.2	277.7	290.4	179.48	
400				1038	353.2	372.6	245.54	0.138
410				1256	441.0	471.5	330.15	0.215
420				1507	545.5	589.3	436.89	0.329
430				1793	667.6	728.3	569.73	0.493
440				2120	808.8	890.9	732.99	0.724
450				2490	971.1	1080	931.36	1.045
460				2910	1156	1297	1169.9	1.485
470					1366	1547	1453.9	2.078
480					1602	1833	1789.0	2.866
490					1868	2159	2181.4	3.899
500					2164	2530	2637.3	5.239
510					2494	2954	3163.3	6.955
520					2861		3766.4	9.131
530					3267		4453.9	11.861
540					3717		5233.5	15.256
550					4216		6113.4	19.438
560					4770		7102.0	25.547
570							8208.6	30.74
580							9443.0	38.19
590							10816	47.09
600							12339	57.64

ENTHALPY OF VAPORIZATION

The molar enthalpy (heat) of vaporization, $\Delta_{vap}H$, of about 700 inorganic and organic substances is tabulated here. Values are given, when available, both at the normal boiling point t_b, referred to a pressure of 101.325 kPa (760 mmHg), and at 25°C. Substances are listed by molecular formula in the modified Hill order (see Preface).

See Reference 4 for a discussion of the accuracy of the data and methods of estimating enthalpy of vaporization at other temperatures.

REFERENCES

1. Chase, M.W., et al., *JANAF Thermochemical Tables, Third Edition, J. Phys. Chem. Ref. Data*, 14, Suppl. 1, 1985.
2. Daubert, T. E., Danner, R. P., Sibul, H. M., and Stebbins, C. C., *Physical and Thermodynamic Properties of Pure Compounds: Data Compilation*, extant 1994 (core with 4 supplements), Taylor & Francis, Bristol, PA (also available as database).
3. Bartels, J., et al., *Landolt-Bornstein Numerical Values and Functions of Physics, Chemistry, Astronomy, Geophysics, and Technology*, Vol. 2, Part 4: Thermal Properties, Springer, Heidelberg, 1961.
4. Majer, V., and Svoboda, V., *Enthalpies of Vaporization of Organic Compounds*, Blackwell Scientific Publications, Oxford, 1985.

Molecular formula	Name	t_b °C	$\Delta_{vap}H$ (t_b) kJ/mol	$\Delta_{vap}H$ (25°C) kJ/mol
Compounds not containing carbon:				
AgBr	Silver bromide	1502	198	
AgCl	Silver chloride	1547	199	
AgI	Silver iodide	1506	143.9	
Al	Aluminum	2519	294	
AlB$_3$H$_{12}$	Aluminum borohydride	44.5	30	
AlBr$_3$	Aluminum tribromide	255	23.5	
AlI$_3$	Aluminum triiodide	382	32.2	
Ar	Argon	-185.85	6.43	
AsBr$_3$	Arsenic tribromide	221	41.8	
AsCl$_3$	Arsenic trichloride	130	35.01	
AsF$_3$	Arsenic trifluoride	57.8	29.7	
AsF$_5$	Arsenic pentafluoride	-52.8	20.8	
AsH$_3$	Arsine	-62.5	16.69	
AsI$_3$	Arsenic triiodide	424	59.3	
Au	Gold	2856	324	
B	Boron	4000	480	
BBr$_3$	Boron tribromide	91	30.5	
BCl$_3$	Boron trichloride	12.65	23.77	23.1
BF$_3$	Boron trifluoride	-101	19.33	
BI$_3$	Boron triiodide	210	40.5	
B$_2$F$_4$	Tetrafluorodiborane(4)	-34	28	
B$_2$H$_6$	Diborane	-92.4	14.28	
B$_4$H$_{10}$	Tetraborane	18	27.1	
B$_5$H$_{11}$	Pentaborane(11)	63	31.8	
Ba	Barium	1897	140	
BeCl$_2$	Beryllium chloride	482	105	
BeI$_2$	Beryllium iodide	487	70.5	
Bi	Bismuth	1564	151	
BiBr$_3$	Bismuth tribromide	453	75.4	
BiCl$_3$	Bismuth trichloride	447	72.61	
BrF	Bromine fluoride	20	25.1	
BrF$_3$	Bromine trifluoride	125.8	47.57	
BrF$_5$	Bromine pentafluoride	40.76	30.6	
BrH	Hydrogen bromide	-66.38		12.69
BrH$_3$Si	Bromosilane	1.9	24.4	
BrIn	Indium bromide (InBr)	656	92	
BrTl	Thallium bromide (TlBr)	819	99.56	
Br$_2$	Bromine (Br$_2$)	58.8	29.96	30.91
Br$_2$Cd	Cadmium bromide	844	115	
Br$_2$H$_2$Si	Dibromosilane	66	31	

Molecular formula	Name	t_b °C	$\Delta_{vap} H$ (t_b) kJ/mol	$\Delta_{vap} H$ (25°C) kJ/mol
Br$_2$Hg	Mercury bromide (HgBr$_2$)	322	58.89	
Br$_2$Pb	Lead bromide (PbBr$_2$)	892	133	
Br$_2$Sn	Tin bromide (SnBr$_2$)	639	102	
Br$_2$Zn	Zinc bromide	697	118	
Br$_3$Ga	Gallium bromide (GaBr$_3$)	279	38.9	
Br$_3$HSi	Tribromosilane	109	34.8	
Br$_3$OP	Phosphoryl bromide	191.7	38	
Br$_3$P	Phosphorus tribromide	172.95	38.8	
Br$_3$Sb	Antimony tribromide	280	59	
Br$_4$Ge	Germanium tetrabromide	186.35	41.4	
Br$_4$Si	Silicon tetrabromide	154	37.9	
Br$_4$Sn	Tin bromide (SnBr$_4$)	205	43.5	
Br$_4$Ti	Titanium bromide (TiBr$_4$)	230	44.37	
Br$_5$Ta	Tantalum bromide (TaBr$_5$)	349	62.3	
Cd	Cadmium	767	99.87	
CdCl$_2$	Cadmium chloride	960	124.3	
CdF$_2$	Cadmium fluoride	1748	214	
CdI$_2$	Cadmium iodide	742	115	
ClF	Chlorine fluoride	-101.1	24	
ClFO$_3$	Perchloryl fluoride	-46.75	19.33	
ClF$_2$P	Phosphorus chloride difluoride	-47.25	17.6	
ClF$_3$	Chlorine trifluoride	11.75	27.53	
ClF$_3$Si	Chlorotrifluorosilane	-70.0	18.7	
ClH	Hydrogen chloride	-85	16.15	9.08
ClH$_3$Si	Chlorosilane	-30.4	21	
ClNO	Nitrosyl chloride	-5.5	25.78	
ClNO$_2$	Nitryl chloride	-15	25.7	
ClO$_2$	Chlorine dioxide (ClO$_2$)	11	30	
ClTl	Thallium chloride (TlCl)	720	102.2	
Cl$_2$	Chlorine (Cl$_2$)	-34.04	20.41	17.65
Cl$_2$Cr	Chromium chloride (CrCl$_2$)	1300	197	
Cl$_2$CrO$_2$	Chromyl chloride (CrO$_2$Cl$_2$)	117	35.1	
Cl$_2$FP	Phosphorus dichloride fluoride	14	24.9	
Cl$_2$F$_2$Si	Dichlorodifluorosilane	-32	21.2	
Cl$_2$H$_2$Si	Dichlorosilane	8.3	25	24.2
Cl$_2$Hg	Mercury chloride (HgCl$_2$)	304	58.9	
Cl$_2$O	Oxygen dichloride	2.2	25.9	
Cl$_2$OS	Thionyl chloride	75.6	31.7	31
Cl$_2$O$_2$S	Sulfuryl chloride	69.4	31.4	30.1
Cl$_2$Pb	Lead chloride (PbCl$_2$)	951	127	
Cl$_2$Sn	Tin chloride (SnCl$_2$)	623	86.8	
Cl$_2$Ti	Titanium chloride (TiCl$_2$)	1500	232	
Cl$_2$Zn	Zinc chloride	732	126	
Cl$_3$Ga	Gallium chloride (GaCl$_3$)	201	23.9	
Cl$_3$HSi	Trichlorosilane	33		25.7
Cl$_3$OP	Phosphoryl chloride	105.5	34.35	38.6
Cl$_3$OV	Vanadium oxytrichloride	127	36.78	
Cl$_3$P	Phosphorus trichloride	75.95	30.5	32.1
Cl$_3$Sb	Antimony trichloride	220.3	45.19	
Cl$_3$Ti	Titanium chloride (TiCl$_3$)	960	124	
Cl$_4$Ge	Germanium tetrachloride	86.55	27.9	
Cl$_4$OW	Tungsten oxide tetrachloride	227.55	67.8	
Cl$_4$Si	Silicon tetrachloride	57.65	28.7	29.7
Cl$_4$Sn	Tin chloride (SnCl$_4$)	114.15	34.9	
Cl$_4$Te	Tellurium chloride (TeCl$_4$)	387	77	
Cl$_4$Th	Thorium chloride (ThCl$_4$)	921	146.4	
Cl$_4$Ti	Titanium chloride (TiCl$_4$)	136.45	36.2	

Molecular formula	Name	t_b °C	$\Delta_{vap} H$ (t_b) kJ/mol	$\Delta_{vap} H$ (25°C) kJ/mol
Cl$_4$V	Vanadium chloride (VCl$_4$)	148	41.4	42.5
Cl$_5$Mo	Molybdenum pentachloride	268	62.8	
Cl$_5$Nb	Niobium chloride (NbCl$_5$)	254.0	52.7	
Cl$_5$Ta	Tantalum chloride (TaCl$_5$)	239.35	54.8	
Cl$_6$W	Tungsten chloride (WCl$_6$)	346.75	52.7	
FH$_3$Si	Fluorosilane	-98.6	18.8	
FLi	Lithium fluoride	1673	147	
FNO	Nitrosyl fluoride	-59.9	19.28	
FNO$_2$	Nitryl fluoride	-72.4	18.05	
FNS	Thionitrosyl fluoride (NSF)	4.8	22.2	
F$_2$	Fluorine (F$_2$)	-188.12	6.62	
F$_2$H$_2$Si	Difluorosilane	-77.8	16.3	
F$_2$O	Oxygen difluoride	-144.75	11.09	
F$_2$OS	Thionyl fluoride	-43.8	21.8	
F$_2$O$_2$	Dioxygen difluoride (FOOF)	-57	19.1	
F$_2$Pb	Lead fluoride (PbF$_2$)	1293	160.4	
F$_2$Zn	Zinc fluoride	1500	190.1	
F$_3$HSi	Trifluorosilane	-95	16.2	
F$_3$N	Nitrogen trifluoride	-128.75	11.56	
F$_3$P	Phosphorus trifluoride	-101.5	16.5	
F$_3$PS	Thiophosphoryl trifluoride	-52.25	19.6	
F$_4$N$_2$	Tetrafluorohydrazine	-74	13.27	
F$_4$S	Sulfur fluoride (SF$_4$)	-40.45	26.44	
F$_4$Se	Selenium fluoride (SeF$_4$)	106	47.2	
F$_4$Th	Thorium fluoride (ThF$_4$)	1680	258	
F$_5$I	Iodine pentafluoride	100.5	41.3	
F$_5$Nb	Niobium fluoride (NbF$_5$)	229	52.3	
F$_5$P	Phosphorus pentafluoride	-84.6	17.2	
F$_5$Ta	Tantalum fluoride (TaF$_5$)	229.2	56.9	
F$_5$V	Vanadium fluoride (VF$_5$)	48.3	44.52	
F$_6$Ir	Iridium fluoride (IrF$_6$)	53	36	
F$_6$Mo	Molybdenum hexafluoride	34	27.2	28
F$_6$S	Sulfur fluoride (SF$_6$)			8.99
F$_6$W	Tungsten fluoride (WF$_6$)	17	27	26.6
Ga	Gallium	2204	254	
GaI$_3$	Gallium iodide (GaI$_3$)	340	56.5	
Ge	Germanium	2833	334	
GeH$_4$	Germane	-88.1	14.06	
Ge$_2$H$_6$	Digermane	30.8	25.1	
Ge$_3$H$_8$	Trigermane	110.5	32.2	
HI	Hydrogen iodide	-35.55	19.76	17.36
HLiO	Lithium hydroxide	1626	188	
HNO$_3$	Nitric acid	83		39.1
HN$_3$	Hydrazoic acid	35.7	30.5	
HNaO	Sodium hydroxide	1388	175	
H$_2$	Hydrogen (H$_2$)	-252.87	0.898	
H$_2$O	Water	100.0	40.65	43.98
H$_2$O$_2$	Hydrogen peroxide	150.2		51.6
H$_2$S	Hydrogen sulfide	-59.55	18.67	14.08
H$_2$S$_2$	Hydrogen sulfide (H$_2$S$_2$)	70.7		33.78
H$_2$Se	Hydrogen selenide	-41.25	19.7	
H$_2$Te	Hydrogen telluride	-2	19.2	
H$_3$N	Ammonia	-33.33	23.33	19.86
H$_3$P	Phosphine	-87.75	14.6	
H$_3$Sb	Stibine	-17	21.3	
H$_4$N$_2$	Hydrazine	113.55	41.8	44.7
H$_4$P$_2$	Diphosphine (P$_2$H$_4$)	63.5	28.8	

Molecular formula	Name	t_b °C	$\Delta_{vap} H$ (t_b) kJ/mol	$\Delta_{vap} H$ (25°C) kJ/mol
H$_4$Si	Silane	-111.9	12.1	
H$_4$Sn	Stannane	-51.8	19.05	
H$_6$Si$_2$	Disilane	-14.3	21.2	
H$_8$Si$_3$	Trisilane	52.9	28.5	
He	Helium	-268.93	0.0829	
Hg	Mercury	356.73	59.11	
HgI$_2$	Mercury iodide (HgI$_2$)	354	59.2	
IIn	Indium iodide (InI)	712	90.8	
ITl	Thallium iodide (TlI)	824	104.7	
I$_2$	Iodine (I$_2$)	184.4	41.57	
I$_2$Pb	Lead iodide (PbI$_2$)	872	104	
I$_2$Sn	Tin iodide (SnI$_2$)	714	105	
I$_3$P	Phosphorus triiodide	227	43.9	
I$_3$Sb	Antimony triiodide	401	68.6	
I$_4$Si	Silicon tetraiodide	287.35	50.2	
I$_4$Sn	Tin iodide (SnI$_4$)	364.35	56.9	
I$_4$Ti	Titanium iodide (TiI$_4$)	377	58.4	
Kr	Krypton	-153.22	9.08	
MoO$_3$	Molybdenum oxide (MoO$_3$)	1155	138	
NO	Nitric oxide	-151.74	13.83	
N$_2$	Nitrogen (N$_2$)	-195.79	5.57	
N$_2$O	Nitrous oxide	-88.48	16.53	
N$_2$O$_4$	Nitrogen tetroxide	21.15	38.12	
Ne	Neon	-246.08	1.71	
O$_2$	Oxygen (O$_2$)	-182.95	6.82	
O$_2$S	Sulfur dioxide	-10.05	24.94	22.92
O$_3$S	Sulfur trioxide	45	40.69	43.14
P	Phosphorus	280.5	12.4	14.2
Pb	Lead	1749	179.5	
S	Sulfur	444.60	45	
STl$_2$	Thallium sulfide (Tl$_2$S)	1367	154	
Se	Selenium	685	95.48	
Te	Tellurium	988	114.1	
Xe	Xenon	-108.04	12.62	

Compounds containing carbon:

Molecular formula	Name	t_b °C	$\Delta_{vap} H$ (t_b) kJ/mol	$\Delta_{vap} H$ (25°C) kJ/mol
CCl$_4$	Tetrachloromethane	76.8	29.82	32.43
CN$_4$O$_8$	Tetranitromethane	126.1	40.74	49.93
CO	Carbon monoxide	-191.5	6.04	
CS$_2$	Carbon disulfide	46	26.74	27.51
CHBr$_3$	Tribromomethane	149.1	39.66	46.05
CHCl$_3$	Trichloromethane	61.17	29.24	31.28
CH$_2$Br$_2$	Dibromomethane	97	32.92	36.97
CH$_2$Cl$_2$	Dichloromethane	40	28.06	28.82
CH$_2$O$_2$	Formic acid	101	22.69	20.10
CH$_3$Br	Bromomethane	3.5	23.91	22.81
CH$_3$Cl	Chloromethane	-24.09	21.40	18.92
CH$_3$I	Iodomethane	42.55	27.34	27.97
CH$_3$NO	Formamide	220		60.15
CH$_3$NO$_2$	Nitromethane	101.19	33.99	38.27
CH$_4$	Methane	-161.48	8.19	
CH$_4$O	Methanol	64.6	35.21	37.43
CH$_5$N	Methylamine	-6.32	25.60	23.37
CH$_6$N$_2$	Methylhydrazine	87.5	36.12	40.37
C$_2$Br$_2$ClF$_3$	1,2-Dibromo-1-chloro-1,2,2-trifluoroethane	93	31.17	35.04
C$_2$Br$_2$F$_4$	1,2-Dibromotetrafluoroethane	47.35	27.03	28.39

Molecular formula	Name	t_b °C	$\Delta_{vap} H (t_b)$ kJ/mol	$\Delta_{vap} H (25°C)$ kJ/mol
C_2ClF_5	Chloropentafluoroethane	-37.95	19.41	
$C_2Cl_3F_3$	1,1,1-Trichlorotrifluoroethane	46.1	26.85	28.08
$C_2Cl_3F_3$	1,1,2-Trichlorotrifluoroethane	47.7	27.04	28.40
C_2Cl_4	Tetrachloroethylene	121.3	34.68	39.68
C_2F_6	Hexafluoroethane	-78.1	16.15	
C_2N_2	Cyanogen	-21.1	23.33	19.75
$C_2HBrClF_3$	1,1,1-Trifluoro-2-bromo-2-chloroethane	50.2	28.08	29.61
C_2HCl_3	Trichloroethylene	87.21	31.40	34.54
$C_2H_2Cl_2$	1,1-Dichloroethylene	31.6	26.14	26.48
$C_2H_2Cl_4$	1,1,2,2-Tetrachloroethane	146.5	37.64	45.71
$C_2H_3Cl_3$	1,1,1-Trichloroethane	74.09	29.86	32.50
$C_2H_3Cl_3$	1,1,2-Trichloroethane	113.8	34.82	40.24
C_2H_3N	Acetonitrile	81.65	29.75	32.94
C_2H_4	Ethylene	-103.77	13.53	
$C_2H_4Br_2$	1,2-Dibromoethane	131.6	34.77	41.73
$C_2H_4Cl_2$	1,1-Dichloroethane	57.4	28.85	30.62
$C_2H_4Cl_2$	1,2-Dichloroethane	83.5	31.98	35.16
$C_2H_4F_2$	1,1-Difluoroethane	-24.95	21.56	19.08
C_2H_4O	Acetaldehyde	20.1	25.76	25.47
C_2H_4O	Ethylene oxide	10.6	25.54	24.75
$C_2H_4O_2$	Acetic acid	117.9	23.70	23.36
$C_2H_4O_2$	Methyl formate	31.7	27.92	28.35
C_2H_5Br	Bromoethane	38.5	27.04	28.03
C_2H_5Cl	Chloroethane	12.3	24.65	
C_2H_5I	Iodoethane	72.5	29.44	31.93
C_2H_5NO	N-Methylformamide	199.51		56.19
C_2H_6	Ethane	-88.6	14.69	5.16
C_2H_6O	Dimethyl ether	-24.8	21.51	18.51
C_2H_6O	Ethanol	78.29	38.56	42.32
C_2H_6S	Dimethyl sulfide	37.33	27.0	27.65
C_2H_6S	Ethanethiol	35.1	26.79	27.30
$C_2H_6S_2$	Dimethyl disulfide	109.8	33.78	37.86
$C_2H_6S_2$	1,2-Ethanedithiol	146.1	37.93	44.68
C_2H_7N	Dimethylamine	6.88	26.40	25.05
C_2H_7NO	Ethanolamine	171	49.83	
$C_2H_8N_2$	1,1-Dimethylhydrazine	63.9	32.55	35.0
$C_2H_8N_2$	1,2-Ethanediamine	117	37.98	44.98
$C_3Cl_2F_6$	1,2-Dichlorohexafluoropropane	34.1	26.28	26.93
$C_3H_3Cl_3O_2$	Methyl trichloroacetate	153.8		48.33
$C_3H_4Cl_2O_2$	Methyl dichloroacetate	142.9	39.28	47.72
$C_3H_4O_2$	2-Oxetanone	162		47.03
C_3H_5Br	3-Bromopropene	70.1	30.24	32.73
$C_3H_5ClO_2$	Methyl chloroacetate	129.5	39.23	46.73
C_3H_5N	Propanenitrile	97.14	31.81	36.03
C_3H_6	Propene	-47.69	18.42	14.24
C_3H_6	Cyclopropane	-32.81	20.05	16.93
$C_3H_6Br_2$	1,2-Dibromopropane	141.9	35.61	41.67
$C_3H_6Br_2$	1,3-Dibromopropane	167.3		47.45
$C_3H_6Cl_2$	1,3-Dichloropropane	120.9	35.18	40.75
C_3H_6O	Acetone	56.05	29.10	30.99
C_3H_6O	Propanal	48	28.31	29.62
C_3H_6O	Methyloxirane	35	27.35	27.89
C_3H_6O	Oxetane	47.6	28.67	29.85
$C_3H_6O_2$	Ethyl formate	54.4	29.91	31.96
$C_3H_6O_2$	Methyl acetate	56.87	30.32	32.29
$C_3H_6O_2$	Propanoic acid	141.15		32.14
C_3H_6S	Thiacyclobutane	95	32.32	35.97

Molecular formula	Name	t_b °C	$\Delta_{vap} H$ (t_b) kJ/mol	$\Delta_{vap} H$ (25°C) kJ/mol
C_3H_7Br	1-Bromopropane	71.1	29.84	32.01
C_3H_7Br	2-Bromopropane	59.5	28.33	30.17
C_3H_7Cl	1-Chloropropane	46.5	27.18	28.35
C_3H_7Cl	2-Chloropropane	35.7	26.30	26.90
C_3H_7I	1-Iodopropane	102.6	32.08	36.25
C_3H_7I	2-Iodopropane	89.5	30.68	34.06
C_3H_7NO	N-Ethylformamide	198		58.44
C_3H_7NO	N,N-Dimethylformamide	153		46.89
C_3H_8	Propane	-42.1	19.04	14.79
C_3H_8O	1-Propanol	97.2	41.44	47.45
C_3H_8O	2-Propanol	82.3	39.85	45.39
$C_3H_8O_2$	2-Methoxyethanol	124.1	37.54	45.17
C_3H_8S	Ethyl methyl sulfide	66.7	29.53	31.85
C_3H_8S	1-Propanethiol	67.8	29.54	31.89
C_3H_8S	2-Propanethiol	52.6	27.91	29.45
$C_3H_8S_2$	1,3-Propanedithiol	172.9		49.66
C_3H_9N	Propylamine	47.22	29.55	31.27
C_3H_9N	Isopropylamine	31.76	27.83	28.36
C_3H_9N	Trimethylamine	2.87	22.94	21.66
$C_3H_{10}N_2$	1,3-Propanediamine	139.8	40.85	50.16
$C_4H_4N_2$	Pyridazine	208		53.47
$C_4H_4N_2$	Pyrimidine	123.8	43.09	49.79
C_4H_4O	Furan	31.5	27.10	27.45
$C_4H_4O_2$	Diketene	126.1	36.80	42.89
C_4H_4S	Thiophene	84.0	31.48	34.70
$C_4H_5Cl_3O_2$	Ethyl trichloroacetate	167.5		50.97
C_4H_5N	Cyclopropanecarbonitrile	135.1	35.55	41.94
C_4H_5N	Pyrrole	129.79	38.75	45.09
C_4H_5NS	4-Methylthiazole	133.3	37.58	43.85
C_4H_6	1,2-Butadiene	10.9	24.02	23.21
C_4H_6	1,3-Butadiene	-4.41	22.47	20.86
C_4H_6	1-Butyne	8.08	24.52	23.35
$C_4H_6Cl_2O_2$	Ethyl dichloroacetate			50.60
C_4H_6S	2,3-Dihydrothiophene	112.1	33.24	37.74
C_4H_6S	2,5-Dihydrothiophene	122.4	34.83	39.95
$C_4H_7ClO_2$	Ethyl chloroacetate	144.3	40.43	49.47
C_4H_7N	Butanenitrile	117.6	33.68	39.33
C_4H_7N	2-Methylpropanenitrile	103.9	32.39	37.13
C_4H_8	1-Butene	-6.26	22.07	20.22
C_4H_8	cis-2-Butene	3.71	23.34	22.16
C_4H_8	trans-2-Butene	0.88	22.72	21.40
C_4H_8	Cyclobutane	12.6	24.19	23.51
$C_4H_8Br_2$	1,4-Dibromobutane	197		53.09
$C_4H_8Cl_2$	1,2-Dichlorobutane	124.1	33.90	39.58
$C_4H_8Cl_2$	1,4-Dichlorobutane	161		46.36
C_4H_8O	2-Butanone	79.59	31.30	34.79
C_4H_8O	Tetrahydrofuran	65	29.81	31.99
$C_4H_8O_2$	1,3-Dioxane	106.1	34.37	39.09
$C_4H_8O_2$	1,4-Dioxane	101.5	34.16	38.60
$C_4H_8O_2$	Ethyl acetate	77.11	31.94	35.60
$C_4H_8O_2$	Methyl propanoate	79.8	32.24	35.85
$C_4H_8O_2$	Propyl formate	80.9	33.61	37.53
$C_4H_8O_2$	Butanoic acid	163.75		40.45
$C_4H_8O_2$	2-Methylpropanoic acid	154.45		35.30
C_4H_8S	Tetrahydrothiophene	121.0	34.66	39.43
C_4H_9Br	1-Bromobutane	101.6	32.51	36.64
C_4H_9Br	1-Bromo-2-methylpropane	91.1	31.33	34.82

Molecular formula	Name	t_b °C	$\Delta_{vap} H$ (t_b) kJ/mol	$\Delta_{vap} H$ (25°C) kJ/mol
C_4H_9Br	2-Bromobutane	91.4	30.77	34.41
C_4H_9Br	2-Bromo-2-methylpropane	73.3	29.23	31.81
C_4H_9Cl	1-Chlorobutane	78.6	30.39	33.51
C_4H_9Cl	1-Chloro-2-methylpropane	68.5	29.22	31.67
C_4H_9Cl	2-Chlorobutane	68.3	29.17	31.53
C_4H_9Cl	2-Chloro-2-methylpropane	50.9	27.55	28.98
C_4H_9I	1-Iodobutane	130.6	34.66	40.63
C_4H_9I	1-Iodo-2-methylpropane	121.1.	33.54	38.83
C_4H_9I	2-Iodobutane	120.1	33.27	38.46
C_4H_9I	2-Iodo-2-methylpropane	100.1	31.43	35.41
C_4H_9N	Pyrrolidine	86.56	33.01	37.52
C_4H_9NO	N-Ethylacetamide	205		64.89
C_4H_9NO	N,N-Dimethylacetamide	165		50.24
C_4H_{10}	Butane	-0.5	22.44	21.02
C_4H_{10}	Isobutane	-11.73	21.30	19.23
$C_4H_{10}O$	1-Butanol	117.73	43.29	52.35
$C_4H_{10}O$	2-Butanol	99.51	40.75	49.72
$C_4H_{10}O$	2-Methyl-2-propanol	82.4	39.07	46.69
$C_4H_{10}O$	2-Methyl-1-propanol	107.89	41.82	50.82
$C_4H_{10}O$	Diethyl ether	34.5	26.52	27.10
$C_4H_{10}O$	Methyl propyl ether	39.1	26.75	27.60
$C_4H_{10}O$	Isopropyl methyl ether	30.77	26.05	26.41
$C_4H_{10}O_2$	2-Ethoxyethanol	135	39.22	48.21
$C_4H_{10}O_2$	1,2-Dimethoxyethane	85	32.42	36.39
$C_4H_{10}S$	1-Butanethiol	98.5	32.23	36.63
$C_4H_{10}S$	2-Butanethiol	85	30.59	33.99
$C_4H_{10}S$	2-Methyl-1-propanethiol	88.5	31.01	34.63
$C_4H_{10}S$	2-Methyl-2-propanethiol	64.3	28.45	30.78
$C_4H_{10}S$	Diethyl sulfide	92.1	31.77	35.80
$C_4H_{10}S$	Methyl propyl sulfide	95.6	32.08	36.24
$C_4H_{10}S$	Isopropyl methyl sulfide	84.8	30.71	34.15
$C_4H_{10}S_2$	1,4-Butanedithiol			55.10
$C_4H_{10}S_2$	Diethyl disulfide	154.1	37.58	45.18
$C_4H_{11}N$	Butylamine	77.00	31.81	35.72
$C_4H_{11}N$	Isobutylamine	67.75	30.61	33.85
$C_4H_{11}N$	sec-Butylamine	62.73	29.92	32.85
$C_4H_{11}N$	tert-Butylamine	44.04	28.27	29.64
$C_4H_{11}N$	Diethylamine	55.5	29.06	31.31
$C_4H_{11}N$	Isopropylmethylamine	50.4	28.71	30.69
$C_5H_2F_6O_2$	Hexafluoroacetylacetone	54.15	27.05	30.58
C_5H_5N	Pyridine	115.23	35.09	40.21
C_5H_6S	2-Methylthiophene	112.6	33.90	38.87
C_5H_6S	3-Methylthiophene	115.5	34.24	39.43
C_5H_7N	trans-3-Pentenenitrile	142.6	37.09	44.77
C_5H_7N	Cyclobutanecarbonitrile	149.6	36.88	44.34
C_5H_8	Spiropentane	39	26.76	27.49
C_5H_8O	Cyclopentanone	130.57	36.35	42.72
C_5H_8O	Cyclopropyl methyl ketone	111.3	34.07	39.41
$C_5H_8O_2$	2,4-Pentanedione	138	34.30	41.77
$C_5H_8O_2$	Methyl cyclopropanecarboxylate	114.9	35.25	41.27
C_5H_9N	Pentanenitrile	141.3	36.09	43.60
C_5H_9N	3-Methylbutanenitrile	127.5	35.10	41.64
C_5H_9N	2,2-Dimethylpropanenitrile	106.1	32.40	37.35
C_5H_{10}	1-Pentene	29.96	25.20	25.47
C_5H_{10}	cis-2-Pentene	36.93		26.86
C_5H_{10}	trans-2-Pentene	36.34		26.76
C_5H_{10}	2-Methyl-1-butene	31.2	25.50	25.86

Molecular formula	Name	t_b °C	$\Delta_{vap} H (t_b)$ kJ/mol	$\Delta_{vap} H (25°C)$ kJ/mol
C_5H_{10}	2-Methyl-2-butene	38.56	26.31	27.06
C_5H_{10}	3-Methyl-1-butene	20.1		23.80
C_5H_{10}	Cyclopentane	49.3	27.30	28.52
$C_5H_{10}Cl_2$	1,2-Dichloropentane	148.3	36.45	43.89
$C_5H_{10}Cl_2$	1,5-Dichloropentane	179		50.71
$C_5H_{10}O$	Cyclopentanol	140.42		57.60
$C_5H_{10}O$	2-Pentanone	102.26	33.44	38.40
$C_5H_{10}O$	3-Pentanone	101.96	33.45	38.52
$C_5H_{10}O$	3-Methyl-2-butanone	94.33	32.35	36.78
$C_5H_{10}O$	3,3-Dimethyloxetanc	80.6	30.85	33.94
$C_5H_{10}O$	Tetrahydropyran	88	31.17	34.58
$C_5H_{10}O_2$	Butyl formate	106.1	36.58	41.11
$C_5H_{10}O_2$	Propyl acetate	101.54	33.92	39.72
$C_5H_{10}O_2$	Isopropyl acetate	88.6	32.93	37.20
$C_5H_{10}O_2$	Ethyl propanoate	99.1	33.88	39.21
$C_5H_{10}O_2$	Methyl butanoate	102.8	33.79	39.28
$C_5H_{10}O_2$	Methyl isobutanoate	92.5	32.61	37.32
$C_5H_{10}O_2$	2-Methylbutanoic acid			46.91
$C_5H_{10}O_3$	Diethyl carbonate			43.60
$C_5H_{10}S$	Cyclopentanethiol	132.1	35.32	41.42
$C_5H_{10}S$	Thiacyclohexane	141.8	35.96	42.58
$C_5H_{11}Br$	1-Bromopentane	129.8	35.01	41.28
$C_5H_{11}Cl$	1-Chloropentane	107.8	33.15	38.24
$C_5H_{11}Cl$	1-Chloro-3-methylbutane	98.9	32.02	36.24
$C_5H_{11}Cl$	2-Chloropentane	97.1.	31.79	36.03
$C_5H_{11}I$	1-Iodopentane	155		45.27
$C_5H_{11}N$	Piperidine	106.22		39.29
C_5H_{12}	Pentane	36.06	25.79	26.43
C_5H_{12}	Isopentane	27.88	24.69	24.85
C_5H_{12}	Neopentane	9.48	22.74	21.84
$C_5H_{12}O$	Butyl methyl ether	70.16	29.55	32.37
$C_5H_{12}O$	Isobutyl methyl ether	58.6	28.02	30.13
$C_5H_{12}O$	sec-Butyl methyl ether	59.1	28.09	30.23
$C_5H_{12}O$	tert-Butyl methyl ether	55.2	27.94	29.82
$C_5H_{12}O$	Ethyl propyl ether	63.21	28.94	31.43
$C_5H_{12}O$	Ethyl isopropyl ether	54.1	28.21	30.08
$C_5H_{12}O$	1-Pentanol	137.98	44.36	57.02
$C_5H_{12}O$	2-Pentanol	119.3	41.40	54.21
$C_5H_{12}O$	3-Pentanol	116.25		54.0
$C_5H_{12}O$	2-Methyl-1-butanol	128		55.16
$C_5H_{12}O$	2-Methyl-2-butanol	102.4	39.04	50.10
$C_5H_{12}O$	3-Methyl-1-butanol	131.1	44.07	55.61
$C_5H_{12}O$	3-Methyl-2-butanol			53.0
$C_5H_{12}O_2$	1-Ethoxy-2-methoxyethane	102.1	34.33	39.83
$C_5H_{12}O_2$	Diethoxymethane	88	31.33	35.65
$C_5H_{12}O_2$	2-Propoxyethanol	149.8	41.40	52.12
$C_5H_{12}S$	Butyl methyl sulfide	123.5	34.47	40.46
$C_5H_{12}S$	tert-Butyl methyl sulfide	99	31.47	35.84
$C_5H_{12}S$	Ethyl propyl sulfide	118.6	34.24	39.97
$C_5H_{12}S$	Ethyl isopropyl sulfide	107.5	32.74	37.78
$C_5H_{12}S$	1-Pentanethiol	126.6	34.88	41.24
$C_5H_{12}S$	2-Methyl-1-butanethiol	119.1	33.79	39.45
$C_5H_{12}S$	2-Methyl-2-butanethiol	99.1	31.37	35.67
$C_5H_{13}N$	Ethylisopropylamine	69.6	29.94	33.13
$C_5H_{13}N$	Pentylamine	104.3	34.01	40.08
C_6ClF_5	Chloropentafluorobenzene	117.96	34.76	41.07
C_6F_6	Hexafluorobenzene	80.26	31.66	35.71

Molecular formula	Name	t_b °C	$\Delta_{vap} H (t_b)$ kJ/mol	$\Delta_{vap} H (25°C)$ kJ/mol
C_6MoO_6	Molybdenum hexacarbonyl	701.2	72.51	
C_6HF_5	Pentafluorobenzene	85.74	32.15	36.27
$C_6H_4Cl_2$	o-Dichlorobenzene	180	39.66	50.21
$C_6H_4Cl_2$	m-Dichlorobenzene	173	38.62	48.58
$C_6H_4Cl_2$	p-Dichlorobenzene	174	38.79	49.0
$C_6H_4F_2$	o-Difluorobenzene	94	32.21	36.18
$C_6H_4F_2$	m-Difluorobenzene	82.6	31.10	34.59
$C_6H_4F_2$	p-Difluorobenzene	89	31.77	35.54
C_6H_5Br	Bromobenzene	156.06		44.54
C_6H_5Cl	Chlorobenzene	131.72	35.19	40.97
C_6H_5F	Fluorobenzene	84.73	31.19	34.58
$C_6H_5NO_2$	Nitrobenzene	210.8		55.01
C_6H_6	Benzene	80.09	30.72	33.83
C_6H_6O	Phenol	181.87	45.69	57.82
C_6H_6S	Benzenethiol	169.1·	39.93	47.56
C_6H_7N	Aniline	184.17	42.44	55.83
C_6H_7N	2-Methylpyridine	129.38	36.17	42.48
C_6H_7N	3-Methylpyridine	144.14	37.35	44.44
C_6H_7N	4-Methylpyridine	145.36	37.51	44.56
C_6H_7N	1-Cyclopentenecarbonitrile			44.98
C_6H_9N	Cyclopentanecarbonitrile			43.43
$C_6H_9NO_3$	Triacetamide			60.41
C_6H_{10}	Cyclohexene	82.98	30.46	33.47
$C_6H_{10}O$	Cyclohexanone	155.43		45.06
$C_6H_{10}O_2$	Methyl cyclobutanecarboxylate	135.5	37.13	44.72
$C_6H_{10}O_4$	Ethylene glycol diacetate	190		61.44
$C_6H_{11}N$	Hexanenitrile	163.65		47.91
C_6H_{12}	Cyclohexane	80.73	29.97	33.01
C_6H_{12}	Methylcyclopentane	71.8	29.08	31.64
C_6H_{12}	Ethylcyclobutane	70.8	28.67	31.24
C_6H_{12}	1-Hexene	63.48		30.61
C_6H_{12}	cis-2-Hexene	68.8		32.19
C_6H_{12}	trans-2-Hexene	67.9		31.60
C_6H_{12}	2-Methyl-1-pentene	62.1		30.48
C_6H_{12}	2-Methyl-2-pentene	67.3		31.60
C_6H_{12}	4-Methyl-1-pentene	53.9		28.71
C_6H_{12}	4-Methyl-cis-2-pentene	56.3		29.48
C_6H_{12}	4-Methyl-trans-2-pentene	58.6		29.97
C_6H_{12}	2,3-Dimethyl-1-butene	55.6		29.18
C_6H_{12}	2,3-Dimethyl-2-butene	73.3	29.64	32.53
C_6H_{12}	2-Ethyl-1-butene	64.7		31.13
$C_6H_{12}Cl_2$	1,2-Dichlorohexane	173		48.16
$C_6H_{12}O$	Cyclohexanol	160.84		62.01
$C_6H_{12}O$	2-Hexanone	127.6	36.35	43.14
$C_6H_{12}O$	3-Hexanone	123.5	35.36	42.47
$C_6H_{12}O$	2-Methyl-3-pentanone	113.5	33.84	39.79
$C_6H_{12}O$	3-Methyl-2-pentanone	117.5	34.16	40.53
$C_6H_{12}O$	4-Methyl-2-pentanone	116.5	34.49	40.61
$C_6H_{12}O$	3,3-Dimethyl-2-butanone	106.1	33.39	37.91
$C_6H_{12}O$	Butyl vinyl ether	94	31.58	36.17
$C_6H_{12}O_2$	Butyl acetate	126.1	36.28	43.86
$C_6H_{12}O_2$	tert-Butyl acetate	95.1	33.07	38.03
$C_6H_{12}O_2$	Propyl propanoate	122.5	35.54	43.45
$C_6H_{12}O_2$	Ethyl butanoate	121.5	35.47	42.68
$C_6H_{12}O_2$	Ethyl 2-methylpropanoate	110.1	33.67	39.83
$C_6H_{12}O_2$	Methyl pentanoate	127.4	35.36	43.10
$C_6H_{12}O_2$	Methyl 2,2-dimethylpropanoate	101.1	33.42	38.76

Molecular formula	Name	t_b °C	$\Delta_{vap} H (t_b)$ kJ/mol	$\Delta_{vap} H (25°C)$ kJ/mol
$C_6H_{12}S$	Cyclohexanethiol	158.9	37.06	44.57
$C_6H_{13}Br$	1-Bromohexane	155.3		45.89
$C_6H_{13}Cl$	1-Chlorohexane	135	35.67	42.83
$C_6H_{13}I$	1-Iodohexane	181		49.75
$C_6H_{13}N$	Cyclohexylamine	134	36.14	43.67
C_6H_{14}	Hexane	68.73	28.85	31.56
C_6H_{14}	2-Methylpentane	60.26	27.79	29.89
C_6H_{14}	3-Methylpentane	63.27	28.06	30.28
C_6H_{14}	2,2-Dimethylbutane	49.73	26.31	27.68
C_6H_{14}	2,3-Dimethylbutane	57.98	27.38	29.12
$C_6H_{14}N_2$	Azopropane			39.88
$C_6H_{14}O$	Methyl pentyl ether	99	32.02	36.85
$C_6H_{14}O$	Butyl ethyl ether	92.3	31.63	36.32
$C_6H_{14}O$	Dipropyl ether	90.08	31.31	35.69
$C_6H_{14}O$	Diisopropyl ether	68.51	29.10	32.12
$C_6H_{14}O$	1-Hexanol	157.6	44.50	61.61
$C_6H_{14}O$	2-Hexanol	138	41.01	58.46
$C_6H_{14}O$	4-Methyl-1-pentanol	151.9	44.46	60.47
$C_6H_{14}O$	2-Methyl-2-pentanol	121.1	39.59	54.77
$C_6H_{14}O_2$	1,1-Diethoxyethane	102.25	36.28	43.20
$C_6H_{14}O_2$	1,2-Diethoxyethane	119.4	36.28	43.20
$C_6H_{14}O_2$	2-Butoxyethanol	168.4		56.59
$C_6H_{14}O_3$	Bis(ethoxymethyl)ether	140.6	36.17	44.69
$C_6H_{14}O_3$	Diethylene glycol dimethyl ether	162	36.17	44.69
$C_6H_{14}S$	Methyl pentyl sulfide	145.1	37.41	45.24
$C_6H_{14}S$	Butyl ethyl sulfide	144.3	37.01	44.51
$C_6H_{14}S$	Dipropyl sulfide	142.9	36.60	44.21
$C_6H_{14}S$	Isopropyl propyl sulfide	132.1	35.11	41.78
$C_6H_{14}S$	Diisopropyl sulfide	120.1	33.80	39.60
$C_6H_{15}N$	Hexylamine	132.8	36.54	45.10
$C_6H_{15}N$	Butylethylamine	107.5	33.97	40.15
$C_6H_{15}N$	Isopropylpropylamine	96.9	32.14	37.23
$C_6H_{15}N$	Triethylamine	89	31.01	34.84
$C_6H_{15}N$	Dipropylamine	109.3	33.47	40.04
$C_6H_{15}N$	Diisopropylamine	83.9	30.40	34.61
$C_7H_3F_5$	2,3,4,5,6-Pentafluorotoluene	117.5	34.75	41.12
$C_7H_5F_3$	(Trifluoromethyl)benzene	102.1	32.63	37.60
C_7H_7F	p-Fluorotoluene	116.6	34.08	39.42
C_7H_8	Toluene	110.63	33.18	38.01
C_7H_8O	o-Cresol	191.04	45.19	
C_7H_8O	m-Cresol	202.27	47.40	61.71
C_7H_8O	p-Cresol	201.98	47.45	
C_7H_8O	Benzyl alcohol	205.31	50.48	
C_7H_8O	Anisole	153.7	38.97	46.90
C_7H_9N	1-Cyclohexenecarbonitrile			53.55
C_7H_9N	2,3-Dimethylpyridine	161.26	39.08	47.73
C_7H_9N	2,4-Dimethylpyridine	158.5	38.53	47.48
C_7H_9N	2,6-Dimethylpyridine	144.1	37.46	45.36
C_7H_9N	3,4-Dimethylpyridine	179.13	39.99	50.53
C_7H_9N	3,5-Dimethylpyridine	172	39.46	49.48
C_7H_9N	Benzylamine	185		60.16
$C_7H_{10}O$	Dicyclopropyl ketone	161		53.70
$C_7H_{11}N$	Cyclohexanecarbonitrile			51.92
C_7H_{12}	1-Methylbicyclo(3,1,0)hexane	93.1	31.07	34.77
C_7H_{14}	Methylcyclohexane	100.93	31.27	35.36
C_7H_{14}	Ethylcyclopentane	103.5	31.96	36.40
C_7H_{14}	cis-1,3-Dimethylcyclopentane	90.8	30.40	34.20

Molecular formula	Name	t_b °C	$\Delta_{vap} H (t_b)$ kJ/mol	$\Delta_{vap} H (25°C)$ kJ/mol
C_7H_{14}	1-Heptene	93.64		35.49
$C_7H_{14}O$	2-Heptanone	151.05		47.24
$C_7H_{14}O$	2,4-Dimethyl-3-pentanone	125.4	34.64	41.51
$C_7H_{14}O$	2,2-Dimethyl-3-pentanone	125.6	36.09	42.34
$C_7H_{14}O_2$	Ethyl 2,2-dimethylpropanoate	118.4	34.51	41.25
$C_7H_{14}O_2$	Methyl hexanoate	149.5	38.55	48.04
$C_7H_{14}O_2$	Ethyl pentanoate	146.1	36.96	47.01
$C_7H_{15}Br$	1-Bromoheptane	179		50.60
$C_7H_{15}Cl$	1-Chloroheptane	159		47.66
C_7H_{16}	Heptane	98.5	31.77	36.57
C_7H_{16}	2-Methylhexane	90.04	30.62	34.87
C_7H_{16}	3-Methylhexane	91.84	30.89	35.06
C_7H_{16}	2,2-Dimethylpentane	79.2	29.23	32.42
C_7H_{16}	2,3-Dimethylpentane	89.78	30.46	34.26
C_7H_{16}	2,4-Dimethylpentane	80.49	29.55	32.88
C_7H_{16}	3,3-Dimethylpentane	86.06	29.62	33.03
C_7H_{16}	3-Ethylpentane	93.5	31.12	35.22
C_7H_{16}	2,2,3-Trimethylbutane	80.86	28.90	32.05
$C_7H_{16}O$	Hexyl methyl ether	126.1	34.93	42.07
$C_7H_{16}O$	Ethyl pentyl ether	117.6	34.41	41.01
$C_7H_{16}O$	Butyl propyl ether	118.1	33.72	40.22
$C_7H_{16}O$	1-Heptanol	176.45		66.81
$C_7H_{17}N$	Heptylamine	156		49.96
C_8F_{18}	Perfluorooctane	105.9	33.38	41.13
$C_8H_8O_2$	Methyl benzoate	199		55.57
C_8H_{10}	Ethylbenzene	136.19	35.57	42.24
C_8H_{10}	o-Xylene	144.5	36.24	43.43
C_8H_{10}	m-Xylene	139.12	35.66	42.65
C_8H_{10}	p-Xylene	138.37	35.67	42.40
$C_8H_{10}O$	2,4-Xylenol	210.98		64.96
$C_8H_{10}O$	2,6-Xylenol	201.07		75.31
$C_8H_{10}O$	3,4-Xylenol	227		85.03
$C_8H_{10}O$	3,5-Xylenol	221.74		82.01
$C_8H_{10}O$	Phenetole	169.81		51.04
$C_8H_{11}N$	N,N-Dimethylaniline	194.15		52.83
$C_8H_{11}N$	2,3,6-Trimethylpyridine	171.6	39.95	50.61
$C_8H_{11}N$	2,4,6-Trimethylpyridine	170.6	39.87	50.33
C_8H_{14}	1-Octyne	126.3	35.83	42.30
C_8H_{14}	2-Octyne	137.6	37.26	44.49
C_8H_{14}	3-Octyne	133.1	36.94	43.92
C_8H_{14}	4-Octyne	131.6	36.0	42.73
$C_8H_{15}N$	Octanenitrile	205.25		56.80
C_8H_{16}	Propylcyclopentane	131	34.70	41.08
C_8H_{16}	Isopropylcyclopentane	126.5	33.56	39.44
C_8H_{16}	1-Ethyl-1-methylcyclopentane	121.6	33.20	38.85
C_8H_{16}	Ethylcyclohexane	131.9	34.04	40.56
C_8H_{16}	1,1-Dimethylcyclohexane	119.6	32.51	37.92
C_8H_{16}	cis-1,2-Dimethylcyclohexane	129.8	33.47	39.70
C_8H_{16}	trans-1,2-Dimethylcyclohexane	123.5	32.96	38.36
C_8H_{16}	cis-1,3-Dimethylcyclohexane	120.1	32.91	38.26
C_8H_{16}	trans-1,3-Dimethylcyclohexane	124.5	33.39	39.16
C_8H_{16}	cis-1,4-Dimethylcyclohexane	124.4	33.28	39.02
C_8H_{16}	trans-1,4-Dimethylcyclohexane	119.4	32.56	37.90
C_8H_{16}	1-Octene	121.29	34.07	40.39
$C_8H_{16}O$	2,2,4-Trimethyl-3-pentanone	135.1	35.64	43.30
$C_8H_{16}O_2$	2-Ethylhexanoic acid			75.60
$C_8H_{16}O_2$	Methyl heptanoate	172		51.62

Molecular formula	Name	t_b °C	$\Delta_{vap} H\ (t_b)$ kJ/mol	$\Delta_{vap} H\ (25°C)$ kJ/mol
$C_8H_{16}O_2$	Ethyl hexanoate	167		51.72
$C_8H_{17}Br$	1-Bromooctane	200		55.77
$C_8H_{17}Cl$	1-Chlorooctane	181.5		52.42
$C_8H_{17}F$	1-Fluorooctane	142.4	40.43	49.65
C_8H_{18}	Octane	125.67	34.41	41.49
C_8H_{18}	2-Methylheptane	117.66	33.26	39.67
C_8H_{18}	3-Methylheptane	118.94	33.66	39.83
C_8H_{18}	4-Methylheptane	117.72	33.35	39.69
C_8H_{18}	2,2-Dimethylhexane	106.86	32.07	37.28
C_8H_{18}	2,3-Dimethylhexane	115.62	33.17	38.78
C_8H_{18}	2,4-Dimethylhexane	109.5˙	32.51	37.76
C_8H_{18}	2,5-Dimethylhexane	109.12	32.54	37.85
C_8H_{18}	3,3-Dimethylhexane	111.97	32.31	37.53
C_8H_{18}	3,4-Dimethylhexane	117.73	33.24	38.97
C_8H_{18}	3-Ethylhexane	118.6	33.59	39.64
C_8H_{18}	3-Ethyl-2-methylpentane	115.66	32.93	38.52
C_8H_{18}	3-Ethyl-3-methylpentane	118.27	32.78	37.99
C_8H_{18}	2,2,3-Trimethylpentane	110	31.94	36.91
C_8H_{18}	2,2,4-Trimethylpentane	99.22	30.79	35.14
C_8H_{18}	2,3,3-Trimethylpentane	114.8	32.12	37.27
C_8H_{18}	2,3,4-Trimethylpentane	113.5	32.36	37.75
C_8H_{18}	2,2,3,3-Tetramethylbutane	106.45		42.90
$C_8H_{18}N_2$	Azobutane			49.31
$C_8H_{18}O$	1-Octanol	195.16		70.98
$C_8H_{18}O$	Dibutyl ether	140.28	36.49	44.97
$C_8H_{18}O$	Di-sec-butyl ether	121.1	34.06	40.84
$C_8H_{18}O$	Di-tert-butyl ether	107.23	32.15	37.61
$C_8H_{18}O_2$	1,2-Dipropoxyethane			50.62
$C_8H_{18}O_3$	Diethylene glycol diethyl ether	188		58.40
$C_8H_{18}S$	Dibutyl sulfide	185		52.96
$C_8H_{18}S$	Diisobutyl sulfide			48.71
$C_8H_{18}S$	Di-tert-butyl sulfide	149.1	33.26	43.76
$C_8H_{19}N$	Dibutylamine	159.6	38.44	49.45
C_9H_7N	Quinoline	237.16		53.90
C_9H_7N	Isoquinoline	243.22		60.26
C_9H_{10}	Cyclopropylbenzene			50.22
C_9H_{10}	Indan	177.97	39.63	48.79
C_9H_{12}	Cumene	152.41		45.13
C_9H_{12}	Propylbenzene	159.24		46.22
C_9H_{12}	1,2,3-Trimethylbenzene	176.12		49.05
C_9H_{12}	1,2,4-Trimethylbenzene	169.38		47.93
C_9H_{12}	Mesitylene	164.74		47.50
$C_9H_{14}O_6$	Triacetin	259		85.74
C_9H_{18}	Butylcyclopentane	156.6	36.16	45.89
C_9H_{18}	Propylcyclohexane	156.7		45.08
C_9H_{18}	Isopropylcyclohexane	154.8		44.02
$C_9H_{18}O$	2-Nonanone	195.3		56.44
$C_9H_{18}O$	5-Nonanone	188.45		53.30
$C_9H_{18}O$	2,6-Dimethyl-4-heptanone	169.4		50.92
$C_9H_{18}O_2$	Methyl octanoate	192.9		56.41
C_9H_{20}	Nonane	150.82	36.91	46.41
C_9H_{20}	2,2,5-Trimethylhexane	124.09	33.65	40.16
C_9H_{20}	2,3,5-Trimethylhexane	131.4	34.43	41.41
C_9H_{20}	3,3-Diethylpentane	146.3	34.61	42.0
C_9H_{20}	2,2,4,4-Tetramethylpentane	122.29	32.51	38.49
$C_9H_{20}O$	1-Nonanol	213.37		76.86
$C_{10}H_{14}$	Butylbenzene	183.31	38.87	51.36

Molecular formula	Name	t_b °C	$\Delta_{vap} H (t_b)$ kJ/mol	$\Delta_{vap} H (25°C)$ kJ/mol
$C_{10}H_{14}$	Isobutylbenzene	172.79		47.86
$C_{10}H_{14}$	*sec*-Butylbenzene			47.98
$C_{10}H_{14}$	*tert*-Butylbenzene	169.1		47.71
$C_{10}H_{19}N$	Decanenitrile			66.84
$C_{10}H_{20}$	Butylcyclohexane	180.9		49.36
$C_{10}H_{20}$	1-Decene	170.5		50.43
$C_{10}H_{22}$	Decane	174.15	38.75	51.38
$C_{10}H_{22}$	2-Methylnonane	167.1	38.23	49.63
$C_{10}H_{22}$	3-Methylnonane	167.9	38.26	49.71
$C_{10}H_{22}$	5-Methylnonane	165.1	38.14	49.34
$C_{10}H_{22}$	2,4-Dimethyloctane	156	36.47	47.13
$C_{10}H_{22}O$	1-Decanol	231.1		81.50
$C_{10}H_{22}S$	1-Decanethiol	240.6		65.48
$C_{11}H_{21}N$	Undecanenitrile	253		71.14
$C_{11}H_{22}$	Pentylcyclohexane	203.7		53.88
$C_{11}H_{24}$	Undecane	195.93		56.43
$C_{11}H_{24}$	2-Methyldecane	189.3	40.25	54.28
$C_{11}H_{24}$	4-Methyldecane	187	40.70	53.76
$C_{11}H_{24}$	2,4,7-Trimethyloctane	168.1	38.22	49.91
$C_{12}H_{16}$	Cyclohexylbenzene	240.1		59.94
$C_{12}H_{22}$	Cyclohexylcyclohexane			57.98
$C_{12}H_{23}N$	Dodecanenitrile	277		76.12
$C_{12}H_{24}$	1-Dodecene	213.8		60.78
$C_{12}H_{26}$	Dodecane	216.32		61.51
$C_{12}H_{26}$	2,2,4,6,6-Pentamethylheptane	177.8		48.97
$C_{12}H_{26}O$	1-Dodecanol	259		91.96
$C_{13}H_{26}O_2$	Methyl dodecanoate	267		77.17
$C_{13}H_{28}$	Tridecane	235.47		66.43
$C_{14}H_{10}$	Phenanthrene	340		75.50
$C_{14}H_{27}N$	Tetradecanenitrile			85.29
$C_{14}H_{30}$	Tetradecane	253.58		71.30
$C_{14}H_{30}O$	1-Tetradecanol	289		102.20
$C_{15}H_{32}$	Pentadecane	270.6		76.11
$C_{16}H_{32}$	1-Hexadecene	284.9		80.25
$C_{16}H_{34}$	Hexadecane	286.86		81.38
$C_{17}H_{36}$	Heptadecane	302.0		86.02

ENTHALPY OF FUSION

This table lists the molar enthalpy (heat) of fusion of about 600 inorganic and organic compounds. Values of $\Delta_{fus} H$ are given in kJ/mol at the normal melting point t_m, except for substances whose triple-point pressures are greater than normal atmospheric pressure (i.e., the sublimation pressure of the solid exceeds 1 atm at temperatures for which the liquid phase does not exist). In such cases the entries are designated by "t" and refer to the triple-point temperature. All values refer to the enthalpy change at equilibrium between the liquid phase and the most stable solid phase at the transition temperature T_m. Substances are listed by molecular formula in the Hill order, with compounds not containing carbon preceding those that do contain carbon.

REFERENCES

1. Chase, M.W., Davies, C.A., Downey, J.R., Frurip, D.J., McDonald, R.A., and Syverud, A.N., *JANAF Thermochemical Tables, Third Edition, J. Phys. Chem. Ref. Data*, Vol. 14, Suppl. 1, 1985.
2. Daubert, T.E., Danner, R.P., Sibul, H.M., and Stebbins, C.C., *Physical and Thermodynamic Properties of Pure Compounds: Data Compilation*, extant 1994 (core with 4 supplements), Taylor & Francis, Bristol, PA.
3. Janz, G.J., et al., *Physical Properties Data Compilations Relevant to Energy Storage. II. Molten Salts,* Natl. Stand. Ref. Data Sys.- Natl. Bur. Standards (U.S.), No. 61, Part 2, 1979.
4. Dinsdale, A.T., *CALPHAD*, 15, 317, 1991.
5. *Landolt-Börnstein, Numerical Values and Functions for Physics, Chemistry, Astronomy, Geophysics, and Technology, Sixth Edition,* Vol. 2, Part 4, Springer-Verlag, Heidelberg, 1961.
6. Gurvich, L.V., Veyts, I.V., and Alcock, C.B., *Thermodynamic Properties of Individual Substances, Fourth Edition*, Vol. 3, CRC Press, Boca Raton, FL, 1994.
7. *Physical Constants of Hydrocarbon and Non-Hydrocarbon Compounds, 2nd Edition*, ASTM Data Series DS 4B, ASTM, Philadelphia, 1988.
8. Trotman-Dickenson, A.F., Executive Editor, *Comprehensive Inorganic Chemistry*, Vol. 1-5, Pergamon Press, Oxford, 1973.

Molecular formula	Name	$t_m/°C$	$\Delta_{fus}H/kJ\ mol^{-1}$
AgBr	Silver(I) bromide	432	9.12
AgCl	Silver(I) chloride	455	13.2
AgI	Silver(I) iodide	558	9.41
AgNO₃	Silver(I) nitrate	212	11.5
Ag₂S	Silver(I) sulfide	825	14.1
Al	Aluminum	660.32	10.71
AlBr₃	Aluminum tribromide	97.5	11.25
AlCl₃	Aluminum trichloride	192.6	35.4
AlF₃	Aluminum trifluoride	2250 t	98
AlI₃	Aluminum triiodide	188.32	15.9
Al₂O₃	Aluminum oxide (corundum)	2054	111.4
Al₂S₃	Aluminum sulfide	1100	55
Am	Americium	1176	14.39
Ar	Argon	-189.3	1.12
As	Arsenic	817 t	24.44
AsBr₃	Arsenic(III) bromide	31.1	11.7
AsCl₃	Arsenic(III) chloride	-16	10.1
AsF₃	Arsenic(III) fluoride	-5.9	10.4
Au	Gold	1064.18	12.55
B	Boron	2075	50.2
BCl₃	Boron trichloride	-107	2.10
BF₃	Boron trifluoride	-126.8	4.20
BHO₂	Metaboric acid	236	14.3
BH₃O₃	Boric acid	170.9	22.3
BN	Boron nitride	2967	81
BNaO₂	Sodium metaborate	966	36.2
B₂O₃	Boron oxide	450	24.56
Ba	Barium	727	7.12
BaBr₂	Barium bromide	857	32.2
BaCl₂	Barium chloride	962	15.85
BaF₂	Barium fluoride	1368	17.8
BaH₂	Barium hydride	1200	25
BaH₂O₂	Barium hydroxide	408	16
BaI₂	Barium iodide	711	26.5
BaO	Barium oxide	1973	46

Molecular formula	Name	$t_m/°C$	$\Delta_{fus}H/kJ$ mol^{-1}
BaO$_4$S	Barium sulfate	1580	40
BaS	Barium sulfide	2230	63
Be	Beryllium	1287	7.895
BeBr$_2$	Beryllium bromide	508	18
BeCl$_2$	Beryllium chloride	415	8.66
BeF$_2$	Beryllium fluoride	552	4.77
BeI$_2$	Beryllium iodide	470	18
BeO	Beryllium oxide	2578	86
BeO$_4$S	Beryllium sulfate	1127	6
Bi	Bismuth	271.40	11.30
BiCl$_3$	Bismuth trichloride	230	10.9
BrF$_5$	Bromine pentafluoride	-60.5	5.67
BrH	Hydrogen bromide	-86.81	2.41
BrIn	Indium(I) bromide	290	15
BrK	Potassium bromide	734	25.5
BrLi	Lithium bromide	552	17.6
BrNa	Sodium bromide	747	26.11
BrNaO$_3$	Sodium bromate	381	28.11
BrRb	Rubidium bromide	682	15.5
BrTl	Thallium(I) bromide	460	16.4
Br$_2$	Bromine	-7.2	10.57
Br$_2$Ca	Calcium bromide	742	29.1
Br$_2$Cd	Cadmium bromide	568	20.9
Br$_2$Fe	Iron(II) bromide	691	50.2
Br$_2$Hg	Mercury(II) bromide	236	17.9
Br$_2$Mg	Magnesium bromide	711	39.3
Br$_2$Pb	Lead(II) bromide	371	16.44
Br$_2$Sr	Strontium bromide	657	10.1
Br$_2$Zn	Zinc bromide	394	16.7
Br$_3$Ga	Gallium(III) bromide	121.5	12.1
Br$_3$In	Indium(III) bromide	420	26
Br$_3$Pu	Plutonium(III) bromide	681	55.2
Br$_3$U	Uranium(III) bromide	727	43.9
Br$_4$Sn	Tin(IV) bromide	31	12
Br$_4$Th	Thorium bromide	679	66.9
Br$_4$Ti	Titanium(IV) bromide	39	12.9
Br$_4$U	Uranium(IV) bromide	519	55.2
Br$_5$Ta	Tantalum(V) bromide	265	45.6
Ca	Calcium	842	8.54
CaCl$_2$	Calcium chloride	775	28.05
CaF$_2$	Calcium fluoride	1418	30
CaH$_2$	Calcium hydride	1000	6.7
CaI$_2$	Calcium iodide	783	41.8
CaO	Calcium oxide	2899	80
CaO$_4$S	Calcium sulfate	1460	28
CaS	Calcium sulfide	2525	70
Cd	Cadmium	321.07	6.19
CdCl$_2$	Cadmium chloride	564	48.58
CdF$_2$	Cadmium fluoride	1110	22.6
CdI$_2$	Cadmium iodide	387	15.3
Ce	Cerium	798	5.46
CeCl$_3$	Cerium(III) chloride	817	54.4
ClCs	Cesium chloride	645	15.9
ClCu	Copper(I) chloride	430	10.2
ClH	Hydrogen chloride	-114.1	2.00
ClI	Iodine chloride	27.38	11.6
ClIn	Indium(I) chloride	211	21.3
ClK	Potassium chloride	771	26.53
ClLi	Lithium chloride	610	19.9

ENTHALPY OF FUSION (continued)

Molecular formula	Name	t_m/°C	$\Delta_{fus}H$/kJ mol^{-1}
ClLiO$_4$	Lithium perchlorate	236	29
ClNa	Sodium chloride	800.7	28.16
ClNaO$_3$	Sodium chlorate	248	22.1
ClRb	Rubidium chloride	715	18.4
ClTl	Thallium(I) chloride	430	15.56
Cl$_2$	Chlorine	-101.5	6.40
Cl$_2$Co	Cobalt(II) chloride	740	45
Cl$_2$Cr	Chromium(II) chloride	814	32.2
Cl$_2$Cu	Copper(II) chloride	630	20.4
Cl$_2$Fe	Iron(II) chloride	677	43.01
Cl$_2$Hg	Mercury(II) chloride	276	19.41
Cl$_2$Mg	Magnesium chloride	714	43.1
Cl$_2$Mn	Manganese(II) chloride	650	30.7
Cl$_2$Ni	Nickel(II) chloride	1009	71.2
Cl$_2$Pb	Lead(II) chloride	501	21.9
Cl$_2$Sn	Tin(II) chloride	247	12.8
Cl$_2$Sr	Strontium chloride	874	17.5
Cl$_3$Fe	Iron(III) chloride	304	43.1
Cl$_3$Ga	Gallium(III) chloride	77.9	11.13
Cl$_3$In	Indium(III) chloride	583	27
Cl$_3$La	Lanthanum chloride	859	43.1
Cl$_3$OP	Phosphorus(V) oxychloride	1	13.1
Cl$_3$P	Phosphorus(III) chloride	-112	7.10
Cl$_3$Sb	Antimony(III) chloride	73.4	12.7
Cl$_4$OW	Tungsten(VI) oxytetrachloride	211	45
Cl$_4$Si	Tetrachlorosilane	-68.85	7.60
Cl$_4$Sn	Tin(IV) chloride	-33	9.20
Cl$_4$Th	Thorium(IV) chloride	770	40.2
Cl$_4$Ti	Titanium(IV) chloride	-25	9.97
Cl$_4$U	Uranium(IV) chloride	590	45
Cl$_4$V	Vanadium(IV) chloride	-25.7	2.30
Cl$_4$Zr	Zirconium(IV) chloride	437 t	50
Cl$_5$Mo	Molybdenum(V) chloride	194	19
Cl$_5$Nb	Niobium(V) chloride	204.7	38.3
Cl$_5$Ta	Tantalum(V) chloride	216	41.6
Cl$_6$W	Tungsten(VI) chloride	275	6.60
Co	Cobalt	1495	16.2
CoF$_2$	Cobalt(II) fluoride	1127	59
Cr	Chromium	1907	21.0
Cr$_2$O$_3$	Chromium(III) oxide	2330	130
Cs	Cesium	28.44	2.09
CsF	Cesium fluoride	703	21.7
Cs$_2$O$_4$S	Cesium sulfate	1005	35.7
Cu	Copper	1084.62	13.26
CuF$_2$	Copper(II) fluoride	836	55
CuO	Copper(II) oxide	1446	11.8
Dy	Dysprosium	1412	11.06
Er	Erbium	1529	19.9
Eu	Europium	822	9.21
FH	Hydrogen fluoride	-83.36	4.58
FK	Potassium fluoride	858	27.2
FLi	Lithium fluoride	848.2	27.09
FNa	Sodium fluoride	996	33.35
FRb	Rubidium fluoride	833	17.3
FTl	Thallium(I) fluoride	326	13.87
F$_2$	Fluorine	-219.6	0.51
F$_2$Fe	Iron(II) fluoride	1100	52
F$_2$HK	Potassium hydrogen fluoride	238.9	6.62
F$_2$Mg	Magnesium fluoride	1263	58.5

Molecular formula	Name	$t_m/°C$	$\Delta_{fus}H/\text{kJ mol}^{-1}$
F_2Pb	Lead(II) fluoride	830	14.7
F_2Sr	Strontium fluoride	1477	28.5
F_3In	Indium(III) fluoride	1170	64
F_3Pu	Plutonium(III) fluoride	1396	59.8
F_4Pu	Plutonium(IV) fluoride	1027	65.3
F_4Th	Thorium(IV) fluoride	1110	44.0
F_4U	Uranium(IV) fluoride	1036	42.7
F_4Zr	Zirconium(IV) fluoride	932 t	64.2
F_5Nb	Niobium(V) fluoride	80	12.2
F_5V	Vanadium(V) fluoride	19.5	49.96
F_6Ir	Iridium(VI) fluoride	44	8.40
F_6Mo	Molybdenum(VI) fluoride	17.5	4.33
F_6Pu	Plutonium(VI) fluoride	52	17.6
F_6S	Sulfur hexafluoride	-50.7 t	5.02
F_6U	Uranium(VI) fluoride	64.0	19.19
F_6W	Tungsten(VI) fluoride	2.3	4.10
Fe	Iron	1538	13.81
FeI_2	Iron(II) iodide	587	45
FeO	Iron(II) oxide	1377	24
FeS	Iron(II) sulfide	1188	31.5
Fe_3O_4	Iron(II,III) oxide	1597	138
Ga	Gallium	29.76	5.59
GaI_3	Gallium(III) iodide	212	12.9
GaSb	Gallium antimonide	712	25.1
Ga_2O_3	Gallium(III) oxide	1807	100
Gd	Gadolinium	1313	10.0
Ge	Germanium	938.25	36.94
HI	Hydrogen iodide	-50.77	2.87
HKO	Potassium hydroxide	406	8.60
HLi	Lithium hydride	688.7	22.59
HLiO	Lithium hydroxide	471.2	20.88
HNO_3	Nitric acid	-41.6	10.5
HNaO	Sodium hydroxide	323	6.60
H_2	Hydrogen	-259.3	0.12
H_2Mg	Magnesium hydride	327	14
H_2O	Water	0.00	6.01
H_2O_2	Hydrogen peroxide	-0.43	12.50
H_2O_2Sr	Strontium hydroxide	535	23
H_2O_4S	Sulfuric acid	10.31	10.71
H_2S	Hydrogen sulfide	-85.5	23.8
H_2Sr	Strontium hydride	1050	23
H_3N	Ammonia	-77.74	5.66
H_3O_2P	Hypophosphorous acid	26.5	9.7
H_3O_3P	Phosphorous acid	74.4	12.8
H_3O_4P	Phosphoric acid	42.4	13.4
H_4IN	Ammonium iodide	551	21
H_4N_2	Hydrazine	1.4	12.6
$H_4N_2O_3$	Ammonium nitrate	210	6.40
Hf	Hafnium	2233	27.2
Hg	Mercury	-38.83	2.29
HgI_2	Mercury(II) iodide	259	18.9
Hg_2I_2	Mercury(I) iodide	290	27
Ho	Holmium	1474	17.0 (est.)
IIn	Indium(I) iodide	364.4	17.26
IK	Potassium iodide	681	24
ILi	Lithium iodide	469	14.6
INa	Sodium iodide	660	23.6
IRb	Rubidium iodide	642	12.5
ITl	Thallium(I) iodide	441.8	14.73

Molecular formula	Name	$t_m/°C$	$\Delta_{fus}H/kJ$ mol^{-1}
I_2	Iodine (rhombic)	113.7	15.52
I_2Mg	Magnesium iodide	634	26
I_2Pb	Lead(II) iodide	410	23.4
I_2Sr	Strontium iodide	538	19.7
I_3In	Indium(III) iodide	207	18.48
I_4Si	Tetraiodosilane	120.5	19.7
I_4Th	Thorium iodide	570	61.4
I_4Ti	Titanium(IV) iodide	150	19.8
I_4U	Uranium(IV) iodide	506	70.7
In	Indium	156.60	3.28
InSb	Indium antimonide	525	25.5
In_2O_3	Indium(III) oxide	1913	105
Ir	Iridium	2446	41.12
K	Potassium	63.38	2.33
KNO_3	Potassium nitrate	337	10.1
K_2O_4S	Potassium sulfate	1069	36.4
K_2S	Potassium sulfide	948	16.15
Kr	Krypton	-157.3	1.37
La	Lanthanum	918	6.20
Li	Lithium	180.5	3.00
$LiNO_3$	Lithium nitrate	253	24.9
Li_2O_3Si	Lithium metasilicate	1201	28
Li_2O_4S	Lithium sulfate	859	7.50
Lu	Lutetium	1663	22 (est.)
Mg	Magnesium	650	8.48
MgO	Magnesium oxide	2826	77
MgO_4S	Magnesium sulfate	1127	14.6
MgS	Magnesium sulfide	2227	63
Mg_2O_4Si	Magnesium orthosilicate	1898	71
Mn	Manganese	1246	12.91
MnO	Manganese(II) oxide	1840	54.4
Mo	Molybdenum	2623	37.48
MoO_3	Molybdenum(VI) oxide	801	48
$NNaO_3$	Sodium nitrate	307	15
NO	Nitric oxide	-163.6	2.30
NO_3Rb	Rubidium nitrate	305	5.60
NO_3Tl	Thallium(I) nitrate	206	9.6
N_2	Nitrogen	-210.0	0.71
N_2O	Nitrous oxide	-90.8	6.54
N_2O_4	Nitrogen tetroxide	-9.3	14.65
Na	Sodium	97.80	2.60
Na_2O	Sodium oxide	1132	48
Na_2O_3Si	Sodium metasilicate	1089	52
Na_2O_4S	Sodium sulfate	884	23.6
Na_2S	Sodium sulfide	1172	19
Nb	Niobium	2477	30
NbO	Niobium(II) oxide	1937	85
NbO_2	Niobium(IV) oxide	1902	92
Nb_2O_5	Niobium(V) oxide	1512	104.3
Nd	Neodymium	1021	7.14
Ne	Neon	-248.5	0.34
Ni	Nickel	1455	17.48
NiS	Nickel(II) sulfide	976	30.1
Np	Neptunium	644	3.20
OSr	Strontium oxide	2532	81
OTl_2	Thallium(I) oxide	579	30.3
OV	Vanadium(II) oxide	1790	63
OZn	Zinc oxide	1975	52.3
O_2	Oxygen	-218.7	0.44

Molecular formula	Name	$t_m/°C$	$\Delta_{fus}H/kJ$ mol^{-1}
O_2Si	Silicon dioxide (cristobalite)	1713	8.51
O_2Zr	Zirconium(IV) oxide	2710	87
O_3S	Sulfur trioxide	16.8	8.60
O_3Tl_2	Thallium(III) oxide	834	53
O_3W	Tungsten(VI) oxide	1472	73
O_3Y_2	Yttrium oxide	2439	105
O_4Os	Osmium(VIII) oxide	41	9.8
O_4SSr	Strontium sulfate	1607	36
O_4STl_2	Thallium(I) sulfate	632	23
O_5P_2	Phosphorus(V) oxide	562	27.2
O_5Ta_2	Tantalum(V) oxide	1785	120
O_5V_2	Vanadium(V) oxide	670	64.5
O_7Re_2	Rhenium(VII) oxide	297	64.2
Os	Osmium	3033	57.85
P	Phosphorus (white)	44.15	0.66
Pa	Protactinium	1572	12.34
Pb	Lead	327.46	4.77
PbS	Lead(II) sulfide	1118	19
Pd	Palladium	1554.9	16.74
Pr	Praseodymium	931	6.89
Pt	Platinum	1768.4	22.17
Pu	Plutonium	640	2.82
Rb	Rubidium	39.31	2.19
Re	Rhenium	3186	60.43
Rh	Rhodium	1964	26.59
Ru	Ruthenium	2334	38.59
S	Sulfur (monoclinic)	115.21	1.72
SSr	Strontium sulfide	2227	63
STl_2	Thallium(I) sulfide	448	12
Sb	Antimony	630.63	19.87
Sc	Scandium	1541	14.1
Se	Selenium	221	6.69
Si	Silicon	1414	50.21
Sm	Samarium	1074	8.62
Sn	Tin (white)	231.93	7.03
Sr	Strontium	777	7.43
Ta	Tantalum	3017	36.57
Tb	Terbium	1356	10.15
Tc	Technetium	2157	33.29
Te	Tellurium	449.51	17.49
Th	Thorium	1750	13.81
Ti	Titanium	1668	14.15
Tl	Thallium	304	4.14
Tm	Thulium	1545	16.84
U	Uranium	1135	9.14
V	Vanadium	1910	21.5
W	Tungsten	3422	52.31
Xe	Xenon	-111.7	1.81
Y	Yttrium	1522	11.4
Yb	Ytterbium	819	7.66
Zn	Zinc	419.53	7.32
Zr	Zirconium	1855	21.00
C	Carbon (graphite)	4492 t	117
$CBaO_3$	Barium carbonate	1555	40
$CBrCl_3$	Bromotrichloromethane	-5.7	2.54
$CCaO_3$	Calcium carbonate	1330	36
CCl_2O	Carbonyl chloride	-127.9	5.74
CCl_4	Tetrachloromethane	-23	3.28
$CHCl_3$	Trichloromethane	-63.6	8.80

Molecular formula	Name	$t_m/°C$	$\Delta_{fus}H/kJ$ mol^{-1}
CHN	Hydrogen cyanide	-13.4	8.41
CH_2Cl_2	Dichloromethane	-95.14	6
CH_2N_2	Cyanamide	44	8.76
CH_2O_2	Formic acid	8.3	12.72
CH_3Br	Bromomethane	-93.7	5.98
CH_4	Methane	-182.4	0.94
CH_4O	Methanol	-97.68	3.18
CH_4S	Methanethiol	-123	5.91
CH_5N	Methylamine	-93.44	6.13
CK_2O_3	Potassium carbonate	898	27.6
CLi_2O_3	Lithium carbonate	723	41
$CMgO_3$	Magnesium carbonate	990	59
CNa_2O_3	Sodium carbonate	858.1	29.7
CO	Carbon monoxide	-205	0.83
CO_2	Carbon dioxide	-56.57 t	9.02
CO_3Sr	Strontium carbonate	1494	40
CO_3Tl_2	Thallium(I) carbonate	272	18.4
CS_2	Carbon disulfide	-111.5	4.40
$C_2Br_2F_4$	1,2-Dibromotetrafluoroethane	-110.4	7.04
$C_2Cl_2F_4$	1,2-Dichlorotetrafluoroethane	-94	1.51
$C_2Cl_3F_3$	1,1,2-Trichlorotrifluoroethane	-35	2.47
$C_2Cl_4F_2$	1,1,2,2-Tetrachloro-1,2-difluoroethane	26	3.7
$C_2HCl_3O_2$	Trichloroacetic acid	57.5	5.88
$C_2H_3ClO_2$	Chloroacetic acid	63	12.28
$C_2H_3Cl_3$	1,1,1-Trichloroethane	-30.4	2.73
$C_2H_3Cl_3$	1,1,2-Trichloroethane	-36.6	11.54
$C_2H_3F_3$	1,1,1-Trifluoroethane	-111.3	6.19
C_2H_4	Ethylene	-169	3.35
$C_2H_4Br_2$	1,2-Dibromoethane	9.93	10.84
$C_2H_4Cl_2$	1,2-Dichloroethane	-35.5	8.83
$C_2H_4O_2$	Acetic acid	16.6	11.54
C_2H_5Cl	Chloroethane	-138.7	4.45
C_2H_6	Ethane	-182.8	2.86
C_2H_6O	Ethanol	-114.1	5.02
C_2H_6O	Dimethyl ether	-141.5	4.94
$C_2H_6O_2$	Ethylene glycol	-13	11.23
C_2H_6S	Ethanethiol	-147.8	4.98
C_2H_6S	Dimethyl sulfide	-98.3	7.99
$C_2H_6S_2$	Dimethyl disulfide	-85	9.19
C_2H_7N	Dimethylamine	-92.2	5.94
$C_2H_8N_2$	1,2-Ethanediamine	11.1	22.58
C_3H_3N	2-Propenenitrile	-83.5	6.23
$C_3H_4O_2$	Propenoic acid	12.3	11.16
$C_3H_5N_3O_9$	Trinitroglycerol	13.5	21.87
C_3H_6	Propene	-185.2	3.00
C_3H_6	Cyclopropane	-127.4	5.44
$C_3H_6Br_2$	1,3-Dibromopropane	-34.2	13.6
$C_3H_6Cl_2$	(±)-1,2-Dichloropropane	-100.5	6.4
C_3H_6O	Acetone	-94.8	5.69
$C_3H_6O_3$	Trioxane	60.2	15.11
C_3H_7Cl	2-Chloropropane	-117.2	7.39
C_3H_7N	Cyclopropylamine	-35.4	13.18
C_3H_8	Propane	-187.6	3.53
C_3H_8O	1-Propanol	-126.1	5.2
C_3H_8O	2-Propanol	-89.5	5.37
$C_3H_8O_3$	Glycerol	18.2	8.48
C_3H_9N	Propylamine	-83	10.97
C_3H_9N	Isopropylamine	-95.14	7.33
C_3H_9N	Trimethylamine	-117.1	6.55

Molecular formula	Name	$t_m/°C$	$\Delta_{fus}H/kJ\ mol^{-1}$
$C_4H_4N_2$	Succinonitrile	54.5	3.92
$C_4H_4O_3$	Succinic anhydride	119	20.41
C_4H_4S	Thiophene	-39.4	4.97
C_4H_5N	Pyrrole	-23.42	7.91
C_4H_6	1,3-Butadiene	-108.9	7.98
C_4H_6	2-Butyne	-32.36	9.23
$C_4H_6O_2$	cis-Crotonic acid	15	12.57
$C_4H_6O_2$	trans-Crotonic acid	72	9.12
$C_4H_6O_2$	γ-Butyrolactone	-43.37	9.57
$C_4H_6O_4$	Succinic acid	188	32.95
$C_4H_6O_4$	Dimethyl oxalate	54.35	21.07
C_4H_8	cis-2-Butene	-138.9	7.58
C_4H_8	Isobutene	-140.4	5.93
C_4H_8O	2-Butanone	-86.67	8.44
C_4H_8O	Tetrahydrofuran	-108.3	8.54
$C_4H_8O_2$	Butanoic acid	-5.7	11.08
$C_4H_8O_2$	Ethyl acetate	-83.6	10.48
$C_4H_8O_2$	1,4-Dioxane	11.8	12.85
C_4H_9Br	(±)-2-Bromobutane	-111.9	6.89
C_4H_9N	Pyrrolidine	-57.84	8.58
C_4H_{10}	Butane	-138.2	4.66
C_4H_{10}	Isobutane	-138.3	4.54
$C_4H_{10}O$	1-Butanol	-89.8	9.28
$C_4H_{10}O$	2-Methyl-2-propanol	25.4	6.79
$C_4H_{10}O$	Diethyl ether	-116.3	7.27
$C_4H_{11}N$	tert-Butylamine	-66.95	0.88
$C_4H_{12}Si$	Tetramethylsilane	-99.04	6.88
C_5H_5N	Pyridine	-41.66	8.28
C_5H_8	1,4-Pentadiene	-148.8	6.14
C_5H_8	2-Methyl-1,3-butadiene	-145.9	4.79
C_5H_8	Cyclopentene	-135.1	3.36
C_5H_8	Spiropentane	-134.6	5.76
$C_5H_8O_3$	4-Oxopentanoic acid	33	9.22
$C_5H_8O_4$	Glutaric acid	97.8	20.9
C_5H_{10}	1-Pentene	-165.2	5.81
C_5H_{10}	cis-2-Pentene	-151.4	7.12
C_5H_{10}	trans-2-Pentene	-140.2	8.36
C_5H_{10}	Cyclopentane	-93.8	0.61
$C_5H_{10}O$	2-Pentanone	-76.9	10.63
$C_5H_{10}O$	3-Pentanone	-39	11.59
$C_5H_{11}N$	Cyclopentylamine	-82.7	8.31
$C_5H_{11}N$	Piperidine	-11.03	14.85
C_5H_{12}	Pentane	-129.7	8.42
C_5H_{12}	Isopentane	-159.9	5.15
C_5H_{12}	Neopentane	-16.6	3.10
$C_5H_{12}O$	1-Pentanol	-78.9	9.83
C_6Cl_6	Hexachlorobenzene	231.8	23.85
C_6F_6	Hexafluorobenzene	5.35	11.59
C_6HF_5	Pentafluorobenzene	-47.3	10.85
C_6HF_5O	Pentafluorophenol	32.8	12.85
$C_6H_3Cl_3$	1,3,5-Trichlorobenzene	63.5	18.2
$C_6H_4ClNO_2$	1-Chloro-3-nitrobenzene	44.4	19.37
$C_6H_4ClNO_2$	1-Chloro-4-nitrobenzene	83.5	20.77
$C_6H_4Cl_2$	o-Dichlorobenzene	-16.7	12.93
$C_6H_4Cl_2$	m-Dichlorobenzene	-24.8	12.64
$C_6H_4Cl_2$	p-Dichlorobenzene	52.7	17.15
$C_6H_4F_2$	m-Difluorobenzene	-69.09	8.58
$C_6H_4O_2$	p-Benzoquinone	115.7	18.53
C_6H_5Br	Bromobenzene	-30.6	10.62

Molecular formula	Name	$t_m/°C$	$\Delta_{fus}H/\text{kJ mol}^{-1}$
C_6H_5Cl	Chlorobenzene	-45.2	9.61
C_6H_5ClO	o-Chlorophenol	9.8	12.52
C_6H_5ClO	m-Chlorophenol	32.6	14.91
C_6H_5ClO	p-Chlorophenol	42.7	14.07
C_6H_5F	Fluorobenzene	-42.21	11.31
C_6H_5I	Iodobenzene	-31.3	9.76
$C_6H_5NO_2$	Nitrobenzene	5.7	11.59
$C_6H_5NO_3$	2-Nitrophenol	44.8	17.44
$C_6H_5NO_3$	3-Nitrophenol	96.8	19.2
$C_6H_5NO_3$	4-Nitrophenol	113.8	18.25
C_6H_6	Benzene	5.53	9.95
$C_6H_6N_2O_2$	o-Nitroaniline	71.2	16.11
$C_6H_6N_2O_2$	m-Nitroaniline	114	23.68
$C_6H_6N_2O_2$	p-Nitroaniline	147	21.1
C_6H_6O	Phenol	40.9	11.29
$C_6H_6O_2$	p-Hydroquinone	172.3	27.11
C_6H_6S	Benzenethiol	-14.9	11.48
C_6H_7N	Aniline	-6.02	10.56
C_6H_7N	2-Methylpyridine	-66.68	9.72
C_6H_7N	3-Methylpyridine	-18.14	14.18
$C_6H_8N_2$	Phenylhydrazine	19.6	16.43
C_6H_{10}	Cyclohexene	-103.5	3.29
$C_6H_{10}O_4$	Adipic acid	153.2	34.85
C_6H_{12}	1-Hexene	-139.7	9.35
C_6H_{12}	2,3-Dimethyl-2-butene	-74.6	5.46
C_6H_{12}	Cyclohexane	6.6	2.63
C_6H_{12}	Methylcyclopentane	-142.5	6.93
$C_6H_{12}O$	Cyclohexanol	25.46	1.76
$C_6H_{12}O$	2-Hexanone	-55.5	14.9
$C_6H_{12}O$	3-Hexanone	-55.5	13.49
C_6H_{14}	Hexane	-95.3	13.08
C_6H_{14}	2-Methylpentane	-153.7	6.27
C_6H_{14}	2,2-Dimethylbutane	-99	0.58
C_6H_{14}	2,3-Dimethylbutane	-128.8	0.80
$C_6H_{14}O$	Dipropyl ether	-126.1	8.83
$C_6H_{14}O$	Diisopropyl ether	-86.8	11.03
C_7F_8	Perfluorotoluene	-65.6	11.58
$C_7H_3F_5$	2,3,4,5,6-Pentafluorotoluene	-29.8	12.99
$C_7H_5ClO_2$	o-Chlorobenzoic acid	140.2	25.73
$C_7H_6O_2$	Benzoic acid	122.4	18.06
C_7H_7NO	Benzamide	129.1	18.49
$C_7H_7NO_2$	p-Nitrotoluene	51.6	16.81
C_7H_8	Toluene	-94.99	6.85
C_7H_8O	o-Cresol	29.8	13.94
C_7H_8O	m-Cresol	11.8	9.41
C_7H_8O	p-Cresol	35.5	11.89
C_7H_8O	Benzyl alcohol	-15.2	8.97
C_7H_9N	p-Methylaniline	43.75	18.22
C_7H_{14}	1-Heptene	-119.7	12.66
C_7H_{14}	Cycloheptane	-8.03	1.88
C_7H_{14}	Methylcyclohexane	-126.6	6.75
C_7H_{16}	Heptane	-90.6	14.16
C_7H_{16}	2-Methylhexane	-118.2	8.87
C_7H_{16}	3-Ethylpentane	-118.6	9.55
C_7H_{16}	2,2-Dimethylpentane	-123.8	5.86
C_7H_{16}	2,4-Dimethylpentane	-119.9	6.69
C_7H_{16}	3,3-Dimethylpentane	-134.9	7.07
C_7H_{16}	2,2,3-Trimethylbutane	-25	2.20
$C_8H_8O_2$	o-Toluic acid	103.7	20.17

Molecular formula	Name	$t_m/°C$	$\Delta_{fus}H/kJ$ mol^{-1}
$C_8H_8O_2$	m-Toluic acid	108.75	15.72
$C_8H_8O_2$	p-Toluic acid	179.6	22.73
$C_8H_8O_2$	Benzeneacetic acid	76.7	14.49
C_8H_{10}	o-Xylene	-25.2	13.61
C_8H_{10}	m-Xylene	-47.8	11.55
C_8H_{10}	p-Xylene	13.2	16.81
$C_8H_{10}O$	2,3-Xylenol	72.8	21.02
$C_8H_{10}O$	2,5-Xylenol	74.8	23.38
$C_8H_{10}O$	2,6-Xylenol	45.7	18.9
$C_8H_{10}O$	3,4-Xylenol	60.8	18.13
$C_8H_{10}O$	3,5-Xylenol	63.6	18
C_8H_{16}	Cyclooctane	14.83	2.41
C_8H_{16}	Ethylcyclohexane	-111.3	8.33
C_8H_{16}	1,1-Dimethylcyclohexane	-33.3	2.06
C_8H_{16}	cis-1,2-Dimethylcyclohexane	-49.9	1.64
C_8H_{16}	trans-1,2-Dimethylcyclohexane	-90	10.49
C_8H_{16}	cis-1,3-Dimethylcyclohexane	-75.6	10.82
C_8H_{16}	trans-1,3-Dimethylcyclohexane	-90.1	9.86
C_8H_{16}	cis-1,4-Dimethylcyclohexane	-87.4	9.31
C_8H_{16}	trans-1,4-Dimethylcyclohexane	-36.9	12.33
$C_8H_{16}O_2$	Octanoic acid	16.3	21.36
C_8H_{18}	Octane	-56.8	20.65
C_8H_{18}	3-Methylheptane	-120	11.38
C_8H_{18}	4-Methylheptane	-121	10.84
C_8H_{18}	2,2,4-Trimethylpentane	-107.3	9.04
C_9H_7N	Quinoline	-14.78	10.66
C_9H_7N	Isoquinoline	26.47	13.54
C_9H_{12}	Propylbenzene	-99.56	9.27
C_9H_{12}	1,2,3-Trimethylbenzene	-25.4	8.37
C_9H_{12}	1,2,4-Trimethylbenzene	-43.8	3.76
C_9H_{12}	1,3,5-Trimethylbenzene	-44.7	9.51
C_9H_{18}	Propylcyclohexane	-94.9	10.37
$C_9H_{18}O$	5-Nonanone	-5.9	24.93
$C_9H_{18}O_2$	Nonanoic acid	12.35	20.28
C_9H_{20}	Nonane	-53.5	15.47
C_9H_{20}	3,3-Diethylpentane	-33.1	10.09
C_9H_{20}	2,2,3,3-Tetramethylpentane	-9.8	2.33
C_9H_{20}	2,2,4,4-Tetramethylpentane	-66.54	9.75
$C_{10}H_7Br$	1-Bromonaphthalene	-0.9	15.16
$C_{10}H_7Cl$	1-Chloronaphthalene	-2.5	12.9
$C_{10}H_8$	Naphthalene	80.2	19.06
$C_{10}H_8O$	1-Naphthol	95	23.33
$C_{10}H_8O$	2-Naphthol	123	17.51
$C_{10}H_{14}$	Butylbenzene	-87.9	11.22
$C_{10}H_{14}$	p-Cymene	-68.9	9.6
$C_{10}H_{14}$	1,2,4,5-Tetramethylbenzene	79.3	21
$C_{10}H_{14}O$	Thymol	51.5	17.27
$C_{10}H_{18}O_4$	Sebacic acid	130.8	40.8
$C_{10}H_{20}$	Butylcyclohexane	-74.73	14.16
$C_{10}H_{20}O_2$	Decanoic acid	31.99	28.02
$C_{10}H_{22}$	Decane	-29.7	28.78
$C_{11}H_{10}$	2-Methylnaphthalene	34.4	11.97
$C_{11}H_{24}$	Undecane	-25.6	22.32
$C_{12}H_9N$	Carbazole	246.2	26.9
$C_{12}H_{10}$	Acenaphthene	93.4	21.54
$C_{12}H_{10}$	Biphenyl	69	18.6
$C_{12}H_{10}N_2$	Azobenzene	68.5	22.04
$C_{12}H_{10}N_2O$	trans-Azoxybenzene	36	17.93
$C_{12}H_{10}O$	Diphenyl ether	26.87	17.22

Molecular formula	Name	$t_m/°C$	$\Delta_{fus}H/kJ\ mol^{-1}$
$C_{12}H_{11}N$	Diphenylamine	52.98	17.86
$C_{12}H_{16}$	Cyclohexylbenzene	7.3	15.3
$C_{12}H_{24}O_2$	Dodecanoic acid	43.22	36.64
$C_{12}H_{26}$	Dodecane	-9.6	36.58
$C_{13}H_{10}$	9H-Fluorene	114.8	19.58
$C_{13}H_{10}O$	Benzophenone	47.88	18.19
$C_{14}H_{10}$	Anthracene	215.0	28.83
$C_{14}H_{10}$	Phenanthrene	99.24	16.46
$C_{14}H_{10}O_2$	Benzil	94.86	23.54
$C_{14}H_{12}$	trans-Stilbene	123	27.4
$C_{14}H_{12}O_2$	α-Phenylbenzeneacetic acid	147	31.27
$C_{14}H_{28}O_2$	Tetradecanoic acid	53.96	45.38
$C_{16}H_{10}$	Fluoranthene	107.8	18.87
$C_{16}H_{10}$	Pyrene	151.2	17.11
$C_{16}H_{32}O_2$	Hexadecanoic acid	61.82	42.04
$C_{16}H_{34}O$	1-Hexadecanol	49.3	34.29
$C_{18}H_{12}$	Chrysene	258.2	26.15
$C_{18}H_{14}$	p-Terphenyl	210.1	35.5
$C_{18}H_{36}O_2$	Stearic acid	68.82	56.59
$C_{18}H_{38}$	Octadecane	28.2	61.39
$C_{19}H_{40}$	Nonadecane	32.1	45.82
$C_{20}H_{12}$	Perylene	274	31.75
$C_{20}H_{42}$	Eicosane	36.8	69.88
$C_{24}H_{12}$	Coronene	437.3	19.2

PRESSURE AND TEMPERATURE DEPENDENCE OF LIQUID DENSITY

This table gives data on the variation of the density of some common liquids with pressure and temperature. The pressure dependence is described to first order by the isothermal compressibility coefficient κ defined as

$$\kappa = -(1/V)\,(\partial V/\partial P)_T$$

where V is the volume, and the temperature dependence by the cubic expansion coefficient α,

$$\alpha = (1/V)\,(\partial V/\partial T)_P$$

Substances are listed by molecular formula in the Hill order. More precise data on the variation of density with temperature over a wide temperature range can be found in Reference 1.

REFERENCES

1. Lide, D. R., and Kehiaian, H. V., *CRC Handbook of Thermophysical and Thermochemical Data*, CRC Press, Boca Raton, FL, 1994.
2. Le Neindre, B., *Effets des Hautes et Très Hautes Pressions*, in *Techniques de l'Ingénieur*, Paris, 1991.
3. *Landolt-Börnstein, Numerical Data and Functional Relationships in Science and Technology, New Series*, IV/4, *High-pressure Properties of Matter*, Springer-Verlag, Heidelberg, 1980.
4. Riddick, J.A., Bunger, W.B., and Sakano, T.K., *Organic Solvents, Fourth Edition*, John Wiley & Sons, New York, 1986.
5. Isaacs, N. S., *Liquid Phase High Pressure Chemistry*, John Wiley, New York, 1981.

Molecular formula	Name	Isothermal Compressibility		Cubic Thermal Expansion	
		$t/°C$	$\kappa \times 10^4/MPa^{-1}$	$t/°C$	$\alpha \times 10^3/°C^{-1}$
Cl_3P	Phosphorus trichloride	20	9.45	20	1.9
H_2O	Water	20	4.591	20	0.206
		25	4.524	25	0.256
		30	4.475	30	0.302
Hg	Mercury	20	0.401	20	1.811
CCl_4	Tetrachloromethane	20	10.50	20	1.14
		40	12.20	40	1.21
		70	15.6	70	1.33
$CHBr_3$	Tribromomethane	50	8.76	25	0.91
$CHCl_3$	Trichloromethane	20	9.96	20	1.21
		50	12.9	50	1.33
CH_2Br_2	Dibromomethane	27	6.85		
CH_2Cl_2	Dichloromethane	25	10.3	25	1.39
CH_3I	Iodomethane	27	10.3	25	1.26
CH_4O	Methanol	20	12.14	20	1.49
		40	13.83	40	1.59
CS_2	Carbon disulfide	20	9.38	20	1.12
		40	10.6	35	1.16
C_2Cl_4	Tetrachloroethylene	25	7.56	25	1.02
C_2HCl_3	Trichloroethylene	25	8.57	25	1.17
$C_2H_2Cl_2$	trans-1,2-Dichloroethylene	25	11.2	25	1.36
$C_2H_4Cl_2$	1,1-Dichloroethane	20	7.97	25	0.93
$C_2H_4Cl_2$	1,2-Dichloroethane	30	8.46	20	1.14
$C_2H_4O_2$	Acetic acid	20	9.08	20	1.08
		80	13.7	80	1.38
C_2H_5Br	Bromoethane	20	11.53	20	1.31
C_2H_5I	Iodoethane	20	9.82	25	1.17
C_2H_6O	Ethanol	20	11.19	20	1.40
		70	15.93	70	1.67
$C_2H_6O_2$	Ethylene glycol	20	3.64	20	0.626
C_3H_6O	Acetone	20	12.62	20	1.46
		40	15.6	40	1.57
C_3H_7Br	1-Bromopropane	0	10.22	25	1.2
C_3H_7Cl	1-Chloropropane	0	12.09	20	1.4
C_3H_7I	1-Iodopropane	0	10.22	25	1.09
C_3H_8O	1-Propanol	0	8.43	0	1.22
C_3H_8O	2-Propanol	40	13.32	40	1.55
$C_3H_8O_2$	1,2-Propanediol	0	4.45	20	0.695

PRESSURE AND TEMPERATURE DEPENDENCE OF LIQUID DENSITY (continued)

Molecular formula	Name	Isothermal Compressibility		Cubic Thermal Expansion	
		$t/°C$	$\kappa \times 10^4/MPa^{-1}$	$t/°C$	$\alpha \times 10^3/°C^{-1}$
$C_3H_8O_2$	1,3-Propanediol	0	4.09	20	0.61
$C_3H_8O_3$	Glycerol	0	2.54	20	0.520
$C_4H_8O_2$	Ethyl acetate	20	11.32	20	1.35
		60	16.2	60	1.54
C_4H_9Br	1-Bromobutane	25	10.26	20	1.13
C_4H_9I	1-Iodobutane	0	7.73	25	1.02
$C_4H_{10}O$	1-Butanol	0	8.10	0	1.12
$C_4H_{10}O$	Diethyl ether	20	18.65	20	1.65
		30	20.85	30	1.72
$C_4H_{10}O_3$	Diethylene glycol	0	3.34	20	0.635
C_5H_{10}	Cyclopentane	20	13.31	20	1.35
$C_5H_{11}Br$	1-Bromopentane	0	8.42	25	1.04
$C_5H_{11}I$	1-Iodopentane	0	7.56		
C_5H_{12}	Pentane	25	21.80	25	1.64
$C_5H_{12}O$	1-Pentanol	0	7.71	0	1.02
C_6H_5Br	Bromobenzene	20	6.46	20	0.86
C_6H_5Cl	Chlorobenzene	20	7.45	20	0.94
$C_6H_5NO_2$	Nitrobenzene	20	4.93	25	0.833
C_6H_6	Benzene	25	9.66	25	1.14
		45	11.28	45	1.21
C_6H_6O	Phenol	60	6.05	60	0.82
C_6H_7N	Aniline	20	4.53	20	0.81
		80	6.32	80	0.91
C_6H_{12}	Cyclohexane	20	11.30	20	1.15
		60	15.2	60	1.29
C_6H_{14}	Hexane	25	16.69	25	1.41
		45	20.27	45	1.52
C_6H_{14}	2-Methylpentane	0	13.97	25	1.43
		0	14.57	25	1.40
C_6H_{14}	2,3-Dimethylbutane	20	17.97	25	1.39
$C_6H_{14}O$	1-Hexanol	25	8.24	25	1.03
$C_6H_{15}NO_3$	Triethanolamine	0	3.61	55	0.53
C_7H_8	Toluene	20	8.96	20	1.05
		50	11.0	50	1.13
C_7H_8O	Anisole	20	6.60	20	0.951
C_7H_{14}	Cycloheptane	20	9.22		
C_7H_{16}	Heptane	25	14.38	25	1.26
C_8H_{10}	o-Xylene	25	8.10	25	0.96
C_8H_{10}	m-Xylene	20	8.46	20	0.99
C_8H_{10}	p-Xylene	25	8.59	25	1.00
C_8H_{16}	Cyclooctane	20	8.03		
C_8H_{18}	Octane	25	12.82	25	1.16
		45	15.06	45	1.23
$C_8H_{18}O$	1-Octanol	25	7.64	25	0.827
C_9H_{12}	Mesitylene	25	8.14	25	0.94
$C_9H_{14}O_6$	Triacetin	0	4.49	25	0.94
C_9H_{20}	Nonane	25	11.75	25	1.08
$C_{10}H_{22}$	Decane	25	10.94	25	1.02
$C_{11}H_{24}$	Undecane	25	10.31	25	0.97
$C_{12}H_{26}$	Dodecane	25	9.88	25	0.93
$C_{13}H_{28}$	Tridecane	25	9.48	25	0.90
$C_{14}H_{30}$	Tetradecane	25	9.10	25	0.87
$C_{15}H_{32}$	Pentadecane	25	8.82		
$C_{16}H_{22}O_4$	Butyl phthalate	0	5.0	25	0.86
$C_{16}H_{34}$	Hexadecane	25	8.57		
		45	9.78		
$C_{19}H_{36}O_2$	Methyl oleate	0	6.18	60	0.85

PROPERTIES OF CRYOGENIC FLUIDS

This table gives physical and thermodynamic properties of eight cryogenic fluids. The properties are:

M	Molar mass in grams per mole
T_t	Triple point temperature in kelvins
P_t	Triple point pressure in kilopascals
$\rho_t(l)$	Liquid density at the triple point in grams per milliliter
$\Delta_{fus}H$ @ T_t	Enthalpy of fusion at the triple point in joules per gram
T_b	Normal boiling point in kelvins at a pressure of 101325 pascals (760 mmHg)
$\Delta_{vap}H$ @ T_b	Enthalpy of vaporization at the normal boiling point in joules per gram
$\rho(l)$ @ T_b	Liquid density at the normal boiling point in grams per milliliter
$\rho(g)$ @ T_b	Vapor density at the normal boiling point in grams per liter
$C_p(l)$ @ T_b	Liquid heat capacity at constant pressure at the normal boiling point in joules per gram kelvin
$C_p(g)$ @ T_b	Vapor heat capacity at constant pressure at the normal boiling point in joules per gram kelvin
T_c	Critical temperature in kelvins
P_c	Critical pressure in megapascals
ρ_c	Critical density in grams per milliliter

In the case of air, the value given for the triple point temperature is the incipient solidification temperature, and the normal boiling point value is the incipient boiling (bubble) point. See Reference 3 for more details.

REFERENCES

1. Youglove, B. A., *J. Phys. Chem. Ref. Data*, 11, Suppl. 1, 1982.
2. Daubert, T. E., Danner, R. P., Sibul, H. M., and Stebbins, C. C., *Physical and Thermodynamic Properties of Pure Compounds: Data Compilation*, extant 1994 (core with 4 supplements), Taylor & Francis, Bristol, PA (also available as database).
3. Sytchev, V. V., et al., *Thermodynamic Properties of Air*, Hemisphere Publishing, New York, 1987.
4. Jacobsen, R. T., Stewart, R. B., and Jahangiri, M., *J. Phys. Chem. Ref. Data*, 15, 735, 1986. [Nitrogen]
5. Stewart, R. B., Jacobsen, R. T., and Wagner, W., *J. Phys. Chem. Ref. Data*, 20, 917, 1991. [Oxygen]
6. McCarty, R. D., *J. Phys. Chem. Ref. Data*, 2, 923, 1973. [Helium] Also, Donnelly, R. J., private communication.
7. Stewart, R. B. and Jacobsen, R. T., *J. Phys. Chem. Ref. Data*, 18, 639, 1989. [Argon]
8. Setzmann, U. and Wagner, W., *J. Phys. Chem. Ref. Data*, 20, 1061, 1991. [Methane]
9. Vargaftik, N. B., *Thermophysical Properties of Liquids and Gases*, 2nd ed., John Wiley, New York, 1975.

Property	Units	Air	N_2	O_2	H_2	He	Ne	Ar	Kr	Xe	CH_4
M	g/mol	28.96	28.014	31.999	2.0159	4.0026	20.180	39.948	83.800	131.290	16.043
T_t	K	59.75	63.15	54.3584	13.8		24.5561	83.8058	115.8	161.4	90.694
P_t	kPa		12.463	0.14633	7.042		50	68.95	72.92	81.59	11.696
$\rho_t(l)$	g/mL	0.959	0.870	1.306	0.0770		1.251	1.417	2.449	2.978	0.4515
$\Delta_{fus}H$ @ T_t	J/g		25.3	13.7	59.5		16.8	28.0	16.3	13.8	58.41
T_b	K	78.67	77.35	90.188	20.28	4.2221	27.07	87.293	119.92	165.10	111.668
$\Delta_{vap}H$ @ T_b	J/g	198.7	198.8	213.1	445	20.7	84.8	161.0	108.4	96.1	510.83
$\rho(l)$ @ T_b	g/mL	0.8754	0.807	1.141	0.0708	0.124901	1.204	1.396	2.418	2.953	0.4224
$\rho(g)$ @ T_b	g/L	3.199	4.622	4.467	1.3390	16.89	9.51	5.79	8.94		1.816
$C_p(l)$ @ T_b	J/g K	1.865	2.042	1.699	9.668	4.545	1.877	1.078	0.533	0.340	3.481
$C_p(g)$ @ T_b	J/g K		1.341	0.980	12.24	9.78		0.570	0.248	0.158	2.218
T_c	K	132.5	126.20	154.581	32.98	5.1953	44.40	150.663	209.40	289.73	190.56
P_c	MPa	3.766	3.390	5.043	1.293	0.227460	2.760	4.860	5.500	5.840	4.592
ρ_c	g/mL	0.316	0.313	0.436	0.031	0.06964	0.484	0.531	0.919	1.110	0.1627

HALOCARBON REFRIGERANTS

Halogen derivatives of the lower aliphatic hydrocarbons, especially fluorocarbons and chlorocarbons, are widely used as refrigerants, solvents, and cleaning agents. This table lists the most important halocarbons in current use or being considered as possible substitutes for those compounds that are potential environmental hazards. The code number for each compound appears in the first column; this number is frequently used in expressions like R 11 or CFC 11 or HFC 134.

In addition to name, molecular formula, CAS Registry Number, and molecular weight, the table lists the following properties:

t_m: normal melting point
t_b: normal boiling point (at 101.324 kPa)
t_c: critical temperature
TLV: Threshold Limit Value, which is the maximum safe concentration in air in the workplace, expressed as the time-weighted average (TWA) in parts per million by volume over an 8-hr workday and 40-hr workweek (see Reference 3). An * following the TLV indicates that the substance in a confirmed or suspected human carcinogen.

REFERENCES

1. *1989 ASHRAE Handbook: Fundamentals*, American Society of Heating, Refrigerating, and Air Conditioning Engineers, Atlanta, GA, 1989.
2. Platzer, B., Polt, A., and Mauer, G., *Thermophysical Properies of Refrigerants*, Springer, Berlin, 1990.
3. 1991-1992 Threshold Limit Values and Biological Exposure Indexes, American Conference of Government Industrial Hygienists, Cincinnati, OH, 1991.

Code no.	Name	Molecular formula	CAS reg. no.	Molecular weight	t_m/°C	t_b/°C	t_c/°C	TLV ppm
10	Tetrachloromethane	CCl_4	56-23-5	153.822	-23	76.8	283.5	5*
11	Trichlorofluoromethane	CCl_3F	75-69-4	137.368	-111.11	23.75	198.1	1000
12	Dichlorodifluoromethane	CCl_2F_2	75-71-8	120.913	-158	-29.8	111.80	1000
12B1	Bromochlorodifluoromethane	$CBrClF_2$	353-59-3	165.365	-159.5	-3.72	153.73	
13	Chlorotrifluoromethane	$CClF_3$	75-72-9	104.459	-181	-81.4	29	
13B1	Bromotrifluoromethane	$CBrF_3$	75-63-8	148.910	-172	-57.89	67.1	
14	Tetrafluoromethane	CF_4	75-73-0	88.005	-183.59	-128.02	-45.5	
20	Trichloromethane	$CHCl_3$	67-66-3	119.377	-63.6	61.17	263.3	10*
21	Dichlorofluoromethane	$CHCl_2F$	75-43-4	102.923	-135	8.95	178.43	10
22	Chlorodifluoromethane	$CHClF_2$	75-45-6	86.468	-157.42	-40.75	96.2	1000
23	Trifluoromethane	CHF_3	75-46-7	70.014	-155.18	-82.1	26.2	
30	Dichloromethane	CH_2Cl_2	75-09-2	84.932	-95.14	40	237	50*
31	Chlorofluoromethane	CH_2ClF	593-70-4	68.478	-133	-9.1		
32	Difluoromethane	CH_2F_2	75-10-5	52.024	-136	-51.69	78.5	
40	Chloromethane	CH_3Cl	74-87-3	50.488	-97.7	-24.09	143.10	50*
41	Fluoromethane	CH_3F	593-53-3	34.033	-141.8	-78.41	44.7	1*
110	Hexachloroethane	C_2Cl_6	67-72-1	236.738	187	187		
111	Pentachlorofluoroethane	C_2Cl_5F	354-56-3	220.284				
112	Tetrachloro-1,2-difluoroethane	$C_2Cl_4F_2$	76-12-0	203.830	26	93	278	500
112a	1,1,1,2-Tetrachloro-2,2-difluoroethane	$C_2Cl_4F_2$	76-11-9	203.830	40.6	91.5		
113	1,1,2-Trichlorotrifluoroethane	$C_2Cl_3F_3$	76-13-1	187.375	-35	47.7	214.2	1000
113a	1,1,1-Trichlorotrifluoroethane	$C_2Cl_3F_3$	354-58-5	187.375	14.2	46.1		
114	1,2-Dichlorotetrafluoroethane	$C_2Cl_2F_4$	76-14-2	170.921	-94	3.8	145.63	1000
114a	1,1-Dichlorotetrafluoroethane	$C_2Cl_2F_4$	374-07-2	170.921	-56.6	4	145.5	1000

HALOCARBON REFRIGERANTS (continued)

Code no.	Name	Molecular formula	CAS reg. no.	Molecular weight	t_m/°C	t_b/°C	t_c/°C	TLV ppm
114B2	1,2-Dibromotetrafluoroethane	$C_2Br_2F_4$	124-73-2	259.824	-110.4	47.35	214.7	
115	Chloropentafluoroethane	C_2ClF_5	76-15-3	154.467	-99.44	-37.95	80.1	1000
116	Hexafluoroethane	C_2F_6	76-16-4	138.012	-100.7	-78.1	20	
120	Pentachloroethane	C_2HCl_5	76-01-7	202.293	-29	159.88		
121	1,1,2,2-Tetrachloro-1-fluoroethane	C_2HCl_4F	354-14-3	185.839	-82.6	116.6		
121a	1,1,1,2-Tetrachloro-2-fluoroethane	C_2HCl_4F	354-11-0	185.840	-95.35	116.5		
122	1,2,2-Trichloro-1,1-difluoroethane	$C_2HCl_3F_2$	354-21-2	169.385		71.9		
122a	1,1,2-Trichloro-1,2-difluoroethane	$C_2HCl_3F_2$	354-15-4	169.385	-140	72.5		
122b	1,1,1-Trichloro-2,2-difluoroethane	$C_2HCl_3F_2$	354-12-1	169.385		73		
123	2,2-Dichloro-1,1,1-trifluoroethane	$C_2HCl_2F_3$	306-83-2	152.931	-107	27.1		
123a	1,2-Dichloro-1,2,2-trifluoroethane	$C_2HCl_2F_3$	354-23-4	152.931		28.2		
124	2-Chloro-1,1,1,2-tetrafluoroethane	C_2HClF_4	2837-89-0	136.476	-12		126.8	
124a	1-Chloro-1,1,2,2-tetrafluoroethane	C_2HClF_4	354-25-6	136.476	-117	-10.2		
125	Pentafluoroethane	C_2HF_5	354-33-6	120.022	-103	-48.5		
130	1,1,2,2-Tetrachloroethane	$C_2H_2Cl_4$	79-34-5	167.849	-43.8	146.5	388.00	1*
131	1,1,2-Trichloro-2-fluoroethane	$C_2H_2Cl_3F$	359-28-4	151.394		102.5	101.2	
132	1,2-Dichloro-1,2-difluoroethane	$C_2H_2Cl_2F_2$	431-06-1	134.940	-155	59		
132b	1,2-Dichloro-1,1-difluoroethane	$C_2H_2Cl_2F_2$	1649-08-7	134.940				
133	1-Chloro-1,2,2-trifluoroethane	$C_2H_2ClF_3$	431-07-2	118.486		17		
133a	2-Chloro-1,1,1-trifluoroethane	$C_2H_2ClF_3$	75-88-7	118.486	-105.5	6.1		
133b	1-Chloro-1,1,2-trifluoroethane	$C_2H_2ClF_3$	421-04-5	118.486		12		
134	1,1,2,2-Tetrafluoroethane	$C_2H_2F_4$	359-35-3	102.032	-89	-19.9	119	
134a	1,1,1,2-Tetrafluoroethane	$C_2H_2F_4$	811-97-2	102.032	-101	-26.5	101.2	
140	1,1,2-Trichloroethane	$C_2H_3Cl_3$	79-00-5	133.404	-36.6	113.8		10
140a	1,1,1-Trichloroethane	$C_2H_3Cl_3$	71-55-6	133.404	-30.4	74.09	272	350
141	1,2-Dichloro-1-fluoroethane	$C_2H_3Cl_2F$	430-57-9	116.950		75.7		
141b	1,1-Dichloro-1-fluoroethane	$C_2H_3Cl_2F$	1717-00-6	116.950	-103.5	32.11		
142	2-Chloro-1,1-difluoroethane	$C_2H_3ClF_2$	338-65-8	100.495		35.1		
142b	1-Chloro-1,1-difluoroethane	$C_2H_3ClF_2$	75-68-3	100.495	-130.8	-9.75	137.14	
143	1,1,2-Trifluoroethane	$C_2H_3F_3$	430-66-0	84.041	-84	5		
143a	1,1,1-Trifluoroethane	$C_2H_3F_3$	420-46-2	84.041	-111.3	-47.55	73.2	
150	1,2-Dichloroethane	$C_2H_4Cl_2$	107-06-2	98.959	-35.5	83.5	288	10*
150a	1,1-Dichloroethane	$C_2H_4Cl_2$	75-34-3	98.959	-96.96	57.4	250	100
151	1-Chloro-2-fluoroethane	C_2H_4ClF	762-50-5	82.505		53.2		
151a	1-Chloro-1-fluoroethane	C_2H_4ClF	1615-75-4	82.505		16.1		
152	1,2-Difluoroethane	$C_2H_4F_2$	624-72-6	66.051		30.7		
152a	1,1-Difluoroethane	$C_2H_4F_2$	75-37-6	66.051	-117	-24.95	113.6	1000
160	Chloroethane	C_2H_5Cl	75-00-3	64.514	-138.7	13.1	187.3	
161	Fluoroethane	C_2H_5F	353-36-6	48.060	-143.2	-37.65	102.16	
218	Perfluoropropane	C_3F_8	76-19-7	188.020	-147.69	-36.65	72.0	
1112a	1,1-Dichloro-2,2-difluoroethylene	$C_2Cl_2F_2$	79-35-5	132.924				
1113	Chlorotrifluoroethylene	C_2ClF_3	79-38-9	116.470	-158	-27.85	106	

HALOCARBON REFRIGERANTS (continued)

Code no.	Name	Molecular formula	CAS reg. no.	Molecular weight	$t_m/°C$	$t_b/°C$	$t_c/°C$	TLV ppm
1114	Tetrafluoroethylene	C_2F_4	116-14-3	100.016	-142.5	-75.95	33.4	
1120	Trichloroethylene	C_2HCl_3	79-01-6	131.388	-84.75	87.21	271.1	50*
1130	cis-1,2-Dichloroethylene	$C_2H_2Cl_2$	156-59-2	96.943	-80	60.19	271.1	200
1130	trans-1,2-Dichloroethylene	$C_2H_2Cl_2$	156-60-5	96.943	-49.8	48.73	243.4	200
1132a	1,1-Difluoroethylene	$C_2H_2F_2$	75-38-7	64.035	-144	-85.7	29.8	
1140	Chloroethylene	C_2H_3Cl	75-01-4	62.499	-153.79	-13.37		5*
1141	Fluoroethylene	C_2H_3F	75-02-5	46.044	-160.5	-72	54.8	
C316	1,2-Dichlorohexafluorocyclobutane	$C_4Cl_2F_6$	356-18-3	232.940				
C317	Chloroheptafluorocyclobutane	C_4ClF_7	377-41-3	216.486				
C318	Perfluorocyclobutane	C_4F_8	115-25-3	200.031	-40.19	-5.99	115.31	

* Confirmed or suspected human carcinogen.

DENSITY AND SPECIFIC VOLUME OF MERCURY

The data in this table have been adjusted to the ITS-90 temperature scale. The uncertainty in density values is 0.0003 g/mL between −20 and −10°C; 0.0001 or less between −10 and 200°C; and 0.0002 between 200 and 300°C.

REFERENCE

Ambrose, D., *Metrologia*, 27, 245, 1990.

$t/°C$	$\rho/(g/mL)$	$v/(mL/kg)$	$t/°C$	$\rho/(g/mL)$	$v/(mL/kg)$	$t/°C$	$\rho/(g/mL)$	$v/(mL/kg)$
−20	13.64461	73.2890	27	13.52869	73.9170	74	13.41423	74.5477
−19	13.64212	73.3024	28	13.52624	73.9304	75	13.41181	74.5612
−18	13.63964	73.3157	29	13.52379	73.9438	76	13.40939	74.5746
−17	13.63716	73.3291	30	13.52134	73.9572	77	13.40697	74.5881
−16	13.63468	73.3424	31	13.51889	73.9705	78	13.40455	74.6016
−15	13.63220	73.3558	32	13.51645	73.9839	79	13.40213	74.6150
−14	13.62972	73.3691	33	13.51400	73.9973	80	13.39971	74.6285
−13	13.62724	73.3824	34	13.51156	74.0107	81	13.39729	74.6420
−12	13.62476	73.3958	35	13.50911	74.0241	82	13.39487	74.6554
−11	13.62228	73.4091	36	13.50667	74.0375	83	13.39245	74.6689
−10	13.61981	73.4225	37	13.50422	74.0509	84	13.39003	74.6824
−9	13.61733	73.4358	38	13.50178	74.0643	85	13.38762	74.6959
−8	13.61485	73.4492	39	13.49934	74.0777	86	13.38520	74.7094
−7	13.61238	73.4625	40	13.49690	74.0911	87	13.38278	74.7229
−6	13.60991	73.4759	41	13.49446	74.1045	88	13.38037	74.7364
−5	13.60743	73.4892	42	13.49202	74.1179	89	13.37795	74.7498
−4	13.60496	73.5026	43	13.48958	74.1313	90	13.37554	74.7633
−3	13.60249	73.5160	44	13.48714	74.1447	91	13.37313	74.7768
−2	13.60002	73.5293	45	13.48470	74.1581	92	13.37071	74.7903
−1	13.59755	73.5427	46	13.48226	74.1715	93	13.36830	74.8038
0	13.59508	73.5560	47	13.47982	74.1850	94	13.36589	74.8173
1	13.59261	73.5694	48	13.47739	74.1984	95	13.36347	74.8308
2	13.59014	73.5827	49	13.47495	74.2118	96	13.36106	74.8443
3	13.58768	73.5961	50	13.47251	74.2252	97	13.35865	74.8579
4	13.58521	73.6095	51	13.47008	74.2386	98	13.35624	74.8714
5	13.58275	73.6228	52	13.46765	74.2520	99	13.35383	74.8849
6	13.58028	73.6362	53	13.46521	74.2655	100	13.35142	74.8984
7	13.57782	73.6495	54	13.46278	74.2789	110	13.3273	75.0337
8	13.57535	73.6629	55	13.46035	74.2923	120	13.3033	75.1693
9	13.57289	73.6763	56	13.45791	74.3057	130	13.2793	75.3052
10	13.57043	73.6896	57	13.45548	74.3192	140	13.2553	75.4413
11	13.56797	73.7030	58	13.45305	74.3326	150	13.2314	75.5778
12	13.56551	73.7164	59	13.45062	74.3460	160	13.2075	75.7147
13	13.56305	73.7297	60	13.44819	74.3594	170	13.1836	75.8519
14	13.56059	73.7431	61	13.44576	74.3729	180	13.1597	75.9895
15	13.55813	73.7565	62	13.44333	74.3863	190	13.1359	76.1274
16	13.55567	73.7698	63	13.44090	74.3998	200	13.1120	76.2659
17	13.55322	73.7832	64	13.43848	74.4132	210	13.0882	76.4047
18	13.55076	73.7966	65	13.43605	74.4266	220	13.0644	76.5440
19	13.54831	73.8100	66	13.43362	74.4401	230	13.0406	76.6838
20	13.54585	73.8233	67	13.43120	74.4535	240	13.0167	76.8241
21	13.54340	73.8367	68	13.42877	74.4670	250	12.9929	76.9650
22	13.54094	73.8501	69	13.42635	74.4804	260	12.9691	77.1064
23	13.53849	73.8635	70	13.42392	74.4939	270	12.9453	77.2484
24	13.53604	73.8769	71	13.42150	74.5073	280	12.9214	77.3909
25	13.53359	73.8902	72	13.41908	74.5208	290	12.8975	77.5341
26	13.53114	73.9036	73	13.41665	74.5342	300	12.8736	77.6779

THERMAL PROPERTIES OF MERCURY
Lev R. Fokin

The first of these tables gives the molar heat capacity at constant pressure of liquid and gaseous mercury as a function of temperature. To convert to specific heat in units of J/g K, divide these values by 200.59, the atomic weight of mercury.

REFERENCE

Douglas, T. B., Ball, A. T., and Ginnings, D. C., *J. Res. Natl. Bur. Stands.*, 46, 334, 1951.

t/°C	C_p/(J/mol K) Liquid	C_p/(J/mol K) Gas	t/°C	C_p/(J/mol K) Liquid	C_p/(J/mol K) Gas	t/°C	C_p/(J/mol K) Liquid	C_p/(J/mol K) Gas
–38.84	28.2746	20.786	140	27.3675	20.786	340	27.1500	20.836
–20	28.1466	20.786	160	27.3090	20.786	356.73	27.1677	20.849
0	28.0190	20.786	180	27.2588	20.790	360	27.1709	20.853
20	27.9002	20.786	200	27.2169	20.790	380	27.1981	20.870
25	27.8717	20.786	220	27.1834	20.794	400	27.2324	20.891
40	27.7897	20.786	240	27.1583	20.794	420	27.2738	20.916
60	27.6880	20.786	260	27.1412	20.799	440	27.3207	20.941
80	27.5952	20.786	280	27.1320	20.807	460	27.3742	20.974
100	27.5106	20.786	300	27.1303	20.815	480	27.4332	21.008
120	27.4349	20.786	320	27.1366	20.824	500	27.4985	21.046

The second table gives the molar heat capacity of solid mercury in its rhombohedral (α-mercury) form.

REFERENCES

1. Busey and Giaque, *J. Am. Chem. Soc.*, 75, 806, 1953.
2. Amitin, Lebedeva, and Paukov, *Rus. J. Phys. Chem.*, 2666, 1979.

t/°C	C_p/J mol^{-1}	t/°C	C_p/J mol^{-1}	t/°C	C_p/J mol^{-1}	t/°C	C_p/J mol^{-1}
–268.99	0.99*	–248.15	12.74	–193.15	23.16	–113.15	26.15
–268.99	0.97**	–243.15	14.78	–183.15	23.76	–93.15	26.69
–268.15	1.6	–233.15	17.90	–173.15	24.24	–73.15	27.28
–263.15	4.6	–223.15	19.94	–153.15	25.00	–53.15	27.96
–258.15	7.6	–213.15	21.40	–133.15	25.61	–38.87	28.5
–253.15	10.33	–203.15	22.42				

*   Superconducting state
** Normal state

The final table gives the cubic thermal expansion coefficient α, the isothermal compressibility coefficient κ_T, and the speed of sound U for liquid mercury as a function of temperature. These properties are defined as follows:

$$\alpha = \frac{1}{v}\left(\frac{\partial v}{\partial T}\right)_p \qquad \kappa_T = -\frac{1}{v}\left(\frac{\partial v}{\partial P}\right)_T \qquad U^2 = \left(\frac{\partial P}{\partial \rho}\right)_s \qquad \rho = v^{-1}$$

where v is the specific volume (given in the table on the preceding page).

REFERENCE

Vukalovich, M. P., et al., *Thermophysical Properties of Mercury*, Moscow Standard Press, 1971.

t/°C	$\alpha \times 10^4$/K^{-1}	$\kappa_T \times 10^6$/bar^{-1} At 1 bar	$\kappa_T \times 10^6$/bar^{-1} At 1000 bar	U/m s^{-1}	t/°C	$\alpha \times 10^4$/K^{-1}	$\kappa_T \times 10^6$/bar^{-1} At 1 bar	$\kappa_T \times 10^6$/bar^{-1} At 1000 bar	U/m s^{-1}
–20	1.818	3.83		1470	120	1.8058	4.513	4.33	1404.7
0	1.8144	3.918	3.78	1460.8	140	1.8074	4.622		1395.4
20	1.8110	4.013	3.87	1451.4	160	1.8100	4.731	4.53	1386.1
40	1.8083	4.109	3.96	1442.0	180	1.8136	4.844		1376.7
60	1.8064	4.207		1432.7	200	1.818	4.96		1367
80	1.8053	4.308	4.14	1423.4	250	1.834	5.26		1344
100	1.8051	4.410		1414.1	300	1.856	5.59		1321

SURFACE TENSION OF COMMON LIQUIDS

The surface tension γ of about 200 liquids is tabulated here as a function of temperature. Values of γ are given in units of millinewtons per meter (mN/m), which is equivalent to dyn/cm in cgs units. The values refer to a nominal pressure of one atmosphere (about 100 kPa) except in cases where the indicated temperature is above the normal boiling point of the substance; in those cases, the applicable pressure is the saturation vapor pressure at the temperature in question.

The uncertainty of the values is 0.1 to 0.2 mN/m or less in most cases. Values at temperatures between the points tabulated can be obtained by linear interpolation to a good approximation.

Substances are listed by molecular formula in the modified Hill order, with substances not containing carbon appearing before those that do contain carbon. A more extensive compilation of surface tension may be found in the Reference.

REFERENCE

Jasper, J. J., *J. Phys. Chem. Ref. Data*, 1, 841, 1972.

Mol. form.	Name	γ in mN/m				
		10°C	25°C	50°C	75°C	100°C
Br_2	Bromine	43.68	40.95	36.40		
Cl_2O_2S	Sulfuryl chloride		28.78			
Cl_3OP	Phosphoryl chloride		32.03	28.85	25.66	
Cl_3P	Phosphorus trichloride		27.98	24.81		
Cl_4Si	Silicon tetrachloride	19.78	18.29	15.80		
H_2O	Water	74.23	71.99	67.94	63.57	58.91
H_4N_2	Hydrazine		66.39			
Hg	Mercury	488.55	485.48	480.36	475.23	470.11
CCl_4	Tetrachloromethane		26.43	23.37	20.31	17.25
CS_2	Carbon disulfide	33.81	31.58	27.87		
$CHBr_3$	Tribromomethane		44.87	41.60	38.33	
$CHCl_3$	Trichloromethane		26.67	23.44	20.20	
CH_2Br_2	Dibromomethane		39.05	35.33	31.61	
CH_2Cl_2	Dichloromethane		27.20			
CH_2O_2	Formic acid		37.13	34.38	31.64	
CH_3I	Iodomethane	32.19	30.34			
CH_3NO	Formamide		57.03	54.92	52.82	50.71
CH_3NO_2	Nitromethane	39.04	36.53	32.33		
CH_4O	Methanol	23.23	22.07	20.14		
CH_5N	Methylamine		19.15			
C_2HCl_5	Pentachloroethane		34.15	31.20	28.26	
$C_2HF_3O_2$	Trifluoroacetic acid		13.53	11.42		
$C_2H_2Cl_4$	1,1,2,2-Tetrachloroethane		35.58	32.41	29.24	26.07
$C_2H_3Cl_3$	1,1,1-Trichloroethane		25.18	22.07		
$C_2H_3Cl_3$	1,1,2-Trichloroethane		34.02	30.65	27.27	23.89
C_2H_3N	Acetonitrile		28.66	25.51		
$C_2H_4Br_2$	1,2-Dibromoethane		39.55	36.25	32.95	
$C_2H_4Cl_2$	1,1-Dichloroethane		24.07			
$C_2H_4Cl_2$	1,2-Dichloroethane		31.86	28.29	24.72	
C_2H_4O	Acetaldehyde	22.54	20.50	17.10		
$C_2H_4O_2$	Acetic acid		27.10	24.61	22.13	
$C_2H_4O_2$	Methyl formate	26.72	24.36	20.43	16.50	12.57
C_2H_5Br	Bromoethane	25.36	23.62			
C_2H_5I	Iodoethane	30.38	28.46	25.24		
$C_2H_5NO_2$	Nitroethane	34.02	32.13	29.00		
C_2H_6O	Ethanol	23.22	21.97	19.89		
C_2H_6OS	Dimethyl sulfoxide		42.92	40.06		
$C_2H_6O_2$	Ethylene glycol		47.99	45.76	43.54	41.31
C_2H_6S	Dimethyl sulfide	25.27	24.06			
C_2H_6S	Ethanethiol		23.08			
$C_2H_6S_2$	Dimethyl disulfide		33.39	30.04		
C_2H_7N	Dimethylamine		26.34			
C_2H_7N	Ethylamine		19.20			

Mol. form.	Name	γ in mN/m				
		10°C	25°C	50°C	75°C	100°C
C_2H_7NO	Ethanolamine		48.32	45.53	42.73	
C_3H_5Br	3-Bromopropene		26.31	23.17		
C_3H_5Cl	3-Chloropropene		23.14			
C_3H_5ClO	Epichlorohydrin	38.40	36.36	32.96	29.56	26.16
C_3H_5N	Propanenitrile		26.75	23.87		
$C_3H_6Cl_2$	1,2-Dichloropropane		28.32	25.22	22.12	
C_3H_6O	Acetone		23.46	20.66		
C_3H_6O	Allyl alcohol	26.63	25.28	23.02	20.77	
$C_3H_6O_2$	Ethyl formate	25.16	23.18			
$C_3H_6O_2$	Methyl acetate	26.66	24.73	21.51		
$C_3H_6O_2$	Propanoic acid		26.20	23.72	21.23	
C_3H_7Br	1-Bromopropane	27.08	25.26	22.21		
C_3H_7Br	2-Bromopropane	25.03	23.25	20.30		
C_3H_7Cl	1-Chloropropane	23.16	21.30			
C_3H_7Cl	2-Chloropropane	20.49	19.16			
$C_3H_7NO_2$	2-Nitropropane	31.02	29.29	26.39		
C_3H_8O	1-Propanol	24.48	23.32	21.38	19.43	
C_3H_8O	2-Propanol	22.11	20.93	18.96	16.98	
$C_3H_8O_2$	2-Methoxyethanol	32.32	30.84	28.38	25.92	23.46
C_3H_8S	1-Propanethiol		24.20	21.02		
C_3H_8S	2-Propanethiol		21.33	18.39		
C_3H_9N	Propylamine		21.75			
C_3H_9N	Trimethylamine		13.41			
$C_4H_4N_2$	Pyridazine	49.51	47.96	45.37	42.78	40.19
$C_4H_4N_2$	Pyrimidine		30.33	27.80	25.28	22.75
C_4H_4S	Thiophene		30.68	27.36		
C_4H_5N	Pyrrole	38.71	37.06	34.31		
$C_4H_6O_3$	Acetic anhydride	34.08	31.93	28.34	24.75	21.16
C_4H_7N	Butanenitrile		26.92	24.33	21.73	
C_4H_8O	2-Butanone		23.97	21.16		
$C_4H_8O_2$	1,4-Dioxane		32.75	29.28	25.80	22.32
$C_4H_8O_2$	Ethyl acetate	25.13	23.39	20.49	17.58	14.68
$C_4H_8O_2$	Methyl propanoate	26.32	24.44	21.29		
$C_4H_8O_2$	Butanoic acid		26.05	23.75	21.45	
C_4H_9Br	1-Bromobutane	27.58	25.90	23.08	20.27	17.45
C_4H_9Cl	1-Chlorobutane	24.85	23.18	20.39		
C_4H_9I	1-Iodobutane	29.79	28.24	25.67	23.09	20.51
C_4H_9N	Pyrrolidine	30.58	29.23	26.98		
$C_4H_{10}O$	1-Butanol	26.28	24.93	22.69	20.44	18.20
$C_4H_{10}O$	2-Butanol	23.74	22.54	20.56	18.57	16.58
$C_4H_{10}O$	2-Methyl-2-propanol		19.96	17.71		
$C_4H_{10}O$	Diethyl ether		16.65			
$C_4H_{10}O_2$	2-Ethoxyethanol		28.35	26.11	23.86	21.62
$C_4H_{10}O_3$	Diethylene glycol		44.77	42.57	40.37	38.17
$C_4H_{10}S$	Diethyl sulfide	26.22	24.57	21.80		
$C_4H_{11}N$	Butylamine		23.44	20.63		
$C_4H_{11}N$	Isobutylamine		21.75	19.02		
$C_4H_{11}N$	*tert*-Butylamine		16.87			
$C_4H_{11}N$	Diethylamine		19.85			
$C_5H_4O_2$	Furfural	45.08	43.09	39.78	36.46	33.14
C_5H_5N	Pyridine		36.56	33.29	30.03	
C_5H_8	Cyclopentene	24.45	22.20			
C_5H_8O	Cyclopentanone	34.45	32.80	30.05	27.30	24.55
C_5H_{10}	1-Pentene	17.10	15.45			
C_5H_{10}	2-Methyl-2-butene	18.61	17.15			
C_5H_{10}	Cyclopentane	24.07	21.88	18.22		
$C_5H_{10}O$	2-Pentanone		23.25	21.62		

Mol. form.	Name	γ in mN/m				
		10°C	25°C	50°C	75°C	100°C
$C_5H_{10}O$	3-Pentanone		24.74	22.13		
$C_5H_{10}O$	Pentanal	26.95	25.44	22.91		
$C_5H_{10}O_2$	Butyl formate	26.05	24.52	21.95	19.39	16.82
$C_5H_{10}O_2$	Propyl acetate	25.48	23.80	21.00	18.20	15.40
$C_5H_{10}O_2$	Isopropyl acetate	23.37	21.76	19.08	16.40	
$C_5H_{10}O_2$	Ethyl propanoate	25.55	23.80	20.88	17.96	
$C_5H_{10}O_2$	Methyl butanoate	26.34	24.62	21.76	18.89	16.03
$C_5H_{11}Cl$	1-Chloropentane	26.01	24.40	21.71	19.02	16.33
$C_5H_{11}N$	Piperidine	30.64	28.91	26.03	23.14	20.26
C_5H_{12}	Pentane	17.15	15.49			
$C_5H_{12}O$	1-Pentanol	26.67	25.36	23.17	20.99	18.80
$C_5H_{12}O$	2-Pentanol	24.96	23.45	20.94	18.43	15.92
$C_5H_{12}O$	3-Methyl-1-butanol	24.94	23.71	21.66	19.61	17.56
$C_5H_{13}N$	Pentylamine		24.69	22.14	19.58	
$C_6H_4Cl_2$	m-Dichlorobenzene	37.15	35.43	32.57	29.70	26.83
C_6H_5Br	Bromobenzene	36.98	35.24	32.34	29.44	26.54
C_6H_5Cl	Chlorobenzene	34.78	32.99	30.02	27.04	24.06
C_6H_5ClO	o-Chlorophenol		39.70	36.89	34.09	31.28
C_6H_5ClO	m-Chlorophenol		41.18	38.66	36.13	33.61
C_6H_5F	Fluorobenzene	28.47	26.66	23.65	20.64	
C_6H_5I	Iodobenzene	40.40	38.71	35.91	33.10	30.29
$C_6H_5NO_2$	Nitrobenzene			40.56	37.66	34.77
C_6H_6	Benzene		28.22	25.00	21.77	
C_6H_6O	Phenol			38.20	35.53	32.86
C_6H_7N	Aniline		42.12	39.41	36.69	
C_6H_7N	2-Methylpyridine		33.00	29.90	26.79	
$C_6H_8N_2$	Adiponitrile		45.45	43.02	40.58	
C_6H_{10}	Cyclohexene	28.01	26.17	23.12		
$C_6H_{10}O$	Cyclohexanone	36.43	34.57	31.46	28.36	25.25
$C_6H_{11}N$	Hexanenitrile		27.37	25.11	22.84	
C_6H_{12}	Cyclohexane	26.43	24.65	21.68		
C_6H_{12}	Methylcyclopentane	23.47	21.72	18.82		
C_6H_{12}	1-Hexene	19.44	17.90	15.33		
$C_6H_{12}O$	Cyclohexanol		32.92	30.50	28.09	25.67
$C_6H_{12}O$	2-Hexanone		25.45	22.72		
$C_6H_{12}O_2$	Butyl acetate	26.48	24.88	22.21	19.54	16.87
$C_6H_{12}O_2$	Isobutyl acetate	24.58	23.06	20.53	17.99	15.46
$C_6H_{12}O_2$	Ethyl butanoate	25.51	23.94	21.33	18.71	16.10
$C_6H_{12}O_3$	Paraldehyde	27.22	25.63	22.97	20.32	17.66
$C_6H_{13}Cl$	1-Chlorohexane	27.28	25.73	23.13	20.54	17.94
$C_6H_{13}N$	Cyclohexylamine		31.22	28.25	25.28	
C_6H_{14}	Hexane	19.42	17.89	15.33		
C_6H_{14}	2-Methylpentane	18.37	16.88	14.39		
C_6H_{14}	3-Methylpentane	19.20	17.61	14.96		
$C_6H_{14}O$	Diisopropyl ether		17.27	14.65		
$C_6H_{14}O$	1-Hexanol		25.81	23.81	21.80	19.80
$C_6H_{14}O_2$	1,1-Diethoxyethane		20.89	18.31	15.74	
$C_6H_{14}O_2$	2-Butoxyethanol	27.36	26.14	24.10	22.06	20.02
$C_6H_{15}N$	Triethylamine		20.22	17.74		
$C_6H_{15}N$	Dipropylamine		22.31	19.75	17.20	
$C_6H_{15}N$	Diisopropylamine		19.14	16.45		
C_7H_5N	Benzonitrile		38.79	35.90	33.00	
C_7H_6O	Benzaldehyde	39.63	38.00	35.27	32.55	29.82
C_7H_8	Toluene	29.71	27.93	24.96	21.98	19.01
C_7H_8O	o-Cresol		36.90	34.38	31.85	29.32
C_7H_8O	m-Cresol		35.69	33.38	31.07	28.76
C_7H_8O	Benzyl alcohol				27.89	24.44

Mol. form.	Name	γ in mN/m				
		10°C	25°C	50°C	75°C	100°C
C_7H_8O	Anisole		35.10	32.09	29.08	
C_7H_9N	N-Methylaniline		36.90	34.47	32.05	
C_7H_9N	2,3-Dimethylpyridine		32.71	30.04	27.36	
C_7H_9N	Benzylamine		39.30	36.27	33.23	
C_7H_{14}	Methylcyclohexane	24.98	23.29	20.46		
C_7H_{14}	1-Heptene	21.29	19.80	17.33	14.85	
$C_7H_{14}O$	2-Heptanone		26.12	23.48		
$C_7H_{14}O_2$	Pentyl acetate	26.67	25.17	22.69	20.20	17.72
$C_7H_{14}O_2$	Heptanoic acid		27.76	25.64		
C_7H_{16}	Heptane	21.12	19.65	17.20	14.75	
C_7H_{16}	3-Methylhexane	20.76	19.31	16.88	14.46	
C_8H_8O	Acetophenone		39.04	36.15	33.27	
$C_8H_8O_2$	Methyl benzoate		37.17	34.25	31.32	
$C_8H_8O_3$	Methyl salicylate	40.98	39.22	36.28	33.35	30.41
C_8H_{10}	Ethylbenzene	30.39	28.75	26.01	23.28	20.54
C_8H_{10}	o-Xylene	31.41	29.76	27.01	24.25	21.50
C_8H_{10}	m-Xylene	30.13	28.47	25.71	22.95	20.19
C_8H_{10}	p-Xylene		28.01	25.32	22.64	19.95
$C_8H_{10}O$	Phenetole		32.41	29.65	26.89	
$C_8H_{11}N$	N,N-Dimethylaniline		35.52	32.90	30.27	
$C_8H_{11}N$	N-Ethylaniline		36.33	33.65	30.98	
C_8H_{16}	Ethylcyclohexane	26.73	25.15	22.51		
C_8H_{18}	Octane	22.57	21.14	18.77	16.39	14.01
C_8H_{18}	2,5-Dimethylhexane	20.77	19.40	17.12	14.84	12.56
$C_8H_{18}O$	1-Octanol	28.30	27.10	25.12		
$C_8H_{19}N$	Dibutylamine		24.12	21.74	19.36	
$C_8H_{19}N$	Diisobutylamine		21.72	19.44	17.16	
C_9H_7N	Quinoline	44.19	42.59	39.94	37.28	34.62
C_9H_{12}	Cumene	29.27	27.69	25.05	22.42	19.78
C_9H_{12}	1,2,4-Trimethylbenzene	30.74	29.20	26.64	24.07	21.51
C_9H_{12}	Mesitylene	28.89	27.55	25.31	23.07	20.82
$C_9H_{18}O$	5-Nonanone		26.28	23.85		
C_9H_{20}	Nonane	23.79	22.38	20.05	17.71	15.37
$C_9H_{20}O$	1-Nonanol	29.03	27.89	26.00	24.10	22.20
$C_{10}H_{12}$	1,2,3,4-Tetrahydronaphthalene		33.17	30.78	28.40	
$C_{10}H_{22}$	Decane	24.75	23.37	21.07	18.77	16.47
$C_{10}H_{22}O$	1-Decanol	29.61	28.51	26.68	24.85	23.02
$C_{11}H_{24}$	Undecane	25.56	24.21	21.96	19.70	17.45
$C_{12}H_{10}O$	Diphenyl ether		26.75	24.80		
$C_{12}H_{27}N$	Tributylamine		24.39	22.32	20.24	
$C_{13}H_{28}$	Tridecane	26.86	25.55	23.37	21.19	19.01
$C_{14}H_{12}O_2$	Benzyl benzoate	44.47	42.82	40.06	37.31	34.55
$C_{14}H_{30}$	Tetradecane	27.43	26.13	23.96	21.78	19.61
$C_{16}H_{34}$	Hexadecane		27.05	24.91	22.78	20.64
$C_{18}H_{38}$	Octadecane		27.87	25.77	23.66	21.55

PERMITTIVITY (DIELECTRIC CONSTANT) OF LIQUIDS
Christian Wohlfarth

The permittivity of a substance (often called the dielectric constant) is the ratio of the electric displacement **D** to the electric field strength **E** when an external field is applied to the substance. The quantity tabulated here is the relative permittivity, which is the ratio of the actual permittivity to the permittivity of a vacuum; it is a dimensionless number. The table gives the static permittivity ε, measured in static fields or at relatively low frequencies where no relaxation effects occur.

This table gives the static permittivity at a specified temperature, usually at 293.15 or 298.15 K. Otherwise, the temperature closest to 293.15 K was chosen or (as is the case for about half the substances included here) a measurement was available at only one temperature, and that value is given in the table. The temperature dependence of the permittivity of many of these substances is given in the table that follows this one.

The static permittivities refer to nominal atmospheric pressure as long as the corresponding temperature is below the normal boiling point. Otherwise, at temperatures above the normal boiling point, the pressure is the saturated vapor pressure of the substance considered.

Substances are listed by molecular formula in the modified Hill order, with substances not containing carbon preceding those that do contain carbon. An asterisk following the substance name indicates that the isomer was not specified in the original reference.

REFERENCES

1. Wohlfarth, Ch., Static Dielectric Constants of Pure Liquids and Binary Liquid Mixtures, *Landolt-Börstein, Numerical Data and Functional Relationships in Science and Technology*, New Series, Editor-in-Chief O. Madelung, Group IV, Macroscopic and Technical Properties of Matter, Volume 6, Springer Verlag, Berlin, Heidelberg, New York, 1991.
2. Marsh, K. N., Ed., *Recommended Reference Materials for the Realization of Physicochemical Properties*, Blackwell Scientific Publications, Oxford, 1987.

Molecular formula	Name	T/K	ε
Ar	Argon	140.00	1.3247
$AlBr_3$	Aluminum tribromide	373.2	3.38
AsH_3	Arsine	200.9	2.40
BBr_3	Boron tribromide	273.2	2.58
B_2H_6	Diborane	180.66	1.8725
B_5H_9	Pentaborane (9)	298.2	21.1
BrF_3	Bromine trifluoride	298.2	106.8
BrF_5	Bromine pentafluoride	297.7	7.91
BrH	Hydrogen bromide	186.8	8.23
BrNO	Nitrosyl bromide	288.4	13.4
Br_2	Bromine	297.9	3.1484
Br_2OS	Thionyl bromide	293.2	9.06
Br_3OV	Vanadium oxytribromide	298.2	3.6
Br_4Ge	Germanium tetrabromide	299.9	2.955
Br_4Sn	Tin tetrabromide	303.45	3.169
$ClFO_3$	Perchloryl fluoride	150.2	2.194
ClF_3	Chlorine trifluoride	293.2	4.394
ClF_5	Chlorine pentafluoride	193.2	4.28
ClH	Hydrogen chloride	158.9	14.3
ClNO	Nitrosyl chloride	285.2	18.2
Cl_2	Chlorine	208.0	2.147
Cl_2F_3P	Phosphorus dichloride trifluoride	228.63	2.8129
Cl_2OS	Thionyl chloride	298.2	8.675
Cl_2OSe	Selenium oxychloride	293.2	46.2
Cl_2O_2S	Sulfuryl chloride	293.2	9.1
Cl_2S	Sulfur chloride (SCl_2)	298.2	2.915
Cl_2S_2	Sulfur chloride (S_2Cl_2)	288.2	4.79
Cl_3F_2P	Phosphorus trichloride difluoride	268.0	2.3752
Cl_3OP	Phosphoryl chloride	293.2	14.1
Cl_3OV	Vanadium oxytrichloride	298.2	3.4
Cl_3P	Phosphorus trichloride	290.2	3.498
Cl_3PS	Phosphorus thiochloride	298.2	4.94
Cl_4FP	Phosphorus tetrachloride fluoride	272.64	2.6499
Cl_4Ge	Germanium tetrachloride	273.2	2.463
Cl_4Si	Silicon tetrachloride	273.2	2.248

Molecular formula	Name	T/K	ε
Cl_4Sn	Tin tetrachloride	273.2	3.014
Cl_4Ti	Titanium tetrachloride	257.4	2.843
Cl_4V	Vanadium tetrachloride	298.2	3.05
Cl_5P	Phosphorus pentachloride	433.2	2.85
Cl_4Pb	Lead tetrachloride	293.2	2.78
Cl_5Sb	Antimony pentachloride	293.0	3.222
DH	Hydrogen-d_1	16.783	1.2690
D_2	Hydrogen-d_2	17.759	1.2896
D_2O	Water-d_2	293.2	79.754
FH	Hydrogen fluoride	273.2	83.6
F_2	Fluorine	53.48	1.4913
F_5I	Iodine pentafluoride	293.2	37.13
F_5IO	Iodine oxypentafluoride	296.2	1.97
F_6S	Sulfur hexafluoride	223.2	1.81
F_6Xe	Xenon hexafluoride	398.2	4.10
F_7I	Iodine heptafluoride	298.2	1.75
$F_{10}S_2$	Disulfur decafluoride	293.2	2.0202
HI	Hydrogen iodide	220.2	3.87
H_2	Hydrogen	13.52	1.2792
H_2O	Water	293.2	80.100
H_2O_2	Hydrogen peroxide	290.2	74.6
H_2S	Hydrogen sulfide	283.2	5.93
H_3N	Ammonia	293.2	16.61
H_4N_2	Hydrazine	298.2	51.7
He	Helium	2.055	1.0555
I_2	Iodine	391.25	11.08
Kr	Krypton	119.80	1.664
Mn_2O_7	Manganese oxide (Mn_2O_7)	293.2	3.28
NO	Nitric oxide		1.997
N_2	Nitrogen	63.15	1.4680
N_2O_3	Nitrogen trioxide	203.2	31.13
N_2O_4	Nitrogen tetroxide	293.2	2.44
Ne	Neon	26.11	1.1907
O_2	Oxygen	54.478	1.5684
O_2S	Sulfur dioxide	298.2	16.3
O_3	Ozone	90.2	4.75
O_3S	Sulfur trioxide	291.2	3.11
P	Phosphorus	307.2	4.096
S	Sulfur	407.2	3.4991
Se	Selenium	510.65	5.44
Xe	Xenon	161.35	1.880
$CBrClF_2$	Bromochlorodifluoromethane	123.2	3.920
$CBrCl_3$	Bromotrichloromethane	293.2	2.405
$CBrF_3$	Bromotrifluoromethane	123.2	3.730
CBr_2Cl_2	Dibromodichloromethane	298.2	2.542
CBr_2F_2	Dibromodifluoromethane	273.2	2.939
CBr_3Cl	Tribromochloromethane	333.2	2.601
CBr_3F	Tribromofluoromethane	293.2	3.00
CBr_3NO_2	Tribromonitromethane	298.2	9.034
$CClF_3$	Chlorotrifluoromethane	123.2	3.010
CCl_2F_2	Dichlorodifluoromethane	123.2	3.500
CCl_2NOP	Phosphorous dichloride isocyanate	293.2	4.16
CCl_2O	Carbonyl chloride	295.2	4.30
CCl_3D	Trichloromethane-d_1	298.2	4.67
CCl_3F	Trichlorofluoromethane	293.2	3.00
CCl_3NO_2	Trichloronitromethane	293.2	7.319
CCl_4	Tetrachloromethane	293.2	2.2379
CF_4	Tetrafluoromethane	126.3	1.685

Molecular formula	Name	T/K	ε
CN_4O_8	Tetranitromethane	293.2	2.317
COS	Carbon oxysulfide	185.0	4.47
COSe	Carbon oxyselenide	283.2	3.47
CO_2	Carbon dioxide	295.0	1.4492
CS_2	Carbon disulfide	293.2	2.6320
$CHBr_3$	Tribromomethane	283.2	4.404
$CHCl_3$	Trichloromethane	293.2	4.8069
CHF_3	Trifluoromethane	294.0	5.2
CHN	Hydrogen cyanide	293.2	114.9
CH_2Br_2	Dibromomethane	283.2	7.77
CH_2Cl_2	Dichloromethane	298.0	8.93
CH_2F_2	Difluoromethane	152.2	53.74
CH_2I_2	Diiodomethane	298.2	5.32
CH_2O_2	Formic acid	298.2	51.1
CH_3Br	Bromomethane	275.7	9.71
CH_3Cl	Chloromethane	295.2	10.0
CH_3ClO_2S	Methanesulfonyl chloride	293.2	34.0
$CH_3Cl_2NO_2S_2$	(Methylsulfonyl)imidosulfurous dichloride	313.2	31.6
CH_3DO	Methan-d_1-ol	297.5	31.68
CH_3F	Fluoromethane	131.0	51.0
CH_3I	Iodomethane	293.2	6.97
CH_3NO	Formamide	293.2	111.0
CH_3NO_2	Methyl nitrite	200.0	20.77
CH_3NO_2	Nitromethane	293.2	37.27
CH_3NO_3	Methyl nitrate	293.2	23.9
CH_4	Methane	91.0	1.6761
CH_4O	Methanol	293.2	33.0
CH_5N	Methylamine	215.2	16.7
$C_2Br_2F_4$	1,2-Dibromotetrafluoroethane	298.2	2.34
$C_2Cl_2F_4$	1,2-Dichlorotetrafluoroethane	273.2	2.4842
$C_2Cl_2O_2$	Oxalyl chloride	294.35	3.470
C_2Cl_3N	Trichloroacetonitrile	292.2	7.85
C_2Cl_4	Tetrachloroethylene	303.2	2.268
$C_2Cl_4F_2$	Tetrachloro-1,2-difluoroethane	308.2	2.52
C_2HBr_3O	Tribromoacetaldehyde	293.2	7.6
C_2HCl_2NO	Dichloromethyl isocyanate	288.2	7.36
C_2HCl_3	Trichloroethylene	301.5	3.390
$C_2HCl_3F_2$	1,2,2-Trichloro-1,1-difluoroethane	303.2	4.01
C_2HCl_3O	Trichloroacetaldehyde	298.2	6.8
$C_2HCl_3O_2$	Trichloroacetic acid	333.2	4.34
C_2HCl_5	Pentachloroethane	298.2	3.716
$C_2HF_3O_2$	Trifluoroacetic acid	293.2	8.42
C_2H_2	Acetylene	195.0	2.4841
C_2H_2BrCl	cis-1-Bromo-2-chloroethene	290.2	7.31
C_2H_2BrCl	trans-1-Bromo-2-chloroethene	290.2	2.50
$C_2H_2Br_2$	cis-1,2-Dibromoethylene	298.2	7.08
$C_2H_2Br_2$	trans-1,2-Dibromoethylene	298.2	2.88
$C_2H_2Br_4$	1,1,2,2-Tetrabromoethane	303.2	6.72
$C_2H_2Cl_2$	1,1-Dichloroethylene	293.2	4.60
$C_2H_2Cl_2$	cis-1,2-Dichloroethylene	298.2	9.20
$C_2H_2Cl_2$	trans-1,2-Dichloroethylene	293.2	2.14
$C_2H_2Cl_2O_2$	Dichloroacetic acid	293.2	8.33
$C_2H_2Cl_4$	1,1,1,2-Tetrachloroethane	207.2	9.22
$C_2H_2Cl_4$	1,1,2,2-Tetrachloroethane	293.2	8.50
$C_2H_2I_2$	cis-1,2-Diiodoethene	345.65	4.46
$C_2H_2I_2$	trans-1,2-Diiodoethene	350.0	3.19
C_2H_3ClO	Acetyl chloride	295.2	15.8
$C_2H_3ClO_2$	Chloroacetic acid	338.2	12.35

Molecular formula	Name	T/K	ε
$C_2H_3Cl_2NO_2$	1,1-Dichloro-1-nitroethane	303.2	16.3
$C_2H_3Cl_3$	1,1,1-Trichloroethane	293.2	7.243
$C_2H_3Cl_3$	1,1,2-Trichloroethane	298.2	7.1937
$C_2H_3F_3O$	2,2,2-Trifluoroethanol	293.2	27.68
$C_2H_3F_3O_2S$	Methyl trifluoromethyl sulfone	293.2	32.0
C_2H_3N	Acetonitrile	293.2	36.64
C_2H_3NO	Methylisocyanate	288.7	21.75
C_2H_4	Ethylene	270.0	1.4833
C_2H_4BrCl	1-Bromo-2-chloroethane	283.2	7.41
$C_2H_4Br_2$	1,2-Dibromoethane	293.2	4.9612
$C_2H_4Cl_2$	1,1-Dichloroethane	298.2	10.10
$C_2H_4Cl_2$	1,2-Dichloroethane	293.2	10.42
$C_2H_4Cl_2O$	Bis(chloromethyl) ether	293.2	3.51
$C_2H_4N_2O_6$	1,2-Ethanediol dinitrate	293.2	28.26
C_2H_4O	Acetaldehyde	291.2	21.0
C_2H_4O	Ethylene oxide	293.2	12.42
C_2H_4OS	Thioacetic acid	298.2	14.30
$C_2H_4O_2$	Methyl formate	288.2	9.20
$C_2H_4O_2$	Acetic acid	293.2	6.20
$C_2H_4O_3S$	Ethylene sulfite	298.2	39.6
C_2H_5Br	Bromoethane	298.2	9.01
C_2H_5Cl	Chloroethane	293.2	9.45
C_2H_5ClO	2-Chloroethanol	293.2	25.80
C_2H_5I	Iodoethane	293.2	7.82
C_2H_5N	Ethyleneimine	298.2	18.3
C_2H_5NO	Acetaldehyde oxime	298.2	4.70
C_2H_5NO	Acetamide	363.7	67.6
C_2H_5NO	N-Methylformamide	293.2	189.0
$C_2H_5NO_2$	Methyl carbamate	328.2	18.48
$C_2H_5NO_2$	Nitroethane	288.2	29.11
$C_2H_5NO_3$	Ethyl nitrate	293.2	19.7
C_2H_6	Ethane	95.0	1.9356
C_2H_6O	Dimethyl ether	258.0	6.18
C_2H_6O	Ethanol	293.2	25.3
C_2H_6OS	Dimethyl sulfoxide	293.2	47.24
$C_2H_6O_2$	Ethylene glycol	293.2	41.4
$C_2H_6O_2S$	Dimethyl sulfone	383.2	47.39
$C_2H_6O_4S$	Dimethyl sulfate	298.2	55.0
C_2H_6S	Dimethyl sulfide	294.2	6.70
C_2H_6S	Ethanethiol	298.2	6.667
$C_2H_6S_2$	1,2-Ethanedithiol	293.2	7.26
$C_2H_6S_2$	Dimethyl disulfide	298.2	9.6
C_2H_7N	Ethylamine	273.2	8.7
C_2H_7NO	Ethanolamine	293.2	31.94
$C_2H_7NO_2S$	N-Methyl methanesulfonamide	298.2	104.4
$C_2H_8N_2$	1,2-Ethanediamine	293.2	13.82
C_3Cl_6O	Hexachloroacetone	291.9	3.925
C_3F_6O	Hexafluoroacetone	202.2	2.104
$C_3N_3O_3P$	Phosphorus triisocyanate	293.2	5.30
$C_3N_3O_3PS$	Phosphorus triisocyanate sulfide	293.2	6.00
$C_3N_3O_4P$	Phosphorus triisocyanate oxide	298.2	4.40
C_3HN	Cyanoacetylene	291.9	72.3
$C_3H_2F_6O$	1,1,1,3,3,3-Hexafluoro-2-propanol	293.2	16.70
$C_3H_2Cl_2O_3$	4,5-Dichloro-1,3-dioxolan-2-one	313.2	31.8
$C_3H_3ClO_3$	4-Chloro-1,3-dioxolan-2-one	313.2	62.0
C_3H_3NO	Vinyl isocyanate	298.2	10.62
C_3H_3N	Propenenitrile	293.2	33.0
$C_3H_3NO_2$	Cyanoacetic acid	277.2	33.4

Molecular formula	Name	T/K	ε
C_3H_4	Allene	269.0	2.025
C_3H_4	Propyne	246.0	3.218
$C_3H_4ClF_3$	3-Chloro-1,1,1-trifluoropropane	295.2	7.32
C_3H_4ClNO	1-Chloro-2-isocyanatoethane	288.2	29.1
$C_3H_4Cl_2O$	1,1-Dichloroacetone	293.2	14.6
$C_3H_4F_4O$	2,2,3,3-Tetrafluoro-1-propanol	298.2	21.03
$C_3H_4N_2O_2$	3-Methylsydnone	313.2	144.0
C_3H_4O	Propargyl alcohol	293.2	20.8
$C_3H_4O_3$	1,3-Dioxolan-2-one	313.2	89.78
C_3H_5Br	3-Bromopropene	293.2	7.0
$C_3H_5BrO_2$	2-Bromopropanoic acid	294.2	11.0
$C_3H_5Br_3$	1,2,3-Tribromopropane	303.2	6.00
C_3H_5Cl	2-Chloropropene	299.25	8.92
C_3H_5Cl	3-Chloropropene	293.2	8.2
$C_3H_5ClN_2O_6$	3-Chloro-1,2-propanediol dinitrate	293.2	17.50
C_3H_5ClO	Epichlorohydrin	293.2	22.6
$C_3H_5ClO_2$	Ethyl chlorocarbonate	308.7	9.736
$C_3H_5ClO_2$	Methyl chloroacetate	293.2	12.0
$C_3H_5Cl_2NO_2S$	Ethoxycarbonylimidosulfurous dichloride	294.2	19.7
$C_3H_5Cl_2NO_3$	1,3-Dichloroisopropyl nitrate	293.2	13.28
$C_3H_5Cl_3$	1,2,3-Trichloropropane	293.2	7.5
C_3H_5I	3-Iodopropene	292.2	6.1
C_3H_5N	Propanenitrile	293.2	29.7
C_3H_5NO	Ethyl isocyanate	293.2	19.7
$C_3H_5NO_2$	Allyl nitrite	298.2	9.12
C_3H_5NS	Ethyl rhodanide	293.2	29.7
C_3H_5NS	Ethyl isothiocyanate	293.2	19.6
$C_3H_5N_3O_9$	Trinitroglycerol	293.2	19.25
C_3H_6	Propene	220.0	2.1365
$C_3H_6Br_2$	1,2-Dibromopropane	283.2	4.60
$C_3H_6Br_2$	1,3-Dibromopropane	293.2	9.482
$C_3H_6ClNO_2$	2-Chloro-2-nitropropane	250.4	31.90
C_3H_6ClNO	N-Methyl-2-chloroacetamide	323.2	92.3
$C_3H_6Cl_2$	1,2-Dichloropropane	293.2	8.37
$C_3H_6Cl_2$	1,3-Dichloropropane	303.2	10.27
$C_3H_6Cl_2$	2,2-Dichloropropane	293.2	11.37
$C_3H_6N_2O_4$	2,2-Dinitropropane	325.1	42.4
$C_3H_6N_2O_6$	1,2-Propanediol dinitrate	293.2	26.80
$C_3H_6N_2O_6N$	1,3-Propanediol dinitrate	293.2	18.97
C_3H_6O	Acetone	293.2	21.01
C_3H_6O	Allyl alcohol	293.2	19.7
C_3H_6O	Propanal	290.2	18.5
$C_3H_6O_2$	Ethyl formate	288.2	8.57
$C_3H_6O_2$	Methyl acetate	288.2	7.07
$C_3H_6O_2$	Propanoic acid	298.2	3.44
$C_3H_6O_3$	Dimethyl carbonate	298.2	3.087
$C_3H_6O_3$	3-Hydroxypropanoic acid	296.2	30.0
$C_3H_6O_3$	1,3,5-Trioxane	338.2	15.55
$C_3H_6O_3S$	1,2-Oxathiolane-2,2-dioxide	308.2	86.3
C_3H_7Br	1-Bromopropane	293.2	8.09
C_3H_7Br	2-Bromopropane	293.2	9.46
C_3H_7Cl	1-Chloropropane	293.2	8.588
C_3H_7ClO	3-Chloro-1-propanol	215.2	36.0
C_3H_7ClO	1-Chloro-2-propanol	153.2	59.0
$C_3H_7ClO_2$	3-Chloro-1,2-propanediol	293.2	31.0
C_3H_7I	1-Iodopropane	293.2	7.07
C_3H_7I	2-Iodopropane	298.2	8.19
C_3H_7NO	N,N-Dimethylformamide	293.2	38.25

Molecular formula	Name	T/K	ε
C_3H_7NO	N-Ethylformamide	298.2	102.7
C_3H_7NO	N-Methylacetamide	303.2	179.0
$C_3H_7NO_2$	Ethyl carbamate	328.2	14.14
$C_3H_7NO_2$	1-Nitropropane	288.2	24.70
$C_3H_7NO_2$	2-Nitropropane	288.2	26.74
$C_3H_7NO_2$	Propyl nitrite	250.0	12.35
$C_3H_7NO_2$	Isopropyl nitrite	260.0	13.92
C_3H_7NS	N,N-Dimethylthioformamide	298.2	47.5
C_3H_8	Propane	293.19	1.6678
C_3H_8O	1-Propanol	293.2	20.8
C_3H_8O	2-Propanol	293.2	20.18
$C_3H_8O_2$	Dimethoxymethane	293.2	2.644
$C_3H_8O_2$	2-Methoxyethanol	298.2	17.2
$C_3H_8O_2$	1,2-Propanediol	303.2	27.5
$C_3H_8O_2$	1,3-Propanediol	293.2	35.1
$C_3H_8O_3$	Glycerol	293.2	46.53
C_3H_8S	1-Propanethiol	288.2	5.937
C_3H_8S	2-Propanethiol	298.2	5.952
$C_3H_8S_2$	1,2-Propanedithiol	293.2	7.24
$C_3H_8S_2$	1,3-Propanedithiol	303.2	8.11
$C_3H_9BO_3$	Trimethylborate	293.2	2.2762
C_3H_9ClSi	Trimethylchlorosilane	273.2	10.21
C_3H_9N	Propylamine	296.2	5.08
C_3H_9N	Isopropylamine	293.2	5.6268
C_3H_9N	Trimethylamine	298.2	2.440
$C_3H_9NO_2S$	N,N-Dimethyl methanesulfonamide	323.2	80.4
$C_3H_9O_3P$	Dimethyl methanephosphate	293.2	22.3
$C_3H_9O_4P$	Trimethyl phosphate	293.2	20.6
C_3H_9PS	Trimethylphosphine sulfide	293.2	71.6
C_4Cl_6	Hexachloro-1,3-butadiene	293.2	2.55
$C_4Cl_6O_3$	Trichloroacetic anhydride	298.2	5.0
$C_4F_6Cl_4$	1,1,3,4-Tetrachlorohexafluorobutane	293.2	2.86
$C_4F_6O_3$	Trifluoroacetic anhydride	298.2	2.7
$C_4H_2Cl_4O_3$	Dichloroacetic anhydride	298.2	15.8
$C_4H_2O_3$	Maleic anhydride	326.2	52.75
$C_4H_3F_7O$	2,2,3,3,4,4,4-Heptafluoro-1-butanol	298.2	14.4
$C_4H_4N_2$	Succinonitrile	298.2	62.6
$C_4H_4N_2$	Pyrazine	323.2	2.80
C_4H_4O	Furan	277.1	2.88
C_4H_4S	Thiophene	293.2	2.739
C_4H_5Cl	2-Chloro-1,3-butadiene	293.2	4.914
$C_4H_5ClO_3$	4-Chloromethyl-1,3-dioxolan-2-one	313.2	97.5
$C_4H_5Cl_3O$	4,4,4-Trichlorobutanal	291.2	10.0
$C_4H_5Cl_3O_2$	Ethyl trichloroacetate	293.2	8.428
C_4H_5N	1H-Pyrrole	293.0	8.00
C_4H_5NO	Allyl isocynate	288.2	15.15
$C_4H_5N_2O_3P$	Diisocyanato ethyl phosphite	298.2	6.93
$C_4H_5N_2O_3PS$	Diisocyanato O-ethyl thionphosphate	293.2	13.4
$C_4H_5N_2O_4P$	Diisocyanato ethyl phosphate	293.2	14.1
C_4H_6	1,3-Butadiene	265.0	2.050
$C_4H_6Cl_2N_2S$	(1-Cyano-1-isopropyl)imido sulfurous dichloride	293.2	21.8
$C_4H_6N_2OS$	2-Methyl-2-(sulfinylamino)-propanenitrile	293.2	16.14
C_4H_6O	Cyclobutanone	298.2	14.27
C_4H_6O	Divinyl ether	288.2	3.94
C_4H_6O	Ethoxyacetylene	298.2	8.05
$C_4H_6O_2$	Methyl acrylate	303.2	7.03
$C_4H_6O_2$	2,3-Butanedione	298.2	4.04
$C_4H_6O_2$	γ-Butyrolactone	293.2	39.0

Molecular formula	Name	T/K	ε
$C_4H_6O_3$	Acetic anhydride	293.2	22.45
$C_4H_6O_3$	4-Methyl-1,3-dioxolan-2-one	293.0	66.14
C_4H_7Br	*trans*-2-Bromo-2-butene	293.2	6.76
C_4H_7Br	*cis*-2-Bromo-2-butene	293.2	5.38
$C_4H_7BrO_2$	Ethyl 2-bromoacetate	303.2	9.75
$C_4H_7BrO_2$	Methyl 2-bromopropanoate	303.2	10.35
$C_4H_7BrO_2$	Methyl 3-bromopropanoate	303.2	5.81
$C_4H_7BrO_2$	2-Bromobutanoic acid	293.2	7.2
$C_4H_7ClO_2$	Propyl chlorocarbonate	293.2	11.2
$C_4H_7ClO_2$	Methyl 2-chloropropanoate	303.2	11.45
C_4H_7N	Butanenitrile	293.2	24.83
C_4H_7N	2-Methylpropanenitrile	293.2	24.42
C_4H_7NO	Pyrrolidine-2-one	298.2	28.18
$C_4H_7NO_2$	3-Methyl oxazolidine-2-one	298.2	77.5
$C_4H_7NO_2$	5-Methyl oxazolidine-2-one	298.2	84.0
C_4H_8	1-Butene	220.0	2.2195
C_4H_8	*cis*-2-Butene	296.0	1.960
C_4H_8	Isobutene	288.7	2.1225
C_4H_8BrCl	3-Bromo-1-chloro-2-methylpropane	303.2	8.90
$C_4H_8Br_2$	1,2-Dibromobutane	293.2	4.74
$C_4H_8Br_2$	1,3-Dibromobutane	293.2	9.14
$C_4H_8Br_2$	1,4-Dibromobutane	303.2	8.68
$C_4H_8Br_2$	2,3-Dibromobutane	293.2	6.36
$C_4H_8Br_2$	*meso*-2,3-Dibromobutane	298.2	6.245
$C_4H_8Br_2$	(±)-2,3-Dibromobutane	298.2	5.758
$C_4H_8Br_2$	1,2-Dibromo-2-methylpropane	293.2	4.1
C_4H_8ClNO	2-Chloro-*N,N*-dimethylacetamide	298.2	49.4
$C_4H_8Cl_2$	1,2-Dichlorobutane	293.2	7.74
$C_4H_8Cl_2$	1,4-Dichlorobutane	308.2	9.30
$C_4H_8Cl_2$	1,2-Dichloro-2-methylpropane	296.0	7.15
$C_4H_8Cl_2O$	Bis(2-chloroethyl) ether	293.2	21.20
$C_4H_8N_2O_6$	1,3-Butanediol dinitrate	293.2	18.85
$C_4H_8N_2O_6$	2,3-Butanediol dinitrate	293.2	28.85
C_4H_8O	Butanal	298.2	13.45
C_4H_8O	2-Butanone	293.2	18.56
C_4H_8O	Tetrahydrofuran	295.2	7.52
C_4H_8OS	Tetrahydrothiophene-*S*-oxide	298.2	42.96
$C_4H_8O_2$	Propyl formate	303.2	6.92
$C_4H_8O_2$	Butanoic acid	287.2	2.98
$C_4H_8O_2$	1,4-Dioxane	293.2	2.2189
$C_4H_8O_2$	Ethyl acetate	293.2	6.0814
$C_4H_8O_2$	2-Methylpropanoic acid	293.2	2.58
$C_4H_8O_2$	Methyl propanoate	293.2	6.200
$C_4H_8O_2S$	Tetrahydrothiophene-*S,S*-dioxide	303.2	43.36
$C_4H_8O_3$	2-Hydroxybutanoic acid	296.2	37.7
$C_4H_8O_3$	3-Hydroxybutanoic acid	296.2	31.5
$C_4H_8O_3$	1,2-Ethanediol monoacetate	303.2	12.95
$C_4H_8O_3$	Ethyl methyl carbonate	293.2	2.985
C_4H_9Br	1-Bromobutane	283.2	7.315
C_4H_9Br	1-Bromo-2-methylpropane	273.2	7.70
C_4H_9Br	2-Bromobutane	298.2	8.64
C_4H_9Br	2-Bromo-2-methylpropane	293.0	10.98
C_4H_9Cl	1-Chlorobutane	293.2	7.276
C_4H_9Cl	1-Chloro-2-methylpropane	293.2	7.027
C_4H_9Cl	2-Chlorobutane	293.2	8.564
C_4H_9Cl	2-Chloro-2-methylpropane	293.2	9.663
$C_4H_9Cl_2NS$	(2-Methyl-2-propyl)imido-sulfurous dichloride	298.2	17.0
C_4H_9I	1-Iodobutane	293.2	6.27

Molecular formula	Name	T/K	ε
C_4H_9I	2-Iodobutane	293.2	7.873
C_4H_9I	2-Iodo-2-methylpropane	283.2	6.65
C_4H_9N	Pyrrolidine	293.0	8.30
C_4H_9NO	N,N-Dimethylacetamide	294.2	38.85
C_4H_9NO	N-Ethylacetamide	293.2	135.0
C_4H_9NO	N-Methylpropanamide	293.2	170.0
C_4H_9NO	N-Isopropylformamide	298.2	65.7
C_4H_9NO	2-Butanone oxime	293.2	3.4
C_4H_9NO	Morpholine	298.2	7.42
C_4H_9NOS	2-Methyl-2-sulfinylaminopropane	298.2	7.22
$C_4H_9NO_2$	tert-Butyl nitrite	298.2	11.47
$C_4H_9NO_2$	Ethyl-N-methyl carbamate	298.2	21.10
$C_4H_9NO_2$	N-(2-Hydroxyethyl)acetamide	298.2	96.6
$C_4H_9NO_2$	Propyl carbamate	338.2	12.06
$C_4H_9NO_3$	Butyl nitrate	293.2	13.10
C_4H_{10}	Butane	295.0	1.7697
C_4H_{10}	Isobutane	295.0	1.7518
$C_4H_{10}Cl_2NO_3P$	Diethyl N,N-dichloroamidophosphate	295.2	12.85
$C_4H_{10}Cl_2NO_3PS$	(Diethoxyphosphinyl)imido-sulfurous dichloride	293.2	19.15
$C_4H_{10}O$	1-Butanol	293.2	17.84
$C_4H_{10}O$	2-Butanol	293.2	17.26
$C_4H_{10}O$	2-Methyl-1-propanol	293.2	17.93
$C_4H_{10}O$	2-Methyl-2-propanol	298.2	12.47
$C_4H_{10}O$	Diethyl ether	293.2	4.2666
$C_4H_{10}O_2$	1,2-Butanediol	298.2	22.4
$C_4H_{10}O_2$	1,3-Butanediol	298.2	28.8
$C_4H_{10}O_2$	1,4-Butanediol	298.2	31.9
$C_4H_{10}O_2$	1,2-Dimethoxyethane	296.7	7.30
$C_4H_{10}O_2$	2-Ethoxyethanol	298.2	13.38
$C_4H_{10}O_2S$	Bis(2-hydroxyethyl) sulfide	293.2	28.61
$C_4H_{10}O_3$	Diethylene glycol	293.2	31.82
$C_4H_{10}O_3S$	Diethyl sulfite	293.2	15.6
$C_4H_{10}O_4$	1,2,3,4-Butanetetrol	393.2	28.2
$C_4H_{10}O_4S$	Diethyl sulfate	293.2	29.2
$C_4H_{10}S$	1-Butanethiol	288.2	5.204
$C_4H_{10}S$	2-Butanethiol	288.2	5.645
$C_4H_{10}S$	2-Methyl-1-propanethiol	298.2	4.961
$C_4H_{10}S$	2-Methyl-2-propanethiol	293.2	5.475
$C_4H_{10}S$	Diethyl sulfide	298.2	5.723
$C_4H_{10}Zn$	Diethyl zinc	293.2	2.55
$C_4H_{11}N$	Butylamine	293.2	4.71
$C_4H_{11}N$	Diethylamine	293.2	3.680
$C_4H_{11}NO_2$	Diethanolamine	293.2	25.75
$C_4H_{11}OP$	Ethyldimethylphosphine oxide	363.2	34.4
$C_4H_{11}O_3P$	Ethanephosphonic acid dimethyl ester	313.2	15.89
$C_4H_{12}Ge$	Tetramethyl germanium	297.2	1.817
$C_4H_{12}NO_3P$	Phosphoramidic acid diethyl ester	313.2	25.3
$C_4H_{12}N_2O$	2-(2-Aminoethylamino)ethanol	293.2	21.81
$C_4H_{12}OSi$	Methoxytrimethylsilane	298.2	3.248
$C_4H_{12}O_2Si$	Dimethoxydimethylsilane	298.2	3.663
$C_4H_{12}O_3Si$	Trimethoxymethylsilane	298.2	4.9
$C_4H_{12}O_4Si$	Tetramethyl silicate	293.2	6.0
$C_4H_{12}Si$	Butylsilane	293.2	2.537
$C_4H_{12}Si$	Isobutylsilane	293.2	2.497
$C_4H_{12}Si$	Diethylsilane	293.2	2.544
$C_4H_{12}Si$	Tetramethylsilane	293.2	1.921
$C_4H_{13}N_3$	Bis(2-amimoethyl)amine	293.2	12.62
C_5FeO_5	Iron pentacarbonyl	293.2	2.602

Molecular formula	Name	T/K	ε
$C_5H_4F_8O$	2,2,3,3,4,4,5,5-Octafluoro-1-pentanol	298.2	15.30
C_5H_4ClN	2-Chloropyridine	298.2	27.32
C_5H_4BrN	2-Bromopyridine	298.2	23.18
$C_5H_4O_2$	Furfural	293.2	42.1
C_5H_5N	Pyridine	293.2	13.260
C_5H_5NO	Pyridine-1-oxide	343.0	35.94
C_5H_6O	2-Methylfuran	293.2	2.76
$C_5H_6O_2$	Furfuryl alcohol	298.2	16.85
$C_5H_7Cl_3O_2$	Propyl trichloroacetate	298.2	8.32
$C_5H_7NO_2$	Ethyl 2-cyanoacetate	263.2	31.62
$C_5H_7N_2O_3P$	Diisocyanato isopropyl phosphite	293.0	7.25
$C_5H_7N_2O_3PS$	Diisocyanato O-isopropyl thionphosphate	288.0	12.8
$C_5H_7N_2O_4P$	Diisocyanato isopropyl phosphate	298.0	13.2
C_5H_8	Cyclopentene	295.0	2.083
C_5H_8	2-Methyl-1,3-butadiene	293.2	2.098
C_5H_8	1,3-Pentadiene*	298.2	2.319
C_5H_8	1,4-Pentadiene	294.0	2.054
$C_5H_8N_2OS$	2-Methyl-2-(sulfinylamino)butanenitrile	293.2	14.70
$C_5H_8N_2O_2$	3-Isopropylsydnone	333.2	66.0
C_5H_8O	Cyclopentanone	298.2	13.58
C_5H_8O	2,3-Dihydropyran	308.2	5.1364
C_5H_8O	2-Methyl-1-butene-3-one	303.2	10.39
$C_5H_8O_2$	2,4-Pentanedione	303.2	26.524
$C_5H_8O_2$	Ethyl acrylate	303.2	6.05
$C_5I_8O_2$	Methyl crotonate	293.2	6.6645
$C_5H_8O_2$	Methyl methacrylate	303.2	6.32
$C_5H_8O_4$	Dimethyl malonate	293.2	9.82
$C_5H_9BrO_2$	2-Bromo-3-methyl butanoic acid	293.2	6.5
$C_5H_9BrO_2$	Ethyl 2-bromopropanoate	293.2	9.4
$C_5H_9ClO_2$	Isobutyl chlorocarbonate	293.2	9.1
$C_5H_9ClO_2$	Methyl 4-chlorobutanoate	303.2	9.51
$C_5H_9ClO_2$	Ethyl 2-chloropropanoate	303.2	11.95
$C_5H_9ClO_2$	Ethyl 3-chloropropanoate	303.2	10.19
$C_5H_9IO_2$	Ethyl 2-iodopropanoate	293.2	8.6
C_5H_9N	2,2-Dimethylpropane nitrile	293.2	21.1
C_5H_9N	Pentanenitrile	293.2	20.04
C_5H_9NO	Butyl isocyanate	293.2	12.290
C_5H_9NO	Isobutyl isocyanate	293.2	11.638
C_5H_9NO	1-Methyl-pyrrolidine-2-one	293.2	32.55
$C_5H_9NO_2$	4,4-Dimethyl-oxazolidine-2-one	333.2	39.2
$C_5H_9NO_2$	4-Ethyl-oxazolidine-2-one	298.2	42.6
$C_5H_9NO_2$	3-Ethyl-oxazolidine-2-one	298.2	66.8
C_5H_{10}	Ethylcyclopropane	293.2	1.933
C_5H_{10}	Cyclopentane	293.2	1.9687
C_5H_{10}	2-Methyl-1-butene	293.2	2.180
C_5H_{10}	2-Methyl-2-butene	296.0	1.979
C_5H_{10}	1-Pentene	293.2	2.011
$C_5H_{10}Br_2$	1,2-Dibromopentane	298.2	4.39
$C_5H_{10}Br_2$	(±)-erythro-2,3-Dibromopentane	298.2	5.430
$C_5H_{10}Br_2$	(±)-threo-2,3-Bibromopentane	298.2	6.507
$C_5H_{10}Br_2$	1,4-Dibromopentane	293.2	9.05
$C_5H_{10}Br_2$	1,5-Dibromopentane	303.2	9.14
$C_5H_{10}Cl_2$	1,2-Dichloropentane	293.2	6.89
$C_5H_{10}Cl_2$	1,5-Dichloropentane	298.2	9.92
$C_5H_{10}N_2O$	1,3-Dimethyl-imidazolidin-2-one	298.2	37.60
$C_5H_{10}O$	Cyclopentanol	288.2	18.5
$C_5H_{10}O$	3-Pentanone	293.2	17.00
$C_5H_{10}O$	2-Pentanone	293.2	15.45

Molecular formula	Name	T/K	ε
$C_5H_{10}O$	3-Methyl-2-butanone	293.2	10.37
$C_5H_{10}O$	2-Methyltetrahydrofuran	298.2	6.97
$C_5H_{10}O$	Tetrahydropyran	293.2	5.66
$C_5H_{10}O$	2,2-Dimethylpropanal	293.2	9.051
$C_5H_{10}O$	Pentanal	293.2	10.00
$C_5H_{10}O_2$	Butyl formate	303.2	6.10
$C_5H_{10}O_2$	Isobutyl formate	293.2	6.41
$C_5H_{10}O_2$	Methyl butanoate	301.2	5.48
$C_5H_{10}O_2$	Propyl acetate	293.2	5.62
$C_5H_{10}O_2$	Ethyl propanoate	293.2	5.76
$C_5H_{10}O_2$	Tetrahydrofurfuryl alcohol	303.2	13.48
$C_5H_{10}O_2$	Pentanoic acid	294.4	2.661
$C_5H_{10}O_2S$	3-Methyl sulfolane	298.2	29.4
$C_5H_{10}O_2S$	Methyl 3-(methylthio)propanoate	303.2	8.66
$C_5H_{10}O_3$	Diethyl carbonate	297.2	2.820
$C_5H_{10}O_3$	Ethyl lactate	303.2	15.4
$C_5H_{10}O_4$	1,2,3-Propanetriol 1-acetate	242.2	38.57
$C_5H_{11}Br$	2-Bromo-2-methylbutane	298.2	9.21
$C_5H_{11}Br$	1-Bromopentane	299.2	6.31
$C_5H_{11}Br$	1-Bromo-3-methylbutane	291.5	6.33
$C_5H_{11}Br$	3-Bromopentane	298.2	8.37
$C_5H_{11}Cl$	1-Chloropentane	293.2	6.654
$C_5H_{11}Cl$	1-Chloro-3-methylbutane	292.0	6.10
$C_5H_{11}Cl$	2-Chloro-2-methylbutane	222.75	12.31
$C_5H_{11}F$	1-Fluoropentane	293.2	3.931
$C_5H_{11}F$	2-Fluoro-2-methylbutane	293.2	5.89
$C_5H_{11}I$	1-Iodopentane	293.2	5.78
$C_5H_{11}I$	2-Iodo-2-methylbutane	293.2	8.192
$C_5H_{11}I$	3-Iodopentane	293.2	7.432
$C_5H_{11}I$	1-Iodo-3-methylbutane	292.2	5.6
$C_5H_{11}N$	Piperidine	293.0	4.33
$C_5H_{11}N$	N-Methylpyrrolidine	298.2	32.2
$C_5H_{11}NO$	N,N-Diethylformamide	293.2	29.6
$C_5H_{11}NO$	N,N-Dimethylpropanamide	293.2	34.6
$C_5H_{11}NO$	2,2-Dimethylpropionamide	298.2	20.13
$C_5H_{11}NO$	N-Ethylpropanamide	298.2	126.8
$C_5H_{11}NO$	N-Propylacetamide	298.2	117.8
$C_5H_{11}NO$	2-Pentanone oxime	293.2	3.3
$C_5H_{11}NO_2$	N-(2-Methoxyethyl)acetamide	298.2	80.7
$C_5H_{11}NO_2$	Pentyl nitrite	298.2	7.21
$C_5H_{11}NO_2$	tert-Pentyl nitrite	298.2	10.88
$C_5H_{11}NO_3$	Pentyl nitrate	291.2	9.0
C_5H_{12}	Isopentane	293.2	1.845
C_5H_{12}	Neopentane	296.0	1.769
C_5H_{12}	Pentane	293.2	1.8371
$C_5H_{12}N_2O$	Tetramethyl urea	293.2	23.10
$C_5H_{12}O$	2,2-Dimethyl-1-propanol	333.2	8.35
$C_5H_{12}O$	2-Methyl-1-butanol	298.2	15.63
$C_5H_{12}O$	2-Methyl-2-butanol	298.2	5.78
$C_5H_{12}O$	3-Methyl-1-butanol	293.2	15.63
$C_5H_{12}O$	3-Methyl-2-butanol	298.2	12.1
$C_5H_{12}O$	1-Pentanol	298.2	15.13
$C_5H_{12}O$	2-Pentanol	298.2	13.71
$C_5H_{12}O$	3-Pentanol	298.2	13.35
$C_5H_{12}O_2$	1,2-Pentanediol	296.8	17.31
$C_5H_{12}O_2$	1,4-Pentanediol	295.7	26.74
$C_5H_{12}O_2$	1,5-Pentanediol	293.2	26.2
$C_5H_{12}O_2$	2,3-Pentanediol	296.9	17.37

Molecular formula	Name	T/K	ε
$C_5H_{12}O_2$	2,4-Pentanediol	294.2	24.69
$C_5H_{12}O_2$	Diethoxymethane	293.2	2.527
$C_5H_{12}O_4$	Tetramethoxymethane	293.2	2.40
$C_5H_{12}O_5$	Xylitol	293.2	40.0
$C_5H_{12}S$	2-Methyl-2-butanethiol	293.2	5.087
$C_5H_{12}S$	1-Pentanethiol	293.2	4.847
$C_5H_{12}S_4$	Tetrakis(methylthio)methane	343.2	2.818
$C_5H_{13}ClNO_3P$	Diethyl N-chloro-N-methylamidophosphate	293.2	8.47
$C_5H_{13}N$	Pentylamine	293.2	4.27
$C_5H_{13}N_3$	1,1,3,3-Tetramethylguanidine	298.2	11.5
$C_5H_{13}O_3P$	Diethyl methanephosphate	313.2	13.405
$C_5H_{14}OSi$	Ethoxytrimethylsilane	298.2	3.013
$C_5H_{15}NSi$	Ethyltrimethylsilazane	303.2	2.2751
$C_5H_{18}B_{10}$	Isopropyl carborane	293.2	45.0
$C_5H_{20}O_5Si_5$	Pentamethyl cyclopentasiloxane	293.2	2.740
C_6F_6	Hexafluorobenzene	298.2	2.029
$C_6F_9Cl_5$	Pentachloro-1,1,2,3,3,4,5,5,6-nonafluorohexane	293.2	2.750
C_6F_{14}	Perfluorohexane	298.2	1.76
$C_6F_{15}OP$	Tris(perfluoroethyl)phosphine oxide	313.2	2.34
$C_6H_3Cl_2NO_2$	2,4-Dichloro-1-nitrobenzene	301.2	13.06
$C_6H_3N_3O_7$	2,4,6-Trinitrophenol	294.2	4.0
$C_6H_4BrClO_2S$	4-Bromobenzene sulfonylchloride	353.2	10.90
$C_6H_4BrCl_2NO_2S_2$	[(4-Bromophenyl)sulfonyl]-imidosulfurous dichloride	351.7	11.08
C_6H_4BrF	1-Bromo-2-fluorobenzene	298.2	4.72
C_6H_4BrF	1-Bromo-3-fluorobenzene	298.2	4.85
C_6H_4BrF	1-Bromo-4-fluorobenzene	298.2	2.60
C_6H_4BrNOS	4-Bromo-N-sulfinylaniline	354.2	3.46
$C_6H_4BrNO_2$	3-Bromonitrobenzene	328.2	20.2
$C_6H_4Br_2$	o-Dibromobenzene	293.2	7.86
$C_6H_4Br_2$	m-Dibromobenzene	293.2	4.81
$C_6H_4Br_2$	p-Dibromobenzene	368.2	2.57
C_6H_4ClF	2-Chlorofluorobenzene	298.2	6.10
C_6H_4ClF	3-Chlorofluorobenzene	298.2	4.96
C_6H_4ClF	4-Chlorofluorobenzene	298.2	3.34
$C_6H_4ClFO_2S$	4-Fluorobenzene sulfonylchloride	313.2	12.65
C_6H_4ClNOS	3-Chloro-N-sulfinylaniline	293.2	7.00
C_6H_4ClNOS	4-Chloro-N-sulfinylaniline	333.2	4.17
$C_6H_4ClNO_2$	o-Chloronitrobenzene	323.2	37.7
$C_6H_4ClNO_2$	m-Chloronitrobenzene	323.2	20.9
$C_6H_4ClNO_2$	p-Chloronitrobenzene	393.2	8.09
$C_6H_4ClNO_4S$	4-Nitrobenzene sulfonylchloride	355.0	3.74
$C_6H_4Cl_2$	o-Dichlorobenzene	293.2	10.12
$C_6H_4Cl_2$	m-Dichlorobenzene	293.2	5.02
$C_6H_4Cl_2$	p-Dichlorobenzene	328.2	2.3943
$C_6H_4Cl_2N_2O_2S$	(3-Nitrophenyl)imidosulfurous dichloride	294.2	21.2
$C_6H_4Cl_2N_2O_4S_2$	[(3-Nitrophenyl)sulfonyl]-imidosulfurous dichloride	348.2	25.55
$C_6H_4Cl_2N_2O_4S_2$	[(4-Nitrophenyl)sulfonyl]-imodosulfurous dichloride	356.7	6.85
$C_6H_4Cl_2O_2S$	4-Chlorobenzene sulfonylchloride	333.2	11.80
$C_6H_4Cl_3NO_2S_2$	[(4-Chlorophenyl)sulfonyl]-imidosulfurous dichloride	341.2	12.50
C_6H_4FI	2-Fluoroiodobenzene	298.2	8.22
C_6H_4FI	m-Fluoroiodobenzene	298.2	4.62
C_6H_4FI	4-Fluoroiodobenzene	298.2	3.12
$C_6H_4F_2$	o-Difluorobenzene	301.2	13.38
$C_6H_4F_2$	m-Difluorobenzene	301.2	5.01
$C_6H_4I_2$	1,2-Diiodobenzene	323.2	5.41
$C_6H_4I_2$	1,3-Diiodobenzene	323.2	4.11
$C_6H_4I_2$	1,4-Diiodobenzene	393.2	2.88
$C_6H_4N_2$	2-Cyanopyridine	303.2	93.77

Molecular formula	Name	T/K	ε
$C_6H_4N_2$	3-Cyanopyridine	323.2	20.54
$C_6H_4N_2$	4-Cyanopyridine	353.2	5.23
$C_6H_4N_2O_2S_2$	N,N'-Disulfinyl-1,2-benzene diamine	339.7	6.99
$C_6H_4N_2O_2S_2$	N,N'-Disulfinyl-1,4-benzene diamine	298.2	7.8
$C_6H_4N_2O_3S$	3-Nitro-N-sulfinylaniline	343.0	19.49
$C_6H_4N_2O_3S$	4-Nitro-N-sulfinylaniline	348.2	14.35
$C_6H_4N_2O_4$	m-Dinitrobenzene	365.2	22.9
C_6H_5Br	Bromobenzene	293.2	5.45
C_6H_5Cl	Chlorobenzene	293.2	5.6895
C_6H_5ClO	o-Chlorophenol	296.2	7.40
C_6H_5ClO	m-Chlorophenol	293.2	6.255
C_6H_5ClO	p-Chlorophenol	314.2	11.18
$C_6H_5ClO_2S$	Benzenesulfonyl chloride	323.2	28.90
C_6H_5ClS	4-Chlorothiophenol	338.2	3.59
$C_6H_5Cl_2OP$	Benzenephosphonic acid dichloride	293.2	26.0
$C_6H_5Cl_2NO_2S_2$	Phenylsulfonylimidosulfurous dichloride	298.2	29.7
C_6H_5F	Fluorobenzene	293.2	5.465
$C_6H_5F_2OP$	Benzenephosphonic acid difluoride	293.2	27.9
C_6H_5I	Iodobenzene	293.2	4.59
C_6H_5NOS	N-Sulfinylaniline	298.2	6.97
$C_6H_5NO_2$	Nitrobenzene	293.0	35.6
$C_6H_5NO_3$	o-Nitrophenol	323.2	16.50
$C_6H_5NO_3$	m-Nitrophenol	373.2	35.45
$C_6H_5NO_3$	p-Nitrophenol	393.2	42.20
C_6H_6	Benzene	293.2	2.2825
C_6H_6BrN	3-Bromoaniline	293.2	13.0
C_6H_6ClN	o-Chloroaniline	293.2	13.40
C_6H_6ClN	m-Chloroaniline	293.2	13.3
$C_6H_6Cl_6$	α-Hexachlorocyclohexane*	429.2	4.77
$C_6H_6N_2O_2$	o-Nitroaniline	353.0	47.3
$C_6H_6N_2O_2$	m-Nitroaniline	398.0	35.6
$C_6H_6N_2O_2$	p-Nitroaniline	428.0	78.5
C_6H_6O	Phenol	303.2	12.40
$C_6H_6O_2$	1,2-Benzenediol	388.2	17.57
$C_6H_6O_2$	1,3-Benzenediol	393.2	13.55
$C_6H_6O_2S$	Methyl thiophene-2-carboxylate	293.2	8.81
$C_6H_6O_3$	Methyl furan-2-carboxylate	293.2	11.01
C_6H_6S	Benzenethiol	303.2	4.26
C_6H_7N	Aniline	293.2	7.06
C_6H_7N	2-Methylpyridine	293.2	10.18
C_6H_7N	3-Methylpyridine	303.0	11.10
C_6H_7N	4-Methylpyridine	293.0	12.2
C_6H_7NO	2-Methylpyridine 1-oxide	323.2	36.4
C_6H_7NO	3-Methylpyridine 1-oxide	318.2	28.26
C_6H_8	1,3-Cyclohexadiene	184.2	2.68
C_6H_8	1,4-Cyclohexadiene	296.0	2.211
C_6H_8	1,3,5-Hexatriene($cis/trans = 0.58$)	297.0	2.276
$C_6H_8N_2$	Phenylhydrazine	293.2	7.15
$C_6H_8N_2$	2,5-Dimethylpyrazine	293.2	2.436
$C_6H_8N_2$	2,6-Dimethylpyrazine	308.2	2.653
$C_6H_8O_2$	1,4-Cyclohexanedione	351.2	4.40
$C_6H_9ClO_2$	cis-Ethyl 3-chlorocrotonate	349.2	7.67
$C_6H_9ClO_2$	$trans$-Ethyl 3-chlorocrotonate	327.2	4.70
$C_6H_9Cl_3O_2$	Butyl trichloroacetate	293.2	7.480
$C_6H_9Cl_3O_2$	Isobutyl trichloroacetate	293.2	7.667
C_6H_9N	Cyclopentanecarbonitrile	293.2	22.68
C_6H_{10}	Cyclohexene	293.2	2.2176
C_6H_{10}	2,3-Dimethyl-1,3-butadiene	293.2	2.102

Molecular formula	Name	T/K	ε
C_6H_{10}	2-Methyl-1,3-pentadiene	298.2	2.422
C_6H_{10}	3-Methyl-1,3-pentadiene	298.2	2.426
C_6H_{10}	4-Methyl-1,3-pentadiene	293.2	2.599
C_6H_{10}	1,5-Hexadiene	294.0	2.125
C_6H_{10}	2,4-Hexadiene	298.2	2.207
C_6H_{10}	cis, cis-2,4-Hexadiene	297.0	2.163
C_6H_{10}	trans, trans-2,4-Hexadiene	297.0	2.123
C_6H_{10}	1-Hexyne	296.0	2.621
$C_6H_{10}N_2O_2$	3-Butylsydnone	298.2	52.8
$C_6H_{10}N_2O_2$	3-Propyl-4-methylsydnone	298.2	66.4
$C_6H_{10}O$	Cyclohexanone	293.0	16.1
$C_6H_{10}O$	Mesityl oxide	273.2	15.6
$C_6H_{10}O$	Butoxyacetylene	298.2	6.62
$C_6H_{10}O_2$	Ethyl methacrylate	303.2	5.68
$C_6H_{10}O_2$	Ethyl crotonate	293.2	5.4
$C_6H_{10}O_2S$	Methyl tetrahydrothiophene-2-carboxylate	293.2	7.30
$C_6H_{10}O_3$	Ethyl acetoacetate	293.2	14.0
$C_6H_{10}O_3$	Propanoic anhydride	293.2	18.30
$C_6H_{10}O_4$	Ethylene glycol diacetate	290.2	7.7
$C_6H_{10}O_4$	Diethyl oxalate	293.2	8.266
$C_6H_{10}O_4$	Monomethyl glutarate	293.2	8.37
$C_6H_{10}O_4$	Dimethyl succinate	293.2	7.19
$C_6H_{11}Br$	Bromocyclohexane	303.2	8.0026
$C_6H_{11}BrO_2$	Ethyl 2-bromobutanoate	303.2	8.57
$C_6H_{11}BrO_2$	Ethyl 2-bromo-2-methylpropanoate	303.2	8.55
$C_6H_{11}BrO_2$	(±)-erythro-2-Acetoxy-3-bromobutane	298.2	7.268
$C_6H_{11}BrO_2$	(±)-threo-2-Acetoxy-3-bromobutane	298.2	7.414
$C_6H_{11}Cl$	Chlorocyclohexane	303.2	7.9505
$C_6H_{11}ClO_2$	3-Methylbutyl chlorocarbonate	293.2	7.8
$C_6H_{11}N$	Hexanenitrile	298.2	17.26
$C_6H_{11}N$	4-Methylpentanenitrile	295.2	17.5
$C_6H_{11}NO$	Cyclohexanone oxime	362.2	3.04
$C_6H_{11}NO_2$	Cyclohexyl nitrite	298.2	9.33
$C_6H_{11}NO_2$	3-Isopropyloxazolidin-2-one	298.2	51.0
$C_6H_{14}O_2$	Hexylene glycol	303.2	23.4
C_6H_{12}	Cyclohexane	293.2	2.0243
C_6H_{12}	Methylcyclopentane	293.2	1.9853
C_6H_{12}	Ethylcyclobutane	293.2	1.965
C_6H_{12}	1-Hexene	294.0	2.077
C_6H_{12}	trans-2-Hexene	295.0	1.978
C_6H_{12}	cis-3-Hexene	296.0	2.069
C_6H_{12}	trans-3-Hexene	293.2	1.954
$C_6H_{12}Br_2$	meso-3,4-Dibromohexane	298.2	4.67
$C_6H_{12}Br_2$	(±)-3,4-Dibromohexane	298.2	6.732
$C_6H_{12}Br_2$	1,6-Dibromohexane	298.2	8.52
$C_6H_{12}ClNO$	2-Chloro-N,N-diethylacetamide	298.2	39.2
$C_6H_{12}Cl_2$	1,6-Dichlorohexane	308.2	8.60
$C_6H_{12}Cl_2O$	Bis(3-chloropropyl) ether	293.2	10.10
$C_6H_{12}N_2O$	1,3-Dimethyl-2-oxohexahydropyrimidine	298.2	36.12
$C_6H_{12}N_3OP$	Tri(1-aziridinyl)phosphine oxide	298.2	14.62
$C_6H_{12}O$	Isobutyl vinyl ether	293.2	3.34
$C_6H_{12}O$	Cyclohexanol	293.2	16.40
$C_6H_{12}O$	cis-2,5-Dimethyltetrahydrofuran	296.2	5.03
$C_6H_{12}O$	2-Hexanone	293.2	14.56
$C_6H_{12}O$	4-Methyl-2-pentanone	293.2	13.11
$C_6H_{12}O$	3,3-Dimethyl-2-butanone	293.2	12.73
$C_6H_{12}O$	1-Methylcyclopentanol	310.1	7.11
$C_6H_{12}O_2$	Diacetone alcohol	298.2	18.2

Molecular formula	Name	T/K	ε
$C_6H_{12}O_2$	Pentyl formate	292.2	5.7
$C_6H_{12}O_2$	Isopentyl formate	288.2	5.44
$C_6H_{12}O_2$	Ethyl butanoate	301.2	5.18
$C_6H_{12}O_2$	Hexanoic acid	298.2	2.600
$C_6H_{12}O_2$	tert-Butylacetic acid	296.2	2.85
$C_6H_{12}O_2$	Butyl acetate	293.2	5.07
$C_6H_{12}O_2$	Isobutyl acetate	293.2	5.068
$C_6H_{12}O_2$	sec-Butyl acetate	293.2	5.135
$C_6H_{12}O_2$	tert-Butyl acetate	293.2	5.672
$C_6H_{12}O_2$	2-Ethylbutanoic acid	296.2	2.72
$C_6H_{12}O_2$	Propyl propanoate	293.2	5.249
$C_6H_{12}O_2$	Methyl pentanoate	293.2	4.992
$C_6H_{12}O_2S$	2,4-Dimethyltetrahydrothiophene-S,S-dioxide	293.2	29.69
$C_6H_{12}O_3$	2-Ethoxyethyl acetate	303.2	7.567
$C_6H_{12}O_3$	2,4,6-Trimethyl-1,3,5-trioxan	292.2	14.70
$C_6H_{12}S$	Cyclohexanethiol	298.2	5.420
$C_6H_{13}Br$	1-Bromohexane	298.2	5.82
$C_6H_{13}Cl$	1-Chlorohexane	293.2	6.104
$C_6H_{13}ClO$	6-Chloro-1-hexanol	242.2	21.6
$C_6H_{13}I$	1-Iodohexane	293.3	5.35
$C_6H_{13}N$	Cyclohexylamine	293.2	4.547
$C_6H_{13}NO$	N-Butylacetamide	293.2	104.0
$C_6H_{13}NO$	N-Isobutylacetamide	293.2	111.0
$C_6H_{13}NO$	N-sec-Butylacetamide	293.2	100.0
$C_6H_{13}NO$	N,N-Diethylacetamide	293.2	32.1
$C_6H_{13}NO$	N,N-Dimethylbutanamide	293.2	29.7
$C_6H_{13}NO$	N-Ethylbutanamide	298.2	107.0
$C_6H_{13}NO$	N-Methylpentanamide	286.0	131.0
$C_6H_{13}NO$	N-Methyl-3-methylbutanamide	299.0	114.0
$C_6H_{13}NO$	N-Methyl-2-methylbutanamide	307.0	123.0
$C_6H_{13}NO$	N-Propylpropanamide	298.2	118.1
C_6H_{14}	2,2-Dimethylbutane	293.2	1.869
C_6H_{14}	2,3-Dimethylbutane	293.2	1.889
C_6H_{14}	2-Methylpentane	293.2	1.886
C_6H_{14}	3-Methylpentane	293.2	1.886
C_6H_{14}	Hexane	293.2	1.8865
$C_6H_{14}Cl_2NO_3PS$	Dipropoxyphosphinylimidosulfurous dichloride	293.2	17.65
$C_6H_{14}N_3OP$	P,P-Bis(1-aziridinyl)-N,N-dimethylphosphinic amide	298.2	8.58
$C_6H_{14}O$	Dipropyl ether	297.0	3.38
$C_6H_{14}O$	Diisopropyl ether	303.2	3.805
$C_6H_{14}O$	Ethyl tert-butyl ether	298.2	7.07
$C_6H_{14}O$	2,2-Dimethyl-1-butanol	293.2	10.5
$C_6H_{14}O$	2-Ethyl-1-butanol	362.2	6.19
$C_6H_{14}O$	1-Hexanol	293.2	13.03
$C_6H_{14}O$	(±)-2-Hexanol	298.2	11.06
$C_6H_{14}O$	3-Hexanol	298.2	9.66
$C_6H_{14}O$	3-Methyl-1-pentanol	298.2	15.2
$C_6H_{14}O$	3-Methyl-3-pentanol	293.2	4.322
$C_6H_{14}OS$	Dipropyl sulfoxide	303.2	30.37
$C_6H_{14}O_2$	2-Butoxyethanol	298.2	9.43
$C_6H_{14}O_2$	1,2-Diethoxyethane	293.2	3.90
$C_6H_{14}O_2$	Hexylene glycol	293.2	25.86
$C_6H_{14}O_2S$	Dipropyl sulfone	303.2	32.62
$C_6H_{14}O_3$	Bis(2-methoyxethyl) ether	298.2	7.23
$C_6H_{14}O_3$	Bis(2-hydroxypropyl) ether	293.2	20.38
$C_6H_{14}O_3$	1,2,6-Hexanetriol	285.3	31.5
$C_6H_{14}O_4$	Triethylene glycol	293.2	23.69
$C_6H_{14}O_6$	Sorbitol	353.2	35.5

Molecular formula	Name	T/K	ε
$C_6H_{14}O_6$	D-Mannitol	443.2	24.6
$C_6H_{14}S$	1-Hexanethiol	293.2	4.436
$C_6H_{15}Al$	Triethyl aluminium	293.2	2.9
$C_6H_{15}B$	Triethylborane	293.2	1.974
$C_{10}H_{22}O_2$	Dipropylamine	293.2	2.923
$C_6H_{15}N$	Hexylamine	293.2	4.08
$C_6H_{15}N$	Triethylamine	293.2	2.418
$C_6H_{15}NO$	2-Dimethylamino-2-methyl-1-propanol	298.2	12.36
$C_6H_{15}OP$	Triethylphosphine oxide	323.2	35.5
$C_6H_{15}O_3P$	Diethyl ethanephosphonate	305.2	10.653
$C_6H_{15}O_3PS$	Triethyl thionphosphate	298.2	10.4
$C_6H_{15}O_4P$	Triethyl phosphate	298.2	13.20
$C_6H_{15}O_4V$	Triethyl orthovanadate	298.2	3.333
$C_6H_{15}PS$	Triethylphosphine sulfide	371.2	39.0
$C_6H_{16}NO_3P$	N,N-Dimethylphosphoramidic acid diethyl ester	293.2	11.75
$C_6H_{16}N_3OP$	P-1-Aziridinyl-N,N,N',N'-tetramethylphosphonic diamide	298.2	14.69
$C_6H_{16}OSi$	Trimethylpropoxysilane	298.2	2.850
$C_6H_{16}OSi$	Trimethylisopropoxysilane	298.2	2.851
$C_6H_{16}O_2Si$	Diethoxydimethylsilane	298.2	3.216
$C_6H_{16}Si$	Triethylsilane	293.2	2.323
$C_6H_{17}N_2O_2P$	N,N,N',N'-Tetramethylphosphordiamic acid ethyl ester	298.2	14.73
$C_6H_{18}N_3OP$	Tris(dimethylamino)phosphine oxide	293.2	31.3
$C_6H_{18}N_3P$	Tris(dimethylamino)phosphine	293.2	3.65
$C_6H_{18}N_3PS$	Tris(dimethylamino)phosphine sulfide	303.2	39.5
$C_6H_{18}N_4$	Triethylenetetramine	293.2	10.76
$C_6H_{18}OSi_2$	Hexamethyldisiloxane	293.2	2.179
$C_6H_{18}O_3Si_3$	Hexamethylcyclotrisiloxane	343.2	2.139
$C_6H_{19}NSi_2$	Hexamethyldisilazane	294.2	2.273
C_7F_{14}	Perfluoromethylcyclohexane	298.2	1.82
C_7F_{16}	Perfluoroheptane	289.2	1.847
$C_7H_3ClF_3NO_2$	2-Chloro-1-trifluoromethyl-5-nitrobenzene	303.2	9.8
$C_7H_3ClF_3NO_2$	4-Chloro-1-trifluoromethyl-3-nitrobenzene	303.2	12.8
$C_7H_3Cl_5$	2,3,4,5,6-Pentachlorotoluene	293.2	4.8
$C_7H_4BrCl_2NOS$	(4-Bromobenzoyl)imidosulfurous dichloride	353.2	7.80
$C_7H_4ClF_3O_2S$	4-(Trifluoromethylsulfonyl)-1-chlorobenzene	333.2	10.3
$C_7H_4ClF_3O_4S_2$	4-(Trifluoromethylsulfonyl)-benzenesulfonyl chloride	298.2	4.72
C_7H_4ClNO	4-Chlorophenyl isocyanate	288.2	3.177
$C_7H_4Cl_2N_2O_3S$	(4-Nitrobenzoyl)imidosulfurous dichloride	366.2	6.02
$C_7H_4Cl_3NOS$	(4-Chlorobenzoyl)imidosulfurous dichloride	298.2	8.25
$C_7H_4F_3NO_2$	1-Trifluoromethyl-3-nitrobenzene	303.2	17.0
$C_7H_4F_3NO_3S_2$	4-(Trifluoromethylsulfonyl)-N-sulfinyl aniline	349.2	15.05
$C_7H_4F_3NO_4S$	4-(Trifluoromethylsulfonyl)-1-nitrobenzene	373.2	3.62
$C_7H_4F_4O_2S$	4-(Trifluoromethylsulfonyl)-1-fluorobenzene	293.2	12.2
C_7H_5BrO	Benzoyl bromide	293.2	21.33
C_7H_5ClO	Benzoyl chloride	293.2	23.0
$C_7H_5Cl_2NOS$	Benzoylimidosulfurous dichloride	313.2	31.6
$C_7H_5Cl_3$	2,4-Dichloro-1-chloromethylbenzene	298.2	6.290
C_7H_5FO	Benzoyl fluoride	293.2	22.7
$C_7H_5F_3$	(Trifluoromethyl)benzene	298.2	9.22
$C_7H_5F_3O_2S$	Trifluoromethylphenyl sulfone	293.2	31.0
$C_7H_5F_3O_2S_2$	4-(Trifluoromethylsulfonyl)-benzenethiol	298.2	28.5
$C_7H_5F_3O_3S$	4-(Trifluoromethylsulfonyl)-phenol	298.2	42.2
C_7H_5N	Benzonitrile	293.2	25.9
C_7H_5NO	Phenyl isocyanate	293.2	8.940
$C_7H_6ClNO_2$	4-Chloro-3-nitrotoluene	301.2	28.07
$C_7H_6Cl_2$	2,6-Dichlorotoluene	301.2	3.36
$C_7H_6Cl_2$	2,4-Dichlorotoluene	301.2	5.68
$C_7H_6Cl_2$	3,4-Dichlorotoluene	301.2	9.39

Molecular formula	Name	T/K	ε
$C_7H_6Cl_2$	(Dichloromethyl)benzene	293.2	6.9
$C_7H_6F_3NO_2S$	4-(Trifluoromethylsulfonyl)aniline	298.2	77.0
C_7H_6O	Benzaldehyde	293.2	17.85
$C_7H_6O_2$	Salicylaldehyde	293.2	18.35
C_7H_7Br	(Bromomethyl)benzene	293.2	6.658
C_7H_7Br	2-Bromotoluene	293.2	4.641
C_7H_7Br	3-Bromotoluene	293.2	5.566
C_7H_7Br	4-Bromotoluene	293.2	5.503
C_7H_7BrO	2-Bromoanisole	303.2	8.96
C_7H_7BrO	4-Bromoanisole	303.2	7.40
C_7H_7BrS	4-Bromo-3-methylthiophenol	303.2	3.70
C_7H_7Cl	Benzyl chloride	293.2	6.854
C_7H_7Cl	o-Chlorotoluene	293.2	4.721
C_7H_7Cl	m-Chlorotoluene	293.2	5.763
C_7H_7Cl	p-Chlorotoluene	293.2	6.25
$C_7H_7Cl_2NO_2S_2$	[(4-Methylphenyl)sulfonyl] imidosulfurous dichloride	327.2	11.08
C_7H_7ClO	4-Chloroanisole	293.2	7.84
$C_7H_7ClO_2S$	p-Toluenesulfonyl chloride	343.2	22.6
$C_7H_7ClO_3S$	4-Methoxybenzenesulfonyl chloride	314.2	27.2
C_7H_7ClS	4-Chlorothioanisole	298.2	5.95
C_7H_7ClS	4-Chloro-2-methylthiophenol	303.2	5.50
C_7H_7F	o-Fluorotoluene	298.2	4.23
C_7H_7F	m-Fluorotoluene	298.2	5.41
C_7H_7F	p-Fluorotoluene	298.2	5.88
C_7H_7I	4-Iodotoluene	308.2	4.4
C_7H_7N	2-Vinylpyridine	293.2	9.126
C_7H_7N	4-Vinylpyridine	293.2	10.50
$C_7H_7NO_2$	Benzyl nitrite	298.2	7.78
$C_7H_7NO_2$	o-Nitrotoluene	293.0	26.26
$C_7H_7NO_2$	m-Nitrotoluene	303.2	24.95
$C_7H_7NO_2$	p-Nitrotoluene	331.2	22.2
$C_7H_7NO_2$	Methyl pyridine-4-carboxylate	293.2	9.51
$C_7H_7NO_2S$	4-Nitrothioanisole	346.0	21.7
$C_7H_7NO_2S$	4-Methyl-N-sulfinyl aniline	293.2	8.75
$C_7H_7NO_3$	2-Nitroanisole	293.2	45.75
$C_7H_7NO_3$	3-Nitroanisole	318.2	25.7
$C_7H_7NO_3$	4-Nitroanisole	338.2	26.95
$C_7H_7NO_3S$	4-Methoxy-N-sulfinylaniline	293.2	11.85
C_7H_8	Toluene	296.35	2.379
$C_7H_8Cl_2N_2S$	Cyanocyclohexenylimidosulfurous dichloride	298.2	23.2
C_7H_8O	Anisole	294.2	4.30
C_7H_8O	Benzyl alcohol	303.2	11.916
C_7H_8O	o-Cresol	298.2	6.76
C_7H_8O	m-Cresol	298.2	12.44
C_7H_8O	p-Cresol	298.2	13.05
$C_7H_8O_2$	2-Methoxyphenol	298.2	11.95
$C_7H_8O_2$	3-Methoxyphenol	298.2	11.59
$C_7H_8O_2$	4-Methoxyphenol	333.7	11.05
$C_7H_8O_2S$	Methyl phenyl sulfone	373.2	37.9
$C_7H_8O_2S$	Ethyl thiophene-2-carboxylate	293.2	7.50
$C_7H_8O_3$	2-Furfuryl acetate	293.2	5.85
$C_7H_8O_3$	Ethyl furan-2-carboxylate	293.2	9.02
C_7H_8S	Benzylthiol	298.2	4.705
C_7H_8S	4-Methylthiophenol	323.2	4.74
C_7H_8S	Thioanisole	303.2	4.88
C_7H_9N	Benzylamine	293.2	5.18
C_7H_9N	2,4-Dimethylpyridine	293.2	9.60
C_7H_9N	2,6-Dimethylpyridine	293.2	7.33

Molecular formula	Name	T/K	ε
C$_7$H$_9$N	2-Ethylpyridine	293.2	8.33
C$_7$H$_9$N	4-Ethylpyridine	293.2	10.98
C$_7$H$_9$N	N-Methylaniline	293.2	5.96
C$_7$H$_9$N	2-Methylaniline	298.2	6.138
C$_7$H$_9$N	3-Methylaniline	298.2	5.816
C$_7$H$_9$N	4-Methylaniline	333.2	5.058
C$_7$H$_9$NO	2-Methoxyaniline	303.2	5.230
C$_7$H$_9$NO	3-Methoxyaniline	298.2	8.76
C$_7$H$_9$NO	4-Methoxyaniline	333.2	7.85
C$_7$H$_9$NO	2,6-Dimethylpyridine-1-oxide	298.2	46.11
C$_7$H$_9$NO$_2$S	N-Methylbenzenesulfonamide	303.2	67.1
C$_7$H$_{10}$N$_2$	1-Methyl-1-phenylhydrazine	292.2	7.3
C$_7$H$_{11}$F$_3$	Trifluoromethylcyclohexane	188.2	11.9
C$_7$H$_{11}$Cl$_3$O$_2$	3-Methyl-1-butyl trichloroacetate	293.2	7.287
C$_7$H$_{12}$	Cycloheptene	295.0	2.265
C$_7$H$_{12}$	1,6-Heptadiene	293.0	2.161
C$_7$H$_{12}$O	2-Methylcyclohexanone	293.2	14.0
C$_7$H$_{12}$O	3-Methylcyclohexanone	293.2	12.4
C$_7$H$_{12}$O	4-Methylcyclohexanone	293.2	12.35
C$_7$H$_{12}$O	Cycloheptanone	298.2	13.16
C$_7$H$_{12}$O$_2$	Butyl acrylate	301.2	5.25
C$_7$H$_{12}$O$_2$	Cyclohexyl formate	293.2	6.47
C$_7$H$_{12}$O$_2$	Cyclohexanecarboxylic acid	304.2	2.67
C$_7$H$_{12}$O$_2$S	Ethyl tetrahydrothiophene-2-carboxylate	293.2	6.18
C$_7$H$_{12}$O$_3$	2-Tetrahydrofurfuryl acetate	293.2	9.65
C$_7$H$_{12}$O$_4$	Monomethyl adipate	293.2	6.69
C$_7$H$_{12}$O$_4$	Dimethyl glutarate	293.2	7.87
C$_7$H$_{12}$O$_4$	Diethyl malonate	304.2	7.550
C$_7$H$_{12}$O$_5$	1,2,3-Propanetriol 1,2-diacetate	244.7	18.16
C$_7$H$_{12}$O$_5$	1,2,3-Propanetriol 1,3-diacetate	288.2	9.80
C$_7$H$_{13}$ClO$_2$	3-Methylbutyl chloroacetate	293.2	7.8
C$_7$H$_{13}$NO	N-Methyl 6-aminohexanoic lactam	293.2	31.0
C$_7$H$_{13}$NO	3-Pentanoneoxime O-vinyl ether	293.2	2.15
C$_7$H$_{13}$NO$_2$	2-Butoxyethyl isocyanate	293.2	9.355
C$_7$H$_{13}$NO$_2$	3-Butyloxazolidine-2-one	298.2	52.4
C$_7$H$_{13}$NO$_2$	3-tert-Butyloxazolidine-2-one	298.2	57.6
C$_7$H$_{13}$NO$_2$	Methyl piperidine-4-carboxylate	293.2	8.24
C$_7$H$_{14}$	Cycloheptane	293.2	2.0784
C$_7$H$_{14}$	3-Ethyl-2-pentene	293.2	2.051
C$_7$H$_{14}$	1-Heptene	293.2	2.092
C$_7$H$_{14}$	Methylcyclohexane	293.2	2.024
C$_7$H$_{14}$	2-Methyl-2-hexene	293.2	2.962
C$_7$H$_{14}$Br$_2$	1,2-Dibromoheptane	298.2	3.77
C$_7$H$_{14}$Br$_2$	2,3-Dibromoheptane	298.2	5.08
C$_7$H$_{14}$Br$_2$	3,4-Dibromoheptane	298.2	4.70
C$_7$H$_{14}$Cl$_2$	1,7-Dichloroheptane	298.2	8.34
C$_7$H$_{14}$O	2-Heptanone	293.2	11.95
C$_7$H$_{14}$O	3-Heptanone	293.2	12.7
C$_7$H$_{14}$O	4-Heptanone	293.2	12.60
C$_7$H$_{14}$O	5-Methyl-2-hexanone	293.2	13.53
C$_7$H$_{14}$O	Heptanal	295.2	9.07
C$_7$H$_{14}$O	Cyclohexyl methanol	333.2	9.70
C$_7$H$_{14}$O	2-Methylcyclohexanol*	293.2	9.375
C$_7$H$_{14}$O	3-Methylcyclohexanol*	293.2	13.79
C$_7$H$_{14}$O	4-Methylcyclohexanol*	293.2	13.45
C$_7$H$_{14}$O$_2$	Methyl hexanoate	293.2	4.615
C$_7$H$_{14}$O$_2$	Ethyl pentanoate	291.2	4.71
C$_7$H$_{14}$O$_2$	2,2-Dimethyl pentanoic acid	296.2	2.58

Molecular formula	Name	T/K	ε
$C_7H_{14}O_2$	Propyl butanoate	293.2	4.3
$C_7H_{14}O_2$	Pentyl acetate	293.2	4.79
$C_7H_{14}O_2$	Isopentyl acetate	293.2	4.72
$C_7H_{14}O_2$	Ethyl 3-methylbutanoate	293.2	4.71
$C_7H_{14}O_2$	Heptanoic acid	288.2	3.04
$C_7H_{14}O_2$	Butyl propanoate	293.2	4.838
$C_7H_{15}Br$	1-Bromoheptane	303.2	5.255
$C_7H_{15}Br$	2-Bromoheptane	295.2	6.46
$C_7H_{15}Br$	3-Bromoheptane	295.2	6.93
$C_7H_{15}Br$	4-Bromoheptane	295.2	6.81
$C_7H_{15}BrO$	1-Bromo-2-ethoxypentane	298.2	6.45
$C_7H_{15}BrO$	2-Bromo-3-ethoxypentane	298.2	6.40
$C_7H_{15}BrO$	3-Bromo-2-ethoxypentane	298.2	8.24
$C_7H_{15}Cl$	1-Chloroheptane	293.2	5.521
$C_7H_{15}Cl$	2-Chloroheptane	295.2	6.52
$C_7H_{15}Cl$	3-Chloroheptane	295.2	6.70
$C_7H_{15}Cl$	4-Chloroheptane	295.2	6.54
$C_7H_{15}I$	1-Iodoheptane	298.2	4.92
$C_7H_{15}I$	3-Iodoheptane	295.2	6.39
$C_7H_{15}NO$	N-Butylpropamide	298.2	100.6
$C_7H_{15}NO$	N,N-Dimethylpentanamide	293.2	26.4
$C_7H_{15}NO$	N,N-Dipropylformamide	293.2	23.5
$C_7H_{15}NO$	N,N-Diisopropylformamide	298.2	24.2
$C_7H_{15}NO$	N-Pentylacetamide	298.2	89.9
$C_7H_{15}NO$	N-Propylbutanamide	313.2	90.8
C_7H_{16}	2,2-Dimethylpentane	293.2	1.915
C_7H_{16}	2,3-Dimethylpentane	293.2	1.929
C_7H_{16}	2,4-Dimethylpentane	293.2	1.902
C_7H_{16}	3,3-Dimethylpentane	291.3	1.9419
C_7H_{16}	3-Ethylpentane	293.2	1.942
C_7H_{16}	Heptane	293.2	1.9209
C_7H_{16}	2-Methylhexane	293.2	1.9221
C_7H_{16}	3-Methylhexane	293.2	1.920
C_7H_{16}	2,2,3-Trimethylbutane	293.2	1.930
$C_7H_{16}O$	Ethyl pentyl ether	296.2	3.6
$C_7H_{16}O$	Ethyl 3-methylbutyl ether	293.2	3.955
$C_7H_{16}O$	2,2-Dimethyl-1-pentanol	293.2	6.020
$C_7H_{16}O$	3-Ethyl-3-pentanol	293.2	3.158
$C_7H_{16}O$	1-Heptanol	293.2	11.75
$C_7H_{16}O$	(±)-2-Heptanol	293.7	9.72
$C_7H_{16}O$	D-(+)-2-Heptanol	293.4	9.67
$C_7H_{16}O$	L-(-)-2-Heptanol	293.3	9.65
$C_7H_{16}O$	3-Heptanol	296.1	7.07
$C_7H_{16}O$	4-Heptanol	296.2	6.18
$C_7H_{16}O$	2-Methyl-2-hexanol	297.0	3.257
$C_7H_{16}O$	3-Methyl-2-hexanol	297.2	4.990
$C_7H_{16}O$	3-Methyl-3-hexanol	298.2	3.248
$C_7H_{16}N_2O$	N,N-Diethyl-N',N'-dimethyl urea	298.2	17.89
$C_7H_{16}O_3$	Triethoxymethane	293.2	4.779
$C_7H_{16}O_7$	Glucoheptitol	293.2	27.4
$C_7H_{16}S$	1-Heptanethiol	293.2	4.194
$C_7H_{17}N$	Heptylamine	293.2	3.81
$C_7H_{17}O_3P$	Diisopropyl methanephosphonate	298.2	7.72
$C_7H_{17}O_3P$	Diethyl propanephosphonate	303.2	9.542
$C_7H_{17}O_3P$	Diethyl 1-methylethylphosphonate	303.2	8.481
$C_7H_{18}OSi$	Butoxytrimethylsilane	298.2	2.758
$C_7H_{18}OSi$	2-Butoxytrimethylsilane	298.2	2.722
$C_7H_{18}O_3Si$	Triethoxymethylsilane	298.2	3.845

Molecular formula	Name	T/K	ε
$C_7H_{19}NSi$	Isobutyltrimethyl silazane	303.2	2.2523
$C_8Cl_6F_{12}$	Hexachloro-1,1,2,3,3,4,5,5,6,7,7,8-dodecafluorooctane	293.2	2.860
C_8F_{16}	Perflouro-1,3-dimethylcyclohexane	298.2	1.91
$C_8H_2Cl_2F_6$	1,2-Dichloro-3,5-bis(trifluoromethyl)benzene	303.2	3.12
$C_8H_3ClF_6$	2-Chloro-1,3-bis(trifluoromethyl)benzene	303.2	3.20
$C_8H_3ClF_6$	4-Chloro-1,3-bis(trifluoromethyl)benzene	303.2	5.44
$C_8H_4F_3NO$	3-Trifluoromethylphenyl isocyanate	293.2	11.665
$C_8H_4F_6$	1,3-Bis(trifluoromethyl)benzene	303.2	5.98
$C_8H_4F_6O_2S_2$	1-Trifluoromethylsulfonyl-4-(trifluoromethylthio)-benzene	298.2	8.34
$C_8H_4F_6O_4S_2$	1,4-Bis(trifluoromethyl sulfonyl)benzene	298.2	4.11
$C_8H_5N_2O_3P$	Phosphorodiisocyanatidous acid phenyl ester	293.2	5.45
$C_8H_5N_2O_3PS$	Phosphorodiisocyanatidothioic acid O-phenyl ester	293.2	9.92
$C_8H_5N_2O_4P$	Phosphorodiisocyanatidic acid phenyl ester	293.2	8.82
C_8H_6	Phenylacetylene	298.2	2.98
$C_8H_6Cl_2$	2,5-Dichlorostyrene	298.2	2.58
$C_8H_6Cl_4$	3,4,5,6-Tetrachloro-1,2-dimethylbenzene	293.2	8.0
$C_8H_6Cl_4$	2,4,5,6-Tetrachloro-1,3-dimethylbenzene	293.2	5.4
C_8H_6O	Phenoxyacetylene	298.2	4.76
$C_8H_7Cl_2NO_2S$	4-Methoxybenzoylimidosulfurous dichloride	298.2	17.7
$C_8H_7Cl_3$	2,5-Dichloro-2-chloroethylbenzene	297.2	5.20
$C_8H_7F_3O_2S$	4-(Trifluoromethylsulfonyl)toluene	313.2	23.4
$C_8H_7F_3O_3S$	4-(Trifluoromethylsulfonyl)anisole	298.2	29.1
C_8H_7N	Phenylacetonitrile	299.2	17.87
$C_8H_7NO_2$	4-Methoxyphenyl isocyanate	333.2	10.26
$C_8H_7NO_4$	Methyl 2-nitrobenzoate	300.1	27.76
C_8H_8	Styrene	293.2	2.4737
C_8H_8O	Acetophenone	298.2	17.44
$C_8H_8O_2$	Benzyl formate	303.2	6.34
$C_8H_8O_2$	Methyl benzoate	302.7	6.642
$C_8H_8O_2$	Phenyl acetate	298.2	5.403
$C_8H_8O_2$	4-Methoxybenzaldehyde	303.2	22.0
$C_8H_8O_2$	2-Hydroxyacetophenone	298.2	21.33
$C_8H_8O_2$	Phenylacetic acid	353.2	3.47
$C_8H_8O_3$	Methyl salicylate	314.4	8.80
C_8H_9Br	1-Bromo-2-ethylbenzene	298.2	5.55
C_8H_9Br	1-Bromo-3-ethylbenzene	298.2	5.56
C_8H_9Br	1-Bromo-4-ethylbenzene	298.2	5.42
C_8H_9BrO	2-Bromophenetole	313.2	7.04
C_8H_9Cl	2-Chloro-ethylbenzene	298.2	4.36
C_8H_9Cl	3-Chloro-ethylbenzene	298.2	5.18
C_8H_9Cl	4-Chloro-ethylbenzene	298.2	5.16
$C_8H_9NO_2$	1-Phenylethyl nitrite	298.2	7.72
$C_8H_9NO_2$	Ethyl 4-pyridinecarboxylate	293.2	8.95
$C_8H_9NO_2$	2-Nitro-ethylbenzene	273.4	21.9
$C_8H_9NO_2$	Methyl 2-aminobenzoate	298.2	21.9
C_8H_{10}	Ethylbenzene	293.2	2.4463
C_8H_{10}	o-Xylene	293.2	2.562
C_8H_{10}	m-Xylene	293.2	2.359
C_8H_{10}	p-Xylene	293.2	2.2735
$C_8H_{10}N_2OS$	N,N-Dimethyl-N'-sulfinyl-1,4-benzenediamine	298.2	34.2
$C_8H_{10}O$	2-Methylanisole	293.2	3.502
$C_8H_{10}O$	3-Methylanisole	293.2	3.967
$C_8H_{10}O$	4-Methylanisole	293.2	3.914
$C_8H_{10}O$	Phenetole	293.2	4.216
$C_8H_{10}O$	1-Phenylethanol	293.2	8.77
$C_8H_{10}O$	2-Phenylethanol	293.2	12.31
$C_8H_{10}O$	2,3-Xylenol	343.2	4.81
$C_8H_{10}O$	2,4-Xylenol	303.2	5.060

Molecular formula	Name	T/K	ε
$C_8H_{10}O$	2,5-Xylenol	338.2	5.36
$C_8H_{10}O$	2,6-Xylenol	313.2	4.90
$C_8H_{10}O$	3,4-Xylenol	333.2	9.02
$C_8H_{10}O$	3,5-Xylenol	323.2	9.06
$C_8H_{10}OS$	4-Methoxythioanisole	298.2	6.9
$C_8H_{10}O_2$	1,2-Dimethoxybenzene	293.2	4.45
$C_8H_{10}O_2$	1,3-Dimethoxybenzene	298.2	5.363
$C_8H_{10}O_2$	1,4-Dimethoxybenzene	333.7	5.60
$C_8H_{10}O_2S$	Ethyl phenyl sulfone	348.2	39.0
$C_8H_{10}O_2S$	Propyl 2-thiophenecarboxylate	293.2	7.03
$C_8H_{10}O_3$	Propyl 2-furancarboxylate	293.2	8.37
$C_8H_{10}O_3$	2-Furfuryl propanoate	293.2	5.45
$C_8H_{10}S$	4-Methylthioanisole	298.2	4.74
$C_8H_{10}S$	Thiophenetole	298.2	4.95
$C_8H_{11}N$	N,N-Dimethylaniline	298.2	4.90
$C_8H_{11}N$	N-Ethylaniline	293.2	5.87
$C_8H_{11}N$	4-Ethylaniline	298.2	4.84
$C_8H_{11}N$	N-Benzylmethylamine	292.2	4.4
$C_8H_{11}N$	2,4,6-Trimethylpyridine	298.2	7.807
$C_8H_{11}NO$	4-Ethoxyaniline	298.2	7.43
$C_8H_{11}NO_2S$	N,N-Dimethylbenzenesulfonamide	323.2	48.6
$C_8H_{11}NO_2S_2$	S,S-Dimethyl-N-(phenylsulfonyl)sulfilimine	381.2	66.0
$C_8H_{12}N_2O_2$	1,6-Diisocyanatohexane	288.2	14.41
$C_8H_{12}O_4$	Diethyl fumarate	296.2	6.56
$C_8H_{12}O_4$	Diethyl maleate	298.2	7.560
C_8H_{14}	cis-Cyclooctene	296.0	2.306
C_8H_{14}	1,2-Dimethylcyclohexene	296.0	2.144
C_8H_{14}	1,3-Dimethylcyclohexene	296.0	2.182
C_8H_{14}	1,7-Octadiene	293.0	2.186
$C_8H_{14}O_2$	Methyl cyclohexanecarboxylate	293.2	4.87
$C_8H_{14}O_2$	Cyclohexyl acetate	293.2	5.08
$C_8H_{14}O_2S$	Propyl tetrahydrothiophene-2-carboxylate	293.2	5.60
$C_8H_{14}O_3$	Butanoic anhydride	293.2	12.8
$C_8H_{14}O_3$	2-Methylpropanoic anhydride	292.2	13.6
$C_8H_{14}O_3$	Tetrahydrofurfuryl propanoate	293.2	8.70
$C_8H_{14}O_4$	Dimethyl adipate	293.2	6.84
$C_8H_{14}O_4$	Diisopropyl oxalate	293.2	6.403
$C_8H_{14}O_4$	Diethyl succinate	293.2	6.098
$C_8H_{14}O_4$	meso-2,3-Diacetoxybutane	298.2	6.644
$C_8H_{14}O_4$	(±)-2,3-Diacetoxybutane	298.2	5.10
$C_8H_{15}NO_2$	N,N-Diethylacetoacetamide	298.2	40.8
$C_8H_{15}NO_2$	Ethyl piperidine-4-carboxylate	293.2	7.60
$C_8H_{15}N$	Octanenitrile	293.2	13.90
C_8H_{16}	Cyclooctane	295.0	2.116
C_8H_{16}	1-Octene	293.2	2.113
C_8H_{16}	cis-3-Octene	298.2	2.062
C_8H_{16}	trans-3-Octene	298.2	2.002
C_8H_{16}	cis-4-Octene	298.2	2.053
C_8H_{16}	trans-4-Octene	298.2	2.004
C_8H_{16}	3-Methyl-2-heptene	293.2	2.436
C_8H_{16}	2,5-Dimethyl-2-hexene	293.2	2.431
C_8H_{16}	2,2,4-Trimethyl-4-pentene	298.2	2.0908
$C_8H_{16}Cl_2$	1,8-Dichlorooctane	298.2	7.64
$C_8H_{16}Br_2$	1,8-Dibromooctane	298.2	7.43
$C_8H_{16}O$	2-Octanone	293.2	9.51
$C_8H_{16}O$	3-Octanone	303.2	10.50
$C_8H_{16}O_2$	Propyl pentanoate	292.2	4.0
$C_8H_{16}O_2$	Butyl butanoate	298.2	4.39

Molecular formula	Name	T/K	ε
$C_8H_{16}O_2$	Ethyl hexanoate	293.2	4.45
$C_8H_{16}O_2$	Octanoic acid	288.2	2.85
$C_8H_{16}O_2$	Hexyl acetate	293.2	4.42
$C_8H_{16}O_2$	2-Ethylhexanoic acid	296.2	2.64
$C_8H_{16}O_2$	Methyl heptanoate	293.2	4.355
$C_8H_{16}O_2$	Pentyl propanoate	293.2	4.552
$C_8H_{16}O_3$	3-Methyl-1-butyl lactate	273.2	11.2
$C_8H_{16}O_2$	Isopentyl propanoate	273.2	5.21
$C_8H_{17}Br$	3-Bromomethylheptane	298.2	6.00
$C_8H_{17}Br$	1-Bromooctane	293.2	5.0957
$C_8H_{17}Br$	2-Bromooctane	293.2	5.44
$C_8H_{17}Br$	3-Bromooctane	293.2	6.08
$C_8H_{17}Br$	4-Bromooctane	293.2	5.95
$C_8H_{17}Cl$	1-Chlorooctane	298.2	5.05
$C_8H_{17}Cl$	2-Chlorooctane	293.2	5.42
$C_8H_{17}Cl$	3-Chlorooctane	293.2	6.17
$C_8H_{17}Cl$	4-Chlorooctane	293.2	5.83
$C_8H_{17}F$	1-Fluorooctane	293.2	3.89
$C_8H_{17}F$	2-Fluorooctane	293.2	4.17
$C_8H_{17}F$	3-Fluorooctane	293.2	4.28
$C_8H_{17}F$	4-Fluorooctane	293.2	4.16
$C_8H_{17}I$	1-Iodooctane	293.2	4.67
$C_8H_{17}I$	2-Iodooctane	293.2	5.48
$C_8H_{17}I$	3-Iodooctane	293.2	5.23
$C_8H_{17}I$	4-Iodooctane	293.2	5.42
$C_8H_{17}NO$	*N,N*-Dimethylhexanamide	293.2	22.7
$C_8H_{17}NO$	*N,N*-Dipropylacetamide	293.2	24.5
$C_8H_{17}NO$	*N,N*-Diisopropylacetamide	298.2	23.2
$C_8H_{17}NO_2$	1-Nitrooctane	293.2	11.46
$C_8H_{17}NO_2$	2-Nitrooctane	293.2	14.0
$C_8H_{17}NO_2$	3-Nitrooctane	293.2	15.85
$C_8H_{17}NO_2$	4-Nitrooctane	293.2	15.7
C_8H_{18}	2,2-Dimethylhexane	293.2	1.9498
C_8H_{18}	2,5-Dimethylhexane	293.95	1.9619
C_8H_{18}	3,3-Dimethylhexane	293.2	1.9645
C_8H_{18}	3,4-Dimethylhexane	292.1	1.9814
C_8H_{18}	3-Ethylhexane	293.2	1.9617
C_8H_{18}	3-Ethyl-3-methylpentane	291.49	1.9869
C_8H_{18}	2-Methylheptane	293.2	1.9519
C_8H_{18}	Octane	293.2	1.948
C_8H_{18}	2,2,4-Trimethylpentane	293.2	1.943
C_8H_{18}	2,2,3-Trimethylpentane	293.2	1.960
C_8H_{18}	2,3,4-Trimethylpentane	293.2	1.9738
C_8H_{18}	2,3,3-Trimethylpentane	293.2	1.9780
$C_8H_{18}Cl_2O_3NPS$	Dibutoxyphosphinylimidosulfurous dichloride	293.2	12.21
$C_8H_{18}N_2S$	Thiodi-*tert*-butylimine	298.2	3.41
$C_8H_{18}O$	Dibutyl ether	293.2	3.0830
$C_8H_{18}O$	2,2-Dimethyl-1-hexanol	293.2	4.50
$C_8H_{18}O$	2-Ethyl-1-hexanol	298.2	7.58
$C_8H_{18}O$	2-Isopropyl-1-pentanol	298.2	8.50
$C_8H_{18}O$	2-Methyl-1-heptanol	293.1	5.16
$C_8H_{18}O$	3-Methyl-1-heptanol	290.3	2.884
$C_8H_{18}O$	4-Methyl-1-heptanol	290.6	4.63
$C_8H_{18}O$	5-Methyl-1-heptanol	290.4	7.68
$C_8H_{18}O$	6-Methyl-1-heptanol	290.3	10.54
$C_8H_{18}O$	2-Methyl-2-heptanol	292.2	3.43
$C_8H_{18}O$	3-Methyl-2-heptanol	289.6	7.47
$C_8H_{18}O$	4-Methyl-2-heptanol	290.0	3.59

Molecular formula	Name	T/K	ε
$C_8H_{18}O$	5-Methyl-2-heptanol	278.5	7.5
$C_8H_{18}O$	6-Methyl-2-heptanol	290.1	6.41
$C_8H_{18}O$	2-Methyl-3-heptanol	293.2	3.260
$C_8H_{18}O$	3-Methyl-3-heptanol	293.2	3.013
$C_8H_{18}O$	4-Methyl-3-heptanol	293.2	3.312
$C_8H_{18}O$	5-Methyl-3-heptanol	293.2	3.832
$C_8H_{18}O$	6-Methyl-3-heptanol	293.2	4.992
$C_8H_{18}O$	2-Methyl-4-heptanol	296.3	3.338
$C_8H_{18}O$	3-Methyl-4-heptanol	290.0	7.46
$C_8H_{18}O$	4-Methyl-4-heptanol	296.2	2.902
$C_8H_{18}O$	3-Methyl-2-isopropyl-1-butanol	298.2	9.83
$C_8H_{18}O$	1-Octanol	293.2	10.30
$C_8H_{18}O$	2-Octanol	293.2	8.13
$C_8H_{18}O$	3-Octanol	293.2	5.55
$C_8H_{18}O$	4-Octanol	293.2	4.48
$C_8H_{18}OS$	Dibutyl sulfoxide	313.2	24.73
$C_8H_{18}O_2$	2-Ethyl-1,3-hexanediol	293.2	18.73
$C_8H_{18}O_2S$	Dibutyl sulfone	323.2	25.72
$C_8H_{18}O_4$	Triethylene glycol dimethyl ether	298.2	7.62
$C_8H_{18}O_5$	Tetraethylene glycol	293.2	20.44
$C_8H_{18}S$	Dibutyl sulfide	298.2	4.29
$C_8H_{18}S$	1-Octanethiol	293.2	3.949
$C_8H_{19}N$	Dibutylamine	293.2	2.765
$C_8H_{19}N$	Octylamine	293.2	3.58
$C_8H_{19}OP$	Dimethylhexylphosphine oxide	323.2	25.8
$C_8H_{20}Ge$	Tetraethylgermane	274.2	1.971
$C_8H_{20}O_2Si$	Dimethyldipropoxysilane	298.2	2.992
$C_8H_{20}O_2Si$	Dimethylbis(1-methylethoxy)silane	298.2	3.050
$C_8H_{20}O_4Si$	2-Ethylhexyl silicate	293.2	4.1
$C_8H_{20}O_4Si$	Tetraethoxysilane	293.2	2.50
$C_8H_{20}Si$	Dimethyldipropylsilane	293.2	2.054
$C_8H_{20}Si$	Tetraethylsilane	293.2	2.090
$C_8H_{20}Sn$	Tetraethyltin	293.2	2.241
$C_8H_{23}NOSi_2$	N-Trimethylsilyl-2-aminoethyl trimethylsilyl ether	293.2	3.045
$C_8H_{23}N_5$	Tetraethylenepentamine	293.2	9.40
$C_8H_{24}O_4Si_4$	Octamethylcyclotetrasiloxane	296.2	2.390
$C_9H_6F_3ClO_2$	Trifluoroethyl 4-chlorobenzoate	298.2	3.96
$C_9H_6N_2O_2$	Toluene-2,4-diisocyanate	293.2	8.433
$C_9H_6N_2O_2$	Toluene-2,6-diisocyanate	293.2	5.147
$C_9H_6O_2$	Coumarin	343.2	34.04
$C_9H_7F_3O_2$	2,2,2-Trifluoroethyl benzoate	298.2	6.98
C_9H_7N	Quinoline	293.2	9.16
C_9H_7N	Isoquinoline	298.2	11.0
C_9H_8O	Cinnamaldehyde	305.8	17.72
$C_9H_8O_4$	Acetylsalicylic acid	333.2	6.55
$C_9H_9Cl_3$	1,2,3-Trichloro-4,5,6-trimethylbenzene	293.2	8.6
$C_9H_9Cl_3$	1,2,5-Trichloro-3,4,6-trimethylbenzene	293.2	6.4
C_9H_9N	2,6-Dimethylbenzonitrile	373.2	16.8
C_9H_{10}	1-Phenylpropene	293.2	2.73
C_9H_{10}	2-Phenylpropene	293.2	2.28
C_9H_{10}	3-Phenylpropene	293.2	2.63
$C_9H_{10}F_3NO_2S$	4-(Trifluoromethylsulfonyl)-N,N-dimethylaniline	432.2	30.1
$C_9H_{10}O_2$	4-Acetylanisole	313.2	17.3
$C_9H_{10}O_2$	Ethyl benzoate	293.2	6.20
$C_9H_{10}O_2$	Benzyl acetate	303.2	5.34
$C_9H_{10}O_2$	Methyl 2-phenylacetate	297.2	5.30
$C_9H_{10}O_2$	Phenyl propanoate	293.2	4.77
$C_9H_{10}O_2$	Methyl 4-methylbenzoate	306.2	4.3

Molecular formula	Name	T/K	ε
$C_9H_{10}O_2S$	Methyl 4-(methylthio)benzoate	357.2	6.39
$C_9H_{10}O_3$	Ethyl salicylate	308.2	8.48
$C_9H_{10}O_3$	Methyl 2-methoxybenzoate	294.2	7.7
$C_9H_{10}OS$	4-Acetylthioanisole	355.2	11.34
$C_9H_{11}Br$	1-Bromo-3-phenylpropane	302.2	5.41
$C_9H_{11}NO$	N,N-Dimethylbenzamide	318.2	20.77
$C_9H_{11}NO$	N-Ethylbenzamide	352.7	42.6
$C_9H_{11}NO_2$	Propyl nicotinate	293.2	9.7
$C_9H_{11}NO_2$	Propyl pyridine-4-carboxylate	293.2	8.24
$C_9H_{11}NO_2$	Ethyl 2-aminobenzoate	298.2	4.14
C_9H_{12}	Propylbenzene	293.2	2.370
C_9H_{12}	Isopropylbenzene	293.2	2.381
C_9H_{12}	1,2,3-Trimethylbenzene	293.2	2.656
C_9H_{12}	1,2,4-Trimethylbenzene	293.2	2.377
C_9H_{12}	Mesitylene	293.2	2.279
C_9H_{12}	2-Ethyl-1-methylbenzene	293.2	2.595
C_9H_{12}	3-Ethyl-1-methylbenzene	293.2	2.365
C_9H_{12}	p-Ethyltoluene	293.2	2.265
$C_9H_{12}O$	2,6-Dimethylanisole	293.2	3.780
$C_9H_{12}O$	3,5-Dimethylanisole	293.2	3.711
$C_9H_{12}O$	Benzyl ethyl ether	298.2	3.90
$C_9H_{12}O$	1-Phenyl-1-propanol	293.2	6.68
$C_9H_{12}O$	1-Phenyl-2-propanol	293.2	9.35
$C_9H_{12}O$	2-Phenyl-2-propanol	303.2	5.61
$C_9H_{12}O$	3-Phenyl-1-propanol	293.2	11.97
$C_9H_{12}O_2S$	1-Butyl thiophene-2-carboxylate	293.2	6.40
$C_9H_{12}O_3$	2-Furfuryl butanoate	293.2	4.93
$C_9H_{12}O_3$	Butyl furan-2-carboxylate	293.2	7.62
$C_9H_{12}S$	3-Phenyl-1-propanethiol	303.2	4.36
$C_9H_{13}N$	3-Phenylpropylamine	303.2	4.83
$C_9H_{13}N$	N-Propylaniline	293.2	5.48
$C_9H_{13}N$	N-Ethylbenzylamine	293.2	4.3
$C_9H_{13}N$	N,N-Dimethyl-o-toluidine	293.2	3.4
$C_9H_{13}N$	N,N-Dimethyl-p-toluidine	293.2	3.9
$C_9H_{14}OSi$	Trimethylphenoxysilane	298.2	3.3953
$C_9H_{14}O_6$	Triacetin	293.6	7.11
$C_9H_{14}Si$	Trimethylphenylsilane	298.2	2.3533
$C_9H_{15}BO_3$	Triallyl borate	293.2	2.38
$C_9H_{16}O_2$	Ethyl cyclohexanecarboxylate	293.2	4.64
$C_9H_{16}O_2$	2-Nonenoic acid	296.2	2.5
$C_9H_{16}O_2$	Cyclohexyl propanoate	293.2	4.82
$C_9H_{16}O_2S$	Butyl tetrahydrothiophene-2-carboxylate	293.2	5.40
$C_9H_{16}O_3$	2-Tetrahydrofurfuryl butanoate	293.2	8.13
$C_9H_{16}O_4$	Diethyl glutarate	303.2	6.659
$C_9H_{16}O_4$	(±)-erythro-2,3-Diacetoxypentane	298.2	6.734
$C_9H_{16}O_4$	(±)-threo-2,3-Diacetoxypentane	298.2	5.228
$C_9H_{17}NO_2$	Propyl piperidine-4-carboxylate	293.2	6.90
$C_9H_{17}N$	1-Cyanooctane	293.2	12.08
$C_9H_{17}N$	2-Cyanooctane	293.2	13.76
$C_9H_{17}N$	3-Cyanooctane	293.2	15.10
$C_9H_{17}N$	4-Cyanooctane	293.2	14.83
C_9H_{18}	1-Nonene	293.2	2.180
C_9H_{18}	4-Ethyl-3-heptene	293.2	2.475
C_9H_{18}	2,6-Dimethyl-2-heptene	293.2	2.606
C_9H_{18}	3,6-Dimethyl-3-heptene	293.2	2.343
$C_9H_{18}Br_2$	1,9-Dibromononane	293.2	7.153
$C_9H_{18}O$	2-Nonanone	295.2	9.14
$C_9H_{18}O$	5-Nonanone	293.2	10.6

Molecular formula	Name	T/K	ε
$C_9H_{18}O$	2,6-Dimethyl-4-heptanone	293.2	9.91
$C_9H_{18}O$	2,2,4,4-Tetramethyl-3-pentanone	287.65	10.0
$C_9H_{18}O_2$	Pentyl butanoate	301.2	4.08
$C_9H_{18}O_2$	2-Methylbutyl butanoate	292.2	4.1
$C_9H_{18}O_2$	3-Methylbutyl butanoate	293.2	4.0
$C_9H_{18}O_2$	2-Methyl-1-propyl pentanoate	292.2	3.8
$C_9H_{18}O_2$	Methyl octanoate	293.2	4.101
$C_9H_{18}O_2$	Heptyl acetate	293.2	4.2
$C_9H_{18}O_2$	2-Ethyl-heptanoic acid	293.2	1.98
$C_9H_{18}O_2$	2-Methyl-octanoic acid	293.2	2.39
$C_9H_{18}O_2$	2-Ethyl-2-methyl hexanoic acid	296.2	2.7
$C_9H_{18}O_2$	Nonanoic acid	294.9	2.475
$C_9H_{18}O_2$	3-Propylhexanoic acid	293.2	2.38
$C_9H_{18}O_2$	2,3,4-Trimethylhexanoic acid	296.2	3.40
$C_9H_{18}O_2$	2,3,5-Trimethylhexanoic acid	296.2	2.60
$C_9H_{19}Br$	1-Bromononane	298.2	4.74
$C_9H_{19}Cl$	1-Chlorononane	293.2	4.803
$C_9H_{19}NO$	N,N-Dibutylformamide	293.2	18.4
$C_9H_{19}NO$	N,N-Diethyl-3-aminopropyl vinyl ether	293.2	2.12
$C_9H_{19}NO$	N,N-Dimethylenanthamide	293.2	20.0
C_9H_{20}	Nonane	293.2	1.9722
C_9H_{20}	4-Methyloctane	293.2	1.967
C_9H_{20}	2-Methyloctane	293.2	1.967
C_9H_{20}	2,6-Dimethylheptane	293.2	1.987
C_9H_{20}	2,4-Dimethylheptane	293.2	1.89
C_9H_{20}	2,5-Dimethylheptane	293.2	1.89
$C_9H_{20}O$	1-Nonanol	293.2	8.83
$C_9H_{20}O$	2-Nonanol	298.2	6.66
$C_9H_{20}O$	3-Nonanol	298.2	4.49
$C_9H_{20}O$	4-Nonanol	298.2	3.69
$C_9H_{20}O$	5-Nonanol	298.2	3.54
$C_9H_{20}N_2O$	Tetraethyl urea	296.8	14.29
$C_9H_{20}S$	2-Methyl-2-octanethiol	293.2	4.069
$C_9H_{21}B$	Tripropylborane	293.2	2.026
$C_9H_{21}N$	Nonylamine	293.2	3.42
$C_9H_{21}N$	Tripropylamine	293.2	2.380
$C_9H_{21}O_4P$	Tripropyl phosphate	293.2	10.93
$C_9H_{21}O_4V$	Tripropyl orthovanadate	298.2	2.961
$C_9H_{21}O_4V$	Triisopropyl orthovanadate	298.2	3.299
$C_9H_{21}PS$	Tripropyl phosphine sulfide	293.2	32.8
$C_9H_{22}OSi$	Trimethylhexoxy silane	298.2	2.690
$C_{10}H_6ClF_5O_2$	Pentafluoropropyl 4-chlorobenzoate	298.2	3.80
$C_{10}H_7Br$	1-Bromonaphthalene	298.2	4.768
$C_{10}H_7Cl$	1-Chloronaphthalene	298.2	5.04
$C_{10}H_7ClF_4O_2$	Tetrafluoropropyl 4-chlorobenzoate	298.2	5.21
$C_{10}H_7F_5O_2$	2,2,3,3,3-Pentafluoropropyl benzoate	298.2	6.86
$C_{10}H_7NO_2$	1-Nitronaphthalene	333.2	19.68
$C_{10}H_8$	Naphthalene	363.2	2.54
$C_{10}H_8F_4O_2$	2,2,3,3-Tetrafluoropropyl benzoate	298.2	9.78
$C_{10}H_8N_2O_2$	1,3-Xylylene diisocyanate	293.2	13.787
$C_{10}H_8N_2O_2$	1,4-Xylylene diisocyanate	323.2	15.438
$C_{10}H_8O$	1-Naphthol	373.0	5.03
$C_{10}H_8O$	2-Naphthol	413.0	4.95
$C_{10}H_9N$	2-Methylquinoline	293.2	7.24
$C_{10}H_9N$	4-Methylquinoline	293.2	9.31
$C_{10}H_9N$	6-Methylquinoline	293.2	8.48
$C_{10}H_9N$	8-Methylquinoline	293.2	6.58
$C_{10}H_9N$	1-Naphthylamine	333.2	5.20

Molecular formula	Name	T/K	ε
$C_{10}H_9N$	2-Naphthylamine	393.0	5.26
$C_{10}H_{10}O_4$	Methyl o-acetylsalicylate	328.9	5.31
$C_{10}H_{10}O_4$	Dimethyl phthalate	293.2	8.66
$C_{10}H_{12}$	Dicyclopentadiene	313.2	2.43
$C_{10}H_{12}$	4-Ethylstyrene	298.2	3.350
$C_{10}H_{12}$	1,2,3,4-Tetrahydronaphthalene	298.2	2.771
$C_{10}H_{12}O$	Tetrahydro-2-naphthol	293.2	11.70
$C_{10}H_{12}O$	4-Isopropylbenzaldehyde	288.2	10.68
$C_{10}H_{12}O_2$	Propyl benzoate	303.2	5.78
$C_{10}H_{12}O_2$	Phenyl butanoate	293.2	4.48
$C_{10}H_{12}O_2$	2-Phenylethyl acetate	297.2	4.93
$C_{10}H_{12}O_2$	4-Allyl-2-methoxyphenol	293.2	9.55
$C_{10}H_{12}O_2$	Ethyl 2-phenylacetate	293.2	5.320
$C_{10}H_{12}O_2$	Benzyl propanoate	303.0	5.11
$C_{10}H_{13}NO_2$	Butyl pyridine-4-carboxylate	293.2	7.80
$C_{10}H_{13}N$	N-Vinyl-N-ethylaniline	293.2	7.8
$C_{10}H_{14}$	Butylbenzene	293.2	2.359
$C_{10}H_{14}$	Isobutylbenzene	293.2	2.318
$C_{10}H_{14}$	sec-Butylbenzene	293.2	2.357
$C_{10}H_{14}$	tert-Butylbenzene	293.2	2.359
$C_{10}H_{14}$	4-Isopropyltoluene	298.2	2.2322
$C_{10}H_{14}$	o-Diethylbenzene	293.2	2.594
$C_{10}H_{14}$	m-Diethylbenzene	293.2	2.369
$C_{10}H_{14}$	p-Diethylbenzene	293.2	2.259
$C_{10}H_{14}$	1-Ethyl-3,5-dimethylbenzene	293.2	2.275
$C_{10}H_{14}$	1,2,3,4-Tetramethylbenzene	296.0	2.538
$C_{10}H_{14}$	1,2,4,5-Tetramethylbenzene	356.0	2.223
$C_{10}H_{14}N_2$	Nicotine	293.2	8.937
$C_{10}H_{14}O$	Butyl phenyl ether	293.2	3.734
$C_{10}H_{14}O$	2-Hydroxy-4-isopropyltoluene	293.2	8.10
$C_{10}H_{14}O$	2-Methyl-1-phenyl-2-propanol	298.2	5.71
$C_{10}H_{14}O$	Thymol	333.2	4.259
$C_{10}H_{14}O_2$	(±)-Camphandione	476.2	16.3
$C_{10}H_{14}O_2S$	Pentyl thiophene-2-carboxylate	293.2	5.85
$C_{10}H_{14}O_3$	Pentyl furan-2-carboxylate	293.2	6.97
$C_{10}H_{14}O_3$	2-Furfuryl pentanoate	293.2	4.73
$C_{10}H_{15}N$	N,N-Diethylaniline	303.2	5.15
$C_{10}H_{16}$	2,6-Dimethyl-2,4,6-octatriene, a-form	298.2	2.5574
$C_{10}H_{16}$	Δ-3-Carene	298.2	2.1988
$C_{10}H_{16}$	(R)-(+)-Limonene	298.2	2.3746
$C_{10}H_{16}$	(±)-Limonene	298.2	2.381
$C_{10}H_{16}$	(S)-(-)-Limonene	298.2	2.3738
$C_{10}H_{16}$	Myrcene	298.2	2.3
$C_{10}H_{16}$	α-Pinene	298.2	2.1787
$C_{10}H_{16}$	(±)-α-Pinene	298.2	2.3695
$C_{10}H_{16}$	(1R)-(+)-α-Pinene	298.2	2.3684
$C_{10}H_{16}$	(1S)-(-)-α-Pinene	298.2	2.3701
$C_{10}H_{16}$	β-Pinene	298.2	2.4970
$C_{10}H_{16}$	α-Terpinene	298.2	2.4526
$C_{10}H_{16}$	γ-Terpinene	298.2	2.2738
$C_{10}H_{16}$	Terpinolene	298.2	2.2918
$C_{10}H_{16}O$	Carvenone	293.2	18.8
$C_{10}H_{16}O$	(±)-Fenchone	294.2	12.8
$C_{10}H_{16}O$	Thujone	273.2	10.8
$C_{10}H_{17}Cl$	(±)-2-Chlorobornane	368.2	5.21
$C_{10}H_{18}$	Decahydronaphthalene	293.2	2.1662
$C_{10}H_{18}$	cis-Decahydronaphthalene	293.2	2.219
$C_{10}H_{18}$	trans-Decahydronaphthalene	293.2	2.184

Molecular formula	Name	T/K	ε
$C_{10}H_{18}$	Pinane	298.2	2.1456
$C_{10}H_{18}O$	1,8-Cineole	298.2	4.57
$C_{10}H_{18}O_2$	Cyclohexyl butanoate	293.2	4.58
$C_{10}H_{18}O_2$	Propyl cyclohexanecarboxylate	293.2	4.44
$C_{10}H_{18}O_2$	6,7-Epoxy-3,7-dimethyl-1-octen-3-ol	297.85	5.78
$C_{10}H_{18}O_2S$	Pentyl tetrahydrothiophene-2-carboxylate	293.2	4.93
$C_{10}H_{18}O_3$	Tetrahydro-2-furfuryl pentanoate	293.2	7.49
$C_{10}H_{18}O_4$	Diethyl adipate	293.2	6.109
$C_{10}H_{19}NO_2$	Butyl piperidine-4-carboxylate	293.2	6.33
$C_{10}H_{20}$	1-Decene	293.2	2.136
$C_{10}H_{20}$	cis-5-Decene	298.2	2.071
$C_{10}H_{20}$	trans-5-Decene	298.2	2.030
$C_{10}H_{20}$	5-Methyl-4-nonene	293.2	2.175
$C_{10}H_{20}$	2,4,6-Trimethyl-3-heptene	293.2	2.293
$C_{10}H_{20}Br_2$	1,10-Dibromodecane	303.2	6.56
$C_{10}H_{20}Cl_2$	1,10-Dichlorodecane	308.2	6.68
$C_{10}H_{20}O$	2-Decanone	287.2	8.3
$C_{10}H_{20}O$	Menthol	309.3	3.90
$C_{10}H_{20}O_2$	2,2-Dimethyloctanoic acid	296.2	2.8
$C_{10}H_{20}O_2$	Octyl acetate	288.2	4.18
$C_{10}H_{20}O_2$	2-Methylheptyl acetate	288.2	4.27
$C_{10}H_{20}O_2$	1-Pentyl pentanoate	305.6	4.076
$C_{10}H_{20}O_2$	3-Methylbutyl pentanoate	292.2	3.6
$C_{10}H_{20}O_2$	3-Methylbutyl 3-methylbutanoate	288.2	4.39
$C_{10}H_{20}O_2$	Methyl nonanoate	293.2	3.943
$C_{10}H_{20}O_3$	2-Methyl-3-hydroxynonanonic acid	296.2	7.45
$C_{10}H_{21}Br$	1-Bromodecane	298.2	4.44
$C_{10}H_{21}Cl$	1-Chlorodecane	293.2	4.581
$C_{10}H_{21}NO$	N,N-Dibutylacetamide	293.2	19.1
$C_{10}H_{21}NO$	N,N-Dimethyloctanamide	293.2	17.4
$C_{10}H_{22}$	Decane	293.2	1.9853
$C_{10}H_{22}$	2,7-Dimethyloctane	293.2	1.98
$C_{10}H_{22}$	4-Propylheptane	293.2	1.9955
$C_{10}H_{22}Cl_2O_3NPS$	Di(pentoxy)phosphinylimidosulfurous dichloride	298.2	11.03
$C_{10}H_{22}O$	Dipentyl ether	298.2	2.798
$C_{10}H_{22}O$	Bis(3-methylbutyl) ether	293.2	2.817
$C_{10}H_{22}O$	1-Decanol	293.2	7.93
$C_{10}H_{22}O$	2-Decanol	298.2	5.82
$C_{10}H_{22}O$	(±)-3-Decanol	298.2	4.05
$C_{10}H_{22}O$	4-Decanol	298.2	3.42
$C_{10}H_{22}O$	5-Decanol	298.2	3.24
$C_{10}H_{22}O$	2,2-Dimethyl-1-octanol	293.2	7.86
$C_{10}H_{22}OS$	Dipentyl sulfoxide	348.2	18.8
$C_{10}H_{22}O_2S$	Dipentyl sulfone	348.2	20.0
$C_{10}H_{22}O_5$	Tetraethylene glycol dimethyl ether	298.2	7.68
$C_{10}H_{22}O_6$	Pentaethylene glycol	293.2	18.16
$C_{10}H_{22}S$	Dipentyl sulfide	298.2	3.826
$C_{10}H_{23}N$	Decylamine	293.2	3.31
$C_{10}H_{23}N$	Bis(3-methylbutyl)amine	291.2	2.5
$C_{10}H_{23}O_3P$	Dimethyl octanephosphonate	293.2	7.45
$C_{10}H_{24}O_2Si$	Dibutoxydimethylsilane	298.2	2.836
$C_{10}H_{24}O_2Si$	Bis(1-methylpropoxy)dimethylsilane	298.2	2.966
$C_{10}H_{24}O_3Si$	Tripropoxymethylsilane	298.2	3.339
$C_{10}H_{24}O_3Si$	Tris(1-methylethoxy)methylsilane	298.2	3.255
$C_{10}H_{27}NO_2Si_2$	Bis(trimethylsilaneoxyethyl)amine	313.2	3.572
$C_{10}H_{30}O_3Si_4$	Decamethyltetrasiloxane	293.2	2.370
$C_{10}H_{30}O_5Si_5$	Decamethylcyclopentasiloxane	293.2	2.50
$C_{11}F_{20}$	Perfluoro-1-methyldecalin	298.2	2.0

Molecular formula	Name	T/K	ε
$C_{11}H_6ClF_7O_2$	2,2,3,3,4,4,4-Heptafluorobutyl 4-chlorobenzoate	298.2	3.56
$C_{11}H_7F_7O_2$	2,2,3,3,4,4,4-Heptafluorobutyl benzoate	298.00	6.08
$C_{11}H_{10}$	1-Methylnaphthalene	293.2	2.915
$C_{11}H_{10}$	2-Methylnaphthalene	313.2	2.747
$C_{11}H_{10}O$	1-Methoxynaphthalene	293.2	4.020
$C_{11}H_{10}O$	2-Methoxynaphthalene	353.2	3.563
$C_{11}H_{12}O_2$	Ethyl trans-cinnamate	293.2	5.63
$C_{11}H_{12}O_3$	Ethyl benzoylacetate	303.2	13.50
$C_{11}H_{14}O_2$	Butyl benzoate	303.2	5.52
$C_{11}H_{14}O_2$	(2-Methylpropyl) benzoate	291.2	5.39
$C_{11}H_{14}O_2$	Phenyl pentanoate	293.2	4.30
$C_{11}H_{14}O_2$	Benzyl butanoate	301.2	4.55
$C_{11}H_{15}Cl$	Chloropentamethylbenzene	293.2	5.8
$C_{11}H_{15}NO_2$	Pentyl pyridine-4-carboxylate	293.2	7.00
$C_{11}H_{16}$	3-tert-Butyl-1-methylbenzene	293.2	2.330
$C_{11}H_{16}$	4-tert-Butyl-1-methylbenzene	293.2	2.250
$C_{11}H_{16}$	3,5-Diethyltoluene	293.2	2.264
$C_{11}H_{16}$	Pentamethylbenzene	334.0	2.358
$C_{11}H_{16}O_2S$	Hexyl thiophene-2-carboxylate	293.2	5.57
$C_{11}H_{16}O_3$	2-Furfuryl hexanoate	293.2	4.54
$C_{11}H_{16}O_3$	Hexyl furan-2-carboxylate	293.2	6.4
$C_{11}H_{17}N$	N-Ethyl-N-propylaniline	293.2	4.90
$C_{11}H_{17}O_3P$	Diethyl 4-methylbenzene phosphonate	303.2	11.18
$C_{11}H_{18}N_2O_2$	2,2,4-Trimethylhexamethylene diisocyanate	293.2	10.731
$C_{11}H_{20}O_2$	Butyl cyclohexanecarboxylate	293.2	4.26
$C_{11}H_{20}O_2$	Cyclohexyl pentanoate	293.2	4.32
$C_{11}H_{20}O_2S$	Hexyl tetrahydrothiophene-2-carboxylate	293.2	4.65
$C_{11}H_{20}O_3$	Tetrahydro-2-furfuryl hexanoate	293.2	6.91
$C_{11}H_{20}O_4$	(±)-erythro-3,4-Diacetoxyheptane	298.2	6.684
$C_{11}H_{20}O_4$	(±)-threo-3,4-Diacetoxyheptane	298.2	5.029
$C_{11}H_{21}NO_2$	Pentyl piperidine-4-carboxylate	293.2	5.95
$C_{11}H_{22}$	1-Undecene	293.2	2.137
$C_{11}H_{22}O$	2-Undecanone	285.3	8.3
$C_7H_{14}O_2$	Methyl hexanoate	293.2	3.785
$C_{11}H_{22}O_2$	Pentyl hexanoate	288.2	4.22
$C_{11}H_{22}O_2$	Nonyl acetate	293.2	3.87
$C_{11}H_{22}N_2O_2$	N,N,N',N'-Tetramethylpimelamide	298.2	37.6
$C_{11}H_{22}O_3$	3-Hydroxy-2-methyl decanoic acid	296.2	5.5
$C_{11}H_{23}Br$	1-Bromoundecane	272.6	4.61
$C_{11}H_{23}NO$	N,N-Dimethylpelargonamide	293.2	15.8
$C_{11}H_{24}$	Undecane	293.2	1.9972
$C_{11}H_{24}O$	1-Undecanol	313.2	5.98
$C_{11}H_{25}N$	Undecylamine	293.2	3.25
$C_{12}F_{27}OP$	Tris(perfluorobutyl)phosphine oxide	293.2	2.21
$C_{12}F_{27}N$	Tris(perfluorobutyl)amine	293.2	2.15
$C_{12}H_4Cl_6$	Hexachlorodiphenyl*	298.6	4.22
$C_{12}H_5Cl_5$	Pentachlorodiphenyl*	282.2	5.26
$C_{12}H_8F_8O_2$	1H,1H,5H-Octafluoropentyl benzoate	293.2	8.19
$C_{12}H_8Br_2N_2S$	Thiobis(p-bromophenyl)imine	298.2	7.7
$C_{12}H_8O$	Dibenzofuran	373.2	3.00
$C_{12}H_{10}$	Biphenyl	348.2	2.53
$C_{12}H_{10}O$	2-Acetonaphthone	333.2	13.03
$C_{12}H_{10}O$	Diphenyl ether	283.2	3.726
$C_{12}H_{10}N_2O$	Azoxybenzene	311.2	5.2
$C_{12}H_{10}OS$	Diphenyl sulfoxide	344.7	16.6
$C_{12}H_{10}O_2S$	Diphenyl sulfone	406.2	21.1
$C_{12}H_{10}N_2S$	Thiobis(N-phenyl)imine	333.2	3.93
$C_{12}H_{10}S$	Diphenyl sulfide	298.2	5.43

Molecular formula	Name	T/K	ε
$C_{12}H_{11}NO$	N-1-naphthylenylacetamide	433.2	24.3
$C_{12}H_{11}N$	Diphenylamine	323.2	3.73
$C_{12}H_{12}$	1,6-Dimethylnaphthalene	293.2	2.7250
$C_{12}H_{12}O$	1-Ethoxynaphthalene	292.2	3.3
$C_{12}H_{14}F_8O_4$	($1H,1H,3H$-Tetrafluoropropyl) 3-methylglutarate	293.2	8.85
$C_{12}H_{14}O_2$	Propyl cinnamate	293.2	5.45
$C_{12}H_{14}O_4$	Diethyl phthalate	293.2	7.86
$C_{12}H_{16}O$	2-Cyclohexylphenol	328.2	3.97
$C_{12}H_{16}O$	4-Cyclohexylphenol	404.2	4.42
$C_{12}H_{16}O_2$	Pentyl benzoate	293.2	5.07
$C_{12}H_{16}O_2$	Phenyl hexanoate	293.2	4.10
$C_{12}H_{16}O_3$	Pentyl salicylate	301.2	6.25
$C_{12}H_{16}O_3$	3-Methylbutyl salicylate	293.12	7.26
$C_{12}H_{17}NO$	N-Butylacetanilide	298.2	11.66
$C_{12}H_{17}NO_2$	Hexyl pyridine-4-carboxylate	293.2	6.75
$C_{12}H_{18}$	Hexamethylbenzene	449.0	2.172
$C_{12}H_{18}$	Hexylbenzene	293.2	2.3
$C_{12}H_{18}$	1,3,5-Triethylbenzene	293.2	2.256
$C_{12}H_{18}N_2O$	N,N-Dipropyl nicotinamide	298.2	2.2770
$C_{12}H_{18}O_2S$	Heptyl thiophene-2-carboxylate	293.2	5.25
$C_{12}H_{18}O_3$	Heptyl furan-2-carboxylate	293.2	5.8
$C_{12}H_{18}O_3$	2-Furfuryl heptanoate	293.2	4.28
$C_{12}H_{20}O_2$	l-Bornyl acetate	303.2	4.46
$C_{12}H_{20}O_6$	1,2,3-Propanetriol tripropanoate	212.7	9.34
$C_{12}H_{22}$	6-Dodecyne	298.2	2.171
$C_{12}H_{22}O$	Cyclododecanone	303.2	11.4
$C_{12}H_{22}O$	Dicyclohexyl ether	293.2	3.45
$C_{12}H_{22}O_2$	Cyclohexyl hexanoate	293.2	4.15
$C_{12}H_{22}O_2$	Pentyl cyclohexanecarboxylate	293.2	4.10
$C_{12}H_{22}O_2S$	Heptyl tetrahydrothiophene-2-carboxylate	293.2	4.53
$C_{12}H_{22}O_2Si_2$	1,2-Bis-(trimethylsiloxy)benzene	298.2	2.9767
$C_{12}H_{22}O_2Si_2$	1,3-Bis-(trimethylsiloxy)benzene	298.2	3.7847
$C_{12}H_{22}O_3$	Tetrahydrofurfuryl heptanoate	293.2	6.33
$C_{12}H_{22}O_6$	Dibutyl tartrate	314.2	9.4
$C_{12}H_{24}$	1-Dodecene	293.2	2.152
$C_{12}H_{24}O_2$	Ethyl decanoate	293.2	3.75
$C_{12}H_{24}O_2$	5,9-Dimethyldecanoic acid	296.2	3.0
$C_{12}H_{24}O_2$	Decyl acetate	293.2	3.75
$C_{12}H_{24}O_2$	Methyl undecanoate	293.2	3.671
$C_{12}H_{25}Br$	1-Bromododecane	298.2	4.07
$C_{12}H_{25}Cl$	1-Chlorododecane	298.2	4.17
$C_{12}H_{25}I$	1-Iodododecane	298.2	3.91
$C_{12}H_{25}NO$	N,N-Dimethyldecanamide	293.2	13.8
$C_{12}H_{25}NO$	N,N-Dipentylacetamide	293.2	15.8
$C_{12}H_{26}$	Dodecane	293.2	2.0120
$C_{12}H_{26}O$	2-Butyl-1-octanol	363.2	3.28
$C_{12}H_{26}O$	2,2-Dimethyl-1-decanol	293.2	3.27
$C_{12}H_{26}O$	1-Dodecanol	303.2	5.82
$C_{12}H_{26}O_7$	Hexaethylene glycol	293.2	16.00
$C_{12}H_{27}BO_3$	Tributyl borate	293.2	2.23
$C_{12}H_{27}N$	Dodecylamine	303.2	3.07
$C_{12}H_{27}N$	Tributylamine	293.2	2.340
$C_{12}H_{27}OP$	Tributyl phosphine oxide	323.2	26.4
$C_{12}H_{27}O_3P$	Diethyl octanephosphonate	305.2	6.291
$C_{12}H_{27}O_3PS$	Tributyl thiophosphate	298.2	6.92
$C_{12}H_{27}O_4P$	Tributyl phosphate	293.2	8.34
$C_{12}H_{27}O_4V$	Tributyl orthovanadate	298.2	2.780
$C_{12}H_{27}O_4V$	Trisisobutyl orthovandate	298.2	2.761

Molecular formula	Name	T/K	ε
$C_{12}H_{27}O_4V$	Tri-*sec*-butyl orthovanadate	298.2	2.699
$C_{12}H_{27}PS$	Tributyl phosphine sulfide	293.2	30.8
$C_{12}H_{28}Ge$	Tetrapropyl germanium	297.2	1.921
$C_{12}H_{28}O_4Si$	Tetrapropoxysilane	298.2	3.21
$C_{12}H_{28}Sn$	Tetrapropyltin	293.2	2.267
$C_{12}H_{30}OSi_2$	Hexaethyl disiloxane	298.2	2.259
$C_{12}H_{30}O_7Si_2$	Hexaethoxy disiloxane	298.2	5.18
$C_{12}H_{36}O_4Si_5$	Dodecamethyl pentasiloxane	293.2	2.476
$C_{12}H_{36}O_6Si_6$	Dodecamethyl cyclohexasiloxane	293.2	2.59
$C_{13}H_9ClN_2OS$	N-Benzoyl-N'-(p-chlorophenyl) thiodiimine	366.2	7.92
$C_{13}H_9FN_2OS$	N-Benzoyl-N'-(p-fluorophenyl) thiodiimine	340.2	7.25
$C_{13}H_{10}O$	Benzophenone	300.2	12.62
$C_{13}H_{10}NO_2P$	Diphenylphosphinic isocyanate	298.2	22.9
$C_{13}H_{10}O_3$	Phenyl salicylate	290.2	6.92
$C_{13}H_{12}$	Diphenylmethane	303.2	2.540
$C_{13}H_{12}O$	Benzyl phenyl ether	313.2	3.748
$C_{13}H_{16}O_2$	Butyl cinnamate	293.2	5.13
$C_{13}H_{18}O_2$	Hexyl benzoate	293.2	4.80
$C_{13}H_{18}O_2$	Phenyl heptanoate	293.2	3.90
$C_{13}H_{18}O_2$	Pentyl 2-phenylacetate	303.2	4.43
$C_{13}H_{19}NO_2$	Heptyl pyridine-4-carboxylate	293.2	6.07
$C_{13}H_{20}$	Heptylbenzene	293.2	2.26
$C_{13}H_{20}O$	α-Ionone	292.4	10.78
$C_{13}H_{20}O$	β-Ionone	297.65	11.66
$C_{13}H_{20}O_2S$	Octyl thiophene-2-carboxylate	293.2	4.93
$C_{13}H_{20}O_3$	2 Furfuryl octanoate	293.2	4.02
$C_{13}H_{20}O_3$	Octyl furan-2-carboxylate	293.2	5.67
$C_{13}H_{24}O$	Cyclotridecanone	303.2	9.34
$C_{13}H_{24}O_2$	Hexyl cyclohexanecarboxylate	293.2	3.94
$C_{13}H_{24}O_2$	Cyclohexyl heptanoate	293.2	3.95
$C_{13}H_{24}O_2S$	Octyl tetrahydrothiophene-2-carboxylate	293.2	4.20
$C_{13}H_{24}O_3$	Tetrahydro-2-furfuryl octanoate	293.2	5.89
$C_{13}H_{24}O_4$	Diethyl azelate	303.2	5.133
$C_{13}H_{25}NO_2$	Heptyl piperidine-4-carboxylate	293.2	5.20
$C_{13}H_{26}$	1-Tridecene	293.2	2.139
$C_{13}H_{26}O$	7-Tridecanone	303.2	7.6
$C_{13}H_{26}O_2$	2,2-Dimethylundecanoic acid	296.2	2.76
$C_{13}H_{26}O_2$	Methyl dodecanoate	293.2	3.539
$C_{13}H_{26}O_2$	Ethyl undecanoate	293.2	3.55
$C_{13}H_{26}O_2$	2,5,9-Trimethyldecanoic acid	296.2	3.04
$C_{13}H_{26}O_3$	3,5,9-Trimethyl-3-hydroxydecanoic acid	296.2	8.75
$C_{13}H_{27}Br$	1-Bromotridecane	281.15	4.19
$C_{13}H_{28}$	5-Butylnonane	293.2	2.0319
$C_{13}H_{28}$	Tridecane	293.2	2.0213
$C_{13}H_{28}O$	1-Tridecanol	333.2	4.02
$C_{13}H_{28}O_4Si_2$	Bis(trimethylsilylmethyl) glutarate	293.2	4.17
$C_{13}H_{30}O_3Si$	Tris(butoxymethyl)silane	298.2	3.138
$C_{13}H_{30}O_3Si$	Tris(*sec*-butoxymethyl)silane	298.2	2.95
$C_{14}H_9Cl_5$	1,1,1-Trichloro-2,2-bis(4-chlorophenyl)ethane	377.2	2.900
$C_{14}H_9F_3O_3N_2S_2$	N-Benzoyl-N'-(p-trifluoro-methylsulfonyl)phenylthiodiimine	343.2	15.9
$C_{14}H_{10}$	Anthracene	502.0	2.649
$C_{14}H_{10}$	Phenanthrene	383.2	2.72
$C_{14}H_{10}ClNO$	1-Chloro-1,1-diphenylmethyl isocyanate	288.2	20.3
$C_{14}H_{10}O_2$	Benzil	368.2	13.04
$C_{14}H_{12}N_2OS$	N-Benzoyl-N'-(p-methylphenyl) thiodiimine	341.2	13.1
$C_{14}H_{12}O_2$	Benzyl benzoate	303.2	5.26
$C_{14}H_{12}O_3$	Benzyl salicylate	301.2	4.12
$C_{14}H_{14}$	1,2-Diphenylethane	331.2	2.47

Molecular formula	Name	T/K	ε
$C_{14}H_{14}$	3,3'-Dimethylbiphenyl	298.15	2.519
$C_{14}H_{14}O$	Dibenzyl ether	293.2	3.821
$C_{14}H_{14}O_3N_2$	4-Azoxyanisole	408.6	5.702
$C_{14}H_{15}N$	Dibenzylamine	293.2	3.446
$C_{14}H_{16}O_2Si$	Dimethyldiphenoxysilane	298.2	3.500
$C_{14}H_{18}O$	Pentyl-3-phenyl-2-propenal*	293.2	11.00
$C_{14}H_{18}O_2$	Pentyl cinnamate	293.2	4.89
$C_{14}H_{20}O_2$	Heptyl benzoate	293.2	4.50
$C_{14}H_{20}O_2$	Phenyl octanoate	293.2	3.72
$C_{14}H_{21}NO_2$	Octyl pyridine-4-carboxylate	293.2	5.80
$C_{14}H_{22}$	Octylbenzene	293.2	2.26
$C_{14}H_{22}O_2S$	Nonyl thiophene-2-carboxylate	293.2	4.77
$C_{14}H_{22}O_3$	Nonyl furan-2-carboxylate	293.2	5.1
$C_{14}H_{26}O_2$	Cyclohexyl octanoate	293.2	3.70
$C_{14}H_{26}O_2$	Heptyl cyclohexanecarboxylate	293.2	3.82
$C_{14}H_{26}O_2S$	Nonyl tetrahydrothiophene-2-carboxylate	293.2	4.02
$C_{14}H_{26}O_3$	Tetrahydro-2-furfuryl pelargonate	293.2	5.67
$C_{14}H_{26}O_4$	Diisobutyl adipate	293.2	5.19
$C_{14}H_{26}O_4$	Diethyl sebacate	303.2	4.995
$C_{14}H_{27}NO_2$	Octyl piperidine-4-carboxylate	293.2	5.05
$C_{14}H_{28}O_2$	Dodecyl acetate	293.2	3.6
$C_{14}H_{28}O_2$	Ethyl laurate	273.2	3.94
$C_{14}H_{28}O_2$	Methyl tridecanoate	293.2	3.442
$C_{14}H_{29}Br$	1-Bromotetradecane	293.2	3.84
$C_{14}H_{30}$	Tetradecane	293.2	2.0343
$C_{14}H_{30}O$	1-Tetradecanol	318.2	4.42
$C_{14}H_{30}OS$	Diheptyl sulfoxide	347.2	13.0
$C_{14}H_{30}O_5Si_2$	Bis(isobutoxymethyl) tetramethyldisiloxane	293.2	4.25
$C_{14}H_{30}O_8$	Heptaethylene glycol	293.2	14.85
$C_{14}H_{31}O_3P$	Diethyl decanephosphonate	293.2	6.06
$C_{14}H_{31}N$	Tetradecylamine	312.55	2.90
$C_{14}H_{42}O_5Si_6$	Tetradecamethyl hexasiloxane	293.2	2.50
$C_{14}H_{42}O_7Si_7$	Tetradecamethyl cycloheptasiloxane	293.2	2.68
$C_{15}F_{33}OP$	Tris(perfluoroisopentyl)phosphine oxide	293.2	2.12
$C_{15}H_9F_3ClNO$	Chlorophenyl(p-trifluoromethylphenyl)methyl isocyanate	303.2	11.40
$C_{15}H_{10}N_2O_2$	1,1'-Diphenylmethane diisocyanate	313.2	7.300
$C_{15}H_{12}F_{16}O_4$	Bis(1H,1H,5H-octafluoropentyl) glutarate	293.2	7.48
$C_{15}H_{12}ClNO$	Chlorophenyl(m-methylphenyl)methyl isocyanate	303.2	18.59
$C_{15}H_{12}ClNO$	Chlorophenyl(p-methylphenyl)methyl isocyanate	288.2	19.4
$C_{15}H_{12}ClNO_2$	Chlorophenyl(p-methoxyphenyl)methyl isocyanate	288.2	24.45
$C_{15}H_{12}O_4$	Phenyl acetylsalicylate	384.2	4.33
$C_{15}H_{13}NO$	N-Vinylbenzanilide	293.2	3.6
$C_{15}H_{14}F_{12}O_6$	Tris(1H,1H,3H-tetrafluoropropyl) carballate	293.2	10.8
$C_{15}H_{14}O_2$	Benzyl 2-phenylacetate	303.2	4.54
$C_{15}H_{16}O_2S$	Tolyl xylyl sulfone*	297.2	23.5
$C_{15}H_{18}O_2$	Hexyl cinnamate	293.2	4.73
$C_{15}H_{22}O_2$	Octyl benzoate	293.2	4.34
$C_{15}H_{23}NO_2$	Nonyl pyridine-4-carboxylate	293.2	5.55
$C_{15}H_{24}O_2S$	Nonyl phenyl sulfone	323.2	17.6
$C_{15}H_{24}O_2S$	Decyl thiophene-2-carboxylate	293.2	4.57
$C_{15}H_{24}O_3$	Decyl furan-2-carboxylate	293.2	4.93
$C_{15}H_{26}O_6$	Tributyrin	282.8	5.72
$C_{15}H_{28}O_2$	Octyl cyclohexanecarboxylate	293.2	3.70
$C_{15}H_{28}O_3$	Tetrahydro-2-furfuryl decanoate	293.2	5.35
$C_{15}H_{30}O_2$	Methyl myristate	293.2	3.352
$C_{15}H_{31}Br$	1-Bromopentadecane	293.35	3.88
$C_{15}H_{32}$	Pentadecane	293.2	2.0391
$C_{15}H_{32}O$	1-Pentadecanol	333.2	3.70

Molecular formula	Name	T/K	ε
$C_{15}H_{33}O_4V$	Tris(1,1-dimethylpropyl) orthovanandate	298.2	2.764
$C_{15}H_{33}N$	Pentadecylamine	313.25	2.85
$C_{15}H_{33}N$	Tris(3-methyl-1-butyl)amine	294.2	2.29
$C_{15}H_{33}PS$	Tripentylphosphine sulfide	293.2	23.3
$C_{15}H_{39}O_3NSi_3$	Tris-(trimethylsilanoxyethyl)amine	293.2	3.333
$C_{16}H_{14}F_{16}O_4$	Bis(1H,1H,5H-octafluoropentyl) 3-methylglutarate	293.2	7.44
$C_{16}H_{18}N_2O_3$	4-Azoxyphenetole	416.2	5.02
$C_{16}H_{22}O_2$	Heptyl cinnamate	293.2	4.58
$C_{16}H_{22}O_4$	Dibutyl phthalate	293.2	6.58
$C_{16}H_{24}O_2$	Nonyl benzoate	293.2	4.20
$C_{16}H_{24}O_2$	Phenyl decanoate	293.2	3.52
$C_{16}H_{25}NO_2$	Decyl pyridine-4-carboxylate	293.2	5.33
$C_{16}H_{26}O_2S$	Undecyl thiophene-2-carboxylate	293.2	4.40
$C_{16}H_{26}O_3$	Undecyl furan-2-carboxylate	303.2	4.73
$C_{16}H_{30}O_2$	Nonyl cyclohexanecarboxylate	293.2	3.60
$C_{16}H_{30}O_2$	Cyclohexyl decanoate	293.2	3.70
$C_{16}H_{30}O_3$	Tetrahydro-2-furfuryl undecanoate	293.2	5.15
$C_{16}H_{32}O_2$	Tetradecyl acetate	293.2	3.41
$C_{16}H_{32}O_2$	Ethyl myristate	293.2	3.50
$C_{16}H_{32}O_2$	Hexadecanoic acid	338.2	2.417
$C_{16}H_{32}O_2$	Methyl pentadecanoate	293.2	3.296
$C_{16}H_{33}Br$	1-Bromohexadecane	298.2	3.68
$C_{16}H_{33}I$	1-Iodohexadecane	293.2	3.57
$C_{16}H_{34}$	Hexadecane	293.2	2.0460
$C_{16}H_{34}$	6-Pentylundecane	293.2	2.0540
$C_{16}H_{34}O$	1-Hexadecanol	333.2	3.69
$C_{16}H_{34}OS$	Dioctyl sulfoxide	360.2	11.8
$C_{16}H_{34}O_2S$	Dioctyl sulfone	348.2	14.4
$C_{16}H_{34}O_2Si$	Trimethylsilylmethyl laurate	293.2	2.92
$C_{16}H_{34}O_5$	Tetraethylene glycol dibutyl ether	298.2	5.15
$C_{16}H_{35}N$	Hexadecylamine	328.35	2.71
$C_{16}H_{35}O_3P$	Diethyl dodecanephosphonate	305.2	5.162
$C_{16}H_{36}Ge$	Tetrabutyl germanium	297.2	2.334
$C_{16}H_{36}O_4Si$	Tetra(1-methylpropoxy)silane	298.2	2.59
$C_{16}H_{36}O_4Si$	Tetra(2-methylpropoxy)silane	298.2	2.81
$C_{16}H_{36}Sn$	Tetrabutyl tin	293.2	9.74
$C_{16}H_{40}O_{10}Si_3$	Octaethoxytrisiloxane	298.2	4.76
$C_{16}H_{48}O_6Si_7$	Hexadecamethyl heptasiloxane	293.2	2.589
$C_{16}H_{48}O_8Si_8$	Hexadecamethyl cycloheptasiloxane	293.2	2.74
$C_{17}H_{12}O_3$	2-Naphthyl salicylate	293.0	6.30
$C_{17}H_{18}F_{16}O_2$	1H,1H,9H-Hexadecafluorononyl 2-ethylhexanoate	293.2	4.66
$C_{17}H_{24}O_2$	Octyl cinnamate	293.2	4.38
$C_{17}H_{26}O_2$	Decyl benzoate	293.2	4.05
$C_{17}H_{27}NO_2$	Undecyl pyridine-4-carboxylate	295.2	5.15
$C_{17}H_{28}O_2S$	Dodecyl thiophene-2-carboxylate	293.2	4.25
$C_{17}H_{32}O_2$	Decyl cyclohexanecarboxylate	293.2	3.52
$C_{17}H_{32}O_3$	Tetrahydro-2-furfuryl laurate	293.2	4.88
$C_{17}H_{34}O$	9-Heptadecanone	328.2	5.43
$C_{17}H_{34}O_2$	Methyl palmitate	313.2	3.124
$C_{17}H_{36}$	Heptadecane	293.2	2.0578
$C_{17}H_{36}N_2O$	Tetrabutyl urea	298.2	9.41
$C_{17}H_{36}O$	1-Heptadecanol	333.2	3.41
$C_{17}H_{36}O_4Si_2$	Bis(trimethylsilylmethyl) azelate	293.2	4.32
$C_{18}H_{10}F_{16}O_4$	Bis(1H,1H,5H-octafluoropentyl) isophthalate	293.2	7.87
$C_{18}H_{10}F_{16}O_4$	Bis(1H,1H,5H-octafluoropentyl) phthalate	293.2	8.67
$C_{18}H_{12}F_{22}O_4$	Bis(1H,1H-undecafluorohexyl) 3-methylglutarate	293.2	4.2
$C_{18}H_{14}O_2$	1,3-Diphenoxybenzene	343.2	3.82
$C_{18}H_{15}NO_4S_2$	Phenylsulfonylimidosulfurous acid diphenyl ester	339.2	23.0

Molecular formula	Name	T/K	ε
$C_{18}H_{26}O_4$	Dipentyl phthalate	293.2	6.00
$C_{18}H_{28}O_2$	Phenyl laurate	293.2	3.28
$C_{18}H_{29}NO_2$	Dodecyl pyridine-4-carboxylate	308.2	4.33
$C_{18}H_{30}O$	1-Phenyl-1-dodecanol	333.2	3.30
$C_{18}H_{30}O_2$	Dodecyl benzoate	293.2	3.90
$C_{18}H_{30}O_2$	Linolenic acid	293.2	2.825
$C_{18}H_{30}O_4$	Dicyclohexyl adipate	308.2	4.84
$C_{18}H_{32}O_2$	Linoleic acid	293.2	2.754
$C_{18}H_{32}O_6$	Tributyl tricarballate	293.2	5.7
$C_{18}H_{34}O_2$	Undecyl cyclohexanecarboxylate	293.2	3.45
$C_{18}H_{34}O_2$	Cyclohexyl laurate	293.2	3.50
$C_{18}H_{34}O_2$	Oleic acid	293.2	2.336
$C_{18}H_{34}O_3$	cis-Ricinoleic acid	373.2	3.547
$C_{18}H_{34}O_3$	trans-Ricinoleic acid	373.2	3.520
$C_{18}H_{34}O_4$	Dibutyl sebacate	293.2	4.54
$C_{18}H_{36}O$	9-Octadecene-1-ol	295.2	3.85
$C_{18}H_{36}O_2$	Cetyl acetate	308.2	3.19
$C_{18}H_{36}O_2$	Methyl heptadecanoate	313.2	3.07
$C_{18}H_{36}O_2$	Ethyl palmitate	303.2	3.07
$C_{18}H_{36}O_2$	Stearic acid	293.2	2.314
$C_{18}H_{37}Br$	1-Bromooctadecane	303.35	3.53
$C_{18}H_{37}NO$	N,N-Dioctylacetamide	293.2	11.5
$C_{18}H_{38}O$	1-Octadecanol	333.2	3.38
$C_{18}H_{38}OS$	Dinonyl sulfoxide	359.2	10.0
$C_{18}H_{38}O_2S$	Dinonyl sulfone	358.2	12.1
$C_{18}H_{39}BO_3$	Trihexyl borate	293.2	2.22
$C_{18}H_{39}N$	1-Octadecylamine	326.35	2.67
$C_{18}H_{39}O_3P$	Diethyl tetradecanephosphonate	305.2	4.631
$C_{19}H_{14}F_6O_4$	2,2,3,3,4,4-Hexafluoro-1,5-pentanediol dibenzoate	293.2	6.73
$C_{19}H_{16}$	Triphenylmethane	367.2	2.46
$C_{19}H_{18}O_3Si$	Methyltriphenoxysilane	298.2	3.628
$C_{19}H_{28}O_2$	Decyl cinnamate	293.2	4.05
$C_{19}H_{32}O$	1-Phenyl-1-tridecanol	333.2	3.20
$C_{19}H_{32}O_2$	Methyl linolenate	293.2	3.466
$C_{19}H_{34}O_2$	Methyl linoleate	293.2	3.355
$C_{19}H_{36}O_2$	Dodecyl cyclohexanecarboxylate	293.2	3.37
$C_{19}H_{36}O_2$	Methyl oleate	293.2	3.211
$C_{19}H_{37}NO$	Octadecyl isocyanate	303.2	5.137
$C_{19}H_{38}O$	10-Nonadecanone	353.2	5.37
$C_{19}H_{38}O_2$	Methyl stearate	313.2	3.021
$C_{19}H_{38}O_4$	1,2,3-Propanetriol monopalmitate	340.3	5.34
$C_{19}H_{40}$	7-Hexyltridecane	293.2	2.0715
$C_{19}H_{40}$	Nonadecane	293.2	2.0706
$C_{19}H_{42}O_3Si$	Trihexoxymethylsilane	298.2	2.853
$C_{20}H_{12}F_{12}O_3$	1H,1H,7H-Dodecafluoroheptyl 2-phenoxybenzoate	293.2	7.60
$C_{20}H_{14}F_8O_4$	Bis(1H,1H,3H-tetrafluoropropyl) 2,2'-biphenyldicarboxylate	293.2	9.7
$C_{20}H_{14}F_{24}O_4$	Bis(1H,1H,7H-dodecafluoroheptyl) 3-methylglutarate	293.2	5.94
$C_{20}H_{18}F_{24}O_2$	Bis(1H,1H,7H-dodecafluoroheptoxy)hexane	293.2	9.4
$C_{20}H_{26}O_5Si_2$	Bis(benzoyloxymethyl)tetramethyl disiloxane	293.2	4.88
$C_{20}H_{30}O_4$	Dihexyl phthalate	293.2	5.62
$C_{20}H_{34}O$	1-Phenyl-1-tetradecanol	333.2	3.10
$C_{20}H_{38}O_2$	Ethyl oleate	301.2	3.17
$C_{20}H_{38}O_3$	cis-Ethyl ricinoleate	373.2	3.475
$C_{20}H_{38}O_3$	trans-Ethyl ricinoleate	373.2	3.775
$C_{20}H_{40}O_2$	Octadecyl acetate	308.2	3.07
$C_{20}H_{40}O_2$	Methyl nonadecanoate	313.2	2.982
$C_{20}H_{40}O_2$	Ethyl stearate	313.2	2.958
$C_{20}H_{42}O$	Didecyl ether	293.2	2.644

Molecular formula	Name	T/K	ε
$C_{20}H_{42}O$	1-Eicosanol	338.2	3.13
$C_{20}H_{42}O_2$	Octadecoxy ethanol	328.2	3.56
$C_{20}H_{43}O_3P$	Dihexyl octanephosphonate	298.2	5.15
$C_{20}H_{43}O_3P$	Diethyl hexadecanephosphonate	305.2	4.28
$C_{20}H_{44}O_4Si$	Tetrapentoxysilane	298.2	2.82
$C_{20}H_{44}Ge$	Tetrapentylgermanium	297.2	2.299
$C_{20}H_{60}O_8Si_9$	Eicosamethyl nonasiloxane	293.2	2.645
$C_{21}H_{14}F_{24}O_6$	Tris(1H,1H,5H-octafluoropentyl) tricarballate	293.2	7.81
$C_{21}H_{21}O_4P$	Tricresyl phosphate*	298.2	6.7
$C_{21}H_{32}O_2$	Dodecyl cinnamate	293.2	3.82
$C_{21}H_{36}O$	1-Phenyl-1-pentadecanol	333.2	3.06
$C_{21}H_{38}O_6$	1,2,3-Propanetriyl hexanoate	293.2	4.476
$C_{21}H_{38}O_6$	Tris(3-methyl-1-butyl)tricarballate	293.2	5.1
$C_{21}H_{42}O_3$	2-Methoxyethyl stearate	323.2	3.387
$C_{21}H_{42}O_4$	1,2,3-Propanetriol monostearate	353.2	4.84
$C_{21}H_{44}O_2$	Octadecoxypropanol	328.2	3.86
$C_{21}H_{45}BO_3$	Triheptyl borate	293.2	2.20
$C_{21}H_{45}OP$	Triheptylphosphine oxide	323.2	30.0
$C_{21}H_{45}PS$	Triheptylphosphine sulfide	293.2	20.4
$C_{22}H_{12}F_{30}O_4$	Bis(1H,1H-pentadecafluorooctyl) 3-methylglutarate	293.2	3.81
$C_{22}H_{30}O_4Si_2$	Bis(trimethylsilylmethyl) 2,2'-biphenyldicarboxylate	293.2	4.50
$C_{22}H_{30}O_5Si_2$	Bis(2-phenylacetoxymethyl)tetramethyldisiloxane	293.2	4.10
$C_{22}H_{34}O_4$	Diheptyl phthalate	293.2	5.22
$C_{22}H_{38}O$	1-Phenyl-1-hexadecanol	333.2	3.00
$C_{22}H_{42}O_2$	cis-13-Docosene carboxylic acid	307.8	2.7028
$C_{22}H_{42}O_2$	Butyl oleate	298.2	4.00
$C_{22}H_{42}O_3$	Isobutyl 12-hydroxy-9-octadecenoate	294.2	4.7
$C_{22}H_{44}O_2$	Butyl stearate	298.2	3.120
$C_{22}H_{44}O_2$	2,9,13,17-Tetramethyloctadecanoic acid	296.2	2.72
$C_{22}H_{45}Br$	1-Bromodocosane	315.85	3.20
$C_{22}H_{46}$	Docosane	293.2	2.0840
$C_{22}H_{46}$	8-Heptylpentadecane	293.2	2.1017
$C_{22}H_{46}O$	1-Docosanol	348.2	2.94
$C_{22}H_{46}O_2$	Eicosanoxyethanol	338.2	3.38
$C_{22}H_{47}O_3P$	Dihexyl decanephosphonate	293.2	4.74
$C_{22}H_{47}O_3P$	Diethyl octadecanephosphonate	305.2	4.052
$C_{23}H_{40}O$	1-Phenyl-1-heptadecanol	333.2	2.98
$C_{23}H_{44}O_2$	Methyl cis-13-docosenecarboxylate	293.2	3.043
$C_{23}H_{44}O_4$	Dinonyl glutarate	293.2	4.0
$C_{23}H_{46}O_2$	2,5,9,13,17-Pentamethyloctadecanoic acid	296.2	2.76
$C_{23}H_{48}O_2$	Eicosanoxypropanol	333.2	3.63
$C_{23}H_{48}O_4$	Dimethoxydidecaoxy methane	293.2	2.39
$C_{24}H_{12}F_{24}O_6$	Tris(1H,1H,5H-octafluoropentyl) hemimellate	293.2	8.98
$C_{24}H_{14}F_{16}O_4$	Bis(1H,1H,5H-octafluoropentyl) 2,2'-biphenyldicarboxylate	293.2	8.0
$C_{24}H_{15}F_{24}O_2$	Bis(1H,1H,7H-dodecafluoroheptyl) 2-phenylsuccinate	293.2	5.60
$C_{24}H_{20}O_4Si$	Tetraphenoxysilane	333.2	3.4915
$C_{24}H_{30}O_4$	Dibenzyl sebacate	298.2	4.61
$C_{24}H_{38}O_4$	Bis(2-ethylhexyl) phthalate	293.2	4.96
$C_{24}H_{38}O_4$	Dioctyl phthalate	293.2	5.22
$C_{24}H_{42}O$	1-Phenyl 1-octadecanol	333.2	2.93
$C_{24}H_{50}O_2$	Docosanoxyethanol	338.2	3.27
$C_{24}H_{51}BO_3$	Trioctyl borate	293.2	2.16
$C_{24}H_{51}O_3P$	Diheptyl decanephosphonate	293.2	4.77
$C_{24}H_{51}PS$	Trioctylphosphine sulfide	293.2	17.2
$C_{24}H_{72}O_{10}Si_{11}$	Tetracosamethyl undecasiloxane	293.2	2.669
$C_{25}H_{16}F_{24}O_4$	Bis(1H,1H,7H-dodecafluoroheptyl) 2-phenylglutarate	293.2	5.34
$C_{25}H_{44}O$	1-Phenyl-1-nonadecanol	333.2	2.88
$C_{25}H_{52}$	9-Octylheptadecane	293.2	2.1061

Molecular formula	Name	T/K	ε
$C_{25}H_{52}O_2$	Docosanoxypropanol	338.2	3.47
$C_{26}H_{42}O_4$	Bis(3,3,5-trimethylhexyl) phthalate	293.2	4.72
$C_{26}H_{50}O_4$	Bis(2-ethylhexyl) sebacate	298.2	4.03
$C_{26}H_{50}O_4$	Dioctyl sebacate	299.2	4.01
$C_{27}H_{50}O_6$	1,2,3-Propanetriyl octanoate	293.2	3.931
$C_{27}H_{57}BO_3$	Trinonyl borate	293.2	2.20
$C_{27}H_{57}OP$	Trinonylphosphine oxide	323.2	15.4
$C_{28}H_{18}F_{32}O_8$	Tetra(1H,1H,5H-octofluoropentyl) 1,2,3,4-cyclo-butanetetracarboxylate	350.2	5.24
$C_{28}H_{56}O_2$	Decyl stearate	313.2	2.81
$C_{28}H_{58}$	10-Nonylnonadecane	293.2	2.102
$C_{29}H_{60}O_4Si_2$	Bis(methyldipentylsilylmethyl)glutarate	293.2	3.19
$C_{30}H_{58}O_4$	Ethylene dimyristate	343.2	2.98
$C_{30}H_{60}O_2$	Tetradecyl palmitate	323.2	2.66
$C_{30}H_{62}$	Triacontane	373.2	1.9112
$C_{30}H_{62}$	Squalane	373.2	1.9106
$C_{30}H_{63}BO_3$	Tridecyl borate	293.2	2.21
$C_{30}H_{63}PS$	Tridecylphosphine sulfide	343.2	11.0
$C_{31}H_{64}$	11-Decylheneicosane	293.2	2.1255
$C_{32}H_{64}O_2$	Tetradecyl stearate	323.2	2.67
$C_{33}H_{62}O_6$	1,2,3-Propanetriol tridecanoate	313.2	3.480
$C_{34}H_{66}$	Tetratriacontadiene*	298.2	2.82
$C_{34}H_{66}O_4$	Ethylene dipalmitate	348.2	2.89
$C_{34}H_{68}O_2$	Cetyl stearate	333.2	2.61
$C_{34}H_{70}$	12-Undecyltricosane	293.2	2.111
$C_{35}H_{68}O_5$	1,2,3-Propanetriol-1,3-dipalmitate	345.2	3.52
$C_{37}H_{76}$	13-Dodecylpentacosane	293.2	2.1203
$C_{38}H_{74}O_4$	Ethylene distearate	353.2	2.79
$C_{39}H_{74}O_6$	Trilaurin	313.2	3.287
$C_{39}H_{76}O_5$	1,2,3-Propanetriol 1,3-distearate	351.2	3.32
$C_{41}H_{84}O_4$	Tetradecoxymethane	293.2	2.44
$C_{48}H_{99}PS$	Trihexadecylphosphine sulfide	378.2	7.7
$C_{51}H_{98}O_6$	Tripalmitin	328.2	2.901
$C_{57}H_{98}O_6$	1,2,3-Propanetriol trilinoleate	293.2	3.470
$C_{57}H_{104}O_6$	*trans, trans, trans*-1,2,3-Propanetriyl 9-octadecenoate	313.2	2.980
$C_{57}H_{104}O_6$	Triolein	293.2	3.109
$C_{57}H_{110}O_6$	Tristearin	353.2	2.740

* Unspecified isomer.

TEMPERATURE DEPENDENCE OF THE PERMITTIVITY (DIELECTRIC CONSTANT) OF LIQUIDS

Christian Wohlfarth

This table describes the temperature dependence of the permittivity (dielectric constant) of about 900 pure liquids (see the preceding table for permittivity values at fixed temperature). The table gives the coefficients of a simple polynomial fitting of permittivity to temperature with an equation of the form

$$\varepsilon(T) = a + bT + cT^2 + dT^3$$

where T is the absolute temperature in K. The temperature range of the fit and the sum of the squares of the deviations divided by the number of points, ESQS/N, are also given. The parameter d was used in only a few cases where the quadratic fit was not satisfactory. When ESQS/N is blank, the number of available data points was equal to the number of fitting parameters.

The coefficients of the fitting equation can be used to calculate dielectric constants within the fitted temperature range. They must not be used for extrapolation without care.

The reader who needs dielectric constant data with more accuracy than can be provided by the fit is referred to Reference 1, which gives the original data together with their literature source.

The static permittivities are given here at nominal atmospheric pressure as long as the corresponding temperature is below the normal boiling point. Otherwise, at temperatures above the normal boiling point, the pressure is the saturated vapor pressure of the substance considered.

Substances are listed by molecular formula in the modified Hill order, with substances not containing carbon preceding those that do contain carbon.

REFERENCES

1. Wohlfarth, Ch., Static Dielectric Constants of Pure Liquids and Binary Liquid Mixtures, *Landolt-Börstein, Numerical Data and Functional Relationships in Science and Technology*, New Series, Editor-in-Chief O. Madelung, Group IV, Macroscopic and Technical Properties of Matter, Volume 6, Springer Verlag Berlin, Heidelberg, New York, 1991.
2. Marsh, K. N., Ed., *Recommended Reference Materials for the Realization of Physicochemical Properties*, Blackwell Scientific Publications, Oxford, 1987.

Molecular formula	Name	a	b	c	d	ESQS/N	Range T/K
Ar	Argon	0.12408E+01	0.68755E-02	-0.45344E-04		0.5213E-04	87.0-149.1
AsH$_3$	Arsine	0.37674E+01	-0.97454E-02	0.14537E-04		0.6566E-04	157.1-200.9
B$_2$H$_6$	Diborane	0.23848E+01	-0.29501E-02	0.64189E-06		0.1731E-06	108.3-180.7
B$_5$H$_9$	Pentaborane(9)	0.40952E+03	-0.24414E+01	0.38225E-02		0.4185E-01	226.2-298.2
BrF$_5$	Bromine pentafluoride	0.11428E+02	-0.11822E-01			0.7246E-04	261.5-297.7
Br$_2$	Bromine	0.32701E+01	-0.12535E-03			0.8507E-01	273.2-327.0
Br$_3$OV	Vanadium oxytribromide	0.61112E+01	-0.84211E-02				203.2-298.2
Br$_4$Ge	Germanium tetrabromide	0.34450E+01	-0.16083E-02			0.3991E-04	299.9-316.0
Br$_4$Sn	Tin tetrabromide	0.50001E+01	-0.60383E-02			0.1724E-05	303.5-316.0
ClFO$_3$	Perchloryl fluoride	0.23808E+01	-0.38629E-03	-0:57143E-05		0.2286E-06	125.2-150.2
ClF$_3$	Chlorine trifluoride	0.96716E+01	-0.18000E-01				273.2-313.2
ClF$_5$	Chlorine pentafluoride	0.78192E+01	-0.20860E-01	0.13132E-04		0.1498E-03	193.2-256.2
ClH	Hydrogen chloride	0.47316E+02	-0.28455E+00	0.48650E-03		0.3369E-01	158.9-258.2
Cl$_2$	Chlorine	0.29440E+01	-0.44649E-02	0.30388E-05		0.3874E-05	208.0-240.0
Cl$_2$F$_3$P	Phosphorus dichloride trifluoride	0.46501E+01	-0.80358E-02				171.7-228.6
Cl$_3$F$_2$P	Phosphorus trichloride difluoride	0.28905E+01	-0.19228E-02				214.9-268.0
Cl$_3$P	Phosphorus trichloride	0.59098E+01	-0.83322E-02			0.5091E-04	290.2-333.0
Cl$_4$FP	Phosphorus tetrachloride fluoride	0.33503E+01	-0.29651E-02				244.1-272.6
Cl$_4$Ge	Germanium tetrachloride	-0.55078E+01	0.64881E-01	-0.13091E-03		0.1299E-03	245.7-273.2
Cl$_4$Si	Silicon tetrachloride	0.58041E+01	-0.27129E-01	0.51678E-04		0.3275E-04	206.7-273.2
Cl$_4$Sn	Tin tetrachloride	0.43951E+01	-0.48805E-02			0.1190E-02	233.7-273.2
Cl$_4$Ti	Titanium tetrachloride	0.33668E+01	-0.19675E-02			0.1033E-03	236.7-257.4
Cl$_5$Sb	Antimony pentachloride	0.45413E+01	-0.45078E-02			0.3465E-04	275.8-320.4
DH	Hydrogen-d_1	0.12837E+01	-0.10700E-02			0.6236E-04	16.8- 22.2
D$_2$	Hydrogen-d_2	0.13063E+01	-0.15143E-02			0.3609E-04	17.8- 22.7
D$_2$O	Water-d_2	0.25232E+03	-0.81067E+00	0.75784E-03		0.7063E-03	277.2-373.2
FH	Hydrogen fluoride	0.50352E+03	-0.19297E+01	0.14372E-02			200.2-273.2
F$_2$	Fluorine	0.14144E+01	0.26387E-02	-0.28356E-04		0.2438E-03	53.5-144.3
F$_5$I	Iodine pentafluoride	0.95184E+02	-0.19800E+00				273.2-313.2
F$_{10}$S$_2$	Sulfur hexafluoride	0.26509E+01	-0.21500E-02			0.9000E-07	263.2-293.2
HI	Hydrogen iodide	0.51557E+03	-0.44552E+01	0.96795E-02			220.2-236.2
H$_2$	Hydrogen	0.13327E+01	-0.51946E-02			0.3687E-04	13.5- 19.4
H$_2$O	Water	0.24921E+03	-0.79069E+00	0.72997E-03		0.1184E-02	273.3-372.2

Molecular formula	Name	a	b	c	d	ESQS/N	Range T/K
H_2O_2	Hydrogen peroxide	0.48511E+03	-0.23145E+01	0.31020E-02		0.3479E-01	233.2-303.2
H_2S	Hydrogen sulfide	0.14736E+02	-0.33675E-01	0.96740E-05		0.1459E-02	212.0-363.2
H_3N	Ammonia	0.66756E+02	-0.24696E+00	0.25913E-03		0.9088E-04	238.2-323.2
H_4N_2	Hydrazine	0.22061E+03	-0.89633E+00	0.11066E-02		0.2980E-03	278.2-323.2
He	Helium	0.10640E+01	-0.35584E-02			0.4670E-06	2.1- 4.2
I_2	Iodine	0.64730E+02	-0.29266E+00	0.39759E-03		0.3503E-03	391.3-440.9
Mn_2O_7	Manganese oxide (Mn_2O_7)	0.37655E+01	-0.16463E-02			0.6028E-05	283.2-312.2
N_2	Nitrogen	0.12550E+01	0.67949E-02	-0.56704E-04		0.2390E-03	63.2-126.2
N_2O_3	Nitrogen trioxide	0.92287E+02	-0.43306E+00	0.65000E-03		0.8000E-05	203.2-243.2
N_2O_4	Nitrogen tetroxide	0.28212E+01	-0.13000E-02			0.6000E-05	253.2-293.2
Ne	Neon	0.12667E+01	-0.29064E-02			0.4057E-08	26.1- 29.0
O_2	Oxygen	0.15434E+01	0.14615E-02	-0.21964E-04		0.1085E-03	54.5-154.0
O_2S	Sulfur dioxide	0.52045E+02	-0.16125E+00	0.11042E-03		0.1637E+00	213.2-449.2
O_3	Ozone	0.86344E+01	-0.54807E-01	0.12596E-03		0.9018E-03	90.2-185.2
P	Phosphorus	0.79018E+00	0.23911E-01	-0.42826E-04		0.2475E-05	307.2-358.2
S	Sulfur	0.51651E+01	-0.77381E-02	0.89120E-05		0.5433E-04	407.2-479.2
Se	Selenium	0.67569E+01	-0.25829E-02			0.4163E-04	510.7-574.6
$CBrClF_2$	Bromochlorodifluoromethane	0.52442E+01	-0.11000E-01			0.1922E-02	123.2-223.2
$CBrCl_3$	Bromotrichloromethane	0.29249E+01	-0.17650E-02			0.3208E-04	273.2-333.2
$CBrF_3$	Bromotrifluoromethane	0.54154E+01	-0.13680E-01				123.2-173.2
CBr_2Cl_2	Dibromodichloromethane	0.32330E+01	-0.23162E-02			0.1126E-06	298.2-333.2
CBr_2F_2	Dibromodifluoromethane	0.67296E+01	-0.22133E-01	0.30213E-04		0.3198E-04	138.5-273.2
CBr_3F	Tribromofluoromethane	0.53203E+01	-0.11061E-01	0.10688E-04		0.8687E-05	205.7-323.2
CBr_3NO_2	Tribromonitromethane	0.16079E+02	-0.23630E-01			0.2011E-03	298.2-328.2
$CClF_3$	Chlorotrifluoromethane	0.43677E+01	-0.11020E-01				123.2-173.2
CCl_2F_2	Dichlorodifluoromethane	0.46984E+01	-0.97600E-02			0.3200E-04	123.2-223.2
CCl_2NOP	Phosphorous dichloride isocyanate	0.69421E+01	-0.95000E-02			0.2222E-04	293.2-333.2
CCl_3F	Trichlorofluoromethane	0.53203E+01	-0.11061E-01	0.10688E-04		0.8687E-05	205.7-323.2
CCl_3NO_2	Trichloronitromethane	0.14403E+02	-0.24178E-01			0.9025E-03	276.2-333.2
CCl_4	Tetrachloromethane	0.28280E+01	-0.20339E-02	0.71795E-07		0.1033E-08	283.2-333.2
CF_4	Tetrafluoromethane	0.20350E+01	-0.27616E-02			0.1176E-05	126.3-141.5
COS	Carbon oxysulfide	0.84702E+01	-0.21488E-01			0.1371E-02	143.0-185.0
COSe	Carbon oxyselenide	0.48740E+01	-0.49425E-02			0.1515E-04	219.2-283.2
CO_2	Carbon dioxide	0.79062E+00	0.10639E-01	-0.28510E-04		0.4024E-04	220.0-300.0
CS_2	Carbon disulfide	0.45024E+01	-0.12054E-01	0.19147E-04		0.7347E-03	154.2-319.4
$CHBr_3$	Tribromomethane	0.71707E+01	-0.98000E-02			0.1502E-03	283.2-343.2
$CHCl_3$	Trichloromethane	0.15115E+02	-0.51830E-01	0.56803E-04		0.5163E-04	218.2-323.2
CHF_3	Trifluoromethane	0.11442E+03	-0.75600E+00	0.13562E-02		0.2121E+00	130.0-263.0
CHN	Hydrogen cyanide	0.37331E+04	-0.23180E+02	0.36963E-01		0.6864E+00	258.2-298.9
CH_2Br_2	Dibromomethane	0.18060E+02	-0.36333E-01				283.2-313.2
CH_2Cl_2	Dichloromethane	0.40452E+02	-0.17748E+00	0.23942E-03		0.4335E-02	184.1-306.0
CH_2F_2	Difluoromethane	0.19428E+03	-0.12939E+01	0.24280E-02		0.1725E-01	152.2-224.2
CH_2O_2	Formic acid	0.14040E+03	-0.24673E+00	-0.17151E-03		0.8184E-01	286.7-358.2
CH_3Br	Bromomethane	0.40580E+02	-0.18418E+00	0.26219E-03		0.1898E-03	194.6-275.7
CH_3Cl	Chloromethane	0.42775E+02	-0.16175E+00	0.17108E-03		0.3439E-01	190.2-392.2
CH_3ClO_2S	Methanesulfonyl chloride	0.10384E+03	-0.33838E+00	0.34156E-03		0.5282E-03	293.2-373.2
CH_3DO	Methan-d_1-ol	0.20839E+03	-0.10318E+01	0.14740E-02		0.5753E-01	176.3-297.5
CH_3F	Fluoromethane	0.11338E+03	-0.63979E+00	0.96983E-03		0.2596E+00	150.0-299.0
CH_3I	Iodomethane	0.24264E+02	-0.93914E-01	0.11926E-03		0.1538E-02	223.2-303.2
CH_3NO	Formamide	0.26076E+03	-0.61145E+00	0.34296E-03		0.2016E-02	278.2-333.2
CH_3NO_2	Methyl nitrite	0.11071E+03	-0.73428E+00	0.14054E-02		0.6757E+00	110.0-260.0
CH_3NO_2	Nitromethane	0.11227E+03	-0.35591E+00	0.34206E-03		0.2039E-01	288.2-343.2
CH_4	Methane	0.15996E+01	0.27434E-02	-0.22086E-04		0.6061E-04	91.0-184.0
CH_4O	Methanol	0.19341E+03	-0.92211E+00	0.12839E-02		0.6618E-01	177.2-293.2
CH_5N	Methylamine	0.34398E+02	-0.73630E-01	-0.41279E-04		0.3172E-01	198.2-258.2
$C_2Cl_2F_4$	1,2-Dichlorotetrafluoroethane	0.36663E+01	-0.42271E-02	-0.36255E-06		0.2399E-07	193.2-273.2
C_2HCl_2NO	Dichloromethyl isocyanate	0.14757E+02	-0.25667E-01				288.2-348.2
C_2HCl_3	Trichloroethylene	0.58319E+01	-0.80828E-02			0.1025E-04	301.5-337.6
$C_2HCl_3F_2$	1,2,2-Trichloro-1,1-difluoroethane	0.75423E+01	-0.11667E-01			0.5000E-04	303.2-333.2
$C_2HCl_3O_2$	Trichloroacetic acid	0.13412E+01	0.90000E-02	-0.24130E-14			333.2-393.2
C_2HCl_5	Pentachloroethane	0.65972E+01	-0.96800E-02			0.5392E-04	298.2-338.2
$C_2HF_3O_2$	Trifluoroacetic acid	0.21652E+02	-0.68146E-01	0.78571E-04		0.2041E-04	263.2-323.2
$C_2H_2Br_4$	1,1,2,2-Tetrabromoethane	0.16246E+02	-0.31500E-01			0.5375E-03	303.2-333.2
$C_2H_2Cl_2O_2$	Dichloroacetic acid	0.11014E+02	-0.10859E-01	0.49242E-05		0.8608E-05	284.0-363.2

Molecular formula	Name	a	b	c	d	ESQS/N	Range T/K
$C_2H_2Cl_4$	1,1,1,2-Tetrachloroethane	0.19606E+02	-0.49847E-01			0.3148E-02	207.2-233.2
$C_2H_3ClO_2$	Chloroacetic acid	0.17310E+02	-0.14674E-01			0.7689E-05	338.2-393.2
$C_2H_3Cl_2NO_2$	1,1-Dichloro-1-nitroethane	0.37576E+02	-0.70400E-01			0.4305E-02	303.2-333.2
$C_2H_3Cl_3$	1,1,1-Trichloroethane	0.27705E+02	-0.10621E+00	0.12424E-03		0.1899E-05	258.2-318.2
$C_2H_3Cl_3$	1,1,2-Trichloroethane	0.17147E+02	-0.33371E-01			0.9082E-05	288.2-318.2
$C_2H_3F_3O$	2,2,2-Trifluoroethanol	0.90593E+02	-0.21421E+00			0.6122E-01	293.2-318.2
C_2H_3N	Acetonitrile	0.29724E+03	-0.15508E+01	0.22591E-02		0.7082E-01	288.2-333.2
C_2H_4	Ethylene	0.13546E+01	0.62614E-02	-0.21374E-04		0.6934E-05	200.0-270.0
C_2H_4BrCl	1-Bromo-2-chloroethane	0.19493E+02	-0.59054E-01	0.58036E-04		0.1224E-03	263.2-363.2
$C_2H_4Br_2$	1,2-Dibromoethane	0.67142E+01	-0.59800E-02			0.2689E-06	293.2-313.2
$C_2H_4Cl_2$	1,1-Dichloroethane	0.24429E+02	-0.48000E-01			0.2250E-03	288.2-318.2
$C_2H_4Cl_2$	1,2-Dichloroethane	0.24404E+02	-0.47892E-01			0.2261E-02	293.2-343.2
C_2H_4O	Ethylene oxide	0.52661E+02	-0.21337E+00	0.25947E-03		0.2614E-03	293.2-243.2
$C_2H_4O_2$	Methyl formate	0.19699E+02	-0.36429E-01				288.2-302.2
$C_2H_4O_2$	Acetic acid	-0.15731E+02	0.12662E+00	-0.17738E-03		0.2367E-02	293.2-363.2
$C_2H_4O_3S$	Ethylene sulfite	0.85483E+02	-0.15400E+00			0.3000E-02	298.2-328.2
C_2H_5Br	Bromoethane	0.28473E+02	-0.85495E-01	0.67971E-04		0.9485E-04	243.2-308.2
C_2H_5Cl	Chloroethane	0.60693E+02	-0.31290E+00	0.47154E-03		0.7954E-03	237.2-293.2
C_2H_5ClO	2-Chloroethanol	0.11155E+03	-0.30149E+00			0.2877E-01	140.2-175.2
C_2H_5I	Iodoethane	0.25598E+02	-0.94367E-01	0.11424E-03		0.3382E-02	183.2-343.2
C_2H_5N	Ethyleneimine	0.61405E+02	-0.14474E+00			0.4298E-02	273.2-298.2
C_2H_5NO	Acetamide	-0.20055E+03	0.15515E+01	-0.22392E-02		0.1046E+00	363.7-448.2
C_2H_5NO	N-Methylformamide	0.10383E+04	-0.43165E+01	0.48398E-02		0.5075E+00	276.2-353.2
$C_2H_5NO_2$	Methyl carbamate	0.36773E+02	-0.55700E-01			0.1998E-02	328.2-368.2
$C_2H_5NO_2$	Nitroethane	0.57406E+02	-0.97657E-01			0.1890E-01	276.2-333.2
C_2H_6	Ethane	0.20185E+01	-0.51493E-03	-0.48148E-05		0.5398E-04	95.0-295.0
C_2H_6O	Dimethyl ether	0.22389E+02	-0.86524E-01	0.91291E-04		0.1023E-01	155.0-258.0
C_2H_6O	Ethanol	0.15145E+03	-0.87020E+00	0.19570E-02	-0.15512E-05	0.1369E+00	163.2-523.2
C_2H_6OS	Dimethyl sulfoxide	0.38478E+02	0.16939E+00	-0.47423E-03		0.3291E-01	288.2-343.2
$C_2H_6O_2$	Ethylene glycol	0.14355E+03	-0.48573E+00	0.46703E-03		0.5617E-02	293.2-423.2
$C_2H_6O_2S$	Dimethyl sulfone	0.10830E+03	-0.15900E+00			0.1875E-03	383.2-398.2
$C_2H_6S_2$	1,2-Ethanedithiol	0.11228E+02	-0.13500E-01			0.2000E-03	293.2-333.2
$C_2H_6S_2$	Dimethyl disulfide	0.19109E+02	-0.32000E-01			0.8889E-03	298.2-323.2
C_2H_7N	Ethylamine	0.30163E+02	-0.79000E-01			0.1420E-01	233.2-273.2
C_2H_7NO	Ethanolamine	0.14890E+03	-0.62491E+00	0.77143E-03		0.5449E-02	253.2-293.2
$C_2H_7NO_2S$	N-Methyl methanesulfonamide	0.54340E+03	-0.22958E+01	0.27619E-02		0.4422E-02	298.2-328.2
$C_2H_8N_2$	1,2-Ethanediamine	0.48922E+02	-0.17021E+00	0.17262E-03		0.1150E-02	273.2-333.2
C_3Cl_6O	Hexachloroacetone	0.76423E+01	-0.15838E-01	0.10618E-04		0.1350E-05	268.7-303.3
C_3F_6O	Hexafluoroacetone	0.34809E+01	-0.92883E-02	0.12282E-04		0.1605E-04	150.6-238.1
C_3HN	Cyanoacetylene	0.91803E+03	-0.49149E+01	0.69104E-02		0.1980E-01	280.8-313.6
$C_3H_2F_6O$	1,1,1,3,3,3-Hexafluoro-2-propanol	0.51961E+03	0.31421E+01	0.48667E-02			283.2-313.2
C_3H_3N	Propenenitrile	0.11109E+03	-0.36806E+00	0.34879E-03		0.3168E-01	233.2-413.2
C_3H_4	Allene	0.26049E+01	-0.44147E-03	-0.63420E-05		0.1019E-04	156.0-269.0
C_3H_4	Propyne	0.60871E+01	-0.11730E-01			0.4998E-03	185.0-246.0
$C_3H_4ClF_3$	3-Chloro-1,1,1-trifluoropropane	0.22361E+02	-0.68840E-01	0.60594E-04		0.2374E-03	275.2-313.2
C_3H_4ClNO	1-Chloro-2-isocyanatoethane	0.64311E+02	-0.12217E+00				288.2-403.2
$C_3H_4N_2O_2$	3-Methylsydnone	0.75278E+03	-0.30204E+01	0.34389E-02		0.5346E-01	313.2-398.2
C_3H_4O	Propargyl alcohol	0.99895E+02	-0.38911E+00	0.40776E-03		0.5853E-02	213.2-293.2
$C_3H_4O_3$	1,3-Dioxolan-2-one	0.20746E+03	-0.37610E+00			0.1402E-01	313.2-343.2
$C_3H_5Br_3$	1,2,3-Tribromopropane	0.11024E+02	-0.16596E-01			0.3222E-03	303.2-358.2
$C_3H_5ClO_2$	Ethyl chlorocarbonate	0.15356E+02	-0.18250E-01			0.1105E-03	308.7-349.2
C_3H_5N	Propanenitrile	0.82222E+02	-0.22937E+00	0.17424E-03		0.4097E-01	213.2-473.2
C_3H_6	Propene	0.29623E+01	-0.37564E-02			0.2492E-06	220.0-250.0
$C_3H_6Br_2$	1,2-Dibromopropane	0.54973E+01	-0.31695E-02			0.1144E-03	283.2-333.2
$C_3H_6Br_2$	1,3-Dibromopropane	0.29193E+02	-0.94450E-01	0.92800E-04		0.7220E-04	293.2-368.2
C_3H_6ClNO	N-Methyl-2-chloroacetamide	0.30695E+03	-0.66475E+00			0.5042E-01	323.2-348.2
$C_3H_6Cl_2$	1,2-Dichloropropane	0.18915E+02	-0.35907E-01			0.1981E-02	281.2-323.2
$C_3H_6Cl_2$	1,3-Dichloropropane	0.21609E+02	-0.37333E-01			0.8000E-03	303.2-333.2
$C_3H_6Cl_2$	2,2-Dichloropropane	0.32421E+02	-0.72188E-01			0.8805E-02	244.7-293.2
C_3H_6O	Acetone	0.88157E+02	-0.34300E+00	0.38925E-03		0.9183E-03	273.2-323.2
C_3H_6O	Allyl alcohol	0.62714E+02	-0.14771E+00	0.37879E-05		0.7439E-03	213.2-303.2
$C_3H_6O_2$	Ethyl formate	0.15884E+02	-0.25333E-01			0.3556E-03	288.2-318.2
$C_3H_6O_2$	Methyl acetate	0.13190E+02	-0.21226E-01			0.2439E-03	276.2-318.2
$C_3H_6O_2$	Propanoic acid	0.18793E+01	0.46841E-02	0.19983E-05		0.4416E-03	289.2-408.2

Molecular formula	Name	a	b	c	d	ESQS/N	Range T/K
$C_3H_6O_3S$	1,2-Oxathiolane-2,2-dioxide	0.24989E+03	-0.72897E+00	0.64286E-03		0.2286E-03	308.2-348.2
C_3H_7Br	1-Bromopropane	0.17769E+02	-0.32599E-01			0.5212E-02	274.2-328.2
C_3H_7Br	2-Bromopropane	0.26195E+02	-0.72995E-01	0.55454E-04		0.9752E-02	186.3-328.2
C_3H_7Cl	1-Chloropropane	0.21214E+02	-0.43130E-01			0.3194E-03	273.2-313.2
C_3H_7ClO	3-Chloro-1-propanol	0.12436E+03	-0.60841E+00	0.92060E-03		0.1844E+00	145.2-215.2
C_3H_7ClO	1-Chloro-2-propanol	-0.19169E+02	0.13605E+01	-0.55567E-02		0.1472E+00	153.2-177.2
C_3H_7I	1-Iodopropane	0.13744E+02	-0.22745E-01			0.4062E-04	293.2-323.3
C_3H_7NO	N,N-Dimethylformamide	0.15364E+03	-0.60367E+00	0.71505E-03		0.3503E-01	213.2-353.2
C_3H_7NO	N-Ethylformamide	0.64764E+03	-0.28499E+01	0.34286E-02		0.3657E-02	298.2-338.2
C_3H_7NO	N-Methylacetamide	0.15975E+04	-0.90451E+01	0.18345E-01	-0.12998E-04	0.3900E-01	303.2-473.2
$C_3H_7NO_2$	Ethyl carbamate	0.32431E+02	-0.65097E-01	0.28571E-04		0.5531E-03	328.2-368.2
$C_3H_7NO_2$	1-Nitropropane	0.94999E+02	-0.38358E+00	0.48480E-03		0.2608E-02	276.2-333.2
$C_3H_7NO_2$	2-Nitropropane	0.60138E+02	-0.11566E+00			0.2164E-02	276.2-303.2
$C_3H_7NO_2$	Propyl nitrite	0.70552E+02	-0.40362E+00	0.66687E-03		0.4240E+00	110.0-310.0
$C_3H_7NO_2$	Isopropyl nitrite	0.74578E+02	-0.38283E+00	0.57071E-03		0.5473E-01	150.0-300.0
C_3H_8	Propane	0.22883E+01	-0.23276E-02	0.84710E-06		0.2970E-04	90.0-300.0
C_3H_8O	1-Propanol	0.98045E+02	-0.36860E+00	0.36422E-03		0.2544E+00	193.2-493.2
C_3H_8O	2-Propanol	0.10416E+03	-0.41011E+00	0.42049E-03		0.1587E+00	193.2-493.2
$C_3H_8O_2$	Dimethoxymethane	0.25877E+01	-0.93019E-03	0.38472E-05		0.6436E-05	170.7-293.2
$C_3H_8O_2$	2-Methoxyethanol	0.11803E+03	-0.58000E+00	0.81001E-03		0.2102E-01	253.6-318.4
$C_3H_8O_2$	1,2-Propanediol	0.24546E+03	-0.15738E+01	0.38068E-02	-0.32544E-05	0.2557E-01	193.2-403.2
$C_3H_8O_2$	1,3-Propanediol	0.11365E+03	-0.36680E+00	0.33766E-03		0.1060E-01	288.2-328.2
$C_3H_8O_3$	Glycerol	0.77503E+02	-0.37984E-01	-0.23107E-03		0.7294E-01	288.2-343.2
C_3H_8S	1-Propanethiol	0.11602E+02	-0.19580E-01			0.2421E-03	273.2-318.2
$C_3H_8S_2$	1,2-Propanedithiol	0.14667E+02	-0.32660E-01	0.25000E-04			293.2-333.2
$C_3H_8S_2$	1,3-Propanedithiol	0.66607E+01	0.31310E-01	-0.87500E-04			303.2-343.2
C_3H_9ClSi	Trimethylchlorosilane	-0.19492E+02	0.29806E+00	-0.69284E-03		0.2640E-02	223.2-273.2
C_3H_9N	Propylamine	0.17719E+02	-0.59022E-01	0.54780E-04		0.3082E-02	204.2-296.2
C_3H_9N	Isopropylamine	0.40429E+02	-0.21441E+00	0.32634E-03		0.7288E-03	213.2-298.2
C_3H_9N	Trimethylamine	0.39745E+01	-0.51331E-02			0.1833E-04	273.2-298.2
$C_3H_9O_3P$	Dimethyl methanephosphate	0.74253E+02	-0.24790E+00	0.24107E-03		0.3143E-03	293.2-373.2
$C_4F_6Cl_4$	1,1,3,4-Tetrachlorohexafluorobutane	0.52319E+01	-0.11882E-01	0.12891E-04		0.4905E-04	203.7-313.2
$C_4H_4N_2$	Succinonitrile	0.17724E+03	-0.54654E+00	0.54046E-03		0.1137E+00	235.5-351.4
C_4H_4O	Furan	0.13636E+01	0.12864E-01	-0.22701E-04		0.9970E-02	187.5-277.1
C_4H_4S	Thiophene	0.32941E+01	-0.19019E-02			0.1433E-04	252.7-293.2
C_4H_5N	1H-Pyrrole	0.12672E+02	-0.14075E-01	-0.62671E-05		0.7743E-04	293.0-356.7
C_4H_5NO	Allyl isocynate	0.34299E+02	-0.66444E-01				288.2-333.2
$C_4H_5N_2O_3P$	Diisocyanato ethyl phosphite	0.11820E+02	-0.16400E-01				298.2-323.2
$C_4H_5N_2O_3PS$	Diisocyanato O-ethyl thionphosphate	0.39572E+02	-0.12459E+00	0.12054E-03		0.6786E-03	293.2-393.2
$C_4H_5N_2O_4P$	Diisocyanato ethyl phosphate	0.24312E+02	-0.35000E-01			0.2500E-02	293.2-353.2
C_4H_6	1,3-Butadiene	0.27674E+01	-0.26738E-02			0.5654E-04	185.0-265.0
$C_4H_6Cl_2N_2S$	(1-Cyano-1-isopropyl)imido sulfurous dichloride	0.38834E+02	-0.58095E-01				293.2-398.2
$C_4H_6N_2OS$	2-Methyl-2-(sulfinylamino)-propanenitrile	0.35288E+02	-0.86192E-01	0.71875E-04		0.6122E-02	293.2-392.2
C_4H_6O	Cyclobutanone	0.43974E+02	-0.15712E+00	0.19264E-03		0.2857E-02	220.2-317.2
$C_4H_6O_2$	Methyl acrylate	0.11968E+02	-0.16500E-01			0.3588E-02	303.2-333.2
$C_4H_6O_2$	2,3-Butanedione	0.46907E+01	-0.22302E-02			0.8205E-04	278.2-348.2
$C_4H_6O_3$	4-Methyl-1,3-dioxolan-2-one	0.15940E+03	-0.39530E+00	0.26284E-03		0.1181E-02	273.0-333.0
$C_4H_7BrO_2$	Ethyl 2-bromoacetate	0.15627E+02	-0.19600E-01			0.3780E-02	303.2-333.2
$C_4H_7BrO_2$	Methyl 2-bromopropanoate	0.19850E+02	-0.31250E-01			0.1250E-02	303.2-343.2
$C_4H_7BrO_2$	Methyl 3-bromopropanoate	0.36001E+01	0.72500E-02			0.2722E-03	303.2-343.2
$C_4H_7ClO_2$	Methyl 2-chloropropanoate	0.22449E+02	-0.36250E-01			0.1389E-03	303.2-343.2
C_4H_7N	Butanenitrile	0.53884E+02	-0.99257E-01			0.1554E-02	293.2-333.2
C_4H_7N	2-Methylpropanenitrile	0.52554E+02	-0.96000E-01			0.3556E-03	293.2-313.2
C_4H_7NO	Pyrrolidine-2-one	0.11054E+03	-0.47945E+00	0.68182E-03		0.8182E-03	298.2-338.2
C_4H_8	1-Butene	0.29354E+01	-0.32580E-02			0.7206E-06	220.0-250.0
C_4H_8	cis-2-Butene	0.28802E+01	-0.31064E-02			0.5256E-04	197.0-296.0
C_4H_8	Isobutene	0.33701E+01	-0.43295E-02			0.8003E-05	220.0-288.7
C_4H_8BrCl	3-Bromo-1-chloro-2-methylpropane	0.32838E+02	-0.10590E+00	0.88889E-04			303.2-333.2
$C_4H_8Br_2$	1,2-Dibromobutane	0.11199E+03	-0.63334E+00	0.91250E-03			293.2-333.2
$C_4H_8Br_2$	1,3-Dibromobutane	0.34031E+02	-0.13254E+00	0.16250E-03			293.2-333.2
$C_4H_8Br_2$	1,4-Dibromobutane	0.20944E+02	-0.55620E-01	0.50000E-04		0.4500E-04	303.2-333.2
$C_4H_8Br_2$	2,3-Dibromobutane	0.23849E+02	-0.96300E-01	0.12500E-03			293.2-333.2

Molecular formula	Name	a	b	c	d	ESQS/N	Range T/K
$C_4H_8Cl_2$	1,2-Dichlorobutane	0.31925E+02	-0.13232E+00	0.17007E-03		0.6993E-03	293.2-356.2
$C_4H_8Cl_2$	1,4-Dichlorobutane	0.59766E+01	0.49300E-01	-0.12500E-03		0.3125E-04	308.2-338.2
$C_4H_8Cl_2$	1,2-Dichloro-2-methylpropane	0.39429E+02	-0.20028E+00	0.30917E-03		0.1837E-01	165.2-296.0
C_4H_8O	2-Butanone	0.15457E+02	0.90152E-01	-0.27100E-03		0.1405E-02	293.2-333.2
C_4H_8O	Tetrahydrofuran	0.30739E+02	-0.12946E+00	0.17195E-03		0.2233E-03	224.2-295.2
C_4H_8OS	Tetrahydrothiophene-S-oxide	0.11842E+03	-0.34315E+00	0.30071E-03		0.5846E-02	298.2-398.2
$C_4H_8O_2$	Butanoic acid	0.15010E+01	0.50046E-02			0.9836E-03	287.2-403.2
$C_4H_8O_2$	1,4-Dioxane	0.27299E+01	-0.17440E-02			0.6960E-07	293.2-313.2
$C_4H_8O_2$	Ethyl acetate	0.15646E+02	-0.44066E-01	0.39137E-04		0.1251E-03	293.2-433.2
$C_4H_8O_2$	Methyl propanoate	0.12798E+02	-0.22540E-01			0.1420E-02	293.2-333.2
$C_4H_8O_2S$	Tetrahydrothiophene-S,S-dioxide	0.10981E+03	-0.30973E+00	0.29809E-03		0.3962E-02	303.2-353.2
C_4H_9Br	1-Bromobutane	0.22542E+02	-0.79306E-01	0.89867E-04		0.1719E-02	183.2-363.2
C_4H_9Br	1-Bromo-2-methylpropane	0.37558E+02	-0.20571E+00	0.35496E-03		0.2884E-01	111.9-273.2
C_4H_9Br	2-Bromobutane	0.18461E+02	-0.32933E-01			0.4880E-05	274.2-328.2
C_4H_9Br	2-Bromo-2-methylpropane	0.35085E+02	-0.14075E+00	0.19960E-03		0.4872E-03	258.0-293.0
C_4H_9Cl	1-Chlorobutane	0.13565E+02	-0.10161E-01	-0.38750E-04		0.4762E-03	273.2-323.2
C_4H_9Cl	1-Chloro-2-methylpropane	0.14945E+02	-0.33747E-01	0.23036E-04		0.4888E-04	273.2-323.2
C_4H_9Cl	2-Chlorobutane	0.30376E+02	-0.11377E+00	0.13429E-03		0.1653E-03	273.2-323.2
C_4H_9Cl	2-Chloro-2-methylpropane	0.35077E+02	-0.12867E+00	0.14304E-03		0.4425E-03	273.2-323.2
C_4H_9I	1-Iodobutane	0.16493E+02	-0.50262E-01	0.52485E-04		0.3476E-05	293.2-323.2
C_4H_9I	2-Iodobutane	0.10883E+02	-0.14680E-02	-0.30000E-04		0.8000E-06	293.2-323.2
C_4H_9I	2-Iodo-2-methylpropane	0.76780E+01	0.69900E-02	-0.37500E-04			283.2-323.2
C_4H_9N	Pyrrolidine	0.38191E+02	-0.15462E+00	0.17941E-03		0.6758E-04	274.0-333.0
C_4H_9NO	N,N-Dimethylacetamide	0.15420E+03	-0.57506E+00	0.61911E-03		0.3225E-01	294.2-433.2
$C_4H_9NO_2$	Ethyl-N-methyl carbamate	0.11477E+03	-0.47568E+00	0.54127E-03		0.1989E-02	298.2-373.2
$C_4H_9NO_2$	N-(2-Hydroxyethyl)acetamide	0.37016E+03	-0.13113E+01	0.13214E-02		0.1190E-02	298.2-348.2
$C_4H_9NO_2$	Propyl carbamate	0.24356E+02	-0.36400E-01			0.9440E-03	338.2-378.2
C_4H_{10}	Butane	0.22379E+01	-0.13884E-02	-0.66711E-06		0.3929E-06	135.0-303.2
C_4H_{10}	Isobutane	0.23295E+01	-0.19953E-02	0.14197E-06		0.2146E-05	115.0-303.2
$C_4H_{10}O$	1-Butanol	0.10578E+03	-0.50587E+00	0.84733E-03	-0.48841E-06	0.1053E-01	193.2-553.2
$C_4H_{10}O$	2-Butanol	0.13850E+03	-0.75146E+00	0.14086E-02	-0.89512E-06	0.4298E-01	172.2-533.2
$C_4H_{10}O$	2-Methyl-1-propanol	0.10762E+03	-0.51398E+00	0.83702E-03	-0.45299E-06	0.8274E-01	173.2-533.2
$C_4H_{10}O$	2-Methyl-2-propanol	0.22541E+03	-0.14990E+01	0.34050E-02	-0.25968E-05	0.2272E-01	298.2-503.2
$C_4H_{10}O$	Diethyl ether	0.79725E+01	-0.12519E-01			0.2822E-02	283.2-301.2
$C_4H_{10}O_2$	1,2-Butanediol	0.63702E+02	-0.13807E+00			0.1069E-01	278.2-323.2
$C_4H_{10}O_2$	1,3-Butanediol	0.72883E+02	-0.14770E+00			0.8115E-02	278.2-323.2
$C_4H_{10}O_2$	1,4-Butanediol	0.13079E+03	-0.46985E+00	0.46320E-03		0.1417E-02	288.2-328.2
$C_4H_{10}O_2$	1,2-Dimethoxyethane	0.48832E+02	-0.24218E+00	0.34413E-03		0.3811E-02	255.6-318.4
$C_4H_{10}O_2S$	Bis(2-hydroxyethyl) sulfide	0.13128E+03	-0.52719E+00	0.60465E-03		0.2327E-02	253.2-333.2
$C_4H_{10}O_3$	Diethylene glycol	0.13973E+03	-0.54725E+00	0.61149E-03		0.1135E-02	288.2-343.2
$C_4H_{10}S$	1-Butanethiol	0.11201E+02	-0.20767E-01			0.8438E-04	273.2-318.2
$C_4H_{10}S$	2-Butanethiol	0.10866E+02	-0.17993E-01			0.6267E-03	273.2-318.2
$C_4H_{10}S$	2-Methyl-2-propanethiol	0.10597E+02	-0.17500E-01			0.1935E-03	283.2-313.2
$C_4H_{11}N$	Butylamine	0.13322E+02	-0.44176E-01	0.50250E-04		0.3162E-03	223.2-333.2
$C_4H_{11}N$	Diethylamine	0.26462E+02	-0.13750E+00	0.20373E-03		0.8237E-03	243.2-323.2
$C_4H_{11}NO_2$	Diethanolamine	0.73435E+02	-0.21377E+00	0.17500E-03		0.7870E-02	273.2-323.2
$C_4H_{11}OP$	Ethyldimethylphosphine oxide	0.76085E+02	-0.11500E+00			0.1389E-01	363.2-383.2
$C_4H_{12}NO_3P$	Phosphoramidic acid diethyl ester	-0.24973E+02	0.40569E+00	-0.78311E-03		0.6669E-02	313.2-383.2
$C_4H_{12}N_2O$	2-(2-Aminoethylamino)ethanol	0.10819E+03	-0.48865E+00	0.65979E-03		0.6209E-01	223.2-323.2
$C_4H_{13}N_3$	Bis(2-amimoethyl)amine	0.57840E+02	-0.23873E+00	0.28841E-03		0.1541E-02	213.2-333.2
C_5H_4ClN	2-Chloropyridine	0.98702E+02	-0.34237E+00	0.34502E-03		0.1033E-02	298.2-398.2
C_5H_4BrN	2-Bromopyridine	0.73391E+02	-0.23678E+00	0.22930E-03		0.2277E-03	298.2-398.2
C_5H_5N	Pyridine	0.43991E+02	-0.15150E+00	0.15925E-03		0.7021E-05	293.2-323.2
C_5H_5NO	Pyridine-1-oxide	0.20878E+02	0.16450E+00	-0.35269E-03		0.9213E-01	343.0-398.0
$C_5H_7N_2O_4P$	Diisocyanato isopropyl phosphate	0.13872E+02	-0.30859E-01	0.28201E-04		0.9850E-05	293.0-398.0
$C_5H_7N_2O_3PS$	Diisocyanato O-isopropyl-thionphosphate	0.41564E+02	-0.14562E+00	0.15873E-03		0.2286E-03	288.0-348.0
$C_5H_7N_2O_4P$	Diisocyanato isopropyl phosphate	0.40521E+02	-0.13936E+00	0.16000E-03			298.0-373.0
C_5H_8	Cyclopentene	0.28177E+01	-0.27597E-02	0.89346E-06		0.2447E-05	171.0-319.0
C_5H_8	2-Methyl-1,3-butadiene	0.28170E+01	-0.23147E-02	-0.43975E-06		0.9675E-05	198.2-293.2
C_5H_8	1,4-Pentadiene	0.29994E+01	-0.34578E-02	0.85300E-06		0.2110E-04	178.0-294.0
$C_5H_8N_2OS$	2-Methyl-2-(sulfinylamino)-butanenitrile	0.37682E+02	-0.10926E+00	0.10536E-03		0.2011E-03	293.2-373.2
$C_5H_8N_2O_2$	3-Isopropylsydnone	0.20651E+03	-0.58628E+00	0.49403E-03		0.4210E-03	333.2-398.2

Molecular formula	Name	a	b	c	d	ESQS/N	Range T/K
C_5H_8O	Cyclopentanone	0.24083E+02	-0.30286E-01	-0.16802E-04		0.2072E-02	219.2-298.2
$C_5H_8O_2$	Ethyl acrylate	0.47827E+02	-0.24394E+00	0.35000E-03		0.2560E-03	303.2-343.2
$C_5H_8O_2$	Methyl crotonate	0.16248E+02	-0.45167E-01	0.42571E-04		0.2684E-05	293.2-343.2
$C_5H_8O_2$	Methyl methacrylate	0.32098E+02	-0.14568E+00	0.20000E-03		0.1120E-03	303.2-343.2
$C_5H_8O_4$	Dimethyl malonate	0.26470E+02	-0.76656E-01	0.67888E-04		0.1435E-01	293.2-433.2
$C_5H_9ClO_2$	Methyl 4-chlorobutanoate	0.17127E+02	-0.25000E-01			0.2689E-02	303.2-343.2
$C_5H_9ClO_2$	Ethyl 2-chloropropanoate	0.25965E+02	-0.46250E-01			0.1389E-03	303.2-343.2
$C_5H_9ClO_2$	Ethyl 3-chloropropanoate	0.21951E+02	-0.38750E-01			0.2722E-03	303.2-343.2
C_5H_9N	2,2-Dimethylpropane nitrile	0.58418E+02	-0.16884E+00	0.14131E-03		0.5891E-03	293.2-453.2
C_5H_9N	Pentanenitrile	0.55793E+02	-0.15750E+00	0.12432E-03		0.5378E-01	182.7-333.2
C_5H_9NO	Butyl isocyanate	0.50564E+02	-0.19487E+00	0.21940E-03		0.4483E-05	293.2-353.2
C_5H_9NO	Isobutyl isocyanate	0.38026E+02	-0.12714E+00	0.12679E-03		0.1083E-02	293.2-353.2
C_4H_9NO	N-Ethylacetamide	0.74494E+03	-0.31400E+01	0.36131E-02		0.6774E-01	213.2-353.2
C_5H_{10}	Cyclopentane	0.24287E+01	-0.15304E-02	-0.13095E-06		0.6057E-08	278.2-313.2
C_5H_{10}	2-Methyl-2-butene	0.26064E+01	-0.19578E-02	-0.53908E-06		0.1541E-05	225.0-296.0
C_5H_{10}	1-Pentene	-0.11438E+01	0.25420E-01	-0.50000E-04			273.2-293.2
$C_5H_{10}Br_2$	1,4-Dibromopentane	0.26443E+02	-0.88640E-01	0.10000E-03			293.2-333.2
$C_5H_{10}Br_2$	1,5-Dibromopentane	0.38192E+02	-0.15648E+00	0.20000E-03		0.2450E-03	303.2-333.2
$C_5H_{10}Cl_2$	1,2-Dichloropentane	0.19016E+02	-0.57954E-01	0.56801E-04		0.5883E-03	293.2-356.2
$C_5H_{10}N_2O$	1,3-Dimethyl-imidazolidin-2-one	0.12262E+03	-0.40480E+00	0.40121E-03		0.3515E-03	298.2-373.2
$C_5H_{10}O$	Cyclopentanol	0.10565E+03	-0.44244E+00	0.48657E-03		0.3078E-02	258.2-323.2
$C_5H_{10}O$	3-Pentanone	0.12690E+02	0.95177E-01	-0.27321E-03		0.1063E+00	233.2-353.2
$C_5H_{10}O$	2-Pentanone	0.40893E+02	-0.10423E+00	0.60557E-04		0.1952E-01	204.2-353.2
$C_5H_{10}O$	3-Methyl-2-butanone	0.30695E+02	-0.10962E+00	0.13810E-03		0.2095E-02	293.2-328.2
$C_5H_{10}O$	Tetrahydropyran	0.19793E+02	-0.76071E-01	0.94852E-04		0.1003E-02	234.0-333.2
$C_5H_{10}O$	2,2-Dimethylpropanal	0.18645E+02	-0.32395E-01	-0.16157E-05		0.8880E-03	280.2-333.2
$C_5H_{10}O_2$	Butyl formate	0.21532E+02	-0.84106E-01	0.10952E-03			288.2-323.2
$C_5H_{10}O_2$	Methyl butanoate	0.38604E+02	-0.19171E+00	0.27128E-03		0.2243E-03	301.2-343.2
$C_5H_{10}O_2$	Propyl acetate	0.17677E+02	-0.61404E-01	0.69196E-04		0.1821E-04	253.2-353.2
$C_5H_{10}O_2$	Pentanoic acid	0.33491E+01	-0.75156E-02	0.17820E-04		0.2526E-03	250.4-343.5
$C_5H_{10}O_2S$	3-Methyl sulfolane	0.53158E+02	-0.93730E-01	0.47275E-04		0.1758E-03	298.2-398.2
$C_5H_{10}O_2S$	Methyl 3-(methylthio)propanoate	0.18059E+02	-0.31000E-01	-0.23141E-13			303.2-343.2
$C_5H_{10}O_3$	Ethyl lactate	0.31225E+02	-0.43531E-01	-0.28571E-04			273.2-373.2
$C_5H_{10}O_4$	1,2,3-Propanetriol 1-acetate	0.10653E+03	-0.26439E+00	-0.62371E-04		0.4024E-01	215.1-242.2
$C_5H_{11}Br$	1-Bromopentane	0.20954E+02	-0.78743E-01	0.98908E-04		0.2956E-02	182.9-328.2
$C_5H_{11}Br$	1-Bromo-3-methylbutane	0.27743E+02	-0.13927E+00	0.22627E-03		0.7417E-02	122.5-291.5
$C_5H_{11}Cl$	1-Chloropentane	0.18626E+02	-0.54719E-01	0.47143E-04		0.1366E-03	273.2-323.2
$C_5H_{11}Cl$	1-Chloro-3-methylbutane	0.22228E+02	-0.93189E-01	0.12991E-03		0.7145E-03	170.7-296.8
$C_5H_{11}Cl$	2-Chloro-2-methylbutane	0.55104E+02	-0.29866E+00	0.47840E-03		0.1025E-03	201.3-222.8
$C_5H_{11}I$	1-Iodopentane	0.15753E+02	-0.50543E-01	0.56401E-04		0.5103E-05	293.2-323.2
$C_5H_{11}N$	Piperidine	0.82317E+01	-0.11229E-01	-0.71429E-05		0.4571E-05	293.0-333.0
$C_5H_{11}NO$	2,2-Dimethylpropanamide	0.10400E+03	-0.46017E+00	0.60000E-03			298.2-328.2
$C_5H_{11}NO$	N-Ethylpropanamide	0.10118E+04	-0.48000E+01	0.61429E-02		0.2714E-01	298.2-328.2
$C_5H_{11}NO$	N-Propylacetamide	0.66775E+03	-0.27964E+01	0.31905E-02		0.4728E-01	298.2-328.2
$C_5H_{11}NO_2$	N-(2-Methoxyethyl)acetamide	0.40999E+03	-0.16529E+01	0.18393E-02		0.3726E-02	298.2-348.2
C_5H_{12}	Isopentane	0.22384E+01	-0.12985E-02	-0.16182E-06		0.4882E-05	142.7-293.2
C_5H_{12}	Neopentane	0.10949E+02	-0.63057E-01	0.10835E-03		0.1509E-03	251.0-296.0
$C_5H_{12}O$	2,2-Dimethyl-1-propanol	0.92350E+02	-0.41870E+00	0.50000E-03			333.2-373.2
$C_5H_{12}O$	2-Methyl-1-butanol	0.14020E+02	0.13948E+00	-0.45000E-03		0.6050E-03	288.2-318.2
$C_5H_{12}O$	2-Methyl-2-butanol	0.11662E+03	-0.69756E+00	0.10920E-02		0.5226E-02	268.2-318.2
$C_5H_{12}O$	3-Methyl-1-butanol	0.79733E+02	-0.31272E+00	0.32014E-03		0.5792E-01	173.2-513.2
$C_5H_{12}O$	1-Pentanol	0.73397E+02	-0.28165E+00	0.28427E-03		0.4917E-01	213.2-513.2
$C_5H_{12}O$	2-Pentanol	0.16437E+03	-0.86506E+00	0.11955E-02		0.7775E-02	273.2-323.2
$C_5H_{12}O$	3-Pentanol	0.12838E+03	-0.60980E+00	0.75000E-03		0.8000E-02	288.2-318.2
$C_5H_{12}O_2$	1,2-Pentanediol	0.18436E+03	-0.10682E+01	0.17037E-02		0.5330E-01	197.2-296.8
$C_5H_{12}O_2$	1,4-Pentanediol	0.13568E+03	-0.59198E+00	0.75398E-03		0.3712E-01	193.2-317.8
$C_5H_{12}O_2$	1,5-Pentanediol	0.11858E+03	-0.45920E+00	0.49341E-03		0.1002E-01	243.2-343.2
$C_5H_{12}O_2$	2,3-Pentanediol	0.95876E+02	-0.46463E+00	0.67434E-03			237.8-296.9
$C_5H_{12}O_2$	2,4-Pentanediol	0.11914E+03	-0.52569E+00	0.69607E-03		0.1773E-01	223.7-294.2
$C_5H_{12}O_2$	Diethoxymethane	0.25294E+01	0.73988E-04	-0.28331E-06		0.1583E-05	227.2-293.2
$C_5H_{12}S$	2-Methyl-2-butanethiol	0.15116E+02	-0.50700E-01	0.56250E-04		0.2000E-06	273.2-333.2
$C_5H_{12}S$	1-Pentanethiol	0.71131E+01	-0.30228E-02	-0.16414E-04		0.5249E-03	273.2-333.2
$C_5H_{13}ClNO_3P$	Diethyl N-chloro-N-methyl-amidophosphate	0.18399E+02	-0.48520E-01	0.50000E-04		0.5000E-05	293.2-353.2

Molecular formula	Name	a	b	c	d	ESQS/N	Range T/K
$C_5H_{13}N$	Pentylamine	0.11274E+02	-0.34965E-01	0.37706E-04		0.1098E-03	223.2-353.2
$C_5H_{18}B_{10}$	Isopropyl carborane	0.77985E+02	-0.11250E+00				293.2-373.2
C_6F_6	Hexafluorobenzene	0.24041E+01	-0.83086E-03	-0.14286E-05		0.1829E-06	298.2-338.2
$C_6F_9Cl_5$	Pentachloro-1,1,2,3,3,4,4,5,5,6-nonafluorohexane	0.53413E+01	-0.13644E-01	0.16309E-04		0.5296E-04	203.2-323.2
$C_6F_{15}OP$	Tris(perfluoroethyl)phosphine oxide	0.34362E+01	-0.35000E-02				313.2-353.2
$C_6H_4Br_2$	o-Dibromobenzene	-0.81849E-02	0.62671E-01	-0.12222E-03			293.2-353.2
$C_6H_4Br_2$	m-Dibromobenzene	0.93214E+01	-0.20273E-01	0.16667E-04			293.2-353.2
C_6H_4ClNOS	3-Chloro-N-sulfinylaniline	0.18958E+02	-0.58646E-01	0.60714E-04		0.3181E-03	293.2-393.2
$C_6H_4ClNO_2$	o-Chloronitrobenzene	0.16800E+03	-0.59708E+00	0.59957E-03		0.4105E-02	323.2-436.2
$C_6H_4ClNO_2$	m-Chloronitrobenzene	0.77193E+02	-0.25118E+00	0.23798E-03		0.2184E-02	323.2-433.2
$C_6H_4Cl_2$	o-Dichlorobenzene	0.13629E+02	0.10622E-02	-0.44444E-04			293.2-353.2
$C_6H_4Cl_2$	m-Dichlorobenzene	0.77565E+01	-0.93333E-02	-0.26880E-14			293.2-353.2
$C_6H_4Cl_2$	p-Dichlorobenzene	0.26999E+01	-0.35325E-03	-0.17619E-05		0.3961E-07	328.2-363.2
$C_6H_4F_2$	o-Difluorobenzene	0.59107E+02	-0.23611E+00	0.27987E-03			273.2-323.2
$C_6H_4F_2$	m-Difluorobenzene	0.14448E+02	-0.46982E-01	0.51948E-04			273.2-323.2
$C_6H_4I_2$	1,2-Diiodobenzene	0.31150E+02	-0.14428E+00	0.20000E-03			323.2-353.2
$C_6H_4N_2$	2-Cyanopyridine	0.45596E+03	-0.17746E+01	0.19105E-02		0.3398E-01	303.2-398.2
$C_6H_4N_2$	3-Cyanopyridine	0.60484E+02	-0.17280E+00	0.15218E-03		0.7473E-04	323.2-398.2
$C_6H_4N_2$	4-Cyanopyridine	0.12533E+02	-0.30115E-01	0.26674E-04		0.6924E-04	353.2-398.2
$C_6H_4N_2O_3S$	3-Nitro-N-sulfinylaniline	0.65583E+02	-0.19656E+00	0.18125E-03		0.1012E-03	343.0-403.0
$C_6H_4N_2O_4$	m-Dinitrobenzene	0.10406E+03	-0.34133E+00	0.32609E-03			365.2-413.2
C_6H_5Br	Bromobenzene	0.94100E+01	-0.12537E-01	-0.31127E-05		0.1294E-03	234.2-333.2
$C_6H_4BrNO_2$	3-Bromonitrobenzene	0.81413E+02	-0.27645E+00	0.27367E-03		0.2068E-02	328.2-413.2
C_6H_5Cl	Chlorobenzene	0.19471E+02	-0.70786E-01	0.82466E-04		0.1658E-01	293.2-430.2
C_6H_5ClO	o-Chlorophenol	0.29755E+02	-0.11256E+00	0.12390E-03		0.7405E-02	296.2-448.2
C_6H_5ClO	p-Chlorophenol	0.31997E+02	-0.94241E-01	0.88392E-04		0.2430E-02	314.2-453.2
$C_6H_5ClO_2S$	Benzenesulfonyl chloride	0.83886E+02	-0.23405E+00	0.19713E-03		0.2732E-02	323.2-473.2
$C_6H_5Cl_2NO_2S_2$	Phenylsulfonylimidosulfurous dichloride	0.57134E+02	-0.92000E-01				298.2-373.2
C_6H_5I	Iodobenzene	0.89442E+01	-0.20008E-01	0.17641E-04		0.6898E-05	243.2-323.2
$C_6H_5NO_2$	Nitrobenzene	0.11212E+03	-0.35211E+00	0.31128E-03		0.1831E+00	278.7-533.0
$C_6H_5NO_3$	o-Nitrophenol	0.33827E+02	-0.62123E-01	0.26774E-04		0.2244E-02	323.2-453.2
$C_6H_5NO_3$	m-Nitrophenol	0.18967E+03	-0.66144E+00	0.66532E-03		0.2395E-01	373.2-458.2
$C_6H_5NO_3$	p-Nitrophenol	0.22901E+03	-0.74264E+00	0.68006E-03		0.2344E-02	393.2-463.2
C_6H_6	Benzene	0.26706E+01	-0.91648E-03	-0.14257E-05		0.2276E-04	293.2-513.2
$C_6H_6N_2O_2$	o-Nitroaniline	0.18900E+03	-0.56977E+00	0.47484E-03		0.3427E-01	353.0-468.0
$C_6H_6N_2O_2$	m-Nitroaniline	0.20352E+03	-0.66582E+00	0.61310E-03		0.1090E-01	398.0-468.0
$C_6H_6N_2O_2$	p-Nitroaniline	0.48673E+03	-0.15040E+01	0.12857E-02		0.3314E-02	428.0-468.0
C_6H_6O	Phenol	0.63391E+02	-0.24988E+00	0.26930E-03		0.1433E-02	303.2-433.2
$C_6H_6O_2$	1,2-Benzenediol	0.74930E+02	-0.22142E+00	0.18919E-03		0.5003E-02	388.2-463.2
$C_6H_6O_2$	1,3-Benzenediol	0.30252E+02	-0.56443E-01	0.35578E-04		0.2084E-03	393.2-463.2
C_6H_6S	Benzenethiol	0.57155E+01	-0.70336E-02	0.73617E-05		0.9383E-06	303.2-358.2
C_6H_7N	Aniline	0.89534E+01	0.38990E-02	-0.36310E-04		0.1847E-01	293.2-413.2
C_6H_7N	2-Methylpyridine	0.34560E+02	-0.11980E+00	0.12500E-03			293.2-333.2
C_6H_7N	3-Methylpyridine	0.19643E+03	-0.11167E+01	0.16667E-02			303.0-333.0
C_6H_7N	4-Methylpyridine	0.33765E+02	-0.10113E+00	0.93860E-04		0.3153E-04	274.0-333.0
C_6H_7NO	2-Methylpyridine 1-oxide	0.11705E+03	-0.35301E+00	0.32000E-03		0.1805E-02	323.2-398.2
C_6H_7NO	3-Methylpyridine 1-oxide	0.59851E+02	-0.12682E+00	0.86622E-04		0.4261E-04	318.2-398.2
C_6H_8	1,4-Cyclohexadiene	0.27459E+01	-0.16975E-02	-0.36461E-06		0.3818E-06	232.0-356.0
C_6H_8	1,3,5-Hexatriene(cis/trans=0.58)	0.40452E+01	-0.95442E-02	0.12064E-04		0.1330E-04	275.0-373.0
C_6H_9N	Cyclopentanecarbonitrile	0.69830E+02	-0.25303E+00	0.31491E-03		0.3209E-02	200.7-293.2
C_6H_{10}	Cyclohexene	0.30598E+01	-0.39841E-02	0.37554E-05		0.5680E-05	141.2-313.2
C_6H_{10}	2,3-Dimethyl-1,3-butadiene	0.26258E+01	-0.17990E-02	0.12035E-06		0.1485E-04	223.2-323.2
C_6H_{10}	4-Methyl-1,3-pentadiene	0.51328E+01	-0.12774E-01	0.14215E-04		0.4665E-04	198.2-323.2
C_6H_{10}	1,5-Hexadiene	0.30014E+01	-0.28668E-02	-0.31026E-06		0.1227E-03	151.0-294.0
C_6H_{10}	cis, cis-2,4-Hexadiene	0.27284E+01	-0.17178E-02	-0.62926E-06		0.3438E-07	234.0-351.0
C_6H_{10}	trans, trans-2,4-Hexadiene	0.26774E+01	-0.16977E-02	-0.55637E-06		0.1285E-05	232.0-353.0
C_6H_{10}	1-Hexyne	0.58591E+01	-0.17099E-01	0.20856E-04		0.3800E-04	184.0-296.0
$C_6H_{10}O$	Cyclohexanone	0.41577E+02	-0.11463E+00	0.92454E-04		0.8572E-02	253.0-423.0
$C_6H_{10}O_2$	Ethyl methacrylate	0.40962E+02	-0.20520E+00	0.29286E-03		0.1997E-02	303.2-343.2
$C_6H_{10}O_4$	Ethylene glycol diacetate	0.25093E+02	-0.95171E-01	0.12224E-03		0.4611E-02	223.2-290.2
$C_6H_{10}O_4$	Diethyl oxalate	0.21938E+02	-0.66226E-01	0.66800E-04		0.2761E-04	293.2-368.2
$C_6H_{10}O_4$	Monomethyl glutarate	0.16779E+02	-0.39839E-01	0.38095E-04		0.4940E-04	293.2-363.2

Molecular formula	Name	a	b	c	d	ESQS/N	Range T/K
$C_6H_{10}O_4$	Dimethyl succinate	0.13551E+02	-0.23109E-01	0.55440E-05		0.4924E-02	293.2-433.2
$C_6H_{11}BrO_2$	Ethyl 2-bromobutanoate	0.49005E+02	-0.23193E+00	0.32500E-03		0.3613E-03	303.2-333.2
$C_6H_{11}BrO_2$	Ethyl 2-bromo-2-methylpropanoate	0.77044E+02	-0.40784E+00	0.60000E-03		0.1250E-03	303.2-333.2
C_6H_{12}	Cyclohexane	0.24293E+01	-0.12095E-02	-0.58741E-06		0.5248E-08	283.2-333.2
C_6H_{12}	Methylcyclopentane	0.21587E+01	-0.22450E-03	-0.12500E-05		0.2813E-07	293.2-323.2
C_6H_{12}	1-Hexene	0.31476E+01	-0.50003E-02	0.46673E-05		0.1654E-04	149.0-294.0
C_6H_{12}	trans-2-Hexene	0.24338E+01	-0.11323E-02	-0.13720E-05		0.5306E-05	157.0-295.0
C_6H_{12}	cis-3-Hexene	0.30691E+01	-0.45458E-02	0.39898E-05		0.1842E-04	155.0-296.0
$C_6H_{12}Br_2$	1,6-Dibromohexane	-0.55185E+01	0.11746E+00	-0.23658E-03		0.1744E-02	274.2-328.2
$C_6H_{12}Cl_2$	1,6-Dichlorohexane	0.11277E+02	0.67200E-02	-0.50000E-04		0.5000E-05	308.2-338.2
$C_6H_{12}N_2O$	1,3-Dimethyl-2-oxohexa-hydropyrimidine	0.12648E+03	-0.43593E+00	0.44574E-03		0.1850E-03	298.2-373.2
$C_6H_{12}O$	Isobutyl vinyl ether	0.48060E+01	-0.50000E-02	-0.41495E-14			293.2-323.2
$C_6H_{12}O$	Cyclohexanol	0.10173E+03	-0.43072E+00	0.47926E-03		0.1373E-01	293.2-423.2
$C_6H_{12}O$	cis-2,5-Dimethyltetrahydrofuran	0.23827E+02	-0.10811E+00	0.15086E-03		0.6833E-03	207.7-302.2
$C_6H_{12}O$	2-Hexanone	0.70378E+02	-0.29385E+00	0.35289E-03		0.5283E-04	243.2-293.2
$C_6H_{12}O$	4-Methyl-2-pentanone	0.36341E+02	-0.97119E-01	0.61896E-04		0.3696E-01	204.2-373.2
$C_6H_{12}O$	3,3-Dimethyl-2-butanone	0.66857E+02	-0.28552E+00	0.34422E-03		0.2113E-03	243.2-293.2
$C_6H_{12}O$	1-Methylcyclopentanol	0.75444E+02	-0.36617E+00	0.47021E-03			310.1-332.8
$C_6H_{12}O_2$	Isopentyl formate	0.29257E+02	-0.14028E+00	0.20000E-03			288.2-323.2
$C_6H_{12}O_2$	Ethyl butanoate	0.48698E+02	-0.25660E+00	0.37237E-03		0.7392E-03	301.2-343.2
$C_6H_{12}O_2$	Hexanoic acid	0.21730E+01	0.14840E-02	-0.16526E-06		0.1204E-04	298.2-433.2
$C_6H_{12}O_2$	Butyl acetate	0.13825E+02	-0.43994E-01	0.48214E-04		0.8571E-05	253.2-353.2
$C_6H_{12}O_2$	Isobutyl acetate	0.14323E+02	-0.46048E-01	0.49286E-04		0.4791E-04	273.2-323.2
$C_6H_{12}O_2$	sec-Butyl acetate	0.12427E+02	-0.32035E-01	0.24286E-04		0.5109E-04	273.2-323.2
$C_6H_{12}O_2$	tert-Butyl acetate	0.55435E+02	-0.30494E+00	0.46107E-03		0.6024E-02	273.2-323.2
$C_6H_{12}O_2S$	2,4-Dimethyltetrahydrothio-phene-S,S-dioxide	0.47328E+02	-0.54486E-01	-0.17599E-04		0.7803E-02	293.2-398.2
$C_6H_{12}O_3$	2-Ethoxyethyl acetate	0.23290E+02	-0.71566E-01	0.65000E-04			303.2-323.2
$C_6H_{12}O_3$	2,4,6-Trimethyl-1,3,5-trioxan	0.86106E+02	-0.36553E+00	0.41440E-03		0.8110E-03	292.2-393.2
$C_6H_{13}Br$	1-Bromohexane	0.15233E+02	-0.44385E-01	0.43039E-04		0.1695E-04	274.2-328.2
$C_6H_{13}Cl$	1-Chlorohexane	0.15994E+02	-0.43647E-01	0.33393E-04		0.5041E-03	273.2-323.2
$C_6H_{13}ClO$	6-Chloro-1-hexanol	-0.73364E+01	0.46377E+00	-0.14202E-02		0.2098E-01	195.2-242.2
$C_6H_{13}I$	1-Iodohexane	0.16685E+02	-0.61309E-01	0.77262E-04		0.2052E-05	293.3-323.2
$C_6H_{13}NO$	N-Butylacetamide	0.70739E+03	-0.37369E+01	0.71585E-02	-0.48716E-05	0.9343E-01	253.2-493.2
$C_6H_{13}NO$	N-Isobutylacetamide	0.84645E+03	-0.46747E+01	0.92801E-02	-0.64581E-05	0.5325E-01	253.2-493.2
$C_6H_{13}NO$	N-sec-Butylacetamide	0.40410E+03	-0.14289E+01	0.13341E-02		0.4917E+00	273.2-493.2
$C_6H_{13}NO$	N-Ethylbutanamide	0.67036E+03	-0.29402E+01	0.35238E-02		0.2014E-01	298.2-328.2
$C_6H_{13}NO$	N-Methylpentanamide	0.81195E+03	-0.35899E+01	0.42135E-02		0.6648E+00	271.0-306.0
$C_6H_{13}NO$	N-Methyl-3-methylbutanamide	0.13713E+04	-0.69930E+01	0.93240E-02			286.0-308.0
$C_6H_{13}NO$	N-Methyl-2-methylbutanamide	0.11578E+04	-0.54895E+01	0.68981E-02		0.1917E+01	307.0-341.0
$C_6H_{13}NO$	N-Propylpropanamide	0.58846E+03	-0.22012E+01	0.20870E-02		0.1640E+01	298.2-328.2
C_6H_{14}	2,2-Dimethylbutane	0.22740E+01	-0.96229E-03	-0.14286E-05		0.5029E-06	273.2-313.2
C_6H_{14}	2,3-Dimethylbutane	0.24305E+01	-0.20081E-02	0.53571E-06		0.6107E-06	273.2-323.2
C_6H_{14}	2-Methylpentane	0.20745E+01	0.50871E-03	-0.39286E-05		0.5857E-06	273.2-323.2
C_6H_{14}	3-Methylpentane	0.24739E+01	-0.23190E-02	0.10714E-05		0.3952E-06	273.2-323.2
C_6H_{14}	Hexane	0.19768E+01	0.70933E-03	-0.34470E-05		0.8445E-05	293.2-473.2
$C_6H_{14}O$	Dipropyl ether	0.14600E+02	-0.72670E-01	0.11742E-03		0.8795E-03	161.0-297.0
$C_6H_{14}O$	2,2-Dimethyl-1-butanol	0.14054E+03	-0.72925E+00	0.97821E-03		0.1192E+00	243.2-393.2
$C_6H_{14}O$	1-Hexanol	0.62744E+02	-0.24214E+00	0.24704E-03		0.4950E-01	233.3-513.2
$C_6H_{14}OS$	Dipropyl sulfoxide	0.84868E+02	-0.23486E+00	0.18198E-03		0.2756E-03	303.2-373.2
$C_6H_{14}O_2$	1,2-Diethoxyethane	0.99099E+01	-0.33403E-01	0.44048E-04		0.4794E-04	223.2-303.2
$C_6H_{14}O_2$	Hexylene glycol	0.14531E+03	-0.65285E+00	0.83503E-03		0.8870E-01	203.2-333.2
$C_6H_{14}O_2S$	Dipropyl sulfone	0.70195E+02	-0.15008E+00	0.86506E-04		0.3052E-03	303.2-398.2
$C_6H_{14}O_3$	Bis(2-methoxyethyl) ether	0.28291E+02	-0.11236E+00	0.14000E-03		0.6228E-04	298.2-333.2
$C_6H_{14}O_3$	Bis(2-hydroxypropyl) ether	0.97446E+02	-0.38029E+00	0.40285E-03		0.4342E-01	213.2-423.2
$C_6H_{14}O_3$	1,2,6-Hexanetriol	0.26127E+03	-0.14552E+01	0.22765E-02		0.1011E+00	261.4-285.3
$C_6H_{14}O_4$	Triethylene glycol	0.91845E+02	-0.33827E+00	0.36062E-03		0.3385E-02	253.2-333.2
$C_6H_{14}S$	1-Hexanethiol	0.11774E+02	-0.37298E-01	0.41875E-04		0.3613E-05	273.2-333.2
$C_{10}H_{22}O_2$	Dipropylamine	0.11376E+02	-0.49796E-01	0.71792E-04		0.2072E-03	243.2-323.2
$C_6H_{15}N$	Hexylamine	0.80244E+01	-0.16627E-01	0.10874E-04		0.7781E-04	253.2-373.2
$C_6H_{15}N$	Triethylamine	0.29205E+01	-0.14007E-02	-0.13469E-05		0.1704E-03	233.2-323.2
$C_6H_{15}O_4P$	Triethyl phosphate	0.61230E+02	-0.26047E+00	0.33333E-03			298.2-333.2
$C_6H_{16}NO_3P$	N,N-Dimethylphosphoramidic acid diethyl ester	0.22012E+02	-0.35000E-01				293.2-393.2

Molecular formula	Name	a	b	c	d	ESQS/N	Range T/K
$C_6H_{18}N_3OP$	Tris(dimethylamino)-phosphine oxide	0.95666E+02	-0.29769E+00	0.26407E-03		0.1112E-01	283.2-363.2
$C_6H_{18}N_4$	Triethylenetetramine	0.50699E+02	-0.21730E+00	0.27582E-03		0.2501E-02	213.2-333.2
$C_6H_{18}OSi_2$	Hexamethyldisiloxane	0.34537E+01	-0.61530E-02	0.61544E-05		0.1541E-06	213.2-313.2
$C_6H_{19}NSi_2$	Hexamethyldisilazane	0.23358E+01	0.16127E-02	-0.62078E-05			294.2-333.2
C_7H_4ClNO	4-Chlorophenyl isocyanate	0.40896E+01	-0.31667E-02				288.2-348.2
C_7H_5BrO	Benzoyl bromide	0.84231E+01	-0.31089E+00	0.32857E-03		0.1429E-04	283.2-313.2
$C_7H_5F_3O_2S$	Trifluoromethylphenyl sulfone	0.96137E+02	-0.30867E+00	0.29464E-03		0.1190E-02	293.2-393.2
$C_7H_5F_3O_3S$	4-(Trifluoromethylsulfonyl)-phenol	0.88287E+02	-0.18630E+00	0.10650E-03		0.5804E-01	298.2-423.2
C_7H_5N	Benzonitrile	0.57605E+02	-0.13354E+00	0.87767E-04		0.6307E-02	273.2-453.2
C_7H_5NO	Phenyl isocyanate	0.17541E+02	-0.29790E-01	0.15476E-05		0.3340E-05	293.2-353.2
C_7H_6O	Benzaldehyde	0.35064E+02	-0.61271E-01	0.16222E-04		0.1256E-02	300.7-345.7
$C_7H_6O_2$	Salicylaldehyde	0.51315E+02	-0.15379E+00	0.14111E-03		0.3279E-02	289.2-453.2
C_7H_7Br	(Bromomethyl)benzene	0.18482E+02	-0.57207E-01	0.57321E-04		0.2730E-03	273.2-323.2
C_7H_7Br	2-Bromotoluene	0.10229E+02	-0.25050E-01	0.20357E-04		0.2511E-04	273.2-323.2
C_7H_7Br	3-Bromotoluene	0.11522E+02	-0.24946E-01	0.15714E-04		0.3025E-04	273.2-323.2
C_7H_7Br	4-Bromotoluene	0.10014E+02	-0.13918E-01	-0.50000E-05			273.2-293.2
C_7H_7BrO	2-Bromoanisole	0.12023E+02	-0.59116E-02	-0.13787E-04		0.4660E-04	303.2-358.2
C_7H_7BrO	4-Bromoanisole	0.74367E+01	0.12648E-01	-0.42128E-04		0.1043E-04	303.2-358.2
C_7H_7BrS	4-Bromo-3-methylthiophenol	0.62893E+01	-0.12330E-01	0.12500E-04			303.2-343.2
C_7H_7Cl	Benzyl chloride	0.17108E+02	-0.45285E-01	0.35000E-04		0.1033E-03	273.2-323.2
C_7H_7Cl	o-Chlorotoluene	0.11507E+02	-0.31148E-01	0.27143E-04		0.6825E-04	273.2-323.2
C_7H_7Cl	m-Chlorotoluene	0.13921E+02	-0.37186E-01	0.31786E-04		0.5979E-04	273.2-323.2
C_7H_7Cl	p-Chlorotoluene	0.20265E+01	0.40060E-01	-0.87500E-04			293.2-333.2
C_7H_7ClO	4-Chloroanisole	0.64019E+01	0.30560E-01	-0.87500E-04			293.2-333.2
C_7H_7ClS	4-Chloro-2-methylthiophenol	0.48451E+01	0.17320E-01	-0.50000E-04			303.2-343.2
$C_7H_7NO_2$	o-Nitrotoluene	0.10420E+03	-0.41726E+00	0.51607E-03		0.6147E-02	273.0-323.0
$C_7H_7NO_2$	m-Nitrotoluene	0.62492E+02	-0.16235E+00	0.12844E-03		0.3424E-01	303.2-403.2
$C_7H_7NO_3$	2-Nitroanisole	0.16684E+03	-0.58196E+00	0.57382E-03		0.3134E-01	293.2-423.2
$C_7H_7NO_3$	3-Nitroanisole	0.65402E+02	-0.16460E+00	0.12560E-03		0.1628E-02	318.2-443.2
$C_7H_7NO_3$	4-Nitroanisole	0.59811E+02	-0.10955E+00	0.36042E-04		0.2498E-01	338.2-443.2
$C_7H_7NO_3S$	4-Methoxy-N-sulfinylaniline	0.33176E+02	-0.10572E+00	0.11250E-03			293.2-333.2
C_7H_8	Toluene	0.32584E+01	-0.34410E-02	0.15937E-05		0.1176E-05	206.7-316.0
C_7H_8O	Anisole	0.10887E+02	-0.32372E-01	0.33629E-04		0.3854E-03	294.2-413.2
C_7H_8O	Benzyl alcohol	0.13661E+03	-0.72127E+00	0.10225E-02		0.5281E-04	303.2-333.2
C_7H_8O	o-Cresol	0.21633E+02	-0.71069E-01	0.70590E-04		0.8393E-03	298.2-453.2
C_7H_8O	m-Cresol	0.81716E+02	-0.35039E+00	0.39878E-03		0.8561E-01	273.7-463.2
C_7H_8O	p-Cresol	0.70253E+02	-0.28870E+00	0.31979E-03		0.4488E-01	298.2-453.2
$C_7H_8O_2$	2-Methoxyphenol	0.31751E+02	-0.88173E-01	0.72953E-04		0.1702E-02	291.2-448.2
$C_7H_8O_2$	3-Methoxyphenol	0.37279E+02	-0.12113E+00	0.11698E-03		0.4976E-03	298.2-432.7
$C_7H_8O_2$	4-Methoxyphenol	0.39483E+02	-0.12142E+00	0.10841E-03		0.1556E-02	333.7-453.2
C_7H_8S	Benzylthiol	0.16628E+02	-0.68276E-01	0.94636E-04		0.2513E-03	298.2-358.2
C_7H_8S	4-Methylthiophenol	0.87052E+01	-0.15347E-01	0.95238E-05			323.2-358.2
C_7H_8S	Thioanisole	0.21841E+02	-0.97630E-01	0.13750E-03			303.2-343.2
C_7H_9N	2,4-Dimethylpyridine	0.25895E+02	-0.73900E-01	0.62500E-04			293.2-333.2
C_7H_9N	2,6-Dimethylpyridine	0.17714E+02	-0.39080E-01	0.12500E-04			293.2-333.2
C_7H_9N	2-Ethylpyridine	0.36397E+02	-0.15070E+00	0.18750E-03			293.2-333.2
C_7H_9N	4-Ethylpyridine	-0.73831E+01	0.14326E+00	-0.27500E-03			293.2-333.2
C_7H_9N	2-Methylaniline	0.10988E+02	-0.18976E-01	0.91958E-05		0.1131E-03	298.2-398.2
C_7H_9N	3-Methylaniline	0.13477E+02	-0.35551E-01	0.33135E-04		0.3507E-04	298.2-398.2
C_7H_9N	4-Methylaniline	0.78897E+01	-0.10196E-01	0.51190E-05		0.1207E-04	333.2-403.2
C_7H_9NO	2-Methoxyaniline	0.79911E+01	-0.92183E-02	0.37879E-06		0.4470E-05	303.2-393.2
C_7H_9NO	3-Methoxyaniline	0.28179E+02	-0.97840E-01	0.11027E-03		0.5895E-03	289.2-393.2
C_7H_9NO	4-Methoxyaniline	0.30149E+02	-0.10523E+00	0.11467E-03		0.3966E-03	333.2-453.2
C_7H_9NO	2,6-Dimethylpyridine-1-oxide	0.22765E+03	-0.90760E+00	0.10011E-02		0.1289E-01	298.2-398.2
$C_7H_9NO_2S$	N-Methylbenzenesulfonamide	0.46369E+02	0.45809E+00	-0.12857E-02		0.6524E-02	303.2-328.2
C_7H_{12}	Cycloheptene	0.32309E+01	-0.42373E-02	0.32572E-05		0.4158E-05	227.0-363.0
C_7H_{12}	1,6-Heptadiene	0.30815E+01	-0.36095E-02	0.16354E-05		0.1583E-04	184.0-293.0
$C_7H_{12}O$	Cycloheptanone	0.17511E+03	-0.11221E+01	0.19417E-02			258.2-298.2
$C_7H_{12}O_2$	Butyl acrylate	0.38296E+02	-0.19109E+00	0.27006E-03		0.3702E-03	301.2-343.2
$C_7H_{12}O_4$	Monomethyl adipate	0.11962E+02	-0.23973E-01	0.20608E-04		0.1881E-03	293.2-433.2
$C_7H_{12}O_4$	Dimethyl glutarate	0.20697E+02	-0.57794E-01	0.48405E-04		0.8804E-03	293.2-433.2
$C_7H_{12}O_4$	Diethyl malonate	0.14809E+02	-0.31207E-01	0.24066E-04		0.6969E-04	304.2-393.2

Molecular formula	Name	a	b	c	d	ESQS/N	Range T/K
$C_7H_{12}O_5$	1,2,3-Propanetriol 1,2-diacetate	0.43342E+02	-0.83764E-01	-0.78234E-04		0.8062E-03	214.9-244.7
$C_7H_{12}O_5$	1,2,3-Propanetriol 1,3-diacetate	0.28321E+02	-0.89073E-01	0.86891E-04		0.7246E-03	257.7-373.7
$C_7H_{13}NO$	N-Methyl 6-aminohexanoic lactam	0.11976E+03	-0.44832E+00	0.49745E-03		0.3082E-02	253.2-393.2
$C_7H_{13}NO_2$	2-Butoxyethyl isocyanate	0.18674E+02	-0.31786E-01	0.21944E-14		0.3306E-05	293.2-353.2
C_7H_{14}	Cycloheptane	0.25136E+01	-0.15089E-02	0.84915E-07		0.1369E-08	278.2-333.2
C_7H_{14}	1-Heptene	0.21755E+01	0.13896E-02	-0.57049E-05		0.5053E-05	273.2-323.2
$C_7H_{14}O$	2-Heptanone	0.38348E+02	-0.12531E+00	0.12005E-03		0.1079E-02	253.2-413.2
$C_7H_{14}O$	4-Heptanone	0.41520E+02	-0.13839E+00	0.13497E-03		0.8429E-03	253.2-393.2
$C_7H_{14}O$	5-Methyl-2-hexanone	0.52353E+02	-0.17695E+00	0.15195E-03		0.3638E-03	293.2-333.2
$C_7H_{14}O$	Cyclohexyl methanol	0.10164E+03	-0.45839E+00	0.54762E-03			333.2-368.2
$C_7H_{14}O$	2-Methylcyclohexanol*	0.17315E+03	-0.98794E+00	0.14634E-02		0.2420E-02	273.2-323.2
$C_7H_{14}O$	3-Methylcyclohexanol*	0.65896E+02	-0.21954E+00	0.14107E-03		0.8784E-02	273.2-323.2
$C_7H_{14}O$	4-Methylcyclohexanol*	0.65021E+02	-0.22896E+00	0.17946E-03		0.5192E-02	273.2-323.2
$C_7H_{14}O_2$	Pentyl acetate	0.12091E+02	-0.36536E-01	0.39732E-04		0.1345E-04	253.2-353.2
$C_7H_{14}O_2$	Heptanoic acid	0.36423E+01	-0.31996E-02	0.39362E-05		0.5705E-04	288.2-423.2
$C_7H_{15}Br$	1-Bromoheptane	0.15289E+02	-0.50621E-01	0.57753E-04		0.4005E-04	203.2-343.2
$C_7H_{15}Cl$	1-Chloroheptane	0.14279E+02	-0.39431E-01	0.32321E-04		0.2586E-03	273.2-323.2
$C_7H_{15}I$	1-Iodoheptane	0.11856E+02	-0.33493E-01	0.34368E-04		0.9935E-05	293.5-323.2
$C_7H_{15}NO$	N-Butylpropamide	0.72659E+03	-0.33492E+01	0.41905E-02		0.9320E-02	298.2-328.2
$C_7H_{15}NO$	N-Pentylacetamide	0.39500E+03	-0.14209E+01	0.13333E-02		0.6259E-02	298.2-328.2
$C_7H_{15}NO$	N-Propylbutanamide	-0.78077E+01	0.12542E+01	-0.30000E-02		0.2813E-01	313.2-328.2
C_7H_{16}	2,2-Dimethylpentane	0.23414E+01	-0.14362E-02	-0.51322E-07		0.1471E-06	153.2-353.2
C_7H_{16}	2,3-Dimethylpentane	0.25637E+01	-0.26328E-02	0.16071E-05		0.5631E-06	273.2-323.2
C_7H_{16}	2,4-Dimethylpentane	0.23979E+01	-0.17436E-02	0.17857E-06		0.6786E-07	273.2-323.2
C_7H_{16}	3,3-Dimethylpentane	0.24007E+01	-0.16802E-02	0.36069E-06		0.4382E-09	291.3-322.2
C_7H_{16}	3-Ethylpentane	0.23771E+01	-0.15140E-02	0.10093E-06		0.3033E-06	163.2-363.2
C_7H_{16}	Heptane	0.24740E+01	-0.22577E-02	0.12428E-05		0.1552E-04	273.2-373.2
C_7H_{16}	2-Methylhexane	0.24759E+01	-0.22535E-02	0.12500E-05		0.1225E-05	293.2-323.2
C_7H_{16}	3-Methylhexane	0.27089E+01	-0.37908E-02	0.37500E-05		0.5155E-06	273.2-323.2
$C_7H_{16}O$	Ethyl 3-methylbutyl ether	0.66541E+01	-0.55450E-02	-0.12500E-04		0.7813E-05	293.2-323.2
$C_7H_{16}O$	2,2-Dimethyl-1-pentanol	0.37318E+02	-0.17095E+00	0.22022E-03		0.3098E-01	283.2-393.2
$C_7H_{16}O$	1-Heptanol	0.60662E+02	-0.24049E+00	0.25155E-03		0.8574E-01	239.2-513.2
$C_7H_{16}O$	(±)-2-Heptanol	0.10050E+03	-0.49793E+00	0.64504E-03		0.5121E-01	207.4-365.1
$C_7H_{16}O$	D-(+)-2-Heptanol	0.10076E+03	-0.50063E+00	0.65038E-03		0.3983E-01	208.0-368.5
$C_7H_{16}O$	L-(-)-2-Heptanol	0.99744E+02	-0.49045E+00	0.62725E-03		0.5421E-01	207.0-352.0
$C_7H_{16}O$	3-Heptanol	0.19586E+03	-0.11465E+01	0.17175E-02		0.1375E+00	247.7-348.8
$C_7H_{16}O$	4-Heptanol	0.28995E+03	-0.18499E+01	0.30109E-02		0.1681E-02	270.0-301.1
$C_7H_{16}O$	3-Methyl-2-hexanol	0.59724E+02	-0.32417E+00	0.47058E-03		0.1191E+00	244.0-372.2
$C_7H_{16}S$	1-Heptanethiol	0.71333E+01	-0.97320E-02	-0.12500E-05		0.2592E-03	273.2-333.2
$C_7H_{17}N$	Heptylamine	0.87794E+01	-0.24363E-01	0.25325E-04		0.4459E-04	253.2-373.2
$C_7H_{17}O_3P$	Diisopropyl methanephosphonate	0.11157E+02	-0.72657E-02	-0.14286E-04			298.2-333.2
$C_8Cl_6F_{12}$	Hexachloro-1,1,2,3,3,4,5,5,6,7,7,8-dodecafluorooctane	0.52291E+01	-0.11867E-01	0.12876E-04		0.4572E-04	203.7-313.2
$C_8H_4F_3NO$	3-Trifluoromethylphenyl isocyanate	0.31492E+02	-0.95578E-01	0.95357E-04		0.9204E-05	293.2-353.2
$C_8H_5N_2O_3PS$	Phosphorodiisocyanatidothioic acid O-phenyl ester	0.25982E+02	-0.74947E-01	0.68750E-04		0.3395E-03	293.2-393.2
$C_8H_5N_2O_4P$	Phosphorodiisocyanatidic acid phenyl ester	0.78505E+01	0.16110E-01	-0.43656E-04		0.1177E-03	283.2-393.2
C_8H_7N	Phenylacetonitrile	0.82175E+02	-0.37416E+00	0.53220E-03			299.2-343.2
$C_8H_7NO_2$	4-Methoxyphenyl isocyanate	0.20780E+02	-0.31571E-01				333.2-403.2
C_8H_8	Styrene	0.44473E+01	-0.11422E-01	0.16000E-04			293.2-313.2
C_8H_8O	Acetophenone	0.26099E+02	0.64048E-02	-0.11905E-03		0.1338E-02	298.2-333.2
$C_8H_8O_2$	Benzyl formate	0.26162E+02	-0.11026E+00	0.14787E-03		0.1278E-02	303.2-358.2
$C_8H_8O_2$	Methyl benzoate	0.17486E+02	-0.51027E-01	0.50222E-04		0.4216E-04	302.7-393.2
$C_8H_8O_2$	Phenyl acetate	0.11327E+02	-0.26707E-01	0.22938E-04		0.4466E-06	298.2-404.2
$C_8H_8O_2$	2-Hydroxyacetophenone	0.42286E+02	-0.69215E-01	-0.35714E-05		0.1131E-04	298.2-368.2
$C_8H_8O_2$	Phenylacetic acid	0.24104E+01	0.30000E-02				353.2-393.2
$C_8H_8O_3$	Methyl salicylate	0.20501E+02	-0.39045E-01	0.68298E-05		0.6832E-02	223.2-398.2
C_8H_9BrO	2-Bromophenetole	0.23146E+02	-0.75753E-01	0.77778E-04		0.4512E-03	313.2-358.2
C_8H_{10}	Ethylbenzene	0.35969E+01	-0.53169E-02	0.47500E-05		0.6125E-08	293.2-323.2
C_8H_{10}	o-Xylene	0.36163E+01	-0.40177E-02	0.14286E-05		0.1086E-05	273.2-323.2
C_8H_{10}	m-Xylene	0.28421E+01	-0.10191E-02	-0.21429E-05		0.2952E-06	273.2-323.2
C_8H_{10}	p-Xylene	0.23140E+01	0.97221E-03	-0.37500E-05		0.4622E-05	293.2-363.2
$C_8H_{10}O$	2-Methylanisole	0.50825E+01	-0.62297E-02	0.28571E-05		0.8114E-06	293.2-333.2

Molecular formula	Name	a	b	c	d	ESQS/N	Range T/K
$C_8H_{10}O$	3-Methylanisole	0.12830E+02	-0.49701E-01	0.66429E-04		0.4023E-05	293.2-333.2
$C_8H_{10}O$	4-Methylanisole	0.86608E+01	-0.23510E-01	0.25000E-04		0.1376E-04	293.2-333.2
$C_8H_{10}O$	Phenetole	-0.15043E+02	0.13752E+00	-0.24500E-03			293.2-313.2
$C_8H_{10}O$	1-Phenylethanol	0.32971E+02	-0.12042E+00	0.12809E-03		0.2179E-02	293.2-423.2
$C_8H_{10}O$	2-Phenylethanol	0.12170E+03	-0.63124E+00	0.87776E-03		0.2728E-01	278.2-333.2
$C_8H_{10}O$	2,3-Xylenol	0.14399E+02	-0.41438E-01	0.39244E-04		0.3926E-04	343.2-433.2
$C_8H_{10}O$	2,4-Xylenol	0.22125E+02	-0.85543E-01	0.96548E-04		0.2163E-04	303.2-363.2
$C_8H_{10}O$	2,5-Xylenol	0.18049E+02	-0.54991E-01	0.51656E-04		0.2805E-05	338.2-455.2
$C_8H_{10}O$	2,6-Xylenol	0.12284E+02	-0.32996E-01	0.29867E-04		0.2147E-03	313.2-453.2
$C_8H_{10}O$	3,4-Xylenol	0.54423E+02	-0.21153E+00	0.22508E-03		0.4946E-02	333.2-453.2
$C_8H_{10}O$	3,5-Xylenol	0.54251E+02	-0.21647E+00	0.23542E-03		0.8998E-02	323.2-453.2
$C_8H_{10}O_2$	1,2-Dimethoxybenzene	0.74604E+01	-0.13445E-01	0.10737E-04		0.1615E-03	293.2-443.2
$C_8H_{10}O_2$	1,3-Dimethoxybenzene	0.11911E+02	-0.30804E-01	0.29643E-04		0.2280E-04	298.2-358.2
$C_8H_{10}O_2$	1,4-Dimethoxybenzene	0.11289E+02	-0.20765E-01	0.11987E-04		0.6627E-02	333.7-463.2
$C_8H_{11}N$	N,N-Dimethylaniline	0.84052E+01	-0.13549E-01	0.62835E-05		0.6669E-03	289.2-453.2
$C_8H_{11}N$	2,4,6-Trimethylpyridine	0.20990E+02	-0.57419E-01	0.44286E-04		0.2100E-04	298.2-358.2
$C_8H_{12}N_2O_2$	1,6-Diisocyanatohexane	0.26715E+02	-0.42696E-01				288.2-403.2
$C_8H_{12}O_4$	Diethyl maleate	0.13953E+02	-0.21969E-01	0.17817E-05		0.9522E-05	298.2-343.2
C_8H_{14}	cis-Cyclooctene	0.31115E+01	-0.32058E-02	0.16713E-05		0.2864E-05	269.0-406.0
C_8H_{14}	1,2-Dimethylcyclohexene	0.26443E+01	-0.17973E-02	0.35815E-06		0.9402E-06	211.0-374.0
C_8H_{14}	1,3-Dimethylcyclohexene	0.29951E+01	-0.34615E-02	0.24026E-05		0.1855E-05	213.0-373.0
C_8H_{14}	1,7-Octadiene	0.28376E+01	-0.17442E-02	-0.16141E-05		0.1505E-04	214.0-293.0
$C_8H_{14}O_4$	Dimethyl adipate	0.11739E+02	-0.17281E-01	0.11447E-05		0.3617E-02	293.2-433.2
$C_8H_{14}O_4$	Diisopropyl oxalate	0.10709E+02	-0.16328E-01	0.56000E-05		0.8000E-06	293.2-368.2
$C_8H_{14}O_4$	Diethyl succinate	0.80213E+01	0.11810E-02	-0.26400E-04			293.2-343.2
C_8H_{16}	Cyclooctane	0.25036E+01	-0.12460E-02	-0.23175E-06		0.2104E-06	295.0-411.0
C_8H_{16}	1-Octene	0.24348E+01	0.34200E-03	-0.50000E-05		0.2520E-04	273.2-323.2
$C_8H_{16}Br_2$	1,8-Dibromooctane	0.94117E+00	0.61520E-01	-0.13333E-03			298.2-328.2
$C_8H_{16}O$	2-Octanone	-0.16219E+02	0.18799E+00	-0.34156E-03		0.1305E-02	293.2-333.2
$C_8H_{16}O_2$	Butyl butanoate	0.79684E+01	-0.12000E-01	0.15266E-13			298.2-318.2
$C_8H_{16}O_2$	Ethyl hexanoate	0.11007E+02	-0.32800E-01	0.35714E-04		0.9524E-04	253.2-353.2
$C_8H_{16}O_2$	Octanoic acid	0.29391E+01	-0.38721E-03			0.2468E-03	288.2-423.2
$C_8H_{16}O_3$	3-Methyl-1-butyl lactate	0.48649E+02	-0.21253E+00	0.27619E-03			273.2-373.2
$C_8H_{16}O_2$	Isopentyl propanoate	0.17665E+02	-0.71718E-01	0.95635E-04			273.2-373.2
$C_8H_{17}Br$	1-Bromooctane	0.12404E+02	-0.35050E-01	0.34542E-04		0.6355E-06	283.2-353.2
$C_8H_{17}Cl$	1-Chlorooctane	0.11346E+02	-0.25120E-01	0.13450E-04		0.1655E-05	274.2-328.2
$C_8H_{17}I$	1-Iodooctane	0.12452E+02	-0.41229E-01	0.50108E-04		0.2450E-04	233.2-313.2
C_8H_{18}	2,5-Dimethylhexane	0.25821E+01	-0.26804E-02	0.19404E-05		0.3626E-09	294.0-323.7
C_8H_{18}	3,4-Dimethylhexane	0.26849E+01	-0.33712E-02	0.32949E-05		0.2463E-07	292.1-323.7
C_8H_{18}	3-Ethyl-3-methylpentane	0.25983E+01	-0.28027E-02	0.24195E-05		0.1888E-07	291.5-324.1
C_8H_{18}	Octane	0.22590E+01	-0.84212E-03	-0.75758E-06		0.5993E-05	233.2-393.2
C_8H_{18}	2,2,4-Trimethylpentane	0.23677E+01	-0.14768E-02	0.94261E-07		0.1993E-06	173.2-373.2
$C_8H_{18}O$	Dibutyl ether	0.65383E+01	-0.16172E-01	0.14969E-04		0.1305E-06	293.2-313.8
$C_8H_{18}O$	2,2-Dimethyl-1-hexanol	0.91244E+01	-0.21785E-01	0.21018E-04		0.1124E-02	283.2-393.2
$C_8H_{18}O$	2-Ethyl-1-hexanol	0.86074E+02	-0.42636E+00	0.55078E-03		0.8583E-01	208.2-318.2
$C_8H_{18}O$	2-Isopropyl-1-pentanol	0.46429E+02	-0.16764E+00	0.13590E-03		0.2950E-02	208.2-318.2
$C_8H_{18}O$	2-Methyl-1-heptanol	0.61698E+02	-0.33647E+00	0.49066E-03		0.1467E-01	235.6-328.1
$C_8H_{18}O$	3-Methyl-1-heptanol	0.84687E+01	-0.33712E-01	0.49793E-04		0.1699E-04	241.2-315.6
$C_8H_{18}O$	4-Methyl-1-heptanol	0.48612E+02	-0.26773E+00	0.39972E-03		0.1079E-01	236.9-332.2
$C_8H_{18}O$	5-Methyl-1-heptanol	0.54581E+02	-0.24772E+00	0.29734E-03		0.4591E-02	235.1-328.2
$C_8H_{18}O$	6-Methyl-1-heptanol	0.57997E+02	-0.23517E+00	0.24663E-03		0.9944E-03	265.1-328.2
$C_8H_{18}O$	3-Methyl-2-heptanol	0.39178E+02	-0.17976E+00	0.24218E-03		0.1914E-02	228.8-328.8
$C_8H_{18}O$	4-Methyl-2-heptanol	0.39715E+02	-0.23115E+00	0.36771E-03		0.1400E-01	239.8-333.2
$C_8H_{18}O$	5-Methyl-2-heptanol	0.68568E+02	-0.40706E+00	0.67433E-03		0.7625E-04	230.2-278.5
$C_8H_{18}O$	6-Methyl-2-heptanol	0.77520E+02	-0.41724E+00	0.59448E-03		0.5279E-01	239.0-328.7
$C_8H_{18}O$	2-Methyl-3-heptanol	-0.59739E+01	0.56700E-01	-0.83125E-04		0.2761E-04	343.2-403.2
$C_8H_{18}O$	3-Methyl-3-heptanol	-0.38440E+01	0.42327E-01	-0.61250E-04		0.7200E-05	343.2-403.2
$C_8H_{18}O$	4-Methyl-3-heptanol	-0.48003E+01	0.50740E-01	-0.75000E-04		0.1445E-04	343.2-403.2
$C_8H_{18}O$	5-Methyl-3-heptanol	0.61967E+01	-0.63750E-02			0.1561E-03	343.2-383.2
$C_8H_{18}O$	6-Methyl-3-heptanol	0.23037E+02	-0.98029E-01	0.12479E-03		0.2760E-02	283.2-383.2
$C_8H_{18}O$	2-Methyl-4-heptanol	0.42102E+00	0.10427E-01	-0.20438E-05		0.4771E-03	229.5-332.7
$C_8H_{18}O$	3-Methyl-4-heptanol	0.33354E+02	-0.14077E+00	0.17750E-03		0.3057E-03	230.1-329.7
$C_8H_{18}O$	3-Methyl-2-isopropyl-1-butanol	0.66911E+02	-0.29898E+00	0.36326E-03		0.3318E-01	208.2-318.2
$C_8H_{18}O$	1-Octanol	0.51647E+02	-0.20371E+00	0.21320E-03		0.6005E-01	258.2-513.2

Molecular formula	Name	a	b	c	d	ESQS/N	Range T/K
$C_8H_{18}O$	2-Octanol	0.63760E+02	-0.27643E+00	0.31075E-03		0.6774E+00	213.2-513.2
$C_8H_{18}O$	3-Octanol	0.12505E+03	-0.70646E+00	0.10245E-02		0.4787E+00	223.2-383.2
$C_8H_{18}O$	4-Octanol	0.51049E+02	-0.26664E+00	0.37280E-03		0.4124E+00	243.2-403.2
$C_8H_{18}OS$	Dibutyl sulfoxide	0.67156E+02	-0.16448E+00	0.92275E-04		0.3399E-02	313.2-393.2
$C_8H_{18}O_2$	2-Ethyl-1,3-hexanediol	0.57919E+02	-0.17128E+00	0.12949E-03		0.2224E-02	233.2-333.2
$C_8H_{18}O_2S$	Dibutyl sulfone	0.66248E+02	-0.16417E+00	0.12001E-03		0.1001E-03	323.2-398.2
$C_8H_{18}O_5$	Tetraethylene glycol	0.83547E+02	-0.31691E+00	0.34689E-03		0.2482E-02	253.2-333.2
$C_8H_{18}S$	1-Octanethiol	0.63667E+01	-0.87920E-02	0.18750E-05		0.6125E-06	273.2-333.2
$C_8H_{19}N$	Dibutylamine	0.52504E+01	-0.10538E-01	0.71485E-05		0.1637E-03	243.2-323.2
$C_8H_{19}N$	1-Octylamine	0.77931E+01	-0.20015E-01	0.19347E-04		0.3729E-04	273.2-373.2
$C_8H_{19}OP$	Dimethylhexylphosphine oxide	0.77256E+02	-0.21595E+00	0.17489E-03		0.1768E-02	323.2-473.2
$C_8H_{23}NOSi_2$	N-Trimethylsilyl-2-amino-ethyl trimethylsilyl ether	0.69934E+01	-0.20076E-01	0.22500E-04		0.1496E-04	293.2-353.2
$C_8H_{23}N_5$	Tetraethylenepentamine	0.40553E+02	-0.16681E+00	0.20659E-03		0.3780E-03	213.2-333.2
$C_8H_{24}O_4Si_4$	Octamethylcyclotetrasiloxane	0.3o286E+01	-0.56885E-02	0.50874E-05			296.2-333.2
$C_9H_6N_2O_2$	Toluene-2,4-diisocyanate	0.22174E+02	-0.66982E-01	0.68571E-04		0.5714E-05	293.2-353.2
$C_9H_6N_2O_2$	2,6-Toluene diisocyanate	0.10834E+02	-0.27987E-01	0.29286E-04		0.4184E-05	293.2-353.2
$C_9H_6O_2$	Coumarin	0.11311E+03	-0.33804E+00	0.31324E-03		0.6791E-02	343.2-423.2
C_9H_7N	Quinoline	0.33432E+02	-0.13497E+00	0.17788E-03		0.3077E-04	258.2-323.2
C_9H_7N	Isoquinoline	0.14412E+03	-0.79935E+00	0.11839E-02		0.1413E-02	298.2-323.2
C_9H_8O	Cinnamaldehyde	0.41837E+02	-0.11060E+00	0.10401E-03		0.2188E-02	305.8-353.5
$C_9H_8O_4$	Acetylsalicylic acid	0.69994E+01	-0.14553E-02			0.3666E-03	333.2-416.2
C_9H_9N	2,6-Dimethylbenzonitrile	0.70509E+02	-0.21399E+00	0.18750E-03		0.1571E-02	373.2-473.2
$C_9H_{10}O_2$	Ethyl benzoate	0.18216E+02	-0.62361E-01	0.72884E-04		0.2257E-04	288.2-343.2
$C_9H_{10}O_2$	Benzyl acetate	0.11727E+02	-0.30869E-01	0.32340E-04		0.1915E-05	303.2-358.2
$C_9H_{10}O_2$	Methyl 2-phenylacetate	0.17413E+02	-0.64222E-01	0.78865E-04		0.5303E-04	297.2-343.2
$C_9H_{10}O_3$	Ethyl salicylate	0.18910E+02	-0.35623E-01	0.46529E-05		0.3547E-02	225.2-321.2
$C_9H_{11}Br$	1-Bromo-3-phenylpropane	0.11360E+02	-0.27471E-01	0.25775E-04		0.2860E-04	302.2-358.2
$C_9H_{11}NO$	N,N-Dimethylbenzamide	0.76725E+02	-0.26908E+00	0.29409E-03		0.2506E-01	318.2-443.2
$C_9H_{11}NO$	N-Ethylbenzamide	-0.20109E+03	0.17866E+01	-0.31065E-02		0.1512E-04	352.7-388.7
C_9H_{12}	Propylbenzene	0.26933E+01	0.21679E-03	-0.44643E-05		0.1304E-04	273.2-323.2
C_9H_{12}	Isopropylbenzene	0.31149E+01	-0.30801E-02	0.19643E-05		0.2107E-06	273.2-323.2
C_9H_{12}	1,2,3-Trimethylbenzene	0.76006E+01	-0.29118E-01	0.41786E-04		0.8694E-04	273.2-323.2
C_9H_{12}	1,2,4-Trimethylbenzene	0.31517E+01	-0.30634E-02	0,14286E-05		0.1390E-05	273.2-323.2
C_9H_{12}	Mesitylene	0.38998E+01	-0.88072E-02	0.11149E-04		0.3592E-04	288.2-358.2
$C_9H_{12}O$	2,6-Dimethylanisole	0.76700E+01	-0.18298E-01	0.17143E-04		0.5851E-05	293.2-333.2
$C_9H_{12}O$	3,5-Dimethylanisole	0.54981E+01	-0.56651E-02	-0.14286E-05		0.2946E-04	293.2-333.2
$C_9H_{12}O$	1-Phenyl-1-propanol	0.44520E+02	-0.21505E+00	0.29443E-03		0.6928E-01	233.2-373.2
$C_9H_{12}O$	1-Phenyl-2-propanol	0.10762E+03	-0.56026E+00	0.76915E-03		0.4330E-01	233.2-373.2
$C_9H_{12}O$	2-Phenyl-2-propanol	0.57072E+02	0.86568E-02	-0.29580E-04		0.9596E-05	303.2-373.2
$C_9H_{12}O$	3-Phenyl-1-propanol	0.94482E+02	-0.45540E+00	0.59307E-03		0.3936E-03	213.2-303.2
$C_9H_{12}S$	3-Phenyl-1-propanethiol	0.82411E+01	-0.15034E-01	0.73617E-05		0.9383E-06	303.2-358.2
$C_9H_{13}N$	3-Phenylpropylamine	0.86941E+01	-0.12750E-01			0.5556E-05	303.2-343.2
$C_9H_{14}O_6$	Triacetin	0.17819E+02	-0.53656E-01	0.57759E-04		0.9260E-03	218.8-303.7
$C_9H_{14}Si$	Trimethylphenylsilane	0.21463E+01	0.32711E-02	-0.86264E-05		0.8667E-06	288.2-323.2
C_9H_{18}	1-Nonene	0.22710E+01	0.15797E-02	-0.64286E-05		0.2152E-05	273.2-323.2
$C_9H_{18}Br_2$	1,9-Dibromononane	0.18931E+02	-0.57764E-01	0.60000E-04			293.2-343.2
$C_9H_{18}O$	2,6-Dimethyl-4-heptanone	0.33178E+02	-0.11290E+00	0.11454E-03		0.1258E-03	273.2-393.2
$C_9H_{18}O_2$	Pentyl butanoate	0.59029E+01	-0.49905E-02	-0.34292E-05		0.3163E-03	301.2-343.2
$C_9H_{18}O_2$	Nonanoic acid	0.25039E+01	0.67274E-03	-0.24180E-05		0.6371E-04	294.9-364.7
$C_9H_{19}Br$	1-Bromononane	0.79870E+01	-0.10488E-01	-0.13450E-05		0.1655E-07	274.2-328.2
$C_9H_{19}Cl$	1-Chlorononane	0.95528E+01	-0.16200E-01	-0.16365E-13			293.2-323.2
C_9H_{20}	Nonane	0.23894E+01	-0.14830E-02	0.14881E-06		0.9003E-05	253.0-393.0
$C_9H_{20}O$	1-Nonanol	0.97467E+02	-0.51103E+00	0.71429E-03		0.4418E-01	288.2-343.2
$C_9H_{20}O$	2-Nonanol	0.10136E+03	-0.55612E+00	0.80000E-03			288.2-308.2
$C_9H_{20}O$	3-Nonanol	0.55214E+02	-0.31920E+00	0.50000E-03			288.2-308.2
$C_9H_{20}O$	4-Nonanol	0.27954E+01	0.30000E-02	-0.52375E-13			288.2-308.2
$C_9H_{20}O$	5-Nonanol	-0.25463E+01	0.35320E-01	-0.50000E-04			288.2-308.2
$C_9H_{20}N_2O$	Tetraethyl urea	0.52820E+02	-0.18790E+00	0.19580E-03		0.1444E-01	205.0-411.4
$C_9H_{20}S$	2-Methyl-2-octanethiol	0.61521E+01	-0.50260E-02	-0.68750E-05		0.1891E-03	273.2-333.2
$C_9H_{21}N$	Nonylamine	0.53575E+01	-0.71982E-02	0.19481E-05		0.1426E-04	293.2-373.2
$C_9H_{21}N$	Tripropylamine	0.33380E+01	-0.86332E-02	0.18322E-04		0.1626E-04	243.2-293.2
$C_9H_{21}O_4P$	Tripropyl phosphate	0.33166E+02	-0.10514E+00	0.10000E-03		0.6400E-04	293.2-373.2
$C_9H_{21}PS$	Tripropyl phosphine sulfide	0.80045E+02	-0.22665E+00	0,22378E-03		0.5040E-02	273.2-373.2

Molecular formula	Name	a	b	c	d	ESQS/N	Range T/K
$C_{10}H_7Br$	1-Bromonaphthalene	0.10561E+02	-0.27671E-01	0.27655E-04		0.2793E-07	293.2-323.2
$C_{10}H_7Cl$	1-Chloronaphthalene	0.84861E+01	-0.12357E-01	0.26899E-05		0.6622E-07	274.2-328.2
$C_{10}H_7NO_2$	1-Nitronaphthalene	0.36267E+02	-0.41283E-01	-0.25595E-04		0.3519E-03	333.2-403.2
$C_{10}H_8N_2O_2$	1,3-Xylylene diisocyanate	0.19264E+02	-0.18354E-01	-0.11310E-05		0.6459E-05	293.2-363.2
$C_{10}H_8N_2O_2$	1,4-Xylylene diisocyanate	0.48117E+02	-0.84299E-01	-0.52143E-04		0.6084E-03	323.2-363.2
$C_{10}H_8O$	1-Naphthol	0.16489E+02	-0.46700E-01	0.42857E-04		0.9143E-05	373.0-453.0
$C_{10}H_8O$	2-Naphthol	0.92865E+01	-0.10500E-01	0.42501E-15			413.0-453.0
$C_{10}H_9N$	2-Methylquinoline	0.11688E+02	-0.78400E-02	-0.25000E-04			293.2-333.2
$C_{10}H_9N$	4-Methylquinoline	0.17788E+02	-0.32580E-01	0.12500E-04			293.2-333.2
$C_{10}H_9N$	6-Methylquinoline	0.21696E+02	-0.63400E-01	0.62500E-04			293.2-333.2
$C_{10}H_9N$	8-Methylquinoline	0.19356E+02	-0.61900E-01	0.62500E-04			293.2-333.2
$C_{10}H_9N$	1-Naphthylamine	0.10577E+02	-0.22114E-01	0.17857E-04		0.4898E-04	333.2-453.2
$C_{10}H_9N$	2-Naphthylamine	0.19722E+02	-0.60679E-01	0.60714E-04		0.1006E-03	393.0-473.0
$C_{10}H_{10}O_4$	Methyl o-acetylsalicylate	0.19579E+02	-0.69970E-01	0.80889E-04		0.4132E-04	328.9-370.7
$C_{10}H_{12}$	Dicyclopentadiene	0.30564E+01	-0.20000E-02	0.82443E-15			313.2-373.2
$C_{10}H_{12}$	1,2,3,4-Tetrahydronaphthalene	0.29172E+01	0.12832E-02	-0.59453E-05		0.1324E-05	298.2-343.2
$C_{10}H_{12}O$	Tetrahydro-2-naphthol	0.98978E+02	-0.48267E+00	0.63008E-03		0.3708E-02	293.2-363.2
$C_{10}H_{12}O_2$	Propyl benzoate	0.10927E+02	-0.20535E-01	0.11745E-04		0.4902E-05	303.2-358.2
$C_{10}H_{12}O_2$	4-Allyl-2-methoxyphenol	0.52377E+02	-0.24380E+00	0.33333E-03			273.2-323.2
$C_{10}H_{12}O_2$	Ethyl 2-phenylacetate	0.13334E+02	-0.39485E-01	0.41440E-04		0.2553E-05	293.2-353.2
$C_{10}H_{12}O_2$	Benzyl propanoate	0.42301E+01	0.13962E-01	-0.36426E-04		0.2788E-03	303.0-358.0
$C_{10}H_{14}$	Isobutylbenzene	0.28055E+01	-0.92614E-03	-0.25000E-05		0.4143E-06	273.2-323.2
$C_{10}H_{14}$	sec-Butylbenzene	0.28348E+01	-0.68586E-03	-0.32143E-05		0.2714E-06	273.2-323.2
$C_{10}H_{14}$	tert-Butylbenzene	0.27924E+01	-0.38350E-03	-0.37500E-05		0.5917E-06	273.2-323.2
$C_{10}H_{14}$	4-Isopropyltoluene	0.25266E+01	-0.25121E-03	-0.24867E-05		0.2735E-05	277.2-333.2
$C_{10}H_{14}$	1,2,3,4-Tetramethylbenzene	0.33822E+01	-0.33630E-02	0.17475E-05		0.2476E-05	273.0-412.0
$C_{10}H_{14}$	1,2,4,5-Tetramethylbenzene	0.26834E+01	-0.10327E-02	-0.73533E-06		0.3294E-06	356.0-430.0
$C_{10}H_{14}N_2$	Nicotine	0.21347E+02	-0.57177E-01	0.50655E-04		0.1625E-04	293.2-363.2
$C_{10}H_{14}O$	2-Methyl-1-phenyl-2-propanol	0.21922E+02	-0.84231E-01	0.99475E-04		0.2998E-02	298.2-423.2
$C_{10}H_{15}N$	N,N-Diethylaniline	0.50773E+01	0.15399E-01	-0.50000E-04		0.3536E-04	303.2-328.2
$C_{10}H_{18}$	Decahydronaphthalene	0.25521E+01	-0.13651E-02	0.16667E-06		0.1190E-08	283.2-353.2
$C_{10}H_{18}$	cis-Decahydronaphthalene	0.25410E+01	-0.11420E-02	0.15092E-06		0.9482E-07	293.2-373.2
$C_{10}H_{18}$	trans-Decahydronaphthalene	0.26615E+01	-0.21241E-02	0.16864E-05		0.9216E-06	293.2-373.2
$C_{10}H_{18}O_4$	Diethyl adipate	0.14824E+02	-0.40749E-01	0.37600E-04			293.2-343.2
$C_{10}H_{20}$	1-Decene	0.19091E+01	0.33442E-02	-0.87500E-05		0.1515E-05	273.2-323.2
$C_{10}H_{20}Br_2$	1,10-Dibromodecane	0.17350E+02	-0.50328E-01	0.48633E-04		0.1554E-05	303.2-368.2
$C_{10}H_{20}Cl_2$	1,10-Dichlorodecane	-0.57423E+01	0.94220E-01	-0.17500E-03		0.2113E-03	308.2-338.2
$C_{10}H_{20}O$	Menthol	0.68202E+01	-0.15894E-01	0.20837E-04		0.4054E-05	309.3-357.6
$C_{10}H_{20}O_2$	Octyl acetate	-0.34691E+01	0.58106E-01	-0.10952E-03			288.2-323.2
$C_{10}H_{20}O_2$	2-Methylheptyl acetate	0.23285E+01	-0.11538E+00	0.17143E-03			288.2-323.2
$C_{10}H_{20}O_2$	1-Pentyl pentanoate	0.77641E+01	-0.14335E-01	0.73740E-05		0.8203E-04	305.6-393.2
$C_{10}H_{20}O_2$	3-Methylbutyl 3-methylbutanoate	0.14698E+02	-0.57726E-01	0.76190E-04			288.2-323.2
$C_{10}H_{21}Br$	1-Bromodecane	0.11202E+02	-0.33491E-01	0.36314E-04		0.1207E-04	274.2-328.2
$C_{10}H_{21}Cl$	1-Chlorodecane	0.68741E+01	-0.12210E-02	-0.22500E-04		0.3613E-05	293.2-323.2
$C_{10}H_{22}$	Decane	0.24054E+01	-0.15445E-02	0.44643E-06		0.1019E-04	253.2-393.2
$C_{10}H_{22}O$	Bis(3-methylbutyl) ether	0.44690E+01	-0.63710E-02	0.25000E-05		0.5513E-05	293.2-323.2
$C_{10}H_{22}O$	1-Decanol	0.47195E+02	-0.20740E+00	0.24942E-03		0.1102E-01	293.2-343.2
$C_{10}H_{22}O$	2-Decanol	0.13621E+03	-0.81000E+00	0.12500E-02			288.2-308.2
$C_{10}H_{22}O$	(±)-3-Decanol	0.52090E+02	-0.31020E+00	0.50000E-03			288.2-308.2
$C_{10}H_{22}O$	4-Decanol	-0.11260E+02	0.93960E-01	-0.15000E-03			288.2-308.2
$C_{10}H_{22}O$	5-Decanol	-0.25832E+01	0.31456E-01	-0.40000E-04			288.2-308.2
$C_{10}H_{22}O$	2,2-Dimethyl-1-octanol	0.69536E+02	-0.34596E+00	0.46250E-03			293.2-333.2
$C_{10}H_{22}O_6$	Pentaethylene glycol	0.66659E+02	-0.23729E+00	0.24524E-03		0.5164E-03	253.2-333.2
$C_{10}H_{23}N$	Decylamine	0.61497E+01	-0.12801E-01	0.10606E-04		0.1286E-04	293.2-373.2
$C_{11}H_{10}$	1-Methylnaphthalene	0.45126E+01	-0.76480E-02	0.75000E-05			293.2-333.2
$C_{11}H_{10}O$	1-Methoxynaphthalene	0.71885E+01	-0.14838E-01	0.13750E-04			293.2-333.2
$C_{11}H_{10}O$	2-Methoxynaphthalene	0.56702E+01	-0.69754E-02	0.28571E-05		0.4571E-07	353.2-373.2
$C_{11}H_{12}O_3$	Ethyl benzoylacetate	0.93644E+01	0.74280E-01	-0.20000E-03			303.2-323.2
$C_{11}H_{14}O_2$	Butyl benzoate	0.77854E+01	-0.34972E-02	-0.13149E-04		0.9117E-04	303.2-358.2
$C_{11}H_{16}$	Pentamethylbenzene	0.30196E+01	-0.22619E-02	0.83831E-06		0.3007E-06	334.0-413.0
$C_{11}H_{18}N_2O_2$	2,2,4-Trimethylhexamethylene diisocyanate	0.32920E+02	-0.10646E+00	0.10500E-03		0.1224E-06	293.2-353.2
$C_{11}H_{22}$	1-Undecene	0.22132E+01	0.13121E-02	-0.53571E-05		0.1462E-05	273.2-323.2
$C_{11}H_{22}O_2$	Pentyl hexanoate	0.83503E+01	-0.18449E-01	0.14286E-04			288.2-323.2

Molecular formula	Name	a	b	c	d	ESQS/N	Range T/K
$C_{11}H_{24}$	Undecane	0.23637E+01	-0.12500E-02	-0.85869E-16			283.2-363.2
$C_{11}H_{25}N$	Undecylamine	0.54945E+01	-0.96161E-02	0.66017E-05		0.3774E-04	293.2-373.2
$C_{12}H_4Cl_6$	Hexachlorodiphenyl*	0.53843E+01	0.85521E-03	-0.15822E-04		0.1952E-04	291.2-303.2
$C_{12}H_5Cl_5$	Pentachlorodiphenyl*	0.92450E+02	-0.60858E+00	0.10617E-02		0.2755E-07	273.2-282.2
$C_{12}H_{10}$	Biphenyl	0.26869E+01	0.63072E-03	-0.30995E-05		0.6116E-05	348.2-428.2
$C_{12}H_{10}O$	2-Acetonaphthone	0.14538E+03	-0.73040E+00	0.10000E-02			333.2-363.2
$C_{12}H_{10}N_2S$	Thio-bis(N-phenyl)imine	0.80384E+01	-0.20660E-01	0.25000E-04			333.2-373.2
$C_{12}H_{11}NO$	N-1-naphthylenylacetamide	0.84739E+02	-0.12391E+00	-0.35714E-04		0.7238E-02	433.2-533.2
$C_{12}H_{16}O_3$	3-Methylbutyl salicylate	0.13129E+02	-0.19190E-01	-0.36060E-05		0.6386E-03	225.2-397.2
$C_{12}H_{18}$	Hexamethylbenzene	0.35710E+01	-0.46912E-02	0.35088E-05			449.0-489.0
$C_{12}H_{20}O_2$	l-Bornyl acetate	0.60791E+01	0.98200E-02	-0.50000E-04			303.2-323.2
$C_{12}H_{22}O$	Cyclododecanone	0.39327E+02	-0.13248E+00	0.13298E-03		0.5512E-03	303.2-423.2
$C_{12}H_{22}O$	Dicyclohexyl ether	0.95324E+01	-0.31740E-01	0.37500E-04			293.2-333.2
$C_{12}H_{24}$	1-Dodecene	0.22581E+01	0.11106E-02	-0.50000E-05		0.1429E-05	273.2-323.2
$C_{12}H_{24}O_2$	Ethyl decanoate	0.70969E+01	-0.15080E-01	0.12500E-04			293.2-353.2
$C_{12}H_{25}Br$	1-Bromododecane	0.86103E+01	-0.20891E-01	0.18994E-04		0.9535E-07	274.2-328.2
$C_{12}H_{25}Cl$	1-Chlorododecane	0.10002E+02	-0.27798E-01	0.27559E-04		0.3442E-04	274.2-328.2
$C_{12}H_{25}I$	1-Iodododecane	0.34641E+01	0.97404E-02	-0.27602E-04		0.5381E-05	293.1-323.0
$C_{12}H_{26}$	Dodecane	0.23697E+01	-0.12200E-02	-0.36375E-16			283.2-363.2
$C_{12}H_{26}O$	2,2-Dimethyl-1-decanol	0.16410E+02	-0.92839E-01	0.16429E-03		0.2333E-02	203.2-293.2
$C_{12}H_{26}O$	1-Dodecanol	0.18518E+02	-0.44859E-01	0.99900E-05		0.9640E-04	303.2-358.2
$C_{12}H_{26}O_7$	Hexaethylene glycol	0.59773E+02	-0.22186E+00	0.24713E-03		0.1306E-02	253.2-333.2
$C_{12}H_{27}N$	Dodecylamine	0.27999E+01	0.44810E-02	-0.11905E-04		0.1756E-04	303.2-373.2
$C_{12}H_{27}N$	Tributylamine	0.19846E+01	0.28108E-02	-0.54545E-05		0.1633E-05	233.2-293.2
$C_{12}H_{27}OP$	Tributyl phosphine oxide	0.94866E+02	-0.29542E+00	0.25788E-03		0.1161E-01	323.2-473.2
$C_{12}H_{27}O_3PS$	Tributyl thiophosphate	0.87018E+01	0.17968E-02	-0.25857E-04		0.7545E-03	298.2-363.2
$C_{12}H_{27}O_4P$	Tributyl phosphate	0.26304E+02	-0.88480E-01	0.92857E-04		0.3314E-04	293.2-373.2
$C_{12}H_{27}PS$	Tributyl phosphine sulfide	0.92150E+02	-0.30005E+00	0.30653E-03		0.5216E-01	273.2-373.2
$C_{12}H_{30}OSi_2$	Hexaethyl disiloxane	0.36559E+01	-0.72406E-02	0.85714E-05			298.2-333.2
$C_{12}H_{36}O_4Si_5$	Dodecamethyl pentasiloxane	0.47894E+01	-0.12286E-01	0.15000E-04		0.1800E-05	293.2-323.2
$C_{13}H_{10}O$	Benzophenone	0.34130E+02	-0.10249E+00	0.10268E-03		0.7143E-04	300.2-420.2
$C_{13}H_{10}O_3$	Phenyl salicylate	0.26545E+02	-0.11180E+00	0.15220E-03		0.5236E-03	290.2-358.2
$C_{13}H_{12}$	Diphenylmethane	0.30638E+01	-0.17286E-02			0.9524E-07	303.2-333.2
$C_{13}H_{18}O_2$	Pentyl 2-phenylacetate	0.16491E+02	-0.66310E-01	0.87500E-04			303.2-343.2
$C_{13}H_{24}O$	Cyclotridecanone	0.10446E+03	-0.58666E+00	0.90000E-03		0.1901E-02	303.2-318.2
$C_{13}H_{26}$	1-Tridecene	0.14154E+01	0.66514E-02	-0.14286E-04		0.2152E-05	273.2-323.2
$C_{13}H_{28}$	Tridecane	0.23731E+01	-0.12000E-02	-0.21841E-15			283.2-363.2
$C_{13}H_{28}O_4Si_2$	Bis(trimethylsilylmethyl) glutarate	0.66133E+01	-0.83333E-02				293.2-413.2
$C_{14}H_9Cl_5$	1,1,1-Trichloro-2,2-bis (4-chlorophenyl)ethane	0.76748E+01	-0.12659E-01				377.2-418.2
$C_{14}H_{10}$	Anthracene	0.20571E+02	-0.69169E-01	0.66667E-04		0.8491E-07	502.0-516.0
$C_{14}H_{10}O_2$	Benzil	-0.23599E+02	0.22715E+00	-0.34667E-03			368.2-393.2
$C_{14}H_{12}O_2$	Benzyl benzoate	0.76856E+01	-0.80000E-02	-0.80361E-15			303.2-358.2
$C_{14}H_{14}$	1,2-Diphenylethane	0.31178E+01	-0.21572E-02	0.59800E-06		0.3427E-05	331.2-451.2
$C_{14}H_{14}O$	Dibenzyl ether	0.80154E+01	-0.20536E-01	0.21250E-04			293.2-333.2
$C_{14}H_{16}O_2Si$	Dimethyldiphenoxysilane	0.51669E+01	-0.77001E-02	0.70156E-05		0.7283E-04	283.2-353.2
$C_{14}H_{18}O$	Pentyl-3-phenyl-2-propenal*	-0.31153E+00	0.11822E+00	-0.27143E-03		0.7273E-02	293.2-333.2
$C_{14}H_{26}O_4$	Diethyl sebacate	0.39143E+02	-0.20965E+00	0.32000E-03			303.2-313.2
$C_{14}H_{29}Br$	1-Bromotetradecane	0.10058E+02	-0.33905E-01	0.43528E-04		0.2320E-03	274.2-328.2
$C_{14}H_{30}$	Tetradecane	0.23832E+01	-0.11900E-02	-0.51229E-16			283.2-363.2
$C_{14}H_{30}O$	1-Tetradecanol	0.12272E+02	-0.24667E-01	-0.13168E-13		0.1481E-04	318.2-358.2
$C_{14}H_{30}O_5Si_2$	Bis(isobutoxymethyl) tetra-methyldisiloxane	0.65712E+01	-0.79167E-02				293.2-413.2
$C_{14}H_{30}O_8$	Heptaethylene glycol	0.52101E+02	-0.18670E+00	0.20345E-03		0.1891E-02	253.2-333.2
$C_{14}H_{31}O_3P$	Diethyl decanephosphonate	0.14446E+02	-0.39313E-01	0.36607E-04		0.7286E-04	293.2-393.2
$C_{15}H_9F_3ClNO$	Chlorophenyl(p-trifluoromethyl-phenyl)methyl isocyanate	0.42140E+02	-0.15199E+00	0.16698E-03		0.1781E-02	288.2-403.2
$C_{15}H_{10}N_2O_2$	1,1'-Diphenylmethane diisocyanate	0.20279E+02	-0.60902E-01	0.62143E-04		0.9611E-05	313.2-353.2
$C_{15}H_{12}ClNO$	Chlorophenyl(m-methylphenyl)-methyl isocyanate	0.19764E+02	0.36339E-01	-0.13228E-03		0.1330E-02	288.2-378.2
$C_{15}H_{14}O_2$	Benzyl 2-phenylacetate	0.11247E+02	-0.33490E-01	0.37500E-04			303.2-343.2
$C_{15}H_{16}O_2S$	Tolyl xylyl sulfone*	0.37584E+02	-0.25882E-01	-0.72420E-04		0.1386E-03	273.2-297.2
$C_{15}H_{26}O_6$	Tributyrin	0.13152E+02	-0.36684E-01	0.36795E-04		0.1298E-06	199.4-282.8

Molecular formula	Name	a	b	c	d	ESQS/N	Range T/K
$C_{15}H_{32}$	Pentadecane	0.23792E+01	-0.11600E-02	-0.71069E-16			283.2-363.2
$C_{15}H_{33}PS$	Tripentylphosphine sulfide	0.71435E+02	-0.23105E+00	0.22960E-03		0.8459E-02	273.2-373.2
$C_{15}H_{39}O_3NSi_3$	Tris-(trimethylsilanoxyethyl)amine	0.60295E+01	-0.12415E-01	0.10952E-04		0.1244E-04	293.2-353.2
$C_{16}H_{22}O_4$	Dibutyl phthalate	0.12444E+02	-0.20000E-01				293.2-333.2
$C_{16}H_{32}O_2$	Ethyl myristate	0.52642E+01	-0.60000E-02	-0.47358E-15		0.1250E-03	293.2-353.2
$C_{16}H_{33}Br$	1-Bromohexadecane	0.58668E+01	-0.73333E-02	-0.52666E-14			298.2-328.2
$C_{16}H_{33}I$	1-Iodohexadecane	0.79531E+01	-0.22859E-01	0.26955E-04		0.3825E-05	293.2-323.2
$C_{16}H_{34}$	Hexadecane	0.23861E+01	-0.11600E-02	0.25555E-15			293.2-363.2
$C_{16}H_{34}O$	1-Hexadecanol	0.85935E+01	-0.14714E-01	-0.45533E-13		0.6122E-05	333.2-363.2
$C_{16}H_{34}O_2Si$	Trimethylsilylmethyl laurate	0.38240E+01	-0.30833E-02				293.2-413.2
$C_{16}H_{36}Sn$	Tetrabutyl tin	0.56115E+02	-0.24812E+00	0.30682E-03			293.2-313.2
$C_{16}H_{48}O_6Si_7$	Hexadecamethyl heptasiloxane	-0.21789E+01	0.36057E-01	-0.67500E-04		0.7813E-05	293.2-323.2
$C_{17}H_{12}O_3$	2-Naphthyl salicylate	0.11229E+02	-0.18857E-01	0.70332E-05		0.3590E-03	293.0-353.0
$C_{17}H_{34}O$	9-Heptadecanone	0.44176E+02	-0.21183E+00	0.28571E-03			328.2-363.2
$C_{17}H_{36}$	Heptadecane	0.23627E+01	-0.10400E-02	-0.10397E-12			293.2-308.2
$C_{17}H_{36}O_4Si_2$	Bis(trimethylsilylmethyl) azelate	0.70077E+01	-0.91667E-02				293.2-413.2
$C_{18}H_{30}O_2$	Linolenic acid	0.33867E+01	-0.19181E-02			0.1227E-05	274.2-368.2
$C_{18}H_{32}O_2$	Linoleic acid	0.32073E+01	-0.15477E-02			0.5542E-06	275.2-368.2
$C_{18}H_{34}O_2$	Oleic acid	0.25385E+01	-0.69448E-03			0.1339E-05	275.2-368.2
$C_{18}H_{36}O$	9-Octadecene-1-ol	0.48099E+01	-0.22728E-02	-0.33404E-05		0.6876E-05	295.2-333.2
$C_{18}H_{36}O_2$	Cetyl acetate	0.47310E+01	-0.50000E-02	0.41338E-14			308.2-348.2
$C_{18}H_{36}O_2$	Ethyl palmitate	0.57938E+01	-0.12294E-01	0.10919E-04			303.2-455.2
$C_{18}H_{36}O_2$	Stearic acid	0.27159E+01	-0.13300E-02			0.6291E-04	293.2-373.2
$C_{18}H_{37}Br$	1-Bromooctadecane	0.46790E+01	-0.30355E-02	-0.24798E-05			303.4-331.6
$C_{18}H_{38}O$	1-Octadecanol	0.73784E+01	-0.12000E-01	-0.22871E-13			333.2-363.2
$C_{19}H_{16}$	Triphenylmethane	0.40201E+01	-0.66507E-02	0.65329E-05		0.1018E-04	367.2-448.2
$C_{19}H_{37}NO$	Octadecyl isocyanate	0.11826E+02	-0.30441E-01	0.27619E-04		0.2361E-05	303.2-363.2
$C_{20}H_{26}O_5Si_2$	Bis(benzoyloxymethyl)tetra-methyl disiloxane	0.77631E+01	-0.98333E-02				293.2-413.2
$C_{20}H_{38}O_2$	Ethyl oleate	0.57033E+01	-0.11223E-01	0.93447E-05		0.1951E-04	301.2-423.2
$C_{20}H_{40}O_2$	Octadecyl acetate	0.44569E+01	-0.45000E-02	0.33923E-14			308.2-348.2
$C_{20}H_{40}O_2$	Ethyl stearate	0.70930E+01	-0.19081E-01	0.19555E-04		0.2137E-04	331.2-440.2
$C_{20}H_{42}O$	Didecyl ether	0.41465E+01	-0.62240E-02	0.37500E-05			293.2-333.2
$C_{20}H_{42}O$	1-Eicosanol	0.21700E+01	0.12497E-01	-0.28571E-04		0.3810E-05	338.2-363.2
$C_{20}H_{42}O_2$	Octadecoxy ethanol	0.67914E+01	-0.12966E-01	0.95238E-05		0.5442E-05	328.2-358.2
$C_{20}H_{60}O_8Si_9$	Eicosamethyl nonasiloxane	0.57840E+01	-0.16568E-01	0.20000E-04		0.1800E-05	293.2-323.2
$C_{21}H_{44}O_2$	Octadecoxypropanol	0.85781E-01	0.30251E-01	-0.57143E-04		0.2286E-05	328.2-348.2
$C_{21}H_{45}OP$	Triheptylphosphine oxide	0.68784E+02	-0.12000E+00				323.2-343.2
$C_{21}H_{45}PS$	Triheptylphosphine sulfide	0.11983E+03	-0.49569E+00	0.53571E-03		0.8363E-02	273.2-343.2
$C_{22}H_{42}O_2$	cis-13-Docosene carboxylic acid	0.32678E+01	-0.18506E-02			0.9529E-05	307.8-343.9
$C_{22}H_{44}O_2$	Butyl stearate	0.73894E+02	-0.46261E+00	0.75500E-03		0.4466E-01	298.2-343.2
$C_{22}H_{45}Br$	1-Bromodocosane	0.19444E+02	-0.94747E-01	0.13714E-03			315.9-333.4
$C_{22}H_{46}O$	1-Docosanol	0.82062E+01	-0.25069E-01	0.28571E-04		0.3810E-05	348.2-373.2
$C_{22}H_{46}O_2$	Eicosanoxyethanol	0.95252E+01	-0.30250E-01	0.35714E-04		0.3929E-05	338.2-363.2
$C_{22}H_{47}O_3P$	Dihexyl decanephosphonate	0.81703E+01	-0.14300E-01	0.89286E-05		0.5029E-04	293.2-373.2
$C_{23}H_{48}O_2$	Eicosanoxypropanol	0.85266E+01	-0.24211E-01	0.28571E-04		0.3810E-05	333.2-358.2
$C_{24}H_{50}O_2$	Docosanoxyethanol	0.77097E+01	-0.21176E-01	0.23810E-04		0.3401E-05	338.2-368.2
$C_{24}H_{51}O_3P$	Diheptyl decanephosphonate	0.98028E+01	-0.23707E-01	0.22321E-04		0.1095E-04	293.2-393.2
$C_{24}H_{51}PS$	Trioctylphosphine sulfide	0.65165E+02	-0.24748E+00	0.28680E-03		0.8228E-02	273.2-353.2
$C_{25}H_{52}O_2$	Docosanoxypropanol	0.54992E+01	-0.60000E-02	-0.63054E-13			338.2-363.2
$C_{27}H_{57}OP$	Trinonylphosphine oxide	0.76825E+02	-0.27965E+00	0.27749E-03		0.1002E-01	323.2-473.2
$C_{28}H_{56}O_2$	Decyl stearate	0.40628E+01	-0.40000E-02	0.11614E-14			313.2-353.2
$C_{29}H_{60}O_4Si_2$	Bis(methyldipentylsilylmethyl)-glutarate	0.43872E+01	-0.40833E-02				293.2-413.2
$C_{30}H_{63}PS$	Tridecylphosphine sulfide	-0.11749E+02	0.15210E+00	-0.25000E-03		0.1250E-03	343.2-373.2
$C_{51}H_{98}O_6$	Tripalmitin	-0.29131E+01	0.32206E-01	-0.44154E-04			328.2-393.2

* Unspecified isomer.

PERMITTIVITY (DIELECTRIC CONSTANT) OF GASES

This table gives the permittivity ε (often called the dielectric constant) of some common gases at a temperature of 20°C and pressure of one atmosphere (101.325 kPa). Values of the permanent dipole moment μ in Debye Units (1 D = 3.33564×10^{-30} C m) are also included.

The density dependence of the permittivity is given by the equation

$$\frac{\varepsilon - 1}{\varepsilon + 2} = \rho_m \left(\frac{4\pi N \alpha}{3} + \frac{4\pi N \mu^2}{9kT} \right)$$

where ρ_m is the molar density, N is Avogadro's number, k is the Boltzmann constant, T is the temperature, and α is the molecular polarizability. Therefore, in regions where the gas can be considered ideal, $\varepsilon - 1$ is approximately proportional to the pressure at constant temperature. For nonpolar gases ($\mu = 0$), $\varepsilon - 1$ is inversely proportional to temperature at constant pressure.

The number of significant figures indicates the accuracy of the values given. The values of ε for air, Ar, H_2, He, N_2, O_2, and CO_2 are recommended as reference values; these are accurate to 1 ppm or better.

The second part of the table gives the permittivity of water vapor in equilibrium with liquid water as a function of temperature (derived from Reference 4).

REFERENCE

1. A. A. Maryott and F. Buckley, *Table of Dielectric Constants and Electric Dipole Moments of Substances in the Gaseous State*, National Bureau of Standards Circular 537, 1953.
2. B. A. Younglove, *J. Phys. Chem. Ref. Data*, 11, Suppl. 1, 1982; 16, 577, 1987 (for data on N_2, H_2, O_2, and hydrocarbons over a range of pressure and temperature).
3. Landolt-Börnstein, *Numerical Data and Functional Relationships in Science and Technology*, New Series, Group IV, Vol. 4, Springer-Verlag, Heidelberg, 1980 (for data at high pressures).
4. G. Birnbaum and S. K. Chatterjee, *J. Appl. Phys.*, 23, 220, 1952 (for data on water vapor).

Mol. form.	Name	ε	μ/D
Compounds not containing carbon			
	Air (dry, CO_2 free)	1.0005364	
Ar	Argon	1.0005172	0
BF_3	Boron trifluoride	1.0011	0
BrH	Hydrogen bromide	1.00279	0.827
ClH	Hydrogen chloride	1.00390	1.109
F_3N	Nitrogen trifluoride	1.0013	0.235
F_6S	Sulfur hexafluoride	1.00200	0
HI	Hydrogen iodide	1.00214	0.448
H_2	Hydrogen	1.0002538	0
H_2S	Hydrogen sulfide	1.00344	0.97
H_3N	Ammonia	1.00622	1.471
He	Helium	1.0000650	0
Kr	Krypton	1.00078	0
NO	Nitric oxide	1.00060	0.159
N_2	Nitrogen	1.0005480	0
N_2O	Nitrous oxide	1.00104	0.161
Ne	Neon	1.00013	0
O_2	Oxygen	1.0004947	0
O_2S	Sulfur dioxide	1.00825	1.633
O_3	Ozone	1.0017	0.534
Xe	Xenon	1.00126	0
Compounds containing carbon			
CF_4	Tetrafluoromethane	1.00121	0
CO	Carbon monoxide	1.00262	0.110
CO_2	Carbon dioxide	1.000922	0

PERMITTIVITY (DIELECTRIC CONSTANT) OF GASES (continued)

Mol. form.	Name	ε	μ/D
CH_3Br	Bromomethane	1.01028	1.822
CH_3Cl	Chloromethane	1.01080	1.892
CH_3F	Fluoromethane	1.00973	1.858
CH_3I	Iodomethane	1.00914	1.62
CH_4	Methane	1.00081	0
C_2H_2	Acetylene	1.00124	0
C_2H_3Cl	Chloroethylene	1.0075	1.45
C_2H_4	Ethylene	1.00134	0
C_2H_5Cl	Chloroethane	1.01325	2.05
C_2H_6	Ethane	1.00140	0
C_2H_6O	Dimethyl ether	1.0062	1.30
C_3H_6	Propene	1.00228	0.366
C_3H_6	Cyclopropane	1.00178	0
C_3H_8	Propane	1.00200	0.084
C_4H_{10}	Butane	1.00258	0
C_4H_{10}	Isobutane	1.00260	0.132

PERMITTIVITY OF SATURATED WATER VAPOR

$t/°C$	ε	$t/°C$	ε
0	1.00007	60	1.00144
10	1.00012	70	1.00213
20	1.00022	80	1.00305
30	1.00037	90	1.00428
40	1.00060	100	1.00587
50	1.00095		

AZEOTROPIC DATA FOR BINARY MIXTURES

Liquid mixtures having an extremum (maximum or minimum) vapor pressure at constant temperature, as a function of composition, are called azeotropic mixtures, or simply azeotropes. Mixtures that do not show a maximum or minimum are called zeotropic. Azeotropes in which the pressure is a maximum are often called positive azeotropes, while pressure-minimum azeotropes are called negative azeotropes. The coordinates of an azeotropic point are the azeotropic temperature t_{az}, pressure P_{az}, and liquid-phase composition, usually expressed as mole fractions. At the azeotropic point, the vapor-phase composition is the same as the liquid-phase composition.

This table gives azeotropic data for a number of binary mixtures at normal atmospheric pressure (P_{az} =101.3 kPa). Component 1 of each mixture is given in bold face. The temperature t_{az} and mole fraction x_1 of component 1 are listed for each choice of component 2.

The components are arranged in a modified Hill order, with substances that do not contain carbon preceding those that do contain carbon.

REFERENCES

1. Lide, D.R., and Kehiaian, H.V., *CRC Handbook of Thermophysical and Thermochemical Data*, CRC Press, Boca Raton, FL, 1994.
2. Horsley, L.H., *Azeotropic Data, III*, American Chemical Society, Washington, D.C., 1973.

Molecular formula	Name	$t_{az}/°C$	x_1
Water H$_2$O			
CHCl$_3$	Trichloromethane	56.1	0.160
CH$_2$O$_2$	Formic acid	107.2	0.427
CH$_3$NO$_2$	Nitromethane	83.6	0.511
CS$_2$	Carbon disulfide	42.6	0.109
C$_2$H$_3$N	Acetonitrile	76.5	0.307
C$_2$H$_5$NO$_2$	Nitroethane	87.2	0.624
C$_2$H$_6$O	Ethanol	78.2	0.096
C$_4$H$_8$O$_2$	Ethyl acetate	70.4	0.312
C$_4$H$_{10}$O	1-Butanol	92.7	0.753
C$_4$H$_{10}$O	2-Butanol	87	0.601
C$_5$H$_5$N	Pyridine	93.6	0.755
C$_5$H$_{11}$N	Piperidine	92.8	0.718
C$_5$H$_{12}$	Pentane	34.6	0.054
C$_6$H$_5$Cl	Chlorobenzene	90.2	0.712
C$_6$H$_6$	Benzene	69.3	0.295
C$_6$H$_6$O	Phenol	99.5	0.981
C$_6$H$_{10}$	Cyclohexene	70.8	0.308
C$_6$H$_{12}$	Cyclohexane	69.5	0.300
C$_6$H$_{14}$	Hexane	61.6	0.221
C$_7$H$_8$	Toluene	84.1	0.444
C$_7$H$_{16}$	Heptane	79.2	0.452
C$_8$H$_{10}$	1,3-Dimethylbenzene	92	0.767
C$_8$H$_{10}$	Ethylbenzene	92	0.744
C$_8$H$_{18}$	Octane	89.6	0.673
C$_8$H$_{18}$O	Dibutyl ether	92.9	0.781
C$_9$H$_{20}$	Nonane	94.8	0.970
C$_{12}$H$_{27}$N	Tributylamine	99.7	0.976
Tetrachloromethane CCl$_4$			
CH$_2$O$_2$	Formic acid	66.7	0.569
CH$_3$NO$_2$	Nitromethane	71.3	0.660
CH$_4$O	Methanol	55.7	0.445
C$_2$H$_3$N	Acetonitrile	65.1	0.566
C$_2$H$_6$O	Ethanol	65.0	0.615
C$_3$H$_6$O	Acetone	56.1	0.047
C$_3$H$_8$O	1-Propanol	73.4	0.820
C$_4$H$_{10}$O	1-Butanol	76.6	0.951
Formic acid CH$_2$O$_2$			
CS$_2$	Carbon disulfide	42.6	0.253
Nitromethane CH$_3$NO$_2$			
CS$_2$	Carbon disulfide	41.2	0.845
Methanol CH$_4$O			
C$_3$H$_6$O	Acetone	55.5	0.198

Molecular formula	Name	$t_{az}/°C$	x_1
$C_3H_6O_2$	Methyl acetate	53.5	0.352
C_5H_{10}	Cyclopentane	38.8	0.263
C_5H_{12}	Pentane	30.9	0.145
$C_5H_{12}O$	*tert*-Butyl methyl ether	51.3	0.315
C_6H_6	Benzene	57.5	0.610
C_6H_{12}	Cyclohexane	53.9	0.601
C_7H_8	Toluene	63.5	0.883
C_7H_{16}	Heptane	59.1	0.769
C_8H_{18}	Octane	62.8	0.881
C_9H_{20}	Nonane	64.1	0.953
	Carbon disulfide CS_2		
C_2H_6O	Ethanol	42.6	0.860
C_3H_6O	Acetone	39.3	0.608
C_3H_8O	1-Propanol	45.7	0.931
$C_4H_8O_2$	Ethyl acetate	46.1	0.974
	Acetonitrile C_2H_3N		
C_2H_6O	Ethanol	72.5	0.469
C_7H_8	Toluene	81.4	0.900
	Acetic acid $C_2H_4O_2$		
$C_4H_8O_2$	1,4-Dioxane	119.5	0.831
C_5H_5N	Pyridine	138.1	0.579
C_6H_6	Benzene	80.1	0.026
C_6H_{12}	Cyclohexane	78.8	0.130
C_6H_{14}	Hexane	68.3	0.084
$C_6H_{15}N$	Triethylamine	163	0.774
C_7H_8	Toluene	100.7	0.375
C_7H_{16}	Heptane	91.7	0.451
C_8H_{10}	Ethylbenzene	114.7	0.774
C_8H_{18}	Octane	105.7	0.688
C_9H_{20}	Nonane	112.9	0.826
	Iodoethane C_2H_5I		
C_6H_{14}	Hexane	64.7	0.420
	Ethanol C_2H_6O		
C_5H_{10}	Cyclopentane	44.7	0.110
C_5H_{12}	Pentane	34.3	0.076
C_6H_6	Benzene	67.9	0.440
C_6H_{12}	Cyclohexane	64.8	0.430
C_6H_{14}	Hexane	58.7	0.332
C_7H_8	Toluene	76.7	0.810
C_8H_{18}	Octane	77	0.898
	Ethylene glycol $C_2H_6O_2$		
C_7H_8	Toluene	110.1	0.034
C_7H_{16}	Heptane	97.9	0.048
$C_8H_{18}O$	Dibutyl ether	139.5	0.125
$C_{10}H_{22}$	Decane	161	0.406
	Dimethyl sulfide C_2H_6S		
C_5H_{12}	Pentane	31.8	0.503
	1,2-Ethanediamine $C_2H_8N_2$		
C_7H_8	Toluene	104	0.406
	Propanenitrile C_3H_5N		
C_6H_{14}	Hexane	63.5	0.134
	Acetone C_3H_6O		
$C_3H_6O_2$	Methyl acetate	55.8	0.544
C_5H_{10}	Cyclopentane	41	0.404
C_6H_{12}	Cyclohexane	53	0.751
	Ethyl formate $C_3H_6O_2$		
C_5H_{12}	Pentane	32.5	0.294
	Methyl acetate $C_3H_6O_2$		
C_6H_{12}	Cyclohexane	55.5	0.801

Molecular formula	Name	$t_{az}/°C$	x_1
C_6H_{14}	Hexane	51.8	0.642
Propanoic acid $C_3H_6O_2$			
C_5H_5N	Pyridine	148.6	0.686
C_7H_{16}	Heptane	97.8	0.027
C_9H_{12}	Propylbenzene	139.5	0.830
1-Nitropropane $C_3H_7NO_2$			
C_3H_8O	1-Propanol	97.0	0.061
C_7H_{16}	Heptane	96.6	0.149
1-Propanol C_3H_8O			
$C_4H_8O_2$	1,4-Dioxane	95.3	0.642
C_6H_6	Benzene	77.1	0.209
C_6H_{12}	Cyclohexane	74.7	0.241
C_7H_{16}	Heptane	84.6	0.470
2-Propanol C_3H_8O			
$C_4H_{11}N$	Butylamine	74.7	0.646
C_5H_{12}	Pentane	35.5	0.071
C_6H_{12}	Cyclohexane	69.4	0.397
C_7H_8	Toluene	80.6	0.773
Ethyl methyl sulfide C_3H_8S			
C_6H_{12}	Methylcyclopentane	65.6	0.664
C_6H_{14}	2,2-Dimethylpentane	66.4	0.908
1-Propanethiol C_3H_8S			
C_6H_{12}	Cyclohexane	67.8	0.978
C_6H_{14}	Hexane	64.4	0.557
$C_6H_{14}O$	Diisopropyl ether	65.9	0.714
Thiophene C_4H_4S			
C_6H_{12}	Cyclohexane	77.9	0.412
C_6H_{14}	Hexane	68.5	0.114
Butanal C_4H_8O			
C_6H_{14}	Hexane	60	0.296
2-Butanone C_4H_8O			
C_4H_9Cl	1-Chlorobutane	77	0.440
$C_4H_{11}N$	Butylamine	74	0.353
C_6H_6	Benzene	78.3	0.460
C_6H_{12}	Cyclohexane	71.8	0.438
C_7H_{16}	Heptane	77	0.764
Butanoic acid $C_4H_8O_2$			
C_5H_5N	Pyridine	163.2	0.912
C_6H_5Cl	Chlorobenzene	131.8	0.035
C_8H_{10}	1,2-Dimethylbenzene	143	0.118
1,4-Dioxane $C_4H_8O_2$			
C_4H_9Br	1-Bromobutane	98	0.580
Ethyl acetate $C_4H_8O_2$			
C_6H_{14}	Hexane	65.2	0.394
Methyl propanoate $C_4H_8O_2$			
C_4H_9Cl	1-Chlorobutane	76.8	0.392
Propyl formate $C_4H_8O_2$			
C_4H_9Cl	1-Chlorobutane	76.1	0.392
C_6H_6	Benzene	78.5	0.440
C_6H_{12}	Cyclohexane	75	0.469
1-Butanol $C_4H_{10}O$			
C_5H_5N	Pyridine	118.6	0.704
C_6H_5Cl	Chlorobenzene	115.3	0.659
C_6H_{10}	Cyclohexene	82	0.055
C_7H_8	Toluene	105.5	0.324
C_7H_{16}	Heptane	93.9	0.229
C_8H_{10}	1,2-Dimethylbenzene	116.8	0.811
$C_8H_{18}O$	Dibutyl ether	117.7	0.892

Molecular formula	Name	$t_{az}/°C$	x_1
	2-Butanol C$_4$H$_{10}$O		
C$_6$H$_6$	Benzene	78.5	0.161
C$_7$H$_{16}$	Heptane	88.1	0.439
	Diethyl ether C$_4$H$_{10}$O		
C$_5$H$_{12}$	Pentane	33.7	0.553
	***tert*-Butyl alcohol C$_4$H$_{10}$O**		
C$_6$H$_6$	Benzene	74.0	0.378
C$_7$H$_{16}$	Heptane	78	0.688
	Methyl propyl ether C$_4$H$_{10}$O		
C$_5$H$_{12}$	Pentane	35.6	0.215
	2-Ethoxyethanol C$_4$H$_{10}$O$_2$		
C$_7$H$_{16}$	Heptane	96.5	0.153
C$_9$H$_{12}$	Propylbenzene	134.6	0.842
	2-Furaldehyde C$_5$H$_4$O$_2$		
C$_7$H$_{16}$	Heptane	98.3	0.055
C$_9$H$_{12}$	Propylbenzene	151.4	0.475
	Pyridine C$_5$H$_5$N		
C$_7$H$_8$	Toluene	110.1	0.249
	Benzene C$_6$H$_6$		
C$_6$H$_{10}$	Cyclohexene	78.9	0.635
C$_6$H$_{12}$	Cyclohexane	77.6	0.538
	Phenol C$_6$H$_6$O		
C$_6$H$_7$N	2-Methylpyridine	185.5	0.752
C$_7$H$_9$N	2,4-Dimethylpyridine	193.4	0.601
C$_9$H$_{12}$	1,3,5-Trimethylbenzene	163.5	0.253
C$_{10}$H$_{22}$	Decane	168	0.449
	Aniline C$_6$H$_7$N		
C$_9$H$_{12}$	1,3,5-Trimethylbenzene	164.4	0.150
C$_{10}$H$_{22}$O	Dipentyl ether	177.5	0.675
C$_{12}$H$_{26}$	Dodecane	180.4	0.821
	2-Methylpyridine C$_6$H$_7$N		
C$_8$H$_{18}$	Octane	121.1	0.470
	Cyclohexanol C$_6$H$_{12}$O		
C$_8$H$_{10}$	1,2-Dimethylbenzene	143	0.147

VISCOSITY OF GASES

The following table gives the viscosity of some common gases as a function of temperature. Unless otherwise noted, the viscosity values refer to a pressure of 100 kPa (1 bar). The notation $P=0$ indicates the low pressure limiting value is given. The difference between the viscosity at 100 kPa and the limiting value is generally less than 1%. Viscosity is given in units of $\mu Pa\ s$; note that $1\ \mu Pa\ s = 10^{-5}$ poise. Substances are listed in the modified Hill order (see Introduction).

		Viscosity in micropascal seconds ($\mu Pa\ s$)						
		100 K	200 K	300 K	400 K	500 K	600 K	Ref.
	Air	7.1	13.3	18.6	23.1	27.1	30.8	1
Ar	Argon	8.0	15.9	22.9	28.8	34.2	39.0	2,8
BF_3	Boron trifluoride		12.3	17.1	21.7	26.1	30.2	13
ClH	Hydrogen chloride			14.6	19.7	24.3		13
F_6S	Sulfur hexafluoride							
	($P=0$)			15.3	19.8	23.9	27.7	10
H_2	Hydrogen ($P=0$)	4.2	6.8	9.0	10.9	12.7	14.4	4
D_2	Deuterium ($P=0$)	5.9	9.6	12.6	15.4	17.9	20.3	11
H_2O	Water			10.0	13.3	17.3	21.4	6
D_2O	Deuterium oxide			11.1	13.7	17.7	22.0	7
He	Helium ($P=0$)	9.7	15.3	20.0	24.4	28.4	32.3	8
Kr	Krypton ($P=0$)	8.8	17.1	25.6	33.1	39.8	45.9	8
NO	Nitric oxide		13.8	19.2	23.8	28.0	31.9	13
N_2	Nitrogen ($P=0$)		12.9	17.9	22.2	26.1	29.6	12
N_2O	Nitrous oxide		10.0	15.0	19.4	23.6	27.4	13
Ne	Neon ($P=0$)	14.4	24.3	32.1	38.9	45.0	50.8	8
O_2	Oxygen ($P=0$)	7.5	14.6	20.8	26.1	30.8	35.1	12
O_2S	Sulfur dioxide		8.6	12.9	17.5	21.7		13
Xe	Xenon ($P=0$)	8.3	15.4	23.2	30.7	37.6	44.0	8
CO	Carbon monoxide	6.7	12.9	17.8	22.1	25.8	29.1	13
CO_2	Carbon dioxide		10.0	15.0	19.7	24.0	28.0	9,10
$CHCl_3$	Chloroform			10.2	13.7	16.9	20.1	13
CH_4	Methane		7.7	11.2	14.3	17.0	19.4	10
CH_4O	Methanol				13.2	16.5	19.6	13
C_2H_2	Acetylene			10.4	13.5	16.5		13
C_2H_4	Ethylene		7.0	10.4	13.6	16.5	19.1	3
C_2H_6	Ethane		6.4	9.5	12.3	14.9	17.3	5
C_2H_6O	Ethanol				11.6	14.5	17.0	13
C_3H_8	Propane			8.3	10.9	13.4	15.8	5
C_4H_{10}	n-Butane			7.5	10.0	12.3	14.6	5
C_4H_{10}	Isobutane			7.6	10.0	12.3	14.6	5
$C_4H_{10}O$	Diethyl ether			7.6	10.1	12.4		13
C_5H_{12}	n-Pentane			6.7	9.2	11.4	13.4	13
C_6H_{14}	n-Hexane				8.6	10.8	12.8	13

REFERENCES

1. K. Kadoya, N. Matsunaga, and A. Nagashima, Viscosity and thermal conductivity of dry air in the gaseous phase, *J. Phys. Chem. Ref. Data*, 14, 947, 1985.
2. B. A. Younglove and H. J. M. Hanley, The viscosity and thermal conductivity coefficients of gaseous and liquid argon, *J. Phys. Chem. Ref. Data*, 15, 1323, 1986.
3. P. M. Holland, B. E. Eaton, and H. J. M. Hanley, A Correlation of the viscosity and thermal conductivity data of gaseous and liquid ethylene, *J. Phys. Chem. Ref. Data*, 12, 917, 1983.
4. M. J. Assael, S. Mixafendi, and W. A. Wakeham, The viscosity and thermal conductivity of normal hydrogen in the limit zero density, *J. Phys. Chem. Ref. Data*, 15, 1315, 1986.
5. B. A. Younglove and J. F. Ely, Thermophysical properties of fluids. II. Methane, ethane, propane, isobutane, and normal butane, *J. Phys. Chem. Ref. Data*, 16, 577, 1987.

6. J. V. Sengers and J. T. R. Watson, Improved international formulations for the viscosity and thermal conductivity of w ter substance, *J. Phys. Chem. Ref. Data*, 15, 1291, 1986.

7. N. Matsunaga and A. Nagashima, Transport properties of liquid and gaseous D_2O over a wide range of temperature and pressure, *J. Phys. Chem. Ref. Data*, 12, 933, 1983.

8. J. Kestin, et al., Equilibrium and transport properties of the noble gases and their mixtures at low density, *J. Phys. Chem. Ref. Data*, 13, 299, 1984.

9. V. Vescovic, et al., The transport properties of carbon dioxide, *J. Phys. Chem. Ref. Data*, 19, 1990.

10. R. D. Trengove and W. A. Wakeham, The viscosity of carbon dioxide, methane, and sulfur hexafluoride in the limit of zero density, *J. Phys. Chem. Ref. Data*, 16, 175, 1987.

11. M. J. Assael, S. Mixafendi, and W. A. Wakeham, The viscosity of normal deuterium in the limit of zero density, *J. Phys. Chem. Ref. Data*, 16, 189, 1987.

12. W. A. Cole and W. A. Wakeham, The viscosity of nitrogen, oxygen, and their binary mixtures in the limit of zero density, *J. Phys. Chem. Ref. Data*, 14, 209, 1985.

13. C. Y. Ho, Ed., *Properties of Inorganic and Organic Fluids*, *CINDAS Data Series on Materials Properties*, Vol. V-1, Hemisphere Publishing Corp., New York, 1988.

VISCOSITY OF LIQUIDS

The absolute viscosity of some common liquids at temperatures between –25 and 100°C is given in this table. Values were derived by fitting experimental data to suitable expressions for the temperature dependence. The substances are arranged by molecular formula in the modified Hill order (see Preface). All values are given in units of millipascal seconds (mPa s); this unit is identical to centipoise (cp).

Viscosity values correspond to a nominal pressure of 1 atmosphere. If a value is given at a temperature above the normal boiling point, the applicable pressure is understood to be the vapor pressure of the liquid at that temperature. A few values are given at a temperature slightly below the normal freezing point; these refer to the supercooled liquid.

The accuracy ranges from 1% in the best cases to 5 to 10% in the worst cases. Additional significant figures are included in the table to facilitate interpolation.

REFERENCES

1. Viswanath, D. S. and Natarajan, G., *Data Book on the Viscosity of Liquids*, Hemisphere Publishing Corp., New York, 1989.
2. Daubert, T. E., Danner, R. P., Sibul, H. M., and Stebbins, C. C., *Physical and Thermodynamic Properties of Pure Compounds: Data Compilation*, extant 1994 (core with 4 supplements), Taylor & Francis, Bristol, PA (also available as database).
3. Ho, C. Y., Ed., *CINDAS Data Series on Material Properties*, Vol. V-1, *Properties of Inorganic and Organic Fluids*, Hemisphere Publishing Corp., New York, 1988.
4. Stephan, K. and Lucas, K., *Viscosity of Dense Fluids*, Plenum Press, New York, 1979.
5. Vargaftik, N. B., *Tables of Thermophysical Properties of Liquids and Gases*, 2nd ed., John Wiley, New York, 1975.

Molecular formula	Name	Viscosity in mPa s					
		–25°C	0°C	25°C	50°C	75°C	100°C
		Compounds not containing carbon					
Br_2	Bromine		1.252	0.944	0.746		
Cl_3HSi	Trichlorosilane		0.415	0.326			
Cl_3P	Phosphorous trichloride	0.870	0.662	0.529	0.439		
Cl_4Si	Tetrachlorosilane			99.4	96.2		
H_2O	Water		1.793	0.890	0.547	0.378	0.282
H_4N_2	Hydrazine			0.876	0.628	0.480	0.384
Hg	Mercury			1.526	1.402	1.312	1.245
NO_2	Nitrogen dioxide		0.532	0.402			
		Compounds containing carbon					
CCl_3F	Trichlorofluoromethane	0.740	0.539	0.421			
CCl_4	Tetrachloromethane		1.321	0.908	0.656	0.494	
CS_2	Carbon disulfide		0.429	0.352			
$CHBr_3$	Tribromomethane			1.857	1.367	1.029	
$CHCl_3$	Trichloromethane	0.988	0.706	0.537	0.427		
CHN	Hydrogen cyanide		0.235	0.183			
CH_2Br_2	Dibromomethane	1.948	1.320	0.980	0.779	0.652	
CH_2Cl_2	Dichloromethane	0.727	0.533	0.413			
CH_2O_2	Formic acid			1.607	1.030	0.724	0.545
CH_3I	Iodomethane		0.594	0.469			
CH_3NO	Formamide		7.114	3.343	1.833		
CH_3NO_2	Nitromethane	1.311	0.875	0.630	0.481	0.383	0.317
CH_4O	Methanol	1.258	0.793	0.544			
CH_5N	Methylamine	0.319	0.231				
$C_2Cl_3F_3$	1,1,2-Trichlorotrifluoro-ethane	1.465	0.945	0.656	0.481		
C_2Cl_4	Tetrachloroethylene		1.114	0.844	0.663	0.535	0.442
C_2HCl_3	Trichloroethylene		0.703	0.545	0.444	0.376	
C_2HCl_5	Pentachloroethane		3.761	2.254	1.491	1.061	
$C_2HF_3O_2$	Trifluoroacetic acid			0.808	0.571		
$C_2H_2Cl_2$	*cis*-1,2-Dichloroethylene	0.786	0.575	0.445			
$C_2H_2Cl_2$	*trans*-1,2-Dichloroethylene	0.522	0.398	0.317	0.261		
$C_2H_2Cl_4$	1,1,1,2-Tetrachloroethane	3.660	2.200	1.437	1.006	0.741	0.570
$C_2H_3ClF_2$	1-Chloro-1,1-difluoro-ethane	0.477	0.376				

Molecular formula	Name	Viscosity in mPa s					
		−25°C	0°C	25°C	50°C	75°C	100°C
C_2H_3ClO	Acetyl chloride			0.368	0.294		
$C_2H_3Cl_3$	1,1,1-Trichloroethane	1.847	1.161	0.793	0.578	0.428	
C_2H_3N	Acetonitrile		0.400	0.369	0.284	0.234	
$C_2H_4Br_2$	1,2-Dibromoethane			1.595	1.116	0.837	0.661
$C_2H_4Cl_2$	1,1-Dichloroethane			0.464	0.362		
$C_2H_4Cl_2$	1,2-Dichloroethane		1.125	0.779	0.576	0.447	
$C_2H_4O_2$	Acetic acid			1.056	0.786	0.599	0.464
$C_2H_4O_2$	Methyl formate		0.424	0.325			
C_2H_5Br	Bromoethane	0.635	0.477	0.374			
C_2H_5Cl	Chloroethane	0.416	0.319				
C_2H_5I	Iodoethane		0.723	0.556	0.444	0.365	
C_2H_5NO	N-Methylformamide		2.549	1.678	1.155	0.824	0.606
$C_2H_5NO_2$	Nitroethane	1.354	0.940	0.688	0.526	0.415	0.337
C_2H_6O	Ethanol	3.262	1.786	1.074	0.694	0.476	
C_2H_6OS	Dimethyl sulfoxide			1.987	1.290		
$C_2H_6O_2$	Ethylene glycol			16.1	6.554	3.340	1.975
C_2H_6S	Dimethyl sulfide		0.356	0.284			
C_2H_6S	Ethanethiol		0.364	0.287			
C_2H_7N	Dimethylamine	0.300	0.232				
C_2H_7NO	Ethanolamine			21.1	8.560	3.935	1.998
C_3H_5Br	3-Bromopropene		0.620	0.471	0.373		
C_3H_5Cl	3-Chloropropene		0.408	0.314			
C_3H_5ClO	Epichlorohydrin	2.492	1.570	1.073	0.781	0.597	0.474
C_3H_5N	Propanenitrile			0.294	0.240	0.202	
C_3H_6O	Acetone	0.540	0.395	0.306	0.247		
C_3H_6O	Allyl alcohol			1.218	0.759	0.505	
C_3H_6O	Propanal			0.321	0.249		
$C_3H_6O_2$	Ethyl formate		0.506	0.380	0.300		
$C_3H_6O_2$	Methyl acetate		0.477	0.364	0.284		
$C_3H_6O_2$	Propanoic acid		1.499	1.030	0.749	0.569	0.449
C_3H_7Br	1-Bromopropane		0.645	0.489	0.387		
C_3H_7Br	2-Bromopropane		0.612	0.458	0.359		
C_3H_7Cl	1-Chloropropane		0.436	0.334			
C_3H_7Cl	2-Chloropropane		0.401	0.303			
C_3H_7I	1-Iodopropane		0.970	0.703	0.541	0.436	0.363
C_3H_7I	2-Iodopropane		0.883	0.653	0.506	0.407	
C_3H_7NO	N,N-Dimethylformamide		1.176	0.794	0.624		
$C_3H_7NO_2$	1-Nitropropane	1.851	1.160	0.798	0.589	0.460	0.374
C_3H_8O	1-Propanol	8.645	3.815	1.945	1.107	0.685	
C_3H_8O	2-Propanol		4.619	2.038	1.028	0.576	
$C_3H_8O_2$	1,2-Propylene glycol		248	40.4	11.3	4.770	2.750
$C_3H_8O_3$	Glycerol			934	152	39.8	14.8
C_3H_8S	1-Propanethiol		0.503	0.385			
C_3H_8S	2-Propanethiol		0.477	0.357	0.280		
C_3H_9N	Propylamine			0.376			
C_3H_9N	Isopropylamine		0.454	0.325			
C_4H_4O	Furan	0.661	0.475	0.361			
C_4H_5N	Pyrrole		2.085	1.225	0.828	0.612	
$C_4H_6O_3$	Acetic anhydride		1.241	0.843	0.614	0.472	0.377
C_4H_7N	Butanenitrile			0.553	0.418	0.330	0.268
C_4H_8O	2-Butanone	0.720	0.533	0.405	0.315	0.249	
C_4H_8O	Tetrahydrofuran	0.849	0.605	0.456	0.359		
$C_4H_8O_2$	1,4-Dioxane			1.177	0.787	0.569	
$C_4H_8O_2$	Ethyl acetate		0.578	0.423	0.325	0.259	
$C_4H_8O_2$	Methyl propionate		0.581	0.431	0.333	0.266	
$C_4H_8O_2$	Propyl formate		0.669	0.485	0.370	0.293	
$C_4H_8O_2$	Butanoic acid		2.215	1.426	0.982	0.714	0.542

VISCOSITY OF LIQUIDS (continued)

Molecular formula	Name	−25°C	0°C	25°C	50°C	75°C	100°C
				Viscosity in mPa s			
$C_4H_8O_2$	2-Methylpropanoic acid		1.857	1.226	0.863	0.639	0.492
$C_4H_8O_2S$	Sulfolane				6.280	3.818	2.559
C_4H_8S	Tetrahydrothiophene			0.973	0.912		
C_4H_9Br	1-Bromobutane		0.815	0.606	0.471	0.379	
C_4H_9Cl	1-Chlorobutane		0.556	0.422	0.329	0.261	
C_4H_9N	Pyrrolidine	1.914	1.071	0.704	0.512		
C_4H_9NO	N,N-Dimethylacetamide			1.956	1.279	0.896	0.661
C_4H_9NO	Morpholine			2.021	1.247	0.850	0.627
$C_4H_{10}O$	1-Butanol	12.19	5.185	2.544	1.394	0.833	0.533
$C_4H_{10}O$	2-Butanol			3.096	1.332	0.698	0.419
$C_4H_{10}O$	2-Methyl-2-propanol			4.312	1.421	0.678	
$C_4H_{10}O$	Diethyl ether		0.283	0.224			
$C_4H_{10}O_3$	Diethylene glycol			30.200	11.130	4.917	2.505
$C_4H_{10}S$	Diethyl sulfide		0.558	0.422	0.331	0.267	
$C_4H_{11}N$	Butylamine		0.830	0.574	0.409	0.298	
$C_4H_{11}N$	Isobutylamine		0.770	0.571	0.367		
$C_4H_{11}N$	Diethylamine			0.319	0.239		
$C_4H_{11}NO_2$	Diethanolamine				109.5	28.7	9.100
$C_5H_4O_2$	Furfural		2.501	1.587	1.143	0.906	0.772
C_5H_5N	Pyridine		1.361	0.879	0.637	0.497	0.409
C_5H_{10}	1-Pentene	0.313	0.241	0.195			
C_5H_{10}	2-Methyl-2-butene		0.255	0.203			
C_5H_{10}	Cyclopentane		0.555	0.413	0.321		
$C_5H_{10}O$	Mesityl oxide	1.291	0.838	0.602	0.465	0.381	0.326
$C_5H_{10}O$	2-Pentanone		0.641	0.470	0.362	0.289	0.238
$C_5H_{10}O$	3-Pentanone		0.592	0.444	0.345	0.276	0.227
$C_5H_{10}O_2$	Butyl formate		0.937	0.644	0.472	0.362	0.289
$C_5H_{10}O_2$	Propyl acetate		0.768	0.544	0.406	0.316	0.255
$C_5H_{10}O_2$	Ethyl propanoate		0.691	0.501	0.380	0.299	0.242
$C_5H_{10}O_2$	Methyl butanoate		0.759	0.541	0.406	0.318	0.257
$C_5H_{10}O_2$	Methyl isobutanoate		0.672	0.488	0.373	0.296	
$C_5H_{11}N$	Piperidine			1.573	0.958	0.649	0.474
C_5H_{12}	Pentane	0.351	0.274	0.224			
C_5H_{12}	Isopentane	0.376	0.277	0.214			
$C_5H_{12}O$	1-Pentanol	25.4	8.512	3.619	1.820	1.035	0.646
$C_5H_{12}O$	2-Pentanol			3.470	1.447	0.761	0.465
$C_5H_{12}O$	3-Pentanol			4.149	1.473	0.727	0.436
$C_5H_{12}O$	2-Methyl-1-butanol			4.453	1.963	1.031	0.612
$C_5H_{12}O$	3-Methyl-1-butanol		8.627	3.692	1.842	1.031	0.631
$C_5H_{13}N$	Pentylamine		1.030	0.702	0.493	0.356	
C_6F_6	Hexafluorobenzene			2.789	1.730	1.151	
$C_6H_4Cl_2$	o-Dichlorobenzene		1.958	1.324	0.962	0.739	0.593
$C_6H_4Cl_2$	m-Dichlorobenzene		1.492	1.044	0.787	0.628	0.525
C_6H_5Br	Bromobenzene		1.560	1.074	0.798	0.627	0.512
C_6H_5Cl	Chlorobenzene	1.703	1.058	0.753	0.575	0.456	0.369
C_6H_5ClO	o-Chlorophenol			3.589	1.835	1.131	0.786
C_6H_5ClO	m-Chlorophenol			4.041			
C_6H_5F	Fluorobenzene		0.749	0.550	0.423	0.338	
C_6H_5I	Iodobenzene		2.354	1.554	1.117	0.854	0.683
$C_6H_5NO_2$	Nitrobenzene		3.036	1.863	1.262	0.918	0.704
C_6H_6	Benzene			0.604	0.436	0.335	
C_6H_6ClN	o-Chloroaniline			3.316	1.913	1.248	0.887
C_6H_6O	Phenol				3.437	1.784	1.099
C_6H_7N	Aniline			3.847	2.029	1.247	0.850
$C_6H_8N_2$	Phenylhydrazine			13.0	4.553	1.850	0.848
C_6H_{10}	Cyclohexene		0.882	0.625	0.467	0.364	
$C_6H_{10}O$	Cyclohexanone			2.017	1.321	0.919	0.671

Molecular formula	Name	\-25°C	0°C	25°C	50°C	75°C	100°C
$C_6H_{11}N$	Hexanenitrile			0.912	0.650	0.488	0.382
C_6H_{12}	Cyclohexane			0.894	0.615	0.447	
C_6H_{12}	Methylcyclopentane	0.927	0.653	0.479	0.364		
C_6H_{12}	1-Hexene	0.441	0.326	0.252	0.202		
$C_6H_{12}O$	Cyclohexanol			57.5	12.3	4.274	1.982
$C_6H_{12}O$	2-Hexanone	1.300	0.840	0.583	0.429	0.329	0.262
$C_6H_{12}O$	4-Methyl-2-pentanone			0.545	0.406		
$C_6H_{12}O_2$	Butyl acetate		1.002	0.685	0.500	0.383	0.305
$C_6H_{12}O_2$	Isobutyl acetate			0.676	0.493	0.370	0.286
$C_6H_{12}O_2$	Ethyl butanoate			0.639	0.453		
$C_6H_{12}O_2$	Diacetone alcohol	28.7	6.621	2.798	1.829	1.648	
$C_6H_{12}O_3$	Paraldehyde			1.079	0.692	0.485	0.362
$C_6H_{13}N$	Cyclohexylamine			1.944	1.169	0.782	0.565
C_6H_{14}	Hexane		0.405	0.300	0.240		
C_6H_{14}	2-Methylpentane		0.372	0.286	0.226		
C_6H_{14}	3-Methylpentane		0.395	0.306			
$C_6H_{14}O$	Dipropyl ether		0.542	0.396	0.304	0.242	
$C_6H_{14}O$	1-Hexanol			4.578	2.271	1.270	0.781
$C_6H_{15}N$	Triethylamine		0.455	0.347	0.273	0.221	
$C_6H_{15}N$	Dipropylamine		0.751	0.517	0.377	0.288	0.228
$C_6H_{15}N$	Diisopropylamine			0.393	0.300	0.237	
$C_6H_{15}NO_3$	Triethanolamine			609	114	31.5	11.7
C_7H_5N	Benzonitrile			1.267	0.883	0.662	0.524
C_7H_7Cl	o-Chlorotoluene		1.390	0.964	0.710	0.547	0.437
C_7H_7Cl	m-Chlorotoluene		1.165	0.823	0.616	0.482	0.391
C_7H_7Cl	p-Chlorotoluene			0.837	0.621	0.483	0.390
C_7H_8	Toluene	1.165	0.778	0.560	0.424	0.333	0.270
C_7H_8O	o-Cresol				3.035	1.562	0.961
C_7H_8O	m-Cresol			12.9	4.417	2.093	1.207
C_7H_8O	Benzyl alcohol			5.474	2.760	1.618	1.055
C_7H_8O	Anisole			1.056	0.747	0.554	0.427
C_7H_9N	N-Methylaniline		4.120	2.042	1.222	0.825	0.606
C_7H_9N	o-Methyl aniline		10.3	3.823	1.936	1.198	0.839
C_7H_9N	m-Methyl aniline		8.180	3.306	1.679	1.014	0.699
C_7H_9N	Benzylamine			1.624	1.080	0.769	0.577
C_7H_{14}	Methylcyclohexane		0.991	0.679	0.501	0.390	0.316
C_7H_{14}	1-Heptene		0.441	0.340	0.273	0.226	
$C_7H_{14}O$	2-Heptanone			0.714	0.407	0.297	
$C_7H_{14}O_2$	Heptanoic acid			3.840	2.282	1.488	1.041
C_7H_{16}	Heptane	0.757	0.523	0.387	0.301	0.243	
C_7H_{16}	3-Methylhexane			0.350			
$C_7H_{16}O$	1-Heptanol			5.810	2.603	1.389	0.849
$C_7H_{16}O$	2-Heptanol			3.955	1.799	0.987	0.615
$C_7H_{16}O$	3-Heptanol				1.957	0.976	0.584
$C_7H_{16}O$	4-Heptanol			4.207	1.695	0.882	0.539
$C_7H_{17}N$	Heptylamine			1.314	0.865	0.600	0.434
C_8H_8	Styrene		1.050	0.695	0.507	0.390	0.310
C_8H_8O	Acetophenone			1.681			0.634
$C_8H_8O_2$	Methyl benzoate			1.857			
$C_8H_8O_3$	Methyl salicylate					1.102	0.815
C_8H_{10}	Ethylbenzene		0.872	0.631	0.482	0.380	0.304
C_8H_{10}	o-Xylene		1.084	0.760	0.561	0.432	0.345
C_8H_{10}	m-Xylene		0.795	0.581	0.445	0.353	0.289
C_8H_{10}	p-Xylene			0.603	0.457	0.359	0.290
$C_8H_{10}O$	Phenetole			1.197	0.817	0.594	0.453
$C_8H_{11}N$	N,N-Dimethylaniline		1.996	1.300	0.911	0.675	0.523
$C_8H_{11}N$	N-Ethylaniline		3.981	2.047	1.231	0.825	0.596

Molecular formula	Name	Viscosity in mPa s					
		−25°C	0°C	25°C	50°C	75°C	100°C
C_8H_{16}	Ethylcyclohexane		1.139	0.784	0.579		
$C_8H_{16}O_2$	Octanoic acid			5.020	2.656	1.654	1.147
C_8H_{18}	Octane		0.700	0.508	0.385	0.302	0.243
$C_8H_{18}O$	1-Octanol			7.288	3.232	1.681	0.991
$C_8H_{18}O$	4-Methyl-3-heptanol		1.904	1.085	0.702	0.497	0.375
$C_8H_{18}O$	5-Methyl-3-heptanol		2.052	1.178	0.762	0.536	0.401
$C_8H_{18}O$	2-Ethyl-1-hexanol		20.7	6.271	2.631	1.360	0.810
$C_8H_{18}O$	Dibutyl ether	1.417	0.918	0.637	0.466	0.356	0.281
$C_8H_{19}N$	Dibutylamine		1.509	0.918	0.619	0.449	0.345
$C_8H_{19}N$	Diisobutylamine		1.115	0.723	0.511	0.384	0.303
C_9H_7N	Quinoline		3.337	1.892	1.201	0.833
C_9H_{10}	Indane		2.230	1.357	0.931	0.692	0.545
C_9H_{12}	Cumene		1.075	0.737	0.547		
$C_9H_{14}O$	Isophorone		4.201	2.329	1.415	0.923	0.638
$C_9H_{18}O$	5-Nonanone			1.199	0.834	0.619	0.484
$C_9H_{18}O_2$	Nonanoic acid			7.011	3.712	2.234	1.475
C_9H_{20}	Nonane		0.964	0.665	0.488	0.375	0.300
$C_9H_{20}O$	1-Nonanol			9.123	4.032		
$C_{10}H_{10}O_4$	Dimethyl phthalate		63.2	14.4	5.309	2.824	1.980
$C_{10}H_{14}$	Butylbenzene			0.950	0.683	0.515	
$C_{10}H_{18}$	cis-Decahydronaphthalene	12.8	5.645	3.042	1.875	1.271	0.924
$C_{10}H_{18}$	trans-Decahydronaphthalene	6.192	3.243	1.948	1.289	0.917	0.689
$C_{10}H_{20}O_2$	Decanoic acid				4.327	2.651	
$C_{10}H_{22}$	Decane	2.188	1.277	0.838	0.598	0.453	0.359
$C_{10}H_{22}O$	1-Decanol			10.9	4.590		
$C_{11}H_{24}$	Undecane		1.707	1.098	0.763	0.562	0.433
$C_{12}H_{10}O$	Diphenyl ether				2.130	1.407	1.023
$C_{12}H_{26}$	Dodecane		2.277	1.383	0.930	0.673	0.514
$C_{13}H_{12}$	Diphenylmethane					1.265	0.929
$C_{13}H_{28}$	Tridecane		2.909	1.724	1.129	0.796	0.594
$C_{14}H_{30}$	Tetradecane			2.128	1.376	0.953	0.697
$C_{16}H_{22}O_4$	Dibutyl phthalate	483	66.4	16.6	6.470	3.495	2.425
$C_{16}H_{34}$	Hexadecane			3.032	1.879	1.260	0.899
$C_{18}H_{38}$	Octadecane				2.487	1.609	1.132

VISCOSITY OF AQUEOUS SOLUTIONS

This table gives the absolute viscosity of aqueous solutions of several common compounds as a function of concentration expresed in mass percent. Viscosity values are in units of millipascal seconds (mPa s), which is equivalent to centipoises (cp).

REFERENCE

Borchers, H., Ed., *Landoldt-Börnstein Numerical Data and Functional Relationships in Physics, Chemistry, Astronomy, Geophysics, and Technology*, 6th ed., Vol.4, Part 1, *Material Values and Mechanical Behavior of Non-Metals*, 1955.

Solute	t/°C	Concentration in mass percent (%)							
		0	5	10	20	40	60	80	100
Ethanol	25	0.89		1.32	1.81	2.35	2.24	1.75	1.07
Glycerol	25	0.89	1.01	1.15	1.54	3.18	8.82	45.9	934
Sucrose	20	1.00	1.13	1.31	1.92	5.98	58.5		
Pyridine	25	0.89		1.14	1.35	1.83	2.19	1.91	0.88
Sodium chloride	20	1.00	1.08	1.19	1.53				
Nitric acid	20	1.00		1.04	1.14	1.55	2.02	1.84	0.88
Sulfuric acid	25	0.89	1.01	1.12	1.40	2.51	5.37	17.4	24.2

THERMAL CONDUCTIVITY OF GASES

This table gives the thermal conductivity of several gases as a function of temperature. Unless otherwise noted, the values refer to a pressure of 100 kPa (1 bar) or to the saturation vapor pressure if that is less than 100 kPa. The notation $P = 0$ indicates the low pressure limiting value is given. In general, the $P = 0$ and $P = 100$ kPa values differ by less than 1%. Units are milliwatts per meter kelvin. Substances are listed in the modified Hill order.

| MF | Name | \multicolumn{6}{c}{Thermal conductivity in mW/m K} | Ref. |
		100 K	200 K	300 K	400 K	500 K	600 K	
	Air	9.4	18.4	26.2	33.3	39.7	45.7	1
Ar	Argon	6.2	12.4	17.9	22.6	26.8	30.6	2,8
BF_3	Boron trifluoride			19.0	24.6			11
ClH	Hydrogen chloride		9.2	14.5	19.5	24.0	28.1	11
F_6S	Sulfur hexafluoride ($P = 0$)			13.0	20.6	27.5	33.8	16
H_2	Hydrogen ($P = 0$)	68.6	131.7	186.9	230.4			4
H_2O	Water			18.7	27.1	35.7	47.1	6
	Deuterium oxide				27.0	36.5	47.6	7
H_2S	Hydrogen sulfide			14.6	20.5	26.4	32.4	11
H_3N	Ammonia			24.4	37.4	51.6	66.8	11
He	Helium ($P = 0$)	75.5	119.3	156.7	190.6	222.3	252.4	8
Kr	Krypton ($P = 0$)	3.3	6.4	9.5	12.3	14.8	17.1	8
NO	Nitric oxide		17.8	25.9	33.1	39.6	46.2	11
N_2	Nitrogen	9.8	18.7	26.0	32.3	38.3	44.0	12
N_2O	Nitrous oxide		9.8	17.4	26.0	34.1	41.8	11
Ne	Neon ($P = 0$)	22.3	37.6	49.8	60.3	69.9	78.7	8
O_2	Oxygen	9.3	18.4	26.3	33.7	41.0	48.1	10
O_2S	Sulfur dioxide			9.6	14.3	20.0	25.6	11
Xe	Xenon ($P = 0$)	2.0	3.6	5.5	7.3	8.9	10.4	8
CCl_2F_2	Dichlorodifluoromethane			9.9	15.0	20.1	25.2	13
CF_4	Tetrafluoromethane ($P = 0$)			16.0	24.1	32.2	39.9	16
CO	Carbon monoxide ($P = 0$)			25.0	32.3	39.2	45.7	14
CO_2	Carbon dioxide		9.6	16.8	25.1	33.5	41.6	9
$CHCl_3$	Trichloromethane			7.5	11.1	15.1		11
CH_4	Methane		22.5	34.1	49.1	66.5	84.1	5,15
CH_4O	Methanol				26.2	38.6	53.0	11
$C_2Cl_2F_4$	1,2-Dichlorotetrafluoro-ethane			10.25	15.7	21.1		13
$C_2Cl_3F_3$	1,1,2-Trichlorotrifluoro-ethane			9.0	13.6	18.3		13
C_2H_2	Acetylene			21.4	33.3	45.4	56.8	11
C_2H_4	Ethylene		11.1	20.5	34.6	49.9	68.6	3
C_2H_6	Ethane		11.0	21.3	35.4	52.2	70.5	5
C_2H_6O	Ethanol			14.4	25.8	38.4	53.2	11
C_3H_6O	Acetone			11.5	20.2	30.6	42.7	11
C_3H_8	Propane			18.0	30.6	45.5	61.9	5
C_4F_8	Perfluorocyclobutane			12.5	19.5			13
C_4H_{10}	Butane			16.4	28.4	43.0	59.1	5
C_4H_{10}	Isobutane			16.1	27.9	42.1	57.6	5
$C_4H_{10}O$	Diethyl ether			15.1	25.0	37.1		11
C_5H_{12}	Pentane			14.4	24.9	37.8	52.7	11
C_6H_{14}	Hexane				23.4	35.4	48.7	11

REFERENCES

1. Kadoya, K. Matsunaga, N., and Nagashima, A., Viscosity and thermal conductivity of dry air in the gaseous phase, *J. Phys. Chem. Ref. Data*, 14, 947, 1985.
2. Younglove, B. A. and Hanley, H. J. M., The viscosity and thermal conductivity coefficients of gaseous and liquid argon, *J. Phys. Chem. Ref. Data*, 15, 1323, 1986.
3. Holland, P. M., Eaton, B. E., and Hanley, H. J. M., A correlation of the viscosity and thermal conductivity data of gaseous and liquid ethylene, *J. Phys. Chem. Ref. Data*, 12, 917, 1983.

THERMAL CONDUCTIVITY OF GASES (continued)

4. Assael, M. J., Mixafendi, S., and Wakeham, W. A., The viscosity and thermal conductivity of normal hydrogen in the limit of zero density, *J. Phys. Chem. Ref. Data*, 15, 1315, 1986.
5. Younglove, B. A. and Ely, J. F., Thermophysical properties of fluids. II. Methane, ethane, propane, isobutane, and normal butane, *J. Phys. Chem. Ref. Data*, 16, 577, 1987.
6. Sengers, J. V. and Watson, J. T. R., Improved international formulations for the viscosity and thermal conductivity of water substance, *J. Phys. Chem. Ref. Data*, 15, 1291, 1986.
7. Matsunaga, N. and Nagashima, A., Transport properties of liquid and gaseous D_2O over a wide range of temperature and pressure, *J. Phys. Chem. Ref. Data*, 12, 933, 1983.
8. Kestin, J. et al., Equilibrium and transport properties of the noble gases and their mixtures at low density, *J. Phys. Chem. Ref. Data*, 13, 229, 1984.
9. Vescovic, V. et al., The transport properties of carbon dioxide, *J. Phys. Chem. Ref. Data*, 19, 1990.
10. Younglove, B. A., Thermophysical properties of fluids. I. Argon, ethylene, parahydrogen, nitrogen, nitrogen trifluoride, and oxygen, *J. Phys. Chem. Ref. Data*, 11, Suppl. 1, 1982.
11. Ho, C. Y., Ed., *Properties of Inorganic and Organic Fluids, CINDAS Data Series on Materials Properties*, Volume V-1, Hemisphere Publishing Corp., New York, 1988.
12. Stephen, K., Krauss, R., and Laesecke, A., Viscosity and thermal conductivity of nitrogen for a wide range of fluid states, *J. Phys. Chem. Ref. Data*, 16, 993, 1987.
13. Krauss, R. and Stephan, K., Thermal conductivity of refrigerants in a wide range of temperature and pressure, *J. Phys. Chem. Ref. Data*, 18, 43, 1989.
14. Millat, J. and Wakeham, W. A., The thermal conductivity of nitrogen and carbon monoxide in the limit of zero density, *J. Phys. Chem. Ref. Data*, 18, 565, 1989.
15. Friend, D. G., Ely, J. F., and Ingham, H., Thermophysical properties of methane, *J. Phys. Chem. Ref. Data*, 18, 583, 1989.
16. Uribe, F. J., Mason, E. A., and Kestin, J., Thermal conductivity of nine polyatomic gases at low density, *J. Phys. Chem. Ref. Data*, 19, 1123, 1990.

THERMAL CONDUCTIVITY OF LIQUIDS

This table gives the thermal conductivity of some common liquids at temperatures between -25 and 100°C. All values are given in units of watts per meter kelvin (W/m K). Values refer to nominal atmospheric pressure (about 100 kPa); when an entry is given at a temperature above the normal boiling point of the substance, the pressure is understood to be the saturation vapor pressure at that temperature.

Substances are arranged by molecular formula in the modified Hill order, with compounds not containing carbon preceding those that do contain carbon.

The values for water, benzene, toluene, heptane, and dimethyl phthalate are particularly well determined and can be used for calibration purposes.

REFERENCES

1. Daubert, T. E., Danner, R. P., Sibul, H. M., and Stebbins, C. C., *Physical and Thermodynamic Properties of Pure Compounds: Data Compilation*, extant 1994 (core with 4 supplements), Taylor & Francis, Bristol, PA (also available as database).
2. Marsh, K. N., Ed., *Recommended Reference Materials for the Realization of Physicochemical Properties,* Blackwell Scientific Publications, Oxford, 1987.

Molecular formula	Name	Thermal conductivity in W/m K					
		–25°C	0°C	25°C	50°C	75°C	100°C
Cl_4Si	Silicon tetrachloride			0.099	0.096		
H_2O	Water		0.5610	0.6071	0.6435	0.6668	0.6791
Hg	Mercury	7.25	7.77	8.25	8.68	9.07	9.43
CCl_4	Tetrachloromethane		0.104	0.099	0.093	0.088	
CS_2	Carbon disulfide		0.154	0.149			
$CHCl_3$	Trichloromethane	0.127	0.122	0.117	0.112	0.107	0.102
CH_2Br_2	Dibromomethane	0.120	0.114	0.108	0.103	0.097	
CH_4O	Methanol	0.214	0.207	0.200	0.193		
C_2Cl_4	Tetrachloroethylene		0.117	0.110	0.104	0.097	0.091
C_2HCl_3	Trichloroethylene	0.133	0.124	0.116	0.108	0.100	
$C_2H_3Cl_3$	1,1,1-Trichloroethane		0.106	0.101	0.096		
C_2H_3N	Acetonitrile	0.208	0.198	0.188	0.178	0.168	
$C_2H_4O_2$	Acetic acid			0.158	0.153	0.149	0.144
C_2H_5Cl	Chloroethane	0.145	0.132	0.119	0.106	0.093	
C_2H_5NO	N-Methylformamide			0.203	0.201	0.199	0.196
C_2H_6O	Ethanol		0.176	0.169	0.162		
$C_2H_6O_2$	Ethylene glycol		0.256	0.256	0.256	0.256	0.256
C_2H_7NO	Ethanolamine			0.299	0.286	0.274	0.261
C_3H_5ClO	Epichlorohydrin	0.142	0.137	0.131	0.125	0.119	0.114
C_3H_6O	Acetone		0.169	0.161			
$C_3H_6O_2$	Methyl acetate	0.174	0.164	0.153	0.143	0.133	0.122
C_3H_7NO	N,N-Dimethylformamide			0.184	0.178	0.171	0.165
C_3H_8O	1-Propanol	0.162	0.158	0.154	0.149	0.145	0.141
C_3H_8O	2-Propanol	0.146	0.141	0.135	0.129	0.124	0.118
$C_3H_8O_2$	1,2-Propanediol		0.202	0.200	0.199	0.198	0.197
$C_3H_8O_3$	Glycerol			0.292	0.295	0.297	0.300
C_3H_9N	Trimethylamine	0.143	0.133				
C_4H_4O	Furan	0.142	0.134	0.126			
C_4H_4S	Thiophene			0.199	0.195	0.191	0.186
C_4H_6	2-Butyne	0.137	0.129	0.121			
C_4H_8O	2-Butanone	0.158	0.151	0.145	0.139	0.133	
C_4H_8O	Tetrahydrofuran	0.132	0.126	0.120	0.114		
$C_4H_8O_2$	1,4-Dioxane			0.159	0.147	0.135	0.123
$C_4H_8O_2$	Ethyl acetate	0.162	0.153	0.144	0.135	0.126	
$C_4H_{10}O$	1-Butanol		0.158	0.154	0.149		
$C_4H_{10}O$	Diethyl ether	0.150	0.140	0.130	0.120	0.110	0.100
C_5H_5N	Pyridine		0.169	0.165	0.161	0.158	
C_5H_8	Cyclopentene	0.143	0.136	0.129			
C_5H_{10}	1-Pentene	0.131	0.124	0.116			
C_5H_{10}	Cyclopentane	0.140	0.133	0.126			
C_5H_{12}	Pentane	0.132	0.122	0.113	0.103	0.095	0.087
$C_5H_{12}O$	1-Pentanol		0.157	0.153	0.149	0.145	
C_6H_5Cl	Chlorobenzene	0.136	0.131	0.127	0.122	0.117	0.112

Molecular formula	Name	\-25°C	0°C	25°C	50°C	75°C	100°C
C_6H_6	Benzene			0.1411	0.1329	0.1247	
C_6H_6O	Phenol				0.156	0.153	0.151
C_6H_{10}	Cyclohexene	0.142	0.136	0.130	0.124	0.118	
$C_6H_{10}O$	Mesityl oxide	0.170	0.163	0.156	0.149	0.142	0.134
C_6H_{12}	Cyclohexane			0.123	0.117	0.111	
C_6H_{12}	1-Hexene	0.137	0.129	0.121	0.113		
$C_6H_{12}O$	Cyclohexanol			0.134	0.131		
$C_6H_{12}O$	2-Hexanone	0.151	0.145	0.139	0.133	0.127	0.121
C_6H_{14}	Hexane	0.137	0.128	0.120	0.111	0.102	0.093
$C_6H_{14}O$	1-Hexanol	0.159	0.154	0.150	0.145	0.141	0.137
C_7H_6O	Benzaldehyde			0.151	0.141	0.131	0.121
C_7H_8	Toluene	0.1461	0.1386	0.1311	0.1236	0.1161	
C_7H_8O	Anisole	0.170	0.163	0.156	0.150	0.143	0.136
C_7H_{16}	Heptane	0.1378	0.1303	0.1228	0.1152	0.1077	
$C_7H_{16}O$	1-Heptanol		0.166	0.159	0.153	0.147	0.141
C_8H_8	Styrene	0.148	0.142	0.137	0.131	0.126	0.120
C_8H_{10}	Ethylbenzene			0.130	0.124	0.118	0.112
C_8H_{10}	o-Xylene			0.131	0.126	0.120	0.114
C_8H_{10}	m-Xylene			0.130	0.124	0.118	0.113
C_8H_{10}	p-Xylene			0.130	0.124	0.118	0.112
C_8H_{18}	Octane	0.143	0.135	0.128	0.120	0.113	0.106
$C_8H_{18}O$	1-Octanol		0.168	0.161	0.154	0.147	0.141
C_9H_{12}	Cumene			0.128	0.120	0.112	0.107
C_9H_{12}	Mesitylene	0.147	0.141	0.136	0.130	0.124	0.118
C_9H_{20}	Nonane	0.144	0.138	0.131	0.124	0.118	0.111
$C_9H_{20}O$	1-Nonanol		0.166	0.161	0.155	0.149	0.143
$C_{10}H_{10}O_4$	Dimethyl phthalate		0.1501	0.1473	0.1443	0.1409	0.1373
$C_{10}H_{14}$	p-Cymene	0.132	0.127	0.122	0.117	0.112	0.107
$C_{10}H_{22}$	Decane	0.144	0.138	0.132	0.126	0.119	0.113
$C_{10}H_{22}O$	1-Decanol			0.162	0.156	0.150	0.145
$C_{11}H_{24}$	Undecane			0.140	0.135	0.129	0.123
$C_{12}H_{10}O$	Diphenyl ether				0.139	0.135	0.131
$C_{12}H_{26}$	Dodecane		0.157	0.152	0.146	0.140	0.135
$C_{12}H_{26}O$	1-Dodecanol			0.146	0.142	0.139	0.135
$C_{13}H_{28}$	Tridecane			0.137	0.132	0.127	0.122
$C_{14}H_{30}$	Tetradecane			0.136	0.131	0.126	0.121
$C_{14}H_{30}O$	1-Tetradecanol				0.167	0.162	0.157
$C_{16}H_{22}O_4$	Dibutyl phthalate	0.144	0.140	0.136	0.133	0.129	0.125
$C_{16}H_{34}$	Hexadecane			0.140	0.135	0.130	0.125
$C_{18}H_{38}$	Octadecane				0.146	0.142	0.137

DIFFUSION IN GASES

This table gives binary diffusion coefficients D_{12} for a number of common gases as a function of temperature. Values refer to atmospheric pressure. The diffusion coefficient is inversely proportional to pressure as long as the gas is in a regime where binary collisions dominate. See Reference 1 for a discussion of the dependence of D_{12} on temperature and composition.

The first part of the table gives data for several gases in the presence of a large excess of air. The remainder applies to equimolar mixtures of gases. Each gas pair is ordered alphabetically according to the most common way of writing the formula. The listing of pairs then follows alphabetical order by the first constituent.

REFERENCES

1. Marrero, T. R., and Mason, E. A., *J. Phys. Chem. Ref. Data*, 1, 1, 1972.
2. Kestin, J., et al., *J. Phys. Chem. Ref. Data*, 13, 229, 1984.

$D_{12}/\text{cm}^2\,\text{s}^{-1}$ for $p = 101.325$ kPa and the Specified T/K

System	200	273.15	293.15	373.15	473.15	573.15	673.15
Large Excess of Air							
Ar-air		0.167	0.148	0.289	0.437	0.612	0.810
CH_4-air			0.106	0.321	0.485	0.678	0.899
CO-air			0.208	0.315	0.475	0.662	0.875
CO_2-air			0.160	0.252	0.390	0.549	0.728
H_2-air		0.668	0.627	1.153	1.747	2.444	3.238
H_2O-air			0.242	0.399	0.638	0.873	1.135
He-air		0.617	0.580	1.057	1.594	2.221	2.933
SF_6-air				0.150	0.233	0.329	0.438
Equimolar Mixture							
Ar-CH_4				0.306	0.467	0.657	0.876
Ar-CO		0.168	0.187	0.290	0.439	0.615	0.815
Ar-CO_2		0.129	0.078	0.235	0.365	0.517	0.689
Ar-H_2		0.698	0.794	1.228	1.876	2.634	3.496
Ar-He	0.381	0.645	0.726	1.088	1.617	2.226	2.911
Ar-Kr	0.064	0.117	0.134	0.210	0.323	0.456	0.605
Ar-N_2		0.168	0.190	0.290	0.439	0.615	0.815
Ar-Ne	0.160	0.277	0.313	0.475	0.710	0.979	1.283
Ar-O_2		0.166	0.189	0.285	0.430	0.600	0.793
Ar-SF_6				0.128	0.202	0.290	0.389
Ar-Xe	0.052	0.095	0.108	0.171	0.264	0.374	0.498
CH_4-H_2			0.782	1.084	1.648	2.311	3.070
CH_4-He			0.723	0.992	1.502	2.101	2.784
CH_4-N_2			0.220	0.317	0.480	0.671	0.890
CH_4-O_2			0.210	0.341	0.523	0.736	0.978
CH_4-SF_6				0.167	0.257	0.363	0.482
CO-CO_2			0.162	0.250	0.384		
CO-H_2	0.408	0.686	0.772	1.162	1.743	2.423	3.196
CO-He	0.365	0.619	0.698	1.052	1.577	2.188	2.882
CO-Kr		0.131	0.581	0.227	0.346	0.485	0.645
CO-N_2	0.133	0.208	0.231	0.336	0.491	0.673	0.878
CO-O_2			0.202	0.307	0.462	0.643	0.849
CO-SF_6				0.144	0.226	0.323	0.432
CO_2-C_3H_8			0.084	0.133	0.209		
CO_2-H_2	0.315	0.552	0.412	0.964	1.470	2.066	2.745
CO_2-H_2O			0.162	0.292	0.496	0.741	1.021
CO_2-He	0.300	0.513	0.400	0.878	1.321		
CO_2-N_2			0.160	0.253	0.392	0.553	0.733
CO_2-N_2O	0.055	0.099	0.113	0.177	0.276		
CO_2-Ne	0.131	0.227	0.199	0.395	0.603	0.847	

DIFFUSION IN GASES (continued)

System	200	273.15	293.15	373.15	473.15	573.15	673.15
CO_2-O_2			0.159	0.248	0.380	0.535	0.710
CO_2-SF_6				0.099	0.155		
D_2-H_2	0.631	1.079	1.219	1.846	2.778	3.866	5.103
H_2-He	0.775	1.320	1.490	2.255	3.394	4.726	6.242
H_2-Kr	0.340	0.601	0.682	1.053	1.607	2.258	2.999
H_2-N_2	0.408	0.686	0.772	1.162	1.743	2.423	3.196
H_2-Ne	0.572	0.982	0.317	1.684	2.541	3.541	4.677
H_2-O_2		0.692	0.756	1.188	1.792	2.497	3.299
H_2-SF_6			0.208	0.649	0.998	1.400	1.851
H_2-Xe		0.513	0.122	0.890	1.349	1.885	2.493
H_2O-N_2			0.242	0.399			
H_2O-O_2			0.244	0.403	0.645	0.882	1.147
He-Kr	0.330	0.559	0.629	0.942	1.404	1.942	2.550
He-N_2	0.365	0.619	0.698	1.052	1.577	2.188	2.882
He-Ne	0.563	0.948	1.066	1.592	2.362	3.254	4.262
He-O_2		0.641	0.697	1.092	1.640	2.276	2.996
He-SF_6			1.109	0.592	0.871	1.190	1.545
He-Xe	0.282	0.478	0.538	0.807	1.201	1.655	2.168
Kr-N_2		0.131	0.149	0.227	0.346	0.485	0.645
Kr-Ne	0.131	0.228	0.258	0.392	0.587	0.812	1.063
Kr-Xe	0.035	0.064	0.073	0.116	0.181	0.257	0.344
N_2-Ne			0.258	0.483	0.731	1.021	1.351
N_2-O_2			0.202	0.307	0.462	0.643	0.849
N_2-SF_6				0.148	0.231	0.328	0.436
N_2-Xe		0.107	0.708	0.188	0.287	0.404	0.539
Ne-Xe	0.111	0.193	0.219	0.332	0.498	0.688	0.901
O_2-SF_6			0.097	0.154	0.238	0.334	0.441

DIFFUSION COEFFICIENTS IN LIQUIDS AT INFINITE DILUTION

This table lists diffusion coefficients D_{AB} at infinite dilution for some binary liquid mixtures. Although values are given to two decimal places, measurements in the literature are often in poor agreement. Therefore most values in the table cannot be relied upon to better than 10%.

Solvents are listed in alphabetical order, as are the solutes within each solvent group.

REFERENCE

Landolt-Börnstein, *Numerical Data and Functional Relationships in Science and Technology*, Sixth Edition, Vol. II/5a, 1969.

Solute	Solvent	$t/°C$	D_{AB} 10^{-5} cm^2 s^{-1}	Solute	Solvent	$t/°C$	D_{AB} 10^{-5} cm^2 s^{-1}
Acetic acid	Acetone	25	3.31	Acetone	Tetrachloromethane	25	1.75
Benzoic acid	Acetone	25	2.62	Benzene	Tetrachloromethane	25	1.42
Formic acid	Acetone	25	3.77	Cyclohexane	Tetrachloromethane	25	1.30
Nitrobenzene	Acetone	20	2.94	Ethanol	Tetrachloromethane	25	1.90
Tetrachloromethane	Acetone	25	3.29	Iodine	Tetrachloromethane	30	1.63
Trichloromethane	Acetone	25	3.64	Trichloromethane	Tetrachloromethane	25	1.66
Water	Acetone	25	4.56	Acetic acid	Toluene	25	2.26
Acetic acid	Benzene	25	2.09	Benzene	Toluene	25	2.54
Aniline	Benzene	25	1.96	Benzoic acid	Toluene	25	1.49
Benzoic acid	Benzene	25	1.38	Cyclohexane	Toluene	25	2.42
Bromobenzene	Benzene	8	1.45	Formic acid	Toluene	25	2.65
2-Butanone	Benzene	30	2.09	Water	Toluene	25	6.19
Chloroethylene	Benzene	8	1.77	Acetone	Trichloromethane	25	2.55
Cyclohexane	Benzene	25	2.25	Benzene	Trichloromethane	25	2.89
Ethanol	Benzene	25	3.02	2-Butanone	Trichloromethane	25	2.13
Formic acid	Benzene	25	2.28	Butyl acetate	Trichloromethane	25	1.71
Heptane	Benzene	25	1.78	Diethyl ether	Trichloromethane	25	2.15
Methanol	Benzene	25	3.80	Ethanol	Trichloromethane	15	2.20
Toluene	Benzene	25	1.85	Ethyl acetate	Trichloromethane	25	2.02
1,2,4-Trichlorobenzene	Benzene	8	1.34	Acetic acid	Water	25	1.29
Trichloromethane	Benzene	25	2.26	Acetone	Water	25	1.28
Adipic acid	1-Butanol	30	0.40	Acetonitrile	Water	15	1.26
Benzene	1-Butanol	25	1.00	Alanine	Water	25	0.91
Biphenyl	1-Butanol	25	0.63	Allyl alcohol	Water	15	0.90
Butyric acid	1-Butanol	30	0.51	Aniline	Water	20	0.92
p-Dichlorobenzene	1-Butanol	25	0.82	Arabinose	Water	20	0.69
Methanol	1-Butanol	30	0.59	Benzene	Water	20	1.02
Oleic acid	1-Butanol	30	0.25	1-Butanol	Water	25	0.56
Propane	1-Butanol	25	1.57	Caprolactam	Water	25	0.87
Water	1-Butanol	25	0.56	Chloroethylene	Water	25	1.34
Benzene	Cyclohexane	25	1.41	Cyclohexane	Water	20	0.84
Tetrachloromethane	Cyclohexane	25	1.49	Diethylamine	Water	20	0.97
Toluene	Cyclohexane	25	1.57	Ethanol	Water	25	1.24
Allyl alcohol	Ethanol	20	0.98	Ethanolamine	Water	25	1.08
Benzene	Ethanol	25	1.81	Ethyl acetate	Water	20	1.00
Iodine	Ethanol	25	1.32	Ethylbenzene	Water	20	0.81
Iodobenzene	Ethanol	20	1.00	Ethylene glycol	Water	25	1.16
3-Methyl-1-butanol	Ethanol	20	0.81	Glucose	Water	25	0.67
Pyridine	Ethanol	20	1.10	Glycerol	Water	25	1.06
Tetrachloromethane	Ethanol	25	1.50	Glycine	Water	25	1.05
Water	Ethanol	25	1.24	Lactose	Water	15	0.38
Acetic acid	Ethyl acetate	20	2.18	Maltose	Water	15	0.38
Acetone	Ethyl acetate	20	3.18	Mannitol	Water	15	0.50
2-Butanone	Ethyl acetate	30	2.93	Methane	Water	25	1.49
Ethyl benzoate	Ethyl acetate	20	1.85	Methanol	Water	15	1.28
Nitrobenzene	Ethyl acetate	20	2.25	3-Methyl-1-butanol	Water	10	0.69
Water	Ethyl acetate	25	3.20	Methylcyclopentane	Water	20	0.85
Benzene	Heptane	25	3.91	Phenol	Water	20	0.89
Toluene	Heptane	25	3.72	1-Propanol	Water	15	0.87
Bromobenzene	Hexane	8	2.60	Propene	Water	25	1.44
2-Butanone	Hexane	30	3.74	Pyridine	Water	25	0.58
Dodecane	Hexane	25	2.73	Raffinose	Water	15	0.33
Iodine	Hexane	25	4.45	Sucrose	Water	25	0.52
Methane	Hexane	25	0.09	Toluene	Water	20	0.85
Propane	Hexane	25	4.87	Urea	Water	25	1.38
Tetrachloromethane	Hexane	25	3.70	Urethane	Water	15	0.80
Toluene	Hexane	25	4.21				

Section 7
Biochemistry

PROPERTIES OF COMMON AMINO ACIDS

This table gives selected properties of 20 α-amino acids commonly found in proteins. The structures of these amino acids are given in a separate table. The compounds are listed in alphabetical order by the three-letter symbols. Dissociation constants refer to aqueous solutions at 25° C.

M_r — Molecular weight
t_m — Melting point
pK_a — Negative of the logarithm of the dissociation constant for the α-COOH group
pK_b — Negative of the logarithm of the dissociation constant for the α-NH_3^+ group
pK_x — Negative of the logarithm of the dissociation constant for any other group present in the molecule
pI — pH at the isoelectronic point
S — Solubility in water at 25° C in units of grams per kilogram of water

Symbol	Name	Mol. form.	M_r	$t_m/°C$	pK_a	pK_b	pK_x	pI	S
Ala	Alanine	$C_3H_7NO_2$	89.09	297	2.34	9.69		6.00	167
Arg	Arginine	$C_6H_{14}N_4O_2$	174.20	238	2.17	9.04	12.48	10.76	181
Asn	Asparagine	$C_4H_8N_2O_3$	132.12	236	2.02	8.80		5.41	25
Asp	Aspartic acid	$C_4H_7NO_4$	133.10	270	1.88	9.60	3.65	2.77	5
Cys	Cysteine	$C_3H_7NO_2S$	121.16	178	1.96	10.28	8.18	5.07	
Gln	Glutamine	$C_5H_{10}N_2O_3$	146.15	185	2.17	9.13		5.65	42
Glu	Glutamic acid	$C_5H_9NO_4$	147.13	249	2.19	9.67	4.25	3.22	
Gly	Glycine	$C_2H_5NO_2$	75.07	290	2.34	9.60		5.97	251
His	Histidine	$C_6H_9N_3O_2$	155.16	277	1.82	9.17	6.00	7.59	43
Ile	Isoleucine	$C_6H_{13}NO_2$	131.17	284	2.36	9.60		6.02	34
Leu	Leucine	$C_6H_{13}NO_2$	131.17	337	2.36	9.60		5.98	23
Lys	Lysine	$C_6H_{14}N_2O_2$	146.19	224—225	2.18	8.95	10.53	9.74	6
Met	Methionine	$C_5H_{11}NO_2S$	149.21	283	2.28	9.21		5.74	56
Phe	Phenylalanine	$C_9H_{11}NO_2$	165.19	284	1.83	9.13		5.48	29
Pro	Proline	$C_5H_9NO_2$	115.13	222	1.99	10.60		6.30	1622
Ser	Serine	$C_3H_7NO_3$	105.09	228	2.21	9.15		5.68	422
Thr	Threonine	$C_4H_9NO_3$	119.12	253	2.09	9.10		5.60	97
Trp	Tryptophan	$C_{11}H_{12}N_2O_2$	204.23	282	2.83	9.39		5.89	12
Tyr	Tyrosine	$C_9H_{11}NO_3$	181.19	344	2.20	9.11	10.07	5.66	0.5
Val	Valine	$C_5H_{11}NO_2$	117.15	292-295	2.32	9.62		5.96	58

REFERENCES

G. D. Fasman, Ed., *Practical Handbook of Biochemistry and Molecular Biology*, CRC Press, Boca Raton, FL, 1989.

E. L. Smith, et al., *Principles of Biochemistry*, 7th ed., McGraw Hill, New York, 1983.

H. J. Hinz, Ed., *Thermodynamic Data for Biochemistry and Biotechnology,* Springer-Verlag, Heidelberg, 1986.

STRUCTURES OF COMMON AMINO ACIDS

Alanine
(Ala, A)

Arginine
(Arg, R)

Asparagine
(Asn, N)

Aspartic acid
(Asp, D)

Cysteine
(Cys, C)

Glutamine
(Gln, Q)

Glutamic acid
(Glu, E)

Glycine
(Gly, G)

Histidine
(His, H)

Isoleucine
(Ile, I)

Leucine
(Leu, L)

Lysine
(Lys, K)

Methionine
(Met, M)

Phenylalanine
(Phe, F)

Proline
(Pro, P)

Serine
(Ser, S)

Threonine
(Thr, T)

Tryptophan
(Trp, W)

Tyrosine
(Tyr, Y)

Valine
(Val, V)

PROPERTIES OF PURINE AND PYRIMIDINE BASES

This table lists some of the important purine and pyrimidine bases that occur in nucleic acids. The pK_a values (negative logarithm of the acid dissociation constant) are given for each ionization stage. The last column gives the aqueous solubility S at the indicated temperature in units of grams per 100 grams of solution.

The numbering system in the rings is:

Purine   Pyrimidine

REFERENCES

1. R. M. C. Dawson, et al., *Data for Biochemical Research*, 3rd Ed., Clarendon Press, Oxford, 1986.
2. S. Budavari, Ed., *The Merck Index*, 11th Ed., Merk and Co., Rahway, NJ., 1989.

Common name	Systematic name	Mol form.	Mol. wt.	pK_a values			S/mass % (temp.)
Pyrimidines							
Cytosine	4-Amino-2-hydroxypyrimidine	$C_4H_5N_3O$	111.10	4.5	12.2		0.76 (25°C)
5-Methylcytosine	4-Amino-2-hydroxy-5-methylpyrimidine	$C_5H_7N_3O$	125.13	4.6	12.4		0.45 (25°C)
5-Hydroxymethyl-cytosine	4-Amino-2-hydroxy-5-hydroxy-methylpyrimidine	$C_5H_7N_3O_2$	141.13	4.3	13		
Uracil	2,4-Dihydroxypyrimidine	$C_4H_4N_2O_2$	112.09	0.5	9.5	>13	0.36 (25°C)
Thymine	5-Methyluracil	$C_5H_6N_2O_2$	126.11	9.9	>13		0.4 (25°C)
Orotic acid	Uracil-6-carboxylic acid	$C_5H_4N_2O_4$	156.10	2.4	9.5	>13	0.18 (18°C)
Purines							
Adenine	6-Aminopurine	$C_5H_5N_5$	135.14	<1	4.1	9.8	0.09 (25°C)
Guanine	2-Amino-6-hydroxypurine	$C_5H_5N_5O$	·151.13	3.3	9.2	12.3	0.004 (40°C)
7-Methylguanine	7-Methyl-2-amino-6-hydroxypurine	$C_6H_7N_5O$	165.16	3.5	9.9		
Isoguanine	6-Amino-2-hydroxypurine	$C_5H_5N_5O$	151.13	4.5	9.0		0.006 (25°C)
Xanthine	2,6-Dioxopurine	$C_5H_4N_4O_2$	152.11	0.8	7.4	11.1	0.05 (20°C)
Hypoxanthine	6-Hydroxypurine	$C_5H_4N_4O$	136.11	2.0	8.9	12.1	0.07 (19°C)
Uric acid	2,6,8-Trihydroxypurine	$C_5H_4N_4O_3$	168.11	5.4	11.3		0.002 (20°C)

THE GENETIC CODE

This table gives the correspondence between a messenger RNA codon and the amino acid which it specifies. The symbols for bases in the codon are:

U: uracil
C: cytosine
A: adenine
G: guanine

The amino acid symbols are given in the table entitled "Structures of Common Amino Acids". A chain-initiating codon is indicated by **init** and a chain-terminating codon by **term**.

Example: UCA codes for **Ser**, UAC codes for **Tyr**, etc.

First position	Second position				Third position
	U	**C**	**A**	**G**	
U	Phe	Ser	Tyr	Cys	U
	Phe	Ser	Tyr	Cys	C
	Leu	Ser	**term**	**term**	A
	Leu	Ser	**term**	Trp	G
C	Leu	Pro	His	Arg	U
	Leu	Pro	His	Arg	C
	Leu	Pro	Gln	Arg	A
	Leu	Pro	Gln	Arg	G
A	Ile	Thr	Asn	Ser	U
	Ile	Thr	Asn	Ser	C
	Ile	Thr	Lys	Arg	A
	Met (**init**)	Thr	Lys	Arg	G
G	Val	Ala	Asp	Gly	U
	Val	Ala	Asp	Gly	C
	Val	Ala	Glu	Gly	A
	Val (**init**)	Ala	Glu	Gly	G

PROPERTIES OF SELECTED FATTY ACIDS

This table gives the systematic names and selected properties of the more important fatty acids of five or more carbon atoms. Compounds are listed first by degree of saturation and, secondly, by number of carbon atoms.

M_r: Molecular weight
t_m: Melting point in °C

S: Aqueous solubility at 20°C in units of grams of solute per 100 grams of water

REFERENCES

1. R. C. M. Dawson, et al., *Data for Biochemical Research*, 3rd ed., Clarendon Press, Oxford, 1986.
2. G. D. Fasman, Ed., *Practical Handbook of Biochemistry and Molecular Biology*, CRC Press, Boca Raton, FL, 1989.

Common name	Systematic name	Mol. form.	M_r	$t_m/°C$	S
		Saturated			
Isovaleric acid	3-Methylbutanoic acid	$C_5H_{10}O_2$	102.13	−29	4.0
Valeric acid	Pentanoic acid	$C_5H_{10}O_2$	102.13	−34	3.2
Caproic acid	Hexanoic acid	$C_6H_{12}O_2$	116.16	−3	0.97
Enanthic acid	Heptanoic acid	$C_7H_{14}O_2$	130.19	−8	0.24
Caprylic acid	Octanoic acid	$C_8H_{16}O_2$	144.21	16	0.068
Pelargonic acid	Nonanoic acid	$C_9H_{18}O_2$	158.24	12	0.026
Capric acid	Decanoic acid	$C_{10}H_{20}O_2$	172.27	32	0.015
Lauric acid	Dodecanoic acid	$C_{12}H_{24}O_2$	200.32	43	0.0055
Tridecylic acid	Tridecanoic acid	$C_{13}H_{26}O_2$	214.35	42	0.0033
Myristic acid	Tetradecanoic acid	$C_{14}H_{28}O_2$	228.38	54	0.0020
Pentadecylic acid	Pentadecanoic acid	$C_{15}H_{30}O_2$	242.40	52	0.0012
Palmitic acid	Hexadecanoic acid	$C_{16}H_{32}O_2$	256.43	62	0.00072
Margaric acid	Heptadecanoic acid	$C_{17}H_{34}O_2$	270.46	61	0.00042
Stearic acid	Octadecanoic acid	$C_{18}H_{36}O_2$	284.48	69	0.00029
Arachidic acid	Eicosanoic acid	$C_{20}H_{40}O_2$	312.54	75	
Phytanic acid	3,7,11,15-Tetramethyl-hexadecanoic acid	$C_{20}H_{40}O_2$	312.54	−65	
Behenic acid	Docosanoic acid	$C_{22}H_{40}O_2$	340.59	81	
Lignoceric acid	Tetracosanoic acid	$C_{24}H_{48}O_2$	368.64	84	
Cerotic acid	Hexacosanoic acid	$C_{26}H_{52}O_2$	396.70	89	
Montanic acid	Octacosanoic acid	$C_{28}H_{56}O_2$	424.75	91	
		Monounsaturated			
Caproleic acid	9-Decenoic acid	$C_{10}H_{18}O_2$	170.25		
Palmitoleic acid	*cis*-9-Hexadecenoic acid	$C_{16}H_{30}O_2$	254.41	0	
Oleic acid	*cis*-9-Octadecenoic acid	$C_{18}H_{34}O_2$	282.47	13	
Vaccenic acid	*trans*-11-Octadecenoic acid	$C_{18}H_{34}O_2$	282.47	7	
Elaidic acid	*trans*-9-Octadecenoic acid	$C_{18}H_{34}O_2$	282.47	44	
Brassidic acid	*trans*-13-Docosenoic acid	$C_{22}H_{42}O_2$	338.57	60	
Erucic acid	*cis*-13-Docosenoic acid	$C_{22}H_{42}O_2$	338.57	34	
Nervonic acid	*cis*-15-Tetracosenoic acid	$C_{24}H_{46}O_2$	366.63	39	
		Diunsaturated			
Linoleic acid	*cis,cis*-9,12-Octadeca-dienoic acid	$C_{18}H_{32}O_2$	280.45	−9	
		Triunsaturated			
Eleosteric acid	*cis,trans,trans*-9,11,13-Octa-decatrienoic acid	$C_{18}H_{30}O_2$	278.44	49	
Linolenic acid	*cis,cis,cis*-9,12,15-Octa-decatrienoic acid	$C_{18}H_{30}O_2$	278.44	−17	
		Tetraunsaturated			
Arachidonic acid	5,8,11,14-Eicosatetraenoic acid	$C_{20}H_{32}O_2$	304.47	−50	

BIOLOGICAL BUFFERS

This table of frequently used buffers gives the pK_a value at 25°C and the useful pH range of each buffer. The buffers are listed in order of increasing pH.

The table is reprinted with permission of Sigma Chemical Company, St. Louis, Mo.

Acronym	Name	Mol. wt.	pK_a	Useful pH range
MES	2-(N-Morpholino)ethanesulfonic acid	195.2	6.1	5.5—6.7
BIS TRIS	Bis(2-hydroxyethyl)iminotris(hydroxymethyl)methane	209.2	6.5	5.8—7.2
ADA	N-(2-Acetamido)-2-iminodiacetic acid	190.2	6.6	6.0—7.2
ACES	2-[(2-Amino-2-oxoethyl)amino]ethanesulfonic acid	182.2	6.8	6.1—7.5
PIPES	Piperazine-N,N'-bis(2-ethanesulfonic acid)	302.4	6.8	6.1—7.5
MOPSO	3-(N-Morpholino)-2-hydroxypropanesulfonic acid	225.3	6.9	6.2—7.6
BIS TRIS PROPANE	1,3-Bis[tris(hydroxymethyl)methylamino]propane	282.3	6.8[a]	6.3—9.5
BES	N,N-Bis(2-hydroxyethyl)-2-aminoethanesulfonic acid	213.2	7.1	6.4—7.8
MOPS	3-(N-Morpholino)propanesulfonic acid	209.3	7.2	6.5—7.9
HEPES	N-(2-Hydroxyethyl)piperazine-N'-(2-ethanesulfonic acid)	238.3	7.5	6.8—8.2
TES	N-Tris(hydroxymethyl)methyl-2-aminoethanesulfonic acid	229.2	7.5	6.8—8.2
DIPSO	3-[N,N-Bis(2-hydroxyethyl)amino]-2-hydroxypropanesulfonic acid	243.3	7.6	7.0—8.2
TAPSO	3-[N-Tris(hydroxymethyl)methylamino)-2-hydroxypropanesulfonic acid	259.3	7.6	7.0—8.2
TRIZMA	Tris(hydroxymethyl)aminomethane	121.1	8.1	7.0—9.1
HEPPSO	N-(2-hydroxyethyl)piperazine-N'-(2-hydroxypropanesulfonic acid)	268.3	7.8	7.1—8.5
POPSO	Piperazine-N,N'-bis(2-hydroxypropanesulfonic acid)	362.4	7.8	7.2—8.5
EPPS	N-(2-Hydroxyethyl)piperazine-N'-(3-propanesulfonic acid)	252.3	8.0	7.3—8.7
TEA	Triethanolamine	149.2	7.8	7.3—8.3
TRICINE	N-Tris(hydroxymethyl)methylglycine	179.2	8.1	7.4—8.8
BICINE	N,N-Bis(2-hydroxyethyl)glycine	163.2	8.3	7.6—9.0
TAPS	N-Tris(hydroxymethyl)methyl-3-aminopropanesulfonic acid	243.3	8.4	7.7—9.1
AMPSO	3-[(1,1-Dimethyl-2-hydroxyethyl)amino]-2-hydroxypropanesulfonic acid	227.3	9.0	8.3—9.7
CHES	2-(N-Cyclohexylamino)ethanesulfonic acid	207.3	9.3	8.6—10.0
CAPSO	3-(Cyclohexylamino)-2-hydroxy-1-propanesulfonic acid	237.3	9.6	8.9—10.3
AMP	2-Amino-2-methyl-1-propanol	89.1	9.7	9.0—10.5
CAPS	3-(Cyclohexylamino)-1-propanesulfonic acid	221.3	10.4	9.7—11.1

[a] $pK_a = 9.0$ for the second dissociation stage.

TYPICAL pH VALUES OF BIOLOGICAL MATERIALS AND FOODS

This table gives typical pH ranges for various biological fluids and common foods. All values refer to 25°C.

Biological Materials

Blood, human	7.35-7.45
Blood, dog	6.9-7.2
Spinal fluid, human	7.3-7.5
Saliva, human	6.5-7.5
Gastric contents, human	1.0-3.0
Duodenal contents, human	4.8-8.2
Feces, human	4.6-8.4
Urine, human	4.8-8.4
Milk, human	6.6-7.6
Bile, human	6.8-7.0

Foods

Apples	2.9-3.3
Apricots	3.6-4.0
Asparagus	5.4-5.8
Bananas	4.5-4.7
Beans	5.0-6.0
Beers	4.0-5.0
Beets	4.9-5.5
Blackberries	3.2-3.6
Bread, white	5.0-6.0
Butter	6.1-6.4
Cabbage	5.2-5.4
Carrots	4.9-5.3
Cheese	4.8-6.4
Cherries	3.2-4.0
Cider	2.9-3.3
Corn	6.0-6.5
Crackers	6.5-8.5
Dates	6.2-6.4
Eggs, fresh white	7.6-8.0
Flour, wheat	5.5-6.5
Gooseberries	2.8-3.0
Grapefruit	3.0-3.3

Grapes	3.5-4.5
Hominy (lye)	6.8-8.0
Jams, fruit	3.5-4.0
Jellies, fruit	2.8-3.4
Lemons	2.2-2.4
Limes	1.8-2.0
Maple syrup	6.5-7.0
Milk, cows	6.3-6.6
Olives	3.6-3.8
Oranges	3.0-4.0
Oysters	6.1-6.6
Peaches	3.4-3.6
Pears	3.6-4.0
Peas	5.8-6.4
Pickles, dill	3.2-3.6
Pickles, sour	3.0-3.4
Pimento	4.6-5.2
Plums	2.8-3.0
Potatoes	5.6-6.0
Pumpkin	4.8-5.2
Raspberries	3.2-3.6
Rhubarb	3.1-3.2
Salmon	6.1-6.3
Sauerkraut	3.4-3.6
Shrimp	6.8-7.0
Soft drinks	2.0-4.0
Spinach	5.1-5.7
Squash	5.0-5.4
Strawberries	3.0-3.5
Sweet potatoes	5.3-5.6
Tomatoes	4.0-4.4
Tuna	5.9-6.1
Turnips	5.2-5.6
Vinegar	2.4-3.4
Water, drinking	6.5-8.0
Wines	2.8-3.8

CHEMICAL COMPOSITION OF THE HUMAN BODY

The elemental composition of the "standard man" of mass 70 kg is given below.

REFERENCES

1. Padikal, T.N., and Fivozinsky, S.P., *Medical Physics Data Book*, *National Bureau of Standards Handbook 138*, U. S. Government Printing Office, Washington, DC, 1981.
2. Snyde, W.S., et al., *Reference Man: Anatomical, Physiological, and Metabolic Characteristics*, Pergamon, New York, 1975.

Element	Amount (g)	Percent of total body mass
Oxygen	43,000	61
Carbon	16,000	23
Hydrogen	7000	10
Nitrogen	1800	2.6
Calcium	1000	1.4
Phosphorus	780	1.1
Sulfur	140	0.20
Potassium	140	0.20
Sodium	100	0.14
Chlorine	95	0.12
Magnesium	19	0.027
Silicon	18	0.026
Iron	4.2	0.006
Fluorine	2.6	0.0037
Zinc	2.3	0.0033
Rubidium	0.32	0.00046
Strontium	0.32	0.00046
Bromine	0.20	0.00029
Lead	0.12	0.00017
Copper	0.072	0.00010
Aluminum	0.061	0.00009
Cadmium	0.050	0.00007
Boron	<0.048	0.00007
Barium	0.022	0.00003
Tin	<0.017	0.00002
Manganese	0.012	0.00002
Iodine	0.013	0.00002
Nickel	0.010	0.00001
Gold	<0.010	0.00001
Molybdenum	<0.0093	0.00001
Chromium	<0.0018	0.000003
Cesium	0.0015	0.000002
Cobalt	0.0015	0.000002
Uranium	0.00009	0.0000001
Beryllium	0.000036	
Radium	$3.1 \cdot 10^{-11}$	

Section 8
Analytical Chemistry

PREPARATION OF SPECIAL ANALYTICAL REAGENTS

Aluminon (qualitative test for aluminum). Aluminon is a trade name for the ammonium salt of aurintricarboxylic acid. Dissolve 1 g of the salt in 1 L of distilled water. Shake the solution well to insure thorough mixing.

Bang's reagent (for glucose estimation). Dissolve 100 g of K_2CO_3, 66 g of KCl and 160 g of $KHCO_3$ in the order given in about 700 mL of water at 30°C. Add 4.4 g of $CuSO_4$ and dilute to 1 L after the CO_2 is evolved. This solution should be shaken only in such a manner as not to allow entry of air. After 24 hours 300 mL are diluted to 1 L with saturated KCl solution, shaken gently and used after 24 hours; 50 mL is equivalent to 10 mg glucose.

Barfoed's reagent (test for glucose). See Cupric acetate.

Baudisch's reagent. See Cupferron.

Benedict's solution (qualitative reagent for glucose). With the aid of heat, dissolve 173 g of sodium citrate and 100 g of Na_2CO_3 in 800 mL of water. Filter, if necessary, and dilute to 850 mL. Dissolve 17.3 g of $CuSO_4 \cdot 5H_2O$ in 100 mL of water. Pour the latter solution, with constant stirring, into the carbonate-citrate solution, and dilute to 1 L.

Benzidine hydrochloride solution (for sulfite determination). Make a paste of 8 g of benzidine hydrochloride ($C_{12}H_8(NH_3)_2 \cdot 2HCl$) and 20 mL of water, add 20 mL of HCl (sp. gr. 1.12) and dilute to 1 L with water. Each mL of this solution is equivalent to 0.00357 g of H_2SO_4.

Bertrand's reagent (glucose estimation). Consists of the following solutions:
1. Dissolve 200 g of Rochelle salt and 150 g of NaOH in sufficient water to make 1 L of solution.
2. Dissolve 40 g of $CuSO_4$ in enough water to make 1 L of solution.
3. Dissolve 50 g of $Fe_2(SO_4)_3$ and 200 g of H_2SO_4 (sp. gr. 1.84) in sufficient water to make 1 L of solution.
4. Dissolve 5 g of $KMnO_4$ in sufficient water to make 1 L of solution.

Bial's reagent (for pentose). Dissolve 1 g of orcinol (5-methyl-1,3-benzenediol) in 500 mL of 30% HCl to which 30 drops of a 10% solution of $FeCl_3$ has been added.

Boutron — Boudet soap solution:
1. Dissolve 100 g of pure castile soap in about 2.5 L of 56% ethanol.
2. Dissolve 0.59 g of $Ba(NO_3)_2$ in 1 L of water.

Adjust the castile soap solution so that 2.4 mL of it will give a permanent lather with 40 mL of solution (b). When adjusted, 2.4 mL of soap solution is equivalent to 220 parts per million of hardness (as $CaCO_3$) for a 40 mL sample. See also Soap solution.

Brucke's reagent (protein precipitation). See Potassium iodide-mercuric iodide.

Clarke's soap solution (estimation of hardness in water).
1. Dissolve 100 g of pure powdered castile soap in 1 L of 80% ethanol and allow to stand over night.
2. Prepare a solution of $CaCl_2$ by dissolving 0.5 g of $CaCO_3$ in HCl (sp. gr. 1.19), neutralize with NH_4OH and make slightly alkaline to litmus, and dilute to 500 mL. One mL is equivalent to 1 mg of $CaCO_3$.

Titrate (1) against (2) and dilute (1) with 80% ethanol until 1 mL of the resulting solution is equivalent to 1 mL of (2) after making allowance for the lather factor (the amount of standard soap solution required to produce a permanent lather in 50 mL of distilled water). One mL of the adjusted solution after subtracting the lather factor is equivalent to 1 mg of $CaCO_3$. See also Soap solution.

Cobalticyanide paper (Rinnmann's test for Zn). Dissolve 4 g of $K_3Co(CN)_6$ and 1 g of $KClO_3$ in 100 mL of water. Soak filter paper in solution and dry at 100°C. Apply drop of zinc solution and burn in an evaporating dish. A green disk is obtained if zinc is present.

Cochineal. Extract 1 g of cochineal for 4 days with 20 mL of alcohol and 60 mL of distilled water. Filter.

Congo red. Dissolve 0.5 g of congo red in 90 mL of distilled water and 10 mL of alcohol.

Cupferron (Baudisch's reagent for iron analysis). Dissolve 6 g of the ammonium salt of N-hydroxy-N-nitrosoaniline (cupferron) in 100 mL of H_2O. Reagent good for 1 week only and must be kept in the dark.

Cupric acetate (Barfoed's reagent for reducing monosaccharides). Dissolve 66 g of cupric acetate and 10 mL of glacial acetic acid in water and dilute to 1 L.

Cupric oxide, ammoniacal; Schweitzer's reagent (dissolves cotton, linen, and silk, but not wool).
1. Dissolve 5 g of cupric sulfate in 100 mL of boiling water, and add sodium hydroxide until precipitation is complete. Wash the precipitate well, and dissolve it in a minimum quantity of ammonium hydroxide.
2. Bubble a slow stream of air through 300 mL of strong ammonium hydroxide containing 50 g of fine copper turnings. Continue for 1 hour.

Cupric sulfate in glycerin-potassium hydroxide (reagent for silk). Dissolve 10 g of cupric sulfate, $CuSO_4 \cdot 5H_2O$, in 100 mL of water and add 5 g of glycerol. Add KOH solution slowly until a deep blue solution is obtained.

Cupron (precipitates copper). Dissolve 5 g of benzoinoxime in 100 mL of 95% ethanol.

Cuprous chloride, acidic (reagent for CO in gas analysis).
1. Cover the bottom of a 2-L flask with a layer of cupric oxide about 0.5 inch deep, suspend a coil of copper wire so as to reach from the bottom to the top of the solution, and fill the flask with hydrochloric acid (sp. gr. 1.10). Shake occasionally. When the solution becomes nearly colorless, transfer to reagent bottles, which should also contain copper wire. The stock bottle may be refilled with dilute hydrochloric acid until either the cupric oxide or the copper wire is used up. Copper sulfate may be substituted for copper oxide in the above procedure.
2. Dissolve 340 g of $CuCl_2 \cdot 2H_2O$ in 600 mL of conc. HCl and reduce the cupric chloride by adding 190 mL of a saturated solution of stannous chloride or until the solution is colorless. The stannous chloride is prepared by treating 300 g of metallic tin in a 500 mL flask with conc. HCl until no more tin goes into solution.
3. (Winkler method). Add a mixture of 86 g of CuO and 17 g of finely divided metallic Cu, made by the reduction of CuO with hydrogen, to a solution of HCl, made by diluting 650 mL of conc. HCl with 325 mL of water. After the mixture has been added slowly and with frequent stirring, a spiral of copper wire is suspended in the bottle, reaching all the way to the bottom. Shake occasionally, and when the solution becomes colorless, it is ready for use.

Cuprous chloride, ammoniacal (reagent for CO in gas analysis).

1. The acid solution of cuprous chloride as prepared above is neutralized with ammonium hydroxide until an ammonia odor persists. An excess of metallic copper must be kept in the solution.

2. Pour 800 mL of acidic cuprous chloride, prepared by the Winkler method, into about 4 L of water. Transfer the precipitate to a 250 mL graduate. After several hours, siphon off the liquid above the 50 mL mark and refill with 7.5% NH_4OH solution which may be prepared by diluting 50 mL of conc. NH_4OH with 150 mL of water. The solution is well shaken and allowed to stand for several hours. It should have a faint odor of ammonia.

Dichlorofluorescein indicator. Dissolve 1 g in 1 L of 70% alcohol or 1 g of the sodium salt in 1 L of water.

Dimethyglyoxime, 0.01 N. Dissolve 0.6 g of dimethylglyoxime (2,3-butanedione oxime) in 500 mL of 95% ethanol. This is an especially sensitive test for nickel, a very definite crimson color being produced.

Diphenylamine (reagent for rayon). Dissolve 0.2 g in 100 mL of concentrated sulfuric acid.

Diphenylamine sulfonate (for titration of iron with $K_2Cr_2O_7$). Dissolve 0.32 g of the barium salt of diphenylamine sulfonic acid in 100 mL of water, add 0.5 g of sodium sulfate and filter off the precipitate of $BaSO_4$.

Diphenylcarbazide. Dissolve 0.2 g of diphenylcarbazide in 10 mL of glacial acetic acid and dilute to 100 mL with 95% ethanol.

Esbach's reagent (estimation of protein). To a water solution of 10 g of picric acid and 20 g of citric acid, add sufficient water to make 1 L of solution.

Eschka's compound. Two parts of calcined ("light") magnesia are thoroughly mixed with 1 part of anhydrous sodium carbonate.

Fehling's solution (reagent for reducing sugars.)

1. Copper sulfate solution. Dissolve 34.66 g of $CuSO_4 \cdot 5H_2O$ in water and dilute to 500 mL.

2. Alkaline tartrate solution. Dissolve 173 g of potassium sodium tartrate (Rochelle salt, $KNaC_4H_4O_6 \cdot 4H_2O$) and 50 g of NaOH in water and dilute when cold to 500 mL.

Mix equal volumes of the two solutions at the time of using.

Ferric-alum indicator. Dissolve 140 g of ferric ammonium sulfate crystals in 400 mL of hot water. When cool, filter, and make up to a volume of 500 mL with dilute nitric acid.

Folin's mixture (for uric acid). To 650 mL of water add 500 g of $(NH_4)_2SO_4$, 5 g of uranium acetate, and 6 g of glacial acetic acid. Dilute to 1 L.

Formaldehyde — sulfuric acid (Marquis' reagent for alkaloids). Add 10 mL of formaldehyde solution to 50 mL of sulfuric acid.

Froehde's reagent. See Sulfomolybdic acid.

Fuchsin (reagent for linen). Dissolve 1 g of fuchsin in 100 mL of alcohol.

Fuchsin — sulfurous acid (Schiff's reagent for aldehydes). Dissolve 0.5 g of fuchsin and 9 g of sodium bisulfite in 500 mL of water, and add 10 mL of HCl. Keep in well-stoppered bottles and protect from light.

Gunzberg's reagent (detection of HCl in gastric juice). Prepare as needed a solution containing 4 g of phloroglucinol (1,3,5-benzenetriol) and 2 g of vanillin in 100 mL of absolute ethanol.

Hager's reagent. See Picric acid.

Hanus solution (for iodine number). Dissolve 13.2 g of resublimed iodine in 1 L of glacial acetic acid which will pass the dichromate test for reducible matter. Add sufficient bromine to double the halogen content, determined by titration (3 mL is about the proper amount). The iodine may be dissolved by the aid of heat, but the solution should be cold when the bromine is added.

Iodine, tincture of. To 50 mL of water add 70 g of I_2 and 50 g of KI. Dilute to 1 L with alcohol.

Iodo-potassium iodide (Wagner's reagent for alkaloids). Dissolve 2 g of iodine and 6 g of KI in 100 mL of water.

Litmus (indicator). Extract litmus powder three times with boiling alcohol, each treatment consuming an hour. Reject the alcoholic extract. Treat residue with an equal weight of cold water and filter; then exhaust with five times its weight of boiling water, cool and filter. Combine the aqueous extracts.

Magnesia mixture (reagent for phosphates and arsenates). Dissolve 55 g of magnesium chloride and 105 g of ammonium chloride in water, barely acidify with hydrochloric acid, and dilute to 1 L. The ammonium hydroxide may be omitted until just previous to use. The reagent, if completely mixed and stored for any period of time, becomes turbid

Magnesium uranyl acetate. Dissolve 100 g of $UO_2(C_2H_3O_2)_2 \cdot 2H_2O$ in 60 mL of glacial acetic acid and dilute to 500 mL. Dissolve 330 g of $Mg(C_2H_3O_2)_2 \cdot 4H_2O$ in 60 mL of glacial acetic acid and dilute to 200 mL. Heat solutions to the boiling point until clear, pour the magnesium solution into the uranyl solution, cool and dilute to 1 L. Let stand over night and filter if necessary.

Marme's reagent. See Potassium-cadmium iodide.

Marquis' reagent. See Formaldehyde-sulfuric acid.

Mayer's reagent (white precipitate with most alkaloids in slightly acid solutions). Dissolve 1.358 g of $HgCl_2$ in 60 mL of water and pour into a solution of 5 g of KI in 10 mL of H_2O. Add sufficient water to make 100 mL.

Methyl orange indicator. Dissolve 1 g of methyl orange in 1 L of water. Filter, if necessary.

Methyl orange, modified. Dissolve 2 g of methyl orange and 2.8 g of xylene cyanole FF in 1 L of 50% alcohol.

Methyl red indicator. Dissolve 1 g of methyl red in 600 mL of alcohol and dilute with 400 mL of water.

Methyl red, modified. Dissolve 0.50 g of methyl red and 1.25 g of xylene cyanole FF in 1 L of 90% alcohol. Or, dissolve 1.25 g of methyl red and 0.825 g of methylene blue in 1 L of 90% alcohol.

Millon's reagent (for albumins and phenols). Dissolve 1 part of mercury in 1 part of cold fuming nitric acid. Dilute with twice the volume of water and decant the clear solution after several hours.

Molisch's reagent. See 1-Naphthol.

1-Naphthol (Molisch's reagent for wool). Dissolve 15 g of 1-naphthol in 100 mL of alcohol or chloroform.

Nessler's reagent (for ammonia). Dissolve 50 g of KI in the smallest possible quantity of cold water (50 mL). Add a saturated solution of mercuric chloride (about 22 g in 350 mL of water will be needed) until an excess is indicated by the formation of a precipitate. Then add 200 mL of 5 N NaOH and dilute to 1 L. Let settle, and draw off the clear liquid.

Nickel oxide, ammoniacal (reagent for silk). Dissolve 5 g of nickel sulfate in 100 mL of water, and add sodium hydroxide solution until nickel hydroxide is completely precipitated. Wash the precipitate well and dissolve in 25 mL of concentrated ammonium hydroxide and 25 mL of water.

Nitron (detection of nitrate radical). Dissolve 10 g of nitron (1,4-diphenyl-3-(phenylamino)-1,2,4-triazolium hydroxide) in 5 mL of glacial acetic acid and 95 mL of water. The solution may be filtered with slight suction through an alumdum crucible and kept in a dark bottle.

1-Nitroso-2-naphthol. Make a saturated solution in 50% acetic acid (1 part of glacial acetic acid with 1 part of water). Does not keep well.

Nylander's solution (carbohydrates). Dissolve 20 g of bismuth subnitrate and 40 g of Rochelle salt in 1 L of 8% NaOH solution. Cool and filter.

Obermayer's reagent (for indoxyl in urine). Dissolve 4 g of $FeCl_3$ in 1 L of HCl (sp. gr. 1.19).

Oxine. Dissolve 14 g of 8-hydroxyquinoline in 30 mL of glacial acetic acid. Warm slightly, if necessary. Dilute to 1 L.

Oxygen absorbent. Dissolve 300 g of ammonium chloride in 1 L of water and add 1 L of concentrated ammonium hydroxide solution. Shake the solution thoroughly. For use as an oxygen absorbent, a bottle half full of copper turnings is filled nearly full with the NH_4Cl-NH_4OH solution and the gas passed through.

Pasteur's salt solution. To 1 L of distilled water add 2.5 g of potassium phosphate, 0.25 g of calcium phosphate, 0.25 g of magnesium sulfate, and 12.00 g of ammonium tartrate.

Pavy's solution (glucose reagent). To 120 mL of Fehling's solution, add 300 mL of NH_4OH (sp. gr. 0.88) and dilute to 1 L with water.

Phenanthroline ferrous ion indicator. Dissolve 1.485 g of 1,10-phenanthroline monohydrate in 100 mL of 0.025 M ferrous sulfate solution.

Phenolphthalein. Dissolve 1 g of phenolphthalein in 50 mL of alcohol and add 50 mL of water.

Phenolsulfonic acid (determination of nitrogen as nitrate). Dissolve 25 g of phenol in 150 mL of conc. H_2SO_4, add 75 mL of fuming H_2SO_4 (15% SO_3), stir well and heat for 2 hours at 100°C.

Phloroglucinol solution (pentosans). Make a 3% phloroglucinol (1,3,5-benzenetriol) solution in alcohol. Keep in a dark bottle.

Phosphomolybdic acid (Sonnenschein's reagent for alkaloids).

    1. Prepare ammonium phosphomolybdate and after washing with water, boil with nitric acid and expel NH_3; evaporate to dryness and dissolve in 2 M nitric acid.

    2. Dissolve ammonium molybdate in HNO_3 and treat with phosphoric acid. Filter, wash the precipitate, and boil with aqua regia until the ammonium salt is decomposed. Evaporate to dryness. The residue dissolved in 10 % HNO_3 constitutes Sonnenschein's reagent.

Phosphoric acid — sulfuric acid mixture. Dilute 150 mL of conc. H_2SO_4 and 100 mL of conc. H_3PO_4 (85%) with water to a volume of 1 L.

Phosphotungstic acid (Schcibicr's reagent for alkaloids).

    1. Dissolve 20 g of sodium tungstate and 15 g of sodium phosphate in 100 mL of water containing a little nitric acid.

    2. The reagent is a 10% solution of phosphotungstic acid in water. The phosphotungstic acid is prepared by evaporating a mixture of 10 g of sodium tungstate dissolved in 5 g of phosphoric acid (sp. gr. 1.13) and enough boiling water to effect solution. Crystals of phosphotungstic acid separate.

Picric acid (Hager's reagent for alkaloids, wool and silk). Dissolve 1 g of picric acid in 100 mL of water.

Potassium antimonate (reagent for sodium). Boil 22 g of potassium antimonate with 1 L of water until nearly all of the salt has dissolved, cool quickly, and add 35 mL of 10% potassium hydroxide. Filter after standing overnight.

Potassium-cadmium iodide (Marme's reagent for alkaloids). Add 2 g of CdI_2 to a boiling solution of 4 g of KI in 12 mL of water, and then mix with 12 mL of saturated KI solution.

Potassium hydroxide (for CO_2 absorption). Dissolve 360 g of KOH in water and dilute to 1 L.

Potassium iodide — mercuric iodide (Brucke's reagent for proteins). Dissolve 50 g of KI in 500 mL of water, and saturate with mercuric iodide (about 120 g). Dilute to 1 L.

Potassium pyrogallate (for oxygen absorption). For mixtures of gases containing less than 28% oxygen, add 100 mL of KOH solution (50 g of KOH to 100 mL of water) to 5 g of pyrogallol. For mixtures containing more than 28% oxygen the KOH solution should contain 120 g of KOH to 100 mL of water.

Pyrogallol, alkaline.

    1. Dissolve 75 g of pyrogallic acid in 75 mL of water.

    2. Dissolve 500 g of KOH in 250 mL of water. When cool, adjust until sp. gr. is 1.55.

    For use, add 270 mL of solution (2) to 30 mL of solution (1).

Rosolic acid (indicator). Dissolve 1 g of rosolic acid in 10 mL of alcohol and add 100 mL of water.

Scheibler's reagent. See Phosphotungstic acid.

Schiff's reagent. See Fuchsin-sulfurous acid.

Schweitzer's reagent. See Cupric oxide, ammoniacal.

Soap solution (reagent for hardness in water). Dissolve 100 g of dry castile soap in 1 L of 80% alcohol (5 parts alcohol to 1 part water). Allow to stand several days and dilute with 70% to 80% alcohol until 6.4 mL produces a permanent lather with 20 mL of standard calcium solution. The latter solution is made by dissolving 0.2 g of $CaCO_3$ in a small amount of dilute HCl, evaporating to dryness and making up to 1 L.

Sodium bismuthate (oxidation of manganese). Heat 20 parts of NaOH nearly to redness in an iron or nickel crucible and add slowly 10 parts of basic bismuth nitrate which has been previously dried. Add 2 parts of sodium peroxide, and pour the brownish-yellow fused mass onto an iron plate to cool. When cold, break up in a mortar, extract with water, and collect on an asbestos filter.

Sodium hydroxide (for CO_2 absorption). Dissolve 330 g of NaOH in water and dilute to 1 L.

Sodium nitroprusside (reagent for hydrogen sulfide and wool). Use a freshly prepared solution of 1 g of sodium nitroferricyanide in 10 mL of water.

Sodium oxalate (primary standard). Dissolve 30 g of the commercial salt in 1 L of water, make slightly alkaline with sodium hydroxide, and let stand until perfectly clear. Filter and evaporate the filtrate to 100 mL. Cool and filter. Pulverize the residue and wash it several times with small volumes of water. The procedure is repeated until the mother liquor is free from sulfate and is neutral to phenolphthalein.

Sodium plumbite (reagent for wool). Dissolve 5 g of sodium hydroxide in 100 mL of water. Add 5 g of litharge (PbO) and boil until dissolved.

Sodium polysulfide. Dissolve 480 g of $Na_2S \cdot 9H_2O$ in 500 mL of water, add 40 g of NaOH and 18 g of sulfur. Stir thoroughly and dilute to 1 L with water.

Sonnenschein's reagent. See Phosphomolybdic acid.

Starch solution.

1. Make a paste with 2 g of soluble starch and 0.01 g of HgI_2 with a small amount of water. Add the mixture slowly to 1 L of boiling water and boil for a few minutes. Keep in a glass stoppered bottle. If other than soluble starch is used, the solution will not clear on boiling; it should be allowed to stand and the clear liquid decanted.

2. A solution of starch which keeps indefinitely is made as follows: Mix 500 mL of saturated NaCl solution (filtered), 80 mL of glacial acetic acid, 20 mL of water and 3 g of starch. Bring slowly to a boil and boil for 2 minutes.

3. Make a paste with 1 g of soluble starch and 5 mg of HgI_2, using as little cold water as possible. Then pour about 200 mL of boiling water on the paste and stir immediately. This will give a clear solution if the paste is prepared correctly and the water actually boiling. Cool and add 4 g of KI. Starch solution decomposes on standing due to bacterial action, but this solution will keep well if stored under a layer of toluene.

Stoke's reagent. Dissolve 30 g of $FeSO_4$ and 20 g of tartaric acid in water and dilute to 1 L. Just before using, add concentrated NH_4OH until the precipitate first formed is redissolved.

Sulfanilic acid (reagent for nitrites). Dissolve 0.5 g of sulfanilic acid in a mixture of 15 mL of glacial acetic acid and 135 mL of recently boiled water.

Sulfomolybdic acid (Froehde's reagent for alkaloids and glucosides). Dissolve 10 g of molybdic acid or sodium molybdate in 100 mL of conc. H_2SO_4.

Tannic acid (reagent for albumin, alkaloids, and gelatin). Dissolve 10 g of tannic acid in 10 mL of alcohol and dilute with water to 100 mL.

Titration mixture. (residual chlorine in water analyasis). Prepare 1 L of dilute HCl (100 mL of HCl (sp. gr. 1.19) in sufficient water to make 1 L). Dissolve 1 g of o-tolidine in 100 mL of the dilute HCl and dilute to 1 L with dilute HCl solution.

Trinitrophenol solution. See Picric acid.

Turmeric tincture (reagent for borates). Digest ground turmeric root with several quantities of water which are discarded. Dry the residue and digest it several days with six times its weight of alcohol. Filter.

Uffelmann's reagent (turns yellow in presence of lactic acid). To a 2% solution of pure phenol in water, add a water solution of $FeCl_3$ until the phenol solution becomes violet in color.

Wagner's reagent. See Iodo-potassium iodide.

Wagner's solution (used in phosphate rock analysis to prevent precipitation of iron and aluminum). Dissolve 25 g of citric acid and 1 g of salicylic acid in water and dilute to 1 L. Use 50 mL of the reagent.

Wij's iodine monochloride solution (for iodine number). Dissolve 13 g of resublimed iodine in 1 L of glacial acetic acid which will pass the dichromate test for reducible matter. Set aside 25 mL of this solution. Pass into the remainder of the solution dry chlorine gas (dried and washed by passing through H_2SO_4 (sp. gr. 1.84)) until the characteristic color of free iodine has been discharged. Now add the iodine solution which was reserved, until all free chlorine has been destroyed. A slight excess of iodine does little or no harm, but an excess of chlorine must be avoided. Preserve in well stoppered, amber colored bottles. Avoid use of solutions which have been prepared for more than 30 days.

Wij's special solution (for iodine number). To 200 mL of glacial acetic acid that will pass the dichromate test for reducible matter, add 12 g of dichloramine T (N,N-dichloro-4-methyl-benzenesulfonamide), and 16.6 g of dry KI (in small quantities with continual shaking until all the KI has dissolved). Make up to 1 L with the same quality of acetic acid used above and preserve in a dark colored bottle.

Zimmermann-Reinhardt reagent (determination of iron). Dissolve 70 g of $MnSO_4 \cdot 4H_2O$ in 500 mL of water, add 125 mL of conc. H_2SO_4 and 125 mL of 85% H_3PO_4, and dilute to 1 L.

Zinc chloride solution, basic (reagent for silk). Dissolve 1000 g of zinc chloride in 850 mL of water, and add 40 g of zinc oxide. Heat until solution is complete.

Zinc uranyl acetate (reagent for sodium). Dissolve 10 g of $UO_2(C_2H_3O_2)_2 \cdot 2H_2O$ in 6 g of 30% acetic acid with heat, if necessary, and dilute to 50 mL. Dissolve 30 g of $Zn(C_2H_3O_2)_2 \cdot H_2O$ in 3 g of 30% acetic acid and dilute to 50 mL. Mix the two solutions, add 50 mg of NaCl, allow to stand overnight and filter.

STANDARD SOLUTIONS OF ACIDS, BASES, AND SALTS

For each compound listed, the last column of this table gives the mass in grams which is contained in 1 liter of a solution whose amount-of-substance concentration divided by the equivalence factor of the compound equals 0.1 mol/L. In the older literature such a solution is often referred to as a "decinormal solution" (0.1 N).

REFERENCE

Compendium of Analytical Nomenclature (IUPAC), Pergamon Press, Oxford, 1978.

Name	Formula	Atomic or molecular weight	Equivalence factor	Mass in grams
Acetic acid	$HC_2H_3O_2$	60.0530	1	6.0053
Ammonia	NH_3	17.0306	1	1.7031
Ammonium ion	NH_4^+	18.0386	1	1.8039
Ammonium chloride	NH_4Cl	53.4916	1	5.3492
Ammonium sulfate	$(NH_4)_2SO_4$	132.1388	1/2	6.6069
Ammonium thiocyanate	NH_4CNS	76.1204	1	7.6120
Barium	Ba	137.34	1/2	6.867
Barium carbonate	$BaCO_3$	197.3494	1/2	9.8675
Barium chloride hydrate	$BaCl_2 \cdot 2H_2O$	244.2767	1/2	12.2138
Barium hydroxide	$Ba(OH)_2$	171.3547	1/2	8.5677
Barium oxide	BaO	153.3394	1/2	7.6670
Bromine	Be	79.909	1	7.9909
Calcium	Ca	40.08	1/2	2.004
Calcium carbonate	$CaCO_3$	100.0894	1/2	5.0045
Calcium chloride	$CaCl_2$	110.9860	1/2	5.5493
Calcium chloride hydrate	$CaCl_2 \cdot 6H_2O$	219.0150	1/2	10.9508
Calcium hydroxide	$Ca(OH)_2$	74.0947	1/2	3.7047
Calcium oxide	CaO	56.0794	1/2	2.8040
Chlorine	Cl	35.453	1	3.5453
Citric acid	$C_6H_8O_7 \cdot H_2O$	210.1418	1/3	7.0047
Cobalt	Co	58.9332	1/2	2.9466
Copper	Cu	63.54	1/2	3.177
Copper oxide (cupric)	CuO	79.5394	1/2	3.9770
Copper sulfate hydrate	$CuSO_4 \cdot 5H_2O$	249.6783	1/2	12.4839
Hydrochloric acid	HCl	36.4610	1	3.6461
Hydrocyanic acid	HCN	27.0258	1	2.7026
Iodine	I	126.9044	1	12.6904
Lactic acid	$C_3H_6O_3$	90.0795	1	9.0080
Malic acid	$C_4H_6O_5$	134.0894	1/2	6.7045
Magnesium	Mg	24.312	1/2	1.2156
Magnesium carbonate	$MgCO_3$	84.3214	1/2	4.2161
Magnesium chloride	$MgCl_2$	95.2180	1/2	4.7609
Magnesium chloride hydrate	$MgCl_2 \cdot 6H_2O$	203.2370	1/2	10.1623
Magnesium oxide	MgO	40.3114	1/2	2.0156
Manganese	Mn	54.938	1/2	2.7469
Manganese sulfate	$MnSO_4$	150.9996	1/2	7.5500
Mercuric chloride	$HgCl_2$	271.4960	1/2	13.5748
Nickel	Ni	58.71	1/2	2.9356
Nitric acid	HNO_3	63.0129	1	6.3013
Oxalic acid	$H_2C_2O_4$	90.0358	1/2	4.5018
Oxalic acid hydrate	$H_2C_2O_4 \cdot 2H_2O$	126.0665	1/2	6.3033
Oxalic acid anhydride	C_2O_3	72.0205	1/2	3.6010
Phosphoric acid	H_3PO_4	97.9953	1/3	3.2665
Potassium	K	39.102	1	3.9102
Potassium bicarbonate	$KHCO_3$	100.1193	1	10.0119
Potassium carbonate	K_2CO_3	138.2134	1/2	6.9106
Potassium chloride	KCl	74.5550	1	7.4555

Name	Formula	Atomic or molecular weight	Equivalence factor	Mass in grams
Potassium cyanide	KCN	65.1199	1	6.5120
Potassium hydroxide	KOH	56.1094	1	5.6109
Potassium oxide	K_2O	94.2034	1/2	4.7102
Potassium tartrate	$K_2H_4C_4O_6$	226.2769	1/2	11.3139
Silver	Ag	107.87	1	10.787
Silver nitrate	$AgNO_3$	169.8749	1	16.9875
Sodium	Na	22.9898	1	2.2990
Sodium bicarbonate	$NaHCO_3$	84.0071	1	8.4007
Sodium carbonate	Na_2CO_3	105.9890	1/2	5.2995
Sodium chloride	NaCl	58.4428	1	5.8443
Sodium hydroxide	NaOH	39.9972	1	3.9997
Sodium oxide	Na_2O	61.9790	1/2	3.0990
Sodium sulfide	Na_2S	78.0436	1/2	3.9022
Succinic acid	$H_2C_4H_4O_4$	118.0900	1/2	5.9045
Sulfuric acid	H_2SO_4	98.0775	1/2	4.9039
Tartaric acid	$C_4H_6O_6$	150.0888	1/2	7.5044
Zinc	Zn	65.37	1/2	3.269
Zinc sulfate hydrate	$ZnSO_4 \cdot 7H_2O$	287.5390	1/2	14.3769

STANDARD SOLUTIONS OF OXIDATION AND REDUCTION REAGENTS

For each reagent listed, the last column of this table gives the mass in grams which is contained in a solution whose amount-of-substance concentration divided by the equivalence factor of the compound equals 0.1 mol/L. The equivalence factor given refers to the most common reactions of the reagent. In the older literature such a solution is often called a "decinormal solution" (0.1 N).

REFERENCE

Compendium of Analytical Nomenclature (IUPAC), Pergamon Press, Oxford, 1978.

Name	Formula	Atomic or molecular weight	Equivalence factor	Mass in grams
Antimony	Sb	121.75	1/2	6.0875
Arsenic	As	74.9216	1/2	3.7461
Arsenic trisulfide	As_2S_3	246.0352	1/4	6.1509
Arsenous oxide	As_2O_3	197.8414	1/4	4.9460
Barium peroxide	BaO_2	169.3388	1/2	8.4669
Barium peroxide hydrate	$BaO_2 \cdot 8H_2O$	313.4615	1/2	15.6730
Calcium	Ca	40.08	1/2	2.004
Calcium carbonate	$CaCO_3$	100.0894	1/2	5.0045
Calcium hypochlorite	$Ca(OCl)_2$	142.9848	1/4	3.5746
Calcium oxide	CaO	56.0794	1/2	2.8040
Chlorine	Cl	35.453	1	3.5453
Chromium trioxide	CrO_3	99.9942	1/3	3.3331
Ferrous ammonium sulfate	$FeSO_4(NH_4)SO_4 \cdot 6H_2O$	392.0764	1	39.2076
Hydroferrocyanic acid	$H_4Fe(CN)_6$	215.9860	1	21.5986
Hydrogen peroxide	H_2O_2	34.0147	1/2	1.7007
Hydrogen sulfide	H_2S	34.0799	1/2	1.7040
Iodine	I	126.9044	1	12.6904
Iron	Fe	55.847	1	5.5847
Iron oxide (ferrous)	FeO	71.8464	1	7.1846
Iron oxide (ferric)	Fe_2O_3	159.6922	1/2	7.9846
Lead peroxide	PbO_2	239.1888	1/2	11.9594
Manganese dioxide	MnO_2	86.9368	1/2	4.3468
Nitric acid	HNO_3	63.0129	1/3	2.1004
Nitrogen trioxide	N_2O_3	76.0116	1/4	1.9002
Nitrogen pentoxide	N_2O_5	108.0104	1/6	1.8001
Oxalic acid	$C_2H_2O_4$	90.0358	1/2	4.5018
Oxalic acid hydrate	$C_2H_2O_4 \cdot 2H_2O$	126.0665	1/2	6.3033
Oxygen	O	15.9994	1/2	0.8000
Potassium dichromate	$K_2Cr_2O_7$	294.1918	1/6	4.9032
Potassium chlorate	$KClO_3$	122.5532	1/6	2.0425
Potassium chromate	K_2CrO_4	194.1076	1/3	6.4733
Potassium ferrocyanide	$K_4Fe(CN)_6$	368.3621	1	36.8362
Potassium ferrocyanide hydrate	$K_4Fe(CN)_6 \cdot 3H_2O$	422.4081	1	42.2408
Potassium iodide	KI	166.0064	1	16.6006
Potassium nitrate	KNO_3	101.1069	1/3	3.3702
Potassium perchlorate	$KClO_4$	138.5526	1/8	1.7319
Potassium permanganate	$KMnO_4$	158.0376	1/5	3.1608
Sodium chlorate	$NaClO_3$	106.4410	1/6	1.7740
Sodium nitrate	$NaNO_3$	84.9947	1/3	2.8332
Sodium thiosulfate hydrate	$Na_2S_2O_3 \cdot 5H_2O$	248.1825	1	24.8183
Stannous chloride	$SnCl_2$	189.5960	1/2	9.4798
Stannous oxide	SnO	134.6894	1/2	6.7345
Sulfur dioxide	SO_2	64.0628	1/2	3.2031
Tin	Sn	118.69	1/2	5.935

ORGANIC ANALYTICAL REAGENTS FOR THE DETERMINATION OF INORGANIC SUBSTANCES
G. Ackermann, L. Sommer, and D. Thorburn Burns

Determination	Reagents	Ref.
Aluminium	Alizarin Red S	Onishi, Part II a, p 28. (5), Snell, *Metals I,* p 587. (7)
	Aluminon	Fries/Getrost, p 16. (2), Onishi, IIa, p 21. (5), Snell, *Metals I,* p 590. (7)
	Aluminon + Cetyltrimethylammonium bromide	Huaxue Shiji, *8*, 85, (1986)
	Chrome Azurol S	Onishi, Part IIa, p 26. (5), Snell, *Metals I,* p 605. (7)
	Chrome Azurol S + Cetyltrimethylammonium bromide	Marczenko, p 133 (3), Snell, *Metals I,* p 606. (7)
	Chromazol KS + Cetylpyridinium bromide	*Analyst, 107,* 428, (1982).
	Eriochrome Cyanine R	Fries/Getrost, p 19 (2), Onishi, Part IIa p 25. (5), Snell, *Metals I,* p 611. (7)
	Eriochrome Cyanine R + Cetyltrimethylammonium bromide	Snell, *Metals I,* p 613. (7), *Analyst, 107,* 1431, (1982).
	8-Hydroxyquinoline	Fries/Getrost, p 22 (2), Marczenko, p 131 (3), Onishi, Part IIa, p 31. (5), Snell, *Metals I , p* 622 (7)
Ammonia	Phenol + Sodium hypochlorite	Boltz, p 210 (1), Marczenko, p 413 (3), Snell, *Nonmetals,* p 604 (9)
Antimony	Brilliant Green	Onishi, Part IIa, p 102. (5), Snell, *Metals I,* p 384. (7)
	Bromopyrogallol Red	*Talanta, 13,* 507, (1966).
	Rhodamine B	Fries/Getrost, p 32, (2), Marczenko, p 141. (3), Onishi, Part IIa, p 93. (5), Snell, *Metals I,* p 404. (7)
	Silver diethyldithiocarbamate	Fries/Getrost, p 36. (2)
Arsenic	Silver diethyldithiocarbamate	Fries/Getrost, p 41. (2), Marczenko, p 153. (3), Onishi, Part IIa, p 153. (5), Snell, *Metals I,* p 370. (7)
Barium	Sulfonazo III	Fries/Getrost, p 46. (2), Snell, *Metals II,* p 1782. (8), Onishi, Part IIa, p 202. (5)
Beryllium	Beryllon II	Snell, *Metals I,* p 667. (7)
	Chrome Azurol S	Marczenko, p 163. (3), Snell, *Metals I,* p 672. (7)
	Chrome Azurol S + Cetyltrimethylammonium bromide	Marczenko, p 164. (3), Snell, *Metals I,* p 673. (7)
	Eriochrome Cyanine R	Snell, *Metals I,* p 675. (7), *Talanta, 31,* 249, (1984).
	Eriochrome Cyanine R + Cetyltrimethylammonium bromide	Zh, *Anal. Khim., 33,* 1298, (1978).
Bismuth	Dithizone	Onishi, Part IIa, p 262. (5), Snell, *Metals I,* p 303. (7)
	Pyrocatechol Violet	Fres. *Z. Anal. Chem., 186,* 418, (1962).
	Pyrocatechol Violet + Cetyltrimethylammonium bromide	Zh. *Anal. Khim., 38,* 216, (1983).
	Thiourea	Onishi, Part IIa, p 260. (5), Snell, *Metals I,* p 317. (7)
	Xylenol Orange	Friez/Getrost, p 57. (2), Marczenko, p 172. (3), Snell, *Metals I,* p 320. (7)
Boron	Azomethine H	Snell, Nonmetals, p 165. (9)
	Carminic acid	Boltz, p 14. (1), Fries/Getrost, p 65. (2), Snell, Nonmetals, p 170. (9), Williams, p 35. (11)
	Curcumin	Boltz, p 8. (1), Fries/Getrost, p 68. (2), Marczenko, p 180. (3), Snell, Nonmetals, p 180. (9), Fres. *Z. Anal. Chem., 323,* 266, (1986).

Determination	Reagents	Ref.
	Methylene Blue	Boltz, p 21. (1), Marczenko, p 183. (3), Snell, Nonmetals, p 205. (9), *Talanta, 31*, 547, (1984).
Bromide	Fluorescein	Boltz, p 48. (1), Snell, Nonmetals, p 276., Fres. Z. *Anal. Chem., 301*, 28 (1980).
	Phenol Red	Boltz, p 44. (1), Marczenko, p 190. (3), Snell, Nonmetals, p 28. (9)
Cadmium	2-(5-Bromo-2-pyridylazo)-5-diethylaminophenol	Marczenko, p 197. (3)
	Cadion	Onishi, Part IIa, p 323. (5)
	Dithizone	Fries/Getrost, p 78. (2), Onishi, Part IIa, p 315. (5), Snell, *Metals I*, p 279. (7), West, p 25. (10).
	4–(2-Pyridylazo)resorcinol	Fres. Z. *Anal. Chem., 310*, 51, (1982).
Calcium	Chlorophosphonazo III	Marczenko, p 207. (3), Snell, *Metals II*, p 1744. (8)
	Glyoxal-bis(2-hydroxyanil)	Fries/Getrost, p 86. (2), Onishi, Part IIa, p 352. (5), Snell, *Metals I*, p 1762. (8)
	Murexide	Onishi, Part IIa, p 357. (5), Snell, *Metals II*, p 1769. (8)
	Phthalein Purple	*Anal. Chim. Acta, 34*, 71 (1966).
Cerium	*N*-benzoyl-*N*-phenylhydroxylamine	*Anal. Chim. Acta, 48*, 155, (1969).
	8-Hydroxyquinoline	Fries/Getrost, p 93. (2), Marczenko, p 220. (3), Onishi, Part IIa, p 383. (7)
Chlorine	*N,N*-Diethyl-1,4-phenylenediamine	Boltz, p 92. (1), Fries/Getrost, p 101. (2), Snell, Nonmetals, p 225. (9), *Analyst, 90*, 187, (1965).
Chromium	1,5-Diphenylcarbazide	Fries/Getrost, p 105. (2), Onishi, Part IIa, p 412. (5), Snell, *Metals I*, p 714. (7), West, p 12. (10)
	4-(2-Pyridylazo)resorcinol	Snell, *Metals I*, p 736. (7), West, p 17. (10)
	4-(2-Pyridylazo)resorcinol + Tetradecyldimethyl-benzylammonium chloride	West, p 17. (10), *Anal. Chim. Acta, 67*, 297, (1973).
	4-(2-Pyridylazo)resorcinol + Hydrogen peroxide	Fres. Z. *Anal. Chem., 304*, 382, (1980).
Cobalt	Nitroso-R salt	Fries/Getrost, p 118. (2), Onishi, Part IIa, p 454. (5), Snell, *Metals I*, p 953. (7)
	1-Nitroso-2-naphthol	Fries/Getrost, p 111. (2), Marczenko, p 246. (3), Snell, *Metals I*, p 947. (5)
	2-Nitroso-1-naphthol	Fries/Getrost, p 113. (2), Onishi, Part IIa, p 459. (5), Snell, *Metals I*, p 949. (7), West, p 45. (10)
	4-(2-Pyridylazo)resorcinol	Snell, *Metals I*, p 969. (7), West, p 44. (10)
	4-(2-Pyridylazo)resorcinol + Diphenylguanidine	Zh. *Anal. Khim., 35*, 1306, (1980).
Copper	Bathocuproine	Fries/Getrost, p 135. (2), Snell, *Metals I*, p 148. (7)
	Bathocuproine disulfonic acid	Fries/Getrost, p 137. (2), West, p 52. (10)
	Dithizone	Marczenko, p 258. (3), Onishi, Part IIa, p 529. (5), Snell, *Metals I*, p 199. (7)
	Neocuproine	Snell, *Metals I*, p 217. (5), West, p 51. (10)
	Cuprizone	Onishi, Part IIa, p 534. (5), Snell, *Metals I*, p 157. (7), West, p 53. (10)
	4-(2-pyridylazo)resorcinol + Tetradecyldimethyl-benzylammonium chloride	*Anal. Chim. Acta, 138*, 321, (1982).
Cyanide	Barbituric Acid + Pyridine	Fries/Getrost, p 153. (2), Snell, Nonmetals, p 653. (9)
	Barbituric Acid + Pyridine-4-carboxylic acid	*Anal. Chim. Acta, 99*, 197, (1978).

Determination	Reagents	Ref.
Fluoride	Alizarin Fluorine blue + Lanthanum(III) ion	Boltz, p 129. (1), Fries/Getrost, p 158. (2), Snell, *Nonmetals,* p 333. (9), Williams, p 354. (11)
	Eriochrome Cyanine R + Zirconium(IV) ion	Boltz, p 119. (1), Snell, *Nonmetals,* p 359. (2), Williams, p 357. (10)
Gallium	Pyrocatechol violet + Diphenylguanidine	Snell, *Metals I,* p 500. (7)
	8-Hydroxyquinoline	Onishi Pt IIa, p 582. (5), Snell, *Metals I,* p 505. (7)
	1-(2-Pyridylazo)-2-naphthol	Snell, *Metals I,* p 512. (7)
	4-(2-Pyridylazo)resorcinol	Snell, *Metals I,* p 513. (7)
	Rhodamine B	Marczenko, p 284. (3), Onishi, Part IIa, p 578. (5), Snell, *Metals I,* p 515. (7)
	Xylenol Orange	Fries/Getrost, p 166. (2), Snell, *Metals I,* p 523. (7)
	Xylenol Orange + 8-Hydroxyquinoline	*Zh. Anal. Khim.,* 26, 75, (1971).
Germanium	Brilliant Green + Molybdate	Snell, *Metals I,* p 562. (7)
	Phenylfluorone	Fries/Getrost, p 168. (2), Marczenko, p 292. (3), Onishi, Part IIa, p 607. (5), Snell, *Metals I,* p 570. (7)
Gold	5-(4-Diethylaminobenzylidene) rhodanine	Fries/Getrost, p 173. (2), Onishi, Part IIa, p 631. (5), Snell, *Metals II,* p 1516. (8)
	Rhodamine B	Fries/Getrost, p 175. (2), Marczenko, p 301. (3), Onishi, Part IIa, p 637. (5), Snell, *Metals II,* p 513. (8)
Hafnium	Arsenazo III	Snell, *Metals II,* p 1184. (8), *Talanta,* 19, 807, (1972).
Indium	Bromopyrogallol Red	Snell, *Metals I,* p 469. (7)
	Chrome Azurol S	Snell, *Metals I,* p 474. (7)
	Chrome Azurol S + Cetyltrimethylammonium bromide	*Anal. Chim. Acta,* 67, 107, (1973).
	Dithizone	Fries/Getrost, p 179. (2), Onishi, Part IIa, p 672. (5), Snell, *Metals I,* p 474. (7)
	8-Hydroxyquinoline	Onishi, Part IIa, p 670. (5), Snell, *Metals I,* p 475. (7)
	1-(2-Pyridylazo)-2-naphthol	Snell, *Metals I,* p 480. (7)
	4-(2-Pyridylazo)resorcinol	Marczenko, p 309. (3), Snell, Metals I, p 480. (7)
Iodide	Neocuproine + Copper(II)	*Anal. Chim. Acta,* 69, 321, (1974).
Iodine	Starch	Boltz, p 162. (1), Marczenko, p 316. (3), Snell, *Nonmetals,* p 307. (9)
Iridium	Rhodamine 6G + Tin(II)	Marczenko, p 323. (3)
	N,N-Dimethyl-4-nitrosoaniline	*Anal. Chem.,* 27, 1776, (1955).
Iron	Bathophenanthroline	Fries/Getrost, p 189. (2), Onishi, Part IIa, p 729. (5), Snell, *Metals I,* p 763. (7)
	Bathophenanthroline disulfonic acid	Fries/Getrost, p 191. (2), Snell, *Metals I,* p 772. (7)
	2,2′-Bipyridyl	Snell, *Metals I,* p 750. (7)
	Chrome Azurol S + Cetyltrimethylammonium bromide	Snell, *Metals I,* p 757. (7), *Coll. Czech. Chem. Comm., 45,* 2656, (1980).
	1,10-Phenanthroline	Fries/Getrost, p 199. (2), Marczenko, p 331. (3), Onishi, Part IIa, p 725. (5), Snell, *Metals I,* p 795. (7)
	1,10-Phenanthroline + Bromothymol Blue	*Zh. Anal. Khim.,* 25, 1348, (1970).
	Ferrozine	Onishi, Part IIa, p 730. (5), Snell, *Metals I,* p 783. (7)
Lanthanum	Arsenazo III	Marczenko, p 468. (3), Snell, *Metals II,* p 1910. (8)

Determination	Reagents	Ref.
Lead	Dithizone	Fries/Getrost, p 207. (2), Onishi, Part IIa, p 824. (5), Snell, *Metals I,* p 2. (7), West, p 34. (10)
	Sodium diethyldithiocarbamate	Fries/Getrost, p 214. (2), Snell, *Metals I,* p 27. (7)
	4-(2-Pyridylazo)resorcinol	Fries/Getrost, p 220. (2), Marczenko, p 347. (3), Snell, *Metals I,* p 34. (7)
Lithium	Thoron	Onishi, Part IIa, p 863. (5), Snell, *Metals II,* p 1726. (8), T*alanta, 30,* 587, (1983).
Magnesium	Eriochrome Black T	Fries/Getrost, p 226. (2), Marczenko, p 355. (3), Onishi, Part IIb, p 13. (6), Snell, *Metals II,* p 1932. (8)
	8-Hydroxyquinoline	Onishi, Part IIb, p 11. (6), Snell, *Metals II,* p 1938. (8)
	8-Hydroxyquinoline + Butylamine	Fries/Getrost, p 228. (2), Snell, *Metals II,* p 1938. (8)
	Titan Yellow	Fries/Getrost, p 234. (2), Marczenko, p 352. (3), Snell, *Metals II,* p 1945. (8)
	Xylidyl Blue	Fries/Getrost, p 231. (2), Onishi, Part IIb, p 14. (6), Snell, *Metals II,* p 1950. (8)
Manganese	Formaldoxime	Fries/Getrost, p 236. (2), Marczenko, p 364. (3), Onishi Part IIb, p 38. (6), Snell, *Metals II,* 1010. (8)
Mercury	Dithizone	Fries/Getrost, p 243. (2), Marczenko, p 373. (3), Onishi, Part IIb, p 66. (6), Snell, *Metals I,* p 107. (7), West, p 29. (10)
	Michler's thioketone	Marczenko, p 375. (3), Snell, *Metals I,* p 126. (7)
	Xylenol Orange	*Talanta, 16,* 1023, (1969)
Molybdenum	Bromopyrogallol Red + Cetylpyridium chloride	West, p 58. (10)
	Phenylfluorone	Snell, *Metals II,* p 1311., *Microchem. J.,* *31,* 56, (1985).
	Toluene-3,4-dithiol	Fries/Getrost, p 251. (2), Marczenko, p 384. (3), Onishi, Part IIb, p 96. (6), Snell, *Metals II,* p 1301. (8)
Nickel	2-(5-Bromo-2-pyridylazo)-5-diethylaminophenol	Marczenko, p 397. (3), *Talanta 28,* 189, (1981).
	Dimethylglyoxime	Fries/Getrost, p 263. (2), Marczenko, p 393. (3), Onishi, Part IIb, p 125. (6), Snell, *Metals I,* p 887. (7)
	Dimethylglyoxime + Oxidant	Fries/Getrost, p 263. (2), Onishi, Part IIb, p 125. (6), Snell, *Metals I,* p 887. (7)
	2,2'-Furildioxime	Marczenko, p 396. (3), Snell, *Metals I,* p 904. (7)
	2-(2-Pyridylazo)-2-naphthol	Snell, *Metals I,* p 910. (7)
	4-(2-Pyridylazo)resorcinol	Snell, *Metals I,* p 911. (7), West, p 39. (10), *Anal. Chim. Acta, 82,* 431, (1976).
Niobium	N-Benzoyl-N-phenylhydroxylamine	Snell, *Metals II,* p 1425. (8)
	Pyrocatechol + EDTA or 2,2'Bipyridyl or 1-(2-thenoyl)-3,3,3,-trifluoroacetone	Snell, *Metals II,* p 1427. (8)
	Bromopyrogallol red	Marczenko, p 407. (3), Snell, *Metals II,* p 1426. (8)
	Bromopyrogallol red + Cetylpyridinium chloride	*Talanta, 32,* 189, (1985).
	4-(2-Pyridylazo)resorcinol	Fries/Getrost, p 274. (2), Marczenko, p 406. (3), Onishi, Part IIb, p 160. (7), Snell, *Metals II,* p 1447. (8)

Determination	Reagents	Ref.
	Sulfochlorophenol S	Onishi, Part IIb, p 161. (7), Snell, *Metals II,* p 1430. (8)
	Xylenol Orange	Onishi, Part IIb, p 164. (7)
Nitrate	Brucine	Boltz, p 227. (1), Fries/Getrost, p 280. (2), Snell, *Nonmetals,* p 546. (9)
	Chromotropic acid	Boltz, p 229. (1), Fries/Getrost, p 281. (2), Snell, *Nonmetals,* p 548. (9), Williams, p 132. (11), *Fres. Z. Anal. Chem., 320,* 490, (1985).
	Sulfanilamide + *N*-(1-Naphthyl)ethylenediamine dihydrochloride	Fries/Getrost, p 279. (2), Snell, *Nonmetals,* p 559. (9)
Nitrite	Sulfanilamide + *N*-(1-Naphthyl)ethylenediamine dihydrochlorine	Boltz, p 241. (1), Snell, N*onmetals, p* 585. (8), A*nalyst, 109,* 1281, (1984).
	Sulfanilic acid + 1-Naphthylamine	Boltz, p 237. (1), Fries/Getrost, p 285. (2), Marczenko, p 419. (3), Snell, *Nonmetals,* p 586. (9)
Osmium	1,5-Diphenylcarbazide	Marczenko, p 428. (3)
Palladium	2-(5 Bromo-2-pyridylazo)-5-diethylaminophenol	*Talanta, 33,* 939, (1986).
	Dithizone	Marczenko, p 440. (3), Onishi, Part IIb, p 227. (6), Snell, *Metals II,* p 1577. (8)
	2-Nitroso-1-naphthol	Fries/Getrost, p 294. (2), Onishi, Part IIb, p 226. (6), Snell, *Metals II,* p 1581. (8)
	4-(2-Pyridylazo)resorcinol	Snell, *Metals II,* p 1583. (8) *Analyst, 107,* 708, (1982).
Phosphate	Rhodamine B + Molybdate	Snell, *Nonmetals,* p 103. (9)
	Malachite Green + Molybdate	Snell, *Nonmetals,* p 12. (9), Analyst, *108,* 361, (1983).
Platinum	Sulfochlorophenolazorhodamine	Onishi, Part IIb, p 253. (6), *Talanta, 34,* 87, (1987).
	Dithizone	Fries/Getrost, p 300. (2), Onishi, Part IIb, p 253. (6), Snell, *Metals II,* p 1534. (8)
	2-Mercaptobenzothiazole	Fries/Getrost, p 302. (2), *Zh. Anal. Khim., 24,* 1172, (1969).
Rare Earths	Arsenazo I	Marczenko, p 470. (3), Onishi, Part IIa, p 785. (5), Snell, *Metals II,* p 1857. (8)
	Arsenazo III	Fries/Getrost, p 309. (2), Marczenko, p 468. (3), Onishi, Part IIa, p 786. (5), Snell, *Metals II,* p 1862. (8)
	Xylenol Orange	Onishi, Part IIa, p 787. (5), Snell, *Metals II,* p 1874. (8)
Rhenium	2,2′-Furildioxime	Fries/Getrost, p 310. (2), Marczenko, p 481. (3), Onishi, Part IIb, p 288. (6), Snell, *Metals II,* p 1659. (8)
Rhodium	1-(2-Pyridylazo)-2-naphthol	Fries/Getrost, p 311. (2), Snell, *Metals II,* p 1553. (8)
Ruthenium	1,10-Phenanthroline	Onishi, Part IIb, p 331. (6), Snell, *Metals II,* p 1623. (8)
	Thiourea	Fries/Getrost, p 318. (2), Onishi, Part IIb, 329. (6), Snell, *Metals II,* p 1626. (8)
	1,4-Diphenylthiosemicarbazide	Marczenko, p 493. (3), Onishi, Part IIb, p 330. (8)
Scandium	Alizarin red S	Fries/Getrost, p 319. (2), Onishi, Part IIb, p 360. (6), Snell, *Metals I,* p 536. (7)
	Arsenazo III	Onishi, Part IIb, p 359. (6), Snell, *Metals I,* p 539. (7)

Determination	Reagents	Ref.
	Chrome Azurol S	Snell, Metals I, p 551. (7), *Anal. Chim. Acta, 159*, 309, (1984).
	Xylenol Orange	Marczenko, p 501. (3), Onishi, Part IIb, p 357. (6), Snell, *Metals I,* p 547. (7)
Selenium	3,3′-Diaminobenzidine	Boltz, p 391. (1), Fries/Getrost, p 323. (2), Marczenko, p 508. (3), Snell, *Nonmetals,* p 490. (9), West, p 4. (10).
	2,3-Diaminonaphthaline	Snell, *Nonmetals,* p 501. (9)
Silver	Dithizone	Fries/Getrost, p 328. (2), Marczenko, p 524. (3), Onishi, Part IIb, p 379. (6), Snell, *Metals I,* p 82. (7)
Sulfate	Eosin + 1,10-Phenanthroline	Snell, *Metals I,* p 93. (7)
Sulfide	Methylthymol blue + Barium (II)	Snell, *Nonmetals,* p 457. (9)
	N,N,-Dimethyl-1,4-phenylenediamine	Boltz, p 483. (1), Fries/Getrost, p 344. (2), Snell, *Nonmetals,* p 400. (9), Williams, p 578. (11)
Sulfite	Pararosaniline + Formaldehyde	Boltz, p 478. (1), Marczenko, p 540. (3), Snell, *Nonmetals,* p 430. (9), Williams, p 591. (11)
Tantalum	Methyl Violet	Marczenko, p 551. (3), Snell, *Metals II,* p 1485. (8)
	4-(2-Pyridylazo)resorcinol	Snell, *Metals II,* p 1488. (8)
	Phenylfluorone	Onishi, Part IIb, p 166. (6), Snell, *Metals II,* p 1486. (8)
Tellurium	Diethyldithiocarbamate	Boltz, p 402. (1), Fries/Getrost, p 348. (2), Snell, *Nonmetals,* p 533. (9), Williams, p 220. (10)
	Bismuthiol II	Boltz, p 401. (1), Marczenko, p 557. (3), Snell, *Nonmetals,* p 524. (9)
Thallium	Brilliant green	Fries/Getrost, p 352. (2), Marczenko, p 567. (3), Onishi, Part IIb, p 426. (6), Snell, *Metals I,* p 45. (7)
	Dithizone	Fries/Getrost, p 355. (2), Onishi, Part IIb, p 426. (6), Snell, *Metals I,* p 54. (7)
	Rhodamine B	Fries/Getrost, p 354. (2), Marczenko, p 566. (3), Onishi, Part IIb, p 424. (6), Snell, *Metals I,* p 63. (7)
Thorium	Arsenazo III	Fries/Getrost, p 360. (2), Marczenko, p 575. (3), Onishi, Part IIb, p 460. (6), Snell, *Metals II,* p 1820. (8)
	Thoron	Marczenko, p 574. (3), Onishi, Part IIb, p 463. (6), Snell, *Metals I,* p 1835. (7)
	Xylenol Orange	Snell, *Metals I,* p 1852. (7)
Tin	Xylenol Orange + Cetyltrimethylammonium bromide	*Talanta, 26*, 499, (1979).
	Pyrocatechol violet (and + Cetyltrimethylammonium bromide)	Marczenko, p 585. (3), Onishi, Part IIb, p 501. (6), Snell, M*etals I, p* 422. (7)
	Gallein	Onishi, Part IIb, p 507, 510. (6), Snell, *Metals I,* p 432. (7)
	Phenylfluorone	Fries/Getrost, p 368. (2), Marczenko, p 582. (3), Onishi, Part IIb, p 497. (6), Snell, *Metals I,* p 444. (7)
	Toluene-3,4-dithiol + Dispersant	Fries/Getrost, p 366. (2), Onishi, Part IIb, p 502. (6), Snell, *Metals I,* p 427. (7)
Titanium	Chromotropic acid	Marczenko, p 593. (3), Onishi, Part IIb, p 551. (6), Snell, *Metals II,* p 1080. (8)

Determination	Reagents	Ref.
	Diantipyrinylmethane	Onishi, Part IIb, p 545. (6), Snell, *Metals II,* 1085. (8)
	Tiron	Fries/Getrost, p 376. (2), Onishi, Part IIb, p 549. (6), Snell, *Metals II,* p 1114. (8)
Tungsten	Pyrocatechol Violet	Snell, Metals II, p 1265. (8)
	Tetraphenylarsonium chloride + Thiocyanate	Onishi, Part IIb, p 596. (6), Snell, *Metals II,* p 1278. (8)
	Toluene-3,5-dithiol	Marczenko, p 605. (3), Onishi, Part IIb, p 590. (6), Snell, *Metals II,* p 1267. (8)
Uranium	Arsenazo III	Marczenko, p 611. (3), Onishi, Part IIb, p 627. (6), Snell, *Metals II,* p 1356. (8)
	2-(5-Bromo-2-pyridylazo)diethylaminophenol	Fries/Getrost, p 388. (2), Onishi, Part IIb, p 625. (6)
	Chlorophosphonazo III	Snell, Metals II, p 1367. (8), *Fres. Z. Anal. Chem., 306,* 110, (1981).
	1-(2-Pyridylazo)-2-naphthol	Fries/Getrost, p 386. (2), Onishi, Part IIb, p 625. (6), Snell, *Metals II,* p 1387. (8)
Vanadium	N-Benzoyl-N-phenylhydroxylamine	Fries/Getrost, p 395. (2), Marczenko, p 625. (3), Snell, *Metals II,* p 1196. (8)
	8-Hydroxyquinoline	Marczenko, p 623. (3), Snell, *Metals II,* 1209. (8)
	4-(2-pyridylazo)resorcinol	Fries/Getrost, p 404. (23), Marczenko, p 628. (3), Onishi, Part IIb, p 625. (6), Snell, *Metals II,* p 1226. (8)
Yttrium	Alizarin Red S	Fries/Getrost, p 406. (2), Onishi, Part IIa, p 784. (5), Snell, *Metals II,* p 1919. (8)
	Arsenazo III	Marczenko, p 468. (3), Onishi, Part IIa, p 786. (5), Snell, *Metals II,* p 1921. (8)
	Xylenol Orange	Fries/Getrost, p 406. (2), Onishi, Part IIa, p 787. (5), Snell, *Metals II,* p 1923. (8)
Zinc	Dithizone	Fries/Getrost, p 408. (2), Marczenko, p 637. (3), Onishi, Part IIb, p 708. (6), Snell, *Metals II,* p 1042. (8)
	1-(2-Pyridylazo)-2-naphthol	Marczenko, p 639. (3), Onishi, Part IIb, p 719. (6), Snell, *Metals II,* p 1056. (8)
	Xylenol Orange	Fries/Getrost, p 417. (2), Snell, *Metals II,* p 1062. (8), Talanta, *26,* 693, (1979).
	Zircon	Fries/Getrost, p 412. (2), Onishi, Part IIb, p 719. (6), Snell, *Metals II,* p 1063. (8), West, p 23. (10)
Zirconium	Alizarin Red S	Fries/Getrost, p 421. (2), Marczenko, p 647. (3), Onishi, Part IIb, p 763. (6), Snell, *Metals II,* p 1136. (8)
	Arsenazo III	Fries/Getrost, p 421. (2), Onishi, Part IIb, p 770. (6), Snell, *Metals II,* p 1143. (8)
	Pyrocatechol Violet	Onishi, Part IIb, p 771. (6), Snell, *Metals II,* p 1149. (8)
	Morin	Fries/Getrost, p 424. (2), Onishi, Part IIb, p 765. (6), Snell, *Metals II,* p 1158. (8)
	Xylenol Orange	Fries/Getrost, p 419. (2), Marczenko, p 648. (3), Onishi, Part IIb, p 767. (6), Snell, *Metals II,* p 1167. (8)

ORGANIC ANALYTICAL REAGENTS FOR THE DETERMINATION OF
INORGANIC SUBSTANCES (continued)

REVIEWS

Sommer, L, Ackermann, G., Thorburn Burns, D., and Savvin, S. B., *Pure and Applied Chem.*, 62, 2147, 1990.
Sommer, L., Ackermann, G., and Thorburn Burns, D., *Pure and Applied Chem.*, 62, 2323, 1990)
Sommer, L., Komarek, J., and Thorburn Burns, D., *Pure and Applied Chem.*, 64, 213, 1992.
Savvin, S. B., *Crit. Rev. Anal. Chem.*, 8, 55, 1979.

MONOGRAPHS

1. Boltz, D. F., and Howell, J. A., *Colorimetric Determination of Nonmetals*, 2nd ed, Wiley, New York, 1978.
2. Fries, J. and Getrost, H., *Organic Reagents for Trace Analysis*, E Merck, Darmstadt, 1977.
3. Marczenko, Z., *Separation and Spectrophotometric Determination of Elements*, Ellis Horwood, Chichester, 1986.
4. Sandell, E. B. and Onishi, H., *Photometric Determination of Traces of Metals. General Aspects, Part I*, 4th ed, J. Wiley, New York, 1978.
5. Onishi, H., *Photometric Determination of Traces of Metals. Part IIa: Individual Metals, Aluminium to Lithium*, 4th ed, J Wiley, New York, 1986.
6. Onishi, H., *Photometric Determination of Traces of Metals. Part IIb: Individual Metals, Magnesium to Zinc*, 4th ed, J. Wiley, New York, 1989.
7. Snell, F. D., *Photometric and Fluorimetric Methods of Analysis. Metals Part 1*, J. Wiley, New York, 1978.
8. Snell, F. D., *Photometric and Fluorimetric Methods of Analysis, Metals Part 2*, J. Wiley, New York, 1978.
9. Snell, F. D., *Photometric and Fluorimetric Methods of Analysis, Nonmetals*, J. Wiley, New York, 1981.
10. West, T. S. and Nürnberg, H. W., Eds., *The Determination of Trace Metals in Natural Waters*, Blackwell, Oxford, 1988.
11. Williams, W. J.,, *Handbook of Anion Determination*, Butterworth, London, 1979.
12. Townshend, A., Burns, D. T., Guilbault, G. G., Lobinski, R., Mavenzenko, Z., Newman, E., and Onishi, H., *Dictionary of Analytical Reagents*, Chapman & Hall, London, 1993.

ACID-BASE INDICATORS

A. K. Covington

Indicator	pK (20°C)	Approx. pH range	Color change	Preparation
Methyl Violet	—	0.0—1.6	Y-B	0.01—0.05% in water
Crystal Violet	—	0.0—1.8	Y-B	0.02% in water
Ethyl Violet	—	0.0—2.4	Y-B	0.1 g in 50 ml 50% v/v methanol-water
Malachite Green	1.3	0.2—1.8	Y-B/G	water
Methyl Green	—	0.2—1.8	Y-B	0.1% in water
2-(p-dimethylaminophenylazo)pyridine	—	0.2—1.8	Y-B	0.1% in ethanol
		4.4—5.6	R-Y	
o-cresolsulfonephthalein (Cresol Red)	8.46	0.4—1.8	Y-R	0.1 g in 26.2 ml 0.01 M NaOH + 222.8 ml water
		7.0—8.8	Y-R	
Quinaldine Red	2.63	1.4—3.2	C-R	1% in ethanol
p-(p-dimethylaminophenylazo)benzoic acid, Na salt (Paramethyl Red)	—	1.0—3.0	R-Y	ethanol
4'-aniline azobenzene m-sulfonic acid, Na salt (Metanil Yellow)	—	1.2—2.4	R-Y	0.01% in water
4-phenylazodiphenylamine	—	1.2—2.6	R-Y	0.01 g in 1 ml 1 M HCl + 50 ml ethanol + 49 ml water
thymolsulfonephthalein (Thymol Blue)	1.65	1.2—2.8	R-Y	0.1 g in 21.5 ml 0.01 M NaOH + 229.5 ml water
		8.0—9.6	R-B	
m-cresolsulfonephthalein (Metacresol Purple)	1.51	1.2—2.8	R-Y	0.1 g in 26.2 ml 0.01 M NaOH + 223.8 ml water
		7.4—9.0	Y-P	
p-(p-anilinophenylazo)benzenesulfonic acid, Na salt (Orange IV)	—	1.4—2.8	R-Y	0.01% in water
4-o-tolylazo-o-toluidine	—	1.4—2.8	O-Y	water
erythrosin, disodium salt	—	2.2—3.6	O-R	0.1 % in water
Benzopurpurine 48	—	2.2—4.2	V-R	0.1 % in water
N,N-dimethyl-p-(m-tolylazo)aniline	—	2.6—4.8	R-Y	0.1% in water
4,4'-bis(2-amino-1-naphthylazo) 2,2'-stilbenedisulfonic acid	—	3.0—4.0	P-R	0.1 g in 5.9 ml 0.05 M NaOH + 94.1 ml water
tetrabromophenolphthaleinethyl ester, K salt	—	3.0—4.2	Y-B	0.1% in water
3',3'',5',5''-tetrabromophenolsulfonephthalein (Bromophenol Blue)	4.10	3.0—4.6	Y-B	0.1 g in 14.9 ml 0.01 M NaOH + 235.1 ml water
2,4-dinitrophenol	4.10	2.0—4.7	C-Y	sat. solution in water
p-dimethylaminoazobenzene	—	2.8—4.4	R-Y	0.1 g in 100 ml 90% v/v ethanol-water
diphenyl bis(1-amino-2-naphthyl-azo-4-sulfonic acid) (Congo Red)	—	3.0—5.0	B-R	0.1% in water
4'-dimethylaminoazobenzene-4-sulfonic acid, Na salt (Methyl Orange)	3.46	3.2—4.4	R-Y	0.01% in water
Ethyl Orange	4.34	3.4—4.8	R-Y	0.05—0.2% in water or aqueous ethanol
4-(4-dimethylamino-1-naphthylazo)-3-methoxybenzenesulfonic acid	—	3.5—4.8	V-Y	0.1% in 60% ethanol-water
3',3'',5',5''-tetrabromo-m-cresol-sulfonephthalein (Bromocresol Green)	4.90	3.8—5.4	Y-B	0.1 g in 14.3 ml 0.01 M NaOH + 235.7 ml water
Resazurin	—	3.8—6.4	O-V	water
4-phenylazo-1-naphthylamine	—	4.0—5.6	R-Y	0.1% in ethanol
Ethyl Red	5.42	4.0—5.8	C-R	0.1 g in 100 ml 50% v/v methanol-water
2-(p-dimethylaminophenylazo)pyridine	—	0.2—1.8	Y-R	0.1% in ethanol
		4.4—5.6	R-Y	
4-(p-ethoxyphenylazo)-m-phenylene-diamine monohydrochloride	—	4.4—5.8	O-Y	0.1% in water

ACID-BASE INDICATORS (continued)

Indicator	pK (20°C)	Approx. pH range	Color change	Preparation
Resorcin Blue	—	4.4—6.2	R-B	0.2% in ethanol
1,2-dihydroxyanthroquinone-3-sulfonic acid (Alizarin Red S)	—	4.6—6.0	Y-R	water
4-dimethylaminoazobenzene-2-carboxylic acid (Methyl Red)	5.00	4.8—6.0	R-Y	0.02 g in 100 ml 60% v/v ethanol-water
Propyl Red	5.48	4.8—6.6	R-Y	ethanol
5,5′-dibromo-o-cresolsulfonephthalein (Bromocresol Purple)	6.40	5.2—6.8	Y-P	0.1 g in 18.5 ml 0.01 M NaOH + 231.5 ml water
3′,3″-dichlorophenolsulfonephthalein (Chlorophenol Red)	6.25	5.2—6.8	Y-R	0.1 g in 23.6 ml 0.01 M NaOH + 226.4 ml water
p-nitropheno	—	5.4—6.6	C-Y	0.1% in water
alizarin	17.15	5.6—7.2	Y-R	0.1% in methanol
	—	11.0—12.4	R-P	
2-(2,4-dinitrophenylazo)-1-naphthol-3,6-disulfonic acid, disodium salt	—	6.0—7.0	Y-B	0.1% in water
3′,3″-dibromothymolsulfonephthalein (Bromothymol Blue)	7.30	6.0—7.6	Y-B	0.1 g in 16 ml 0.01 M NaOH + 234 ml water
6,8-dinitro-2,4-(1H)quinazolinedione	—	6.4—8.0	C-Y	25 g in 115 ml 1 M NaOH + 50 ml water at 100° C
Brilliant Yellow	—	6.6—7.8	Y-R	1% in water
phenolsulfonephthalein (Phenol Red)	8.00	6.6—8.0	Y-R	0.1 g in 28.2 ml 0.01 M NaOH + 221.8 ml water
3-amino-6-dimethylamino-2-methyl-phenazinium chloride (Neutral Red)	—	6.8—8.0	R-A	0.01 g in 100 ml 50%v/v ethanol-water
m-nitrophenol	8.35	6.8—8.6	C-Y	0.3% in water
o-cresolsulfonephthalein (Cresol Red)	—	0.0—1.0	R-Y	0.1 g in 26.2 ml 0.01 M NaOH + 223.8 ml water
	8.46	7.0—8.8	Y-R	
Turmaric (Curcumin)	—	7.4—8.6	Y-R	ethanol
m-cresolsulfonephthalein (Metacresol Purple)	1.5	1.2—2.8	R-Y	0.1 g in 26.2 ml 0.01 M NaOH + 223.8 ml water
	8.3	7.4—9.0	Y-P	
4,4′-bis(4-amino-1-naphthylazo)-2,2′-stilbenedisulfonic acid	—	8.0—9.0	B-R	0.1 g in 5.9 ml 0.05 M NaOH + 94.1 ml water
thymolsulfonephthalein (Thymol Blue)	1.65	1.2—2.8	R-Y	0.1 g in 21.5 ml 0.01 M NaOH + 228.5 ml water
	9.20	8.0—9.6		
o-cresolphthalein	—	8.2—9.8	C-R	0.04% in ethanol
p-naphtholbenzene	—	8.2—10.0	O-B	1% in dil. alkali
di(p-dioxydiphenyl)phthalide (Phenolphthalein)	9.5	8.2—10.0	C-Pk	0.05 g in 100 ml 50% v/v ethanol-water
ethyl-bis(2,4—dimethylphenyl)ethanoate	—	8.4—9.6	C-B	sat. solution in 50% acetone-alcohol
thymolphthalein	—	9.4—10.6	C-B	0.04 g in 100 ml in 50% v/v ethanol-water
5-(p-nitrophenylazo)salicylic acid, Na salt (Alizarin Yellow R)	—	10.1—12.0	Y-R	0.01% in water
p-(2,4-dihydroxyphenylazo)benzene-sulfonic acid, Na salt	—	11.4—12.6	Y-O	0.1% in water
5,5′-indigodisulfonic acid, disodium salt	—	11.4—13.0	B-Y	water
2,4,6-trinitrotoluene	—	11.5—13.0	C-O	0.1—0.5% in ethanol
1,3,5-trinitrobenzene	—	12.0—14.0	C-O	0.1—0.5% in ethanol
Clayton Yellow	—	12.2—13.2	Y-A	0.1% in water

Note: C = colorless; Pk = pink; R = red; O = orange; A = amber; Y = yellow; B = blue; B/G = blue/ green; V = violet; P = purple.

Useful reference: E. Bishop, Ed., *Indicators*, Pergamon, Oxford, 1972.

FLUORESCENT INDICATORS

Jack DeMent

Fluorescent indicators are substances which show definite changes in fluorescence with change in pH. Some fluorescent materials are not suitable for indicators since their change in fluorescence is too gradual. Fluorescent indicators find greatest utility in the titration of opaque, highly turbid or deeply colored solutions. A long wavelength ultraviolet ("black light") lamp in a dimly lighted room provides the best environment for titrations involving fluorescent indicators, although bright daylight is sometimes sufficient to evoke a response in the bright green, yellow and orange fluorescent indicators. Titrations are carried out in non-fluorescent glassware. One should check the glassware prior to use to make certain that it does not fluoresce due to the wavelengths of light involved in the titration. The meniscus of the liquid in the burette can be followed when a few particles of an insoluble fluorescent solid are dropped onto its surface.

In this table the indicators are arranged by approximate pH range covered. In the case of some of the dyestuffs the end point may vary slightly with the source or manufacturer.

pH 0 to 2

Indicator	C.I.	From pH	To pH
Benzoflavine	—	0.3, yellow fl.	1.7, green fl.
3,6-Dioxyphthalimide	—	0, blue fl.	2.4, green fl.
Eosine YS	768	0, yellow colored	3.0, yellow fl.
Erythrosine	772	0, yellow colored	3.6, yellow fl.
Esculin	—	1.5, colorless	2, blue fl.
4-Ethoxyacridone	—	1.2, green fl.	3.2, blue fl.
3,6-Tetramethyldiaminooxanthone	—	1.2, green fl.	3.4, blue fl.

pH 2 to 4

Indicator	C.I.	From pH	To pH
Chromotropic acid	—	3.5, colorless	4.5, blue fl.
Fluorescein	766	4, colorless	4.5, green fl.
Magdala Red	—	3.0, purple colored	4.0, fl.
α-Naphthylamine	—	3.4, colorless	4.8, blue fl.
β-Naphthylamine	—	2.8, colorless	4.4, violet fl.
Phloxine	774	3.4, colorless	5.0, bright yellow fl.
Salicylic acid	—	2.5, colorless	3.5, blue fl.

pH 4 to 6

Indicator	C.I.	From pH	To pH
Acridine	788	4.9, green fl.	5.1, violet colored
Dichlorofluorescein	—	4.0, colorless	5.0, green fl.
3,6-Dioxyxanthone	—	5.4, colorless	7.6, blue-violet fl.
Erythrosine	772	4.0, colorless	4.5, yellow-green fl.
β-Methylesculetin	—	4.0, colorless	6.2, blue fl.
Neville-Winther acid	—	6.0, colorless	6.5, blue fl.
Resorufin	—	4.4, yellow fl.	6.4, weak orange fl.
Quininic acid	—	4.0, yellow colored	5.0, blue fl.
Quinine [first end point]	—	5.0, blue fl.	6.1, violet fl.

pH 6 to 8

Indicator	C.I.	From pH	To pH
Acid R Phosphine	—	(claimed for range pH 6.0–7.0)	
Brilliant Diazol Yellow	—	6.5, colorless	7.5, violet fl.
Cleves acid	—	6.5, colorless	7.5, green fl.
Coumaric acid	—	7.2, colorless	9.0, green fl.
3,6-Dioxyphthalic dinitrile	—	5.8, blue fl.	8.2, green fl.
Magnesium 8-hydroxyquinolinate	—	6.5, colorless	7.5, golden fl.
β-Methylumbelliferone	—	7.0, colorless	7.5, blue fl.
1-Naphthol-4-sulfonic acid	—	6.0, colorless	6.5, blue fl.
Orcinaurine	—	6.5, colorless	8.0, green fl.
Patent Phosphine	789	(for the range pH 6.0–7.0, green-yellow fl.)	
Thioflavine	816	(for the region pH 6.5–7.0, yellow fl.)	
Umbelliferone	—	6.5, colorless	7.6, blue fl.

pH 8 to 10

Indicator	C.I.	From pH	To pH
Acridine Orange	788	8.4, orange colored	10.4, green fl.
Ethoxyphenylnaphthostilbazonium chloride	—	9, green fl.	11, non-fl.
G Salt	—	9.0, dull blue fl.	9.5, bright blue fl.
Naphthazol derivatives	—	8.2, colorless	10.0, yellow or green fl.
α-Naphthionic acid	—	9, blue fl.	11, green fl.
2-Naphthol-3,6-disulfonic acid	—	9.5, dark blue fl.	Light blue fl. at higher pH
β-Naphthol	—	8.6, colorless	Blue fl. at higher pH
α-Naphtholsulfonic acid	—	8.0, dark blue fl.	9.0, bright violet fl.
1,4-Naphtholsulfonic acid	—	8.2, dark blue fl.	Light blue fl. at higher pH
Orcinsulfonphthalein	—	8.6, yellow colored	10.0 fl.
Quinine [second end point]	—	9.5, violet fl.	10.0, colorless
R-Salt	—	9.0, dull blue fl.	9.5, bright blue fl.
Sodium 1-naphthol-2-sulfonate	—	9.0, dark blue fl.	10.0, bright violet fl.

pH 10 to 12

Indicator	C.I.	From pH	To pH
Coumarin	—	9.8, deep green fl.	12, light green fl.
Eosine BN	771	10.5, colorless	14.0, yellow fl.
Papaverine (permanganate oxidized)	—	9.5, yellow fl.	11.0, blue fl.
Schaffers Salt	—	5.0, violet fl.	11.0, green-blue fl.
SS-Acid (sodium salt)	—	10.0, violet fl.	12.0, yellow colored

pII 12 to 14

Indicator	C.I.	From pH	To pH
Cotarnine	—	12.0, yellow fl.	13.0, white fl.
α-Naphthionic acid	—	12, blue fl.	13, green fl.
β-Naphthionic acid	—	12, blue fl.	13, violet fl.

CONVERSION FORMULAS FOR CONCENTRATION OF SOLUTIONS

A = Weight per cent of solute
B = Molecular weight of solvent
E = Molecular weight of solute
F = Grams of solute per liter of solution
G = Molality
M = Molarity
N = Mole fraction
R = Density of solution in grams per milliliter

Concentration of solute—SOUGHT	Concentration of solute—GIVEN				
	A	N	G	M	F
A	—	$\dfrac{100N \times E}{N \times E + (1 - N)B}$	$\dfrac{100G \times E}{1000 + G \times E}$	$\dfrac{M \times E}{10R}$	$\dfrac{F}{10R}$
N	$\dfrac{\dfrac{A}{E}}{\dfrac{A}{E} + \dfrac{100 - A}{B}}$	—	$\dfrac{B \times G}{B \times G + 1000}$	$\dfrac{B \times M}{M(B - E) + 1000R}$	$\dfrac{B \times F}{F(B - E) + 1000R \times E}$
G	$\dfrac{1000A}{E(100 - A)}$	$\dfrac{1000N}{B - N \times B}$	—	$\dfrac{1000M}{1000R - (M \times E)}$	$\dfrac{1000F}{E(1000R - F)}$
M	$\dfrac{10R \times A}{E}$	$\dfrac{1000R \times N}{N \times E + (1 - N)B}$	$\dfrac{1000R \times G}{1000 + E \times G}$	—	$\dfrac{F}{E}$
F	$10AR$	$\dfrac{1000R \times N \times E}{N \times E + (1 - N)B}$	$\dfrac{1000R \times G \times E}{1000 + G \times E}$	$M \times E$	—

ELECTROCHEMICAL SERIES

Petr Vanýsek

There are three tables for this electrochemical series. Each table lists standard reduction potentials, $E°$ values, at 298.15 K (25°C), and at a pressure of 101.325 kPa (1 atm). Table 1 is an alphabetical listing of the elements, according to the symbol of the elements. Thus, data for silver (Ag) precedes those for aluminum (Al). Table 2 lists only those reduction reactions which have $E°$ values positive in respect to the standard hydrogen electrode. In Table 2, the reactions are listed in the order of increasing positive potential, and they range from 0.0000 V to +3.4 V. Table 3 lists only those reduction potentials which have $E°$ negative with respect to the standard hydrogen electrode. In Table 3, the reactions are listed in the order of decreasing potential and range from 0.0000 V to −4.10 V. The reliability of the potentials is not the same for all the data. Typically, the values with fewer significant figures have lower reliability. The values of reduction potentials, in particular those of less common reactions, are not definite; they are subject to occasional revisions.

Abbreviations: bipy = 2,2′-dipyridine, or bipyridine; en = ethylenediamine; phen = 1,10-phenanthroline.

REFERENCES

1. G. Milazzo, S. Caroli, and V. K. Sharma, *Tables of Standard Electrode Potentials*, Wiley, Chichester, 1978.
2. A. J. Bard, R. Parsons, and J. Jordan, *Standard Potentials in Aqueous Solutions*, Marcel Dekker, New York, 1985.
3. S. G. Bratsch, *J. Phys. Chem. Ref. Data*, 18, 1—21, 1989.

TABLE 1
Alphabetical Listing

Reaction	$E°$/V	Reaction	$E°$/V
$Ac^{3+} + 3\,e \rightleftharpoons Ac$	−2.20	$Al(OH)_4^- + 3\,e \rightleftharpoons Al + 4\,OH^-$	−2.328
$Ag^+ + e \rightleftharpoons Ag$	0.7996	$H_2AlO_3^- + H_2O + 3\,e \rightleftharpoons Al + 4\,OH^-$	−2.33
$Ag^{2+} + e \rightleftharpoons Ag^+$	1.980	$AlF_6^{3-} + 3\,e \rightleftharpoons Al + 6\,F^-$	−2.069
$Ag(ac) + e \rightleftharpoons Ag + (ac)^-$	0.643	$Am^{4+} + e \rightleftharpoons Am^{3+}$	2.60
$AgBr + e \rightleftharpoons Ag + Br^-$	0.07133	$Am^{2+} + 2\,e \rightleftharpoons Am$	−1.9
$AgBrO_3 + e \rightleftharpoons Ag + BrO_3^-$	0.546	$Am^{3+} + 3\,e \rightleftharpoons Am$	−2.048
$Ag_2C_2O_4 + 2\,e \rightleftharpoons 2\,Ag + C_2O_4^{2-}$	0.4647	$Am^{3+} + e \rightleftharpoons Am^{2+}$	−2.3
$AgCl + e \rightleftharpoons Ag + Cl^-$	0.22233	$As + 3\,H^+ + 3\,e \rightleftharpoons AsH_3$	−0.608
$AgCN + e \rightleftharpoons Ag + CN^-$	−0.017	$As_2O_3 + 6\,H^+ + 6\,e \rightleftharpoons 2\,As + 3\,H_2O$	0.234
$Ag_2CO_3 + 2\,e \rightleftharpoons 2\,Ag + CO_3^{2-}$	0.47	$HAsO_2 + 3\,H^+ + 3\,e \rightleftharpoons As + 2\,H_2O$	0.248
$Ag_2CrO_4 + 2\,e \rightleftharpoons 2\,Ag + CrO_4^{2-}$	0.4470	$AsO_2^- + 2\,H_2O + 3\,e \rightleftharpoons As + 4\,OH^-$	−0.68
$AgF + e \rightleftharpoons Ag + F^-$	0.779	$H_3AsO_4 + 2\,H^+ + 2\,e^- \rightleftharpoons HAsO_2 + 2\,H_2O$	0.560
$Ag_4[Fe(CN)_6] + 4\,e \rightleftharpoons 4\,Ag + [Fe(CN)_6]^{4-}$	0.1478	$AsO_4^{3-} + 2\,H_2O + 2\,e \rightleftharpoons AsO_2^- + 4\,OH^-$	−0.71
$AgI + e \rightleftharpoons Ag + I^-$	−0.15224	$At_2 + 2\,e \rightleftharpoons 2\,At^-$	0.3
$AgIO_3 + e \rightleftharpoons Ag + IO_3^-$	0.354	$Au^+ + e \rightleftharpoons Au$	1.692
$Ag_2MoO_4 + 2\,e \rightleftharpoons 2\,Ag + MoO_4^{2-}$	0.4573	$Au^{3+} + 2\,e \rightleftharpoons Au^+$	1.401
$AgNO_2 + e \rightleftharpoons Ag + 2\,NO_2^-$	0.564	$Au^{3+} + 3\,e \rightleftharpoons Au$	1.498
$Ag_2O + H_2O + 2\,e \rightleftharpoons 2\,Ag + 2\,OH^-$	0.342	$Au^{2+} + e^- \rightleftharpoons Au^+$	1.8
$Ag_2O_3 + H_2O + 2\,e \rightleftharpoons 2\,AgO + 2\,OH^-$	0.739	$AuOH^{2+} + H^+ + 2\,e \rightleftharpoons Au^+ + H_2O$	1.32
$Ag^{3+} + 2\,e \rightleftharpoons Ag^+$	1.9	$AuBr_2^- + e \rightleftharpoons Au + 2\,Br^-$	0.959
$Ag^{3+} + e \rightleftharpoons Ag^{2+}$	1.8	$AuBr_4^- + 3\,e \rightleftharpoons Au + 4\,Br^-$	0.854
$Ag_2O_2 + 4\,H^+ + e \rightleftharpoons 2\,Ag + 2\,H_2O$	1.802	$AuCl_4^- + 3\,e \rightleftharpoons Au + 4\,Cl^-$	1.002
$2\,AgO + H_2O + 2\,e \rightleftharpoons Ag_2O + 2\,OH^-$	0.607	$Au(OH)_3 + 3\,H^+ + 3\,e \rightleftharpoons Au + 3\,H_2O$	1.45
$AgOCN + e \rightleftharpoons Ag + OCN^-$	0.41	$H_2BO_3^- + 5\,H_2O + 8\,e \rightleftharpoons BH_4^- + 8\,OH^-$	−1.24
$Ag_2S + 2\,e \rightleftharpoons 2\,Ag + S^{2+}$	−0.691	$H_2BO_3^- + H_2O + 3\,e \rightleftharpoons B + 4\,OH^-$	−1.79
$Ag_2S + 2\,H^+ + 2\,e \rightleftharpoons 2\,Ag + H_2S$	−0.0366	$H_3BO_3 + 3\,H^+ + 3\,e \rightleftharpoons B + 3\,H_2O$	−0.8698
$AgSCN + e \rightleftharpoons Ag + SCN^-$	0.08951	$B(OH)_3 + 7\,H^+ + 8\,e \rightleftharpoons BH_4^- + 3\,H_2O$	−0.481
$Ag_2SeO_3 + 2\,e \rightleftharpoons 2\,Ag + SeO_3^{2-}$	0.3629	$Ba^{2+} + 2\,e \rightleftharpoons Ba$	−2.912
$Ag_2SO_4 + 2\,e \rightleftharpoons 2\,Ag + SO_4^{2-}$	0.654	$Ba^{2+} + 2\,e \rightleftharpoons Ba(Hg)$	−1.570
$Ag_2WO_4 + 2\,e \rightleftharpoons 2\,Ag + WO_4^{2-}$	0.4660	$Ba(OH)_2 + 2\,e \rightleftharpoons Ba + 2\,OH^-$	−2.99
$Al^{3+} + 3\,e \rightleftharpoons Al$	−1.662	$Be^{2+} + 2\,e \rightleftharpoons Be$	−1.847
$Al(OH)_3 + 3\,e \rightleftharpoons Al + 3\,OH^-$	−2.31	$Be_2O_3^{2-} + 3\,H_2O + 4\,e \rightleftharpoons 2\,Be + 6\,OH^-$	−2.63

TABLE 1
Alphabetical Listing (continued)

Reaction	$E°/V$	Reaction	$E°/V$
p–benzoquinone $+ 2 H^+ + 2 e \rightleftharpoons$ hydroquinone	0.6992	$HClO_2 + 3 H^+ + 4 e \rightleftharpoons Cl^- + 2 H_2O$	1.570
$Bi^+ + e \rightleftharpoons Bi$	0.5	$ClO_2^- + H_2O + 2 e \rightleftharpoons ClO^- + 2 OH^-$	0.66
$Bi^{3+} + 3 e \rightleftharpoons Bi$	0.308	$ClO_2^- + 2 H_2O + 4 e \rightleftharpoons Cl^- + 4 OH^-$	0.76
$Bi^{3+} + 2 e \rightleftharpoons Bi^+$	0.2	$ClO_2(aq) + e \rightleftharpoons ClO_2^-$	0.954
$Bi + 3 H^+ + 3 e \rightleftharpoons BiH_3$	−0.8	$ClO_3^- + 2 H^+ + e \rightleftharpoons ClO_2 + H_2O$	1.152
$BiCl_4^- + 3 e \rightleftharpoons Bi + 4 Cl^-$	0.16	$ClO_3^- + 3 H^+ + 2 e \rightleftharpoons HClO_2 + H_2O$	1.214
$Bi_2O_3 + 3 H_2O + 6 e \rightleftharpoons 2 Bi + 6 OH^-$	−0.46	$ClO_3^- + 6 H^+ + 5 e \rightleftharpoons 1/2 Cl_2 + 3 H_2O$	1.47
$Bi_2O_4 + 4 H^+ + 2 e \rightleftharpoons 2 BiO^+ + 2 H_2O$	1.593	$ClO_3^- + 6 H^+ + 6 e \rightleftharpoons Cl^- + 3 H_2O$	1.451
$BiO^+ + 2 H^+ + 3 e \rightleftharpoons Bi + H_2O$	0.320	$ClO_3^- + H_2O + 2 e \rightleftharpoons ClO_2^- + 2 OH^-$	0.33
$BiOCl + 2 H^+ + 3 e \rightleftharpoons Bi + Cl^- + H_2O$	0.1583	$ClO_3^- + 3 H_2O + 6 e \rightleftharpoons Cl^- + 6 OH^-$	0.62
$Bk^{4+} + e \rightleftharpoons Bk^{3+}$	1.67	$ClO_4^- + 2 H^+ + 2 e \rightleftharpoons ClO_3^- H_2O$	1.189
$Bk^{2+} + 2 e \rightleftharpoons Bk$	−1.6	$ClO_4^- + 8 H^+ + 7 e \rightleftharpoons 1/2 Cl_2 + 4 H_2O$	1.39
$Bk^{3+} + e \rightleftharpoons Bk^{2+}$	−2.8	$ClO_4^- + 8 H^+ + 8 e \rightleftharpoons Cl^- + 4 H_2O$	1.389
$Br_2(aq) + 2 e \rightleftharpoons 2 Br^-$	1.0873	$ClO_4^- + H_2O + 2 e \rightleftharpoons ClO_3^- + 2 OH^-$	0.36
$Br_2(\ell) + 2 e \rightleftharpoons 2 Br^-$	1.066	$Cm^{4+} + e \rightleftharpoons Cm^{3+}$	3.0
$HBrO + H^+ + 2 e \rightleftharpoons Br^- + H_2O$	1.331	$Cm^{3+} + 3 e \rightleftharpoons Cm$	−2.04
$HBrO + H^+ + e \rightleftharpoons 1/2 Br_2(aq) + H_2O$	1.574	$Co^{2+} + 2 e \rightleftharpoons Co$	−0.28
$HBrO + H^+ + e \rightleftharpoons 1/2 Br_2(\ell) + H_2O$	1.596	$Co^{3+} + e \rightleftharpoons Co^{2+}$	1.92
$BrO^- + H_2O + 2 e \rightleftharpoons Br^- + 2 OH^-$	0.761	$[Co(NH_3)_6]^{3+} + e \rightleftharpoons [Co(NH_3)_6]^{2+}$	0.108
$BrO_3^- + 6 H^+ + 5 e \rightleftharpoons 1/2 Br_2 + 3 H_2O$	1.482	$Co(OH)_2 + 2 e \rightleftharpoons Co + 2 OH^-$	−0.73
$BrO_3^- + 6 H^+ + 6 e \rightleftharpoons Br^- + 3 H_2O$	1.423	$Co(OH)_3 + e \rightleftharpoons Co(OH)_2 + OH^-$	0.17
$BrO_3^- + 3 H_2O + 6 e \rightleftharpoons Br^- + 6 OH^-$	0.61	$Cr^{2+} + 2 e \rightleftharpoons Cr$	−0.913
$(CN)_2 + 2 H^+ + 2 e \rightleftharpoons 2 HCN$	0.373	$Cr^{3+} + e \rightleftharpoons Cr^{2+}$	−0.407
$2 HCNO + 2 H^+ + 2 e \rightleftharpoons (CN)_2 + 2 H_2O$	0.330	$Cr^{3+} + 3 e \rightleftharpoons Cr$	−0.744
$(CNS)_2 + 2 e \rightleftharpoons 2 CNS^-$	0.77	$Cr_2O_7^{2-} + 14 H^+ + 6 e \rightleftharpoons 2 Cr^{3+} + 7 H_2O$	1.232
$CO_2 + 2 H^+ + 2 e \rightleftharpoons HCOOH$	−0.199	$CrO_2^- + 2 H_2O + 3 e \rightleftharpoons Cr + 4 OH^-$	−1.2
$Ca^+ + e \rightleftharpoons Ca$	−3.80	$HCrO_4^- + 7 H^+ + 3 e \rightleftharpoons Cr^{3+} + 4 H_2O$	1.350
$Ca^{2+} + 2 e \rightleftharpoons Ca$	−2.868	$CrO_2 + 4 H^+ + e \rightleftharpoons Cr^{3+} \cdot 2H_2O$	1.48
$Ca(OH)_2 + 2 e \rightleftharpoons Ca + 2 OH^-$	−3.02	$Cr_2O_5 + 10 H^+ + e \rightleftharpoons CrO_2 \cdot 5H_2O$	1.3
Calomel electrode, 1 molal KCl	0.2800	$CrO_4^{2-} + 4 H_2O + 3 e \rightleftharpoons Cr(OH)_3 + 5 OH^-$	−0.13
Calomel electrode, 1 molar KCl (NCE)	0.2801	$Cr(OH)_3 + 3 e \rightleftharpoons Cr + 3 OH^-$	−1.48
Calomel electrode, 0.1 molar KCl	0.3337	$Cs^+ + e \rightleftharpoons Cs$	−3.026
Calomel electrode, saturated KCl (SCE)	0.2412	$Cu^+ + e \rightleftharpoons Cu$	0.521
Calomel electrode, saturated NaCl (SSCE)	0.2360	$Cu^{2+} + e \rightleftharpoons Cu^+$	0.153
$Cd^{2+} + 2 e \rightleftharpoons Cd$	−0.4030	$Cu^{2+} + 2 e \rightleftharpoons Cu$	0.3419
$Cd^{2+} + 2 e \rightleftharpoons Cd(Hg)$	−0.3521	$Cu^{2+} + 2 e \rightleftharpoons Cu(Hg)$	0.345
$Cd(OH)_2 + 2 e \rightleftharpoons Cd(Hg) + 2 OH^-$	−0.809	$Cu^{3+} + e \rightleftharpoons Cu^{2+}$	2.4
$CdSO_4 + 2 e \rightleftharpoons Cd + SO_4^{2-}$	−0.246	$Cu_2O_3 + 6 H^+ + e \rightleftharpoons Cu^{2+} + 3 H_2O$	2.0
$Cd(OH)_4^{2-} + 2 e \rightleftharpoons Cd + 4 OH^-$	−0.658	$Cu^{2+} + 2 CN^- + e \rightleftharpoons [Cu(CN)_2]^-$	1.103
$CdO + H_2O + 2 e \rightleftharpoons Cd + 2 OH^-$	−0.783	$CuI_2^- + e \rightleftharpoons Cu + 2 I^-$	0.00
$Ce^{3+} + 3 e \rightleftharpoons Ce$	−2.336	$Cu_2O + H_2O + 2 e \rightleftharpoons 2 Cu + 2 OH^-$	−0.360
$Ce^{3+} + 3 e \rightleftharpoons Ce(Hg)$	−1.4373	$Cu(OH)_2 + 2 e \rightleftharpoons Cu + 2 OH^-$	−0.222
$Ce^{4+} + e \rightleftharpoons Ce^{3+}$	1.72	$2 Cu(OH)_2 + 2 e \rightleftharpoons Cu_2O + 2 OH^- + H_2O$	−0.080
$CeOH^{3+} + H^+ + e \rightleftharpoons Ce^{3+} + H_2O$	1.715	$2 D^+ + 2 e \rightleftharpoons D_2$	−0.013
$Cf^{4+} + e \rightleftharpoons Cf^{3+}$	3.3	$Dy^{2+} + 2 e \rightleftharpoons Dy$	−2.2
$Cf^{3+} + e \rightleftharpoons Cf^{2+}$	−1.6	$Dy^{3+} + 3 e \rightleftharpoons Dy$	−2.295
$Cf^{3+} + 3 e \rightleftharpoons Cf$	−1.94	$Dy^{3+} + e \rightleftharpoons Dy^{2+}$	−2.6
$Cf^{2+} + 2 e \rightleftharpoons Cf$	−2.12	$Er^{2+} + 2 e \rightleftharpoons Er$	−2.0
$Cl_2(g) + 2 e \rightleftharpoons 2 Cl^-$	1.35827	$Er^{3+} + 3 e \rightleftharpoons Er$	−2.331
$HClO + H^+ + e \rightleftharpoons 1/2 Cl_2 + H_2O$	1.611	$Er^{3+} + e \rightleftharpoons Er^{2+}$	−3.0
$HClO + H^+ + 2 e \rightleftharpoons Cl^- + H_2O$	1.482	$Es^{3+} + e \rightleftharpoons Es^{2+}$	−1.3
$ClO^- + H_2O + 2 e \rightleftharpoons Cl^- + 2 OH^-$	0.81	$Es^{3+} + 3 e \rightleftharpoons Es$	−1.91
$ClO_2 + H^+ + e \rightleftharpoons HClO_2$	1.277	$Es^{2+} + 2 e \rightleftharpoons Es$	−2.23
$HClO_2 + 2 H^+ + 2 e \rightleftharpoons HClO + H_2O$	1.645	$Eu^{2+} + 2 e \rightleftharpoons Eu$	−2.812
$HClO_2 + 3 H^+ + 3 e \rightleftharpoons 1/2 Cl_2 + 2 H_2O$	1.628	$Eu^{3+} + 3 e \rightleftharpoons Eu$	−1.991

Reaction	$E°/V$	Reaction	$E°/V$
$Eu^{3+} + e \rightleftharpoons Eu^{2+}$	−0.36	$Ho^{3+} + 3e \rightleftharpoons Ho$	−2.33
$F_2 + 2H^+ + 2e \rightleftharpoons 2HF$	3.053	$Ho^{3+} + e \rightleftharpoons Ho^{2+}$	−2.8
$F_2 + 2e \rightleftharpoons 2F^-$	2.866	$I_2 + 2e \rightleftharpoons 2I^-$	0.5355
$F_2O + 2H^+ + 4e \rightleftharpoons H_2O + 2F^-$	2.153	$I_2^- + 2e \rightleftharpoons 3I^-$	0.536
$Fe^{2+} + 2e \rightleftharpoons Fe$	−0.447	$H_3IO_6 + 2e \rightleftharpoons IO_3 + 3OH^-$	0.7
$Fe^{3+} + 3e \rightleftharpoons Fe$	−0.037	$H_5IO_6 + H^+ + 2e \rightleftharpoons IO_3^- + 3H_2O$	1.601
$Fe^{3+} + e \rightleftharpoons Fe^{2+}$	0.771	$2HIO + 2H^+ + 2e \rightleftharpoons I_2 + 2H_2O$	1.439
$HFeO_4^- + 2H^+ + 3e \rightleftharpoons Fe_2O_3 + H_2O$	2.09	$HIO + H^+ + 2e \rightleftharpoons I^- + H_2O$	0.987
$HFeO_4^- + 4H^+ + 3e \rightleftharpoons FeOOH + 2H_2O$	2.08	$IO^- + H_2O + 2e \rightleftharpoons I^- + 2OH^-$	0.485
$HFeO_4^- + 8H^+ + 3e \rightleftharpoons Fe^{3+} + 4H_2O$	2.07	$2IO_3^- + 12H^+ + 10e \rightleftharpoons I_2 + 6H_2O$	1.195
$Fe_2O_3 + 4H^+ + 2e \rightleftharpoons 2FeOH^+ + H_2O$	0.16	$IO_3^- + 6H^+ + 6e \rightleftharpoons I^- + 3H_2O$	1.085
$[Fe(CN)_6]^{3-} + e \rightleftharpoons [Fe(CN)_6]^{4-}$	0.358	$IO_3^- + 2H_2O + 4e \rightleftharpoons IO^- + 4OH^-$	0.15
$FeO_4^{2-} + 8H^+ + 3e \rightleftharpoons Fe^{3+} + 4H_2O$	2.20	$IO_3^- + 3H_2O + 6e \rightleftharpoons IO^- + 6OH^-$	0.26
$[Fe(bipy)_2]^{3+} + e \rightleftharpoons Fe(bipy)_2^{2+}$	0.78	$In^+ + e \rightleftharpoons In$	−0.14
$[Fe(bipy)_3]^{3+} + e \rightleftharpoons Fe(bipy)_3^{2+}$	1.03	$In^{2+} + e \rightleftharpoons In^+$	−0.40
$Fe(OH)_3 + e \rightleftharpoons Fe(OH)_2 + OH^-$	−0.56	$In^{3+} + e \rightleftharpoons In^{2+}$	−0.49
$[Fe(phen)_3]^{3+} + e \rightleftharpoons [Fe(phen)_3]^{2+}$	1.147	$In^{3+} + 2e \rightleftharpoons In^+$	−0.443
$[Fe(phen)_3]^{3+} + e \rightleftharpoons [Fe(phen)_3]^{2+}$ (1 molar H_2SO_4)	1.06	$In^{3+} + 3e \rightleftharpoons In$	−0.3382
$[Ferricinium]^+ + e \rightleftharpoons$ ferrocene	0.400	$In(OH)_3 + 3e \rightleftharpoons In + 3OH^-$	−0.99
$Fm^{3+} + e \rightleftharpoons Fm^{2+}$	−1.1	$In(OH)_4^- + 3e \rightleftharpoons In + 4OH^-$	−1.007
$Fm^{3+} + 3e \rightleftharpoons Fm$	−1.89	$In_2O_3 + 3H_2O + 6e \rightleftharpoons 2In + 6OH^-$	−1.034
$Fm^{2+} + 2e \rightleftharpoons Fm$	−2.30	$Ir^{3+} + 3e \rightleftharpoons Ir$	1.156
$Fr^+ + e \rightleftharpoons Fr$	−2.9	$[IrCl_6]^{2-} + e \rightleftharpoons [IrCl_6]^{3-}$	0.8665
$Ga^{3+} + 3e \rightleftharpoons Ga$	−0.549	$[IrCl_6]^{3-} + 3e \rightleftharpoons Ir + 6Cl^-$	0.77
$Ga^+ + e \rightleftharpoons Ga$	−0.2	$Ir_2O_3 + 3H_2O + 6e \rightleftharpoons 2Ir + 6OH^-$	0.098
$GaOH^{2+} + H^+ + 3e \rightleftharpoons Ga + H_2O$	−0.498	$K^+ + e \rightleftharpoons K$	−2.931
$H_2GaO^-_3 + H_2O + 3e \rightleftharpoons Ga + 4OH^-$	−1.219	$La^{3+} + 3e \rightleftharpoons La$	−2.379
$Gd^{3+} + 3e \rightleftharpoons Gd$	−2.279	$La(OH)_3 + 3e \rightleftharpoons La + 3OH^-$	−2.90
$Ge^{2+} + 2e \rightleftharpoons Ge$	0.24	$Li^+ + e \rightleftharpoons Li$	−3.0401
$Ge^{4+} + 4e \rightleftharpoons Ge$	0.124	$Lr^{3+} + 3e \rightleftharpoons Lr$	−1.96
$Ge^{4+} + 2e \rightleftharpoons Ge^{2+}$	0.00	$Lu^{3+} + 3e \rightleftharpoons Lu$	−2.28
$GeO_2 + 2H^+ + 2e \rightleftharpoons GeO + H_2O$	−0.118	$Md^{3+} + e \rightleftharpoons Md^{2+}$	−0.1
$H_2GeO_3 + 4H^+ + 4e \rightleftharpoons Ge + 3H_2O$	−0.182	$Md^{3+} + 3e \rightleftharpoons Md$	−1.65
$2H^+ + 2e \rightleftharpoons H_2$	0.00000	$Md^{2+} + 2e \rightleftharpoons Md$	−2.40
$H_2 + 2e \rightleftharpoons 2H^-$	−2.23	$Mg^+ + e \rightleftharpoons Mg$	−2.70
$HO_2 + H^+ + e \rightleftharpoons H_2O_2$	1.495	$Mg^{2+} + 2e \rightleftharpoons Mg$	−2.372
$2H_2O + 2e \rightleftharpoons H_2 + 2OH^-$	−0.8277	$Mg(OH)_2 + 2e \rightleftharpoons Mg + 2OH^-$	−2.690
$H_2O_2 + 2H^+ + 2e \rightleftharpoons 2H_2O$	1.776	$Mn^{2+} + 2e \rightleftharpoons Mn$	−1.185
$Hf^{4+} + 4e \rightleftharpoons Hf$	−1.55	$Mn^{3+} + 3 \rightleftharpoons Mn^{2+}$	1.5415
$HfO^{2+} + 2H^+ + 4e \rightleftharpoons Hf + H_2O$	−1.724	$MnO_2 + 4H^+ + 2e \rightleftharpoons Mn^{2+} + 2H_2O$	1.224
$HfO_2 + 4H^+ + 4e \rightleftharpoons Hf + 2H_2O$	−1.505	$MnO_4^- + e \rightleftharpoons MnO_4^{2-}$	0.558
$HfO(OH)_2 + H_2O + 4e \rightleftharpoons Hf + 4OH^-$	−2.50	$MnO_4^- + 4H^+ + 3e \rightleftharpoons MnO_2 + 2H_2O$	1.679
$Hg^{2+} + 2e \rightleftharpoons Hg$	0.851	$MnO_4^- + 8H^+ + 5e \rightleftharpoons Mn^{2+} + 4H_2O$	1.507
$2Hg^{2+} + 2e \rightleftharpoons Hg_2^{2+}$	0.920	$MnO_4^- + 2H_2O + 3e \rightleftharpoons MnO_2 + 4OH^-$	0.595
$Hg_2^{2+} + 2e \rightleftharpoons 2Hg$	0.7973	$MnO_4^{2-} + 2H_2O + 2e \rightleftharpoons MnO_2 + 4OH^-$	0.60
$Hg_2(ac)_2 + 2e \rightleftharpoons 2Hg + 2(ac)^-$	0.51163	$Mn(OH)_2 + 2e \rightleftharpoons Mn + 2OH^-$	−1.56
$Hg_2Br_2 + 2e \rightleftharpoons 2Hg + 2Br^-$	0.13923	$Mn(OH)_3 + e \rightleftharpoons Mn(OH)_2 + OH^-$	0.15
$Hg_2Cl_2 + 2e \rightleftharpoons 2Hg + 2Cl^-$	0.26808	$Mn_2O_3 + 6H + e \rightleftharpoons Mn^{2+} + 3H_2O$	1.485
$Hg_2HPO_4 + 2e \rightleftharpoons 2Hg + HPO_4^{2-}$	0.6359	$Mo^{3+} + 3e \rightleftharpoons Mo$	−0.200
$Hg_2I_2 + 2e \rightleftharpoons 2Hg + 2I^-$	−0.0405	$MoO_2 + 4H^+ + e \rightleftharpoons Mo + 4H_2O$	−0.152
$Hg_2O + H_2O + 2e \rightleftharpoons 2Hg + 2OH^-$	0.123	$H_3Mo_7O_{24}^{3-} + 45H^+ + 42e \rightleftharpoons 7Mo + 24H_2O$	0.082
$HgO + H_2O + 2e \rightleftharpoons Hg + 2OH^-$	0.0977	$MoO_3 + 6H^+ + 6e \rightleftharpoons Mo + 3H_2O$	0.075
$Hg(OH)_2 + 2H^+ + 2e \rightleftharpoons Hg + 2H_2O$	1.034	$N_2 + 2H_2O + 6H^+ + 6e \rightleftharpoons 2NH_4OH$	0.092
$Hg_2SO_4 + 2e \rightleftharpoons 2Hg + SO_4^{2-}$	0.6125	$3N_2 + 2H^+ + 2e \rightleftharpoons 2HN_3$	−3.09
$Ho^{2+} + 2e \rightleftharpoons Ho$	−2.1	$N_5^+ + 3H^+ + 2e \rightleftharpoons 2NH_4^+$	1.275

TABLE 1
Alphabetical Listing (continued)

Reaction	$E°$/V	Reaction	$E°$/V
$N_2O + 2 H^+ + 2 e \rightleftharpoons N_2 + H_2O$	1.766	$H_2P_2^- + e \rightleftharpoons P + 2 OH^-$	−1.82
$H_2N_2O_2 + 2 H^+ + 2 e \rightleftharpoons N_2 + 2 H_2O$	2.65	$H_3PO_2 + H^+ + 3 e \rightleftharpoons P + 2 H_2O$	−0.508
$N_2O_4 + 2 e \rightleftharpoons 2 NO_2^-$	0.867	$H_3PO_3 + 2 H^+ + 2 e \rightleftharpoons H_3PO_2 + H_2O$	−0.499
$N_2O_4 + 2 H^+ + 2 e \rightleftharpoons 2 NHO_2$	1.065	$H_3PO_3 + 3 H^+ + 3 e \rightleftharpoons P + 3 H_2O$	−0.454
$N_2O_4 + 4 H^+ + 4 e \rightleftharpoons 2 NO + 2 H_2O$	1.035	$HPO_3^{2-} + 2 H_2O + 2 e \rightleftharpoons H_2PO_2^- + 3 OH^-$	−1.65
$2 NH_3OH^+ + H^+ + 2 e \rightleftharpoons N_2H_5^+ + 2 H_2O$	1.42	$HPO_3^{2-} + 2 H_2O + 3 e \rightleftharpoons P + 5 OH^-$	−1.71
$2 NO + 2 H^+ + 2 e \rightleftharpoons N_2O + H_2O$	1.591	$H_3PO_4 + 2 H^+ + 2 e \rightleftharpoons H_3PO_3 + H_2O$	−0.276
$2 NO + H_2O + 2 e \rightleftharpoons N_2O + 2 OH^-$	0.76	$PO_4^{3-} + 2 H_2O + 3 e \rightleftharpoons HPO_3^{2-} + 3 OH^-$	−1.05
$HNO_2 + H^+ + e \rightleftharpoons NO + H_2O$	0.983	$Pa^{3+} + 3 e \rightleftharpoons Pa$	−1.34
$2 HNO_2 + 4 H^+ + 4 e \rightleftharpoons H_2N_2O_2 + 2 H_2O$	0.86	$Pa^{4+} + 4 e \rightleftharpoons Pa$	−1.49
$2 HNO_2 + 4 H^+ + 4 e \rightleftharpoons N_2O + 3 H_2O$	1.297	$Pa^{4+} + e \rightleftharpoons Pa^{3+}$	−1.9
$NO_2^- + H_2O + 3 e \rightleftharpoons NO + 2 OH^-$	−0.46	$Pb^{2+} + 2 e \rightleftharpoons Pb$	−0.1262
$2 NO_2^- + 2 H_2O + 4 e \rightleftharpoons N_2^{2-} + 4 OH^-$	−0.18	$Pb^{2+} + 2 e \rightleftharpoons Pb(Hg)$	−0.1205
$2 NO_2^- + 3 H_2O + 4 e \rightleftharpoons N_2O + 6 OH^-$	0.15	$PbBr_2 + 2 e \rightleftharpoons Pb + 2 Br^-$	−0.284
$NO_3^- + 3 H^+ + 2 e \rightleftharpoons HNO_2 + H_2O$	0.934	$PbCl_2 + 2 e \rightleftharpoons Pb + 2 Cl^-$	−0.2675
$NO_3^- + 4 H^+ + 3 e \rightleftharpoons NO + 2 H_2O$	0.957	$PbF_2 + 2 e \rightleftharpoons Pb + 2 F^-$	−0.3444
$2 NO_3^- + 4 H^+ + 2 e \rightleftharpoons N_2O_4 + 2 H_2O$	0.803	$PbHPO_4 + 2 e \rightleftharpoons Pb + HPO_4^{2-}$	−0.465
$NO_3^- + H_2O + 2 e \rightleftharpoons NO_2^- + 2 OH^-$	0.01	$PbI_2 + 2 e \rightleftharpoons Pb + 2 I^-$	−0.365
$2 NO_3^- + 2 H_2O + 2 e \rightleftharpoons N_2O_4 + 4 OH^-$	−0.85	$PbO + H_2O + 2 e \rightleftharpoons Pb + 2 OH^-$	−0.580
$Na^+ + e \rightleftharpoons Na$	−2.71	$PbO_2 + 4 H^+ + 2 e \rightleftharpoons Pb^{2+} + 2 H_2O$	1.455
$Nb^{3+} + 3 e \rightleftharpoons Nb$	−1.099	$HPbO_2^- + H_2O + 2 e \rightleftharpoons Pb + 3 OH^-$	−0.537
$NbO_2 + 2 H^+ + 2 e \rightleftharpoons NbO + H_2O$	−0.646	$PbO_2 + H_2O + 2 e \rightleftharpoons PbO + 2 OH^-$	0.247
$NbO_2 + 4 H^+ + 4 e \rightleftharpoons Nb + 2 H_2O$	−0.690	$PbO_2 + SO_4^{2-} + 4 H^+ + 2 e \rightleftharpoons PbSO_4 + 2 H_2O$	1.6913
$NbO + 2 H^+ + 2 e \rightleftharpoons Nb + H_2O$	−0.733	$PbSO_4 + 2 e \rightleftharpoons Pb + SO_4^{2-}$	−0.3588
$Nb_2O_5 + 10 H^+ + 10 e \rightleftharpoons 2 Nb + 5 H_2O$	−0.644	$PbSO_4 + 2 e \rightleftharpoons Pb(Hg) + SO_4^{2-}$	−0.3505
$Nd^{3+} + 3 e \rightleftharpoons Nd$	−2.323	$Pd^{2+} + 2 e \rightleftharpoons Pd$	0.951
$Nd^{2+} + 2 e \rightleftharpoons Nd$	−2.1	$[PdCl_4]^{2-} + 2 e \rightleftharpoons Pd + 4 Cl^-$	0.591
$Nd^{3+} + e \rightleftharpoons Nd^{2+}$	−2.7	$[PdCl_6]^{2-} + 2 e \rightleftharpoons [PdCl_4]^{2-} + 2 Cl^-$	1.288
$Ni^{2+} + 2 e \rightleftharpoons Ni$	−0.257	$Pd(OH)_2 + 2 e \rightleftharpoons Pd + 2 OH^-$	0.07
$Ni(OH)_2 + 2 e \rightleftharpoons Ni + 2 OH^-$	−0.72	$Pm^{2+} + 2 e \rightleftharpoons Pm$	−2.2
$NiO_2 + 4 H^+ + 2 e \rightleftharpoons Ni^{2+} + 2 H_2O$	1.678	$Pm^{3+} + 3 e \rightleftharpoons Pm$	−2.30
$NiO_2 + 2 H_2O + 2 e \rightleftharpoons Ni(OH)_2 + 2 OH^-$	−0.490	$Pm^{3+} + e \rightleftharpoons Pm^{2+}$	−2.6
$No^{3+} + e \rightleftharpoons No^{2+}$	1.4	$Po^{4+} + 2 e \rightleftharpoons Po^{2+}$	0.9
$No^{3+} + 3 e \rightleftharpoons No$	−1.20	$Po^{4+} + 4 e \rightleftharpoons Po$	0.76
$No^{2+} + 2 e \rightleftharpoons No$	−2.50	$Pr^{4+} + e \rightleftharpoons Pr^{3+}$	3.2
$Np^{3+} + 3 e \rightleftharpoons Np$	−1.856	$Pr^{2+} + 2 e \rightleftharpoons Pr$	−2.0
$Np^{4+} + e \rightleftharpoons Np^{3+}$	0.147	$Pr^{3+} + 3 e \rightleftharpoons Pr$	−2.353
$NpO_2 + H_2O + H^+ + e \rightleftharpoons Np(OH)_3$	−0.962	$Pr^{3+} + e \rightleftharpoons Pr^{2+}$	−3.1
$O_2 + 2 H^+ + 2 e \rightleftharpoons H_2O_2$	0.695	$Pt^{2+} + 2 e \rightleftharpoons Pt$	1.18
$O_2 + 4 H^+ + 4 e \rightleftharpoons 2 H_2O$	1.229	$[PtCl_4]^{2-} + 2 e \rightleftharpoons Pt + 4 Cl^-$	0.755
$O_2 + H_2O + 2 e \rightleftharpoons HO_2^- + OH^-$	−0.076	$[PtCl_6]^{2-} + 2 e \rightleftharpoons [PtCl_4]^{2-} + 2 Cl^-$	0.68
$O_2 + 2 H_2O + 2 e \rightleftharpoons H_2O_2 + 2 OH^-$	−0.146	$Pt(OH)_2 + 2 e \rightleftharpoons Pt + 2 OH^-$	0.14
$O_2 + 2 H_2O + 4 e \rightleftharpoons 4 OH^-$	0.401	$PtO_3 + 2 H^+ + 2 e \rightleftharpoons PtO_2 + H_2O$	1.7
$O_3 + 2 H^+ + 2 e \rightleftharpoons O_2 + H_2O$	2.076	$PtO_3 + 4 H^+ + 2 e \rightleftharpoons Pt(OH)_2^{2+} + H_2O$	1.5
$O_3 + H_2O + 2 e \rightleftharpoons O_2 + 2 OH^-$	1.24	$PtOH^+ + H^+ + 2 e \rightleftharpoons Pt + H_2O$	1.2
$O(g) + 2 H^+ + 2 e \rightleftharpoons H_2O$	2.421	$PtO_2 + 4 H^+ + 2 e \rightleftharpoons PtO + 2 H_2O$	1.01
$OH + e \rightleftharpoons OH^-$	2.02	$PtO_2 + 4 H^+ + 4 e \rightleftharpoons Pt + 2 H_2O$	1.00
$HO_2^- + H_2O + 2 e \rightleftharpoons 3 OH^-$	0.878	$Pu^{3+} + 3 e \rightleftharpoons Pu$	−2.031
$OsO_4 + 8 H^+ + 8 e \rightleftharpoons Os + 4 H_2O$	0.838	$Pu^{4+} + e \rightleftharpoons Pu^{3+}$	1.006
$OsO_4 + 4 H^+ + 4 e \rightleftharpoons OsO_2 + 2 H_2O$	1.02	$Pu^{5+} + e \rightleftharpoons Pu^{4+}$	1.099
$[Os(bipy)_2]^{3+} + e \rightleftharpoons [Os(bipy)_2]^{2+}$	0.81	$PuO_2(OH)_2 + 2 H^+ + 2 e \rightleftharpoons Pu(OH)_4$	1.325
$[Os(bipy)_3]^{3+} + e \rightleftharpoons [Os(bipy)_3]^{2+}$	0.80	$PuO_2(OH)_2 + H^+ + e \rightleftharpoons PuO_2OH + H_2O$	1.062
$P(red) + 3 H^+ + 3 e \rightleftharpoons PH_3(g)$	−0.111	$Ra^{2+} + 2 e \rightleftharpoons Ra$	−2.8
$P(white) + 3 H^+ + 3 e \rightleftharpoons PH_3(g)$	−0.063	$Rb^+ + e \rightleftharpoons Rb$	−2.98
$P + 3 H_2O + 3 e \rightleftharpoons PH_3(g) + 3 OH^-$	−0.87	$Re^{3+} + 3 e \rightleftharpoons Re$	0.300

TABLE 1
Alphabetical Listing (continued)

Reaction	$E°/V$	Reaction	$E°/V$
$ReO_4^- + 4\,H^- + 3\,e \rightleftharpoons ReO_2 + 2\,H_2O$	0.510	$SiO_2 \text{ (quartz)} + 4\,H^+ + 4\,e \rightleftharpoons Si + 2\,H_2O$	0.857
$ReO_2 + 4\,H^+ + 4\,e \rightleftharpoons Re + 2\,H_2O$	0.2513	$SiO_3^{2-} + 3\,H_2O + 4\,e \rightleftharpoons Si + 6\,OH^-$	−1.697
$ReO_4^- + 2\,H^+ + e \rightleftharpoons ReO_3 + H_2O$	0.768	$Sm^{3+} + e \rightleftharpoons Sm^{2+}$	−1.55
$ReO_4^- + 4\,H_2O + 7\,e \rightleftharpoons Re + 8\,OH^-$	−0.584	$Sm^{3+} + 3\,e \rightleftharpoons Sm$	−2.304
$ReO_4^- + 8\,H^+ + 7\,e \rightleftharpoons Re + 4\,H_2O$	0.368	$Sm^{2+} + 2\,e \rightleftharpoons Sm$	−2.68
$Rh^+ + e \rightleftharpoons Rh$	0.600	$Sn^{2+} + 2\,e \rightleftharpoons Sn$	−0.1375
$Rh^+ + 2\,e \rightleftharpoons Rh$	0.600	$Sn^{4+} + 2\,e \rightleftharpoons Sn^{2+}$	0.151
$Rh^{3+} + 3\,e \rightleftharpoons Rh$	0.758	$Sn(OH)_3^+ + 3\,H^+ + 2\,e \rightleftharpoons Sn^{2+} + 3\,H_2O$	0.142
$[RhCl_6]^{3-} + 3\,e \rightleftharpoons Rh + 6\,Cl^-$	0.431	$SnO_2 + 4\,H^+ + 2\,e^- \rightleftharpoons Sn^{2+} + 2\,H_2O$	−0.094
$RhOH^{2+} + H^+ + 3\,e \rightleftharpoons Rh + H_2O$	0.83	$SnO_2 + 4\,H^+ + 4\,e \rightleftharpoons Sn + 2\,H_2O$	−0.117
$Ru^{2+} + 2\,e \rightleftharpoons Ru$	0.455	$SnO_2 + 3\,H^+ + 2\,e \rightleftharpoons SnOH^+ + H_2O$	−0.194
$Ru^{3+} + e \rightleftharpoons Ru^{2+}$	0.2487	$SnO_2 + 2\,H_2O + 4\,e \rightleftharpoons Sn + 4\,OH^-$	−0.945
$RuO_2 + 4\,H^+ + 2\,e \rightleftharpoons Ru^{2+} + 2\,H_2O$	1.120	$HSnO_2^- + H_2O + 2\,e \rightleftharpoons Sn + 3\,OH^-$	−0.909
$RuO_4^- + e \rightleftharpoons RuO_4^{2-}$	0.59	$Sn(OH)_6^{2-} + 2\,e \rightleftharpoons HSnO_2^- + 3\,OH^- + H_2O$	−0.93
$RuO_4 + e \rightleftharpoons RuO_4^-$	1.00	$Sr^+ + e \rightleftharpoons Sr$	−4.10
$RuO_4 + 6\,H^+ + 4\,e \rightleftharpoons Ru(OH)_2^{2+} + 2\,H_2O$	1.40	$Sr^{2+} + 2\,e \rightleftharpoons Sr$	−2.899
$RuO_4 + 8\,H^+ + 8\,e \rightleftharpoons Ru + 4\,H_2O$	1.038	$Sr^{2+} + 2\,e \rightleftharpoons Sr(Hg)$	−1.793
$[Ru(bipy)_3]^{3+} + e^- \rightleftharpoons [Ru(bipy)_3]^{2+}$	1.24	$Sr(OH)_2 + 2\,e \rightleftharpoons Sr + 2\,OH^-$	−2.88
$[Ru(H_2O)_6]^{3+} + e^- \rightleftharpoons [Ru(H_2O)_6]^{2+}$	0.23	$Ta_2O_5 + 10\,H^+ + 10\,e \rightleftharpoons 2\,Ta + 5\,H_2O$	−0.750
$[Ru(NH_3)_6]^{3+} + e^- \rightleftharpoons [Ru(NH_3)_6]^{2+}$	0.10	$Ta^{3+} + 3\,e \rightleftharpoons Ta$	−0.6
$[Ru(en)_3]^{3+} + e^- \rightleftharpoons Ru(en)_3^{2+}$	0.210	$Tc^{2+} + 2\,e \rightleftharpoons Tc$	0.400
$[Ru(CN)_6]^{3-} + e^- \rightleftharpoons [Ru(CN)_6]^{4-}$	0.86	$TcO_4^- + 4\,H^+ + 3\,e \rightleftharpoons TcO_2 + 2\,H_2O$	0.782
$S + 2\,e \rightleftharpoons S^{2-}$	−0.47627	$Tc^{3+} + e \rightleftharpoons Tc^{2+}$	0.3
$S + 2\,H^+ + 2\,e \rightleftharpoons H_2S(aq)$	0.142	$TcO_4^- + 8\,H^+ + 7\,e \rightleftharpoons Tc + 4\,H_2O$	0.472
$S + H_2O + 2\,e \rightleftharpoons SH^- \, OH^-$	−0.478	$Tb^{4+} + e \rightleftharpoons Tb^{3+}$	3.1
$2\,S + 2\,e \rightleftharpoons S_2^{2-}$	−0.42836	$Tb^{3+} + 3\,e \rightleftharpoons Tb$	−2.28
$S_2O_6^{2-} + 4\,H^+ + 2\,e \rightleftharpoons 2\,H_2SO_3$	0.564	$Te + 2\,e \rightleftharpoons Te^{2-}$	−1.143
$S_2O_8^{2-} + 2\,e \rightleftharpoons 2\,SO_4^{2-}$	2.010	$Te + 2\,H^+ + 2\,e \rightleftharpoons H_2Te$	−0.793
$S_2O_8^{2-} + 2\,H^+ + 2\,e \rightleftharpoons 2\,HSO_4^-$	2.123	$Te^{4+} + 4\,e \rightleftharpoons Te$	0.568
$S_4O_6^{2-} + 2\,e \rightleftharpoons 2\,S_2O_3^{2-}$	0.08	$TeO_2 + 4\,H^+ + 4\,e \rightleftharpoons Te + 2\,H_2O$	0.593
$2\,H_2SO_3 + H^+ + 2\,e \rightleftharpoons HS_2O_4^- + 2\,H_2O$	−0.056	$TeO_3^{2-} + 3\,H_2O + 4\,e \rightleftharpoons Te + 6\,OH^-$	−0.57
$H_2SO_3 + 4\,H^+ + 4\,e \rightleftharpoons S + 3\,H_2O$	0.449	$TeO_4^- + 8\,H^+ + 7\,e \rightleftharpoons Te + 4\,H_2O$	0.472
$2\,SO_3^{2-} + 2\,H_2O + 2\,e \rightleftharpoons S_2O_4^{2-} + 4\,OH^-$	−1.12	$H_6TeO_6 + 2\,H^+ + 2\,e \rightleftharpoons TeO_2 + 4\,H_2O$	1.02
$2\,SO_3^{2-} + 3\,H_2O + 4\,e \rightleftharpoons S_2O_3^{2-} + 6\,OH^-$	−0.571	$Th^{4+} + 4\,e \rightleftharpoons Th$	−1.899
$SO_4^{2-} + 4\,H^+ + 2\,e \rightleftharpoons H_2SO_3 + H_2O$	0.172	$ThO_2 + 4\,H^+ + 4\,e \rightleftharpoons Th + 2\,H_2O$	−1.789
$2\,SO_4^{2-} + 4\,H^+ + 2\,e \rightleftharpoons S_2O_6^{2-} + H_2O$	−0.22	$Th(OH)_4 + 4\,e \rightleftharpoons Th + 4\,OH^-$	−2.48
$SO_4^{2-} + H_2O + 2\,e \rightleftharpoons SO_3^{2-} + 2\,OH^-$	−0.93	$Ti^{2+} + 2\,e \rightleftharpoons Ti$	−1.630
$Sb + 3\,H^+ + 3\,e \rightleftharpoons SbH_3$	−0.510	$Ti^{3+} + e \rightleftharpoons Ti^{2+}$	−0.9
$Sb_2O_3 + 6\,H^+ + 6\,e \rightleftharpoons 2\,Sb + 3\,H_2O$	0.152	$TiO_2 + 4\,H^+ + 2\,e \rightleftharpoons Ti^{2+} + 2\,H_2O$	−0.502
$Sb_2O_5 \text{ (senarmontite)} + 4\,H^+ + 4\,e \rightleftharpoons Sb_2O_3 + 2\,H_2O$	0.671	$Ti^{3+} + 3\,e \rightleftharpoons Ti$	−1.37
$Sb_2O_5 \text{ (valentinite)} + 4\,H^+ + 4\,e \rightleftharpoons Sb_2O_3 + 2\,H_2O$	0.649	$TiOH^{3+} + H^+ + e \rightleftharpoons Ti^{3+} + H_2O$	−0.055
$Sb_2O_5 + 6\,H^+ + 4\,e \rightleftharpoons 2\,SbO^+ + 3\,H_2O$	0.581	$Tl^+ + e \rightleftharpoons Tl$	−0.336
$SbO^+ + 2\,H^+ + 3\,e \rightleftharpoons Sb + 2\,H_2O$	0.212	$Tl^+ + e \rightleftharpoons Tl(Hg)$	−0.3338
$SbO_2^- + 2\,H_2O + 3\,e \rightleftharpoons Sb + 4\,OH^-$	−0.66	$Tl^{3+} + 2\,e \rightleftharpoons Tl^+$	1.252
$SbO_3^- + H_2O + 2\,e \rightleftharpoons SbO_2^- + 2\,OH^-$	−0.59	$Tl^{3+} + 3\,e \rightleftharpoons Tl$	0.741
$Sc^{3+} + 3\,e \rightleftharpoons Sc$	−2.077	$TlBr + e \rightleftharpoons Tl + Br^-$	−0.658
$Se + 2\,e \rightleftharpoons Se^{2-}$	−0.924	$TlCl + e \rightleftharpoons Tl + Cl^-$	−0.5568
$Se + 2\,H^+ + 2\,e \rightleftharpoons H_2Se(aq)$	−0.399	$TlI + e \rightleftharpoons Tl + I^-$	−0.752
$H_2SeO_3 + 4\,H^+ + 4\,e \rightleftharpoons Se + 3\,H_2O$	0.74	$Tl_2O_3 + 3\,H_2O + 4\,e \rightleftharpoons 2\,Tl^+ + 6\,OH^-$	0.02
$Se + 2\,H^+ + 2\,e \rightleftharpoons H_2Se$	−0.082	$TlOH + e \rightleftharpoons Tl + OH^-$	−0.34
$SeO_3^{2-} + 3\,H_2O + 4\,e \rightleftharpoons Se + 6\,OH^-$	−0.366	$Tl(OH)_3 + 2\,e \rightleftharpoons TlOH + 2\,OH^-$	−0.05
$SeO_4^{2-} + 4\,H^+ + 2\,e \rightleftharpoons H_2SeO_3 + H_2O$	1.151	$Tl_2SO_4 + 2\,e \rightleftharpoons Tl + SO_4^{2-}$	−0.4360
$SeO_4^{2-} + H_2O + 2\,e \rightleftharpoons SeO_3^{2-} + 2\,OH^-$	0.05	$Tm^{3+} + e \rightleftharpoons Tm^{2+}$	−2.2
$SiF_6^{2-} + 4\,e \rightleftharpoons Si + 6\,F^-$	−1.24	$Tm^{3+} + 3\,e \rightleftharpoons Tm$	−2.319
$SiO + 2\,H^+ + 2\,e \rightleftharpoons Si + H_2O$	−0.8	$Tm^{2+} + 2\,e \rightleftharpoons Tm$	−2.4

TABLE 1
Alphabetical Listing (continued)

Reaction	$E°$/V	Reaction	$E°$/V
$U^{3+} + 3\,e \rightleftharpoons U$	−1.798	$2\,WO_3 + 2\,H^+ + 2\,e \rightleftharpoons W_2O_5 + H_2O$	−0.029
$U^{4+} + e \rightleftharpoons U^{3+}$	−0.607	$H_4XeO_6 + 2\,H^+ + 2\,e \rightleftharpoons XeO_3 + 3\,H_2O$	2.42
$UO_2^+ + 4\,H^+ + e \rightleftharpoons U^{4+} + 2\,H_2O$	0.612	$XeO_3 + 6\,H^+ + 6\,e \rightleftharpoons Xe + 3\,H_2O$	2.10
$UO_2^{2+} + e \rightleftharpoons UO^+_2$	0.062	$XeF + e \rightleftharpoons Xe + F^-$	3.4
$UO_2^{2+} + 4\,H^+ + 2\,e \rightleftharpoons U^{4+} + 2\,H_2O$	0.327	$Y^{3+} + 3\,e \rightleftharpoons Y$	−2.372
$UO_2^{2+} + 4\,H^+ + 6\,e \rightleftharpoons U + 2\,H_2O$	−1.444	$Yb^{3+} + e \rightleftharpoons Yb^{2+}$	−1.05
$V^{2+} + 2\,e \rightleftharpoons V$	−1.175	$Yb^{3+} + 3\,e \rightleftharpoons Yb$	−2.19
$V^{3+} + e \rightleftharpoons V^{2+}$	−0.255	$Yb^{2+} + 2\,e \rightleftharpoons Yb$	−2.76
$VO^{2+} + 2\,H^+ + e \rightleftharpoons V^{3+} + H_2O$	0.337	$Zn^{2+} + 2\,e \rightleftharpoons Zn$	−0.7618
$VO_2^+ + 2\,H^+ + e \rightleftharpoons VO^{2+} + H_2O$	0.991	$Zn^{2+} + 2\,e \rightleftharpoons Zn(Hg)$	−0.7628
$V_2O_5 + 6\,H^+ + 2\,e \rightleftharpoons 2\,VO^{2+} + 3\,H_2O$	0.957	$ZnO_2^{2-} + 2\,H_2O + 2\,e \rightleftharpoons Zn + 4\,OH^-$	−1.215
$V_2O_5 + 10\,H^+ + 10\,e \rightleftharpoons 2\,V + 5\,H_2O$	−0.242	$ZnSO_4 + 7\,H_2O + 2\,e = Zn(Hg) + SO_4^{2-}$	−0.7993
$V(OH)_4^+ + 2\,H^+ + e \rightleftharpoons VO^{2+} + 3\,H_2O$	1.00	(Saturated $ZnSO_4$)	
$V(OH)_4^+ + 4\,H^+ + 5\,e \rightleftharpoons V + 4\,H_2O$	−0.254	$ZnOH^+ + H^+ + 2\,e \rightleftharpoons Zn + H_2O$	−0.497
$[V(phen)_3]^{3+} + e \rightleftharpoons [V(phen)_3]^{2+}$	0.14	$Zn(OH)_4^{2-} + 2\,e \rightleftharpoons Zn + 4\,OH^-$	−1.199
$W^{3+} + 3\,e \rightleftharpoons W$	0.1	$Zn(OH)_2 + 2\,e \rightleftharpoons Zn + 2\,OH^-$	−1.249
$W_2O_5 + 2\,H^+ + 2\,e \rightleftharpoons 2\,WO_2 + H_2O$	−0.031	$ZnO + H_2O + 2\,e \rightleftharpoons Zn + 2\,OH^-$	−1.260
$WO_2 + 4\,H^+ + 4\,e \rightleftharpoons W + 2\,H_2O$	−0.119	$ZrO_2 + 4\,H^+ + 4\,e \rightleftharpoons Zr + 2\,H_2O$	−1.553
$WO_3 + 6\,H^+ + 6\,e \rightleftharpoons W + 3\,H_2O$	−0.090	$ZrO(OH)_2 + H_2O + 4\,e \rightleftharpoons Zr + 4\,OH^-$	−2.36
$WO_3 + 2\,H^+ + 2\,e \rightleftharpoons WO_2 + H_2O$	0.036	$Zr^{4+} + 4\,e \rightleftharpoons Zr$	−1.45

TABLE 2
Reduction Reactions Having $E°$ Values More Positive than that of the Standard Hydrogen Electrode

Reaction	$E°$/V	Reaction	$E°$/V
$2\,H^+ + 2\,e \rightleftharpoons H_2$	0.00000	$Sn(OH)_3^+ + 3\,H^+ + 2\,e \rightleftharpoons Sn^{2+} + 3\,H_2O$	0.142
$CuI_2^- + e \rightleftharpoons Cu + 2\,I^-$	0.00	$Np^{4+} + e \rightleftharpoons Np^{3+}$	0.147
$Ge^{4+} + 2\,e \rightleftharpoons Ge^{2+}$	0.00	$Ag_4[Fe(CN)_6] + 4\,e \rightleftharpoons 4\,Ag + [Fe(CN)_6]^{4-}$	0.1478
$NO_3^- + H_2O + 2\,e \rightleftharpoons NO_2^- + 2\,OH^-$	0.01	$IO_3^- + 2\,H_2O + 4\,e \rightleftharpoons IO^- + 4\,OH^-$	0.15
$Tl_2O_3 + 3\,H_2O + 4\,e \rightleftharpoons 2\,Tl^+ + 6\,OH^-$	0.02	$Mn(OH)_3 + e \rightleftharpoons Mn(OH)_2 + OH^-$	0.15
$SeO_4^{2-} + H_2O + 2\,e \rightleftharpoons SeO_3^{2-} + 2\,OH^-$	0.05	$2\,NO_2^- + 3\,H_2O + 4\,e \rightleftharpoons N_2O + 6\,OH^-$	0.15
$WO_3 + 2\,H^+ + 2\,e \rightleftharpoons WO_2 + H_2O$	0.036	$Sn^{4+} + 2\,e \rightleftharpoons Sn^{2+}$	0.151
$UO_2^{2+} + e = UO_2^-$	0.062	$Sb_2O_3 + 6\,H^+ + 6\,e \rightleftharpoons 2\,Sb + 3\,H_2O$	0.152
$Pd(OH)_2 + 2\,e \rightleftharpoons Pd + 2\,OH^-$	0.07	$Cu^{2+} + e \rightleftharpoons Cu^+$	0.153
$AgBr + e \rightleftharpoons Ag + Br^-$	0.07133	$BiOCl + 2\,H^+ + 3\,e \rightleftharpoons Bi + Cl^- + H_2O$	0.1583
$MoO_3 + 6\,H^+ + 6\,e \rightleftharpoons Mo + 3\,H_2O$	0.075	$BiCl_4^- + 3\,e \rightleftharpoons Bi + 4\,Cl^-$	0.16
$S_4O_6^{2-} + 2\,e \rightleftharpoons 2\,S_2O_3^{2-}$	0.08	$Fe_2O_3 + 4\,H^+ + 2\,e \rightleftharpoons 2\,FeOH^+ + H_2O$	0.16
$H_3Mo_7O_{24}^{3-} + 45\,H^+ + 42\,e \rightleftharpoons 7\,Mo + 24\,H_2O$	0.082	$Co(OH)_3 + e \rightleftharpoons Co(OH)_2 + OH^-$	0.17
$AgSCN + e \rightleftharpoons Ag + SCN^-$	0.8951	$SO_4^{2-} + 4\,H^+ + 2\,e \rightleftharpoons H_2SO_3 + H_2O$	0.172
$N_2 + 2\,H_2O + 6\,H^+ + 6\,e \rightleftharpoons 2\,NH_4OH$	0.092	$Bi^{3+} + 2\,e \rightleftharpoons Bi^+$	0.2
$HgO + H_2O + 2\,e \rightleftharpoons Hg + 2\,OH^-$	0.0977	$[Ru(en)_3]^{3+} + e \rightleftharpoons [Ru(en)_3]^{2+}$	0.210
$Ir_2O_3 + 3\,H_2O + 6\,e \rightleftharpoons 2\,Ir + 6\,OH^-$	0.098	$SbO^+ + 2\,H^+ + 3\,e \rightleftharpoons Sb + 2\,H_2O$	0.212
$2\,NO + 2\,e \rightleftharpoons N_2O_2^{2-}$	0.10	$AgCl + e \rightleftharpoons Ag + Cl^-$	0.22233
$[Ru(NH_3)_6]^{3+} + e \rightleftharpoons [Ru(NH_3)_6]^{2+}$	0.10	$[Ru(H_2O)_6]^{3+} + e \rightleftharpoons [Ru(H_2O)_6]^{2+}$	0.23
$W^{3+} + 3\,e \rightleftharpoons W$	0.1	$As_2O_3 + 6\,H^+ + 6\,e \rightleftharpoons 2\,As + 3\,H_2O$	0.234
$[Co(NH_3)_6]^{3+} + e \rightleftharpoons [Co(NH_3)_6]^{2+}$	0.108	Calomel electrode, saturated NaCl (SSCE)	0.2360
$Hg_2O + H_2O + 2\,e \rightleftharpoons 2\,Hg + 2\,OH^-$	0.123	$Ge^{2+} + 2\,e \rightleftharpoons Ge$	0.24
$Ge^{4+} + 4\,e \rightleftharpoons Ge$	0.124	$Ru^{3+} + e \rightleftharpoons Ru^{2+}$	0.24
$Hg_2Br_2 + 2\,e \rightleftharpoons 2\,Hg + 2\,Br^-$	0.13923	Calomel electrode, saturated KCl	0.2412
$Pt(OH)_2 + 2\,e \rightleftharpoons Pt + 2\,OH^-$	0.14	$PbO_2 + H_2O + 2\,e \rightleftharpoons PbO + 2\,OH^-$	0.247
$[V(phen)_3]^{3+} + e \rightleftharpoons [V(phen)_3]^{2+}$	0.14	$HAsO_2 + 3\,H^+ + 3\,e \rightleftharpoons As + 2\,H_2O$	0.248
$S + 2\,H^+ + 2\,e \rightleftharpoons H_2S(aq)$	0.142	$Ru^{3+} + e \rightleftharpoons Ru^{2+}$	0.2487

TABLE 2
Reduction Reactions Having $E°$ Values More Positive than that of the Standard Hydrogen Electrode (continued)

Reaction	$E°$/V	Reaction	$E°$/V
$ReO_2 + 4 H^+ + 4 e \rightleftharpoons Re + 2 H_2O$	0.2513	$[PdCl_4]^{2-} + 2 e \rightleftharpoons Pd + 4 Cl^-$	0.591
$IO_3^- + 3 H_2O + 6 e \rightleftharpoons I^- + OH^-$	0.26	$TeO_2 + 4 H^+ + 4 e \rightleftharpoons Te + 2 H_2O$	0.593
$Hg_2Cl_2 + 2 e \rightleftharpoons 2 Hg + 2 Cl^-$	0.26808	$MnO_4^- + 2 H_2O + 3 e \rightleftharpoons MnO_2 + 4 OH^-$	0.595
Calomel electrode, 1 molal KCl	0.2800	$Rh^{2+} + 2 e \rightleftharpoons Rh$	0.600
Calomel electrode, 1 molar KCl (NCE)	0.2801	$Rh^+ + e \rightleftharpoons Rh$	0.600
$At_2 + 2 e \rightleftharpoons 2 At^-$	0.3	$MnO_4^{2-} + 2 H_2O + 2 e \rightleftharpoons MnO_2 + 4 OH^-$	0.60
$Re^{3+} + 3 e \rightleftharpoons Re$	0.300	$2 AgO + H_2O + 2 e \rightleftharpoons Ag_2O + 2 OH^-$	0.607
$Tc^{3+} + e \rightleftharpoons Tc^{2+}$	0.3	$BrO_3^- + 3 H_2O + 6 e \rightleftharpoons Br^- + 6 OH^-$	0.61
$Bi^{3+} + 3 e \rightleftharpoons Bi$	0.308	$UO_2^+ + 4 H^+ + e \rightleftharpoons U^{4+} + 2 H_2O$	0.612
$BiO^+ + 2 H^+ + 3 e \rightleftharpoons Bi + H_2O$	0.320	$Hg_2SO_4 + 2 e \rightleftharpoons 2 Hg + SO_4^{2-}$	0.6125
$UO_2^{2+} + 4 H^+ + 2 e \rightleftharpoons U^{4+} + 2 H_2O$	0.327	$ClO_3^- + 3 H_2O + 6 e \rightleftharpoons Cl^- + 6 OH^-$	0.62
$ClO_3^- + H_2O + 2 e \rightleftharpoons ClO_2^- + 2 OH^-$	0.33	$Hg_2HPO_4 + 2 e \rightleftharpoons 2 Hg + HPO_4^{2-}$	0.6359
$2 HCNO + 2 H^+ + 2 e \rightleftharpoons (CN)_2 + 2 H_2O$	0.330	$Ag(ac) + e \rightleftharpoons Ag + (ac)^-$	0.643
Calomel electrode, 0.1 molar KCl	0.3337	$Sb_2O_5(valentinite) + 4 H^+ + 4 e \rightleftharpoons Sb_2O_3 + 2 H_2O$	0.649
$VO^{2+} + 2 H^+ + e \rightleftharpoons V^{3+} + H_2O$	0.337	$Ag_2SO_4 + 2 e \rightleftharpoons 2 Ag + SO_4^{2-}$	0.654
$Cu^{2+} + 2 e \rightleftharpoons Cu$	0.3419	$ClO_2^- + H_2O + 2 e \rightleftharpoons ClO^- + 2 OH^-$	0.66
$Ag_2O + H_2O + 2 e \rightleftharpoons 2 Ag + 2 OH^-$	0.342	$Sb_2O_5(senarmontite) + 4 H^+ + 4 e \rightleftharpoons Sb_2O_5 + 2 H_2O$	0.671
$Cu^{2+} + 2 e \rightleftharpoons Cu(Hg)$	0.345	$[PtCl_6]^{2-} + 2 e \rightleftharpoons [PtCl_4]^{2-} + 2 Cl^-$	0.68
$AgIO_3 + e \rightleftharpoons Ag + IO_3^-$	0.354	$O_2 + 2 H^+ + 2 e \rightleftharpoons H_2O_2$	0.695
$[Fe(CN)_6]^{3-} + e \rightleftharpoons [Fe(CN)_6]^{4-}$	0.358	$p-benzoquinone + 2 H^+ + 2 e \rightleftharpoons hydroquinone$	0.6992
$ClO_4^- + H_2O + 2 e \rightleftharpoons ClO_3^- + 2 OH^-$	0.36	$H_3IO_6 + 2 e \rightleftharpoons IO_3^- + 3 OH^-$	0.7
$Ag_2SeO_3 + 2 e \rightleftharpoons 2 Ag + SeO_3^{2-}$	0.3629	$Ag_2O_3 + H_2O + 2 e \rightleftharpoons 2 AgO + 2 OH^-$	0.739
$ReO_4^- + 8 H^+ + 7 e \rightleftharpoons Re + 4 H_2O$	0.368	$Tl^{3+} + 3 e \rightleftharpoons Tl$	0.741
$(CN)_2 + 2 H^+ + 2 e \rightleftharpoons 2 HCN$	0.373	$[PtCl_4]^{2-} + 2 e \rightleftharpoons Pt + 4 Cl^-$	0.755
$[Ferricinium]^+ + e \rightleftharpoons ferrocene$	0.400	$Rh^{3+} + 3 e \rightleftharpoons Rh$	0.758
$Tc^{2+} + 2 e \rightleftharpoons Tc$	0.400	$ClO_2^- + 2 H_2O + 4 e \rightleftharpoons Cl^- + 4 OH^-$	0.76
$O_2 + 2 H_2O + 4 e \rightleftharpoons 4 OH^-$	0.401	$2 NO + H_2O + 2 e \rightleftharpoons N_2O + 2 OH^-$	0.76
$AgOCN + e \rightleftharpoons Ag + OCN^-$	0.41	$Po^{4+} + 4 e \rightleftharpoons Po$	0.76
$[RhCl_6]^{3-} + 3 e \rightleftharpoons Rh + 6 Cl^-$	0.431	$BrO^- + H_2O + 2 e \rightleftharpoons Br^- + 2 OH^-$	0.761
$Ag_2CrO_4 + 2 e \rightleftharpoons 2 Ag + CrO_4^{2-}$	0.4470	$ReO_4^- + 2 H^+ + e \rightleftharpoons ReO_3 + H_2O$	0.768
$H_2SO_3 + 4 H^+ + 4 e \rightleftharpoons S + 3 H_2O$	0.449	$(CNS)_2 + 2 e \rightleftharpoons 2 CNS^-$	0.77
$Ru^{2+} + 2 e \rightleftharpoons Ru$	0.455	$[IrCl_6]^{3-} + 3 e \rightleftharpoons Ir + 6 Cl^-$	0.77
$Ag_2MoO_4 + 2 e \rightleftharpoons 2 Ag + MoO_4^{2-}$	0.4573	$Fe^{3+} + e \rightleftharpoons Fe^{2+}$	0.771
$Ag_2C_2O_4 + 2 e \rightleftharpoons 2 Ag + C_2O_4^{2-}$	0.4647	$AgF + e \rightleftharpoons Ag + F^-$	0.779
$Ag_2WO_4 + 2 e \rightleftharpoons 2 Ag + WO_4^{2-}$	0.4660	$[Fe(bipy)_2]^{3+} + e \rightleftharpoons [Fe(bipy)_2]^{2+}$	0.78
$Ag_2CO_3 + 2 e \rightleftharpoons 2 Ag + CO_3^{2-}$	0.47	$TcO_4^- + 4 H^+ + 3 e \rightleftharpoons TcO_2 + 2 H_2O$	0.782
$TcO_4^- + 8 H^+ + 7 e \rightleftharpoons Tc + 4 H_2O$	0.472	$Hg_2^{2+} + 2 e \rightleftharpoons 2 Hg$	0.7973
$TeO_4^- + 8 H^+ + 7 e \rightleftharpoons Te + 4 H_2O$	0.472	$Ag^+ + e \rightleftharpoons Ag$	0.7996
$IO^- + H_2O + 2 e \rightleftharpoons I^- + 2 OH^-$	0.485	$[Os(bipy)_3]^{3+} + e \rightleftharpoons [Os(bipy)_3]^{2+}$	0.80
$NiO_2 + 2 H_2O + 2 e \rightleftharpoons Ni(OH)_2 + 2 OH^-$	0.490	$2 NO_3^- + 4 H^+ + 2 e \rightleftharpoons N_2O_4 + 2 H_2O$	0.803
$Bi^+ + e \rightleftharpoons Bi$	0.5	$[Os(bipy)_2]^{3+} + e \rightleftharpoons [Os(bipy)_2]^{2+}$	0.81
$ReO_4^- + 4 H^+ + 3 e \rightleftharpoons ReO_2 + 2 H_2O$	0.510	$RhOH^{2+} + H + 3 e \rightleftharpoons Rh + H_2O$	0.83
$Hg_2(ac)_2 + 2 e \rightleftharpoons 2 Hg + 2(ac)^-$	0.51163	$OsO_4 + 8 H^+ + 8 e \rightleftharpoons Os + 4 H_2O$	0.838
$Cu^+ + e \rightleftharpoons Cu$	0.521	$ClO^- + H_2O + 2 e \rightleftharpoons Cl^- + 2 OH^-$	0.841
$I_2 + 2 e \rightleftharpoons 2 I^-$	0.5355	$Hg^{2+} + 2 e \rightleftharpoons Hg$	0.851
$I_3^- + 2 e \rightleftharpoons 3 I^-$	0.536	$AuBr_4^- + 3 e \rightleftharpoons Au + 4 Br^-$	0.854
$AgBrO_3 + e \rightleftharpoons Ag + BrO_3^-$	0.546	$SiO_2(quartz) + 4 H^+ + 4 e \rightleftharpoons Si + 2 H_2O$	0.857
$MnO_4^- + e \rightleftharpoons MnO_4^{2-}$	0.558	$2 HNO_2 + 4 H^+ + 4 e \rightleftharpoons H_2N_2O_2 + H_2O$	0.86
$H_3AsO_4 + 2 H^+ + 2 e \rightleftharpoons HAsO_2 + 2 H_2O$	0.560	$[Ru(CN)_6]^{3-} + e^- \rightleftharpoons [Ru(CN)_6]^{4-}$	0.86
$S_2O_6^{2-} + 4 H^+ + 2 e \rightleftharpoons 2 H_2SO_3$	0.564	$[IrCl_6]^{2-} + e \rightleftharpoons [IrCl_6]^{3-}$	0.8665
$AgNO_2 + e \rightleftharpoons Ag + NO_2^-$	0.564	$N_2O_4 + 2 e \rightleftharpoons 2 NO_2^-$	0.867
$Te^{4+} + 4 e \rightleftharpoons Te$	0.568	$HO_2^- + H_2O + 2 e \rightleftharpoons 3 OH^-$	0.878
$Sb_2O_5 + 6 H^+ + 4 e \rightleftharpoons 2 SbO^+ + 3 H_2O$	0.581	$Po^{4+} + 2 e \rightleftharpoons Po^{2+}$	0.9
$RuO_4^- + e \rightleftharpoons RuO_4^{2-}$	0.59	$2 Hg^+ + 2 e \rightleftharpoons Hg_2^{2+}$	0.920

TABLE 2
Reduction Reactions Having $E°$ Values More Positive than that of the Standard Hydrogen Electrode
(continued)

Reaction	$E°/V$	Reaction	$E°/V$
$NO_3^- + 3 H^+ + 2 e \rightleftharpoons HNO_2 + H_2O$	0.934	$Cl_2(g) + 2 e \rightleftharpoons 2 Cl^-$	1.35827
$Pd^{2+} + 2 e \rightleftharpoons Pd$	0.951	$ClO_4^- + 8 H^+ + 8 e \rightleftharpoons Cl^- + 4 H_2O$	1.389
$ClO_2(aq) + e \rightleftharpoons ClO_2^-$	0.954	$ClO_4^- + 8 H^+ + 7 e \rightleftharpoons 1/2 Cl_2 + 4 H_2O$	1.39
$NO_3^- + 4 H^+ + 3 e \rightleftharpoons NO + 2 H_2O$	0.957	$No^{3+} + e \rightleftharpoons No^{2+}$	1.4
$V_2O_5 + 6 H^+ + 2 e \rightleftharpoons 2 VO^{2+} + 3 H_2O$	0.957	$RuO_4 + 6 H^+ + 4 e \rightleftharpoons Ru(OH)_2^{2+} + 2 H_2O$	1.40
$AuBr_2^- + e \rightleftharpoons Au + 2 Br^-$	0.959	$Au^{3+} + 2 e \rightleftharpoons Au^+$	1.401
$HNO_2 + H^+ + e \rightleftharpoons NO + H_2O$	0.983	$2 NH_3OH^+ + H^+ + 2 e \rightleftharpoons N_2H_5^+ + 2 H_2O$	1.42
$HIO + H^+ + 2 e \rightleftharpoons I^- + H_2O$	0.987	$BrO_3^- + 6 H^+ + 6 e \rightleftharpoons Br^- + 3 H_2O$	1.423
$VO_2^+ + 2 H^+ + e \rightleftharpoons VO^{2+} + H_2O$	0.991	$2 HIO + 2 H^+ + 2 e \rightleftharpoons I_2 + 2 H_2O$	1.439
$PtO_2 + 4 H^+ + 4 e \rightleftharpoons Pt + 2 H_2O$	1.00	$Au(OH)_3 + 3 H^+ + 3 e \rightleftharpoons Au^- + 3 H_2O$	1.45
$RuO_4 + e \rightleftharpoons RuO_4^-$	1.00	$3 IO_3^- + 6 H^+ + 6 e \rightleftharpoons Cl^- + 3 H_2O$	1.451
$V(OH)_4^+ + 2 H^+ + e \rightleftharpoons VO^{2+} + 3 H_2O$	1.00	$PbO_2 + 4 H^+ + 2 e \rightleftharpoons Pb^{2+} + 2 H_2O$	1.455
$AuCl_4^- + 3 e \rightleftharpoons Au + 4 Cl^-$	1.002	$ClO_3^- + 6 H^+ + 5 e \rightleftharpoons 1/2 Cl_2 + 3 H_2O$	1.47
$Pu^{4+} + e \rightleftharpoons Pu^{3+}$	1.006	$CrO_2 + 4 H^+ + e \rightleftharpoons Cr^{3+} + 2 H_2O$	1.48
$PtO_2 + 4 H^+ + 2 e \rightleftharpoons PtO + 2 H_2O$	1.01	$BrO_3^- + 6 H^+ + 5 e \rightleftharpoons 1/2 Br_2 + 3 H_2O$	1.482
$OsO_4 + 4 H + 4 e \rightleftharpoons OsO_2 + 2 H_2O$	1.02	$HClO + H^+ + 2 e \rightleftharpoons Cl^- + H_2O$	1.482
$H_6TeO_6 + 2 H^+ + 2 e \rightleftharpoons TeO_2 + 4 H_2O$	1.02	$Mn_2O_3 + 6 H + e \rightleftharpoons Mn^{2+} + 3 H_2O$	1.485
$[Fe(bipy)_3]^{3+} + e \rightleftharpoons [Fe(bipy)_3]^{2+}$	1.03	$HO_2 + H^+ + e \rightleftharpoons H_2O_2$	1.495
$Hg(OH)_2 + 2 H^+ + 2 e \rightleftharpoons Hg + 2 H_2O$	1.034	$Au^{3+} + 3 e \rightleftharpoons Au$	1.498
$N_2O_4 + 4 H^+ + 4 e \rightleftharpoons 2 NO + 2 H_2O$	1.035	$PtO_3 + 4 H^+ + 2 e \rightleftharpoons Pt(OH)_2^{2+} + H_2O$	1.5
$RuO_4 + 8 H^+ + 8 e \rightleftharpoons Ru + 4 H_2O$	1.038	$MnO_4^- + 8 H^+ + 5 e \rightleftharpoons Mn^{2+} + 4 H_2O$	1.507
$[Fe(phen)_3]^{3+} + e \rightleftharpoons [Fe(phen)_3]^{2+}$ (1 molar H_2SO_4)	1.06	$Mn^{3+} + e \rightleftharpoons Mn^{2-}$	1.5415
$PuO_2(OH)_2 + H^+ + e \rightleftharpoons PuO_2OH + H_2O$	1.062	$HClO_2 + 3 H^+ + 4 e \rightleftharpoons Cl^- + 2 H_2O$	1.570
$N_2O_4 + 2 H^+ + 2 e \rightleftharpoons 2 HNO_2$	1.065	$HBrO + H^+ + e \rightleftharpoons 1/2 Br_2(aq) + H_2O$	1.574
$Br_2(\ell) + 2 e \rightleftharpoons 2 Br^-$	1.066	$2 NO + 2 H^+ + 2 e \rightleftharpoons N_2O + H_2O$	1.591
$IO_3^- + 6 H^+ + 6 e \rightleftharpoons I^- + 3 H_2O$	1.085	$Bi_2O_4 + 4 H^+ + 2 e \rightleftharpoons 2 BiO^+ + 2 H_2O$	1.593
$Br_2(aq) + 2 e \rightleftharpoons 2 Br^-$	1.0873	$HBrO + H^+ + e \rightleftharpoons 1/2 Br_2(l) + H_2O$	1.596
$Pu^{5+} + e \rightleftharpoons Pu^{4+}$	1.099	$H_5IO_6 + H^+ + 2 e \rightleftharpoons IO_3^- + 3 H_2O$	1.601
$Cu^{2+} + 2 CN^- + e \rightleftharpoons [Cu(CN)_2]^-$	1.103	$HClO + H^+ + e \rightleftharpoons 1/2 Cl_2 + H_2O$	1.611
$RuO_2 + 4 H^+ + 2 e \rightleftharpoons Ru^{2+} + 2 H_2O$	1.120	$HClO_2 + 3 H^+ + 3 e \rightleftharpoons 1/2 Cl_2 + 2 H_2O$	1.628
$[Fe(phen)_3]^{3+} + e \rightleftharpoons [Fe(phen)_3]^{2+}$	1.147	$HClO_2 + 2 H^+ + 2 e \rightleftharpoons HClO + H_2O$	1.645
$SeO_4^{2-} + 4 H^+ + 2 e \rightleftharpoons H_2SeO_3 + H_2O$	1.151	$Bk^{4+} + e \rightleftharpoons Bk^{3+}$	1.67
$ClO_3^- + 2 H^+ + e \rightleftharpoons ClO_2 + H_2O$	1.152	$NiO_2 + 4 H^+ + 2 e \rightleftharpoons Ni^{2+} + 2 H_2O$	1.678
$Ir^{3+} + 3 e \rightleftharpoons Ir$	1.156	$MnO_4^- + 4 H^+ + 3 e \rightleftharpoons MnO_2 + 2 H_2O$	1.679
$Pt^{2+} + 2 e \rightleftharpoons Pt$	1.18	$PbO_2 + SO_4^{2-} + 4 H^+ + 2 e \rightleftharpoons PbSO_4 + 2 H_2O$	1.6913
$ClO_4^- + 2 H^+ + 2 e \rightleftharpoons ClO_3^- + H_2O$	1.189	$Au^+ + e \rightleftharpoons Au$	1.692
$2 IO_3^- + 12 H^+ + 10 e \rightleftharpoons I_2 + 6 H_2O$	1.195	$PtO_3 + 2 H^+ + 2 e \rightleftharpoons PtO_2 + H_2O$	1.7
$PtOH^+ + H^+ + 2 e \rightleftharpoons Pt + H_2O$	1.2	$CeOH^{3+} + H^+ + e \rightleftharpoons Ce^{3+} + H_2O$	1.715
$ClO_3^- + 3 H^+ + 2 e \rightleftharpoons HClO_2 + H_2O$	1.214	$Ce^{4+} + e \rightleftharpoons Ce^{3+}$	1.72
$MnO_2 + 4 H^+ + 2 e \rightleftharpoons Mn^{2+} + 2 H_2O$	1.224	$N_2O + 2 H^+ + 2 e \rightleftharpoons N_2 + H_2O$	1.766
$O_2 + 4 H^+ + 4 e \rightleftharpoons 2 H_2O$	1.229	$H_2O_2 + 2 H^+ + 2 e \rightleftharpoons 2 H_2O$	1.776
$Cr_2O_7^{2-} + 14 H^+ + 6 e \rightleftharpoons 2 Cr^{3+} + 7 H_2O$	1.232	$Ag^{3+} + e \rightleftharpoons Ag^{2+}$	1.8
$O_3 + H_2O + 2 e \rightleftharpoons O_2 + 2 OH^-$	1.24	$Au^{2+} + e^- \rightleftharpoons Au^+$	1.8
$[Ru(bipy)_3]^{3+} + e \rightleftharpoons [Ru(bipy)_3]^{2+}$	1.24	$Ag_2O_2 + 4 H^+ + e \rightleftharpoons 2 Ag + 2 H_2O$	1.802
$Tl^{3+} + 2 e \rightleftharpoons Tl^+$	1.252	$Co^{3+} + e \rightleftharpoons Co^{2-}$ (2 molar H_2SO_4)	1.83
$N_2H_5^+ + 3 H^+ + 2 e \rightleftharpoons 2 NH_4^+$	1.275	$Ag^{3+} + 2 e \rightleftharpoons Ag^+$	1.9
$ClO_2 + H^+ + e \rightleftharpoons HClO_2$	1.277	$Co^{3+} + e \rightleftharpoons Co^{2+}$	1.92
$[PdCl_6]^{2-} + 2 e \rightleftharpoons [PdCl_4]^{2-} + 2 Cl^-$	1.288	$Ag^{2+} + e \rightleftharpoons Ag^+$	1.980
$2 HNO_2 + 4 H^+ + 4 e \rightleftharpoons N_2O + 3 H_2O$	1.297	$Cu_2O_3 + 6 H^+ + e \rightleftharpoons Cu^{2+} + 3 H_2O$	2.0
$Cr_2O_5 + 10 H^+ + 6 e \rightleftharpoons CrO_2 + 5 H_2O$	1.3	$S_2O_8^{2-} + 2 e \rightleftharpoons 2 SO_4^{2-}$	2.010
$AuOH^{2+} + H^+ + 2 e \rightleftharpoons Au^+ + H_2O$	1.32	$OH + e \rightleftharpoons OH^-$	2.02
$PuO_2(OH)_2 + 2 H^- + 2 e \rightleftharpoons Pu(OH)_4$	1.325	$HFeO_4^- + 8 H^+ + 3 e \rightleftharpoons Fe^{3+} + 4 H_2O$	2.07
$HBrO + H^+ + 2 e \rightleftharpoons Br^- + H_2O$	1.331	$O_3 + 2 H^+ + 2 e \rightleftharpoons O_2 + H_2O$	2.076
$HCrO_4^- + 7 H^+ + 3 e \rightleftharpoons Cr^{3+} + 4 H_2O$	1.350	$HFeO_4^- + 4 H^+ + 3 e \rightleftharpoons FeOOH + 2 H_2O$	2.08

TABLE 2
Reduction Reactions Having $E°$ Values More Positive than that of the Standard Hydrogen Electrode (continued)

Reaction	$E°/V$	Reaction	$E°/V$
$HFeO_4^- + 2\,H^+ + 3\,e \rightleftharpoons Fe_2O_3 + H_2O$	2.09	$H_2N_2O_2 + 2\,H^+ + 2\,e \rightleftharpoons N_2 + 2\,H_2O$	2.65
$XeO_3 + 6\,H^+ + 6\,e \rightleftharpoons Xe + 3\,H_2O$	2.10	$F_2 + 2\,e \rightleftharpoons 2\,F^-$	2.866
$S_2O_8^{2-} + 2\,H^+ + 2\,e \rightleftharpoons 2\,HSO_4^-$	2.123	$Cm^{4+} + e \rightleftharpoons Cm^{3+}$	3.0
$F_2O + 2\,H^+ + 4\,e \rightleftharpoons H_2O + 2\,F^-$	2.153	$F_2 + 2\,H^+ + 2\,e \rightleftharpoons 2\,HF$	3.053
$FeO_4^{2-} + 8\,H^+ + 3\,e \rightleftharpoons Fe^{3+} + 4\,H_2O$	2.20	$Tb^{4+} + e \rightleftharpoons Tb^{3+}$	3.1
$Cu^{3+} + e \rightleftharpoons Cu^{2+}$	2.4	$Pr^{4+} + e \rightleftharpoons Pr^{3+}$	3.2
$H_4XeO_6 + 2\,H^+ + 2\,e \rightleftharpoons XeO_3 + 3\,H_2O$	2.42	$Cf^{4+} + e \rightleftharpoons Cf^{3+}$	3.3
$O(g) + 2\,H^+ + 2\,e \rightleftharpoons H_2O$	2.421	$XeF + e \rightleftharpoons Xe + F^-$	3.4
$Am^{4+} + e \rightleftharpoons Am^{3+}$	2.60		

TABLE 3
Reduction Reactions Having $E°$ Values More Negative than that of the Standard Hydrogen Electrode

Reaction	$E°/V$	Reaction	$E°/V$
$2\,H^+ + 2\,e \rightleftharpoons H_2$	0.00000	$Cu(OH)_2 + 2\,e \rightleftharpoons Cu + 2\,OH^-$	−0.222
$2\,D^+ + 2\,e \rightleftharpoons D_2$	−0.013	$V_2O_5 + 10\,H^+ + 10\,e \rightleftharpoons 2\,V + 5\,H_2O$	−0.242
$AgCN + e \rightleftharpoons Ag + CN^-$	−0.017	$CdSO_4 + 2\,e \rightleftharpoons Cd + SO_4^{2-}$	−0.246
$2\,WO_3 + 2\,H^+ + 2\,e \rightleftharpoons W_2O_5 + H_2O$	−0.029	$V(OH)_4^+ + 4\,H^+ + 5\,e \rightleftharpoons V + 4\,H_2O$	−0.254
$W_2O_5 + 2\,H^+ + 2\,e \rightleftharpoons 2\,WO_2 + H_2O$	−0.031	$V^{3+} + e \rightleftharpoons V^{2+}$	−0.255
$Ag_2S + 2\,H^+ + 2\,e \rightleftharpoons 2\,Ag + H_2S$	−0.0366	$Ni^{2+} + 2\,e \rightleftharpoons Ni$	−0.257
$Fe^{3+} + 3\,e \rightleftharpoons Fe$	−0.037	$PbCl_2 + 2\,e \rightleftharpoons Pb + 2\,Cl^-$	−0.2675
$Hg_2I_2 + 2\,e \rightleftharpoons 2\,Hg + 2\,I^-$	−0.0405	$H_3PO_4 + 2\,H^+ + 2\,e \rightleftharpoons H_3PO_3 + H_2O$	−0.276
$Tl(OH)_3 + 2\,e \rightleftharpoons TlOH + 2\,OH^-$	−0.05	$Co^{2+} + 2\,e \rightleftharpoons Co$	−0.28
$TiOH^{3+} + H^+ + e \rightleftharpoons Ti^{3+} + H_2O$	−0.055	$PbBr_2 + 2\,e \rightleftharpoons Pb + 2\,Br^-$	−0.284
$2\,H_2SO_3 + H^+ + 2\,e \rightleftharpoons HS_2O_4^- + 2\,H_2O$	−0.056	$Tl^+ + e \rightleftharpoons Tl(Hg)$	−0.3338
$P(white) + 3\,H^+ + 3\,e \rightleftharpoons PH_3(g)$	−0.063	$Tl^+ + e \rightleftharpoons Tl$	−0.336
$O_2 + H_2O + 2\,e \rightleftharpoons HO_2^- + OH^-$	−0.076	$In^{3+} + 3\,e \rightleftharpoons In$	−0.3382
$2\,Cu(OH)_2 + 2\,e \rightleftharpoons Cu_2O + 2\,OH^- + H_2O$	−0.080	$TlOH + e \rightleftharpoons Tl + OH^-$	−0.34
$Se + 2\,H^+ + 2\,e \rightleftharpoons H_2Se$	−0.082	$PbF_2 + 2\,e \rightleftharpoons Pb + 2\,F^-$	−0.3444
$WO_3 + 6\,H^+ + 6\,e \rightleftharpoons W + 3\,H_2O$	−0.090	$PbSO_4 + 2\,e \rightleftharpoons Pb(Hg) + SO_4^{2-}$	−0.3505
$SnO_2 + 4\,H^+ + 2\,e \rightleftharpoons Sn^{2+} + 2\,H_2O$	−0.094	$Cd^{2+} + 2\,e \rightleftharpoons Cd(Hg)$	−0.3521
$Md^{3+} + e \rightleftharpoons Md^{2+}$	−0.1	$PbSO_4 + 2\,e \rightleftharpoons Pb + SO_4^{2-}$	−0.3588
$P(red) + 3\,H^+ + 3\,e \rightleftharpoons PH_3(g)$	−0.111	$Cu_2O + H_2O + 2\,e \rightleftharpoons 2\,Cu + 2\,OH^-$	−0.360
$SnO_2 + 4\,H^+ + 4\,e \rightleftharpoons Sn + 2\,H_2O$	−0.117	$Eu^{3+} + e \rightleftharpoons Eu^{2+}$	−0.36
$GeO_2 + 2\,H^+ + 2\,e \rightleftharpoons GeO + H_2O$	−0.118	$PbI_2 + 2\,e \rightleftharpoons Pb + 2\,I^-$	−0.365
$WO_2 + 4\,H^+ + 4\,e \rightleftharpoons W + 2\,H_2O$	−0.119	$SeO_3^{2-} + 3\,H_2O + 4\,e \rightleftharpoons Se + 6\,OH^-$	−0.366
$Pb^{2+} + 2\,e \rightleftharpoons Pb(Hg)$	−0.1205	$Se + 2\,H^+ + 2\,e \rightleftharpoons H_2Se(aq)$	−0.399
$Pb^{2+} + 2\,e \rightleftharpoons Pb$	−0.1262	$In^{2+} + e \rightleftharpoons In^+$	−0.40
$CrO_4^{2-} + 4\,H_2O + 3\,e \rightleftharpoons Cr(OH)_3 + 5\,OH^-$	−0.13	$Cd^{2+} + 2\,e \rightleftharpoons Cd$	−0.4030
$Sn^{2-} + 2\,e \rightleftharpoons Sn$	−0.1375	$Cr^{3+} + e \rightleftharpoons Cr^{2+}$	−0.407
$In^+ + e \rightleftharpoons In$	−0.14	$2\,S + 2\,e \rightleftharpoons S_2^{2-}$	−0.42836
$O_2 + 2\,H_2O + 2\,e \rightleftharpoons H_2O_2 + 2\,OH^-$	−0.146	$Tl_2SO_4 + 2\,e \rightleftharpoons Tl + SO_4^{2-}$	−0.4360
$MoO_2 + 4\,H^+ + e \rightleftharpoons Mo + 4\,H_2O$	−0.152	$In^{3+} + 2\,e \rightleftharpoons In^+$	−0.443
$AgI + e \rightleftharpoons Ag + I^-$	−0.15224	$Fe^{2+} + 2\,e \rightleftharpoons Fe$	−0.447
$2\,NO_2^- + 2\,H_2O + 4\,e \rightleftharpoons N_2O_2^2 + 4\,OH^-$	−0.18	$H_3PO_3 + 3\,H^+ + 3\,e \rightleftharpoons P + 3\,H_2O$	−0.454
$H_2GeO_3 + 4\,H^+ + 4\,e \rightleftharpoons Ge + 3\,H_2O$	−0.182	$Bi_2O_3 + 3\,H_2O + 6\,e \rightleftharpoons 2\,Bi + 6\,OH^-$	−0.46
$SnO_2 + 3\,H^+ + 2\,e \rightleftharpoons SnOH^+ + H_2O$	−0.194	$NO_2^- + H_2O + e \rightleftharpoons NO + 2\,OH$	−0.46
$CO_2 + 2\,H^+ + 2\,e \rightleftharpoons HCOOH$	−0.199	$PbHPO_4 + 2\,e \rightleftharpoons Pb + HPO_4^{2-}$	−0.465
$Mo^{3+} + 3\,e \rightleftharpoons Mo$	−0.200	$S + 2\,e \rightleftharpoons S^{2-}$	−0.47627
$Ga^+ + e \rightleftharpoons Ga$	−0.2	$S + H_2O + 2\,e \rightleftharpoons HS^- + OH^-$	−0.478
$2\,SO_2^{2-} + 4\,H^+ + 2\,e \rightleftharpoons S_2O_6^{2-} + H_2O$	−0.22	$B(OH)_3 + 7\,H^+ + 8\,e \rightleftharpoons BH_4^- + 3\,H_2O$	−0.481

TABLE 3
Reduction Reactions Having $E°$ Values More Negative than that of the Standard Hydrogen Electrode
(continued)

Reaction	$E°/V$	Reaction	$E°/V$
$In^{3+} + e \rightleftharpoons In^{2+}$	−0.49	$SnO_2 + 2 H_2O + 4 e \rightleftharpoons Sn + 4 OH^-$	−0.945
$ZnOH^+ + H^+ + 2 e \rightleftharpoons Zn + H_2O$	−0.497	$In(OH)_3 + 3 e \rightleftharpoons In + 3 OH^-$	−0.99
$GaOH^{2+} + H^+ + 3 e \rightleftharpoons Ga + H_2O$	−0.498	$NpO_2 + H_2O + H^+ + e \rightleftharpoons Np(OH)_3$	−0.962
$H_3PO_3 + 2 H^+ + 2 e \rightleftharpoons H_3PO_2 + H_2O$	−0.499	$In(OH)_4^- + 3 e \rightleftharpoons In + 4 OH^-$	−1.007
$TiO_2 + 4 H^+ + 2 e \rightleftharpoons Ti^{2+} + 2 H_2O$	−0.502	$In_2O_3 + 3 H_2O + 6 e \rightleftharpoons 2 In + 6 OH^-$	−1.034
$H_3PO_2 + H^+ + e \rightleftharpoons P + 2 H_2O$	−0.508	$PO_4^{3-} + 2 H_2O + 2 e \rightleftharpoons HPO_3^{2-} + 3 OH^-$	−1.05
$Sb + 3 H^+ + 3 e \rightleftharpoons SbH_3$	−0.510	$Yb^{3+} + e \rightleftharpoons Yb^{2+}$	−1.05
$HPbO_2^- + H_2O + 2 e \rightleftharpoons Pb + 3 OH^-$	−0.537	$Nb^{3+} + 3 e \rightleftharpoons Nb$	−1.099
$Ga^{3+} + 3 e \rightleftharpoons Ga$	−0.549	$Fm^{3+} + e \rightleftharpoons Fm^{2+}$	−1.1
$TlCl + e \rightleftharpoons Tl + Cl^-$	−0.5568	$2 SO_3^{2-} + 2 H_2O + 2 e \rightleftharpoons S_2O_4^{2-} + 2 OH^-$	−1.12
$Fe(OH)_3 + e \rightleftharpoons Fe(OH)_2 + OH^-$	−0.56	$Te + 2 e \rightleftharpoons Te^{2-}$	−1.143
$TeO_3^{2-} + 3 H_2O + 4 e \rightleftharpoons Te + 6 OH^-$	−0.57	$V^{2+} + 2 e \rightleftharpoons V$	−1.175
$2 SO_3^{2-} + 3 H_2O + 4 e \rightleftharpoons S_2O_3^{2-} + 6 OH^-$	−0.571	$Mn^{2+} + 2 e \rightleftharpoons Mn$	−1.185
$PbO + H_2O + 2 e \rightleftharpoons Pb + 2 OH^-$	−0.580	$Zn(OH)_4^{2-} + 2 e \rightleftharpoons Zn + 40 H^-$	−1.199
$ReO_2 + 4 H_2O + 7 e \rightleftharpoons Re + 8 OH^-$	−0.584	$CrO_2 + 2 H_2O + 3 e \rightleftharpoons Cr + 4 OH^-$	−1.2
$SbO_3^- + H_2O + 2 e \rightleftharpoons SbO_2^- + 2 OH^-$	−0.59	$No^{3+} + 3 e \rightleftharpoons No$	−1.20
$Ta^{3+} + 3 e \rightleftharpoons Ta$	−0.6	$ZnO_2^- + 2 H_2O + 2 e \rightleftharpoons Zn + 4 OH^-$	−1.215
$U^{4+} + e \rightleftharpoons U^{3+}$	−0.607	$H_2GaO_3^- + H_2O + 3 e \rightleftharpoons Ga + 4 OH^-$	−1.219
$As + 3 H^+ + 3 e \rightleftharpoons AsH_3$	−0.608	$H_2BO_3^- + 5 H_2O + 8 e \rightleftharpoons BH_4^- + 8 OH^-$	−1.24
$Nb_2O_5 + 10 H^+ + 10 e \rightleftharpoons 2 Nb + 5 H_2O$	−0.644	$SiF_6^{2-} + 4 e \rightleftharpoons Si + 6 F^-$	−1.24
$NbO_2 + 2 H^+ + 2 e \rightleftharpoons NhO + H_2O$	−0.646	$Zn(OH)_2 + 2 e \rightleftharpoons Zn + 2 OH^-$	−1.249
$Cd(OH)_4^{2-} + 2 e \rightleftharpoons Cd + 4 OH^-$	−0.658	$ZnO + H_2O + 2 e \rightleftharpoons Zn + 2 OH^-$	−1.260
$TlBr + e \rightleftharpoons Tl + Br^-$	−0.658	$Es^{3+} + e \rightleftharpoons Es^{2+}$	−1.3
$SbO_2^- + 2 H_2O + 3 e \rightleftharpoons Sb + 4 OH^-$	−0.66	$Pa^{3+} + 3 e \rightleftharpoons Pa$	−1.34
$AsO_2^- + 2 H_2O + 3 e \rightleftharpoons As + 4 OH^-$	−0.68	$Ti^{3+} + 3 e \rightleftharpoons Ti$	−1.37
$NbO_2 + 4 H^+ + 4 e \rightleftharpoons Nb + 2 H_2O$	−0.690	$Ce^{3+} + 3 e \rightleftharpoons Ce(Hg)$	−1.4373
$Ag_2S + 2 e \rightleftharpoons 2 Ag + S^{2-}$	−0.691	$UO_2^{2+} + 4 H^+ + 6 e \rightleftharpoons U + 2 H_2O$	−1.444
$AsO_4^{3-} + 2 H_2O + 2 e \rightleftharpoons AsO_2^- + 4 OH^-$	−0.71	$Zr^{4+} + 4 e \rightleftharpoons Zr$	−1.45
$Ni(OH)_2 + 2 e \rightleftharpoons Ni + 2 OH^-$	−0.72	$Cr(OH)_3 + 3 e \rightleftharpoons Cr + 3 OH^-$	−1.48
$Co(OH)_2 + 2 e \rightleftharpoons Co + 2 OH^-$	−0.73	$Pa^{4+} + 4 e \rightleftharpoons Pa$	−1.49
$NbO + 2 H^+ + 2 e \rightleftharpoons Nb + H_2O$	−0.733	$HfO_2 + 4 H^+ + 4 e \rightleftharpoons Hf + 2 H_2O$	−1.505
$H_2SeO_3 + 4 H^+ + 4 e \rightleftharpoons Se + 3 H_2O$	−0.74	$Hf^{4+} + 4 e \rightleftharpoons Hf$	−1.55
$Cr^{3+} + 3 e \rightleftharpoons Cr$	−0.744	$Sm^{3+} + e \rightleftharpoons Sm^{2+}$	−1.55
$Ta_2O_5 + 10 H^+ + 10 e \rightleftharpoons 2 Ta + 5 H_2O$	−0.750	$ZrO_2 + 4 H^+ + 4 e \rightleftharpoons Zr + 2 H_2O$	−1.553
$TlI + e \rightleftharpoons Tl + I^-$	−0.752	$Mn(OH)_2 + 2 e \rightleftharpoons Mn + 2 OH^-$	−1.56
$Zn^{2+} + 2 e \rightleftharpoons Zn$	−0.7618	$Ba^{2+} + 2 e \rightleftharpoons Ba(Hg)$	−1.570
$Zn^{2+} + 2 e \rightleftharpoons Zn(Hg)$	−0.7628	$Bk^{2+} + 2 e \rightleftharpoons Bk$	−1.6
$CdO + H_2O + 2 e \rightleftharpoons Cd + 2 OH^-$	−0.783	$Cf^{3+} + e \rightleftharpoons Cf^{2+}$	−1.6
$Te + 2 H^+ + 2 e \rightleftharpoons H_2Te$	−0.793	$Ti^{2+} + 2 e \rightleftharpoons Ti$	−1.630
$ZnSO_4 \cdot 7H_2O + 2 e \rightleftharpoons Zn(Hg) + SO_4^{2-}$	−0.7993	$Md^{3+} + 3 e \rightleftharpoons Md$	−1.65
(Saturated $ZnSO_4$)		$HPO_3^{2-} + 2 H_2O + 2 e \rightleftharpoons H_2PO_2^- + 3 OH^-$	−1.65
$Bi + 3 H^+ + 3 e \rightleftharpoons BiH_3$	−0.8	$Al^{3+} + 3 e \rightleftharpoons Al$	−1.662
$SiO + 2 H^+ + 2 e \rightleftharpoons Si + H_2O$	−0.8	$SiO_3^{2-} + H_2O + 4 e \rightleftharpoons Si + 6 OH^-$	−1.697
$Cd(OH)_2 + 2 e \rightleftharpoons Cd(Hg) + 2 OH^-$	−0.809	$HPO_3^{2-} + 2 H_2O + 3 e \rightleftharpoons P + 5 OH^-$	−1.71
$2 H_2O + 2 e \rightleftharpoons H_2 + 2 OH^-$	−0.8277	$HfO^{2+} + 2 H^+ + 4 e \rightleftharpoons Hf + H_2O$	−1.724
$2 NO_3^- + 2 H_2O + 2 e \rightleftharpoons N_2O_4 + 4 OH^-$	−0.85	$ThO_2 + 4 H^+ + 4 e \rightleftharpoons Th + 2 H_2O$	−1.789
$H_3BO_3 + 3 H^+ + 3 e \rightleftharpoons B + 3 H_2O$	−0.8698	$H_2BO_3^- + H_2O + 3 e \rightleftharpoons B + 4 OH^-$	−1.79
$P + 3 H_2O + 3 e \rightleftharpoons PH_3(g) + 3 OH^-$	−0.87	$Sr^{2+} + 2 e \rightleftharpoons Sr(Hg)$	−1.793
$Ti^{3+} + e \rightleftharpoons Ti^{2+}$	−0.9	$U^{3+} + 3 e \rightleftharpoons U$	−1.798
$HSnO_2^- + H_2O + 2 e \rightleftharpoons Sn + 3 OH^-$	−0.909	$H_2PO_2^- + e \rightleftharpoons P + 2 OH^-$	−1.82
$Cr^{2+} + 2 e \rightleftharpoons Cr$	−0.913	$Be^{2+} + 2 e \rightleftharpoons Be$	−1.847
$Se + 2 e \rightleftharpoons Se^{2-}$	−0.924	$Np^{3+} + 3 e \rightleftharpoons Np$	−1.856
$SO_4^{2-} + H_2O + 2 e \rightleftharpoons SO_3^{2-} + 2 OH^-$	−0.93	$Fm^{3+} + 3 e \rightleftharpoons Fm$	−1.89
$Sn(OH)_6^{2-} + 2 e \rightleftharpoons HSnO_2^- + 3 OH^- + H_2O$	−0.93	$Th^{4+} + 4 e \rightleftharpoons Th$	−1.899

TABLE 3
Reduction Reactions Having $E°$ Values More Negative than that of the Standard Hydrogen Electrode
(continued)

Reaction	$E°$/V	Reaction	$E°$/V
$Am^{2+} + 2e \rightleftharpoons Am$	−1.9	$ZrO(OH)_2 + H_2O + 4e \rightleftharpoons Zr + 4 OH^-$	−2.36
$Pa^{4+} + e \rightleftharpoons Pa^{3+}$	−1.9	$Mg^{2+} + 2e \rightleftharpoons Mg$	−2.372
$Es^{3+} + 3e \rightleftharpoons Es$	−1.91	$Y^{3+} + 3e \rightleftharpoons Y$	−2.372
$Cf^{3+} + 3e \rightleftharpoons Cf$	−1.94	$La^{3+} + 3e \rightleftharpoons La$	−2.379
$Lr^{3+} + 3e \rightleftharpoons Lr$	−1.96	$Tm^{2+} + 2e \rightleftharpoons Tm$	−2.4
$Eu^{3+} + 3e \rightleftharpoons Eu$	−1.991	$Md^{2+} + 2e \rightleftharpoons Md$	−2.40
$Er^{2+} + 2e \rightleftharpoons Er$	−2.0	$Th(OH)_4 + 4e \rightleftharpoons Th + 4 OH^-$	−2.48
$Pr^{2+} + 2e \rightleftharpoons Pr$	−2.0	$HfO(OH)_2 + H_2O + 4e \rightleftharpoons Hf + 4 OH^-$	−2.50
$Pu^{3+} + e \rightleftharpoons Pu$	−2.031	$No^{2+} + 2e \rightleftharpoons No$	−2.50
$Cm^{3+} + 3e \rightleftharpoons Cm$	−2.04	$Dy^{3+} + e \rightleftharpoons Dy^{2+}$	−2.6
$Am^{3+} + 3e \rightleftharpoons Am$	−2.048	$Pm^{3+} + e \rightleftharpoons Pm^{2+}$	−2.6
$AlF_6^{3-} + 3e \rightleftharpoons Al + 6 F^-$	−2.069	$Be_2O_3^{2-} + 3 H_2O + 4e \rightleftharpoons 2 Be + 6 OH^-$	−2.63
$Sc^{3+} + 3e \rightleftharpoons Sc$	−2.077	$Sm^{2+} + 2e \rightleftharpoons Sm$	−2.68
$Ho^{2+} + 2e \rightleftharpoons Ho$	−2.1	$Mg(OH)_2 + 2e \rightleftharpoons Mg + 2 OH^-$	−2.690
$Nd^{2+} + 2e \rightleftharpoons Nd$	−2.1	$Nd^{3+} + e \rightleftharpoons Nd^{2+}$	−2.7
$Cf^{2+} + 2e \rightleftharpoons Cf$	−2.12	$Mg^+ + e \rightleftharpoons Mg$	−2.70
$Yb^{3+} + 3e \rightleftharpoons Yb$	−2.19	$Na^+ + e \rightleftharpoons Na$	−2.71
$Ac^{3+} + 3e \rightleftharpoons Ac$	−2.20	$Yb^{2+} + 2e \rightleftharpoons Yb$	−2.76
$Dy^{2+} + 2e \rightleftharpoons Dy$	−2.2	$Bk^{3+} + e \rightleftharpoons Bk^{2+}$	−2.8
$Tm^{3+} + e \rightleftharpoons Tm^{2+}$	−2.2	$Ho^{3+} + e \rightleftharpoons Ho^{2+}$	−2.8
$Pm^{2+} + 2e \rightleftharpoons Pm$	−2.2	$Ra^{2+} + 2e \rightleftharpoons Ra$	−2.8
$Es^{2+} + 2e \rightleftharpoons Es$	−2.23	$Eu^{2+} + 2e \rightleftharpoons Eu$	−2.812
$H_2 + 2e \rightleftharpoons 2 H^-$	−2.23	$Ca^{2+} + 2e \rightleftharpoons Ca$	−2.868
$Gd^{3+} + 3e \rightleftharpoons Gd$	−2.279	$Sr(OH)_2 + 2e \rightleftharpoons Sr + 2 OH^-$	−2.88
$Tb^{3+} + 3e \rightleftharpoons Tb$	−2.28	$Sr^{2+} + 2e \rightleftharpoons Sr$	−2.89
$Lu^{3+} + 3e \rightleftharpoons Lu$	−2.28	$Fr^+ + e \rightleftharpoons Fr$	−2.9
$Dy^{3+} + 3e \rightleftharpoons Dy$	−2.295	$La(OH)_3 + 3e \rightleftharpoons La + 3 OH^-$	−2.90
$Am^{3+} + e \rightleftharpoons Am^{2+}$	−2.3	$Ba^{2+} + 2e \rightleftharpoons Ba$	−2.912
$Fm^{2+} + 2e \rightleftharpoons Fm$	−2.30	$K^+ + e \rightleftharpoons K$	−2.931
$Pm^{3+} + 3e \rightleftharpoons Pm$	−2.30	$Rb^+ + e \rightleftharpoons Rb$	−2.98
$Sm^{3+} + 3e \rightleftharpoons Sm$	−2.304	$Ba(OH)_2 + 2e \rightleftharpoons Ba + 2 OH^-$	−2.99
$Al(OH)_3 + 3e \rightleftharpoons Al + 3 OH^-$	−2.31	$Er^{3+} + e \rightleftharpoons Er^{2+}$	−3.0
$Tm^{3+} + 3e \rightleftharpoons Tm$	−2.319	$Ca(OH)_2 + 2e \rightleftharpoons Ca + 2 OH^-$	−3.02
$Nd^{3+} + 3e \rightleftharpoons Nd$	−2.323	$Cs^+ + e \rightleftharpoons Cs$	−3.026
$Al(OH)_4^- + 3e \rightleftharpoons Al + 4 OH^-$	−2.328	$Li^+ + e \rightleftharpoons Li$	−3.0401
$H_2AlO_3^- + H_2O + 3e \rightleftharpoons Al + 4 OH^-$	−2.33	$3 N_2 + 2 H^+ + 2e \rightleftharpoons 2 HN_3$	−3.09
$Ho^{3+} + 3e \rightleftharpoons Ho$	−2.33	$Pr^{3+} + e \rightleftharpoons Pr^{2+}$	−3.1
$Er^{3+} + 3e \rightleftharpoons Er$	−2.331	$Ca^+ + e \rightleftharpoons Ca$	−3.80
$Ce^{3+} + 3e \rightleftharpoons Ce$	−2.336	$Sr^+ + e \rightleftharpoons Sr$	−4.10
$Pr^{3+} + 3e \rightleftharpoons Pr$	−2.353		

REDUCTION AND OXIDATION POTENTIALS FOR CERTAIN ION RADICALS

Petr Vanýsek

There are two tables for ion radicals. The first table lists reduction potentials for organic compounds which produce anion radicals during reduction, a process described as $A + e^- 1 A^{-\cdot}$. The second table lists oxidation potentials for organic compounds which produce cation radicals during oxidation, a process described as $A 1 A^{+\cdot} + e^-$. To obtain reduction potential for a reverse reaction, the sign for the potential is changed.

Unlike the table of the Electrochemical Series, which lists *standard* potentials, values for radicals are experimental values with experimental conditions given in the second column. Since the measurements leading to potentials for ion radicals are very dependent on conditions, an attempt to report standard potentials for radicals would serve no useful purpose. For the same reason, the potentials are also reported as experimental values, usually a half-wave potential ($E_{1/2}$ in polarography) or a peak potential (E_p in cyclic voltammetry). Unless otherwise stated, the values are reported vs. SCE (saturated calomel electrode). To obtain a value vs. normal hydrogen electrode, 0.241 V has to be added to the SCE values. All the ion radicals chosen for inclusion in the tables result from electrochemically reversible reactions. More detailed data on ion radicals can be found in the *Encyclopedia of Electrochemistry of Elements,* (A. J. Bard, Ed.), Vol. XI and XII in particular, Marcel Dekker, New York, 1978.

Abbreviations are: CV — cyclic voltammetry; DMF — *N,N*-Dimethylformamide; E swp — potential sweep; $E°$ — standard potential; E_p — peak potential; $E_{p/2}$ — half-peak potential; $E_{1/2}$ — half wave potential; M — mol/L; MeCN — acetonitrile; pol — polarography; rot Pt dsk — rotated Pt disk; SCE — saturated calomel electrode; TBABF$_4$ — tetrabutylammonium tetrafluoroborate; TBAI — tetrabutylammonium iodide; TBAP — tetrabutylammonium perchlorate; TEABr — tetraethylammonium bromide; TEAP — tetraethylammonium perchlorate; THF — tetrahydrofuran; TPACF$_3$SO$_3$ — tetrapropylammonium trifluoromethanesulfite; TPAP — tetrapropylammonium perchlorate; and wr — wire.

Reduction Potentials (Products are Anion Radicals)

Substance	Conditions/electrode/technique	Potential V (vs. SCE)
Acetone	DMF, 0.1 M TEABr/Hg/pol	$E_{1/2} = -2.84$
1-Naphthyphenylacetylene	DMF, 0.03 M TBAI/Hg/pol	$E_{1/2} = -1.91$
1-Naphthalenecarboxyaldehyde	-/Hg/pol	$E_{1/2} = -0.91$
2-Naphthalenecarboxyaldehyde	-/Hg/pol	$E_{1/2} = -0.96$
2-Phenanthrenecarboxaldehyde	-/Hg/pol	$E_{1/2} = -1.00$
3-Phenanthrenecarboxaldehyde	-/Hg/pol	$E_{1/2} = -0.94$
9-Phenanthrenecarboxaldehyde	-/Hg/pol	$E_{1/2} = -0.83$
1-Anthracenecarboxaldehyde	-/Hg/pol	$E_{1/2} = -0.75$
1-Pyrenecarboxaldehyde	-/Hg/pol	$E_{1/2} = -0.76$
2-Pyrenecarboxaldehyde	-/Hg/pol	$E_{1/2} = -1.00$
Anthracene	DMF, 0.1 M TBAP/Pt dsk/CV	$E_p = -2.00$
	DMF, 0.5 M TBABF$_4$/Hg/CV	$E_{1/2} = -1.93$
	MeCN, 0.1 M TEAP/Hg/CV	$E_{1/2} = -2.07$
	DMF, 0.1 M TBAI/Hg/pol	$E_{1/2} = -1.92$
9,10-Dimethylanthracene	DMF, 0.1 M TBAP/Pt/CV	$E_p = -2.08$
	MeCN, 0.1 M TBAP/Pt/CV	$E_p = -2.10$
1-Phenylanthracene	DMF, 0.5 M TBABF$_2$/Hg/CV	$E_{1/2} = -1.91$
	DMF, 0.1 M TBAI/Hg/pol	$E_{1/2} = -1.878$
2-Phenylanthracene	DMF, 0.1 M TBAI/Hg/pol	$E_{1/2} = -1.875$
8-Phenylanthracene	DMF, 0.5 M TBABF$_4$/Hg/CV	$E_{1/2} = -1.91$
9-Phenylanthracene	DMF, 0.5 M TBABF$_4$/Hg/CV	$E_{1/2} = -1.93$
	DMF, 0.1 M TBAI/Hg/pol	$E_{1/2} = -1.863$
1,8-Diphenylanthracene	DMF, 0.5 M TBABF$_4$/Hg/CV	$E_{1/2} = -1.88$
1,9-Diphenylanthracene	DMF, 0.1 M TBAI/Hg/pol	$E_{1/2} = -1.846$
1,10-Diphenylanthracene	DMF, 0.1 M TBAI/Hg/pol	$E_{1/2} = -1.786$
8,9-Diphenylanthracene	DMF, 0.5 M TBABF$_4$/Hg/CV	$E_{1/2} = -1.90$
9,10-Diphenylanthracene	MeCN, 0.1 M TBAP/rot Pt/E swp	$E_{1/2} = -1.83$
	DMF, 0.1 M TBAI/Hg/pol	$E_{1/2} = -1.835$
1,8,9-Triphenylanthracene	DMF, 0.5 M TBABF$_4$/Hg/CV	$E_{1/2} = -1.85$
1,8,10-Triphenylanthracene	DMF, 0.5 M TBABF$_4$/Hg/CV	$E_{1/2} = -1.81$
9,10-Dibiphenylanthracene	MeCN, 0.1 M TBAP/rot Pt/E swp	$E_{1/2} = -1.94$
Benz(a)anthracene	MeCN, 0.1 M TEAP/Hg/CV	$E_{1/2} = -2.11$
	MeCN, 0.1 M TEAP/Hg/pol	$E_{1/2} = -2.40$[a]
Azulene	DMF, 0.1 M TBAI/Hg/pol	$E_{1/2} = -1.10$[c]
Annulene	DMF, 0.5 M TBAP 0°C/Hg/pol	$E_{1/2} = -1.23$
Benzaldehyde	DMF, 0.1 M TBAP/Hg/pol	$E_{1/2} = -1.67$
Benzil	DMSO, 0.1 M TBAP/Hg/pol	$E_{1/2} = -1.04$

Substance	Conditions/electrode/technique	Potential V (vs. SCE)
Benzophenone	-/Hg/pol	$E_{1/2} = -1.80$
	DMF/Pt dsk/CV	$E° = -1.72$
Chrysene	MeCN, 0.1 M TEAP/Hg/pol	$E_{1/2} = -2.73$[a]
Fluoranthrene	DMF, 0.1 M TBAP/Pt dsk/CV	$E_p = -1.76$
Cyclohexanone	DMF, 0.1 M TEABr/Hg/pol	$E_{1/2} = -2.79$
5,5-Dimethyl-3-phenyl-2-cyclohexen-1-one	DMF, 0.5 M/Hg/pol	$E_{1/2} = -1.71$
1,2,3-Indanetrione hydrate (ninhydrin)	DMF, 0.2 M NaNO$_3$/Hg/pol	$E_{1/2} = -0.039$
Naphthacene	DMF, 0.1 M TBAI/Hg/pol	$E_{1/2} = -1.53$
Naphthalene	DMF, 0.1 M TBAP/Pt dsk/CV	$E_p = -2.55$
	DMF, 0.5 M TBABF$_4$/Hg/CV	$E_{1/2} = -2.56$
	DMF, MeCN, 0.1 M TEAP/Hg/CV	$E_{1/2} = -2.63$
	DMF, 0.1 M TBAI/Hg/pol	$E_{1/2} = -2.50$
1-Phenylnaphthalene	DMF, 0.5 M TBABF$_4$/Hg/CV	$E_{1/2} = -2.36$
1,2-Diphenylnaphthalene	DMF, 0.5 M TBABF$_4$/Hg/CV	$E_{1/2} = -2.25$
Cyclopentanone	DMF, 0.1 M TEABr/Hg/pol	$E_{1/2} = -2.82$
Phenanthrene	MeCN, 0.1 M TBAP/Pt wr/CV	$E_{1/2} = -2.47$
	MeCN, 0.1 M TEAP/Hg/pol	$E_{1/2} = -2.88$[a]
Pentacene	THF, 0.1 M TBAP/rot Pt dsk/E swp	$E_{1/2} = -1.40$
Perylene	MeCN, 0.1 M TEAP/Hg/CV	$E_{1/2} = -1.73$
1,3-Diphenyl-1,3-propanedione	DMSO, 0.2 M TBAP/Hg/CV	$E_{1/2} = -1.42$
2,2-Dimethyl-1,3-diphenyl-1,3 propanedione	DMSO, TBAP/Hg/CV	$E_{1/2} = -1.80$
Pyrene	DMF, 0.1 M TBAP/Pt/CV	$E_p = -2.14$
	MeCN, 0.1 M TEAP/Hg/pol	$E_{1/2} = -2.49$[a]
Diphenylsulfone	DMF, TEABr	$E_{1/2} = -2.16$
Triphenylene	MeCN, 0.1 M TEAP/Hg/pol	$E_{1/2} = -2.87$[a]
9,10-Anthraquinone	DMF, 0.5 M TBAP, 20°/Pt dsk/CV	$E_{1/2} = -1.01$
1,4-Benzoquinone	MeCN, 0.1 M TEAP/Pt/CV	$E_p = -0.54$
1,4-Naphthohydroquinone, dipotassium salt	DMF, 0.5 M TBAP, 20°/Pt dsk/CV	$E_{1/2} = -1.55$
Rubrene	DMF, 0.1 M TBAP/Pt dsk/CV	$E_p = -1.48$
	DMF, 0.1 M TBAI/Hg/pol	$E_{1/2} = -1.410$
Benzocyclooctatetraene	THF, 0.1 M TBAP/Hg/pol	$E_{1/2} = -2.13$
sym-Dibenzocyclooctatetraene	THF, 0.1 M TBAP/Hg/pol	$E_{1/2} = -2.29$
Ubiquinone-6	MeCN, 0.1 M TEAP/Pt/CV	$E_p = -1.05$[e]
(9-Phenyl-fluorenyl)$^+$	10.2 M H$_2$SO$_4$/Hg/CV	$E_p = -0.01$[b]
(Triphenylcyclopropenyl)$^+$	MeCN, 0.1 M TEAP/Hg/CV	$E_p = -1.87$
(Triphenylmethyl)$^+$	MeCN, 0.1 M TBAP/Hg/pol	$E_{1/2} = 0.27$
	H$_2$SO$_4$, 10.2 M/Hg/CV	$E_p = -0.58$[b]
(Tribiphenylmethyl)$^+$	MeCN, 0.1 M TBAP/Hg/pol	$E_{1/2} = 0.19$
(Tri-4-t-butyl-5-phenylmethyl)$^+$	MeCN, 0.1 M TBAP/Hg/pol	$E_{1/2} = 0.13$
(Tri-4-isopropylphenylmethyl)$^+$	MeCN, 0.1 M TBAP/Hg/pol	$E_{1/2} = 0.07$
(Tri-4-methylphenylmethyl)$^+$	MeCN, 0.1 M TBAP/Hg/pol	$E_{1/2} = 0.05$
(Tri-4-cyclopropylphenylmethyl)$^+$	MeCN, 0.1 M TBAP/Hg/pol	$E_{1/2} = 0.01$
(Tropylium)$^+$	MeCN, 0.1 M TBAP/Hg/pol	$E_{1/2} = -0.17$
	DMF, 0.15 M TBAI/Hg/pol	$E_{1/2} = -1.55$
	DMF, 0.15 M TBAI/Hg/pol	$E_{1/2} = -1.55$
	DMF, 0.15 M TBAI/Hg/pol	$E_{1/2} = -1.57$
	DMF, 0.15 M TBAI/Hg/pol	$E_{1/2} = -1.60$
	DMF, 0.15 M TBAI/Hg/pol	$E_{1/2} = -1.87$
	DMF, 0.15 M TBAI/Hg/pol	$E_{1/2} = -1.96$
	DMF, 0.15 M TBAI/Hg/pol	$E_{1/2} = -2.05$

Oxidation Potentials (Products are Cation Radicals)

Substance	Conditions/electrode/technique	Potential V (vs. SCE)
Anthracene	CH_2Cl_2, 0.2 M TBABF$_4$, −70°C/Pt dsk/CV	$E_p = +0.73$[d]
9,10-Dimethylanthracene	MeCN, 0.1 M LiClO$_4$/Pt wr/CV	$E_p = +1.0$
9,10-Dipropylanthracene	MeCN, 0.1 M TEAP/Pt/CV	$E_p = +1.08$
1,8-Diphenylanthracene	CH_2Cl_2, 0.2 M TPrACF$_3$SO$_3$/rot Pt wr/E swp	$E_{1/2} = +1.34$
8,9-Diphenylanthracene	CH_2Cl_2, 0.2 M TPrACF$_3$SO$_3$/rot Pt wr/E swp	$E_{1/2} = +1.30$
9,10-Diphenylanthracene	MeCN/Pt/CV	$E_p = +1.22$
Perylene	MeCN, 0.1 M TBAP/Pt/CV	$E_p = +1.34$
Pyrene	DMF, 0.1 M TBAP/Pt dsk/CV	$E_p = +1.25$
Rubrene	DMF, 0.1 M TBAP/Pt dsk/CV	$E_p = +1.10$
Tetracene	CH_2Cl_2, 0.2 M TBABF$_4$, −70°C/Pt wr/CV	$E_p = +0.35$[d]
1,4-Dithiabenzene	MeCN, 0.1 M TEAP/Pt dsk/rot	$E_{1/2} = +0.69$
1,4-Dithianaphthalene	MeCN, 0.1 M TEAP/Pt dsk/rot	$E_{1/2} = +0.80$
Thianthrene	0.1 M TPAP/Pt/CV	$E_{1/2} = +1.28$

[a] vs 0.01 M Ag/AgClO$_4$
[b] vs. Hg/Hg$_2$SO$_4$, 17 M H$_2$SO$_4$
[c] vs Hg pool
[d] vs Ag/saturated AgNO$_3$
[e] vs Ag/0.01 M Ag$^+$

pH SCALE FOR AQUEOUS SOLUTIONS

A. K. Covington

The pH value is the negative decadic logarithm of the (relative) ion activity of the hydrogen ion in the solution.

$$pH = -\log a_H \tag{1}$$

This is only a notional definition since Equation 1 involves a single ion activity, which is immeasurable, and has to be attained through a nonthermodynamic assumption such as that described in Equation 5 below. In terms of substance concentration, molarity, Equation 1 may be rewritten

$$pH = -\log (c_H y_H/c^o) \tag{2}$$

where c^o is an arbitrary constant representing the standard state condition and equal to 1 mol dm^{-3}, c_H is the concentration of hydrogen ion and y_H is the single ion activity of the hydrogen ion. In terms of molality, Equation 1 may be rewritten

$$pH = -\log (m_H \gamma_H/m^o) \tag{3}$$

where m^o is an arbitrary constant representing the standard state condition and equal to 1 mol kg^{-1}, m_H is the concentration of hydrogen ion and γ_H is the single ion activity of the hydrogen ion. For most purposes the difference between these two scales can be ignored for dilute aqueous solutions; the difference is 0.001 at 25° C and 0.02 at 100° C. Arising from the nonexperimental determinability of single ion activities, the definition and determination of pH have an operational basis, and depend on the assignment of pH values to a standard solution (or solutions) together with the determination of pH difference by a cell with liquid junction called the operational cell.

The Operational Definition of pH Difference[1,2]

The electromotive force, EMF, $E(X)$ of the cell with liquid junction:

$$\text{Reference electrode} \mid \text{KCl (aq., concentrated)} \parallel \text{Solution X} \mid H_2 \mid \text{Pt} \tag{I}$$

is measured, and likewise that, $E(S)$, of the cell:

$$\text{Reference electrode} \mid \text{KCl (aq., concentrated)} \parallel \text{Solution S} \mid H_2 \mid \text{Pt} \tag{II}$$

The temperature of both cells (I and II) must be equal and uniform throughout, and the hydrogen gas pressures identical. The two bridge solutions may be any molality of KCl not less than 3.5 mol kg^{-1} provided they are the same.

The pH of the solution X, pH(X), is then related to the assigned pH of the solution S, pH(S) by the definition:

$$pH(X) = pH(S) + [E(S) - E(X)]/[(RT/F) \ln 10] \tag{4}$$

where R is the gas constant, T the thermodynamic temperature, F the Faraday constant. The quantity $k = (RT/F) \ln 10$ is called the slope factor whose values are given as a function of temperature in Table 1. As a consequence of this definition any difference in liquid junction potential between cells I and II is subsumed into the value of pH(X).

The pH Scale

The pH scale at a particular temperature is defined by Equation 4 as a straight line, on the plot of pH against $E(X)$, having a slope of k drawn through the pH value assigned to the Reference Value Standard (RVS) solution (as given in Table 2) and the value of $E(S)$ for cell II when it contains the Reference Value Standard solution. The solution chosen for the RVS is 0.05 mol kg^{-1} aqueous potassium hydrogen phthalate. The procedure by which pH(RVS) values have been assigned to the Reference Value Standard (RVS) is the cell III without transference:[1,3]

$$\text{Pt(Pd)} \mid H_2 \text{ (g, } p=1 \text{ atm} = 101\ 325 \text{ Pa)} \mid \text{RVS, Cl}^- \mid \text{AgCl} \mid \text{Ag} \tag{III}$$

The palladised-platinum hydrogen electrode is used to reduce the catalytised chemical reduction of the phthalate by hydrogen gas. The calculation involves a non-thermodynamic assumption, the Bates-Guggenheim Convention, for the single ion activity of the chloride ion[1,2] as

$$\log (\gamma_{Cl})^o = -A(I/m^o)^{1/2}/[1 + 1.5\ (I/m^o)^{1/2}] \tag{5}$$

where I is the ionic strength = $(1/2)\Sigma m_i z_i^2 = 0.0534$ mol kg^{-1} for the RVS solution and A is a known function of temperature (Table 1).

To prepare the RVS solution, dry the sample at 110° C for 2 h before use. The water should have a conductivity of less than 0.1 mS m^{-1}. The required solution contains 10.211 g kg^{-1} water. It can be prepared on a volume basis by dissolving 10.138 g potassium hydrogen phthalate in water and making up to 1 L at 20° C. This solution is 0.04964 mol/L with a density of 1.00300 g/L at 20° C.

Primary Standards[2]

pH values may be assigned by the cell without transference method (cell III) to six other buffer solutions which meet certain criteria of reproducibility of preparation and properties. These solutions are called primary pH standards (PS), and details and the pH(PS) values assigned to them are given in Table 3. When these PS solutions are used in the operational cell I, the experimental value of the slope will not be in accord with the slope factor values of Table 1, and, moreover, the experimental value could change if additional primary solutions were to be defined. Hence the pH value determined for an unknown solution can be slightly dependent (±0.02) on the choice of primary standard.[2,4,5] Some useful data for standard buffers are given in Table 5.

Operational Standards[2,6]

Operational standards (OS) are also defined which are traceable to the Reference Value Standard (RVS). Values are assigned by means of the operational cells I and II where the liquid junctions are the free diffusion type reproducibly formed in 1 mm vertical capillary tubes. These operational standards are not restricted in number provided certain preparation criteria are met, and pH(OS) values for 16 solutions are given in Table 4.[2] These OS represent an alternative procedure and are in no way to be regarded as inferior to the primary standards. As a consequence of their definition, all pH(OS) values fall on the line with slope given by the slope factor value for the appropriate temperature in Table 1. Any difference in liquid junction potential between the solutions of cells I and II and KCl is subsumed into the assigned value of pH(OS).

Measurement of pH. Choice of Standard Reference Solution

1a. If pH is not required to better than ±0.05 any standard reference solution may be selected.

1b. If pH is required to ±0.002 and interpretation in terms of hydrogen ion concentration or activity is desired, choose a standard reference solution, pH(PS) or pH(OS), to match X as closely as possible in terms of pH, composition and ionic strength.

2. Alternatively, a bracketting procedure may be adopted whereby two standard reference solutions are chosen whose pH values, pH(S1), pH(S2) are on either side of pH(X). Then if the corresponding potential difference measurements are $E(S1)$, $E(S2)$, $E(X)$, then pH(X) is obtained from

$$pH(X) = pH(S1) + [E(X) - E(S1)]/ \%k$$

where $\%k = 100[E(S2) - E(S1)]/[pH(S2) - pH(S1)]$ is the apparent percentage slope. This procedure is very easily done on some pH meters simply by adjusting downwards the slope factor control with the electrodes in S2. The purpose of the bracketting procedure is to compensate for deficiencies in the electrodes and measuring system.

Information to be Given about the Measurement of pH(X)

The standard solutions selected for calibration of the pH meter system should be reported with the measurement as follows,

1. System calibrated with pH(RVS) = ... at ...K.
2. System calibrated with two primary standards pH(PS1) = ... and pH(PS2) = ... at ∴ K.
3. System calibrated with two operational standards pH(OS1) =..... and pH(OS2) = ... at .... K.

Interpretation of pH(X) in Terms of Hydrogen Ion Concentration

The operationally defined pH has no simple interpretation in terms of hydrogen ion concentration but the mean ionic activity coefficient of a typical 1:1 electrolyte can be substituted into equation 2 or 3 to obtain hydrogen ion concentration subject to an uncertainty of 3.9% in concentration corresponding to 0.02 in pH.

REFERENCES

1. R. G. Bates, *Measurement of pH. Theory and Practice*, 2nd ed., John Wiley & Sons,, New York, 1973.
2. A. K. Covington, R. G. Bates, and R. A. Durst, *Pure Appl. Chem.*, 57, 531, 1985.
3. H. P. Butikofer and A. K. Covington, *Anal. Chim. Acta*, 108, 179, 1979.
4. R. G. Bates, *Crit. Rev. Anal. Chem.*, 10, 247, 1981.
5. A. K. Covington, *Anal. Chim. Acta*, 127, 1, 1981.
6. A. K. Covington and M. J. Rebelo, *Anal. Chim. Acta*, 200, 245, 1987.
7. V. E. Bower and R. G. Bates, *J. Res. Natl. Bur. Stand.*, 39, 263, 1954.

pH SCALE FOR AQUEOUS SOLUTIONS (continued)

TABLE 1
Standard EMF, Slope Factor and Debye-Huckel Constant A
(Unit Weight of Solvent) as Functions of Temperature

Temperature/°C	$E°$/mV[7]	Slope Factor k/mV	A^1
0	236.55	54.199	0.4918
5	234.13	55.191	0.4952
10	231.42	56.183	0.4988
15	228.57	57.175	0.5026
20	225.57	58.167	0.5066
25	222.34	59.159	0.5108
30	219.04	60.152	0.5150
35	215.65	61.144	0.5196
40	212.08	62.136	0.5242
45	208.35	63.128	0.5291
50	204.49	64.120	0.5341
55	200.56	65.112	0.5393
60	196.49	66.104	0.5448
70	187.82	68.088	0.5562
80	178.73	70.073	0.5685
90	169.52	72.057	0.5817
95	165.11	73.049	0.5886

TABLE 2
Values of pH(RVS) for the Reference Value Standard of 0.05 mol kg^{-1} Potassium Hydrogen Phthalate at Various Temperatures

t/°C	pH(RVS)	t/°C	pH(RVS)	t/°C	pH(RVS)
0	4.000	35	4.018	65	4.097
5	3.998	37	4.022	70	4.116
10	3.997	40	4.027	75	4.137
15	3.998	45	4.038	80	4.159
20	4.001	50	4.050	85	4.183
25	4.005	55	4.064	90	4.21
30	4.011	60	4.080	95	4.24

pH SCALE FOR AQUEOUS SOLUTIONS (continued)

TABLE 3
Values of pH(PS) for Primary Standard Reference Solutions

Primary ref. standard	t/°C															
	0	5	10	15	20	25	30	35	37	40	50	60	70	80	90	95
Saturated (at 25° C) Potassium hydrogen tartrate	—	—	—	—	—	3.557	3.552	3.549	3.548	3.547	3.549	3.560	3.580	3.610	3.650	3.674
0.1 mol/kg Potassium dihydrogen citrate	3.863	3.840	3.820	3.802	3.788	3.776	3.766	3.759	3.756	3.754	3.749	—	—	—	—	—
0.025 mol/kg Disodium hydrogen phosphate +0.025 mol/kg Potassium dihydrogen phosphate	6.984	6.951	6.923	6.900	6.881	6.865	6.853	6.844	6.841	6.838	6.833	6.836	6.845	6.859	6.876	6.886
0.03043 mol/kg Disodium hydrogen phosphate +0.008695 mol/kg Potassium dihydrogen phosphate	7.534	7.500	7.472	7.448	7.429	7.413	7.400	7.389	7.386	7.380	7.367					
0.01 mol/kg Disodium tetraborate	9.464	9.395	9.332	9.276	9.225	9.180	9.139	9.102	9.088	9.068	9.011	8.962	8.921	8.884	8.850	8.833
0.025 mol/kg Sodium hydrogen carbonate +0.025 mol/kg sodium carbonate	10.317	10.245	10.179	10.118	10.062	10.012	9.966	9.926	9.910	9.889	9.828	—	—	—	—	—

Note: Based on an uncertainty of ±0.2 mV in determined $(E-E^\theta)$, the uncertainty is ±0.003 in pH in the range 0—50° C.

pH SCALE FOR AQUEOUS SOLUTIONS (continued)

TABLE 4

pH (OS) Values for Operational Reference Solutions

Operational standard ref. solution	t/°C														
	0	5	10	15	20	25	30	37	40	50	60	70	80	90	95
0.1 mol/kg Potassium tetroxalate[a]	—	—	—	—	1.475	1.479	1.483	1.490	1.493	1.503	1.513	1.52	1.53	1.53	1.53
0.05 mol/kg potassium tetroxalate[a]	—	—	1.638	1.642	1.644	1.646	1.648	1.649	1.650	1.653	1.660	1.671	1.689	1.72	1.73
0.05 mol/kg sodium hydrogen diglycolate[b]	—	3.466	3.470	3.476	3.484	3.492	3.502	3.519	3.527	3.558	3.595	—	—	—	—
Saturated (at 25° C) potassium hydrogen tartrate	—	—	—	—	—	3.556	3.549	3.544	3.542	3.544	3.553	3.570	3.596	3.627	3.649
0.05 mol/kg Potassium hydrogen phthalate (RVS)	4.000	3.998	3.997	3.998	4.000	4.005	4.011	4.022	4.027	4.050	4.080	4.115	4.159	4.21	4.24
0.1 mol/dm³ Acetic acid + 0.1 mol/dm³ sodium acetate	4.664	4.657	4.652	4.647	4.645	4.644	4.643	4.647	4.650	4.663	4.684	4.713	4.75	4.80	4.83
0.01 mol/dm³ Acetic acid + 0.1 mol/dm³ sodium acetate	4.729	4.722	4.717	4.714	4.712	4.713	4.715	4.722	4.726	4.743	4.768	4.800	4.839	4.88	4.91
0.02 mol/kg Piperazine phosphate[c]	—	6.477	6.419	6.364	6.310	6.259	6.209	6.143	6.116	6.030	5.952	—	—	—	—
0.025 mol/kg Disodium hydrogen phosphate + 0.025 mol/kg potassium dihydrogen phosphate	6.961	6.935	6.912	6.891	6.873	6.857	6.843	6.828	6.823	6.814	6.817	6.830	6.85	6.90	6.92
0.03043 mol/kg Disodium hydrogen phosphate + 0.008695 mol/kg potassium disodium phosphate	7.506	7.482	7.460	7.441	7.423	7.406	7.390	7.369	—	—	—	—	—	—	—
0.04 mol/kg Disodium hydrogen phosphate + 0.01 mol/kg potassium dihydrogen phosphate	—	7.512	7.488	7.466	7.445	7.428	7.414	7.404	—	—	—	—	—	—	—
0.05 mol/kg Tris hydrochloride + 0.01667 mol/kg Tris[d]	8.399	8.238	8.083	7.933	7.788	7.648	7.513	7.332	7.257	7.018	6.794	—	—	—	—
0.05 mol/kg Disodium tetraborate (Na₂B₄O₇)	9.475	9.409	9.347	9.288	9.233	9.182	9.134	9.074	9.051	8.983	8.932	8.898	8.88	8.84	8.89
0.01 mol/kg Disodium tetraborate (Na₂B₄O₇)	9.451	9.388	9.329	9.275	9.225	9.179	9.138	9.086	9.066	9.009	8.965	8.932	8.91	8.90	8.89

TABLE 4
pH(OS) Values for Operational Standard Reference Solutions (continued)

Operational standard ref. solution	0	5	10	15	20	25	30	37	40	50	60	70	80	90	95
												$t/°C$			
0.025 mol/kg Sodium hydrogen carbonate + 0.025 mol/kg sodium carbonate	10.273	10.212	10.154	10.098	10.045	9.995	9.948	9.889	9.866	9.800	9.753	9.728	9.725	9.75	9.77
Saturated (at 20° C) calcium hydroxide	13.360	13.159	12.965	12.780	12.602	12.431	12.267	12.049	11.959	11.678	11.423	11.192	10.984	10.80	10.71

Note: Uncertainty is ±0.003 in pH between 0 and 60° C rising to ±0.01 above 70° C.

a Potassium trihydrogen dioxalate ($KH_3C_4O_8$).
b Sodium hydrogen 2,2′-oxydiethanoate.
c $C_4H_{10}N_2 \cdot H_3PO_4$.
d 2-Amino-2-(hydroxymethyl)-1,3 propanediol or tris(hydroxymethyl)aminomethane.

TABLE 5
Useful Data on Some Standard Buffer Solutions

	Molecular formula	Molality (mol/kg)	Relative molar mass	Density at 20° C (g/cm³)	Molarity at 20° C (mol/L)	Mass of 1 L at 20° C (g)	Mass tolerance for ±0.001 pH[a] (g)	Mass tolerance expressed as a percentage (%)
Potassium tetraoxalate	$KH_3C_4O_8 \cdot 2H_2O$	0.1	254.1913	1.0091	0.09875	25.1017	0.07	0.27
Potassium tetraoxalate	$KH_3C_4O_8 \cdot 2H_2O$	0.05	254.1913	1.0038	0.04965	12.6202	0.034	0.26
Disodium hydrogen orthophosphate	Na_2HPO_4	0.025	141.9588	1.0038	0.02492	3.5379	0.02	0.56
Potassium dihydrogen orthophosphate	KH_2PO_4	0.025	136.0852	1.0038		3.3912	0.02	0.58
Disodium tetraborate	$Na_2B_4O_7 \cdot 10H_2O$	0.05	381.367	1.0075	0.04985	19.0117	0.9	4.73
Disodium tetraborate	$Na_2B_4O_7 \cdot 10H_2O$	0.01	381.367	1.0001	0.009981	3.8064	0.19	0.49
Sodium carbonate	Na_2CO_3	0.025	105.9887	1.0021	0.02494	2.6428	0.017	0.064
Sodium hydrogen carbonate	$NaHCO_3$	0.025	84.0069			2.0947	0.013	0.62

a Calculated from known dilution value of solution.

PRACTICAL pH MEASUREMENTS ON NATURAL WATERS

A. K. Covington and W. Davison

(1) Dilute solutions and freshwater including 'acid-rain' samples ($I < 0.02$ mol kg^{-1})

Major problems could be encountered due to errors associated with the liquid junction. It is recommended that either a free diffusion junction is used or it is verified that the junction is working correctly using dilute solutions as follows. For commercial electrodes calibrated with IUPAC aqueous RVS or PS standards, the pH(X) of dilute solutions should be within ±0.02 of those given in Table 1. The difference in determined pH(X) between a stirred and unstirred dilute solution should be < 0.02. The characteristics of glass electrodes are such that below pH 5 the readings should be stable within 2 min, but for pH 5 to 8, 8 or so minutes may be necessary to attain stability. Interpretation of pH(X) measured in this way in terms of activity of hydrogen ion, a_{H+} is subject[1] to an uncertainty of ±0.02 in pH.

(2) Seawater

Measurements made by calibration of electrodes with IUPAC aqueous RVS or PS standards to obtain pH(X) are perfectly valid. However, the interpretation of pH(X) in terms of the activity of hydrogen ion is complicated by the non zero residual liquid junction potential as well as by systematic differences between electrode pairs, principally attributable to the reference electrode. For 35‰ salinity seawater ($S = 0.035$) a_{H+} calculated from pH(X) is typically 12% too low. Special seawater pH scales have been devised to overcome this problem:
(i) The total hydrogen ion scale, pH$_T$, is defined in terms of the sum of free and complexed (total) hydrogen ion concentrations, where

$$^{T}C_H = [H^+] + [HSO_4^-] + [HF].$$

$$\text{So, pH}_T = - \log {}^{T}C_H$$

Calibration of the electrodes with a buffer having a composition similar to that of seawater, to which pH$_T$ has been assigned, results in values of pHT(X) (Tables 2, 3) which are accurately interpretable in terms of $^{T}C_H$.
(ii) The free hydrogen ion scale, pH$_F$, is defined, and fully interpretable, in terms of the concentration of free hydrogen ions.

$$\text{pH}_F = - \log [H^+]$$

Values of pH$_F$ as a function of temperature have been assigned to the same set of pH$_T$ seawater buffers, and so alternatively can be used for calibration (Tables 2, 3) [2,3]

(3) Estuarine water

Prescriptions for seawater scale buffers are available for a range of salinities. Reliable estuarine pH measurements can be made by calibrating with a buffer of the same salinity as the sample. However, these buffers are difficult to prepare and their use presumes prior knowledge of salinity of the sample. Interpretable measurements of estuarine pH can be made by calibration with IUPAC aqueous RVS or PS standards if the electrode pair is additionally calibrated using a 20‰ salinity seawater buffer.[4] The difference between the assigned pH$_{SWS}$ of the seawater buffer and its measured pH(X) value using RVS or PS standards is

$$\Delta pH = pH_{SWS} - pH(X)$$

Values of ΔpH should be in the range of 0.08 to 0.18. It empirically corrects for differences between the two pH scales and for measurement errors associated with the electrode pair. The pH(X) of samples measured using IUPAC aqueous buffers, can be converted to pH$_T$ or pH$_F$ using the appropriate measured ΔpH:

$$pH_T = pH(X) - \Delta pH$$
$$\text{or } pH_F = pH(X) - \Delta pH$$

This simple procedure is appropriate to pH measurement at salinities from 2‰ to 35‰. For salinities lower than 2‰ the procedures for freshwaters should be adopted.

REFERENCES

1. Davison, W. and Harbinson, T. R., *Analyst*, 113, 709, 1988.
2. Culberson, C. H., in *Marine Electrochemistry*, Whitfield, M. and Jagner, D., Eds., Wiley, 1981.
3. Millero, F. J., *Limnol. Oceanogr.*, 31, 839, 1986.
4. Covington, A. K., Whalley, P. D., Davison, W., and Whitfield, M., in *The Determination of Trace Metals in Natural Waters*, West, T. S. and Nurnberg, H. W., Eds., Blackwell, Oxford, 1988.
5. Koch, W. F., Marinenko, G., and Paule, R. C., *J. Res. NBS*, 91, 33, 1986.

Table 1
pH of Dilute Solutions at 25°C, Degassed and Equilibrated with Air, Suitable as Quality Control Standards

	Ionic strength mmol kg^{-1}	Concentration(x) mmol kg^{-1}	pH $p_{CO_2} = 0$	pH $p_{CO_2} =$ air
Potassium hydrogen phthalate	10.7	10	4.12	4.12
	1.1	1	4.33	4.33
$xKH_2PO_4 + xNa_2HPO_4$	9.9	2.5	7.07	7.05
$xKH_2PO_4 + 3.5xNa_2HPO_4$	10	0.87	7.61	7.58
$Na_2B_4O_7 \cdot 10H_2O$	10	5	9.20	—
HCl	0.1	0.1	4.03	4.03
SRM2694-I[a]	—	—	4.30	—
SRM2694-II[a]	—	—	3.59	—

Note: The pH of solutions near to pH 4 is virtually independent of temperature over the range of 5 to 30°C.

[a] Simulated rainwater samples are available (Reference 5) from NIST containing sulfate, nitrate, chloride, fluoride, sodium, potassium, calcium and magnesium

Table 2
Composition of Seawater Buffer of Salinity $S = 35‰$ at 25°C
(Reference 3)

Solute	mol dm^{-3}	mol kg^{-1}	g kg^{-1}	g dm^{-3}
NaCl	0.3666	0.3493	20.416	20.946
Na_2SO_4	0.02926	0.02788	3.96	4.063
KCl	0.01058	0.01008	0.752	0.772
$CaCl_2$	0.01077	0.01026	1.139	1.169
$MgCl_2$	0.05518	0.05258	5.006	5.139
Tris	0.06	0.05717	6.926	7.106
Tris · HCl	0.06	0.05717	9.010	9.244

Tris = tris(hydroxymethyl)aminomethane $(HOCH_2)_3CNH_2$.
A 20‰ buffer is made by diluting the 35‰ in the ratio 20:35.

Table 3
Assigned Values of 20‰ and 35‰ Buffers on Free and Total Hydrogen Ion Scales. Calculated from Equations Provided by Millero (Reference 3)

Temp (°C)	pH$_T$ $S = 20‰$	pH$_T$ $S = 35‰$	pH$_F$ $S = 20‰$	pH$_F$ $S = 35‰$
5	8.683	8.718	8.759	8.81
10	8.513	8.542	8.597	8.647
15	8.351	8.374	8.442	8.491
20	8.195	8.212	8.292	8.341
25	8.045	8.057	8.149	8.197
30	7.901	7.908	8.011	8.059
35	7.762	7.764	7.879	7.926

BUFFER SOLUTIONS GIVING ROUND VALUES OF pH AT 25°C

A pH	A x	B pH	B x	C pH	C x	D pH	D x	E pH	E x
1.00	67.0	2.20	49.5	4.10	1.3	5.80	3.6	7.00	46.6
1.10	52.8	2.30	45.8	4.20	3.0	5.90	4.6	7.10	45.7
1.20	42.5	2.40	42.2	4.30	4.7	6.00	5.6	7.20	44.7
1.30	33.6	2.50	38.8	4.40	6.6	6.10	6.8	7.30	43.4
1.40	26.6	2.60	35.4	4.50	8.7	6.20	8.1	7.40	42.0
1.50	20.7	2.70	32.1	4.60	11.1	6.30	9.7	7.50	40.3
1.60	16.2	2.80	28.9	4.70	13.6	6.40	11.6	7.60	38.5
1.70	13.0	2.90	25.7	4.80	16.5	6.50	13.9	7.70	36.6
1.80	10.2	3.00	22.3	4.90	19.4	6.60	16.4	7.80	34.5
1.90	8.1	3.10	18.8	5.00	22.6	6.70	19.3	7.90	32.0
2.00	6.5	3.20	15.7	5.10	25.5	6.80	22.4	8.00	29.2
2.10	5.10	3.30	12.9	5.20	28.8	6.90	25.9	8.10	26.2
2.20	3.9	3.40	10.4	5.30	31.6	7.00	29.1	8.20	22.9
		3.50	8.2	5.40	34.1	7.10	32.1	8.30	19.9
		3.60	6.3	5.50	36.6	7.20	34.7	8.40	17.2
		3.70	4.5	5.60	38.8	7.30	37.0	8.50	14.7
		3.80	2.9	5.70	40.6	7.40	39.1	8.60	12.2
		3.90	1.4	5.80	42.3	7.50	40.9	8.70	10.3
		4.00	0.1	5.90	43.7	7.60	42.4	8.80	8.5
						7.70	43.5	8.90	7.0
						7.80	44.5	9.00	5.7
						7.90	45.3		
						8.00	46.1		

F pH	F x	G pH	G x	H pH	H x	I pH	I x	J pH	J x
8.00	20.5	9.20	0.9	9.60	5.0	10.90	3.3	12.00	6.0
8.10	19.7	9.30	3.6	9.70	6.2	11.00	4.1	12.10	8.0
8.20	18.8	9.40	6.2	9.80	7.6	11.10	5.1	12.20	10.2
8.30	17.7	9.50	8.8	9.90	9.1	11.20	6.3	12.30	12.8
8.40	16.6	9.60	11.1	10.00	10.7	11.30	7.6	12.40	16.2
8.50	15.2	9.70	13.1	10.10	12.2	11.40	9.1	12.50	20.4
8.60	13.5	9.80	15.0	10.20	13.8	11.50	11.1	12.60	25.6
8.70	11.6	9.90	16.7	10.30	15.2	11.60	13.5	12.70	32.2
8.80	9.6	10.00	18.3	10.40	16.5	11.70	16.2	12.80	41.2
8.90	7.1	10.10	19.5	10.50	17.8	11.80	19.4	12.90	53.0
9.00	4.6	10.20	20.5	10.60	19.1	11.90	23.0	13.00	66.0
9.10	2.0	10.30	21.3	10.70	20.2	12.00	26.9		
		10.40	22.1	10.80	21.2				
		10.50	22.7	10.90	22.0				
		10.60	23.3	11.00	22.7				
		10.70	23.8						
		10.80	24.25						

A. 25 ml of 0.2 molar KCl + x ml of 0.2 molar HCl.
B. 50 ml of 0.1 molar potassium hydrogen phthalate + x ml of 0.1 molar HCl.
C. 50 ml of 0.1 molar potassium hydrogen phthalate + x ml of 0.1 molar NaOH.
D. 50 ml of 0.1 molar potassium dihydrogen phosphate + x ml of 0.1 molar NaOH.
E. 50 ml of 0.1 molar tris(hydroxymethyl)aminomethane + x ml of 0.1 M HCl.
F. 50 ml of 0.025 molar borax + x ml of 0.1 molar HCl.
G. 50 ml of 0.025 molar borax + x ml of 0.1 molar NaOH.
H. 50 ml of 0.05 molar sodium bicarbonate + x ml of 0.1 molar NaOH.
I. 50 ml of 0.05 molar disodium hydrogen phosphate + x ml of 0.1 molar NaOH.
J. 25 ml of 0.2 molar KCl + x ml of 0.2 molar NaOH.

Final volume of mixtures = 100 ml.

REFERENCES

1. Bower, V.E., and Bates, R.G., *J. Res. Natl. Bur. Stand.*, 55, 197, 1955 (A–D).
2. Bates, R.G., and Bower, V.E., *Anal. Chem.*, 28, 1322, 1956 (E–J).

DISSOCIATION CONSTANTS OF INORGANIC ACIDS AND BASES

The data in this table are presented as values of pK_a, defined as the negative logarithm of the acid dissociation constant K_a for the reaction

$$BH \rightleftharpoons B^- + H^+$$

Thus $pK_a = -\log K_a$, and the hydrogen ion concentration $[H^+]$ can be calculated from

$$K_a = \frac{[H^+][B^-]}{[BH]}$$

In the case of bases, the entry in the table is for the conjugate acid; e.g., ammonium ion for ammonia. The OH⁻ concentration in the system

$$NH_3 + H_2O \rightleftharpoons NH_4^+ + OH^-$$

can be calculated from the equation

$$K_b = K_{water} / K_a = \frac{[OH^-][NH_4^+]}{[NH_3]}$$

where $K_{water} = 1.01 \times 10^{-14}$ at 25 °C. Note that $pK_a + pK_b = pK_{water}$.

All values refer to dilute aqueous solutions at zero ionic strength at the temperature indicated. The table is arranged alphabetically by compound name.

REFERENCE

1. Perrin, D. D., *Ionization Constants of Inorganic Acids and Bases in Aqueous Solution, Second Edition*, Pergamon, Oxford, 1982.

Name	Formula	Step	$t/°C$	pK_a
Aluminum(III) ion	Al^{+3}		25	5.0
Ammonia	NH_3		25	9.25
Arsenic acid	H_3AsO_4	1	25	2.26
		2	25	6.76
		3	25	11.29
Arsenious acid	H_2AsO_3		25	9.29
Barium(II) ion	Ba^{+2}		25	13.4
Boric acid	H_3BO_3	1	20	9.27
		2	20	>14
Calcium(II) ion	Ca^{+2}		25	12.6
Carbonic acid	H_2CO_3	1	25	6.35
		2	25	10.33
Chlorous acid	$HClO_2$		25	1.94
Chromic acid	H_2CrO_4	1	25	0.74
		2	25	6.49
Cyanic acid	$HCNO$		25	3.46
Germanic acid	H_2GeO_3	1	25	9.01
		2	25	12.3
Hydrazine	N_2H_4		25	8.1
Hydrazoic acid	HN_3		25	4.6
Hydrocyanic acid	HCN		25	9.21
Hydrofluoric acid	HF		25	3.20
Hydrogen peroxide	H_2O_2		25	11.62
Hydrogen selenide	H_2Se	1	25	3.89
		2	25	11.0
Hydrogen sulfide	H_2S	1	25	7.05
		2	25	19
Hydrogen telluride	H_2Te	1	18	2.6
		2	25	11
Hydroxylamine	NH_2OH		25	5.94
Hypobromous acid	$HBrO$		25	8.55

Name	Formula	Step	$t/°C$	pK_a
Hypochlorous acid	$HClO$		25	7.40
Hypoiodous acid	HIO		25	10.5
Iodic acid	HIO_3		25	0.78
Lithium ion	Li^+		25	13.8
Magnesium(II) ion	Mg^{+2}		25	11.4
Nitrous acid	HNO_2		25	3.25
Perchloric acid	$HClO_4$		20	-1.6
Periodic acid	HIO_4		25	1.64
Phosphoric acid	H_3PO_4	1	25	2.16
		2	25	7.21
		3	25	12.32
Phosphorous acid	H_3PO_3	1	20	1.3
		2	20	6.70
Pyrophosphoric acid	$H_4P_2O_7$	1	25	0.91
		2	25	2.10
		3	25	6.70
		4	25	9.32
Selenic acid	H_2SeO_4	2	25	1.7
Selenious acid	H_2SeO_3	1	25	2.62
		2	25	8.32
Silicic acid	H_4SiO_4	1	30	9.9
		2	30	11.8
		3	30	12
		4	30	12
Sodium ion	Na^+		25	14.8
Strontium(II) ion	Sr^{+2}		25	13.2
Sulfamic acid	NH_2SO_3H		25	1.05
Sulfuric acid	H_2SO_4	2	25	1.99
Sulfurous acid	H_2SO_3	1	25	1.85
		2	25	7.2
Telluric acid	H_2TeO_4	1	18	7.68
		2	18	11.0
Tellurous acid	H_2TeO_3	1	25	6.27
		2	25	8.43
Tetrafluoroboric acid	HBF_4		25	0.5
Thiocyanic acid	$HSCN$		25	-1.8
Water	H_2O		25	13.995

DISSOCIATION CONSTANTS OF ORGANIC ACIDS AND BASES

This table lists the acid-base dissociation constants of over 600 organic compounds, including many amino acids. All data apply to dilute aqueous solutions and are presented in the form of pK_a, which is the negative of the logarithm of the acid dissociation constant K_a. See the preceding table, "Dissociation Constants of Inorganic Acids and Bases", for further details on notation.

Compounds are listed by molecular formula in the Hill order.

REFERENCES

1. Perrin, D. D., *Dissociation Constants of Organic Bases in Aqueous Solution*, Butterworths, London, 1965; Supplement, 1972.
2. Serjeant, E. P., and Dempsey, B., *Ionization Constants of Organic Acids in Aqueous Solution*, Pergamon, Oxford, 1979.
3. Albert, A., "Ionization Constants of Heterocyclic Substances", in *Physical Methods in Heterocyclic Chemistry*, Katritzky, A. R., Ed., Academic Press, New York, 1963.
4. Sober, H. A., Ed., *CRC Handbook of Biochemistry*, CRC Press, Cleveland, Ohio, 1968.
5. Perrin, D. D., Dempsey, B., and Serjeant, E. P., *pK_a Prediction for Organic Acids and Bases*, Chapman & Hall, London, 1981.
6. Dawson, R. M. C., Elliot, D. C., Elliot, W. H., and Jones, K. M., *Data for Biochemical Research*, Oxford Science Publications, Oxford, 1986.

Molecular formula	Name	Step	$T/°C$	pK_a
CH_2O_2	Formic acid		20	3.75
CH_4N_2O	Urea		21	0.10
CH_5N	Methylamine		25	10.63
$C_2HCl_3O_2$	Trichloroacetic acid		25	0.70
$C_2H_2Cl_2O_2$	Dichloroacetic acid		25	1.48
$C_2H_2O_3$	Glyoxylic acid		25	3.18
$C_2H_2O_4$	Oxalic acid	1	25	1.23
		2	25	4.19
$C_2H_3BrO_2$	Bromoacetic acid		25	2.69
$C_2H_3ClO_2$	Chloroacetic acid		25	2.85
$C_2H_3IO_2$	Iodoacetic acid		25	3.12
C_2H_4OS	Thioacetic acid		25	3.33
$C_2H_4O_2$	Acetic acid		25	4.76
$C_2H_4O_3$	Glycolic acid		25	3.83
C_2H_5N	Ethyleneimine		25	8.01
C_2H_5NO	Acetamide		25	0.63
$C_2H_5NO_2$	Glycine	1	25	2.35
		2	25	9.78
$C_2H_6O_2$	Ethylene glycol		25	14.22
$C_2H_7AsO_2$	Cacodylic acid	1	25	1.57
		2	25	6.27
C_2H_7N	Dimethylamine		25	10.68
C_2H_7N	Ethylamine		25	10.70
C_2H_7NO	Ethanolamine		25	9.50
$C_2H_7NO_3S$	Taurine	1	25	1.5
		2	25	9.061
C_2H_7NS	Cysteamine	1	25	8.35
		2	25	10.81
$C_2H_8N_2$	1,2-Ethanediamine	1	0	10.712
		2	0	7.564
$C_3H_3NO_2$	Cyanoacetic acid		25	2.45
C_3H_3NS	Thiazole		20	2.44
$C_3H_4N_2$	1H-Imidazole		25	6.953
$C_3H_4N_2S$	2-Thiazolamine		20	5.36
$C_3H_4O_2$	Acrylic acid		25	4.25
$C_3H_4O_3$	Pyruvic acid		25	2.39
$C_3H_4O_4$	Malonic acid	1	25	2.83
		2	25	5.69
$C_3H_5ClO_2$	2-Chloropropanoic acid		25	2.83
$C_3H_5ClO_2$	3-Chloropropanoic acid		25	3.98
$C_3H_6N_6$	Melamine		25	5.00
$C_3H_6O_2$	Propanoic acid		25	4.86
$C_3H_6O_3$	3-Hydroxypropanoic acid		25	4.51
$C_3H_6O_3$	Lactic acid		100	3.08
$C_3H_6O_4$	Glyceric acid		25	3.52

Molecular formula	Name	Step	$T/°C$	pK_a
C_3H_7N	Azetidine		25	11.29
$C_3H_7NO_2$	L-Alanine	1	25	2.34
		2	25	9.87
$C_3H_7NO_2$	β-Alanine	1	25	3.55
		2	25	10.24
$C_3H_7NO_2$	Methylglycine	1	25	2.21
		2	25	10.12
$C_3H_7NO_2S$	Cysteine	1	25	1.92
		2	25	8.37
		3	25	10.70
$C_3H_7NO_3$	Serine	1	25	2.19
		2	25	9.21
$C_3H_7NO_5S$	l-Cysteic acid	1	25	1.3
		2	25	1.9
		3	25	8.70
$C_3H_7N_3O_2$	Glycocyamine		25	2.82
$C_3H_8O_3$	Glycerol		25	14.15
C_3H_9N	Propylamine		20	10.60
C_3H_9N	Trimethylamine		25	9.80
C_3H_9NO	1-Amino-2-methoxyethane		10	9.89
C_3H_9NO	Trimethylamine oxide		25	4.65
$C_3H_{10}N_2$	1,2-Propanediamine	1	25	9.82
		2	25	6.61
$C_3H_{10}N_2$	1,3-Propanediamine	1	10	10.94
		2	10	9.03
$C_3H_{11}N_3$	1,2,3-Triaminopropane	1	20	9.59
		2	20	7.95
$C_4H_4N_2$	Pyrazine		27	0.65
$C_4H_4N_2$	Pyridazine		20	2.24
$C_4H_4N_2O_3$	Barbituric acid		25	4.01
$C_4H_4N_2O_5$	Alloxanic acid		25	6.64
$C_4H_4N_4O_2$	5-Nitropyrimidinamine		20	0.35
$C_4H_4O_4$	trans-Fumaric acid	1	18	3.03
		2	18	4.44
$C_4H_4O_4$	Maleic acid	1	25	1.83
		2	25	6.07
$C_4H_4O_5$	Oxaloacetic acid	1	25	2.22
		2	25	3.89
		3	25	13.03
$C_4H_5N_3$	2-Pyrimidinamine		20	3.45
$C_4H_6N_2$	1-Methylimidazol		25	6.95
$C_4H_6N_4O_3$	Allantoin		25	8.96
$C_4H_6O_2$	3-Butenoic acid		25	4.34
$C_4H_6O_2$	trans-Crotonic acid		25	4.69
$C_4H_6O_3$	Acetoacetic acid		18	3.58
$C_4H_6O_3$	2-Oxobutanoic acid		25	2.50
$C_4H_6O_4$	Methylmalonic acid	1	25	3.07
		2	25	5.76
$C_4H_6O_4$	Succinic acid	1	25	4.16
		2	25	5.61
$C_4H_6O_5$	Malic acid	1	25	3.40
		2	25	5.11
$C_4H_6O_6$	α-Tartaric acid	1	25	2.98
		2	25	4.34
$C_4H_6O_6$	meso-Tartaric acid	1	25	3.22
		2	25	4.82
$C_4H_6O_8$	Dihydroxytartaric acid		25	1.92
$C_4H_7ClO_2$	2-Chlorobutanoic acid		RT	2.86
$C_4H_7ClO_2$	3-Chlorobutanoic acid		RT	4.05
$C_4H_7ClO_2$	4-Chlorobutanoic acid		RT	4.52

Molecular formula	Name	Step	$T/°C$	pK_a
$C_4H_7NO_2$	4-Cyanobutanoic acid		25	2.42
$C_4H_7NO_3$	N-Acetylglycine		25	3.669
$C_4H_7NO_4$	Aspartic acid	1	25	1.99
		2	25	3.90
		3	25	9.90
$C_4H_7N_3O$	Creatinine	1	25	4.83
		2		9.2
$C_4H_8N_2O_3$	Asparagine	1	20	2.17
		2	20	8.80
$C_4H_8N_2O_3$	N-Glycylglycine		25	3.139
$C_4H_8O_2$	Butanoic acid		25	4.83
$C_4H_8O_2$	2-Methylpropanoic acid		25	4.88
$C_4H_8O_3$	3-Hydroxybutanoic acid		25	4.70
$C_4H_8O_3$	4-Hydroxybutanoic acid		25	4.72
C_4H_9N	Pyrrolidine		25	11.27
C_4H_9NO	Morpholine		25	8.33
$C_4H_9NO_2$	2-Aminobutanoic acid	1	25	2.29
		2	25	9.83
$C_4H_9NO_2$	4-Aminobutanoic acid	1	25	4.031
		2	25	10.556
$C_4H_9NO_2$	N,N-Dimethylglycine		25	9.89
$C_4H_9NO_2$	2-Methylalanine	1	25	2.36
		2	25	10.21
$C_4H_9NO_2S$	Homocysteine	1	25	2.22
		2	25	8.87
		3	25	10.86
$C_4H_9NO_3$	Homoserine	1	25	2.71
		2	25	9.62
$C_4H_9NO_3$	DL-Methoxyalanine		25	2.037
$C_4H_9NO_3$	Threonine	1	25	2.09
		2	25	9.10
$C_4H_9N_3O_2$	Creatine	1	25	2.63
		2	25	14.3
$C_4H_{10}N_2$	Piperazine	1	23	9.83
		2	23	5.56
$C_4H_{10}N_2O$	2,4-Diaminobutanoic acid	1	25	1.85
		2	25	8.24
		3	25	10.44
$C_4H_{11}N$	Butylamine		20	10.77
$C_4H_{11}N$	sec-Butylamine		25	10.56
$C_4H_{11}N$	tert-Butylamine		25	10.68
$C_4H_{11}N$	Diethylamine		40	11.02
$C_4H_{12}N_2$	1,4-Butanediamine	1	20	10.80
		2	20	9.35
$C_4H_{12}O_2$	1,2-Dimethylaminoethane	1	25	10.40
		2	25	8.26
C_5H_4BrN	3-Bromopyridine		25	2.84
C_5H_4ClN	3-Chloropyridine		25	2.84
$C_5H_4N_4$	Purine	1	20	2.30
		2	20	8.96
$C_5H_4N_4O_3$	Uric acid		12	3.89
C_5H_5N	Pyridine		25	5.25
C_5H_5NO	4-Pyridinol	1	20	3.20
		2	20	11.12
C_5H_5NO	2(1H)-Pyridinone	1	20	0.75
		2	20	11.65
$C_5H_5N_5$	1H-Purin-6-amine	1	25	4.12
		2	25	9.83
$C_5H_6N_2$	2-Methylpyrazine		27	1.45
$C_5H_6N_2$	2-Pyridinamine		20	6.82

Molecular formula	Name	Step	$T/°C$	pK_a
$C_5H_6N_2$	4-Pyridinamine		25	9.114
$C_5H_6O_4$	1,1-Cyclopropanedicarboxylic acid	1	25	1.82
		2	25	7.43
$C_5H_6O_4$	Itaconic acid	1	25	3.85
		2	25	5.45
$C_5H_6O_4$	Mesaconic acid	1	25	3.09
		2	25	4.75
$C_5H_6O_5$	2-Oxoglutaric acid	1	25	2.47
		2	25	4.68
$C_5H_7NO_3$	L-2-Pyrrolidone-5-carboxylic aci		25	3.32
$C_5H_7N_3$	Methylaminopyrazine		25	3.39
$C_5H_7N_3$	2,5-Pyridinediamine		20	6.48
$C_5H_8N_2$	2,4-Dimethylimidazol		25	8.36
$C_5H_8O_4$	Dimethylmalonic acid		25	3.15
$C_5H_8O_4$	Glutaric acid	1	25	4.31
		2	25	5.41
$C_5H_8O_4$	Methylsuccinic acid	1	25	4.13
		2	25	5.64
$C_5H_9NO_2$	Proline	1	25	1.952
		2	25	10.64
$C_5H_9NO_3$	5-Aminolevulinic acid	1	25	4.05
		2	25	8.90
$C_5H_9NO_3$	trans-4-Hydroxyproline	1	25	1.818
		2	25	9.662
$C_5H_9NO_3$	8-Hydroxypurine	1	20	2.56
		2	20	8.26
$C_5H_9NO_4$	L-Glutamic acid	1	25	2.13
		2	25	4.31
$C_5H_9N_3$	Histamine	1	25	6.04
		2	25	9.75
$C_5H_{10}N_2O_2$	Diaminopimelic acid	1	25	1.8
		2	25	2.2
		3	25	8.8
		4	25	9.9
$C_5H_{10}N_2O_3$	L-Glutamine	1	25	2.17
		2	25	9.13
$C_5H_{10}N_2O_3$	Glycylalanine		25	3.15
$C_5H_{10}N_2O_4$	Glycylserine	1	25	2.98
		2	25	8.38
$C_5H_{10}O_2$	2-Methylbutanoic acid		25	4.80
$C_5H_{10}O_2$	3-Methylbutanoic acid		25	4.77
$C_5H_{10}O_2$	Pentanoic acid		25	4.84
$C_5H_{10}O_2$	Trimethylacetic acid		25	5.03
$C_5H_{11}N$	N-Methylpyrrolidine		25	10.32
$C_5H_{11}N$	Piperidine		25	11.123
$C_5H_{11}NO_2$	5-Aminopentanoic acid	1	25	4.27
		2	25	10.766
$C_5H_{11}NO_2$	Betaine		0	1.83
$C_5H_{11}NO_2$	Norvaline	1	25	2.32
		2	25	9.81
$C_5H_{11}NO_2$	N-Propylglycine	1	25	2.35
		2	25	10.19
$C_5H_{11}NO_2$	Valine	1	25	2.29
		2	25	9.74
$C_5H_{11}NO_2S$	Methionine	1	25	2.13
		2	25	9.27
$C_5H_{12}N_2O_2$	Ornithine	1	25	1.705
		2	25	8.69
		3	25	10.76

Molecular formula	Name	Step	$T/°C$	pK_a
$C_5H_{13}N$	1-Amino-2,2-dimethylpropane		25	10.15
$C_5H_{13}N$	Diethylmethylamine		25	10.35
$C_5H_{13}N$	3-Methyl-1-butanamine		25	10.60
$C_5H_{13}N$	2-Methyl-2-butanamine		19	10.85
$C_5H_{13}N$	3-Pentanamine		17	10.59
$C_5H_{13}N$	Pentylamine		25	10.63
$C_5H_{14}NO$	Choline		25	13.9
$C_5H_{14}N_2$	Cadaverine	1	25	10.05
		2	25	10.93
$C_6H_3ClN_4$	6-Chloropteridine		20	3.68
$C_6H_3N_3O_7$	Picric acid		25	0.38
$C_6H_4Cl_2O$	2,3-Dichlorophenol		25	7.44
$C_6H_4N_2O_3$	5-Hydroxylysine		25	2.13
$C_6H_4N_2O_5$	2,4-Dinitrophenol		15	3.96
$C_6H_4N_2O_5$	3,6-Dinitrophenol		15	5.15
$C_6H_4N_4$	Pteridine		20	4.05
$C_6H_5Br_2N$	3,5-Dibromoaniline		25	2.34
C_6H_5ClO	2-Chlorophenol		25	8.49
C_6H_5ClO	3-Chlorophenol		25	8.85
C_6H_5ClO	4-Chlorophenol		25	9.18
$C_6H_5Cl_2N$	2,4-Dichloroaniline		22	2.05
C_6H_5NO	2-Pyridinecarboxaldehyde		20	3.80
$C_6H_5NO_2$	Nitrobenzene		0	3.98
$C_6H_5NO_2$	Picolinic acid	1	25	1.07
		2	25	5.25
$C_6H_5NO_2$	3-Pyridinecarboxylic acid		25	4.85
$C_6H_5NO_2$	4-Pyridinecarboxylic acid		25	4.96
$C_6H_5NO_3$	2-Nitrophenol		25	7.17
$C_6H_5NO_3$	3-Nitrophenol		25	8.28
$C_6H_5NO_3$	4-Nitrophenol		25	7.15
$C_6H_5N_5O$	2-Amino-4-hydroxypteridine	1	20	2.27
		2	20	7.96
$C_6H_5N_5O_2$	Xanthopterin	2	20	6.59
		3	20	9.31
C_6H_6BrN	2-Bromoaniline		25	2.53
C_6H_6BrN	3-Bromoaniline		25	3.58
C_6H_6BrN	4-Bromoaniline		25	3.86
C_6H_6ClN	2-Chloroaniline		25	2.65
C_6H_6ClN	3-Chloroaniline		25	3.46
C_6H_6ClN	4-Chloroaniline		25	4.15
C_6H_6FN	2-Fluoroaniline		25	3.20
C_6H_6FN	3-Fluoroaniline		25	3.50
C_6H_6FN	4-Fluoroaniline		25	4.65
C_6H_6IN	2-Iodoaniline		25	2.60
$C_6H_6N_2O$	2-Pyridinecarboxaldehyde, oxime	1	20	3.59
		2	20	10.18
$C_6H_6N_2O_2$	2-Nitroaniline		25	-0.26
$C_6H_6N_2O_2$	3-Nitroaniline		25	2.466
$C_6H_6N_2O_2$	4-Nitroaniline		25	1.0
C_6H_6O	Phenol		20	9.89
$C_6H_6O_2$	Hydroquinone		20	10.35
$C_6H_6O_2$	Pyrocatcchol		20	9.85
$C_6H_6O_2$	Resorcinol		25	9.81
$C_6H_6O_3S$	Benzenesulfonic acid		25	0.70
$C_6H_6O_6$	cis-Aconitic acid		25	1.95
$C_6H_6O_6$	trans-Aconitic acid	1	25	2.80
		2	25	4.46
C_6H_7N	Aniline		25	4.63
C_6H_7N	2-Methylpyridine		20	5.97

Molecular formula	Name	Step	$T/°C$	pK_a
C_6H_7N	3-Methylpyridine		20	5.68
C_6H_7N	4-Methylpyridine		20	6.02
C_6H_7NO	Methoxypyridine		25	6.47
$C_6H_7NO_3S$	o-Aminobenzenesulfonic acid		25	2.48
$C_6H_7NO_3S$	m-Aminobenzenesulfonic acid		25	3.73
$C_6H_7NO_3S$	p-Aminobenzenesulfonic acid		25	3.24
$C_6H_8N_2$	N-Methylpyridinamine		20	9.65
$C_6H_8O_6$	Ascorbic acid	1	24	4.10
		2	16	11.79
$C_6H_8O_7$	Citric acid	1	20	3.14
		2	20	4.77
		3	20	6.39
$C_6H_8O_7$	Isocitric acid	1	25	3.29
		2	25	4.71
		3	25	6.40
$C_6H_9NO_6$	γ-Carboxyglutamic acid	1	25	1.7
		2	25	3.2
		3	25	4.75
		4	25	9.9
$C_6H_9N_3$	4,6-Dimethylpyrimidinamine		20	4.82
$C_6H_9N_3O_2$	Histidine	1	25	1.80
		2	25	6.04
		3	25	9.33
$C_6H_{10}O_3$	2-Oxo-3-methylpentanoic acid		25	2.3
$C_6H_{10}O_4$	Adipic acid	1	25	4.43
		2	25	5.41
$C_6H_{10}O_4$	3-Methylglutaric acid		25	4.24
$C_6H_{11}NO_2$	l-Pipecolic acid	1	25	2.28
		2	25	10.72
$C_6H_{11}NO_3$	Adipamic acid		25	4.63
$C_6H_{11}NO_4$	2-Aminoadipic acid	1	25	2.14
		2	25	4.21
		3	25	9.77
$C_6H_{11}N_3O_4$	Glycylasparagine	1	25	2.942
		2	18	8.44
$C_6H_{11}N_3O_4$	N-(N-Glycylglycyl)glycine	1	25	3.225
		2	25	8.09
$C_6H_{12}N_2O_4S_2$	Cystine	1	35	2.1
		2	35	8.0
$C_6H_{12}O_2$	Hexanoic acid		25	4.85
$C_6H_{12}O_2$	4-Methylpentanoic acid		18	4.84
$C_6H_{13}N$	Cyclohexylamine		24	10.66
$C_6H_{13}N$	1,2-Dimethylpyrrolidine		26	10.20
$C_6H_{13}N$	1-Methylpiperidine		25	10.08
$C_6H_{13}NO_2$	6-Aminohexanoic acid	1	25	4.373
		2	25	10.804
$C_6H_{13}NO_2$	Isoleucine	1	25	2.32
		2	25	9.76
$C_6H_{13}NO_2$	L-Leucine	1	25	2.328
		2	25	9.744
$C_6H_{13}NO_2$	Norleucine	1	25	2.335
		2	25	9.83
$C_6H_{13}N_3O_3$	Citrulline	1	25	2.43
		2	25	9.69
$C_6H_{14}N_2$	cis-1,2-Cyclohexanediamine	1	20	9.93
		2	20	6.13
$C_6H_{14}N_2$	trans-1,2-Cyclohexanediamine	1	20	9.94
		2	20	6.47
$C_6H_{14}N_2$	2,5-Dimethylpiperazine	1	25	9.66
		2	25	5.20

Molecular formula	Name	Step	$T/^\circ C$	pK_a
$C_6H_{14}N_2O_2$	Lysine	1	25	2.16
		2	25	9.06
		3	25	10.54
$C_6H_{14}N_4O_2$	Arginine	1	25	1.82
		2	25	8.99
		3	25	12.48
$C_6H_{15}N$	3-Amino-3-methylpentane		16	11.01
$C_6H_{15}N$	Diisopropylamine		25	11.05
$C_6H_{15}N$	Hexylamine		25	10.56
$C_6H_{15}N$	Triethylamine		25	10.75
$C_6H_{16}N_2$	Hexamethylenediamine	1	0	11.857
		2	0	10.762
$C_7H_5BrO_2$	2-Bromobenzoic acid		25	2.84
$C_7H_5BrO_2$	3-Bromobenzoic acid		25	3.86
$C_7H_5ClO_2$	2-Chlorobenzoic acid		25	2.92
$C_7H_5ClO_2$	3-Chlorobenzoic acid		25	3.82
$C_7H_5ClO_2$	4-Chlorobenzoic acid		25	3.98
$C_7H_5IO_2$	2-Iodobenzoic acid		25	2.85
$C_7H_5IO_2$	3-Iodobenzoic acid		25	3.80
$C_7H_5NO_3S$	Saccharin		18	11.68
$C_7H_5NO_4$	Dinicotinic acid		25	2.80
$C_7H_5NO_4$	Dipicolinic acid	1	25	2.16
		2	25	4.76
$C_7H_5NO_4$	Lutidinic acid		25	2.15
$C_7H_5NO_4$	2-Nitrobenzoic acid		18	2.16
$C_7H_5NO_4$	3-Nitrobenzoic acid		25	3.47
$C_7H_5NO_4$	4-Nitrobenzoic acid		25	3.41
$C_7H_5NO_4$	Quinolinic acid	1	25	2.43
		2	25	4.78
$C_7H_6N_2$	Benzimidazole		25	5.532
$C_7H_6N_4O$	6-Hydroxy-4-methylpteridine	1	20	4.08
		2	20	6.41
$C_7H_6O_2$	Benzoic acid		25	4.19
$C_7H_6O_3$	o-Hydroxybenzoic acid	1	19	2.97
		2	18	13.40
$C_7H_6O_3$	m-Hydroxybenzoic acid	1	19	4.06
		2	19	9.92
$C_7H_6O_3$	p-Hydroxybenzoic acid	1	19	4.48
		2	19	9.32
$C_7H_6O_4$	2,5-Dihydroxybenzoic acid	1	25	2.97
$C_7H_6O_4$	3,4-Dihydroxybenzoic acid	1	25	4.48
		2	25	8.83
		3	25	12.6
$C_7H_6O_4$	3,5-Dihydroxybenzoic acid	1	25	4.04
$C_7H_6O_4$	Dihydroxymalic acid		25	1.92
$C_7H_6O_5$	Gallic acid		25	4.41
$C_7H_6O_5$	2,4,6-Trihydroxybenzoic acid		25	1.68
$C_7H_7NO_2$	2-Aminobenzoic acid	1	25	2.108
		2	25	4.946
$C_7H_7NO_2$	3-Aminobenzoic acid	2	25	4.78
$C_7H_7NO_2$	4-Animobenzoic acid	1	25	2.501
		2	25	4.874
$C_7H_8N_4O_2$	Theobromine		18	7.89
C_7H_8O	o-Cresol		25	10.20
C_7H_8O	m-Cresol		25	10.01
C_7H_8O	p-Cresol		25	10.17
C_7H_9N	Benzylamine		25	9.33
C_7H_9N	2,3-Dimethylpyridine		25	6.57
C_7H_9N	2,4-Dimethylpyridine		25	6.99

Molecular formula	Name	Step	$T/°C$	pK_a
C_7H_9N	2,5-Dimethylpyridine		25	6.40
C_7H_9N	2,6-Dimethylpyridine		25	6.65
C_7H_9N	3,4-Dimethylpyridine		25	6.46
C_7H_9N	3,5-Dimethylpyridine		25	6.15
C_7H_9N	2-Ethylpyridine		25	5.89
C_7H_9N	N-Methylaniline		25	4.84
C_7H_9N	o-Methylaniline		25	4.44
C_7H_9N	m-Methylaniline		25	4.73
C_7H_9N	p-Methylaniline		25	5.08
C_7H_9NO	o-Anisidine		25	4.52
C_7H_9NO	m-Anisidine		25	4.23
C_7H_9NO	p-Anisidine		25	5.34
C_7H_9NS	4-Methylthioaniline		25	4.35
$C_7H_9N_5$	2-Dimethylaminopurine	1	20	4.00
		2	20	10.24
$C_7H_{11}N_3O_2$	l-1-Methylhistidine	1	25	1.69
		2	25	6.48
		3	25	8.85
$C_7H_{11}N_3O_2$	l-3-Methylhistidine	1	25	1.92
		2	25	6.56
		3	25	8.73
$C_7H_{12}O_2$	Cyclohexanecarboxylic acid		25	4.90
$C_7H_{12}O_4$	Heptanedioic acid	1	25	4.71
		2	25	5.58
$C_7H_{13}NO$	3-Acetylpiperidine		25	3.18
$C_7H_{13}NO_4$	α-Ethylglutamic acid	1	25	3.846
		2	25	7.838
$C_7H_{14}O_2$	Heptanoic acid		25	4.89
$C_7H_{15}N$	1,2-Dimethylpiperidine		25	10.22
$C_7H_{15}N$	1-Ethylpiperidine		23	10.45
$C_7H_{15}NO_3$	Carnitine		25	3.80
$C_7H_{17}N$	2-Heptanamine		19	10.7
$C_7H_{17}N$	Heptylamine		25	10.67
$C_8H_6N_2$	Cinnoline		20	2.37
$C_8H_6N_2$	Quinazoline		20	3.43
$C_8H_6N_2$	Quinoxaline		20	0.56
$C_8H_6N_2O$	5-Hydoxyquinazoline	1	20	3.62
		2	20	7.41
$C_8H_6O_4$	o-Phthalic acid	1	25	2.89
		2	25	5.51
$C_8H_6O_4$	m-Phthalic acid	1	25	3.54
		2	18	4.60
$C_8H_6O_4$	p-Phthalic acid	1	25	3.51
		2	16	4.82
$C_8H_6O_4$	Terephthalic acid		25	3.51
$C_8H_7ClO_2$	2-Chlorophenylacetic acid		25	4.07
$C_8H_7ClO_2$	3-Chlorophenylacetic acid		25	4.14
$C_8H_7ClO_2$	4-Chlorophenylacetic acid		25	4.19
$C_8H_7ClO_3$	2-Chlorophenoxyacetic acid		25	3.05
$C_8H_7ClO_3$	3-Chlorophenoxyacetic acid		25	3.10
$C_8H_7NO_4$	2-Nitrophenylacetic acid		25	4.00
$C_8H_7NO_4$	3-Nitrophenylacetic acid		25	3.97
$C_8H_7NO_4$	4-Nitrophenylacetic acid		25	3.85
$C_8H_8N_2$	2-Methylbenzimidazole		25	6.19
$C_8H_8O_2$	Phenylacetic acid		18	4.28
$C_8H_8O_2$	o-Toluic acid		25	3.91
$C_8H_8O_2$	m-Toluic acid		25	4.27
$C_8H_8O_2$	p-Toluic acid		25	4.36
$C_8H_8O_3$	DL-Mandelic acid		25	3.85

Molecular formula	Name	Step	$T/°C$	pK_a
$C_8H_8O_4$	Homogentisic acid		25	4.40
$C_8H_9NO_2$	2-(Methylamino)benzoic acid		25	5.34
$C_8H_9NO_2$	3-(Methylamino)benzoic acid		25	5.10
$C_8H_9NO_2$	4-(Methylamino)benzoic acid		25	5.04
$C_8H_9NO_2$	Phenylglycine	1	25	1.83
		2	25	4.39
$C_8H_{10}BrN$	4-Bromo-N,N-dimethylaniline		25	4.23
$C_8H_{10}ClN$	3-Chloro-N,N-dimethylaniline		20	3.83
$C_8H_{10}ClN$	4-Chloro-N,N-dimethylaniline		20	4.39
$C_8H_{10}N_2O_2$	N,N-Dimethyl-3-nitroaniline		25	2.62
$C_8H_{11}N$	N,N-Dimethylaniline		25	5.15
$C_8H_{11}N$	N-Ethylaniline		24	5.12
$C_8H_{11}N$	Phenylethylamine		25	9.84
$C_8H_{11}N$	2,4,6-Trimethylpyridine		25	7.43
$C_8H_{11}NO$	o-Phenetidine		28	4.43
$C_8H_{11}NO$	m-Phenetidine		25	4.18
$C_8H_{11}NO$	p-Phenetidine		28	5.20
$C_8H_{11}NO$	Tyramine	1	25	9.74
		2	25	10.52
$C_8H_{11}NO_2$	Dopamine	1	25	8.9
		2	25	10.6
$C_8H_{11}NO_3$	Noradrenaline	1	25	8.64
		2	25	9.70
$C_8H_{12}N_2O_3$	Veronal		25	7.43
$C_8H_{14}O_4$	Octanedioic acid		25	4.52
$C_8H_{16}N_2O_3$	N-Glycylleucine		25	3.18
$C_8H_{16}N_2O_3$	N-Leucylglycine	1	25	3.25
		2	25	8.2
$C_8H_{16}N_2O_4S_2$	Homocystine	1	25	1.59
		2	25	2.54
		3	25	8.52
		4	25	9.44
$C_8H_{16}O_2$	Octanoic acid		25	4.89
$C_8H_{17}N$	2,2,4-Trimethylpiperidine		30	11.04
$C_8H_{19}N$	Dibutylamine		21	11.25
$C_8H_{19}N$	N-Methyl-2-heptanamine		17	10.99
$C_8H_{19}N$	Octylamine		25	10.65
C_9H_6BrN	3-Bromoquinoline		25	2.69
$C_9H_7ClO_2$	o-Chlorocinnamic acid		25	4.23
$C_9H_7ClO_2$	m-Chlorocinnamic acid		25	4.29
$C_9H_7ClO_2$	p-Chlorocinnamic acid		25	4.41
C_9H_7N	Isoquinoline		20	5.42
C_9H_7N	Quinoline		20	4.90
C_9H_7NO	7-Isoquinolinol	1	20	5.68
		2	20	8.90
C_9H_7NO	3-Quinolinol	1	20	4.28
		2	20	8.08
C_9H_7NO	8-Quinolinol	1	20	5.017
		2	25	9.812
$C_9H_7NO_3$	o-Cyanophenoxyacetic acid		25	2.98
$C_9H_7NO_3$	m-Cyanophenoxyacetic acid		25	3.03
$C_9H_7NO_3$	p-Cyanophenoxyacetic acid		25	2.93
$C_9H_8N_2$	1-Isoquinolinamine		20	7.59
$C_9H_8N_2$	3-Quinolinamine		20	4.91
$C_9H_8O_2$	cis-Cinnamic acid		25	3.89
$C_9H_8O_2$	$trans$-Cinnamic acid		25	4.44
$C_9H_9ClO_2$	3-(2-Chlorophenyl)propanoic acid		25	4.58
$C_9H_9ClO_2$	3-(3-Chlorophenyl)propanoic acid		25	4.59
$C_9H_9ClO_2$	3-(4-Chlorophenyl)propanoic acid		25	4.61

Molecular formula	Name	Step	$T/°C$	pK_a
$C_9H_9I_2NO_3$	3,5-Diiodotyrosine	1	25	2.12
		2	25	5.32
		3	25	9.48
$C_9H_9NO_3$	Hippuric acid		25	3.62
$C_9H_9NO_4$	3-(2-Nitrophenyl)propanoic acid		25	4.50
$C_9H_9NO_4$	3-(4-Nitrophenyl)propanoic acid		25	4.47
$C_9H_{10}INO_3$	3-Iodotyrosine	1	25	2.2
		2	25	8.7
		3	25	9.1
$C_9H_{10}N_2$	2-Ethylbenzimidazole		25	6.18
$C_9H_{10}O_2$	Mesitylenic acid		25	4.32
$C_9H_{10}O_2$	α-Phenylpropanoic acid		25	4.64
$C_9H_{10}O_2$	β-Phenylpropanoic acid		25	4.37
$C_9H_{11}N$	N-Allylaniline		25	4.17
$C_9H_{11}N$	1-Aminoindane		22	9.21
$C_9H_{11}NO_2$	Phenylalanine	1	25	2.20
		2	25	9.31
$C_9H_{11}NO_3$	Tyrosine	1	25	2.20
		2	25	9.21
		3	25	10.46
$C_9H_{11}NO_4$	L-3,4-Dihydroxyphenylalanine	1	25	2.32
		2	25	8.72
		3	25	9.96
		4	25	11.79
$C_9H_{12}N_2O_2$	Tyrosineamide		25	7.33
$C_9H_{13}NO_3$	D-Adrenaline	1	25	8.66
		2	25	9.95
$C_9H_{14}N_4O_3$	N-β-Alanylhistidine	1	20	2.73
		2	20	6.87
		3	20	9.73
$C_9H_{14}N_4O_3$	L-Carnosine	1	25	2.62
		2	25	6.66
		3	25	9.24
$C_9H_{18}O_2$	Nonanic acid		25	4.96
$C_9H_{19}N$	1-Butylpiperidine		23	10.47
$C_9H_{19}N$	2,2,6,6-Tetramethylpiperidine		25	11.07
$C_9H_{21}N$	Nonylamine		25	10.64
$C_{10}H_7NO_2$	8-Quinolinecarboxylic acid		25	1.82
$C_{10}H_8O$	1-Naphthol		25	9.34
$C_{10}H_8O$	2-Naphthol		25	9.51
$C_{10}H_9N$	2-Methylquinoline		20	5.83
$C_{10}H_9N$	4-Methylquinoline		20	5.67
$C_{10}H_9N$	5-Methylquinoline		20	5.20
$C_{10}H_9N$	α-Naphthylamine		25	3.92
$C_{10}H_9N$	β-Naphthylamine		25	4.16
$C_{10}H_9NO$	1-Amino-6-hydroxynaphthalene		25	3.97
$C_{10}H_9NO$	6-Methoxyquinoline		20	5.03
$C_{10}H_{10}O_2$	o-Methylcinnamic acid		25	4.50
$C_{10}H_{10}O_2$	m-Methylcinnamic acid		25	4.44
$C_{10}H_{10}O_2$	p-Methylcinnamic acid		25	4.56
$C_{10}H_{12}N_2$	Tryptamine		25	10.2
$C_{10}H_{12}N_2O$	5-Hydroxytryptamine	1	25	9.8
		2	25	11.1
$C_{10}H_{12}O_2$	4-Phenylbutanoic acid		25	4.76
$C_{10}H_{12}O_3$	2-(m-Anisyl)propanoic acid		25	4.65
$C_{10}H_{12}O_3$	2-(p-Anisyl)propanoic acid		25	4.69
$C_{10}H_{12}O_3$	2-(o-Anisyl)propanonic acid		25	4.80
$C_{10}H_{14}N_2$	Nicotine	1	25	8.02
		2	25	3.12

Molecular formula	Name	Step	$T/°C$	pK_a
$C_{10}H_{15}N$	N,N-Diethylaniline		22	6.61
$C_{10}H_{15}NO$	d-Ephedrine		10	10.139
$C_{10}H_{15}NO$	l-Ephedrine		10	9.958
$C_{10}H_{17}N_3O_6S$	l-Glutathione	1	25	2.12
		2	25	3.59
		3	25	8.75
		4	25	9.65
$C_{10}H_{17}N_5O_6$	Tetraglycylglycine	1	20	3.10
		2	20	8.02
$C_{10}H_{18}N_4O_5$	L-Argininosuccinic acid	1	25	1.62
		2	25	2.70
		3	25	4.26
		4	25	9.58
$C_{10}H_{19}N$	Bornylamine		25	10.17
$C_{10}H_{19}N$	Neobornylamine		25	10.01
$C_{10}H_{21}N$	Butylcyclohexylamine		25	11.23
$C_{10}H_{23}N$	Decylamine		25	10.64
$C_{11}H_8N_2$	Perimidine		20	6.35
$C_{11}H_8O_2$	1-Naphthoic acid		25	3.70
$C_{11}H_8O_2$	2-Naphthoic acid		25	4.17
$C_{11}H_{11}N$	Methyl-1-naphthylamine		27	3.67
$C_{11}H_{12}N_2O_2$	Tryptophan	1	25	2.46
		2	25	9.41
$C_{11}H_{16}N_2O_2$	Pilocarpine		30	6.87
$C_{11}H_{25}N$	Undecylamine		25	10.63
$C_{12}H_8N_2$	1,10-Phenanthroline		25	4.84
$C_{12}H_{11}N$	2-Aminobiphenyl		22	3.82
$C_{12}H_{11}N$	2-Benzylpyridine		25	5.13
$C_{12}H_{11}N$	Diphenylamine		25	0.79
$C_{12}H_{11}N_3$	4-Aminoazobenzene		25	2.82
$C_{12}H_{12}N_2$	p-Benzidine	1	30	4.66
		2	30	3.57
$C_{12}H_{13}N$	N,N-Dimethyl-1-naphthylamine		25	4.83
$C_{12}H_{13}N$	N,N-Dimethyl-2-naphthylamine		25	4.566
$C_{12}H_{27}N$	Dodecylamine		25	10.63
$C_{13}H_9N$	Acridine		20	5.58
$C_{13}H_9N$	Phenanthridine		20	5.58
$C_{13}H_{10}N_2$	2-Phenylbenzimidazole	1	25	5.23
		2	25	11.91
$C_{13}H_{10}O_2$	2-Phenylbenzoic acid		25	3.46
$C_{13}H_{12}N_2O$	4-(p-Aminobenzoyl)aniline	1	25	2.93
$C_{13}H_{13}N$	4-Benzylaniline		25	2.17
$C_{13}H_{29}N$	Tridecylamine		25	10.63
$C_{14}H_{12}O_2$	Diphenylacetic acid		25	3.94
$C_{14}H_{15}N_3$	4-Dimethylaminoazobenzene		25	3.226
$C_{14}H_{31}N$	Tetradecylamine		25	10.62
$C_{15}H_{11}I_4NO_4$	L-Thyroxine	1	25	2.2
		2	25	6.45
		3	25	10.1
$C_{15}H_{33}N$	Pentadecylamine		25	10.61
$C_{16}H_{35}N$	Hexadecylamine		25	10.63
$C_{17}H_{19}NO_3$	Morphine		25	8.21
$C_{18}H_{21}NO_3$	Codeine		25	8.21
$C_{18}H_{39}N$	Octadecylamine		25	10.60
$C_{20}H_{21}NO_4$	Papaverine		25	6.40
$C_{20}H_{24}N_2O_2$	Quinine	1	25	8.52
		2	25	4.13
$C_{21}H_{22}N_2O_2$	Strychnine		25	8.26
$C_{23}H_{26}N_2O_4$	Brucine	1	25	8.28

DENSITY, REFRACTIVE INDEX, FREEZING POINT DEPRESSION, AND VISCOSITY OF AQUEOUS SOLUTIONS

This table gives properties of aqueous solutions of 45 common substances as a function of concentration. All data refer to a temperature of 20°C. The properties are:

Mass %: Mass of solute divided by total mass of solution, expressed as percent.
m Molality (moles of solute per kg of water).
c Molarity (moles of solute per liter of solution).
ρ Density of solution in g/cm³.
n Index of refraction, relative to air, at a wavelength of 589 nm (sodium D line); the index of pure water at 20°C is 1.3330.
Δ Freezing point depression relative to pure water.
η Absolute (dynamic) viscosity in mPa s (equal to centipoise, cP); the viscosity of pure water at 20°C is 1.002 mPa s.

Density data over a wider range of temperatures and pressures (and for other compounds) may be found in Reference 2.

REFERENCES

1. Wolf, A.V., Aqueous Solutions and Body Fluids, Hoeber, 1966.
2. Söhnel, O., and Novotny, P., *Densities of Aqueous Solutions of Inorganic Substances*, Elsevier, Amsterdam, 1985.

Solute	Mass %	m/mol kg⁻¹	c/mol L⁻¹	ρ/g cm⁻³	n	Δ/°C	η/mPa s
Acetic acid	0.5	0.084	0.083	0.9989	1.3334	0.16	1.012
CH₃COOH	1.0	0.168	0.166	0.9996	1.3337	0.32	1.022
	2.0	0.340	0.333	1.0011	1.3345	0.63	1.042
	3.0	0.515	0.501	1.0025	1.3352	0.94	1.063
	4.0	0.694	0.669	1.0038	1.3359	1.26	1.084
	5.0	0.876	0.837	1.0052	1.3366	1.58	1.105
	6.0	1.063	1.006	1.0066	1.3373	1.90	1.125
	7.0	1.253	1.175	1.0080	1.3381	2.23	1.143
	8.0	1.448	1.345	1.0093	1.3388	2.56	1.162
	9.0	1.647	1.515	1.0107	1.3395	2.89	1.186
	10.0	1.850	1.685	1.0121	1.3402	3.23	1.210
	11.0	2.058	1.856	1.0134	1.3409	3.57	1.231
	12.0	2.271	2.028	1.0147	1.3416	3.91	1.253
	13.0	2.488	2.200	1.0161	1.3423	4.26	1.275
	14.0	2.711	2.372	1.0174	1.3430	4.61	1.298
	15.0	2.939	2.545	1.0187	1.3437	4.97	1.320
	16.0	3.172	2.718	1.0200	1.3444	5.33	1.341
	17.0	3.411	2.891	1.0213	1.3451	5.69	1.360
	18.0	3.655	3.065	1.0225	1.3458	6.06	1.380
	19.0	3.906	3.239	1.0238	1.3465	6.43	1.405
	20.0	4.163	3.414	1.0250	1.3472	6.81	1.431
	22.0	4.697	3.764	1.0275	1.3485	7.57	1.478
	24.0	5.259	4.116	1.0299	1.3498	8.36	1.525
	26.0	5.851	4.470	1.0323	1.3512	9.17	1.572
	28.0	6.476	4.824	1.0346	1.3525	10.00	1.613
	30.0	7.137	5.180	1.0369	1.3537	10.84	1.669
	32.0	7.837	5.537	1.0391	1.3550	11.70	1.715
	34.0	8.579	5.896	1.0413	1.3562	12.55	1.762
	36.0	9.367	6.255	1.0434	1.3574	13.38	1.812
	38.0	10.207	6.615	1.0454	1.3586		1.852
	40.0	11.102	6.977	1.0474	1.3598		1.912
	42.0	12.059	7.339	1.0493	1.3610		1.960
	44.0	13.084	7.701	1.0510	1.3621		2.007
	46.0	14.186	8.065	1.0528	1.3632		2.052
	48.0	15.372	8.429	1.0545	1.3642		2.110
	50.0	16.653	8.794	1.0562	1.3653		2.158
	52.0	18.041	9.159	1.0577	1.3663		2.212
	54.0	19.549	9.525	1.0592	1.3673		2.265
	56.0	21.194	9.890	1.0605	1.3682		2.308

Solute	Mass %	m/mol kg^{-1}	c/mol L^{-1}	ρ/g cm^{-3}	n	Δ/°C	η/mPa s
	58.0	22.997	10.256	1.0618	1.3691		2.360
	60.0	24.979	10.620	1.0629	1.3700		2.409
	62.0	27.170	10.986	1.0640	1.3708		2.456
	64.0	29.605	11.351	1.0650	1.3716		2.502
	66.0	32.326	11.715	1.0659	1.3724		2.553
	68.0	35.387	12.080	1.0668	1.3732		2.594
	70.0	38.857	12.441	1.0673	1.3738		2.629
	72.0	42.821	12.801	1.0676	1.3745		2.662
	74.0	47.396	13.159	1.0678	1.3751		2.687
	76.0	52.734	13.517	1.0680	1.3757		2.714
	78.0	59.042	13.874	1.0681	1.3762		2.720
	80.0	66.611	14.228	1.0680	1.3767		2.720
	82.0	75.863	14.580	1.0677	1.3770		2.696
	84.0	87.427	14.930	1.0673	1.3773		2.658
	86.0	102.296	15.275	1.0666	1.3774		2.596
	88.0	122.120	15.619	1.0658	1.3774		2.511
	90.0	149.875	15.953	1.0644	1.3771		2.386
	92.0	191.507	16.284	1.0629	1.3766		2.240
	94.0	260.894	16.602	1.0606	1.3759		2.036
	96.0	399.667	16.911	1.0578	1.3748		1.813
	98.0	815.987	17.198	1.0538	1.3734		1.535
	100.0		17.447	1.0477	1.3716		1.223
Acetone	0.5	0.087	0.086	0.9975	1.3334	0.16	1.013
$(CH_3)_2CO$	1.0	0.174	0.172	0.9968	1.3337	0.32	1.024
	1.5	0.262	0.257	0.9961	1.3341	0.48	1.035
	2.0	0.351	0.343	0.9954	1.3344	0.65	1.047
	2.5	0.441	0.428	0.9947	1.3348	0.81	1.059
	3.0	0.533	0.513	0.9940	1.3352	0.97	1.072
	3.5	0.624	0.599	0.9933	1.3355	1.13	1.085
	4.0	0.717	0.684	0.9926	1.3359	1.30	1.099
	4.5	0.811	0.769	0.9919	1.3363	1.46	1.112
	5.0	0.906	0.853	0.9912	1.3366	1.63	1.125
	6.0	1.099	1.023	0.9899	1.3373	1.96	1.150
	7.0	1.296	1.191	0.9886	1.3381	2.29	1.174
	8.0	1.497	1.360	0.9874	1.3388	2.62	1.198
	9.0	1.703	1.528	0.9861	1.3395	2.95	1.221
	10.0	1.913	1.696	0.9849	1.3402	3.29	1.244
Ammonia	0.5	0.295	0.292	0.9960	1.3332	0.55	1.009
NH_3	1.0	0.593	0.584	0.9938	1.3335	1.14	1.015
	2.0	1.198	1.162	0.9895	1.3339	2.32	1.029
	3.0	1.816	1.736	0.9853	1.3344	3.53	1.043
	4.0	2.447	2.304	0.9811	1.3349	4.78	1.057
	5.0	3.090	2.868	0.9770	1.3354	6.08	1.071
	6.0	3.748	3.428	0.9730	1.3359	7.43	1.085
	7.0	4.420	3.983	0.9690	1.3365	8.95	1.099
	8.0	5.106	4.533	0.9651	1.3370	10.34	1.113
	9.0	5.807	5.080	0.9613	1.3376	11.90	1.127
	10.0	6.524	5.622	0.9575	1.3381	13.55	1.141
	11.0	7.257	6.160	0.9538	1.3387	15.29	1.155
	12.0	8.007	6.695	0.9502	1.3393	17.13	1.169
	13.0	8.774	7.226	0.9466	1.3398	19.07	1.182
	14.0	9.558	7.753	0.9431	1.3404	21.13	1.195
	15.0	10.362	8.275	0.9396	1.3410	23.32	1.207
	16.0	11.184	8.794	0.9361	1.3416	25.63	1.218
	17.0	12.026	9.310	0.9327	1.3422	28.09	1.228
	18.0	12.889	9.823	0.9294	1.3428	30.70	1.237
	19.0	13.773	10.332	0.9261	1.3434	33.47	1.245

DENSITY, REFRACTIVE INDEX, FREEZING POINT DEPRESSION, AND VISCOSITY OF AQUEOUS SOLUTIONS (continued)

Solute	Mass %	m/mol kg^{-1}	c/mol L^{-1}	ρ/g cm^{-3}	n	Δ/°C	η/mPa s
	20.0	14.679	10.837	0.9228	1.3440	36.42	1.254
	22.0	16.561	11.838	0.9164	1.3453	43.36	1.268
	24.0	18.542	12.826	0.9102	1.3465	51.38	1.280
	26.0	20.630	13.801	0.9040	1.3477	60.77	1.288
	28.0	22.834	14.764	0.8980	1.3490	71.66	
	30.0	25.164	15.713	0.8920	1.3502	84.06	
Ammonium	0.5	0.094	0.093	0.9998	1.3340	0.32	0.999
chloride	1.0	0.189	0.187	1.0014	1.3349	0.64	0.996
NH$_4$Cl	2.0	0.382	0.376	1.0045	1.3369	1.27	0.992
	3.0	0.578	0.565	1.0076	1.3388	1.91	0.988
	4.0	0.779	0.756	1.0107	1.3407	2.57	0.985
	5.0	0.984	0.948	1.0138	1.3426	3.25	0.982
	6.0	1.193	1.141	1.0168	1.3445	3.94	0.979
	7.0	1.407	1.335	1.0198	1.3464	4.66	0.976
	8.0	1.626	1.529	1.0227	1.3483	5.40	0.974
	9.0	1.849	1.726	1.0257	1.3502	6.16	0.972
	10.0	2.077	1.923	1.0286	1.3521	6.95	0.970
	11.0	2.311	2.121	1.0315	1.3540	7.76	0.969
	12.0	2.549	2.320	1.0344	1.3559	8.60	0.969
	13.0	2.793	2.521	1.0373	1.3578	9.47	0.969
	14.0	3.043	2.722	1.0401	1.3596		0.969
	15.0	3.299	2.924	1.0429	1.3615		0.970
	16.0	3.561	3.128	1.0457	1.3634		0.971
	17.0	3.829	3.332	1.0485	1.3652		0.972
	18.0	4.104	3.537	1.0512	1.3671		0.973
	19.0	4.385	3.744	1.0540	1.3689		0.975
	20.0	4.674	3.951	1.0567	1.3708		0.978
	22.0	5.273	4.368	1.0621	1.3745		0.986
	24.0	5.903	4.789	1.0674	1.3782		0.996
Barium	0.5	0.024	0.024	1.0026	1.3337	0.12	1.009
chloride	1.0	0.049	0.048	1.0070	1.3345	0.23	1.016
BaCl$_2$	2.0	0.098	0.098	1.0159	1.3360	0.46	1.026
	3.0	0.149	0.148	1.0249	1.3375	0.69	1.037
	4.0	0.200	0.199	1.0341	1.3391	0.93	1.049
	5.0	0.253	0.251	1.0434	1.3406	1.18	1.062
	6.0	0.307	0.303	1.0528	1.3422	1.44	1.075
	7.0	0.361	0.357	1.0624	1.3438	1.70	1.087
	8.0	0.418	0.412	1.0721	1.3454	1.98	1.101
	9.0	0.475	0.468	1.0820	1.3470	2.27	1.114
	10.0	0.534	0.524	1.0921	1.3487	2.58	1.129
	12.0	0.655	0.641	1.1128	1.3520	3.22	1.161
	14.0	0.782	0.763	1.1342	1.3555	3.92	1.195
	16.0	0.915	0.889	1.1564	1.3591	4.69	1.234
	18.0	1.054	1.019	1.1793	1.3627		1.277
	20.0	1.201	1.156	1.2031	1.3664		1.325
	22.0	1.355	1.297	1.2277	1.3703		1.378
	24.0	1.517	1.444	1.2531	1.3741		1.437
	26.0	1.687	1.597	1.2793	1.3781		1.503
Calcium	0.5	0.045	0.045	1.0024	1.3342	0.22	1.015
chloride	1.0	0.091	0.091	1.0065	1.3354	0.44	1.028
CaCl$_2$	2.0	0.184	0.183	1.0148	1.3378	0.88	1.050
	3.0	0.279	0.277	1.0232	1.3402	1.33	1.078
	4.0	0.375	0.372	1.0316	1.3426	1.82	1.110
	5.0	0.474	0.469	1.0401	1.3451	2.35	1.143
	6.0	0.575	0.567	1.0486	1.3475	2.93	1.175
	7.0	0.678	0.667	1.0572	1.3500	3.57	1.208
	8.0	0.784	0.768	1.0659	1.3525	4.28	1.242

Solute	Mass %	m/mol kg^{-1}	c/mol L^{-1}	ρ/g cm^{-3}	n	Δ/°C	η/mPa s
	9.0	0.891	0.872	1.0747	1.3549	5.04	1.279
	10.0	1.001	0.976	1.0835	1.3575	5.86	1.319
	11.0	1.114	1.083	1.0923	1.3600	6.74	1.362
	12.0	1.229	1.191	1.1014	1.3625	7.70	1.408
	13.0	1.346	1.301	1.1105	1.3651	8.72	1.457
	14.0	1.467	1.413	1.1198	1.3677	9.83	1.508
	15.0	1.590	1.526	1.1292	1.3704	11.01	1.564
	16.0	1.716	1.641	1.1386	1.3730	12.28	1.625
	17.0	1.846	1.759	1.1482	1.3757	13.65	1.691
	18.0	1.978	1.878	1.1579	1.3784	15.11	1.764
	19.0	2.114	1.999	1.1677	1.3812	16.70	1.843
	20.0	2.253	2.122	1.1775	1.3839	18.30	1.930
	22.0	2.541	2.374	1.1976	1.3895	21.70	2.127
	24.0	2.845	2.634	1.2180	1.3951	25.30	2.356
	26.0	3.166	2.902	1.2388	1.4008	29.70	2.645
	28.0	3.504	3.179	1.2600	1.4066	34.70	3.000
	30.0	3.862	3.464	1.2816	1.4124	41.00	3.467
	32.0	4.240	3.759	1.3036	1.4183	49.70	4.035
	34.0	4.642	4.062	1.3260	1.4242		4.820
	36.0	5.068	4.375	1.3488	1.4301		5.807
	38.0	5.522	4.698	1.3720	1.4361		7.321
	40.0	6.007	5.030	1.3957	1.4420		8.997
Copper	0.5	0.031	0.031	1.0033	1.3339	0.08	1.017
sulfate	1.0	0.063	0.063	1.0085	1.3348	0.14	1.036
$CuSO_4$	2.0	0.128	0.128	1.0190	1.3367	0.26	1.084
	3.0	0.194	0.194	1.0296	1.3386	0.37	1.129
	4.0	0.261	0.261	1.0403	1.3405	0.48	1.173
	5.0	0.330	0.329	1.0511	1.3424	0.59	1.221
	6.0	0.400	0.399	1.0620	1.3443	0.70	1.276
	7.0	0.472	0.471	1.0730	1.3462	0.82	1.336
	8.0	0.545	0.543	1.0842	1.3481	0.93	1.400
	9.0	0.620	0.618	1.0955	1.3501	1.05	1.469
	10.0	0.696	0.694	1.1070	1.3520	1.18	1.543
	11.0	0.774	0.771	1.1186	1.3540	1.31	1.620
	12.0	0.854	0.850	1.1304	1.3560	1.45	1.701
	13.0	0.936	0.930	1.1424	1.3581	1.60	1.790
	14.0	1.020	1.013	1.1545	1.3601	1.75	1.889
	15.0	1.106	1.097	1.1669	1.3622		2.004
	16.0	1.193	1.182	1.1796	1.3644		2.136
	17.0	1.283	1.270	1.1926	1.3666		2.285
	18.0	1.375	1.360	1.2059	1.3689		2.449
Ethanol	0.5	0.109	0.108	0.9973	1.3333	0.20	1.023
CH_3CH_2OH	1.0	0.219	0.216	0.9963	1.3336	0.40	1.046
	2.0	0.443	0.432	0.9945	1.3342	0.81	1.095
	3.0	0.671	0.646	0.9927	1.3348	1.23	1.140
	4.0	0.904	0.860	0.9910	1.3354	1.65	1.183
	5.0	1.142	1.074	0.9893	1.3360	2.09	1.228
	6.0	1.385	1.286	0.9878	1.3367	2.54	1.279
	7.0	1.634	1.498	0.9862	1.3374	2.99	1.331
	8.0	1.887	1.710	0.9847	1.3381	3.47	1.385
	9.0	2.147	1.921	0.9833	1.3388	3.96	1.442
	10.0	2.412	2.131	0.9819	1.3395	4.47	1.501
	11.0	2.683	2.341	0.9805	1.3403	5.00	1.563
	12.0	2.960	2.551	0.9792	1.3410	5.56	1.627
	13.0	3.243	2.759	0.9778	1.3417	6.13	1.694
	14.0	3.534	2.967	0.9765	1.3425	6.73	1.761
	15.0	3.830	3.175	0.9752	1.3432	7.36	1.826
	16.0	4.134	3.382	0.9739	1.3440	8.01	1.890

Solute	Mass %	m/mol kg^{-1}	c/mol L^{-1}	ρ/g cm^{-3}	n	Δ/°C	η/mPa s
	17.0	4.446	3.589	0.9726	1.3447	8.69	1.955
	18.0	4.765	3.795	0.9713	1.3455	9.40	2.019
	19.0	5.092	4.000	0.9700	1.3462	10.14	2.081
	20.0	5.427	4.205	0.9687	1.3469	10.92	2.142
	22.0	6.122	4.613	0.9660	1.3484	12.60	2.259
	24.0	6.855	5.018	0.9632	1.3498	14.47	2.370
	26.0	7.626	5.419	0.9602	1.3511	16.41	2.476
	28.0	8.441	5.817	0.9571	1.3524	18.43	2.581
	30.0	9.303	6.212	0.9539	1.3535	20.47	2.667
	32.0	10.215	6.601	0.9504	1.3546	22.44	2.726
	34.0	11.182	6.987	0.9468	1.3557	24.27	2.768
	36.0	12.210	7.370	0.9431	1.3566	25.98	2.803
	38.0	13.304	7.747	0.9392	1.3575	27.62	2.829
	40.0	14.471	8.120	0.9352	1.3583	29.26	2.846
	42.0	15.718	8.488	0.9311	1.3590	30.98	2.852
	44.0	17.055	8.853	0.9269	1.3598	32.68	2.850
	46.0	18.490	9.213	0.9227	1.3604	34.36	2.843
	48.0	20.036	9.568	0.9183	1.3610	36.04	2.832
	50.0	21.706	9.919	0.9139	1.3616	37.67	2.813
	52.0	23.515	10.266	0.9095	1.3621	39.20	2.789
	54.0	25.481	10.607	0.9049	1.3626	40.65	2.754
	56.0	27.626	10.945	0.9004	1.3630	42.06	2.701
	58.0	29.975	11.278	0.8958	1.3634	43.49	2.632
	60.0	32.559	11.605	0.8911	1.3638	44.93	2.547
	62.0	35.415	11.930	0.8865	1.3641	46.28	2.479
	64.0	38.589	12.250	0.8818	1.3644	47.52	2.415
	66.0	42.135	12.565	0.8771	1.3647	48.64	2.347
	68.0	46.125	12.877	0.8724	1.3650	49.52	2.281
	70.0	50.648	13.183	0.8676	1.3652		2.214
	72.0	55.816	13.486	0.8629	1.3654		2.148
	74.0	61.779	13.783	0.8581	1.3655		2.082
	76.0	68.736	14.077	0.8533	1.3657		2.015
	78.0	76.958	14.366	0.8485	1.3657		1.948
	80.0	86.824	14.649	0.8436	1.3658		1.881
	82.0	98.883	14.928	0.8387	1.3657		1.808
	84.0	113.957	15.197	0.8335	1.3656		1.741
	86.0	133.337	15.464	0.8284	1.3655		1.674
	88.0	159.178	15.724	0.8232	1.3653		1.606
	90.0	195.355	15.980	0.8180	1.3650		1.542
	92.0	249.620	16.225	0.8125	1.3646		1.475
	94.0	340.062	16.466	0.8070	1.3642		1.407
	96.0	520.946	16.697	0.8013	1.3636		1.342
	98.0		16.920	0.7954	1.3630		1.273
	100.0		17.133	0.7893	1.3614		1.203
Ethylene glycol $(CH_2OH)_2$	0.5	0.081	0.080	0.9988	1.3335	0.15	1.010
	1.0	0.163	0.161	0.9995	1.3339	0.30	1.020
	2.0	0.329	0.322	1.0007	1.3348	0.61	1.048
	3.0	0.498	0.484	1.0019	1.3358	0.92	1.074
	4.0	0.671	0.646	1.0032	1.3367	1.24	1.099
	5.0	0.848	0.809	1.0044	1.3377	1.58	1.125
	6.0	1.028	0.972	1.0057	1.3386	1.91	1.153
	7.0	1.213	1.136	1.0070	1.3396	2.26	1.182
	8.0	1.401	1.299	1.0082	1.3405	2.62	1.212
	9.0	1.593	1.464	1.0095	1.3415	2.99	1.243
	10.0	1.790	1.628	1.0108	1.3425	3.37	1.277
	12.0	2.197	1.959	1.0134	1.3444	4.16	1.348
	14.0	2.623	2.292	1.0161	1.3464	5.01	1.424
	16.0	3.069	2.626	1.0188	1.3484	5.91	1.500
	18.0	3.537	2.962	1.0214	1.3503	6.89	1.578

Solute	Mass %	m/mol kg^{-1}	c/mol L^{-1}	ρ/g cm^{-3}	n	Δ/°C	η/mPa s
	20.0	4.028	3.300	1.0241	1.3523	7.93	1.661
	24.0	5.088	3.981	1.0296	1.3564	10.28	1.843
	28.0	6.265	4.669	1.0350	1.3605	13.03	2.047
	32.0	7.582	5.364	1.0405	1.3646	16.23	2.280
	36.0	9.062	6.067	1.0460	1.3687	19.82	2.537
	40.0	10.741	6.776	1.0514	1.3728	23.84	2.832
	44.0	12.659	7.491	1.0567	1.3769	28.32	3.166
	48.0	14.872	8.212	1.0619	1.3811	33.30	3.544
	52.0	17.453	8.939	1.0670	1.3851	38.81	3.981
	56.0	20.505	9.671	1.0719	1.3892	44.83	4.475
	60.0	24.166	10.406	1.0765	1.3931	51.23	5.026
Ferric	0.5	0.031	0.031	1.0025	1.3344	0.21	1.024
chloride	1.0	0.062	0.062	1.0068	1.3358	0.39	1.047
$FeCl_3$	2.0	0.126	0.125	1.0153	1.3386	0.75	1.093
	3.0	0.191	0.189	1.0238	1.3413	1.15	1.139
	4.0	0.257	0.255	1.0323	1.3441	1.56	1.187
	5.0	0.324	0.321	1.0408	1.3468	2.00	1.238
	6.0	0.394	0.388	1.0493	1.3496	2.48	1.292
	7.0	0.464	0.457	1.0580	1.3524	2.99	1.350
	8.0	0.536	0.526	1.0668	1.3552	3.57	1.412
	9.0	0.610	0.597	1.0760	1.3581	4.19	1.480
	10.0	0.685	0.669	1.0853	1.3611	4.85	1.553
	12.0	0.841	0.817	1.1040	1.3670	6.38	1.707
	14.0	1.004	0.969	1.1228	1.3730	8.22	1.879
	16.0	1.174	1.126	1.1420		10.45	2.080
	18.0	1.353	1.289	1.1615		13.08	2.311
	20.0	1.541	1.457	1.1816		16.14	2.570
	24.0	1.947	1.810	1.2234		23.79	3.178
	28.0	2.398	2.189	1.2679		33.61	4.038
	32.0	2.901	2.595	1.3153		49.16	5.274
	36.0	3.468	3.030	1.3654			7.130
	40.0	4.110	3.496	1.4176			9.674
D-Fructose	0.5	0.028	0.028	1.0002	1.3337	0.05	1.015
$C_6H_{12}O_6$	1.0	0.056	0.056	1.0021	1.3344	0.10	1.028
	2.0	0.113	0.112	1.0061	1.3358	0.21	1.054
	3.0	0.172	0.168	1.0101	1.3373	0.32	1.080
	4.0	0.231	0.225	1.0140	1.3387	0.43	1.106
	5.0	0.292	0.283	1.0181	1.3402	0.54	1.134
	6.0	0.354	0.340	1.0221	1.3417	0.66	1.165
	7.0	0.418	0.399	1.0262	1.3431	0.78	1.198
	8.0	0.483	0.458	1.0303	1.3446	0.90	1.232
	9.0	0.549	0.517	1.0344	1.3461	1.03	1.270
	10.0	0.617	0.576	1.0385	1.3476	1.16	1.309
	11.0	0.686	0.637	1.0427	1.3492	1.29	1.349
	12.0	0.757	0.697	1.0469	1.3507	1.43	1.391
	13.0	0.829	0.759	1.0512	1.3522	1.57	1.435
	14.0	0.904	0.820	1.0554	1.3538	1.71	1.483
	15.0	0.980	0.882	1.0597	1.3554	1.86	1.533
	16.0	1.057	0.945	1.0640	1.3569	2.01	1.587
	17.0	1.137	1.008	1.0684	1.3585	2.16	1.643
	18.0	1.218	1.072	1.0728	1.3601	2.32	1.703
	19.0	1.302	1.136	1.0772	1.3618	2.48	1.768
	20.0	1.388	1.201	1.0816	1.3634	2.64	1.837
	22.0	1.566	1.332	1.0906	1.3667	3.05	1.986
	24.0	1.753	1.465	1.0996	1.3700	3.43	2.154

Solute	Mass %	m/mol kg^{-1}	c/mol L^{-1}	ρ/g cm^{-3}	n	Δ/°C	η/mPa s
	26.0	1.950	1.600	1.1089	1.3734	3.82	2.348
	28.0	2.159	1.738	1.1182	1.3768	4.20	2.562
	30.0	2.379	1.878	1.1276	1.3803		2.817
	32.0	2.612	2.020	1.1372	1.3839		3.112
	34.0	2.859	2.164	1.1469	1.3874		3.462
	36.0	3.122	2.312	1.1568	1.3911		3.899
	38.0	3.402	2.461	1.1668	1.3948		4.418
	40.0	3.700	2.613	1.1769	1.3985		5.046
	42.0	4.019	2.767	1.1871	1.4023		5.773
	44.0	4.361	2.925	1.1975	1.4062		6.644
	46.0	4.728	3.084	1.2080	1.4101		7.753
	48.0	5.124	3.247	1.2187	1.4141		9.060
D-Glucose	0.5	0.028	0.028	1.0001	1.3337	0.05	1.010
$C_6H_{12}O_6$	1.0	0.056	0.056	1.0020	1.3344	0.11	1.021
	2.0	0.113	0.112	1.0058	1.3358	0.21	1.052
	3.0	0.172	0.168	1.0097	1.3373	0.32	1.083
	4.0	0.231	0.225	1.0136	1.3387	0.43	1.113
	5.0	0.292	0.282	1.0175	1.3402	0.55	1.145
	6.0	0.354	0.340	1.0214	1.3417	0.67	1.179
	7.0	0.418	0.398	1.0254	1.3432	0.79	1.214
	8.0	0.483	0.457	1.0294	1.3447	0.91	1.250
	9.0	0.549	0.516	1.0334	1.3462	1.04	1.289
	10.0	0.617	0.576	1.0375	1.3477	1.17	1.330
	11.0	0.686	0.636	1.0416	1.3492	1.30	1.372
	12.0	0.757	0.697	1.0457	1.3508	1.44	1.416
	13.0	0.829	0.758	1.0498	1.3523	1.59	1.462
	14.0	0.904	0.819	1.0540	1.3539	1.73	1.512
	15.0	0.980	0.881	1.0582	1.3555	1.88	1.566
	16.0	1.057	0.944	1.0624	1.3571	2.03	1.625
	17.0	1.137	1.007	1.0667	1.3587	2.19	1.688
	18.0	1.218	1.070	1.0710	1.3603	2.35	1.757
	19.0	1.302	1.134	1.0753	1.3619	2.52	1.829
	20.0	1.388	1.199	1.0797	1.3635	2.70	1.904
	22.0	1.566	1.329	1.0884	1.3668	3.07	2.063
	24.0	1.753	1.462	1.0973	1.3702	3.48	2.242
	26.0	1.950	1.597	1.1063	1.3736	3.90	2.458
	28.0	2.159	1.734	1.1154	1.3770	4.34	2.707
	30.0	2.379	1.873	1.1246	1.3805	4.79	2.998
	32.0	2.612	2.014	1.1340	1.3840		3.324
	34.0	2.859	2.158	1.1434	1.3876		3.704
	36.0	3.122	2.304	1.1529	1.3912		4.193
	38.0	3.402	2.452	1.1626	1.3949		4.786
	40.0	3.700	2.603	1.1724	1.3986		5.493
	42.0	4.019	2.756	1.1823	1.4024		6.288
	44.0	4.361	2.912	1.1924	1.4062		7.235
	46.0	4.728	3.071	1.2026	1.4101		8.454
	48.0	5.124	3.232	1.2130	1.4141		9.883
	50.0	5.551	3.396	1.2235	1.4181		11.884
	52.0	6.013	3.562	1.2342	1.4222		14.489
	54.0	6.516	3.732	1.2451	1.4263		17.916
	56.0	7.064	3.905	1.2562	1.4306		22.886
	58.0	7.665	4.081	1.2676	1.4349		29.389
	60.0	8.326	4.261	1.2793	1.4394		37.445
Glycerol	0.5	0.055	0.054	0.9994	1.3336	0.07	1.011
$CH_2OHCHOHCH_2OH$	1.0	0.110	0.109	1.0005	1.3342	0.18	1.022

Solute	Mass %	m/mol kg^{-1}	c/mol L^{-1}	ρ/g cm^{-3}	n	Δ/°C	η/mPa s
	2.0	0.222	0.218	1.0028	1.3353	0.41	1.048
	3.0	0.336	0.327	1.0051	1.3365	0.63	1.074
	4.0	0.452	0.438	1.0074	1.3376	0.85	1.100
	5.0	0.572	0.548	1.0097	1.3388	1.08	1.127
	6.0	0.693	0.659	1.0120	1.3400	1.32	1.157
	7.0	0.817	0.771	1.0144	1.3412	1.56	1.188
	8.0	0.944	0.883	1.0167	1.3424	1.81	1.220
	9.0	1.074	0.996	1.0191	1.3436	2.06	1.256
	10.0	1.207	1.109	1.0215	1.3448	2.32	1.291
	12.0	1.481	1.337	1.0262	1.3472	2.88	1.365
	14.0	1.768	1.568	1.0311	1.3496	3.47	1.445
	16.0	2.068	1.800	1.0360	1.3521	4.09	1.533
	18.0	2.384	2.035	1.0409	1.3547	4.76	1.630
	20.0	2.715	2.271	1.0459	1.3572	5.46	1.737
	24.0	3.429	2.752	1.0561	1.3624	7.01	1.988
	28.0	4.223	3.242	1.0664	1.3676	8.77	2.279
	32.0	5.110	3.742	1.0770	1.3730	10.74	2.637
	36.0	6.108	4.252	1.0876	1.3785	12.96	3.088
	40.0	7.239	4.771	1.0984	1.3841	15.50	3.653
	44.0	8.532	5.300	1.1092	1.3897		4.443
	48.0	10.024	5.838	1.1200	1.3954		5.413
	52.0	11.764	6.385	1.1308	1.4011		6.666
	56.0	13.820	6.944	1.1419	1.4069		8.349
	60.0	16.288	7.512	1.1530	1.4129		10.681
	64.0	19.305	8.092	1.1643	1.4189		13.657
	68.0	23.075	8.680	1.1755	1.4249		18.457
	72.0	27.923	9.277	1.1866	1.4310		27.625
	76.0	34.387	9.884	1.1976	1.4370		40.57
	80.0	43.436	10.498	1.2085	1.4431		59.9
	84.0	57.009	11.121	1.2192	1.4492		84.3
	88.0	79.632	11.753	1.2299	1.4553		147.5
	92.0	124.878	12.392	1.2404	1.4613		384
	96.0	260.615	13.039	1.2508	1.4674		780
	100.0		13.694	1.2611	1.4735		
Hydrochloric	0.5	0.138	0.137	1.0007	1.3341	0.49	1.008
acid	1.0	0.277	0.275	1.0031	1.3353	0.99	1.015
HCl	2.0	0.560	0.553	1.0081	1.3376	2.08	1.029
	3.0	0.848	0.833	1.0130	1.3399	3.28	1.044
	4.0	1.143	1.117	1.0179	1.3422	4.58	1.059
	5.0	1.444	1.403	1.0228	1.3445	5.98	1.075
	6.0	1.751	1.691	1.0278	1.3468	7.52	1.091
	7.0	2.064	1.983	1.0327	1.3491	9.22	1.108
	8.0	2.385	2.277	1.0377	1.3515	11.10	1.125
	9.0	2.713	2.574	1.0426	1.3538	13.15	1.143
	10.0	3.047	2.873	1.0476	1.3561	15.40	1.161
	11.0	3.390	3.176	1.0526	1.3584	17.85	1.180
	12.0	3.740	3.481	1.0576	1.3607	20.51	1.199
	13.0	4.098	3.789	1.0626	1.3630		1.219
	14.0	4.465	4.099	1.0676	1.3653		1.239
	15.0	4.840	4.413	1.0726	1.3676		1.261
	16.0	5.224	4.729	1.0777	1.3700		1.282
	17.0	5.617	5.049	1.0828	1.3723		1.304
	18.0	6.020	5.370	1.0878	1.3746		1.326
	19.0	6.433	5.695	1.0929	1.3769		1.350
	20.0	6.857	6.023	1.0980	1.3792		1.374
	22.0	7.736	6.687	1.1083	1.3838		1.426
	24.0	8.661	7.362	1.1185	1.3884		1.483
	26.0	9.636	8.049	1.1288	1.3930		1.547
	28.0	10.666	8.748	1.1391	1.3976		1.620

Solute	Mass %	m/mol kg^{-1}	c/mol L^{-1}	ρ/g cm^{-3}	n	Δ/°C	η/mPa s
	30.0	11.754	9.456	1.1492	1.4020		1.705
	32.0	12.907	10.175	1.1594	1.4066		1.799
	34.0	14.129	10.904	1.1693	1.4112		1.900
	36.0	15.427	11.642	1.1791	1.4158		2.002
	38.0	16.810	12.388	1.1886	1.4204		2.105
	40.0	18.284	13.140	1.1977	1.4250		
Lithium	0.5	0.119	0.118	1.0012	1.3341	0.42	1.019
chloride	1.0	0.238	0.237	1.0041	1.3351	0.84	1.037
LiCl	2.0	0.481	0.476	1.0099	1.3373	1.72	1.072
	3.0	0.730	0.719	1.0157	1.3394	2.68	1.108
	4.0	0.983	0.964	1.0215	1.3415	3.73	1.146
	5.0	1.241	1.211	1.0272	1.3436	4.86	1.185
	6.0	1.506	1.462	1.0330	1.3457	6.14	1.226
	7.0	1.775	1.715	1.0387	1.3478	7.56	1.269
	8.0	2.051	1.971	1.0444	1.3499	9.11	1.313
	9.0	2.333	2.230	1.0502	1.3520	10.79	1.360
	10.0	2.621	2.491	1.0560	1.3541	12.61	1.411
	12.0	3.217	3.022	1.0675	1.3583	16.59	1.522
	14.0	3.840	3.564	1.0792	1.3625	21.04	1.647
	16.0	4.493	4.118	1.0910	1.3668		1.787
	18.0	5.178	4.683	1.1029	1.3711		1.942
	20.0	5.897	5.260	1.1150	1.3755		2.128
	22.0	6.653	5.851	1.1274	1.3799		2.341
	24.0	7.449	6.453	1.1399	1.3844		2.600
	26.0	8.288	7.069	1.1527	1.3890		2.925
	28.0	9.173	7.700	1.1658	1.3936		3.318
	30.0	10.109	8.344	1.1791	1.3983		3.785
Magnesium	0.5	0.053	0.053	1.0022	1.3343	0.26	1.024
chloride	1.0	0.106	0.106	1.0062	1.3356	0.52	1.046
MgCl$_2$	2.0	0.214	0.213	1.0144	1.3381	1.06	1.091
	3.0	0.325	0.322	1.0226	1.3406	1.65	1.139
	4.0	0.438	0.433	1.0309	1.3432	2.30	1.188
	5.0	0.553	0.546	1.0394	1.3457	3.01	1.241
	6.0	0.670	0.660	1.0479	1.3483		1.298
	7.0	0.791	0.777	1.0564	1.3508		1.358
	8.0	0.913	0.895	1.0651	1.3534		1.423
	9.0	1.039	1.015	1.0738	1.3560		1.493
	10.0	1.167	1.137	1.0826	1.3587		1.570
	12.0	1.432	1.387	1.1005	1.3641		1.745
	14.0	1.710	1.645	1.1189	1.3695		1.956
	16.0	2.001	1.911	1.1372	1.3749		2.207
	18.0	2.306	2.184	1.1553	1.3804		2.507
	20.0	2.626	2.467	1.1742	1.3859		2.867
	22.0	2.962	2.758	1.1938	1.3915		3.323
	24.0	3.317	3.060	1.2140	1.3972		3.917
	26.0	3.690	3.371	1.2346	1.4030		4.694
	28.0	4.085	3.692	1.2555	1.4089		5.709
	30.0	4.501	4.022	1.2763	1.4148		7.017
Magnesium	0.5	0.042	0.042	1.0033	1.3340	0.10	1.027
sulfate	1.0	0.084	0.084	1.0084	1.3350	0.19	1.054
MgSO$_4$	2.0	0.170	0.169	1.0186	1.3371	0.36	1.112
	3.0	0.257	0.256	1.0289	1.3391	0.52	1.177
	4.0	0.346	0.345	1.0392	1.3411	0.69	1.249
	5.0	0.437	0.436	1.0497	1.3431	0.87	1.328
	6.0	0.530	0.528	1.0602	1.3451	1.05	1.411
	7.0	0.625	0.623	1.0708	1.3471	1.24	1.498

DENSITY, REFRACTIVE INDEX, FREEZING POINT DEPRESSION, AND VISCOSITY OF AQUEOUS SOLUTIONS (continued)

Solute	Mass %	m/mol kg^{-1}	c/mol L^{-1}	ρ/g cm^{-3}	n	Δ/°C	η/mPa s
	8.0	0.722	0.719	1.0816	1.3492	1.43	1.593
	9.0	0.822	0.817	1.0924	1.3512	1.64	1.702
	10.0	0.923	0.917	1.1034	1.3532	1.85	1.829
	12.0	1.133	1.122	1.1257	1.3572	2.31	2.104
	14.0	1.352	1.336	1.1484	1.3613	2.86	2.412
	16.0	1.582	1.557	1.1717	1.3654	3.67	2.809
	18.0	1.824	1.788	1.1955	1.3694		3.360
	20.0	2.077	2.027	1.2198	1.3735		4.147
	22.0	2.343	2.275	1.2447	1.3776		5.199
	24.0	2.624	2.532	1.2701	1.3817		6.498
	26.0	2.919	2.800	1.2961	1.3858		8.066
Maltose	0.5	0.015	0.015	1.0003	1.3337	0.03	1.016
$C_{12}H_{22}O_{11}$	1.0	0.030	0.029	1.0023	1.3345	0.06	1.030
	2.0	0.060	0.059	1.0063	1.3359	0.11	1.060
	3.0	0.090	0.089	1.0104	1.3374	0.17	1.092
	4.0	0.122	0.119	1.0144	1.3389	0.23	1.126
	5.0	0.154	0.149	1.0184	1.3404	0.29	1.162
	6.0	0.186	0.179	1.0224	1.3420	0.35	1.200
	7.0	0.220	0.210	1.0265	1.3435	0.42	1.239
	8.0	0.254	0.241	1.0305	1.3450	0.48	1.281
	9.0	0.289	0.272	1.0345	1.3466	0.55	1.325
	10.0	0.325	0.303	1.0385	1.3482	0.62	1.372
	11.0	0.361	0.335	1.0425	1.3497	0.69	1.422
	12.0	0.398	0.367	1.0465	1.3513	0.77	1.474
	13.0	0.437	0.399	1.0505	1.3529	0.84	1.530
	14.0	0.476	0.431	1.0545	1.3546	0.92	1.588
	15.0	0.516	0.464	1.0585	1.3562	1.00	1.649
	16.0	0.556	0.497	1.0629	1.3578	1.08	1.715
	17.0	0.598	0.530	1.0672	1.3595	1.17	1.784
	18.0	0.641	0.564	1.0716	1.3612	1.25	1.859
	19.0	0.685	0.597	1.0759	1.3628	1.34	1.940
	20.0	0.730	0.631	1.0801	1.3644	1.43	2.021
	22.0	0.824	0.700	1.0894	1.3678	1.64	2.216
	24.0	0.923	0.770	1.0984	1.3714	1.85	2.463
	26.0	1.026	0.842	1.1080	1.3749	2.08	2.753
	28.0	1.136	0.914	1.1171	1.3785	2.34	3.066
	30.0	1.252	0.988	1.1269	1.3821	2.62	3.427
	32.0	1.375	1.063	1.1367	1.3858	2.93	3.918
	34.0	1.505	1.139	1.1463	1.3896	3.25	4.447
	36.0	1.643	1.216	1.1561	1.3935	3.60	5.050
	38.0	1.791	1.295	1.1663	1.3974	3.99	5.832
	40.0	1.948	1.375	1.1769	1.4013	4.41	6.926
	42.0	2.116	1.457	1.1878	1.4051	4.88	8.191
	44.0	2.295	1.540	1.1979	1.4094	5.35	9.649
	46.0	2.489	1.624	1.2084	1.4136		11.473
	48.0	2.697	1.710	1.2194	1.4177		14.148
	50.0	2.921	1.797	1.2304	1.4217		17.786
	52.0	3.165	1.886	1.2416	1.4260		22.034
	54.0	3.429	1.976	1.2528	1.4308		28.757
	56.0	3.718	2.068	1.2638	1.4350		38.226
	58.0	4.034	2.159	1.2740	1.4394		49.298
	60.0	4.382	2.253	1.2855	1.4440		
Methanol	0.5	0.157	0.156	0.9973	1.3331	0.28	1.022
CH_3OH	1.0	0.315	0.311	0.9964	1.3332	0.56	1.040
	2.0	0.637	0.621	0.9947	1.3334	1.14	1.070

Solute	Mass %	m/mol kg^{-1}	c/mol L^{-1}	ρ/g cm^{-3}	n	$\Delta/°C$	η/mPa s
	3.0	0.965	0.930	0.9930	1.3336	1.75	1.100
	4.0	1.300	1.238	0.9913	1.3339	2.37	1.131
	5.0	1.643	1.544	0.9896	1.3341	3.02	1.163
	6.0	1.992	1.850	0.9880	1.3343	3.71	1.196
	7.0	2.349	2.155	0.9864	1.3346	4.41	1.229
	8.0	2.714	2.459	0.9848	1.3348	5.13	1.264
	9.0	3.087	2.762	0.9832	1.3351	5.85	1.297
	10.0	3.468	3.064	0.9816	1.3354	6.60	1.329
	11.0	3.858	3.365	0.9801	1.3356	7.36	1.360
	12.0	4.256	3.665	0.9785	1.3359	8.14	1.389
	13.0	4.664	3.964	0.9770·	1.3362	8.93	1.418
	14.0	5.081	4.262	0.9755	1.3365	9.72	1.446
	15.0	5.508	4.560	0.9740	1.3367	10.53	1.474
	16.0	5.945	4.856	0.9725	1.3370	11.36	1.501
	17.0	6.393	5.152	0.9710	1.3373	12.23	1.528
	18.0	6.851	5.447	0.9695	1.3376	13.13	1.554
	19.0	7.321	5.740	0.9680	1.3379	14.06	1.579
	20.0	7.803	6.034	0.9666	1.3381	15.02	1.604
	22.0	8.803	6.616	0.9636	1.3387	16.98	1.652
	24.0	9.856	7.196	0.9606	1.3392	19.04	1.697
	26.0	10.966	7.771	0.9576	1.3397	21.23	1.735
	28.0	12.138	8.341	0.9545	1.3402	23.59	1.769
	30.0	13.376	8.908	0.9514	1.3407	25.91	1.795
	32.0	14.688	9.470	0.9482	1.3411	28.15	1.814
	34.0	16.078	10.028	0.9450	1.3415	30.48	1.827
	36.0	17.556	10.580	0.9416	1.3419	32.97	1.835
	38.0	19.129	11.127	0.9382	1.3422	35.60	1.839
	40.0	20.807	11.669	0.9347	1.3425	38.60	1.837
	42.0	22.601	12.205	0.9311	1.3427	41.50	1.831
	44.0	24.523	12.734	0.9273	1.3429	44.50	1.821
	46.0	26.587	13.259	0.9235	1.3430	47.80	1.805
	48.0	28.810	13.777	0.9196	1.3431	51.20	1.785
	50.0	31.211	14.288	0.9156	1.3431	54.50	1.761
	52.0	33.812	14.792	0.9114	1.3431	58.10	1.735
	54.0	36.639	15.290	0.9072	1.3430	62.00	1.708
	56.0	39.723	15.783	0.9030	1.3429	66.00	1.676
	58.0	43.101	16.269	0.8987	1.3428	70.00	1.641
	60.0	46.816	16.749	0.8944·	1.3426	74.50	1.600
	62.0	50.923	17.224	0.8901	1.3425	79.30	1.553
	64.0	55.486	17.690	0.8856	1.3422	84.40	1.503
	66.0	60.586	18.148	0.8810	1.3419	89.60	1.456
	68.0	66.323	18.598	0.8763	1.3415	96.30	1.413
	70.0	72.826	19.040	0.8715	1.3411		1.368
	72.0	80.257	19.476	0.8667	1.3407		1.318
	74.0	88.831	19.904	0.8618	1.3402		1.267
	76.0	98.835	20.324	0.8568	1.3397		1.218
	78.0	110.657	20.737	0.8518	1.3391		1.172
	80.0	124.844	21.144	0.8468	1.3385		1.128
	82.0	142.183	21.539	0.8416	1.3379		1.074
	84.0	163.858	21.931	0.8365	1.3372		1.023
	86.0	191.725	22.311	0.8312	1.3365		0.970
	88.0	228.881	22.684	0.8259	1.3357		0.916
	90.0	280.899	23.045	0.8204	1.3348		0.861
	92.0	358.926	23.396	0.8148	1.3339		0.806
	94.0	488.972	23.732	0.8089	1.3328		0.749
	96.0	749.064	24.072	0.8034	1.3316		0.695
	98.0		24.396	0.7976	1.3304		0.639

Solute	Mass %	m/mol kg^{-1}	c/mol L^{-1}	ρ/g cm^{-3}	n	Δ/°C	η/mPa s
	100.0		24.710	0.7917	1.3290		0.586
Nitric acid	0.5	0.080	0.079	1.0009	1.3336	0.28	1.004
HNO$_3$	1.0	0.160	0.159	1.0037	1.3343	0.56	1.005
	2.0	0.324	0.320	1.0091	1.3356	1.12	1.007
	3.0	0.491	0.483	1.0146	1.3368	1.70	1.010
	4.0	0.661	0.648	1.0202	1.3381	2.32	1.014
	5.0	0.835	0.814	1.0257	1.3394	2.96	1.018
	6.0	1.013	0.982	1.0314	1.3407	3.63	1.022
	7.0	1.194	1.152	1.0370	1.3421	4.33	1.027
	8.0	1.380	1.324	1.0427	1.3434	5.05	1.032
	9.0	1.570	1.498	1.0485	1.3447	5.81	1.038
	10.0	1.763	1.673	1.0543	1.3460	6.60	1.044
	11.0	1.961	1.851	1.0602	1.3474	7.42	1.051
	12.0	2.164	2.030	1.0660	1.3487	8.27	1.058
	13.0	2.371	2.212	1.0720	1.3500	9.15	1.066
	14.0	2.583	2.395	1.0780	1.3514	10.08	1.075
	15.0	2.801	2.580	1.0840	1.3527	11.04	1.084
	16.0	3.023	2.768	1.0901	1.3541	12.04	1.094
	17.0	3.250	2.958	1.0963	1.3555	13.08	1.105
	18.0	3.484	3.149	1.1025	1.3569	14.16	1.116
	19.0	3.723	3.343	1.1087	1.3582	15.30	1.128
	20.0	3.967	3.539	1.1150	1.3596		1.141
	22.0	4.476	3.937	1.1277	1.3624		1.169
	24.0	5.011	4.344	1.1406	1.3652		1.199
	26.0	5.576	4.760	1.1536	1.3680		1.233
	28.0	6.172	5.185	1.1668	1.3708		1.271
	30.0	6.801	5.618	1.1801	1.3736		1.311
	32.0	7.468	6.060	1.1934	1.3763		1.354
	34.0	8.175	6.512	1.2068	1.3790		1.400
	36.0	8.927	6.971	1.2202	1.3817		1.450
	38.0	9.727	7.439	1.2335	1.3842		1.504
	40.0	10.580	7.913	1.2466	1.3867		1.561
Phosphoric	0.5	0.051	0.051	1.0010	1.3335	0.12	1.010
acid	1.0	0.103	0.102	1.0038	1.3340	0.24	1.020
H$_3$PO$_4$	2.0	0.208	0.206	1.0092	1.3349	0.46	1.050
	3.0	0.316	0.311	1.0146	1.3358	0.69	1.079
	4.0	0.425	0.416	1.0200	1.3367	0.93	1.108
	5.0	0.537	0.523	1.0254	1.3376	1.16	1.138
	6.0	0.651	0.631	1.0309	1.3385	1.38	1.169
	7.0	0.768	0.740	1.0363	1.3394	1.62	1.200
	8.0	0.887	0.850	1.0418	1.3403	1.88	1.232
	9.0	1.009	0.962	1.0474	1.3413	2.16	1.267
	10.0	1.134	1.075	1.0531	1.3422	2.45	1.303
	11.0	1.261	1.189	1.0589	1.3431	2.72	1.341
	12.0	1.392	1.304	1.0647	1.3441	3.01	1.382
	13.0	1.525	1.420	1.0705	1.3450	3.38	1.424
	14.0	1.661	1.538	1.0765	1.3460	3.76	1.469
	15.0	1.801	1.657	1.0825	1.3470	4.08	1.516
	16.0	1.944	1.777	1.0885	1.3480	4.45	1.565
	17.0	2.090	1.899	1.0947	1.3489	4.82	1.616
	18.0	2.240	2.022	1.1009	1.3500	5.25	1.671
	19.0	2.394	2.147	1.1071	1.3510	5.72	1.727
	20.0	2.551	2.273	1.1135	1.3520	6.23	1.788
	22.0	2.878	2.529	1.1263	1.3540	7.38	1.914
	24.0	3.223	2.791	1.1395	1.3561	8.69	2.049
	26.0	3.585	3.059	1.1528	1.3582	10.12	2.198
	28.0	3.968	3.333	1.1665	1.3604	11.64	2.365
	30.0	4.373	3.614	1.1804	1.3625	13.23	2.553

Solute	Mass %	m/mol kg^{-1}	c/mol L^{-1}	ρ/g cm^{-3}	n	Δ/°C	η/mPa s
	32.0	4.802	3.901	1.1945	1.3647	14.94	2.766
	34.0	5.257	4.194	1.2089	1.3669	16.81	3.001
	36.0	5.740	4.495	1.2236	1.3691	18.85	3.260
	38.0	6.254	4.803	1.2385	1.3713	21.09	3.544
	40.0	6.803	5.117	1.2536	1.3735	23.58	3.856
Potassium	0.5	0.042	0.042	1.0018	1.3336	0.15	1.000
bromide	1.0	0.085	0.084	1.0054	1.3342	0.29	0.998
KBr	2.0	0.171	0.170	1.0127	1.3354	0.59	0.994
	3.0	0.260	0.257	1.0200	1.3366	0.88	0.990
	4.0	0.350	0.345	1.0275	1.3379	1.18	0.985
	5.0	0.442	0.435	1.0350	1.3391	1.48	0.981
	6.0	0.536	0.526	1.0426	1.3403	1.78	0.977
	7.0	0.633	0.618	1.0503	1.3416	2.10	0.974
	8.0	0.731	0.711	1.0581	1.3429	2.42	0.970
	9.0	0.831	0.806	1.0660	1.3441	2.74	0.967
	10.0	0.934	0.903	1.0740	1.3454	3.07	0.964
	11.0	1.039	1.000	1.0821	1.3467	3.42	0.961
	12.0	1.146	1.099	1.0903	1.3481	3.76	0.958
	13.0	1.256	1.200	1.0986	1.3494	4.12	0.956
	14.0	1.368	1.302	1.1070	1.3507	4.49	0.953
	15.0	1.483	1.406	1.1155	1.3521	4.86	0.951
	16.0	1.601	1.512	1.1242	1.3535	5.25	0.949
	17.0	1.721	1.619	1.1330	1.3548	5.64	0.948
	18.0	1.845	1.727	1.1419	1.3562	6.04	0.946
	19.0	1.971	1.838	1.1509	1.3577	6.46	0.945
	20.0	2.101	1.950	1.1601	1.3591	6.88	0.944
	22.0	2.370	2.179	1.1788	1.3620	7.76	0.943
	24.0	2.654	2.416	1.1980	1.3650	8.70	0.943
	26.0	2.952	2.661	1.2179	1.3680	9.68	0.944
	28.0	3.268	2.914	1.2383	1.3711	10.72	0.947
	30.0	3.601	3.175	1.2593	1.3743	11.82	0.952
	32.0	3.954	3.445	1.2810	1.3776	12.98	0.959
	34.0	4.329	3.724	1.3033	1.3809		0.968
	36.0	4.727	4.012	1.3263	1.3843		0.979
	38.0	5.150	4.311	1.3501	1.3878		0.993
	40.0	5.602	4.620	1.3746	1.3914		1.010
Potassium	0.5	0.067	0.067	1.0014	1.3337	0.23	1.000
chloride	1.0	0.135	0.135	1.0046	1.3343	0.46	0.999
KCl	2.0	0.274	0.271	1.0110	1.3357	0.92	0.999
	3.0	0.415	0.409	1.0174	1.3371	1.38	0.998
	4.0	0.559	0.549	1.0239	1.3384	1.85	0.997
	5.0	0.706	0.691	1.0304	1.3398	2.32	0.996
	6.0	0.856	0.835	1.0369	1.3411	2.80	0.994
	7.0	1.010	0.980	1.0434	1.3425	3.29	0.992
	8.0	1.166	1.127	1.0500	1.3438	3.80	0.990
	9.0	1.327	1.276	1.0566	1.3452	4.30	0.989
	10.0	1.490	1.426	1.0633	1.3466	4.81	0.988
	11.0	1.658	1.579	1.0700	1.3479	5.33	0.989
	12.0	1.829	1.733	1.0768	1.3493	5.88	0.990
	13.0	2.004	1.890	1.0836	1.3507	6.45	0.992
	14.0	2.184	2.048	1.0905	1.3521		0.994
	15.0	2.367	2.208	1.0974	1.3535		0.997
	16.0	2.555	2.370	1.1043	1.3549		0.999
	17.0	2.747	2.534	1.1114	1.3563		1.001
	18.0	2.944	2.701	1.1185	1.3577		1.004
	19.0	3.146	2.869	1.1256	1.3592		1.007
	20.0	3.353	3.039	1.1328	1.3606		1.012
	22.0	3.783	3.386	1.1474	1.3635		1.024

Solute	Mass %	m/mol kg^{-1}	c/mol L^{-1}	ρ/g cm^{-3}	n	Δ/°C	η/mPa s
	24.0	4.236	3.742	1.1623	1.3665		1.040
Potassium	0.5	0.090	0.089	1.0025	1.3340	0.30	1.010
hydroxide	1.0	0.180	0.179	1.0068	1.3350	0.61	1.019
KOH	2.0	0.364	0.362	1.0155	1.3369	1.24	1.038
	3.0	0.551	0.548	1.0242	1.3388	1.89	1.058
	4.0	0.743	0.736	1.0330	1.3408	2.57	1.079
	5.0	0.938	0.929	1.0419	1.3427	3.36	1.102
	6.0	1.138	1.124	1.0509	1.3445	4.14	1.126
	7.0	1.342	1.322	1.0599	1.3464	4.92	1.151
	8.0	1.550	1.524	1.0690	1.3483		1.177
	9.0	1.763	1.729	1.0781	1.3502		1.205
	10.0	1.980	1.938	1.0873	1.3520		1.233
	11.0	2.203	2.150	1.0966	1.3539		1.264
	12.0	2.431	2.365	1.1059	1.3558		1.294
	13.0	2.663	2.584	1.1153	1.3576		1.327
	14.0	2.902	2.806	1.1246	1.3595		1.361
	15.0	3.145	3.032	1.1341	1.3614		1.397
	16.0	3.395	3.261	1.1435	1.3632		1.436
	17.0	3.651	3.494	1.1531	1.3651		1.477
	18.0	3.913	3.730	1.1626	1.3670		1.521
	19.0	4.181	3.970	1.1722	1.3688		1.568
	20.0	4.456	4.213	1.1818	1.3707		1.619
	22.0	5.027	4.711	1.2014	1.3744		1.732
	24.0	5.629	5.223	1.2210	1.3781		1.861
	26.0	6.262	5.750	1.2408	1.3818		2.006
	28.0	6.931	6.293	1.2609	1.3854		2.170
	30.0	7.639	6.851	1.2813	1.3889		2.357
	32.0	8.388	7.426	1.3020	1.3923		2.570
	34.0	9.182	8.017	1.3230	1.3957		2.820
	36.0	10.026	8.626	1.3444	1.3993		3.111
	38.0	10.924	9.253	1.3661	1.4030		3.460
	40.0	11.882	9.896	1.3881	1.4068		3.879
	42.0	12.907	10.558	1.4104	1.4106		4.389
	44.0	14.004	11.239	1.4331	1.4143		5.013
	46.0	15.183	11.938	1.4560	1.4179		5.781
	48.0	16.453	12.654	1.4791	1.4214		6.726
	50.0	17.824	13.389	1.5024	1.4247		7.892
Potassium	0.5	0.030	0.030	1.0019	1.3337	0.11	0.999
iodide	1.0	0.061	0.061	1.0056	1.3343	0.22	0.997
KI	2.0	0.123	0.122	1.0131	1.3357	0.43	0.991
	3.0	0.186	0.184	1.0206	1.3370	0.64	0.986
	4.0	0.251	0.248	1.0282	1.3384	0.86	0.981
	5.0	0.317	0.312	1.0360	1.3397	1.08	0.976
	6.0	0.385	0.377	1.0438	1.3411	1.30	0.969
	7.0	0.453	0.443	1.0517	1.3425	1.53	0.963
	8.0	0.524	0.511	1.0598	1.3440	1.77	0.957
	9.0	0.596	0.579	1.0679	1.3454	2.01	0.951
	10.0	0.669	0.648	1.0762	1.3469	2.26	0.946
	12.0	0.821	0.790	1.0931	1.3498	2.77	0.937
	14.0	0.981	0.937	1.1105	1.3529	3.30	0.929
	16.0	1.147	1.088	1.1284	1.3560	3.87	0.921
	18.0	1.322	1.244	1.1469	1.3593	4.46	0.915
	20.0	1.506	1.405	1.1659	1.3626	5.09	0.910
	22.0	1.699	1.571	1.1856	1.3661	5.76	0.905
	24.0	1.902	1.744	1.2060	1.3696	6.46	0.901
	26.0	2.117	1.922	1.2270	1.3733	7.21	0.898

Solute	Mass %	m/mol kg^{-1}	c/mol L^{-1}	ρ/g cm^{-3}	n	Δ/°C	η/mPa s
	28.0	2.343	2.106	1.2487	1.3771	8.01	0.895
	30.0	2.582	2.297	1.2712	1.3810	8.86	0.892
	32.0	2.835	2.495	1.2944	1.3851	9.76	0.891
	34.0	3.103	2.700	1.3185	1.3893	10.72	0.890
	36.0	3.388	2.913	1.3434	1.3936	11.73	0.890
	38.0	3.692	3.134	1.3692	1.3981	12.81	0.893
	40.0	4.016	3.364	1.3959	1.4027	13.97	0.897
Potassium nitrate KNO_3	0.5	0.050	0.050	1.0014	1.3335	0.17	0.999
	1.0	0.100	0.099	1.0045	1.3339	0.33	0.996
	2.0	0.202	0.200	1.0108	1.3349	0.64	0.990
	3.0	0.306	0.302	1.0171	1.3358	0.94	0.986
	4.0	0.412	0.405	1.0234	1.3368	1.22	0.983
	5.0	0.521	0.509	1.0298	1.3377	1.50	0.980
	6.0	0.631	0.615	1.0363	1.3386	1.76	0.977
	7.0	0.744	0.722	1.0428	1.3396	2.02	0.975
	8.0	0.860	0.830	1.0494	1.3405	2.27	0.973
	9.0	0.978	0.940	1.0560	1.3415	2.52	0.971
	10.0	1.099	1.051	1.0627	1.3425	2.75	0.970
	12.0	1.349	1.277	1.0762	1.3444		0.970
	14.0	1.610	1.509	1.0899	1.3463		0.972
	16.0	1.884	1.747	1.1039	1.3482		0.976
	18.0	2.171	1.991	1.1181	1.3502		0.982
	20.0	2.473	2.240	1.1326	1.3521		0.990
	22.0	2.790	2.497	1.1473	1.3541		0.999
	24.0	3.123	2.759	1.1623	1.3561		1.010
Potassium permanganate $KMnO_4$	0.5	0.032	0.032	1.0017		0.11	1.001
	1.0	0.064	0.064	1.0051		0.22	1.000
	1.5	0.096	0.096	1.0085		0.32	0.999
	2.0	0.129	0.128	1.0118		0.43	0.998
	2.5	0.162	0.161	1.0152			0.996
	3.0	0.196	0.193	1.0186			0.995
	3.5	0.230	0.226	1.0220			0.994
	4.0	0.264	0.260	1.0254			0.992
	4.5	0.298	0.293	1.0288			0.991
	5.0	0.333	0.327	1.0322			0.989
	5.5	0.368	0.360	1.0356			0.987
	6.0	0.404	0.394	1.0390			0.985
Potassium sulfate K_2SO_4	0.5	0.029	0.029	1.0022	1.3336	0.14	1.006
	1.0	0.058	0.058	1.0062	1.3343	0.26	1.011
	2.0	0.117	0.116	1.0143	1.3355	0.50	1.021
	2.5	0.147	0.146	1.0183	1.3362	0.61	1.027
	3.0	0.177	0.176	1.0224	1.3368	0.73	1.033
	3.5	0.208	0.206	1.0265	1.3374	0.84	1.039
	4.0	0.239	0.237	1.0306	1.3380	0.95	1.045
	4.5	0.270	0.267	1.0347	1.3386	1.06	1.051
	5.0	0.302	0.298	1.0388	1.3393	1.17	1.058
	6.0	0.366	0.360	1.0470	1.3405		1.072
	7.0	0.432	0.424	1.0553	1.3417		1.087
	8.0	0.499	0.488	1.0637	1.3428		1.102
	9.0	0.568	0.554	1.0721	1.3440		1.117
	10.0	0.638	0.620	1.0806	1.3452		1.132
1-Propanol $CH_3CH_2CH_2OH$	1.0	0.168	0.166	0.9963	1.3339	0.31	1.051
	2.0	0.340	0.331	0.9946	1.3348	0.61	1.100
	3.0	0.515	0.496	0.9928	1.3357	0.93	1.152

Solute	Mass %	m/mol kg^{-1}	c/mol L^{-1}	ρ/g cm^{-3}	n	Δ/°C	η/mPa s
	4.0	0.693	0.660	0.9911	1.3366	1.24	1.208
	5.0	0.876	0.823	0.9896	1.3376	1.57	1.267
	6.0	1.062	0.987	0.9882	1.3385	1.91	1.325
	7.0	1.252	1.149	0.9868	1.3394	2.26	1.387
	8.0	1.447	1.312	0.9855	1.3404	2.61	1.449
	9.0	1.646	1.474	0.9842	1.3414	2.99	1.514
	10.0	1.849	1.635	0.9829	1.3423	3.36	1.577
	12.0	2.269	1.958	0.9804	1.3442	4.09	1.710
	14.0	2.709	2.278	0.9779	1.3460	4.91	1.849
	16.0	3.169	2.595	0.9749	1.3477	5.78	1.986
	18.0	3.652	2.911	0.9719	1.3494	6.67	2.106
	20.0	4.160	3.223	0.9686	1.3510	7.76	2.218
	24.0	5.254	3.838	0.9612	1.3539	9.12	2.432
	28.0	6.471	4.441	0.9533	1.3566	10.17	2.612
	32.0	7.830	5.033	0.9452	1.3592	10.66	2.765
	36.0	9.359	5.613	0.9370	1.3614		2.900
	40.0	11.093	6.182	0.9288	1.3635		3.010
	60.0	24.958	8.860	0.8875	1.3734		3.186
	80.0	66.556	11.275	0.8470	1.3812		2.822
	100.0		13.368	0.8034	1.3852		2.227
2-Propanol	1.0	0.168	0.166	0.9960	1.3338	0.30	1.056
$CH_3CHOHCH_3$	2.0	0.340	0.331	0.9939	1.3346	0.60	1.112
	3.0	0.515	0.495	0.9920	1.3355	0.93	1.166
	4.0	0.693	0.659	0.9902	1.3364	1.26	1.225
	5.0	0.876	0.822	0.9884	1.3373	1.61	1.287
	6.0	1.062	0.985	0.9871	1.3382	1.96	1.352
	7.0	1.252	1.148	0.9855	1.3392	2.32	1.417
	8.0	1.447	1.310	0.9843	1.3400	2.68	1.485
	9.0	1.646	1.472	0.9831	1.3410	3.06	1.553
	10.0	1.849	1.633	0.9816	1.3420	3.48	1.629
	12.0	2.269	1.955	0.9793	1.3439	4.43	1.794
	14.0	2.709	2.276	0.9772	1.3459	5.29	1.970
	16.0	3.169	2.596	0.9751	1.3478	6.36	2.160
	18.0	3.652	2.913	0.9725	1.3496	7.40	2.352
	20.0	4.160	3.227	0.9696	1.3514	8.52	2.550
	40.0	11.093	6.191	0.9302	1.3642		
	60.0	24.958	8.809	0.8824	1.3717		
	80.0	66.556	11.103	0.8341	1.3765		
	100.0		13.058	0.7848	1.3742		
Silver	0.5	0.030	0.030	1.0027	1.3336	0.10	1.003
nitrate	1.0	0.059	0.059	1.0070	1.3342	0.20	1.005
$AgNO_3$	2.0	0.120	0.120	1.0154	1.3352	0.40	1.009
	3.0	0.182	0.181	1.0239	1.3363	0.59	1.013
	4.0	0.245	0.243	1.0327	1.3374	0.78	1.016
	5.0	0.310	0.307	1.0417	1.3385	0.96	1.020
	6.0	0.376	0.371	1.0506	1.3396	1.15	1.024
	7.0	0.443	0.437	1.0597	1.3407	1.33	1.027
	8.0	0.512	0.503	1.0690	1.3419	1.51	1.031
	9.0	0.582	0.571	1.0785	1.3431	1.69	1.035
	10.0	0.654	0.641	1.0882	1.3443	1.87	1.039
	12.0	0.803	0.783	1.1079	1.3467	2.21	1.049
	14.0	0.958	0.930	1.1284	1.3493	2.55	1.060
	16.0	1.121	1.083	1.1496	1.3519	2.86	1.072
	18.0	1.292	1.241	1.1715	1.3546		1.086
	20.0	1.472	1.406	1.1942	1.3574		1.101
	22.0	1.660	1.577	1.2177	1.3602		1.117
	24.0	1.859	1.755	1.2420	1.3632		1.135

DENSITY, REFRACTIVE INDEX, FREEZING POINT DEPRESSION, AND VISCOSITY OF AQUEOUS SOLUTIONS (continued)

Solute	Mass %	m/mol kg^{-1}	c/mol L^{-1}	ρ/g cm^{-3}	n	$\Delta/°C$	η/mPa s
	26.0	2.068	1.940	1.2672	1.3662		1.154
	28.0	2.289	2.132	1.2933	1.3694		1.176
	30.0	2.523	2.332	1.3204	1.3726		1.200
	32.0	2.770	2.541	1.3487	1.3760		1.227
	34.0	3.033	2.758	1.3780	1.3795		1.257
	36.0	3.311	2.985	1.4087	1.3832		1.290
	38.0	3.608	3.223	1.4407	1.3871		1.326
	40.0	3.925	3.472	1.4743	1.3911		1.366
Sodium	0.5	0.061	0.061	1.0008	1.3337	0.22	1.021
acetate	1.0	0.123	0.122	1.0034	1.3344	0.43	1.040
CH$_3$COONa	2.0	0.249	0.246	1.0085	1.3358	0.88	1.080
	3.0	0.377	0.371	1.0135	1.3372	1.34	1.124
	4.0	0.508	0.497	1.0184	1.3386	1.82	1.171
	5.0	0.642	0.624	1.0234	1.3400	2.32	1.222
	6.0	0.778	0.752	1.0283	1.3414	2.85	1.278
	7.0	0.918	0.882	1.0334	1.3428	3.40	1.337
	8.0	1.060	1.013	1.0386	1.3442	3.98	1.401
	9.0	1.206	1.145	1.0440	1.3456	4.57	1.468
	10.0	1.354	1.279	1.0495	1.3470		1.539
	12.0	1.662	1.552	1.0607	1.3498		1.688
	14.0	1.984	1.829	1.0718	1.3526		1.855
	16.0	2.322	2.112	1.0830	1.3554		2.054
	18.0	2.676	2.400	1.0940	1.3583		2.284
	20.0	3.047	2.694	1.1050	1.3611		2.567
	22.0	3.438	2.993	1.1159	1.3639		2.948
	24.0	3.849	3.297	1.1268	1.3666		3.400
	26.0	4.283	3.606	1.1377	1.3693		3.877
	28.0	4.741	3.921	1.1488	1.3720		4.388
	30.0	5.224	4.243	1.1602	1.3748		4.940
Sodium	0.5	0.060	0.060	1.0018	1.3337	0.20	1.015
bicarbonate	1.0	0.120	0.120	1.0054	1.3344	0.40	1.028
NaHCO$_3$	1.5	0.181	0.180	1.0089	1.3351	0.59	1.042
	2.0	0.243	0.241	1.0125	1.3357	0.78	1.057
	2.5	0.305	0.302	1.0160	1.3364	0.98	1.071
	3.0	0.368	0.364	1.0196	1.3370	1.16	1.086
	3.5	0.432	0.426	1.0231	1.3377	1.35	1.102
	4.0	0.496	0.489	1.0266	1.3383	1.54	1.118
	4.5	0.561	0.552	1.0301	1.3390	1.72	1.134
	5.0	0.627	0.615	1.0337	1.3396	1.90	1.151
	5.5	0.693	0.679	1.0372	1.3403	2.08	1.168
	6.0	0.760	0.743	1.0408	1.3409	2.26	1.185
Sodium	0.5	0.049	0.049	1.0021	1.3337	0.17	1.004
bromide	1.0	0.098	0.098	1.0060	1.3344	0.34	1.007
NaBr	2.0	0.198	0.197	1.0139	1.3358	0.69	1.012
	3.0	0.301	0.298	1.0218	1.3372	1.04	1.017
	4.0	0.405	0.400	1.0298	1.3386	1.39	1.022
	5.0	0.512	0.504	1.0380	1.3401	1.76	1.028
	6.0	0.620	0.610	1.0462	1.3415	2.14	1.034
	7.0	0.732	0.717	1.0546	1.3430	2.53	1.040
	8.0	0.845	0.826	1.0630	1.3445	2.93	1.046
	9.0	0.961	0.937	1.0716	1.3460	3.34	1.053
	10.0	1.080	1.050	1.0803	1.3475	3.77	1.060
	11.0	1.201	1.164	1.0892	1.3491	4.21	1.068
	12.0	1.325	1.281	1.0981	1.3506	4.67	1.077
	13.0	1.452	1.399	1.1072	1.3522	5.15	1.086
	14.0	1.582	1.519	1.1164	1.3538	5.65	1.096

Solute	Mass %	m/mol kg^{-1}	c/mol L^{-1}	ρ/g cm^{-3}	n	Δ/°C	η/mPa s
	15.0	1.715	1.641	1.1257	1.3554	6.18	1.107
	16.0	1.851	1.765	1.1352	1.3570	6.74	1.119
	17.0	1.991	1.891	1.1448	1.3587	7.32	1.131
	18.0	2.133	2.020	1.1546	1.3604		1.144
	19.0	2.280	2.150	1.1645	1.3621		1.159
	20.0	2.430	2.283	1.1745	1.3638		1.174
	22.0	2.741	2.555	1.1951	1.3673		1.207
	24.0	3.069	2.837	1.2163	1.3708		1.244
	26.0	3.415	3.129	1.2382	1.3745		1.287
	28.0	3.780	3.431	1.2608	1.3783		1.336
	30.0	4.165	3.744	1.2842	1.3822		1.395
	32.0	4.574	4.069	1.3083	1.3862		1.465
	34.0	5.007	4.406	1.3333	1.3903		1.546
	36.0	5.467	4.755	1.3592	1.3946		1.639
	38.0	5.957	5.119	1.3860	1.3990		1.745
	40.0	6.479	5.496	1.4138	1.4035		1.866
Sodium carbonate Na$_2$CO$_3$	0.5	0.047	0.047	1.0034	1.3341	0.22	1.025
	1.0	0.095	0.095	1.0086	1.3352	0.43	1.049
	2.0	0.193	0.192	1.0190	1.3375	0.75	1.102
	3.0	0.292	0.291	1.0294	1.3397	1.08	1.159
	4.0	0.393	0.392	1.0398	1.3419	1.42	1.222
	5.0	0.497	0.495	1.0502	1.3440	1.77	1.292
	6.0	0.602	0.600	1.0606	1.3462	2.13	1.367
	7.0	0.710	0.707	1.0711	1.3483		1.448
	8.0	0.820	0.816	1.0816	1.3504		1.538
	9.0	0.933	0.927	1.0922	1.3525		1.638
	10.0	1.048	1.041	1.1029	1.3547		1.754
	11.0	1.166	1.156	1.1136	1.3568		1.884
	12.0	1.287	1.273	1.1244	1.3589		2.028
	13.0	1.410	1.392	1.1353	1.3610		2.186
	14.0	1.536	1.514	1.1463	1.3631		2.361
	15.0	1.665	1.638	1.1574	1.3652		2.551
Sodium chloride NaCl	0.1	0.017	0.017	0.9989	1.3332	0.06	1.004
	0.2	0.034	0.034	0.9997	1.3333	0.12	1.006
	0.3	0.051	0.051	1.0004	1.3335	0.18	1.008
	0.4	0.069	0.069	1.0011	1.3337	0.24	1.009
	0.5	0.086	0.086	1.0018	1.3339	0.30	1.011
	1.0	0.173	0.172	1.0053	1.3347	0.59	1.020
	1.5	0.261	0.259	1.0089	1.3356	0.89	1.028
	2.0	0.349	0.346	1.0125	1.3365	1.19	1.036
	2.5	0.439	0.435	1.0160	1.3374	1.49	1.044
	3.0	0.529	0.523	1.0196	1.3383	1.79	1.052
	3.5	0.621	0.613	1.0232	1.3391	2.10	1.060
	4.0	0.713	0.703	1.0268	1.3400	2.41	1.068
	4.5	0.806	0.793	1.0304	1.3409	2.73	1.076
	5.0	0.901	0.885	1.0340	1.3418	3.05	1.085
	6.0	1.092	1.069	1.0413	1.3435	3.70	1.104
	7.0	1.288	1.256	1.0486	1.3453	4.38	1.124
	8.0	1.488	1.445	1.0559	1.3470	5.08	1.145
	9.0	1.692	1.637	1.0633	1.3488	5.81	1.168
	10.0	1.901	1.832	1.0707	1.3505	6.56	1.193
	11.0	2.115	2.029	1.0781	1.3523	7.35	1.220
	12.0	2.333	2.229	1.0857	1.3541	8.18	1.250
	13.0	2.557	2.432	1.0932	1.3558	9.04	1.283
	14.0	2.785	2.637	1.1008	1.3576	9.94	1.317
	15.0	3.020	2.845	1.1085	1.3594	10.89	1.352
	16.0	3.259	3.056	1.1162	1.3612	11.89	1.388

Solute	Mass %	m/mol kg^{-1}	c/mol L^{-1}	ρ/g cm^{-3}	n	Δ/°C	η/mPa s
	17.0	3.505	3.270	1.1240	1.3630	12.94	1.424
	18.0	3.756	3.486	1.1319	1.3648	14.04	1.463
	19.0	4.014	3.706	1.1398	1.3666	15.22	1.507
	20.0	4.278	3.928	1.1478	1.3684	16.46	1.557
	21.0	4.548	4.153	1.1558	1.3702	17.78	1.614
	22.0	4.826	4.382	1.1640	1.3721	19.18	1.676
	23.0	5.111	4.613	1.1721	1.3739	20.67	1.745
	24.0	5.403	4.847	1.1804	1.3757		1.821
	25.0	5.704	5.085	1.1887	1.3776		1.902
	26.0	6.012	5.326	1.1972	1.3795		1.990
Sodium	0.5	0.126	0.125	1.0039	1.3344	0.43	1.027
hydroxide	1.0	0.253	0.252	1.0095	1.3358	0.86	1.054
NaOH	2.0	0.510	0.510	1.0207	1.3386	1.74	1.112
	3.0	0.773	0.774	1.0318	1.3414	2.64	1.176
	4.0	1.042	1.043	1.0428	1.3441	3.59	1.248
	5.0	1.316	1.317	1.0538	1.3467	4.57	1.329
	6.0	1.596	1.597	1.0648	1.3494	5.60	1.416
	7.0	1.882	1.883	1.0758	1.3520	6.69	1.510
	8.0	2.174	2.174	1.0869	1.3546	7.87	1.616
	9.0	2.473	2.470	1.0979	1.3572	9.12	1.737
	10.0	2.778	2.772	1.1089	1.3597	10.47	1.882
	11.0	3.090	3.080	1.1199	1.3623	11.89	2.039
	12.0	3.409	3.393	1.1309	1.3648	13.42	2.201
	13.0	3.736	3.711	1.1419	1.3673	15.04	2.376
	14.0	4.070	4.036	1.1530	1.3697	16.76	2.568
	15.0	4.412	4.365	1.1640	1.3722		2.789
	16.0	4.762	4.701	1.1751	1.3746		3.043
	17.0	5.121	5.041	1.1861	1.3770		3.344
	18.0	5.488	5.387	1.1971	1.3793		3.698
	19.0	5.865	5.739	1.2082	1.3817		4.119
	20.0	6.250	6.096	1.2192	1.3840		4.619
	22.0	7.052	6.827	1.2412	1.3885		5.765
	24.0	7.895	7.579	1.2631	1.3929		7.100
	26.0	8.784	8.352	1.2848	1.3971		8.744
	28.0	9.723	9.145	1.3064	1.4012		10.832
	30.0	10.715	9.958	1.3277	1.4051		13.517
	32.0	11.766	10.791	1.3488	1.4088		16.844
	34.0	12.880	11.643	1.3697	1.4123		20.751
	36.0	14.064	12.512	1.3901	1.4156		25.290
	38.0	15.324	13.398	1.4102	1.4186		30.461
	40.0	16.668	14.300	1.4299	1.4215		36.312
Sodium	0.5	0.059	0.059	1.0016	1.3336	0.20	1.004
nitrate	1.0	0.119	0.118	1.0050	1.3341	0.40	1.007
NaNO$_3$	2.0	0.240	0.238	1.0117	1.3353	0.79	1.012
	3.0	0.364	0.359	1.0185	1.3364	1.18	1.018
	4.0	0.490	0.483	1.0254	1.3375	1.56	1.025
	5.0	0.619	0.607	1.0322	1.3387	1.94	1.032
	6.0	0.751	0.734	1.0392	1.3398	2.32	1.040
	7.0	0.886	0.862	1.0462	1.3409	2.70	1.049
	8.0	1.023	0.991	1.0532	1.3421	3.08	1.059
	9.0	1.164	1.123	1.0603	1.3432	3.46	1.069
	10.0	1.307	1.256	1.0674	1.3443	3.84	1.081
	12.0	1.604	1.527	1.0819	1.3466	4.60	1.107
	14.0	1.915	1.806	1.0967	1.3489	5.37	1.138
	18.0	2.583	2.387	1.1272	1.3536	6.98	1.215
	20.0	2.941	2.689	1.1429	1.3559	7.81	1.263
	24.0	3.715	3.318	1.1752	1.3607	9.52	1.377

Solute	Mass %	m/mol kg^{-1}	c/mol L^{-1}	ρ/g cm^{-3}	n	Δ/°C	η/mPa s
	28.0	4.575	3.981	1.2085	1.3654	11.28	1.522
	30.0	5.042	4.326	1.2256	1.3678		1.609
	34.0	6.061	5.044	1.2610	1.3726		1.818
	40.0	7.844	6.200	1.3175	1.3802		2.226
Sodium	0.5	0.035	0.035	1.0027	1.3338	0.17	1.013
sulfate	1.0	0.071	0.071	1.0071	1.3345	0.32	1.026
Na$_2$SO$_4$	2.0	0.144	0.143	1.0161	1.3360	0.61	1.058
	3.0	0.218	0.217	1.0252	1.3376	0.87	1.091
	4.0	0.293	0.291	1.0343	1.3391	1.13	1.126
	5.0	0.371	0.367	1.0436	1.3406	1.36	1.163
	6.0	0.449	0.445	1.0526	1.3420	1.56	1.202
	7.0	0.530	0.523	1.0619	1.3435		1.244
	8.0	0.612	0.603	1.0713	1.3449		1.289
	9.0	0.696	0.685	1.0808	1.3464		1.337
	10.0	0.782	0.768	1.0905	1.3479		1.390
	11.0	0.870	0.852	1.1002	1.3494		1.447
	12.0	0.960	0.938	1.1101	1.3509		1.508
	13.0	1.052	1.025	1.1201	1.3524		1.574
	14.0	1.146	1.114	1.1301	1.3539		1.646
	15.0	1.242	1.204	1.1402	1.3553		1.725
	16.0	1.341	1.296	1.1503	1.3567		1.812
	17.0	1.442	1.389	1.1604	1.3581		1.905
	18.0	1.545	1.483	1.1705	1.3595		2.005
	19.0	1.651	1.579	1.1806	1.3608		2.112
	20.0	1.760	1.677	1.1907	1.3620		2.227
	22.0	1.986	1.875	1.2106	1.3643		2.481
Sodium	0.5	0.032	0.032	1.0024	1.3340	0.14	1.012
thiosulfate	1.0	0.064	0.064	1.0065	1.3351	0.28	1.023
Na$_2$S$_2$O$_3$	2.0	0.129	0.128	1.0148	1.3371	0.57	1.044
	3.0	0.196	0.194	1.0231	1.3392	0.84	1.066
	4.0	0.264	0.261	1.0315	1.3413	1.09	1.090
	5.0	0.333	0.329	1.0399	1.3434	1.34	1.115
	6.0	0.404	0.398	1.0483	1.3454	1.59	1.141
	7.0	0.476	0.468	1.0568	1.3475	1.83	1.169
	8.0	0.550	0.539	1.0654	1.3496	2.06	1.199
	9.0	0.626	0.611	1.0740	1.3517	2.30	1.231
	10.0	0.703	0.685	1.0827	1.3538	2.55	1.267
	12.0	0.862	0.835	1.1003	1.3581	3.06	1.345
	14.0	1.030	0.990	1.1182	1.3624	3.60	1.435
	16.0	1.205	1.150	1.1365	1.3667	4.17	1.537
	18.0	1.388	1.315	1.1551	1.3711	4.76	1.657
	20.0	1.581	1.485	1.1740	1.3756	5.37	1.798
	22.0	1.784	1.660	1.1932	1.3801		1.958
	24.0	1.997	1.841	1.2128	1.3847		2.141
	26.0	2.222	2.027	1.2328	1.3893		2.356
	28.0	2.460	2.219	1.2532	1.3940		2.596
	30.0	2.711	2.417	1.2739	1.3987		2.903
	32.0	2.976	2.621	1.2950	1.4035		3.305
	34.0	3.258	2.831	1.3164.	1.4084		3.792
	36.0	3.558	3.047	1.3382	1.4132		4.359
	38.0	3.876	3.269	1.3603	1.4181		5.011
	40.0	4.216	3.498	1.3827	1.4229		5.758
Strontium	0.5	0.032	0.032	1.0027	1.3339	0.16	1.012
chloride	1.0	0.064	0.064	1.0071	1.3348	0.31	1.021
SrCl$_2$	2.0	0.129	0.128	1.0161	1.3366	0.62	1.039
	3.0	0.195	0.194	1.0252	1.3384	0.93	1.057

Solute	Mass %	m/mol kg^{-1}	c/mol L^{-1}	ρ/g cm^{-3}	n	Δ/°C	η/mPa s
	4.0	0.263	0.261	1.0344	1.3402	1.26	1.076
	5.0	0.332	0.329	1.0437	1.3421	1.61	1.096
	6.0	0.403	0.399	1.0532	1.3440	1.98	1.116
	7.0	0.475	0.469	1.0628	1.3459	2.38	1.136
	8.0	0.549	0.541	1.0726	1.3478	2.80	1.157
	9.0	0.624	0.615	1.0825	1.3498	3.25	1.180
	10.0	0.701	0.689	1.0925	1.3518	3.74	1.204
	12.0	0.860	0.843	1.1131	1.3558	4.81	1.258
	14.0	1.027	1.002	1.1342	1.3599	6.03	1.317
	16.0	1.202	1.167	1.1558	1.3641	7.41	1.383
	18.0	1.385	1.338	1.1780	1.3684	8.98	1.460
	20.0	1.577	1.515	1.2008	1.3728	10.74	1.549
	22.0	1.779	1.699	1.2241	1.3772	12.74	1.650
	24.0	1.992	1.890	1.2481	1.3817	14.99	1.765
	26.0	2.216	2.087	1.2728	1.3864		1.897
	28.0	2.453	2.293	1.2983	1.3911		2.056
	30.0	2.703	2.507	1.3248	1.3961		2.245
	32.0	2.968	2.730	1.3523	1.4013		2.527
	34.0	3.250	2.962	1.3811	1.4067		2.846
	36.0	3.548	3.205	1.4114	1.4124		3.206
Sucrose	0.5	0.015	0.015	1.0002	1.3337	0.03	1.015
$C_{12}H_{22}O_{11}$	1.0	0.030	0.029	1.0021	1.3344	0.06	1.028
	2.0	0.060	0.059	1.0060	1.3359	0.11	1.055
	3.0	0.090	0.089	1.0099	1.3373	0.17	1.084
	4.0	0.122	0.118	1.0139	1.3388	0.23	1.114
	5.0	0.154	0.149	1.0178	1.3403	0.29	1.146
	6.0	0.186	0.179	1.0218	1.3418	0.35	1.179
	7.0	0.220	0.210	1.0259	1.3433	0.42	1.215
	8.0	0.254	0.241	1.0299	1.3448	0.49	1.254
	9.0	0.289	0.272	1.0340	1.3463	0.55	1.294
	10.0	0.325	0.303	1.0381	1.3478	0.63	1.336
	11.0	0.361	0.335	1.0423	1.3494	0.70	1.381
	12.0	0.398	0.367	1.0465	1.3509	0.77	1.429
	13.0	0.437	0.399	1.0507	1.3525	0.85	1.480
	14.0	0.476	0.431	1.0549	1.3541	0.93	1.534
	15.0	0.516	0.464	1.0592	1.3557	1.01	1.592
	16.0	0.556	0.497	1.0635	1.3573	1.10	1.653
	17.0	0.598	0.530	1.0678	1.3589	1.19	1.719
	18.0	0.641	0.564	1.0722	1.3606	1.27	1.790
	19.0	0.685	0.598	1.0766	1.3622	1.37	1.865
	20.0	0.730	0.632	1.0810	1.3639	1.47	1.945
	22.0	0.824	0.700	1.0899	1.3672	1.67	2.124
	24.0	0.923	0.771	1.0990	1.3706	1.89	2.331
	26.0	1.026	0.842	1.1082	1.3741	2.12	2.573
	28.0	1.136	0.914	1.1175	1.3776	2.37	2.855
	30.0	1.252	0.988	1.1270	1.3812	2.64	3.187
	32.0	1.375	1.063	1.1366	1.3848	2.94	3.762
	34.0	1.505	1.139	1.1464	1.3885	3.27	4.052
	36.0	1.643	1.216	1.1562	1.3922	3.63	4.621
	38.0	1.791	1.295	1.1663	1.3960	4.02	5.315
	40.0	1.948	1.375	1.1765	1.3999	4.45	6.162
	42.0	2.116	1.456	1.1868	1.4038	4.93	7.234
	44.0	2.295	1.539	1.1972	1.4078		8.596
	46.0	2.489	1.623	1.2079	1.4118		10.301
	48.0	2.697	1.709	1.2186	1.4159		12.515
	50.0	2.921	1.796	1.2295	1.4201		15.431
	60.0	4.382	2.255	1.2864	1.4419		58.49
	70.0	6.817	2.755	1.3472	1.4654		481

Solute	Mass %	m/mol kg^{-1}	c/mol L^{-1}	ρ/g cm^{-3}	n	Δ/°C	η/mPa s
	80.0	11.686	3.299	1.4117	1.4906		
	84.0	15.337	3.530	1.4383	1.5010		
Sulfuric acid	0.5	0.051	0.051	1.0016	1.3336	0.21	1.010
H_2SO_4	1.0	0.103	0.102	1.0049	1.3342	0.42	1.019
	2.0	0.208	0.206	1.0116	1.3355	0.80	1.036
	3.0	0.315	0.311	1.0183	1.3367	1.17	1.059
	4.0	0.425	0.418	1.0250	1.3379	1.60	1.085
	5.0	0.537	0.526	1.0318	1.3391	2.05	1.112
	6.0	0.651	0.635	1.0385	1.3403	2.50	1.136
	7.0	0.767	0.746	1.0453	1.3415	2.95	1.159
	8.0	0.887	0.858	1.0522	1.3427	3.49	1.182
	9.0	1.008	0.972	1.0591	1.3439	4.08	1.206
	10.0	1.133	1.087	1.0661	1.3451	4.64	1.230
	11.0	1.260	1.204	1.0731	1.3463	5.25	1.256
	12.0	1.390	1.322	1.0802	1.3475	5.93	1.282
	13.0	1.524	1.441	1.0874	1.3488	6.67	1.309
	14.0	1.660	1.563	1.0947	1.3500	7.49	1.337
	15.0	1.799	1.685	1.1020	1.3513	8.35	1.367
	16.0	1.942	1.810	1.1094	1.3525	9.26	1.399
	17.0	2.088	1.936	1.1169	1.3538	10.23	1.434
	18.0	2.238	2.064	1.1245	1.3551	11.29	1.470
	19.0	2.392	2.193	1.1321	1.3563	12.43	1.508
	20.0	2.549	2.324	1.1398	1.3576	13.64	1.546
	22.0	2.876	2.592	1.1554	1.3602	16.48	1.624
	24.0	3.220	2.866	1.1714	1.3628	19.85	1.706
	26.0	3.582	3.147	1.1872	1.3653	24.29	1.797
	28.0	3.965	3.435	1.2031	1.3677	29.65	1.894
	30.0	4.370	3.729	1.2191	1.3701	36.21	2.001
	32.0	4.798	4.030	1.2353	1.3725	44.76	2.122
	34.0	5.252	4.339	1.2518	1.3749	55.28	2.255
	36.0	5.735	4.656	1.2685	1.3773		2.392
	38.0	6.249	4.981	1.2855	1.3797		2.533
	40.0	6.797	5.313	1.3028	1.3821		2.690
	42.0	7.383	5.655	1.3205	1.3846		2.872
	44.0	8.011	6.005	1.3386	1.3870		3.073
	46.0	8.685	6.364	1.3570	1.3895		3.299
	48.0	9.411	6.734	1.3759	1.3920		3.546
	50.0	10.196	7.113	1.3952	1.3945		3.826
	52.0	11.045	7.502	1.4149	1.3971		4.142
	54.0	11.969	7.901	1.4351	1.3997		4.499
	56.0	12.976	8.312	1.4558	1.4024		4.906
	58.0	14.080	8.734	1.4770	1.4050		5.354
	60.0	15.294	9.168	1.4987	1.4077		5.917
	62.0	16.635	9.608	1.5200			
	64.0	18.126	10.063	1.5421			
	66.0	19.792	10.529	1.5646			
	68.0	21.666	11.006	1.5874			
	70.0	23.790	11.494	1.6105			
	72.0	26.218	11.994	1.6338			
	74.0	29.019	12.505	1.6574			
	76.0	32.287	13.026	1.6810			
	78.0	36.149	13.554	1.7043			
	80.0	40.783	14.088	1.7272			
	82.0	46.447	14.623	1.7491			
	84.0	53.528	15.153	1.7693			
	86.0	62.631	15.671	1.7872			
	88.0	74.769	16.170	1.8022			
	90.0	91.762	16.649	1.8144			

Solute	Mass %	m/mol kg^{-1}	c/mol L^{-1}	ρ/g cm^{-3}	n	Δ/°C	η/mPa s
	92.0	117.251	17.109	1.8240			
	94.0	159.734	17.550	1.8312			
	96.0	244.698	17.966	1.8355			
	98.0	499.592	18.346	1.8361			
	100.0		18.663	1.8305			
Urea	0.5	0.084	0.083	0.9995	1.3337	0.16	1.007
$(NH_2)_2CO$	1.0	0.168	0.167	1.0007	1.3344	0.31	1.010
	2.0	0.340	0.334	1.0033	1.3358	0.62	1.012
	3.0	0.515	0.502	1.0058	1.3372	0.93	1.017
	4.0	0.694	0.672	1.0085	1.3387	1.24	1.025
	5.0	0.876	0.842	1.0111	1.3401	1.55	1.033
	6.0	1.063	1.013	1.0138	1.3416	1.88	1.041
	7.0	1.253	1.185	1.0165	1.3431	2.22	1.049
	8.0	1.448	1.358	1.0192	1.3446	2.56	1.057
	9.0	1.647	1.531	1.0220	1.3461	2.91	1.065
	10.0	1.850	1.706	1.0248	1.3476	3.26	1.074
	11.0	2.058	1.882	1.0276	1.3491	3.61	1.083
	12.0	2.270	2.059	1.0304	1.3506	3.95	1.091
	13.0	2.488	2.236	1.0332	1.3521	4.30	1.100
	14.0	2.710	2.415	1.0360	1.3537	4.66	1.109
	15.0	2.938	2.594	1.0388	1.3552	5.02	1.119
	16.0	3.171	2.775	1.0417	1.3568	5.40	1.130
	17.0	3.410	2.956	1.0445	1.3583	5.79	1.141
	18.0	3.655	3.139	1.0473	1.3599	6.19	1.153
	19.0	3.906	3.322	1.0502	1.3614	6.59	1.165
	20.0	4.163	3.506	1.0530	1.3629	7.00	1.178
	22.0	4.696	3.878	1.0586	1.3661	7.81	1.205
	24.0	5.258	4.253	1.0643	1.3692	8.64	1.235
	26.0	5.850	4.632	1.0699	1.3723	9.52	1.266
	28.0	6.475	5.014	1.0756	1.3754	10.45	1.298
	30.0	7.136	5.401	1.0812	1.3785	11.40	1.332
	32.0	7.835	5.791	1.0869	1.3817	12.34	1.371
	34.0	8.577	6.185	1.0926	1.3848	13.27	1.413
	36.0	9.366	6.584	1.0984	1.3881	14.20	1.459
	38.0	10.205	6.988	1.1044	1.3913	15.11	1.509
	40.0	11.100	7.397	1.1106	1.3947	15.99	1.565
	42.0	12.057	7.812	1.1171	1.3982	16.83	1.629
	44.0	13.082	8.234	1.1239	1.4018	17.62	1.700
	46.0	14.183	8.665	1.1313	1.4056		1.780
Zinc sulfate	0.5	0.031	0.031	1.0034	1.3339	0.08	1.021
$ZnSO_4$	1.0	0.063	0.062	1.0085	1.3348	0.15	1.040
	2.0	0.126	0.126	1.0190	1.3366	0.28	1.081
	3.0	0.192	0.191	1.0296	1.3384	0.41	1.126
	4.0	0.258	0.258	1.0403	1.3403	0.53	1.175
	5.0	0.326	0.326	1.0511	1.3421	0.65	1.227
	6.0	0.395	0.395	1.0620	1.3439	0.77	1.283
	7.0	0.466	0.465	1.0730	1.3457	0.89	1.341
	8.0	0.539	0.537	1.0842	1.3475	1.01	1.403
	9.0	0.613	0.611	1.0956	1.3494	1.14	1.470
	10.0	0.688	0.686	1.1071	1.3513	1.27	1.545
	11.0	0.766	0.762	1.1188	1.3532	1.41	1.627
	12.0	0.845	0.840	1.1308	1.3551	1.55	1.716
	13.0	0.926	0.920	1.1429	1.3570	1.71	1.814
	14.0	1.008	1.002	1.1553	1.3590	1.89	1.918
	15.0	1.093	1.085	1.1679	1.3610	2.09	2.031
	16.0	1.180	1.170	1.1806	1.3630	2.31	2.152

ION PRODUCT OF WATER SUBSTANCE
William L. Marshall and E. U. Franck

Pressure (bars)	Temperature (°C)								
	0	25	50	75	100	150	200	250	300
Saturated vapor	14.938	13.995	13.275	12.712	12.265	11.638	11.289	11.191	11.406
250	14.83	13.90	13.19	12.63	12.18	11.54	11.16	11.01	11.14
500	14.72	13.82	13.11	12.55	12.10	11.45	11.05	10.85	10.86
750	14.62	13.73	13.04	12.48	12.03	11.36	10.95	10.72	10.66
1,000	14.53	13.66	12.96	12.41	11.96	11.29	10.86	10.60	10.50
1,500	14.34	13.53	12.85	12.29	11.84	11.16	10.71	10.43	10.26
2,000	14.21	13.40	12.73	12.18	11.72	11.04	10.57	10.27	10.08
2,500	14.08	13.28	12.62	12.07	11.61	10.92	10.45	10.12	9.91
3,000	13.97	13.18	12.53	11.98	11.53	10.83	10.34	9.99	9.76
3,500	13.87	13.09	12.44	11.90	11.44	10.74	10.24	9.88	9.63
4,000	13.77	13.00	12.35	11.82	11.37	10.66	10.16	9.79	9.52
5,000	13.60	12.83	12.19	11.66	11.22	10.52	10.00	9.62	9.34
6,000	13.44	12.68	12.05	11.53	11.09	10.39	9.87	9.48	9.18
7,000	13.31	12.55	11.93	11.41	10.97	10.27	9.75	9.35	9.04
8,000	13.18	12.43	11.82	11.30	10.86	10.17	9.64	9.24	8.93
9,000	13.04	12.31	11.71	11.20	10.77	10.07	9.54	9.13	8.82
10,000	12.91	12.21	11.62	11.11	10.68	9.98	9.45	9.04	8.71

Pressure (bars)	Temperature (°C)								
	350	400	450	500	600	700	800	900	1000
Saturated vapor	12.30	—	—	—	—	—	—	—	—
250	11.77	19.43	21.59	22.40	23.27	23.81	24.23	24.59	24.93
500	11.14	11.88	13.74	16.13	18.30	19.29	19.92	20.39	20.80
750	10.79	11.17	11.89	13.01	15.25	16.55	17.35	17.93	18.39
1,000	10.54	10.77	11.19	11.81	13.40	14.70	15.58	16.22	16.72
1,500	10.22	10.29	10.48	10.77	11.59	12.50	13.30	13.97	14.50
2,000	9.98	9.98	10.07	10.23	10.73	11.36	11.98	12.54	12.97
2,500	9.79	9.74	9.77	9.86	10.18	10.63	11.11	11.59	12.02
3,000	9.61	9.54	9.53	9.57	9.78	10.11	10.49	10.89	11.24
3,500	9.47	9.37	9.33	9.34	9.48	9.71	10.02	10.35	10.62
4,000	9.34	9.22	9.16	9.15	9.23	9.41	9.65	9.93	10.13
5,000	9.13	8.99	8.90	8.85	8.85	8.95	9.11	9.30	9.42
6,000	8.96	8.80	8.69	8.62	8.57	8.61	8.72	8.86	8.97
7,000	8.81	8.64	8.51	8.42	8.34	8.34	8.40	8.51	8.64
8,000	8.68	8.50	8.36	8.25	8.13	8.10	8.13	8.21	8.38
9,000	8.57	8.37	8.22	8.10	7.95	7.89	7.89	7.95	8.12
10,000	8.46	8.25	8.09	7.96	7.78	7.70	7.68	7.70	7.85

Data in this table were calculated from the equation, $\log_{10} K_w^* = A + B/T + C/T^2 + D/T^3 + (E + F/T + G/T^2) \log_{10} \rho_w^*$, where $K_w^* = K_w/(\text{mol kg}^{-1})$, and $\rho_w^* = \rho_w/(\text{g cm}^{-3})$. The parameters are:

$A = -4.098$

$B = -3245.2$ K

$C = +2.2362 \times 10^5$ K^2

$D = -3.984 \times 10^7$ K^3

$E = +13.957$

$F = 1262.3$ K

$G = +8.5641 \times 10^5$ K^2

Reprinted with permission from W. L. Marshall and E. U. Franck, *J. Phys. Chem. Ref. Data*, 10, 295, 1981.

IONIZATION CONSTANT OF NORMAL AND HEAVY WATER

This table gives the ionization constant in molality terms for H_2O and D_2O at temperatures from 0 to 100°C at the saturated vapor pressure. The quantity tabulated is $-\log K_W$, where K_W is defined by

$$K_W = m_+ \times m_-$$

and m_+ and m_- are the molalities, in mol/kg of water, for H^+ and OH^-, respectively.

REFERENCES

1. W.L. Marshall and E.U. Franck, *J. Phys. Chem. Ref. Data*, 10, 295, 1981.
2. R.E. Mesmer and D.L. Herting, *J. Solution Chem.*, 7, 901, 1978.

	$-\log K_W$	
$t/°C$	H_2O	D_2O
0	14.938	15.972
5	14.727	15.743
10	14.528	15.527
15	14.340	15.324
20	14.163	15.132
25	13.995	14.951
30	13.836	14.779
35	13.685	14.616
40	13.542	14.462
45	13.405	14.316
50	13.275	14.176
55	13.152	14.044
60	13.034	13.918
65	12.921	13.798
70	12.814	13.683
75	12.712	13.574
80	12.613	13.470
85	12.520	13.371
90	12.428	13.276
95	12.345	13.186
100	12.265	13.099

SOLUBILITY OF SELECTED GASES IN WATER

L. H. Gevantman

The values in this table are taken almost exclusively from the International Union of Pure and Applied Chemistry "Solubility Data Series". Unless noted, they comprise evaluated data fitted to a smoothing equation. The data at each temperature are then derived from the smoothing equation which expresses the mole fraction solubility X_1 of the gas in solution as:

$$\ln X_1 = A + B/T^* + C \ln T^*$$

where

$$T^* = T/100 \text{ K}$$

All values refer to a partial pressure of the gas of 101.325 kPa (one atmosphere).

The equation constants, the standard deviation for $\ln X_1$ (except where noted), and the temperature range over which the equation applies are given in the column headed Equation constants. There are two exceptions. The equation for methane has an added term, DT^*. The equation for H_2Se and H_2S takes the form,

$$\ln X_1 = A + B/T + C \ln T + DT$$

where T is the temperature in kelvin.

Solubilities given for those gases which react with water, namely ozone, nitrogen oxides, chlorine and its oxides, carbon dioxide, hydrogen sulfide, hydrogen selenide and sulfur dioxide, are recorded as bulk solubilities; i.e., all chemical species of the gas and its reaction products with water are included.

Gas	T/K	Solubility (X_1)	Equation constants	Ref.
Hydrogen (H_2)	288.15	1.510×10^{-5}	$A = -48.1611$	1
$M_r = 2.01588$	293.15	1.455×10^{-5}	$B = 55.2845$	
	298.15	1.411×10^{-5}	$C = 16.8893$	
	303.15	1.377×10^{-5}	Std. dev. $= \pm 0.54\%$	
	308.15	1.350×10^{-5}	Temp.range = 273.15—353.15	
Deuterium (D_2)	283.15	$1.675 \times 10^{-5} \pm 0.57\%$	Averaged experimental	1
$M_r = 4.0282$	288.15	$1.595 \times 10^{-5} \pm 0.57\%$	values	
	293.15	$1.512 \times 10^{-5} \pm 0.78\%$	Temp. range = 278.15—303.15	
	298.15	$1.460 \times 10^{-5} \pm 0.52\%$		
	303.15	$1.395 \times 10^{-5} \pm 0.37\%$		
Helium (He)	288.15	7.123×10^{-6}	$A = -41.4611$	2
$A_r = 4.0026$	293.15	7.044×10^{-6}	$B = 42.5962$	
	298.15	6.997×10^{-6}	$C = 14.0094$	
	303.15	6.978×10^{-6}	Std. dev. $= \pm 0.54\%$	
	308.15	6.987×10^{-6}	Temp.range = 273.15—348.15	
Neon (Ne)	288.15	8.702×10^{-6}	$A = -52.8573$	2
$A_r = 20.1797$	293.15	8.395×10^{-6}	$B = 61.0494$	
	298.15	8.152×10^{-6}	$C = 18.9157$	
	303.15	7.966×10^{-6}	Std. dev. $= \pm 0.47\%$	
	308.15	7.829×10^{-6}	Temp.range = 273.15—348.15	
Argon (Ar)	288.15	3.025×10^{-5}	$A = -57.6661$	3
$A_r = 39.948$	293.15	2.748×10^{-5}	$B = 74.7627$	
	298.15	2.519×10^{-5}	$C = 20.1398$	
	303.15	2.328×10^{-5}	Std. dev. $= \pm 0.26\%$	
	308.15	2.169×10^{-5}	Temp.range = 273.15—348.15	
Krypton (Kr)	288.15	5.696×10^{-5}	$A = -66.9928$	4
$A_r = 83.80$	293.15	5.041×10^{-5}	$B = 91.0166$	
	298.15	4.512×10^{-5}	$C = 24.2207$	

Gas	T/K	Solubility (X_1)	Equation constants	Ref.
	303.15	4.079×10^{-5}	Std. dev. = ±0.32%	
	308.15	3.725×10^{-5}	Temp.range = 273.15—353.15	
Xenon (Xe) $A_r = 131.29$	288.15	10.519×10^{-5}	$A = -74.7398$	4
	293.15	9.051×10^{-5}	$B = 105.210$	
	298.15	7.890×10^{-5}	$C = 27.4664$	
	303.15	6.961×10^{-5}	Std. dev. = ±0.35%	
	308.15	6.212×10^{-5}	Temp.range = 273.15—348.15	
Radon-222(^{222}Rn) $A_r = 222$	288.15	2.299×10^{-4}	$A = -90.5481$	
	293.15	1.945×10^{-4}	$B = 130.026$	
	298.15	1.671×10^{-4}	$C = 35.0047$	
	303.15	1.457×10^{-4}	Std. dev. = ±1.02%	
	308.15	1.288×10^{-4}	Temp.range = 273.15—373.15	
Oxygen (O_2) $M_r = 31.9988$	288.15	2.756×10^{-5}	$A = -66.7354$	5
	293.15	2.501×10^{-5}	$B = 87.4755$	
	298.15	2.293×10^{-5}	$C = 24.4526$	
	303.15	2.122×10^{-5}	Std. dev. = ±0.36%	
	308.15	1.982×10^{-5}	Temp.range = 273.15—348.15	
Ozone (O_3) $M_r = 47.9982$	293.15	$1.885 \times 10^{-6} \pm 10\%$ pH = 7.0	Experimental value derived from Henry's Law Constant	5
Nitrogen (N_2) $M_r = 28.0134$	288.15	1.386×10^{-5}	$A = -67.3877$	6
	293.15	1.274×10^{-5}	$B = 86.3213$	
	298.15	1.183×10^{-5}	$C = 24.7981$	
	303.15	1.108×10^{-5}	Std. dev. = ±0.72%	
	308.15	1.047×10^{-5}	Temp.range = 273.15—348.15	
Nitrous oxide (N_2O) $M_r = 44.0129$	288.15	5.948×10^{-4}	$A = -60.7467$	7
	293.15	5.068×10^{-4}	$B = 88.8280$	
	298.15	4.367×10^{-4}	$C = 21.2531$	
	303.15	3.805×10^{-4}	Std. dev. = ±1.2%	
	308.15	3.348×10^{-4}	Temp.range = 273.15—313.15	
Nitric oxide (NO) $M_r = 30.0061$	288.15	4.163×10^{-5}	$A = -62.8086$	7
	293.15	3.786×10^{-5}	$B = 82.3420$	
	298.15	3.477×10^{-5}	$C = 22.8155$	
	303.15	3.222×10^{-5}	Std. dev. = ±0.76%	
	308.15	3.012×10^{-5}	Temp.range = 273.15—358.15	
Carbon monoxide (CO) $M_r = 28.0104$	288.15	2.095×10^{-5}	Derived from Henry's Law Constant Equation	8
	293.15	1.918×10^{-5}	Std. dev. = ±0.043%	
	298.15	1.774×10^{-5}	Temp.range = 273.15—328.15	
	303.15	1.657×10^{-5}		
	308.15	1.562×10^{-5}		
Carbon dioxide (CO_2) $M_r = 44.0098$	288.15	8.21×10^{-4}	Derived from Henry's Law Constant Equation	9
	293.15	7.07×10^{-4}	Std. dev. = ±1.1%	
	298.15	6.15×10^{-4}	Temp.range = 273.15—353.15	
	303.15	5.41×10^{-4}		
	308.15	4.80×10^{-4}		
Hydrogen selenide (H_2Se) $M_r = 80.976$	288.15	1.80×10^{-3}	$A = 9.15$	10
	298.15	1.49×10^{-3}	$B = 974$	
	308.15	1.24×10^{-3}	$C = -3.542$	
			$D = 0.0042$	

Gas	T/K	Solubility (X_1)	Equation constants	Ref.
			Std. dev. = $\pm 2.3 \times 10^{-5}$	
			Temp. range = 288.15—343.15	
Hydrogen sulfide (H$_2$S)	288.15	2.335×10^{-3}	$A = -24.912$	10
$M_r = 34.082$	293.15	2.075×10^{-3}	$B = 3477$	
	298.15	1.85×10^{-3}	$C = 0.3993$	
	303.15	1.66×10^{-3}	$D = 0.0157$	
	308.15	1.51×10^{-3}	Std. dev. = $\pm 6.5 \times 10^{-5}$	
			Temp. range = 283.15—603.15	
Sulfur dioxide (SO$_2$)	288.15	3.45×10^{-2}	$A = -25.2629$	11
$M_r = 64.0648$	293.15	2.90×10^{-2}	$B = 45.7552$	
	298.15	2.46×10^{-2}	$C = 5.6855$	
	303.15	2.10×10^{-2}	Std. dev. = $\pm 1.8\%$	
	308.15	1.80×10^{-2}	Temp.range = 278.15—328.15	
Chlorine (Cl$_2$)	283.15	$2.48 \times 10^{-3} \pm 2\%$	Experimental data	11
$M_r = 70.9054$	293.15	$1.88 \times 10^{-3} \pm 2\%$	Temp.range = 283.15—333.15	
	303.15	$1.50 \times 10^{-3} \pm 2\%$		
	313.15	$1.23 \times 10^{-3} \pm 2\%$		
Chlorine monoxide (Cl$_2$O)	273.15	$5.25 \times 10^{-1} \pm 1\%$	Experimental data	11
$M_r = 86.9048$	276.61	$4.54 \times 10^{-1} \pm 1\%$	Temp. range = 273.15—293.15	
	283.15	$4.273 \times 10^{-1} \pm 1\%$		
	293.15	$3.353 \times 10^{-1} \pm 1\%$		
Chlorine dioxide (ClO$_2$)	288.15	2.67×10^{-2}	$A = 7.9163$	11
$M_r = 67.4515$	293.15	2.20×10^{-2}	$B = 0.4791$	
	298.15	1.823×10^{-2}	$C = 11.0593$	
	303.15	1.513×10^{-2}	Std. dev. = $\pm 4.6\%$	
	308.15	1.259×10^{-2}	Temp.range = 283.15—333.15	
Methane (CH$_4$)	288.15	3.122×10^{-5}	$A = -115.6477$	12
$M_r = 16.0428$	293.15	2.806×10^{-5}	$B = 155.5756$	
	298.15	2.552×10^{-5}	$C = 65.2553$	
	303.15	2.346×10^{-5}	$D = -6.6170$	
	308.15	2.180×10^{-5}	Std. dev. = $\pm 0.056\%$	
			Temp.range = 273.15—328.15	
Ethane (C$_2$H$_6$)	288.15	4.556×10^{-5}	$A = -90.8225$	13
$M_r = 30.0696$	293.15	3.907×10^{-5}	$B = 126.9559$	
	298.15	3.401×10^{-5}	$C = 34.7413$	
	303.15	3.002×10^{-5}	Std. dev. = $\pm 0.13\%$	
	308.15	2.686×10^{-5}	Temp.range = 273.15—323.15	
Propane (C$_3$H$_8$)	288.15	3.813×10^{-5}	$A = -102.044$	14
$M_r = 44.097$	293.15	3.200×10^{-5}	$B = 144.345$	
	298.15	2.732×10^{-5}	$C = 39.4740$	
	303.15	2.370×10^{-5}	Std. dev. = $\pm 0.012\%$	
	308.15	2.088×10^{-5}	Temp.range = 273.15—347.15	
Butane (C$_4$H$_{10}$)	288.15	3.274×10^{-5}	$A = -102.029$	14
$M_r = 58.123$	293.15	2.687×10^{-5}	$B = 146.040$	
	298.15	2.244×10^{-5}	$C = 38.7599$	
	303.15	1.906×10^{-5}	Std. dev. = $\pm 0.026\%$	
	308.15	1.645×10^{-5}	Tcmp.range = 273.15—349.15	
2-Methyl propane (Isobutane)	288.15	2.333×10^{-5}	$A = -129.714$	14

SOLUBILITY OF SELECTED GASES IN WATER (continued)

Gas	T/K	Solubility (X_1)	Equation constants	Ref.
(C_4H_{10})	293.15	1.947×10^{-5}	$B = 183.044$	
$M_r = 58.123$	298.15	1.659×10^{-5}	$C = 53.4651$	
	303.15	1.443×10^{-5}	Std. dev. = ±0.034%	
	308.15	1.278×10^{-5}	Temp.range = 278.15—318.15	

REFERENCES

1. C. L. Young, Ed., *IUPAC Solubility Data Series*, Vol. 5/6, Hydrogen and Deuterium, Pergamon Press, Oxford, England, 1981.
2. H. L. Clever, Ed., *IUPAC Solubility Data Series*, Vol. 1, Helium and Neon, Pergamon Press, Oxford, England, 1979.
3. H. L. Clever, Ed., *IUPAC Solubility Data Series*, Vol. 4, Argon, Pergamon Press, Oxford, England, 1980.
4. H. L. Clever, Ed., *IUPAC Solubility Data Series*, Vol. 2, Krypton, Xenon and Radon, Pergamon Press, Oxford, England, 1979.
5. R. Battino, Ed., *IUPAC Solubility Data Series*, Vol. 7, Oxygen and Ozone, Pergamon Press, Oxford, England, 1981.
6. R. Battino, Ed., *IUPAC Solubility Data Series*, Vol. 10, Nitrogen and Air, Pergamon Press, Oxford, England, 1982.
7. C. L. Young, Ed., *IUPAC Solubility Data Series*, Vol. 8, Oxides of Nitrogen, Pergamon Press, Oxford, England, 1981.
8. R. W. Cargill, Ed., *IUPAC Solubility Data Series*, Vol. 43, Carbon Monoxide, Pergamon Press, Oxford, England, 1990.
9. R. Crovetto, Evaluation of Solubility Data for the System CO_2-H_2O, *J. Phys. Chem. Ref. Data*, 20, 575, 1991.
10. P. G. T. Fogg and C. L. Young, Eds., *IUPAC Solubility Data Series*, Vol. 32, Hydrogen Sulfide, Deuterium Sulfide, and Hydrogen Selenide, Pergamon Press, Oxford, England, 1988.
11. C. L. Young, Ed., *IUPAC Solubility Data Series*, Vol. 12, Sulfur Dioxide, Chlorine, Fluorine and Chlorine Oxides, Pergamon Press, Oxford, England, 1983.
12. H. L. Clever and C. L. Young, Eds., *IUPAC Solubility Data Series*, Vol. 27/28, Methane, Pergamon Press, Oxford, England, 1987.
13. W. Hayduk, Ed., *IUPAC Solubility Data Series*, Vol. 9, Ethane, Pergamon Press, Oxford, England, 1982.
14. W. Hayduk, Ed., *IUPAC Solubility Data Series*, Vol. 24, Propane, Butane and 2-Methylpropane, Pergamon Press, Oxford, England, 1986.

SOLUBILITY OF CARBON DIOXIDE IN WATER AT VARIOUS TEMPERATURES AND PRESSURES

The solubility of CO_2 in water, expressed as mole fraction of CO_2 in the liquid phase, is given for pressures up to atmospheric and temperatures of 0 to 100°C. Note that 1 standard atmosphere equals 101.325 kPa. The references give data over a wider range of temperature and pressure. The estimated accuracy is about 2%.

REFERENCES

1. Carroll, J. J., Slupsky, J. D., and Mather, A. E., *J. Phys. Chem. Ref. Data*, 20, 1201, 1991.
2. Fernandez-Prini, R. and Crovetto, R., *J. Phys. Chem. Ref. Data*, 18, 1231, 1989.
3. Crovetto, R., *J. Phys. Chem. Ref. Data*, 20, 575, 1991

$1000 \times$ mole fraction of CO_2 in liquid phase

$t/°C$	Partial pressure of CO_2 in kPa						
	5	10	20	30	40	50	100
0	0.067	0.135	0.269	0.404	0.538	0.671	1.337
5	0.056	0.113	0.226	0.338	0.451	0.564	1.123
10	0.048	0.096	0.191	0.287	0.382	0.477	0.950
15	0.041	0.082	0.164	0.245	0.327	0.409	0.814
20	0.035	0.071	0.141	0.212	0.283	0.353	0.704
25	0.031	0.062	0.123	0.185	0.247	0.308	0.614
30	0.027	0.054	0.109	0.163	0.218	0.271	0.541
35	0.024	0.048	0.097	0.145	0.193	0.242	0.481
40	0.022	0.043	0.087	0.130	0.173	0.216	0.431
45	0.020	0.039	0.078	0.117	0.156	0.196	0.389
50	0.018	0.036	0.071	0.107	0.142	0.178	0.354
55	0.016	0.033	0.065	0.098	0.131	0.163	0.325
60	0.015	0.030	0.060	0.090	0.121	0.150	0.300
65	0.014	0.028	0.056	0.084	0.112	0.140	0.279
70	0.013	0.026	0.052	0.079	0.105	0.131	0.261
75	0.012	0.025	0.049	0.074	0.099	0.123	0.245
80	0.012	0.023	0.047	0.070	0.093	0.116	0.232
85	0.011	0.022	0.044	0.067	0.089	0.111	0.221
90	0.011	0.021	0.042	0.064	0.085	0.106	0.211
95	0.010	0.020	0.041	0.061	0.082	0.102	0.203
100	0.010	0.020	0.039	0.059	0.079	0.098	0.196

AQUEOUS SOLUBILITY AND HENRY'S LAW CONSTANTS OF ORGANIC COMPOUNDS

The solubility in water of about 580 organic compounds, including many compounds of environmental interest, is tabulated here. Values are given at 25°C or at the nearest temperature to this where data are available. In some cases solubility values are given at other temperatures as well.

Solubility of solids is defined as the concentration of the compound in a solution that is in equilibrium with the solid phase at the specified temperature and one atmosphere pressure. For liquids whose water mixtures separate into two phases, the solubility given here is the concentration of the compound in the water-rich phase at equilibrium. In the case of gases (i.e., compounds whose vapor pressure at the specified temperature exceeds one atmosphere) the solubility is defined here as the concentration in the water phase when the partial pressure of the compound above the solution is 101.325 kPa (1 atm). Values for gases are marked with an asterisk.

All solubility values are expressed as mass percent of solute, $S = 100\,w_2$, where the mass fraction w_2 is given by $w_2 = m_2/(m_1 + m_2)$. In these equations m_2 is the mass of solute and m_1 the mass of water. This quantity is related to other common measures of solubility as follows:

Molality: $m_2 = 1000w_2/M_2(1-w_2)$

Mole fraction: $x_2 = (w_2/M_2)/\{(w_2/M_2) + (1-w_2)/M_1\}$

Mass of solute per 100 g of H$_2$O: $r_2 = 100w_2/(1-w_2)$

Here M_2 is the molar mass of the solute and $M_1 = 18.015$ g/mol is the molar mass of water. For small values of S the amount of substance concentration c_2 in moles per liter is approximately $10S/M_2$.

Data have been selected from evaluated sources wherever possible, in particular the *IUPAC Solubility Data Series* (References 1-4). The primary source for each value is listed in the column following the solubility values. The user is cautioned that wide variations of data are found in the literature for the lower solubility compounds.

The table also contains values of the Henry's Law constant k_H, which provides a measure of the partition of a substance between the atmosphere and the aqueous phase. Here k_H is defined as the limit of p_2/c_2 as the concentration approaches zero, where p_2 is the partial pressure of the solute above the solution and c_2 is the concentration in the solution at equilibrium (other formulations of Henry's Law are often used; see Reference 5). The values of k_H listed here are based on direct experimental measurement whenever available, but many of them are simply calculated as the ratio of the pure compound vapor pressure to the solubility. This approximation is reliable only for compounds of very low solubility. In fact, values of k_H found in the literature frequently differ by a factor of two or three, and variations over an order of magnitude are not unusual (Reference 5). Therefore the data given here should be taken only as a rough indication of the true Henry's Law constant, which is difficult to measure precisely.

All values of k_H refer to 25°C. If the vapor pressure of the compound at 25°C is greater than one atmosphere, it can be assumed that the k_H value has been calculated as $101.325/c_2$ kPa m^3/mol. The source of the Henry's Law data is given in the last column. The air-water partition coefficient (i.e., ratio of air concentration to water concentration when both are expressed in the same units) is equal to k_H/RT or $k_H/2.48$ in the units used here.

Compounds are listed by molecular formula following the Hill convention. To locate a compound by name or CAS Registry Number when the molecular formula is not known, use the table "Physical Constants of Organic Compounds" in Section 3 and its indexes to determine the molecular formula.

* Indicates a value of S for a gas at a partial pressure of 101.325 kPa (1 atm) in equilibrium with the solution.

REFERENCES

1. *Solubility Data Series, International Union of Pure and Applied Chemistry, Vol. 15,* Pergamon Press, Oxford, 1982.
2. *Solubility Data Series, International Union of Pure and Applied Chemistry, Vol. 20,* Pergamon Press, Oxford, 1985.
3. *Solubility Data Series, International Union of Pure and Applied Chemistry, Vol. 37,* Pergamon Press, Oxford, 1988.
4. *Solubility Data Series, International Union of Pure and Applied Chemistry, Vol. 38,* Pergamon Press, Oxford, 1988.
5. Mackay, D., and Shiu, W.Y., *J. Phys. Chem. Ref. Data,* 10, 1175, 1981.
6. Pearlman, R.S., and Yalkowsky, S.H., *J. Phys. Chem. Ref. Data,* 13, 975, 1984.
7. Shiu, W.Y., and Mackay, D., *J. Phys. Chem. Ref. Data,* 15, 911, 1986.
8. Varhanickova, D., Lee, S.C., Shiu, W. Y., and Mackay, D., *J. Chem. Eng. Data,* 40, 620, 1995.
9. Miller, M.M., Ghodbane, S., Wasik, S. P., Tewari, Y.B., and Martire, D.E., *J. Chem. Eng. Data,* 29, 184, 1984.
10. Riddick, J.A., Bunger, W.B., and Sakano, T.K., *Organic Solvents, Fourth Edition,* John Wiley & Sons, New York, 1986.
11. Mackay, D., Shiu, W.Y., and Ma, K.C., *Illustrated Handbook of Physical-Chemical Properties and Environmental Fate for Organic Chemicals,* Vol. I, Lewis Publishers/CRC Press, Boca Raton, FL, 1992.
12. Mackay, D., Shiu, W.Y., and Ma, K.C., *Illustrated Handbook of Physical-Chemical Properties and Environmental Fate for Organic Chemicals,* Vol. II, Lewis Publishers/CRC Press, Boca Raton, FL, 1992.
13. Mackay, D., Shiu, W.Y., and Ma, K.C., *Illustrated Handbook of Physical-Chemical Properties and Environmental Fate for Organic Chemicals,* Vol. III, Lewis Publishers/CRC Press, Boca Raton, FL, 1993.
14. Horvath, A.L., *Halogenated Hydrocarbons,* Marcel Dekker, New York, 1982.
15. Howard, P.H., *Handbook of Environmental Fate and Exposure Data for Organic Chemicals, Vol. I,* Lewis Publishers/CRC Press, Boca Raton, FL, 1989.
16. Howard, P.H., *Handbook of Environmental Fate and Exposure Data for Organic Chemicals, Vol. II,* Lewis Publishers/CRC Press, Boca Raton, FL, 1990.
17. Banergee, S., Yalkowsky, S.H., and Valvani, S.C., *Environ. Sci. Technol.,* 14, 1227, 1980.
18. Gevantman, L., in *CRC Handbook of Chemistry and Physics, 77th Edition,* p. 6-3, CRC Press, Boca Raton, FL, 1996.
19. Wilhelm, E., Battino, R., and Wilcock, R.J., *Chem. Rev.,* 77, 219, 1977.
20. Stephenson, R.M., *J. Chem. Eng. Data,* 37, 80, 1992.
21. Stephenson, R.M., Stuart, J., and Tabak, M., *J. Chem. Eng. Data,* 29, 287, 1984.

Mol. Form.	Name	$t/^\circ C$	S/mass %	Ref.	k_H/kPa m^3 mol^{-1}	Ref.
CBrF$_3$	Bromotrifluoromethane	25	0.032*	14		
CBr$_3$F	Tribromofluoromethane	25	0.040	14		
CBr$_4$	Tetrabromomethane	30	0.024	14		
CClF$_3$	Chlorotrifluoromethane	25	0.009*	10	6.9	13
CCl$_2$F$_2$	Dichlorodifluoromethane	20	0.028*	5	41	13
CCl$_3$F	Trichlorofluoromethane	20	0.11	5	10.2	13
CCl$_4$	Tetrachloromethane	25	0.065	20	2.99	13
CF$_4$	Tetrafluoromethane	25	0.00187*	19		
CHBr$_3$	Tribromomethane	25	0.30	5	0.047	13
CHClF$_2$	Chlorodifluoromethane	25	0.30*	10	3.0	13
CHCl$_2$F	Dichlorofluoromethane	25	0.95*	10		
CHCl$_3$	Trichloromethane	25	0.80	20	0.43	13
CHF$_3$	Trifluoromethane	25	0.09*	14		
CHI$_3$	Triiodomethane	25	0.012	14		
CH$_2$BrCl	Bromochloromethane	25	1.7	10	0.18	13
CH$_2$Br$_2$	Dibromomethane	25	1.14	14	0.086	13
CH$_2$ClF	Chlorofluoromethane	25	1.05*	14		
CH$_2$Cl$_2$	Dichloromethane	25	1.73	20	0.30	13
CH$_2$I$_2$	Diiodomethane	30	0.124	10	0.032	13
CH$_3$Br	Bromomethane	20	1.80*	5	0.63	13
CH$_3$Cl	Chloromethane	25	0.535*	5	0.98	13
CH$_3$F	Fluoromethane	30	0.177*	5		
CH$_3$I	Iodomethane	20	1.4	10	0.54	13
CH$_3$NO$_2$	Nitromethane	25	11.1	10		
CH$_4$	Methane	25	0.00227*	18	67.4	5
CO	Carbon monoxide	25	0.00276*	18		
CO$_2$	Carbon dioxide	25	0.1501	18		
CS$_2$	Carbon disulfide	20	0.210	10		
C$_2$ClF$_5$	Chloropentafluoroethane	25	0.006*	10	260	13
C$_2$Cl$_2$F$_4$	1,2-Dichlorotetrafluoroethane	25	0.013*	10	127	13
C$_2$Cl$_3$F$_3$	1,1,2-Trichlorotrifluoroethane	25	0.017	10	32	13
C$_2$Cl$_4$	Tetrachloroethylene	25	0.026	20	1.73	13
C$_2$Cl$_4$F$_2$	1,1,2,2-Tetrachloro-1,2-difluoroethane	25	0.012	10		
C$_2$Cl$_6$	Hexachloroethane	25	0.0050	5	0.85	13
C$_2$F$_4$	Tetrafluoroethylene	25	0.0158*	19		
C$_2$HCl$_3$	Trichloroethylene	25	0.11	5	1.03	13
C$_2$HCl$_5$	Pentachloroethane	25	0.048	5	0.25	13
C$_2$H$_2$	Acetylene	25	0.1081*	19		
C$_2$H$_2$Br$_4$	1,1,2,2-Tetrabromoethane	30	0.0651	10		
C$_2$H$_2$Cl$_2$	1,1-Dichloroethylene	25	0.040	5	2.62	13
C$_2$H$_2$Cl$_2$	cis-1,2-Dichloroethylene	25	0.35	5	0.46	13
C$_2$H$_2$Cl$_2$	trans-1,2-Dichloroethylene	25	0.63	5	0.96	13
C$_2$H$_2$Cl$_4$	1,1,1,2-Tetrachloroethane	25	0.11	5	0.24	13
C$_2$H$_2$Cl$_4$	1,1,2,2-Tetrachloroethane	25	0.30	5	0.026	13
C$_2$H$_3$Cl	Chloroethylene	25	0.27*	5	2.68	13
C$_2$H$_3$Cl$_3$	1,1,1-Trichloroethane	25	0.072	5	1.76	13
C$_2$H$_3$Cl$_3$	1,1,2-Trichloroethane	25	0.442	5	0.092	13
C$_2$H$_4$	Ethylene	25	0.01336*	19	21.7	5
C$_2$H$_4$BrCl	1-Bromo-2-chloroethane	30	0.683	10		
C$_2$H$_4$Br$_2$	1,2-Dibromoethane	25	0.17	5	0.066	13
C$_2$H$_4$Cl$_2$	1,1-Dichloroethane	25	0.51	5	0.63	13
C$_2$H$_4$Cl$_2$	1,2-Dichloroethane	25	0.75	20	0.14	13
C$_2$H$_4$O$_2$	Methyl formate	25	23	10		
C$_2$H$_5$Br	Bromoethane	20	0.91	10	1.23	13
C$_2$H$_5$Cl	Chloroethane	20	0.57*	5	1.02	13
C$_2$H$_5$F	Fluoroethane	25	0.216*	14		
C$_2$H$_5$I	Iodoethane	30	3.88	10	0.52	13
C$_2$H$_5$NO	Acetamide	20	40.8	10		
C$_2$H$_5$NO$_2$	Nitroethane	25	4.68	10		
C$_2$H$_6$	Ethane	25	0.00568*	18	50.6	5

Mol. Form.	Name	$t/°C$	S/mass %	Ref.	k_H/kPa m^3 mol^{-1}	Ref.
C_2H_6O	Dimethyl ether	24	35.3*	10	0.077	13
C_2H_6OS	Dimethyl sulfoxide	25	25.3	10		
C_2H_6S	Dimethyl sulfide	25	2	10		
C_3Cl_6	Hexachloropropene	25	0.00170	14		
C_3F_6	Perfluoropropene	25	0.0194*	14		
C_3F_8	Perfluoropropane	15	0.0015*	14		
C_3H_3N	2-Propenenitrile	20	7.35	10		
C_3H_4	Propyne	25	0.364*	5	1.11	5
$C_3H_4Cl_2$	cis-1,3-Dichloropropene	25	0.27	5	0.24	5
$C_3H_4Cl_2$	trans-1,3-Dichloropropene	20	0.28	5	0.18	5
$C_3H_4Cl_2$	2,3-Dichloropropene	25	0.215	5	0.36	5
C_3H_4O	Acrolein	20	20.8	10		
C_3H_5Cl	3-Chloropropene	25	0.33	5	1.10	5
C_3H_5ClO	Epichlorohydrin	20	6.58	10	0.003	13
$C_3H_5Cl_3$	1,2,3-Trichloropropane	20	0.19	10	0.038	13
C_3H_5N	Propanenitrile	25	10.3	10		
C_3H_6	Propene	25	0.0200*	5	21.3	5
C_3H_6	Cyclopropane	25	0.0484*	19		
$C_3H_6Br_2$	1,2-Dibromopropane	25	0.143	10		
$C_3H_6Cl_2$	1,2-Dichloropropane	25	0.247	10	0.29	13
$C_3H_6N_6O_6$	Hexahydro-1,3,5-trinitro-1,3,5-triazine	25	0.0060	17		
C_3H_6O	Propanal	25	30.6	10		
C_3H_6O	Methyloxirane	20	40.5	10	0.0087	13
$C_3H_6O_2$	Ethyl formate	25	11.8	10		
$C_3H_6O_2$	Methyl acetate	20	24.5	10		
C_3H_7Br	1-Bromopropane	30	0.230	10	3.8	13
C_3H_7Br	2-Bromopropane	18	0.286	10	1.27	13
C_3H_7Cl	1-Chloropropane	20	0.271	10	1.41	13
C_3H_7Cl	2-Chloropropane	12	0.342	10		
C_3H_7F	1-Fluoropropane	14	0.386*	14		
C_3H_7F	2-Fluoropropane	15	0.366	14		
C_3H_7I	1-Iodopropane	30	0.104	10	0.93	13
C_3H_7I	2-Iodopropane	20	0.140	10		
$C_3H_7NO_2$	1-Nitropropane	25	1.50	10		
$C_3H_7NO_2$	2-Nitropropane	25	1.71	10		
C_3H_8	Propane	25	0.00669*	18	71.6	5
$C_3H_8O_2$	Dimethoxymethane	16	24.4	10		
C_4F_8	Perfluorocyclobutane	21	0.014*	14		
$C_4H_4N_2$	Succinonitrile	25	11.5	10		
C_4H_4O	Furan	25	1	10	0.54	13
C_4H_5N	Methylacrylonitrile	20	2.57	10		
C_4H_5N	Pyrrole	25	4.5	10		
C_4H_6	1,3-Butadiene	25	0.0735*	5	20.7	13
C_4H_6	1-Butyne	25	0.287*	5	1.91	5
C_4H_6O	trans-2-Butenal	20	15.6	10		
$C_4H_6O_2$	Methacrylic acid	20	8.9	10		
$C_4H_6O_2$	trans-Crotonic acid	25	9	10		
$C_4H_6O_2$	Vinyl acetate	20	2.0	10		
$C_4H_6O_2$	Methyl acrylate	25	4.94	10		
C_4H_7Cl	1-Chloro-2-methylpropene	25	0.916	5	0.12	5
C_4H_7N	Butanenitrile	20	3.3	10		
C_4H_8	1-Butene	25	0.0222*	5	25.6	13
C_4H_8	Isobutene	25	0.0263*	5	21.6	13
$C_4H_8Cl_2O$	Bis(2-chloroethyl) ether	25	1.03	20	0.003	13
C_4H_8O	cis-Crotonyl alcohol	20	16.6	10		
C_4H_8O	Ethyl vinyl ether	20	0.9	10		
C_4H_8O	Butanal	25	7.1	10		
C_4H_8O	Isobutanal	20	9.1	10		
C_4H_8O	2-Butanone	25	25.9	20		
$C_4H_8O_2$	2-Methylpropanoic acid	20	22.8	10		

Mol. Form.	Name	$t/°C$	$S/mass\ \%$	Ref.	$k_H/kPa\ m^3\ mol^{-1}$	Ref.
$C_4H_8O_2$	Propyl formate	22	2.05	10		
$C_4H_8O_2$	Ethyl acetate	25	8.08	10		
C_4H_9Br	1-Bromobutane	30	0.0608	10	1.2	13
C_4H_9Cl	1-Chlorobutane	20	0.11	10	1.54	13
C_4H_9Cl	1-Chloro-2-methylpropane	12	0.092	10		
C_4H_9I	1-Iodobutane	17	0.012	10	1.87	13
C_4H_{10}	Butane	25	0.00724*	18	95.9	5
C_4H_{10}	Isobutane	25	0.00535*	18	120	5
$C_4H_{10}O$	Diethyl ether	25	6.04	10	0.088	13
$C_4H_{10}O$	1-Butanol	0	10.4	1		
		25	7.4	1		
		50	6.4	1		
$C_4H_{10}O$	2-Butanol	10	23.9	1		
		25	18.1	1		
		50	14.0	1		
$C_4H_{10}O$	2-Methyl-1-propanol	0	11.5	1		
		25	8.1	1		
		50	6.5	1		
$C_4H_{10}S$	1-Butanethiol	20	0.0597	10		
$C_4H_{11}NO_2$	Diethanolamine	20	95.4	10		
$C_4H_{12}Si$	Tetramethylsilane	25	0.00196	10		
$C_5H_4O_2$	Furfural	20	8.2	10		
C_5H_6	1,3-Cyclopentadiene	25	0.068	3		
$C_5H_7NO_2$	Ethyl cyanoacetate	20	25.9	10		
C_5H_8	1,4-Pentadiene	25	0.056	3	12	5
C_5H_8	2-Methyl-1,3-butadiene	25	0.061	3	7.78	5
		50	0.076*	3		
C_5H_8	1-Pentyne	25	0.157	3	2.5	5
C_5H_8	Cyclopentene	25	0.054	3	6.56	13
$C_5H_8O_2$	Ethyl acrylate	25	1.50	10		
$C_5H_8O_2$	Methyl methacrylate	20	1.56	10		
$C_5H_8O_2$	2,4-Pentanedione	20	16.6	10		
C_5H_{10}	1-Pentene	25	0.0148	3	40.3	5
C_5H_{10}	cis-2-Pentene	25	0.0203	3	22.8	5
C_5H_{10}	3-Methyl-1-butene	25	0.013*	3	54.7	5
C_5H_{10}	2-Methyl-2-butene	25	0.041	3		
C_5H_{10}	Cyclopentane	25	0.0157	3	19.1	13
$C_5H_{10}O$	2-Pentanone	25	5.5	20		
$C_5H_{10}O$	3-Pentanone	25	4.72	20		
$C_5H_{10}O$	2-Methyltetrahydrofuran	25	13.9	10	0.67	13
$C_5H_{10}O_2$	Pentanoic acid	20	2.4	10		
$C_5H_{10}O_2$	3-Methylbutanoic acid	21	4.1	10		
$C_5H_{10}O_2$	Isobutyl formate	22	1.0	10		
$C_5H_{10}O_2$	Propyl acetate	20	2.3	10		
$C_5H_{10}O_2$	Isopropyl acetate	20	2.9	10		
$C_5H_{10}O_2$	Ethyl propanoate	20	1.92	10		
$C_5H_{11}Br$	1-Bromopentane	25	0.0127	10		
$C_5H_{11}Cl$	1-Chloropentane	25	0.02	10	2.37	13
C_5H_{12}	Pentane	25	0.0041	3	128	13
C_5H_{12}	Isopentane	25	0.00485	3	479	13
C_5H_{12}	Neopentane	25	0.00332*	3	220	13
$C_5H_{12}O$	1-Pentanol	0	3.1	1		
		25	2.20	1		
		50	1.8	1		
$C_5H_{12}O$	2-Pentanol	25	4.3	21		
$C_5H_{12}O$	3-Pentanol	25	5.6	21		
$C_5H_{12}O$	2-Methyl-1-butanol	25	3.0	3		
$C_5H_{12}O$	3-Methyl-1-butanol	25	2.7	1		
$C_5H_{12}O$	2-Methyl-2-butanol	25	11.0	1		
$C_5H_{12}O$	3-Methyl-2-butanol	25	5.6	1		

Mol. Form.	Name	$t/°C$	S/mass %	Ref.	k_H/kPa m^3 mol^{-1}	Ref.
$C_5H_{12}O$	2,2-Dimethyl-1-propanol	25	3.5	1		
$C_5H_{12}O$	Methyl *tert*-butyl ether	25	3.62	20	0.070	13
C_6Cl_6	Hexachlorobenzene	25	0.0000005	2	0.131	11
C_6HCl_5	Pentachlorobenzene	25	0.000055	2	0.085	11
C_6HCl_5O	Pentachlorophenol	25	0.0010	2		
$C_6H_2Br_4$	1,2,4,5-Tetrabromobenzene	25	0.00000434	2		
$C_6H_2Cl_4$	1,2,3,4-Tetrachlorobenzene	25	0.0000433	2	0.144	11
$C_6H_2Cl_4$	1,2,3,5-Tetrachlorobenzene	25	0.000346	2	0.59	11
$C_6H_2Cl_4$	1,2,4,5-Tetrachlorobenzene	25	0.0000606	2	0.122	11
$C_6H_2Cl_4O$	2,3,4,6-Tetrachlorophenol	25	1.8	2		
$C_6H_2Cl_4O_2$	3,4,5,6-Tetrachloro-1,2-benzenediol	25	0.071	8		
$C_6H_3Br_3$	1,2,4-Tribromobenzene	25	0.0010	2		
$C_6H_3Br_3$	1,3,5-Tribromobenzene	25	0.0000789	2		
$C_6H_3Br_3O$	2,4,6-Tribromophenol	15	0.0007	2		
$C_6H_3Cl_3$	1,2,3-Trichlorobenzene	25	0.00309	2	0.242	11
$C_6H_3Cl_3$	1,2,4-Trichlorobenzene	25	0.00379	2	0.277	11
$C_6H_3Cl_3$	1,3,5-Trichlorobenzene	25	0.000655	2	1.1	11
$C_6H_3Cl_3O$	2,4,5-Trichlorophenol	25	0.1	2		
$C_6H_3Cl_3O$	2,4,6-Trichlorophenol	25	0.04	2		
$C_6H_3Cl_3O_2$	3,4,5-Trichloro-1,2-benzenediol	25	0.051	8		
C_6H_4BrCl	1-Bromo-2-chlorobenzene	25	0.0124	2		
C_6H_4BrCl	1-Bromo-3-chlorobenzene	25	0.0118	2		
C_6H_4BrCl	1-Bromo-4-chlorobenzene	25	0.00442	2		
C_6H_4BrI	1-Bromo-4-iodobenzene	25	0.000794	2		
$C_6H_4Br_2$	*o*-Dibromobenzene	25	0.00748	2		
$C_6H_4Br_2$	*m*-Dibromobenzene	25	0.0064	2		
$C_6H_4Br_2$	*p*-Dibromobenzene	25	0.0020	2		
$C_6H_4Br_2O$	2,4-Dibromophenol	25	0.2	2		
C_6H_4ClI	1-Chloro-2-iodobenzene	25	0.00689	2		
C_6H_4ClI	1-Chloro-3-iodobenzene	25	0.00674	2		
C_6H_4ClI	1-Chloro-4-iodobenzene	25	0.00311	2		
$C_6H_4Cl_2$	*o*-Dichlorobenzene	0	0.0142	2		
		25	0.0147	2	0.244	11
		50	0.0212	2		
$C_6H_4Cl_2$	*m*-Dichlorobenzene	10	0.0103	2		
		25	0.0106	2	0.376	11
		50	0.0165	2		
$C_6H_4Cl_2$	*p*-Dichlorobenzene	10	0.00512	2		
		25	0.00829	2	0.160	11
		50	0.0167	2		
$C_6H_4Cl_2O$	2,4-Dichlorophenol	20	0.45	2		
$C_6H_4Cl_2O_2$	3,5-Dichloro-1,2-benzenediol	25	0.78	8		
$C_6H_4Cl_2O_2$	4,5-Dichloro-1,2-benzenediol	25	1.19	8		
$C_6H_4F_2$	*o*-Difluorobenzene	25	0.114	2		
$C_6H_4F_2$	*m*-Difluorobenzene	25	0.114	2		
$C_6H_4F_2$	*p*-Difluorobenzene	25	0.122	2		
$C_6H_4I_2$	*o*-Diiodobenzene	25	0.00192	2		
$C_6H_4I_2$	*m*-Diiodobenzene	25	0.000185	2		
$C_6H_4I_2$	*p*-Diiodobenzene	25	0.000893	2		
C_6H_5Br	Bromobenzene	10	0.0387	2		
		25	0.0445	2	0.21	5
		40	0.0516	2		
C_6H_5BrO	*p*-Bromophenol	25	1.86	2		
C_6H_5Cl	Chlorobenzene	10	0.0387	2		
		25	0.0495	2	0.38	11
		50	0.0882	2		
C_6H_5ClO	*o*-Chlorophenol	25	2.0	2		
C_6H_5ClO	*m*-Chlorophenol	25	2.2	2		
C_6H_5ClO	*p*-Chlorophenol	25	2.7	2		
C_6H_5F	Fluorobenzene	27	0.154	2	0.70	11

Mol. Form.	Name	$t/°C$	S/mass %	Ref.	k_H/kPa m^3 mol^{-1}	Ref.
C_6H_5I	Iodobenzene	10	0.0193	2		
		25	0.0226	2	0.078	11
		45	0.0279	2		
$C_6H_5NO_2$	Nitrobenzene	25	0.21	17		
C_6H_6	Benzene	10	0.178	3		
		25	0.177	3	0.557	11
		50	0.208	3		
C_6H_6ClN	o-Chloroaniline	25	0.876	10		
C_6H_6O	Phenol	25	8.66	10		
C_6H_7N	Aniline	25	3.38	10	14	15
C_6H_8	1,4-Cyclohexadiene	25	0.08	3	1.03	13
$C_6H_8N_2$	Adiponitrile	20	0.80	16		
$C_6H_8O_4$	Dimethyl maleate	25	8.0	10		
C_6H_{10}	1,5-Hexadiene	25	0.017	3		
C_6H_{10}	1-Hexyne	25	0.036	3	4.14	13
C_6H_{10}	Cyclohexene	25	0.016	3	4.57	13
$C_6H_{10}O$	Cyclohexanone	25	8.8	20		
$C_6H_{10}O$	Mesityl oxide	20	2.89	10		
$C_6H_{10}O_3$	Ethyl acetoacetate	25	12	10		
$C_6H_{11}NO$	Caprolactam	25	84.0	10		
C_6H_{12}	1-Hexene	25	0.0053	3	41.8	5
C_6H_{12}	trans-2-Hexene	25	0.0067	3		
C_6H_{12}	2-Methyl-1-pentene	25	0.0078	3	28.1	5
C_6H_{12}	4-Methyl-1-pentene	25	0.0048	3	63.2	5
C_6H_{12}	2,3-Dimethyl-1-butene	30	0.046	3		
C_6H_{12}	Cyclohexane	25	0.0058	3	19.4	13
C_6H_{12}	Methylcyclopentane	25	0.0043	3	36.7	5
$C_6H_{12}O$	1-Hexen-3-ol	25	2.52	1		
$C_6H_{12}O$	4-Hexen-2-ol	25	3.81	1		
$C_6H_{12}O$	Cyclohexanol	10	4.62	1		
		25	3.8	1		
		40	3.30	1		
$C_6H_{12}O$	Butyl vinyl ether	20	0.3	10		
$C_6H_{12}O$	2-Hexanone	20	1.75	10		
$C_6H_{12}O$	4-Methyl-2-pentanone	25	1.7	10		
$C_6H_{12}O_2$	Hexanoic acid	20	0.958	10		
$C_6H_{12}O_2$	Butyl acetate	20	0.68	10		
$C_6H_{12}O_2$	Isobutyl acetate	20	0.63	10		
$C_6H_{12}O_2$	sec-Butyl acetate	20	0.62	10		
$C_6H_{12}O_2$	Ethyl butanoate	20	0.49	10		
C_6H_{14}	Hexane	25	0.0011	3	183	13
		60	0.00136	3		
C_6H_{14}	2-Methylpentane	25	0.00137	3	176	13
C_6H_{14}	3-Methylpentane	25	0.00129	3	170	13
C_6H_{14}	2,2-Dimethylbutane	25	0.0021	3	199	13
C_6H_{14}	2,3-Dimethylbutane	25	0.0021	3	144	13
$C_6H_{14}O$	1-Hexanol	0	0.79	1		
		25	0.60	1		
		50	0.51	1		
$C_6H_{14}O$	2-Hexanol	25	1.4	1		
$C_6H_{14}O$	3-Hexanol	25	1.6	1		
$C_6H_{14}O$	Dipropyl ether	25	0.49	10	0.26	13
$C_6H_{14}O$	2-Methyl-1-pentanol	25	0.81	1		
$C_6H_{14}O$	4-Methyl-1-pentanol	25	0.76	1		
$C_6H_{14}O$	2-Methyl-2-pentanol	25	3.2	1		
$C_6H_{14}O$	3-Methyl-2-pentanol	25	1.9	1		
$C_6H_{14}O$	4-Methyl-2-pentanol	27	1.5	1		
$C_6H_{14}O$	2-Methyl-3-pentanol	25	2.0	1		
$C_6H_{14}O$	3-Methyl-3-pentanol	25	4.3	1		
$C_6H_{14}O$	2-Ethyl-1-butanol	25	1.0	1		

Mol. Form.	Name	$t/°C$	S/mass %	Ref.	k_H/kPa m^3 mol^{-1}	Ref.
$C_6H_{14}O$	2,2-Dimethyl-1-butanol	25	0.8	1		
$C_6H_{14}O$	Diisopropyl ether	20	1.2	10	0.26	13
$C_6H_{14}O$	2,3-Dimethyl-2-butanol	25	4.2	1		
$C_6H_{14}O$	3,3-Dimethyl-2-butanol	25	2.4	1		
$C_6H_{14}O_2$	1,1-Diethoxyethane	25	5	10		
$C_6H_{15}N$	Dipropylamine	20	2.5	10		
$C_6H_{15}N$	Triethylamine	20	5.5	10		
$C_7H_4Cl_4O$	2,3,4,6-Tetrachloro-5-methylphenol	25	0.00061	2		
$C_7H_5Cl_3$	(Trichloromethyl)benzene	5	0.0053	10		
$C_7H_5Cl_3O$	2,4,6-Trichloro-3-methylphenol	25	0.0112	2		
C_7H_5N	Benzonitrile	25	0.2	10		
C_7H_5NO	Benzoxazole	20	0.834	6		
$C_7H_6Cl_2$	(Dichloromethyl)benzene	30	0.025	10		
$C_7H_6Cl_2O$	2,6-Dichloro-4-methylphenol	25	0.0673	2		
$C_7H_6Cl_2O$	2,4-Dichloro-6-methylphenol	25	0.0283	2		
$C_7H_6N_2$	1H-Benzimidazole	20	0.201	6		
$C_7H_6N_2$	1H-Indazole	20	0.0827	6		
C_7H_6O	Benzaldehyde	20	0.3	10		
$C_7H_6O_2$	Salicylaldehyde	86	1.68	10		
C_7H_7Br	p-Bromotoluene	25	0.011	2		
C_7H_7Cl	(Chloromethyl)benzene	20	0.0493	10		
C_7H_7ClO	4-Chloro-2-methylphenol	25	0.68	2		
C_7H_7ClO	4-Chloro-3-methylphenol	25	0.40	2		
C_7H_7ClO	2-Chloro-6-methylphenol	25	0.36	2		
$C_7H_7NO_3$	2-Nitroanisole	30	0.619	10		
C_7H_8	Toluene	5	0.063	3		
		25	0.053	3	0.680	13
C_7H_8	1,3,5-Cycloheptatriene	25	0.064	3	0.47	13
C_7H_8	1,6-Heptadiyne	25	0.125	3		
C_7H_8O	Benzyl alcohol	20	0.08	10		
C_7H_8O	o-Cresol	40	3.08	10		
C_7H_8O	m-Cresol	40	2.51	10		
C_7H_8O	p-Cresol	40	2.26	10		
C_7H_8O	Anisole	25	0.19	20	0.025	13
C_7H_9N	o-Methylaniline	20	1.66	10		
C_7H_9N	p-Methylaniline	21	7.35	10		
C_7H_{12}	1-Heptyne	25	0.0094	3	4.47	13
C_7H_{12}	Cycloheptene	25	0.0066	3	4.9	13
C_7H_{12}	1-Methylcyclohexene	25	0.0052	3		
C_7H_{14}	1-Heptene	25	0.032	3	40.3	13
C_7H_{14}	$trans$-2-Heptene	25	0.015	3	42.2	13
C_7H_{14}	Cycloheptane	25	0.0030	3	9.59	13
C_7H_{14}	Methylcyclohexane	25	0.00151	3	43.3	13
		50	0.0019	3		
C_7H_{14}	Ethylcyclopentane	20	0.012	3		
$C_7H_{14}O$	2-Heptanone	25	0.43	10		
$C_7H_{14}O$	3-Heptanone	20	1.43	10		
$C_7H_{14}O$	2,4-Dimethyl-3-pentanone	20	0.59	10		
$C_7H_{14}O_2$	Pentyl acetate	20	0.17	10		
$C_7H_{14}O_2$	Isopentyl acetate	20	0.2	10		
$C_7H_{14}O_2$	Ethyl 3-methylbutanoate	20	0.2	10		
C_7H_{16}	Heptane	0	0.0003	3		
		25	0.00024	3	209	13
		40	0.00025	3		
C_7H_{16}	2-Methylhexane	25	0.00025	3	346	5
C_7H_{16}	3-Methylhexane	25	0.00026	3	249	13
C_7H_{16}	2,2-Dimethylpentane	25	0.00044	3	318	5
C_7H_{16}	2,3-Dimethylpentane	25	0.00052	3	175	5
C_7H_{16}	2,4-Dimethylpentane	25	0.00042	3	323	13
C_7H_{16}	3,3-Dimethylpentane	25	0.00059	3	186	5

Mol. Form.	Name	$t/°C$	S/mass %	Ref.	k_H/kPa m^3 mol^{-1}	Ref.
C$_7$H$_{16}$O	1-Heptanol	10	0.25	1		
		25	0.174	1		
		50	0.12	1		
C$_7$H$_{16}$O	2-Heptanol	30	0.33	1		
C$_7$H$_{16}$O	3-Heptanol	25	0.43	1		
C$_7$H$_{16}$O	4-Heptanol	25	0.47	1		
C$_7$H$_{16}$O	2-Methyl-2-hexanol	25	1.0	1		
C$_7$H$_{16}$O	5-Methyl-2-hexanol	25	0.49	1		
C$_7$H$_{16}$O	3-Methyl-3-hexanol	25	1.2	1		
C$_7$H$_{16}$O	2,3-Dimethyl-2-pentanol	25	1.5	1		
C$_7$H$_{16}$O	2,4-Dimethyl-2-pentanol	25	1.3	1		
C$_7$H$_{16}$O	3-Ethyl-3-pentanol	25	1.7	1		
C$_7$H$_{16}$O	2,2-Dimethyl-3-pentanol	25	0.82	1		
C$_7$H$_{16}$O	2,3-Dimethyl-3-pentanol	25	1.6	1		
C$_7$H$_{16}$O	2,4-Dimethyl-3-pentanol	25	0.70	1		
C$_7$H$_{16}$O	2,3,3-Trimethyl-2-butanol	40	2.2	1		
C$_8$H$_4$F$_6$	1,3-Bis(trifluoromethyl)benzene	25	0.0041	2		
C$_8$H$_6$N$_2$	Quinoxaline	50	54	6		
C$_8$H$_6$S	Benzo[b]thiophene	20	0.0130	6		
C$_8$H$_7$ClO$_3$	3-Chloro-4-hydroxy-5-methoxybenzaldehyde	25	0.093	8		
C$_8$H$_7$ClO$_3$	2-Chloro-4-hydroxy-5-methoxybenzaldehyde	25	0.013	8		
C$_8$H$_7$Cl$_3$O	2,4,6-Trichloro-3,5-dimethylphenol	25	0.00050	2		
C$_8$H$_7$N	Indole	20	0.187	6		
C$_8$H$_8$	Styrene	25	0.025	4	0.3	15
		50	0.046	4	0.30	13
C$_8$H$_8$N$_2$	2-Methyl-1H-benzimidazole	20	0.145	6		
C$_8$H$_8$O	Acetophenone	25	0.55			
C$_8$H$_8$O$_2$	Methyl benzoate	20	0.21	10		
C$_8$H$_8$O$_3$	Methyl salicylate	30	0.74	10		
C$_8$H$_8$O$_3$	4-Hydroxy-3-methoxybenzaldehyde	25	0.247	8		
C$_8$H$_9$ClO	4-Chloro-2,5-dimethylphenol	25	0.89	2		
C$_8$H$_9$ClO	4-Chloro-2,6-dimethylphenol	25	0.52	2		
C$_8$H$_9$ClO	4-Chloro-3,5-dimethylphenol	25	0.34	2		
C$_8$H$_{10}$	Ethylbenzene	0	0.020	4		
		25	0.0169	4	0.887	11
		40	0.0200	4		
C$_8$H$_{10}$	o-Xylene	25	0.0173	4	0.565	13
		45	0.021	4		
C$_8$H$_{10}$	m-Xylene	0	0.0203	4		
		25	0.016	4	0.730	13
		40	0.022	4		
C$_8$H$_{10}$	p-Xylene	0	0.0160	4		
		25	0.018	4	0.578	13
		40	0.022	4		
C$_8$H$_{10}$O	Phenetole	25	0.12	10		
C$_8$H$_{10}$O	2,4-Xylenol	25	0.787	10		
C$_8$H$_{10}$O	3,5-Xylenol	29	0.62	10		
C$_8$H$_{12}$	4-Vinylcyclohexene	25	0.005	4		
C$_8$H$_{14}$	1-Octyne	25	0.0024	4	7.87	13
C$_8$H$_{16}$	1-Octene	25	0.00027	4	96.3	13
C$_8$H$_{16}$	Cyclooctane	25	0.00079	4	10.7	13
C$_8$H$_{16}$	Ethylcyclohexane	40	0.00066	4		
C$_8$H$_{16}$	cis-1,2-Dimethylcyclohexane	25	0.00060	4	36	5
C$_8$H$_{16}$	trans-1,4-Dimethylcyclohexane	25	0.000384	4	88.2	5
C$_8$H$_{16}$	Propylcyclopentane	25	0.00020	4	90.2	5
C$_8$H$_{16}$	1,1,3-Trimethylcyclopentane	25	0.00037	4	159	5
C$_8$H$_{16}$O	2-Octanone	25	0.113	10		
C$_8$H$_{16}$O$_2$	Octanoic acid	25	0.0798	10		

Mol. Form.	Name	$t/°C$	S/mass %	Ref.	k_H/kPa m^3 mol^{-1}	Ref.
$C_8H_{16}O_2$	Hexyl acetate	20	0.02	10		
$C_8H_{16}O_2$	sec-Hexyl acetate	20	0.13	10		
$C_8H_{16}O_2$	Isobutyl isobutanoate	20	0.5	10		
$C_8H_{17}Cl$	3-(Chloromethyl)heptane	20	0.01	10		
C_8H_{18}	Octane	25	0.000071	4	311	13
		50	0.00010	4		
C_8H_{18}	3-Methylheptane	25	0.000079	4	376	5
C_8H_{18}	2,2,4-Trimethylpentane	25	0.00022	4	307	13
C_8H_{18}	2,3,4-Trimethylpentane	25	0.00018	4	206	13
$C_8H_{18}O$	Dibutyl ether	20	0.03	10	0.48	13
$C_8H_{18}O$	1-Octanol	25	0.054	1		
$C_8H_{18}O$	2-Octanol	25	0.4	1		
$C_8H_{18}O$	2-Methyl-2-heptanol	30	0.25	1		
$C_8H_{18}O$	2-Ethyl-1-hexanol	25	0.01	1		
$C_8H_{19}N$	Dibutylamine	20	0.47	10		
$C_8H_{19}N$	2-Ethylhexylamine	20	0.25	10		
$C_8H_{20}Si$	Tetraethylsilane	25	0.0000325	10		
C_9H_7N	Quinoline	20	0.633	6		
C_9H_7N	Isoquinoline	20	0.452	6		
C_9H_9N	3-Methyl-1H-indole	20	0.050	6		
C_9H_{10}	Indan	25	0.010	4		
$C_9H_{10}O_2$	Ethyl benzoate	25	0.083	20		
C_9H_{12}	1,8-Nonadiyne	25	0.0125	4		
C_9H_{12}	Propylbenzene	25	0.0055	4	1.04	11
C_9H_{12}	Isopropylbenzene	25	0.0056	4	1.47	11
C_9H_{12}	o-Ethyltoluene	25	0.0093	5	0.529	13
C_9H_{12}	p-Ethyltoluene	25	0.0094	5	0.500	13
C_9H_{12}	1,2,3-Trimethylbenzene	25	0.0069	4	0.343	11
C_9H_{12}	1,2,4-Trimethylbenzene	25	0.0056	4	0.569	11
C_9H_{12}	1,3,5-Trimethylbenzene	25	0.00489	4	0.781	11
$C_9H_{14}O_6$	Triacetin	25	5.8	10		
C_9H_{16}	1-Nonyne	25	0.00072	4		
C_9H_{18}	1,1,3-Trimethylcyclohexane	25	0.000177	4	105	13
$C_9H_{18}O$	Diisobutyl ketone	25	0.043	10		
$C_9H_{18}O_2$	Nonanoic acid	20	0.0284	10		
C_9H_{20}	Nonane	25	0.000017	4	333	13
		50	0.000022	4		
C_9H_{20}	4-Methyloctane	25	0.0000115	4	1000	5
C_9H_{20}	2,2,5-Trimethylhexane	25	0.00008	4	246	13
$C_9H_{20}O$	3,5-Dimethyl-4-heptanol	15	0.072	1		
$C_9H_{20}O$	1-Nonanol	25	0.014	1		
$C_9H_{20}O$	2-Nonanol	15	0.026	1		
$C_9H_{20}O$	3-Nonanol	15	0.032	1		
$C_9H_{20}O$	4-Nonanol	15	0.0026	1		
$C_9H_{20}O$	5-Nonanol	15	0.0032	1		
$C_{10}H_7Cl$	1-Chloronaphthalene	25	0.00224	5		
$C_{10}H_7Cl$	2-Chloronaphthalene	25	0.00117	5		
$C_{10}H_8$	Naphthalene	10	0.0019	4		
		25	0.0031	4	0.0430	5
		50	0.0082	4		
$C_{10}H_9N$	3-Methylisoquinoline	20	0.092	6		
$C_{10}H_{10}O_4$	Dimethyl phthalate	25	0.40	15		
$C_{10}H_{14}$	Butylbenzene	25	0.0015	4	1.33	11
$C_{10}H_{14}$	sec-Butylbenzene	25	0.0014	4	1.89	11
$C_{10}H_{14}$	tert-Butylbenzene	25	0.0032	4	1.28	11
$C_{10}H_{14}$	Isobutylbenzene	25	0.0010	4	3.32	11
$C_{10}H_{14}$	p-Cymene	25	0.00234	4	0.80	5
$C_{10}H_{14}$	1,2,4,5-Tetramethylbenzene	25	0.000348	4	2.55	11
$C_{10}H_{16}$	d-Limonene	0	0.00097	4		
		25	0.00138	4		

Mol. Form.	Name	$t/°C$	S/mass %	Ref.	k_H/kPa m^3 mol^{-1}	Ref.
$C_{10}H_{16}O$	Camphor	20	0.01	10		
$C_{10}H_{18}$	trans-Decahydronaphthalene	25	0.000089	4	3	13
$C_{10}H_{20}$	Pentylcyclopentane	25	0.0000115	4	185	5
$C_{10}H_{20}$	1-Decene	25	0.00057	4		
$C_{10}H_{22}$	Decane	0	0.0000015	4	479	13
$C_{10}H_{22}O$	Diisopentyl ether	20	0.02	10		
$C_{10}H_{22}O$	1-Decanol	25	0.0037	1		
$C_{11}H_{10}$	1-Methylnaphthalene	25	0.0028	4	0.0450	12
$C_{11}H_{10}$	2-Methylnaphthalene	25	0.0025	4	0.051	12
$C_{11}H_{16}$	Pentylbenzene	25	0.00105	5	1.69	11
$C_{12}Cl_{10}$	Decachlorobiphenyl	25	0.00000000012	7	0.0208	7
$C_{12}HCl_9$	2,2',3,3',4,5,5',6,6'-Nonachlorobiphenyl	25	0.0000000018	7		
$C_{12}H_2Cl_8$	2,2',3,3'5,5',6,6'-Octachlorobiphenyl	25	0.0000003	7	0.0381	7
$C1_2H_3Cl_7$	2,2',3,3',4,4',6-Heptachlorobiphenyl	25	0.0000002	7	0.0054	7
$C_{12}H_4Cl_6$	2,2',3,3',4,4'-Hexachlorobiphenyl	25	0.00000006	7	0.0119	7
$C_{12}H_4Cl_6$	2,2',4,4',6,6'-Hexachlorobiphenyl	25	0.00000007	7	0.818	7
$C_{12}H_4Cl_6$	2,2',3,3',6,6'-Hexachlorobiphenyl	25	0.00000008	7		
$C_{12}H_5Cl_5$	2,3,4,5,6-Pentachlorobiphenyl	25	0.0000008	7		
$C_{12}H_5Cl_5$	2,2',4,5,5'-Pentachlorobiphenyl	25	0.000001	7	0.0355	7
$C_{12}H_6Cl_4$	2,3,4,5-Tetrachlorobiphenyl	25	0.000002	7		
$C_{12}H_6Cl_4$	2,2',4',5-Tetrachlorobiphenyl	25	0.0000016	9		
$C_{12}H_7Cl_3$	2,4,5-Trichlorobiphenyl	25	0.000014	7	0.0243	7
$C_{12}H_7Cl_3$	2,4,6-Trichlorobiphenyl	25	0.00002	7	0.0495	7
$C_{12}H_8Cl_2$	2,5-Dichlorobiphenyl	25	0.0002	7	0.0201	7
$C_{12}H_8Cl_2$	2,6-Dichlorobiphenyl	25	0.00014	7		
$C_{12}H_8O$	Dibenzofuran	25	0.000656	6	0.011	12
$C_{12}H_8S$	Dibenzothiophene	25	0.000103	6		
$C_{12}H_9Cl$	2-Chlorobiphenyl	25	0.00055	7	0.0701	7
$C_{12}H_9N$	Carbazole	22	0.000120	6		
$C_{12}H_{10}$	Acenaphthene	0	0.00015	4		
		25	0.00038	4	0.012	12
		50	0.00092	4		
$C_{12}H_{10}$	Biphenyl	0	0.000272	4		
		25	0.00072	4	0.028	5
		50	0.0022	4		
$C_{12}H_{10}N_2O$	N-Nitrosodiphenylamine	25	0.0035	17		
$C_{12}H_{10}O$	Diphenyl ether	25	0.00180	6	0.027	13
$C_{12}H_{12}$	1-Ethylnaphthalene	25	0.00101	4	0.039	12
$C_{12}H_{12}$	2-Ethylnaphthalene	25	0.00080	4	0.078	12
$C_{12}H_{12}$	1,3-Dimethylnaphthalene	25	0.0008	4		
$C_{12}H_{12}$	1,4-Dimethylnaphthalene	25	0.00114	4		
$C_{12}H_{12}$	1,5-Dimethylnaphthalene	25	0.00031	4		
$C_{12}H_{12}$	2,3-Dimethylnaphthalene	25	0.00025	4		
$C_{12}H_{12}$	2,6-Dimethylnaphthalene	25	0.00017	4		
$C_{12}H_{18}$	Hexylbenzene	25	0.00021	4		
$C_{12}H_{26}$	Dodecane	25	0.00000037	4	750	5
$C_{12}H_{26}O$	1-Dodecanol	25	0.0004	1		
$C_{12}H_{27}O_4P$	Tributyl phosphate	25	0.039	10		
$C_{13}H_9N$	Acridine	25	0.00466	6		
$C_{13}H_9N$	Benzo[f]quinoline	25	0.0079	6		
$C_{13}H_{10}$	9H-Fluorene	0	0.00007	4		
		25	0.00019	4	0.0079	12
		50	0.00063	4		
$C_{13}H_{12}$	Diphenylmethane	25	0.000141	4	0.001	12
$C_{13}H_{14}$	1,4,5-Trimethylnaphthalene	25	0.00021	4		
$C_{14}H_{10}$	Anthracene	0	0.0000022	4		
		25	0.0000062	4	0.0040	12
$C_{14}H_{10}$	Phenanthrene	10	0.000050	4		
		25	0.00011	4	0.0032	12
		50	0.00041	4		

Mol. Form.	Name	$t/°C$	S/mass %	Ref.	k_H/kPa m^3 mol^{-1}	Ref.
$C_{14}H_{12}$	*trans*-Stilbene	25	0.000029	4	0.040	12
$C_{14}H_{14}$	1,2-Diphenylethane	25	0.00044	6	0.017	12
$C_{14}H_{14}O$	Dibenzyl ether	35	0.0040	10		
$C_{14}H_{30}$	Tetradecane	25	0.000012	5		
$C_{14}H_{30}O$	1-Tetradecanol	25	0.000031	1		
$C_{15}H_{12}$	2-Methylanthracene	25	0.0000030	4		
$C_{15}H_{12}$	9-Methylanthracene	25	0.000026	4		
$C_{15}H_{12}$	1-Methylphenanthrene	25	0.0000269	4		
$C_{15}H_{32}O$	1-Pentadecanol	25	0.000010	1		
$C_{16}H_{10}$	Fluoranthene	25	0.000024	4	0.0010	12
$C_{16}H_{10}$	Pyrene	25	0.0000132	4	0.00092	12
		50	0.00009	4		
$C_{16}H_{14}$	9,10-Dimethylanthracene	25	0.0000056	4		
$C_{16}H_{22}O_4$	Dibutyl phthalate	25	0.00112	15		
$C_{16}H_{34}O$	1-Hexadecanol	25	0.000003	1		
$C_{17}H_{12}$	11H-Benzo[a]fluorene	25	0.0000045	4		
$C_{17}H_{12}$	11H-Benzo[b]fluorene	25	0.0000002	4		
$C_{18}H_{12}$	Benz[a]anthracene	25	0.0000011	4	0.00058	12
$C_{18}H_{12}$	Chrysene	25	0.00000019	4	0.006	12
$C_{18}H_{12}$	Naphthacene	25	0.00000006	4	0.000004	12
$C_{18}H_{12}$	Triphenylene	25	0.0000041	4	0.00001	12
$C_{18}H_{12}N_2$	2,2'-Biquinoline	24	0.000102	6		
$C_{18}H_{34}O_4$	Dibutyl sebacate	20	0.004	10		
$C_{19}H_{14}$	5-Methylchrysene	27	0.0000062	4		
$C_{19}H_{14}$	9-Methylbenz[a]anthracene	27	0.0000066	4		
$C_{19}H_{14}$	10-Methylbenz[a]anthracene	25	0.0000055	4		
$C_{20}H_{12}$	Benzo[a]pyrene	25	0.00000038	4	0.000046	12
$C_{20}H_{12}$	Perylene	25	0.00000004	4	0.000003	12
$C_{20}H_{12}$	Benzo[e]pyrene	20	0.00000046	4	0.00002	12
$C_{20}H_{13}N$	13H-Dibenzo[a,i]carbazole	24	0.00000104	6		
$C_{20}H_{14}$	1,2-Dihydrobenz[j]aceanthrylene	25	0.00000036	6		
$C_{20}H_{42}$	Eicosane	25	0.00000019	4		
$C_{21}H_{13}N$	Dibenz[a,j]acridine	25	0.000016	6		
$C_{21}H_{16}$	1,2-Dihydro-3-methylbenz[j]aceanthrylene	25	0.00000022	6		
$C_{22}H_{12}$	Benzo[ghi]perylene	25	0.000000026	4	0.000075	12
$C_{22}H_{14}$	Picene	25	0.00000025	4		
$C_{22}H_{14}$	Benzo[b]triphenylene	25	0.0000027	4		
$C_{22}H_{14}$	Dibenz[a,h]anthracene	25	0.00000006	4		
$C_{22}H_{14}$	Dibenz[a,j]anthracene	25	0.0000012	4		
$C_{22}H_{44}O_2$	Butyl stearate	25	0.2	10		
$C_{24}H_{12}$	Coronene	25	0.000000014	4		

AQUEOUS SOLUBILITY OF INORGANIC COMPOUNDS AT VARIOUS TEMPERATURES

The solubility of over 300 common inorganic compounds in water is tabulated here as a function of temperature. Solubility is defined as the concentration of the compound in a solution that is in equilibrium with a solid phase at the specified temperature. In this table the solid phase is generally the most stable crystalline phase at the temperature in question. An asterisk * on solubility values in adjacent columns indicates that the solid phase changes between those two temperatures (usually from one hydrated phase to another or from a hydrate to the anhydrous solid). In such cases the slope of the solubility vs. temperature curve may show a discontinuity.

All solubility values are expressed as mass percent of solute, $100 \cdot w_2$, where

$$w_2 = m_2/(m_1 + m_2)$$

and m_2 is the mass of solute and m_1 the mass of water. This quantity is related to other common measures of solubility as follows:

Molality: $m_2 = 1000 w_2/M_2(1-w_2)$
Mole fraction: $x_2 = (w_2/M_2)/\{(w_2/M_2) + (1-w_2)/M_1\}$
Mass of solute per 100 g of H_2O: $r_2 = 100 w_2/(1-w_2)$

Here M_2 is the molar mass of the solute and $M_1 = 18.015$ g/mol is the molar mass of water.

The data in the table have been derived from the references indicated; in many cases the data have been refitted or interpolated in order to present solubility at rounded values of temperature. Where available, values were taken from the IUPAC *Solubility Data Series* (Reference 1) or the related papers in the *Journal of Physical and Chemical Reference Data* (References 2 to 5), which present carefully evaluated data.

The solubility of sparingly soluble compounds that do not appear in this table may be calculated from the data in the table "Solubility Product Constants". Solubility of inorganic gases may be found in the table "Solubility of Selected Gases in Water".

Compounds are listed alphabetically by chemical formula in the most commonly used form (e.g., NaCl, NH_4NO_3, etc.).

REFERENCES

1. *Solubility Data Series*, International Union of Pure and Applied Chemistry. Volumes 1 to 53 were published by Pergamon Press, Oxford, from 1979 to 1994; subsequent volumes were published by Oxford University Press, Oxford. The number following the colon is the volume number in the series.
2. Clever, H.L., and Johnston, F.J., *J. Phys. Chem. Ref. Data*, 9, 751, 1980.
3. Marcus, Y., *J. Phys. Chem. Ref. Data*, 9, 1307, 1980.
4. Clever, H.L., Johnson, S.A., and Derrick, M.E., *J. Phys. Chem. Ref. Data*, 14, 631, 1985.
5. Clever, H.L., Johnson, S.A., and Derrick, M.E., *J. Phys. Chem. Ref. Data*, 21, 941, 1992.
6. Söhnel, O., and Novotny, P., *Densities of Aqueous Solutions of Inorganic Substances*, Elsevier, Amsterdam, 1985.
7. Krumgalz, B.S., *Mineral Solubility in Water at Various Temperatures*, Israel Oceanographic and Limnological Research Ltd., Haifa, 1994.
8. Potter, R.W., and Clynne, M.A., *J. Research U.S. Geological Survey*, 6, 701, 1978; Clynne, M.A., and Potter, R.W., *J. Chem. Eng. Data*, 24, 338, 1979.
9. Marshal, W.L., and Slusher, R., *J. Phys. Chem.*, 70, 4015, 1966; Knacke, O., and Gans, W., *Zeit. Phys. Chem.*, NF, 104, 41, 1977.
10. Stephen, H., and Stephen, T., *Solubilities of Inorganic and Organic Compounds, Vol. 1*, Macmillan, New York, 1963.

AQUEOUS SOLUBILITY OF INORGANIC COMPOUNDS AT VARIOUS TEMPERATURES (continued)

Compound	0°C	10°C	20°C	25°C	30°C	40°C	50°C	60°C	70°C	80°C	90°C	100°C	Ref.
AgBrO3				0.193							1.32		7
AgClO2	0.17	0.31	0.47	0.55	0.64	0.82	1.02	1.22	1.44	1.66	1.88	2.11	7
AgClO3				15									7
AgClO4	81.6	83.0	84.2	84.8	85.3	86.3	86.9	87.5	87.9	88.3	88.6	88.8	6
AgNO2	0.155			0.413									7
AgNO3	55.9	62.3	67.8	70.1	72.3	76.1	79.2	81.7	83.8	85.4	86.7	87.8	6
Ag2SO4	0.56	0.67	0.78	0.83	0.88	0.97	1.05	1.13	1.20	1.26	1.32	1.39	6
AlCl3	30.84	30.91	31.03	31.10	31.18	31.37	31.60	31.87	32.17	32.51	32.90	33.32	7
Al(ClO4)3	54.9										64.4		7
AlF3	0.25	0.34	0.44	0.50	0.56	0.68	0.81	0.96	1.11	1.28	1.45	1.64	7
Al(NO3)3	37.0	38.2	39.9	40.8	42.0	44.5	47.3	50.4	53.8*			61.5*	6
Al2(SO4)3	27.5			27.8	28.2	29.2	30.7	32.6	34.9	37.6	40.7	44.2	7
As2O3	1.19	1.48	1.80	2.01	2.27	2.86	3.43	4.11	4.89	5.77	6.72	7.71	10
BaBr2	47.6	48.5	49.5	50.0	50.4	51.4	52.5	53.5	54.5	55.5	56.6	57.6	6
Ba(BrO3)2	0.285	0.442	0.656	0.788	0.935	1.30	1.74	2.27	2.90	3.61	4.40	5.25	1:14
Ba(C2H3O2)2	37.0			44.2									7
BaCl2	23.30	24.88	26.33	27.03	27.70	29.00	30.27	31.53	32.81	34.14	35.54	37.05	8
Ba(ClO2)2	30.5			31.3								44.7	7
Ba(ClO3)2	16.90	21.23	23.66	27.50	29.43	33.16	36.69	40.05	43.04	45.90	48.70	51.17	1:14
Ba(ClO4)2	67.30	70.96	74.30	75.75	77.05	79.23	80.92	82.21	83.16	83.88	84.43	84.90	7
BaF2	0.158	0.158		0.161									7
BaI2	62.5	64.7	67.3	68.8	69.1	69.5	70.1	70.7	71.3	72.0	72.7	73.4	6
Ba(IO3)2	0.0182	0.0262	0.0342	0.0396	0.045*	0.058*	0.073	0.090	0.109	0.131	0.156	0.182	1:14
Ba(NO2)2	31.1	36.6	41.8	44.3	46.8	51.6	56.2	60.5	64.6	68.5	72.1	75.6	10
Ba(NO3)2	4.7	6.3	8.2	9.3	10.2	12.4	14.7	17.0	19.3	21.5	23.5	25.5	6
Ba(OH)2	1.67			4.68	8.4	19	33	52	74	101			7
BaS	2.79	4.78	6.97	8.21	9.58	12.67	16.18	20.05	24.19	28.55	33.04	37.61	7
Ba(SCN)2				62.6									1:26
BaSO3				0.0011									7
BeCl2	40.5			41.7									7
Be(ClO4)2				59.5									7
BeSO4	26.69	27.58	28.61	29.22	29.90	31.51	33.39	35.50	37.78	40.21	42.72	45.28	7
CaBr2	55	56	59	61	63	68	71	73					10
CaCl2	36.70	39.19	42.13	44.83*	49.12*	52.85*	56.05*	56.73	57.44	58.21	59.04	59.94	8
Ca(ClO3)2	63.2	64.2	65.5	66.3	67.2	69.0	71.0	73.2	75.5*	77.4*	77.7	78.0	1:14
Ca(ClO4)2				65.3									7
CaF2	0.0013			0.0016									10
CaI2	64.6	66.0	67.6	68.3	69.0	70.8	72.4	74.0	76.0	78.0	79.6	81.0	7
Ca(IO3)2	0.082	0.155	0.243	0.305	0.384*	0.517*	0.590	0.652	0.811*	0.665*	0.668		1:14
Ca(NO2)2	38.6	39.5	44.5	48.6									7
Ca(NO3)2	50.1	53.1	56.7	59.0	60.9	65.4	77.8	78.1	78.2	78.3	78.4	78.5	6
CaSO3				0.0054	0.0049	0.0041	0.0035	0.0030	0.0026	0.0023	0.0020	0.0019	1:26
CaSO4	0.174	0.191	0.202	0.205	0.208	0.210	0.207	0.201	0.193	0.184	0.173	0.163	9

AQUEOUS SOLUBILITY OF INORGANIC COMPOUNDS AT VARIOUS TEMPERATURES (continued)

Compound	0°C	10°C	20°C	25°C	30°C	40°C	50°C	60°C	70°C	80°C	90°C	100°C	Ref.
$CdBr_2$	36.0	43.0	49.9	53.4	56.4	60.3*	60.3*	60.5	60.7	60.9	61.3	61.6	6
CdC_2O_4				0.0060									5
$CdCl_2$	47.2	50.1	53.2	54.6	56.3*	57.3*	57.5	57.8	58.1	58.51	58.98	59.5	6
$Cd(ClO_4)_2$				58.7								66.9	7
CdF_2					3.76								5
CdI_2	44.1	44.9	45.8	46.3	46.8	47.9	49.0	50.2	51.5	52.7	54.1	55.4	6
$Cd(IO_3)_2$				0.091									5
$Cd(NO_3)_2$	55.4	57.1	59.6	61.0	62.8	66.5	70.6	86.1	86.5	86.8	87.1	87.4	6
$CdSO_4$	43.1	43.1	43.2	43.4	43.6	44.1	43.5	42.5	41.4	40.2	38.5	36.7	6
$CdSeO_4$	42.04	40.59	39.02	38.18	37.29	35.35	33.15	30.65	27.84	24.69	21.24	17.49	5
$Ce(NO_3)_3$	57.99	59.80	61.89	63.05	64.31*	67.0*	68.6	71.1*	74.9*	79.2	80.9	83.1	1:13
$CoCl_2$	30.30	32.60	34.87	35.99	37.10	39.27	41.38	43.46	45.50	47.51	49.51	51.50	7
$Co(ClO_4)_2$	50.0			53.0									7
CoF_2				1.4									7
CoI_2	58.00	61.78	65.35	66.99	68.51	71.17	73.41	75.29	76.89	78.28	79.52	80.70	7
$Co(NO_2)_2$	0.076			0.49									7
$Co(NO_3)_2$	45.5	47.0	49.4	50.8	52.4	56.0	60.1	62.6	64.9	67.7			6
$CoSO_4$	19.9	23.0	26.1	27.7	29.2	32.3	34.4	35.9	35.5	33.2	30.6	27.8	6
$Co(SCN)_2$				50.7									7
CrO_3	62.2	62.3	62.6	62.8	63.0	63.5	64.1	64.7	65.5	66.2	67.1	67.9	6
$CsBr$				55.2									7
$CsBrO_3$	1.16	1.93	3.01	3.69	4.46	6.32	8.60	11.32	14.45	17.96	21.83	25.98	1:30
$CsCl$	61.83	63.48	64.96	65.64	66.29	57.50	68.60	69.61	70.54	71.40	72.21	72.96	1:47
$CsClO_3$	2.40	3.87	5.94	7.22	8.69	12.15	16.33	21.14	26.45	32.10	37.89	43.42	1:30
$CsClO_4$	0.79	1.01	1.51	1.96	2.57	4.28	6.55	9.29	12.41	15.80	19.39	23.07	7
CsI	30.9	37.2	43.2	45.9	48.6	53.3	57.3	60.7	63.6	65.9	67.7	69.2	6
$CsIO_3$	1.08	1.58	2.21	2.59	3.02	3.96	5.06	6.29	7.70	9.20	10.79	12.45	1:30
$CsNO_3$	8.46	13.0	18.6	21.8	25.1	32.0	39.0	45.7	51.9	57.3	62.1	66.2	6
$CsOH$					75								7
Cs_2SO_4	62.6	63.4	64.1	64.5	64.8	65.5	66.1	66.7	67.3	67.8	68.3	68.8	6
$CuBr_2$				55.8									7
$CuCl_2$	40.8	41.7	42.6	43.1	43.7	44.8	46.0	47.2	48.5	49.9	51.3	52.7	6
$Cu(ClO_4)_2$	54.3				59.3								7
CuF_2				0.075									7
$Cu(NO_3)_2$	45.2	49.8	56.3	59.2	61.1	62.0	63.1	64.5	65.9	67.5	69.2	71.0	6
$CuSO_4$	12.4	14.4	16.7	18.0	19.3	22.2	25.4	28.8	32.4	36.3	40.3	43.5	6
$CuSeO_4$	10.6			16.0									7
$Dy(NO_3)_3$	58.79	59.99	61.49	62.35	63.29	65.43	68.04	71.58	77.75				1:13
$Er(NO_3)_3$	61.58	63.15	64.84	65.75	66.69	68.70	70.96	73.64					1:13
$Eu(NO_3)_3$	55.2	56.7	58.5	59.4	60.4	62.5	64.6						1:13
$FeBr_2$				54.6								64.8*	7
$FeCl_2$	33.2*			39.4*								48.7*	7
$FeCl_3$	42.7	44.9	47.9	47.7	51.6	74.8	76.7	84.6	84.3	84.3	84.4	84.7	6

AQUEOUS SOLUBILITY OF INORGANIC COMPOUNDS AT VARIOUS TEMPERATURES (continued)

Compound	0°C	10°C	20°C	25°C	30°C	40°C	50°C	60°C	70°C	80°C	90°C	100°C	Ref.
$Fe(ClO_4)_2$	63.39			67.76									7
FeF_3				5.59									7
$Fe(NO_3)_3$	40.15			46.57									7
$Fe(NO_3)_2$	41.44			46.67									7
$FeSO_4$	13.5	17.0	20.8	22.8	24.8	28.8	32.8	35.5	33.6	30.4	27.1	24.0	6
$Gd(NO_3)_3$	56.3	57.7	59.2	60.1	61.0	62.9	65.2	67.9	71.5				1:13
HIO_3	73.45	74.10	74.98	75.48	76.03	77.20	78.46	79.78	81.13	82.48	83.82	85.14	1:30
H_3BO_3	2.61	3.57	4.77	5.48	6.27	8.10	10.3	12.9	15.9	19.3	23.1	27.3	6
$HgBr_2$	0.26	0.37	0.52	0.61	0.72	0.96	1.26	1.63	2.08	2.61	3.23	3.95	4
$Hg(CN)_2$	6.57	7.83	9.33	10.2	11.1	13.1	15.5	18.2	21.2	24.6	28.3	32.3	6
$HgCl_2$	4.24	5.05	6.17	6.81	7.62	9.53	12.02	15.18	19.16	24.06	29.90	36.62	4
HgI_2			0.0041	0.0055	0.0072	0.0122	0.0199						4
$Hg(SCN)_2$				0.070									4
Hg_2Cl_2				0.0004									3
$Hg_2(ClO_4)_2$	73.8			79.8*								85.3*	7
Hg_2SO_4	0.038	0.043	0.048	0.051	0.054	0.059	0.065	0.070	0.076	0.082	0.088	0.093	4
$Ho(NO_3)_3$				63.8									1:13
KBF_4	0.28	0.34	0.45	0.55	0.75	1.38	2.09	2.82	3.58	4.34	5.12	5.90	10
KBr	35.0	37.3	39.4	40.4	41.4	43.2	44.8	46.2	47.6	48.8	49.8	50.8	6
$KBrO_3$	2.97	4.48	6.42	7.55	8.79	11.57	14.71	18.14	21.79	25.57	29.42	33.28	1:30
$KC_2H_3O_2$	68.40	70.29	72.09	72.92	73.70	75.08	76.27	77.31	78.22	79.04	79.80	80.55	7
KCl	21.74	23.61	25.39	26.22	27.04	28.59	30.04	31.40	32.66	33.86	34.99	36.05	1:47
$KClO_3$	3.03	4.67	6.74	7.93	9.21	12.06	15.26	18.78	22.65	26.88	31.53	36.65	1:30
$KClO_4$	0.70	1.10	1.67	2.04	2.47	3.54	4.94	6.74	8.99	11.71	14.94	18.67	6
KF	30.90	39.8	47.3	50.41	53.2					60.0			7
$KHCO_3$	18.62	21.73	24.92	26.6	28.13	31.32	34.46	37.51	40.45				6
$KHSO_4$	27.1	29.7	32.3	33.6	35.0	37.8	40.5	43.4	46.2	49.02	51.82	54.6	6
KH_2PO_4	11.74	14.91	18.25	19.97	21.77	25.28	28.95	32.76	36.75	40.96	45.41	50.12	1:31
KI	56.0	57.6	59.0	59.7	60.4	61.6	62.8	63.8	64.8	65.7	66.6	67.4	6
KIO_3	4.53	5.96	7.57	8.44	9.34	11.09	13.22	15.29	17.41	19.58	21.78	24.03	1:30
KIO_4	0.16	0.22	0.37	0.51	0.70	1.24	1.96	2.83	3.82	4.89	6.02	7.17	7
$KMnO_4$	2.74	4.12	5.96	7.06	8.28	11.11	14.42	18.16					6
KNO_2	73.7	74.6	75.3	75.7	76.0	76.7	77.4	78.0	78.5	79.1	79.6	80.1	6
KNO_3	12.0	17.6	24.2	27.7	31.3	38.6	45.7	52.2	58.0	63.0	67.3	70.8	6
KOH	48.7	50.8	53.2	54.7	56.1	57.9	58.6	59.5	60.6	61.8	63.1	64.6	6
$KSCN$	63.8	66.4	69.1	70.4	71.6	74.1	76.5	78.9	81.1	83.3	85.3	87.3	6
K_2CO_3	51.3	51.7	52.3	52.7	53.1	54.0	54.9	56.0	57.2	58.4	59.6	61.0	6
K_2CrO_4	37.1	38.1	38.9	39.4	39.8	40.5	41.3	41.9	42.6	43.2	43.8	44.3	6
$K_2Cr_2O_7$	4.30	7.12	10.9	13.1	15.5	20.8	26.3	31.7	36.9	41.5	45.5	48.9	6
K_2HAsO_4	48.5*			63.6*								79.8*	7
K_2HPO_4	57.0	59.1	61.5	62.7	64.1	67.7*		72.7*					1:31
K_2MoO_4				64.7							66.5		7
K_2SO_3	51.30	51.39	51.49	51.55	51.62	51.76	51.93	52.11	52.32	52.54	52.79	53.06	1:26

Compound	0°C	10°C	20°C	25°C	30°C	40°C	50°C	60°C	70°C	80°C	90°C	100°C	Ref.
K_2SO_4	7.11	8.46	9.95	10.7	11.4	12.9	14.2	15.5	16.7	17.7	18.6	19.3	6
$K_2S_2O_3$	49.0*			62.3*							75.7*		7
$K_2S_2O_5$	22.1	26.7	31.1	33.1	35.2	39.0	42.6	46.0	49.1	52.0	54.6		1:26
K_2SeO_3	68.4*			68.5*								68.5*	7
K_2SeO_4	52.70	52.93	53.17	53.30	53.43	53.70	53.99	54.30	54.61	54.94	55.26	55.60	7
K_3AsO_4	51.5*			55.6*								73*	7
$K_3Fe(CN)_6$	23.9	27.6	31.1	32.8	34.3	37.2	39.6	41.7	43.5	45.0	46.1	47.0	6
K_3PO_4	44.3			51.4									7
$K_4Fe(CN)_6$	12.5	17.3	22.0	23.9	25.6	29.2	32.5	35.5	38.2	40.6	41.4	43.1	6
$LaCl_3$	49.0	48.5	48.6	48.9	49.3	50.5	52.1	54.0	56.3	58.9	61.7		6
$La(NO_3)_3$	55.0	56.9	58.9	60.0	61.1	63.6	66.3	69.9*	74.1*				6
$LiBr$	58.4	60.1	62.7	64.4	65.9	67.8	68.3	69.0	69.8	70.7	71.7	72.8	6
$LiBrO_3$	61.03	62.62	64.44	65.44	66.51	68.90	71.58*	73.24*	74.43	75.66	76.93	78.32	1:30
$LiC_2H_3O_2$	23.76	26.49	29.42	31.02	32.72	36.48	40.55	45.15	49.93	54.91	60.04	65.26	7
$LiCl$	40.45	42.46*	45.29*	45.81	46.25	47.30	48.47	49.78	51.27	52.98	54.98*	56.34*	1:47
$LiClO_3$	73.2	75.6*	80.8*	82.1	83.4	85.9*	87.1*	88.2	89.6	91.3	93.4	95.7	1:30
$LiClO_4$	30.1	32.6	35.5	37.0	38.6	41.9	45.5	49.2	53.2	57.2	61.3	71.4	6
LiF	0.120	0.126	0.131	0.134									7
LiH_2PO_4	55.8												7
LiI	59.4	60.5	61.7	62.3	63.0	64.3	65.8	67.3	68.8	81.3	81.7	82.6	6
$LiIO_3$				43.8									1:30
$LiNO_2$	41	45	49	51	53	56	60	63	66	68			10
$LiNO_3$	34.8	37.6	42.7	50.5	57.9	50.1	62.2	64.0	65.7	67.2	68.5	69.7	6
$LiOH$	10.8	10.8	11.0	11.1	11.3	11.7	12.2	12.7	13.4	14.2	15.1	16.1	6
$LiSCN$				54.5									7
Li_2CO_3	1.54	1.43	1.33	1.28	1.24	1.15	1.07	0.99	0.92	0.85	0.78	0.72	7
$Li_2C_2O_4$				5.87									7
Li_2HPO_3	9.07	8.40	7.77	7.47	7.18	6.64	6.16	5.71	5.30	4.91	4.53	4.16	7
Li_2SO_4	26.3	25.9	25.6	25.5	25.3	25.0	24.8	24.5	24.3	24.0	23.8	23.6	6
Li_3PO_4				0.027									1:31
$Lu(NO_3)_3$				71.1									1:13
$MgBr_2$	49.3	49.8	50.3	50.6	50.9	51.5	52.1	52.8	53.5	54.2	55.0	55.7	6
$Mg(BrO_3)_2$	43.0	45.2	48.0	49.4	51.0	54.3	57.9	61.6	65.3	69.0*	70.9*	71.7	1:14
$Mg(C_2H_3O_2)_2$	36.18	37.55	38.92	39.61									7
MgC_2O_4				0.038									7
$MgCl_2$	33.96	34.85	35.58	35.90	36.20	36.77	37.34	37.97	38.71	39.62	40.75	42.15	8
$Mg(ClO_3)_2$	53.35	54.40	56.81	58.66	60.91*	65.46*	67.33	69.27	71.01	72.44	73.48		1:14
$Mg(ClO_4)_2$	47.8	48.7	49.6	50.1	50.5	51.3	52.1						7
$MgCrO_4$	32.06*			35.39*						67.0			6
$Mg_2P_2O_7$				58.9									7
MgF_2				0.013									7
MgI_2	54.7	56.1	58.2	59.4	60.8	63.9	65.0	65.0	65.0	65.0	65.1	65.2	6
$Mg(IO_3)_2$	3.19*	6.70*	7.92	8.52	9.11	10.45	11.99	13.7	15.6	17.6	19.6		1:14

AQUEOUS SOLUBILITY OF INORGANIC COMPOUNDS AT VARIOUS TEMPERATURES (continued)

Compound	0°C	10°C	20°C	25°C	30°C	40°C	50°C	60°C	70°C	80°C	90°C	100°C	Ref.
Mg(NO2)2				47									7
Mg(NO3)2	38.4	39.5	40.8	41.6	42.4	44.1	45.9	47.9	50.0	52.2	70.6	72.0	6
MgSO3	0.32	0.37	0.46	0.52	0.61	0.87*	0.85*	0.76	0.69	0.64	0.62	0.60	1:26
MgSO4	18.2	21.7	25.1	26.3	28.2	30.9	33.4	35.6	36.9	35.9	34.7	33.3	6
MgS2O3	30.7												7
MgSeO4	31.4*			35.7*								47*	7
MnBr2	56.00	57.72	59.39	60.19	60.96	62.41	63.75	65.01	66.19	67.32	68.42	69.50	7
MnCl2	38.7	40.6	42.5	43.6	44.7	47.0	49.4	54.1	54.7	55.2	55.7	56.1	6
MnF2	0.80*			1.01*							0.34	0.48	7
Mn(IO3)2				0.27									7
Mn(NO3)2	50.5			61.7									7
MnSO4	34.6	37.3	38.6	38.9	38.9	37.7	36.3	34.6	32.8	30.8	28.8	26.7	6
NH4Br	37.5	40.2	42.7	43.9	45.1	47.3	49.4	51.3	53.0	54.6	56.1	57.4	7
NH4Cl	22.92	25.12	27.27	28.34	29.39	31.46	33.50	35.49	37.46	39.40	41.33	43.24	1:47
NH4ClO4	10.8	14.1	17.8	19.7	21.7	25.8	29.8	33.6	37.3	40.7	43.8	46.6	6
NH4F	41.7	43.2	44.7	45.5	46.3	47.8	49.3	50.9	52.5	54.1			7
NH4HCO3	10.6	13.7	17.6	19.9	22.4	27.9	34.2	41.4	49.3	58.1	67.6	78.0	7
NH4H2AsO4	25.2	29.0	32.7	34.5	36.3	39.7	43.1	46.2	49.3	52.2	55.0		7
NH4H2PO4	17.8	22.0	26.4	28.8	31.2	36.2	41.6	47.2	53.0	59.2	65.7	72.4	7
NH4I	60.7	62.1	63.4	64.0	64.6	65.8	66.8	67.8	68.7	69.6	70.4	71.1	6
NH4IO3				3.70	4.20	5.64	7.63						1:30
NH4NO2	55.7	59.0	64.9	68.8	70.3	74.3	77.7	80.8	83.4	85.8	88.2	90.3	7
NH4NO3	54.0	60.1	65.5	68.0					81.1				6
NH4SCN				64.4									7
(NH4)2C2O4	2.31	3.11	4.25	4.94	5.73	7.56	9.73	12.2	15.1	18.3	21.8	25.7	7
(NH4)2HPO4	36.4	38.2	40.0	41.0	42.0	44.1	46.2	48.5	50.9	53.3	55.9	58.6	1:26
(NH4)2S2O5	65.5	67.9	69.8	70.5	71.3	72.3	72.9	73.1					7
(NH4)2S2O8	37.00	40.45	43.84	45.49	47.11	50.25	53.28	56.23	59.13	62.00			1:26
(NH4)2SO3	32.2	34.9	37.7	39.1	40.6	43.7	47.0	50.6	54.5	58.9			7
(NH4)2SO4	41.3	42.1	42.9	43.3	43.8	44.7	45.6	46.6	47.5	48.5	49.5	50.5	1:26
(NH4)2SeO3	49.0	51.1	53.4	54.7	56.0	58.9	62.0	65.4	69.1				6
(NH4)2SeO4				54.02									7
(NH4)3PO4				15.5									7
NaBr	44.4	45.9	47.7	48.6	49.6	51.6	53.7	54.1	54.3	54.5	54.7	54.9	6
NaBrO3	20.0	23.22	26.65	28.28	29.86	32.83	35.55	38.05	40.37	42.52			1:30
NaCHO2	30.8	37.9	45.7	48.7	50.6	52.0	53.5	55.0					6
NaC2H3O2	26.5	28.8	31.8	33.5	35.5	39.9	45.1	58.3	59.3	60.5	61.7	62.9	6
NaCl	26.28	26.32	26.41	26.45	26.52	26.67	26.84	27.03	27.25	27.50	27.78	28.05	1:47
NaClO	22.7			44.4									7
NaClO2				97.0*				95.3*					7
NaClO3	44.27	46.67	49.3	50.1	51.2	53.6	55.5	57.0	58.5	60.5	63.3	67.1	1:30
NaClO4	61.9	64.1	66.2	67.2	68.3	70.4	72.5	74.1	74.7	75.4	76.1	76.7	6
NaF	3.52	3.72	3.89	3.97	4.05	4.20	4.34	4.46	4.57	4.66	4.75	4.82	6

AQUEOUS SOLUBILITY OF INORGANIC COMPOUNDS AT VARIOUS TEMPERATURES (continued)

Compound	0°C	10°C	20°C	25°C	30°C	40°C	50°C	60°C	70°C	80°C	90°C	100°C	Ref.
NaHCO$_3$	6.48	7.59	8.73	9.32	9.91	11.13	12.40	13.70	15.02	16.37	17.73	19.10	7
NaHSO$_4$				22.2								33.3	10
NaH$_2$PO$_4$	36.54	41.07	46.00	48.68	51.54	57.89*	61.7*	62.3*	65.9	68.7			1:31
NaI	61.2	62.4	63.9	64.8	65.7	67.7	69.8	72.0	74.7	74.8	74.9	75.1	6
NaIO$_3$	2.43	4.40	7.78*	8.65*	9.60	11.67	13.99	16.52	19.25*	21.1*	22.9	24.7	1:30
NaIO$_4$				12.62									7
NaNO$_2$	41.9	43.4	45.1	45.9	46.8	48.7	50.7	52.8	55.0	57.2	59.5	61.8	6
NaNO$_3$	42.2	44.4	46.6	47.7	48.8	51.0	53.2	55.3	57.5	59.6	61.7	63.8	6
NaOH	30	39	46	50	53	58	63	67	71	74	76	79	10
NaSCN		52.9	57.1	60.2	62.7	63.5	64.2	65.0	65.9	66.9	67.9	69.0	6
Na$_2$B$_4$O$_7$	1.23	1.71	2.50	3.07	3.82	6.02	9.7	14.9	17.1	19.9	23.5	28.0	6
Na$_2$CO$_3$	6.44	10.8	17.9	23.5	28.7	32.8	32.2	31.7	31.3	31.1	30.9	30.9	6
Na$_2$C$_2$O$_4$	2.62	2.95	3.30	3.48	3.65	4.00	4.36	4.71	5.06	5.41	5.75	6.08	6
Na$_2$CrO$_4$	22.6	32.3	44.6	46.7	46.9	48.9	51.0	53.4	55.3	55.5	55.8	56.1	6
Na$_2$Cr$_2$O$_7$	62.1	63.1	64.4	65.2	66.1	68.0	70.1	72.3	74.6	77.0	79.6	80.7	6
Na$_2$HAsO$_4$	5.6*			29.3*								67*	7
Na$_2$HPO$_4$	1.66	4.19	7.51	10.55	16.34*	35.17*	44.64*	45.20	46.81	48.78	50.52	51.53	1:31
Na$_2$MoO$_4$	30.6	38.8	39.4	39.4	39.8	40.3	41.0	41.7	42.6	43.5	44.5	45.5	6
Na$_2$S	11.1	13.2	15.7	17.1	18.6	22.1	26.7	28.1	30.2	33.0	36.4	41.0	6
Na$_2$SO$_3$	12.0	16.1	20.9	23.5	26.3*	27.3*	25.9	24.8	23.7	22.8	22.1	21.5	1:26
Na$_2$SO$_4$			16.13	21.94	29.22*	32.35*	31.55	30.90	30.39	30.02	29.79	29.67	8
Na$_2$S$_2$O$_3$	33.1	36.3	40.6	43.3	45.9	52.0	62.3	65.7	68.8	69.4	70.1	71.0	6
Na$_2$S$_2$O$_5$		38.4	39.5	40.0	40.6	41.8	43.0	44.2	45.5	46.8	48.1	49.5	1:26
Na$_2$SeO$_3$				47.3*								45*	7
Na$_2$SeO$_4$	11.7			36.9*								42.1*	7
Na$_2$WO$_4$	41.6	41.9	42.3	42.6	42.9	43.6	44.4	45.3	46.2	47.3	48.4	49.5	6
Na$_3$PO$_4$	4.28	7.30	10.8	12.6	14.1	15.6	22.9	28.4	32.4	37.6	40.4	43.5	6
Na$_2$P$_2$O$_7$	2.23	3.28	4.81	6.62	7.00	10.10	14.38	20.07	27.31	36.03	32.37	30.67	6
NdCl$_3$	49.0	49.3	49.7	50.0	50.4	51.2	52.2	53.3	54.5	55.8	57.1	58.5	1:13
Nd(NO$_3$)$_3$	55.76	57.49	59.37	60.38	61.43	63.69	66.27	69.47					6
NiCl$_2$	34.7	36.1	38.5	40.3	41.7	42.1	43.2	45.0	46.1	46.2	46.4	46.6	6
Ni(ClO$_4$)$_2$	51.1			52.8									7
NiF$_2$				2.50							2.52		7
NiI$_2$	55.40	57.68	59.78	60.69	61.50	62.80	63.73	64.38	64.80	65.09	65.30		7
Ni(NO$_3$)$_2$	44.1	46.0	48.4	49.8	51.3	54.6	58.3	61.0	63.1	65.6	67.9	69.0	7
NiSO$_4$	21.4	24.4	27.4	28.8	30.3*	32.0*	34.1	35.8	37.7	39.9	42.3	44.8	6
Ni(SCN)$_2$				35.48									7
NiSeO$_4$	21.6		26.2*									45.6*	7
PbBr$_2$	0.449	0.620	0.841	0.966	1.118	1.46	1.89	1.93	2.24	2.60	2.99	3.42	2
PbCl$_2$	0.66	0.81	0.98	1.07	1.17	1.39	1.64						2
Pb(ClO$_4$)$_2$				81.5									7
PbF$_2$		0.0603	0.0649	0.0670	0.0693								2
PbI$_2$	0.041	0.052	0.067	0.076	0.086	0.112	0.144	0.187	0.243	0.315			2

AQUEOUS SOLUBILITY OF INORGANIC COMPOUNDS AT VARIOUS TEMPERATURES (continued)

Compound	0°C	10°C	20°C	25°C	30°C	40°C	50°C	60°C	70°C	80°C	90°C	100°C	Ref.
Pb(IO3)2				0.0025									7
Pb(NO3)2	28.46	32.13	35.67	37.38	39.05	42.22	45.17	47.90	50.42	52.72	54.82	56.75	2
PbSO4	0.0033	0.0038	0.0042	0.0044	0.0047	0.0052	0.0058						2
PrCl3	48.0	48.1	48.6	49.0	49.5	50.8	52.3	54.1	56.1	58.3			6
Pr(NO3)3	57.50	59.20	61.16	62.24	63.40*	65.7*	67.8	70.2	73.4				1:13
RbBr	47.4	50.1	52.6	53.8	54.9	57.0	58.8	60.6	62.1	63.5	64.8	65.9	6
RbBrO3	0.97	1.55	2.36	2.87	3.45	4.87	6.64	8.78	11.29	14.15	17.32	20.76	1:30
RbCl	43.58	45.65	47.53	48.42	49.27	50.86	52.34	53.67	54.92	56.08	57.16	58.15	1:47
RbClO3	2.10	3.38	5.14	6.22	7.45	10.35	13.85	17.93	22.53	27.57	32.96	38.60	1:30
RbClO4	1			1.5								17	7
RbF			75										7
RbHCO3			53.7										7
RbI	55.8	58.6	61.1	62.3	63.4	65.4	67.2	68.8	70.3	71.6	72.7	73.8	6
RbIO3	1.09	1.53	2.07	2.38	2.74	3.52	4.41	5.42	6.52	7.74	9.00	10.36	1:30
RbNO3	16.4	25.0	34.6	39.4	44.2	53.1	60.8	67.2	72.2	76.1	79.0	81.2	6
RbOH					63.4								7
Rb2CrO4	38.27			43.26									7
Rb2SO4	27.3	30.0	32.5	33.7	34.8	36.9	38.7	40.3	41.8	43.0	44.1	44.9	6
SbCl3	85.7			90.8									7
SbF3	79.4			83.1									7
Sc(NO3)3	57.0	59.3	61.6	62.8	63.9	66.2	68.5		70.8	74.2			1:13
Sm(NO3)3	54.83	56.33	58.08	59.05	60.08	62.38	65.05*	68.1*					1:13
SmCl3		48.0	48.2	48.4	48.6	49.2	50.0						7
SnCl2	46	64											6
SnI2			0.97									3.87	7
SrBr2	46.0	48.3	50.6	51.7	52.9	55.2	57.6	59.9	62.3	64.6	66.8	69.0	6
Sr(BrO3)2	18.53	22.00	25.39	27.02	28.59	31.55	34.21	36.57	38.64*	40.2*	40.8	41.0	1:14
SrCl2	31.94	32.93	34.43	35.37	36.43	38.93	41.94	45.44*	46.81*	47.69	48.70	49.87	8
Sr(ClO2)2	13.0	13.6	14.1	14.3	14.5	14.9	15.3	15.6	15.9				7
Sr(ClO3)2	63.29	63.42	63.64	63.77	63.93	64.29	64.70	65.16	65.65	66.18	66.74	67.31	1:14
Sr(ClO4)2	70.04*			75.35*		78.44*							7
SrF2	0.011			0.021									7
SrI2	62.5	62.8	63.5	63.9	64.5	65.8	67.3	69.0	70.8	72.7	74.7	79.2	6
Sr(IO3)2	0.102	0.126	0.152	0.165	0.179	0.206	0.233	0.259	0.284	0.307	0.328	0.346	1:14
Sr(MnO4)2	2.5												7
Sr(NO2)2					41.9	44.3	47.9	48.4	48.9			58.6	7
Sr(NO3)2	28.2	34.6	41.0	44.5	47.0	47.4				49.5	50.1	50.7	6
Sr(OH)2	0.9			2.2									7
SrSO3				0.0015									1:26
SrSO4				0.0135									7
SrS2O3	8.8	13.2	17.7	20.0	22.2	26.8							7
Tb(NO3)3			60.6	61.02									1:13
Tl2SO4	2.65	3.56	4.61	5.19	5.80	7.09	8.46	9.89	11.33	12.77	14.18	15.53	6

AQUEOUS SOLUBILITY OF INORGANIC COMPOUNDS AT VARIOUS TEMPERATURES (continued)

Compound	0°C	10°C	20°C	25°C	30°C	40°C	50°C	60°C	70°C	80°C	90°C	100°C	Ref.
Tm(NO₃)₃				67.9									1:13
UO₂(NO₃)₂	49.52	51.82	54.42	55.85	57.55	61.59	67.07	67.9	72.5				1:55
Y(NO₃)₃	55.57	56.93	58.75	59.86	61.11*	63.3*	64.9						1:13
Yb(NO₃)₃				70.5									1:13
ZnBr₂	79.3	80.1	81.8	83.0	84.1	85.6	85.8	86.1	86.3	86.6	86.8	87.1	6
ZnC₂O₄		0.0010	0.0019	0.0026									5
ZnCl₂		76.6	79.0	80.3	81.4	81.8	82.4	83.0	83.7	84.4	85.2	86.0	6
Zn(ClO₄)₂	44.29*			46.27*			48.70						7
ZnF₂				1.53									5
ZnI₂	81.1	81.2	81.3	81.4	81.5	81.7	82.0	82.3	82.6	83.0	83.3	83.7	6
Zn(IO₃)₂			0.58	0.64	0.69	0.77	0.82						5
Zn(NO₃)₂	47.8	50.8	54.4	54.6	58.5	79.1	80.1	87.5	89.9				6
ZnSO₃			0.1786	0.1790	0.1794	0.1803	0.1812						5
ZnSO₄	29.1	32.0	35.0	36.6	38.2	41.3	43.0	42.1	41.0	39.9	38.8	37.6	6
ZnSeO₄	33.06	34.98	37.38	38.79	40.34								5

SOLUBILITY PRODUCT CONSTANTS

The solubility product constant K_{sp} is a useful parameter for calculating the aqueous solubility of sparingly soluble compounds under various conditions. It may be determined by direct measurement or calculated from the standard Gibbs energies of formation $\Delta_f G°$ of the species involved at their standard states. Thus if $K_{sp} = [M^+]^m [A^-]^n$ is the equilibrium constant for the reaction

$$M_m A_n(s) \rightleftharpoons mM^+(aq) + nA^- (aq),$$

where $M_m A_n$ is the slightly soluble substance and M^+ and A^- are the ions produced in solution by the dissociation of $M_m A_n$, then the Gibbs energy change is

$$\Delta G° = m \, \Delta_f G° (M^+,aq) + n \, \Delta_f G° (A^-,aq) - \Delta_f G° (M_m A_n, s)$$

The solubility product constant is calculated from the equation

$$\ln K_{sp} = -\Delta G°/RT$$

The first table below gives selected values of K_{sp} at 25°C. Many of these have been calculated from standard state thermodynamic data in References 1 and 2; other values are taken from publications of the IUPAC Solubility Data Project (References 3 to 7).

The above formulation is not convenient for treating sulfides because the S^{-2} ion is usually not present in significant concentrations (see Reference 8). This is due to the hydrolysis reaction

$$S^{-2} + H_2O \rightleftharpoons HS^- + OH^-$$

which is strongly shifted to the right except in very basic solutions. Furthermore, the equilibrium constant for this reaction, which depends on the second ionization constant of H_2S, is poorly known. Therefore it is more useful in the case of sulfides to define a different solubility product K_{spa} based on the reaction

$$M_m S_n(s) + 2H^+ \rightleftharpoons mM^+ + nH_2S (aq)$$

Values of K_{spa}, taken from Reference 8, are given for several sulfides in the auxiliary table following the main table. Additional discussion of sulfide equilibria may be found in References 7 and 9.

REFERENCES

1. Wagman, D.D., Evans, W.H., Parker, V.B., Schumm, R.H., Halow, I., Bailey, S.M., Churney, K.L., and Nuttall, R L., *The NBS Tables of Chemical Thermodynamic Properties, J. Phys. Chem. Ref. Data*, Vol. 11, Suppl. 2, 1982.
2. Garvin, D., Parker, V.B., and White, H.J., *CODATA Thermodynamic Tables*, Hemisphere, New York, 1987.
3. *Solubility Data Series* (53 Volumes), International Union of Pure and Applied Chemistry, Pergamon Press, Oxford, 1979—1992.
4. Clever, H.L., and Johnston, F.J., *J. Phys. Chem. Ref. Data*, 9, 751, 1980.
5. Marcus, Y., *J. Phys. Chem. Ref. Data*, 9, 1307, 1980.
6. Clever, H.L., Johnson, S.A., and Derrick, M.E., *J. Phys. Chem. Ref. Data*, 14, 631, 1985.
7. Clever, H.L., Johnson, S.A., and Derrick, M.E., *J. Phys. Chem. Ref. Data*, 21, 941, 1992.
8. Myers, R.J., *J. Chem. Educ.*, 63, 687, 1986.
9. Licht, S., *J. Electrochem. Soc.*,135, 2971, 1988.

Compound	Formula	K_{sp}
Aluminum phosphate	$AlPO_4$	$9.84 \cdot 10^{-21}$
Barium bromate	$Ba(BrO_3)_2$	$2.43 \cdot 10^{-4}$
Barium carbonate	$BaCO_3$	$2.58 \cdot 10^{-9}$
Barium chromate	$BaCrO_4$	$1.17 \cdot 10^{-10}$
Barium fluoride	BaF_2	$1.84 \cdot 10^{-7}$
Barium hydroxide octahydrate	$Ba(OH)_2 \cdot 8H_2O$	$2.55 \cdot 10^{-4}$
Barium iodate	$Ba(IO_3)_2$	$4.01 \cdot 10^{-9}$
Barium iodate monohydrate	$Ba(IO_3)_2 \cdot H_2O$	$1.67 \cdot 10^{-9}$
Barium molybdate	$BaMoO_4$	$3.54 \cdot 10^{-8}$
Barium nitrate	$Ba(NO_3)_2$	$4.64 \cdot 10^{-3}$
Barium selenate	$BaSeO_4$	$3.40 \cdot 10^{-8}$
Barium sulfate	$BaSO_4$	$1.08 \cdot 10^{-10}$
Barium sulfite	$BaSO_3$	$5.0 \cdot 10^{-10}$
Beryllium hydroxide	$Be(OH)_2$	$6.92 \cdot 10^{-22}$
Bismuth arsenate	$BiAsO_4$	$4.43 \cdot 10^{-10}$

Compound	Formula	K_{sp}
Bismuth iodide	BiI_3	$7.71 \cdot 10^{-19}$
Cadmium arsenate	$Cd_3(AsO_4)_2$	$2.2 \cdot 10^{-33}$
Cadmium carbonate	$CdCO_3$	$1.0 \cdot 10^{-12}$
Cadmium fluoride	CdF_2	$6.44 \cdot 10^{-3}$
Cadmium hydroxide	$Cd(OH)_2$	$7.2 \cdot 10^{-15}$
Cadmium iodate	$Cd(IO_3)_2$	$2.5 \cdot 10^{-8}$
Cadmium oxalate trihydrate	$CdC_2O_4 \cdot 3H_2O$	$1.42 \cdot 10^{-8}$
Cadmium phosphate	$Cd_3(PO_4)_2$	$2.53 \cdot 10^{-33}$
Calcium carbonate (calcite)	$CaCO_3$	$3.36 \cdot 10^{-9}$
Calcium fluoride	CaF_2	$3.45 \cdot 10^{-11}$
Calcium hydroxide	$Ca(OH)_2$	$5.02 \cdot 10^{-6}$
Calcium iodate	$Ca(IO_3)_2$	$6.47 \cdot 10^{-6}$
Calcium iodate hexahydrate	$Ca(IO_3)_2 \cdot 6H_2O$	$7.10 \cdot 10^{-7}$
Calcium molybdate	$CaMoO_4$	$1.46 \cdot 10^{-8}$
Calcium oxalate monohydrate	$CaC_2O_4 \cdot H_2O$	$2.32 \cdot 10^{-9}$
Calcium phosphate	$Ca_3(PO_4)_2$	$2.07 \cdot 10^{-33}$
Calcium sulfate	$CaSO_4$	$4.93 \cdot 10^{-5}$
Calcium sulfate dihydrate	$CaSO_4 \cdot 2H_2O$	$3.14 \cdot 10^{-5}$
Calcium sulfite hemihydrate	$CaSO_3 \cdot 0.5H_2O$	$3.1 \cdot 10^{-7}$
Cesium perchlorate	$CsClO_4$	$3.95 \cdot 10^{-3}$
Cesium periodate	$CsIO_4$	$5.16 \cdot 10^{-6}$
Cobalt(II) arsenate	$Co_3(AsO_4)_2$	$6.80 \cdot 10^{-29}$
Cobalt(II) hydroxide (blue)	$Co(OH)_2$	$5.92 \cdot 10^{-15}$
Cobalt(II) iodate dihydrate	$Co(IO_3)_2 \cdot 2H_2O$	$1.21 \cdot 10^{-2}$
Cobalt(II) phosphate	$Co_3(PO_4)_2$	$2.05 \cdot 10^{-35}$
Copper(I) bromide	$CuBr$	$6.27 \cdot 10^{-9}$
Copper(I) chloride	$CuCl$	$1.72 \cdot 10^{-7}$
Copper(I) cyanide	$CuCN$	$3.47 \cdot 10^{-20}$
Copper(I) iodide	CuI	$1.27 \cdot 10^{-12}$
Copper(I) thiocyanate	$CuSCN$	$1.77 \cdot 10^{-13}$
Copper(II) arsenate	$Cu_3(AsO_4)_2$	$7.95 \cdot 10^{-36}$
Copper(II) iodate monohydrate	$Cu(IO_3)_2 \cdot H_2O$	$6.94 \cdot 10^{-8}$
Copper(II) oxalate	CuC_2O_4	$4.43 \cdot 10^{-10}$
Copper(II) phosphate	$Cu_3(PO_4)_2$	$1.40 \cdot 10^{-37}$
Europium(III) hydroxide	$Eu(OH)_3$	$9.38 \cdot 10^{-27}$
Gallium(III) hydroxide	$Ga(OH)_3$	$7.28 \cdot 10^{-36}$
Iron(II) carbonate	$FeCO_3$	$3.13 \cdot 10^{-11}$
Iron(II) fluoride	FeF_2	$2.36 \cdot 10^{-6}$
Iron(II) hydroxide	$Fe(OH)_2$	$4.87 \cdot 10^{-17}$
Iron(III) hydroxide	$Fe(OH)_3$	$2.79 \cdot 10^{-39}$
Iron(III) phosphate dihydrate	$FePO_4 \cdot 2H_2O$	$9.91 \cdot 10^{-16}$
Lanthanum iodate	$La(IO_3)_3$	$7.50 \cdot 10^{-12}$
Lead(II) bromide	$PbBr_2$	$6.60 \cdot 10^{-6}$
Lead(II) carbonate	$PbCO_3$	$7.40 \cdot 10^{-14}$
Lead(II) chloride	$PbCl_2$	$1.70 \cdot 10^{-5}$
Lead(II) fluoride	PbF_2	$3.3 \cdot 10^{-8}$
Lead(II) hydroxide	$Pb(OH)_2$	$1.43 \cdot 10^{-20}$
Lead(II) iodate	$Pb(IO_3)_2$	$3.69 \cdot 10^{-13}$
Lead(II) iodide	PbI_2	$9.8 \cdot 10^{-9}$
Lead(II) selenate	$PbSeO_4$	$1.37 \cdot 10^{-7}$
Lead(II) sulfate	$PbSO_4$	$2.53 \cdot 10^{-8}$
Lithium carbonate	Li_2CO_3	$8.15 \cdot 10^{-4}$
Lithium fluoride	LiF	$1.84 \cdot 10^{-3}$
Lithium phosphate	Li_3PO_4	$2.37 \cdot 10^{-11}$
Magnesium carbonate	$MgCO_3$	$6.82 \cdot 10^{-6}$
Magnesium carbonate trihydrate	$MgCO_3 \cdot 3H_2O$	$2.38 \cdot 10^{-6}$
Magnesium carbonate pentahydrate	$MgCO_3 \cdot 5H_2O$	$3.79 \cdot 10^{-6}$
Magnesium fluoride	MgF_2	$5.16 \cdot 10^{-11}$
Magnesium hydroxide	$Mg(OH)_2$	$5.61 \cdot 10^{-12}$
Magnesium oxalate dihydrate	$MgC_2O_4 \cdot 2H_2O$	$4.83 \cdot 10^{-6}$

Compound	Formula	K_{sp}
Magnesium phosphate	$Mg_3(PO_4)_2$	$1.04 \cdot 10^{-24}$
Manganese(II) carbonate	$MnCO_3$	$2.24 \cdot 10^{-11}$
Manganese(II) iodate	$Mn(IO_3)_2$	$4.37 \cdot 10^{-7}$
Manganese(II) oxalate dihydrate	$MnC_2O_4 \cdot 2H_2O$	$1.70 \cdot 10^{-7}$
Mercury(I) bromide	Hg_2Br_2	$6.40 \cdot 10^{-23}$
Mercury(I) carbonate	Hg_2CO_3	$3.6 \cdot 10^{-17}$
Mercury(I) chloride	Hg_2Cl_2	$1.43 \cdot 10^{-18}$
Mercury(I) fluoride	Hg_2F_2	$3.10 \cdot 10^{-6}$
Mercury(I) iodide	Hg_2I_2	$5.2 \cdot 10^{-29}$
Mercury(I) oxalate	$Hg_2C_2O_4$	$1.75 \cdot 10^{-13}$
Mercury(I) sulfate	Hg_2SO_4	$6.5 \cdot 10^{-7}$
Mercury(I) thiocyanate	$Hg_2(SCN)_2$	$3.2 \cdot 10^{-20}$
Mercury(II) bromide	$HgBr_2$	$6.2 \cdot 10^{-20}$
Mercury(II) iodide	HgI_2	$2.9 \cdot 10^{-29}$
Neodymium carbonate	$Nd_2(CO_3)_3$	$1.08 \cdot 10^{-33}$
Nickel(II) carbonate	$NiCO_3$	$1.42 \cdot 10^{-7}$
Nickel(II) hydroxide	$Ni(OH)_2$	$5.48 \cdot 10^{-16}$
Nickel(II) iodate	$Ni(IO_3)_2$	$4.71 \cdot 10^{-5}$
Nickel(II) phosphate	$Ni_3(PO_4)_2$	$4.74 \cdot 10^{-32}$
Palladium(II) thiocyanate	$Pd(SCN)_2$	$4.39 \cdot 10^{-23}$
Potassium hexachloroplatinate	K_2PtCl_6	$7.48 \cdot 10^{-6}$
Potassium perchlorate	$KClO_4$	$1.05 \cdot 10^{-2}$
Potassium periodate	KIO_4	$3.71 \cdot 10^{-4}$
Praseodymium hydroxide	$Pr(OH)_3$	$3.39 \cdot 10^{-24}$
Radium iodate	$Ra(IO_3)_2$	$1.16 \cdot 10^{-9}$
Radium sulfate	$RaSO_4$	$3.66 \cdot 10^{-11}$
Rubidium perchlorate	$RbClO_4$	$3.00 \cdot 10^{-3}$
Scandium fluoride	ScF_3	$5.81 \cdot 10^{-24}$
Scandium hydroxide	$Sc(OH)_3$	$2.22 \cdot 10^{-31}$
Silver(I) acetate	$AgCH_3COO$	$1.94 \cdot 10^{-3}$
Silver(I) arsenate	Ag_3AsO_4	$1.03 \cdot 10^{-22}$
Silver(I) bromate	$AgBrO_3$	$5.38 \cdot 10^{-5}$
Silver(I) bromide	$AgBr$	$5.35 \cdot 10^{-13}$
Silver(I) carbonate	Ag_2CO_3	$8.46 \cdot 10^{-12}$
Silver(I) chloride	$AgCl$	$1.77 \cdot 10^{-10}$
Silver(I) chromate	Ag_2CrO_4	$1.12 \cdot 10^{-12}$
Silver(I) cyanide	$AgCN$	$5.97 \cdot 10^{-17}$
Silver(I) iodate	$AgIO_3$	$3.17 \cdot 10^{-8}$
Silver(I) iodide	AgI	$8.52 \cdot 10^{-17}$
Silver(I) oxalate	$Ag_2C_2O_4$	$5.40 \cdot 10^{-12}$
Silver(I) phosphate	Ag_3PO_4	$8.89 \cdot 10^{-17}$
Silver(I) sulfate	Ag_2SO_4	$1.20 \cdot 10^{-5}$
Silver(I) sulfite	Ag_2SO_3	$1.50 \cdot 10^{-14}$
Silver(I) thiocyanate	$AgSCN$	$1.03 \cdot 10^{-12}$
Strontium arsenate	$Sr_3(AsO_4)_2$	$4.29 \cdot 10^{-19}$
Strontium carbonate	$SrCO_3$	$5.60 \cdot 10^{-10}$
Strontium fluoride	SrF_2	$4.33 \cdot 10^{-9}$
Strontium iodate	$Sr(IO_3)_2$	$1.14 \cdot 10^{-7}$
Strontium iodate monohydrate	$Sr(IO_3)_2 \cdot H_2O$	$3.77 \cdot 10^{-7}$
Strontium iodate hexahydrate	$Sr(IO_3)_2 \cdot 6H_2O$	$4.55 \cdot 10^{-7}$
Strontium sulfate	$SrSO_4$	$3.44 \cdot 10^{-7}$
Thallium(I) bromate	$TlBrO_3$	$1.10 \cdot 10^{-4}$
Thallium(I) bromide	$TlBr$	$3.71 \cdot 10^{-6}$
Thallium(I) chloride	$TlCl$	$1.86 \cdot 10^{-4}$
Thallium(I) chromate	Tl_2CrO_4	$8.67 \cdot 10^{-13}$
Thallium(I) hydroxide	$Tl(OH)_3$	$1.68 \cdot 10^{-44}$
Thallium(I) iodate	$TlIO_3$	$3.12 \cdot 10^{-6}$
Thallium(I) iodide	TlI	$5.54 \cdot 10^{-8}$
Thallium(I) thiocyanate	$TlSCN$	$1.57 \cdot 10^{-4}$
Tin(II) hydroxide	$Sn(OH)_2$	$5.45 \cdot 10^{-27}$

Compound	Formula	K_{sp}
Yttrium carbonate	$Y_2(CO_3)_3$	$1.03 \cdot 10^{-31}$
Yttrium fluoride	YF_3	$8.62 \cdot 10^{-21}$
Yttrium hydroxide	$Y(OH)_3$	$1.00 \cdot 10^{-22}$
Yttrium iodate	$Y(IO_3)_3$	$1.12 \cdot 10^{-10}$
Zinc arsenate	$Zn_3(AsO_4)_2$	$2.8 \cdot 10^{-28}$
Zinc carbonate	$ZnCO_3$	$1.46 \cdot 10^{-10}$
Zinc carbonate monohydrate	$ZnCO_3 \cdot H_2O$	$5.42 \cdot 10^{-11}$
Zinc fluoride	ZnF_2	$3.04 \cdot 10^{-2}$
Zinc hydroxide	$Zn(OH)_2$	$3 \cdot 10^{-17}$
Zinc iodate dihydrate	$Zn(IO_3)_2 \cdot 2H_2O$	$4.1 \cdot 10^{-6}$
Zinc oxalate dihydrate	$ZnC_2O_4 \cdot 2H_2O$	$1.38 \cdot 10^{-9}$
Zinc selenide	$ZnSe$	$3.6 \cdot 10^{-26}$
Zinc selenite monohydrate	$ZnSeO_3 \cdot H_2O$	$1.59 \cdot 10^{-7}$

Sulfides

Compound	Formula	K_{spa}
Cadmium sulfide	CdS	$8 \cdot 10^{-7}$
Copper(II) sulfide	CuS	$6 \cdot 10^{-16}$
Iron(II) sulfide	FeS	$6 \cdot 10^{2}$
Lead(II) sulfide	PbS	$3 \cdot 10^{-7}$
Manganese(II) sulfide (green)	MnS	$3 \cdot 10^{7}$
Mercury(II) sulfide (red)	HgS	$4 \cdot 10^{-33}$
Mercury(II) sulfide (black)	HgS	$2 \cdot 10^{-32}$
Silver(I) sulfide	Ag_2S	$6 \cdot 10^{-30}$
Tin(II) sulfide	SnS	$1 \cdot 10^{-5}$
Zinc sulfide (sphalerite)	ZnS	$2 \cdot 10^{-4}$
Zinc sulfide (wurtzite)	ZnS	$3 \cdot 10^{-2}$

SOLUBILITY CHART

Abbreviations: **W**, soluble in water; **A**, insoluble in water but soluble in acids; **a**, insoluble in water and only sparingly soluble in acids; **I**, insoluble in water and acids; **w**, sparingly soluble in water but soluble in acids; **d**, decomposes in water. * Indicates two modifications of the salt

No.	Anion	Al	NH$_4$	Sb	Ba	Bi	Cd	Ca	Cr	Co	Cu	Au(I)	Au(II)	H	Fe(II)	Fe(III)
1	Acetate —$(C_2H_3O_2)$	W $Al(-)_3$	W $NH_4(-)$		W $Ba(-)_2$	W $Bi(-)_3$	W $Cd(-)_2$	W $Ca(-)_2$	W $Cr(-)_3$	W $Co(-)_2$	W $Cu(-)_2$	W	W	W $C_2H_4O_2$	W $Fe(-)_2$	W $Fe_2(-)_6$
2	Arsenate —(AsO_4)	a $Al(-)$	W $(NH_4)_3(-)$	A $Sb(-)$	w $Ba_3(-)_2$	A $Bi(-)$	A $Cd_3(-)_2$	w $Ca_3(-)_2$		A $Co_3(-)_2$	A $Cu_3(-)_2$			W H_3AsO_4	A $Fe_3(-)_2$	A $Fe(-)$
3	Arsenite —(AsO_3)	$Al(-)$	W NH_4AsO_2	A $Sb(-)$							w $Cu_3(-)_2$					
4	Benzoate —$(C_7H_5O_2)$	W $Al(-)_3$	W $NH_4(-)$		W $Ba(-)_2$	A $Bi(-)_3$	W $Cd(-)_2$	W $Ca(-)_2$	W $Cr(-)_3$	W $Co_3H_6(-)_4$	w $Cu(-)_2$			W $C_7H_6O_2$	W $Fe(-)_2$	A $Fe(-)_3$
5	Bromide —Br	W $AlBr_3$	W NH_4Br	d $SbBr_3$	W $BaBr_2$	d $BiBr_3$	W $CdBr_2$	W $CaBr_2$	W $CrBr_3$	A $CoBr_2$	W $CuBr_2$	w $AuBr$	W $AuBr_3$	W HBr	W $FeBr_2$	W $FeBr_3$
6	Carbonate —(CO_3)		W $(NH_4)_2CO_3$		A $BaCO_3$	d	A $CdCO_3$	w $CaCO_3$	W $CrCO_3$	A $CoCO_3$	A $CuCO_3$				w $FeCO_3$	
7	Chlorate —(ClO_3)	W $Al(-)_3$	W $NH_4(-)$		W $Ba(-)_2$	W $Bi(-)_3$	W $Cd(-)_2$	W $Ca(-)_2$	W(I)* $Cr(-)_3$	W $Co(-)_2$	W $Cu(-)_2$			W $HClO_3$	W $Fe(-)_2$	W $Fe(-)_3$
8	Chloride —Cl	W $AlCl_3$	W NH_4Cl	W $SbCl_3$	W $BaCl_2$	d $BiCl_3$	W $CdCl_2$	W $CaCl_2$	I $CrCl_3$	W $CoCl_2$	W $CuCl_2$	w $AuCl$	W $AuCl_3$	W HCl	W $FeCl_2$	W $FeCl_3$
9	Chromate —(CrO_4)	W $Al_2(-)_3$	W $(NH_4)_2(-)$		A $Ba(-)$		A $Cd(-)$	W $Ca(-)$	W $Cr_2(-)_3$	A $Co(-)$	W $Cu(-)$			W (H_2CrO_4)	W $Fe(-)$	W $Fe_2(-)_3$
10	Citrate —$(C_6H_5O_7)$	W $Al(-)$	W $(NH_4)_3(-)$		w $Ba_3(-)_2$	A $Bi(-)$	A $Cd_3(-)_2$	w $Ca_3(-)_2$		w $Co_3(-)_2$				W $C_6H_8O_7$	W $Fe_3(-)_2$	W $Fe(-)$
11	Cyanide —(CN)		W NH_4CN		W $Ba(CN)_2$		A $Cd(CN)_2$	W $Ca(CN)_2$	W $Cr(CN)_3$	A $Co(CN)_2$	A $Cu(CN)_2$	w $AuCN$	W $Au(CN)_3$	W HCN	a $Fe(CN)_2$	
12	Ferricy'de —$[Fe(CN)_6]$	w $Al(-)_3$	W $(NH_4)_3(-)$		W $Ba_3(-)_2$	W $Bi(CN)_3$	A $Cd_3(-)_2$	W $Ca_3(-)_2$		I $Co_3(-)_2$	I $Cu_3(-)_2$			W $H_3(-)$	I $Fe_3(-)_2$	a $Fe(-)$
13	Ferrocy'de —$[Fe(CN)_6]$	w $Al_4(-)_3$	W $(NH_4)_4(-)$		A $Ba_2(-)$		A $Cd_2(-)$	W $Ca_2(-)$		I $Co_2(-)$	I $Cu_2(-)$			W $H_4(-)$	I $Fe_2(-)$	w $Fe_4(-)_3$
14	Fluoride —F	W AlF_3	W NH_4F	W SbF_3	W BaF_2	W BiF_3	W CdF_2	A CaF_2	W(a)* CrF_3	W CoF_2	W CuF_2			W HF	W FeF_2	w FeF_3
15	Formate —(CHO_2)	W $Al(-)_3$	W $NH_4(-)$		W $Ba(-)_2$		W $Cd(-)_2$	W $Ca(-)_2$	W $Cr(-)_3$	W $Co(-)_2$	W $Cu(-)_2$			W CH_2O_2	W $Fe(-)_2$	W $Fe(-)_3$
16	Hydroxide —(OH)	A $Al(OH)_3$	W NH_4OH		W $Ba(OH)_2$	A $Bi(OH)_3$	A $Cd(OH)_2$	W $Ca(OH)_2$	A $Cr(OH)_3$	A $Co(OH)_2$	A $Cu(OH)_2$	W $AuOH$	A $Au(OH)_3$		A $Fe(OH)_2$	A $Fe(OH)_3$
17	Iodide —I	W AlI_3	W NH_4I	d SbI_3	W BaI_2	d BiI_3	W CdI_2	W CaI_2	W CrI_3	W CoI_2	W CuI	a AuI	a AuI_3	W HI	W FeI_2	w FeI_3
18	Nitrate —(NO_3)	W $Al(NO_3)_3$	W NH_4NO_3		W $Ba(NO_3)_2$	A $Bi(NO_3)_3$	W $Cd(NO_3)_2$	W $Ca(NO_3)_2$	W $Cr(NO_3)_3$	W $Co(NO_3)_2$	W $Cu(NO_3)_2$			W HNO_3	W $Fe(NO_3)_2$	W $Fe(NO_3)_3$
19	Oxalate —(C_2O_4)	A $Al_2(-)_3$	W $(NH_4)_2(-)$		A $Ba(-)$	A $Bi_2(-)_3$	A $Cd(-)$	A $Ca(-)$	a $Cr(-)$	A $Co(-)$	A $Cu(-)$			W $C_2H_2O_4$	A $Fe(-)$	A $Fe_2(-)_3$
20	Oxide —O	a Al_2O_3		w Sb_2O_3	W BaO	A Bi_2O_3	A CdO	W CaO	a Cr_2O_3	A CoO	A CuO	w Au_2O	A Au_2O_3	W H_2O_2	A FeO	w Fe_2O_3
21	Phosphate —(PO_4)	I $AlPO_4$	W $NH_4H_2PO_4$		A $Ba_3(PO_4)_2$	A $BiPO_4$	A $Cd_3(PO_4)_2$	A $Ca_3(PO_4)_2$	A $Cr_2(PO_4)_2$	A $Co_3(PO_4)_2$	A $Cu_3(PO_4)_2$		H_3PO_4	I $Fe_3(PO_4)_2$		$FePO_4$
22	Silicate, —(SiO_3)	W $Al_2(-)_3$					A $Cd(-)$	W Ca_2SiO_3		W Co_2SiO_3				W H_2SiO_3	W Fe_2SiO_4	
23	Sulfate —(SO_4)	W $Al_2(SO_4)_3$	W $(NH_4)_2SO_4$	A $Sb_2(SO_4)_3$	A $BaSO_4$	d $Bi_2(SO_4)_3$	W $CdSO_4$	w $CaSO_4$	W(I)* $Cr_2(SO_4)_3$	W $CoSO_4$	W $CuSO_4$			W H_2SO_4	W $FeSO_4$	w $Fe_2(SO_4)_3$
24	Sulfide —S	d Al_2S_3	W $(NH_4)_2S$	A Sb_2S_3	w BaS	A Bi_2S_3	A CdS	W CaS	d Cr_2S_3	A CoS	A CuS	I Au_2S	I Au_2S_3	W H_2S	A FeS	d Fe_2S_3
25	Tartrate —$(C_4H_4O_6)$	w $Al_2(-)_3$	W $(NH_4)_2(-)$	W $Sb_2(-)_3$	W $Ba(-)$	W $Bi_2(-)_3$	A $Cd(-)$	W $Ca(-)$		W $Co(-)$	W $Cu(-)$			W $C_4H_6O_6$	W $Fe(-)$	W $Fe_2(-)_3$

SOLUBILITY CHART (continued)

No.	Anion	Al	Pb	NH₄	Mg	Mn	Sb	Ba	Hg(I)	Bi	Hg(II)	Cd	Ni	Ca	K	Pt	Co	Ag	Cu	Na	Au(I)	Sn(IV)	Au(II)	Sn(II)	H	Sr	Fe(II)	Fe(III)	Zn
26	Thiocy'te —(CNS)			W NH₄CNS				W Ba(CNS)₂						W Ca(CNS)			W Co(CNS)₂		d CuCNS						W CNSH		W Fe(CNS)₂	W Fe(CNS)₃	
1	Acetate —(C₂H₃O₂)		W Pb(—)₂	W	W Mg(—)₂	W Mn(—)₂			w Hg₂(—)₂		w Hg(—)		W Ni(—)₂	W	W K(—)			w Ag(—)		W Na(—)		W Sn(—)₄		d Sn(—)₂		W Sr(—)₂			W Zn(—)₂
2	Arsenate —(AsO₄)		A PbH(—)		A Mg₃(—)₂	w MnH(—)			A Hg₃(—)		w Hg₃(—)		A Ni₃(—)₂	A	W K₃AsO₄			A Ag₃(—)		W Na₃(—)				A Sn₃H(—)		w SrH(—)			A Zn₃(—)₂
3	Arsenite —(AsO₃)		w		W Mg₃(—)₂				A Hg₃(—)		w Hg₃(—)		A Ni₃H₃(—)₄		W K₃AsO₃			A Ag₃(—)		W Na₂H(—)						A Sr₃(—)₂			W Zn₃(—)₂
4	Benzoate —(C₇H₅O₂)		w Pb(—)₂	W	W Mg(—)₂	W			A Hg₂(—)₂		w Hg(—)		W		W			w Ag(—)		W Na(—)						W Sr₃(—)₂			W Zn(—)₂
5	Bromide		W PbBr₂	W	W MgBr₂	W MnBr₂			A HgBr		W HgBr₂		W NiBr₂		W KBr			a AgBr		W NaBr		W SnBr₄		W SnBr₂		W SrBr₂			W ZnBr₂
6	Carbonate		A PbCO₃		W MgCO₃	A MnCO₃			A Hg₂CO₃		W Hg(—)₂		W NiCO₃		W K₂CO₃			A Ag₂CO₃		W Na₂CO₃						w SrCO₃			W ZnCO₃
7	Chlorate —(ClO₃)		W Pb(—)₂	W	W Mg(—)₂	W Mn(—)₂			W Hg(—)		W Hg(—)₂		W Ni(—)₂		W K(—)			W Ag(—)		W Na(—)				W Sn(—)₂		W Sr(—)₂			W Zn(—)₂
8	Chloride		W PbCl₂	W	W MgCl₂	W MnCl₂			a HgCl		W HgCl₂		W NiCl₂		W KCl	PtCl₄		a AgCl		W NaCl		W SnCl₄		W SnCl₂		W SrCl₂			W ZnCl₂
9	Chromate —(CrO₄)		A PbCrO₄	W	W MgCrO₄				A Hg₂(—)		w Hg(—)		A Ni(—)		W K₂(—)			A Ag₂(—)		W Na₂(—)				A Sn(—)		A Sr(—)			w Zn(—)
10	Citrate —(C₆H₅O₇)		W Pb₃(—)₂	W	W Mg₃(—)₂				w				W Ni₃(—)₂		W K₃(—)			w Ag(—)		W Na₃(—)						W Sr(—)			A Zn(—)
11	Cyanide		A Pb(CN)₂	W	W Mg(CN)₂	w MnH(—)			HgCN		W Hg(CN)₂		a Ni(CN)₂	W	W KCN	I Pt(CN)₂		a AgCN		W NaCN				a Sn₂(—)		A SrH(—)			A Zn(CN)₂
12	Ferricy'de —Fe(CN)₆		W Pb₃(—)₂	W	W Mg₃(—)₂	A Mn₃(—)					A Hg₃(—)₂		I Ni₃(—)₂		W K₃(—)			I Ag₃(—)		W Na₃(—)						W Sr(CN)₂			W Zn₃(—)₂
13	Ferrocy'de —Fe(CN)₆		a Pb₂(—)	W	W Mg₂(—)						I Hg₂(—)		Ni₂(—)		W K₄(—)			I Ag₄(—)		W Na₄(—)						W Sr₂(—)			I Zn₃(—)
14	Fluoride		w PbF₂	W	w MgF₂	A MnF₂			d HgF		d HgF₂		w NiF₂	w	W KF	w PtF₄		W AgF		W NaF		W SnF₄		SnF₂		A SrF₂			W ZnF₂
15	Formate —(CHO₂)		w Pb(—)₂	W	W Mg(—)₂	W Mn(—)₂			w Hg(—)		Hg₂(—)		W Ni(—)₂		W K(—)			w Ag(—)		W Na(—)						W Sr₂(—)			W Zn(—)₂
16	Hydroxide		w Pb(OH)₂	W	A Mg(OH)₂	A Mn(OH)₂			Hg(—)		w Hg(OH)₂		A Ni(OH)₂	W	W KOH	Pt(OH)₄		Ag(—)		W NaOH		Sn(OH)₄		A Sn(OH)₂		W Sr(OH)₂			A Zn(OH)₂
17	Iodide		W PbI₂	W	W MgI₂	W MnI₂			A HgI		I HgI₂		W NiI₂		W KI	I PtI		I AgI		W NaI		d SnI₄		SnI₂		W SrI₂			I ZnI₂
18	Nitrate		W Pb(NO₃)₂	W	W Mg(NO₃)₂	W Mn(NO₃)₂			W HgNO₃		W Hg(NO₃)₂		W Ni(NO₃)₂		W KNO₃	W Pt(NO₃)₄		W AgNO₃		W NaNO₃				d Sn(NO₃)₂		W Sr(NO₃)₂			W Zn(NO₃)₂
19	Oxalate —(C₂O₄)		A Pb(—)	W	w Mg(—)	w Mn(—)			a Hg₂(—)		Hg(—)		A Ni(—)		W K₂(—)			a Ag₂(—)		W Na₂(—)				Sn(—)		A Sr(—)			W Zn(—)
20	Oxide		w PbO		A MgO	A MnO			Hg₂O		w HgO		A NiO		W K₂O	A PtO		a Ag₂O		d Na₂O		A SnO₂		A SnO		w SrO			w ZnO
21	Phosphate		A Pb₃(PO₄)₂	W	A Mg₃(PO₄)₂	I Mn₃(PO₄)₂			Hg₃PO₄		Hg₃(PO₄)₂		A Ni₃(PO₄)₂		W K₃PO₄			A Ag₃PO₄		W Na₃PO₄				Sn₃(PO₄)₂		A Sr₃(PO₄)₂			A Zn₃(PO₄)₂
22	Silicate —(SiO₃)		w Pb(—)	W	A Mg(—)	Mn(—)									W K₂(—)					W Na₂(—)						A Sr(—)			A Zn(—)
23	Sulfate		w PbSO₄	W	W MgSO₄	W MnSO₄			w Hg₂SO₄		d HgSO₄		W NiSO₄		W K₂SO₄	W Pt(SO₄)₂		w Ag₂SO₄		W Na₂SO₄		W Sn(SO₄)₂		W SnSO₄		W SrSO₄			W ZnSO₄

SOLUBILITY CHART (continued)

No.		Pb	Mg	Mn	Hg (I)	Hg (II)	Ni	K	Pt	Ag	Na	Sn (IV)	Sn (II)	Sr	Zn
24	Sulfide	A PbS	d MgS	A MnS	I Hg_2S	I HgS	A NiS	W K_2S	I PtS	A Ag_2S	W Na_2S	A SnS_2	A SnS	W SrS	A ZnS
25	Tartrate $-(C_4H_4O_6)$	A Pb(—)	w Mg(—)	w Mn(—)	I Hg_2(—)	I	A Ni(—)	W K_2(—)	I	A Ag_2(—)	W Na_2(—)	A	A Sn(—)	W Sr(—)	A Zn(—)
26	Thiocy'te	w $Pb(CNS)_2$	W $Mg(CNS)_2$	W $Mn(CNS)_2$	A HgCNS	w $Hg(CNS)_2$		W KCNS		I AgCNS	NaCNS			$Sr(CNS)_2$	$Zn(CNS)_2$

REDUCTIONS OF WEIGHINGS IN AIR TO VACUO

When the mass M of a body is determined in air, a correction is necessary for the buoyancy of the air. The following table is computed for an air density of 0.0012 g/cm³. The corrected mass is equal to $M + kM/1000$. Values of k are given in the table.

Density of body weighted (g/cm³)	Correction factor, k		
	Pt Ir weights	Brass weights	Quartz or Al weights
0.5	+2.34	+2.26	+1.95
0.6	+1.94	+1.86	+1.55
0.7	+1.66	+1.57	+1.26
0.75	+1.55	+1.46	+1.15
0.80	+1.44	+1.36	+1.05
0.85	+1.36	+1.27	+0.96
0.90	+1.28	+1.19	+0.88
0.95	+1.21	+1.12	+0.81
1.00	+1.14	+1.06	+0.75
1.1	+1.04	+0.95	+0.64
1.2	+0.94	+0.86	+0.55
1.3	+0.87	+0.78	+0.47
1.4	+0.80	+0.71	+0.40
1.5	+0.75	+0.66	+0.35

Density of body weighted (g/cm³)	Correction factor, k		
	Pt Ir weights	Brass weights	Quartz or Al weights
1.6	+0.69	+0.61	+0.30
1.7	+0.65	+0.56	+0.25
1.8	+0.62	+0.52	+0.21
1.9	+0.58	+0.49	+0.18
2.0	+0.54	+0.46	+0.15
2.5	+0.43	+0.34	+0.03
3.0	+0.34	+0.26	-0.05
4.0	+0.24	+0.16	-0.15
6.0	+0.14	+0.06	-0.25
8.0	+0.09	+0.01	-0.30
10.0	+0.06	-0.02	-0.33
15.0	+0.03	-0.06	-0.37
20.0	+0.004	-0.08	-0.39
22.0	-0.001	-0.09	-0.40

SOLVENTS FOR LIQUID CHROMATOGRAPHY

Solvent	Viscosity (mPa·s; 20 °C)	UV cutoff (nm)	Refractive index (20 °C)	Normal boiling point (°C)	Dielectric constant (20 °C)
Acetic acid	1.31(15)		1.372	117.9	6.15
Acetone	0.30(25)	330	1.359	56.1	20.7(25)
Acetonitrile	0.34(25)	190	1.344	81.6	37.5
Benzene	0.65	278	1.501	80.1	2.284
1-Butanol	2.95	215	1.399	117.7	17.8
2-Butanol	4.21	260	1.397	99.5	15.8(25)
n-Butyl acetate	0.73	254	1.394	126.1	
n-Butyl chloride	0.47(15)	220	1.402	78.4	
Carbon tetrachloride	0.97	263	1.460	76.8	2.238
Chlorobenzene	0.80	287	1.525	131.7	2.708
Chloroform	0.58	245	1.446	61.2	4.806
Cyclohexane	0.98	200	1.426	80.7	2.023
Cyclopentane	0.44	200	1.406	49.3	1.965
o-Dichlorobenzene	1.32(25)	295	1.551	180.0	9.93(25)
N,N-Dimethylacetamide	2.14	268	1.438	166.0	37.8
Dimethylformamide	0.92	268	1.430	153.0	36.7
Dimethyl sulfoxide	2.20	286	1.478	189.0	4.7
Dioxane	1.44(15)	215	1.422	101.4	2.209(25)
2-Ethoxyethanol	2.05	210	1.408	135.0	
Ethyl acetate	0.46	256	1.372	77.1	6.02(25)
Ethyl ether	0.24	218	1.352	34.4	4.335
Glyme (ethylene glycol dimethyl ether)	0.46(25)	220	1.380	93.0	
Heptane	0.42	200	1.388	98.4	1.92
Hexadecane	3.34	200	1.434	286.9	
Hexane	0.31	200	1.375	68.7	1.890
Isobutyl alcohol	4.70(15)	200	1.396	107.9	15.8(25)
Methanol	0.55	205	1.328	64.6	32.63(25)
2-Methoxyethanol	1.72	210	1.402	124.1	16.9
2-Methoxyethyl acetate		254	1.402	143.0	
Methylene chloride	0.45(15)	233	1.424	39.8	9.08
Methylethylketone	0.42(15)	329	1.379	79.6	18.5
Methylisoamylketone		330	1.406	144.0	
Methylisobutylketone	0.54(25)	334	1.396	116.5	
N-Methyl-2-pyrrolidone	1.67(25)	285	1.488	202.0	32.0
Nonane	0.72	200	1.405	150.8	1.972
Pentane	0.24	200	1.357	36.1	1.84
Petroleum ether	0.30	226		30—60	
β-Phenethylamine		285	1.529(25)	197—198	
1-Propanol	2.26	210	1.386	97.2	20.1(25)
2-Propanol	2.86(15)	205	1.377	82.2	18.3(25)
Propylene carbonate			1.419	240.0	
Pyridine	0.95	330	1.510	115.3	12.3(25)
Tetrachloroethylene	0.93(15)	295	1.506	121.2	
Tetrahydrofuran	0.55	212	1.407	64.8	7.6
Tetramethyl urea		265	1.449(25)	175.2	23.0
Toluene	0.59	284	1.497	110.6	2.379(25)
Trichloroethylene	0.57	273	1.477	87.2	3.4(16)
1,2,2-Trichloro-1,2,2-trifluoroethane	0.71	231	1.356(25)	47.6	
2,2,4-Trimethylpentane	0.50	215	1.391	99.2	1.94
Water	1.00	<190	1.333	100.0	78.54
o-Xylene	0.81	288	1.505	144.4	2.568
p-Xylene		290	1.496	138.4	2.270

Reprinted from T. J. Bruno and P. D. N. Svoronos, *CRC Handbook of Basic Tables for Chemical Analysis*, CRC Press, Boca Raton, FL, 1989, pg. 89.

PROPERTIES OF CARRIER GASES FOR GAS CHROMATOGRAPHY

Carrier gas	Density (kg/m³)	Thermal conductivity × 10⁻² (W/[m·K])	Thermal conductivity differences			Viscosity × 10⁻⁵ (Pa·s)	Heat capacity (J/[kg·K])	Relative molecular mass
			δλ (He)	δλ (N₂)	δλ (Ar)			
Hydrogen	0.08988	19.71	3.97	16.96	17.81	0.876 (20.7 °C) 1.086 (129.4 °C) 1.381 (299.0 °C)	14,112.7	2.016
Helium	0.17847	15.74	—	12.99	13.84	1.941 (20.0 °C) 2.281 (100.0 °C) 2.672 (200.0 °C)	5,330.6	4.003
Methane	0.71680	3.74	−12.00	0.99	1.84	1.087 (20.0 °C) 1.331 (100.0 °C) 1.605 (200.5 °C)	2,217.2	16.04
Oxygen	1.42904	2.85	−12.89	0.10	0.95	2.018 (19.1 °C) 2.568 (127.7 °C) 3.017 (227.0 °C)	915.3	32.00
Nitrogen	1.25055	2.75	−12.99	—	0.85	1.781 (27.4 °C) 2.191 (127.2 °C) 2.559 (226.7 °C)	1,030.5	28.016
Carbon monoxide	1.25040	2.67	−13.07	−0.08	0.77	1.753 (21.7 °C) 2.183 (126.7 °C) 2.548 (227.0 °C)	1,030.7	28.01
Ethane	1.35660	2.44	−13.30	−0.31	0.54	0.901 (17.2 °C) 1.143 (100.4 °C) 1.409 (200.3 °C)	1,614.0	30.07
Ethene	1.26040	2.30	−13.44	−0.45	0.40	1.008 (20.0 °C) 1.257 (100.0 °C) 1.541 (200.0 °C)	—	28.05
Propane	2.00960	2.03	−13.71	−0.72	0.13	0.795 (17.9 °C) 1.009 (100.4 °C) 1.253 (199.3 °C)	—	44.09
Argon	1.78370	1.90	−13.84	−0.85	—	2.217 (20.0 °C) 2.695 (100.0 °C) 3.223 (200.0 °C)	523.7	39.94
Carbon dioxide	1.97690	1.83	−13.91	−0.92	−0.07	1.480 (20.0 °C) 1.861 (99.1 °C) 2.221 (182.4 °C)	836.6	44.01
n-Butane	2.51900	1.82	−13.92	−0.93	−0.08	0.840 (14.7 °C)	—	58.12
Sulfur hexafluoride	6.50 (20°C)	1.63	−14.11	−1.12	−0.27	1.450 (21.1 °C)	674.0	146.05

Note: Values refer to a pressure of 101 kPa (760 torr). The reference temperatures for the quantities are: density, 0°C, thermal conductivity, 48.9°C (120°F); heat capacity, 15°C.

Reprinted from T. J. Bruno and P. D. N. Svoronos, *CRC Handbook of Basic Tables for Chemical Analysis.* CRC Press, Boca Raton, FL, 1989, pg 4.

SOLVENTS FOR ULTRAVIOLET SPECTROPHOTOMETRY

Solvent	Wavelength cutoff (nm)	Dielectric constant (20°C)	
Acetic acid	260	6.15	
Acetone	330	20.7	(25°C)
Acetonitrile	190	37.5	
Benzene	280	2.284	
2-Butanol	260	15.8	(25°C)
n-Butyl acetate	254		
Carbon disulfide	380	2.641	
Carbon tetrachloride	265	2.238	
1-Chlorobutane	220	7.39	(25°C)
Chloroform[a]	245	4.806	
Cyclohexane	210	2.023	
1,2-Dichloroethane	226	10.19	(25°C)
1,2-Dimethoxyethane	240		
N,N-Dimethylacetamide	268	59	(83°C)
N,N-Dimethylformamide	270	36.7	
Dimethylsulfoxide	265	4.7	
1,4-Dioxane	215	2.209	(25°C)
Diethyl ether	218	4.335	
Ethanol	210	24.30	(25°C)
2-Ethoxyethanol	210		
Ethyl acetate	255	6.02	(25°C)
Glycerol	207	42.5	(25°C)
n-Hexadecane	200	2.06	(25°C)
n-Hexane	210	1.890	
Methanol	210	32.63	(25°C)
2-Methoxyethanol	210	16.9	
Methyl cyclohexane	210	2.02	(25°C)
Methyl ethyl ketone	330	18.5	
Methyl isobutyl ketone	335		
2-Methyl-1-propanol	230	1	
N-Methyl-2-pyrrolidone	285	32.0	
Pentane	210	1.844	
n-Pentyl acetate	212		
1-Propanol	210	20.1	(25°C)
2-Propanol	210	18.3	(25°C)
Pyridine	330	12.3	(25°C)
Tetrachloroethylene[b]	290		
Tetrahydrofuran	220	7.6	
Toluene	286	2.379	(25°C)
1,1,2-Trichloro-1,2,2-trifluoroethane	231		
2,2,4-Trimethylpentane	215	1.936	(25°C)
o-Xylene	290	2.568	
m-Xylene	290	2.374	
p-Xylene	290	2.270	
Water		78.54	(25°C)

[a] Stabilized with ethanol to avoid phosgene formation.
[b] Stabilized with thymol (isopropyl meta-cresol).

Reprinted from T. J. Bruno and P. D. N. Svoronos, *CRC Handbook of Basic Tables for Chemical Analysis*, CRC Press, Boca Raton, FL, 1989, pg. 212.

13C CHEMICAL SHIFTS OF USEFUL NMR SOLVENTS

The following table gives the expected carbon-13 chemical shift(s) for various useful NMR solvents in parts per million (ppm). In some solvents, slight changes can occur with change of concentration.[1,2]

REFERENCES

1. **Silverstein, R. M., Bassler, G. C., and Morrill, T. C.,** *Spectrometric Identification of Organic Compounds,* John Wiley & Sons, New York, 1981.
2. **Rahman, A-U.,** *Nuclear Magnetic Resonance. Basic Principles,* Springer-Verlag, New York, 1986.

13C CHEMICAL SHIFTS OF USEFUL NMR SOLVENTS

Solvent	Formula	Chemical shift (ppm)
Acetone-d_6	$(CD_3)_2C=O$	29.2 (CD_3) 204.1 ($>C=O$)
Acetonitrile-d_3	$CD_3C\equiv N$	1.3 (CD_3) 117.1 ($C\equiv N$)
Benzene-d_6	C_6D_6	128.4
Carbon disulfide	CS_2	192.3
Carbon tetrachloride	CCl_4	96.0
Chloroform-d_3	$CDCl_3$	77.05
Cyclohexane-d_{12}	C_6D_{12}	27.5
Dichloromethane-d_2	CD_2Cl_2	53.6
Dimethylformamide-d_7	$(CD_3)_2NCOD$	31 (CD_3) 36 (CD_3) 162.4 ($DC=O$)
Dimethylsulfoxide-d_6	$(CD_3)_2S=O$	39.6
Dioxane-d_8	$C_4D_8O_2$	67.4
Methanol-d_4	CD_3OD	49.3
Nitromethane-d_3	CD_3-NO_2	57.3
Pyridine-d_5	C_5D_5N	123.9 (C–3) 135.9 (C–4) 150.2 (C–2)
1,1,2,2-tetrachloroethane-d_2	$CDCl_2-CDCl_2$	75.5
Tetrahydrofuran-d_8	C_4D_8O	25.8 (C–2) 67.9 (C–1)
Trichlorofluoromethane	$CFCl_3$	117.6
Water (heavy)	D_2O	—

Reprinted from T. J. Bruno and P. D. N. Svoronos, *CRC Handbook of Basic Tables for Chemical Analysis*, CRC Press, Boca Raton, FL, 1989, pg. 330.

MASS SPECTRAL PEAKS OF COMMON ORGANIC SOLVENTS

The strongest peaks in the mass spectra of 200 important solvents are listed in this table. The *e/m* value for each peak is followed by the relative intensity in parentheses, with the strongest peak assigned an intensity of 100. The peaks for each compound are listed in order of decreasing intensity. Solvents are listed in alphabetical order by common name.

Data on the physical properties of the same compounds may be found in Section 15 in the table "Properties of Common Laboratory Solvents".

REFERENCES

1. NIST/EPA/NIH Mass Spectral Database, National Institute of Standards and Technology, Gaithersburg, MD, 20899.
2. Lide, D.R., and Milne, G.W.A., Editors, *Handbook of Data on Organic Compounds, Third Edition*, CRC Press, Boca Raton, FL, 1994. (Also available as a CD ROM database.)

MASS SPECTRAL PEAKS OF COMMON ORGANIC SOLVENTS (continued)

e/m (intensity)

Compound										
Acetal (1,1-Diethoxyethane)	44(100)	43(92)	29(77)	31(76)	45(74)	27(52)	72(48)	73(23)	28(17)	46(15)
Acetic acid	43(100)	45(87)	60(57)	15(42)	42(14)	29(13)	14(13)	28(7)	18(6)	16(6)
Acetone	43(100)	15(34)	58(23)	27(9)	14(9)	42(8)	26(7)	29(5)	28(5)	39(4)
Acetonitrile	41(100)	40(46)	39(13)	14(9)	38(6)	28(4)	26(4)	25(3)	42(2)	27(2)
Acetylacetone	43(100)	85(31)	100(20)	27(12)	42(10)	29(10)	41(7)	39(7)	31(5)	26(5)
Acrylonitrile	53(100)	26(85)	52(79)	51(34)	27(13)	50(8)	25(7)	38(5)	54(3)	37(3)
Adiponitrile	41(100)	68(50)	54(42)	40(21)	55(20)	27(17)	39(16)	28(13)	52(7)	42(6)
Allyl alcohol	57(100)	31(34)	29(32)	28(31)	58(25)	39(22)	27(20)	30(16)	32(14)	26(11)
Allylamine	30(100)	56(80)	28(76)	57(33)	39(21)	29(20)	27(18)	26(13)	41(8)	18(8)
2-Aminoisobutanol	58(100)	41(18)	18(17)	42(13)	28(11)	56(10)	30(10)	29(8)	43(6)	59(5)
Benzal chloride	125(100)	127(32)	160(14)	89(13)	162(9)	63(9)	126(8)	62(7)	105(5)	39(5)
Benzaldehyde	51(100)	77(81)	50(55)	106(44)	105(43)	52(26)	78(16)	39(13)	27(10)	74(8)
Benzene	78(100)	77(20)	52(19)	51(17)	50(15)	39(12)	79(6)	76(5)	74(4)	38(4)
Benzonitrile	103(100)	76(34)	50(13)	104(9)	75(7)	51(7)	77(5)	52(4)	39(4)	74(3)
Benzyl chloride	91(100)	126(20)	65(14)	92(9)	39(9)	63(8)	128(6)	45(6)	89(5)	125(3)
Bromochloromethane	49(100)	130(67)	128(52)	51(31)	93(23)	81(20)	79(20)	95(17)	132(16)	47(8)
Bromoform (Tribromomethane)	173(100)	171(50)	175(49)	93(22)	91(22)	79(18)	81(17)	94(13)	92(13)	254(11)
Butyl acetate	43(100)	56(34)	41(17)	27(16)	29(15)	73(11)	61(10)	28(7)	55(6)	39(6)
Butyl alcohol	31(100)	56(81)	41(62)	43(60)	27(50)	42(31)	29(31)	28(17)	39(16)	55(12)
sec-Butyl alcohol	45(100)	31(22)	27(22)	59(20)	29(18)	43(13)	41(12)	44(8)	18(8)	28(5)
tert-Butyl alcohol	59(100)	31(33)	41(22)	43(18)	29(13)	27(11)	57(10)	42(4)	60(3)	28(3)
Butylamine	30(100)	73(10)	28(5)	41(3)	27(3)	18(3)	44(2)	42(2)	31(2)	29(2)
tert-Butylamine	58(100)	41(21)	42(15)	18(9)	30(8)	15(8)	39(7)	57(6)	28(6)	59(4)
Butyl methyl ketone	43(100)	58(60)	57(17)	100(16)	29(15)	41(13)	85(8)	27(8)	71(7)	59(5)
p-tert-Butyltoluene	133(100)	105(38)	41(23)	148(18)	93(16)	91(14)	115(13)	134(11)	39(11)	116(10)
γ-Butyrolactone	28(100)	42(74)	29(48)	27(33)	41(27)	56(25)	86(24)	26(18)	85(10)	39(10)
Caprolactam	55(100)	113(87)	30(81)	56(66)	84(60)	85(57)	42(51)	41(33)	28(26)	43(17)
Carbon disulfide	76(100)	32(22)	44(17)	78(9)	38(6)	28(5)	77(3)	64(1)	46(1)	39(1)
Carbon tetrachloride	117(100)	119(98)	121(31)	82(24)	47(23)	84(16)	35(14)	49(8)	28(8)	36(6)
1-Chloro-1,1-difluoroethane	65(100)	45(31)	85(14)	31(10)	64(8)	44(7)	35(6)	26(6)	87(5)	81(4)
Chlorobenzene	112(100)	77(63)	114(33)	51(29)	50(14)	75(8)	113(7)	78(5)	76(5)	28(4)
Chloroform	83(100)	85(64)	47(35)	35(19)	48(16)	49(12)	87(10)	37(6)	50(5)	84(4)
Chloropentafluoroethane	85(100)	69(61)	31(38)	87(32)	50(17)	35(8)	119(6)	66(4)	100(3)	47(3)
Cumene (Isopropylbenzene)	105(100)	120(25)	77(13)	51(12)	79(10)	106(9)	39(9)	27(8)	103(6)	91(5)
Cyclohexane	56(100)	84(71)	41(70)	27(37)	55(36)	39(35)	42(30)	69(23)	28(18)	43(14)
Cyclohexanol	57(100)	44(68)	41(68)	39(51)	32(40)	43(38)	31(32)	42(22)	67(18)	82(16)
Cyclohexanone	55(100)	42(85)	41(34)	27(33)	98(31)	39(27)	69(26)	70(20)	43(14)	28(14)
Cyclohexylamine	56(100)	43(23)	28(17)	99(10)	70(8)	57(6)	30(6)	93(5)	54(4)	41(4)
Cyclopentane	42(100)	70(30)	55(29)	41(29)	39(22)	27(15)	40(7)	29(5)	28(4)	43(3)
Cyclopentanone	55(100)	28(50)	84(42)	41(38)	56(29)	27(24)	39(19)	42(15)	26(9)	29(7)
p-Cymene (1-Methyl-4-isopropyl-benzene)	119(100)	91(42)	134(33)	39(27)	41(20)	117(18)	65(18)	77(17)	27(16)	120(15)
cis-Decalin	67(100)	81(87)	41(81)	138(67)	96(62)	82(62)	39(50)	55(45)	27(44)	95(42)
trans-Decalin	41(100)	68(91)	67(88)	82(67)	27(65)	96(61)	95(55)	138(51)	81(51)	29(51)

MASS SPECTRAL PEAKS OF COMMON ORGANIC SOLVENTS (continued)

Compound	e/m (intensity)									
Diacetone alcohol	43(100)	59(41)	58(17)	101(10)	41(9)	31(9)	83(6)	56(6)	55(6)	29(6)
1,2-Dibromoethane	27(100)	107(77)	109(72)	26(24)	28(10)	81(5)	79(5)	25(5)	95(4)	93(4)
Dibromofluoromethane	111(100)	113(98)	192(29)	43(16)	41(16)	190(15)	194(14)	81(9)	79(9)	122(7)
Dibromomethane	174(100)	93(96)	95(84)	172(53)	176(50)	91(11)	81(9)	79(9)	94(5)	65(5)
1,2-Dibromotetrafluoroethane	179(100)	181(97)	129(34)	131(33)	100(17)	31(13)	260(12)	50(8)	69(7)	262(6)
Dibutylamine	86(100)	72(52)	30(48)	44(40)	29(31)	57(24)	41(21)	73(15)	28(15)	43(13)
o-Dichlorobenzene	146(100)	148(64)	111(38)	75(23)	113(12)	74(12)	50(11)	150(10)	73(9)	147(7)
1,1-Dichloroethane (Ethylidene dichloride)	63(100)	27(71)	65(31)	26(19)	83(11)	85(7)	61(7)	35(6)	98(5)	62(5)
1,2-Dichloroethane (Ethylene dichloride)	62(100)	27(91)	49(40)	64(32)	26(31)	63(19)	98(14)	51(13)	61(12)	100(9)
1,1-Dichloroethylene	61(100)	96(61)	98(38)	63(32)	26(16)	60(15)	62(7)	25(7)	100(6)	35(6)
cis-1,2-Dichloroethylene	61(100)	96(73)	98(47)	63(32)	26(30)	60(21)	25(13)	35(12)	62(9)	100(8)
trans-1,2-Dichloroethylene	61(100)	96(67)	98(43)	26(34)	63(32)	60(24)	25(15)	62(10)	100(7)	47(7)
Dichloroethyl ether	93(100)	63(74)	27(38)	95(32)	65(24)	31(9)	49(4)	28(4)	94(3)	62(3)
Dichloromethane (Methylene chloride)	49(100)	84(64)	86(39)	51(31)	47(14)	48(8)	88(6)	50(3)	85(2)	83(2)
1,2-Dichloropropane	63(100)	62(71)	27(57)	41(49)	39(32)	65(31)	76(27)	64(25)	49(13)	77(12)
1,2-Dichlorotetrafluoroethane	85(100)	135(52)	87(33)	137(17)	101(9)	31(9)	103(6)	100(6)	50(5)	69(4)
Diethanolamine	30(100)	74(82)	28(77)	56(69)	18(50)	42(46)	29(36)	27(34)	45(30)	43(19)
Diethylamine	30(100)	58(81)	44(28)	73(18)	29(18)	28(17)	72(12)	42(11)	27(11)	59(4)
Diethyl carbonate	29(100)	45(70)	31(53)	27(39)	91(24)	28(15)	63(11)	26(10)	30(6)	43(5)
Diethylene glycol	45(100)	75(23)	31(20)	44(16)	27(14)	76(12)	29(12)	43(11)	42(9)	41(4)
Diethylene glycol dimethyl ether (Diglyme)	59(100)	58(43)	31(34)	29(32)	45(28)	28(19)	89(15)	43(9)	27(5)	60(4)
Diethylene glycol monoethyl ether (Carbitol)	45(100)	59(56)	72(37)	73(22)	60(14)	31(13)	75(11)	44(9)	104(8)	103(7)
Diethylene glycol monoethyl ether acetate	43(100)	29(51)	31(42)	45(40)	59(24)	72(18)	44(10)	73(9)	42(9)	30(6)
Diethylene glycol monomethyl ether	45(100)	31(42)	59(41)	29(38)	28(32)	58(21)	43(14)	27(13)	44(11)	32(10)
Diethylenetriamine	44(100)	73(59)	30(35)	19(18)	56(16)	28(16)	27(16)	42(11)	99(8)	43(8)
Diethyl ether	31(100)	29(63)	59(40)	27(35)	45(33)	74(23)	15(17)	43(9)	28(9)	26(9)
Diisobutyl ketone (Isovalerone)	57(100)	85(82)	41(46)	43(39)	58(33)	28(30)	26(30)	39(22)	42(12)	142(11)
Diisopropyl ether	45(100)	43(39)	87(15)	41(12)	59(10)	27(8)	39(4)	69(3)	42(3)	31(3)
N,N-Dimethylacetamide	44(100)	87(69)	43(46)	45(23)	42(19)	72(15)	15(11)	30(8)	28(5)	88(4)
Dimethylamine	44(100)	45(81)	18(32)	28(30)	43(19)	42(15)	15(9)	46(5)	41(5)	27(5)
Dimethyl disulfide	94(100)	45(63)	79(59)	46(38)	47(26)	15(18)	48(14)	61(12)	64(11)	96(9)
N,N-Dimethylformamide	73(100)	44(86)	42(36)	30(22)	28(20)	29(8)	43(7)	72(6)	58(5)	74(4)
Dimethyl sulfoxide	63(100)	78(70)	15(40)	45(35)	29(16)	61(13)	46(12)	31(11)	48(10)	47(10)
1,4-Dioxane	28(100)	88(31)	58(24)	31(17)	15(17)	27(15)	30(13)	43(11)	31(11)	26(9)
1,3-Dioxolane	73(100)	29(56)	44(53)	45(28)	28(21)	43(20)	27(13)	30(13)	74(5)	42(3)
Dipentene	68(100)	93(50)	67(44)	94(22)	39(22)	107(18)	92(18)	53(18)	136(16)	79(16)
Epichlorohydrin	57(100)	27(39)	29(32)	49(25)	31(22)	62(18)	28(16)	92(1)	53(18)	79(16)

MASS SPECTRAL PEAKS OF COMMON ORGANIC SOLVENTS (continued)

e/m (intensity)

Compound										
Ethanolamine (Glycinol)	30(100)	18(30)	28(15)	42(7)	31(6)	17(6)	61(5)	15(5)	43(3)	29(3)
Ethyl acetate	43(100)	29(46)	27(33)	45(32)	61(28)	28(25)	42(18)	73(11)	88(10)	70(10)
Ethyl acetoacetate	43(100)	29(24)	88(18)	28(16)	85(14)	27(12)	42(11)	60(9)	130(6)	45(6)
Ethyl alcohol	31(100)	45(44)	46(18)	27(18)	29(15)	43(14)	30(6)	42(3)	19(3)	14(3)
Ethylamine	30(100)	28(32)	44(20)	45(19)	27(13)	15(10)	42(9)	29(8)	41(5)	40(5)
Ethylbenzene	91(100)	106(31)	51(14)	39(10)	77(8)	65(8)	105(7)	92(7)	78(7)	27(6)
Ethyl bromide (Bromoethane)	108(100)	110(97)	29(62)	27(51)	28(35)	26(14)	93(6)	32(6)	95(5)	81(5)
Ethyl chloride (Cloroethane)	64(100)	28(91)	29(84)	27(75)	66(32)	26(28)	49(25)	51(8)	63(6)	65(4)
Ethylene carbonate	29(100)	44(62)	43(54)	88(40)	30(16)	28(11)	45(7)	58(6)	42(6)	73(4)
Ethylenediamine (1,2-Ethanediamine)	30(100)	18(13)	42(6)	43(5)	27(5)	44(4)	29(4)	17(4)	15(4)	41(3)
Ethylene glycol	31(100)	33(35)	29(13)	32(11)	43(6)	27(5)	28(4)	62(3)	30(3)	44(2)
Ethylene glycol diethyl ether	31(100)	59(71)	29(58)	45(43)	27(33)	74(27)	43(15)	15(14)	28(12)	44(10)
Ethylene glycol dimethyl ether	45(100)	60(13)	29(13)	90(7)	58(6)	31(5)	28(5)	43(4)	59(3)	46(2)
Ethylene glycol monobutyl ether	57(100)	45(38)	29(35)	41(31)	87(16)	27(12)	56(11)	31(9)	75(7)	28(7)
Ethylene glycol monoethyl ether (Cellosolve)	31(100)	29(52)	59(50)	27(27)	45(26)	72(14)	43(14)	15(14)	28(8)	26(6)
Ethylene glycol monoethyl ether acetate	43(100)	31(34)	59(31)	72(28)	44(25)	29(24)	45(12)	27(11)	15(11)	87(7)
Ethylene glycol monomethyl ether	45(100)	31(15)	29(14)	28(11)	47(9)	76(6)	43(6)	58(4)	46(4)	27(4)
Ethylene glycol monomethyl ether acetate	43(100)	45(48)	58(42)	29(10)	42(4)	31(4)	73(3)	27(3)	59(2)	26(2)
Ethyl formate	31(100)	28(73)	27(51)	29(38)	45(34)	26(17)	74(11)	43(9)	47(8)	56(4)
Furan	68(100)	39(64)	40(9)	38(9)	42(6)	29(6)	37(5)	69(4)	34(2)	67(1)
Furfural	39(100)	96(55)	95(52)	38(38)	29(35)	37(29)	40(11)	97(9)	50(7)	42(7)
Furfuryl alcohol	98(100)	41(65)	39(59)	81(55)	53(53)	97(51)	42(49)	69(39)	70(36)	45(10)
Glycerol	61(100)	43(90)	31(57)	44(54)	29(38)	18(32)	27(12)	42(11)	60(10)	70(18)
Heptane	43(100)	41(56)	29(49)	57(47)	71(45)	56(27)	42(26)	39(23)	70(18)	69(41)
1-Heptanol	41(100)	70(87)	56(86)	31(78)	29(70)	55(67)	27(65)	42(54)	39(27)	86(14)
Hexane	57(100)	43(78)	41(77)	29(61)	27(57)	56(45)	42(39)	39(27)	28(16)	69(27)
1-Hexanol (Caproyl alcohol)	56(100)	43(78)	31(74)	41(71)	27(64)	29(59)	55(58)	42(53)	39(37)	31(10)
Hexylene glycol	59(100)	43(61)	56(25)	45(17)	41(16)	57(13)	42(13)	85(11)	61(10)	55(17)
Hexyl methyl ketone	43(100)	58(79)	41(56)	59(52)	71(49)	27(46)	29(36)	39(27)	57(18)	86(2)
Isobutyl acetate	43(100)	56(26)	73(15)	41(10)	29(5)	71(3)	57(3)	39(3)	27(3)	86(2)
Isobutyl alcohol	43(100)	33(73)	31(72)	41(66)	42(60)	27(43)	29(18)	39(17)	28(8)	74(6)
Isobutylamine	30(100)	28(9)	41(6)	73(5)	27(5)	39(4)	29(3)	15(3)	58(2)	56(2)
Isopentyl acetate	43(100)	70(49)	55(38)	61(15)	42(15)	41(14)	27(12)	87(11)	29(10)	73(9)
Isophorone	82(100)	39(20)	138(17)	54(13)	27(12)	41(10)	53(8)	83(7)	29(7)	55(6)
Isopropyl acetate	43(100)	61(17)	41(14)	87(9)	59(8)	27(8)	42(7)	39(4)	45(3)	44(2)
Isopropyl alcohol	45(100)	43(19)	27(17)	29(12)	41(7)	31(6)	19(6)	42(5)	44(4)	59(3)
Isoquinoline	129(100)	102(26)	51(20)	128(18)	50(11)	130(10)	75(10)	76(9)	103(8)	74(7)
d-Limonene (Citrene)	68(100)	93(50)	67(49)	41(22)	94(21)	79(21)	39(21)	136(20)	53(19)	121(16)
2,6-Lutidine (2,6-Dimethylpyridine)	107(100)	39(39)	106(29)	66(22)	92(18)	65(18)	38(12)	27(11)	79(9)	63(9)

MASS SPECTRAL PEAKS OF COMMON ORGANIC SOLVENTS (continued)

e/m (intensity)

Compound										
Mesitylene (1,3,5-Trimethyl-benzene)	105(100)	120(64)	119(15)	77(13)	39(11)	106(9)	91(9)	51(8)	27(7)	121(6)
Mesityl oxide	55(100)	83(89)	43(73)	29(42)	98(36)	39(32)	27(28)	53(11)	41(10)	56(5)
Methyl acetate	43(100)	74(52)	42(19)	59(17)	44(8)	32(8)	29(6)	31(4)		75(2)
Methylal (Dimethoxymethane)	45(100)	75(61)	29(59)	31(13)	30(6)	15(6)	47(5)	76(2)	46(2)	44(2)
Methyl alcohol	31(100)	32(67)	15(42)	30(6)	14(10)	28(12)	30(9)	13(6)	12(3)	16(2)
Methylamine	30(100)	31(87)	28(56)	29(19)	32(15)	15(12)	27(9)			
Methyl benzoate	105(100)	77(81)	51(45)	136(24)	50(18)	139(5)	106(8)	78(6)	39(5)	27(5)
Methylcyclohexane	83(100)	55(82)	41(60)	98(44)	42(35)	56(30)	27(29)	39(27)	69(23)	70(22)
Methyl ethyl ketone	43(100)	72(24)	29(19)	27(12)	57(7)	42(5)	26(4)	28(3)	44(2)	39(2)
N-Methylformamide	59(100)	30(54)	28(34)	29(13)	58(8)	15(7)	60(3)	41(3)	28(3)	31(2)
Methyl formate	31(100)	29(63)	32(34)	60(28)	30(7)	28(7)	44(2)	18(2)	61(1)	59(1)
Methyl iodide (Iodomethane)	142(100)	127(38)	141(14)	15(13)	139(5)	140(4)	128(3)	14(1)	13(1)	71(0)
Methyl isobutyl ketone	43(100)	58(84)	29(65)	41(55)	57(44)	27(42)	39(31)	85(19)	100(14)	42(14)
Methyl isopentyl ketone	43(100)	58(34)	27(14)	41(13)	15(13)	57(11)	39(9)	71(8)	59(8)	29(8)
2-Methylpentane	43(100)	42(53)	41(35)	27(31)	71(29)	39(20)	29(18)	57(11)	15(10)	70(7)
4-Methyl-2-pentanol	45(100)	43(47)	69(30)	41(27)	27(19)	39(13)	29(12)	87(11)	84(10)	57(10)
Methyl pentyl ketone	43(100)	58(60)	71(14)	41(11)	27(11)	59(9)	39(8)	29(8)	42(5)	114(4)
Methyl propyl ketone	43(100)	41(17)	86(12)	42(12)	27(11)	39(8)	71(7)	58(7)	45(7)	44(3)
N-Methyl-2-pyrrolidone	99(100)	44(89)	98(80)	42(60)	41(38)	43(17)	28(17)	71(13)	39(11)	70(10)
Morpholine	57(100)	29(100)	87(69)	23(69)	30(38)	56(33)	86(28)	31(28)	27(12)	15(7)
Nitrobenzene	77(100)	51(59)	123(42)	50(25)	30(15)	65(14)	39(10)	93(9)	74(7)	78(6)
Nitroethane	29(100)	30(12)	28(11)	26(9)	27(8)	43(5)	41(5)	14(5)	15(3)	46(2)
Nitromethane	30(100)	61(64)	46(39)	28(30)	45(8)	44(7)	29(7)	60(5)	43(4)	
1-Nitropropane	43(100)	27(93)	41(90)	39(34)	30(25)	44(20)	42(20)	26(20)	28(13)	54(12)
2-Nitropropane	43(100)	41(73)	27(71)	39(30)	30(18)	15(11)	42(9)	42(9)	26(8)	38(6)
Octane	43(100)	57(30)	85(25)	41(25)	71(19)	29(17)	56(14)	70(10)	42(10)	27(10)
1-Octanol	41(100)	56(85)	55(81)	31(69)	27(69)	29(68)	42(62)	70(53)	70(10)	69(48)
Pentachloroethane	167(100)	165(91)	117(90)	119(89)	83(58)	169(54)	130(43)	132(42)	60(40)	85(37)
Pentamethylene glycol (1,5-Pentanediol)	31(100)	56(85)								
Pentane	43(100)	42(55)	41(67)	57(59)	55(51)	44(45)	29(37)	43(31)	68(29)	27(26)
1-Pentanol (Amyl alcohol)	42(100)	70(72)	41(45)	27(42)	29(26)	39(19)	57(13)	28(9)	15(9)	72(8)
Pentyl acetate (Amyl acetate)	43(100)	70(90)	55(65)	41(56)	31(47)	29(41)	27(26)	57(22)	28(22)	43(21)
2-Picoline (2-Methylpyridine)	93(100)	66(41)	42(52)	28(51)	61(50)	55(41)	73(21)	41(20)	29(14)	69(11)
α-Pinene	93(100)	92(30)	39(31)	92(20)	78(19)	51(19)	65(16)	38(13)	50(12)	52(11)
β-Pinene	93(100)	41(64)	39(24)	41(23)	77(22)	91(21)	27(21)	79(18)	121(13)	53(10)
Piperidine (Hexahydropyridine)	84(100)	85(53)	69(47)	39(33)	27(31)	79(20)	91(21)	77(18)	94(13)	91(13)
Propanenitrile	28(100)	54(63)	56(46)	57(43)	28(41)	29(37)	44(34)	42(30)	30(30)	43(25)
Propyl acetate	43(100)	61(19)	26(20)	27(17)	52(11)	55(10)	51(9)	15(9)	53(7)	25(7)
Propyl alcohol	31(100)	27(19)	31(18)	27(15)	42(11)	73(9)	41(9)	29(9)	59(5)	39(5)
Propylamine	30(100)	28(13)	29(18)	59(11)	42(3)	60(7)	41(7)	28(7)	43(3)	32(3)
Propylbenzene	91(100)	120(21)	59(8)	27(7)	41(5)	42(3)	39(3)	29(3)	26(3)	18(3)
1,2-Propylene glycol	45(100)	18(46)	92(10)	38(10)	65(9)	78(6)	51(6)	27(5)	63(4)	105(3)

MASS SPECTRAL PEAKS OF COMMON ORGANIC SOLVENTS (continued)

e/m (intensity)

Compound										
Pseudocumene (1,2,4-Trimethyl-benzene)	105(100)	120(56)	119(17)	77(15)	39(15)	51(11)	91(10)	27(10)	106(9)	79(7)
Pyridine	79(100)	52(62)	51(31)	50(19)	78(11)	53(7)	39(7)	80(6)	27(3)	77(2)
Pyrrole	67(100)	41(58)	39(58)	40(51)	28(42)	38(20)	37(12)	66(7)	68(5)	27(3)
Pyrrolidine	43(100)	28(52)	70(33)	71(26)	42(22)	41(20)	27(16)	39(15)	29(10)	30(9)
2-Pyrrolidone	85(100)	42(43)	41(36)	28(33)	30(29)	56(16)	84(14)	40(12)	27(12)	29(9)
Quinoline	129(100)	51(28)	76(25)	128(24)	44(24)	50(20)	32(19)	75(18)	74(12)	103(11)
Styrene	104(100)	103(41)	78(32)	51(28)	77(23)	105(12)	50(12)	52(11)	39(11)	102(10)
Sulfolane	41(100)	28(94)	56(82)	55(72)	120(37)	27(32)	39(19)	29(17)	26(11)	48(5)
α-Terpinene	121(100)	93(85)	136(43)	91(40)	77(34)	39(33)	27(33)	79(27)	41(26)	43(18)
1,1,1,2-Tetrachloro-2,2-difluoroethane	167(100)	169(96)	117(85)	119(82)	171(31)	85(29)	121(26)	82(14)	47(14)	101(13)
Tetrachloro-1,2-difluoroethane	101(100)	103(64)	167(54)	169(52)	117(19)	119(18)	171(17)	105(11)	31(11)	132(9)
1,1,1,2-Tetrachloroethane	131(100)	133(96)	117(76)	119(73)	95(34)	135(31)	131(23)	97(23)	61(19)	60(18)
1,1,2,2-Tetrachloroethane	83(100)	85(63)	95(11)	87(10)	168(8)	133(8)	131(8)	96(8)	61(8)	60(8)
Tetrachloroethylene	166(100)	164(82)	131(71)	129(71)	168(45)	94(38)	47(31)	96(24)	133(20)	59(17)
Tetraethylene glycol	45(100)	89(10)	44(8)	43(6)	31(6)	29(6)	27(6)	101(5)	75(5)	28(5)
Tetrahydrofuran	42(100)	41(52)	27(33)	72(29)	71(27)	39(24)	43(22)	29(22)	40(13)	15(10)
1,2,3,4-Tetrahydronaphthalene	104(100)	132(53)	91(43)	51(17)	39(17)	131(15)	117(15)	115(14)	78(13)	77(13)
Tetrahydropyran	41(100)	28(64)	56(57)	45(57)	29(51)	27(49)	85(47)	86(42)	39(28)	55(23)
Tetramethylsilane	73(100)	43(14)	45(12)	74(8)	29(7)	15(5)	75(4)	44(4)	42(4)	31(4)
Toluene	91(100)	92(73)	39(20)	65(14)	63(11)	51(11)	50(7)	27(6)	93(5)	90(5)
o-Toluidine	106(100)	107(83)	77(17)	79(13)	39(12)	53(10)	52(10)	54(9)	51(9)	28(9)
Triacetin	43(100)	103(44)	145(34)	116(17)	115(13)	44(10)	86(9)	28(8)	73(7)	42(7)
Tributylamine	142(100)	100(19)	143(11)	29(8)	185(7)	57(6)	44(6)	41(6)	30(5)	86(4)
1,1,1-Trichloroethane	97(100)	99(64)	61(58)	26(31)	27(24)	117(19)	63(19)	119(18)	35(17)	62(11)
1,1,2-Trichloroethane	97(100)	83(95)	99(62)	85(60)	61(58)	26(23)	96(21)	63(19)	27(17)	98(15)
Trichloroethylene	95(100)	130(90)	132(85)	60(65)	97(64)	35(40)	134(27)	47(26)	62(21)	59(13)
Trichlorofluoromethane	101(100)	103(66)	66(13)	105(11)	35(11)	47(9)	31(8)	82(4)	68(4)	37(4)
1,1,2-Trichlorotrifluoroethane	101(100)	151(68)	103(64)	85(45)	31(45)	153(44)	35(20)	66(19)	47(18)	87(14)
Triethanolamine	118(100)	56(69)	45(60)	42(56)	44(27)	43(25)	41(14)	116(8)	57(8)	86(7)
Triethylamine	86(100)	30(68)	58(37)	28(24)	29(23)	27(19)	44(18)	101(17)	42(16)	56(8)
Triethylene glycol	45(100)	58(11)	89(9)	31(8)	29(8)	75(7)	44(7)	43(7)	27(7)	28(5)
Triethyl phosphate	99(100)	81(71)	155(56)	82(45)	45(45)	109(44)	127(41)	43(24)	125(16)	111(14)
Trimethylamine	58(100)	59(47)	30(29)	42(26)	44(17)	15(14)	28(10)	18(10)	43(8)	57(7)
Trimethylene glycol (1,3-Propanediol)	28(100)	58(93)	31(76)	57(70)	29(40)	27(26)	45(24)	43(23)	19(18)	30(17)
Trimethyl phosphate	110(100)	109(35)	79(34)	95(25)	80(23)	15(20)	140(18)	47(10)	31(7)	139(5)
Veratrole	138(100)	95(65)	77(48)	123(44)	52(42)	41(33)	65(30)	51(29)	39(19)	63(17)
o-Xylene	91(100)	106(40)	39(21)	105(17)	51(17)	77(15)	27(12)	65(10)	92(8)	79(8)
m-Xylene	91(100)	106(65)	105(29)	39(18)	51(15)	77(14)	27(10)	92(8)	79(8)	78(8)
p-Xylene	91(100)	106(62)	105(30)	51(16)	39(16)	77(13)	27(11)	92(7)	78(7)	65(7)

Section 9
Molecular Structure and Spectroscopy

BOND LENGTHS IN CRYSTALLINE ORGANIC COMPOUNDS

The following table gives average interatomic distances for bonds between the elements H, B, C, N, O, F, Si, P, S, Cl, As, Se, Br, Te, and I as determined from X-ray and neutron diffraction measurements on organic crystals. The table has been derived from an analysis of high-precision structure data on about 10,000 crystals contained in the 1985 version of the Cambridge Structural Database, which is maintained by the Cambridge Crystallographic Data Center. The explanation of the columns is:

Column 1: Specification of elements in the bond, with coordination number given in parentheses, and bond type (single, double, etc.). For carbon, the hybridization state is given.

Column 2: Substructure in which the bond is found. The target bond is set in boldface. Where X is not specified, it denotes any element type. C# indicates any sp^3 carbon atom, and C* denotes an sp^3 carbon whose bonds, in addition to those specified in the linear formulation, are to C and H atoms only.

Column 3: d is the unweighted mean in Å units of all the values for that bond length found in the sample.

Column 4: m is the median in Å units of all values.

Column 5: σ is the standard deviation in the sample.

Column 6: q_1 is the lower quartile for the sample (i.e., 25% of values are less than q_1 and 75% exceed it).

Column 7: q_2 is the upper quartile for the sample.

Column 8: n is number of observations in the sample.

Column 9: Notes refer to the footnotes in Appendix 1.

References to special cases are given in a shorthand form and listed in Appendix 2. Further information on the method of analysis of the data may be found in the reference cited below.

The table is reprinted with permission of the authors, the Royal Society of Chemistry, and the International Union of Crystallography.

REFERENCE

Frank H. Allen, Olga Kennard, David G. Watson, Lee Brammer, A. Guy Orpen, and Robin Taylor, *J. Chem. Soc. Perkin Trans. II*, S1—S19, 1987.

Bond	Substructure	d	m	σ	q_1	q_u	n	Note
As(3)–As(3)	X₂–**As–As**–X₂	2.459	2.457	0.011	2.456	2.466	8	
As–B	see CUDLOC (2.065), CUDLUI (2.041)							
As–Br	see CODDEE, CODDII (2.346—3.203)							
As(4)–C	X₃–**As–C**H₃	1.903	1.907	0.016	1.893	1.916	12	
	(X)₂(C,O,S=)**As–C**sp^3	1.927	1.929	0.017	1.921	1.937	16	
	As–Car in Ph₄As⁺	1.905	1.909	0.012	1.897	1.912	108	
	(X)₂(C,O,S=)**As–C**ar	1.922	1.927	0.016	1.908	1.934	36	
As(3)–C	X₂–**As–C**sp^3	1.963	1.965	0.017	1.948	1.978	6	
	X₂–**As–C**ar	1.956	1.956	0.015	1.944	1.964	41	
As(3)–Cl	X₂–**As–Cl**	2.268	2.256	0.039	2.247	2.281	10	
As(6)–F	in **As**F₆⁻	1.678	1.676	0.020	1.659	1.695	36	
As(3)–I	see OPIMAS (2.579, 2.590)							
As(3)–N(3)	X₂–**As–N**–X₂	1.858	1.858	0.029	1.839	1.873	19	
As(4)=N(2)	see TPASSN (1.837)							
As(4)–O	(X)₂(O=)**As–O**H	1.710	1.712	0.017	1.695	1.726	6	
As(3)–O	see ASAZOC, PHASOC01 (1.787—1.845)							
As(4)=O	X₃–**As=O**	1.661	1.661	0.016	1.652	1.667	9	
As(3)=P(3)	see BELNIP (2.350, 2.362)							†
As(3)–P(3)	see BUTHAZ10 (2.124)							†
As(3)–S	X₂–**As–S**	2.275	2.266	0.032	2.247	2.298	14	
As(4)=S	X₃–**As=S**	2.083	2.082	0.004	2.080	2.086	9	
As(3)–Se(2)	see COSDIX, ESEARS (2.355—2.401)							†
As(3)–Si(4)	see BICGEZ, MESIAD (2.351—2.365)							†
As(3)–Te(2)	see ETEARS (2.571, 2.576)							†
B(n)–B(n)	$n = 5$—7 in boron cages	1.775	1.773	0.031	1.763	1.786	688	
B(4)–B(4)	see CETTAW (2.041)							
B(4)–B(3)	see COFVOI (1.698)							
B(3)–B(3)	X₂–**B–B**–X₂	1.701	1.700	0.014	1.691	1.712	8	
B(6)–Br		1.967	1.971	0.014	1.954	1.979	7	†
B(4)–Br		2.017	2.008	0.031	1.990	2.044	15	†
B(n)–C	$n = 5$—7: **B–C** in cages	1.716	1.717	0.020	1.707	1.728	96	
	$n = 3$—4: **B–C**sp^3 not cages	1.597	1.599	0.022	1.585	1.611	29	1
	$n = 4$: **B–C**ar	1.606	1.607	0.012	1.596	1.615	41	
	$n = 4$: **B–C**ar in Ph₄B⁻	1.643	1.643	0.006	1.641	1.645	16	
B(n)–C	$n = 3$: **B–C**ar	1.556	1.552	0.015	1.546	1.566	24	
B(n)–Cl	**B**(5)–**Cl** and **B**(3)–**Cl**	1.751	1.751	0.011	1.743	1.761	14	
	B(4)–**Cl**	1.833	1.833	0.013	1.821	1.843	22	
B(4)–F	**B–F** (B neutral)	1.366	1.368	0.017	1.356	1.375	25	
	B⁻–**F** in BF₄⁻	1.365	1.372	0.029	1.352	1.390	84	
B(4)–I	see TMPBTI (2.220, 2.253)							
B(4)–N(3)	X₃–**B–N**(=C)(X)	1.611	1.617	0.013	1.601	1.625	8	
	in pyrazaboles	1.549	1.552	0.015	1.536	1.560	10	

Bond	Substructure	d	m	σ	q_1	q_u	n	Note
B(3)–N(3)	X_2–**B**–**N**–C_2: all coplanar	1.404	1.404	0.014	1.389	1.408	40	2
	for $\tau(BN) > 30°$ see BOGSUL, BUSHAY, CILRUK (1.434—1.530)							
	S_2–**B**–**N**–X_2	1.447	1.443	0.013	1.435	1.470	14	
B(4)–O	**B**$^-$–**O** in BO_4^-	1.468	1.468	0.022	1.453	1.479	24	
	for neutral B–O see Note 3							3
B(3)–O(2)	X_2–**B**–**O**–X	1.367	1.367	0.024	1.349	1.382	35	
B(n)–P	$n = 4$: **B**–**P**	1.922	1.927	0.027	1.900	1.954	10	
	$n = 3$: see BUPSIB10 (1.892, 1.893)							
B(4)–S	**B(4)**–**S(3)**	1.930	1.927	0.009	1.925	1.934	10	
	B(4)–**S(2)**	1.896	1.896	0.004	1.893	1.899	6	
B(3)–S	**N**–**B**–S_2	1.806	1.806	0.010	1.799	1.816	28	
	(=X–)(N–)**B**–**S**	1.851	1.854	0.013	1.842	1.859	10	
Br–Br	see BEPZEB, TPASTB	2.542	2.548	0.015	2.526	2.551	4	
Br–C	**Br**–**C***	1.966	1.967	0.029	1.951	1.983	100	4
	Br–**C**sp^3 (cyclopropane)	1.910	1.910	0.010	1.900	1.914	8	
	Br–**C**sp^2	1.883	1.881	0.015	1.874	1.894	31	4
	Br–**C**ar (mono-Br + m,p-Br$_2$)	1.899	1.899	0.012	1.892	1.906	119	4
	Br–**C**ar (o-Br$_2$)	1.875	1.872	0.011	1.864	1.884	8	4
$^-$Br(2)–Cl	see TEACBR (2.362—2.402)							†
Br–I	see DTHIBR10 (2.646), TPHOSI (2.695)							
Br–N	see NBBZAM (1.843)							
Br–O	see CIYFOF	1.581	1.581	0.007	1.574	1.587	4	
Br–P	see CISTED (2.366)							
Br–S(2)	see BEMLIO (2.206)							†
Br–S(3)	see CIWYIQ (2.435, 2.453)							†
Br–S(3)$^+$	see THINBR (2.321)							†
Br–Se	see CIFZUM (2.508, 2.619)							
Br–Si	see BIZJAV (2.284)							
Br–Te	In **Br**$_6$**Te**$^{2-}$ see CUGBAH (2.692—2.716)							
	Br–**Te(4)** see BETUTE10 (3.079, 3.015)							
	Br–**Te(3)** see BTUPTE (2.835)							
Csp^3–Csp^3	C#–**CH**$_2$–**CH**$_3$	1.513	1.514	0.014	1.507	1.523	192	
	(C#)$_2$–**CH**–**CH**$_3$	1.524	1.526	0.015	1.518	1.534	226	
	(C#)$_3$–**C**–**CH**$_3$	1.534	1.534	0.011	1.527	1.541	825	
	C#–**CH**$_2$–**CH**$_2$–C#	1.524	1.524	0.014	1.516	1.532	2 459	
	(C#)$_2$–**CH**–**CH**$_2$–C#	1.531	1.531	0.012	1.524	1.538	1 217	
	(C#)$_3$–**C**–**CH**$_2$–C#	1.538	1.539	0.010	1.533	1.544	330	
	(C#)$_2$–**CH**–**CH**–(C#)$_2$	1.542	1.542	0.011	1.536	1.549	321	
	(C#)$_3$–**C**–**CH**–(C#)$_2$	1.556	1.556	0.011	1.549	1.562	215	
	(C#)$_3$–**C**–**C**–(C#)$_3$	1.588	1.580	0.025	1.566	1.610	21	
	C*–**C*** (overall)	1.530	1.530	0.015	1.521	1.539	5 777	5,6
	in cyclopropane (any subst.)	1.510	1.509	0.026	1.497	1.523	888	7
	in cyclobutane (any subst.)	1.554	1.553	0.021	1.540	1.567	679	8
	in cyclopentane (C,H-subst.)	1.543	1.543	0.018	1.532	1.554	1 641	
	in cyclohexane (C,H-subst.)	1.535	1.535	0.016	1.525	1.545	2 814	
	cyclopropyl-**C*** (exocyclic)	1.518	1.518	0.019	1.505	1.531	366	7
	cyclobutyl-**C*** (exocyclic)	1.529	1.529	0.016	1.519	1.539	376	8
	cyclopentyl-**C*** (exocyclic)	1.540	1.541	0.017	1.527	1.549	956	
	cyclohexyl-**C*** (exocyclic)	1.539	1.538	0.016	1.529	1.549	2 682	
	in cyclobutene (any subst.)	1.573	1.574	0.017	1.566	1.586	25	8
	in cyclopentene (C,H-subst.)	1.541	1.539	0.015	1.532	1.549	208	
	in cyclohexene (C,H-subst.)	1.541	1.541	0.020	1.528	1.554	586	
	in oxirane (epoxide)	1.466	1.466	0.015	1.458	1.474	249	9
	in aziridine	1.480	1.481	0.021	1.465	1.496	67	9
	in oxetane	1.541	1.541	0.019	1.527	1.557	16	
	in azetidine	1.548	1.543	0.018	1.536	1.558	22	
	oxiranyl-**C*** (exocyclic)	1.509	1.507	0.018	1.497	1.519	333	9
	aziridinyl-**C*** (exocyclic)	1.512	1.512	0.018	1.496	1.526	13	9
Csp^3–Csp^2	**CH**$_3$–**C**=C	1.503	1.504	0.011	1.497	1.509	215	
	C#–**CH**$_2$–**C**=C	1.502	1.502	0.013	1.494	1.510	483	
	(C#)$_2$–**CH**–**C**=C	1.510	1.510	0.014	1.501	1.518	564	
	(C#)$_3$–**C**–**C**=C	1.522	1.522	0.016	1.511	1.533	193	
Csp^3–Csp^2	**C***–**C**=C (overall)	1.507	1.507	0.015	1.499	1.517	1 456	5
	C*–**C**=C (endocyclic)							
	in cyclopropene	1.509	1.508	0.016	1.500	1.516	20	10
	in cyclobutene	1.513	1.512	0.018	1.500	1.525	50	8
	in cyclopentene	1.512	1.512	0.014	1.502	1.521	208	
	in cyclohexene	1.506	1.505	0.016	1.495	1.516	391	
	in cyclopentadiene	1.502	1.503	0.019	1.490	1.515	18	
	in cyclohexa-1,3-diene	1.504	1.504	0.017	1.491	1.517	56	
	C*–**C**=C (exocyclic):							
	cyclopropenyl-**C***	1.478	1.475	0.012	1.470	1.485	7	10
	cyclobutenyl-**C***	1.489	1.483	0.015	1.479	1.496	11	8

Bond	Substructure	d	m	σ	q_1	q_u	n	Note
	cyclopentenyl-C*	1.504	1.506	0.012	1.495	1.512	115	
	cyclohexenyl-C*	1.511	1.511	0.013	1.502	1.519	292	
	C*–CH=O in aldehydes	1.510	1.510	0.008	1.501	1.518	7	
	(C*)₂–C=O							
	in ketones	1.511	1.511	0.015	1.501	1.521	952	11
	in cyclobutanone	1.529	1.530	0.016	1.514	1.545	18	
	in cyclopentanone	1.514	1.514	0.016	1.505	1.523	312	
	acyclic and 6+ rings	1.509	1.509	0.016	1.499	1.519	626	
	C*–COOH in carboxylic acids	1.502	1.502	0.014	1.495	1.510	176	
	C*–COO⁻ in carboxylate anions	1.520	1.521	0.011	1.516	1.528	57	
	C*–C(=O)(–OC*)							
	in acyclic esters	1.497	1.496	0.018	1.484	1.509	553	12
	in β-lactones	1.519	1.519	0.020	1.500	1.538	4	13
	in γ-lactones	1.512	1.512	0.015	1.501	1.521	110	12
	in δ-lactones	1.504	1.502	0.013	1.495	1.517	27	12
	cyclopropyl (C)–C=O in ketones, acids							
	and esters	1.486	1.485	0.018	1.474	1.497	105	7
	C*–C(=O)(–NH₂) in acyclic amides	1.514	1.512	0.016	1.506	1.526	32	14
	C*–C(=O)(–NHC*) in acyclic amides	1.506	1.505	0.012	1.498	1.515	78	14
	C*–C(=O)[–N(C*)₂] in acyclic amides	1.505	1.505	0.011	1.496	1.517	15	14
Csp³–Car	CH₃–Car	1.506	1.507	0.011	1.501	1.513	454	
	C#–CH₂–Car	1.510	1.510	0.009	1.505	1.516	674	
	(C#)₂–CH–Car	1.515	1.515	0.011	1.508	1.522	363	
	(C#)₃–C–Car	1.527	1.530	0.016	1.517	1.539	308	
	C*–Car (overall)	1.513	1.513	0.014	1.505	1.521	1 813	
	cyclopropyl (C)–Car	1.490	1.490	0.015	1.479	1.503	90	7
Csp³–Csp¹	C*–C≡C	1.466	1.465	0.010	1.460	1.469	21	15
	C#–C≡C	1.472	1.472	0.012	1.464	1.481	88	15
	C*–C≡N	1.470	1.469	0.013	1.463	1.479	106	7b
	cyclopropyl (C)–C≡N	1.444	1.447	0.010	1.436	1.451	38	7
Csp²–Csp²	C=C–C=C							
	(conjugated)	1.455	1.455	0.011	1.447	1.463	30	16,18
	(unconjugated)	1.478	1.476	0.012	1.470	1.479	8	17,18
	(overall)	1.460	1.460	0.015	1.450	1.470	38	
	C=C–C=C–C=C	1.443	1.445	0.013	1.431	1.454	29	18
	C=C–C=C (endocyclic in TCNQ)	1.432	1.433	0.012	1.424	1.441	280	19
	C=C–C(=O)(–C*)							
	(conjugated)	1.464	1.462	0.018	1.453	1.476	211	16,18
	(unconjugated)	1.484	1.486	0.017	1.475	1.497	14	17,18
	(overall)	1.465	1.462	0.018	1.453	1.478	226	
	C=C–C(=O)–C=C							
	in benzoquinone (C,H-subst. only)	1.478	1.476	0.011	1.469	1.488	28	
	in benzoquinone (any subst.)	1.478	1.478	0.031	1.464	1.498	172	
	non-quinonoid	1.456	1.455	0.012	1.447	1.464	28	
	C=C–COOH	1.475	1.476	0.015	1.461	1.488	22	
	C=C–COOC*	1.488	1.489	0.014	1.478	1.497	113	
	C=C–COO⁻	1.502	1.499	0.017	1.488	1.510	11	
	HOOC–COOH	1.538	1.537	0.007	1.535	1.541	9	
	HOOC–COO⁻	1.549	1.552	0.009	1.546	1.553	13	
	⁻OOC–COO⁻	1.564	1.559	0.022	1.554	1.568	9	
	formal Csp²–Csp² single bond in selected							
	non-fused heterocycles:							
	in 1H-pyrrole (C3–C4)	1.412	1.410	0.016	1.401	1.427	29	
	in furan (C3–C4)	1.423	1.423	0.016	1.412	1.433	62	
	in thiophene (C3–C4)	1.424	1.425	0.015	1.415	1.433	40	
	in pyrazole (C3–C4)	1.410	1.412	0.016	1.400	1.418	20	
	in isoxazole (C3–C4)	1.425	1.425	0.016	1.413	1.438	9	
	in furazan (C3–C4)	1.428	1.427	0.007	1.422	1.435	6	
	in furoxan (C3–C4)	1.417	1.417	0.006	1.412	1.422	14	
Csp²–Car	C=C–Car							
	(conjugated)	1.470	1.470	0.015	1.463	1.480	37	16,18
Csp²–Car		1.488	1.490	0.012	1.480	1.496	87	17,18
	(overall)	1.483	1.483	0.015	1.472	1.494	124	
	cyclopropenyl (C=C)–Car	1.447	1.448	0.006	1.441	1.452	8	10
	Car–C(=O)–C*	1.488	1.489	0.016	1.478	1.500	84	
	Car–C(=O)–Car	1.480	1.481	0.017	1.468	1.494	58	
	Car–COOH	1.484	1.485	0.014	1.474	1.491	75	
	Car–C(=O)(–OC*)	1.487	1.487	0.012	1.480	1.494	218	
	Car–COO⁻	1.504	1.509	0.014	1.495	1.512	26	
	Car–C(=O)–NH₂	1.500	1.503	0.020	1.498	1.510	19	
	Car–C=N–C#							
	(conjugated)	1.476	1.478	0.014	1.466	1.486	27	16
	(unconjugated)	1.491	1.490	0.008	1.485	1.496	48	17
	(overall)	1.485	1.487	0.013	1.481	1.493	75	

Bond	Substructure	d	m	σ	q_1	q_u	n	Note
	in indole (C3–C3a)	1.434	1.434	0.011	1.428	1.439	40	
Csp^2–Csp^1	C=C–C≡C	1.431	1.427	0.014	1.425	1.441	11	7b
	C=C–C≡N in TCNQ	1.427	1.427	0.010	1.420	1.433	280	19
Car–Car	in biphenyls (*ortho* subst. all H)	1.487	1.488	0.007	1.484	1.493	30	
	(⩾1 non-H *ortho*-subst.)	1.490	1.491	0.010	1.486	1.495	212	
Car–Csp^1	Car–C≡C	1.434	1.436	0.006	1.430	1.437	37	
	Car–C≡N	1.443	1.444	0.008	1.436	1.448	31	
Csp^1–Csp^1	C≡C–C≡C	1.377	1.378	0.012	1.374	1.384	21	
Csp^2=Csp^2	C*–CH=CH$_2$	1.299	1.300	0.027	1.280	1.311	42	
	(C*)$_2$–C=CH$_2$	1.321	1.321	0.013	1.313	1.328	77	
	C*–CH=CH–C*							
	(cis)	1.317	1.318	0.013	1.310	1.323	106	
	(trans)	1.312	1.311	0.011	1.304	1.320	19	
	(overall)	1.316	1.317	0.015	1.309	1.323	127	
	(C*)$_2$–C=CH–C*	1.326	1.328	0.011	1.319	1.334	168	
	(C*)$_2$–C=C–(C*)$_2$	1.331	1.330	0.009	1.326	1.334	89	
	(C*,H)$_2$–C=C–(C*,H)$_2$ (overall)	1.322	1.323	0.014	1.315	1.331	493	5
	in cyclopropene (any subst.)	1.294	1.288	0.017	1.284	1.302	10	10
	in cyclobutene (any subst.)	1.335	1.335	0.019	1.324	1.347	25	8
	in cyclopentene (C,H-subst.)	1.323	1.324	0.013	1.314	1.331	104	
	in cyclohexene (C,H-subst.)	1.326	1.325	0.012	1.318	1.334	196	
	C=C=C (allenes, any subst.)	1.307	1.307	0.005	1.303	1.310	18	
	C=C–C=C (C,H subst., conjugated)	1.330	1.330	0.014	1.322	1.338	76	16
	C=C–C=C–C=C (C,H subst., conjugated)	1.345	1.345	0.012	1.337	1.350	58	16
	C=C–Car (C,H subst., conjugated)	1.339	1.340	0.011	1.334	1.346	124	16
	C=C in cyclopenta-1,3-diene (any subst.)	1.341	1.341	0.017	1.328	1.356	18	
	C=C in cyclohexa-1,3-diene (any subst.)	1.332	1.332	0.013	1.323	1.341	56	
	in C=C–C=O							
	(C,H subst., conjugated)	1.340	1.340	0.013	1.332	1.348	211	16,18
	(C,H subst., unconjugated)	1.331	1.330	0.008	1.326	1.339	14	17,18
	(C,H subst., overall)	1.340	1.339	0.013	1.332	1.348	226	
	in cyclohexa-2,5-dien-1-ones	1.329	1.327	0.011	1.321	1.335	28	
	in *p*-benzoquinones							
	(C*,H subst.)	1.333	1.337	0.011	1.325	1.338	14	
	(any subst.)	1.349	1.339	0.030	1.330	1.364	86	
	in TCNQ							
	(endocyclic)	1.352	1.353	0.010	1.345	1.358	142	19
	(exocyclic)	1.392	1.391	0.017	1.379	1.405	139	19
	C=C–OH in enol tautomers	1.362	1.360	0.020	1.349	1.370	54	
	in heterocycles (any subst.):							
	1*H*-pyrrole (C2–C3, C4–C5)	1.375	1.377	0.018	1.361	1.388	58	
	furan (C2–C3, C4–C5)	1.341	1.342	0.021	1.329	1.351	125	
	thiophene (C2–C3, C4–C5)	1.362	1.359	0.025	1.346	1.377	60	
	pyrazole (C4–C5)	1.369	1.372	0.019	1.362	1.383	20	
	imidazole (C4–C5)	1.360	1.361	0.014	1.352	1.367	44	
	isoxazole (C4–C5)	1.341	1.336	0.012	1.331	1.355	9	
	indole (C2–C3)	1.364	1.363	0.012	1.355	1.371	40	
Car≃Car	in phenyl rings with C*,H subst. only							
	H–C≃C–H	1.380	1.381	0.013	1.372	1.388	2 191	
	C*–C≃C–H	1.387	1.388	0.010	1.382	1.393	891	
	C*–C≃C–C*	1.397	1.397	0.009	1.392	1.403	182	
	C≃C (overall)	1.384	1.384	0.013	1.375	1.391	3 264	
	F–C≃C–F	1.372	1.374	0.011	1.366	1.380	84	4
	Cl–C≃C–Cl	1.388	1.389	0.014	1.380	1.398	152	4
	in naphthalene (D_{2h}, any subst.)							
	C1–C2	1.364	1.364	0.014	1.356	1.373	440	
	C2–C3	1.406	1.406	0.014	1.397	1.415	218	
	C1–C8a	1.420	1.419	0.012	1.412	1.426	440	
	C4a–C8a	1.422	1.424	0.011	1.417	1.429	109	
Car≃Car	in anthracene (D_{2h}, any subst.)							
	C1–C2	1.356	1.356	0.009	1.350	1.360	56	
	C2–C3	1.410	1.410	0.010	1.401	1.416	34	
	C1–C9a	1.430	1.430	0.006	1.426	1.434	56	
	C4a–C9a	1.435	1.436	0.007	1.429	1.440	34	
	C9–C9a	1.400	1.402	0.009	1.395	1.406	68	
	in pyridine (C,H subst.)	1.379	1.381	0.012	1.371	1.387	276	20
	(any subst.)	1.380	1.380	0.015	1.371	1.389	537	20
	in pyridinium cation							
	(N$^+$–H; C,H subst. on C)							
	C2–C3	1.373	1.375	0.012	1.368	1.380	30	
	C3–C4	1.379	1.380	0.011	1.371	1.388	30	
	(N$^+$–X; C,H subst. on C)							
	C2–C3	1.373	1.372	0.019	1.362	1.382	151	
	C3–C4	1.383	1.385	0.019	1.372	1.394	151	

Bond	Substructure	d	m	σ	q_1	q_u	n	Note
	in pyrazine (H subst. on C)	1.379	1.377	0.010	1.370	1.388	10	
	(any subst. on C)	1.405	1.405	0.024	1.388	1.420	60	
	in pyrimidine (C,H subst. on C)	1.387	1.389	0.018	1.379	1.400	28	
$Csp^1{\equiv}Csp^1$	X–C≡C–X	1.183	1.183	0.014	1.174	1.193	119	15
	C,H–C≡C–C,H	1.181	1.181	0.014	1.173	1.192	104	15
	in C≡C–C(sp^2,ar)	1.189	1.193	0.010	1.181	1.195	38	15
	in C≡C–C≡C	1.192	1.192	0.010	1.187	1.197	42	15
	in CH≡C–C#	1.174	1.174	0.011	1.167	1.180	42	15
Csp^3–Cl	Omitting 1,2-dichlorides:							
	C–CH$_2$–Cl	1.790	1.790	0.007	1.783	1.795	13	4
	C$_2$–CH–Cl	1.803	1.802	0.003	1.800	1.807	8	4
	C$_3$–C–Cl	1.849	1.856	0.011	1.837	1.858	5	4
	X–CH$_2$–Cl (X = C,H,N,O)	1.790	1.791	0.011	1.783	1.797	37	4
	X$_2$–CH–Cl (X = C,H,N,O)	1.805	1.803	0.014	1.800	1.812	26	4
	X$_3$–C–Cl (X = C,H,N,O)	1.843	1.838	0.014	1.835	1.858	7	4
	X$_2$–C–Cl$_2$ (X = C,H,N,O)	1.779	1.776	0.015	1.769	1.790	18	4
	X–C–Cl$_3$ (X = C,H,N,O)	1.768	1.765	0.011	1.761	1.776	33	4
	Cl–CH(–C)–CH(–C)–Cl	1.793	1.793	0.013	1.786	1.800	66	4
	Cl–C(–C$_2$)–C(–C$_2$)–Cl	1.762	1.760	0.010	1.757	1.765	54	4
	cyclopropyl–Cl	1.755	1.756	0.011	1.749	1.763	64	
Csp^2–Cl	C=C–Cl (C,H,N,O subst. on C)	1.734	1.729	0.019	1.719	1.748	63	4
	C=C–Cl$_2$ (C,H,N,O subst. on C)	1.720	1.716	0.013	1.708	1.729	20	4
	Cl–C=C–Cl	1.713	1.711	0.011	1.705	1.720	80	4
Car–Cl	Car–Cl (mono-Cl + m,p-Cl$_2$)	1.739	1.741	0.010	1.734	1.745	340	4
	Car–Cl (o-Cl$_2$)	1.720	1.720	0.010	1.713	1.717	364	4
Csp^1–Cl	see HCLENE10 (1.634, 1.646)							
Csp^3–F	Omitting 1,2-difluorides							
	C–CH$_2$–F and C$_2$-CH–F	1.399	1.399	0.017	1.389	1.408	25	4
	C$_3$–C–F	1.428	1.431	0.009	1.421	1.435	11	4
	(C*,H)$_2$-C–F$_2$	1.349	1.347	0.012	1.342	1.356	58	4
	C*–C–F$_3$	1.336	1.334	0.007	1.330	1.344	12	4
	F–C*–C*–F	1.371	1.374	0.007	1.362	1.375	26	4
	X$_3$–C–F (X = C,H,N,O)	1.386	1.389	0.033	1.373	1.408	70	4
	X$_2$–C–F$_2$ (X = C,H,N,O)	1.351	1.349	0.013	1.342	1.356	58	4
	X–C–F$_3$ (X = C,H,N,O)	1.322	1.323	0.015	1.314	1.332	309	4
	F–C(–X)$_2$–C(–X)$_2$–F (X = C,H,N,O)	1.373	1.374	0.009	1.362	1.377	30	4
	F–C(–X)$_2$–NO$_2$ (X = any subst.)	1.320	1.319	0.009	1.312	1.327	18	
Csp^2–F	C=C–F (C,H,N,O subst. on C)	1.340	1.340	0.013	1.334	1.346	34	4
Car–F	Car–F (mono-F + m,p-F$_2$)	1.363	1.362	0.008	1.357	1.368	38	4
	Car–F (o-F$_2$)	1.340	1.340	0.009	1.336	1.344	167	4
Csp^3–H	C–C–H$_3$ (methyl)	1.059	1.061	0.030	1.039	1.083	83	21
	C$_2$–C–H$_2$ (primary)	1.092	1.095	0.013	1.088	1.099	100	21
	C$_3$–C–H (secondary)	1.099	1.097	0.004	1.095	1.103	14	21
	C$_{2,3}$–C–H (primary and secondary)	1.093	1.095	0.012	1.089	1.100	118	21
	X–C–H$_3$ (methyl)	1.066	1.074	0.028	1.049	1.087	160	21
	X$_2$–C–H$_2$ (primary)	1.092	1.095	0.012	1.088	1.099	230	21
	X$_3$–C–H (secondary)	1.099	1.099	0.007	1.095	1.103	117	21
	X$_{2,3}$–C–H (primary and secondary)	1.094	1.096	0.011	1.091	1.100	348	21
Csp^2–H	C–C=C–H	1.077	1.079	0.012	1.074	1.085	14	21
Car–H	Car–H	1.083	1.083	0.011	1.080	1.087	218	21
Csp^3–I	C*–I	2.162	2.159	0.015	2.149	2.179	15	4
Car–I	Car–I	2.095	2.095	0.015	2.089	2.104	51	4
Csp^3–N(4)	C*–NH$_3{}^+$	1.488	1.488	0.013	1.482	1.495	298	
	(C*)$_2$–NH$_2{}^+$	1.494	1.493	0.016	1.484	1.503	249	
	(C*)$_3$–NH$^+$	1.502	1.502	0.015	1.491	1.512	509	
	(C*)$_4$–N$^+$	1.510	1.509	0.020	1.496	1.523	319	
	C*–N$^+$ (overall)	1.499	1.498	0.018	1.488	1.510	1 370	
Csp^3–N(3)	C*–N$^+$ in N-subst. pyridinium	1.485	1.484	0.009	1.477	1.490	32	
	C*–NH$_2$ (Nsp^3: pyramidal)	1.469	1.470	0.010	1.462	1.474	19	22
	(C*)$_2$–NH (Nsp^3: pyramidal)	1.469	1.467	0.012	1.461	1.477	152	5,22
	(C*)$_3$–N (Nsp^3: pyramidal)	1.469	1.468	0.014	1.460	1.476	1 042	5,22
	C*–Nsp^3 (overall)	1.469	1.468	0.014	1.460	1.476	1 201	
	Csp^3–Nsp^3							
	in aziridine	1.472	1.471	0.016	1.464	1.482	134	
	in azetidine	1.484	1.481	0.018	1.472	1.495	21	
	in tetrahydropyrrole	1.475	1.473	0.016	1.464	1.483	66	
	in piperidine	1.473	1.473	0.013	1.460	1.479	240	
	Csp^3–Nsp^2 (N planar) in:							23
	acyclic amides C*–NH–C=O	1.454	1.451	0.011	1.446	1.461	78	14
	β-lactams C*–N(–X)–C=O (endo)	1.464	1.465	0.012	1.458	1.475	23	13
	γ-lactams							
	C*–NH–C=O (endo)	1.457	1.458	0.011	1.449	1.465	20	13
	C*–N(–C*)–C=O (endo)	1.462	1.461	0.010	1.453	1.466	15	13
	C*–N(–C*)–C=O (exo)	1.458	1.456	0.014	1.448	1.465	15	13

Bond	Substructure	d	m	σ	q_1	q_u	n	Note
	δ-lactams							
	C*–NH–C=O (endo)	1.478	1.472	0.016	1.467	1.491	6	14
	C*–N(–C*)–C=O (endo)	1.479	1.476	0.007	1.475	1.482	15	14
	C*–N(–C*)–C=O (exo)	1.468	1.471	0.009	1.462	1.477	15	14
	nitro compounds (1,2-dinitro omitted):							
	C–**CH$_2$–NO$_2$**	1.485	1.483	0.020	1.478	1.502	8	
	C$_2$–**CH–NO$_2$**	1.509	1.509	0.011	1.502	1.511	12	
	C$_3$–**C–NO$_2$**	1.533	1.533	0.013	1.530	1.539	17	
	C$_2$–**C–(NO$_2$)$_2$**	1.537	1.536	0.016	1.525	1.550	19	
	1,2-dinitro: NO$_2$–**C*–C***–NO$_2$	1.552	1.550	0.023	1.536	1.572	32	
Csp^3–N(2)	**C#–N=N**	1.493	1.493	0.020	1.477	1.506	54	
	C*–N=C–Car	1.465	1.468	0.011	1.461	1.472	75	
Csp^2–N(3)	**C=C–NH$_2$** Nsp^2 planar	1.336	1.344	0.017	1.317	1.348	10	23
	C=C–NH–C# Nsp^2 planar	1.339	1.340	0.016	1.327	1.351	17	23
	C=C–N–(C#)$_2$							
	Nsp^2 planar	1.355	1.358	0.014	1.341	1.363	22	23
	Nsp^3 pyramidal	1.416	1.418	0.018	1.397	1.432	18	22
Csp^2–Nsp^2 (N planar) in:								23
	acyclic amides							
	NH$_2$–C=O	1.325	1.323	0.009	1.318	1.331	32	14
	C*–NH–C=O	1.334	1.333	0.011	1.326	1.343	78	14
	(C*)$_2$–N–C=O	1.346	1.342	0.011	1.339	1.356	5	14
	β-lactams **C*–NH–C=O**	1.385	1.388	0.019	1.374	1.396	23	13
	γ-lactams							
	C*–NH–C=O	1.331	1.331	0.011	1.326	1.337	20	13
	C*–N(–C*)–C=O	1.347	1.344	0.014	1.335	1.359	15	13
	δ-lactams							
	C*–NH–C=O	1.334	1.334	0.006	1.330	1.339	6	14
	C*–N(–C*)–C=O	1.352	1.353	0.010	1.344	1.356	15	14
	peptides **C#–N(–X)–C(–C#)(=O)**	1.333	1.334	0.013	1.326	1.340	380	24
	ureas							
	(NH$_2$)$_2$–C=O	1.334	1.334	0.008	1.329	1.339	48	25,26
	(C#–NH)$_2$–C=O	1.347	1.345	0.010	1.341	1.354	26	25
	[(C#)$_n$–N]$_2$–C=O	1.363	1.359	0.014	1.354	1.370	40	25,27
	thioureas	1.346	1.343	0.023	1.328	1.361	192	
	(X$_2$N)$_2$–C=S							
	imides							
	[C#–C(=O)]$_2$–NH	1.376	1.377	0.012	1.369	1.383	64	
	[C#–C(=O)]$_2$–N–C#	1.389	1.383	0.017	1.376	1.404	38	
	[Csp^2–C(=O)]$_2$–N–C#	1.396	1.396	0.010	1.389	1.403	46	
	[Csp^2–C(=O)]$_2$–N–Csp^2	1.409	1.406	0.020	1.391	1.419	28	
	guanidinium **[C–(NH$_2$)$_3$]$^+$** (unsubst.)	1.321	1.320	0.008	1.314	1.327	39	
	(any subst.)	1.328	1.325	0.015	1.317	1.333	140	
	in heterocyclic systems (any subst.)							
	1H-pyrrole (N1–C2, N1–C5)	1.372	1.374	0.016	1.363	1.384	58	
	indole (N1–C2)	1.370	1.370	0.012	1.364	1.377	40	
	pyrazole (N1–C5)	1.357	1.359	0.012	1.347	1.365	20	
	imidazole (N1–C2)	1.349	1.349	0.018	1.338	1.358	44	
	imidazole (N1–C5)	1.370	1.370	0.010	1.365	1.377	44	
Csp^2–N(2)	in imidazole (N3–C4)	1.376	1.377	0.011	1.369	1.384	44	
Car–N(4)	**Car–N$^+$–(C,H)$_3$**	1.465	1.466	0.007	1.461	1.470	23	
Car–N(3)	**Car–NH$_2$**							
	(Nsp^2: planar)	1.355	1.360	0.020	1.340	1.372	33	23
	(Nsp^3: pyramidal)	1.394	1.396	0.011	1.385	1.403	25	22
	(overall)	1.375	1.377	0.025	1.363	1.394	98	28
Car–N(3)	**Car–NH–C#**							
	(Nsp^2: planar)	1.353	1.353	0.007	1.347	1.359	16	23
	(Nsp^3: pyramidal)	1.419	1.423	0.017	1.412	1.432	8	22
	(overall)	1.380	1.364	0.032	1.353	1.412	31	28
	Car–N–(C#)$_2$							
	(Nsp^2: planar)	1.371	1.370	0.016	1.363	1.382	41	23
	(Nsp^3: pyramidal)	1.426	1.425	0.011	1.421	1.431	22	22
	(overall)	1.390	1.385	0.030	1.366	1.420	69	28
	in indole (N1–C7a)	1.372	1.372	0.007	1.367	1.376	40	
	Car–NO$_2$	1.468	1.469	0.014	1.460	1.476	556	
Car–N(2)	**Car–N=N**	1.431	1.435	0.020	1.422	1.442	26	
Csp^2=N(3)	in furoxan ($^+$N2=C3)	1.316	1.316	0.009	1.311	1.324	14	
Csp^2=N(2)	**Car–C=N–C#**	1.279	1.279	0.008	1.275	1.285	75	
	(C,H)$_2$–C=N–OH in oximes	1.281	1.280	0.013	1.273	1.288	67	
	S–C=N–X	1.302	1.302	0.021	1.285	1.319	36	
	in pyrazole (N2=C3)	1.329	1.331	0.014	1.315	1.339	20	
	in imidazole (C2=N3)	1.313	1.314	0.011	1.307	1.319	44	
	in isoxazole (N2=C3)	1.314	1.315	0.009	1.305	1.320	9	

Bond	Substructure	d	m	σ	q_1	q_u	n	Note
	in furazan (N2=C3, C4=N5)	1.298	1.299	0.006	1.294	1.303	12	
	in furoxan (C4=N5)	1.304	1.306	0.008	1.300	1.308	14	
$Car \simeq N(3)$	$C \simeq N^+ - H$ (pyrimidinium)	1.335	1.334	0.015	1.325	1.342	30	
	$C \simeq N^+ - C^*$ (pyrimidinium)	1.346	1.346	0.010	1.340	1.352	64	
	$C \simeq N^+ - O^-$ (pyrimidinium)	1.362	1.359	0.013	1.353	1.369	56	
$Car \simeq N(2)$	$C \simeq N$ (pyridine)	1.337	1.338	0.012	1.330	1.344	269	
	$C \simeq N$ (pyrazine)	1.336	1.335	0.022	1.319	1.347	120	
	$C \simeq N \simeq C$ (pyrimidine)	1.339	1.338	0.015	1.333	1.342	28	
	$N \simeq C \simeq N$ (pyrimidine)	1.333	1.335	0.013	1.326	1.337	28	
	$C \simeq N$ (pyrimidine) (overall)	1.336	1.337	0.014	1.331	1.339	56	
	in any 6-membered N-containing aromatic ring:							
	$H-C \simeq N \simeq C-H$	1.334	1.334	0.014	1.327	1.341	146	
	$H-C \simeq N \simeq C-C^*$	1.339	1.341	0.013	1.336	1.345	38	
	$C^*-C \simeq N \simeq C-C^*$	1.345	1.345	0.008	1.342	1.348	24	
	$C \simeq N \simeq C$ (overall)	1.336	1.337	0.014	1.329	1.344	204	
$Csp^1 \equiv N(2)$	$X-S-N \equiv C^-$ (isothiocyanide)	1.144	1.147	0.006	1.140	1.148	6	
$Csp^1 \equiv N(1)$	$C^*-C \equiv N$	1.136	1.137	0.010	1.131	1.142	140	
	$C=C-C \equiv N$ in TCNQ	1.144	1.144	0.008	1.139	1.149	284	19
	$Car-C \equiv N$	1.138	1.138	0.007	1.133	1.143	31	
	$X-C \equiv N$	1.144	1.141	0.012	1.138	1.151	10	
	$(S-C \equiv N)^-$	1.155	1.156	0.012	1.147	1.165	14	
$Csp^3-O(2)$	in alcohols							
	CH_3-OH	1.413	1.414	0.018	1.395	1.425	17	
	$C-CH_2-OH$	1.426	1.426	0.011	1.420	1.431	75	
	$C_2-CH-OH$	1.432	1.431	0.011	1.425	1.439	266	
	C_3-C-OH	1.440	1.440	0.012	1.432	1.449	106	
	C^*-OH (overall)	1.432	1.431	0.013	1.424	1.441	464	
	in dialkyl ethers							29
	CH_3-O-C^*	1.416	1.418	0.016	1.405	1.426	110	
	$C-CH_2-O-C^*$	1.426	1.424	0.011	1.418	1.435	34	
	$C_2-CH-O-C^*$	1.429	1.430	0.010	1.420	1.437	53	
	$C_3-C-O-C^*$	1.452	1.450	0.011	1.445	1.458	39	
	C^*-O-C^* (overall)	1.426	1.425	0.019	1.414	1.437	236	5
	in aryl alkyl ethers							29
	$CH_3-O-Car$	1.424	1.424	0.012	1.417	1.431	616	
	$C-CH_2-O-Car$	1.431	1.430	0.013	1.422	1.438	188	
	$C_2-CH-O-Car$	1.447	1.446	0.020	1.435	1.466	58	
	$C_3-C-O-Car$	1.470	1.469	0.018	1.456	1.483	55	
	$C^*-O-Car$ (overall)	1.429	1.427	0.018	1.419	1.436	917	
	in alkyl esters of carboxylic acids							12,29
	$CH_3-O-C(=O)-C^*$	1.448	1.449	0.010	1.442	1.455	200	
	$C-CH_2-O-C(=O)-C^*$	1.452	1.453	0.009	1.445	1.458	32	
	$C_2-CH-O-C(=O)-C^*$	1.460	1.460	0.010	1.454	1.465	78	
	$C_3-C-O-C(=O)-C^*$	1.477	1.475	0.008	1.472	1.484	6	
	$C^*-O-C(=O)-C^*$ (overall)	1.450	1.451	0.014	1.442	1.459	314	
	in alkyl esters of α,β-unsaturated acids:							
	$C^*-O-C(=O)-C=C$ (overall)	1.453	1.452	0.013	1.444	1.459	112	
	in alkyl esters of benzoic acid							
	$C^*-O-C(=O)-C$(phenyl) (overall)	1.454	1.454	0.012	1.446	1.463	219	
	in ring systems							
	oxirane (epoxides) (any subst.)	1.446	1.446	0.014	1.438	1.456	498	9
	oxetane (any subst.)	1.463	1.460	0.015	1.451	1.474	16	
	tetrahydrofuran (C,H subst.)	1.442	1.441	0.017	1.430	1.451	154	
$Csp^3-O(2)$	tetrahydropyran (C,H subst.)	1.441	1.442	0.015	1.431	1.451	22	
	β-lactones: $C^*-O-C(=O)$	1.492	1.494	0.010	1.481	1.501	4	16
	γ-lactones: $C^*-O-C(=O)$	1.464	1.464	0.012	1.455	1.473	110	12
	δ-lactones: $C^*-O-C(=O)$	1.461	1.464	0.017	1.452	1.473	27	12
	O–C–O system in *gem*-diols, and pyranose and furanose sugars:							30,31
	$HO-C^*-OH$	1.397	1.401	0.012	1.388	1.405	18	
	$C_5-O_5-C_1-O_1H$ in pyranoses							
	O_1 axial (α):							
	C_5-O_5	1.439	1.440	0.008	1.432	1.445	29	
	O_5-C_1	1.427	1.426	0.012	1.421	1.432	29	
	C_1-O_1	1.403	1.400	0.012	1.391	1.412	29	
	O_1 equatorial (β):							
	C_5-O_5	1.435	1.436	0.008	1.429	1.440	17	
	O_5-C_1	1.430	1.431	0.010	1.424	1.436	17	
	C_1-O_1	1.393	1.393	0.007	1.386	1.399	17	
	$\alpha + \beta$ (overall):							
	C_5-O_5	1.439	1.440	0.008	1.432	1.446	60	
	O_5-C_1	1.430	1.429	0.012	1.421	1.436	60	
	C_1-O_1	1.401	1.399	0.011	1.392	1.407	60	

Bond	Substructure	d	m	σ	q_1	q_u	n	Note
	C_4–O_4–C_1–O_1H in furanoses							
	(overall values)							
	C_4–O_4	1.442	1.446	0.012	1.436	1.449	18	
	O_4–C_1	1.432	1.432	0.012	1.421	1.443	18	
	C_1–O_1	1.404	1.405	0.013	1.397	1.409	18	
	C_5–O_5–C_1–O_1–C^* in pyranoses							
	O_1 axial (α):							
	C_5–O_5	1.439	1.438	0.010	1.433	1.446	67	
	O_5–C_1	1.417	1.417	0.009	1.410	1.424	67	
	C_1–O_1	1.409	1.409	0.014	1.401	1.417	67	
	O_1–C^*	1.435	1.435	0.013	1.427	1.443	67	
	O_1 equatorial (β):							
	C_5–O_5	1.434	1.435	0.006	1.429	1.439	39	
	O_5–C_1	1.424	1.424	0.008	1.418	1.431	39	
	C_1–O_1	1.390	1.390	0.011	1.381	1.400	39	
	O_1–C^*	1.437	1.438	0.013	1.428	1.445	39	
	α + β (overall):							
	C_5–O_5	1.436	1.436	0.009	1.431	1.442	126	
	O_5–C_1	1.419	1.419	0.011	1.412	1.426	126	
	C_1–O_1	1.402	1.403	0.016	1.391	1.413	126	
	O_1–C^*	1.436	1.436	0.013	1.428	1.445	126	
	C_4–O_4–C_1–O_1–C^* in furanoses							
	(overall values)							
	C_4–O_4	1.443	1.445	0.013	1.429	1.453	23	
	O_4–C_1	1.421	1.418	0.012	1.413	1.431	23	
	C_1–O_1	1.410	1.409	0.014	1.401	1.420	23	
	O_1–C^*	1.439	1.437	0.014	1.429	1.449	23	
	Miscellaneous:							
	$C\#$–O–SiX_3	1.416	1.416	0.017	1.405	1.428	29	
	C^*–O–SO_2–C	1.465	1.461	0.014	1.454	1.475	33	
Csp^2–O(2)	in enols: $C=C$–OH	1.333	1.331	0.017	1.324	1.342	53	
	in enol esters: $C=C$–O–C^*	1.354	1.353	0.016	1.341	1.363	40	
	in acids:							
	C^*–$C(=O)$–OH	1.308	1.311	0.019	1.298	1.320	174	
	$C=C$–$C(=O)$–OH	1.293	1.295	0.019	1.279	1.307	22	
	Car–$C(=O)$–OH	1.305	1.311	0.020	1.291	1.317	75	
	in esters:							
	C^*–$C(=O)$–O–C^*	1.336	1.337	0.014	1.328	1.346	551	12,29
	$C=C$–$C(=O)$–O–C^*	1.332	1.331	0.011	1.324	1.339	112	
	Car–$C(=O)$–O–C^*	1.337	1.335	0.013	1.329	1.344	219	12
	C^*–$C(=O)$–O–$C=C$	1.362	1.359	0.018	1.351	1.374	26	
	C^*–$C(=O)$–O–$C=C$	1.407	1.405	0.017	1.394	1.420	26	
	C^*–$C(=O)$–O–Car	1.360	1.359	0.011	1.355	1.367	40	12
	in anhydrides: $O=C$–O–$C=O$	1.386	1.386	0.011	1.379	1.393	70	
	in ring systems:							
	furan (O1–C2, O1–C5)	1.368	1.369	0.015	1.359	1.377	125	
	isoxazole (O1–C5)	1.354	1.354	0.010	1.345	1.360	9	
	β-lactones: C^*–$C(=O)$–O–C^*	1.359	1.359	0.013	1.348	1.371	4	13
	γ-lactones: C^*–$C(=O)$–O–C^*	1.350	1.349	0.012	1.342	1.359	110	12
	δ-lactones: C^*–$C(=O)$–O–C^*	1.339	1.339	0.016	1.332	1.347	27	12
Car–O(2)	in phenols: Car–OH	1.362	1.364	0.015	1.353	1.373	551	
	in aryl alkyl ethers: Car–O–C^*	1.370	1.370	0.011	1.363	1.377	920	29,32
Car–O(2)	in diaryl ethers: Car–O–Car	1.384	1.381	0.014	1.375	1.391	132	
	in esters: Car–O–$C(=O)$–C^*	1.401	1.401	0.010	1.394	1.408	40	12
Csp^2=O(1)	in aldehydes and ketones:							
	C^*–$CH=O$	1.192	1.192	0.005	1.188	1.197	7	
	$(C^*)_2$–$C=O$	1.210	1.210	0.008	1.206	1.215	474	5
	$(C\#)_2$–$C=O$							
	in cyclobutanones	1.198	1.198	0.007	1.194	1.204	12	
	in cyclopentanones	1.208	1.208	0.007	1.203	1.212	155	
	in cyclohexanones	1.211	1.211	0.009	1.207	1.216	312	
	$C=C$–$C=O$	1.222	1.222	0.010	1.216	1.229	225	
	$(C=C)_2$–$C=O$	1.233	1.229	0.010	1.226	1.242	28	
	Car–$C=O$	1.221	1.218	0.014	1.212	1.229	85	
	$(Car)_2$–$C=O$	1.230	1.226	0.015	1.220	1.238	66	
	$C=O$ in benzoquinones	1.222	1.220	0.013	1.211	1.231	86	
	delocalized double bonds in carboxylate anions:							
	H–$C\simeq O_2^-$ (formate)	1.242	1.243	0.012	1.234	1.252	24	
	C^*–$C\simeq O_2^-$	1.254	1.253	0.010	1.247	1.261	114	
	$C=C$–$C\simeq O_2^-$	1.250	1.248	0.017	1.238	1.261	52	
	Car–$C\simeq O_2^-$	1.255	1.253	0.010	1.249	1.262	22	
	$HOOC$–$C\simeq O_2^-$ (hydrogen oxalate)	1.243	1.247	0.015	1.232	1.256	26	
	$^-O_2\simeq C$–$C\simeq O_2^-$ (oxalate)	1.251	1.251	0.007	1.248	1.254	18	

Bond	Substructure	d	m	σ	q_1	q_u	n	Note
	in carboxylic acids (X–COOH)							
	C*–C(=O)–OH	1.214	1.214	0.019	1.203	1.224	175	
	C=C–C(=O)–OH	1.229	1.226	0.017	1.218	1.237	22	
	Car–C(=O)–OH	1.226	1.223	0.020	1.211	1.241	75	
	in esters:							
	C*–C(=O)–O–C*	1.196	1.196	0.010	1.190	1.202	551	12
	C=C–C(=O)–O–C*	1.199	1.198	0.009	1.193	1.203	113	
	Car–C(=O)–O–C*	1.202	1.201	0.009	1.196	1.207	218	12
	C*–C(=O)–O–C=C	1.190	1.190	0.014	1.184	1.198	26	
	C*–C(=O)–O–Car	1.187	1.188	0.011	1.181	1.195	40	12
	in anhydrides: O=C–O–C=O	1.187	1.187	0.010	1.184	1.193	70	
	in β-lactones: C*–C(=O)–O–C*	1.193	1.193	0.006	1.187	1.198	4	13
	γ-lactones: C*–C(=O)–O–C*	1.201	1.202	0.009	1.196	1.206	109	12
	δ-lactones: C*–C(=O)–O–C*	1.205	1.207	0.008	1.201	1.209	27	12
	in amides:							
	NH_2–C(–C*)=O	1.234	1.233	0.012	1.225	1.243	32	14
	(C*–)(C*,H–)N–C(–C*)=O	1.231	1.231	0.012	1.224	1.238	378	14
	β-lactams: C*–NH–C=O	1.198	1.200	0.012	1.193	1.204	23	13
	γ-lactams:							
	C*–NH–C=O	1.235	1.235	0.008	1.232	1.240	20	13
	C*–N(–C*)–C=O	1.225	1.226	0.011	1.217	1.233	15	13
	δ-lactams:							
	C*–NH–C=O	1.240	1.241	0.003	1.237	1.243	6	14
	C*–N(–C*)–C=O	1.233	1.233	0.007	1.229	1.239	15	14
	in ureas:							
	$(NH_2)_2$–C=O	1.256	1.256	0.007	1.249	1.261	24	25,26
	(C#–NH)$_2$–C=O	1.241	1.237	0.011	1.235	1.245	13	25
	$[(C\#)_n$–N]$_2$–C=O	1.230	1.230	0.007	1.224	1.234	20	25,27
Csp^3–P(4)	C_3–P^+–C*	1.800	1.802	0.015	1.790	1.812	35	33
	C_2–P(=O)–CH_3	1.791	1.790	0.006	1.786	1.795	10	
	C_2–P(=O)–CH_2–C	1.806	1.806	0.009	1.801	1.813	45	
	C_2–P(=O)–CH–C_2	1.821	1.821	0.009	1.815	1.828	15	
	C_2–P(=O)–C–C_3	1.841	1.842	0.008	1.835	1.847	14	
	C_2–P(=O)–C* (overall)	1.813	1.811	0.017	1.800	1.822	84	
Csp^3–P(3)	C_2–P–C*	1.855	1.857	0.019	1.840	1.870	23	
Car–P(4)	C_3–P^+–Car	1.793	1.792	0.011	1.786	1.800	276	
	C_2–P(=O)–Car	1.801	1.802	0.011	1.796	1.807	98	
	Ph_3–P=N^+=P–Ph_3	1.795	1.795	0.008	1.789	1.800	197	
Car–P(3)	C_2–P–Car	1.836	1.837	0.010	1.830	1.844	102	
	(N≃)$_2$P–Car (P≃N aromatic)	1.795	1.793	0.011	1.788	1.803	43	
Csp^3–S(4)	C*–SO_2–C (C* = CH_3 excluded)	1.786	1.782	0.018	1.774	1.797	75	
	C*–SO_2–C (overall)	1.779	1.778	0.020	1.764	1.790	94	
	C*–SO_2–O–X	1.745	1.744	0.009	1.738	1.754	7	34
	C*–SO_2–N–X_2	1.758	1.756	0.018	1.746	1.773	17	34
Csp^3–S(3)	C*–S(=O)–C (C* = CH_3 excluded)	1.818	1.814	0.024	1.802	1.829	69	
	C*–S(=O)–C (overall)	1.809	1.806	0.025	1.793	1.820	88	
	CH_3–S^+–X_2	1.786	1.787	0.007	1.779	1.792	21	
	C*–S^+–X_2 (C* = CH_3 excluded)	1.823	1.820	0.016	1.812	1.834	18	
	C*–S^+–X_2 (overall)	1.804	1.794	0.025	1.788	1.820	41	
Csp^3–S(2)	C*–SH	1.808	1.805	0.010	1.800	1.819	6	
	CH_3–S–C*	1.789	1.787	0.008	1.784	1.794	9	
Csp^3–S(2)	C–CH_2–S–C*	1.817	1.816	0.013	1.808	1.824	92	
	C_2–CH–S–C*	1.819	1.819	0.011	1.811	1.825	32	
	C_3–C–S–C*	1.856	1.860	0.011	1.854	1.863	26	
	C*–S–C* (overall)	1.819	1.817	0.019	1.809	1.827	242	
	in thiirane	1.834	1.835	0.025	1.810	1.858	4	9
	in thietane: see ZCMXSP (1.817, 1.844)							
	in tetrahydrothiophene	1.827	1.826	0.018	1.811	1.837	20	
	in tetrahydrothiopyran	1.823	1.821	0.014	1.812	1.832	24	
	C–CH_2–S–S–X	1.823	1.820	0.014	1.813	1.832	41	
	C_3–C–S–S–X	1.863	1.865	0.015	1.848	1.878	11	
	C*–S–S–X (overall)	1.833	1.828	0.022	1.818	1.848	59	
Csp^2–S(2)	C=C–S–C*	1.751	1.755	0.017	1.740	1.764	61	
	C=C–S–C=C (in tetrathiafulvalene)	1.741	1.741	0.011	1.733	1.750	88	
	C=C–S–C=C (in thiophene)	1.712	1.712	0.013	1.703	1.722	60	
	O=C–S–C#	1.762	1.759	0.018	1.747	1.778	20	
Car–S(4)	Car–SO_2–C	1.763	1.764	0.009	1.756	1.769	96	
	Car–SO_2–O–X	1.752	1.750	0.008	1.749	1.756	27	
	Car–SO_2–N–X_2	1.758	1.759	0.013	1.749	1.765	106	35
Car–S(3)	Car–S(=O)–C	1.790	1.790	0.010	1.783	1.798	41	
	Car–S^+–X_2	1.778	1.779	0.010	1.771	1.787	10	
Car–S(2)	Car–S–C*	1.773	1.774	0.009	1.765	1.779	44	
	Car–S–Car	1.768	1.767	0.010	1.762	1.774	158	
	Car–S–Car (in phenothiazine)	1.764	1.764	0.008	1.760	1.769	48	

Bond	Substructure	d	m	σ	q_1	q_u	n	Note
	Car-S-S-X	1.777	1.777	0.012	1.767	1.785	47	
Csp^1-S(2)	N≡C-S-X	1.679	1.683	0.026	1.645	1.698	10	
Csp^1-S(1)	(N≡C-S)⁻	1.630	1.630	0.014	1.619	1.641	14	
Csp^2=S(1)	(C*)₂-C=S: see IPMUDS (1.599)							
	$(Car)_2$-C=S: see CELDOM (1.611)							
	(X)₂-C=S (X = C,N,O,S)	1.671	1.675	0.024	1.656	1.689	245	
	X₂N-C(=S)-S-X	1.660	1.660	0.016	1.648	1.674	38	
	(X₂N)₂-C=S (thioureas)	1.681	1.684	0.020	1.669	1.693	96	
	N-C(≃S)₂	1.720	1.721	0.012	1.709	1.731	20	
Csp^3-Se	C#-Se	1.970	1.967	0.032	1.948	1.998	21	
Csp^2-Se(2)	C=C-Se-C=C (in tetraselenafulvalene)	1.893	1.895	0.013	1.882	1.902	32	
Car-Se(3)	Ph₃-Se⁺	1.930	1.929	0.006	1.924	1.936	13	
Csp^3-Si(5)	C#-Si⁻-X₄	1.874	1.876	0.015	1.859	1.884	9	
Csp^3-Si(4)	CH₃-Si-X₃	1.857	1.857	0.018	1.848	1.869	552	
	C*-Si-X₃ (C* = CH₃ excluded)	1.888	1.887	0.023	1.872	1.905	124	
	C*-Si-X₃ (overall)	1.863	1.861	0.024	1.850	1.875	681	
Car-Si(4)	Car-Si-X₃	1.868	1.868	0.014	1.857	1.878	178	
Csp^1-Si(4)	C≡C-Si-X₃	1.837	1.840	0.012	1.824	1.849	8	
Csp^3-Te	C#-Te	2.158	2.159	0.030	2.128	2.177	13	
Car-Te	Car-Te	2.116	2.115	0.020	2.104	2.130	72	
Csp^2=Te	see CEDCUJ (2.044)							
Cl-Cl	see PHASCL (2.306, 2.227)							
Cl-I	see CMBIDZ (2.563), HXPASC (2.541, 2.513), METAMM (2.552), BQUINI (2.416, 2.718)							
Cl-N	see BECTAE (1.743—1.757), BOGPOC (1.705)							
Cl-O(1)	in ClO₄⁻	1.414	1.419	0.026	1.403	1.431	252	
Cl-P	(N≃)₂P-Cl (N≃P aromatic)	1.997	1.994	0.015	1.989	2.004	46	
	Cl-P (overall)	2.008	2.001	0.035	1.986	2.028	111	
Cl-S	Cl-S (overall)	2.072	2.079	0.023	2.047	2.091	6	
	see also longer bonds in CILSAR (2.283), BIHXIZ (2.357), CANLUY (2.749)							
Cl-Se	see BIRGUE10, BIRHAL10, CTCNSE (2.234—2.851)							
Cl-Si(4)	Cl-Si-X₃ (monochloro)	2.072	2.075	0.009	2.066	2.078	5	
	Cl₂-Si-X₂ and Cl₃-Si-X	2.020	2.012	0.015	2.007	2.036	5	
Cl-Te	Cl-Te in range 2.34—2.60	2.520	2.515	0.034	2.493	2.537	22	36
	see also longer bonds in BARRIV, BOJPUL, CETUTE, EPHTEA, OPNTEC10 (2.73—2.94)							
F-N(3)	F-N-C₂ and F₂-N-C	1.406	1.404	0.016	1.395	1.416	9	
F-P(6)	in hexafluorophosphate, PF₆⁻	1.579	1.587	0.025	1.563	1.598	72	
F-P(3)	(N≃)₂P-F (N≃P aromatic)	1.495	1.497	0.016	1.481	1.510	10	
F-S	43 observations in range 1.409—1.770 in a wide variety of environments; F-S(6) in							
	F₂-SO₂-C₂ (see FPSULF10, BETJOZ)	1.640	1.646	0.011	1.626	1.649	6	
	F-S(4) in F₂-S(=O)-N (see BUDTEZ)	1.527	1.528	0.004	1.524	1.530	24	37
F-Si(6)	in SiF₆²⁻	1.694	1.701	0.013	1.677	1.703	6	
F-Si(5)	F-Si⁻-X₄	1.636	1.639	0.035	1.602	1.657	10	
F-Si(4)	F-Si-X₃	1.588	1.587	0.014	1.581	1.599	24	
F-Te	see CUCPIZ (F-Te(6) = 1.942, 1.937), FPHTEL (F-Te(4) = 2.006)							
H-N(4)	X₃-N⁺-H	1.033	1.036	0.022	1.026	1.045	87	21
H-N(3)	X₂-N-H	1.009	1.010	0.019	0.997	1.023	95	21
H-O(2)	in alcohols C*-O-H	0.967	0.969	0.010	0.959	0.974	63	21
	C#-O-H	0.967	0.970	0.010	0.959	0.974	73	21
	in acids O=C-O-H	1.015	1.017	0.017	1.001	1.031	16	21,38
I-I	in I₃⁻	2.917	2.918	0.011	2.907	2.927	6	
I-N	see BZPRIB, CMBIDZ, HMTITI, HMTNTI, IFORAM, IODMAM (2.042—2.475)							
I-O	X-I-O (see BZPRIB, CAJMAB, IBZDAC11)	2.144	2.144	0.028	2.127	2.164	6	
	for IO₆⁻ see BOVMEE (1.829—1.912)							
I-P(3)	see CEHKAB (2.490—2.493)							†
I-S	see DTHIBR10 (2.687), ISUREA10 (2.629), BZTPPI (3.251)							
I-Te(4)	I-Te-X₃	2.926	2.928	0.026	2.902	2.944	8	
N(4)-N(3)	X₃-N⁺-N⁰-X₂ (N⁰ planar)	1.414	1.414	0.005	1.412	1.418	13	
N(3)-N(3)	(C)(C,H)-Nₐ-N_b-(C)(C,H)							5,39
	Nₐ, N_b pyramidal	1.454	1.452	0.021	1.444	1.457	44	40
	Nₐ pyramidal, N_b planar	1.420	1.420	0.015	1.407	1.433	68	40
	Nₐ, N_b planar	1.401	1.401	0.018	1.384	1.418	40	40
	overall	1.425	1.425	0.027	1.407	1.443	139	
N(3)-N(2)	in pyrazole (N1-N2)	1.366	1.366	0.019	1.350	1.375	20	
	in pyridazinium (N1⁺≃N2)	1.350	1.349	0.010	1.345	1.361	7	

Bond	Substructure	d	m	σ	q_1	q_u	n	Note
$N(2) \simeq N(2)$	$N \simeq N$ (aromatic) in pyridazine							
	with C,H as *ortho* substituents	1.304	1.300	0.019	1.287	1.326	6	
	with N,Cl as *ortho* substituents	1.368	1.373	0.011	1.362	1.375	9	
$N(2)=N(2)$	$C\#-N=N-C\#$							
	cis	1.245	1.244	0.009	1.239	1.252	21	
	trans	1.222	1.222	0.006	1.218	1.227	6	
	(overall)	1.240	1.241	0.012	1.230	1.251	27	
	$Car-N=N-Car$	1.255	1.253	0.016	1.247	1.262	13	
	$X-N=N=N$ (azides)	1.216	1.226	0.028	1.202	1.237	19	
$N(2)=N(1)$	$X-N=N=N$ (azides)	1.124	1.128	0.015	1.114	1.137	19	
$N(3)-O(2)$	$(C,H)_2-N-OH$ (Nsp^2: planar)	1.396	1.394	0.012	1.390	1.401	28	
	$C_2-N-O-C$							
	(Nsp^3: pyramidal)	1.463	1.465	0.012	1.457	1.468	22	
	(Nsp^2: planar)	1.397	1.394	0.011	1.388	1.409	12	
	in furoxan (N2-O1)	1.438	1.436	0.009	1.430	1.447	14	
$N(3)-O(1)$	$(C\simeq)_2N^+-O^-$ in pyridine *N*-oxides	1.304	1.299	0.015	1.291	1.316	11	
	in furoxan ($^+N2-O6^-$)	1.234	1.234	0.008	1.228	1.240	14	
$N(2)-O(2)$	in oximes							
	$(C\#)_2-C=N-OH$	1.416	1.418	0.006	1.416	1.420	7	
	$(H)(Csp^2)-C=N-OH$	1.390	1.390	0.011	1.380	1.401	20	
	$(C\#)(Csp^2)-C=N-OH$	1.402	1.403	0.010	1.393	1.410	18	
	$(Csp^2)_2-C=N-OH$	1.378	1.377	0.017	1.365	1.393	16	
	$(C,H)_2-C=N-OH$ (overall)	1.394	1.395	0.018	1.379	1.408	67	
	in furazan (O1-N2, O1-N5)	1.385	1.383	0.013	1.378	1.392	12	
	in furoxan (O1-N5)	1.380	1.380	0.011	1.370	1.388	14	
	in isoxazole (O1-N2)	1.425	1.425	0.010	1.417	1.434	9	
$N(3)=O(1)$	in nitrate ions NO_3^-	1.239	1.240	0.020	1.227	1.251	105	
	in nitro groups							
	C^*-NO_2	1.212	1.214	0.012	1.206	1.221	84	
	$C\#-NO_2$	1.210	1.210	0.011	1.203	1.218	251	
	$Car-NO_2$	1.217	1.218	0.011	1.211	1.215	1 116	
	$C-NO_2$ (overall)	1.218	1.219	0.013	1.210	1.226	1 733	
$N(3)-P(4)$	$X_2-P(=X)-NX_2$							
	Nsp^2: planar	1.652	1.651	0.024	1.634	1.670	205	
	Nsp^3: pyramidal	1.683	1.683	0.005	1.680	1.686	6	
	(overall)	1.662	1.662	0.029	1.639	1.682	358	
	subsets of this group are:							
	$O_2-P(=S)-NX_2$	1.628	1.624	0.015	1.615	1.634	9	
	$C-P(=S)-(NX_2)_2$	1.691	1.694	0.018	1.678	1.703	28	
	$O-P(=S)-(NX_2)_2$	1.652	1.654	0.014	1.642	1.664	28	
	$P(=O)-(NX_2)_3$	1.663	1.668	0.026	1.640	1.679	78	
$N(3)-P(3)$	$-NX-P(-X)-NX-P(-X)-$ (P_2N_2 ring)	1.730	1.721	0.017	1.716	1.748	20	
	$-NX-P(=S)-NX-P(=S)-$ (P_2N_2 ring)	1.697	1.697	0.015	1.690	1.703	44	
	in *P*-substituted phosphazenes:							
	$(N\simeq)_2P-N$ (amino)	1.637	1.638	0.014	1.625	1.651	16	
	(aziridinyl)	1.672	1.674	0.010	1.665	1.676	15	
$N(2)-P(4)$	$Ph_3-P=N^+=P-Ph_3$	1.571	1.573	0.013	1.563	1.580	66	
$N(2)-P(3)$	$Ph_3-P=N-C,S$	1.599	1.597	0.018	1.580	1.615	7	
$N(2) \simeq P(3)$	$N \simeq P$ aromatic							
	in phosphazenes	1.582	1.582	0.019	1.571	1.594	126	
	in $P \simeq N \simeq S$	1.604	1.606	0.009	1.594	1.612	36	
$N(3)-S(4)$	$C-SO_2-NH_2$	1.600	1.601	0.012	1.591	1.610	14	35
	$C-SO_2-NH-C\#$	1.633	1.633	0.019	1.615	1.652	47	35
	$C-SO_2-N-C(\#)_2$	1.642	1.641	0.024	1.623	1.659	38	35
$N(3)-S(2)$	$C-S-NX_2$ Nsp^2: planar	1.710	1.707	0.019	1.698	1.722	22	23
	(for Nsp^3 pyramidal see MODIAZ: 1.765)							
	$X-S-NX_2$ Nsp^2: planar	1.707	1.705	0.012	1.699	1.715	30	23
$N(2)-S(2)$	$C=N-S-X$	1.656	1.663	0.027	1.632	1.677	36	
$N(2) \simeq S(2)$	$N \simeq S$ aromatic in $P \simeq N \simeq S$	1.560	1.558	0.011	1.554	1.563	37	
$N(2)=S(2)$	$N=S$ in $N=S=N$ and $N=S=S$	1.541	1.546	0.022	1.521	1.558	37	
$N(3)-Se$	see COJCUZ (1.830), DSEMOR10 (1.846, 1.852), MORTRS10 (1.841)							
$N(2)-Se$	see SEBZQI (1.805), NAPSEZ10 (1.809, 1.820)							
$N(2)=Se$	see CISMUM (1.790, 1.791)							
$N(3)-Si(5)$	see DMESIP01, BOJLER, CASSAQ, CASYOK, CECXEN, CINTEY, CIPBUY, FMESIB, MNPSIL, PNPOSI (1.973—2.344)							
$N(3)-Si(4)$	$X_3-Si-NX_2$ (overall)	1.748	1.746	0.022	1.735	1.757	170	
	subsets of this group are:							
	$X_3-Si-NHX$	1.714	1.719	0.014	1.702	1.727	16	
	$X_3-Si-NX-Si-X_3$ acyclic	1.743	1.744	0.016	1.731	1.755	45	
	N-Si-N in 4-membered rings	1.742	1.742	0.009	1.735	1.748	53	
	N-Si-N in 5-membered rings	1.741	1.742	0.019	1.726	1.749	33	
$N(2)-Si(4)$	$X_3-Si-N^--Si-X_3$	1.711	1.712	0.019	1.693	1.729	15	

Bond	Substructure	d	m	σ	q_l	q_u	n	Note
N–Te	see ACLTEP (2.402), BIBLAZ (1.980), CESSAU (2.023)							
O(2)–O(2)	C*–O–O–C*,H							
	$\tau(OO) = 70$—$85°$	1.464	1.464	0.009	1.458	1.472	12	
	$\tau(OO)$ ca. 180°	1.482	1.480	0.005	1.478	1.486	5	
	overall	1.469	1.471	0.012	1.461	1.478	17	
	O=C–O–O–C=O see ACBZPO01 (1.446), CEYLUN (1.452), CIMHIP (1.454)							
	Si–O–O–Si	1.496	1.499	0.005	1.490	1.499	10	
O(2)–P(5)	X–P–(OX)$_4$							41
	trigonal bipyramidal:							
	axial	1.689	1.685	0.024	1.675	1.712	20	
	equatorial	1.619	1.622	0.024	1.604	1.628	20	
	square pyramidal	1.662	1.661	0.020	1.649	1.673	28	
O(2)–P(4)	C–O–P($\simeq$O)$_3{}^{2-}$	1.621	1.622	0.007	1.615	1.628	12	
	(H–O)$_2$–P($\simeq$O)$_2{}^-$	1.560	1.561	0.009	1.555	1.566	16	
	(C–O)$_2$–P($\simeq$O)$_2{}^-$	1.608	1.607	0.013	1.599	1.615	16	
	(C#–O)$_3$–P=O	1.558	1.554	0.011	1.550	1.564	30	
	(Car–O)$_3$–P=O	1.587	1.588	0.014	1.572	1.599	19	
	X–O–P(=O)–(C,N)$_2$	1.590	1.585	0.016	1.577	1.601	33	
	(X–O)$_2$–P(=O)–(C,N)	1.571	1.572	0.013	1.563	1.579	70	
O(2)–P(3)	(N$\simeq$)$_2$P–O–C (N $\simeq$ P aromatic)	1.573	1.573	0.011	1.563	1.584	16	
O(1)=P(4)	C–O–P($\simeq$O)$_3{}^{2-}$ (delocalized)	1.513	1.512	0.008	1.508	1.518	42	
	(H–O)$_2$–P($\simeq$O)$_2{}^-$ (delocalized)	1.503	1.503	0.005	1.499	1.508	16	
	(C–O)$_2$–P($\simeq$O)$_2{}^-$ (delocalized)	1.483	1.485	0.008	1.474	1.490	16	
	(C–O)$_3$–P=O	1.449	1.448	0.007	1.446	1.452	18	
	C$_3$–P=O	1.489	1.486	0.010	1.481	1.496	72	
	N$_3$–P=O	1.461	1.462	0.014	1.449	1.470	26	
	(C)$_2$(N)–P=O	1.487	1.489	0.007	1.479	1.493	5	
	(C,N)$_2$(O)–P=O	1.467	1.465	0.007	1.462	1.472	33	
	(C,N)(O)$_2$–P=O	1.457	1.458	0.009	1.454	1.462	35	
O(2)–S(4)	C–O–SO$_2$–C	1.577	1.576	0.015	1.566	1.584	41	
	C–O–SO$_2$–CH$_3$	1.569	1.569	0.013	1.556	1.582	7	
	C–O–SO$_2$–Car	1.580	1.578	0.015	1.571	1.588	27	
O(1)=S(4)	C–SO$_2$–C	1.436	1.437	0.010	1.431	1.442	316	42
	X–SO$_2$–NX$_2$	1.428	1.428	0.010	1.422	1.434	326	
	C–SO$_2$–N–(C,H)$_2$	1.430	1.430	0.009	1.425	1.435	206	
	C–SO$_2$–O–C	1.423	1.423	0.008	1.418	1.428	82	
	in SO$_4{}^{2-}$	1.472	1.473	0.013	1.463	1.481	104	
O(1)=S(3)	C–S(=O)–C	1.497	1.498	0.013	1.489	1.505	90	5
O–Se	see BAPPAJ, BIRGUE10, BIRHAL10, CXMSEO, DGLYSE, SPSEBU (1.597 for O=Se to 1.974 for O–Se)							
O(2)–Si(5)	(X–O)$_3$–Si–(N)(C)	1.663	1.658	0.023	1.650	1.665	21	
O(2)–Si(4)	X$_3$–Si–O–X (overall)	1.631	1.630	0.022	1.617	1.646	191	
O(2)–Si(4)	subsets of this group are:							
	X$_3$–Si–O–C#	1.645	1.647	0.012	1.634	1.652	29	
	X$_3$–Si–O–Si–X$_3$	1.622	1.625	0.014	1.614	1.631	70	
	X$_3$–Si–O–O–Si–X$_3$	1.680	1.676	0.008	1.673	1.688	10	
O(2)–Te(6)	(X–O)$_6$–Te	1.927	1.927	0.020	1.908	1.942	16	
O(2)–Te(4)	(X–O)$_2$–Te–X$_2$	2.133	2.136	0.054	2.078	2.177	12	
P(4)–P(4)	X$_3$–P–P–X$_3$	2.256	2.259	0.025	2.243	2.277	6	
P(4)–P(3)	see CECHEX (2.197), COZPIQ (2.249)							
P(3)–P(3)	X$_2$–P–P–X$_2$	2.214	2.210	0.022	2.200	2.224	41	
P(4)=P(4)	see BUTSUE (2.054)							
P(3)=P(3)	see BALXOB (2.034)							
P(4)=S(1)	C$_3$–P=S	1.954	1.952	0.005	1.950	1.957	13	
	(N,O)$_2$(C)–P=S	1.922	1.924	0.014	1.913	1.927	26	
	(N,O)$_3$–P=S	1.913	1.914	0.014	1.906	1.921	50	
P(4)=Se(1)	X$_3$–P=Se	2.093	2.099	0.019	2.075	2.108	12	
P(3)–Si(4)	X$_2$–P–Si–X$_3$: 3- and 4-rings excluded (see BOPFER, BOPFIV, CASTOF10, COZVIW: 2.201—2.317)	2.264	2.260	0.019	2.249	2.283	22	
P(4)=Te(1)	see MOPHTE (2.356), TTEBPZ (2.327)							
S(2)–S(2)	C–S–S–C							
	$\tau(SS) = 75$—$105°$	2.031	2.029	0.015	2.021	2.038	46	
	$\tau(SS) = 0$—$20°$	2.070	2.068	0.022	2.057	2.077	28	
	(overall)	2.048	2.045	0.026	2.028	2.068	99	
	in polysulphide chain–S–S–S–	2.051	2.050	0.022	2.037	2.065	126	
S(2)–S(1)	X–N=S–S	1.897	1.896	0.012	1.887	1.908	5	
S–Se(4)	see BUWZUO (2.264, 2.269)							
S–Se(2)	X–Se–S (any)	2.193	2.195	0.015	2.174	2.207	9	
S(2)–Si(4)	X$_3$–Si–S–X	2.145	2.138	0.020	2.130	2.158	19	

Bond	Substructure	d	m	σ	q_1	q_u	n	Note
S(2)–Te	X–S–Te (any)	2.405	2.406	0.022	2.383	2.424	10	
	X=S–Te (any)	2.682	2.686	0.035	2.673	2.694	28	
Se(2)–Se(2)	X–Se–Se–X	2.340	2.340	0.024	2.315	2.361	15	
Se(2)–Te(2)	see BAWFUA, BAWGAH (2.524—2.561)							†
Si(4)–Si(4)	X_3–Si–Si–X_3 3-membered rings excluded: see CIHRAM (2.511)	2.359	2.359	0.012	2.349	2.366	42	
Te–Te	see CAHJOK (2.751, 2.704)							

Appendix 1. (Footnotes to Table)

1. Sample dominated by B–CH₃. For longer bonds in B⁻–CH₃ see LITMEB10 [B(4)–CH₃ = 1.621—1.644Å].
2. $p(\pi)$–$p(\pi)$ Bonding with Bsp^2 and Nsp^2 coplanar (τBN = 0 $\pm$ 15°) predominates. See G. Schmidt, R. Boese, and D. Bläser, *Z. Naturforsch.*, 1982, **37b**, 1230.
3. 84 observations range from 1.38 to 1.61 Å and individual values depend on substituents on B and O. For a discussion of borinic acid adducts see S. J. Rettig and J. Trotter, *Can. J. Chem.*, 1982, **60**, 2957.
4. See M. Kaftory in 'The Chemistry of Functional Groups. Supplement D: The Chemistry of Halides, Pseudohalides, and Azides' eds. S. Patai and Z. Rappoport, Wiley: New York, 1983, Part 2, ch. 24.
5. Bonds which are endocyclic or exocyclic to any 3- or 4-membered rings have been omitted from all averages in this section.
6. The overall average given here is for Csp^3–Csp^3 bonds which carry only C or H substituents. The value cited reflects the relative abundance of each 'substitution' group. The 'mean of means' for the 9 subgroups is 1.538 (σ = 0.022) Å.
7. See F. H. Allen, (*a*) *Acta Crystallogr.*, 1980, **B36**, 81; (*b*) 1981, **B37**, 890.
8. See F. H. Allen, *Acta Crystallogr.*, 1984, **B40**, 64.
9. See F. H. Allen, *Tetrahedron*, 1982, **38**, 2843.
10. See F. H. Allen, *Tetrahedron*, 1982, **38**, 645.
11. Cyclopropanones and cyclobutanones excluded.
12. See W. B. Schweizer and J. D. Dunitz, *Helv. Chim. Acta*, 1982, **65**, 1547.
13. See L. Norskov-Lauritsen, H.-B. Bürgi, P. Hoffmann, and H. R. Schmidt, *Helv. Chim. Acta*, 1985, **68**, 76.
14. See P. Chakrabarti and J. D. Dunitz, *Helv. Chim. Acta*, 1982, **65**, 1555.
15. See J. L. Hencher in 'The Chemistry of the C≡C Triple Bond,' ed. S. Patai, Wiley, New York, 1978, ch. 2.
16. Conjugated: torsion angle about central C–C single bond is 0 $\pm$ 20° (*cis*) or 180 $\pm$ 20° (*trans*).
17. Unconjugated: torsion angle about central C–C single bond is 20—160°.
18. Other conjugative substituents excluded.
19. TCNQ is tetracyanoquinodimethane.
20. No difference detected between C2≃C3 and C3≃C4 bonds.
21. Derived from neutron diffraction results only.
22. Nsp^3: pyramidal; mean valence angle at N is in range 108—114°.
23. Nsp^2: planar; mean valence angle at N is ⩾117.5°.
24. Cyclic and acyclic peptides.
25. See R. H. Blessing, *J. Am. Chem. Soc.*, 1983, **105**, 2776.
26. See L. Lebioda, *Acta Crystallogr.*, 1980, **B36**, 271.
27. n = 3 or 4, *i.e.* tri- or tetra-substituted ureas.
28. Overall value also includes structures with mean valence angle at N in the range 115—118°.
29. See F. H. Allen and A. J. Kirby, *J. Am. Chem. Soc.*, 1984, **106**, 6197.
30. See A. J. Kirby, 'The Anomeric Effect and Related Stereoelectronic Effects at Oxygen,' Springer, Berlin, 1983.
31. See B. Fuchs, L. Schleifer, and E. Tartakovsky, *Nouv. J. Chim.*, 1984, **8**, 275.
32. See S. C. Nyburg and C. H. Faerman, *J. Mol. Struct.*, 1986, **140**, 347.
33. Sample dominated by P–CH₃ and P–CH₂–C.
34. Sample dominated by C* = methyl.
35. See A. Kalman, M. Czugler, and G. Argay, *Acta Crystallogr.*, 1981, **B37**, 868.
36. Bimodal distribution resolved into 22 'short' bonds and 5 longer outliers.
37. All 24 observations come from BUDTEZ.
38. 'Long' O–H bonds in centrosymmetric O – – – H – – – O H-bonded dimers are excluded.
39. N–N bond length also dependent on torsion angle about N–N bond and on nature of substituent C atoms; these effects are ignored here.
40. N pyramidal has average angle at N in range 100—113.5°; N planar has average angle of ⩾117.5°.
41. See R. R. Holmes and J. A. Deiters, *J. Amer. Chem. Soc.*, 1977, **99**, 3318.
42. No detectable variation in S=O bond length with type of C-substituent.

Appendix 2.

Short-form references to individual CSD entries cited by reference code in the Table. A full list of CSD bibliographic entries is given in SUP 56701.

ACBZPO01	*J. Am. Chem. Soc.*, 1975, **97**, 6729.		BIBLAZ	*Zh. Strukt. Khim.*, 1981, **22**, 118.
ACLTEP	*J. Organomet. Chem.*, 1980, **184**, 417.		BICGEZ	*Z. Anorg. Allg. Chem.*, 1982, **486**, 90.
ASAZOC	*Dokl. Akad. Nauk SSSR*, 1979, **249**, 120.		BIHXIZ	*J. Chem. Soc., Chem. Commun.*, 1982, 982.
BALXOB	*J. Am. Chem. Soc.*, 1981, **103**, 4587.		BIRGUE10	*Z. Naturforsch., Teil B*, 1983, **38**, 20.
BAPPAJ	*Inorg. Chem.*, 1981, **20**, 3071.		BIRHAL10	*Z. Naturforsch., Teil B*, 1982, **37**, 1410.
BARRIV	*Acta Chem. Scand., Ser. A*, 1981, **35**, 443.		BIZJAV	*J. Organomet. Chem.*, 1982, **238**, C1.
BAWFUA	*Cryst. Struct. Commun.*, 1981, **10**, 1345.		BOGPOC	*Z. Naturforsch., Teil B*, 1982, **37**, 1402.
BAWGAH	*Cryst. Struct. Commun.*, 1981, **10**, 1353.		BOGSUL	*Z. Naturforsch., Teil B*, 1982, **37**, 1230.
BECTAE	*J. Org. Chem.*, 1981, **46**, 5048, 1981.		BOJLER	*Z. Anorg. Allg. Chem.*, 1982, **493**, 53.
BELNIP	*Z. Naturforsch., Teil B*, 1982, **37**, 299.		BOJPUL	*Acta Chem. Scand., Ser. A*, 1982, **36**, 829.
BEMLIO	*Chem. Ber.*, 1982, **115**, 1126.		BOPFER	*Chem. Ber.*, 1983, **116**, 146.
BEPZEB	*Cryst. Struct. Commun.*, 1982, **11**, 175.		BOPFIV	*Chem. Ber.*, 1983, **116**, 146.
BETJOZ	*J. Am. Chem. Soc.*, 1982, **104**, 1683.		BOVMEE	*Acta Crystallogr., Sect. B*, 1982, **38**, 1048.
BETUTE10	*Acta Chem. Scand., Ser. A*, 1976, **30**, 719.		BQUINI	*Acta Crystallogr., Sect. B*, 1979, **35**, 1930.

BTUPTE	*Acta Chem. Scand., Ser. A*, 1975, **29**, 738.		CUGBAH	*Acta Crystallogr., Sect. C*, 1985, **41**, 476.
BUDTEZ	*Z. Naturforsch., Teil B*, 1983, **38**, 454.		CXMSEO	*Acta Crystallogr., Sect. B*, 1973, **29**, 595.
BUPSIB10	*Z. Anorg. Allg. Chem.*, 1981, **474**, 31.		DGLYSE	*Acta Crystallogr., Sect. B*, 1975, **31**, 1785.
BUSHAY	*Z. Naturforsch., Teil. B*, 1983, **38**, 692.		DMESIP01	*Acta Crystallogr., Sect. C*, 1984, **40**, 895.
BUTHAZ10	*Inorg. Chem.*, 1984, **23**, 2582.		DSEMOR10	*J. Chem. Soc., Dalton Trans.*, 1980, 628.
BUTSUE	*J. Chem. Soc., Chem. Commun.*, 1983, 862.		DTHIBR10	*Inorg. Chem.*, 1971, **10**, 697.
BUWZUO	*Acta Chem. Scand., Ser A*, 1983, **37**, 219.		EPHTEA	*Inorg. Chem.*, 1980, **19**, 2487.
BZPRIB	*Z. Naturforsch., Teil B*, 1981, **36**, 922.		ESEARS	*J. Chem. Soc. C*, 1971, 1511.
BZTPPI	*Inorg. Chem.*, 1978, **17**, 894.		ETEARS	*J. Chem. Soc. C*, 1971, 1511.
CAHJOK	*Inorg. Chem.*, 1983, **22**, 1809.		FMESIB	*J. Organomet. Chem.*, 1980, **197**, 275.
CAJMAB	*Chem. Z*, 1983, **107**, 169.		FPHTEL	*J. Chem. Soc., Dalton Trans.*, 1980, 2306.
CANLUY	*Tetrahedron Lett.*, 1983, **24**, 4337.		FPSULF10	*J. Am. Chem. Soc.*, 1982, **104**, 1683.
CASSAQ	*J. Struct. Chem.*, 1983, **2**, 101.		HCLENE10	*Acta Crystallogr., Sect. B*, 1982, **38**, 3139.
CASTOF10	*Acta Crystallogr., Sect. C*, 1984, **40**, 1879.		HMTITI	*Acta Crystallogr., Sect. B*, 1975, **31**, 1505.
CASYOK	*J. Struct. Chem.*, 1983, **2**, 107.		HMTNTI	*Z. Anorg. Allg. Chem.*, 1974, **409**, 237.
CECHEX	*Z. Anorg. Allg. Chem.*, 1984, **508**, 61.		HXPASC	*J. Chem. Soc., Dalton Trans.*, 1975, 1381.
CECXEN	*J. Struct. Chem.*, 1983, **2**, 207.		IBZDAC11	*J. Chem. Soc., Dalton Trans.*, 1979, 854.
CEDCUJ	*J. Org. Chem.*, 1983, **48**, 5149.		IFORAM	*Monatsh. Chem.*, 1974, **105**, 621.
CEHKAB	*Z. Naturforsch., Teil B*, 1984, **39**, 139.		IODMAM	*Acta Crystallogr., Sect. B*, 1977, **33**, 3209.
CELDOM	*Acta Crystallogr., Sect. C*, 1984, **40**, 556.		IPMUDS	*Acta Crystallogr., Sect. B*, 1973, **29**, 2128.
CESSAU	*Acta Crystallogr., Sect. C*, 1984, **40**, 653.		ISUREA10	*Acta Crystallogr., Sect. B*, 1972, **28**, 643.
CETTAW	*Chem. Ber.*, 1984, **117**, 1089.		LITMEB10	*J. Am. Chem. Soc.*, 1975, **97**, 6401.
CETUTE	*Acta Chem. Scand., Ser A*, 1975, **29**, 763.		MESIAD	*Z. Naturforsch., Teil B*, 1980, **35**, 789.
CEYLUN	*Izv. Akad. Nauk SSSR, Ser. Khim.*, 1983, 2744.		METAMM	*Acta Crystallogr.*, 1964, **17**, 1336.
CIFZUM	*Acta Chem. Scand., Ser A*, 1984, **38**, 289.		MNPSIL	*J. Am. Chem. Soc.*, 1969, **91**, 4134.
CIHRAM	*Angew. Chem., Int. Ed. Engl.*, 1984, **23**, 302.		MODIAZ	*J. Heterocycl. Chem.*, 1980, **17**, 1217.
CILRUK	*J. Chem. Soc., Chem. Commun.*, 1984, 1023.		MOPHTE	*Acta Chem. Scand., Ser. A*, 1980, **34**, 333.
CILSAR	*J. Chem. Soc., Chem. Commun.*, 1984, 1021.		MORTRS10	*J. Chem. Soc., Dalton Trans.*, 1980, 628.
CIMHIP	*Acta Crystallogr., C*, 1984, **40**, 1458.		NAPSEZ10	*J. Am. Chem. Soc.*, 1980, **102**, 5070.
CINTEY	*Dokl. Akad. Nauk SSSR*, 1984, **274**, 615.		NBBZAM	*Z. Naturforsch., Teil B*, 1977, **32**, 1416.
CIPBUY	*J. Struct. Chem.*, 1983, **2**, 281.		OPIMAS	*Aust. J. Chem.*, 1977, **30**, 2417.
CISMUM	*Z. Naturforsch., Teil B*, 1984, **39**, 485.		OPNTEC10	*J. Chem. Soc., Dalton Trans.*, 1982, 251.
CISTED	*Z. Anorg. Allg. Chem.*, 1984, **511**, 95.		PHASCL	*Acta Crystallogr., Sect. B*, 1981, **37**, 1357.
CIWYIQ	*Inorg. Chem.*, 1984, **23**, 1946.		PHASOC01	*Aust. J. Chem.*, 1975, **28**, 15.
CIYFOF	*Inorg. Chem.*, 1984, **23**, 1790.		PNPOSI	*J. Am. Chem. Soc.*, 1968, **90**, 5102.
CMBIDZ	*J. Org. Chem.*, 1979, **44**, 1447.		SEBZQI	*J. Chem. Soc., Chem. Commun.*, 1977, 325.
CODDEE	*Z. Naturforsch., Teil B*, 1984, **39**, 1257.		SPSEBU	*Acta Chem. Scand., Ser. A*, 1979, **33**, 403.
CODDII	*Z. Naturforsch., Teil B*, 1984, **39**, 1257.		TEACBR	*Cryst. Struct. Commun.*, 1974, **3**, 753.
COFVOI	*Z. Naturforsch., Teil B*, 1984, **39**, 1027.		THINBR	*J. Am. Chem. Soc.*, 1970, **92**, 4002.
COJCUZ	*Chem. Ber.*, 1984, **117**, 2686.		TMPBTI	*Acta Crystallogr., Sect. B*, 1975, **31**, 1116.
COSDIX	*Z. Naturforsch., Teil B*, 1984, **39**, 1344.		TPASSN	*J. Chem. Soc., Dalton Trans.*, 1977, 514.
COZPIQ	*Chem. Ber.*, 1984, **117**, 2063.		TPASTB	*Cryst. Struct. Commun.*, 1976, **5**, 39.
COZVIW	*Z. Anorg. Allg. Chem.*, 1984, **515**, 7.		TPHOSI	*Z. Naturforsch., Teil B*, 1979, **34**, 1064.
CTCNSE	*J. Am. Chem. Soc.*, 1980, **102**, 5430.		TTEBPZ	*Z. Naturforsch., Teil B*, 1979, **34**, 256.
CUCPIZ	*J. Am. Chem. Soc.*, 1984, **106**, 7529.		ZCMXSP	*Cryst. Struct. Commun.*, 1977, **6**, 93.
CUDLOC	*J. Cryst. Spectrosc.*, 1985, **15**, 53.			
CUDLUI	*J. Cryst. Spectrosc.*, 1985, **15**, 53.			

BOND LENGTHS AND ANGLES IN GAS-PHASE MOLECULES

This table is reprinted from *Kagaku Benran, 3rd Edition*, Vol. II, pp. 649—661 (1984), with permission of the publisher, Maruzen Company, LTD. (Copyright 1984 by the Chemical Society of Japan). Translation was carried out by Kozo Kuchitsu.

Internuclear distances and bond angles are represented in units of Å (1 Å = 10^{-10} m) and degrees, respectively. The same but inequivalent atoms are discriminated by subscripts a, b, etc. In some molecules ax for axial and eq for equatorial are also used. All measurements were made in the gas phase. The methods used are abbreviated as follows. UV: ultraviolet (including visible) spectroscopy; IR: infrared spectroscopy; R: Raman spectroscopy; MW: microwave spectroscopy; ED: electron diffraction; NMR: nuclear magnetic resonance; LMR: laser magnetic resonance; EPR: electron paramagnetic resonance; MBE: molecular beam electric resonance. If two methods were used jointly for structure determination, they are listed together, as (ED, MW). If the numerical values listed refer to the equilibrium values, they are specified by r_e and θ_e. In other cases the listed values represent various average values in vibrational states; it is frequently the case that they represent the r_s structure derived from several isotopic species for MW or the r_g structure (i.e., the average internuclear distances at thermal equilibrium) for ED. These internuclear distances for the same atom pair with different definitions may sometimes differ as much as 0.01 Å. Appropriate comments are made on the symmetry and conformation in the equilibrium structure.

In general, the numerical values listed in the following tables contain uncertainties in the last digits. However, for certain molecules such as diatomic molecules, with experimental uncertainties of the order of 10^{-5} Å or smaller, numerical values are listed to four decimal places.

REFERENCES

1. L. E. Sutton, ed., *Tables of Interatomic Distances and Configuration in Molecules and Ions*, The Chemical Society Special Publication, No. 11, 18, The Chemical Society (London) (1958, 1965).
2. K.-H. Hellwege, ed., *Landolt-Börnstein Numerical Data and Functional Relations in Science and Technology*, New Series, II/7, J. H. Callomon, E. Hirota, K. Kuchitsu, W. J. Lafferty, A. G. Maki, C. S. Pote, with assistance of I. Buck and B. Starck, *Structure Data of Free Polyatomic Molecules*, Springer-Verlag (1976).
3. K. P. Huber and G. Herzberg, *Molecular Spectra and Molecular Structure IV. Constants of Diatomic Molecules*, Van Nostrand Reinhold Co., London (1979).
4. B. Starck, *Microwave Catalogue and Supplements*.
5. B. Starck, *Electron Diffraction Catalogue and Supplements*.

STRUCTURES OF ELEMENTS AND INORGANIC COMPOUNDS
Compounds are Arranged in Alphabetical Order by their Chemical Formulas
(Lengths in Å and Angles in Degrees)

Compound	Structure				Method
AgBr	Ag—Br (r_e)	2.3931			MW
AgCl	Ag—Cl (r_e)	2.2808			MW
AgF	Ag—F (r_e)	1.9832			MW
AgH	Ag—H (r_e)	1.617			UV
AgI	Ag—I (r_e)	2.5446			MW
AgO	Ag—O (r_e)	2.0030			UV
AlBr	Al—Br (r_e)	2.295			UV
AlCl	Al—Cl (r_e)	2.1301			MW
AlF	Al—F (r_e)	1.6544			MW
AlH	Al—H (r_e)	1.6482			UV
AlI	Al—I (r_e)	2.5371			MW
AlO	Al—O (r_e)	1.6176			UV
Al_2Br_6	Br_a, Br_b, Br_a / Al, Al / Br_a, Br_b, Br_a		Al—Br_a	2.22	ED
			Al—Br_b	2.38	
			$\angle Br_b AlBr_b$	82	
			$\angle Br_a AlBr_a$	118	
			(D_{2h})		
Al_2Cl_6	Cl_a, Cl_b, Cl_a / Al, Al / Cl_a, Cl_b, Cl_a		Al—Cl_a	2.04	ED
			Al—Cl_b	2.24	
			$\angle Cl_b AlCl_b$	87	
			$\angle Cl_a AlCl_a$	122	
			(D_{2h})		
$AsBr_3$	As—Br	2.324	$\angle BrAsBr$	99.6	ED
$AsCl_3$	As—Cl	2.165	$\angle ClAsCl$	98.6	ED, MW
AsF_3	As—F	1.710	$\angle FAsF$	95.9	ED, MW

Compound	Structure						Method
AsF$_5$	As—F$_a$ 1.711		As—F$_b$ 1.656 (D$_{3h}$)				
AsH$_3$	As—H (r_e) 1.511		∠HAsH (θ_e) 92.1				MW, IR
AsI$_3$	As—I 2.557		∠IAsI 100.2				ED
AuH	Au—H (r_e) 1.5237						UV
BBr$_3$	B—Br 1.893		(D$_{3h}$)				ED
BCl$_3$	B—Cl 1.742		(D$_{3h}$)				ED
BF	B—F (r_e) 1.2626						UV
BF$_2$H	B—H 1.189		B—F 1.311		∠FBF 118.3		MW
BF$_2$OH	B—F 1.32		B—O 1.34		O—H 0.941		MW
	∠FBF 118		∠FBO 123		∠BOH 114.1		
BF$_3$	B—F 1.313		(D$_{3h}$)				ED, IR
BH	B—H (r_e) 1.2325						UV
BH$_3$PH$_3$	B—P 1.937		B—H 1.212		P—H 1.399		MW
	∠PBH 103.6		∠BPH 116.9		∠HBH 114.6		
	∠HPH 101.3		staggered form				
BI$_3$	B—I 2.118		(D$_{3h}$)				ED
BN	B—N (r_e) 1.281						UV
BO	B—O (r_e) 1.2045						EPR
BO$_2$	B—O 1.265		linear				UV
BS	B—S 1.6091						UV
B$_2$H$_6$			B—H$_a$ 1.19				IR, ED
			B—H$_b$ 1.33				
			B···B 1.77				
			∠H$_a$BH$_a$ 122				
			∠H$_b$BH$_b$ 97				
B$_3$H$_3$O$_3$	B—O 1.376		∠BOB≅∠OBO 120				ED
B$_3$H$_6$N$_3$	B—N 1.435		B—H 1.26		N—H 1.05		ED
	∠NBN 118		∠BNB 121		(C$_2$)		
BaH	Ba—H (r_e) 2.2318						UV
BaO	Ba—O (r_e) 1.9397						MW
BaS	Ba—S (r_e) 2.5074						MBE
BeF	Be—F (r_e) 1.3609						UV
BeH	Be—H (r_e) 1.3431						UV
BeO	Be—O (r_e) 1.3308						UV
BiBr	Bi—Br (r_e) 2.6095						MW
BiBr$_3$	Bi—Br 2.63		∠BrBiBr 90		(C$_{3v}$)		ED
BiCl	Bi—Cl (r_e) 2.4716						MW
BiCl$_3$	Bi—Cl 2.423		∠ClBiCl 100		(C$_{3v}$)		ED
BiF	Bi—F (r_e) 2.0516						MW
BiH	Bi—H (r_e) 1.805						UV
BiI	Bi—I (r_e) 2.8005						MW
BiO	Bi—O (r_e) 1.934						UV
BrCN	C—N (r_e) 1.157		C—Br (r_e) 1.790				IR
BrCl	Br—Cl (r_e) 2.1361						MW
BrF	Br—F (r_e) 1.7590						MW
BrF$_3$			Br—F$_a$ 1.810		Br—F$_b$ 1.721		MW
			∠F$_a$BrF$_b$ 86.2		(C$_{2v}$)		
BrF$_5$	Br—F (average) 1.753						ED, MW
	(Br—F$_{eq}$) – (Br—F$_{ax}$) = 0.069						
	∠F$_{ax}$BrF$_{eq}$ 85.1		(C$_{4v}$)				
BrO	Br—O (r_e) 1.7172						MW
Br$_2$	Br—Br (r_e) 2.2811						R
CBr$_4$	C—Br 1.935		(T$_d$)				ED
CCl	C—Cl 1.6512						UV

Compound	Structure						Method
$CClF_3$	C—Cl	1.752		C—F	1.325	∠FCF 108.6	ED, MW
CCl_3F	C—Cl	1.754		C—F	1.362	∠ClCCl 111	MW
						(C_{3v})	
CCl_4	C—Cl		1.767		(T_d)		ED
CF	C—F (r_e)		1.2718				EPR
CF_3I	C—I	2.138		C—F	1.330	∠FCF 108.1	ED, MW
CF_4	C—F		1.323		(T_d)		ED
CH	C—H (r_e)		1.1199				UV
CI_4	C—I		2.15		(T_d)		ED
CN	C—N (r_e)		1.1718				MW
CO	C—O (r_e)		1.1283				MW
$COBr_2$	C—O		1.178		C—Br	1.923	ED, MW
	∠BrCBr		112.3				
COClF	C—F	1.334		C—O	1.173	C—Cl 1.725	ED, MW
	∠FCCl	108.8		∠ClCO	127.5		
$COCl_2$	C—O		1.179		C—Cl	1.742	ED, MW
	∠ClCCl		111.8				
COF_2	C—F		1.3157		C—O	1.172	ED, MW
	∠FCF		107.71				
CO_2	C—O (r_e)		1.1600				IR
CP	C—P (r_e)		1.562				UV
CS	C—S (r_e)		1.5349				MW
CS_2	C—S (r_e)		1.5526				IR
C_2	C—C (r_e)		1.2425				UV
C_3O_2	C—O		1.163		C—C	1.289	ED
	linear (large-amplitude bending vibration)						
CaH	Ca—H (r_e)		2.002				UV
CaO	Ca—O (r_e)		1.8221				UV
CaS	Ca—S (r_e)		2.3178				UV
CdH	Cd—H (r_e)		1.781				EPR
$CdBr_2$	Cd—Br		2.35		linear		ED
$CdCl_2$	Cd—Cl		2.24		linear		ED
CdI_2	Cd—I		2.56		linear		ED
ClCN	C—Cl (r_e)		1.629		C—N (r_e)	1.160	MW
ClF	Cl—F (r_e)		1.6283				MW
ClF_3	F_a—Cl—F_a			$Cl—F_a$	1.698	$Cl—F_b$ 1.598	MW
				∠F_aClF_b	87.5	(C_{2v})	
	F_b						
ClO	Cl—O (r_e)		1.5696				MW, UV
ClOH	O—Cl		1.690	O—H	0.975	∠HOCl 102.5	MW, IR
ClO_2	Cl—O		1.470		∠OClO	117.38	MW
$ClO_3(OH)$	O_a—Cl		1.407		O_b—Cl	1.639	ED
	∠O_aClO_a		114.3		∠O_aClO_b	104.1	

$$\begin{array}{c} H \\ | \\ O_b \\ | \\ Cl \\ O_a \diagdown O_a \\ O_a \end{array}$$

Compound	Structure						Method
Cl_2	Cl—Cl (r_e)		1.9878				UV
Cl_2O	Cl—O		1.6959		∠ClOCl	110.89	MW
CoH	Co—H (r_e)		1.542				UV
$Cr(CO)_6$	C—O		1.16		Cr—C	1.92	ED
	∠CrCO		180				
CrO	Cr—O (r_e)		1.615				UV
CsBr	Cs—Br (r_e)		3.0723				MW
CsCl	Cs—Cl (r_e)		2.9063				MW
CsF	Cs—F (r_e)		2.3454				MW
CsH	Cs—H (r_e)		2.4938				UV
CsI	Cs—I (r_e)		3.3152				MW

Compound	Structure						Method
CsOH	Cs—O (r_e)	2.395		O—H (r_e)	0.97		MW
CuBr	Cu—Br (r_e)	2.1734					MW
CuCl	Cu—Cl (r_e)	2.0512					MW
CuF	Cu—F (r_e)	1.7449					MW
CuH	Cu—H (r_e)	1.4626					UV
CuI	Cu—I (r_e)	2.3383					MW
FCN	C—F	1.262		C—N	1.159		MW
FOH	O—H	0.96	O—F	1.442	∠HOF	97.2	MW
F_2	F—F (r_e)	1.4119					R
$Fe(CO)_5$	Fe—C (average)	1.821					ED
	$(Fe—C)_{eq} - (Fe—C)_{ax}$		0.020				
	C—O (average)	1.153		(D_{3h})			
GaBr	Ga—Br (r_e)	2.3525					MW
GaCl	Ga—Cl (r_e)	2.2017					MW
GaF	Ga—F (r_e)	1.7744					MW
GaF_3	Ga—F	1.88		(D_{3h})			ED
GaI	Ga—I (r_e)	2.5747					MW
GaI_3	Ga—I	2.458		(D_{3h})			ED
GdI_3	Gd—I	2.841		∠IGdI	108	(C_{3v})	ED
$GeBrH_3$	Ge—H	1.526		Ge—Br	2.299		MW, IR
	∠HGeH	106.2					
$GeBr_4$	Ge—Br	2.272		(T_d)			ED
$GeClH_3$	Ge—H	1.537		Ge—Cl	2.150		IR, MW
	∠HGeH	111.0					
$GeCl_2$	Ge—Cl	2.183		∠ClGeCl	100.3		ED
$GeCl_4$	Ge—Cl	2.113		(T_d)			ED
$GeFH_3$	Ge—H	1.522		Ge—F	1.732		MW, IR
	∠HGeH	113.0					
GeF_2	Ge—F (r_e)	1.7321		∠FGeF (θ_e)	97.17		MW
GeH	Ge—H (r_e)	1.5880					UV
GeH_4	Ge—H	1.5251		(T_d)			IR, R
GeO	Ge—O (r_e)	1.6246					MW
GeS	Ge—S (r_e)	2.0121					MW
GeSe	Ge—Se (r_e)	2.1346					MW
GeTe	Ge—Te (r_e)	2.3402					MW
Ge_2H_6	Ge—H	1.541		Ge—Ge	2.403		ED
	∠HGeH	106.4		∠GeGeH	112.5		
HBr	H—Br (r_e)	1.4145					MW
HCN	C—H (r_e)	1.0655		C—N (r_e)	1.1532		MW, IR
				linear			
HCNO	H—C	1.027	C—N	1.161	N—O	1.207	MW
				linear			
HCl	H-Cl (r_e)	1.2746					MW
HF	H—F (r_e)	0.9169					MW
HI	H—I (r_e)	1.6090					MW
HNCO	N—H	0.986	N—C	1.209	C—O	1.166	MW
	∠HNC	128.0					
HNCS	N—H	0.989	N—C	1.216	C—S	1.561	MW
	∠HNC	135.0	∠NCS	180			
HNO	N—H	1.063	N—O	1.212	∠HNO	108.6	UV

HNO_2 O$_a$ N—O$_b$H

	s-trans conformer		s-cis conformer	MW
O$_b$—H	0.958		0.98	
N—O$_b$	1.432		1.39	
N—O$_a$	1.170		1.19	
∠O$_a$NO$_b$	110.7		114	
∠NO$_b$H	102.1		104	

HNO_3

				MW
O$_c$—H	0.96	N—O$_c$	1.41	
N—O$_a$	1.20	N—O$_b$	1.21	
∠HO$_c$N	102.2	∠O$_c$NO$_a$	113.9	
∠O$_c$NO$_b$	115.9	planar		

Compound	Structure						Method
HNSO	N—H	1.029		N—S	1.512	S—O 1.451	MW
	∠HNS	115.8		∠NSO	120.4		
				planar			
H_2	H—H (r_e)	0.7414					UV
H_2O	O—H (r_e)	0.9575			∠HOH (θ_e)	104.51	MW, IR
H_2O_2	O—O	1.475			∠OOH	94.8	IR
	dihedral angle of internal rotation				119.8	(C_2)	
H_2S	H—S (r_e)	1.3356			∠HSH (θ_e)	92.12	MW, IR
H_2SO_4	H_b			O—H	0.97	S—O_a 1.574	MW
	O_b			S—O_c	1.422	∠H_aO_aS 108.5	
	O_a H_a			∠O_aSO_b	101.3	∠O_cSO_d 123.3	
	S			∠O_aSO_c	108.6	∠O_aSO_d 106.4	
	O_d O_c		dihedral angle between the H_aO_aS and O_aSO_c planes		20.8		
			dihedral angle between the H_aO_aS and O_aSO_b planes		90.9		
			dihedral angle between the H_aSO_b and O_cSO_d planes		88.4	(C_2)	
H_2S_2	S—S	2.055		S—H	1.327	∠SSH 91.3	ED, MW
	dihedral angle of internal rotation		90.6	(C_2)			
$HfCl_4$	Hf—Cl	2.33		(T_d)			ED
$HgCl_2$	Hg—Cl	2.252		linear			ED
HgH	Hg—H (r_e)	1.7404					UV
HgI_2	Hg—I	2.553		linear			ED
IBr	I—Br (r_e)	2.4691					MW
ICN	C—I	1.995		C—N	1.159		MW
ICl	I—Cl (r_e)	2.3210					MW
IF_5	I—F (average)	1.860		$(I—F)_{eq} - (I—F)_{ax}$	0.03		ED, MW
	∠$F_{ax}IF_{eq}$	82.1		(C_{4v})			
IO	I—O (r_e)	1.8676					MW
I_2	I—I (r_e)	2.6663					R
InBr	In—Br (r_e)	2.5432					MW
InCl	In—Cl (r_e)	2.4012					MW
InF	In—F (r_e)	1.9854					MW
InH	In—H (r_e)	1.8376					UV
InI	In—I (r_e)	2.7537					MW
IrF_6	Ir—F	1.830		(O_h)			ED
KBr	K—Br (r_e)	2.8208					MW
KCl	K—Cl (r_e)	2.6667					MW
KF	K—F (r_e)	2.1716					MW
KH	K—H (r_e)	2.244					UV
KI	K—I (r_e)	3.0478					MW
KOH	O—H	0.91		K—O	2.212 linear		MW
K_2	K—K (r_e)	3.9051					UV
KrF_2	Kr—F	1.89		linear			ED
LiBr	Li—Br (r_e)	2.1704					MW
LiCl	Li—Cl (r_e)	2.0207					MW
LiF	Li—F (r_e)	1.5639					MW
LiH	Li—H (r_e)	1.5949					MW
LiI	Li—I (r_e)	2.3919					MW
Li_2	Li—Li (r_e)	2.6729					UV
Li_2Cl_2	Li			Li—Cl	2.23		ED
	Cl Cl			Cl—Cl	3.61		
	Li			∠ClLiCl	108		
$LuCl_3$	Lu—Cl	2.417		∠ClLuCl	112	(C_{3v})	ED
MgF	Mg—F (r_e)	1.7500					UV
MgH	Mg—H (r_e)	1.7297					UV
MgO	Mg—O (r_e)	1.749					UV
MnH	Mn—H (r_e)	1.7308					UV
$Mo(CO)_6$	Mo—C	2.063		C—O	1.145	(O_h)	ED
$MoCl_4O$	Mo—Cl	2.279		Mo—O	1.658		ED
	∠ClMoCl	87.2		(C_{4v})			

Compound	Structure					Method
MoF_6	Mo—F	1.820		(O_h)		ED
$NClH_2$	N—H	1.017		N—Cl	1.748	MW, IR
	∠HNCl	103.7		∠HNH	107	
NCl_3	N—Cl	1.759		∠ClNCl	107.1	ED
NF_2	N—F	1.3528		∠FNF	103.18	MW
NH_2	N—H	1.024		∠HNH	103.3	UV
NH_2CN	N—H	1.00		N_a—C	1.35	MW
	C—N_b	1.160		∠HNH	114	
	angle between the NH_2 plane and the N—C bond				142	

$$\overset{\displaystyle H}{\underset{\displaystyle H}{\Big\backslash}} N_a - C \equiv N_b$$

NH_2NO_2	N—N	1.427		N—H	1.005	MW
	∠HNH	115.2		∠ONO	130.1	
	dihedral angle between the NH_2 and NNO_2 planes				128.2	
NH_3	N—H (r_e)	1.012		∠HNH (θ_e)	106.7	IR
NH_4Cl	N—H	1.22	N—Cl	2.54	(C_{3v})	ED
NF_2CN	F_2N_b—C≡N_a		C—N_a	1.158	C—N_b 1.386	MW
	N_b—F	1.399	∠N_aCN_b	174		
	∠CN_bF	105.4	∠FN_bF	102.8		
NH	N—H (r_e)	1.0362				LMR
NH_2OH	N—H	1.02		N—O	1.453	O—H 0.962	MW
	∠HNH	107		∠HNO	103.3	∠NOH 101.4	
	The bisector of H—N—H angle is *trans* to the O—H bond					
NO	N—O (r_e)	1.1506				IR
NOCl	N—Cl	1.975		N—O	1.14	∠ONCl 113	MW
NOF	O—N	1.136		N—F	1.512	∠FNO 110.1	MW
NO_2	N—O	1.193		∠ONO	134.1		MW
NO_2Cl	N—Cl	1.840		N—O	1.202		MW
	∠ONO	130.6		(C_{2v})		
NO_2F	N—O	1.1798		N—F	1.467		MW
	∠ONO	136		(C_{2v})		
NS	N—S (r_e)	1.4940				IR
N_2	N—N (r_e)	1.0977				UV
N_2H_4	N—H	1.021		N—N	1.449	ED, MW
	∠HNH	106.6 (assumed)		∠NNH_a	112	
	∠NNH_b 106	dihedral angle of internal rotation		91		
	H_a: the H atom closer to the C_2 axis, H_b: the H atom farther from the C_2 axis					
N_2O	N—N (r_e)	1.1284		N—O (r_e)	1.1841	MW, IR
N_2O_3			N_a—N_b	1.864	N_a—O_a 1.142	MW
			N_b—O_b	1.202	N_b—O_c 1.217	
			∠$O_aN_aN_b$	105.05		
			∠$N_aN_bO_b$	112.72		
			∠$N_aN_bO_c$	117.47		

$$\overset{\displaystyle O_a \qquad O_b}{\underset{\displaystyle O_c}{N_a - N_b}}$$

N_2O_4			N—N	1.782	N—O 1.190	ED
			∠ONO	135.4	(D_{2h})	

$$\overset{\displaystyle O \qquad O}{\underset{\displaystyle O \qquad O}{N - N}}$$

NaBr	Na—Br (r_e)	2.5020				MW
NaCl	Na—Cl (r_e)	2.3609				MW
NaF	Na—F (r_e)	1.9260				MW
NaH	Na—H (r_e)	1.8873				UV
NaI	Na—I (r_e)	2.7115				MW
Na_2	Na—Na (r_e)	3.0789				UV
$NbCl_5$	Nb—Cl_{eq} 2.241		Nb—Cl_{ax} 2.338 (D_{3h})		ED	
NbO	Nb—O (r_e)	1.691				UV
$Ni(CO)_4$	Ni—C 1.838		C—O	1.141	(T_d)	ED
NiH	Ni—H (r_e)	1.476				UV
NpF_6	Np—F	1.981		(O_h)		ED
OCS	C—O (r_e)	1.1578		C—S (r_e)	1.5601	MW

Compound	Structure						Method
OCSe	C—O	1.159		C—Se	1.709		MW
OF	O—F (r_e)	1.3579					LMR
OF$_2$	O—F (r_e)	1.4053		∠FOF (θ_e)	103.07	(C_{2v})	MW
O(SiH$_3$)$_2$	Si—H	1.486		Si—O	1.634		ED
	∠SiOSi	144.1					
O$_2$	O—O (r_e)	1.2074					MW
O$_2$F$_2$	O—O	1.217		F—O	1.575		MW
	∠OOF	109.5	dihedral angle of internal rotation	87.5	(C_2)		
O$_3$	O—O (r_e)	1.2716		∠OOO (θ_e)	117.47	(C_{2v})	MW
OsF$_6$	Os—F	1.831		(O$_h$)			ED
OsO$_4$	Os—O	1.712		(T$_d$)			ED
PBr$_3$	P—Br	2.220		∠BrPBr	101.0		ED
PCl$_3$	P—Cl	2.039		∠ClPCl	100.27		ED
PCl$_5$	Cl$_a$ Cl$_b$ / Cl$_b$—P / Cl$_a$ Cl$_b$		P—Cl$_a$	2.124	P—Cl$_b$ (D$_{3h}$)	2.020	ED
PF	P—F (r_e)	1.5896					UV
PF$_3$	P—F	1.570		∠FPF	97.8		ED, MW
PF$_5$	P—F$_{ax}$	1.577	P—F$_{eq}$	1.534	(D$_{3h}$)		ED
PH	P—H (r_e)	1.4223					LMR
PH$_2$	P—H	1.418		∠HPH	91.70		UV
PH$_3$	P—H	1.4200		∠HPH	93.345		MW
PN	N—P (r_e)	1.4909					MW
PO	O—P (r_e)	1.4759					UV
POCl$_3$	P—O	1.449		P—Cl	1.993		ED
	∠ClPCl	103.3					
POF$_3$	P—O	1.436	P—F	1.524	∠FPF	101.3	ED, MW
P$_2$	P—P (r_e)	1.8931					UV
P$_2$F$_4$	P—F	1.587		P—P	2.281		ED
	∠PPF	95.4		∠FPF	99.1		
	The two PF$_2$ planes are *trans* to each other (the *gauche* conformer is less than 10%)						
P$_4$	P—P	2.21		(T$_d$)			ED
P$_4$O$_6$	P—O	1.638	∠POP	126.4	(T$_d$)		ED
PbH	Pb—H (r_e)	1.839					UV
PbO	Pb—O (r_e)	1.9218					MW
PbS	Pb—S (r_e)	2.2869					MW
PbSe	Pb—Se (r_e)	2.4022					MW
PbTe	Pb—Te (r_e)	2.5950					MW
PrI$_3$	Pr—I	2.904	∠IPrI	113	(C_{3v})		ED
PtO	Pt—O (r_e)	1.7273					UV
PuF$_6$	Pu—F	1.971		(O$_h$)			ED
RbBr	Rb—Br (r_e)	2.9447					MW
RbCl	Rb—Cl (r_e)	2.7869					MW
RbF	Rb—F (r_e)	2.2703					MW
RbH	Rb—H (r_e)	2.367					UV
RbI	Rb—I (r_e)	3.1768					MW
RbOH	Rb—O	2.301	O—H	0.957	linear		MW
ReClO$_3$	Re—O	1.702		Re—Cl	2.229		MW
	∠ClReO	109.4		(C_{3v})			
ReF$_6$	Re—F	1.832		(O$_h$)			ED
RuO$_4$	Ru—O	1.706		(T$_d$)			ED
SCSe	C—Se	1.693		C—S	1.553		MW
SCTe	C—S	1.557		C—Te	1.904		MW
SCl$_2$	S—Cl	2.006	∠ClSCl	103.0	(C_{2v})		ED
SF	S—F (r_e)	1.6006					MW
SF$_2$	S—F	1.5921		∠FSF	98.20		MW
SF$_6$	S—F	1.561		(O$_h$)			ED

Compound	Structure						Method
SO	S—O (r_e)	1.4811					MW
SOCl$_2$	S—O	1.44		S—Cl	2.072		MW
	∠ClSCl	97.2		∠OSCl	108.0		
SOF$_2$	S—O	1.420		S—F	1.583		ED
	∠OSF	106.2		∠FSF	92.2		
SOF$_4$	F$_b$ F$_b$		S—O	1.403	S—F$_a$	1.575	ED
	F$_a$—S—F$_a$		S—F$_b$	1.552	∠OSF$_a$	90.7	
	O		∠OSF$_b$	124.9	∠F$_a$SF$_b$	89.6	
			∠F$_b$SF$_b$	110.2	(C$_{2v}$)		
SO$_2$	S—O (r_e)	1.4308		∠OSO (θ_e)	119.329		MW
SO$_2$Cl$_2$	S—O	1.404	S—Cl	2.011	∠OSO	123.5	ED
	∠ClSCl	100.0	(C$_{2v}$)				
SO$_2$F$_2$	S—O	1.397	S—F	1.530	∠OSO	123	ED
	∠FSF	97	(C$_{2v}$)				
SO$_3$	S—O	1.4198		(D$_{3h}$)			IR
S(SiH$_3$)$_2$	Si—H	1.494	Si—S	2.136	∠SiSSi	97.4	ED
S$_2$	S—S (r_e)	1.8892					R
S$_2$Br$_2$	S—Br	2.24		S—S	1.98		ED
	∠SSBr	105	dihedral angle of internal rotation		83.5		
S$_2$Cl$_2$	S—Cl	2.057		S—S	1.931		ED
	∠SSCl	108.2	dihedral angle of internal rotation		84.1	(C$_2$)	
S$_2$O$_2$	S—O	1.458		S—S	2.025	∠OSS	112.8	MW
					planar *cis* form		
S$_8$			S—S	2.07			ED
			∠SSS	105			
			(D$_{4d}$)				
SbCl$_3$	Sb—Cl	2.333		∠ClSbCl	97.2		ED
SbH$_3$	Sb—H	1.704		∠HSbH	91.6		MW
SeF	Se—F	1.742					MW
SeF$_6$	Se—F	1.69		(O$_h$)			ED
SeO	Se—O (r_e)	1.6393					MW
SeOF$_2$	Se—O	1.576		Se—F	1.730		MW
	∠OSeF	104.82		∠FSeF	92.22		
SeO$_2$	Se—O (r_e)	1.6076		∠OSeO (θ_e)	113.83		MW
SeO$_3$	Se—O	1.69		(D$_{3h}$)			ED
Se$_2$	Se—Se (r_e)	2.1660					UV
Se$_6$	Se—Se	2.34		∠SeSeSe	102		ED
	six-membered ring with chair conformation						
SiBrF$_3$	Si—F	1.560		Si—Br	2.153		MW
	∠FSiBr	108.5		(C$_{3v}$)			
SiBrH$_3$	Si—H	1.485		Si—Br	2.210		MW
	∠HSiBr	107.8		(C$_{3v}$)			
SiClH$_3$	Si—H	1.482		Si—Cl	2.048		MW
	∠HSiCl	107.9		(C$_{3v}$)			
SiCl$_4$	Si—Cl	2.019		(T$_d$)			ED
SiF	Si—F	1.6008					UV
SiFH$_3$	Si—H	1.484		Si—F	1.593		MW, IR
	∠HSiH	110.63		(C$_{3v}$)			
SiF$_2$	Si—F (r_e)	1.590		∠FSiF (θ_e)	100.8		MW
SiF$_3$H	Si—H (r_e)	1.4468		Si—F (r_e)	1.5624		MW
	∠HSiF (θ_e)	110.64					
SiF$_4$	Si—F	1.553		(T$_d$)			ED
SiH	Si—H (r_e)	1.5201					UV
SiH$_3$I	Si—H	1.485		Si—I	2.437		MW
	∠HSH	107.8					
SiH$_4$	Si—H	1.4798		(T$_d$)			IR
SiN	N—Si (r_e)	1.572					UV
SiO	Si—O (r_e)	1.5097					MW

Compound	Structure					Method
SiS	Si—S (r_e)	1.9293				MW
SiSe	Se—Si (r_e)	2.0583				MW
Si$_2$	Si—Si (r_e)	2.246				UV
Si$_2$Cl$_6$	Si—Si	2.32	Si—Cl	2.009		ED
	∠ClSiCl	109.7				
Si$_2$F$_6$	Si—Si	2.317	Si—F	1.564		ED
	∠FSiF	108.6				
Si$_2$H$_6$	Si—H	1.492	Si—Si	2.331		ED
	∠SiSiH	110.3	∠HSiH	108.6		
			staggered form (assumed)			
SnCl$_4$	Sn—Cl	2.280	(T$_d$)			ED
SnH	Sn—H (r_e)	1.7815				UV
SnH$_4$	Sn—H	1.711	(T$_d$)			R, IR
SnO	Sn—O	1.8325				MW
SnS	S—Sn (r_e)	2.2090				MW
SnSe	Se—Sn (r_e)	2.3256				MW
SnTe	Sn—Te (r_e)	2.5228				MW
SrH	Sr—H (r_e)	2.1455				UV
SrO	Sr—O (r_e)	1.9198				MW
SrS	S—Sr (r_e)	2.4405				UV
TaCl$_5$	Ta—Cl$_{eq}$	2.227	Ta—Cl$_{ax}$	2.369	(D$_{3h}$)	ED
TaO	Ta—O (r_e)	1.6875				UV
TeF$_6$	Te—F	1.815	(O$_h$)			ED
Te$_2$	Te—Te (r_e)	2.5574				UV
ThCl$_4$	Th—Cl	2.58	(T$_d$)			ED
ThF$_4$	Th—F	2.14	(T$_d$)			ED
TlBr	Tl—Br (r_e)	2.6182				MW
TlCl	Tl—Cl (r_e)	2.4848				MW
TlF	Tl—F (r_e)	2.0844				MW
TlH	Tl—H (r_e)	1.870				UV
TlI	Tl—I (r_e)	2.8137				MW
TiBr$_4$	Ti—Br	2.339	(T$_d$)			ED
TiCl$_4$	Ti—Cl	2.170	(T$_d$)			ED
TiO	Ti—O (r_e)	1.620				UV
TiS	Ti—S (r_e)	2.0825				UV
UF$_6$	U—F	1.996	(O$_h$)			ED
V(CO)$_6$	V—C	2.015	C—O	1.138		ED
	(O$_h$, involving dynamic Jahn-Teller effect)					
VCl$_3$O	V—O	1.570	V—Cl	2.142		ED, MW
	∠ClVCl	111.3				
VCl$_4$	V—Cl	2.138	(T$_d$, involving dynamic Jahn-Teller effect)			ED
VF$_5$	V—F (average)	1.71				ED
VO	V—O (r_e)	1.5893				UV
W(CO)$_6$	W—C	2.059	C—O	1.149	(O$_h$)	ED
WClF$_5$			W—Cl	2.251		MW
			W—F (average)	1.836		
			∠F$_a$WF$_b$	88.7		
WF$_4$O	W—O	1.666	W—F	1.847		ED
	∠FWF	86.2	(C$_{4v}$)			
WF$_6$	W—F	1.832	(O$_h$)			ED
XeF$_2$	Xe—F	1.977	linear			IR
XeF$_4$	Xe—F	1.94	(D$_{4h}$)			ED
XeF$_6$	Xe—F	1.890	(large-amplitude bending vibration around the O$_h$ structure)			ED
XeO$_4$	Xe—O	1.736	(T$_d$)			ED
ZnH	Zn—H (r_e)	1.5949				UV
ZrCl$_4$	Zr—Cl	2.32	(T$_d$)			ED
ZrF$_4$	Zr—F	1.902	(T$_d$)			ED
ZrO	Zr—O (r_e)	1.7116				UV

For WClF$_5$:

```
      Cl  F_b
       |  /
 F_b—W—F_b
      / |
   F_b  F_a
```

STRUCTURES OF ORGANIC MOLECULES
Compounds are Arranged in Alphabetical Order by Chemical Name; Cross References are Given for Common Synonyms (Lengths in Å and Angles in Degrees)

Compound	Structure			Method
Acetaldehyde		C_a—O	1.210	ED, MW
		C_b—H	1.107	
		C_a—H	1.128	
	C_a—C_b 1.515	$\angle C_bC_aO$	124.1	
	$\angle HC_bH$ 109.8	$\angle C_bC_aH$	115.3	
Acetamide	C—O 1.220	C—N	1.380	ED
CH_3CONH_2	C—C 1.519	N—H	1.022	
	C—H 1.124	$\angle NCO$	122.0	
	$\angle CCN$ 115.1			
Acetic acid		C—C	1.520	ED
		C—O_a	1.214	
		C—O_b	1.364	
	C—H 1.10	$\angle CCO_a$	126.6	
	$\angle CCO_b$ 110.6			
Acetone	C—C 1.520	C—O	1.213	ED, MW
$(CH_3)_2CO$	C—H 1.103	$\angle CCC$	116.0	
	$\angle HCH$ 108.5	symmetry axis of each methyl group		
		is tilted 2 from the C—C bond		
Acetonitrile	C—H 1.107	C—C	1.468	ED, MW
CH_3CN	C—N 1.159	$\angle CCH$	109.7	
Acetonitrile oxide	C—C 1.442	C—N	1.169	MW
CH_3CNO	N—O 1.217	(C_{3v})		
Acetyl chloride	C—H 1.105	C—O	1.187	ED, MW
CH_3COCl	C—C 1.506	C—Cl	1.798	
	$\angle HCH$ 108.6	$\angle OCCl$	121.2	
	$\angle CCCl$ 111.6			
Acetyl cyanide → Pyruvonitrile				
Acetylene	C—H (r_e) 1.060	C—C (r_e)	1.203	IR
HC≡CH				
Acrolein → Acrylaldehyde				
Acrylaldehyde		C_b—C_c	1.484	ED, MW
		C_a—C_b	1.345	
		C_c—O	1.217	
		C_a—H	1.10	
		C_c—H	1.13	
	$\angle C_aC_bC_c$ 120.3	$\angle C_bC_cO$	123.3	
	$\angle HC_cC_b$ 114	other CCH angles (average)	122	
	planar s-trans form			
Acrylonitrile		C_a—C_b	1.343	ED, MW
		C_b—C_c	1.438	
		C_c—N	1.167	
		C_a—H	1.114	
		$\angle C_aC_bC_c$	121.7	
	$\angle C_bC_cN$ 178	$\angle HCC$	120	
Acryloyl chloride	C—H 1.086 (assumed)	C_b—C_c	1.48	MW
	C_c—Cl 1.82	C_a—C_b	1.35	
		C_c—O	1.19	
		$\angle C_aC_bH$	120 (assumed)	
		$\angle C_bC_aH$	121.5 (assumed)	
		$\angle C_aC_bC_c$	123	
		$\angle C_bC_cCl$	116	
		$\angle C_bC_cO$	127	

Compound	Structure					Method

Allene
$CH_2=C=CH_2$

C—C	1.3084	C—H	1.087	IR
∠HCH	118.2			

Allyl chloride

Cl

$H_2C_a=C_bH—C_cH_2$

∠CCCl   109.6

cis conformer   C—Cl   1.811   MW
  ∠CCCl   115.2
skew conformer   C—Cl   1.809
CCCCl   dihedral angle of internal rotation   122.4

Aniline
$C_6H_5NH_2$

C—C	1.392	C—N	1.431	MW
N—H	0.998	∠HNH	113.9	

dihedral angle between the NH_2 plane and the N—C bond   140.6

Azetidine

$CH_2—CH_2$
 | |
$CH_2—NH$

N—H   1.03
∠CCC   86.9

C—N   1.482   ED
C—C   1.553
C—H   1.107
∠CNC   92.2
∠CCN   85.8

dihedral angle between the CCC and CNC planes   147

Aziridine

H_a
H_b N H_b
  C — C
H_c H_c

N—H   1.016   MW
N—C   1.475
C—C   1.481
C—H   1.084
∠CNC   60.3
∠H_aNC   109.3

∠H_bCH_c 115.7   ∠H_bCC 117.8
∠H_bCN 118.3   ∠H_cCC 119.3
∠H_cCN 114.3

Azomethane
$CH_3N=NCH_3$

C—N	1.482	N—N	1.247	ED
∠CNN	112.3	*trans* conformer		

Benzene
C_6H_6

C—C	1.399	C—H	1.101	ED, IR

p-Benzoquinone

H   H
  $C_b=C_b$
$O=C_a$ $C_a=O$
  $C_b=C_b$
H   H

C_a—O   1.225   ED
C_b—C_b 1.344
C_a—C_b 1.481
∠$C_bC_aC_b$ 118.1

Biacetyl
$CH_3COCOCH_3$

C—O	1.215	C—C (average)	1.524	ED
C—H	1.108	∠CCO	119.5	
∠CCC	116.2	*trans* conformer		

Bicyclo[1.1.0]butane

H_a
C_a
H_b H_b
  C_a
C_b C_b
H_a
H_c H_c

C_a—C_a 1.497   MW
C_a—C_b 1.498
C_a—H_a 1.071
C_b—H_b, C_b—H_c 1.093
∠$H_bC_bH_c$ 115.6

∠$C_bC_aH_a$ 130.4   ∠$C_aC_aC_a$ 60.0
∠$C_aC_aH_a$ 128.4   dihedral angle between the two $C_aC_aC_b$ planes   121.7

Bicyclo[2.2.1]hepta-2,5-diene

H_2
C_c
C_aH
HC_b C_bH
C_a
HC_b H C_bH

C_a—C_b 1.535   ED
C_b—C_b 1.343
C_a—C_c 1.573
C—H   1.12
∠$C_aC_cC_a$ 94

dihedral angle between the two $C_aC_bC_bC_a$ planes   115.6
(C_{2v})

Bicyclo[2.2.1]heptane
C_7H_{12}

See the preceding molecule for the labels of the C atoms   ED

C_a—C_b 1.54 C_b—C_b 1.56
C_a—C_c 1.56   C—C (average)   1.549
∠$C_aC_cC_a$ 93.1   dihedral angle between the two $C_aC_bC_bC_a$ planes   113.1

Compound	Structure			Method
Bicyclo[2.2.0]hexa-2,5-diene	H C_a HC_b C_bH HC_b C_bH C_a H	C_b—C_b C_a—C_a C_a—C_b dihedral angle between the two $C_aC_bC_bC_a$ planes	1.345 1.574 1.524 117.3	ED
Bicyclo[2.2.2]octane	$HC_a(C_bH_2C_bH_2)_3C_aH$ C_b—C_b 1.55 C—C (average) 1.542 large-amplitude torsional motion about the D_{3h} symmetry axis	C_a—C_b ∠$C_aC_bC_b$	1.54 109.7	ED
Bicyclo[1.1.1]pentane C_5H_8	C—C 1.557	∠CCC	74.2	ED
Bicyclo[2.1.0]pentane	C_bH_2—C_aH C_b C_a C_cH_2 H_2 H Dihedral angle between the $C_aC_aC_bC_b$ and $C_cC_aC_c$ planes	C_a—C_a C_b—C_b C_a—C_b C_a—C_c 	1.536 1.565 1.528 1.507 112.7	MW
Biphenyl	(structure: two fused benzene rings)	C—C (intra-ring) (inter-ring) torsional dihedral angle between the two rings	1.396 1.49 ~40	ED
4,4′-Bipyridyl	C—C, C—N (intra-ring) N (structure: two pyridine rings) N	1.375 C—C (inter-ring) 1.465 torsional dihedral angle between the two rings ~37		ED
Bis(cyclopentadienyl) beryllium $(C_5H_5)_2Be$	Be—(cyclopentadienyl plane) 1.470, 1.92 C—C 1.423 (C_{5v}) (The Be atom has two equilibrium positions)			ED
Bis(cyclopentadienyl) iron → Ferrocene				
Bis(cyclopentadienyl) lead $(C_5H_5)_2Pb$	C—C 1.430 dihedral angle between the two C_5H_5 planes 40~50 (The two rings are not parallel.)	Pb—C	2.79	ED
Bis(cyclopentadienyl) manganese $(C_5H_5)_2Mn$	Mn—C 2.383 C—C 1.429 (D_{5h})			ED
Bis(cyclopentadienyl) nickel $(C_5H_5)_2Ni$	Ni—C 2.196	C—C	1.430 (D_{5h})	ED
Bis(cyclopentadienyl) ruthenium $(C_5H_5)_2Ru$	C—C 1.439	Ru—C	2.196	ED
Bis(cyclopentadienyl) tin $(C_5H_5)_2Sn$	C—C 1.431 C—H 1.14	Sn—C (D_{5h})	2.71	ED
Bis(trifluoromethyl) peroxide CF_3OOCF_3	O—O 1.42 C—F 1.320 ∠FCF 109.0	C—O ∠COO COOC dihedral angle of internal rotation	1.399 107 123	ED
Borine carbonyl BH_3CO	B—H 1.194 C—O 1.131 ∠BCO 180	B—C ∠HBH (C_{3v})	1.540 113.9	MW
Bromobenzene	Br C_a HC_b C_bH HC_c C_cH C_d H	C—H C_c—C_d C_b—C_c C—Br C_a—C_b ∠$C_bC_aC_b$	1.072 1.401 1.375 1.85 1.42 117.4	MW
Bromoform $CHBr_3$	C—Br 1.924 ∠BrCBr 111.7	C—H (C_{3v})	1.11	ED, MW
Bromoiodoacetylene $IC≡CBr$	C—I 1.972 C—Br 1.795	C—C	1.206	ED

Compound	Structure					Method

1,3-Butadiene

C_aH_2
$C_bH—C_bH$
C_aH_2

$C_b—C_b$	1.467	ED
$C_a—C_b$	1.349	
C—H (average)	1.108	
∠CCC	124.4	

∠C_bC_aH 120.9   *anti* conformer (C_{2h})

1,3-Butadiyne $HC_a≡C_b—C_b≡C_aH$

C—H	1.09	ED
$C_a—C_b$	1.218	
$C_b—C_b$	1.384	
linear		

Butane $CH_3CH_2CH_2CH_3$

C—C	1.531	C—H	1.117	ED
∠CCC	113.8	∠CCH	111.0	

trans conformer   54%   dihedral angle for the *gauche* conformer   65

2-Butanone → Ethyl methyl ketone

Butatriene $H_2C_a=C_b=C_b=C_aH_2$

		C—H	1.08	ED
$C_a—C_b$	1.32	$C_b—C_b$	1.28 (D_{2h})	

2-Butene $C_aH_3—C_bH=C_bH—C_aH_3$

		ED	
$C_a—C_b$	*cis* conformer	1.506	*trans* conformer 1.508
$C_b—C_b$	1.346	1.347	
∠$C_aC_bC_b$	125.4	123.8	

3-Buten-1-yne → Vinylacetylene

tert-Butyl chloride $(CH_3)_3CCl$

C—H	1.102	C—C	1.528	ED, MW
C—Cl	1.828	∠CCCl	107.3	
∠CCH	110.8	∠CCC	111.6	

tert-Butyl cyanide → Pivalonitrile

2-Butyne $C_aH_3—C_b≡C_b—C_aH_3$

C—H	1.116	ED	
$C_b—C_b$	1.214	$C_a—C_b$	1.468
∠C_bC_aH	110.7		

Compound			Method
Carbon C_2	C—C (r_e)	1.3119	UV
Carbon C_3	C—C	1.277 linear	UV

Carbon suboxide → Tricarbon dioxide

Carbon tetrabromide CBr_4	C—Br	1.935 (T_d)	ED
Carbon tetrachloride CCl_4	C—Cl	1.767 (T_d)	ED
Carbon tetrafluoride CF_4	C—F	1.323 (T_d)	ED
Carbon tetraiodide CI_4	C—I	2.15 (T_d)	ED

Carbonyl cyanide $CO(CN)_2$

C—O	1.209	C—C	1.466	ED, MW
C—N	1.153	∠CCC	115	
∠CCN	180			

Chloroacetylene $HC≡CCl$

C—H	1.0550	C—C	1.2033	MW
C—Cl	1.6368			

Chlorobenzene C_6H_5Cl

C—C	1.400	C—Cl	1.737	ED
C—H	1.083	∠CC(Cl)C	121.7	
∠CC(H)C	120			

Chlorobromoacetylene $ClC≡CBr$

Cl—C	1.636	C—C	1.206	ED
C—Br	1.784			

Chlorocyanoacetylene $ClC≡CCN$

C—Cl	1.624	C—C	1.205	ED
C—CN	1.362	C—N	1.160	

Chloroethane → Ethyl chloride

2-Chloroethanol $ClCH_2CH_2OH$

C—O	1.413	C—C	1.519	ED
C—Cl	1.801	C—H	1.093	
O—H	1.033 ∠CCCl 110.7 ∠CCO 113.8			

fraction of the *gauche* conformer at 37C is 92 ~ 94%,
dihedral angle of internal rotation   62.4

Chloroethylene → Vinyl chloride

Chloroform $CHCl_3$

C—H	1.100	C—Cl	1.758	MW
∠ClCCl	111.3	(C_{3v})		

Compound	Structure					Method
Chloroiodoacetylene	C—Cl	1.63		C—I	1.99	MW
ClC≡CI	C—C	1.209 (assumed)				
Chloromethane → Methyl chloride						
3-Chloropropene → Allyl chloride						
Cyanamide	N_a—C	1.346		C—N_b	1.160	MW
$H_2N_aCN_b$	N—H	1.00		∠HNH	114	
	dihedral angle between the NH$_2$ plane and the N—C bond				142	
Cyanoacetylene	C_b—H	1.058		C_a—C_b	1.205	MW
H—C_b≡C_a—C_c≡N	C_a—C_c	1.378		C_c—N	1.159	
Cyanocyclopropane	C—C (ring)	1.513		C—C_a	1.472	MW
$C_3H_5C_aN$	C—H	1.107		C_a—N	1.157	
	∠HCH	114.6		∠C_aCH	119.6	
Cyanogen	C—N	1.163		C—C	1.393	ED
(CN)$_2$					linear	
Cyclobutane	C—H	1.113		C—C	1.555	ED
(CH$_2$)$_4$	dihedral angle between the two CCC planes 145					
Cyclobutanone				C_a—C_b	1.527	MW
	C_bH_2			C_b—C_c	1.556	
	C_cH_2 C_a=O			∠$C_bC_aC_b$	93.1	
	C_bH_2			∠$C_aC_bC_c$	88.0	
Cyclobutene	H_2C_a—C_aH_2	C_b—C_b	1.342	C_a—C_a	1.566	MW
	HC$_b$=C$_b$H	C_a—C_b	1.517	C_a—H	1.094	
		C_b—H	1.083			
	∠$C_aC_bC_b$	94.2		∠C_bC_aH	133.5	
	∠C_aC_aH	114.5		∠$C_aC_aC_b$	85.8	
	∠HC_aH	109.2		dihedral angle between the CH$_2$ plane and the C_a—C_a bond 135.8		
Cyclohexane	C—C	1.536		C—H	1.119	ED
C_6H_{12}	∠CCC	111.3		chair form		
Cyclohexene				C_a—C_a	1.334	ED
	HC$_a$=C$_a$H			C_a—C_b	1.50	
	H_2C_b C_bH_2			C_b—C_c	1.52	
	C_cH_2—C_cH_2			C_c—C_c	1.54	
	∠$C_aC_aC_b$	123.4		∠$C_aC_bC_c$	112.0	
	∠$C_bC_cC_c$	110.9		(C$_2$)	half-chair form	
Cyclooctatetraene				C_a—C_b	1.476	ED
				C—H	1.100	
				C_a—C_a, C_b—C_b	1.340	
				∠$C_bC_aC_a$, ∠$C_aC_bC_b$	126.1	
	dihedral angle between the $C_aC_aC_aC_a$ and $C_aC_bC_bC_a$ planes				136.9	
	tub form (D$_{2d}$)					
1,3-Cyclopentadiene				C_a—C_b	1.509	MW
	C_aH_2			C_b—C_c	1.342	
	HC$_b$ C_bH			C_c—C_c	1.469	
	HC$_c$—C_cH			∠$C_aC_bC_c$	109.3	
	∠$C_bC_cC_c$	109.4		∠$C_cC_aC_b$	102.8	
Cyclopentadienylindium	In			In—C	2.621	ED
	HC—CH			C—C	1.426	
	HC CH			(C$_{5v}$)		
	CH					
Cyclopentane	C—H	1.114		C—C	1.546	ED
(CH$_2$)$_5$	∠HCH	111.7				
	(The out-of-plane vibration of the C atoms is essentially free pseudorotation; average value of the displacements of the C atoms from the molecular plane 0.43)					

Compound	Structure					Method

Cyclopentene

C_aH_2
H_2C_b C_bH_2
$C_cH=C_cH$

C_a—C_b	1.546	ED
C_b—C_c	1.519	
C_c—C_c	1.342	
$\angle C_bC_aC_b$	104.0	
$\angle C_aC_bC_c$	103.0	

$\angle C_bC_cC_c$ 110.0

dihedral angle between the $C_bC_aC_b$ and $C_bC_cC_cC_b$ planes   151.2

Cyclopropane
$(CH_2)_3$

C—C	1.512	C—H	1.083	R
$\angle$HCH	114.0			

Cyclopropanone

H_2C_b
   $C_a=O$
H_2C_b

C—H	1.086	C_a—C_b	1.475	MW
C_b—C_b	1.575	C_a—O	1.191	
$\angle C_aC_bC_b$	57.7			
$\angle HC_bH$	114			

dihedral angle between the CH_2 plane and the C_b—C_b bond   151

Cyclopropene

C_aH_2
HC_b═══C_bH

C_b—C_b	1.304	ED
C_a—C_b	1.519	
C_b—H	1.077	
$\angle C_cC_bH$	133	
C_a—H	1.112		
$\angle HC_aH$	118		

Decalin
$C_{10}H_{18}$

C—C (average)	1.530	C—H (average)	1.113	ED
$\angle$CCC (average)	111.4			

Dewer benzene → Bicyclo[2.2.0] hexa-2,5-diene

Diacetylene → 1,3-Butadiyne

1,4-Diazabicyclo[2.2.2]octane

CH_2CH_2
N CH_2CH_2—N
CH_2CH_2

C—N	1.472	ED
C—C	1.562	
$\angle$NCC	110.2	
$\angle$CNC	108.7	

large-amplitude torsional motion about the D_{3h} symmetry axis

2,3-Diaza-1,3-butadiene → Formaldehyde azine

Diazirine

CH_2
 N‖N

C—H	1.09	MW
C—N	1.482	
N—N	1.228	
$\angle$HCH	117	

Diazoacetonitrile

H
C_b
C_a N_b
N_a N_c

C_b—N_b	1.280	MW
N_b—N_c	1.132	
C_a—N_a	1.165	
C—H	1.082	
C_a—C_b	1.424	

$\angle C_aC_bH$ 117 $\angle C_aC_bN_b$ 119.5

Diazomethane
CH_2N_2

C—H	1.075	C—N	1.32	MW, IR
N—N	1.12	$\angle$HCH	126.0 linear	

1,2-Dibromoethane
CH_2BrCH_2Br

C—C	1.506	C—Br	1.950	ED
C—H	1.108	$\angle$CCBr	109.5	

$\angle$CCH   110   fraction of the *trans* conformer at 25°C   95%

Dibromomethane
CH_2Br_2

C—H	1.08	C—Br	1.924	ED
$\angle$HCBr	109	$\angle$BrCBr	113.2	

2,2′-Dichlorobiphenyl
C_6H_4Cl—C_6H_4Cl

C—C	1.398	C—C inter-ring	1.495	ED
C—Cl	1.732	C—H	1.10	
$\angle$CCCl	121.4	$\angle$CCH	126	

dihedral angle between the two aromatic rings 74 (defined to be 0 for that of the *cis* conformer)

trans-1,4-Dichlorocyclohexane
$C_6H_{10}Cl_2$

C—H	1.102	C—Cl	1.810	ED
C—C	1.530	$\angle$CCC	111.5	
$\angle$CCCl (*ee*)	108.6	$\angle$CCCl (*aa*)	110.6	
$\angle$HCCl (*ee*)	111.5	$\angle$HCCl (*aa*)	107.6	
ee 49%		*aa* 51%	e: equatorial, a: axial	

Compound	Structure				Method
1,1-Dichloroethane	C—Cl	1.766	C—C	1.540	MW
$CHCl_2CH_3$	∠ClCCl	112.0	∠CCCl	111.0	
1,2-Dichloroethane	C—C	1.531	C—Cl	1.790	ED
CH_2ClCH_2Cl	C—H	1.11	∠CCCl	109.0	
	∠CCH	113			
	fraction of the *trans* conformer at room temperature 73%, that of the *gauche* conformer 27%				
1,1-Dichloroethylene	C—C	1.32 (assumed)	C—Cl	1.73	MW
$CH_2=CCl_2$	∠ClCC	123	(C_{2v})		
cis-1,2-Dichloroethylene	C—Cl	1.718	C—C	1.354	ED
CHCl=CHCl	∠ClCC	123.8			
Dichloromethane	C—H (r_e)	1.087	C—Cl (r_e)	1.765	MW, IR
CH_2Cl_2	∠HCH (θ_e)	111.5	∠ClCCl (θ_e)	112.0	
1,1-Difluoroethane	C—C	1.498	C—H (average)	1.081	ED
CH_3CHF_2	C—F	1.364	∠CCH (average)	111.0	
	∠CCF	110.7	dihedral angle between the two CCF planes	118.9	
1,2-Difluoroethane	C—F	1.389	C—C	1.503	ED
CH_2FCH_2F	C—H	1.103	∠CCF	110.3	
	∠CCH	111	dihedral angle of internal rotation	109	
	fraction of the *gauche* conformer at 22°C 94%				
1,1-Difluoroethane	C—C	1.340	C—F	1.315	ED, MW
$CH_2=CF_2$	C—H	1.091	∠CCF	124.7	
	∠CCH	119.0			
cis-1,2-Difluoroethylene	C—C	1.33	C—F	1.342	ED, MW
CHF=CHF	C—H	1.099	∠CCF	122.0	
	∠CCH	124.1			
Difluoromethane	C—H	1.093	C—F	1.357	MW
CH_2F_2	∠HCH	113.7	∠FCF	108.3	
Dimethoxymethane			C_a—O	1.432	ED
			C_b—O	1.382	
			C—H (average)	1.108	
	∠COC	114.6	∠OCO	114.3	
	∠OCH	110.3			
Dimethylacetylene → 2-Butyne					
Dimethylamine	C—H	1.106	N—H	1.00	ED
$(CH)_2NH$	C—N	1.455	∠CNC	111.8	
	∠CNH	107	∠NCH	112	
	∠HCH	107			
Dimethylberyllium	Be—C	1.698	C—H	1.127	ED
$(CH_3)_2Be$	∠BeCH	113.9	CBeC linear		
Dimethylcadmium	C—Cd	2.112	∠HCH	108.4	R
$(CH_3)_2Cd$					
Dimethyl carbonate	$C_aH_3O_a$		C_b—O_b	1.209	ED
			C_b—O_a	1.34	
			C_a—O_a	1.42	
	∠$O_aC_bO_a$	107	∠$C_bO_aC_a$	114.5	
Dimethylcyanamide	C_b—N_b	1.161	C_b—N_a	1.338	ED
$(C_aH_3)_2N_a$—$C_b≡N_b$	C_a—N_a	1.463	∠C_aNC_a	115.5	
	∠C_aNC_b	116.0			
1,2-Dimethyldiborane			B—B	1.799	ED
			B—C	1.580	
			B—H_b	1.358 (*cis*), 1.365 (*trans*)	
	B—H_t	1.24			
	∠BBC	122.6 (*cis*), 121.8 (*trans*)			
Dimethyl diselenide	C—H	1.13	C—Se	1.95	ED
$(CH_3)_2Se_2$	Se—Se	2.326	∠CSeSe	98.9	
	∠HCSe	108	dihedral angle between the CSeSe and SeSeC planes	88	

Compound	Structure				Method
Dimethyl disulfide	C—S	1.816	S—S	2.029	ED
$(CH_3)_2S_2$	C—H	1.105	∠SSC	103.2	
	∠SCH	111.3	CSSC dihedral angle of internal rotation	85	
S,S'-Dimethyl dithiocarbonate	$C_aH_3SC_b SC_aH_3$		C_b—O	1.206	ED
	‖		C_b—S	1.777	
	O		C_a—S	1.802	
	∠OCS	124.9	∠CSC	99.3	
			syn-syn conformer		
Dimethyl ether	C—O	1.416	C—H	1.121	ED
$(CH_3)_2O$	∠COC	112	∠HCH	108	
Dimethylglyoxal → Biacetyl					
N,N'-Dimethylhydrazine	N—N	1.42	C—N	1.46	ED
$CH_3NH—NHCH_3$	N—H	1.03	C—H	1.12	
	∠NNC	112	CNNC dihedral angle of internal rotation	90	
Dimethylmercury	C—Hg	2.083	C—H	1.160 (assumed)	ED
$(CH_3)_2Hg$	Hg⋯H	2.71			
Dimethylphosphine	C—P	1.848	P—H	1.419	MW
$(CH_3)_2PH$	∠CPC	99.7	∠CPH	97.0	
Dimethyl selenide	C—H	1.093	Se—C	1.943	MW
$(CH_3)_2Se$	∠CSeC	96.2	∠SeCH	108.7	
	∠HCH	110.3			
Dimethyl sulfide	C—S	1.807	C—H	1.116	ED, MW
$(CH_3)_2S$	∠CSC	99.05	∠HCH	109.3	
Dimethyl sulfone	C—H	1.114	S—O	1.435	ED
$(CH_3)_2SO_2$	S—C	1.771	∠CSC	102	
	∠OSO	121			
Dimethyl sulfoxide	C—H	1.081	C—S	1.799	MW
$(CH_3)_2SO$	S—O	1.485	∠CSC	96.6	
	∠CSO	106.7	∠HCH	110.3	
	dihedral angle between the SCC plane and the S—O bond			115.5	
Dimethylzinc	Zn—C	1.929	∠HCH	107.7	R
$(CH_3)_2Zn$					
1,4-Dioxane	C—C	1.523	C—O	1.423	ED
CH_2CH_2	C—H	1.112	∠COC	112.45	
O⟨⟩O	∠CCO	109.2	chair form		
CH_2CH_2					
Ethanal → Acetaldehyde					
Ethane	C—C	1.5351	C—H	1.0940	MW
C_2H_6	∠CCH	111.17	staggered conformation		
Ethanethiol	$C_bH_3—C_aH_2—SH$		C_a—H	1.090	MW
	C_b—H	1.093	C_a—C_b	1.530	
	C_a—S	1.829	S—H	1.350	
	∠C_bC_aH	109.6	∠C_aC_bH	109.7	
	∠C_bC_aS	108.3	∠C_aSH	96.4	
Ethanol	$C_bH_3C_aH_2OH$		C—C	1.512	MW
	C—O	1.431	O—H	0.971	
	C_a—H	1.10	C_b—H	1.09	
	∠CCO	107.8	∠COH	105	
	∠C_bC_aH	111	∠C_aC_bH	110	
	staggered conformation				
Ethyl chloride	H_b Cl		C—C	1.528	ED, MW
			C—Cl	1.802	
	H_b—C_b—C_a—H_a		C—H	1.103	
	H_b H_a		C_a—H_a=C_b—H_b (assumed)		
			∠CCCl	110.7	
	∠$H_bC_bH_b$	109.8	∠$H_aC_aH_a$	109.2	
	∠$C_bC_aH_a$	110.6			
Ethylene	C—H	1.087	C—C	1.339	MW
$CH_2=CH_2$	∠CCH	121.3			

Compound	Structure					Method
Ethylenediamine	C—N	1.469		C—C	1.545	ED
$H_2NCH_2CH_2NH_2$	C—H	1.11		∠CCN	110.2	
	gauche conformer	dihedral angle between the NCC and CCN planes			64	
Ethylene dibromide → 1,2-Dibromoethane						
Ethylene dichloride → 1,2-Dichloroethane						
Ethyleneimine → Aziridine						
Ethylene oxide	C—C	1.466		C—H	1.085	MW
CH_2	C—O	1.431		∠HCH	116.6	
⟍O	dihedral angle between the NH_2 plane and the N—C bond				158.0	
CH_2						
Ethylene sulfide → Thiirane						
Ethyl methyl ether	C—O (average)	1.418		C—C	1.520	ED
$C_2H_5OCH_3$	C—H (average)	1.118		∠COC	111.9	
	∠OCC	109.4		∠HCH	109.0	
	fraction of the *trans* conformer at 20°C	80%				
Ethyl methyl ketone				C—C (average)	1.518	ED
				C_c—O	1.219	
				C—H (average)	1.102	
				∠$C_aC_bC_c$	113.5	
	∠C_bC_cO, ∠C_dC_cO	121.9		*trans* conformer	95%	
Ethyl methyl sulfide	C—S (average)	1.813		C—C	1.536	ED
$C_2H_5SCH_3$	C—H	1.111		∠CSC	97	
	∠SCC	114.0		∠HCH	110	
	fraction of the *gauche* conformer at 20°C	75%				
Ferrocene	C—C	1.440		C—H	1.104	ED
$(C_5H_5)_2Fe$	Fe—C	2.064		(D_{5h})		
Fluoroform	C—H	1.098		C—F	1.332	MW
CHF_3	∠FCF	108.8		(C_{3v})		
Formaldehyde	C—H	1.116		C—O	1.208	MW
H_2CO	∠HCH	116.5				
Formaldehyde azine	$H_2C=N—N=CH_2$	N—N	1.418			ED
	C—N	1.277		C—H	1.094	
	∠CNN	111.4		∠HCN	120.7	
	fraction of the *trans* conformer at –30°C	91%				
Formaldehyde dimethylacetal → Dimethoxy-methane						
Formaldoxime				C—H_a	1.085	MW
				C—H_b	1.086	
				C—N	1.276	
	N—O	1.408		O—H_c	0.956	
	∠H_bCN	115.6		∠CNO	110.2	
	∠H_aCN	121.8		∠NOH_c	102.7	
Formamide				C—H_a	1.125	ED, MW
				N—H	1.027	
				C—N	1.368	
	C—O	1.212		∠NCO	125.0	
	∠CNH (average)	119.2				
Formic acid				C—O_a	1.202	MW
				C—O_b	1.343	
	O_b—H	0.972		C—H	1.097	
	∠HCO_a	124.1		∠O_aCO_b	124.9	
	∠CO_bH	106.3		planar		

Compound	Structure				Method

Formic acid dimer

$O_a \cdots O_b$	2.703	ED
$C - O_a$	1.220	
$C - O_b$	1.323	
$\angle O_a C O_b$	126.2	
$\angle C O_a O_b$	108.5	

Formyl radical

$C - H$	1.110	$C - O$	1.1712	MW
$\angle HCO$	127.43			

Fulvene

$C_a - C_d$	1.349	MW
$C_a - C_b$	1.470	
$C_b - C_c$	1.355	
$C_c - C_c$	1.476	
$C_b - H$	1.078	
$C_c - H$	1.080	
$\angle C_b C_a C_b$	106.6	
$\angle C_b C_c C_c$	109	
$\angle C_b C_c H$	126.4	

$C_d - H$	1.13
$\angle C_a C_b C_c$	107.7
$\angle C_a C_b H$	124.7
$\angle HC_d H$	117

2-Furaldehyde

$C_a - C_e$	1.458	MW
$C_e - O_b$	1.250	
$C_e - H$	1.088	
$\angle C_e C_a C_b$	133.9	

$\angle C_a C_e H$	116.9	$\angle C_a C_e O$	121.6

trans conformer (with respect to the O_a and O_b atoms)

Furan

$C_b - C_b$	1.431	MW
$C_a - C_b$	1.361	
$C_a - O$	1.362	
$C_a - H_a$	1.075	
$C_b - H_b$	1.077	

$\angle C_a C_b C_b$	106.1	$\angle C_a O C_a$	106.6
$\angle C_b C_a O$	110.7	$\angle O C_a H_a$	115.9
$\angle C_b C_b H_b$	128.0		

Furfural → 2-Furaldehyde

Glycolaldehyde

$C_b - O_b$	1.209	MW
$C_a - O_a$	1.437	
$C_a - C_b$	1.499	
$O_a - H_a$	1.051	
$C_b - H_c$	1.102	
$C_a - H_b$	1.093	

$\angle C_a C_b O_b$	122.7	$\angle C_a C_b H_c$	115.3
$\angle C_b C_a O_a$	111.5	$\angle C_a O_a H_a$	101.6
$\angle C_b C_a H_b$	109.2	$\angle H_b C_a H_b$	107.6
$\angle H_b C_a O_a$	109.7		

Glyoxal
CHOCHO

$C - C$	1.526	$C - O$	1.212	ED, UV
$C - H$	1.132	$\angle CCO$	121.2	
$\angle HCO$	112	*trans* conformer	(C_{2h} (assumed))	

Hexachloroethane
Cl_3CCCl_3

$C - C$	1.56	$C - Cl$	1.769	ED
$\angle CCCl$	110.0			

2,4-Hexadiyne

$C_aH_3 - C_b \equiv C_c - C_c \equiv C_b - C_aH_3$

				ED
$C_a - C_b$	1.450	$C_b - C_c$	1.208	
$C_c - C_c$	1.377	$C_a - H$	1.09	

Hexafluoroethane
F_3CCF_3

$C - C$	1.545	$C - F$	1.326	ED
$\angle CCF$	109.8	staggered conformation		

Hexafluoropropene
$CF_2=CFCF_3$

average value of the C=C and C—F distances			1.329	ED
$C - C$	1.513	$\angle CCC$	127.8	
$\angle FCC (CF_2)$	124	$\angle FCC (CF)$	120	
$\angle FCC (CF_3)$	110			

Compound	Structure				Method

1,3,5-Hexatriene — ED

$H_2C_a=C_bH—C_cH=C_cH—C_bH=C_aH_2$

$C_a—C_b$	1.337	$C_b—C_c$	1.458
$C_c—C_c$	1.368	$\angle C_aC_bC_c$	121.7
$\angle C_bC_cC_c$	124.4		

Iminocyanide radical — UV
$H\dot{N}CN$

N—H	1.034	N···N	2.470
$\angle HNC$	116.5	$\angle NCN$	~180

Iodocyanoacetylene — MW
$I—C_a\equiv C_b—C_c\equiv N$

$I—C_a$	1.985	$C_a—C_b$	1.207
$C_b—C_c$	1.370	$C_c—N$	1.160

Isobutane — ED, MW
$(C_bH_3)_3C_aH$

$C_a—H$	1.122	$C_b—H$	1.113
$C_a—C_b$	1.535	$\angle C_bC_aC_b$	110.8
$\angle C_aC_bH$	111.4		

Isobutylene → 2-Methylpropene

Ketene — MW
$CH_2=C=O$

C—C	1.317	C—O	1.161
C—H	1.080	$\angle HCH$	123.0

Malononitrile — MW
$C_aH_2(C_bN)_2$

C—H	1.091	C—C	1.480
C—N	1.147	$\angle CCC$	110.4
$\angle HCH$	108.4	$\angle CCN$	176.6

(The two N atoms are bent away from each other in the plane of $C_b—C_a—C_b$.)

Methane — MW
CH_4

C—H (r_e)	1.0870	(T_d)	

Methanethiol — MW
CH_3SH

C—H	1.09	C—S	1.819
S—H	1.34	$\angle HSC$	96.5
$\angle HCH$	109.8		

angle between the CH_3 symmetry axis and the C—S bond 2.2.
(The axis of the CH_3 group is tilted away from the H atom with respect to the C—S bond.)

Methanol — MW
CH_3OH

C—H	1.0936	C—O	1.4246
O—H	0.9451	$\angle HCH$	108.63
$\angle COH$	108.53		

angle between the CH_3 symmetry axis and the C—O bond
(The axis of the CH_3 group is tilted away from the H atom
with respect to the C—O bond.) 3.27

Methyl radical — UV
$\cdot CH_3$

C—H	1.08	planar	

N-Methylacetamide — ED

H_3C_a ... H / $C_b—N$ / O C_cH_3

		$C_a—C_b$	1.520
		$N—C_c$	1.469
		C—H	1.107
$C_b—N$	1.386	$C_b—O$	1.225
$\angle C_bNC_c$	119.7		
$\angle NC_bO$	121.8		
$\angle C_aC_bN$	114.1		

Methylacetylene → Propyne

Methylal → Dimethoxymethane

Methylamine — MW
CH_3NH_2

N—H	1.010	C—N	1.471
C—H	1.099	$\angle NHN$	107.1
$\angle HNC$	110.3	$\angle HCH$	108.0

dihedral angle between the CH_3 symmetry axis and the
C—N bond (The axis of the CH_3 group is tilted away
from the NH_2 group with respect to the C—N bond.) 2.9

Methyl azide — ED

CH_3 / $N_a—N_b—N_c$

		C—H	1.09
		$C—N_a$	1.468
		$N_a—N_b$	1.216
$N_b—N_c$	1.113	$\angle CN_aN_b$	116.8
NNN linear			

Methyl bromide — MW, IR
CH_3Br

C—H (r_e)	1.086	C—Br (r_e)	1.933
$\angle HCH$ (θ_e)	111.2	(C_{3v})	

Methyl chloride — MW, IR
CH_3Cl

C—H	1.090	C—Cl	1.785
$\angle HCH$	110.8		

Compound	Structure					Method
Methyldiazirine	$CH_3CH \diagdown \!\!\!\!\!\begin{array}{c} N \\ \| \\ N \end{array}$	C—N	1.481	C—C	1.501	MW
		N—N	1.235	∠NCN	49.3	
		dihedral angle between the CNN plane and the C—C bond			122.3	
Methylene :CH$_2$	C—H	1.078	∠HCH		130	LMR
Methylenecyclopropane	C_cH_2 | $C_b{=}C_aH_2$ | C_cH_2	C$_a$—C$_b$			1.332	MW
		C$_b$—C$_c$			1.457	
		C$_c$—C$_c$			1.542	
	C$_c$—H	1.09	∠C$_c$C$_b$C$_c$		63.9	
	∠HC$_a$H	114.3	∠HC$_c$H		113.5	
	dihedral angle between the C$_c$H$_2$ plane and the C$_c$—C$_c$ bond			150.8		
3-Methyleneoxetane	$\begin{array}{c} C_cH_2 \\ O \diagdown\!\!\diagup\;\; C_b = C_aH_2 \\ C_cH_2 \end{array}$	C$_b$—C$_c$			1.52	MW
		C$_c$—O			1.45	
		C$_a$—C$_b$			1.33	
	C—H	1.09 (assumed)	∠C$_c$C$_b$C$_c$		87	
	∠HC$_c$H	114 (assumed)	∠HC$_a$H		120 (assumed)	
Methyl fluoride CH$_3$F	C—H (r_e)	1.095				MW, IR
	C—F (r_e)	1.382	∠HCH (θ_e)		110.45 (C_{3v})	
Methyl formate	$\begin{array}{c} C_aH_3 \quad\quad O_b \\ \diagdown\quad\diagup\!\!\| \\ O_a - C_b \\ \diagdown \\ H_b \end{array}$	C$_a$—H			1.08	ED
		C$_b$—O$_b$			1.206	
		C—O (average)			1.393	
		C$_b$—H			1.101 (assumed)	
	∠O$_a$C$_b$O$_b$	127	∠COC		114	
	∠O$_a$C$_a$H	110				
Methylgermane CH$_3$GeH$_3$	C—H	1.083	Ge—H		1.529	MW
	C—Ge	1.945	∠HCH		108.4	
	∠HGeH	109.3				
Methyl hypochlorite CH$_3$OCl	C—H	1.103	O—Cl		1.674	MW
	O—C	1.389	∠HCH		109.6	
	∠COCl	112.8				
Methylidyne radical :ĊH	C—H (r_e)	1.1198				UV
Methylidyne phosphide HCP	H—C (r_e)	1.0692	C—P (r_e)		1.5398	MW
Methyl iodide CH$_3$I	C—H (r_e)	1.084	C—I (r_e)		2.132	MW, IR
	∠HCH (θ_e)	111.2	(C_{3v})			
Methyl isocyanide	$C_aH_3{-}N{\equiv}C_b$	C$_a$—H	1.102	C$_a$—N	1.424	MW
	N—C$_b$	1.166	∠NC$_a$H		109.12	
Methylketene	$\begin{array}{c} C_cH_3 \\ \diagdown \\ C_b = C_a = O \\ \diagup \\ H \end{array}$	O—C$_a$			1.171	MW
		C$_b$—C$_c$			1.518	
		C$_c$—H			1.10	
	C$_a$—C$_b$	1.306	C$_b$—H		1.083	
	∠OC$_a$C$_b$	180.5	∠C$_a$C$_b$C$_c$		122.6	
	∠C$_a$C$_b$H	113.7	∠C$_c$C$_b$H		123.7	
	∠HCH	109.2				
Methylmercury chloride CH$_3$HgCl	Hg—Cl	2.282	C—H		1.15	MW,
	Hg—C	1.99	(C_{3v})			NMR
Methyl nitrate	$\begin{array}{c} H_a \quad H_a \;\; O_a \\ \diagdown\;\;\diagup\quad\diagup \\ C \!-\! N \\ \diagup\quad\diagdown\quad\diagdown \\ H_b \quad O \quad O_b \end{array}$	C—H$_a$			1.10	MW
		C—H$_b$			1.09	
		C—O			1.437	
		O—N			1.402	
	N—O$_a$	1.205	N—O$_b$		1.208	
	∠OCH$_a$	110	∠OCH$_b$		103	
	∠CON	112.7	∠ONO$_a$		118.1	
	∠ONO$_b$	112.4				

Compound	Structure					Method
Methylphosphine CH_3PH_2	C—P	1.858	C—H	1.094		ED
2-Methylpropane → Isobutane						
2-Methylpropene	C_aH_3 H_c $C_b=C_c$ C_aH_3 H_c		C_a—H	1.119		ED, MW
			C_c—H_c	1.10		
			C_a—C_b	1.508		
			C_b—C_c	1.342		
	∠HC_aC_b (average)	111.4	∠$H_cC_cH_c$	118.5		
	∠$C_aC_bC_a$	115.6	∠$C_aC_bC_c$	122.2		
	∠HC_aH	107.9	∠C_bC_cH	121		
Methylsilane CH_3SiH_3	C—H	1.093	C—Si	1.867		MW
	Si—H	1.485	∠HCH	107.7		
	∠HSiH	108.3	(C_{3v})			
Methylstannane CH_3SnH_3	C—Sn	2.143	Sn—H	1.700		MW
			(C_{3v})			
Methyl thiocyanate	C_aH_3 $S—C_b—N$		S—C_b	1.684	S—C_a	1.824	MW
			C_b—N	1.170	C—H	1.081	
	∠C_aSC_b	99.0	∠HCH	110.6	∠HCS	108.3	
Naphthalene	(ring structure)		C_a—C_b	1.37		ED
			C_b—C_b	1.41		
			C_a—C_c	1.42		
			C_c—C_c	1.42		
			C—C (average)	1.40		
			∠$C_aC_cC_c$	119.4		
Neopentane $C(CH_3)_4$	C—C	1.537	C—H	1.114		ED
	∠CCH	112				
Nickelocene → Bis (cyclopentadienyl) nickel						
Nitromethane CH_3NO_2	C—H	1.088 (assumed)	C—N	1.489		MW
	N—O	1.224	∠NCH	107		
	∠ONO	125.3				
N-Nitrosodimethylamine $(CH_3)_2NNO$	N—O	1.235	N—N	1.344		ED
	C—N	1.461	∠ONN	113.6		
	∠CNC	123.2	∠CNN	116.4		
Nitrosomethane CH_3NO	C—N	1.49	N—O	1.22		MW
	C—H	1.084	∠CNO	112.6		
	∠NCH	109.0				
Norbornane → Bicyclo[2.2.1]heptane						
Norbornadiene → Bicyclo[2.2.1]hepta-2,5-diene						
1,2,5-Oxadiazole	(ring structure)		O—N	1.380	∠NON	110.4	MW
			C—N	1.300	∠ONC	105.8	
			C—C	1.421	∠CCN	109.0	
			C—H	1.076	∠CCH	130.2	
			∠NCH	120.9	planar		
1,3,4-Oxadiazole	(ring structure)		O—C	1.348	∠COC	102.0	MW
			C—N	1.297	∠OCN	113.4	
			N—N	1.399	∠CNN	105.6	
			C—H	1.075	∠OCH	118.1	
			∠NCH	128.5	planar		
Oxalic acid	(structure)		C—C	1.544		ED
			C—O_a	1.205		
			C—O_b	1.336		
			O_b—H	1.05		
			∠CCO_a	123.1		
			∠O_aCO_b	125.0		
			∠CO_bH	104		

Compound	Structure			Method

Oxalyl dichloride

	C—O	1.182		ED
	C—C	1.534		
	C—Cl	1.744		
∠CCO	124.2	∠CCCl	111.7	

fraction of the *trans* conformer at 0°C 68%, that of the gauche conformer 32%

Oxetane

	C—O	1.448		MW
	C—C	1.546		
	C—H (average)	1.090		
	∠COC	92		
∠CCC	85	∠OCC	92	
∠HCH (average)	109.9			

Oxirane → Ethylene oxide

Phenol

C—C (average)	1.397	MW
C_b—H	1.084	
C_c—H	1.076	
C_d—H	1.082	
C_a—O	1.364	
O—H	0.956	
∠COH	109.0	

Phosphirane

C—P	1.867	P—H	1.43	MW
C—C	1.502	C—H	1.09	
∠CPC	47.4	∠HCH	114.4	
∠HPC	95.2	∠CCH	118	

dihedral angle between the PCC plane and the PH bond 95.7

Piperazine

C—C	1.540	ED
C—N	1.467	
C—H	1.110	
∠CNC	109.0	∠CCN 110.4
		(C_{2h})

Pivalonitrile
$(C_cH_3)_3C_b$—C_a≡N

C_a—C_b	1.495	C_a—N	1.159	MW
C_b—C_c	1.536	∠$C_cC_bC_c$	110.5

Propadiene → Allene

Propane
C_3H_8

C—C	1.532	C—H	1.107	ED
∠CCC	112	∠HCH	107	

Propenal → Acrylaldehyde

Propene

C_a—H_a	1.104	ED, MW
C_a—C_b	1.341	
C_c—H_d	1.117	
C_b—C_c	1.506	
∠$C_bC_aH_{a,b,c}$	121.3	
∠$C_bC_cH_d$ 110.7	∠$C_aC_bC_c$	124.3

1-Propenyl chloride
CH_3—C_bH=C_aH—Cl C_a—Cl   1.728   MW
∠C_bC_aCl 121.9   *trans* conformer

Propiolaldehyde
H_aC_a≡C_b—C_cH_cO

C_a—C_b	1.211	C_a—H_a	1.085	ED, MW
C_c—H_c	1.130	C_b—C_c	1.453
∠C_bC_cO	124.2	C_c—O	1.214
∠$C_aC_bC_c$	178.6	∠$C_bC_cH_c$	113.7
		planar	

Propylene → Propene

Propylene oxide
$C_aH_3C_bH$——C_cH_2

C_a—C_b	1.51	MW
∠$C_aC_bC_c$	121.0	

dihedral angle between the C_bC_cO plane
and the C_aC_b bond   123.8

Compound	Structure					Method

Propynal → Propiolaldehyde

Propyne

H_3C_c—C_b≡C_aH

C_c—C_b	1.459	C_c—H	1.105	MW
C_a—H	1.056	C_b—C_a	1.206	
		∠HC_cC_b	110.2	

Pyrazine

N⬡N

C—C	1.339	C—N	1.403	ED
C—H	1.115	∠CCN	115.6	
∠CCH	123.9			

Pyridazine

N—C_a	1.341		ED, MW
C_a—C_b	1.393		
N—N	1.330		
C_b—C_b	1.375		

∠NCC	123.7	∠NNC	119.3

Pyridine

N—C_a	1.340	MW
C_b—C_c	1.394	
C_b—H_b	1.081	
C_a—C_b	1.395	
C_a—H_a	1.084	
C_c—H_c	1.077	

∠C_aNC_a	116.8	∠NC_aC_b	123.9
∠$C_aC_bC_c$	118.5	∠$C_bC_cC_b$	118.3
∠NC_aH_a	115.9	∠$C_cC_bH_b$	121.3

Pyrimidine

N—C	1.340	C—C	1.393	ED
∠NCN	127.6	∠CNC	115.5	
(C_{2v} assumed)				

Pyrrole

N—Ca	1.370	MW
C_b—C_b	1.417	
C_a—C_b	1.382	
N—H	0.996	
C_a—H_a	1.076	

C_b—H_b	1.077	∠C_aNC_a	109.8
∠NC_aC_b	107.7	∠$C_aC_bC_b$	107.4
∠NC_aH_a	121.5	∠C_bC_bH	127.1

Pyruvonitrile

C_aH_3—C_b(=O)—C_c≡N

C—H	1.12		ED, MW
C—N	1.17		
C—O	1.208		
C_b—C_c	1.477		

C_a—C_b	1.518	∠HCH	109.2
∠C_aC_bO	124.5	∠$C_aC_bC_c$	114.2
∠CCN	179		

Ruthenocene → Bis (cyclopentadienyl) ruthenium

Silacyclobutane

CH_2—CH_2
CH_2—SiH_2

Si—C	1.892		ED
C—C	1.600		
Si—H	1.47		

C—H	1.14	∠CSiC	80.7
∠SiCC	84.8	∠CCC	99.8
dihedral angle between the CCC and CSiC planes			146

Spiropentane

H_2C_b ⟍ C_bH_2 / C_a / H_2C_b ⟍ C_bH_2

C_b—C_b	1.52		ED
C_a—C_b	1.47		
C—H	1.09		
∠$C_bC_aC_b$	62		

∠HCH	118	(D_{2d})

Compound	Structure					Method
Succinonitrile	CH₂CN | CH₂CN	C—C C—N ∠CCC	1.561 1.161 110.4	C—C(N) C—H	1.465 1.09	ED

fraction of the *anti* conformer at 170 C 74%,
   dihedral angle of CCCC for the *gauche* conformer   75

Compound	Structure					Method
Tetrachloroethylene CCl₂=CCl₂	C—Cl ∠ClCCl	1.718 115.7		C—C	1.354	ED
Tetracyanoethylene (CN)₂C=C(CN)₂	C—N C=C	1.162 1.357		C—C ∠CC=C	1.435 121.1	ED
Tetrafluoro-1,3-dithietane	S F₂C CF₂ S ∠FCS	113.7		C—S C—F ∠CSC (D₂ₕ assumed)	1.785 1.314 83.2	ED
Tetrafluoroethylene CF₂=CF₂	C—C ∠CCF	1.31 123.8		C—F (D₂ₕ assumed)	1.319	ED
Tetrahydrofuran CH₂CH₂ | O CH₂CH₂	C—H C—C	1.115 1.536		C—O	1.428	ED

The skeletal bending vibration of the molecular plane is essentially free
   pseudorotation

Compound	Structure					Method
Tetrahydropyran	H₂ C H₂C CH₂ H₂C CH₂ O	C—O C—C C—H ∠COC ∠OCC	1.420 1.531 1.116 111.5 111.8			ED
	∠CCC (C) chair form	108	∠CCC (O)	111		
Tetrahydrothiophene CH₂CH₂ | S CH₂CH₂	C—S C—C ∠SCC	1.839 1.536 106.1	C—H ∠CSC ∠CCC	1.120 93.4 105.0		
Tetramethylgermane (CH₃)₄Ge	Ge—C ∠GeCH	1.945 108	C—H (Tₐ excluding the H atoms)	1.12		ED
Tetramethyllead (CH₃)₄Pb	Pb—C	2.238	(Tₐ excluding the H atoms)			ED
Tetramethylsilane (CH₃)₄Si	C—H ∠HCH	1.115 109.8	C—Si (Tₐ excluding the H atoms)	1.875		ED
Tetramethylstannane (CH₃)₄Sn	C—Sn C—H	2.144 1.12	(Tₐ excluding the H atoms)			ED

Compound	Structure					Method
1,2,5-Thiadiazole	S N N HC — CH	S—N C—N C—C C—H	1.631 1.328 1.420 1.079	∠NSN ∠CCN ∠CCH planar	99.6 113.8 126.2	MW
1,3,4-Thiadiazole	S HC CH N — N	S—C N—N C—N C—H ∠NCH	1.721 1.371 1.302 1.08 123.5	∠CSC ∠SCN ∠CCN ∠SCH planar	86.4 114.6 112.2 121.9	MW
Thietane	CH₂ — CH₂ | | CH₂ — S	C—S C—C C—H (average)	1.847 1.549 1.100	∠HCH (average)	112	ED, MW
	∠CSC	76.8				

dihedral angle between the CCC and CSC planes   154

Compound	Structure					Method
Thiirane	H₂C S H₂C	C—C C—H C—S	1.484 1.083 1.815	∠HCH ∠CSC ∠CCS	116 48.3 65.9	MW

dihedral angle between the CH₂ plane and the
   C—C bond   152

Compound	Structure					Method
Thioformaldehyde	C—S	1.611		C—H	1.093	MW
CH_2S	∠HCH	116.9				
Thioformamide				N—H_a	1.002	MW
				N—H_b	1.007	
				C—N	1.358	
	C—S	1.626		C—H_c	1.10	
	∠H_aNH_b	121.7		∠H_aNC	117.9	
	∠H_bNC	120.4		∠NCS	125.3	
	∠NCH_c	108		∠SCH_c	127	
Thiolane → Tetrahydrothiophene						
Thiophene				C_a—H_a	1.078	MW
				C_b—H_b	1.081	
				C_a—S	1.714	
				C_a—C_b	1.370	
				C_b—C_b	1.423	
				∠C_aSC_a	92.2	
	∠SC$_a$C$_b$	115.5		∠$C_a$$C_b$$C_b$	112.5	
	∠SC$_a$H$_a$	119.9		∠$C_b$$C_b$$H_b$	124.3	
Toluene	CH_3	C—C (ring)	1.399	C—CH_3	1.524	ED
		C—H (average)	1.11			
		the difference between the C—H(CH_3) and				
		C—H(ring): about 0.01				
1,1,1-Tribromoethane	C—Br	1.93		C—H	1.095 (assumed)	MW
CH_3CBr_3	C—C	1.51 (assumed)		∠CCBr	108	
	∠BrCBr	111		∠CCH	109.0 (assumed)	
Tribromomethane → Bromoform						
Tri-*tert*-butyl methane	C_a—C_b	1.611		C—H	1.111	ED
$HC_a[C_b(C_cH_3)_3]_3$	C_b—C_c	1.548		∠$C_a$$C_b$$C_c$	113.0	
Tricarbon dioxide	C—O	1.163		C—C	1.289	ED
OCCCO	linear (with a large-amplitude bending vibration)					
Trichloroacetonitrile	C—N	1.165		C—C	1.460	ED
CCl_3CN	C—Cl	1.763		∠ClCCl	110.0	
1,1,1-Trichloroethane	C—H	1.090		C—C	1.541	MW
CH_3CCl_3	C—Cl	1.771		∠HCH	110.0	
	∠CCH	108.9		∠ClCCl	109.4	
	∠CCCl	109.6				
Trichloro(methyl)germane	Ge—Cl	2.132		Ge—C	1.89	ED, MW
CH_3GeCl_3	C—H	1.103 (assumed)		∠ClGeCl	106.4	
	∠GeCH	110.5 (assumed)				
Trichloro(methyl)silane	C—Si	1.876		Si—Cl	2.021	MW
CH_3SiCl_3				(C_{3v})		
Trichloro(methyl)stannane	Sn—Cl	2.304		Sn—C	2.10	ED
CH_3SnCl_3	C—H	1.100		∠CSnCl	113.9	
	∠ClSnCl	104.7		∠SnCH	108	
Triethylenediamine → 1,4-Diazabicyclo [2.2.2]octane						
Trifluoroacetic acid				C—F	1.325	ED
	CF_3C			C—C	1.546	
	O_bH			C—O_a	1.192	
	C—O_b	1.35		O—H	0.96 (assumed)	
	∠CCO_a	126.8		∠CCO_b	111.1	
	∠CCF	109.5				
1,1,1-Trifluoroethane	C—C	1.494		C—F	1.340	ED
CH_3CF_3	C—H	1.081		∠CCF	119.2	
	∠CCH	112				
Trifluoromethane → Fluoroform						

Compound	Structure				Method
1,1,1-Trifluoro-2,2,2-trichloroethane	C—C	1.54	C—F	1.33	MW
CF$_3$CCl$_3$	C—Cl	1.77	∠CCF	110	
	∠CCCl	109.6	staggered conformation		
Trimethylaluminium	C—H	1.113	Al—C	1.957	ED
(CH$_3$)$_3$Al	∠AlCH	111.7	∠CAlC	120	
Trimethylamine	C—N	1.458	C—H	1.100	ED
(CH$_3$)$_3$N	∠CNC	110.9	∠HCH	110	
Trimethylarsine	C—As	1.979	C—H		ED
(CH$_3$)$_3$As	∠AsCH	111.4	∠CAsC	98.8	
Trimethylbismuth	Bi—C	2.263	C—H	1.07	ED
(CH$_3$)$_3$Bi	∠CBiC	97.1			
Trimethylborane	C—B	1.578	C—H	1.114	ED
(CH$_3$)$_3$B	∠CBC	120.0	∠BCH	112.5	
Trimethyleneimine → Azetidine					
Trimethylphosphine	C—P	1.847	C—H	1.091	ED
(CH$_3$)$_3$P	∠CPC	98.6	∠PCH	110.7	

1,3,5-Trioxane

C—O	1.422			MW
∠OCO	112.2			
∠COC	110.3			

Triphenylamine	C—C	1.392	C—N	1.42	ED
(C$_6$H$_5$)$_3$N	∠CNC	116	(C$_3$)		

torsional dihedral angle of the two phenyl rings 47 (defined to be 0
when the symmetry axis is contained in the phenyl planes)

Tropone

C$_a$—O	1.23		ED
C$_a$—C$_b$	1.45		
C$_b$—C$_c$	1.36		
C$_c$—C$_d$	1.46		
C$_d$—C$_d$	1.34		
∠C$_b$C$_a$C$_b$	122		
∠C$_a$C$_b$C$_c$	133		

∠C$_b$C$_c$C$_d$	126	∠C$_c$C$_d$C$_d$	130
		(C$_{2v}$)	

Vinylacetylene

C$_b$—C$_c$	1.434		ED, MW
C$_a$—C$_b$	1.344		
C$_c$—C$_d$	1.215		
C$_a$—H$_a$	1.11		
C$_d$—H$_d$	1.09		
∠C$_a$C$_b$C$_c$	123.1		

∠C$_b$C$_c$C$_d$	178	∠H$_a$C$_a$C$_b$	119
∠H$_b$C$_a$C$_b$	122	∠H$_c$C$_b$C$_a$	122
∠C$_c$C$_d$H$_d$	182		

Vinyl chloride

C—C	1.342		ED, MW
C—Cl	1.730		
C—H	1.09		

∠CCCl	122.5	∠CCH$_a$	124
∠CCH$_b$	120	∠CCH$_c$	121.1

DIPOLE MOMENTS OF MOLECULES IN THE GAS PHASE

David R. Lide

This table gives values of the electric dipole moment for about 500 gas-phase molecules. When available, values determined by microwave spectroscopy, molecular beam electric resonance, and other spectroscopic techniques were selected; otherwise, the values come from measurements of the dielectric constant in the gas phase. Compounds are listed by molecular formula in the modified Hill order; compounds not containing carbon are listed first, followed by compounds containing carbon.

The dipole moment μ is given in debye units (D). The conversion factor to SI units is $1\ D = 3.33564 \times 10^{-30}$ C m. The column headed Acc. indicates the estimated accuracy of the value through the following codes:

A $\pm 1\%$ for $\mu > 1.0$ D; ± 0.01 D for $\mu < 1.0$ D
B $\pm 2\%$ for $\mu > 1.0$ D; ± 0.02 D for $\mu < 1.0$ D
C $\pm 5\%$ for $\mu > 1.0$ D; ± 0.05 D for $\mu < 1.0$ D
D $\pm 10\%$ for $\mu > 1.0$ D; ± 0.10 D for $\mu < 1.0$ D
Q   Questionable value
i   Indicates the value may be ambiguous because of the presence of rotational isomers

Dipole moments of individual rotational isomers are given when they have been measured. The isomers are designated as *gauche*, *trans-trans*, etc. The meaning of these designations can be found in the references. Other information, such as the components of the dipole moment in the molecular framework and the variation with vibrational state and isotopic species, may also be found in the references.

REFERENCES

1. R. D. Nelson, D. R. Lide, and A. A. Maryott, *Selected Values of Electric Dipole Moments for Molecules in the Gas Phase*, Natl. Stand. Ref. Data Ser. — Natl. Bur. Stnds. 10, 1967.
2. K. H. Hellwege, Ed., *Landolt-Börnstein, Numerical Data and Functional Relationships in Science and Technology*, Group II, Vol. 6, Molecular Constants, Springer-Verlag, Heidelberg, 1974.
3. K. H. Hellwege, Ed., *Landolt-Börnstein, Numerical Data and Functional Relationships in Science and Technology*, Group II, Vol. 14, Subvolume a, Springer-Verlag, Heidelberg, 1982.
4. K. H. Hellwege, Ed., *Landolt-Börnstein, Numerical Data and Functional Relationships in Science and Technology*, Group II, Vol. 14, Subvolume b, Springer-Verlag, Heidelberg, 1983.
5. R. Takashi, et. al., *J. Mol. Spec.*, 138, 450, 1989.
6. M. Hayashi and T. Inagusa, *J. Mol. Spec.*, 138, 135, 1989.
7. F. Scappini, A. C. Fantoni, and W. Caminati, *J. Mol. Spec.*, 120, 101, 1986.

Mol. form.	Name	μ/D	Acc.	Ref.
	Compounds not containing carbon			
AgCl	Silver chloride	5.70	C	2
AgF	Silver fluoride	6.22	C	2
AgI	Silver iodide	5.10	C	4
AlF	Aluminum fluoride (AlF)	1.53	D	1
$AsCl_3$	Arsenic trichloride	1.59	C	1
AsF_3	Arsenic trifluoride	2.59	B	1
AsH_3	Arsine	0.20	C	1
BF	Fluoroborane (BF)	0.5	Q	2
B_4H_{10}	Tetraborane	0.486	A	3
B_5H_9	Pentaborane	2.13	B	1
B_6H_{10}	Hexaborane	2.50	B	3
BaO	Barium oxide	7.954	A	2
BaS	Barium sulfide	10.86	A	2
BrCl	Bromine chloride	0.519	A	3
BrF	Bromine fluoride	1.42	A	3
BrF_5	Bromine pentafluoride	1.51	D	1
BrH	Hydrogen bromide	0.827	A	3
BrH_3Si	Bromosilane	1.319	A	3
BrI	Iodine bromide	0.726	A	4

Mol. form.	Name	μ/D	Acc.	Ref.
BrK	Potassium bromide	10.628	A	2
BrLi	Lithium bromide	7.268	A	2
BrNO	Nitrosyl bromide	1.8	Q	1
BrNa	Sodium bromide	9.118	A	2
BrO	Bromine oxide (BrO)	1.76	B	2
BrTl	Thallium bromide (TlBr)	4.49	A	2
CaCl	Calcium monochloride	3.6	Q	4
ClCs	Cesium chloride	10.387	A	2
ClF	Chlorine fluoride	0.888	A	2
ClFO$_3$	Perchloryl fluoride	0.023	A	3
ClF$_3$	Chlorine trifluoride	0.6	D	1
ClGeH$_3$	Chlorogermane	2.13	A	1
ClH	Hydrogen chloride	1.109	A	3
ClHO	Hypochlorous acid (HOCl)	1.3	Q	2
ClH$_3$Si	Chlorosilane	1.31	A	1
ClI	Iodine chloride	1.24	B	2
ClIn	Indium chloride	3.79	C	2
ClK	Potassium chloride	10.269	A	2
ClLi	Lithium chloride	7.129	A	2
ClNO$_2$	Nitryl chloride	0.53	A	1
ClNS	Thionitrosyl chloride (NSCl)	1.87	A	2
ClNa	Sodium chloride	9.001	A	2
ClO	Chlorine oxide (ClO)	1.239	A	2
ClRb	Rubidium chloride	10.510	A	2
ClTl	Thallium chloride (TlCl)	4.543	A	2
Cl$_2$H$_2$Si	Dichlorosilane	1.17	B	1
Cl$_2$OS	Thionyl chloride	1.45	B	1
Cl$_2$O$_2$S	Sulfuryl chloride	1.81	B	1
Cl$_2$S	Sulfur dichloride	0.36	A	3
Cl$_3$FSi	Trichlorofluorosilane	0.49	A	2
Cl$_3$HSi	Trichlorosilane	0.86	A	2
Cl$_3$N	Nitrogen trichloride	0.39	A	3
Cl$_3$OP	Phosphorous oxychloride	2.54	B	2
Cl$_3$P	Phosphorous trichloride	0.56	B	2
CsF	Cesium fluoride	7.884	A	2
CsNa	Cesium sodium (CsNa)	4.75	C	2
CuF	Copper fluoride (CuF)	5.77	C	2
FGa	Gallium fluoride (GaF)	2.45	B	2
FGeH$_3$	Fluorogermane	2.33	C	2
FH	Hydrogen fluoride	1.826	A	2
FHO	Hypofluorous acid (HOF)	2.23	C	3
FH$_3$Si	Fluorosilane	1.298	A	3
FI	Iodine fluoride	1.95	A	3
FIn	Indium fluoride (InF)	3.40	B	2
FK	Potassium fluoride	8.585	A	2
FLi	Lithium fluoride	6.326	A	3
FNO	Nitrosyl fluoride	1.730	A	3
FNO$_2$	Nitryl flouride	0.466	A	2
FNS	Thionitrosyl fluoride (NSF)	1.902	A	2
FNa	Sodium fluoride	8.156	A	2
FRb	Rubidium fluoride	8.546	A	2
FS	Sulfur fluoride (SF)	0.794	B	3
FTl	Thallium fuoride (TlF)	4.228	A	2
F$_2$Ge	Germanium fluoride (GeF$_2$)	2.61	A	2
F$_2$HN	Difluoramine	1.92	A	1
F$_2$H$_2$Si	Difluorosilane	1.55	A	1

Mol. form.	Name	μ/D	Acc.	Ref.
F_2N_2	*cis*-Difluorodiazine	0.16	A	1
F_2O	Oxygen difluoride	0.297	A	1
F_2OS	Thionyl fluoride	1.63	A	1
F_2O_2	Dioxygen difluoride (FOOF)	1.44	C	1
F_2O_2S	Sufuryl fluoride	1.12	B	1
F_2S	Sulfur difluoride	1.05	C	2
F_2Si	Difluorosilylene (SiF$_2$)	1.23	B	2
F_3HSi	Trifluorosilane	1.27	B	1
$F_3H_3Si_2$	1,1,1-Trifluorodisilane	2.03	C	3
F_3N	Nitrogen trifluoride	0.235	A	1
F_3NO	Trifluoramine oxide	0.039	A	3
F_3OP	Phosphorous oxyfluoride	1.868	A	3
F_3P	Phosphorous trifluoride	1.03	A	1
F_3PS	Thiophosphoryl trifluoride	0.64	B	1
F_4N_2	Tetrafluorohydrazine (*gauche*)	0.26	B	1
F_4S	Sulfur tetrafluoride	0.632	A	1
F_4Se	Selenium tetrafluoride	1.78	C	2
F_5I	Iodine pentafluoride	2.18	C	1
GeO	Germanium oxide (GeO)	3.282	A	2
GeS	Germanium sulfide (GeS)	2.00	C	2
HI	Hydrogen iodide	0.448	A	2
HLi	Lithium hydride	5.884	A	2
HLiO	Lithium hydroxide	4.754	A	3
HN	Imidogen (NH)	1.39	C	3
HNO	Nitrosyl hydride (HNO)	1.62	B	3
HNO_2	Nitrous acid (HONO) (*cis*)	1.423	A	2
	Nitrous acid (HONO) (*trans*)	1.855	A	2
HNO_3	Nitric acid	2.17	A	1
HN_3	Hydrazoic acid	1.70	C	3
HO	Hydroxyl (OH)	1.668	A	3
HS	Mercapto (SH)	0.758	A	3
H_2O	Water	1.854	A	3
H_2O_2	Hydrogen peroxide	1.573	A	3
H_2S	Hydrogen sulfide	0.97	A	1
H_3N	Ammonia	1.471	A	3
H_3NO	Hydroxylamine (NH$_2$OH)	0.59	C	2
H_3P	Phosphine	0.574	A	3
H_3Sb	Stibine	0.12	C	1
H_4N_2	Hydrazine	1.75	C	1
H_6OSi_2	Disiloxane	0.24	B	1
ILi	Lithium iodide	7.428	A	2
INa	Sodium iodide	9.236	A	2
IO	Iodine oxide (IO)	2.45	B	2
ITl	Thallium iodide (TlI)	4.61	B	2
KLi	Lithium potassium (LiK)	3.45	C	2
KNa	Potassium sodium (KNa)	2.73	C	3
LiNa	Lithium sodium (LiNa)	0.463	A	2
LiO	Lithium oxide (LiO)	6.84	A	2
LiRb	Lithium rubidium (LiRb)	4.00	C	2
NO	Nitric oxide	0.159	A	2
NO_2	Nitrogen dioxide	0.316	A	1
NS	Nitrogen sulfide (NS)	1.81	A	2
NP	Phosphorous nitride (PN)	2.747	A	2
N_2O	Nitrous oxide	0.161	A	3
N_2O_3	Nitrogen trioxide	2.122	A	2
NaRb	Rubidium sodium (RbNa)	3.1	D	2

Mol. form.	Name	μ/D	Acc.	Ref.
PbO	Lead oxide (PbO)	4.64	D	2
OS	Sulfur oxide (SO)	1.55	A	1
OS_2	Sulfur oxide (SSO)	1.47	B	1
OSi	Silicon monoxide (SiO)	3.098	A	2
OSn	Tin oxide (SnO)	4.32	C	2
OSr	Strontium oxide	8.900	A	2
O_2S	Sulfur dioxide	1.633	A	3
O_2Se	Selenium dioxide	2.62	B	2
O_3	Ozone	0.534	A	3
PbS	Lead sulfide (PbS)	3.59	C	2
SSi	Silicon monosulfide (SiS)	1.73	C	2
SSn	Tin sulfide (SnS)	3.18	C	2

Compounds containing carbon

Mol. form.	Name	μ/D	Acc.	Ref.
$CBrF_3$	Bromotrifluoromethane	0.65	C	1
CBr_2F_2	Dibromodifluoromethane	0.66	C	1
$CClF_3$	Chlorotrifluoromethane	0.50	A	1
CClN	Cyanogen chloride	2.833	A	3
CCl_2F_2	Dichlorodifluoromethane	0.51	C	1
CCl_2O	Phosgene	1.17	A	1
CCl_3F	Trichlorofluoromethane	0.46	B	2
CF	Fluoromethylidene	0.645	A	3
CFN	Cyanogen fluoride	2.120	A	3
CF_2	Difluoromethylene	0.47	B	3
CF_2O	Carbonyl fluoride	0.95	A	1
CF_3I	Trifluoroiodomethane	1.048	A	3
CO	Carbon monoxide	0.110	A	3
COS	Carbon oxysulfide	0.715	A	3
COSe	Carbon oxyselenide	0.73	B	1
CS	Carbon sulfide	1.958	A	2
CSe	Carbon selenide	1.99	B	3
CH	Methylidyne	1.46	Q	2
$CHBr_3$	Tribromomethane (bromoform)	0.99	B	1
$CHClF_2$	Chlorodifluoromethane	1.42	B	1
$CHCl_2F$	Dichlorofluoromethane	1.29	B	1
$CHCl_3$	Trichloromethane (chloroform)	1.04	B	2
CHF_3	Trifluoromethane (fluoroform)	1.651	A	3
CHN	Hydrogen cyanide	2.984	A	3
CHN	Hydrogen isocyanide (HNC)	3.05	C	3
CHNO	Fulminic acid (HCNO)	3.099	A	5
CHNO	Isocyanic acid (HNCO)	1.6	Q	2
CH_2Br_2	Dibromomethane	1.43	B	1
CH_2ClF	Chlorofluoromethane	1.82	B	1
CH_2Cl_2	Dichloromethane	1.60	B	1
CH_2F_2	Difluoromethane	1.978	A	3
CH_2N_2	Cyanamide	4.27	C	1
CH_2N_2	Diazomethane	1.50	A	1
CH_2O	Formaldehyde	2.332	A	3
CH_2O_2	Formic acid	1.41	A	1
CH_3BO	Carbonyltrihydroboron (BH_3CO)	1.70	B	3
CH_3Br	Bromomethane (methyl bromide)	1.822	A	3
CH_3Cl	Chloromethane (methyl chloride)	1.892	A	2
CH_3F	Fluoromethane (methyl fluoride)	1.858	A	3
CH_3I	Iodomethane (methyl iodide)	1.62	B	1
CH_3NO	Formamide	3.73	B	1

Mol. form.	Name	μ/D	Acc.	Ref.
CH_3NO_2	Nitromethane	3.46	A	1
CH_3N_3	Methyl azide	2.17	B	2
CH_4O	Methanol	1.70	A	1
CH_4S	Methanethiol (methyl mercaptan)	1.52	C	1
CH_5N	Methylamine	1.31	B	1
CH_6OSi	Methoxysilane	1.15	B	2
CH_6Si	Methylsilane	0.735	A	1
C_2ClF_3	Chlorotrifluoroethylene	0.40	D	1
C_2ClF_5	Chloropentafluoroethane	0.52	C	1
$C_2Cl_2F_4$	1,2-Dichlorotetrafluoroethane	0.5	Q	1
C_2F_3N	Trifluoroacetonitrile	1.262	A	3
C_2HBr	Bromoacetylene	0.234	A	3
C_2HCl	Chloroacetylene	0.44	A	1
C_2HCl_5	Pentachloroethane	0.92	C	1
C_2HF	Fluoroacetylene	0.721	A	3
C_2HF_3	Trifluoroethylene	1.32	B	2
$C_2HF_3O_2$	Trifluoroacetic acid	2.28	D	1
$C_2H_2Cl_2$	1,1-Dichloroethylene	1.34	A	1
$C_2H_2Cl_2$	cis-1,2-Dichloroethylene	1.90	B	1
$C_2H_2Cl_2O$	Chloroacetyl chloride	2.23	Ci	1
$C_2H_2Cl_4$	1,1,2,2-Tetrachloroethane	1.32	Ci	1
$C_2H_2F_2$	1,1-Difluoroethylene	1.368	A	3
$C_2H_2F_2$	cis-1,2-Difluoroethylene	2.42	A	1
C_2H_2O	Ketene	1.422	A	3
$C_2H_2O_2$	Glyoxal (cis)	4.8	C	2
C_2H_3Br	Bromoethylene (vinyl bromide)	1.42	B	1
C_2H_3Cl	Chloroethylene (vinyl chloride)	1.45	B	1
$C_2H_3ClF_2$	1-Chloro-1,1-difluoroethane	2.14	B	1
C_2H_3ClO	Acetyl chloride	2.72	C	1
$C_2H_3Cl_3$	1,1,1-Trichloroethane	1.755	A	2
C_2H_3F	Fluoroethylene (vinyl fluoride)	1.468	A	6
C_2H_3FO	Acetyl fluoride	2.96	A	1
$C_2H_3F_3$	1,1,1-Trifluoroethane	2.347	A	3
C_2H_3N	Acetonitrile	3.924	A	3
C_2H_3NO	Methylisocyanate	2.8	Q	1
C_2H_4ClF	1-Chloro-1-fluoroethane	2.068	A	3
$C_2H_4Cl_2$	1,1-Dichloroethane	2.06	B	1
$C_2H_4F_2$	1,1-Difluoroethane	2.27	B	1
$C_2H_4F_2$	1,2-Difluoroethane (gauche)	2.67	C	2
C_2H_4O	Acetaldehyde	2.750	A	3
C_2H_4O	Ethylene oxide (oxirane)	1.89	A	1
$C_2H_4O_2$	Acetic acid	1.70	B	2
$C_2H_4O_2$	Methyl formate	1.77	B	1
$C_2H_4O_2$	Glycolaldehyde (hydroxyacetylene)	2.73	B	2
C_2H_5Br	Bromoethane (ethyl bromide)	2.03	A	1
C_2H_5Cl	Chloroethane (ethyl chloride)	2.05	A	1
C_2H_5ClO	2-Chloroethanol	1.78	Ci	1
C_2H_5F	Fluoroethane (ethyl fluoride)	1.94	B	1
C_2H_5I	Iodoethane (ethyl iodide)	1.91	B	1
C_2H_5N	Ethyleneimine	1.90	A	1
C_2H_5NO	Acetamide	3.76	Bi	1
C_2H_5NO	N-Methylformamide	3.83	Bi	1
$C_2H_5NO_2$	Nitroethane	3.23	A	2
C_2H_6O	Dimethyl ether	1.30	A	1
C_2H_6O	Ethanol (average)	1.69	Bi	1
	Ethanol (trans)	1.44	B	2

Mol. form.	Name	μ/D	Acc.	Ref.
	Ethanol (*gauche*)	1.68	B	3
C_2H_6OS	Dimethyl sulfoxide	3.96	A	1
$C_2H_6O_2$	Ethylene glycol	2.28	Ci	1
C_2H_6S	Dimethyl sulfide	1.554	A	3
C_2H_6S	Ethanethiol (*trans*)	1.58	C	3
	Ethanethiol (*gauche*)	1.61	C	3
C_2H_7N	Dimethylamine	1.01	B	2
C_2H_7N	Ethylamine	1.22	Ci	1
$C_2H_8N_2$	1,2-Ethanediamine	1.99	Ci	1
C_3HN	Cyanoacetylene	3.724	A	3
C_3H_2O	Propynal (propargyl aldehyde)	2.47	B	1
$C_3H_3Cl_2F$	1,1-Dichloro-2-fluoropropene	2.43	A	3
$C_3H_3F_3$	3,3,3-Trifluoropropene	2.45	B	1
C_3H_3N	Propenenitrile (acrylonitrile)	3.87	B	1
C_3H_3NO	Oxazole	1.503	B	3
C_3H_3NO	Isoxazole	2.90	B	2
C_3H_4	Propyne (methylacetylene)	0.784	A	3
C_3H_4	Cyclopropene	0.45	A	1
$C_3H_4F_2$	1,1-Difluoro-1-propene	0.889	A	2
$C_3H_4N_2$	Imidazole	3.8	D	2
C_3H_4O	Propenal (*trans*)	3.12	B	1
C_3H_4O	Propargyl alcohol	1.13	C	2
C_3H_4O	Cyclopropanone	2.67	C	2
$C_3H_4O_2$	Vinyl formate	1.49	A	1
$C_3H_4O_2$	2-Oxetanone (beta-propiolactone)	4.18	A	1
$C_3H_4O_2$	3-Oxetanone	0.887	A	2
C_3H_5Br	3-Bromopropene (allyl bromide)	1.9	Q	1
C_3H_5Cl	*cis*-1-Chloropropene	1.67	C	1
C_3H_5Cl	*trans*-1-Chloropropene	1.97	C	1
C_3H_5Cl	2-Chloropropene	1.647	A	3
C_3H_5Cl	3-Chloropropene (allyl chloride)	1.94	Ci	1
C_3H_5F	*cis*-1-Fluoropropene	1.46	B	1
C_3H_5F	*trans*-1-Fluoropropene	1.9	Q	1
C_3H_5F	2-Fluoropropene	1.61	B	1
C_3H_5F	3-Fluoropropene (*cis*)	1.76	A	1
	3-Fluoropropene (*gauche*)	1.94	A	1
C_3H_5N	Propanenitrile	4.05	A	3
C_3H_6	Propene (propylene)	0.366	A	1
$C_3H_6Cl_2$	1,3-Dichloropropane	2.08	Bi	1
C_3H_6O	Acetone	2.88	A	1
C_3H_6O	Allyl alcohol (*average*)	1.60	Ci	1
	Allyl alcohol (*gauche*)	1.55	C	3
C_3H_6O	Propanal (*cis*)	2.52	B	1
C_3H_6O	Methyloxirane	2.01	A	1
C_3H_6O	Oxetane (trimethylene oxide)	1.94	A	1
C_3H_6O	Methyl vinyl ether (*cis*)	0.96	B	2
$C_3H_6O_2$	Ethyl formate (*trans*)	1.98	A	2
	Ethyl formate (*gauche*)	1.81	A	2
$C_3H_6O_2$	Methyl acetate	1.72	Ci	1
$C_3H_6O_2$	Propanoic acid (*average*)	1.75	Ci	1
	Propanoic acid (*cis*)	1.46	C	2
$C_3H_6O_2$	1,3-Dioxolane	1.19	C	3
$C_3H_6O_3$	Trioxane	2.08	A	1
C_3H_6S	Thiacyclobutane (trimethylene sulfide)	1.85	C	1
C_3H_7Br	1-Bromopropane (propyl bromide)	2.18	Ci	1
C_3H_7Br	2-Bromopropane (isopropyl bromide)	2.21	C	1

Mol. form.	Name	μ/D	Acc.	Ref.
C_3H_7Cl	1-Chloropropane (propyl chloride)	2.05	Bi	1
C_3H_7Cl	2-Chloropropane (isopropyl chloride)	2.17	C	1
C_3H_7F	1-Fluoropropane (*trans*)	2.05	B	1
	1-Fluoropropane (*gauche*)	1.90	C	1
C_3H_7F	2-Fluoropropane	1.96	B	3
C_3H_7I	1-Iodopropane (propyl iodide)	2.04	Ci	1
C_3H_7N	Allylamine	1.2	Q	1
C_3H_7N	Cyclopropylamine	1.19	A	2
C_3H_7N	Propyleneimine (*cis*)	1.77	C	2
	Propyleneimine (*trans*)	1.57	B	2
C_3H_7NO	N,N-Dimethylformamide	3.82	Bi	1
$C_3H_7NO_2$	1-Nitropropane	3.66	Bi	1
$C_3H_7NO_2$	2-Nitropropane	3.73	B	1
C_3H_8	Propane	0.084	A	1
C_3H_8O	1-Propanol (*trans*)	1.55	B	2
	1-Propanol (*gauche*)	1.58	B	2
C_3H_8O	2-Propanol (*trans*)	1.58	B	2
C_3H_8O	Ethyl methyl ether (*trans*)	1.17	B	3
$C_3H_8O_2$	2-Methoxyethanol (*gauche*)	2.36	B	2
C_3H_8S	Ethyl methyl sulfide (*trans*)	1.56	B	3
C_3H_8S	1-Propanethiol (*trans*)	1.60	C	3
	1-Propanethiol (*gauche*)	1.683	A	3
C_3H_8S	2-Propanethiol (*trans*)	1.61	B	3
	2-Propanethiol (*gauche*)	1.53	B	3
C_3H_9N	Propylamine	1.17	Ci	1
C_3H_9N	Isopropylamine	1.19	C	3
C_3H_9N	Trimethylamine	0.612	A	1
C_4H_4	1-Buten-3-yne (vinylacetylene)	0.22	B	3
$C_4H_4N_2$	Pyridazine	4.22	A	2
$C_4H_4N_2$	Pyrimidine	2.334	A	2
C_4H_4O	Furan	0.66	A	1
$C_4H_4O_2$	Diketene	3.53	B	1
C_4H_4S	Thiophene	0.55	A	1
C_4H_5N	Pyrrole	1.74	B	2
C_4H_5N	Methylacrylonitrile	3.69	C	1
C_4H_6	1,2-Butadiene (methylallene)	0.403	A	1
C_4H_6	1-Butyne (ethylacetylene)	0.80	C	1
C_4H_6	Cyclobutene	0.132	A	1
C_4H_6O	*trans*-2-Butenal (*trans*-crotonaldehyde)	3.67	Bi	1
C_4H_6O	2-Methylpropenal (methacrolein)	2.68	Ci	1
C_4H_6O	Divinyl ether	0.78	C	2
C_4H_6O	Cyclobutanone	2.89	A	2
C_4H_6O	2,3-Dihydrofuran	1.32	B	2
$C_4H_6O_2$	*gamma*-Butyrolactone	4.27	A	3
$C_4H_6O_3$	Acetic anhydride	2.8	Q	1
C_4H_6S	2,5-Dihydrothiophene	1.75	A	3
C_4H_7N	Butanenitrile	4.07	Bi	1
C_4H_7N	2-Methylpropanenitrile (isobutyronitrile)	4.29	B	3
C_4H_8	1-Butene (*cis*)	0.438	A	2
	1-Butene (*skew*)	0.359	A	2
C_4H_8	*cis*-2-Butene	0.253	A	2
C_4H_8	Isobutene	0.503	A	1
C_4H_8	Methylcyclopropane	0.139	A	2
$C_4H_8Cl_2$	1,4-Dichlorobutane	2.22	Ci	1
C_4H_8O	Butanal (butyraldehyde)	2.72	Bi	1
C_4H_8O	2-Butanone (methyl ethyl ketone)	2.78	A	2

Mol. form.	Name	μ/D	Acc.	Ref.
C_4H_8O	Tetrahydrofuran	1.75	B	2
$C_4H_8O_2$	1,3-Dioxane	2.06	B	2
$C_4H_8O_2$	Ethyl acetate	1.78	Ci	1
C_4H_9Br	1-Bromobutane (butyl bromide)	2.08	Ci	1
C_4H_9Br	2-Bromobutane (*sec*-butyl bromide)	2.23	Ci	1
C_4H_9Cl	1-Chlorobutane (butyl chloride)	2.05	Bi	1
C_4H_9Cl	1-Chloro-2-methylpropane (isobutyl chloride)	2.00	Ci	1
C_4H_9Cl	2-Chlorobutane (*sec*-butyl chloride)	2.04	Ci	1
C_4H_9Cl	2-Chloro-2-methylpropane (*tert*-butyl chloride)	2.13	B	1
C_4H_9I	2-Iodobutane (*sec*-butyl iodide)	2.12	Ci	1
C_4H_9NO	Morpholine	1.55	B	3
C_4H_{10}	Isobutane	0.132	A	1
$C_4H_{10}O$	1-Butanol	1.66	Bi	1
$C_4H_{10}O$	2-Methyl-1-propanol (isobutyl alcohol)	1.64	C	1
$C_4H_{10}O$	Diethyl ether	1.15	Bi	1
$C_4H_{10}O$	Methyl propyl ether (*trans-trans*)	1.107	A	3
$C_4H_{10}S$	2-Methyl-2-propanethiol (*tert*-butyl mercaptan)	1.66	B	3
$C_4H_{10}S$	Diethyl sulfide	1.54	Ci	1
$C_4H_{11}N$	Butylamine	1.0	Q	1
$C_4H_{11}N$	Diethylamine	0.92	Ci	1
C_5H_5N	Pyridine	2.215	A	3
C_5H_6	1-Penten-3-yne	0.66	B	2
C_5H_6	*cis*-3-Penten-1-yne	0.78	B	2
C_5H_6	*trans*-3-Penten-1-yne	1.06	C	2
C_5H_6	2-Methyl-1-buten-3-yne	0.513	B	2
C_5H_6	1,3-Cyclopentadiene	0.419	A	1
C_5H_6O	2-Methylfuran	0.65	C	2
C_5H_6O	3-Methylfuran	1.03	B	2
C_5H_6S	2-Methylthiophene	0.674	A	2
C_5H_6S	3-Methylthiophene	0.95	C	2
C_5H_7N	Cyclobutanecarbonitrile	4.11	A	3
C_5H_8	2-Methyl-1,3-butadiene (isoprene)	0.25	A	1
C_5H_8	*cis*-1,3-Pentadiene	0.500	B	2
C_5H_8	*trans*-1,3-Pentadiene	0.585	A	2
C_5H_8	Cyclopentene	0.20	B	1
C_5H_8	1-Pentyne (*trans*)	0.842	A	2
	1-Pentyne (*gauche*)	0.77	C	2
C_5H_8O	Cyclopentanone	3.3	Q	1
C_5H_8O	Cyclopropyl methyl ketone	2.62	D	2
C_5H_9N	Pentanenitrile	4.12	Bi	1
C_5H_9N	2,2-Dimethylpropanenitrile (*tert*-butyl cyanide)	3.95	A	1
C_5H_{10}	1-Pentene	0.5	Q	1
C_5H_{10}	3-Methyl-1-butene (*trans*)	0.320	A	3
	3-Methyl-1-butene (*gauche*)	0.398	A	3
$C_5H_{10}O$	Tetrahydropyran (*chair form*)	1.74	B	2
$C_5H_{10}O$	2,2-Dimethylpropanal (pivaldehyde)	2.66	B	2
$C_5H_{10}O_3$	Diethyl carbonate	1.10	Ci	1
$C_5H_{10}S$	Thiacyclohexane	1.781	A	3
$C_5H_{11}Br$	1-Bromopentane (amyl bromide)	2.20	Ci	1
$C_5H_{11}Cl$	1-Chloropentane (amyl chloride)	2.16	Ci	1
$C_5H_{11}N$	Piperidine (*equatorial*)	0.82	B	3
	Piperidine (*axial*)	1.19	B	3
C_5H_{12}	Isopentane	0.13	C	1
$C_6H_4ClNO_2$	*o*-Chloronitrobenzene	4.64	B	1
$C_6H_4ClNO_2$	*m*-Chloronitrobenzene	3.73	B	1
$C_6H_4ClNO_2$	*p*-Chloronitrobenzene	2.83	B	1

Mol. form.	Name	μ/D	Acc.	Ref.
$C_6H_4Cl_2$	o-Dichlorobenzene	2.50	B	1
$C_6H_4Cl_2$	m-Dichlorobenzene	1.72	C	1
$C_6H_4FNO_2$	p-Fluoronitrobenzene	2.87	B	1
$C_6H_4F_2$	o-Difluorobenzene	2.46	B	2
$C_6H_4F_2$	m-Difluorobenzene	1.51	A	2
C_6H_5Br	Bromobenzene	1.70	B	1
C_6H_5Cl	Chlorobenzene	1.69	B	1
C_6H_5ClO	p-Chlorophenol	2.11	C	1
C_6H_5F	Fluorobenzene	1.60	C	1
C_6H_5I	Iodobenzene	1.70	C	1
$C_6H_5NO_2$	Nitrobenzene	4.22	B	1
C_6H_6	Fulvene	0.424	A	2
C_6H_6O	Phenol	1.224	A	3
C_6H_7N	Aniline	1.13	B	3
C_6H_7N	2-Methylpyridine	1.85	B	2
C_6H_7N	4-Methylpyridine	2.70	A	2
C_6H_{10}	Cyclohexene (half-chair form)	0.332	A	2
C_6H_{10}	1-Hexyne (butylacetylene)	0.83	Ci	1
C_6H_{10}	3,3-Dimethyl-1-butyne (tert-butylacetylene)	0.661	A	1
$C_6H_{10}O$	Cyclohexanone	2.87	B	2
$C_6H_{11}Cl$	Chlorocyclohexane (equatorial)	2.44	C	7
	Chlorocyclohexane (axial)	1.91	B	7
$C_6H_{11}F$	Fluorocyclohexane (equatorial)	2.11	B	2
	Fluorocyclohexane (axial)	1.81	B	2
$C_6H_{12}O_2$	Pentyl formate	1.90	Ci	1
$C_6H_{12}O_3$	Paraldehyde	1.43	C	1
$C_6H_{14}O$	Dipropyl ether	1.21	Ci	1
$C_6H_{14}O$	Diisopropyl ether	1.13	Di	1
$C_6H_{15}N$	Triethylamine	0.66	Ci	1
$C_7H_5F_3$	(Trifluoromethyl)benzene	2.86	B	1
C_7H_5N	Benzonitrile	4.18	B	1
C_7H_7Cl	o-Chlorotoluene	1.56	C	1
C_7H_7Cl	p-Chlorotoluene	2.21	B	1
C_7H_7F	o-Fluorotoluene	1.37	C	1
C_7H_7F	m-Fluorotoluene	1.82	B	2
C_7H_7F	p-Fluorotoluene	2.00	C	1
C_7H_8	Toluene	0.375	A	3
C_7H_8O	Benzyl alcohol	1.71	C	1
C_7H_8O	Anisole (methyl phenyl ether)	1.38	C	1
$C_7H_{14}O_2$	Pentyl acetate	1.75	Ci	1
$C_7H_{15}Br$	1-Bromoheptane	2.16	Ci	1
C_8H_8O	Acetophenone (methyl phenyl ketone)	3.02	B	1
C_8H_{10}	Ethylbenzene	0.59	C	1
C_8H_{10}	o-Xylene	0.640	A	2
$C_8H_{10}O$	Phenetole (ethyl phenyl ether)	1.45	Di	1
$C_8H_{11}N$	N,N-Dimethylaniline	1.68	D	1
$C_8H_{18}O$	Dibutyl ether	1.17	Ci	1
C_9H_7N	Quinoline	2.29	C	1
C_9H_7N	Isoquinoline	2.73	C	1
$C_9H_{10}O_2$	Ethyl benzoate	2.00	Ci	1
C_9H_{12}	Cumene (isopropylbenzene)	0.79	Q	1
$C_{10}H_8$	Azulene	0.80	B	1
$C_{10}H_{14}$	tert-Butylbenzene	0.83	Q	1
$C_{12}H_{10}$	Acenaphthene	0.85	Q	1
$C_{12}H_{10}O$	Diphenyl ether	1.3	Q	1

STRENGTHS OF CHEMICAL BONDS*
J. A. Kerr

The strength of a chemical bond, $D°(R\text{-}X)$, often known as the bond dissociation energy, is defined as the standard enthalpy change of the reaction in which the bond is broken: $RX \rightarrow R + X$. It is given by the thermochemical equation, $D°(R\text{-}X) = \Delta_f H°(R) + \Delta_f H°(X) - \Delta_f H°(RX)$. Some authors list bond strengths at a temperature of absolute zero but here the values at 298 K are given because more thermodynamic data are available for this temperature. Bond strengths or bond dissociation energies are not equal to, and may differ considerably from, mean bond energies determined solely from thermochemical data on atoms and molecules.

Table 1
BOND STRENGTHS IN DIATOMIC MOLECULES

These have usually been measured spectroscopically or by mass spectrometric analysis of hot gases effusing a Knudsen cell. Excellent accounts of these and other methods are given in (i) *Dissociation Energies and Spectra of Diatomic Molecules*, by A. G. Gaydon, 3rd. ed., Chapman & Hall, London, 1968 and (ii) "Mass Spectrometric Determination of Bond Energies of High-Temperature Molecules", K. A. Gingerich, *Chimia*, 26, 619, 1972. The errors quoted in the table are those given in the original paper or review article. The references have been chosen primarily as a key to the literature. It should not be assumed that the author referred to was responsible for the value quoted, as the reference may be to a review article.

Bond strengths reported at a temperature of absolute zero, $D°_0$, have been converted to $D°_{298}$ by the use of enthalpy functions taken mainly from the JANAF Thermochemical Tables, Third Edition, *J. Phys. Chem. Ref. Data*, 14, Suppl. 1, 1985, wherever possible. For most bonds, however, where these data are not available, the conversion has been made by the approximate relation:

$$D°_{298} = D°_0 + (3/2)RT$$

The list below does not include the increasing number of bond strengths of diatomic molecules now being calculated by *ab initio* methods. The Table has been arranged in an alphabetical order of the atoms.

Molecule	$D°_{298}$/kJ mol^{-1}	Ref.	Molecule	$D°_{298}$/kJ mol^{-1}	Ref.	Molecule	$D°_{298}$/kJ mol^{-1}	Ref.
Ag-Ag	160.3 ± 3.4	296	Al-Au	325.9 ± 6.3	116	As-F	410	192
Ag-Al	183.7 ± 9.2	71	Al-Br	429 ± 6	176	As-Ga	209.6 ± 1.2	75
Ag-Au	202.9 ± 9.2	3	Al-Cl	511.3 ± 0.8	294	As-H	274.0 ± 2.9	26
Ag-Bi	193 ± 42	230	Al-Cu	227.1 ± 9.8	19	As-I	296.6 ± 28.0	308
Ag-Br	293 ± 29	112	Al-D	290.8	230	As-In	201	281
Ag-Cl	341.4	151	Al-F	663.6 ± 6.3	72	As-N	489 ± 2	292
Ag-Cu	174.1 ± 9.2	128	Al-H	284.9 ± 6.3	72	As-O	481 ± 8	245
Ag-D	226.8	192	Al-I	369.9 ± 2.1	250	As-P	433.5 ± 12.6	129
Ag-Dy	130 ± 19	190	Al-Kr	5.89 ± 0.81	39	As-S	379.5 ± 6.3	245
Ag-Eu	129.7 ± 12.6	58	Al-Li	175.7 ± 14.6	152	As-Sb	330.5 ± 5.4	87
Ag-F	354.4 ± 16.3	112	Al-N	297 ± 96	112	As-Se	96	272
Ag-Ga	180 ± 15	40	Al-Ni	225 ± 5	18	As-Tl	198.3 ± 14.6	284
Ag-Ge	174.5 ± 20.9	127	Al-O	511 ± 3	49,66	At-At	~80	81
Ag-H	215.1 ± 8	202	Al-P	216.7 ± 12.6	72	Au-Au	224.7 ± 1.5	31,234
Ag-Ho	123.4 ± 16.7	54	Al-Pd	254.4 ± 12.1	56	Au-B	367.8 ± 10.5	137
Ag-I	234 ± 29	112	Al-S	373.6 ± 7.9	352	Au-Ba	254.8 ± 10.0	127
Ag-In	166.5 ± 4.9	11	Al-Sb	216.3 ± 5.9	277	Au-Be	285 ± 8	112
Ag-Li	177.4 ± 6.3	260	Al-Se	337.6 ± 10.0	352	Au-Bi	297 ± 8.4	127
Ag-Mn	100 ± 21	230	Al-Si	229.3 ± 30.1	44	Au-Ca	243	127
Ag-Na	138.1 ± 8.4	275,282	Al-Te	267.8 ± 10.0	352	Au-Ce	339 ± 21	127
Ag-Nd	<209	206	Al-U	326 ± 29	115	Au-Cl	343 ± 9.6	112
Ag-O	220.1 ± 20.9	271	Al-Xe	7.43 ± 0.69	39	Au-Co	222 ± 17	127
Ag-S	217.1	329	Ar-Ar	4.73 ± 0.04	172	Au-Cr	213 ± 17	127
Ag-Se	202.5	329	Ar-He	3.89	230	Au-Cs	255 ± 3.3	37
Ag-Si	177.8 ± 10.0	303	Ar-Hg	6.15	230	Au-Cu	228.0 ± 5.0	31,127
Ag-Sn	136.0 ± 20.9	2	Ar-I	10.0	36	Au-D	318.4	192
Ag-Te	195.8	329	Ar-K	4.2	192	Au-Dy	259 ± 21	190
Al-Al	133 ± 6	110	As-As	382.0 ± 10.5	231	Au-Eu	241.0 ± 10.5	58
Al-Ar	5.34 ± 0.78	39	As-Cl	448	72	Au-Fe	187.0 ± 16.7	205
Al-As	202.9 ± 7.1	278,286	As-D	270.3	192	Au-Ga	234 ± 38	127

* Revised to December 1995.

Table 1
BOND STRENGTHS IN DIATOMIC MOLECULES (continued)

Molecule	$D°_{298}$/kJ mol⁻¹	Ref.	Molecule	$D°_{298}$/kJ mol⁻¹	Ref.	Molecule	$D°_{298}$/kJ mol⁻¹	Ref.
Au-Ge	274.1 ± 5.0	127	Ba-Br	362.8 ± 8.4	97,186,214	Br-Li	418.8 ± 4	341
Au-H	292.0 ± 8	202	Ba-Cl	436.0 ± 8.4	184,186	Br-Mg	≤327.2	192
Au-Ho	267.4 ± 16.7	54,234	Ba-D	≤193.7	192	Br-Mn	314.2 ± 9.6	112
Au-In	286.0 ± 5.7	11	Ba-F	587.0 ± 6.7	94,183	Br-N	276 ± 21	112
Au-La	336.4 ± 20.9	119	Ba-H	176 ± 14.6	112	Br-Na	367.4 ± 0.8	341,354
Au-Li	284.5 ± 6.7	260	Ba-I	320.8 ± 6.3	178	Br-Ni	360 ± 13	112
Au-Lu	332.2 ± 16.7	124	Ba-O	561.9 ± 13.4	271	Br-O	235.1 ± 0.4	72
Au-Mg	243 ± 42	230	Ba-Pd	221.8 ± 5.0	117	Br-Pb	247 ± 38	112
Au-Mn	185.4 ± 12.6	322	Ba-Rh	259.4 ± 25.1	117	Br-Rb	380.7 ± 4	341
Au-Na	215.1 ± 12.6	282	Ba-S	400.0 ± 18.8	62	Br-Sb	314 ± 59	112
Au-Nd	299.2 ± 20.9	119	Be-Be	59	32,79	Br-Sc	444 ± 63	230
Au-Ni	255 ± 21	127	Be-Br	381 ± 84	230	Br-Se	297 ± 84	230
Au-O	221.8 ± 20.9	271	Be-Cl	388.3 ± 9.2	101,181,358	Br-Si	367.8 ± 10.0	99
Au-Pb	130 ± 42	230	Be-D	203.05	192	Br-Sn	≥552	270
Au-Pd	155 ± 21	127	Be-F	577 ± 42	72,101	Br-Sr	333.0 ± 9.2	186
Au-Pr	310 ± 21	127	Be-H	200.0 ± 1.3	59	Br-Th	364	177
Au-Rb	243 ± 2.9	37	Be-O	434.7 ± 13.4	271	Br-Ti	439	230
Au-Rh	231.0 ± 29	58	Be-S	372 ± 59	112	Br-Tl	333.9 ± 1.7	25
Au-S	418 ± 25	123	Bi-Bi	200.4 ± 7.5	290,307	Br-U	377.4 ± 6.3	243
Au-Sc	280.3 ± 16.7	120	Bi-Br	267.4 ± 4.2	68	Br-V	439 ± 42	230
Au-Se	243.1	329	Bi-Cl	301 ± 4	70	Br-W	329.3	208
Au-Si	305.4 ± 5.9	131	Bi-D	283.7	249	Br-Y	485 ± 84	230
Au-Sn	254.8 ± 7.1	217	Bi-F	367 ± 13	375	Br-Zn	142 ± 29	230
Au-Sr	264 ± 42	230	Bi-Ga	159 ± 17	280	C-C	607 ± 21	72
Au-Tb	289.5 ± 33.5	144,234	Bi-H	≤283.3	249	C-Ce	444 ± 13	220
Au-Te	317.6	329	Bi-I	218.0 ± 4.6	69	C-Cl	397 ± 29	267
Au-U	318 ± 29	115	Bi-In	153.6 ± 1.7	304	C-D	341.4	192
Au-V	240.6 ± 12.1	165	Bi-Li	154.0 ± 5.0	262,288	C-F	552	182
Au-Y	307.1 ± 8.4	170	Bi-O	337.2 ± 12.6	271	C-Ge	460 ± 21	112
B-B	297 ± 21	72	Bi-P	280 ± 13	129	C-H	338.4 ± 1.2	192
B-Br	396	29	Bi-Pb	141.8 ± 14.6	307	C-Hf	540 ± 25	336
B-C	448 ± 29	230	Bi-S	315.5 ± 4.6	351	C-I	209 ± 21	112
B-Ce	305 ± 21	230	Bi-Sb	251 ± 4	228	C-Ir	632 ± 4	164
B-Cl	536	17	Bi-Se	280.3 ± 5.9	351	C-La	462 ± 20	274
B-D	341.0 ± 6.3	230	Bi-Sn	210.0 ± 8.4	127	C-Mo	481 ± 15.9	160
B-F	757	240	Bi-Te	232.2 ± 11.3	351	C-N	754.3 ± 10	38
B-H	340	287	Bi-Tl	121 ± 13	76	C-Nb	569 ± 13.0	160
B-I	220.5 ± 0.8	297	Br-Br	192.807	1	C-O	1076.5 ± 0.4	72
B-Ir	514.2 ± 17.2	357	Br-C	280 ± 21	112	C-Os	≥594	118
B-La	339 ± 63	230	Br-Ca	310.9 ± 9.2	302	C-P	513.4 ± 8	330
B-N	389 ± 21	72	Br-Cd	159 ± 96	112	C-Pt	598 ± 5.9	164,355
B-O	808.8 ± 20.9	271	Br-Cl	217.53 ± 0.29	50	C-Rh	580.0 ± 3.8	312
B-P	346.9 ± 16.7	139	Br-Co	331 ± 42	230	C-Ru	616.2 ± 10.5	313
B-Pd	329.3 ± 20.9	357	Br-Cr	328.0 ± 24.3	112	C-S	714.1 ± 1.2	63,333
B-Pt	477.8 ± 16.7	251	Br-Cs	389.1 ± 4	269,341	C-Sc	≤444 ± 21	140
B-Rh	475.7 ± 20.9	357	Br-Cu	331 ± 25	112	C-Se	590.4 ± 5.9	323
B-Ru	446.9 ± 20.9	357	Br-D	370.74	192	C-Si	451.5	83,363
B-S	580.7 ± 9.2	350	Br-F	280 ± 12	197	C-Tc	565 ± 29	305
B-Sc	276 ± 63	230	Br-Fe	247 ± 96	112	C-Th	453 ± 17	159,336
B-Se	461.9 ± 14.6	350	Br-Ga	444 ± 17	72	C-Ti	423 ± 29	155,336
B-Si	288.3	363	Br-Ge	255 ± 29	112	C-U	454.8 ± 15.1	158,162
B-Te	354.4 ± 20.1	350	Br-H	366.35	192	C-V	427 ± 23.8	160
B-Th	297	132	Br-Hg	72.8 ± 4	72	C-Y	418 ± 14	318
B-Ti	276 ± 63	230	Br-I	179.1 ± 0.4	112,291	C-Zr	561 ± 25	336
B-U	322 ± 33	230	Br-In	414 ± 21	72	Ca-Ca	≤46	371,373
B-Y	293 ± 63	230	Br-K	379.9 ± 0.8	341,354	Ca-Cl	409 ± 9	252

Table 1
BOND STRENGTHS IN DIATOMIC MOLECULES (continued)

Molecule	$D°_{298}$/kJ mol⁻¹	Ref.	Molecule	$D°_{298}$/kJ mol⁻¹	Ref.	Molecule	$D°_{298}$/kJ mol⁻¹	Ref.
Ca-D	≤169.9	192	Cl-Rb	427.6 ± 8	342	Cu-Ho	142 ± 21	190
Ca-F	527 ± 21	94,180	Cl-S	277.0	209	Cu-I	197 ± 21	112
Ca-H	167.8	112	Cl-Sb	360 ± 50	112	Cu-In	187.4 ± 7.9	11
Ca-I	284.7 ± 8.4	178	Cl-Sc	331	362	Cu-Li	192.9 ± 8.8	260
Ca-Li	84.9 ± 8.4	371	Cl-Se	322	230	Cu-Mn	158.6 ± 17	207
Ca-O	402.1 ± 16.7	271	Cl-Si	406	361	Cu-Na	176.1 ± 16.7	283
Ca-S	337.6 ± 18.8	62,192	Cl-Sm	≥423 ± 13	374	Cu-Ni	205.9 ± 17	207
Cd-Cd	7.36	235	Cl-Sn	414 ± 17	230	Cu-O	269.0 ± 20.9	271
Cd-Cl	208.4	192	Cl-Sr	406 ± 13	184,186	Cu-S	276	329
Cd-F	305 ± 21	28	Cl-Ta	544	20	Cu-Se	251	329
Cd-H	69.0 ± 0.4	112	Cl-Th	489	244	Cu-Si	221.3 ± 6.3	303
Cd-I	97.23	200	Cl-Ti	494	230	Cu-Sn	169.5 ± 6.7	2,221
Cd-In	138	230	Cl-Tl	372.8 ± 2.1	25	Cu-Tb	193 ± 19	190
Cd-O	235.6 ± 83.7	149,271	Cl-U	452 ± 8	242	Cu-Te	278.7	1
Cd-S	208.4 ± 20.9	149	Cl-V	477 ± 63	230	D-D	443.533	192
Cd-Se	127.6 ± 25.1	149	Cl-W	423 ± 42	230	D-F	576.6	192
Cd-Te	100.0 ± 15.1	149	Cl-Xe	6.7	192	D-Ga	<272.8	237
Ce-Ce	245.2	119	Cl-Y	527 ± 84	230	D-Ge	≤322	192
Ce-F	582 ± 42	230	Cl-Yb	~322	107	D-H	439.433	192
Ce-Ir	586	141	Cl-Zn	228.9 ± 19.7	65	D-Hg	42.05	192
Ce-N	519 ± 21	138	Cm-O	736	321	D-In	246.0	192
Ce-O	795 ± 8	88	Co-Co	167 ± 25	203	D-Li	240.1892 ± 0.0046	194,339
Ce-Os	506 ± 33	118	Co-Cu	167 ± 17	127	D-Mg	135.1	192
Ce-Pd	322.2	55	Co-F	435 ± 63	230	D-Ni	≤302.9	192
Ce-Pt	556	141	Co-Ge	234 ± 21	127	D-Pt	≤350.2	192
Ce-Rh	548	141	Co-H	226 ± 42	347	D-S	351	192
Ce-Ru	531 ± 25	118	Co-I	285 ± 21	230	D-Si	302.5	192
Ce-S	569	21	Co-O	384.5 ± 13.4	271	D-Sr	≥275.7	192
Ce-Se	494.5 ± 14.6	255	Co-S	331	329	D-Zn	88.7	192
Ce-Te	189.4 ± 12.8	236	Co-Si	276 ± 17	356	Dy-F	531	381
Cl-Cl	242.580 ± 0.004	192	Cr-Cr	142.9 ± 5.4	188	Dy-O	607 ± 17	88
Cl-Co	389	238	Cr-Cu	155 ± 21	207	Dy-S	414 ± 42	230
Cl-Cr	366.1 ± 24.3	112	Cr-F	444.8 ± 19.7	211	Dy-Se	322 ± 42	230
Cl-Cs	448 ± 8	269,342	Cr-Ge	154 ± 7	189	Dy-Te	234 ± 42	230
Cl-Cu	382.8 ± 4.6	154	Cr-H	190.3 ± 7.0	46	Er-F	565 ± 17	381
Cl-D	436.47	192	Cr-I	287.0 ± 24.3	112	Er-O	615 ± 13	88
Cl-Eu	~326	107	Cr-N	377.8 ± 18.8	144,334	Er-S	418 ± 42	230
Cl-F	256.23	192,264	Cr-O	429.3 ± 29.3	271	Er-Se	326 ± 42	230
Cl-Fe	~351	112	Cr-Pb	105 ± 2	189	Er-Te	238 ± 42	230
Cl-Ga	481 ± 13	72	Cr-S	331	85	Eu-Eu	33.5 ± 17	58
Cl-Ge	~431	192	Cr-Sn	141 ± 3	189	Eu-F	544	226
Cl-H	431.62	192	Cs-Cs	43.919 ± 0.010	368	Eu-Li	66.9 ± 2.9	259
Cl-Hg	100 ± 8	112	Cs-F	519 ± 8	269	Eu-O	479 ± 10	88
Cl-I	211.3 ± 0.4	112	Cs-H	175.364	376	Eu-Rh	233.9 ± 33	58
Cl-In	439 ± 8	72	Cs-Hg	8	192	Eu-S	362.3 ± 13.0	255,327
Cl-K	433.0 ± 8	342	Cs-I	337.2 ± 2.1	269,340	Eu-Se	301 ± 14.6	22,171,255
Cl-Li	469 ± 13	72	Cs-Na	63.2 ± 1.3	78	Eu-Te	243 ± 14.6	22,255
Cl-Mg	327.6 ± 2.1	98,184,358	Cs-O	295.8 ± 62.8	271	F-F	158.78	192
Cl-Mn	360.7 ± 9.6	112	Cs-Rb	49.57 ± 0.01	167	F-Ga	577 ± 14.6	254
Cl-N	333.9 ± 9.6	47	Cu-Cu	176.52 ± 2.38	127,306	F-Gd	590.4 ± 27.2	378
Cl-Na	412.1 ± 8	342	Cu-D	270.3	192	F-Ge	485 ± 21	91
Cl-Ni	372 ± 21	112	Cu-Dy	142 ± 21	190	F-H	569.87 ± 0.06	377
Cl-O	269.1	192,301	Cu-F	413.4 ± 13	92	F-Hf	650 ± 15	14
Cl-P	289 ± 42	230	Cu-Ga	215.9 ± 15.1	40	F-Hg	~180	192
Cl-Pb	301 ± 29	112	Cu-Ge	208.8 ± 21	257	F-Ho	540	381
Cl-Ra	343 ± 75	112	Cu-H	277.8	202,300	F-I	≤271.5	6,30,51,67

Table 1
BOND STRENGTHS IN DIATOMIC MOLECULES (continued)

Molecule	$D°_{298}$/kJ mol^{-1}	Ref.	Molecule	$D°_{298}$/kJ mol^{-1}	Ref.	Molecule	$D°_{298}$/kJ mol^{-1}	Ref.
F-In	506 ± 14.6	254	Gd-O	719 ± 10	88	Hg-O	220.9 ± 33.1	149
F-K	497.5 ± 2.5	15	Gd-S	526.8 ± 10.5	109,325	Hg-Rb	8.4	192
F-La	598 ± 42	230	Gd-Se	431 ± 14.6	22	Hg-S	217.1 ± 22.2	149
F-Li	577 ± 21	72	Gd-Te	343 ± 14.6	22	Hg-Se	144.3 ± 30.1	149
F-Lu	333.5	195	Ge-Ge	263.6 ± 7.1	222	Hg-Te	≤142	230
F-Mg	461.9 ± 5.0	94,183	Ge-H	≤321.7	227	Hg-Tl	4	174
F-Mn	423.4 ± 14.6	212	Ge-Ni	290.3 ± 10.9	315	Ho-Ho	84 ± 17	54
F-Mo	464.8	185	Ge-O	659.4 ± 12.6	271	Ho-O	611 ± 17	88
F-N	343	192	Ge-Pd	254.7 ± 10.5	314	Ho-S	428.4 ± 14.6	325
F-Na	519	192	Ge-S	551.0 ± 2.5	80,86	Ho-Se	335 ± 17	22
F-Nd	545.2 ± 12.6	379	Ge-Sc	271.0 ± 11	218	Ho-Te	259 ± 17	22
F-Ni	430 ± 20	77	Ge-Se	489 ± 4	266	I-I	151.088	192,348
F-O	222 ± 17	52	Ge-Si	301 ± 21	112	I-In	331	360
F-P	439 ± 96	112	Ge-Te	456 ± 13	106	I-K	325.1 ± 0.8	340,354
F-Pb	356 ± 8	383	Ge-Y	279.8 ± 11.4	219	I-Li	345.2 ± 4.2	340
F-Pm	540 ± 42	230	H-H	435.990	192	I-Mg	~285	23
F-Pr	582 ± 46	230	H-Hg	39.844	192	I-Mn	282.8 ± 9.6	112
F-Pu	538.5 ± 29	213	H-I	298.407	192	I-N	159 ± 17	230
F-Rb	494 ± 21	72	H-In	243.1	192	I-Na	304.2 ± 2.1	340,354
F-Ru	402	175	H-K	174.576	193,376	I-Ni	293 ± 21	112
F-S	342.7 ± 5.0	27,187,215	H-Li	238.049 ± 0.004	367	I-O	249.0 ± 1.3	301
F-Sb	439 ± 96	112	H-Mg	126.4 ± 2.9	12,13,93	I-Pb	193 ± 4	320
F-Sc	589.1 ± 13	380	H-Mn	234 ± 29	112	I-Rb	318.8 ± 2.1	340
F-Se	339 ± 42	230	H-N	≤339	192	I-Si	293	192
F-Si	552.7 ± 2.1	102	H-Na	185.69 ± 0.25	258,298	I-Sn	234 ± 42	230
F-Sm	565	226	H-Ni	252.3 ± 8	202	I-Sr	269.9 ± 5.9	223
F-Sn	466.5 ± 13	383	H-O	427.6	192	I-Te	192 ± 42	230
F-Sr	541.8 ± 6.7	94,183	H-P	297	192	I-Ti	310 ± 42	230
F-Ta	573 ± 13	239	H-Pb	≤157	192	I-Tl	272 ± 8	23
F-Tb	561 ± 42	230	H-Pd	234 ± 25	347	I-Zn	108.29	200
F-Th	652	246	H-Pt	≤335	192	I-Zr	305	225
F-Ti	569 ± 33	382	H-Rb	167 ± 21	112	In-In	100 ± 8	230
F-Tl	445.2 ± 19.2	25	H-Rh	247 ± 21	347	In-Li	92.5 ± 14.6	152
F-Tm	510	226	H-Ru	234 ± 21	347	In-O	<320.1 ± 41.8	271
F-U	659.0 ± 10.5	148,241	H-S	344.3 ± 12.1	198	In-P	197.9 ± 8.4	278
F-V	590 ± 63	230	H-Sc	~180	309	In-S	289 ± 17	61
F-W	548 ± 63	230	H-Se	314.47 ± 0.96	114	In-Sb	151.9 ± 10.5	73
F-Xe	15.77	299,346	H-Si	≤299.2	192	In-Se	247 ± 17	61
F-Y	605.0 ± 20.9	380	H-Sn	264 ± 17	112	In-Te	218 ± 17	61
F-Yb	≥521.3 ± 9.6	16,107,374	H-Sr	163 ± 8	112	Ir-La	577 ± 13	169
F-Zn	368 ± 63	230	H-Te	268 ± 2.1	111	Ir-O	414.6 ± 42.3	271
F-Zr	616 ± 15	14	H-Ti	204.6 ± 8.8	45	Ir-Si	462.8 ± 20.9	357
Fe-Fe	75 ± 17	310	H-Tl	188 ± 8	112	Ir-Th	573	125
Fe-Ge	210.9 ± 29	204	H-V	208.7 ± 7.0	46	Ir-Ti	422 ± 13	273
Fe-H	180 ± 25	347	H-Yb	159 ± 38	112	Ir-Y	456.1 ± 16.7	170
Fe-O	390.4 ± 17.2	271	H-Zn	85.8 ± 2.1	112	K-K	54.63 ± 0.02	5,248
Fe-S	322	85	He-He	3.8	192	K-Kr	4.6	192
Fe-Si	297 ± 25	356	He-Hg	6.61	230	K-Li	82.0 ± 4.2	96,385
Ga-Ga	112.1 ± 7	317	Hf-C	548 ± 63	230	K-Na	65.994 ± 0.008	33,385
Ga-H	<274.1	237	Hf-N	536 ± 29	144,229	K-O	277.8 ± 20.9	271
Ga-I	339 ± 9.6	112	Hf-O	801.7 ± 13.4	271	K-Xe	5.0	192
Ga-Li	133.1 ± 14.6	152	Hg-Hg		190,191	Kr-Kr	5.23	42,192
Ga-O	353.5 ± 41.8	271	Hg-I	34.69 ± 0.96	365	Kr-O	<8	230
Ga-P	229.7 ± 12.6	122	Hg-K	8.24 ± 0.21	230	Kr-Xe	5.505 ± 0.002	7
Ga-Sb	192.0 ± 12.6	279	Hg-Li	13.8	192	La-La	247 ± 21	362
Ga-Te	251 ± 25	353	Hg-Na	9.2	192,386	La-N	519 ± 42	230

Table 1
BOND STRENGTHS IN DIATOMIC MOLECULES (continued)

Molecule	$D°_{298}$/kJ mol⁻¹	Ref.	Molecule	$D°_{298}$/kJ mol⁻¹	Ref.	Molecule	$D°_{298}$/kJ mol⁻¹	Ref.
La-O	799 ± 4	88	Nd-O	703 ± 13	88	P-Th	550.2 ± 42	127
La-Pt	502 ± 21	256	Nd-S	471.5	22	P-Tl	209 ± 13	277
La-Rh	527 ± 17	57	Nd-Se	385 ± 17	22,147,255	P-U	297 ± 21	230
La-S	573.2 ± 1.7	199,338	Nd-Te	305 ± 17	22	P-W	305 ± 4	142
La-Se	477 ± 17	22,255	Ne-Ne	3.93	344	Pb-Pb	86.6 ± 0.8	130,288
La-Te	381 ± 17	22,145	Ni-Ni	203.26 ± 0.96	253	Pb-S	346.0 ± 1.7	351
La-Y	202.1	362	Ni-O	382.0 ± 16.7	271	Pb-Sb	161.5 ± 10.5	384
Li-Li	110.21 ± 4	359,372	Ni-Pt	273.7 ± 0.3	345	Pb-Se	302.9 ± 4	351
Li-Mg	67.4 ± 6.3	370	Ni-S	344.3	85	Pb-Te	251 ± 13	351
Li-Na	87.181 ± 0.001	95,104	Ni-Si	318 ± 17	356	Pd-Pd	100 ± 15	311
Li-O	333.5 ± 8.4	271	Ni-V	206.3 ± 0.2	332	Pd-Si	261 ± 12	316
Li-Pb	78.7 ± 7.9	262	Np-O	718.4 ± 41.8	271	Pd-Y	238 ± 17	295
Li-S	312.5 ± 7.5	216	O-O	498.36 ± 0.17	35,192	Pm-S	423 ± 63	230
Li-Sb	172.8 ± 10.0	263	O-Os	575	179	Pm-Se	339 ± 63	230
Li-Sm	49.0 ± 4.2	259	O-P	599.1 ± 12.6	271	Pm-Te	255 ± 63	230
Li-Tm	69.0 ± 3.3	259	O-Pa	788.3 ± 17.2	224	Po-Po	187.0	192
Li-Yb	37.2 ± 2.9	259	O-Pb	382.0 ± 12.6	271	Pr-S	492.5 ± 4.6	105
Lu-Lu	142 ± 33	230	O-Pd	380.7 ± 83.7	271	Pr-Se	446.4 ± 23.0	146,255
Lu-O	678 ± 8	88	O-Pm	674 ± 63	230	Pr-Te	326 ± 42	230
Lu-Pt	402 ± 33	133	O-Pr	753 ± 13	88	Pt-Pt	357.3 ± 15.1	163
Lu-S	507.1 ± 14.6	108,325	O-Pt	391.6 ± 41.8	271	Pt-Si	501.2 ± 18.0	357
Lu-Se	418 ± 17	22	O-Pu	715.9 ± 33.9	271	Pt-Th	552	125
Lu-Te	326 ± 17	22	O-Rh	255 ± 84	34	Pt-Ti	397 ± 13	166
Mg-Mg	8.552 ± 0.004	247,371	O-Re	626.8 ± 83.7	271	Pt-Y	474.0 ± 12.1	163
Mg-O	363.2 ± 12.6	271	O-Rh	405.0 ± 41.8	271	Rb-Rb	48.898 ± 0.005	4
Mg-S	234	62	O-Ru	528.4 ± 41.8	271	Rh-Rh	285.3 ± 20.9	53,285
Mn-Mn	25.9	206	O-S	521.7 ± 4.2	271	Rh-Sc	443.9 ± 10.5	168
Mn-O	402.9 ± 41.8	210,271	O-Sb	434.3 ± 41.8	271	Rh-Si	395.0 ± 18.0	357
Mn-S	301 ± 17	369	O-Sc	681.6 ± 11.3	271	Rh-Th	515 ± 21	121
Mn-Se	239.3 ± 9.2	331	O-Se	464.8 ± 21.3	271,324	Rh-Ti	390.8 ± 14.6	53
Mo-Mo	406 ± 21	161	O-Si	799.6 ± 13.4	271	Rh-U	519 ± 17	121
Mo-Nb	456 ± 25	156	O-Sm	565 ± 13	88	Rh-V	364 ± 29	127
Mo-O	560.2 ± 20.9	271	O-Sn	531.8 ± 12.6	271	Rh-Y	445.2 ± 10.5	168
N-N	945.33 ± 0.59	192	O-Sr	425.5 ± 16.7	271	Ru-Si	397.1 ± 20.9	357
N-O	630.57 ± 0.13	192	O-Ta	799.1 ± 12.6	271	Ru-Th	591.6 ± 42	126
N-P	617.1 ± 20.9	64,143	O-Tb	711 ± 13	88	Ru-V	414 ± 29	127
N-Pu	473 ± 63	230	O-Te	376.1 ± 20.9	271	S-S	425.30	192
N-S	464 ± 21	230	O-Th	878.6 ± 12.1	271	S-Sb	378.7	103
N-Sb	301 ± 50	112	O-Ti	672.4 ± 9.2	271	S-Sc	477 ± 13	338,349
N-Sc	469 ± 84	230	O-Tm	502 ± 13	88	S-Se	371.1 ± 6.7	82
N-Se	370 ± 11	333	O-U	759.4 ± 13.4	271	S-Si	623	192
N-Si	470 ± 15	293	O-V	626.8 ± 18.8	10,271	S-Sm	389	105
N-Ta	611 ± 84	230	O-W	672.0 ± 41.8	271	S-Sn	464 ± 3.3	80
N-Th	577.4 ± 33.1	136,144	O-Xe	36.4	230	S-Sr	339	41
N-Ti	476.1 ± 33.1	144,335	O-Y	719.6 ± 11.3	196,271	S-Tb	515 ± 42	230
N-U	531.4 ± 2.1	134	O-Yb	397 ± 17	88	S-Te	339 ± 21	80
N-V	477.4 ± 17.2	100,144	O-Zn	159 ± 4	48	S-Ti	418 ± 3	89,276
N-Xe	23.0	173	O-Zr	776.1 ± 13.4	271	S-Tm	368 ± 42	230
N-Y	481 ± 63	230	P-P	489.5 ± 10.5	143	S-U	522.6 ± 9.6	338
N-Zr	564.8 ± 25.1	135,144	P-Pt	≤416.7 ± 17	328	S-V	450	90,192
Na-Na	73.60 ± 0.25	364	P-Rh	353.1 ± 17	328	S-Y	528.4 ± 10.5	337
Na-O	256.1 ± 16.7	271	P-S	444 ± 8	84	S-Yb	167	230
Na-Rb	63.25	366	P-Sb	356.9	233	S-Zn	205 ± 13	74,149
Nb-Nb	510 ± 10.0	157	P-Se	363.6 ± 10.0	84	S-Zr	575.3 ± 16.7	338
Nb-O	771.5 ± 25.1	271	P-Si	363.6	326	Sb-Sb	299.2 ± 6.3	73,232
Nd-Nd	<163	230	P-Te	297.9 ± 10.0	84	Sb-Te	277.4 ± 3.8	289,343

Table 1
BOND STRENGTHS IN DIATOMIC MOLECULES (continued)

Molecule	$D°_{298}$/kJ mol^{-1}	Ref.	Molecule	$D°_{298}$/kJ mol^{-1}	Ref.	Molecule	$D°_{298}$/kJ mol^{-1}	Ref.
Sb-Tl	126.8 ± 10.5	9,277	Se-V	347 ± 21	230	Te-Tm	276 ± 42	230
Sc-Sc	162.8 ± 21	128	Se-Y	435 ± 13	230	Te-Y	339 ± 13	230
Sc-Se	385 ± 17	230	Se-Zn	170.7 ± 25.9	74,149	Te-Zn	117.6 ± 18.0	149
Sc-Si	228.7 ± 14	218	Si-Si	326.8 ± 10.0	44	Th-Th	≤289	132
Sc-Te	289 ± 17	230	Si-Te	452 ± 8	192,265	Ti-Ti	141.4 ± 21	201
Se-Se	332.6 ± 0.4	82,351	Si-Y	258.8 ± 17.3	219	Tl-Tl	64.4 ± 17	8
Se-Si	548	192	Sm-Te	272.4 ± 14.6	255	U-U	222 ± 21	230
Se-Sm	331.0 ± 14.6	255	Sn-Sn	187.1 ± 0.3	268	V-V	269.2 ± 0.2	332
Se-Sn	401.2 ± 5.9	60	Sn-Te	359.8	192	Xe-Xe	6.53 ± 0.29	43
Se-Sr	~285	24	Sr-Sr	15.5 ± 0.4	113	Y-Y	159 ± 21	230
Se-Tb	423 ± 42	230	Tb-Tb	131.4 ± 25.1	234	Yb-Yb	20.5 ± 17	153
Se-Te	291.6 ± 4	80,82,150	Tb-Te	339 ± 42	230	Zn-Zn	29	319
Se-Ti	381 ± 42	230	Te-Te	259.8 ± 5.0	261			
Se-Tm	276 ± 42	230	Te-Ti	289 ± 17	230			

REFERENCES

1. Abbasov, A. S., Azizov, T. Kh., Alleva, N. A., Aliev, I. Ya, Mustafaev, F. M., and Mamedov, A. N., *Zh. Fiz. Khim.*, 50, 2172, 1976.
2. Ackerman, M., Drowart, J., Stafford, F. E., and Verhaegen, G., *J. Chem. Phys.*, 36, 1557, 1962.
3. Ackerman, M., Stafford, F. E., and Drowart, J., *J. Chem. Phys.*, 33, 1784, 1960.
4. Amiot, C., *J. Chem. Phys.*, 93, 8591, 1990.
5. Amiot, C., *J. Mol. Spectrosc.*, 147, 370, 1991.
6. Appelman, E. H. and Clyne, M. A. A., *J. Chem. Soc. Faraday Trans. 1*, 71, 2072, 1975.
7. Balakrishnan, A., Jones, W. J., Mahajan, C. G., and Stoicheff, B. P., *Chem. Phys. Lett.*, 155, 43, 1989.
8. Balducci, G. and Piacente, V., *J. Chem. Soc. Chem. Commun.*, 1287, 1980.
9. Balducci, G., Ferro, D., and Piacente, V., *High Temp. Sci.*, 14, 207, 1981.
10. Balducci, G., Gigli, G., and Guido, M., *J. Chem. Phys.*, 79, 5616, 1983.
11. Balducci, G., Nunzio, P. E., Gigli, G., and Guido, M., *J. Chem. Phys.*, 90, 406, 1989.
12. Balfour, W. J. and Cartwright, H. M., *Astron. Astrophys. Suppl. Ser.*, 26, 389, 1976.
13. Balfour, W. J. and Lingren, B., *Can. J. Chem.*, 56, 767, 1978.
14. Barkovskii, N. V., Tsirel'nikov, V. I., Emel'yanov, A. M., and Khodeev, Yu. S., *Teplofiz. Vys. Temp.*, 29, 474, 1991.
15. Barrow, R. F. and Caunt, A. D., *Proc. R. Soc. London Ser. A*, 219, 120, 1953.
16. Barrow, R. F. and Chojnicki, A. H., *J. Chem. Soc. Faraday Trans. 2*, 71, 728, 1975.
17. Barrow, R. F., *Trans. Faraday Soc.*, 56, 952, 1960.
18. Behm, J. M., Arrington, C. A., and Morse, M. D., *J. Chem. Phys.*, 99, 6409, 1993.
19. Behm, J. M., Arrington, C. A., Langenberg, J. D., and Morse, M. D., *J. Chem. Phys.*, 99, 6394, 1993.
20. Behrens, R. G. and Feber, R. C., *J. Less-Common Met.*, 75, 281, 1980.
21. Bergman, C. and Gingerich, K. A., *J. Phys. Chem.*, 76, 2332, 1972.
22. Bergman, C., Coppens P., Drowart, J., and Smoes, S., *Trans. Faraday Soc.*, 66, 800, 1970.
23. Berkowitz, J. and Chupka, W. A., *J. Chem. Phys.*, 45, 1287, 1966.
24. Berkowitz, J. and Chupka, W. A., *J. Chem. Phys.*, 45, 4289, 1966.
25. Berkowitz, J. and Walter, T., *J. Chem. Phys.*, 49, 1184, 1968.
26. Berkowitz, J., *J. Chem. Phys.*, 89, 7065, 1988.
27. Berneike ,W., Kreuttle, U., and Neuert, H., *Chem. Phys. Lett.*, 76, 525, 1980.
28. Besenbruch, G., Kana'an, A. S., and Margrave, J. L., *J. Phys. Chem.*, 69, 3174, 1965.
29. Bharate, N. S., Bhartiya, J. B., and Behere, S. H., *Proc. Indian Natl. Sci. Acad. Part A*, 57, 419, 1991.
30. Birks, J. W., Gabelnick, S. D., and Johnston, H. S., *J. Mol. Spectrosc.*, 57, 23, 1975.
31. Bishea, G. A. and Morse, M. D., *Chem. Phys. Lett.*, 171, 430, 1990.
32. Bondybey, V. E., *Chem. Phys. Lett.*, 109, 436, 1984.
33. Breford, E. J. and Engelke, F., *J. Chem. Phys.*, 71, 1994, 1979.
34. Brewer, L. and Rosenblatt, G. M., *Adv. High Temp. Sci.*, 2, 1, 1969.
35. Brix, P. and Herzberg, G., *J. Chem. Phys.*, 21, 2240, 1953.
36. Burns, G., LeRoy, L. J., Morris, D. J., and Blake, J. A., *Proc. R. Soc. London Ser. A*, 316, 81, 1970.
37. Busse, V. B. and Weil, K. G., *Ber. Bunsenges. Phys. Chem.*, 85, 309, 1981.
38. Calculated from $D_f H\infty$(CN) = 435.1 ±10 kJ mol^{-1} (Table 4), but see also Costes, M., Naulin, C., and Dorthe, G., *Astron. Astrophys.*, 232, 270, 1990.
39. Callender, C. L., Mitchell, S. A., and Hackett, P. A., *J. Chem. Phys.*, 90, 5252, 1989.
40. Carbonel, M., Bergman, C., and Laffite, M., *Colloq. Int. Cent. Nat. Rech. Sci.*, 201, 311, 1972.

Table 1
BOND STRENGTHS IN DIATOMIC MOLECULES (continued)

41. Cater, E. D. and Johnson, F. W., *J. Chem. Phys.*, 47, 5353, 1967.
42. Chashchina, G. I. and Shreider, E. Ya., *Zh. Prikl. Spektrosk.*, 21, 696, 1974.
43. Chashchina, G. I. and Shreider, E. Ya., *Zh. Prikl. Spektrosk.*, 25, 163, 1976.
44. Chatillon, C., Allibert, M., and Pattoret, A., *C. R. Acad. Sci. Ser. C*, 280, 1505, 1975.
45. Chen, Y. M., Clemmer, D. E., and Armentrout, P. B., *J. Chem. Phys.*, 95, 1228, 1991.
46. Chen, Y.-M., Clemmer, D. E., and Armentrout, P. B., *J. Chem. Phys.*, 98, 4929, 1993.
47. Clarke, T. C. and Clyne, M. A. A., *Trans. Faraday Soc.*, 66, 877, 1970.
48. Clemmer, D. E., Daltaska, N. F., and Armentrout, P. B., *J. Chem. Phys.*, 95, 7263, 1991.
49. Clemmer, D. E., Weber, M. E., and Armentrout, P. B., *J. Phys. Chem.*, 96, 10888, 1992.
50. Clyne, M. A. A. and McDermid, I. S., *Faraday Discuss. Chem. Soc.*, 67, 316, 1979.
51. Clyne, M. A. A. and McDermid, I. S., *J. Chem. Soc. Faraday Trans. 2*, 72, 2252, 1976.
52. Clyne, M. A. A. and Watson, R. T., *Chem. Phys. Lett.*, 12, 344, 1971.
53. Cocke, D. L. and Gingerich, K. A., *J. Chem. Phys.*, 60, 1958, 1974.
54. Cocke, D. L. and Gingerich, K. A., *J. Phys. Chem.*, 75, 3264, 1971.
55. Cocke, D. L. and Gingerich, K. A., *J. Phys. Chem.*, 76, 2332, 1972.
56. Cocke, D. L., Gingerich, K. A., and Chang, C. A., *J. Chem. Soc. Faraday Trans. 1*, 72, 268, 1976.
57. Cocke, D. L., Gingerich, K. A., and Kordis, J., *High Temp. Sci.*, 5, 474, 1973.
58. Cocke, D. L., Gingerich, K. A., and Kordis, J., *High Temp. Sci.*, 7, 61, 1975.
59. Colin, R. and De Greef, D., *Can. J. Phys.*, 53, 2142, 1975.
60. Colin, R. and Drowart, J., *Trans. Faraday Soc.*, 60, 673, 1964.
61. Colin, R. and Drowart, J., *Trans. Faraday Soc.*, 64, 2611, 1968.
62. Colin, R., Goldfinger, P., and Jeunehomme, M., *Trans. Faraday Soc.*, 60, 306, 1964.
63. Coppens, P., Reynaert, J. C., and Drowart, J., *J. Chem. Soc. Faraday Trans. 2*, 75, 292, 1979.
64. Coquart, B. and Prudhomme, J. C., *J. Mol. Spectrosc.*, 87, 75, 1981.
65. Corbett, J. D. and Lynde, R. A., *Inorg. Chem.*, 6, 2199, 1967.
66. Costes, M., Naulin, C., Dorthe, G., Vaucamps, C., Nouchi, G., *Faraday Discuss. Chem. Soc.*, 84, 75, 1987.
67. Coxon, J. A., *Chem. Phys. Lett.*, 33, 136, 1975.
68. Cubicciotti, D., *Inorg. Chem.*, 7, 208, 1968.
69. Cubicciotti, D., *Inorg. Chem.*, 7, 211, 1968.
70. Cubicciotti, D., *J. Phys. Chem.*, 71, 3066, 1967.
71. Cuthill, A. M., Fabian, D. J., and Shu-Shou-Shen, S., *J. Phys. Chem.*, 77, 2008, 1973.
72. Darwent, B. de B., *Bond Dissociation Energies in Simple Molecules*, NSRDS-NBS-31, National Bureau of Standards, Washington, D. C., 1970.
73. De Maria, G., Drowart, J., and Inghram, M. G., *J. Chem. Phys.*, 31, 1076, 1959.
74. De Maria, G., Goldfinger, P., Malaspina, L., and Piacente, V., *Trans. Faraday Soc.*, 61, 2146, 1965.
75. De Maria, G., Malaspina, L., and Piacente, V., *J. Chem. Phys.*, 52, 1019, 1970.
76. De Maria, G., Malaspina, L., and Piacente, V., *J. Chem. Phys.*, 56, 1978, 1972.
77. Devore, T. C., McQuaid, M., and Gole, J. L., *High Temp. Sci.*, 30, 83, 1990.
78. Diemer, U., Weickenmeier, H., Wahl, M., and Demtroeder, W., *Chem. Phys. Lett.*, 104, 489, 1984.
79. Drowart, J. and Goldfinger, P., *Angew. Chem.*, 6, 581, 1967.
80. Drowart, J. and Goldfinger, P., *Q. Rev. (London)*, 20, 545, 1966.
81. Drowart, J. and Honig, R. E., *J. Phys. Chem.*, 61, 980, 1957.
82. Drowart, J. and Smoes, S., *J. Chem. Soc. Faraday Trans. 2*, 73, 1755, 1977.
83. Drowart, J., De Maria, G., and Inghram, M. G., *J. Chem. Phys.*, 29, 1015, 1958.
84. Drowart, J., Myers, C. E., Szwarc, R., Vander Auwera-Mahieu, A., and Uy, O. M., *High Temp. Sci.*, 5, 482, 1973.
85. Drowart, J., Pattoret, A., and Smoes, S., *Proc. Br. Ceramic Soc.*, No. 8, 67, 1967.
86. Drowart, J., Smets, J., Reynaert, J. C., and Coppens, P., *Adv. Mass Spectrom.*, 7A, 647, 1978.
87. Drowart, J., Smoes, S, and Vander Auwera-Mahieu, A., *J. Chem. Thermodyn.*, 10, 453, 1978.
88. Dulick, M., Murad, E., and Barrow, R. F., *J. Chem. Phys.*, 85, 385, 1986.
89. Edwards, J. G., Franklin, H. F., and Gilles, P. W., *J. Chem. Phys.*, 54, 545, 1971.
90. Edwards, J. G., *J. Chem. Phys.*, 96, 866, 1992
91. Ehlert, T. C. and Margrave, J. L., *J. Chem. Phys.*, 41, 1066, 1964.
92. Ehlert, T. C. and Wang, J. S., *J. Phys. Chem.*, 81, 2069, 1977.
93. Ehlert, T. C., Hilmer, R. M., and Beauchamp, E. A., *J. Inorg. Nucl. Chem.*, 30, 3112, 1968.
94. Engelke, F., *Chem. Phys.*, 39, 279, 1979.
95. Engelke, F., Ennen, G., and Meiwes, K. H., *Chem. Phys.*, 66, 391, 1982.
96. Engelke, F., Hage, H., and Sprick, U., *Chem. Phys.*, 88, 443, 1984.
97. Estler, C. and Zare, R. N., *Chem. Phys.*, 28, 253, 1978.
98. Farber, M. and Srivastava, R. D., *Chem. Phys. Lett.*, 42, 567, 1976.

Table 1
BOND STRENGTHS IN DIATOMIC MOLECULES (continued)

99. Farber, M. and Srivastava, R. D., *High Temp. Sci.*, 12, 21, 1980.
100. Farber, M. and Srivastava, R. D., *J. Chem. Soc. Faraday Trans. 1*, 69, 390, 1973.
101. Farber, M. and Srivastava, R. D., *J. Chem. Soc. Faraday Trans. 1*, 70, 1581, 1974.
102. Farber, M. and Srivastava, R. D., *J. Chem. Soc. Faraday Trans. 1*, 74, 1089, 1978.
103. Faure, F. M., Mitchell, M. J., and Bartlett, R. W., *High Temp. Sci.*, 4, 181, 1972.
104. Fellows, C. E., *J. Chem. Phys.*, 94, 5855, 1991.
105. Fenochka, B. V. and Gorkienko, S. P., *Zh. Fiz. Khim*, 47, 2445, 1973.
106. Ferro, D., Piacente, V., and Bardi, G., *High Temp. Sci.*, 21, 69, 1986.
107. Filippenko, N. V., Morozov, E. V., Giricheva, N. L., and Krasnev, K. S., *Izv. Vyssh. Ucheb. Zaved Khim.Technol.*, 15, 1416, 1972.
108. Franzen, H. and Hariharan, A. V., *J. Chem. Phys.*, 70, 4907, 1979.
109. Fries, J. A. and Cater, E. D., *J. Chem. Phys.*, 68, 3978, 1978.
110. Fu, Z., Lemire, G. W., Bishea, G. A., and Morse, M. D., *J. Chem. Phys.*, 93, 8420, 1990.
111. Gal, J. F., Maria, P. C., and Decouzon, M., *Int. J. Mass Spectrom. Ion Processes*, 93, 87, 1989.
112. Gaydon, A. G., *Dissociation Energies and Spectra of Diatomic Molecules, 3rd ed.*, Chapman & Hall, London, 1968.
113. Gerber, G. and Moeller, R., *Contrib. Symp. At. Surf. Phys.*, 168, 1982.
114. Gibson, S. T., Greene, J. P., and Berkowitz, J., *J. Chem. Phys.*, 85, 4815, 1986.
115. Gingerich, K. A. and Blue, G. D., *J. Chem. Phys.*, 47, 5447, 1967.
116. Gingerich, K. A. and Blue, G. D., *J. Chem. Phys.*, 59, 186, 1973.
117. Gingerich, K. A. and Choudary, U. V., *J. Chem. Phys.*, 68, 3265, 1978.
118. Gingerich, K. A. and Cocke, D. L., *Inorg. Chim. Acta*, 28, L171, 1978.
119. Gingerich, K. A. and Finkbeiner, H. C., *J. Chem. Phys.*, 54, 2621, 1971.
120. Gingerich, K. A. and Finkbeiner, H. C., *Proc. 9th Rare Earth Res. Conf.*, 2, 795, 1971.
121. Gingerich, K. A. and Gupta, S. K., *J. Chem. Phys.*, 69, 505, 1978.
122. Gingerich, K. A. and Piacente, V., *J. Chem. Phys.*, 54, 2498, 1971.
123. Gingerich, K. A., *Chem. Commun.*, 580, 1970.
124. Gingerich, K. A., *Chem. Phys. Lett.*, 13, 262, 1972.
125. Gingerich, K. A., *Chem. Phys. Lett.*, 23, 270, 1973.
126. Gingerich, K. A., *Chem. Phys. Lett.*, 25, 523, 1974.
127. Gingerich, K. A., *Chem. Soc. Faraday, Symp.*, No.14, 109, 1980.
128. Gingerich, K. A., *Chimia*, 26, 619, 1972.
129. Gingerich, K. A., Cocke, D. L., and Kordis, J., *J. Phys. Chem.*, 78, 603, 1974.
130. Gingerich, K. A., Cocke, D. L., and Miller, F., *J. Chem. Phys.*, 64, 4027, 1976.
131. Gingerich, K. A., Haque, R., and Kingcade, J. E., *Thermochim. Acta*, 30, 61, 1979.
132. Gingerich, K. A., *High Temp. Sci.*, 1, 258, 1969.
133. Gingerich, K. A., *High Temp. Sci.*, 3, 415, 1971.
134. Gingerich, K. A., *J. Chem. Phys.*, 47, 2192, 1967.
135. Gingerich, K. A., *J. Chem. Phys.*, 49, 14, 1968.
136. Gingerich, K. A., *J. Chem. Phys.*, 49, 19, 1968.
137. Gingerich, K. A., *J. Chem. Phys.*, 54, 2646, 1971.
138. Gingerich, K. A., *J. Chem. Phys.*, 54, 3720, 1971.
139. Gingerich, K. A., *J. Chem. Phys.*, 56, 4239, 1972.
140. Gingerich, K. A., *J. Chem. Phys.*, 74, 6407, 1981.
141. Gingerich, K. A., *J. Chem. Soc. Faraday Trans. 2*, 70, 471, 1974.
142. Gingerich, K. A., *J. Phys. Chem.*, 68, 768, 1964.
143. Gingerich, K. A., *J. Phys. Chem.*, 73, 2734, 1969.
144. Gingerich, K. A., *NBS Spec. Publ. (U. S.)*, 561, 289, 1979.
145. Gordienko, S. P. and Fenochka, B. V., *Izv. Akad. Nauk. SSSR Neorg. Mater.*, 18, 1811, 1982.
146. Gordienko, S. P., Fenochka, B. V., Viksman, G. Sh., Klockkova, L. A., and Mikhlina, T. M., *Izv. Akad. Nauk. SSSR Neorg. Mater.*, 18, 18, 1982.
147. Gordienko, S. P., *Izv. Akad. Nauk. SSSR Neorg. Mater.*, 20, 1472, 1984.
148. Gorokhov, L. N., Smirnov, V. K., and Khodeev, Yu. S., *Zh. Fiz. Khim.*, 58, 1603, 1984.
149. Grade, M. and Hirschwald, W., *Ber. Bunsenges. Phys. Chem.*, 86, 899, 1982.
150. Grade, M., Wienecke, J., Rosinger, W., and Hirschwald, W., *Ber. Bunsenges. Phys. Chem.*, 87, 355, 1983.
151. Graeber, P. and Weil, K. G., *Ber. Bunsenges. Phys. Chem.*, 76, 417, 1972.
152. Guggi, D. J., Neubert, A., and Zmbov, K. F., *Conf. Int. Thermodyn. Chim. [C. R.] 4th*, 3, 124, 1975.
153. Guido, M. and Balducci, G., *J. Chem. Phys.*, 57, 5611, 1972.
154. Guido, M., Gigli, G., and Balducci, G., *J. Chem. Phys.*, 57, 3731, 1972.
155. Gupta, S. K. and Gingerich, K. A., *High Temp. - High Pressures*, 12, 273, 1980.
156. Gupta, S. K. and Gingerich, K. A., *J. Chem. Phys.*, 69, 4318, 1978.

Table 1
BOND STRENGTHS IN DIATOMIC MOLECULES (continued)

157. Gupta, S. K. and Gingerich, K. A., *J. Chem. Phys.*, 70, 5350, 1979.
158. Gupta, S. K. and Gingerich, K. A., *J. Chem. Phys.*, 71, 3072, 1979.
159. Gupta, S. K. and Gingerich, K. A., *J. Chem. Phys.*, 72, 2795, 1980.
160. Gupta, S. K. and Gingerich, K. A., *J. Chem. Phys.*, 74, 3584, 1981.
161. Gupta, S. K., Atkins, R. M., and Gingerich, K. A., *Inorg. Chem.*, 17, 3211, 1978.
162. Gupta, S. K., Kingcade, J. E., and Gingerich, K. A., *Adv. Mass Spectrom.*, 8A, 445, 1980.
163. Gupta, S. K., Nappi, B. M., and Gingerich, K. A., *Inorg. Chem.*, 20, 966, 1981.
164. Gupta, S. K., Nappi, B. M., and Gingerich, K. A., *J. Phys. Chem.*, 85, 971, 1981.
165. Gupta, S. K., Pelino, M., and Gingerich, K. A., *J. Chem. Phys.*, 70, 2044, 1979.
166. Gupta, S. K., Pelino, M., and Gingerich, K. A., *J. Phys. Chem.*, 83, 2335, 1979.
167. Gustavsson, T., Amiot, C., Verges, J., *Mol. Phys.*, 64, 279, 1988.
168. Haque, R. and Gingerich, K. A., *J. Chem. Thermodyn.*, 12, 439, 1980.
169. Haque, R., Pelino, M., and Gingerich, K. A., *J. Chem. Phys.*, 71, 2929, 1979.
170. Haque, R., Pelino, M., and Gingerich, K. A., *J. Chem. Phys.*, 73, 4045, 1980.
171. Hariharan, A. V. and Eick, H. A., *J. Chem. Thermodyn.*, 6, 373, 1974.
172. Herman, P. R., La Rocque, P. E., and Stoicheff, B., *J. Chem. Phys.*, 89, 4535, 1988.
173. Herman, R. and Herman, L., *J. Phys. Radium.*, 24, 73, 1963.
174. Herzberg, G., *Molecular Spectra and Molecular Structure. I. Spectra of Diatomic Molecules*, 2nd ed.,Van Nostrand, New York, 1950.
175. Hildenbrand, D. L. and Lau, K. H., *J. Chem. Phys.*, 89, 5825, 1988.
176. Hildenbrand, D. L. and Lau, K. H., *J. Chem. Phys.*, 91, 4909, 1989.
177. Hildenbrand, D. L. and Lau, K. H., *J. Chem. Phys.*, 93, 5983, 1990.
178. Hildenbrand, D. L. and Lau, K. H., *J. Chem. Phys.*, 96, 3830, 1992.
179. Hildenbrand, D. L. and Lau, K. H., *J. Phys. Chem.*, 96, 2325, 1992.
180. Hildenbrand, D. L. and Murad, E., *J. Chem. Phys.*, 44, 1524, 1966.
181. Hildenbrand, D. L. and Theard, L. P., *J. Chem. Phys.*, 50, 5350, 1969.
182. Hildenbrand, D. L., *Chem. Phys. Lett.*, 32, 523, 1975.
183. Hildenbrand, D. L., *J. Chem. Phys.*, 48, 3657, 1968.
184. Hildenbrand, D. L., *J. Chem. Phys.*, 52, 5751, 1970.
185. Hildenbrand, D. L., *J. Chem. Phys.*, 65, 614, 1976.
186. Hildenbrand, D. L., *J. Chem. Phys.*, 66, 3526, 1977.
187. Hildenbrand, D. L., *J. Phys. Chem.*, 77, 897, 1973.
188. Hilpert, K. and Ruthardt, K., *Ber. Bunsenges. Phys. Chem.*, 91, 724, 1987.
189. Hilpert, K. and Ruthardt, K., *Ber. Bunsenges. Phys. Chem.*, 93, 1070, 1989.
190. Hilpert, K., *Ber. Kernforschungsanlage Juelich, JUEL-1744*, 272, 1981.
191. Hilpert, K., *J. Chem. Phys.*, 77, 1425, 1982.
192. Huber, K. P. and Herzberg, G., *Molecular Spectra and Molecular Structure Constants of Diatomic Molecules*, Van Nostrand, New York, 1979
193. Hussein, K., Effantin, C., D'Incan, J., Verges, J., and Barrow, R. F., *Chem. Phys. Lett.*, 124, 105, 1986.
194. Ihle, H. R. and Wu, C. H., *J. Chem. Phys.*, 63, 1605, 1975.
195. Ishwar, N. B., Varma, M. P., and Jha, B. L., *Acta Phys. Pol. A*, A61, 503, 1982.
196. Ishwar, N. B., Varma, M. P., and Jha, B. L., *Indian J. Pure Appl. Phys.*, 20, 992, 1982.
197. Jeyagopal, T., Rajavel, S. R. K., Ramakrishnan, M., and Rajamanickam, N., *Acta Phys. Hung.*, 68, 145, 1990.
198. Johns, J. W. C. and Ramsey, D. A., *Can. J. Phys.*, 39, 210, 1961.
199. Jones, R. W. and Gole, J. L., *Chem. Phys.*, 20, 311, 1977.
200. Jordan, K. J., Lipson, R. H., McDonald, N. A., and Le Roy, R. J., *J. Phys. Chem.*, 96, 4778, 1992.
201. Kant, A. and Lin, S.-S., *J. Chem. Phys.*, 51, 1644, 1969.
202. Kant, A. and Moon, K. A., *High Temp. Sci.*, 11, 55, 1979.
203. Kant, A. and Strauss, B. H., *J. Chem. Phys.*, 41, 3806, 1964.
204. Kant, A. and Strauss, B., *J. Chem. Phys.*, 49, 3579, 1968.
205. Kant, A., *J. Chem. Phys.*, 49, 5144, 1968.
206. Kant, A., Lin, S.-S, and Strauss, B., *J. Chem. Phys.*, 49, 1983, 1968.
207. Kant, A., Strauss, B., and Lin, S.-S., *J. Chem. Phys.*, 52, 2384, 1970.
208. Kaposi, O., *Magy. Kem. Foly.*, 83, 356, 1977.
209. Kaufel, R., Vahl, G., Nunkwitz, R., and Baumgaertel, H., *Z. Anorg. Allg. Chem.*, 481, 207, 1981.
210. Kazenas ,E., Tagirov, V. K., and Zviadadze, G. N., *Izv. Akad. Nauk. SSSR Met.*, 58, 1984.
211. Kent, R. A. and Margrave, J. L., *J. Am. Chem. Soc.*, 87, 3582, 1965.
212. Kent, R. A., Ehlert, T. C., and Margrave, J. L., *J. Am. Chem. Soc.*, 86, 5090, 1964.
213. Kent, R. A., *J. Am. Chem. Soc.*, 90, 5657, 1968.
214. Khitrov, A. N., Ryabova, V. G., and Gurvich, L. V., *Teplofiz. Vys. Tempo.*, 11, 1126, 1973.
215. Kiang, T. and Zare, R. N., *J. Am. Chem. Soc.*, 102, 4024, 1980.

216. Kimura, H., Asano, M., and Kubo, K., *J. Nucl. Mater.*, 97, 259, 1981.

217. Kingcade, J. E. Jr. and Gingerich, K. A., *J. Chem. Phys.*, 84, 3432, 1986.

218. Kingcade, J. E. Jr. and Gingerich, K. A., *J. Chem. Soc. Faraday Trans. 2*, 85, 195, 1989.

219. Kingcade, J. E. Jr., and Gingerich, K. A., *J. Chem. Phys.*, 84, 4574, 1986.

220. Kingcade, J. E., Cocke, D. L., and Gingerich, K. A., *High Temp. Sci.*, 16, 89, 1983,

221. Kingcade, J. E., Dufner, D. C., Gupta, S. K., and Gingerich, K. A., *High Temp. Sci.*, 10, 213, 1978.

222. Kingcade, J. E., Nagarathna, H. M., Shim, I., and Gingerich, K. A., *J. Phys. Chem.*, 90, 2830, 1986.

223. Kleinschmidt, P. D. and Hildenbrand, D. L., *J. Chem. Phys.*, 68, 2819, 1978.

224. Kleinschmidt, P. D. and Ward, J. W., *J. Less-Common. Met.*, 121, 61, 1986.

225. Kleinschmidt, P. D., Cubicciotti, D., and Hildenbrand, D. L., *J. Electrochem. Soc.*, 125, 1543, 1978; *Proc. Electrochem. Soc.*, 78, 217, 1978.

226. Kleinschmidt, P. D., Lau, K. H., and Hildenbrand, D. L., *J. Chem. Phys.*, 74, 653, 1981.

227. Klynning, L. and Lindgren, B., *Arkiv. Fysik.*, 32, 575, 1966.

228. Kohl, F. J. and Carlson, K. D., *J. Am. Chem. Soc.*, 90, 4814, 1968.

229. Kohl, F. J. and Stearns, C. A., *J. Phys. Chem.*, 78, 273, 1974.

230. Kondratiev, V. N., *Bond Dissociation Energies, Ionization Potentials and Electron Affinities*, Mauka Publishing House, Moscow, 1974.

231. Kordis, J. and Gingerich, K. A., *J. Chem. Eng. Data*, 18, 135, 1973.

232. Kordis, J. and Gingerich, K. A., *J. Chem. Phys.*, 58, 5141, 1973.

233. Kordis, J. and Gingerich, K. A., *J. Phys. Chem.*, 76, 2336, 1972.

234. Kordis, J., Gingerich, K. A., and Seyse, R. J., *J. Chem. Phys.*, 61, 5114, 1974.

235. Kowalski, A., Czaikowski, M., and Breckenridge, W. H., *Chem. Phys. Lett.*, 119, 368, 1985.

236. Koyama, T. and Yamawaki, M., *J. Nucl. Mater.*, 152, 30, 1988.

237. Kronekvist, M., Lagerqvist, A., and Neuhaus, H., *J. Mol. Spectrosc.*, 39, 516, 1971.

238. Kulkarni, M. P. and Dadape, V. V., *High Temp. Sci.*, 3, 277, 1971.

239. Lau, K. H. and Hildenbrand, D. L., *J. Chem. Phys.*, 71, 1572, 1979.

240. Lau, K. H. and Hildenbrand, D. L., *J. Chem. Phys.*, 72, 4928, 1980.

241. Lau, K. H. and Hildenbrand, D. L., *J. Chem. Phys.*, 76, 2646, 1982.

242. Lau, K. H. and Hildenbrand, D. L., *J. Chem. Phys.*, 80, 1312, 1984.

243. Lau, K. H. and Hildenbrand, D. L., *J. Chem. Phys.*, 86, 2949, 1987.

244. Lau, K. H. and Hildenbrand, D. L., *J. Chem. Phys.*, 92, 6124, 1990.

245. Lau, K. H., Brittain, R. D., and Hildenbrand, D. L., *Chem. Phys. Lett.*, 81, 227, 1981; *J. Phys. Chem.*, 86, 4429, 1982.

246. Lau, K. H., Brittain, R. D., and Hildenbrand, D. L., *J. Chem. Phys.*, 90, 1158, 1989.

247. Li, K. C. and Stwalley, W. C., *J. Chem. Phys.*, 59, 4423, 1973.

248. Li, L., Lyyra, A. M., Luh, W. T., and Stwalley, W. C., *J. Chem. Phys.*, 93, 8452, 1990.

249. Lindgren, B. and Nilsson, Ch., *J. Mol. Spectrosc.*, 55, 407, 1975.

250. Martin, E. and Barrow, R. F., *Phys. Scr.*, 17, 501, 1978.

251. McIntyre, N. S., Vander Auwera-Mahieu, A., and Drowart, J., *Trans. Faraday Soc.*, 64, 3006, 1968.

252. Menendez, M., Garay, M., Verdasco, E., and Gonzalez, U. A., *J. Chem. Phys.*, 99, 2760, 1993.

253. Morse, M. D., Hansen, G. P., Langridge-Smith, P. R. R., Zheng, L. S., Geusic, M. E., Michalopoulos, D. L., and Smalley, R. E., *J. Chem. Phys.*, 80, 5400, 1984.

254. Murad, E., Hildenbrand, D. L., and Main, R. P., *J. Chem. Phys.*, 45, 263, 1966.

255. Nagai, S., Shinmei, M., and Yokokawa, T., *J. Inorg. Nucl. Chem.*, 36, 1904, 1974.

256. Nappi, B. M. and Gingerich, K. A., *Inorg. Chem.*, 20, 522, 1981.

257. Neckel, A. and Sodeck, G., *Monatsch. Chem.*, 103, 367, 1972.

258. Nedelec, O. and Giroud, M., *J. Chem. Phys.*, 79, 2121, 1983.

259. Neubert, A. and Zmbov, K. F., *Chem. Phys.*, 76, 469, 1983.

260. Neubert, A. and Zmbov, K. F., *J. Chem. Soc. Faraday Trans. 1*, 70, 2219, 1974.

261. Neubert, A., *High Temp. Sci.*, 10, 213, 1978.

262. Neubert, A., Ihle, H. R., and Gingerich, K. A., *J. Chem. Phys.*, 73, 1406, 1980.

263. Neubert, A., Zmbov, K. F., Gingerich, K. A., and Ihle, H. R., *J. Chem. Phys.*, 77, 5218, 1982.

264. Nordine, P. C., *J. Chem. Phys.*, 61, 224, 1974.

265. O'Hare, P. A. G., *J. Phys. Chem. Ref. Data*, 22, 1455, 1993.

266. O'Hare, P. A. G., Susman, S., and Volin, K.J., *J. Chem. Thermodyn.*, 21, 827, 1989.

267. Ovcharenko, I. E., Ya, Kuzyankow, Y., and Tatevaskii, V. M., *Opt. Spectrosk.*, 19, 528, 1965.

268. Pak, K., Cai, M. F., Dzugan, T. P., and Bondybey, V. E., *Faraday Discuss. Chem. Soc.*, 86, 153, 1988.

269. Parks, E. K. and Wexler, S., *J. Phys. Chem.*, 88, 4492, 1984.

270. Parr, T. P., Behrens, R., Freedman, A., and Heron, R. R., *Chem. Phys. Lett.*, 56, 71, 1978.

271. Pedley, J. B. and Marshall, E. M., *J. Phys. Chem. Ref. Data*, 12, 967, 1984.

272. Pelevin, O. V., Mil'vidskii, M. G., Belyaev, A. I., and Khotin, B. A., *Izv. Akad. Nauk. SSSR Neorg. Mater.*, 2, 924, 1966.

273. Pelino, M. and Gingerich, K. A., *J. Chem. Phys.*, 90, 1286, 1989.

Table 1
BOND STRENGTHS IN DIATOMIC MOLECULES (continued)

274. Pelino, M., and Gingerich, K. A., *J. Chem. Phys.*, 93, 1581, 1989.
275. Pelino, M., Piacente, V., and Ascenzo, G., *Thermochim. Acta,* 31, 383, 1979.
276. Pelino, M., Viswanadham, P., and Edwards, J. G., *J. Phys. Chem.*, 83, 2964, 1979.
277. Piacente, V. and Balducci, G., *Adv. Mass Spectrom.*, 7A, 626, 1978.
278. Piacente, V. and Balducci, G., *Dyn. Mass Spectrom.*, 4, 295, 1976.
279. Piacente, V. and Balducci, G., *High Temp. Sci.*, 6, 254, 1974.
280. Piacente, V. and Desideri, A., *J. Chem. Phys.*, 57, 2213, 1972.
281. Piacente, V. and Gigli, R., *J. Chem. Phys.*, 77, 4790, 1982.
282. Piacente, V. and Gingerich, K. A., *High Temp. Sci.*, 9, 189, 1977.
283. Piacente, V. and Gingerich, K. A., *Z. Naturforsch. Teil A*, 28, 316, 1973.
284. Piacente, V. and Malaspina, L., *J. Chem. Phys.*, 56, 1780, 1972.
285. Piacente, V., Balducci, G., and Bardi, G. , J. *Less-Commun. Metals.*, 37, 123, 1974.
286. Piacente, V., *J. Chem. Phys.*, 70, 5911, 1979.
287. Pianalto, F. S., O'Brien, L. C., Keller, P. C., and Bernath, P. F., *J. Mol. Spectrosc.*, 129, 348, 1988.
288. Pitzer, K. S., *J. Chem. Phys.*, 74, 3078, 1981.
289. Porter, R. F. and Spencer, C. W. J., *J. Chem. Phys.*, 32, 943, 1960.
290. Prasad, R., Venugopal, V., and Sood, D. D., *J. Chem. Thermodyn.*, 9, 593, 1977.
291. Rajamanickam, N., Palaniselvam, K., Rajavel, S. R. K., Rajesh, M., and Sureshkumar, G., *Acta Phys. Hung.*, 70, 141, 1991.
292. Rajamanickam, N., Senthilkumar, R. N., Ganesan, S., Gopalakrishnan, N., Rajkumer, J., Jegadesan, V., and Dandapani, C., *Acta Phys. Hung.*, 70, 71, 1991.
293. Rajamanickan, N., *Acta Ciencia Indica Phys.,* 14, 18, 1988.
294. Ram, R. S., Rai, S. B., Ram, R. S., Upadhya, K. N., J. *Chim. Phys. Phys-Chim. Biol.*, 76, 560, 1979.
295. Ramakrishnan, E. S., Shim, I., and Gingerich, K. A., *J. Chem. Soc. Faraday Trans. 2,* 80, 395, 1984.
296. Ran, Q., Schmude, R. W., Gingerich, K. A., Wilhite, D. W., and Kincade, J. E., *J. Phys. Chem.*, 97, 8535, 1993.
297. Rao, P. S. and Rao, T. V. R., *J. Quant. Spectrosc. Radiat. Transfer*, 27, 207, 1982.
298. Rao, S. P. and Rao, T. V. R., *Acta Ciencia Indica Phys.*, 7, 58, 1981.
299. Rao, T. V. R., Reddy, R. R., and Rao, P. S., *Indian J. Pure Appl. Phys.*, 19, 1219, 1981.
300. Rao, V. M., Rao, M. L. P., and Rao, P. T., *J. Quant. Spectrosc. Radiat. Transfer*, 25, 547, 1981.
301. Reddy, R. R., Rao, T. V. R., and Reddy, A. S. R., *Indian J. Pure Appl. Phys.*, 27, 243, 1989.
302. Reddy, R. R., Reddy, A. S. R., and Rao, T. V. R., *Acta Phys. Slovaca*, 36, 273, 1986.
303. Riekert, G., Lamparter, P., and Steeb, S., *Z. Metallkd.,* 72, 765, 1981.
304. Riekert, G., Rainer-Harbach, G., Lamparter, P., and Steeb, S., *Z. Metallkd.*, 76, 406, 1981.
305. Rinehart, G. H. and Behrens, R. G., *J. Phys. Chem.*, 83, 2052, 1979.
306. Rohlfing, E. A. and Valentine, J. J., *J. Chem. Phys.*, 84, 6560, 1986.
307. Rovner, L., Drowart, A., and Drowart, J., *Trans. Faraday Soc.*, 63, 2910, 1967.
308. Rusin, A. D., Zhukov, E., Agamirova, L. M., and Kalinnikov, V. T., *Zh. Neorg. Khim.*, 24, 1457, 1979.
309. Scott, P. R. and Richards, W. G., *J. Phys. B*, 7, 1679, 1974.
310. Shim, I. and Gingerich, K. A., *J. Chem. Phys.*, 77, 2490, 1982.
311. Shim, I. and Gingerich, K. A., *J. Chem. Phys.,* 80, 5107, 1984.
312. Shim, I., and Gingerich, K. A., *J. Chem. Phys.*, 81, 5937, 1984.
313. Shim, I., Finkbeiner, H. C., and Gingerich, K. A., *J. Phys. Chem.*, 91, 3171, 1987.
314. Shim, I., Kingcade, J. E. Jr., and Gingerich, K. A., *J. Chem. Phys.*, 85, 6629, 1986.
315. Shim, I., Kingcade, J. E. Jr., and Gingerich, K. A., *J. Chem. Phys.*, 89, 3104, 1988.
316. Shim, I., Kingcade, J. E. Jr., and Gingerich, K. A., *Z. Phys. D - Atoms Molecules and Clusters*, 7, 261, 1987.
317. Shim, I., Mandix, K., and Gingerich, K. A., *J. Phys. Chem.*, 95, 5435, 1991.
318. Shim, I., Pelino, M., and Gingerich, K. A., *J. Chem. Phys.,* 97, 9240, 1992.
319. Siegel, B., *Q. Rev. (London)*, 19, 77, 1965.
320. Simons, J. W., Oldenberg, R. C., and Baughaim, S. L., *J. Phys. Chem.*, 91, 3840, 1987.
321. Smith, P. K. and Peterson, D. E., *J .Chem. Phys.*, 52, 4963, 1970.
322. Smoes, S. and Drowart, J., *Chem. Commun.,* p.534, 1968.
323. Smoes, S. and Drowart, J., *J. Chem. Soc. Faraday Trans. 2*, 73, 1746, 1977.
324. Smoes, S. and Drowart, J., *J. Chem. Soc. Faraday Trans. 2*, 80, 1171, 1984.
325. Smoes, S., Coppens, P., Bergman, C., and Drowart, J., *Trans. Faraday Soc.*, 65, 682, 1969.
326. Smoes, S., Depiere, D., and Drowart, J., *Rev. Int. Hautes Temp. Refractaires Paris*, 9, 171, 1972.
327. Smoes, S., Drowart, J., Welter, J. M., *J. Chem. Thermodyn.*, 9, 275, 1977; *Adv. Mass Spectrom.*, 7A, 622, 1978.
328. Smoes, S., Huguet, R., and Drowart, J., *Z. Naturforsch. Teil A*, 26, 1934, 1971.
329. Smoes, S., Mandy, F., Vander Auwera-Mahieu, A., and Drowart, J., *Bull. Soc. Chim. Belg.*, 81, 45, 1972.
330. Smoes, S., Myers, C. E., and Drowart, J., *Chem. Phys. Lett.*, 8, 10, 1971.
331. Smoes, S., Pattje, W. R., and Drowart, J., *High Temp. Sci.*, 10, 109, 1978.

Table 1
BOND STRENGTHS IN DIATOMIC MOLECULES (continued)

332. Spain, E. M. and Morse, M. D., *Int. J. Mass Spectrom. Ion Processes*, 102, 183, 1990.
333. Sreedhara Murthy, N., *Indian J. Phys.*, 62B, 92, 1988.
334. Srivastara, R. D. and Farber, M., *High Temp. Sci.*, 5, 489, 1973.
335. Stearns, C. A. and Kohl, F. J., *High Temp. Sci.*, 2, 146, 1970.
336. Stearns, C. A. and Kohl, F. J., *High Temp. Sci.*, 6, 284, 1974.
337. Steiger, R. A. and Cater, E. D., *High Temp. Sci.,* 7, 204, 1975.
338. Steiger, R. P. and Cater, E. D., *High Temp. Sci.*, 7, 288, 1975.
339. Stwalley, W. C., Way, K. R., and Velasco, R., *J. Chem. Phys.,* 60, 3611, 1974.
340. Su, T.-M. R. and Riley, S. J., *J. Chem. Phys.*, 71, 3194, 1979.
341. Su, T.-M. R. and Riley, S. J., *J. Chem. Phys.*, 72, 1614, 1980.
342. Su, T.-M. R. and Riley, S. J., *J. Chem. Phys.*, 72, 6632, 1980.
343. Sullivan, C. L., Zehe, M. J., and Carlson, K. D., *High Temp. Sci.*, 6, 80, 1974.
344. Tanaka, Y., Yushina, K., and Freeman, D. E., *J. Chem. Phys.*, 59, 564, 1973.
345. Taylor, S., Spain, E. M., and Morse, M. D., *J. Chem. Phys.*, 92, 2698, 1990.
346. Tellinghuisen, J., Tisone, G. C., Hoffmann, J. M., and Hays, A. K., *J. Chem. Phys.*, 64, 4796, 1976.
347. Tolbert, M. A. and Beauchamp, J. L., *J. Phys. Chem.*, 90, 5015, 1986.
348. Tromp, J. W., LeRoy, R. J., Gerstenkorn, S., and Luc, P., *J. Mol. Spectrosc.*, 100, 82, 1983.
349. Tuenge, R. T., Laabs, F., and Franzen, H. F., *J. Chem. Phys.*, 65, 2400, 1976.
350. Uy, O. M. and Drowart, J., *High Temp. Sci.,* 2, 293, 1970.
351. Uy, O. M. and Drowart, J., *Trans. Faraday Soc.*, 65, 3221, 1969.
352. Uy, O. M. and Drowart, J., *Trans. Faraday Soc.*, 67, 1293, 1971.
353. Uy, O. M., Muenow, D. W., Ficalora, P, J., and Margrave, J. L., *Trans. Faraday Soc.*, 64, 2998, 1968.
354. Van Veen, N. J. A., DeVries, M., and DeVries, A. E., *Chem. Phys. Lett.*, 64, 213, 1979.
355. Vander Auwera-Mahieu, A. and Drowart, J., *Chem. Phys. Lett.*, 1, 311, 1967.
356. Vander Auwera-Mahieu, A., McIntyre, N. S., and Drowart, J., *Chem. Phys. Lett.*, 4, 198, 1969.
357. Vander Auwera-Mahieu, A., Peeters, R., McIntyre, N. S., and Drowart, J., *Trans. Faraday Soc.*, 66, 809, 1970.
358. Varma, M. P., Ishwar, N. B., and Jha, B. L., *Indian J. Pure Appl. Phys.*, 20, 828, 1982.
359. Velasco, R., Ottinger, C., and Zare, R. N., *J. Chem. Phys.,* 51, 5522, 1969.
360. Vempati, S. N. and Jones, W. E., *J. Mol. Spectrosc.*, 127, 232, 1988.
361. Venkataramanaiah, M. and Lakshman, S. V. J., *J. Quant. Spectrosc. Radiat. Transfer*, 26, 11, 1981.
362. Verhaegen, G., Smoes, S., and Drowart, J., *J. Chem. Phys.*, 40, 239, 1964.
363. Verhaegen, G., Stafford, F. E., and Drowart, J., *J. Chem. Phys.*, 40, 1622, 1964.
364. Verma, K. K., Vu, T. H., and Stwalley, W. C., *J. Mol. Spectrosc.,* 85, 131, 1980.
365. Viswanathan, K. S. and Tellinghuisen, J., *J. Mol. Spectrosc.*, 98, 185, 1983.
366. Wang, Y. C., Kajitani, M., Kasahara, S., Bata, M., Ishikawa, K., and Kato, H., *J. Chem. Phys.*, 95, 6229, 1991.
367. Way, K. R. and Stwalley, W. C., *J. Chem. Phys.*, 59, 5298, 1973.
368. Weickenmeier, W., Diemer, U., Wahl, M., Raab, M., Demtroeder, W., and Mueller, W., *J. Phys.Chem.*, 82, 5354, 1985.
369. Wiedemeier, H. and Gilles, P. W., *J. Chem. Phys.*, 42, 2765, 1965.
370. Wu, C. H. and Ihle, H. R., *Adv. Mass Spectrom.*, 8A, 374, 1980.
371. Wu, C. H., Ihle, H. R., and Gingerich, K. A., *Int. J. Mass Spectrom. Ion Phys.*, 47, 235, 1983.
372. Wu, C. H., *J. Chem. Phys.*, 65, 3181, 1976; 65, 2040, 1976.
373. Wyss, J. C., *J. Chem. Phys.*, 71, 2949, 1979.
374. Yokozeki, A. and Menzinger, M., *Chem. Phys.*, 14, 427, 1976.
375. Yoo, R. K., Ruscic, B., and Berkowitz, J., *Chem. Phys.*, 166, 215, 1992.
376. Zemke, W. T. and Stwalley, W. C., *Chem. Phys. Lett.*, 143, 84, 1988.
377. Zemke, W. T., Stwalley, W. C., Langhoff, S. R., Valderrama, G. L., and Berry, M. J., *J. Chem. Phys.*, 95, 7846, 1991.
378. Zmbov, K .F. and Margrave, J. L., *J. Inorg. Nucl. Chem.*, 29, 59, 1967.
379. Zmbov, K. F. and Margrave, J. L., *J. Chem. Phys.,* 45, 3167, 1966.
380. Zmbov, K. F. and Margrave, J. L., *J. Chem. Phys.*, 47, 3122, 1967.
381. Zmbov, K. F. and Margrave, J. L., *J. Phys. Chem.*, 70, 3379, 1966.
382. Zmbov, K. F. and Margrave, J. L., *J. Phys. Chem.*, 71, 2893, 1967.
383. Zmbov, K. F., Hastie, J. W., and Margrave, J. L., *Trans. Faraday Soc.*, 64, 861, 1968.
384. Zmbov, K. F., Neubert, A., and Ihle, H. R., *Z. Naturforsch., A*, 36A, 914, 1981.
385. Zmbov, K. F., Wu, C. H., and Ihle, H. R., *J. Chem. Phys.*, 67, 4603, 1977.
386. Zollweg, R. J., *Contrib. Pap. Int. Conf. Phenom. Ioniz. Gases.* 11th, 402, 1973.

Table 2
ENTHALPY OF FORMATION OF GASEOUS ATOMS FROM ELEMENTS IN THEIR STANDARD STATES

For elements that are diatomic gases in their standard states these are readily obtained from the bond strength. For elements that are crystalline in their standard states they are derived from vapor pressure data.

Atom	$\Delta_f H°_{298}$/kJ mol^{-1}	Ref.	Atom	$\Delta_f H°_{298}$/kJ mol^{-1}	Ref.	Atom	$\Delta_f H°_{298}$/kJ mol^{-1}	Ref.
			Hf	619 ± 4	1	Re	774 ± 6.3	1
Ag	284.9 ± 0.8	2	Hg	61.38 ± 0.04	2	Rh	556 ± 4	1
Al	330.0 ± 4.0	2	I	106.76 ± 0.04	2	Ru	650.6 ± 6.3	1
As	302.5 ± 13	1	In	243 ± 4	1	S	277.17 ± 0.15	2
Au	368.2 ± 2.1	1	Ir	669 ± 4	1	Sb	264.4 ± 2.5	1
B	565 ± 5	2	K	89.0 ± 0.8	2	Sc	377.8 ± 4	1
Ba	177.8 ± 4	1	Li	159.3 ± 1.0	2	Se	227.2 ± 4	1
Be	324 ± 5	2	Mg	147.1 ± 0.8	2	Si	450 ± 8	2
Bi	209.6 ± 2.1	1	Mn	283.3 ± 4	1	Sn	301.2 ± 1.5	2
Br	111.87 ± 0.12	2	Mo	658.1 ± 2.1	1	Sr	163.6 ± 2.1	1
C	716.68 ± 0.45	2	N	472.68 ± 0.40	2	Ta	782.0 ± 2.5	1
Ca	177.8 ± 0.8	2	Na	107.5 ± 0.7	2	Te	196.6 ± 2.1	1
Cd	111.80 ± 0.20	2	Nb	721.3 ± 4	1	Th	602 ± 6	2
Ce	423 ± 13	1	Ni	430.1 ± 2.1	1	Ti	473 ± 3	2
Cl	121.301 ± 0.008	2	O	249.18 ± 0.10	2	Tl	182.21 ± 0.4	1
Co	428.4 ± 4	1	Os	787 ± 6.3	1	U	533 ± 8	2
Cr	397 ± 4	1	P	316.5 ± 1.0	2	V	514.2 ± 1.3	1
Cs	76.5 ± 1.0	2	Pb	195.2 ± 0.8	2	W	849.8 ± 4	1
Cu	337.4 ± 1.2	2	Pd	$376.6 + 2.1$	1	Y	424.7 ± 2.1	1
Er	317.1 ± 4	1	Pt	565.7 ± 1.3	1	Yb	152.09 ± 0.8	1
F	79.38 ± 0.30	2	Pu	364.4 ± 17	1	Zn	130.40 ± 0.40	2
Ge	372 ± 3	2	Rb	80.9 ± 0.8	2	Zr	608.8 ± 4	1
H	217.998 ± 0.006	2						

REFERENCES

1. Brewer, L. and Rosenblatt, G. M., *Adv. High Temp. Chem.*, 2, 1, 1969.
2. Cox, J. D., Wagman, D. D., and Medvedev, V. A., Eds., *CODATA Key Values for Thermodynamics*, Hemisphere Publishing Corporation, New York, 1989.

Table 3
BOND STRENGTHS IN POLYATOMIC MOLECULES

The values below refer to a temperature of 298 K and have mostly been determined by kinetic methods (see (i) S. W. Benson, *J. Chem. Educ.*, 42, 502, 1965, (ii) J. A. Kerr, *Chem. Rev.*, 66, 465, 1966 and (iii) D. F. McMillen and D. M. Golden, *Ann. Rev. Phys. Chem.*, 33, 493, 1982, for a full description of the methods). An increasing number of bond strengths are being determined from gas-phase acidity cycles and from photoionization mass spectrometry (see J. Berkowitz, G. B. Ellison and D. Gutman, *J. Phys. Chem.*, 98, 2744, 1994).

Bond strengths in polyatomic molecules are notoriously difficult to measure accurately since the mechanisms of the kinetic systems involved in many of the measurements are seldom straightforward. Thus much controversy has taken place in the literature over the past 15 years concerning C-H bond strengths in simple alkanes, although these now appear to have been resolved. Other examples illustrating the difficulties involved are concerned with the C-H bond strengths in ethene and methanol or the corresponding enthalpies of formation of the vinyl and hydroxymethyl radicals and changes to the recommendations could well arise.

Some of the bond strengths have been calculated from the enthalpies of formation of the species involved according to the equations:

$$D°(R–X) = \Delta_f H°(R) + \Delta_f H°(X) – \Delta_f H°(RX)$$
$$D°(R–R) = 2 \Delta_f H°(R) – \Delta_f H°(RR)$$

The enthalpies of formation of the radicals are taken from Table 4 and for the molecules from the appropriate References following Table 3.

An attempt has been made to list all the important values obtained by methods that are considered to be valid. The references are intended to serve as a guide to the literature.

Table 3
BOND STRENGTHS IN POLYATOMIC MOLECULES (continued)

Bond	D°_{298}/kJ mol^{-1}	Ref.	Bond	D°_{298}/kJ mol^{-1}	Ref.
H-CH	424.0 ± 4.2	1	H-C(CH$_3$)$_2$CN	361.9 ± 8.4	57
H-CH$_2$	462.0 ± 4.0	1	H-CH$_2$NH$_2$	390.4 ± 8.4	57
H-CH$_3$	438.9 ± 0.4	14	H-CH$_2$NHCH$_3$	364 ± 8	57
H-CCH	556.1 ± 2.9	3,34	H-CH$_2$N(CH$_3$)$_2$	351 ± 8	57
H-CHCH$_2$	465.3 ± 3.4	34,74	H-CHO	368.5 ± 1.0	21
H-C$_2$H$_5$	423.0 ± 1.6	14	H-CHCO	440.6 ± 8.8	14
H-Cycloprop-2-en-1-yl	379.1 ± 17	57	H-COCH$_3$	373.8 ± 1.5	60
H-CH$_2$CCH	374.0 ± 8	57	H-COCHCH$_2$	364.4 ± 4.2	57
H-CH$_2$CHCH$_2$	361.9 ± 8.8	14	H-COC$_2$H$_5$	365.7 ± 4	57
H-Cyclopropyl	444.8 ± 1.3	57	H-COC$_6$H$_5$	363.6 ± 4	57
H-n-C$_3$H$_7$	420.0 ± 2.5	20,67	H-COCF$_3$	380.7 ± 8	57
H-i-C$_3$H$_7$	412.5 ± 1.6	14	H-CH$_2$CHO	394.6 ± 9.2	14
H-CH$_2$CCCH$_3$	364.8 ± 8	57	H-CH$_2$COCH$_3$	411.3 ± 7.5	57
H-CH(CH$_3$)CCH	347.7 ± 9.2	57	H-CH(CH$_3$)COCH$_3$	386.2 ± 5.9	57
H-Cyclobutyl	403.8 ± 4	57	H-CH$_2$OCH$_3$	389 ± 4	57
H-Cyclopropylmethyl	407.5 ± 6.7	57	H-CH(CH$_3$)OC$_2$H$_5$	383.7 ± 1.7	49
H-CH(CH$_3$)CHCH$_2$	345.2 ± 5.4	57	H-Tetrahydrofuran-2-yl	385 ± 4	57
H-CH$_2$CHCHCH$_3$	358.2 ± 6.3	57	H-2-Furylmethyl	361.9 ± 8	57
H-CH$_2$C(CH$_3$)CH$_2$	358.2 ± 4	81,85	H-CH$_2$OH	401.92 ± 0.63	14
H-s-C$_4$H$_9$	410.9 ± 2.1	14	H-CH(CH$_3$)OH	389 ± 4	57
H-t-C$_4$H$_9$	403.8 ± 1.6	14	H-CH(OH)CHCH$_2$	341.4 ± 7.5	57
H-Cyclopenta-1,3-dien-5-yl	297.5 ± 6.3	57	H-C(CH$_3$)$_2$OH	381 ± 4	57
H-Spiropentyl	413.4 ± 4	57	H-CH$_2$OCOC$_6$H$_5$	419.2 ± 5.4	57
H-Cyclopent-1-en-3-yl	344.3 ± 4	57	H-COOCH$_3$	387.9 ± 4	57
H-CH$_2$CHCHCHCH$_2$	347 ± 13	57	H-CH$_2$F	423.8 ± 4	70
H-CH(C$_2$H$_3$)$_2$	319.7	57,82	H-CHF$_2$	431.8 ± 4	70
H-CH(CH$_3$)CCCH$_3$	365.3 ± 11.3	57	H-CF$_3$	446.4 ± 4	57
H-C(CH$_3$)$_2$CCH	338.9 ± 9.6	57	H-CHFCl	421.7 ± 5.4	87
H-C(CH$_3$)$_2$CHCH$_2$	323.0 ± 6.3	57	H-CF$_2$Cl	421.3 ± 8.3	58
H-Cyclopentyl	403.5 ± 2.5	20,67	H-CHFCl$_2$	413.8 ± 5.0	87
H-CH$_2$C(CH$_3$)$_3$	418 ± 8	57	H-CH$_2$Cl	421.7 ± 4.2	87
H-C(CH$_3$)$_2$C$_2$H$_5$	404.0 ± 6.3	1,67,79	H-CHCl$_2$	411.7 ± 5.0	87
H-C$_6$H$_5$	465.3 ± 3.4	14	H-CCl$_3$	392.5 ± 2.5	44
H-Cyclohexa-1,3-dien-5-yl	305 ± 21	57	H-CH$_2$Br	425.1 ± 4.2	87
H-Cyclohexa-1,4-dien-3-yl	305.4 ± 8.4	42	H-CHBr$_2$	417.2 ± 7.5	87
H-Cyclohexyl	399.6 ± 4	57	H-CBr$_3$	401.7 ± 6.7	57
H-C(CH$_3$)$_2$CCCH$_3$	344.3 ± 11.3	57	H-CH$_2$I	431 ± 8	57
H-CH$_2$C(CH$_3$)C(CH$_3$)$_2$	326.4 ± 4.6	57	H-CHI$_2$	431 ± 8	57
H-C(CH$_3$)$_2$C(CH$_3$)CH$_2$	319.2 ± 4.6	57	H-CHCF$_2$	448 ± 8	80
H-CH$_2$C$_6$H$_5$	370.3 ± 6.3	14	H-CFCHF	448 ± 8	80
H-Cyclohepta-1,3,5-trien-7-yl	305.4 ± 8	57	H-CFCF$_2$	452 ± 8	80
H-Norbornyl	404.6 ± 10.5	57	H-CH$_2$CF$_3$	446.4 ± 4.6	57
H-Cycloheptyl	387.0 ± 4	57	H-CF$_2$CH$_3$	416.3 ± 10.5	57
H-CH(CH$_3$)C$_6$H$_5$	357.3 ± 6.3	57	H-C$_2$F$_5$	429.7 ± 2.1	57
H-Inden-1-yl	351 ± 13	57	H-CFCFCl	444 ± 8	80
H-C(CH$_3$)$_2$C$_6$H$_5$	353.1 ± 6.3	57	H-CHClCF$_3$	425.9 ± 6.3	57
H-1-Naphthylmethyl	356.1 ± 6.3	57	H-CClCFCl	439 ± 8	80
H-CH(C$_6$H$_5$)$_2$	340.6	73	H-CClCH$_2$	>433.5	74
H-9,10-Dihydroanthracen-9-yl	315.1 ± 6.3	57	H-CClCHCl	435 ± 8	80
H-C(CH$_3$)(C$_6$H$_5$)$_2$	339 ± 8	57	H-CCl$_2$CHCl$_2$	393 ± 8	57
H-9-Anthracenylmethyl	342.3 ± 6.3	57	H-C$_2$Cl$_5$	397 ± 8	57
H-9-Phenanthrenylmethyl	356.1 ± 6.3	57	H-CClBrCF$_3$	404.2 ± 6.3	57
H-CN	527.6 ± 1.7	14	H-n-C$_3$F$_7$	435 ± 8	57
H-CH$_2$CN	392.9 ± 8.4	14	H-i-C$_3$F$_7$	433.5 ± 2.5	35
H-CH$_2$NC	380.7 ± 8.8	14	H-CHClCHCH$_2$	370.7 ± 5.9	57
H-CH(CH$_3$)CN	376.1 ± 9.6	57	H-C$_6$F$_5$	476.6	57

Table 3
BOND STRENGTHS IN POLYATOMIC MOLECULES (continued)

Bond	D°_{298}/kJ mol^{-1}	Ref.	Bond	D°_{298}/kJ mol^{-1}	Ref.
H-CH$_2$Si(CH$_3$)$_3$	415.1 ± 4	89	CH$_3$-CH$_3$	376.0 ± 2.1	1,67,76
H-CSH	399.6 ± 5.0	14	CH$_3$-CH$_2$CCH	318.0 ± 8	57
H-CH$_2$SH	392.9 ± 8.4	14	CH$_3$-CH$_2$CCCH$_3$	308.4 ± 6.3	57
H-CH$_2$SCH$_3$	384.9 ± 5.9	45	CH$_3$-CH(CH$_3$)CCH	305.4	57
H-NH$_2$	452.7 ± 1.3	14	CH$_3$-C(CH$_3$)CCH$_2$	320.1 ± 9.2	57
H-NHCH$_3$	418.4 ± 10.5	57	CH$_3$-CH$_2$CHCHCH$_3$	305.0 ± 3.3	57
H-N(CH$_3$)$_2$	382.8 ± 8	57	CH$_3$-CH$_2$C(CH$_3$)CH$_2$	301.2 ± 3.3	81
H-NHC$_6$H$_5$	368.2 ± 8	57	CH$_3$-t-C$_4$H$_9$	425.9 ± 8	57
H-N(CH$_3$)C$_6$H$_5$	366.1 ± 8	57	CH$_3$-CH(CH$_3$)CCCH$_3$	320.9 ± 6.3	57
H-NO	≤207.1	23	CH$_3$-C(CH$_3$)$_2$CCH	295.8 ± 6.3	57
H-NO$_2$	327.6 ± 2.1	57	n-C$_3$H$_7$-CH$_2$CCH	306.3 ± 6.3	57
H-NF$_2$	316.7 ± 10.5	57	CH$_3$-C(CH$_3$)$_2$CHCH$_2$	284.9 ± 6.3	57
H-NHNH$_2$	366.1	41	n-C$_3$H$_7$-CH$_2$CHCH$_2$	295.8	86
H-NH$_3$	385 ± 21	57	CH$_3$-C$_6$H$_5$	317.1 ± 6.3	57
H-OH	498 ± 4	57	CH$_3$-C(CH$_3$)$_2$CCCH$_3$	303.3 ± 6.3	57
H-OCH$_3$	436.0 ± 3.8	14	CHCCH$_2$-s-C$_4$H$_9$	300.0 ± 6.3	57
H-OC$_2$H$_5$	437.7 ± 3.4	14	CH$_3$-CH$_2$C$_6$H$_5$	332.2 ± 4	57
H-OC(CH$_3$)$_3$	439.7 ± 4	57	CH$_3$-CH(CH$_3$)C$_6$H$_5$	312.1 ± 6.3	57
H-OCH$_2$C(CH$_3$)$_3$	428.0 ± 6.3	57	C$_2$H$_5$-CH$_2$C$_6$H$_5$	294.1 ± 4	57
H-OC$_6$H$_5$	361.9 ± 8	57	CH$_3$-1-Naphthylmethyl	305.0 ± 6.3	57
H-O$_2$H	369.0 ± 4.2	78	CH$_3$-C(CH$_3$)$_2$C$_6$H$_5$	308.4 ± 6.3	57
H-O$_2$CH$_3$	370.3 ± 2.1	50	CHCCH$_2$-CH$_2$C$_6$H$_5$	256.9 ± 8	57
H-O$_2$C(CH$_3$)$_3$	374.0 ± 0.8	43	n-C$_3$H$_7$-CH$_2$C$_6$H$_5$	292.9 ± 4	57
H-OCOCH$_3$	442.7 ± 8	57	CH$_3$-9-Anthracenylmethyl	282.8 ± 6.3	57
H-OCOC$_2$H$_5$	445.2 ± 8	57	CH$_3$-9-Phenanthrenylmethyl	305.0 ± 6.3	57
H-OCO-n-C$_3$H$_7$	443.1 ± 8	57	CH$_3$-CH(C$_6$H$_5$)$_2$	301 ± 8	57
H-ONO	327.6 ± 2.1	13	CH$_3$-C(CH$_3$)(C$_6$H$_5$)$_2$	289 ± 8	57
H-ONO$_2$	423.4 + 2.1	13	CH$_3$-CN	509.6 ± 8	57
H-SiH	351	57	C$_2$H-CN	602 ± 4	63
H-SiH$_2$	268	57	C$_2$H$_5$-CH$_2$NH$_2$	332.2 ± 8	57
H-SiH$_3$	384.1 ± 2.0	75	CH$_3$-CH$_2$CN	336.4 ± 4	83
H-SiH$_2$CH$_3$	374.9	89	C$_2$H$_5$-CH$_2$CN	321.7 ± 7.1	57
H-SiH(CH$_3$)$_2$	374.0	89	CH$_3$-CH(CH$_3$)CN	329.7 ± 8	57
H-Si(CH$_3$)$_3$	377.8	57,89	C$_2$H$_5$-CH$_2$CN	321.7 ± 7.1	57
D-Si(CH$_3$)$_3$	389 ± 7.1	33	CH$_3$-C(CH$_3$)$_2$CN	312.5 ± 6.7	57
H-SiH$_2$C$_6$H$_5$	369.0	57,89	CH$_3$-C(CH$_3$)(CN)C$_6$H$_5$	250.6	57
H-SiF$_3$	418.8	57,89	C$_6$H$_5$CH$_2$-CH$_2$NH$_2$	284.5 ± 8	57
H-SiCl$_3$	382.0	57,89	C$_6$H$_5$CH$_2$-C$_5$H$_4$N	362.8	73
H-Si$_2$H$_5$	361.1	57	CN-CN	536 ± 4	29
H-Si(CH$_3$)$_2$Si(CH$_3$)$_3$	356.9 ± 8.4	42	CH$_3$-2-Furylmethyl	314 ± 8	57
H-Si(Si(CH$_3$)$_3$)$_3$	330.5 ± 8.4	42	C$_6$H$_5$CH$_2$-COCH$_2$C$_6$H$_5$	273.6 ± 8	57
H-PH$_2$	351.0 ± 2.1	14	CH$_3$CO-COCH$_3$	282.0 ± 9.6	57
H-SH	381.6 ± 2.9	59	C$_6$H$_5$CH$_2$-COOH	280	57
H-SCH$_3$	365.3 ± 2.5	59	C$_6$H$_5$CO-COC$_6$H$_5$	277.8	57
H-SC$_6$H$_5$	348.5 ± 8	57	(C$_6$H$_5$)$_2$CH-COOH	248.5 ± 13	57
H-SO	172.8	90	CF$_3$-COC$_6$H$_5$	308.8 ± 8	57
H-GeH$_3$	349.0 ± 8	14	CF$_2$=CF$_2$	319.2 ± 13	92
H-GeH$_2$I	331 ± 8	61	CH$_2$F-CH$_2$F	368 ± 8	46
H-Ge(CH$_3$)$_3$	339 ± 8	32	CH$_3$-CF$_3$	423.4 ± 4.6	72
H-AsH$_2$	319.2 ± 0.8	14	CF$_3$-CF$_3$	413.0 ± 10.5	57
H-SeH$_2$	334.93 ± 0.75	14	C$_6$F$_5$-C$_6$F$_5$	487.9 ± 24.7	71
H-Sn(n-C$_4$H$_9$)$_3$	308.4 ± 8.4	17	CH$_3$-BF$_2$	~473	57
H-SbH$_2$	288.3 ± 2.1	14	C$_6$H$_5$-BCl$_2$	~510	57
H-TeH	277.0 ± 5.0	14	CH$_2$CHCH$_2$-Si(CH$_3$)$_3$	293	57
HCĊH	965 ± 8	1,22,67	s-C$_4$H$_9$-Si(CH$_3$)$_3$	414	57
H$_2$C=CH$_2$	733 ± 8	1,67	CH$_3$-NHC$_6$H$_5$	298.7 ± 8	57

Table 3
BOND STRENGTHS IN POLYATOMIC MOLECULES (continued)

Bond	D°_{298}/kJ mol^{-1}	Ref.	Bond	D°_{298}/kJ mol^{-1}	Ref.
$C_6H_5CH_2$-NH_2	297.5 ± 4	57	F-CH_3	472	1,55
CH_3-$N(CH_3)C_6H_5$	296.2 ± 8	57	F-CN	469.9 ± 5.0	57
$C_6H_5CH_2$-$NHCH_3$	287.4 ± 8	57	F-COF	535 ± 12	16
$C_6H_5CH_2$-$N(CH_3)_2$	259.8 ± 8	57	F-CHFCl	465.3 ± 9.6	87
CH_2=N_2	<175	52	F-CF_2Cl	490 ± 25	38
CH_3-N_2CH_3	219.7	11	F-$CFCl_2$	462.3 ± 10.0	87
C_2H_5-$N_2C_2H_5$	209.2	11	F-CF_2CH_3	522.2 ± 8	57
i-C_3H_7-N_2i-C_3H_7	198.7	11	F-C_2F_5	530.5 ± 7.5	57
n-C_4H_9-N_2n-C_4H_9	209.2	11	Cl-CN	421.7 ± 5.0	57
i-C_4H_9-N_2i-C_4H_9	205.0	11	Cl-COC_6H_5	310 ± 13	57
s-C_4H_9-N_2s-C_4H_9	195.4	11	Cl-CSCl	265.3 ± 2.1	64
t-C_4H_9-N_2t-C_4H_9	182.0	11	Cl-CF_3	360.2 ± 3.3	26
$C_6H_5CH_2$-$N_2CH_2C_6H_5$	157.3	11	Cl-CHFCl	354.4 ± 11.7	87
CF_3-N_2CF_3	231.0	11	Cl-CF_2Cl	346.0 ± 13.4	87
CH_3-NO	167.4 ± 3.3	57	Cl-$CFCl_2$	305 ± 8	37
i-C_3H_7-NO	152.7 ± 13	57	Cl-CH_2Cl	350.2 ± 0.8	87
t-C_4H_9-NO	165.3 ± 6.3	57	Cl-$CHCl_2$	338.5 ± 4.2	87
C_6H_5-NO	212.5 ± 4	57	Cl-CCl_3	305.9 ± 7.5	57
NC-NO	120.5 ± 10.5	40	Cl-C_2F_5	346.0 ± 7.1	26
CF_3-NO	179.1 ± 8	57	Cl-CF_2CF_2Cl	326 ± 8	57
C_6F_5-NO	208.4 ± 4	57	Cl-$SiCl_3$	464	89
CCl_3-NO	134 ± 13	57	Br-CH_3	292.9 ± 5.0	36
t-C_4H_9-NOt-C_4H_9	121	19	Br-C_6H_5	336.8 ± 8	57
CH_3-NO_2	254.4	57	Br-CN	367.4 ± 5.0	57
$CH_2C(CH_3)$-NO_2	245.2	57	Br-CH_2COCH_3	261.5	91
i-C_3H_7-NO_2	246.9	57	Br-COC_6H_5	268.6	11
t-C_4H_9-NO_2	244.8	57	Br-CHF_2	289 ± 8	57
C_6H_5-NO_2	298.3 ± 4	57	Br-CF_3	295.4 ± 13	57
$C(NO_2)_3$-NO_2	169.5 ± 4	57	Br-CF_2CH_3	287.0 ± 5.4	69
CH_3-$OC(CH_3)CH_2$	277.4	91	Br-C_2F_5	287.4 ± 6.3	57
CH_3-OC_6H_5	238 ± 8	66	Br-n-C_3F_7	278.2 ± 10.5	57
CH_3-$OCH_2C_6H_5$	280.3	25	Br-i-C_3F_7	274.1 ± 4.6	57
C_2H_5-OC_6H_5	264 ± 6.3	57	Br-$CH_2C_6F_5$	225 ± 6	48
CH_2CHCH_2-OC_6H_5	208.4 ± 8	57	Br-$CHClCF_3$	274.9 ± 6.3	57
O=CO	532.2 ± 0.4	28	Br-CCl_3	231.4 ± 4	57
CH_3-O_2	135.6 ± 2.9	56	Br-$CClBrCF_3$	251.0 ± 6.3	57
C_2H_5-O_2	147.2 ± 6.3	56	Br-CH_2Br	296.7 ± 1.3	87
CH_2CHCH_2-O_2	76.2 ± 2.1	56	Br-$CHBr_2$	292.0 ± 8	87
i-C_3H_7-O_2	157.9 ± 7.5	56	Br-CBr_3	235.1 ± 7.5	57
t-C_4H_9-O_2	153.6 ± 7.9	56	Br-NO_2	82.0 ± 7.1	51
$C_6H_5CH_2$-O_2CCH_3	280 ± 8	57	Br-NF_2	≤ 222	24
$C_6H_5CH_2$-$O_2CC_6H_5$	289	11	I-$CHCH_2$	259.0 ± 4.2	18
CH_3-O_2SCH_3	279.5	57	I-n-C_4H_9	205.0 ± 4	57
CH_2CHCH_2-O_2SCH_3	207.5	57	I-Norbornyl	261.5 ± 10.5	62
$C_6H_5CH_2$-O_2SCH_3	221.3	57	I-CN	305 ± 4	29
CF_3-O_2CF_3	361.5	9	I-CF_3	223.8 ± 2.9	2
CH_2Cl-O_2	120.9	56	I-CF_2CH_3	218.0 ± 4.2	57
$CHCl_2$-O_2	105.9	56	I-CH_2CF_3	235.6 ± 4	57
CCl_3-O_2	82.4 ± 3.8	56	I-C_2F_5	218.8 ± 2.9	2
CH_3-SH	312.5 ± 4.2	59	I-n-C_3F_7	208.4 ± 4.2	57
t-C_4H_9-SH	286.2 ± 6.3	57	I-i-C_3F_7	215.1 ± 2.9	2
C_6H_5-SH	361.9 ± 8	57	I-n-C_4F_9	205.0 ± 4.2	65
CH_3-SCH_3	307.9 ± 3.3	59	I-$C(CF_3)_3$	206	31
CH_3-SC_6H_5	290.4 ± 8	57	I-C_6H_5	273.6 ± 8	57
$C_6H_5CH_2$-SCH_3	256.9 ± 8	57	I-C_6F_5	~277.0	57
S-CS	430.5 ± 13	57	C_5H_5-FeC_5H_5	381 ± 13	53

Table 3
BOND STRENGTHS IN POLYATOMIC MOLECULES (continued)

Bond	D°_{298}/kJ mol^{-1}	Ref.	Bond	D°_{298}/kJ mol^{-1}	Ref.
CH_3-$ZnCH_3$	285 ± 17	57	Cl-NF_2	~134	1,68
C_2H_5-ZnC_2H_5	238 ± 17	57	I-NO	77.8 ± 0.4	39
CH_3-$Ga(CH_3)_2$	264 ± 17	57	I-NO_2	76.6 ± 4	88
C_2H_5-$Ga(C_2H_5)_2$	209 ± 17	57	HO-OH	213 ± 4	57
CH_3-$Ge(CH_3)_3$	347 ± 17	57	HO-$OCH_2C(CH_3)_3$	193.7 ± 7.9	57
CH_3-$As(CH_3)_2$	280 ± 17	57	CH_3O-OCH_3	157.3 ± 8	57
CH_3-$CdCH_3$	251 ± 17	57	C_2H_5O-OC_2H_5	158.6 ± 4	57
CH_3-$In(CH_3)_2$	205 ± 17	57	n-C_3H_7O-On-C_3H_7	155.2 ± 4	57
CH_3-$Sn(CH_3)_3$	297 ± 17	57	i-C_3H_7O-Oi-C_3H_7	157.7 ± 4	57
C_2H_5-$Sn(C_2H_5)_3$	264 ± 17	57	s-C_4H_9O-Os-C_4H_9	152.3 ± 4	57
CH_3-$Sb(CH_3)_2$	255 ± 17	57	t-C_4H_9O-Ot-C_4H_9	159.0 ± 4	57
C_2H_5-$Sb(C_2H_5)_2$	243 ± 17	57	$C_2H_5C(CH_3)_2O$-$OC(CH_3)_2C_2H_5$	164.4 ± 4	57
CH_3-$HgCH_3$	255 ± 17	57	$(CH_3)_3CCH_2O$-$OCH_2C(CH_3)_3$	152.3 ± 4	57
C_2H_5-HgC_2H_5	205 ± 17	57	CF_3O-OCF_3	193.3	57
CH_3-$Tl(CH_3)_2$	167 ± 17	57	$(CF_3)_3CO$-$OC(CF_3)_3$	148.5 ± 4.6	57
CH_3-$Pb(CH_3)_3$	238 ± 17	57	t-C_4H_9O-$OSi(CH_3)_3$	197	57
C_2H_5-$Pb(C_2H_5)_3$	230 ± 17	57	SF_5O-OSF_5	155.6	57
CH_3-$Bi(CH_3)_2$	218 ± 17	57	t-C_4H_9O-$OGe(C_2H_5)_3$	192	57
CO-$Cr(CO)_5$	155 ± 8	54	t-C_4H_9O-$OSn(C_2H_5)_3$	192	57
CO-$Fe(CO)_4$	172 ± 8	54	$FClO_2$-O	244.3	11
CO-$Mo(CO)_5$	167 ± 8	54	CF_3O-O_2CF_3	126.8 ± 8	57
CO-$W(CO)_5$	192 ± 8	54	SF_5O-O_2SF_5	126.8	57
BH_3-BH_3	146	11	CH_3CO_2-O_2CCH_3	127.2 ± 8	57
NH_2-NH_2	275.3	57	$C_2H_5CO_2$-$O_2CC_2H_5$	127.2 ± 8	57
NH_2-$NHCH_3$	268.2 ± 8	57	n-$C_3H_7CO_2$-O_2Cn-C_3H_7	127.2 ± 8	57
NH_2-$N(CH_3)_2$	246.9 ± 8	57	O-SO	552 ± 8	28
NH_2-NHC_6H_5	218.8 ± 8	57	F-OCF_3	182.0 ± 2.1	27
ON-NO_2	40.6 ± 2.1	57	HO-Cl	251 ± 13	47
O_2N-NO_2	56.9	57	O-ClO	247 ± 13	28
NF_2-NF_2	88 ± 4	57	HO-Br	234 ± 13	47
O-N_2	167	1,12	HO-I	234 ± 13	47
O-NO	305	1,12	O=PF_3	544 ± 21	47
O-NO_2	208.7 ± 1.1	30	O=PCl_3	510 ± 21	47
HO-NO	206.3	57	O=PBr_3	498 ± 21	47
HO-NO_2	206.7	57	HO-$Si(CH_3)_3$	536	57
HO_2-NO_2	96 ± 8	57	HS-SH	276 ± 8	57
CH_3O-NO	174.9 ± 3.8	8,10	CH_3S-SCH_3	272.8 ± 3.8	59
C_2H_5O-NO	175.7 ± 5.4	7,10	F-SF_5	420 ± 10	84
CH_3COO_2-NO_2	118.8 ± 3.0	15	I-SH	206.7 ± 8	57
n-C_3H_7O-NO	167.8 ± 7.5	10	I-SO	180	57
i-C_3H_7O-NO	171.5 ± 5.4	6,10	I-SCH_3	206.3 ± 7.1	77
n-C_4H_9O-NO	177.8 ± 6.3	10	I-$Si(CH_3)_3$	322	89
i-C_4H_9O-NO	175.7 ± 6.3	10	H_3Si-SiH_3	310	57,89
s-C_4H_9O-NO	173.6 ± 3.3	4,10	$(CH_3)_3Si$-$Si(CH_3)_3$	336.8	57,89
t-C_4H_9O-NO	171.1 ± 3.3	5,10	$(C_6H_5)_3Si$-$Si(C_6H_5)_3$	368 ± 29	57,89
HO-$NCHCH_3$	207.9	11			

REFERENCES

1. A value calculated from one of the thermochemical equations above, taking enthalpy data from Table 4.
2. Ahonkhai, S. I. and Whittle, E., *Int. J. Chem. Kinet.,* 16, 543, 1984.
3. Baldwin, D. P., Buntine, M. A., and Chandler, D.W., *J. Chem. Phys.,* 93, 6578, 1990.
4. Batt, L. and McCulloch, R. D., *Int. J. Chem. Kinet.,* 8, 911, 1976.
5. Batt, L. and Milne, R. T. , *Int. J. Chem. Kinet.,* 8, 59, 1976.
6. Batt, L. and Milne, R. T. , *Int. J. Chem. Kinet.,* 9, 141, 1977.
7. Batt, L. and Milne, R. T. , *Int. J. Chem. Kinet.,* 9, 549, 1977.
8. Batt, L. and Milne, R. T., and McCulloch, R. D., *Int. J. Chem. Kinet.,* 9, 567, 1977.

Table 3
BOND STRENGTHS IN POLYATOMIC MOLECULES (continued)

9. Batt, L. and Walsh, R., *Int. J. Chem. Kinet.*, 15, 605, 1983.

10. Batt, L., Christie, K., Milne, R. T., and Summers, A. J., *Int. J. Chem. Kinet.*, 6, 877, 1974.

11. Benson, S. W. and O'Neal, H. E., *Kinetic data on Gas Phase Unimolecular Reactions*, National Bureau of Standards, NSRDS-NBS, Washington, D. C., 21, 1970.

12. Benson, S. W., *J. Chem. Educ.*, 42, 502, 1965.

13. Benson, S. W., *Thermochemical Kinetics, 2nd ed.*, John Wiley & Sons, New York, 1976.

14. Berkowitz, J., Ellison, G. B., and Gutman, D., *J. Phys. Chem.*, 98, 2744, 1994.

15. Bridier, I., Caralp, F., Loirat, H., Lesclaux, R., Veyret, B., Becker, K. H., Reimer, A., and Zabel, F., *J. Phys. Chem.*, 95, 3594, 1991.

16. Buckley, T. J., Johnson, R. D., Huie, R. E., Zhang, Z., Kuo, S. C., and Klemm, B., *J. Phys. Chem.*, 99, 4879, 1995

17. Burkey, T. J., Majewski, M., and Griller, D., *J. Am. Chem. Soc.*, 108, 2218, 1986.

18. Cao, J. R., Zhang, J. M., Zhong, X., Huang, Y. H., Fang, W. Q., Wu, X. J., and Zhu, Q. H., *Chem. Phys.*, 138, 377, 1989.

19. Carmichael, P. J., Gowenlock, B. G., and Johnson, C. A. F., *J. Chem. Soc. Perkin Trans. 2*, 1853, 1973.

20. Castelhano, A. L. and Griller, D., *J. Am. Chem. Soc.*, 104, 3655, 1982.

21. Chuang, M.-C., Foltz, M. F., and Moore, C. B., *J. Chem. Phys.*, 87, 3855, 1987.

22. Chupka, W. A. and Liftshitz, C., *J. Chem. Phys.*, 48, 1109, 1968.

23. Clement, M. J. Y. and Ramsay, D. A., *Can. J. Phys.*, 39, 205, 1961.

24. Clyne, M. A. A. and Connor, J., *J. Chem. Soc. Faraday Trans. 1*, 68, 1220, 1972.

25. Colussi, A. J., Zabel, F., and Benson, S. W., *Int. J. Chem. Kinet.*, 9, 161, 1977.

26. Coomber, J. W. and Whittle, E., *Trans. Faraday Soc.*, 63, 2656, 1967.

27. Czarnarski, J., Castellano, E., and Schumacher, H. J., *Chem. Comm.*, p. 1255, 1968.

28. Darwent, D. deB., *Bond Dissociation Energies in Simple Molecules*, National Bureau of Standards, NSRDS-NBS, 31 Washington, D. C., 1970.

29. Davis, D. D. and Okabe, H., *J. Chem. Phys.*, 49, 5526, 1968.

30. Davis, H. F., Kim, B., Johnston, H. S., and Lee, Y. T., *J. Chem. Phys.*, 97, 2172, 1993.

31. Dobychin, S. L., Mashendzhinov, V. I., Mishin, V. I., Semenov, V. N., and Shpak, V. S., *Doklady Akademii Nauk SSSR*, 312, 1166 1991.

32. Doncaster, A. M. and Walsh, R., *J. Phys. Chem.*, 83, 578, 1979.

33. Ellul, E., Potzinger, P., Reimann, B., and Camilleri, P., *Ber. Bunsenges. Phys. Chem.*, 85, 407, 1981.

34. Ervin, K. M., Gronert, S., Barlow, S. E., Gilles, M. K., Harrison, A. G., Bierbaum, V. M., DePuy, C. H., Lineberger, W. C., and Ellison, G. B., *J. Am. Chem. Soc.*, 112, 5750, 1990.

35. Evans, B. S., Weeks, I., and Whittle, E., *J. Chem. Soc. Faraday Trans. 1*, 79, 1471, 1983.

36. Ferguson, K. C., Okafo, E. N., and Whittle, E., *J. Chem. Soc. Faraday Trans. 1*, 69, 295, 1973.

37. Foon, R. and Tait, K. B., *J. Chem. Soc. Faraday Trans. 1*, 68, 104, 1972.

38. Foon, R. and Tait, K. B., *J. Chem. Soc. Faraday Trans. 1*, 68, 1121, 1972.

39. Forte, E., Hippler, H., and van den Bergh, H., *Int. J. Chem. Kinet.*, 13, 1227, 1981.

40. Gowenlock, B. G., Jonhson, C. A. F., Keary, C. M., and Pfaf, J., *J. Chem. Soc. Perkin Trans. 2*, 71, 351, 1975.

41. Grela, M. A. and Colussi, A. J., *Int. J. Chem. Kinet.*, 20, 713, 1988.

42. Griller, D. and Wayner, D. D. M., *Pure Appl. Chem.*, 61, 717, 1989.

43. Heneghan, S. P. and Benson, S. W., *Int. J. Chem. Kinet.*, 15, 815, 1983.

44. Hudgens, J. W., Johnson, R. D., Timonen, R. S., Seetula, J. A., and Gutman, D., *J. Phys. Chem.*, 95, 4400, 1991.

45. Jefferson, A., Nicovich, J. M., and Wine, P. H., *J. Phys. Chem.*, 98, 7128, 1994.

46. Kerr, J. A. and Timlin, D. M., *Int. J. Chem. Kinet.*, 3, 427, 1971.

47. Kerr, J. A., *Chem. Rev.*, 66, 465, 1966.

48. Kominar, R. J., Krech, M. J., and Price, S. J. W., *Can. J. Chem.*, 58, 1906, 1980.

49. Kondo, O. and Benson, S. W., *Int. J. Chem. Kinet.*, 16, 949, 1984.

50. Kondo, O. and Benson, S. W., *J. Phys. Chem.*, 88, 6675, 1984.

51. Kreutter, K. D., Nicovich, J. M., and Wine, P. H., *J. Phys. Chem.*, 95, 4020, 1991.

52. Laufer, A. H. and Okabe, H., *J. Am. Chem. Soc.*, 93, 4137, 1971.

53. Lewis, K. E. and Smith, G. P., *J. Am. Chem. Soc.*, 106, 4650, 1984.

54. Lewis, K. E., Golden, D. M., and Smith, G. P., *J. Am. Chem. Soc.*, 106, 3905, 1984.

55. Lias, S. G., Bartmess, J. E., Liebman, J. F., Holmes, J. L., Levin, R. D., and Mallard, W. G., *J. Phys. Chem. Ref. Data 17*, Suppl. 1, 1988.

56. Lightfoot, P. D., Cox, R. A., Crowley, J. N., Destriau, M., Hayman G. D., Jenkin, M. E., Mootrgat, G. K., and Zabel, F., *Atmos. Environ.*, 26A, 1805, 1992.

57. McMillen, D. F. and Golden, D. M., *Ann. Rev. Phys. Chem.*, 33, 493, 1982.

58. Miyokawa, K. and Tschuikow-Roux, E., *J. Phys. Chem.*, 96, 7328, 1992.

59. Nicovich, J. M., Kreutter, K. D., van Dijk, C. A., and Wine, P. H., *J. Phys. Chem.*, 96, 2518, 1992.

60. Niiranen, J. T., Gutman, D., and Krasnoperov, L. N., *J. Phys. Chem.*, 96, 5881, 1992.

61. Noble, P. N. and Walsh, R., *Int. J. Chem. Kinet.*, 15, 561, 1983.

62. O'Neal, H. E., Bagg, J. W., and Richardson, W. H., *Int. J. Chem. Kinet.*, 2, 493, 1970.

Table 3
BOND STRENGTHS IN POLYATOMIC MOLECULES (continued)

63. Okabe, H. and Dibeler, V. H., *J. Chem. Phys.*, 59, 2430, 1973.
64. Okabe, H., *J. Chem. Phys.,* 66, 2058, 1977.
65. Okafo, E. N. and Whittle, E., *Int. J. Chem. Kinet.*, 7, 287, 1975.
66. Paul, S. and Back, M. H., *Can J. Chem.*, 53, 3330, 1975.
67. Pedley, J. B. and Rylance, J., "Sussex - N.P.L. Computer Analysed Thermochemical Data; Organic and Organometallic Compounds", University of Sussex, 1977.
68. Petry, R. C., *J. Am. Chem. Soc.*, 89, 4600, 1967.
69. Pickard, J. M. and Rodgers, A. S., *Int. J. Chem. Kinet.*, 9, 759, 1977.
70. Pickard, J. M. and Rodgers, A. S., *Int. J. Chem. Kinet.,* 15, 569, 1983.
71. Price, S. J. W. and Sapiano, H. J., *Can. J. Chem.*, 57, 1468, 1979.
72. Rogers, A. S. and Ford, W. G. F., *Int. J. Chem. Kinet.*, 5, 965, 1973.
73. Rossi, M., McMillen, D. F., and Golden, D. M., *J. Phys. Chem.*, 88, 5031, 1984.
74. Russell, J. J., Senkan, S. M., Seetula, J. A., and Gutman, D., *J. Phys. Chem.*, 93, 5184, 1989.
75. Seetula, J. A., Feng, Y., Gutman, D., Seakins, P. W., and Pilling, M. J., *J. Phys. Chem.*, 95, 1658, 1991.
76. Seetula, J. A., Russell, J. J., and Gutman, D., *J. Am. Chem. Soc.,* 112, 1347, 1990.
77. Shum, L. G. S. and Benson, S. W., *Int. J. Chem.*, 15, 433, 1983.
78. Shum, L. G. S. and Benson, S. W., *J. Phys. Chem.*, 87, 3479, 1983.
79. Stein, S. E., SRD Thermochemical Database, 25. N.I.S.T. Structures and Properties Database and Estimation Program, U.S. Department of Commerce, 1992.
80. Steinkruger, F. J. and Rowland, F. S., *J. Phys. Chem.,* 85, 136, 1981.
81. Trenwith, A. B. and Wrigley, S. P., *J. Chem. Soc. Faraday Trans. 1*, 73, 817, 1977.
82. Trenwith, A. B., *J. Chem. Soc. Faraday Trans. 1*, 78, 3131, 1982.
83. Trenwith, A. B., *J. Chem. Soc. Faraday Trans. 1*, 79, 2755, 1983.
84. Tsang, W. and Herron, J. T., *J. Chem. Phys.,* 96, 4272, 1992.
85. Tsang, W., *Int. J. Chem. Kinet.,* 5, 929, 1973.
86. Tsang, W., *Int. J. Chem. Kinet.*, 10, 1119, 1978.
87. Tschuikow-Roux, E. and Paddison, S., *Int. J. Chem. Kinet.*, 19, 15, 1987.
88. van den Bergh, H. and Troe, J., *J. Chem. Phys.*, 64, 736, 1976.
89. Walsh, R., *Acc. Chem. Res.,* 14, 246, 1981.
90. White, J. N. and Gardiner, W. C., *Chem. Phys. Lett.*, 58, 470, 1978.
91. Zabel, F., Benson, S. W., and Golden, D. M., *Int. J. Chem. Kinet.*, 10, 295, 1978.
92. Zmbov, K. F., Uy, O. M., and Margrave, J. L., *J. Am. Chem. Soc.*, 90, 5090, 1968.

Table 4
ENTHALPIES OF FORMATION OF FREE RADICALS

The enthalpies of formation of the free radicals are related to the corresponding bond strengths by the equations

$$D°(R\text{-}X) = \Delta_f H°(R) + \Delta_f H°(X) - \Delta_f H°(RX)$$

or

$$D°(R\text{-}R) = 2\Delta_f H°(R) + \Delta_f H°(RR)$$

For an excellent review of the methods of determining the enthalpies of formation of free radicals the reader is referred to "Thermochemistry of Free Radicals" by H. E. O'Neal and S. W. Benson in *Free Radicals*, Kochi, J. K., Ed., John Wiley & Sons, New York, 1973, 275.

Radical	$\Delta_f H°_{298}$/kJ mol^{-1}	Ref.	Radical	$\Delta_f H°_{298}$/kJ mol^{-1}	Ref.
CH	596.4 ± 1.2	5	C_2H_5	120.9 ± 1.6	9
CH_2(triplet)	390.4 ± 4	13,23	Cycloprop-2-en-1-yl	439.7 ± 17.2	34
CH_2(singlet)	428.3 ± 4	13	CH≡CCH$_2$	340.6 ± 8.4	34
CH_3	146.4 ± 0.4	9	CH_2=CHCH$_2$	170.7 ± 8.8	9
CH≡C	566.1 ± 2.9	4,19	Cyclopropyl	279.9 ± 1.1	34
CH_2=CH	300.0 ± 3.4	19,51	n-C_3H_7	97.5 ± 2.5	14

Table 4
ENTHALPIES OF FORMATION OF FREE RADICALS (continued)

Radical	$\Delta_f H^°_{298}$/kJ mol⁻¹	Ref.	Radical	$\Delta_f H^°_{298}$/kJ mol⁻¹	Ref.
i-C₃H₇	90.0 ± 1.6	9	CH₃COCHCH₃	-70.3 ± 7.1	34
CH₃C≡CCH₂	293.7	34	CH₃OCH₂	-11.7 ± 5.0	34
CH₂=CHCHCH₃	125.5 ± 6.3	34	C₂H₅OCHCH₃	-84.5	28
CH≡CCHCH₃	295.0 ± 9.2	34	Tetrahydrofuran-2-yl	-18.0 ± 6.3	34
Cyclobutyl	214.2 ± 4.2	34	CH₂OH	-17.07 ± 3.3	9
Cyclopropylmethyl	213.8 ± 6.7	34	CH₃CHOH	-63.6 ± 4.2	34
CH₂=C(CH₃)CH₂	127.2 ± 5.4	34	CH₂=CHCHOH	0.0	34
CH₃CH=CHCH₂	125.5 ± 6.3	34	(CH₃)₂COH	-111.3 ± 4.6	34
s-C₄H₉	67.4 ± 2.1	9	COOH	-223.0	34
t-C₄H₉	51.5 ± 1.6	52,54	COOCH₃	-169.0 ± 4	34
Cyclopenta-1,3-dien-5-yl	242.3 ± 6.3	34	C₆H₅COOCH₂	-69.9 ± 8.4	34
Spiropentyl	380.7 ± 4	34	CHF	163 ± 13	48
Cyclopent-1-en-3-yl	160.7 ± 4	34	CH₂F	-31.8 ± 8.4	45
CH₂=CHCH=CHCH₂	205 ± 13	34	FCO	-152.1 ± 12	12
(C₂H₃)₂CH	205 ± 13	34	CHF₂	-238.9 ± 4	45
CH₃C≡CCHCH₃	272.8 ± 9.6	34	CF₂	-194.1 ± 9.2	43
CH≡CC(CH₃)₂	257.3 ± 8.4	34	CF₃	-467.4 ± 15.1	34
CH₂=CHC(CH₃)₂	77.4 ± 6.3	34	CH₂Cl	121.8 ± 4.2	59
Cyclopentyl	107.1 ± 2.5	14	CHFCl	-60.7 ± 10.0	59
(CH₃)₃CCH₂	36.4 ± 8	34	CF₂Cl	-279.1 ± 8.3	35
C₂H₅C(CH₃)₂	32.2 ± 6.3	57	ClCO	-21.8	40
C₆H₅	330.1 ± 3.4	9	CHCl₂	98.3 ± 5.0	59
Cyclohexa-1,3-dien-5-yl	197 ± 21	34	CFCl₂	-89.1 ± 10.0	59
Cyclohexyl	58.2 ± 4	34	CCl₂	239	42
CH₃C≡CC(CH₃)₂	221.8 ± 9.6	34	CCl₃	71.1 ± 2.5	25
(CH₃)₂C=C(CH₃)CH₂	39.8 ± 6.3	34	CH₂Br	169.0 ± 4.2	59
CH₂=C(CH₃)C(CH₃)₂	37.7 ± 6.3	34	CHBr₂	188.3 ± 9.2	59
C₆H₅CH₂	202.5 ± 6.3	9	CBr₃	207.1 ± 8	59
Cyclohepta-1,3,5-trien-7-yl	271.1 ± 8	34	CH₂I	230.1 ± 6.7	34
CH₃CH₂CH₂C(CH₃)₂	3.4 ± 8.4	55	CHI₂	333.9 ± 9.2	34
Norbornyl	136.4 ± 10.5	34	CH₃CF₂	-302.5 ± 8	34
Cycloheptyl	51.1 ± 4	34	CF₃CH₂	-517.1 ± 5.0	34
C₆H₅CHCH₃	169.0	34	C₂F₅	-892.9 ± 4	34
C₆H₅C(CH₃)₂	134.7	34	CCl=CH₂	>251	51
1-Naphthylmethyl	252.7	34	CHCl₂CCl₂	23.4 ± 8	34
(C₆H₅)₂CH	289	50	CF₂ClCF₂	-686 ± 17	34
9,10-Dihydroanthracen-9-yl	256.9 ± 6.3	34	C₂Cl₅	35.2 ± 8	34
9-Anthracenylmethyl	337.6	34	C₆F₅	-547.7 ± 8	34
9-Phenanthrenylmethyl	311.3	34	(CH₃)₃SiCH₂	-34.7	34
CH₂CN	243.1 ± 11.3	9	CS	278.5 ± 3.8	47
CH₂NC	326.4 ± 11.3	9	HCS	300.4 ± 8.4	9
CH₃CHCN	209.2 ± 9.6	34	CH₂SH	151.9 ± 8.4	9
(CH₃)₂CCN	166.5 ± 8.4	34	CH₃SCH₂	136.8 ± 5.9	26
C₆H₅C(CH₃)CN	248.5	34	NH	352.3 ± 9.6	46
CH₂NH₂	149.4 ± 8	34	NH₂	168.7 ± 1.3	9
CH₃NHCH₂	126 ± 8	34	NF₂	34 ± 4	34
(CH₃)₂NCH₂	109 ± 8	34	N₂H₃	243.5	22
CN	441.4 ± 4.6	9	N₃	469 ± 21	34
CHO	41.8 ± 0.8	16	CH₂NH	104.6 ± 13	22
CHCO	175.3 ± 8.4	9	CH₃NH	177.4 ± 8	34
CH₃CO	-10.0 ± 1.2	41	(CH₃)₃N	145.2 ± 8	34
CH₂=CHCO	72.4	34	C₆H₅NH	237.2 ± 8	34
C₂H₅CO	-42.7 ± 4.2	34	C₆H₅NCH₃	233.5 ± 8	34
C₆H₅CO	109.2 ± 8	34	NCO	159	15
CH₂CHO	10.5 ± 9.2	9	CH₃N₂	215.5 ± 7.5	2
CH₃COCH₂	-23.9 ± 10.9	34	C₂H₅N₂	187.4 ± 10.5	2

Table 4
ENTHALPIES OF FORMATION OF FREE RADICALS (continued)

Radical	$\Delta_f H°_{298}$/kJ mol^{-1}	Ref.	Radical	$\Delta_f H°_{298}$/kJ mol^{-1}	Ref.
i-C$_3$H$_7$N$_2$	158.6 ± 9.2	2	SiH$_3$	200.5 ± 2.5	53
OH	39.3	34	CH$_3$Si	310	62
CH$_3$O	17.2 ± 3.8	9	CH$_3$SiH	213.0 ± 14.6	60
C$_2$H$_5$O	-15.5 ± 3.4	9	CH$_3$SiH$_2$	152.7	34,62
n-C$_3$H$_7$O	-41.4	34	(CH$_3$)$_2$Si	109	62
i-C$_3$H$_7$O	-52.3	34	(CH$_3$)$_2$SiH	59.8	34,62
n-C$_4$H$_9$O	-62.8	34	(CH$_3$)$_3$Si	-3.3	34,62
s-C$_4$H$_9$O	-69.5 ± 3.3	34	C$_6$H$_5$Si(CH$_3$)$_2$	163	42
t-C$_4$H$_9$O	-90.8	34	(C$_6$H$_5$)$_2$SiCH$_3$	326	42
C$_6$H$_5$O	47.7	34	(C$_6$H$_5$)$_3$Si	486.2	42
CF$_3$O	-655.6	7	SiF	-19.3	34,62
FO	109 ± 4	8	SiF$_2$	-587.9	34,62
ClO	101.6 ± 0.1	1	SiF$_3$	-1025	34,62
BrO	125	61	SiCl	195.8	34,62
IO	107	6,49	SiCl$_2$	-163.6	34,62
HO$_2$	14.6	56	SiCl$_3$	-318	34,62
CH$_3$O$_2$	10.4 ± 3.1	29	SiH$_3$SiH	269.9 ± 14.6	60
C$_2$H$_5$O$_2$	-28.7 ± 6.5	29	Si$_2$H$_5$	223.0	34,62
CH$_2$=CHCH$_2$O$_2$	87.9 ± 5.5	29	PH$_2$	138.5 ± 2.5	9
i-C$_3$H$_7$O$_2$	-68.9 ± 8.1	29	HS	143.0 ± 2.8	39
t-C$_4$H$_9$O$_2$	-105.0 ± 8.1	29	CH$_3$S	124.6 ± 1.8	39
HOCH$_2$O$_2$	-162.1 ± 2.1	29	C$_6$H$_5$S	229.7 ± 8	34
CF$_3$O$_2$	-614.0	29	SO	5.0	15
CF$_2$ClO$_2$	-406.5	29	HSO	-4	30
CFCl$_2$O$_2$	-213.7	29	HSO$_2$	-222	10
CH$_2$ClO$_2$	9.2	29	CH$_3$SO$_2$	-239.3	42
CHCl$_2$O$_2$	1.6	29	HOSO$_2$	-385	32
CCl$_3$O$_2$	-11.3 ± 4.6	29	SO$_3$	-395.7	61
CH$_3$CHClO$_2$	-54.7 ± 3.7	27	SF$_4$	-746 ± 12	58
CH$_3$CO$_2$	-207.5 ± 4	34	SF$_5$	-879.9 ± 20	58
CH$_3$COO$_2$	-172 ± 20	11	CH$_3$S$_2$	68.6 ± 8	24
C$_2$H$_5$CO$_2$	-228.5 ± 4	34	C$_2$H$_5$S$_2$	43.5 ± 8	24
n-C$_3$H$_7$CO$_2$	-249.4 ± 4	34	i-C$_3$H$_7$S$_2$	13.8 ± 8	24
FO$_2$	26.1	31,44	t-C$_4$H$_9$S$_2$	-19.3 ± 8	24
ClO$_2$	97.5	3,38	HOCS$_2$	110.5	36
OClO	95.6	20,37	GeH$_3$	222 ± 8	9
NO$_3$	73.7 ± 1.4	18	AsH$_2$	167.8 ± 1.3	9
sym-ClO$_3$	232.6	17	HSe	144.8 ± 2.1	9
SiH	377	34,62	SbH$_2$	215.5 ± 2.5	9
SiH$_2$	269.0 ± 1.3	21,33	HTe	158.6 ± 5.0	9

REFERENCES

1. Abramowitz, S. and Chase, M. W., *Pure Appl. Chem.*, 63, 1448, 1991.
2. Acs, G. and Peter, A., *Int. J. Chem. Kinet.*, 19, 929, 1987.
3. Baer, S., Hippler, H., Rabu, R., Siefke, M., Seitzinger, N., and Troe, J., *J. Chem. Phys.*, 95, 6463, 1991.
4. Baldwin, D. P., Buntine, M. A., and Chandler, D. W., *J. Chem. Phys.*, 93, 6578, 1990.
5. Based on D(C-H), see Table 1 and D$_f$H°(C) and D$_f$H°(H), see Table 2.
6. Based on D(I-O), see Table 1 and D$_f$H°(I) and D$_f$H°(O) see Table 2.
7. Batt, L. and Walsh, R., Int. J. Chem. Kinet., 14, 933, 1982.
8. Benson, S. W.,*Thermochemical Kinetics, 2nd ed.*, John Wiley & Sons, New York, 1976.
9. Berkowitz, J., Ellison, G. B., and Gutman, D., *J. Phys. Chem.,* 98, 2744, 1994.
10. Boyd, R. J., Gupta, A., Langler, R. F., Lownie, S. P., and Pincock, J. A., *Can. J. Chem.*, 58, 331, 1980.
11. Bridier, I., Caralp, F., Loirat, H., Lesclaux, R., Veyret, B., Becker, K. H., Reimer, A., and Zabel, F., *J. Phys. Chem.*, 95, 3594, 1991.
12. Buckley, T. J., Johnson, R. D., Huie, R. E., Zhang, Z., Kuo, S. C., and Klemm, B., *J. Phys. Chem.*, 99, 4879, 1995.
13. Bunker, P. R. and Sears, T. J., *J. Chem. Phys.,* 83, 4866, 1985.

Table 4
ENTHALPIES OF FORMATION OF FREE RADICALS (continued)

14. Castelhano, A. L. and Griller, D., *J. Am. Chem. Soc.*, 104, 3655, 1982; value adjusted to $D_fH°(CH_3) = 146$ kJ mol^{-1} and error limits assigned here.
15. Chase, M. W. Jr., Davies, C. A., Downey, J.R. Jr., Frurip, D. J., McDonald, R. A.. and Syverud, A. N., *J. Phys. Chem. Ref. Data,* 14, Suppl. 1, 1985.
16. Chuang, M.-C., Foltz, M. F., and Moore, C. B., *J. Chem. Phys.*, 87, 3855, 1987.
17. Colussi, A. J., *J. Phys. Chem.*, 94, 8922, 1990.
18. Davis, H. F., Kim, B., Johnston, H. S., and Lee, Y. T., *J. Phys. Chem.,* 97, 2172, 1993.
19. Ervin, K. M., Gronert, S., Barlow, S. E., Gilles, M. K., Harrison, A. G., Bierbaum, V. M., DePuy, C. H., Lineberger, W. C., and Ellison, G. B., *J. Am. Chem. Soc,* 112, 5750, 1990.
20. Flesch, R., E. Rühl, K. Hottmann, and H. Baumgartel, *J. Phys. Chem.*, 97, 837 (1993).
21. Frey, H. M., Walsh, R., and Watts, I. M., *J. Chem. Soc. Chem. Comm.*, 1189, 1986.
22. Grela, M. A. and Colussi, A. J., *Int. J. Chem. Kinet.*, 20, 713, 1988.
23. Gurvich, l.V., Veyts, I.V., Alcock, C.B., *Thermodynamic Properties of Individual Substances, 4th ed.*, Hemisphere, New York, 1991, Vol. 2.
24. Howari, J. A. Griller, D., and Lossing, F. P., *J. Am. Chem. Soc.*, 108, 3273, 1986.
25. Hudgens, J. W., Johnson, R. D., Timonen, R. S., Seetula, J. A., and Gutman, D., *J. Phys. Chem.*, 95, 4400, 1991.
26. Jefferson, A., Nicovich, J. M., and Wine, P. H., *J. Phys. Chem.*, 98, 7128, 1994.
27. Knyazev, V. D., Bencsura, A., Dubinsky, I. A., Gutman, D., Melius, C. F., and Senkan, S. M., *J. Phys. Chem.*, 99, 230, 1995.
28. Kondo, O. and Benson, S. W., *Int. J. Chem. Kinet.*, 16, 949, 1984.
29. Lightfoot, P. D., Cox, R. A., Crowley, J. N., Destriau, M., Hayman, G.D., Jenkin, M. E., Moortgat, G. K., and Zabel, F., *Atmos Environ.*, 26A, 1805, 1992.
30. Lovejoy, E. R., Wang, N. S., and Howard, C. J., *J. Phys. Chem.,* 91, 5749, 1987.
31. Lyman, J. L. and Holland, R., *J. Phys. Chem.*, 92, 7232, 1988.
32. Margitan, J. J., *J. Phys. Chem.*, 88, 3314, 1984.
33. Martin, J. G., Ring, M. A., and O'Neal, H. E., *Int. J. Chem. Kinet.*, 19, 715, 1987.
34. McMillen, D. F. and Golden, D. M., *Ann. Rev. Phys. Chem.*, 33, 493, 1982.
35. Miyokawa, K. and Tschuikow-Roux, E., *J. Phys. Chem.,* 96, 7328, 1992.
36. Murrells, T. P., Lovejoy, E.R., and Ravishankara, A. R., *J. Phys. Chem.*, 80, 4065, 1984.
37. Nickolaisen, S. L., R. R. Friedl, and S. P. Sander, *J. Phys. Chem.*, 98, 155 (1994).
38. Nicovich, J. M., Kreutter, K. D., Shockelford, C. J., and Wine, P. H., *Chem. Phys. Lett.*, 179, 367, 1991.
39. Nicovich, J. M., Kreutter, K. D., van Dijk, C. A., and Wine, P. H., *J. Phys. Chem.*, 96, 2518, 1992.
40. Nicovich, J.M., Kreutter, K. D., and Wine, P.H., *J. Chem. Phys.*, 92, 3539 (1990).
41. Niiranen, J. T., Gutman, D., and Krasnoperov, L. N., *J. Phys. Chem.*, 96, 5881, 1992.
42. O'Neal, H. E. and Benson, S. W., in *Free Radicals*, Kochi, J. K., Ed., John Wiley & Sons, New York, 1973, 275.
43. Okafo, E. N. and Whittle, E., *J. Chem. Soc. Faraday Soc.*, 1, 70, 1366, 1974.
44. Pagsberg, P., Ratajczak, E., Sillesen, A., and Jodkowski, J. T., *Chem. Phys. Lett.*, 141, 88, 1987.
45. Pickard, J. M. and Rodgers, A. S., *Int. J. Chem. Kinet.*, 15, 569, 1983.
46. Piper, L. G., *J. Chem. Phys.*, 70, 3417, 1979.
47. Prinslow, D. A. and Armentrout, P. B., *J. Chem. Phys.,* 94, 3563, 1991.
48. Pritchard, G. O., Nilson, W. B., and Kirtman, B., *Int. J. Chem. Kinet.*, 16, 1637, 1984.
49. Reddy, R. R., Rao, T. V. R., and Reddy, A. S. R., *Indian J. Pure Appl. Phys.*, 27, 243, 1989.
50. Rossi, M., McMillen, D. F., and Golden, D. M., *J. Phys. Chem.*, 88, 5031, 1984.
51. Russell, J. R., Senkan, S. M., Seetula, J. A., and Gutman, D., *J. Phys. Chem.*, 93, 5184, 1989.
52. Seakins, P. W., Pilling, M. J., Niiranen, J. T., Gutman, D., and Krasnoperov, L. N., *J. Phys. Chem.*, 96, 9847, 1992.
53. Seetula, J. A., Feng, Y., Gutman, D., Seakins, P. W., and Pilling, M. J., *J. Phys. Chem.,* 95, 1658, 1991.
54. Seetula, J. A., Russell, J. A., and Gutman, D., *J. Am. Chem. Soc.*, 112, 1347, 1990.
55. Seres, L., Gorgenyi, M., and Farkas, J., *Int. J. Chem. Kinet.*, 15, 1133, 1983.
56. Shum, L. G. S. and Benson, S. W., *J. Phys. Chem.,* 87, 3479, 1983.
57. Stein, S. E., SRD Thermochemical Database, 25. N.I.S.T. Structures and Properties Database and Estimation Program, U.S. Department of Commerce, 1992.
58. Tsang, W. and Herron, J. T., *J. Chem. Phys.*, 96, 4272, 1992.
59. Tschuikow-Roux, E. and Paddison, S., *Int. J. Chem. Kinet.*, 19, 15, 1987.
60. Vanderwielen, A. J., Ring, M. A., and O'Neal, H. E., *J. Am. Chem. Soc.,* 97, 993, 1975.
61. Wagman, D. D., Evans, W. H., Parker, V. B., Schumm, R. H., Halow, I., Bailey, S. M., Churney, K. L., and Nuttall, R.L., *J. Phys. Chem. Ref. Data*, 11, Suppl. 2, 1978.
62. Walsh, R., *Acc. Chem. Res.*, 14, 246, 1981.

Table 5
BOND STRENGTHS OF SOME ORGANIC MOLECULES

Bond strengths at 298 K expressed in kJ/mol for some organic molecules of the general formula R-X are presented below. Some are experimental values taken from the preceding tables; the remainder are calculated from the enthalpies of formation of the radicals, listed in Table 4 above, and the enthalpies of formation of the parent compounds from sources indicated by the references below. The table also includes bond strengths for the inorganic molecules, hydrogen, the hydrogen halides, water and ammonia.

	H	F	Cl	Br	I	OH	NH$_2$	CH$_3$O	CH$_3$	CH$_3$CO	NO	CF$_3$	CCl$_3$
H	435.990	570.3	431.62	366.35	298.407	498	449	436	439	374	207	446	393
CH$_3$	439	472	349[c]	293	239[c]	387[i]	354[i]	348[i]	376[i]	339[i]	167	423	362[d]
C$_2$H$_5$	423	463[f]	353[d]	294[d]	235[d]	395[i]	354[i]	355[i]	372[i]	338[i]	—	—	—
i-C$_3$H$_7$	413	463[f]	356[i]	300[i]	238[i]	402[i]	359[i]	360[i]	371[i]	328[i]	153	—	—
t-C$_4$H$_9$	404	—	355[i]	296[i]	230[i]	403[i]	358[i]	353[i]	365[i]	317[i]	165	—	—
C$_6$H$_5$	465	524[i]	399[i]	337[i]	274[i]	465[i]	427[i]	415[i]	425[i]	391[i]	213	—	—
C$_6$H$_5$CH$_2$	370	—	303[i]	248[i]	207[i]	340[i]	—	—	332	274[i]	—	—	—
CCl$_3$	393	435[a]	288[c]	225[h]	167[e]	—	—	—	362[d]	—	134	335[d]	292[d]
CF$_3$	446	545[c]	360[c]	295[c]	224	—	—	—	427[d]	—	179	413	335[d]
C$_2$F$_5$	430	531[d]	346[d]	287[g]	219	—	—	—	—	—	—	—	—
CH$_3$CO	374	499[i]	340[i]	278[i]	209[i]	447[i]	399[i]	403	339[i]	282	—	—	—
CN	528	470	422	367	305	—	—	—	—	—	121	—	—
C$_6$F$_5$	477	477[b]	385[b]	337	277[b]	448[b]	—	—	444[b]	—	208[b]	—	—

REFERENCES

[a] Chen, S. S., Wilhoit, R. C., and Zwolinski, B. J., *J. Phys. Chem. Ref. Data*, 5, 571, 1976.

[b] Choo, K. Y., Mendenhall, G. D., Golden, D.M., and Benson, S. W., *Int. J. Chem. Kinet.*, 6, 813, 1974.

[c] Kolesov, V. P., *Russ. Chem. Rev.*, 47, 599, 1978.

[d] Kolesov, V. P. and Papina, T. S., *Russ. Chem. Rev.*, 52, 425, 1983.

[e] Kudchadker, S. A. and Kudchadker, A. P., *J. Phys. Chem. Ref. Data*, 1, 1285, 1978.

[f] Lias, S. G., Bartmess, J. E., Liebman, J. F., Holmes, J. L., Levin, R. D., and Mallard, W. G., *J. Phys. Chem. Ref. Data*, 17, Suppl. No. 1, 1988.

[g] McMillen, D. F. and Golden, D. M., *Ann. Rev. Phys. Chem.*, 33, 493, 1982.

[h] Mendenhall, G. D., Golden, D. M., and Benson, S. W., *J. Phys. Chem.*, 77, 2707, 1973.

[i] Pedley, J. B. and Rylance, J., "Sussex.N.P.L. Computer Analysed Thermochemical Data: Organic and Organometallic Compounds", University of Sussex, 1977.

[j] Wu, E.-C. and Rodgers, A. S., *J. Am. Chem. Soc.*, 98, 6112, 1976.

ELECTRONEGATIVITY

Electronegativity is a parameter originally introduced by Pauling which describes, on a relative basis, the tendency of an atom in a molecule to attract bonding electrons. While electronegativity is not a precisely defined molecular property, the electronegativity difference between two atoms provides a useful measure of the polarity and ionic character of the bond between them. This table gives the electronegativity X, on the Pauling scale, for the most common oxidation state. Other scales are described in the references.

REFERENCES

1. Pauling, L., *The Nature of the Chemical Bond, Third Edition*, Cornell University Press, Ithaca, New York, 1960.
2. Allen, L.C., *J. Am. Chem. Soc.*, 111, 9003, 1989.
3. Allred, A.L., *J. Inorg. Nucl. Chem.*, 17, 215, 1961.

Z	Symbol	X	Z	Symbol	X	Z	Symbol	X
1	H	2.20	33	As	2.18	65	Tb	—
2	He	—	34	Se	2.55	66	Dy	1.22
3	Li	0.98	35	Br	2.96	67	Ho	1.23
4	Be	1.57	36	Kr	—	68	Er	1.24
5	B	2.04	37	Rb	0.82	69	Tm	1.25
6	C	2.55	38	Sr	0.95	70	Yb	—
7	N	3.04	39	Y	1.22	71	Lu	1.0
8	O	3.44	40	Zr	1.33	72	Hf	1.3
9	F	3.98	41	Nb	1.6	73	Ta	1.5
10	Ne	—	42	Mo	2.16	74	W	1.7
11	Na	0.93	43	Tc	2.10	75	Re	1.9
12	Mg	1.31	44	Ru	2.2	76	Os	2.2
13	Al	1.61	45	Rh	2.28	77	Ir	2.2
14	Si	1.90	46	Pd	2.20	78	Pt	2.2
15	P	2.19	47	Ag	1.93	79	Au	2.4
16	S	2.58	48	Cd	1.69	80	Hg	1.9
17	Cl	3.16	49	In	1.78	81	Tl	1.8
18	Ar	—	50	Sn	1.96	82	Pb	1.8
19	K	0.82	51	Sb	2.05	83	Bi	1.9
20	Ca	1.00	52	Te	2.1	84	Po	2.0
21	Sc	1.36	53	I	2.66	85	At	2.2
22	Ti	1.54	54	Xe	2.60	86	Rn	—
23	V	1.63	55	Cs	0.79	87	Fr	0.7
24	Cr	1.66	56	Ba	0.89	88	Ra	0.9
25	Mn	1.55	57	La	1.10	89	Ac	1.1
26	Fe	1.83	58	Ce	1.12	90	Th	1.3
27	Co	1.88	59	Pr	1.13	91	Pa	1.5
28	Ni	1.91	60	Nd	1.14	92	U	1.7
29	Cu	1.90	61	Pm	—	93	Np	1.3
30	Zn	1.65	62	Sm	1.17	94	Pu	1.3
31	Ga	1.81	63	Eu	—			
32	Ge	2.01	64	Gd	1.20			

FORCE CONSTANTS FOR BOND STRETCHING

Representative force constants (f) for stretching of chemical bonds are listed in this table. Except where noted, all force constants are derived from values of the harmonic vibrational frequencies ω_e. Values derived from the observed vibrational fundamentals ν, which are noted by a, are lower than the harmonic force constants, typically by 2 to 3% in the case of heavy atoms (often by 5 to 10% if one of the atoms is hydrogen). Values are given in the SI unit newton per centimeter (N/cm), which is identical to the commonly used cgs unit mdyn/Å.

REFERENCES

1. Huber, K. P., and Herzberg, G., *Molecular Spectra and Molecular Structure. IV. Constants of Diatomic Molecules*, Van Nostrand Reinhold, New York, 1979.
2. Shimanouchi, T., *The Molecular Force Field*, in Eyring, H., Henderson, D., and Yost, W., Eds., *Physical Chemistry: An Advanced Treatise*, Vol. IV, Academic Press, New York, 1970.
3. Tasumi, M., and Nakata, M., *Pure and Appl. Chem.*, 57, 121—147, 1985.

Bond	Molecule	f/(N/cm)	Note	Bond	Molecule	f/(N/cm)	Note
H-H	H_2	5.75			OCS	7.44	
Be-H	BeH	2.27		C-N	CN	16.29	
B-H	BH	3.05			HCN	18.78	
C-H	CH	4.48			CH_3CN	18.33	
	CH_4	5.44	b		CH_3NH_2	5.12	a,c
	C_2H_6	4.83	a,b,c	C-P	CP	7.83	
	CH_3CN	5.33	b	Si-Si	Si_2	2.15	
	CH_3Cl	5.02	a,b,c	Si-O	SiO	9.24	
	$CCl_2=CH_2$	5.57	b	Si-F	SiF	4.90	
	HCN	6.22		Si-Cl	SiCl	2.63	
N-H	NH	5.97		N-N	N_2	22.95	
O-H	OH	7.80			N_2O	18.72	
	H_2O	8.45		N-O	NO	15.95	
P-H	PH	3.22			N_2O	11.70	
S-H	SH	4.23		P-P	P_2	5.56	
	H_2S	4.28		P-O	PO	9.45	
F-H	HF	9.66		O-O	O_2	11.77	
Cl-H	HCl	5.16			O_3	5.74	a
Br-H	HBr	4.12		S-O	SO	8.30	
I-H	HI	3.14			SO_2	10.33	a
Li-H	LiH	1.03		S-S	S_2	4.96	
Na-H	NaH	0.78		F-F	F_2	4.70	
K-H	KH	0.56		Cl-F	ClF	4.48	
Rb-H	RbH	0.52		Br-F	BrF	4.06	
Cs-H	CsH	0.47		Cl-Cl	Cl_2	3.23	
C-C	C_2	12.16		Br-Cl	BrCl	2.82	
	$CCl_2=CH_2$	8.43		Br-Br	Br_2	2.46	
	C_2H_6	4.50	a,c	I-I	I_2	1.72	
	CH_3CN	5.16		Li-Li	Li_2	0.26	
C-F	CF	7.42		Li-Na	LiNa	0.21	
	CH_3F	5.71	a,c	Na-Na	Na_2	0.17	
C-Cl	CCl	3.95		Li-F	LiF	2.50	
	CH_3Cl	3.44	a,c	Li-Cl	LiCl	1.43	
	$CCl_2=CH_2$	4.02	b	Li-Br	LiBr	1.20	
C-Br	CH_3Br	2.89	a,c	Li-I	LiI	0.97	
C-I	CH_3I	2.34	a,c	Na-F	NaF	1.76	
C-O	CO	19.02		Na-Cl	NaCl	1.09	
	CO_2	16.00		Na-Br	NaBr	0.94	
	OCS	16.14		Na-I	NaI	0.76	
	CH_3OH	5.42	a,c	Be-O	BeO	7.51	
C-S	CS	8.49		Mg-O	MgO	3.48	
	CS_2	7.88		Ca-O	CaO	3.61	

[a] Derived from fundamental frequency, without anharmonicity correction.
[b] Average of symmetric and antisymmetric (or degenerate) modes.
[c] Calculated from Local Symmetry Force Field (see Reference 2).

FUNDAMENTAL VIBRATIONAL FREQUENCIES OF SMALL MOLECULES

This table lists the fundamental vibrational frequencies of selected three-, four-, and five-atom molecules. Both stable molecules and transient free radicals are included. The data have been taken from evaluated sources. In general, the selected values are based on gas-phase infrared, Raman, or ultraviolet spectra; when these were not available, liquid-phase or matrix-isolation spectra were used.

Molecules are grouped by structural type. Within each group, related molecules appear together for convenient comparison.

The vibrational modes are described by their approximate character in terms of stretching, bending, deformation, etc. However, it should be emphasized that most such descriptions are only approximate, and that the true normal mode usually involves a mixture of motions. Abbreviations are:

sym.	symmetric
antisym.	antisymmetric
str.	stretch
deform.	deformation
scis.	scissors
rock.	rocking
deg.	degenerate

In the case of free radicals, strong interactions may exist between the electronic and bending vibrational motions. Details can be found in References 3 and 4. The references should be consulted for information on the accuracy of the data and for data on other molecules not listed here.

All fundamental frequencies (more precisely, wavenumbers) are given in units of cm^{-1}

XY$_2$ Molecules
Point groups D$_{\infty h}$(linear) and C$_{2v}$(bent)

Molecule	Structure	Sym. str.	Bend	Antisym. str.
CO_2	Linear	1333	667	2349
CS_2	Linear	658	397	1535
C_3	Linear	1224	63	2040
CNC	Linear		321	1453
NCN	Linear	1197	423	1476
BO_2	Linear	1056	447	1278
BS_2	Linear	510	120	1015
KrF_2	Linear	449	233	590
XeF_2	Linear	515	213	555
$XeCl_2$	Linear	316		481
H_2O	Bent	3657	1595	3756
D_2O	Bent	2671	1178	2788
F_2O	Bent	928	461	831
Cl_2O	Bent	639	296	686
O_3	Bent	1103	701	1042
H_2S	Bent	2615	1183	2626
D_2S	Bent	1896	855	1999
SF_2	Bent	838	357	813
SCl_2	Bent	525	208	535
SO_2	Bent	1151	518	1362
H_2Se	Bent	2345	1034	2358
D_2Se	Bent	1630	745	1696
NH_2	Bent	3219	1497	3301
NO_2	Bent	1318	750	1618
NF_2	Bent	1075	573	942
ClO_2	Bent	945	445	1111
CH_2	Bent		963	
CD_2	Bent		752	
CF_2	Bent	1225	667	1114
CCl_2	Bent	721	333	748
CBr_2	Bent	595	196	641
SiH_2	Bent	2032	990	2022
SiD_2	Bent	1472	729	1468
SiF_2	Bent	855	345	870

Molecule	Structure	Sym. str.	Bend	Antisym. str.
$SiCl_2$	Bent	515		505
$SiBr_2$	Bent	403		400
GeH_2	Bent	1887	920	1864
$GeCl_2$	Bent	399	159	374
SnF_2	Bent	593	197	571
$SnCl_2$	Bent	352	120	334
$SnBr_2$	Bent	244	80	231
PbF_2	Bent	531	165	507
$PbCl_2$	Bent	314	99	299
ClF_2	Bent	500		576

XYZ Molecules
Point Groups $C_{\infty v}$ (linear) and C_s (bent)

Molecule	Structure	XY str.	Bend	YZ str.
HCN	Linear	3311	712	2097
DCN	Linear	2630	569	1925
FCN	Linear	1077	451	2323
ClCN	Linear	744	378	2216
BrCN	Linear	575	342	2198
ICN	Linear	486	305	2188
CCN	Linear	1060	230	1917
CCO	Linear	1063	379	1967
HCO	Bent	2485	1081	1868
HCC	Linear	3612		1848
OCS	Linear	2062	520	859
NCO	Linear	1270	535	1921
NNO	Linear	2224	589	1285
HNB	Linear	3675		2035
HNC	Linear	3653		2032
HNSi	Linear	3583	523	1198
HBO	Linear		754	1817
FBO	Linear		500	2075
ClBO	Linear	676	404	1958
BrBO	Linear	535	374	1937
FNO	Bent	766	520	1844
ClNO	Bent	596	332	1800
BrNO	Bent	542	266	1799
HNF	Bent		1419	1000
HNO	Bent	2684	1501	1565
HPO	Bent	2095	983	1179
HOF	Bent	3537	886	1393
HOCl	Bent	3609	1242	725
HOO	Bent	3436	1392	1098
FOO	Bent	579	376	1490
ClOO	Bent	407	373	1443
BrOO	Bent			1487
HSO	Bent		1063	1009
NSF	Bent	1372	366	640
NSCl	Bent	1325	273	414
HCF	Bent		1407	1181
HCCl	Bent		1201	815
HSiF	Bent	1913	860	834
HSiCl	Bent		808	522
HSiBr	Bent	1548	774	408

FUNDAMENTAL VIBRATIONAL FREQUENCIES OF SMALL MOLECULES (continued)

Symmetric XY$_3$ Molecules
Point Groups D$_{3h}$ (planar) and C$_{3v}$ (pyramidal)

Molecule	Structure	Sym. str.	Sym. deform.	Deg. str.	Deg. deform.
NH$_3$	Pyram.	3337	950	3444	1627
ND$_3$	Pyram.	2420	748	2564	1191
PH$_3$	Pyram.	2323	992	2328	1118
AsH$_3$	Pyram.	2116	906	2123	1003
SbH$_3$	Pyram.	1891	782	1894	831
NF$_3$	Pyram.	1032	647	907	492
PF$_3$	Pyram.	892	487	860	344
AsF$_3$	Pyram.	741	337	702	262
PCl$_3$	Pyram.	504	252	482	198
PI$_3$	Pyram.	303	111	325	79
AsI$_3$	Pyram.	219	94	224	71
AlCl$_3$	Pyram.	375	183	595	150
SO$_3$	Planar	1065	498	1391	530
BF$_3$	Planar	888	691	1449	480
BH$_3$	Planar		1125	2808	1640
CH$_3$	Planar		606	3161	1396
CD$_3$	Planar		453	2369	1029
CF$_3$	Pyram.	1090	701	1260	510
SiF$_3$	Pyram.	830	427	937	290

Linear XYYX Molecules
Point Group D$_{\infty h}$

Molecule	Sym. XY str.	Antisym. XY str.	YY str.	Bend	Bend
C$_2$H$_2$	3374	3289	1974	612	730
C$_2$D$_2$	2701	2439	1762	505	537
C$_2$N$_2$	2330	2158	851	507	233

Planar X$_2$YZ Molecules
Point Group C$_{2v}$

Molecule	Sym.XY str.	YZ str.	YX$_2$ scis.	Antisym. XY str.	YX$_2$ rock	YX$_2$ wag
H$_2$CO	2783	1746	1500	2843	1249	1167
D$_2$CO	2056	1700	1106	2160	990	938
F$_2$CO	965	1928	584	1249	626	774
Cl$_2$CO	567	1827	285	849	440	580
O$_2$NF	1310	822	568	1792	560	742
O$_2$NCl	1286	793	370	1685	408	652

Tetrahedral XY$_4$ Molecules
Point Group T$_d$

Molecule	Sym. str.	Deg. deform.(e)	Deg. str.(f)	Deg. deform.(f)
CH$_4$	2917	1534	3019	1306
CD$_4$	2109	1092	2259	996
CF$_4$	909	435	1281	632
CCl$_4$	459	217	776	314

FUNDAMENTAL VIBRATIONAL FREQUENCIES OF SMALL MOLECULES (continued)

Molecule	Sym. str.	Deg. deform.(e)	Deg. str.(f)	Deg. deform.(f)
CBr_4	267	122	672	182
CI_4	178	90	555	125
SiH_4	2187	975	2191	914
SiD_4	1558	700	1597	681
SiF_4	800	268	1032	389
$SiCl_4$	424	150	621	221
GeH_4	2106	931	2114	819
GeD_4	1504	665	1522	596
$GeCl_4$	396	134	453	172
$SnCl_4$	366	104	403	134
$TiCl_4$	389	114	498	136
$ZrCl_4$	377	98	418	113
$HfCl_4$	382	102	390	112
RuO_4	885	322	921	336
OsO_4	965	333	960	329

REFERENCES

1. T. Shimanouchi, Tables of Molecular Vibrational Frequencies, Consolidated Volume I, Natl. Stand. Ref. Data Ser. Natl. Bur. Stand. (U.S.), 39, 1972.
2. T. Shimanouchi, Tables of Molecular Vibrational Frequencies, Consolidated Volume II, *J. Phys. Chem. Ref. Data*, 6, 993, 1977.
3. G. Herzberg, *Electronic Spectra and Electronic Structure of Polyatomic Molecules*, D. Van Nostrand Co., Princeton, 1966.
4. M. E. Jacox, Ground state vibrational energy levels of polyatomic transient molecules, *J. Phys. Chem. Ref. Data*, 13, 945, 1984.

INFRARED CORRELATION CHARTS

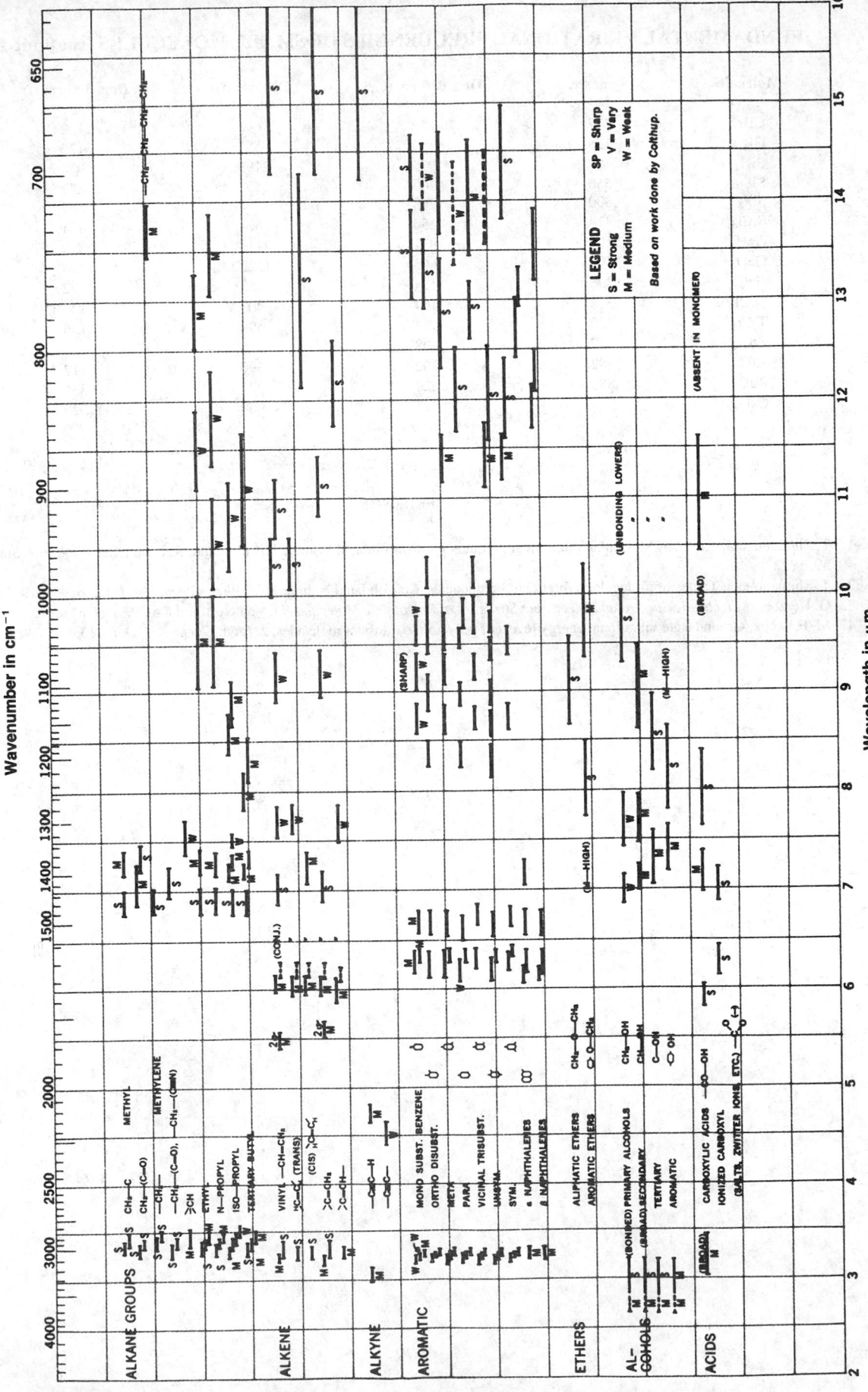

INFRARED CORRELATION CHARTS (continued)

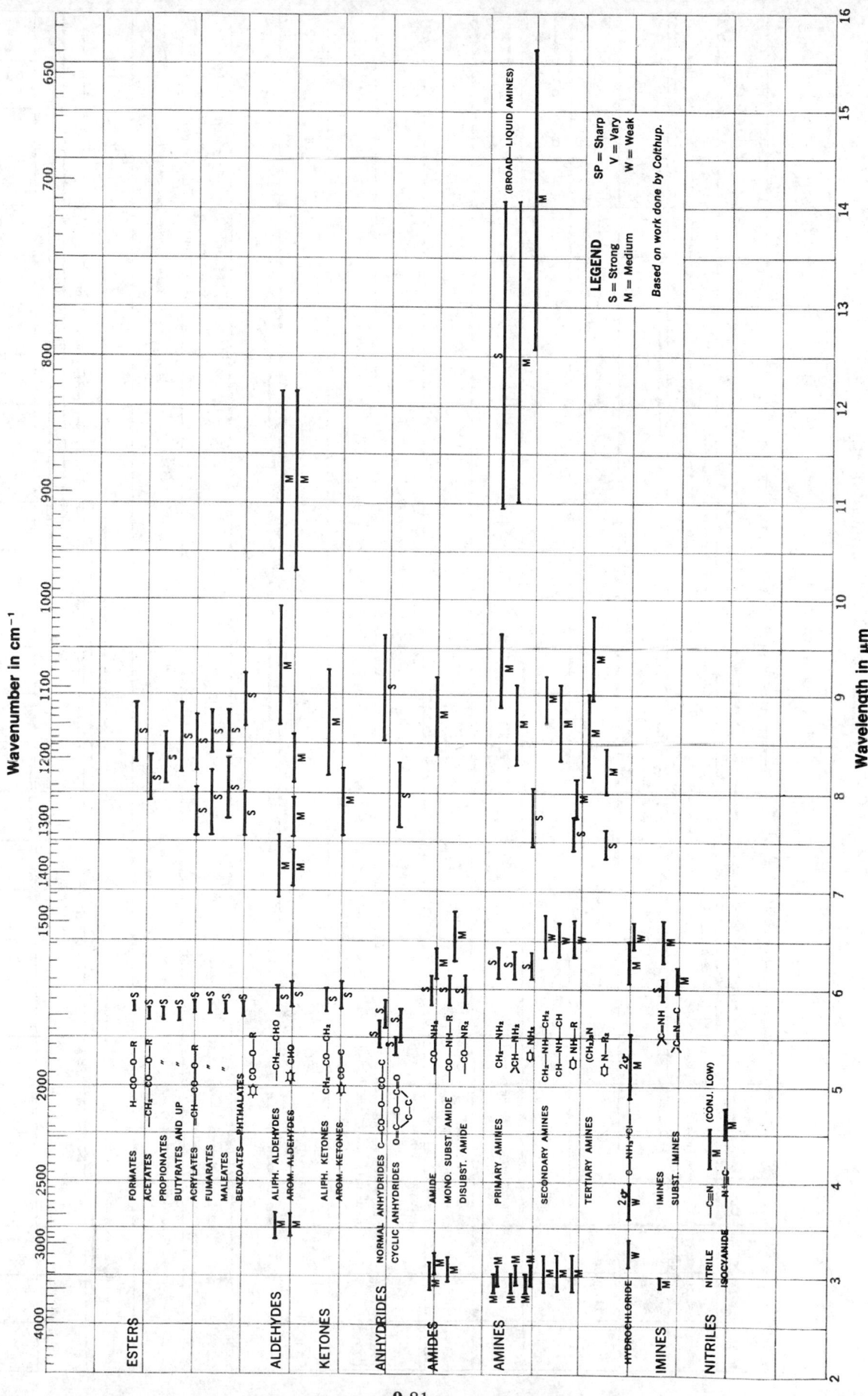

INFRARED CORRELATION CHARTS (continued)

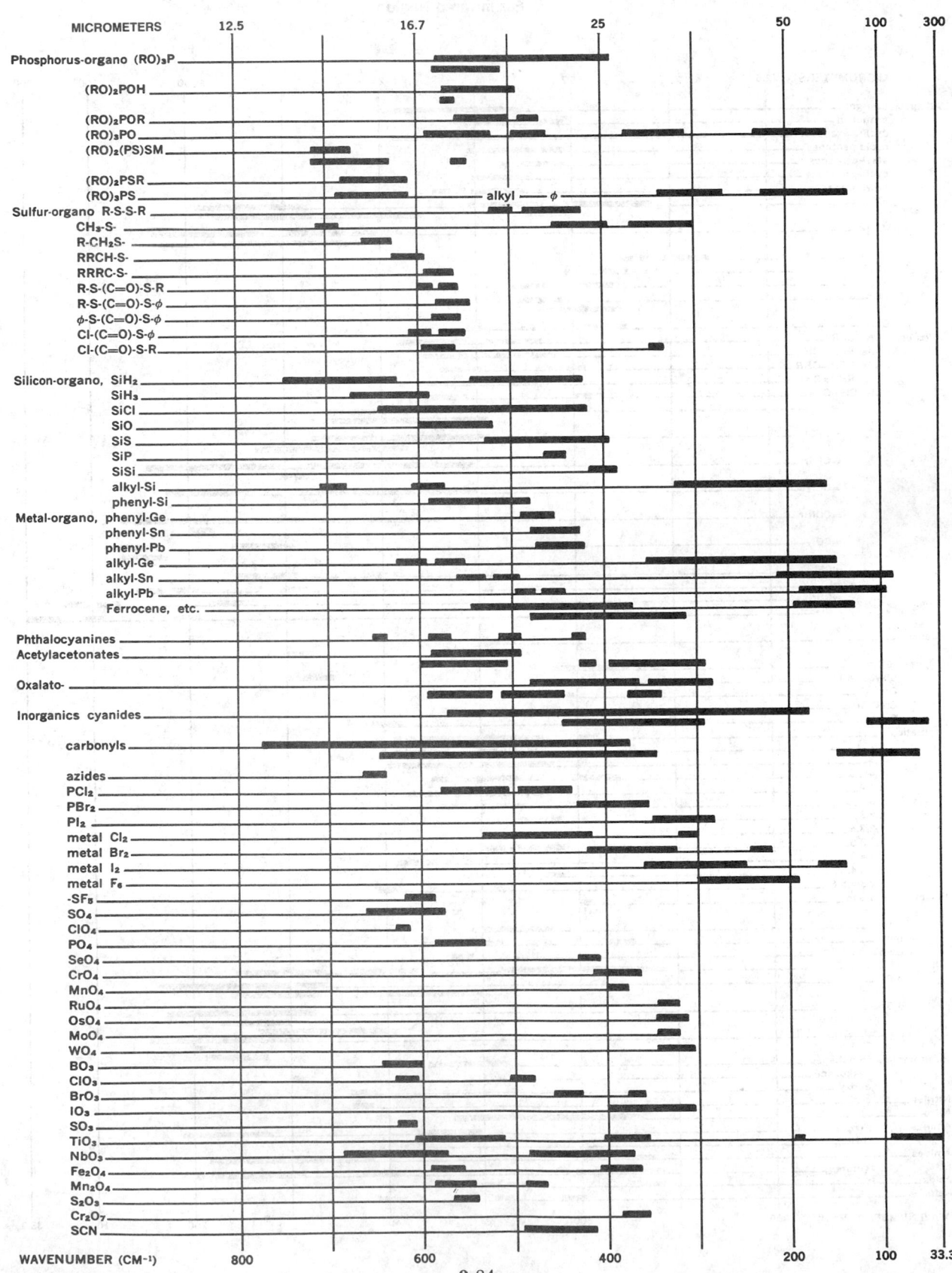

INFRARED CORRELATION CHART (Low Frequencies)

MICROMETERS

| | 12.5 | 16.7 | 25 | 50 | 100 | 300 |

Phosphorus-organo (RO)₃P
(RO)₂POH
(RO)₂POR
(RO)₃PO
(RO)₂(PS)SM
(RO)₂PSR
(RO)₃PS   alkyl — φ
Sulfur-organo R-S-S-R
CH₃-S-
R-CH₂S-
RRCH-S-
RRRC-S-
R-S-(C=O)-S-R
R-S-(C=O)-S-φ
φ-S-(C=O)-S-φ
Cl-(C=O)-S-φ
Cl-(C=O)-S-R

Silicon-organo, SiH₂
SiH₃
SiCl
SiO
SiS
SiP
SiSi
alkyl-Si
phenyl-Si
Metal-organo, phenyl-Ge
phenyl-Sn
phenyl-Pb
alkyl-Ge
alkyl-Sn
alkyl-Pb
Ferrocene, etc.

Phthalocyanines
Acetylacetonates

Oxalato-

Inorganics cyanides

carbonyls

azides
PCl₂
PBr₂
PI₂
metal Cl₂
metal Br₂
metal I₂
metal F₆
-SF₅
SO₄
ClO₄
PO₄
SeO₄
CrO₄
MnO₄
RuO₄
OsO₄
MoO₄
WO₄
BO₃
ClO₃
BrO₃
IO₃
SO₃
TiO₃
NbO₃
Fe₂O₄
Mn₂O₄
S₂O₃
Cr₂O₇
SCN

WAVENUMBER (CM⁻¹)

| 800 | 600 | 400 | 200 | 100 | 33.3 |

NUCLEAR SPINS, MOMENTS, AND OTHER DATA RELATED TO NMR SPECTROSCOPY

This table presents the following data relevant to nuclear magnetic resonance spectroscopy:

Z: Atomic number

Isotope: Element symbol and mass number

Abundance: Natural abundance of the isotope in percent. An * indicates a radioactive nuclide; if no value is given, the nuclide is not present in nature or its abundance is highly variable.

I: Nuclear spin

ν: Resonant frequency in megahertz for an applied field H_o of 1 tesla (in cgs units, 10 kilogauss)

Relative sensitivity: Sensitivity relative to ^1H (=1) assuming an equal number of nuclei and constant temperature. Values were calculated from the expressions:

$$\text{For constant } H_o: 0.0076508(\mu/\mu_N)^3(I+1)/I^2$$
$$\text{For constant } \nu: 0.23871(\mu/\mu_N)(I+1)$$

μ: Nuclear magnetic moment in units of the nuclear magneton μ_N

Q: Nuclear quadrupole moment in units of femtometers squared (1 $fm^2 = 10^{-2}$ barn)

The table includes all stable nuclides of non-zero spin for which spin and magnetic moment values have been measured, as well as selected radioactive nuclides of current or potential interest. At least one isotope is included for each element through $Z = 95$ for which data are available. See Reference 2 for a complete listing of spins and moments.

The assistance of P. Pyykkö in providing recent data on nuclear quadrupole moments is gratefully acknowledged.

REFERENCES

1. Raghavan, P., "Table of Nuclear Moments", *At. Data Nuc. Data Tables,* 42, 189, 1989.
2. Holden, N. E., "Table of the Isotopes", in Lide, D. R., Ed., *CRC Handbook of Chemistry and Physics,* 74th Ed., CRC Press, Boca Raton, FL, 1993.
3. Pyykkö, P., *Z. Naturforsch.*, 47a, 189, 1992.

Z	Isotope	Abundance %	I	ν/MHz for H_o = 1 T	Relative sensitivity Const. H_o	Const. ν	μ/μ_N	Q/fm²
1	^{1}n	*	1/2	29.1639	0.32139	0.6850	-1.91304275	
1	^{1}H	99.985	1/2	42.5764	1.00000	1.0000	+2.7928474	
1	^{2}H	0.015	1	6.53573	0.00965	0.4093	+0.8574382	+0.2860
1	^{3}H	*	1/2	45.4137	1.21354	1.0666	+2.9789625	
2	^{3}He	0.0001	1/2	32.4352	0.44212	0.7618	-2.1276248	
3	^{6}Li	7.5	1	6.2660	0.00850	0.3925	+0.8220467	-0.082
3	^{7}Li	92.5	3/2	16.5478	0.29355	1.9433	+3.256427	-4.01
4	^{9}Be	100	3/2	5.986	0.01389	0.7029	-1.1779	+5.288
5	^{10}B	19.9	3	4.5751	0.01985	1.7193	+1.800645	+8.459
5	^{11}B	80.1	3/2	13.6626	0.16522	1.6045	+2.688649	+4.059
6	^{13}C	1.10	1/2	10.7081	0.01591	0.2515	+0.7024118	
7	^{14}N	99.634	1	3.0776	0.00101	0.1928	+0.4037610	+2.02
7	^{15}N	0.366	1/2	4.3172	0.00104	0.1014	-0.2831888	
8	^{17}O	0.038	5/2	5.7741	0.02910	1.5822	-1.89379	-2.558
9	^{19}F	100	1/2	40.0765	0.83400	0.9413	+2.628868	
10	^{21}Ne	0.27	3/2	3.3630	0.00246	0.3949	-0.661797	+10.155
11	^{23}Na	100	3/2	11.2686	0.09270	1.3233	+2.217522	+10.89
12	^{25}Mg	10.00	5/2	2.6082	0.00268	0.7147	-0.85545	+19.94
13	^{27}Al	100	5/2	11.1028	0.20689	3.0424	+3.641507	+14.03
14	^{29}Si	4.67	1/2	8.4653	0.00786	0.1988	-0.55529	
15	^{31}P	100	1/2	17.2510	0.06652	0.4052	+1.13160	
16	^{33}S	0.75	3/2	3.2716	0.00227	0.3842	+0.6438212	-6.78
17	^{35}Cl	75.77	3/2	4.1764	0.00472	0.4905	+0.8218743	-8.165
17	^{37}Cl	24.23	3/2	3.4764	0.00272	0.4083	+0.6841236	-6.435
18	^{37}Ar	*	3/2	5.818	0.01276	0.6833	+1.145	
18	^{39}Ar	*	7/2	2.8	0.00617	1.3964	-1.3	
19	^{39}K	93.2581	3/2	1.9893	0.00051	0.2336	+0.3914662	+6.01
19	^{40}K	0.0117	4	2.4737	0.00523	1.5493	-1.298100	-7.49
19	^{41}K	6.7302	3/2	1.0919	0.00008	0.1282	+0.2148701	+7.33

Z	Isotope	Abundance %	I	ν/MHz for $H_o = 1$ T	Relative sensitivity Const. H_o	Const. ν	μ/μ_N	Q/fm^2
20	^{43}Ca	0.135	7/2	2.8688	0.00642	1.4150	-1.31726	-4.08
21	^{45}Sc	100	7/2	10.3588	0.30244	5.1093	+4.756487	-22
22	^{47}Ti	7.3	5/2	2.4040	0.00210	0.6587	-0.78848	+29
22	^{49}Ti	5.5	7/2	2.4047	0.00378	1.1861	-1.10417	+24
23	^{50}V	0.250	6	4.2504	0.05571	5.5904	+3.345689	+21
23	^{51}V	99.750	7/2	11.2130	0.38360	5.5306	+5.1487057	-5.2
24	^{53}Cr	9.501	3/2	2.4114	0.00091	0.2832	-0.47454	-15
25	^{55}Mn	100	5/2	10.5760	0.17881	2.8980	+3.46872	+33
26	^{57}Fe	2.1	1/2	1.3815	0.00003	0.0324	+0.0906230	
27	^{59}Co	100	7/2	10.077	0.27841	4.9702	+4.627	+42
28	^{61}Ni	1.140	3/2	3.8113	0.00359	0.4476	-0.75002	+16.2
29	^{63}Cu	69.17	3/2	11.2979	0.09342	1.3268	+2.22329	-22.0
29	^{65}Cu	30.83	3/2	12.1027	0.11484	1.4213	+2.38167	-20.4
30	^{67}Zn	4.1	5/2	2.6693	0.00287	0.7314	+0.875479	+15.0
31	^{69}Ga	60.108	3/2	10.2475	0.06971	1.2034	+2.01659	+17.0
31	^{71}Ga	39.892	3/2	13.0204	0.14300	1.5291	+2.56227	+10.0
32	^{73}Ge	7.73	9/2	1.4897	0.00141	1.1546	-0.8794677	-17.3
33	^{75}As	100	3/2	7.3148	0.02536	0.8590	+1.439475	+31.4
34	^{77}Se	7.63	1/2	8.1566	0.00703	0.1916	+0.535042	
35	^{79}Br	50.69	3/2	10.7039	0.07945	1.2570	+2.106400	+33.1
35	^{81}Br	49.31	3/2	11.5381	0.09951	1.3550	+2.270562	+27.6
36	^{83}Kr	11.5	9/2	1.6442	0.00190	1.2744	-0.970669	+25.3
37	^{85}Rb	72.165	5/2	4.1253	0.01061	1.1304	+1.35303	+27.4
37	^{87}Rb	27.835	3/2	13.9807	0.17703	1.6418	+2.75124	+13.2
38	^{87}Sr	7.00	9/2	1.8524	0.00272	1.4358	-1.093603	+33.5
39	^{89}Y	100	1/2	2.0949	0.00012	0.0492	-0.1374154	
40	^{91}Zr	11.22	5/2	3.9747	0.00949	1.0891	-1.30362	-20.6
41	^{93}Nb	100	9/2	10.4520	0.48821	8.1011	+6.1705	-32
42	^{95}Mo	15.92	5/2	2.7874	0.00327	0.7638	-0.9142	-2.2
42	^{97}Mo	9.55	5/2	2.8462	0.00349	0.7799	-0.9335	+25.5
43	^{99}Tc	*	9/2	9.63	0.38174	7.4633	+5.6847	-12.9
44	^{99}Ru	12.7	5/2	1.9553	0.00113	0.5358	-0.6413	+7.9
44	^{101}Ru	17.0	5/2	2.192	0.00159	0.6005	-0.7188	+45.7
45	^{103}Rh	100	1/2	1.3476	0.00003	0.0317	-0.08840	
46	^{105}Pd	22.33	5/2	1.957	0.00113	0.5364	-0.642	+66.0
47	^{107}Ag	51.839	1/2	1.7330	0.00007	0.0407	-0.1136796	
47	^{109}Ag	48.161	1/2	1.9924	0.00010	0.0468	-0.1306906	
48	^{111}Cd	12.80	1/2	9.0689	0.00966	0.2130	-0.5948861	
48	^{113}Cd	12.22	1/2	9.4868	0.01106	0.2228	-0.6223009	
49	^{113}In	4.3	9/2	9.3652	0.35121	7.2588	+5.5289	+79.9
49	^{115}In	95.7	9/2	9.3854	0.35348	7.2744	+5.5408	+81
50	^{115}Sn	0.34	1/2	14.0074	0.03561	0.3290	-0.91883	
50	^{117}Sn	7.68	1/2	15.2606	0.04605	0.3584	-1.00104	
50	^{119}Sn	8.59	1/2	15.9656	0.05273	0.3750	-1.04728	
51	^{121}Sb	57.36	5/2	10.2549	0.16302	2.8100	+3.3634	-36
51	^{123}Sb	42.64	7/2	5.5530	0.04659	2.7389	+2.5498	-49
52	^{123}Te	0.908	1/2	11.2346	0.01837	0.2639	-0.7369478	
52	^{125}Te	7.139	1/2	13.5451	0.03220	0.3181	-0.8885051	
53	^{127}I	100	5/2	8.5776	0.09540	2.3504	+2.813273	-78.9
54	^{129}Xe	26.4	1/2	11.8601	0.02162	0.2786	-0.7779763	
54	^{131}Xe	21.2	3/2	3.5158	0.00282	0.4129	+0.6918619	-12.0
55	^{133}Cs	100	7/2	5.6232	0.04838	2.7735	+2.582025	-0.37
56	^{135}Ba	6.592	3/2	4.2581	0.00500	0.5001	+0.837943	+16.0
56	^{137}Ba	11.23	3/2	4.7633	0.00700	0.5594	+0.937365	+24.5
57	^{138}La	* 0.0902	5	5.6614	0.09404	5.3188	+3.713646	+45
57	^{139}La	99.9098	7/2	6.0610	0.06058	2.9895	+2.7830455	+20
58	^{137}Ce	*	3/2	4.62	0.00641	0.5431	0.91	

Z	Isotope	Abundance %	I	ν/MHz for $H_0 = 1$ T	Relative sensitivity Const. H_0	Const. ν	μ/μ_N	Q/fm^2
58	^{139}Ce	*	3/2	4.62	0.00641	0.5431	0.91	
58	^{141}Ce	*	7/2	2.37	0.00364	1.1708	1.09	
59	^{141}Pr	100	5/2	13.0355	0.33483	3.5720	+4.2754	-5.9
60	^{143}Nd	12.18	7/2	2.319	0.00339	1.1440	-1.065	-63
60	^{145}Nd	8.30	7/2	1.429	0.00079	0.7047	-0.656	-33
61	^{143}Pm	*	5/2	11.6	0.23510	3.1748	3.8	
61	^{147}Pm	*	7/2	5.7	0.04940	2.7928	+2.6	+70
62	^{147}Sm	15.0	7/2	1.7747	0.00152	0.8753	-0.8149	-26
62	^{149}Sm	13.8	7/2	1.4631	0.00085	0.7216	-0.6718	+9.4
63	^{151}Eu	47.8	5/2	10.5854	0.17929	2.9006	+3.4718	+90.3
63	^{153}Eu	52.2	5/2	4.6744	0.01544	1.2809	+1.5331	+241
64	^{155}Gd	14.80	3/2	1.317	0.00015	0.1546	-0.2591	+127
64	^{157}Gd	15.65	3/2	1.727	0.00033	0.2028	-0.3399	+135
65	^{159}Tb	100	3/2	10.23	0.06945	1.2019	+2.014	+143.2
66	^{161}Dy	18.9	5/2	1.4653	0.00048	0.4015	-0.4806	+247
66	^{163}Dy	24.9	5/2	2.0507	0.00130	0.5619	+0.6726	+265
67	^{165}Ho	100	7/2	9.0881	0.20423	4.4825	+4.173	+358
68	^{167}Er	22.95	7/2	1.2281	0.00050	0.6057	-0.5639	+357
69	^{169}Tm	100	1/2	3.531	0.00057	0.0829	-0.2316	
70	^{171}Yb	14.3	1/2	7.5259	0.00552	0.1768	+0.49367	
70	^{173}Yb	16.12	5/2	2.0730	0.00135	0.5680	-0.67989	+280
71	^{175}Lu	97.41	7/2	4.8624	0.03128	2.3983	+2.2327	+349
71	^{176}Lu	* 2.59	7	3.451	0.03975	6.0516	+3.169	+497
72	^{177}Hf	18.606	7/2	1.7281	0.00140	0.8524	+0.7935	+336
72	^{179}Hf	13.629	9/2	1.0856	0.00055	0.8414	-0.6409	+379
73	^{180}Ta	0.012	9	4.04	0.10251	11.3862	4.77	
73	^{181}Ta	99.988	7/2	5.1625	0.03744	2.5463	+2.3705	+317
74	^{183}W	14.3	1/2	1.7956	0.00008	0.0422	+0.1177847	
75	^{185}Re	37.40	5/2	9.717	0.13870	2.6627	+3.1871	+218
75	^{187}Re	*62.60	5/2	9.817	0.14300	2.6900	+3.2197	+207
76	^{187}Os	1.6	1/2	0.9856	0.00001	0.0231	+0.06465189	
76	^{189}Os	16.1	3/2	3.3535	0.00244	0.3938	+0.659933	+85.6
77	^{191}Ir	37.3	3/2	0.766	0.00003	0.0899	+0.1507	+81.6
77	^{193}Ir	62.7	3/2	0.832	0.00004	0.0977	+0.1637	+75.1
78	^{195}Pt	33.8	1/2	9.2920	0.01039	0.2182	+0.60952	
79	^{197}Au	100	3/2	0.7406	0.00003	0.0870	+0.145746	+54.7
80	^{199}Hg	16.87	1/2	7.7121	0.00594	0.1811	+0.5058855	
80	^{201}Hg	13.18	3/2	2.8468	0.00149	0.3343	-0.5602257	+38.6
81	^{203}Tl	29.524	1/2	24.7310	0.19598	0.5809	+1.6222579	
81	^{205}Tl	70.476	1/2	24.9742	0.20182	0.5866	+1.6382146	
82	^{207}Pb	22.1	1/2	9.0338	0.00955	0.2122	+0.59258	
83	^{209}Bi	100	9/2	6.9628	0.14433	5.3967	+4.1106	-50
84	^{209}Po	*	1/2	11.7	0.02096	0.2757	+0.77	
86	^{211}Rn	*	1/2	9.16	0.00997	0.2152	+0.601	
87	^{223}Fr	*	3/2	5.94	0.01362	0.6982	+1.17	+117
88	^{223}Ra	*	3/2	1.3746	0.00017	0.1614	+0.2705	+119
88	^{225}Ra	*	1/2	11.187	0.01814	0.2627	-0.7338	
89	^{227}Ac	*	3/2	5.6	0.01131	0.6564	+1.1	+170
90	^{229}Th	*	5/2	1.40	0.00042	0.3843	+0.46	+430
91	^{231}Pa	*	3/2	10.2	0.06903	1.1995	2.01	-172
92	^{235}U	* 0.7200	7/2	0.83	0.00015	0.4082	-0.38	+493.6
93	^{237}Np	*	5/2	9.57	0.13264	2.6234	+3.14	+388.6
94	^{239}Pu	*	1/2	3.09	0.00038	0.0727	+0.203	
95	^{243}Am	*	5/2	4.91	0.01788	1.3451	+1.61	+421

PROTON NMR CHEMICAL SHIFTS FOR CHARACTERISTIC ORGANIC STRUCTURES

The chart below summarizes the range of chemical shifts for protons in several classes or organic compounds and substituent groups. The chemical shifts δ are given in parts per million relative to tetramethylsilane.

REFERENCE

Mohacsi, E., *J. Chem. Edu.*, 41, 38, 1964 (with permission).

¹³C-NMR ABSORPTIONS OF MAJOR FUNCTIONAL GROUPS

The table below lists the ¹³C chemical shifts in parts per million (ppm), in descending order with the corresponding functional groups. A series of typical simple compounds for every family is added to illustrate the correlations above. The shifts for the carbons of interest are given in parentheses — either for each carbon as it appears from left to right in the formula, or by italics.[1-3]

REFERENCES

1. **Yoder, C. H. and Schaeffer, C. D., Jr.,** *Introduction to Multinuclear NMR: Theory and Application,* Benjamin/Cummings, Menlo Park, CA, 1987.
2. **Silverstein, R. M., Bassler, G. C., and Morrill, T. C.,** *Spectrometric Identification of Organic Compounds,* John Wiley & Sons, New York, 1981.
3. **Brown, D. W.,** A short set of ¹³NMR Correlation Tables, *J. Chem. Educ.,* 62, 209, 1985.

¹³C-NMR ABSORPTIONS OF MAJOR FUNCTIONAL GROUPS

δ (ppm)	Group	Family	Example (δ of italicized carbon)	
220—165	>C=O	Ketones	$(CH_3)_2CO$	(206.0)
			$(CH_3)_2CHCOCH_3$	(212.1)
		Aldehydes	CH_3CHO	(199.7)
		α,β-Unsaturated carbonyls	$CH_3CH=CHCHO$	(192.4)
			$CH_2=CHCOCH_3$	(169.9)
		Carboxylic acids	HCO_2H	(166.0)
			CH_3CO_2H	(178.1)
		Amides	$HCONH_2$	(165.0)
			CH_3CONH_2	(172.7)
		Esters	$CH_3CO_2CH_2CH_3$	(170.3)
			$CH_2=CHCO_2CH_3$	(165.5)
140—120	>C=C<	Aromatic	C_6H_6	(128.5)
		Alkenes	$CH_2=CH_2$	(123.2)
			$CH_2=CHCH_3$	(115.9,136.2)
			$CH_2=CHCH_2Cl$	(117.5.133.7)
			$CH_3CH=CHCH_2CH_3$	(132.7)
125—115	–C≡N	Nitriles	$CH_3-C≡N$	(117.7)
80—70	–C≡C–	Alkynes	$HC≡CH$	(71.9)
			$CH_3C≡CH_3$	(73.9)
70—45	–C–O	Esters	$CH_3OOCH_2CH_3$	(57.6,67.9)
		Alcohols	$HOCH_3$	(49.0)
			$HOCH_2CH_3$	(57.0)
40—20	–C–NH₂	Amines	CH_3NH_2	(26.9)
			$CH_3CH_2NH_2$	(35.9)
30—15	–S–CH₃	Sulfides (thioethers)	$C_6H_5-S-CH_3$	15.6
30—(–2.3)	–C–H	Alkanes, cycloalkanes	CH_4	(–2.3)
			CH_3CH_3	(5.7)
			$CH_3CH_2CH_3$	(15.8,16.3)
			$CH_3CH_2CH_2CH_3$	(13.4,25.2)
			$CH_3CH_2CH_2CH_2CH_3$	(13.9,22.8,34.7)
			Cyclohexane	(26.9)

Reprinted from T. J. Bruno and P. D. N. Svoronos, *CRC Handbook of Basic Tables for Chemical Analysis,* CRC Press, Boca Raton, FL, 1989, pg. 329.

Section 10
Atomic, Molecular, and Optical Physics

LINE SPECTRA OF THE ELEMENTS

Edited by Joseph Reader and Charles H. Corliss*
National Bureau of Standards

These tables were prepared under the auspices of the Committee on Line Spectra of the Elements of the National Academy of Sciences—National Research Council. They contain the outstanding spectral lines of neutral (I), singly ionized (II), doubly ionized (III), triply ionized (IV), and quadruply ionized (V) atoms. Listed are lines that appear in emission from the vacuum ultraviolet to the far infrared. For most atoms these lines were selected from much larger lists in such a way as to include the stronger observed lines in each spectral region. In a few cases prominent monoxide band heads are also given.

The data were compiled by the following contributors, whose initials are given in the headings of the tables that they prepared:

- J. G. Conway—Lawrence Berkeley Laboratory
- C. H. Corliss—National Bureau of Standards
- R. D. Cowan—Los Alamos Scientific Laboratory
- C. R. Cowley—University of Michigan
- Henry M. and Hannah Crosswhite—Argonne National Laboratory
- S. P. Davis—University of California, Berkeley
- V. Kaufman—National Bureau of Standards
- R. L. Kelly—Naval Postgraduate School
- J. F. Kielkopf—University of Louisville
- W. C. Martin—National Bureau of Standards
- T. K. McCubbin—Pennsylvania State University
- L. J. Radziemski—Los Alamos Scientific Laboratory
- J. Reader—National Bureau of Standards
- C. J. Sansonetti—National Bureau of Standards
- G. V. Shalimoff—Lawrence Berkeley Laboratory
- R. W. Stanley—Purdue University
- J. O. Stoner, Jr.—University of Arizona
- H. H. Stroke—New York University
- D. R. Wood—Wright State University
- E. F. Worden—Lawrence Livermore Laboratory
- J. J. Wynne—International Business Machines Corporation
- R. Zalubas—National Bureau of Standards

The literature references are collected at the end of the entire set of tables.

All wavelengths are given in Angstroms. Below 2000 Å the wavelengths are in vacuum; above 2000 Å the wavelengths are in air. Wavelengths given to three decimal places have an uncertainty of less than 0.001 Å and are therefore suitable for the calibration of most spectrographs. In the air region, the elements used most commonly for calibration purposes are Ne, Ar, Kr, Fe, Th, and Hg; in the vacuum region, the most common are C, N, O, Si, Cu.

A large number of the lines for neutral and singly ionized atoms were extracted from the National Bureau of Standards (NBS) Tables of Spectral-Line Intensities.[1] The intensities of these lines represent quantitative estimates of relative line strengths that take account of varying detection sensitivity at different wavelengths. They are on a linear scale. For nearly all of the other lines the intensities represent qualitative estimates of the relative strengths of lines not greatly separated in wavelength. Because different observers frequently use different scales for their intensity estimates, these intensities are useful only as a rough indication of the appearance of a spectrum. In some cases the intensity scale is not intended to be linear. In the tables of first and second spectra the intensities of the lines of the singly ionized atom relative to those of the neutral atom should be used with caution, inasmuch as the concentration of ions in a light source depends greatly on the excitation conditions.

Descriptive symbols used in the tables have the following meanings:

- c—complex
- d—line consists of two unresolved lines
- h—hazy
- l—shaded to longer wavelengths
- s—shaded to shorter wavelengths
- p—perturbed by a close line
- b—band head
- r—easily reversed
- w—wide

ACTINIUM (Ac)
Z = 89

Ac I and II
Ref. 193 — J.G.C.

Intensity		Wavelength (Air)	
8	h	2100.00	II
20		2712.50	II
10		2726.23	II
10	h	2760.18	II
10	h	2781.56	II
20		2797.59	II
20		2806.76	II
8		2833.47	II
150	h	2847.16	II
8		2895.20	II
30		2896.82	II
30		2923.02	II
200		2994.17	II
500		3043.30	II
200		3069.36	II
100		3078.07	II
100		3086.04	II
100		3087.37	II
200		3112.83	II
100		3120.16	II
500	s	3153.09	II
600	s	3154.41	II
200	s	3164.81	II
300	s	3230.59	II
150	s	3237.70	II
500		3260.91	II
100	s	3318.01	II
200	s	3383.53	II
200		3413.84	II
500		3417.77	II
500	s	3481.16	II
200		3489.53	II
100		3529.24	II
100		3534.63	II
200	s	3554.99	II
1000	s	3565.59	II
100		3694.88	II
300	s	3756.67	II
200		3799.82	II
2000	s	3863.12	II
100		3914.47	II
400		4061.60	II
3000		4088.44	II
3000	s	4168.40	II
100		4179.98	I
20		4183.12	I
20		4194.40	I
300	s	4209.69	II
300		4359.13	II
20	1	4384.53	I
1000	1	4386.41	II
20		4396.71	I
20		4462.73	I
1000	1	4507.20	II
500		4605.45	II
10		4716.58	I
400	s	4720.16	II
300		4812.22	II
100		4945.18	II
100		4958.23	II
100		4960.87	II
150		5446.38	II
300	1	5732.05	II
400		5758.97	II
1000		5910.85	II
600	1	6164.75	II
200	1	6167.83	II
400		6242.83	II
20		6359.86	I
20	1	6691.27	I
6		7290.40	I
6	1	7866.10	I

Ac III
Ref. 193 — J.G.C.

Intensity		Wavelength (Air)	
1000	h	2626.44	III
50	h	2682.90	III
2000	h	2952.55	III
2000	h	3392.78	III
3000		3487.59	III
2000	h	4413.09	III
3000	h	4569.87	III
8	h	5193.21	III

Ac IV
Ref. 193 — J.G.C.

Intensity		Wavelength (Air)	
20	h	2062.00	IV
30	h	2502.12	IV
100	h	2558.08	IV
5	h	2790.83	IV
50	h	2793.90	IV
20	1	3224.7	IV

ALUMINUM (Al)
Z = 13

Al I and II
Ref. 81, 89, 144, 227, 228, 282 — E.F.W.

Intensity		Wavelength (Vacuum)	
40		1177.43	II
50		1191.812	II
150		1350.18	II
800		1539.830	II
100		1569.385	II
125		1596.059	II
150		1625.627	II
100		1644.235	II
100		1644.809	II
1000		1670.787	II
100		1686.250	II
800		1719.440	II
500		1721.244	II
900		1721.271	II
500		1724.952	II
900		1724.984	II
350		1760.104	II
300		1761.975	II
290		1763.00	I
500		1763.869	II
700		1763.952	II
450		1765.64	I

---

* Charles H. Corliss is now retired.

Aluminum (Cont.)

Intensity		Wavelength	Spectrum
300		1765.815	II
450		1766.38	I
400		1767.731	II
450		1769.14	I
600		1828.588	II
400		1832.837	II
250		1834.808	II
300		1855.929	II
700		1858.026	II
120		1859.980	II
1000		1862.311	II
200		1929.978	II
150		1931.048	II
200		1932.377	II
400		1934.503	II
150		1934.713	II
150		1936.907	II
220		1939.261	II
700		1990.531	II

Air

Intensity		Wavelength	Spectrum
150		2016.052	II
150		2016.234	II
100		2016.368	II
200		2074.008	II
700		2094.264	II
150		2094.744	II
300		2094.791	II
100		2095.104	II
200		2095.141	II
60		2150.70	I
60		2181.00	I
400		2269.10	I
120		2269.22	I
60		2312.49	I
70		2313.53	I
90		2317.48	I
60		2319.06	I
140		2321.56	I
460		2367.05	I
110		2367.61	I
110		2368.11	I
180		2369.30	I
140		2370.22	I
70		2370.73	I
160		2372.07	I
850		2373.12	I
170		2373.35	I
110		2373.57	I
60		2378.40	I
60		2513.30	I
240		2567.98	I
480		2575.10	I
60		2575.40	I
80		2631.55	II
110		2637.70	II
150		2652.48	I
200		2660.39	I
160		2669.17	II
650		2816.19	II
90		2837.96	I
90		2840.10	I
150		3041.28	II
360		3050.07	I
60		3054.68	I
450		3057.14	I
90		3064.29	I
60		3066.14	I
150		3074.64	II
4500	rS	3082.153	I
7200	r	3092.710	I
1800	r	3092.839	I
150		3428.92	II
70		3439.35	I
150		3443.64	I
70		3444.86	I
70		3458.22	I
60		3479.81	I
60		3482.63	I
450		3586.56	II
360		3587.07	II
290		3587.45	II
220		3651.06	II
110		3651.10	II
150		3654.98	II
290		3655.00	II
450		3900.68	II
60		3932.00	I
4500	r	3944.006	I
9000	r	3961.520	I
110		3995.86	II
290		4226.81	II
150		4585.82	II
110		4588.19	II
550		4666.80	II
110		4898.76	II
110		4902.77	II
150		5280.21	II
70		5107.52	I
290		5283.77	II
150		5285.85	II
110		5312.32	II
220		5316.07	II
150		5371.84	II
180		5557.06	I
110		5557.95	I
450		5593.23	II
110		5853.62	II
220		5971.94	II
290		6001.76	II
220		6001.88	II
450		6006.42	II
150		6061.11	II
290		6068.43	II
110		6068.53	II
450		6073.23	II
110		6181.57	II
150		6181.68	II
290		6182.28	II
220		6182.45	II
450	h	6183.42	II
450		6201.52	II
360		6201.70	II
290		6226.18	II
360		6231.78	II
450		6243.36	II
450		6335.74	II
360		6696.02	I
230		6698.67	I
60		7083.97	I
70		7084.64	I
110		7361.57	I
140		7362.30	I
60		7606.16	I
90		7614.82	I
230		7835.31	I
290		7836.13	I
60		7993.05	I
90		8003.19	I
70		8065.97	I
110		8075.35	I
290		8640.70	II
360		8772.87	I
450		8773.90	I
110		8828.91	I
180		8841.28	I
90		8912.90	I
140		8923.56	I
60		9089.91	I
70		9139.95	I
150		9290.65	II
110		9290.75	II
150		10076.29	II
110		10768.36	I
140		10782.04	I
110		10872.98	I
230		10891.73	I
450		11253.19	I
570		11254.88	I
570		13123.41	I
450		13150.76	I
230		16718.96	I
300		16750.56	I
140		16763.36	I
300		21093.04	I
360		21163.75	I

Al III
Ref. 127 — E.F.W.

Intensity	Wavelength	Spectrum
	Vacuum	
70	486.884	III
30	486.912	III
250	511.138	III
150	511.191	III
500	560.317	III
200	560.433	III
100	670.068	III
200	671.118	III
500	695.829	III
400	696.217	III
200	725.683	III
300	726.915	III
400	855.034	III
500	856.746	III
400	892.024	III
50	893.887	III
450	893.897	III
10	1162.59	III
5	1162.62	III
100	1352.81	III
5	1352.82	III
70	1352.86	III
600	1379.67	III
800	1384.13	III
700	1605.766	III
100	1611.814	III
800	1611.874	III
1000	1854.716	III
600	1862.790	III
300	1935.840	III
15	1935.863	III
200	1935.949	III

Air

Intensity		Wavelength	Spectrum
110		2399.00	III
285		2762.77	III
220		2762.87	III
450		2906.93	III
360		3348.52	III
290		3350.88	III
870		3601.63	III
550		3601.93	III
750		3612.36	III
450		3702.11	III
550		3713.12	III
110	h	3980.14	III
110		4082.45	III
150		4088.61	III
110	h	4142.37	III
650		4149.92	III
650		4150.17	III
110	h	4364.64	III
650		4479.89	III
650		4479.97	III
760		4512.56	III
550		4528.94	III
870		4529.19	III
110		4701.15	III
150		4701.41	III
110	h	4904.10	III
110	h	5151.01	III
110	h	5163.89	III
1200		5696.60	III
1000		5722.73	III
110		6055.21	III
220	h	7635.37	III
150		7660.26	III
220		7681.97	III
360		7881.79	III
150		7882.52	III
290		7905.51	III
290	h	8243.59	III
360	h	8275.11	III
290		9571.52	III
360		9605.99	III

Al IV
Ref. 8, 146 — E.F.W.

Intensity	Wavelength	Spectrum
	Vacuum	
400	124.03	IV
700	129.73	IV
800	160.07	IV
700	161.69	IV
500	1027.34	IV
800	1042.17	IV
700	1048.52	IV
500	1058.90	IV
500	1061.43	IV
600	1064.89	IV
500	1066.57	IV
600	1069.44	IV
400	1105.74	IV
600	1118.82	IV
500	1125.61	IV
400	1136.82	IV
400	1198.50	IV
400	1220.55	IV
900	1237.19	IV
600	1240.21	IV
700	1240.86	IV
700	1248.79	IV
900	1257.62	IV
800	1264.18	IV
1000	1272.76	IV
400	1337.90	IV
500	1376.62	IV
500	1388.79	IV
600	1431.94	IV
700	1441.82	IV
800	1447.51	IV
600	1457.96	IV
700	1486.89	IV
800	1494.79	IV
400	1519.07	IV
800	1537.54	IV
500	1550.19	IV
1000	1557.25	IV
500	1559.03	IV
700	1564.16	IV
900	1582.04	IV
800	1584.46	IV
400	1589.28	IV
400	1606.65	IV
400	1617.81	IV
600	1627.54	IV
500	1636.82	IV
800	1639.06	IV
1000	1818.56	IV
700	1881.16	IV

Air

Intensity	Wavelength	Spectrum
400	2515.87	IV
500	3208.20	IV
500	3267.21	IV
600	3285.13	IV
400	3344.46	IV
500	3473.54	IV
900	3492.23	IV
800	3508.46	IV
500	3511.28	IV
700	3517.56	IV
400	3527.03	IV
500	3541.08	IV

Al V
Ref. 6 — E.F.W.

Intensity	Wavelength	Spectrum
	Vacuum	
300	103.80	V
400	103.88	V
250	104.07	V
250	104.18	V
600	107.95	V
300	108.06	V
300	108.11	V
250	118.50	V
900	125.53	V
800	126.07	V
800	130.41	V
1000	130.85	V
900	131.00	V
900	131.44	V
500	132.63	V
1000	278.69	V
900	281.39	V
250	1068.26	V
500	1088.67	V
300	1090.14	V
300	1150.30	V
350	1165.42	V
250	1168.48	V
500	1287.70	V
400	1330.06	V
400	1350.52	V
400	1363.35	V
600	1369.20	V
300	1373.70	V
400	1445.87	V
300	1455.26	V
600	1475.64	V
300	1486.05	V
700	1508.37	V
1000	1526.14	V
500	1539.12	V
300	1577.90	V
350	1589.87	V

AMERICIUM (Am)
Z = 95

Am I and II
Ref. 92 — J.G.C.

Intensity		Wavelength	Spectrum
		Air	
100	s	2706.35	II
100	s	2728.69	II
200	s	2756.55	II
100	l	2812.10	II
200	s	2812.92	II
1000	l	2815.28	II
100	s	2815.98	II
100	l	2831.24	II
5000	s	2832.26	II
100	l	2833.95	II
100	l	2861.92	II
100	l	2866.20	II
1000	l	2888.51	II
100	l	2893.29	II
200	l	2899.56	II
100	l	2909.86	II
200	l	2911.13	II
1000	s	2920.59	II
200	l	2927.53	II
200	l	2936.99	II
100	l	2939.08	II
500	l	2950.39	II
100	s	2957.05	II
100	l	2958.39	II
100	l	2963.02	II
1000	l	2966.71	II
1000	l	2969.29	II
1000	s	2987.24	II
500	l	2993.51	II
1000	s	3004.25	II

Americium (Cont.)

Intensity		Wavelength	
500	s	3027.99	II
100	l	3028.86	II
500	l	3038.36	II
200	s	3053.69	II
2000	s	3120.49	II
200	s	3161.83	II
100	s	3167.86	II
100	l	3203.26	II
500	s	3282.32	II
200	l	3286.67	II
100	l	3343.87	I
500	l	3362.55	I
200	l	3395.01	II
200	s	3419.66	II
200		3446.19	I
1000	l	3452.10	II
5000	l	3483.31	II
5000		3510.13	I
1000	l	3530.95	I
200	s	3562.68	II
5000		3569.16	I
100	s	3596.07	II
500		3603.41	I
5000		3673.12	I
100	l	3684.57	II
1000	l	3696.42	II
100	l	3707.86	II
5000	l	3777.50	II
5000	l	3926.25	II
1000	l	3952.58	II
100	s	4020.25	I
100	s	4035.81	I
500	l	4036.37	I
5000	s	4089.29	II
100	l	4089.32	II
100	s	4140.96	I
1000	s	4188.12	II
1000		4265.55	I
5000		4289.26	I
200	s	4309.65	II
2000	s	4324.57	II
2000	s	4441.36	II
5000	l	4509.45	II
5000	l	4575.59	II
1000	l	4593.31	II
100	l	4649.12	I
100	l	4653.45	I
5000	l	4662.79	I
2000	l	4681.65	I
2000	l	4699.70	II
1000	l	4706.80	I
2000	l	4872.22	II
200	l	4990.79	I
100	s	5000.21	I
1000	l	5020.96	II
200	l	5215.99	II
1000	o	5402.62	I
1000	l	5424.70	I
1000	l	5584.21	II
1000	s	5598.13	I
10000	l	6054.64	I
1000	l	6405.11	I
500	l	6544.16	I
500	s	6955.58	I

ANTIMONY (Sb)
Z = 51

Sb I and II
Ref. 167, 194 — L.J.R. and J.R.

Intensity		Wavelength	
		Vacuum	
1		691.20	II
1		764.43	II
1		814.85	II
1		849.39	II
2		855.08	II
4		876.84	II
4		921.07	II
6		983.57	II
6		1001.13	II
6		1009.43	II
6		1052.21	II
8		1056.27	II
8		1057.32	II
6		1073.81	II
6		1230.30	II
8		1274.98	II
8		1327.40	II
6		1358.04	II
8		1384.70	II
6		1407.83	II
10		1430.76	I
8		1436.49	II
10	h	1464.19	I
20	r	1486.57	I
40	h	1491.36	I
50	r	1512.57	I
120	r	1532.74	I
80	r	1535.06	I
6		1565.51	II
8		1576.11	II
7		1581.36	II
80	r	1599.96	I
10		1606.98	II
200	w	1612.8	I
100	w	1623.3	I
50	h	1651.20	I
20		1657.04	II
100	w	1662.6	I
50		1698.85	I
80	r	1716.93	I
150	r	1717.45	I
150	r	1723.43	I
100	r	1736.19	I
8		1736.43	II
50		1757.79	I
100	h	1765.76	I
100		1780.87	I
100	r	1788.24	I
150		1800.18	I
50	r	1810.50	I
80	r	1814.20	I
100		1829.50	I
60		1858.89	I
50	r	1868.17	I
300	r	1871.15	I
150	r	1882.56	I
70		1891.28	I
70		1899.39	I
100		1927.08	I
200	r	1950.39	I
80	h	1964.3	I
60		1986.05	I
6		1990.60	II
		Air	
50		2024.00	I
60	r	2029.49	I
70	r	2039.77	I
150	r	2049.57	I
50		2063.43	I
1000	r	2068.33	I
100		2079.56	I
50	r	2098.41	I
80	r	2118.48	I
100	r	2127.34	I
50	r	2137.05	I
100	r	2139.69	I
10		2141.80	II
50	r	2141.83	I
100	r	2144.86	I
50		2158.91	I
1500	r	2175.81	I
250	r	2179.19	I
6		2179.25	II
200	r	2201.32	I
300	r	2208.45	I
6		2208.50	II
150	r	2220.73	I
100		2221.98	I
120	r	2224.93	I
6		2225.15	II
300	r	2262.51	I
120		2288.98	I
150	r	2293.44	I
300	r	2306.46	I
2500	r	2311.47	I
150		2315.89	I
400	r	2373.67	I
300	h	2383.64	I
100		2395.22	I
150		2422.13	I
250		2426.35	I
400	r	2445.51	I
400		2478.32	I
150		2480.44	I
8		2480.46	II
100		2510.54	I
2000	r	2528.52	I
15		2528.54	II
10		2567.75	II
150		2574.06	I
1500	r	2598.05	I
500	r	2598.09	I
300	r	2612.51	I
200	r	2652.60	I
12		2656.55	II
300	r	2670.64	I
200	r	2682.76	I
120		2692.23	I
150	r	2718.90	I
400	r	2769.95	I
12		2851.09	II
100		2851.11	I
1000	r	2877.92	I
12		2966.10	II
15		2980.96	II
500	r	3029.83	I
12		3034.01	II
12		3040.67	II
600	r	3232.52	I
20		3241.28	II
700	r	3267.51	I
12		3383.09	II
100		3383.15	I
15		3498.46	II
12		3520.47	II
25		3637.80	II
250		3637.83	I
20		3722.78	II
200	r	3722.79	I
20		3850.22	II
200		4033.55	I
20		4033.56	II
20		4133.63	II
15		4140.54	II
15		4195.17	II
20		4219.07	II
20		4314.32	II
12		4344.83	II
12		4411.42	II
12		4446.48	II
12		4506.92	II
15		4514.50	II
30		4596.90	II
20		4599.09	II
15		4604.77	II
30		4647.32	II
20		4675.74	II
40		4711.26	II
12		4735.44	II
20		4757.81	II
20		4765.36	II
12		4766.91	II
30		4784.03	II
20		4802.01	II
20		4832.82	II
20		4877.24	II
15		4947.40	II
15		5044.56	II
12		5166.32	II
15		5176.55	II
20		5238.94	II
20		5354.24	II
15		5464.08	II
30	h	5490.32	I
40	h	5556.10	I
15		5568.13	II
30	l	5602.19	I
100	l	5632.02	I
30		5639.75	II
60	h	5830.34	I
15		5895.09	II
100		6005.21	II
20		6053.41	II
30		6079.80	II
50		6130.04	II
20		6154.94	II
12		6302.76	II
20		6611.49	I
30		6647.44	II
15		6688.01	II
6		6806.67	II
30	h	7648.28	I
80		7844.44	I
200		7924.65	I
40	h	7969.55	I
60		8411.69	I
150		8572.64	I
100		8619.55	I
30	h	8682.7	I
30		9132.21	I
30		9518.68	I
30		9866.78	I
400		9949.14	I
200		10078.49	I
300		10261.01	I
50		10364.33	I
50	h	10488.3	I
200		10585.60	I
1000		10677.41	I
800		10741.94	I
80		10794.11	I
600		10839.52	I
200		10868.58	I
400		10879.55	I
300		11012.79	I
40		11079.95	I
30	h	11084.98	I
50	h	11104.84	I
50	h	11108.52	I
30		11189.61	I
150		11266.23	I
30		11863.37	I
1		11957.7	I
5		12116.06	I
2		12276.6	I
2		12466.75	I

Sb III
Ref. 164 — L.J.R. and J.R.

Intensity	Wavelength	
	Vacuum	
10	691.18	III
10	698.69	III
15	722.86	III
8	724.81	III
15	732.33	III
15	999.62	III
40	1011.94	III
10	1056.58	III
40	1065.90	III
20	1069.93	III
20	1070.43	III
5	1073.76	III
30	1075.82	III
5	1078.10	III
20	1084.06	III
10	1098.34	III
10	1135.43	III
30	1151.49	III
40	1157.74	III
12	1166.96	III
50	1205.20	III
50	1210.64	III
20	1306.69	III
8	1379.58	III
20	1404.18	III
10	1429.57	III
15	1673.89	III
3	1710.23	III
15	1711.84	III
15	1725.33	III
12	1762.30	III
12	1839.32	III
10	1946.13	III
	Air	
3	2054.10	III
2	2091.85	III
5	2127.00	III
5	2507.71	III
15	2590.13	III
1	2614.20	III
12	2617.17	III
1	2617.63	III
20	2669.39	III
5	2785.87	III
20	2790.27	III
50	3336.61	III
50	3504.07	III
15	3519.06	III
15	3533.45	III
40	3559.18	III
40	3566.25	III
30	3738.90	III
40	4265.09	III
50	4352.16	III
30	4591.89	III
30	4692.91	III
1	5247.71	III
1	5690.8	III
1	5717.3	III
3	5845.5	III
5	6246.7	III
3	6287.6	III

Sb IV
Ref. 386 — L.J.R.

Intensity	Wavelength	
	Vacuum	
861.5	SB IV	1
873.5	SB IV	1
888.3	SB IV	1
891.2	SB IV	1
1087.6	SB IV	1
1099.3	SB IV	1
1115.1	SB IV	1
1120.4	SB IV	1
1145.9	SB IV	1
1151.5	SB IV	1
1171.4	SB IV	1
1192.9	SB IV	1
1199.1	SB IV	1
1499.2	SB IV	1

Sb V
Ref. 406 — L.J.R.

Intensity	Wavelength	
	Vacuum	
3	699.22	V
1	746.06	V
6	831.00	V
1	898.02	V
8	1104.32	V
12	1226.00	V
12	1505.70	V
12	1524.47	V

ARGON (Ar)
Z = 18

Ar I and II
Ref. 190, 203, 204, 219, — E.F.W.

Intensity		Wavelength	

Vacuum

Intensity		Wavelength	Spectrum
30		487.227	II
50		490.650	II
30		490.701	II
30		519.327	II
30		542.912	II
200		543.203	II
70		547.461	II
70		556.817	II
70		573.362	II
30		576.736	II
70		580.263	II
30		583.437	II
70		597.700	II
30		602.858	II
30		612.372	II
500		661.867	II
30		664.562	II
200		666.011	II
1000		670.946	II
3000		671.851	II
70		676.242	II
30		677.952	II
30		679.218	II
200		679.401	II
200		718.090	II
3000		723.361	II
500		725.548	II
70		730.930	II
200		740.269	II
200		744.925	II
70		745.322	II
20		802.859	I
100		806.471	I
60		806.869	I
30		807.218	I
40		807.653	I
50		809.927	I
120		816.232	I
70		816.464	I
80		820.124	I
120		825.346	I
120		826.365	I
150		834.392	I
100		835.002	I
100		842.805	I
180		866.800	I
150		869.754	I
180	r	876.058	I
180	r	879.947	I
150		894.310	I
1000		919.781	II
1000		932.054	II
1000	r	1048.220	I
500	r	1066.660	I

Air

Intensity		Wavelength	Spectrum
5		2420.456	II
10		2516.789	II
10		2534.709	II
15		2562.087	II
25		2891.612	II
200		2942.893	II
100		2979.050	II
50		3033.508	II
50		3093.402	II
8		3200.37	I
20		3243.689	II
25		3293.640	II
20		3307.228	II
7		3319.34	I
25		3350.924	II
7		3373.47	I
25		3376.436	II
25		3388.531	II
7		3393.73	I
7		3461.07	I
70		3476.747	II
20		3478.232	II
50		3491.244	II
100		3491.536	II
70		3509.778	II
70		3514.388	II
70		3545.596	II
70		3545.845	II
7		3554.306	I
100		3559.508	II
100		3561.030	II
70		3576.616	II
25		3581.608	II

Intensity	Wavelength	Spectrum
50	3582.355	II
70	3588.441	II
7	3606.522	I
25	3622.138	II
20	3639.833	II
35	3718.206	II
70	3729.309	II
50	3737.889	II
150	3765.270	II
50	3766.119	II
20	3770.369	I
20	3770.520	II
25	3780.840	II
25	3803.172	II
50	3809.456	II
7	3834.679	I
70	3850.581	II
35	3868.528	II
35	3925.719	II
50	3928.623	II
25	3932.547	II
70	3946.097	II
7	3947.505	I
35	3948.979	I
20	3979.356	II
35	3994.792	II
50	4013.857	II
50	4033.809	II
20	4035.460	II
150	4042.894	II
200	4072.005	II
70	4072.385	II
25	4076.628	II
35	4079.574	II
25	4082.387	II
150	4103.912	II
300	4131.724	II
35	4156.086	II
400	4158.590	I
50	4164.180	I
35	4179.297	II
50	4181.884	I
100	4190.713	I
50	4191.029	I
200	4198.317	I
400	4200.674	I
25	4218.665	II
25	4222.637	II
25	4226.988	II
100	4228.158	II
100	4237.220	II
25	4251.185	I
200	4259.362	I
100	4266.286	I
70	4266.527	II
150	4272.169	I
550	4277.528	II
20	4282.898	II
100	4300.101	I
25	4300.650	II
70	4309.239	II
200	4331.200	II
50	4332.030	II
100	4333.561	II
50	4335.338	I
25	4345.168	I
800	4348.064	II
50	4352.205	II
25	4362.066	II
50	4367.832	II
200	4370.753	II
70	4371.329	II
50	4375.954	II
150	4379.667	II
50	4385.057	II
70	4400.097	II
200	4400.986	II
400	4426.001	II
150	4430.189	II
50	4430.996	II
50	4433.838	II
20	4439.461	II
35	4448.879	II
100	4474.759	II
200	4481.811	II
100	4510.733	I
20	4522.323	I
20	4530.552	II
400	4545.052	II
20	4564.405	II
400	4579.350	II
400	4589.898	II
15	4596.097	I
550	4609.567	II
7	4628.441	I
35	4637.233	II
400	4657.901	II
15	4702.316	I
20	4721.591	II
550	4726.868	II
50	4732.053	II

Intensity	Wavelength	Spectrum
300	4735.906	II
800	4764.865	II
550	4806.020	II
150	4847.810	II
50	4865.910	II
800	4879.864	II
70	4889.042	II
20	4904.752	II
35	4933.209	II
200	4965.080	II
50	5009.334	II
70	5017.163	II
70	5062.037	II
20	5090.495	II
100	5141.783	II
70	5145.308	II
5	5151.391	I
15	5162.285	I
25	5165.773	II
20	5187.746	I
20	5216.814	II
7	5221.271	I
5	5421.352	I
10	5451.652	I
25	5495.874	I
5	5506.113	I
25	5558.702	I
10	5572.541	I
35	5606.733	I
20	5650.704	I
10	5739.520	I
5	5834.263	I
10	5860.310	I
15	5882.624	I
25	5888.584	I
50	5912.085	I
15	5928.813	I
5	5942.669	I
7	5987.302	I
5	5998.999	I
5	6025.150	I
7	6032.127	I
70	6043.223	I
35	6052.723	I
10	6059.372	I
7	6098.803	I
10	6105.635	I
100	6114.923	II
10	6145.441	I
7	6170.174	I
150	6172.278	II
10	6173.096	I
10	6212.503	I
5	6215.938	I
25	6243.120	II
7	6296.872	I
15	6307.657	I
7	6369.575	I
20	6384.717	I
70	6416.307	I
25	6483.082	II
15	6538.112	I
15	6604.853	I
25	6638.221	II
20	6639.740	II
50	6643.698	II
5	6660.676	I
5	6664.051	I
25	6666.359	II
100	6677.282	I
35	6684.293	II
150	6752.834	I
5	6756.163	I
15	6766.612	I
20	6861.269	II
150	6871.289	I
5	6879.582	I
10	6888.174	I
50	6937.664	I
7	6951.478	I
7	6960.250	I
10000	6965.431	I
150	7030.251	I
10000	7067.218	I
100	7068.736	I
25	7107.478	I
25	7125.820	I
1000	7147.042	I
15	7158.839	I
70	7206.980	I
15	7265.170	I
7	7270.664	I
2000	7272.936	I
35	7311.716	I
25	7316.005	I
5	7350.814	I
70	7353.293	I
200	7372.118	I
20	7380.426	II
10000	7383.980	I
20	7392.980	I
15	7412.337	I
10	7425.294	I

Intensity	Wavelength	Spectrum
25	7435.368	I
10	7436.297	I
20000	7503.869	I
15000	7514.652	I
25000	7635.106	I
15000	7723.761	I
10000	7724.207	I
10	7891.075	I
20000	7948.176	I
20000	8006.157	I
25000	8014.786	I
7	8053.308	I
20000	8103.693	I
35000	8115.311	I
10000	8264.522	I
20	8392.27	I
15000	8408.210	I
20000	8424.648	I
15000	8521.442	I
7	8605.776	I
4500	8667.944	I
20	8771.860	II
180	8849.91	I
20	9075.394	I
35000	9122.967	I
550	9194.638	I
15000	9224.499	I
400	9291.531	I
1600	9354.220	I
25000	9657.786	I
4500	9784.503	I
180	10052.06	I
30	10332.72	I
100	10467.177	II
1600	10470.054	I
13	10478.034	I
180	10506.50	I
200	10673.565	I
11	10681.773	I
7	10683.034	II
30	10733.87	I
30	10759.16	I
7	10812.896	II
11	11078.869	I
30	11106.46	I
12	11441.832	I
400	11488.109	I
200	11668.710	I
12	11719.488	I
200	12112.326	I
50	12139.738	I
50	12343.393	I
200	12402.827	I
200	12439.321	I
100	12456.12	I
200	12487.663	I
150	12702.281	I
30	12733.418	I
12	12746.232	I
200	12802.739	I
50	12933.195	I
500	12956.659	I
200	13008.264	I
200	13213.99	I
200	13228.107	I
100	13230.90	I
500	13272.64	I
1000	13313.210	I
1000	13367.111	I
30	13499.41	I
1000	13504.191	I
11	13573.617	I
30	13599.333	I
400	13622.659	I
200	13678.550	I
1000	13718.577	I
10	13825.715	I
10	13907.478	I
200	14093.640	I
100	15046.50	I
25	15172.69	I
10	15329.34	I
30	15989.49	I
30	16519.86	I
500	16940.58	I
12	18427.76	I
50	20616.23	I
30	20986.11	I
20	23133.20	I
20	23966.52	I

Ar III
Ref. 367, 372, 373, 375 — E.F.W.

Intensity	Wavelength	Spectrum

Vacuum

Intensity	Wavelength	Spectrum
12	769.15	III

Argon (Cont.)

Intensity	Wavelength	
10	871.10	III
9	875.53	III
12	878.73	III
8	879.62	III
9	883.18	III
10	887.40	III
7	1669.67	III
7	1673.42	III
7	1675.48	III
9	1914.40	III
7	1915.56	III

Air

Intensity	Wavelength	
10	2125.16	III
15	2133.87	III
10	2138.59	III
10	2148.73	III
15	2166.19	III
10	2168.26	III
20	2170.23	III
25	2177.22	III
8	2184.06	III
10	2188.22	III
15	2192.06	III
7	2248.73	III
10	2279.10	III
7	2281.22	III
7	2282.21	III
12	2293.03	III
10	2300.85	III
15	2302.17	III
9	2317.00	III
15	2317.47	III
12	2318.04	III
10	2319.13	III
10	2319.37	III
9	2345.17	III
7	2351.67	III
9	2360.26	III
10	2395.63	III
12	2399.15	III
10	2413.20	III
7	2415.61	III
10	2418.82	III
12	2423.52	III
12	2423.93	III
7	2443.69	III
8	2472.95	III
7	2476.10	III
12	2488.86	III
7	2631.90	III
10	2654.63	III
8	2674.02	III
9	2678.38	III
10	2724.84	III
7	2762.23	III
7	2842.88	III
8	2855.29	III
9	2884.12	III
10	3010.02	III
12	3024.05	III
12	3054.82	III
10	3064.77	III
10	3078.15	III
7	3110.41	III
7	3127.90	III
25	3285.85	III
20	3301.88	III
15	3311.25	III
7	3323.59	III
25	3336.13	III
20	3344.72	III
15	3358.49	III
7	3361.28	III
15	3391.85	III
7	3417.49	III
9	3424.25	III
8	3438.04	III
9	3471.32	III
20	3480.55	III
12	3499.67	III
15	3503.58	III
8	3511.12	III
20	3795.37	III
10	3858.32	III
7	3907.84	III
8	3960.53	III
6	4023.60	III
5	4146.70	III

Ar IV
Ref. 367, 368, 374 —
E.F.W.

Intensity	Wavelength	
	Vacuum	
4	396.87	IV
4	398.55	IV
6	623.77	IV

Intensity	Wavelength	
10	683.28	IV
7	688.39	IV
12 p	689.01	IV
6	699.41	IV
8	700.28	IV
4	754.20	IV
5	761.47	IV
5	800.57	IV
10	801.09	IV
10	801.41	IV
5	801.91	IV
15	840.03	IV
20	843.77	IV
25	850.60	IV
5	900.36	IV
9	901.17	IV

Air

Intensity	Wavelength	
4	2299.72	IV
8	2447.71	IV
12	2513.28	IV
6	2518.40	IV
9	2525.69	IV
12	2562.17	IV
10	2568.07	IV
7	2569.53	IV
12	2599.47	IV
10	2608.06	IV
7	2608.44	IV
12	2615.68	IV
6	2619.98	IV
12	2621.36	IV
12	2624.92	IV
15	2640.34	IV
9	2682.63	IV
14	2757.92	IV
10	2776.26	IV
12	2784.47	IV
14	2788.96	IV
7	2797.11	IV
16	2809.44	IV
10	2830.25	IV
6	2874.40	IV
12	2913.00	IV
11	2926.33	IV
6	3037.98	IV
8	3077.40	IV

Ar V
Ref. 414, 421 — E.F.W.

Intensity	Wavelength	
	Vacuum	
3	336.56	V
3	337.56	V
6	338.00	V
2	338.43	V
2	339.01	V
3	339.89	V
3	350.88	V
2	436.67	V
5	446.00	V
8	446.95	V
4	447.53	V
18	449.06	V
4	449.49	V
3	458.12	V
2	458.98	V
6 p	461.23	V
3	462.42	V
7	463.94	V
3	522.09	V
5	524.19	V
6	527.69	V
2	554.50	V
5	558.48	V
3	635.12	V
3	705.35	V
5	709.20	V
4	715.60	V
3	715.65	V
2	725.11	V
4	822.16	V
5	827.05	V
3	827.35	V
4 p	834.88	V
2	836.13	V

ARSENIC (As)
Z = 33

As I and II
Ref. 168, 197 — R.L.K.

Intensity		Wavelength	
		Vacuum	
165		761.24	II
165		802.83	II
340		1021.96	II
340		1082.35	II
500		1139.40	II
615		1149.31	II
555		1181.51	II
555		1189.87	II
615		1196.38	II
615		1196.56	II
340		1207.44	II
800		1211.17	II
800		1218.10	II
340		1223.15	II
760		1241.31	II
965		1243.08	II
870		1245.67	II
800		1258.58	II
965		1263.77	II
800		1266.34	II
800		1267.59	II
715		1280.99	II
715		1287.54	II
715		1305.70	II
340		1307.74	II
760		1333.15	II
965		1341.55	II
760		1355.93	II
965		1369.77	II
800		1373.65	II
1000		1375.07	II
760		1375.78	II
800		1394.64	II
800		1400.31	II
500		1448.59	II
500		1558.88	II
500		1570.99	II
100	r	1593.60	I
500		1660.55	II
100		1758.60	I
170		1806.15	I
340		1860.34	II
1000	r	1890.42	I
500		1912.94	II
800	r	1937.59	I
585	r	1972.62	I
170	r	1990.35	I
100		1991.13	I
100	r	1995.43	I

Air

Intensity		Wavelength	
230	r	2003.34	I
100	r	2009.19	I
100		2013.32	I
100		2112.99	I
100		2144.08	I
135		2165.52	I
350	r	2288.12	I
350	r	2349.84	I
100	r	2370.77	I
135	r	2381.18	I
170	r	2456.53	I
340		2602.00	II
170	r	2780.22	I
300		2830.359	II
300		2831.164	II
100	r	2860.44	I
300		2884.406	II
615		2959.572	II
300		3003.819	II
300		3116.516	II
340		3842.60	II
715		4190.082	II
615		4197.40	II
615		4242.982	II
500		4315.657	II
500		4323.867	II
500		4336.64	II
715		4352.145	II
425		4352.864	II
375		4371.17	II
615		4427.106	II
615		4431.562	II
715		4458.469	II
340		4461.075	II
715		4466.348	II
500		4474.46	II
800		4494.230	II
850		4507.659	II
615		4539.74	II
715		4543.483	II
615		4602.427	II
340		4629.787	II
340		4707.586	II
340		4730.67	II
340		4888.557	II
100		5068.98	I
340		5105.58	II

Intensity	Wavelength	
500	5107.55	II
100	5121.34	I
100	5141.63	I
425	5231.38	II
500	5331.23	II
100	5408.13	I
135	5451.32	I
340	5497.727	II
425	5558.09	II
425	5651.32	II
425	6110.07	II
500	6170.27	II
300	6511.74	II
300	7092.27	II
300	7102.72	II
340	7990.53	II
300	8174.51	II
100	8428.91	I
100	8564.71	I
100	8654.14	I
135	8821.73	I
100	8869.66	I
135	9267.28	I
200	9300.61	I
230	9597.95	I
290	9626.70	I
230	9833.76	I
100	9886.05	I
140	9900.55	I
170	9915.71	I
290	9923.05	I
100	10010.63	I
290	10024.04	I
100	10453.09	I
100	10575.02	I
170	10614.07	I

As III
Ref. 163 — R.L.K.

Intensity	Wavelength	
	Vacuum	
185	849.9	III
185	866.3	III
510	871.7	III
325	889.0	III
325	927.5	III
325	953.6	III
325	937.2	III
325	963.8	III
120	1172.2	III
185	1209.3	III

Air

Intensity	Wavelength	
80	2926.3	III
185	2982.0	III
325	3922.6	III
185	4037.2	III

As IV
Ref. 244 — R.L.K.

Intensity	Wavelength	
	Air	
150	2253.1	IV
200	2263.2	IV
200	2301.0	IV
250	2417.5	IV
150	2446.1	IV
250	2454.0	IV
200	2461.4	IV
150	3108.8	IV

As V
Ref. 280 — R.L.K.

Intensity	Wavelength	
	Vacuum	
25	600.7	V
40	616.0	V
120	715.5	V
150	734.8	V
60	737.2	V
250	987.7	V
250	1029.5	V
40	1051.6	V
60	1056.6	V

ASTATINE (At)
Z = 85

At I
Ref. 188 — E.F.W.

Astatine (Cont.)

Intensity	Wavelength	
	Air	
8	2162.25	I
10	2244.01	I

BARIUM (Ba)
Z = 56

Ba I and II
Ref. 1, 252, 277, 279 — J.J.W.

Intensity	Wavelength	
	Vacuum	
200	1486.72	II
400	1504.01	II
300	1554.38	II
200	1572.73	II
	1573.92	II
	1630.40	II
100	1674.51	II
400	1694.37	II
	1697.16	II
	1761.75	II
	1771.03	II
	1786.93	II
100	1904.15	II
500	1924.70	II
	1985.60	II
300	1999.54	II
	Air	
	2009.20	II
400	2023.95	II
	2052.68	II
	2054.57	II
500	2214.7	II
800	2245.61	II
1000	2254.73	II
1400	2304.24	II
2000	2335.27	II
190	2347.58	II
60	2528.51	II
8 h	2596.64	I
100	2634.78	II
8	2702.63	I
18	2771.36	II
15	2785.28	I
100 r	3071.58	I
10 h	3108.21	I
8	3132.60	I
8 h	3135.72	I
10	3137.70	I
10	3155.34	I
10	3155.67	I
12	3158.05	I
12 h	3158.54	I
25	3165.60	I
15 h	3173.69	I
30	3183.16	I
15	3183.96	I
10	3193.91	I
25 h	3203.70	I
30	3221.63	I
40	3222.19	I
50	3261.96	I
60 r	3262.34	I
40	3281.50	I
15	3281.77	I
50	3322.80	I
80 h	3356.80	I
60 r	3377.08	I
20	3377.39	I
70 r	3420.32	I
25	3421.01	I
30 h	3421.48	I
40	3463.74	I
200 r	3501.11	I
80 h	3524.97	I
30 h	3531.35	I
80 h	3544.66	I
20 h	3547.68	I
100	3552.45	II
200	3567.73	II
100	3576.28	II
30	3577.62	I
80 h	3579.67	I
200	3596.57	II
40	3630.64	I
40 h	3636.83	I
20 h	3688.47	I
400	3735.75	II
200	3816.69	II
200	3842.80	II
100	3854.76	II
20	3889.33	I
1400 l	3891.78	II
20	3892.65	I
40	3909.91	I
500	3914.73	II
50	3935.72	I
20	3937.87	I
200	3939.67	II
500	3949.51	II
80	3993.40	I
30	3995.66	I
300	4036.26	II
200	4083.77	II
30 h	4084.86	I
1500 h	4130.66	II
20	4132.43	I
200	4166.00	II
500	4216.04	II
800	4267.95	II
100	4283.10	I
300	4287.80	I
200	4297.60	II
800	4309.32	II
20 h	4323.00	I
600	4325.73	II
200	4326.74	II
300	4329.62	II
80	4350.33	I
60	4402.54	I
400	4405.23	II
60 h	4488.98	I
40	4431.89	I
50 h	4493.64	I
40	4505.92	I
200	4509.63	II
60 h	4523.17	I
130	4524.93	II
65000	4554.03	II
40	4573.85	I
80	4579.64	I
30	4599.75	I
20 h	4619.92	I
25 h	4628.33	I
300	4644.10	II
30	4673.62	I
35	4691.62	I
20	4700.43	I
800	4708.94	II
40	4726.44	I
800	4843.46	II
300	4847.14	II
200	4850.84	II
30 h	4877.65	II
400	4899.97	II
15	4902.90	I
20000	4934.09	II
8	4947.35	I
1000	4957.15	II
300	4997.81	II
1000	5013.00	II
20 h	5159.94	I
20	5267.03	I
800	5361.35	II
1000	5391.60	II
200	5421.05	II
100	5424.55	I
200	5428.79	II
300	5480.30	II
200	5519.05	I
1000 r	5535.48	I
20 h	5620.40	I
10	5680.18	I
400	5777.62	I
800	5784.18	II
100	5800.23	I
20	5805.69	I
150	5826.28	I
2800	5853.68	II
15	5907.64	I
100	5971.70	I
800	5981.25	II
100	5997.09	I
300	5999.85	II
100	6019.47	I
200	6063.12	I
300	6110.78	I
400	6135.83	II
20000	6141.72	II
150	6341.68	I
500	6378.91	I
90	6450.85	I
150	6482.91	I
12000	6496.90	II
300	6498.76	I
150	6527.31	I
3000	6595.33	I
150	6654.10	I
1500	6675.27	I
1800	6693.84	I
1000	6769.62	II
600	6865.69	I
300 h	6867.85	I
1000	6874.09	II
6000	7059.94	I
2400 hS	7120.33	I
600	7195.24	I
600 hL	7228.84	I
3000	7280.30	I
1200	7392.41	I
300	7417.53	I
900 hL	7459.78	I
600	7488.08	I
450 hL	7636.90	I
600 hL	7642.91	I
1800	7672.09	I
1200	7780.48	I
180 h	7839.57	I
1500	7905.75	I
600	7911.34	I
900	8210.24	I
1800 h	8559.97	I
100	8710.74	II
100	8737.71	II
300 h	8799.76	I
300	8860.98	I
450	8914.99	I
300	9219.69	I
300	9308.08	I
300 h	9324.58	I
1500	9370.06	I
300	9455.92	I
450	9589.37	I
900	9608.88	I
300 h	9645.72	I
1500 hL	9830.37	I
900	10001.08	I
600	10032.10	I
1200 h	10233.23	I
300	10471.26	I
120 hL	10791.25	I
180 hL	11012.69	I
150 h	11114.42	I
240	11303.04	I
120 h	11697.45	I
120	13207.30	I
120	13810.50	I
120	14077.90	I
120	15000.40	I
120	20712.00	I
150	25515.70	I
150	29223.90	I

Ba III
Ref. III — J.J.W.

Intensity	Wavelength	
	Vacuum	
5	403.82	III
2	407.12	III
7	420.12	III
4	423.84	III
9	448.95	III
8	456.96	III
14	555.48	III
14	587.57	III
18	647.27	III
9	653.36	III
15	743.12	III
12	1097.41	III
15	1113.67	III
11	1116.01	III
14	1133.05	III
12	1151.76	III
12	1170.62	III
13	1207.29	III
11	1218.92	III
12	1224.55	III
12	1288.53	III
11	1299.18	III
12	1307.40	III
12	1308.87	III
12	1315.72	III
12	1334.01	III
11	1354.71	III
12	1369.53	III
11	1416.61	III
12	1478.85	III
12	1510.68	III
12	1514.22	III
12	1565.61	III
12	1566.12	III
12	1574.55	III
12	1596.80	III
12	1610.95	III
12	1615.78	III
12	1711.53	III
12	1861.74	III
12	1883.92	III
11	1974.76	III
	Air	
10	2001.30	III
15	2008.40	III
13	2022.45	III
10	2038.84	III
12	2070.43	III
12	2071.68	III
10	2076.00	III
12	2081.35	III
10	2134.87	III
16	2156.37	III
10	2160.76	III
20	2230.33	III
30	2280.68	III
35	2323.51	III
60	2331.10	III
25	2476.73	III
25	2505.07	III
40	2512.28	III
40	2523.83	III
25	2530.92	III
50	2559.54	III
25	2570.48	III
40	2681.89	III
30	2745.78	III
25	2938.95	III
25	2960.05	III
30	2962.48	III
20	3014.22	III
30	3043.42	III
40	3079.14	III
30	3103.92	III
30	3119.22	III
30	3152.70	III
25	3195.17	III
25	3235.04	III
25	3281.65	III
20	3286.79	III
50	3368.18	III
30	3369.68	III
25	3649.18	III
25	3926.85	III
25	3993.06	III
18 p	4053.71	III
15	4697.44	III
10	5049.55	III
10	5097.54	III
12 p	5102.25	III
10	5134.54	III
10	5998.00	III
13	6101.99	III
10	6377.11	III
10	6383.76	III
8	6526.17	III
8	7095.49	III
8	8308.69	III
8	9521.76	III

Ba IV
Ref. 78 — J.J.W.

Intensity	Wavelength	
	Vacuum	
40000	794.89	IV
50000	923.74	IV

Ba V
Ref. 259—J.R.

Intensity	Wavelength	
	Vacuum	
15	612.55	V
5	658.11	V
5	681.09	V
300	719.86	V
150	721.85	V
100	760.45	V
1000	766.87	V
100	783.61	V
100	816.41	V
40	875.69	V
300	877.41	V
15	892.28	V
200	946.26	V

BERKELIUM (Bk)
Z = 97

Bk I and II
Ref.53, 339—J.G.C.

Intensity	Wavelength	
	Air	
10000 s	2748.02	II
10000 s	2827.57	II
10000 l	2872.11	II
10000 s	2878.57	II
10000 l	2884.77	II
10000 l	2889.80	II
10000 s	2893.66	II
10000 s	2910.65	II
10000 s	2926.49	
10000	2927.91	
10000 l	2941.71	II
10000 l	2951.76	II
10000 l	2969.13	II
10000 l	2987.76	II
10000 l	3178.47	II
10000	3239.72	I
10000 s	3247.26	II
10000	3252.19	I
10000 s	3263.47	II
10000 l	3288.75	I

Intensity	Wavelength	Ion
10000	3289.35	I
10000 s	3302.35	II
10000	3335.26	I
10000 s	3387.45	II
10000	3408.28	I
10000 l	3412.13	II
10000	3426.95	I
10000 l	3432.62	I
10000 l	3437.47	I
10000	3442.66	I
10000 s	3461.24	II
10000 l	3453.90	I
10000	3464.13	II
10000 s	3472.02	II
10000 s	3477.62	II
10000	3528.72	I
10000 l	3531.40	I
10000 l	3535.73	I
10000 s	3542.19	II
10000	3553.60	I
10000 s	3555.88	I
10000	3556.52	I
10000	3565.41	I
10000 l	3567.25	II
10000	3590.32	I
10000	3595.88	I
10000	3601.12	I
10000 s	3603.20	II
10000	3604.78	I
10000 l	3608.49	I
10000	3609.61	I
10000	3611.03	I
10000 l	3611.93	I
10000	3613.91	I
10000	3616.62	I
10000	3619.37	I
10000 s	3621.81	I
10000	3627.61	I
10000	3633.28	I
10000 l	3637.05	I
10000	3640.26	I
10000	3640.93	II
10000 l	3675.59	I
10000 s	3681.22	II
10000	3684.43	I
10000 s	3685.21	I
10000 l	3686.74	I
10000	3692.73	I
10000	3695.37	I
10000	3703.28	I
10000 s	3704.02	I
10000	3705.26	I
10000 s	3711.14	II
10000	3712.93	I
10000	3725.39	I
10000	3739.92	I
10000	3743.05	I
10000	3745.40	I
10000	3750.08	I
10000	3751.91	I
10000	3757.35	I
10000	3757.85	I
10000 s	3771.06	II
10000	3780.72	I
10000	3781.17	I
10000	3785.38	I
10000	3788.21	I
10000	3791.42	I
10000	3796.21	I
10000	3797.12	
10000	3798.63	I
10000	3801.08	II
10000 s	3802.35	I
10000	3802.47	I
10000	3815.29	I
10000 s	3823.10	II
10000 s	3824.08	II
10000	3825.19	I
10000 s	3825.84	II
10000	3827.41	I
10000	3830.55	I
10000 l	3831.57	II
10000 s	3833.48	
10000 s	3835.97	II
10000	3842.19	I
10000 l	3846.62	I
10000	3847.63	I
10000	3855.03	I
10000 l	3859.89	I
10000 l	3877.94	II
10000	3880.11	I
10000 l	3882.60	I
10000 l	3894.55	II
10000 s	3906.09	I
10000 l	3912.16	II
10000 l	3916.37	II
10000	3921.42	I
10000 l	3928.05	I
10000 l	4147.13	II
10000 s	4189.69	II
10000 s	4197.44	II
10000 l	4329.58	I
10000	4351.50	I
10000	4363.64	I
10000	4423.01	I
10000	4466.46	I
10000	4685.70	
10000 l	4765.40	II
10000 s	5056.73	I
10000 l	5118.24	I
10000 s	5135.53	II
10000	5170.61	I
10000	5197.55	I
10000	5212.53	I
10000	5271.95	I
10000	5392.03	I
10000	5394.24	I
10000 l	5404.62	I
10000	5449.63	I
10000 l	5467.47	I
10000 s	5484.58	I
10000	5512.22	II
10000 l	5537.93	I
10000	5556.80	I
10000 s	5557.09	I
10000	5581.21	I
10000	5656.54	I
10000	5659.03	I
10000	5702.24	I
10000	5910.71	I
10000 l	7040.85	I
10000	7107.85	I
10000 l	7176.22	I
10000	7249.26	
10000	7252.50	I
10000 l	7257.21	I
10000	7306.94	I
10000	7394.26	
10000 l	7511.26	I
10000	7551.12	I
10000	7579.77	I
10000	7729.93	I
10000 s	7903.90	I
10000 s	9319.30	I
10000 s	9429.13	I
10000 l	9801.18	I
10000 l	9862.39	II
10000 l	9879.29	I
10000 l	9892.38	I
10000	10126.20	I
10000	10186.58	I
10000	10292.44	I
10000 l	10527.71	I
10000 l	10570.53	I
10000 l	11293.14	I
10000 l	11500.30	I
10000 s	11575.34	I
10000 s	11793.09	I
10000 l	12159.05	I
10000	13061.13	I
10000	13498.36	I
10000 s	14196.93	I
10000	15136.10	I
10000	18352.31	I
10000 l	19273.87	I
10000 s	19653.22	I
10000 s	23902.85	I
10000 s	24192.62	I

BERYLLIUM (Be)
Z = 4

Be I and II
Ref. 15, 44, 115, 134, 135, 198, 335 — J.O.S.

Intensity	Wavelength	Ion

Vacuum

Intensity	Wavelength	Ion
	82.58	II
	83.66	II
	89.16	I
	89.80	II
	90.04	II
	90.21	I
	90.67	I
	91.06	I
	91.36	II
	91.74	II
	92.19	I
	92.61	II
	93.14	II
	93.42	II
	93.93	II
	94.78	II
	95.76	II
	96.29	I
	97.24	I
	97.44	I
	97.86	I
	97.97	I
	98.12	I
	98.37	I
	98.66	I
	98.94	I
	99.19	I
	100.86	I
	101.20	I
	102.13	I
	102.49	II
	104.40	II
	104.67	I
	105.80	I
	107.26	I
	107.38	I
	714.0	II
	725.71	II
	743.58	II
8	775.37	II
20	842.06	II
	865.3	II
2	925.25	II
10	943.56	II
10	973.27	II
	981.4	II
	1020.1	II
8	1026.93	II
5	1036.32	II
15	1048.23	II
20	1143.03	II
	1155.9	II
60	1197.19	II
	1426.12	I
	1491.76	II
20	1512.30	II
60	1512.43	II
100	1661.49	II
15	1776.12	II
20	1776.34	II
	1907.	I
	1909.0	II
	1912.	I
	1919.	I
5	1929.67	I
10	1943.68	I
	1956.	I
50	1964.59	I
5	1985.13	I
	1997.95	I
	1997.98	I
60	1998.01	I

Air

Intensity	Wavelength	Ion
	2033.25	I
	2033.28	I
	2033.38	I
	2055.90	I
	2056.01	I
	2125.57	I
	2125.68	I
	2145.	I
	2174.99	I
	2175.10	I
	2273.5	II
	2324.6	II
	2337.0	I
2	2348.61	I
16	2350.66	I
20	2350.71	I
	2350.83	I
35	2413.34	II
35	2413.46	II
100	2453.84	II
16	2480.6	I
5	2494.54	I
20	2494.58	I
100	2494.73	I
60	2507.43	II
	2617.99	II
5	2618.13	II
20	2650.45	I
100	2650.55	I
6	2650.62	I
40	2650.69	I
100	2650.76	I
3	2697.46	II
2	2697.58	II
10	2728.88	II
10	2738.05	I
20	2764.2	II
	2898.13	I
	2898.19	I
10	2898.25	I
10	2986.06	I
20	2986.42	I
	3019.3	I
	3019.49	I
	3019.53	I
	3019.60	I
	3046.52	II
	3046.69	II
	3090.3	I
	3110.81	I
	3110.92	I
	3110.99	I
	3120.	I
480	3130.42	II
320	3131.07	II
	3136.	I
	3150.	I
	3160.6	I
	3163.	I
	3168.	I
	3180.7	II
	3187.	I
20	3193.81	I
20	3197.10	II
30	3197.15	II
20	3208.60	I
	3220.	I
60	3229.63	I
2	3233.52	II
10	3241.62	II
30	3241.83	II
15	3269.02	I
100	3274.58	II
30	3274.67	II
30	3282.91	I
30	3321.01	I
30	3321.09	I
220	3321.34	I
20	3345.43	I
60	3367.63	I
	3405.6	II
5	3451.37	I
300	3455.18	I
20	3476.56	I
300	3515.54	I
10	3555.	I
100	3736.30	I
700	3813.45	I
40	3865.13	I
80	3865.42	I
	3865.51	I
6	3865.72	I
100	3866.03	I
100	4253.05	I
60	4253.76	I
300	4360.66	II
500	4360.99	II
400	4407.94	I
	4526.6	I
	4548.	I
12	4572.66	I
700	4673.33	II
1000	4673.42	II
6	4709.37	I
200	4828.16	II
40	4849.16	I
2 h	4858.22	II
80	5087.75	I
8	5218.12	II
20	5218.33	II
3	5255.86	II
64	5270.28	II
500	5270.81	II
20	5403.04	II
20	5410.21	II
	5558.	I
	6229.11	I
10	6279.43	II
16	6279.73	II
30	6473.54	I
60	6547.89	II
60	6558.36	II
30	6564.52	II
2 h	6636.44	II
1	6756.72	II
2	6757.13	II
30	6786.56	I
1 h	6884.22	I
6 h	6884.44	I
100	6982.75	I
6 h	7154.40	I
40 h	7154.65	I
100	7209.13	I
3	7401.20	II
2	7401.43	II
10	7551.90	I
10 h	7618.68	I
20 h	7618.88	I
60	8090.06	I
5 h	8158.99	I
10 h	8159.24	I
4	8254.07	I
10 h	8287.07	I
30	8547.36	I
60	8547.67	I
300	8801.37	I
6	8882.18	I
40	9190.45	I
20 h	9243.92	I
1 h	9343.89	II
40	9392.74	I
2	9476.43	II
16	9477.03	II
20	9847.32	I
10 h	9895.63	I
20 h	9895.96	I
80	9939.78	I
16	10095.52	II

Beryllium (Cont.)

Intensity		Wavelength	
20		10095.73	II
60		10119.92	II
80		10331.03	I
30		11066.46	I
		11173.	II
1		11173.73	II
120		11496.39	I
2	h	11625.16	II
		11659.	II
2		11660.25	II
100		12095.36	II
30		12098.18	II
100		14643.92	I
60		14644.75	I
200		16157.72	I
80		17855.38	I
120		17856.63	I
100		18143.54	I
160		31775.05	I
200		31778.70	I

Be III
Ref. 73, 102, 175 — J.O.S.

Intensity		Wavelength	
		Vacuum	
1	h	76.10	III
2		76.48	III
3		78.53	III
4		78.66	III
1	h	78.92	III
5		81.89	III
10		82.38	III
20		83.20	III
30		84.76	III
50		88.31	III
100		100.25	III
3		509.99	III
2		549.31	III
6		582.08	III
4		661.32	III
8		675.59	III
4		725.59	III
7		746.23	III
2		767.75	III
1		1114.69	III
2		1213.12	III
1		1214.32	III
2		1362.25	III
1		1401.52	III
10		1421.26	III
5		1422.86	III
1		1435.17	III
2		1440.77	III
2	h	1754.69	III
3		1917.03	III
60	h	1954.97	III

Intensity		Wavelength	
		Air	
75	h	2076.94	III
60	h	2080.38	III
25		2118.56	III
15	h	2122.27	III
15	h	2127.20	III
5		2137.25	III
5		2191.57	III
100		3720.36	III
		3720.92	III
		3722.98	III
90	h	4249.14	III
2		4485.52	III
100	h	4487.30	III
1		4495.09	III
140	h	4497.8	III
140	h	6142.01	III

Be IV
Ref. 171 — J.O.S.

Intensity	Wavelength	
	Vacuum	
	58.13	IV
	58.57	IV
	59.32	IV
	60.74	IV
	64.06	IV
	75.93	IV

BISMUTH (Bi)
Z = 83

Bi I and II
Ref. 1, 357-359 — C.H.C.

Intensity		Wavelength	
		Vacuum	
15		1058.88	II
20		1085.47	II
10		1099.20	II
8		1163.19	II
8		1167.06	II
10		1225.43	II
15		1232.78	II
10		1241.05	II
10		1265.35	II
15		1283.73	II
10		1306.18	II
20		1325.46	II
20		1329.47	II
20		1350.07	II
25		1372.61	II
15		1376.02	II
20		1393.92	II
45		1436.83	II
25		1447.94	II
50		1455.11	II
25		1462.14	II
35		1486.93	II
20		1502.50	II
40		1520.57	II
40		1533.17	II
30		1536.77	II
35		1538.06	II
20		1563.67	II
40		1573.70	II
60		1591.79	II
25		1601.58	II
40		1609.70	II
40		1611.38	II
20		1652.81	II
20		1749.29	II
80		1777.11	II
60		1787.47	II
70		1791.93	II
70		1823.80	II
100		1902.41	II
9000		1954.53	I
7000		1960.13	I
25		1989.35	II

Intensity		Wavelength	
		Air	
7000		2021.21	I
9000		2061.70	I
45	h	2068.9	II
4600		2110.26	I
2500		2133.63	I
15		2143.40	II
15		2143.46	II
60		2186.9	II
40	h	2214.0	II
360		2228.25	I
1700		2230.61	I
340		2276.58	I
16		2368.12	II
12		2368.25	II
190		2400.88	I
10		2501.0	II
25		2515.69	I
70		2524.49	I
20	h	2544.5	II
700		2627.91	I
12		2693.0	II
280	c	2696.76	I
20		2713.3	II
140	d	2730.50	I
360		2780.52	I
15		2803.42	II
11		2803.70	II
12		2805.3	II
140	c	2809.62	I
4000		2897.98	I
15		2936.7	II
3200		2938.30	I
20		2950.4	II
12		2963.4	II
2800		2989.03	I
700		2993.34	I
2400		3024.64	I
60		3034.87	I
9000	c	3067.72	I
140		3076.66	I
550	c	3397.21	I
10		3430.83	II
12		3431.23	II
500	c	3510.85	I
380	c	3596.11	I
12		3654.2	II
70	h	3792.5	II
12		3811.1	II
20		3815.8	II
10		3845.8	II
30		3863.9	II
40	h	4079.1	II
10		4097.2	II
140		4121.53	I
140		4121.86	I
75	h	4259.4	II
25		4272.0	II
70	h	4301.7	II
12	h	4339.8	II
25	h	4340.5	II
12	h	4379.4	II
25	h	4476.8	II
60	h	4705.3	II
600	c	4722.52	I
30		4730.3	II
20		4749.7	II
12		4908.2	II
10		4916.6	II
12		4969.7	II
20		4993.6	II
10		5091.6	II
50	h	5124.3	II
60	h	5144.3	II
20		5201.5	II
75	h	5209.2	II
40	h	5270.3	II
10		5397.8	II
10	c	5552.35	I
3		5599.41	I
20		5655.2	II
40	h	5719.2	II
6		5742.55	I
12		5818.3	II
20		5860.2	II
20		5973.0	II
15		6059.1	II
15		6128.0	II
6		6134.82	I
3		6475.73	I
3		6476.24	I
15		6497.7	II
10		6577.2	II
40	h	6600.2	II
50	h	6808.6	II
4	h	6991.12	I
12		7033.	II
2		7036.15	I
10	h	7381.	II
2		7502.33	I
10	h	7637.	II
10		7750.	II
3		7838.70	I
2		7840.33	I
20		7965.	II
12	h	8050.	II
15		8328.	II
15		8388.	II
30		8532.	II
2		8544.54	I
1		8579.74	I
25		8653.	II
2		8754.88	I
3		8761.54	I
25		8863.	II
2		8907.81	I
2000	d	9657.04	I
40		9827.78	I
20		10104.5	I
15		10138.8	I
20		10300.6	I
20		10536.19	I
50		11072.44	I
15		11551.6	I
1500	d	11710.37	I
40		11999.49	I
200		12165.08	I
10		12374.64	I
200		12690.04	I
100		12817.8	I
200		14330.5	I
50		16001.5	I
60		22551.6	I

Bi III
Ref. 359 — C.H.C.

Intensity		Wavelength	
		Vacuum	
1		590.73	III
5		670.76	III
4		775.16	III
6		803.65	III
7		920.93	III
6		925.48	III
25		1039.99	III
50	h	1045.76	III
30		1051.81	III
20		1139.01	III
15		1145.91	III
50		1224.64	III
40		1326.84	III
60		1346.12	III
35		1423.33	III
35		1423.52	III
60	h	1461.00	III
60	h	1606.40	III
20	h	1691.5	III
20		1834.32	III
10		1863.9	III
10		1912.12	III
10		1988.26	III

Intensity		Wavelength	
		Air	
20		2020.75	III
20		2021.15	III
10		2073.22	III
14		2073.37	III
15		2103.42	III
30		2213.55	III
75	h	2414.6	III
10		2437.6	III
30	h	2847.4	III
80	h	2855.6	III
35		3115.0	III
40	h	3451.0	III
40		3473.8	III
35		3485.5	III
15		3540.8	III
45		3613.4	III
50		3695.32	III
50		3695.68	III
12		4224.6	III
25		4327.8	III
30		4560.84	III
30		4561.54	III
40	h	4797.4	III
45	h	5079.3	III
12		6623.4	III
10	h	7381.	III
12		7551.	III
25		7598.	III
10	h	7637.	III
40		8008.	III
50		8070.	III
20		8100.	III
15		8671.	III
20		8934.	III

Bi IV
Ref. 360 — C.H.C.

Intensity	Wavelength	
	Vacuum	
6	420.7	IV
6	431.2	IV
6	790.5	IV
6	790.6	IV
8	792.5	IV
10	820.3	IV
9	822.9	IV
12	824.9	IV
15	872.6	IV
8	876.8	IV
9	916.7	IV
12	923.9	IV
15	943.3	IV
9	967.6	IV
8	968.8	IV
8	989.8	IV
24	1103.4	IV
7	1128.8	IV
6	1138.6	IV
6	1139.8	IV
7	1149.7	IV
60	1317.0	IV
30	1910.0	IV

Intensity	Wavelength	
	Air	
30	2093.	IV
100	2311.	IV
100	2326.	IV
100	2376.	IV
100	2629.	IV
100	2677.	IV
100	2767.	IV
100	2772.	IV
100	2786.	IV
100	2842.	IV
100	2924.	IV
100	2933.	IV
100	2936.	IV
100	3012.	IV
100	3042.	IV
100	3239.	IV
100	3643.	IV
100	3682.	IV
100	3734.	IV
100	3868.	IV
30	4342.	IV

Intensity	Wavelength	Spectrum
30	5347.	IV

Bi V
Ref. 361 — C.H.C.

Intensity	Wavelength (Vacuum)	
1	355.77	V
1	369.52	V
1	429.78	V
1	435.63	V
2	488.39	V
1	492.72	V
3	563.62	V
2	678.87	V
6	686.88	V
1	706.54	V
5	730.71	V
10	738.17	V
6	849.86	V
5	855.68	V
15 d	864.45	V
6	880.17	V
6	929.81	V
15 d	1139.46	V

BORON (B)
Z = 5

B I and II
Ref. 66, 104, 171, 222 — R.L.K.

Intensity	Wavelength (Vacuum)	
70	693.95	II
40	731.36	II
40	731.44	II
110	882.54	II
110	882.68	II
40	984.67	II
110	1081.88	II
110	1082.07	II
110	1230.16	II
220	1362.46	II
70	1600.46	I
120	1600.73	I
160	1623.58	II
110	1623.77	II
220	1624.02	II
70	1624.16	II
160	1624.34	II
100	1663.04	I
150	1666.87	I
200	1667.29	I
150	1817.86	I
200	1818.37	I
300	1825.91	I
300	1826.41	I
110	1842.81	II

Intensity	Wavelength (Air)	
250	2066.38	I
250	2066.65	I
100	2066.93	I
300	2067.19	I
500	2088.91	I
500	2089.57	I
70	2220.30	II
40	2323.03	II
40	2328.67	II
40	2393.20	II
220	2395.05	II
40	2459.69	II
40	2459.90	II
1000	2496.77	I
1000	2497.73	I
160	2918.08	II
110	3032.26	II
70	3179.33	II
110	3323.18	II
110	3323.60	II
450	3451.29	II
285	4121.93	II
110	4194.79	II
110	4472.10	II
110	4472.85	II
70	4784.21	II
110	4940.38	II
110	6080.44	II
70	6285.47	II
70	7030.20	II
70	7031.90	II
70	8668.57	I
20	8667.22	I
800	11660.04	I
570	11662.47	I
125	15629.08	I
200	16240.38	I
250	16244.67	I
235	18994.33	I

B III
Ref. 69, 221 — R.L.K.

Intensity	Wavelength (Vacuum)	
150	518.24	III
75	518.27	III
40	411.80	III
20	510.77	III
40	510.85	III
110	677.00	III
160	677.14	III
40	758.48	III
70	758.67	III
20	1953.83	III

Intensity	Wavelength (Air)	
550	2065.78	III
450	2067.23	III
160	2077.09	III
40	2234.09	III
70	2234.59	III
40	4242.98	III
70	4243.61	III
220	4487.05	III
360	4497.73	III
110	7835.25	III
70	7841.41	III

B IV
Ref. 74 — R.L.K.

Intensity	Wavelength (Vacuum)	
10	52.68	IV
30	60.31	IV
160	344.0	IV
450	385.0	IV
285	418.7	IV
70	1112.2	IV
450	1168.9	IV
70	1170.9	IV

Intensity	Wavelength (Air)	
70	2524.7	IV
160	2530.3	IV
450	2821.68	IV
70	2824.57	IV
285	2825.85	IV

B V
Ref. 94 — R.L.K.

Intensity	Wavelength (Vacuum)	
30	41.00	V
	48.59	V
	194.37	V
	262.37	V
	512.53	V
	749.74	V

BROMINE (Br)
Z = 35

Br I and II
Ref. 122, 124, 240, 248, 316
G.V.S.

Intensity	Wavelength (Vacuum)	
300	711.68	II
250	815.48	II
350	856.19	II
1000	889.23	II
500	896.64	II
500	905.99	II
300	922.56	II
1000	948.97	II
500	984.93	II
500	1012.10	II
1000	1015.54	II
500	1037.02	II
1000	1049.00	II
450	1064.76	II
500	1071.87	II
250	1101.50	I
300	1134.59	I
250	1136.29	I
250	1177.23	I
400	1178.90	I
1000	1189.28	I
250	1189.38	I
1000	1189.50	I
500	1198.37	I
800	1209.76	I
1000	1210.73	I
750	1216.01	I
1000	1221.13	I
900	1221.87	I
1000	1223.24	I
1200	1224.41	I
1200	1226.90	I
750	1228.05	I
7500	1232.43	I
1200	1243.90	I
800	1249.59	I
1500	1251.66	I
1000	1255.80	I
1500	1259.20	I
1200	1261.66	I
1200	1266.20	I
1000	1279.48	I
1000	1286.26	I
3000	1309.91	I
3000	1316.74	I
1000	1317.37	I
2000	1317.70	I
12000	1384.60	I
3000	1449.90	I
50000	1488.45	I
30000	1531.74	I
25000	1540.65	I
30000	1574.84	I
20000	1576.39	I
25000	1582.31	I
75000	1633.40	I

Intensity		Wavelength (Air)	
350		2285.17	II
350		2287.60	II
500	h	2317.30	II
400		2336.93	II
350		2386.45	II
500		2386.70	II
300		2388.69	II
450		2388.96	II
500		2389.69	II
350		2392.21	II
400		2392.42	II
300		2488.50	II
300	h	2495.22	II
450		2521.70	II
400		2541.48	II
400		2556.92	II
350		2690.17	II
400		2713.77	II
350	h	2746.52	II
300	h	2807.55	II
400	h	2893.40	II
400	h	2917.18	II
400	h	2967.21	II
500	h	2972.26	II
300		2981.86	II
300	h	2985.87	II
300	h	2986.53	II
300	h	3016.48	II
350		3423.82	II
300	h	3606.80	II
350		3714.30	II
1200		3815.65	I
350		3834.69	II
300		3871.21	II
400		3891.63	II
300		3901.24	II
300		3914.20	II
500		3914.38	II
350		3919.51	II
400		3924.09	II
300		3929.55	II
350		3939.69	II
350		3950.61	II
500		3980.38	II
1500		3992.36	I
300		4024.04	II
300		4135.66	II
300		4140.20	II
400		4179.63	II
300		4193.45	II
1000		4223.89	II
300		4236.89	II
300		4291.39	II
2000		4365.14	I
1000		4365.60	II
1500		4425.14	I
10000		4441.74	I
10000		4472.61	I
20000		4477.72	I
1000		4490.42	I
3000		4513.44	I
15000		4525.59	I
300		4529.60	II
500		4542.92	II
3000		4575.74	I
300		4601.36	II
2500		4614.58	I
350		4622.70	II
300		4642.02	II
300		4651.98	II
500		4678.70	II
400		4693.17	II
500		4704.85	II
350		4719.76	II
400		4720.36	II
300		4728.20	II
400		4735.41	II
400		4742.64	II
2500		4752.28	I
350		4766.00	II
400		4779.40	II
4000		4780.31	I
1600		4785.19	I
500		4785.50	II
300		4802.33	II
500		4816.70	II
300		4818.46	II
350		4844.81	II
350		4848.75	II
400		4921.12	II
400		4928.79	II
450		4930.66	II
300		4945.51	II
4000		4979.76	I
300		5038.74	II
300		5054.64	II
400		5164.38	II
300		5180.01	II
500		5182.35	II
300		5193.90	II
500		5238.23	II
300		5272.68	II
350		5304.10	II
400		5330.57	II
500		5332.05	II
1200		5395.48	I
400		5422.78	II
350		5424.99	II
300		5435.07	II
1200		5466.22	I
350		5478.47	II
300		5488.79	II
300		5495.06	II
500		5506.69	II
350		5589.94	II
300		5718.71	II
300		5830.78	II
1800		5852.08	I
1600		5940.48	I
2400		6122.14	I
40000		6148.60	I
300		6161.74	II
2000		6177.39	I
1500		6335.48	I
60000		6350.73	I
400		6352.94	II
2500		6410.32	I
1800		6483.56	I
1000		6514.62	I
20000		6544.57	I
1500		6548.09	I
50000	c	6559.80	I
1000		6571.31	I
1800		6579.14	I
20000		6582.17	I
1500		6620.47	I
50000	c	6631.62	I
20000		6682.28	I
10000		6692.13	I
8000		6728.28	I
2000		6760.06	I
2000		6779.48	I
2200		6786.74	I
6500		6790.04	I
1600	c	6791.48	I
1800		6861.15	I
10000		7005.19	I
2000		7260.45	I
10000		7348.51	I
40000		7512.96	I
1600		7591.61	I
1800		7595.07	I
2000		7616.41	I
30000		7803.02	I
1200		7827.23	I
2500	s	7881.45	I
2500		7881.57	I
2500		7925.81	I
30000	c	7938.68	I
3000		7947.94	I
3000		7950.18	I

Bromine (Cont.)

Intensity		Wavelength	
8000		7978.44	I
10000		7978.57	I
30000		7989.94	I
2000		8026.35	I
2500		8026.54	I
30000		8131.52	I
1000	c	8152.65	I
10000		8153.75	I
25000		8154.00	I
5000		8246.86	I
15000		8264.96	I
75000	c	8272.44	I
20000		8334.70	I
10000		8343.70	I
1200		8384.04	I
40000		8446.55	I
4000		8477.45	I
1500		8513.38	I
1000		8557.73	I
1000		8566.28	I
20000		8638.66	I
4000		8698.53	I
10000	c	8793.47	I
15000		8819.96	I
25000		8825.22	I
4000		8888.98	I
30000		8897.62	I
6000		8932.40	I
1800		8949.39	I
9000		8964.00	I
350		9024.42	II
30000		9166.06	I
15000		9173.63	I
20000		9178.16	I
40000		9265.42	I
15000		9320.86	I
300		9434.04	II
6000		9793.48	I
10000		9896.40	I
3000		10140.08	I
6000		10237.74	I
1000		10299.62	I
1500		10377.65	I
30000		10457.96	I
1000		10742.14	I
3000		10755.92	I
1700		13217.17	I
1800		14354.57	I
1250		14888.70	I
1800		16731.19	I
1200		18568.31	I
3500		19733.62	I
1000		20281.73	I
1000		20624.67	I
1200		21787.24	I
4000		22865.65	I
1000		23513.15	I
500		28346.50	I
500		30380.85	I
600		31630.13	I
120		32693.90	I
150		34181.87	I
150		38345.75	I
120		39964.36	I

Br III
Ref. 246, 250 — G.V.S.

Intensity		Wavelength	
		Vacuum	
450		611.1	III
300		620.4	III
500		665.54	III
500		677.19	III
300		677.8	III
450		687.68	III
400		690.2	III
350		696.99	III
300		727.0	III
300		736.4	III
250		769.63	III
250		817.79	III
350		949.0	III
400		960.4	III
450		984.9	III
250		1313.5	III
250		1402.9	III
		Air	
400	h	2293.44	III
300		2313.29	III
300		2462.39	III
300		2482.60	III
350		2499.25	III
350		2529.49	III
350	h	2551.09	III
350	h	2570.83	III
300	h	2573.17	III
400	h	2584.99	III
500		2589.14	III
300	h	2594.48	III
400		2595.98	III
450	h	2606.20	III
350		2608.15	III
500	h	2613.13	III
350	h	2616.26	III
500		2626.52	III
350	h	2629.23	III
350		2639.60	III
350	h	2671.53	III
350	h	2735.83	III
300		2770.50	III
300	h	2785.28	III
300		2804.16	III
400		2926.96	III
300		2936.22	III
350		2969.00	III
400		2994.04	III
500		3020.76	III
300		3033.63	III
350		3036.45	III
500		3074.42	III
350		3091.94	III
350		3117.29	III
300		3147.81	III
400		3174.08	III
300		3321.08	III
450		3333.07	III
500		3349.64	III
300		3385.25	III
450		3447.36	III
400		3487.58	III
300		3506.47	III
450		3517.36	III
500		3540.16	III
300		3551.08	III
500		3562.43	III
450		3600.71	III
250		3693.53	III
450		3820.26	III
200		3903.95	III
350		4506.53	III
200		4519.74	III
150		5175.87	III
100		5446.80	III
100		7192.8	III
100		7673.1	III

Br IV
Ref. 139, 142, 243, 249
G.V.S.

Intensity		Wavelength	
		Vacuum	
700		379.73	IV
700		400.37	IV
1000		545.43	IV
1000		559.76	IV
1000		569.19	IV
1000		576.59	IV
1000		585.10	IV
1000		586.71	IV
1000		597.51	IV
1000		600.09	IV
1000		601.27	IV
1000		607.03	IV
1000		617.85	IV
1000		619.87	IV
1000		630.14	IV
1000		642.23	IV
1000		661.53	IV
1000		683.51	IV
1000		697.72	IV
1000		715.39	IV
1000		731.00	IV
1000		800.12	IV
1000		813.66	IV
900		1274.82	IV
1000		1703.51	IV
		Air	
1000		2133.79	IV
1000		2145.02	IV
1000		2257.21	IV
1000		2272.73	IV
1000		2307.40	IV
1000		2408.16	IV
1000		2411.58	IV
700		2491.14	IV
1000		2581.19	IV
600		2661.40	IV
700		2820.87	IV
1000		2842.88	IV
1100	h	2907.71	IV
500		3041.18	IV
500		3380.56	IV

Br V
Ref. 42 — G.V.S.

Intensity	Wavelength	
	Vacuum	
600	468.37	V
800	482.11	V
900	531.97	V
1000	547.90	V
700	549.77	V
800	621.03	V
800	632.22	V
700	645.44	V
400	652.64	V
800	657.54	V
800	679.62	V
700	812.95	V
1000	850.81	V
150	855.27	V
600	1041.60	V
1000	1069.15	V
500	1080.54	V
900	1112.13	V
1000	1143.56	V
150	1429.75	V
400	1442.60	V
150	1470.35	V

# CADMIUM (Cd)
Z = 48

Cd I and II
Ref. 44, 285, 296 — R.D.C.

Intensity		Wavelength	
		Vacuum	
100		1256.00	II
150		1296.43	II
100		1326.50	II
150		1370.91	II
200		1514.26	II
200		1571.58	II
100		1668.60	II
50		1702.47	II
50		1724.41	II
100		1785.84	II
100		1827.70	II
300		1922.23	II
100		1943.54	II
40		1965.54	II
30		1986.89	II
200		1995.43	II
		Air	
100		2007.49	II
50		2032.45	II
75		2036.23	II
150		2096.00	II
1000	r	2144.41	II
50		2155.06	II
100		2187.29	II
1000		2194.56	II
1000		2265.02	II
1500	r	2288.022	I
1000		2312.77	II
200		2321.07	II
40		2376.82	II
50		2418.69	II
50		2469.73	II
40		2487.93	II
3		2491.00	I
40		2495.58	II
10		2508.91	I
50		2509.11	II
30		2516.22	II
15	h	2518.59	I
25	h	2525.196	I
50		2544.613	I
50		2551.98	II
25		2553.465	I
3		2565.789	I
500		2572.93	II
50		2580.106	I
3		2584.87	I
30		2592.026	I
25	h	2602.048	I
50		2628.979	I
40		2632.190	I
75		2639.420	I
40		2659.23	II
50	h	2660.325	I
25		2668.20	II
50		2672.62	II
100		2677.540	I
25		2677.748	I
100		2707.00	II
75		2712.505	I
50		2733.820	I
1000		2748.54	II
100	h	2763.894	I
50	h	2764.230	I
50		2774.958	I
30		2823.19	II
200		2836.900	I
25		2856.46	II
100		2868.180	I
200	r	2880.767	I
50	r	2881.224	I
200		2914.67	II
50		2927.87	II
200		2929.27	II
1000	r	2980.620	I
200	r	2981.362	I
50		2981.845	I
50		3030.60	II
150		3080.822	I
25		3081.48	II
30		3082.593	I
100		3092.34	II
200		3133.167	I
50		3146.79	II
150		3250.33	II
300		3252.524	I
300		3261.055	I
50		3343.21	II
50		3385.49	II
30		3388.88	II
800		3403.652	I
50		3417.49	II
50		3442.42	II
100		3464.43	II
1000		3466.200	I
800		3467.655	I
25		3483.08	II
150		3495.44	II
25		3499.952	I
100		3524.11	II
100		3535.69	II
1000		3610.508	I
800		3612.873	I
60		3614.453	I
20		3649.558	I
10		3981.926	I
100		4029.12	II
200		4134.77	II
50		4141.49	II
100		4285.08	II
8		4306.672	I
100		4412.41	II
3		4412.989	I
1000		4415.63	II
30		4440.45	II
8		4662.352	I
200		4678.149	I
30		4744.69	II
300		4799.912	I
50		4881.72	II
50		5025.50	II
1000	h	5085.822	I
6		5154.660	I
100		5268.01	II
100		5271.60	II
1000		5337.48	II
1000		5378.13	II
200		5381.89	II
40		5843.30	II
50		5880.22	II
300		6099.142	I
100		6111.49	I
100		6325.166	I
30		6330.013	I
400		6354.72	II
500		6359.98	II
2000		6438.470	I
400		6464.94	II
25		6567.65	II
500		6725.78	II
100		6759.19	II
30		6778.116	I
50		7237.01	II
100		7284.38	II
1000		7345.670	I
50		8066.99	II
5		8200.309	I
20		9292	I
15		11655	I
35		14491	I
80		15712	I
55	d	19125	I
25		24378	I
35		25455	I

Cd III
Ref. 296 — R.D.C.

Intensity	Wavelength	
	Vacuum	
8	677.39	III

Cadmium (Cont.)

Intensity	Wavelength	
15	684.58	III
10	720.70	III
5	1383.60	III
15	1392.10	III
10	1396.78	III
25	1416.28	III
5	1420.29	III
8	1420.54	III
5	1432.86	III
20	1446.08	III
25	1447.55	III
30	1455.74	III
8	1466.14	III
5	1471.97	III
15	1491.81	III
5	1511.01	III
10	1511.65	III
5	1513.13	III
10	1523.55	III
15	1528.40	III
30	1529.30	III
5	1532.10	III
50	1545.17	III
25	1547.57	III
10	1550.07	III
20	1550.45	III
15	1550.89	III
5	1552.18	III
5	1556.48	III
15	1560.66	III
5	1566.03	III
15	1568.98	III
10	1582.39	III
40	1601.59	III
20	1604.87	III
20	1606.64	III
10	1607.28	III
15	1608.91	III
10	1609.61	III
15	1612.51	III
15	1625.27	III
25	1628.54	III
20	1651.87	III
25	1655.63	III
30	1678.15	III
10	1699.70	III
40	1707.16	III
10	1721.93	III
40	1722.95	III
30	1725.66	III
25	1739.00	III
5	1745.69	III
40	1747.67	III
15	1748.15	III
30	1768.82	III
40	1773.06	III
30	1789.19	III
75	1793.40	III
15	1796.10	III
5	1800.57	III
40	1823.41	III
50	1844.66	III
40	1851.13	III
20	1851.37	III
40	1855.85	III
200	1856.67	III
150	1874.08	III
15	1886.49	III
15	1903.48	III
25	1909.98	III
15	1910.57	III
10	1939.59	III
15	1988.81	III

Air

Intensity	Wavelength	
20	2000.60	III
15	2004.07	III
15	2016.12	III
40	2039.83	III
50	2045.61	III
10	2061.25	III
75	2087.91	III
10	2097.45	III
5	2100.47	III
50	2111.60	III
5	2188.13	III
5	2218.43	III
5	2224.43	III
7	2418.24	III
10	2426.36	III
25	2499.81	III
15	2618.81	III
5	2630.56	III
20	2766.99	III
30	2805.59	III
10	3035.72	III

Cd IV
Ref. 353, 399 — R.D.C.

Intensity	Wavelength	
	Vacuum	
50	427.01	IV
20	437.88	IV
50	447.85	IV
60	480.90	IV
10	489.49	IV
70	493.00	IV
70	495.13	IV
70	498.14	IV
70	498.53	IV
80	504.09	IV
70	504.20	IV
70	504.50	IV
80	506.31	IV
60	508.01	IV
50	508.95	IV
70	509.55	IV
25	509.81	IV
70	511.40	IV
80	513.00	IV
70	514.50	IV
60	519.42	IV
15	520.97	IV
80	524.41	IV
70	524.47	IV
50	524.77	IV
70	525.10	IV
60	525.19	IV
70	527.07	IV
50	530.79	IV
80	531.09	IV
80	531.51	IV
70	534.29	IV
70	536.77	IV
50	537.24	IV
60	540.90	IV
70	541.74	IV
80	542.60	IV
80	546.55	IV
40	548.01	IV
20	548.33	IV
15	548.90	IV
25	551.27	IV
20	552.90	IV
60	553.06	IV
80	554.05	IV
25	560.26	IV
10	564.16	IV
60	567.01	IV
20	1062.23	IV
150	1118.16	IV
30	1126.00	IV
20	1134.08	IV
15	1139.04	IV
20	1154.64	IV
10	1155.73	IV
100	1164.65	IV
20	1165.78	IV
40	1167.30	IV
20	1179.73	IV
15	1183.07	IV
100	1183.40	IV
40	1194.13	IV
20	1195.63	IV
30	1196.47	IV
20	1198.93	IV
15	1215.38	IV
20	1223.52	IV
20	1246.06	IV
15	1246.56	IV
15	1249.94	IV
30	1266.47	IV
15	1274.41	IV
20	1285.63	IV
20	1287.58	IV
20	1299.46	IV
40	1304.36	IV
30	1306.07	IV
15	1316.89	IV
15	1321.85	IV
20	1325.55	IV
30	1340.97	IV
15	1346.15	IV
30	1354.78	IV
20	1358.11	IV
30	1362.55	IV
60	1370.48	IV
30	1380.98	IV
20	1397.65	IV
30	1403.68	IV
15	1406.58	IV
60	1418.89	IV
15	1429.83	IV
20	1447.54	IV
20	1452.63	IV
15	1465.97	IV
15	1466.67	IV
15	1482.95	IV
20	1491.79	IV
20	1570.20	IV
20	1598.73	IV
20	1600.42	IV
15	1622.87	IV

CALCIUM (Ca)
Z = 20

Ca I and II
Ref. 70, 150, 270 — J.J.W. and H.H.S.

Intensity	Wavelength	
	Vacuum	
24	1341.89	II
12	1342.54	II
20	1433.75	II
12	1432.50	II
20	1553.18	II
32	1554.64	II
4	1642.80	II
20	1643.77	II
36	1644.44	II
60	1649.86	II
32	1651.99	II
12	1673.86	II
20	1680.05	II
2	1680.13	II
8	1691.78	II
16	1698.18	II
20	1807.34	II
40	1814.50	II
4	1814.65	II
60	1840.06	II
40	1838.01	II
20	1843.09	II
40	1850.69	II
	Air	
	2103.24	II
	2112.76	II
	2113.15	II
	2128.75	II
	2131.51	II
	2132.30	II
2	2150.80	I
	2197.79	II
5	2200.73	I
	2208.61	II
6	2275.46	I
8	2398.56	I
7	2721.65	I
9	2994.96	I
8	2997.31	I
8	2999.64	I
9	3000.86	I
10	3006.86	I
9	3009.21	I
2	3024.94	I
2	3034.54	I
2	3045.74	I
3	3055.32	I
2	3071.57	I
2	3076.95	I
2	3080.79	I
2	3099.30	I
10	3125.18	II
5	3136.02	I
6	3140.79	I
7	3150.75	I
170	3158.87	II
180	3179.33	II
5	3180.52	I
150	3181.28	II
7	3209.96	I
8	3215.17	I
6	3215.34	I
9	3225.90	I
6	3226.15	I
5	3274.67	I
6	3286.07	I
10	3308.02	II
20	3316.51	II
10	3344.51	I
10	3347.04	II
11	3350.21	I
9	3350.36	I
12	3361.92	I
9	3362.14	I
10	3452.66	II
20	3461.87	II
20	3468.48	I
11	3474.76	I
10	3485.16	II
13	3487.60	I
10	3495.16	II
15	3624.11	I
17	3630.75	I
14	3630.97	I
20	3644.41	I
14	3644.77	I
8	3644.99	I
5	3675.29	I
6	3678.21	I
30	3683.70	II
40	3694.11	II
10	3694.36	II
170	3706.03	II
180	3736.90	II
10	3739.38	II
6	3748.35	I
8	3750.29	I
9	3753.34	I
20	3755.67	II
30	3758.39	II
9	3870.48	I
11	3872.54	I
11	3872.56	I
12	3875.78	I
12	3875.80	I
6	3889.10	I
6	3923.48	I
230	3933.66	II
9	3935.29	I
6	3946.04	I
15	3948.90	I
17	3957.05	I
220	3968.47	II
8	3972.57	I
18	3973.71	I
50	4097.10	II
15	4098.53	I
15	4098.57	I
60	4109.82	II
30	4110.28	II
40	4206.18	II
50	4220.07	II
50	4226.73	I
15	4240.46	I
24	4283.01	I
22	4289.36	I
22	4298.99	I
25	4302.53	I
23	4307.74	I
22	4318.65	I
20	4355.08	I
25	4425.44	I
26	4434.96	I
25	4435.69	I
30	4454.78	I
28	4455.89	I
20	4456.61	I
20	4472.04	II
10	4479.23	II
20	4489.18	II
23	4526.94	I
22	4578.55	I
23	4581.40	I
23	4581.47	I
24	4585.87	I
24	4585.96	I
20	4685.27	I
30	4716.74	II
40	4721.03	II
40	4799.97	II
25	4878.13	I
70	5001.48	II
80	5019.97	II
40	5021.14	II
23	5041.62	I
25	5188.85	I
22	5261.71	I
23	5262.24	I
22	5264.24	I
24	5265.56	I
25	5270.27	I
60	5285.27	II
70	5307.22	II
50	5339.19	II
27	5349.47	I
23	5512.98	I
25	5581.97	I
27	5588.76	I
24	5590.12	I
26	5594.47	I
25	5598.49	I
24	5601.29	I
24	5602.85	I
30	5857.45	I
10	5922.72	II
10	5923.69	II
27	6102.72	I
29	6122.22	I
22	6161.29	I
30	6162.17	I
22	6163.76	I
24	6166.44	I
26	6169.06	I
28	6169.56	I
35	6439.07	I
30	6449.81	I
22	6455.60	I
80	6456.87	II
34	6462.57	I
29	6471.66	I

Calcium (Cont.)

Intensity	Wavelength	
32	6493.78	I
28	6499.65	I
23	6572.78	I
30	6717.69	I
33	7148.15	I
31	7202.19	I
	7291.47	II
	7323.89	II
33	7326.15	I
30	7575.81	II
60	7581.11	II
80	7601.30	II
20	7602.32	II
40	7820.78	II
60	7843.38	II
20	8017.50	II
20	8020.50	II
70	8133.05	II
100	8201.72	II
110	8248.80	II
70	8254.73	II
14	8256.67	I
10	8338.04	I
12	8339.12	I
10	8352.39	I
11	8357.17	I
130	8498.02	II
170	8542.09	II
10	8633.95	I
160	8662.14	II
12	8842.61	I
15	8909.18	I
100	8912.07	II
110	8927.36	II
12	8967.47	I
16	9099.10	I
13	9105.62	I
12	9108.82	I
10	9171.14	I
110	9213.90	II
90	9312.00	II
100	9319.56	II
110	9320.65	II
25	9416.97	I
10	9456.80	I
10	9534.88	I
11	9548.38	I
100	9567.97	II
110	9599.24	II
80	9601.82	II
10	9604.28	I
12	9663.65	I
10	9664.41	I
14	9676.30	I
14	9688.67	I
13	9701.94	I
80	9854.74	II
110	9890.63	II
90	9931.39	II
100	10223.04	II
20	10343.81	I
13	10838.97	I
13	10861.58	I
13	10863.87	I
14	10869.50	I
14	10879.87	I
20	11838.99	II
10	11949.72	II
25	12816.04	I
24	12823.86	I
25	12905.10	I
30	13033.57	I
21	13086.44	I
24	13134.95	I
20	16150.77	I
22	16157.36	I
21	16197.04	I
20	18925.47	I
24	18970.14	I
30	19046.14	I
48	19309.20	I
49	19452.99	I
47	19505.72	I
50	19776.79	I
35	19853.10	I
34	19862.22	I
23	19917.19	I
24	19933.70	I
	21389.00	II
	21428.90	II
20	22607.93	I
25	22624.93	I
30	22651.23	I

Ca III
Ref. 25, 16 — J.J.W. and H.H.S.

Intensity	Wavelength	
	Vacuum	
6	296.96	III

Intensity	Wavelength	
9	403.72	III
7	409.95	III
5	439.69	III
5	633.59	III
5	685.41	III
5	697.55	III
5	699.09	III
5	699.89	III
6	701.39	III
5	727.66	III
8	740.55	III
6	746.25	III
5	747.98	III
5	779.61	III
5	800.30	III
5	809.93	III
5	817.06	III
6	821.57	III
6	840.56	III
6	1020.07	III
5	1034.65	III
5	1187.30	III
8	1188.61	III
8	1188.61	III
5	1190.86	III
10	1262.65	III
11	1278.39	III
10	1281.55	III
12	1286.52	III
12	1298.04	III
11	1317.70	III
10	1328.95	III
11	1335.13	III
10	1360.01	III
11	1385.43	III
11	1397.69	III
13	1453.16	III
12	1459.79	III
11	1461.88	III
15	1463.34	III
16	1484.87	III
12	1496.88	III
11	1506.88	III
20	1545.29	III
15	1555.53	III
18	1562.47	III
13	1571.27	III
13	1586.13	III
10	1762.26	III
10	1783.93	III
10	1794.22	III
12	1800.21	III
13	1807.89	III
14	1812.15	III
11	1813.59	III
12	1830.06	III
10	1860.43	III
14	1870.26	III
14	1872.37	III
10	1894.12	III
11	1910.10	III
12	1935.72	III
10	1939.68	III
11	1943.01	III
12	1948.26	III
10	1953.55	III
10	1958.97	III
13	1964.61	III
13	1967.94	III
12	1972.82	III
10	1977.01	III
10	1978.55	III
11	1981.19	III

Intensity	Wavelength	
	Air	
12	2033.36	III
12	2041.53	III
13	2078.92	III
13	2098.49	III
15	2114.41	III
17	2123.03	III
14	2129.19	III
14	2133.96	III
13	2140.36	III
16	2152.43	III
12	2171.57	III
12	2276.52	III
15	2312.08	III
14	2497.74	III
15	2541.50	III
13	2587.15	III
12	2590.41	III
15	2620.82	III
15	2634.14	III
12	2686.72	III
16	2687.76	III
15	2704.86	III
14	2771.28	III
15	2791.59	III
16	2813.88	III
17	2866.54	III

Intensity	Wavelength	
18	2869.95	III
19	2881.78	III
21	2899.79	III
19	2924.33	III
20	2988.63	III
18	2989.27	III
15	3028.59	III
19	3119.67	III
15	3367.79	III
19	3372.67	III
18	3537.77	III
15	4081.77	III
15	4153.57	III
15	4164.31	III
15	4184.20	III
18	4207.24	III
17	4233.74	III
16	4240.74	III
15	4284.39	III
20	4302.81	III
15	4329.19	III
16	4333.57	III
15	4358.38	III
19	4399.59	III
17	4406.29	III
17	4431.30	III
19	4499.88	III
18	4516.59	III
18	4572.12	III
11	4708.83	III
11	4716.27	III
10	4859.17	III
10	5008.95	III
10	5050.07	III
10	5231.82	III
11	5247.37	III
13	5271.98	III
10	5301.32	III
11	5321.29	III
10	5328.06	III
11	5570.58	III
10	5579.06	III
13	6069.98	III
10	6173.22	III
12	6213.98	III
11	6294.89	III
11	6370.11	III
10	6387.55	III
12	6424.51	III
12	6485.35	III
10	6538.78	III
10	6542.24	III
10	7308.69	III
10	7843.06	III
12	7898.46	III
10 1	8217.20	III

Ca IV
Ref. 150 — J.J.W. and H.H.S.

Intensity	Wavelength	
	Vacuum	
150	249.41	IV
150	250.15	IV
150	251.35	IV
250	296.55	IV
200	299.32	IV
200	318.09	IV
50	318.39	IV
120	321.59	IV
250	329.12	IV
150	329.39	IV
200	331.44	IV
250	331.99	IV
235	332.53	IV
150	332.81	IV
200	338.83	IV
150	339.79	IV
200	340.29	IV
200	341.29	IV
200	341.46	IV
250 c	342.45	IV
100	343.19	IV
200	343.44	IV
250	343.93	IV
200	344.96	IV
215	345.13	IV
250	374.74	IV
600	434.57	IV
100	437.27	IV
250	437.77	IV
200	438.93	IV
750	443.82	IV
50	445.02	IV
500	450.57	IV
50	454.55	IV
250	456.98	IV
250	461.09	IV
150	565.46	IV
750	656.00	IV
500	669.70	IV

Ca V
Ref. 150 — J.J.W. and H.H.S.

Intensity	Wavelength	
	Vacuum	
200	190.36	V
250	190.46	V
250	196.97	V
300	199.55	V
250	200.51	V
265	257.98	V
165	260.45	V
400	267.77	V
300	270.31	V
200	271.14	V
250	272.27	V
200	272.98	V
400	280.99	V
300	284.98	V
450 c	286.96	V
500	322.17	V
250	322.76	V
300	323.22	V
250	324.48	V
250	325.28	V
300	330.94	V
200	333.44	V
200	333.57	V
300	334.55	V
250 c	335.34	V
200	336.55	V
200	337.54	V
250	338.06	V
200	343.64	V
450	352.92	V
250	356.25	V
250	377.18	V
200	387.08	V
750	425.00	V
500	558.60	V
400	637.93	V
300	643.12	V
400	646.57	V
250	647.88	V
250	651.55	V
300	656.76	V

CALIFORNIUM (Cf)
Z = 98

Cf I and II
Ref. 52, 331 — J.G.C.

Intensity	Wavelength	
	Air	
10000	2739.31	
10000 s	2759.10	
10000	2774.52	
10000 l	2852.03	
10000 s	2855.24	
10000	3298.14	
10000	3352.71	
10000 l	3367.79	
10000	3392.22	I
10000	3481.07	
10000 l	3513.47	I
10000	3531.49	I
10000	3540.98	I
10000	3598.77	I
10000	3605.32	I
10000	3612.11	II
10000	3617.49	I
10000 s	3626.76	II
10000	3659.46	
10000	3662.70	I
10000 s	3699.49	
10000 l	3722.11	II
10000	3739.35	I
10000	3785.61	
10000 l	3789.04	II
10000 s	3893.23	II
10000 l	3993.57	II
10000	4035.45	
10000	4099.12	I
10000	4242.38	I
10000	4329.03	I
10000	4335.22	I
10000	5173.96	I
10000	5179.08	I
10000	5219.24	I
10000 s	5279.01	
10000 s	5320.09	
10000 s	5339.13	
10000	5408.88	I
10000	5726.05	I
10000	6622.83	I
10000	6631.26	I

Californium (Cont.)

Intensity		Wavelength	
10000		6677.90	I
10000		6894.59	I
10000		6927.10	II
10000	1	7074.52	I
10000	s	7307.90	I
10000		8141.29	I
10000		8241.77	I
10000		8333.85	II
10000		8423.49	II
10000		8568.83	II
10000	1	9228.52	
10000		9337.70	
10000		9649.51	
10000	s	10308.41	
10000	1	10568.83	
10000	s	10614.84	
10000		11300.19	
10000		11681.85	
10000		11941.33	
10000	1	12183.05	
10000	s	12352.72	
10000		12437.48	
10000	1	12789.41	
10000	1	13329.98	
10000	s	13362.98	
10000		13376.89	
10000	1	13474.44	
10000	1	14772.49	
10000	s	15281.32	
10000		15587.12	
10000		15675.92	
10000		16759.06	
10000	s	17626.25	
10000	1	18718.69	
10000	h	19068.71	
10000	1	19336.96	
10000	1	19576.84	
10000	1	20393.38	
10000	s	20869.98	

CARBON (C)
Z = 6

C I and II
Ref. 211   R.L.K.

Intensity		Wavelength	

Vacuum

9		595.022	II
30		687.053	II
50		687.345	II
10		858.092	II
20		858.559	II
30		903.624	II
60		903.962	II
150		904.142	II
30		904.480	II
9		1009.86	II
10		1010.08	II
10		1010.37	II
80		1036.337	II
150		1037.018	II
150		1157.910	I
150		1158.019	I
150		1158.035	I
150		1188.992	I
150		1189.447	I
200		1189.631	I
300		1193.009	I
300		1193.031	I
300		1193.240	I
300		1193.264	I
100		1193.393	I
150		1193.649	I
150		1193.679	I
100		1194.064	I
100		1194.488	I
100		1261.552	I
250		1277.245	I
250		1277.282	I
300		1277.513	I
300		1277.550	I
200		1280.333	I
100		1311.363	I
9		1323.951	II
120		1329.578	I
120		1329.600	I
150		1334.532	II
300		1335.708	II
100		1354.288	I
150		1355.84	I
120		1364.164	I
100		1459.032	I
200		1463.336	I
120		1467.402	I
150		1481.764	I
150		1560.310	I
400		1560.683	I
400		1560.708	I
100		1561.341	I
400		1561.438	I
150		1656.266	I
120		1656.928	I
300		1657.008	I
120		1657.380	I
120		1657.907	I
150		1658.122	I
500		1751.823	I
1000		1930.905	I

Air

800		2478.56	I
250		2509.12	II
350		2512.06	II
250	h	2574.83	II
350	1	2741.28	II
250		2746.49	II
1000		2836.71	II
800		2837.60	II
800	h	2992.62	II
350		3876.19	II
350		3876.41	II
350		3876.66	II
570		3918.98	II
800		3920.69	II
250		4074.52	II
350	1	4075.85	II
800		4267.02	II
1000		4267.26	II
200		4771.75	I
200		4932.05	I
200		5052.17	I
350		5132.94	II
350		5133.28	II
350		5143.49	II
570		5145.16	II
400		5151.09	II
300		5380.34	I
250		5648.07	II
350		5662.47	II
570		5889.77	II
350		5891.59	II
200		6001.13	I
250		6006.03	I
110		6007.18	I
150		6010.68	I
300		6013.22	I
250		6014.84	I
800		6578.05	II
570		6582.88	II
200		6587.61	I
250		6783.90	II
250		7113.18	I
250		7115.19	I
250		7115.63	II
200		7116.99	I
350		7119.90	II
800		7231.12	II
1000		7236.42	II
200		7860.89	I
200		8058.62	I
520		8335.15	I
250		9061.43	I
200		9062.47	I
200		9078.28	I
250		9088.51	I
450		9094.83	I
300		9111.80	I
800		9405.73	I
150		9603.03	I
250		9620.80	I
300		9658.44	I
200		10683.08	I
300		10691.25	I
12		11619.29	I
23		11628.83	I
13		11658.85	I
47		11659.68	I
24		11669.63	I
85		11748.22	I
142		11753.32	I
114		11754.76	I
11		11777.54	I
17		11892.91	I
30		11895.75	I
26		12614.10	I
20		13502.27	I
38		14399.65	I
16		14403.25	I
61		14420.12	I
12		14429.03	I
13		14442.24	I
12		16559.66	I
50		16890.38	I
10		17338.56	I
11		17448.60	I
13		18139.80	I
23		19721.99	I

C III
Ref. 22, 211 — R.L.K.

Intensity		Wavelength	

Vacuum

250		371.69	III
250		371.75	III
150		371.78	III
500		386.203	III
200		450.734	III
400		459.46	III
500		459.52	III
570		459.63	III
250		511.522	III
250		535.288	III
300		538.080	III
350		538.149	III
400		538.312	III
350		574.281	III
800		977.03	III
370		1174.93	III
350		1175.26	III
330		1175.59	III
500		1175.71	III
350		1175.99	III
370		1176.37	III

Air

250		2162.94	III
800		2296.87	III
150		2697.95	III
110	1	2724.85	III
150	1	2725.30	III
150	1	2725.90	III
200		2982.11	III
150		4056.06	III
200		4067.94	III
250		4068.91	III
250		4070.26	III
150		4162.86	III
250	h	4186.90	III
200		4325.56	III
600		4647.42	III
520		4650.25	III
375		4651.47	III
200		4665.86	III
450		5695.92	III
150		5826.42	III
150		6744.38	III
150	h	7037.25	III
150		7612.65	III
300	h	8196.48	III
150		8332.99	III
300		8500.32	III

C IV
Ref. 66, 211 — R.L.K.

Intensity		Wavelength	

Vacuum

250		244.91	IV
200		289.14	IV
250		289.23	IV
570		312.42	IV
500		312.46	IV
650		384.03	IV
700		384.18	IV
400		419.52	IV
500		419.71	IV
1000		1548.202	IV
900		1550.774	IV

Air

200	1	2524.41	IV
300	s	2529.98	IV
200	w	4658.30	IV
250		5801.33	IV
200		5811.98	IV
90	w	7726.2	IV

C V
Ref. 211 — R.L.K.

Intensity		Wavelength	

Vacuum

110		34.973	V
450		40.268	V
110		227.19	V
160		248.66	V
160		248.74	V

Air

40		2270.91	V
5		2277.25	V
20		2277.92	V
5		4943.88	V
5		4944.56	V

CERIUM (Ce)
Z = 58

Ce I and II
Ref. 1 — C.H.C.

Intensity	Wavelength	

Air

130	2462.97	II
110	2518.51	II
200	2548.68	II
340	2651.01	II
120	2696.07	II
120	2706.88	II
120	2723.38	II
110	2741.96	II
100	2750.89	II
150	2761.42	II
120	2784.27	II
100	2785.35	II
100	2790.53	II
140	2791.42	II
270	2830.90	II
100	2833.31	II
250	2874.14	II
110	2908.42	II
100	2918.67	II
120	2955.94	II
110	2964.80	II
100	2972.58	II
400	2976.91	II
150	2977.46	II
120	2980.41	II
250	2990.87	II
110	2994.42	II
320	2995.64	II
400	3008.79	II
370	3017.20	II
210	3037.73	II
200	3051.98	II
350	3055.24	II
320	3056.78	II
680	3063.01	II
320	3083.67	II
250	3084.44	II
200	3090.37	II
370	3103.38	II
200	3107.47	II
320	3110.28	II
300	3111.17	II
220	3127.53	II
200	3130.33	II
240	3130.87	II
200	3144.60	II
290	3145.28	II
290	3146.41	II
290	3164.15	II
290	3169.18	II
290	3171.61	II
480	3183.52	II
240	3186.13	II
200	3190.34	II
710	3194.83	II
200	3199.28	II
990	3201.71	II
200	3218.38	II
710	3218.94	II
880	3221.17	II
330	3225.67	II
710	3227.11	II
240	3229.36	II
480	3231.24	II
710	3234.16	II
330	3234.89	II
390	3236.74	II
390	3243.37	II
200	3246.67	II
200	3260.98	II
200	3263.88	II
990	3272.25	II
330	3274.86	II
200	3279.84	II
330	3285.22	II
240	3295.28	II
200	3296.88	II
220	3300.15	II
240	3304.84	II
200	3312.22	II
240	3314.72	II
200	3317.80	II
200	3334.46	II
240	3341.87	II
330	3343.86	II
440	3344.76	II
200	3355.02	II

Cerium (Cont.)

Int.	λ		Int.	λ		Int.	λ		Int.	λ	
240	3357.22	II	370	3857.64	II	670	4081.22	II	700	4349.79	II
200	3360.54	II	200	3862.46	II	910	4083.23	II	560	4352.71	II
240	3366.55	II	200	3868.13	II	450	4085.23	II	910	4364.66	II
200	3371.18	II	270	3874.68	II	250	4087.36	II	350	4373.82	II
200	3373.46	II	620	3876.97	II	230	4088.85	II	530	4375.92	II
200	3373.73	II	1100	3878.36	II	450	4101.77	II	910	4382.17	II
480	3377.13	II	1500	3882.45	II	250	4105.00	II	700	4386.84	II
200	3383.68	II	1000	3889.98	II	510	4107.42	II	310	4388.01	II
200	3404.91	II	210	3890.75	II	200	4110.38	II	1700	4391.66	II
240	3405.98	II	210	3890.98	II	250	4111.39	II	200	4398.79	II
290	3417.45	II	620	3895.11	II	420	4115.37	II	510	4399.20	II
600	3422.71	II	590	3896.80	II	250	4117.01	II	350	4410.64	II
390	3426.21	II	490	3898.27	II	200	4117.29	II	350	4410.76	II
290	3441.21	II	270	3898.94	II	200	4117.59	II	310	4416.90	II
480	3476.84	II	200	3903.34	II	770	4118.14	II	980	4418.78	II
240	3482.35	II	250	3904.34	II	250	4119.02	II	200	4423.68	II
710	3485.05	II	200	3906.92	II	310	4119.79	II	310	4427.07	II
210	3507.94	II	770	3907.29	II	310	4119.88	II	480	4427.92	II
600	3517.38	II	560	3908.41	II	450	4120.83	II	310	4428.44	II
210	3520.52	II	390	3908.54	II	510	4123.24	II	650	4429.27	II
330	3521.88	II	270	3909.31	II	510	4123.49	II	480	4444.39	II
210	3526.68	II	230	3912.19	II	980	4123.87	II	450	4444.70	II
600	3534.05	II	980	3912.44	II	510	4124.79	II	770	4449.34	II
770	3539.08	II	390	3915.52	II	980	4127.37	II	620	4450.73	II
210	3545.60	II	390	3916.14	II	250	4127.74	II	2400	4460.21	II
290	3546.19	II	230	3917.64	II	200	4128.07	II	450	4461.14	II
240	3552.73	II	770	3918.28	II	530	4130.71	II	420	4463.41	II
420	3555.00	II	480	3919.81	II	480	4131.10	II	280	4467.54	II
1200	3560.80	II	590	3921.73	II	2700	4133.80	II	1400	4471.24	II
210	3576.23	II	560	3923.11	II	270	4135.44	II	450	4472.72	II
1000	3577.45	II	450	3924.64	II	270	4137.47	II	700	4479.36	II
330	3590.60	II	770	3931.09	II	2000	4137.65	II	700	4483.90	II
390	3607.63	II	310	3931.37	II	270	4138.10	II	840	4486.91	II
550	3609.69	II	230	3931.83	II	210	4138.35	II	250	4497.85	II
420	3613.70	II	310	3933.73	II	770	4142.40	II	100	4506.41	I
440	3622.15	II	560	3938.09	II	390	4144.49	II	110	4515.86	II
380	3623.74	II	770	3940.34	II	670	4145.00	II	100	4519.59	II
440	3623.84	II	310	3940.97	II	480	4146.23	II	770	4523.08	II
200	3631.19	II	2000	3942.15	II	280	4148.90	II	840	4527.35	II
350	3646.97	II	2700	3942.75	II	420	4149.79	II	840	4528.47	II
260	3647.75	II	770	3943.89	II	980	4149.94	II	110	4532.49	II
260	3647.95	II	310	3947.97	II	420	4150.91	II	110	4539.07	II
420	3653.11	II	3100	3952.54	II	1400	4151.97	II	840	4539.75	II
660	3653.67	II	340	3953.66	II	230	4153.13	II	210	4544.96	II
310	3654.97	II	310	3955.36	II	450	4159.03	II	250	4551.30	II
1800	3655.85	II	230	3956.06	II	310	4163.52	II	650	4560.28	II
440	3659.23	II	980	3956.28	II	1300	4165.61	II	310	4560.96	II
350	3659.97	II	230	3958.27	II	620	4166.88	II	2100	4562.36	II
880	3660.64	II	230	3958.87	II	250	4167.80	II	420	4565.84	II
880	3667.98	II	770	3960.91	II	320	4169.77	II	1100	4572.28	II
220	3672.18	I	390	3964.50	II	320	4169.88	II	420	4582.50	II
350	3672.79	II	770	3967.05	II	340	4176.70	II	130	4591.12	II
220	3679.42	II	450	3971.68	II	340	4181.08	II	840	4593.93	II
300	3694.91	II	270	3972.07	II	340	4185.33	II	420	4606.40	II
220	3704.98	II	270	3975.07	II	3500	4186.60	II	420	4624.90	II
1000	3709.29	II	770	3978.65	II	530	4187.32	II	1700	4628.16	II
1000	3709.93	II	560	3980.88	II	560	4193.09	II	170	4632.32	I
1400	3716.37	II	560	3982.89	II	370	4193.28	II	110	4650.51	I
420	3718.19	II	310	3983.29	II	370	4193.87	II	130	4654.29	II
420	3718.38	II	770	3984.68	II	630	4196.34	II	110	4669.50	II
210	3719.80	II	370	3989.44	II	280	4198.00	II	150	4680.13	II
420	3725.68	II	700	3992.39	II	280	4198.67	II	270	4684.61	II
490	3728.02	II	370	3992.91	II	840	4198.72	II	200	4714.00	II
800	3728.42	II	910	3993.82	II	240	4201.24	II	100	4714.81	II
320	3748.06	II	2800	3999.24	II	910	4202.94	II	110	4725.09	II
250	3751.45	II	230	4001.56	II	270	4209.41	II	100	4733.52	II
200	3755.43	II	910	4003.77	II	370	4214.04	II	310	4737.28	II
300	3762.98	II	370	4005.64	II	310	4217.59	II	100	4739.53	II
680	3764.12	II	210	4007.59	II	1500	4222.60	II	160	4747.17	II
200	3765.04	II	2700	4012.39	II	770	4227.75	II	110	4757.84	II
300	3768.76	II	910	4014.90	II	390	4231.74	II	100	4768.77	II
210	3770.76	II	250	4015.88	II	240	4234.21	II	230	4773.94	II
300	3771.60	II	200	4019.04	II	200	4236.02	II	110	4822.55	I
250	3776.61	II	240	4022.27	II	980	4239.92	II	140	4847.77	I
620	3781.62	II	840	4024.49	II	390	4242.72	II	180	4882.46	II
440	3782.52	II	240	4025.15	II	310	4245.89	II	110	4943.44	I
200	3783.58	II	840	4028.41	II	310	4245.98	II	130	4971.50	II
860	3786.63	II	250	4030.34	II	390	4246.72	II	130	4994.63	I
520	3788.75	II	840	4031.34	II	1100	4248.68	II	210	5009.10	I
300	3792.32	II	340	4037.67	II	390	4253.37	II	100	5011.77	II
2500	3801.52	II	2100	4040.76	II	620	4255.79	II	120	5022.87	II
800	3803.09	II	910	4042.58	II	200	4263.43	II	120	5037.78	II
1000	3808.11	II	230	4045.21	II	620	4270.19	II	120	5040.85	I
490	3809.21	II	620	4046.34	II	390	4270.72	II	180	5044.02	II
250	3812.20	II	210	4051.43	II	200	4278.86	II	120	5071.78	I
490	3815.85	II	210	4051.99	II	280	4285.37	II	240	5075.35	II
470	3817.46	II	700	4053.51	II	200	4288.66	II	470	5079.68	II
300	3819.02	II	450	4054.99	II	200	4289.44	II	130	5112.70	I
470	3823.90	II	280	4062.22	II	2000	4289.94	II	160	5117.17	II
470	3830.55	II	230	4062.94	II	200	4296.07	II	170	5129.57	I
490	3831.08	II	280	4067.28	II	1500	4296.67	II	110	5147.57	II
490	3834.55	II	420	4068.84	II	420	4296.78	II	100	5149.99	I
270	3836.10	II	1100	4071.81	II	590	4299.36	II	280	5159.69	I
1100	3838.54	II	270	4072.92	II	770	4300.33	II	280	5161.48	I
200	3843.76	II	1800	4073.48	II	420	4305.14	II	190	5174.55	I
220	3846.52	II	210	4073.74	II	770	4306.72	II	370	5187.46	II
250	3848.10	II	1500	4075.71	II	390	4309.74	II	210	5191.66	II
860	3848.59	II	1500	4075.85	II	560	4320.72	II	190	5211.92	I
860	3853.15	II	210	4076.24	II	310	4330.45	II	260	5223.46	II
1200	3854.18	II	420	4077.47	II	310	4332.71	II	180	5229.75	I
1200	3854.31	II	530	4078.32	II	240	4336.23	II	140	5232.92	II
620	3855.29	II	270	4078.52	II	980	4337.77	II	260	5245.92	I
390	3857.02	II	270	4080.44	II	340	4339.31	II	130	5265.71	II

Cerium (Cont.)

Intensity	Wavelength	Ion
340	5274.23	II
130	5296.56	I
130	5328.08	I
190	5330.54	II
450	5353.53	II
300	5393.40	II
150	5397.64	T
280	5409.23	II
110	5420.38	I
140	5449.24	I
140	5468.37	I
140	5472.29	II
260	5512.08	II
110	5556.25	I
170	5564.97	I
130	5565.97	I
100	5595.88	I
240	5601.28	I
190	5655.14	I
240	5669.96	I
120	5677.75	I
120	5692.94	I
300	5696.99	I
370	5699.23	I
240	5719.03	I
140	5773.12	I
120	5788.15	I
120	5812.92	I
230	5940.86	I
11	6000.18	I
55	6001.90	I
55	6005.86	I
15	6006.20	I
55	6006.82	I
19	6007.37	I
75	6013.42	I
23	6016.59	I
110	6024.20	I
15	6027.16	I
11	6031.26	I
23	6033.58	II
35	6034.20	II
23	6034.41	I
35	6035.49	II
110	6043.39	II
28	6045.42	I
55	6047.40	I
19	6051.80	II
23	6057.50	I
35	6058.00	I
23	6066.75	I
19	6069.46	I
35	6069.48	I
35	6072.00	I
35	6076.61	I
17	6077.16	I
17	6080.37	I
17	6081.28	I
19	6088.86	I
19	6088.96	I
35	6093.19	I
45	6098.34	II
11	6099.80	I
28	6108.74	II
15	6118.56	I
17	6118.90	I
45	6123.67	I
19	6132.00	II
19	6132.18	I
11	6135.45	I
23	6139.03	I
15	6142.92	I
35	6143.36	II
23	6146.43	I
19	6147.84	I
23	6151.72	I
19	6159.82	I
19	6162.14	I
19	6165.45	I
19	6175.28	I
35	6186.17	I
15	6187.97	I
15	6195.23	I
19	6195.53	I
19	6198.05	I
35	6208.98	I
11	6216.82	I
35	6228.94	I
19	6229.13	I
23	6232.45	II
28	6237.45	I
13	6238.71	I
11	6241.87	I
13	6242.91	I
15	6253.65	I
13	6257.99	I
15	6264.27	I
45	6272.05	II
15	6276.47	I
35	6295.58	I
28	6299.51	II
23	6300.21	I
13	6306.64	I

Intensity	Wavelength	Ion
35	6310.01	I
15	6335.40	I
11	6337.21	I
13	6340.70	I
35	6343.95	II
35	6371.11	II
28	6386.84	I
23	6393.02	II
11	6395.16	I
11	6425.29	II
35	6430.07	I
19	6434.39	I
23	6436.40	I
19	6446.12	I
35	6458.03	I
19	6466.88	II
28	6467.39	I
35	6473.72	I
17	6490.97	I
11	6503.27	II
11	6507.16	II
23	6513.59	II
19	6517.31	I
19	6551.70	I
45	6555.65	I
23	6579.10	I
15	6606.35	I
15	6606.86	II
22	6612.06	I
10	6623.00	I
30	6628.93	I
13	6650.89	I
22	6652.72	II
10	6661.41	I
13	6665.59	I
10	6675.54	II
15	6686.60	I
26	6700.66	I
35	6704.27	I
13	6704.52	II
10	6706.04	II
15	6728.71	I
15	6729.57	I
15	6744.70	II
10	6746.90	I
30	6774.28	TT
35	6775.59	I
10	6778.20	I
18	6807.81	I
10	6808.82	I
15	6818.23	I
10	6829.73	II
13	6847.25	I
12	6856.55	I
10	6893.66	I
10	6898.45	II
30	6924.81	I
10	6939.45	I
19	6973.50	II
10	6983.82	II
30	6986.02	I
12	7054.51	I
11	7058.68	II
11	7060.00	I
35	7061.74	II
11	7064.49	I
35	7086.35	II
11	7105.04	II
11	7115.08	II
10	7124.73	I
16	7141.42	I
19	7150.23	II
10	7151.67	I
16	7155.25	I
16	7156.99	II
10	7189.40	II
10	7191.72	I
11	7201.56	II
16	7201.89	I
10	7203.55	I
12	7210.67	I
19	7217.36	I
16	7235.71	II
22	7238.36	II
12	7241.73	I
25	7252.75	I
12	7262.64	I
11 h	7277.90	I
11	7296.17	I
19	7301.42	II
19	7313.45	II
25	7329.91	I
16	7334.68	II
12	7343.44	I
25	7397.77	I
11	7401.27	I
12	7417.94	II
11	7424.70	C
12	7433.08	I
11	7438.56	I
12	7444.44	I
10	7472.41	I
16	7486.57	II

Intensity	Wavelength	Ion
11	7527.46	I
11	7527.68	I
10	7533.73	I
10	7551.25	I
12	7562.44	I
10	7562.86	I
10 h	7563.60	I
10	7603.10	I
25	7616.11	II
12	7646.08	I
10	7647.88	I
12	7682.47	I
25	7689.17	II
10	7732.33	I
16	7748.35	I
10	7797.70	I
12	7842.59	I
22	7844.94	II
16	7850.02	II
16	7851.18	II
22	7857.54	II
12	7864.49	I
10	7866.04	I
16	7898.96	II
11	7913.52	I
10	7927.30	C
10	7927.72	I
10	7934.50	II
30	8025.56	II
16	8070.71	I
10	8094.43	I
16	8120.36	I
10	8241.55	II
12	8261.09	I
16	8418.23	II
11	8495.82	I
12	8539.08	II
10	8612.64	I
10 h	8647.66	I
11 h	8702.38	II
25	8772.14	II
12	8810.84	I
30	8891.20	II

Ce III
Ref. 136, 305 — J.R.

Intensity	Wavelength	
	Vacuum	
100	840.24	III
20	844.11	III
40	845.02	III
20	847.88	III
200	851.18	III
200	852.63	III
200	853.47	III
60	853.78	III
200	855.16	III
400	858.30	III
200	860.15	III
200	862.25	III
40	868.74	III
20	869.51	III
40	869.84	III
20	871.15	III
40	871.27	III
20	880.68	III
60	881.75	III
60	884.04	III
20	885.22	III
30	888.39	III
80	892.75	III
20	899.32	III
100	912.77	III
40	937.04	III
40	999.26	III
20	1025.25	III
20	1025.29	III
20	1026.28	III
40	1029.37	III
20	1034.55	III
20	1041.14	III
100	1042.74	III
50	1051.61	III
70	1057.40	III
100	1057.66	III
100	1058.46	III
50	1062.99	III
30	1063.26	III
50	1063.51	III
100	1067.76	III
100	1068.69	III
20	1070.54	III
200	1072.79	III
40	1073.69	III
30	1079.35	III
20	1080.82	III
30	1088.70	III
30	1090.03	III
20	1092.48	III

Intensity	Wavelength	
20	1099.25	III
40	1100.71	III
20	1107.09	III
20	1111.19	III
20	1116.30	III
20	1125.58	III
20	1129.73	III
20	1132.74	III
100	1142.55	III
20	1192.41	III
50	1201.87	III
20	1204.05	III
20	1719.43	III
20	1796.89	III
20	1836.66	III
20	1836.99	III
30	1862.32	III
30	1950.36	III
20	1990.54	III
	Air	
100	2033.34	III
100	2057.65	III
100	2077.87	III
200	2083.32	III
100	2089.96	III
400	2109.07	III
100	2122.55	III
500	2136.95	III
1000	2151.44	III
3000	2166.88	III
2000	2169.48	III
5000	2180.64	III
1000	2183.71	III
3000	2203.15	III
3000	2218.11	III
5000	2222.01	III
5000	2225.08	III
3000	2227.84	III
3000	2228.05	III
5000	2242.29	III
2000	2249.25	III
2000	2264.85	III
2000	2268.20	III
2000	2287.82	III
2000	2298.70	III
3000	2300.65	III
4000	2302.09	III
5000	2317.34	III
10000	2318.64	III
5000	2324.31	III
2000	2337.66	III
5000	2350.10	III
2000	2362.54	III
2000	2367.77	TTT
10000	2372.34	III
5000	2377.07	III
5000	2377.48	III
10000	2380.12	III
3000	2382.28	III
3000	2385.06	III
5000	2395.04	III
3000	2406.15	III
4000	2408.08	III
2000	2410.26	III
5000	2415.60	III
2000	2417.01	III
2000	2423.02	III
3000	2428.64	III
5000	2430.24	III
10000	2431.45	III
15000	2439.80	III
3000	2441.55	III
2000	2444.78	III
10000	2454.32	III
10000	2469.95	III
3000	2471.66	III
5000	2477.25	III
8000	2479.44	III
3000	2479.51	III
10000	2483.82	III
10000	2497.50	III
3000	2503.56	III
2000	2504.43	III
20000	2531.99	III
3000	2539.27	III
3000	2557.49	III
4000	2577.67	III
2000	2578.30	III
2000	2584.71	III
10000	2603.59	III
2000	2607.96	III
2000	2615.79	III
2000	2649.38	III
2000	2662.81	III
3000	2719.30	III
2000	2730.04	III
3000	2743.71	III
4000	2748.90	III
4000	2754.87	III
4000	2768.28	III
3000	2849.40	III

Cerium (Cont.)

Intensity	Wavelength	Spectrum
2000	2861.39	III
4000	2907.05	III
10000	2923.81	III
5000	2925.26	III
10000	2931.54	III
2000	2948.53	III
5000	2973.72	III
10000	3022.75	III
50000	3031.58	III
95000	3055.59	III
20000	3056.56	III
40000	3057.23	III
20000	3057.58	III
40000	3085.10	III
20000	106.98	III
30000	3110.53	III
30000	3121.56	III
20000	3141.29	III
20000	3143.96	III
20000	3147.06	III
20000	3228.57	III
3000	3234.20	III
4000	3267.76	III
3000	3267.94	III
20000	3353.29	III
10000	3395.77	III
4000	3398.91	III
30000	3427.36	III
40000	3443.63	III
30000	3454.39	III
40000	3459.39	III
60000	3470.92	III
50000	3497.81	III
60000	3504.64	III
500	3514.41	III
50000	3544.07	III
3000	3784.29	III
800	3936.80	III
300	3957.10	III
500	4169.42	III
300	4191.70	III
500	4194.83	III
300	4213.26	III
300	4217.13	III
400	4284.77	III
300	4304.71	III
600	4346.35	III
400	4389.97	III
600	4448.32	III
500	4485.27	III
1000	4521.92	III
1000	4535.73	III
300	4576.90	III
500	4627.60	III
300	4766.07	III
500	4976.45	III
500	5650.97	III
1000	5664.20	III
500	5691.08	III
300	5710.59	III
500	5749.47	III
500	5949.83	III
2000	5962.22	III
500	5962.71	III
400	5979.56	III
1000	5983.40	III
3000	6002.63	III
10000	6032.54	III
10000	6060.91	III
500	6061.79	III
500	6097.35	III
500	6098.87	III
500	6135.10	III
300	6287.79	III
500	6308.16	III
300	6341.75	III
1000	6944.94	III
700	7739.04	III
300	7758.27	III
500	7826.80	III
300	7948.64	III
500	7960.31	III
500	7991.01	III
400	8030.80	III
300	8084.12	III
400	8177.33	III
300	8186.03	III
300	8222.16	III
300	9056.53	III
300	9079.58	III
400	9328.20	III
300	9367.03	III
300	9567.37	III
400	10458.37	III
400	10494.42	III
300	10534.36	III
400	10684.46	III
15	12756.96	III
12	12821.62	III
80	15847.58	III
80	15956.79	III
12	15960.59	III
87	16128.75	III
42	18579.82	III
38	19141.29	III
27	19377.15	III
26	19466.14	III
20	19498.14	III
55	19524.18	III
30	20685.63	III
12	21380.23	III

Ce IV
Ref. 166 — J.R.

Intensity	Wavelength	
	Vacuum	
2	447.58	IV
1	443.11	IV
8	558.92	IV
8	571.59	IV
40	741.19	IV
30	754.60	IV
12	755.75	IV
6	975.20	IV
5	1009.31	IV
2	1022.12	IV
9	1057.67	IV
1	1059.64	IV
50	1289.41	IV
75	1332.16	IV
75	1372.72	IV
2	1577.60	IV
1	1572.62	IV
15	1641.58	IV
20	1775.30	IV
20	1779.03	IV
35	1914.75	IV
10	1937.21	IV
	Air	
100	2000.42	IV
35	2003.11	IV
100	2009.94	IV
3	2433.50	IV
5	2445.50	IV

Ce V
Ref. 261 — J.R.

Intensity	Wavelength	
	Vacuum	
100	365.66	V
300	399.36	V
150	404.21	V
200	482.96	V
100	552.13	V

CESIUM (Cs)
Z = 55
Cs I and II

Ref. 82, 154, 155, 200, 263, 325— C.J.S.

Intensity		Wavelength	Spectrum
		Vacuum	
250		591.04	II
2000		639.36	II
500		668.39	II
15000		718.14	II
15000		808.76	II
15000		813.84	II
35000		901.27	II
40000		926.66	II
3		1656.15	II
7		1689.46	II
2		1691.83	II
7		1717.64	II
7		1718.97	II
3		1727.79	II
3		1736.77	II
8		1807.83	II
8		1813.75	II
7		1815.16	II
18		1840.50	II
13		1859.16	II
7		1864.83	II
7		1876.72	II
18		1883.93	II
5		1914.61	II
29		1935.19	II
		Air	
72		2024.97	II
46		2028.32	II
180		2080.05	II
160		2087.20	II
100		2102.22	II
160		2205.52	II
190		2220.53	II
120		2245.81	II
1600		2267.65	II
750		2273.84	II
220		2285.41	II
250		2286.15	II
330		2315.69	II
1400		2332.46	II
240		2375.86	II
220		2379.27	II
1000		2392.86	II
370		2425.17	II
220		2543.93	II
630		2596.99	II
450		2609.43	II
170		2628.86	II
300		2699.18	II
180		2789.78	II
150		2793.31	II
680	w	2816.92	II
150		2829.04	II
140	c	2866.36	II
1600		2931.08	II
290		3151.19	II
520	c	3265.91	II
650	w	3267.12	II
850	c	3271.63	II
570		3368.57	II
150		3514.05	II
110	w	3615.01	II
170	w	3680.11	II
130		3732.56	II
1100	w	3785.44	II
2700		3805.12	II
520	c	3861.50	II
2100	c	3876.15	I
600	c	3888.61	I
3400		3896.99	II
4200		3959.51	II
2500		3965.20	II
2100		3974.25	II
8000		4039.85	II
2000		4068.78	II
320	w	4151.27	II
2000		4213.14	II
1900		4232.20	II
980	c	4234.41	II
14000		4264.70	II
18000	w	4277.13	II
5100		4288.38	II
1900		4300.65	II
7600		4363.30	II
3700		4373.04	II
960		4384.44	II
3900	w	4405.26	II
12000		4501.55	II
20000		4526.74	II
4100		4538.97	II
1000	c	4555.28	I
460		4593.17	I
100000		4603.79	II
4200	c	4616.17	II
2800		4646.52	II
990		4670.29	II
1500		4732.99	II
7000		4763.64	II
1900		4786.38	II
25000		4830.19	II
19000		4870.04	II
4900	c	4880.05	II
37000		4952.85	II
8200		4972.60	II
27000		5043.80	II
2800	c	5059.87	II
2900		5096.60	II
6500	c	5209.58	II
75000		5227.04	II
29000		5249.38	II
11000		5274.05	II
1300		5306.60	II
10000	c	5349.13	II
22000		5370.99	II
1200		5402.78	II
2900		5419.67	II
60	c	5465.94	I
37		5502.88	I
39000		5563.02	II
100		5635.21	I
210	c	5664.02	I
27		5745.72	I
4500		5814.16	II
24000		5831.16	II
59	c	5838.83	I
300		5845.14	I
51000		5925.63	II
1400	c	5984.39	II
640		6010.49	I
86		6034.09	I
760		6076.72	II
9800		6128.61	II
1000		6213.10	I
170		6217.60	I
320	c	6354.55	I
2000		6419.52	II
8300		6495.53	II
10000	w	6536.44	II
490		6586.51	I
530	c	6628.01	II
97		6628.66	I
8800		6646.57	II
3300	c	6723.28	I
9600		6724.47	II
200		6824.65	I
300		6870.45	I
37000		6955.50	II
4800		6973.30	I
16000		6979.67	II
980		6983.49	I
2300		7130.54	II
13000	w	7149.54	II
1600		7160.90	II
630		7188.37	II
1100		7206.04	II
790		7228.53	I
960		7248.88	II
130		7279.90	I
1100		7279.96	I
780		7369.36	II
550		7437.78	II
440	w	7523.39	II
2600	c	7608.90	I
910	w	7651.95	II
310	h	7746.98	II
2400		7852.52	II
3300		7943.88	I
22000		7997.44	II
2100		8012.98	II
3500		8015.73	I
8200	c	8047.13	II
1300		8078.50	II
510		8078.94	I
4500		8079.04	I
59000	cr	8521.13	I
680		8521.62	II
1700		8608.31	II
420	w	8695.60	II
15000	c	8761.41	I
840		8775.42	II
340	w	8857.39	II
61000	cr	8943.47	I
18000		9172.32	I
5200		9208.53	I
540	c	9212.36	II
3300	c	9220.75	II
310	c	9718.11	II
630	c	9932.91	II
1400		9994.79	II
19000		10024.36	I
4800		10123.41	I
26000		10123.60	I
3000	c	10176.02	II
2700		10379.66	II
1300		10480.93	II
4700		10504.51	II
1900	w	10807.88	II
610	w	11324.34	II
530	c	11496.56	II
410	c	11704.18	II
330	c	11797.93	II
710	c	11840.84	II
1100		12604.29	II
2000	c	12735.52	II
400	c	12746.34	II
850		13406.83	II
2900		13424.31	I
38000	c	13588.29	I
8400		13602.56	I
4200	c	13692.91	II
5700		13758.81	I
1400	c	13868.82	II
880	c	14482.21	II
55000	c	14694.91	I
350		14906.91	II
1900	c	15293.80	II
1600	c	15356.61	II
620		15445.47	II
940	c	15735.48	II
1100	c	16426.14	II
820		16535.63	I
1500		17012.32	I
340		18160.84	II
430		18179.35	II
710		18221.36	II
390		18222.40	II
510		18404.77	II
310		18407.23	II
590		18509.52	II
610		18509.85	II
100	c	18742.18	II
530	c	18921.76	II
150	c	18986.48	II
180		19924.85	II
390		20110.77	II
760		20138.47	I
200		20301.54	II
220		20443.87	II
92	w	21103.42	II
79	c	22344.98	II
78	c	22448.98	II
880		22811.86	I
1100		23037.98	I
3900		23344.47	I

Cesium (Cont.)

Intensity		Wavelength	
320	c	23408.41	II
180		24132.52	II
4400		24251.21	I
850		24374.96	I
120	c	24528.78	II
140		24810.89	II
170		25189.83	II
900		25220.37	II
240		25733.29	II
890	d	25763.51	II
500		25764.73	I
120		26448.83	II
340		26503.26	II
190		26727.33	II
17		26954.56	II
22		26978.68	II
23		27187.98	II
9		28649.38	II
680	c	29310.06	I
7		29542.08	II
2800		30103.27	I
9		30487.16	II
610	c	30953.06	I
1100		34900.13	I
190		36131.00	I
2	c	39177.28	I
2	d	39421.25	I
1		39424.11	I

Cs III
Ref. 78, 200, 201—C.J.S.

Intensity		Wavelength	
		Vacuum	
10000		614.01	III
2000		638.17	III
2500		666.25	III
5000		691.60	III
3500		703.89	III
20000		721.79	III
20000		722.20	III
5000		731.56	III
12000		740.29	III
7500		830.39	III
15000		920.35	III
25000	c	1054.79	III
17	c	1673.99	III
12	c	1705.25	III
10		1801.83	III
20	c	1822.40	III
11		1823.93	III
12		1824.70	III
12		1841.80	III
25		1915.50	III
25	c	1923.29	III
12		1961.33	III
17		1996.56	III
		Air	
710		2035.11	III
120		2056.43	III
330		2076.43	III
540		2077.30	III
410		2088.68	III
210		2101.63	III
200		2141.47	III
1000		2316.88	III
230		2325.95	III
390		2340.49	III
1600		2455.81	III
1600		2477.57	III
890		2485.45	III
410		2495.07	III
1400		2525.67	III
430		2573.05	III
16000		2596.86	III
390		2610.12	III
6200		2630.51	III
370		2700.32	III
710		2701.20	III
390		2776.44	III
270		2810.87	III
630		2845.70	III
3100		2859.32	III
200		2893.85	III
180		2921.13	III
3200		2976.86	III
210		3001.28	III
1700		3066.59	III
1100	c	3149.36	III
1400		3152.36	III
8400		3268.32	III
1300		3315.51	III
550		3340.60	III
430		3344.02	III
1200		3349.46	III
400		3463.45	III
580		3476.83	III
480		3559.82	III
7200		3597.45	III
1300		3608.31	III
2300		3618.19	III
300	c	3641.34	III
520		3651.08	III
4800		3661.40	III
640		3699.50	III
430		3837.46	III
2900		3888.37	III
2700		3925.60	III
680	c	4001.70	III
3100		4006.55	III
420		4006.78	III
520		4043.42	III
370		4403.86	III
1200		4410.22	III
940		4425.60	III
530		4471.48	III
1200		4506.72	III
590		4522.86	III
420	h	4620.61	III
210		4665.52	III
140		4851.59	III
370		5035.72	III
230		5380.79	III
140		5950.14	III
110		5979.97	III
150		6043.99	III
870		6079.86	III
330		6150.42	III
450		6242.96	III
510		6456.33	III
400		6753.12	III
1900	c	7219.60	III

Cs IV
Ref. 259 — J.R.

Intensity	Wavelength	
	Vacuum	
35	707.20	IV
60	759.57	IV
5	778.21	IV
500	824.80	IV
350	828.86	IV
400	868.18	IV
1000	874.84	IV
400	896.92	IV
400	923.02	IV
600	986.14	IV
300	995.14	IV
60	1019.13	IV
550	1068.95	IV
200	1282.66	IV

CHLORINE (Cl)
Z = 17
Cl I and II
Ref 238, 239 — L.J.R.

Intensity		Wavelength	
		Vacuum	
350		559.305	II
400		571.904	II
800		574.406	II
500		586.24	II
700		618.057	II
600		619.982	II
800		620.298	II
700		626.735	II
800		635.881	II
1000		636.626	II
1000		650.894	II
1000		659.811	II
1300		661.841	II
2000		663.074	II
1500		682.053	II
1500		687.656	II
1500		693.594	II
2000		725.271	II
2500		728.951	II
2000		777.562	II
5000		787.580	II
5000		788.740	II
5000		793.342	II
6000		839.297	II
8000		839.599	II
7000	p	841.41	II
5000		851.691	II
2000		888.026	II
2000		893.549	II
2000		961.499	II
30		969.92	I
40		978.284	I
25		998.372	I
25		998.432	I
75		1002.346	I
150		1013.664	I
90		1025.553	I
6000		1063.831	II
3000		1067.945	II
9000		1071.036	II
6000		1071.767	II
5000		1075.230	II
5000		1079.080	II
200		1084.667	I
200		1085.171	I
250		1085.304	I
400		1088.06	I
350		1090.271	I
250		1090.982	I
250		1092.437	I
400		1094.769	I
350		1095.148	I
350		1095.662	I
400		1095.797	I
250		1096.810	I
300		1097.369	I
200		1098.068	I
200		1099.523	I
500		1107.528	I
800		1139.214	II
800		1167.148	I
3000		1179.293	I
1200		1188.774	I
900		1201.353	I
3000		1335.726	I
10000		1347.240	I
5000		1351.657	I
12000		1363.447	I
2500		1373.116	I
20000		1379.528	I
25000		1389.693	I
20000		1389.957	I
12000		1396.527	I
500		1441.470	II
500		1528.569	II
500		1542.942	II
500		1558.144	II
500		1565.050	II
500		1857.488	II
450	h	1997.307	II
		Air	
450		2032.116	II
350	h	2088.583	II
350	h	2091.458	II
170		2427.476	II
360		2434.07	II
340		2498.53	II
470		2502.74	II
260		2546.96	II
500		2549.88	II
460		2564.84	II
320		2603.31	II
950		2658.72	II
750		2676.95	II
1200		2688.04	II
410		2912.05	II
950		2996.65	II
500		3006.06	II
950		3057.96	II
1300		3071.32	II
1400		3092.19	II
1200		3123.72	II
1900		3315.43	II
1200		3329.10	II
2500		3353.35	II
20		3726.54	I
1200		3749.96	II
1000		3781.17	II
1500		3798.76	II
1900		3805.18	II
1300		3809.46	II
1700		3820.20	II
2800		3827.59	II
4500		3833.35	II
2500		3843.20	II
3100		3845.37	II
3900		3845.65	II
1500		3845.80	II
10000		3850.99	II
7900		3851.37	II
1200		3851.65	II
25000		3860.83	II
4400		3860.99	II
1000		3861.37	II
1500		3913.87	II
1100		3916.63	II
20		3944.82	I
20		4104.79	I
10000	h	4132.50	II
65		4209.67	I
50		4226.42	I
60		4264.58	I
100		4363.27	I
100		4369.50	I
5000		4372.93	II
100		4379.90	I
100		4389.75	I
90		4390.40	I
90		4403.03	I
100		4438.49	I
90		4475.30	I
1500		4489.91	II
100		4526.19	I
80		4600.98	I
40		4623.938	I
50		4654.040	I
80		4661.207	I
45		4691.523	I
40		4721.255	I
45		4740.729	II
4300		4768.65	II
13000		4781.32	II
99000		4794.55	II
29000		4810.06	II
16000		4819.47	II
81000		4896.77	II
47000		4904.78	II
26000		4917.73	II
10000		4995.48	II
26000		5078.26	II
30		5099.789	I
56000		5217.94	II
23000		5221.36	II
15000		5392.12	II
99000		5423.23	II
10000		5423.51	II
19000		5443.37	II
10000		5444.21	II
5600		5457.02	II
40		5532.162	I
50	d	5796.305	I
45		5799.914	I
30		5856.742	I
100	d	5948.58	I
50		6019.812	I
35		6082.61	I
1900		6094.69	II
160		6114.43	I
200		6140.245	I
160		6194.757	I
160		6398.66	I
150		6434.833	I
150		6531.43	I
1400		6661.67	II
150		6678.43	I
1300		6686.02	II
1200		6713.41	II
150		6840.29	I
300		6932.903	I
300		6981.886	I
600		7086.814	I
7500		7256.62	I
5000		7414.11	I
550		7462.370	I
550		7489.47	I
700		7492.118	I
11000		7547.072	I
2300		7672.42	I
450		7702.828	I
7000		7717.581	I
10000		7744.97	I
2200		7769.16	I
650		7771.09	I
2200		7821.36	I
1700		7830.75	I
3000		7878.22	I
220		7893.34	I
2300		7899.31	I
1800		7915.08	I
3000		7924.645	I
2100		7933.89	I
1700		7935.012	I
650		7952.52	I
1500		7974.72	I
1300		7976.97	I
600		7980.60	I
2900		7997.85	I
2200		8015.61	I
1100		8023.33	I
400		8051.07	I
1700		8084.51	I
2200		8085.56	I
3000		8086.67	I
1300		8087.73	I
250		8094.67	I
2500		8194.42	I
2200		8199.13	I
2200		8200.21	I
800		8203.78	I
18000		8212.04	I
3000		8220.45	I
20000		8221.74	I
18000		8333.31	I
1000		8360.71	II
560		8361.84	II
99900		8375.94	I
180		8382.61	II
100		8392.02	II
400		8406.199	I
15000		8428.25	I
2200		8467.34	I
2200		8550.44	I
20000		8575.24	I
750		8578.02	I
75000		8585.97	I
450		8628.54	I
300		8641.71	I
3500		8686.26	I
2200		8912.92	I
3000		8948.06	I
2000		9038.982	I
2500		9045.43	I
1000		9069.656	I
2000		9073.17	I
7500		9121.15	I

Chlorine (Cont.)

Cl I and II

Intensity	Wavelength	
3000	9191.731	I
500	9197.596	I
4000	9288.86	I
1500	9393.862	I
3500	9452.10	I
500	9486.964	I
1000	9584.801	I
3500	9592.22	I
250	9632.509	I
1000	9702.439	I
250	9744.426	I
200	9807.057	I
400	9875.970	I
331	10392.549	I
38	10432.83	II
10	10506.62	II
14	10509.12	II
19	10512.46	II
25	10514.17	II
9	10801.47	II
5	10885.42	II
1	10955.71	II
300	11123.05	I
231	11392.62	I
269	11409.69	I
1000	11436.33	I
180	11720.56	I
195	11866.76	I
172	12021.7	I
350	13243.8	I
310	13296.0	I
550	13346.8	I
525	13821.7	I
148	14369.7	I
294	14931.7	I
269	15108.0	I
381	15465.1	I
169	15467.6	I
1094	15520.3	I
1487	15730.1	I
193	15818.4	I
2780	15869.7	I
277	15883.3	I
342	15928.9	I
735	15960.0	I
283	15970.5	I
129	16077.6	I
259	16198.5	I
227	19370.3	I
717	19755.3	I
185	19766.8	I
227	20199.4	I
85	20370.1	I
100	24470.0	I
	39603.7	I
	39615.3	I
	39716.0	I
	39744.0	I
	39750.9	I
	39875.3	I
	39881.0	I
	39985.7	I
	40085.5	I
	40089.5	I
	40171.0	I
	40310.3	I
	40335.4	I
	40532.2	I

Cl III
Ref. 28, 30 — L.J.R.

Intensity	Wavelength	
	Vacuum	
100	406.27	III
400	411.37	III
400	411.81	III
600	556.23	III
700	556.61	III
700	557.12	III
600	2965.56	III
600	3104.46	III
800	3139.34	III
900	3191.45	III
700	3289.80	III
700	3320.57	III
800	3329.06	III
900	3340.42	III
800	3392.89	III
800	3393.45	III
900	3530.03	III
800	3560.68	III
900	3602.10	III
800	3612.85	III
700	3622.69	III
700	3656.95	III
700	3670.28	III
700	3682.05	III
600	3705.45	III
600	3707.34	III
800	3720.45	III
800	3748.81	III
500	3779.35	III
500	3925.87	III
700	3991.50	III
600	4018.50	III
600	4059.07	III
500	4104.23	III
500	4106.83	III
400	4370.91	III
500	4608.21	III
300	4703.14	III
100	4863.75	III
10	4971.64	III
700	561.53	III
700	561.68	III
700	561.74	III
500	606.35	III
400	621.28	III
300	670.38	III
300	673.13	III
100	936.28	III
500	1005.28	III
600	1008.78	III
700	1015.02	III
600	1822.50	III
500	1828.40	III
500	1901.61	III
500	1983.61	III
	Air	
400	2006.84	III
700	2253.07	III
500	2268.95	III
500	2278.34	III
700	2283.93	III
600	2323.50	III
500	2336.45	III
600	2340.64	III
600	2359.67	III
600	2370.37	III
700	2416.42	III
600	2447.14	III
600	2448.58	III
500	2486.91	III
500	2532.48	III
600	2580.67	III
500	2603.59	III
500	2632.67	III
500	2633.18	III
600	2665.54	III
700	2710.37	III

Cl IV
Ref. 11, 28, 30, 31 — L.J.R.

Intensity	Wavelength	
	Vacuum	
300	319.62	IV
200	331.84	IV
400	437.83	IV
400	464.86	IV
800	486.17	IV
800	534.73	IV
700	535.67	IV
600	536.15	IV
900	537.61	IV
600	538.12	IV
500	549.22	IV
400	550.02	IV
700	552.02	IV
600	553.30	IV
700	554.62	IV
500	601.50	IV
500	604.59	IV
400	608.90	IV
400	612.07	IV
400	653.70	IV
400	745.21	IV
400	831.43	IV
500	834.84	IV
500	834.97	IV
400	840.81	IV
600	840.93	IV
	865.3	IV
500	973.21	IV
600	977.56	IV
400	977.90	IV
700	984.95	IV
400	985.75	IV
300	1537.21	IV
200	1539.30	IV
200	1545.19	IV
200	1549.15	IV
200	1622.86	IV
	Air	
400	2701.36	IV
500	2724.03	IV
500	2751.23	IV
400	2770.64	IV
700	2782.47	IV
400	2835.4	IV
500	3063.13	IV
600	3076.68	IV
200	3167.87	IV

Cl V
Ref. 11, 28, 30, 85, 233 — L.J.R.

Intensity	Wavelength	
	Vacuum	
300	287.33	V
300	373.78	V
400	390.15	V
500	392.43	V
300	536.53	V
400	537.01	V
300	537.46	V
500	538.03	V
400	538.68	V
800	542.23	V
600	542.30	V
400	542.87	V
1000	545.11	V
600	546.33	V
1000	547.63	V
400	633.19	V
400	635.32	V
400	681.92	V
400	683.17	V
400	688.93	V
	715.55	V
	716.19	V
400	883.13	V
400	894.34	V
100	894.91	V
	914.5	V

CHROMIUM (Cr)
Z = 24

Cr I and II
Ref. 1 — C.H.C.

Intensity	Wavelength	
	Vacuum	
19000	2055.52	II
14000	2061.49	II
8900	2065.42	II
80 h	2364.71	I
130	2383.33	I
140	2408.62	I
170	2496.31	I
110	2502.53	I
190	2504.31	I
50	2508.11	I
60	2508.98	I
40	2513.62	I
110	2516.92	I
80	2518.71	I
390	2519.52	I
190	2527.12	I
40	2530.45	I
70	2534.34	II
50	2545.64	I
160	2549.54	I
40	2553.06	I
80	2557.15	I
130	2560.69	I
150	2571.74	I
100	2577.65	I
50	2588.20	I
380	2591.85	I
35	2603.57	I
35	2622.86	I
22	2625.32	I
18	2626.60	I
18	2629.82	I
35	2642.12	I
250	2653.59	II
250	2658.59	II
70	2661.73	II
320	2663.42	II
70	2663.68	II
440	2666.02	II
280	2668.71	II
350	2671.81	II
280	2672.83	II
1800	2677.16	II
35	2678.16	I
320	2678.79	II
18	2680.34	II
230	2687.09	II
60	2688.04	I
55	2688.29	II
26	2690.26	I
280	2691.04	II
35	2693.52	II
35	2697.91	II
180	2698.41	II
180	2698.69	II
18	2700.60	I
110	2701.99	I
18	2702.53	I
70	2703.48	I
35	2703.55	II
18	2703.86	II
60	2705.43	I
35	2708.79	II
140	2709.31	II
45	2712.31	II
55	2716.18	I
45	2717.51	II
170	2718.43	II
18	2722.75	II
420 h	2724.04	II
45	2726.51	I
280 h	2727.26	I
170 h	2731.91	I
70	2736.47	I
70	2739.38	II
95	2740.10	I
95	2741.07	II
95	2742.03	I
250	2742.17	II
35	2743.64	I
110 h	2746.21	II
330	2748.29	I
390	2748.98	II
45	2750.73	II
280	2751.60	I
110 h	2751.87	II
35	2752.88	I
22	2754.28	I
22	2754.90	I
22	2755.27	I
150	2756.75	I
350	2757.10	I
60	2757.72	II
80	2758.98	II
45	2759.39	II
90 h	2759.73	II
750	2761.76	II
22	2762.59	I
80 h	2763.06	I
750	2764.35	II
22	2766.54	I
250 h	2767.54	I
18	2769.92	II
45	2771.45	I
22	2778.06	II
80	2779.14	I
610	2780.30	I
70	2780.70	I
35	2785.70	II
35	2787.63	I
90	2787.84	II
55	2792.16	II
70	2798.67	II
80	2800.77	II
60	2812.01	II
45	2818.36	II
180	2822.01	II
22	2822.37	I
180	2826.75	II
70	2830.47	II
2500	2834.26	II
45	2835.63	II
55	2836.48	II
110	2838.79	II
1700	2840.02	II
22	2843.25	II
45	2846.02	I
1200	2849.29	I
120	2849.84	II
55	2851.36	II
55	2853.22	II
880	2855.07	II
90	2855.68	II
70	2856.77	II
610	2857.40	II
440	2858.91	II
790	2860.93	II
750	2862.57	II
55	2865.11	II
610	2865.33	II
90	2866.74	II
480	2867.10	II
210	2867.65	II
110	2870.44	II
160	2871.63	I
90	2873.48	II
320	2873.82	II
230	2875.99	II
180	2876.24	II
70	2877.98	II
120	2878.45	I
95	2879.27	II
30	2880.26	I
170	2881.14	I
55	2887.00	I
700	2888.74	II
55	2889.29	I
55	2889.82	II
370	2891.42	I
190	2893.25	II
55	2894.17	I
55	2896.46	II

Chromium (Cont.)

Int.		λ (Å)	Sp.	Int.		λ (Å)	Sp.	Int.		λ (Å)	Sp.	Int.		λ (Å)	Sp.
210		2896.75	I	30		3208.59	II	220		3641.83	I	40		3993.97	I
55	d	2897.67	II	170		3209.18	II	45		3646.16	I	160		4001.44	I
		2897.73	II	140		3217.40	II	85		3648.53	I	120		4012.47	II
90		2898.54	II	30		3229.20	I	220		3649.00	I	30		4014.67	I
80		2899.21	I	28		3234.06	II	170		3653.91	I	85		4022.26	I
55		2899.48	II	65		3237.73	I	220		3656.26	I	70		4025.01	I
26		2903.97	II	120		3245.54	I	45		3662.84	I	120		4026.17	I
55		2904.68	I	130		3251.84	I	130		3663.21	I	85		4027.10	I
180		2905.49	I	130		3257.82	I	45		3665.98	I	85		4030.68	I
260		2909.05	I	95		3259.98	I	95		3666.64	I	190		4039.10	I
260		2910.90	I	30		3295.43	II	55		3668.03	I	160		4048.78	I
250		2911.14	I	24		3307.02	II	65		3676.32	I	120		4058.77	I
45		2911.68	II	55		3324.06	II	40		3677.68	II	40		4065.72	I
60		2913.73	I	28		3326.59	I	55		3677.89	II	85		4066.94	I
22		2915.23	II	30		3328.35	II	40		3679.82	I	35		4074.86	I
22		2915.46	II	30		3329.05	I	19		3681.69	I	40		4076.06	I
90		2921.24	II	95		3336.33	II	120		3685.55	I	40		4077.09	I
60		2921.82	II	130		3339.80	II	130		3686.80	I	40		4077.68	I
60		2927.08	II	110		3342.59	II	130		3687.25	I	40		4104.87	I
80		2928.15	II	30		3343.34	I	75		3687.54	I	40		4109.58	I
95		2928.30	II	95		3346.02	I	19		3688.46	I	40		4120.61	I
26		2929.44	II	95		3346.74	II	75		3712.95	II	40		4121.82	I
35		2930.85	II	95		3347.84	II	40		3716.53	I	35		4122.16	I
26		2932.70	II	65		3349.01	I	130		3730.81	I	40		4123.39	I
55		2933.97	II	55		3349.32	I	150		3732.03	I	140		4126.52	I
90		2935.14	II	30		3351.60	I	95		3742.97	I	35		4127.30	I
45		2940.22	II	55		3351.97	I	480		3743.58	I	40		4127.64	I
60		2946.84	II	55	h	3353.03	I	570		3743.88	I	40		4131.36	I
55		2953.36	II			3353.13	II	85		3744.49	I	30		4152.78	I
45		2953.71	II	170		3358.50	II	55		3748.61	I	120		4153.82	I
55		2961.73	II	160		3360.30	II	340		3749.00	I	85		4161.42	I
45		2966.05	II	65		3361.77	II	50		3757.17	I	140		4163.62	I
480		2967.64	I	55		3362.21	I	230		3757.66	I	70		4165.52	I
480		2971.11	I	430		3368.05	II	60		3758.04	I	40		4169.84	I
210		2971.91	II	30		3376.40	I	24		3767.43	I	35		4170.20	I
480		2975.48	I	55		3378.34	I	260		3768.24	I	40		4172.77	I
30		2976.72	II	30		3379.17	I	95		3768.73	I	170		4174.80	I
190		2979.74	II	30		3379.37	II	95		3788.86	I	30		4175.94	I
350		2980.79	I	95		3379.83	II	95		3790.45	I	170		4179.26	I
110		2985.32	II	140		3382.68	II	130		3791.38	I	35		4184.90	I
480		2985.85	I	95		3391.43	II	130		3792.14	I	30		4186.36	I
1500		2986.00	I	55		3392.99	II	120		3793.29	I	35		4190.13	I
2100		2986.47	I	70		3393.84	II	130		3793.86	I	85		4191.27	I
660		2988.65	I	55		3394.30	II	85		3794.61	I	35		4192.10	I
160		2989.19	II	30		3402.40	II	140		3797.13	I	85		4193.66	I
480		2991.89	I	170		3403.32	II	200		3797.72	I	70		4194.95	I
230		2994.07	I	360		3408.76	II	530		3804.80	I	40		4197.23	I
300		2995.10	I	210		3421.21	II	110		3806.83	I	85		4198.52	I
700		2996.58	I	270		3422.74	II	110		3807.93	I	60		4203.59	I
210		2998.79	I	140		3433.31	II	180		3815.43	I	40		4204.47	I
1100		3000.89	I	270		3433.60	I	70		3818.48	I	35		4208.36	I
750		3005.06	I	55		3434.11	I	180		3819.56	I	110		4209.37	I
140		3013.03	I	160		3436.19	I	70		3823.52	I	40		4209.76	I
710		3013.71	I	70		3441.12	I	130		3826.42	I	40		4211.35	I
710		3014.76	I	140		3441.44	I	130		3830.03	I	40		4216.36	I
1400		3014.92	I	30		3443.79	I	380		3841.28	I	85		4217.63	I
710		3015.19	I	170		3445.62	I	190		3848.98	I	40		4221.57	I
2800		3017.57	I	30		3447.02	I	140		3849.36	I	40		4222.73	I
430		3018.50	I	170		3447.43	I	290		3850.04	I	40		4238.96	I
240		3018.82	I	70		3447.76	I	140		3852.22	I	60		4240.70	I
430		3020.67	I	190		3453.33	I	190		3854.22	I	20000		4254.35	I
2800		3021.56	I	40		3453.74	I	110		3855.29	I	70		4255.50	I
1100		3024.35	I	130		3455.60	I	140		3855.57	I	60		4261.35	I
85		3026.65	II	100		3460.43	I	260		3857.63	I	110		4263.14	I
170		3029.16	I	65		3465.25	I	70		3874.53	I	30		4271.06	I
710		3030.24	I	40		3467.02	I	660		3883.29	I	40		4272.91	I
140		3031.35	I	70		3467.72	I	50		3883.66	I	16000		4274.80	I
28		3032.93	II	45		3469.59	I	570		3885.22	I	85		4280.40	I
390		3034.19	I	16		3472.76	I	380		3886.79	I	10000		4289.72	I
550		3037.04	I	24		3472.91	I	60		3891.93	I	40		4291.96	I
80		3039.78	I	40		3473.61	I	260		3894.04	I	85		4295.76	I
550		3040.85	I	70		3481.30	I	40		3897.65	I	70		4297.74	I
		3040.91	II	55		3481.54	I	35		3902.11	I	35		4300.51	I
55		3041.74	II	55		3494.97	I	360		3902.92	I	50		4301.18	I
110		3050.14	II	40		3495.38	II	60		3903.16	I	30		4305.45	I
710		3053.88	I	80		3510.54	I	960		3908.76	I	35		4319.64	I
24		3059.52	II	40		3511.84	II	120	Hd	3911.82	I	60		4325.08	I
85		3065.07	I	120		3550.64	I			3912.00	I	780		4337.57	I
28		3067.16	II	80		3558.52	I	120		3915.84	I	1100		4339.45	I
85		3073.68	II	130		3566.16	I	190		3916.24	I	380		4339.72	I
55		3077.83	I	130		3573.64	I	35		3917.60	I	60		4340.13	I
28		3095.86	I	80		3574.04	I	1900		3919.16	I	1900		4344.51	I
28		3109.34	I	330	h	3574.80	I	600		3921.02	I	70		4346.83	I
28		3110.86	I			3574.94	I	30		3926.65	I	380		4351.05	I
240		3118.65	II	19000		3578.69	I	600		3928.64	I	2300		4351.77	I
45		3119.25	I	160	h	3584.33	I	410		3941.49	I	570		4359.63	I
40		3119.71	I	130		3585.30	II	30		3951.10	II	70		4363.13	I
430		3120.37	II	17000		3593.49	I	40		3952.40	I	530		4371.28	I
28		3122.60	II	350		3601.67	I	35		3953.16	I	70		4373.25	I
470		3124.94	II	40		3602.57	I	1900		3963.69	I	110		4374.16	I
		3125.02	II	85		3603.74	I	120		3969.06	I	70		4375.33	I
120		3128.70	II			3603.78	II	1600		3969.75	I	50		4381.11	I
590		3132.06	II	13000		3605.33	I	85		3971.26	I	530		4384.98	I
140		3136.68	II	40		3608.40	I	1600		3976.66	I	60		4387.50	I
140		3147.23	II	40		3609.48	I	85		3978.68	I	70		4391.75	I
85		3148.44	I	40		3610.05	I	40		3979.80	I	60		4403.50	I
100		3155.15	I	70		3612.61	I	85		3981.23	I	24		4410.30	I
100		3163.76	I	85		3615.64	I	960		3983.91	I	60		4411.09	I
240		3180.70	II	130		3632.84	I	190		3984.34	I	35		4412.25	I
30		3181.43	II	350		3636.59	I	160		3989.99	I	50		4413.87	I
65	h	3188.01	I	630		3639.80	I	960		3991.12	I	60		4424.28	I
220		3197.08	II	85		3640.39	I	160		3991.67	I	24		4428.50	I
24		3198.11	I	70		3641.47	I	190		3992.84	I	50		4430.49	I

Chromium (Cont.)

Intensity		Wavelength		
50		4432.18		I
110		4458.54		I
30		4459.74		I
30		4465.36		I
30		4482.88		I
40		4488.05		I
50		4489.47		I
60		4492.31		I
660		4496.86		I
50		4498.73		I
70		4500.30		I
50		4501.11		I
22		4501.79		I
24		4506.85		I
95		4511.90		I
12		4514.37		I
35		4514.53		I
24		4521.14		I
24		4526.11		I
380		4526.47		I
70	d	4527.34		I
		4527.47		I
24		4529.85		I
380		4530.74		I
50		4535.15		I
240		4535.72		I
40		4539.79		I
240		4540.50		I
240		4540.72		I
35		4541.07		I
19		4541.51		I
24		4542.62		I
140		4544.62		I
24		4545.34		I
600		4545.96		I
50		4556.17		I
22		4558.66		II
19		4564.17		I
120		4565.51		I
95		4569.64		I
120		4571.68		I
22		4575.12		I
360		4580.06		I
24		4586.14		I
360		4591.39		I
70		4595.59		I
50		4600.10		I
480		4600.75		I
50		4601.02		I
240		4613.37		I
600		4616.14		I
70		4619.55		I
85		4621.96		I
70		4622.49		I
24		4622.76		I
550		4626.19		I
24		4632.18		I
40		4637.18		I
50		4637.77		I
50	d	4639.52		I
		4639.70		I
1600		4646.17		I
24		4646.81		I
24		4648.13		I
24		4648.87		I
35		4649.46		I
570		4651.28		I
840		4652.16		I
35		4654.74		I
19		4656.19		I
40		4663.33		I
70		4663.83		I
95		4664.80		I
35		4665.90		I
22		4666.22		I
70		4666.51		I
50		4669.34		I
40		4680.54		I
19		4680.87		I
70		4689.37		I
60		4693.95		I
24		4695.15		I
60		4697.06		I
240	d	4698.46		I
		4698.62	CR	I
35		4700.61		I
190		4708.04		I
240		4718.43		I
50		4723.10		I
50		4724.42		I
50		4727.15		I
24		4729.72		I
120		4730.71		I
140		4737.35		I
19		4745.31		I
70		4752.08		I
340		4756.11		I
50		4764.29		I
22		4766.63		I
30		4767.86		I
190		4789.32		I
95		4792.51		I

Intensity		Wavelength		
120		4801.03		I
110		4829.38		I
14		4836.86		I
17		4861.20		I
70		4861.84		I
140		4879.80		I
35		4885.78		I
19		4885.96		I
130		4887.01		I
19		4888.53		I
35		4903.24		I
260		4922.27		I
110		4936.33		I
70		4942.50		I
110		4954.81		I
35		4964.93		I
60		5013.32		I
17		5051.90		I
17		5065.91		I
40		5067.71		I
40		5072.92		I
30		5110.75		I
17		5113.13		I
17		5123.46		I
50		5139.65		I
14		5144.67		I
70		5166.23		I
35		5177.43		I
70		5184.59		I
70		5192.00		I
12		5193.49		I
85		5196.44		I
35		5200.19		I
5300		5204.52		I
8400		5206.04		I
11000		5208.44		I
19		5214.13		I
30		5221.75		I
85		5224.94		I
12		5226.89		I
19		5238.97		I
30		5243.40		I
290		5247.56		I
60		5254.92		I
60		5255.13		I
19		5261.75		I
530		5264.15		I
30		5265.16		I
180		5265.72		I
35		5272.01		I
30		5273.44		I
95	h	5275.17		I
35	h	5275.69		I
70	h	5276.03		I
19		5280.29		I
10		5287.19		I
340		5296.69		I
70	h	5297.36		I
660		5298.27		I
85		5300.75		I
17		5304.21		I
24		5312.88		I
24		5318.78		I
340	h	5328.34		I
70	h	5329.17		I
17	h	5329.72		I
14		5340.44		I
10		5344.76		I
780		5345.81		I
380		5348.32		I
30		5386.98		I
22		5387.57		I
10		5390.39		I
40		5400.61		I
22		5405.00		I
1400		5409.79		I
12		5442.41		I
19		5463.97		I
19		5480.50		I
24		5628.64		I
7		5642.36		I
12	h	5649.37		I
24		5664.04		I
7	h	5681.20		I
7	h	5682.48		I
24		5694.73		I
40		5698.33		I
24		5702.31		I
12		5712.64		I
24		5712.78		I
7		5719.82		I
7		5746.43		I
7		5753.69		I
12	h	5781.20		I
6	h	5781.81		I
24	h	5783.11		I
30	h	5783.93		I
24	h	5785.00		I
19		5785.82		I
60	h	5787.99		I
180		5791.00		I
35		6330.10		I

Intensity		Wavelength	
22		6362.87	I
19		6661.08	I
11		6669.26	I
5	h	6881.62	I
10	h	6882.38	I
21	h	6883.03	I
27	h	6924.13	I
17	h	6925.20	I
30	h	6978.48	I
11	h	6979.82	I
7		7185.52	I
6	h	7236.20	I
85		7355.90	I
130		7400.21	I
150		7462.31	I
11	h	7942.04	I
5	h	8163.18	I
9		8348.28	I
6		8450.26	I
3		8455.24	I
6		8548.86	I
40		8947.15	I
19		8976.83	I

Cr III
Ref. 412 — C.H.C.

Intensity	Wavelength	
	Vacuum	
20	969.26	III
40	1000.86	III
40	1001.04	III
30	1002.96	III
50	1017.14	III
50	1017.31	III
50	1017.57	III
30	1028.33	III
60	1030.47	III
30	1030.89	III
50	1033.23	III
50	1033.45	III
100	1033.69	III
50	1035.93	III
100	1036.03	III
30	1040.17	III
40	1040.53	III
40	1045.06	III
40	1045.14	III
60	1059.13	III
60	1060.15	III
60	1061.04	III
50	1062.68	III
30	1064.32	III
30	1064.43	III
50	1066.23	III
80	1068.41	III
30	1100.61	III
30	1101.43	III
30	1102.88	III
30	1117.19	III
30	1132.75	III
50	1136.67	III
50	1161.43	III
30	1187.65	III
60	1206.38	III
80	1209.13	III
80	1211.12	III
40	1221.07	III
40	1221.90	III
30	1225.65	III
30	1228.65	III
30	1231.88	III
50	1232.96	III
40	1236.20	III
40	1238.51	III
50	1252.61	III
40	1259.02	III
40	1261.86	III
30	1262.34	III
35	1263.61	III
35	1264.21	III
40	1287.05	III
30	1455.27	III
40	1584.60	III
30	1603.19	III
30	1679.25	III
30	1690.28	III
60	1692.89	III
60	1696.94	III
60	1701.48	III
80	1707.43	III
40	1707.78	III
45	1762.81	III
30	1766.92	III
30	1769.17	III
30	1827.26	III
	AIR	
60	2036.39	III

Intensity	Wavelength	
50	2039.63	III
80	2047.23	III
100	2113.73	III
100	2113.83	III
50	2114.26	III
50	2114.53	III
100	2114.87	III
100	2117.53	III
80	2123.53	III
80	2139.11	III
100	2141.15	III
80	2144.15	III
50	2147.16	III
50	2147.56	III
50	2148.65	III
50	2149.48	III
50	2152.76	III
100	2157.17	III
50	2163.86	III
60	2166.25	III
100	2170.70	III
50	2183.71	III
100	2185.01	III
50	2190.09	III
100	2190.76	III
100	2191.58	III
100	2197.89	III
100	2198.62	III
100	2203.22	III
60	2208.70	III
200	2226.72	III
100	2231.81	III
100	2233.81	III
200	2235.91	III
150	2237.59	III
150	2244.10	III
80	2251.45	III
50	2257.92	III
100	2273.30	III
80	2275.43	III
100	2276.38	III
80	2277.47	III
150	2284.44	III
50	2289.23	III
80	2290.66	III
60	2295.55	III
50	2309.99	III
80	2314.63	III
100	2319.07	III
150	2324.88	III
60	2340.51	III
50	2456.83	III
100	2472.88	III
100	2479.77	III
100	2483.06	III
60	2488.26	III
80	2506.41	III
80	2530.99	III
80	2537.73	III
80	2544.37	III
50	2545.17	III
80	2564.76	III
80	2616.50	III
100	2626.08	III
100	2640.73	III
50	2647.50	III
40	2655.28	III
40	2916.57	III

Cr IV
Ref. 379, 412 — C.H.C.

Intensity	Wavelength	
	Vacuum	
50	575.05	IV
30	576.24	IV
30	576.62	IV
30	595.09	IV
50	612.64	IV
40	613.75	IV
40	614.03	IV
40	614.90	IV
30	615.34	IV
30	615.60	IV
50	616.82	IV
40	618.23	IV
40	619.13	IV
100	620.66	IV
60	621.36	IV
40	622.09	IV
30	623.54	IV
40	625.04	IV
40	625.99	IV
100	629.26	IV
50	629.74	IV
80	630.30	IV
30	632.62	IV
30	637.34	IV
50	637.55	IV
50	638.13	IV

Intensity		Wavelength	
30		638.54	IV
100		666.55	IV
75		667.30	IV
40		677.55	IV
40		687.12	IV
50		688.46	IV
100		693.92	IV
50		695.21	IV
50		705.98	IV
30		712.90	IV
80		1055.89	IV
60		1057.85	IV
30		1367.39	IV
40		1375.05	IV
70		1401.82	IV
100		1417.42	IV
30		1485.05	IV
80		1595.04	IV
90	d	1595.59	IV
100		1658.08	IV
120		1672.66	IV
90		1686.07	IV
100		1690.88	IV
80		1725.26	IV
90		1727.07	IV
100		1732.04	IV
40		1733.98	IV
80		1734.16	IV
50		1739.19	IV
70		1746.88	IV
80		1747.13	IV
110		1755.64	IV
120		1758.51	IV
100		1769.64	IV
100		1777.82	IV
40		1791.09	IV
140		1802.72	IV
130		1812.41	IV
60		1819.23	IV
30		1826.21	IV
30		1826.86	IV
100		1840.14	IV
50		1851.89	IV
100		1863.11	IV
140		1873.89	IV
35		1883.16	IV
40		1937.63	IV
30		1946.59	IV
140		1967.18	IV
120		1972.07	IV
40		1990.25	IV

Air

Intensity		Wavelength	
50	d	2042.91	IV
40		2055.73	IV
70		2299.21	IV
90		2299.59	IV
100		2316.85	IV
40		2324.06	IV
50		2360.40	IV
70		2405.15	IV
60		2423.32	IV

Cr V
Ref. 380 — C.H.C.

Intensity   Wavelength

Vacuum

Intensity	Wavelength	
100	438.62	V
100	464.02	V
50	469.64	V
50	825.60	V
50	968.70	V
50	1045.04	V
60	1060.65	V
50	1112.45	V
60	1114.35	V
100	1116.48	V
80	1117.56	V
50	1118.16	V
150	1121.07	V
150	1127.63	V
100	1193.95	V
80	1196.04	V
50	1210.50	V
50	1259.99	V
100	1263.50	V
150	1465.86	V
50	1481.65	V
50	1482.76	V
50	1484.67	V
100	1489.71	V
150	1497.97	V
170	1519.03	V
220	1579.70	V
170	1591.72	V
150	1603.19	V
60	1638.50	V
50	1639.40	V
200	1837.44	V

COBALT (Co)
Z = 27

Co I and II
Ref. 1, 125, 276 — C.R.C.

Intensity   Wavelength

Vacuum

Intensity		Wavelength	
20		1265.93	II
40		1271.94	II
20		1276.90	II
30		1293.97	II
25		1295.53	II
30		1295.86	II
20		1297.10	II
80		1299.58	II
30		1302.39	II
20		1306.76	II
80		1306.95	II
40		1311.12	II
40		1311.86	II
30		1315.42	II
30		1316.09	II
20		1318.19	II
30		1318.60	II
20		1319.84	II
20		1409.33	II
20		1466.21	II
20		1471.87	II
30		1472.90	II
30		1475.81	II
20		1484.26	II
30		1486.50	II
20		1509.23	II
20		1590.54	II
20		1595.77	II
20		1599.30	II
20		1693.34	II
8		1706.05	II
10		1723.01	II
15	h	1740.55	II
15		1743.39	II
8		1754.21	II
20	d	1808.01	II
10		1837.56	II
15		1839.37	II
1500		1842.34	I
1800		1847.89	I
1800		1852.71	I
2400		1855.05	I
1500		1878.28	I
10		1917.62	II
1800		1936.58	I
1500		1946.79	I
30		1950.09	II
1500		1951.90	I
1800		1954.22	I
1800		1955.17	I
30		1957.42	II
1500		1958.55	I
1500		1961.59	I
1500	h	1968.69	I
1500	h	1968.93	I
3000		1970.71	I
1800		1971.16	I
1800	h	1972.52	I
1500		1973.85	I
1800		1976.97	I
2400	h	1980.89	I
1500		1989.80	I
1800		1990.34	I
1500	l	1998.49	I

Air

Intensity		Wavelength	
20		2000.79	II
1500		2002.32	I
900		2008.04	I
50		2011.51	II
1200	h	2014.58	I
900		2016.17	I
50		2022.35	II
40		2025.76	II
50		2027.04	II
900		2031.96	I
30		2036.58	II
1500		2039.95	I
1200		2041.11	I
20		2049.17	II
40		2058.82	II
40		2063.78	II
50		2065.54	II
1500	h	2077.76	I
900		2085.67	I
900		2087.55	I
900		2089.35	I
900		2093.40	I
900		2094.86	I
900		2095.77	I
1200		2097.51	I
1500		2104.73	I
1500		2106.80	I
900		2108.98	I
30		2111.44	II
900	s	2117.68	I
20		2128.79	II
900		2137.78	I
900		2138.97	I
900		2163.03	I
20		2164.44	II
30		2173.33	II
1100		2174.60	I
20		2181.99	II
30		2187.01	II
20		2190.68	II
20		2192.50	II
200		2193.60	II
20		2200.40	II
40	p	2202.95	II
200		2256.73	II
150		2260.00	II
200		2283.52	II
1000		2286.15	II
200		2291.98	II
300	d	2293.38	II
300		2301.40	II
800	d	2307.85	II
2600		2309.02	I
500		2311.60	II
500		2314.05	II
300		2314.96	II
200	p	2317.06	II
2400		2323.14	I
300	p	2324.31	II
200	d	2326.11	II
500		2326.47	II
1400		2335.99	I
1600		2338.67	I
200		2347.39	II
1600		2352.85	I
200	d	2353.41	II
2000		2353.42	I
500		2363.80	II
400		2378.62	II
1400		2380.48	I
200		2381.76	II
300	p	2383.45	II
1400		2384.86	I
200		2386.36	II
500		2388.92	II
200		2397.38	II
1100	d	2402.06	I
200	p	2404.16	II
5300		2407.25	I
5300		2411.62	I
1600		2412.76	I
4800		2414.46	I
4800		2415.30	I
300		2417.65	II
4100		2424.93	I
3300		2432.21	I
2900		2436.66	I
2400		2439.05	I
200		2442.63	II
200	d	2446.03	II
200	p	2447.69	II
200		2450.00	II
200		2464.20	II
200		2486.44	II
200		2498.82	II
570		2504.52	I
500		2506.46	II
360		2506.88	I
200		2511.16	II
860		2517.87	I
500		2519.63	II
4300		2521.36	I
200	h	2524.65	II
300		2524.97	II
500		2528.62	II
2900		2528.97	I
200	p	2530.09	II
720		2530.13	I
860		2532.18	I
200	d	2533.82	II
2900		2535.96	I
860		2536.49	I
300		2541.94	II
1700		2544.25	I
200		2546.74	II
340		2548.34	I
310		2553.37	I
310		2555.07	I
300		2559.41	II
200		2560.03	II
960		2562.15	I
500		2564.04	II
1100		2567.35	I
960		2574.35	I
800		2580.32	II
300	d	2582.22	II
500		2587.22	II
500		2587.52	II
200		2588.91	II
100	p	2605.71	II
100		2612.50	II
100		2614.36	II
100	p	2628.77	II
100		2632.26	II
100		2636.07	II
310		2646.42	I
770		2648.64	I
100		2653.72	II
100		2663.53	II
200		2666.73	II
100		2675.85	II
100		2684.45	II
100		2702.02	II
200		2706.62	II
200		2707.35	II
190		2715.99	I
100		2727.78	II
80		2734.54	II
190		2745.10	II
100		2753.22	I
190		2764.19	I
100		2766.70	II
100		2774.97	II
100		2791.00	II
100		2793.73	II
150		2815.56	I
80		2835.63	II
80		2847.35	II
80		2871.22	II
190		2886.44	I
100		2918.38	II
100		2930.24	II
100		2954.73	II
690		2987.16	I
690		2989.59	I
20		3008.86	II
60		3022.59	II
30		3035.13	II
3100		3044.00	I
1700		3061.82	I
20		3352.79	II
80		3387.70	II
1100		3388.17	I
2200		3395.38	I
11000		3405.12	I
4500		3409.18	I
6700		3412.34	I
2200		3412.63	I
30		3415.77	II
2700		3417.16	I
50		3423.84	II
2500		3431.58	I
4500		3433.04	I
1600		3442.93	I
8800		3443.64	I
50		3446.39	II
4100		3449.17	I
2100		3449.44	I
21000		3453.50	I
1000		3455.23	I
5100		3462.80	I
5100		3465.80	I
8000		3474.02	I
1900		3483.41	I
4800		3489.40	I
2400		3495.69	I
40		3497.33	II
50		3501.72	II
9600		3502.28	I
7000		3506.32	I
50		3507.77	II
2900		3509.84	I
1400		3510.43	I
4800		3512.64	I
3800		3513.48	I
10		3514.23	II
30		3517.50	II
4800		3518.35	I
1300		3520.08	I
2700		3521.57	I
3800		3523.43	I
60		3523.51	II
6400		3526.85	I
2700		3529.03	I
7300		3529.81	I
1900		3533.36	I
40		3535.92	II
50		3545.03	II
20		3555.93	II
1100		3560.89	I
80		3561.07	II
20		3566.98	II
8800		3569.38	I
50		3574.95	II

Cobalt (Cont.)

Intensity		Wavelength	
1600		3574.96	I
60		3575.32	II
2500		3575.36	I
60		3577.96	II
1000		3585.16	I
6700		3587.19	I
1900		3594.87	I
1600		3602.08	I
100		3621.21	II
1000		3627.81	I
80		3643.61	II
10		3656.75	II
60		3681.35	II
5		3695.32	II
5		3714.73	II
1100		3745.50	I
20		3754.69	II
1400		3842.05	I
6900		3845.47	I
5500		3873.12	I
2800		3873.96	I
7900		3894.08	I
20	h	3911.40	II
1500		3935.97	I
80	h	3963.10	II
40	h	3976.74	II
10		3983.02	II
6000		3995.31	I
970		3997.91	I
350		4020.90	I
10	h	4036.14	II
20	h	4037.37	II
4		4040.02	II
370		4045.39	I
5	h	4050.23	II
10	h	4052.40	II
20	h	4062.73	II
2	h	4064.50	II
350		4066.37	I
5	h	4074.34	II
830		4092.39	I
1		4096.57	II
550		4110.54	I
2800		4118.77	I
4400		4121.32	I
3		4130.88	II
3		4145.13	II
3	d	4160.67	II
1		4181.13	II
90		4190.71	I
1		4208.61	II
30	s	4244.25	II
8		4272.33	II
20	h	4288.25	II
3		4328.86	II
2		4384.26	II
2	h	4396.94	II
3		4413.91	II
90		4469.56	I
10		4482.50	I
2	h	4489.12	II
4	h	4497.44	II
10	d	4500.54	II
0		4516.65	II
690		4530.96	I
2	h	4533.22	II
0	h	4537.95	II
90		4549.66	I
1	h	4559.29	II
140		4565.59	I
1		4569.26	II
190		4581.60	I
5		4616.30	II
120		4629.38	I
25	h	4660.66	II
85		4663.41	I
110		4792.86	I
10	d	4831.16	II
100		4840.27	I
150		4867.88	I
80	h	4964.18	II
10	h	4970.05	II
10	h	4990.47	II
20	h	4995.98	II
35		5146.74	I
50		5212.71	I
50		5230.22	I
45		5235.21	I
50		5247.93	I
26		5266.30	I
45		5266.49	I
26		5268.52	I
45	h	5280.65	I
26		5301.06	I
50		5342.71	I
26		5343.39	I
50		5352.05	I
26		5353.48	I
35		5369.58	I
45		5483.34	I
17		5530.77	I
17		5647.22	I
17		5991.88	I
17		6082.44	I
17		6282.63	I
45		6450.24	I
21		6455.00	I
15		6563.42	I
15		6632.44	I
14		6814.94	I
14		6872.40	I
21		7052.89	I
45		7084.99	I
8		7417.38	I
8		7712.68	I
7		7908.71	I
9		7987.38	I
13		8007.27	I
9		8093.96	I
9		8372.84	I
4		8575.35	I
3		8819.15	I

Co III
Ref. 291 — C.R.C.

Intensity	Wavelength	
	Vacuum	
1000	1696.01	III
800	1697.99	III
500	1702.79	III
1000	1707.35	III
500	1707.95	III
500	1723.97	III
400	1745.67	III
500	1755.98	III
5000	1760.35	III
500	1769.96	III
500	1773.22	III
5000	1773.57	III
500	1774.42	III
1000	1777.14	III
2000	1780.05	III
3000	1782.97	III
500	1784.06	III
1000	1787.08	III
1000	1789.07	III
500	1790.26	III
500	1791.28	III
500	1798.06	III
500	1805.54	III
400	1811.47	III
400	1821.26	III
400	1821.69	III
400	1821.77	III
1000	1823.08	III
400	1825.36	III
750	1825.95	III
2000	1830.09	III
2000	1831.44	III
750	1831.92	III
400	1832.20	III
5000	1835.00	III
1000	1837.63	III
500	1846.16	III
500	1852.92	III
400	1854.39	III
400	1854.76	III
1000	1861.78	III
2000	1863.83	III
400	1864.19	III
500	1871.87	III
1000	1881.70	III
500	1895.37	III
500	1919.12	III
500	1928.57	III
500	1940.15	III
500	1953.94	III
500	1959.41	III
400	1989.60	III
	Air	
100	2001.09	III
200	2011.62	III
200	2013.88	III
100	2031.81	III
200	2053.11	III
100	2056.70	III
10	2062.17	III
10	2079.74	III
15	2088.58	III
10	2090.51	III
10	2097.63	III
10	2134.15	III
10	2452.16	III
20	2811.75	III
10	2888.31	III
10	2933.27	III
10	2978.01	III
20	2991.89	III
25	3010.92	III
10	3116.68	III
2	3151.40	III
2	3180.64	III
20	3232.11	III
2	3249.24	III
20	3259.68	III
2	3269.23	III
10	3287.68	III
15	3305.38	III
10	3451.25	III
2	3526.24	III
1	3634.21	III
1	3636.31	III
2	3667.52	III
2	3677.23	III
3	3680.74	III
1	3762.50	III
15	3782.27	III

Co IV
Ref. 236 — C.R.C.

Intensity	Wavelength	
	Vacuum	
81	606.79	IV
74	607.59	IV
55	608.24	IV
66	609.16	IV
70	609.21	IV
64	609.28	IV
43	610.04	IV
37	610.25	IV
24	610.79	IV

Co V
Ref. 100, 159 — C.R.C.

Intensity	Wavelength	
	Vacuum	
20	355.52	V
18	355.88	V
12	356.06	V
4	1006.86	V
10	1007.51	V
15	1009.02	V
10	1010.94	V
10	1013.80	V
1	1017.43	V
10	1018.36	V
10	1021.14	V
1	1028.08	V
3	1226.31	V
8	1228.19	V
15	1231.73	V
2	1234.55	V
20	1236.95	V
2	1239.85	V
8	1246.91	V
6	1258.61	V
5	1263.28	V
20	1270.70	V
20	1272.23	V
2	1275.52	V
50	1277.01	V
30	1281.63	V
15	1284.00	V
28	1286.95	V
25	1295.55	V
40	1295.87	V
35	1301.12	V
50	1345.67	V
6	1351.22	V
15	1353.42	V
40	1355.20	V
30	1357.67	V
20	1361.32	V
30	1362.46	V
30	1364.17	V
30	1368.24	V
4	1369.30	V
10	1371.01	V
30	1373.09	V
30	1375.20	V
25	1378.12	V
10	1379.05	V
10	1380.21	V
32	1389.11	V
15	1459.77	V
35	1468.98	V
30	1476.65	V
25	1482.62	V
20	1482.91	V
25	1486.02	V
20	1488.73	V

COPPER (Cu)
Z = 29

Cu I and II

Ref. 273, 290—V.K.

Intensity	Wavelength	
	Vacuum	
80	685.141	II
100	709.313	II
100	718.179	II
150	724.489	II
200	735.520	II
250	736.032	II
80	779.295	II
100	797.455	II
150	810.998	II
200	813.883	II
300	826.996	II
150	848.808	II
250	851.303	II
250	858.487	II
400	861.994	II
400	865.390	II
250	869.336	II
150	873.263	II
200	876.723	II
250	877.012	II
200	877.555	II
500	878.699	II
100	884.133	II
250	885.847	II
600	886.943	II
600	890.567	II
500	892.414	II
800	893.678	II
400	894.227	II
600	896.759	II
400	896.976	II
600	901.073	II
400	906.113	II
800	914.213	II
600	922.019	II
500	924.239	II
400	935.232	II
600	935.898	II
600	943.335	II
600	945.525	II
500	945.965	II
200	954.383	II
250	956.290	II
400	958.154	II
200	960.414	II
250	968.042	II
200	974.759	II
250	977.567	II
100	987.657	II
250	992.953	II
300	1004.055	II
300	1008.569	II
300	1008.728	II
300	1010.269	II
250	1012.597	II
500	1018.707	II
500	1027.831	II
250	1028.328	II
200	1030.263	II
600	1036.470	II
600	1039.348	II
600	1039.582	II
800	1044.519	II
800	1044.744	II
500	1049.755	II
600	1054.690	II
400	1055.797	II
600	1056.955	II
400	1058.799	II
600	1059.096	II
600	1060.634	II
600	1063.005	II
200	1065.782	II
200	1066.134	II
500	1069.195	II
300	1073.745	II
200	1088.395	II
300	1094.402	II
250	1097.053	II
150	1119.947	II
200	1142.640	II
300	1144.856	II
100	1250.048	II
150	1265.506	II
300	1275.572	II
150	1282.455	II
150	1287.468	II
150	1298.395	II
300	1308.297	II
300	1314.337	II
100	1320.686	II
100	1326.395	II
150	1350.594	II
250	1351.837	II
150	1355.305	II
300	1358.773	II
200	1359.009	II

Copper (Cont.)

Int.		λ	Sp.
200		1362.600	II
250		1367.951	II
200		1371.840	II
100		1393.128	II
100		1398.642	II
150		1402.777	II
150		1407.169	II
100		1414.898	II
250		1418.426	II
250		1421.759	II
200		1427.829	II
400		1430.243	II
250		1434.904	II
150		1436.236	II
150		1442.139	II
200		1445.984	II
200		1449.058	II
250		1450.304	II
200		1452.294	II
300		1458.002	II
250		1459.412	II
200		1463.752	II
400		1463.838	II
200		1466.070	II
400		1470.697	II
200		1472.395	II
250		1473.978	II
200		1474.935	II
150		1476.059	II
200		1481.544	II
200		1485.328	II
750		1488.831	II
300		1492.834	II
250		1493.366	II
250		1495.430	II
350		1496.687	II
150		1503.368	II
250		1504.757	II
200		1505.388	II
300		1508.632	II
350		1510.506	II
200		1512.465	II
200		1513.366	II
500		1514.492	II
200		1517.631	II
500		1519.492	II
600		1519.837	II
200		1520.540	TT
200		1524.860	II
150		1525.764	II
500		1531.856	II
300		1532.131	II
250		1533.986	II
250		1535.002	II
500		1537.559	II
200		1540.239	II
300		1540.389	II
300		1540.588	II
750		1541.703	II
400		1544.677	II
100		1547.958	II
300		1550.653	II
300		1551.389	II
500		1552.646	II
250		1553.896	II
400		1555.134	II
500		1555.703	II
300		1558.345	II
400		1565.924	II
400		1566.415	II
100		1569.416	II
300		1579.492	II
300		1580.626	II
400		1581.995	II
500		1583.682	II
400		1590.165	II
600		1593.556	II
400		1598.402	II
400		1602.388	II
200		1604.848	II
300		1605.281	II
400		1606.834	II
250		1608.639	II
150		1610.296	II
200		1617.915	II
600		1621.426	II
400		1622.428	II
250		1630.268	II
100		1636.605	II
250		1649.458	II
30	r	1655.32	I
200		1656.322	II
200		1660.001	II
300		1663.002	II
100		1672.776	II
30		1688.09	I
30		1691.08	I
30	r	1703.84	I
50	r	1713.36	I
150		1717.721	II
50	r	1725.66	I
100		1736.551	II
50	r	1741.57	I
150		1753.281	II

Int.		λ	Sp.
200	r	1774.82	I
100	r	1825.35	I
250		1929.751	II
250		1944.597	II
100		1946.493	II
200		1957.518	II
150		1970.495	II
150		1977.027	II
500		1979.956	II
300		1989.855	II

Air

Int.		λ	Sp.
250		1999.698	II
270		2035.854	II
250		2037.127	II
350		2043.802	II
300		2054.980	II
100		2078.663	II
110		2098.398	II
320		2104.797	II
300		2112.100	II
320		2117.310	II
350		2122.980	II
350		2126.044	II
420		2134.341	II
900		2135.981	II
400		2148.984	II
150		2161.320	II
1300	r	2165.09	I
250		2174.982	II
1600	r	2178.94	I
700		2179.410	II
1700	r	2181.72	I
700		2189.630	II
900		2192.268	II
400		2195.683	II
1700	r	2199.58	I
1300	r	2199.75	I
100		2200.509	II
200		2209.806	II
750		2210.268	II
1600	r	2214.58	I
250		2215.106	II
1000	r	2215.65	I
750		2218.108	II
2100	r	2225.70	I
150		2226.700	II
1600	r	2227.78	I
350		2228.868	TT
2500	r	2230.08	I
1100	r	2238.45	I
900		2242.618	II
2300	r	2244.26	I
1000		2247.002	II
1300	r	2260.53	I
2200	r	2263.08	I
150		2263.786	II
200		2276.258	II
100		2286.645	II
2500	r	2293.84	I
170		2294.368	II
1000		2303.12	I
150		2369.890	II
2500	r	2392.63	I
120		2403.337	II
1500		2406.66	I
1000	r	2441.64	I
100		2485.792	II
2000	r	2492.15	I
150		2506.273	II
120		2526.593	II
300		2544.805	II
100		2571.756	II
150		2590.529	II
200		2600.270	II
2500	r	2618.37	I
200		2666.304	II
750		2689.300	II
700		2700.962	II
650		2703.184	II
700		2713.508	II
650		2718.778	II
300		2721.677	II
120		2737.342	II
270		2745.271	II
2500	r	2766.37	I
800		2769.669	II
200		2791.795	II
170		2799.528	II
100		2810.804	II
1250	r	2824.37	I
350		2837.368	II
100		2857.748	II
600		2877.100	II
270		2884.196	II
2500	r	2961.16	I
100		2986.335	II
2000		2997.36	I
2000		3010.84	I
2500		3036.10	I
2500		3063.41	I
1400		3073.80	I
1500		3093.99	I

Int.		λ	Sp.
1250		3099.93	I
2000		3108.60	I
1400	h	3126.11	I
1500		3194.10	I
1400		3208.23	I
1500	h	3243.16	I
10000	r	3247.54	I
10000	r	3273.96	I
1400	h	3282.72	I
400		3290.418	II
1500	h	3290.54	I
110		3300.881	II
250		3301.229	II
2500	h	3307.95	I
200		3316.276	II
1500		3337.84	I
150		3338.648	II
200		3365.648	II
450		3370.454	II
300		3374.952	II
200		3380.712	II
100		3384.945	II
1250	h	3483.76	I
1250		3524.23	I
2000		3530.38	I
1400		3599.13	I
1400		3602.03	I
1000		3686.555	II
150		3786.270	II
170		3797.849	II
100		3818.879	II
140		3826.921	II
160		3864.127	II
280		3884.131	II
150		3892.924	II
170		3903.177	II
140		3920.654	II
120		3933.268	II
120		3987.024	II
150		3993.302	II
140		4003.476	II
1250		4022.63	I
100		4032.647	II
600		4043.484	II
500		4043.751	II
2000		4062.64	I
120		4068.106	II
500		4131.363	II
200		4143.017	II
300		4153.623	II
500		4161.140	II
370		4164.284	II
400		4171.851	II
500		4179.512	II
500		4211.866	II
320		4230.449	II
200		4255.635	II
950		4275.11	I
300		4279.962	II
500		4292.470	II
400		4365.370	II
100		4444.831	II
400		4506.002	II
150		4516.049	II
150		4541.032	II
500		4555.920	II
100		4596.906	II
120		4649.271	II
2000		4651.12	I
120		4661.363	II
320		4671.702	II
300		4673.577	II
450		4681.994	II
100		4758.433	II
400		4812.948	II
120		4851.262	II
300		4854.988	II
100		4873.304	II
150		4901.427	II
1000		4909.734	II
500		4918.376	II
200		4926.424	II
900		4931.698	II
120		4943.026	II
700		4953.724	II
500		4985.506	II
400		5006.801	II
350		5009.851	II
400		5012.620	II
350		5021.279	II
200		5039.016	II
300		5047.348	II
900		5051.793	II
400		5058.910	II
500		5065.459	II
450		5067.094	II
350		5072.302	II
450		5088.277	II
420		5093.816	II
350		5100.067	II
1500		5105.54	I
250		5124.476	II
2000		5153.24	I
100		5158.093	II

Int.	λ	Sp.
100	5183.367	II
2500	5218.20	I
100	5269.991	II
100	5276.525	II
1650	5292.52	I
100	5368.383	II
1500	5700.24	I
1500	5782.13	I
150	5805.989	II
100	5833.515	II
200	5897.971	II
120	5937.577	II
400	5941.196	II
100	5993.260	II
650	6000.120	II
100	6023.264	II
250	6072.218	II
150	6080.343	II
150	6099.990	II
160	6107.412	II
300	6114.493	II
600	6150.384	II
750	6154.222	II
500	6172.037	II
550	6186.884	II
400	6188.676	II
300	6198.092	II
470	6204.261	II
450	6208.457	II
750	6216.939	II
700	6219.844	II
500	6261.848	II
1000	6273.349	II
350	6288.696	II
900	6301.009	II
550	6305.972	II
400	6312.492	II
120	6326.466	II
400	6373.268	II
750	6377.840	II
400	6403.384	II
850	6423.884	II
200	6442.965	II
750	6448.559	II
170	6466.246	II
950	6470.168	II
750	6481.437	TT
400	6484.421	II
220	6517.317	II
400	6530.083	II
120	6551.286	II
200	6577.080	II
750	6624.292	II
800	6641.396	II
450	6660.962	II
100	6770.362	II
300	6806.216	II
400	6809.647	II
320	6823.202	II
250	6844.157	II
320	6868.791	II
270	6872.231	II
270	6879.404	II
220	6937.553	II
150	6952.871	II
150	6977.572	II
200	7022.860	II
300	7194.896	II
400	7326.008	II
300	7331.694	II
250	7382.277	II
1000	7404.354	II
270	7434.156	II
500	7562.015	II
700	7652.333	II
1000	7664.648	II
150	7681.788	II
450	7744.097	II
800	7778.738	II
750	7805.184	II
1500	7807.659	II
1000	7825.654	II
350	7860.577	II
300	7890.567	II
700	7902.553	II
1500	7933.13	I
400	7944.438	II
400	7972.033	II
1200	7988.163	II
2000	8092.63	I
500	8277.560	II
800	8283.160	II
250	8503.396	II
750	8511.061	II
200	8609.134	II
500	9813.213	II
250	9827.978	II
200	9830.798	II
600	9861.280	II
600	9864.137	II
200	9883.969	II
550	9916.419	II
500	9917.954	II
550	9925.594	II

Copper (Cont.)

Intensity	Wavelength	
450	9938.998	II
500	9960.354	II
450	10006.588	II
550	10022.969	II
550	10038.093	II
650	10054.938	II
450	10080.354	II

Cu III
Ref. 295—V.K.

Intensity		Wavelength	
		Vacuum	
75		542.90	III
200		615.67	III
150		616.03	III
150		687.98	III
150		715.53	III
125		730.38	III
250		788.07	III
250		788.46	III
250		791.36	III
150		801.14	III
100		829.34	III
40		1048.88	III
50		1186.80	III
50		1200.96	III
300		1219.30	III
200		1244.38	III
100		1279.14	III
200		1312.39	III
300		1332.97	III
200		1339.48	III
150		1363.08	III
300	r	1376.79	III
200	r	1377.49	III
150		1423.48	III
300	r	1481.23	III
200		1543.46	III
500	r	1593.75	III
1000	r	1642.21	III
300		1679.14	III
400		1702.10	III
500		1722.37	III
600		1741.37	III
200		1768.86	III
200		1840.91	III
100		1971.95	III
		Air	
200		2013.22	III
150		2157.28	III
100		2299.47	III
500		2368.17	III
400		2391.74	III
800		2405.50	III
700		2412.34	III
2000		2444.44	III
500		2468.41	III
1000		2482.36	III
700		2486.46	III
500		2508.49	III
500		2522.38	III
500		2538.66	III
400		2566.37	III
400		2573.33	III
500		2609.32	III
200		2643.92	III
200		2696.38	III
20		2751.33	III
100		2812.94	III
100		2978.87	III
75		3548.87	III
100		3639.42	III
500		3702.92	III
800		3744.70	III
400		3748.27	III
600		3752.06	III
1000		3776.97	III
800		3790.80	III
600		3804.13	III
600		3809.18	III
300		3881.68	III
150		3953.81	III
100		4090.49	III
200		4283.40	III
500		4351.97	III
1000		4352.80	III
500		4355.24	III
500		4370.84	III
500		4371.40	III
500		4373.43	III
1000		4377.11	III
200		4386.42	III
150		4927.41	III
400		5094.28	III
200		5168.97	III
400		5208.34	III
600		5219.21	III
200		5268.59	III
400		5317.78	III
300		5369.79	III
350		5418.48	III
250	d	5494.94	III
50		5573.94	III
100		5609.00	III
75		5702.12	III
100		5768.56	III
100		5850.72	III
200		5965.25	III
30		6100.87	III
50		6369.27	III
20		6512.54	III
20		6644.13	III
50		6793.20	III

Cu IV
Ref. 199—V.K.

Intensity		Wavelength	
		Vacuum	
30		360.86	IV
20		374.40	IV
30		405.24	IV
80		406.45	IV
40		413.45	IV
70		443.68	IV
80		451.16	IV
80		463.72	IV
80		484.53	IV
90		497.00	IV
90		504.60	IV
70		509.38	IV
40		519.51	IV
40		540.65	IV
60		550.92	IV
20		584.85	IV
60		1056.13	IV
30		1074.72	IV
50		1091.65	IV
30		1105.50	IV
25		1119.43	IV
40	p	1152.18	IV
60		1227.44	IV
70		1228.87	IV
70		1258.69	IV
90		1274.84	IV
90		1293.46	IV
90		1309.41	IV
70		1321.17	IV
100	d	1340.08	IV
100		1350.42	IV
100		1362.05	IV
100		1372.14	IV
100		1377.82	IV
100		1388.80	IV
90		1405.49	IV
90		1415.27	IV
90		1434.34	IV
90		1449.69	IV
80		1466.18	IV
70		1482.77	IV
80		1499.81	IV
90		1515.28	IV
90		1535.12	IV
80		1551.12	IV
90		1567.35	IV
80		1583.47	IV
70		1595.12	IV
80	p	1608.14	IV
90		1639.75	IV
20		1650.16	IV
70		1704.37	IV
30		1797.99	IV
70		1817.56	IV
70		1819.23	IV
60		1837.04	IV
80		1849.62	IV
30		1867.24	IV
30		1918.71	IV
40		1966.31	IV

Cu V
Ref. 324—V.K.

Intensity		Wavelength	
		Vacuum	
9	h	258.95	V
49		271.33	V
49		283.97	V
22		293.41	V
56		299.64	V
65		305.83	V
51		312.51	V
66		321.05	V
74		326.57	V
82		333.56	V
81		339.88	V
81		346.00	V
86		355.41	V
77		363.96	V
65		370.63	V
74		377.76	V
70		387.40	V
51		396.06	V
25		406.94	V
13		1097.10	V
42		1106.24	V
77		1113.22	V
67		1121.20	V
76		1128.80	V
59		1133.86	V
63		1142.38	V
54		1149.06	V
72		1157.54	V
64		1167.35	V
77		1176.53	V
84		1183.63	V
76		1192.54	V
83		1201.22	V
77		1204.90	V
71		1214.36	V
70		1221.34	V
76		1230.11	V
80		1239.73	V
79		1246.99	V
65		1253.07	V
70		1260.24	V
77		1269.35	V
78		1278.20	V
68		1286.13	V
67		1292.08	V
73		1299.22	V
65		1309.72	V
55		1318.89	V
65		1323.28	V
64		1329.22	V

CURIUM (Cm)
Z = 96

Cm I and II
Ref. 51, 332—J.G.C.

Intensity		Wavelength	
		Air	
10000		2462.76	II
10000		2617.17	II
10000		2636.42	II
10000		2651.17	II
10000		2653.80	II
10000		2725.68	II
10000		2736.89	II
10000		2748.04	II
10000		2784.83	II
10000		2811.62	II
10000		2824.20	II
10000		2833.58	II
10000		2899.90	II
10000		2912.97	II
10000		2928.92	II
10000		2996.18	II
10000		2999.39	I
10000	b	3014.87	II
10000		3044.85	II
10000		3109.69	I
10000		3116.41	I
10000		3137.16	I
10000		3147.33	I
10000		3155.10	I
10000		3158.60	I
10000		3169.98	II
10000		3177.55	I
10000		3179.10	I
10000		3186.41	I
10000		3188.11	I
10000		3207.12	II
10000		3207.71	I
10000		3209.89	II
10000		3209.94	I
10000		3210.05	II
10000		3220.76	II
10000		3224.23	I
10000		3225.11	I
10000		3226.41	II
10000		3230.28	I
10000		3230.35	II
10000		3236.74	I
10000		3238.55	II
10000		3242.66	II
10000		3246.25	I
10000		3252.68	I
10000		3265.81	I
10000		3280.45	I
10000		3296.71	II
10000		3304.85	I
10000		3317.14	I
10000		3374.70	I
10000		3452.92	I
10000		3458.34	I
10000		3510.28	I
10000		3522.36	I
10000		3524.94	I
10000		3542.06	I
10000		3547.02	I
10000		3547.92	I
10000		3561.44	I
10000		3572.95	II
10000		3600.62	I
10000		3639.94	I
10000		3664.34	I
10000		3709.43	I
10000		3729.00	I
10000		3732.35	I
10000		3747.86	I
10000		3763.05	I
10000		3775.75	I
10000		3816.30	I
10000		3825.14	I
10000		3833.32	I
10000		3837.59	I
10000		3842.00	I
10000		3849.92	I
10000		3854.11	I
10000		3900.25	I
10000		3904.06	II
10000		3908.24	II
10000		3936.67	I
10000		3942.03	I
10000		3944.15	I
10000		3948.68	I
10000		3953.36	I
10000		3964.83	I
10000		3995.10	I
10000		4016.17	I
10000		4031.76	I
10000		4048.29	I
10000		4049.65	I
10000		4113.29	I
10000		4129.71	I
10000		4207.66	II
10000		4211.62	I
10000		4266.45	I
10000		4293.00	I
10000		4330.82	I
10000		4345.69	I
10000		4447.77	I
10000		4459.16	I
10000		4608.40	I
10000		5846.07	I
10000		5952.41	I
10000		6058.90	I
10000		6243.35	I
10000		6376.71	I
10000		6510.16	I
10000		6554.41	I
10000		6640.17	I
10000		6663.25	I
10000		6686.87	I
10000		6706.85	I
10000		6726.68	I
10000		6793.15	I
10000		7162.69	I
10000		7577.80	I
10000		7673.79	I
10000		7720.47	I
10000		8392.37	I
10000		9293.25	I
10000		9567.08	I
10000		9657.12	I
10000		10310.83	I
10000		10351.73	I
10000		10424.49	I
10000		10508.11	I
10000		10542.98	I
10000		10792.25	I
10000		10897.45	I
10000		11507.45	I
10000		11707.73	I
10000		11780.95	I
10000		11834.28	I
10000		12017.85	I
10000		12394.16	I
10000		12454.98	I
10000		12464.99	I
10000		13004.56	I
10000		13258.18	I
10000		13289.84	I
10000		13344.62	I
10000		13480.54	I
10000		13590.01	I
10000		13644.77	I
10000		13789.52	I
10000		13840.18	I
10000		13908.46	I
10000		13964.14	I
10000		14235.27	I
10000		14334.52	I
10000		14563.41	I
10000		14580.23	I
10000		15018.13	I
10000		15222.27	I
10000		15642.59	I

Curium (Cont.)

Intensity	Wavelength	
10000	15757.23	I
10000	15793.31	I
10000	16008.41	I
10000	17148.22	I
10000	17453.18	I
10000	17619.28	I
10000	18069.02	I
10000	19572.62	I
10000	19975.98	I
10000	20526.32	I
10000	20853.49	I
10000	20911.52	I
10000	20968.11	I
10000	21241.06	I
10000	21393.23	I

DYSPROSIUM (Dy)
Z = 66

Dy I and II
Ref. 1—C.H.C.

Intensity	Wavelength (Air)	
260	2356.91	II
65	2381.95	
130	2387.36	II
150	2392.15	
180	2402.29	II
240	2410.01	II
150	2422.75	II
260	2439.84	II
90	2455.15	II
110	2459.99	II
90	2471.40	II
110	2480.93	II
170	2490.61	II
90	2510.31	II
170	2513.55	II
170	2517.61	II
130	2543.81	II
90	2545.12	II
150	2552.29	II
180	2557.94	II
90	2560.21	II
90	2566.25	II
220	2585.30	I
90	2591.56	II
75	2592.54	II
130	2600.16	II
130	2600.76	II
75	2608.69	II
370	2623.69	I
440	2634.80	II
110	2642.15	I
110	2645.35	II
110	2667.94	I
55	2676.84	II
50	2677.34	II
85	2689.31	II
85	2692.83	II
55	2709.01	II
55	2727.17	II
85	2729.50	II
55	2735.79	I
40	2739.30	II
85	2740.70	II
220	2755.75	II
55	2757.08	II
70	2766.50	II
70	2772.42	II
110	2772.61	II
40	2779.58	II
55	2791.44	II
120	2800.33	II
110	2800.53	II
110	2801.41	II
300	2816.39	II
140	2825.42	II
140	2862.70	I
190	2877.88	II
110	2884.28	II
120	2885.53	I
120	2890.74	II
120	2900.82	II
110	2904.62	II
190	2906.39	II
390	2913.95	II
110	2934.31	II
250	2934.52	II
110	2941.05	II
140	2944.56	II
150	2947.06	II
150	2947.21	II
250	2948.31	II
170	2950.33	II
110	2952.12	II
140	2953.70	II
220	2964.60	I
110	2977.42	II

Intensity	Wavelength	
110	2985.97	II
220	3015.68	II
390	3026.16	II
210	3029.81	II
610	3038.28	II
280	3043.13	II
210	3047.56	II
280	3060.64	II
390	3062.62	II
220	3066.99	II
330	3071.91	II
280	3073.54	II
220	3078.68	II
280	3101.93	II
220	3103.24	II
410	3109.76	II
330	3128.41	II
830	3135.38	II
360	3140.64	II
500	3141.14	II
220	3143.83	II
250	3146.16	II
1200	3156.52	II
670	3162.83	II
1000	3169.99	II
400	3177.89	II
220	3178.37	II
200	3184.79	II
330	3186.38	II
240	3187.68	II
330	3193.30	II
240	3206.40	II
220	3207.12	II
290	3208.85	II
470	3215.19	II
830	3216.63	II
240	3221.49	II
290	3223.28	II
240	3225.08	II
330	3225.95	II
490	3235.89	II
290	3236.69	II
490	3245.12	II
200	3248.36	II
1200	3251.27	II
200	3252.19	II
290	3256.26	II
200	3266.21	II
240	3269.11	II
200	3272.73	II
890	3280.09	II
490	3282.77	II
200	3287.94	II
200	3293.88	II
200	3296.30	II
200	3305.40	II
200	3305.51	II
240	3306.19	II
440	3308.79	II
1100	3308.88	II
510	3312.72	II
780	3316.32	II
240	3317.12	II
1000	3319.88	II
270	3326.19	II
780	3341.00	II
270	3341.88	II
200	3347.83	II
270	3352.69	II
510	3353.58	II
240	3359.46	II
510	3368.11	II
5300	3385.02	II
210	3386.57	II
610	3388.85	II
210	3391.96	II
3800	3393.57	II
1300	3396.16	II
380	3407.16	II
5300	3407.80	II
420	3408.14	II
1300	3413.78	II
530	3414.82	II
780	3419.63	II
530	3425.06	II
420	3429.44	II
1900	3434.37	II
330	3438.94	II
560	3440.93	II
1300	3441.45	II
3800	3445.57	II
830	3446.99	II
440	3449.89	II
2700	3454.32	II
440	3454.51	II
1300	3456.56	II
4400	3460.97	II
720	3468.43	II
560	3471.14	II
560 d	3471.53	II
380	3473.70	II
1300	3477.07	II
4400	3494.49	II
560	3496.34	II

Intensity	Wavelength	
400	3497.81	II
830	3498.71	II
400	3501.50	II
830	3504.53	II
830	3505.45	II
1300	3506.81	II
560	3517.26	II
4400	3523.98	II
22000	3531.70	II
4400	3534.96	II
5500	3536.02	II
4400	3538.52	II
400	3539.37	II
1700	3542.33	II
400	3544.20	II
400	3544.35	II
1400	3546.83	II
330	3548.19	II
4400	3550.22	II
2200	3551.62	II
440 h	3558.23	II
440	3559.30	II
2200	3563.15	II
560	3563.69	II
780	3573.83	II
1400	3574.15	II
4400	3576.24	II
1700	3576.81	II
830	3577.98	II
440	3580.04	II
400	3584.42	II
3300	3585.06	II
1400	3585.78	II
560	3586.11	II
360	3590.07	II
1100	3591.41	II
560	3591.81	II
560	3592.11	II
1800	3595.04	II
400	3596.06	II
560	3600.38	II
360	3602.82	II
1800	3606.12	II
440	3618.51	II
560	3620.16	II
470	3624.27	II
1100	3629.42	II
4000	3630.24	II
440	3632.78	II
400	3635.27	II
360	3637.28	II
1100	3640.25	II
400	3643.92	II
11000	3645.40	II
360	3645.86	II
1000	3648.78	II
700	3664.62	II
400	3666.84	I
990	3672.30	II
420	3672.70	II
400	3673.14	II
1400	3674.08	II
2200	3676.59	II
640	3678.51	I
820	3684.85	I
1300	3685.78	I
4700	3694.81	II
370	3697.31	II
990	3698.21	II
540	3701.63	II
330	3707.40	II
440	3707.57	II
440	3708.22	II
420	3710.07	II
330	3711.66	II
1600	3724.45	II
300	3728.00	I
930	3739.24	I
1200	3747.82	II
1400	3753.51	II
1400	3753.75	II
1200	3757.05	I
4700	3757.37	II
640	3767.63	I
330	3771.11	I
640	3773.05	I
370	3774.71	I
420	3781.47	I
330	3785.41	II
3300	3786.18	II
1600	3788.44	II
700	3791.87	II
510	3804.14	II
580	3806.27	II
470	3812.27	I
470	3813.67	II
1400	3816.76	II
700	3825.68	II
2300	3836.50	II
370	3840.89	I
1400	3841.31	II
330	3842.00	II
330	3844.36	I

Intensity	Wavelength	
420	3846.34	II
420	3847.02	I
330	3849.39	II
1200	3853.03	II
420	3858.40	I
370	3866.58	II
560	3868.45	II
1600	3868.81	I
300	3869.42	II
820	3869.86	II
7000	3872.11	II
1200	3873.99	II
470	3879.11	II
300	3881.99	II
5800	3898.53	II
540	3914.87	II
540	3915.59	II
540 d	3917.29	I
320	3923.38	II
420	3927.86	I
540	3930.14	I
2100	3931.52	II
320	3932.22	II
370	3933.00	II
320	3934.21	II
420	3936.70	I
540	3942.53	II
10000	3944.68	II
420	3946.93	II
540	3950.39	II
420	3954.55	II
800	3957.79	II
370	3962.59	I
320	3967.51	I
14000	3968.39	II
2700	3978.57	II
1400	3981.92	II
1600	3983.65	II
800	3984.21	II
540	3991.32	II
1600	3996.69	II
8000	4000.45	II
420	4005.84	I
320	4006.07	I
540	4011.29	II
540	4013.82	I
540	4014.70	II
370	4023.71	I
420	4027.78	II
520 d	4028.32	II
520	4032.47	II
420	4033.65	II
420	4036.32	II
320	4041.98	II
12000	4045.97	I
1600	4050.56	II
520	4055.14	II
2500	4073.12	II
7400	4077.96	II
370	4085.34	I
390	4096.10	I
3900	4103.30	II
860	4103.87	I
1500	4111.34	II
490	4124.63	II
390	4128.24	II
350	4129.12	I
990	4129.42	II
350	4130.35	I
390	4133.85	I
470	4141.50	II
1200	4143.10	II
990	4146.06	I
5700	4167.97	I
370	4171.93	I
930	4183.72	I
12000	4186.82	I
320	4190.94	I
2200	4191.64	I
6800	4194.84	I
320	4195.19	II
800	4198.02	I
680	4201.30	I
680	4202.24	I
230	4205.06	I
370	4206.54	II
440	4211.24	I
16000	4211.72	I
1800	4213.18	I
3700	4215.16	I
4400	4218.09	I
4400	4221.11	I
540	4222.21	I
2700	4225.16	I
680	4232.02	I
680	4239.85	I
440	4245.91	I
440	4256.33	II
250	4276.69	I
370 d	4294.93	II
	4295.04	I
1000	4308.63	II
320	4325.86	I
200	4358.44	II

Dysprosium (Cont.)

Intensity		Wavelength	Type
320		4374.24	II
320		4374.76	II
540		4409.38	II
150		4444.58	I
740		4449.70	II
110		4455.60	II
250		4468.14	II
100	d	4527.58	I
		4527.76	II
100		4541.66	II
140		4565.09	I
420		4577.78	I
2100		4589.36	I
990		4612.26	I
50		4613.83	I
50		4614.82	I
60		4617.26	II
140		4620.03	II
50		4662.72	I
110		4664.66	II
85		4673.60	II
50		4682.03	II
50		4689.75	II
95		4698.68	II
85		4721.22	I
70		4727.13	II
170		4731.84	II
40		4745.73	II
60		4754.99	II
50		4760.04	II
60		4771.94	I
50		4774.80	I
120	h	4775.79	I
75		4786.92	II
95		4791.29	I
29		4800.64	I
50		4807.94	I
40		4810.28	I
50		4812.80	I
75		4819.04	I
85		4824.96	I
75		4828.88	I
50		4829.68	II
70		4832.38	I
35		4833.75	II
75		4841.75	I
40		4856.24	II
40		4868.05	II
40		4875.93	I
85		4880.16	I
40		4884.55	I
95		4888.08	I
40		4889.33	II
75		4890.10	II
50		4893.68	I
24		4899.24	I
55		4916.41	I
50		4922.22	II
65		4923.16	II
480		4957.34	II
24		4959.59	I
28		4973.57	I
40		4985.52	I
50		5003.87	I
55		5004.28	II
24		5010.60	I
24		5017.98	II
70		5022.12	I
30		5024.03	I
24		5024.54	I
40		5027.87	I
50		5033.00	I
160		5042.63	I
24		5047.25	I
50	h	5050.21	I
30		5053.35	I
24		5055.46	I
95		5070.68	I
120		5077.67	I
80		5090.38	II
80		5110.32	I
130	h	5120.04	I
30		5135.02	I
190		5139.60	II
40		5161.03	II
40		5164.12	II
50		5165.34	I
110		5169.69	II
20		5172.90	II
80		5185.30	I
40		5188.45	II
290		5192.86	II
95		5197.66	II
50		5246.94	II
70		5259.88	I
130		5260.56	I
55	B1	5263.3	D
65		5267.11	I
50		5272.25	II
50		5275.29	II
50		5279.70	II
55		5282.07	I
28		5284.99	II
40		5297.82	II
160		5301.58	I
40		5309.02	II
50		5324.69	I
24		5337.43	II
65		5340.30	I
30		5352.11	I
30		5368.20	II
20		5385.23	II
85		5389.58	II
40		5395.57	I
20	h	5398.26	D
24		5399.93	I
50		5404.19	I
80		5419.13	I
70		5423.32	I
30		5424.27	I
40		5426.70	II
30		5443.34	II
95		5451.11	I
30		5455.47	II
24		5469.10	II
28		5496.83	I
28		5502.79	I
24		5506.52	I
24		5515.41	I
30		5528.01	I
65		5547.27	I
40	d	5600.65	II
24		5605.53	I
30		5613.23	I
20		5627.49	I
100		5639.50	I
55	h	5645.99	I
80		5652.01	I
24		5685.58	I
28	h	5693.67	D
24	h	5694.10	D
28	Cw	5694.54	D
28		5698.72	II
24		5702.91	I
70		5718.46	I
28	h	5725.84	D
55	h	5728.64	D
24	h	5738.73	D
50		5740.20	I
55		5745.53	I
24		5750.48	I
24		5758.79	I
80	h	5832.01	D
55	h	5833.85	D
40	h	5834.86	D
28	h	5844.41	D
24		5845.65	D
40	h	5848.05	D
40		5855.56	D
55	h	5868.11	II
40		5915.16	II
20		5924.56	II
70		5945.80	I
50	l	5964.46	I
120		5974.49	I
24		5984.86	I
140		5988.56	I
24	h	6005.75	D
24	h	6006.54	D
24	h	6006.97	D
30		6008.94	I
65		6010.82	I
24		6017.26	I
24		6030.98	I
24	B1	6042.49	D
24		6058.18	I
30		6085.06	I
140		6088.26	I
24		6127.15	I
24		6133.64	I
24		6158.28	I
100		6168.43	I
20		6196.23	II
270		6259.09	I
30		6260.36	I
14		6343.32	I
40		6386.80	I
24		6396.60	II
50		6421.92	I
13	h	6436.55	I
8		6460.83	I
10		6468.58	II
11		6474.91	I
20		6483.59	II
28		6486.59	I
20		6558.02	I
160		6579.37	I
14		6594.14	II
15		6643.37	I
22		6658.36	I
29		6661.64	I
75		6667.86	I
10		6700.64	II
29		6747.93	I
10		6757.62	I
45		6765.89	I
12		6818.20	I
180		6835.42	I
80		6852.96	I
22		6856.46	I
22		6888.83	I
15		6897.97	II
65		6899.32	II
22		6906.53	II
15		6929.55	I
29		6950.28	II
11		6951.42	I
40		6958.08	I
13	h	6982.44	I
13		6991.30	I
45		6998.10	I
20		7017.42	I
35		7055.95	II
24		7075.14	II
17		7109.26	II
11		7120.81	II
13		7175.11	II
11		7213.27	I
17	h	7230.04	I
13		7250.01	I
17		7345.13	II
11		7370.23	II
20		7376.04	I
11		7407.59	I
24		7412.37	I
55		7426.86	II
20		7457.05	II
17		7516.61	II
55		7543.73	I
17	h	7553.00	I
27		7559.78	I
40		7562.96	II
20	h	7577.46	II
27	h	7591.30	I
13	h	7611.55	I
11	h	7617.70	I
35	h	7641.09	I
17		7645.86	I
13		7646.64	I
80		7662.36	I
11		7666.78	II
35		7715.33	I
45		7729.76	II
20		7751.62	I
35		7812.06	I
27		7909.38	I
11		7968.63	I
12		7982.85	I
13		8147.29	I
27		8198.77	II
100		8201.57	II
11		8218.62	II
20		8265.53	I
35		8326.10	I
35		8392.01	II
12		8405.85	II
20		8416.64	II
24		8438.58	II
11		8630.12	I
27		8655.94	II
17		8657.68	II
17		8678.49	II
11		8696.83	II
11	h	8715.95	II
20		8750.40	II
12		8780.83	I
45		8791.39	II
13		8833.08	II
24		8850.37	II

EINSTEINIUM (Es)
Z = 99

Es I and II
Ref. 333—J.G.C.

Intensity		Wavelength	
		Air	
300	l	2694.32	II
100	l	2703.84	I
10000	s	2708.66	II
100	s	2716.02	II
1000	s	2724.57	II
1000	l	2765.76	II
3000	l	2787.10	II
1000	l	2796.11	II
3000	l	2815.15	I
100	s	2885.84	II
100	s	2886.44	I
3000		2907.03	
100		3003.28	
1000	l	3065.40	
3000		3135.25	I
1000	l	3154.27	II
10000		3413.17	
300	l	3423.12	I
100	l	3424.28	
10000	s	3428.48	I
300		3437.31	
100		3437.34	
3000	s	3445.25	
300	l	3446.93	
3000		3452.36	
100		3453.16	
300	s	3470.77	
3000		3484.59	I
300		3494.30	
10000	s	3498.11	I
10000	l	3514.33	I
10000	l	3521.38	
10000	s	3523.49	I
300	l	3528.58	
10000		3536.01	
10000	s	3547.75	II
300		3549.97	
10000	l	3555.34	I
100		3555.53	
3000	s	3556.65	
3000		3560.92	
1000	s	3575.68	
100		3578.56	
300	s	3579.38	
1000	s	3582.95	
3000		3590.28	
1000		3595.47	
10000	s	3602.43	II
300		3605.58	
1000	s	3606.75	
100	l	3624.52	
3000	l	3631.09	
10000		3632.87	
100	l	3634.41	
1000		3651.94	
10000	l	3670.01	II
100	l	3672.32	
100	s	3713.56	
1000	l	3720.56	
300		3722.32	
10000	l	3728.55	II
100	h	3737.47	
300		3776.27	
10000	l	3792.99	
10000	l	3801.49	
300	s	3929.10	
3000		3930.77	
100		3957.19	
100		3995.35	
300		4077.71	
10000	s	4082.24	
3000	l	4107.59	
1000	l	4176.94	
100	s	4496.25	
100	l	4631.66	
300	l	4650.86	
1000		4789.93	
300	h	4802.17	
1000	h	4802.21	
3000	h	4958.29	
10000	s	5052.08	
1000	s	5102.93	
100	s	5155.82	
10000	s	5161.74	I
10000	l	5204.40	I
3000	L	5615.51	
100	S	6539.71	ES

ERBIUM (Er)
Z = 68

Er I and II
Ref. 1—C.H.C.

Wavelength			Intensity
		Air	
110		2358.51	II
100		2386.58	II
120		2387.17	II
110		2396.38	III
140		2446.39	II
100		2537.02	II
110		2547.28	II
290		2586.73	II
110		2587.04	II
130		2592.57	II
120		2595.03	II
140		2624.18	II
490		2670.26	II
330		2672.25	II
100		2675.35	II
270		2739.31	II
310		2750.19	II
230		2755.01	II
610		2755.63	II
510		2770.02	II
230		2778.97	II
230		2802.53	II
310		2804.35	II
410		2820.19	II
270	d	2833.91	II

10-26

Erbium (Cont.)

Intensity	Wavelength	Species
390	2838.71	II
270	2848.37	II
250	2855.41	II
310	2859.84	II
310	2896.96	II
390	2897.52	II
1000	2904.47	II
210	2909.58	II
1500	2910.36	II
270	2915.62	II
350	2929.27	II
270	2945.28	II
230	2946.62	II
1500	2964.52	II
410	2968.76	II
210	2974.47	II
230	2975.68	II
270	2983.80	II
1200	3002.41	II
310	3002.65	II
230	3012.47	II
230	3016.84	II
290	3025.95	II
270	3028.27	II
370	3031.31	II
310	3036.22	II
210	3054.42	II
230	3066.22	II
450	3070.74	II
560	3072.53	II
610	3073.34	II
210	3078.87	II
720	3082.08	II
610	3084.02	II
370	3099.19	II
230	3106.78	II
310 d	3113.43	II
	3113.54	II
770	3122.72	II
290	3132.52	II
470	3132.77	II
410	3141.10	II
	3141.15	II
250	3144.33	II
410	3154.29	II
870	3181.92	II
410	3183.42	II
250	3185.25	II
310	3200.58	II
230	3205.15	II
270	3214.44	II
870	3220.73	II
610	3223.31	II
210	3227.16	II
2300	3230.58	II
250	3232.03	II
330	3237.98	II
330	3249.34	II
560	3259.05	II
	3259.11	II
2700	3264.78	II
430	3267.10	II
	3267.18	II
330	3269.41	II
250	3278.22	II
720	3279.33	II
720	3280.22	II
470	3286.77	II
330 d	3303.88	II
	3303.95	II
370	3305.56	II
2300	3312.42	II
560	3316.39	II
770	3323.19	II
290	3329.66	II
770	3332.70	II
370	3337.25	II
290	3337.79	II
250	3340.03	II
290	3341.84	II
1300	3346.04	II
470	3350.06	II
350	3350.26	II
1400	3364.08	II
1400 d	3368.02	II
	3368.13	I
450	3370.55	II
7700	3372.71	II
970	3374.17	II
290	3381.32	II
230	3382.06	I
1700	3385.08	II
450	3389.74	II
2300	3392.00	II
350	3396.07	II
290	3396.84	II
390	3401.83	II
350	3417.63	II
490	3428.39	II
770	3441.13	II
390	3442.68	I
490	3469.51	I
970	3471.71	II
610	3479.41	II
970	3485.85	II
350	3486.82	II
350	3496.86	II
6700	3499.10	II
610	3502.78	I
390	3508.38	II
490	3514.89	II
390	3518.18	II
610	3524.91	II
410	3539.59	I
310	3548.26	II
820	3549.84	II
310	3553.20	II
1500	3558.02	I
510	3558.71	II
1000	3559.90	II
310	3565.17	I
920	3570.75	II
310	3578.24	I
1000	3580.52	II
370	3586.60	I
610	3590.76	I
410	3595.84	I
610	3599.50	II
1000	3599.83	II
510	3604.90	II
410	3607.42	I
3100	3616.56	II
510	3617.85	II
510	3618.92	II
720	3628.04	I
310	3629.37	I
1000	3633.54	II
510	3634.67	I
1600	3638.68	I
900	3645.94	II
520	3650.41	II
360	3652.58	II
500	3652.87	II
360	3664.45	I
470	3669.02	II
500	3682.70	II
320	3684.01	I
380	3684.28	II
7900	3692.65	II
450	3696.25	II
380	3697.68	I
540	3700.72	II
520	3707.64	II
520	3712.39	II
320	3719.35	I
1300	3729.52	II
450	3731.26	II
540	3738.16	II
340	3741.10	II
900	3742.64	II
900	3747.43	I
540	3756.05	I
410	3781.01	II
1800	3786.84	II
560	3787.86	II
560	3791.83	II
500	3792.79	I
560	3797.06	II
1600	3810.33	I
3600	3830.48	II
540	3849.91	I
320	3851.60	II
680	3855.90	I
540	3858.39	II
7500	3862.85	I
1500	3880.61	II
1200	3882.89	II
400	3890.61	II
4200	3892.68	I
5200	3896.23	II
810	3902.76	II
250	3903.98	I
250	3904.56	II
1200	3905.40	I
11000	3906.31	II
280	3918.05	I
210	3918.35	II
280	3921.88	II
810	3932.25	II
3200	3937.01	I
2100	3938.63	II
3200	3944.42	I
550	3948.06	I
250	3951.48	I
320	3956.42	I
280	3966.35	I
2700	3973.04	II
3200	3973.58	I
1400	3974.72	II
280	3976.73	I
810	3977.02	I
1100	3982.33	I
280	3987.53	I
810	3987.66	I
230	3991.15	I
230	4004.05	I
14000	4007.96	I
230	4008.18	II
280	4009.16	II
1100	4012.58	I
350	4015.57	II
3000	4020.51	I
450	4021.55	I
230	4043.01	II
1000	4046.96	I
280	4048.34	II
200	4049.49	II
940	4055.47	II
550	4059.51	I
690	4059.78	II
420	4077.88	I
550	4081.24	II
3500	4087.63	I
210	4092.90	I
1100	4098.10	I
350	4100.56	II
320	4116.36	I
320	4118.55	I
600	4131.50	I
550	4142.91	II
6900	4151.11	I
280	4189.98	II
1000	4190.70	I
130	4205.32	I
1400	4218.43	I
200	4220.99	I
320	4230.20	II
140	4234.78	I
200	4251.94	II
140	4276.48	II
690	4286.56	I
320	4298.91	I
320	4301.60	II
140	4303.81	II
110	4319.94	II
130	4328.81	I
110	4331.36	I
140	4340.92	I
190	4348.34	I
110	4369.39	II
160	4382.17	I
300	4384.70	II
300	4386.40	I
100	4403.17	II
810	4409.34	I
180	4418.70	I
570	4419.61	II
110	4422.51	II
320	4424.57	I
370	4426.77	I
110	4437.66	I
100	4459.24	II
100	4473.50	II
130	4496.39	I
200	4500.75	II
130	4522.74	I
160	4563.26	II
1000	4606.61	I
160	4630.88	I
110	4640.60	II
110	4665.44	II
310	4673.16	I
570	4675.62	II
150	4679.06	II
230	4722.69	I
150	4729.05	I
130	4751.52	II
170	4759.65	II
190	4820.35	II
140	4857.44	I
150	4872.09	II
210	4900.08	I
210	4934.11	II
130	4944.36	I
180	4951.74	II
130	4976.42	I
250	5007.25	I
140	5028.33	I
120	5028.91	I
200	5035.94	I
210	5042.05	I
130	5043.86	I
130	5044.89	I
130	5077.59	II
120	5124.56	I
130	5127.41	I
120	5131.53	I
130	5133.83	I
170	5164.77	II
130	5172.78	I
160	5188.90	II
150	5206.52	I
60	5212.91	II
30	5215.13	II
30	5218.26	II
45	5229.34	II
140	5255.93	II
22	5256.47	II
27	5257.02	II
35	5264.77	II
80	5272.91	I
55	5277.71	I
27	5279.34	II
45	5302.30	II
55	5333.06	I
27	5333.33	II
27	5334.23	II
22	5343.94	II
30	5344.50	II
90	5348.06	I
45	5350.47	I
35	5368.85	I
35	5395.87	II
60	5414.63	II
18	5422.81	II
18 h	5451.30	I
35	5454.27	II
180	5456.62	I
35	5462.43	II
90	5468.32	I
18	5477.47	I
80	5485.97	II
27	5497.44	II
27	5516.02	I
80	5593.46	I
45 d	5601.14	I
	5601.32	I
45 h	5609.94	I
60	5611.82	I
70	5622.01	I
80	5626.53	II
30	5636.20	I
90	5640.36	I
22	5641.42	I
22 h	5658.63	II
70	5664.95	I
45	5665.44	II
55	5675.48	II
14	5695.53	II
27	5710.87	II
55	5717.48	I
70	5719.55	I
55	5726.97	I
22	5733.43	II
22	5736.56	I
22	5736.94	I
100	5739.19	I
35	5740.61	I
60	5748.65	I
55	5752.53	I
70	5757.63	II
290	5762.80	I
70	5769.92	I
45	5782.82	I
70	5784.66	I
22	5791.15	II
70	5800.79	I
22	5806.10	I
430	5826.79	I
45	5835.84	I
100	5850.07	I
120	5855.31	I
140	5872.35	I
120	5881.14	I
27	5886.30	II
27	5902.08	II
55	5906.06	I
45	5909.24	I
35	5933.50	I
22	5946.37	I
55	5968.68	I
27	5975.49	I
35	6006.79	II
22	6008.75	II
55	6014.83	I
35	6015.74	II
70	6022.56	I
22	6032.12	II
22	6045.63	II
22	6048.14	II
45	6054.85	I
70	6061.25	I
60	6076.45	II
35	6116.01	I
35	6125.32	I
30	6170.06	II
27	6183.21	II
360	6221.02	I
35	6230.90	I
55	6262.56	I
45	6267.93	I
60	6268.87	I
35	6274.94	I
30	6286.86	I
45	6299.42	I
130	6308.77	I
55	6326.13	I
22	6347.16	II
45	6388.19	I
22	6432.53	I
27	6485.87	I
55	6492.35	I

Erbium (Cont.)

Intensity		Wavelength	
22		6541.57	I
60		6583.48	I
70		6601.11	I
27		6721.91	I
70		6759.87	I
22		6762.92	I
27		6773.37	I
35		6790.92	I
22		6825.44	I
22		6825.98	I
70		6848.10	I
55		6865.13	I
27		6879.98	I
22		7001.40	I
12		7058.55	I
12	h	7065.04	I
11		7070.99	II
18		7101.27	I
8		7109.67	I
11		7155.40	II
5		7161.91	I
14	h	7197.00	I
7	h	7264.82	II
7	h	7283.95	I
14		7329.73	II
18		7355.37	I
11		7356.34	I
18		7428.67	I
55		7459.55	I
9		7460.42	I
120		7469.51	I
22		7532.34	I
6	h	7539.18	I
27		7556.26	I
6	h	7574.21	I
5	h	7590.51	I
11		7597.33	I
6		7607.23	I
11		7613.52	I
6		7623.48	I
16	h	7645.67	I
8		7650.63	I
22		7654.45	II
12	h	7658.05	I
22		7659.25	I
4		7665.64	I
35		7680.01	I
9		7722.14	I
8		7726.19	II
11	h	7747.44	I
22		7754.63	I
4		7762.16	I
9		7796.69	I
35		7797.47	I
9		7838.80	I
11		7844.00	I
16		7847.55	I
5		7875.36	I
5		7879.36	I
18		7899.55	I
8		7913.08	I
35		7921.85	I
30		7937.84	I
8		7952.93	I
12		7964.51	I
8		7979.03	II
8		7980.87	I
5		8023.03	I
12		8035.91	I
12		8181.85	I
35		8312.82	I
18		8328.57	II
5		8367.58	II
55		8409.90	I
11		8466.18	II
35		8472.42	I
14		8517.71	II
18		8521.37	II
22		8768.64	I
11	h	8776.63	II
9		8866.84	II

Er III
Ref. 301—J.R.

Intensity	Wavelength	
	Air	
2	2165.26	III
3	2190.77	III
10	2198.15	III
1	2223.98	III
4	2232.35	III
60	2235.28	III
8	2245.60	III
2	2255.95	III
80	2269.36	III
600	2277.65	III
100	2309.19	III
40	2358.69	III
10	2358.79	III
50	2359.33	III

Intensity	Wavelength	
200	2367.64	III
20	2375.50	III
10	2377.07	III
80	2381.25	III
20	2381.40	III
40	2381.75	III
6	2391.96	III
60	2393.08	III
5	2393.63	III
250	2396.40	III
10	2398.91	III
2	2402.75	III
80	2404.58	III
100	2410.47	III
200	2419.81	III
200	2422.47	III
40	2431.51	III
60	2464.60	III
2	2492.04	III
100	2508.59	III
2	2531.03	III
100	2532.36	III
8	2536.76	III
80	2540.91	III
3	2543.31	III
10	2545.95	III
50	2557.22	III
80	2570.74	III
40	2580.02	III
2	2589.55	III
80	2590.72	III
20	2591.56	III
200	2591.83	III
20	2598.39	III
3	2599.18	III
100	2603.62	III
40	2604.91	III
25	2614.53	III
30	2617.64	III
2	2618.40	III
2	2618.94	III
8	2625.19	III
20	2626.37	III
4	2637.52	III
200	2637.77	III
5	2651.49	III
25	2683.10	III
400	2723.29	III
100	2738.53	III
500	2739.27	III
8	2741.41	III
80	2746.03	III
6	2752.20	III
80	2756.20	III
400	2759.23	III
150	2761.92	III
60	2762.66	III
15	2767.11	III
100	2768.72	III
60	2772.07	III
60	2774.80	III
20	2775.55	III
2	2780.60	III
80	2783.11	III
500	2792.54	III
10	2804.10	III
100	2805.87	III
6	2808.44	III
50	2824.75	III
150	2830.34	III
1	2831.95	III
8	2833.03	III
8	2845.29	III
60	2846.08	III
6	2849.63	III
1	2869.52	III
8	2878.24	III
1	2955.93	III
1	2958.63	III
1000	3055.10	III
1000	3070.40	III
500	3100.40	III
1500	3166.25	III
3	3172.47	III
1	3173.45	III
50	3175.74	III
400	3214.95	III
2000	3301.13	III
8	3341.00	III
200	3480.54	III
8	3592.96	III
600	3715.67	III
200	3739.43	III
4000	3816.78	III
600	3962.87	III
40	4009.70	III
2	4088.58	III
1000	4288.18	III
40000	4290.06	III
300	4338.24	III
20000	4386.86	III
30	4612.93	III
15000	4735.56	III
2000	4783.12	III

Intensity	Wavelength	
8	4876.07	III
8000	5903.30	III

EUROPIUM (Eu)
Z = 63

Eu I and II
Ref. 1—C.H.C.

Intensity		Wavelength	
		Air	
21		2499.39	II
26		2554.78	II
26		2559.18	II
160		2564.17	II
110		2568.17	II
26		2574.76	II
230		2577.14	II
26		2581.86	II
26		2604.61	I
30		2635.50	II
1000		2638.77	II
380		2641.27	II
40		2653.61	II
640		2668.34	II
110		2673.41	II
250		2678.29	II
250		2685.66	II
550		2692.03	II
700		2701.14	II
800		2701.90	II
240		2705.28	II
180		2709.99	I
700		2716.98	I
70		2723.96	I
4200		2727.78	II
190		2729.33	II
380		2729.44	II
50		2731.37	I
40		2732.61	I
80		2735.25	I
160		2740.62	II
70		2743.28	I
120		2744.26	II
40		2745.61	I
70		2747.29	II
80		2747.83	I
90		2752.17	II
480		2781.89	II
1900		2802.84	II
220		2811.75	II
30		2813.08	II
3400		2813.94	II
550		2816.18	II
2000		2820.78	II
400	Cw	2828.72	II
120		2829.30	II
140		2833.26	II
80		2843.96	II
60		2852.05	II
260		2859.67	II
280		2862.57	II
25		2864.42	II
60		2876.06	II
100		2878.87	I
80		2887.85	II
200		2892.54	I
140		2893.01	I
360		2893.83	I
3200		2906.68	II
160		2908.99	I
30		2917.44	II
850		2925.04	II
60		2947.29	II
200	Cw	2952.68	II
30		2958.91	I
35		2959.47	II
260		2960.21	II
300		2991.33	II
35		2995.22	II
40		3006.26	II
35		3022.15	I
30		3040.77	II
320	Cw	3054.94	II
120		3058.98	I
35		3069.11	II
35		3076.07	II
220		3077.36	II
35		3089.35	II
120		3097.45	II
320		3106.18	I
950		3111.43	I
120		3130.73	II
40		3132.16	I
45		3149.88	II
85		3173.61	II
40		3185.54	I
420		3210.57	I
1000		3212.81	I
420		3213.75	I
45		3235.13	I

Intensity		Wavelength	
95		3241.40	I
45		3246.03	I
45		3247.32	II
100		3247.55	I
100		3266.39	II
150		3272.77	II
210		3277.78	II
150		3301.95	II
45		3304.50	II
140		3308.02	II
140		3313.33	II
65		3319.89	II
95		3321.86	II
85		3322.26	I
950		3334.33	I
45		3338.75	II
110		3350.40	I
40		3351.56	II
40		3354.38	II
45		3367.64	II
140		3369.06	II
65		3380.25	II
75		3390.78	II
190		3391.99	II
280		3396.58	II
45		3419.84	II
65		3423.09	II
150		3425.02	II
45		3426.44	II
45		3435.05	II
65		3435.20	II
40		3435.72	II
45		3440.82	II
150		3441.00	II
45		3445.18	II
85		3457.05	I
45		3457.56	II
130		3461.38	II
85		3467.88	I
75		3477.07	I
75	h	3505.30	II
470	Cw	3521.09	II
75		3531.15	II
45		3532.23	II
65		3538.08	II
150		3542.15	II
85		3543.85	II
45		3549.71	II
180		3552.52	II
75		3589.27	I
45		3591.31	II
150		3603.20	II
75		3611.57	II
45		3616.15	II
95		3622.54	II
95		3632.18	II
45		3673.19	II
45		3674.63	II
45		3678.26	II
6400		3688.42	II
60		3710.87	II
95		3713.45	II
95		3714.90	II
35		3716.94	II
35		3717.69	II
40		3719.16	I
20000	Cw	3724.94	II
45		3729.68	II
45		3729.74	II
21		3732.20	I
45		3738.03	II
350		3741.31	II
100		3743.56	II
260		3761.12	II
95		3765.93	II
40		3774.10	I
60		3781.40	II
40		3788.76	II
45		3791.50	II
130		3799.01	II
70		3801.36	II
95		3807.54	II
120		3811.33	I
120		3815.50	II
39000	Cw	3819.67	II
120		3826.68	II
140		3844.23	II
190		3865.57	I
45		3872.72	II
70		3877.27	II
150		3884.75	I
23		3896.78	I
23		3900.18	I
70		3900.51	I
28000	Cw	3907.10	II
45		3915.24	II
45		3916.00	I
230		3917.29	I
23		3917.70	I
40		3918.52	I
100		3919.09	II
40		3928.87	II
32000	Cw	3930.48	II
55		3941.56	II

Intensity		Wavelength	
30	h	3942.21	II
60		3942.94	II
120		3943.08	II
30		3944.59	II
30		3945.67	II
30		3949.13	II
60		3949.60	I
45		3950.76	II
55		3951.33	II
60		3955.75	I
40		3957.92	II
30		3963.61	I
120		3964.90	II
150		3966.59	II
45		3967.18	I
30000	Cw	3971.96	II
60		3978.42	I
30		3979.63	II
55	h	3986.60	I
40		3988.24	II
30		3993.93	II
55		3995.98	II
60		4003.71	II
180		4011.69	II
150		4017.58	II
120		4039.19	I
45	h	4078.24	I
120		4085.38	II
75		4096.80	II
60		4106.88	I
90	h	4112.04	II
45		4119.30	II
75		4127.28	I
33000	Cw	4129.70	II
30		4136.59	II
40		4137.07	I
30		4141.02	II
60		4141.72	II
30		4151.52	II
45		4151.64	II
30		4157.72	I
110		4172.80	II
30		4175.16	II
110		4182.22	I
40		4196.18	II
60000	Cw	4205.05	II
45		4221.08	II
40		4223.88	II
90	h	4227.40	II
75		4229.33	I
75		4237.45	II
90		4237.51	II
45		4238.69	II
45		4244.74	I
45		4247.06	II
45		4253.80	II
30		4270.24	I
150		4298.73	I
90		4329.36	II
75		4329.97	I
60		4330.61	II
40		4331.18	I
90		4337.68	I
240		4355.09	II
27		4361.57	II
55		4369.47	II
45	h	4372.20	II
75		4383.17	II
90		4387.88	I
21	h	4405.27	II
55		4407.07	II
18		4419.66	II
120		4434.81	II
14000	Cw	4435.56	II
75	h	4464.97	II
24		4485.15	II
3000		4522.57	II
45	h	4535.59	I
11000		4594.03	I
21		4602.63	I
9800		4627.22	I
8300		4661.88	I
30		4713.59	I
27		4740.50	I
45		4792.59	I
40	h	4829.30	I
60		4830.33	I
40	h	4840.47	I
60	h	4849.64	I
110		4867.62	I
40	h	4884.05	I
90		4894.68	I
60		4900.86	I
150		4907.18	I
180		4911.40	I
55		4953.52	I
55		4960.21	I
55		4962.55	I
45		4975.76	I
180		5013.17	I
170		5022.91	I
110		5029.54	I
90		5033.55	I
75		5067.95	I

Intensity		Wavelength	
75	h	5092.69	I
90	h	5096.44	I
170		5114.37	I
90		5124.77	I
170		5129.10	I
90		5130.08	I
210		5133.52	I
270		5160.07	I
210		5166.70	I
60		5193.74	I
200		5199.85	I
110		5200.96	I
120		5206.44	I
750		5215.10	I
300		5223.49	I
120		5239.24	I
200		5266.40	I
390		5271.96	I
110		5272.48	I
150		5282.82	I
55		5287.25	I
60		5289.25	I
120		5291.26	I
60		5293.68	I
120		5294.64	I
90		5303.85	I
30	h	5350.41	I
75	h	5351.69	I
40		5352.84	I
90		5355.10	I
540		5357.61	I
60		5360.83	I
120		5361.61	I
110		5376.94	I
120		5392.94	I
450		5402.77	I
45		5405.33	I
45		5411.86	I
55		5421.07	I
90		5426.94	I
40		5443.56	I
380		5451.51	I
260		5452.94	I
40		5457.62	I
90		5472.32	I
120		5488.65	I
45		5495.20	I
15		5500.83	I
120		5510.52	I
30		5526.63	I
30		5533.25	I
30		5542.54	I
200		5547.44	I
150		5570.33	I
200		5577.14	I
75		5579.63	I
120		5580.03	I
90		5586.24	I
75		5586.83	I
18		5592.25	I
18		5599.80	I
18		5605.86	I
40		5618.81	I
60		5622.44	I
75		5632.54	I
210		5645.80	I
15		5651.11	I
60		5673.85	I
27		5681.10	I
27		5684.24	I
60		5730.87	I
60		5739.00	I
330		5765.20	I
180		5783.69	I
15		5792.72	I
60		5800.27	I
170		5818.74	II
600	Cw	5830.98	I
27		5845.77	I
27		5860.97	I
15		5864.77	I
90		5872.98	II
15		5895.31	I
27		5902.97	I
12		5909.94	I
75		5915.74	I
12		5925.30	I
27		5926.52	I
45		5942.72	I
27		5953.49	I
27		5953.84	II
30		5954.28	I
90		5963.76	I
330		5966.07	II
480	Cw	5967.10	I
15	h	5968.43	I
30		5971.69	I
170		5972.75	I
15		5980.47	I
27		5983.14	I
27		5983.78	I
240		5992.83	I
60		6004.36	I
15	h	6005.61	I

Intensity		Wavelength	
60	h	6012.20	I
110		6012.50	I
60		6015.58	I
420		6018.15	I
60		6023.15	I
170		6029.00	I
60		6044.66	I
420		6049.51	II
140		6057.36	I
90		6075.58	I
30		6077.38	I
240		6083.84	I
240		6099.35	I
60		6108.15	I
120		6118.78	I
60		6124.67	I
330		6173.05	II
110		6178.76	I
260	Cw	6188.13	I
140		6195.07	I
15	h	6207.60	I
15		6230.51	I
90	h	6233.73	I
55		6250.47	I
240		6262.25	I
55		6266.95	I
15	h	6285.95	I
60		6291.34	I
170		6299.77	I
230		6303.41	II
24	h	6313.78	I
15		6318.58	I
75		6335.82	I
120	Cw	6350.04	I
60		6355.89	I
60		6369.25	I
55		6382.73	I
75		6383.86	I
120	Cw	6400.93	I
40		6406.11	I
180		6410.04	I
140		6411.32	I
55		6428.29	I
830		6437.64	II
18		6439.93	I
120		6457.96	I
12		6470.70	I
18		6483.02	I
45		6501.53	I
60		6519.59	I
15		6522.72	I
8	h	6549.12	I
75		6567.87	I
45		6593.79	I
18	h	6603.55	I
1400		6645.11	II
26		6685.21	I
95		6693.96	I
7	h	6701.06	I
12		6710.45	I
30		6744.88	I
30	h	6782.54	I
14	h	6787.48	I
140		6802.72	I
35		6816.06	I
11	h	6834.30	I
17		6840.93	I
17	h	6844.83	I
14	h	6847.04	I
360		6864.54	I
21		6898.21	I
60	h	6903.67	I
14	h	6910.17	I
30	h	6914.82	I
120		7040.20	I
12		7074.54	I
330		7077.10	II
100		7106.48	I
6		7164.66	I
30		7175.55	I
570		7194.81	II
570		7217.55	II
11	h	7224.68	I
15		7258.72	I
30		7262.77	I
11	h	7281.53	I
6	h	7297.56	I
540		7301.17	II
11		7310.46	I
12		7313.63	I
55	Cw	7336.18	I
4		7346.25	I
4		7356.65	I
11		7362.25	I
55	Cw	7369.60	I
720		7370.22	II
4		7387.36	I
12		7389.16	I
11	h	7404.41	I
300		7426.57	II
21	h	7436.59	I
8		7470.53	I
5	h	7491.00	I
50	Cw	7528.70	I

Intensity		Wavelength	
5	h	7533.02	I
6		7547.32	I
150		7583.91	I
60	Cw	7742.57	I
70		7746.19	I
8	h	7803.32	I
8		7818.21	I
35		7887.99	I
7		8015.47	I
24	Cw	8209.80	I
15	Cw	8226.81	I
6	h	8464.71	I
21	Cw	8642.67	I
7		8727.77	I
6		8782.46	I
12	Cw	8790.88	I
18		8870.30	I

Eu III
Ref. 312—J.R.

Intensity	Wavelength	
	Air	
10	2073.40	III
10	2093.50	III
30	2124.69	III
10	2167.12	III
10	2173.59	III
10	2184.68	III
10	2190.59	III
10	2194.81	III
20	2211.85	III
20	2212.63	III
20	2214.66	III
10	2215.34	III
30	2217.23	III
30	2219.33	III
20	2219.42	III
10	2223.13	III
10	2235.17	III
10	2240.14	III
10	2261.88	III
20	2265.74	III
10	2269.39	III
20	2276.85	III
40	2291.62	III
20	2304.37	III
10	2311.92	III
10	2327.69	III
10	2334.56	III
10	2336.96	III
10	2339.84	III
10	2343.10	III
10	2346.83	III
10	2347.64	III
10	2350.38	III
200	2350.51	III
10	2352.28	III
10	2357.87	III
10	2359.08	III
10	2360.65	III
20	2363.76	III
20	2368.04	III
20	2374.08	III
10	2375.20	III
4000	2375.46	III
10	2376.42	III
20	2377.23	III
10	2381.81	III
10	2383.62	III
20	2387.29	III
20	2389.11	III
150	2389.98	III
20	2391.11	III
20	2391.90	III
10	2392.59	III
10	2394.66	III
20	2395.62	III
20	2398.79	III
20	2401.00	III
20	2402.34	III
20	2404.08	III
10	2406.14	III
20	2407.30	III
20	2408.32	III
10	2409.63	III
20	2410.08	III
40	2412.02	III
20	2412.96	III
20	2413.26	III
10	2413.41	III
10	2419.11	III
10	2419.25	III
10	2419.58	III
30	2422.00	III
10	2422.90	III
10	2425.33	III
50	2425.68	III
40	2427.67	III
10	2429.32	III
10 d	2429.66	III
10	2430.04	III

Europium (Cont.)

Intensity		Wavelength		Intensity		Wavelength	
10		2431.49	III	10		2928.91	III
10		2431.76	III	20	c	2931.00	III
10		2432.55	III	10	c	2950.20	III
10		2433.65	III	20		2956.74	III
10		2434.19	III	10		2956.90	III
100	d	2435.14	III	10	c	2972.30	III
20		2436.39	III	30	c	2982.29	III
10		2436.77	III	20	c	3000.11	III
10		2438.83	III	20		3006.37	III
10		2440.26	III	20	c	3013.28	III
50		2440.67	III	10	c	3018.43	III
1000		2444.38	III	20		3022.08	III
4000		2445.99	III	50	c	3022.69	III
30	h	2446.43	III	20	c	3023.40	III
20		2448.57	III	100	c	3023.93	III
20		2451.24	III	10		3025.32	III
10		2451.73	III	10	c	3026.09	III
30		2455.22	III	200	c	3026.79	III
10		2461.79	III	50	c	3029.92	III
30		2463.30	III	20		3031.24	III
30		2464.47	III	40	c	3032.84	III
40		2470.51	III	20	c	3036.98	III
10		2474.94	III	10		3038.64	III
10		2476.24	III	10		3039.05	III
20		2476.45	III	20	c	3039.98	III
10		2477.78	III	10		3054.07	III
20		2480.02	III	10	c	3054.97	III
20		2483.29	III	20		3076.43	III
10		2486.92	III	10	c	3089.09	III
10		2488.91	III	10		3105.25	III
10		2490.50	III	10		3109.67	III
10		2491.08	III	10		3129.31	III
10		2492.48	III	10		3142.54	III
10		2496.92	III	50	c	3171.00	III
10		2499.17	III	10		3178.08	III
2000		2513.76	III	20	h	3178.87	III
20		2517.94	III	50	c	3183.78	III
200		2522.14	III	10	h	3191.46	III
10		2539.14	III	20	c	3194.34	III
10		2548.30	III	10		3206.30	III
20		2548.59	III	10		3208.95	III
10		2554.50	III	10	h	3213.84	III
10		2558.07	III	10		4837.98	III
10		2560.36	III	50		6666.35	III
10		2594.71	III	30		7221.84	III
20		2594.76	III	20		7690.44	III
10		2596.34	III	10		8079.07	III
10		2604.44	III				
30		2608.34	III				
30		2610.09	III				
50	c	2616.11	III				
20		2616.26	III				
10		2616.33	III				
10		2616.35	III				
10		2620.79	III				
20		2623.33	III				
10		2626.98	III				
20	c	2628.46	III				
10		2628.82	III				
10		2631.98	III				
30	c	2642.27	III				
20		2645.22	III				
20	c	2650.93	III				
10		2653.19	III				
10		2655.09	III				
10		2662.24	III				
20	c	2666.86	III				
20	c	2668.21	III				
40	c	2676.09	III				
20	c	2683.21	III				
20		2686.13	III				
20	c	2687.74	III				
40	c	2693.51	III				
10		2694.80	III				
20		2699.87	III				
20	c	2700.78	III				
10	c	2708.25	III				
20	c	2708.84	III				
10		2712.08	III				
50	c	2720.67	III				
20	c	2725.54	III				
10		2743.94	III				
10		2752.68	III				
20		2755.12	III				
10		2757.75	III				
20	c	2760.21	III				
10		2761.72	III				
10		2766.26	III				
20	c	2768.38	III				
10	c	2768.54	III				
10		2769.71	III				
20	c	2780.48	III				
20		2792.51	III				
10		2808.09	III				
10		2817.58	III				
20	c	2839.56	III				
20		2844.99	III				
10	c	2848.44	III				
20		2850.39	III				
10	c	2892.60	III				
40		2912.23	III				
40	c	2912.64	III				
10		2913.04	III				

FLUORINE (F)
Z = 9

F I and II
Ref. 169, 224—G.V.S.

Intensity	Wavelength	
	Vacuum	
30	375.30	II
30	380.90	II
40	407.04	II
50	430.91	II
40	431.55	II
40	435.64	II
70	457.18	II
40	471.95	II
60	472.00	II
50	472.71	II
40	473.02	II
90	484.60	II
50	513.64	II
70	514.94	II
70	546.85	II
60	547.87	II
50	548.32	II
40	548.52	II
90	605.67	II
80	606.29	II
100	606.80	II
70	606.92	II
80	607.47	II
90	608.06	II
15	780.39	I
10	780.52	I
10	782.38	I
12	791.88	I
10	792.54	I
10	794.42	I
150	806.96	I
125	809.60	I
500	951.87	I
1000	954.83	I
750	955.55	I
500	958.52	I
20	972.40	I
350	973.90	I
100	976.22	I
40	976.51	I
100	977.75	I
40	1129.76	II
40	1327.06	II
50	1328.11	II

Intensity		Wavelength	
40		1333.59	II
50		1343.60	II
40		1344.04	II
50		1400.61	II
40		1407.14	II
60		1493.09	II
50		1493.24	II
40		1493.31	II
40		1702.13	II
40		1744.75	II
50		1745.55	II
60		1747.39	II
		Air	
100		2556.11	II
100		2871.40	II
120		3059.99	II
140		3153.49	II
170		3202.76	II
140		3264.08	II
140		3414.65	II
150		3416.45	II
140		3416.80	II
160		3417.00	II
160		3472.96	II
150		3473.31	II
170		3474.78	II
190		3501.39	II
200		3501.45	II
200		3501.57	II
180		3502.84	II
200		3502.96	II
210		3503.11	II
170		3505.37	II
200		3505.52	II
220		3505.63	II
160		3522.89	II
150		3536.87	II
160		3541.77	II
160		3590.52	II
6		3594.10	I
170		3598.69	II
180		3601.39	II
190		3602.84	II
12		3668.17	I
180		3704.53	II
160		3710.35	II
160		3739.57	II
140		3805.83	II
270		3847.09	II
260		3849.99	II
250		3851.67	II
5		3898.48	I
190		3898.83	II
180		3901.93	II
170		3903.82	II
8		3930.69	I
5		3934.26	I
5		3948.56	I
150		3972.04	II
160		3972.67	II
170		3974.78	II
240		4024.73	II
220		4025.01	II
230		4025.49	II
160		4083.91	II
190		4103.07	II
170		4103.22	II
200		4103.51	II
180		4103.71	II
170		4103.87	II
170		4109.16	II
160		4116.54	II
150		4119.21	II
140		4207.15	II
170	h	4225.16	II
150	h	4244.12	II
200		4246.23	II
190		4246.39	II
180		4246.59	II
170		4246.77	II
160		4246.84	II
170	h	4275.36	II
160	h	4277.53	II
160	h	4278.93	II
200		4299.17	II
160		4446.53	II
170		4446.72	II
180		4447.19	II
140		4734.38	II
170		4859.39	II
160		4933.26	II
6		4960.65	I
140		5002.00	II
150		5173.25	II
15		5230.41	I
12		5279.01	I
18		5540.52	I
12		5552.43	I
10		5577.33	I
160		5589.27	II
20		5624.06	I

Intensity		Wavelength	
12		5626.93	I
15		5659.15	I
40		5667.53	I
90		5671.67	I
18		5689.14	I
25		5700.82	I
25		5707.31	I
12		5950.15	I
25		5959.19	I
70		5965.28	I
50		5994.43	I
150		6015.83	I
80		6038.04	I
900		6047.54	I
100		6080.11	I
800		6149.76	I
400		6210.87	I
13000		6239.65	I
140		6247.90	II
10000		6348.51	I
8000		6413.65	I
450		6569.69	I
300		6580.39	I
400		6650.41	I
1800		6690.48	I
400		6708.28	I
7000		6773.98	I
1500		6795.53	I
9000		6834.26	I
50000		6856.03	I
8000		6870.22	I
15000		6902.48	I
6000		6909.82	I
4000		6966.35	I
45000		7037.47	I
30000		7127.89	I
130		7179.90	II
15000		7202.36	I
130	h	7211.79	II
1000		7309.03	I
15000		7311.02	I
700		7314.30	I
5000		7331.96	I
10000		7398.69	I
4000		7425.65	I
2200		7482.72	I
2500		7489.16	I
900		7514.92	I
5000		7552.24	I
5000		7573.38	I
7000		7607.17	I
18000		7754.70	I
15000		7800.21	I
300		7879.18	I
500		7898.59	I
350		7936.31	I
300		7956.32	I
80		8016.01	II
1000		8040.93	I
900		8075.52	I
350		8077.52	I
350		8126.56	I
600		8129.26	I
300		8159.51	I
600		8179.34	I
300		8191.24	I
350		8208.63	I
2500		8214.73	I
3000		8230.77	I
500		8232.19	I
1500		8274.62	I
2000		8298.58	I
600		8302.40	I
900		8807.58	I
1000		8900.92	I
300		8912.78	I
350		9025.49	I
400		9042.10	I
350		9178.68	I
200		9433.67	I
25		9505.30	I
12		9662.04	I
25		9734.34	I
15		9822.11	I
12		9902.65	I
80	h	10047.98	II
15		10285.45	I
20		10862.31	I

F III
Ref. 225—G.V.S.

Intensity		Wavelength	
		Vacuum	
50	h	230.12	III
50		255.72	III
60		255.77	III
70		255.86	III
70		261.71	III
60		261.75	III
80		263.81	III

Fluorine (Cont.)

Intensity	Wavelength	
70	279.69	III
80	315.22	III
70	315.54	III
60	315.75	III
100	429.51	III
110	430.15	III
80	430.22	III
90	464.29	III
100	465.11	III
120	508.39	III
120	567.69	III
110	567.75	III
80	630.14	III
90	630.20	III
120	656.12	III
130	656.87	III
140	658.33	III
80	1219.03	III
80	1266.87	III
90	1267.71	III
70	1297.54	III
70	1359.92	III
110	1498.93	III
120	1502.01	III
110	1504.18	III
140	1504.79	III
130	1506.30	III
110	1506.77	III
100	1553.02	III
110	1557.59	III
100	1563.73	III
100	1565.54	III
100	1623.40	III
100	1650.76	III
130	1670.39	III
140	1677.40	III
100	1716.99	III
120	1770.09	III
150	1770.67	III
110	1772.93	III
140	1773.36	III
160	1791.65	III
110	1803.03	III
100	1804.70	III
170	1805.90	III
110	1839.30	III
120	1839.97	III
110	1840.14	III
80	1900.76	III

Air

Intensity	Wavelength	
100	2027.44	III
120	2030.32	III
120	2217.17	III
120	2452.07	III
130	2464.85	III
130	2470.29	III
120	2478.73	III
150	2484.37	III
120	2542.77	III
120	2580.04	III
130	2583.81	III
120	2593.23	III
130	2595.53	III
140	2599.28	III
130	2625.01	III
140	2629.70	III
120	2656.44	III
130	2755.55	III
160	2759.63	III
120	2788.15	III
160	2811.45	III
140	2833.99	III
150	2835.63	III
150	2860.33	III
120	2862.86	III
140	2887.58	III
150	2889.45	III
120	2905.30	III
140	2913.29	III
160	2916.34	III
140	2932.49	III
140	2994.28	III
120 h	2997.21	III
130	2997.53	III
120	2999.47	III
130	3039.25	III
120	3039.75	III
160	3042.80	III
150	3049.14	III
140	3113.62	III
160	3115.70	III
180	3121.54	III
140	3124.79	III
140	3134.23	III
140	3146.99	III
180	3174.17	III
170	3174.76	III
120	3214.00	III
140 h	4420.30	III
120 h	4427.35	III
120 h	4432.32	III

Intensity	Wavelength	
140 h	4479.99	III
150	5012.54	III
160	5110.99	III
140	5753.17	III
120	5761.20	III
150	6091.82	III
140	6125.50	III
130	6233.57	III
140	6363.05	III
120	7336.77	III
130	7354.94	III

F IV
Ref. 68, 226—G.V.S.

Intensity	Wavelength	
	Vacuum	
30	169.79	IV
30	169.84	IV
30	171.07	IV
40	176.37	IV
40	181.52	IV
40	181.57	IV
30	187.24	IV
50	196.39	IV
60	196.45	IV
50	199.76	IV
50	199.80	IV
50	199.85	IV
50	199.93	IV
50	200.00	IV
70	200.09	IV
60	201.01	IV
70	201.06	IV
60	201.10	IV
80	201.16	IV
60	201.22	IV
90	208.25	IV
70	213.85	IV
70	214.06	IV
70	220.77	IV
60	226.94	IV
50	227.10	IV
60	233.22	IV
50	233.39	IV
70	239.86	IV
70	240.02	IV
90	240.08	IV
70	240.15	IV
70	240.28	IV
70	240.37	IV
100	251.03	IV
60	270.23	IV
140	419.65	IV
150	420.05	IV
160	420.73	IV
150	430.76	IV
130	490.57	IV
160	491.00	IV
50	497.38	IV
60	497.83	IV
70	498.80	IV
140	570.64	IV
140	571.30	IV
150	571.39	IV
160	572.66	IV
140	676.12	IV
130	677.15	IV
150	677.22	IV
130	678.99	IV
160	679.21	IV

Intensity	Wavelength	
	Air	
40	2171.44	IV
50	2298.29	IV
40	2451.58	IV
50	2456.92	IV
40	2820.74	IV
50	2826.13	IV

F V
Ref. 68, 226—G.V.S.

Intensity	Wavelength	
	Vacuum	
40	134.54	V
40	147.95	V
50	148.00	V
40	152.51	V
40	158.54	V
40	162.27	V
40	163.50	V
50	163.56	V
90	165.98	V
100	166.18	V
40	174.70	V
50	178.43	V

Intensity	Wavelength	
40	178.59	V
40	182.98	V
40	186.72	V
40	186.79	V
50	186.84	V
40	186.97	V
40	187.01	V
60	190.57	V
70	190.84	V
40	191.97	V
40	205.55	V
100	464.37	V
110	465.37	V
120	465.98	V
100	466.99	V
90	506.16	V
100	508.08	V
70	513.97	V
60	514.08	V
80	524.59	V
90	525.29	V
100	526.30	V
70	647.67	V
100	647.77	V
110	647.87	V
70	647.97	V
130	654.03	V
110	657.23	V
140	657.33	V
60	757.04	V
60	1082.31	V
70	1088.39	V

Intensity	Wavelength	
	AIR	
10	2229.18	V
20	2252.72	V
20	2450.63	V
10	2461.33	V
10	2693.98	V
10	2702.30	V
10	2703.96	V
20	2707.17	V

FRANCIUM (Fr)
Z = 87

Fr I
Ref. 408 — C.H.C.

Intensity	Wavelength	
	Air	
	7177.	I

GADOLINIUM (Gd)
Z = 64

Gd I and II
Ref. 1—C.H.C.

Intensity	Wavelength	
	Air	
100	2468.22	II
55	2471.58	II
35	2485.47	II
70	2487.46	II
110	2488.72	II
55	2493.29	II
35	2496.35	II
45	2499.04	II
28	2543.68	II
28	2586.13	II
28	2661.50	II
70	2720.50	II
430	2750.22	II
460	2764.08	II
40	2768.51	II
320	2769.81	II
230	2770.17	II
21	2770.98	II
45	2778.76	I
45	2779.14	II
440	2781.40	II
70	2787.68	I
390	2791.96	II
100	2794.66	II
930	2796.93	II
60	2808.38	II
750	2809.72	II
160	2810.93	II
45	2814.01	II
300	2833.75	II
35	2836.69	II
70	2837.00	II
560	2840.23	II
140	2841.33	II
40	2853.91	II

Intensity	Wavelength	
60	2856.52	II
19	2859.78	II
120	2862.48	II
60	2865.06	II
40	2866.33	II
40	2871.75	II
460	2881.33	II
40	2882.13	II
130	2885.60	II
35	2907.44	II
170	2910.53	II
60	2913.08	II
45	2918.52	II
95	2923.32	II
35	2924.25	II
35	2928.34	II
35	2947.80	II
70	2948.01	II
35	2952.43	I
35	2955.60	II
70	2960.93	II
130	2963.60	II
80	2965.43	II
29	2972.74	II
560	2980.15	II
35	2983.74	II
40	2991.52	II
95	2993.04	II
1200	2999.04	II
370	3002.86	II
100	3005.09	II
2100	3010.13	II
130	3012.19	II
1900	3027.60	II
120	3028.98	II
2100	3032.84	II
1600	3034.05	II
130	3043.01	I
160	3046.48	I
280	3053.57	II
100	3059.92	I
1000	3068.64	II
560	3072.56	II
640	3076.92	II
150	3077.08	II
2100	3081.99	II
140	3084.01	II
280	3089.95	II
140	3092.06	II
460	3098.64	II
190	3098.90	II
3500	3100.50	II
120	3101.18	II
230	3101.91	II
580	3102.55	II
130	3108.36	II
170	3111.19	I
160	3113.17	II
120	3118.60	II
120	3119.01	I
510	3119.94	II
100	3120.18	II
370	3123.99	II
120	3124.25	II
130	3128.56	II
130	3130.81	II
100	3133.09	II
460	3133.85	II
210	3135.03	II
190	3136.93	I
190	3137.30	I
120	3138.71	I
230	3143.13	II
930	3145.00	II
370	3145.52	II
230	3146.88	II
980	3156.53	II
200	3158.63	I
140	3160.69	II
980	3161.37	II
220	3190.28	I
220	3199.30	I
160	3199.58	I
110	3203.41	I
690	3223.74	II
	3223.78	I
110	3225.46	II
160	3226.32	II
220	3232.78	I
100	3250.19	II
110	3259.25	II
540	3266.73	I
250	3267.64	II
140	3268.34	II
110	3274.18	II
110	3279.53	II
100	3281.61	II
250	3282.25	I
	3282.30	II
430	3291.48	I
370	3292.21	II
430	3294.08	I
330	3313.73	II
200	3315.59	II

Gadolinium (Cont.)

Int	λ		Int	λ		Int	λ		Int	λ	
430	3330.34	II	6100	3646.19	II	1200	3957.67	II	1700	4251.73	II
1400	3331.38	II	310	3649.44	II	750	3959.44	II	860	4253.37	II
830	3332.13	II	450	3650.95	II		3959.52	GD II	650	4253.61	II
1100	3336.18	II	620	3652.54	II	220	3963.66	II	810	4260.12	I
590	3345.98	II	3900	3654.62	II	590	3966.28	I	1600	4262.09	I
200	3350.10	II	3100	3656.15	II	590	3968.26	II	650	4266.60	I
5400	3350.47	II	210	3658.19	I	750	3969.00	I	470	4267.00	I
220	3357.61	I	1400	3662.26	II	270	3969.29	II	300	4274.17	I
270	3358.43	II	2700	3664.60	II	450	3971.75	II	910	4280.49	II
4300	3358.62	II	2000	3671.20	II	390	3972.71	I	430	4285.82	I
780	3360.71	II	1000	3674.05	I	590	3973.98	II	300	4286.12	I
5400	3362.23	II	350	3679.21	I	300	3974.81	I	540	4296.08	II
270	3364.24	II	2000	3684.13	I	750	3979.33	I	220	4297.17	II
200	3365.59	II	720	3686.33	II	450	3987.33	II	430	4299.29	I
220	3374.69	II	3100	3687.74	II	470	3987.84	I	1100	4306.34	I
220	3379.76	II	210	3694.03	II	320	3992.69	I	260	4309.29	I
220	3380.52	II	2000	3697.73	II	220	3993.21	II	1800	4313.84	I
1100	3392.53	II	1300	3699.73	II	650	3994.16	II	520	4314.40	I
540	3395.12	II	2700	3712.70	II	700	3996.32	II	520	4316.05	II
220 d	3397.22	I	2000	3713.57	I	320	3997.76	II	370	4320.52	I
	3397.32	I	1400	3716.36	II	470	4001.26	II	750	4321.11	II
200	3399.41	II	2000	3717.48	I	260	4004.94	II		4321.20	I
540	3399.99	II	1800 d	3719.45	II	320	4008.33	I	2600 d	4325.57	II
540	3402.07	II		3719.53	II	300	4008.91	II		4325.69	I
200	3406.92	I	250	3722.01	II	300	4013.80	II	1900	4327.12	I
1100 d	3407.56	II	430	3725.47	II	200	4015.58	I	370	4329.58	I
	3407.61	II	1500	3730.84	II	300	4017.25	II	340	4330.61	II
250	3409.30	II	270	3732.32	I	430	4017.71	I	240	4331.38	I
220	3411.02	I	230	3732.45	II	300	4019.73	I	450	4341.28	II
220	3413.27	II	230	3732.67	I	300	4022.33	II	910	4342.18	II
1400	3416.95	II	510	3733.08	II	1100	4023.14	I	1000	4344.30	II
1400	3418.73	II	490	3739.76	I	810	4023.35	I	2200	4346.46	I
6900	3422.47	II	330	3740.02	II	220	4027.61	I	910	4346.62	I
390	3422.75	II	4500	3743.47	II	1100	4028.15	I	220	4347.31	II
1100	3423.90	I	620	3744.83	I	860	4030.88	I	300	4369.77	II
	3423.92	II	230	3757.74	II	700	4033.49	I	970	4373.83	I
830	3424.59	II	1000	3757.94	I	340	4035.40	I	280	4392.06	I
390	3425.93	II	1400	3758.31	II	260	4036.84	I	1400	4401.86	I
220	3428.47	II	820	3759.00	II	1400	4037.33	II	520	4403.14	II
690	3432.99	II	620	3760.71	II	700	4037.90	II	260	4406.67	II
1700	3439.21	II	290	3760.92	II	410	4043.71	I	260	4408.25	II
830	3439.78	II	870	3762.20	I	1600	4045.01	I	220	4409.25	I
2700	3439.99	II	210	3763.33	II	270	4046.84	II	520	4411.16	I
390	3449.62	II	370	3764.20	II	270	4047.09	I	860	4414.16	I
1400	3450.38	II	870	3767.04	II	270	4049.20	I	700	4414.73	I
1100	3451.23	II	8700	3768.39	II	1300	4049.43	II	340	4419.03	II
540	3454.14	II	620	3769.45	II	2200	4049.86	II	1400	4422.41	I
880	3454.90	II	1400	3770.69	II	270	4050.37	I	1100	4430.63	I
200	3455.27	I	250	3771.26	I	810	4053.29	II	240 d	4436.10	I
200	3457.05	II	210	3773.45	I	2600	4053.64	I		4436.22	II
220	3461.95	II	210	3776.83	I	810	4054.72	I	300	4464.74	I
220	3463.00	II	1000	3782.34	II	2600	4058.22	I	300	4466.55	II
2700	3463.98	II	2900	3783.05	I	650	4059.88	I		4466.60	I
330	3466.95	II	1100	3787.56	II	270	4061.30	II	520	4467.08	I
1700	3467.27	II	200	3790.63	I	650	4062.59	II	700	4474.13	I
1700	3468.99	II	770	3791.17	II	1900	4063.39	II	860	4476.12	I
1400	3473.22	II	490	3792.39	II	540	4063.59	II	220	4478.80	II
2200	3481.28	II	5100	3796.37	II	260	4066.04	I	280	4481.06	II
1700	3481.80	II	720	3801.29	II	520	4068.35	I	220	4483.33	II
490	3482.60	II	210	3804.39	I	260	4068.74	I	220	4484.70	I
220	3486.20	I	210	3805.09	II	750	4070.29	II	280	4486.90	I
980	3491.95	II	560	3805.52	II		4070.39	II	500	4497.13	I
1700	3494.40	II	3700	3813.97	II	650	4073.20	II	220	4497.32	I
1400	3505.51	II	430	3814.74	II	300	4073.76	II	430	4506.21	I
780	3512.22	II	770	3816.64	II	1300	4078.44	II	140	4506.33	II
1100	3512.50	II	430	3818.75	II	2800	4078.70	I	140	4514.50	II
830	3513.65	I	350	3826.05	II	520	4083.70	I	1100	4519.66	I
980	3524.20	II	230	3827.33	II	1500	4085.56	II	300	4522.82	II
430	3528.54	II	230	3829.46	II	260	4087.69	II	150	4524.12	I
540	3542.77	II	370	3831.80	II	650	4090.41	I	910	4537.81	I
4300	3545.80	II	210	3832.97	I	1100	4092.71	I	220	4540.02	II
3900	3549.36	II	330	3834.99	II	260	4093.72	I	300	4542.03	I
1400	3557.05	II	970	3836.91	II	260	4094.48	II	240	4548.00	I
540	3558.19	II	1000	3839.64	II	2600	4098.61	II	120	4558.08	II
430	3558.47	II	1200	3842.20	II	520	4098.90	II	130	4573.81	I
200	3564.05	II	1400	3843.28	I	650	4100.26	I	260	4575.91	I
690	3571.93	II	1400	3844.54	II	390	4111.44	II	280	4579.59	I
330	3574.74	II	3300	3850.69	II	2200	4130.37	II	410	4581.29	II
390	3578.36	II	5100	3850.77	II	270	4131.48	II	130	4582.53	II
980	3581.91	II	4300	3852.45	II	1100	4132.28	II	410	4583.07	I
5400	3584.96	II	470	3855.96	II	750	4134.16	I	160	4586.99	I
540	3590.47	II	250	3863.05	II	410	4137.10	II	220	4596.98	II
1100	3592.71	II	1600	3866.99	I	280	4148.86	I	320	4597.91	II
200	3593.44	II	250	3873.57	I	540	4162.73	II	410	4598.90	I
540	3600.96	II	220	3875.46	II	280	4163.09	II	340	4601.05	II
1100	3604.87	I	1500	3894.70	II	280	4167.16	II	240	4602.93	I
270	3605.26	II	450	3895.79	II		4167.27	I	520	4614.50	I
250	3605.66	II	750	3902.40	II	2400	4175.54	I	140	4624.42	I
830	3608.75	II	300	3902.71	I	2400	4184.25	II	430	4636.64	I
830	3610.76	II	240	3904.29	I	2200	4190.78	I	110	4639.00	II
220	3610.91	II	450	3905.65	I	750	4191.07	II	170	4640.04	I
540	3613.39	II	2200	3916.51	II	750	4191.63	I	170	4646.00	I
270 d	3614.21	II	450	3923.25	II	450	4197.68	II	170	4647.64	I
	3614.42	I	1200	3934.79	I	590	4204.86	II	170 d	4648.59	I
430	3617.16	II		3934.82	II	1300	4212.00	II		4648.70	I
390	3620.46	II	220	3935.38	I	970	4215.02	II	430	4653.54	I
270	3624.89	II	450	3941.80	I	650	4217.20	II	140 h	4670.87	I
250	3629.51	II	590	3942.63	I	320	4225.03	I	170	4679.18	I
330	3634.76	II	270	3943.24	I	4800	4225.85	I	260	4680.04	I
220	3639.05	II	220	3943.62	I	220	4227.14	II	430	4683.33	I
250	3640.18	II	1400	3945.54	I	220	4229.80	II	140	4688.12	I
330	3641.39	II	300	3952.00	II	650	4238.78	II	700	4694.33	I
870	3645.62	II	590	3953.37	I	200	4246.57	II	170	4695.49	I

Gadolinium (Cont.)

Intensity		Wavelength	
430		4697.42	I
170		4703.13	I
200		4709.78	I
110		4721.46	I
150		4728.47	II
220		4732.60	II
260		4735.75	I
410		4743.65	I
110		4745.82	I
320		4758.70	I
110		4760.74	I
130		4763.82	I
470		4767.24	I
180		4781.92	I
300		4784.62	I
110		4786.75	I
140		4801.05	II
220		4807.45	I
320		4821.69	I
130		4835.26	I
110		4848.10	I
110		4862.59	I
170		4865.02	II
120		4871.50	I
280		4934.12	I
220		4938.61	I
110		4952.47	I
130		4958.79	I
65		5010.82	II
55		5011.74	I
750		5015.04	I
55		5023.13	II
65		5031.29	II
75		5039.09	I
65		5050.88	II
55		5073.74	I
55		5082.80	I
95		5092.25	II
65		5096.06	II
130		5098.38	II
55		5100.94	II
910		5103.45	I
180		5108.91	II
120		5125.56	II
65		5130.28	II
65		5135.59	I
75		5136.04	I
85		5140.84	II
75		5141.50	I
75		5142.68	I
860		5155.84	I
55		5156.76	II
75		5158.48	I
75		5163.70	I
55		5164.54	II
190		5176.28	II
55		5187.24	II
55		5187.88	I
55		5191.08	II
410		5197.77	I
55		5210.49	II
85		5217.48	I
280		5219.40	I
75		5220.30	II
130		5233.93	I
65		5246.87	I
320		5251.18	I
120		5252.14	II
85		5254.75	I
140		5255.80	I
65		5268.78	I
55		5272.91	I
55		5282.48	I
280		5283.08	I
280		5301.67	I
220		5302.76	I
55		5306.70	I
280		5307.30	I
130		5321.50	I
280		5321.78	I
110		5327.32	I
65		5328.30	I
170		5333.30	I
55		5337.53	I
300		5343.00	I
85		5345.13	I
75		5345.68	I
200		5348.67	I
300		5350.38	I
240		5353.26	I
55		5361.66	I
95		5365.38	I
95		5369.92	I
150		5370.63	I
85		5389.50	I
85		5413.20	I
85		5415.69	I
65		5453.46	I
55		5583.68	II
55	d	5591.85	I
190		5617.91	I
65		5629.55	I
110		5632.25	I
260		5643.24	I
55	B1	5680.89	G
390		5696.22	I
95		5701.35	I
65		5709.42	I
120		5733.86	II
85		5746.36	I
85	d	5754.17	I
75		5776.02	I
240		5791.38	I
65	h	5796.80	I
55		5802.92	I
55	Hs	5807.72	I
55	h	5809.22	I
55		5815.85	II
65	Hs	5819.51	G
55		5840.47	II
220		5851.63	I
55		5855.24	II
280		5856.22	I
55		5860.73	II
65		5877.26	II
55		5886.46	I
55		5904.07	II
110		5904.56	I
170		5911.45	II
65		5913.55	II
55		5916.77	I
85		5930.29	I
85		5936.84	I
65		5937.71	I
55	h	5940.95	G
55	h	5942.78	G
55		5951.60	II
55		5956.48	II
85		5977.25	I
110	h	5988.02	I
85		5999.08	I
65		6000.96	G
75	h	6001.87	G
55		6004.57	II
55		6008.71	I
55		6021.13	I
55		6080.65	II
430		6114.07	I
55		6180.42	II
110	B1	6182.68	G
110	B1	6200.86	G
110	B1	6211.71	G
110	B1	6220.93	G
55	B1	6231.62	G
75	B1	6241.66	G
55	b	6252.12	G
55	B1	6262.64	G
45	b	6273.00	G
85		6289.73	II
30		6292.87	I
75		6305.15	II
30		6309.11	II
27		6317.19	I
40		6331.35	I
17		6333.75	I
17		6336.34	I
27		6346.65	II
27	h	6351.72	I
17		6363.23	I
40		6380.95	II
17		6382.19	II
22		6408.55	I
22		6422.42	II
17		6424.52	I
19	h	6470.29	I
15		6480.11	II
40	h	6538.15	I
22		6549.25	I
55		6564.78	I
10		6568.00	II
10		6573.80	I
30		6591.60	I
15		6593.42	I
10		6610.04	II
50		6634.36	II
35		6640.08	I
10		6642.76	I
30		6643.98	I
10		6646.85	I
10		6653.55	I
10		6679.56	II
35		6681.23	II
10		6692.86	I
10		6704.18	II
14		6718.14	II
17		6727.83	II
85		6730.73	I
50		6752.67	II
14		6753.91	II
14		6783.39	I
26		6786.33	II
10		6787.18	I
12		6814.56	I
26		6816.49	I
17		6820.90	I
100		6828.25	I
35		6846.60	II
30		6857.13	II
15		6864.25	I
21		6887.63	II
14		6900.73	II
100		6916.57	I
21		6920.62	II
15		6924.99	II
21		6926.49	I
17		6945.98	II
15		6957.74	II
15		6959.24	II
14		6964.33	I
15		6971.66	II
12		6976.35	II
10		6978.27	II
26	h	6980.86	I
50		6985.89	II
10		6988.75	II
75		6991.92	I
21		6993.18	I
60		6996.76	II
17		7000.75	II
45		7006.16	II
10		7016.60	I
21		7037.26	II
14		7051.00	II
13		7054.62	II
10		7058.02	II
10		7068.09	II
18		7071.00	I
18		7073.63	I
14		7098.11	I
14		7098.73	I
10	h	7116.77	II
21		7118.86	II
35		7122.57	I
13		7135.73	II
18		7147.31	II
13		7158.28	I
170		7168.37	I
21		7172.26	II
28		7189.57	II
13		7197.08	II
13		7201.41	II
10		7228.02	I
25		7233.45	I
14		7252.70	II
28		7262.66	I
14		7291.35	I
21		7313.28	I
18		7324.89	II
14		7373.81	I
14		7376.41	I
13		7377.27	II
13		7380.28	I
13		7394.90	II
13		7430.19	I
35		7441.85	I
40		7464.36	I
55		7562.97	I
10		7563.19	II
10		7588.20	I
10		7611.78	I
21		7621.96	I
21		7650.32	I
25		7672.56	I
10		7676.06	I
13		7694.45	I
80		7733.50	I
35		7749.30	I
10		7755.97	I
10		7766.48	II
11	h	7844.87	I
10		7845.80	I
35		7846.35	II
35		7856.93	I
14		7869.72	I
25		7930.25	II
13		8077.59	I
18		8146.15	I
11		8218.08	I
10		8275.42	I
10		8349.73	I
11		8398.30	I
10		8445.47	I
13	h	8527.88	I
21		8668.63	I
11		8770.36	I
13		8784.85	I
10		8795.76	I
21	h	8832.06	II
14	h	8849.14	I
18	h	8867.31	I

Gd III
Ref. 46, 137, 151—J.F.K.

Intensity	Wavelength	
	Vacuum	
600	1813.47	III
900	1946.26	III
1100	1974.34	III
2200	1975.24	III
	Air	
900	2008.79	III
3400	2018.07	III
1800	2027.82	III
800	2046.02	III
500	2057.79	III
1500	2080.08	III
1800	2098.20	III
1300	2125.68	III
1400	2148.03	III
1700	2176.84	III
1700	2223.95	III
1700	2236.73	III
1200	2239.84	III
1500	2243.75	III
1700	2250.18	III
1300	2257.05	III
1300	2292.51	III
1000	2300.38	III
1200	2303.72	III
1500	2307.03	III
1100	2313.50	III
1700	2313.56	III
1000	2315.09	III
1400	2323.12	III
1700	2323.18	III
2200	2323.78	III
1900	2329.35	III
2100	2335.01	III
1600	2336.02	III
1900	2338.97	III
1600	2339.88	III
2100	2342.74	III
2500	2346.52	III
2800	2359.31	III
1900	2360.87	III
2300	2361.91	III
1400	2362.38	III
2100	2363.26	III
1600	2365.22	III
2000	2373.38	III
1300	2374.29	III
1300	2381.38	III
1600	2307.02	III
2000	2388.77	III
1200	2393.86	III
1200	2397.34	III
1200	2405.03	III
1300	2408.41	III
1200	2409.35	III
1500	2466.84	III
1100	2469.14	III
1300	2499.53	III
1600	2520.38	III
1600	2534.11	III
1300	2536.10	III
1600	2551.56	III
2200	2553.90	III
2100	2554.04	III
2500	2563.33	III
2100	2564.46	III
1000	2565.04	III
2400	2565.95	III
1800	2569.27	III
1800	2573.57	III
2000	2576.06	III
1300	2576.15	III
1400	2578.13	III
1600	2578.76	III
1700	2583.62	III
2000	2588.21	III
2000	2588.46	III
1300	2595.81	III
1800	2609.77	III
1200	2619.40	III
1200	2621.52	III
1400	2623.52	III
1400	2625.48	III
2000	2628.10	III
1300	2628.99	III
2400	2629.83	III
2100	2632.30	III
1800	2633.32	III
1400	2635.71	III
1600	2636.44	III
1700	2637.15	III
2100	2637.97	III
2100	2638.06	III
2200	2640.53	III
1600	2641.65	III
2100	2643.71	III
1600	2644.52	III
1800	2646.04	III
1800	2646.84	III
1600	2651.48	III
2000	2655.59	III
1900	2656.55	III
2200	2660.83	III

Gadolinium (Cont.)

Intensity	Wavelength	
1800	2675.75	III
1800	2679.44	III
1700	2680.63	III
1800	2682.52	III
1500	2683.91	III
1600	2692.78	III
1900	2692.86	III
1500	2694.43	III
2800	2697.39	III
1500	2702.91	III
2800	2703.28	III
1600	2704.53	III
1800	2717.35	III
2700	2727.89	III
450	2751.24	III
1800	2833.83	III
9000	2904.73	III
1800	2918.40	III
9500	2955.53	III
1000	2975.42	III
1000	2984.10	III
1000	3116.59	III
2500	3118.04	III
4000	3176.66	III
400	3253.53	III
400	3330.34	III
400	3371.05	III
400	3402.97	III
450	3624.90	III
250	3700.47	III
300	3831.73	III
300	3910.24	III
300	4016.91	III
600	4177.26	III
400	4279.96	III
300	4314.28	III
300	4445.91	III
600	4684.25	III
600	4715.06	III
600	4782.79	III
250	4976.72	III
5000	5091.70	III
300	5124.06	III
1800	5347.95	III
3000	5365.96	III
1100	5412.62	III
4000	5553.30	III
3000	5587.88	III
3000	5658.98	III
1800	5786.96	III
1500	5862.09	III
1500	5987.85	III
5000	14332.88	III
2000	17474.78	III
800	19996.34	III
800	21259.44	III
600	22493.33	III

Gd IV
Ref. 152—J.F.K.

Intensity	Wavelength	
	Vacuum	
1000	967.92	IV
1000	983.42	IV
1000	987.10	IV
1000	987.91	IV
1000	995.04	IV
1000	995.80	IV
1000	996.49	IV
1000	999.24	IV
1000	1000.36	IV
1000	1002.73	IV
1000	1004.46	IV
1000	1005.66	IV
1000	1006.55	IV
1200	1007.24	IV
1200	1063.84	IV
500	1228.37	IV
500	1307.23	IV
500	1313.29	IV
500	1316.71	IV
600	1321.42	IV
500	1330.79	IV
1100	1393.24	IV
1600	1476.98	IV
1500	1705.03	IV
1600	1706.01	IV
2000	1736.24	IV
1500	1815.32	IV
400	1997.89	IV
	Air	
800	2049.28	IV
800	2061.30	IV
800	2070.40	IV
800	2076.66	IV
1000	2094.29	IV
1000	2296.89	IV
1000	2352.66	IV
800	2379.17	IV
900	2385.65	IV
500	2390.07	IV
700	2392.30	IV
700	2393.29	IV
700	2395.76	IV
500	2396.22	IV
600	2396.27	IV
1400	2397.87	IV
700	2402.70	IV
500	2412.21	IV
900	2419.26	IV
500	2439.84	IV
600	2440.38	IV
600	2468.60	IV

GALLIUM (Ga)
Z = 31

Ga I and II
Ref. 19, 132, 195, 281—L.J.R.

Intensity		Wavelength	
		Vacuum	
2		829.60	II
2		958.67	II
1		960.57	II
2		969.19	II
2		998.52	II
3		1002.95	II
5		1012.38	II
3		1019.10	II
5		1023.80	II
8		1033.69	II
1		1113.87	II
3		1119.25	II
5		1130.81	II
1		1167.62	II
2		1173.78	II
3		1186.81	II
1		1227.13	II
5		1286.38	II
5		1327.81	II
20		1414.44	II
5		1449.49	II
2		1463.65	II
3		1473.73	II
3		1483.52	II
3		1485.95	II
3		1495.21	II
3		1504.41	II
3		1505.01	II
5		1514.57	II
3		1515.19	II
8		1535.40	II
5		1536.37	II
1		1536.91	II
3		1669.83	II
5		1695.85	II
5		1799.42	II
10		1813.98	II
15		1845.30	II
		Air	
20		2091.34	II
1		2218.04	I
1		2255.03	I
1		2259.23	I
2		2294.19	I
1		2297.87	I
3		2338.24	I
1		2338.60	I
3		2371.29	I
2		2377.53	II
4		2418.69	I
5		2438.88	II
6		2450.08	I
7		2500.19	I
3		2500.71	I
5		2513.55	II
3		2514.15	II
2		2551.26	II
3		2552.87	II
4		2555.28	II
5		2607.47	I
8		2624.82	I
10		2632.66	I
3		2659.87	I
10		2665.05	I
8		2691.29	I
20		2700.47	II
3		2719.66	I
15		2780.15	II
6		2874.24	I
1		2886.45	II
		2893.65	GA II
2		2910.77	II
6		2943.64	I
6		2944.17	I
3		2969.41	II
1		2971.01	II
3		2971.60	II
5		2974.77	II
1		2992.84	II
2		3011.90	II
1		3158.18	II
4		3374.94	II
1		3375.95	II
2		3436.66	II
3		3446.46	II
2		3447.26	II
5		3470.34	II
1		3471.46	II
1		3472.52	II
2		3583.60	II
1		3693.93	II
2		3705.85	II
4		3734.85	II
9		3924.39	II
10		4032.99	I
10		4172.04	I
4		4251.11	II
15		4251.16	II
10	h	4254.04	II
4	h	4255.64	II
5		4255.70	II
10		4255.77	II
40		4262.00	II
3		5218.21	II
1		5338.3	II
2		5353.49	I
2		5360.6	II
1		5363.5	II
3		5416.8	II
1		5421.6	II
1		5425.6	II
8		6334.2	II
2000		6396.56	I
1000		6413.44	I
5		6419.4	II
3		6456.3	II
1		7000.0	II
3	h	7051.24	I
5	h	7106.82	I
1	h	7116.3	I
2	h	7172.9	I
5	h	7193.6	I
7		7198.7	II
10	h	7251.4	I
3	h	7289.6	I
5	h	7349.3	I
20	h	7403.0	I
30	h	7464.0	I
6	h	7556.6	I
10	h	7620.5	I
50	h	7734.77	I
2		7793.0	II
100	h	7800.01	I
4	h	7801.6	I
15	h	8002.55	I
20	h	8074.25	I
3	h	8167.5	I
5	h	8171.6	I
100	h	8311.86	I
200	h	8386.49	I
10	h	8389.30	I
7	h	8415.51	I
10	h	8419.91	I
20	h	8808.75	I
30	h	8813.56	I
20	h	8856.37	I
30	h	8944.33	I
200	h	9492.92	I
200	h	9493.12	I
300	h	9589.36	I
20	h	9594.25	I
60	h	10898.10	I
100	h	10905.95	I
10		10968.27	I
20		11103.51	I
400		11949.12	I
200		12109.78	I
40		12885.05	I
50		13057.50	I
50		14982.75	I
60		14996.64	I
20		17757.91	I
10		17868.96	I
60		22016.81	I
70		22568.71	I

Ga III
Ref. 141—L.J.R.

Intensity	Wavelength	
	Vacuum	
50	620.00	III
40	622.01	III
90	806.51	III
90	817.30	III
50	828.70	III
80	1085.00	III
60	1105.61	III
90	1150.27	III
90	1267.16	III
80	1293.46	III
60	1295.36	III
60	1323.15	III
70	1353.92	III
90	1495.07	III
50	1534.46	III
	Air	
90	2417.70	III
90	2423.98	III
15	2424.36	III
50	3521.77	III
80	3581.19	III
100	3589.34	III
10	3731.10	III
10	3806.60	III
100	4380.69	III
150	4381.76	III
100	4863.00	III
150	4993.78	III
10	5808.28	III
20	5848.25	III
15	5993.51	III

Ga IV
Ref. 141, 143—L.J.R.

Intensity	Wavelength	
	Vacuum	
14	294.53	IV
61	295.67	IV
41	304.99	IV
4	422.12	IV
25	423.18	IV
16	439.92	IV
67	1137.06	IV
70	1156.10	IV
70	1163.60	IV
75	1170.58	IV
48	1171.71	IV
68	1185.23	IV
40	1186.06	IV
73	1190.89	IV
73	1193.02	IV
75	1195.02	IV
69	1201.54	IV
72	1206.89	IV
63	1216.15	IV
50	1228.03	IV
60	1236.38	IV
60	1238.59	IV
45	1241.81	IV
75	1245.53	IV
83	1258.77	IV
81	1264.66	IV
82	1267.15	IV
81	1279.24	IV
80	1285.33	IV
82	1295.86	IV
83	1299.46	IV
82	1303.53	IV
80	1309.68	IV
80	1314.82	IV
85	1338.09	IV
77	1347.03	IV
76	1351.06	IV
74	1364.63	IV
60	1395.54	IV
77	1402.55	IV
70	1405.32	IV
73	1465.87	IV

Ga V
Ref. 2, 62, 140—L.J.R.

Intensity	Wavelength	
	Vacuum	
5	290.53	V
1	296.13	V
5	296.82	V
30	298.44	V
20	299.47	V
30	300.01	V
25	300.57	V
10	300.78	V
20	301.19	V
30	302.86	V
20	303.84	V
30	307.03	V
30	308.26	V
15	309.64	V
30	311.79	V
25	312.41	V
30	313.68	V

Gallium (Cont.)

Intensity	Wavelength		Spectrum
15	315.95		V
20	316.48		V
40	319.41		V
12	320.53		V
40	322.31		V
50	322.99		V
30	323.10		V
40	324.25		V
40	324.95		V
40	326.14		V
30	326.77		V
30	328.65		V
5	336.61		V
20	878.17		V
40	973.21		V
10	977.89		V
15	979.60		V
20	984.95		V
40	989.75		V
90	1014.47		V
90	1019.71		V
20	1033.55		V
30	1038.76		V
30	1047.50		V
120	1050.48		V
80	1054.56		V
90	1058.12		V
80	1066.69		V
35	1068.59		V
30	1069.45		V
60	1069.60		V
55	1071.19		V
45	1071.41		V
80	1073.77		V
90	1078.83		V
110	1079.60		V
60	1080.99		V
250	1085.01		V
80	1087.37		V
40	1090.53		V
90	1091.71		V
100	1094.36		V
80	1095.10		V
70	1101.62		V
160	1102.83		V
140	1103.03		V
60	1104.93		V
75	1105.62		V
70	1106.17		V
40	1115.55		V
80	1118.34		V
55	1123.18		V
80	1123.66		V
120	1126.40		V
80	1127.75		V
130	1128.10		V
120	1128.53		V
100	1129.94		V
80	1131.43		V
40	1133.91		V
130	1136.07		V
65	1138.20		V
60	1144.30		V
50	1145.70		V
30	1148.42		V
45	1150.09		V
130	1150.23		V
120	1156.51		V
35	1157.74		V
25	1169.40		V
40	1178.95		V
80	1213.17		V
30	1265.45		V
30	1276.85		V
15	1283.64		V
10	1311.35		V

Intensity	Wavelength		Spectrum
500	1181.65		II
200	1188.73		II
100	1189.62		II
300	1191.26		II
50	1191.72		II
500	1237.059		II
500	1261.905		II
100	1264.710		II
100	1380.42		II
50	1392.26		II
200	1401.24		II
200	1538.091		II
500	1576.855		II
75	1581.070		II
100	1602.486		II
3	1615.57	r	I
2	1624.130	r	I
2	1630.173	r	I
3	1636.31	r	I
2	1638.96		I
4	1639.730	r	I
2	1647.531		I
200	1649.194		II
2	1651.528		I
4	1651.955	r	I
3	1661.345		I
4	1663.539	r	I
10	1665.275	h	I
4	1667.802		I
3	1670.608	r	I
100	1691.090	r	I
200	1716.784	r	I
100	1739.102	h	I
100	1742.195		I
50	1746.065		I
200	1750.043		I
100	1758.279		I
100	1764.185	h	I
100	1765.284	h	I
50	1766.433	h	I
200	1774.176		I
200	1785.046		I
100	1793.071	h	I
75	1801.432	h	I
200	1841.328	h	I
200	1842.410	h	I
100	1844.410	h	I
100	1845.872	h	I
100	1846.958	h	I
200	1853.134		I
500	1860.086	r	I
100	1865.052		I
300	1874.256	r	I
100	1895.197		I
500	1904.702	r	I
50	1908.434	h	I
30	1912.409		I
300	1917.592	r	I
100	1923.467	h	I
500	1929.826	r	I
10	1934.048	h	I
100	1937.483	r	I
500	1938.008		II
100	1938.300	r	II
500	1938.891		II
30	1944.116	s	I
200	1944.731		I
200	1955.115		I
500	1962.013		I
30	1963.373	h	I
30	1965.383		I
200	1970.880		I
200	1979.274		II
300	1987.849	h	I
300	1988.267		I
500	1998.887	r	I

GERMANIUM (Ge)
Z = 32

Ge I and II
Ref. 5, 119, 293, 340—C.H.C.

Intensity	Wavelength		Spectrum
	Vacuum		
1	822.97		II
3	835.08		II
10	850.50		II
10	862.234		II
15	875.493		II
15	905.977		II
20	920.554		II
50	999.101		II
100	1016.638		II
100	1075.072		II
300	1085.51		II
200	1098.71		II
500	1106.74		II
500	1120.46		II
200	1164.27		II
500	1181.19		II

Intensity	Wavelength		Spectrum
	Air		
50	2007.04		II
200	2011.29		I
1700	2019.068		I
2400	2041.712	r	I
1600	2043.770	r	I
420	2054.461		I
220	2057.238	h	I
750	2065.215	r	I
2600	2068.656	r	I
420	2086.021		I
2000	2094.258	r	I
240	2105.824		I
95	2124.744	h	I
50	2186.451	h	I
100	2197.62		II
340	2198.714	r	I
100	2205.85		II
15	2220.375		I
18	2256.001		I
18	2314.201		I
24	2327.918		I
15	2359.233		I
20	2379.144		I
10	2389.472		I
15	2397.895		I
130	2417.367		I
30	2436.412		I
100	2478.66		II
90	2497.962		I
500	2500.54		II
70	2533.230		I
3	2556.298		I
28	2589.188		I
500	2592.534		I
8	2644.184		I
1200	2651.172		I
500	2651.568		I
6	2691.341		I
200	2704.03		II
850	2709.624		I
400	2729.78		II
40	2740.426		I
650	2754.588		I
50	2770.30		II
75	2772.35		II
70	2793.925		I
80	2829.008		I
1000	2831.843		II
50	2834.28		II
75	2839.68		II
1000	2845.527		II
75	2853.97		II
750	3039.067		I
600	3067.021		I
20	3124.816		I
50	3186.72		II
100	3221.64		II
110	3269.081		II
50	3312.56		II
75	3323.64		II
100	3455.72		II
300	3499.21		II
30	3845.11		II
70	4226.562		I
10	4685.829		I
75	4689.87	h	II
50	4690.02	h	II
1000	4741.806		II
1000	4814.608		II
50	4824.097		II
100	5131.752		II
200	5178.648		II
3	5194.583		I
6	5265.291		I
6	5513.263		I
8	5564.741		I
8	5607.010		I
6	5616.135		I
7	5621.249		I
8	5655.96		I
6	5664.226		I
5	5664.842		I
9	5691.954		I
6	5701.776		I
5	5717.877		I
6	5801.029		I
9	5802.093		I
1000	5893.389		II
500	6021.041		II
150	6078.39		II
50	6267.14		II
150	6268.07		II
100	6268.34		II
75	6283.452		II
100	6336.377		II
100	6484.181		II
6	6557.488		I
50	6780.51		II
50	7049.369		II
6	7130.12		I
30	7145.390		II
7	7330.38		I
5	7353.334		I
7	7384.208		I
6	7402.64		I
7	7511.57		I
5	7776.20		I
10	7833.575		I
7	7837.63		I
6	7853.77		I
7	7878.12		I
5	7962.26		I
5	7983.33		I
10	8031.039		I
6	8044.165		I
5	8095.29		I
5	8225.22		I
7	8226.09		I
10	8256.013		I
5	8264.15		I
5	8280.09		I
6	8281.04		I
8	8367.81		I
7	8391.70		I
5	8396.36		I
5	8429.42		I
10	8482.21		I
8	8506.70		I
5	8507.66		I
8	8564.89		I
6	8599.27		I
6	8652.42		I
5	8669.60		I
9	8700.60		I
5	8712.90		I
6	8734.78		I
6	8789.88		I
5	9068.785		I
5	9095.957		I
6	9398.868		I
20	9474.993		II
20	9475.645		II
4	9492.559		I
7	9625.664		I
5	10039.436		I
4	10200.952		I
10	10382.427		I
10	10404.913		I
8	10734.068		I
8	10947.416		I
10	11125.130		I
230	11252.83		I
24	11293.40		I
33	11318.13		I
55	11459.05		I
150	11483.77		I
175	11614.81		I
600	11714.76		I
10	11839.77		I
55	11917.01		I
10	12025.64		I
10	12055.49		I
30	12061.41		I
45	12065.76		I
1300	12069.20		I
30	12198.88		I
20	12207.73		I
60	12286.75		I
55	12338.76		I
1050	12391.58		I
48	12540.41		I
15	12636.80		I
150	12676.58		I
40	12681.28		I
115	12800.66		I
175	12836.38		I
12	12847.92		I
170	12955.73		I
15	13028.64		I
235	13107.61		I
20	13492.28		I
42	13534.85		I
28	13724.48		I
42	14116.70		I
42	14297.15		I
40	14569.84		I
12	14667.52		I
470	14822.38		I
16	14921.97		I
15	15001.75		I
13	15041.21		I
20	15504.34		I
14	16424.77		I
12	16626.64		I
70	16699.29		I
150	16759.79		I
135	17214.34		I
16	18428.30		I
35	18495.54		I
10	18764.11		I
70	18811.86		I
62	19279.24		I
28	20673.64		I
4	21518.30		I
9	22091.84		I
5	23921.92		I

Ge III
Ref. 341—C.H.C.

Intensity	Wavelength	Spectrum
	Vacuum	
2	542.90	III
2	663.77	III
3	670.88	III
2	680.28	III
2	952.76	III
12	988.96	III
15	995.72	III
10	996.50	III
15	1011.21	III
10	1012.31	III
8	1032.62	III
12	1040.99	III
12	1058.91	III
40	1088.45	III
10	1137.92	III
12	1150.55	III

Germanium (Cont.)

Intensity	Wavelength	
8	1159.15	III
8	1159.62	III
8	1160.79	III
10	1173.78	III
8	1212.47	III
4	1323.24	III
10	1525.32	III
2	1527.15	III
9	1600.09	III
6	1883.26	III
2	1978.22	III

Air

Intensity	Wavelength	
2	2019.22	III
4	2022.25	III
3	2062.14	III
15	2100.05	III
15	2102.42	III
25	2104.45	III
3	2922.86	III
25	3197.56	III
35	3211.86	III
25	3214.95	III
40	3255.05	III
20	3259.90	III
5	3369.57	III
20	3414.27	III
40	3434.03	III
8	3464.59	III
40	3489.08	III
2	3724.51	III
15	3884.78	III
200	4178.96	III
12	4245.41	III
200	4260.85	III
150	4291.71	III
10	4674.36	III
10	5016.88	III
18	5134.75	III
5	5229.37	III
3	5256.61	III

Ge IV
Ref. 341—C.H.C.

Vacuum

Intensity	Wavelength	
1	440.11	IV
1	441.95	IV
3	847.80	IV
3	868.30	IV
8	915.00	IV
8	936.70	IV
4	938.90	IV
1	1073.44	IV
20	1188.99	IV
20	1229.81	IV
2	1494.89	IV
6	1500.61	IV
3	1648.14	IV

Air

Intensity	Wavelength	
2	2293.0	IV
2	2343.37	IV
15	2445.38	IV
15	2445.71	IV
30	2488.25	IV
20	2542.44	IV
5	2631.78	IV
3	2698.08	IV
15	2717.44	IV
30	2736.09	IV
30	2788.61	IV
5	3071.84	IV
60	3554.19	IV
50	3676.65	IV

Ge V
Ref. 342—C.H.C.

Vacuum

Intensity	Wavelength	
700	294.51	V
1000	295.64	V
200	304.98	V
20	621.52	V
35	716.26	V
50	724.21	V
35	733.54	V
35	735.35	V
35	741.52	V
60	746.88	V
40	750.26	V
35	755.84	V
60	760.05	V
60	958.51	V
300	971.35	V
150	984.92	V
200	988.13	V
300	990.66	V
300	1004.38	V
300	1016.66	V
250	1038.40	V
900	1045.71	V
400	1050.05	V
300	1054.59	V
300	1068.43	V
400	1069.13	V
700	1072.66	V
600	1086.65	V
500	1087.85	V
800	1089.49	V
300	1092.09	V
1000	1116.94	V
300	1122.01	V
700	1163.39	V
300	1165.26	V
200	1176.69	V
700	1222.30	V

GOLD (Au)
Z = 79

Au I and II
Ref. 38, 72, 234—C.H.C.

Vacuum

Intensity	Wavelength	
	925.72	II
	946.03	II
	950.39	II
20	957.78	II
3	967.94	II
3	974.47	II
2	982.24	II
	1062.67	II
8	1066.96	II
	1085.00	II
8	1090.78	II
5	1094.92	II
20	1103.31	II
3	1166.76	II
5	1210.86	II
20	1224.57	II
40 h	1305.34	I
25	1310.47	I
100 h	1328.37	I
3	1336.26	II
10 h	1338.37	I
10 h	1342.80	I
20	1350.09	I
20	1350.84	I
22	1351.74	I
25	1352.82	I
25	1354.14	I
30	1355.79	I
35	1357.86	I
40	1360.51	I
6	1362.33	II
20	1362.47	I
8	1363.15	II
50	1363.98	I
25	1364.15	I
10	1364.74	II
60	1368.62	I
35	1368.98	I
70	1374.82	I
50	1375.76	I
30	1378.87	I
8	1380.53	II
80	1382.75	I
50	1385.33	I
20	1389.14	I
60	1392.27	I
6	1393.80	II
50	1402.12	I
30	1405.12	II
70	1407.38	I
100	1408.45	I
	1410.69	II
	1415.22	II
80	1429.19	I
50	1435.79	I
20	1436.61	II
10	1468.85	II
10	1469.17	II
10	1469.28	II
100	1481.76	I
25	1486.55	II
20	1532.82	I
20	1532.86	I
10	1562.04	II
200	1587.16	I
12	1593.41	II
70	1598.24	I
12	1611.11	II
	1616.65	AU II
2	1622.83	II
100	1624.34	I
2	1632.53	II
50	1639.90	I
150	1646.67	I
10	1656.99	II
100	1665.76	I
25	1673.59	II
7	1694.38	II
2	1698.65	II
200	1699.34	I
30	1700.69	II
10	1720.04	II
25	1725.75	II
45	1740.52	II
10	1749.80	II
35	1756.15	II
60	1783.22	II
35	1793.31	II
35	1800.58	II
25	1823.24	II
100	1879.83	I
20	1919.64	I
20	1921.64	II
45	1942.31	I
25	1951.93	I
30	1978.19	I

Air

Intensity	Wavelength	
25	2000.81	II
11000	2012.00	I
2600	2021.38	I
50	2044.54	II
150	2082.09	II
35	2095.13	II
20	2098.14	II
60	2110.68	II
30	2125.29	II
15	2126.63	I
20	2170.75	I
35	2188.81	II
25	2201.32	II
35	2215.63	II
45	2228.88	II
30	2231.18	II
25	2240.16	II
70	2248.56	II
80	2263.62	II
18	2263.88	II
25	2277.52	II
25	2283.30	II
25	2291.40	II
45	2304.69	II
25	2314.55	II
20	2315.75	II
25	2340.06	II
180	2352.65	I
20	2376.28	I
120	2387.75	I
2600	2427.95	I
60	2533.52	II
16	2544.19	I
45	2552.67	II
20	2589.25	I
30	2590.04	I
50	2616.40	II
20	2627.02	II
250	2641.48	I
3400	2675.95	I
20	2687.63	II
20	2688.16	II
30	2688.71	I
80	2700.89	I
1100	2748.25	I
20	2748.71	II
100	2780.82	I
30	2800.93	I
1000	2802.04	II
300	2819.79	II
100	2822.55	II
30	2823.13	II
100 h	2825.44	II
30	2833.03	II
300	2837.85	II
100	2846.92	II
30	2856.74	II
3	2872.36	I
300	2883.45	I
3	2886.96	I
10	2888.40	I
300	2891.96	I
100	2893.25	II
3	2905.74	I
30	2905.90	I
100	2907.04	II
300	2913.52	II
3	2914.82	I
300	2918.24	II
16	2932.19	I
30	2940.67	I
100	2954.22	II
30	2973.33	II
10	2990.27	II
300	2994.80	II
10	3002.65	I
3	3005.85	I
10	3024.67	I
320	3029.20	I
30	3033.25	I
300	3065.42	I
10	3102.63	I
10 h	3117.01	I
100	3122.50	II
1600	3122.78	I
30	3126.86	II
30	3127.03	I
10	3164.88	I
10	3172.35	II
30	3191.76	I
100	3194.72	I
30	3200.37	I
30	3204.74	I
1	3221.86	I
30	3225.25	I
300	3230.63	I
10	3253.94	I
10 h	3265.10	I
30 h	3267.07	I
10	3271.63	I
10	3273.47	I
300	3308.30	I
300	3309.64	I
100	3320.12	I
100	3355.15	I
30 h	3368.44	I
10	3381.90	I
100 h	3391.31	I
100	3395.40	I
30 h	3440.36	I
100	3467.21	I
30	3471.61	I
30	3509.04	I
10	3510.82	I
3	3523.34	II
3	3545.61	I
30	3553.57	I
300	3557.36	I
10	3558.22	I
30	3565.97	I
30 h	3584.37	I
300	3586.73	I
30 h	3588.79	I
30	3598.06	I
100	3611.57	I
30 h	3614.00	I
30	3622.74	I
100 h	3631.31	I
10	3633.22	II
30	3634.53	I
10	3635.12	I
300	3637.90	I
3	3639.87	I
100 h	3645.02	I
10	3649.09	I
100	3650.74	I
10	3653.53	I
3	3654.69	I
10	3655.30	I
10	3656.90	I
30	3706.55	II
100	3709.62	I
10	3766.61	I
10	3770.76	I
100	3796.01	I
30	3801.92	I
30	3804.01	II
10	3821.85	I
10	3825.70	I
100	3874.73	I
30	3880.25	I
30	3889.48	I
100	3892.26	I
400	3897.86	I
30	3901.09	I
300	3909.38	I
100	3927.69	I
10 h	3959.10	I
30	3966.23	I
30	3976.65	I
30	3979.68	I
30	3991.37	I
3	4012.57	I
10	4016.07	II
400	4040.93	I
30	4052.79	II
700	4065.07	I
10	4076.35	II
3	4083.28	II
100	4084.10	I
30	4101.70	I
30	4128.59	I
30	4201.13	I
30 h	4227.88	I
100	4241.80	I
200	4315.11	I
30	4361.04	II
10	4420.61	II
120 h	4437.27	I

Gold (Cont.)

Intensity		Wavelength	Spectrum
250		4488.25	I
900	h	4607.51	I
100	h	4620.56	I
1		4663.92	I
3		4663.97	I
10		4694.69	I
3		4760.17	II
500		4792.58	I
100		4811.60	I
10		4822.96	I
30	h	4950.82	I
30		5064.59	I
30	h	5108.84	I
100		5147.44	I
300		5230.26	I
100	h	5261.76	I
100		5655.77	I
100	h	5721.36	I
300		5837.37	I
100	h	5862.93	I
300	h	5956.96	I
30	h	5962.68	I
600		6278.17	I
100		6562.68	I
30		6652.89	I
600		7510.73	I
10		8145.06	I
10		9254.28	I

Au III
Ref. 72, 393, 395 — R.D.C.

Intensity		Wavelength	Spectrum

Vacuum

Intensity		Wavelength	Spectrum
30		779.73	III
30		788.78	III
50		799.93	III
40		811.83	III
50		817.95	III
40		820.06	III
80		833.16	III
100		843.44	III
100		845.14	III
80		855.49	III
80		859.90	III
80		863.42	III
80		901.03	III
80		910.45	III
80		924.02	III
200		945.10	III
100		1040.63	III
80		1044.49	III
80		1046.81	III
100	h	1239.96	III
100		1278.51	III
100		1314.84	III
200		1336.72	III
180		1341.68	III
100		1348.89	III
150		1350.32	III
150		1355.61	III
150		1356.13	III
80		1362.06	III
500		1365.40	III
200		1367.17	III
180		1377.73	III
150		1378.69	III
150		1379.98	III
125		1380.53	III
200		1381.36	III
300		1385.79	III
100		1389.41	III
180		1391.46	III
180		1396.00	III
100		1402.91	III
225		1409.50	III
250		1413.80	III
100		1414.27	III
80		1415.54	III
100		1417.09	III
125		1417.39	III
150		1427.42	III
300		1428.93	III
250		1430.06	III
275		1433.37	III
250		1435.81	III
80		1436.12	III
300		1439.12	III
200		1441.21	III
150		1446.37	III
80		1446.69	III
250		1448.42	III
250		1454.95	III
100		1464.72	III
150		1471.28	III
80		1473.32	III
100		1474.73	III
150		1481.10	III
300		1487.15	III
250		1487.91	III
200		1489.47	III
250		1500.37	III
200		1502.47	III
200		1503.74	III
80		1540.26	III
100		1542.00	III
80		1542.25	III
100		1548.50	III
80		1554.61	III
80		1562.33	III
80		1562.41	III
200		1567.54	III
80		1571.94	III
200		1574.85	III
200		1579.44	III
150		1584.10	III
200		1589.56	III
80		1589.68	III
150		1593.41	III
200		1600.51	III
250		1617.16	III
100		1617.78	III
500		1621.93	III
300	d	1629.13	III
250		1638.88	III
100		1644.17	III
250		1652.74	III
250		1664.77	III
100		1668.11	III
125		1673.93	III
1000		1693.94	III
150		1697.09	III
200		1698.98	III
200		1700.00	III
200		1702.25	III
100		1707.53	III
250		1710.16	III
200		1715.69	III
100		1716.71	III
300		1717.83	III
500		1727.31	III
100	d	1733.17	III
300		1738.48	III
150		1744.39	III
500		1746.10	III
500		1756.92	III
500		1761.95	III
300		1767.42	III
100		1774.42	III
800		1775.17	III
200		1776.40	III
100		1780.57	III
300		1786.11	III
150		1792.65	III
500		1793.76	III
200		1801.98	III
400		1805.24	III
100		1809.81	III
400		1821.17	III
400		1844.89	III
150		1848.83	III
80		1850.15	III
500		1861.80	III
150		1871.92	III
150		1918.28	III
100		1932.04	III
100		1935.42	III
200		1948.79	III
100		1958.47	III
400		1989.63	III
150		1996.85	III

Air

Intensity	Wavelength	Spectrum
300	2083.09	III
80	2085.45	III
100	2159.08	III
80	2167.33	III
200	2172.20	III
100	2184.11	III
500	2188.97	III
300	2322.27	III
100	2382.40	III
150	2402.71	III
150	2405.12	III
100	3227.99	III
100	3309.86	III

HAFNIUM (Hf)
Z = 72

Hf I and II
Ref. 1—C.H.C.

Intensity	Wavelength	Spectrum

Air

Intensity	Wavelength	Spectrum
6200	2012.78	II
8500	2028.18	II
1200	2096.18	II
540	2210.82	II
320	2254.01	II
160	2255.15	II
250	2266.83	II
620	2277.16	II
230	2321.14	II
580	2322.47	II
300	2323.25	II
120	2324.50	II
300	2324.89	II
200	2332.97	II
200	2337.33	II
230	2343.32	II
320	2347.44	II
540	2351.22	II
110	2353.02	I
90	2365.98	II
250	2380.30	II
100	2381.00	II
170	2393.18	II
450	2393.36	II
670	2393.83	II
130	2400.78	II
70	2404.56	II
540	2405.42	II
130	2406.44	II
370	2410.14	II
90	2413.33	II
55	2415.96	II
320	2417.69	II
120	2425.98	II
45	2428.75	I
120	2428.99	II
130	2433.57	II
45	2434.74	II
35	2444.99	I
390	2447.25	II
140	2449.44	II
35	2452.30	II
110	2453.34	II
450	2460.49	II
70	2463.97	II
430	2464.19	II
90	2465.06	II
35	2465.67	I
140	2467.97	II
210	2469.18	II
100	2473.92	II
55	2481.44	II
55	2482.65	I
55	2487.16	I
290	2496.99	II
580	2512.69	II
580	2513.03	II
130	2515.48	II
890	2516.88	II
340	2531.19	II
200	2537.33	II
110	2548.20	II
320	2551.40	II
130	2559.19	II
250	2563.61	II
890	2571.67	II
320	2573.90	II
320	2576.82	II
300	2578.14	II
320	2582.54	II
130	2591.33	II
390	2606.37	II
450	2607.03	II
120	2608.45	I
230	2613.60	II
450	2622.74	II
160	2637.00	I
1100	2638.71	II
1100	2641.41	II
160	2642.75	I
670	2647.29	II
100	2651.16	II
160	2657.84	II
210	2661.88	II
290	2683.35	II
670	2705.61	I
110	2706.73	II
210	2712.42	II
140	2713.84	I
250	2718.59	II
120	2730.85	I
710	2738.76	II
200	2743.64	I
360	2751.81	II
450	2761.63	I
160	2766.96	I
170	2773.02	I
980	2773.36	II
180	2774.02	II
390	2779.37	I
100	2789.50	II
140	2789.73	II
230	2808.00	II
230	2813.86	II
170	2814.48	II
230	2817.68	I
140	2818.94	I
230	2819.74	II
1200	2820.22	II
490	2822.68	II
180	2833.28	I
110	2834.13	I
410	2845.83	I
270	2849.21	II
270	2850.96	I
180	2851.21	II
180	2860.56	I
760	2861.01	II
760	2861.70	II
2100	2866.37	I
130	2869.82	II
150	2876.33	II
210	2887.14	I
100	2887.54	I
800	2889.62	I
1800	2898.26	I
130	2898.71	II
1200	2904.41	I
890	2904.75	I
140	2909.91	II
2000	2916.48	I
580	2918.58	I
320	2919.59	II
180	2924.62	I
490	2929.63	II
450	2929.90	I
710	2937.80	II
2000	2940.77	I
160	2944.71	I
1200	2950.68	I
1100	2954.20	I
540	2958.02	I
120	2961.80	II
1400	2964.88	I
620	2966.93	I
140	2967.23	II
710	2968.81	II
110	2973.37	I
890	2975.88	II
150	2979.28	I
1100	2980.81	I
210	2982.72	I
170	3000.10	II
800	3005.56	I
1100	3012.90	II
540	3016.78	I
1100	3016.94	II
900	3018.31	I
1200	3020.53	I
140	3025.29	II
410	3031.16	II
110	3046.08	II
710	3050.76	I
1100	3057.02	I
130	3063.78	I
130	3064.68	II
850	3067.41	I
2100	3072.88	I
170	3074.10	I
250	3074.79	I
150	3080.66	II
430	3080.84	I
200	3096.76	I
340	3101.40	II
710	3109.12	II
130	3110.87	II
130	3119.98	I
710	3131.81	I
850	3134.72	II
130	3137.51	I
170	3139.65	II
120	3140.76	II
220	3145.32	II
220	3148.41	I
120	3151.63	I
450	3156.63	I
270	3159.82	I
710	3162.61	II
450	3168.39	I
890	3172.94	I
450	3176.86	II
220	3181.01	I
120	3181.15	I
130	3189.62	I
360	3193.53	II
670	3194.19	II
200	3196.93	I
130	3199.99	II
310	3206.11	I
180	3210.98	I
180	3217.30	II
180	3220.61	I
130	3230.06	I
130	3239.44	I
130	3243.35	I
360	3247.66	I
220	3249.53	I
890	3253.70	II
270	3255.28	II
120	3262.47	I
180	3273.66	II
270	3279.98	II

Intensity		Wavelength	Spectrum
160		3291.05	I
210		3306.12	I
120		3309.19	I
340		3310.27	I
670		3312.86	I
180		3317.99	II
130		3328.21	II
890		3332.73	I
370		3352.06	II
130		3356.78	I
230		3358.91	I
180		3360.06	I
140		3366.68	I
180		3378.93	I
140		3384.14	II
230		3384.70	II
170		3386.21	I
800		3389.83	II
230		3392.81	I
230		3394.59	II
140		3394.98	II
230		3397.26	I
230		3397.60	I
2300		3399.80	II
170		3400.21	I
180		3402.51	I
140		3407.76	II
230		3410.17	II
230		3417.34	I
410		3419.18	I
140		3427.44	I
200		3428.37	II
250		3438.24	II
140		3438.43	I
100		3441.84	I
100		3452.31	I
140		3462.64	II
140		3467.60	I
710		3472.40	I
200		3478.99	II
480		3479.28	II
250		3495.75	II
250		3497.16	I
980		3497.49	I
100		3498.98	I
1200		3505.23	II
150		3513.28	I
130		3518.75	II
980		3523.02	I
100		3530.87	I
100		3531.23	I
980		3535.54	II
760		3536.62	I
180		3548.81	I
540		3552.70	II
150		3554.00	I
1300		3561.66	II
150		3564.31	I
270		3567.36	I
1100		3569.04	II
150		3579.90	I
110		3583.28	I
210		3597.42	II
540		3599.87	I
110		3615.04	I
800		3616.89	I
110		3617.68	I
110		3624.00	II
320		3630.87	II
100		3635.43	I
800		3644.36	II
320		3649.10	I
200		3651.84	I
140		3661.05	II
220		3665.35	II
100		3668.21	I
200		3672.27	I
480		3675.74	I
2200		3682.24	I
280		3696.51	I
100		3698.40	II
240		3699.72	II
340		3701.15	II
100		3704.92	I
120		3705.40	II
1000		3717.80	I
650		3719.28	II
140		3726.49	I
160		3729.10	I
460		3733.79	I
160		3737.88	II
120		3739.04	I
100		3744.98	II
400		3746.80	I
140		3753.22	I
100		3765.05	I
100		3765.56	I
170		3766.92	II
200		3768.25	I
1400		3777.64	I
1400		3785.46	I
650		3793.37	II
100		3798.66	I
850	d	3800.38	I
140		3806.07	II
320		3811.78	I
100		3817.20	II
100		3819.38	I
1300		3820.73	I
140		3829.67	I
280		3830.47	I
800		3849.18	I
140		3849.52	II
600		3858.31	I
230		3860.91	I
200		3872.55	II
160		3877.10	II
380		3880.82	II
200		3882.52	I
150		3883.77	II
200		3889.23	I
200		3889.33	I
620		3899.94	I
620		3918.09	II
200		3923.90	II
120		3926.42	I
150		3927.57	I
110		3929.54	II
320		3931.38	I
120		3935.65	II
120		3939.04	I
410		3951.83	I
160		3968.01	I
150	B1	3970.05	H
200		3973.48	I
180		4032.27	I
100		4047.96	II
230		4062.84	I
140		4066.21	I
180		4083.35	I
540		4093.16	II
110		4104.23	I
140		4106.58	I
110		4113.53	II
110		4118.60	I
150		4127.80	II
140		4145.76	I
150		4162.36	II
110		4162.69	I
1100		4174.34	I
120		4190.95	I
160		4206.58	II
190		4209.10	I
170		4228.08	I
170		4232.44	II
120	B1	4252.08	H
170		4260.98	I
200		4263.39	I
170		4272.85	II
320		4294.79	I
120		4318.14	I
160		4330.27	I
180		4336.66	II
150		4350.51	II
250		4356.33	I
110		4367.90	II
180		4370.97	II
120		4417.35	II
160		4417.91	I
200		4438.04	I
140		4457.34	I
140		4461.18	I
140		4540.93	I
250		4565.94	I
500	d	4598.80	I
230		4620.86	I
210		4655.19	I
120		4699.01	I
160		4782.74	I
310		4800.50	I
130		4859.24	I
120		4975.25	I
95		5018.20	I
15		5021.75	I
55		5040.82	II
95		5047.45	I
55	b	5074.74	H
30		5079.65	II
55	b	5093.88	H
15		5112.13	I
19		5128.53	II
30		5157.96	I
55		5167.42	I
75		5170.18	I
230		5181.86	I
30		5186.84	I
30		5187.75	II
110		5243.99	I
55		5247.10	II
25		5260.44	II
30		5264.95	II
55		5275.04	I
22		5286.09	I
120		5294.87	I
45		5298.06	II
30		5307.82	I
45		5309.68	I
55		5311.60	II
12		5324.26	II
9		5346.30	II
110		5354.73	I
110		5373.86	I
40		5389.34	I
19		5391.36	II
19		5404.47	I
28		5424.02	I
12		5435.78	I
40		5438.74	I
14		5444.07	II
75		5452.92	I
30		5463.38	II
15		5497.30	I
15		5510.12	I
15		5510.45	I
19		5524.35	II
45		5538.02	I
28		5538.26	I
230		5550.60	I
230		5552.12	I
55		5575.86	I
14		5600.77	I
95		5613.27	I
25		5614.01	I
8		5628.27	I
19		5650.83	I
40	B1	5698.03	H
25		5713.28	I
160		5719.18	I
25	B1	5720.16	H
12		5748.72	I
14		5767.18	II
12		5809.50	II
19		5817.47	I
25		5842.23	II
25		5845.87	I
19		5847.77	I
22		5883.66	I
15		5926.47	I
60		5933.69	I
75		5974.28	I
25		5974.72	I
60		5978.66	I
25		5992.96	I
45	c	6016.79	I
28	b	6021.12	H
25	b	6043.19	H
25		6054.17	I
95		6098.67	I
95		6185.13	I
55		6210.70	I
28		6216.82	I
45		6238.58	I
60		6248.95	II
22	h	6299.54	I
25	h	6311.85	I
19		6318.33	I
30		6338.10	I
19	h	6380.19	I
60		6386.23	I
19	h	6409.52	I
15	h	6556.50	I
28		6587.23	I
45		6644.60	II
19		6647.06	II
11		6659.40	I
30		6713.48	I
17		6754.61	II
11		6769.95	I
85		6789.27	I
160		6818.94	I
15		6826.56	I
13		6850.07	I
35		6858.70	I
45		6911.40	I
10		6926.19	I
19		6979.59	I
21		6980.91	II
7		7019.25	I
7		7030.33	II
7		7035.13	I
11		7061.90	I
15		7062.87	I
160		7063.83	I
11	h	7094.40	I
15		7100.54	I
55		7119.52	I
570		7131.81	I
650		7237.10	I
410		7240.87	I
6		7262.62	I
75		7320.05	I
16		7321.76	I
6		7356.10	I
6		7365.28	I
20		7390.70	I
6		7423.69	I
25		7437.56	I
13		7463.86	I
7		7484.56	I
15		7556.37	I
75		7562.93	I
15		7564.22	I
11		7576.95	I
11		7592.96	I
13		7608.59	I
360		7624.40	I
20		7645.64	I
110		7740.17	I
8		7743.57	I
5		7757.89	II
40		7790.90	I
7		7796.81	I
35		7814.55	I
310		7845.35	I
7		7846.56	I
130		7920.71	I
29		7938.06	I
250		7994.73	I
7		8010.58	I
25		8056.52	I
25		8080.32	I
16		8173.89	I
130		8204.58	I
7		8248.81	I
55		8276.95	I
13		8305.91	II
25		8344.25	I
5		8380.06	I
5		8382.98	I
35		8460.01	I
150		8546.48	I
160		8640.06	I
40		8711.24	I
65		9004.73	I

Hf III
Ref. 404 — R.L.K.

Intensity		Wavelength	
		Vacuum	
20		1449.83	III
30		1507.82	III
50		1683.95	III
50		1756.91	III
60		1843.64	III
60		1870.58	III
50		1874.81	III
150		1885.15	III
100		1991.44	III
		Air	
100		2037.76	III
300		2070.94	III
150	h	2085.33	III
200	h	2099.30	III
200	h	2110.31	III
100	h	2119.69	III
200		2155.66	III
200		2183.50	III
200		2195.44	III
100		2213.54	III
200		2234.59	III
200	h	2313.44	III
300		2336.47	III
150	h	2355.48	III
100		2373.30	III
120		2377.57	III
250		2383.540	III
400		2461.74	III
2000		2495.16	III
1000		2515.16	III
100		2534.33	III
400	h	2560.74	III
300	h	2567.46	III
200		2687.22	III
500		2753.60	III
100	h	3060.08	III
200	h	3279.67	III
100	h	3741.94	III

Hf IV
Ref. 369, 425 — R.L.K.

Intensity	Wavelength	
	Vacuum	
40	520.04	IV
50	569.19	IV
100	596.56	IV
100	600.90	IV
100	603.16	IV
200	618.27	IV
100	620.19	IV
100	633.58	IV
100	643.05	IV
200	644.54	IV
400	647.39	IV
600	665.65	IV
100	671.36	IV
200	673.49	IV

Hafnium (Cont.)

Intensity	Wavelength	Spectrum
15	1305.24	IV
12	1357.40	IV
40	1390.39	IV
50	1491.67	IV
35	1528.82	IV
15	1560.18	IV
25	1572.03	IV
100	1717.21	IV
20	1718.57	IV

Air

Intensity	Wavelength	Spectrum
100	2054.46	IV
7	2014.06	IV
20	7751.29	IV
10	7267.58	IV

Hf V
Ref. 410 — R.L.K.

Intensity	Wavelength	Spectrum
	Vacuum	
220	545.41	V
180	600.00	V
100	816.81	V
100	830.69	V
100	836.74	V
100	846.87	V
100	856.32	V
100	861.80	V
135	865.16	V
270	867.25	V
180	875.88	V
135	877.87	V
135	880.37	V
100	880.85	V
180	885.58	V
135	885.80	V
135	894.24	V
100	894.41	V
180	896.14	V
100	896.47	V
135	899.70	V
180	901.54	V
135	901.92	V
135	904.95	V
135	909.70	V
135	913.68	V
135	918.48	V
180	919.10	V
270	921.67	V
135	928.01	V
135	931.50	V
135	947.12	V
245	951.62	V
180	960.12	V
180	964.74	V
160	971.51	V
135	974.62	V
120	984.64	V
135	991.50	V
100	1078.42	V
100	1079.92	V
160	1092.76	V
135	1097.28	V
135	1137.49	V
160	1201.76	V
135	1208.88	V
135	1224.62	V
135	1227.98	V
135	1230.21	V
270	1232.03	V
200	1233.59	V
160	1237.42	V
100	1238.85	V
160	1239.53	V
160	1244.46	V
100	1259.25	V
440	1396.66	V
270	1400.09	V
160	1401.70	V
135	1405.77	V
370	1407.17	V
370	1408.38	V
270	1412.28	V
270	1413.51	V
160	1421.96	V
220	1422.53	V
370	1433.43	V
370	1437.27	V
500	1437.73	V
370	1445.40	V
270	1457.91	V
270	1719.32	V
550	1729.08	V
750	1731.83	V
750	1733.96	V
440	1741.74	V
1000	1749.11	V
1000	1750.19	V
500	1760.89	V
370	1765.62	V
270	1774.02	V
135	1792.39	V

HELIUM (He)
Z = 2

He I and II
Ref. 16, 94, 173, 183, 317
W.C.M.

Intensity	Wavelength	Spectrum
	Vacuum	
15	231.454	II
20	232.584	II
30	234.347	II
50	237.331	II
100	243.027	II
300	256.317	II
1000	303.780	II
500	303.786	II
10	320.293	I
2	505.500	I
3	505.684	I
4	505.912	I
5	506.200	I
7	506.570	I
10	507.058	I
15	507.718	I
20	508.643	I
25	509.998	I
35	512.098	I
50	515.616	I
100	522.213	I
400	537.030	I
1000	584.334	I
50	591.412	I
5	958.70	II
6	972.11	II
8	992.36	II
15	1025.27	II
30	1084.94	II
35	1215.09	II
50	1215.17	II
120	1640.34	II
180	1640.47	II

Air

Intensity	Wavelength	Spectrum
7	2385.40	II
9	2511.20	II
50	2577.6	I
1	2723.19	I
12	2733.30	II
2	2763.80	I
10	2818.2	I
4	2829.08	I
10	2945.11	I
40	3013.7	I
20	3187.74	I
3	3202.96	II
15	3203.10	II
1	3354.55	I
2	3447.59	I
1	3587.27	I
3	3613.64	I
2	3634.23	I
3	3705.00	I
1	3732.86	I
10	3819.607	I
1	3819.76	I
500	3888.65	I
20	3964.729	I
1	4009.27	I
50	4026.191	I
5	4026.36	I
12	4120.82	I
2	4120.99	I
3	4143.76	I
10	4387.929	I
3	4437.55	I
200	4471.479	I
25	4471.68	I
6	4685.4	II
30	4685.7	II
30	4713.146	I
4	4713.38	I
20	4921.931	I
100	5015.678	I
10	5047.74	I
5	5411.52	II
500	5875.62	I
100	5875.97	I
8	6560.10	II
100	6678.15	I
3	6867.48	I
200	7065.19	I
30	7065.71	I
50	7281.35	I
1	7816.15	I
2	8361.69	I
2	9063.27	I
2	9210.34	I
10	9463.61	I
4	9516.60	I
3	9526.17	I
1	9529.27	I
1	9603.42	I
3	9702.60	I
6	10027.73	I
2	10031.16	I
15	10123.6	II
1	10138.50	I
10	10311.23	I
2	10311.54	I
3	10667.65	I
300	10829.09	I
1000	10830.25	I
2000	10830.34	I
9	10913.05	I
3	10917.10	I
4	11626.4	II
30	11969.12	I
20	12527.52	I
50	12784.99	I
20	12790.57	I
7	12845.96	I
10	12968.45	I
2	12984.89	I
12	15083.64	I
200	17002.47	I
1	18555.55	I
6	18636.8	II
500	18685.34	I
200	18697.23	I
100	19089.38	I
20	19543.08	I
1000	20581.30	I
80	21120.07	I
10	21121.43	I
20	21132.03	I
3	30908.5	II
4	40478.90	I

HOLMIUM (Ho)
Z = 67

Ho I and II
Ref. 1 — C.H.C.

Intensity		Wavelength	Spectrum
		Air	
170		2502.91	II
80		2508.53	II
110		2513.55	II
95		2518.73	II
170		2533.80	I
130		2536.86	II
80		2556.84	I
80		2567.73	II
80		2586.52	I
60		2591.05	II
95		2592.99	I
190		2605.86	II
110		2610.51	II
95		2613.99	II
60		2625.20	II
80		2640.09	I
80		2640.30	II
60		2649.68	II
80		2666.24	II
70		2689.03	II
210		2713.65	II
230		2733.95	II
270		2750.35	II
110	c	2759.35	II
110		2766.85	II
270		2769.89	II
110		2772.83	II
140		2777.10	II
140		2794.41	II
100		2799.99	II
100		2806.72	II
160	c	2809.99	II
220		2811.36	II
180		2812.00	II
190		2814.74	II
300		2824.20	II
140		2826.64	II
270	c	2831.69	II
210		2834.99	II
110		2835.85	II
110		2844.18	II
100		2844.68	II
270		2849.10	II
100		2861.23	II
250		2861.49	II
150		2862.72	II
210		2871.99	II
230		2874.06	II
160		2874.43	II
360		2880.26	II
460		2880.98	II
340		2894.99	II
160		2895.62	II
170		2900.84	II
570	c	2909.41	II
170		2915.82	II
300		2919.62	II
110		2925.35	II
160		2926.09	II
300	c	2928.30	II
220		2942.05	II
300		2944.49	II
250	c	2953.11	II
390		2973.00	II
410	c	2979.63	II
180		2981.46	II
140		2985.48	II
410		2987.64	II
250		2990.27	II
110		2995.86	II
320	c	3008.10	II
220		3014.60	II
270		3038.69	II
480	c	3049.38	II
410	c	3054.00	II
500	c	3057.45	II
230		3074.30	II
500		3082.34	II
910		3084.36	II
430		3086.54	II
200		3108.31	II
200	c	3109.91	II
760		3118.50	II
300	c	3130.99	II
200		3134.39	II
300	c	3144.36	II
200		3156.18	II
270		3156.97	II
200	c	3159.67	II
580	c	3166.62	II
390	D1	3171.72	II
810		3173.78	II
390		3174.84	II
270	c	3176.97	II
810	c	3181.50	II
390		3183.84	II
270	Cw	3184.48	II
200		3186.37	I
390	c	3197.83	II
390		3201.76	II
200		3206.86	II
270	c	3210.41	II
200	c	3221.42	II
320		3233.34	II
200		3236.90	II
200		3237.40	II
200	c	3257.45	II
390	c	3278.15	II
270		3279.25	II
980	c	3281.97	II
390		3288.46	II
270	c	3290.96	II
200	c	3305.16	II
200		3319.87	II
230		3320.25	II
200		3331.93	II
630	c	3337.23	II
390	c	3338.86	II
980	c	3343.58	II
200		3344.47	II
360		3350.49	II
320		3352.10	II
320	Cw	3353.55	II
320		3354.58	II
320		3357.91	II
320		3364.27	II
290		3370.87	II
230		3374.16	II
290	c	3390.75	II
320		3394.60	II
8100	c	3398.98	II
810	c	3410.26	II
1200		3421.63	II
3200		3453.14	II
390	c	3410.65	II
1400	c	3414.90	II
5400		3416.46	II
2000	c	3425.34	II
2000		3428.13	II
630	c	3429.18	II
320	c	3432.10	II
390		3449.35	I
810	c	3455.70	II
16000	c	3456.00	II
1600		3461.97	II
360		3467.07	II
810	c	3473.91	II
5400	c	3474.26	II
6300		3484.84	II
490		3489.58	II
580	c	3493.09	II

Holmium (Cont.)

Intensity		Wavelength	Spectrum
2500	c	3494.76	II
810	c	3498.88	II
410	c	3506.95	II
320		3509.37	II
810		3510.73	I
4100	c	3515.59	II
410	c	3519.94	II
630		3540.76	II
1600		3546.05	II
1100	c	3556.78	II
410		3560.15	II
410	c	3573.24	II
630	c	3574.80	II
810		3579.12	I
410		3580.75	II
410		3581.83	II
630	c	3592.23	II
1100	Cw	3598.77	II
340		3599.48	I
540	c	3600.95	II
340		3613.31	II
410		3618.43	I
430	c	3626.69	II
490		3627.25	II
430	c	3631.76	II
430	c	3638.30	II
1600	c	3662.29	I
430		3662.99	I
720		3666.65	I
1400		3667.97	I
320		3669.05	II
450		3669.52	II
450	c	3674.77	II
720		3679.19	I
670		3679.70	I
720		3682.65	I
430		3685.16	II
580		3690.65	I
340		3691.95	I
410		3700.04	I
490	c	3702.35	II
320		3709.76	I
430		3712.88	I
450		3720.72	I
1100		3731.40	I
360		3732.09	I
810		3736.35	I
3200	Cw	3748.17	II
320	c	3753.73	II
340		3769.09	I
320		3788.08	II
8900	c	3796.75	II
8900	c	3810.73	II
490		3811.86	I
900	c	3813.25	II
300		3821.73	II
390		3829.27	I
320	Cw	3831.9	II
410	c	3835.35	II
1300	Cw	3837.51	II
410	c	3842.05	II
1100		3843.86	II
490	c	3846.73	II
300		3849.88	I
320		3852.40	II
1800	c	3854.07	II
390	Cw	3856.94	II
720		3857.72	I
2700	c	3861.68	II
540		3862.62	I
360		3872.05	II
320	c	3874.09	II
630		3874.68	II
540		3881.61	II
3000	c	3888.96	II
490		3890.42	I
13000	c	3891.02	II
540		3896.76	II
290		3902.23	II
320		3904.44	I
1300	Cw	3905.68	II
320		3911.80	I
320		3919.45	I
320	c	3936.44	II
220		3938.85	I
320	Cw	3940.53	II
220		3950.56	I
580		3955.73	I
230	c	3959.51	II
490		3959.68	I
220		3975.88	I
390	c	3976.93	I
220	Cw	3985.71	II
220		3993.73	II
380		3999.58	I
160	Cw	4002.59	II
( )		4003.39	I
110		4013.50	I
320		4014.20	I
160	c	4018.09	II
160	c	4022.76	II
160	c	4023.94	II
110		4025.39	I
320		4027.21	I
270		4028.86	I
180	c	4031.80	I
220		4037.62	I
220	c	4038.87	II
2700		4040.81	I
5400	c	4045.44	II
220	c	4047.52	I
8100		4053.93	I
540		4054.48	II
270		4057.55	I
220		4060.31	I
1700		4065.09	II
170		4067.57	I
720		4068.05	I
270		4071.83	I
270		4073.13	I
290		4073.51	I
120	c	4080.23	II
230		4083.67	I
140		4085.09	I
170		4087.35	I
200		4087.59	I
140		4091.64	I
120		4094.78	I
230		4100.22	I
8900		4103.84	I
120		4105.04	I
270		4106.50	I
100		4107.36	I
2900		4108.62	I
300		4112.00	I
100		4112.72	I
270		4116.73	I
1500		4120.20	I
1300		4125.65	I
4300		4127.16	I
300		4134.54	I
1500		4136.22	I
130		4139.34	I
230		4142.19	I
290		4148.97	I
980	Cw	4152.61	II
8100		4163.03	I
160		4172.23	I
2500		4173.23	I
540		4194.35	I
100		4198.08	I
130		4203.21	I
100		4211.30	II
290		4222.29	I
290		4223.47	I
2000		4227.04	I
390		4229.52	II
130	h	4231.24	I
290		4243.78	I
1300	Cw	4254.43	I
130	c	4258.61	II
490		4264.05	I
300		4266.04	I
100		4273.63	II
200		4311.04	I
250		4330.64	II
300		4337.13	II
100	Cw	4346.84	II
1300		4350.73	I
290		4356.73	II
140		4363.93	II
170		4379.14	II
180	c	4384.83	II
150		4400.55	II
120		4401.24	II
180		4403.27	I
200		4420.56	II
130		4444.63	I
100		4473.59	II
300		4477.64	II
120		4484.57	II
140		4510.82	I
100	c	4526.14	II
170		4530.08	II
170	c	4531.28	I
130	c	4531.65	II
170		4534.58	I
200		4562.52	I
120	Cw	4609.32	II
130		4613.37	I
100		4618.84	I
100	c	4628.22	I
290		4629.10	II
200	c	4649.77	II
130	c	4661.33	II
140		4674.62	II
70		4701.17	II
80		4701.69	II
130		4709.84	II
65		4711.39	I
130	c	4717.52	I
35		4728.72	II
35		4738.00	II
290		4742.04	II
35	c	4749.09	II
35		4751.40	I
100	c	4757.01	I
35		4762.39	II
35		4763.57	II
55		4777.48	II
30		4779.42	I
70	c	4781.19	I
65		4782.92	I
55	c	4786.29	I
35		4791.48	II
35		4795.92	II
45	h	4798.87	I
27		4812.92	II
55		4832.31	II
30		4833.32	I
30		4855.54	II
45		4860.39	I
27	c	4889.67	II
30		4892.35	I
35		4896.44	II
55		4906.99	II
45		4922.73	I
55	c	4934.89	I
290		4939.01	I
27	c	4946.80	II
45		4948.18	II
65	c	4959.42	II
35		4961.03	II
55	Cw	4966.73	II
250	c	4967.21	II
220		4979.97	I
35	c	4988.96	I
90		4995.05	I
35	c	5012.42	I
55		5013.28	II
65		5026.53	I
30		5028.17	I
55		5032.95	II
65	c	5037.60	I
130		5042.37	I
35		5044.73	I
30		5051.44	II
30		5054.92	II
35	c	5060.75	I
65		5074.34	I
80		5093.07	I
140		5127.81	I
55		5129.27	II
130		5142.59	II
110		5143.22	II
160		5149.59	II
90	c	5167.88	I
130	c	5182.11	I
55		5187.85	I
90		5190.11	II
18		5195.23	I
45		5221.54	I
35	c	5244.47	I
65		5251.82	I
55		5275.48	I
90		5301.25	I
35		5319.24	I
35		5319.65	I
80		5330.11	I
90		5359.99	I
55		5381.40	I
30		5384.56	I
30		5384.97	I
18	h	5393.85	I
70		5403.17	I
100		5407.08	I
14		5413.62	II
16		5434.39	II
18		5435.87	I
30		5445.39	I
18		5449.8	II
30	h	5451.90	I
14		5454.0	II
30	c	5498.57	I
30		5504.51	I
27		5515.56	II
18		5516.45	II
30		5534.33	I
27		5553.14	I
35	c	5560.94	I
35	b	5563.6	H
70		5566.52	I
18		5573.96	II
35	B1	5584.7	I
55	b	5591.1	H
55	B1	5592.3	I
30	b	5607.1	H
27		5613.64	I
45	b	5626.4	H
65		5627.60	I
30		5628.24	II
55		5640.62	I
70	Bs	5655.9	H
65		5658.9	H
140		5659.58	I
70	c	5671.84	I
65		5674.70	I
140	c	5691.47	I
70	Bs	5696.3	H
140	c	5696.57	I
27		5734.02	I
45		5736.4	H
55		5739.24	I
22		5749.58	I
30		5766.64	
27	b	5803.8	H
45	b	5819.2	H
27	h	5821.90	I
22		5839.47	I
45	b	5849.4	H
140	c	5860.28	I
27	h	5864.42	I
45		5870.85	I
27	b	5879.6	H
70	c	5882.99	I
35	c	5892.56	I
22		5904.29	I
70		5921.76	I
30	c	5933.71	I
70	Cw	5948.03	I
45		5955.98	I
70		5972.76	I
90		5973.52	I
22		5981.43	I
230	c	5982.90	I
55		6002.04	I
27		6005.33	I
35		6021.43	I
16		6038.97	I
27		6050.71	I
45		6060.31	I
120		6081.79	I
70	Cw	6133.60	I
35		6156.38	I
27		6156.58	I
55		6191.68	I
70		6208.65	I
18		6234.17	I
45	c	6255.75	I
70	c	6305.36	I
22		6306.68	I
30		6321.94	
30	c	6354.35	I
30	c	6372.59	I
14	h	6373.86	I
22	h	6413.41	I
27	c	6471.77	I
13		6479.17	I
11		6515.30	I
11	h	6538.99	I
70		6550.97	I
15		6560.08	I
35	d	6600.58	I
260		6604.94	I
55		6607.47	I
13		6628.35	I
120		6628.99	I
15		6632.24	I
9	h	6652.98	I
15		6662.52	I
19	c	6680.46	I
24	c	6681.62	I
15	h	6682.02	I
55	Cw	6694.32	I
15	Cw	6722.34	I
40		6745.05	I
13		6766.74	I
28	c	6774.68	I
55	c	6785.43	I
13		6793.7	I
13	Cw	6811.04	I
15	Cw	6820.38	I
24		6821.64	I
17	c	6825.72	I
8	h	6826.62	
8	h	6852.97	
17	Cw	6865.85	I
9		6883.36	I
13		6888.50	I
15	c	6892.96	I
17		6897.95	I
15	h	6903.80	I
15	Cw	6913.47	I
9		6916.70	I
40	Cw	6939.49	I
45	Cw	6950.39	I
13	H1	6955.3	I
19		6976.7	II
10		6985.11	
9		6994.38	
14	h	7000.71	
10		7079.07	
12		7098.58	
9		7242.08	I
9		7250.60	I
14		7308.55	
25		7341.43	
18		7389.40	
5	h	7496.20	I
10	h	7510.74	I
140		7555.09	I
18		7589.20	I
25		7591.87	I
9	h	7593.64	I
7	h	7594.35	I

Holmium (Cont.)

Intensity		Wavelength	Spectrum
12		7602.31	II
16		7605.35	I
12		7617.05	I
14		7627.98	I
40	c	7628.42	I
9	c	7641.14	I
4		7648.16	I
14	c	7653.80	I
12	c	7667.30	I
20		7690.43	I
50	c	7693.15	I
40	Cw	7715.06	I
16		7719.05	I
16	h	7738.98	I
8	c	7752.01	I
60	Cw	7815.48	I
40	Cw	7823.63	I
8	h	7879.22	I
60		7894.64	I
10	h	8464.66	I
10	h	8482.67	I
50		8512.94	I
20		8545.61	II
18		8601.84	II
40		8670.19	I
8	h	8697.32	I
16	h	8805.48	II
20	c	8834.49	I
90		8915.98	II

HYDROGEN (H)
Z = 1

H I
Ref. 214 — W.C.M.

Intensity	Wavelength	Spectrum
	Vacuum	
15	926.226	I
20	930.748	I
30	937.803	I
50	949.743	I
100	972.537	I
300	1025.722	I
1000	1215.668	I
500	1215.674	I
	Air	
5	3835.384	I
6	3889.049	I
8	3970.072	I
15	4101.74	I
30	4340.47	I
80	4861.33	I
120	6562.72	I
180	6562.852	I
5	9545.97	I
7	10049.4	I
12	10938.1	I
20	12818.1	I
40	18751.0	I
5	21655.3	I
8	26251.5	I
15	40511.6	I
4	46525.1	I
6	74578	I
3	123685	I

INDIUM (In)
Z = 49

In I and II
Ref. 1, 132, 348—350 — C.H.C.

Intensity		Wavelength	Spectrum
		Vacuum	
2		1648.00	I
1	h	1676.16	I
5	h	1711.54	I
2	h	1741.23	I
1	h	1758.49	I
		Air	
10		2103.89	II
10		2166.88	II
2		2179.90	I
2		2182.40	I
2		2187.40	I
2		2190.84	I
15		2195.67	II
2		2197.41	I
2		2202.24	I
50		2205.28	II
3		2211.09	I
5		2230.70	I
3		2241.66	I
30		2255.79	II
10		2259.99	I
5		2278.20	I
40		2281.64	II
2		2283.75	I
2		2298.33	I
2		2298.70	I
2		2302.49	I
100	c	2306.05	II
25		2306.86	I
3		2309.32	I
2		2309.75	I
90	d	2313.21	II
2		2315.09	I
30		2323.40	II
5		2324.41	I
3		2324.92	I
70	d	2327.95	II
3		2332.76	I
80	h	2334.57	II
10		2340.19	I
8		2345.90	I
5		2346.56	I
50	d	2350.75	II
5		2358.70	I
15		2378.14	I
10		2379.00	I
110	d	2382.63	II
40		2389.54	I
40		2393.18	II
10		2399.18	I
50	h	2406.47	II
50		2408.76	II
50		2419.06	II
50		2419.20	II
70	h	2427.20	II
20		2429.86	I
10		2430.99	I
50		2432.73	II
60		2442.63	II
100		2447.90	II
60		2453.23	II
60		2460.08	I
30	h	2468.02	I
70		2486.15	II
110	d	2488.62	II
90		2488.95	II
80		2498.59	II
100		2499.60	II
90		2500.99	II
60		2508.16	II
110	d	2512.31	II
100		2521.37	I
10		2522.98	I
70		2553.56	II
160	d	2554.44	II
1100		2560.15	I
70		2565.13	II
70	d	2598.75	II
200		2601.76	I
50		2604.04	II
90	d	2654.70	II
100	d	2662.63	II
140	d	2668.65	II
140	d	2674.56	II
80		2683.12	II
1600		2710.26	I
300		2713.94	I
130	d	2749.75	II
700		2753.88	I
40		2775.37	I
60		2798.76	II
90	d	2818.97	II
180	c	2836.92	I
30	c	2858.14	I
80		2865.68	II
120	d	2890.18	II
1100		2932.63	I
100		2941.05	II
20	c	2957.01	II
60	d	2966.17	II
110	c	2999.40	II
8000		3039.36	I
8	d	3051.15	I
110	d	3099.80	II
180		3101.8	II
130	c	3138.60	II
80		3142.75	II
130	d	3146.70	II
150		3155.77	II
100	c	3158.40	II
90	c	3176.30	II
90	d	3198.11	II
13000		3256.09	I
3000		3258.56	I
90	c	3338.50	II
75	c	3376.59	II
100	c	3404.28	II
110	d	3438.40	II
180	c	3693.91	II
95	c	3708.13	II
380	w	3716.14	II
120	c	3718.30	II
160	c	3718.72	II
160	c	3723.40	II
170	w	3795.61	II
230	c	3799.21	II
250	c	3834.65	II
200	c	3842.18	II
100		3889.78	II
100	c	3902.07	II
60	d	3922.12	II
65	c	3934.40	II
250	w	3962.35	II
120	c	4004.66	II
140	d	4013.92	II
410	w	4056.94	II
17000		4101.76	I
140	c	4205.14	II
100	c	4213.04	II
110	d	4219.66	II
150	c	4372.87	II
150	c	4500.78	II
18000		4511.31	I
110	c	4549.01	II
140	c	4570.85	II
180	w	4578.02	II
180	w	4578.40	II
140	c	4616.08	II
170	c	4617.17	II
250	c	4620.14	II
150	w	4620.70	II
170	c	4627.30	II
140	c	4637.04	II
380	c	4638.16	II
220	c	4644.58	II
360	c	4655.62	II
320	w	4656.74	II
190	c	4681.11	II
450	w	4684.8	II
3		4878.59	I
90	d	4907.06	II
70	h	4924.93	II
150	c	4973.77	II
80	h	5109.36	II
100	w	5115.14	II
140	c	5117.40	II
270	c	5120.80	II
200	w	5121.75	II
80	d	5129.85	II
240	c	5175.42	II
140	c	5184.44	II
30		5254.32	I
12		5262.74	I
150	c	5309.45	II
80		5411.41	II
140	c	5418.45	II
220	w	5436.70	II
130	c	5497.50	II
140	c	5507.08	II
320	w	5513.00	II
250	w	5523.28	II
130	c	5536.50	II
190	w	5555.45	II
240	c	5576.90	II
200	w	5636.70	II
160	c	5708.50	II
50		5709.91	I
100	c	5721.80	II
50		5727.68	I
210	c	5853.15	II
490	w	5903.4	II
260	w	5915.4	II
120	c	5918.78	II
130	c	6062.9	II
250	c	6095.95	II
210	c	6108.66	II
180	w	6115.9	II
230	w	6128.7	II
240	w	6129.4	II
320	w	6132.1	II
150	c	6140.0	II
90		6143.23	II
140	c	6148.10	II
190	w	6149.5	II
80		6161.15	II
180	w	6162.45	II
100	c	6224.28	II
280	w	6228.3	II
140	c	6231.1	II
270	w	6304.8	II
290	w	6362.3	II
300	w	6469.0	II
210	c	6541.20	II
190	c	6751.88	II
180	c	6765.9	II
100	c	6783.72	II
8	h	6847.44	I
320	c	6891.5	II
4	h	6900.13	I
380	w	7182.9	II
180	c	7255.0	II
210	c	7276.5	II
180	c	7303.4	II
320	c	7350.6	II
100	c	7632.7	II
100	c	7682.9	II
210	c	7740.7	II
100	c	7776.96	II
180	c	7789.0	II
70	c	7806.8	II
70	c	7814.5	II
90	c	7840.9	II
20	h	8050.78	I
240	c	8227.0	II
30	h	8238.66	I
15	h	8314.92	I
50	c	8434.55	II
30		8678.95	I
20		8682.63	I
50		8700.25	I
100	w	8813.5	II
80	c	8832.6	II
40		8894.47	I
10		9170.08	I
120	c	9197.7	II
120	c	9202.0	II
220	w	9213.0	II
160	d	9241.1	II
40	h	9349.83	I
60	h	9370.27	I
20		9427.99	I
100		9977.86	I
200		10257.03	I
60	h	10717.42	I
100	h	10744.31	I
20		11334.72	I
10		11731.48	I
9		12912.59	I
9		13429.96	I
5		13824.48	I
6		14316.25	I
3		14419.20	I
6		14668.66	I
7		14719.08	I
2		16504.31	I
6		22291.06	I
7		23879.13	I

In III
Ref. 351 — C.H.C.

Intensity	Wavelength	Spectrum
	Vacuum	
7	685.31	III
5	691.62	III
1	782.17	III
10	882.24	III
10	890.84	III
10	915.87	III
2	917.45	III
5	926.83	III
30	1403.08	III
30	1434.85	III
20	1487.70	III
20	1494.14	III
10	1524.78	III
20	1530.21	III
30	1532.95	III
100	1625.42	III
20	1642.28	III
20	1702.53	III
100	1748.83	III
2	1767.88	III
1	1810.71	III
30	1842.41	III
40	1850.30	III
15	1862.98	III
	Air	
30	2154.08	III
2	2154.42	III
10	2199.52	III
5	2232.18	III
20	2261.26	III
5	2266.26	III
5	2272.41	III
5	2272.84	III
10	2300.90	III
100	2527.41	III
50	2725.52	III
80	2726.15	III
100	2982.80	III
100	3008.08	III
30	3008.82	III
30	3293.55	III
8	3350.91	III
5	3562.32	III
100	3852.82	III
100	4023.77	III
150	4032.32	III
50	4062.30	III
100	4071.57	III
100	4072.93	III
100	4252.68	III
40	4509.58	III
200	5248.77	III

Indium (Cont.)

Intensity	Wavelength	
100	5645.15	III
40	5723.17	III
100	5819.50	III
200	6197.72	III

In IV
Refs. 352, 435, 436 — J.R.

Intensity	Wavelength	
	Vacuum	
622	472.71	IV
689	479.39	IV
709	498.62	IV
61	945.74	IV
85	954.67	IV
87	973.50	IV
86	991.60	IV
89	1024.68	IV
85	1024.79	IV
88	1031.45	IV
82	1031.98	IV
80	1054.43	IV
84	1063.03	IV
83	1068.25	IV
82	1069.82	IV
86	1077.64	IV
90	1082.10	IV
83	1086.33	IV
82	1096.81	IV
84	1097.18	IV
85	1116.10	IV
80	1124.06	IV
90	1131.46	IV
85	1144.43	IV
80	1145.41	IV
89	1146.62	IV
83	1154.11	IV
84	1154.60	IV
90	1157.71	IV
90	1157.82	IV
85	1159.78	IV
88	1176.50	IV
85	1191.58	IV
83	1204.87	IV
90	1206.55	IV
88	1221.50	IV
85	1221.90	IV
85	1233.58	IV
87	1235.84	IV
90	1373.20	IV
88	1398.77	IV
81	1412.09	IV

In V
Ref. 353 — C.H.C.

Intensity	Wavelength	
	Vacuum	
6	368.67	V
6	370.10	V
10	372.82	V
10	372.94	V
2	374.95	V
6	375.84	V
6	376.07	V
10	376.79	V
17	378.61	V
3	379.24	V
9	380.27	V
11	381.56	V
9	382.14	V
11	382.76	V
10	383.05	V
17	386.21	V
10	386.70	V
3	388.66	V
14	388.91	V
11	390.03	V
11	390.92	V
9	392.29	V
9	392.46	V
1	393.60	V
25	393.89	V
11	395.74	V
3	397.73	V
10	399.79	V
9	400.05	V
25	400.57	V
25	402.39	V
3	405.33	V
9	407.28	V
3	407.36	V
9	407.95	V
9	417.43	V
2	418.45	V
2	423.16	V

IODINE (I)

Z = 53
I I and II
Ref. 124, 153, 176, 184
L.J.R.

Intensity	Wavelength	
	Vacuum	
2	655.80	II
6	659.00	II
8	663.98	II
8	664.52	II
8	665.06	II
150	665.70	II
1000	719.55	II
1000	722.98	II
1000	798.16	II
1200	834.10	II
600	847.80	II
1500	873.49	II
1000	875.94	II
2000	879.84	II
1500	881.88	II
1000	891.00	II
1000	893.17	II
1200	1000.57	II
1000	1003.35	II
4000	1018.58	II
10000	1034.66	II
1500	1054.74	II
2000	1066.34	II
3000	1075.21	II
5000	1105.00	II
2500	1111.16	II
1500	1117.22	II
3500	1125.25	II
2000	1131.50	II
1200	1139.75	II
10000	1139.80	II
1500	1154.67	II
1000	1159.87	II
10000	1160.56	II
20000	1166.48	II
1500	1167.05	II
5000	1175.84	II
10000	1178.65	II
15000	1187.34	II
10000	1190.85	II
15	1195.29	I
5000	1198.88	II
7000	1200.22	II
200	1218.41	I
20000	1220.89	II
600	1224.05	I
600	1224.08	I
500	1228.89	I
20000	1234.06	II
600	1251.34	I
2500	1259.15	I
3000	1259.51	I
800	1261.27	I
600	1267.57	I
600	1267.60	I
1500	1275.26	I
3000	1289.40	I
10000	1300.34	I
3000	1302.98	I
3000	1313.95	I
3000	1317.54	I
2000	1330.19	I
20000	1336.52	II
5000	1355.10	I
3000	1357.97	I
5000	1360.97	I
3000	1361.11	I
2500	1367.71	I
2500	1368.22	I
4000	1383.23	I
3000	1390.75	I
2000	1392.90	I
2000	1400.01	I
8000	1425.49	I
5000	1446.26	I
5000	1453.18	I
5000	1457.39	I
5000	1457.47	I
10000	1457.98	I
2500	1458.79	I
4000	1459.15	I
2500	1465.83	I
1000	1485.92	I
5000	1492.89	I
5000	1507.04	I
5000	1514.68	I
15000	1518.05	I
2500	1526.45	I
5000	1593.58	I
5000	1617.60	I
2500	1640.78	I
15000	1702.07	I
12000	1782.76	I
5000	1799.09	I
75000	1830.38	I
15000	1844.45	I

Intensity		Wavelength	
		Air	
2000		2061.63	I
100		2408.01	II
100		2419.18	II
100		2494.74	II
100		2533.60	II
200		2534.27	II
1000		2566.24	II
2000		2582.79	II
300		2593.46	II
200	c	2688.98	II
500		2730.12	II
20		2765.15	II
200		2808.59	II
1500		2878.63	II
1000		2993.87	II
5000		3078.75	II
200		3161.03	II
1000		3175.07	II
300		3355.53	II
250	c	3424.99	II
300	c	3497.41	II
500		3526.90	II
200		3742.14	II
200		4102.23	I
200		4129.21	I
100	d	4134.15	I
500	d	4321.84	I
300		4452.86	II
200		4599.77	II
300	c	4632.45	II
500	d	4666.48	II
1000		4675.53	II
250		4763.31	I
1000		4862.32	I
200		4916.94	I
1000		4986.92	I
400		5065.37	II
10000		5119.29	I
200		5149.73	II
3000	c	5161.20	II
300		5176.19	II
600		5216.27	II
500	d	5228.97	II
1000		5234.57	I
3000		5245.71	II
500		5269.36	II
400		5299.78	II
400		5322.80	II
10000		5338.22	II
5000	c	5345.15	II
1000		5369.86	II
800	c	5405.42	II
800	c	5407.36	II
600	c	5427.06	I
3000		5435.83	II
1000		5438.00	II
2000	c	5464.62	II
800		5491.50	II
1000	c	5496.94	II
1000		5504.72	II
600	c	5522.06	II
600	c	5598.52	II
1000		5600.32	II
1500		5612.89	II
10000		5625.69	II
1000		5678.08	II
2000	c	5690.91	II
500		5702.05	II
4000	c	5710.53	II
1000		5738.27	II
1000		5760.72	II
1000	d	5764.33	I
500	c	5774.83	II
500		5787.22	II
2000		5894.03	I
5000		5950.25	II
300		5984.86	I
2000	d	6024.08	I
500		6068.93	II
2000	c	6074.98	II
1000		6082.43	I
2000	c	6127.49	II
800		6191.88	I
1000		6204.86	II
500		6213.10	I
800		6244.48	I
900	c	6257.49	II
1000		6293.98	I
500		6313.13	I
800		6330.37	I
400		6333.50	I
2000		6337.85	I
1000		6339.44	I
500		6359.16	I
1000		6566.49	I
2000		6583.75	I
1000		6585.27	I
5000		6619.66	I
500		6661.11	I
600		6665.96	II
500	c	6697.29	I
300		6718.83	II
400		6732.03	I
4000		6812.57	II
1000		6958.78	II
500		6989.78	I
200	c	7085.21	II
500		7120.05	I
1200		7122.05	I
2000		7142.06	I
1000		7164.79	I
400	d	7191.66	I
700		7227.30	I
1000		7236.78	I
500		7237.84	I
500		7351.35	II
5000		7402.06	I
1000		7410.50	I
500		7416.48	I
5000		7468.99	I
500	c	7490.52	I
2000		7554.18	I
500	d	7556.65	I
2000	c	7700.20	I
500		7798.98	II
600		7897.98	I
500		7969.48	I
1000		8003.63	I
99000		8043.74	I
300	d	8065.70	I
1000		8090.76	I
800	c	8169.38	I
500	d	8222.57	I
4000		8240.05	I
10000	c	8393.30	I
150		8414.60	II
1000		8486.11	I
1500	c	8664.95	I
500	c	8700.80	I
250	d	8748.22	I
1000		8853.24	I
2000		8853.80	I
3000		8857.50	I
1000	d	8898.50	I
400		8964.69	I
400		8993.13	I
5000		9022.40	I
15000		9058.33	I
1000		9098.86	I
12000		9113.91	I
600		9128.03	I
30		9195.30	II
600		9227.74	I
1000		9335.05	I
4000		9426.71	I
3000		9427.15	I
10	c	9480.33	II
2000		9598.22	I
2000		9649.61	I
3000	d	9653.06	I
5000		9731.73	I
500		10003.05	I
750		10131.16	I
1000		10238.82	I
400		10375.20	I
400		10391.74	I
6		10405.49	II
5000		10466.54	I
1		11084.68	II
400		11236.56	I
350		11558.46	I
320		11778.34	I
450		11996.86	I
300		12033.69	I
150		12304.58	I
60		13149.16	I
140		13958.27	I
200		14287.02	I
100		14460.00	I
225		15032.57	I
105		15528.65	I
150		16037.33	I
15		18275.71	I
20		18348.52	I
15		18982.41	I
35		19070.17	I
110		19105.12	I
50		19370.02	I
10		20648.69	I
220		22183.03	I
150		22226.53	I
30		22309.21	I
32		24420.82	I
12		27365.42	I
9		27573.05	I
10		30361.93	I
8		30383.88	I
10		34295.73	I
9		34513.11	I
3		40228.54	I
2		41633.80	I

Iodine (Cont.)

I III
Ref. 20, 21, 161 — L.J.R.

Intensity	Wavelength	
	Vacuum	
6	666.81	III
8	705.11	III
7	784.64	III
7	784.80	III
8	795.52	III
5	865.97	III
5	920.38	III
6	961.17	III
6	1078.58	III
8	1094.20	III
4	1244.66	III
8	1252.35	III
5	1306.93	III
	Air	
1	2224.43	III
1	2238.12	III
3	2249.31	III
2	2309.38	III
3	2340.85	III
3	2350.43	III
2	2353.46	III
4	2367.74	III
2	2371.45	III
3	2372.45	III
4	2376.47	III
4	2387.12	III
3	2392.01	III
2	2403.06	III
2	2403.63	III
2	2414.85	III
2	2418.49	III
2	2418.85	III
2	2423.91	III
5	2426.12	III
3	2434.88	III
2	2462.50	III
3	2466.69	III
3	2466.99	III
6	2475.36	III
4	2489.27	III
2	2493.21	III
2	2494.27	III
3	2495.16	III
2	2496.07	III
3	2501.41	III
2	2516.82	III
6	2519.75	III
4	2521.72	III
3	2531.99	III
2	2537.56	III
7	2545.71	III
4	2640.77	III
4	2642.11	III
6	2652.25	III
2	2818.48	III
2	2839.44	III
4	2864.67	III
4	2885.15	III
3	2910.98	III
3	2917.35	III
2	2931.11	III
2	3005.68	III
3	3069.23	III
3	3153.88	III
3	3170.14	III
3	3181.66	III
3	3210.14	III
4	3213.49	III
4	3224.93	III
2	3300.47	III
2	3479.53	III
3	3546.92	III
3	3613.81	III
2	3754.40	III
2	3754.55	III
3	3963.16	III
3	4077.14	III

I IV
Ref. 21, 58 — L.J.R.

Intensity	Wavelength	
	Vacuum	
5	601.86	IV
6	612.46	IV
4	615.17	IV
4	654.22	IV
4	654.56	IV
7	919.28	IV
	Air	
5	2249.30	IV
4	2340.84	IV
7	2361.13	IV
5	2367.75	IV
6	2372.45	IV
7	2376.46	IV
4	2385.28	IV
8	2387.11	IV
6	2392.00	IV
4	2403.05	IV
2	2418.45	IV
3	2423.89	IV
9	2426.10	IV
6	2434.85	IV
3	2466.68	IV
3	2466.96	IV
8	2475.35	IV
4	2485.51	IV
5	2489.24	IV
4	2493.20	IV
2	2501.38	IV
3	2513.74	IV
8	2519.74	IV
6	2521.72	IV
4	2531.98	IV
5	2537.54	IV
8	2545.67	IV
4	2640.77	IV
5	2642.11	IV
8	2652.23	IV
3	2818.45	IV
6	2864.68	IV
4	2910.97	IV
5	2917.33	IV
4	3069.17	IV
4	3170.11	IV
4	3181.64	IV
4	3210.12	IV
6	3213.48	IV
6	3224.90	IV
4	3546.90	IV

IV
Ref. 84 — L.J.R.

Intensity	Wavelength	
	Vacuum	
30	363.78	V
36	380.74	V
45	565.53	V
50	607.57	V

• IRIDIUM (Ir)
Z = 77

Ir I and II
Ref. 1 — C.H.C.

Intensity	Wavelength	
	Air	
9900	2010.65	I
8700	2022.35	I
15000	2033.57	I
6200	2052.22	I
5000	2060.64	I
3700	2083.22	I
3100	2085.74	I
17000	2088.82	I
14000	2092.63	I
2700	2112.68	I
1800	2119.54	I
2000	2125.44	I
4500	2126.81	II
2000	2127.52	I
4500	2127.94	I
3700	2148.22	I
2500	2150.54	I
3500	2152.68	II
2900	2155.81	I
7900	2158.05	I
2100	2162.88	I
5800	2169.42	II
4500	2175.24	I
2700	2178.17	I
1600	2187.43	II
1100	2190.38	II
740	2191.64	I
910	2208.09	II
1300	2220.37	I
790	2221.07	I
2500	2242.68	II
620	2245.76	II
2100	2253.38	I
	2253.49	I
2100	2255.10	I
1400	2255.81	I
350	2258.51	I
1400	2258.86	I
830	2264.61	I
1100	2266.33	I
1000	2268.90	I
660	2280.00	I
950	2281.02	II
660	2281.91	I
330	2284.60	I
330	2295.08	I
790	2298.05	I
	2298.16	I
460	2299.53	I
910	2300.50	I
2700	2304.22	I
410	2305.47	I
210	2307.27	I
910	2308.93	I
460	2315.38	I
410	2321.45	I
410	2321.58	I
210	2327.98	I
540	2333.30	I
210	2333.84	I
580	2334.50	I
1600	2343.18	I
740	2343.61	I
100	2352.62	I
580	2355.00	I
230	2357.53	II
410	2358.16	I
500	2360.73	I
2500	2363.04	I
370	2368.04	II
3500	2372.77	I
290	2375.09	II
250	2377.28	I
250	2377.98	I
500	2379.38	I
540	2381.62	I
210	2383.17	I
120	2386.58	II
1300	2386.89	I
2500	2390.62	I
2700	2391.18	I
230	2407.59	I
290	2409.37	I
290	2410.17	I
290	2410.73	I
540	2413.31	I
370	2415.86	I
620	2418.11	I
120	2424.32	I
120	2424.66	I
210	2424.89	I
370	2424.99	I
290	2425.66	I
170	2426.53	II
540	2427.61	I
540	2431.24	I
1300	2431.94	I
170	2432.36	I
100	2432.58	I
270	2435.14	I
250	2445.34	I
250	2447.76	I
190	2448.23	I
910	2452.81	I
1300	2455.61	I
230	2455.87	I
210	2457.03	I
210	2457.23	I
120	2465.09	I
870	2467.30	I
3300	2475.12	I
210	2478.11	I
2100	2481.18	I
100	2485.38	I
620	2493.08	I
210	2496.27	I
250	2502.63	I
4100	2502.98	I
170	2504.37	I
120	2505.74	I
120	2507.63	I
170	2509.71	I
170	2511.94	I
170	2512.58	II
210	2513.71	I
120	2515.36	I
40	2524.88	II
170	2525.05	I
120	2532.52	I
990	2533.13	I
1100	2534.46	I
580	2537.22	I
170	2537.68	I
100	2541.48	I
580	2542.02	I
40	2542.80	II
7900	2543.97	I
150	2545.54	I
790	2546.03	I
120	2547.20	I
120	2547.69	I
210	2551.40	I
190	2554.40	I
210	2555.35	I
170	2555.88	I
150	2563.28	I
910	2564.18	I
210	2569.88	I
100	2570.62	I
230	2572.70	I
740	2577.26	I
100	2578.71	I
35	2579.49	II
740	2592.06	I
740	2599.04	I
150	2602.04	I
190	2604.55	I
190	2607.52	I
700	2608.25	I
1800	2611.30	I
210	2614.98	I
330	2617.78	I
210	2619.88	I
70	2623.64	II
250	2625.32	I
100	2626.76	I
700	2634.17	I
170	2635.27	I
250	2639.42	I
3500	2639.71	I
210	2644.19	I
170	2653.76	I
100	2656.81	I
1800	2661.98	I
350	2662.63	I
2700	2664.79	I
140	2668.99	I
520	2669.91	I
520	2671.84	I
330	2673.61	I
120	2676.83	I
110	2684.04	I
270	2692.34	I
3000	2694.23	I
110	2704.03	I
160	2712.74	I
140	2744.00	I
330	2772.46	I
250	2775.55	I
520	2781.29	I
330	2785.22	I
540	2797.35	I
1600	2797.70	I
380	2798.18	I
410	2800.62	I
680	2823.18	I
1200	2824.45	I
110	2833.24	II
110	2835.66	I
820	2836.40	I
160	2837.33	I
1100	2839.16	I
820	2840.22	I
160	2842.28	I
3800	2849.72	I
110	2863.84	I
380	2875.60	I
380	2875.98	I
270	2877.68	I
140	2879.41	I
820	2882.64	I
650	2897.15	I
260	2901.95	I
260	2904.80	I
200	2907.24	I
440	2916.36	I
230	2918.57	I
4400	2924.79	I
1200	2934.64	I
880	2936.68	I
250	2938.47	I
190	2939.27	I
140	2940.54	I
2700	2943.15	I
230	2946.97	I
200	2949.76	I
1200	2951.22	I
150	2962.99	I
200	2974.95	I
440	2980.65	I
150	2985.80	I
190	2990.62	I
300	2996.08	I
180	2997.41	I
220	3002.25	I
600	3003.63	I
160	3011.69	I
120	3016.43	I
270	3017.31	I
140	3019.23	I
110	3025.82	I
380	3029.36	I
330	3039.26	I
35	3042.65	II
300	3047.16	I
300	3049.44	I
300	3057.28	I

Intensity	Wavelength		Intensity	Wavelength	
1600	3068.89	I	30	4548.48	I
190	3069.09	I	13	4550.78	I
190	3069.71	I	35	4568.09	I
170	3076.69	I	18	4570.02	I
320	3083.22	I	18	4604.48	I
240	3086.44	I	75	4616.39	I
390	3088.04	I	26	4656.18	I
510	3100.29	I	17 h	4668.99	I
510	3100.45	I	21	4708.88	I
340	3120.76	I	50	4728.86	I
200	3121.78	I	21	4731.86	I
3400	3133.32	I	26	4756.46	I
190	3150.61	I	13	4757.96	I
190	3154.74	I	65	4778.16	I
190	3159.15	I	30	4795.67	I
140	3168.18	I	10	4807.14	I
490	3168.88	I	21	4809.47	I
370	3177.58	I	10	4840.77	I
170	3180.35	I	17	4845.38	I
370	3198.92	I	50	4938.09	I
610	3212.12	I	26	4970.48	I
370	3219.51	I	25	4999.74	I
5100	3220.78	I	25	5002.74	I
100	3221.28	I	17	5009.17	l
300	3229.28	I	30	5014.98	I
100	3230.76	I	17	5046.06	I
470	3241.52	I	30	5123.66	I
200	3262.01	I	20	5177.95	I
390	3266.44	I	22	5238.92	I
160	3277.28	I	12	5340.74	I
100	3287.59	I	35	5364.32	I
160	3310.52	I	75	5449.50	I
200	3322.60	I	30	5454.50	I
130	3334.16	I	7	5469.40	I
560	3368.48	I	10	5620.04	I
660	3437.02	I	45	5625.55	I
100	3437.50	I	10	5828.55	I
410	3448.97	I	10	5882.30	I
3200	3513.64	I	7	5887.36	I
220	3515.95	I	35	5894.06	I
410	3522.03	I	7	6026.10	I
160	3557.17	I	12	6067.83	I
320	3558.99	I	20	6110.67	I
1200	3573.72	I	12	6288.28	I
320	3594.39	I	7	6334.44	I
220	3609.77	I	5	6624.73	I
190	3617.21	I	10	6686.08	I
160	3626.29	I	5	6830.01	I
660	3628.67	I	5	6929.88	I
220	3636.20	I	4	7183.71	I
300	3661.71	I	6	7834.32	I
300	3664.62	I			
320	3674.98	I			
200	3687.08	I			
140	3725.38	I			
200	3731.36	II			
130	3738.53	I			
530	3747.20	I			
120	3793.79	I			
3100	3800.12	I			
230	3817.24	I			

IRON (Fe)
Z = 26

Fe I and II
Ref. 56, 63, 105, 138, 174, 278
— H.M.C. and H.C.

Intensity	Wavelength	
	Vacuum	

Intensity	Wavelength		Intensity	Wavelength		Intensity	Wavelength	
170	3865.64	I	12	1055.27	II	18	1788.07	II
480	3902.51	I	15	1068.36	II	30	1934.538	I
480	3915.38	I	15	1071.60	II	25	1937.269	I
400	3934.84	I	15	1096.89	II	50	1946.988	I
120	3946.27	I	12	1099.12	II	25	1951.571	I
590	3976.31	I	18	1112.09	II	30	1952.59	I
460	3992.12	I	12	1121.99	II	30	1953.005	I
180	4020.03	I	12	1122.86	II	60	1957.823	I
350	4033.76	I	12	1128.07	II	60	1960.144	I
130	4040.08	I	12	1130.43	II	30	1961.25	I
370	4069.92	I	15	1133.41	II	50	1962.111	I
150	4070.68	I	12	1133.68	II	12	1963.11	II
100	4092.61	I	12	1138.64	II			
140	4115.78	I	18	1144.95	II		Air	
23	4127.92	I	12	1147.41	II			
27	4155.70	I	15	1148.29	II	100	2084.122	I
15	4166.04	I	12	1151.16	II	50	2157.794	I
90	4172.56	I	12	1267.44	II	15	2162.02	II
35	4182.47	I	12	1272.00	II	40	2166.773	I
15 h	4183.21	I	12	1371.02	II	300	2178.118	I
18	4185.66	I	12	1563.79	II	250	2186.486	I
23	4197.54	I	12	1580.62	II	60	2186.892	I
27	4217.76	I	18	1608.46	II	120	2187.195	I
13	4220.80	I	12	1618.47	II	250	2191.839	I
75	4259.11	I	15	1621.68	II	150	2196.043	I
27	4265.30	I	15	1629.15	II	80	2200.390	I
260	4268.10	I	15	1631.12	II	80	2200.724	I
23	4286.62	I	18	1635.40	II	15	2208.41	II
75	4301.60	I	15	1636.32	II	20	2213.65	II
55	4310.59	I	15	1639.40	II	12	2218.26	II
220	4311.50	I	12	1641.76	II	20	2220.38	II
18	4351.30	I	12	1647.16	II	25	2245.58	II
18	4352.56	I	12	1670.74	II	50	2250.790	I
18	4392.59	I	12	1702.04	II	60	2251.874	I
160	4399.47	I	12	1761.38	II	25	2255.77	II
65	4403.78	I	20	1785.26	II	300	2259.511	I
110	4426.27	I	20	1786.74	II	60	2264.389	I
15	4450.18	I				80	2267.085	I
55	4478.48	I				80	2267.469	I
16	4495.35	I				50	2270.862	I
11 h	4496.03	I				150	2272.070	I
55	4545.68	I				150	2276.026	I

Intensity	Wavelength		Intensity	Wavelength	
80	2279.937	I	1000	2382.04	II
150	2284.086	I	20	2382.90	II
150	2287.250	I	20	2383.06	II
300	2292.524	I	60	2383.25	II
80	2294.41	I	50	2384.39	II
200	2297.787	I	40	2388.37	II
600	2298.169	I	300	2388.63	II
80	2299.220	I	200	2389.973	I
300	2300.142	I	30	2390.10	II
50	2301.684	I	20	2390.77	II
100	2303.424	I	15	2391.48	II
150	2303.581	* I	20	2392.58	II
120	2308.999	I	40	2395.42	II
150	2313.104	I	1000	2395.62	II
200	2320.358	I	15	2396.72	II
100	2327.40	II	300	2399.24	II
15	2327.88	II	20	2400.05	II
100	2331.31	II	15	2401.29	II
15	2331.97	II	50	2404.43	II
300	2332.80	II	800	2404.88	II
200	2338.01	II	250	2406.66	II
600	2343.49	II	80	2406.97	II
80	2343.96	II	300	2410.52	II
150	2344.28	II	200	2411.07	II
25	2344.98	II	50	2411.81	II
50	2345.34	II	150	2413.31	II
200	2348.11	II	20	2416.45	II
250	2348.30	II	80	2417.87	II
50	2351.20	II	15	2418.44	II
15	2351.67	II	60	2420.396	I
25	2352.31	II	60	2422.69	II
30	2353.47	II	60	2423.089	I
15	2353.68	II	40	2423.21	II
50	2354.48	II	150	2424.14	II
40	2354.89	II	15	2424.39	II
200	2359.12	II	30	2424.59	II
15	2359.59	II	30	2428.29	II
150	2360.00	II	120	2428.36	II
120	2360.29	II	25	2428.80	II
30	2360.51	II	25	2429.03	II
40	2362.02	II	20	2429.39	II
60	2363.86	II	30	2429.86	II
200	2364.83	II	120	2430.08	II
80	2365.76	II	25	2431.02	II
25	2366.59	II	80	2432.26	II
80	2368.59	II	60	2432.87	II
80	2369.456	I	25	2434.06	II
80	2369.95	II	20	2434.24	II
25	2370.50	II	20	2434.65	II
120	2371.430	I	50	2434.73	II
300	2373.624	I	50	2434.95	II
150	2373.74	II	25	2436.62	II
120	2374.518	I	60	2438.182	I
60	2375.19	II	150	2439.30	II
120	2376.43	II	150	2439.74	I
20	2378.13	II	80	2440.11	II
80	2379.27	II	40	2440.42	II
20	2379.41	II	30	2442.37	II
40	2380.20	II	100	2442.57	I
120	2380.76	II	60	2443.71	II
150	2381.835	I	250	2443.872	I
			100	2444.51	II
			50	2445.11	II
			50	2445.212	I
			100	2445.57	II
			40	2445.80	II
			50	2446.11	II
			30	2446.47	II
			40	2447.20	II
			25	2447.33	II
			60	2447.709	I
			30	2447.75	II
			25	2449.96	II
			25	2450.20	II
			100	2453.476	I
			20	2453.98	II
			30	2454.58	II
			15	2455.71	II
			15	2455.90	II
			15	2457.09	II
			1500	2457.598	I
			150	2458.78	II
			40	2458.97	II
			60	2460.44	II
			80	2461.28	II
			100	2461.86	II
			100	2462.181	I
			1500	2462.647	I
			50	2463.29	II
			50	2463.730	I
			40	2464.01	II
			40	2464.90	II
			800	2465.149	I
			50	2465.91	II
			15	2466.50	II
			60	2466.67	II
			60	2466.43	II
			60	2467.732	I
			15	2468.29	II
			600	2468.879	I
			60	2469.51	II
			25	2470.41	II
			80	2470.67	II

Int.	λ	Sp.	Int.	λ	Sp.	Int.	λ	Sp.	Int.	λ	Sp.
80	2470.965	I	50	2537.14	II	250	2718.436	I	250	2957.364	I
800	2472.336	I	50	2538.20	II	4000	2719.027	I	80	2959.99	I
40	2472.43	II	40	2538.50	II	100	2719.420	I	150	2965.254	I
40	2472.60	II	100	2538.80	II	50	2720.197	I	1500	2966.898	I
1000	2472.895	I	100	2538.91	II	1500	2720.903	I	120	2969.36	I
200	2473.16	I	150	2538.99	II	400	2723.578	I	800	2970.099	I
50	2473.32	II	50	2539.357	I	30	2724.88	II	15	2970.52	II
30	2474.05	II	200	2540.66	II	150	2724.953	I	1200	2973.132	I
600	2474.814	I	600	2540.972	I	80	2726.05	I	500	2973.235	I
50	2475.12	II	80	2541.10	II	50	2726.235	I	600	2981.445	I
40	2475.54	II	60	2541.84	II	25	2727.38	II	1000	2983.570	I
15	2476.26	II	300	2542.10	I	80	2727.54	II	60	2984.77	I
60	2476.657	I	25	2542.78	II	200	2728.020	I	50	2984.82	II
25	2477.34	II	60	2543.38	II	50	2728.820	I	13	2985.54	II
60	2478.57	II	250	2543.92	I	80	2728.90	II	1000	2994.427	I
120	2479.480	I	150	2544.70	I	40	2730.73	II	250	2994.502	I
1200	2479.776	I	40	2544.97	II	1000	2733.581	I	500	2999.512	I
100	2480.16	II	40	2545.22	II	60	2734.005	I	120	3000.451	I
15	2481.05	II	800	2545.978	I	50	2734.268	I	800	3000.948	I
80	2482.12	II	40	2546.44	II	500	2735.475	I	60	3001.655	I
25	2482.32	II	80	2546.67	II	50	2735.612	I	15	3002.64	II
100	2482.66	II	80	2546.87	I	500	2737.310	I	200	3007.282	I
10000	2483.271	I	100	2548.74	II	120	2737.83	I	500	3008.14	I
300	2483.533	I	80	2549.08	II	400	2739.55	II	120	3009.569	I
1000	2484.185	I	80	2549.39	II	250	2742.254	I	60	3017.627	I
60	2484.24	II	60	2549.46	II	800	2742.405	I	60	3018.983	I
30	2484.44	II	600	2549.613	I	200	2743.20	II	60	3020.01	II
50	2485.990	I	40	2549.77	II	150	2743.565	I	500	3020.491	I
800	2486.373	I	60	2550.03	II	200	2744.068	I	1500	3020.639	I
100	2486.691	I	25	2550.15	II	80	2744.527	I	600	3021.073	I
100	2487.066	I	50	2550.68	II	300	2746.48	II	500	3024.032	I
120	2487.370	I	40	2560.28	II	100	2749.32	II	150	3025.638	I
4000	2488.143	I	25	2562.09	II	500	2749.48	II	500	3025.842	I
100	2488.945	I	400	2562.53	II	1200	2750.140	I	80	3030.148	I
80	2489.48	II	200	2563.48	II	20	2751.13	II	60	3031.214	I
1000	2489.750	I	60	2566.91	II	20	2752.15	II	60	3034.484	I
50	2489.83	II	25	2570.52	II	80	2753.29	II	40	3036.96	II
50	2489.913	I	30	2570.85	II	50	2753.69	I	800	3037.389	I
3000	2490.644	I	150	2574.36	II	150	2754.032	I	80	3041.637	I
100	2490.71	II	50	2575.74	I	100	2754.426	I	800	3047.604	I
60	2490.86	II	300	2576.691	I	30	2754.89	II	600	3057.446	I
2000	2491.155	I	25	2576.86	II	800	2755.73	II	1000	3059.086	I
100	2491.40	II	60	2577.92	II	250	2756.328	I	250	3067.244	I
25	2492.34	II	50	2582.30	I	100	2757.316	I	120	3075.719	I
100	2493.18	II	100	2582.58	II	50	2759.81	I	120	3091.577	I
500	2493.26	II	1500	2584.54	I	120	2761.780	I	80	3098.189	I
60	2494.000	I	650	2585.88	II	150	2761.81	II	100	3099.895	I
50	2494.251	I	90	2588.00	I	150	2762.026	I	100	3099.968	I
100	2495.87	I	90	2591.54	II	120	2762.772	I	60	3100.303	I
600	2496.533	I	30	2592.78	II	120	2763.109	I	100	3100.665	I
50	2497.82	II	60	2593.51	I	20	2763.66	II	12	3154.20	II
150	2498.90	I	90	2593.73	II	25	2765.13	II	80	3175.445	I
40	2500.92	II	650	2598.37	II	80	2766.910	I	150	3184.895	I
1000	2501.132	I	2000	2599.40	II	250	2767.522	I	250	3191.659	I
40	2501.31	II	300	2599.57	I	50	2769.30	I	500	3193.226	I
50	2501.693	I	20	2605.34	II	25	2769.35	II	800	3193.299	I
60	2502.39	II	20	2605.42	II	300	2772.07	I	12	3196.08	II
40	2503.33	II	60	2605.657	I	50	2773.23	I	200	3196.928	I
60	2503.87	II	300	2606.51	II	20	2774.69	II	80	3199.500	I
80	2506.09	II	800	2606.827	I	15	2776.91	II	60	3200.47	I
40	2506.80	II	650	2607.09	II	60	2778.07	I	50	3205.398	I
500	2507.900	I	20	2611.07	II	600	2778.220	I	50	3211.67	I
30	2508.34	II	600	2611.87	II	40	2779.30	II	100	3211.88	I
50	2508.753	I	320	2613.82	II	50	2783.69	II	13	3213.31	II
1000	2510.835	I	320	2617.62	II	30	2785.19	II	200	3214.011	I
120	2511.76	II	250	2618.018	I	3000	2788.10	I	200	3214.396	I
80	2512.275	I	20	2619.07	II	20	2793.89	II	60	3215.938	I
400	2512.365	I	90	2620.41	II	200	2797.78	I	50	3217.377	I
50	2514.38	II	20	2620.69	II	30	2799.29	II	80	3219.583	I
80	2516.570	I	40	2621.67	II	400	2804.521	I	60	3219.766	I
50	2517.13	II	400	2623.53	II	1500	2806.98	I	300	3222.045	I
300	2517.661	I	50	2625.49	II	2500	2813.287	I	600	3225.78	I
800	2518.102	I	200	2625.67	II	300	2823.276	I	13	3227.73	II
60	2519.05	II	150	2628.29	II	600	2825.56	I	80	3227.796	I
150	2519.629	I	20	2630.07	II	50	2825.687	I	20	3230.42	II
40	2521.09	II	250	2631.05	II	120	2828.808	I	80	3233.05	I
30	2521.82	II	250	2631.32	II	25	2831.56	II	50	3233.967	I
50	2522.480	I	50	2631.61	II	1500	2832.436	I	120	3234.613	I
4000	2522.849	I	100	2632.237	I	120	2835.950	I	300	3236.222	I
200	2523.66	I	300	2635.809	I	200	2838.119	I	100	3239.433	I
500	2524.293	I	50	2641.646	I	30	2839.51	II	80	3244.187	I
100	2525.02	I	200	2643.998	I	20	2839.80	II	80	3246.005	I
200	2525.39	II	60	2664.66	II	15	2840.65	II	60	3254.36	I
25	2526.07	II	30	2666.64	II	200	2843.631	I	80	3265.046	I
300	2526.29	II	300	2666.812	I	1000	2843.977	I	50	3265.617	I
2000	2527.435	I	60	2666.965	I	100	2845.594	I	50	3271.000	I
30	2527.70	II	600	2679.062	I	15	2848.11	II	50	3280.26	I
800	2529.135	I	500	2684.75	II	15	2848.32	II	150	3286.75	I
25	2529.23	II	400	2689.212	I	800	2851.797	I	120	3305.97	I
80	2529.31	I	60	2692.60	II	30	2856.91	II	200	3306.343	I
250	2529.55	II	50	2696.28	I	25	2858.34	II	400	3355.227	I
150	2529.836	I	200	2699.106	I	50	2869.307	I	80	3355.517	I
40	2530.11	II	60	2703.99	II	50	2872.334	I	60	3369.546	I
200	2530.687	I	80	2706.012	I	80	2874.172	I	120	3370.783	I
120	2533.63	II	400	2706.582	I	50	2894.504	I	50	3378.678	I
60	2533.80	II	60	2708.571	I	120	2912.157	I	50	3380.110	I
100	2534.42	II	20	2709.05	II	120	2929.007	I	60	3383.978	I
120	2535.49	II	200	2711.655	I	1200	2936.903	I	12	3388.13	II
400	2535.607	I	80	2714.41	II	60	2941.343	I	50	3392.304	I
60	2536.67	II	50	2716.22	II	12	2944.40	II	150	3392.651	I
200	2536.792	I	50	2716.257	I	1000	2947.876	I	150	3399.333	I
200	2536.80	II	50	2717.786	I	60	2950.24	I	80	3404.353	I
50	2536.84	II	50	2717.87	II	600	2953.940	I	500	3407.458	I

Iron (Cont.)

Intensity	Wavelength	Spectrum
250	3413.131	I
60	3424.284	I
500	3427.119	I
60	3428.748	I
6000	3440.606	I
2500	3440.989	I
1000	3443.876	I
200	3445.149	I
15	3453.61	II
1200	3465.860	I
2000	3475.450	I
500	3476.702	I
2500	3490.574	I
500	3497.840	I
250	3513.817	I
300	3521.261	I
400	3526.040	I
100	3526.166	I
60	3526.237	I
60	3526.381	I
60	3526.467	I
100	3533.199	I
200	3536.556	I
300	3541.083	I
250	3542.075	I
80	3553.739	I
400	3554.925	I
200	3556.878	I
400	3558.515	I
1000	3565.379	I
1200	3570.097	I
800	3570.25	I
120	3571.996	I
100	3573.393	I
60	3573.829	I
60	3573.888	I
4000	3581.19	I
150	3582.199	I
150	3584.660	I
120	3584.929	I
300	3585.319	I
150	3585.705	I
200	3586.103	I
400	3586.984	I
100	3594.633	I
150	3603.204	I
200	3605.454	I
500	3606.680	I
1500	3608.859	I
250	3610.16	I
60	3612.068	I
150	3617.788	I
1500	3618.768	I
200	3621.462	I
150	3622.004	I
150	3623.19	I
100	3631.096	I
1200	3631.463	I
60	3632.041	I
100	3638.298	I
200	3640.389	I
80	3643.717	I
1500	3647.842	I
250	3649.506	I
80	3650.279	I
200	3651.467	I
120	3670.024	I
150	3670.089	I
100	3676.311	I
150	3677.629	I
1500	3679.913	I
200	3682.242	I
120	3683.054	I
150	3684.107	I
120	3685.998	I
500	3687.456	I
120	3689.477	I
150	3694.008	I
120	3695.051	I
150	3701.086	I
80	3704.462	I
1200	3705.566	I
60	3707.041	I
150	3707.821	I
300	3707.919	I
600	3709.246	I
120	3716.442	I
8000	3719.935	I
1500	3722.563	I
120	3724.377	I
60	3725.491	I
60	3727.093	I
500	3727.619	I
150	3732.396	I
1200	3733.317	I
5000	3734.864	I
120	3735.324	I
6000	3737.131	I
100	3738.306	I
400	3743.362	I
80	3743.47	I
6000	3745.561	I
1200	3745.899	I
3000	3748.262	I
80	3748.964	I
3000	3749.485	I
1500	3758.232	I
400	3760.05	I
1500	3763.788	I
400	3765.54	I
600	3767.191	I
60	3776.452	I
250	3785.95	I
100	3786.68	I
250	3787.880	I
250	3790.092	I
150	3794.34	I
400	3795.002	I
120	3797.518	I
250	3798.511	I
400	3799.547	I
200	3805.345	I
80	3806.696	I
600	3812.964	I
60	3813.059	I
1500	3815.840	I
2500	3820.425	I
150	3821.179	I
80	3824.306	I
2500	3824.444	I
1500	3825.880	I
1200	3827.823	I
1000	3834.222	I
120	3839.257	I
500	3840.437	I
800	3841.047	I
120	3843.256	I
80	3846.800	I
200	3849.96	I
120	3850.817	I
2500	3856.372	I
150	3859.212	I
10000	3859.911	I
150	3865.523	I
60	3867.215	I
250	3872.501	I
150	3873.761	I
250	3878.018	I
2000	3878.573	I
4000	3886.282	I
200	3887.048	I
300	3888.513	I
800	3895.656	I
1200	3899.707	I
400	3902.945	I
250	3906.479	I
80	3916.731	I
600	3920.258	I
1200	3922.911	I
1200	3927.920	I
2000	3930.296	I
60	3948.774	I
60	3949.953	I
50	3951.164	I
50	3952.601	I
60	3956.454	I
250	3956.68	I
60	3966.614	I
100	3969.257	I
80	3977.741	I
40	3981.771	I
50	3983.956	I
60	3994.114	I
200	3997.392	I
40	3998.053	I
400	4005.241	I
60	4009.713	I
80	4014.53	I
100	4021.867	I
50	4040.638	I
4000	4045.813	I
1500	4063.594	I
50	4066.975	I
50	4067.977	I
1200	4071.737	I
40	4076.629	I
40	4100.737	I
40	4107.489	I
150	4118.544	I
40	4127.608	I
400	4132.058	I
80	4134.676	I
40	4136.997	I
200	4143.415	I
800	4143.869	I
40	4153.898	I
50	4154.500	I
60	4156.799	I
50	4172.744	I
50	4174.912	I
60	4175.635	I
50	4177.593	I
120	4181.754	I
50	4184.891	I
120	4187.038	I
120	4187.795	I
80	4191.430	I
40	4195.329	I
150	4198.304	I
40	4199.095	I
300	4202.029	I
40	4203.984	I
80	4206.696	I
80	4210.343	I
400	4216.153	I
100	4219.360	I
50	4222.212	I
50	4225.956	I
200	4227.423	I
11	4233.17	II
100	4233.602	I
250	4235.936	I
50	4238.809	I
50	4247.425	I
200	4250.118	I
300	4250.787	I
40	4258.315	I
800	4260.473	I
250	4271.153	I
1200	4271.759	I
1200	4282.402	I
80	4291.462	I
250	4299.234	I
1200	4307.901	I
150	4315.084	I
1500	4325.761	I
80	4352.734	I
80	4369.771	I
800	4375.929	I
3000	4383.544	I
1200	4404.750	I
300	4415.122	I
600	4427.299	I
400	4461.652	I
120	4466.551	I
80	4476.017	I
80	4482.169	I
200	4482.252	I
50	4489.739	I
50	4528.613	I
11	4583.83	II
30	4647.433	I
30	4736.771	I
50	4859.741	I
120	4871.317	I
60	4872.136	I
30	4878.208	I
100	4890.754	I
250	4891.492	I
30	4903.309	I
150	4918.992	I
500	4920.502	I
12	4923.92	II
1500	4957.597	I
11	4990.50	II
80	5001.862	I
18	5001.91	II
11	5004.20	II
30	5005.711	I
100	5006.117	I
60	5012.067	I
30	5014.941	I
12	5018.43	II
11	5030.64	II
25	5030.77	I
12	5035.71	II
150	5041.755	I
30	5049.819	I
30	5051.634	I
25	5074.748	I
18	5100.73	II
15	5100.95	II
150	5110.357	I
40	5133.69	I
40	5139.251	I
100	5139.462	I
11	5144.36	II
12	5149.46	II
25	5151.910	I
30	5162.27	I
80	5166.281	I
2500	5167.487	I
80	5168.897	I
12	5169.03	II
500	5171.595	I
50	5191.454	I
80	5192.343	I
200	5194.941	I
30	5204.582	I
25	5215.179	I
150	5216.274	I
18	5216.85	II
60	5226.862	I
1000	5227.150	I
13	5227.49	II
250	5232.939	I
13	5247.95	II
13	5251.23	II
18	5260.26	II
11	5264.18	II
100	5266.555	I
1200	5269.537	I
800	5270.357	I
30	5281.789	I
60	5283.621	I
25	5302.299	I
11	5306.18	II
13	5316.23	II
150	5324.178	I
800	5328.038	I
300	5328.531	I
100	5332.899	I
14	5339.59	II
80	5339.928	I
500	5341.023	I
25	5364.87	I
40	5367.47	I
50	5369.96	I
400	5371.489	I
60	5383.37	I
14	5387.06	II
40	5393.167	I
12	5395.86	II
300	5397.127	I
15	5402.06	II
60	5404.12	I
250	5405.774	I
30	5410.91	I
60	5415.20	I
60	5424.07	I
30	5427.83	II
250	5429.695	I
13	5429.99	II
100	5434.523	I
200	5446.871	I
25	5455.45	I
120	5455.609	I
16	5465.93	II
20	5466.94	II
16	5482.31	II
14	5493.83	II
25	5497.516	I
20	5501.464	I
18	5506.20	II
30	5506.778	I
12	5510.78	II
12	5529.06	II
13	5544.76	II
30	5569.618	I
60	5572.841	I
120	5586.755	I
200	5615.644	I
20	5624.541	I
12	5645.40	II
50	5662.515	I
20	5762.990	I
11	5783.63	II
30	5862.353	I
13	5885.02	II
16	5902.82	II
30	5914.114	I
14	5955.70	II
30	5986.956	I
18	5961.71	II
30	5962.4	II
13	5965.63	II
40	6065.482	I
30	6102.159	I
40	6136.614	I
40	6137.694	I
30	6147.73	II
20	6149.24	II
15	6175.16	II
40	6191.558	I
30	6213.429	I
30	6219.279	I
40	6230.726	I
20	6238.37	II
20	6246.317	I
80	6247.56	II
30	6252.554	I
15	6305.32	II
12	6331.97	II
15	6383.75	II
20	6393.602	I
30	6399.999	I
20	6411.647	I
20	6416.90	II
20	6421.349	I
30	6430.844	I
20	6446.43	II
200	6456.38	II
60	6494.981	I
20	6516.05	II
20	6546.239	I
20	6592.913	I
40	6677.989	I
15	6855.18	I
15	6945.21	I
20	7067.44	II
15	7130.94	I
25	7164.443	I
80	7187.313	I

Iron (Cont.)

Intensity	Wavelength	Spectrum
30	7207.381	I
12	7224.51	II
50	7307.97	II
40	7320.70	II
20	7376.46	II
30	7445.746	I
20	7462.38	II
40	7495.059	I
60	7511.045	I
15	7586.04	I
15	7711.71	II
30	7780.59	I
40	7832.22	I
80	7937.131	I
60	7945.984	I
80	7998.939	I
60	8046.047	I
50	8085.176	I
150	8220.41	I
120	8327.053	I
20	8331.908	I
120	8387.770	I
30	8468.404	I
15	8514.069	I
60	8661.898	I
150	8688.621	I
12	8793.38	I
12	8824.23	I
20	8866.96	I
15	8999.56	I
15	10216.32	I
13	10469.65	I
21	11119.80	I
14	11374.08	I
52	11422.32	I
87	11439.12	I
91	11593.59	I
255	11607.57	I
160	11638.26	I
230	11689.98	I
160	11783.26	I
580	11882.84	I
225	11884.08	I
1030	11973.05	I
15	12638.71	I
14	12879.76	I
17	13565.04	I
30	14236.25	I
24	14285.11	I
14	14292.38	I
16	14308.69	I
96	14400.56	I
20	14442.28	I
72	14512.23	I
50	14555.06	I
14	14565.95	I
40	14826.43	I
37	15051.77	I
28	15207.55	I
94	15294.58	I
16	15335.40	I
30	15621.67	I
25	15631.97	I
14	15723.59	I
41	15769.42	I
28	15813.13	I
13	16444.82	I
20	16486.69	I
105	18856.65	I
47	18987.01	I
25	19113.68	I
22	19791.88	I
14	22380.82	I
21	22619.85	I
38	26222.04	I
17	26659.22	I

Fe III
Ref. 71, 101 — J.R.

Intensity · Wavelength

Vacuum

Intensity		Wavelength	Spectrum
6		728.81	III
5		730.00	III
5		737.71	III
5		739.26	III
9		807.55	III
8		807.86	III
8		808.84	III
8	p	811.28	III
10		813.38	III
8		838.05	III
10		844.28	III
9		845.41	III
8	w	847.42	III
8		859.72	III
8	p	861.76	III
10	p	861.83	III
8		873.46	III
9		890.76	III
10		891.17	III

Intensity		Wavelength	Spectrum
8		891.44	III
8		899.42	III
10		950.33	III
10		981.37	III
10	w	983.88	III
8		985.82	III
9		991.23	III
9		1017.25	III
8		1017.74	III
8		1018.29	III
8		1032.12	III
8		1063.87	III
9		1122.53	III
9		1124.88	III
8		1128.02	III
10	h	1505.17	III
10	h	1538.63	III
12	h	1550.20	III
10	h	1601.21	III
10		1869.83	III
12		1877.99	III
10		1882.05	III
12		1886.76	III
13		1890.67	III
11		1893.98	III
20		1895.46	III
10	s	1907.58	III
19		1914.06	III
15		1915.08	III
15		1922.79	III
10	p	1926.01	III
18		1926.30	III
15		1930.39	III
14		1931.51	III
14		1937.34	III
10	l	1938.90	III
14	s	1943.48	III
12		1945.34	III
10		1950.33	III
12		1951.01	III
11		1952.65	III
13		1953.32	III
10		1953.49	III
10	w	1954.22	III
11		1958.58	III
13		1960.32	III
15		1987.50	III
14		1991.61	III
13		1994.07	III
12		1995.56	III
12		1996.42	III

Air

Intensity		Wavelength	Spectrum
10		2061.55	III
12		2068.24	III
14		2078.99	III
10		2084.35	III
12		2090.14	III
15		2097.48	III
12		2097.69	III
12		2103.80	III
10		2107.32	III
15		2151.78	III
12		2157.71	III
12		2158.47	III
10		2161.27	III
12		2166.95	III
12		2171.04	III
15		2174.66	III
12		2180.41	III
10	p	2208.85	III
10		2221.83	III
10		2229.27	III
10		2232.43	III
10		2232.69	III
10		2235.91	III
10		2238.16	III
12	p	2241.54	III
12		2261.59	III
10		2267.42	III
10		2293.06	III
15		2295.86	III
10	p	2317.70	III
10		2319.22	III
10	p	2321.71	III
10		2326.95	III
10	p	2336.77	III
10		2338.96	III
8		2389.53	III
8		2438.17	III
8	p	2582.37	III
8		2595.62	III
8		2617.15	III
9		2645.39	III
10	h	2695.13	III
9	h	2695.34	III
8	h	2700.02	III
8	h	2701.13	III
8		2773.31	III
10	p	2813.24	III
8	p	2895.08	III
9	p	2902.47	III

Intensity		Wavelength	Spectrum
12		2904.43	III
8	p	2905.80	III
10		2907.50	III
12		2907.70	III
8		2923.90	III
8		2948.39	III
8		2963.23	III
12		3001.62	III
12	h	3007.28	III
15		3013.17	III
10	p	3136.43	III
10		3174.09	III
10		3175.99	III
10		3178.01	III
13		3266.88	III
11		3276.08	III
10		3288.81	III
9		3305.22	III
9		3339.39	III
9		3499.59	III
9		3500.28	III
10		3501.76	III
10		3586.04	III
11		3600.94	III
11		3603.88	III
16		3954.33	III
11		3968.72	III
9		3969.49	III
10	w	3979.42	III
10		4035.42	III
11		4053.11	III
10		4081.00	III
10		4120.90	III
11		4122.02	III
11		4122.78	III
15		4137.76	III
13		4139.48	III
9		4140.48	III
9		4154.96	III
18		4164.73	III
9		4164.92	III
13		4166.84	III
13		4174.26	III
9		4210.67	III
11		4222.27	III
13		4235.56	III
9		4238.62	III
12		4243.75	III
12	h	4273.40	III
12		4279.72	III
14	h	4286.16	III
16	h	4296.85	III
18	h	4304.78	III
20	h	4310.36	III
9		4323.68	III
9	h	4372.04	III
9	h	4372.14	III
11	h	4372.31	III
14	h	4372.53	III
18	h	4372.81	III
9		4395.76	III
12		4419.60	III
9		4431.02	III
9		5111.07	III
9		5127.35	III
12		5156.12	III
10		5199.08	III
10		5235.66	III
18		5243.31	III
13	l	5260.34	III
9		5272.37	III
14		5272.98	III
15		5276.48	III
16		5282.30	III
12		5284.83	III
11		5298.12	III
12		5299.93	III
14	w	5302.60	III
10		5306.76	III
9		5310.88	III
10		5322.74	III
11		5346.88	III
12		5353.77	III
12		5363.76	III
10		5368.06	III
11	l	5375.47	III
11		5719.88	III
9		5744.19	III
10		5756.38	III
18		5833.93	III
9		5848.76	III
10		5854.62	III
9		5876.26	III
15		5891.91	III
9		5898.68	III
9		5918.96	III
10	p	5920.13	III
18	p	5929.69	III
10		5952.31	III
14		5953.62	III
9		5968.48	III
12		5979.32	III
9	h	5981.01	III

Intensity		Wavelength	Spectrum
12	h	5989.08	III
18		5999.54	III
9		6031.02	III
16		6032.59	III
13		6036.56	III
11		6048.72	III
11		6054.18	III
9		6056.36	III
9		6149.99	III
9		6169.74	III
9		6185.26	III
7		6186.56	III
7		6194.79	III
6		6195.43	III
6		6201.37	III
5	s	6203.04	III
5		6259.81	III
6	p	6294.50	III
5		6357.81	III
5	h	7317.63	III
6	h	7320.14	III
5	w	7921.17	III
5	w	8230.88	III
5	w	8231.79	III
9	w	8235.45	III
8	w	8236.75	III
6	w	8238.98	III
5		8563.49	III

Fe IV
Ref. 382 — J.R.

Intensity · Wavelength

Vacuum

Intensity	Wavelength	Spectrum
10	502.42	IV
11	506.69	IV
11	505.35	IV
17	525.69	IV
15	526.29	IV
10	526.57	IV
13	526.63	IV
10	530.91	IV
10	531.78	IV
10	535.55	IV
14	536.61	IV
10	536.74	IV
15	537.10	IV
13	537.26	IV
14	537.79	IV
13	537.94	IV
10	538.44	IV
10	544.20	IV
10	546.22	IV
10	548.80	IV
11	550.32	IV
10	551.77	IV
13	552.14	IV
11	552.74	IV
10	554.26	IV
10	555.66	IV
10	572.88	IV
10	576.76	IV
10	579.76	IV
14	607.53	IV
13	608.80	IV
10	609.65	IV
12	1425.73	IV
13	1431.43	IV
12	1473.20	IV
12	1489.53	IV
12	1495.18	IV
13	1526.60	IV
13	1530.26	IV
14	1532.63	IV
13	1532.91	IV
15	1533.86	IV
13	1533.95	IV
14	1536.58	IV
12	1538.29	IV
13	1542.16	IV
14	1542.70	IV
12	1546.40	IV
12	1552.35	IV
12	1552.71	IV
12	1562.46	IV
13	1566.26	IV
14	1568.27	IV
12	1570.18	IV
12	1570.42	IV
12	1571.24	IV
12	1577.20	IV
12	1577.76	IV
12	1590.62	IV
13	1591.51	IV
13	1592.05	IV
12	1596.67	IV
13	1598.01	IV
12	1600.50	IV
13	1600.58	IV
13	1601.67	IV

Iron (Cont.)

Intensity	Wavelength	
12	1602.08	IV
13	1603.18	IV
13	1603.73	IV
13	1604.88	IV
13	1605.68	IV
15	1605.97	IV
13	1606.98	IV
17	1609.10	IV
14	1609.83	IV
13	1610.47	IV
13	1611.20	IV
13	1613.64	IV
15	1614.02	IV
13	1614.64	IV
13	1615.00	IV
12	1615.61	IV
16	1616.68	IV
14	1617.68	IV
14	1619.02	IV
12	1620.91	IV
13	1621.16	IV
14	1621.57	IV
13	1623.38	IV
13	1623.53	IV
15	1626.47	IV
14	1626.90	IV
13	1628.54	IV
13	1630.18	IV
17	1631.08	IV
12	1632.08	IV
14	1632.40	IV
13	1634.01	IV
12	1638.07	IV
12	1638.30	IV
14	1639.40	IV
16	1640.04	IV
14	1640.16	IV
15	1641.87	IV
12	1642.88	IV
15	1647.09	IV
15	1651.58	IV
15	1652.90	IV
13	1653.41	IV
13	1656.11	IV
15	1656.65	IV
12	1657.82	IV
12	1658.43	IV
14	1660.10	IV
12	1661.57	IV
13	1662.32	IV
13	1662.52	IV
13	1663.54	IV
13	1668.09	IV
12	1669.61	IV
14	1671.04	IV
12	1672.86	IV
13	1673.68	IV
14	1675.66	IV
12	1676.78	IV
12	1677.12	IV
13	1681.36	IV
12	1681.95	IV
15	1687.69	IV
15	1698.88	IV
12	1700.40	IV
12	1704.93	IV
13	1709.81	IV
15	1711.41	IV
14	1712.76	IV
12	1717.11	IV
14	1717.90	IV
14	1718.16	IV
12	1718.42	IV
14	1719.46	IV
14	1722.71	IV
14	1724.06	IV
12	1724.26	IV
16	1725.63	IV
13	1761.08	IV
12	1764.92	IV
12	1767.36	IV
13	1792.10	IV
13	1796.93	IV
12	1805.32	IV
12	1820.42	IV
13	1827.98	IV
12	1840.24	IV
12	1860.42	IV
12	1869.64	IV
12	1874.23	IV

Fe V
Ref. 381 — J.R.

Intensity	Wavelength	
	Vacuum	
300	361.28	V
300	365.43	V
300	365.86	V
300	374.24	V
300	374.87	V
300	375.98	V
300	379.59	V
300	380.31	V
300	381.27	V
300	384.96	V
300	384.97	V
300	385.03	V
300	385.11	V
300	385.25	V
300	385.26	V
300	385.30	V
300	385.75	V
300	385.88	V
350	386.16	V
300	386.74	V
300	386.78	V
300	386.85	V
300	386.88	V
350	386.88	V
400	387.20	V
400	387.50	V
300	387.62	V
400	387.76	V
400	387.78	V
300	387.98	V
300	388.61	V
300	388.82	V
300	390.11	V
300	390.19	V
300	390.78	V
300	391.94	V
300	392.06	V
300	392.38	V
300	392.50	V
300	392.51	V
300	392.70	V
300	392.91	V
300	393.27	V
300	393.72	V
300	393.73	V
300	393.91	V
300	393.97	V
300	394.04	V
300	394.64	V
300	395.15	V
300	395.79	V
400	395.90	V
300	399.84	V
300	400.11	V
300	400.51	V
300	400.52	V
300	400.63	V
300	401.04	V
300	401.64	V
300	401.86	V
300	402.87	V
300	403.06	V
400	404.62	V
400	405.50	V
800	407.42	V
600	407.44	V
400	407.49	V
500	407.75	V
400	409.71	V
400	410.20	V
600	411.55	V
300	415.01	V
300	416.66	V
300	416.84	V
700	417.39	V
700	418.04	V
500	418.47	V
300	420.56	V
700	421.06	V
500	421.78	V
300	422.28	V
500	422.31	V
300	423.23	V
500	426.06	V
500	426.11	V
300	426.83	V
350	426.97	V
300	434.42	V
300	439.22	V
300	444.70	V
300	445.44	V
300	446.04	V
300	458.16	V
300	486.17	V
400	1317.86	V
300	1318.35	V
300	1320.41	V
300	1321.34	V
300	1321.49	V
400	1323.27	V
400	1330.40	V
300	1345.61	V
400	1359.01	V
300	1361.28	V
300	1361.45	V
600	1361.82	V
300	1363.08	V
300	1363.64	V
300	1365.57	V
700	1373.59	V
600	1373.67	V
300	1374.12	V
500	1376.34	V
300	1376.46	V
500	1378.56	V
300	1385.68	V
800	1387.94	V
400	1397.97	V
600	1400.24	V
800	1402.39	V
400	1406.67	V
500	1406.82	V
400	1407.25	V
300	1409.03	V
300	1409.22	V
600	1409.45	V
400	1415.20	V
300	1418.12	V
600	1420.46	V
800	1430.57	V
800	1440.53	V
300	1440.79	V
400	1442.22	V
800	1446.62	V
700	1448.85	V
400	1449.93	V
300	1455.56	V
700	1456.16	V
500	1459.83	V
400	1460.73	V
500	1462.63	V
700	1464.68	V
500	1465.38	V
400	1466.65	V
500	1469.00	V
300	1475.60	V
500	1479.47	V
300	1554.22	V

KRYPTON (Kr)
Z = 36

Kr I and II
Ref. 61, 121, 123, 147, 208, 232
— E.F.W.

Intensity		Wavelength	
		Vacuum	
60		729.40	II
200		761.18	II
100		763.98	II
60		766.20	II
200		771.03	II
60	p	773.69	II
200		782.10	II
100		783.72	II
60		818.15	II
60		830.38	II
100		844.06	II
60		864.82	II
60		868.87	II
200		884.14	II
1000		886.30	II
400		891.01	II
200		911.39	II
2000		917.43	II
50		945.44	I
50		946.54	I
20		951.06	I
50		953.40	I
50		963.37	I
2000		964.97	II
100		1001.06	I
100		1003.55	I
100		1030.02	I
200		1164.87	I
650		1235.84	I
		Air	
100	h	2464.77	II
60		2492.48	II
80	h	2712.40	II
100		2833.00	II
100	h	3607.88	II
200		3631.889	II
250		3653.928	II
80		3665.324	I
150		3669.01	II
100		3679.559	I
80		3686.182	II
300	h	3718.02	II
200		3718.595	II
150		3721.350	II
200		3741.638	II
150		3744.80	II
80		3754.245	II
500		3778.089	II
500		3783.095	II
150	h	3875.44	II
150		3906.177	II
200		3920.081	II
100		3994.840	II
100	h	3997.793	II
300		4057.037	II
300		4065.128	II
500		4088.337	II
250		4098.729	II
100		4109.248	II
250		4145.122	II
150		4250.580	II
1000		4273.969	I
100		4282.967	I
600		4292.923	II
200		4300.49	II
500	h	4317.81	II
400		4318.551	I
1000		4319.579	I
150	h	4322.98	II
100		4351.359	I
3000		4355.477	II
500		4362.641	I
200		4369.69	II
800		4376.121	I
300	h	4386.54	II
200		4399.965	I
100		4425.189	I
500		4431.685	II
600		4436.812	II
600		4453.917	I
800		4463.689	I
800		4475.014	II
400	h	4489.88	II
600		4502.353	I
400	h	4523.14	II
200	h	4556.61	II
800		4577.209	II
300		4582.978	II
150	h	4592.80	II
500		4615.292	II
1000		4619.166	II
800		4633.885	II
2000		4658.876	II
500		4680.406	II
100		4691.301	II
200		4694.360	II
3000		4739.002	II
300		4762.435	II
1000		4765.744	II
300		4811.76	II
300		4825.18	II
800		4832.077	II
700		4846.612	II
150		4857.20	II
300		4945.59	II
200		5022.40	II
250		5086.52	II
400	h	5125.73	II
500		5208.32	II
200		5308.66	II
500		5333.41	II
200		5468.17	II
500		5562.224	I
2000		5570.288	I
80		5580.386	I
100		5649.561	I
400		5681.89	II
200	h	5690.35	II
100		5832.855	I
3000		5870.914	I
200		5992.22	II
60		5993.849	I
60		6056.125	I
300		6420.18	II
100		6421.026	I
200		6456.288	I
150		6570.07	II
60		6699.228	I
100		6904.678	I
250		7213.13	II
100		7224.104	I
80		7287.258	I
400		7289.78	II
400		7407.02	I
60		7425.541	I
200		7435.78	II
100		7486.862	I
300		7524.46	II
1000		7587.411	I
2000		7601.544	I
150		7641.16	II
1000		7685.244	I
1200		7694.538	I
250		7735.69	II
150		7746.827	I
800		7854.821	I
200		7913.423	I
180		7928.597	I
200		7933.22	II
120		7973.62	II
100		7982.401	I
1500		8059.503	I
4000		8104.364	I

Intensity		Wavelength	
6000		8112.899	I
60		8132.967	I
3000		8190.054	I
200		8202.72	II
80		8218.365	I
3000		8263.240	I
100		8272.353	I
5000		8298.107	I
1500		8281.050	I
100		8412.430	I
3000		8508.870	I
150		8764.110	I
6000		8776.748	I
2000		8928.692	I
500		9238.48	II
500	hL	9293.82	II
200	h	9320.99	II
300		9361.95	II
100		9362.082	I
200	h	9402.82	II
200	h	9470.93	II
500		9577.52	II
500	h	9605.80	II
400	h	9619.61	II
200		9663.34	II
200	h	9711.60	II
2000		9751.758	I
500		9803.14	II
500		9856.314	I
1000		10221.46	II
100		11187.108	I
200		11257.711	I
150		11259.126	I
500		11457.481	I
150		11792.425	I
1500		11819.377	I
600		11997.105	I
160		12077.224	I
100		12861.892	I
1100		13177.412	I
1000		13622.415	I
2400		13634.220	I
800		13658.394	I
200		13711.036	I
600		13738.851	I
150		13974.027	I
550		14045.657	I
140		14104.298	I
180		14402.22	I
2000		14426.793	I
100		14517.84	I
1600		14734.436	I
550		14762.672	I
450		14765.472	I
400		14961.894	I
120		15005.307	I
140		15209.526	I
1700		15239.615	I
130		15326.480	I
1500		15334.958	I
700		15372.037	I
200		15474.026	I
180		15681.02	I
120		15820.09	I
200		16726.513	I
2000		16785.128	I
1000		16853.488	I
2400		16890.441	I
1600		16896.753	I
1800		16935.806	I
600		17098.771	I
700		17367.606	I
120		17404.443	I
150		17616.854	I
650		17842.737	I
700		18002.229	I
2600		18167.315	I
100		18399.786	I
150		18580.896	I
300		18696.294	I
170		18785.460	I
200		18797.703	I
140		20209.878	I
300		20423.964	I
140		20446.971	I
600		21165.471	I
1800		21902.513	I
120		22485.775	I
180		23340.416	I
120		24260.506	I
180		24292.221	I
600		25233.820	I
180		28610.55	I
1000		28655.72	I
150		28769.71	I
140		28822.49	I
300		29236.69	I
300		30663.54	I
300		30979.16	I
500		39300.6	I
1100		39486.52	I
220		39557.25	I
100		39572.60	I
1400		39588.4	I
1100		39589.6	I
500		39954.8	I
300		39966.6	I
1300		40306.1	I
250		40685.16	I

Kr III
Ref. 208, 366, 390, 421 — E.F.W.

Intensity		Wavelength	
		Vacuum	
30		467.35	III
30		540.86	III
30		565.64	III
30		569.16	III
30		571.98	III
30		579.83	III
30		585.14	III
30		585.96	III
30		593.70	III
30		594.10	III
30		596.41	III
40		600.17	III
30		603.67	III
30		605.86	III
35		606.47	III
50		611.12	III
35		616.72	III
40		621.45	III
45		622.80	III
50		625.02	III
30		625.76	III
45		628.59	III
50		630.04	III
35		633.09	III
50		639.98	III
60		646.41	III
50		651.20	III
50		659.72	III
30		664.86	III
40		672.34	III
35		672.85	III
35		676.57	III
35		680.13	III
35		683.68	III
45		686.25	III
45		687.98	III
45		691.93	III
50		695.61	III
30		698.05	III
50		708.36	III
50		714.00	III
100	p	722.04	III
30		746.70	III
60		785.97	III
50		837.66	III
50		854.73	III
60		862.58	III
40		870.84	III
50		876.08	III
75		897.81	III
50		987.29	III
30		1158.74	III
6		1638.82	III
6		1914.09	III
		Air	
40		2393.94	III
40		2494.11	III
30		2563.25	III
60		2639.76	III
30		2680.32	III
40		2681.19	III
30		2841.00	III
30		2851.16	III
50		2870.61	III
100		2892.18	III
30		2909.17	III
50		2952.56	III
60		2992.22	III
50		3022.30	III
80		3024.45	III
50		3046.93	III
30		3056.72	III
60		3063.13	III
40		3097.16	III
60		3112.25	III
30		3120.61	III
100		3124.39	III
60		3141.35	III
100		3189.11	III
80		3191.21	III
40		3239.52	III
40		3240.44	III
300		3245.69	III
150		3264.81	III
100		3268.48	III
30		3271.65	III
30		3285.89	III
30		3304.75	III
50		3311.47	III
200		3325.75	III
60		3330.76	III
50		3342.48	III
100		3351.93	III
40		3374.96	III
100		3439.46	III
70		3474.65	III
100		3488.59	III
200		3507.40	III
100		3564.23	III
30		3641.34	III
30		3690.65	III
40	h	3868.70	III
50		4067.37	III
40		4131.33	III
40		4154.46	III
20	h	5016.45	III
10		5501.43	III
10	h	6037.17	III
10	h	6078.38	III
10		6310.22	III

Kr IV
Ref. 366, 409, 417 — E.F.W.

Intensity	Wavelength	
	Vacuum	
	793.44	IV
	794.11	IV
7	805.76	IV
18	816.82	IV
22	842.04	IV
	Air	
3	2237.34	IV
6	2291.26	IV
3	2329.3	IV
4	2336.75	IV
4	2348.27	IV
3	2358.5	IV
3	2388.05	IV
4	2416.9	IV
3	2428.04	IV
5	2442.68	IV
4	2451.7	IV
6	2459.74	IV
5	2474.06	IV
4	2517.0	IV
5	2518.02	IV
6	2519.38	IV
5	2524.5	IV
5	2546.0	IV
6	2547.0	IV
4	2558.08	IV
3	2586.9	IV
5	2606.17	IV
10	2609.5	IV
8	2615.3	IV
7	2621.11	IV
3	2730.55	IV
8	2748.18	IV
6	2774.70	IV
3	2829.60	IV
3	2836.08	IV
5	2853.0	IV
3	2859.3	IV
3	3142.01	IV
6	3224.99	IV
3	3261.70	IV
3	3809.30	IV
5	3860.58	IV
5	3934.29	IV

Kr V
Ref. 409, 421 — E.F.W.

Intensity	Wavelength	
	Vacuum	
150	472.16	V
100	484.39	V
250	496.25	V
120	500.77	V
200	507.20	V
60	548.04	V
120	637.87	V
	690.86	V
	691.75	V
600	708.85	V
	810.70	V

LANTHANUM (La)
Z = 57

La I and II

Ref. 1 — C.H.C.

Intensity		Wavelength	
		Air	
240		2187.87	II
770		2256.76	II
200		2319.44	II
400		2610.34	II
420		2808.39	II
130		2885.14	II
160		2893.07	II
110		2950.50	II
180		3104.59	II
130		3142.76	II
510		3245.13	II
260		3249.35	II
550		3265.67	II
800		3303.11	II
1500		3337.49	II
870		3344.56	II
200		3376.33	II
1500		3380.91	II
130		3452.18	II
180		3453.17	I
200		3574.43	I
320		3628.83	II
120		3637.15	II
170	d	3641.53	I
		3641.66	II
1000		3645.42	II
390		3650.18	II
170		3662.08	II
120		3704.54	I
320		3705.82	II
550		3713.54	II
140		3714.87	II
270		3715.53	II
2400		3759.08	II
120		3780.67	II
3700		3790.83	II
3900		3794.78	II
190		3835.08	II
600		3840.72	II
120		3846.00	II
1600		3849.02	II
130		3854.91	II
3400		3871.64	II
1700		3886.37	II
1300		3916.05	II
1100		3921.54	II
160		3927.56	I
2200		3929.22	II
180		3936.22	II
9000		3949.10	II
4400		3988.52	II
3600		3995.75	II
180		4015.39	I
250		4025.88	II
2800		4031.69	II
140		4037.21	I
3000		4042.91	II
320		4050.08	II
220		4060.33	I
160		4064.79	I
850		4067.39	II
110		4076.71	II
2800		4077.35	II
120		4079.18	I
5500		4086.72	II
180		4089.61	I
280		4099.54	II
110		4104.87	I
4400		4123.23	II
110		4137.04	I
550		4141.74	II
1100		4151.97	II
220		4152.78	II
100		4160.26	I
280		4187.32	I
280		4192.36	II
1500		4196.55	II
240		4204.04	II
300		4217.56	II
200		4230.95	II
1600		4238.38	II
140		4249.99	II
320		4263.59	II
480		4269.50	II
240		4275.64	II
300		4280.27	I
600		4286.97	II
600		4296.05	II
120		4300.44	II
440		4322.51	II
4600		4333.74	II
550		4354.40	II
110		4364.67	II
110	b1	4371.97	L
110	b1	4375.84	L
110		4378.10	II
280		4383.44	II
100		4385.20	II

Lanthanum (Cont.)

Intensity		Wavelength	
220	b1	4418.24	L
160	b1	4423.17	L
160		4423.90	I
260		4427.55	II
100	b1	4428.10	L
2000		4429.90	II
160	b1	4432.98	L
100	b1	4438.01	L
100		4452.15	I
100		4455.80	II
850		4522.37	II
170		4525.31	II
420		4526.12	II
400		4558.46	II
110		4559.29	II
160		4567.91	I
200		4570.02	I
400		4574.88	II
200		4580.06	II
160		4605.78	II
410		4613.39	II
410		4619.88	II
110		4645.28	II
540		4655.50	II
360		4662.51	II
230		4663.76	II
200		4668.91	II
160		4671.83	II
230		4692.50	II
140		4703.28	II
170		4716.44	II
140		4719.94	II
230		4728.42	II
500		4740.28	II
390		4743.09	II
320		4748.73	II
160		4766.89	I
160		4804.04	II
160		4809.01	II
200		4824.06	II
320		4860.91	II
850		4899.92	II
1000		4920.98	II
1000		4921.79	II
140		4934.83	II
110		4946.47	II
370		4949.77	I
340		4970.39	II
370		4986.83	II
140		4991.28	II
720		4999.47	II
140		5046.88	I
210		5050.57	I
170		5056.46	I
200		5106.23	I
470		5114.56	II
470		5122.99	II
450		5145.42	I
180		5156.74	II
180		5157.43	II
290		5158.69	I
120		5163.62	II
580		5177.31	I
850		5183.42	II
260		5188.22	II
170		5204.15	II
720		5211.86	I
520		5234.27	I
340		5253.46	I
110		5259.39	II
370		5271.19	I
140		5290.84	II
370		5301.98	II
140		5302.62	II
180		5303.55	II
110		5340.67	II
110		5357.86	I
130		5377.09	II
140		5380.99	II
500		5455.15	I
470		5501.34	I
110	b1	5602.50	L
160		5631.22	I
240		5648.25	I
130		5657.72	I
180		5740.66	I
160		5744.41	I
160		5761.84	I
160		5769.07	II
370		5769.34	I
320		5789.24	I
450		5791.34	I
220		5797.58	II
160		5805.78	II
140		5821.99	I
320		5930.62	I
720		6249.93	I
260	d	6262.30	II
180		6296.09	II
160		6320.39	II
110		6325.91	I
170		6390.48	II
450		6394.23	I
210		6410.99	I

Intensity		Wavelength	
250		6455.99	I
110		6526.99	II
130		6543.16	I
140		6578.51	I
180		6709.50	I
120		6774.26	II
13	b	7011.22	L
75		7023.67	I
26		7032.05	I
26	b	7040.84	L
110		7045.96	I
13	b	7054.80	L
160		7066.23	II
65		7068.37	I
21	b1	7070.79	L
13		7076.38	I
21	b1	7085.40	L
26	b1	7101.02	L
10	h	7116.8	II
19	b1	7131.58	L
10		7149.77	I
40	h	7158.08	I
50		7161.25	I
10		7162.60	L
21		7219.91	I
10	b	7257.16	L
26		7270.09	I
10		7270.30	I
110	cw	7282.34	II
10		7320.91	I
110	cw	7334.18	I
65		7345.34	I
50	b1	7379.71	L
85	b1	7380.08	L
35		7382.73	I
110	b1	7403.52	L
210	b1	7403.75	L
50	b	7411.34	L
65	b1	7434.28	L
110	b1	7434.36	L
30	b	7442.92	L
50	h	7463.08	I
50	b1	7465.25	L
95	b1	7465.48	L
75	cw	7483.50	II
40	b1	7496.50	L
95	b1	7496.78	L
50		7498.83	L
30	b	7506.79	L
19	b1	7528.21	L
50	b1	7528.39	L
30		7533.59	I
85		7539.23	I
35	b1	7560.09	L
35	b1	7592.26	L
19	h	7612.94	II
19	b	7624.99	L
21		7664.34	I
15	h	7841.80	I
21	b	7876.87	L
75	b1	7877.22	L
75	b1	7910.19	L
150	b1	7910.54	L
50	b	7944.61	L
110	b1	7944.95	L
40		7964.83	I
35	b	7979.34	L
75	b1	7979.70	L
35	h	8001.89	I
21	b	8014.43	L
65	b1	8014.79	L
30	b	8019.48	L
35	hc	8051.39	I
75		8086.05	I
15	b	8122.20	L
15	b	8159.02	L
7	h	8203.38	I
50		8247.44	I
13	h	8316.04	I
85		8324.69	I
95		8346.53	I
8	h	8379.80	I
8	b	8453.55	L
8	h	8467.62	I
26		8476.48	I
13	h	8507.37	I
13	h	8513.57	I
8	h	8514.65	II
17	b	8526.59	L
17	c	8543.46	I
65		8545.44	I
15	b	8563.54	L
9	h	8590.94	I
9	b	8600.81	L
7	h	8624.22	I
15		8638.47	I
19	hw	8672.11	I
40		8674.43	I
13	h	8720.41	I
35		8748.38	I
19		8818.93	I
35		8825.82	I
21		8839.63	I

La III
Ref. 220, 309 — J.R.

Intensity	Wavelength	
Vacuum		
3	744.19	III
10	753.03	III
1	786.64	III
200	787.14	III
1	796.03	III
400	796.99	III
1	797.20	III
10	835.03	III
30	845.62	III
1	850.73	III
1	860.39	III
2	860.88	III
5	865.04	III
2000	870.40	III
30	872.43	III
1000	882.34	III
20	882.72	III
200	929.71	III
400	942.86	III
30	967.69	III
10	974.33	III
50	979.99	III
10	980.29	III
200	1058.63	III
1000	1072.59	III
5000	1076.91	III
50000	1081.61	III
95000	1099.73	III
5000	1100.70	III
30	1208.80	III
30	1212.29	III
200	1236.54	III
100	1253.99	III
2000	1255.63	III
100	1259.55	III
100	1322.42	III
5000	1330.04	III
10000	1349.18	III
5000	1459.49	III
2000	1466.44	III
10000	1523.79	III
500	1528.55	III
5000	1536.17	III
200	1923.34	III
500	1938.57	III
Air		
60	2216.07	III
20	2238.36	III
25	2258.61	III
5	2260.30	III
250	2297.74	III
400	2379.37	III
10	2387.99	III
20	2392.49	III
100	2476.60	III
50	2478.65	III
2	2513.43	III
4	2588.87	III
2	2604.83	III
400	2651.50	III
100	2682.34	III
150	2684.76	III
110	2897.88	III
160	2904.58	III
7	2950.84	III
10	2953.77	III
40	2992.10	III
4	3006.19	III
15	3009.22	III
100	3075.17	III
4	3085.38	III
25	3093.03	III
15	3096.26	III
200	3111.97	III
50	3116.74	III
1000	3171.63	III
1500	3171.71	III
50	3172.69	III
20	3196.84	III
70	3289.11	III
15	3301.48	III
35	3327.66	III
500	3517.09	III
600	3517.72	III
3	4129.24	III
5	4137.43	III
200	4482.97	III
300	4499.05	III
5	5145.73	III
8	5158.41	III
6	5467.81	III
55	5491.90	III
2	5511.72	III
1	5518.19	III
45	5529.54	III
1	5744.09	III
200	5778.14	III
2	5813.45	III
3	5875.63	III
55	5888.62	III
3	5932.71	III
2	6017.11	III
20	6055.84	III
35	6119.25	III
120	6141.99	III
55	6220.00	III
60	6348.21	III
3	8114.42	III
2	8135.96	III
250	8252.60	III
100	8275.39	III
200	8287.75	III
250	8321.11	III
300	8583.45	III
120	9184.38	III
100	9212.63	III
80	9923.99	III
140	10284.79	III
20	10370.34	III
12	10937.90	III
	13894.47	III
	14096.18	III
	17898.09	III

La IV
Ref. 79 — J.R.

Intensity		Wavelength	
Vacuum			
100		344.12	IV
7000		453.50	IV
10000		463.14	IV
15000		499.54	IV
40000		552.02	IV
30000		631.26	IV
25		724.92	IV
15		733.29	IV
10		797.03	IV
10	c	980.03	IV
50		1039.30	IV
60	p	1062.09	IV
75		1158.35	IV
50		1164.29	IV
400		1230.90	IV
75	p	1260.79	IV
300		1261.12	IV
150		1283.19	IV
2000		1302.31	IV
1200		1333.53	IV
3500		1334.96	IV
750		1352.76	IV
25000		1368.04	IV
8000		1377.49	IV
3000		1394.32	IV
5000		1414.58	IV
7000		1432.55	IV
7000		1441.63	IV
7500		1462.15	IV
20000		1463.47	IV
7500		1467.54	IV
15000		1507.87	IV
5000		1527.19	IV
2500		1575.92	IV
1500		1583.61	IV
750		1585.11	IV
750		1637.42	IV
750	d	1645.21	IV
1000		1664.84	IV
750		1684.17	IV
2000	p	1767.65	IV
4000		1808.66	IV
1000		1851.81	IV
1500		1852.77	IV
750		1879.79	IV
1000		1881.57	IV
800		1889.22	IV
1000		1891.47	IV
5000		1902.97	IV
800		1907.44	IV
1500	c	1950.80	IV
1200	c	1957.57	IV
Air			
3000		2012.42	IV
750		2037.43	IV
2000	c	2066.50	IV
3000	c	2073.18	IV
1500		2143.23	IV
4000	c	2197.45	IV
1000	w	2221.12	IV
900		2227.34	IV
3000		2244.95	IV
7500	w	2265.91	IV
2000		2315.89	IV
750		2348.36	IV
750		2355.31	IV
2000	c	2407.10	IV

Lanthanum (Cont.)

Intensity		Wavelength	
25000	w	2417.58	IV
1200	p	2443.92	IV
18000	c	2502.81	IV
15000		2515.02	IV
50000		2532.75	IV
900	d	2535.76	IV
45000		2582.05	IV
18000	c	2591.30	IV
95000	w	2597.50	IV
5000	c	2608.01	IV
70000	w	2662.75	IV
50000	w	2848.30	IV
12000	c	2863.30	IV
30000	c	2962.58	IV
70000	w	3009.51	IV
90000	c	3056.68	IV
3500		3522.28	IV
2000		3650.40	IV
2000	p	4270.76	IV
1500	w	4549.80	IV
500		4836.89	IV

La V
Ref. 78 — J.R.

Intensity	Wavelength	
	Vacuum	
2	389.03	V
400	390.72	V
1	398.53	V
30	399.34	V
350	405.10	V
50	416.13	V
3	421.55	V
50	423.07	V
400	424.78	V
1000	432.11	V
2500	435.28	V
700	436.14	V
700	436.84	V
300	437.11	V
700	437.55	V
20	444.01	V
10	444.07	V
1250	450.40	V
600	457.30	V
1000	463.85	V
150	476.67	V
5000	482.16	V
200	482.43	V
2000	483.30	V
7000	498.08	V
4000	499.03	V
10000	503.58	V
40	508.15	V
1500	525.71	V
12000	526.76	V
10000	531.07	V
15000	533.23	V
4000	540.20	V
6000	544.80	V
8000	547.44	V
3000	570.90	V
2500	593.18	V
750	597.70	V
2000	600.01	V
5000	600.24	V
700	611.70	V
500	617.60	V

LEAD (Pb)
Z = 82

Pb I and II
Ref. 64, 274, 283, 329, 330 — D.R.W.

Intensity		Wavelength	
		Vacuum	
2		846.04	II
2	h	849.88	II
3		855.57	II
3		863.00	II
6		873.71	II
2		877.96	II
8		889.68	II
3		896.30	II
5		926.44	II
2		958.76	II
2		960.21	II
3		965.36	II
10		967.23	II
9		972.56	II
8		982.17	II
10		986.71	II
10		995.89	II
6		1001.81	II
10		1016.61	II
10		1049.82	II
10		1050.77	II
10		1060.66	II
9		1065.58	II
10		1103.94	II
10		1108.43	II
10		1109.84	II
10		1119.57	II
10		1121.36	II
10		1133.14	II
4		1145.91	II
10		1203.63	II
10		1231.20	II
10		1331.65	II
10		1335.20	II
10		1348.37	II
10		1433.96	II
3		1449.35	II
10		1512.42	II
10		1671.53	II
10		1682.15	II
20		1726.75	II
2		1740.00	I
2		1766.64	I
2		1794.67	I
10		1796.670	II
5		1812.97	I
10		1822.050	II
4		1868.76	I
10		1904.77	I
7		1921.471	II
4		1972.44	I
2		1977.88	I
2		1991.60	I
2		1992.31	I
		Air	
5	r	2022.02	I
5		2050.88	I
8	r	2053.28	I
6		2111.758	I
10		2115.066	I
500	r	2170.00	I
7		2175.580	I
7		2187.888	I
8		2189.603	I
10		2203.534	II
20		2237.425	I
20		2246.86	I
25		2246.89	I
150		2332.418	I
180		2388.797	I
550	r	2393.792	I
140		2399.597	I
320	r	2401.940	I
320	r	2411.734	I
150	r	2443.829	I
160	r	2446.181	I
130	r	2476.378	I
8	c	2526.69	II
8	c	2576.60	II
80	r	2577.260	I
2	c	2608.38	II
500	r	2613.655	I
900	r	2614.175	I
160		2628.262	I
4		2634.256	II
10		2657.094	I
700		2663.154	I
10		2697.541	I
25000	r	2801.995	I
100		2822.58	I
14000	r	2823.189	I
35000	r	2833.053	I
6		2840.557	II
14000	r	2873.311	I
3	c	2887.30	II
3		2914.442	II
2	c	2947.43	II
3	c	2948.53	II
15		2966.460	I
15		2972.991	I
15		2980.157	I
4		2986.876	II
10	c	3016.39	II
150		3118.894	I
600		3220.528	I
100		3229.613	I
400		3240.186	I
200		3262.355	I
35000		3572.729	I
50000	r	3639.568	I
20000		3671.491	I
70000	r	3683.462	I
10		3713.982	II
25000		3739.935	I
15000		4019.632	I
95000		4057.807	I
14000		4062.136	I
5		4110.76	II
4		4113.35	II
10		4152.82	II
10		4157.814	I
10000		4168.033	I
9	c	4242.14	II
20	c	4244.92	II
7		4293.82	II
6		4296.65	II
200		4340.413	I
10		4352.74	II
20	c	4386.46	II
10		4579.051	II
10		4582.27	II
1000		5005.416	I
100		5006.572	I
50		5042.58	II
10		5070.58	II
10		5074.53	II
10		5076.35	I
50		5089.484	I
20		5090.01	I
10		5107.242	I
10		5111.64	II
2000		5201.437	I
10		5367.64	II
10		5372.099	I
10	c	5544.25	II
20	c	5608.85	II
40		5692.346	I
200		5895.624	I
2000		6001.862	I
9	c	6009.58	II
500		6011.667	I
8	c	6041.17	II
500		6059.356	I
40		6075.74	II
40		6081.409	II
50		6110.520	I
10		6159.89	II
100		6235.266	I
50	c	6660.20	II
10		6892.11	I
5		7128.94	I
20		7193.60	II
20000		7228.965	I
5		7304.68	I
8		7330.15	I
10		7346.676	I
10		7558.97	II
10		7632.56	II
4		7732.96	II
20		7809.259	I
5		7817.97	I
6		7829.01	I
5		7896.737	I
2	d	8156.91	II
10		8168.001	I
6		8191.886	I
5		8217.711	I
8		8255.61	I
40		8272.690	I
6		8333.34	II
10		8395.68	II
20		8409.384	I
10		8478.492	I
8		8532.17	I
7	c	8544.55	II
7		8709.90	II
5		8719.39	II
5		8722.810	I
10		8857.457	I
10		9050.82	II
10		9063.43	II
2	d	9245.28	II
8		9293.476	I
5		9384.35	I
5		9385.89	I
15		9438.05	I
15		9604.297	I
6		9608.73	I
15		9674.351	I
200		10290.458	I
5		10434.32	I
100		10498.965	I
50		10649.249	I
5		10759.41	I
7		10759.74	I
15		10886.688	I
40		10969.53	I
6		11059.22	I
3		11333.08	I
2	d	11479.49	II
2		11488.76	I
5		11627.91	II
1		12561.37	I
		13495.3	I
		13498.2	I
		13512.6	I
		14722.8	I
		14742.1	I
		14743.0	I
		15314.8	I
		15327.6	I
		15331.0	I
		15349.6	I
		38831.1	I
		38950.1	I
		38958.6	I
		39039.4	I

Pb III
Ref. 54, 256, 297 — D.R.W.

Intensity	Wavelength	
	Vacuum	
1	961.01	III
3	1030.5	III
12	1048.9	III
4	1069.2	III
3	1074.7	III
4	1118.67	III
4	1167.0	III
4	1250.6	III
1	1266.9	III
20	1553.1	III
1	1610.1	III
4	1711.23	III
	Air	
10	3043.85	III
4	3089.08	III
4	3102.74	III
10	3137.81	III
10	3176.50	III
5	3242.84	III
1	3530.17	III
7	3589.87	III
7	3689.31	III
3	3706.02	III
5	3728.69	III
12	3854.08	III
8	3951.92	III
3	4031.16	III
3	4094.54	III
2	4128.11	III
8	4272.66	III
6	4499.34	III
7	4571.21	III
1	4596.45	III
6	4761.12	III
4	4798.50	III
1	4826.86	III
2	4855.06	III
3	5065.12	III
4	5191.56	III
5	5523.97	III
3	5779.41	III
6	5857.96	III

Pb IV
Ref. 106 — D.R.W.

Intensity	Wavelength	
	Vacuum	
8	475.36	IV
7	478.35	IV
10	496.38	IV
12	499.94	IV
9	515.07	IV
14	529.78	IV
20	570.16	IV
8	573.90	IV
8	584.52	IV
10	648.50	IV
9	656.10	IV
10	761.09	IV
18	802.07	IV
12	802.82	IV
10	812.59	IV
8	822.07	IV
10	827.41	IV
12	832.60	IV
8	840.99	IV
8	842.88	IV
12	845.94	IV
18	857.64	IV
8	859.02	IV
16	862.33	IV
14	870.44	IV
12	879.96	IV
7	880.35	IV
14	884.96	IV
14	884.99	IV
16	890.72	IV
12	908.51	IV
12	917.90	IV
10	922.12	IV
12	922.49	IV
7	924.52	IV
10	927.64	IV
14	932.20	IV
8	937.00	IV
7	952.85	IV

Lead (Cont.)

Intensity	Wavelength	
8	1012.44	IV
14	1028.61	IV
20	1032.05	IV
16	1041.24	IV
18	1044.14	IV
15	1056.53	IV
12	1072.09	IV
7	1079.88	IV
18	1080.81	IV
20	1084.17	IV
6	1089.94	IV
7	1099.47	IV
6	1115.30	IV
20	1116.08	IV
18	1137.84	IV
8	1142.77	IV
14	1144.93	IV
20	1189.95	IV
8	1267.55	IV
8	1290.82	IV
10	1291.10	IV
20	1313.05	IV
8	1323.92	IV
12	1343.06	IV
16	1388.94	IV
6	1397.02	IV
18	1400.26	IV
10	1404.34	IV
7	1510.76	IV
14	1535.71	IV
8	1798.39	IV
8	1893.19	IV
12	1959.34	IV
16	1973.16	IV

Air

Intensity	Wavelength	
10	2042.58	IV
12	2049.34	IV
12	2079.22	IV
8	2151.96	IV
15	2154.01	IV
12	2177.46	IV
16	2359.53	IV
4	2864.24	IV
4	2864.50	IV
16	2417.61	IV
4	2978.14	IV
4	3052.56	IV
4	3221.17	IV
4	3962.48	IV
4	4049.80	IV
10	4496.15	IV
16	4534.60	IV
8	4605.40	IV
2	5914.54	IV

Pb V
Ref. 106 — D.R.W.

Intensity	Wavelength	
	Vacuum	
2	367.40	V
2	372.53	V
2	387.87	V
2	394.38	V
5	424.64	V
3	431.03	V
3	436.60	V
2	438.47	V
6	438.91	V
4	453.45	V
3	461.70	V
3	496.20	V
4	694.42	V
8	696.20	V
20	703.73	V
4	706.29	V
6	707.66	V
5	730.85	V
12	749.46	V
10	752.52	V
10	755.80	V
6	762.76	V
10	765.87	V
18	767.45	V
18	769.49	V
14	771.42	V
14	782.79	V
10	787.05	V
15	797.02	V
5	799.80	V
18	809.63	V
5	812.32	V
8	814.10	V
8	820.09	V
5	825.52	V
8	829.32	V
5	851.98	V
20	863.97	V
10	867.10	V
6	880.50	V
18	883.90	V
14	888.37	V
14	894.40	V
12	896.08	V
8	915.09	V
14	915.71	V
12	918.09	V
12	920.28	V
12	920.66	V
6	940.74	V
8	946.20	V
6	950.93	V
12	954.35	V
4	954.95	V
10	955.28	V
4	964.38	V
6	989.14	V
8	1005.42	V
10	1051.26	V
4	1059.26	V
10	1088.86	V
9	1096.52	V
6	1104.79	V
4	1121.33	V
10	1137.50	V
4	1152.36	V
12	1157.88	V
14	1185.43	V
11	1233.50	V
10	1248.47	V
8	1635.75	V
2	1802.87	V
2	1843.00	V
2	1888.67	V
2	1897.02	V
2	1914.33	V
4	1919.74	V
5	1957.96	V
2	1998.58	V
10	1998.83	V

Air

Intensity	Wavelength	
8	2078.45	V
10	2142.55	V
10	2167.97	V
20	2259.01	V
10	2276.66	V
8	2301.49	V
15	2424.81	V
4	4809.36	V
5	6650.99	V
4	6753.20	V

LITHIUM (Li)
Z = 3

Li I and II
Ref. 3, 15, 17, 18, 37, 44, 112, 284, 321, 335 — J.D.S.

Intensity	Wavelength	
	Vacuum	
	125.5	II
	136.5	II
	140.5	II
	167.21	II
	168.74	II
	171.58	II
	178.02	II
	199.28	II
	207.5	II
	456.	II
	483.	II
	540.	II
	729.	II
	800.	II
	820.	II
	861.	II
	905.5	II
	917.5	II
	936.	II
	945.	II
	965.	II
	972.	II
	988.	II
	1018.	II
	1032.	II
	1036.	II
	1093.	II
	1103.	II
	1109.	II
	1116.	II
	1132.1	II
	1141.	II
	1166.4	II
	1198.09	II
	1215.	II
	1238.	II
	1253.8	II
	1420.89	II
	1424.	II
3	1492.93	II
5	1492.97	II
1	1493.04	II
	1555.	II
3	1653.08	II
5	1653.13	II
1	1653.21	II
	1681.66	II
	1755.33	II

Air

Intensity	Wavelength	
	2009.	II
	2039.	I
	2068.	II
	2131.	II
	2164.	II
	2173.4	I
	2183.	II
	2214.	II
	2222.	II
	2237.	II
H	2249.21	II
	2286.09	II
	2302.57	II
	2303.33	II
	2304.59	I
	2304.92	I
	2305.36	I
	2305.83	I
	2306.29	I
	2306.82	I
	2307.44	I
	2308.97	I
	2309.88	I
	2310.94	I
	2312.11	I
	2313.49	I
	2315.08	I
	2316.95	I
	2319.18	I
	2321.88	I
	2325.11	I
4	2329.02	I
4	2329.84	II
1	2333.94	I
	2336.88	II
	2336.91	II
3	2337.00	II
5	2340.15	I
2	2348.22	I
	2358.93	I
	2373.54	II
	2381.54	II
	2383.20	II
1	2394.39	I
	2402.33	II
3	2410.84	II
	2425.43	I
	2429.81	II
10	2460.2	I
	2475.06	II
	2506.94	II
	2508.78	II
	2518.	I
	2539.49	II
24	2551.7	II
	2559	II
15	2562.31	I
	2605.08	II
	2640.	I
2	2657.29	II
3	2657.30	II
	2674.46	II
0	2728.24	II
5	2728.29	II
2	2728.32	II
3	2730.47	II
1	2730.55	II
5	2741.20	I
	2766.99	II
	2790.31	II
	2801.	I
	2846.	I
	2868.	I
	2895.	I
2	2934.02	II
2	2934.07	II
5	2934.12	II
1	2934.25	II
	2968.	I
3	3029.12	II
3	3029.14	II
	3144.	I
3	3155.31	II
4	3155.33	II
1	3196.26	II
9	3196.33	II
4	3196.36	II
5	3199.33	II
2	3199.43	II
17	3232.66	I
	3249.87	II
	3306.28	II
	3393.	
	3488.	I
	3579.8	I
	3618.	I
	3662.	I
	3684.32	II
1	3714.00	II
5	3714.16	II
6 d	3714.27	II
8	3714.29	II
7 d	3714.40	II
10	3714.41	II
1	3714.51	II
0	3714.58	II
3	3718.7	I
6	3794.72	I
20	3915.30	I
20	3915.35	I
10	3985.48	I
10	3985.54	I
40	4132.56	I
40	4132.62	I
	4196.	I
20	4273.07	I
20	4273.13	I
5	4325.42	II
5	4325.47	II
1	4325.54	II
	4516.45	II
	4590.	
13	4602.83	I
13	4602.89	I
	4607.34	II
0	4671.51	II
6	4671.65	II
2	4671.70	II
3	4678.06	II
1	4678.29	II
	4760.	I
	4763.	II
	4788.36	II
	4843.0	II
4	4881.32	II
4	4881.39	II
1	4881.49	II
8	4971.66	I
8	4971.75	I
	5037.92	II
	5095.	
	5114.	
	5190.	
	5271.	I
	5315.	I
	5395.	I
	5440.	I
600 c	5483.55	II
600 c	5485.65	II
320	6103.54	I
320	6103.65	I
3600	6707.76	I
3600	6707.91	I
48	8126.23	I
48	8126.45	I
	8517.37	II
	9581.42	II
	10120.	II
	12232.	I
	12782.	I
	13566.	I
	17552.	I
	18697.	I
	19290.	I
	24467.	I
	40475.	I

Li III
Ref. 335 — J.O.S.

Intensity	Wavelength	
	Vacuum	
	102.9	III
	103.4	III
	104.1	III
	105.5	III
	108.0	III
	113.9	III
	135.0	III
	540.0	III
	729.1	III

LUTETIUM (Lu)
Z = 71

Lu I and II
Ref. 1 — C.H.C.

Intensity	Wavelength
	Air

Intensity		Wavelength		
1700	h	2195.54	II	
95		2276.94	II	
190		2297.41	II	
1300		2392.19	II	
120		2399.14	II	
80		2419.21	II	
55		2430.26	II	
130		2459.64	II	
80		2469.27	II	
21	h	2481.72	II	
370		2536.95	II	
40		2546.87	II	
20		2549.44	I	
20		2549.72	I	
35		2561.80	II	
930		2571.23	II	
1700		2578.79	II	
80	h	2582.13	II	
1800		2613.40	II	
18000		2615.42	II	
1800		2619.26	II	
90		2657.05		
2700		2657.80	II	
90	h	2677.25	I	
570	h	2685.08	I	
90	h	2685.54	I	
4200		2701.71	II	
90	h	2715.91	I	
180	d	2719.09	I	
480	h	2728.95	I	
75	c	2738.17	II	
3600		2754.17	II	
750	h	2765.74	I	
2700		2796.63	II	
35		2821.23	II	
270	c	2834.35	II	
330		2845.13	I	
3000		2847.51	II	
570	h	2885.14	I	
6300		2894.84	II	
4500		2900.30	II	
300		2903.05	I	
9000		2911.39	II	
270	h	2949.73	I	
1200		2951.69	II	
60		2955.78	II	
4200		2963.32	II	
2400		2969.82	II	
1800		2989.27	I	
3000		3020.54	II	
120		3027.29	II	
2100		3056.72	II	
7500		3077.60	II	
390		3080.11	I	
5100	h	3081.47	I	
3000		3118.43	I	
2400		3171.36	I	
100		3183.73	II	
260		3191.80	II	
1400		3198.12	II	
4800		3254.31	II	
3800		3278.97	I	
7600		3281.74	I	
6200		3312.11	I	
7600		3359.56	I	
6200		3376.50	I	
950		3385.50	I	
160	h	3391.55	I	
1400		3396.82	I	
4100		3397.07	II	
4800		3472.48	II	
8300	c	3507.39	II	
1600		3508.42	I	
4800		3554.43	II	
4800		3567.84	I	
340		3596.34	I	
800		3623.99	II	
680		3636.25	I	
2600		3647.77	I	
60		3684.32	I	
60		3710.95	I	
110		3756.70	I	
110		3756.79	I	
30	h	3786.18	I	
150		3800.67	I	
75	h	3829.07	I	
2700		3841.18	I	
75		3843.61	I	
95		3853.29	I	
40		3874.61	I	
530		3876.65	II	
29		3911.77	I	
50		3918.86	I	
35	h	3926.62	I	
480		3968.46	I	
50		3991.38	I	
670		4054.45	I	
75	B1	4096.13	L	
35	h	4107.44	I	
95	h	4112.67	I	
310		4122.49	I	
3100		4124.73	I	
150	c	4131.79	I	

Intensity		Wavelength		
460		4154.08	I	
24		4158.98	I	
1600		4184.25	II	
150		4277.50	I	
250		4281.03	I	
330	d	4295.97	I	
		4296.09	I	
150		4309.57	I	
75		4332.72	I	
29		4341.98	II	
65	h	4420.96	I	
190	c	4430.48	I	
35		4438.79	I	
190		4450.81	I	
50	h	4471.55	I	
60	h	4498.85	I	
3300		4518.57	I	
24	b	4560.95	L	
24	b	4575.31	L	
85	c	4605.39	I	
95	h	4645.47	I	
100	h	4648.21	I	
95	h	4648.85	I	
65	b	4654.03	L	
1000		4658.02	I	
85	h	4659.03	I	
630	B1	4661.75	L	
310	B1	4672.31	L	
420	B1	4684.16	L	
270	B1	4695.46	L	
190	B1	4708.00	L	
30		4716.70	I	
65	b	4720.86	L	
65	h	4726.20	I	
100	B1	4735.00	L	
75	B1	4749.11	L	
40	B1	4764.22	L	
150		4785.42	II	
85		4815.05	I	
50	c	4839.62	II	
18		4865.36	II	
460		4904.88	I	
180		4942.34	I	
800		4994.13	II	
800		5001.14	I	
55	h	5057.60	I	
140		5134.05	I	
2700		5135.09	I	
130	B1	5170.11	L	
170		5196.61	I	
90		5206.47	I	
40		5304.40	I	
80		5349.12	I	
500		5402.57	I	
140	c	5421.90	I	
100		5437.88	I	
35		5453.57	I	
2100		5476.69	II	
9		5664.89	II	
14		5713.49	II	
550		5736.55	I	
55		5775.40	I	
80		5800.59	I	
40	h	5860.79	I	
9		5866.30	I	
690	Cw	5983.9	II	
140		5997.13	I	
1400		6004.52	I	
35	h	6041.66	I	
440		6055.03	I	
11		6140.41	I	
150		6159.94	II	
160		6199.66	II	
2100		6221.87	II	
35		6228.14	II	
80		6235.36	II	
160		6242.34	II	
16	h	6248.80	I	
70	h	6345.35	I	
18	h	6354.85	I	
9		6365.79	I	
16		6366.00	I	
22		6441.14	I	
11		6444.89	II	
1100		6463.12	II	
29		6477.67	I	
55	c	6523.18	I	
35	Cw	6611.28	II	
		6611.58	LU	II
		6611.80	LU	II
		6611.95	LU	II
		6612.04	LU	II
11		6619.15	II	
23	c	6677.14	I	
9	h	6735.76	I	
30	c	6793.77	I	
11		6826.59	II	
45		6917.31	I	
8		6943.96	II	
23		7031.24	I	
14	c	7096.34	I	
45		7125.84	II	
9		7142.79	I	

Intensity		Wavelength		
7		7143.10	I	
8		7165.94	II	
14	Ch	7237.98	I	
5		7409.70	II	
11	c	7441.52	I	
8		7456.96	II	
7	Ch	7640.08	I	
7	Cw	7758.30	I	
7	h	7815.9	I	
9	c	8178.16	I	
17		8382.08	I	
35		8459.19	II	
10	d	8478.50	I	
29	c	8508.08	I	
35	c	8610.98	I	

Lu III
Ref. 148 — J.R.

Intensity	Wavelength	
	Vacuum	
1	677.34	III
7	691.05	III
30	700.25	III
50	714.89	III
100	738.76	III
200	755.03	III
3	755.16	III
500	810.73	III
100	830.53	III
2000	832.28	III
10	972.66	III
2	991.26	III
100	996.44	III
400	1001.18	III
1	1022.40	III
100	1029.83	III
200	1030.33	III
100	1031.54	III
50	1056.53	III
20	1061.99	III
3	1092.84	III
200	1187.34	III
50	1228.7	III
200	1277.53	III
30	1283.41	III
100	1331.93	III
1000	1854.57	III
	Air	
40	2050.72	III
1500	2065.35	III
1500 c	2070.56	III
100	2083.34	III
200	2099.44	III
1000	2236.14	III
2000	2236.22	III
500 c	2381.59	III
300	2563.51	III
4500 c	2603.35	III
200	2721.65	III
2000	2772.55	III
20	2781.16	III
10	2788.37	III
500	2800.90	III
20 p	2993.21	III
1000	3057.86	III
200	4251.44	III
300	4271.91	III
200	4490.00	III
400	4956.43	III
10	5046.12	III
150 c	5145.86	III
70	5419.42	III
60	5519.88	III
5	5526.80	III
5	5748.71	III
70	5786.46	III
80	5869.71	III
60	5889.76	III
300	6197.96	III
600	6198.13	III
100	7309.95	III
200	7310.25	III
50	7534.27	III
70 1	7936.45	III
3	8008.59	III

Lu IV
Ref. 310 — J.R.

Intensity	Wavelength	
	Vacuum	
400	876.80	IV
100	902.06	IV
300	1015.18	IV
20	1136.17	IV

Intensity		Wavelength	
50		1189.27	IV
20	p	1194.59	IV
60		1213.08	IV
15		1220.74	IV
20		1223.75	IV
20		1240.07	IV
20		1248.10	IV
40		1266.27	IV
100		1272.42	IV
20		1273.02	IV
40		1274.77	IV
20		1276.54	IV
50		1289.38	IV
60		1310.08	IV
50		1323.02	IV
15		1331.04	IV
800		1333.79	IV
300		1334.94	IV
50		1338.20	IV
100		1339.49	IV
300		1342.58	IV
200		1351.68	IV
300		1353.74	IV
200		1355.85	IV
20		1359.67	IV
100		1363.24	IV
50		1363.37	IV
20		1367.34	IV
20		1373.54	IV
15		1375.36	IV
100		1376.02	IV
30		1379.56	IV
15		1383.18	IV
15		1389.85	IV
60		1390.07	IV
200		1390.30	IV
50		1390.69	IV
40		1392.38	IV
40		1397.18	IV
30		1401.32	IV
100		1401.46	IV
200		1406.64	IV
100		1407.00	IV
250		1407.04	IV
20		1420.32	IV
100		1421.59	IV
200		1429.08	IV
400		1429.38	IV
40		1430.80	IV
100		1440.62	IV
100		1448.14	IV
200		1452.33	IV
30		1462.65	IV
100		1483.79	IV
200		1493.24	IV
400		1511.26	IV
200		1521.06	IV
100		1522.21	IV
100		1537.77	IV
30		1549.35	IV
20		1551.59	IV
20		1562.06	IV
30		1592.55	IV
15		1594.92	IV
100	c	1607.72	IV
50	c	1631.65	IV
20		1684.50	IV
60	c	1693.67	IV
400	c	1721.42	IV
100		1725.14	IV
100		1735.79	IV
100		1736.78	IV
100		1741.74	IV
50	c	1743.84	IV
40	c	1752.60	IV
200		1759.61	IV
300		1772.08	IV
600		1772.57	IV
200		1782.45	IV
100		1797.52	IV
20	c	1901.63	IV
100		1983.92	IV
20		1990.52	IV
40		1996.18	IV
		Air	
100		2003.18	IV
100		2020.94	IV
20		2071.10	IV
100		2081.09	IV
400	c	2085.70	IV
600	c	2086.47	IV
400	c	2092.16	IV
100		2103.63	IV
1000	c	2104.41	IV
200		2107.85	IV
1000	c	2108.31	IV
100	c	2127.43	IV

Lu V
Ref. 401 — J.R.

Lutetium (Cont.)

Intensity		Wavelength	
		Vacuum	
40		555.44	V
100		563.72	V
50		601.54	V
60		614.23	V
40		628.79	V
50	p	637.44	V
40		637.53	V
50		663.29	V
60		850.06	V
100		861.92	V
70		866.93	V
40		870.84	V
50		875.89	V
50		876.45	V
100		880.32	V
40		884.21	V
70		886.16	V
50		886.32	V
50		886.44	V
100		891.81	V
60		895.01	V
40		895.15	V
40		898.42	V
100		914.72	V
40		918.26	V
50		920.92	V
40		921.32	V
60		921.90	V
40		922.73	V
80		925.79	V
50		927.22	V
40		947.80	V
50	w	1420.02	V
50		1432.50	V
100		1432.77	V
200		1441.76	V
40		1443.64	V
100		1448.14	V
40		1449.32	V
100		1450.36	V
100		1450.69	V
100		1452.64	V
200		1453.35	V
100		1454.38	V
100		1455.21	V
100		1460.11	V
40	c	1467.81	V
200		1468.99	V
50		1469.45	V
100		1471.20	V
400		1472.12	V
200		1473.71	V
40		1475.77	V
200		1485.58	V
50		1709.02	V
40		1728.90	V
50	c	1775.92	V
40	c	1777.68	V
40	c	1784.71	V
100	c	1786.25	V
60	c	1787.58	V
60	c	1793.85	V
60		1809.73	V
60		1814.24	V

MAGNESIUM (Mg)
Z = 12

Mg I and II
Ref. 49, 83, 103, 217, 269, 315, 335 — J.O.S.

Vacuum

Intensity	Wavelength	
	184.05	II
	184.31	II
	184.68	II
	184.81	II
	185.26	II
	185.59	II
	185.98	II
	186.47	II
	186.84	II
	187.19	II
	187.38	II
	188.54	II
	188.91	II
	189.01	II
	189.23	II
	189.37	II
	191.30	II
	191.56	II
	191.65	II
	192.40	II
	192.55	II
	192.84	II
	193.09	II
	193.31	II
	193.40	II
	193.64	II
	197.76	II
	199.31	II
	200.29	I
	202.00	II
	202.27	II
	202.51	II
	202.94	II
	203.15	I
	203.42	II
	203.53	II
	204.22	I
	209.09	II
	209.43	II
	209.84	I
	213.53	I
	215.12	I
	215.31	I
	215.45	I
	215.66	I
	215.79	I
	216.22	I
	216.36	I
	216.68	I
	217.21	I
	217.37	I
	218.19	I
	218.34	I
	218.42	I
	218.74	I
	219.04	I
	219.28	I
	220.03	I
	220.33	I
	222.03	I
	222.67	I
	223.45	I
	223.74	I
	225.18	I
	225.54	I
	226.26	I
	247.14	II
	248.47	II
	884.70	II
	884.72	II
	907.38	II
	907.41	II
8	946.70	II
9	946.77	II
14	1025.96	II
12	1026.11	II
25	1239.94	II
20	1240.40	II
6	1248.51	II
8	1249.93	II
8	1271.24	II
9	1271.94	II
8	1272.72	II
11	1273.43	II
11	1306.71	II
12	1307.88	II
12	1308.28	II
14	1309.44	II
14	1365.45	II
	1365.54	II
15	1367.26	II
15	1367.70	II
18	1369.42	II
20	1476.00	II
25	1478.01	II
20	1480.89	II
30	1482.90	II
	1625.22	I
	1625.50	I
	1625.81	I
	1626.16	I
	1626.36	I
	1626.56	I
	1626.79	I
	1627.02	I
	1627.27	I
	1627.53	I
	1627.82	I
	1628.12	I
	1628.46	I
	1628.80	I
	1629.21	I
	1629.59	I
	1630.52	I
	1631.62	I
	1632.93	I
	1634.52	I
	1636.48	I
	1638.90	I
	1641.97	I
	1645.93	I
1	1651.16	I
2	1658.31	I
5	1668.43	I
10	1683.41	I
15	1707.06	I
40	1734.84	II
50	1737.62	II
20	1747.80	I
40	1750.65	II
50	1753.46	II
30	1827.93	I

Air

Intensity	Wavelength	
9	2025.82	I
3	2329.58	II
6	2449.57	II
1	2557.23	I
1	2560.94	I
1	2562.26	I
1	2564.94	I
1	2570.91	I
1	2572.25	I
10	2574.94	I
3	2577.89	I
2	2580.59	I
4	2584.22	I
15	2585.56	I
20	2588.28	I
6	2591.89	I
6	2593.23	I
9	2595.97	I
10	2602.50	I
8	2603.85	I
9	2606.62	I
5	2613.36	I
4	2614.73	I
14	2617.51	I
13	2628.66	I
6	2630.05	I
28	2632.87	I
3	2644.80	I
7	2646.21	I
10	2649.06	I
6	2660.76	II
5	2660.82	II
7	2668.12	I
	2669.55	I
	2672.46	I
3	2693.72	I
5	2695.18	I
6	2698.14	I
8	2731.99	I
10	2733.49	I
12	2736.53	I
5	2765.22	I
7	2768.34	I
38	2776.69	I
32	2778.27	I
90	2779.83	I
8	2781.29	I
32	2781.42	I
36	2782.97	I
13	2790.79	II
1000	2795.53	II
16	2798.06	II
600	2802.70	II
3	2809.76	I
2	2811.11	I
1	2811.78	I
12	2846.72	I
14	2848.34	I
16	2851.66	I
6000	2852.13	I
2	2902.92	I
4	2906.36	I
3	2915.45	I
2	2928.75	II
3	2936.54	II
10	2936.74	I
12	2938.47	I
13	2942.00	I
1	2967.87	II
1	2971.70	II
20	3091.08	I
22	3092.99	I
14	3096.90	I
9	3104.71	II
8	3104.81	II
6	3168.98	II
6	3172.71	II
7	3175.78	II
2	3197.62	I
17	3329.93	I
6	3332.15	I
9	3336.68	I
7	3535.04	II
8	3538.86	II
7	3549.52	II
8	3553.37	II
140	3829.30	I
300	3832.30	I
500	3838.29	I
8	3848.24	II
1	3848.91	I
7	3850.40	II
2	3853.96	I
1	3854.96	I
2	3858.86	I
3	3878.31	I
2	3891.91	I
2	3893.30	I
3	3895.57	I
4	3903.86	I
6	3938.40	I
1	3984.21	I
8	3986.75	I
2	4054.69	I
10	4057.50	I
3	4075.06	I
2	4081.83	I
4	4165.10	I
15	4167.27	I
20	4351.91	I
6	4354.53	I
6	4380.38	I
9	4384.64	II
10	4390.59	II
8	4428.00	II
9	4433.99	II
5	4436.49	II
4	4436.60	II
14	4481.16	II
13	4481.33	II
6	4534.29	II
28	4571.10	I
3	4621.30	I
7	4702.99	I
10	4730.03	I
6	4739.59	II
5	4739.71	II
7	4851.10	II
75	5167.33	I
220	5172.68	I
400	5183.61	I
8	5264.21	II
7	5264.37	II
1	5345.98	I
9	5401.54	II
2	5509.60	I
6	5528.41	I
30	5711.09	I
5	5785.31	I
4	5785.56	I
7	5916.43	II
6	5918.16	II
10	6318.72	I
9	6319.24	I
7	6319.49	I
10	6346.74	II
9	6346.96	II
11	6545.97	II
5	6620.44	II
6	6620.57	II
2	6630.83	I
7	6781.45	II
8	6787.85	II
7	6812.86	II
8	6819.27	II
4	6894.90	I
6	6965.40	I
8	7060.41	I
10	7193.17	I
10	7291.06	I
5	7387.00	I
12	7387.69	I
4	7580.76	II
20	7657.60	I
19	7659.15	I
17	7659.90	I
8	7690.16	I
15	7691.55	I
1	7722.61	I
1	7746.34	I
1	7759.30	I
5	7786.50	II
4	7790.98	II
3	7811.14	I
12	7877.05	II
2	7881.67	I
13	7896.37	II
7	7930.81	I
3	8047.73	I
5	8049.85	I
7	8054.23	I
10	8098.72	I
9	8115.22	II
8	8120.43	II
1	8154.64	I
2	8159.13	I
10	8209.84	I
20	8213.03	I
10	8213.99	II
7	8222.92	II
7	8233.19	II
11	8234.64	II

Intensity	Wavelength	
7	8303.31	I
9	8305.60	I
10	8310.26	I
15	8346.12	I
2	8466.48	I
5	8468.84	I
7	8473.69	I
10	8710.18	I
12	8712.69	I
13	8717.83	I
10	8734.99	II
17	8736.02	I
11	8745.66	II
14	8806.76	I
10	8824.32	II
11	8835.08	II
20	8923.57	I
7	8989.03	I
9	8991.69	I
10	8997.16	I
14	9218.25	II
13	9244.27	II
12	9246.50	I
30	9255.78	I
10	9327.54	II
10	9340.54	I
25	9414.96	I
17	9429.81	I
19	9432.76	I
20	9438.78	I
8	9502.45	I
7	9503.11	I
5	9503.43	I
12	9631.89	II
11	9632.43	II
15	9953.20	I
15	9983.20	I
17	9986.47	I
18	9993.21	I
14	10092.16	II
5	10391.76	II
6	10392.23	II
35	10811.08	I
11	10914.23	II
7	10915.27	II
10	10951.78	II
25	10953.32	I
27	10957.30	I
28	10965.45	I
15	11032.10	I
14	11033.66	I
5	11255.93	II
4	11256.35	II
45	11828.18	I
30	12083.66	I
28	14877.62	I
35	15024.99	I
30	15040.24	I
25	15047.70	I
6	15740.71	I
8	15748.99	I
10	15765.84	I
30	17108.66	I
5	26392.90	I

Mg III
Ref. 4, 83, 177 — J.O.S.

Intensity	Wavelength	
	Vacuum	
	106.30	III
	106.92	III
	108.08	III
	110.16	III
	114.32	III
	126.50	III
15	170.80	III
15	171.39	III
15	182.24	III
12	182.97	III
20	186.51	III
20	187.20	III
10	188.53	III
100	231.73	III
80	234.26	III
10	1274.83	III
11 h	1280.70	III
12	1391.27	III
15	1393.39	III
10	1431.14	III
16	1572.71	III
12	1586.24	III
13	1687.09	III
13	1697.28	III
10	1722.04	III
22	1738.84	III
12	1747.56	III
18	1748.93	III
15	1772.98	III
20	1783.25	III
14	1794.58	III
15	1800.66	III
13	1858.19	III
12	1879.49	III
10	1908.50	III
12	1923.90	III
13	1930.67	III
11	1937.84	III
	Air	
15	2039.55	III
15	2055.49	III
25	2064.90	III
15	2085.90	III
20	2091.96	III
13	2097.93	III
15	2112.77	III
16	2134.06	III
20	2177.70	III
20	2395.15	III
15	2467.75	III
10	2490.54	III
10	2529.19	III
12	3299.05	III
13	3306.39	III
12	3335.90	III
11	3342.58	III
12	3361.41	III
10	3381.24	III
11	3382.90	III
11	3387.37	III
10	3706.74	III
10	4916.00	III
10	5839.82	III
15	6256.75	III

Mg IV
Ref. 7, 128, 129 — J.O.S.

Intensity	Wavelength	
	Vacuum	
40	118.16	IV
80 p	118.81	IV
70	123.59	IV
240	124.65	IV
300	129.86	IV
300	132.81	IV
400	146.95	IV
300	147.41	IV
300	147.54	IV
350	180.07	IV
400	180.62	IV
400	180.80	IV
350	181.34	IV
4000	320.99	IV
3000	323.31	IV
40	800.41	IV
150	857.29	IV
30	866.74	IV
50	919.03	IV
30	929.78	IV
40	1008.76	IV
30	1026.41	IV
250	1037.41	IV
80	1044.37	IV
60	1055.76	IV
300	1210.99	IV
300	1342.19	IV
800	1346.57	IV
300	1346.68	IV
600	1352.05	IV
900	1384.46	IV
500	1385.77	IV
800	1387.53	IV
300	1404.68	IV
1000	1409.36	IV
500	1437.53	IV
1000	1437.64	IV
300	1447.42	IV
300	1459.54	IV
400	1459.62	IV
400	1481.51	IV
350	1490.45	IV
300	1495.50	IV
300	1607.11	IV
500	1683.02	IV
400	1698.81	IV
300	1844.17	IV
	Air	
12 p	2518.40	IV
4	2534.79	IV

Mg V
Ref. 128 — J.O.S.

Intensity	Wavelength	
	Vacuum	
5	251.58	V
35	276.58	V
10	312.30	V
20	351.09	V
18	352.20	V
30	353.09	V
15	353.30	V
18	354.22	V
20	355.33	V

MANGANESE (Mn)
Z = 25

Mn I and II
Ref. 1, 126 — C.H.C.

Intensity	Wavelength	
	Vacuum	
20	1726.47	II
30	1732.70	II
50	1733.55	II
40	1734.49	II
30	1737.93	II
20	1740.16	II
20	1742.00	II
30	1853.27	II
20	1857.92	II
50	1902.95	II
20	1907.84	II
30	1911.41	II
20 d	1914.68	II
100	1915.10	II
20	1918.64	II
30	1919.64	II
80	1921.25	II
20	1923.07	II
20	1923.34	II
30	1925.52	II
50	1926.59	II
30	1931.40	II
20	1945.15	II
20	1947.93	II
20	1950.14	II
30	1953.23	II
20 d	1954.81	II
30	1959.25	II
20	1969.24	II
30	1994.23	II
9700	1996.06	I
14000	1999.51	I
	Air	
18000	2003.85	I
50	2037.31	II
40	2037.64	II
40	2039.97	II
30	2076.21	II
1500	2092.16	I
20	2097.46	II
20	2102.50	II
1700	2109.58	I
30	2113.96	II
290	2208.81	I
540	2213.85	I
770	2221.84	I
20	2373.36	II
20	2427.38	II
50	2427.72	II
30	2427.94	II
30	2437.37	II
20	2437.88	II
30	2452.49	II
50	2499.00	II
30	2507.60	II
20	2516.60	II
30	2516.74	II
20	2521.66	II
20	2530.72	II
20	2531.80	II
50	2532.78	II
75	2533.06	I
50	2533.33	II
30	2534.10	II
80	2534.22	II
100	2535.66	II
30	2535.98	II
100	2537.92	II
50	2541.11	II
80	2542.92	II
50	2543.45	II
100	2548.75	II
50	2551.85	II
30	2553.27	II
75	2556.57	II
3u	2556.89	II
50	2557.54	II
95	2558.59	II
30	2559.41	II
150	2563.65	II
30	2565.22	II
580	2572.76	I
480	2575.51	I
12000	2576.10	II
550	2584.31	I
30	2588.97	II
45	2589.71	II
250	2592.94	I
6200	2593.73	II
250	2595.76	I
95	2598.90	II
40	2602.14	I
30	2602.72	II
45	2603.72	II
4300	2605.69	II
190	2610.20	II
500	2618.14	II
140	2622.90	I
150	2624.04	I
40	2624.80	II
200	2625.58	II
95	2626.64	I
30	2630.26	I
60	2630.57	I
190	2632.35	II
130	2638.17	II
80	2639.84	II
27	2650.99	II
60	2655.91	II
30	2666.77	II
45	2667.00	I
30	2667.03	II
110	2672.59	II
55	2673.37	II
55	2674.43	II
30	2676.33	I
45	2680.34	II
30	2680.68	II
30	2681.25	II
40	2681.72	I
45	2683.02	I
23	2683.75	I
55	2684.55	II
55	2685.94	II
110	2688.25	II
85	2692.66	I
27	2693.19	II
55	2695.36	II
27	2698.97	II
85	2701.00	II
50	2701.17	II
160	2701.70	II
100	2703.98	II
130	2705.74	II
80	2707.53	II
110	2708.45	II
45	2709.96	II
80	2710.33	II
110	2711.58	II
30	2716.80	II
30	2717.53	II
30	2719.01	II
50	2719.74	II
30	2722.10	II
30	2724.46	II
55	2728.61	II
30	2738.86	I
45 h	2760.93	I
30	2771.44	I
30 h	2776.23	I
30	2780.00	I
55	2789.20	I
60	2790.36	I
60	2791.08	I
6200	2794.82	I
5100	2798.27	I
220	2799.84	I
3700	2801.06	I
70	2804.10	I
60	2806.14	I
55	2808.02	I
110	2809.11	I
60	2812.84	I
70	2813.47	I
60	2815.02	II
30	2816.33	II
85	2817.97	I
40	2818.77	I
55	2821.45	I
55	2822.55	I
80	2830.79	I
27	2836.31	I
60	2870.08	II
30	2872.94	II
80	2879.49	I
40	2882.90	I
70	2886.68	II
160	2889.58	II
55	2892.39	II
50	2898.70	II
80	2900.16	II
40	2907.22	I
140 h	2914.60	I
190 h	2925.57	I

Manganese (Cont.)

Intensity		Wavelength	Species
27		2928.68	I
1100		2933.06	II
27		2934.02	I
1500		2939.30	II
250	h	2940.39	I
		2940.48	I
60		2941.04	I
1900		2949.20	II
40		3007.66	I
40		3011.16	I
40		3011.38	I
40		3014.67	I
60		3016.45	I
30		3019.92	II
70		3022.75	I
55		3031.06	II
30		3035.35	II
95		3040.60	I
27		3042.73	I
85		3043.36	I
330		3044.57	I
120		3045.59	I
200		3047.04	I
40		3048.86	I
30		3050.65	II
250		3054.36	I
140		3062.12	I
170		3066.02	I
170		3070.27	I
160		3073.13	I
90		3079.63	I
50		3081.33	I
73		3082.05	I
40		3097.06	I
40		3110.68	I
60	h	3148.18	I
90	h	3161.04	I
140	h	3178.50	I
220		3212.88	I
65		3216.95	I
1000		3228.09	I
300		3230.72	I
850		3236.78	I
330		3243.78	I
650		3248.52	I
100		3251.14	I
310		3252.95	I
65		3254.04	I
310		3256.14	I
220		3258.41	I
180		3260.23	I
180		3264.71	I
65		3296.88	I
65		3298.22	I
65		3320.69	I
70		3330.67	I
200		3330.78	II
100		3336.39	II
30		3365.02	II
30		3400.12	II
50		3438.97	II
720		3441.99	II
50		3460.03	II
360		3460.33	II
360	h	3474.04	II
		3474.13	II
290		3482.91	II
180		3488.68	II
140		3495.84	II
50		3496.81	II
100		3497.54	II
360		3531.85	I
		3532.00	I
1100		3532.12	I
1300		3547.80	I
1100		3548.03	I
390		3548.20	I
2200		3569.49	I
720		3569.80	I
		3570.04	I
1400		3577.88	I
720		3586.54	I
290		3595.12	I
420		3607.54	I
420		3608.49	I
360		3610.30	I
290		3619.28	I
220		3623.79	I
140		3629.74	I
100		3660.40	I
70		3670.52	I
70		3676.96	I
50		3682.09	I
280		3693.67	I
180		3696.57	I
70		3701.73	I
210		3706.08	I
130		3718.93	I
55		3728.89	I
130		3731.93	I
260		3790.22	I
55		3799.26	I
110		3800.55	I

Intensity		Wavelength	Species
55		3801.91	I
3200		3806.72	I
700		3809.59	I
55		3810.69	I
90		3816.75	I
2100		3823.51	I
390		3823.89	I
200		3829.68	I
480		3833.86	I
1300		3834.36	I
350		3839.78	I
670		3841.08	I
350		3843.98	I
65		3918.32	I
120		3926.47	I
65		3952.84	I
55		3975.89	I
65		3977.08	I
130		3982.58	I
150		3985.24	I
190		3986.83	I
150		3987.10	I
1500		4018.10	I
150		4026.44	I
27000		4030.76	I
19000		4033.07	I
11000		4034.49	I
1500		4035.73	I
55		4038.73	I
5600		4041.36	I
210	d	4045.13	I
		4045.21	I
1100		4048.76	I
80		4049.00	I
55		4051.73	I
65		4052.47	I
150		4055.21	I
1900		4055.54	I
210		4057.95	I
1100		4058.93	I
150		4059.39	I
730		4061.74	I
730		4063.53	I
80		4065.08	I
80		4068.00	I
290		4070.28	I
730		4079.24	I
730		4079.42	I
1100		4082.94	I
1100		4083.63	I
65		4089.94	I
55		4105.36	I
200		4110.90	I
150		4131.12	I
120		4135.04	I
80		4141.06	I
55		4147.53	I
80		4148.80	I
150		4176.60	I
120		4189.99	I
65		4201.76	I
65		4211.75	I
370		4235.14	I
510		4235.29	I
190		4239.72	I
290		4257.66	I
290		4265.92	I
270		4281.10	I
65		4284.08	I
65		4312.55	I
50		4323.63	II
45		4374.95	I
45		4381.70	I
55		4411.88	I
350		4414.88	I
55		4419.78	I
210		4436.35	I
800		4451.59	I
160		4453.00	I
130		4455.01	I
160		4455.32	I
110		4455.82	I
55		4457.04	I
210		4457.55	I
270		4458.26	I
55		4460.38	I
150		4461.08	I
510		4462.02	I
290		4464.68	I
200		4470.14	I
130		4472.79	I
40		4479.40	I
170		4490.08	I
240		4498.90	I
240		4502.22	I
80		4605.36	I
80		4626.54	I
35		4671.69	I
50		4701.16	I
160		4709.72	I
180		4727.48	I
130		4739.11	I
1000		4754.04	I

Intensity		Wavelength	Species
180		4761.53	I
750		4762.38	I
300		4765.86	I
500		4766.43	I
940		4783.42	I
1000		4823.52	I
25		4844.32	I
35		4965.48	I
19		5004.91	I
30		5074.79	I
60		5117.94	I
50		5150.89	I
50		5196.59	I
85		5255.32	I
160		5341.00	I
19		5349.88	I
95		5377.63	I
95		5394.67	I
50		5399.49	I
95		5407.42	I
35		5413.69	I
85		5420.36	I
35		5432.55	I
12		5457.47	I
60		5470.64	I
40		5481.40	I
30		5505.87	I
50		5516.77	I
40		5537.76	I
21		5551.98	I
8		5567.76	I
7		5573.01	I
8		5573.68	I
7		5738.29	I
7		5780.19	I
7		5816.84	I
140		6013.50	I
200		6016.64	I
290		6021.80	I
7		6384.67	I
17		6440.97	I
24		6491.71	I
14	h	6942.52	I
12		6989.96	I
14		7069.84	I
12		7184.25	I
10		7247.82	I
24	h	7283.82	I
35	h	7302.89	I
50		7326.51	I
12		7680.20	I
10		7712.42	I
10	h	7764.72	I
10	h	8670.92	I
12	h	8672.06	I
10	h	8673.97	I
12	h	8701.05	I
17	h	8703.76	I
30	h	8740.93	I

Mn III
Ref. 385 — C.H.C.

Intensity		Wavelength	Species
		Vacuum	
20		892.39	III
20		1108.16	III
30		1183.30	III
25	w	1183.86	III
30		1198.49	III
30		1219.80	III
100		1228.97	III
500		1283.58	III
400		1287.59	III
300		1291.62	III
1000		1360.72	III
800		1365.20	III
400		1369.43	III
300		1371.65	III
300		1596.95	III
500	h	1609.17	III
1000		1614.14	III
2000		1620.60	III
300		1623.91	III
400		1629.12	III
500		1633.80	III
250		1647.46	III
400		1653.57	III
400		1804.06	III
300		1806.47	III
300		1811.02	III
400		1877.62	III
300		1885.21	III
500		1941.28	III
250	w	1942.89	III
800		1943.21	III
500		1952.36	III
1000		1952.52	III
300		1956.61	III
250		1962.04	III

Intensity		Wavelength	Species
500		1978.95	III
400		1982.76	III
400		1989.59	III
		Air	
300		2022.19	III
1000	w	2027.83	III
500	w	2028.14	III
300		2044.57	III
400		2048.93	III
500		2049.68	III
300		2056.80	III
500		2066.38	III
1000		2069.02	III
900		2077.38	III
300		2078.13	III
800		2084.23	III
600		2090.05	III
300		2090.25	III
500		2094.14	III
500		2094.78	III
500		2097.93	III
500		2099.97	III
300		2123.25	III
1000		2169.78	III
700		2174.15	III
900		2176.87	III
800		2181.86	III
800	w	2184.87	III
600		2185.13	III
400		2211.95	III
600		2212.42	III
800		2215.21	III
900		2220.55	III
1000		2227.42	III
100		3287.49	III
100		3540.52	III
150		3601.72	III
100		3616.00	III
100		4246.17	III
200		5079.20	III
150		5100.03	III
100		5117.03	III
100		5252.23	III
100		5365.59	III
150		5454.07	III
200		5474.68	III
100		5671.12	III
200		5946.65	III
100	s	6213.11	III
200		6231.21	III
100		6238.64	III
100		6273.71	III

Mn IV
Ref. 433 — C.H.C.

Intensity		Wavelength	Species
		Vacuum	
60		579.79	IV
60		581.44	IV
60		581.65	IV
60		585.21	IV
90		1242.25	IV
90		1244.50	IV
85		1247.73	IV
95		1251.93	IV
95		1257.28	IV
90		1264.41	IV
70		1603.60	IV
70		1611.10	IV
75		1653.83	IV
70		1656.39	IV
70		1659.25	IV
75		1664.73	IV
80	b	1667.00	IV
70		1670.08	IV
75		1691.68	IV
75		1693.15	IV
80		1698.30	IV
75		1698.70	IV
70		1699.06	IV
75		1707.43	IV
65		1718.67	IV
75	b	1720.52	IV
75		1720.74	IV
75		1721.41	IV
65		1722.94	IV
75		1724.83	IV
85	b	1742.10	IV
85		1751.59	IV
75		1759.82	IV
70		1762.71	IV
75		1762.94	IV
85	d	1766.27	IV
75		1767.09	IV
65		1772.11	IV
75		1773.51	IV
75		1782.21	IV
75		1786.02	IV

Manganese (Cont.)

Intensity	Wavelength	
75	1787.04	IV
75	1787.38	IV
75	1788.64	IV
75	1790.44	IV
80	1795.65	IV
80	1795.79	IV
60	1907.03	IV
75	1910.25	IV
65	1997.54	IV

Mn V
Ref. 405 — C.H.C.

Intensity	Wavelength	
	Vacuum	
300	404.36	V
380	406.02	V
300	406.40	V
600	410.30	V
600	410.60	V
480	410.98	V
400	411.32	V
460	412.74	V
460	413.75	V
600	415.62	V
650	415.98	V
350	419.80	V
600	428.59	V
500	429.05	V
400	433.54	V
600	435.67	V
350	436.16	V
500	436.18	V
450	438.74	V
350	439.35	V
1000	441.72	V
850	442.49	V
400	467.32	V
300	474.82	V

MERCURY (198) (Hg)
Z = 80

Hg I and II (198)
Ref. 43, 50, 69, 145, 229, 242 — R.W.S.

Intensity	Wavelength	
	Vacuum	
80	1250.564	I
8	1259.242	I
100	1268.825	I
5	1307.751	I
20	1402.619	I
10	1435.503	I
1000	1849.492	I
	Air	
60	2262.210	II
20	2302.065	I
20	2345.440	I
100	2378.325	I
20	2380.004	I
40	2399.349	I
20	2399.729	I
20	2446.900	I
15	2464.064	I
40	2481.999	I
30	2482.713	I
40	2483.821	I
90	2534.769	I
15000	2536.506	I
25	2563.861	I
25	2576.290	I
250	2652.043	I
400	2653.683	I
100	2655.130	I
50	2698.831	I
80	2752.783	I
20	2759.710	I
40	2803.471	I
30	2804.438	I
750	2847.675	II
50	2856.939	I
150	2893.598	I
150	2916.227	II
60	2925.413	I
1200	2967.283	I
300	3021.500	I
120	3023.476	I
30	3025.608	I
50	3027.490	I
400	3125.670	I
320	3131.551	I
320	3131.842	I
80	3341.481	I
2800	3650.157	I
300	3654.839	I
80	3662.883	I
240	3663.281	I
30	3701.432	I
35	3704.170	I
30	3801.660	I
20	3901.867	I
60	3906.372	I
200	3983.839	II
1800	4046.572	I
150	4077.838	I
40	4108.057	I
250	4339.224	I
400	4347.496	I
4000	4358.337	I
80	4916.068	I
1100	5460.753	I
160	5675.922	I
240	5769.598	I
280	5790.663	I
20	6072.713	I
30	6234.402	I
160	6716.429	I
250	6907.461	I
240	11287.407	I

MERCURY (NATURAL) (Hg)
Z = 80

Hg I and II (nat.)
Ref. 34, 45, 90, 117, 133, 189, 235, 304, 327, 328 — R.W.S.

Intensity	Wavelength	
	Vacuum	
400	893.08	II
300	915.83	II
150	923.39	II
200	940.80	II
100	962.74	II
50	969.13	II
800	1099.26	II
80	1250.58	I
8	1259.24	I
100	1268.82	I
5	1307.75	I
300	1307.93	II
400	1321.71	II
400	1331.74	II
80	1350.07	II
200	1361.27	II
20	1402.62	I
200	1414.43	II
10	1435.51	I
15	1619.46	II
120	1623.95	II
20	1628.25	II
150	1649.94	II
50	1653.64	II
200	1672.41	II
100	1702.73	II
100	1707.40	II
120	1727.18	II
250	1732.14	II
20	1775.68	I
40	1783.70	II
30	1796.22	II
200	1796.90	II
60	1798.74	II
30	1803.89	II
40	1808.29	II
400	1820.34	II
5	1832.74	I
1000	1849.50	I
160	1869.23	II
300	1870.55	II
200	1875.54	II
20	1900.28	II
30	1927.60	II
300	1942.27	II
100	1972.94	II
200	1973.89	II
150	1987.98	II
	Air	
90	2026.97	II
90	2052.93	II
70	2148.00	II
5	2247.55	I
60	2262.23	II
20	2302.06	I
15	2323.20	I
5	2340.57	I
20	2345.43	I
20	2352.48	I
100	2378.32	I
20	2380.00	I
40	2399.38	I
20	2399.73	I
10	2400.49	I
60	2407.35	II
50	2414.13	II
5	2441.06	I
20	2446.90	I
15	2464.06	I
40	2482.00	I
30	2482.72	I
40	2483.82	I
90	2534.77	I
15000	2536.52	I
25	2563.86	I
25	2576.29	I
5	2578.91	I
15	2625.19	I
5	2639.78	I
250	2652.04	I
400	2653.69	I
100	2655.19	I
5	2674.91	I
20	2698.83	I
50	2699.38	I
80	2705.36	II
80	2752.78	I
20	2759.71	I
40	2803.46	I
30	2804.43	I
2	2805.34	I
2	2806.77	I
150	2814.93	II
750	2847.68	II
50	2856.94	I
150	2893.60	I
150	2916.27	II
60	2925.41	I
150	2935.94	II
400	2947.08	II
1200	2967.28	I
300	3021.50	I
120	3023.47	I
30	3025.61	I
50	3027.49	I
400	3125.67	I
320	3131.55	I
320	3131.84	I
400	3208.20	II
400	3264.06	II
80	3341.48	I
100	3385.25	II
400	3451.69	II
200	3549.42	II
2800	3650.15	I
300	3654.84	I
80	3662.88	I
240	3663.28	I
30	3701.44	I
35	3704.17	I
30	3801.66	I
100	3806.38	II
20	3901.87	I
60	3906.37	I
100	3918.92	II
200	3983.96	II
1800	4046.56	I
150	4077.83	I
40	4108.05	I
250	4339.22	I
400	4347.49	I
4000	4358.33	I
100	4398.62	II
90	4660.28	II
80	4855.72	II
5	4883.00	I
5	4889.91	I
80	4916.07	I
5	4970.37	I
5	4980.64	I
20	5102.70	I
40	5120.64	I
100	5128.45	II
20	5137.94	I
5	5290.74	I
60	5316.78	I
5	5354.05	I
60	5384.63	I
1100	5460.74	I
30	5549.63	I
160	5675.86	I
240	5769.60	I
100	5789.66	I
280	5790.66	I
140	5803.78	I
60	5859.25	I
60	5871.73	II
20	5871.98	I
20	6072.72	I
1000	6149.50	II
30	6234.40	I
80	6521.13	II
160	6716.43	I
250	6907.52	I
250	7081.90	I
200	7091.86	I
40	7346.37	II
100	7485.87	II
20	7728.82	I
100	7944.66	II
2000	10139.75	I
240	11287.40	I
120	13209.95	I
140	13426.57	I
60	13468.38	I
80	13505.58	I
500	13570.21	I
450	13673.51	I
200	13950.55	I
500	15295.82	I
100	16881.48	I
400	16920.16	I
300	16942.00	I
500	17072.79	I
400	17109.93	I
20	17116.75	I
20	17198.67	I
20	17213.20	I
70	17329.41	I
30	17436.18	I
50	18130.38	I
40	19700.17	I
	22493.28	I
250	23253.07	I
	32148.06	I
	36303.03	I

Hg III
Ref. 343 — C.H.C.

Intensity	Wavelength	
	Vacuum	
3	621.44	III
2	679.68	III
2	878.59	III
1	886.48	III
1	988.89	III
2	1009.29	III
5	1068.03	III
2	1161.95	III
9	1681.40	III
15	1759.75	III
1	1894.77	III
	Air	
7	2314.15	III
4	2380.55	III
8	2431.65	III
5	2480.56	III
7	2484.50	III
2	2612.92	III
4	2617.97	III
3	2670.49	III
70	2724.43	III
6	2769.22	III
3	2844.76	III
15	3090.05	III
5	3283.02	III
12	3312.28	III
8	3389.01	III
5	3450.77	III
3	3500.35	III
4	3538.88	III
5	3557.24	III
15	3803.51	III
70	4122.07	III
10	4140.34	III
100	4216.74	III
15	4470.58	III
12	4552.84	III
50	4797.01	III
10	4869.85	III
80	4973.57	III
30	5210.82	III
6	5695.71	III
25	6220.35	III
35	6418.98	III
40	6501.38	III
10	6584.26	III
6	6610.12	III
30	6709.29	III
12	7517.46	III
7	7808.10	III
25	7946.75	III
50	7984.51	III
5	8151.64	III

MOLYBDENUM (Mo)
Z = 42

Mo I and II
Ref. I — C.H.C.

Intensity	Wavelength AIR	
19000	2015.11	II
40000	2020.30	II
21000	2038.44	II
17000	2045.98	II
4800	2081.68	II
2400	2089.52	II
2200	2092.50	II
4000	2093.11	II
2700	2100.84	II
1500	2104.29	II
1400	2108.02	II
400	2269.69	II
160	2304.25	II
160	2306.97	II
130	2325.94	I
240	2330.46	I
110	2332.12	II
190	2340.47	I
190	2341.59	II
80	2352.61	I
80	2355.22	I
80	2355.42	II
70	2364.37	I
50	2366.09	I
140	2372.27	I
100	2380.41	I
150	2383.52	I
110	2389.20	I
140	2403.61	II
80	2404.66	II
140	2405.86	I
40	2408.39	I
40	2412.84	II
120	2413.01	II
70	2415.33	I
80	2417.96	II
65	2419.01	II
80	2420.18	II
70	2424.00	II
65	2430.43	I
65	2435.96	II
65	2440.28	II
40	2461.81	II
50	2466.68	II
50	2466.97	II
50	2468.78	II
30	2470.04	II
150 h	2471.97	I
70	2477.57	II
70 h	2481.81	I
65	2482.57	II
40	2484.75	II
40 h	2485.31	I
24	2496.24	II
85	2498.28	II
40	2500.44	II
65	2502.84	II
50	2511.80	II
65	2515.08	II
70	2527.14	II
50	2530.34	II
70	2532.31	II
440	2538.46	II
50	2539.44	II
110 h	2540.45	I
330	2542.67	II
40	2543.61	II
330	2548.22	II
110 h	2550.85	I
65	2555.42	II
40	2556.75	II
80	2558.88	II
65	2562.08	II
85	2564.34	II
40	2566.26	II
250	2567.05	I
20	2571.45	II
320	2572.34	I
50	2574.42	II
40	2576.56	II
40	2578.36	II
250	2582.16	I
30	2585.95	II
65	2588.78	II
40	2591.77	II
250	2593.70	II
100	2595.40	I
40	2597.38	II
250	2602.80	II
40	2605.08	II
40	2605.93	II

Intensity	Wavelength	
250	2607.37	I
190	2611.20	I
290	2613.08	I
130	2615.39	I
400	2616.78	I
70	2619.34	II
140	2621.07	I
320	2627.55	I
160	2628.74	I
440	2629.85	I
330	2636.67	II
250	2638.30	I
720	2638.76	II
410	2640.99	I
600	2644.35	II
370	2646.49	I
640	2649.46	I
480	2653.35	II
560 h	2655.03	I
290	2658.11	I
640	2660.58	II
110	2665.10	I
55	2671.83	II
720	2672.84	II
250	2673.27	II
1000	2679.85	I
95	2681.36	II
640	2683.23	II
880	2684.14	II
560	2687.99	II
30	2692.61	II
55	2695.22	II
30	2696.83	II
55	2699.41	II
140	2701.03	I
480	2701.42	II
30	2701.87	II
30	2704.93	II
40	2710.19	II
30	2711.49	II
50	2712.35	II
190	2713.51	II
290	2717.35	II
110	2724.41	I
180	2725.15	I
85	2726.97	II
140	2729.68	II
80	2730.20	II
330	2732.88	II
250	2733.39	I
160	2736.96	II
80 h	2737.88	II
50	2738.60	II
40	2741.32	II
55	2741.62	II
240	2743.07	I
290	2746.30	II
320	2751.47	I
110	2756.07	II
65 d	2758.63	II
20	2760.53	II
190	2761.53	I
220	2763.62	II
110	2766.26	I
240	2769.76	II
160	2773.78	II
190	2774.39	II
1700	2775.40	II
130	2777.74	I
65	2777.86	II
880	2780.04	II
400	2784.99	II
180	2787.83	I
40	2791.54	II
240 d	2797.93	I
220	2801.47	I
400	2807.76	II
28	2812.58	II
24	2814.67	II
1700	2816.15	II
220	2817.44	II
50	2822.03	II
240	2826.54	I
80	2827.74	II
40	2831.44	II
30	2832.07	II
80	2834.39	II
80	2835.33	II
160	2842.15	II
24	2843.73	II
220	2844.39	I
1700	2848.23	II
160	2849.38	I
370	2853.23	II
50	2856.00	II
24	2863.20	II
370	2863.81	II
160	2864.31	I
140	2864.66	I
40	2865.62	II
220	2866.69	II
40	2868.11	II
40	2868.32	II

Intensity	Wavelength	
1700	2871.51	II
85	2872.88	II
220	2879.05	II
65	2888.15	II
95	2891.28	II
1300	2890.99	II
190	2892.81	II
950	2894.45	II
140	2897.63	II
70	2900.80	II
290	2903.07	II
160	2905.27	I
80	2907.12	II
600	2909.12	II
1100	2911.92	II
55	2913.81	II
120	2918.83	II
1300	2923.39	II
140	2924.32	II
65	2927.54	II
50	2930.06	II
1100	2930.50	II
55	2930.77	II
800	2934.30	II
65	2935.20	II
120	2937.66	I
40	2938.30	II
95	2940.10	II
110	2941.22	II
140	2944.21	I
150	2944.82	II
140	2945.66	I
	2945.95	II
190	2946.01	I
140	2946.42	I
140	2946.69	II
95	2947.28	II
95	2955.84	II
240	2956.06	II
70	2956.90	II
95	2960.24	II
140	2962.89	I
250	2963.79	II
50	2964.96	II
210	2965.27	II
70	2971.91	II
250	2972.61	II
80	2975.40	II
180	2978.28	I
120	2981.52	I
110	2987.92	I
160	2988.68	I
190	2989.80	I
95	2992.84	II
50	2993.52	II
190	3002.21	I
40	3004.46	II
130	3013.39	I
140 h	3013.76	I
250	3025.00	I
95	3027.77	II
100	3036.31	I
300	3041.70	I
150	3046.80	I
210	3047.31	I
210	3055.32	I
100	3060.78	II
160	3061.59	I
800	3064.28	I
250	3065.04	II
100	3068.00	I
250	3070.90	I
800	3074.37	I
85	3077.66	II
150	3079.88	I
210	3080.41	I
800	3085.62	I
270	3087.62	II
100	3089.12	I
100	3089.71	I
190	3092.07	II
560	3094.66	I
110	3099.93	I
110	3100.88	I
560	3101.34	I
1400	3112.12	I
290	3122.00	II
14000	3132.59	I
110	3138.72	II
220	3147.35	I
220	3152.82	II
55	3155.64	II
6000	3158.16	I
120	3164.53	I
8700	3170.35	I
95	3172.03	II
160	3172.74	II
370	3183.03	I
120	3184.57	I
370	3185.10	I
180	3185.71	I
120 d	3187.59	II
7600	3193.97	I

Intensity	Wavelength	
290	3195.96	I
120	3198.85	I
40	3201.50	II
330	3205.22	I
880	3205.88	I
3000	3208.83	I
240	3210.97	I
560	3215.07	I
350	3221.74	I
880	3228.22	I
600	3229.79	I
1100	3233.14	I
950	3237.08	I
65	3240.71	II
950	3256.21	I
300	3262.63	I
480	3264.40	I
800	3270.90	I
240	3285.02	I
320	3285.36	I
1100	3289.02	I
950	3290.82	I
190	3292.31	II
320	3305.56	I
320	3307.12	I
100	3313.62	II
190	3320.90	II
640	3323.95	I
360	3325.67	I
360	3327.30	I
240	3340.17	I
1300	3344.75	I
95	3346.40	II
320	3347.02	I
1600	3358.12	I
250	3361.37	I
950	3363.78	I
950	3379.97	I
320	3382.48	I
1900	3384.62	I
130	3395.36	II
640	3404.34	I
1300	3405.94	I
240	3418.52	I
250	3420.04	I
250	3422.31	I
380	3434.79	I
320	3435.45	I
640	3437.22	I
250	3438.87	I
250	3441.44	I
250	3443.26	I
130	3446.08	II
3200	3447.12	I
640	3449.07	I
300	3451.75	I
250	3452.60	I
950	3456.39	I
640	3460.78	I
320	3466.83	I
250	3467.85	I
320	3469.22	I
240	3485.93	I
800	3504.41	I
240	3505.32	I
560	3508.12	I
480	3521.41	I
240	3524.98	I
640	3537.28	I
320	3542.17	I
520	3558.10	I
400	3563.14	I
300	3566.05	I
240	3570.65	I
320	3573.88	I
1400	3581.89	I
200	3590.74	I
210	3598.88	I
270	3602.94	I
210	3608.37	I
200	3623.23	I
1400	3624.46	I
330	3626.18	I
28	3635.14	II
1000	3635.43	I
400	3657.35	I
540	3664.81	I
290	3666.72	I
590	3672.82	I
1300	3680.60	I
45	3684.22	II
65	3688.31	II
240	3690.59	I
180	3692.64	II
1400	3694.94	I
220	3702.03	I
220	3715.65	I
500	3727.69	I
330 d	3732.71	I
240	3742.28	I
80	3744.37	II
360	3770.45	I
220	3779.77	I

Molybdenum (Cont.)

Intensity		Wavelength		Intensity		Wavelength		Intensity		Wavelength		Intensity		Wavelength	
360		3781.59		35		5092.16	I	50		5634.86	I	13		7063.34	I
250		3797.30	I	40		5095.89	I	230		5650.13	I	13		7081.22	I
29000		3798.25	I	100		5096.65	I	23		5673.63	I	110		7109.87	I
290		3801.84	I	130		5097.52	I	55		5674.47	I	27		7134.08	I
520		3826.70	I	35		5098.03	I	40		5677.89	I	150		7242.50	I
940		3828.87	I	130		5109.71	I	35		5682.89	I	40		7245.85	I
1700		3833.75	I	80		5114.97	I	460		5689.14	I	22		7267.62	I
380		3847.25	I	35		5116.97	I	23		5699.28	I	17		7300.19	I
29000		3864.11	I	29		5123.83	I	80		5705.72	I	13		7348.49	I
580		3869.08	I	150		5145.38	I	23		5711.80	I	13		7361.65	I
580		3886.82	I	110		5147.39	I	210		5722.74	I	10	h	7364.41	I
380		3901.77	I	80		5163.19	I	23		5728.77	I	40		7391.36	I
19000		3902.96	I	100		5167.76	I	26	d	5729.45	I	10		7434.10	I
65		3941.48	II	160	d	5171.08	I			5729.59	I	13		7447.34	I
230		3943.04	I			5171.25	I	620		5751.40	I	13	h	7452.85	I
270		4056.01	I	230	h	5172.94	I	23		5774.55	I	140		7485.74	I
1400		4062.08	I	160	h	5174.18	I	40		5779.36	I	13		7504.47	I
2300		4069.88	I	40		5191.44	I	23	h	5783.33	I	11		7572.64	I
1300		4081.44	I	110		5200.17	I	520		5791.85	I	11	h	7595.16	I
940		4084.38	I	50		5200.74	I	23	h	5795.77	I	11		7601.84	I
250		4102.15	I	26		5210.44	I	26		5800.46	I	17	h	7656.76	I
730		4107.47	I	50		5211.86	I	35		5802.67	I	13		7679.49	I
630		4120.10	I	80		5219.40	I	23		5825.20	I	27		7720.77	I
2900		4143.55	I	65		5231.06	I	23		5835.59	I	17		7829.65	I
230		4148.94	I	26		5232.36	I	20		5839.99	I	15		7854.45	I
250		4155.28	I	100		5234.26	I	20	h	5848.86	I	11		7923.15	I
200		4157.40	I	460	h	5238.20	I	55	h	5849.73	I	15		7986.60	I
200		4178.27	I	230	h	5240.88	I	50	h	5851.52	I	22	h	8245.06	I
480		4185.82	I	110	h	5242.81	I	520		5858.27	I	40	h	8328.44	I
2500		4188.32	I	100		5245.51	I	20		5861.38	I	45	h	8389.32	I
250		4194.56	I	150		5259.04	I	50		5869.33	I	45	h	8483.39	I
1500		4232.59	I	16		5260.17	I	26		5876.59	I				
270		4269.28	I	65		5261.14	I	820		5888.33	I				
890		4276.91	I	20		5268.95	I	23		5892.29	I				
1200		4277.24	I	35		5271.80	I	50	h	5893.38	I				
1400		4288.64	I	35		5276.28	I	20		5898.78	I				
680		4292.13	I	65		5279.65	I			5898.82	MO I				
890		4293.21	I	210		5280.86	I	40		5901.47	I				
360		4293.88	I	20		5283.84	I	40	h	5926.36	I				
840		4326.14	I	55		5292.08	I	160	h	5928.88	I				
250		4326.74	I	35		5293.46	I	40		5988.17	I				
230		4350.34	I	55		5295.47	I	35		6025.49	I				
230		4369.04	I	20		5306.26	I	16		6027.26	I				
1900		4381.64	I	55		5313.89	I	1300		6030.66	I				
2500		4411.57	I	35		5315.04	I	20		6047.83	I				
210		4423.62	I	20		5319.89	I	20		6054.81	I				
990		4434.95	I	20		5324.47	I	20		6079.58	I				
200		4442.20	T	35		5327.06	I	10		6081.27	I				
340		4449.74	I	20		5352.35	I	40		6101.87	I				
480		4457.36	T	80		5354.88	I	10		6130.63	I				
630		4474.56	I	35		5355.51	I	10		6197.66	T				
230		4491.28	I	65		5356.48	I	20		6217.89	I				
120		4504.90	I	560	H1	5360.56	I	10		6264.27	I				
140		4512.15	I	110	H1	5364.28	I	10		6265.88	I				
230		4517.13	I	35	H1	5367.11	I	15		6290.74	I				
230		4524.34	I	35		5372.40	I	13		6301.75	I				
120		4529.40	I	26		5388.69	I	11		6323.54	I				
400		4536.80	I	65		5394.52	I	40		6357.22	I				
110		4558.11	I	35		5397.38	I	16		6389.11	T				
210		4576.50	I	50		5400.47	I	11		6391.12	I				
170		4595.16	I	35		5405.79	I	35		6401.07	I				
360		4609.88	I	35		5406.39	I	26		6409.11	I				
100		4621.38	I	40		5417.38	I	10		6412.39	I				
460		4626.47	I	23		5426.89	I	100		6424.37	I				
100		4627.48	I	55		5435.68	I	20		6446.34	I				
220		4662.76	I	65		5437.75	I	20		6471.20	I				
130		4671.90	I	40		5450.51	I	20		6473.99	I				
130		4688.22	I	35		5456.46	I	10		6493.13	I				
640		4707.26	I	26		5460.53	I	23		6519.84	I				
150		4708.22	I	23		5465.57	I	15	h	6611.20	I				
220		4717.92	I	35		5475.90	I	230		6619.13	I				
100		4729.14	I	35		5490.28	I	10		6624.57	I				
700		4731.44	I	20		5492.17	I	50		6650.38	I				
100		4750.39	I	26	h	5493.80	I	13		6659.68	I				
770		4760.19	I	26		5498.49	I	18		6690.47	I				
150		4776.34	I	50		5501.54	I	110		6733.98	I				
100		4796.52	I	23		5501.87	I	21		6746.08	I				
410		4819.25	I	26	h	5503.54	I	50		6746.27	I				
410		4830.51	I	7800		5506.49	I	35		6753.97	I				
360		4868.00	I	23		5520.04	I	13		6763.50	I				
110		4950.62	I	26		5520.68	I	10	h	6799.88	I				
150		4957.54	I	40		5526.52	I	10		6802.62	I				
210		4979.12	I	40		5526.97	I	10		6812.03	I				
110		4999.91	I	5200		5533.05	I	13		6825.63	I				
20		5010.81	I	40		5539.41	I	18	d	6828.87	I				
180		5014.60	I	50		5543.12	I			6829.05	I				
26		5016.78	I	40		5544.49	I	40		6838.88	I				
20		5019.85	I	55		5556.28	I	16		6848.92	I				
80		5029.00	I	26		5556.72	I	21		6886.28	I				
65		5030.78	I	20		5564.05	I	16		6892.36	I				
23		5038.91	I	40		5568.62	I	10		6898.01	I				
26		5046.52	I	26		5569.48	I	10		6898.98	I				
100		5047.71	I	2500		5570.45	I	13		6908.20	I				
50		5055.00	I	35		5575.19	I	35		6914.01	I				
35		5058.07	I	20		5591.58	I	13		6934.10	I				
200		5059.88	I	40		5602.76	I	10		6947.39	I				
35		5062.52	I	23		5608.62	I	10		6960.64	I				
29		5064.64	I	23		5609.23	I	16		6978.71	I				
35		5079.87	I	100		5610.93	I	26		6988.94	I				
100		5080.02	I	23		5613.07	I	12		6999.13	I				
35		5081.26	I	20		5618.45	I	16		7001.60	I				
40		5090.97	I	23		5619.38	I	22		7037.98	I				
35		5091.34	I	330		5632.47	I	22		7060.21	I				

Mo III
Ref. 420 — C.H.C.

Intensity	Wavelength	
	Vacuum	
50	1166.07	III
100	1169.33	III
50	1173.67	III
50	1209.60	III
50	1225.46	III
50	1230.34	III
50	1234.63	III
30	1236.10	III
100	1254.93	III
100	1262.21	III
100	1263.74	III
100	1274.37	III
50	1274.94	III
100	1276.40	III
200	1277.40	III
200	1277.58	III
100	1278.06	III
200	1278.40	III
150	1281.90	III
150	1283.60	III
100	1258.52	III
100	1286.42	III
50	1288.07	III
50	1288.25	III
100	1290.49	III
100	1299.82	III
50	1305.58	III
50	1437.37	III
50	1452.38	III
50	1534.86	III
50	1751.22	III
50	1760.57	III
50	1807.60	III
100	1854.73	III
	Air	
75	2165.19	III
75	2170.57	III
50	2172.46	III
75	2179.37	III
100	2184.37	III
100	2211.02	III
50	2223.19	III
100	2253.18	III
150	2269.71	III
50	2275.47	III
50	2275.64	III
200	2294.97	III
80	2304.26	III
50	2326.75	III
150	2330.93	III
100	2359.76	III
80	2386.96	III
50	2403.61	III
70	2412.71	III
50	2422.18	III
200	2506.19	III
90	2597.13	III
50	2756.06	III
100	2807.74	III
125	2947.32	III

Molybdenum (Cont.)

Intensity	Wavelength	
75	2983.94	III
80	3254.70	III
200	3271.69	III

Mo IV
Ref. 383 — C.H.C.

Intensity		Wavelength	

Vacuum

10		857.75	IV
10		859.72	IV
25		865.24	IV
20		865.53	IV
20		863.63	IV
50		867.92	IV
10		878.43	IV
100		884.19	IV
10		884.82	IV
60		886.05	IV
50		891.74	IV
40		894.80	IV
15		895.41	IV
20		1819.50	IV
30		1821.59	IV
30		1850.69	IV
20		1877.88	IV
80		1926.26	IV
100		1929.24	IV
20		1949.44	IV
80		1971.06	IV
20		1991.41	IV

Air

70		2010.92	IV
20		2023.78	IV
25		2055.64	IV
50		2060.38	IV
15		2091.89	IV
10		2113.78	IV
5		2140.33	IV

NEODYMIUM (Nd)
Z = 60

Nd I and II
Ref. 1 — C.H.C.

Intensity		Wavelength	

Air

75		2702.46	
75		2704.54	
75		2764.98	I
60		2785.79	I
50		2863.95	
50		2921.26	
55		2962.88	II
65		2963.58	II
80		2993.20	II
40		2994.73	
95		3007.97	II
95		3014.19	II
95		3018.35	II
80		3026.47	II
50		3038.98	II
50		3043.29	
50		3051.11	II
80		3052.15	II
140		3056.71	II
130		3069.73	II
65	d	3071.43	II
		3071.50	II
160		3075.38	II
95		3079.38	II
95		3080.94	II
95		3092.73	
240		3092.92	II
140		3098.48	II
55		3099.52	II
130		3105.43	II
95		3106.18	II
65		3108.01	II
260		3115.18	II
190		3116.15	II
50		3119.75	II
160		3123.06	II
190		3124.58	II
290		3133.60	II
220		3134.90	II
100		3137.24	II
170		3141.46	II
170		3142.44	H
100		3144.55	II
100		3144.82	II
100		3148.51	II
100		3149.29	II
100		3149.51	II

100		3162.62	II
100		3175.99	II
50		3181.54	II
50		3188.73	II
50		3200.62	II
150		3203.47	II
85		3211.00	II
100		3217.12	II
50		3222.62	I
50		3228.04	II
60		3234.62	
40		3237.91	II
100		3254.08	II
50		3256.91	II
220		3259.24	II
100		3260.66	II
220		3265.12	II
50		3265.38	II
170		3267.25	II
100		3273.18	II
320		3275.22	II
50		3281.49	II
50		3282.78	II
290		3285.10	II
100		3286.62	II
50		3289.52	
100		3290.65	II
70		3293.84	II
70		3294.68	II
70		3298.61	
300		3300.16	II
200		3312.75	II
200		3325.90	II
410		3328.28	II
250		3331.57	II
290		3334.48	II
290		3339.07	II
320		3353.59	II
200		3355.93	II
270		3364.96	II
290		3393.63	II
120	h	3484.88	I
200		3527.53	II
290		3543.35	II
200		3555.77	II
410		3560.75	II
340		3568.87	II
470		3587.51	II
300		3592.59	II
340		3598.02	II
300		3600.91	II
320		3609.79	II
370		3615.82	II
300		3618.96	II
300		3631.02	II
340		3634.30	II
240		3637.00	II
240		3637.23	II
240		3640.24	II
240		3645.78	II
340		3648.20	II
240		3649.46	II
240		3650.42	II
410		3653.15	II
240		3654.16	II
470		3662.26	II
540		3665.18	II
540		3672.36	II
580		3673.54	II
240		3678.18	II
1200		3685.80	II
440		3687.30	II
410		3689.69	II
300		3694.81	II
410		3697.56	II
240		3702.84	II
240		3704.95	II
200		3712.81	II
470		3713.70	II
370		3714.20	II
640	d	3714.73	II
250		3715.04	II
200		3715.39	II
470		3715.68	II
410		3718.54	II
410		3721.35	II
220		3722.42	II
780		3723.50	II
410		3724.87	II
250		3726.90	II
710		3728.13	II
470		3730.58	II
270		3732.78	II
1000	d	3735.54	II
		3735.60	II
440		3737.10	II
1000		3738.06	II
270		3741.42	II
200		3749.85	II
320		3750.31	II
580		3752.49	II
370		3752.67	II
250		3754.83	II

370		3755.60	II
510		3757.82	II
930		3758.95	II
300		3759.79	II
930		3763.47	II
300		3766.59	II
510		3769.65	II
1400		3775.50	II
250		3776.34	II
710		3779.47	II
580		3780.40	II
510		3781.32	II
300		3783.78	II
2400		3784.25	II
270		3784.73	II
340		3791.50	II
340		3795.45	II
240		3799.55	II
370		3801.12	II
200		3801.38	II
340		3802.30	II
1200		3803.47	II
200		3804.10	II
2500		3805.36	II
340		3805.55	II
470		3807.23	II
540		3808.77	II
440		3809.06	II
580		3810.49	II
240		3811.06	II
270		3811.77	II
200		3812.53	II
710		3814.73	II
240		3819.70	II
410		3822.47	II
1200		3826.42	II
240		3828.00	II
540		3828.85	II
440		3829.16	II
510		3830.47	II
740		3836.54	II
340		3837.91	II
1700		3838.98	II
340		3839.51	II
410	d	3841.82	II
		3841.88	II
1700	d	3848.24	II
		3848.31	II
1500		3848.52	II
470		3850.22	II
2400	d	3851.66	II
		3851.74	II
340		3858.55	II
270		3860.94	II
300		3862.52	II
3700	d	3863.33	II
		3863.40	II
240		3866.52	II
220		3866.81	II
850		3869.07	II
240		3875.74	II
470		3875.87	II
1100		3878.58	II
1000		3879.55	II
780		3880.38	II
1200		3880.78	II
200		3881.59	II
540		3887.87	II
370	h	3889.66	II
1300		3889.93	II
1300		3890.58	II
1300		3890.94	II
580		3891.51	II
470		3892.06	II
810		3894.63	II
270		3896.13	II
440		3897.63	II
2000		3900.21	II
1300		3901.84	II
1700		3905.89	II
200		3907.70	II
510		3907.84	II
2000		3911.16	II
850		3912.23	II
340		3913.69	II
440		3915.13	II
610		3915.95	II
340		3917.65	II
220		3919.92	II
1100		3920.96	II
510		3927.10	II
200		3929.26	II
610		3934.82	II
410		3936.11	II
510		3938.86	II
2000		3941.51	II
2000		3951.16	II
810		3952.20	II
320		3952.87	II
320		3953.52	II
240		3957.45	II
590		3958.00	II
510		3962.21	II

1400		3963.12	II
270		3963.90	II
1100		3973.30	II
740		3973.69	II
740		3976.85	II
740		3979.49	II
320		3982.36	II
470		3986.25	II
1400		3990.10	II
1000		3991.74	II
1100		3994.68	II
410		4000.50	II
540		4004.02	II
410		4007.43	II
3700		4012.25	II
540		4012.70	II
370		4018.81	II
1000		4020.87	II
1000		4021.34	II
1000		4021.78	II
1200		4023.00	II
340		4024.78	II
410		4030.47	II
1200		4031.82	II
270		4038.12	II
3000		4040.80	II
200		4041.06	II
410		4043.59	II
410		4048.81	II
850		4051.15	II
850		4059.96	II
4700		4061.09	II
1100		4069.28	II
710		4075.12	II
470		4075.28	II
240		4077.62	II
470		4080.23	II
240		4085.82	II
270		4096.13	II
220		4098.18	II
200		4106.59	II
1400		4109.08	II
2500		4109.46	II
510		4110.48	II
300	h	4113.83	II
410		4123.88	II
470		4133.36	II
510		4135.33	II
3000		4156.08	II
510		4156.26	II
340		4160.57	II
410		4168.00	II
810		4175.61	II
2400		4177.32	II
200		4178.64	II
640		4179.59	II
250		4184.98	II
470		4205.60	II
470		4211.29	II
290		4220.25	II
440		4227.73	II
1300		4232.38	II
250		4234.19	II
290	h	4235.24	II
290		4239.84	II
2000		4247.38	II
850		4252.44	II
290		4254.29	
410		4261.84	II
340		4266.71	II
240		4270.56	II
340		4272.79	II
340		4275.09	II
470		4282.44	II
240		4282.57	II
710		4284.52	II
270		4297.80	II
5400		4303.58	II
340		4304.45	II
200		4307.78	II
470		4314.52	II
1100		4325.76	II
510		4327.93	II
540		4338.70	II
680		4351.29	II
850		4358.17	II
240		4366.38	II
340		4368.64	II
470	d	4374.93	II
		4375.04	II
710		4385.66	II
250		4390.66	II
540		4400.83	II
510		4411.06	II
580		4446.39	II
1400		4451.57	II
200		4451.99	II
300		4456.40	II
740		4462.99	II
410		4501.82	II
200		4506.59	II
170		4513.34	II
250		4516.36	II

Neodymium (Cont.)

Intensity		Wavelength	Ion
120		4527.25	I
340		4541.27	II
340		4542.61	II
100		4556.14	II
170		4559.67	I
340		4563.22	II
200		4578.89	II
200		4579.32	II
100		4586.62	I
200		4597.02	II
100		4603.82	I
100		4609.87	I
300		4621.94	I
100		4627.98	I
510		4634.24	I
340		4641.10	I
250		4645.77	II
200		4646.40	I
300		4649.67	I
200		4654.73	I
130		4670.56	II
170		4680.74	II
310		4683.45	II
110		4684.04	I
110		4690.35	I
190		4696.44	I
130		4703.57	II
470		4706.54	II
140		4706.96	I
190		4709.71	II
190		4715.59	II
240		4719.02	I
190		4724.35	II
140		4731.77	I
120		4779.46	I
170		4789.41	II
120		4797.15	II
240		4811.34	II
140		4820.34	II
350		4825.48	II
130		4832.28	II
110		4849.06	II
280		4859.02	II
190		4866.74	I
350		4883.81	I
140		4889.10	II
220		4890.70	II
240		4891.07	II
280		4896.93	I
120		4901.53	I
210		4901.84	I
110		4902.03	II
190		4913.41	I
170		4914.37	II
330		4920.68	II
470		4924.53	I
260		4944.83	I
290		4954.78	I
290		4959.13	II
150		4961.39	II
250		4989.94	II
150		5033.52	II
110		5063.73	II
360		5076.59	II
150	h	5089.84	II
360		5092.80	II
180		5102.39	II
150	d	5105.21	II
		5105.35	I
360		5107.59	II
340		5123.79	II
680		5130.60	II
170		5132.33	II
170		5165.14	II
130		5181.17	II
120		5182.60	II
500		5191.45	II
630		5192.62	II
330		5200.12	II
310		5212.37	II
150		5213.23	I
130		5225.05	II
130		5228.43	II
450		5234.20	II
250		5239.79	II
720		5249.59	II
200		5250.82	II
360		5255.51	II
120		5269.48	II
590		5273.43	II
150		5276.88	II
110		5291.67	I
680		5293.17	II
160		5302.28	II
110		5306.47	II
220		5311.46	II
500		5319.82	II
180		5356.98	II
290		5361.47	II
150		5371.94	II
110		5385.90	II
160		5431.53	II
110		5451.12	II

Intensity		Wavelength	Ion
170		5485.70	II
35		5501.47	I
45		5525.72	I
90		5533.82	I
55		5535.27	II
55		5543.24	I
55		5548.47	II
55		5561.17	I
27		5575.50	I
27		5576.70	I
27		5577.50	I
27		5587.61	I
240		5594.43	II
55		5601.43	I
45		5601.92	I
220		5620.54	I
65		5635.76	I
45		5639.54	I
35		5653.57	I
70		5668.87	II
65		5669.77	I
140	d	5675.97	I
55		5676.33	I
220		5688.53	II
23		5689.51	I
30		5701.57	I
130		5702.24	II
80		5706.21	II
160		5708.28	II
80		5718.12	II
65		5726.83	II
100		5729.29	I
23		5734.55	I
70		5740.86	II
55		5749.19	I
27		5749.66	I
23		5767.33	I
45		5776.12	I
45		5784.96	I
45		5788.22	I
45		5800.09	I
160		5804.02	II
80		5811.57	II
45		5813.89	I
27		5820.37	I
70		5825.87	II
30		5826.74	I
80		5842.39	II
30		5844.66	I
23		5845.95	I
55		5858.91	I
35		5867.08	I
30		5868.90	I
27		5871.04	I
30		5883.29	I
23		5886.24	I
30		5887.91	I
27		5921.22	I
27		5955.87	I
30		5994.76	I
27		5996.47	I
45		6007.67	I
35		6031.27	II
27		6033.29	I
45		6034.24	II
55		6066.03	I
27		6071.70	I
30		6073.97	I
23	d	6133.47	II
27		6149.28	I
27		6155.06	I
35		6157.83	II
23		6166.67	II
35		6170.49	II
45		6178.59	I
27		6183.91	II
27		6208.24	I
45		6223.39	I
27		6226.50	I
23		6238.50	II
35		6244.08	I
23		6257.49	I
27		6258.73	II
23		6277.29	II
27		6285.79	I
23		6292.84	II
23		6297.07	I
55		6310.49	I
27		6341.51	II
23		6382.07	II
65		6385.20	II
35		6485.69	I
45		6630.14	I
35		6637.96	I
45		6650.57	II
30		6655.67	I
25		6737.79	II
40		6740.11	II
25		6742.54	I
30		6790.37	II
30		6804.00	II
25		6846.72	II
40		6900.43	II

Intensity		Wavelength	Ion
24		6941.39	II
17	h	7010.80	II
8		7018.85	II
17		7020.92	II
17		7024.58	II
10		7033.21	I
35		7037.30	II
7		7052.14	II
7		7054.74	II
40		7066.89	II
8		7082.93	II
12	h	7089.71	II
12	h	7092.09	
12	h	7092.74	II
12	h	7092.94	
17	h	7093.98	I
20	h	7095.42	I
29		7129.35	II
12	h	7142.04	II
10		7143.72	
8		7151.03	II
6		7153.09	I
6	h	7185.01	
10		7189.09	II
24		7189.42	II
20		7192.01	II
10		7199.00	II
8	h	7227.01	I
15		7236.54	II
7	h	7261.64	II
9		7285.29	II
9		7288.56	II
6		7291.38	II
7		7298.72	II
12		7316.81	II
7	h	7321.43	I
7		7323.12	II
6		7334.54	I
6		7357.10	I
6		7374.04	II
7		7381.79	II
9		7401.31	I
10		7406.62	II
6		7411.20	II
10		7418.18	II
9		7427.41	II
9		7448.71	II
5		7481.28	II
12		7511.16	II
17		7513.73	II
7	h	7514.44	II
7		7516.02	II
9		7526.45	II
12		7528.99	II
10		7538.26	II
5		7540.97	II
7		7547.00	II
5		7577.54	II
7		7587.65	II
6		7590.75	II
6		7603.73	II
5		7605.92	II
5		7614.72	I
9		7639.79	II
8		7646.00	II
6		7663.52	II
12		7696.56	II
6		7718.20	II
4		7743.90	II
4		7748.92	II
10		7750.95	II
6		7773.06	II
7		7792.22	II
6		7796.40	II
8		7797.32	II
5		7798.32	II
10		7808.47	II
7		7818.83	II
5		7825.20	II
12		7863.04	II
5	h	7872.03	I
7		7886.60	II
4	h	7896.50	II
9		7900.40	II
5	h	7906.03	I
12		7917.01	II
10		7925.03	II
5		7947.93	II
10		7949.68	II
5	h	7955.38	II
12		7958.95	I
12		7965.73	II
15		7982.09	II
12		7982.68	II
12		8000.76	II
9		8007.70	I
4	h	8020.07	II
8		8026.35	
10		8043.24	I
8		8051.33	II
5		8064.00	II
10		8099.17	I

Intensity		Wavelength	Ion
10		8120.93	II
12		8122.07	II
12		8141.75	II
12		8143.27	II
7	h	8164.97	I
8		8172.56	II
9		8179.83	II
9		8182.41	II
4		8185.58	II
7	h	8205.38	II
10		8231.52	II
4		8248.76	II
5	h	8249.68	II
4	h	8262.80	II
7	h	8266.72	II
4		8272.79	II
4	h	8302.74	II
10		8307.72	II
6		8324.50	II
4		8332.01	II
12		8346.36	II
4		8375.16	II
4		8375.33	II
4		8394.71	II
7		8400.85	II
5	h	8456.87	II
4		8530.53	II
5		8582.03	II
5		8591.53	II
7		8594.87	II
8	c	8643.43	II
5		8667.07	II
5		8677.48	
6		8691.29	II
6		8695.07	II
6		8712.82	II
6		8715.03	I
17		8839.10	II

NEON (Ne)
Z = 10

Ne I and II
Ref. 56, 58, 118, 150, 230 — S.P.D.

Intensity		Wavelength	Ion
		Vacuum	
90		352.956	II
60		354.962	II
90		361.433	II
60		362.455	II
150		405.854	II
120		407.138	II
200		445.040	II
300		446.256	II
250		446.590	II
180		447.815	II
150		454.654	II
200		455.274	II
10		456.275	II
120		456.348	II
90		456.896	II
1000		460.728	II
500		462.391	II
35		587.213	I
35		589.179	I
35		589.911	I
70		591.830	I
100		595.920	I
75		598.706	I
35		598.891	I
70		600.036	I
170		602.726	I
170		615.628	I
170		618.672	I
120		619.102	I
200		626.823	I
200		629.739	I
1000		735.896	I
400		743.720	I
60		993.88	II
70		1068.65	II
90		1131.72	II
100		1131.85	II
90		1229.83	II
90		1418.38	II
90		1428.58	II
90		1436.09	II
120		1681.68	II
180		1688.36	II
100		1888.11	II
100		1889.71	II
200		1907.49	II
500		1916.08	II
300		1930.03	II
200		1938.83	II
100	c	1945.46	II
		AIR	

Neon (Cont.)

Intensity		Wavelength	
80		2007.01	II
80		2025.56	II
150		2085.47	II
180		2096.11	II
120		2096.25	II
80	p	2562.12	II
90	w	2567.12	II
80		2623.11	II
80		2629.89	II
90	w	2636.07	II
80		2638.29	II
80		2644.10	II
80		2762.92	II
90		2792.02	II
80		2794.22	II
100		2809.48	II
80		2906.59	II
80		2906.82	II
90		2910.06	II
90		2910.41	II
80		2911.14	II
80		2915.12	II
80		2925.62	II
80	w	2932.10	II
80		2940.65	II
90		2946.04	II
150		2955.72	II
150		2963.24	II
150		2967.18	II
100		2973.10	II
15		2974.72	I
100		2979.46	II
12		2982.67	I
150		3001.67	II
120	p	3017.31	II
300		3027.02	II
300		3028.86	II
100		3030.79	II
120		3034.46	II
100		3035.92	II
100		3037.72	II
100		3039.59	II
100		3044.09	II
100		3045.56	II
120		3047.56	II
100		3054.34	II
100		3054.68	II
100		3059.11	II
100		3062.49	II
100		3063.30	II
100		3070.89	II
100		3071.53	II
100		3075.73	II
120		3088.17	II
100		3092.09	II
120		3092.90	II
100		3094.01	II
100		3095.10	II
100		3097.13	II
100		3117.98	II
120		3118.16	II
10		3126.199	I
300		3141.33	II
100		3143.72	II
100	p	3148.68	II
100		3164.43	II
100		3165.65	II
100		3188.74	II
120		3194.58	II
500		3198.59	II
60		3208.96	II
120		3209.36	II
120		3213.74	II
150		3214.33	II
150		3218.19	II
120		3224.82	II
120		3229.57	II
200		3230.07	II
120		3230.42	II
120		3232.02	II
150		3232.37	II
100		3243.40	II
100		3244.10	II
100		3248.34	II
100		3250.36	II
150		3297.73	II
150		3309.74	II
300		3319.72	II
1000		3323.74	II
150		3327.15	II
100		3329.16	II
200		3334.84	II
150		3344.40	II
300		3345.45	II
150		3345.83	II
200		3355.02	II
120		3357.82	II
200		3360.60	II
120		3362.16	II
100		3362.71	II
120		3367.22	II
12		3369.808	I
40		3369.908	I

Intensity		Wavelength	
100		3371.80	II
500		3378.22	II
150		3388.42	II
120		3388.94	II
300		3392.80	II
100		3404.82	II
120		3406.95	II
100		3413.15	II
120		3416.91	II
120		3417.69	II
50		3417.904	I
15		3418.006	I
120		3428.69	II
60		3447.703	I
50		3454.195	I
100		3456.61	II
100		3459.32	II
25		3460.524	I
30		3464.339	I
30		3466.579	I
60		3472.571	I
150		3479.52	II
200		3480.72	II
200		3481.93	II
25		3498.064	I
30		3501.216	I
25		3515.191	I
150		3520.472	I
120		3542.85	II
120		3557.80	II
100		3561.20	II
250		3568.50	II
100		3574.18	II
200		3574.61	II
50		3593.526	I
30		3593.640	I
15		3600.169	I
20		3633.665	I
150		3643.93	II
200		3664.07	II
20		3682.243	I
12		3685.736	I
200		3694.21	II
10		3701.225	I
150		3709.62	II
250		3713.08	II
250		3727.11	II
800		3766.26	II
1000		3777.13	II
100		3818.43	II
120		3829.75	II
150		4219.74	II
100		4233.85	II
120		4250.65	II
120		4369.86	II
70		4379.40	II
150		4379.55	II
100		4385.06	II
200		4391.99	II
150		4397.99	II
150		4409.30	II
100		4413.22	II
100		4421.39	II
100	p	4428.52	II
100	p	4428.63	II
150	p	4430.90	II
150	p	4430.94	II
120		4457.05	II
100		4522.72	II
10		4537.754	I
10		4540.380	I
100		4569.06	II
15		4704.395	I
12		4708.862	I
10		4710.067	I
10		4712.066	I
15		4715.347	I
10		4752.732	I
12		4788.927	I
10		4790.22	I
10		4827.344	I
10		4884.917	I
4		5005.159	I
10		5037.751	I
10		5144.938	I
25		5330.778	I
20		5341.094	I
8		5343.283	I
60		5400.562	I
5		5562.766	I
10		5656.659	I
5		5719.225	I
12		5748.298	I
80		5764.419	I
12		5804.450	I
40		5820.156	I
500		5852.488	I
100		5872.828	I
100		5881.895	I
60		5902.462	I
60		5906.429	I
100		5944.834	I
100		5965.471	I

Intensity	Wavelength	
100	5974.627	I
120	5975.534	I
80	5987.907	I
100	6029.997	I
100	6074.338	I
80	6096.163	I
60	6128.450	I
100	6143.063	I
120	6163.594	I
250	6182.146	I
150	6217.281	I
150	6266.495	I
60	6304.789	I
7	6328.165	I
100	6334.428	I
120	6382.992	I
200	6402.246	I
150	6506.528	I
60	6532.882	I
150	6598.953	I
70	6652.093	I
90	6678.276	I
20	6717.043	I
100	6929.467	I
90	7024.050	I
100	7032.413	I
50	7051.292	I
80	7059.107	I
100	7173.938	I
150	7213.20	II
150	7235.19	II
100	7245.167	I
150	7343.94	II
40	7472.439	I
90	7488.871	I
100	7492.10	II
150	7522.82	II
80	7535.774	I
60	7544.044	I
100	7724.628	I
120	7740.74	II
300	7839.055	I
120	7926.20	II
400	7927.118	I
700	7936.996	I
2000	7943.181	I
2000	8082.458	I
100	8084.34	II
1000	8118.549	I
600	8128.911	I
3000	8136.406	I
2500	8259.379	I
100	8264.81	II
2500	8266.077	I
800	8267.117	I
6000	8300.326	I
100	8315.00	II
1500	8365.749	I
100	8372.11	II
8000	8377.606	I
1000	8417.159	I
4000	8418.427	I
1500	8463.358	I
800	8484.444	I
5000	8495.360	I
600	8544.696	I
1000	8571.352	I
4000	8591.259	I
6000	8634.647	I
3000	8647.041	I
15000	8654.383	I
4000	8655.522	I
100	8668.26	II
5000	8679.492	I
5000	8681.921	I
2000	8704.112	I
4000	8771.656	I
12000	8780.621	I
10000	8783.753	I
500	8830.907	I
7000	8853.867	I
1000	8865.306	I
1000	8865.755	I
3000	8919.501	I
2000	8988.57	I
100	9079.46	II
6000	9148.67	I
6000	9201.76	I
4000	9220.06	I
2000	9221.58	I
2000	9226.69	I
1000	9275.52	I
200	9287.56	II
6000	9300.85	I
1500	9310.58	I
3000	9313.97	I
6000	9326.51	I
2000	9373.31	I
5000	9425.38	I
3000	9459.21	I
5000	9486.68	I
5000	9534.16	I
3000	9547.40	I

Intensity	Wavelength	
120	9577.01	II
1000	9665.42	I
100	9808.86	II
800	10295.42	I
2000	10562.41	I
1500	10798.07	I
2000	10844.48	I
3000	11143.020	I
3500	11177.528	I
1600	11390.434	I
1100	11409.134	I
3000	11522.746	I
1500	11525.020	I
950	11536.344	I
500	11601.537	I
1200	11614.081	I
300	11688.002	I
2000	11766.792	I
1500	11789.044	I
500	11789.889	I
1000	11984.912	I
3000	12066.334	I
800	12459.389	I
1000	12689.201	I
1100	12912.014	I
700	13219.241	I
800	15230.714	I
400	17161.930	I
400	18035.80	I
1000	18083.21	I
350	18221.11	I
250	18227.02	I
2500	18276.68	I
2000	18282.62	I
1200	18303.97	I
250	18359.12	I
1200	18384.85	I
2000	18389.95	I
1000	18402.84	I
1200	18422.39	I
300	18458.65	I
400	18475.79	I
900	18591.55	I
1600	18597.70	I
350	18618.96	I
550	18625.16	I
1200	21041.295	I
750	21708.145	I
300	22247.35	I
350	22428.13	I
2250	22530.40	I
400	22661.81	I
600	23100.51	I
1000	23260.30	I
1050	23373.00	I
850	23565.36	I
3500	23636.52	I
300	23701.64	I
1100	23709.2	I
1800	23951.42	I
600	23956.46	I
1000	23978.12	I
200	24098.54	I
500	24161.42	I
600	24249.64	I
1500	24365.05	I
800	24371.60	I
400	24447.85	I
700	24459.4	I
300	24776.46	I
550	24928.88	I
250	25161.69	I
650	25524.37	I
125	28386.21	I
150	30200.	I
250	33173.09	I
450	33352.35	I
1300	33901.	I
2200	33912.10	I
600	34131.31	I
100	34471.44	I
120	35834.78	I

Ne III
Ref. 365, 371, 402 — R.L.K.

Intensity	Wavelength	
	Vacuum	
20	251.14	III
20	251.56	III
20	251.73	III
40	267.06	III
40	267.52	III
20	267.71	III
40	283.18	III
160	283.21	III
110	283.69	III
40	283.89	III
220	301.12	III
220	313.05	III
220	313.68	III
40	313.95	III

Neon (Cont.)

Intensity		Wavelength	
220		379.31	III
285		488.10	III
220		488.87	III
450		489.50	III
70		489.64	III
220		490.31	III
360		491.05	III
20		1255.03	III
110		1255.68	III
160		1257.19	III

Air

200	2086.96	III
300	2089.43	III
240	2092.44	III
400	2095.54	III
200	2161.22	III
300	2163.77	III
200	2180.89	III
200	2209.35	III
200	2211.85	III
240	2213.76	III
300	2216.07	III
240	2263.21	III
200	2264.91	III
300	2412.73	III
240	2412.94	III
200	2413.78	III
200	2473.40	III
800	2590.04	III
600	2593.60	III
400	2595.68	III
300	2610.03	III
240	2613.41	III
200	2615.87	III
200	2638.70	III
200	2641.07	III
600	2677.90	III
500	2678.64	III

Ne IV
Ref. 69, 364, 388, 400, 413, 430, — R.L.K.

Intensity	Wavelength	

Vacuum

15	151.82	IV
15	152.23	IV
15	158.65	IV
15	158.82	IV
80	172.62	IV
80	177.16	IV
150	186.58	IV
100	194.28	IV
100	208.48	IV
100	208.73	IV
80	208.90	IV
150	212.56	IV
140	223.24	IV
120	223.60	IV
140	234.32	IV
120	234.70	IV
50	357.83	IV
200	358.72	IV
125	387.14	IV
100	388.22	IV
150	421.61	IV
140	469.77	IV
200	469.82	IV
180	469.87	IV
140	469.92	IV
120	521.74	IV
140	521.82	IV
80	541.13	IV
100	542.07	IV
150	543.89	IV

Air

65	2018.44	IV
110	2022.19	IV
30	2203.88	IV
10	2220.81	IV
250	2258.02	IV
175	2262.08	IV
110	2264.54	IV
550	2285.79	IV
30	2293.14	IV
250	2293.49	IV
250	2363.28	IV
110	2365.49	IV
250	2350.84	IV
450	2352.52	IV
700	2357.96	IV
250	2362.68	IV
350	2372.16	IV
65	2384.20	IV
350	2384.95	IV

Ne V
Ref. 69, 388, 389, 400, 413 — R.L.K.

Intensity	Wavelength	

Vacuum

66	119.01	V
200	122.52	V
66	125.12	V
45	131.99	V
50	132.04	V
150	140.76	V
150	140.79	V
100	142.44	V
100	142.50	V
150	142.72	V
100	143.27	V
150	143.34	V
150	147.13	V
66	151.23	V
120	151.42	V
45	154.50	V
100	164.02	V
100	164.14	V
500	173.93	V
400	357.96	V
500	358.47	V
500	359.38	V
1000	365.59	V
800	416.20	V
250	480.41	V
150	481.28	V
250	481.36	V
500	482.99	V
400	568.42	V
250	569.76	V
500	569.83	V
250	572.11	V
800	572.34	V

Air

75	2227.42	V
110	2232.41	V
65	2245.48	V
65	2259.57	V
65	2263.39	V
250	2265.71	V

NEPTUNIUM (Np)
Z = 93
Np I and II
Ref. 93 = J.G.C.

Intensity		Wavelength	

Air

300		3481.93	I
300	h	3501.50	I
300	l	3986.89	I
300	s	5044.66	I
300	l	5601.70	I
300	l	5652.75	I
300	l	5784.39	I
300	l	5878.04	I
300	s	6011.22	I
300		6056.09	I
300	s	6073.90	I
300	s	6080.05	I
300	l	6120.49	I
300		6188.59	I
300	l	6200.00	I
300	s	6215.90	I
300	l	6317.84	I
300	l	6341.38	I
300	l	6566.11	I
300	l	6720.68	I
300	s	6751.32	I
300	s	6795.21	I
300	l	6802.62	I
300	l	6805.81	I
300	s	6816.44	I
300	l	6865.45	I
300	s	6907.13	I
300	h	6912.91	I
1000	s	6930.31	I
300	l	6963.63	I
3000	s	6972.09	I
300		7014.02	I
300	l	7018.91	I
300	s	7039.14	I
300	s	7080.01	I
300	l	7174.83	I
300	l	7184.93	I
300	l	7284.28	I
300	l	7292.29	I
300	l	7332.52	I

300	s	7370.60	I
300	l	7381.03	I
300	l	7381.65	I
300	l	7402.70	I
300	s	7512.22	I
300	l	7515.15	I
300	l	7546.05	I
300	l	7624.83	I
300		7626.85	I
300	s	7681.01	I
300		7685.25	I
1000	l	7735.14	I
300	l	7761.61	I
1000	l	7765.75	I
300	s	7776.07	I
300		7787.46	I
1000	l	7791.38	I
300	l	7851.44	I
300	l	7887.88	I
300	l	7901.71	I
300	l	7975.98	I
300	l	8080.32	I
300	s	8124.59	I
300		8155.11	I
300	l	8167.42	I
300	l	8183.06	I
300	l	8188.61	I
300	l	8247.82	I
300	l	8287.11	I
300	s	8287.75	I
300	l	8306.22	I
300	s	8313.66	I
1000	l	8339.12	I
300		8356.79	I
300	l	8367.11	I
3000		8372.88	I
3000		8529.96	I
1000	s	8696.23	I
1000	s	8906.02	I
1000		8942.70	I
1000	s	9004.75	I
1000	l	9006.31	I
10000	l	9016.18	I
3000	l	9141.30	I
3000		9379.33	I
3000	l	9468.66	I
3000	l	9679.13	I
3000	l	9930.55	I
10000	l	10091.99	I
10000	s	10817.45	I
10000	l	11695.15	I
10000	l	11776.64	I
10000	s	12148.18	I
10000	s	12377.42	I
10000	l	12407.99	I
10000	l	13834.33	I

NICKEL (Ni)
Z = 28
Ni I and II
Ref. 1,294 = C.H.C.

Intensity	Wavelength	

Vacuum

500	1317.22	II
400	1335.20	II
500	1370.14	II
1000	1741.55	II
500	1748.28	II

Air

1000	2165.55	II
2000	2169.10	II
2000	2174.67	II
1500	2175.15	II
500	2177.09	II
400	2177.36	II
400	2179.35	II
800	2180.47	II
800	2184.60	II
2500	2185.50	II
3000	2192.09	II
600	2201.41	II
5000	2205.55	II
4000	2206.72	II
6000	2216.48	II
800	2220.40	II
500	2221.06	II
900	2222.96	II
500	2242.68	II
500	2253.85	II
1000	2264.46	II
2000	2270.21	II
800	2277.28	II
400	2278.32	II
800	2278.77	II
500	2287.65	II
1600	2289.98	I

400	2296.55	II
400	2297.14	II
630	2300.78	I
1000	2303.00	II
2000	2310.96	I
1700	2312.34	I
1400	2313.66	I
1400	2313.98	I
1000	2316.04	II
1400	2317.16	I
500	2319.75	II
2600	2320.03	I
1900	2321.38	I
240	2322.68	I
1400	2325.79	I
940	2329.96	I
500	2334.58	II
460	2337.49	I
160	2337.82	I
500	2341.20	II
1200	2345.54	I
190	2346.63	I
400	2347.52	I
160	2360.63	I
200	2362.06	I
1000	2375.42	II
240	2386.58	I
1000	2394.52	II
2000	2416.13	II
240	2419.31	I
85	2421.23	I
70	2423.33	I
70	2423.66	I
70	2424.03	I
500	2437.89	II
85	2453.99	I
160	2472.06	I
85	2476.87	I
500	2510.87	II
500	2565.92	II
500	2606.26	II
500	2609.94	II
500	2615.06	II
45	2696.49	I
150	2798.65	I
250	2821.29	I
500	2864.02	II
50	2865.50	I
60	2907.46	I
25	2914.01	I
500	2943.91	I
570	2981.65	I
250	2984.13	I
500	2992.60	I
1000	2994.46	I
4000	3002.49	I
2200	3003.63	I
3700	3012.00	I
350	3019.14	I
120	3031.87	I
1700	3037.94	I
150	3045.01	I
3500	3050.82	I
1500	3054.32	I
1900	3057.64	I
500	3064.62	I
420	3080.76	I
260	3097.12	I
210	3099.12	I
2600	3101.55	I
1300	3101.88	I
220	3105.47	I
270	3114.12	I
2900	3134.11	I
55	3145.72	I
55	3181.74	I
100	3184.37	I
55	3195.57	I
150	3197.11	I
55	3202.14	I
180	3214.06	I
180	3217.83	I
100	3221.27	I
150	3221.65	I
210	3225.02	I
1100	3232.96	I
290	3234.65	I
600	3243.06	I
100	3248.46	I
120	3250.74	I
100	3271.12	I
120	3282.70	I
400	3292.87	II
500	3297.60	II
400	3305.71	II
660	3315.66	I
330	3320.26	I
310	3322.31	I
2000	3331.88	II
400	3335.64	II
500	3338.09	II
500	3348.84	II
500	3349.24	II

Intensity		Wavelength	Spectrum
600		3358.68	II
330		3361.56	I
500		3363.45	II
330		3365.77	I
330		3366.17	I
65		3366.81	I
65		3367.89	I
2900		3369.57	I
400		3371.99	I
260		3374.22	I
130		3374.64	I
500		3378.97	II
3300		3380.57	I
240		3380.85	I
1300		3391.05	I
3300		3392.99	I
500		3401.05	II
130		3409.58	I
330		3413.48	I
330		3413.94	I
8200		3414.76	I
1600		3423.71	I
2600		3433.56	I
990		3437.28	I
4800		3446.26	I
1300		3452.89	I
5000		3458.47	I
5000		3461.65	I
200		3467.50	I
240		3469.49	I
1600		3472.54	I
550		3483.77	I
130		3485.89	I
5500		3492.96	I
660		3500.85	I
65		3502.60	I
55		3507.69	I
2600		3510.34	I
260		3513.93	I
6600		3515.05	I
660		3519.77	I
8200		3524.54	I
110		3527.98	I
330		3548.18	I
55		3551.53	I
65		3561.75	I
5000		3566.37	I
990		3571.87	I
130		3587.93	I
1300		3597.70	I
1300		3610.46	I
530		3612.74	I
6600		3619.39	I
130		3624.73	I
200		3664.10	I
130		3669.24	I
180		3670.43	I
260		3674.15	I
160		3688.42	I
80		3693.93	I
120		3722.48	I
150		3736.81	I
60		3739.23	I
600		3775.57	I
700		3783.53	I
700		3807.14	I
110		3831.69	I
1200		3858.30	I
30		3889.67	I
35		3972.17	I
110		3973.56	I
110		4401.55	I
85		4459.04	I
18		4462.46	I
55		4470.48	I
35		4592.53	I
18		4600.37	I
65		4605.00	I
18		4606.23	I
75		4648.66	I
23		4686.22	I
110		4714.42	I
22		4715.78	I
30		4756.52	I
15		4763.95	I
45		4786.54	I
22		4807.00	I
22	h	4829.03	I
19		4831.18	I
45		4855.41	I
30		4866.27	I
17		4873.44	I
40		4904.41	I
22		4918.36	I
16		4935.83	I
45		4980.16	I
45		4984.13	I
500		4992.02	II
16	h	5000.34	I
18		5012.46	I
50		5017.59	I
100		5035.37	I
16		5048.85	I

Intensity		Wavelength	Spectrum
100		5080.52	I
65		5081.11	I
26	h	5084.08	I
18		5099.32	I
26	h	5099.95	I
21		5115.40	I
18	h	5129.38	I
23		5137.08	I
23	h	5142.77	I
40	h	5146.48	I
40	h	5155.76	I
16		5168.66	I
13		5176.56	I
8		5435.87	I
180		5476.91	I
6		5510.00	I
6		5578.73	I
9		5587.86	I
13		5592.28	I
9		5614.79	I
5	h	5625.33	I
4		5649.70	I
5		5664.02	I
12		5682.20	I
8		5695.00	I
23		5709.56	I
10		5711.90	I
10		5715.09	I
16		5754.68	I
8		5760.85	I
10		5857.76	I
10		5892.88	I
10		6108.12	I
10		6176.81	I
10		6191.18	I
13		6256.36	I
10		6314.66	I
16		6643.64	I
22		6767.77	I
9		6772.32	I
10		6914.56	I
5		7110.90	I
26		7122.20	I
6		7182.00	I
5		7197.02	I
5		7261.93	I
5		7291.45	I
4		7385.24	I
16		7393.60	I
16		7409.35	I
5		7414.51	I
23		7422.28	I
13		7522.76	I
9		7525.12	I
19		7555.60	I
8		7574.05	I
23		7617.00	I
9		7619.21	I
16		7714.32	I
5	h	7715.58	I
19		7727.61	I
19		7748.89	I
10		7788.94	I
13		7797.59	I
2		7917.44	I
1000		8096.75	II
500		8114.21	II
700		8121.48	II
2		8809.42	I
9		8862.55	I
500	w	9900.92	II

Ni III
Ref. 422 — C.H.C.

Intensity	Wavelength Vacuum	Spectrum
100	625.68	III
500	630.71	III
200	637.54	III
200	662.37	III
150	663.57	III
500	676.94	III
200	700.17	III
300	713.33	III
300	713.38	III
500	718.48	III
200	721.26	III
300	722.09	III
250	725.20	III
500	729.82	III
250	730.11	III
400	731.70	III
300	732.16	III
200	738.26	III
300	747.99	III
200	749.68	III
300	750.05	III
200	752.02	III
300	757.80	III

Intensity		Wavelength	Spectrum
250		758.73	III
250		758.27	III
400		770.22	III
200		772.04	III
500		778.81	III
200	d	785.02	III
300		788.04	III
200		788.30	III
200		805.01	III
500		811.57	III
500		826.14	III
200		826.50	III
500		842.14	III
400		845.24	III
300		847.43	III
200		857.09	III
300		860.64	III
300		862.88	III
300		863.22	III
300		867.51	III
200		869.70	III
200		870.84	III
300		973.79	III
400		979.59	III
200		1428.87	III
200		1434.31	III
200		1451.50	III
200		1604.54	III
300		1652.87	III
200		1653.12	III
250		1656.13	III
200		1661.79	III
400		1687.90	III
1000		1692.51	III
200		1707.35	III
200		1707.43	III
800		1709.90	III
650		1715.30	III
500		1719.46	III
200		1721.26	III
400		1722.28	III
250		1733.13	III
500		1738.25	III
300		1739.78	III
300		1741.96	III
550		1747.01	III
300		1752.43	III
400		1753.01	III
800		1764.69	III
500		1767.94	III
2000		1769.64	III
400		1776.07	III
200		1788.30	III
250		1790.40	III
200		1790.93	III
200		1791.64	III
200		1794.90	III
300		1807.24	III
200		1811.69	III
300		1819.28	III
800		1823.06	III
400		1830.01	III
200		1830.08	III
650		1847.28	III
800		1854.15	III
300		1858.75	III
200		1930.43	III
200		1952.54	III

Ni IV
Ref. 415 — C.H.C.

Intensity	Wavelength Vacuum	Spectrum
33	392.68	IV
32	393.24	IV
49	424.40	IV
57	444.21	IV
67	469.67	IV
65	471.24	IV
65	485.42	IV
66	536.28	IV
67	537.96	IV
58	1345.72	IV
69	1357.07	IV
76	1398.19	IV
74	1411.45	IV
69	1419.58	IV
74	1421.22	IV
70	1427.45	IV
67	1428.93	IV
67	1435.24	IV
70	1438.82	IV
73	1449.01	IV
76	1452.22	IV
70	1455.42	IV
69	1472.63	IV
68	1476.82	IV
73	1482.25	IV
67	1489.53	IV

Intensity	Wavelength	Spectrum
72	1489.83	IV
69	1493.01	IV
74	1493.67	IV
68	1498.71	IV
71	1498.77	IV
72	1498.90	IV
67	1499.97	IV
70	1512.74	IV
70	1516.66	IV
73	1520.63	IV
75	1525.31	IV
74	1527.68	IV
74	1527.80	IV
76	1534.71	IV
73	1537.25	IV
69	1538.93	IV
75	1543.41	IV
74	1546.23	IV
68	1548.04	IV
69	1557.28	IV
67	1560.18	IV

Ni V
Ref. 416 — C.H.C.

Intensity	Wavelength Vacuum	Spectrum
29	304.02	V
55	315.24	V
56	315.71	V
63	336.79	V
68	343.93	V
78	347.34	V
70	347.46	V
67	347.72	V
71	348.10	V
69	350.77	V
69	353.59	V
72	354.18	V
76	354.42	V
68	354.49	V
68	355.61	V
70	355.78	V
65	357.37	V
69	358.57	V
68	358.58	V
66	359.47	V
69	365.62	V
70	370.62	V
67	371.31	V
68	371.76	V
67	373.60	V
72	377.68	V
70	393.91	V
66	394.31	V
66	395.24	V
41	400.59	V

# NIOBIUM (Nb)
Z = 41

Nb I and II
Ref. I = C.H.C.

Intensity	Wavelength Air	Spectrum
3300	2029.32	II
3000	2032.99	II
2000	2109.42	II
1700	2125.21	II
1100	2126.54	II
1500	2131.18	II
370	2295.68	II
280	2302.08	II
170	2376.40	II
110	2387.09	II
140	2387.52	II
45	2388.27	II
160	2398.48	II
55	2405.34	II
55	2405.85	II
140	2412.46	II
160	2416.99	II
140	2418.69	II
75	2433.80	II
40	2435.95	II
35	2436.33	I
45	2437.42	II
40	2442.14	II
28	2442.68	II
65	2451.87	II
65	2453.95	II
55	2458.09	II
65	2462.89	I
35	2466.73	I
55	2469.08	I
110	2477.38	II
65	2478.29	II

Niobium (Cont.)

Int	λ	Sp	Int	λ	Sp	Int	λ	Sp	Int	λ	Sp
65	2479.94	II	140	3099.19	II	250	3544.65	I	870	4192.07	I
35	2483.88	II	150	3111.45	I	300	3550.45	I	870	4195.09	I
110	2504.65	I	270	3127.53	II	250	3554.52	I	1300	4195.66	I
110	2511.00	II	1500	3130.79	II	1000	3554.66	I	310	4198.51	I
110	2521.40	II	390	3145.40	II	630	3563.50	I	350	4201.52	I
390	2544.80	II	140	3151.87	I	630	3563.62	I	870	4205.31	I
110	2551.38	II	1200	3163.40	II	1500	3575.85	I	350	4214.73	I
130	2556.94	II	150	3175.78	II	200	3577.72	I	420	4217.94	I
130	2562.41	II	390	3180.29	II	5000	3580.27	I	420	4229.15	I
130	2565.41	I	200	3187.49	I	500	3584.97	I	250	4255.44	I
100	2569.03	I	300	3191.10	II	750	3589.11	I	770	4262.05	I
110	2571.33	II	150	3191.43	II	500	3589.36	I	420	4266.02	I
200	2578.74	I	1000	3194.98	II	500	3593.97	I	290	4270.69	I
390	2583.99	II	120	3203.35	II	500	3602.56	I	400	4286.99	I
390	2590.94	II	300	3206.34	II	300	3619.51	II	580	4299.60	I
270	2592.20	I	390	3215.60	II	200	3621.03	I	580	4300.99	I
130	2616.48	I	800	3225.48	II	420	3639.33	I	120	4309.56	I
130	2623.51	II	140	3229.56	II	250	3649.85	I	390	4311.27	I
130	2627.44	I	400	3236.40	II	400	3650.81	I	120	4312.45	I
130	2628.49	I	200	3247.47	II	200	3651.19	II	350	4326.33	I
200	2642.24	II	120	3248.94	II	630	3659.61	II	120	4327.38	I
320	2646.26	II	160	3249.52	I	900	3660.37	I	390	4331.37	I
330	2647.50	I	320	3254.07	II	220	3664.70	I	140	4342.82	I
240	2649.52	I	230	3260.56	II	270	3669.01	I	140	4348.65	I
330	2654.45	I	160	3263.37	II	1500	3674.78	I	110	4349.03	I
310	2656.08	II	160	3264.59	I	330	3697.85	I	290	4351.57	I
160	2657.62	I	120	3270.47	I	3300	3711.34	I	210	4368.43	I
110	2665.25	II	100	3270.76	I	480	3713.01	I	140	4377.96	I
110	2666.59	II	200	3272.07	I	2700	3716.99	I	130	4388.36	I
110	2667.30	II	160	3277.67	I	270	3726.24	I	160	4392.69	I
130	2668.29	I	200	3283.46	II	2700	3738.42	I	330	4410.21	I
400	2671.93	II	230	3285.66	I	670	3739.80	I	190	4419.44	I
200	2673.57	II	200	3287.59	I	270	3740.73	II	230 c	4437.22	I
200	2675.94	II	160	3287.92	I	1700	3741.78	I	290	4447.18	I
130	2687.15	I	160	3292.02	II	250	3742.39	I	140	4456.80	I
160	2691.77	II	320	3296.01	I	210	3753.18	I	140	4457.42	I
1000	2697.06	II	160	3299.61	I	530	3755.77	I	140	4469.71	I
320	2698.86	II	120	3304.83	I	350	3763.49	I	140	4471.29	I
320	2702.20	II	120	3308.05	I	250	3765.08	I	140	4472.53	I
150	2702.52	II	120	3310.47	I	530	3766.13	I	150	4503.04	I
470	2716.62	II	400	3312.60	I	870	3771.85	I	530	4523.41	I
470	2721.98	II	200	3315.22	I	1700	3781.01	I	480	4546.82	I
310	2733.26	II	200	3318.98	I	1300	3787.06	I	370	4564.53	I
110	2737.09	II	120	3319.26	I	3500	3790.15	I	720	4573.08	I
200	2746.91	I	120	3319.58	II	2700	3791.21	I	480	4581.62	I
200	2748.85	I	240	3326.62	I	270	3798.12	I	1200	4606.77	I
190	2753.01	I	170	3329.36	I	2700	3801.30	I	170	4616.17	I
280	2758.61	I	110	3332.16	I	670	3802.92	I	450	4630.11	I
240	2768.13	II	130	3341.60	II	530	3803.88	I	450	4648.95	I
310	2773.20	I	1300	3341.97	I	670	3804.74	I	110	4649.27	I
270	2780.24	II	1300	3343.71	I	530	3810.49	I	450	4663.83	I
130	2782.36	I	130	3346.93	I	530	3811.03	I	340	4666.24	I
110	2793.05	II	1700	3349.06	I	210	3815.51	I	240	4667.22	I
190	2827.08	II	420	3349.52	I	210	3818.86	II	580	4672.09	I
150	2836.24	I	340	3354.74	I	670	3819.15	I	530	4675.37	I
110	2840.94	I	130	3357.04	I	350	3824.88	I	110	4678.48	I
250	2841.15	II	1700	3358.42	I	250	3835.18	I	320	4685.14	I
280	2842.65	II	130	3365.58	II	210	3836.45	I	130 c	4706.14	I
160	2846.28	II	340	3366.96	I	290	3845.90	I	260	4708.29	I
110	2851.45	I	130	3369.16	II	350	3858.95	I	150	4713.50	I
240	2861.09	II	170	3371.33	I	270	3863.38	I	110 c	4733.89	I
100	2864.32	I	350	3374.92	I	530	3867.92	I	220	4749.70	I
100	2865.61	II	270	3380.41	I	870	3877.56	I	110	4816.38	I
500	2868.52	II	130	3380.86	I	670	3878.82	I	110 c	4848.37	I
800	2875.39	II	170	3386.24	II	1100	3883.14	I	130 c	4967.78	I
270	2876.95	II	350	3392.34	I	670	3885.44	I	110	4973.14	I
530	2877.03	II	170	3395.93	I	210	3885.68	I	190	4988.97	I
100	2880.72	II	120	3399.40	I	580	3886.07	I	85	5000.95	I
570	2883.18	II	230	3405.41	I	210	3891.30	I	65	5002.25	I
280	2888.83	II	130	3406.13	I	670	3908.97	I	40	5013.27	I
470	2897.81	II	270	3408.38	I	530	3914.70	I	230	5017.75	I
400	2899.24	II	230	3408.68	II	670	3920.20	I	40	5019.51	I
470	2908.24	II	180	3409.19	II	520	3937.44	I	150	5026.36	I
670	2910.59	II	230	3412.94	II	250	3943.67	I	40	5030.13	I
470	2911.74	II	180	3415.97	I	250	3965.69	I	210	5039.04	I
1100	2927.81	II	180	3423.76	I	910 d	3966.09	I	40	5047.96	I
110	2931.47	II	230	3425.42	II	210	3971.85	I	170	5058.01	I
870	2941.54	II	130	3425.85	I	1100	4032.52	I	65	5059.35	I
110 h	2945.88	II	230	3426.57	II	250	4039.53	I	130	5065.25	I
110	2946.12	II	230	3427.45	I	16000 c	4058.94	I	40	5077.40	I
110	2946.90	II	130	3429.04	I	210	4059.51	I	750	5078.96	I
1100	2950.88	II	180	3432.70	II	350	4060.79	I	40 c	5094.41	I
400	2972.57	II	180	3440.59	II	210	4068.26	I	420	5095.30	I
320	2974.10	II	170	3463.81	I	12000	4079.73	I	170	5100.16	I
210	2977.68	II	180	3465.86	I	270	4084.86	I	170	5120.30	I
200	2982.11	II	130	3469.44	I	440	4100.40	I	85	5121.80	I
330	2990.26	II	100	3471.19	I	6700	4100.92	I	85	5127.66	I
470	2994.73	II	140	3473.02	I	310	4116.90	I	40	5133.34	I
140	3024.74	II	290	3478.69	I	5300	4123.81	I	210	5134.75	I
350	3028.44	II	200	3479.56	II	670	4129.43	I	75	5140.58	I
300	3032.77	II	100	3484.05	II	770	4129.93	I	75	5147.54	I
100	3044.76	II	230	3491.03	I	2300	4137.10	I	40	5150.64	I
150	3048.10	I	200	3497.81	I	440	4139.44	I	75	5152.63	I
110	3053.09	I	500	3498.63	I	2700	4139.71	I	250	5160.33	I
100	3055.52	II	460	3507.96	I	350	4143.21	I	250	5164.38	I
220	3064.53	II	200	3510.26	II	870	4150.12	I	230	5180.31	I
110	3069.68	II	200	3515.42	II	4400	4152.58	I	110	5186.98	I
100	3070.90	II	200	3517.67	II	870	4163.47	I	190	5189.20	I
110	3071.56	II	200	3520.06	I	4400	4163.66	I	170	5193.08	I
100	3073.24	II	2000	3535.30	I	4000	4164.66	I	150	5195.84	I
400	3076.87	II	1300	3537.48	II	3500	4168.13	I	65	5203.22	I
110	3080.35	II	250	3540.96	II	310	4184.44	I	35	5205.13	I
1800	3094.18	II	500	3544.02	I	1200	4190.88	I	85 c	5219.10	I

Niobium (Cont.)

Intensity		Wavelength	
65		5225.16	I
150		5232.81	I
85	c	5237.43	I
29		5240.39	I
150	d	5251.62	I
		5251.81	I
75		5253.03	I
85		5253.93	I
50		5269.92	I
270		5271.53	I
25		5272.48	I
130	c	5276.20	I
29	c	5279.43	I
50		5285.26	I
35		5296.34	I
50		5315.55	I
17		5317.01	I
250		5318.60	I
50		5319.49	I
75		5334.87	I
25		5336.81	I
50		5340.80	I
25		5343.58	I
460		5344.17	I
340		5350.74	I
40		5353.28	I
25		5355.31	I
40		5355.70	I
29		5359.19	I
17		5362.01	I
40		5375.27	I
40		5381.34	I
17		5388.30	I
21		5395.86	I
29		5396.33	I
29		5411.24	I
21		5416.30	I
65		5422.44	I
21		5431.26	I
110		5437.27	I
19		5448.31	I
19		5456.19	I
40		5458.04	I
19	h	5468.10	I
40		5481.00	I
13		5483.09	I
19		5483.49	I
13		5491.06	I
17		5499.53	I
40		5504.58	I
17		5509.12	I
35	c	5512.82	I
17		5517.39	I
50		5523.57	I
25		5541.47	I
85		5551.35	I
29		5563.00	I
17	c	5571.44	I
35	c	5576.16	I
35		5578.29	I
50		5586.97	I
17	c	5590.95	I
13		5594.89	I
17	c	5599.59	I
40		5603.52	I
13		5603.93	I
25		5628.26	I
65		5629.17	I
35	c	5635.42	I
170		5642.11	I
35		5645.30	I
17		5654.14	I
130		5664.71	I
170		5665.63	I
17		5666.86	I
65	Cw	5671.02	I
85		5671.91	I
25		5677.47	I
25		5693.09	I
35	d	5697.90	I
		5698.03	I
40		5706.16	I
85		5706.48	I
29		5709.33	I
17		5715.59	I
65		5716.35	I
25		5725.66	I
130		5729.19	I
21		5737.36	I
13		5738.20	I
85		5751.44	I
110		5760.34	I
65		5764.99	I
29		5771.08	I
50	c	5776.07	I
17		5780.34	I
85		5787.54	I
17		5789.79	I
50		5794.24	I
50		5804.03	I
29	h	5815.33	I
110		5819.43	I
35		5820.62	I

Intensity		Wavelength	
75		5834.90	I
25		5838.15	I
130	d	5838.64	I
50		5842.47	I
17		5846.09	I
65		5866.47	I
35		5874.70	I
17		5877.79	I
40		5893.44	I
190	Cw	5900.62	I
40	c	5903.80	I
29		5927.41	I
40	c	5934.16	I
40		5957.70	I
150		5983.22	I
65		5986.08	I
85	Cw	5997.93	I
50		6029.75	I
50		6031.84	I
50		6045.50	I
25		6048.72	I
29		6056.65	I
29		6107.71	I
40		6142.51	I
50		6148.13	I
50		6164.32	I
29		6213.06	I
75		6221.96	I
40	c	6251.76	I
21		6260.77	I
85	c	6430.46	I
50	c	6433.22	I
17		6497.84	I
65		6544.61	I
15		6574.73	I
19	Cw	6591.00	I
19		6606.16	I
19		6607.28	I
35		6614.15	I
19		6626.98	I
210	Cw	6660.84	I
150	Cw	6677.33	I
65		6701.20	I
130	c	6723.62	I
75		6739.88	I
25		6795.31	I
85		6828.11	I
25	c	6849.35	I
19		6870.92	I
40		6876.36	I
25	c	6902.89	I
35		6908.07	I
40		6918.32	I
17		6946.07	I
17		6972.49	I
25		6986.09	I
85		6990.32	I
17	c	6996.11	I
21		7023.48	I
17		7038.04	I
190	c	7046.81	I
8		7066.41	I
8		7075.23	I
40	c	7098.94	I
17	Cw	7102.01	I
19		7119.31	I
15		7122.95	I
35		7126.17	I
17		7130.06	I
130		7159.43	I
17		7191.37	I
19	c	7208.94	I
50		7252.35	I
15		7274.81	I
13		7317.03	I
17	c	7323.92	I
29	Cw	7328.38	I
65	c	7353.16	I
190	Cw	7372.50	I
13		7419.83	I
15		7436.02	I
19		7478.20	I
65		7515.93	I
29	c	7519.77	I
170	c	7574.58	I
17	c	7583.21	I
13		7639.81	I
13		7647.71	I
25		7703.33	I
75	c	7726.68	I
25		7757.31	I
6		7787.11	I
13	Cw	7873.41	I
35		7885.31	I
25		7938.89	I
8		7954.76	I
40		8135.20	I
13	Cw	8240.00	I
29	Cw	8320.93	I
29		8346.08	I
10		8350.04	I
17		8439.77	I
17	Cw	8475.98	I

Intensity		Wavelength	
25		8526.99	I
13	c	8547.25	I
17	c	8560.54	I
17		8575.87	I
21	c	8697.55	I
21		8740.96	I
21		8767.97	I
29	Cw	8815.56	I
35		8905.78	I

Nb III
Ref. 392 — C.H.C.

Intensity		Wavelength	
		Vacuum	
60		1314.56	III
50		1319.15	III
60		1431.92	III
60		1433.39	III
50		1435.26	III
80		1445.43	III
80		1445.98	III
80		1447.09	III
50		1448.50	III
60		1451.63	III
100		1456.68	III
80		1484.73	III
50		1486.79	III
100		1495.94	III
80		1498.02	III
80		1499.45	III
50		1501.53	III
100		1501.99	III
60		1505.03	III
50		1509.71	III
50		1512.34	III
50		1513.25	III
80		1513.81	III
50		1517.38	III
100		1524.91	III
60		1532.98	III
60		1537.50	III
50		1566.92	III
50		1570.19	III
50		1586.82	III
100		1590.21	III
80		1598.86	III
80		1604.72	III
80		1639.51	III
100		1682.77	III
60		1684.40	III
100		1705.44	III
100		1707.14	III
50		1739.30	III
50		1758.63	III
60		1763.72	III
50		1808.70	III
50		1863.13	III
100		1892.92	III
100		1938.84	III
60	h	1979.07	III
50		1985.15	III
50		1997.11	III
		Air	
50	h	2007.28	III
50		2032.47	III
50		2060.29	III
80	h	2130.24	III
60		2206.01	III
60		2240.31	III
60		2244.19	III
60		2265.63	III
80		2273.92	III
100		2275.23	III
80		2279.36	III
100		2281.51	III
80		2284.40	III
100		2290.36	III
60		2304.78	III
50		2309.92	III
100		2313.30	III
50		2330.22	III
100		2338.09	III
80		2344.12	III
90		2349.21	III
80		2355.54	III
100		2362.06	III
80		2362.50	III
80		2365.70	III
100		2372.73	III
100		2387.41	III
80		2388.23	III
50		2404.23	III
80		2404.89	III
100		2413.94	III
60		2414.50	III
100		2421.91	III
50		2437.74	III

Intensity	Wavelength	
60	2446.45	III
100	2456.99	III
50	2460.34	III
50	2460.45	III
50	2463.72	III
80	2468.72	III
60	2469.39	III
80	2475.87	III
50	2486.02	III
60	2488.74	III
60	2493.02	III
100	2499.73	III
60	2508.53	III
50	2511.95	III
100	2545.64	III
80	2557.94	III
50	2567.44	III
60	2593.75	III
80	2598.86	III
50	2628.67	III
80	2633.17	III
80	2657.99	III
50	2937.71	III
80	3001.84	III
80	3142.26	III
60	3266.11	III

Nb IV
Ref. 407 — C.H.C.

Intensity	Wavelength	
	Vacuum	
12	542.38	IV
12	543.09	IV
12	545.21	IV
10	559.94	IV
10	566.22	IV
18	981.27	IV
60	993.54	IV
18	996.16	IV
50	1002.76	IV
400	1005.72	IV
500	1007.05	IV
500	1010.19	IV
45	1030.27	IV
100	1116.08	IV
150	1120.02	IV
50	1447.48	IV
40	1473.43	IV
40	1487.23	IV
60	1502.30	IV
60	1524.36	IV
50	1534.06	IV
40	1635.68	IV
40	1910.70	IV
60	1922.41	IV
60	1978.22	IV
	Air	
40	2027.50	IV
65	2032.53	IV
40	2034.67	IV
40	2068.62	IV
55	2084.07	IV
35	2093.12	IV
50	2093.30	IV
20	2122.68	IV
25	2130.23	IV
45	2146.36	IV
18	2249.98	IV

Nb V
Ref. 431 — C.H.C.

Intensity	Wavelength	
	Vacuum	
80	464.55	V
80	468.32	V
60	753.01	V
80	763.77	V
80	774.02	V
40	1007.02	V
50	1044.90	V
70	1212.21	V
100	1258.87	V
40	1267.60	V
100	1758.33	V
100	1877.34	V

NITROGEN (N)
Z = 7

N I and II
Ref. 213 — R.L.K.

Intensity	Wavelength

Nitrogen (Cont.)

Vacuum

Intensity	Wavelength	
285	644.634	II
360	644.837	II
450	645.178	II
140	647.50	I
360	660.286	II
170	671.016	II
285	671.386	II
150	671.630	II
160	671.773	II
170	672.001	II
350	692.70	I
285	746.984	II
650	775.965	II
90	885.67	I
90	909.697	I
80	910.278	I
40	910.645	I
450	915.612	II
450	915.962	II
550	916.012	II
650	916.701	II
90	953.415	I
100	953.655	I
130	953.970	I
130	963.990	I
115	964.626	I
70	965.041	I
90	1067.614	I
60	1068.612	I
450	1083.990	II
600	1084.580	II
430	1085.546	II
650	1085.701	II
175	1097.237	I
115	1098.095	I
115	1098.260	I
105	1100.360	I
40	1100.465	I
90	1101.291	I
360	1134.165	I
385	1134.415	I
410	1134.980	I
105	1143.65	I
130	1163.884	I
60	1164.206	I
105	1164.325	I
270	1167.448	I
105	1168.334	I
60	1168.417	I
195	1168.536	I
230	1176.510	I
105	1176.630	I
195	1177.695	I
410	1199.550	I
385	1200.223	I
360	1200.710	I
175	1225.026	I
160	1225.37	I
130	1228.41	I
160	1228.79	I
360	1243.179	I
315	1243.306	I
290	1310.540	I
250	1310.95	I
230	1319.00	I
315	1319.68	I
115	1326.57	I
115	1327.92	I
360	1411.94	I
700	1492.625	I
490	1492.820	I
640	1494.675	I
775	1742.729	I
700	1745.252	I

Air

Intensity	Wavelength	
160	2095.53	II
70	2096.20	II
110	2096.86	II
110	2130.18	II
160	2142.78	II
160	2206.09	II
160	2286.69	II
110	2288.44	II
220	2316.49	II
160	2316.69	II
285	2317.05	II
160	2461.27	II
110	2496.83	II
70	2496.97	II
110	2520.22	II
160	2520.79	II
220	2522.23	II
110	2590.94	II
160	2709.84	II
110	2799.22	II
110	2823.64	II
160	2885.27	II
220	3006.83	II
360	3437.15	II
285	3838.37	II
360	3919.00	II
450	3955.85	II
1000	3995.00	II
360	4035.08	II
550	4041.31	II
360	4043.53	II
140	4099.94	I
185	4109.95	I
285	4176.16	II
285	4227.74	II
285	4236.91	II
220	4237.05	II
450	4241.78	II
285	4432.74	II
650	4447.03	II
360	4530.41	II
550	4601.48	II
450	4607.16	II
360	4613.87	II
450	4621.39	II
870	4630.54	II
550	4643.08	II
285	4788.13	II
450	4803.29	II
180	4847.38	I
285	4895.11	II
160	4914.94	I
210	4935.12	I
160	4950.23	I
350	4963.98	I
285	4987.37	II
450	4994.36	II
650	5001.48	II
360	5002.70	II
870	5005.15	II
550	5007.32	II
450	5010.62	II
360	5016.39	II
360	5025.66	II
550	5045.10	II
185	5281.20	I
140	5292.68	I
450	5495.67	II
285	5535.36	II
650	5666.63	II
550	5676.02	II
870	5679.56	II
450	5686.21	II
450	5710.77	II
285	5747.30	II
700	5752.50	I
240	5764.75	I
265	5829.54	I
235	5854.04	I
360	5927.81	II
550	5931.78	II
285	5940.24	II
650	5941.65	II
285	5952.39	II
160	5999.43	I
210	6008.47	I
285	6167.76	II
360	6379.62	II
185	6411.65	I
210	6420.64	I
210	6423.02	I
210	6428.32	I
185	6437.68	I
235	6440.94	I
185	6457.90	I
300	6468.44	I
750	6482.05	II
360	6482.70	I
300	6483.75	I
265	6481.71	I
325	6484.80	I
160	6491.22	I
210	6499.54	I
185	6506.31	I
750	6610.56	II
185	6622.54	I
185	6636.94	I
235	6644.96	I
185	6646.50	I
235	6653.46	I
210	6656.51	I
185	6722.62	I
210	7398.64	I
160	7406.12	I
265	7406.24	I
685	7423.64	I
785	7442.29	I
900	7468.31	I
185	7608.80	I
450	7762.24	II
400	8184.87	I
400	8188.02	I
250	8200.36	I
300	8210.72	I
570	8216.34	I
400	8223.14	I
400	8242.39	I
550	8438.74	II
500	8567.74	I
570	8594.00	I
650	8629.24	I
500	8655.89	I
220	8676.08	II
700	8680.28	I
650	8683.40	I
500	8686.15	I
110	8687.43	II
110 h	8699.00	II
500	8703.25	I
160 h	8710.54	II
570	8711.70	I
500	8718.83	I
250	8728.89	I
200	8747.36	I
500	9386.80	I
570	9392.79	I
250	9460.68	I
200	9863.33	I
160 h	9865.41	II
110 h	9868.21	II
160 h	9887.39	II
220 h	9891.09	II
160 h	9961.86	II
220 h	9969.34	II
285 h	10023.27	II
220 h	10035.45	II
220 h	10065.15	II
160 h	10070.12	II
250	10105.13	I
300	10108.89	I
350	10112.48	I
400	10114.64	I
110 h	10126.27	II
250	10539.57	I
200	12074.51	I
380	12186.82	I
225	12288.97	I
290	12328.76	I
310	12381.65	I
180	12438.40	I
510	12461.25	I
920	12469.62	I
500	13429.61	I
840	13581.33	I
180	13587.73	I
180	13602.27	I
290	13624.18	I
250	14757.07	I
100	14868.87	I
160	14966.60	I
180	15582.27	I
120 s	17516.58	I
100 l	17584.86	I
100	17878.26	I

N III
Ref. 66, 213 = R.L.K.

Vacuum

Intensity		Wavelength	
500		257.95	III
650		258.50	III
700		259.19	III
800		260.09	III
800		261.28	III
500		262.91	III
500		265.23	III
500		265.27	III
500		268.70	III
150		314.715	III
200		314.850	III
90		314.877	III
600		323.26	III
500		338.35	III
500		340.20	III
500		351.98	III
120		362.833	III
150		362.881	III
150		362.946	III
90		362.985	III
300		374.204	III
350		374.441	III
500		387.48	III
250		451.869	III
300		452.226	III
500		684.996	III
570		685.513	III
650		685.816	III
500		686.335	III
500		763.336	III
570		764.359	III
250		771.544	III
300		771.901	III
350		772.385	III
200		772.891	III
150		772.975	III
650		979.842	III
700		979.919	III
900		989.790	III
700		991.514	III
1000		991.579	III
500		1183.031	III
570		1184.550	III
150		1387.371	III
250		1729.945	III
570		1747.848	III
350		1751.218	III
650		1751.657	III
150		1804.486	III
200		1805.669	III
150		1846.42	III
350		1885.06	III
400		1885.546	III
200		1907.99	III
150		1919.55	III
150		1919.77	III
300		1920.65	III
150		1920.84	III
200		1921.30	III

Air

Intensity		Wavelength	
200		2064.01	III
250		2064.42	III
120		2068.68	III
90		2071.09	III
90		2117.59	III
90		2121.50	III
90		2147.31	III
200		2188.20	III
150		2188.38	III
250	w	2682.18	III
90		2689.20	III
120		3367.34	III
90		3754.67	III
120		3771.05	III
90		3938.52	III
150		3998.63	III
200		4003.58	III
250		4097.33	III
200		4103.43	III
120		4195.76	III
150		4200.10	III
90		4332.91	III
120		4345.68	III
300		4379.11	III
90		4510.91	III
120		4514.86	III
90		4634.14	III
120		4640.64	III
90		4858.82	III
150		4867.15	III
90		5314.35	III
200		5320.82	III
150		5327.18	III
90		6454.11	III
120		6467.02	III

N IV
Ref. 108, 212 = R.L.K.

Vacuum

Intensity		Wavelength	
400		181.75	IV
400		191.7	IV
400		192.9	IV
500		196.87	IV
500		197.23	IV
500		202.60	IV
500		205.94	IV
500		205.97	IV
500		206.03	IV
500		217.20	IV
500	d	217.90	IV
500	d	223.4	IV
800	w	225.12	IV
800	w	225.21	IV
600	w	234.12	IV
600	w	234.20	IV
600	w	234.25	IV
550		236.07	IV
500		237.99	IV
500	w	238.7	IV
600		238.80	IV
500	w	239.62	IV
900		247.20	IV
500	w	248.43	IV
500	w	248.46	IV
500	w	248.48	IV
600		260.45	IV
650		270.99	IV
250		283.42	IV
300		283.48	IV
350		283.58	IV
600		285.56	IV
600	w	297.7	IV
700		297.82	IV
650		300.32	IV
90		303.123	IV

Nitrogen (Cont.)

Intensity		Wavelength	
500		303.28	IV
150		315.053	IV
120		322.5u3	IV
150		322.570	IV
200		322.724	IV
120		323.175	IV
300		335.050	IV
500	w	351.93	IV
700		353.06	IV
500		420.77	IV
650		463.74	IV
570		765.148	IV
520		921.992	IV
500		922.519	IV
480		923.057	IV
520		921.992	IV
500		922.519	IV
520		924.283	IV
1000		955.335	IV
150	w	1036.16	IV
90		1078.71	IV
90		1188.01	IV
1000		1718.55	IV

Air

Intensity		Wavelength	
90		2080.34	IV
90	w	2318.09	IV
150		2477.69	IV
250		2645.65	IV
300		2646.18	IV
350		2646.96	IV
90		3078.25	IV
90		3463.37	IV
570		3478.71	IV
500		3482.99	IV
400		3484.96	IV
90		3747.54	IV
150		4057.76	IV
90		4606.33	IV
150		6380.77	IV

N V
Ref. 66, 107, 318 = R.L.K.

Intensity		Wavelength	

Vacuum

52		166.947	V
52		186.069	V
62		186.153	V
90		209.303	V
90		247.561	V
120		247.706	V
150		266.196	V
200		266.379	V
90		713.518	V
150		713.860	V
150		748.195	V
200		748.291	V
1000		1238.821	V
900		1242.804	V
90		1549.336	V
200	1	1616.33	V
350	1	1619.69	V
90	w	1860.37	V

Air

60	1	2859.16	V
90	1	2974.52	V
150	w	2980.78	V
250	w	2981.31	V
60	w	2998.43	V
350		4603.73	V
250		4619.98	V
200	w	4944.56	V
60	w	7618.46	V

OSMIUM (Os)
Z = 76

Os I and II
Ref. 1 = C.H.C.

Intensity	Wavelength	

Air

9600	2001.45	I
13000	2003.73	I
9000	2004.78	
17000	2010.15	I
29000	2018.14	I
29000	2020.26	
14000	2022.76	I
14000	2028.23	I
18000	2034.44	I
26000	2045.36	I
8600	2058.69	I
	2058.78	I
13000	2061.69	I
7800	2067.21	II
4200	2070.67	II
7200	2076.95	I
7200	2078.09	
14000	2079.97	I
2900	2082.54	I
2900	2089.03	I
2900	2089.21	I
6000	2097.60	I
5500	2100.63	I
2100	2117.66	I
4800	2117.96	I
6600	2119.79	
1900	2123.84	I
5300	2137.11	I
2400	2149.97	
2600	2154.59	I
1300	2157.44	I
1200	2158.53	I
2400	2161.00	
3100	2166.90	I
1100	2167.75	I
2100	2171.65	I
960	2184.68	I
840	2194.39	II
760	2202.49	I
600	2227.98	I
1100	2234.61	I
1300	2252.15	I
2000	2255.85	II
1400	2264.60	I
360	2268.28	I
960	2270.17	I
1400	2282.26	II
840	2283.67	I
570	2289.32	I
380	2297.31	I
660	2308.31	I
190	2313.75	II
550	2320.18	I
310	2323.98	I
660	2324.24	I
330	2326.99	I
310	2334.56	I
720	2336.80	II
430	2338.63	I
290	2340.69	I
430	2343.74	I
260	2345.75	I
430	2347.38	I
230	2350.23	II
360	2352.99	I
120	2355.28	II
240	2356.92	I
240	2357.25	I
310	2362.41	I
900	2362.77	I
500	2367.35	II
290	2369.24	I
500	2370.70	I
480	2371.18	I
95	2375.06	II
2600	2377.03	I
260	2377.61	I
900	2379.39	I
240	2384.62	I
1700	2387.29	I
330	2394.29	I
290	2395.36	I
1100	2395.88	I
220	2396.78	I
960	2401.13	I
260	2402.23	I
200	2403.54	I
330	2403.85	I
95	2405.08	II
290	2405.45	I
200	2405.96	I
360	2408.67	I
240	2410.98	I
290	2414.52	I
530	2417.99	I
530	2418.53	I
95	2420.02	II
200	2423.07	II
70	2424.02	II
500	2424.56	I
1400	2424.97	I
240	2426.81	I
70	2427.90	II
380	2431.19	I
380	2431.61	I
360	2446.02	I
900	2450.74	I
530	2451.73	I
530	2453.90	I
110	2454.91	II
530	2456.46	I
1800	2461.42	I
110	2468.90	II
290	2472.28	I
290	2474.78	I
900	2476.84	I
360	2482.43	I
530	2486.24	II
4500	2488.55	I
290	2491.02	I
290	2491.69	I
360	2492.42	I
2600	2498.41	I
330	2499.92	I
330	2502.29	I
500	2504.39	I
260	2504.51	I
35	2507.18	II
70	2509.71	II
660	2512.87	I
2400	2513.25	I
660	2515.04	I
500	2517.92	I
660	2518.44	I
200	2519.29	I
330	2519.79	I
200	2532.44	I
780	2538.00	II
240	2538.10	I
1000	2542.51	I
30	2548.83	II
310	2554.46	I
190	2563.16	II
600	2566.49	I
290	2566.88	I
480	2568.83	I
340	2571.78	I
150	2578.32	I
130	2580.03	II
360	2581.05	I
740	2581.96	I
1000	2590.76	I
200	2591.98	I
170	2596.00	II
210	2609.20	I
380	2609.56	I
400	2610.78	I
470	2612.63	I
1800	2613.06	I
800	2619.94	I
230	2620.62	I
530	2621.82	I
380	2628.48	I
27	2631.22	II
3800	2637.13	I
1900	2644.11	I
340	2646.89	I
380	2647.73	I
380	2649.34	I
490	2656.68	I
1900	2658.60	I
640	2659.83	I
380	2661.18	I
40	2664.29	II
580	2674.57	I
400	2674.88	I
2100	2689.82	I
510	2699.59	I
580	2706.70	I
3000	2714.64	I
580	2715.36	I
1300	2720.04	I
850	2721.86	I
580	2730.61	I
40	2731.36	II
580	2732.80	I
690	2761.42	I
470	2763.27	I
340	2765.04	I
960	2770.71	I
300	2776.91	I
740	2782.55	I
40	2783.88	II
640	2786.31	I
230	2793.99	I
230	2794.19	I
530	2796.73	I
320	2804.07	I
2800	2806.91	I
470	2808.94	I
420	2813.84	I
740	2814.20	I
300	2815.78	I
420	2829.27	I
230	2837.42	I
470	2838.17	I
5100	2838.63	I
740	2841.60	I
2300	2844.40	I
420	2846.39	I
420	2848.25	I
1500	2850.76	I
1500	2860.96	I
35	2863.37	II
360	2874.96	I
300	2878.40	I
35	2879.39	II
30	2880.20	II
260	2896.06	I
9600	2909.06	I
2100	2912.33	I
530	2917.26	I
2100	2919.79	I
300	2925.57	I
360	2929.51	I
510	2931.28	I
260	2934.64	I
200	2942.85	I
1100 h	2948.23	I
1400	2949.53	I
210 d	2949.81	I
300	2961.01	I
530	2962.15	I
450	2964.06	I
740	2970.97	I
450	2977.64	I
510	2982.90	I
340	2983.49	I
260	2997.65	I
330	3013.07	I
570	3017.25	I
4400	3018.04	I
480	3019.38	I
1100	3030.70	I
2900	3040.90	I
210	3043.50	I
120	3042.74	II
230	3049.46	I
210	3050.39	I
8600	3058.66	I
290	3060.30	I
570	3062.19	I
210	3069.94	I
360	3074.08	I
290	3074.96	I
290	3077.44	I
1100	3077.72	I
360	3078.11	I
230	3078.38	I
230	3090.08	I
270	3093.59	I
310	3101.53	I
360	3105.99	I
310	3108.98	I
620	3109.38	I
250	3111.09	I
310	3118.33	I
480	3131.12	I
250	3152.67	I
290	3153.61	I
3100	3156.25	I
250	3156.78	I
310	3166.51	I
180	3173.93	II
420	3178.06	I
230	3181.88	I
230	3185.33	I
310	3186.98	I
310	3189.46	I
310	3194.23	I
150	3213.31	II
1900	3232.06	I
290	3238.63	I
190	3241.04	I
120	3248.00	I
190	3254.91	I
190	3256.92	I
190	3260.30	I
3100	3262.29	I
380	3262.75	I
3100	3267.94	I
620	3269.21	I
190	3272.16	I
530	3275.20	I
330	3277.97	I
190	3288.84	I
1200	3290.26	I
7600	3301.56	I
250	3306.23	I
620	3310.91	I
120	3315.42	I
250	3324.33	I
310	3327.42	I
960	3336.15	I
110	3351.74	I
120	3353.91	I
230	3357.97	I
250	3361.15	I
190	3364.12	I
120	3370.20	I
960	3370.59	I
160	3372.08	I
120	3378.68	I
310	3384.00	I
190	3385.94	I
620	3387.84	I
120	3401.17	I
620	3401.86	I
250	3402.51	I
120	3408.76	I
120	3412.74	I

Osmium (Cont.)

Int	Wavelength		Int	Wavelength		Int	Wavelength	
120	3421.69	I	120	4184.13	I	170	5780.82	I
150	3427.67	I	320	4189.91	I	40	5800.60	I
250	3440.60	I	180	4201.45	I	8	5842.49	I
120	3444.46	I	250	4202.06	I	110	5857.76	I
160	3445.55	I	1200	4211.86	I	28	5860.64	I
310	3449.20	I	120	4213.86	I	11	5882.92	I
120	3458.38	I	100	4215.16	I	11	5903.98	I
120	3465.44	I	170	4233.46	I	11	5906.84	I
120	3478.53	I	4900	4260.85	I	7	5908.95	I
120	3482.11	I	100	4264.75	I	7	5981.36	I
120	3487.46	I	120	4269.61	I	11	5983.22	I
120	3490.33	I	100	4285.90	I	65	5996.00	I
160	3498.54	I	560	4293.95	I	20	6015.79	I
250	3501.16	I	560	4311.40	I	7	6054.63	I
620	3504.66	I	110	4326.25	I	20	6144.53	I
440	3512.99	I	340	4328.68	I	11	6158.03	I
310	3518.72	I	100	4338.75	I	35	6227.70	I
120	3520.00	I	100	4351.53	I	7	6241.70	I
480	3523.64	I	210	4365.67	I	22	6269.41	I
120	3526.04	I	110	4370.66	I	11	6274.94	I
1200	3528.60	I	520	4394.86	I	11	6286.83	I
230	3530.06	I	160	4397.26	I	9	6398.86	I
230	3532.80	I	160	4402.74	I	22	6403.15	I
120	3533.41	I	4900	4420.47	I	9	6448.13	I
230	3542.71	I	100	4432.41	I	6	6520.85	I
960	3559.79	I	290	4436.32	I	7	6528.87	I
1200	3560.86	I	100	4439.64	I	7	6533.14	I
120	3562.34	I	230	4447.35	I	11	6538.30	I
310	3569.78	I	120	4484.76	I	11	6576.83	I
120	3574.08	I	110	4548.66	I	8	6614.56	I
120	3587.32	I	540	4550.41	I	4	6615.43	I
620	3598.11	I	140	4551.30	I	7	6661.81	I
190	3601.83	I	170	4616.78	I	27	6729.56	I
95	3604.48	II	170	4631.83	I	18	6791.53	I
250	3616.57	I	140	4663.82	I	14	6806.61	I
120	3619.43	I	670	4793.99	I	5	6878.70	I
450	3640.33	I	110	4865.60	I	4	6901.58	I
230	3654.49	I	55	5031.83	I	11	6956.02	I
330	3656.90	I	45	5039.12	I	6	6984.95	I
120	3666.31	I	35	5072.88	I	15	7060.67	I
480	3670.89	I	35	5074.77	I	22	7145.54	I
120	3675.45	I	35	5079.09	I	10	7149.89	I
250	3689.06	I	90	5103.50	I	4	7184.10	I
190	3703.25	I	55	5110.81	I	10	7206.33	I
120	3706.56	I	22	5122.23	I	5	7209.96	I
120	3709.14	I	22	5145.54	I	9	7251.16	I
230	3713.73	I	140	5149.74	I	6	7253.49	I
210	3719.52	I	28	5152.01	I	6	7375.07	I
230	3720.13	I	28	5168.98	I	9	7407.95	I
180	3746.47	I	40	5193.52	I	26	7602.95	I
3700	3752.52	I	270	5202.63	I	4	7701.46	I
100	3757.12	I	35	5203.23	I	7	7789.96	I
130	3766.30	I	20	5250.46	I	7	7852.17	I
120	3768.14	I	45	5255.82	I	6	7981.20	I
120	3774.40	I	55	5265.15	I	7	8041.29	I
110	3774.62	I	20	5283.89	I			
120	3776.25	I	20	5295.65	I			
290	3776.99	I	40	5298.78	I			
2100	3782.20	I	13	5302.58	I			
620	3790.14	I	18	5336.23	I			
180	3790.73	I	11	5346.03	I			
370	3793.91	I	13	5352.25	I			
250	3836.06	I	110	5376.79	I			
150	3840.30	I	16	5403.43	I			
150	3841.29	I	13	5412.14	I			
190	3849.94	I	120	5416.34	I			
230	3857.09	I	45	5416.69	I			
230	3865.47	I	28	5417.51	I			
730	3876.77	I	16	5441.82	I			
250	3881.86	I	55	5443.31	I			
140	3900.39	I	22	5446.93	I			
190	3901.71	I	11	5447.76	I			
100	3930.00	I	20	5449.37	I			
250	3938.59	I	20	5453.40	I			
100	3949.78	I	22	5457.30	I			
200	3961.02	I	28	5470.00	I			
1000	3963.63	I	13	5474.58	I			
100	3964.96	I	13	5475.13	I			
150	3969.67	I	9	5477.27	I			
110 h	3975.44	I	16	5481.85	I			
730	3977.23	I	22	5509.33	I			
100	3988.18	I	9	5516.01	I			
150	4003.48	I	270	5523.53	I			
100	4004.02	I	22	5546.82	I			
150	4005.16	I	9	5549.79	I			
160	4018.26	I	13	5552.88	I			
100	4037.84	I	11	5560.62	I			
280	4041.92	I	16	5580.66	I			
160	4048.05	I	80	5584.44	I			
960	4066.69	I	8	5600.50	I			
250	4070.86	I	35	5620.08	I			
190	4071.56	I	9	5637.41	I			
230	4074.68	I	22	5642.56	I			
490	4091.82	I	28	5645.25	I			
120	4100.30	I	7	5648.98	I			
1200	4112.02	I	9	5660.21	I			
180	4124.60	I	7	5674.38	I			
180	4128.96	I	28	5680.88	I			
2500	4135.78	I	11	5709.37	I			
150	4137.84	I	170	5721.93	I			
180	4172.57	I	8	5737.89	I			
1200	4173.23	I	8	5739.72	I			
620	4175.63	I	22	5765.05	I			

OXYGEN (O)
Z = 8
OI and II
Ref. 66, 69, 209, 210, 215 —
R.L.K.

Intensity	Wavelength		Intensity	Wavelength	
	Vacuum				
250	537.83	II	300	1306.029	I
300	538.26	II		**Air**	
220	539.09	II	30 d	2283.42	II
200	539.55	II	30 d	2284.89	II
150	539.85	II	110	2293.32	II
150	644.148	II	200	2300.35	II
200	672.95	II	30 d	2313.05	II
150	673.77	II	30 d	2316.12	II
70	685.544	I	30 d	2316.79	II
900	718.484	II	50 d	2319.68	II
600	718.562	II	30 d	2322.15	II
70	744.794	I	30 d	2339.31	II
70	770.793	I	110	2411.60	II
90	771.056	I	80	2425.55	II
70	775.321	I	250	2433.56	II
70	791.973	I	80 d	2436.06	II
300	796.66	II	80	2444.26	II
90	804.267	I	300	2445.55	II
70	804.848	I	300	2733.34	II
70	805.295	I	110	2747.46	II
80	805.810	I	265	2972.29	I
240	832.762	II	160	3122.62	II
450	833.332	II	220	3129.44	II
600	834.467	II	450	3134.82	II
40	877.879	I	285	3138.44	II
80	922.008	I	220	3270.98	II
90	935.193	I	220	3273.52	II
40	948.686	I	220	3277.69	II
90	971.738	I	360	3287.59	II
40	976.448	I	160	3305.15	II
160	988.773	I	160	3306.60	II
40	990.204	I	220	3377.20	II
250	1025.762	I	285	3390.25	II
90	1027.431	I	220	3407.38	II
160	1039.230	I	160	3409.84	II
60	1040.942	I	285	3470.81	II
40	1152.152	I	220	3712.75	II
900	1302.168	I	285	3727.33	II
600	1304.858	I	160	3739.92	II
			360	3749.49	II
			160	3803.14	II
			120	3823.41	I
			450	3911.96	II
			160	3919.29	II
			185	3947.29	I
			160	3947.48	I
			140	3947.59	I
			220	3954.37	II
			100	3954.61	I
			450	3973.26	II
			220	3982.20	II
			160	4069.90	II
			285	4072.16	II
			450	4075.87	II
			80 d	4083.91	II
			50 d	4087.14	II
			150 d	4089.27	II
			110	4097.24	II
			220	4105.00	II
			285	4119.22	II
			160	4132.81	II
			50	4146.06	II
			220	4153.30	II
			285	4185.46	II
			450	4189.79	II
			80	4233.27	I
			50 d	4253.74	II
			50 d	4253.98	II
			50 d	4275.47	II
			50 d	4303.78	II
			285	4317.14	II
			160	4336.86	II
			220	4345.56	II
			285	4349.43	II
			220	4366.90	II
			100	4368.25	I
			220	4395.95	II
			450	4414.91	II
			285	4416.98	II
			160	4448.21	II
			160	4452.38	II
			50	4465.45	II
			50 d	4466.28	II
			50	4467.83	II
			50	4469.41	II
			360	4590.97	II
			285	4596.17	II
			80 d	4609.39	II
			160	4638.85	II
			360	4641.81	II
			450	4649.14	II
			160	4650.84	II
			360	4661.64	II
			285	4676.23	II
			220	4699.21	II
			285	4705.36	II
			160	4924.60	II
			220	4943.06	II
			135	5329.10	I
			160	5329.68	I
			190	5330.74	I
			90	5435.18	I
			110	5435.78	I

Oxygen (Cont.)

Intensity		Wavelength	
135		5436.86	I
120		5577.34	I
160		5958.39	I
190		5958.58	I
80		5995.28	I
160		6046.23	I
190		6046.44	I
110		6046.49	I
100		6106.27	I
400		6155.98	I
450		6156.77	I
490		6158.18	I
80		6256.83	I
100		6261.55	I
100		6366.34	I
100		6374.32	I
320		6453.60	I
360		6454.44	I
400		6455.98	I
80		6604.91	I
100		6653.83	I
360		7001.92	I
450		7002.23	I
210		7156.70	I
400		7254.15	I
450		7254.45	I
320		7254.53	I
210		7476.44	I
100		7477.24	I
120		7479.08	I
120		7480.67	I
100		7706.75	I
870		7771.94	I
810		7774.17	I
750		7775.39	I
80		7886.27	I
100		7943.15	I
100		7947.17	I
235		7947.55	I
210		7950.80	I
185		7952.16	I
110		7981.94	I
135		7982.40	I
190		7986.98	I
135		7987.33	I
250		7995.07	I
400		8221.82	I
265		8227.65	I
265		8230.02	I
325		8233.00	I
120		8235.35	I
120		8426.16	I
810		8446.25	I
1000		8446.36	I
935		8446.76	I
325		8820.43	I
160	d	9057.01	I
120		9118.29	I
80		9134.71	I
80		9150.14	I
80		9151.48	I
235		9156.01	I
450		9260.81	I
490		9260.84	I
450		9260.94	I
400		9262.58	I
540		9262.67	I
590		9262.77	I
490		9265.94	I
640		9266.01	I
185		9399.19	I
120		9481.16	I
120	d	9482.88	I
235		9487.43	I
140		9492.71	I
265		9497.97	I
160		9499.30	I
235		9505.59	I
210		9521.96	I
120		9523.36	I
120		9523.96	I
100		9528.28	I
100		9622.13	I
120		9625.29	I
160		9677.38	I
80		9694.66	I
65		9694.91	I
235		9741.50	I
235		9760.65	I
120		9909.05	I
140		9936.98	I
120		9940.41	I
160		9995.31	I
120	d	10421.18	I
590		11286.34	I
640		11286.91	I
490		11287.02	I
490		11287.32	I
490		11295.10	I
540		11297.68	I
590		11302.38	I
265		11358.69	I
490		12464.02	I
450		12570.04	I
120		12990.77	I
160		13076.91	I
700		13163.89	I
750		13164.85	I
640		13165.11	I
160		16212.06	I
120		17966.70	I
590		18021.21	I
120		18041.48	I
120		18042.19	I
120		18046.23	I
140		18229.23	I
540		18243.63	I
140		26173.56	I

O III
Ref. 23, 66, 210 — R.L.K.

Intensity		Wavelength	
		Vacuum	
80	d	264.34	III
110		264.48	III
110		266.97	III
150		266.98	III
150		267.03	III
150		277.38	III
80		295.62	III
110		295.66	III
120		295.72	III
150		303.41	III
150		303.46	III
140		303.52	III
160		303.62	III
160		303.69	III
250		303.80	III
200		305.60	III
250		305.66	III
190		305.70	III
300		305.77	III
190		305.84	III
450		320.979	III
300		328.45	III
250		328.74	III
300		345.31	III
110		355.14	III
90		355.33	III
80		355.47	III
200		359.02	III
190		359.22	III
150		359.38	III
210		373.80	III
200		374.00	III
300		374.08	III
190		374.16	III
200		374.33	III
210		374.44	III
450		395.558	III
300		434.98	III
800		507.391	III
900		507.683	III
1000		508.182	III
1000		525.795	III
700		597.818	III
1000		599.598	III
110		609.70	III
160		610.04	III
200		610.75	III
100		610.85	III
800		702.332	III
800		702.822	III
900		702.899	III
1000		703.850	III
600		832.927	III
780		833.742	III
600		835.096	III
800		835.292	III
160		1476.89	III
285		1590.01	III
160		1591.33	III
220		1760.12	III
110		1760.42	III
220		1763.22	III
220		1764.48	III
750		1767.78	III
550		1768.24	III
360		1771.67	III
110		1773.00	III
110		1773.85	III
220		1779.16	III
160		1781.03	III
160		1784.85	III
220		1789.66	III
110		1848.26	III
110		1856.62	III
285		1872.78	III
285		1872.87	III
285		1874.94	III
160		1920.04	III
110		1920.75	III
110		1921.52	III
220		1923.49	III
110		1923.82	III
110		1926.94	III

Air (O III)

Intensity		Wavelength	
360		2013.27	III
160		2026.96	III
220		2045.67	III
160		2052.74	III
200	d	2390.44	III
80		2394.33	III
80		2422.84	III
80	d	2438.83	III
200		2454.99	III
200		2558.06	III
80		2687.53	III
110		2695.49	III
80		2959.68	III
250		2983.78	III
80		3017.63	III
80		3023.45	III
80		3043.02	III
200		3047.13	III
110		3059.30	III
80		3121.71	III
110		3132.86	III
80		3238.57	III
200		3260.38	III
300		3265.46	III
80		3267.31	III
80		3312.30	III
110		3340.74	III
80		3444.10	III
80		3455.12	III
80		3698.70	III
80		3702.75	III
80		3703.37	III
110		3707.24	III
110		3715.08	III
110		3744.00	III
150		3754.67	III
80		3757.21	III
250		3759.87	III
110		3791.26	III
200		3961.59	III
110		5592.37	III

O IV
Ref. 36, 66 = R.L.K.

Intensity	Wavelength	
	Vacuum	
150	195.86	IV
200	196.01	IV
110	207.18	IV
150	207.24	IV
140	233.46	IV
150	233.50	IV
110	233.52	IV
200	233.56	IV
110	233.60	IV
90	238.36	IV
180	238.57	IV
110	252.56	IV
110	252.95	IV
150	253.08	IV
300	260.39	IV
250	260.56	IV
300	279.63	IV
375	279.94	IV
110	285.71	IV
150	285.84	IV
200	306.62	IV
150	306.88	IV
700	553.330	IV
775	554.075	IV
850	554.514	IV
700	555.261	IV
580	608.398	IV
640	609.829	IV
270	616.952	IV
150	617.005	IV
200	617.036	IV
520	624.617	IV
580	625.130	IV
640	625.852	IV
200	779.734	IV
315	779.821	IV
360	779.912	IV
200	779.997	IV
640	787.711	IV
520	790.109	IV
700	790.199	IV
200	802.200	IV
160	802.255	IV
130	921.296	IV
160	921.366	IV
200	923.367	IV
130	923.433	IV
200	1338.612	IV
130	1342.992	IV
230	1343.512	IV

Air

Intensity		Wavelength	
200		2449.372	IV
200		2450.040	IV
200		2493.44	IV
200		2493.77	IV
200		2507.73	IV
230		2509.19	IV
200		2517.2	IV
160		2836.26	IV
160		2921.45	IV
460		3063.42	IV
410		3071.61	IV
160		3209.66	IV
230		3348.08	IV
270		3349.11	IV
160		3354.27	IV
200		3375.40	IV
130		3378.06	IV
360		3381.20	IV
360		3385.52	IV
270		3396.79	IV
360		3403.52	IV
230		3409.66	IV
410		3411.69	IV
230		3413.64	IV
200		3489.83	IV
160		3492.24	IV
230		3560.39	IV
270		3563.33	IV
315	w	3725.93	IV
360		3729.03	IV
410		3736.85	IV
230		3744.89	IV

O V
Ref. 24, 66 = R.L.K.

Intensity		Wavelength	
		Vacuum	
80		124.616	V
110		135.523	V
80		138.109	V
110		139.029	V
80		151.447	V
110		151.477	V
150		151.546	V
80		164.574	V
110		164.657	V
80		164.709	V
80		166.235	V
150		167.99	V
110		170.219	V
450		172.169	V
250		185.745	V
375		192.751	V
450		192.799	V
520		192.906	V
80		193.003	V
200		194.593	V
80		202.161	V
80		202.224	V
80		202.283	V
80		202.334	V
150		202.393	V
110		203.78	V
150		203.82	V
100		203.85	V
200		203.89	V
100		203.94	V
300		207.794	V
150		215.040	V
200		215.103	V
250		215.245	V
250		216.018	V
520		220.352	V
80		227.372	V
80		227.469	V
150		227.511	V
80		227.549	V
80		227.634	V
80		227.689	V
150		231.823	V
110		248.459	V
110		286.448	V
1000		629.730	V
230		681.272	V
700		758.678	V
640		759.441	V
580		760.228	V
775		760.445	V
640		761.128	V
700		762.003	V
520		774.518	V
640		1371.292	V
160	w	1506.72	V
315	w	1643.68	V
160		1707.996	V

Air

Intensity		Wavelength	
1000		2781.01	V
920		2786.99	V
775		2789.85	V
200		2941.33	V
210		2941.65	V
160		3144.66	V
100		4123.99	V
230	w	4930.27	V
130		5597.91	V
130		6500.24	V

PALLADIUM (Pd)
Z = 46

Pd I and II
Ref. 1, 287 — C.H.C.

Intensity		Wavelength	
		Air	
50		2162.27	II
50		2182.35	II
50		2212.15	II
100	r	2231.59	II
200	r	2296.53	II
50		2351.32	II
50		2362.31	II
75		2367.92	II
60		2372.16	II
50		2388.29	II
60		2414.73	II
75		2418.72	II
75		2424.49	II
100		2426.87	II
100		2430.94	II
100		2433.11	II
100		2435.32	II
150		2446.17	II
75		2446.72	II
1100		2447.91	I
80		2448.15	II
100		2457.29	II
60		2457.76	II
150		2469.29	II
80		2470.06	II
100		2471.18	II
50		2472.55	II
1700		2476.42	I
250		2486.52	II
300		2488.92	II
75		2489.61	II
200		2498.81	II
150		2505.73	II
50		2514.47	II
80		2534.57	II
50	h	2539.44	II
150		2551.84	II
150		2565.51	II
100		2569.56	II
60		2593.24	II
50		2628.24	II
70		2635.92	II
150		2658.75	II
1900		2763.09	I
150	h	2776.85	II
100	h	2787.92	II
50	h	2800.64	II
50	h	2807.59	II
200		2854.59	II
100	h	2871.37	II
100	h	2878.01	II
520		2922.49	I
50		2980.63	II
650		3002.65	I
45		3009.78	I
1500		3027.91	I
1100		3065.31	I
2600		3114.04	I
270		3142.81	I
11000		3242.70	I
2700		3251.64	I
3500		3258.78	I
460		3287.25	I
3600		3302.13	I
5000		3373.00	I
24000		3404.58	I
13000		3421.24	I
5000		3433.45	I
6400		3441.40	I
7700		3460.77	I
10000		3481.15	I
2000		3489.77	I
12000		3516.94	I
12000		3553.08	I
4500		3571.16	I
20000		3609.55	I
20000		3634.70	I
5500		3690.34	I
1400		3718.91	I
1500		3799.19	I
1500		3832.29	I
2200		3894.20	I
1500		3958.64	I
290		4087.34	I
90		4169.84	I
2500		4212.95	I
180		4473.59	I
55	h	4788.18	I
45	h	4817.51	I
35		4875.43	I
55		5110.81	I
75		5117.02	I
160		5163.84	I
55		5234.86	I
120		5295.63	I
18		5312.57	I
15		5345.10	I
35		5395.24	I
55		5542.80	I
35		5547.02	I
27		5619.44	I
15		5642.69	I
14		5655.42	I
75		5670.07	I
11		5690.14	I
55	h	5695.09	I
18		5736.61	I
23		6774.54	I
65		6784.52	I
4	h	6833.42	I
11		7016.44	I
13	h	7310.06	I
75		7368.12	I
27		7391.92	I
16		7486.90	I
120		7764.03	I
27		7786.67	I
45		7915.80	I
18		7961.08	I
55		8132.82	I
45		8300.83	I
9	h	8353.58	I
18	h	8532.74	I
16	h	8599.10	I
65		8761.35	I

Pd III
Ref. 424 — L.J.R.

Intensity		Wavelength	
		Vacuum	
10		688.74	III
20		689.46	III
50		689.54	III
50		695.91	III
200		705.49	III
150		707.80	III
100		709.89	III
100		717.90	III
100		719.47	III
200		727.72	III
150		738.79	III
100		756.85	III
100		757.41	III
500		763.06	III
500		766.42	III
200		772.11	III
200		776.51	III
2000		781.02	III
200		784.99	III
200		787.31	III
200		787.95	III
200		789.58	III
500		794.08	III
500		797.52	III
500		800.03	III
500		800.10	III
500		803.67	III
500		825.35	III
500		840.58	III
500		856.47	III
500		864.04	III
500		880.59	III
500		888.84	III
1000		889.29	III
300		947.78	III
300		965.52	III
100		1505.40	III
200		1517.18	III
200	h	1526.88	III
100		1542.63	III
200	h	1545.95	III
300		1596.89	III
200	h	1606.10	III
150		1630.84	III
50		1679.73	III
100		1704.33	III
50		1706.40	III
200		1719.86	III
500		1741.62	III
400		1758.19	III
4000		1782.55	III
400		1804.91	III
400		1843.49	III
1500		1851.59	III
2000		1852.27	III
1000		1859.21	III
1500		1874.63	III
2000		1885.83	III
1000		1887.40	III
1500		1891.34	III
4000		1914.62	III
1000		1930.33	III
2000		1941.64	III
400		1951.56	III
300		1972.29	III
300		1977.53	III
		Air	
800		2002.16	III
1000		2004.47	III
500		2055.11	III
500		2149.82	III
500		2177.55	III
500		2177.63	III
100		2291.45	III
100		2452.42	III
100		2633.22	III

PHOSPHORUS (P)
Z = 15

P I and II
Ref. 182 = R.L.K.

Intensity	Wavelength	
	Vacuum	
10	810.24	II
10	865.44	II
20	1249.32	II
20	1301.87	II
20	1304.47	II
15	1304.68	II
35	1305.48	II
25	1309.87	II
60	1310.70	II
15	1372.033	I
15	1373.500	I
10	1374.732	I
15	1377.080	I
15	1377.937	I
25	1379.429	I
25	1381.469	I
15	1381.637	I
30	1452.89	II
80	1532.51	II
120	1535.90	II
80	1536.39	II
120	1542.29	II
140	1671.070	I
100	1671.510	I
180	1671.680	I
140	1672.035	I
140	1672.474	I
600	1674.591	I
600	1679.695	I
140	1685.976	I
100	1694.028	I
100	1694.486	I
100	1706.376	I
100	1707.553	I
600	1774.951	I
500	1782.838	I
400	1787.656	I
140	1834.801	I
140	1847.165	I
100	1849.820	I
140	1851.194	I
100	1852.069	I
500	1858.886	I
400	1859.393	I
140	1864.348	I
180	1905.481	I
140	1906.403	I
280	1907.665	I
	Air	
280	2023.489	I
180	2024.516	I
400	2032.432	I
400	2033.477	I
400	2135.465	I
400	2136.182	I
400	2149.145	I
280	2152.940	I
500	2154.080	I
180	2235.732	I
100	2484.19	II
750	2533.976	I
950	2535.603	I
750	2553.262	I
500	2554.915	I
150	2606.06	II
100	2626.18	II
90	2636.76	II
150	3308.92	II
125	3419.34	II
100	3425.00	II
100	4178.48	II
200	4288.60	II
200	4385.35	II
400	4420.71	II
100	4452.46	II
150	4463.00	II
120	4467.98	II
200	4475.26	II
200	4499.24	II
120	4530.81	II
120	4554.83	II
120	4558.07	II
120	4581.71	II
500	4588.04	II
500	4589.86	II
600	4602.08	II
300	4626.70	II
300	4658.31	II
200	4864.42	II
150	4927.20	II
500	4943.53	II
300	4954.39	II
300	4969.71	II
100	5079.381	I
100	5098.221	I
100	5100.974	I
140	5109.628	I
140	5154.844	I
180	5162.290	I
150	5191.41	II
300	5253.52	II
140	5293.539	I
400	5296.13	II
250	5316.07	II
300	5344.85	II
180	5345.851	I
100	5364.631	I
250	5378.20	II
300	5386.88	II
200	5409.72	II
400	5425.91	II
100	5428.094	I
400	5450.74	II
140	5458.305	I
125	5461.20	II
180	5477.672	I
140	5477.860	I
140	5478.267	I
200	5483.55	II
200	5499.73	II
200	5507.19	II
100	5514.774	I
100	5516.997	I
200	5541.14	II
200	5583.27	II
250	5588.34	II
100	5727.71	II
500	6024.18	II
400	6034.04	II
500	6043.12	II
250	6055.50	II
100	6057.86	II
350	6087.87	II
180	6097.690	I
350	6165.59	II
500	6199.024	I
180	6210.499	I
100	6232.29	II
200	6367.27	II
140	6375.681	I
100	6388.579	I
250	6435.32	II
130	6436.31	II
600	6459.99	II
600	6503.46	II
600	6507.97	II
150	6713.28	II
100	6717.411	I
100	7102.200	I
100	7158.367	I
180	7165.465	I
180	7175.102	I
180	7176.660	I
250	7845.63	II
100	8046.801	I
140	8278.058	I
100	8367.856	I
140	8531.475	I
140	8613.835	I
180	8637.578	I
400	8741.529	I
100	8872.174	I
140	9153.34	I
180	9175.819	I
950	9193.85	I
600	9278.88	I
1250	9304.94	I
500	9323.50	I
100	9327.13	I
140	9372.09	I
950	9435.069	I
950	9441.86	I
600	9452.83	I
100	9481.84	I
100	9492.12	I
1250	9493.56	I
180	9521.78	I
1700	9525.73	I
1500	9545.18	I
280	9556.81	I
1700	9563.439	I
280	9593.50	I
750	9609.04	I
180	9625.80	I
180	9628.42	I
400	9638.939	I
140	9675.41	I
500	9676.24	I
180	9706.533	I
1500	9734.750	I
280	9736.680	I
1500	9750.77	I

Phosphorus (Cont.)

Intensity	Wavelength	
100	9760.77	I
100	9776.85	I
100	9779.11	I
600	9790.21	I
1700	9796.85	I
280	9834.80	I
400	9903.68	I
280	9976.67	I
229	10084.27	I
174	10432.66	I
132	10455.87	I
458	10511.58	I
962	10529.52	I
1235	10581.57	I
415	10596.90	I
435	10681.40	I
265	10813.13	I
134	10932.72	I
103	10967.37	I
180	11160.05	I
764	11183.23	I
402	11186.75	I
76	13438.43	I
86	13485.19	I
479	14241.64	I
150	14272.75	I
256	14307.83	I
173	14430.50	I
135	14470.62	I
98	14646.42	I
714	15711.52	I
228	15962.53	I
296	16254.77	I
203	16292.97	I
1627	16482.92	I
588	16590.07	I
225	16613.05	I
221	16738.68	I
419	16803.39	I
471	17112.48	I
104	17223.28	I
289	17286.91	I
145	17359.00	I
299	17423.67	I
95	17665.68	I
186	18007.63	I
106	18518.90	I
92	18881.16	I
125	20841.62	I
124	23038.83	I
287	23844.97	I
98	26134.44	I
118	27959.52	I
188	28049.42	I
127	28154.52	I
132	28284.16	I
98	28288.69	I
311	29097.16	I
92	31483.21	I
91	32270.90	I
146	35551.93	I
90	35582.27	I
192	35802.53	I
146	36417.43	I

P III
Ref. 180 = R.L.K.

Intensity	Wavelength	
	Vacuum	
90	471.146	III
90	484.278	III
120	498.180	III
200	569.853	III
200	581.831	III
200	844.646	III
150	845.038	III
250	845.664	III
300	847.669	III
200	848.016	III
120	848.465	III
150	848.639	III
250	852.686	III
350	855.624	III
200	859.406	III
500	859.652	III
250	859.729	III
300	913.971	III
300	917.120	III
350	918.665	III
250	921.849	III
200	997.999	III
250	1003.598	III
500	1334.808	III
650	1344.327	III
300	1344.845	III
250	1380.463	III
150	1381.089	III
350	1502.228	III
250	1504.663	III
150	1618.632	III
200	1618.907	III
	Air	
200	2611.147	III
300	2632.713	III
200	2680.133	III
250	2895.241	III
250	3186.186	III
300	3219.307	III
150	3233.536	III
400	3233.602	III
200	3556.546	III
200	3577.526	III
200	3904.812	III
250	3914.314	III
300	3957.641	III
350	3978.307	III
200	4057.440	III
400	4059.312	III
300	4080.084	III
500	4222.195	III
350	4246.720	III
200	4428.171	III
200	4463.668	III
250	4479.776	III
150	6083.409	III
150	6409.204	III
150	6484.440	III
150	6486.381	III
150	6992.690	III
150	8113.528	III

P IV
Ref. 336 – R.L.K.

Intensity	Wavelength	
	Vacuum	
90	282.301	IV
90	304.996	IV
120	359.293	IV
150	359.899	IV
120	361.514	IV
150	361.629	IV
120	371.299	IV
150	371.504	IV
200	372.001	IV
500	388.318	IV
120	414.604	IV
200	414.999	IV
250	415.805	IV
250	444.245	IV
300	445.158	IV
250	568.038	IV
350	629.008	IV
400	629.914	IV
500	631.779	IV
350	648.482	IV
300	756.510	IV
300	776.353	IV
650	823.179	IV
700	824.730	IV
800	827.932	IV
250	847.019	IV
350	849.799	IV
200	850.392	IV
700	877.476	IV
1000	950.655	IV
570	1025.563	IV
500	1028.096	IV
570	1030.517	IV
500	1033.111	IV
500	1035.517	IV
570	1118.551	IV
200	1206.422	IV
200	1335.705	IV
500	1366.695	IV
400	1372.674	IV
350	1377.282	IV
500	1484.507	IV
400	1487.788	IV
300	1489.098	IV
250	1862.762	IV
120	1862.893	IV
200	1863.580	IV
650	1888.523	IV
200	1910.183	IV
120	1985.682	IV
150	1985.851	IV
200	1986.114	IV
150	1987.022	IV
	Air	
200	2477.823	IV
150	2478.070	IV
250	2478.256	IV
250	2605.506	IV
400	2644.295	IV
300	2724.764	IV
400	2728.770	IV
200	2729.120	IV
500	2739.309	IV
250	2739.872	IV
200	2740.223	IV
200	2961.242	IV
650	3347.736	IV
570	3364.467	IV
400	3371.122	IV
200	3413.543	IV
200	3733.393	IV
300	4249.656	IV
250	4540.288	IV
250	4541.112	IV
150	4548.056	IV
200	4548.449	IV
150	5235.499	IV
150	5989.774	IV
150	6142.605	IV
150	6713.939	IV
120	6715.906	IV
200	7443.657	IV

P V
Ref. 179 = R.L.K.

Intensity	Wavelength	
	Vacuum	
80	255.59	V
50	255.67	V
110	310.58	V
150	311.34	V
300	328.47	V
250	328.78	V
150	347.23	V
200	348.20	V
110	378.56	V
250	389.50	V
300	390.70	V
150	410.03	V
375	475.60	V
110	534.63	V
80	534.99	V
520	542.57	V
600	544.92	V
450	673.90	V
450	865.45	V
600	871.39	V
250	997.62	V
150	1000.38	V
900	1117.98	V
700	1128.01	V
150	1379.62	V
250	1385.05	V
375	1447.83	V
450	1610.50	V
	Air	
200	2180.29	V
150	2186.42	V
375	2424.40	V
450	2440.93	V
200	2441.24	V
300	2961.00	V
450	2978.55	V
700	3175.09	V
520	3204.04	V
150	4083.18	V
110	4094.95	V
110	5156.72	V

PLATINUM (Pt)
Z = 78
Pt I and II
Ref. 1, 288 — C.H.C.

Intensity		Wavelength	
		Vacuum	
30		1621.66	II
30		1723.13	II
30		1751.70	II
50	r	1777.09	II
30		1781.86	II
30		1879.09	II
40		1883.05	II
50		1889.52	II
50		1911.70	II
30		1929.25	II
30		1929.68	II
30		1939.80	II
30		1949.90	II
30		1983.74	II
		Air	
40		2014.93	II
3200		2030.63	I
4400		2032.41	I
100		2036.46	II
40		2041.57	II
5500		2049.37	I
1500		2067.50	I
3000		2084.59	I
1000		2103.33	I
30		2115.57	II
950		2128.61	I
30		2130.69	II
1900		2144.23	I
100		2144.24	II
600		2165.17	I
1500		2174.67	I
30		2190.32	II
400		2202.22	I
50	h	2202.58	II
320		2222.61	I
50	h	2233.11	II
150		2249.30	I
30	h	2240.99	II
100		2245.52	II
30		2251.52	II
30	h	2251.92	II
190		2268.84	I
30	h	2271.72	II
280		2274.38	I
50	h	2287.50	II
30		2288.20	II
150		2289.27	I
150		2292.40	I
240		2308.04	I
50		2310.96	II
90		2315.50	I
220		2318.29	I
100		2326.10	I
170		2340.18	I
280		2357.10	I
180		2368.28	I
50		2377.28	II
130		2383.64	I
40		2386.81	I
120		2389.53	I
35		2396.17	I
70		2401.87	I
200		2403.09	I
100		2418.06	I
50		2424.87	II
80		2428.04	I
50		2428.20	I
25		2429.10	I
180		2436.69	I
650		2440.06	I
60		2450.97	I
440		2467.44	I
35		2471.01	I
1000		2487.17	I
25		2488.74	II
200		2490.12	I
160		2495.82	I
240		2498.50	I
50		2505.93	I
120		2508.50	I
50		2514.07	I
60		2515.03	I
240		2515.58	I
140		2524.30	I
40		2529.41	I
50		2536.49	I
160		2539.20	I
18		2549.46	I
50		2552.25	I
50		2596.00	I
70		2603.14	I
30		2616.76	II
50		2619.57	I
30		2625.34	II
1100		2628.03	I
130		2639.35	I
1000		2646.89	I
500		2650.86	I
20		2658.17	I
2800		2659.45	I
40		2674.57	I
440		2677.15	I
200		2698.43	I
2000		2702.40	I
1600		2705.89	I
60		2713.13	I
1300		2719.04	I
130		2729.92	I
1800		2733.96	I
70		2738.48	I
70		2747.61	I
80		2753.86	I
200		2754.92	I
30		2769.84	I
500		2771.67	I
40		2773.24	?
20		2774.00	i
50		2774.77	II
50		2793.27	I
100		2794.21	II
40	h	2799.98	II
140		2803.24	I
10		2808.51	I
50		2818.25	I
30	h	2822.27	II
1400		2830.30	I
70		2834.71	I
16		2853.11	I
80	h	2860.68	II
40	h	2865.05	II
40	h	2875.85	II
100	h	2877.52	II
25		2888.20	I
25		2893.22	I
600		2893.86	I
300		2897.87	I
60		2905.90	I
120		2912.26	I
120		2913.54	I
70		2919.34	I
30		2921.38	I

Platinum (Cont.)

Intensity		Wavelength	Spectrum
1700		2929.79	I
30		2942.76	I
30		2944.75	I
25		2959.10	I
60		2960.75	I
1800		2997.97	I
35		3001.17	II
220		3002.27	I
30		3017.88	I
30	h	3031.22	II
130		3036.45	I
800		3042.64	I
3200		3064.71	I
30		3071.94	I
130		3100.04	I
320		3139.39	I
140		3156.56	I
120		3200.71	I
320		3204.04	I
30		3230.29	I
20		3233.42	I
20		3250.36	I
40		3251.98	I
160		3255.92	I
25		3268.42	I
25		3281.97	I
120		3290.22	I
500		3301.86	I
60		3315.05	I
35		3323.80	I
340		3408.13	I
35		3427.93	I
60		3483.43	I
160		3485.27	I
120		3628.11	I
70		3638.79	I
70		3643.17	I
50		3663.10	I
80		3671.99	I
80		3674.04	I
35		3699.91	I
18		3706.53	I
20		3801.05	
80		3818.69	I
40		3900.73	I
110		3922.96	I
35		3948.40	I
100		3966.36	I
20		3996.57	I
110		4118.69	I
80		4164.56	I
40		4197.43	I
18		4327.06	I
18		4391.83	I
80		4442.55	I
14		4445.55	I
25		4498.76	I
12		4520.90	I
35		4552.42	I
12		4879.53	I
14		5044.04	I
30		5059.48	I
35		5227.66	I
40		5301.02	I
12		5368.99	I
12		5390.79	I
14		5475.77	I
14		5478.50	I
6		5763.57	I
20		5840.12	I
8		5844.84	I
6		6026.04	I
7		6318.37	I
8		6326.58	I
9		6523.45	I
10		6710.42	I
20		6760.02	I
60		6842.60	I
20		7113.73	I
10		8224.74	I

PLUTONIUM (Pu)
Z = 94
Pu I and II
Ref. 91 — J.G.C.

Intensity		Wavelength	Spectrum
		Air	
10000		2781.40	II
10000		2784.48	II
10000		2806.11	II
10000		2815.77	II
10000		2897.97	II
10000		2898.94	II
10000		2904.25	II
10000		2904.94	II
10000		2910.40	II
10000		2918.00	II
10000		2918.80	II
10000		2926.08	II
10000		2928.25	II
10000		2929.71	II
10000		2930.98	II
10000		2932.32	II
10000		2933.30	II
10000		2938.54	II
10000		2938.95	II
10000		2941.39	II
10000		2945.26	II
10000		2946.00	II
10000		2950.06	II
10000		2951.62	II
10000		2954.46	II
10000		2963.47	II
10000		2966.84	II
10000		2967.54	II
10000		2972.50	II
10000		2977.81	II
10000		2978.37	II
10000		2980.23	II
10000		2981.23	II
10000		2986.95	II
10000		2988.21	II
10000		2991.31	II
10000		2996.40	II
10000		3000.31	II
10000		3009.57	II
10000		3028.85	II
10000		3042.61	II
10000		3043.12	II
10000		3060.32	II
10000		3069.32	II
10000		3091.33	II
10000		3091.94	II
10000		3092.59	II
10000		3104.12	II
10000		3105.04	II
10000		3106.03	
10000		3123.87	II
10000		3159.21	II
10000		3161.73	II
10000		3163.18	II
10000		3174.49	II
10000		3179.41	II
10000		3185.12	II
10000		3187.60	II
10000		3189.23	II
10000		3193.54	
10000		3193.55	II
10000		3194.56	II
10000		3198.47	II
10000		3200.23	II
10000		3201.00	II
10000		3201.66	II
10000		3204.48	II
10000		3206.80	II
10000		3207.97	II
10000		3215.08	I
10000		3216.15	II
10000		3220.94	II
10000		3224.87	II
10000		3231.86	II
10000		3232.24	
10000		3232.63	II
10000		3241.39	II
10000		3242.96	II
10000		3243.40	II
10000		3244.16	I
10000		3245.25	II
10000		3245.71	
10000		3246.35	II
10000		3247.50	
10000		3247.56	II
10000		3252.08	I
10000		3260.54	II
10000		3265.17	
10000		3273.11	II
10000		3274.71	II
10000		3275.24	I
10000		3292.56	I
10000		3293.61	I
10000		3296.91	I
10000		3297.87	I
10000		3298.47	II
10000		3301.76	I
10000		3306.55	I
10000		3306.66	I
10000		3307.66	II
10000		3308.75	I
10000		3312.65	II
10000		3315.34	II
10000		3316.96	II
10000		3320.61	I
10000		3320.84	I
10000		3323.48	
10000		3327.19	I
10000		3330.11	
10000		3331.52	II
10000		3332.34	I
10000		3333.03	
10000		3337.71	II
10000		3338.40	II
10000		3338.94	I
10000		3347.87	II
10000		3349.63	I
10000		3351.82	II
10000		3356.61	II
10000		3358.41	II
10000		3358.84	II
10000		3362.26	II
10000		3365.20	I
10000		3365.66	
10000		3368.86	I
10000		3370.64	II
10000		3371.19	I
10000		3375.80	I
10000		3376.76	II
10000		3376.94	II
10000		3377.37	II
10000		3379.51	I
10000		3381.82	I
10000		3381.97	II
10000		3382.70	I
10000		3390.33	II
10000		3391.41	II
10000		3393.67	I
10000		3394.32	I
10000		3418.88	II
10000		3465.10	II
10000		3473.64	II
10000		3483.20	I
10000		3585.87	II
10000		3632.21	II
10000		3699.19	II
10000		3720.59	I
10000		3725.98	I
10000		3726.11	II
10000		3726.79	II
10000		3732.03	II
10000		3744.78	I
10000		3753.63	I
10000		3755.94	I
10000		3757.82	I
10000		3758.34	I
10000		3774.38	I
10000		3776.71	I
10000		3792.22	I
10000		3799.37	I
10000		3805.93	I
10000		3811.40	I
10000		3812.30	II
10000		3827.57	I
10000		3835.52	I
10000		3836.96	I
10000		3838.92	I
10000		3842.10	I
10000		3851.01	I
10000		3851.85	I
10000		3878.54	I
10000		3895.89	I
10000		3928.53	I
10000		3975.43	II
10000		4097.12	I
10000		4101.96	I
10000		4105.95	II
10000		4111.07	I
10000		4114.91	I
10000		4128.12	I
10000		4129.93	II
10000		4133.01	I
10000		4135.97	I
10000		4140.04	I
10000		4141.20	II
10000		4151.09	I
10000		4151.45	I
10000		4155.46	I
10000		4159.39	II
10000		4167.77	I
10000		4170.95	I
10000		4178.28	II
10000		4189.90	II
10000		4190.06	II
10000		4196.20	II
10000		4206.48	I
10000		4208.23	I
10000		4221.87	I
10000		4224.20	I
10000		4229.77	II
10000		4254.76	I
10000		4261.88	I
10000		4269.77	I
10000		4273.34	II
10000		4281.17	I
10000		4289.08	II
10000		4337.18	II
10000		4352.71	II
10000		4367.41	I
10000		4379.91	II
10000		4385.35	II
10000		4393.93	I
10000		4404.90	I
10000		4441.65	II
10000		4468.54	II
10000		4472.79	II
10000		4493.78	II
10000		4504.91	II
10000		4536.15	II
10000		4735.40	I
10000		4989.34	I
10000		5269.86	I
10000		5381.02	I
10000		5498.50	I
10000		5510.72	I
10000		5537.59	I
10000		5549.62	I
10000		5590.54	I
10000		5592.33	I
10000		5712.39	I
10000		5770.26	I
10000		5839.05	I
10000		5983.35	I
10000		6012.78	I
10000		6192.80	I
10000		6304.66	I
10000		6449.75	I
10000		6486.71	I
10000		6488.86	I
10000		6488.89	I
10000		6535.27	I
10000		6544.21	I
10000		6608.95	I
10000		6672.72	I
10000		6784.66	I
10000		6880.16	I
10000		6891.38	I
10000		7059.23	I
10000		7068.90	I
10000		7092.46	I
10000		7116.88	I
10000		7141.66	I
10000		7177.14	I
10000		7231.09	I
10000		7258.06	I
10000		7322.23	I
10000		7325.97	I
10000		7331.81	I
10000		7431.18	I
10000		7447.99	I
10000		7507.80	I
10000		7526.93	I
10000		7547.45	I
10000		7564.50	I
10000		7571.87	I
10000		7572.93	I
10000		7609.77	I
10000		7689.40	I
10000		7758.20	I
10000		7798.54	I
10000		7953.17	I
10000		8102.54	I
10000		8130.86	I
10000		8309.61	I
10000		8435.47	I
10000		8476.13	I
10000		8495.75	I
10000		8597.26	I
10000		8665.02	I
10000		8691.94	I
3000		8729.82	I
3000		8836.16	I
3000		9533.07	I
3000		10046.75	I
3000		11114.82	I
3000		12144.46	I
3000		12231.22	I
3000		15377.31	I
3000		16397.38	I

POLONIUM (Po)
Z = 84
Po I and II
Ref. 47, 48 — E.F.W.

Intensity		Wavelength	Spectrum
		Air	
250	w	2139.02	I
300	h	2203.80	
300		2220.67	I
200		2222.13	I
200		2284.22	
250		2344.61	I
250		2421.72	I
300		2426.09	I
1500	w	2450.08	I
700		2483.94	I
700		2490.53	I
200	h	2502.18	
300		2534.95	I
300		2557.33	I
1500	w	2558.01	I
400		2562.31	
300		2578.80	
400		2587.64	I
200		2637.01	
300		2645.36	I
700	h	2663.33	
200		2671.67	I
600		2761.92	I
400		2800.26	I
250		2824.11	I
300		2866.01	I
400		2919.31	I
600		2958.92	I
2500	w	3003.21	I
450		3069.31	I
200		3115.95	

Potassium (Cont.)

Intensity		Wavelength	
400		3189.02	I
600		3240.24	I
250		3286.38	I
600		3328.60	I
300		3489.79	
200		3493.65	
400		3588.33	
200		3671.36	
500		3861.93	I
200		4051.98	
1200		4170.52	I
250		4236.13	
200	h	4415.58	
800		4493.21	I
350		4611.44	I
200		4867.12	
400		4876.24	I
450		4946.81	
350		5323.23	I
300		5744.85	I
600		7962.62	I
300		8433.87	I
500		8618.26	I
250		9227.87	

POTASSIUM (K)
Z = 19
K I and II
Ref. 59, 76, 172, 268 — L.J.R.

Intensity	Wavelength	
	Vacuum	
5	261.20	II
25	441.81	II
5	465.08	II
	469.50	II
10	476.03	II
30	495.14	II
30	600.77	II
25	607.93	II
30	612.62	II
3	1725.0	II
	Air	
6	2190.00	II
4	2210.53	II
5	2265.04	II
6	2550.02	II
4	2743.55	II
	2992.12	I
	2992.42	I
	3034.76	I
	3034.92	I
5	3062.18	II
4	3101.79	I
3	3102.04	I
6	3105.00	II
5	3190.07	II
7	3217.16	I
6	3217.62	I
4	3220.60	II
5	3290.65	II
6	3345.32	II
6	3373.60	II
6	3380.62	II
6	3384.86	II
6	3404.24	II
7	3440.05	II
11	3446.37	I
10	3447.38	I
6	3481.11	II
7	3530.75	II
5	3608.88	II
6	3618.49	II
4	3626.42	II
3	3648.84	I
4	3648.98	I
6	3681.54	II
5	3716.60	II
5	3721.34	II
5	3739.13	II
5	3744.42	II
6	3767.36	II
6	3783.19	II
6	3800.14	II
6	3816.56	II
7	3817.50	II
5	3873.74	II
4	3878.62	II
8	3897.92	II
5	3923.00	II
5	3926.36	II
6	3942.53	II
6	3955.21	II
6	3966.72	II
6	3972.58	II
6	3995.10	II
7	4001.24	II
5	4012.10	II
6	4042.59	II
18	4044.14	I
17	4047.21	I
5	4093.69	II
6	4114.99	II

Intensity	Wavelength	
7	4134.72	II
7	4149.19	II
8	4186.24	II
7	4222.97	II
7	4225.67	II
7	4263.40	II
7	4305.00	II
7	4309.10	II
5	4340.03	II
7	4388.16	II
5	4466.65	II
6	4505.33	II
5	4595.65	II
8	4608.45	II
10	4641.88	I
11	4642.37	I
5	4659.38	II
4	4740.91	I
6	4744.35	I
5	4753.93	I
7	4757.39	I
5	4786.49	I
7	4791.05	I
6	4799.75	I
8	4804.35	I
9	4829.23	II
7	4849.86	I
8	4856.09	I
8	4863.48	I
9	4869.76	I
8	4942.02	I
6	4943.29	II
9	4950.82	I
9	4956.15	I
10	4965.03	I
8	5005.60	II
7	5056.27	II
10	5084.23	I
11	5097.17	I
11	5099.20	I
12	5112.25	I
5	5310.24	II
12	5323.28	I
13	5339.69	I
12	5342.97	I
14	5359.57	I
6	5470.13	II
5	5642.73	II
4	5772.32	II
16	5782.38	I
17	5801.75	I
15	5812.15	I
17	5831.89	I
2	5969.64	II
8	6120.27	II
6	6246.59	II
7	6307.29	II
5	6427.96	II
2	6595.00	II
19	6911.08	I
12	6936.28	I
20	6938.77	I
7	6964.18	I
12	6964.67	I
25	7664.90	I
24	7698.96	I
5	7955.37	I
4	7956.83	I
7	8078.11	I
6	8079.62	I
9	8250.18	I
8	8251.74	I
3	8390.22	I
	8391.44	I
2	8417.54	I
1	8420.00	I
11	8503.45	I
10	8505.11	I
4	8763.96	I
3	8767.05	I
13	8902.19	I
12	8904.02	I
5	8923.31	I
4	8925.44	I
7	9347.24	I
3	9349.25	I
6	9351.59	I
15	9595.70	I
14	9597.83	I
6	9949.67	I
5	9954.14	I
9	10479.63	I
5	10482.15	I
8	10487.15	I
17	11019.87	I
16	11022.67	I
17	11690.21	I
16	11769.62	I
17	11772.83	I
	12432.24	I
	12522.11	I
	13377.86	I
	13397.09	I
	15163.08	I

Intensity	Wavelength	
	15168.40	I
	40158.37	I

K III
Ref. 60, 76 — L.J.R.

Intensity	Wavelength	
	Vacuum	
2	325.28	III
5	327.60	III
25	330.68	III
30	341.92	III
15	348.00	III
30	379.12	III
25	380.48	III
30	382.23	III
15	398.63	III
20	402.10	III
30	406.48	III
40	408.96	III
50	413.79	III
30	414.87	III
30	416.00	III
30	417.54	III
30	418.62	III
75	434.72	III
50	435.68	III
75	444.34	III
75	448.60	III
75	466.79	III
100	470.09	III
75	471.57	III
45	474.92	III
40	479.18	III
10	482.11	III
10	482.41	III
75	497.10	III
10	514.94	III
50	520.61	III
25	523.79	III
40	529.80	III
15	539.71	III
15	546.12	III
20	708.84	III
20	765.31	III
30	765.64	III
35	778.53	III
20	872.31	III
10	873.86	III
15	874.04	III
	Air	
6	2550.02	III
5	2635.11	III
1	2736.96	III
5	2689.90	III
1	2898.90	III
5	2938.45	III
1	2948.94	III
5	2986.20	III
6	2992.24	III
6	3052.07	III
5	3056.84	III
6	3201.95	III
6	3209.34	III
6	3278.79	III
6	3289.06	III
6	3322.40	III
6	3364.22	III
6	3420.82	III
4	3421.83	III
6	3468.32	III
6	3481.11	III
5	3513.88	III
1	3885.50	III

K IV
Ref. 32, 76, 86, 150, 160, 314, 322 — L.J.R.

Intensity	Wavelength	
	Vacuum	
150	271.82	IV
100	273.06	IV
	279.88	IV
300	340.46	IV
150	340.74	IV
300	354.93	IV
150	356.26	IV
300	359.73	IV
200	359.91	IV
250	362.08	IV
150	362.15	IV
150	363.02	IV
300	375.96	IV
300	379.88	IV
250	380.48	IV
200	381.70	IV
300	382.23	IV
150	382.49	IV
200	382.65	IV
300	382.91	IV
250	384.10	IV
200	386.61	IV
250	388.92	IV
250	389.07	IV

Intensity	Wavelength	
250	390.42	IV
300	390.57	IV
200	391.46	IV
200	392.47	IV
500	393.14	IV
400	400.21	IV
300	402.91	IV
250	403.97	IV
150	404.41	IV
250	408.08	IV
150	417.28	IV
200	442.30	IV
300	443.57	IV
200	445.61	IV
250	446.83	IV
750	448.60	IV
400	456.33	IV
250	523.00	IV
200	526.45	IV
150	527.62	IV
750	646.19	IV
500	737.14	IV
500	741.95	IV
500	745.26	IV
400	746.35	IV
300	749.99	IV
150	754.19	IV
400	754.67	IV

K V
Ref. 32, 75, 76, 150, 322 — L.J.R.

Intensity	Wavelength	
	Vacuum	
100	214.35	V
150	282.35	V
150	293.33	V
300	294.84	V
200	296.17	V
200	297.06	V
200	300.25	V
200	300.50	V
200	311.24	V
250	312.77	V
200	315.18	V
250	327.38	V
250	389.07	V
200	349.50	V
500	372.15	V
200	372.46	V
200	372.77	V
300	375.96	V
250	377.76	V
300	379.12	V
300	387.80	V
250	390.11	V
250	395.40	V
200	398.36	V
200	398.88	V
200	399.75	V
250	415.05	V
200	415.79	V
400	422.18	V
300	425.16	V
500	425.59	V
250	438.02	V
200	449.71	V
200	452.90	V
250	455.67	V
400	456.33	V
200	482.71	V
200	483.75	V
750	580.32	V
250	585.51	V
500	586.32	V
250	602.27	V
400	603.43	V
250	638.67	V
300	687.50	V
300	720.43	V
400	724.42	V
600	731.86	V
150	770.29	V
150	771.46	V
	1035.60	V

PRASEODYMIUM (Pr)
Z = 59
Pr I and II
Ref. 1 — C.H.C.

Intensity		Wavelength	
		Air	
25		2558.58	II
25		2578.27	I
30		2579.31	I
40	h	2598.04	II
25		2608.92	II
25		2615.75	II
25		2648.48	II
30		2654.75	II
25		2666.70	II
20		2672.52	II
30		2685.19	II
45		2685.70	II

Intensity		Wavelength	Ion
50		2698.92	II
60		2700.38	II
30		2702.25	II
100	h	2707.37	II
20		2714.16	II
60		2720.17	II
30		2721.90	II
50		2726.50	II
12		2731.78	II
25		2733.12	II
50		2734.30	II
25		2737.90	II
40		2742.12	II
25		2744.66	II
20		2746.28	II
60		2760.35	II
50		2769.60	II
50	d	2775.94	II
		2776.03	II
40		2778.80	II
50		2783.31	II
30		2789.05	II
35		2792.51	II
50		2802.05	II
20		2823.17	II
20		2824.14	II
20		2828.29	II
20		2842.98	I
20		2844.01	II
20		2850.62	I
25		2853.99	II
30		2865.64	II
50		2881.60	I
30		2882.31	II
30		2884.89	II
30		2943.97	II
30		2967.58	II
30		2971.13	II
40	d	2971.40	II
		2971.46	II
50		2984.98	II
30		2986.18	II
30		2990.22	II
110		3082.11	II
100		3111.34	II
140		3121.58	II
140		3163.73	II
270		3168.24	II
160		3172.31	II
110		3191.42	II
200	d	3195.99	II
110		3199.04	II
100		3207.89	II
190		3219.48	II
100		3234.27	II
100		3245.48	II
140		3355.67	II
110		3394.62	II
110		3465.74	II
200		3584.21	II
130		3611.94	II
170		3630.96	II
100		3645.55	II
250		3645.66	II
250		3646.30	II
100		3648.30	II
150	c	3660.36	II
100		3661.62	II
370		3668.83	II
250		3687.03	II
150		3687.19	II
100		3689.71	II
150		3698.06	II
230		3706.75	II
170	c	3711.10	II
290		3714.05	II
120	c	3733.03	II
210	c	3734.41	II
250		3735.76	II
190		3736.49	II
410		3739.18	II
150		3740.99	II
120		3743.98	II
190		3750.98	II
140		3759.60	II
120		3760.08	II
680		3761.87	II
230		3764.77	II
230		3768.94	II
170	c	3772.82	II
170		3774.06	II
140		3777.62	II
170		3780.66	II
150		3785.46	II
150		3786.86	II
210		3792.51	II
190		3794.93	II
680		3800.30	II
290		3804.84	II
140		3809.18	II
390		3811.84	II
1300	h	3816.02	II
120		3817.66	II
680		3818.28	II

Intensity		Wavelength	Ion
120		3819.14	II
310		3821.80	II
150	c	3823.18	II
120		3826.67	II
960		3830.72	II
140		3834.93	II
480		3840.99	II
270		3842.34	II
150	c	3844.54	II
580		3846.59	II
1200		3850.79	II
720	c	3851.55	II
960		3852.80	II
120		3858.25	II
110		3859.14	II
480	c	3865.45	II
210		3867.52	II
210		3870.72	II
480		3876.19	II
1700	c	3877.18	II
270		3879.20	II
680		3880.47	II
440	c	3885.19	II
440	c	3889.34	II
120	c	3891.71	II
190		3897.25	II
210		3898.84	II
250		3902.45	II
770	c	3908.05	II
630		3912.90	II
310		3913.55	II
210		3914.76	II
1300	c	3918.85	II
420		3919.63	II
250		3920.53	II
960		3925.47	II
480		3927.46	II
370		3929.29	II
370		3935.82	II
250		3938.30	II
730	c	3947.63	II
900	c	3949.43	II
900	c	3953.51	II
380		3956.75	II
190		3959.44	I
470		3962.45	II
560		3964.26	II
1600	c	3964.81	II
560	c	3966.57	II
500		3971.16	II
320		3971.67	II
620	c	3972.14	II
320		3974.85	II
1300	c	3989.68	II
230		3991.91	II
340		3992.16	II
1600		3994.79	II
270		3995.83	II
560	c	3997.04	II
230		3997.96	II
320		3999.12	II
620	c	4000.17	II
730		4004.70	II
1900		4008.69	II
620		4010.60	II
730		4015.39	II
620		4020.96	II
470		4022.71	II
360		4025.54	II
230		4026.83	II
230		4029.00	II
360	c	4029.72	II
730	c	4031.75	II
230		4032.47	II
960		4033.83	II
230		4034.33	II
230		4038.22	II
730		4038.45	II
470		4039.34	II
1300		4044.81	II
230		4045.70	II
230		4046.63	II
340		4047.08	II
450		4051.13	II
2200		4054.88	II
2200		4056.54	II
450		4058.80	II
230		4062.22	II
3400		4062.81	II
210		4068.80	II
500	c	4079.77	II
500	c	4080.98	II
790		4081.85	II
500		4083.34	II
200	c	4087.21	II
560		4096.82	II
380		4098.40	II
2900	c	4100.72	II
270	c	4113.89	II
1700	c	4118.46	II
250		4129.15	II
340		4130.77	II
200		4133.61	II
1500	c	4141.22	II

Intensity		Wavelength	Ion
2700		4143.11	II
270	c	4146.50	II
270		4148.44	II
200		4156.50	II
1700	c	4164.16	II
270		4168.04	II
230		4169.45	II
620		4171.82	II
730		4172.25	II
250		4175.32	II
250		4175.62	II
200		4178.63	II
5200		4179.39	II
2500		4189.48	II
560	c	4191.60	II
290		4201.17	II
2500	c	4206.32	II
500		4208.32	II
320		4211.86	II
320		4217.81	II
3800		4222.93	II
3800		4225.35	II
320		4233.11	II
320	c	4236.15	II
270		4240.02	II
960		4241.01	II
340		4243.51	II
840	c	4247.63	II
500		4254.40	II
270	c	4263.78	II
320		4269.09	II
790		4272.27	II
470	c	4280.07	II
790	c	4282.42	II
450	c	4298.98	II
290		4303.61	II
1500		4305.76	II
210		4323.55	II
270		4329.41	II
1300		4333.97	II
200		4335.74	II
360		4338.70	II
620	Cw	4344.30	II
470	c	4347.49	II
340		4350.40	II
450		4354.91	II
410	c	4359.79	II
1200		4368.33	II
320		4371.62	II
270		4396.08	II
170		4403.60	II
100		4405.12	II
430		4405.83	II
1700		4408.82	II
410		4413.77	II
160		4419.04	II
190		4419.65	II
160	c	4421.22	II
160		4424.58	II
1200	c	4429.13	II
110		4432.28	II
730		4449.83	II
140		4451.90	II
140		4454.68	II
100		4465.97	II
960		4468.66	II
140	c	4477.26	II
1100		4496.46	II
790		4510.15	II
200	c	4517.58	II
340	c	4534.15	II
340		4535.92	II
200		4563.12	II
140		4612.08	II
270	c	4628.74	II
140		4632.28	I
140		4635.68	I
200		4639.55	I
110	c	4643.49	II
140		4646.05	II
200	c	4651.50	II
140		4664.65	II
270	c	4672.09	II
180		4687.80	I
290		4695.77	I
140	Cw	4708.07	II
140		4709.52	I
180		4730.67	I
250		4736.69	I
100		4744.16	I
150		4746.92	II
100		4762.72	II
110		4783.35	II
110		4906.99	I
140		4914.02	I
200		4924.60	I
140		4936.00	I
320		4939.74	I
160		4940.30	I
380		4951.37	I
110		4975.75	I
120		5018.59	I
200		5019.76	I
200		5026.96	I

Intensity		Wavelength	Ion
100		5033.38	I
270		5034.41	II
110		5043.83	I
320		5045.52	I
160		5053.40	I
180		5087.12	I
360		5110.38	II
560		5110.76	II
410		5129.52	II
270		5133.44	I
270		5135.14	II
100		5139.81	I
100	c	5152.30	II
200		5161.74	II
620		5173.90	II
200		5191.32	II
120		5194.43	I
150		5195.11	II
200		5195.31	II
360		5206.55	II
150		5207.90	II
360		5219.05	II
560		5220.11	II
110		5227.97	I
680		5259.73	II
180		5263.88	II
340	c	5292.02	II
340		5292.62	II
230		5298.09	II
430		5322.76	II
200		5352.40	II
16		5501.50	I
40		5508.79	II
65		5509.15	II
16	c	5511.63	II
55		5513.58	II
28		5515.12	II
13		5519.38	II
20	c	5520.31	II
45	c	5522.79	II
28	c	5524.15	I
28	c	5525.91	II
16	c	5527.93	I
13		5530.21	I
45		5531.16	I
150		5535.17	II
28		5538.37	I
20		5538.78	II
55		5545.01	II
20		5548.33	II
11		5553.42	II
22		5561.46	II
45	c	5562.06	I
13		5565.52	I
13		5566.91	II
45		5571.83	II
11		5574.61	II
11		5578.81	I
13		5582.35	II
11		5584.02	II
22		5594.92	I
22		5597.29	II
13		5601.30	II
90		5605.65	II
13		5606.68	I
28		5608.93	II
55		5610.22	II
11		5620.06	II
20		5620.26	I
45	c	5621.89	II
110		5623.05	II
90		5624.45	II
11	h	5633.03	I
22		5636.46	II
55	c	5638.79	II
16		5640.37	II
16	Cw	5643.16	I
22		5645.41	II
35		5654.23	II
55		5659.84	II
35	h	5661.57	I
16		5662.19	II
65	c	5668.46	II
45		5669.55	II
35		5669.99	II
16		5674.14	II
16		5677.03	II
55		5681.89	II
13		5685.60	II
16		5686.52	I
22	h	5687.17	II
65		5688.44	II
22		5689.21	II
55	h	5690.97	II
22		5695.90	II
65		5704.38	I
40		5711.63	II
22		5713.83	II
16		5716.08	II
45		5719.08	II
45	d	5719.63	II
		5719.80	II
11		5728.38	I

Int.		Wavelength	Sp.
40		5731.88	II
20		5747.13	II
11		5747.74	I
11		5747.95	II
22		5753.02	II
90		5756.17	II
16		5759.40	II
22		5760.20	I
22		5769.16	II
16		5769.79	II
45		5773.16	II
11		5775.91	II
16		5777.29	II
90		5779.28	I
65	c	5785.28	II
65		5786.17	II
16	h	5788.29	II
16		5788.92	II
16		5790.86	II
45		5791.36	II
22		5792.95	I
40		5810.58	II
16		5813.55	II
160	d	5815.17	II
		5815.33	II
55		5818.57	II
40		5820.62	II
16	h	5821.36	I
55		5822.59	II
90		5823.72	II
45		5830.94	II
40		5835.13	I
35	c	5844.65	II
40		5844.98	II
65		5847.13	II
65	c	5850.64	II
45		5852.63	II
11	c	5854.44	I
45		5856.07	II
55		5856.90	II
90		5859.68	II
80		5868.83	II
22		5873.83	II
35		5874.72	I
35		5878.10	I
35		5879.04	I
80		5879.25	II
35	c	5884.72	I
55		5892.23	II
22		5894.22	II
40		5903.11	II
45		5904.45	II
40		5908.67	II
11		5915.31	I
11		5915.97	I
40		5920.76	I
40		5930.66	II
16		5936.33	II
160		5939.90	II
65		5940.72	II
22		5941.65	I
35		5947.16	II
22	c	5949.76	I
55		5951.27	II
20		5951.76	II
90		5956.60	II
		5956.70	I
13		5959.25	I
20		5962.18	I
28		5963.00	I
110		5967.82	II
13	c	5976.95	I
13		5978.88	I
65		5981.19	II
40		5986.14	I
45	c	5987.14	I
		5987.29	II
13		5991.27	I
13	c	5994.89	I
11		5996.06	I
29		6002.44	II
90		6006.33	II
13		6008.54	I
55		6016.48	II
150		6017.80	II
28	c	6019.85	I
150		6025.72	II
35		6042.87	II
55		6046.66	II
35		6049.26	II
28		6050.04	II
11		6050.88	I
140		6055.13	II
13		6067.27	II
13		6085.81	I
28		6086.16	II
65		6087.52	II
20		6090.38	II
28		6093.09	II
18		6096.28	I
22		6106.72	II
18		6109.08	I
65		6114.38	II
22	c	6118.02	I

Int.		Wavelength	Sp.
22	c	6122.15	I
35		6141.51	II
65		6148.23	I
		6148.24	II
22		6157.82	II
13		6159.10	II
190		6161.18	II
18		6165.38	I
270		6165.94	II
55		6182.34	II
13		6187.96	I
35		6197.45	II
35		6200.81	II
13		6205.63	II
13		6210.59	I
22		6212.73	I
18		6218.06	I
20	h	6236.80	I
20	h	6241.05	I
45		6244.35	II
35		6255.10	II
40		6262.55	II
18		6264.54	II
22	c	6274.66	II
		6274.81	II
40		6278.68	II
110		6281.28	II
18	c	6289.02	I
11	c	6298.01	I
11		6302.05	I
35		6302.35	II
16		6304.05	I
35		6305.23	II
11	h	6318.13	II
45	c	6322.36	I
22	h	6343.88	I
28		6347.11	II
18	c	6350.98	I
22	c	6357.20	I
55	c	6359.03	II
11	h	6363.62	II
16		6377.61	I
16		6378.59	I
11		6389.57	I
18	c	6391.99	I
40		6393.18	I
45		6397.96	II
10	h	6410.69	I
55		6411.23	II
40		6413.68	II
10		6415.43	I
45		6429.63	II
45		6431.84	II
7	h	6442.78	II
9	h	6443.91	II
16	c	6453.44	I
9		6454.84	II
9		6456.18	I
9	h	6460.19	I
18		6467.72	II
9	h	6475.26	II
35	Cw	6478.02	II
45		6486.55	I
9	h	6486.97	II
40	h	6491.75	I
9		6493.49	I
11		6494.89	I
22	c	6497.11	I
18		6498.94	II
22		6500.72	I
9		6504.09	I
8		6517.14	I
16		6518.79	II
8		6534.52	I
16		6540.47	I
7	h	6553.30	I
22		6564.62	II
45		6566.77	II
7		6571.03	I
6		6578.00	I
6		6584.56	II
9	h	6593.74	II
11		6595.48	I
15		6609.86	I
55		6616.67	I
11		6618.34	II
7	h	6631.00	I
13	h	6632.06	I
14		6647.12	I
75		6656.83	II
55		6673.41	II
75		6673.78	II
5	h	6687.51	I
4	h	6699.25	I
13		6736.79	I
35	c	6747.09	I
19	c	6749.19	I
7	c	6784.99	II
55	Cw	6798.60	I
11		6811.76	II
17	Cw	6812.87	II
13		6814.04	II
9		6817.61	I
35	Cw	6827.60	II

Int.		Wavelength	Sp.
19		6830.50	II
9		6844.39	I
9	h	6845.47	II
9		6846.59	II
17	c	6850.46	II
11		6852.77	I
11	c	6870.44	I
7		6884.66	I
8		6892.71	I
8	h	6970.38	I
8	c	6980.12	I
40		7021.51	II
10		7024.53	I
13		7042.40	I
10		7044.45	II
7		7051.07	I
10		7079.99	I
11	c	7095.18	I
20		7114.55	I
10	h	7116.90	I
11		7118.24	II
7		7137.33	I
10	h	7159.88	I
7		7167.77	II
7	h	7189.95	I
10	c	7208.85	II
24		7227.70	II
13		7231.53	I
7	c	7243.26	I
7	c	7259.21	I
7	c	7287.61	I
7	h	7289.19	I
7	Cw	7324.42	I
7		7328.47	I
7		7344.86	I
16		7407.56	II
20	h	7451.74	II
11	h	7495.59	I
6	h	7499.42	I
14		7541.02	II
6		7574.86	I
20		7645.66	II
7		7704.98	I
16		7721.84	II
6	h	7786.16	II
6	Cw	7841.27	I
14		7871.67	I
6		7881.09	I
6	Cw	7888.56	II
6		7915.19	II
6		8031.92	I
6		8055.43	I
14		8067.44	I
10	Cw	8122.78	II
11		8141.10	I
5		8181.34	II
5	c	8211.93	I
6		8289.93	I
6		8379.84	I
6	h	8427.82	I
6	h	8605.27	II
10		8714.59	II

Pr III

Ref. 306, 308 — J.R.

Intensity		Wavelength	
		Vacuum	
25		1008.61	III
50		1021.35	III
25		1026.18	III
100		1029.03	III
50		1038.29	III
50		1042.96	III
25		1043.80	III
25		1044.03	III
25		1046.20	III
150		1047.24	III
100		1049.09	III
50		1052.63	III
25		1061.60	III
25		1066.03	III
25	p	1068.85	III
25		1069.88	III
25		1084.42	III
25		1088.66	III
100		1104.84	III
25		1108.82	III
30		1352.70	III
25		1881.22	III
		Air	
50		2031.10	III
100		2033.30	III
25		2043.12	III
100		2052.30	III
50		2052.87	III
50		2053.85	III
200		2058.59	III
50		2064.08	III
25		2075.08	III
200		2090.75	III
200		2093.49	III
25		2096.85	III
50		2096.94	III

Intensity		Wavelength	
10	w	2148.14	III
10	w	2194.24	III
10	w	2197.25	III
10	w	2205.48	III
10	w	2206.26	III
10	w	2214.45	III
10	w	2215.25	III
10	w	2217.12	III
10	w	2223.23	III
10	w	2230.35	III
10	w	2237.26	III
10		2239.06	III
10		2239.42	III
10	w	2242.15	III
10	w	2284.62	III
10		2307.59	III
10		2307.77	III
10		2308.41	III
10		2311.29	III
10		2311.44	III
10		2314.18	III
10	w	2315.46	III
10		2318.15	III
10		2318.36	III
10		2318.64	III
10		2318.82	III
10		2318.97	III
10	w	2319.40	III
10		2320.41	III
10	w	2328.56	III
10	w	2336.13	III
10		2365.52	III
10		2368.78	III
10		2369.08	III
10	w	2378.06	III
10	w	2378.97	III
10		2395.44	III
10		2399.70	III
10	w	2405.56	III
10		2408.19	III
10		2409.80	III
10		2412.40	III
10		2417.69	III
10		2418.95	III
10		2426.14	III
10		2426.85	III
10	w	2430.32	III
10	w	2434.18	III
10		2434.39	III
10		2435.91	III
10		2436.89	III
10		2438.63	III
10	w	2444.93	III
10		2445.49	III
10		2446.77	III
10		2448.16	III
10		2452.02	III
10	w	2452.81	III
10	w	2452.85	III
10		2454.60	III
10		2454.82	III
10	w	2459.77	III
10	w	2460.72	III
10		2462.18	III
10		2462.90	III
10		2468.20	III
10		2468.97	III
10	w	2473.42	III
10	w	2478.32	III
10		2479.98	III
10		2481.02	III
10		2483.30	III
10		2483.99	III
10	w	2484.60	III
10	w	2485.16	III
10	w	2488.72	III
10		2491.97	III
10	w	2494.20	III
10	w	2495.37	III
10	w	2495.51	III
10		2499.97	III
20	w	2587.71	III
40	w	2624.91	III
20	w	2644.62	III
20	w	2656.88	III
20	w	2667.51	III
70	w	2679.47	III
40	w	2710.30	III
20	w	2718.65	III
100	w	2724.03	III
20	l	2841.94	III
70	s	2910.61	III
50	s	2911.77	III
100	l	2914.49	III
50	s	2930.19	III
70	s	2942.43	III
70	s	2953.58	III
90	w	2954.40	III
90	s	2964.85	III
150	s	2968.83	III
80	s	2969.41	III
150	l	2976.86	III
150	s	2977.06	III
500	s	2980.54	III

Intensity		Wavelength	
100	l	2981.65	III
150	c	2982.42	III
500	s	2985.82	III
150	s	2997.12	III
70	l	2998.79	III
150	l	3000.46	III
150	s	3003.20	III
60	l	3006.47	III
150	s	3008.04	III
150	l	3010.61	III
90	s	3014.60	III
100	l	3015.13	III
90	s	3016.26	III
70	s	3021.77	III
70	s	3025.26	III
100	l	3029.38	III
100	l	3033.31	III
90	l	3034.25	III
70		3040.02	III
60	c	3040.94	III
60	l	3041.78	III
100	s	3042.35	III
100	l	3045.81	III
70	l	3046.98	III
120	l	3050.30	III
70	s	3055.30	III
150	l	3058.90	III
150	l	3066.71	III
70	s	3078.68	III
150	l	3080.20	III
50	w	3248.39	III
90	s	3280.92	III
90	l	3292.58	III
90	s	3296.10	III
100	l	3306.14	III
60	s	3333.26	III
90	s	3340.58	III
500	s	3341.43	III
70	s	3341.68	III
70	s	3345.38	III
70	l	3345.44	III
50	w	3351.07	III
50	s	3353.87	III
250	s	3354.91	III
500	l	3357.56	III
500	l	3359.41	III
100	l	3364.52	III
50		3364.88	III
50	s	3365.80	III
200	s	3367.35	III
500	s	3367.58	III
100	l	3371.92	III
100	s	3377.14	III
50	s	3379.13	III
150	s	3380.21	III
150	s	3381.26	III
300	s	3381.84	III
100	s	3391.08	III
1000	w	3394.22	III
600	d	3396.07	III
300	s	3396.62	III
300	l	3397.46	III
50	l	3402.97	III
500	c	3413.21	III
300	s	3415.15	III
150	s	3420.07	III
300	l	3422.22	III
50	c	3426.27	III
500	l	3427.02	III
300	l	3436.36	III
150	l	3440.62	III
50	l	3445.29	III
70	s	3454.05	III
180		3653.58	III
60		3817.25	III
60	l	3861.80	III
150		3980.51	III
200		4000.20	III
90		4018.36	III
180		4029.60	III
90		4142.46	III
120		4144.48	III
90		4147.85	III
90		4172.15	III
150		4179.77	III
180		4184.18	III
240		4197.01	III
120		4219.45	III
180		4231.45	III
180		4275.07	III
120		4286.32	III
90		4298.27	III
90		4301.73	III
90		4316.34	III
60		4354.28	III
90		4379.82	III
90		4381.47	III
120		4404.71	III
120		4421.10	III
120		4431.85	III
150	w	4447.93	III
180	w	4450.14	III
120	w	4451.00	III

Intensity		Wavelength	
120		4461.02	III
200		4462.81	III
300	w	4500.31	III
450	w	4612.02	III
600	w	4625.18	III
120		4654.16	III
600	w	4713.70	III
300	w	4725.55	III
270		4728.21	III
300		4747.11	III
300	w	4771.83	III
450	w	4775.30	III
600		4857.39	III
150		5208.51	III
150		5261.68	III
1000		5264.44	III
1500		5284.70	III
1500		5299.99	III
1500		5340.02	III
100		5427.70	III
100		5581.74	III
150		5646.80	III
600		5765.27	III
1500		5844.41	III
200		5947.98	III
7000	w	5956.05	III
900	w	5998.94	III
1500	w	6053.01	III
900		6071.09	III
9000	w	6090.02	III
5000		6160.24	III
1500	w	6161.22	III
100		6195.05	III
2000		6195.63	III
200		6310.36	III
100		6361.65	III
300		6429.26	III
300		6444.74	III
600		6500.04	III
300		6501.49	III
200		6578.90	III
100		6616.46	III
600		6706.70	III
100		6727.63	III
200		6827.96	III
100		6854.63	III
200		6857.30	III
1000		6866.80	III
1000		6899.06	III
500		6903.52	III
7000		6910.14	III
150	w	6934.55	III
500		6970.96	III
100		6979.83	III
5000		7030.39	III
100		7075.21	III
4500		7076.62	III
100		7083.99	III
500		7112.53	III
100	w	7165.64	III
250		7231.62	III
100	w	7238.26	III
250		7240.21	III
100	w	7262.32	III
150	w	7340.69	III
350		7343.70	III
200		7349.75	III
300	w	7350.61	III
100	w	7355.52	III
2000		7426.48	III
4000		7429.47	III
100		7463.96	III
250		7487.40	III
200		7493.20	III
100		7511.17	III
500		7529.11	III
100	w	7549.20	III
150	w	7588.64	III
300		7596.41	III
100	w	7625.63	III
100	w	7648.34	III
100	w	7670.65	III
200	w	7674.65	III
500		7742.34	III
250	w	7745.59	III
500		7754.31	III
100	w	7755.48	III
3000		7781.98	III
200	w	7814.74	III
1500		7866.14	III
1000		7888.12	III
400		7897.09	III
1000		7914.00	III
100		7923.16	III
1000	w	7972.75	III
250		8001.14	III
3000		8102.90	III
250		8119.54	III
400		8132.23	III
100		8138.34	III
150	w	8235.33	III
250	w	8244.89	III
100		8409.10	III
100	w	8494.99	III

Intensity		Wavelength	
200	w	8567.63	III
5000	w	8602.74	III
500		8691.58	III
500		8771.38	III
1000		8854.05	III
125		8886.17	III
100		8908.70	III
250		9099.98	III
250		9131.90	III
200		9222.32	III
250	w	9265.56	III
125		9320.54	III
250		9334.33	III
175		9377.44	III
175		9388.56	III
175		9549.77	III
100		9579.74	III
150		9802.98	III
175		9806.37	III
500	w	9991.16	III
500		10031.10	III
500		10160.33	III
500		10238.63	III
500		10301.58	III
500		10324.59	III
500		10716.58	III

Pr IV

Ref. 337, 338 — J.R.

Intensity	Wavelength	
	Vacuum	
20	718.23	IV
30	721.34	IV
60	722.41	IV
30	722.58	IV
20	726.04	IV
50	730.37	IV
30	731.77	IV
30	734.86	IV
20	735.04	IV
20	736.19	IV
50	736.32	IV
100	737.17	IV
100	741.45	IV
20	743.15	IV
20	743.89	IV
20	746.14	IV
40	763.16	IV
20	764.00	IV
300	1226.40	IV
2000	1228.59	IV
500	1230.69	IV
400	1238.19	IV
200	1249.35	IV
200	1255.64	IV
200	1261.27	IV
400	1268.32	IV
300	1270.58	IV
1000	1275.10	IV
200	1275.40	IV
1000	1278.65	IV
200	1279.34	IV
1000	1287.44	IV
300	1290.93	IV
1000	1292.30	IV
5000	1293.22	IV
5000	1295.28	IV
400	1296.50	IV
200	1298.26	IV
300	1298.54	IV
200	1304.71	IV
300	1306.86	IV
200	1308.08	IV
200	1310.71	IV
500	1314.96	IV
300	1315.28	IV
300	1316.96	IV
500	1320.10	IV
1000	1320.70	IV
5000	1321.36	IV
500	1322.51	IV
500	1326.38	IV
5000	1333.57	IV
300	1335.96	IV
500	1339.29	IV
1000	1340.74	IV
200	1341.32	IV
300	1344.23	IV
1000	1347.07	IV
1000	1352.81	IV
500	1354.35	IV
5000	1354.66	IV
2000	1360.64	IV
1000	1364.81	IV
2000	1365.77	IV
400	1368.90	IV
5000	1374.41	IV
1000	1382.62	IV
200	1384.23	IV
200	1385.91	IV
300	1394.11	IV
500	1397.11	IV
1000	1399.31	IV
1000	1400.96	IV

Intensity		Wavelength	
400		1410.90	IV
1000		1424.36	IV
500		1426.59	IV
5000		1435.56	IV
200		1459.95	IV
200		1461.76	IV
1000		1474.91	IV
200		1477.32	IV
500		1485.88	IV
500		1503.35	IV
400		1516.86	IV
200		1520.71	IV
2000		1520.98	IV
400		1522.46	IV
500		1553.62	IV
500		1559.49	IV
500		1570.13	IV
200		1572.80	IV
5000		1574.55	IV
5000		1575.10	IV
3000		1578.38	IV
500		1585.10	IV
300		1613.00	IV
400		1613.65	IV
1000		1618.03	IV
2000		1622.30	IV
300		1634.77	IV
400		1676.08	IV
200		1688.49	IV
200		1713.53	IV
500		1732.86	IV
300		1762.86	IV
1000		1766.88	IV
1000		1771.14	IV
500		1841.08	IV
10000		1884.87	IV
400		1951.23	IV
200		1954.61	IV
		Air	
200		2025.06	IV
1000		2039.15	IV
200		2047.05	IV
200		2050.73	IV
200		2058.48	IV
2000		2083.23	IV
500		2100.42	IV
1000		2154.31	IV
300		2193.37	IV
1000		2205.13	IV
300		2265.70	IV
200	c	2334.46	IV
200	c	2339.08	IV
500	c	2376.09	IV
2000	c	2378.98	IV
1000	c	2379.66	IV
500	c	2427.07	IV
500	c	2428.13	IV
500	c	2438.57	IV
500	c	2455.64	IV
500	c	2705.19	IV
200	c	2708.01	IV
200	c	2753.47	IV
200	c	2767.60	IV

Pr V

Ref. 149 — J.R.

Intensity	Wavelength	
	Vacuum	
200	843.78	V
7000	865.90	V
5000	869.17	V
80	869.66	V
1000	896.65	V
750	922.29	V
250	1234.07	V
250	1342.78	V
200	1958.09	V
400	1958.20	V
	AIR	
300	2246.06	V
300	2246.20	V

PROMETHIUM (Pm)
Z = 61

Pm I and II
Ref. 196, 260 — C.H.C.

Intensity		Wavelength	
		Air	
40	w	2502.12	II
40		2608.24	II
150		2632.00	II
70		2638.46	II
100		2671.05	II
50	w	2787.72	II
40		2808.05	II
100	h	2820.10	II
100	w	2840.82	II
150	w	2841.86	II
200	c	2857.46	II
100		3004.59	II
100		3008.85	II
300		3072.41	II
150		3086.02	II

Promethium (Cont.)

Int.		λ		Int.		λ		Int.		λ		Int.		λ	
120		3090.19	II	600		4075.84	II	400		4645.94	I	120		5556.88	II
150		3091.86	II	200	r	4085.31	I	600	r	4647.03	I	150		5558.39	II
150		3108.11	II	500		4086.10	II	600	r	4650.42	I	200		5561.73	II
100		3115.36	II	250		4140.46	II	500		4650.52	I	800		5576.02	II
100		3117.22	II	200		4185.74	II	400	r	4653.41	I	200		5641.29	II
100		3118.76	II	300		4192.92	II	400		4654.50	I	200		5730.81	I
35		3162.23	I	200		4194.70	II	500	r	4655.05	I	200		5768.16	II
35		3168.82	I	200		4222.15	II	300		4659.38	I	200		5776.99	I
100		3172.77	II	300	r	4264.32	I	500	r	4660.79	I	500		5823.93	II
35		3222.04	I	300	r	4284.37	I	300		4663.26	I	300	c	5868.79	II
35		3238.55	I	600		4297.78	II	600	r	4663.46	I	200	c	5875.31	II
60		3239.62	II	200		4303.89	II	400		4665.19	I	100	c	5878.76	II
75		3296.63	I	200	r	4305.64	I	500	r	4671.23	I	150		5899.76	II
60		3311.76	II	400		4318.80	I	400		4671.76	I	250		5904.71	I
50		3313.38	I	250		4325.92	II	500	r	4674.42	I	100		5905.90	I
50		3329.22	I	200		4332.05	II	200		4677.46	I	125		5914.96	I
75		3331.57	I	300		4336.54	II	500	r	4677.92	I	250	c	5927.17	II
100		3354.45	I	200		4337.48	II	400		4678.09	I	150		5939.66	I
100		3358.14	II	300		4342.12	II	700	r	4682.92	I	400	c	5946.49	II
80		3360.21	II	200		4347.72	I	500	r	4696.80	I	800		5956.42	I
80		3364.44	II	350	r	4363.92	I	200		4699.51	I	200		5956.69	I
300		3366.03	I	300	r	4369.64	I	250		4722.06	II	100	c	5960.08	II
90		3377.68	II	200		4381.88	II	300		4727.06	I	150		5963.00	II
100		3391.28	II	400	r	4388.49	I	900		4728.36	I	400		5967.89	I
100		3408.06	II	200		4388.76	II	400	r	4728.68	I	200		5979.73	I
500		3427.40	II	400	r	4409.40	I	800	r	4734.27	I	200		5984.82	I
120		3441.15	II	500	r	4412.47	I	200	r	4737.99	I	100	c	5987.13	II
400		3449.80	II	1000		4417.96	II	100		4739.08	II	400		5997.12	I
250		3460.25	II	400		4432.51	II	200		4739.78	I	200		6027.11	II
200		3462.91	II	250	r	4435.86	I	350	r	4745.13	I	300		6030.06	I
200		3480.61	II	300	r	4436.55	I	500	r	4757.73	I	400		6031.32	I
150		3497.13	II	300	r	4438.68	I	800	r	4759.00	I	500		6043.39	I
100		3514.85	II	500		4445.41	II	700	r	4762.57	I	150	c	6052.57	II
200		3546.81	I	600		4446.90	II	700	r	4773.46	I	100	c	6067.00	II
100		3559.43	II	800		4453.95	II	900	r	4781.29	I	500		6069.06	I
200		3565.31	II	200		4459.97	II	250		4794.59	I	100		6076.40	II
150		3580.10	II	250	r	4468.16	I	200		4795.43	I	200		6085.41	I
200		3610.76	II	200		4471.48	II	700	r	4798.98	I	900		6100.21	I
200		3629.84	II	300		4473.23	II	900	r	4801.36	I	400		6106.40	I
300		3634.20	II	200		4477.46	II	700	r	4809.54	I	100		6114.90	II
300		3659.39	II	350	r	4478.58	I	900	r	4811.96	I	400	h	6151.76	I
300	r	3669.22	I	300	r	4481.60	I	400	r	4817.12	I	100		6159.53	II
200		3674.85	I	300	r	4485.05	I	400	r	4827.72	I	400		6163.16	I
200		3678.51	II	300	r	4490.50	I	800	r	4837.66	I	100		6184.52	II
300	r	3679.85	I	250		4492.05	II	400	r	4838.92	I	200		6208.91	I
200		3687.65	II	600		4500.15	II	300		4839.62	I	500		6229.64	I
400		3689.79	II	350	r	4500.33	I	200		4844.01	I	400		6237.79	I
300		3692.50	II	250		4506.84	I	350		4852.73	I	100		6263.25	II
300		3697.50	II	100		4509.38	II	400	r	4860.62	I	400		6272.69	I
300	r	3697.63	I	100		4513.56	II	700	r	4860.74	I	400		6286.06	I
400		3702.63	II	200		4517.31	I	300	r	4865.30	I	500		6308.29	I
800		3711.72	II	200		4523.32	I	500	r	4865.72	I	100		6314.20	II
200		3715.75	II	600		4525.20	II	400		4869.80	I	700		6323.84	I
200		3721.72	II	250	r	4526.12	I	700	r	4872.42	I	500		6390.31	I
500		3726.01	I	250	r	4526.76	I	500	r	4887.02	I	100		6429.64	II
200		3738.43	I	400	r	4527.70	I	700	r	4892.52	I	500	h	6431.93	I
300		3740.68	I	800		4529.21	II	400		4900.30	I	100		6436.57	II
300		3742.52	II	300		4540.06	I	300	r	4904.28	I	400		6487.61	I
300	r	3742.97	I	300		4541.42	I	400	r	4918.28	I	400		6510.34	I
500		3745.86	II	450	r	4541.75	I	600	r	4932.99	I	500		6517.25	I
300		3747.09	II	500	r	4544.08	I	700	r	4959.46	I	200		6519.43	II
500		3750.09	II	200		4545.17	I	100		4971.40	II	1000	d	6520.45	I
200		3761.68	I	400	r	4549.78	I	500	r	4997.10	I	500		6542.20	I
300		3765.75	I	300	r	4554.03	I	200	r	5030.80	I	100	h	6558.48	II
300		3775.42	I	200		4554.63	I	300	r	5058.31	I	100		6586.39	II
300		3780.77	I	500	r	4555.34	I	100		5067.35	II	100		6592.29	II
200		3781.43	I	200		4556.06	I	150		5080.52	II	900		6598.15	I
400		3795.66	II	300		4557.03	I	150		5089.35	II	800		6598.66	I
250		3806.06	II	300		4559.21	I	200		5092.42	I	700		6606.37	I
300	r	3809.20	I	100		4564.83	II	400	r	5094.83	I	800	w	6625.23	I
400		3810.93	I	300		4568.14	I	200		5096.18	I	100	h	6625.54	II
200		3819.26	II	200		4570.37	I	150		5097.30	II	700		6649.81	I
300		3820.53	II	300	r	4572.15	I	400	r	5100.77	I	400		6659.05	II
300		3839.52	I	400	r	4575.27	I	250		5121.47	II	100		6661.25	II
200		3842.88	II	300		4578.28	I	400	r	5127.34	I	500		6661.68	I
300		3842.98	II	200	r	4578.41	I	200		5129.75	I	400		6663.76	I
250		3845.38	II	300	r	4579.48	I	400	r	5145.13	I	800	c	6667.51	I
300		3874.03	I	300		4581.14	I	500	r	5146.30	I	700	h	6677.47	I
800		3877.62	II	300	r	4585.49	I	400		5153.86	II	200		6680.89	II
300	r	3885.79	I	200		4593.82	I	300		5169.71	II	500		6685.55	I
250		3890.97	I	400	r	4595.82	I	500		5171.58	II	500		6685.68	I
1000		3892.15	II	800		4597.55	I	300		5194.05	II	150		6690.09	II
300		3898.73	I	500	r	4600.25	I	500		5208.09	II	600		6700.33	I
400		3899.78	II	400	r	4602.96	I	150		5215.96	II	100		6706.27	II
250		3909.50	II	400		4604.59	I	250		5225.12	II	700		6714.67	I
1000		3910.26	II	600	r	4605.66	I	500		5236.26	II	500		6717.26	I
1000		3919.10	II	500	r	4609.85	I	300		5236.66	II	500		6720.71	I
800		3936.48	II	100		4615.87	II	400		5246.33	II	700		6727.50	I
300		3944.21	II	600		4617.02	I	150		5262.42	II	600		6743.71	I
300	r	3954.76	I	200		4618.40	I	150		5270.64	II	900		6749.91	I
1000		3957.74	II	400	r	4618.49	I	200		5293.92	II	900		6750.48	I
500		3980.74	II	500	r	4619.75	I	100		5308.86	II	200		6756.45	II
300		3995.05	II	500		4621.57	I	150		5318.58	II	300		6772.29	II
1000	r	3998.96	II	500		4623.31	I	200		5410.45	II	400		6778.78	I
500		4009.96	II	700	r	4623.68	I	200		5424.54	II	100		6783.09	II
200		4012.72	II	900		4624.41	I	180		5424.79	II	100		6796.87	II
250		4014.20	II	500	r	4625.29	I	150		5429.04	II	200		6811.68	II
200		4019.34	II	400	r	4627.60	I	100		5467.64	II	800		6833.30	I
250		4028.20	II	200		4630.93	I	150		5495.45	II	400		6848.37	I
200		4045.36	II	600	r	4633.45	I	100		5516.42	II	50		6858.58	II
300		4051.54	II	400	r	4640.96	I	180		5534.96	II				
600	r	4055.20	II	700	r	4643.36	I	200		5537.38	I				
200	r	4056.56	I	700	r	4643.76	I	800		5546.08	II				

PROTACTINIUM (Pa)
Z = 91

Pa I and II
Ref. 96 — J.G.C.

Intensity		Wavelength	
		Air	
3000	h	2466.85	
3000	h	2492.85	
3000		2599.16	II
3000		2699.22	II
3000		2822.79	II
3000	l	2832.14	
3000		2870.01	
3000	h	2871.42	II
3000	h	2891.14	II
3000	h	2906.93	
3000	l	3011.10	II
3000	s	3033.59	II
3000	l	3071.24	II
3000	h	3083.19	
3000	l	3093.23	II
3000	l	3126.23	II
3000	l	3146.28	II
3000	l	3170.89	II
3000	l	3171.54	II
3000		3204.16	
3000	l	3240.58	II
3000		3274.46	II
3000	l	3332.69	II
3000	s	3346.66	II
3000	l	3394.49	
3000	l	3452.82	II
3000		3504.97	I
3000	s	3530.65	II
3000		3570.56	I
3000		3571.82	I
3000		3618.07	
10000		3636.52	I
3000		3702.74	I
3000		3752.67	I
3000		3873.35	I
3000		3931.83	I
3000	s	3952.62	II
10000	l	3957.85	II
3000	s	3970.07	II
3000		3981.82	I
10000		3982.23	I
3000	l	4012.96	II
3000	l	4018.21	II
3000		4030.16	II
3000	s	4046.93	II
10000	s	4056.20	II
10000	s	4070.40	II
3000		4117.62	I
3000	l	4176.18	II
10000	l	4217.23	II
10000	s	4248.08	II
3000	s	4291.34	II
3000		4400.77	
3000		4436.13	
3000	s	4601.43	II
3000		4628.19	
3000		4820.34	
3000	s	4861.49	
3000	l	6035.78	I
3000		6162.56	I
3000		6216.35	
3000	l	6358.61	I
3000		6379.25	I
3000	l	6438.97	I
3000	h	6792.75	I
10000		6945.72	I
3000		6960.09	I
3000	h	6961.78	I
3000	s	6992.73	I
3000		7076.27	I
3000	h	7100.94	I
10000	s	7114.89	I
3000	h	7171.55	I
3000		7227.13	I
3000		7318.79	I
10000	l	7368.25	I
3000	h	7471.89	I
10000	h	7493.15	I
3000	h	7558.26	I
10000	h	7608.20	I
10000		7626.79	I
10000	s	7635.18	I
10000		7669.34	I
3000		7679.20	I
10000	h	7749.19	I
3000		7872.95	I
3000	l	7945.56	I
10000		8039.34	I
10000	h	8099.84	I
10000		8199.04	I
10000		8271.87	I
3000	s	8358.98	I
3000	s	8369.60	I
3000	h	8441.04	I
10000	h	8532.66	I
10000	s	8572.96	I
3000	h	8639.91	I
3000	h	8653.51	I
10000		8735.27	I
3000		10594.38	
3000		10923.32	I
3000		11646.78	
10000		11791.73	I
3000		12279.01	
3000		13234.09	
10000		13522.40	
10000		14344.76	I
3000		18478.61	I

RADIUM (Ra)
Z = 88

Ra I and II
Ref. 253, 254 — E.F.W.

Intensity		Wavelength	
		Air	
8		2369.73	II
8		2460.55	II
10		2475.50	II
8		2586.61	II
10		2643.73	II
20		2708.96	II
10		2795.21	II
30		2813.76	II
10		3033.44	II
5		3101.80	I
100		3649.55	II
200		3814.42	II
8		4194.09	II
8		4244.72	II
100		4340.64	II
20		4436.27	II
30		4533.11	II
8		4641.29	I
100		4682.28	II
8		4699.28	I
100		4825.91	I
10		4856.07	I
10		4859.41	II
10		4927.53	II
10		5097.56	I
10		5205.93	I
10		5283.28	I
10		5320.29	I
10		5399.80	I
20		5400.23	I
20		5406.81	I
8		5482.13	I
10		5501.98	I
10		5553.57	I
20		5555.85	I
10		5616.66	I
50		5660.81	I
20		5813.63	II
30		6200.30	I
10	p	6336.90	I
20		6446.20	I
20		6487.32	I
10		6593.34	II
10		6719.32	II
20		6980.22	I
20		7118.50	I
50		7141.21	I
20		7225.16	I
10		7310.27	I
20		7838.12	I
50		8019.70	II
6		8177.31	I
5		8335.07	I
5		9932.21	I
50		4508.48	I
50		4577.72	I
50		4609.38	I
30		4721.76	I
6		5722.58	I
10		6061.92	I
6		6200.75	I
6		6380.45	I
10		6557.49	I
10		6606.43	I
15		6627.23	I
6		6669.60	I
8		6704.28	I
20		6751.81	I
6		6806.79	I
8		6836.95	I
8		6837.57	I
10		6891.16	I
10		6998.90	I
200		7055.42	I
100		7268.11	I
20		7291.00	I
6		7320.98	I
10		7419.04	I
300		7450.00	I
8		7470.89	I
8		7483.13	I
8		7514.13	I
8		7516.92	I
6		7523.93	I
6		7597.55	I
8		7601.28	I
10		7657.48	I
10		7738.43	I
20		7746.64	I
100		7809.82	I
20		8049.00	I
100		8099.51	I
6		8173.84	I
100		8270.96	I
8		8314.51	I
6		8349.74	I
10		8381.05	I
10		8487.48	I
10		8494.89	I
20		8520.95	I
100		8600.07	I
10		8639.76	I
15		8675.83	I
10		8807.75	I
50		9327.02	I
6		9948.57	I
5		10106.13	I

RADON (Rn)
Z = 86

Rn I
Ref. 251 — E.F.W.

Intensity	Wavelength	
	Air	
5	3514.60	I
10	3739.89	I
20	3753.65	I
10	3917.20	I
10	3941.72	I
10	3952.36	I
10	4226.06	I
80	4307.76	I
7	4335.78	I
100	4349.60	I
40	4435.05	I
50	4459.25	I

RHENIUM (Re)
Z = 75

Re I and II
Ref. 1 — C.H.C.

Intensity		Wavelength	
		Air	
25000		2003.53	I
16000		2017.87	I
27000		2049.08	I
4200		2074.70	I
3700		2083.92	I
10000		2085.59	I
4700		2092.41	II
9800		2097.12	I
2700		2109.22	I
3400		2139.04	II
1600		2142.74	II
		2142.97	I
3700		2156.67	I
4900		2167.94	I
3400		2176.21	I
4200	c	2214.26	II
2200		2214.58	I
1700		2226.42	I
920		2235.44	I
440		2255.73	I
860		2256.19	I
2000		2264.39	I
2100		2274.62	I
5200	c	2275.25	II
1600		2281.62	I
2900		2287.51	I
2700		2294.49	I
390		2298.09	II
390		2299.77	I
610		2302.99	I
680		2306.54	I
230		2312.97	I
220		2313.34	I
220		2319.19	I
370		2320.16	I
800		2322.49	I
300		2328.66	I
270		2334.33	I
270		2335.73	I
220		2336.10	I
270		2337.95	I
860		2344.78	I
230		2349.39	I
220	d	2350.46	I
680		2352.07	I
210	d	2353.95	I
250		2356.50	I
200		2365.32	I
1200		2365.90	I
570		2367.68	I
180		2368.53	II
520		2369.27	II
220		2370.76	II
210		2371.52	I
150		2373.48	II
320		2375.07	I
75		2378.53	II
370		2379.77	I
180		2386.90	II
340		2388.57	I
230		2393.65	I
320		2394.37	I
320		2396.79	I
200		2397.31	I
210	d	2400.72	I
		2400.89	I
210		2401.68	I
75		2403.04	II
1500		2405.06	I
740		2405.60	I
320		2406.70	I
270		2410.37	I
60		2418.20	II
1200		2419.81	I
300		2421.73	I
300		2421.88	I
60		2423.84	II
2500		2428.58	I
490		2431.54	I
420		2432.18	I
340	c	2441.47	I
230		2442.51	I
250		2444.94	I
610		2446.98	I
85		2449.03	II
85		2449.52	II
610		2449.71	I
200		2455.83	II
390		2461.20	I
800	c	2461.84	II
200		2467.57	II
120		2467.85	I
150	c	2469.36	II
120		2470.61	II
75		2471.05	II
150		2473.72	II
160		2475.17	II
75		2477.43	II
200		2479.02	I
1200		2483.92	I
390		2485.81	I
980		2487.33	I
75		2490.16	II
200		2492.84	I
370		2496.04	I
200		2498.22	I
370		2501.72	I
570		2502.35	II
230		2504.60	II
270		2505.94	I
1800	c	2508.99	I
570		2520.01	I
540		2521.50	I
150		2534.10	II
370		2534.80	I
570		2540.51	I
740	d	2544.74	I
370		2545.48	I
160		2550.09	II
300		2552.02	I
150	c	2553.59	II
370		2554.63	II
1000		2556.51	I
250		2559.08	I
340		2564.19	I
540		2568.64	II
370		2571.81	II
380		2586.79	I
290		2599.86	I
290		2603.89	I
660		2608.50	II
610	d	2611.54	I
160	c	2616.72	II
200		2622.76	I
310		2635.83	II
550		2636.64	I
190		2637.01	II
90		2641.02	II
270		2642.75	I
65		2648.46	II
270		2649.05	I
660		2651.90	I

Intensity		Wavelength	
400		2654.12	I
220		2663.63	I
940		2674.34	I
220		2688.53	I
1300		2715.47	I
200		2731.56	II
220		2732.21	I
610		2733.04	II
110	h	2753.64	II
220		2758.00	I
210		2763.79	I
200		2766.39	I
310		2767.74	I
220		2768.85	I
220		2769.32	I
350		2770.42	I
550		2783.57	I
220		2791.29	I
120		2803.28	II
220		2814.68	I
75		2819.78	II
880		2819.95	I
310		2834.08	I
200		2837.55	I
200		2840.35	I
220		2843.00	I
270		2850.98	I
240		2867.19	I
200		2875.28	I
200		2883.44	I
2900		2887.68	I
130	c	2888.06	II
490		2896.01	I
830	c	2902.48	I
210		2905.58	I
550		2909.82	I
65	h	2916.73	II
830		2927.42	I
270		2930.61	I
440		2943.14	I
130	h	2957.91	II
270		2962.27	I
720		2965.11	I
1500		2965.76	I
90		2968.98	II
310		2976.29	I
210		2978.15	I
220		2980.82	I
220		2982.19	I
220		2988.47	I
1800		2992.36	I
5500		2999.60	I
350		3001.14	I
220		3004.14	I
200		3006.42	I
500		3016.02	I
300		3016.49	I
380		3030.45	I
240		3047.25	I
200		3058.78	I
1600		3067.40	I
320		3069.94	I
260		3071.16	I
200		3072.96	I
550		3082.43	I
340		3088.76	I
200		3093.64	I
200		3095.06	I
700		3100.67	I
140		3103.06	II
700		3108.81	I
340		3110.86	I
340	c	3118.19	I
340		3121.36	I
420		3128.94	I
260		3134.02	I
250		3141.38	I
440		3151.64	I
330		3153.79	I
360	c	3158.31	I
220		3164.52	I
700		3168.37	I
220		3174.61	I
440		3177.71	I
260		3178.61	I
600		3182.87	I
1100		3184.76	I
1100		3185.57	I
260		3190.78	I
260		3192.36	I
200		3194.50	I
220		3198.58	I
1100	c	3204.25	I
380		3235.94	I
600		3258.85	I
600		3259.55	I
200		3261.56	I
300		3268.89	I
200		3294.83	I
280		3296.70	I
280		3296.99	I
280		3301.60	I
240		3302.23	I
320		3303.21	II

Intensity		Wavelength	
280		3303.75	I
240		3313.95	I
600		3322.48	I
200		3331.52	I
2000		3338.18	I
1600		3342.24	I
810		3344.32	I
320		3346.20	I
240	d	3356.33	I
200		3358.02	I
200		3362.74	I
240		3377.74	I
320		3379.06	II
320		3379.70	I
200		3385.76	I
240		3389.43	I
200		3390.25	I
4000		3399.30	I
650		3404.72	I
650		3405.89	I
240		3408.67	I
320		3409.83	I
320		3417.77	I
810		3419.41	I
8000		3424.62	I
400		3426.19	I
300		3427.61	I
320		3437.71	I
400		3449.37	I
16000	c	3451.88	I
240		3453.50	I
55000	c	3460.46	I
40000	c	3464.73	I
400		3467.96	I
240		3476.44	I
400		3480.38	I
320		3480.85	I
240		3482.23	I
560		3503.06	I
100	c	3512.28	I
320		3516.65	I
320		3517.33	I
120		3534.82	I
320		3537.46	I
160		3539.33	I
240		3549.89	I
160		3551.29	I
160		3553.65	I
160		3558.94	I
160		3568.23	I
240		3570.26	I
360		3579.12	I
810	c	3580.15	II
650		3580.97	I
810		3583.02	I
160		3596.39	I
160		3610.49	I
320		3617.08	I
160		3621.46	I
160		3625.91	I
140		3637.06	I
810		3637.84	I
440		3651.97	I
120		3669.78	I
320		3670.53	I
860	c	3689.50	I
1500	c	3691.48	I
100		3697.71	I
520		3703.24	I
100		3705.02	I
240		3709.93	I
360	c	3717.28	I
4000		3725.76	I
140		3731.87	I
140		3732.28	I
240	c	3735.01	I
810		3735.31	I
910		3740.10	I
140		3740.41	I
130		3742.26	II
300	Cw	3745.44	I
140		3766.48	I
120		3768.26	I
140		3777.66	I
700		3787.52	I
160		3796.59	I
160		3797.59	I
190		3807.74	I
120		3815.66	I
120		3836.30	I
240		3869.94	I
240		3875.26	I
240		3876.86	I
100		3908.21	I
130		3913.92	I
380	c	3917.27	I
550		3929.85	I
140		3936.90	I
110		3944.72	I
180		3945.91	I
280		3961.04	I
350	c	3962.48	I
100		4004.93	I
140		4022.96	I

Intensity		Wavelength	
100		4023.31	I
110	c	4029.63	I
220		4033.31	I
110		4037.49	I
200		4048.99	I
240		4081.43	I
140		4104.42	I
240	c	4110.89	I
190		4121.64	I
240	Cw	4133.42	I
1800		4136.45	I
700		4144.36	I
140		4149.96	I
160		4170.40	I
220		4182.90	I
220		4183.06	I
650		4221.08	I
3600	c	4227.46	I
150		4241.39	I
260	c	4257.60	I
120	c	4291.17	I
200		4304.40	I
200		4332.25	I
40		4357.98	II
380		4358.59	I
190		4367.58	I
140		4391.34	I
360	Cw	4394.38	I
110	Cw	4406.40	I
180		4415.82	I
150		4475.08	I
120		4478.39	I
120	c	4507.04	I
2600		4513.31	I
260		4516.64	I
500		4522.73	I
120		4523.88	I
120		4529.95	I
100		4545.17	I
120		4580.68	I
120		4605.73	I
100		4621.38	I
190	c	4791.42	I
2200	Cw	4889.14	I
220		4923.90	I
40		5058.56	I
70		5096.50	I
20		5120.32	I
25		5161.65	I
40	c	5178.89	I
20		5181.74	I
35		5234.31	I
50		5248.86	I
1300		5270.95	I
1600	Cw	5275.56	I
100		5278.24	I
30		5305.56	I
20		5317.28	I
35		5321.28	I
50		5327.46	I
20		5331.90	I
20		5332.76	I
20		5333.85	I
35		5369.48	I
50	c	5369.80	I
100	c	5377.10	I
25		5431.90	I
14		5437.03	I
14		5447.92	I
25		5460.64	I
14	h	5520.05	I
25		5521.10	I
50	c	5532.68	I
50	c	5563.24	I
25		5573.47	I
25		5584.72	I
10	h	5607.21	I
12	h	5612.27	I
100		5667.08	I
25		5711.43	I
18		5716.95	I
110	c	5752.93	I
110	Cw	5776.83	I
18		5791.60	I
10	c	5815.92	I
550		5834.31	I
10		5919.86	I
60		5943.91	I
10		5950.21	I
18	h	5969.77	I
10		5989.99	I
18	h	5995.73	I
30		6114.22	I
35	c	6145.81	I
50		6146.82	I
18		6203.24	I
25		6217.97	I
30	Cw	6229.24	I
35	Cw	6243.24	I
35	d	6260.02	I
		6260.24	I
18	c	6271.37	I
18		6278.76	I
10		6286.41	I

Intensity		Wavelength	
10		6303.42	I
200		6307.70	I
200		6321.90	I
80	d	6350.75	I
16	h	6382.94	I
14		6411.47	I
50		6511.47	I
14		6515.25	I
12		6544.91	I
35	c	6577.11	I
40	Cw	6592.52	I
100	Cw	6605.19	I
30	c	6623.91	I
10	c	6637.25	I
27	Cw	6652.39	I
15	h	6683.28	I
9	c	6711.30	I
30		6751.22	I
5		6761.19	I
180	c	6813.41	I
260		6829.90	I
85		6971.53	I
35	Cw	7006.63	I
65	Cw	7024.15	I
65	Cw	7246.67	I
13	Cw	7292.72	I
40	Cw	7578.73	I
13		7611.89	I
7	Cw	7620.25	I
50	Cw	7640.94	I
65	Cw	7912.94	I
35	Cw	7980.77	I
40		8417.13	I
29	Cw	8527.73	I

RHODIUM (Rh)
Z = 45

Rh I and II
Ref. 1 — C.H.C.

Intensity		Wavelength	
		Air	
150		2276.21	II
140		2288.57	I
110		2309.82	I
55		2318.36	I
95		2319.10	I
95		2321.73	I
350		2322.58	I
140		2326.47	I
80		2328.64	I
190		2334.77	II
55		2345.41	I
55		2352.47	I
55		2359.18	I
300		2361.92	I
110		2368.34	I
270		2382.89	I
230		2383.40	I
40		2384.65	I
270		2386.14	II
80		2407.88	I
27		2408.19	I
27		2410.25	I
80		2415.84	II
55		2418.64	I
45		2419.75	I
45		2420.18	II
65		2420.98	II
75		2423.94	I
65		2427.11	II
130		2427.68	I
230		2429.52	I
40		2431.85	II
40		2432.66	I
18		2437.08	I
110		2437.90	I
330		2440.34	I
50	h	2444.27	I
65		2448.84	I
50		2449.04	I
75		2450.56	I
30		2455.70	II
65		2458.90	II
90		2461.04	II
30		2463.61	I
75		2470.39	I
90		2471.47	I
30		2472.51	I
130		2473.09	I
15		2475.64	II
15		2477.54	II
25		2482.04	I
50		2483.33	I
150		2487.47	II
100		2490.77	II
30		2492.30	I
75	h	2494.51	I
15		2499.02	I
40		2500.58	I
130		2502.46	I

Rhodium (Cont.)

Intensity	Wavelength	Species
15	2503.84	II
300	2504.29	II
40	2505.10	II
150	2505.67	I
350	2509.70	I
50	2510.66	II
300	2511.03	II
75	2513.36	II
200	2515.75	I
130	2520.53	II
13	2525.99	I
13	2531.74	I
50	2532.66	
13	2533.59	
50	2534.07	
110	2536.71	
110	2537.04	II
30	2539.72	
40	2544.22	
350	2545.70	I
13	2548.60	
550	2555.36	I
25	2558.62	I
50	2565.79	I
45	2566.04	I
25	2566.92	II
50	2567.28	
25	2574.66	
25	2575.75	I
13	2576.23	
40	2587.29	II
30	2598.07	
30	2603.32	II
75	2606.44	II
75	2613.60	
150	2622.58	I
230	2625.88	I
100	2630.42	I
40	2634.99	I
30	2638.74	II
75	2643.00	I
110	2647.28	I
400	2652.66	I
30	2659.01	I
30	2671.06	I
65	2676.11	I
25	2680.28	I
100	2680.63	I
30	2681.78	I
30 h	2686.50	
30 h	2686.91	
50	2694.31	I
400	2703.73	I
40	2705.63	II
40	2707.23	I
75	2714.41	I
100	2715.31	II
75	2717.51	I
180	2718.54	I
65	2720.14	I
30	2720.52	I
160	2728.94	I
40	2736.76	I
75	2741.75	I
50	2767.73	I
100	2771.51	I
50	2778.06	I
75	2779.54	I
130	2783.03	I
25	2791.16	I
75	2796.63	I
150	2826.43	I
180	2826.68	I
30	2827.31	I
75	2834.12	I
45	2835.44	I
75	2836.69	I
50	2856.16	I
50 d	2860.68	I
	2860.76	I
280	2862.94	I
65	2864.40	I
50	2871.35	I
30	2873.62	I
110	2878.66	I
75	2880.76	I
140	2882.37	I
75	2885.97	I
75	2889.11	I
75	2889.84	I
65	2899.96	I
25	2904.81	I
160	2907.21	I
65	2910.17	II
75	2912.62	I
90	2915.42	I
30	2923.10	I
180	2924.02	I
130	2929.11	I
130	2931.94	I
30	2955.41	I
230	2968.66	I
25	2974.03	I
160	2977.68	I
450	2986.20	I
90	2986.99	I
50	2987.45	I
110	3004.46	I
50	3019.54	I
130	3023.91	I
50	3028.43	I
30	3045.77	I
30	3046.76	I
25	3057.89	I
65	3067.30	I
180	3083.96	I
29	3087.42	I
70	3114.91	I
140	3121.76	I
240	3123.70	I
35	3130.79	I
95	3137.71	I
45	3151.36	I
45	3152.60	I
130	3155.78	I
70	3179.73	I
80	3185.59	I
140	3189.05	I
470	3191.19	I
190	3197.13	I
70	3214.32	I
80	3237.66	I
520	3263.14	I
520	3271.61	I
2300	3280.55	I
110	3281.70	I
2300	3283.57	I
280	3289.14	I
45	3289.64	I
210	3294.28	I
45	3296.72	I
260	3300.46	I
4200	3323.09	I
60	3331.09	I
45	3331.24	I
330	3338.54	I
70	3342.90	I
80	3344.20	I
60	3359.90	I
280	3360.80	I
60	3362.18	I
420	3368.38	I
45	3369.68	I
1100	3372.25	I
110	3377.14	I
80	3377.71	I
110	3385.78	I
5600	3396.82	I
820	3399.70	I
160	3406.55	I
820	3412.27	I
60	3420.16	I
330	3421.22	I
120 d	3424.38	I
8200	3434.89	I
1400	3440.53	I
35	3442.63	I
120	3447.74	I
60	3448.58	I
120	3450.29	I
60	3451.15	I
400	3455.22	I
60	3455.42	I
180	3457.07	I
220	3457.93	I
5900	3462.04	I
180	3469.62	I
4700	3470.66	I
120	3472.25	I
4700	3474.78	I
2100	3478.91	I
95	3484.04	I
80	3491.07	I
110	3494.44	I
1200	3498.73	I
5900	3502.52	I
60	3505.41	I
2800	3507.32	I
60	3511.78	I
60	3513.10	I
60	3519.54	I
8800	3528.02	I
880 d	3538.14	I
	3538.26	I
280	3541.91	I
1200	3543.95	I
1800	3549.54	I
240	3564.13	I
1200	3570.18	I
4700	3583.10	I
120	3583.53	I
4700	3596.19	I
5900	3597.15	I
310	3605.86	I
3100	3612.47	I
240	3614.78	I
200	3620.46	I
1800	3626.59	I
95	3627.80	I
310	3639.51	I
350	3654.87	I
8200	3657.99	I
280	3661.86	I
1300	3666.22	I
180	3666.91	I
140	3674.76	l
560	3681.04	I
1900	3690.70	I
9400	3692.36	I
60	3694.95	I
940	3695.52	I
280	3698.26	I
380	3698.60	I
7600	3700.91	I
940	3713.02	I
60	3713.43	I
45	3714.83	I
16	3724.94	I
650	3735.28	I
420	3737.27	I
420	3744.17	I
1200	3748.22	I
240	3754.12	I
380	3754.27	I
490	3755.58	I
1000	3760.40	I
2300	3765.08	I
490	3769.97	I
70	3775.72	I
380	3778.13	I
1000	3788.47	I
1300	3792.18	I
3800	3793.22	I
4900	3799.31	I
760	3805.92	I
1300	3806.76	I
45	3809.50	I
95	3812.45	I
470	3815.01	I
760	3816.47	I
1300	3818.19	I
3800	3822.26	I
2300	3828.48	I
2000	3833.89	I
45	3834.75	I
5900	3856.52	I
490	3870.01	I
70	3872.39	I
380	3877.34	I
70	3888.34	I
29	3904.22	I
23	3912.83	I
120	3913.71	I
240	3922.19	I
2000	3934.23	I
45	3934.98	I
50	3935.84	I
590	3942.72	I
95	3958.24	I
3800	3958.86	I
45	3964.54	II
380	3975.31	I
240	3984.40	I
240	3995.61	I
380	3996.15	I
120	4023.14	I
60	4048.41	I
23	4049.04	I
40	4053.44	I
23	4056.34	I
70	4077.57	I
560	4082.78	I
19	4084.28	I
45	4087.79	I
60	4088.50	I
140	4097.52	I
45	4107.49	I
70	4116.33	I
120	4119.68	I
1100	4121.68	I
1500	4128.87	I
2100	4135.27	I
240	4154.37	I
330	4196.50	I
70	4206.62	I
3300	4211.14	I
29	4230.20	I
40	4244.41	I
60	4273.43	I
60	4278.60	I
820	4288.71	I
70	4296.77	I
23	4342.44	I
45	4373.04	I
4200	4374.80	I
95	4379.92	I
23	4433.32	I
35	4492.47	I
29	4503.78	I
23	4528.72	I
16	4544.27	I
35	4548.73	I
40	4551.64	I
19	4560.89	I
16	4565.19	I
130	4569.00	I
14	4571.31	I
29	4608.12	I
14	4619.91	I
23	4643.18	I
150	4675.03	I
19	4721.00	I
70	4745.11	I
12	4755.58	I
23	4810.49	I
21	4842.43	I
45	4843.99	I
60	4851.63	I
60	4963.71	I
60	4977.75	I
40	4979.18	I
14	5085.52	I
70	5090.63	I
23	5120.69	I
19	5130.76	I
60	5155.54	I
14	5157.09	I
40	5158.69	I
60	5175.97	I
12	5177.27	I
35	5184.19	I
95	5193.14	I
16	5206.95	I
16	5211.52	I
19	5212.73	I
16	5214.79	I
19	5222.66	I
19	5230.62	I
45	5237.16	I
9	5237.80	I
14	5269.27	I
11 h	5280.12	I
14	5292.14	I
14	5314.79	I
40 h	5329.74	I
14 h	5331.08	I
9	5349.31	I
130	5354.40	I
23	5356.47	I
45	5379.10	I
95	5390.44	I
23 h	5404.73	I
60 h	5424.07	I
19	5424.72	I
19 h	5425.45	I
12	5439.58	I
12 h	5441.36	I
9 h	5444.32	I
35 h	5445.23	I
23 h	5468.11	I
35 h	5470.85	I
12	5476.12	I
12	5481.42	I
16	5484.23	I
9	5504.65	I
29	5535.04	I
21 l	5544.58	I
160	5599.42	I
7	5607.71	I
16	5608.35	I
5	5632.77	I
9	5659.62	I
40	5686.38	I
9 h	5702.47	I
6	5727.30	I
29	5792.66	I
9	5795.79	I
9	5803.34	I
40	5806.91	I
6	5821.84	I
35	5831.58	I
7	5907.31	I
9	5918.54	I
7	5941.46	I
130	5983.60	I
9	5991.19	I
35	6102.72	I
6	6116.15	I
8	6128.06	I
8	6186.89	I
14	6199.99	I
16	6253.72	I
5	6276.66	I
8	6277.46	I
6	6293.38	I
29	6319.53	I
12	6414.72	I
16	6510.41	I
19	6519.70	I
9	6627.80	I
19	6630.16	I
40	6752.35	I
9	6796.65	I
13	6827.33	I
11	6857.68	I
20	6879.94	I

Rhodium (Cont.)

Intensity		Wavelength	
65		6965.67	I
8		6972.91	I
16		6979.15	I
16		7001.58	I
11		7038.76	I
18		7101.64	I
15		7104.45	I
6		7142.55	I
9		7219.06	I
18		7268.18	I
35		7270.82	I
12		7271.94	I
5		7273.03	I
9		7375.57	I
5	h	7386.64	I
9		7430.80	I
18	h	7442.39	I
7		7446.77	I
12		7475.74	I
12		7495.24	I
8		7542.02	I
11		7557.67	I
8		7577.22	I
11		7690.05	I
18		7772.90	I
29		7791.61	I
55		7824.91	I
15		7830.05	I
15		7846.50	I
21		8029.91	I
11	h	8036.09	I
29		8045.36	I
7		8063.50	I
15		8136.20	I
7	h	8193.67	I
5	h	8369.67	I
8		8425.59	I

Rh III
Ref. 396 — L.J.R.

Intensity		Wavelength	
		Vacuum	
10		746.28	III
30		759.54	III
50		813.44	III
30		826.01	III
30		843.63	III
30		849.08	III
40		852.70	III
40		854.77	III
40		859.89	III
40		861.34	III
50		863.78	III
50		865.22	III
50		870.40	III
80		882.51	III
100		925.75	III
150		937.28	III
100		976.12	III
500		991.62	III
400		992.48	III
500	d	1009.60	III
200		1012.22	III
200		1015.17	III
100		1050.00	III
100		1058.97	III
200		1073.87	III
100		1100.58	III
100		1113.79	III
100		1768.43	III
150		1784.24	III
200		1784.94	III
150		1796.50	III
200		1816.03	III
1000		1832.05	III
500		1859.85	III
100		1874.70	III
800		1880.66	III
500		1884.91	III
500		1887.36	III
700		1888.62	III
800		1901.32	III
500		1910.16	III
600		1919.37	III
500		1927.07	III
700		1931.79	III
500		1954.25	III
400		1965.16	III
500		1994.26	III
		Air	
400		2005.14	III
800		2013.71	III
500		2017.47	III
500		2028.53	III
800		2036.72	III
600		2037.61	III
1000		2040.18	III
3000		2048.67	III
2000		2064.11	III
800		2076.84	III
1000		2118.53	III
1000		2118.63	III
1000		2139.44	III
1000		2152.23	III
3000		2158.17	III
3000		2163.19	III
3000		2167.33	III
100		2207.00	III
100		2230.66	III
50		2250.84	III
30		2374.84	III
20		2470.65	III
50		3006.43	III
50		3052.44	III
50		3310.69	III
1		3852.98	III

RUBIDIUM (Rb)
Z = 37

Rb I and II
Ref. 12, 130, 241, 257, 264 — J.R.

Intensity		Wavelength	
		Vacuum	
10		474.88	II
40		481.118	II
90		497.430	II
20		508.434	II
150		513.266	II
300		530.173	II
75		533.801	II
40		542.887	II
200		555.036	II
2500		589.419	II
1500		643.878	II
3000		697.049	II
6000		711.187	II
10000		741.456	II
1000		1604.12	II
200		1644.96	II
200		1707.52	II
600		1716.85	II
5000		1760.50	II
200		1803.47	II
500		1809.68	II
500		1865.33	II
500		1889.42	II
500		1954.24	II
300		1956.54	II
200		1971.42	II
500		1983.19	II
		Air	
300		2042.23	II
300		2052.21	II
500		2052.80	II
2000		2068.92	II
1000		2071.50	II
10000		2075.95	II
1000		2090.29	II
200		2108.06	II
300		2116.50	II
1000		2125.25	II
400		2129.82	II
200		2143.10	II
30000		2143.83	II
200		2190.36	II
600		2197.99	II
600		2198.26	II
300		2207.86	II
10000		2217.08	II
200		2223.79	II
400		2237.72	II
500		2250.65	II
200		2251.43	II
800		2254.19	II
200		2254.55	II
500		2263.54	II
500		2263.94	II
500		2286.82	II
5000		2291.71	II
300		2298.80	II
250		2333.01	II
2000		2333.39	II
350		2353.11	II
300		2353.96	II
400		2356.97	II
300		2358.04	II
300		2364.27	II
200		2364.32	II
200		2365.15	II
300		2367.51	II
200		2373.21	II
2000		2385.34	II
250		2405.94	II
400		2434.17	II
800		2459.14	II
50000		2472.20	II
300		2484.56	II
700		2484.70	II
2000		2496.38	II
200		2502.67	II
250		2514.18	II
1000		2524.24	II
200		2594.56	II
400		2623.76	II
400		2645.58	II
1000		2684.10	II
1000		2711.76	II
250		2741.01	II
500		2812.15	II
350		2838.51	II
750		2873.88	II
1000		3051.36	II
2		3082.02	I
250		3088.58	II
10		3112.57	I
3		3113.06	I
5000	c	3148.90	II
25		3157.54	I
5		3158.26	II
1200		3161.00	II
50		3227.98	I
6		3229.16	I
2000		3270.99	II
1500		3321.49	II
1200		3340.55	II
60		3348.72	I
75		3350.82	I
750		3353.89	II
1200		3393.03	II
750		3415.58	II
1000		3434.18	II
1500		3461.50	II
3000		3521.39	II
3000	1	3531.55	II
1000		3541.15	II
100		3587.05	I
40		3591.57	I
5000		3600.60	II
10000		3600.64	II
600	c	3639.80	II
400	c	3646.26	II
350		3647.56	II
1000	c	3662.74	II
900	c	3663.81	II
350		3666.72	II
300		3675.66	II
2500	c	3699.58	II
350		3746.33	II
3500		3796.81	II
2500		3801.90	II
1000		3826.66	II
450		3860.74	II
250		3907.29	II
500		3922.20	II
2500	1	3926.44	II
25000		3940.51	II
1000	c	3978.15	II
1700		4029.49	II
2500	c	4083.88	II
2000		4104.28	II
1700	c	4136.11	II
3500		4193.08	II
1000		4201.80	I
500		4215.53	II
90000		4244.40	II
500		4266.58	II
250	c	4270.25	II
15000		4273.14	II
2500	c	4287.97	II
1500		4293.97	II
500	c	4306.26	II
1000		4346.96	II
2500		4377.12	II
300		4440.10	II
1000		4469.47	II
400	c	4493.92	II
700		4519.04	II
3000		4530.34	II
500	1	4533.79	II
400		4540.74	II
20000		4571.77	II
3000	c	4622.42	II
350	c	4631.89	II
10000		4648.57	II
500		4659.28	II
1000		4730.45	II
1000		4755.30	II
400	c	4757.82	II
30000		4775.95	II
5000	c	4782.83	II
300	c	4855.34	II
1500	c	4885.59	II
2		5087.987	I
2		5132.471	I
10		5150.134	I
10000		5152.08	II
300		5164.58	II
1		5165.023	I
2		5165.142	I
1		5169.65	I
15		5195.278	I
2		5233.968	I
20		5260.034	I
1		5260.228	I
200		5270.51	II
3		5322.380	I
40		5362.601	I
4		5390.568	I
75		5431.532	I
3		5431.830	I
500		5512.55	II
5000		5522.78	II
6		5578.788	I
5000	c	5635.99	II
40		5647.774	I
20		5653.750	I
3000	d	5699.15	II
60		5724.121	I
3		5724.614	I
200		5739.64	II
75		6070.755	I
200		6135.27	II
30	c	6159.626	I
1000		6199.08	II
75	c	6206.309	I
300		6269.40	II
120	c	6298.325	I
5		6299.224	I
10000		6458.33	II
1000		6555.62	II
5000		6560.81	II
3000	1	6775.07	II
100	1	7279.997	II
300	c	7316.52	II
150		7408.173	I
200	1	7618.933	I
300		7757.651	I
60		7759.436	I
90000	c	7800.27	I
5	1	7925.26	I
4		7925.54	I
45000	c	7947.60	I
40		8271.41	I
30		8271.71	I
2000		8603.96	II
40	1	8868.512	I
30		8868.852	I
300		8978.88	II
300		9021.77	II
3		9224.64	I
2		9234.25	I
500	c	9246.41	II
300		9338.87	II
200	w	9373.50	II
300		9391.36	II
1000		9479.32	II
700	1	9493.72	II
30	1	9522.65	I
5		9523.05	I
20	1	9540.18	I
300		9612.99	II
300		9671.54	II
2000	c	9689.05	II
200		9776.06	II
200		9934.76	II
35	1	10075.282	I
30	1	10075.708	I
100		13235.17	I
20		13442.81	I
30		13443.57	I
75		13665.01	I
1000		14752.41	I
800		15288.43	I
150		15289.48	I
20		22529.65	I
10		22932.47	I
4		27314.31	I
2		27905.37	I

Rb III
Ref. 258, 262 — J.R.

Intensity		Wavelength	
		Vacuum	
30		465.85	III
35	p	482.43	III
30	p	482.47	III
500		482.83	III
300		484.84	III
500		489.66	III
100		489.96	III
600		493.48	III
-50		497.82	III
100		500.28	III
30		508.33	III
400		516.79	III
800		533.64	III
1200		535.86	III
1200		556.19	III
500		558.36	III

Rubidium (Cont.)

Intensity		Wavelength	
700		564.77	III
1500		566.71	III
1000		572.82	III
1500		576.65	III
2500		579.63	III
1500		581.26	III
500		582.34	III
800		586.77	III
100		591.42	III
900		593.65	III
1000		594.94	III
1300		595.88	III
1200		598.49	III
450		602.09	III
50		605.51	III
500		607.28	III
400		613.31	III
500		619.67	III
20		620.83	III
100		622.24	III
250		630.06	III
500		645.67	III
20		674.81	III
5000		769.04	III
2500		815.28	III

Air

Intensity		Wavelength	
100		2153.21	III
250		2164.59	III
100		2268.00	III
150		2300.12	III
500		2304.14	III
150		2304.45	III
250		2312.46	III
200		2337.07	III
100		2341.90	III
200		2345.37	III
100		2349.81	III
150		2380.44	III
100		2381.29	III
150		2418.46	III
300		2561.86	III
100		2573.71	III
100		2577.07	III
200		2586.83	III
1000		2631.75	III
350		2636.83	III
100		2656.68	III
100		2713.86	III
500		2798.86	III
150		2800.27	III
500		2807.58	III
100		2845.44	III
150		2869.77	III
500		2903.69	III
150		2949.62	III
100		2951.01	III
2000		2956.07	III
500	l	2967.45	III
150		2968.13	III
500		2970.74	III
250		2987.40	III
350		3023.61	III
200		3039.62	III
200		3041.48	III
250		3070.70	III
500		3086.84	III
100		3098.49	III
500		3111.36	III
250	s	3114.82	III
120		3118.92	III
100		3169.34	III
200		3222.60	III
500		3286.41	III
100		3330.16	III
200		3346.92	III
250		3439.26	III
100		3492.68	III

Rb IV
Ref. 109 — J.R.

Intensity　　　Wavelength

Vacuum

Intensity	Wavelength	
10	595.18	IV
25	663.76	IV
25	716.24	IV
20	733.41	IV
50	740.85	IV
20	749.86	IV
20	753.75	IV
10	771.54	IV
25	776.89	IV
9	817.92	IV
15	850.18	IV
10	988.00	IV

RUTHENIUM (Ru)
Z = 44

Ru I and II
Ref. 1 — C.H.C.

Intensity　　　Wavelength

Air

Intensity		Wavelength	
2400		2076.43	I
2600		2083.77	I
2400		2090.89	I
690		2255.52	I
290		2259.53	I
780		2272.09	I
240		2278.19	I
780		2279.57	I
170		2285.38	I
290		2302.54	I
480		2317.80	I
150		2322.01	I
120		2334.96	II
240		2340.69	I
190	h	2342.85	II
190		2349.34	I
310		2351.33	I
170		2357.91	II
140		2360.56	I
170		2370.17	I
240		2375.27	I
80		2375.63	II
160		2392.42	I
95		2396.71	II
780		2402.72	II
150		2407.92	II
55		2410.89	I
55		2414.82	II
130		2420.82	I
55		2422.92	I
45		2429.60	I
65		2432.93	I
30		2447.45	I
30		2450.58	I
65		2454.92	I
180		2455.53	II
150		2456.44	II
370		2456.57	II
65	h	2458.62	I
55		2462.94	I
85		2464.70	I
30		2474.04	I
110		2475.41	I
100		2476.88	I
280		2478.93	II
28		2481.11	II
30		2489.91	I
18		2491.78	I
65		2493.69	II
85		2494.02	I
45		2494.48	II
85		2495.69	II
65		2496.56	I
140		2498.42	II
140		2498.57	II
85		2499.78	I
260		2507.01	II
130		2508.27	I
110		2509.07	I
110		2512.81	I
110		2513.32	II
110		2517.32	II
150		2535.59	II
65		2543.25	II
280		2544.22	I
120		2546.67	I
280		2549.48	I
550		2549.58	I
130		2560.26	I
120		2560.83	I
110		2563.15	I
160		2568.77	I
100		2570.97	I
100		2578.57	I
100		2579.53	I
100		2589.57	I
170		2591.12	I
120		2592.02	I
100		2593.70	I
110		2594.85	I
370		2609.06	I
830		2612.07	I
100		2615.09	I
220		2631.30	I
220		2635.86	I
170		2636.67	I
110		2640.33	I
460		2642.96	I
110		2647.32	I
110		2651.29	I
330		2651.84	I
28		2656.25	II
400		2659.62	I

Intensity	Wavelength	
23	2661.17	II
330	2661.61	II
200	2664.76	I
30	2667.40	II
690	2678.76	II
220	2686.29	I
28	2687.50	II
	2688.16	II
330	2692.06	II
110	2701.34	I
110	2702.83	I
170	2709.20	I
200	2712.41	II
690	2719.52	I
130	2722.65	I
140	2725.47	II
310	2734.35	II
1800	2735.72	I
170	2739.22	I
130	2744.45	I
35	2747.97	II
75	2752.45	II
75	2752.77	II
260	2763.42	I
35	2765.44	II
90	2768.93	II
100	2778.38	II
110	2787.83	II
140	2802.81	I
35	2806.74	II
350	2810.03	I
1700	2810.55	I
350	2818.36	I
110	2822.03	I
200	2827.87	I
400	2829.16	I
130	2834.00	I
150	2840.54	I
35	2841.68	II
640	2854.07	I
180	2860.02	I
420	2861.41	I
550	2866.64	I
110	2868.31	I
1800	2874.98	I
220	2879.76	I
55	2882.12	II
130	2883.60	I
740	2886.54	I
180	2892.56	I
110	2901.94	I
140	2905.65	I
370	2908.88	I
1100	2916.26	I
150	2919.61	I
35	2927.54	II
180	2945.67	II
180	2946.99	I
370	2949.50	I
150	2954.49	I
18	2963.40	II
550	2965.16	I
170	2965.55	II
140	2976.59	II
550	2976.92	I
45	2977.23	II
75	2979.96	II
1400	2988.95	I
35	2991.62	II
110	2993.27	I
460	2994.96	I
440	3006.59	I
330	3017.24	I
310	3020.88	I
240	3033.45	I
200	3040.31	I
220	3042.48	I
110	3045.71	I
110	3048.78	I
150	3054.94	I
390	3064.84	I
170	3089.14	I
120	3089.80	I
330	3096.57	I
120	3097.60	I
830	3099.28	I
740	3100.84	I
120	3125.96	I
120	3153.82	I
290	3159.92	I
200	3168.52	I
60	3177.05	II
180	3186.04	I
240	3188.34	I
240	3189.98	I
180	3196.59	I
180	3223.27	I
110	3226.37	I
100	3227.88	I
220	3228.53	I
220	3238.53	I
120	3241.24	I
120	3243.50	I

Intensity		Wavelength	
280		3260.35	I
120	d	3264.55	I
120		3266.44	I
200		3268.21	I
200		3273.08	I
200		3274.71	I
100		3277.57	I
490		3294.11	I
370		3301.59	I
220		3306.17	I
290		3315.23	I
290		3316.39	I
100		3325.00	I
120		3335.69	I
930		3339.55	I
240		3341.66	I
200		3361.15	I
370		3368.45	I
100		3371.86	I
130		3374.65	I
120		3378.02	I
100		3379.60	I
130		3380.18	I
130		3385.14	I
130		3388.71	I
100		3389.50	I
370		3392.54	I
310		3401.74	I
310		3409.28	I
3100		3417.35	I
4900		3428.31	I
490		3430.77	I
310		3432.74	I
6400		3436.74	I
260		3438.37	I
220		3440.20	I
260		3473.75	I
240		3481.30	I
8300		3498.94	I
640		3514.49	I
330		3519.64	I
200		3528.68	I
240		3532.81	I
390		3537.95	I
790		3539.37	I
200		3541.63	I
690		3570.59	I
200		3574.58	I
390		3587.20	I
6400		3589.22	I
6900		3593.02	I
6400		3596.18	I
1300		3599.76	I
350		3625.20	I
370		3626.74	I
3100		3634.93	I
210		3637.47	I
200		3640.64	I
290		3650.32	I
310		3654.40	I
6200		3661.35	I
830		3663.37	I
650		3669.49	I
240		3678.32	I
260		3696.59	I
410		3717.00	I
260		3719.33	I
550		3726.10	I
8700		3726.93	I
11000		3728.03	I
7100		3730.43	I
280		3737.40	I
410		3739.46	I
3500		3742.28	I
870		3742.78	I
280		3744.22	I
410		3744.40	I
2800		3745.59	I
760		3753.54	I
310		3755.09	I
870		3755.93	I
1200		3759.84	I
370		3760.03	I
600		3761.51	I
600		3767.35	I
1500		3777.59	I
460		3781.18	I
600		3782.74	I
3900		3786.06	I
6000		3790.51	I
240		3794.92	I
760		3798.05	I
7600		3798.90	I
7600		3799.35	I
310		3800.26	I
310		3808.68	I
600		3812.72	I
760		3817.27	I
760		3819.03	I
650		3822.09	I
550		3824.93	I
760		3831.80	I
220		3835.05	I

Intensity	Wavelength		
310	3838.07		I
930	3839.70		I
480	3846.68		I
760	3850.43		I
480	3856.46		I
1300	3857.55		I
220	3860.72		I
650	3862.69		I
1300	3867.84		I
260	3873.52		I
650	3892.21		I
760	3909.08		I
260	3920.92		I
1500	3923.47		I
3300	3925.92		I
600	3931.76		I
310	3933.55		I
760	3945.57		I
460	3950.21		I
310	3952.68		I
460	3964.90		I
600	3978.44		I
600	3979.42		I
870	3984.86		I
280	3995.98		I
1500	4022.16		I
600	4023.83		I
310	4039.21		I
1400	4051.40		I
710	4054.05		I
370	4064.46		I
200	4067.61		I
760	4068.37		I
200	4073.00		I
980	4076.73		I
6000	4080.60		I
310	4085.43		I
930	4097.79		I
350	4101.74		I
1900	4112.74		I
2000	4144.16		I
650	4145.74		I
260	4146.77		I
870	4167.51		I
550	4197.58		I
550	4198.88		I
7600	4199.90		I
1500	4206.02		I
5400	4212.06		I
760	4214.44		I
930	4217.27		I
370	4220.68		I
550	4230.31		I
760	4241.05		I
760	4243.06		I
370	4246.73		I
310	4258.99		I
760	4284.33		I
220	4293.28		I
260	4294.79		I
550	4295.93		I
3700	4297.71		I
930	4307.60		I
370	4318.43		I
550	4319.87		I
550	4342.07		I
350	4349.70		I
710	4354.13		I
870	4361.21		I
2400	4372.21		I
870	4385.39		I
1300	4385.65		I
1700	4390.44		I
1600	4410.03		I
160	4421.46		I
330	4428.46		I
460	4439.76		I
440	4449.34		I
1100	4460.04		I
190	4473.93		I
150	4480.45		I
350	4498.14		I
120	4510.10		I
220	4516.89		I
220	4517.82		I
110	4520.95		I
170	4547.33		I
110	4547.85		I
5400	4554.51		I
110	4559.98		I
1700	4584.44		I
110	4591.10		I
150	4592.52		I
330	4599.08		I
170	4635.69		I
200	4645.09		I
720	4647.61		I
290	4654.32		I
290	4681.79		I
190	4684.02		I
290	4690.11		I
1400	4709.48		I
140	4731.33		I

Intensity	Wavelength		
120	4733.52		I
500	4757.84		I
260	4815.52		I
120	4844.56		I
550	4869.15		I
160	4895.60		I
470	4903.05		I
120	4907.89		I
260	4921.07		I
180	4938.43		I
160	4968.90		I
160	4980.35		I
120	4992.74		I
160	5011.23		I
90	5014.95		I
90	5026.18		I
65	5028.16		I
35	5040.35		I
35	5040.74		I
65	5047.31		I
450	5057.33		I
21	5062.64		I
90	5072.97		I
120	5076.32		I
200	5093.83		I
80	5107.07		I
24	5123.73		I
55	5127.26		I
65	5133.89		I
530	5136.55		I
170	5142.76		I
250	5147.24		I
110	5151.07		I
55	5153.20		I
500	5155.14		I
55	5160.00		I
920	5171.03		I
180	5195.02		I
80	5199.87		I
45	5202.12		I
45	5213.43		I
65	5223.55		I
40	5242.38		I
55	5251.67		I
40	5257.07		I
40	5266.47		I
40	5266.83		I
40	5280.82		I
130	5284.08		I
40	5291.16		I
80	5304.86		I
260	5309.27		I
13	5315.33		I
40	5332.93		I
45	5334.70	h	I
110	5335.93		I
130	5361.77		I
65	5377.84		I
65	5385.88		I
110	5401.04	h	I
40	5401.39		I
40	5418.86		I
55	5427.59		I
26	5439.21	l	I
13	5452.71		I
80	5454.82	h	I
90	5456.13		I
13	5475.18	h	I
55	5479.40		I
26	5480.30		I
80	5484.32		I
18	5484.64		I
26	5496.69		I
13	5501.02		I
130	5510.71		I
20	5512.37		I
8	5517.86		I
12	5521.78		I
12	5530.99		I
24	5540.66		I
12	5556.52		I
90	5559.75		I
11	5569.03		I
21	5578.40		I
21	5603.14		I
8	5603.55		I
13	5606.73		I
11	5629.79		I
290	5636.24		I
11	5641.66		I
7	5649.56		I
7	5653.30		I
11	5665.20		I
16	5679.63		I
180	5699.05		I
13	5724.82		I
13	5725.73		I
16	5745.99		I
16	5747.47		I
11	5752.02		I
11	5756.83		I
11	5767.92		I
16	5804.39		I

Intensity	Wavelength		
65	5814.98		I
8	5828.06		I
16	5833.21	h	I
55	5919.34		I
80	5921.45		I
21	5926.87		I
26	5932.38		I
8	5936.65		I
8	5951.15		I
21	5973.38	h	I
8	5974.17		I
16	5988.67		I
35	5993.65		I
18	6116.77		I
26	6199.42		I
26	6225.20		I
9	6284.49		I
18	6295.22		I
13	6330.62		I
9	6336.12		I
9	6363.41	h	I
9	6376.45		I
16	6390.23		I
8	6417.57		I
26	6444.84	h	I
8	6496.44		I
11	6528.74		I
4	6560.45		I
4	6593.74		I
9	6618.20		I
21	6663.14		I
55	6690.00		I
11	6707.52		I
15	6718.30		I
15	6730.45		I
7	6756.54		I
21	6766.95		I
30	6775.02		I
13	6787.23		I
8	6813.51		I
15	6823.88		I
21	6824.17		I
7	6831.52		I
26	6911.48		I
110	6923.23		I
26	6982.01		I
26	7027.98		I
9	7086.06		I
12	7087.35		I
4	7141.72		I
6	7219.26		I
35	7238.92		I
7	7266.96		I
8	7323.56		I
16	7393.93		I
18	7468.91		I
12	7475.40		I
26	7485.79		I
70	7499.75		I
7	7532.07		I
26	7559.61		I
5	7612.94		I
18	7621.50		I
18	7722.87		I
5	7729.91		I
22	7791.86		I
4	7797.89		I
4	7806.92		I
3	7813.43		I
4	7829.81		I
5	7833.39	h	I
6	7841.90	h	I
30	7847.80		I
80	7881.49		I
16	7890.37		I
16	7924.43		I
5	7948.15		I
9	7967.84		I
9	8112.47		I
18	8264.96		I
11	8348.98		I
6	8352.94		I
4	8435.77		I
11	8473.64		I
11	8483.56		I
22	8710.84		I
14	8724.98		I
9	8777.36		I

Ru III
Ref. 423 — C.H.C.

Intensity	Wavelength	
	Vacuum	
250	850.09	III
200	850.30	III
50	851.22	III
50	852.49	III
150	856.32	III
50	867.48	III
250	919.74	III

Intensity	Wavelength		
50	921.78		III
250	928.08		III
150	937.16		III
500	940.09		III
250	940.68		III
50	941.85		III
50	942.63		III
50	943.06		III
150	945.68		III
100	946.05		III
100	947.14		III
100	949.83		III
50	950.35		III
100	950.45		III
50	952.59		III
50	957.06		III
50	957.18		III
50	961.58		III
250	961.68		III
100	962.56		III
500	966.54		III
250	967.09		III
150	967.85		III
150	967.92		III
150	971.83		III
250	972.40		III
100	973.54		III
150	973.78		III
750	974.14		III
250	974.46		III
250	977.51		III
100	978.18		III
900	979.43		III
500	981.35		III
250	983.81		III
250	983.91		III
250	985.55		III
250	986.84		III
200	987.87		III
250	991.67		III
250	992.75		III
900	994.56		III
250	995.30		III
200	1000.78		III
300	1001.65		III
250	1004.29		III
500	1009.13		III
900	1009.87		III
500	1014.68		III
100	1018.72		III
100	1019.33		III
100	1020.77		III
200	1080.00		III
100	1184.37		III
800	1190.51		III
500	1200.07		III
100	1204.57		III
200	1204.88		III
500	1207.17		III
500	1209.77		III
300	1211.31		III
200	1232.57		III
100	1653.77	h	III
200	1699.84		III
100	1715.97	h	III
200	1759.49		III
200	1880.95	h	III
100	1883.56	h	III
200	1899.04	h	III
100	1899.42	h	III
100	1908.31		III
500	1941.35		III
100	1981.82		III
100	1982.10		III
200	1989.22		III
200	1993.32		III
100	1997.55		III
	Air		
200	2005.71		III
100	2006.46		III
500	2009.28		III
100	2011.17		III
50	2011.56		III
50	2011.66		III
100	2015.20		III
50	2018.58		III
100	2044.59		III

SAMARIUM (Sm)
Z = 62

Sm I and II
Ref. 1 — C.H.C.

Intensity	Wavelength
	Air
45	2610.07
90	2640.27

Samarium (Cont.)

Intensity		Wavelength	Spectrum
35		2649.17	
45		2657.68	
70		2662.42	
120		2675.15	
100		2688.60	
45		2690.90	II
130		2693.34	
45		2693.74	
60		2696.08	
85		2707.96	
50		2732.42	
35		2739.87	
29		2762.28	
35		2764.18	
85		2767.85	II
60		2774.77	
85		2776.11	
85		2779.23	II
85		2786.64	
150		2789.38	II
130	h	2796.70	
85		2807.36	
150		2809.50	
120		2810.86	II
85		2817.20	II
29		2820.96	II
220		2830.94	
60		2840.30	
60		2847.49	II
60		2851.35	
120		2866.09	II
70		2868.40	II
70		2881.34	
85		2881.68	
60		2883.09	
45		2889.06	
60		2891.34	
85	d	2907.88	
		2907.99	II
130		2910.28	II
85		2937.48	II
70		2943.49	II
150		2953.19	II
85		2962.74	II
160		2969.02	II
100		2983.43	II
60		2991.57	II
100		3021.01	
150		3034.84	II
100		3039.13	II
120		3046.93	II
150		3067.54	
120		3071.29	II
100		3086.45	II
120		3096.88	II
100	h	3102.30	II
250		3106.52	II
220		3110.20	II
200		3117.72	II
270		3136.30	II
150		3139.97	II
150		3147.19	II
180		3152.10	II
410		3152.52	II
150		3162.30	II
360		3169.88	II
180		3178.12	II
720		3183.92	II
310		3187.01	II
430		3187.22	II
360		3187.79	II
360		3193.01	II
360		3196.18	II
150		3201.80	II
150		3204.90	II
360		3207.18	II
180		3208.17	II
600		3211.73	II
150		3214.12	II
270		3215.26	II
530		3216.85	II
600		3218.61	II
150		3219.43	II
270		3226.84	II
180		3228.50	II
270		3228.78	II
720		3230.56	II
360		3231.53	II
150		3231.95	II
430		3233.68	II
720		3236.64	II
150		3237.89	II
720		3239.66	II
530		3241.16	II
180		3241.59	II
180		3242.04	II
150		3244.69	II
240		3249.75	II
720		3250.37	II
360		3253.40	II
270		3253.94	II
850		3254.38	II
110		3255.63	II
360		3262.28	II
430		3264.94	II
180		3270.49	II
180		3270.68	II
430	d	3272.48	II
		3272.60	II
430		3272.81	II
430		3273.48	II
430		3276.75	II
270		3280.84	II
180		3285.46	II
430		3286.23	II
720	d	3290.28	II
		3290.39	II
180		3290.65	II
240		3293.37	II
360		3295.44	II
430		3295.81	II
720		3298.10	II
170		3300.98	II
340		3301.68	II
340		3304.52	II
340		3305.18	II
1700		3306.39	II
170		3306.61	II
850		3307.02	II
340		3309.52	II
850		3310.66	II
600		3312.42	II
410		3316.58	II
430		3320.16	II
110		3320.59	II
1200		3321.18	II
340		3323.77	II
340		3325.26	II
170		3325.48	II
340		3327.88	II
170		3333.64	II
170		3336.12	II
850		3340.58	II
240		3343.49	II
110		3343.64	II
240		3344.35	II
170		3347.30	II
240		3348.68	II
220		3350.88	II
410	d	3354.18	II
		3354.30	II
170		3354.72	II
1200		3365.86	II
150		3367.27	II
340		3368.57	II
340		3369.46	II
170		3370.59	II
340		3371.21	II
150		3376.48	II
1200		3382.40	II
510		3384.66	II
150		3384.86	
150		3387.66	II
410		3389.32	II
150		3391.11	II
410		3396.19	II
150		3397.76	II
150		3399.84	II
600		3402.46	II
210		3403.09	II
850		3408.68	II
270		3418.15	II
430		3418.51	II
170		3419.77	II
120		3424.78	II
170		3426.20	II
170		3433.68	II
150		3437.10	II
170		3438.06	II
240		3440.50	II
170		3453.56	II
170		3459.20	II
120		3459.42	II
240		3461.13	II
120		3464.07	II
170		3467.87	II
130		3473.96	II
130		3479.53	II
130		3480.26	II
170		3480.56	II
170		3487.41	II
170		3493.61	II
220		3499.84	II
340		3511.23	II
310		3530.60	II
220		3532.57	II
270		3535.65	II
240		3554.15	II
510		3559.10	II
220		3566.84	II
4200		3568.27	II
270		3577.79	II
390		3580.94	II
310		3583.39	II
4200		3592.60	II
340		3601.69	II
1700		3604.28	II
3400		3609.49	II
240		3620.58	II
1700		3621.23	II
240		3623.32	II
850		3627.01	II
850		3631.13	II
3400		3634.29	II
240		3634.93	II
410		3638.77	II
360		3645.29	II
300		3645.39	II
660		3649.53	II
340		3650.19	II
340		3656.22	II
2200		3661.36	II
220		3662.69	II
340		3667.93	II
340		3670.66	II
2200		3670.84	II
340		3677.79	II
270		3681.73	II
270		3688.42	II
270		3692.22	II
1100		3693.99	II
480		3706.75	II
480		3706.98	II
480		3708.41	II
930		3708.65	II
480		3711.54	II
350		3712.76	II
930		3718.88	II
930		3721.85	II
420		3724.90	II
1600		3728.47	II
2100		3731.26	II
1600		3735.98	II
800		3737.14	II
320		3737.48	II
2900		3739.12	II
		3739.20	II
800		3741.29	II
1200		3743.87	II
930		3745.46	I
		3745.60	II
480		3747.62	II
800		3755.28	II
800		3756.41	I
1200		3757.53	II
450		3758.45	II
660		3758.97	II
350		3760.04	II
1900		3760.69	II
660		3762.59	II
1100		3764.37	II
480		3767.36	II
480		3767.76	II
370	d	3773.33	I
		3773.42	II
1100		3778.14	II
660		3780.76	II
420		3780.93	II
320		3787.20	II
1500		3788.12	II
1600		3793.97	II
420		3797.28	II
1600		3797.73	II
500		3799.54	II
800		3800.89	II
320		3805.63	II
420		3808.46	II
320		3809.75	II
320		3809.88	II
420		3810.43	II
500		3812.07	II
480		3813.63	II
420		3814.63	II
930	d	3820.82	II
530		3824.18	II
1600		3826.20	II
530		3830.29	II
1100		3831.50	II
530		3833.83	II
560		3834.48	I
560		3834.60	II
370		3835.72	II
500		3838.94	II
400		3840.45	II
1600		3843.50	II
530		3847.51	II
640		3848.78	II
420		3851.88	II
530		3853.30	I
2700		3854.21	II
480		3854.56	I
800		3855.90	II
480		3857.91	II
400		3858.74	I
660		3862.05	II
350		3862.23	II
320		3865.24	II
800		3871.78	II
400		3875.19	II
560		3875.54	II
800		3880.77	II
450		3881.38	II
450		3881.79	II
320		3882.50	II
3700		3885.29	II
660		3889.16	II
		3889.22	II
610		3890.08	II
320		3891.21	II
400		3894.05	II
1600		3896.98	II
1300		3903.42	II
620		3917.44	II
2500		3922.40	II
1900		3928.28	II
470		3935.76	II
1300		3941.87	II
620		3943.24	II
500		3946.51	II
740		3948.11	II
470		3951.89	I
370		3959.53	II
1500		3963.00	II
620		3966.04	II
470		3967.68	II
740		3970.53	II
1500		3971.40	II
620		3974.66	I
960		3976.27	II
1000		3976.43	II
960		3979.20	II
740		3983.14	II
740		3986.68	II
370		3987.43	II
1500		3990.00	II
		3990.02	I
740		3993.31	II
280		4003.46	II
470		4007.48	II
280		4019.98	II
880		4023.23	II
740		4035.11	II
590		4041.68	II
740		4042.72	II
880		4042.90	II
240		4044.11	II
560		4045.05	II
440		4046.16	II
740		4047.16	II
210		4048.62	II
590		4049.81	II
440		4058.87	II
560		4063.54	II
280		4064.32	II
1400		4064.58	II
810		4066.74	II
710		4068.33	II
810		4075.84	II
280		4076.65	II
240		4080.56	II
410		4082.60	II
280		4083.58	II
220		4084.40	II
1000		4092.27	II
290		4094.05	II
240		4104.13	II
810		4107.28	II
		4107.39	II
410		4109.40	II
280		4110.19	II
410		4113.90	II
1900		4118.55	II
410		4121.36	II
280		4122.51	II
710		4123.96	II
280		4129.23	II
250		4135.14	II
320		4147.71	II
810		4149.83	II
1200		4152.21	II
530		4153.33	II
560		4155.22	II
810		4169.48	II
410		4171.57	II
440		4178.02	II
530		4181.10	II
210		4183.33	I
530		4183.76	II
1000		4188.13	II
410		4191.93	II
270		4199.45	II
650		4202.92	II
1100		4203.05	II
660		4206.13	II
270		4206.62	II
660		4210.35	II
740		4220.66	II
1000		4225.33	II
740		4229.70	II
620		4234.57	II
1200		4236.74	II
500		4237.66	II
620		4244.70	II
210		4249.55	II
250		4251.78	II

Samarium (Cont.)

Int.	λ		Spec.	Int.	λ		Spec.	Int.	λ		Spec.	Int.	λ		Spec.
2100	4256.39		II	290	4606.51		II	70	5773.77		I	19	7283.33		II
210	4258.58		II	290	4615.44		II	60	5778.33		I	16	7288.92		II
1300	4262.68		II	470	4615.69		II	45	5779.24		I	13	7290.23		I
500	4265.08		II	150	4630.21		II	45	5781.93		I	26	7300.72	h	II
1200	4279.68		II	880	4642.24		II	70	5786.98	d	II	13	7327.08		II
	4279.75		II	290	4645.40		I	60	5788.38		I	13	7332.65		I
240	4279.94		II	290	4646.68		II	60	5800.52		I	8	7338.04		I
2200	4280.79		II	240	4648.16		II	65	5802.84		I	26	7347.30		I
710	4282.21		I	380	4649.49		I	45	5814.89		I	26	7376.69		II
470	4282.83		I	150	4655.13		II	45	5831.02		II	13	7393.98		II
240	4283.50		I	290	4663.56		I	45	5836.37		II	30	7444.56		I
350	4286.64		II	740	4669.40		II	35	5860.78		I	26	7445.41		I
350	4292.18		II	620	4669.65		II	65	5867.79		I	26	7453.03	d	II
1600	4296.74		I	470	4670.75	d	I	45	5868.61		I	13	7470.76		I
320	4304.94		II		4670.83		I	35	5871.06		I	26	7481.99		II
880	4309.01		II	1100	4674.60		II	50	5874.21		I	23	7502.39	h	II
240	4312.85		I	680	4676.91		II	45	5897.39		II	10	7517.00	h	II
1900	4318.94		II	210	4681.55		I	50	5898.96		I	23	7541.42	h	II
470	4319.53		I	370	4687.18		II	35	5938.90		II	9	7544.74		I
590	4323.28		II	370	4688.73		I	65	5965.71		I	10	7546.57		I
240	4324.46		I	130	4693.63		II	35	5968.82	h	II	12	7560.03	h	II
1800	4329.02		II	120	4699.34		II	35	5984.29		I	19	7562.94		II
440	4330.02		I	530	4704.40		II	50	6045.00		I	23	7570.95		II
1300	4334.15		II	270	4713.06		II	45	6045.39		I	23	7572.29		II
880	4336.14		I	130	4715.26		II	50	6070.06		I	19	7578.09		II
560	4345.86		II	730	4716.10		I	45	6084.12		I	30	7585.85		II
1100	4347.80		II	270	4717.07		I	35	6091.40	h	I	23	7588.31		II
560	4350.46		II	210	4717.72		II	45	6110.66		II	10	7598.01		I
560	4352.10		II	190	4718.33		II	45	6159.56	h	I	23	7607.48	d	II
560	4360.72		II	270	4719.84		II	45	6246.76		II		7607.74		I
220	4361.07		II	130	4726.02		II	45	6256.54		I	12	7613.94		II
810	4362.04		II	770	4728.42		I	45	6256.66		II	10	7631.77	h	II
440	4362.91		I	470	4745.68		II	100	6267.28		II	23	7637.94		II
220	4363.45		II	150	4750.72		I	50	6291.82		II	45	7645.09		II
500	4368.03		II	730	4760.27		I	35	6307.06		II	12	7645.82		I
210	4369.92		II	110	4770.20		I	70	6327.47		II	19	7648.02		II
440	4373.46		II	110	4774.15		II	45	6426.64		II	10	7655.78		II
320	4374.98		II	190	4777.85		II	45	6472.34		II	19	7667.20		II
880	4378.24		II	580	4783.10		I	35	6484.52		II	8	7672.49		II
530	4380.42		I	350	4785.86		I	35	6498.67		II	10	7678.79	h	II
290	4384.29		II	160	4789.96		I	50	6542.76		II	10	7695.78	h	I
1600	4390.86		II	230	4791.58		II	140	6569.31		II	23	7712.04		II
210	4393.35		I	430	4815.81		II	35	6570.67	h	II	30	7728.56		II
290	4397.34		I	130	4829.57		II	40	6585.21	h	II	30	7736.26		II
410	4401.17		I	970	4841.70		I	110	6589.72		II	30	7749.30		II
810	4403.06	d	II	310	4844.21		II	40	6601.83		II	23	7755.20		II
	4403.13		I	140	4847.76		II	95	6604.56		II	10	7794.50		I
410	4403.36		II	270	4848.32		I	40	6632.28	h	II	10	7801.54		I
520	4409.33		II	120	4854.36		II	50	6671.51		I	8	7812.75	h	II
290	4411.58		I	210	4883.77		I	70	6679.21		II	16	7820.15		II
380	4417.58		II	730	4883.97		I	70	6693.55		II	10	7831.40		II
470	4419.33		I	170	4904.97		I	40	6723.07	d	I	40	7835.08	w	II
1500	4420.53		II	630	4910.40		I	120	6731.84	d	II	26	7837.27		II
960	4421.14		II	350	4913.25		II	70	6734.06	d	II	10	7844.82		II
2900	4424.34		II	430	4918.99		II	40	6734.81	d	II	6	7859.53		I
470	4429.66		I	110	4924.04		I	55	6741.47		II	19	7863.65		II
1600	4433.88		II	120	4938.10		II	40	6778.61	h	II	10	7880.01	h	II
1800	4434.32		II	170	4948.63		II	60	6790.00		II	16	7895.96		I
530	4441.81		I	120	4952.37		II	95	6794.20		II	26	7914.96		II
440	4442.28		I	170	4961.94		II	55	6844.71		II	90	7928.14		II
710	4444.26		II	170	4975.98		I	75	6856.03		II	9	7931.92		I
710	4445.15		I	140	5028.44		II	120	6860.93		I	19	7937.09		II
1300	4452.73		II	400	5044.28		I	40	6862.82		II	16	7948.12		II
250	4452.95		I	200	5052.76		II	30	6950.51		II	19	8001.61	w	II
1200	4454.63		II	170	5069.46		II	120	6955.29		II	19	8014.92	w	II
1000	4458.52		II	540	5071.20		I	90	7020.44		II	23	8025.12		II
250	4459.29		I	170	5100.22		II	13	7036.73		II	23	8026.32	w	II
2200	4467.34		II		5100.39		I	90	7039.22		II	16	8032.03		II
810	4470.89		I	260	5103.09		II	90	7042.24		II	40	8048.70		II
470	4472.43		II	140	5104.48		II	13	7049.15		II	16	8065.16		I
620	4473.02		II	140	5116.70		II	90	7051.52		II	45	8068.46		II
740	4478.66		II	510	5117.16		II	16	7054.97		II	9	8117.16	w	II
370	4499.11		I	350	5122.14		I	19	7074.67		I	9	8125.12		II
370	4499.48		II	360	5155.03		II	90	7082.37		II	26	8161.82		II
240	4503.38		I	250	5172.74		I	40	7085.52	d	II	19	8195.50	w	II
180	4505.05		II	470	5175.42		I	26	7088.30		I	6	8206.30		I
120	4511.33		I	250	5200.59		I	16	7091.16		I	26	8218.76	w	II
560	4511.83		II	260	5251.92		I	30	7095.50		I	9	8230.33		I
440	4515.09		II	400	5271.40		I	16	7096.33		I	16	8240.98		II
880	4519.63		II	250	5282.91		I	30	7104.54		I	19	8289.26	w	II
440	4523.04		II	190	5320.60		I	19	7106.23		I	10	8300.88		II
	4523.18		I	110	5341.29		I	26	7115.96		I	40	8305.79	w	II
650	4523.91		II	140	5368.36		I	23	7117.51		II	10	8315.45		I
290	4533.80		I	130	5405.23		I	26	7119.81	h	II	19	8348.68	w	II
270	4536.51		II	220	5453.00		I	12*	7122.40		II	19	8383.71		I
710	4537.95		II	140	5466.72		I	23	7125.11	h	II	19	8387.77		II
150	4538.53		II	230	5493.72		I	13	7131.80		I	30	8432.64	w	II
290	4540.19		II	80	5512.10		I	10	7136.01		I	19	8473.54	w	II
380	4542.06		II	230	5516.09		I	12	7139.39		II	45	8485.99	w	II
810	4543.95		II	50	5548.95		I	40	7143.98	d	II	30	8510.90	w	II
100	4544.83		II	140	5550.40		I	85	7149.60	d	II	23	8543.22		II
410	4552.66		II	45	5573.42		I	10	7172.67		I	23	8617.03	w	II
270	4554.45		II	35	5588.20		I	10	7189.57		I	23	8632.82	w	II
240	4560.43		II	50	5600.86		II	9	7210.95		I	12	8677.81	w	II
470	4566.21		II	50	5621.79		I	23	7213.82		I	13	8706.32		II
590	4577.69		II	70	5626.01		I	26	7218.09	d	II	45	8708.43	w	II
290	4581.58		I	85	5644.10		I	13	7220.07		I	30	8717.89	w	II
440	4581.73		I	140	5659.86		I	13	7237.02		I	30	8758.28	w	II
560	4584.83		II	120	5696.73		I	60	7240.90		II	16	8780.59	w	II
290	4591.82		II	85	5706.20		I	9	7257.11		I	23	8788.83	w	II
380	4593.54		II	35	5710.93		I	9	7261.52	d	II	26	8859.76		II
560	4595.29		II	50	5732.95		I	13	7279.25		I	95	8913.66		II
240	4596.74		I	50	5743.35		II	26	7281.47		II				
220	4604.18		II	45	5759.52		II	8	7282.21		I				

SCANDIUM (Sc)
Z = 21

Sc I and II
Ref. 1, 88 — C.H.C.

Intensity		Wavelength	
		Air	

Intensity		Wavelength	Spectrum
65		2429.16	I
110		2438.62	I
560		2545.22	II
2900		2552.37	II
560		2555.82	II
2300		2560.25	II
1100		2563.21	II
40		2611.22	II
19		2684.23	II
120		2692.78	I
360		2706.77	I
210		2707.95	I
580		2711.35	I
30		2819.54	II
35		2822.15	II
60		2826.68	II
340		2965.86	I
1200		2974.01	I
1400		2980.75	I
340		2988.95	I
2200		3015.36	I
2700		3019.34	I
360		3030.76	I
30		3039.93	II
70		3045.72	II
85		3052.93	II
120	h	3056.31	I
130		3065.11	II
45		3139.75	II
990		3251.32	II
1500		3255.69	I
4400		3269.91	I
5500		3273.63	I
110	d	3343.28	II
270		3352.05	II
9900		3353.73	II
65	d	3357.30	II
2000		3359.68	II
1700		3361.27	II
1700		3361.94	II
4000		3368.95	II
6600		3372.15	II
90		3416.68	I
130		3418.51	I
65		3419.36	I
200		3429.21	I
200		3429.48	I
270		3431.36	I
530		3435.56	I
90		3439.41	I
65		3440.18	I
65		3448.49	I
270		3457.45	I
180		3462.19	I
130	d	3469.65	I
110		3471.13	I
200		3498.91	I
2700		3535.73	II
6600		3558.55	II
6100		3567.70	II
13000		3572.53	II
9900		3576.35	II
7700		3580.94	II
4000		3589.64	II
4000		3590.48	II
28000		3613.84	II
110		3617.43	I
20000		3630.75	II
13000		3642.79	II
6600		3645.31	II
110		3646.90	I
5300		3651.80	II
110		3664.25	II
290		3666.54	II
55		3675.26	II
40		3678.35	II
75	h	3717.10	I
270		3833.07	II
610		3843.03	II
90		3894.97	I
20000		3907.49	I
23000		3911.81	I
45		3923.51	II
4400		3933.38	I
45		3952.27	I
45		3989.06	II
5500		3996.61	I
530		4014.49	II
20000		4020.40	I
20000		4023.69	I
220		4030.67	I
140		4031.39	I
100		4034.23	I

Intensity		Wavelength	Spectrum
220		4043.80	I
200		4046.48	I
2700		4047.79	I
120		4049.95	I
5500		4054.55	I
220		4056.59	I
160	h	4074.97	I
160		4078.57	I
6100		4082.40	I
200		4086.67	I
400		4087.16	I
40	h	4093.13	I
65		4094.85	I
55	h	4098.35	I
65		4100.33	I
440	h	4133.00	I
530	h	4140.30	I
65	h	4147.40	I
720		4152.36	I
55	h	4154.72	
90	Hd	4161.88	I
1100		4165.19	I
65	h	4171.56	I
45	h	4186.45	I
65	h	4187.62	I
75		4205.20	I
65		4212.34	I
45		4212.49	I
75	h	4216.10	I
110	h	4218.26	I
110	h	4219.73	I
40		4221.88	I
90	d	4225.59	I
180		4231.93	I
200		4233.61	I
100		4237.82	I
400		4238.05	I
90		4239.57	I
100		4246.12	I
15000		4246.83	II
55		4283.56	I
290		4294.77	II
350		4305.71	II
4200		4314.09	II
3300		4320.74	II
2400		4325.01	II
28		4348.53	I
180		4354.61	II
110		4358.64	I
55		4359.08	I
28		4364.92	I
2000		4374.46	II
130		4384.81	II
45	h	4389.60	I
1100		4400.37	II
880		4415.56	II
28		4420.66	II
45		4431.36	II
65		4542.55	I
90		4544.68	I
120	h	4557.24	I
160	h	4573.99	I
65	h	4592.94	I
65	h	4598.45	I
55	h	4604.72	I
45		4609.53	I
45		4609.95	I
350		4670.40	II
40	h	4680.49	I
50		4698.29	II
120		4706.97	I
120		4709.34	I
200		4728.77	I
490		4729.23	I
40	h	4732.30	I
590		4734.10	I
60		4735.08	I
690		4737.65	I
790		4741.02	I
1200		4743.81	I
200		4753.16	I
220		4779.35	I
90		4791.50	I
100		4827.28	I
100		4833.67	I
170		4839.44	I
40		4840.47	I
80		4847.68	I
80		4852.68	I
140	BLd	4857.79	S
		4858.09	SCO
80		4906.67	I
90		4909.76	I
90		4922.84	I
90		4934.25	I
45		4935.74	I
70		4941.33	I
170		4954.06	I
120		4973.66	I
150		4980.37	I
80		4983.45	I
140		4991.92	I
80		5018.39	I
70		5020.14	I

Intensity		Wavelength	Spectrum
80		5021.51	I
530		5031.02	II
55		5032.74	I
250		5064.32	I
80		5068.86	I
530		5070.23	I
250		5075.81	I
2100		5081.56	I
1200		5083.72	I
1100		5085.55	I
750		5086.95	I
390		5087.14	I
270		5089.89	I
45		5092.46	I
390		5096.73	I
620		5099.23	I
370		5101.12	I
180		5109.06	I
150		5112.86	I
320		5116.69	I
70	b	5133.68	S
45	b	5171.06	S
390		5210.52	I
45		5211.28	I
280		5219.67	I
350		5239.82	II
280		5258.33	I
35		5284.97	I
210		5285.76	I
35		5301.94	I
22		5318.35	II
70		5331.77	I
14		5334.23	II
95		5339.41	I
120		5341.05	I
95		5342.96	I
350		5349.30	I
120		5349.71	I
60		5350.30	I
210		5355.75	I
530		5356.10	I
14		5357.19	II
270		5375.35	I
370		5392.08	I
45		5416.12	I
45		5425.57	I
45		5429.41	I
35		5432.94	I
55		5433.23	I
45		5438.22	I
55		5439.03	I
55	h	5442.60	I
270		5446.20	I
18		5447.39	I
120		5451.34	I
30		5455.21	I
18		5465.20	I
55		5468.40	I
60		5472.19	I
18		5474.64	I
750		5481.99	I
530		5484.62	I
570		5514.22	I
16		5515.39	I
660		5520.50	I
45		5526.06	I
660		5526.82	II
55		5541.04	I
30		5546.40	I
18		5550.40	I
5		5552.25	II
35		5553.59	I
16		5561.10	I
70		5564.86	I
18		5571.24	I
14		5579.76	I
110		5591.33	I
35	h	5593.38	I
22		5604.19	I
22		5631.02	I
80		5640.98	II
45		5646.36	I
16		5647.60	I
55		5649.56	I
250		5657.88	II
60		5658.34	II
55		5667.16	II
70		5669.04	II
1500		5671.81	I
95		5684.20	II
1200		5686.84	I
1100		5700.21	I
190		5708.61	I
880		5711.75	I
230		5717.28	I
180		5724.08	I
55	B1	5736.85	S
55	B1	5764.45	S
95		5772.74	S
55	B1	5775.32	S
70	B1	5809.84	S
70	B1	5811.60	S

Intensity		Wavelength	Spectrum
95	B1	5847.73	S
70	B1	5849.07	S
70	b	5887.38	S
35	B1	5918.04	S
30		5919.11	I
60	B1	5928.10	S
35		5961.49	I
60	B1	5968.25	S
35		5969.19	I
90		5988.42	I
160	B1	6017.07	S
60		6026.18	I
620	B1	6036.17	S
490	B1	6064.31	S
440		6072.65	S
620	B1	6079.30	S
320	B1	6101.87	S
370	B1	6109.93	S
370	B1	6115.97	S
180	b	6148.70	S
150	b	6153.93	S
150	b	6188.09	S
150	b	6192.90	S
620		6210.68	I
90		6239.41	I
320		6239.78	I
120		6245.63	II
110		6249.96	I
250		6258.96	I
60		6262.25	I
55		6276.31	I
45		6279.76	II
18		6300.70	II
750		6305.67	I
26		6309.90	II
16		6320.85	II
26		6344.83	I
60		6378.82	I
55	B1	6408.41	S
90		6413.35	I
26	b	6437.08	S
55	B1	6446.24	S
26	b	6457.78	S
35		6485.40	S
26	b	6495.90	S
55	b	6525.62	S
22	b	6535.30	S
45	h	6557.84	S
35	b	6566.88	S
18	h	6575.85	S
60		6604.60	II
26	B1	6609.99	S
18	B1	6617.94	S
18	B1	6645.08	S
22	B1	6654.42	S
26	B1	6661.01	S
18	b	6700.48	S
18	b	6705.93	S
65		6737.87	I
35		6739.40	I
35		6817.08	I
50		6819.52	I
29		6829.54	I
50		6835.03	I
5	b	6963.12	S
5	B1	6990.68	S
5	B1	7025.72	S
8	b	7035.77	S
5	b	7072.37	S
5	b	7094.38	S
12	h	7138.14	I
14		7169.13	I
12		7257.57	I
8		7275.57	I
3	h	7300.62	I
12	h	7524.13	I
14	h	7553.96	I
15	h	7574.44	I
11		7617.45	I
14	h	7665.72	I
30		7697.73	I
18		7729.72	I
55	h	7741.17	I
5	h	7750.37	I
5		7752.72	I
6	h	7771.06	I
15		7785.17	I
8		7794.68	I
30		7800.44	I
11		7821.64	I
11	h	8196.98	I
15		8241.13	I
19	h	8761.40	I
11	h	8774.8	
15	h	8794.72	I
15	h	8823.8	I
30	h	8834.35	I
70		20616.32	
30		20985.81	
400		22051.86	I
150		22065.05	I

Scandium (Cont.)

Sc III
Ref. 323 — C.H.C.

Intensity	Wavelength	
	Vacuum	
10	730.60	III
15	731.65	III
15	1148.24	III
20	1154.52	III
20	1162.44	III
25	1168.61	III
10	1168.88	III
80	1598.00	III
180	1603.06	III
150	1610.19	III
40	1895.44	III
60	1912.62	III
90	1993.89	III
	Air	
160	2010.42	III
50	2012.26	III
350	2699.07	III
230	2734.05	III
10	2831.75	III
80	4061.21	III
100	4068.66	III
40	4309.47	III
10	4740.95	III
15	4780.87	III
50	4992.89	III
60	5032.09	III
80	6256.01	III
60	6307.60	III
90	7449.16	III
70	7548.15	III
70	7868.65	III
35	8814.29	III
50	8829.78	III
30	8865.89	III
15	8881.58	III

Sc IV
Ref. 298 — C.H.C.

Intensity	Wavelength	
	Vacuum	
8	220.28	IV
15	289.85	IV
15	296.31	IV
15	299.04	IV
10	371.16	IV
9	438.80	IV
8	557.50	IV
8	584.83	IV
9	617.08	IV
8	761.43	IV
8	769.70	IV
10	785.12	IV
8	789.00	IV
8	791.71	IV
8	861.24	IV
8	861.30	IV
8	890.87	IV
8	1219.40	IV
9	1228.20	IV
9	1424.66	IV
9	1444.10	IV
8	1489.64	IV
8	1514.96	IV
8	1535.76	IV
9	1543.86	IV
9	1549.55	IV
15	1550.80	IV
8	1555.72	IV
9	1563.81	IV
10	1574.92	IV
9	1583.41	IV
8	1584.64	IV
8	1592.23	IV
8	1660.71	IV
10	1665.92	IV
8	1746.23	IV
	Air	
10	2056.06	IV
8	2078.93	IV
12	2118.97	IV
9	2164.43	IV
11	2185.43	IV
11	2205.46	IV
14	2222.22	IV
11	2271.33	IV
9	2464.45	IV
8	2520.93	IV
11	2586.93	IV
9	2595.17	IV
8	2678.01	IV
8	2723.52	IV
8	2773.04	IV
8 d	4594.42	IV
8 d	4639.96	IV
8	5501.74	IV
9	5620.72	IV
10	5706.82	IV
14	5771.63	IV
9	6548.03	IV

Sc V
Ref. 150 — C.H.C.

Intensity	Wavelength	
	Vacuum	
150	179.42	V
350	180.14	V
200	180.82	V
200	180.96	V
50	181.55	V
200	182.39	V
300	228.56	V
100	230.85	V
40	243.82	V
500	243.87	V
400	246.42	V
400	250.98	V
500	252.85	V
500	253.73	V
50	255.38	V
300	255.64	V
200	257.16	V
150	258.24	V
40	258.81	V
50	260.05	V
400	281.00	V
900	283.91	V
800	284.45	V
600	288.29	V
900	289.59	V
1000 d	291.93	V
800	293.25	V
400	296.17	V
700	300.00	V
400	375.05	V
100	378.68	V
200	388.68	V
400	395.32	V
200	399.50	V
1000	573.36	V
600	587.94	V

SELENIUM (Se)
Z = 34

Se I and II
Ref. 80, 181, 216, 275 — R.L.K.

Intensity	Wavelength	
	Vacuum	
285	828.5	II
360	832.7	II
285	906.6	II
360	912.9	II
360	1013.4	II
360	1014.0	II
450	1033.6	II
450	1049.6	II
360	1057.4	II
285	1097.8	II
360	1141.8	II
220	1156.0	II
285	1156.9	II
285	1168.5	II
450	1192.3	II
220	1205.7	II
220	1234.9	II
285	1291.0	II
285	1308.9	II
100	1405.4	I
100	1406.4	I
100	1406.6	I
120	1435.3	I
120	1435.8	I
100	1444.8	I
100	1446.8	I
100	1447.0	I
150	1449.2	I
120	1456.3	I
150	1500.9	I
250	1530.4	I
150	1531.3	I
200	1531.8	I
120	1547.1	I
120	1560.3	I
150	1575.3	I
150	1577.6	I
150	1577.9	I
150	1579.5	I
200	1580.0	I
150	1587.5	I
150	1593.2	I
250	1606.5	I
100	1610.7	I
100	1611.3	I
200	1617.4	I
150	1621.2	I
100	1622.7	I
120	1626.2	I
150	1643.4	I
250	1671.2	I
250	1675.3	I
250	1690.7	I
250	1793.3	I
300	1795.3	I
300	1855.2	I
250	1858.8	I
400	1898.6	I
350	1913.8	I
300	1919.2	I
500	1960.9	I
150	1995.1	I
	Air	
500	2039.8	I
500	2074.8	I
500	2164.2	I
150	2332.8	I
600	2413.5	I
300	2548.0	I
220	3038.7	II
220	3041.3	II
285	4070.2	II
360	4175.3	II
450	4180.9	II
120	4328.7	I
100	4330.3	I
285	4382.9	II
285	4446.0	II
220	4449.2	II
285	4467.6	II
500	4730.8	I
400	4739.0	I
300	4742.2	I
285	4840.6	II
360	4845.0	II
450	5227.5	II
360	5305.4	II
100	5365.5	I
120	5369.9	I
110	5374.1	I
285	5522.4	II
285	5566.9	II
285	5866.3	II
450	6056.0	II
200	6325.6	I
360	6444.2	II
285	6490.5	II
285	6535.0	II
150	6831.3	I
120	6990.690	I
100	6991.792	I
200	7010.809	I
150	7013.875	I
300	7062.065	I
200	7575.1	I
250	7583.4	I
150	7592.2	I
120	7606.8	I
300	8001.0	I
200	8036.4	I
120	8060.9	I
120	8065.3	I
120	8081.1	I
150	8093.2	I
150	8094.7	I
180	8149.3	I
150	8152.0	I
200	8157.7	I
180	8163.1	I
150	8182.9	I
100	8185.0	I
120	8194.6	I
150	8440.47	I
150	8450.38	I
150	8742.33	I
300	8918.86	I
100	8969.69	I
200	9001.97	I
200	9038.61	I
80	9083.14	I
120	9088.79	I
80	9140.83	I
60	9181.88	I
60	9271.12	I
100	9432.50	I
60	9825.58	I
200	10217.25	I
377	10307.45	I
900	10327.26	I
640	10386.36	I
124	10650.30	I
125	11934.56	I
275	11946.87	I
100	11947.92	I
105	11952.27	I
170	11952.64	I
100	11966.04	I
205	11972.93	I
115	11973.07	I
315	14817.93	I
410	14917.47	I
500	15151.44	I
115	15469.06	I
320	15471.00	I
265	15520.97	I
395	15618.40	I
115	15620.38	I
360	16659.44	I
505	16813.78	I
165	16817.76	I
205	16866.54	I
115	16972.71	I
235	21374.24	I
680	21442.56	I
415	21473.48	I
270	21716.36	I
240	21730.60	I
105	23133.66	I
150	23388.85	I
110	23628.17	I
265	24148.18	I
170	24159.23	I
185	24204.44	I
375	24385.99	I
160	24413.67	I
225	24471.17	I
255	25017.51	I
510	25127.43	I

Se III
Ref. 9, 247 — R.L.K.

Intensity	Wavelength	
	Vacuum	
220	709.2	III
220	709.4	III
220	720.6	III
360	724.3	III
285	726.4	III
220	737.2	III
220	741.9	III
285	777.3	III
220	790.8	III
360	843.0	III
220	879.2	III
285	953.7	III
220	954.4	III
220	954.7	III
160	974.1	III
360	974.8	III
285	1079.6	III
360	1099.1	III
450	1119.2	III
	Air	
285	2057.5	III
285	2767.2	III
220	2773.8	III
285	3379.8	III
450	3387.2	III
450	3413.9	III
285	3428.4	III
450	3457.8	III
360	3543.6	III
285	3570.2	III
450	3637.6	III
360	3711.7	III
450	3738.7	III
285	3743.0	III
450	3800.9	III
360	4046.7	III
220	4083.2	III
450	4169.1	III
220	4637.9	III
285	6303.8	III

Se IV
Ref. 245 — R.L.K.

Intensity	Wavelength	
	Vacuum	
285	636.0	IV
285	654.2	IV
360	652.7	IV
450	670.1	IV
285	671.9	IV
220	722.8	IV
285	734.6	IV
450	746.4	IV
285	759.0	IV
285	776.5	IV
285	803.8	IV

Selenium (Cont.)

Intensity	Wavelength	
360	959.6	IV
450	996.7	IV
220	1307.2	IV
285	1314.4	IV

Air

Intensity	Wavelength	
220	2090.0	IV
285	2136.6	IV
160	2165.2	IV
160	2166.6	IV
360	2665.5	IV
285	2724.3	IV
160	2951.6	IV

Se V
Ref. 245 — R.L.K.

Intensity	Wavelength	

Vacuum

Intensity	Wavelength	
285	596.0	V
285	601.0	V
220	608.7	V
360	613.0	V
285	614.3	V
450	759.1	V
285	785.8	V
285	804.3	V
360	808.7	V
220	814.8	V
220	820.7	V
360	830.3	V
450	839.5	V
360	845.8	V
360	1094.7	V
220	1151.0	V
450	1227.6	V

SILICON (Si)
Z = 14

Si I and II
Ref. 170, 237, 292 — L.J.R.

Intensity		Wavelength	

Vacuum

Intensity		Wavelength	
10	h	805.10	II
20	h	820.52	II
20	h	843.72	II
40	h	845.77	II
10		850.14	II
100		889.72	II
200		892.00	II
10		899.41	II
20		901.74	II
10		913.01	II
20		913.85	II
20		929.81	II
100		989.87	II
200		992.68	II
25		1020.70	II
50		1023.69	II
30		1057.05	II
15		1057.50	II
20	h	1127.44	II
40	h	1127.91	II
100		1190.42	II
200		1193.28	II
250		1194.50	II
100		1197.39	II
10	h	1216.12	II
20		1223.91	II
20		1224.25	II
10		1224.97	II
50		1226.81	II
20		1226.89	II
40		1226.99	II
100		1227.60	II
10		1228.44	II
25		1228.62	II
150		1228.75	II
200		1229.39	II
10		1235.92	II
100		1246.74	II
150		1248.43	II
100		1250.09	II
150		1250.43	II
200		1251.16	II
10		1255.28	I
40		1256.49	I
50		1258.80	I
1000		1260.42	II
2000		1264.73	II
200		1265.02	II
100		1304.37	II
50	h	1305.59	II
200		1309.27	II
20	h	1309.46	II
100		1346.87	II
100		1348.54	II
150		1350.06	II
20		1350.52	II
20		1350.66	II
100		1352.64	II
100		1353.72	II
10	h	1409.07	II
20	h	1410.22	II
10	h	1416.97	II
15	h	1474.65	II
15		1484.87	II
90	h	1485.02	II
30		1485.22	II
100	h	1485.51	II
100	h	1509.10	II
50	h	1512.07	II
30	p	1513.57	II
60	p	1516.91	II
500		1526.72	II
1000		1533.45	II
10		1562.45	II
15		1562.85	II
10		1563.77	II
50		1573.87	I
50		1574.82	I
50		1592.41	I
150		1594.55	I
50		1594.93	I
30		1597.95	I
100		1622.87	I
30		1625.71	I
300		1629.43	I
200		1629.92	I
75		1631.13	I
50		1633.98	I
30	h	1653.35	I
30		1664.52	I
50		1666.37	I
100		1667.62	I
100		1668.52	I
100		1672.59	I
200		1675.20	I
30		1682.68	I
30		1686.82	I
50	h	1689.29	I
30	h	1690.79	I
50		1693.29	I
50		1695.51	I
200		1696.20	I
200		1697.94	I
50		1700.42	I
30		1700.63	I
30		1702.86	I
50		1704.43	I
10	h	1710.83	II
20	h	1711.30	II
30		1743.88	I
50		1747.40	I
30	h	1753.11	I
50		1763.66	I
40		1765.03	I
30	h	1765.60	I
30		1766.06	I
30		1770.63	I
100	h	1770.92	I
100	h	1776.83	I
50	h	1783.23	I
100	h	1799.12	I
150		1808.00	II
50	h	1809.09	I
500	h	1814.07	I
200		1816.92	II
10		1817.45	II
50		1822.45	I
200		1836.51	I
30	h	1838.01	I
100	h	1841.15	I
200		1841.44	I
200		1843.77	I
300		1845.51	I
100		1846.10	I
400		1847.47	I
200		1848.14	I
100		1848.74	I
500		1850.67	I
30	h	1851.79	I
200		1852.46	I
50		1853.15	I
20		1869.32	II
15		1870.23	II
100		1873.10	I
500	h	1874.84	I
100		1875.81	I
200		1881.85	I
200		1887.70	I
200		1893.25	I
1000	h	1901.33	I
100	h	1902.46	II
50	h	1904.66	I
50	h	1910.62	II
50		1941.67	II
15		1944.59	II
10		1949.33	II
100		1949.56	II
100		1954.97	I
30		1984.43	I
50		1991.85	I

Air

Intensity		Wavelength	
30		2010.97	I
50		2054.83	I
50		2058.65	II
50		2059.01	II
40		2061.19	I
30		2065.52	I
200		2072.02	II
200		2072.70	II
30	h	2103.21	I
30		2114.63	I
100		2124.12	I
10	h	2133.99	II
30	h	2136.40	II
50	h	2136.56	II
50	h	2147.91	I
110		2207.98	I
115		2210.89	I
110		2211.74	I
120		2216.67	I
120		2218.06	I
50		2218.91	I
35		2291.03	I
55		2303.06	I
30		2334.40	II
30		2334.61	II
10		2344.20	II
10	h	2349.54	II
20		2350.17	II
20		2353.09	II
100	h	2356.30	II
30	h	2357.18	II
50	h	2357.97	II
10	h	2360.20	II
30		2366.97	II
20		2374.26	II
10	h	2428.45	II
300		2435.15	I
65		2438.77	I
65		2443.36	I
70		2452.12	I
425		2506.90	I
375		2514.32	I
500		2516.113	I
350		2519.202	I
425		2524.108	I
450		2528.509	I
110		2532.381	I
30		2563.679	I
85		2568.641	I
45		2577.151	I
190		2631.282	I
10	h	2682.21	II
1000		2881.579	I
10	h	2887.51	II
300		2904.28	II
500		2905.69	II
55		2970.355	I
150		2987.645	I
50		3006.739	I
75		3020.004	I
20	h	3021.55	II
100	h	3030.00	II
20	h	3041.57	II
30	h	3042.19	II
100	h	3043.69	II
10	h	3043.85	II
10	h	3045.77	II
50	h	3048.30	II
150	h	3053.18	II
150		3188.97	II
50		3192.25	II
150		3193.09	II
50		3194.21	II
50		3194.69	II
100		3195.41	II
200		3199.51	II
20		3202.49	II
100	h	3203.87	II
200	h	3210.03	II
75		3214.66	II
15	h	3217.99	II
10		3220.44	II
20		3223.01	II
300		3333.14	II
500		3339.82	II
100	h	3853.66	II
500	h	3856.02	II
200	h	3862.60	II
300		3905.523	I
10	h	3955.74	II
10	h	3977.46	II
15	h	3991.77	II
10	h	3998.01	II
20	h	4075.45	II
15	h	4076.78	II
70		4102.936	I
300	h	4128.07	II
500	h	4130.89	II
10	h	4183.35	II
100	h	4190.72	II
50		4198.13	II
100		4621.42	II
150		4621.72	II
50		4782.991	I
35		4792.212	I
80		4792.324	I
15	h	4883.20	II
20	h	4906.99	II
20	h	4932.80	II
30		4947.607	I
40		5006.061	I
1000		5041.03	II
1000		5055.98	II
100		5181.90	II
100	h	5185.25	II
200	h	5192.86	II
500	h	5202.41	II
30	h	5295.19	II
100		5405.34	II
15	h	5417.24	II
15		5428.92	II
15		5432.89	II
100	h	5438.62	II
20	h	5447.26	II
15		5454.49	II
100		5456.45	II
500		5466.43	II
500		5466.87	II
100	h	5469.21	II
40		5493.23	I
200	h	5496.45	II
35		5517.535	I
100	h	5540.74	II
150	h	5576.66	II
30		5622.221	I
100	h	5632.97	II
200	h	5639.48	II
90		5645.611	I
150	h	5660.66	II
80		5665.554	I
1000	h	5669.56	II
30		5681.44	II
120		5684.484	I
300	h	5688.81	II
100		5690.425	I
90		5701.105	I
200	h	5701.37	II
100	h	5706.37	II
160		5708.397	I
45		5747.667	I
45		5753.625	I
45		5754.220	I
45		5762.977	I
70		5772.145	I
70		5780.384	I
30	h	5785.73	II
90		5793.071	I
30	h	5794.90	II
100		5797.859	I
150	h	5800.47	II
200		5806.74	II
30		5827.80	II
50		5846.13	II
10	h	5867.48	II
300	h	5868.40	II
40		5873.764	I
150		5915.22	II
200		5948.545	I
500		5957.56	II
500		5978.93	II
10	h	6067.45	II
20	h	6080.06	II
10	h	6086.67	II
90		6125.021	I
85		6131.574	I
90		6131.850	I
100		6142.487	I
100		6145.015	I
160		6155.134	I
160		6237.320	I
40		6238.287	I
125		6243.813	I
125		6244.468	I
180		6254.188	I
45		6331.954	I
1000		6347.10	II
1000		6371.36	II
45		6526.609	I
45		6527.199	I
45		6555.462	I
50	h	6660.52	II
15		6665.00	II
100		6671.88	II
20		6699.38	II
50	h	6717.04	II
100		6721.853	I
30		6741.64	I
20	h	6750.28	II
30		6818.45	II
50		6829.82	II
30		6848.568	I
80		6976.523	I

Silicon (Cont.)

Intensity	Wavelength	
180	7003.567	I
180	7005.883	I
30	7017.28	I
90	7017.646	I
250	7034.903	I
70	7164.69	I
200	7165.545	I
70	7184.89	I
65	7193.58	I
30	7193.90	I
100	7226.206	I
100	7235.326	I
60	7235.82	I
180	7250.625	I
160	7275.294	I
40	7282.81	I
400	7289.173	I
55	7290.26	I
35	7373.00	I
375	7405.774	I
200	7409.082	I
40	7415.35	I
275	7415.946	I
425	7423.497	I
85	7424.60	I
100	7680.267	I
40	7742.71	I
30	7800.008	I
400	7848.80	II
500	7849.72	II
30	7849.967	I
90	7918.386	I
120	7932.349	I
140	7944.001	I
35	7970.306	I
35	8035.619	I
70	8093.241	I
35	8230.642	I
40	8443.982	I
40	8501.547	I
60	8502.221	I
40	8536.165	I
120	8556.780	I
50	8648.462	I
40	8728.011	I
75	8742.451	I
100	8752.009	I
35	8790.389	I
100	9412.72	II
100	9413.506	I
30	10371.269	I
120	10585.141	I
120	10603.431	I
120	10660.975	I
30	10694.251	I
30	10727.408	I
60	10749.384	I
30	10784.550	I
80	10786.856	I
140	10827.091	I
60	10843.854	I
30	10868.79	I
130	10869.541	I
30	10882.802	I
30	10885.336	I
80	10979.308	I
30	10982.061	I
80	11017.965	I
13	11187.60	I
12	11289.84	I
12	11611.09	I
370	11984.19	I
220	11991.57	I
440	12031.51	I
150	12103.53	I
120	12270.68	I
11	13176.90	I
190	15888.39	I
40	15960.04	I
95	16060.03	I
20	16094.80	I
60	16163.71	I
11	16215.68	I
16	16381.55	I
29	16680.77	I
28	17327.29	I
26	18722.90	I
15	19385.94	I
48	19432.97	I
13	19493.38	I
110	19722.50	I
31	19928.88	I
12	20917.13	I
21	21354.24	I
12	22062.71	I

Si III
Ref. 320 — L.J.R.

Intensity	Wavelength	
	Vacuum	
8	566.61	III
6	652.22	III
8	653.33	III
5	673.48	III
5	800.07	III
9	823.41	III
5	883.40	III
7	939.09	III
9	967.95	III
10	993.52	III
13	994.79	III
7	997.39	III
7	1005.37	III
7	1031.16	III
8	1033.92	III
7	1037.05	III
6	1083.22	III
14	1108.37	III
16	1109.97	III
18	1113.23	III
6	1140.55	III
7	1141.58	III
6	1142.28	III
8	1144.31	III
6	1144.96	III
8	1145.11	III
7	1145.18	III
6	1155.00	III
6	1155.96	III
7	1158.10	III
6	1160.26	III
8	1161.58	III
5	1174.37	III
6	1174.43	III
8	1178.00	III
30	1206.50	III
30	1206.53	III
9	1207.52	III
10	1210.46	III
7	1235.43	III
6	1280.35	III
17	1294.54	III
14	1296.73	III
15	1298.89	III
18	1298.96	III
14	1301.15	III
16	1303.32	III
13	1312.59	III
8	1341.47	III
7	1342.39	III
6	1343.39	III
7	1361.60	III
6	1362.37	III
6	1363.47	III
8	1365.26	III
7	1367.05	III
5	1369.44	III
5	1373.03	III
5	1387.99	III
13	1417.24	III
6	1433.69	III
8	1435.77	III
7	1436.17	III
5	1441.73	III
6	1447.20	III
5	1457.25	III
12	1500.24	III
10	1501.19	III
9	1501.87	III
6	1506.06	III
7	1673.51	III
9	1842.55	III

Intensity	Wavelength	
	Air	
5	2176.89	III
6	2295.48	III
10	2296.87	III
8	2300.93	III
10	2308.19	III
11	2449.48	III
6	2483.20	III
25	2541.82	III
10	2546.09	III
14	2559.21	III
11	2640.79	III
14	2655.51	III
9	2817.11	III
7	2831.49	III
5	2839.62	III
5	2959.15	III
5	2980.52	III
5	3013.09	III
6	3034.73	III
8	3037.29	III
9	3040.93	III
7	3043.93	III
5	3045.08	III
7	3068.24	III
25	3086.24	III
6	3086.46	III
20	3093.42	III
5	3093.65	III
16	3096.83	III
6	3126.27	III
7	3147.37	III
8	3161.61	III
16	3185.13	III
13	3186.02	III
14	3196.50	III
15	3210.55	III
7	3216.25	III
12	3230.50	III
14	3233.95	III
15	3241.62	III
7	3253.40	III
5	3253.74	III
7	3254.80	III
12	3258.66	III
6	3270.46	III
10	3276.26	III
7	3279.26	III
15	3486.91	III
9	3525.94	III
8	3569.67	III
20	3590.47	III
8 h	3622.54	III
5 h	3639.45	III
6 h	3645.12	III
7 h	3681.40	III
5 h	3682.15	III
20 c	3791.41	III
25	3796.15	III
30	3806.54	III
7	3842.46	III
20	3924.47	III
6 h	3947.49	III
6	3963.84	III
5	3981.24	III
5 h	4101.86	III
8	4102.42	III
5 h	4115.50	III
9	4338.50	III
8	4341.40	III
8 h	4377.63	III
6 h	4405.90	III
8 h	4406.72	III
6	4494.05	III
30	4552.62	III
8	4554.00	III
25	4567.82	III
20	4574.76	III
7	4619.66	III
7	4638.28	III
8	4665.87	III
9	4683.02	III
7	4683.80	III
16	4716.65	III
7	4730.52	III
8	4800.43	III
15	4813.33	III
16	4819.72	III
18	4828.97	III
10 h	5091.42	III
7 h	5113.76	III
8 h	5114.12	III
5	5197.26	III
6	5451.46	III
7	5473.05	III
7	5704.60	III
8	5716.29	III
20	5739.73	III
10 h	5898.74	III
7	6314.46	III
6 h	6524.36	III
6 h	6831.56	III
7 h	6851.65	III
5 h	7461.89	III
8 h	7462.62	III
9 h	7466.32	III
12 h	7612.36	III
9 h	8102.86	III
11 h	8103.45	III
7 h	8190.43	III
6 h	8191.16	III
8 h	8191.68	III
9 h	8262.15	III
5 h	8265.64	III
8 h	8269.32	III
5 h	8271.38	III
6 h	8271.94	III

Si IV
Ref. 319 — L.J.R.

Intensity	Wavelength	
	Vacuum	
4	457.82	IV
3	458.16	IV
2	515.12	IV
3	516.35	IV
2	645.76	IV
5	749.94	IV
7	815.05	IV
8	818.13	IV
8	1066.63	IV
8	1122.49	IV
10	1128.34	IV
15	1393.76	IV
12	1402.77	IV
1	1634.61	IV
6	1722.53	IV
5	1727.38	IV

Intensity	Wavelength	
	Air	
3	2120.18	IV
4	2127.47	IV
5 h	2287.04	IV
2 h	2328.56	IV
2	2366.76	IV
3	2370.99	IV
2	2482.82	IV
1	2485.38	IV
7	2517.51	IV
1	2672.19	IV
4	2675.12	IV
4	2675.25	IV
1	2677.57	IV
3 h	2723.81	IV
3 h	2895.13	IV
2 h	2904.47	IV
1 h	2971.52	IV
7	3149.56	IV
9	3165.71	IV
1 h	3244.19	IV
8	3762.44	IV
6	3773.15	IV
1	4031.39	IV
2 h	4038.06	IV
10	4088.85	IV
9	4116.10	IV
7 h	4212.41	IV
3	4314.10	IV
5	4328.18	IV
2 h	4403.73	IV
1 h	4411.65	IV
1 h	4611.27	IV
3 h	4628.62	IV
9 h	4631.24	IV
10 h	4654.32	IV
3 h	4656.92	IV
1 h	4667.14	IV
2 h	4673.30	IV
1 h	4947.45	IV
3	4950.11	IV
2 h	5304.97	IV
1 h	5309.49	IV
5	6667.56	IV
7	6701.21	IV
3 h	6998.36	IV
6 h	7047.94	IV
4 h	7068.41	IV
2 h	7630.50	IV
4 h	7654.56	IV
4 h	7678.75	IV
5 h	7718.79	IV
6 h	7723.82	IV
2 h	7725.64	IV
1 h	7730.47	IV
1 h	7752.91	IV
1 h	8240.61	IV
2 h	8957.25	IV
1 h	9018.16	IV

Si V
Ref. 87 — L.J.R.

Intensity	Wavelength	
	Vacuum	
1	78.61	V
1	78.90	V
2	80.81	V
2	81.11	V
10	85.18	V
6	85.58	V
4	90.45	V
4	90.85	V
15	96.44	V
10	97.14	V
2	98.21	V
20	117.86	V
20	118.97	V

SILVER (Ag)
Z = 47

Ag I and II
Ref. 13, 99, 255, 286, 289 — C.H.C.

Intensity	Wavelength	
	Vacuum	
25	730.83	II

Silver (Cont.)

Intensity		Wavelength	
30		752.80	II
15		1005.32	II
10		1065.49	II
12		1072.23	II
250		1074.22	II
150		1107.03	II
150		1112.46	II
60		1195.83	II
50		1223.33	II
50		1240.80	II
50		1246.87	II
55		1256.81	II
55		1257.55	II
50		1266.63	II
70		1273.67	II
65		1297.51	II
85		1311.20	II
55		1313.81	II
50		1314.61	II
60		1323.84	II
60		1342.09	II
50		1342.57	II
70		1346.62	II
50		1353.54	II
150		1364.50	II
100		1396.00	II
100		1410.93	II
90		1419.72	II
95		1432.60	II
100		1464.72	II
50		1466.23	II
50	r	1507.37	I
100	r	1515.63	I
50	r	1548.58	I
100		1555.16	II
100		1644.50	II
60		1651.52	I
50		1652.10	I
120		1682.82	II
10		1708.11	I
50		1709.27	I
125		1736.44	II
10	h	1766.14	I
75		1790.37	I
20		1847.71	I
100		1967.38	II

Air

Intensity		Wavelength	
150		2015.96	II
150		2033.98	II
200		2061.17	I
100		2069.85	I
80	r	2113.82	II
60		2145.60	II
15		2170.00	I
50		2186.76	II
60		2229.53	II
100	r	2246.43	II
75	r	2248.74	II
75		2280.03	II
30	h	2309.56	I
10	h	2312.60	I
70	r	2317.05	II
80	r	2320.29	II
70	r	2324.68	II
80	r	2331.40	II
70		2357.92	II
50	h	2375.02	I
75		2411.41	II
90	r	2413.23	II
100	r	2437.81	II
80		2447.93	II
80		2473.84	II
60		2506.63	II
50	h	2575.63	I
60		2660.49	II
60		2721.77	I
75		2767.54	II
100	h	2824.39	I
10	h	2926.77	I
20	h	2938.42	I
20		3099.10	I
30	h	3130.02	I
10	h	3170.58	I
90		3180.70	II
15	h	3215.67	I
10		3225.15	I
15		3233.18	I
100		3267.35	II
55000	r	3280.68	I
10		3305.67	I
28000	r	3382.89	I
10	h	3403.78	I
30		3469.16	I
70		3475.82	II
80		3495.28	II
20	h	3501.92	I
20		3508.03	I
15	h	3513.38	I
10		3521.12	I
50		3542.61	I
10	h	3547.16	I
10	h	3557.01	I
20	h	3586.67	I
10	h	3623.49	I
50	h	3624.68	I
75		3682.46	II
30		3682.50	I
80		3683.34	II
50	h	3709.20	I
10	h	3727.42	I
20	h	3753.14	I
200		3810.94	I
50		3811.78	I
100	h	3840.74	I
15		3847.85	I
50	h	3907.41	I
50		3909.31	II
50	h	3914.40	I
70		3920.10	II
10	h	3928.01	I
10	h	3940.43	I
10	h	3942.97	I
60		3949.43	II
100	h	3981.58	I
70		3985.19	II
10	h	3992.15	I
100	h	4055.48	I
10	h	4083.43	I
80		4085.91	II
100		4185.48	II
90	h	4210.96	I
100		4212.82	I
50		4311.07	I
20		4396.23	I
50	h	4476.04	I
20	h	4556.0	I
30	h	4615.69	I
80		4620.04	II
50		4620.46	II
60	h	4668.48	I
30	h	4677.60	I
100		4788.40	II
20	h	4796.2	I
30	h	4847.82	I
100		4874.10	I
20		4888.21	I
10	h	4917.5	I
10		4935.75	I
20	h	4992.89	I
80		5027.35	II
15	h	5123.50	I
1000		5209.08	I
10	h	5333.62	I
1000		5465.50	I
100		5471.55	I
20		5475.38	I
20	h	5545.67	I
10	h	5559.58	I
100		5667.34	I
10	h	6083.78	I
10	h	6268.50	I
20		6621.08	I
20		7359.96	I
320		7687.78	I
25		8005.4	II
15		8254.7	II
500		8273.52	I
20		8324.4	II
15		8379.5	II
25		8403.8	II
15		8492.5	II
30	h	8645.70	I
10	h	8704.85	I
12		8747.6	II
15		9000.9	II
10		12551.0	I
60		16819.5	I
20		17416.7	I
15		18307.9	I
15		18382.3	I

Ag III
Ref. 363, 387, 398 — R.D.C.

Intensity	Wavelength	
	Vacuum	
200	709.80	III
200	713.85	III
100	717.73	III
200	718.53	III
300	726.96	III
350	730.04	III
150	730.28	III
150	730.94	III
200	736.57	III
100	738.13	III
200	740.98	III
200	742.29	III
150	748.30	III
150	755.73	III
100	758.27	III
200	767.19	III
250	768.33	III
150	769.61	III
350	776.38	III
200	782.91	III
150	785.76	III
200	789.08	III
250	792.35	III
200	796.54	III
250	797.91	III
400	799.41	III
300	808.88	III
150	816.12	III
180	822.39	III
200	838.11	III
120	1373.22	III
120	1374.76	III
110	1404.93	III
120	1413.90	III
120	1414.29	III
120	1428.61	III
120	1452.74	III
200	1456.41	III
100	1471.44	III
300	1489.01	III
150	1515.08	III
100	1524.23	III
120	1527.04	III
150	1541.14	III
150	1550.89	III
130	1553.04	III
130	1587.41	III
120	1589.28	III
100	1613.79	III
130	1619.14	III
100	1634.46	III
100	1652.24	III
130	1653.60	III
300	1654.43	III
700	1656.18	III
150	1657.10	III
100	1661.54	III
130	1670.75	III
150	1676.14	III
100	1681.07	III
500	1693.51	III
200	1705.06	III
150	1708.86	III
130	1717.68	III
200	1722.27	III
150	1726.76	III
250	1728.14	III
200	1747.34	III
120	1749.64	III
150	1750.89	III
750	1751.03	III
100	1760.57	III
150	1762.62	III
150	1768.70	III
100	1771.81	III
100	1783.85	III
100	1791.70	III
100	1792.69	III
150	1793.90	III
150	1802.24	III
150	1802.26	III
100	1802.77	III
300	1808.23	III
250	1816.83	III
150	1822.45	III
350	1828.83	III
250	1832.33	III
120	1832.50	III
100	1834.31	III
250	1836.10	III
150	1838.64	III
400	1840.14	III
100	1846.96	III
120	1849.93	III
150	1856.33	III
120	1858.91	III
100	1860.39	III
100	1860.64	III
350	1867.12	III
150	1868.10	III
100	1872.55	III
400	1873.45	III
250	1880.36	III
300	1889.57	III
400	1916.92	III
600	1917.08	III
200	1925.30	III
150	1946.32	III
100	1948.44	III
700	1957.62	III
120	1959.27	III
100	1960.86	III
400	1966.89	III
600	1975.92	III
500	1977.03	III
150	1981.87	III
200	1987.02	III
130	1995.16	III
600	2000.24	III

Air

Intensity	Wavelength	
300	2007.30	III
200	2011.49	III
150	2013.65	III
150	2041.33	III
150	2053.17	III
150	2053.83	III
200	2056.99	III
200	2081.04	III
300	2146.47	III
150	2149.19	III
600	2161.89	III
150	2166.21	III
150	2211.23	III
100	2238.40	III
500	2246.51	III
100	2286.50	III
700	2310.04	III
100	2386.85	III
300	2395.69	III
100	2469.62	III
100	2562.87	III

SODIUM (Na)
Z = 11

Na I and II
Ref. 268, 334 — T.K.M.

Intensity		Wavelength	
		Vacuum	
160		300.15	II
160		300.20	II
90		301.32	II
100		301.44	II
60		302.45	II
300		372.08	II
350		376.38	II
60		1293.97	II
50		1327.74	II
45		1347.54	II
90		1374.69	II
90		1404.68	II
45		1495.21	II
45		1497.73	II
80		1506.41	II
60		1506.91	II
70		1513.10	II
60		1519.63	II
60		1657.92	II
90		1776.57	II
60		1783.04	II
80		1787.19	II
45		1708.85	II
80		1798.41	II
45		1801.26	II
90		1807.09	II
60		1808.38	II
50		1821.70	II
45		1833.87	II
80		1835.22	II
45		1837.89	II
60		1841.82	II
70		1845.02	II
45		1850.15	II
70		1851.19	II
80		1853.17	II
45		1866.45	II
45		1873.37	II
60		1875.08	II
160		1881.91	II
50		1885.09	II
45		1885.74	II

Air

Intensity	Wavelength	
80	2228.53	II
80	2303.58	II
300	2315.65	II
130	2393.28	II
100	2401.01	II
300	2420.99	II
300	2424.73	II
200	2439.14	II
250	2441.50	II
200	2448.72	II
200	2452.18	II
1000	2493.15	II
300	2502.84	II
450	2506.30	II
600	2515.46	II
600	2531.54	II
20	2543.84	I
10	2543.87	II
550	2586.31	II
70	2593.87	I
35	2593.92	II
600	2594.96	II
850	2611.81	II
300	2627.41	II

Sodium (Cont.)

Intensity	Wavelength		Intensity	Wavelength		Intensity	Wavelength		Intensity	Wavelength	
850	2661.00	II	1	4249.41	I	50	11197.21	I	10	1995.68	III
350	2666.46	II	2	4252.52	I	400	11381.45	I			
1000	2671.83	II	15	4273.64	I	1000	11403.78	I		Air	
850	2678.09	II	20	4276.79	I	400	12679.17	I			
200	2680.34	I	2	4287.84	I	60	14767.48	I	10	2004.21	III
100	2680.43	I	3	4291.01	I	100	14779.73	I	11	2005.22	III
650	2808.71	II	250	4292.48	II	60	16373.85	I	11	2008.47	III
850	2809.52	II	250	4292.86	II	100	16388.85	I	15	2011.87	III
600	2829.87	II	250	4308.81	II	400	18465.25	I	11	2014.17	III
800	2839.56	II	250	4309.04	II	50	22056.04	I	12	2017.03	III
1000	2841.72	II	250	4320.91	II	25	22083.67	I	12	2028.56	III
400	2852.81	I	30	4321.40	I	60	23348.41	I	12	2031.13	III
200	2853.01	I	40	4324.62	I	100	23379.13	I	11	2035.90	III
650	2856.51	II	250	4337.29	II				12	2041.66	III
800	2859.49	II	3	4341.49	I		Na III		12	2043.29	III
750	2871.28	II	250	4344.11	II	Ref. 178, 205, 207 — T.K.M.			10	2044.82	III
650	2872.95	II	5	4344.74	I				10	2045.44	III
900	2881.15	II	200	4368.60	II	Intensity	Wavelength		11	2051.48	III
850	2886.26	II	200	4375.22	II		Vacuum		10	2060.36	III
2	2893.62	I	200	4387.49	II				15	2066.60	III
700	2893.95	II	40	4390.03	I	5	183.95	III	13	2082.91	III
900	2901.14	II	250	4392.81	II	5 h	189.35	III	15	2140.72	III
800	2904.72	II	60	4393.34	I	5	193.80	III	14	2144.54	III
1100	2904.92	II	200	4405.12	II	5 h	194.04	III	15	2202.83	III
1100	2917.52	II	5	4419.88	I	5 h	194.17	III	15	2225.93	III
1100	2919.05	II	8	4423.25	I	5	194.29	III	30	2230.33	III
1200	2919.85	II	200	4446.70	II	6	194.68	III	16	2232.19	III
1300	2920.95	II	200	4447.41	II	6	195.53	III	20 h	2246.70	III
1000	2923.49	II	200	4454.74	II	6	202.15	III	14	2251.47	III
750	2930.88	II	200	4455.23	II	6	202.19	III	15	2278.42	III
850	2934.08	II	200	4457.21	II	8	202.49	III	13	2285.66	III
950	2937.74	II	200	4474.63	II	5 d	202.71	III	15	2309.99	III
800	2945.70	II	200	4478.80	II	7 d	202.72	III	18	2386.99	III
950	2947.50	II	200	4481.67	II	8	202.76	III	17	2394.03	III
1200	2951.24	II	200	4490.15	II	8 p	203.06	III	15	2406.59	III
1100	2952.40	II	200	4490.87	II	8	203.28	III	25	2459.31	III
850	2960.12	II	60	4494.18	I	8	203.33	III	18	2468.85	III
500	2970.73	II	100	4497.66	I	10	207.30	III	20	2474.73	III
600	2974.24	II	200	4499.62	II	10 c	215.34	III	25	2497.03	III
750	2974.99	II	200	4506.97	II	12	215.86	III	17	2510.26	III
1000	2977.13	II	200	4519.21	II	12	216.12	III	15	2530.25	III
1100	2979.66	II	200	4524.98	II	15	229.87	III	14	2542.80	III
1100	2980.63	II	200	4533.32	II	12	230.59	III			
1300	2984.19	II	10	4541.63	I	50 c	250.52	III		Na IV	
550	3004.15	II	15	4545.19	I	30	251.37	III	Ref. 206 — T.K.M.		
750	3007.44	II	200	4551.53	II	25	266.90	III			
750	3009.14	II	160	4590.92	II	70	267.65	III	Intensity	Wavelength	
600	3015.40	II	120	4664.811	I	50	267.87	III		Vacuum	
550	3053.67	II	200	4668.560	I	50	268.63	III			
550	3055.35	II	160	4722.23	II	20 p	272.08	III	4	136.551	IV
550	3056.16	II	160	4731.10	II	20	272.45	III	4	136.854	IV
550	3057.38	II	160	4741.67	II	100	378.14	III	4	139.961	IV
550	3057.95	II	20	4747.941	I	70	380.10	III	7	142.232	IV
550	3058.72	II	30	4751.822	I	7	1336.76	III	6	142.359	IV
700	3060.25	II	160	4768.79	II	7	1337.36	III	8	146.064	IV
800	3061.35	II	100	4788.79	II	8	1340.67	III	7	146.302	IV
500	3064.38	II	200	4978.541	I	9 d	1342.39	III	9	150.298	IV
500	3066.22	II	400	4982.813	I	10	1342.73	III	7	150.543	IV
500	3066.54	II	40	5148.838	I	11	1355.29	III	7 c	150.64	IV
550	3074.33	II	80	5153.402	I	12	1361.90	III	8	150.687	IV
550	3078.32	II	100	5191.65	II	11	1372.34	III	7	151.299	IV
550	3080.25	II	80	5208.55	II	10	1420.89	III	7	155.083	IV
550	3087.06	II	70	5400.46	II	10	1444.19	III	7	155.240	IV
550	3092.04	II	90	5414.55	II	12	1449.31	III	7	155.448	IV
550	3092.73	II	280	5682.633	I	11	1562.87	III	8	155.510	IV
650	3094.45	II	70	5688.193	I	10	1565.29	III	8	156.537	IV
650	3095.55	II	560	5688.205	I	10	1598.18	III	12	162.448	IV
500	3103.58	II	80000	5889.950	I	11	1688.94	III	10	163.190	IV
500	3104.40	II	40000	5895.924	I	10	1699.29	III	12	168.411	IV
500	3113.69	II	120	6154.225	I	10	1711.12	III	10	168.546	IV
1700	3124.42	II	240	6160.747	I	11	1728.27	III	10	190.445	IV
600	3125.21	II	60	6175.25	II	10	1731.11	III	10	199.772	IV
600	3129.38	II	70	6199.26	II	10	1755.48	III	10 c	205.49	IV
2500	3135.48	II	70	6234.68	II	15	1807.07	III	10	319.644	IV
1700	3137.86	II	80	6260.01	II	10	1810.77	III	10	360.76	IV
950	3145.71	II	80	6274.74	II	11	1811.67	III	12	408.684	IV
2000	3149.28	II	70	6361.15	II	10	1816.81	III	10	409.614	IV
2000	3163.74	II	70	6366.41	II	10 d	1835.22	III	15	410.372	IV
700	3175.09	II	90	6514.21	II	10	1838.94	III	10	411.334	IV
1000	3179.06	II	80	6524.68	II	11	1844.36	III	13	412.242	IV
1700	3189.79	II	130	6530.70	II	12 d	1847.53	III	10	1580.50	IV
1600	3212.19	II	130	6544.04	II	10 d	1847.59	III	11	1582.18	IV
700	3234.93	II	130	6545.75	II	15	1849.56	III	10	1582.33	IV
1500	3257.96	II	80	6552.43	II	12	1850.38	III	11 d	1583.98	IV
650	3260.21	II	20	7373.23	I	10	1855.92	III	12	1584.14	IV
950	3274.22	II	10	7373.49	I	10	1856.71	III	10 d	1586.99	IV
1700	3285.60	II	50	7809.78	I	10	1861.21	III	12 d	1587.05	IV
1700	3301.35	II	25	7810.24	I	10	1880.66	III	10	1613.95	IV
1200	3302.37	I	4400	8183.256	I	10 d	1887.39	III	11	1615.92	IV
600	3302.98	II	800	8194.790	I	20 d	1887.47	III	12	1618.57	IV
1500	3304.96	II	8800	8194.824	I	15 d	1890.75	III	11	1655.47	IV
1000	3318.04	II	100	8649.92	I	15	1900.16	III	15 c	1701.97	IV
950	3327.69	II	60	8650.89	I	10	1918.45	III	10	1702.41	IV
50	3426.86	I	25	8942.96	I	11	1923.96	III	11	1960.76	IV
1500	3533.05	II	40	9153.88	I	14	1926.26	III	11	1965.08	IV
1200	3631.27	II	60	9465.94	I	12	1927.24	III	10	1967.60	IV
850	3711.07	II	80	9961.28	I	12	1932.74	III			
300	4113.70	II	20	10566.00	I	13	1933.89	III		Air	
250	4123.08	II	60	10572.28	I	10	1943.52	III			
250	4233.26	II	200	10746.44	I	12	1946.43	III	10	2018.39	IV
6	4238.99	I	80	10749.29	I	12	1950.91	III	12 d	2106.33	IV
250	4240.90	II !	120	10834.87	I	14	1951.24	III	10	2114.53	IV
10	4242.08	I	35	11190.19	I	10	1977.16	III			
						13	1985.57	III			

Sodium (Cont.)

Intensity	Wavelength	
10	2155.76	IV

Na V
Ref. 299 — T.K.M.

Intensity	Wavelength	
	Vacuum	
100	106.28	V
100	106.30	V
100	106.40	V
100	106.49	V
200 c	107.93	V
200	108.02	V
200 c	110.82	V
200	110.88	V
100	111.51	V
300 c	112.01	V
100 h	114.70	V
100	114.74	V
100	117.99	V
400	117.99	V
100	120.04	V
400	125.18	V
400	125.22	V
500	125.29	V
300	125.43	V
300	125.46	V
200	125.90	V
100	126.21	V
200	126.56	V
100	126.61	V
400	127.44	V
400	127.47	V
400	128.03	V
400	128.05	V
200	130.68	V
300	131.35	V
200	131.41	V
300 h	131.64	V
500	133.16	V
400	133.39	V
200	134.27	V
300	135.79	V
300	135.85	V
200	138.81	V
300	138.92	V
400	148.64	V
300	148.86	V
400	151.13	V
300	157.21	V
300	163.62	V
800	307.15	V
1000	308.26	V
800	332.55	V
900	333.91	V
800	360.32	V
800	360.37	V
1000	400.72	V
500	445.05	V
600	445.19	V
600	459.90	V
850	461.05	V
1000	463.26	V

STRONTIUM (Sr)
Z = 38

Sr I and II
Ref. 1, 218, 279, 313 — J.J.W.

Intensity	Wavelength	
	Air	
1400	2152.84	II
1400	2165.96	II
160	2428.10	I
120	2569.47	I
200	2931.83	I
300	3301.73	I
300	3329.99	I
400	3351.25	I
300	3366.33	I
650	3380.71	II
950	3464.46	II
120	3474.89	II
300 h	3940.80	I
600	3969.26	I
300	3970.04	I
1300	4030.38	I
300	4032.38	I
46000	4077.71	II
200	4161.80	II
32000	4215.52	II
340	4305.45	II
350 h	4438.04	I
	4526.10	II
	4585.91	II
65000	4607.33	I
3200	4722.28	I

Intensity	Wavelength	
2200	4741.92	I
1400	4784.32	I
4800	4811.88	I
3600	4832.08	I
500	4855.04	I
600	4868.70	I
3000	4872.49	I
600	4876.06	I
2000	4876.32	I
1000	4891.98	I
8000	4962.26	I
1300	4967.94	I
800 h	5156.07	I
1400	5222.20	I
2000	5225.11	I
2000	5229.27	I
2800	5238.55	I
4800	5256.90	I
	5303.13	II
350 h	5329.82	I
	5379.13	II
	5385.45	II
1500	5450.84	I
7000	5480.84	I
1100	5486.12	I
3500	5504.17	I
2600	5521.83	I
2000	5534.81	I
2000	5540.05	I
250 h	5543.36	I
	5622.94	II
	5650.54	II
	5723.70	II
	5819.00	II
200 h	5970.10	I
250 h	6345.75	I
250 h	6363.94	I
350 h	6369.96	I
1000	6380.75	I
900 h	6386.50	I
600 h	6388.24	I
9000	6408.47	I
250	6446.68	I
250 h	6465.79	I
	6483.17	II
5500	6504.00	I
	6509.20	II
1000	6546.79	I
1700	6550.26	I
3000	6617.26	I
800	6643.54	I
1800	6791.05	I
4800	6878.38	I
1200	6892.59	I
5500	7070.10	I
60	7153.09	I
250 h	7167.24	I
200	7232.27	I
2500	7309.41	I
500	7621.50	I
400 h	7673.06	I
50 hL	7850.00	I
30 h	7866.90	I
20 hL	7874.00	I
200 h	8422.80	I
120	8505.69	II
200	8688.91	II
30	8719.56	II
40	9170.00	I
30	9204.50	I
20	9283.90	I
100	9294.10	I
15	9306.60	I
30	9319.20	I
60	9380.45	I
40 h	9411.25	I
400 h	9448.95	I
600	9596.00	I
300	9624.70	I
100	9638.10	I
100 h	9647.70	II
300	10036.66	II
1000	10327.31	II
7	10872.70	II
200	10914.88	II
10	10984.00	I
13	11224.57	II
700	11241.25	I
100	12014.76	II
20	12236.20	I
60	12445.90	II
20	12479.60	I
40	12495.00	I
15	12652.20	I
75	12974.70	I
100	13123.80	II
15	13522.80	I
15	17140.90	I
30	17170.50	I
50	17447.40	I
4	17626.00	I
30	17743.00	I
15	19759.60	I
230	20261.40	I

Intensity	Wavelength	
120	20700.70	I
40	20764.50	I
15	20778.70	I
30	26023.60	I

Sr III
Ref. 231, 265 — J.J.W.

Intensity	Wavelength	
	Vacuum	
25	307.18	III
50	316.11	III
50	321.61	III
125	330.67	III
500	351.62	III
75	358.80	III
250	363.49	III
150	371.21	III
1000	437.24	III
1875	491.79	III
1250	507.04	III
3750	514.38	III
2500	562.75	III
20	968.37	III
20	975.78	III
25	992.98	III
50	1025.23	III
35	1044.91	III
20	1057.74	III
25	1060.20	III
20	1098.77	III
35	1125.49	III
20	1140.24	III
20	1168.27	III
20	1182.09	III
50	1236.23	III
20	1940.58	III
30 p	1958.44	III
30	1966.92	III
	Air	
25	2068.63	III
50	2099.59	III
25	2114.31	III
30	2118.48	III
50	2119.52	III
50	2133.12	III
30	2142.80	III
20	2145.74	III
30	2178.91	III
30	2180.14	III
50	2190.88	III
50	2203.86	III
50	2219.50	III
50	2220.05	III
50	2267.03	III
100	2273.71	III
50	2277.87	III
30	2310.33	III
50	2314.95	III
50	2334.79	III
100	2340.13	III
50	2404.17	III
30	2410.52	III
50	2454.03	III
100	2486.52	III
50	2503.59	III
30	2599.10	III
35	2622.69	III
30	2642.96	III
30	2648.51	III
35	2654.66	III
40	2722.47	III
50	2786.00	III
50	2821.42	III
30	2874.86	III
30	2929.34	III
30	2983.00	III
100	3002.61	III
200	3012.32	III
100	3021.73	III
30	3059.83	III
50	3061.43	III
30	3104.25	III
50	3182.61	III
100	3235.39	III
30	3302.72	III
50	3430.76	III
30	3874.26	III
30	3936.40	III
30	3936.72	III
30	3958.75	III
30	4094.03	III
30	4097.02	III
30	4105.63	III
35	4335.80	III
30	5071.09	III
30	5130.34	III
35	5158.26	III
40	5257.71	III
30	5262.21	III

Intensity	Wavelength	
30	5288.32	III
30	5391.03	III
40	5443.48	III
30	5463.90	III
30	5664.66	III
30	5689.72	III

Sr IV
Ref. 110 — J.J.W.

Intensity	Wavelength	
	Vacuum	
12	284.31	IV
12	291.09	IV
12	291.19	IV
12	293.22	IV
15	298.12	IV
15	300.12	IV
12	300.27	IV
12	301.67	IV
20	378.53	IV
75	392.44	IV
50	393.00	IV
45	394.90	IV
50	396.22	IV
40	399.92	IV
35	403.85	IV
35	406.94	IV
30	412.93	IV
40	413.07	IV
40	415.32	IV
30	419.78	IV
25	430.21	IV
30	430.65	IV
25	442.73	IV
25	471.76	IV
25	484.20	IV
25 p	508.14	IV
25	534.19	IV
200	664.43	IV
100	710.35	IV
20	1189.21	IV
30	1244.14	IV
20 p	1244.75	IV
20 p	1244.87	IV
20 p	1257.78	IV
20	1268.62	IV
20	1331.13	IV
30	1347.90	IV
20	1361.15	IV
25	1408.67	IV
20	1592.74	IV
25	1677.03	IV
20	1705.16	IV
20	1724.23	IV
25	1729.53	IV
20	1732.12	IV
20	1777.25	IV
20	1994.61	IV
	Air	
20	2104.38	IV
20	2117.90	IV
20	2217.99	IV
20	2230.41	IV
20	2240.49	IV
20	2253.38	IV
50	2346.97	IV
20	2357.34	IV
20	2438.93	IV
25	2441.41	IV
30	2482.79	IV
25	2483.57	IV
18	2500.57	IV
20	2508.02	IV
20	2534.03	IV
18	2548.02	IV
40	2555.60	IV
40	2571.04	IV
25	2571.58	IV
15	2589.34	IV
25	2620.35	IV
20	2621.16	IV
20	2642.16	IV
15	2830.53	IV
9	2934.60	IV
10	3019.29	IV
9	3266.52	IV
9	3566.43	IV
9	3741.05	IV
9	4298.57	IV
9	4685.08	IV

Sr V
Ref. 109 — J.J.W.

Intensity	Wavelength	
	Vacuum	

Strontium (Cont.)

Intensity	Wavelength	
10	517.28	V
6	540.51	V
25	578.01	V
30	624.93	V
25	642.23	V
50	649.21	V
20	659.15	V
25	660.94	V
9	669.93	V
35	686.23	V
6	715.79	V
12	747.82	V
9	862.32	V

SULFUR (S)
Z = 16

S I and II
Ref. 144, 209, 210, 266—R.L.K.

Intensity	Wavelength	
	Vacuum	
40	906.9	II
40	910.5	II
40	912.7	II
40	937.4	II
40	937.7	II
20	996.0	II
20	1000.5	II
20	1014.4	II
20	1019.5	II
20	1096.6	II
40	1102.3	II
20	1131.0	II
20	1131.6	II
40	1234.1	II
40	1250.5	II
110	1253.8	II
110	1259.5	II
275	1270.782	I
250	1277.216	I
280	1295.653	I
275	1302.337	I
235	1302.863	I
235	1303.110	I
245	1303.430	I
260	1305.883	I
265	1310.194	I
355	1316.542	I
290	1316.618	I
375	1323.515	I
355	1326.643	I
775	1381.552	I
710	1385.510	I
960	1388.435	I
640	1389.154	I
775	1392.588	I
1000	1396.112	I
300	1409.337	I
510	1425.030	I
425	1433.280	I
300	1436.968	I
300	1448.229	I
425	1472.972	I
550	1473.995	I
300	1474.380	I
355	1481.665	I
485	1483.039	I
300	1483.233	I
330	1485.622	I
390	1487.150	I
680	1666.688	I
640	1687.530	I
710	1807.311	I
680	1820.343	I
640	1826.245	I
710	1900.286	I
550	1914.698	I
	Air	
20	2629.1	II
40	2670.0	II
40	2847.7	II
285	3867.6	II
285	3902.0	I
360	3933.3	II
450	4120.8	I
280	4142.3	II
360	4145.1	II
450	4153.1	II
450	4162.7	II
450	4694.1	I
285	4695.4	I
160	4696.2	I
280	4716.2	II
450	4815.5	II
360	4924.1	II
450	4925.3	II
285	4993.5	I
360	5428.6	II
650	5432.8	II
1000	5453.8	II
1000	5473.6	II
1000	5509.7	II
280	5564.6	II
1000	5606.1	II
450	5640.0	II
450	5640.3	II
280	5647.0	II
650	5659.9	II
450	5664.7	II
160	5706.1	I
450	5819.2	II
450	6052.7	I
280	6286.4	II
450	6287.1	II
450	6305.5	II
450	6312.7	II
280	6384.9	II
280	6397.3	II
280	6398.0	II
360	6413.7	II
160	6743.6	I
285	6748.8	I
450	6757.2	I
450	7579.0	I
450	7629.8	I
285	7686.1	I
450	7696.7	I
1000	7924.0	I
160	7928.8	I
285	7930.3	I
450	7931.7	I
450	7967.4	I
450	7967.4	II
450	8314.7	I
450	8314.0	II
450	8585.6	I
285	8680.5	I
450	8694.7	I
360	8874.5	I
110	8882.5	I
220	8884.2	I
160	9035.9	I
450	9212.9	I
450	9228.1	I
450	9237.5	I
285	9413.5	I
285	9421.9	I
285	9437.1	I
650	9649.9	I
450	9672.3	I
450	9680.8	I
450	9693.7	I
285	9697.3	I
285	9739.7	I
110	9741.9	I
285	9932.3	I
285	9949.8	I
285	9958.9	I
285	10455.5	I
70	10456.8	I
285	10459.5	I

S III
Ref. 209, 210 — R.L.K.

Intensity	Wavelength	
	Vacuum	
70	729.5	III
110	732.42	III
70	735.2	III
70	738.5	III
70	789.0	III
70	796.7	III
70	824.9	III
70	836.3	III
285	1077.1	III
70	1194.0	III
70	1201.0	III
	Air	
110	2460.5	III
110	2489.6	III
160	2496.2	III
160	2499.1	III
220	2508.2	III
70	2636.9	III
220	2665.4	III
70	2680.5	III
110	2691.8	III
110	2702.8	III
220	2718.9	III
110	2721.4	III
220	2726.8	III
220	2731.1	III
110	2741.0	III
285	2756.9	III
110	2775.2	III
160	2785.5	III
70	2797.4	III
70	2856.0	III
110	2863.5	III
160	2904.3	III
70	2964.8	III
160	2986.0	III
70	3234.2	III
70	3324.9	III
110	3497.3	III
160	3632.0	III
70	3662.0	III
110	3709.4	III
160	3717.8	III
160	3838.3	III
160	3928.6	III
360	4253.6	III
110	4285.0	III
70	4332.7	III

S IV
Ref. 29, 202, 209 — R.L.K.

Intensity	Wavelength	
	Vacuum	
20	519.3	IV
20	520.1	IV
40	520.8	IV
20	522.0	IV
20	522.5	IV
20	551.2	IV
40	652.5	IV
40	653.0	IV
70	653.6	IV
40	654.0	IV
70	655.6	IV
20	655.9	IV
110	657.3	IV
40	660.9	IV
160	661.4	IV
40	663.7	IV
40	664.8	IV
70	666.1	IV
110	744.9	IV
110	748.4	IV
110	750.2	IV
110	753.8	IV
40	798.3	IV
70	800.5	IV
70	804.0	IV
70	809.7	IV
110	816.0	IV
160	1062.7	IV
160	1073.0	IV
70	1073.5	IV
110	1108.4	IV
20	1110.9	IV
20	1624.0	IV
20	1629.2	IV
	Air	
20	2387.0	IV
40	2398.9	IV
110	3097.5	IV
40	3117.7	IV

S V
Ref. 29 — R.L.K.

Intensity	Wavelength	
	Vacuum	
5	437.4	V
5	438.2	V
5	439.6	V
40	658.3	V
70	659.8	V
110	663.2	V
5	676.2	V
5	677.3	V
20	678.1	V
40	680.3	V
110	680.9	V
40	681.6	V
5	686.2	V
5	686.9	V
5	689.8	V
5	691.7	V
20	693.5	V
285	786.5	V
160	849.2	V
110	852.2	V
220	854.8	V
110	857.9	V
110	860.5	V
20	883.6	V
20	884.5	V
5	885.8	V
20	900.9	V
5	902.8	V
20	905.9	V

TANTALUM (Ta)
Z = 73

Ta I and II
Ref. 1 — C.H.C.

Intensity		Wavelength	
		Air	
1100		2140.13	II
1500		2146.87	II
740		2150.62	II
600		2165.01	II
740		2178.03	II
1200		2182.71	II
540		2193.20	II
1100		2193.88	II
1500		2196.03	II
1500		2199.67	II
500		2207.14	II
1400	d	2210.03	II
		2210.19	II
420		2215.60	II
1400		2239.48	II
240		2248.48	II
480		2249.79	II
1200		2250.76	II
260		2254.86	II
440		2255.77	II
360		2256.51	II
500		2258.71	II
840		2261.42	II
260		2261.62	II
990		2262.30	II
220		2269.56	II
740		2271.85	II
990		2272.59	II
200		2279.85	I
320		2282.19	II
130		2285.02	II
790		2285.25	II
600		2286.59	II
240		2287.27	II
990		2289.16	II
180		2292.54	II
160		2295.18	
160		2301.47	II
440		2302.24	II
440		2302.93	II
300		2303.49	II
100		2308.46	II
440		2312.60	II
420		2315.46	II
260		2319.16	II
100		2331.29	II
690		2331.98	II
550		2332.19	II
110		2334.13	II
180		2334.88	II
140		2335.75	II
300		2338.28	II
200		2340.94	II
200		2341.61	II
130		2343.64	II
100		2346.42	II
90		2351.99	II
170		2353.86	II
120		2355.22	II
170		2356.05	II
140		2356.90	II
250		2357.30	I
170		2359.16	II
260		2361.09	I
160		2362.78	II
130		2363.32	II
600		2364.24	II
50		2367.24	II
150		2369.32	II
300		2370.76	II
320		2371.58	I
100		2373.94	II
70		2375.91	I
150		2378.31	II
440		2381.13	II
240		2381.52	II
170		2383.72	II
240		2384.28	II
130		2385.73	I
1400		2387.06	II
80		2388.37	II
160		2389.11	II
70		2396.30	I
110		2399.15	I
20		2399.92	II
2400		2400.63	II
140		2402.13	II
100		2403.68	II
130		2406.55	I

Tantalum (Cont.)

Int		λ		Int		λ		Int		λ		Int		λ	
130		2408.26	II	470		2720.76	I	100		3885.20	I	90		5901.91	I
120		2414.32	I	470		2727.44	II	210		3918.51	I	30		5916.51	I
240		2415.21	II	410		2727.78	I	140		3922.78	I	90		5918.95	I
320		2416.89	II	310		2736.25	II	140		3922.92	I	15		5925.90	I
220		2417.86	II	210		2739.26	II	210		3970.10	I	15		5930.62	I
150		2418.77	II	210		2743.59	I	210		3996.17	I	23		5931.05	I
140		2421.03	I	510		2746.68	I	100		3999.28	I	20		5931.68	I
150		2421.85	II	1200		2748.78	I	190		4006.84	I	18		5935.54	I
170		2423.48	II	860		2749.83	I	190		4026.94	I	130		5939.76	I
130		2425.91	II	410		2752.49	II	140		4029.94	I	240		5944.02	I
360		2427.64	I	1000		2758.31	I	120		4040.87	I	25		5951.78	I
360		2429.71	II	430		2761.68	II	410		4061.40	I	18		5960.13	I
170		2431.06	II	770		2775.88	I	210		4064.63	I	190	c	5997.23	I
480		2432.70	II	390		2787.69	I	100		4067.24	I	25	h	6009.89	I
130		2433.59	II	680		2796.34	I	310		4067.91	I	25		6015.90	I
130		2436.51	II	680		2797.76	II	120		4105.02	I	100		6020.72	I
110		2437.07	I	380		2802.07	I	210		4129.38	I	250		6045.39	I
110		2438.64	II	430		2806.30	I	230		4136.20	I	100		6047.25	I
200		2439.91	I	510		2806.58	I	230		4147.89	I	25		6053.70	I
130		2442.39	I	260		2817.10	II	210		4175.21	I	30		6090.82	I
100		2444.13	II	260		2842.82	I	100		4177.92	I	18		6092.06	I
100		2447.17	I	640		2844.25	I	130		4181.15	I	100		6101.58	I
100		2454.48	I	290		2844.46	II	300		4205.88	I	25		6140.07	I
100		2458.68	I	290	c	2845.35	I	120		4206.40	I	65		6144.56	I
100		2460.55	I	560		2848.52	I	130		4245.35	I	30		6152.54	I
160		2463.82	II	1500		2850.49	I	130		4268.26	I	130		6154.50	I
130		2466.99	II	1900		2850.98	I	160	c	4302.98	I	40		6158.84	I
130		2467.37	II	220		2858.44	II	110		4355.14	I	15		6170.46	I
380		2470.90	II	360		2861.98	I	100		4378.82	I	15		6189.66	I
120		2471.38	I	310		2868.65	I	150		4386.07	I	15		6193.11	I
120		2472.13	I	470		2871.42	I	110		4398.45	I	25		6208.37	I
150		2473.13	I	270		2873.36	I	180		4402.50	I	40		6249.79	I
120		2473.31	II	260		2873.56	I	130		4415.74	I	150		6256.68	I
600		2474.62	I	210		2874.17	I	360	c	4510.98	I	150		6268.70	I
120		2475.33	I	380		2880.02	I	190		4530.85	I	50		6278.34	I
200		2476.67	II	770		2891.84	I	130		4551.95	I	65		6281.33	I
150		2478.22	I	260		2899.04	I	170		4565.85	I	15		6287.36	I
120		2481.86	II	560		2902.05	I	340		4574.31	I	40		6287.91	I
100		2482.10	I	210		2914.12	I	260		4619.51	I	40		6289.34	I
100		2482.58	II	310		2915.49	I	130		4669.14	I	50		6309.06	I
100		2484.04	II	410		2925.19	I	450		4681.88	I	150		6309.58	I
500		2484.95	I	310		2932.70	I	130		4691.90	I	25	c	6312.22	I
120		2486.70	I	1700		2933.55	I	150		4740.16	I	75		6325.08	I
600		2488.70	II	470		2940.06	I	220		4756.51	I	50		6332.91	I
500		2490.46	I	1200		2940.22	I	120		4768.98	I	65		6341.17	I
600		2504.45	I	240		2942.14	I	220		4812.75	I	30		6346.02	I
600		2507.45	I	510		2951.92	I	110		4920.11	I	75		6356.16	I
240		2512.65	I	340		2953.56	I	100		4921.27	I	65		6360.84	I
1200	d	2526.35	I	1500		2963.32	I	110		4926.00	I	40		6373.06	I
600		2532.12	II	770		2965.13	II	150		4936.42	I	15		6379.07	I
240		2545.49	II	770		2965.54	I	200		5037.37	I	90		6389.45	I
240		2546.80	I	340		2969.47	I	100		5067.87	I	23	h	6392.21	I
460	d	2551.07	I	430		2975.56	I	110		5115.84	I	65		6428.60	I
460		2554.62	II	210		3011.88	I	100		5141.62	I	250		6430.79	I
240		2555.05	I	1800		3012.54	II	100		5143.69	I	13		6437.36	I
1200		2559.43	I	290	d	3027.48	I	330		5156.56	I	40		6444.61	I
460		2562.10	I	290		3042.06	II	110		5212.74	I	30		6445.87	I
340		2571.51	II	530		3049.56	I	110	d	5218.45	I	200		6450.36	I
430		2573.54	I	530		3069.24	I			5218.66	I	20		6455.83	I
390		2573.79	I	360		3077.24	I	140		5341.05	I	30		6459.92	I
600		2577.37	II	560		3103.25	I	200		5402.51	I	380		6485.37	I
340		2577.78	I	380		3124.97	I	130		5419.19	I	18	h	6502.43	I
210		2580.16	I	380		3130.58	I	18		5500.68	I	65		6505.52	I
340		2584.03	II	270		3132.64	I	20		5505.66	I	100		6514.39	I
430		2593.08	I	320		3170.29	I	15	c	5516.27	I	100		6516.10	I
410		2593.66	II	270		3173.59	I	90		5518.91	I	25	h	6561.60	I
310		2594.25	II	200		3176.29	I	9		5521.15	I	25	Cw	6564.26	I
560		2595.26	I	600		3180.95	I	10		5523.98	I	100		6574.84	I
310	l	2596.45	II	240		3184.55	I	13		5528.36	I	10		6585.13	I
220		2600.14	I	200		3198.67	I	10	l	5545.20	I	15	c	6587.16	I
600		2603.49	II	200		3213.91	II	20		5548.32	I	110		6611.95	I
1400		2608.63	I	300		3223.83	I	15		5584.02	I	75		6621.30	I
210		2609.00	I	230		3229.24	I	75		5598.75	I	15	Cw	6662.24	I
310	d	2611.34	I	200		3242.05	I	15	Cw	5599.52	I	100		6673.73	I
340		2615.46	I	200		3242.83	I	100		5605.50	I	180		6675.53	I
310		2615.66	I	210		3274.95	II	9	c	5617.71	I	30		6684.00	I
1200		2635.58	II	1100		3311.16	I	40		5620.68	I	15		6693.61	I
470		2636.67	I	210		3317.93	I	13		5628.20	I	15		6706.46	I
860		2636.90	I	680		3318.84	I	20		5635.71	I	25		6709.39	I
510		2646.22	I	330	d	3330.99	II	40		5640.18	I	10		6714.44	I
600		2646.37	I	230		3358.47	I	150		5645.91	I	15	h	6723.61	I
2400		2647.47	I	640		3371.54	I	130		5664.90	I	75	c	6740.73	I
270		2651.22	II	360		3385.05	I	30		5688.25	I	40		6754.91	I
2600		2653.27	I	230		3398.33	I	40		5699.24	I	13	c	6755.85	I
1900		2656.61	I	450		3406.94	I	15		5704.31	I	13		6770.37	I
1500		2661.34	I	230		3463.77	I	25		5706.28	I	75		6771.74	I
220		2665.60	II	490		3480.52	I	30		5715.24	I	40	Cw	6774.25	I
220		2668.07	I	380		3497.85	I	8		5716.53	I	40	Cw	6788.99	I
600		2668.62	I	240		3503.87	I	23		5746.71	I	13	c	6790.06	I
770		2675.90	II	490		3511.04	I	30		5755.81	I	13		6799.27	I
270		2680.06	I	200		3513.61	I	15	h	5761.61	I	40	c	6810.46	I
220		2680.66	II	750		3607.41	I	25		5766.56	I	160	c	6813.25	I
600		2684.28	I	980		3626.62	I	30	c	5767.91	I	20		6819.36	I
1500		2685.17	II	500		3642.06	I	10		5771.93	I	18	c	6824.96	I
340		2691.31	I	100		3686.18	I	130		5776.77	I	13		6832.00	I
260		2692.40	I	100		3689.73	I	25		5780.02	I	15		6850.83	I
470		2694.52	II	130		3731.02	I	90		5780.71	I	15		6865.13	I
240		2696.81	I	140		3736.76	I	130		5811.10	I	210		6866.23	I
1000		2698.30	I	130		3746.36	I	25		5816.51	I	180		6875.27	I
470		2706.69	I	110		3754.52	I	45	c	5843.94	I	40		6877.49	I
310		2709.27	II	110		3777.10	I	13		5849.68	I	15		6896.77	I
1200		2710.13	I	110		3792.02	I	15		5866.61	I	40		6900.55	I
2600		2714.67	I	210		3833.74	II	240		5877.36	I	150		6902.10	I
240		2717.18	I	100		3848.05	I	130		5882.30	I	140		6927.38	I

Tantalum (Cont.)

Intensity		Wavelength	Spectrum
140		6928.54	I
8	h	6939.33	I
20	c	6946.87	I
65		6951.26	I
45		6953.88	I
180		6966.13	I
8		6969.49	I
8	c	6971.31	I
9		6971.53	I
23		6983.52	I
110	d	6995.22	I
		6995.49	I
20		7000.21	I
40		7005.07	I
75		7006.96	I
50		7025.03	I
13		7031.51	I
40		7039.07	I
15	h	7081.30	I
20		7085.40	I
23		7093.02	I
8		7108.05	I
15	c	7117.52	I
20		7121.27	I
40		7125.72	I
150		7148.63	I
110		7172.90	I
13	c	7174.91	I
13		7191.35	I
8		7233.45	I
30	h	7250.27	I
11	h	7264.82	I
6		7272.29	I
30		7276.96	I
5		7277.54	I
9		7286.36	I
13	c	7296.32	I
140		7301.74	I
20		7319.84	I
11	c	7322.17	I
13		7325.95	I
11	Cw	7340.19	I
160		7346.41	I
140	c	7352.86	I
100		7356.96	I
90	Cw	7369.09	I
160		7407.89	I
11	c	7435.19	I
23		7440.17	I
30		7467.75	I
23		7486.01	I
30		7520.56	I
6		7569.23	I
9		7590.22	I
6		7649.62	I
11	h	7722.02	I
11		7763.11	I
9	c	7779.67	I
20	c	7842.76	I
100		7882.37	I
30		7950.19	I
5		7952.07	I
6		7998.75	I
6		8022.09	I
75		8026.50	I
5		8029.04	I
15		8039.08	I
8		8053.93	I
15		8068.98	I
5		8100.11	I
13		8128.76	I
9		8158.54	I
5	c	8180.74	I
13	c	8248.95	I
20	d	8264.85	I
75		8281.62	I
11		8389.06	I
5		8415.73	I
25	Cw	8447.62	I
11	h	8550.49	I
15		8575.92	I
10	Cw	8595.84	I

Ta IV
Ref. 411 — R.L.K.

Intensity	Wavelength	
	Vacuum	
10	763.14	IV
32	934.41	IV
67	999.34	IV
65	1063.53	IV
68	1067.17	IV
71	1074.47	IV
71	1086.39	IV
72	1094.60	IV
68	1100.13	IV
79	1116.10	IV
71	1118.83	IV
78	1136.17	IV
66	1138.26	IV
75	1149.72	IV
75	1150.42	IV
75	1151.92	IV
76	1172.51	IV
85	1175.51	IV
80	1189.28	IV
78	1192.52	IV
80	1192.67	IV
78	1211.94	IV
80	1212.68	IV
85	1213.09	IV
70	1214.66	IV
85	1215.53	IV
80	1220.73	IV
80	1220.96	IV
90	1223.73	IV
88	1238.13	IV
95	1240.06	IV
87	1258.34	IV
94	1264.91	IV
98	1272.42	IV
94	1275.48	IV
86	1275.94	IV
92	1308.51	IV
85	1311.35	IV
87	1315.58	IV
81	1325.19	IV
92	1332.38	IV
86	1343.30	IV
75	1350.46	IV
92	1365.88	IV
79	1376.62	IV
78	1388.80	IV
91	1398.78	IV
93	1413.40	IV
79	1430.11	IV
83	1441.54	IV
91	1454.32	IV
92	1464.41	IV
93	1469.82	IV
90	1495.25	IV
95	1514.19	IV
70	1525.69	IV
82	1565.97	IV
82	1584.64	IV
84	1594.91	IV
85	1607.70	IV
84	1631.65	IV
79	1639.82	IV
84	1668.76	IV
84	1676.45	IV
85	1712.16	IV
85	1716.13	IV
82	1753.90	IV
82	1759.04	IV
79	1763.03	IV
83	1865.92	IV
84	1901.63	IV
84	1907.66	IV
84	1924.75	IV
77	1940.25	IV
79	1985.68	IV
82	1989.44	IV
	Air	
85	2055.75	IV
83	2079.27	IV
75	2111.53	IV
90	2199.58	IV
90	2207.64	IV
68	2697.42	IV
10	3076.06	IV

Ta V
Ref. 426 — R.L.K.

Intensity	Wavelength	
	Vacuum	
20	478.29	V
60	493.07	V
200	841.31	V
1000	890.87	V
500	947.30	V
100	990.29	V
200	1066.64	V
200	1140.49	V
500	1213.42	V
100	1242.98	V
5000	1392.56	V
7000	1709.10	V

TECHNETIUM (Tc)
Z = 43
Tc I and II
Ref. 35 — C.H.C.

Intensity		Wavelength	
		Air	
15		2106.23	II
20		2116.44	II
15		2119.41	I
30		2156.27	I
30		2185.39	I
30		2189.06	I
40		2193.35	I
10		2266.22	II
10		2282.12	I
10		2282.71	I
50		2285.45	I
100		2298.08	I
30		2416.22	I
50		2423.23	I
20		2424.54	I
20		2435.82	I
10		2436.99	I
80	w	2463.69	I
20		2465.09	I
30		2466.87	I
20		2475.11	I
50		2480.70	I
50		2483.22	I
20		2486.50	I
25		2492.72	I
20		2493.43	I
100		2496.77	II
30		2510.17	I
80		2529.34	II
500		2543.23	II
60		2544.81	II
50		2547.92	II
50		2558.61	II
50		2567.01	II
30		2575.06	II
80		2576.28	II
40		2577.86	II
300	H1	2578.79	I
200		2589.86	I
20	w	2590.19	I
100		2592.82	I
20		2597.19	II
500		2608.86	I
1000	c	2609.99	II
1500		2614.23	I
1000		2615.87	I
30		2618.28	I
200		2634.91	II
80		2636.36	I
30		2641.26	I
100		2642.37	I
40		2644.50	II
1000	c	2647.01	II
300	c	2649.21	I
100		2652.35	II
30		2653.57	I
100		2654.31	I
120		2660.88	I
100		2662.30	I
80		2681.19	II
60		2683.14	I
80		2683.89	I
80		2693.11	I
50		2696.64	I
70		2702.27	I
40		2702.96	II
100		2707.90	II
1000		2708.78	I
30		2715.20	I
30		2723.55	I
1000		2726.69	I
30	c	2728.47	I
500		2730.53	I
300		2732.87	I
150		2736.23	I
60	c	2736.83	II
100		2737.97	I
20	c	2738.83	II
100		2755.76	I
100		2762.13	I
200		2762.34	I
60		2765.95	I
500		2766.89	I
20		2777.31	II
150		2778.91	I
25		2781.22	I
1000		2782.05	I
500		2785.59	I
40		2788.89	I
500		2789.25	I
100		2794.53	I
80		2795.65	I
200		2795.78	II
1000		2802.81	I
150		2803.02	I
500		2808.36	I
50	c	2809.65	II
500		2811.61	II
30		2814.86	I
40		2819.46	I
100		2821.35	II
200		2828.04	I
60		2831.18	II
50		2840.38	II
60		2845.04	I
10		2846.39	II
60		2849.20	I
150		2850.96	I
500	h	2857.13	I
2000	c	2859.11	I
500		2864.49	I
100		2868.09	I
1000		2887.73	I
100		2888.46	I
30		2889.20	II
200		2893.16	I
150		2893.45	I
200		2894.32	I
1000		2896.34	I
40		2903.81	I
1000		2913.15	I
500		2921.91	I
20	c	2923.34	II
1000		2928.20	I
80		2933.89	I
200		2955.93	I
200		2973.65	I
100		2979.34	I
150		2985.36	I
100		3010.83	I
300		3017.23	I
150		3021.56	I
100		3022.66	I
200		3023.68	I
80		3025.26	I
300	w	3026.89	I
50		3033.16	I
80		3034.57	I
40		3036.88	I
20		3037.90	II
100		3038.23	I
40		3042.64	I
100	h	3051.55	I
40		3052.47	I
80		3062.11	I
200		3062.36	I
300		3064.67	I
100	c	3066.60	I
120	c	3068.34	I
80		3076.24	I
150		3089.34	I
1000		3099.10	I
200		3099.52	I
60		3108.25	I
40		3109.15	I
60		3115.98	I
80		3119.17	I
40		3119.66	I
700		3122.64	I
1500		3131.23	I
40	c	3150.26	I
300		3161.67	I
3000		3173.30	I
200		3180.30	I
2000		3182.37	I
2000		3183.11	I
800	c	3195.20	II
40	w	3197.53	I
300	c	3202.83	I
1000		3212.02	II
40		3220.74	I
60		3230.02	I
1000		3237.02	II
100		3241.84	I
500		3244.19	I
300		3252.05	I
40		3256.10	I
40		3261.94	I
100		3287.14	I
30		3298.84	II
100		3300.77	I
80		3305.89	I
200		3310.65	I
150		3313.65	I
200		3325.55	I
150	c	3327.10	I
100		3330.77	I
50		3332.47	I
60		3350.56	I
50	c	3350.83	I
400		3366.75	I
40		3386.67	I
50		3392.23	I
300		3394.18	I
60		3396.90	I
40		3397.83	I
300		3398.33	I
200		3402.10	I
200	c	3403.93	I
80		3405.33	I
80		3407.28	I
50		3408.33	I
40		3411.80	I
40	c	3418.20	I
100		3419.10	I

Int		λ	Sp	Int		λ	Sp	Int		λ	Sp	Int		λ	Sp
60		3427.85		200		3797.44	I	200		4633.15	I	300		5629.94	I
40		3431.75	I	1000		3797.77	I	3000		4637.50	I	1500		5642.13	I
200		3434.70	I	200		3814.67	I	500		4643.28	I	800		5644.94	I
40		3435.68	I	300		3816.89	I	2000		4648.33	I	100		5656.00	I
150		3437.44	I	300		3824.47	I	2000	c	4660.21	I	60		5672.15	I
80		3438.73	I	500		3828.54	I	2000		4669.30	I	200		5687.30	I
200	c	3443.47	I	200		3830.35	I	400		4672.17	I	200		5689.05	I
200		3451.05	I	200		3832.45	I	200		4678.90	I	700		5725.31	I
200		3456.85	I	600		3832.82	I	400		4689.36	I	500	c	5771.47	I
400		3457.24	I	1500		3837.56	I	300		4694.28	I	100		5794.65	I
40		3457.60	I	800		3841.31	I	1000		4706.92	I	80		5799.85	I
5000	c	3466.28	I	800		3845.97	I	200		4714.22	I	100	c	5814.24	I
150		3470.51	I	500		3847.60	I	2000		4717.77	I	200		5831.48	I
80		3475.18	I	300		3851.22	I	500	c	4719.02	I	150	c	5836.33	I
1000		3475.59	I	500	c	3856.73	I	4000	c	4719.28	I	150		5923.36	I
60		3484.62	I	200		3863.07	I	200	c	4736.51	I	1000	c	5924.47	I
1000	c	3486.23	I	400		3864.11	I	10000		4740.61	I	200		5926.29	I
100		3490.30	I	1000		3868.24	I	500		4749.61	I	600	c	5931.93	I
400		3493.39	I	200		3875.66	I	1000		4752.72	I	60		6032.36	I
500		3494.62	I	500	c	3879.16	I	200		4762.36	I	60		6047.99	I
40		3499.14	I	600	c	3880.72	I	4000		4771.54	I	200		6065.09	I
1000		3500.70	I	300	w	3892.10	I	200		4773.89	I	800		6085.23	I
200		3501.24	I	200		3893.22	I	200		4783.92	I	300		6099.39	I
800	c	3502.70	I	600		3899.83	I	500		4785.60	I	500	c	6102.96	I
100		3507.19	I	300		3919.38	I	200	c	4790.48	I	1000		6120.68	I
100		3508.27	I	300	c	3923.66	I	250		4791.62	I	1000		6130.80	I
100		3510.91	I	200		3927.57	I	300		4799.98	I	150	c	6132.23	I
800		3525.83	I	200		3933.70	I	100		4805.69	I	100		6184.70	I
300		3526.18	I	4000	c	3946.57	I	100		4809.42	I	800		6192.66	I
100		3529.83	I	2000		3947.09	I	500		4816.79	I	600	c	6244.18	I
150		3534.88	I	200		3955.73	I	10000		4820.74	I	100		6312.18	I
500		3535.51	I	300		3979.64	I	300		4831.35	I	100		6354.86	I
300		3538.12	I	500		3980.35	I	1000		4834.37	I	100		6356.73	I
800		3538.68	I	10000	c	3984.97	I	1000		4835.39	I	80		6389.87	I
2000	c	3541.77	I	400		3987.78	I	100		4841.36	I	100		6408.83	I
6000	c	3549.72	I	300		3994.04	I	20000		4853.59	I	1000		6455.90	I
4000	c	3550.64	I	2000		3994.51	I	100		4857.21	I	600	c	6461.93	I
300		3559.75	I	200		3996.97	I	100	c	4862.19	I	100		6470.27	I
800		3560.32	I	300		4004.69	I	10000		4866.73	I	200	c	6491.68	I
100		3565.22	I	500		4007.14	I	200		4870.77	I	200		6526.82	I
800		3568.85	I	1000		4012.00	I	100		4888.70	I	150		6579.24	I
100		3570.65	I	400		4016.68	I	150	c	4890.88	I	500	c	6625.57	I
100		3575.42	I	600		4017.22	I	8000		4891.92	I	300	c	6673.66	I
1000		3580.06	I	2000		4020.76	I	150		4892.49	I	100		6687.10	I
600		3581.26	I	20000	c	4031.63	I	1000		4908.51	I	80		6786.00	I
800		3582.08	I	1000		4039.25	I	2000		4909.57	I	70		6798.63	I
2000		3582.63	I	200		4041.78	I	500		4913.02	I	60		6856.90	I
4000		3587.94	I	10000	c	4049.11	I	150		4914.70	I	150		7002.37	I
200		3593.47	I	500		4051.95	I	200	c	4920.67	I	100		7016.57	I
300		3594.57	I	200		4053.18	I	300		4923.60	I	500	c	7086.18	I
1000	c	3595.66	I	200	c	4056.08	I	400		4948.06	I	60		7093.12	I
1000	c	3607.32	I	400		4083.54	I	5000		4976.34	I	200		7141.28	I
200		3607.62	I	10000		4088.71	I	400		4995.00	I	200	c	7157.62	I
2000		3608.27	I	200		4093.69	I	200		5002.67	I	70		7256.08	I
200		3618.94	I	15000		4095.67	I	100		5005.74	I	100		7322.38	I
1000	c	3627.36	I	1000		4110.22	I	200	c	5014.52	I	80		7329.14	I
200		3630.39	I	10000		4115.08	I	500		5026.24	I	100		7396.80	I
3000	a	3635.15	I	600		4119.27	I	300		5026.79	I	100	c	7402.61	I
10000	c	3636.07	I	8000		4124.22	I	150		5027.89	I	200		7405.36	I
1000		3638.22	I	1000		4128.27	I	80		5032.45	I	60		7427.15	I
200		3638.85	I	300		4134.81	I	300		5055.27	I	150		7434.12	I
900		3639.38	I	300		4139.12	I	60		5058.33	I	600		7452.49	I
400		3640.23	I	800		4139.85	I	500		5060.69	I	60		7461.59	I
1000	c	3648.04	I	400		4141.27	I	500		5090.74	I	80		7534.95	I
600		3651.47	I	6000		4144.95	I	5000		5096.28	I	800		7540.26	I
1000	c	3658.59	I	3000		4145.08	I	200	c	5103.24	I	80		7543.39	I
400	c	3661.45	I	200		4147.62	I	500		5104.32	I	200		7574.02	I
200		3664.92	I	10000		4165.61	I	200		5109.81	I	500		7579.26	I
1000		3679.15	I	500		4167.42	I	100		5120.60	I	90		7624.53	I
300		3680.32	I	1000		4169.68	I	500		5139.26	I	100		7684.45	I
5000		3684.74	I	4000		4170.27	I	500		5150.63	I	500		7697.37	I
300		3692.76	I	5000		4172.53	I	2000		5161.81	I	80	c	7698.19	I
800		3703.83	I	1000		4176.28	I	2000		5174.81	I	800	c	7793.04	I
300		3704.80	I	800		4186.51	I	100		5206.56	I	60		7798.28	I
200		3706.70	I	300		4218.61	I	200		5225.55	I	60		7816.74	I
200		3707.63	I	10000	c	4238.19	I	200	c	5260.22	I	800		7817.72	I
200		3708.26	I	20000		4262.27	I	200	c	5261.44	I	100		7856.38	I
1000		3712.26	I	1000		4262.69	I	1000		5275.51	I	200		7861.44	I
300		3712.82	I	800		4274.97	II	800		5285.07	I	400	d	7871.25	I
500		3715.94	I	800		4278.90	I	100		5305.31	I	60	c	7874.76	I
10000		3718.86	I	30000		4297.06	I	400		5314.96	I	70		7965.45	I
1500		3723.67	I	400	c	4336.86	I	600		5320.20	I	500		7999.73	I
2000		3724.40	I	400		4358.49	I	200		5334.79	I	200		8126.55	I
5000		3726.35	I	200		4359.26	I	500	c	5353.48	I	200		8170.55	I
200		3727.36	I	1000		4429.59	I	200		5356.63	I	150		8205.27	I
400	c	3729.18	I	1000		4481.53	I	300		5358.65	I	100		8206.49	I
500		3731.74	I	3000		4487.06	I	200		5360.14	I	150		8211.31	I
300		3737.42	I	400		4495.03	I	500	h	5375.20	I	500	c	8237.08	I
400		3745.01	I	1000		4515.98	I	150	c	5423.05	I	200		8308.15	I
1000		3746.15	I	10000		4522.84	I	200		5447.40	I	200		8309.16	I
5000		3746.84	I	2000		4539.53	I	500	c	5451.90	I	60		8315.50	I
1000		3752.13	I	400		4542.09	I	100		5455.95	I	100		8531.06	I
4000		3754.37	I	400		4552.01	I	300		5471.96	I	100		8543.61	I
1000		3758.54	I	800		4552.85	I	70		5483.01	I	100	c	8707.21	I
2000		3761.81	I	1000		4557.05	I	60		5485.37	I	100	c	8737.93	I
5000		3768.77	I	2000		4564.54	I	80	c	5506.89	I	200	c	8829.82	I
3000		3771.03	I	1000		4578.45	I	150		5524.11	I				
500		3777.27	I	1000		4593.35	I	100		5528.23	I				
2000		3779.37	I	300	c	4609.16	I	200		5541.94	I				
3000	c	3780.68	I	1000		4616.86	I	80		5543.63	I				
500		3784.06	I	200	c	4622.69	I	100		5550.53	I				
200		3786.06	I	300		4624.96	I	3000	c	5589.02	I				
500		3791.28	I	1000		4630.57	I	200		5602.23	I				
300		3791.73	I					2000	c	5620.45	I				

TELLURIUM (Te)
Z = 52

Te I and II
Ref. 1, 344—347—C.H.C.

Tellurium (Cont.)

Vacuum

Intensity	Wavelength	
6	799.60	II
8	802.28	II
6	942.62	II
6	1003.73	II
6	1007.80	II
5	1014.27	II
5	1022.79	II
6	1057.00	II
8	1059.51	II
6	1068.86	II
8	1077.66	II
6	1090.11	II
6	1144.04	II
5	1153.10	II
10	1161.42	II
10	1174.34	II
12	1175.79	II
9	1208.54	II
5	1213.00	II
9	1220.98	II
9	1253.62	II
9	1270.52	II
7	1274.76	II
8	1306.53	II
10	1324.92	II
7	1336.42	II
7	1345.20	II
9	1363.24	II
8	1366.73	II
10	1374.80	II
6	1395.22	II
6	1439.52	II
6	1465.25	II
7	1489.56	II
8	1607.99	II
10	1608.41	II
10	1613.15	II
6	1638.91	II
5	1655.4	I
5	1688.5	I
6	1700.0	I
6	1701.58	II
5	1708.0	I
6	1751.0	I
5	1759.4	I
5	1775.0	I
6	1795.7	I
6	1796.3	I
10	1822.4	I
6	1825.5	I
6	1850.6	I
6	1852.1	I
6	1853.8	I
8	1857.2	I
6	1860.4	I
3	1962.88	II
7	1994.83	I

Air

Intensity	Wavelength	
6	2000.2	I
26000	2002.02	I
8	2070.9	I
6500	2081.16	I
18000	2142.81	I
3200	2147.25	I
360	2159.85	I
9	2208.74	I
10	2255.49	I
500	2259.02	I
10	2265.52	I
20	2373.06	II
1200	2383.26	I
1500	2385.78	I
20	2387.82	II
10	2401.63	II
10	2436.47	II
50	2438.69	II
120	2530.72	I
20	2567.82	II
10	2574.96	II
5	2576.10	II
7	2579.24	II
10	2591.12	II
10	2592.85	II
10	2605.72	II
5	2621.92	II
10	2624.86	II
20	2627.96	II
20	2641.89	II
20	2648.48	II
100	2649.66	II
40	2657.70	II
80	2661.10	II
110	2677.13	I
20	2711.58	II
6	2769.65	I
10	2841.17	II
10	2846.15	II
100	2858.29	II
20	2861.00	II
40	2868.82	II
150	2895.41	II
30	2919.89	II
50	2942.11	II
50	2946.68	II
70	2967.29	II
20	2973.67	II
50	2975.90	II
15	2997.04	II
15	3012.02	II
50	3017.58	II
20	3023.31	II
70	3047.00	II
20	3052.46	II
10	3063.16	II
15	3073.56	II
8	3104.44	II
10	3132.58	II
20	3160.66	II
100	3175.14	I
10	3189.83	II
5	3211.21	II
60	3256.80	II
30	3268.77	II
30	3282.63	II
40	3321.92	II
40	3323.11	II
60	3329.22	II
60	3352.10	II
60	3362.79	II
25	3374.10	II
150	3406.79	II
20	3419.63	II
50	3442.25	II
40	3455.12	II
20	3456.88	II
20	3480.32	II
40	3483.67	II
20	3486.11	II
50	3521.11	II
50	3552.19	II
100	3611.78	II
50	3617.57	II
40	3644.46	II
20	3679.26	II
30	3725.66	II
40	3797.22	II
20	3800.92	II
20	3905.67	II
20	3918.54	II
30	3931.49	II
20	3947.98	II
40	3969.22	II
25	3975.94	II
20	3981.77	II
50	4006.52	II
20	4011.69	II
30	4029.73	II
40	4047.17	II
30	4048.88	II
15	4073.48	II
30	4101.04	II
70	4127.32	II
30	4163.55	II
100	4169.77	II
30	4179.29	II
25	4211.31	II
80	4225.73	II
30	4246.47	II
20	4251.15	II
100	4261.11	II
30	4264.36	II
60	4273.43	II
80	4285.85	II
40	4320.90	II
30	4361.28	II
150	4364.00	II
30	4377.12	II
75	4385.10	II
60	4396.00	II
170	4478.63	II
80	4537.07	II
100	4557.78	II
70	4630.62	II
100	4641.12	II
180	4654.37	II
200	4686.91	II
100	4696.38	II
100	4706.53	II
100	4766.05	II
70	4771.56	II
100	4784.87	II
100	4827.14	II
150	4831.28	II
150	4842.90	II
130	4865.12	II
200	4866.24	II
80	4885.22	II
80	4904.44	II
60	4961.88	II
60	5000.82	II
8	5083.0	I
7	5148.7	I
50	5449.84	II
50	5487.95	II
150	5576.35	II
150	5649.26	II
100	5666.20	II
200	5708.12	II
7	5733.5	I
150	5755.85	II
8	5789.1	I
50	5936.15	II
100	5974.68	II
8 d	6273.5	I
8 h	6349.7	I
50	6367.13	II
8	6405.9	I
7 h	6456.7	I
8 h	6613.4	I
10	6648.58	II
8	6660.2	I
8 h	6690.0	I
10 h	6790.0	I
20 h	6837.6	I
20 h	6854.7	I
10	7016.06	II
10	7039.13	II
15 h	7191.1	I
10	7236.62	II
20 h	7263.5	I
8	7280.9	I
10	7289.26	II
10	7445.39	II
12	7460.98	II
15	7468.75	II
10	7481.26	II
10	7556.8	I
6	7688.61	II
15	7759.1	I
8	7818.79	II
8	7861.61	II
15	7921.69	II
15	7943.14	II
10	7950.34	II
20	7972.9	I
6	8056.15	II
30 h	8061.4	I
10 h	8082.5	I
10	8122.66	II
8	8130.39	II
8	8154.47	II
20	8186.44	II
6	8190.94	II
10	8251.5	I
15	8273.53	II
10	8276.6	I
10	8291.1	I
15	8355.8	I
10	8372.12	II
7	8469.8	I
8	8492.2	I
8	8500.8	I
12	8521.4	I
10	8535.68	II
12	8575.78	II
10	8604.63	II
8	8621.68	II
7 h	8632.1	I
15	8672.95	II
12	8701.09	I
10	8733.81	II
205	8758.18	I
12	8831.52	I
18	8851.15	I
6	8897.92	II
81	9004.37	I
18	9043.39	I
12	9196.80	I
15	9206.78	I
17	9207.64	I
12	9469.00	I
5660	9722.74	I
185	9785.54	I
109	9842.30	I
532	9868.92	I
118	9902.61	I
689	9956.30	I
37	9959.93	I
325	9977.13	I
136	9979.31	I
45	9985.85	I
5950	10051.41	I
4097	10091.01	I
104	10099.57	I
279	10106.05	I
381	10118.08	I
296	10151.06	I
397	10300.56	I
205	10323.05	I
745	10493.57	I
197	10509.86	I
1880	10918.34	I
298	11007.80	I
10200	11089.56	I
508	11163.74	I
6620	11487.23	I
280	11978.96	I
188	12566.24	I
389	12589.19	I
161	12805.50	I
400	13104.18	I
1580	13247.75	I
483	13316.63	I
217	14037.09	I
144	14072.53	I
434	14335.74	I
220	14417.46	I
1050	14513.51	I
129	14554.68	I
1480	15452.45	I
2430	15546.23	I
3760	16403.90	I
1960	17303.54	I
2780	18291.59	I
394	18777.30	I
269	19623.52	I
239	20147.54	I
1020	21043.73	I
464	21602.50	I
37	21799.64	I
74	22555.29	I
48	22755.66	I
27	23294.94	I
17	23978.70	I
25	24059.04	I
13	26428.62	I
38	26539.17	I
15	26553.74	I
7	27179.26	I

TERBIUM (Tb)
Z = 65

Tb I and II
Ref. 1 — C.H.C.

Air

Intensity	Wavelength	
29	2577.73	II
110	2584.61	II
29	2590.31	II
29	2591.42	II
24	2592.64	II
55	2597.71	II
40	2602.93	II
110	2608.57	II
40	2616.90	
130	2628.69	II
55	2655.96	II
50	2661.40	II
24	2661.64	II
55	2667.64	II
50	2668.86	II
140	2669.29	II
40	2674.13	II
40	2674.69	II
29	2678.15	II
40	2683.97	II
35	2687.82	II
50	2691.90	II
35	2693.05	II
55	2693.41	II
35	2695.46	II
50	2696.83	II
190	2704.07	II
130	2736.24	II
160	2759.47	II
270	2769.53	II
130	2784.49	II
180	2800.51	II
250	2802.75	II
250	2809.30	II
180	2812.64	II
190	2852.14	II
110	2857.68	II
230	2886.29	II
160	2894.45	II
320	2897.44	II
160	2898.86	II
110	2901.54	I.
110	2910.30	II
160	2914.75	II
160	2915.30	II
190	2915.60	II
120	2916.24	II
120	2918.89	II
120	2924.16	II
120	2924.53	II
160	2932.89	II
150	2940.05	II
250	2956.21	II
170	2968.87	II
170	2977.78	II
110	2987.03	II
110	2988.57	II

Terbium (Cont.)

Intensity		Wavelength	Species
130		2996.00	II
110		2999.03	II
130		3005.52	II
170		3009.30	II
230		3010.59	II
230		3016.18	II
130		3019.17	II
170		3020.29	II
110		3023.43	II
170		3027.33	II
230		3031.60	II
230		3044.96	II
190		3051.13	II
130		3053.24	II
460		3053.55	II
130		3062.78	II
230		3064.09	II
110		3065.69	II
230		3067.20	II
270		3069.03	II
460		3070.05	II
270		3072.60	II
670		3078.86	II
480		3082.36	II
120		3086.78	II
250		3088.43	II
480		3089.58	II
230		3102.54	II
480		3102.96	II
290		3117.89	II
290		3119.62	II
230		3121.94	II
230		3123.05	II
160		3124.54	II
110		3131.35	II
250		3134.26	II
440		3139.64	II
190		3140.06	II
230		3145.22	II
150		3146.67	II
310		3147.04	II
310		3147.15	II
310		3148.71	II
120		3155.62	II
130		3162.42	II
290		3162.93	II
190		3165.74	II
380		3167.52	II
140		3168.32	II
230		3169.84	II
190		3173.76	II
380		3174.66	II
380		3180.54	II
140		3183.88	II
480		3187.26	II
290		3188.03	II
190		3194.69	II
380		3195.60	II
480		3199.56	II
1100		3218.93	II
1200		3219.98	II
250		3230.03	II
250		3231.06	II
210	d	3239.60	II
250		3240.00	II
480		3252.32	II
250		3262.97	II
230	d	3263.87	II
230		3264.90	II
400		3266.40	II
250		3274.14	II
250		3274.33	II
210		3277.32	II
760		3280.31	II
760		3281.40	II
520		3283.10	II
1000		3285.04	II
310		3287.55	II
310		3291.56	II
1500		3293.07	II
210		3295.33	II
310		3298.66	II
210		3304.95	II
420	d	3307.44	II
210		3308.51	II
210		3314.38	II
340	d	3321.15	II
420		3322.28	II
210		3323.38	II
210		3323.89	II
3800		3324.40	II
520		3329.08	II
210		3334.48	II
250		3336.70	II
310		3338.03	II
250		3339.00	II
210		3347.27	II
210		3348.07	II
760		3349.42	II
320		3362.25	II
760		3364.93	II
230	d	3370.61	II
320		3371.50	II
520		3372.36	II

Intensity		Wavelength	Species
460	d	3372.72	II
520		3375.03	II
320		3378.73	II
520		3378.86	II
320		3382.80	II
210		3390.60	II
380		3391.28	II
270		3398.35	II
320		3399.10	II
270		3400.53	II
210	d	3400.86	II
420		3402.33	II
210		3410.40	II
210		3410.68	II
520		3413.76	II
270		3416.24	II
400	d	3420.34	II
210		3430.61	II
320		3433.26	II
270		3439.72	II
520		3440.37	II
320		3444.58	II
210		3446.40	II
270		3449.46	II
810		3454.06	II
380		3460.38	II
230	d	3462.97	II
620		3468.03	II
270		3471.73	II
270		3472.37	II
810	d	3472.79	II
210		3473.00	II
380		3480.17	II
230		3483.04	II
230		3483.69	II
290	d	3489.51	II
210	d	3492.00	II
270		3494.21	II
270		3495.36	II
810		3500.84	II
570		3507.45	II
5700		3509.17	II
380		3510.10	II
320		3513.10	II
570		3519.76	II
1300		3523.66	II
380		3525.14	II
440		3525.61	II
440		3536.32	II
570		3537.94	II
1100		3540.24	II
810		3543.89	II
310	d	3551.03	
320		3551.96	II
460	d	3558.77	
3200		3561.74	II
480		3562.90	II
570		3565.74	
810		3567.35	II
4200		3568.52	II
1600		3568.98	II
320		3572.07	II
1100		3579.20	II
710		3585.03	II
570		3587.44	II
810		3596.38	II
440		3598.06	II
1600		3600.44	II
320		3604.90	II
320		3611.33	II
320		3614.63	II
320	d	3615.66	II
320		3616.58	II
380		3617.86	II
380		3619.73	II
810		3625.54	II
570		3626.50	II
380		3629.44	II
670		3633.29	II
670		3638.46	II
670		3641.66	II
440		3647.06	II
570		3647.75	II
2300		3650.40	II
810		3654.88	II
2000		3658.88	II
450		3663.12	II
3800		3676.35	II
300		3677.89	II
810		3682.26	II
320		3688.15	II
610		3691.15	II
300		3692.95	II
450		3693.58	I
320		3696.85	II
450		3700.12	I
4700		3702.76	II
300		3703.12	I
2400		3703.92	II
370		3709.30	II
1000	d	3711.76	II
300		3719.45	II
650		3729.91	II
430		3732.39	II

Intensity		Wavelength	Species
430		3743.09	
650		3745.04	I
870		3747.17	II
870		3747.34	II
1100		3755.24	II
430		3757.44	II
430	d	3757.90	II
650		3759.35	I
350		3761.14	I
1700		3765.14	I
2100		3776.49	II
330		3779.22	
600		3783.53	I
410	d	3787.22	II
410		3789.92	I
390		3792.20	I
600		3793.55	
330		3801.80	II
760	d	3806.85	II
1500		3830.26	I
540		3833.42	I
920	d	3842.50	II
370	d	3845.61	II
3700		3848.73	II
450	d	3869.75	II
3500	w	3874.17	II
330		3883.34	I
480		3888.22	I
490		3894.64	I
330		3895.99	I
330		3896.58	II
330		3897.89	I
2400		3899.20	II
1600		3901.33	I
480		3908.06	I
380		3909.14	I
330		3909.55	I
650		3915.43	I
480		3919.52	II
300		3922.10	II
480		3922.74	II
760		3925.45	II
650		3935.24	II
810	d	3939.52	II
650		3946.89	II
350	d	3958.36	II
2200	d	3976.84	II
1800		3981.87	II
300		3983.85	II
350		3999.40	II
350	d	4002.19	II
970		4002.59	II
1900		4005.47	II
300		4010.04	I
760		4012.75	II
330		4013.26	I
370		4019.14	II
540		4020.47	II
220		4022.88	I
370		4024.77	II
520		4031.66	II
870		4032.28	I
2100		4033.03	II
350		4036.22	I
210		4038.86	I
300		4051.86	II
300		4052.87	II
430		4054.12	I
410		4060.37	I
220		4060.87	II
1300		4061.58	I
220		4063.89	II
390		4066.22	II
260		4075.22	I
390		4081.24	I
210		4086.06	I
210		4092.19	I
260		4094.37	II
260		4094.49	I
260		4103.90	II
650		4105.37	I
300		4112.50	I
260		4119.92	I
280		4143.51	I
1100		4144.81	II
350		4158.53	I
240		4169.09	I
240		4169.32	I
240		4171.05	I
240		4172.60	I
240		4172.82	I
260		4173.47	I
240		4186.21	I
300		4187.16	I
390		4196.74	I
450		4201.00	II
650		4203.74	I
600		4206.49	I
300	Cw	4213.50	I
300		4214.42	II
480		4215.09	I
300		4217.56	I
260		4219.16	I
260		4224.28	I

Intensity		Wavelength	Species
480	Cw	4226.45	II
260		4231.89	I
480		4232.82	I
300		4235.35	I
370		4255.24	I
480		4258.23	II
260		4263.66	I
650		4266.34	I
330		4269.69	I
220		4275.21	I
760	Cw	4278.52	II
300		4285.13	II
300		4289.70	I
370		4298.36	I
300		4299.90	I
240		4302.95	I
240		4307.18	I
450		4310.42	I
300		4311.56	I
370		4313.25	I
2200		4318.83	I
600		4322.23	I
600		4325.83	II
3000		4326.43	I
240		4328.90	I
600		4332.12	I
870		4336.43	I
600		4337.64	I
1700		4338.41	I
700		4340.62	I
430	Cw	4342.53	I
430	d	4353.20	II
280		4356.09	I
870		4356.81	I
280		4360.16	I
220		4367.30	II
220		4372.02	I
330		4382.45	I
300		4388.23	I
260		4390.91	I
200		4416.27	II
140		4420.19	I
350		4423.10	I
110		4432.72	I
240		4436.12	I
110		4439.38	I
240		4448.04	I
110		4467.69	I
430		4493.07	I
45	d	4509.04	II
150	h	4511.52	I
45		4512.96	II
75		4514.31	II
45		4519.72	II
45		4525.01	II
45		4529.76	
45	h	4531.83	II
45		4534.13	I
45		4537.14	I
45		4537.23	I
110		4549.07	I
45		4549.72	II
110		4550.45	I
110		4556.46	I
55		4562.24	II
110		4563.69	II
30		4564.85	II
55		4573.19	II
210		4578.69	II
65		4584.84	II
65		4591.56	II
45		4592.38	I
45	h	4604.10	II
30		4611.96	I
45	h	4615.92	II
27		4617.49	I
30		4619.36	II
75	d	4626.32	II
95		4626.94	II
65		4632.07	I
65	h	4636.59	I
30		4636.99	II
85		4641.00	II
210		4641.98	II
260	Cw	4645.31	II
80		4647.23	I
60		4658.38	I
20		4658.73	
80		4662.79	I
50	c	4665.45	I
40		4669.40	I
80		4676.90	I
70	c	4681.87	I
50		4682.52	I
25	c	4682.79	II
80		4688.63	II
80		4693.11	II
30	h	4693.39	II
200		4702.41	II
110		4707.94	II
40	w	4716.07	II
40		4728.16	II
60	Cw	4734.20	I
80		4739.93	I

Terbium (Cont.)

Intensity		Wavelength		Intensity		Wavelength	
70		4747.80	I	40	c	5686.48	I
410	Cw	4752.53	II	85	c	5747.58	II
40		4758.44	II	24		5762.66	I
40		4760.19	II	24		5785.18	II
30		4762.37	II	75		5795.64	I
25		4764.47	II	75		5803.13	II
35		4778.36	II	65		5815.36	I
35		4778.80	II	29		5842.97	I
180		4786.78	I	65		5851.07	I
40	Cw	4789.91	II	65		5870.62	I
30		4801.87	II	35		5898.84	I
100		4813.77	I	24		5902.40	I
60		4837.59	II	35		5904.71	I
25		4840.39	I	65	c	5920.78	I
30	c	4842.69	II	50	c	5939.38	I
30		4844.89	II	35		5940.17	I
30		4854.81	I	24		5951.17	I
20		4856.54	II	75		5967.34	II
30		4858.87	II	29		6038.97	I
80		4875.57	II	29		6039.38	I
25		4876.12	I	24		6104.29	II
80		4881.15	II	24	c	6292.43	I
29		4894.33		35		6331.68	II
95		4915.90	I	24		6334.91	II
35		4924.09	I	24		6446.87	II
35		4926.83	I	35	Cw	6518.68	I
50		4928.93	I	24	c	6574.04	II
65		4931.79	I	35		6581.82	I
29		4970.99	II	30		6607.17	II
29		4971.42	I	90		6677.94	II
29		4973.04	I	40	Cw	6702.61	I
29		4980.16	II	20	c	6706.79	II
29		4980.56	I	30		6785.12	II
85		4993.82	II	130		6794.58	II
50		4995.84	II	40		6874.18	II
55		4997.95	I	55		6896.37	II
29		5006.10	II	45	h	6899.95	I
50		5022.16	I	40		6901.98	I
29		5024.24	II	9		7005.99	II
29		5024.65	I	17		7082.85	II
50		5033.12	I	11		7089.22	II
50	w	5042.06	II	11		7112.69	I
55		5054.30	I	10		7187.48	I
55		5065.79	I	10	h	7195.89	II
110		5078.25	I	65		7204.28	I
24		5080.05	II	19	h	7234.98	I
24		5081.11	I	40		7257.73	I
75		5089.12	II	17		7311.57	I
24		5089.66	I	45		7348.88	II
24		5101.09	I	10		7398.27	I
24		5108.56	I	15	h	7424.24	II
35		5118.39	I	10	h	7429.62	II
24		5120.18	I	9		7472.15	I
50	w	5131.69	I	22		7484.54	I
50	w	5141.08	II	9		7495.45	I
50		5147.58	I	45		7496.12	I
24		5164.27	I	17		7499.69	II
29		5170.13	I	27		7511.40	I
24		5170.61	I	9		7519.77	II
50		5176.51	I	6		7557.59	II
50		5179.97	I	27		7582.03	II
50		5184.59	I	27		7587.49	I
85		5186.13	I	45		7590.24	I
50		5188.48	I	65		7596.44	I
50		5198.86	I	17	h	7601.18	II
35	w	5202.77	I	17		7616.01	II
40		5204.55	I	22	h	7624.05	I
40		5207.97	I	30		7627.81	I
40		5214.28	I	9	h	7639.05	II
40		5221.99	I	8		7672.72	II
120		5228.12	I	8	h	7694.74	II
40		5235.11	I	22	h	7706.16	II
75		5248.71	I	22	h	7726.97	II
75	w	5262.11	II	30		7737.63	I
24		5275.03	I	22		7793.20	II
75		5281.05	I	8		7807.33	II
65		5304.72	I	16		7832.91	II
29		5308.19	I	30		7855.79	II
29		5309.46	I	15		7864.99	I
110		5319.23	I	6	h	7885.70	I
35		5331.04	I	6		7913.11	I
65	w	5337.90	I	27		7927.90	II
35	d	5338.59	I	13		7955.31	I
24		5347.83	II	11		7998.05	I
160		5354.88	I	17		8001.04	I
75		5369.72	I	13	h	8010.16	I
75		5375.98	I	30		8025.42	II
29	d	5402.06	II	6		8053.80	I
29		5413.65	I	19		8067.35	II
29		5416.20	I	30		8085.06	II
50		5424.10	II	27		8164.17	II
29	c	5426.43	I	13		8171.70	I
35		5443.38	I	65		8194.82	I
29		5457.00	I	95		8212.57	I
55		5459.81	I	11		8214.33	I
29	w	5470.34	II	8		8259.08	I
24		5481.45	I	40		8450.06	II
55		5509.61	I	8	h	8465.80	II
50		5514.54	I	13		8502.70	I
65		5524.12	I	30	h	8511.80	I
24	c	5525.62	I	45		8583.45	II
35		5565.93	I	30		8603.40	I
29	c	5638.80	I	9		8678.25	I
29	c	5685.74	II	65		8765.74	II

Tb IV
Ref. 302 — J.R.

Intensity	Wavelength	
	Vacuum	
30	1176.58	IV
30	1192.01	IV
70	1200.58	IV
50	1213.94	IV
80	1221.22	IV
500	1235.04	IV
1000	1259.40	IV
300	1301.48	IV
300	1308.30	IV
600	1311.70	IV
700	1315.12	IV
500	1325.56	IV
1000	1327.67	IV
100	1367.56	IV
400	1367.71	IV
700	1369.64	IV
1000	1373.86	IV
400	1376.46	IV
200	1378.23	IV
300	1381.00	IV
100	1382.83	IV
20	1389.92	IV
200	1516.17	IV
50	1530.10	IV
5000	1595.39	IV
2000	1633.19	IV
300	1649.38	IV
400	1654.75	IV
400	1667.58	IV
200	1672.55	IV
400	1681.98	IV
5	1684.46	IV
100	1685.37	IV
400	1691.95	IV
10	1695.23	IV
300	1698.36	IV
30	1701.60	IV
50	1704.79	IV
20	1705.05	IV
3	1943.94	IV
50	1970.90	IV
	Air	
2000	2027.79	IV
200	2029.22	IV
400	2048.88	IV
200	2078.83	IV
1000	2089.98	IV
1000	2332.54	IV
100	2436.01	IV

THALLIUM (Tl)
Z = 81

Tl I and II
Ref. 1, 195, 348, 354 — C.H.C.

Intensity		Wavelength		Intensity		Wavelength	
		Vacuum		5		1685.40	I
3		650.90	II	1		1728.	I
5	r	670.87	II	10	r	1792.76	II
4		674.10	II	12	r	1814.85	II
15	r	696.30	II	3	h	1847.	I
5	r	709.23	II	8		1892.72	II
10	r	817.18	II	25	r	1908.64	II
5	r	836.34	II			Air	
8	r	1018.85	II	100	r	2007.56	I
10	r	1049.73	II	2		2209.75	I
8	r	1050.30	II	100	r	2210.71	I
5	r	1074.97	II	3		2287.6	I
10	r	1130.17	II	30		2298.04	II
15	r	1162.55	II	140		2315.98	I
10	r	1167.43	II	900	h	2379.69	I
10	r	1183.41	II	8		2451.83	II
12	r	1194.84	II	6		2469.03	II
8		1231.81	II	1		2508.2	I
5	r	1246.00	II	20		2530.86	II
15	r	1307.50	II	700		2580.14	I
8	r	1310.20	II	60		2608.99	I
25	r	1321.71	II	80		2665.57	I
8	r	1330.40	II	420		2709.23	I
10	r	1373.52	II	50	h	2710.67	I
1		1423.2	I	4400	d	2767.87	I
8	r	1489.65	I	280		2826.16	I
5		1490.50	II	10		2849.80	II
10	r	1499.30	II	2800		2918.32	I
10	r	1507.82	II	440		2921.52	I
15	r	1561.58	II	5		3029.01	II
10	r	1568.57	II	20		3091.56	II
7	r	1593.26	II	15		3185.51	II
5	h	1616.	I	15		3186.56	II
1		1650.2	I	15		3187.74	II
				1200		3229.75	I
				15		3291.01	II
				12		3319.91	II
				12		3321.04	II
				8		3322.25	II
				15		3369.15	II
				8		3381.00	II
				6		3381.80	II
				6	d	3460.48	II
				20000		3519.24	I
				5000		3529.43	I
				8		3540.08	II
				9		3560.68	II
				5		3567.67	II
				12000	Cw	3775.72	I
				8		3793.95	II
				10		3832.30	II
				6		3869.15	II
				10		3887.15	II
				8		4223.05	II
				20		4274.98	II
				40		4306.80	II
				2		4359.9	I
				8		4490.77	II
				20		4737.05	II
				15		4981.35	II
				25		5078.54	II
				25		5152.14	II
				6		5181.95	II
				6		5183.10	II
				18000		5350.46	I
				15	d	5384.85	II
				7		5409.92	II
				10		5410.97	II
				25		5949.48	II
				10		6179.98	II
				8	d	6239.03	II
				10		6378.32	II
				16	h	6549.84	I
				6	h	6713.80	I
				10		6966.5	II
				3		7493.6	I
				2		7678.93	I
				10		7815.80	I
				8		8130.0	I
				20		8373.6	I
				8		8445.8	II
				10		8474.27	I
				8		8632.9	II
				10		8664.1	II
				4		8850.4	I
				5		8976.75	I
				3		9038.4	I
				20		9130.	II
				20		9130.5	I
				2	h	9183.1	I
				4		9225.	II
				2	h	9252.6	I
				3		9254.	II
				40		9509.4	I
				10		9863.4	I
				20		9930.4	I
				2		9937.4	I
				30		10011.9	I
				40		10488.80	I
				5		11101.61	I
				4		11483.7	I
				1000		11512.82	I

Thallium (Cont.)

Intensity	Wavelength	
5	11592.9	I
15	12491.8	I
150	12736.4	I
700	13013.2	I

Tl III
Ref 355 — C.H.C.

Intensity	Wavelength	

Vacuum

7	1231.57	III
10	1266.33	III
4	1332.36	III
10	1477.14	III
4	1506.37	III
8	1558.67	III
8	1660.05	III

Air

6	3163.53	III
3	3300.80	III
9	3456.34	III
4	3507.41	III
6	3933.05	III
2	3946.02	III
7	4109.85	III
4	4155.75	III
6	4269.81	III
2	4380.57	III
4	5086.99	III
4	5362.40	III
2	5499.4	III
5	5927.8	III
4	8001.	III

Tl IV
Ref. 356 — C.H.C.

Intensity	Wavelength	

Vacuum

7	531.26	IV
10	570.49	IV
4	597.01	IV
1	868.99	IV
3	912.74	IV
8	917.31	IV
30	1028.69	IV
20	1034.73	IV
20	1036.61	IV
10	1049.48	IV
10	1057.56	IV
20	1068.04	IV
20	1070.47	IV
30	1079.68	IV
5	1079.70	IV
2	1092.90	IV
4	1094.41	IV
6	1099.60	IV
4	1125.52	IV
5	1139.30	IV
3	1144.07	IV
3	1225.45	IV
6	1273.03	IV
3	1304.55	IV
6	1323.66	IV
7	1337.10	IV
7	1358.56	IV
5	1374.62	IV
7	1377.75	IV
8	1404.60	IV
5	1412.93	IV
6	1434.72	IV
5	1449.37	IV
5	1883.2	IV
3	1974.6	IV

THORIUM (Th)
Z = 90

Th I and II
Ref. 1, 97, 98, 434 — J.G.C. and R.Z.

Intensity	Wavelength	

Air

100	2326.926	II
190	2377.84	II
90	2404.504	II
100	2413.409	II
30	2439.433	I
5	2532.894	I
150	2547.901	II
500	2565.593	II
270	2566.588	II
200	2576.688	II
230	2589.059	II
230	2597.047	II
230	2600.882	II
100	2609.855	II
230	2618.91	II
270	2623.448	II
270	2625.737	II
55	2628.812	II
270	2641.488	II
170	2650.583	II
150	2658.663	II
50	2680.692	I
360	2684.288	II
480	2692.415	II
100	2695.553	II
270	2703.958	II
170	2708.176	II
230	2721.691	II
170	2722.380	II
250	2729.327	II
250	2732.808	II
28	2735.834	II
520	2747.156	II
100	2749.530	II
410	2752.166	II
130	2760.391	II
100	2765.123	II
270	2768.841	II
200	2770.816	II
70	2773.951	II
50	2774.066	II
70	2778.706	II
35	2791.496	I
90	2794.255	II
70	2797.737	II
55	2799.114	II
110	2807.827	II
180	2808.998	II
70	2814.319	II
45	2816.071	II
100	2819.322	II
100	2820.336	II
100	2822.025	II
170	2826.855	II
70	2830.442	II
800	2832.315	II
1200	2837.295	II
320	2842.812	II
100	2848.084	I
270	2851.260	II
50	2854.342	I
30	2860.490	II
220	2861.42	II
35	2868.461	I
70	2869.916	II
550	2870.406	II
30	2878.657	I
40	2882.511	II
320	2884.289	II
360	2885.049	II
360	2887.817	II
40	2892.172	II
250	2899.720	II
45	2903.167	II
50	2908.506	I
200	2910.594	II
90	2911.320	II
90	2912.009	II
140	2919.840	II
250	2925.050	II
250	2928.254	II
50	2931.281	I
55	2934.135	II
100	2936.086	I
35	2940.589	II
340	2942.860	II
100	2943.729	I
150	2949.068	II
80	2950.438	II
35	2955.849	II
170	2957.580	II
28	2959.853	I
28	2963.607	I
270	2968.686	II
110	2971.481	II
220	2974.011	II
50	2976.104	I
55	2980.334	II
160	2985.243	II
360	2988.232	II
150	2991.062	II
110	2996.986	II
50	3002.686	I
30	3004.248	I
180	3008.497	II
50	3010.736	I
20	3018.644	I
50	3021.056	I
150	3026.575	II
40	3030.487	I
370	3034.065	II
170	3035.110	II
85	3038.598	II
130	3045.564	II
420	3049.092	II
30	3056.692	I
220	3061.699	II
450	3067.729	II
370	3072.114	II
670	3078.828	II
480	3080.217	II
240	3088.470	II
130	3090.093	II
50	3093.711	I
140	3097.266	II
200	3102.664	II
510	3108.296	II
50	3115.538	I
100	3116.263	I
510	3119.526	II
510	3122.963	II
370	3124.387	II
480	3125.507	II
150	3131.070	II
100	3136.216	I
420	3139.306	II
420	3142.835	II
310	3146.044	II
150	3150.455	II
310	3154.300	II
50	3157.221	I
30	3161.364	I
140	3166.099	II
110	3169.328	II
420	3175.726	II
270	3179.048	II
1100	3180.193	II
310	3184.948	II
770	3188.233	II
85	3191.221	II
55	3192.585	I
55	3195.689	I
30	3202.520	I
170	3210.308	II
55	3214.380	I
560	3221.292	II
30	3223.168	I
560	3229.009	II
110	3230.868	II
480	3235.84	II
590	3238.116	II
240	3241.108	II
110	3244.448	I
280	3251.915	II
910	3256.274	II
180	3257.366	I
910	3262.668	II
180	3267.003	II
110	3272.027	II
30	3278.733	I
50	3281.048	I
130	3285.752	I
620	3287.789	II
910	3291.739	II
620	3292.520	II
240	3297.832	II
240	3301.650	I
480	3304.238	I
130	3309.365	I
50	3314.790	I
30	3318.390	I
510	3321.450	II
390	3324.752	II
840	3325.120	II
55	3328.255	II
250	3330.476	I
620	3334.604	II
620	3337.870	II
30	3342.073	II
180	3346.557	II
310	3348.768	I
980	3351.228	II
310	3354.179	II
620	3358.602	II
75	3361.738	II
390	3367.819	II
250	3374.974	I
390	3378.573	II
130	3380.859	I
310	3385.531	II
310	3386.501	II
110	3387.920	I
1300	3392.035	II
200	3396.727	I
250	3398.544	II
200	3405.558	I
250	3413.012	I
50	3417.497	I
390	3421.210	I
270	3423.989	I
50	3428.622	I
980	3433.998	II
770	3435.976	II
340	3438.949	II
110	3442.578	I
50	3446.547	I
130	3451.702	I
50	3457.068	I
340	3462.850	II
130	3465.924	II
390	3468.219	II
1300	3469.920	II
170	3471.218	I
250	3479.173	II
70	3480.052	I
200	3486.552	I
100	3489.184	I
270	3493.518	II
70	3496.810	I
130	3498.621	I
70	3503.786	I
50	3506.645	I
110	3511.157	I
140	3518.404	I
70	3521.059	I
70	3526.633	I
140	3531.450	I
670	3539.587	II
180	3544.018	I
170	3549.595	I
140	3551.401	I
200	3555.013	I
530	3559.451	II
110	3563.375	I
70	3569.820	I
200	3576.557	I
100	3583.101	I
170	3589.750	I
270	3592.780	I
270	3598.120	I
390	3601.034	II
170	3608.377	I
980	3609.445	II
200	3612.427	I
480	3615.133	II
400	3617.118	II
270	3623.970	II
390	3625.627	II
140	3632.831	I
270	3635.943	I
70	3638.644	I
210	3642.248	I
170	3649.735	I
50	3658.808	I
100	3659.629	I
220	3663.202	I
140	3668.140	I
280	3669.968	I
700	3675.567	II
150	3682.486	I
50	3688.658	I
100	3690.624	I
170	3692.566	I
180	3698.105	I
50	3703.229	I
340	3706.767	I
280	3711.305	II
590	3719.435	I
50	3719.836	I
770	3721.825	II
110	3727.902	I
50	3733.672	I
1300	3741.183	II
310	3747.539	I
650	3752.569	II
140	3757.694	I
110	3765.240	I
180	3770.056	I
50	3772.649	I
85	3776.271	I
50	3780.966	I
340	3785.600	II
100	3789.167	I
85	3795.386	I
50	3800.197	I
590	3803.075	I
50	3807.273	I
340	3813.068	II
50	3818.685	I
75	3825.133	I
450	3828.384	I
70	3836.584	I
840	3839.746	II
280	3841.960	II
100	3846.887	I
85	3852.135	I
390	3854.511	II
140	3859.840	II
450	3863.405	II
100	3869.663	I
210	3875.374	I
140	3879.644	I
100	3886.915	I
340	3895.419	I
50	3901.661	I
110	3903.102	I
170	3905.186	II
50	3908.750	I
85	3911.909	I
50	3916.417	I

Int.	λ	Sp.	Int.	λ	Sp.	Int.	λ	Sp.	Int.	λ	Sp.
110	3919.023	I	25	4452.565	I	70	5277.501	II	8	6317.185	I
140	3925.093	I	85	4458.002	I	15	5281.069	I	21	6327.278	I
590	3929.669	II	220	4465.341	II	10	5294.397	I	35	6342.860	I
200	3932.911	I	30	4475.221	I	30	5297.743	I	50	6355.911	II
140	3937.040	II	75	4482.169	I	30	5307.466	II	14	6369.140	I
50	3942.072	I	50	4489.664	I	35	5312.002	I	40	6376.931	I
50	3948.030	I	110	4498.940	I	20	5317.494	I	30	6411.899	I
200	3948.964	II	55	4505.216	I	60	5325.145	II	24	6413.615	I
50	3952.760	I	280	4510.527	II	50	5326.976	I	15	6437.762	I
110	3959.300	I	70	4521.194	I	20	5330.080	I	15	6450.005	I
390	3967.392	I	22	4530.319	I	60	5343.581	I	60	6457.283	I
200	3972.155	I	40	4535.255	I	14	5351.126	I	50	6462.614	I
150	3980.089	I	30	4545.915	II	30	5358.707	I	5	6466.717	I
110	3991.730	I	70	4555.812	I	20	5369.281	I	14	6490.738	I
530	3994.549	II	40	4563.660	I	30	5378.836	I	5	6501.992	I
50	3998.061	I	65	4570.972	I	70	5390.466	II	20	6512.364	I
240	4003.309	II	50	4588.426	I	50	5392.572	I	5	6522.044	I
250	4007.021	I	75	4595.421	I	20	5399.175	I	50 h	6531.342	I
220	4008.210	I	26	4612.554	II	24	5417.486	I	6	6554.160	I
220	4009.056	I	30	4621.163	I	60	5425.678	II	5	6558.876	I
280	4012.495	I	140	4631.761	II	50	5435.893	II	3	6565.070	I
4200	4019.129	I	140	4631.761	II	40	5449.479	II	5	6577.215	I
210	4025.656	II	30	4641.254	I	30	5462.615	II	24	6583.907	I
140	4027.009	I	140	4651.558	II	15	5470.759	I	24	6588.540	I
250	4030.842	I	23	4663.202	I	24	5484.147	II	24	6593.940	I
250	4036.047	I	50	4669.984	I	10	5496.137	I	24	6605.416	II
240	4036.565	II	65	4676.056	I	19	5504.302	I	24	6619.946	II
240	4041.204	II	50	4686.195	I	35	5509.994	I	24	6619.947	II
55	4048.287	I	140	4694.091	II	12	5524.584	I	21	6644.650	II
110	4050.887	I	50	4703.990	I	50	5539.262	I	6	6658.678	II
140	4059.253	I	20	4712.841	I	70	5539.911	II	30	6662.269	I
250	4063.407	I	90	4723.438	I	35	5548.176	I	6	6674.697	I
300	4069.201	II	30	4729.128	I	50	5558.342	I	3	6683.367	I
100	4069.461	I	190	4740.529	II	60	5564.203	II	8	6692.724	II
55	4075.503	I	140	4752.414	II	40	5573.354	I	5	6697.712	I
110	4081.368	I	13	4766.600	I	60	5587.026	I	16	6727.459	I
85	4085.434	I	50	4778.294	I	24	5595.064	I	3	6735.126	I
700	4086.520	II	20	4786.531	I	50	5604.515	II	5	6742.884	I
70	4088.726	I	40	4789.387	I	35	5615.320	I	20	6756.453	I
700	4094.747	II	45	4808.134	I	7	5630.297	I	6	6765.677	I
150	4100.341	I	20	4819.193	I	70	5639.746	II	15	6778.313	I
270	4105.330	II	26	4822.855	I	7	5648.991	I	15	6780.413	I
840	4108.421	II	40	4826.700	I	12	5657.925	I	6	6791.236	I
240	4112.754	I	45	4831.121	I	15	5667.128	I	3	6798.747	I
280	4115.758	I	50	4840.843	I	20	5677.053	I	3	6809.511	I
1100	4116.713	II	30	4848.362	I	10	5685.192	I	5	6823.509	I
30	4123.600	I	15	4852.868	I	65	5700.918	II	11	6829.036	I
200	4127.411	I	40	4858.333	I	95	5707.103	II	14	6834.925	I
110	4131.002	I	280	4863.163	II	50	5720.183	I	4	6862.873	I
340	4132.753	II	40	4872.917	I	30	5732.975	II	5	6866.367	I
200	4134.067	I	26	4878.733	I	24	5742.084	II	8	6874.754	I
220	4140.235	II	45	4894.955	I	30	5749.388	II	20	6889.303	II
250	4142.701	II	20	4907.209	I	70	5760.551	I	3	6908.988	I
200	4148.182	II	240	4919.816	II	15	5773.946	I	24	6911.227	I
450	4149.986	II	18	4927.780	I	20	5789.645	I	5	6916.129	I
110	4158.535	I	40	4939.642	I	35	5804.141	I	5	6936.652	I
140	4165.766	I	60	4947.575	II	19	5815.422	II	35	6943.611	I
620	4178.060	II	50	4954.659	II	10	5832.370	I	5	6954.657	I
250	4179.714	II	30	4965.731	I	10	5845.919	I	15	6965.947	I
30	4184.138	I	35	4975.950	II	15	5854.121	I	3	6981.086	I
130	4193.017	I	24	4985.372	I	15	5868.373	I	55	6989.656	I
620	4208.890	II	50	5002.097	I	10	5878.933	I	24	6993.038	II
130	4210.923	I	50	5002.097	I	8	5891.451	I	18	7000.806	I
28	4214.828	I	50	5015.889	II	15	5899.844	I	18	7000.806	I
55	4220.065	I	260	5017.255	II	20	5914.387	II	3	7015.319	I
55	4227.387	I	20	5029.892	I	19	5925.893	II	10	7018.569	I
30	4230.824	I	24	5039.230	I	10	5937.162	I	3	7026.462	I
85	4235.463	I	50	5044.719	I	10	5944.648	I	7	7036.281	I
20	4241.094	I	240	5049.796	II	8	5957.587	I	30	7045.795	II
30	4247.989	II	85	5055.347	II	30	5973.665	I	15	7053.619	II
110	4253.538	I	70	5058.562	II	85	5989.044	II	6	7060.654	I
70	4256.254	I	110	5067.974	I	24	5994.129	I	24	7075.333	II
110	4260.333	I	30	5081.446	I	21	6007.072	I	30	7084.171	I
28	4269.942	I	50	5090.051	I	30	6015.426	II	24	7089.339	II
280	4273.357	II	50	5098.043	II	17	6021.036	I	10	7100.512	II
480	4277.313	II	40	5101.130	I	17	6037.698	I	3	7109.861	I
700	4282.042	II	50	5110.867	II	24	6044.431	II	11	7124.562	I
28	4288.669	I	30	5115.044	I	10	6053.381	I	3	7132.613	I
55	4297.306	I	10	5125.950	I	5	6061.536	I	5	7148.560	I
85	4299.839	I	20	5134.746	I	30	6077.106	I	10	7154.954	I
100	4307.176	I	95	5143.267	II	30	6087.262	II	30	7168.896	I
200	4309.991	II	120	5148.211	II	24	6099.083	I	15	7173.373	I
55	4315.254	I	50	5151.612	I	30	6104.580	I	40	7191.132	II
110	4318.416	I	50	5154.243	I	40	6112.837	I	7	7200.046	I
30	4325.274	I	85	5158.604	I	30	6120.571	II	35	7208.006	I
28	4330.844	I	70	5160.730	I	10	6124.480	I	11	7212.69	I
130	4337.277	I	20	5168.922	I	14	6151.993	I	10	7217.755	II
85	4342.256	II	50	5176.961	I	10	6161.354	I	11	7218.054	I
130	4344.326	II	35	5183.990	II	60	6169.822	I	3	7242.355	I
55	4349.072	I	50	5190.871	II	50	6182.622	I	5	7255.354	I
55	4354.484	I	50 h	5195.814	I	12	6191.906	I	3	7270.558	I
85	4359.372	I	50	5198.800	I	24	6193.858	II	5	7284.904	I
85	4365.930	I	95	5199.164	I	12	6203.493	I	7	7298.143	I
85	4374.123	I	50	5211.230	I	12	6224.528	I	11	7305.405	II
1300	4381.860	II	95	5216.596	II	24 h	6234.856	I	5	7315.067	I
1100	4391.110	II	50	5218.528	II	8	6240.954	I	7	7324.808	I
55	4392.974	I	35	5219.110	I	10	6257.424	I	5	7339.606	I
55	4401.580	I	110	5231.160	I	21	6261.063	I	8	7341.152	I
85	4408.882	I	85	5233.225	I	21	6261.418	I	5	7361.349	I
210	4412.741	II	85	5233.229	II	50	6274.116	II	5	7384.175	I
28	4416.845	I	95	5247.654	II	50	6274.117	II	18	7385.501	I
50	4422.048	I	10	5255.573	I	30	6279.172	II	3	7393.431	II
250	4432.963	II	35	5258.360	I	8	6291.192	I	5	7402.252	I
140	4440.866	II	12	5266.710	I	10	6303.251	II	3	7418.550	I

Intensity	Wavelength	
21	7428.940	I
10	7430.254	I
3	7444.749	I
2	7462.993	I
10	7481.355	I
2	7493.427	I
50	7525.508	II
7	7549.314	I
18	7567.740	I
12	7585.69	I
12	7585.792	I
4	7598.204	I
2	7607.824	I
5	7627.176	I
3	7636.176	I
30	7647.380	I
4	7658.324	I
7	7676.219	I
21	7685.305	I
4	7710.269	I
4	7728.951	I
10	7731.72	II
4	7771.948	I
15	7787.79	II
15	7788.937	I
5	7798.360	I
4	7810.625	I
21	7817.771	I
8	7834.459	II
15	7847.540	I
4	7864.023	I
12	7865.95	I
6	7886.284	I
11	7900.31	I
4	7937.732	I
11	7941.72	I
5	7954.594	I
7	7972.598	I
24	7978.974	I
11	7987.97	I
5	7993.680	I
2	8014.502	I
5	8024.253	II
11	8032.433	I
11	8062.64	I
5	8085.220	I
5	8093.626	I
5	8129.407	I
11	8138.477	I
18	8143.139	I
12	8159.729	I
10	8163.125	II
7	8169.788	I
15	8186.914	I
12	8203.199	II
2	8231.408	I
5	8252.395	I
3	8259.512	I
18	8275.629	I
3	8292.529	I
15	8320.857	I
30	8330.451	I
4	8358.726	I
6	8387.104	II
18	8403.767	II
15	8416.729	I
12	8421.227	I
21	8446.509	I
5	8464.230	I
18	8478.360	I
5	8510.621	I
5	8516.557	I
5	8539.795	I
6	8554.946	I
11	8573.122	I
12	8591.838	I
3	8621.325	I
3	8638.363	I
10	8665.487	I
4	8668.116	I
2	8701.127	I
5	8709.236	I
8	8732.401	II
18	8748.033	I
15	8758.244	I
8	8775.573	I
4	8792.058	I
3	8804.590	I
5	8841.185	I
18	8842.073	II
15	8868.834	I
4	8875.233	I
2	8907.038	I
5	8955.848	I
15	8957.97	II
40	8967.641	I
3	8987.408	I
5	9031.819	I
25	9048.252	I
5	9063.953	I
6	9094.831	I
3	9107.225	I
2	9118.140	I
3	9170.825	I
10	9203.963	I
10	9266.208	I
10	9276.276	I
10	9289.563	I
4	9317.722	I
7	9340.706	I
15	9399.085	I
10	9431.603	I
15	9461.030	I
15	9474.882	I
15	9495.501	I
15	9497.191	I
10	9505.392	I
10	9561.24	I
8	9613.689	II
12	9632.647	I
10	9664.700	I
4	9676.106	I
15	9700.564	I
15	9746.46	I
10	9812.70	I
10	9826.45	I
20	9833.42	I
15	10039.364	I
15	10089.138	I
15	10133.56	II
15	10419.57	II
15	10556.45	I
15	10723.92	II
20	10726.93	I
20	10942.24	II
15	11051.90	I
30	11230.259	I
20	11354.719	I
15	11703.46	I
15	11864.25	I
15	11940.64	I
20	11984.67	II
15	12018.72	I
20	12127.30	I
20	12194.16	II
15	12206.89	I
20	12231.94	I
15	12338.00	I
15	12477.30	I
20	12646.54	I
15	12866.64	I
15	12940.65	II
15	12959.82	I
15	13145.90	II
15	13565.67	I
15	14090.25	I
15	14168.67	I
15	14424.54	I
15	14618.98	I
15	14654.91	I
15	14940.49	II
15	15240.24	II
15	15429.78	I
15	15831.75	I
20	17208.22	II
15	17307.66	I
15	17381.91	I
15	17481.04	I
15	17584.52	I
15	17936.43	II
15	18811.88	I
15	19145.60	II
15	19338.98	II
10	19774.30	II
10	20634.36	I
10	20692.06	II
10	22264.35	II

Th III
Ref. 157 — J.G.C.

Intensity	Wavelength	
	Vacuum	
100	1888.12	III
	Air	
50	2149.18	III
50	2162.82	III
50	2199.74	III
50	2206.62	III
50	2291.59	III
100	2301.18	III
100	2319.42	III
80	2324.68	III
150	2335.50	III
100	2340.58	III
100	2363.06	III
50	2368.91	III
100	2371.42	III
80	2381.47	III
100	2391.48	III
200	2413.50	III
50	2424.54	III
200	2427.94	III
200	2431.68	III
200	2441.24	III
100	2463.66	III
50	2473.93	III
100	2501.08	III
60	2512.69	III
50	2514.31	III
100	4555.73	III
50	4589.28	III
100	5376.13	III
50	5447.18	III
50	6242.95	III
50	6599.39	III
50	7461.59	III
50	8105.14	III

Th IV
Ref. 156, 165 — J.G.C.

Intensity	Wavelength	
	Vacuum	
4	797.53	IV
1	835.55	IV
30	846.91	IV
1	854.02	IV
30	882.39	IV
12	886.66	IV
100	1565.85	IV
70	1682.22	IV
30	1684.01	IV
150	1707.37	IV
200	1959.02	IV
	Air	
200	2002.34	IV
100	2066.70	IV
20	2143.91	IV
30	2146.81	IV
1	2242.11	IV
2	2261.26	IV
5	2296.81	IV
100	2693.99	IV
2	4937.09	IV
4	4952.52	IV
3	5420.38	IV
2	6711.87	IV
3	6740.37	IV
50	6901.16	IV
	9839.25	IV
	10875.05	IV

THULIUM (Tm)
Z = 69

Tm I and II
Ref. 1 — C.H.C.

Intensity	Wavelength	
	Air	
360	2284.79	II
120	2329.77	II
70	2340.92	II
120	2363.91	II
45	2365.96	II
160	2367.11	II
150	2383.68	II
110	2388.95	II
450	2409.02	II
110	2412.44	II
120	2421.65	II
450	2426.17	II
140	2445.47	II
770	2480.13	II
150	2481.15	II
130	2487.52	II
250	2491.60	II
100	2499.54	II
130	2507.15	II
1300	2509.08	II
200	2520.87	II
250	2522.17	II
180	2524.11	II
130	2527.02	I
110	2527.42	II
120	2542.66	II
360	2552.76	I
540	2561.65	II
150	2563.86	II
430	2588.27	II
170 h	2596.49	I
110	2601.09	I
220	2606.02	II
810	2607.06	II
730	2624.33	II
210	2640.76	II
130	2646.45	II
160	2650.27	II
190	2658.48	II
250	2660.09	II
140	2668.20	II
310	2679.57	II
170	2697.50	II
540	2721.19	II
200	2744.08	II
270	2779.55	II
350	2785.07	II
680	2794.60	II
730	2797.27	II
250	2818.47	II
250	2827.02	II
580	2827.92	II
200	2831.55	II
310	2844.67	II
200	2854.17	I
200	2860.12	II
200	2861.74	II
1600	2869.23	II
630	2890.94	II
210	2918.27	II
270	2925.65	II
680	2926.74	II
630	2935.99	II
350	2951.26	II
430	2965.86	II
490	2973.22	I
540	2981.48	II
350	2986.52	II
630	2990.54	II
200	2993.26	II
230	3013.71	II
430	3014.65	II
1500	3015.30	II
270	3017.09	II
330	3026.07	II
280	3042.35	II
340 d	3046.76	II
320	3050.73	II
340	3056.07	II
580	3073.08	II
360	3081.12	I
740	3098.60	II
7400	3131.26	II
2300	3133.89	II
230	3144.90	II
230	3146.16	II
1900	3151.04	II
1500	3157.34	II
450	3172.65	I
2300	3172.83	II
380	3173.58	II
230	3195.33	II
320	3210.56	II
320	3210.82	II
320	3212.01	II
230	3231.51	II
470	3235.44	II
1200	3236.81	II
1600	3240.23	II
2300	3241.54	II
320	3246.96	I
420	3247.46	II
1900	3258.05	II
400	3261.65	II
320	3264.10	II
1600	3266.64	II
1200	3267.40	II
790	3268.99	II
1100	3276.81	II
1200	3283.40	II
1200	3285.61	II
2300	3291.00	II
2000	3302.46	II
210	3306.01	II
210	3306.91	II
210	3308.01	II
1200	3309.80	II
640	3310.59	II
400	3316.88	II
210	3318.65	II
230	3349.99	I
230	3354.86	II
4000	3362.61	II
490	3374.50	II
420 d	3384.99	II
1700	3397.50	II
420	3399.95	II
850	3410.05	I
340	3412.59	I
340	3416.59	I
6400	3425.08	II
950	3425.63	II
340	3429.33	I
850	3429.96	II
420	3431.19	II
4900	3441.50	II
4900	3453.66	II
8500	3462.20	II
210	3467.51	I

Thulium (Cont.)

Intensity	Wavelength		Spectrum
340	3476.69		I
340	3480.98		I
340	3481.75		II
420	3487.38		I
210	3492.58		II
340	3499.95		I
250	3513.02		II
250	3517.60		I
250	3534.85		II
1700	3535.52		II
490	3536.21		II
850	3536.58		II
420	3537.91		I
210	3555.82		I
420	3557.79		II
340	3560.92		I
420	3563.88		I
490	3565.91		II
1300	3566.47		II
420	3567.36		I
280	3574.06		II
280	3586.07		I
2100	3608.77		II
250	3609.53		II
380	3638.41		I
950	3643.65		II
240	3647.72		II
600	3653.61		II
500	3665.81		II
1100	3668.09		II
410	3677.98		II
450	3678.85	d	II
410	3694.74		II
4800	3700.26		II
3800	3701.36		II
330	3704.85		II
7700	3717.91		I
890	3725.06		II
2400	3734.12		II
5000	3744.06		I
1700	3751.81		I
310	3756.86		II
6000	3761.33		II
4800	3761.91		II
260	3783.55		II
380	3795.16		II
7100	3795.75		II
770	3798.54		I
240	3798.75		II
600	3807.72		I
380	3810.72		II
550	3817.39		II
290	3826.39		I
1300	3838.20		II
290	3840.87		I
8900	3848.02		II
140	3857.84		II
6800	3883.13		I
1800	3883.44		II
5400	3887.35		I
440	3890.53		II
440	3896.62		I
680	3900.79		II
3500	3916.48		I
120	3928.66		II
570	3929.58		II
1500	3949.27		I
1500	3958.10		II
440	3995.58		II
1800	3996.52		II
220	4024.23		I
380	4044.47		I
10000	4094.19		I
9500	4105.84		I
120	4132.69		II
1100	4138.33		I
120	4149.14		I
120	4158.60		I
8800	4187.62		I
520	4199.92		II
6000	4203.73		I
220	4206.00		II
380	4222.67		I
3000	4242.15		II
270	4271.71		I
150	4298.36		I
2700	4359.93		I
1400	4386.43		I
200	4394.42		I
120	4395.96		I
140	4396.50		I
55	4437.40		II
80	4442.74		I
50	4447.58		I
120	4454.03		I
80	4459.99		I
50	4467.98		I
540	4481.26		II
80	4489.70		I
150	4519.60		I
260	4522.57		II
180	4529.38		II
80	4532.15		I
110	4548.60		I

Intensity	Wavelength		Spectrum
40	4556.68		II
40	4561.86		II
80	4564.68		I
40	4567.11		II
95	4596.63		I
270	4599.02		I
35	4601.29		II
55	4603.43		II
40	4604.85		I
50	4613.97		I
40	4614.47		II
300	4615.94		II
35	4619.06		II
40	4621.72		I
80	4626.33		II
95	4626.56		II
40	4626.97		I
110	4634.26		II
40	4642.96		II
95	4643.12		I
35	4644.58		I
120	4655.09		I
35	4666.70		II
35	4671.99		II
35	4675.10		I
80	4675.31		I
40	4677.86		II
160	4681.92		I
70	4685.11		I
120	4691.11		I
110	4724.26		I
680	4733.34		I
35	4750.75		II
70	4759.90		I
27	4789.92		II
27	4807.48		I
35	4808.68		I
35	4813.50		I
27	4826.99		II
27	4828.97		I
80	4831.20		II
35	4835.75		I
27	4851.76	d	I
	4851.90		II
19	4872.28		II
27	4879.19		I
27	4891.64		I
24	4909.74		I
55	4923.83		I
140	4957.18		I
40	4970.87		II
27	4971.26		I
40	4975.12		II
50	4978.90		I
40	4980.68		II
55	4989.32		II
27	4993.79		II
19	4994.72		II
35	5001.02		I
27	5001.59		I
160	5009.77		II
35	5014.56		II
27	5017.87		II
160	5034.22		II
27	5041.00	h	II
22	5043.50		I
35	5045.41		I
27	5060.42		II
150	5060.90		I
27	5062.25		I
27	5065.88		I
80	5066.67		I
27	5072.42		I
27	5076.36		I
27	5077.18		I
35	5085.09		I
40	5107.53		I
95	5113.97		I
50	5114.55		II
22	5120.67		I
22	5140.28		II
40	5149.40		II
19	5182.68		I
40	5185.25		I
14	5204.51		II
80	5213.38		I
22	5228.23		II
14	5260.93		II
24	5267.34		II
40	5291.14		I
40	5294.32		I
35	5300.21		I
35	5302.69		I
55	5305.87		II
650	5307.12		I
16	5322.79		II
35	5338.90		I
	5339.03		I
80	5346.49		II
27	5372.98		II
14	5391.96		II
27	5400.46		II
27	5402.23		I
14	5405.98		II

Intensity	Wavelength		Spectrum
14	5461.95		II
14	5464.14		I
14	5465.54		II
16	5500.30		II
14	5526.82		II
24	5528.34		I
14	5539.03		II
27	5566.00		I
22	5581.37		I
14	5586.65		II
14	5589.94		II
14	5606.64		I
270	5631.41		I
40	5642.60		I
27	5645.40		I
70	5658.30		I
520	5675.84		I
14	5683.59		I
40	5684.76		II
14	5696.42		II
35	5709.97		II
22	5715.79		I
14	5733.81		II
11	5737.20	d	II
	5737.25		II
14	5738.92	h	II
27	5758.02		I
55	5760.20		I
190	5764.29		I
5	5778.82		II
19	5782.36		II
22	5784.46		II
11	5799.97		II
14	5811.09		II
14	5816.46	h	I
35	5838.76		II
240	5895.63		I
35	5899.47		I
24	5901.57		I
8	5912.58		I
11	5931.70		I
27	5935.90		I
140	5971.26		I
27	5975.02		I
11	5984.87		I
19	6025.44		I
11	6067.78		II
16	6131.53		I
14	6175.29		I
14	6181.41		II
14	6299.46		II
27	6352.66		I
22	6401.44		I
8	6430.94		II
14	6440.54		I
200	6460.26		I
14	6490.70		I
14	6519.78		I
8	6575.54		I
95	6604.96		I
8	6627.25		I
35	6657.72		I
11	6658.64		I
11	6692.93		I
30	6721.36		I
9	6726.34		I
9	6727.94		II
18	6739.22		I
9	6767.48	h	I
9	6777.93		I
110	6779.77		I
14	6782.00	h	I
18	6788.52		I
13	6820.27	h	I
14	6826.95		I
14	6829.12		II
23	6831.09		I
120	6844.26		I
80	6845.76		I
18	6854.12		I
6	6898.56		I
6	6915.86		I
10	6937.37		I
5	6949.54		I
5	6976.69	h	II
5	7010.79		I
6	7014.31	h	II
10	7017.90		I
6	7029.40	h	I
12	7034.34		I
10	7056.43		II
5	7060.97		I
6	7079.78		II
10	7106.14		I
5	7231.33	h	I
5	7233.74		II
4	7257.72		I
17	7272.62		I
8	7284.30		I
11	7286.16	h	I
14	7310.51		I
11	7336.63		II
14	7432.18		I
5	7434.51		II

Intensity		Wavelength	Spectrum
5		7439.95	II
75		7481.08	I
75		7490.20	I
10	h	7507.28	I
14		7545.78	I
140		7558.33	I
17	h	7580.61	I
20	h	7593.74	I
17		7595.07	II
5		7629.85	I
5	h	7648.76	II
17		7655.00	I
4		7660.32	I
7	h	7666.24	I
8		7676.04	II
8	h	7701.46	I
80		7731.53	I
4	h	7778.27	I
12	h	7782.35	I
8	h	7785.51	I
		7785.90	I
17		7803.93	I
4		7829.22	I
40		7856.08	I
3		7861.67	I
5		7918.10	I
55		7927.51	I
110		7930.84	I
6		7971.56	I
11	h	7985.93	I
14	h	8014.77	I
95		8017.90	I
3	h	8021.33	I
14		8194.19	I
5		8294.52	I
7		8365.75	I
7		8460.79	II
27		8472.01	II
7	h	8546.07	II
11		8565.73	II

Tm III
Ref. 307 — J.R.

Intensity	Wavelength (Air)	Spectrum
500	2099.11	III
500	2107.10	III
200	2136.67	III
200	2156.29	III
800	2182.98	III
300	2183.91	III
5000	2185.94	III
100	2212.25	III
300	2230.86	III
400	2231.25	III
200	2243.34	III
400	2243.98	III
200	2246.68	III
200	2269.39	III
1000	2276.91	III
100	2280.08	III
100	2281.27	III
100	2282.86	III
200	2282.98	III
200	2286.57	III
400	2287.21	III
500	2294.73	III
20000	2296.21	III
200	2297.43	III
100	2304.64	III
400	2304.82	III
5000	2305.03	III
20000	2311.16	III
5000	2312.72	III
200	2314.88	III
400	2317.35	III
500	2320.96	III
200	2322.83	III
100	2323.71	III
100	2323.77	III
100	2324.43	III
500	2324.62	III
5000	2326.19	III
100	2327.02	III
300	2327.25	III
6000	2328.50	III
6000	2329.29	III
200	2330.87	III
3000	2331.80	III
400	2335.01	III
1000	2338.36	III
500	2341.74	III
300	2342.04	III
100	2344.59	III
500	2345.61	III
300	2347.43	III
400	2353.10	III
100	2355.65	III
3000	2357.05	III

Thulium (Cont.)

Intensity	Wavelength		Intensity	Wavelength	
1000	2361.23	III	100	2974.85	III
500	2363.97	III	1000	2998.28	III
1000	2375.32	III	100	3048.11	III
700	2375.83	III	700	3078.87	III
400	2389.52	III	200	3120.15	III
4000	2406.63	III	200	3277.26	III
500	2435.31	III	200	3407.73	III
500	2457.86	III	100	3415.40	III
500	2471.23	III	100	3415.96	III
30000	2489.44	III	100	3436.93	III
200	2496.25	III	400	3467.93	III
2000	2504.71	III	200	3529.29	III
3000	2519.78	III	200	3533.28	III
10000	2552.46	III	100	3537.47	III
500	2557.90	III	300	3562.41	III
1000	2574.52	III	100	3563.42	III
500	2574.98	III	200	3587.74	III
100	2581.84	III	600	3617.96	III
100	2585.48	III	1000	3629.09	III
300	2589.20	III	100	3706.11	III
500	2608.96	III	100	3799.41	III
300	2609.66	III	300	3998.84	III
500	2617.22	III	200	4021.92	III
500	2618.78	III	200	4026.03	III
1000	2621.12	III	700	4032.13	III
400	2621.35	III	100	4076.15	III
400	2622.31	III	200	4335.47	III
100	2627.09	III	500	4385.41	III
100	2628.83	III			
300	2634.66	III			
200	2636.68	III			
200	2637.30	III			
100	2640.32	III			
500	2643.58	III			
100	2645.05	III			
200	2649.27	III			
100	2650.82	III			
100	2654.05	III			
100	2656.30	III			
500	2661.51	III			
1000	2663.00	III			
500	2664.76	III			
500	2664.88	III			
200	2665.05	III			
1000	2666.93	III			
200	2668.59	III			
100	2668.66	III			
200	2669.18	III			
100	2671.42	III			
100	2675.30	III			
1000	2676.64	III			
500	2676.91	III			
100	2678.28	III			
100	2680.49	III			
5000	2682.32	III			
300	2682.64	III			
300	2687.14	III			
300	2695.69	III			
400	2698.21	III			
1000	2699.49	III			
1000	2699.80	III			
100	2703.63	III			
200	2703.68	III			
100	2704.93	III			
2000	2707.03	III			
300	2707.19	III			
200	2707.44	III			
500	2707.60	III			
1000	2709.74	III			
200	2710.79	III			
1000	2713.38	III			
200	2715.81	III			
300	2717.56	III			
100	2718.02	III			
3000	2719.47	III			
3000	2724.44	III			
4000	2727.56	III			
200	2728.13	III			
1000	2731.38	III			
300	2732.11	III			
400	2737.98	III			
800	2744.74	III			
400	2745.99	III			
500	2752.46	III			
500	2753.20	III			
400	2756.15	III			
800	2765.98	III			
700	2769.92	III			
200	2772.64	III			
100	2777.43	III			
400	2781.12	III			
2000	2806.77	III			
300	2821.12	III			
200	2849.52	III			
700	2882.02	III			
100	2899.29	III			
100	2912.33	III			
400	2921.08	III			
200	2947.02	III			
1000	2947.72	III			
500	2953.18	III			
100	2966.15	III			
500	2966.85	III			
400	2972.61	III			

TIN (Sn)
Z = 50

Sn I and II
Ref. 187, 191 — C.H.C.

Intensity		Wavelength		Intensity		Wavelength		Intensity		Wavelength	
		Vacuum									
1		899.92	II	50	h	1977.6	I	550	r	3262.34	I
2		917.40	II	80		1984.20	I	50		3283.21	II
1		935.63	II	15		1991.88	I	110		3330.62	I
3		945.83	II	20		1994.98	I	60		3351.97	II
4		954.50	II					2		3407.48	II
7		985.13	II			Air		10		3472.46	II
4		997.21	II					7		3537.57	II
2		1016.26	II	25		2008.05	I	11		3575.45	II
4		1040.78	II	30		2015.76	I	3		3582.39	II
1		1041.32	II	30		2025.98	I	2		3620.08	II
3		1062.10	II	50		2040.66	I	6		3620.54	II
8		1108.19	II	20		2040.90	I	40		3655.78	I
4		1159.05	II	50		2054.03	I	6		3715.23	II
10		1161.43	II	70		2058.31	I	280	r	3801.02	I
3		1162.94	II	20		2064.00	I	4		3841.44	II
4		1180.51	II	80		2068.58	I	1		4294.65	II
9		1219.07	II	100		2072.89	I	40		4524.74	I
13		1223.70	II	100		2073.08	I	1		4579.13	II
11		1243.00	II	25		2080.62	I	1		4580.29	II
20		1290.86	II	30		2091.58	I	2		4877.22	II
20		1316.59	II	40		2094.35	I	3		4944.31	II
25		1400.52	II	200		2096.39	I	20		4979.73	I
20		1475.15	II	100		2100.93	I	2		5071.14	II
9		1489.22	II	100	r	2113.93	I	2		5072.67	II
7		1609.47	II	50		2121.26	I	20		5174.54	I
10	r	1737.21	I	25		2140.73	I	10		5332.36	II
15	r	1751.46	I	20		2141.43	I	20		5561.95	II
10	h	1753.3	I	15		2148.46	I	25		5588.92	II
7		1758.00	II	1		2148.63	II	2		5596.20	II
20	r	1764.98	I	40	r	2148.73	I	500		5631.71	I
20		1773.40	I	20	r	2151.43	I	15		5753.59	II
30	r	1790.75	I	30		2151.54	II	1		5797.20	II
80	r	1804.60	I	80		2171.32	I	15		5799.18	II
15		1811.34	I	150	r	2194.49	I	50		5925.44	I
30		1813.04	I	300	r	2199.34	I	100		5970.30	I
40	r	1815.74	I	400	r	2209.65	I	150		6037.70	I
25		1819.31	I	4		2209.67	II	200		6054.86	+
120	r	1823.00	I	40		2211.05	I	250		6069.00	I
9		1831.89	II	80	r	2231.72	I	100		6073.46	I
50	r	1848.75	I	400	r	2246.05	I	6		6077.48	II
30		1852.00	I	6		2246.07	II	5		6079.70	II
200	r	1860.32	I	60		2251.17	I	400		6149.71	I
20		1861.42	I	30		2267.19	I	200		6154.60	I
20		1865.52	I	400	r	2268.91	I	150		6171.50	I
30		1865.99	I	20		2282.26	I	100		6310.78	I
15		1873.29	I	200	r	2286.68	I	40		6354.35	I
30		1882.64	I	600	r	2317.23	I	70		6453.50	II
80		1886.05	I	300	r	2334.80	I	8		6761.45	II
100		1891.40	I	1000	r	2354.84	I	25		6844.05	II
20		1897.29	I	20		2357.39	I	20		7191.40	II
12		1899.91	II	3		2360.34	II	10		7387.79	I
50		1909.30	I	22		2368.33	II	20	h	7398.6	I
40		1911.61	I	60		2380.72	I	1		7408.62	II
20		1913.52	I	4		2384.54	II	30		7685.30	I
80		1925.31	I	100		2408.15	I	13		7741.80	II
20		1926.77	I	800	r	2421.70	I	100		7754.97	I
15		1927.95	I	1000	r	2429.49	I	3		7904.00	II
40	h	1928.9	I	1		2433.52	II	100	h	8030.5	I
25		1933.17	I	15		2448.98	II	30	h	8039.3	I
20		1942.69	I	60		2455.24	I	200		8114.09	I
150		1952.15	I	20		2476.40	I	30	h	8121.0	I
15		1960.21	I	300		2483.39	I	30		8349.35	I
30		1971.46	I	13		2483.48	II	80		8357.04	I
				10		2486.99	II	300		8422.72	I
				200		2495.70	I	400		8552.60	I
				5		2522.61	II	50	h	8681.7	I
				90		2523.92	I	30		9018.95	I
				80	h	2531.17	I	50	h	9410.86	I
				400		2546.55	I	80	h	9415.37	I
				40	h	2558.01	I	150		9616.40	I
				500	r	2571.58	I	50		9741.1	I
				200		2594.42	I	100	h	9742.8	I
				50	h	2636.94	I	300	h	9805.38	I
				200		2661.24	I	500		9850.52	I
				2		2664.93	II	25		10456.47	I
				700	r	2706.51	I	11		10807.58	I
				2		2727.82	II	54		10894.00	I
				20		2761.78	I	70		11191.85	I
				150		2779.81	I	56		11277.66	I
				80		2785.03	I	17		11336.97	I
				60		2787.96	I	200		11454.59	I
				60		2812.59	I	200		11616.26	I
				80		2813.58	I	76		11670.77	I
				2		2825.52	II	25		11694.45	I
				1400	r	2839.99	I	258		11739.78	I
				1		2846.42	II	96		11825.18	I
				200		2850.62	I	106		11835.82	I
				1000	r	2863.32	I	254		11932.99	I
				1		2912.80	II	48		12009.50	I
				200		2913.54	I	111		12313.24	I
				6		2919.82	II	33		12335.6	I
				3		2991.00	II	42		12530.87	I
				7		2994.44	II	42		12536.5	I
				700	r	3009.14	I	37		12788.2	I
				1		3012.18	II	89		12888.5	I
				8		3023.94	II	187		12981.1	I
				200		3032.80	I	20		13000.3	I
				850	r	3034.12	I	187		13018.5	I
				12		3047.50	II	68		13081.5	I
				6		3094.69	II	378		13460.2	I
				60		3141.84	I	144		13608.2	I
				550	r	3175.05	I	13		15018.2	I
				40		3218.71	I	30		15464.2	I
								20		17000.5	I

Tin (Cont.)

Intensity	Wavelength	
10	17807.5	I
20	20622.2	I
40	20861.7	I
8	21686.2	I
4	22131.7	I
3	22997.2	I
4	24327.2	I
4	24738.2	I

Sn III
Ref. 423 — C.H.C.

Intensity	Wavelength	
	Vacuum	
100	753.01	III
50	760.62	III
75	775.79	III
50	784.68	III
200	910.92	III
50	1010.92	III
50	1048.84	III
1000	1139.29	III
1000	1158.33	III
200	1161.09	III
100	1161.58	III
100	1180.62	III
1000	1184.25	III
200	1189.99	III
200	1204.06	III
2000	1210.52	III
100	1215.10	III
100	1218.14	III
100	1230.17	III
100	1231.38	III
500	1243.63	III
1000	1259.92	III
40	1276.31	III
1000	1305.97	III
1000	1327.34	III
200	1334.70	III
200	1346.05	III
1000	1347.65	III
200	1369.71	III
1000	1386.74	III
500	1410.61	III
200	1449.77	III
1000	1570.36	III
50	1674.29	III
500	1811.71	III
500	1941.86	III
50	1955.52	III

Sn IV
Ref. 423 — C.H.C.

Intensity	Wavelength	
	Vacuum	
50	605.23	IV
50	619.04	IV
50	628.73	IV
50	908.22	IV
500	956.25	IV
500	1019.72	IV
1000	1044.49	IV
100	1058.37	IV
50	1058.59	IV
1000	1073.41	IV
200	1087.50	IV
300	1096.92	IV
50	1103.24	IV
000	1119.34	IV
200	1120.68	IV
1000	1314.55	IV
1000	1437.52	IV
100	1532.90	IV

Sn V
Ref. 399, 423 — C.H.C.

Intensity	Wavelength	
	Vacuum	
120	355.14	V
150	361.01	V
100	372.55	V
200	1089.35	V
100	1132.79	V
200	1160.74	V
100	1176.26	V
100	1189.92	V
100	1205.72	V
2000	1251.38	V
100	1283.81	V
200	1294.36	V
100	1302.20	V

TITANIUM (Ti)
Z = 22

Ti I and II
Ref. 1 — C.H.C.

Intensity	Wavelength		Intensity	Wavelength		Intensity	Wavelength	
	Air							
140	2272.61	I	280	2884.11	II	310	3272.08	II
180	2273.28	I	65	2888.93	II	200	3278.29	II
130	2276.70	I	55	2891.07	II	260	3278.92	II
190	2279.96	I	55	2905.66	I	220	3282.33	II
150	2299.85	I	30	2909.92	II	530	3287.66	II
140	2302.73	I	450	2912.08	I	290	3292.08	I
190	2305.67	I	340	2928.34	I	170	3299.41	I
65	2380.81	I	15	2931.03	II	170	3306.88	I
50	2384.52	I	180	2933.55	I	220	3308.39	I
55	2418.36	I	26	2935.96	II	220	3308.81	II
75	2421.30	I	150	2937.32	I	260	3309.50	I
95	2424.24	I	1100	2942.00	I	60	3309.73	I
40	2428.23	I	1300	2948.26	I	110	3312.69	I
35	2433.22	I	30	2954.58		840	3314.42	I
19	2434.10	I	1600	2956.13	I		3314.52	I
35	2440.21	II	170	2956.80	I	290	3315.32	II
65	2440.98	I	30	2958.77		330	3318.02	II
24	2450.44	II	26	2959.71	I	550	3321.70	II
24	2504.54	I	35	2959.99	I	2900	3322.94	II
75	2517.43	II	170	2965.71	I	380	3326.76	II
40	2519.04	I	190	2967.22	I	2100	3329.46	II
140	2520.54	I	26	2968.23	I	550	3332.11	II
75	2524.64	II	75	2970.38	I	1800	3335.20	II
360	2525.60	II	30	2974.93	I	1100	3340.34	I
29	2527.98	I	170	2983.31	I	5700	3341.88	I
210	2529.85	I	35	3000.87	I	120	3342.15	I
190	2531.25	II	120	3017.19	II	260	3343.77	II
190	2534.62	II	140	3029.73	II	330	3346.73	II
130	2535.87	II	110	3046.68	II	4300	3349.04	II
190	2541.92	I	130	3056.74	II	12000	3349.41	II
65	2555.99	II	130	3057.40	II	120	3352.94	I
110	2571.03	II	170	3058.09	II	4100	3354.64	I
50	2572.65	II	85	3059.74	II	290	3358.28	I
50	2580.82	I	1300 d	3066.22	II	290	3360.99	I
35	2590.26	I		3066.35	II	7200	3361.21	II
190	2593.64	I	70	3071.24	II		3361.26	I
65	2596.58	I	600	3072.11	II	120	3361.84	I
270	2599.92	I	1100	3072.97	II	1100	3370.44	I
340	2605.15	I	1600	3075.22	II	4300	3371.45	I
510	2611.28	I	2300	3078.64	II	140	3372.21	II
75	2611.48	I	3600	3088.02	II	5700	3372.80	II
300	2619.94	I	180	3089.40	II	60	3374.35	II
170	2631.54	I	180	3097.19	II	2900 d	3377.48	I
170	2632.42	I	180	3100.67	I		3377.58	I
640	2641.10	I	230	3103.80	II	290	3379.22	I
800	2644.26	I	230	3105.08	II	1400	3380.28	II
950	2646.64	I	260	3106.23	II	170	3382.31	I
30	2649.30	I	70	3106.81	I	5700	3383.76	II
15	2654.93	I	50	3110.67	II	170	3385.66	I
35	2657.19	I	50	3112.48	I	1400	3385.95	I
85	2661.97	I	140	3117.67	II	1400	3387.84	II
95	2669.60	I	720	3119.72	I	60	3388.76	II
130	2679.93	I		3119.80	II	140	3390.68	I
26	2684.80	I	190	3123.07	I	140	3392.71	I
30	2685.14	I	240	3130.80	II	1100	3394.58	II
65	2688.82	I	140	3141.54	I	60	3398.63	I
26	2716.25	II	95	3141.67	I	60	3402.42	II
85	2725.07	I	220	3143.76	II	60	3407.20	II
75	2727.42	I	240	3148.04	II	95	3409.81	II
21	2731.13	I	240	3152.25	II	60	3439.30	I
40	2731.58	I	240	3154.20	II	890	3444.31	II
170	2733.26	I	240	3155.67	II	60	3452.47	II
55	2735.29	I	500	3161.20	II	180	3456.39	II
40	2735.61	I	780	3161.77	II	600	3461.50	II
85	2739.81	I	1000	3162.57	II	95	3467.26	I
250	2742.32	I	1600	3168.52	II	600	3477.18	II
40	2749.06	I	2400	3186.45	I	60	3478.92	I
65	2757.40	I	1000	3190.87	II	240	3480.53	I
95	2758.08	I	3100	3191.99	I	60	3485.69	II
15	2761.29	II	50	3197.52	II	480	3489.74	II
250	2802.50	I	3800	3199.92	I	480	3491.05	II
55	2805.70	I	780	3202.54	II	60	3495.75	I
30	2806.50	II	50	3203.44	II	95	3499.10	I
40	2809.17	I	240	3203.83	I	890	3504.89	II
75	2810.30	II	50	3204.87	I	120	3506.64	I
30	2812.98	I	110	3213.14	II	600	3510.84	II
30	2817.40	I	260	3214.24	I	60	3520.25	II
65	2817.84	I	190	3214.75	II	310	3535.41	II
	2817.87	II	1100	3217.06	II	190	3547.03	I
65	2828.04	I	110	3217.94	I	120	3573.74	II
	2828.15	II	260	3218.27	II	60	3574.24	I
130	2832.16	II	110	3219.21	I	60	3587.13	II
190	2841.94	II	110	3221.38	I	240	3596.05	II
110	2851.10	II	1300	3222.84	II	190	3598.72	I
40	2853.93	I	220	3223.52	I	600	3610.16	I
95	2862.32	II	240	3224.24	I	190	3624.82	II
55	2868.74	II	140	3226.13	I	95	3635.20	I
180	2877.44	II	530	3228.60	II	4800	3635.46	I
			780	3229.19	II	120	3637.97	I
			530	3229.42	II	190	3641.33	II
			110	3231.32	II	6600	3642.68	I
			240	3232.28	II	180	3646.20	I
			6600	3234.52	II	7200	3653.50	I
			220	3236.12	II	290	3654.59	I
			5200	3236.57	II	660	3658.10	I
			4100	3239.04	II	120	3659.76	II
			220	3239.66	II	380	3660.63	I
			2600	3241.99	II	190	3662.24	II
			1200	3248.60	II	380	3668.97	I
			950	3251.91	II	600	3671.67	I
			1200	3252.91	II	3100	3685.20	II
			1200	3254.25	II	120	3685.96	I
			1200	3261.60	II	95	3687.35	I
			310	3271.65	II	600	3689.91	I

Titanium (Cont.)

Intensity	Note	Wavelength	Spectrum
140		3694.45	I
30		3698.18	I
60		3698.43	I
60		3700.08	I
120		3702.29	I
190		3704.30	I
140		3706.23	II
50		3707.53	I
290		3709.96	I
30		3715.40	I
450		3717.40	I
140		3721.64	II
330		3722.57	I
600		3724.57	I
380		3725.16	I
2900		3729.82	I
50		3735.67	I
60		3738.90	I
3300		3741.06	I
330		3741.64	II
160		3748.10	I
5200		3752.86	I
600		3753.64	I
140		3757.69	II
3300		3759.30	II
2900		3761.32	II
50		3761.89	II
60		3766.45	I
600		3771.66	I
30		3776.06	II
840		3786.04	I
120		3789.30	I
70		3795.90	I
60		3798.31	I
70		3818.22	I
60		3822.03	I
240		3828.19	I
95		3833.68	I
95		3836.78	I
60		3846.45	I
130		3853.05	I
130		3853.73	I
170		3858.14	I
240		3866.44	I
170		3868.40	I
120		3873.21	I
260		3875.26	I
170		3882.15	I
170		3882.33	I
500		3882.89	I
60	h	3888.02	I
70		3889.95	I
200	h	3895.25	I
85		3898.49	I
530		3900.54	II
180		3900.96	I
2600		3904.78	I
110	h	3911.19	I
500		3913.46	II
500		3914.34	I
24		3914.74	I
35		3919.82	I
290		3921.42	I
1100		3924.53	I
110		3926.32	I
890		3929.88	I
35		3932.02	II
70		3934.24	I
1100		3947.78	I
4500		3948.67	I
4500		3956.34	I
5200		3958.21	I
950		3962.85	I
950		3964.27	I
4800		3981.76	I
570		3982.48	I
60		3984.33	I
35		3985.25	I
60		3985.59	I
5700		3989.76	I
35		3994.70	I
7800		3998.64	I
70		3999.36	I
70		4002.49	I
70		4003.81	I
35		4005.97	I
70		4008.06	I
950		4008.93	I
190		4009.66	I
70		4012.39	II
180		4013.58	I
70		4015.38	I
35		4016.28	I
120	h	4017.77	I
140		4021.83	I
1200		4024.57	I
40		4025.14	II
190	h	4026.54	I
40		4027.48	I
40		4028.34	II
190	h	4030.51	I
40		4033.91	I
30		4034.91	I
110		4035.83	I
35	h	4040.32	I
290		4055.02	I
85		4057.62	I
85		4058.14	I
410		4060.26	I
200		4064.22	I
200		4065.10	I
840		4078.47	I
40		4079.72	I
290		4082.46	I
85		4099.17	I
220		4112.71	I
85		4122.17	I
40		4123.31	I
85		4123.57	I
130		4127.54	I
40		4129.17	I
40		4131.25	I
140		4137.29	I
85		4143.05	I
170		4150.96	I
85		4159.64	I
70		4163.65	II
35		4164.14	I
40		4166.32	I
85		4169.35	I
120		4171.03	I
40		4171.90	II
35		4183.30	I
360		4186.12	I
40		4188.69	I
70		4200.75	I
85		4203.46	I
35		4211.73	I
40		4224.79	I
40		4227.65	I
130		4237.89	I
85		4249.12	I
130		4256.04	I
70		4258.54	I
70		4261.60	I
330		4263.13	I
35		4265.71	I
40		4266.22	I
70		4270.14	I
85		4272.43	I
240		4274.58	I
120		4276.43	I
120		4278.23	I
30		4278.81	I
110		4281.38	I
220		4282.71	I
160		4284.99	I
840		4286.01	I
840		4287.40	I
30		4288.16	I
950		4289.07	I
120		4290.23	II
840		4290.94	I
120		4291.14	I
140		4294.12	II
840		4295.76	I
2000		4298.66	I
200		4299.23	I
200		4299.64	I
200		4300.05	II
2900		4300.56	I
4100		4301.09	I
85		4301.93	II
6000		4305.92	I
180		4307.90	II
35		4308.50	I
40		4311.65	I
85		4312.87	II
85		4314.35	I
1200		4314.80	I
360		4318.64	I
180		4321.66	I
190		4325.13	I
160		4326.36	I
30		4334.84	I
160		4337.92	II
24		4344.29	II
70		4346.11	I
35		4354.06	I
95		4360.49	I
24		4368.94	I
95		4369.68	I
60		4372.38	I
30		4388.08	I
170		4393.92	I
330		4395.04	II
60		4399.77	II
240		4404.28	I
60		4404.90	I
30		4405.68	I
60		4416.54	I
220		4417.28	I
60		4417.72	II
120		4421.76	I
120		4422.82	I
24		4424.39	I
30		4425.83	I
120		4426.06	I
890		4427.10	I
21		4430.02	I
85		4430.37	I
50		4431.28	I
30		4432.60	I
24		4433.58	I
170		4434.00	I
70		4436.59	I
30		4438.23	I
130		4440.35	I
50		4441.27	I
230		4443.80	II
24		4444.27	I
840		4449.15	I
30		4450.49	II
550		4450.90	I
840		4453.32	I
290		4453.71	I
950		4455.33	I
1100		4457.43	I
21		4462.09	I
70		4463.38	I
95		4463.54	I
290		4465.81	I
240		4468.50	II
240		4471.24	I
95		4474.85	I
95		4479.70	II
50		4480.59	I
530		4481.26	I
95		4482.69	I
19		4488.32	II
260		4489.09	I
24		4492.55	I
40		4495.01	I
240		4496.15	I
24		4497.73	I
200		4501.27	II
40		4503.78	I
21		4506.36	I
50		4511.17	I
780		4512.74	I
19		4515.62	I
1000		4518.03	I
95		4518.70	I
1000		4522.80	I
780		4527.31	I
6000		4533.24	I
240		4533.97	II
3600		4534.78	I
2400		4535.58	I
1200		4535.92	I
1200		4536.05	I
24		4537.23	I
24		4539.10	I
720		4544.69	I
950		4540.77	I
240		4549.63	II
950		4552.46	I
24		4555.08	I
720		4555.49	I
19		4557.86	I
19		4558.11	I
60		4559.92	I
50		4562.63	I
35		4563.43	I
110		4563.77	II
35		4570.91	I
240		4571.98	II
19		4585.84	I
24		4589.95	II
60		4599.23	I
21		4609.37	I
950		4617.27	I
24		4619.52	I
480		4623.09	I
190		4629.34	I
50	d	4634.87	I
60		4637.88	I
240		4639.37	I
220		4639.67	I
190		4639.95	I
140		4645.19	I
120		4650.02	I
24		4656.04	I
720		4656.47	I
840		4667.59	I
70		4675.12	I
950		4681.92	I
21		4686.92	I
24		4690.80	I
190		4691.34	I
40		4693.68	I
24		4696.94	I
190		4698.76	I
120		4710.19	I
24		4715.30	I
65		4722.62	I
65		4723.17	I
55		4731.16	I
45		4733.43	I
18		4734.68	I
22		4742.11	I
170		4742.79	I
22		4747.68	I
310		4758.12	I
310		4759.28	I
45		4766.33	I
28		4769.77	I
65		4778.26	I
45		4781.72	I
110		4792.49	I
45		4796.22	I
35		4797.98	I
110		4799.80	I
28		4805.10	II
110		4805.43	I
45		4808.53	I
22		4811.08	I
40		4812.25	I
200		4820.42	I
22		4825.46	I
40		4836.13	I
470		4840.87	I
65		4848.47	I
290		4856.01	I
35		4864.18	I
200		4868.26	I
250		4870.14	I
28		4880.91	I
45		4882.35	I
400		4885.08	I
380		4899.91	I
320		4913.62	I
55		4915.24	I
130		4919.87	I
180		4921.77	I
55		4925.41	I
30		4926.16	I
150		4928.34	I
30		4937.74	I
95		4938.29	I
30		4941.58	I
21		4948.19	I
21		4958.25	I
55		4964.75	I
21		4966.04	I
65		4968.58	I
75		4973.05	I
120		4975.35	I
65		4977.74	I
120		4978.20	I
5800		4981.73	I
150		4989.15	I
4600		4991.07	I
30		4995.08	I
140		4997.10	I
4000		4999.51	I
230		5001.01	I
3600		5007.21	I
120		5009.65	I
230		5013.30	I
3200	d	5014.19	I
		5014.24	I
580		5016.17	I
840		5020.03	I
840		5022.87	I
580		5024.84	I
300		5025.58	I
1200		5035.91	I
840		5036.47	I
740		5038.40	I
1200		5039.95	I
75		5040.62	I
85		5043.59	I
35		5044.27	I
55		5045.41	I
26		5048.21	I
110		5052.87	I
21		5054.08	I
110		5062.11	I
35		5064.07	I
1400		5064.66	I
95		5065.99	I
35	h	5068.33	I
65		5069.35	I
130		5071.48	I
40		5085.34	I
130		5087.07	I
21		5103.15	I
55		5109.44	I
190		5113.44	I
270		5120.42	I
30		5129.15	II
270		5145.47	I
230		5147.48	I
210		5152.20	I
21	B1	5166.86	I
1100		5173.75	I
40		5186.34	I
85		5188.70	II
30		5189.58	I
1300		5192.98	I
85	h	5194.04	I
65		5201.10	I
120		5206.08	I
75		5207.87	I
65		5208.42	I

Intensity		Wavelength	
1400		5210.39	I
65		5212.29	I
150		5219.71	I
95		5222.69	I
85		5223.64	I
250		5224.32	I
95		5224.56	I
190		5224.95	I
65		5226.56	II
120		5238.58	I
21		5246.15	I
55		5246.57	I
75		5247.31	I
21		5250.95	I
110		5252.11	I
75		5255.83	I
55		5259.99	I
55		5263.50	I
150		5265.98	I
40		5282.39	I
140		5283.45	I
35		5284.39	I
26		5288.81	
65		5295.79	I
120		5297.26	I
65		5298.44	I
26		5336.81	II
17		5341.50	I
75		5351.08	I
26		5366.65	I
55		5369.64	I
40		5389.18	I
55		5389.99	I
17		5396.60	I
85		5397.09	I
35		5404.02	I
110		5409.61	I
40		5426.26	I
75		5429.15	I
26		5436.73	I
17		5438.32	I
40		5446.64	I
11	B1	5448.34	T
30		5448.90	I
21		5449.16	I
35		5453.65	I
55		5460.51	I
75		5471.21	I
35		5472.70	I
40	h	5473.55	I
85		5474.23	I
30		5474.46	I
120	h	5477.71	I
110		5481.43	I
75		5481.87	I
85	h	5488.20	I
150		5490.15	I
26		5490.84	I
110		5503.90	I
40		5511.78	I
340		5512.53	I
270		5514.35	I
320		5514.54	I
26		5530.49	I
110		5565.49	I
13		5579.16	
21	h	5582.98	
30	h	5585.68	
65	B1	5597.85	T
55	B1	5629.28	T
17		5635.84	
250		5644.14	I
75		5648.58	I
26	B1	5661.55	T
190		5662.16	I
75		5662.91	I
21		5673.42	I
130		5675.44	I
30	h	5679.94	I
95		5689.47	I
75		5702.68	I
35		5708.23	I
65		5711.88	I
40	h	5713.92	I
95		5715.13	I
55		5716.48	I
35		5720.48	I
85		5739.51	I
40		5740.02	I
19		5741.22	I
21		5752.84	I
19		5756.86	I
40	h	5762.27	I
55	h	5766.35	I
75	h	5774.05	I
30		5780.78	I
75	h	5785.98	I
65	H1	5804.26	I
21	B1	5814.96	T
40		5823.71	I
21	h	5841.18	I
21		5852.34	I
400		5866.46	I
65		5880.31	I

Intensity		Wavelength	
21	h	5888.68	
230		5899.32	I
55		5903.33	I
120		5918.55	I
150		5922.12	I
75		5937.82	I
120		5941.76	I
300		5953.17	I
200		5965.84	I
270		5978.56	I
340		5999.04	I
65		5999.68	I
21		6012.73	
110		6064.63	I
120		6085.23	I
120		6091.17	I
40		6092.81	I
40	h	6098.67	I
35	h	6121.01	I
120		6126.22	I
19		6138.38	I
30		6146.62	I
21		6149.74	I
30	B1	6162.23	T
35		6186.15	I
95	h	6215.28	I
75	h	6220.49	I
65	h	6221.41	I
380		6258.10	I
380		6258.70	I
300		6261.10	I
65		6303.75	I
55		6312.24	I
26		6318.03	I
30		6336.10	I
35		6366.35	I
11		6419.10	I
17		6497.69	I
19		6508.14	I
55		6546.28	I
65		6554.23	I
11	h	6554.83	I
75		6556.07	I
19	h	6565.62	I
14	h	6575.18	I
35		6599.11	I
18	B1	6651.46	T
18	h	6666.55	I
22	h	6667.74	I
9		6668.39	I
18		6677.18	I
22	b	6691.21	T
26		6716.68	I
16	B1	6723.95	T
80		6743.12	I
22		6745.52	I
18		6844.64	I
18		6860.39	I
35		6861.47	I
9		6873.94	I
12		6913.19	I
14	h	6933.15	I
14	h	6943.70	I
23		6996.63	I
15		7004.66	I
14		7008.35	I
14		7010.44	I
14	h	7035.86	I
40		7038.80	I
14		7050.65	I
40	B1	7054.51	T
23		7069.11	I
23		7072.05	I
45	B1	7087.89	T
30	b	7124.9	T
40	B1	7125.61	T
26		7138.91	I
26		7167.13	I
23		7171.53	I
55		7189.89	I
26	b	7203.64	T
260		7209.44	I
60		7216.20	I
130		7244.86	I
130		7251.72	I
19		7263.40	I
19		7266.29	I
19	b	7269.05	T
15		7315.56	I
26		7318.39	I
120		7344.72	I
11		7352.16	I
90		7357.74	I
60		7364.11	I
26		7440.60	I
9		7474.94	I
26		7489.61	I
19		7496.12	I
12		7580.55	I
9	B1	7589.62	T
15		7614.50	I
23		7654.44	I
11	B1	7705.21	T

Intensity		Wavelength	
30		7949.17	I
26	h	7961.58	I
60		7978.88	I
9		7979.07	I
30		7996.53	I
7	h	8003.55	
55		8024.84	I
30		8068.24	I
8		8267.62	
14	h	8306.31	I
9	h	8307.41	I
9	h	8311.76	I
8	h	8312.85	I
12		8334.37	I
14		8353.15	I
75		8364.24	I
100		8377.85	I
100		8382.54	I
55		8382.82	I
75		8396.87	I
120		8412.36	I
19		8416.98	I
15		8424.41	I
170		8426.52	I
490		8434.94	I
240		8435.70	I
40		8438.93	I
40		8450.89	I
9	h	8457.10	I
19	h	8467.15	I
45		8468.50	I
15		8496.04	I
19	h	8518.05	I
40		8518.32	I
14		8539.38	I
40		8548.12	I
9		8569.77	I
9	h	8598.18	I
90		8675.39	I
45		8682.99	I
23		8692.33	I
19		8734.69	I
23		8766.64	I
15	h	8778.71	I

Ti III
Ref. 378 — C.H.C.

Intensity		Wavelength	
		Vacuum	
6		1282.48	III
6		1286.23	III
15		1286.36	III
10		1289.30	III
10		1291.62	III
10		1293.23	III
15		1294.70	III
10		1295.88	III
20		1298.66	III
20		1298.97	III
12		1327.59	III
10		1420.04	III
10		1420.44	III
10		1421.63	III
10		1421.77	III
12		1422.40	III
10		1424.14	III
23		1455.19	III
10		1498.70	III
		Air	
10		2199.22	III
12		2237.77	III
10		2327.02	III
15		2331.35	III
15		2331.66	III
15		2334.34	III
17		2339.00	III
18		2346.79	III
18		2374.99	III
22		2413.99	III
25		2516.05	III
24		2527.84	III
23		2540.06	III
24		2563.44	III
23		2565.42	III
22		2567.56	III
15		2576.47	III
15		2580.46	III
10		2692.16	III
12		2701.96	III
22		2984.75	III
12	d	3354.71	III
12		3872.50	III
12		3881.21	III
12		3893.63	III
10		3896.33	III
15		3915.47	III
12		3921.38	III
10		3921.61	III
12		3922.95	III
10		3924.86	III
10		4060.21	III
10		4119.14	III
11		4215.52	III
11		4269.84	III
11		4296.70	III
10		4348.04	III
11		4433.91	III
10		4540.22	III
15		4549.84	III
10	d	4555.46	III
15	d	4572.20	III
10		4649.45	III
12		4652.86	III
10		4874.00	III
10		4950.10	III
10		4971.19	III
10		5083.80	III
14		5147.31	III
12	d	5226.28	III
11		5247.49	III
17		5278.12	III
10		5278.70	III
12		5298.43	III
16		5301.20	III
15		5306.88	III
10		5395.69	III
12		5533.01	III
10		5817.44	III
12		6611.38	III
18		6621.58	III
10		6629.37	III
14		6647.47	III
18		6667.99	III
15		6674.19	III
14		6707.76	III
12		6724.80	III
16		6734.10	III
15		6862.26	III
12		6874.35	III
10		6896.12	III
12		7015.38	III
10		7071.93	III
20		7072.64	III
18		7084.57	III
15		7124.13	III
11		7171.79	III
10		7175.92	III
10		7217.50	III
9		7225.55	III
12		7270.67	III
14		7316.30	III
10		7316.68	III
12		7379.96	III
10		7408.13	III
10		7457.85	III
15		7506.87	III
17		7507.68	III
10		7523.85	III
12		7544.29	III
9		7566.25	III
10	h	8172.21	III
9	h	8173.37	III
9	h	8178.00	III
10		8182.42	III
9	h	8192.68	III
9	h	8194.75	III
9	h	8263.67	III
15	h	8267.32	III
10		8338.54	III
12		8394.20	III
20		8466.87	III
5		8699.85	III
3		9017.10	III

Ti IV
Ref. 428 — C.H.C.

Intensity	Wavelength	
	Vacuum	
10	776.76	IV
18	779.07	IV
16	781.73	IV
8	1183.64	IV
10	1195.21	IV
18	1451.74	IV
20	1467.34	IV
12	1469.19	IV
	Air	
20	2067.56	IV
18	2103.16	IV
10	2359.14	IV
10	2359.50	IV

Titanium (Cont.)

Intensity		Wavelength	
8		2541.79	IV
10		2546.88	IV
5		2862.60	IV
6		2929.96	IV
14		2937.33	IV
12		2957.31	IV
15		3541.36	IV
17		3576.44	IV
10		3581.39	IV
13		4131.22	IV
14		4133.78	IV
10		4397.33	IV
9		4403.45	IV
15		4618.11	IV
20		5398.93	IV
8		5470.98	IV
18		5492.51	IV
10		5517.72	IV
14		5877.79	IV
15		5885.96	IV
7		5891.15	IV
6		6231.62	IV
17		6246.65	IV
11		6247.74	IV
15		6292.41	IV
12		6913.85	IV
15		6978.51	IV
9		7491.37	IV
8		7494.77	IV
5	h	7652.12	IV
8	h	7706.85	IV

Ti V
Ref. 427 — C.H.C.

Intensity		Wavelength	
		Vacuum	
12		225.35	V
10		228.91	V
17		252.96	V
7		323.36	V
7		461.41	V
8		474.69	V
8		483.99	V
10	d	488.58	V
15		498.26	V
14		502.08	V
7		502.71	V
12		504.66	V
7		506.47	V
8		513.37	V
7		523.05	V
12		524.58	V
13		526.57	V
8		529.32	V
10		535.84	V
10		535.89	V
8	d	540.14	V
8		541.46	V
9		541.71	V
7		543.10	V
7		543.34	V
7		1128.55	V
8		1192.35	V
9		1198.66	V
9		1222.36	V
10		1230.36	V
11		1239.96	V
10		1241.67	V
7		1246.13	V
8		1268.49	V
8		1306.11	V
8		1411.31	V
9		1675.15	V
8		1687.16	V
11		1717.40	V
8		1759.76	V
7		1771.45	V
10		1841.49	V
7		1864.45	V
7		1881.89	V
7		1920.16	V
7		1988.75	V

TUNGSTEN (W)
Z = 74

W I and II
Ref. 1 — C.H.C.

Intensity		Wavelength	
		Air	
5800		2001.71	II
13000		2008.07	II
5100		2009.98	II
4100		2010.23	II
4100		2014.23	II
7300		2026.08	II
15000		2029.98	II
2700		2035.03	II
5300		2049.63	II
2300		2065.57	II
3400		2071.21	II
2200		2075.59	II
9700		2079.11	II
3600		2088.19	II
2200		2089.14	II
1700		2090.48	I
6100		2094.75	II
2400		2098.53	II
2200		2100.67	II
1500		2101.54	I
1500		2106.18	II
1300		2110.34	II
2100		2118.87	II
2400		2121.59	II
850		2153.56	II
850		2157.80	II
1500		2166.32	II
480		2182.90	I
440		2194.52	II
1300		2204.48	II
460		2248.75	II
460		2249.80	I
180		2270.24	I
95		2271.37	I
510		2277.58	I
160		2284.91	I
320		2285.17	I
530	d	2294.49	I
		2294.54	II
270		2298.33	I
240		2303.83	II
240		2306.59	I
340		2309.02	I
440		2313.17	I
220		2314.17	I
190		2315.02	II
460		2321.63	I
290		2326.09	II
390	d	2326.56	I
		2326.70	I
75		2328.31	II
130		2333.77	II
210		2341.37	I
75		2349.26	II
120	d	2350.37	II
320		2354.61	I
60		2358.81	II
580		2360.44	I
850		2363.07	I
60		2364.22	II
510		2374.47	I
210		2382.99	I
670		2384.82	I
240		2389.08	I
120		2390.37	II
120		2392.93	II
730		2397.09	II
560		2397.73	I
560		2397.98	I
75		2404.24	II
1700	d	2405.58	I
		2405.69	I
75		2411.54	II
320		2414.01	I
610		2415.68	II
50		2419.34	II
50		2421.01	II
870		2424.21	I
190		2427.49	II
170		2429.39	II
580		2431.08	I
630		2433.98	I
60		2435.01	II
1800		2435.96	I
250		2436.62	I
580		2444.06	I
160		2446.39	II
270		2448.39	I
270		2451.35	I
780		2451.48	II
870		2452.00	I
430		2454.72	I
630		2454.98	I
780		2455.51	I
780		2456.53	I
1100		2459.30	I
270		2460.16	I
480		2462.79	I
270		2464.30	I
230		2466.52	II
1400		2466.85	I
75		2470.80	II
480		2472.51	I
1200		2474.15	I
290		2477.80	II
870		2480.13	I
390		2480.96	I
1500		2481.44	I
480	d	2482.10	I
		2482.21	I
29		2484.40	II
580		2484.74	I
390		2487.50	I
270		2488.77	II
390		2489.23	II
75		2492.93	II
630		2495.26	I
230		2496.64	II
95		2497.48	II
140		2499.69	II
40		2500.11	II
		2501.90	II
680		2504.70	I
270		2506.02	I
24		2508.00	II
250		2510.17	I
75		2510.47	II
60		2518.14	II
310		2520.46	I
780		2521.32	I
270		2522.04	II
430		2523.41	I
780		2527.76	I
50		2533.64	I
		2534.82	II
580		2545.34	I
1200		2547.14	I
50		2549.09	II
40		2550.10	II
780		2550.38	I
2700		2551.35	I
450		2553.82	I
410		2554.86	II
580		2555.09	II
		2555.21	I
310		2556.75	I
290		2560.12	I
730		2561.97	I
230		2563.16	II
110		2563.91	II
530		2571.44	II
170	d	2572.24	II
		2572.35	II
75		2573.95	II
190		2579.26	II
290		2580.34	I
870		2580.49	I
40		2581.20	II
390		2584.39	I
390		2589.17	II
170		2591.49	II
110		2598.74	II
370		2601.96	I
75		2602.51	II
75		2603.02	II
270		2603.54	I
680		2606.39	I
320		2607.38	I
370		2608.32	I
970		2613.08	I
480		2613.82	I
230		2615.12	I
70		2615.44	II
210		2619.18	I
400		2620.25	I
400		2622.21	I
400		2625.22	I
210		2628.26	I
400		2632.48	I
400		2632.70	I
810		2633.13	I
290		2636.54	I
400	d	2638.62	I
		2638.75	I
210		2645.69	I
650		2646.18	I
400		2646.73	I
75		2647.74	II
40		2653.42	II
80		2653.57	II
1600		2656.54	I
400		2657.38	I
400	d	2658.04	II
		2658.18	I
810		2662.84	I
260		2664.97	I
75		2666.49	II
210		2669.30	I
810		2671.47	I
80		2673.59	II
650		2677.28	I
160	d	2677.79	II
		2677.91	I
400		2678.88	I
2100		2681.42	I
290		2683.35	I
210		2691.09	I
650		2695.67	I
210		2697.71	II
650		2699.59	I
400		2700.01	I
40		2701.48	II
160		2702.11	II
210		2702.52	I
400		2706.58	I
400		2708.59	I
400	d	2708.80	I
		2708.93	I
80		2709.58	II
40		2710.78	II
400		2715.50	I
80		2716.32	II
80		2718.04	II
2100		2718.91	I
320		2719.33	I
210		2719.86	I
2600		2724.35	I
210		2724.62	I
400		2725.03	I
		2725.06	I
80		2729.62	II
75		2740.79	II
650		2748.84	I
40	d	2760.74	II
80		2761.59	II
400		2762.34	I
400		2764.27	II
210		2768.98	I
400		2769.74	I
810		2770.88	I
210		2773.70	I
810		2774.00	I
810		2774.48	I
160		2776.50	II
40		2778.69	II
210		2787.98	I
340		2791.96	I
810		2792.70	I
80		2799.03	II
400		2799.93	I
160	d	2801.05	II
		2801.17	I
130		2805.92	II
40		2812.25	II
810		2818.06	I
160		2822.57	II
260		2829.82	I
1600		2831.38	I
810		2833.63	I
210		2835.64	I
400		2841.57	I
810		2848.02	I
650		2856.03	I
650		2866.06	I
230		2878.72	I
610		2879.11	I
610		2879.40	I
440		2896.01	I
1500		2896.44	I
230		2910.48	I
270		2911.00	I
360		2918.25	I
50		2918.63	II
360		2923.10	I
230		2923.54	I
230		2925.13	I
690		2935.00	I
2400		2944.40	I
2400		2946.99	I
480		2947.39	I
210		2952.29	II
440		2964.52	I
480		2977.11	I
		2977.21	I
730	d	2979.71	I
		2979.86	I
400		2993.61	I
240		2995.26	I
190		3009.09	I
360		3013.79	I
520		3016.47	I
770		3017.44	I
110		3024.50	II
210		3024.93	I
310	d	3026.67	I
		3026.79	I
160		3033.56	I
160		3034.19	I
160		3039.31	I
440	d	3041.73	I
		3041.86	I
270		3043.80	I
440		3046.44	I
110		3048.66	I
810		3049.69	I
110		3064.93	I
180		3073.28	I
110		3077.52	II
180	d	3084.83	I
		3084.91	I
370		3093.50	I
240		3107.23	I
240		3108.02	I

Tungsten (Cont.)

Intensity		Wavelength	
230		3117.57	I
260		3120.18	I
160		3133.88	I
130		3141.42	I
65		3149.85	II
290		3163.42	I
130		3164.44	I
130		3165.38	I
320		3176.60	I
130		3179.06	I
190		3181.82	I
130		3184.05	I
130		3184.42	I
65		3189.24	II
390		3191.57	I
390		3198.84	I
520		3207.25	I
140		3208.28	I
1000		3215.56	I
140		3221.21	I
140		3221.91	I
190		3232.49	I
140		3237.09	I
140		3242.03	I
140		3252.29	I
210		3254.36	I
140		3259.43	I
210		3259.66	I
210	d	3266.62	I
		3266.77	I
150		3281.94	I
150		3293.71	I
730		3300.82	I
440		3311.38	I
440		3326.20	I
440		3331.69	I
150		3354.45	I
150		3371.04	I
390		3373.75	I
150		3412.96	I
150		3413.53	I
150		3422.42	I
150		3427.71	I
230		3429.59	I
240		3443.00	I
160		3477.94	I
400		3495.24	I
160		3508.73	I
160		3510.02	I
160		3526.85	I
160		3535.54	I
160		3537.45	I
650		3545.22	I
160		3568.04	I
240		3570.65	I
80		3572.48	II
160		3575.22	I
80		3592.42	II
240		3606.06	I
80		3613.79	II
1900		3617.52	I
160		3622.34	I
130		3627.24	I
320		3631.94	I
240		3641.41	II
80		3646.52	II
80		3657.59	II
160		3675.55	I
650		3682.08	I
400		3683.30	I
		3683.39	I
160		3683.93	I
570		3688.06	I
810		3707.92	I
60		3716.08	II
100		3719.39	I
50		3736.22	II
120		3741.71	I
510		3757.92	I
680		3760.13	I
1000		3768.45	I
120		3769.21	I
120		3769.86	I
340		3773.71	I
1000		3780.77	I
170		3792.76	I
290		3809.22	I
190		3810.38	I
260		3810.79	I
1400		3817.48	I
110		3829.13	I
1100		3835.06	I
290		3838.51	I
730		3846.22	I
250		3847.49	I
27		3851.57	II
150		3855.55	I
150		3859.30	I
180		3864.34	I
1800		3867.99	I
250		3872.84	I
110		3874.41	I
730		3881.41	I
110		3892.72	I

Intensity		Wavelength	
140	h	3897.91	I
150		3935.03	I
120		3936.97	I
120		3947.58	I
120		3952.52	I
120		3952.90	I
160		3953.15	I
200		3955.30	I
160		3965.14	I
130		3968.59	I
150	h	3970.80	I
130		3979.29	I
130		3980.64	I
250		3983.29	I
8600		4008.75	I
540		4015.22	I
170	h	4016.52	I
220		4019.23	I
130		4022.12	I
180		4028.79	I
180		4036.86	I
140		4039.85	I
140	h	4044.28	I
910		4045.59	I
180		4064.79	I
150		4069.79	I
730		4069.95	I
340		4070.61	I
100		4071.93	I
5000		4074.36	I
150		4082.96	I
130		4088.33	I
100		4095.69	I
1000		4102.70	I
150		4109.75	I
100		4111.82	I
150		4118.05	I
100		4118.19	I
100		4120.85	I
100		4125.16	I
150		4126.80	I
100		4133.48	I
540		4137.46	I
150		4138.02	I
110		4142.25	I
140		4145.16	I
110		4145.95	I
160		4154.66	I
160		4170.53	I
450		4171.17	I
160		4204.40	I
220		4207.05	I
110		4215.38	I
250		4219.37	I
110		4222.04	I
150		4234.34	I
290		4241.44	I
540		4244.36	I
290		4259.35	I
200		4260.29	I
200		4263.30	I
1400		4269.38	I
110		4269.77	I
220		4274.55	I
160		4275.49	I
160		4276.74	I
110		4282.34	I
110		4286.01	I
110		4294.10	I
4100		4294.61	I
2200		4302.11	I
160		4306.87	I
110		4307.64	I
110		4332.13	I
100		4347.00	I
150		4355.17	I
100		4361.81	I
150		4364.78	I
100	d	4365.95	I
		4366.07	I
150		4372.52	I
200		4378.48	I
180		4384.85	I
100		4403.95	I
200		4408.28	I
130		4412.19	I
160		4436.90	I
140		4460.49	I
140		4466.34	I
140		4466.74	I
640		4484.19	I
160		4504.86	I
130		4512.88	I
120		4513.25	I
150		4543.54	I
150		4546.47	I
150		4551.82	I
140		4570.64	I
170		4588.73	I
140		4599.94	I
140		4609.89	I
160		4613.30	I
100		4642.53	I
130		4657.42	I

Intensity		Wavelength	
640		4659.87	I
640		4680.51	I
100		4693.72	I
140		4757.54	I
790		4843.81	I
380		4886.90	I
220		4982.59	I
330		5006.15	I
220		5015.30	I
820		5053.28	I
210		5054.60	I
210		5069.12	I
120		5071.74	I
770		5224.66	I
27		5500.49	I
27		5503.44	I
10		5508.61	I
220		5514.68	I
15		5531.38	I
15		5537.72	I
13		5568.09	I
13		5604.31	I
11		5631.27	I
27		5631.94	I
65		5648.37	I
35		5660.72	I
27		5674.39	I
13		5676.60	I
15		5676.90	I
15		5697.79	I
55		5735.09	I
13		5749.24	I
11		5756.10	I
13		5793.06	I
13		5796.49	I
45		5804.85	I
13	d	5806.05	I
		5806.24	I
13		5833.61	I
13		5838.97	I
17		5845.27	I
28		5851.58	I
11		5856.61	I
22		5864.63	I
11		5874.22	I
13		5880.21	I
13		5891.61	I
13		5901.20	I
40		5902.64	I
13		5928.58	I
55		5947.57	I
13		5953.96	I
13		5956.19	I
27		5960.83	I
55		5965.86	I
27		5972.51	I
20		5978.86	I
20		5983.82	I
13		6009.01	I
55		6012.78	I
40		6021.52	I
20		6028.32	I
20		6043.31	I
13		6049.92	I
13		6065.08	I
22		6081.44	I
13		6111.66	I
13		6115.52	I
22		6128.25	I
13		6143.94	I
20		6153.72	I
20		6154.87	I
20		6203.51	I
20		6254.28	I
27		6285.88	I
45		6292.02	I
20		6303.21	I
13		6386.47	I
35		6404.21	I
40		6445.12	I
11		6508.05	I
15		6532.39	I
13		6538.11	I
13		6563.20	I
20		6573.93	I
11		6607.13	I
11		6609.05	I
17		6611.62	I
11		6621.74	I
13		6678.42	I
15		6693.08	I
5		6746.56	I
5		6764.45	I
7		6805.31	I
9		6814.92	I
9		6820.27	I
8		6828.43	I
4		6853.74	I
4		6876.01	I
5		6908.29	I
9		6934.23	I
8		6964.12	I
13		6984.27	I
8		6993.27	I

Intensity		Wavelength	
4		6994.06	I
8		7017.88	I
3		7028.68	I
3		7098.22	I
4	h	7111.18	I
15		7140.52	I
9		7162.64	I
5	h	7191.33	I
5		7198.62	I
11		7200.16	I
5		7216.35	I
4		7226.06	I
8		7237.12	I
5		7274.47	I
10		7278.24	I
15		7285.81	I
15		7296.55	I
3		7298.25	I
7		7385.08	I
4		7451.39	I
3		7456.37	I
8		7483.35	I
7		7504.13	I
10		7509.00	I
3		7520.66	I
9		7537.45	I
9		7550.48	I
17		7569.92	I
5		7582.88	I
3		7612.18	I
17		7614.15	I
3		7631.29	I
3		7654.81	I
13		7688.97	I
4		7701.01	I
5		7761.16	I
3		7776.73	I
11		7784.15	I
7		7808.96	I
2		7823.82	I
4		7863.47	I
2		7867.04	I
4		7880.40	I
5		7886.48	I
9		7940.92	I
3		7957.06	I
22		8017.19	I
7		8054.89	I
22		8055.64	I
5		8060.38	I
13		8123.82	I
5		8143.19	I
3		8165.72	I
5		8210.22	I
4		8322.05	I
10		8338.08	I
4		8348.81	I
7		8358.72	I
3		8382.94	I
4		8402.60	I
4		8475.14	I
27		8585.11	I
10		8594.42	I
8		8613.27	I
3		8614.50	I
13		8865.53	I

URANIUM (U)
Z = 92

U I and II
Ref. 1, 303 — J.G.C.

Intensity	Wavelength (Air)	
440	2565.41	II
340	2569.71	II
340	2591.25	II
610	2635.53	II
470	2645.47	II
340	2669.17	II
470	2683.28	II
320	2691.04	II
370	2706.95	II
370	2733.97	II
470	2754.16	II
340	2762.85	II
390	2770.04	II
410	2784.45	II
830	2793.94	II
870	2802.56	II
630	2807.05	II
440	2808.98	II
630	2817.96	II
870	2821.12	II
390	2824.37	II
680	2828.90	II
920	2832.06	II
360	2837.19	II
460	2839.89	II

Intensity	Wavelength	Spectrum
360	2842.09	II
360	2849.48	II
390	2860.47	II
970	2865.68	II
340	2870.97	II
490	2882.74	II
460	2887.25	II
410	2888.26	II
1200	2889.62	II
320	2894.14	II
410	2894.51	II
780	2906.80	II
780	2908.28	II
320	2914.25	II
360	2914.63	II
440 p	2921.68	II
320	2927.38	II
490	2928.60	II
580	2931.41	II
440	2932.61	II
340	2933.86	II
530 p	2940.37	II
1300	2941.92	II
830	2943.90	II
340	2948.09	II
390	2954.77	II
580	2956.06	II
460	2965.03	II
580	2967.94	II
580	2971.06	II
410	2976.35	II
320	2982.74	II
530	2984.61	II
410	2992.72	II
360	3007.91	II
320	3021.22	II
630	3022.21	II
320	3024.51	II
320	3028.19	II
630	3031.99	II
490	3033.18	II
490	3044.16	II
580	3050.20	II
630	3057.91	II
460	3061.62	II
630	3062.54	II
580	3072.78	II
580	3093.01	II
320	3095.75	II
320	3098.01	II
580	3102.39	II
460	3104.15	II
970	3111.62	II
530	3119.35	II
680	3124.95	II
530	3139.61	II
410	3144.97	II
490	3145.56	II
680	3149.24	II
530	3153.11	II
340	3176.21	II
340	3177.33	II
340	3206.05	II
730	3229.50	II
680	3232.16	II
440	3244.22	II
340	3265.79	II
440	3270.12	II
440	3288.21	II
730	3291.33	II
1100	3305.89	II
390	3337.79	II
440	3341.66	II
390	3357.84	I
730	3390.38	I
340	3394.77	II
580	3424.56	II
580	3435.49	I
360	3453.55	II
320	3454.23	II
320	3457.05	II
320	3457.71	II
360	3459.92	I
320	3462.22	I
460	3463.55	I
630	3466.30	I
390	3472.52	I
320	3473.43	I
360	3480.36	I
680	3482.49	II
1600	3489.37	I
390	3493.33	II
340	3494.00	I
320	3494.84	II
530	3496.41	II
630	3500.08	I
320	3504.01	I
320	3505.07	II
780	3507.34	I
320	3508.84	II
1600	3514.61	I
390	3509.66	II
320	3513.67	I
390	3519.96	II
390	3531.11	II
630	3533.57	II
320	3534.33	I
530	3540.47	II
320	3542.57	I
390	3547.19	II
320	3549.20	I
1200	3550.82	II
320	3552.17	II
680	3555.32	I
320	3561.41	II
1200	3561.80	I
390	3563.66	I
2300	3566.59	I
530	3569.08	I
320	3574.76	I
360	3577.92	I
630	3578.72	II
360	3581.84	II
3200	3584.88	I
320	3590.50	I
390	3591.74	I
460	3593.52	II
460	3605.27	I
360	3606.32	II
320	3616.33	I
320	3616.76	II
320	3620.08	I
320	3622.70	I
390	3623.06	II
460	3630.73	II
840	3638.20	I
310	3640.76	II
420	3644.24	I
310	3645.03	II
660	3651.54	I
490	3652.06	I
960	3659.15	I
2800	3670.07	II
380	3678.75	II
540	3691.92	II
330	3693.70	II
540	3700.57	II
1100	3701.52	II
350	3713.55	I
300	3717.42	II
350	3718.11	II
350	3729.82	II
350	3732.62	II
350	3733.07	II
600	3738.04	II
300	3744.25	II
680	3746.42	II
350	3747.14	II
950	3748.68	II
600	3751.17	I
350	3752.66	II
350	3755.48	II
490	3758.35	I
350	3759.24	II
330	3763.26	I
490	3764.57	II
430	3766.89	I
330	3769.53	II
540	3773.43	I
300	3776.48	I
380	3780.71	I
1900	3782.84	II
430	3783.84	II
570	3793.10	II
380	3793.24	I
380	3793.57	II
380	3808.92	I
380	3809.22	II
1900	3811.99	I
380	3813.79	II
380	3814.06	II
750	3826.51	II
2000	3831.46	II
1200	3839.63	I
490 p	3848.60	II
620	3854.22	I
2400	3854.64	II
4900	3859.57	II
490	3861.17	II
1900	3865.92	II
380	3866.80	II
1500	3871.03	I
620	3874.04	II
620	3878.08	II
1000	3881.45	II
490	3882.36	II
380	3883.28	II
2200	3890.36	II
620	3892.68	II
490	3894.12	I
490	3896.77	II
620	3899.78	II
410	3902.55	II
460	3904.30	II
380	3906.45	I
330	3911.67	II
380	3915.88	II
330	3926.21	I
330	3926.72	I
430	3930.98	II
2000	3932.02	II
490	3935.38	II
330	3940.48	II
1200	3943.82	II
300	3948.44	I
300	3953.58	II
360	3954.67	II
350	3964.21	I
600	3966.52	II
1200	3985.79	II
460	3990.42	II
380	3992.53	II
350	3998.24	II
350	4004.06	II
430	4005.21	I
570	4017.72	II
300	4018.99	II
1000	4042.75	I
520	4044.41	II
410	4047.61	I
1600	4050.04	II
540	4051.91	II
300	4054.30	II
430	4058.19	II
880	4062.54	II
520	4067.75	II
410	4071.12	II
300	4074.48	II
330	4076.69	II
330	4080.60	II
2200	4090.13	II
460	4093.03	II
380	4106.38	II
810	4116.10	II
410	4124.73	II
410	4128.34	II
460	4141.22	II
880	4153.97	I
380	4156.65	I
350	4163.68	II
1400	4171.59	II
300	4189.27	II
350	4222.37	I
1000	4241.67	II
520	4244.37	II
680	4341.69	II
430 h	4355.74	I
430	4362.05	I
330	4393.59	I
600	4472.33	II
240	4515.28	II
620	4543.63	II
300	4620.21	I
240	4627.07	II
210	4631.62	I
220	4646.60	II
140	4666.85	II
100	4671.40	II
170	4689.07	II
100	4702.51	II
160	4722.72	II
120	4731.59	II
100	4755.74	II
150	4756.81	I
100	4772.70	II
100	4860.99	II
110	5008.21	II
170	5027.38	I
70	5117.24	II
80	5160.32	II
55	5164.14	I
55	5184.57	II
45	5204.31	II
45	5247.75	II
45	5257.04	II
70	5280.38	I
55	5386.19	II
80	5475.70	II
70	5480.26	II
70	5481.20	II
45	5482.53	II
160	5492.95	II
70	5527.82	II
70	5564.17	I
45	5581.59	II
55	5620.78	I
70	5780.59	II
70	5798.53	II
45	5836.02	I
55	5837.68	II
230	5915.39	I
55	5971.50	I
100	5976.32	I
45	5997.31	I
28	6017.38	II
55	6051.74	II
45	6067.22	II
90	6077.29	I
28	6087.34	II
40	6171.86	I
35	6175.39	I
28	6280.18	II
28	6359.29	I
55	6372.46	I
28	6378.52	II
28	6392.77	I
90	6395.42	I
110	6449.16	I
35	6464.98	I
90	6826.92	I
35	6876.74	II
23	7074.79	I
27	7101.61	I
30	7128.90	I
16	7147.89	I
16	7254.45	I
23	7425.50	I
45	7533.93	I
16	7619.35	I
50	7881.94	I
18	7970.46	I
16	8174.66	I
18	8262.06	I
16	8318.35	I
16	8337.50	II
18	8381.87	I
16	8441.21	I
35	8445.39	I
18	8450.03	I
16	8570.52	I
75	8607.95	I
23	8691.28	I
18	8710.76	I
18	8753.69	I
30	8757.76	I
16	8951.96	I
16	8989.92	I
10	9093.67	I
10	9139.56	I
10	9201.51	I
10	9265.34	I
10	9276.44	I
10	9385.90	I
10	9653.26	I
10	9819.00	I
10	9819.05	I
10	9868.36	I
10	9932.76	I
10	9964.11	I
50	10157.91	I
50	10259.55	I
100	10554.93	I
50	10799.78	I
25	10823.93	I
25	11095.77	I
75	11167.84	I
50	11294.13	I
100	11384.13	I
25	11410.43	I
50	11503.38	I
25	11568.81	I
20	11784.72	II
100	11859.42	I
100	11908.83	I
100	12250.46	I
25	13088.28	I
100	13185.16	I
75	13306.23	I
100	13961.58	I
50	16906.00	I
50	17451.11	I
50	18136.65	I
50	18366.96	I
75	18634.43	I
25	19029.39	I
10	20201.13	I
10	20271.41	I
10	20374.13	I
10	20517.29	I
10	20690.64	I
10	20772.19	I
10	21008.38	I
10	21099.98	I
10	21112.14	I
20	21144.90	II
10	21674.51	I
10	21693.38	I
75	21910.22	I
10	22110.73	I
10	23156.76	I
10	23948.19	I
10	29557.07	I

VANADIUM (V)
Z = 23

V I and II
Ref. 1 — C.H.C.

Intensity	Wavelength	
	Air	
2100	2092.44	I

Vanadium (Cont.)

Int	λ			Int	λ			Int	λ			Int	λ	
40	2384.00	II		120	2917.37	II		560	3592.53	I		150	4035.63	II
40	2384.28	I		210	2919.99	II		270	3593.33	II		120	4042.64	I
60	2386.96	I		380	2920.38	II		110	3606.69	I		360	4050.96	I
60	2388.92	I		710	2923.62	I		110	3639.02	I		360	4051.35	I
75	2390.87	I		2400	2924.02	II		110	3644.71	I		280	4057.07	I
75	2391.26	I		1700	2924.64	II		250	3663.59	I		130	4057.82	I
85	2392.90	I		710	2930.81	II		250	3667.74	I		230	4063.93	I
70	2397.78	I		210	2934.40	II		110	3669.41	II		230	4071.54	I
70	2398.27	I		110	2935.87	I		170	3671.20	I		1100	4090.58	I
70	2399.96	I		900	2941.37	II		280	3673.40	I		180	4092.41	I
120	2406.75	I		450	2941.49	II		280	3675.70	I		1800	4092.69	I
110	2407.90	I		230 d	2942.33	I		170	3676.68	I		120	4093.50	I
120	2415.33	I		230	2943.20	II		300	3680.11	I		890	4095.49	I
120	2416.75	I		1100	2944.57	II		570	3683.13	I		2800	4099.80	I
100	2420.12	I		110	2946.53	I		190	3686.26	I		590	4102.16	I
100	2421.06	I		230	2949.63	I		470	3687.47	I		230	4104.40	I
100	2421.98	I		300	2950.35	II		1300	3688.07	I		260	4104.78	I
110	2428.28	I		640	2952.08	I		1000	3690.28	I		2800	4105.17	I
110	2435.52	I		120	2954.33	I		1500	3692.22	I		120	4108.22	I
140	2501.61	I		260	2957.52	II		450	3695.34	I		2300	4109.79	I
150	2506.90	I		410	2962.77	I		1000	3695.86	I		8900	4111.78	I
240	2507.78	I		600	2968.38	II		3800	3703.58	I		120	4112.33	I
180	2511.65	I		120	2972.25	II		1800	3704.70	I		230	4113.52	I
180	2511.95	I		120	2976.20	II		570	3705.04	I		4300	4115.18	I
180	2517.14	I		380	2976.52	II		130	3708.72	I		1800	4116.47	I
240	2519.62	I		240	2977.54	I		320	3715.47	II		180	4118.18	I
410	2526.22	I		260	3001.20	II		250	3727.34	II		180	4118.64	I
210	2527.90	II		140	3014.82	II		280	3732.76	II		230	4119.46	I
120	2528.47	II		180	3016.78	II		150	3734.43	I		180	4120.54	I
150	2528.84	II		270	3033.45	II		230	3745.80	II		180	4123.19	I
240	2530.18	I		290	3033.82	II		210	3750.87	I		2000	4123.57	I
110	2549.28	II		230	3043.12	I		210	3770.97	II		120	4124.07	I
120	2552.65	I		230	3043.56	I		270	3778.68	I		3100	4128.07	I
210	2562.13	I		230	3044.94	I		520	3790.32	I		120	4128.86	I
110	2564.82	I		230	3048.22	II		1100	3794.96	I		3100	4132.02	I
230	2574.02	I		170	3050.89	I		570	3799.91	I		2300	4134.49	I
140	2630.67	II		180	3053.39	II		570	3803.47	I		150	4159.69	I
130	2642.21	II		450	3053.65	I		190	3806.80	I		100	4174.01	I
150	2645.26	I		1200	3056.33	I		300	3807.50	I		230	4179.42	I
140	2651.90	I		1400	3060.46	I		520	3808.52	I		150	4182.59	I
150	2656.22	I		140	3063.25	II		230	3809.60	I		180	4189.84	I
180	2661.42	I		2400	3066.38	I		1000	3813.49	I		180	4191.56	I
290	2672.00	II		200	3067.12	II		140 d	3817.84	I		230	4209.86	I
380	2677.80	II		140	3069.64	I		1300	3818.24	I		120	4226.62	I
270	2678.57	II		170	3073.82	I		230	3819.96	I		360	4232.46	I
380	2679.32	II		100	3075.27	I		230	3821.49	I		180	4232.95	I
180	2682.87	II		150	3082.11	I		570	3822.01	I		180	4234.00	I
180	2683.09	II		3800	3093.11	II		450	3822.89	I		120	4235.76	I
1100	2687.96	II		200	3094.20	II		300	3823.21	I		100	4257.37	I
170	2688.72	II		180	3100.94	II		1700	3828.56	I		120	4259.31	I
150	2689.88	II		3000	3102.30	II		280	3834.22	I		120	4262.16	I
230	2690.24	II		2600	3110.71	II		160	3839.00	I		560	4268.64	I
240	2690.79	II		2000	3118.38	II		110	3839.38	I		460	4271.55	I
120	2696.99	I		380	3121.14	II		570	3840.44	I		460	4276.96	I
120	2697.74	I		150	3122.90	II		2600	3840.75	I		430	4284.06	I
680	2700.94	II		1500	3125.28	II		110	3841.89	I		330	4291.82	I
380	2702.19	II		260	3126.22	II		380	3844.44	I		220	4296.11	I
530	2706.17	II		530	3130.27	II		320	3847.33	I		170	4297.68	I
150	2706.70	II		410	3133.33	II		110	3849.32	I		170	4298.03	I
110	2707.86	II		210	3134.93	II		1200	3855.37	I		170	4306.21	I
170	2711.74	II		150	3136.51	II		3000	3855.84	I		140	4307.18	I
120	2714.20	II		150	3139.74	II		150	3862.22	I		170	4309.80	I
640	2715.69	II		200	3142.48	II		130	3863.87	I		460	4330.02	I
150	2722.56	I		150	3145.34	II		1300	3864.86	I		510	4332.82	I
240	2728.64	II		3200	3183.41	I		230	3867.60	I		760	4341.01	I
180	2731.35	I		5300	3183.98	I		170	3871.08	I		1000	4352.87	I
100	2739.71	II		3800	3185.40	I		1500	3875.08	I		130	4354.98	I
140	2753.40	II		410	3187.71	II		420	3875.90	I		150	4355.94	I
140	2765.67	II		530	3188.51	II		570	3876.09	I		150	4368.04	I
140	2777.73	II		750	3190.68	II		130	3878.71	II		140 d	4373.23	I
120	2803.47	II		530	3198.01	I		700	3890.18	I		120	4375.30	I
120	2846.57	I		750	3202.38	I		460	3892.86	I		12000	4379.24	I
110	2847.57	II		450	3205.58	I		280 h	3898.02	I		100	4380.55	I
140	2852.87	I		450	3207.41	I		140	3899.13	II		7000	4384.72	I
140	2854.34	II		410	3212.43	I		140 h	3900.18	I		4800	4389.97	I
200	2855.22	I		210	3217.11	II		140 h	3901.15	I		3600	4395.23	I
180	2859.97	I		150	3237.87	II		2400	3902.25	I		1400	4400.58	I
240	2864.36	I		140	3254.77	II		100	3906.75	I		2300	4406.64	I
170	2866.59	I		140	3263.24	I		700	3909.89	I		2800	4407.64	I
210	2868.10	I		1100	3267.70	II		100	3910.79	I		3600	4408.20	I
140	2869.13	II		900	3271.12	II		220	3912.21	I		4600	4408.51	I
210	2870.55	I		750	3276.12	II		140	3914.33	II		140	4412.14	I
110	2877.69	II		110	3279.84	II		100	3916.41	II		640	4416.47	I
110	2879.16	II		140	3298.14	I		100	3920.49	I		120	4419.94	I
350	2880.03	II		110	3329.86	I		100	3921.90	I		640	4421.57	I
380	2882.50	II		110	3365.55	I		230	3922.43	I		460	4426.00	I
380	2884.78	II		110	3377.62	I		240	3924.66	I		120	4427.31	I
140	2888.25	II		170	3400.40	I		150	3925.24	I		310	4428.52	I
380	2889.62	II		110	3425.07	I		200	3927.93	I		230	4429.80	I
900	2891.64	II		110	3485.92	II		260	3930.02	I		430	4436.14	I
530	2892.44	II		210	3504.44	II		150	3931.34	I		640	4437.84	I
900	2892.66	II		560	3517.30	II		260	3934.01	I		830	4441.68	I
1400	2893.32	II		150	3520.02	II		150	3935.14	I		640	4444.21	I
360	2896.21	II		110	3524.72	II		100	3936.28	I		610	4452.01	I
110	2899.60	I		230	3529.74	I		150	3943.66	I		410	4457.48	I
360	2903.08	II		230	3530.77	II		100	3950.23	I		120	4457.76	I
150	2906.13	I		560	3533.68	I		140	3951.97	II		1000	4459.76	I
900	2906.46	II		110	3543.50	I		100	3973.64	II		2000	4460.29	I
490	2907.47	II		560	3545.20	II		540	3990.57	I		610	4462.36	I
2400	2908.82	II		110	3553.27	I		260	3992.80	I		120	4468.01	I
710	2910.02	II		560	3556.50	II		430	3998.73	I		380	4469.71	I
530	2910.39	II		110	3566.18	I		170	4005.71	II		120	4474.04	I
560	2911.06	II		560	3589.76	II		120	4023.39	II		200	4474.71	I
380	2914.93	I		490	3592.02	II		120	4031.83	I		380	4488.89	I

Vanadium (Cont.)

Intensity	Wavelength		
100	4496.06		I
120	4501.95		I
140	4524.22		I
360	4545.39		I
100	4549.65		I
280	4560.71		I
200	4571.78		I
510	4577.17		I
140	4578.73		I
640	4580.40		I
830	4586.36		I
170	4591.22		I
1300	4594.11		I
100	4606.15		I
30	4609.65		I
25	4611.74		I
230	4619.77		I
65	4624.41		I
50	4626.48		I
100	4635.18		I
65	4640.07		I
65	4640.74		I
130	4646.40		I
30	4648.89		I
30	4666.14		I
160	4670.49		I
24	4684.45		I
35	4686.92		I
55	4706.16		I
80	4706.57		I
80	4710.56		I
65	4714.12		I
35	4715.89		I
55	4717.69		I
40	4721.51		I
40	4722.86		I
40	4729.53		I
27	4730.38		I
27	4742.63		I
24	4746.63		I
40	4748.52		I
45	4750.98		I
35	4751.56		I
40	4753.93		I
65	4757.48		I
55	4766.63		I
130	4776.36		I
	4776.52		I
110	4786.51		I
130	4796.92		I
19	4799.77		I
130	4807.53		I
130	4827.45		I
150	4831.64		I
120	4832.43		I
19	4833.02		I
19	4848.81		I
320	4851.48		I
35	4862.61		I
480	4864.74		I
21	4871.26		I
620	4875.48		I
55	4880.56		I
740	4881.56		I
27	4891.60		I
21	4894.21		I
55	4900.62		I
95	4904.29	d	I
	4904.34		I
85	4925.65		I
35	4932.03		I
23	4966.12		I
70	5002.33		I
85	5014.62		I
28	5051.63		I
35	5064.12		I
35	5105.14		I
110	5128.53		I
110	5138.42		I
25	5139.53		I
70	5148.72		I
40	5159.35		I
23	5169.94		I
70	5176.77		I
20	5192.01		I
110	5192.99		I
23	5193.62		I
110	5194.83		I
55	5195.36		I
20	5206.61		I
40	5216.59		I
35	5225.77		I
35	5233.75		I
110	5234.07		I
20	5240.20		I
110	5240.87		I
17	5260.98		I
40	5353.41		I
35	5383.43		I
40	5385.14		I
14	5388.30		I
11	5397.87		I
100	5401.93		I
140	5415.26		I

Intensity	Wavelength		
28	5418.09		I
50	5424.66		I
40	5434.18		I
11	5437.66		I
17	5458.12		I
13	5471.33		
25	5487.22		
85	5487.92		I
25	5489.94		I
28	5504.87		I
70	5507.75		I
14	5511.18		I
23	5545.93		I
70	5547.07		I
35	5558.75		I
28	5561.66		I
140	5584.50		I
23	5586.00		I
100	5592.42		I
28	5601.38		I
70	5604.94		I
13	5624.20		I
200	5624.60		I
70	5624.89		I
55	5626.01		I
400	5627.64		I
13	5632.46		I
10	5633.90		I
13	5635.51		I
85	5646.11		I
110	5657.44		I
110	5668.36		I
310	5670.85		I
20	5683.22		I
1200	5698.52		I
920	5703.56		I
570	5706.98		I
11	5708.95		I
11	5716.21	h	I
70	5725.64		I
850	5727.03		I
170	5727.66		I
230	5731.25		I
40	5734.01		I
230	5737.06		I
110	5743.45		I
17	5747.70		I
40	5748.87		I
17	5752.74		I
17	5761.41		I
70	5772.42		I
35	5776.64		I
11	5782.61		I
11	5783.50		I
40	5784.38	h	I
55	5786.16	h	I
23	5788.56		I
35	5807.14	h	I
23	5817.06		I
35	5817.53	h	I
55	5830.72	h	I
85	5846.90	h	I
11	5850.32		I
40	5924.57		I
28	5978.91		I
20	5980.78		I
28	6002.31		I
55	6002.63		I
28	6016.12		I
20	6025.41		I
450	6039.73		I
100	6058.14		I
20	6067.26		I
480	6081.44		I
1300	6090.22		I
28	6106.98		I
280	6111.67		I
600	6119.52		I
20	6128.34		I
280	6135.38		I
180	6150.15		I
85	6170.36		I
23	6189.35		I
450	6199.19		I
130	6213.87		I
450	6216.37		I
28	6218.31	h	I
130	6224.50		I
430	6230.74		I
100	6233.20		I
55	6240.13		I
170	6242.81		I
710	6243.10		I
280	6251.82		I
85	6256.90		I
85	6258.57		I
55	6261.22		I
85	6266.32		I
130	6268.82		I
170	6274.65		I
17	6282.33	h	I
200	6285.16		I
200	6292.83		I
170	6296.49		I

Intensity		Wavelength	
28	h	6311.50	I
14		6324.06	I
70		6326.84	I
55		6339.09	I
50		6349.48	I
14		6355.58	I
50		6357.30	I
25		6358.82	I
35		6361.27	l
23		6379.36	I
14		6393.28	I
35		6430.47	I
23		6431.63	I
14		6433.18	I
11		6435.16	I
70		6452.34	I
11		6488.05	I
55		6504.17	I
110		6531.43	I
28		6543.51	I
17		6558.02	I
11		6565.88	I
50		6605.97	I
15		6607.83	I
10		6623.54	I
50		6624.85	I
13		6633.26	I
13		6643.79	I
8		6693.66	I
8		6708.07	I
65	c	6753.00	I
10		6760.12	I
50	c	6766.49	I
40		6784.98	I
15		6786.32	I
26		6812.40	I
9	c	6829.94	I
15		6832.44	I
12		6839.58	I
12		6841.90	I
10	c	6870.88	I
8		6871.56	I
7		6894.00	I
12		6974.50	I
21		7026.07	I
7		7063.69	I
11	h	7092.08	I
6		7102.58	I
24		7148.15	I
7		7151.36	I
7		7182.08	I
14		7264.29	I
8		7321.44	I
40		7338.92	I
35		7356.54	I
11		7358.66	I
24		7361.39	I
12		7362.49	I
24		7363.16	I
9		7385.91	I
6	h	7393.49	
12	h	7485.90	I
12	h	7488.08	I
12	h	7492.44	
12	h	7578.75	I
9	h	7591.24	I
14	h	7596.92	I
12	h	7598.28	I
24		7624.81	I
5		7701.37	I
8		7704.81	I
8	h	7851.18	I
14	B1	7865.51	VO
12		7896.40	I
14	h	7898.81	
24		7937.92	I
29	c	8027.39	I
14		8028.13	I
14	h	8035.38	
14	h	8045.71	
12		8051.89	
14		8093.48	I
8		8102.44	I
12		8108.59	I
9	h	8109.07	I
120	Cw	8116.80	I
11	h	8136.79	I
29		8144.59	I
9		8154.55	I
70	c	8161.07	I
14		8171.35	I
7		8180.21	I
35		8186.71	I
24		8187.33	I
29		8198.87	I
35		8203.07	I
24		8241.61	I
29	c	8253.51	I
29		8255.88	I
5	h	8280.39	I
19		8282.37	I
8		8324.42	I
14	h	8331.23	I
14		8342.03	I

Intensity		Wavelength	
7		8402.81	I
12		8499.52	I
6		8534.49	I
6	B1	8624.86	VO
60	c	8919.85	I
29	c	8932.93	I
12		8971.62	I

V III
Ref. 394 — C.H.C.

Intensity		Wavelength	
		Vacuum	
25		616.09	III
50		633.94	III
40		635.41	III
100		864.27	III
75		948.84	III
500		1006.46	III
500		1149.94	III
400		1157.18	III
300		1160.77	III
500		1252.11	III
400		1254.01	III
500		1287.87	III
400		1289.42	III
300		1290.77	III
400		1313.35	III
300		1313.27	III
500		1331.99	III
500		1335.12	III
1000		1643.03	III
1000		1650.14	III
300		1668.03	III
300		1670.66	III
300		1679.19	III
1000		1694.78	III
400		1721.98	III
300		1724.63	III
500	d	1751.68	III
500		1757.73	III
1000		1760.07	III
300		1773.43	III
400		1770.02	III
500		1779.72	III
400		1784.44	III
1000		1788.26	III
500		1793.82	III
1000		1794.60	III
300		1796.77	III
500		1798.15	III
300		1802.55	III
500		1804.13	III
1000		1812.19	III
400		1831.15	III
400		1831.64	III
300		1845.07	III
300		1850.69	III
400		1852.01	III
500		1854.42	III
300		1855.06	III
500		1856.64	III
300		1864.51	III
300		1878.68	III
400		1880.41	III
300		1895.01	III
400		1899.81	III
500		1902.23	III
300		1934.00	III
		Air	
500		2232.91	III
400		2241.53	III
1000		2292.86	III
400		2314.18	III
500		2318.06	III
400		2319.00	III
500		2323.82	III
2500		2330.42	III
500		2331.75	III
400		2334.20	III
500		2343.10	III
500		2358.73	III
500		2366.31	III
2500		2371.06	III
1000		2382.46	III
500		2393.58	III
500		2404.18	III
500		2516.14	III
250		2521.16	III
250		2521.55	III
250		2548.21	III
150		2554.22	III
150		2563.32	III
250		2593.05	III
250		2595.10	III
100		3679.86	III
50	h	3705.35	III

Vanadium (Cont.)

Intensity	Wavelength	
40	4714.89	III
50	6597.20	III

V IV
Ref. 397 — C.H.C.

Intensity	Wavelength	
	Vacuum	
200	677.34	IV
60	678.74	IV
50	679.65	IV
500	684.37	IV
100	684.45	IV
100	691.53	IV
50	693.13	IV
400	737.85	IV
150	750.11	IV
60	1226.52	IV
50	1308.06	IV
80	1355.13	IV
60	1395.00	IV
50	1414.41	IV
80	1419.58	IV
100	1426.65	IV
60	1520.14	IV
80	1601.92	IV
80	1611.88	IV
80	1806.18	IV
60	1809.85	IV
100	1817.68	IV
200	1825.84	IV
300	1861.56	IV
500	1939.06	IV
400	1951.43	IV
300	1963.10	IV
500	1997.72	IV
200	1999.32	IV
	Air	
100	2002.48	IV
50 h	2088.74	IV
50 h	2146.83	IV
100 h	2155.34	IV
500	2268.30	IV
50 h	2421.32	IV
50 h	2433.53	IV
50 h	2446.80	IV
50 h	2450.87	IV
50 h	2556.92	IV
50 h	2570.72	IV
80 h	2624.21	IV
80 h	2645.54	IV
50 h	2655.41	IV
50 h	2656.87	IV
50	3284.56	IV
60	3334.79	IV
50	3448.41	IV
50	3496.42	IV
80 h	3514.25	IV
50 h	5130.78	IV
50 h	5262.16	IV
60 h	5352.32	IV
40 h	5940.12	IV

V V
Ref. 432 — C.H.C.

Intensity	Wavelength	
	Vacuum	
18	224.91	V
20	225.46	V
17	227.88	V
16	239.41	V
19	239.48	V
20	251.66	V
18	252.44	V
18	285.98	V
20	286.84	V
17	312.39	V
35	483.01	V
25	484.51	V
20	820.86	V
30	829.48	V
15	962.03	V
15	1142.74	V
25	1157.58	V
20	1490.11	V
100	1680.20	V
50	1716.72	V
15	1724.99	V
25	1792.99	V
30	1811.42	V

Air (V)

Intensity	Wavelength	
20 d	2319.66	V
20	2577.90	V
15	2775.82	V
15	3617.97	V
12 d	3746.36	V
20	4200.32	V
15 w	4930.53	V
8	5356.07	V
7	6628.80	V
3	7595.51	V

XENON (Xe)
Z = 54

Xe I and II
Ref. 33, 116, 118, 120, 232 — S.P.D.

Intensity	Wavelength	
	Vacuum	
350	740.41	II
350	803.07	II
600	880.80	II
350	885.54	II
600	925.87	II
250	935.40	II
800	972.77	II
700	976.68	II
500	1032.44	II
700	1037.68	II
1100	1041.31	II
1000	1048.27	II
1200	1051.92	II
2000	1074.48	II
600	1083.86	II
1200	1100.43	II
600	1158.47	II
250	1169.63	II
800 p	1183.05	II
250	1192.04	I
600	1244.76	II
250	1250.20	I
1000	1295.59	I
600	1469.61	I
	Air	
200	2864.73	II
150 h	2895.22	II
400	2979.32	II
100 h	3017.43	II
300	3128.87	II
200 h	3366.72	II
2	3400.07	I
2	3418.37	I
2	3420.00	I
3	3442.66	I
100 h	3461.26	II
4	3469.81	I
4	3472.36	I
5	3506.74	I
10	3549.86	I
10	3554.04	I
15	3610.32	I
8	3613.06	I
6	3633.06	I
10	3669.91	I
40	3685.90	I
40	3693.49	I
100 l	3907.91	II
100	4037.59	II
200 l	4057.46	II
100 h	4098.89	II
200 l	4158.04	II
1000 h	4180.10	II
500 h	4193.15	II
300 h	4208.48	II
100 h	4209.47	II
300 h	4213.72	II
100	4215.60	II
300 h	4223.00	II
400 h	4238.25	II
500 h	4245.38	II
100 l	4251.57	II
500 h	4296.40	II
500 h	4310.51	II
1000 l	4330.52	II
200 h	4369.20	II
100 l	4373.78	II
500 l	4393.20	II
500 l	4395.77	II
200 l	4406.88	II
150 l	4416.07	II
500 h	4448.13	II
1000 h	4462.19	II
500 l	4480.86	II
100 l	4521.86	II
600	4734.152	I
150	4792.619	I
500	4807.02	I
400	4829.71	I
300	4843.29	I
500	4916.51	I
500	4923.152	I
200 l	4971.71	II
400	4972.71	II
300	4988.77	II
100 l	4991.17	II
200	5028.280	I
200	5044.92	II
1000	5080.62	II
300	5122.42	II
100	5125.70	II
100	5178.82	II
300	5188.04	II
400	5191.37	II
100	5192.10	II
500	5260.44	II
500	5261.95	II
2000	5292.22	II
300	5309.27	II
1000	5313.87	II
2000	5339.33	II
200	5363.20	II
200	5368.07	II
500	5372.39	II
100	5392.80	I
3000	5419.15	II
800	5438.96	II
300	5445.45	II
200	5450.45	II
400	5460.39	II
1000	5472.61	II
100 l	5494.86	II
200	5525.53	II
600	5531.07	II
100	5566.62	I
300	5616.67	II
300	5659.38	II
600	5667.56	II
150	5670.91	II
100	5695.75	I
200	5699.61	II
200	5716.10	II
500	5726.91	II
500	5751.03	II
300	5758.65	II
300	5776.39	II
100	5815.96	II
300	5823.89	I
150	5824.80	I
100	5875.02	I
300	5893.29	II
100	5894.99	I
200	5905.13	II
100	5934.17	I
500	5945.53	II
300	5971.13	II
2000	5976.46	II
200	6008.92	II
1000	6036.20	II
2000	6051.15	II
600	6093.50	II
1500	6097.59	II
400	6101.43	II
100	6115.08	II
100	6146.45	II
150	6178.30	I
120	6179.66	I
300	6182.42	I
500	6194.07	II
100	6198.26	I
100	6220.02	II
500	6270.82	II
400	6277.54	II
100	6284.41	II
100	6286.01	I
250	6300.86	II
500	6318.06	I
400	6343.96	II
600	6356.35	II
200	6375.28	II
100	6397.99	II
300	6469.90	I
150	6472.84	I
120	6487.76	I
100	6498.72	I
200 h	6504.18	I
300	6512.83	II
200	6528.65	II
100	6533.16	I
1000	6595.01	II
100	6595.56	I
400	6597.25	II
100	6598.84	II
150	6668.92	I
300	6694.32	II
200	6728.01	I
150	6788.71	II
100	6790.37	II
1000	6805.74	II
200	6827.32	I
100	6872.11	I
300	6882.16	I
80	6910.22	II
100	6925.53	I
800 h	6942.11	II
100	6976.18	I
2000	6990.88	II
150	7082.15	II
500	7119.60	I
50 s	7147.50	II
200	7149.03	II
500	7164.83	II
100	7284.34	II
200	7301.80	II
200	7339.30	II
100	7386.00	I
150	7393.79	I
300	7548.45	II
200	7584.68	I
80	7618.57	II
500	7642.02	I
100	7643.91	I
200	7670.66	II
60	7787.04	II
100	7802.65	I
100	7881.32	I
300	7887.40	I
500	7967.34	I
100	8029.67	I
200	8057.26	I
150	8061.34	I
100	8101.98	I
150 h	8151.80	II
100	8171.02	I
700	8206.34	I
10000	8231.635	I
500	8266.52	I
7000	8280.116	I
2000	8346.82	I
100	8347.24	II
2000	8409.19	I
50 h	8515.19	II
200	8576.01	I
50 h	8604.23	II
250	8648.54	I
100	8692.20	I
200	8696.86	I
50 h	8716.19	II
300	8739.39	I
100	8758.20	I
5000	8819.41	I
300	8862.32	I
200	8908.73	I
200	8930.83	I
1000	8952.25	I
100	8981.05	I
200	8987.57	I
400	9045.45	I
500	9162.65	I
100	9167.52	I
100	9374.76	I
200	9513.38	I
50 h	9591.35	II
150	9685.32	I
50 l	9698.68	II
100	9718.16	I
2000	9799.70	I
3000	9923.19	I
100	10838.37	I
90	11742.01	I
375	12235.24	I
100	12257.76	I
300	12590.20	I
2500	12623.391	I
250	13544.15	I
2000	13657.055	I
1250	14142.444	I
800	14240.96	I
375	14364.99	I
140	14660.81	I
3000	14732.806	I
100	15099.72	I
2500	15418.394	I
150	15557.13	I
250	15979.54	I
100	16039.90	I
1000	16053.28	I
125	16554.49	I
1500	16728.15	I
1500	17325.77	I
350	18788.13	I
150	20187.19	I
3000	20262.242	I
250	21470.09	I
1250	23193.33	I
110	23279.54	I
1800	24824.71	I
175	25145.84	I
2000	26269.08	I
2500	26510.86	I
250	28381.54	I
750	28582.25	I
300	29384.41	I
150	29448.06	I

Xenon (Cont.)

Intensity	Wavelength	
100	29649.58	I
100	29813.62	I
600	30253.14	I
1500	30475.46	I
100	30504.12	I
500	30794.18	I
6000	31069.23	I
125	31336.01	I
550	31607.91	I
100	32293.08	I
1800	32739.26	I
3500	33666.69	I
150	34014.67	I
450	34335.27	I
170	34744.00	I
5000	35070.25	I
110	35246.92	I
250	36209.21	I
150	36231.74	I
450	36508.36	I
850	36788.83	I
140	38685.98	I
175	38737.82	I
270	38939.60	I
120	39955.14	I

Xe III
Ref. 33, 384, 391, 429 — R.L.K.

Intensity	Wavelength	
	Vacuum	
8	657.8	III
8	660.1	III
9	673.8	III
9	674.0	III
9	676.6	III
10	694.0	III
20	698.5	III
12	705.1	III
10	721.2	III
15	731.0	III
10	733.3	III
15	742.6	III
10	756.0	III
10	761.5	III
10	769.1	III
25	779.1	III
15	792.9	III
12	796.1	III
15	802.0	III
25	823.2	III
30	824.9	III
25	853.0	III
15	889.3	III
20	894.0	III
20	896.0	III
10	965.5	III
35	1003.4	III
35	1017.7	III
10	1047.8	III
12	1066.4	III
30	1130.3	III
25	1232.1	III
	Air	
80	2668.98	III
100	2717.33	III
30	2814.45	III
40	2815.91	III
30	2827.45	III
40	2847.65	III
30	2862.40	III
80 w	2871.10	III
60 w	2871.24	III
30	2871.7	III
30	2896.62	III
50	2906.6	III
40	2911.89	III
80 w	2912.36	III
40	2940.2	III
60	2945.2	III
40	2947.5	III
40	2948.1	III
80 w	2970.47	III
40	2992.87	III
30	3004.25	III
100	3023.81	III
40	3083.5	III
50	3091.1	III
30	3106.46	III
100 w	3138.3	III
80 c	3150.82	III
40	3185.2	III
100	3242.86	III
80	3268.98	III
30	3287.82	III
80 w	3301.55	III
40	3331.6	III

Intensity	Wavelength	
30	3358.0	III
80	3384.12	III
60	3444.2	III
70	3454.2	III
100 w	3458.7	III
40	3468.22	III
80	3522.83	III
50	3542.3	III
50	3552.1	III
40	3561.4	III
100	3579.7	III
80	3583.6	III
100 w	3595.4	III
100	3606.06	III
40	3607.0	III
100 w	3615.9	III
40	3623.1	III
600	3624.08	III
50	3676.67	III
40	3776.3	III
300	3781.02	III
100	3841.5	III
200	3877.8	III
60	3880.5	III
500	3922.55	III
300	3950.59	III
200	4050.07	III
60	4060.4	III
100	4109.1	III
100	4145.7	III
30	4285.9	III
50	4434.2	III
100 w	4462.1	III
100 w	4569.1	III
100 w	4570.1	III
100 w	4641.4	III
30	4673.7	III
60	4683.57	III
30	4723.60	III
100 w	4757.3	III
40	4869.5	III
60	5239.0	III
30	5367.1	III
50	5401.0	III
40	5524.4	III
60	6205.97	III
25	6221.7	III
60	6238.2	III
60	6259.05	III

YTTERBIUM (Yb)
Z = 70

Yb I and II
Ref. 1 = C.H.C.

Intensity	Wavelength	
	Air	
2500	2116.67	II
3000	2126.74	II
370	2161.60	II
850	2185.71	II
640	2224.46	II
140	2320.81	II
50	2362.89	II
170	2390.74	II
18	2398.02	II
28	2421.35	II
25	2447.26	II
28	2460.25	II
460	2464.50	I
14	2484.89	II
70	2502.02	II
28	2505.48	II
11	2508.07	II
140	2512.06	II
18	2516.35	II
50	2522.44	II
65	2537.65	II
270	2538.67	II
14	2550.06	II
70	2552.15	II
55	2552.70	II
21	2565.57	II
28	2571.36	II
13	2573.15	II
18	2596.16	II
28	2596.32	II
21	2615.26	II
100	2617.01	II
55	2634.31	II
45	2639.45	II
85	2641.89	II
110	2644.31	II
28	2646.44	II
28	2647.46	II
28	2648.80	II
50	2649.79	II
28	2650.73	II

Intensity	Wavelength	
990	2653.75	II
35	2656.12	II
21	2659.27	II
200	2665.04	II
55	2668.75	II
390	2671.96	I
390	2672.66	II
21	2680.40	II
14	2683.42	II
70	2684.75	II
25	2687.98	II
28	2695.43	II
14	2696.62	II
18	2700.80	II
21	2708.84	II
65	2710.54	II
25	2711.78	II
55	2712.66	II
170	2718.35	II
28	2722.20	II
70	2732.74	II
21	2734.09	II
55	2741.71	II
55	2747.58	II
18	2748.04	II
230	2748.66	II
1300	2750.48	II
85	2751.45	II
21	2759.00	II
65	2760.78	II
65	2761.37	II
35	2764.41	II
85	2771.32	II
170	2776.28	II
100	2784.66	II
18	2787.96	II
45	2793.28	II
25	2794.44	II
21	2795.07	II
18	2795.29	II
35	2797.80	II
100	2798.21	II
45	2799.38	II
50	2800.00	II
35	2800.06	II
14	2810.02	II
65	2814.53	II
28	2816.32	II
140	2821.15	II
100	2824.97	II
190	2830.99	II
18	2832.20	II
28	2834.97	II
14	2842.59	II
230 h	2847.18	II
100	2848.44	II
21	2849.34	II
360	2851.13	II
55	2851.86	II
21	2853.41	II
18	2853.68	II
55	2854.14	II
45	2854.49	II
45	2858.33	II
45	2858.46	II
100	2859.39	II
430	2859.80	II
55	2860.39	II
140	2861.21	II
100	2861.34	II
200	2867.06	II
25	2870.06	II
45	2873.49	I
28	2885.97	II
70	2886.26	II
200	2888.04	II
3600	2891.38	II
45	2893.62	II
28	2896.90	II
85	2899.70	II
21	2902.41	II
21	2902.92	II
28	2906.88	II
28	2908.33	II
35	2909.19	II
55	2909.48	II
85	2911.52	II
18	2912.86	II
170	2914.21	II
140	2915.28	II
18	2916.43	II
280	2919.35	II
55	2921.12	II
45	2924.24	II
25	2927.85	II
35	2934.36	I
55	2935.11	II
21	2937.19	II
45	2939.53	II
45	2940.52	II
28	2942.04	II
140	2945.91	II
45	2946.30	II
18	2946.76	II

Intensity	Wavelength	
28	2950.33	II
45	2955.32	II
18	2957.63	II
65	2962.52	II
21	2963.26	II
45	2963.46	II
130	2964.76	II
2000	2970.56	II
45	2982.49	II
21	2982.66	II
28	2983.70	II
200	2983.99	II
90	2985.08	II
35	2985.88	II
45	2990.37	II
65	2991.87	II
28	2993.94	II
170	2994.80	II
28	2995.86	II
70	3000.46	II
25	3002.61	II
310	3005.77	II
100	3009.39	II
65	3010.62	II
55	3014.43	II
160	3017.56	II
160	3026.67	II
920	3031.11	II
55	3034.64	II
25	3037.99	II
55	3039.67	II
80	3042.65	II
21	3044.00	II
45	3046.48	II
35	3047.05	II
45	3063.12	II
21	3063.67	II
110	3065.04	II
18	3076.01	II
100	3089.10	II
70	3093.87	II
28	3100.74	I
45	3101.36	II
28	3102.07	II
55	3107.76	II
170	3107.90	II
85	3115.34	II
55	3116.70	II
190	3117.01	II
50	3136.76	II
230	3140.94	II
80	3141.73	II
80	3145.06	II
28	3145.54	II
28	3153.18	II
90	3153.88	II
50	3155.18	II
28	3162.29	I
70	3163.80	II
50	3165.21	II
120	3169.06	II
120	3180.92	II
390	3192.88	II
70	3198.65	II
240	3201.16	II
80	3217.18	II
50	3218.32	II
50	3225.88	II
45	3239.20	II
35	3239.58	I
35	3246.06	II
130	3261.51	II
18000	3289.37	I
130	3305.25	I
140	3305.73	II
50	3315.10	II
80	3319.41	I
50	3333.06	II
240	3337.17	II
280 d	3342.93	II
	3343.07	II
	3346.50	
80	3347.54	II
50	3351.09	II
	3351.26	
100	3352.49	II
100	3362.44	II
50	3363.64	II
240	3375.48	II
50	3376.62	II
28	3382.54	
140	3387.50	I
50	3390.25	II
28	3390.42	II
50	3391.10	II
50 h	3394.44	II
50	3401.01	II
35	3404.10	II
50	3412.45	I
140	3418.39	I
360	3426.04	I
80	3428.46	II
240	3431.11	I
45	3434.61	II

Ytterbium (Cont.)

Intensity		Wavelength		Intensity		Wavelength		Intensity		Wavelength		Intensity		Wavelength	
50		3438.71	II	220		5074.34	I	65		1810.88	III	100		2438.27	III
100		3438.85	II	50		5076.74	I	60		1826.41	III	20		2439.31	III
35		3443.59		20		5135.98	II	20		1826.77	III	20		2440.43	III
35		3446.89	II	14		5147.02	II	30		1838.01	III	10		2458.64	III
85		3452.40	I	20		5184.15	II	30		1847.30	III	10		2464.59	III
500		3454.08	II	60		5196.08	I	30		1849.24	III	200		2490.42	III
190	d	3458.29	II	85		5211.60	I	10		1849.42	III	20		2491.69	III
		3458.39	I	35		5240.51	II	75		1852.36	III	40		2506.25	III
360		3460.27	I	100		5244.11	I	75		1852.94	III	300		2516.82	III
35		3462.34	II	40		5257.49	II	90		1854.80	III	15		2522.07	III
2400		3464.37	I	150	h	5277.04	I	80		1857.16	III	20		2529.14	III
500		3476.30	II	35		5279.53	II	100		1863.32	III	40		2550.39	III
500		3478.84	II	17		5300.94	II	5		1864.85	III	300		2555.29	III
50		3482.56	II	170		5335.15	II	10		1867.23	III	100		2560.56	III
85		3485.76	II	30	d	5345.66	II	10		1867.63	III	10		2561.66	III
85		3488.43	II			5345.83	II	10		1868.19	III	100		2566.78	III
100	Hw	3495.90	II	60		5347.22	II	5		1868.92	III	2000		2567.61	III
85		3507.83	II	30	h	5351.29	I	10		1870.07	III	1000		2579.57	III
50		3517.00	I	150		5352.95	II	15		1870.83	III	100		2588.62	III
230		3520.29	II	30		5358.64	II	10		1871.15	III	20		2592.69	III
50		3545.72	II	30		5363.66	I	200		1872.03	III	500		2597.23	III
100		3549.82	II	17		5389.84	II	800		1873.91	III	800		2599.14	III
35		3559.03	i	14		5432.71	II	100		1875.41	III	30		2609.14	III
200		3560.33	II	40		5449.27	II	75		1875.92	III	600		2621.11	III
170		3560.70	II	14		5478.50	II	70		1880.30	III	300		2627.07	III
50	h	3563.94	II	60		5481.92	I	80		1884.22	III	30		2635.37	III
85		3570.57	II	40		5505.49	I	70		1885.07	III	500		2638.06	III
50		3572.50	II	17		5524.54	I	70		1887.22	III	300		2640.48	III
50		3574.58	II	85	h	5539.05	I	10		1890.34	III	1000		2642.56	III
360		3585.47	II	2400		5556.47	I	100		1890.87	III	100		2643.62	III
130		3606.48	II	35		5562.09	I	15		1892.42	III	1000		2651.74	III
50		3610.23	II	20		5568.11	I	5		1895.50	III	700		2652.25	III
70		3611.30	II	20		5586.36	I	100		1896.18	III	100		2659.98	III
200		3619.80	II	40		5588.45	II	10		1897.57	III	70		2664.89	III
110		3634.52		60		5651.98	II	500		1898.25	III	2000		2666.13	III
240		3637.76	II	7		5686.53	II	7		1906.74	III	2000		2666.99	III
70		3648.15	I	220		5719.99	I	10		1908.50	III	30		2673.33	III
90		3655.73	I	10		5749.91	II	100		1909.66	III	500		2677.39	III
240		3669.69	II	10	h	5755.89	I	70		1910.86	III	500		2691.01	III
50		3670.69	II	27		5771.66	II	20		1920.53	III	30		2708.04	III
140		3675.08	II	10		5803.44	I	10		1926.76	III	400		2712.32	III
50		3690.56	II	10		5819.41	II	70		1928.09	III	500		2749.91	III
32000		3694.19	II	35		5833.99	II	15		1930.63	III	200		2755.94	III
70		3698.60	II	35		5837.14	II	55		1942.59	III	200		2756.76	III
70		3700.58	I	27		5854.51	I	40		1950.34	III	100		2765.50	III
50		3710.34	II	8		5897.21	II	15		1962.80	III	300		2788.24	III
60		3724.21	II	20		5908.36	II	80		1967.13	III	600		2795.60	III
180		3734.69	I	17		5989.33	I	20		1969.47	III	400		2803.32	III
550		3770.10	I	40		5991.51	II	30		1969.73	III	1000		2803.43	III
80		3774.32	I	10		6052.88	II	10		1973.96	III	10		2807.22	III
60	h	3791.74	I	10		6054.57	I	10		1974.18	III	50		2808.51	III
170		3839.91	I	60		6152.57	II	25		1976.46	III	600		2816.92	III
340		3872.85	I	30		6246.97	II	2		1981.74	III	1000		2818.72	III
340		3900.85	I	60		6274.78	II	5		1983.88	III	15		2826.01	III
50		3904.81	II	14		6308.15	II	25		1984.62	III	300		2842.96	III
140		3911.27	I	35	h	6400.35	I	7		1985.74	III	400		2875.86	III
32000		3987.99	I	35	h	6417.91	I	80		1986.43	III	600		2898.30	III
930		3990.88	I	20		6432.73	II	80	h	1989.82	III	1000		2906.31	III
50		4007.36	I	17	h	6463.15	II	50		1991.14	III	300		2928.97	III
70		4052.28	I	340		6489.06	I	45		1995.05	III	50		2977.84	III
85		4077.28	II	20		6643.55	I	55		1997.28	III	800		2998.00	III
440		4089.68	I	180		6667.82	I	55		1997.66	III	2000		3029.49	III
120	h	4119.25	II	15		6678.17	I	500		1998.82	III	100		3031.62	III
70		4135.09	II	25		6727.61	II					30		3040.65	III
470		4149.07	I	25		6768.70	I			Air		3000		3092.50	III
120		4174.56	I	690		6799.60	I					20		3102.18	III
340		4180.81	II	18		6934.05	II	20		2054.80	III	4000		3126.01	III
150	d	4218.56	II	20		6999.88	II	10		2066.49	III	1000		3138.58	III
		4218.69	I	10		7043.78	II	10		2073.64	III	100		3151.44	III
120		4231.97	I	9	h	7244.41	I	30		2078.05	III	70		3179.34	III
70		4277.74	I	8	h	7305.22	I	10		2087.37	III	800		3191.35	III
120		4305.97	I	10	h	7313.05	I	50		2087.98	III	50		3216.27	III
70		4316.95	II	16	h	7350.04	I	20		2091.23	III	2000		3228.58	III
60	h	4393.69	I	25		7448.28	I	20		2092.26	III	2000		3325.51	III
60	h	4430.21	I	30	h	7527.46	I	10		2094.77	III	50		3358.25	III
440		4439.19	I	750		7699.48	I	80		2095.31	III	20		3364.30	III
85	h	4482.42	I	7		7895.08	I	15		2096.79	III	2000		3384.01	III
85		4515.16	II	70	h	8922.56	II	30		2098.36	III	150		3392.56	III
35		4553.58	II					10		2106.71	III	100		3397.66	III
85	h	4563.95	I					50		2109.54	III	80		3432.94	III
640		4576.21	I			Yb III		20		2119.18	III	40		3456.18	III
200		4582.36	I			Ref. 40, 192 — J.R.		20		2198.14	III	150		3463.51	III
70		4589.21	I					80		2202.27	III	20		3469.98	III
140		4590.83	I	Intensity		Wavelength		300		2240.11	III	300		3550.87	III
40		4598.36	II					100		2244.28	III	200		3613.89	III
35		4683.81	II			Vacuum		200		2257.03	III	30		3659.84	III
40		4684.27	I					100		2262.26	III	30		3663.74	III
190		4726.08	II	5		968.46	III	200		2265.67	III	200		3664.74	III
170	h	4781.87	I	20		973.16	III	150		2282.99	III	20		3675.78	III
170		4786.61	II	10		994.56	III	100		2283.99	III	400		3711.91	III
35		4816.43	I	10		1560.66	III	300		2305.32	III	20		3879.98	III
40		4820.24	II	80		1561.42	III	100		2309.27	III	10		3882.58	III
35		4836.96	II	30		1669.60	III	200		2314.49	III	20		3887.17	III
40		4837.46	I	50		1670.78	III	200		2337.97	III	150		3896.55	III
17		4851.15	II	50		1719.82	III	40		2361.08	III	15		3912.75	III
40	h	4894.60	I	60		1739.18	III	200		2365.43	III	20		3913.23	III
27		4912.36	I	70		1762.80	III	50		2367.46	III	500		3931.23	III
710		4935.50	I	80	h	1765.21	III	30		2369.99	III	100		3985.56	III
24		4937.22	II	65		1775.29	III	20		2377.22	III	10		3991.74	III
140		4966.90	I	70		1779.74	III	50		2403.95	III	10		3997.67	III
24		5009.52	II	70		1781.31	III	20		2410.04	III	2000		4028.14	III
17		5067.30	II	20		1793.70	III	60		2412.33	III	10		4033.03	III
30		5067.80	I	60		1798.85	III	10		2429.18	III	20		4074.53	III
70		5069.14	I					20		2433.43	III	20		4090.67	III

Ytterbium (Cont.)

Intensity		Wavelength		Intensity	Wavelength		Intensity	Wavelength		Intensity		Wavelength	
20		4098.23	III	300	1491.57	IV	480	2974.59	I	2000		4174.14	I
15		4121.06	III	400	1765.03	IV	30	2980.55	II	8000		4177.54	II
10		4150.04	III	200	1776.18	IV	750	2984.26	I	120		4199.28	II
15		4153.11	III	200	1778.20	IV	70	2995.26	I	380		4204.70	I
100		4162.72	III	200	1779.34	IV	140	2996.94	I	80		4213.02	I
60		4172.95	III	300	1789.71	IV	70	3005.26	I	40		4213.54	I
30		4194.34	III	800	1791.06	IV	55	3018.95	I	160		4217.80	I
100		4194.95	III	200	1801.67	IV	130	3021.73	I	280	h	4220.63	I
10		4198.74	III	250	1809.63	IV	90	3022.28	I	80		4224.25	I
300		4213.64	III	600	1813.84	IV	26	3026.49	II	600		4235.73	II
10		4220.83	III	400	1816.07	IV	30	3036.59	II	2200		4235.94	I
15		4231.07	III	250	1817.58	IV	45	3044.84	I	300		4251.20	I
20		4289.64	III	300	1819.02	IV	190	3045.37	I	360	h	4302.30	I
40		4301.14	III	200	1824.22	IV	22	3047.11	I	2800		4309.63	II
20		4304.01	III				60	3055.57	II	50		4316.30	I
15		4350.80	III		Air		60	3086.85	II	110		4330.78	I
10		4380.07	III				55	3091.70 h	I	30		4337.29	I
100		4517.58	III	300	2106.48	IV	22	3093.76	II	60		4344.65	I
40		4639.14	III	900	2116.65	IV	95	3095.88	II	440	h	4348.79	I
10		4834.93	III	500	2121.29	IV	45	3111.81	I	60		4352.33	I
15		5054.94	III	250	2122.84	IV	55	3112.04	II	60		4352.70	I
20		5256.85	III	800	2123.32	IV	22	3114.28	I	120		4357.73	I
20		5331.54	III	600	2125.72	IV	60	3128.77	II	800		4358.73	II
15		5740.83	III	200	2129.65	IV	80	3129.93	II	120		4366.03	I
10	d	5949.02	III	500	2135.21	IV	95	3135.17	II	12000		4374.94	II
20		5973.05	III	300	2137.58	IV	110	3173.06	II	150	h	4375.61	I
40		6055.85	III	500	2138.35	IV	220	3179.41	II	80		4379.33	I
100		6214.22	III	200	2138.53	IV	70	3191.31	I	30		4385.48	I
200		6328.52	III	800	2139.99	IV	2300	3195.62	II	100		4387.74	I
10		6365.88	III	400	2141.04	IV	2200	3200.27	II	30		4394.01	I
150		6378.33	III	200	2142.20	IV	2200	3203.32	II	30		4394.67	I
25		6466.33	III	300	2143.42	IV	3900	3216.69	II	1800		4398.02	II
20		6985.15	III	300	2143.89	IV	6200	3242.28	II	890		4422.59	II
10		7037.04	III	20000	2144.77	IV	310	3280.91	II	80		4437.34	I
15		7157.72	III	400	2148.10	IV	19	3308.47	II	100		4443.66	I
10		7311.02	III	300	2148.52	IV	4700	3327.89	II	130		4446.63	I
10		7399.98	III	15000	2154.18	IV	55	3340.38	I	20		4465.27	I
80		7410.01	III	250	2165.55	IV	160	3362.00	II	40		4473.89	I
15		7456.86	III	300	2169.12	IV	85	3388.59	I	170		4475.72	I
70		7664.41	III	200	2172.16	IV	45	3397.04	I	180		4476.96	I
80		7892.39	III	300	2177.53	IV	85	3412.47	I	160		4477.45	I
20		7893.10	III	400	2183.32	IV	200	3448.82	II	110		4487.28	I
100		7971.46	III	270	2186.13 h	IV	70	3450.95	I	300		4487.47	I
20		8056.02	III	270	2187.17	IV	110	3467.88	II	30		4491.75	I
10		8117.44	III	90	2189.90 h	IV	170	3485.73	I	25		4492.42	I
10		8326.86	III	120	2193.34	IV	1700	3496.09	II	500		4505.95	I
30		8327.88	III	150	2198.27	IV	80	3521.53	I	50		4513.58	I
20		8400.01	III	90	2224.64	IV	45	3546.01	II	80		4514.01	I
30		8489.90	III	150	2231.28	IV	3900	3549.01	II	40	h	4522.05	I
200		10110.60	III	90	2233.30	IV	130	3551.80	I	890		4527.25	I
100		10830.36	III	90	2244.20	IV	540	3552.69	I	440		4527.80	I
				140	2331.36	IV	170	3558.76	I	100		4544.32	I

Yb IV
Ref. 40, 311 — J.R.

Intensity	Wavelength						
	Vacuum				190	3571.43	I
					260	3576.05	I
200	828.96	IV			3300	3584.52	II

YTTRIUM (Y)
Z = 39
Y I and II
Ref. 1 — C.H.C.

Intensity	Wavelength		
300	3587.75	I	
100	3589.69	I	
2800	3592.92	I	
10000	3600.73	II	
6200	3601.92	II	
7800	3611.05	II	
4300	3620.94	I	
1900	3628.71	II	
7800	3633.12	II	
3000	3664.61	II	
45	3668.49	II	
170	3692.53	I	
13000	3710.30	II	
60	3718.12	I	
60	3738.61	I	
1200	3747.55	II	
50	3749.89	I	
10000	3774.33	II	
1400	3776.56	II	
50	3782.30	II	
7400	3788.70	II	
1300	3818.35	II	
4000	3832.88	II	
70	3847.87	II	
80	3876.82	I	
480	3878.28	II	
30	3887.77	I	
60	3904.59	I	h
50	3918.25	I	h
60	3930.11	I	h
240	3930.66	II	
4400	3950.36	II	
150	3951.60	II	
60	3955.09	I	h
3600	3982.60	II	
40	3987.50	I	
940	4039.83	I	
2400	4047.64	I	
9400	4077.38	I	
90	4081.22	I	h
2000	4083.71	I	
9900	4102.38	I	
60	4106.39	I	h
80	4110.81	I	
320	4124.92	II	
8900	4128.31	I	
7500	4142.85	I	
100	4157.63	I	h
2400	4167.52	I	

Yb IV — Vacuum (cont.)

Intensity	Wavelength	
200	870.35	IV
300	902.46	IV
300	927.01	IV
300	936.22	IV
400	943.04	IV
400	946.20	IV
200	975.21	IV
1000	1050.24	IV
1000	1054.46	IV
400	1092.51	IV
200	1109.96	IV
200	1110.55	IV
5000	1134.43	IV
300	1136.24	IV
500	1166.01	IV
600	1185.58	IV
200	1290.24	IV
600	1305.58	IV
900	1316.04	IV
200	1326.32	IV
800	1326.36	IV
200	1340.06	IV
300	1345.36	IV
900	1350.26	IV
200	1353.43	IV
400	1356.15	IV
200	1361.75	IV
300	1365.88	IV
300	1369.72	IV
400	1375.42	IV
300	1376.66	IV
200	1384.41	IV
250	1393.93	IV
350	1398.77	IV
400	1407.05	IV
300	1413.14	IV
400	1416.15	IV
400	1417.72	IV
200	1423.99	IV
300	1430.29	IV
200	1440.61	IV
400	1477.92	IV

Yttrium (Y) — Air (cont.)

Intensity		Wavelength	
350		2243.06	II
50		2354.20	I
30		2373.83	
50		2385.24	
25		2413.93	II
560		2422.20	II
60		2460.61	II
25		2490.42	I
12		2540.28	
14		2547.57	
10		2550.17	
20		2681.65	I
60		2694.21	I
26		2695.39	I
95		2723.00	I
22		2730.08	I
22		2734.85	II
70		2742.53	I
140	h	2760.10	I
30		2785.21	II
12		2785.59	II
12		2791.20	I
30		2800.11	II
26		2813.64	I
18		2818.86	I
45		2822.56	I
22		2825.37	II
45		2826.38	II
70		2854.43	II
26		2856.30	II
11		2857.87	II
95		2886.48	I
18		2897.29	II
14		2898.82	II
160		2919.05	I
18	h	2930.03	II
390		2948.40	I
350		2964.96	I
18		2973.91	II

Intensity		Wavelength	
100		4559.37	I
30		4564.39	I
60	h	4573.56	I
35		4581.32	I
30		4581.77	I
130		4596.55	I
95		4604.80	I
40		4613.00	I
2000		4643.70	I
200	h	4658.32	I
70		4658.89	I
85		4667.47	I
60		4670.82	I
2000		4674.84	I
60		4678.35	I
260		4682.32	II
85		4689.77	I
180		4696.81	I
35		4708.85	I
60		4725.85	I
170		4728.53	I
60	h	4732.37	I
85		4741.40	I
160		4752.79	I
410		4760.98	I
17		4780.18	I
120		4781.04	I
160		4786.58	II
170		4786.89	I
180		4799.30	I
50		4804.31	I
70		4804.81	I
85	B1	4817.38	YO
140	B1	4818.20	YO
140		4819.64	I
120		4822.13	I
190		4823.31	II
60		4839.15	I
770		4839.87	I
550		4845.68	I
410		4852.69	I
120		4854.25	I
890		4854.87	II
50		4856.70	I
330		4859.84	I
30		4879.65	I
1900		4883.69	II
50		4886.28	I
40		4886.65	I
95		4893.44	I
1100		4900.12	II
100		4906.11	I

Yttrium (Cont.)

Intensity		Wavelength	
45		4909.00	I
150		4921.87	I
35		4930.93	I
45		4950.66	I
120		4974.30	I
120		4982.13	II
100		5006.97	I
75		5070.21	I
75		5072.19	I
1100		5087.42	II
30		5088.18	I
210		5119.11	II
450		5123.21	II
180		5135.20	I
120		5196.43	II
960		5200.41	II
1500		5205.72	II
180		5240.81	I
60		5289.82	II
45		5320.78	II
75		5380.62	I
220		5402.78	II
24		5417.03	I
90		5424.37	I
190		5438.24	I
710		5466.46	I
100		5468.47	I
90		5473.39	II
90		5480.74	II
60		5493.17	I
35		5495.59	I
240		5497.41	II
300		5503.45	I
250		5509.90	II
60		5513.64	I
120		5521.63	I
		5521.70	II
24		5526.76	I
740		5527.54	I
35		5541.63	I
120		5544.50	I
		5544.61	II
90		5546.02	II
75		5556.43	I
60		5567.75	I
180		5577.42	I
24		5581.08	I
620		5581.87	I
21		5590.96	I
21		5594.12	I
120		5606.33	I
15		5623.91	I
560		5630.13	I
24		5632.25	I
21		5632.89	I
120		5644.69	I
120		5648.47	I
740		5662.94	II
90		5675.27	I
18		5693.63	I
160		5706.73	I
24		5720.61	I
75		5728.89	II
150	B1	5730.12	YO
21		5732.09	I
90		5743.85	I
18	B1	5746.93	YO
24	B1	5764.22	YO
75		5765.64	I
35		5773.95	I
100		5781.69	II
15	B1	5800.00	YO
15	B1	5818.58	YO
30		5821.87	I
21		5832.27	I
9	b	5838.07	YO
15		5858.83	YO
15		5871.83	I
24		5876.14	YO
24		5879.96	I
24	b	5893.94	YO
35		5902.96	I
24	b	5912.19	YO
24	b	5931.10	YO
90	B1	5939.08	YO
45		5945.72	I
24		5950.02	I
75	b	5956.41	YO
1300	B1	5972.04	YO
50		5981.86	I
1000	B1	5987.64	YO
740	B1	6003.60	YO
120		6004.65	I
120		6009.19	I
620	B1	6019.87	YO
120		6023.41	I
500	B1	6036.60	YO
420	B1	6053.81	YO
130	B1	6072.78	YO
50		6088.00	I
210	B1	6089.35	YO
160	B1	6096.78	YO
130	B1	6107.82	YO
130	B1	6114.73	YO

Intensity		Wavelength	
75	B1	6127.38	YO
1400	B1	6132.06	YO
120		6135.04	I
150		6138.43	I
1100	B1	6148.36	YO
120		6151.72	YO
820	B1	6165.08	YO
560	B1	6182.23	YO
1200		6191.73	I
590	B1	6199.82	YO
450	B1	6217.96	YO
300		6222.59	I
270	B1	6236.21	YO
45		6251.05	I
120	B1	6275.01	YO
60	b	6295.06	YO
24	b	6316.20	YO
24	b	6338.10	YO
15	b	6359.48	YO
15		6369.87	YO
75		6402.01	I
1000		6435.04	I
24		6437.18	I
18	h	6501.23	YO
18	h	6518.33	YO
18	h	6535.84	YO
90		6538.60	I
12	h	6553.84	YO
70		6557.39	I
12	h	6572.58	
35		6576.85	I
23		6584.87	I
95		6613.75	II
14		6622.49	I
19	h	6636.49	I
40		6650.61	I
21		6664.40	I
150		6687.58	I
14	h	6691.83	I
7		6694.75	I
16	h	6699.26	I
70		6700.71	I
35		6713.20	I
40		6735.99	I
190		6793.71	I
70	h	6795.41	II
12	h	6803.15	I
21		6815.16	I
14		6832.49	II
45		6845.24	I
14		6858.24	II
29		6887.22	I
21		6896.00	II
9		6908.26	I
14		6933.52	I
24	h	6950.31	I
10		6951.68	II
10		6958.04	I
24		6979.88	I
13	h	7008.97	I
10		7009.93	I
19	h	7035.18	I
29		7052.94	I
13	h	7054.28	I
9		7075.13	I
11		7127.92	I
35		7191.66	I
10	h	7195.93	I
35		7264.17	II
9	h	7293.08	I
9	h	7330.62	I
5		7332.96	II
50		7346.46	I
11	h	7398.77	I
29		7450.30	II
17		7494.88	I
7	h	7536.71	I
35		7563.13	I
8	h	7617.72	I
19	h	7622.94	I
7		7652.89	I
5		7689.49	I
8	h	7698.00	I
19		7719.89	I
19		7724.08	I
13		7788.42	I
13		7796.32	I
6		7802.52	I
17		7812.16	I
29		7855.52	I
110		7881.90	II
10	h	7999.33	I
9		8329.61	I
24		8344.43	I
8	h	8365.64	I
17		8450.36	I
8	h	8528.94	I
95		8800.62	I
19	h	8835.85	II

Y III
Ref. 77 — J.R.

Intensity		Wavelength	
		Vacuum	
1		643.68	III
4		646.69	III
6		653.87	III
10		656.98	III
25		668.74	III
40		671.98	III
100		691.72	III
4		693.85	III
200		695.20	III
9		727.91	III
4		728.47	III
2		728.83	III
20		729.73	III
600		730.49	III
15		732.70	III
800		734.36	III
15		770.78	III
10		771.79	III
20		804.26	III
5000		805.20	III
75		806.18	III
150		808.97	III
7000		809.92	III
100		855.64	III
60		857.82	III
25		984.23	III
15		987.96	III
15000		989.21	III
25000		996.37	III
20		999.19	III
150		1000.56	III
25		1003.35	III
1000		1006.58	III
1200		1007.86	III
120		1077.52	III
500		1081.35	III
75	p	1084.63	III
350	p	1088.39	III
250		1095.25	III
25		1095.87	III
150		1103.21	III
3000		1289.74	III
2500		1306.96	III
5000		1314.51	III
1500		1316.10	III
4000		1334.04	III
8		1549.08	III
15	p	1553.81	III
30		1635.14	III
75		1640.43	III
200		1779.80	III
600		1786.05	III
		Air	
10		2041.93	III
5		2042.07	III
1500		2060.58	III
4000		2068.98	III
10000		2127.98	III
16000		2191.16	III
8000		2200.76	III
8000		2206.03	III
150		2261.41	III
80		2261.57	III
10000		2284.34	III
3		2319.92	III
10000		2327.31	III
50000		2367.23	III
40000		2414.64	III
100	p	2710.30	III
90	h	2710.54	III
5		2780.11	III
70		2791.44	III
20		2803.27	III
100		2807.00	III
90000		2817.00	III
6000		2867.67	III
6000		2913.41	III
1500		2917.74	III
1600		2918.56	III
15		2940.53	III
99000		2946.01	III
20	p	2948.48	III
6000		2970.42	III
1400		3013.93	III
1500		3018.85	III
3		3267.10	III
25	l	3276.80	III
500		3866.96	III
3000		3900.74	III
4000		3914.58	III
3800		4039.60	III
3000		4040.11	III
120	h	4121.61	III
2000	c	4737.62	III
7500		5102.88	III
1300		5120.40	III
10000		5238.10	III
3000		5263.58	III
4000		5383.64	III
6000		5562.81	III
600		5567.27	III
4000		5572.24	III
400		5595.48	III
3000		5602.08	III
2000	h	7254.58	III
9000		7558.71	III
6000		7864.53	III
8000		7916.71	III
400		7989.41	III
10000		7991.43	III
8000		8171.41	III
4000		8645.09	III
10000		8796.21	III
8000		9116.59	III

Y IV
Ref. 265 — J.R.

Intensity	Wavelength	
	Vacuum	
3	211.80	IV
3	214.51	IV
6	215.97	IV
6	217.39	IV
12	221.71	IV
3	222.18	IV
6	222.98	IV
2	228.84	IV
20	228.94	IV
3	229.78	IV
3	235.17	IV
25	235.77	IV
1	242.12	IV
30	242.30	IV
3	244.14	IV
3	263.72	IV
150	264.64	IV
30	272.40	IV
150	273.03	IV
10	278.60	IV
900	355.86	IV
300	370.42	IV
500	386.82	IV
600	425.03	IV
300	473.10	IV

Y V
Ref. 419 — J.R.

Intensity	Wavelength	
	Vacuum	
5	289.18	V
50	299.99	V
3	312.89	V
200	313.35	V
40	320.47	V
40	321.69	V
150	325.58	V
200	326.57	V
175	328.34	V
50	330.40	V
900	333.09	V
500	333.80	V
100	335.12	V
400	335.14	V
500	336.62	V
500	339.02	V
200	340.02	V
5	340.42	V
500	344.59	V
100	349.65	V
2	349.75	V
10	351.36	V
100	353.98	V
100	355.56	V
300	372.05	V
400	379.96	V
200	397.77	V
300	403.45	V
1	408.81	V
200	409.31	V
200	415.03	V
100	418.18	V
150	418.59	V
50	419.79	V
300	420.74	V
1	427.87	V
50	430.75	V
100	437.66	V
3	441.62	V
30	442.96	V

Yttrium (Cont.)

Intensity		Wavelength	
2		451.97	V
15		455.84	V
85		457.84	V
4000		584.98	V
2000		630.97	V

ZINC (Zn)
Z = 30

Zn I and II
Ref. 39, 55, 113, 131, 185, 186 — R.D.C.

Intensity		Wavelength	
		Vacuum	
60		1193.23	II
60		1277.31	II
60	d	1366.68	II
		1404.12	I
60		1410.44	II
60		1439.09	II
60		1445.04	II
50		1456.91	II
		1457.57	I
60		1477.02	II
50		1514.76	II
90		1572.99	II
		1589.57	I
60		1617.68	II
60		1658.25	II
50		1713.25	II
60		1715.76	II
80	d	1735.61	II
60		1736.89	II
50		1737.90	II
75		1747.12	II
80	c	1762.19	II
75		1774.04	II
80		1790.76	II
100		1797.64	II
100	d	1811.05	II
80		1816.48	II
80		1831.38	II
100	d	1833.57	II
70		1836.01	II
75		1836.65	II
75		1847.56	II
100		1864.12	II
100		1866.08	II
100		1872.13	II
75		1894.26	II
60		1901.52	II
60		1914.81	II
100	d	1918.96	II
70		1920.27	II
100	d	1929.67	II
60		1945.58	II
60		1951.91	II
80		1953.00	II
75		1954.87	II
80		1964.54	II
100		1969.40	II
100		1982.11	II
70		1985.61	II
100		1986.99	II
50		1993.37	II
50		1996.92	II
		Air	
100		2011.94	II
500		2025.48	II
60		2039.31	II
500		2062.00	II
200		2064.23	II
120		2079.08	I
50		2079.93	II
60		2087.33	I
80		2096.93	I
300		2099.94	II
200		2102.18	II
150		2104.42	I
75		2122.74	II
800	r	2138.56	II
75		2147.42	II
60		2210.18	II
50		2273.15	II
1000		2501.99	II
150		2515.81	I
50		2527.96	II
1000		2557.95	II
50		2567.80	II
50		2567.98	II
100	h	2569.87	I
100		2582.44	I
300		2582.49	I
200		2608.56	I
300		2608.64	I
200		2670.53	I
300		2684.16	I
300		2712.49	I
200		2756.45	I
300		2770.86	I
300		2770.98	I
400		2800.87	I
100		2801.06	I
5		2801.17	I
100		2801.96	II
100		2902.30	II
125		3018.36	I
200		3035.78	I
200		3072.06	I
150		3075.90	I
100		3171.45	II
100		3172.23	II
300		3196.31	II
100		3197.10	II
500	r	3282.33	I
50		3299.42	II
800		3302.58	I
700	r	3302.94	I
75		3306.01	II
800		3345.02	I
500		3345.57	I
150		3345.94	I
5		3799.00	I
50		3806.34	II
100		3840.29	II
50		3883.34	I
15		3965.43	I
10		4113.21	I
25		4292.88	I
25		4298.33	I
35		4629.81	I
300		4680.14	I
400		4722.15	I
400		4810.53	I
800		4911.62	II
500		4924.03	II
7		5068.66	I
15		5069.58	I
200		5181.98	I
8		5308.65	I
7		5310.24	I
7		5311.02	I
4		5772.10	I
4		5775.50	I
10		5777.11	I
500		5894.33	II
500		6021.18	II
500		6102.49	II
100		6111.53	II
500		6214.61	II
8		6237.90	I
8		6239.17	I
1000	h	6362.34	I
10		6479.10	I
15		6928.32	I
8		6938.47	I
3		6943.20	I
200		7478.8	II
300		7588.5	II
100		7612.9	II
300		7732.5	II
200		7757.9	II
10		7799.36	I
100		11054.25	I
100		13053.63	I
100		13150.59	I
20		13196.61	I
100		14038.70	I
20		15680.29	I
20		16483.45	I
20		16491.98	I
20		16505.23	I
5		24044.16	I
10		24375.02	I

Zn III
Ref. 376, 377 — R.D.C.

Intensity	Wavelength	
	Vacuum	
1000	677.63	III
750	677.96	III
200	713.90	III
100	1432.15	III
200	1456.72	III
100	1464.20	III
100	1465.75	III
300	1473.41	III
100	1489.26	III
100	1490.96	III
200	1498.79	III
300	1499.42	III
300	1500.42	III
300	1505.92	III
300	1515.85	III
30	1533.09	III
300	1552.30	III
200	1552.94	III
80	1553.11	III
200	1560.79	III
150	1562.55	III
200	1581.53	III
100	1582.06	III
100	1598.52	III
100	1600.87	III
100	1619.61	III
150	1622.51	III
200	1629.19	III
200	1639.33	III
150	1644.82	III
100	1651.74	III
200	1673.05	III
100	1688.59	III
80	1695.40	III
80	1706.65	III
150	1749.63	III
50	1753.84	III
100	1767.69	III
80	1839.32	III

Zn IV
Ref. 370, 377 — R.D.C.

Intensity	Wavelength	
	Vacuum	
30	412.67	IV
30	423.42	IV
50	423.54	IV
200	425.90	IV
200	428.54	IV
80	428.79	IV
150	429.30	IV
200	430.59	IV
150	431.54	IV
120	431.62	IV
10	434.41	IV
150	435.02	IV
150	435.76	IV
80	436.25	IV
30	436.38	IV
50	436.82	IV
50	441.15	IV
20	441.52	IV
20	441.56	IV
100	441.70	IV
200	442.39	IV
150	444.39	IV
100	444.46	IV
100	446.58	IV
30	447.85	IV
10	449.13	IV
200	449.98	IV
200	450.99	IV
30	451.62	IV
20	452.80	IV
80	456.67	IV
150	457.32	IV
200	466.93	IV
150	468.43	IV
200	472.09	IV
200	472.66	IV
200	473.02	IV
200	473.51	IV
200	474.56	IV
50	475.78	IV
200	476.42	IV
200	478.65	IV
200	478.90	IV
200	482.10	IV
10	482.68	IV
10	485.48	IV
10	489.19	IV
10	490.96	IV
10	493.37	IV
10	496.72	IV
100	497.70	IV
15	1193.29	IV
15	1203.44	IV
8	1212.71	IV
5	1214.14	IV
25	1224.35	IV
20	1227.62	IV
40	1228.63	IV
30	1231.46	IV
30	1237.26	IV
50	1239.12	IV
3	1246.26	IV
15	1247.01	IV
50	1249.69	IV
30	1253.67	IV
30	1257.31	IV
8	1259.68	IV
500	1265.74	IV
5	1269.15	IV
100	1272.21	IV
200	1272.98	IV
30	1275.78	IV
25	1278.51	IV
100	1280.47	IV
100	1291.83	IV
100	1292.49	IV
10	1294.32	IV
100	1296.62	IV
100	1296.73	IV
20	1301.88	IV
500	1306.66	IV
200	1318.00	IV
200	1320.74	IV
200	1321.22	IV
200	1322.33	IV
200	1322.43	IV
200	1326.74	IV
200	1329.11	IV
100	1333.32	IV
150	1344.08	IV
200	1347.98	IV
200	1349.90	IV
50	1352.27	IV
100	1356.20	IV
200	1357.82	IV
100	1363.43	IV
200	1363.95	IV
40	1368.17	IV
200	1369.53	IV
50	1370.42	IV
100	1375.33	IV
50	1375.98	IV
200	1377.65	IV
50	1391.24	IV
80	1393.07	IV
50	1394.54	IV
30	1400.14	IV
15	1403.98	IV
30	1409.40	IV
5	1410.33	IV
30	1419.60	IV
50	1427.79	IV
20	1438.58	IV
80	1455.65	IV
200	1459.98	IV
100	1476.43	IV
100	1481.25	IV
100	1529.84	IV
50	1533.60	IV

ZIRCONIUM (Zr)
Z = 40

Zr I and II
Ref. 1 — C.H.C.

Intensity	Wavelength	
	Air	
60	2374.42	I
60	2384.17	I
50	2388.01	I
50	2389.21	I
45	2405.52	I
60	2419.41	II
150	2449.85	II
21	2457.44	II
75	2487.29	II
45	2496.48	II
180	2532.46	II
90	2539.65	I
220	2542.10	II
45	2550.51	I
220	2550.74	II
45	2556.43	I
60	2567.45	II
570	2567.64	II
1600	2568.87	II
2100	2571.39	II
75	2583.40	II
130	2589.07	II
22	2589.65	I
45	2609.43	I
150	2630.91	II
80	2635.42	I
210	2639.09	II
70	2643.40	II
55	2647.78	I
110	2650.38	II
70	2658.69	I
180	2667.80	II
55	2669.49	II
120	2670.96	II
1800	2678.63	II
35	2681.76	II
90	2687.75	I
90	2692.60	II
22	2692.92	I
160	2693.53	II
180	2694.06	II
70	2695.43	II

Zirconium (Cont.)

Int.	λ	Sp.	Int.	λ	Sp.	Int.	λ	Sp.	Int.	λ	Sp.
95	2699.60	II	880	3182.86	II	960	3698.17	II	200	4496.97	II
750	2700.13	II	540	3191.21	I	720	3709.26	II	550	4507.12	I
280	2711.51	II	210	3191.90	II	270	3731.26	II	610	4535.75	I
140	2712.42	II	540	3212.01	I	560	3745.98	II	490	4542.22	I
140	2714.26	II	760	3214.19	II	880	3751.60	II	200	4553.01	I
1300	2722.61	II	110	3222.47	II	480	3764.39	I	200	4555.13	I
140	2725.47	I	200	3228.81	II	480	3766.72	I	140	4555.52	I
800	2726.49	II	630	3231.69	II	340	3766.82	II	490	4575.52	I
490	2732.72	II	630	3234.12	I	720	3780.54	I	100	4582.29	I
1400	2734.86	II	110	3236.58	II	560	3791.40	I	140	4590.55	I
110	2740.51	II	760	3241.05	II	210	3817.58	II	350	4602.57	I
140	2741.55	II	320	3250.39	I	560	3822.41	I	140	4604.42	I
1100	2742.56	II	200	3254.28	I	2200	3835.96	I	210	4626.41	I
660	2745.86	II	200	3260.11	I	1300	3836.76	II	700	4633.98	I
660	2752.21	II	190	3269.66	I	550	3843.02	II	210	4644.83	I
530	2758.81	II	150	3271.13	II	550	3847.01	II	260	4683.42	I
200 d	2768.73	II	540	3272.22	II	550	3849.25	I	2300	4687.80	I
	2768.85	II	1000	3273.05	II	2900	3863.87	I	510	4688.45	I
170 d	2774.04	I	1300	3279.26	II	770	3864.34	I	110	4707.79	I
	2774.16	II	320 d	3282.73	I	990	3877.60	I	1900	4710.08	I
200	2790.14	I	880	3284.71	II	200	3879.05	I	160	4711.92	I
120	2792.04	I	140	3285.88	II	1500	3885.42	I	120	4717.62	I
160	2796.90	II	150	3288.80	II	2900	3890.32	I	210	4719.12	I
110	2799.15	II	540	3305.15	II	2000	3891.38	I	300	4732.33	I
180	2810.91	II	880	3306.28	II	400	3900.52	I	1400	4739.48	I
620	2814.90	I	150	3313.70	II	310	3915.94	II	190	4762.78	I
390	2818.74	II	210	3314.50	II	610	3921.79	I	870	4772.31	I
530	2825.56	II	150	3319.02	II	1200	3929.53	I	210	4784.92	I
110	2833.91	II	380	3322.99	II	200	3934.12	II	160	4788.67	I
710	2837.23	I	380	3326.80	II	200	3934.79	I	260	4805.87	I
120	2839.34	II	380	3334.25	II	940	3958.22	II	140	4809.47	I
130	2843.52	II	210	3334.62	II	490	3966.66	I	190	4815.04	I
660	2844.58	II	190	3338.41	II	990	3968.26	I	700	4815.63	I
210	2848.19	II	760	3340.56	II	660	3973.50	I	280	4824.29	I
350	2848.52	I	380	3344.79	II	200	3975.29	I	190	4828.04	I
350	2851.97	II	130	3353.66	I	200 h	3981.60	I	110	4838.78	I
340	2869.81	II	180	3354.39	II	770	3991.13	II	210	4851.36	I
490	2875.98	I	760	3356.09	II	770	3998.97	II	160	4866.06	I
120	2892.26	I	540	3357.26	II	200	4007.60	I	110	4881.24	I
160	2905.23	II	180	3359.96	II	200	4012.25	I	110	4883.60	I
300	2915.99	II	150	3360.46	I	400	4023.98	I	100	4994.76	I
110	2916.64	II	150	3363.82	II	770	4024.92	I	30	5011.46	I
270	2918.24	II	150	3367.82	II	990	4027.20	I	250	5046.58	I
320	2926.99	II	150	3370.59	I	240	4028.95	I	85	5060.39	I
160	2934.61	II	180	3373.42	II	400	4029.68	II	360	5064.91	I
160	2936.31	II	380	3374.73	II	490	4030.04	I	110	5065.22	I
320	2948.94	II	110	3376.27	II	400	4035.89	I	100	5070.26	I
210	2951.48	II	150	3377.46	II	240	4042.22	I	75	5073.98	I
320	2955.78	II	570	3387.87	II	610	4043.58	I	470	5078.25	I
320	2960.87	I	760	3388.30	II	490	4044.56	I	85	5085.26	I
320	2962.68	II	5700	3391.98	II	400	4045.61	II	50	5112.27	II
320	2968.96	II	570	3393.12	II	610	4048.67	II	140	5115.24	I
120	2969.19	I	160	3396.33	II	200	4050.33	II	50	5120.42	I
230	2969.63	II	380	3399.35	II	200	4050.48	I	85	5133.40	I
130	2976.61	II	570	3404.83	II	770	4055.03	I	300	5155.45	I
320	2978.05	II	760	3410.25	II	600	4055.71	I	200	5158.00	I
230	2979.18	II	380	3414.66	I	330	4061.53	I	35	5158.67	I
160	2981.02	II	1000	3430.53	II	1500	4064.16	I	75	5160.99	I
820	2985.39	I	380	3437.14	II	2000	4072.70	I	85	5165.96	I
320	3003.74	II	4700	3438.23	II	310	4074.93	I	17	5178.99	I
100	3005.37	I	600	3447.36	I	200	4076.53	I	100	5183.70	I
160	3005.50	I	200	3455.91	I	240	4078.31	I	30	5187.03	I
820	3011.75	I	410	3457.56	II	2000	4081.22	I	100	5191.60	II
100	3013.32	II	200	3458.93	II	200	4108.40	I	100	5201.15	I
160	3019.84	II	820	3463.02	II	400	4121.46	I	85	5209.30	I
350	3020.47	II	600	3471.19	I	1200	4149.20	II	85	5224.93	I
500	3028.04	II	200	3478.79	I	200	4152.64	I	30	5243.47	I
880	3029.52	I	1200	3479.39	II	290	4156.24	II	120	5277.41	I
180	3030.92	II	1300	3481.15	II	400	4161.21	II	75	5280.05	I
350 d	3036.39	II	760	3483.54	II	400	4166.36	I	60	5294.82	I
100	3045.83	I	4100	3496.21	II	200	4183.32	I	120	5296.79	I
690	3054.84	II	350	3505.48	II	660	4187.56	I	60	5301.97	I
100	3060.11	II	820	3505.67	II	400	4194.76	I	110	5311.40	I
100	3064.63	II	1000	3509.32	I	610	4199.09	I	25	5321.26	I
110	3085.34	I	200	3510.46	II	610	4201.46	I	22	5330.84	I
110	3094.80	I	2000	3519.60	I	610	4208.98	II	12	5338.43	I
250	3095.07	II	440	3525.81	II	200	4211.88	II	30	5350.09	II
110	3095.82	I	440	3533.22	I	400	4213.86	I	30	5350.35	II
280	3099.23	II	210	3535.16	I	2000	4227.76	I	25	5350.90	I
690	3106.58	II	630	3542.62	II	200	4236.06	I	25	5351.92	I
110	3108.37	I	1800	3547.68	I	2000	4239.31	I	75	5362.56	I
210	3110.88	II	210	3549.74	I	770	4240.34	I	12	5363.35	I
350	3120.74	I	630	3550.34	I	770	4241.20	I	17	5369.39	I
320	3125.92	II	1800	3551.95	II	1200	4241.69	I	20	5382.37	I
500	3129.18	II	2100	3556.60	II	310	4268.02	I	270	5385.14	I
500	3129.76	II	1100	3566.10	I	550	4282.20	I	30	5386.65	I
140	3131.11	I	210	3568.88	I	550	4294.79	I	17	5391.18	I
350	3132.07	I	2100	3572.47	II	310	4302.89	I	17	5395.88	I
110	3133.23	I	210	3573.08	II	550	4341.13	I	25	5405.13	I
350	3133.48	II	1100	3575.79	I	1000	4347.89	I	85	5407.62	I
180	3136.96	I	1300	3576.85	II	290	4359.74	II	17	5413.93	I
690	3138.68	II	880	3586.29	I	310	4360.81	I	20	5421.86	I
140	3139.80	I	440	3587.98	II	350	4366.45	I	15	5426.36	I
180	3148.82	I	3500	3601.19	I	240	4379.78	II	25	5428.42	I
290	3155.67	II	690	3611.89	II	190	4413.04	I	25	5437.76	I
150	3157.00	II	1100	3613.10	II	240	4420.46	I	15	5440.41	I
320	3157.82	I	1100	3614.77	II	120	4427.24	I	35	5448.57	I
540	3164.31	II	1100	3623.86	I	160	4431.49	I	10	5474.92	I
150	3165.45	II	320	3634.15	I	140	4443.00	II	10	5477.40	I
880	3165.97	II	260	3661.20	I	110	4457.43	I	35	5478.33	I
150	3166.26	II	1100	3663.65	I	110	4466.91	I	35	5480.83	I
190	3178.09	II	390	3671.27	II	110	4470.31	I	10	5481.16	I
190	3181.58	II	800	3674.72	II	190	4470.56	I	30	5486.09	I
150	3181.92	II	390	3697.46	II				140	5502.12	I

Zirconium (Cont.)

Intensity		Wavelength	Species
25		5507.87	I
30		5517.11	I
10		5518.05	I
75		5528.41	I
20		5532.30	I
45		5537.46	I
50		5545.32	I
22	B1	5551.75	z
25	B1	5553.17	z
12		5612.11	I
120		5620.14	I
35		5623.53	I
25	B1	5629.02	z
25	B1	5629.58	z
160		5664.51	I
20		5666.28	I
120		5680.90	I
15		5685.42	I
30		5708.89	I
75	B1	5718.21	z
120		5735.70	I
35	B1	5748.17	z
17	B1	5778.57	z
160		5797.74	I
30		5847.32	I
50		5868.27	I
110		5869.50	I
340		5879.80	I
85		5885.62	I
50		5901.09	I
30	B1	5908.61	z
140		5925.13	I
100		5935.20	I
110		5955.35	I
30	B1	5977.80	z
100		5984.23	I
17		5995.37	I
50		6001.05	I
30		6025.36	I
85		6032.61	I
170		6045.85	I
100		6049.24	I
140		6062.84	I
50		6120.83	I
170		6121.91	I
85		6124.84	I
680		6127.44	I
340		6134.55	I
100		6140.46	I
440		6143.20	I
30		6155.61	I
75		6157.71	I
25		6160.20	I
35		6189.40	I
60		6192.96	I
85		6213.05	I
100		6214.69	I
170	B1	6226.51	z
100		6257.26	I
50	b	6261.05	z
35		6267.06	I
45	B1	6292.84	z
120		6299.66	I
15		6304.34	I
300		6313.02	I
30		6314.71	I
50		6321.35	I
22		6340.36	I
50	B1	6345.10	z
75		6345.22	I
75	B1	6378.56	z
35		6407.00	I
50	b	6412.39	z
12		6426.17	I
35		6434.33	I
60		6445.74	I
20		6451.62	I
20		6457.63	I
110		6470.21	I
60	B1	6473.79	z
11		6484.35	I
110		6489.64	I
22		6493.10	I
50		6503.26	I
50		6506.36	I
50	B1	6508.15	z
30	B1	6542.90	z
35		6550.54	I
30		6569.43	I
20		6576.56	I
30	b	6578.06	z
50		6591.99	I
10		6596.71	I
10		6598.84	I
50		6603.27	I
15		6620.56	I
11		6678.01	II
22		6688.18	I
11		6702.12	I
17		6709.61	I
27		6717.88	I
40		6752.73	I
75		6762.38	I
85		6769.16	I
27		6772.89	I
15		6787.15	II
35		6790.85	I
45		6828.78	I
45		6832.89	I
13		6845.33	I
17		6846.34	I
100		6846.97	I
27		6849.26	I
13		6852.56	t
120		6888.29	I
29		6900.59	I
20		6904.36	I
29		6907.37	I
20		6916.87	I
16		6932.38	I
29		6948.46	I
150		6953.84	I
60		6966.44	I
10		6975.91	I
150		6990.84	I
80		6994.32	I
10		7005.46	I
100		7027.40	I
25		7057.36	I
14		7057.96	I
140		7087.30	I
25		7089.43	I
35		7094.46	I
50		7095.59	I
540		7097.70	I
280		7102.91	I
170		7103.72	I
140		7111.68	I
40		7112.82	I
18		7113.52	I
12		7132.05	I
16		7140.74	I
12		7144.47	I
590		7169.09	I
50		7201.62	I
12		7258.17	I
35		7264.76	I
20		7306.21	I
25		7311.62	I
35		7313.72	I
90		7318.08	I
10		7327.82	I
50		7335.97	I
50		7343.96	I
20		7373.50	I
25		7383.63	I
14		7400.90	I
10		7411.39	I
10		7422.75	I
10		7433.10	I
110		7439.86	I
18		7467.57	I
16		7479.58	I
14	h	7515.70	I
12		7517.95	I
20	h	7540.62	I
20		7544.59	I
29		7551.46	I
40		7554.70	I
25		7558.45	I
12		7560.09	I
12		7562.12	I
80		7607.15	I
14		7612.08	I
20		7621.17	I
29		7658.60	I
18		7690.83	I
14		7704.27	I
10		7708.42	I
10		7766.55	I
12		7816.32	I
110		7819.35	I
35		7822.94	I
40		7826.72	I
90		7849.35	I
35		7869.99	I
14		7876.25	I
16		7882.18	I
10		7897.98	I
16		7908.46	I
20		7940.47	I
160		7944.61	I
80		7956.66	I
80		7959.98	I
20		7963.63	I
160		8005.27	I
25		8046.05	I
16		8053.06	I
20		8055.29	I
20		8055.76	I
60		8058.08	I
150		8063.09	I
790		8070.08	I
10		8114.28	I
20		8120.17	I
390		8132.99	I
20		8152.58	I
12		8188.77	I
40		8194.73	I
60		8201.37	I
280		8212.53	I
20		8240.37	I
40		8283.81	I
140		8305.90	I
14		8332.44	I
50		8370.23	I
120		8389.41	I
70		8414.00	I
50		8453.17	I
50		8464.65	I
40		8498.44	I
18		8584.21	I
10		8734.86	I
12		8749.68	I
10		8786.23	I
16		8804.90	I
70		8836.09	I
60		8899.52	I

Zr III
Ref. 403 — J.R.

Intensity	Wavelength	
	Vacuum	
25	687.64	III
50	690.39	III
25	819.59	III
30	820.21	III
35	823.69	III
25	829.50	III
30	850.61	III
25	859.56	III
30	864.86	III
25	868.64	III
25	868.99	III
25	919.59	III
30	1320.81	III
25	1375.13	III
30	1378.93	III
40	1403.48	III
35	1420.12	III
25	1420.87	III
25	1465.44	III
25	1563.24	III
40	1593.59	III
100	1612.38	III
50	1620.62	III
75	1631.31	III
50	1638.33	III
35	1675.06	III
35	1675.75	III
40	1703.36	III
40	1725.03	III
40	1754.30	III
35	1759.12	III
30	1764.75	III
30	1771.96	III
40	1773.90	III
100	1779.51	III
40	1783.35	III
200	1790.19	III
150	1793.56	III
125	1798.13	III
75	1800.03	III
25	1801.67	III
100	1805.26	III
35	1831.89	III
40	1850.06	III
40	1853.38	III
30	1859.12	III
25	1860.47	III
25	1861.77	III
75	1864.06	III
30	1865.45	III
35	1877.00	III
50	1892.07	III
65	1914.25	III
75	1921.96	III
75	1932.54	III
50	1934.32	III
40	1935.20	III
75	1936.48	III
75	1936.67	III
80	1937.27	III
200	1940.25	III
100	1941.08	III
25	1946.12	III
80	1946.61	III
50	1946.99	III
100	1953.95	III
50	1961.32	III
100	1962.01	III
40	1962.92	III
85	1966.22	III
25	1967.81	III
60	1974.99	III
40	1983.14	III
50	1989.83	III
30	1990.95	III
30	1994.46	III
	Air	
45	2000.23	III
100	2006.82	III
30	2013.30	III
35	2016.63	III
30	2021.52	III
60	2026.78	III
100	2035.42	III
50	2036.92	III
75	2056.13	III
40	2058.73	III
75	2060.83	III
50	2061.47	III
125	2070.43	III
50	2074.12	III
100	2077.92	III
100	2080.99	III
30	2081.81	III
25	2085.35	III
200	2086.78	III
40	2089.50	III
40	2097.03	III
40	2102.30	III
50	2103.16	III
25	2104.23	III
40	2113.98	III
35	2114.10	III
40	2125.06	III
35	2137.90	III
35	2138.45	III
25	2139.85	III
40	2159.24	III
40	2162.20	III
100	2175.80	III
100	2191.15	III
35	2192.05	III
60	2206.33	III
40	2206.97	III
30	2231.00	III
50	2245.36	III
25	2251.14	III
40	2257.83	III
30	2281.43	III
100	2301.60	III
75	2308.12	III
35	2405.81	III
40	2406.21	III
75	2420.65	III
25	2438.70	III
50	2444.58	III
100	2448.86	III
100	2593.64	III
250	2620.56	III
50	2621.28	III
60	2628.26	III
200	2643.79	III
100	2656.46	III
150	2664.26	III
100	2682.16	III
75	2686.28	III
70	2690.49	III
60	2698.31	III
50	2709.05	III
45	2715.76	III
40	2720.07	III
75	2735.76	III
25	2775.23	III
40	2836.18	III
40	3278.86	III

Zr IV
Ref. 362 — J.R.

Intensity	Wavelength	
	Vacuum	
15	478.97	IV
60	480.66	IV
60	497.23	IV
60	500.22	IV
4	500.34	IV
4	584.65	IV
15	585.42	IV
2	586.42	IV
60	588.89	IV
100	589.74	IV
600	628.66	IV
500	633.56	IV
200	633.63	IV
8	712.49	IV
90	754.39	IV
90	760.15	IV
150	846.40	IV
300	863.65	IV
500	864.59	IV
200	881.30	IV

Zirconium (Cont.)

Intensity	Wavelength		Intensity		Wavelength		Intensity		Wavelength		Intensity		Wavelength	
200	882.59	IV					400		836.57	V	500	p	1323.81	V
100	1099.76	IV			**Zr V**		3000		841.40	V	300		1332.06	V
150	1100.00	IV			Ref. 418 — J.R.		300	p	852.87	V	300		1337.34	V
9000	1183.97	IV					700		853.68	V	300		1355.21	V
9000	1201.77	IV	Intensity		Wavelength		200		873.11	V	500		1355.98	V
200	1212.71	IV					300		885.68	V	300		1361.39	V
100	1213.01	IV			Vacuum		300		900.48	V	500		1376.54	V
10000	1219.86	IV					500		906.66	V	200		1396.79	V
500	1285.89	IV	300		292.19	V	500		915.30	V	300		1410.03	V
500	1290.56	IV	500		304.01	V	400		923.10	V	200		1413.40	V
70	1291.70	IV	300		305.24	V	400		940.41	V	300		1460.05	V
500	1417.70	IV	400	p	368.18	V	200		949.70	V	300		1486.90	V
500	1440.65	IV	200		519.25	V	200		978.06	V	300		1491.33	V
500	1441.06	IV	200		536.50	V	200		980.70	V	200		1520.47	V
1000	1469.47	IV	200	p	674.13	V	200		984.18	V	200		1550.12	V
10000	1546.17	IV	200		675.53	V	300		995.59	V	500		1633.03	V
10	1596.29	IV	300		679.39	V	400		1002.48	V	500		1654.46	V
10000	1598.95	IV	200		688.27	V	400		1038.69	V	700		1725.02	V
150	1605.26	IV	200	p	703.03	V	200		1044.41	V	500		1786.20	V
5000	1607.95	IV	300		717.24	V	300		1047.77	V	200	p	1790.81	V
90	1609.50	IV	200		740.33	V	400		1068.55	V	500		1806.09	V
35	1818.06	IV	2000		740.61	V	300		1072.25	V	500		1860.48	V
500	1836.15	IV	500		742.74	V	300		1083.45	V	600		1860.86	V
500	1846.37	IV	300		752.48	V	200		1087.05	V	500		1878.33	V
200	1848.03	IV	300		764.58	V	200		1093.54	V	400		1926.24	V
70	1851.91	IV	500		766.20	V	200		1108.79	V	500		1927.43	V
			200		773.80	V	300		1194.24	V	400		1934.88	V
	Air		200		775.58	V	300		1200.76	V				
			200		779.21	V	200		1233.91	V		**Air**		
20	2045.12	IV	400		784.50	V	300		1238.93	V				
40	2047.15	IV	400		797.23	V	300		1253.61	V	500		2009.29	V
10000	2091.49	IV	10000		800.00	V	200		1259.70	V	600		2028.54	V
10000	2092.36	IV	10000		806.89	V	400		1260.91	V	600		2132.42	V
10000	2163.68	IV	200		809.75	V	500		1265.38	V	200		2150.18	V
10000	2286.67	IV	10000		812.05	V	300		1295.81	V	200		2336.94	V
200	2473.75	IV	500		822.06	V	400		1302.80	V				
300	2476.71	IV	300	p	823.46	V	500		1303.93	V				
5	2572.32	IV	200		823.78	V	400		1306.76	V				
400	2573.66	IV	200		825.13	V	200		1315.14	V				
400	2583.32	IV	200		830.59	V	400		1320.74	V				

REFERENCES

1. **Meggers, W. F., Corliss, C. H., and Scribner, B. F.,** *Natl. Bur. Stand. (U.S.) Monogr.,* 145, Washington, D.C., 1975.
2. **Aksenov, V. P. and Ryabtsev, A. N.,** *Opt. Spectrosc.,* 37, 860, 1970.
3. **Andersen, N., Bickel, W. S., Carriveau, G. W., Jensen, K., and Veje, E.,** *Phys. Scr.,* 4, 113, 1971.
4. **Andersson, E. and Johannesson, G. A.,** *Phys. Scr.,* 3, 203, 1971.
5. **Andrew, K. L. and Meissner, K. W.,** *J. Opt. Soc. Am.,* 49, 146, 1959.
6. **Artru, M. C. and Brillet, W. U. L.,** *J. Opt. Soc. Am.,* 64, 1063, 1974.
7. **Artru, M. C. and Kaufman, V.,** *J. Opt. Soc. Am.,* 62, 949, 1972.
8. **Artru, M. C. and Kaufman, V.,** *J. Opt. Soc. Am.,* 65, 594, 1975.
9. **Badami, J. S. and Rao, K. R.,** *Proc. R. Soc. London,* 140(A), 387, 1933.
10. **Baird, K. M. and Smith, D. S.,** *J. Opt. Soc. Am.,* 48, 300, 1958.
11. **Bashkin, S. and Martinson, I.,** *J. Opt. Soc. Am.,* 61, 1686, 1971.
12. **Beacham, J. R.,** Ph.D. thesis, Purdue University, 1970.
13. **Benschop, H., Joshi, Y. N., and van Kleef, T. A. M.,** *Can. J. Phys.,* 53, 700, 1975.
14. **Berry, H. G., Bromander, J., and Buchta, R.,** *Phys. Scr.,* 1, 181, 1970.
15. **Berry, H. G., Bromander, J., Martinson, I., and Buchta, R.,** *Phys. Scr.,* 3, 63, 1971.
16. **Berry, H. G., Desesquelles, J., and Dufay, M.,** *Phys. Rev. Sect. A.,* 6, 600, 1972.
17. **Berry, H. G., Desesquelles, J., and Dufay, M.,** *Nucl. Instrum. Methods,* 110, 43, 1973.
18. **Berry, H. G., Pinnington, E. H., and Subtil, J. L.,** *J. Opt. Soc. Am.,* 62, 767, 1972.
19. **Bidelman, W. P. and Corliss, C. H.,** *Astrophys. J.,* 135, 968, 1962.
20. **Bloch, L. and Bloch, E.,** *Ann. Phys.* (Paris), 10(11), 141, 1929.
21. **Bloch, L., Bloch, E., and Felici, N.,** *J. Phys. Radium,* 8, 355, 1937.
22. **Bockasten, K.,** *Ark. Fys.,* 9, 457, 1955.
23. **Bockasten, K., Hallin, R., Johansson, K. B., and Tsui, P.,** *Phys. Lett.* (Netherlands), 8, 181, 1964.
24. **Bockasten, K. and Johansson, K. B.,** *Ark. Fys.,* 38, 563, 1969.
25. **Borgstrom, A.,** *Ark. Fys.,* 38, 243, 1968.
26. **Borgstrom, A.,** *Phys. Scr.,* 3, 157, 1971.
27. **Bowen, I. S.,** *Phys. Rev.,* 29, 231, 1927.
28. **Bowen, I. S.,** *Phys. Rev.,* 31, 34, 1928.
29. **Bowen, I. S.,** *Phys. Rev.,* 39, 8, 1932.
30. **Bowen, I. S.,** *Phys. Rev.,* 45, 401, 1934.
31. **Bowen, I. S.,** *Phys. Rev.,* 46, 377, 1934.
32. **Bowen, I. S.,** *Phys. Rev.,* 46, 791, 1934.
33. **Boyce, J. C.,** *Phys. Rev.,* 49, 730, 1936.
34. **Boyce, J. C. and Robinson, H. A.,** *J. Opt. Soc. Am.,* 26, 133, 1936.
35. **Bozman, W. R., Meggers, W. F., and Corliss, C. H.,** *J. Res. Natl. Bur. Stand. Sect. A,* 71, 547, 1967.
36. **Bromander, J.,** *Ark. Fys.,* 40, 257, 1969.
37. **Bromander, J. and Buchta, R.,** *Phys. Scr.,* 1, 184, 1970.
38. **Brown, C. M. and Ginter, M. L.,** *J. Opt. Soc. Am.,* 68, 243, 1978.
39. **Brown, C. M., Tilford, S. G., and Ginter, M. L.,** *J. Opt. Soc. Am.,* 65, 1404, 1975.
40. **Bryant, B. W.,** *Johns Hopkins Spectroscopic Report* No. 21, 1961.
41. **Buchet, J. P., Buchet-Poulizac, M. C., Berry, H. G., and Drake, G. W. F.,** *Phys. Rev. Sect. A,* 7, 922, 1973.
42. **Budhiraja, C. J. and Joshi, Y. N.,** *Can. J. Phys.,* 49, 391, 1971.

43. Burns, K. and Adams, K. B., *J. Opt. Soc. Am.*, 42, 56, 1952.
44. Burns, K. and Adams, K. B., *J. Opt. Soc. Am.*, 46, 94, 1956.
45. Burns, K., Adams, K. B., and Longwell, J., *J. Opt. Soc. Am.*, 40, 339, 1950.
46. Callahan, W. R., Ph.D. thesis, Johns Hopkins University, 1962.
47. Charles, G. W., *J. Opt. Soc. Am.*, 56, 1292, 1966.
48. Charles, G. W., Hunt, D. J., Pish, G., and Timma, D. L., *J. Opt. Soc. Am.*, 45, 869, 1955.
49. Codling, K., *Proc. Phys. Soc.*, 77, 797, 1961.
50. Comite Consultatif Pour La Definition du Metre, *J. Phys. Chem. Ref. Data*, 3, 852, 1974.
51. Conway, J. G., Blaise, J., and Verges, J., *Spectrochim. Acta Part B*, 31, 31, 1976.
52. Conway, J. G., Worden, E. F., Blaise, J., and Verges, J., *Spectrochim. Acta Part B*, 32, 97, 1977.
53. Conway, J. G., Worden, E. F., Blaise, J., Camus, P., and Verges, J., *Spectrochim. Acta Part B*, 32, 101, 1977.
54. Crooker, A. M., *Can. J. Res. Sect. A*, 14, 115, 1936.
55. Crooker, A. M. and Dick, K. A., *Can. J. Phys.*, 46, 1241, 1968.
56. Crosswhite, H. M., *J. Res. Natl. Bur. Stand. Sect. A*, 79, 17, 1975.
58. Crosswhite, H. M. and Dieke, G. H., *American Institute of Physics Handbook*, Section 7, 1972.
59. de Bruin, T. L., *Z. Phys.*, 38, 94, 1926.
60. de Bruin, T. L., *Z. Phys.*, 53, 658, 1929.
61. de Bruin, T. L., Humphreys, C. J., and Meggers, W. F., *J. Res. Natl. Bur. Stand.*, 11, 409, 1933.
62. Dick, K. A., *J. Opt. Soc. Am.*, 64, 702, 1973.
63. Dobbie, J. C., *Ann. Solar Phys. Observ.* (Cambridge), 5, 1, 1938.
64. Earls, L. T. and Sawyer, R. A., *Phys. Rev.*, 47, 115, 1935.
65. Edlen, B., *Z. Phys.*, 85, 85, 1933.
66. Edlen, B., *Nova Acta Reglae Soc. Sci. Ups.*, (IV) 9, No. 6, 1934.
67. Edlen, B., *Z. Phys.*, 93, 726, 1935.
68. Edlen, B., *Z. Phys.*, 94, 47, 1935.
69. Edlen, B., *Rep. Prog. Phys.*, 26, 181, 1963.
70. Edlen, B. and Risberg, P., *Ark. Fys.*, 10, 553, 1956.
71. Edlen, B. and Swings, P., *Astrophys. J.*, 95, 532, 1942.
72. Ehrhardt, J. C. and Davis, S. P., *J. Opt. Soc. Am.*, 61, 1342, 1971.
73. Eidelsberg, M., *J. Phys. B*, 5, 1031, 1972.
74. Eidelsberg, M., *J. Phys. B*, 7, 1476, 1974.
75. Ekberg, J. O. and Svensson, L. A., *Phys. Scr.*, 2, 283, 1970.
76. Ekefors, E., *Z. Phys.*, 71, 53, 1931.
77. Epstein, G. L. and Reader, J., *J. Opt. Soc. Am.*, 65, 310, 1975.
78. Epstein, G. L. and Reader, J., *J. Opt. Soc. Am.*, 66, 590, 1976.
79. Epstein, G. L. and Reader, J., unpublished.
80. Eriksson, K. B. S., *Phys. Lett. A.*, 41, 97, 1972.
81. Eriksson, K. B. S. and Isberg, H. B. S., *Ark. Fys.*, 23, 527, 1963.
82. Eriksson, K. B. S. and Wenaker, I., *Phys. Scr.*, 1, 21, 1970.
83. Esteva, J. M. and Mehlman, G., *Astrophys. J.*, 193, 747, 1974.
84. Even-Zohar, M. and Fraenkel, B. S., *J. Phys. B*, 5, 1596, 1972.
85. Fawcett, B. C., *J. Phys. B*, 3, 1732, 1970.
86. Fawcett, B. C., Culham Laboratory Report ARU-R4, 1971.
87. Ferner, E., *Ark. Mat. Astron. Fys.*, 28(A), 4, 1941.
88. Fischer, R. A., Knopf, W. C., and Kinney, F. E., *Astrophys. J.*, 130, 683, 1959.
89. Fowler, A., *Report on Series in Line Spectra*, Fleetway Press, London, 1922.
90. Fowles, G. R., *J. Opt. Soc. Am.*, 44, 760, 1954.
91. Fred, M., *Argonne Natl. Lab.*, unpublished, 1977.
92. Fred, M. and Tomkins, F. S., *J. Opt. Soc. Am.*, 47, 1076, 1957.
93. Fred, M., Tomkins, F. S., Blaise, J. E., Camus, P., and Verges, J., Argonne National Laboratory Report No. 76-68, 1976.
94. Garcia, J. D. and Mack, J. E., *J. Opt. Soc. Am.*, 55, 654, 1965.
96. Giacchetti, A., *Argonne Natl. Lab.*, unpublished, 1975.
97. Giacchetti, A., Blaise, J., Corliss, C. H., and Zalubas, R., *J. Res. Natl. Bur. Stand. Sect. A*, 78, 247, 1974.
98. Giacchetti, A., Stanley, R. W., and Zalubas, R., *J. Opt. Soc. Am.*, 69, 474, 1970.
99. Gilbert, W. P., *Phys. Rev.*, 47, 847, 1935.
100. Gilroy, H. T., *Phys. Rev.*, 38, 2217, 1931.
101. Glad, S., *Ark. Fys.*, 10, 291, 1956.
102. Goldsmith, S., *J. Phys. B*, 2, 1075, 1969.
103. Goorvitch, D., Mehlman-Balloffet, G., and Valero, F. P. J., *J. Opt. Soc. Am.*, 60, 1458, 1970.
104. Goorvitch, D. and Valero, F. P. J., *Astrophys. J.*, 171, 643, 1972.
105. Green, L. C., *Phys. Rev.*, 55, 1209, 1939.
106. Gutman, F., *Diss. Abstr. Int. B*, 31, 363, 1970.
107. Hallin, R., *Ark. Fys.*, 31, 511, 1966.
108. Hallin, R., *Ark. Fys.*, 32, 201, 1966.
109. Hansen, J. E. and Persson, W., *J. Opt. Soc. Am.*, 64, 696, 1974.
110. Hansen, J. E. and Persson, W., *Phys. Scr.*, 13, 166, 1976.
111. Hellintin, P., *Phys. Scr.*, 13, 155, 1976.
112. Herzberg, G. and Moore, H. R., *Can. J. Phys.*, 37, 1293, 1959.
113. Hetzler, C. W., Boreman, R, W., and Burns, K., *Phys. Rev.*, 48, 656, 1935.
114. Holmstrom, J. E. and Johansson, L., *Ark. Fys.*, 40, 133, 1969.
115. Hontzeas, S., Martinson, I., Erman, P., and Buchta, R., *Nucl. Instrum. Methods*, 110, 51, 1973.
116. Humphreys, C. J., *J. Res. Natl. Bur. Stand.*, 22, 19, 1939.
117. Humphreys, C. J., *J. Opt. Soc. Am.*, 43, 1027, 1953.
118. Humphreys, C. J., *J. Phys. Chem. Ref. Data*, 2, 519, 1973.
119. Humphreys, C. J. and Andrew, K. L., *J. Opt. Soc. Am.*, 54, 1134, 1964.
120. Humphreys, C. J. and Meggers, W. F., *J. Res. Natl. Bur. Stand.*, 10, 139, 1933.
121. Humphreys, C. J. and Paul, E., Jr., *J. Opt. Soc. Am.*, 60, 200, 1970.
122. Humphreys, C. J. and Paul, E., Jr., *J. Opt. Soc. Am.*, 62, 432, 1972.
123. Humphreys, C. J., Paul, E., Jr., Cowan, R. D., and Andrew, K. L., *J. Opt. Soc. Am.*, 57, 855, 1967.
124. Humphreys, C. J., Paul, E., Jr., and Minnhagen, L., *J. Opt. Soc. Am.*, 61, 110, 1971.
125. Iglesias, L., Inst. of Optics, Madrid, unpublished, 1977.

126. Iglesias, L. and Velasco, R., *Publ. Inst. Opt. Madrid*, No. 23, 1964.
127. Isberg, B., *Ark. Fys.*, 35, 551, 1967.
128. Johannesson, G. A., Lundstrom, T., and Minnhagen, L., *Phys. Scr.*, 6, 129, 1972.
129. Johannesson, G. A. and Lundstrom, T., *Phys. Scr.*, 8, 53, 1973.
130. Johansson, I., *Ark. Fys.*, 20, 135, 1961.
131. Johansson, I. and Contreras, R., *Ark. Fys.*, 37, 513, 1968.
132. Johansson, I. and Litzen, U., *Ark. Fys.*, 34, 573, 1967.
133. Johansson, I. and Svensson, K. F., *Ark. Fys.*, 16, 353, 1960.
134. Johansson, L., *Ark. Fys.*, 20, 489, 1961.
135. Johansson, L., *Ark. Fys.*, 23, 119, 1963.
136. Johansson, S. and Litzen, U., *Phys. Scr.*, 6, 139, 1972.
137. Johansson, S. and Litzen, U., *Phys. Scr.*, 8, 43, 1973.
138. Johansson, S. and Litzen, U., *Phys. Scr.*, 10, 121, 1974.
139. Joshi, Y. N., St. Francis Xavier Univ., Nova Scotia, unpublished.
140. Joshi, Y. N., Bhatia, K. S., and Jones, W. E., *Sci. Light Tokyo*, 21, 113, 1972.
141. Joshi, Y. N., Bhatia, K. S., and Jones, W. E., *Spectrochim. Acta Part B*, 28, 149, 1973.
142. Joshi, Y. N. and Budhiraja, C. J., *Can. J. Phys.*, 49, 670, 1971.
143. Joshi, Y. N. and van Kleef, T. A. M., *Can. J. Phys.*, 52, 1891, 1974.
144. Kaufman, V., *Natl. Bur. Stand.*, unpublished.
145. Kaufman, V., *J. Opt. Soc. Am.*, 52, 866, 1962.
146. Kaufman, V., Artru, M. C., and Brillet, W. U. L., *J. Opt. Soc. Am.*, 64, 197, 1974.
147. Kaufman, V. and Humphreys, C. J., *J. Opt. Soc. Am.*, 59, 1614, 1969.
148. Kaufman, V. and Sugar, J., *J. Opt. Soc. Am.*, 61, 1693, 1971.
149. Kaufman, V. and Sugar, J., *J. Res. Natl. Bur. Stand. Sect. A*, 71, 583, 1967.
150. Kelly, R. L. and Palumbo, L. J., *Naval Research Laboratory Report 7599, Washington, D. C.*, 1973.
151. Kielkopf, J. F., *Univ. of Louisville*, unpublished, 1975.
152. Kielkopf, J. F., *Univ. of Louisville*, unpublished, 1976.
153. Kiess, C. C. and Corliss, C. H., *J. Res. Natl. Bur. Stand. Sect. A*, 63, 1, 1959.
154. Kleiman, H., *J. Opt. Soc. Am.*, 52, 441, 1962.
155. Eriksson, K. B., Johansson, I., and Norlen, G., *Ark. Fys.*, 28, 233, 1964.
156. Klinkenberg, P. F. A., *Physica*, 15, 774, 1949.
157. Klinkenberg, P. F. A., *Physica*, 16, 618, 1950.
158. Krishnamurty, S. G., *Proc. Phys. Soc. London*, 48, 277, 1936.
159. Kruger, P. G. and Gilroy, H. T., *Phys. Rev.*, 48, 720, 1935.
160. Kruger, P. G. and Pattin, H. S., *Phys. Rev.*, 52, 621, 1937.
161. Lacroute, P., *Ann. Phys. (Paris)*, 3, 5, 1935.
162. Lang, R. J., *Phys. Rev.*, 30, 762, 1927.
163. Lang, R. J., *Phys. Rev.*, 32, 737, 1928.
164. Lang, R. J., *Phys. Rev.*, 35, 445, 1930.
165. Lang, R. J., *Can. J. Res. Sect. A*, 14, 43, 1936.
166. Lang, R. J., *Can. J. Res. Sect. A*, 14, 127, 1936.
167. Lang, R. J. and Vestine, E. H., *Phys. Rev.*, 42, 233, 1932.
168. Li, H. and Andrew, K. L., *J. Opt. Soc. Am.*, 61, 96, 1971.
169. Liden, K., *Ark. Fys.*, 1, 229, 1949.
170. Litzen, U., *Ark. Fys.*, 28, 239, 1965.
171. Litzen, U., *Phys. Scr.*, 1, 251, 1970.
172. Litzen, U., *Phys. Scr.*, 1, 253, 1970.
173. Litzen, U., *Phys. Scr.*, 2, 103, 1970.
174. Litzen, U. and Verges, J., *Phys. Scr.*, 13, 240, 1976.
175. Lofstrand, B., *Phys. Scr.*, 8, 57, 1973.
176. Luc-Koenig, E., Morillon, C., and Verges, J., *Phys. Scr.*, 12, 199, 1975.
177. Lundstrom, T., *Phys. Scr.*, 7, 62, 1973.
178. Lundstrom, T. and Minnhagen, L., *Phys. Scr.*, 5, 243, 1972.
179. Magnusson, C. E. and Zetterberg, P. O., *Phys. Scr.*, 10, 177, 1974.
180. Magnusson, C. E., and Zetterberg, P. O., *Phys. Scr.*, 15, 237, 1977.
181. Martin, D. C., *Phys. Rev.*, 48, 938, 1935.
182. Svendenius, N., *Phys. Scr.*, 22, 240, 1980.
183. Martin, W. C., *J. Res. Natl. Bur. Stand. Sect. A*, 64, 19, 1960.
184. Martin, W. C. and Corliss, C. H., *J. Res. Natl. Bur. Stand. Sect. A*, 64, 443, 1960.
185. Martin, W. C. and Kaufman, V., *J. Res. Natl. Bur. Stand. Sect. A*, 74, 11, 1970.
186. Martin, W. C. and Kaufman, V., *J. Opt. Soc. Am.*, 60, 1096, 1970.
187. McCormick, W. W. and Sawyer, R. A., *Phys. Rev.*, 54, 71, 1938.
188. McLaughlin, R., *J. Opt. Soc. Am.*, 54, 965, 1964.
189. McLennan, J. C., McLay, A. B., and Crawford, M. F., *Proc. R. Soc. London Ser. A*, 134, 41, 1931.
190. Meissner, K. W., *Z. Phys.*, 39, 172, 1926.
191. Meggers, W. F., *J. Res. Natl. Bur. Stand.*, 24, 153, 1940.
192. Meggers, W. F. and Corliss, C. H., *J. Res. Natl. Bur. Stand. Sect. A*, 70, 63, 1966.
193. Meggers, W. F., Fred, M., and Tomkins, F. S., *J. Res. Natl. Bur. Stand.*, 58, 297, 1957.
194. Meggers, W. F. and Humphreys, C. J., *J. Res. Natl. Bur. Stand.*, 28, 463, 1942.
195. Meggers, W. F. and Murphy, R. J., *J. Res. Natl. Bur. Stand.*, 48, 334, 1952.
196. Meggers, W. F., Scribner, B. F., and Bozman, W. R., *J. Res. Natl. Bur. Stand.*, 46, 85, 1951.
197. Meggers, W. F., Shenstone, A. G., and Moore, C. E., *J. Res. Natl. Bur. Stand.*, 45, 346, 1950.
198. Mehlman, G. and Esteva, J. M., *Astrophys. J.*, 188, 191, 1974.
199. Meinders, E., *Physica*, 84(C), 117, 1976.
200. Sansonetti, C. J., Dissertation, Purdue University, 1981.
201. Sansonetti, C. J., *Natl. Bur. Stand. (U.S.)*, unpublished.
202. Millikan, R. A. and Bowen, I. S. *Phys. Rev.*, 25, 600, 1925.
203. Minnhagen, L., *J. Opt. Soc. Am.*, 61, 1257, 1925.
204. Minnhagen, L., *J. Opt. Soc. Am.*, 63, 1185, 1973.
205. Minnhagen, L., *Phys. Scr.*, 11, 38, 1975.
206. Minnhagen, L., *J. Opt. Soc. Am.*, 66, 659, 1976.
207. Minnhagen, L. and Nietsche, H., *Phys. Scr.*, 5, 237, 1972.
208. Minnhagen, L., Strihed, H., and Petersson, B., *Ark. Fys.*, 39, 471, 1969.
209. Moore, C. E., *Natl. Bur. Stand. (U.S.) Circ.*, 488, 1950.
210. Moore, C. E., *Revised Multiplet Table*, Princeton University Observatory No. 20, 1945.

211. **Moore, C. E.**, National Standard Reference Data Series — National Bureau of Standards 3, Sect. 3, 1970.
212. **Moore, C. E.**, National Standard Reference Data Series — National Bureau of Standards 3, Sect. 4, 1971.
213. **Moore, C. E.**, National Standard Reference Data Series — National Bureau of Standards 3, Sect. 5, 1975.
214. **Moore, C. E.**, National Standard Reference Data Series — National Bureau of Standards 3, Sect. 6, 1972.
215. **Moore, C. E.**, *National Standard Reference Data Series — National Bureau of Standards 3*, Sect. 7, 1975.
216. **Morillon, C. and Verges, J.**, *Phys. Scr.*, 10, 227, 1974.
217. **Newsom, G. H.**, *Astrophys. J.*, 166, 243, 1971.
218. **Newsom, G. H., O'Connor, S., and Learner, R. C. M.**, *J. Phys. B*, 6, 2162, 1973.
219. **Norlen, G.**, *Phys. Scr.*, 8, 249, 1973.
220. **Odabasi, H.**, *J. Opt. Soc. Am.*, 57, 1459, 1967.
221. **Olme, A.**, *Ark. Fys.*, 40, 35, 1969.
222. **Olme, A.**, *Phys. Scr.*, 1, 256, 1970.
223. **Johansson, S., and Litzen, U.**, *J. Opt. Soc. Am.*, 61, 1427, 1971.
224. **Palenius, H. P.**, *Ark. Fys.*, 39, 15, 1969.
225. **Palenius, H. P.**, *Phys. Scr.*, 1, 113, 1970.
226. **Palenius, H. P.**, *Univ. of Lund, Sweden*, unpublished.
227. **Paschen, F.**, *Ann. Phys.*, Series 5, 12, 509, 1932.
228. **Paschen, F. and Ritschl, R.**, *Ann. Phys.*, Series 5, 18, 867, 1933.
229. **Peck, E. R., Khanna, B. N., and Anderholm, N. C.** *J. Opt. Soc. Am.*, 52, 53, 1962.
230. **Persson, W.**, *Phys. Scr.*, 3, 133, 1971.
231. **Persson, W. and Valind S.**, *Phys. Scr.*, 5, 187, 1972.
232. **Petersson, B.**, *Ark. Fys.*, 27, 317, 1964.
233. **Phillips, L. W. and Parker, W. L.**, *Phys. Rev.*, 60, 301, 1941.
234. **Platt, J. R. and Sawyer, R. A.**, *Phys. Rev.*, 60, 866, 1941.
235. **Plyer, E. K., Blaine, L. R., and Tidwell, E.**, *J. Res. Natl. Bur. Stand.*, 55, 279, 1955.
236. **Poppe, R., van Kleef, T. A. M., and Raassen, A. J. J.**, *Physica*, 77, 165, 1974.
237. **Radziemski, L. J., Jr. and Andrew, K. L.**, *J. Opt. Soc. Am.*, 55, 474, 1965.
238. **Radziemski, L. J., Jr. and Kaufman, V.**, *J. Opt. Soc. Am.*, 59, 424, 1969.
239. **Radziemski, L. J., Jr. and Kaufman, V.**, *J. Opt. Soc. Am.*, 64, 366, 1974.
240. **Ramanadham, R. and Rao, K. R.**, *Indian J. Phys.*, 18, 317, 1944.
241. **Ramb, R.**, *Ann. Phys.*, 10, 311, 1931.
242. **Rank, D. H., Bennett, J. M., and Bennett, H. E.**, *J. Opt. Soc. Am.*, 40, 477, 1950.
243. **Rao, A. S. and Krishnamurty, S. G.**, *Proc. Phys. Soc. London*, 46, 531, 1943.
244. **Rao, K. R.**, *Proc. R. Soc. London, Ser. A*, 134, 604, 1932.
245. **Rao, K. R. and Badami, J. S.**, *Proc. R. Soc. London Ser. A*, 131, 154, 1931.
246. **Rao, K. R. and Krishnamurty, S. G.**, *Proc. R. Soc. London Ser. A*, 161, 38, 1937.
247. **Rao, K. R. and Murti, S. G. K.**, *Proc. R. Soc. London Ser. A*, 145, 681, 1934.
248. **Rao, Y. B.**, *Indian J. Phys.*, 32, 497, 1958.
249. **Rao, Y. B.**, *Indian J. Phys.*, 33, 546, 1959.
250. **Rao, Y. B.**, *Indian J. Phys.*, 35, 386, 1961.
251. **Rasmussen, E.**, *Z. Phys.*, 80, 726, 1933.
252. **Rasmussen, E.**, *Z. Phys.*, 83, 404, 1933.
253. **Rasmussen, E.**, *Z. Phys.*, 86, 24, 1934.
254. **Rasmussen, E.**, *Z. Phys.*, 87, 607, 1934.
255. **Rasmussen, E.**, *Phys. Rev.*, 57, 840, 1940.
256. **Rau, A. S. and Narayan, A. L.**, *Z. Phys.*, 59, 687, 1930.
257. **Reader, J.**, *J. Opt. Soc. Am.*, 65, 286, 1975.
258. **Reader, J.**, *J. Opt. Soc. Am.*, 65, 988, 1975.
259. **Reader, J.**, *J. Opt Soc. Am.*, 73, 349, 1983.
260. **Reader, J. and Davis, S.**, *J. Res. Natl. Bur. Stand. Sect. A*, 71, 587, 1967, and unpublished.
261. **Reader, J. and Ekberg, J. O.**, *J. Opt. Soc. Am.*, 62, 464, 1972.
262. **Reader, J. and Epstein, G. L.**, *J. Opt. Soc. Am.*, 62, 1467, 1972.
263. **Reader, J. and Epstein, G. L.**, *J. Opt. Soc. Am.*, 65, 638, 1975.
264. **Reader, J. and Epstein, G. L.**, *Natl. Bur. Stand.*, unpublished.
265. **Reader, J., Epstein, G. L., and Ekberg, J. O.**, *J. Opt. Soc. Am.*, 62, 273, 1972.
266. **Kaufman, V.**, *Phys. Scr.*, 26, 439, 1982.
267. **Ricard, R., Givord, M., and George, F.**, *C. R. Acad. Sci. Paris*, 205, 1229, 1937.
268. **Risberg, P.**, *Ark. Fys.*, 10, 583, 1956.
269. **Risberg, G.**, *Ark. Fys.*, 28, 381, 1965.
270. **Risberg, G.**, *Ark. Fys.*, 37, 231, 1968.
271. **Robinson, H. A.**, *Phys. Rev.*, 49, 297, 1936.
272. **Robinson, H. A.**, *Phys. Rev.*, 50, 99, 1936.
273. **Ross, C. B., Jr.**, Doctoral dissertation, Purdue University, 1969.
274. **Ross, C. B., Wood, D. R., and Scholl, P. S.**, *J. Opt. Soc. Am.*, 66, 36, 1976.
275. **Ruedy, J. E. and Gibbs, R. C.**, *Phys. Rev.*, 46, 880, 1934.
276. **Russell, H. N., King, R. B., and Moore, C. E.**, *Phys. Rev.*, 58, 407, 1940.
277. **Russell, H. N. and Moore, C. E.**, *J. Res. Natl. Bur. Stand.*, 55, 299, 1955.
278. **Russell, H. N., Moore, C. E., and Weeks, D. W.**, *Trans. Am. Philos. Soc.*, 34(2), 111, 1944.
279. **Saunders, F., Schneider, E., and Buckingham, E.**, *Proc. Natl. Acad. Sci.*, 20, 291, 1934.
280. **Sawyer, R. A. and Humphreys, C. J.**, *Phys. Rev.*, 32, 583, 1928.
281. **Sawyer, R. A. and Lang, R. J.**, *Phys. Rev.*, 34, 712, 1929.
282. **Sawyer, R. A. and Paschen, F.** *Ann. Phys.*, 84(4), 1, 1927.
283. **Scholl, P. S.**, M.S. thesis, Wright State Univ., 1975.
284. **Schurmann, D.**, *Z. Phys.*, 17, 4, 1975.
285. **Seguier, J.**, *C. R. Acad. Sci. Paris*, 256, 1703, 1963.
286. **Shenstone, A. G.**, *Phys. Rev.*, 31, 317, 1928.
287. **Shenstone, A. G.**, *Phys. Rev.*, 32, 30, 1928.
288. **Shenstone, A. G.**, *Trans. R. Soc. London*, 237(A), 57, 1938.
289. **Shenstone, A. G.**, *Phys. Rev.*, 57, 894, 1940.
290. **Shenstone, A. G.** *Philos. Trans. R. Soc. London Ser. A*, 241, 297, 1948.
291. **Shenstone, A. G.**, *Can. J. Phys.*, 38, 677, 1960.
292. **Shenstone, A. G.**, *Proc. R. Soc. London*, 261(A), 153, 1961.

293. Shenstone, A. G., *Proc. R. Soc. London*, 276(A), 293, 1963.
294. Shenstone, A. G., *J. Res. Natl. Bur. Stand. Sect. A*, 74, 801, 1970.
295. Shenstone, A. G., *J. Res. Natl. Bur. Stand. Sect. A*, 79, 497, 1975.
296. Shenstone, A. G. and Pittenger, J. T., *J. Opt. Soc. Am.*, 39, 219, 1949.
297. Smith, S., *Phys. Rev.*, 36, 1, 1930.
298. Smitt, R., *Phys. Scr.*, 8, 292, 1973.
299. Soderqvist, J., *Ark. Mat. Astronom. Fys.*, 32(A), 1, 1946.
300. Sommer, L. A., *Ann. Phys.*, 75, 163, 1924.
301. Spector, N., *J. Opt. Soc. Am.*, 63, 358, 1973.
302. Spector, N. and Sugar, J., *J. Opt. Soc. Am.*, 66, 436, 1976.
303. Steinhaus, D. W., Radziemski, L. J., Jr., and Blaise, J., *Los Alamos Sci. Lab.*, unpublished, 1975.
304. Subbaraya, T. S., *Z. Phys.*, 78, 541, 1932.
305. Sugar, J., *J. Opt. Soc. Am.*, 55, 33, 1965.
306. Sugar, J., *J. Res. Natl. Bur. Stand. Sect. A*, 73, 333, 1969.
307. Sugar, J., *J. Opt. Soc. Am.*, 60, 454, 1970.
308. Sugar, J., *J. Res. Natl. Bur. Stand. Sect. A*, 78, 555, 1974.
309. Sugar, J. and Kaufman, V., *J. Opt. Soc. Am.*, 55, 1283, 1965.
310. Sugar, J. and Kaufman, V., *J. Opt. Soc. Am.*, 62, 562, 1972.
311. Sugar, J., Kaufman, V., and Spector, N., *J. Res. Natl. Bur. Stand.*, Sect. A, 83, 233, 1978.
312. Sugar, J. and Spector, N., *J. Opt. Soc. Am.*, 64, 1484, 1974.
313. Sullivan, F. J. *Univ. Pittsburgh Bull.*, 35, 1, 1938.
314. Svensson, L. A. and Ekberg, J. O., *Ark. Fys.*, 37, 65, 1968.
315. Swensson, J. W. and Risberg, G., *Ark. Fys.*, 31, 237, 1966.
316. Tech. J. L., *J. Res. Natl. Bur. Stand. Sect. A*, 67, 505, 1963.
317. Tech, J. L. and Ward, J. F., *Phys. Rev. Lett.*, 27, 367, 1971.
318. Tilford, S. G., *J. Opt. Soc. Am.*, 53, 1051, 1963.
319. Toresson, Y. G., *Ark. Fys.*, 17, 179, 1960.
320. Toresson, Y. G., *Ark. Fys.*, 18, 389, 1960.
321. Toresson, Y. G. and Edlen, B., *Ark. Fys.*, 23, 117, 1963.
322. Tsien, W. Z., *Chin. J. Phys.*, Peiping, 3, 117, 1939.
323. van Deurzen, C. H. H., Conway, J., and Davis, S. P., *J. Opt. Soc. Am.*, 63, 158, 1973.
324. van Kleef, T. A. M., Raassen, A. J. J., and Joshi, Y. N., *Physica*, 84(C), 401, 1976.
325. Sansonetti, C. J., Andrew, K. L., and Verges, J., *J. Opt. Soc. Am.*, 71, 423, 1981.
326. Wheatley, M. A. and Sawyer, R. A., *Phys. Rev.*, 61, 591, 1942.
327. Wilkinson, P. G., *J. Opt. Soc. Am.*, 45, 862, 1955.
328. Wilkinson, P. G. and Andrew, K. L., *J. Opt. Soc. Am.*, 53, 710, 1963.
329. Wood, D. and Andrew, K. L., *J. Opt. Soc. Am.*, 58, 818, 1968.
330. Wood, D. R., Ross, C. B., Scholl, P. S., and Hoke, M., *J. Opt. Soc. Am.*, 64, 1159, 1974.
331. Worden, E. F. and Conway, J. G., *Lawrence Livermore Lab.*, unpublished, 1977.
332. Worden, E. F., Hulet, E. K., Gutmacher, R. G., Conway, J. G., *At. Data Nucl. Data Tables*, 18, 459, 1976.
333. Worden, E. F., Lougheed, R. W., Gutmacher, R. G., and Conway, J. G., *J. Opt. Soc. Am.*, 64, 77, 1974.
334. Wu, C. M., Ph.D. thesis, University of British Columbia, 1971.
335. Zaidel, A. N., Prokofev, V. K., Raiskii, S. M., Slavnyi, V. A., and Schreider, E. Y., *Tables of Spectral Lines*, 3rd ed., Plenum, New York, 1970.
336. Zetterberg, P. O. and Magnusson, C. E., *Phys. Scr.*, 15, 189, 1977.
337. Sugar, J., *J. Opt. Soc. Am.*, 55, 1058, 1965.
338. Sugar, J., *J. Opt. Soc. Am.*, 61, 727, 1971.
339. Worden, E. F., and Conway, J. G., *At. Data Nucl. Data Tables*, 22, 329, 1978.
340. Kaufman, V. and Edlen, B., *J. Phys. Chem. Ref. Data*, 3, 825, 1974.
341. Lang, R. J., *Phys. Rev.*, 34, 697, 1929.
342. Ryabtsev, A. N., *Opt. Spectros.*, 39, 455, 1975.
343. Foster, E. W., *Proc. R. Soc. London*, 200(A), 429, 1950.
344. Morillon, C. and Verges, J., *Phys. Scr.*, 12, 129, 1975.
345. Ruedy, J. E., *Phys. Rev.*, 41, 588, 1932.
346. McLennan, J. C., McLay, A. B., and McLeod, J. H., *Philos. Mag.*, 4, 486, 1927.
347. Handrup, M. B. and Mack, J. E., *Physica*, 30, 1245, 1964.
348. Clearman, H. E., *J. Opt. Soc. Am.*, 42, 373, 1952.
349. Paschen, F., *Ann. Physik*, 424, 148, 1938.
350. Paschen, F. and Campbell, J. S., *Ann. Phys.*, 31(5), 29, 1938.
351. Nodwell, R., *Univ. of British Columbia, Vancouver*, unpublished, 1955.
352. Gibbs, R. C. and White, H. E., *Phys. Rev.*, 31, 776, 1928.
353. Green, M., *Phys. Rev.*, 60, 117, 1941.
354. Ellis, C. B. and Sawyer, R. A., *Phys. Rev.*, 49, 145, 1936.
355. McLennan, J. C., McLay, A. B., and Crawford, M. F., *Proc. R. Soc. London Ser. A*, 125, 50, 1929.
356. Mack, J. E. and Fromer, M., *Phys. Rev.*, 48, 346, 1935.
357. Humphreys, C. J. and Paul, E., *U.S. Nav. Ord. Lab.*, Navord Rep. 4589, 25, 1956.
358. Walters, F. M., *Sci. Pap. Bur. Stand.*, 17, 161, 1921.
359. Crawford, M. F. and McLay, A. B., *Proc. R. Soc. London Ser. A*, 143, 540, 1934.
360. McLay, A. B. and Crawford, M. F., *Phys. Rev.*, 44, 986, 1933.
361. Schoepfle, G. K., *Phys. Rev.*, 47, 232, 1935.
362. Acquista, N., and Reader, J., *J. Opt. Soc. Am.*, 70, 789, 1980.
363. Benschop, H., Joshi, Y. N., and Van Kleef, T. A. M., *Can. J. Phys.*, 53, 498, 1975.
364. Bockasten, K., Hallin, R., and Hughes, T. P., *Proc. Phys. Soc.*, 81, 522, 1963.
365. Boyce, J. C., *Phys. Rev.*, 46, 378, 1934.
366. Boyce, J. C., *Phys. Rev.*, 47, 718, 1935.
367. Boyce, J. C., *Phys. Rev.*, 48, 396, 1935.
368. Boyce, J. C., *Phys. Rev.*, 49, 351, 1936.
369. Corliss, C. H. and Meggers, W. F., *J. Res. Natl. Bur. Stand.*, 61, 269, 1958.
370. Crooker, A. M. and Dick, K. A., *Can. J. Phys.*, 42, 766, 1964.
371. De Bruin, T. L., *Z. Physik*, 77, 505, 1932.
372. De Bruin, T. L., *Proc. Roy. Acad. Amsterdam*, 36, 727, 1933.
373. De Bruin, T. L., *Zeeman Verhandelingen*, (The Hague), 1935, p. 415.
374. De Bruin, T. L., *Physica*, 3, 809, 1936.
375. De Bruin, T. L., *Proc. Roy. Acad. Amsterdam*, 40, 339, 1937.

376. Dick, K. A., *Can. J. Phys.*, 46, 1291, 1968.
377. Dick, K. A., unpublished, 1978.
378. Edlen, B. and Swensson, J. W., *Phys. Scr.*, 12, 21, 1975.
379. Ekberg, J. O., *Phys. Scr.*, 7, 55, 1973.
380. Ekberg, J. O., *Phys. Scr.*, 7, 59, 1973.
381. Ekberg, J. O., *Phys. Scr.*, 12, 42, 1975.
382. Ekberg, J. O. and Edlen, B., *Phys. Scr.*, 18, 107, 1978.
383. Eliason, A. Y., *Phys. Rev.*, 43, 745, 1933.
384. Gallardo, M., Massone, C. A., Tagliaferri, A. A., Garavaglia, M., and Persson, W., *Phys. Scr.*, to be published, 1979.
385. Garcia-Riquelme, O., *Optica Pura Y Aplicada*, 1, 53, 1968.
386. Gibbs, R. C., Vieweg, A. M., and Gartlein, C. W., *Phys. Rev.*, 34, 406, 1929.
387. Gilbert, W. P., *Phys. Rev.*, 48, 338, 1935.
388. Goldsmith, S. and Kaufman, A. S., *Proc. Phys. Soc.*, 81, 544, 1963.
389. Hermansdorfer, H., *J. Opt. Soc. Am.*, 62, 1149, 1972.
390. Humphreys, C. J., *Phys. Rev.*, 47, 712, 1935.
391. Humphreys, C. J., *J. Res. Natl. Bur. Stand.*, 16, 639, 1936.
392. Iglesias, L., *J. Opt. Soc. Am.*, 45, 856, 1955.
393. Iglesias, L., *J. Res. Natl. Bur. Stand.*, 64A, 481, 1960.
394. Iglesias, L., *Anales Fisica Y Quimica*, 58A, 191, 1962.
395. Iglesias, L., *J. Res. Natl. Bur. Stand.*, 70A, 465, 1966.
396. Iglesias, L., *Can. J. Phys.*, 44, 895, 1966.
397. Iglesias, L., *J. Res. Natl. Bur. Stand.*, 72A, 295, 1968.
398. Joshi, Y. N., *Can. Spectrosc.*, 15, 96, 1970.
399. Joshi, Y. N. and Van Kleef, T. A. M., *Can. J. Phys.*, 55, 714, 1977.
400. Kaufman, A. S., Hughes, T. P., and Williams, R. V., *Proc. Phys. Soc.*, 76, 17, 1960.
401. Kaufman, V. and Sugar, J., *J. Opt. Soc. Am.*, 68, 1529, 1978.
402. Keussler, V., *Z. Physik*, 85, 1, 1933.
403. Kiess, C. C., *J. Res. Natl. Bur. Stand.*, 56, 167, 1956.
404. Klinkenberg, P. F. A., Van Kleef, T. A. M., and Noorman, P. E., *Physica*, 27, 1177, 1961.
405. Kovalev, V. I., Romanos, A. A., and Ryabtsev, A. N., *Opt. Spectrosc.*, 43, 10, 1977.
406. Lang, R. J., *Proc. Natl. Acad. Sci.*, 13, 341, 1927.
407. Lang, R. J., *Zeeman Verhandelingen*, (The Hague), 44, 1935.
408. Liberman, S., et al., *C. R. Acad. Sci. (Paris)*, 286, 253, 1978.
409. Livingston, A. E., *J. Phys.*, B9, L215, 1976.
410. Meijer, F. G., *Physica*, 72, 431, 1974.
411. Meijer, F. G. and Metsch, B. C., *Physica*, 94C, 259, 1978.
412. Moore, F. L., thesis, Princeton, 1949.
413. Paul, F. W. and Polster, H. D., *Phys. Rev.*, 59, 424, 1941.
414. Phillips, L. W. and Parker, W. L., *Phys. Rev.*, 60, 301, 1941.
415. Poppe, R., *Physica*, 81C, 351, 1976.
416. Raassen, A. J. J., Van Kleef, T. A. M., and Metsch, B. C., *Physica*, 84C, 133, 1976.
417. Rao, A. B. and Krishnamurty, S. G., *Proc. Phys. Soc. (London)*, 51, 772, 1939.
418. Reader, J. and Acquista, N., *J. Opt. Soc. Am.*, 69, 239, 1979.
419. Reader, J. and Epstein, G. L., *J. Opt. Soc. Am.*, 62, 619, 1972.
420. Rico, F. R., *Anales, Real Soc. Esp. Fis. Quim.*, 61, 103, 1965.
421. Schonheit, E., *Optik*, 23, 409, 1966.
422. Shenstone, A. G., *J. Opt. Soc. Am.*, 44, 749, 1954.
423. Shenstone, A. G., unpublished, 1958.
424. Shenstone, A. G., *J. Res. Natl. Bur. Stand.*, 67A, 87, 1963.
425. Sugar, J. and Kaufman, V., *J. Opt. Soc. Am.*, 64, 1656, 1974.
426. Sugar, J. and Kaufman, V., *Phys. Rev.*, C12, 1336, 1975.
427. Svensson, L. A., *Phys. Scr.*, 13, 235, 1976.
428. Swensson, J. W. and Edlen, B., *Phys. Scr.*, 9, 335, 1974.
429. Tagliaferri, A. A., Gallego Lluesma, E., Garavaglia, M., Gallardo, M., and Massone, C. A., *Optica Pura Y Aplica*, 7, 89.
430. Tilford, S. G. and Giddings, L. E., *Astrophys. J.*, 141, 1222, 1965.
431. Trawick, M. W., *Phys. Rev.*, 46, 63, 1934.
432. Van Deurzen, C. H. H., *J. Opt. Soc. Am.*, 67, 476, 1977.
433. Yarosewick, S. L. and Moore, F. L., *J. Opt. Soc. Am.*, 57, 1381, 1967.
434. Zalubas, R., unpublished, 1979.
435. Bhatia, K. S., Jones, W. E., and Crooker, A. M., *Can. J. Phys.*, 50, 2421, 1972.
436. van Kleef, A. M. and Joshi, Y. N., *Phys. Scr.*, 24, 557, 1981.

NIST ATOMIC TRANSITION PROBABILITY TABLES

J.R. Fuhr and W.L. Wiese

These tables substantially update and enlarge our earlier tables in this *Handbook*. The new tables contain critically evaluated atomic transition probabilities for about 9000 selected lines of all elements for which reliable data are available on an absolute scale. The material is largely for neutral and singly ionized spectra, but also includes a number of prominent lines of more highly charged ions of important elements.

Many of the data are obtained from comprehensive compilations of the Data Center on Atomic Transition Probabilities at the National Institute of Standards and Technology (formerly the National Bureau of Standards). Specifically, data have been taken from three recent comprehensive critical compilations on C, N and O,[1] on Sc through Mn,[2] and Fe through Ni.[3] Material from earlier compilations for the elements H through Ne[4] and Na through Ca[5] was supplemented by more recent material taken directly from the original literature. For the highly charged ions, some of the data were derived from studies of the systematic behavior of transition probabilities.[6-8] Most of the original literature is cited in the above tables and in recent bibliographies[9,10]; for lack of space, individual literature references are not cited here.

The wavelength range for the neutral species is normally the visible spectrum or shorter wavelengths; only the very prominent near infrared lines are included. For the higher ions, most of the strong lines are located in the far UV. The tabulation is limited to electric dipole — including intercombination — lines and comprises essentially the fairly strong transitions with estimated uncertainties of 50% or less. With the exception of hydrogen, helium, and the alkalis, most transitions are between states with low principal quantum numbers.

The transition probability, A, is given in units of 10^8 s^{-1} and is listed to as many digits as is consistent with the indicated accuracy. The power of 10 is indicated by the E notation (i.e., E-02 means 10^{-2}). Generally, the estimated uncertainties of the A-values are ±25 to 50% for two-digit numbers, ±10 to 25% for three-digit numbers and $\pm1\%$ or better for four- and five-digit numbers.

Each transition is identified by the wavelength, λ, in angstroms; and the statistical weights, g_i and g_k, of the lower (i) and upper (k) states [the product $g_k A$ (or $g_i f$) is needed for many applications]. Whenever the wavelengths of individual lines within a multiplet are extremely close, only an average wavelength for the multiplet as well as the multiplet A-value are given, and this is indicated by an asterisk (*) to the left of the wavelength. This also has been done when the transition probability for an entire multiplet has been taken from the literature and values for individual lines cannot be determined because of insufficient knowledge of the coupling of electrons. The wavelength data have been taken either from recent compilations or from the original literature cited in bibliographies published by the Atomic Energy Levels Data Center[11,12] at the National Institute of Standards and Technology. Wavelength values are consistent with those given in the table "Line Spectra of the Elements", which appears elsewhere in this *Handbook*.

The transition probabilities for hydrogen and hydrogen-like ions are known precisely. Because of the hydrogen degeneracy, a "transition" is actually the sum of all fine-structure transitions between the principal quantum numbers listed in the transition column; therefore, the special hydrogen table which appears below gives weighted average A-values.

In addition to the transition probability A, the atomic oscillator strength f and the line strength S are often used in the literature. The conversion factors between these quantities are (for electric-dipole transitions):

$$g_i f = 1.499 \times 10^{-8} \lambda^2 g_k A = 303.8 \lambda^{-1} S$$

where λ is in angstroms, A is in 10^8 s^{-1}, and S is in atomic units, which are
$a_0^2 e^2 = 7.188 \times 10^{-59}$ m^2 C^2.

After the special table for hydrogen, the tables for other elements appear in alphabetical sequence by element name (not symbol). Within each element, the tables are ordered by increasing ionization stage (e.g., Al I, Al II, etc.).

REFERENCES

1. Wiese, W.L., Fuhr, J.R., and Deters, T.M., *Atomic Transition Probabilities of Carbon, Nitrogen and Oxygen, J. Phys. Chem. Ref. Data, Monograph 7*, 1996.
2. Martin, G.A., Fuhr, J.R., and Wiese, W.L., *Atomic Transition Probabilities—Scandium through Manganese, J. Phys. Chem. Ref. Data, 17, Suppl. 3*, 1988.
3. Fuhr, J.R., Martin, G.A., and Wiese, W.L., *Atomic Transition Probabilities—Iron through Nickel, J. Phys. Chem. Ref. Data, 17, Suppl. 4*, 1988.
4. Wiese, W.L., Smith, M.W., and Glennon, B.M., *Atomic Transition Probabilities (H through Ne—A Critical Data Compilation)*, National Standard Reference Data Series, National Bureau of Standards 4, Vol. I, U.S. Government Printing Office, Washington, D.C., 1966.
5. Wiese, W.L., Smith, M.W., and Miles, B.M., *Atomic Transition Probabilities (Na through Ca—A Critical Data Compilation)*, National Standard Reference Data Series, National Bureau of Standards 22, Vol. II, U. S. Government Printing Office, Washington, D.C., 1969.
6. Wiese, W.L. and Weiss, A.W., *Phys. Rev.*, 175, 50, 1968.
7. Smith, M.W. and Wiese,M.L., *Astrophys. J., Suppl. Ser.*, 23, No. 196, 103, 1971.
8. Martin, G.A., and Wiese,W.L., *J. Phys. Chem. Ref. Data*, 5, 537, 1976.
9. Fuhr, J.R., Miller, B.J., and Martin, G.A., *Bibliography on Atomic Transition Probabilities (1914 through October 1997)*, National Bureau of Standards Special Publication 505, 1978; Miller, B.J., Fuhr, J.R., and Martin, G.A., *Bibliography on Atomic Transition Probabilities (November 1977 through February 1980)*, National Bureau of Standards Special Publication 505, Supplement 1, 1980.
10. Wiese, W.L., Reports on Astronomy, *Trans. Int. Astron. Union*, 18A, 116—123, 1982; 19A, 122—138, 1985.; 20A, 117—123, 1988, Reidel, D., Ed., Kluwer, Dordrecht, Holland.
11. Moore, C.E., *Bibliography on the Analyses of Optical Atomic Spectra*, National Bureau of Standards Special Publication 306—Section 1, 1968; Sections 2—4, 1969.

12. Hagan, L. and Martin, W.C., *Bibliography on Atomic Energy Levels and Spectra (July 1968 through June 1971)*, National Bureau of Standards Special Publication 363, 1972; Hagan, L., *Bibliography on Atomic Energy Levels and Spectra (July 1971 through June 1975)*, National Bureau of Standards Special Publication 363, Supplement 1, 1977; Zalubas, R. and Albright, A., *Bibliography on Atomic Energy Levels and Spectra (July 1975 through June 1979)*, National Bureau of Standards Special Publication 363, Supplement 2, 1980; Musgrove, A. and Zalubas, R., *Bibliography on Atomic Energy Levels and Spectra (July 1979 through December 1983)*, National Bureau of Standards Special Publication 363, Supplement 3, 1985.

13. Younger, S.M. and Weiss, A., *J. Res. Natl. Bur. Stand.*, 79A, 629, 1975.

Transition Probabilities for Allowed Lines of Hydrogen

λ Å	Weights g_i	Weights g_k	A 10^8 s^{-1}	λ Å	Weights g_i	Weights g_k	A 10^8 s^{-1}	λ Å	Weights g_i	Weights g_k	A 10^8 s^{-1}
Hydrogen				3664.68	8	1568	4.022E-06	8598.40	18	392	9.211E-05
H I				3666.10	8	1458	4.826E-06	8665.02	18	338	1.343E-04
912.768	2	1800	5.167E-06	3667.68	8	1352	5.830E-06	8750.48	18	288	2.021E-04
912.839	2	1682	6.122E-06	3669.46	8	1250	7.096E-06	8862.79	18	242	3.156E-04
912.918	2	1568	7.297E-06	3671.48	8	1152	8.707E-06	9014.91	18	200	5.156E-04
913.006	2	1458	8.753E-06	3673.76	8	1058	1.078E-05	9229.02	18	162	8.905E-04
913.104	2	1352	1.057E-05	3676.36	8	968	1.347E-05	9545.97	18	128	1.651E-03
913.215	2	1250	1.286E-05	3679.35	8	882	1.700E-05	10049.4	18	98	3.358E-03
913.339	2	1152	1.578E-05	3682.81	8	800	2.172E-05	10938.1	18	72	7.783E-03
913.480	2	1058	1.952E-05	3686.83	8	722	2.809E-05	12818.1	18	50	2.201E-02
913.641	2	968	2.438E-05	3691.55	8	648	3.685E-05	16407.2	32	288	1.620E-04
913.826	2	882	3.077E-05	3697.15	8	578	4.910E-05	16806.5	32	242	2.556E-04
914.039	2	800	3.928E-05	3703.85	8	512	6.658E-05	17362.1	32	200	4.235E-04
914.286	2	722	5.077E-05	3711.97	8	450	9.210E-05	18174.1	32	162	7.459E-04
914.576	2	648	6.654E-05	3721.94	8	392	1.303E-04	18751.0	18	32	8.986E-02
914.919	2	578	8.858E-05	3734.37	8	338	1.893E-04	19445.6	32	128	1.424E-03
915.329	2	512	1.200E-04	3750.15	8	288	2.834E-04	21655.3	32	98	3.041E-03
915.824	2	450	1.657E-04	3770.63	8	242	4.397E-04	26251.5	32	72	7.711E-03
916.429	2	392	2.341E-04	3797.90	8	200	7.122E-04	27575	50	288	1.402E-04
917.181	2	338	3.393E-04	3835.38	8	162	1.216E-03	28722	50	242	2.246E-04
918.129	2	288	5.066E-04	3889.05	8	128	2.215E-03	30384	50	200	3.800E-04
919.351	2	242	7.834E-04	3970.07	8	98	4.389E-03	32961	50	162	6.908E-04
920.963	2	200	1.263E-03	4101.73	8	72	9.732E-03	37395	50	128	1.388E-03
923.150	2	162	2.143E-03	4340.46	8	50	2.530E-02	40511.5	32	50	2.699E-02
926.226	2	128	3.869E-03	4861.32	8	32	8.419E-02	43753	72	288	1.288E-04
930.748	2	98	7.568E-03	6562.80	8	18	4.410E-01	46525	50	98	3.253E-03
937.803	2	72	1.644E-02	8392.40	18	800	1.517E-05	46712	72	242	2.110E-04
949.743	2	50	4.125E-02	8413.32	18	722	1.964E-05	51273	72	200	3.688E-04
972.537	2	32	1.278E-01	8437.96	18	648	2.580E-05	59066	72	162	7.065E-04
1025.72	2	18	5.575E-01	8467.26	18	578	3.444E-05	74578	50	72	1.025E-02
1215.67	2	8	4.699E+00	8502.49	18	512	4.680E-05	75004	72	128	1.561E-03
3662.26	8	1800	2.847E-06	8545.39	18	450	6.490E-05	123680	72	98	4.561E-03
3663.40	8	1682	3.374E-06								

For hydrogen-like ions of nuclear charge Z, the following scaling laws hold:

$$A_Z = Z^4 A_{\text{Hydrogen}}; f_Z = f_H; S_Z = Z^{-2} S_H$$
$$(\text{For wavelengths, } \lambda_Z = Z^{-2} \lambda_H)$$

For very highly charged hydrogen-like ions, starting at about $Z > 25$, relativistic corrections[13] must be applied.

Transition Probabilities for Other Elements

λ Å	Weights g_i	g_k	A 10^8 s^{-1}	λ Å	Weights g_i	g_k	A 10^8 s^{-1}	λ Å	Weights g_i	g_k	A 10^8 s^{-1}
Aluminum				1384.1	4	2	9.1E+00	*4761	2	6	2.55E-01
Al I				1605.8	2	4	1.22E+01	5172	2	4	3.95E-02
2263.5	2	4	6.6E-01	1611.8	4	4	2.42E+00	5551	4	6	3.85E-02
2269.1	4	6	7.9E-01	1611.9	4	6	1.45E+01	5687	4	4	6.0E-03
2269.2	4	4	1.3E-01	1854.7	2	4	5.40E+00				
2367.1	2	4	7.2E-01	1862.8	2	2	5.33E+00	**Argon**			
2373.1	4	6	8.6E-01	*1935.9	10	14	1.22E+01	**Ar I**			
2373.4	4	4	1.4E-01	3601.6	6	4	1.34E+00	1048.22	1	3	5.36E+00
2568.0	2	4	2.3E-01	3601.9	4	4	1.49E-01	1066.66	1	3	1.29E+00
2575.1	4	6	2.8E-01	3612.4	4	2	1.5E+00	3406.18	3	1	3.9E-03
2575.4	4	4	4.4E-02					3461.08	3	5	6.7E-04
2652.5	2	2	1.33E-01	**Al X**				3554.30	5	5	2.7E-03
2660.4	4	2	2.64E-01	39.925	1	3	2.22E+03	3563.29	1	3	1.2E-03
3082.2	2	4	6.3E-01	51.979	1	3	4.8E+03	3567.66	5	7	1.1E-03
3092.7	4	6	7.4E-01	55.227	1	3	5.2E+03	3572.30	3	1	5.1E-03
3092.8	4	4	1.2E-01	55.272	3	5	7.2E+03	3606.52	3	1	7.6E-03
3944.0	2	2	4.93E-01	55.376	5	7	9.5E+03	3632.68	3	5	6.6E-04
3961.5	4	2	9.8E-01	59.107	3	5	4.6E+03	3634.46	3	3	1.3E-03
6696.0	2	4	1.69E-02	332.78	1	3	5.6E+01	3643.12	3	5	2.4E-04
6698.7	2	2	1.69E-02	394.83	3	1	8.3E+01	3649.83	3	1	8.0E-03
7835.3	4	6	5.7E-02	395.36	3	5	1.2E+02	3659.53	3	3	4.4E-04
7836.1	6	8	6.2E-02	397.76	1	3	1.7E+01	3670.67	3	5	3.1E-04
				400.43	3	3	1.3E+01	3675.23	3	3	4.9E-04
Al II				401.12	5	5	3.6E+01	3770.37	1	3	7.0E-04
1047.9	1	3	3.6E-01	403.55	3	1	4.9E+01	3834.68	3	1	7.5E-03
1048.6	3	5	4.8E-01	406.31	5	3	1.9E+01	3894.66	3	3	5.7E-04
1539.8	3	5	8.8E+00	670.06	3	5	9.8E+00	3947.50	5	5	5.6E-04
1670.8	1	3	1.46E+01	2535	1	3	3.8E-01	3948.98	5	3	4.55E-03
1719.4	1	3	6.79E+00					4044.42	3	5	3.33E-03
1764.0	5	5	9.8E+00	**Al XI**				4045.96	3	3	4.1E-04
1772.8	1	3	9.5E+00	*36.675	2	6	1.5E+03	4054.53	3	3	2.7E-04
1777.0	5	7	1.7E+01	39.091	2	4	2.6E+03	4158.59	5	5	1.40E-02
*1819.0	15	15	5.6E+00	39.180	4	6	3.1E+03	4164.18	5	3	2.88E-03
1855.9	1	3	8.32E-01	39.530	2	2	1.8E+02	4181.88	1	3	5.61E-03
1858.0	3	3	2.48E+00	39.623	4	2	3.7E+02	4190.71	5	5	2.80E-03
1862.3	5	3	4.12E+00	48.298	2	4	3.09E+03	4191.03	1	3	5.39E-03
1931.0	3	1	1.08E+01	48.338	2	2	3.08E+03	4198.32	3	1	2.57E-02
1990.5	3	5	1.47E+01	52.299	2	4	8.1E+03	4200.67	5	7	9.67E-03
2816.2	3	1	3.83E+00	52.446	4	6	9.6E+03	4251.18	5	3	1.11E-03
4663.1	5	3	5.3E-01	52.458	4	4	1.6E+03	4259.36	3	1	3.98E-02
6226.2	1	3	6.2E-01	54.217	2	4	4.8E+02	4266.29	3	5	3.12E-03
6231.8	3	5	8.4E-01	54.388	4	2	9.6E+02	4272.17	3	3	7.97E-03
6243.4	5	7	1.1E+00	*99.083	2	6	2.2E+02	4300.10	3	5	3.77E-03
6335.7	5	3	1.4E-01	103.6	2	4	4.2E+02	4333.56	3	5	5.68E-03
6823.4	3	3	3.4E-01	103.8	4	6	5.0E+02	4335.34	3	3	3.87E-03
6837.1	5	3	5.7E-01	*141.6	2	6	4.07E+02	4345.17	3	3	2.97E-03
6920.3	3	1	9.6E-01	150.31	2	4	8.5E+02	4363.79	3	3	1.2E-03
7042.1	3	5	5.9E-01	150.61	4	6	9.9E+02	4424.00	1	3	7.3E-05
7056.7	3	3	5.8E-01	157.0	2	2	1.3E+02	4510.73	3	1	1.18E-02
7471.4	5	7	9.4E-01	157.4	4	2	2.6E+02	4522.32	1	3	8.98E-04
				*205.0	2	6	6.3E+01	4544.75	3	3	8.3E-04
Al III				*308.6	2	6	9.9E+01	4554.32	3	5	3.8E-04
*560.36	2	6	4.0E-01	*341.3	6	2	1.3E+02	4584.96	3	5	1.6E-04
695.83	2	4	7.4E-01	550.05	2	4	8.55E+00	4586.61	3	3	2.3E-04
696.22	2	2	7.2E-01	568.12	2	2	7.73E+00	4587.21	3	1	4.9E-03
*1352.8	10	14	4.40E+00	1997	2	4	1.07E+00	4589.29	3	5	6.2E-05
1379.7	2	2	4.59E+00	2069	2	2	9.7E-01	4596.10	3	3	9.47E-04

λ Å	g_i	g_k	A 10^8 s^{-1}	λ Å	g_i	g_k	A 10^8 s^{-1}	λ Å	g_i	g_k	A 10^8 s^{-1}
4628.44	3	5	3.83E-04	5473.46	5	3	2.0E-03	6025.15	5	3	9.0E-03
4642.15	3	5	9.6E-04	5490.12	5	5	8.5E-04	6043.22	5	7	1.47E-02
4647.49	3	3	1.2E-03	5492.09	3	1	5.6E-03	6052.73	3	5	1.9E-03
4702.32	3	3	1.09E-03	5495.87	7	9	1.69E-02	6064.76	5	7	5.8E-04
4746.82	3	1	3.6E-03	5506.11	5	7	3.6E-03	6081.25	3	3	7.5E-04
4752.94	3	3	4.5E-03	5524.96	7	7	1.7E-03	6085.86	3	3	9.0E-05
4768.68	3	5	8.6E-04	5528.97	1	3	1.2E-03	6090.79	1	3	3.0E-03
4798.74	7	9	8.8E-04	5534.49	5	3	2.7E-03	6098.81	3	3	5.2E-03
4835.97	7	9	9.3E-04	5540.87	7	5	4.1E-04	6101.16	3	3	3.3E-03
4836.70	3	5	1.02E-03	5552.77	3	3	7.9E-04	6104.58	3	1	3.4E-03
4876.26	3	5	7.8E-03	5558.70	3	5	1.42E-02	6105.64	3	5	1.21E-02
4886.29	7	9	1.2E-03	5559.66	3	5	2.2E-03	6113.46	3	5	4.7E-04
4887.95	3	3	1.3E-02	5572.54	5	7	6.6E-03	6119.66	3	3	5.1E-04
4894.69	3	1	1.8E-02	5574.22	3	5	4.6E-03	6121.86	3	5	1.3E-04
4921.04	5	7	5.9E-04	5581.87	7	5	5.6E-04	6127.42	5	3	1.1E-03
4937.72	7	5	3.6E-04	5588.72	5	5	1.5E-03	6128.73	3	5	8.6E-04
4956.75	7	9	1.8E-03	5597.48	5	7	4.2E-03	6145.44	5	7	7.6E-03
4989.95	5	7	1.1E-03	5606.73	3	3	2.20E-02	6155.24	5	3	5.1E-03
5032.03	7	5	8.2E-04	5618.01	3	3	2.1E-03	6165.12	5	5	9.89E-04
5048.81	3	5	4.6E-03	5620.92	3	1	3.6E-03	6170.17	5	5	5.0E-03
5054.18	3	3	4.5E-03	5623.78	5	5	1.4E-03	6173.10	3	5	6.7E-03
5056.53	3	1	5.7E-03	5635.58	3	5	9.6E-04	6179.41	5	3	6.6E-04
5060.08	7	9	3.7E-03	5637.33	1	3	9.1E-04	6212.50	5	7	3.9E-03
5070.99	5	5	2.6E-03	5639.12	1	3	2.1E-03	6215.94	5	5	5.7E-03
5073.08	3	5	5.9E-04	5641.39	3	5	8.7E-04	6230.93	5	5	1.2E-04
5078.03	7	7	4.7E-04	5648.69	5	3	1.2E-03	6243.40	3	1	1.3E-03
5087.09	5	7	1.6E-03	5650.70	3	1	3.20E-02	6244.73	3	5	2.0E-04
5104.74	3	5	8.7E-04	5659.13	5	5	2.6E-03	6248.41	3	5	6.8E-04
5118.21	5	7	2.7E-03	5681.90	5	7	2.0E-03	6278.65	5	7	2.0E-04
5127.80	5	5	3.3E-04	5683.73	5	5	2.0E-03	6296.87	3	5	9.0E-03
5151.39	3	1	2.39E-02	5700.87	5	7	5.9E-03	6307.66	5	5	6.0E-03
5152.30	3	5	1.1E 03	5712.51	1	3	8.7E-04	6309.14	3	3	7.6E-04
5162.29	3	3	1.90E-02	5739.52	3	5	8.7E-03	6364.89	3	1	5.6E-03
5177.54	7	5	2.4E-03	5772.11	5	7	2.0E-03	6369.58	5	3	4.2E-03
5192.72	7	7	1.2E-04	5773.99	5	5	1.1E-03	6384.72	3	3	4.21E-03
5194.02	3	1	7.8E-03	5783.54	3	5	8.1E-04	6416.31	3	5	1.16E-02
5210.49	7	7	1.1E-03	5789.48	5	5	4.6E-04	6431.56	5	3	5.1E-04
5214.77	5	3	2.1E-03	5790.40	5	3	3.4E-04	6466.55	1	3	1.5E-03
5216.28	5	3	1.3E-03	5802.08	5	3	4.2E-03	6481.14	1	3	9.4E-04
5221.27	7	9	8.8E-03	5843.77	3	5	3.3E-04	6513.85	3	3	5.4E-04
5241.09	5	5	1.3E-03	5882.62	3	1	1.23E-02	6538.11	7	7	1.1E-03
5246.24	5	7	1.2E-03	5888.58	7	9	1.29E-02	6596.12	7	5	2.3E-04
5249.20	5	5	7.9E-04	5916.58	5	3	5.9E-04	6598.68	5	5	3.6E-04
5252.79	5	7	5.4E-03	5927.11	7	7	3.7E-04	6604.02	7	5	2.8E-03
5254.47	3	5	3.6E-03	5928.81	5	3	1.12E-02	6604.85	5	7	1.3E-04
5286.07	5	7	9.6E-04	5940.86	1	3	1.2E-03	6632.09	3	3	5.3E-04
5290.00	5	3	9.0E-04	5942.67	5	5	1.8E-03	6656.88	3	3	3.1E-04
5309.52	5	5	1.2E-03	5943.89	7	5	3.6E-04	6660.68	3	1	7.8E-03
5317.73	5	7	2.6E-03	5949.26	3	3	1.5E-03	6664.05	5	5	1.5E-03
5373.50	3	5	2.7E-03	5964.48	1	3	7.7E-04	6677.28	3	1	2.36E-03
5393.27	5	5	9.6E-04	5968.32	3	3	1.8E-03	6684.73	3	5	3.9E-04
5410.48	5	7	2.0E-03	5971.60	3	1	1.1E-02	6698.47	3	3	2.5E-04
5421.35	7	5	6.0E-03	5981.90	5	7	1.2E-03	6698.88	5	3	1.6E-03
5439.99	3	3	1.9E-03	5987.30	7	7	1.2E-03	6719.22	1	3	2.4E-03
5442.24	7	7	9.3E-04	5988.13	3	5	6.1E-04	6722.88	5	7	3.2E-04
5451.65	3	5	4.7E-03	5994.66	3	5	2.6E-04	6752.84	3	5	1.93E-02
5457.42	5	3	3.6E-03	5999.00	5	5	1.4E-03	6754.37	3	3	2.1E-03
5459.65	7	7	3.8E-04	6005.73	5	3	1.4E-03	6756.10	5	5	3.6E-03
5467.16	5	5	7.6E-04	6013.68	7	5	1.4E-03	6766.61	5	3	4.0E-03

λ Å	Weights g_i	g_k	A 10^8 s^{-1}	λ Å	Weights g_i	g_k	A 10^8 s^{-1}	λ Å	Weights g_i	g_k	A 10^8 s^{-1}
6779.93	1	3	1.21E-03	8014.79	5	5	9.28E-02	13622.4	3	5	7.3E-02
6818.29	3	1	2.0E-03	8037.23	1	3	3.59E-03	13678.5	3	5	6.2E-02
6827.25	5	3	2.4E-03	8046.13	3	1	1.12E-02	14093.6	1	3	4.3E-02
6851.88	3	5	6.7E-04	8053.31	5	3	8.6E-03	14739.1	5	7	8.8E-04
6871.29	3	3	2.78E-02	8066.60	5	5	1.4E-03	15046.4	1	3	5.2E-02
6879.59	3	5	1.8E-03	8103.69	3	3	2.5E-01	15172.3	1	3	1.3E-02
6887.10	5	7	1.3E-03	8115.31	5	7	3.31E-01	15329.6	5	5	1.2E-03
6888.17	3	5	2.5E-03	8264.52	3	3	1.53E-01	15555.5	5	7	9.8E-05
6925.01	3	3	1.2E-03	8384.73	5	7	2.4E-03	15734.9	5	3	2.9E-04
6937.67	3	1	3.08E-02	8408.21	3	5	2.23E-01	15816.8	5	3	8.7E-04
6951.46	5	5	2.2E-03	8424.65	3	5	2.15E-01	15989.3	1	3	1.9E-02
6960.23	5	5	2.4E-03	8490.30	3	5	9.6E-04	16122.7	5	3	3.9E-04
6965.43	5	3	6.39E-02	8521.44	3	3	1.39E-01	16180.0	5	5	1.2E-03
6992.17	3	1	7.5E-03	8605.78	5	5	1.04E-02	16264.1	3	3	3.0E-04
7030.25	7	5	2.67E-02	8620.46	1	3	9.2E-03	16520.1	3	5	2.6E-03
7067.22	5	5	3.80E-02	8667.94	1	3	2.43E-02	16739.8	3	5	3.1E-03
7068.73	5	3	2.0E-02	8761.69	3	5	9.5E-03	16940.4	5	5	2.5E-03
7086.70	1	3	1.5E-03	8784.61	3	1	2.4E-03	20317.0	1	3	1.6E-03
7107.48	5	5	4.5E-03	8799.08	5	3	4.6E-03	20616.5	5	5	3.9E-03
7125.83	3	3	6.0E-03	8962.19	3	3	1.6E-03	20812.0	5	7	7.6E-04
7147.04	5	3	6.25E-03	9075.42	3	1	1.2E-02	21332.2	3	3	3.2E-04
7158.83	3	1	2.1E-02	9122.97	5	3	1.89E-01	21534.9	3	5	1.1E-03
7162.57	1	3	5.8E-04	9194.64	3	3	1.76E-02	22039.2	3	1	1.2E-03
7206.98	5	3	2.48E-02	9224.50	3	5	5.03E-02	22077.4	5	3	1.4E-03
7229.93	5	5	6.6E-04	9291.53	3	1	3.26E-02	23133.4	3	3	1.7E-03
7265.17	3	3	1.7E-02	9354.22	3	3	1.06E-02	23844.8	9	7	1.1E-02
7270.66	7	7	1.1E-02	9657.78	3	3	5.43E-02	23967.5	3	1	3.6E-03
7272.93	3	3	1.83E-02	9784.50	3	5	1.47E-02				
7285.44	5	3	1.2E-03	10470.05	1	3	9.8E-03	**Ar II**			
7311.72	3	3	1.7E-02	10478.0	3	3	2.44E-02	2317.7	6	4	1.4E-01
7316.01	3	3	9.6E-03	10950.7	5	3	3.96E-03	2891.6	4	2	1.82E-01
7350.78	3	1	1.2E-02	11078.9	5	5	8.3E-03	2942.9	4	4	5.3E-01
7353.32	5	7	9.6E-03	11393.7	3	1	2.22E-02	2979.1	2	2	4.16E-01
7372.12	7	9	1.9E-02	11441.8	5	3	1.39E-02	3033.5	2	4	9.9E-02
7383.98	3	5	8.47E-02	11467.5	3	5	3.69E-03	3139.0	6	6	5.2E-01
7392.97	5	3	7.2E-03	11488.11	3	3	1.9E-03	3169.7	4	6	4.9E-01
7412.33	3	5	3.9E-03	11668.7	5	5	3.76E-02	3181.0	6	4	3.7E-01
7422.26	3	5	6.6E-04	11719.5	5	3	9.52E-03	3212.5	4	4	5.2E-02
7425.29	5	7	3.1E-03	12026.6	1	3	4.2E-03	3221.6	6	6	1.8E-02
7435.33	5	5	9.0E-03	12112.2	7	7	3.1E-02	3226.0	4	4	2.1E-02
7436.25	7	5	2.7E-03	12139.8	3	3	4.5E-02	3243.7	4	2	1.06E+00
7471.17	3	3	2.2E-04	12343.7	5	7	2.0E-02	3249.8	2	4	6.3E-01
7484.24	3	5	3.4E-03	12402.9	3	3	1.1E-01	3263.6	2	4	1.55E-01
7503.84	3	1	4.45E-01	12439.2	3	5	4.9E-02	3281.7	2	2	4.2E-01
7510.42	5	5	4.5E-03	12456.1	5	3	8.9E-02	3430.4	6	8	6.2E-02
7514.65	3	1	4.02E-01	12487.6	7	5	1.1E-01	3454.1	6	4	3.14E-01
7618.33	3	5	2.9E-03	12554.4	7	5	1.2E-03	3466.3	8	6	3.0E-02
7628.86	3	5	2.9E-03	12702.4	3	3	7.1E-02	3476.7	6	6	1.25E+00
7635.11	5	5	2.45E-01	12733.6	5	5	1.1E-01	3491.2	4	4	1.79E+00
7670.04	5	3	2.8E-03	12746.3	3	3	2.0E-02	3491.5	6	8	2.31E+00
7704.81	5	7	6.3E-04	12802.7	5	5	5.7E-02	3509.8	2	2	2.55E+00
7723.76	5	3	5.18E-02	12933.3	3	1	1.0E-01	3514.4	4	6	1.36E+00
7724.21	1	3	1.17E-01	12956.6	3	3	7.4E-02	3520.0	6	6	5.2E-01
7798.55	3	5	8.7E-04	13008.5	5	3	8.9E-02	3521.3	8	8	2.27E-01
7868.20	1	3	3.50E-03	13214.7	3	1	8.1E-02	3535.3	2	4	5.7E-01
7891.08	5	5	9.5E-03	13273.1	5	7	1.5E-01	3548.5	4	4	8.7E-01
7916.45	3	3	1.2E-03	13313.4	3	5	1.3E-01	3550.0	6	6	2.6E-02
7948.18	1	3	1.86E-01	13504.0	5	7	1.1E-01	3556.9	2	2	5.0E-01
8006.16	3	5	4.90E-02	13599.2	5	5	2.2E-02	3559.5	6	8	2.88E+00

λ Å	Weights		A 10⁸ s⁻¹	λ Å	Weights		A 10⁸ s⁻¹	λ Å	Weights		A 10⁸ s⁻¹
	g_i	g_k			g_i	g_k			g_i	g_k	
3565.0	2	4	5.5E-01	4228.2	4	6	1.31E-01	6483.1	4	2	1.06E-01
3576.6	6	8	2.75E+00	4237.2	4	4	1.12E-01	6638.2	6	4	1.37E-01
3581.6	2	4	1.76E+00	4266.5	6	6	1.64E-01	6639.7	4	2	1.69E-01
3582.4	4	6	2.53E+00	4277.5	6	4	8.0E-01	6643.7	10	8	1.47E-01
3588.4	8	10	3.03E+00	4282.9	4	2	1.32E-01	6666.4	2	2	8.8E-02
3605.9	4	6	4.4E-02	4300.6	6	6	5.7E-02	6684.3	8	6	1.07E-01
3656.0	6	6	7.6E-02	4331.2	4	4	5.74E-01	6756.6	4	4	2.0E-02
3682.5	4	2	1.7E-02	4332.0	4	2	1.92E-01	6863.5	6	6	2.5E-02
3709.9	4	4	4.7E-02	4348.1	6	8	1.17E+00	7233.5	2	4	3.7E-02
3717.2	6	8	5.2E-02	4352.2	2	2	2.12E-01	7380.4	4	4	5.6E-02
3729.3	6	4	4.80E-01	4362.1	4	6	5.5E-02	7589.3	6	4	1.07E-01
3746.9	4	6	2.1E-02	4370.8	4	4	6.6E-01				
3763.5	8	6	1.78E-01	4371.3	6	4	2.21E-01	**Ar III**			
3766.1	4	4	7.4E-02	4376.0	4	2	2.05E-01	769.15	5	3	6.0E+00
3777.5	2	2	1.1E-01	4379.7	2	2	1.00E+00	871.10	5	3	1.59E+00
3780.8	8	8	7.7E-01	4383.8	4	4	1.1E-02	875.53	3	1	3.74E+00
3786.4	8	6	1.5E-02	4400.1	4	4	1.60E-01	878.73	5	5	2.79E+00
3799.4	6	4	1.7E-01	4401.0	8	6	3.04E-01	879.62	3	3	9.2E-01
3808.6	6	6	1.0E-02	4412.9	6	8	6.1E-02	883.18	1	3	1.22E+00
3826.8	6	6	2.81E-01	4420.9	2	4	3.1E-02	887.40	3	5	9.0E-01
3841.5	4	2	2.69E-01	4426.0	4	6	8.17E-01	3024.1	5	7	2.6E+00
3844.7	6	8	4.8E-02	4430.2	2	4	5.69E-01	3027.2	5	5	6.4E-01
3845.4	6	4	1.6E-02	4431.0	6	6	1.09E-01	3054.8	3	5	1.9E+00
3850.6	4	4	3.87E-01	4460.6	4	6	1.5E-02	3064.8	3	3	1.0E+00
3868.5	4	6	1.4E+00	4474.8	4	2	2.90E-01	3078.2	1	3	1.4E+00
3872.1	4	4	1.5E-01	4481.8	6	6	4.55E 01	3285.9	5	7	2.0E+00
3875.3	4	2	8.2E-02	4491.0	6	4	4.6E-02	3301.9	5	5	2.0E+00
3880.3	2	2	2.32E-01	4530.5	6	4	2.1E-02	3311.3	5	3	2.0E+00
3891.4	2	2	4.3E-02	4545.1	4	4	4.71E-01	3336.1	7	9	2.0E+00
3892.0	6	4	6.3E-02	4579.4	2	2	8.0E-01	3344.7	5	7	1.8E+00
3900.6	4	6	7.2E-02	4589.9	4	6	6.64E 01	3352.1	7	7	2.2E-01
3911.6	2	4	7.7E-02	4598.8	4	4	6.7E-02	3358.5	3	5	1.6E+00
3914.8	4	4	3.7E-02	4609.6	6	8	7.89E-01	3361.3	5	5	3.0E-01
3928.6	2	4	2.44E-01	4637.2	6	6	7.1E-02	3472.6	5	7	2.0E-01
3931.2	2	4	2.0E-02	4657.9	4	2	8.92E-01	3480.6	7	7	1.6E+00
3932.5	4	4	9.3E-01	4726.9	4	4	5.88E-01	3499.7	3	3	1.3E+00
3944.3	8	6	4.1E-02	4732.1	6	4	6.7E-02	3500.6	3	5	2.6E-01
3952.7	4	4	2.08E-01	4735.9	6	4	5.80E-01	3502.7	5	3	4.3E-01
3958.4	6	4	3.8E-02	4764.9	2	4	6.4E-01	3503.6	5	5	1.2E+00
3968.4	6	6	4.8E-02	4806.0	6	6	7.80E-01	3511.7	7	5	2.6E-01
3979.4	4	2	9.8E-01	4847.8	4	2	8.49E-01				
3988.2	6	6	4.1E-02	4879.9	4	6	8.23E-01	**Ar IV**			
3992.1	4	6	1.6E-02	4889.0	2	2	1.9E-01	840.03	4	2	2.73E+00
4013.9	8	8	1.05E-01	4904.8	6	8	3.7E-02	843.77	4	4	2.70E+00
4031.4	4	2	7.5E-02	4933.2	4	4	1.44E-01	850.60	4	6	2.63E+00
4035.5	4	6	4.4E-02	4965.1	2	4	3.94E-01				
4038.8	6	8	1.2E-02	4972.2	2	2	9.7E-02	**Ar VI**			
4042.9	4	4	4.06E-01	5009.3	4	6	1.51E-01	292.15	2	2	6.9E+01
4045.7	4	4	1.6E-02	5017.2	4	6	2.07E-01	294.05	4	2	1.36E+02
4052.9	2	4	6.7E-01	5017.6	4	4	1.1E-02				
4065.1	4	4	1.1E-02	5062.0	2	4	2.23E-01	**Ar VII**			
4072.0	6	6	5.8E-01	5141.8	6	8	8.1E-02	*250.41	9	3	2.78E+02
4079.6	6	4	1.19E-01	5145.3	4	6	1.06E-01	*477.54	9	15	9.92E+01
4082.4	6	6	2.9E-02	5176.2	6	6	1.7E-02	585.75	1	3	7.83E+01
4112.8	4	4	1.1E-02	6103.5	2	2	1.7E-02	*637.30	9	9	6.7E+01
4128.6	8	6	1.4E-02	6114.9	10	8	2.00E-01				
4131.7	4	2	8.5E-01	6138.7	6	4	1.2E-02	**Ar VIII**			
4178.4	6	4	1.2E-02	6172.3	8	6	2.00E-01	158.92	2	4	1.1E+02
4202.0	2	4	2.1E-02	6243.1	8	6	3.0E-02	159.18	2	2	1.11E+02

λ Å	Weights g_i	Weights g_k	A 10^8 s^{-1}	λ Å	Weights g_i	Weights g_k	A 10^8 s^{-1}	λ Å	Weights g_i	Weights g_k	A 10^8 s^{-1}
229.44	2	2	1.12E+02	2860.4	2	2	5.5E-01	5777.6	5	7	6.5E-01
230.88	4	2	2.21E+02	2898.7	4	2	9.9E-02	5784.0	3	5	2.1E-01
337.09	4	4	1.2E+01					5800.2	5	5	9.9E-02
337.26	6	4	1.0E+02	**Barium**				5805.7	7	7	1.1E-02
338.22	4	2	1.1E+02	**Ba I**				5826.3	5	3	5.6E-01
519.43	2	4	6.3E+01	2409.2	1	3	8.6E-04	5907.6	3	5	1.5E-02
526.46	4	6	7.2E+01	2414.1	1	3	1.5E-03	5971.7	5	5	1.8E-01
526.87	4	4	1.2E+01	2420.1	1	3	2.3E-03	5997.1	3	3	2.7E-01
700.24	2	4	2.55E+01	2427.4	1	3	5.6E-03	6019.5	3	1	1.4E+00
713.81	2	2	2.4E+01	2432.5	1	3	7.2E-03	6063.1	5	3	5.7E-01
				2438.8	1	3	1.4E-03	6083.4	3	1	1.1E-01
Ar IX				2444.6	1	3	4.5E-03	6110.8	7	5	5.5E-01
48.739	1	3	1.69E+03	2452.4	1	3	8.1E-04	6129.2	3	1	6.0E-02
				2473.2	1	3	4.6E-03	6341.7	5	7	1.9E-01
Ar XIII				2500.2	1	3	1.5E-02	6450.9	3	5	1.1E-01
162.96	5	3	3.4E+02	2543.2	1	3	4.1E-02	6482.9	5	7	4.4E-01
*163.08	9	3	5.3E+02	2596.6	1	3	1.2E-01	6498.8	7	7	8.6E-01
184.90	5	5	1.66E+02	2646.5	1	3	1.1E-01	6527.3	5	5	5.9E-01
186.38	1	3	8.8E+01	2702.6	1	3	2.5E-02	6595.3	3	3	3.9E-01
*207.89	9	9	9.5E+01	2739.2	1	3	9.1E-03	6675.3	5	3	1.9E-01
*245.10	9	15	3.7E+01	2785.3	1	3	2.8E-02	6693.8	7	5	2.8E-01
				3071.6	1	3	4.1E-01	6865.7	5	5	2.3E-02
Ar XIV				3501.1	1	3	1.9E-01	7059.9	7	9	7.1E-01
180.29	2	4	4.5E+01	3889.3	1	3	8.8E-03	7120.3	3	5	2.1E-01
183.41	2	2	1.69E+02	3909.9	3	5	4.9E-01	7195.2	1	3	2.4E-01
187.95	4	4	1.97E+02	3935.7	5	7	4.7E-01	7280.3	5	7	5.3E-01
191.35	4	2	7.5E+01	3937.9	5	5	1.1E-01	7392.4	3	3	5.0E-01
194.39	2	2	4.6E+01	3993.4	7	9	5.5E-01	7417.5	7	5	2.5E-02
203.35	4	2	7.8E+01	3995.7	7	7	8.8E-02	7488.1	7	7	1.0E-01
				4132.4	1	3	7.1E-03	7528.2	5	5	2.7E-02
Ar XV				4239.6	5	3	2.4E-01	7672.1	3	5	3.1E-01
25.05	1	3	1.7E+04	4242.6	3	5	5.6E-02	7780.5	5	5	1.3E-01
221.10	1	3	9.55E+01	4264.4	1	3	1.5E-01	7905.8	5	3	6.3E-01
*265.3	9	9	8.1E+01	4283.1	5	7	6.4E-01	7911.3	1	3	2.98E-03
				4323.0	3	5	1.5E-01	8147.7	5	5	6.3E-02
Ar XVI				4325.2	5	7	7.1E-02	9645.6	7	5	1.1E-01
*23.52	2	6	1.43E+04	4332.9	3	3	1.5E-01	9704.3	3	1	1.6E-01
*24.96	6	10	4.4E+04	4350.3	3	5	6.0E-01	9821.5	3	1	5.5E-02
353.88	2	4	1.5E+01	4402.5	3	5	2.7E-01	10370.3	3	5	1.3E-02
389.11	2	2	1.1E+01	4406.8	5	5	1.0E-01	10649.1	5	5	2.7E-02
1268	2	4	1.9E+00	4431.9	1	3	1.2E+00	11075.7	3	3	3.6E-05
1401	2	2	1.4E+00	4467.1	5	7	6.6E-02	11303.1	5	3	1.2E-03
2975	2	4	9.0E-02	4489.0	5	7	4.2E-01	11373.8	3	1	1.3E-01
3514	4	6	6.5E-02	4493.6	5	5	3.6E-01	14158.4	9	7	2.0E-03
				4505.9	3	3	1.1E+00	14723.2	3	5	8.6E-03
Arsenic				4523.2	5	5	9.6E-01	14999.9	5	3	2.8E-03
As I				4573.9	3	1	1.21E+00	17123.7	7	7	3.3E-03
1890.4	4	6	2.0E+00	4579.6	5	5	7.0E-01	17187.1	3	1	2.7E-02
1937.6	4	4	2.0E+00	4591.8	5	5	1.6E-02	20563.9	5	7	2.6E-03
1972.6	4	2	2.0E+00	4599.7	3	1	4.07E-01				
2288.1	6	4	2.8E+00	4605.0	3	1	7.7E-02	**Ba II**			
2344.0	2	4	3.5E-01	4619.9	1	3	9.3E-02	1413.4	6	8	1.7E-02
2349.8	4	2	3.1E+00	4628.3	5	3	6.0E-02	1417.1	4	6	3.8E-02
2369.7	4	4	6.0E-01	4673.6	7	5	6.5E-02	1444.9	4	6	8.1E-02
2370.8	4	6	4.2E-01	4691.6	5	3	1.6E+00	1461.5	6	8	8.7E-02
2456.5	6	4	7.2E-02	4700.4	3	3	2.4E-01	1487.0	4	6	1.4E-01
2492.9	4	2	1.2E-01	4726.4	5	3	4.6E-01	1503.9	6	8	1.5E-01
2745.0	2	4	2.6E-01	5519.1	3	5	5.0E-01	1554.4	4	6	2.6E-01
2780.2	4	4	7.8E-01	5535.5	1	3	1.19E+00	1572.7	6	8	2.4E-01

λ Å	g_i	g_k	A 10^8 s⁻¹	λ Å	g_i	g_k	A 10^8 s⁻¹	λ Å	g_i	g_k	A 10^8 s⁻¹
1573.9	6	6	1.6E-02	5428.8	6	4	2.3E-02	2061.7	4	6	9.9E-01
1630.4	2	2	1.7E-02	5480.3	8	6	1.8E-02	2110.3	4	2	9.1E-01
1674.5	4	6	2.2E-01	5784.2	2	4	2.0E-01	2177.3	4	2	2.6E-02
1694.4	6	8	2.1E-01	5853.7	4	4	4.8E-02	2228.3	4	4	8.9E-01
1697.2	6	6	1.7E-02	5981.3	4	6	1.6E-01	2230.6	4	6	2.6E+00
1761.8	4	4	3.9E-03	5999.9	4	4	2.6E-02	2276.6	4	4	2.5E-01
1771.0	4	2	3.4E-02	6135.8	2	2	8.5E-02	2515.7	4	6	4.3E-02
1786.9	6	4	4.4E-02	6141.7	6	4	3.7E-01	2627.9	4	4	4.7E-02
1892.7	2	4	9.0E-02	6363.2	6	4	2.9E-03	2696.8	4	6	6.4E-02
1904.2	4	6	1.1E-02	6372.9	4	4	6.7E-04	2780.5	4	2	3.09E-01
1906.8	2	2	5.1E-02	6378.9	4	2	9.9E-02	2798.7	6	6	3.6E-02
1924.7	6	8	3.1E-02	6457.7	6	4	3.0E-03	2898.0	4	2	1.53E+00
1954.2	4	6	1.3E-01	6496.9	4	2	3.32E-01	2938.3	6	4	1.23E+00
1955.1	4	4	1.8E-02	7556.8	6	4	1.6E-03	2989.0	4	4	5.5E-01
1970.2	4	2	6.7E-02	7678.2	8	6	6.6E-04	2993.3	4	6	1.6E-01
1985.6	2	4	2.5E-01	8710.7	6	8	8.0E-01	3024.6	6	6	8.8E-01
1999.5	2	4	1.0E-01	8737.7	4	6	9.3E-01	3067.7	4	2	2.07E+00
2009.2	2	2	8.6E-02					3076.7	4	4	3.5E-02
2052.7	4	6	2.0E-01	**Beryllium**				3397.2	6	4	1.81E-01
2054.6	4	4	2.9E-02	**Be I**				3402.9	6	6	1.6E-02
2080.0	4	2	1.0E-01	1491.8	1	3	1.3E-02	3510.9	6	4	6.8E-02
2153.9	2	4	5.3E-01	1661.5	1	3	2.0E-01	3596.1	2	4	1.98E-01
2200.9	2	2	2.0E-01	2348.6	1	3	5.55E+00	3888.2	2	2	6.9E-02
2232.8	4	6	2.9E-01	*2494.7	9	15	1.6E+00	4121.5	2	2	1.64E-01
2235.4	4	4	4.4E-02	*2650.6	9	9	4.24E+00	4308.5	2	4	1.6E-02
2286.0	4	2	1.3E-01	4572.7	3	5	7.9E-01	4493.0	2	4	1.5E-02
2528.5	2	4	7.1E-01					4722.5	4	2	1.17E-01
2634.8	4	6	7.6E-01	**Be II**				6134.8	4	4	1.8E-02
2641.4	4	4	1.2E-01	1197.1	2	2	4.7E-01				
2647.3	2	2	2.0E-01	1197.2	4	2	9.4E-01	**Boron**			
2771.4	4	2	4.0E-01	1512.3	2	4	9.2E+00	**B I**			
3816.7	4	6	2.3E-03	1512.4	4	6	1.1E+01	1378.6	2	4	3.50E+00
3842.8	6	8	2.2E-03	1776.1	2	2	1.4E+00	1378.9	2	2	1.40E+01
3891.8	2	4	1.67E+00	1776.3	4	2	2.9E+00	1378.9	4	4	1.75E+01
4024.1	6	4	5.3E-03	*2453.8	2	6	1.42E-01	1379.2	4	2	7.0E+00
4057.5	8	6	1.2E-02	3046.5	2	4	4.8E-01	1465.5	2	4	3.34E+00
4130.7	4	6	1.80E+00	3046.7	4	6	5.9E-01	1465.7	4	4	6.7E+00
4166.0	4	4	3.7E-01	3130.4	2	4	1.14E+00	1465.8	6	4	1.00E+01
4216.0	2	4	5.8E-02	3131.1	2	2	1.15E+00	1825.9	2	4	1.76E+00
4287.8	2	2	2.4E-02	3241.6	2	2	1.41E-01	1826.4	4	6	2.11E+00
4325.7	4	6	5.9E-02	3241.8	4	2	2.8E-01	2088.9	2	4	2.8E-01
4329.6	4	4	8.8E-03	3274.6	2	4	1.9E-01	2089.6	4	6	3.3E-01
4405.2	4	2	3.9E-02	3274.7	2	2	1.9E-01	2496.8	2	2	8.64E-01
4470.7	6	4	1.4E-02	4360.7	2	4	9.2E-01	2497.7	4	2	1.73E+00
4509.6	8	6	1.2E-02	4361.0	4	6	1.1E+00				
4524.9	2	2	7.2E-01	*5255.9	2	6	2.56E-02	**Bromine**			
4554.0	2	4	1.17E+00	5270.3	2	2	3.30E-01	**Br I**			
4708.9	2	4	9.7E-02	5270.8	4	2	6.6E-01	1488.5	4	4	1.2E+00
4843.5	4	6	9.3E-02	6279.4	2	4	1.2E-01	1540.7	4	4	1.4E+00
4847.1	2	2	4.1E-02	6279.7	4	6	1.43E-01	1574.8	2	4	2.0E-01
4850.8	4	4	1.4E-02	6756.7	2	2	5.1E-02	1576.4	4	6	2.1E-02
4900.0	4	2	7.75E-01	6757.1	4	2	1.02E-01	1633.4	2	4	8.1E-02
4934.1	2	2	9.55E-01	7401.2	2	4	3.0E-02	4365.1	2	4	7.5E-03
4997.8	4	2	6.1E-02	7401.4	2	2	3.0E-02	4425.1	4	2	4.2E-03
5185.0	2	4	1.8E-02					4441.7	6	4	7.5E-03
5361.4	4	6	4.8E-02	**Bismuth**				4472.6	4	4	9.3E-03
5391.6	6	8	5.2E-02	**Bi I**				4477.7	6	8	1.3E-02
5413.6	6	6	8.4E-04	1954.5	4	6	1.2E+00	4513.4	6	4	2.8E-03
5421.1	6	6	1.9E-03	2021.2	4	4	6.0E-02				

λ Å	Weights g_i	g_k	A 10^8 s^{-1}	λ Å	Weights g_i	g_k	A 10^8 s^{-1}	λ Å	Weights g_i	g_k	A 10^8 s^{-1}
4525.6	6	6	7.2E-03	3361.9	5	7	2.23E-01	6493.8	3	5	4.4E-01
4575.7	4	4	1.6E-02	3624.1	1	3	2.12E-01	6499.7	5	5	8.1E-02
4614.6	4	6	5.4E-03	3630.8	3	5	2.97E-01				
4979.8	4	4	2.6E-03	3631.0	3	3	1.53E-01	**Ca II**			
5245.1	2	4	3.1E-03	3644.4	5	7	3.55E-01	1341.9	2	4	1.5E-02
5345.4	2	4	7.6E-03	3644.8	5	5	9.4E-02	1342.5	2	2	1.5E-02
7348.5	4	6	1.2E-01	3870.5	3	5	7.2E-02	1649.9	2	4	3.2E-03
7513.0	6	4	1.2E-01	3957.1	3	3	9.8E-02	1652.0	2	2	3.1E-03
7803.0	2	4	5.3E-02	3973.7	5	3	1.75E-01	1673.9	2	4	2.24E-01
7938.7	6	6	1.9E-01	4092.6	3	5	1.1E-01	1680.1	4	6	2.65E-01
8131.5	2	4	3.8E-02	4094.9	5	7	1.2E-01	1680.1	4	4	4.41E-02
8343.7	2	2	2.2E-01	4098.5	7	9	1.3E-01	1807.3	2	4	3.54E-01
8446.6	4	4	1.2E-01	4108.5	5	7	9.0E-01	1814.5	4	6	4.2E-01
8638.7	6	4	9.7E-02	4226.7	1	3	2.18E+00	1814.7	4	4	7.0E-02
				4283.0	3	5	4.34E-01	1843.1	2	2	1.6E-01
Br II				4289.4	1	3	6.0E-01	1850.7	4	2	3.08E-01
4704.9	5	7	1.1E+00	4299.0	3	3	4.66E-01	2103.2	2	4	8.2E-01
4785.5	5	5	9.4E-01	4302.5	5	5	1.36E+00	2112.8	4	6	9.7E-01
4816.7	5	3	1.1E+00	4307.7	3	1	1.99E+00	2113.2	4	4	1.6E-01
				4318.7	5	3	7.4E-01	2197.8	2	2	3.1E-01
Cadmium				4355.1	5	7	1.9E-01	2208.6	4	2	6.2E-01
Cd I				4425.4	1	3	4.98E-01	3158.9	2	4	3.1E+00
2288.0	1	3	5.3E+00	4435.0	3	5	6.7E-01	3179.3	4	6	3.6E+00
2836.9	1	3	2.8E-01	4435.7	3	3	3.42E-01	3181.3	4	4	5.8E-01
2880.8	3	5	4.2E-01	4454.8	5	7	8.7E-01	3706.0	2	2	8.8E-01
2881.2	3	3	2.4E-01	4455.9	5	5	2.0E-01	3736.9	4	2	1.7E+00
2980.6	5	7	5.9E-01	4526.9	5	3	4.1E-01	3933.7	2	4	1.47E+00
2981.4	5	5	1.5E-01	4578.6	3	5	1.76E-01	3968.5	2	2	1.4E+00
3261.1	1	3	4.06E-03	4581.4	5	7	2.09E-01				
3403.7	1	3	7.7E-01	4585.9	7	9	2.29E-01	**Ca III**			
3466.2	3	5	1.2E+00	4685.3	3	5	8.0E-02	357.97	1	3	8.8E+02
3467.7	3	3	6.7E-01	4878.1	5	7	1.88E-01	439.69	1	3	1.9E-01
3610.5	5	7	1.3E+00	5041.6	5	3	3.3E-01	490.55	1	3	1.6E-02
3612.9	5	5	3.5E-01	5188.9	3	5	4.0E-01				
4140.5	3	5	4.7E-02	5261.7	3	3	1.5E-01	**Ca V**			
4662.4	3	5	5.5E-02	5262.2	3	1	6.0E-01	558.60	5	3	2.2E+01
4678.1	1	3	1.3E-01	5264.2	5	5	9.1E-02	637.93	5	3	3.9E+00
4799.9	3	3	4.1E-01	5265.6	5	3	4.4E-01	643.12	3	1	9.1E+00
5085.8	5	3	5.6E-01	5270.3	7	5	5.0E-01	646.57	5	5	6.9E+00
6438.5	3	5	5.9E-01	5582.0	5	7	6.0E-02	647.88	3	3	2.3E+00
				5588.8	7	7	4.9E-01	651.55	1	3	2.9E+00
Cd II				5590.1	3	5	8.3E-02	656.76	3	5	2.1E+00
2144.4	2	4	2.8E+00	5594.5	5	5	3.8E-01				
2265.0	2	2	3.0E+00	5598.5	3	3	4.3E-01	**Ca VII**			
2572.9	2	2	1.7E+00	5601.3	7	5	8.6E-02	550.20	5	5	1.8E+01
2748.5	4	2	2.8E+00	5602.9	5	3	1.4E-01	624.39	1	3	3.3E+00
4415.6	4	6	1.4E-02	5857.5	3	5	6.6E-01	630.54	3	5	4.5E+00
				6102.7	1	3	9.6E-02	630.79	3	3	2.2E+00
Calcium				6122.2	3	3	2.87E-01	639.15	5	7	5.7E+00
Ca I				6161.3	5	5	3.3E-02	640.41	5	5	1.3E+00
2275.5	1	3	3.01E-01	6162.2	5	3	3.54E-01				
2995.0	1	3	3.67E-01	6163.8	3	3	5.6E-02	**Ca VIII**			
2997.3	3	5	2.41E-01	6166.4	3	1	2.2E-01	182.71	2	2	1.6E+02
2999.6	3	3	2.79E-01	6169.1	5	3	1.7E-01	184.16	4	2	3.2E+02
3000.9	3	1	1.58E+00	6169.6	7	5	1.9E-01				
3006.9	5	5	7.5E-01	6439.1	7	9	5.3E-01	**Ca IX**			
3009.2	5	3	4.30E-01	6449.8	3	5	9.0E-02	163.23	5	3	3.76E+02
3344.5	1	3	1.51E-01	6462.6	5	7	4.7E-01	371.89	1	3	8.8E+01
3350.2	3	5	1.78E-01	6471.7	7	7	5.9E-02	373.81	3	5	1.16E+02

λ Å	g_i	g_k	A 10^8 s^{-1}	λ Å	g_i	g_k	A 10^8 s^{-1}	λ Å	g_i	g_k	A 10^8 s^{-1}
378.08	5	7	1.5E+02	1261.00	3	3	4.42E-01	4817.37	3	3	8.76E-04
395.03	3	5	2.2E+02	1261.12	3	5	3.71E-01	4826.80	5	3	6.28E-04
466.24	1	3	1.12E+02	1261.43	5	3	7.06E-01	4932.05	3	1	6.02E-02
498.01	3	5	2.49E+01	1261.55	5	5	1.27E+00	5023.84	7	9	1.81E-03
506.18	5	5	7.2E+01	1274.11	5	7	1.03E-02	5039.06	7	9	4.73E-03
515.57	5	3	3.75E+01	1277.25	1	3	1.27E+00	5041.48	3	5	5.25E-03
				1277.28	3	5	1.73E+00	5041.79	5	7	3.28E-03
Ca X				1277.51	3	3	9.12E-01	5052.17	3	5	2.60E-02
110.96	2	4	2.9E+02	1277.55	5	7	2.31E+00	5380.34	3	3	1.86E-02
111.20	2	2	2.92E+02	1277.72	5	5	6.35E-01	5545.05	3	3	3.04E-03
151.84	2	2	2.3E+02	1277.95	5	3	5.56E-02	5668.94	3	3	2.35E-02
153.02	4	2	4.5E+02	1279.23	5	7	1.10E-01	5793.12	7	5	3.44E-03
206.57	4	4	2.9E+01	1279.89	3	5	3.08E-01	5794.47	5	5	6.44E-04
206.75	6	4	2.6E+02	1280.14	1	3	3.11E-01	5800.23	3	3	1.04E-03
207.39	4	2	2.8E+02	1280.33	5	5	5.77E-01	5800.60	5	3	3.04E-03
411.70	2	4	8.3E+01	1280.40	3	3	1.73E-01	5805.20	3	1	4.12E-03
419.75	4	6	9.5E+01	1280.60	3	1	8.22E-01	6001.12	5	5	3.22E-03
420.47	4	4	1.6E+01	1280.85	5	3	3.33E-01	6006.02	7	5	1.79E-02
557.76	2	4	3.50E+01	1328.83	1	3	7.95E-01	6007.18	3	3	5.34E-03
574.01	2	2	3.2E+01	1329.09	3	1	2.41E+00	6010.68	3	1	2.13E-02
				1329.58	5	5	1.79E+00	6013.17	7	5	1.79E-02
Ca XI				1329.60	5	3	1.00E+00	6013.21	7	9	4.35E-03
30.448	1	3	6.2E+03	1355.84	5	7	1.04E+00	6014.83	5	3	1.60E-02
30.867	1	3	4.9E+04	1364.16	5	5	1.57E-01	6016.45	5	7	3.86E-03
35.212	1	3	2.0E+03	1431.60	5	7	2.11E+00	6587.61	3	3	5.09E-02
				1432.10	5	5	2.01E+00	6655.52	3	3	5.03E-03
Ca XII				1432.53	5	3	2.11E+00	6828.12	3	5	9.89E-03
140.05	4	2	3.7E+02	1459.03	5	3	4.76E-01	7111.47	3	5	2.17E-02
147.27	2	2	1.6E+02	1463.34	5	7	1.88E+00	7113.18	7	9	2.47E-02
				1467.40	5	3	5.49E-01	7115.17	5	7	2.19E-02
Ca XV				1468.41	5	3	3.90E-02	7115.18	3	1	4.43E-02
141.69	5	3	4.08E+02	1470.09	5	7	1.37E-02	7116.99	7	5	3.26E-02
*142.23	9	3	6.3E+02	1472.23	5	3	8.01E-03	7119.66	5	3	3.12E-02
161.00	5	5	1.9E+02	1481.76	5	5	3.92E-01	7860.88	5	5	1.53E-02
				1560.31	1	3	6.57E-01	8058.62	5	5	1.09E-02
Ca XVII				1561.34	5	5	2.94E-01	8335.15	3	1	3.51E-01
19.558	1	3	3.8E+04	1561.44	5	7	1.18E+00	9061.44	3	5	7.31E-02
21.198	3	5	4.9E+04	1656.27	3	5	8.58E-01	9062.49	1	3	9.48E-02
192.82	1	3	1.21E+02	1656.93	1	3	1.13E+00	9078.29	3	3	7.07E-02
218.82	3	5	2.76E+01	1657.01	5	5	2.52E+00	9088.51	3	1	3.00E-01
223.02	1	3	3.44E+01	1657.38	3	3	8.64E-01	9094.83	5	5	2.28E-01
228.72	3	3	2.37E+01	1657.91	3	1	3.43E+00	9111.81	5	3	1.35E-01
232.83	5	5	6.5E+01	1658.12	5	3	1.44E+00	9405.73	3	5	2.91E-01
244.06	5	3	3.28E+01	1751.83	1	3	9.07E-01	9603.03	1	3	3.06E-02
				1763.91	1	3	3.59E-02	9620.78	3	3	8.62E-02
Ca XVIII				1765.37	1	3	1.04E-02	9658.43	5	3	1.25E-01
*18.71	2	6	2.31E+04	1930.90	5	3	3.51E+00				
*19.74	6	10	7.0E+04	2478.56	1	3	3.40E-01	**C II**			
302.19	2	4	2.0E+01	2902.23	1	3	4.32E-03	687.345	4	6	2.84E+01
344.76	2	2	1.3E+01	2903.27	3	3	1.29E-02	858.092	2	2	1.18E+00
				2905.00	5	3	2.15E-02	858.559	4	2	2.35E+00
Carbon				4371.37	3	3	1.27E-02	903.623	2	4	6.85E+00
C I				4762.31	1	3	3.37E-03	903.962	2	2	2.74E+01
945.191	1	3	3.79E+00	4762.53	3	5	2.72E-03	904.142	4	4	3.42E+01
945.338	3	3	1.14E+01	4766.67	3	3	2.36E-03	904.480	4	2	1.37E+01
945.579	5	3	1.89E+01	4770.03	3	1	1.07E-02	1009.86	2	4	5.71E+00
1193.24	5	7	1.22E+00	4771.74	5	5	7.97E-03	1010.08	4	4	1.14E+01
1260.74	1	3	5.32E-01	4775.90	5	3	4.84E-03	1010.37	6	4	1.71E+01
1260.93	3	1	1.70E+00	4812.92	1	3	4.03E-04	1036.34	2	2	7.61E+00

λ Å	Weights g_i	g_k	A 10^8 s⁻¹	λ Å	Weights g_i	g_k	A 10^8 s⁻¹	λ Å	Weights g_i	g_k	A 10^8 s⁻¹
1037.02	4	2	1.52E+01	7046.25	4	2	3.20E-01	5695.92	3	5	4.27E-01
1323.91	4	4	4.33E+00	7053.09	4	4	3.19E-01	5858.34	3	1	1.34E-01
1323.95	6	6	4.49E+00	7063.68	4	6	3.17E-01	5863.25	3	3	3.35E-02
1334.53	2	4	2.37E+00	7112.48	2	4	2.94E-01	5871.68	5	3	1.00E-01
1335.71	4	6	2.84E+00	7113.04	4	6	3.15E-01	5880.56	5	5	1.99E-02
2091.14	2	4	1.00E-01	7115.63	6	8	3.60E-01	5894.07	7	5	1.11E-01
2091.19	4	6	1.69E-01	7119.76	4	4	1.17E-01	6727.48	1	3	1.12E-01
2091.65	6	8	2.41E-01	7119.91	8	10	4.19E-01	6731.04	3	5	1.50E-01
2093.16	6	6	7.20E-02	7125.72	6	6	1.02E-01	6742.15	3	3	8.32E-02
2173.85	2	4	2.31E-01	7132.47	6	4	8.33E-03	6744.39	5	7	1.99E-01
2174.17	2	2	2.31E-01	7134.10	8	8	5.93E-02	6762.17	5	5	4.95E-02
2509.13	2	4	4.53E-01	7231.33	2	4	3.52E-01	6773.39	5	3	5.47E-03
2511.74	4	4	9.04E-02	7236.42	4	6	4.22E-01	6851.18	3	5	7.60E-03
2512.06	4	6	5.42E-01	7237.17	4	4	7.03E-02	6853.68	5	7	5.64E-03
2727.31	2	4	6.63E-02	8028.85	2	2	1.71E-02	6857.24	3	3	3.79E-02
2728.72	4	4	3.31E-01	8037.73	2	4	4.26E-02	6862.69	5	5	3.51E-02
2729.21	2	2	2.65E-01	8039.40	4	2	8.51E-02	6868.78	5	3	1.26E-02
2730.63	4	2	1.32E-01	8048.31	4	6	1.36E-02	6872.04	7	7	4.46E-02
5132.95	2	4	3.89E-01	8062.10	4	6	3.04E-02	6881.10	7	5	7.80E-03
5133.28	4	6	2.80E-01	8062.80	6	4	4.56E-02	7353.88	5	3	3.09E-02
5137.26	2	2	1.55E-01	8076.64	6	6	7.05E-02	7707.43	3	5	1.30E-01
5139.17	4	4	1.24E-01	9238.30	4	6	3.34E-02	7771.76	3	1	1.77E-01
5143.49	4	2	7.73E-01	9251.01	2	4	2.77E-02	7780.41	3	3	1.76E-01
5145.16	6	6	6.49E-01	9863.06	2	4	5.56E-02	7796.00	3	5	1.75E-01
5151.08	6	4	4.16E-01	9870.78	4	6	9.31E-02	8500.32	1	3	1.01E-01
5640.55	2	4	9.89E-02	9882.68	6	8	1.33E-01	9593.32	3	3	5.32E-03
5648.07	4	4	1.97E-01					9651.47	5	5	1.57E-02
5662.46	6	4	2.93E-01	**C III**				9696.48	5	7	7.53E-03
5818.31	2	2	3.38E-02	310.170	1	3	6.56E+00	9696.54	3	5	7.12E-03
5822.98	2	4	3.38E-03	386.203	1	3	3.46E+01	9699.57	7	9	8.47E-03
5823.18	4	2	3.38E-02	459.466	1	3	5.91E+01	9701.10	1	3	4.40E-02
5827.85	4	4	2.16E-02	459.514	3	5	7.97E+01	9705.41	3	5	5.93E-02
5836.37	6	4	4.22E-02	459.627	5	7	1.06E+02	9706.44	3	3	3.29E-02
5843.62	6	6	1.20E-02	574.281	3	5	6.24E+01	9715.09	5	7	7.88E-02
5856.06	8	6	5.31E-02	977.020	1	3	1.767E+01	9717.75	5	5	1.97E-02
6095.29	2	4	4.20E-01	1174.93	3	5	3.293E+00	9718.79	5	3	2.19E-03
6098.51	4	6	5.03E-01	1175.26	1	3	4.385E+00				
6102.56	4	4	8.37E-02	1175.59	3	3	3.287E+00	**C IV**			
6578.05	2	4	3.63E-01	1175.71	5	5	9.856E+00	*312.43	2	6	4.63E+01
6582.88	2	2	3.62E-01	1175.99	3	1	1.313E+01	*384.13	6	10	1.76E+02
6724.56	2	4	3.17E-02	1176.37	5	3	5.468E+00	1548.19	2	4	2.65E+00
6727.07	2	2	6.34E-02	1247.38	3	1	2.082E+00	1550.77	2	2	2.64E+00
6727.26	4	6	2.96E-02	2296.87	3	5	1.376E+00	5801.31	2	4	3.17E-01
6731.07	4	4	5.06E-02	2849.05	3	1	1.95E-01	5811.97	2	2	3.16E-01
6733.58	4	2	6.32E-02	3703.70	3	3	5.90E-01				
6734.00	6	8	1.80E-02	4325.56	3	5	1.24E-01	**C V**			
6738.61	6	6	7.23E-02	4647.42	3	5	7.26E-01	34.9728	1	3	2.554E+03
6742.43	6	4	4.41E-02	4650.25	3	3	7.25E-01	40.2678	1	3	8.873E+03
6750.54	8	8	1.08E-01	4651.02	3	5	2.28E-01	*227.19	3	9	1.363E+02
6755.16	8	6	2.38E-02	4651.47	3	1	7.24E-01	247.315	1	3	1.278E+02
6779.94	4	6	2.56E-01	4652.05	1	3	3.04E-01	*248.71	9	15	4.247E+02
6780.59	2	4	1.52E-01	4659.06	3	3	2.27E-01	*260.19	9	3	6.680E+01
6783.91	6	8	3.65E-01	4663.64	3	1	9.05E-01	267.267	3	5	3.947E+02
6787.21	2	2	3.04E-01	4665.86	5	5	6.78E-01	*2273.9	3	9	5.646E-01
6791.47	4	4	1.94E-01	4673.95	5	3	3.75E-01	3526.66	1	3	1.663E-01
6798.10	4	2	6.04E-02	5244.66	1	3	5.30E-02	8420.72	3	5	6.898E-02
6800.69	6	6	1.09E-01	5253.58	3	3	1.58E-01	*8433.2	3	9	6.868E-02
6812.28	6	4	1.80E-02	5272.52	5	3	2.61E-01	8448.12	3	1	6.832E-02

λ Å	Weights g_i	g_k	A 10^8 s^{-1}	λ Å	Weights g_i	g_k	A 10^8 s^{-1}	λ Å	Weights g_i	g_k	A 10^8 s^{-1}
8449.19	3	3	6.829E-02	7256.6	6	4	1.5E-01	3530.0	6	8	1.8E+00
				7414.1	6	4	4.7E-02	3560.7	4	6	1.7E+00
Cesium				7547.1	4	4	1.2E-01	3602.1	6	8	1.7E+00
Cs I				7717.6	4	4	3.0E-02	3612.9	4	6	1.2E+00
3203.5	2	4	7.6E-06	7745.0	2	4	6.3E-02	3720.5	4	6	1.7E+00
3205.3	2	4	7.9E-06	7769.2	6	6	6.0E-02				
3207.5	2	4	8.5E-06	7821.4	6	8	9.8E-02	**Chromium**			
3210.0	2	4	9.4E-06	7830.8	4	4	9.7E-02	**Cr I**			
3212.8	2	4	1.19E-05	7878.2	6	6	1.8E-02	1999.95	9	9	1.4E+00
3216.2	2	4	1.49E-05	7899.3	4	6	5.1E-02	2383.30	9	11	4.1E-01
3220.1	2	4	1.7E-05	7924.6	2	4	2.1E-02	2389.21	3	5	2.3E-01
3220.2	2	2	1.07E-07	7935.0	6	8	3.9E-02	2408.60	9	7	6.7E-01
3224.8	2	4	2.0E-05	7997.9	4	4	2.1E-02	2408.72	7	5	2.9E-01
3225.0	2	2	1.43E-07					2492.57	3	5	4.5E-01
3230.5	2	4	2.5E-05	**Cl II**				2495.08	3	3	2.7E-01
3230.7	2	2	1.97E-07	3329.1	5	7	1.5E+00	2496.30	5	7	5.6E-01
3237.4	2	4	2.8E-05	3522.1	7	7	1.4E+00	2502.55	7	9	2.2E-01
3237.6	2	2	2.63E-07	3798.8	5	7	1.6E+00	2504.31	7	9	4.5E-01
3245.9	2	4	3.45E-05	3805.2	7	9	1.8E+00	2508.11	5	5	2.1E-01
3246.2	2	2	3.7E-07	3809.5	3	5	1.5E+00	2508.97	5	3	3.8E-01
3256.7	2	4	4.25E-05	3851.0	5	7	1.8E+00	2527.11	9	9	5.3E-01
3257.1	2	2	7.0E-07	3851.4	5	5	1.6E+00	2549.55	3	3	4.8E-01
3270.5	2	4	5.6E-05	3854.7	3	5	2.2E+00	2560.70	5	5	4.3E-01
3271.0	2	2	9.8E-07	3861.9	5	7	2.4E+00	2571.74	7	5	6.4E-01
3288.6	2	4	1.0E-04	3868.6	7	9	2.7E+00	2577.66	7	7	2.6E-01
3289.3	2	2	2.7E-06	3913.9	9	9	8.2E 01	2591.84	9	7	6.5E-01
3313.1	2	4	1.6E-04	3990.2	5	7	8.4E-01	2620.48	5	3	1.9E-01
3314.0	2	2	5.2E-06	4132.5	5	5	1.6E+00	2673.64	3	3	1.8E-01
3347.5	2	4	2.2E-04	4276.5	9	7	7.6E-01	2701.99	9	11	2.1E-01
3348.8	2	2	1.1E-05	4768.7	3	5	7.7E-01	2726.50	5	7	7.5E-01
3397.9	2	4	4.0E-04	4781.3	5	7	1.0E+00	2731.90	5	5	7.8E-01
3400.0	2	2	2.4E-05	4794.6	5	7	1.04E+00	2736.46	5	3	7.5E-01
3476.8	2	4	6.6E-04	4810.1	5	5	9.9E-01	2752.85	3	3	8.7E-01
3480.0	2	2	6.6E-05	4819.5	5	3	1.00E+00	2757.09	5	5	6.8E-01
3611.4	2	4	1.5E-03	4904.8	5	7	8.1E-01	2761.74	5	3	6.8E-01
3617.3	2	2	2.5E-04	4917.7	3	5	7.5E-01	2764.36	7	7	3.7E-01
3876.1	2	4	3.8E-03	5078.3	7	7	7.7E-01	2769.90	7	5	1.1E+00
3888.6	2	2	9.7E-04	5219.1	3	9	8.6E-01	2780.70	9	7	1.4E+00
4555.3	2	4	1.88E-02	5392.1	5	7	1.0E+00	2879.27	5	7	2.1E-01
4593.2	2	2	8.0E-03					2887.00	3	5	2.7E-01
8521.1	2	4	3.276E-01	**Cl III**				2889.22	9	9	6.6E-01
8943.5	2	2	2.87E-01	2298.5	4	4	4.2E+00	2893.25	7	7	5.2E-01
				2340.6	6	6	4.2E+00	2894.17	1	3	3.3E-01
Chlorine				2370.4	8	6	2.8E+00	2896.76	5	5	3.0E-01
Cl I				2531.8	2	4	4.4E+00	2905.48	3	1	1.3E+00
1188.8	4	6	2.33E+00	2532.5	4	6	5.3E+00	2909.05	5	3	6.8E-01
1188.8	4	4	2.71E-01	2577.1	4	6	4.3E+00	2910.89	7	5	3.4E-01
1201.4	2	4	2.39E+00	2580.7	6	8	4.7E+00	2911.15	9	7	2.6E-01
1335.7	4	2	1.74E+00	2601.2	2	4	4.6E+00	2967.64	7	9	3.9E-01
1347.2	4	4	4.19E+00	2603.6	4	6	5.0E+00	2971.10	5	7	7.1E-01
1351.7	2	2	3.23E+00	2609.5	6	8	5.7E+00	2975.48	3	5	8.9E-01
1363.4	2	4	7.5E-01	2617.0	8	10	6.6E+00	2980.78	1	3	5.10E-01
4323.3	4	4	1.1E-02	2661.6	4	6	3.4E+00	2988.64	5	7	5.2E-01
4363.3	4	6	6.8E-03	2665.5	6	8	4.8E+00	2991.88	3	1	3.0E+00
4379.9	4	4	1.4E-02	2691.5	4	4	3.5E+00	2994.06	5	5	2.5E-01
4389.8	6	8	1.4E-02	2710.4	4	6	3.5E+00	2995.09	5	5	4.3E-01
4526.2	4	4	5.1E-02	3340.4	6	6	1.5E+00	2996.57	5	3	2.0E+00
4601.0	2	2	4.2E-02	3392.9	4	4	1.9E+00	2998.78	5	3	4.07E-01
4661.2	2	4	1.2E-02	3393.5	6	6	1.9E+00	3000.88	7	5	1.6E+00

λ Å	Weights g_i	g_k	A 10^8 s^{-1}	λ Å	Weights g_i	g_k	A 10^8 s^{-1}	λ Å	Weights g_i	g_k	A 10^8 s^{-1}
3005.06	9	7	9.2E-01	4432.77	15	15	4.9E-01	2857.40	6	8	2.8E-01
3013.72	3	5	8.3E-01	4443.72	3	1	4.5E-01	2860.92	2	4	6.9E-01
3015.20	1	3	1.63E+00	4482.88	3	3	3.0E-01	2862.57	8	8	6.3E-01
3020.67	3	3	1.5E+00	4490.55	9	7	3.9E-01	2866.72	4	4	1.2E+00
3021.58	9	11	2.91E+00	4492.31	5	3	4.47E-01	2867.09	4	4	1.1E+00
3024.36	5	5	1.27E+00	4495.28	9	7	2.0E-01	2867.65	2	2	1.1E+00
3029.17	5	3	3.8E-01	4500.29	7	7	2.1E-01	2870.43	6	6	1.3E+00
3030.25	7	7	1.1E+00	4506.84	13	11	2.7E-01	2873.81	4	2	8.8E-01
3031.35	5	3	3.1E-01	4540.72	11	11	3.14E-01	2880.86	6	4	7.9E-01
3034.19	7	7	3.5E-01	4564.17	11	13	5.1E-01	2898.53	10	12	1.2E+00
3037.05	9	9	5.4E-01	4595.60	13	13	4.7E-01	2921.81	8	10	9.0E-01
3040.84	7	5	7.4E-01	4622.47	7	7	4.1E-01	2930.83	2	4	1.1E+00
3053.87	9	7	7.97E-01	4663.33	3	3	2.0E-01	2935.12	6	8	1.8E+00
3148.44	9	11	5.6E-01	4665.90	3	3	3.0E-01	2953.34	2	2	1.8E+00
3155.16	11	13	5.7E-01	4689.38	7	5	2.3E-01	2966.03	10	8	5.4E-01
3163.76	13	15	6.0E-01	4698.46	9	7	2.2E-01	2971.90	14	14	2.0E+00
3237.73	9	9	1.3E+00	4708.02	11	9	4.31E-01	2979.73	12	12	1.8E+00
3238.09	11	11	2.0E+00	4718.43	13	11	3.4E-01	2985.32	10	10	2.2E+00
3578.68	7	9	1.48E+00	4730.69	7	5	3.83E-01	2989.18	8	8	2.2E+00
3593.48	7	7	1.50E+00	4737.33	9	7	3.38E-01	3118.64	2	4	1.7E+00
3605.32	7	5	1.62E+00	4741.09	3	5	2.2E-01	3120.36	4	6	1.5E+00
3639.80	13	11	1.8E+00	4752.07	13	13	6.2E-01	3122.59	12	12	4.4E-01
3743.89	13	13	7.61E-01	4756.09	11	9	4.0E-01	3128.69	4	4	8.1E-01
3757.66	7	7	4.13E-01	4792.49	7	5	2.6E-01	3136.68	6	6	6.4E-01
3768.24	5	5	5.10E-01	4801.02	9	7	3.06E-01	4588.22	8	6	1.2E-01
3804.80	9	9	6.9E-01	4816.13	9	9	1.8E-01				
3963.69	13	15	1.3E+00	4870.79	7	9	3.5E-01	**Cr V**			
3969.75	11	13	1.2E+00	4887.01	9	11	3.2E-01	434.306	9	9	1.5E+00
3983.90	7	9	1.05E+00	4922.28	11	13	4.0E-01	436.351	9	7	2.4E+01
3991.12	5	7	1.07E+00	4966.80	3	1	3.0E-01	436.601	7	5	2.1E+01
4001.44	9	11	6.8E-01	5204.51	5	3	5.09E-01	437.420	7	7	1.4E+01
4039.10	15	15	6.7E-01	5206.02	5	5	5.14E-01	437.655	5	5	1.3E+01
4048.78	13	13	6.4E-01	5208.42	5	7	5.06E-01	441.056	5	3	2.3E+01
4058.78	11	11	6.7E-01	5243.38	5	3	2.19E-01	456.357	1	3	9.5E+00
4065.71	9	11	3.5E-01	5297.37	7	9	3.88E-01	456.637	3	1	3.3E+01
4165.52	11	13	7.5E-01	5297.99	7	7	3.0E-01	456.743	3	3	9.1E+00
4204.48	13	11	3.1E-01	5328.36	9	11	6.2E-01	457.028	5	5	2.7E+01
4254.33	7	9	3.15E-01	5329.17	9	9	2.25E-01	457.504	5	3	1.2E+01
4263.15	15	17	6.4E-01	5783.11	3	3	2.1E-01	464.015	9	7	3.6E+01
4274.81	7	7	3.07E-01	5783.89	5	5	2.02E-01	469.634	5	5	2.3E+01
4275.98	11	11	2.2E-01	5787.97	5	7	2.35E-01	1106.25	7	9	1.2E+01
4280.42	13	15	4.7E-01					1121.07	7	9	2.1E+01
4289.73	7	5	3.16E-01	**Cr II**				1127.63	9	11	3.5E+01
4291.97	7	5	2.4E-01	2653.57	4	6	3.5E-01	1465.86	5	3	1.1E+01
4297.75	11	13	4.9E-01	2658.59	2	4	5.8E-01	1481.65	3	1	1.0E+01
4298.05	9	9	2.6E-01	2666.02	6	8	5.9E-01	1519.03	5	7	9.5E+00
4300.52	9	7	1.9E-01	2668.71	4	2	1.4E+00	1579.70	7	9	8.6E+00
4301.19	11	9	2.6E-01	2671.80	6	4	1.0E+00				
4302.78	11	11	2.5E-01	2672.83	8	6	5.5E-01	**Cr VI**			
4319.66	5	3	1.8E-01	2744.97	4	6	8.5E-01	161.687	6	6	1.7E+02
4337.25	5	7	2.0E-01	2787.61	6	6	1.5E+00	168.088	4	6	2.0E+02
4373.65	9	9	2.8E-01	2822.38	14	16	2.3E+00	201.007	4	4	2.5E+03
4376.80	13	13	3.2E-01	2835.63	10	12	2.0E+00	201.224	4	6	1.8E+02
4413.86	7	5	2.7E-01	2840.01	10	12	2.7E+00	201.388	6	4	2.7E+02
4422.70	5	5	2.7E-01	2843.24	8	10	6.4E-01	201.606	6	6	2.6E+03
4424.29	9	7	2.1E-01	2849.83	6	8	9.2E-01	202.442	6	4	1.0E+03
4429.93	3	3	2.4E-01	2851.35	8	10	2.2E+00	202.739	4	2	1.2E+03
4432.16	1	3	1.8E-01	2856.77	4	6	4.3E-01	226.241	6	8	7.2E+02

λ Å	Weights g_i	g_k	A $10^8\,\mathrm{s}^{-1}$	λ Å	Weights g_i	g_k	A $10^8\,\mathrm{s}^{-1}$	λ Å	Weights g_i	g_k	A $10^8\,\mathrm{s}^{-1}$
227.202	4	6	6.6E+02	270	3	1	1.7E+02	346.5	6	8	2.5E+02
				276.4	5	7	2.2E+02				
Cr X				277	1	3	2.1E+02	**Cr XV**			
216.72	6	8	9.0E+02	279.32	3	5	3.5E+02	18.497	1	3	1.62E+05
223.86	4	2	7.7E+02	286	3	1	4.6E+02	18.782	1	3	2.8E+04
224.74	4	4	7.6E+02	328.29	1	3	1.86E+02	19.015	1	3	6.3E+02
226.24	4	6	7.3E+02	345	7	9	1.74E+02	20.863	1	3	6.0E+03
227.42	4	4	5.2E+02					21.153	1	3	5.6E+03
227.50	4	6	1.8E+01	**Cr XIV**				102	3	3	1.6E+02
228.63	6	4	8.1E+01	*38.036	2	6	2.47E+02	102.18	5	3	7.0E+02
228.71	6	6	4.5E+02	39.796	2	4	3.05E+02	103	3	1	3.8E+02
231.21	2	4	1.2E+02	40.018	4	6	3.6E+02	105	7	5	5.3E+02
232.96	4	4	4.4E+02	40.782	2	4	3.9E+02	111.27	3	3	1.7E+02
242.20	2	4	5.0E+01	40.800	2	2	3.9E+02				
244.19	4	6	5.8E+01	41.556	2	4	4.5E+02	**Cr XVI**			
395.984	4	4	2.4E+01	41.788	4	6	5.3E+02	17.073	4	6	1.2E+04
398.150	6	6	2.1E+01	44.597	2	4	7.1E+02	17.242	2	4	8.6E+04
				44.869	4	6	8.3E+02	17.299	4	4	2.5E+04
Cr XI				46.125	4	2	3.1E+02	17.372	4	4	1.4E+05
214.31	5	7	1.4E+01	46.468	2	4	6.6E+02	17.438	4	2	1.1E+05
226.45	5	7	6.0E+02	46.527	2	2	6.7E+02	17.514	2	4	1.1E+05
232	3	1	4.1E+02	48.300	4	6	5.9E+02	17.587	2	4	2.0E+04
235.53	5	7	5.5E+02	48.338	6	8	6.3E+02	17.656	2	2	2.0E+04
240.76	1	3	4.8E+02	50.821	2	4	1.2E+03	19.442	4	2	9.9E+03
250.28	5	7	1.0E+01	51.172	4	6	1.4E+03	19.714	2	2	1.1E+04
366.491	3	3	1.2E+01	51.180	4	4	2.3E+02				
366.942	3	1	3.0E+01	52.321	4	6	1.0E+03	**Cr XVII**			
374.927	5	5	2.3E+01	52.363	6	8	1.1E+03	16.31	5	3	9.6E+03
422.083	3	5	1.0E+01	53.760	2	2	3.0E+02	16.32	5	7	3.2E+04
				54.164	4	2	5.9E+02	16.37	3	1	9.7E+04
Cr XII				60.699	4	6	2.05E+03	16.44	5	7	1.3E+05
216	4	6	2.4E+02	60.756	6	8	2.19E+03	16.59	3	1	5.7E+04
218	6	8	2.4E+02	63.324	2	4	1.07E+03	16.65	5	5	1.1E+04
239	2	2	1.6E+02	63.539	2	2	1.13E+03	16.66	1	3	1.8E+05
244.70	2	4	3.0E+02	68.594	2	4	1.98E+03	16.68	5	7	6.8E+04
247	4	2	2.4E+02	69.213	4	6	2.31E+03	16.80	5	7	4.4E+04
247	2	2	3.3E+02	69.247	4	4	3.8E+02	16.97	1	3	2.63E+04
248	6	8	1.4E+02	86.060	4	6	5.3E+03	16.97	3	3	1.5E+04
250	6	8	3.5E+02	86.169	6	8	5.9E+03	17.968	5	3	8.6E+03
250	6	6	2.2E+02	86.185	6	6	3.9E+02	18.336	5	3	1.7E+04
251.52	4	6	3.4E+02	101.05	6	4	4.4E+02	18.336	5	5	1.6E+04
252	4	6	2.0E+02	101.42	4	2	4.83E+02	18.389	1	3	9.2E+03
256	2	2	1.5E+02	104.4	4	6	3.0E+02				
259	2	4	3.2E+02	104.5	6	8	3.1E+02	**Cr XVIII**			
269	2	2	2.1E+02	109.8	2	4	2.3E+02	95.77	4	2	3.08E+02
300.32	2	2	1.4E+02	110.4	4	6	2.8E+02	102.32	4	4	1.54E+02
305.81	4	4	2.76E+02	118.3	4	2	2.1E+02	104.98	6	4	8.7E+02
309	4	2	2.7E+02	125.2	4	6	5.0E+02	106.84	4	2	3.4E+02
309	6	6	1.6E+02	125.3	6	8	5.4E+02	110.41	4	2	7.9E+02
311.55	4	2	1.6E+02	148.5	2	4	2.18E+02	112.27	4	2	4.24E+02
324	4	6	2.2E+02	149.1	2	2	2.1E+02	119.62	2	2	3.2E+02
327	6	8	2.2E+02	157.1	2	4	3.3E+02	123.87	6	4	3.9E+02
332.06	6	4	1.4E+02	*158.4	4	6	3.7E+02	125.51	4	4	3.4E+02
				187.02	4	6	9.3E+02	128.10	6	6	2.8E+02
Cr XIII				187.30	6	8	9.6E+02	136.52	4	2	1.66E+02
49.59	1	3	9.9E+02	189.1	2	2	2.13E+02	139.87	4	4	1.49E+02
67.01	1	3	1.67E+03	191.0	4	2	4.11E+02	140.82	4	2	2.66E+02
228	5	7	1.8E+02	222.9	4	2	2.2E+02	155.46	2	2	2.84E+02
267.73	5	7	1.9E+02	346.3	4	6	2.4E+02	157.40	4	4	2.83E+02

λ Å	Weights g_i	Weights g_k	A $10^8\,s^{-1}$	λ Å	Weights g_i	Weights g_k	A $10^8\,s^{-1}$	λ Å	Weights g_i	Weights g_k	A $10^8\,s^{-1}$
Cr XIX				14.04	3	5	1.2E+05	2467.69	6	8	7.0E-02
14.73	3	3	7.1E+04	14.24	1	3	1.41E+05	2470.27	10	12	1.5E-01
14.80	1	3	1.3E+05					2476.64	10	8	2.2E-01
14.81	5	3	3.4E+04	**Cr XXII**				2504.52	10	8	1.8E-01
14.84	5	7	1.3E+05	2.190	4	2	1.7E+06	2511.02	10	10	9.2E-01
109.64	3	3	2.46E+02	2.191	2	2	2.5E+06	2521.36	10	8	3.0E+00
110.37	5	3	6.0E+02	2.198	4	4	4.5E+06	2528.97	8	6	2.8E+00
113.97	5	3	5.5E+02	2.199	2	4	2.3E+06	2530.13	6	6	7.1E-02
118.31	3	1	3.29E+02	2.202	4	6	1.6E+06	2535.96	6	4	1.9E+00
118.67	5	3	2.1E+02	2.203	4	2	1.3E+06	2536.50	8	8	3.0E-01
118.83	3	3	1.35E+02	13.149	2	4	1.29E+05	2544.25	4	2	3.0E+00
126.30	1	3	1.56E+02	13.292	4	6	1.54E+05	2562.12	4	4	3.9E-01
126.33	5	5	4.35E+02					2567.34	6	6	3.0E-01
130.99	7	5	2.9E+02	**Cr XXIII**				2574.35	8	8	1.7E-01
134.89	3	1	1.98E+02	1.7632	1	3	3.68E+05	2685.34	6	8	5.5E-02
138.15	3	1	1.75E+02	1.8557	1	3	8.97E+05	3017.55	8	6	6.9E-02
138.45	5	5	1.71E+02	2.095	3	1	3.5E+06	3044.00	10	10	1.9E-01
140.92	5	3	1.38E+02	2.101	1	3	2.0E+06	3048.89	6	4	7.5E-02
143.57	3	1	7.2E+02	2.101	5	5	7.9E+05	3061.82	8	8	1.6E-01
163.94	5	5	3.1E+02	2.102	3	5	2.1E+06	3072.34	6	6	1.5E-01
179.18	3	1	1.45E+02	2.103	3	5	1.2E+06	3086.78	4	4	1.9E-01
				2.104	1	3	1.4E+06	3354.37	8	6	1.1E-01
Cr XX				2.105	3	3	9.6E+05	3367.11	10	8	6.0E-02
14.13	2	4	1.1E+05	2.106	3	3	2.0E+06	3385.22	8	6	1.1E-01
14.26	4	6	1.3E+05	2.107	5	5	2.3E+06	3388.16	6	4	2.4E-01
128.42	4	4	3.8E+02	2.107	3	5	3.3E+06	3395.37	6	8	2.9E-01
131.31	6	4	1.27E+02	2.109	5	3	1.7E+06	3405.12	10	10	1.0E+00
133.82	2	4	8.3E+01	2.113	3	5	5.9E+05	3409.17	8	8	4.2E-01
135.26	4	2	2.41E+02	2.119	3	1	2.7E+05	3412.34	8	10	6.1E-01
140.75	4	4	1.35E+02	2.129	3	1	5.1E+05	3412.63	10	8	1.2E-01
148.99	6	4	1.75E+02	2.1818	1	3	3.37E+06	3414.74	4	4	8.8E-02
156.00	2	4	8.4E+01	2.1923	1	3	2.34E+05	3417.15	6	6	3.2E-01
167.97	6	6	1.12E+02					3431.58	8	6	1.1E-01
180.85	4	4	1.6E+02	**Cobalt**				3433.05	4	4	1.0E+00
				Co I				3442.92	6	4	1.2E-01
Cr XXI				2287.80	8	8	8.6E-01	3443.64	8	8	6.9E-01
12.97	3	1	4.8E+04	2295.22	10	8	2.2E-01	3449.17	6	6	7.6E-01
12.98	5	5	3.9E+04	2309.03	10	10	5.6E-01	3449.44	10	10	1.8E-01
13.02	3	5	3.8E+04	2323.13	8	8	5.0E-01	3453.51	10	12	1.1E+00
13.02	5	7	3.9E+04	2325.53	6	8	1.1E-01	3455.24	4	2	1.9E-01
13.08	1	3	5.2E+04	2335.98	6	6	5.1E-01	3462.80	4	6	7.9E-01
13.22	3	1	4.6E+04	2338.66	4	4	7.7E-01	3465.79	10	12	9.2E-02
13.34	3	5	5.2E+04	2353.36	8	10	1.5E-01	3474.02	6	8	5.6E-02
13.49	1	3	9.0E+04	2355.48	6	8	1.3E-01	3483.41	8	10	5.5E-02
13.53	3	3	6.6E+04	2358.18	4	6	1.4E-01	3489.40	8	6	1.3E+00
13.55	3	5	1.2E+05	2365.06	10	10	1.3E-01	3491.32	4	4	5.0E-02
13.65	5	7	1.5E+05	2371.85	6	8	7.3E-02	3495.68	4	6	4.9E-01
13.66	3	1	1.2E+05	2384.86	10	8	2.4E-01	3502.28	10	8	8.0E-01
13.67	5	5	3.9E+04	2392.03	6	6	4.0E-01	3502.63	6	6	5.2E-02
13.68	3	3	8.2E+04	2402.06	8	6	5.1E-01	3506.32	8	6	8.2E-01
13.75	5	3	4.5E+04	2407.25	10	12	3.6E+00	3509.84	6	8	3.2E-01
13.75	5	5	9.5E+04	2412.76	4	6	6.5E-01	3512.64	6	4	1.0E+00
13.76	1	3	1.51E+05	2414.46	6	8	3.4E+00	3513.48	8	10	7.8E-02
13.78	5	7	1.7E+05	2415.29	4	6	3.6E+00	3518.34	6	4	1.6E+00
13.84	5	7	2.59E+05	2424.93	10	10	3.2E+00	3521.58	10	8	1.8E+00
13.87	3	5	8.5E+04	2432.21	8	8	2.6E+00	3523.42	4	2	9.8E-01
13.92	3	5	8.5E+04	2436.66	6	6	2.6E+00	3526.85	10	10	1.3E-01
13.93	5	7	4.2E+04	2439.04	4	4	2.7E+00	3529.03	6	8	8.8E-02
13.95	5	5	3.8E+04	2460.80	4	6	1.2E-01	3529.82	8	10	4.6E-01

λ Å	Weights g_i	g_k	A 10^8 s^{-1}	λ Å	Weights g_i	g_k	A 10^8 s^{-1}	λ Å	Weights g_i	g_k	A 10^8 s^{-1}
3533.36	4	6	9.1E-02	**Copper**				3868.8	17	17	3.1E+00
3560.89	4	4	2.3E-01	**Cu I**				3967.5	17	19	8.7E-01
3564.95	6	8	7.0E-02	*2024.3	2	6	9.8E-02	4046.0	17	15	1.5E+00
3569.37	8	8	1.6E+00	2165.1	2	4	5.1E-01	4103.9	13	11	1.7E+00
3574.97	6	6	1.5E-01	2178.9	2	4	9.13E-01	4186.8	17	17	1.32E+00
3575.36	8	8	9.6E-02	2181.7	2	2	1.0E+00	4194.8	17	17	7.2E-01
3585.15	8	8	7.1E-02	2225.7	2	2	4.6E-01	4211.7	17	19	2.08E+00
3587.19	6	6	1.4E+00	2244.3	2	4	1.19E-02	4218.1	15	15	1.85E+00
3594.87	6	6	9.2E-02	2441.6	2	2	2.0E-01	4221.1	15	17	1.52E+00
3602.08	4	4	1.0E-01	2492.2	2	4	3.11E-02	4225.2	13	15	4.5E+00
3704.06	6	8	1.2E-01	2618.4	6	4	3.07E-01	4268.3	15	15	3.6E-02
3745.49	8	8	7.5E-02	2766.4	4	4	9.6E-02	4276.7	13	13	7.3E-01
3842.05	8	6	1.3E-01	2824.4	6	6	7.8E-02	4292.0	15	15	5.8E-02
3845.47	8	10	4.6E-01	2961.2	6	8	3.76E-02	4577.8	17	19	2.2E-02
3861.16	6	4	1.4E-01	3063.4	4	4	1.55E-02	4589.4	17	15	1.3E-01
3873.12	10	8	1.2E-01	3194.1	4	4	1.55E-02	4612.3	17	15	8.2E-02
3873.95	8	6	1.0E-01	3247.5	2	4	1.39E+00	5077.7	17	17	5.7E-03
3881.87	6	4	8.2E-02	3274.0	2	2	1.37E+00	5301.6	17	15	1.1E-02
3894.07	6	8	6.9E-01	3337.8	6	8	3.8E-03	5547.3	17	17	2.7E-03
3894.98	4	2	8.8E-02	4022.6	2	4	1.90E-01	5639.5	17	19	4.7E-03
3935.96	8	10	6.2E-01	4062.6	4	6	2.10E-01	5974.5	17	17	4.0E-03
3995.31	8	10	2.5E-01	4249.0	2	2	1.95E-01	5988.6	17	15	5.3E-03
3997.90	6	8	7.0E-02	4275.1	6	8	3.45E-01	6010.8	15	15	2.6E-02
4092.39	8	8	5.7E-02	4480.4	2	2	3.0E-02	6088.3	15	13	3.5E-02
4110.53	6	6	5.5E-02	4509.4	4	2	2.75E-01	6168.4	15	17	2.5E-02
4118.77	6	8	1.6E-01	4530.8	4	2	8.4E-02	6259.1	17	19	8.5E-03
4121.32	8	10	1.9E-01	4539.7	6	4	2.12E-01	6579.4	17	15	7.5E-03
5146.75	8	8	1.5E-01	4587.0	8	6	3.20E-01				
5212.70	10	10	1.9E-01	4651.1	10	8	3.80E-01	**Erbium**			
5265.79	6	8	5.0E-02	4704.6	8	8	5.5E-02	**Er I**			
5280.63	10	8	2.8E-01	5105.5	6	4	2.0E-02	3862.9	13	13	2.5E+00
5352.05	12	10	2.7E-01	5153.2	2	4	6.0E-01	4008.0	13	15	2.6E+00
5477.09	6	8	6.8E-02	5218.2	4	6	7.5E-01	4151.1	13	11	1.8E+00
5483.96	8	10	7.3E-02	5220.1	4	4	1.50E-01				
6082.43	10	10	5.4E-02	5292.5	8	8	1.09E-01	**Europium**			
6455.00	8	10	9.0E-02	5700.2	4	4	2.4E-03	**Eu I**			
7838.12	8	10	5.4E-02	5782.1	4	2	1.65E-02	2372.9	8	6	1.9E-01
8093.93	12	10	2.0E-01					2375.3	8	8	2.0E-01
8372.79	10	10	8.7E-02	**Cu II**				2379.7	8	10	2.0E-01
				2489.7	5	5	1.5E-02	2619.3	8	10	7.0E-03
Co II				2544.8	9	7	1.1E+00	2643.8	8	8	6.6E-03
2286.15	11	13	3.3E+00	2689.3	7	7	4.1E-01	2659.4	8	10	1.2E-02
2307.85	9	11	2.6E+00	2701.0	5	5	6.7E-01	2682.6	8	6	1.2E-02
2311.61	7	9	2.8E+00	2703.2	3	3	1.2E+00	2710.0	8	10	1.4E-01
2314.05	5	7	2.8E+00	2713.5	5	5	6.8E-01	2724.0	8	8	1.2E-01
2314.97	3	5	2.7E+00	2769.7	7	7	6.1E-01	2731.4	8	8	3.1E-02
2330.36	5	3	1.32E+00					2732.6	8	6	3.7E-02
2344.28	3	3	1.5E+00	**Dysprosium**				2735.3	8	10	4.7E-02
2353.41	7	7	1.9E+00	**Dy I**				2738.6	8	10	1.3E-02
2363.80	9	9	2.1E+00	2862.7	17	15	6.5E-02	2743.3	8	6	1.1E-01
2378.62	11	9	1.9E+00	2964.6	17	17	6.5E-02	2745.6	8	6	5.0E-02
2383.45	9	7	1.8E+00	3147.7	15	17	1.1E-01	2747.8	8	8	5.2E-02
2388.92	11	11	2.8E+00	3263.2	15	13	1.4E-01	2772.9	8	6	1.0E-02
2389.54	5	3	1.5E+00	3511.0	15	13	3.1E-01	2878.9	8	10	2.8E-02
2404.17	3	3	1.5E+00	3571.4	15	13	2.0E-01	2892.5	8	8	1.0E-01
2417.66	9	9	8.5E-01	3757.1	17	19	3.0E+00	2893.0	8	6	1.0E-01

λ Å	Weights g_i	g_k	A 10^8 s^{-1}	λ Å	Weights g_i	g_k	A 10^8 s^{-1}	λ Å	Weights g_i	g_k	A 10^8 s^{-1}
2909.0	8	10	6.9E-02	7425.7	4	2	3.4E-01	2754.6	5	3	1.1E+00
2958.9	8	6	1.6E-02	7482.7	4	4	5.6E-02	3039.1	5	3	2.8E+00
3059.0	8	8	3.8E-02	7489.2	2	2	1.1E-01	3124.8	5	5	3.1E-02
3067.0	8	10	9.1E-03	7514.9	2	2	5.2E-02	3269.5	5	3	2.9E-01
3106.2	8	10	5.5E-02	7552.2	4	6	7.8E-02	4226.6	1	3	2.1E-01
3111.4	8	10	3.0E-01	7573.4	2	4	1.0E-01	4685.8	1	3	9.5E-02
3168.3	8	10	6.9E-02	7607.2	4	4	7.0E-02				
3185.5	8	10	5.8E-03	7754.7	4	6	3.82E-01	**Ge II**			
3210.6	8	8	1.1E-01	7800.2	2	4	2.1E-01	999.10	2	4	1.9E+00
3212.8	8	8	2.9E-01					1016.6	4	6	2.1E+00
3213.8	8	6	1.8E-01	**Gallium**				1017.1	4	4	3.5E-01
3235.1	8	10	1.0E-02	**Ga I**				1055.0	2	2	6.9E-01
3241.4	8	8	2.3E-02	2195.4	2	2	1.9E-02	1075.1	4	2	1.3E+00
3246.0	8	6	1.4E-02	2199.7	4	2	3.3E-02	1237.1	2	4	1.9E+01
3247.6	8	8	2.3E-02	2214.4	4	6	1.2E-02	1261.9	4	6	2.2E+01
3322.3	8	6	3.5E-02	2235.9	4	2	4.3E-02	1264.7	4	4	3.5E-01
3334.3	8	6	3.4E-01	2255.0	2	2	3.1E-02	1602.5	2	2	3.4E+00
3350.4	8	10	1.5E-02	2259.2	4	6	3.1E-02	1649.2	4	2	6.5E+00
3353.7	8	8	5.8E-03	2294.2	2	4	7.0E-02	4741.8	2	4	4.6E-01
3457.1	8	8	8.4E-03	2297.9	4	2	5.8E-02	4814.6	4	6	5.1E-01
3467.9	8	8	1.0E-02	2338.2	4	6	9.8E-02	4824.1	4	4	8.6E-02
3589.3	8	6	6.9E-03	2371.3	2	2	5.7E-02	5131.8	4	6	1.9E+00
4594.0	8	10	1.4E+00	2418.7	4	2	1.0E-01	5178.5	6	6	1.3E-01
4627.2	8	8	1.3E+00	2450.1	2	4	2.8E-01	5178.6	6	8	2.0E+00
4661.9	8	6	1.3E+00	2500.2	4	6	3.4E-01	5893.4	2	4	9.2E-01
5645.8	8	6	5.4E-03	2659.9	2	2	1.2E-01	6021.0	2	2	8.4E-01
5765.2	8	8	1.1E-02	2719.7	4	2	2.3E-01	6336.4	2	2	4.4E-01
6018.2	8	10	8.5E-03	2874.2	2	4	1.2E+00	6484.2	4	2	8.5E-01
6291.3	8	6	1.8E-03	2943.6	4	6	1.4E+00				
6864.5	8	10	5.8E-03	2944.2	4	4	2.7E-01	**Gold**			
7106.5	8	8	2.6E-03	4033.0	2	2	4.9E-01	**Au I**			
				4172.0	4	2	9.2E-01	2427.95	2	4	1.99E+00
Fluorine								2675.95	2	2	1.64E+00
F I				**Ga II**				3122.78	6	4	1.90E-01
806.96	4	6	3.3E+00	829.60	1	3	2.2E-01	6278.30	4	2	3.4E-02
809.60	2	4	2.8E+00	1414.4	1	3	1.88E+01				
951.87	4	2	2.6E+00					**Helium**			
954.83	4	4	5.77E+00	**Germanium**				**He I**			
955.55	2	2	5.1E+00	**Ge I**				510.00	1	3	4.6224E-01
958.52	2	4	1.3E+00	1944.7	3	1	7.0E-01	512.10	1	3	7.3174E-01
6239.7	6	4	2.5E-01	1955.1	3	3	2.8E-01	515.62	1	3	1.2582E+00
6348.5	4	4	1.8E-01	1988.3	5	3	2.5E-01	522.21	1	3	2.4356E+00
6413.7	2	4	1.1E-01	1998.9	5	5	5.5E-01	537.03	1	3	5.6634E+00
6708.3	6	4	1.4E-02	2041.7	1	3	1.1E+00	584.33	1	3	1.7989E+01
6774.0	6	6	1.0E-01	2065.2	3	3	8.5E-01	*2677.1	3	9	4.4174E-03
6795.5	4	2	5.2E-02	2068.7	3	5	1.2E+00	*2696.1	3	9	6.0234E-03
6834.3	4	4	2.1E-01	2086.0	3	5	4.0E-01	*2723.2	3	9	8.4996E-03
6856.0	6	8	4.94E-01	2094.3	5	7	9.7E-01	*2763.8	3	9	1.2508E-02
6870.2	2	2	3.8E-01	2105.8	5	5	1.7E-01	*2829.1	3	9	1.9389E-02
6902.5	4	6	3.2E-01	2256.0	5	5	3.2E-02	*2945.1	3	9	3.2006E-02
6909.8	2	4	2.2E-01	2417.4	5	5	9.6E-01	*3187.7	3	9	5.6361E-02
6966.4	4	2	1.1E-01	2498.0	1	3	1.3E-01	3231.3	1	3	5.1015E-03
7037.5	4	4	3.0E-01	2533.2	3	3	1.0E-01	3258.3	1	3	6.9627E-03
7127.9	2	2	3.8E-01	2589.2	5	3	5.1E-02	3296.8	1	3	9.8432E-03
7309.0	6	8	4.7E-01	2592.5	3	5	7.1E-01	3354.6	1	3	1.4537E-02
7311.0	4	2	3.9E-01	2651.2	5	5	2.0E+00	3447.6	1	3	2.2691E-02
7314.3	4	6	4.8E-01	2651.6	1	3	8.5E-01	*3554.4	9	15	7.5971E-03
7332.0	6	4	3.1E-01	2691.3	3	3	6.1E-01	*3563.0	9	3	4.8362E-03
7398.7	6	6	2.85E-01	2709.6	3	1	2.8E+00	*3587.3	9	15	1.8107E-02

λ Å	g_i	g_k	A 10^8 s^{-1}	λ Å	g_i	g_k	A 10^8 s^{-1}	λ Å	g_i	g_k	A 10^8 s^{-1}
3613.6	1	3	3.8022E-02	**In II**				2276.03	9	7	1.7E-01
*3634.2	9	15	2.6062E-02	2941.1	3	1	1.4E+00	2277.11	7	5	3.7E+01
*3652.0	9	3	9.7444E-03					2287.25	5	3	3.4E-01
*3705.0	9	15	3.9528E-02	**Iodine**				2292.52	7	9	4.3E-02
*3819.6	9	15	6.4351E-02	**I I**				2294.41	3	1	6.1E-01
3833.6	3	5	9.6470E-03	1782.8	4	4	2.71E+00	2300.14	5	7	8.0E-02
*3867.5	9	3	2.4465E-02	1830.4	4	6	1.6E-01	2301.68	1	3	1.3E-01
3871.8	3	5	1.3386E-02					2303.42	1	3	9.4E-02
*3888.7	3	9	9.4746E-02	**Iridium**				2303.58	3	5	7.6E-02
3926.5	3	5	1.9371E-02	**Ir I**				2309.00	3	5	1.5E-01
3935.9	3	1	7.4475E-03	2475.12	10	10	2.1E-01	2313.10	5	7	1.4E-01
3964.7	1	3	6.9507E-02	2502.98	10	12	3.2E-01	2320.36	7	9	1.2E-01
4009.3	3	5	2.9612E-02	2639.71	10	10	4.7E-01	2371.43	5	5	5.2E-02
4024.0	3	1	1.1281E-02	2661.98	10	10	2.5E-01	2373.62	7	7	6.7E-02
*4026.2	9	15	1.1600E-01	2664.79	10	8	4.0E-01	2374.52	1	3	2.9E-01
*4120.8	9	3	4.4529E-02	2694.23	10	12	4.8E-01	2381.83	3	5	5.4E-02
4143.8	3	5	4.8812E-02	2849.72	10	10	2.2E-01	2389.97	5	7	5.0E-02
4169.0	3	1	1.8298E-02	2853.31	10	10	2.0E-03	2462.18	7	5	1.5E-01
4387.9	3	5	8.9889E-02	2882.64	10	8	7.2E-02	2462.65	9	9	5.8E-01
4437.6	3	1	3.2689E-02	2924.79	10	12	1.42E-01	2479.78	5	5	1.8E+00
*4471.5	9	15	2.4578E-01	2934.64	8	10	2.0E-01	2483.27	9	11	4.9E+00
*4713.2	9	3	9.5209E-02	2951.22	10	8	2.8E-02	2488.14	7	9	4.7E+00
4921.9	3	5	1.9863E-01	3003.63	8	10	5.9E-02	2490.64	5	7	3.8E+00
5015.7	1	3	1.3372E-01	3168.88	8	10	5.47E-02	2491.15	3	5	3.0E+00
5047.7	3	1	6.7712E-02	3220.78	10	8	2.4E-01	2501.13	9	7	6.8E-01
*5875.7	9	15	7.0703E-01	3558.99	6	8	1.5E-02	2510.83	7	5	1.3E+00
6678.2	3	5	6.3705E-01	3573.72	8	10	5.4E-02	2518.10	5	3	1.9E+00
*7065.2	9	3	2.7853E-01	3617.21	6	8	2.0E-02	2522.85	9	9	2.9E+00
7281.4	3	1	1.8299E-01	3628.67	8	8	2.8E-02	2524.29	3	1	3.4E+00
*8361.7	3	9	3.8126E-03	3661.71	8	10	4.0E-02	2527.43	7	7	1.9E+00
*9463.6	3	9	5.6868E-03	3734.77	8	8	2.7E-02	2529.13	5	5	9.8E-01
9603.4	1	3	5.8286E-03	4033.76	8	10	2.7E-02	2535.61	1	3	9.7E-01
*9702.6	9	3	8.6511E-03	4069.92	6	8	3.6E-02	2540.97	3	5	9.2E-01
*10311	9	15	1.9945E-02	4913.35	12	12	3.3E-02	2545.98	5	7	6.7E-01
*10668	9	3	1.4471E-02	4939.24	10	12	2.5E-03	2549.61	7	9	3.6E-01
*10830	3	9	1.0216E-01					2584.54	11	13	4.6E-01
*10913	15	21	1.9801E-02	**Iron**				2606.83	9	11	4.2E-01
10917	5	7	1.6083E-02	**Fe I**				2618.02	7	7	4.0E-01
*10997	15	9	1.4253E-03	1934.54	9	7	2.5E-01	2623.53	7	9	3.3E-01
11013	1	3	9.2496E-03	1937.27	9	7	2.2E-01	2656.15	13	15	2.8E-01
11045	3	5	1.8457E-02	1940.66	7	5	2.6E-01	2669.49	11	13	1.7E-01
11226	3	1	1.1168E-02	2084.12	9	7	3.7E-01	2679.06	11	11	1.9E-01
*11969	9	15	3.4781E-02	2102.35	7	7	8.8E-02	2719.03	9	7	1.4E+00
*12528	3	9	7.0932E-03	2112.97	1	3	1.9E-01	2720.90	7	5	1.1E+00
12756	5	3	1.2754E-03	2132.02	9	9	7.6E-02	2723.58	5	3	6.4E-01
*12785	15	21	4.1339E-02	2145.19	7	7	5.7E-02	2733.58	11	9	8.6E-01
12791	5	7	3.2475E-02	2153.01	5	5	6.9E-02	2735.48	9	7	6.2E-01
*12846	9	3	2.7317E-02	2161.58	3	5	5.0E-02	2737.31	3	3	8.5E-01
12968	3	5	3.3615E-02	2166.77	9	7	2.7E+00	2742.41	5	5	6.3E-01
*12985	15	9	2.7292E-03	2171.30	5	7	5.1E-02	2744.07	1	3	3.5E-01
				2173.21	3	5	8.3E-02	2750.14	7	7	3.9E-01
Indium				2176.84	1	3	1.0E-01	2756.33	3	5	2.0E-01
In I				2191.20	1	3	7.3E-02	2788.10	11	13	6.3E-01
2560.2	2	4	4.0E-01	2191.84	5	5	1.2E+00	2894.50	5	5	6.2E-01
2710.3	4	6	4.0E-01	2196.04	3	3	1.2E+00	2899.42	5	3	5.9E-01
3039.4	2	4	1.3E+00	2200.72	3	5	2.8E-01	2920.69	5	5	5.2E-02
3256.1	4	6	1.3E+00	2259.51	9	11	7.0E-02	2923.29	11	11	1.6E+00
4101.8	2	2	5.6E-01	2267.08	7	5	7.1E-02	2925.36	7	9	1.8E-01
4511.3	4	2	1.02E+00	2272.07	7	9	3.8E-02	2929.01	7	5	7.3E-02

λ Å	Weights g_i	g_k	A 10^8 s^{-1}	λ Å	Weights g_i	g_k	A 10^8 s^{-1}	λ Å	Weights g_i	g_k	A 10^8 s^{-1}
2936.90	9	9	1.3E-01	3215.94	5	5	8.0E-01	3442.36	5	5	4.55E-02
2941.34	5	3	5.6E-02	3217.38	11	9	2.2E-01	3443.88	5	3	6.2E-02
2947.88	7	7	2.0E-01	3219.58	7	9	6.2E-01	3445.15	5	7	2.8E-01
2953.94	5	5	1.89E-01	3222.07	11	11	3.3E-01	3447.28	5	5	9.1E-02
2954.65	5	7	1.0E-01	3225.79	11	13	8.8E-01	3450.33	3	3	2.0E-01
2957.36	3	3	1.77E-01	3227.80	9	7	1.4E+00	3476.70	1	3	5.4E-01
2965.25	1	3	1.16E-01	3228.25	5	3	4.5E-01	3477.85	3	1	4.2E-02
2966.90	9	11	2.72E-01	3229.99	9	11	4.5E-01	3485.34	5	3	1.4E-01
2969.36	3	1	3.66E-02	3230.21	5	5	1.9E-01	3495.29	9	7	9.46E-02
2973.13	5	7	1.35E-01	3230.96	7	5	3.9E-01	3497.10	7	7	1.4E-01
2973.24	7	9	1.83E-01	3233.05	13	15	5.4E-01	3505.07	5	3	9.9E-02
2980.53	7	7	2.2E-01	3233.97	9	9	2.0E-01	3506.50	5	5	7.1E-02
2981.45	7	5	6.54E-02	3246.96	5	3	9.9E-02	3508.49	9	11	5.7E-02
2983.57	9	7	2.80E-01	3248.20	7	7	2.2E-01	3510.44	1	3	4.4E-02
2987.29	9	7	6.6E-02	3253.60	7	9	1.8E-01	3516.56	7	5	3.7E-02
2990.39	9	11	3.9E-01	3254.36	11	13	5.1E-01	3521.84	3	5	9.6E-02
2994.43	7	5	4.4E-01	3257.59	7	5	1.4E-01	3523.31	5	3	7.6E-02
2996.39	3	5	1.6E-01	3265.62	7	5	3.8E-01	3524.08	7	5	7.5E-02
2999.51	11	11	2.3E-01	3268.23	3	3	5.9E-02	3524.24	5	7	4.2E-02
3000.95	5	3	6.42E-01	3271.00	5	3	6.6E-01	3527.79	9	9	2.0E-01
3008.14	3	1	1.07E+00	3280.26	9	11	5.4E-01	3529.82	3	3	7.6E-01
3009.09	13	11	6.7E-02	3282.89	3	5	3.0E-01	3536.56	5	7	7.8E-01
3009.57	9	9	1.7E-01	3284.59	5	5	5.4E-02	3537.73	5	3	1.1E-01
3011.48	7	9	4.7E-01	3290.99	3	5	6.0E-02	3537.90	11	11	8.4E-02
3015.92	11	9	5.9E-02	3292.02	7	9	6.1E-01	3540.12	7	9	1.2E-01
3016.18	5	3	8.5E-02	3292.59	3	3	2.6E-01	3541.08	9	11	6.2E-01
3017.63	3	3	6.82E-02	3298.13	3	5	8.1E-02	3542.08	7	9	7.4E-01
3018.98	7	7	1.3E-01	3305.97	5	7	4.7E-01	3543.67	3	5	1.8E-01
3021.07	7	7	4.56E-01	3306.36	3	5	6.1E-01	3548.02	5	3	9.7E-02
3024.03	3	5	4.88E-02	3307.23	13	13	2.0E-01	3552.11	3	5	4.5E-02
3025.84	1	3	3.48E-01	3314.74	5	7	6.9E-01	3552.83	5	5	1.5E-01
3026.46	5	5	1.1E-01	3322.47	9	11	6.2E-02	3553.74	11	9	8.1E-01
3031.63	3	3	1.5E-01	3323.74	5	5	3.0E-01	3556.88	9	11	4.4E-01
3037.39	3	5	3.2E-01	3328.87	11	11	2.7E-01	3559.50	3	3	1.9E-01
3042.02	3	5	4.9E-02	3337.66	11	9	5.7E-02	3560.70	7	9	6.5E-02
3042.66	5	7	5.7E-02	3347.93	5	5	4.0E-02	3565.38	7	9	3.8E-01
3047.60	5	7	2.84E-01	3354.06	1	3	7.7E-02	3567.03	5	7	6.5E-02
3053.07	3	5	1.5E-01	3355.23	9	9	3.2E-01	3568.42	5	3	5.3E-02
3057.45	11	9	4.4E-01	3369.55	9	9	2.4E-01	3568.82	7	9	5.6E-02
3059.09	7	9	1.7E-01	3370.78	11	11	3.3E-01	3570.10	9	11	6.77E-01
3067.24	9	7	3.4E-01	3380.11	7	7	2.4E-01	3572.00	11	11	2.4E-01
3068.17	5	3	9.8E-02	3383.98	7	7	9.3E-02	3573.39	5	7	7.5E-02
3075.72	7	5	2.9E-01	3392.65	7	7	2.6E-01	3576.76	11	9	9.6E-02
3083.74	5	3	3.0E-01	3394.58	5	3	9.9E-02	3578.38	1	3	6.3E-02
3091.58	3	1	5.4E-01	3399.33	5	5	3.8E-01	3581.19	11	13	1.02E+00
3098.19	11	11	1.1E-01	3402.26	13	13	2.8E-01	3582.20	13	11	2.5E-01
3100.67	7	7	1.4E-01	3406.44	3	5	3.0E-01	3583.33	1	3	2.3E-01
3119.49	11	9	8.2E-02	3407.46	7	9	5.8E-01	3585.32	7	7	1.3E-01
3120.43	9	7	8.9E-02	3410.17	3	5	4.7E-01	3585.71	9	9	3.75E-02
3156.27	7	7	5.4E-01	3411.35	9	9	5.5E-02	3586.98	5	5	1.6E-01
3160.66	9	9	1.9E-01	3413.13	5	7	3.6E-01	3591.48	1	3	6.0E-02
3161.95	11	13	1.2E-01	3417.84	3	3	5.1E-01	3592.67	7	5	4.0E-02
3166.44	9	7	1.14E-01	3418.51	3	1	1.3E+00	3594.63	9	9	2.7E-01
3168.85	5	7	5.7E-02	3424.28	7	7	2.0E-01	3595.30	5	5	5.4E-02
3175.45	11	11	1.3E-01	3425.01	9	7	2.8E-01	3597.02	5	3	1.7E-01
3176.36	5	3	9.2E-02	3427.12	7	9	5.5E-01	3599.62	11	9	1.8E-01
3196.93	9	11	9.0E-01	3428.19	5	5	2.1E-01	3603.20	11	11	2.6E-01
3199.53	9	9	2.6E-01	3428.75	7	5	2.7E-01	3603.82	3	3	1.7E-01
3205.40	3	3	1.2E+00	3440.99	7	5	8.4E-02	3605.45	9	9	6.4E-01

λ Å	g_i	g_k	A 10^8 s^{-1}	λ Å	g_i	g_k	A 10^8 s^{-1}	λ Å	g_i	g_k	A 10^8 s^{-1}
3606.68	11	13	8.2E-01	3724.38	5	7	1.3E-01	3820.43	11	9	6.68E-01
3608.86	3	5	8.14E-01	3726.93	5	5	4.6E-01	3821.18	11	13	7.0E-01
3610.16	13	13	4.8E-01	3727.09	9	7	2.0E-01	3821.83	5	5	7.8E-02
3610.70	5	3	7.1E-02	3727.62	7	5	2.25E-01	3825.88	9	7	5.98E-01
3612.07	11	13	7.5E-02	3730.39	9	11	1.3E-01	3827.82	7	5	1.05E+00
3613.45	7	7	6.7E-02	3730.95	5	7	3.8E-02	3833.31	9	9	4.69E-02
3615.19	3	3	5.8E-02	3732.40	5	5	2.8E-01	3834.22	7	5	4.53E-01
3617.79	5	7	6.5E-01	3733.32	3	3	6.2E-02	3836.33	5	5	3.7E-01
3618.77	5	7	7.3E-01	3734.86	11	11	9.02E-01	3839.26	9	9	2.8E-01
3621.46	9	11	5.1E-01	3735.32	9	9	2.4E-01	3839.61	3	5	3.9E-01
3622.00	7	7	5.1E-01	3737.13	7	9	1.42E-01	3840.44	5	3	4.70E-01
3623.19	13	13	7.4E-02	3738.31	11	13	3.8E-01	3841.05	5	3	1.3E+00
3624.06	5	3	5.4E-02	3740.24	7	9	1.4E-01	3843.26	9	7	4.7E-01
3630.35	9	7	7.6E-02	3742.62	9	9	1.0E-01	3845.17	3	3	6.8E-02
3631.46	7	9	5.17E-01	3743.36	5	3	2.60E-01	3845.69	5	7	4.9E-02
3632.04	3	5	4.8E-01	3744.10	5	3	3.6E-01	3846.00	9	7	4.3E-02
3632.55	11	9	5.2E-02	3745.56	5	7	1.15E-01	3846.41	11	9	1.9E-01
3635.19	5	3	1.4E-01	3745.90	1	3	7.33E-02	3846.80	7	7	6.6E-01
3637.86	9	9	5.5E-02	3746.93	7	7	2.2E-01	3849.96	3	1	6.06E-01
3638.30	7	9	2.6E-01	3748.26	3	5	9.15E-02	3856.37	7	5	4.64E-02
3640.39	9	11	3.8E-01	3749.48	9	9	7.64E-01	3859.21	13	11	8.5E-02
3644.80	7	5	7.8E-02	3753.61	7	5	9.3E-02	3859.91	9	9	9.70E-02
3645.82	1	3	5.7E-01	3756.94	11	11	2.4E-01	3865.52	3	3	1.55E-01
3647.84	9	11	2.92E-01	3757.45	5	3	1.2E-01	3867.22	5	5	3.4E-01
3649.51	11	9	4.2E-01	3758.23	7	7	6.34E-01	3871.75	11	11	6.7E-02
3650.03	7	7	9.9E-02	3760.05	13	15	4.47E-02	3872.50	5	5	1.05E-01
3651.47	7	9	6.2E-01	3760.53	3	5	4.8E-02	3873.76	11	9	8.0E-02
3655.46	5	5	1.0E-01	3763.79	5	5	5.44E-01	3878.02	7	7	7.72E-02
3659.52	9	9	5.8E-02	3765.54	13	15	9.8E-01	3878.57	5	3	6.6E-02
3667.25	9	7	1.4E-01	3766.67	5	3	9.7E-02	3883.28	7	7	1.6E-01
3669.15	9	7	7.4E-02	3767.19	3	3	6.40E-01	3884.36	11	9	3.5E-02
3669.52	9	7	3.0E-01	3768.03	3	1	8.4E-02	3885.51	3	5	5.8E-02
3670.09	11	13	7.6E-02	3774.82	3	3	4.7E-02	3886.28	7	7	5.30E-02
3674.77	5	3	6.7E-02	3778.51	7	5	1.2E-01	3887.05	9	9	3.52E-02
3676.31	9	11	4.63E-02	3781.94	5	7	3.7E-02	3888.51	5	5	2.6E-01
3677.31	5	7	3.1E-01	3785.95	11	13	4.2E-02	3888.82	5	3	`2.7E-01
3677.63	7	5	8.0E-01	3786.19	5	5	1.2E-01	3891.93	3	3	4.0E-01
3678.86	3	5	4.1E-02	3787.16	5	5	1.0E-01	3893.39	11	11	1.3E-01
3682.24	5	5	1.7E+00	3787.88	3	5	1.29E-01	3895.66	3	1	9.40E-02
3684.11	9	7	3.4E-01	3789.82	9	7	3.9E-02	3900.52	7	7	7.5E-02
3686.00	9	11	2.6E-01	3791.73	5	3	6.3E-02	3902.95	7	7	2.14E-01
3686.26	3	1	1.2E-01	3793.87	3	3	7.4E-02	3903.90	9	9	9.6E-02
3687.46	11	9	8.01E-02	3794.34	9	11	3.8E-02	3906.75	5	7	6.7E-02
3688.48	7	9	6.9E-02	3795.00	5	7	1.15E-01	3907.93	7	5	6.7E-02
3690.73	11	11	2.7E-01	3799.55	7	9	7.32E-02	3909.66	3	5	5.3E-02
3694.01	5	7	6.8E-01	3801.68	5	7	6.6E-02	3909.83	3	3	6.5E-02
3697.43	7	7	2.1E-01	3802.00	11	13	3.5E-02	3914.27	3	3	5.4E-02
3698.60	5	7	3.8E-02	3802.28	5	5	5.0E-02	3916.73	13	11	1.2E-01
3699.15	5	7	4.5E-02	3804.01	11	9	4.7E-02	3919.07	9	9	3.9E-02
3701.09	7	9	4.8E-01	3805.35	9	11	9.8E-01	3925.20	1	3	5.7E-02
3702.03	3	1	3.5E-01	3806.22	3	3	2.3E-01	3931.12	5	7	4.5E-02
3703.69	9	11	5.3E-02	3806.70	11	11	5.4E-01	3941.28	5	5	8.4E-02
3703.82	1	3	1.2E-01	3807.54	3	5	8.0E-02	3942.44	3	5	9.0E-02
3704.46	11	9	1.3E-01	3808.73	9	9	3.54E-02	3946.99	9	11	4.4E-02
3709.25	9	7	1.56E-01	3810.76	5	3	2.0E-01	3948.77	11	9	2.2E-01
3711.41	3	5	7.3E-02	3813.88	13	11	8.7E-02	3949.14	3	3	3.9E-02
3718.41	7	7	5.3E-02	3815.84	9	7	1.3E+00	3949.95	7	5	5.9E-02
3719.93	9	11	1.62E-01	3817.64	11	11	8.3E-02	3951.16	3	5	3.6E-01
3722.56	5	5	4.97E-02	3819.50	7	5	4.6E-02	3952.60	11	11	4.1E-02

λ Å	Weights g_i	g_k	A 10^8 s^{-1}	λ Å	Weights g_i	g_k	A 10^8 s^{-1}	λ Å	Weights g_i	g_k	A 10^8 s^{-1}
3953.15	7	9	3.7E-02	4109.07	1	3	4.5E-02	4250.79	7	7	1.0E-01
3955.34	3	3	1.4E-01	4109.80	3	3	1.6E-01	4260.47	11	11	3.2E-01
3955.96	3	3	5.7E-02	4112.96	11	13	1.4E-01	4267.83	1	3	9.4E-02
3956.45	13	11	2.1E-01	4114.45	5	5	4.7E-02	4268.75	5	3	4.2E-02
3957.02	5	7	1.6E-01	4118.54	11	13	5.8E-02	4271.15	7	9	1.82E-01
3960.28	5	7	4.2E-02	4126.18	11	11	3.9E-02	4271.76	9	11	2.28E-01
3963.10	3	5	1.7E-01	4127.61	1	3	1.3E-01	4282.40	7	5	1.1E-01
3967.42	9	7	2.3E-01	4132.06	5	7	1.2E-01	4300.83	5	5	4.7E-02
3967.96	7	9	6.3E-02	4132.90	3	5	9.4E-02	4305.45	5	3	6.0E-02
3969.26	9	7	2.3E-01	4134.68	5	7	1.8E-01	4307.90	7	9	3.4E-01
3970.39	3	1	3.5E-01	4137.00	3	5	2.2E-01	4315.08	5	5	7.7E-02
3971.32	11	9	5.7E-02	4137.42	5	7	6.1E-02	4325.76	5	7	5.0E-02
3973.65	5	7	6.6E-02	4142.63	3	5	7.4E-02	4327.09	5	5	7.8E-02
3976.61	3	5	1.8E-01	4143.87	7	9	1.5E-01	4352.73	3	5	3.9E-02
3977.74	5	5	7.0E-02	4149.37	11	13	3.6E-02	4369.77	9	9	7.2E-02
3981.77	9	9	3.9E-02	4150.25	3	3	7.1E-02	4383.54	9	11	5.00E-01
3983.96	9	7	7.6E-02	4153.90	7	9	2.3E-01	4387.89	3	3	3.9E-02
3985.39	5	5	6.7E-02	4154.80	9	11	1.5E-01	4388.41	7	7	1.3E-01
3989.86	5	7	5.0E-02	4156.80	5	5	1.9E-01	4401.29	7	7	5.9E-02
3996.97	9	9	6.7E-02	4158.79	3	5	1.6E-01	4404.75	7	9	2.75E-01
3997.39	9	11	1.5E-01	4170.90	5	5	6.1E-02	4415.12	5	7	1.19E-01
3998.05	11	9	6.6E-02	4172.12	7	5	9.7E-02	4422.57	3	3	8.8E-02
4003.76	3	3	7.1E-02	4175.64	3	5	1.6E-01	4430.61	3	1	7.45E-02
4005.24	7	5	2.04E-01	4181.75	5	7	3.6E-01	4433.22	5	3	2.3E-01
4006.31	11	9	4.7E-02	4182.38	5	5	4.9E-02	4438.34	3	1	7.9E-02
4007.27	7	5	4.2E-02	4184.89	5	5	1.1E-01	4442.34	5	5	3.76E-02
4009.71	3	5	5.2E-02	4187.04	7	5	2.15E-01	4443.19	1	3	1.1E-01
4014.53	11	11	2.4E-01	4187.79	9	7	1.52E-01	4446.83	3	3	5.3E-02
4017.15	9	11	4.5E-02	4191.68	1	3	4.8E-02	4447.72	3	3	5.11E-02
4021.87	7	9	1.0E-01	4196.21	7	7	9.8E-02	4454.38	5	5	3.8E-02
4024.72	7	9	8.9E-02	4198.30	11	9	8.03E-02	4455.03	9	7	3.9E-02
4031.96	3	5	7.1E-02	4198.64	5	5	1.3E-01	4466.55	5	7	1.2E-01
4040.64	5	7	4.4E-02	4199.09	9	11	6.1E-01	4469.37	5	7	2.6E-01
4044.61	5	3	1.1E-01	4200.09	7	7	4.0E-02	4481.61	3	3	4.2E-02
4045.81	9	9	8.63E-01	4200.92	7	9	4.2E-02	4484.22	7	9	7.0E-02
4054.87	5	3	1.6E-01	4202.03	9	9	8.22E-02	4485.67	3	3	1.1E-01
4058.22	9	7	4.9E-02	4203.67	7	9	8.6E-02	4528.61	7	9	5.44E-02
4059.73	5	3	8.1E-02	4203.94	13	13	1.3E-01	4533.13	3	1	3.7E-02
4062.44	3	3	2.2E-01	4205.54	5	5	3.6E-02	4547.85	5	7	7.6E-02
4063.59	7	7	6.8E-01	4207.13	5	3	4.3E-02	4619.29	7	5	4.7E-02
4065.40	3	1	1.9E-01	4210.34	3	3	1.7E-01	4669.17	5	3	4.0E-02
4067.98	9	9	1.7E-01	4213.65	3	1	1.9E-01	4673.16	5	7	4.6E-02
4070.77	7	5	1.3E-01	4217.55	3	5	2.3E-01	4678.85	7	9	7.4E-02
4071.74	5	5	7.65E-01	4219.36	11	13	3.8E-01	4704.95	3	1	8.1E-02
4073.76	5	3	1.6E-01	4220.34	3	1	1.9E-01	4736.77	9	11	4.9E-02
4074.79	9	9	4.8E-02	4222.21	7	7	5.77E-02	4789.65	5	5	7.2E-02
4076.63	9	9	1.9E-01	4224.17	9	11	1.3E-01	4859.74	5	3	1.3E-01
4078.35	5	3	4.2E-02	4224.51	3	5	7.1E-02	4871.32	7	5	2.2E-01
4079.18	5	5	5.1E-02	4225.45	5	7	1.7E-01	4872.14	3	3	2.4E-01
4079.84	1	3	6.3E-02	4226.42	3	3	3.7E-02	4878.21	1	3	9.1E-02
4080.21	3	1	2.4E-01	4233.60	3	5	1.85E-01	4890.75	5	5	2.1E-01
4082.44	3	3	3.8E-02	4235.94	9	9	1.88E-01	4891.49	9	7	2.9E-01
4084.49	11	9	1.1E-01	4238.81	7	9	2.2E-01	4892.87	3	3	4.8E-02
4085.00	3	5	4.2E-02	4240.37	5	3	5.7E-02	4903.31	3	5	4.7E-02
4085.30	7	7	1.1E-01	4245.26	1	3	8.3E-02	4917.23	5	3	6.1E-02
4085.98	7	5	5.0E-02	4246.08	7	5	5.7E-02	4918.01	1	3	4.0E-02
4088.57	5	3	3.9E-02	4247.43	9	11	2.0E-01	4918.99	7	7	1.7E-01
4098.18	7	7	6.8E-02	4248.22	3	5	3.5E-02	4920.50	11	9	3.5E-01
4107.49	5	3	2.5E-01	4250.12	5	7	2.08E-01	4930.31	3	3	4.1E-02

λ Å	g_i	g_k	A 10^8 s^{-1}	λ Å	g_i	g_k	A 10^8 s^{-1}	λ Å	g_i	g_k	A 10^8 s^{-1}
4969.92	3	3	1.8E-01	5679.02	5	7	3.6E-02	2370.50	4	4	1.4E-01
4973.10	3	3	1.0E-01	5686.53	9	11	4.4E-02	2373.74	10	10	3.3E-01
4978.60	5	3	1.1E-01	5691.51	3	1	6.2E-02	2375.19	4	2	9.8E-01
4988.95	7	7	4.9E-02	5705.99	7	9	6.7E-02	2379.27	8	8	1.5E-01
4991.27	5	7	8.2E-02	5717.85	1	3	5.0E-02	2380.76	6	8	3.1E-01
5001.86	9	7	3.9E-01	5753.12	3	5	7.0E-02	2382.04	10	12	3.8E+00
5004.04	5	3	3.5E-02	5762.99	5	7	1.0E-01	2382.90	12	14	2.2E-01
5014.94	7	5	3.0E-01	5816.36	9	11	3.7E-02	2383.25	6	6	3.4E-01
5022.24	5	3	2.6E-01	5905.67	5	3	1.2E-01	2384.39	4	4	2.3E-01
5074.75	9	11	1.5E-01	5927.80	5	3	5.1E-02	2388.37	10	12	2.2E-01
5090.78	7	5	2.0E-01	5930.17	5	7	1.6E-01	2388.63	8	8	1.0E+00
5109.65	3	5	5.4E-02	6020.17	7	9	1.1E-01	2390.10	14	16	5.5E+00
5121.64	5	5	7.9E-02	6024.07	9	11	1.3E-01	2390.77	6	6	9.3E-01
5125.11	9	7	2.6E-01	6055.99	7	9	7.0E-02	2395.42	6	4	3.3E-01
5133.69	11	13	2.7E-01	6170.49	5	5	1.3E-01	2395.62	8	10	2.5E+00
5137.38	11	9	1.1E-01	6336.84	3	3	4.9E-02	2399.24	6	6	1.4E+00
5159.06	5	3	1.3E-01	6338.90	5	3	4.8E-02	2400.06	12	14	5.2E+00
5162.27	11	11	2.4E-01	6400.00	7	9	5.5E-02	2401.29	6	8	2.5E+00
5184.26	5	7	3.5E-02	6411.65	5	7	3.5E-02	2404.43	4	2	7.1E-01
5208.59	7	5	5.2E-02	6419.98	7	7	1.3E-01	2404.89	6	8	1.7E+00
5232.94	9	11	1.4E-01	6469.21	3	3	9.0E-02	2406.66	4	4	1.6E+00
5263.30	5	5	5.2E-02	6495.78	3	3	6.0E-02	2410.52	4	6	1.5E+00
5266.55	7	9	8.6E-02	6496.46	5	5	8.5E-02	2411.07	2	2	2.4E+00
5283.62	7	7	8.0E-02	6569.23	7	9	6.5E-02	2413.31	2	4	1.1E+00
5302.30	3	5	6.3E-02	6633.76	7	7	3.6E-02	2416.45	8	10	1.6E+00
5324.18	9	9	1.5E-01	6733.16	3	1	3.9E-02	2418.44	6	8	1.6E+00
5339.93	5	7	7.0E-02	6841.35	5	7	3.6E-02	2423.21	4	6	1.4E+00
5353.39	9	7	4.8E-02	7130.94	3	5	4.3E-02	2428.36	8	10	2.7E+00
5364.87	5	7	5.5E-01					2432.87	14	14	3.2E+00
5367.47	7	9	5.8E-01	**Fe II**				2434.06	8	6	7.0E-01
5369.96	9	11	4.7E-01	1144.94	10	12	4.8E+00	2434.24	8	10	2.0E+00
5373.71	7	9	3.5E-02	1635.40	8	6	2.4E+00	2434.73	12	12	3.2E+00
5383.37	11	13	5.6E-01	1641.76	6	4	1.8E+00	2439.30	12	14	2.8E+00
5389.48	7	7	1.3E-01	1647.16	6	6	5.2E-01	2445.11	12	12	1.9E+00
5398.29	5	5	9.8E-02	2208.41	10	10	1.8E+00	2445.80	4	6	1.5E+00
5400.50	9	9	1.8E-01	2213.66	14	14	4.4E-01	2446.47	12	14	2.9E-01
5410.91	7	9	4.8E-01	2218.27	8	10	1.9E+00	2447.20	6	6	1.2E+00
5415.20	11	13	5.6E-01	2327.40	6	4	5.9E-01	2453.98	8	10	7.3E-01
5424.07	13	15	5.0E-01	2331.31	10	8	2.9E-01	2455.71	8	8	1.0E+00
5432.95	5	5	4.1E-02	2332.80	8	6	1.5E+00	2458.78	10	12	2.7E+00
5445.04	11	11	2.0E-01	2338.01	4	4	1.1E+00	2458.97	6	4	2.0E+00
5463.27	9	9	3.2E-01	2343.49	10	8	1.7E+00	2460.44	10	12	5.3E-01
5466.39	9	7	7.5E-02	2343.96	8	6	2.9E-01	2461.28	6	8	2.6E+00
5473.90	7	7	5.5E-02	2344.28	2	4	8.2E-01	2461.86	8	10	2.6E+00
5480.87	3	1	1.2E-01	2348.11	10	8	5.1E-01	2466.52	2	4	2.1E+00
5487.74	7	5	8.6E-02	2348.30	6	6	1.2E+00	2469.51	8	6	2.8E+00
5554.89	9	9	8.7E-02	2351.67	6	6	1.7E+00	2472.61	8	10	3.7E+00
5569.62	5	3	2.1E-01	2352.31	2	4	4.2E+00	2475.12	4	6	3.9E+00
5572.84	7	5	2.1E-01	2353.68	8	8	1.3E+00	2475.54	6	8	3.5E+00
5576.09	3	1	2.1E-01	2354.89	6	4	2.4E-01	2481.05	12	12	1.9E-01
5586.76	9	7	1.9E-01	2360.00	10	10	2.4E-01	2484.44	8	8	2.3E+00
5598.30	5	5	1.8E-01	2360.29	8	6	5.9E-01	2492.34	10	12	1.6E-01
5615.64	11	9	1.7E-01	2362.02	8	8	1.3E-01	2493.26	14	16	3.4E+00
5624.54	5	5	5.3E-02	2363.86	8	10	5.1E-01	2501.31	2	2	1.4E+00
5633.97	11	13	8.7E-02	2364.83	8	8	6.1E-01	2503.87	10	10	2.4E+00
5638.27	9	7	4.0E-02	2365.77	6	6	2.1E+00	2508.34	8	10	2.7E+00
5650.01	3	5	5.0E-02	2366.59	6	6	9.9E-02	2533.63	12	12	1.3E+00
5655.18	7	9	5.3E-02	2368.60	6	4	5.9E-01	2534.42	8	8	1.2E+00
5658.82	7	7	3.6E-02	2369.95	10	12	5.7E+00	2535.36	6	4	3.3E+00

λ Å	Weights g_i	g_k	A 10^8 s^{-1}	λ Å	Weights g_i	g_k	A 10^8 s^{-1}	λ Å	Weights g_i	g_k	A 10^8 s^{-1}
2535.49	10	8	5.4E-01	2592.78	14	16	2.1E+00	2712.39	10	12	1.3E-01
2536.67	12	12	4.0E-01	2593.72	2	4	1.3E-01	2714.41	8	6	5.5E-01
2537.14	10	10	1.4E+00	2594.96	8	8	1.0E-01	2716.22	6	6	1.1E+00
2538.20	14	12	1.2E+00	2598.37	8	6	1.3E+00	2716.56	14	12	1.6E+00
2538.50	8	6	3.3E-01	2599.40	10	10	2.2E+00	2717.87	16	14	1.4E+00
2538.80	12	10	8.2E-01	2604.05	8	8	1.1E-01	2718.64	10	8	1.3E+00
2538.91	10	8	7.8E-01	2605.04	6	8	2.1E-01	2719.30	6	8	3.7E-01
2538.99	14	12	1.2E+00	2605.34	4	4	1.6E+00	2722.06	8	8	1.1E-01
2540.52	2	2	1.5E+00	2605.42	6	6	2.6E-01	2722.74	6	8	7.8E-01
2541.10	8	6	7.3E-01	2605.90	4	2	1.2E+00	2724.88	6	6	9.7E-01
2541.84	8	6	7.7E-01	2606.51	6	6	1.8E+00	2727.38	12	10	3.2E-01
2542.73	2	2	1.9E+00	2607.09	6	4	1.7E+00	2727.54	6	4	8.5E-01
2543.38	10	12	4.4E-01	2609.13	8	10	3.0E-01	2728.91	8	10	8.8E-02
2543.43	6	4	7.1E-01	2609.87	8	8	1.8E-01	2730.73	4	4	2.5E-01
2544.97	4	6	4.0E-01	2611.87	8	8	1.1E+00	2732.94	8	6	7.8E-01
2545.22	8	10	3.3E-01	2613.82	4	2	2.0E+00	2739.55	8	8	1.9E+00
2545.44	8	10	1.4E-01	2617.62	6	6	4.4E-01	2741.40	6	6	1.7E-01
2546.67	8	8	6.2E-01	2619.07	10	10	2.7E-01	2743.20	2	4	1.8E+00
2547.34	8	8	2.0E-01	2620.17	6	6	1.3E-01	2746.48	4	6	1.9E+00
2548.33	4	6	2.0E-01	2620.70	8	8	3.3E-01	2746.98	6	6	1.6E+00
2548.59	10	10	1.9E-01	2621.67	2	2	4.9E-01	2749.18	4	4	1.1E+00
2548.74	4	2	1.7E+00	2623.11	14	14	1.1E-01	2749.32	6	8	2.1E+00
2548.92	12	10	4.8E-01	2623.73	6	6	2.2E-01	2749.49	2	2	1.1E+00
2549.08	10	8	1.5E+00	2625.49	12	14	2.2E+00	2753.29	10	12	1.2E+00
2549.40	4	4	1.3E+00	2625.67	8	10	3.4E-01	2754.91	8	6	8.4E-01
2549.46	6	6	8.0E-01	2626.50	4	6	3.4E-01	2755.73	8	10	2.1E+00
2549.77	8	6	2.5E-01	2628.29	2	4	8.6E-01	2761.81	2	4	1.1E-01
2550.03	10	10	1.2E+00	2629.59	6	8	6.2E-01	2762.34	6	6	3.7E-01
2550.15	8	10	4.0E-01	2630.07	4	6	5.7E-01	2763.66	14	12	1.3E+00
2550.68	12	12	8.9E-01	2631.05	4	6	7.7E-01	2765.13	10	8	1.2E+00
2551.21	10	8	3.2E-01	2631.32	6	8	6.0E-01	2767.50	12	14	1.9E+00
2555.07	6	8	1.8E-01	2631.61	10	12	5.3E-01	2769.36	12	14	1.6E+00
2555.45	4	6	2.5E-01	2633.20	6	4	1.7E+00	2774.69	2	4	2.4E-01
2557.51	10	8	1.3E-01	2636.69	4	4	1.2E-01	2776.91	8	8	3.0E-01
2559.77	6	8	2.4E-01	2637.50	6	6	5.2E-01	2779.30	10	8	7.6E-01
2559.92	6	8	2.4E-01	2637.64	2	4	8.3E-01	2779.91	2	4	2.3E-01
2560.28	4	4	1.5E+00	2639.56	2	2	1.1E+00	2780.04	2	2	2.9E-01
2562.09	4	2	1.5E+00	2642.01	6	6	3.6E-01	2783.69	12	10	7.0E-01
2562.54	8	6	1.5E+00	2649.47	6	8	1.8E-01	2785.19	12	10	1.0E+00
2563.48	6	4	1.3E+00	2650.48	6	8	1.6E+00	2787.24	8	6	1.3E-01
2566.22	8	10	2.5E+00	2654.63	4	4	7.7E-01	2793.89	10	12	9.6E-02
2566.40	8	6	2.1E+00	2658.25	8	8	3.2E-01	2796.63	10	10	1.0E-01
2566.91	4	2	1.1E+00	2662.56	2	2	9.6E-01	2799.29	10	8	1.1E-01
2568.41	2	4	4.4E-01	2664.66	8	10	1.5E+00	2809.78	8	8	1.6E-01
2569.78	2	4	1.2E+00	2666.64	6	8	1.7E+00	2817.09	6	4	2.1E-01
2570.53	6	8	1.2E+00	2667.22	4	6	9.2E-01	2831.56	4	6	5.8E-01
2570.85	8	6	1.7E+00	2669.93	2	4	4.7E-01	2833.09	6	6	2.7E-01
2573.21	8	10	1.4E-01	2671.40	2	4	5.6E-01	2835.71	4	6	3.1E-01
2574.36	6	4	1.6E+00	2682.51	8	10	7.0E-01	2838.22	4	2	4.2E-01
2576.86	10	12	1.1E+00	2683.00	4	6	6.4E-01	2839.51	10	8	9.9E-01
2577.92	2	2	1.3E+00	2684.75	8	10	1.4E+00	2839.80	8	10	4.1E-01
2582.41	6	8	2.4E-01	2692.60	10	12	1.2E+00	2840.65	2	4	5.3E-01
2582.58	4	4	7.7E-01	2697.33	4	4	2.7E-01	2840.76	10	12	1.1E-01
2585.63	10	10	3.6E-01	2697.46	4	2	1.8E+00	2844.96	2	2	4.5E-01
2585.88	10	8	8.1E-01	2699.20	4	4	6.6E-01	2847.77	4	4	3.3E-01
2587.95	8	10	1.4E+00	2703.99	8	8	1.2E+00	2848.11	6	6	7.0E-01
2588.18	2	2	1.6E-01	2707.13	4	6	8.5E-01	2848.32	6	4	1.1E+00
2590.55	4	6	9.1E-02	2709.05	4	6	3.7E-01	2855.69	8	10	1.0E-01
2591.54	6	6	5.1E-01	2711.84	12	14	3.8E-01	2856.38	6	8	2.7E-01

NIST ATOMIC TRANSITION PROBABILITY TABLES (continued)

λ (Å)	Weights g_i	Weights g_k	A (10^8 s^{-1})
2856.91	8	8	8.7E-01
2857.17	6	8	9.5E-02
2872.39	10	8	1.5E-01
2873.40	8	10	3.4E-01
2875.35	8	10	9.5E-02
2883.71	12	14	1.0E-01
2884.77	6	8	1.4E-01
2895.22	8	10	8.0E-02
2897.27	6	4	1.4E-01
2944.40	4	2	4.6E-01
2947.66	6	4	2.0E-01
2949.18	10	8	2.0E-01
2959.84	8	6	1.6E-01
2964.63	2	2	9.3E-02
2969.93	8	6	1.8E-01
2982.06	4	6	2.1E-01
2984.82	6	6	3.6E-01
2985.55	2	4	1.8E-01
2997.30	6	8	8.3E-02
3002.65	4	6	1.4E-01
3036.96	6	6	1.6E-01
3048.99	4	4	2.8E-01
3062.23	12	10	1.2E-01
3071.12	2	4	1.9E-01
3076.44	4	6	2.8E-01
3077.17	14	12	1.1E-01
3078.68	6	8	4.2E-01
3135.36	6	6	8.4E-02
3154.20	10	10	1.5E-01
3167.86	8	8	1.3E-01
3177.54	8	8	8.1E-02
3179.50	6	8	9.9E-02
5247.95	4	6	1.7E+00
5506.20	12	14	1.4E+00
5961.71	10	12	7.7E-01

Fe III

λ (Å)	g_i	g_k	A (10^8 s^{-1})
1843.4	9	7	4.8E+00
1844.3	7	5	4.9E+00
1846.9	5	3	5.5E+00
1854.38	3	1	5.7E+00
1865.20	7	7	6.1E+00
1893.98	11	9	5.5E+00
1896.80	13	11	5.0E+00
1904.3	5	5	5.7E+00
1907.58	15	13	5.3E+00
1915.08	13	15	6.0E+00
1922.79	11	13	5.5E+00
1930.39	9	11	5.1E+00
1931.51	9	11	5.3E+00
1937.35	7	9	5.1E+00
1943.48	5	7	5.0E+00
1950.33	13	15	5.5E+00
1951.01	11	11	5.3E+00
1952.65	9	9	4.9E+00
1953.32	7	7	5.1E+00
1987.50	13	13	4.9E+00

Fe VII

λ (Å)	g_i	g_k	A (10^8 s^{-1})
150.807	5	7	1.3E+03
150.852	7	9	1.3E+03
151.023	9	11	1.6E+03
151.046	7	7	2.2E+02
151.145	9	9	2.1E+02
151.432	5	7	2.2E+02
151.512	5	5	5.3E+02
151.675	7	7	3.9E+02
151.782	9	9	2.4E+02
154.307	3	1	8.9E+02
154.335	5	7	1.2E+03
154.363	3	3	4.2E+02
154.565	5	3	3.5E+02
154.650	5	5	8.8E+02
154.848	1	3	7.7E+02
154.921	3	5	9.7E+02
154.941	3	3	2.4E+02
154.949	5	7	1.0E+03
155.994	9	11	1.8E+03
158.481	9	9	2.3E+02
165.087	1	3	6.9E+02
165.919	7	5	2.8E+03
166.365	9	7	2.9E+03
173.441	9	9	3.6E+03
176.744	9	9	2.7E+03
176.928	7	7	2.4E+03
177.172	5	5	1.5E+03
235.221	5	3	1.7E+02
240.053	3	1	1.3E+02
243.379	9	7	2.1E+02

Fe VIII

λ (Å)	g_i	g_k	A (10^8 s^{-1})
112.472	4	4	3.6E+02
112.486	6	6	4.3E+02
116.196	4	6	4.5E+02
117.197	6	8	3.8E+02
167.486	4	4	3.0E+03
168.172	6	6	3.1E+03
168.545	6	4	2.0E+03
168.929	4	2	2.1E+03
185.213	6	8	1.0E+03
186.601	4	6	9.4E+02

Fe X

λ (Å)	g_i	g_k	A (10^8 s^{-1})
76.822	2	2	1.8E+03
77.865	4	6	1.6E+03
100.026	8	10	2.6E+03
101.733	6	8	1.8E+03
101.846	4	6	1.7E+03
102.095	10	12	2.9E+03
102.192	10	12	2.9E+03
102.829	4	6	2.1E+03
103.319	6	8	2.6E+03
103.724	6	8	1.7E+03
104.638	8	10	2.1E+03
174.534	4	6	1.8E+03

λ (Å)	g_i	g_k	A (10^8 s^{-1})
175.266	2	4	1.72E+03

Fe XI

λ (Å)	g_i	g_k	A (10^8 s^{-1})
72.166	5	7	2.9E+03
72.310	5	5	1.5E+03
72.635	5	7	1.6E+03
91.394	5	7	2.6E+03
91.472	7	9	2.5E+03
91.63	3	5	2.3E+03
91.63	7	9	3.4E+03
91.63	5	7	2.8E+03
91.733	9	11	4.1E+03
92.81	9	11	3.7E+03
92.87	11	13	3.9E+03
93.433	9	11	3.2E+03
179.762	5	7	1.67E+03

Fe XII

λ (Å)	g_i	g_k	A (10^8 s^{-1})
65.905	4	4	2.0E+03
66.526	6	8	1.7E+03
66.960	4	6	1.6E+03
67.164	4	2	1.1E+03
67.821	4	6	1.4E+03
68.382	2	4	1.7E+03
80.541	6	6	8.7E+02
81.943	6	4	1.4E+03
82.226	4	2	1.9E+03
84.48	4	6	4.5E+03
84.48	8	10	4.9E+03
84.52	10	12	5.2E+03
84.52	6	8	4.0E+03
84.85	6	8	2.3E+03
85.14	8	10	3.4E+03
85.477	10	12	4.6E+03
186.880	6	8	1.0E+03
192.394	4	2	9.0E+02
193.509	4	4	9.1E+02
195.119	4	6	8.6E+02

Fe XIII

λ (Å)	g_i	g_k	A (10^8 s^{-1})
62.353	1	3	2.0E+03
62.46	5	7	1.2E+03
62.699	3	5	2.3E+03
63.188	5	7	3.9E+03
64.139	1	3	2.1E+03
74.845	5	5	1.0E+03
75.892	5	3	7.7E+02
76.117	5	3	2.1E+03
78.452	9	11	6.3E+03
84.270	7	9	5.5E+03
107.384	7	5	1.8E+03

Fe XIV

λ (Å)	g_i	g_k	A (10^8 s^{-1})
58.963	2	4	2.7E+03
59.579	4	6	3.1E+03
69.176	4	6	5.6E+02
69.386	2	4	7.6E+02

λ Å	g_i	g_k	A 10^8 s^{-1}	λ Å	g_i	g_k	A 10^8 s^{-1}	λ Å	g_i	g_k	A 10^8 s^{-1}
69.66	2	2	8.9E+02	248	3	1	5.4E+02	12.264	1	3	5.9E+04
69.66	6	6	1.3E+03	284.160	1	3	2.28E+02	12.526	1	3	3.0E+03
70.251	6	4	8.1E+02					12.681	1	3	3.5E+03
70.613	4	2	1.7E+03	**Fe XVI**				13.823	1	3	3.3E+04
72.80	10	12	7.9E+03	31.041	2	4	5.2E+02	13.891	1	3	3.4E+03
76.022	4	6	6.6E+03	31.242	4	6	6.1E+02	15.015	1	3	2.28E+05
76.152	6	8	7.0E+03	32.166	2	4	6.8E+02	15.262	1	3	6.0E+04
91.009	6	4	5.1E+02	32.192	2	2	6.7E+02	16.777	1	3	8.29E+03
91.273	4	2	5.6E+02	32.433	2	4	7.7E+02	17.054	1	3	9.33E+03
188	4	6	2.7E+02	32.652	4	6	9.1E+02	41.37	9	11	4.8E+03
190	6	8	2.8E+02	34.857	2	4	1.23E+03	49.427	3	3	4.0E+03
207	2	2	2.1E+02	35.106	4	6	1.44E+03	50.26	7	9	6.0E+03
211.316	2	4	3.6E+02	35.333	4	6	6.4E+02	58.76	9	11	1.2E+04
213	4	2	2.8E+02	35.368	6	8	6.8E+02				
214	2	2	4.0E+02	36.01	4	2	5.0E+02	**Fe XIX**			
216	6	8	1.7E+02	36.749	2	4	1.1E+03	13.413	5	3	1.3E+04
217	6	8	4.0E+02	36.803	2	2	1.2E+03	13.426	5	7	4.8E+03
217	6	6	2.6E+02	37.096	4	6	1.0E+03	13.47	3	1	1.5E+05
219	2	4	4.8E+02	37.138	6	8	1.07E+03	13.520	5	7	2.0E+05
219	4	6	2.4E+02	39.827	2	4	2.1E+03	13.56	3	5	1.0E+04
219.123	4	6	3.9E+02	40.153	4	6	2.5E+03	13.68	3	1	8.0E+04
220	4	4	3.2E+02	40.161	4	4	4.1E+02	13.69	5	7	2.3E+04
221	4	6	5.9E+02	40.199	4	6	1.7E+03	13.700	1	3	2.7E+05
226	2	4	3.9E+02	40.245	6	8	1.8E+03	13.71	5	5	2.2E+04
234	2	2	2.8E+02	41.91	2	2	4.72E+02	13.738	5	7	1.0E+04
264.787	4	4	3.38E+02	42.30	4	2	9.2E+02	13.796	5	7	7.0E+04
265	4	4	1.5E+02	46.661	4	6	3.46E+03	13.83	5	5	1.4E+04
266	6	4	1.7E+02	46.718	6	8	3.7E+03	13.934	1	3	4.51E+04
268	6	6	2.1E+02	50.350	2	4	1.86E+03	13.961	3	3	2.0E+04
268	4	2	3.3E+02	50.555	2	2	1.98E+03	14.668	5	7	1.1E+04
270.524	4	2	2.1E+02	54.142	2	4	3.41E+03	14.671	5	3	1.1E+04
274.203	2	2	1.8E+02	54.728	4	6	4.16E+03	14.929	3	3	1.2E+04
280	4	6	2.8E+02	54.769	4	4	6.97E+02	14.966	5	3	2.5E+04
283	6	8	2.7E+02	62.879	2	2	1.05E+03	14.995	5	5	2.2E+04
288.45	6	4	1.6E+02	63.719	4	2	2.18E+03	15.015	1	3	1.4E+04
				66.263	4	6	9.39E+03	16.668	3	1	1.1E+04
Fe XV				66.368	6	8	1.00E+04				
38.95	1	3	1.69E+03	66.392	6	6	6.69E+02	**Fe XX**			
52.911	1	3	2.94E+03	76.502	6	4	6.7E+02	12.67	6	6	1.0E+04
59.404	3	5	3.4E+03	76.796	4	2	7.72E+02	12.69	4	6	1.2E+04
63.959	5	7	1.6E+03	80.192	4	6	5.2E+02	12.73	4	2	4.0E+04
65.370	1	3	3.2E+02	80.270	6	8	5.4E+02	12.77	4	4	2.1E+05
65.612	3	3	9.8E+02	85.587	2	4	4.0E+02	12.78	4	2	6.9E+04
66.238	5	3	1.6E+03	86.133	4	6	4.8E+02	12.78	2	4	1.4E+05
68.860	9	11	9.2E+03	96.256	4	6	8.7E+02	12.79	6	4	1.7E+04
69.7	3	1	1.9E+03	96.348	6	8	9.3E+02	12.82	4	4	1.1E+05
69.942	3	5	7.4E+03	117.2	2	4	3.93E+02	12.88	6	4	2.7E+04
69.989	5	7	7.9E+03	117.7	2	2	3.9E+02	12.89	4	4	4.4E+04
70.052	7	9	8.8E+03	123.4	2	4	5.9E+02	12.90	4	2	6.2E+03
70.224	1	3	4.13E+03	124.5	4	6	7.0E+02	12.90	4	6	1.4E+05
70.53	7	5	2.6E+02	144.06	4	6	1.6E+03	12.92	2	4	1.7E+04
70.59	7	7	1.7E+03	144.25	6	8	1.6E+03	12.93	4	6	1.6E+05
73.199	7	9	8.8E+03	148	4	2	6.5E+02	12.93	2	2	1.2E+04
73.473	5	7	6.2E+03	266.7	4	6	3.9E+02	12.98	2	2	6.7E+04
233.857	5	7	2.2E+02	267.0	6	8	4.3E+02	12.99	6	6	5.1E+04
235	1	3	2.5E+02					13.00	6	4	1.1E+04
243	1	3	2.4E+02	**Fe XVII**				13.01	2	4	3.0E+04
243	5	7	2.3E+02	11.023	1	3	2.1E+04	13.03	4	2	8.6E+04
243.790	3	5	4.2E+02	12.123	1	3	8.0E+04	13.07	6	4	8.2E+03

λ Å	Weights g_i	Weights g_k	A 10^8 s^{-1}	λ Å	Weights g_i	Weights g_k	A 10^8 s^{-1}	λ Å	Weights g_i	Weights g_k	A 10^8 s^{-1}
13.13	2	4	8.9E+04	13.14	3	1	2.0E+04	11.325	3	5	1.7E+05
13.24	4	4	1.2E+04	13.41	1	3	7.3E+03	11.338	3	3	9.3E+04
13.28	4	4	6.1E+03					11.429	3	1	1.7E+05
13.70	4	6	1.1E+04	**Fe XXII**				11.433	3	3	1.2E+05
13.71	2	2	9.9E+03	9.002	4	6	5.5E+04	11.441	5	7	2.2E+05
13.78	4	4	1.0E+04	9.006	6	8	5.7E+04	11.445	5	5	5.6E+04
13.79	6	6	1.2E+04	9.006	6	6	5.3E+04	11.485	3	5	1.40E+05
13.83	4	2	9.8E+03	9.163	4	6	6.9E+04	11.491	5	3	5.9E+04
13.90	4	2	1.2E+04	9.183	6	8	8.3E+04	11.519	5	5	1.16E+05
13.98	6	4	1.6E+04	9.241	4	6	5.1E+04	11.520	1	3	2.16E+05
13.99	4	2	2.2E+04	11.748	4	4	1.2E+05	11.524	5	7	2.3E+05
14.05	4	4	1.7E+04	11.748	4	6	1.6E+05	11.593	5	7	3.58E+05
14.23	2	2	6.3E+03	11.748	4	2	1.8E+05	11.613	3	5	1.0E+05
				11.763	2	4	1.6E+05	11.615	3	3	4.4E+04
Fe XXI				11.789	2	2	2.6E+05	11.691	5	7	7.7E+04
8.53	3	1	1.8E+04	11.789	6	8	1.2E+05	11.698	5	5	7.3E+04
8.53	3	5	6.1E+03	11.797	2	4	1.7E+05	11.737	3	5	1.8E+05
8.53	3	3	1.5E+04	11.823	6	4	7.9E+04	11.898	1	3	2.03E+05
8.56	5	7	2.0E+04	11.837	6	8	2.3E+05				
8.56	1	3	2.1E+04	11.837	6	6	1.7E+05	**Fe XXIV**			
8.56	5	3	6.5E+03	11.886	4	6	1.3E+05	1.8523	2	2	1.0E+05
8.64	5	7	1.5E+04	11.898	2	4	8.2E+04	1.8552	2	4	4.82E+06
8.65	5	7	3.9E+04	11.922	4	6	1.8E+05	1.8563	4	2	2.43E+06
8.66	5	5	4.4E+03	11.976	6	8	5.9E+04	1.8572	2	2	3.06E+06
8.74	1	3	2.5E+04	12.027	2	4	6.9E+04	1.858	2	4	1.2E+05
9.42	3	1	4.3E+04	12.045	6	8	2.4E+05	1.8614	4	4	6.24E+06
9.42	3	3	3.3E+04	12.045	4	4	9.7E+04	1.8626	2	4	3.16E+06
9.44	3	5	1.7E+04	12.053	4	6	6.1E+04	1.8627	2	2	5.47E+06
9.45	1	3	5.2E+04	12.077	2	4	1.0E+05	1.8637	2	2	1.91E+06
9.46	5	3	1.5E+04	12.077	4	6	2.4E+05	1.8655	4	6	2.14E+06
9.47	5	7	4.9E+04	12.095	6	6	7.8E+04	1.8672	4	2	1.63E+06
9.47	5	5	6.1E+03	12.193	2	4	7.2E+04	1.8678	4	4	3.5E+05
9.52	3	3	8.1E+03	12.193	4	6	9.9E+04	1.8721	4	6	3.2E+05
9.58	5	5	5.2E+03	12.325	2	2	1.5E+05	1.8721	2	2	2.0E+05
9.59	5	5	1.0E+04					1.8730	2	4	1.5E+05
9.67	1	3	5.7E+04	**Fe XXIII**				1.8739	4	4	8.3E+04
9.68	5	7	4.0E+03	7.733	5	7	3.0E+04	1.891	2	2	9.7E+04
9.74	5	3	5.3E+03	7.849	5	7	4.9E+04	1.897	4	2	9.8E+04
12.02	1	3	1.3E+04	8.307	1	3	4.8E+04	8.231	2	4	6.10E+04
12.13	3	3	1.8E+04	8.529	1	3	4.3E+04	8.316	4	6	7.07E+04
12.18	5	7	2.2E+04	8.550	3	5	6.0E+04	10.619	2	4	7.28E+04
12.19	5	3	6.4E+03	8.552	3	3	3.2E+04	10.663	2	2	7.51E+04
12.21	3	1	1.5E+05	8.614	5	7	7.7E+04	11.030	2	4	1.84E+05
12.21	3	3	1.2E+05	8.664	3	3	4.4E+04	11.171	4	6	2.18E+05
12.25	1	3	2.1E+05	8.669	5	7	6.1E+04				
12.28	5	3	5.2E+04	8.672	1	3	6.8E+04	**Fe XXV**			
12.30	5	7	2.1E+05	8.752	5	7	1.2E+05	1.4607	1	3	2.54E+05
12.36	3	3	3.6E+04	8.764	5	7	4.6E+04	1.4945	1	3	5.05E+05
12.37	5	7	3.1E+05	8.814	3	5	6.2E+04	1.5730	1	3	1.24E+06
12.38	5	7	6.9E+03	10.902	5	5	5.3E+04	1.5749	1	3	1.5E+05
12.47	5	7	5.8E+04	10.910	3	1	6.7E+04	1.778	3	3	8.7E+04
12.47	5	3	1.3E+04	10.927	5	7	6.0E+04	1.782	3	1	4.69E+06
12.49	5	7	1.3E+04	10.934	3	5	5.4E+04	1.787	1	3	2.57E+06
12.53	5	5	1.5E+04	10.979	1	3	7.9E+04	1.787	5	5	1.19E+06
12.57	1	3	7.2E+04	11.018	1	3	4.9E+04	1.788	3	5	2.68E+06
12.73	5	5	8.2E+03	11.086	3	1	6.5E+04	1.788	3	5	1.63E+06
12.95	3	5	6.2E+03	11.165	3	5	6.7E+04	1.789	1	3	1.78E+06
13.00	1	3	7.2E+03	11.255	3	3	3.7E+04	1.790	3	3	1.23E+06
13.03	5	5	1.3E+04	11.298	1	3	1.3E+05	1.791	3	5	4.10E+06

λ Å	g_i	g_k	A 10^8 s^{-1}	λ Å	g_i	g_k	A 10^8 s^{-1}	λ Å	g_i	g_k	A 10^8 s^{-1}
1.791	3	3	2.59E+06	4619.2	4	6	8.1E-01				
1.792	3	1	4.92E+06	4633.9	4	6	7.1E-01	**Magnesium**			
1.792	5	5	2.81E+06	4658.9	6	4	6.5E-01	**Mg I**			
1.793	3	1	2.67E+06	4739.0	6	6	7.6E-01	2025.8	1	3	8.4E-01
1.794	5	3	2.22E+06	4762.4	2	4	4.2E-01	*2779.8	9	9	5.2E+00
1.797	3	5	8.8E+05	4765.7	4	6	6.7E-01	*2850.0	9	15	2.3E-01
1.798	3	3	1.0E+05	4811.8	2	4	1.7E-01	2852.1	1	3	4.95E+00
1.800	1	3	8.6E+04	4825.2	2	4	1.9E-01	*3094.9	9	15	5.2E-01
1.802	3	1	4.1E+05	4832.1	4	2	7.3E-01	3329.9	1	3	3.3E-02
1.810	3	1	5.9E+05	5208.3	4	4	1.4E-01	3332.2	3	3	9.7E-02
1.8502	1	3	4.57E+06	5308.7	4	6	2.4E-02	3336.7	5	3	1.6E-01
1.8593	1	3	4.42E+05	7407.0	6	6	7.0E-02	*3835.3	9	15	1.68E+00
10.038	3	3	8.08E+04					4703.0	3	5	2.55E-01
				Lead				5167.3	1	3	1.16E-01
Krypton				**Pb I**				5172.7	3	3	3.46E-01
Kr I				2022.0	1	3	5.2E-02	5183.6	5	3	5.75E-01
1164.9	1	3	3.16E+00	2053.3	1	3	1.2E-01	5528.4	3	5	1.99E-01
1235.8	1	3	3.12E+00	2170.0	1	3	1.5E+00				
4274.0	5	5	2.6E-02	2401.9	3	3	1.9E-01	**Mg II**			
4351.4	3	1	3.2E-02	2446.2	3	3	2.5E-01	1239.9	2	4	1.4E-02
4362.6	5	3	8.4E-03	2476.4	3	5	2.8E-01	1240.4	2	2	1.4E-02
4376.1	3	1	5.6E-02	2577.3	5	3	5.0E-01	*2660.8	10	14	3.8E-01
4400.0	3	5	2.0E-02	2613.7	3	3	2.7E-01	2790.8	2	4	4.0E+00
4410.4	3	3	4.4E-03	2614.2	3	5	1.9E+00	2795.5	2	4	2.6E+00
4425.2	3	3	9.7E-03	2628.3	5	3	3.1E-02	2797.9	4	4	7.9E-01
4453.9	3	5	7.8E-03	2657.1	3	5	9.8E-04	2798.1	4	6	4.8E+00
4463.7	3	3	2.3E-02	2663.2	5	5	7.1E-01	2802.7	2	2	2.6E+00
4502.4	3	5	9.2E-03	2802.0	5	7	1.6E+00	2928.8	2	2	1.2E+00
5562.2	5	5	2.8E-03	2823.2	5	5	2.6E-01	2936.5	4	2	2.3E+00
5570.3	5	3	2.1E-02	2833.1	1	3	5.8E-01	*3104.8	10	14	8.1E-01
5649.6	1	3	3.7E-03	2873.3	5	5	3.7E-01	3848.2	6	4	2.8E-02
5870.9	3	5	1.8E-02	3572.7	5	3	9.9E-01	3848.3	4	4	3.0E-03
6904.7	3	5	1.3E-02	3639.6	3	3	3.4E-01	3850.4	4	2	3.0E-02
7224.1	3	5	1.4E-02	3671.5	5	3	4.4E-01	*4481.2	10	14	2.23E+00
7587.4	3	1	5.1E-01	3683.5	3	1	1.5E+00	9218.3	2	4	3.6E-01
7601.5	5	5	3.1E-01	3739.9	5	5	7.3E-01	9244.3	2	2	3.6E-01
7685.2	3	1	4.9E-01	4019.6	5	7	3.5E-02				
7694.5	5	3	5.6E-02	4057.8	5	3	8.9E-01	**Mg IV**			
7854.8	1	3	2.3E-01	4062.1	5	3	9.2E-01	320.99	4	2	1.2E+02
8059.5	1	3	1.9E-01	4168.0	5	5	1.2E-02	323.31	2	2	5.9E+01
8104.4	5	5	1.3E-01	5005.4	1	3	2.7E-01	1219.0	6	6	5.9E+00
8112.9	5	7	3.6E-01	5201.4	1	3	1.9E-01	1375.5	4	4	4.5E+00
8190.1	3	5	1.1E-01	7229.0	5	3	8.9E-03	1459.6	6	4	4.6E+00
8263.2	3	5	3.5E-01					1495.5	4	6	6.4E+00
8281.1	3	3	1.9E-01	**Lithium**				1510.7	4	4	6.7E+00
8298.1	3	3	3.2E-01	**Li I**				1683.0	6	8	5.8E+00
8508.9	3	3	2.4E-01	*2741.2	2	6	1.3E-02	1698.8	4	6	3.9E+00
8776.7	3	5	2.7E-01	*3232.7	2	6	1.17E-02	1893.9	6	6	2.8E+00
8928.7	5	3	3.7E-01	*4602.9	6	10	2.23E-01				
				*6103.6	6	10	6.860E-01	**Mg VI**			
Kr II				*6707.8	2	6	3.691E-01	*269.92	10	6	3.1E+02
4250.6	4	4	1.2E-01					*292.53	6	6	9.0E+01
4292.9	4	4	9.6E-01	**Lutetium**				*314.64	6	2	1.8E+02
4355.5	6	8	1.0E+00	**Lu I**				*349.15	10	10	6.1E+01
4431.7	2	2	1.8E+00	3376.5	4	4	2.23E+00	*387.94	6	10	1.3E+01
4436.8	2	4	6.6E-01	3567.8	4	6	5.9E-01	399.29	4	2	2.8E+01
4577.2	6	8	9.6E-01	3620.3	6	4	1.1E-02	400.68	4	4	2.8E+01
4583.0	6	4	7.6E-01	3841.2	6	6	2.5E-01	403.32	4	6	2.7E+01
4615.3	4	4	5.4E-01	4518.6	4	4	2.1E-01				

λ Å	Weights		A 10^8 s^{-1}	λ Å	Weights		A 10^8 s^{-1}	λ Å	Weights		A 10^8 s^{-1}
	g_i	g_k			g_i	g_k			g_i	g_k	
Mg VII				3007.65	6	8	1.8E-01	3773.86	12	12	2.5E-01
277.01	3	3	9.5E+01	3011.38	8	10	3.1E-01	3800.55	6	8	2.7E-01
278.41	5	3	1.5E+02	3016.45	10	12	2.9E-01	3806.72	10	12	5.9E-01
280.74	5	3	2.0E+02	3043.36	8	8	5.9E-01	3823.51	8	10	5.21E-01
319.02	5	5	8.9E+01	3044.57	10	8	5.7E-01	3823.89	6	6	2.31E-01
*366.42	9	9	4.4E+01	3045.59	10	10	6.7E-01	3833.87	4	4	3.14E-01
*433.04	9	15	1.6E+01	3045.80	8	10	1.7E-01	3834.37	6	8	4.29E-01
1334.3	5	5	5.3E+00	3047.03	12	12	6.1E-01	3839.78	2	2	4.64E-01
1410.0	5	5	2.57E+00	3054.36	8	6	4.6E-01	3841.07	4	6	3.3E-01
1487.0	3	5	3.02E+00	3070.27	6	6	1.9E-01	3843.99	2	4	2.11E-01
1487.9	5	7	3.66E+00	3073.18	4	4	3.7E-01	3889.46	12	14	3.1E-01
				3082.71	14	14	2.9E-01	3898.37	6	8	1.7E-01
Mg VIII				3110.68	6	8	2.7E-01	3899.34	4	6	2.4E-01
*74.976	6	10	4.3E+03	3113.80	12	10	2.6E-01	3924.08	2	4	9.4E-01
315.02	4	4	1.2E+02	3118.10	4	6	1.7E-01	3926.48	6	8	5.4E-01
*342.29	10	6	6.3E+01	3122.88	10	10	1.9E-01	3951.98	2	2	3.1E-01
353.86	4	4	3.89E+01	3126.85	8	6	2.3E-01	3952.84	6	6	4.1E-01
356.00	6	4	5.7E+01	3132.28	10	10	2.1E-01	3975.88	2	4	1.8E-01
*428.52	10	10	3.24E+01	3132.79	8	8	2.7E-01	3982.16	4	2	3.5E-01
*434.62	6	10	1.6E+01	3175.58	8	10	1.8E-01	3982.58	6	4	2.3E-01
*489.33	6	6	3.9E+01	3201.11	4	6	2.2E-01	3982.90	6	4	5.5E-01
*686.92	6	10	9.4E+00	3228.09	10	12	6.4E-01	3991.60	2	2	2.1E-01
				3230.23	10	12	1.9E-01	4011.91	8	8	2.3E-01
Mg IX				3230.72	8	8	3.5E-01	4018.11	10	8	2.54E-01
62.751	1	3	2.87E+03	3240.88	6	4	2.2E-01	4030.76	6	8	1.7E-01
*67.189	9	15	6.20E+03	3243.78	6	6	5.3E-01	4033.07	6	6	1.65E-01
*71.965	9	3	1.22E+03	3251.13	4	2	2.3E-01	4034.49	6	4	1.58E-01
72.312	3	5	4.43E+03	3252.95	4	4	1.8E-01	4041.36	10	10	7.87E-01
77.737	3	1	3.92E+02	3256.14	4	6	5.0E-01	4048.75	6	4	7.5E-01
368.07	1	3	5.27E+01	3258.41	2	2	9.7E-01	4052.48	6	8	3.8E-01
438.69	3	1	7.9E+01	3260.24	2	4	3.8E-01	4055.55	8	8	4.31E-01
*443.74	9	9	4.19E+01	3267.79	14	14	3.5E-01	4058.94	4	2	7.25E-01
749.55	3	5	8.2E+00	3268.72	6	8	3.3E-01	4061.74	8	6	1.9E-01
1639.8	3	5	2.1E+00	3270.35	12	12	2.6E-01	4063.53	6	6	1.69E-01
2814.2	1	3	3.35E-01	3273.02	10	10	2.7E-01	4065.08	12	14	2.5E-01
				3298.23	6	4	2.8E-01	4066.24	10	8	2.2E-01
Mg X				3303.28	4	4	1.9E-01	4070.28	2	2	2.3E-01
57.876	2	4	2.09E+03	3463.66	8	8	3.2E-01	4079.42	2	4	3.8E-01
57.920	2	2	2.09E+03	3470.01	6	8	2.4E-01	4082.95	4	6	2.95E-01
63.152	2	4	5.6E+03	3511.83	12	12	2.7E-01	4083.63	6	8	2.8E-01
63.295	4	6	6.7E+03	3535.30	10	10	1.7E-01	4089.94	8	10	1.7E-01
609.79	2	4	7.53E+00	3559.81	6	6	2.1E-01	4105.37	10	8	1.7E-01
624.94	2	2	7.01E+00	3577.87	10	8	9.4E-01	4135.03	12	12	3.0E-01
2212.5	2	4	9.64E-01	3595.11	6	4	1.8E-01	4141.06	10	10	2.6E-01
2278.7	2	2	8.82E-01	3601.27	12	10	2.3E-01	4148.80	8	8	2.3E-01
5918.7	2	4	3.20E-02	3607.53	8	8	2.3E-01	4176.61	14	12	2.4E-01
6229.6	4	6	3.30E-02	3608.49	6	6	3.6E-01	4189.99	12	10	2.0E-01
				3610.30	4	4	4.2E-01	4201.78	10	8	2.3E-01
Mg XI				3635.70	10	8	2.1E-01	4235.30	8	6	9.17E-01
7.310	1	3	1.15E+04	3660.40	12	14	9.1E-01	4239.74	4	2	3.9E-01
7.473	1	3	2.27E+04	3675.67	6	8	2.2E-01	4257.67	2	2	3.7E-01
7.850	1	3	5.50E+04	3676.96	10	12	7.3E-01	4265.93	4	4	4.92E-01
9.169	1	3	1.97E+05	3680.15	12	10	1.9E-01	4281.10	6	6	2.3E-01
				3682.09	8	10	7.6E-01	4411.87	12	10	2.6E-01
Manganese				3684.87	6	8	2.6E-01	4414.89	8	6	2.93E-01
Mn I				3706.08	12	14	1.4E+00	4419.77	10	8	2.1E-01
2794.82	6	8	3.7E+00	3718.92	10	12	9.6E-01	4436.36	6	4	4.37E-01
2798.27	6	6	3.6E+00	3731.94	8	10	1.0E+00	4451.58	8	8	7.98E-01
2801.08	6	4	3.7E+00	3771.44	14	14	1.9E-01	4453.01	4	2	5.44E-01

λ Å	Weights g_i	g_k	A 10^8 s^{-1}	λ Å	Weights g_i	g_k	A 10^8 s^{-1}	λ Å	Weights g_i	g_k	A 10^8 s^{-1}
4455.82	4	6	1.7E-01	1285.10	5	7	1.1E+01	2751.47	7	9	2.54E-01
4457.04	6	4	2.34E-01	1333.87	7	9	1.0E+01	2756.26	5	3	1.18E-01
4457.55	6	6	4.27E-01					2761.53	9	11	2.06E-01
4458.26	6	8	4.62E-01	**Mercury**				2763.02	3	1	4.44E-01
4461.09	8	8	1.7E-01	**Hg I**				2766.25	3	5	1.17E-01
4462.03	8	10	7.00E-01	2536.52	1	3	8.00E-02	2787.83	9	7	2.85E-01
4464.68	6	6	4.39E-01	2652.04	3	5	3.88E-01	2792.96	5	3	1.53E-01
4470.14	4	4	3.00E-01	2655.13	3	5	1.1E-01	2798.02	7	5	1.22E-01
4472.79	2	2	4.35E-01	2752.78	1	3	6.10E-02	2801.47	5	7	1.24E-01
4479.40	8	10	3.4E-01	2856.94	3	1	1.1E-02	2825.68	5	7	2.53E-01
4490.08	2	4	2.49E-01	2893.60	3	3	1.6E-01	2826.75	7	7	4.23E-01
4498.90	4	6	2.49E-01	2925.4	5	3	7.7E-02	2876.54	9	9	2.84E-01
4502.22	6	8	1.86E-01	2967.3	1	3	4.5E-01	2886.60	11	11	4.74E-01
4605.37	10	12	3.6E-01	3021.50	5	7	5.09E-01	2906.06	3	3	8.04E-01
4626.54	12	14	3.6E-01	3023.48	5	5	9.4E-02	2913.52	5	3	1.38E-01
4709.71	8	8	1.72E-01	3027.49	5	5	2.0E-02	2915.38	5	3	7.31E-01
4727.46	6	6	1.7E-01	3125.66	3	5	6.56E-01	2918.84	5	3	3.79E-01
4739.11	4	4	2.40E-01	3341.48	5	3	1.68E-01	2930.39	1	3	1.91E-01
4754.05	6	8	3.03E-01	3650.15	5	7	1.3E+00	2936.50	11	11	2.33E-01
4761.53	2	4	5.35E-01	3654.83	5	5	1.8E-01	2945.43	7	7	3.66E-01
4762.38	8	10	7.83E-01	4046.56	1	3	2.1E-01	2945.66	3	3	4.08E-01
4765.86	4	6	4.1E-01	4077.81	3	1	4.0E-02	2946.01	5	5	1.68E-01
4766.43	6	8	4.6E-01	4108.1	3	1	3.0E-02	2951.45	9	9	1.43E-01
4783.43	8	8	4.01E-01	4339.22	3	5	2.88E-02	2959.48	9	11	1.75E-01
4823.53	10	8	4.99E-01	4347.50	3	5	8.4E-02	2972.96	5	3	2.69E-01
6013.48	4	6	1.72E-01	4358.34	3	3	5.57E-01	2977.27	9	7	3.28E-01
6021.79	8	6	3.32E-01	4916.07	3	1	5.8E-02	2978.28	7	5	1.50E-01
				5025.64	3	3	2.7E-04	2983.04	1	3	2.82E-01
Mn II				5460.75	5	3	4.87E-01	2987.92	3	5	8.43E-01
2593.72	7	7	2.6E+00	5769.59	3	5	2.36E-01	2988.23	5	7	4.28E-01
2605.68	7	5	2.7E+00	6234.4	1	3	5.3E-03	2988.68	7	9	1.61E-01
2933.05	5	3	2.0E+00	6716.4	1	3	4.3E-03	2989.80	9	7	9.27E-01
2939.31	5	5	1.9E+00	6907.5	3	5	2.8E-02	3000.24	9	9	1.40E-01
2949.20	5	7	1.9E+00	7728.8	1	3	9.7E-03	3000.44	5	5	1.25E-01
3441.99	9	7	4.3E-01	10139.79	3	1	2.71E-01	3000.85	5	7	2.58E-01
3460.32	7	5	3.2E-01					3001.43	5	5	2.31E-01
3474.13	5	3	1.5E-01	**Molybdenum**				3007.71	7	5	1.90E-01
3482.90	5	5	2.0E-01	**Mo I**				3013.39	7	5	6.06E-01
3488.68	3	3	2.5E-01	2616.79	3	5	7.34E-01	3016.78	9	9	2.75E-01
				2621.06	7	7	1.16E-01	3025.00	5	5	8.49E-01
Mn VI				2628.96	3	3	2.81E-01	3036.31	3	5	5.81E-01
307.999	9	9	3.7E+01	2629.85	5	7	7.75E-01	3041.70	13	11	5.94E-01
309.440	9	7	5.7E+01	2631.50	1	3	2.54E-01	3046.80	13	11	1.63E-01
309.579	7	5	4.4E+01	2638.30	5	5	7.57E-01	3047.31	11	9	5.01E-01
310.058	7	7	3.4E+01	2640.98	7	5	1.20E+00	3055.32	9	7	4.29E-01
310.182	5	5	2.8E+01	2644.36	5	7	1.96E-01	3057.56	7	5	2.64E-01
311.748	5	3	5.7E+01	2649.46	7	9	9.84E-01	3061.59	7	5	4.41E-01
320.598	3	5	1.5E+01	2655.02	9	7	4.08E-01	3064.27	13	13	8.46E-01
320.681	1	3	2.2E+01	2658.11	7	5	6.43E-01	3065.04	13	13	3.08E-01
320.874	3	1	7.8E+01	2665.09	7	9	1.32E-01	3069.51	5	5	1.52E-01
320.979	3	5	2.2E+01	2679.85	9	11	1.31E+00	3069.96	11	11	2.72E-01
321.176	5	5	6.0E+01	2684.16	9	9	4.18E-01	3070.89	9	11	1.87E-01
321.541	5	3	2.7E+01	2706.11	3	5	2.03E-01	3074.37	11	11	1.42E+00
325.146	9	7	1.3E+02	2710.74	3	3	1.57E-01	3079.88	9	11	9.55E-01
328.431	5	5	4.4E+01	2725.15	3	5	2.79E-01	3080.40	7	9	3.61E-01
328.558	3	5	1.2E+01	2728.71	3	3	1.26E-01	3081.16	3	5	2.35E-01
329.043	1	3	1.1E+01	2733.39	5	7	2.95E-01	3085.62	9	9	1.63E+00
1236.23	5	3	1.3E+01	2743.71	1	3	2.47E-01	3089.13	11	9	1.53E-01
1255.77	3	1	1.2E+01	2745.38	13	11	1.29E-01	3089.71	5	7	2.34E-01

λ Å	Weights g_i	g_k	A 10^8 s^{-1}	λ Å	Weights g_i	g_k	A 10^8 s^{-1}	λ Å	Weights g_i	g_k	A 10^8 s^{-1}
3094.66	7	7	1.63E+00	3285.35	9	7	4.49E-01	3456.52	3	3	2.96E-01
3099.92	9	7	1.45E-01	3287.38	5	5	1.38E-01	3460.22	5	3	2.77E-01
3100.88	7	9	1.20E+00	3289.01	9	9	5.08E-01	3460.78	9	7	6.03E-01
3101.34	5	5	1.92E+00	3290.82	7	5	5.44E-01	3465.84	3	1	9.99E-01
3106.34	7	5	2.21E-01	3305.56	5	3	1.74E-01	3466.19	9	7	2.11E-01
3117.54	13	13	1.89E-01	3305.91	7	9	3.06E-01	3466.96	7	7	1.52E-01
3123.03	3	3	2.81E-01	3307.13	7	9	1.25E-01	3467.85	5	7	2.63E-01
3125.03	5	3	1.98E-01	3312.33	7	5	1.62E-01	3469.22	5	3	6.96E-01
3132.59	7	9	1.79E+00	3323.95	9	7	2.82E-01	3469.63	13	15	1.51E-01
3135.90	9	11	3.68E-01	3325.13	5	3	2.26E-01	3470.92	3	5	2.91E-01
3136.75	9	11	1.57E-01	3325.67	5	5	1.72E-01	3475.03	3	3	4.68E-01
3142.75	3	5	4.10E-01	3327.30	1	3	2.88E-01	3479.42	7	5	2.26E-01
3147.35	13	11	2.41E-01	3336.56	9	9	1.64E-01	3483.67	7	7	1.13E-01
3155.19	7	7	2.75E-01	3340.16	5	3	1.20E-01	3483.83	7	5	1.41E-01
3158.17	7	7	4.63E-01	3344.73	3	5	6.04E-01	3489.43	7	7	3.27E-01
3170.34	7	7	1.37E+00	3346.83	11	11	1.13E-01	3504.41	7	9	8.06E-01
3171.38	5	7	2.03E-01	3347.00	3	3	2.72E-01	3505.31	7	9	2.25E-01
3175.59	13	11	8.40E-01	3358.12	5	7	7.59E-01	3508.11	9	9	1.59E-01
3179.78	11	13	2.33E-01	3361.37	9	9	1.38E-01	3510.77	13	13	4.75E-01
3183.03	11	9	3.98E-01	3363.78	5	7	2.74E-01	3517.55	11	11	5.41E-01
3184.58	7	5	2.77E-01	3363.87	5	7	1.39E-01	3518.21	3	3	3.64E-01
3185.10	7	7	2.54E-01	3373.81	3	3	2.03E-01	3521.38	9	9	1.39E-01
3185.71	5	3	6.10E-01	3375.22	7	7	1.38E-01	3521.41	9	11	6.06E-01
3188.10	7	9	3.45E-01	3375.65	7	9	1.56E-01	3524.65	5	3	3.10E-01
3188.41	5	7	4.40E-01	3378.19	3	1	1.88E-01	3524.98	7	9	2.25E-01
3192.79	9	11	1.88E-01	3378.46	13	13	3.75E-01	3538.92	11	11	2.24E-01
3193.98	7	5	1.53E+00	3379.96	5	5	4.11E-01	3540.57	5	3	4.46E-01
3194.88	9	11	1.75E-01	3382.48	3	3	2.66E-01	3542.17	7	5	4.93E-01
3195.96	9	7	4.10E-01	3384.61	7	9	7.32E-01	3552.71	9	7	3.64E-01
3197.18	1	3	1.47E-01	3385.87	9	11	3.30E-01	3555.64	3	3	3.46E-01
3198.85	15	13	7.22E-01	3389.79	5	7	1.85E-01	3558.09	5	7	5.43E-01
3200.89	3	5	1.82E-01	3392.17	9	9	1.97E-01	3563.75	1	3	1.53E-01
3205.22	1	3	4.27E-01	3393.65	11	11	2.08E-01	3566.05	9	9	2.67E-01
3205.43	9	11	2.55E-01	3404.33	7	7	2.10E-01	3566.74	7	7	1.43E-01
3205.89	9	9	5.35E-01	3413.37	11	11	1.25E-01	3570.64	15	15	7.18E-01
3208.84	7	5	2.77E-01	3415.27	9	9	1.83E-01	3573.88	3	5	3.58E-01
3210.97	7	5	6.94E-01	3415.61	7	9	1.29E-01	3580.54	13	11	5.49E-01
3214.44	9	7	2.01E-01	3416.14	9	11	2.45E-01	3581.88	11	13	3.81E-01
3215.07	3	5	4.20E-01	3418.52	5	3	1.41E-01	3584.25	3	3	1.73E-01
3216.78	15	13	2.10E-01	3419.69	7	7	1.15E-01	3585.57	7	5	3.95E-01
3221.73	3	1	1.41E+00	3420.04	5	5	3.28E-01	3588.95	7	7	1.18E-01
3228.21	5	7	3.85E-01	3422.31	9	9	2.52E-01	3590.74	7	9	2.23E-01
3229.79	9	11	1.44E-01	3425.13	11	11	2.29E-01	3595.55	5	5	2.32E-01
3233.14	13	13	6.33E-01	3427.90	11	13	4.09E-01	3598.88	13	11	5.67E-01
3237.06	7	9	2.95E-01	3434.79	7	7	1.75E-01	3600.73	9	9	2.07E-01
3244.47	5	3	2.80E-01	3435.45	15	15	1.50E+00	3601.88	7	9	1.15E-01
3247.61	5	5	1.71E-01	3437.21	11	9	8.06E-01	3602.94	5	7	2.96E-01
3249.93	5	3	1.87E-01	3438.87	1	3	2.34E-01	3604.07	9	7	3.25E-01
3251.65	3	5	3.05E-01	3441.87	5	3	1.34E-01	3610.61	5	3	1.78E-01
3256.21	5	3	6.89E-01	3442.66	3	3	2.94E-01	3611.99	7	7	1.16E-01
3256.72	3	3	1.31E-01	3445.03	7	9	1.53E-01	3615.16	7	9	1.96E-01
3259.16	11	13	1.62E-01	3445.26	7	5	2.96E-01	3623.22	11	9	5.58E-01
3262.63	7	9	3.62E-01	3445.80	9	9	1.14E-01	3624.46	9	11	5.27E-01
3264.40	11	9	5.42E-01	3447.12	9	11	8.75E-01	3624.62	5	7	1.37E-01
3265.14	5	7	2.60E-01	3447.29	5	3	1.79E-01	3638.20	5	3	3.51E-01
3266.16	9	11	1.95E-01	3449.07	7	9	1.52E-01	3638.21	5	3	3.33E-01
3270.90	7	7	3.59E-01	3449.85	5	7	1.65E-01	3640.62	7	5	1.94E-01
3276.07	11	9	1.18E-01	3452.60	7	7	2.48E-01	3647.84	7	7	2.11E-01
3285.03	1	3	1.41E-01	3456.15	5	5	3.60E-01	3648.70	7	5	1.15E-01

λ Å	Weights g_i	g_k	A 10^8 s^{-1}	λ Å	Weights g_i	g_k	A 10^8 s^{-1}	λ Å	Weights g_i	g_k	A 10^8 s^{-1}
3654.58	3	3	1.80E-01	3834.64	3	5	1.20E-01	4536.80	13	15	5.03E-01
3657.36	5	7	2.03E-01	3846.18	7	7	1.26E-01	4598.23	1	3	1.47E-01
3658.13	9	9	1.86E-01	3847.25	3	1	2.41E-01	4624.23	9	9	1.32E-01
3659.36	7	9	6.70E-01	3848.30	9	9	1.26E-01	4633.08	3	5	2.35E-01
3660.92	3	5	1.34E-01	3851.99	11	9	1.78E-01	4649.06	3	1	1.25E-01
3662.15	7	9	1.45E-01	3864.10	7	7	6.24E-01	4652.24	5	7	1.55E-01
3662.99	11	11	3.48E-01	3866.69	3	5	1.74E-01	4686.08	3	3	1.72E-01
3663.27	7	5	2.30E-01	3867.67	5	3	2.22E-01	4688.21	13	15	1.54E-01
3664.81	11	13	9.54E-01	3869.08	5	3	1.35E-01	4707.25	7	9	3.63E-01
3664.88	1	3	1.92E-01	3874.15	7	5	1.67E-01	4718.86	5	5	2.17E-01
3669.34	9	7	2.16E-01	3902.95	7	5	6.17E-01	4723.05	9	9	1.23E-01
3672.81	9	11	1.95E-01	3909.54	9	7	1.13E-01	4731.44	9	11	4.49E-01
3672.82	9	9	1.13E-01	3911.94	5	7	1.15E-01	4758.50	11	9	3.01E-01
3676.23	3	1	5.22E-01	3915.43	5	5	1.40E-01	4760.18	11	13	4.67E-01
3680.68	11	11	2.96E-01	3916.43	5	3	1.78E-01	4764.11	9	7	2.16E-01
3681.72	9	7	1.68E-01	3919.55	11	13	2.24E-01	4811.05	13	11	4.36E-01
3683.01	3	5	1.20E-01	3955.48	13	11	1.71E-01	4819.25	11	9	2.71E-01
3687.96	5	7	2.12E-01	3973.76	11	13	4.39E-01	4830.51	9	7	4.07E-01
3688.97	11	9	3.26E-01	3977.90	9	7	1.35E-01	4858.39	13	11	1.24E-01
3690.59	11	9	2.07E-01	3980.20	5	3	2.70E-01	4868.02	7	5	3.11E-01
3694.94	5	7	6.36E-01	3991.85	11	9	1.29E-01	5037.18	9	7	1.14E-01
3696.04	11	11	3.59E-01	4010.13	5	3	4.38E-01	5044.36	7	5	1.31E-01
3698.07	7	5	1.48E-01	4021.01	9	11	2.65E-01	5047.70	3	1	2.61E-01
3708.55	7	9	1.28E-01	4051.18	13	11	1.36E-01	5163.18	9	11	2.03E-01
3715.75	9	7	2.38E-01	4062.08	11	9	1.96E-01	5171.06	5	7	1.84E-01
3718.48	5	7	1.34E-01	4069.88	13	11	3.25E-01	5172.94	5	5	4.11E-01
3720.25	7	9	2.86E-01	4076.19	9	9	1.16E-01	5174.18	5	3	5.83E-01
3725.55	7	7	1.60E-01	4084.37	9	7	1.94E-01	5191.45	7	9	1.62E-01
3727.68	9	11	1.51E-01	4102.15	5	3	1.22E-01	5238.21	7	9	3.74E-01
3728.30	7	5	1.55E-01	4107.46	7	5	2.02E-01	5240.87	7	7	3.89E-01
3728.50	7	9	2.20E-01	4120.09	13	15	6.05E-01	5242.80	7	5	2.01E-01
3733.02	7	7	1.45E-01	4131.92	9	11	1.56E-01	5261.53	5	7	1.13E-01
3733.41	13	13	2.80E-01	4148.98	9	11	1.56E-01	5280.85	5	5	1.28E-01
3735.62	11	11	1.66E-01	4157.40	13	11	2.17E-01	5355.52	9	9	1.21E-01
3742.28	7	7	1.56E-01	4157.90	9	11	1.60E-01	5356.46	11	11	2.11E-01
3747.19	5	7	3.07E-01	4185.82	11	13	3.82E-01	5360.51	9	11	6.19E-01
3748.48	9	11	3.95E-01	4188.32	11	13	3.32E-01	5364.28	9	9	2.26E-01
3755.10	3	5	1.41E-01	4194.56	11	11	2.70E-01	5460.50	5	3	3.46E-01
3755.16	9	9	2.48E-01	4232.59	9	11	3.17E-01	5493.76	7	5	2.13E-01
3758.52	9	9	1.22E-01	4240.83	5	5	1.68E-01	5506.49	5	7	3.61E-01
3759.60	9	7	1.82E-01	4246.02	11	13	2.00E-01	5533.03	5	5	3.72E-01
3760.88	9	9	2.16E-01	4251.88	13	11	1.76E-01	5570.44	5	3	3.30E-01
3768.73	9	9	2.88E-01	4254.95	7	9	2.01E-01	5849.71	3	3	3.02E-01
3769.99	7	9	2.46E-01	4269.28	11	11	1.36E-01	5851.50	3	5	1.55E-01
3777.72	13	11	1.66E-01	4276.91	7	9	2.85E-01	5893.36	5	5	2.60E-01
3788.25	7	9	2.87E-01	4277.24	9	11	1.35E-01	5895.93	5	7	3.12E-01
3794.43	9	9	1.22E-01	4317.92	15	15	1.28E-01	5926.37	7	7	1.63E-01
3797.47	7	5	1.48E-01	4325.80	3	3	1.84E-01	5928.88	7	9	5.32E-01
3798.25	7	9	6.90E-01	4326.14	5	7	2.56E-01	7154.11	9	9	3.45E-01
3801.84	9	7	3.16E-01	4340.74	5	7	1.23E-01				
3805.99	5	5	2.44E-01	4381.63	13	13	2.93E-01	**Neodymium**			
3819.78	9	11	1.47E-01	4382.41	11	13	3.83E-01	**Nd II**			
3824.78	5	7	1.40E-01	4409.94	13	13	1.38E-01	3780.4	16	18	1.4E-01
3827.15	7	7	1.94E-01	4411.69	11	11	2.63E-01	3805.4	14	16	6.9E-01
3828.88	7	7	1.35E-01	4434.95	9	9	2.51E-01	3807.2	10	12	4.9E-02
3830.81	5	5	1.83E-01	4446.42	11	11	1.90E-01	3863.3	8	10	1.5E-01
3831.07	7	9	1.20E-01	4457.35	7	7	1.28E-01	3941.5	10	10	6.1E-01
3832.11	9	9	3.05E-01	4474.57	5	5	2.10E-01	3951.2	12	12	6.0E-01
3833.75	9	9	1.70E-01	4491.65	11	11	2.09E-01	3973.3	18	18	6.3E-01

λ Å	g_i	g_k	A 10^8 s^{-1}	λ Å	g_i	g_k	A 10^8 s^{-1}	λ Å	g_i	g_k	A 10^8 s^{-1}
3979.5	10	12	2.7E-01	3464.3	5	5	6.7E-03	6402.2	5	7	5.14E-01
3990.1	16	16	5.2E-01	3466.6	1	3	1.3E-02	6506.5	3	5	3.00E-01
4012.3	18	20	5.5E-01	3472.6	5	7	1.7E-02	6532.9	1	3	1.08E-01
4061.1	16	18	4.4E-01	3498.1	3	5	5.1E-03	6599.0	3	3	2.32E-01
4106.6	14	16	6.8E-02	3501.2	3	3	1.2E-02	6602.9	3	3	5.9E-03
4109.5	14	16	3.7E-01	3510.7	5	3	2.2E-03	6652.1	3	1	2.9E-03
4133.4	14	12	1.5E-01	3515.2	3	5	6.9E-03	6678.3	3	5	2.33E-01
4156.1	12	14	3.4E-01	3520.5	3	1	9.3E-02	6717.0	3	3	2.17E-01
4205.6	18	16	1.8E-01	3593.5	3	5	9.9E-03	6721.1	3	3	4.9E-04
4284.5	18	18	8.5E-02	3593.6	3	3	6.6E-03	6929.5	3	5	1.74E-01
4303.6	8	10	4.7E-01	3600.2	3	3	4.3E-03	7024.1	3	3	1.89E-02
4325.8	16	16	1.6E-01	3633.7	3	1	1.1E-02	7032.4	5	3	2.53E-01
4358.2	14	14	1.5E-01	3682.2	3	5	1.6E-03	7051.3	3	3	3.0E-02
4382.7	12	10	4.0E-02	3685.7	3	3	3.9E-03	7059.1	3	5	6.8E-02
4400.8	10	10	6.8E-02	3701.2	3	5	2.2E-03	7173.9	3	5	2.87E-02
4451.6	12	14	2.5E-01	4536.3	3	3	5.0E-03	7245.2	3	3	9.35E-02
4456.4	16	18	6.4E-02	4702.5	3	3	2.1E-03	7304.8	1	3	2.55E-03
4463.0	14	16	1.8E-01	4708.9	3	3	4.2E-02	7438.9	1	3	2.31E-02
4958.1	12	10	1.2E-02	4955.4	3	3	3.3E-03	7472.4	3	3	4.0E-02
5130.6	22	20	1.6E-01	5113.7	3	3	1.0E-02	7535.8	3	3	4.3E-01
5192.6	20	18	1.7E-01	5120.5	3	3	5.6E-03	7937.0	5	5	7.8E-03
5249.6	18	16	1.8E-01	5154.4	3	3	1.9E-02	8082.5	3	3	1.2E-03
5276.9	12	10	1.2E-01	5191.3	3	3	1.3E-02	8118.5	3	3	4.9E-02
5293.2	16	14	1.2E-01	5326.4	3	3	6.8E-03	8128.9	3	5	7.2E-03
5302.3	20	18	1.1E-01	5333.3	3	3	5.3E-03	8259.4	5	5	2.03E-02
5311.5	14	12	1.1E-01	5341.1	3	3	1.1E-01	8571.4	3	3	5.5E-02
5319.8	12	10	1.6E-01	5400.6	3	1	9.0E-03	8582.9	3	5	1.00E-02
5357.0	18	16	1.8E-01	5418.6	3	3	5.2E-03	8647.0	5	5	3.91E-02
5371.9	20	20	5.1E-02	5433.7	3	3	2.83E-03	8681.9	3	3	2.1E-01
5485.7	18	18	5.7E-02	5652.6	3	3	8.9E-03	8767.5	3	3	1.1E-03
5594.4	16	16	7.0E-02	5662.5	3	3	6.9E-03	8771.7	3	3	1.6E-01
5620.6	18	18	1.3E-01	5852.5	3	1	6.82E-01	8783.8	3	5	3.13E-01
5688.5	14	14	5.9E-02	5868.4	3	3	1.4E-02	8865.3	3	3	9.4E-03
5718.1	16	16	8.7E-02	5881.9	5	3	1.15E-01	9201.8	3	3	9.1E-02
5726.8	10	10	5.6E-02	5913.6	3	3	4.8E-02	9433.0	3	3	1.1E-03
5740.9	12	12	7.2E-02	5939.3	5	3	2.00E-03	9486.7	3	3	2.5E-02
5804.0	10	10	4.6E-02	5944.8	5	5	1.13E-01	9534.2	3	3	6.3E-02
5865.1	16	18	1.3E-02	5961.6	3	3	3.3E-02	10621	3	3	2.4E-03
6051.9	12	10	1.1E-02	5975.5	5	3	3.51E-02	11409	3	3	4.2E-02
				6030.0	3	3	5.61E-02	11525	3	3	8.4E-02
Neon				6046.1	3	3	2.26E-03	11767	3	3	6.9E-02
Ne I				6074.3	3	1	6.03E-01	12459	3	3	1.5E-02
615.63	1	3	3.8E-01	6096.2	3	5	1.81E-01				
618.67	1	3	9.3E-01	6118.0	5	3	6.09E-03	**Ne II**			
619.10	1	3	3.3E-01	6128.5	3	3	6.7E-03	*357.03	6	10	3.8E+01
626.82	1	3	7.4E-01	6143.1	5	5	2.82E-01	*361.77	6	2	1.6E+01
629.74	1	3	4.8E-01	6150.3	3	3	1.5E-02	*406.28	6	10	1.8E+01
735.90	1	3	6.11E+00	6163.6	1	3	1.46E-01	*446.37	6	6	4.07E+01
743.72	1	3	4.86E-01	6217.3	5	3	6.37E-02	460.73	4	2	4.7E+01
3369.8	5	5	1.0E-03	6266.5	1	3	2.49E-01	462.39	2	2	2.3E+01
3369.9	5	3	7.6E-03	6273.0	3	3	9.7E-03	1907.5	4	2	2.8E-01
3375.6	5	3	2.2E-03	6293.7	3	3	6.39E-03	1916.1	4	4	6.9E-01
3417.9	3	5	9.2E-03	6304.8	3	5	4.16E-02	1930.0	2	2	5.7E-01
3418.0	3	3	2.2E-03	6328.2	5	3	3.39E-02	1938.8	2	4	1.3E-01
3423.9	3	3	1.0E-03	6330.9	3	3	2.3E-02	2858.0	6	6	7.9E-01
3447.7	5	5	2.1E-02	6334.4	5	5	1.61E-01	2870.0	6	6	1.7E-01
3450.8	5	3	4.9E-03	6351.9	1	3	3.45E-03	2873.0	6	4	3.8E-01
3454.2	3	1	3.7E-02	6383.0	3	3	3.21E-01	2876.3	4	6	7.8E-01
3460.5	1	3	7.0E-03	6401.1	3	3	1.39E-02	2876.5	6	4	3.3E-01

λ Å	Weights g_i	Weights g_k	A 10^8 s^{-1}	λ Å	Weights g_i	Weights g_k	A 10^8 s^{-1}	λ Å	Weights g_i	Weights g_k	A 10^8 s^{-1}
2878.1	2	2	6.9E-02	3309.7	4	2	3.1E-01	3568.5	6	8	1.4E+00
2888.4	4	6	7.0E-02	3310.5	4	4	6.9E-02	3571.2	4	4	6.3E-01
2891.5	4	4	6.1E-02	3311.3	4	2	2.6E-01	3574.2	6	6	1.0E-01
2897.0	6	8	5.2E-02	3314.7	6	6	4.4E-02	3574.6	4	6	1.3E+00
2906.8	2	4	5.5E-01	3319.7	4	2	1.6E+00	3590.4	4	6	3.6E-02
2910.1	4	2	1.7E+00	3320.2	8	6	2.1E-01	3594.2	4	2	1.3E+00
2910.4	2	4	5.9E-01	3323.7	4	4	1.6E+00	3612.3	2	4	2.6E-01
2916.2	6	4	9.6E-02	3327.2	4	4	9.1E-02	3628.0	4	4	6.0E-01
2925.6	2	2	5.6E-01	3329.2	8	8	8.8E-01	3632.7	4	4	1.3E-01
2933.7	6	6	6.9E-02	3330.7	6	6	3.9E-02	3643.9	4	4	3.2E-01
2955.7	6	4	1.2E+00	3334.8	6	8	1.8E+00	3644.9	2	4	9.9E-01
3001.7	4	4	8.7E-01	3336.1	4	6	1.1E+00	3659.9	4	6	6.7E-02
3017.3	6	4	3.5E-01	3344.4	2	2	1.5E+00	3664.1	6	4	7.0E-01
3027.0	6	6	1.4E+00	3345.5	6	4	1.4E+00	3679.8	4	2	3.2E-01
3028.7	4	2	8.5E-01	3345.8	4	4	2.2E-01	3694.2	6	6	1.0E+00
3028.9	2	4	4.7E-01	3353.6	4	2	1.2E-01	3697.1	2	2	2.8E-01
3034.5	6	8	3.1E+00	3355.0	4	6	1.3E+00	3701.8	4	6	2.7E-01
3037.7	4	4	2.1E+00	3356.3	6	6	2.0E-01	3709.6	4	2	1.1E+00
3045.6	2	2	2.5E+00	3357.8	6	6	5.0E-01	3713.1	4	6	1.3E+00
3047.6	4	6	1.8E+00	3360.3	2	4	8.6E-01	3721.8	4	6	2.0E-01
3054.7	2	4	9.4E-01	3360.6	2	4	8.2E-01	3726.9	4	4	1.2E-01
3092.9	6	6	1.3E+00	3362.9	4	2	3.5E-01	3727.1	2	4	9.8E-01
3097.1	8	8	1.3E+00	3371.8	4	6	2.2E-01	3734.9	4	4	1.9E-01
3118.0	8	6	4.2E-02	3374.1	4	4	3.0E-01	3744.6	2	4	2.6E-01
3134.1	6	4	2.6E-01	3378.2	2	2	1.7E+00	3751.2	2	2	1.8E-01
3140.4	8	6	2.4E-01	3379.3	2	2	3.0E-01	3753.8	4	6	4.5E-01
3151.1	6	6	4.8E-02	3386.2	4	6	5.5E-02	3766.3	4	6	2.9E-01
3154.8	8	6	1.8E-02	3388.4	4	6	2.2E+00	3777.1	2	4	4.2E-01
3164.4	8	8	1.6E-01	3390.6	2	4	7.7E-02	3800.0	4	4	3.7E-01
3165.7	6	6	1.2E-01	3392.8	2	4	4.4E-01	3818.4	2	4	6.1E-01
3173.6	6	4	4.5E-02	3404.8	4	6	1.9E+00	3829.8	4	6	8.4E-01
3176.1	4	6	6.0E-02	3407.0	6	8	2.3E+00	3942.3	4	6	1.0E-02
3187.6	4	6	1.4E-02	3411.4	4	2	6.1E-01				
3188.7	6	6	3.9E-01	3413.2	4	4	1.8E+00	**Ne V**			
3190.9	4	6	1.5E-01	3414.9	4	6	1.8E-02	*142.61	9	9	6.7E+02
3194.6	4	4	5.2E-01	3416.9	6	6	6.4E-01	*143.32	9	15	1.2E+03
3198.6	6	8	1.7E+00	3417.7	6	8	1.6E+00	147.13	5	7	1.5E+03
3198.9	4	4	2.3E-01	3438.9	2	2	1.4E+00	151.23	5	5	3.38E+02
3209.0	8	8	1.6E-01	3440.7	2	2	3.5E-01	154.50	1	3	7.0E+02
3209.4	2	4	6.0E-01	3453.1	4	4	4.6E-01	*167.69	9	9	1.5E+02
3213.7	2	4	1.7E+00	3454.8	4	4	1.6E+00	*358.93	9	3	2.1E+02
3214.3	4	6	2.2E+00	3456.6	2	4	9.6E-01	365.59	5	3	1.35E+02
3218.2	8	10	3.6E+00	3457.1	4	6	9.9E-02	*482.15	9	9	3.01E+01
3224.8	6	8	3.5E+00	3459.3	6	6	1.6E+00	*571.04	9	15	1.0E+01
3229.5	8	8	1.3E-01	3475.2	4	4	1.2E-02	2259.6	3	5	1.9E+00
3229.6	8	10	3.6E+00	3477.6	4	6	4.3E-01	2265.7	5	7	2.4E+00
3230.1	6	6	1.8E+00	3481.9	4	2	1.4E+00				
3230.4	4	6	1.4E-01	3503.6	2	2	2.0E+00	**Ne VII**			
3232.0	6	4	2.7E-01	3522.7	4	2	2.3E-02	97.502	1	3	1.07E+03
3232.4	4	4	1.6E+00	3538.0	4	2	7.6E-01	*115.46	9	3	4.8E+02
3243.4	6	6	2.3E-01	3539.9	4	4	3.6E-02	116.69	3	5	1.6E+03
3244.1	6	8	1.5E+00	3542.2	6	4	6.0E-01	127.66	3	1	1.9E+02
3248.1	4	4	2.4E-01	3542.9	4	6	1.2E+00	465.22	1	3	4.09E+01
3255.4	6	4	3.8E-02	3546.2	2	4	6.3E-02	558.61	3	5	8.11E+00
3263.4	2	4	3.9E-01	3551.6	2	4	3.7E-02	559.95	1	3	1.07E+01
3269.9	4	6	5.1E-01	3557.8	2	2	1.9E-01	561.38	3	3	7.99E+00
3270.8	6	4	5.7E-02	3561.2	4	6	2.1E-01	561.73	5	5	2.39E+01
3297.7	6	6	4.3E-01	3565.8	4	4	6.2E-01	562.99	3	1	3.17E+01

λ Å	Weights g_i	g_k	A 10^8 s^{-1}	λ Å	Weights g_i	g_k	A 10^8 s^{-1}	λ Å	Weights g_i	g_k	A 10^8 s^{-1}
564.53	5	3	1.31E+01	2313.98	5	5	5.0E+00	4752.43	3	3	2.0E-01
				2317.16	7	5	3.8E+00	4756.52	9	9	1.5E-01
Ne VIII				2320.03	9	11	6.9E+00	4786.54	11	11	1.8E-01
*88.09	2	6	8.4E+02	2321.38	5	7	5.6E+00	4812.00	3	1	9.5E-02
*98.208	6	10	2.77E+03	2324.65	7	9	1.8E-01	4829.03	5	7	1.9E-01
770.41	2	4	5.90E+00	2325.79	7	9	3.5E+00	4831.18	9	7	1.6E-01
780.32	2	2	5.69E+00	2329.96	5	3	5.3E+00	4838.64	9	7	2.2E-01
2820.7	2	4	7.20E-01	2345.54	9	7	2.2E+00	4855.41	5	5	5.7E-01
2860.1	2	2	6.88E-01	2346.63	7	5	5.5E-01	4904.41	5	3	6.2E-01
				2347.51	9	9	2.2E-01	4912.03	3	3	1.5E-01
Nickel				2348.73	7	7	2.2E-01	4913.97	1	3	2.2E-01
Ni I				2419.31	7	5	2.0E-01	4918.36	9	7	2.3E-01
1963.85	7	7	1.1E-01	2943.91	7	5	1.1E-01	4935.83	7	5	2.4E-01
1976.87	7	9	1.1E+00	2981.65	5	3	2.8E-01	4937.34	9	9	1.2E-01
1981.61	5	5	1.3E-01	3002.48	7	7	8.0E-01	4953.20	5	5	1.2E-01
1990.25	5	7	8.3E-01	3003.62	5	5	6.9E-01	4980.17	9	11	1.9E-01
2007.01	5	5	1.7E-01	3012.00	5	5	1.3E+00	5000.34	7	7	1.4E-01
2007.69	7	7	9.0E-02	3037.93	7	7	2.8E-01	5012.46	7	7	1.1E-01
2014.25	3	5	9.3E-01	3050.82	7	9	6.0E-01	5017.58	11	11	2.0E-01
2025.40	7	5	2.3E-01	3054.31	5	5	4.0E-01	5035.37	7	9	5.7E-01
2026.62	9	7	2.4E-01	3057.64	3	3	1.0E+00	5042.20	3	5	1.4E-01
2047.35	7	5	1.8E-01	3064.62	5	7	1.1E-01	5048.85	7	7	1.6E-01
2052.04	9	9	9.7E-02	3101.56	5	7	6.3E-01	5080.53	9	11	3.2E-01
2055.50	5	3	3.3E-01	3101.88	5	7	4.9E-01	5081.11	7	9	5.7E-01
2059.92	7	5	2.1E-01	3134.11	3	5	7.3E-01	5082.35	3	3	2.5E-01
2060.20	5	3	2.3E-01	3225.02	5	3	9.3E-02	5084.08	7	9	3.1E-01
2064.39	3	1	4.0E-01	3369.56	9	7	1.8E-01	5099.95	7	7	2.9E-01
2069.52	5	5	1.1E-01	3380.57	5	3	1.3E+00	5115.40	11	9	2.2E-01
2085.57	5	5	2.6E+00	3392.98	7	7	2.4E-01	5129.37	7	5	1.2E-01
2089.09	7	5	9.7E-02	3414.76	7	9	5.5E-01	5155.14	5	5	1.1E-01
2095.13	5	7	1.1E-01	3423.71	3	3	3.3E-01	5155.76	5	7	2.9E-01
2114.43	5	5	9.7E-02	3433.56	7	7	1.7E-01	5176.57	5	5	1.8E-01
2121.40	7	5	2.8E-01	3446.26	5	5	4.4E-01	5371.33	7	7	1.6E-01
2124.80	5	3	3.8E-01	3452.88	5	7	9.8E-02	5476.91	1	3	9.5E-02
2147.80	5	3	4.7E-01	3458.46	3	5	6.1E-01	5637.12	3	3	1.1E-01
2157.83	5	3	4.1E-01	3461.66	7	9	2.7E-01	5664.02	5	7	1.1E-01
2158.31	7	5	6.9E-01	3472.55	5	7	1.2E-01	5695.00	3	3	1.7E-01
2161.04	5	5	1.3E-01	3483.77	5	3	1.4E-01	6086.29	3	5	1.1E-01
2173.54	5	3	1.5E-01	3492.96	5	3	9.8E-01	6175.42	3	3	1.7E-01
2174.48	3	1	8.9E-01	3510.33	3	1	1.2E+00	7122.24	5	7	2.1E-01
2182.38	7	5	1.3E-01	3515.05	5	7	4.2E-01	7381.94	9	11	9.7E-02
2183.91	5	5	1.2E-01	3524.54	7	5	1.0E+00	7422.30	7	5	1.8E-01
2190.22	5	5	3.0E-01	3566.37	5	5	5.6E-01	7727.66	7	7	1.1E-01
2197.35	3	3	7.8E-01	3597.70	3	3	1.4E-01				
2201.59	5	3	7.3E-01	3619.39	5	7	6.6E-01	**Ni II**			
2221.94	5	3	2.2E-01	4027.67	5	7	1.3E-01	2165.55	10	10	2.4E+00
2244.46	5	5	3.8E-01	4295.88	9	7	1.7E-01	2169.10	8	8	1.58E+00
2253.57	7	7	1.9E-01	4401.54	9	11	3.8E-01	2174.67	8	10	1.43E+00
2254.81	9	9	9.6E-02	4462.46	3	5	1.7E-01	2175.15	6	6	1.77E+00
2258.15	7	5	1.7E-01	4470.48	5	7	1.9E-01	2184.61	4	4	2.90E+00
2259.56	5	3	2.0E-01	4600.37	5	3	2.6E-01	2201.41	4	6	1.3E+00
2261.42	9	7	9.1E-02	4604.99	9	7	2.3E-01	2206.72	6	8	1.66E+00
2287.32	3	5	1.8E-01	4606.23	5	3	1.0E-01	2216.48	10	12	3.4E+00
2289.98	9	7	2.1E+00	4648.66	11	9	2.4E-01	2220.40	6	8	2.3E+00
2293.11	5	5	3.8E-01	4686.22	5	5	1.4E-01	2222.96	10	10	9.8E-01
2300.77	7	7	7.5E-01	4701.54	9	9	1.4E-01	2224.86	8	8	1.55E+00
2302.97	3	3	4.5E-01	4714.42	13	11	4.6E-01	2226.33	6	6	1.3E+00
2307.35	5	7	1.6E-01	4715.78	7	7	2.0E-01	2253.85	4	6	1.98E+00
2312.34	7	7	5.5E+00	4732.47	7	9	9.3E-02	2264.46	6	8	1.43E+00

λ Å	Weights g_i	g_k	A 10^8 s^{-1}	λ Å	Weights g_i	g_k	A 10^8 s^{-1}	λ Å	Weights g_i	g_k	A 10^8 s^{-1}
2270.21	8	10	1.56E+00	168	6	8	3.2E+02	292	5	7	2.2E+02
2278.77	8	6	2.8E+00	182	2	2	2.5E+02				
2287.09	6	4	2.8E+00	185.23	2	4	4.2E+02	**Ni XVIII**			
2296.55	8	8	1.98E+00	187	4	6	1.2E+02	24.881	2	4	8.6E+02
2297.14	6	4	2.70E+00	187	4	2	3.3E+02	25.070	4	6	9.9E+02
2297.49	4	2	3.0E+00	188	2	2	4.7E+02	26.02	2	4	1.26E+03
2298.27	6	6	2.8E+00	190	6	8	2.0E+02	26.020	2	4	1.1E+03
2303.00	8	6	2.9E+00	192	6	8	4.54E+02	26.046	2	2	1.1E+03
2316.04	10	8	2.88E+00	192	6	6	3.1E+02	26.218	4	6	1.5E+03
2334.58	8	8	8.0E-01	194	4	6	2.8E+02	27.98	4	6	1.0E+03
2375.42	6	8	6.6E-01	194	2	4	5.5E+02	27.982	2	4	2.0E+03
2394.52	8	10	1.70E+00	194	2	2	1.1E+02	28.018	6	8	1.1E+03
2416.13	6	8	2.1E+00	194	4	4	3.5E+02	28.220	4	6	2.33E+03
2437.89	8	10	5.4E-01	194.04	4	6	4.6E+02	29.383	4	6	1.58E+03
2510.87	8	10	5.8E-01	195.27	4	4	9.5E+01	29.422	6	8	1.69E+03
				196	4	6	6.7E+02	29.779	2	4	1.9E+03
Ni III				197	4	6	1.5E+02	29.829	2	2	1.9E+03
1692.51	11	13	7.9E+00	197	4	2	1.2E+02	31.845	4	6	2.7E+03
1709.90	9	11	6.3E+00	199	2	4	4.9E+02	31.890	6	8	3.0E+03
1719.46	5	7	6.0E+00	206	2	2	3.7E+02	32.034	2	4	3.4E+03
1722.28	3	5	5.9E+00	217	4	4	1.1E+02	32.340	4	6	4.0E+03
1724.52	3	1	6.7E+00	218.391	2	4	9.5E+01	36.990	4	6	5.5E+03
1741.96	9	7	5.7E+00	223.119	2	2	1.3E+02	37.049	6	8	5.9E+03
1752.43	7	5	5.5E+00	231	4	4	1.6E+02	41.015	2	4	2.97E+03
1760.56	5	3	6.5E+00	232.475	4	4	4.07E+02	41.218	2	2	3.2E+03
1769.64	11	11	6.2E+00	233	6	4	2.4E+02	43.814	2	4	5.5E+03
1823.06	9	9	5.6E+00	235	4	2	3.8E+02	44.365	4	6	6.8E+03
				235	6	6	2.5E+02	44.405	4	4	1.14E+03
Ni XIV				236	4	4	1.2E+02	52.615	4	6	1.5E+04
164.13	6	8	1.2E+03	237.875	4	2	2.6E+02	52.720	6	8	1.6E+04
168	2	4	2.4E+02	238	6	4	1.3E+02	52.745	6	6	1.06E+03
168.12	4	2	8.5E+02	239.550	2	2	2.6E+02	59.950	6	4	9.6E+02
169.69	4	4	9.8E+02	245	4	4	1.4E+02	60.212	4	2	1.1E+03
170.50	4	4	7.1E+02	245	4	6	3.2E+02	63.512	4	6	7.9E+02
171.37	4	6	9.4E+02	249	6	8	3.3E+02	63.589	6	8	8.5E+02
172.16	6	6	4.7E+02	249	6	4	1.2E+02	69.075	4	6	8.0E+02
172.80	6	4	1.4E+02	250	4	2	1.6E+02	76.254	4	6	1.38E+03
177.28	4	4	5.6E+02	254	6	4	1.8E+02	76.359	6	8	1.47E+03
178	2	4	8.9E+01					99.275	2	4	1.0E+03
181	4	6	7.4E+01	**Ni XVII**				100.4	4	6	1.2E+03
182.14	4	2	1.5E+02	30.919	1	3	2.77E+03	114.46	4	6	2.5E+03
196	4	2	3.8E+01	42.855	1	3	4.75E+03	114.74	6	8	2.7E+03
288.894	4	4	4.6E+01	54.451	9	11	1.5E+04				
292.399	6	6	3.6E+01	55.361	1	3	6.7E+03	**Ni XIX**			
				57.348	7	9	1.4E+04	9.140	1	3	3.1E+04
Ni XV				197.39	1	3	1.6E+02	9.153	1	3	5.2E+03
50.249	5	7	6.8E+03	199.87	3	5	2.1E+02	9.977	1	3	1.1E+05
60.890	9	11	1.0E+04	204	3	3	1.8E+02	10.110	1	3	9.4E+04
64.635	7	9	9.6E+03	205	3	1	2.4E+02	10.283	1	3	4.7E+03
163.64	5	7	5.6E+01	206	1	3	3.0E+02	10.433	1	3	5.1E+03
173.73	5	7	7.6E+02	207.50	5	7	2.5E+02	11.539	1	3	4.8E+04
175	3	1	5.7E+02	215.89	3	5	4.8E+02	11.599	1	3	6.3E+03
179.28	5	7	7.5E+02	216	1	3	2.7E+02	12.435	1	3	3.66E+05
181	1	3	6.8E+02	217	5	7	2.4E+02	12.656	1	3	1.0E+05
269	3	1	5.3E+01	227	5	5	1.6E+02	13.779	1	3	1.23E+04
278.386	5	5	4.3E+01	249.180	1	3	2.75E+02	14.043	1	3	1.31E+04
				281.50	3	1	2.1E+02	40.7	3	3	6.4E+03
Ni XVI				282	3	1	2.4E+02	40.7	3	1	8.4E+03
166	4	6	3.1E+02	284	5	3	1.5E+02	41.132	7	9	9.4E+03

λ Å	g_i	g_k	A 10^8 s^-1	λ Å	g_i	g_k	A 10^8 s^-1	λ Å	g_i	g_k	A 10^8 s^-1
Ni XXI				103.53	4	2	4.17E+02	1.539	1	3	2.6E+06
11.13	3	3	1.7E+04	104.64	2	2	4.7E+02	1.539	3	5	2.6E+06
11.23	5	3	1.7E+04	106.68	4	2	3.67E+02	1.540	3	3	1.7E+06
11.239	5	7	5.7E+04	113.14	4	4	1.65E+02	1.541	3	5	5.5E+06
11.28	3	1	2.2E+05	118.52	2	4	1.5E+02	1.542	3	3	3.6E+06
11.318	5	7	2.8E+05	122.72	6	4	2.17E+02	1.542	5	5	3.5E+06
11.48	3	1	1.1E+05	134.73	6	6	1.44E+02	1.544	5	3	3.2E+06
11.48	1	3	4.0E+05	135.47	4	4	8.0E+01	1.546	3	5	1.6E+06
11.517	5	7	1.4E+05	137.01	4	4	2.6E+02	1.547	3	3	2.1E+05
11.539	5	7	1.2E+05	138.80	4	6	7.2E+01	1.549	1	3	2.0E+05
11.67	1	3	8.0E+04	153.47	2	2	1.27E+02	1.551	3	1	8.2E+05
11.72	3	3	2.3E+04	159.69	2	4	8.9E+01	1.558	3	1	6.5E+05
12.454	5	3	3.3E+04					1.5883	1	3	6.02E+06
12.472	3	3	1.8E+04	**Ni XXV**				1.5963	1	3	7.70E+05
12.502	5	5	2.8E+04	9.30	3	1	9.3E+04				
				9.31	5	7	8.2E+04	**Nitrogen**			
Ni XXII				9.32	3	5	7.8E+04	**N I**			
72.52	4	2	2.84E+02	9.34	1	3	1.1E+05	1163.88	6	6	7.52E-01
84.06	6	4	1.2E+03	9.42	3	1	9.0E+04	1164.00	4	6	1.27E-02
84.24	4	2	5.6E+02	9.49	3	5	8.9E+04	1164.21	6	4	5.17E-02
85.86	4	2	4.9E+02	9.60	1	3	1.8E+05	1164.32	4	4	6.94E-01
88.00	4	2	1.2E+03	9.63	3	5	2.4E+05	1167.45	6	8	1.29E+00
95.95	2	2	4.4E+02	9.64	3	3	1.3E+05	1168.42	6	4	4.24E-02
98.16	4	4	5.2E+02	9.71	3	1	2.3E+05	1168.54	4	6	1.24E+00
98.58	4	4	2.45E+02	9.71	3	3	1.8E+05	1176.51	6	4	9.22E-01
100.60	6	6	3.9E+02	9.74	5	7	3.0E+05	1176.63	4	4	1.02E-01
101.31	6	4	4.83E+02	9.75	3	5	1.3E+05	1177.69	4	2	1.02E+00
103.31	4	2	2.66E+02	9.76	1	3	3.03E+05	1199.55	4	6	4.01E+00
106.04	4	4	2.36E+02	9.76	5	3	7.5E+04	1200.22	4	4	3.99E+00
106.16	4	2	5.1E+02	9.78	5	7	2.9E+05	1200.71	4	2	3.98E+00
124.31	2	2	3.7E+02	9.86	5	7	4.8E+05	1310.54	4	6	8.42E-01
126.32	4	4	3.3E+02	9.87	3	5	2.03E+05	1316.29	4	6	1.42E-02
				9.92	5	5	1.3E+05	1492.63	6	4	3.13E+00
Ni XXIII				9.94	5	7	1.29E+05	1492.82	4	4	3.51E-01
87.66	3	3	2.8E+02	9.97	3	5	2.5E+05	1494.68	4	2	3.72E+00
88.11	5	3	8.3E+02	10.08	1	3	2.80E+05	3822.03	2	2	3.70E-02
90.49	3	3	1.77E+02					3830.43	4	4	4.67E-02
90.96	5	3	2.5E+02	**Ni XXVI**				3834.22	4	2	1.89E-02
91.83	5	3	7.5E+02	1.5930	4	2	3.4E+06	4099.94	2	4	3.48E-02
92.32	3	1	4.39E+02	1.5935	2	2	4.0E+06	4109.95	4	6	3.90E-02
100.42	1	3	2.1E+02	1.5973	4	4	8.1E+06	4113.97	4	4	6.62E-03
102.08	5	5	5.3E+02	1.5977	2	4	4.4E+06	4137.64	2	4	2.80E-03
103.23	3	3	2.4E+02	1.5982	2	2	7.3E+06	4143.43	4	4	6.09E-03
103.67	5	5	1.78E+02	1.5996	2	2	2.7E+06	4151.48	6	4	1.01E-02
104.70	3	1	2.94E+02	1.6005	4	6	2.7E+06	4249.87	4	2	2.59E-02
106.02	5	5	2.87E+02	1.6036	4	2	2.1E+06	4264.00	6	4	2.26E-02
108.27	7	5	3.32E+02	9.390	2	4	2.59E+05	4356.29	6	8	5.10E-02
111.23	3	1	2.26E+02	9.535	4	6	2.96E+05	4385.54	2	2	8.84E-03
111.78	5	3	2.19E+02					4392.41	4	2	1.76E-02
111.86	1	3	1.7E+02	**Ni XXVII**				4435.43	2	4	7.51E-03
112.55	3	1	1.0E+03	1.2534	1	3	3.35E+05	4442.45	4	4	3.81E-02
128.87	5	5	4.02E+02	1.2824	1	3	6.38E+05	4669.89	4	4	7.49E-03
133.54	3	3	1.86E+02	1.3500	1	3	1.63E+06	4914.94	2	2	8.08E-03
137.55	3	1	2.53E+02	1.3516	1	3	2.4E+05	4935.12	4	2	1.76E-02
				1.531	3	3	2.0E+05	5199.84	2	2	1.87E-02
Ni XXIV				1.534	3	1	6.9E+06	5201.61	2	4	1.87E-02
101.13	6	4	1.63E+02	1.537	5	5	2.3E+06	5281.20	6	6	2.45E-03
102.11	4	4	5.4E+02	1.537	1	3	3.7E+06	5344.05	6	6	6.10E-04
103.43	2	4	1.3E+02	1.538	3	5	3.9E+06	5356.62	4	6	1.41E-03

λ Å	Weights g_i	g_k	A 10^8 s^{-1}	λ Å	Weights g_i	g_k	A 10^8 s^{-1}	λ Å	Weights g_i	g_k	A 10^8 s^{-1}
5367.01	4	4	1.07E-03	9810.01	4	2	5.30E-02	1085.55	5	5	9.47E-01
5372.61	2	4	8.34E-04	9814.02	6	8	6.56E-03	1085.70	5	7	3.87E+00
5378.27	2	2	1.66E-03	9822.75	6	6	4.95E-02	3408.13	3	1	2.19E-01
6606.18	4	6	8.87E-04	9834.61	6	4	4.50E-02	3437.14	3	1	2.07E+00
6622.54	6	6	7.93E-03	9863.33	8	8	9.62E-02	3593.60	3	5	1.21E-01
6626.99	2	4	2.20E-03	9872.15	8	6	2.97E-02	3609.10	3	3	1.41E-01
6636.94	4	4	1.40E-02	9883.38	2	2	2.93E-02	3615.86	3	1	1.53E-01
6644.96	8	6	3.49E-02	9905.52	4	2	3.11E-03	3829.80	3	5	2.42E-01
6646.50	2	2	2.18E-02	9909.22	2	4	7.58E-03	3838.37	5	5	6.98E-01
6653.46	6	4	2.74E-02	9931.47	4	4	3.64E-02	3842.19	1	3	3.06E-01
6656.51	4	2	2.17E-02	9947.07	6	8	1.08E-02	3847.40	3	3	2.22E-01
6926.67	4	6	7.75E-03	9965.75	4	6	7.60E-03	3855.10	3	1	8.82E-01
6945.18	6	6	1.83E-02	9968.51	6	4	4.50E-03	3856.06	5	3	3.71E-01
6951.60	2	4	1.03E-02	9980.42	4	6	8.10E-03	3919.00	3	3	6.76E-01
6960.50	4	4	4.67E-03	9997.73	8	8	9.20E-03	3955.85	3	5	1.31E-01
6973.07	2	2	3.83E-03					3995.00	3	5	1.35E+00
6979.18	6	4	9.83E-03	**N II**				4114.33	3	3	1.42E-03
6982.03	4	2	2.04E-02	474.891	5	5	9.66E+00	4124.08	3	5	3.20E-01
7423.64	2	4	5.95E-02	475.647	1	3	1.17E+01	4133.67	5	5	5.30E-01
7442.30	4	4	1.24E-01	475.698	3	5	1.58E+01	4145.77	7	5	7.36E-01
7468.31	6	4	1.93E-01	475.757	3	3	8.75E+00	4374.99	3	5	5.55E-03
7898.98	6	4	2.82E-01	475.803	5	7	2.10E+01	4447.03	3	5	1.14E+00
7899.28	4	4	3.28E-02	475.884	5	5	5.25E+00	4459.94	3	1	1.12E-01
7915.42	4	2	3.13E-01	508.697	5	5	1.91E+00	4465.53	3	3	2.36E-02
8184.86	4	6	8.58E-02	510.758	5	7	1.87E+01	4477.68	5	3	8.85E-02
8188.01	2	4	1.27E-01	513.849	5	5	1.24E+01	4488.09	5	5	1.30E-02
8200.36	2	2	4.95E-02	529.355	1	3	7.23E+00	4507.56	7	5	1.00E-01
8210.72	4	4	4.84E-02	529.413	3	1	2.43E+01	4564.76	3	5	1.41E-02
8216.34	6	6	2.23E-01	529.491	3	3	6.75E+00	4601.48	3	5	2.35E-01
8223.13	4	2	2.64E-01	529.637	3	5	4.92E+00	4607.15	1	3	3.26E-01
8242.39	6	4	1.36E-01	529.722	5	3	1.03E+01	4613.87	3	3	2.26E-01
8567.74	2	4	4.92E-02	529.867	5	5	1.94E+01	4621.39	3	1	9.55E-01
8594.00	2	2	2.09E-01	533.511	1	3	2.39E+01	4630.54	5	5	7.72E-01
8629.24	4	4	2.66E-01	533.581	3	5	3.20E+01	4643.09	5	5	4.51E-01
8655.88	4	2	1.05E-01	533.650	3	3	1.66E+01	4654.53	3	5	2.43E-02
8680.28	6	8	2.46E-01	533.729	5	7	4.13E+01	4667.21	3	3	2.99E-02
8683.40	4	6	1.80E-01	533.815	5	5	9.19E+00	4674.91	3	1	1.05E-01
8686.15	2	4	1.09E-01	547.818	5	3	2.16E+00	4694.27	1	3	1.23E-01
8703.25	2	2	2.10E-01	559.762	1	3	1.14E+01	4695.90	3	5	1.29E-01
8711.70	4	4	1.28E-01	574.650	5	7	3.60E+01	4697.64	3	3	3.06E-02
8718.84	6	6	6.75E-02	582.156	5	5	2.85E+01	4698.55	3	1	3.67E-01
8728.90	4	2	3.76E-1	635.197	1	3	2.33E-01	4700.03	5	7	1.05E-01
8747.37	6	4	1.04E-02	644.634	1	3	1.21E+01	4702.50	5	5	9.15E-02
9028.92	2	2	3.02E-01	644.837	3	3	3.64E+01	4704.25	5	3	2.13E-01
9045.88	6	8	2.80E-01	645.178	5	3	6.07E+01	4706.40	7	9	6.09E-02
9049.49	6	6	1.88E-02	660.286	5	3	3.69E+01	4709.58	7	7	1.82E-01
9049.89	4	6	2.60E-01	671.016	3	5	2.47E+00	4712.07	7	5	1.46E-01
9060.48	2	4	2.95E-01	671.386	5	5	7.40E+00	4718.38	9	9	3.02E-02
9187.45	6	6	2.44E-01	671.411	1	3	3.04E+00	4721.58	9	7	7.75E-02
9187.86	4	6	1.76E-02	671.630	3	3	2.27E+00	4774.24	3	5	3.24E-02
9207.59	6	4	2.70E-02	671.773	3	1	9.85E+00	4779.72	3	3	2.52E-01
9208.00	4	4	2.33E-01	672.001	5	3	3.87E+00	4781.19	5	7	2.05E-02
9386.81	2	4	2.24E-01	745.841	1	3	1.25E+01	4788.14	5	5	2.52E-01
9392.79	4	6	2.63E-01	746.984	5	3	3.85E+01	4793.65	5	3	7.77E-02
9460.68	4	4	3.98E-02	748.369	5	3	3.83E+00	4803.29	7	7	3.18E-01
9776.90	2	4	1.18E-02	775.965	5	5	3.08E+00	4810.30	7	5	4.75E-02
9786.78	4	6	1.13E-02	915.612	1	3	4.38E+00	4860.17	3	5	1.61E-02
9788.29	2	2	2.99E-02	915.962	3	1	1.32E+01	4987.38	3	1	7.48E-01
9798.56	4	4	2.75E-02	1083.99	1	3	2.18E+00	4991.24	3	5	3.54E-01

λ Å	Weights g_i	g_k	A 10^8 s^{-1}	λ Å	Weights g_i	g_k	A 10^8 s^{-1}	λ Å	Weights g_i	g_k	A 10^8 s^{-1}
4994.36	5	7	2.62E-01	5551.92	7	7	2.00E-01	2978.84	2	4	1.66E-01
4994.37	3	3	7.60E-01	5552.68	5	3	1.50E-01	2983.64	4	4	8.24E-01
4997.22	3	3	1.96E-01	5565.26	7	5	3.97E-02	3342.76	2	2	3.80E-01
5001.13	3	5	9.76E-01	5666.63	3	5	3.74E-01	3353.98	2	4	7.66E-01
5001.47	5	7	1.05E+00	5676.02	1	3	2.96E-01	3354.32	4	6	5.51E-01
5002.70	1	3	8.45E-02	5679.56	5	7	5.25E-01	3355.46	4	2	7.51E-01
5005.15	7	9	1.16E+00	5686.21	3	3	1.94E-01	3358.78	2	2	3.05E-01
5005.30	5	5	6.51E-02	5710.77	5	5	1.24E-01	3360.98	4	4	2.44E-01
5007.33	3	5	7.89E-01	5730.66	5	3	1.34E-02	3365.80	4	2	1.52E+00
5010.62	3	3	2.19E-01	5747.30	3	5	3.40E-02	3367.36	6	6	1.27E+00
5011.31	5	3	5.84E-01	5767.45	3	3	2.44E-02	3374.07	6	4	8.13E-01
5012.04	7	7	5.19E-01	5893.15	5	7	2.88E-01	3745.95	2	4	1.90E-01
5016.38	5	5	1.62E-01	5897.25	3	5	2.16E-01	3752.63	2	2	6.67E-02
5023.05	7	5	3.61E-01	5899.83	1	3	1.60E-01	3754.69	4	4	3.78E-01
5025.66	7	7	1.07E-01	5927.81	1	3	3.22E-01	3762.60	4	4	4.24E-02
5040.71	7	5	3.78E-03	5931.78	3	5	4.27E-01	3771.03	6	4	5.59E-01
5045.10	5	3	3.42E-01	5940.24	3	3	2.26E-01	3771.36	6	4	8.28E-02
5073.59	3	3	2.59E-02	5941.65	5	7	5.54E-01	3792.97	8	6	1.03E-01
5168.05	3	5	3.06E-01	5952.39	5	5	1.27E-01	3934.50	2	4	7.49E-01
5170.16	3	3	6.54E-01	5960.91	5	3	1.34E-02	3938.51	4	6	8.96E-01
5171.27	3	1	8.71E-01	6065.00	3	5	2.21E-03	3942.88	4	4	1.49E-01
5171.47	5	7	5.81E-01	6284.32	5	3	7.74E-02	4097.36	2	4	8.70E-01
5172.34	3	5	6.01E-01	6379.62	3	3	6.11E-02	4103.39	2	2	8.67E-01
5172.97	1	3	5.01E-01	6482.05	3	3	3.01E-01	4195.74	2	4	9.37E-01
5173.39	5	7	7.36E-01	6610.56	5	7	6.34E-01	4200.07	4	6	1.12E+00
5174.46	5	5	5.07E-01	6857.03	5	3	2.53E-01	4215.77	4	4	1.85E-01
5175.89	7	9	8.93E-01	6869.58	5	5	2.51E-01	4318.78	2	4	5.40E-02
5176.57	5	3	2.17E-01	6887.83	5	7	2.49E-01	4321.22	2	2	1.08E-01
5177.06	3	3	5.00E-01	7762.24	5	5	8.74E-02	4321.39	4	6	5.03E-02
5179.34	7	9	8.67E-01	8438.74	1	3	2.24E-01	4325.43	4	4	8.60E-02
5179.52	9	11	1.07E+00	8831.75	1	3	8.42E-03	4327.69	6	8	3.06E-02
5180.36	5	5	4.28E-01	8855.30	3	3	2.51E-02	4327.88	4	2	1.07E-01
5183.20	7	7	2.88E-01	8893.29	5	3	4.12E-02	4332.95	6	6	1.23E-01
5184.96	7	7	3.20E-01					4337.01	6	4	7.47E-02
5185.09	5	3	7.11E-02	**N III**				4345.81	8	8	1.82E-01
5186.21	7	5	5.76E-02	374.198	2	4	9.89E+01	4351.11	8	6	4.01E-02
5190.38	9	9	1.77E-01	451.871	2	2	1.03E+01	4510.88	2	4	2.84E-01
5191.96	7	5	4.25E-02	452.227	4	2	2.05E+01	4510.96	4	6	4.77E-01
5199.50	9	7	1.51E-02	684.998	2	4	9.63E+00	4514.85	6	8	6.80E-01
5313.42	3	3	1.41E-01	685.515	2	2	3.83E+01	4518.14	2	2	5.65E-01
5320.20	5	3	4.20E-01	685.817	4	4	4.54E+01	4523.56	4	4	3.61E-01
5320.96	3	5	2.52E-01	686.336	4	2	1.95E+01	4530.86	4	2	1.12E-01
5327.76	5	5	4.65E-02	763.334	2	2	9.58E+00	4534.58	6	6	2.01E-01
5338.73	5	7	1.85E-01	764.351	4	2	1.85E+01	4547.30	6	4	3.33E-02
5340.21	7	5	2.59E-01	771.545	2	4	8.19E+00	4634.13	2	4	6.36E-01
5351.23	7	7	3.67E-01	771.901	4	4	1.64E+01	4640.64	4	6	7.60E-01
5383.72	3	5	3.31E-03	772.384	6	4	2.45E+01	4641.85	4	4	1.26E-01
5452.07	1	3	8.89E-02	772.889	6	4	2.09E+01	4858.70	2	4	4.35E-01
5454.22	3	1	3.34E-01	772.955	4	2	2.34E+01	4858.98	4	6	4.66E-01
5462.58	3	3	1.00E-01	979.832	4	4	8.84E+00	4861.27	6	8	5.32E-01
5478.09	3	5	4.75E-02	979.905	6	6	9.21E+00	4867.12	4	4	1.73E-01
5480.05	5	3	1.30E-01	989.799	2	4	4.18E+00	4867.17	8	10	6.18E-01
5495.65	5	5	2.40E-01	991.511	4	4	8.17E-01	4873.60	6	6	1.50E-01
5526.23	3	5	2.13E-01	991.577	4	6	4.97E+00	4881.78	6	4	1.22E-02
5530.24	5	7	4.04E-01	1747.85	2	4	1.28E+00	4884.14	8	8	8.71E-02
5535.35	7	9	6.04E-01	1751.22	4	4	2.48E-01	4896.58	8	6	5.86E-03
5535.38	3	3	4.53E-01	1751.66	4	6	1.51E+00	5260.86	2	2	2.80E-02
5540.06	3	1	6.03E-01	2972.55	2	2	6.67E-01	5270.57	2	4	6.95E-02
5543.47	5	5	3.51E-01	2977.33	4	2	3.32E-01	5272.68	4	2	1.39E-01

λ Å	Weights g_i	g_k	A 10^8 s^{-1}	λ Å	Weights g_i	g_k	A 10^8 s^{-1}	λ Å	Weights g_i	g_k	A 10^8 s^{-1}
5282.43	4	4	2.21E-02	4769.86	5	3	2.50E-02	935.193	5	5	1.33E+00
5297.75	4	6	4.93E-02	4786.92	7	7	8.79E-02	1028.16	1	3	4.22E-01
5298.95	6	4	7.38E-02	4796.66	7	5	1.53E-02	1152.15	5	5	5.28E+00
5314.36	6	6	1.14E-01	5200.41	3	5	2.67E-01	1217.65	1	3	2.06E+00
5320.87	6	8	5.68E-01	5204.28	5	7	3.55E-01	1302.17	5	3	3.41E+00
5327.19	4	6	5.29E-01	5205.15	1	3	1.97E-01	1304.86	3	3	2.03E+00
5352.46	6	6	3.72E-02	5226.70	3	3	1.46E-01	1306.03	1	3	6.76E-01
6365.84	2	2	2.18E-01	5245.60	5	5	8.66E-02	3823.41	7	7	6.63E-03
6394.75	2	4	2.15E-01	5272.35	5	3	9.48E-03	3823.87	5	3	1.87E-03
6445.34	2	4	8.89E-02	5288.25	5	3	3.22E-02	3824.35	5	5	5.19E-03
6450.79	2	2	1.77E-01	5736.93	3	5	1.84E-01	3825.02	3	3	5.59E-03
6454.08	4	6	1.49E-01	5776.31	1	3	1.85E-02	3825.19	5	7	8.31E-04
6463.09	4	4	1.13E-01	5784.76	3	1	5.51E-02	3855.01	5	5	1.63E-02
6467.02	6	8	2.11E-01	5795.09	3	3	1.37E-02	3947.29	5	7	4.91E-03
6468.57	4	2	3.52E-02	5812.31	3	5	1.36E-02	3947.48	5	5	4.88E-03
6478.76	6	6	6.31E-02	5826.43	5	3	2.25E-02	3947.59	5	3	4.87E-03
6487.84	6	4	1.05E-02	5843.84	5	5	4.01E-02	3951.93	3	1	3.10E-03
7371.51	4	4	3.53E-02	6380.75	1	3	1.42E-01	3952.98	5	3	1.29E-03
7404.54	6	6	3.61E-02	7103.24	1	3	6.28E-02	3953.00	1	3	1.03E-03
8307.51	2	4	1.65E-02	7109.35	3	5	8.46E-02	3954.52	3	5	7.73E-04
8344.95	2	2	6.52E-02	7111.28	3	3	4.70E-02	3954.61	5	5	2.32E-03
8386.39	4	4	8.03E-02	*7116.8	9	15	1.12E-01	3997.95	5	3	2.41E-02
8424.56	4	2	3.17E-02	7122.98	5	7	1.12E-01	4217.09	3	1	5.44E-03
				7127.25	5	5	2.80E-02	4222.77	5	3	2.26E-03
N IV				7127.25	5	3	3.11E-03	4222.82	1	3	1.81E-03
247.205	1	3	1.19E+02	9165.07	3	5	4.23E-02	4233.27	5	5	4.04E-03
*283.52	9	15	3.05E+02	9182.16	5	7	4.45E-02	4368.19	3	1	7.56E-03
*322.64	9	3	8.99E+01	9222.99	7	9	4.95E-02	4368.24	3	5	7.59E-03
335.047	3	5	1.845E+02	9247.04	5	5	7.66E-03	4967.38	3	5	4.43E-03
387.356	3	1	2.55E+01	9311.55	7	7	5.36E-03	4967.88	5	7	8.44E-03
765.147	1	3	2.320E+01					4968.79	7	9	1.27E-02
*923.16	9	9	1.759E+01	**N V**				5019.29	5	5	7.13E-03
955.334	3	1	2.919E+01	*209.29	2	6	1.21E+02	5020.22	7	5	9.98E-03
1718.55	3	5	2.321E+00	*247.66	6	10	4.26E+02	5329.11	3	5	9.48E-03
2649.88	3	3	1.07E+00	1238.82	2	4	3.40E+00	5329.69	5	7	1.81E-02
3052.20	1	3	1.33E-01	1242.80	2	2	3.37E+00	5330.74	7	9	2.71E-02
3059.60	3	3	3.95E-01	4603.74	2	4	4.14E-01	5435.18	3	5	7.74E-03
3075.19	5	3	6.48E-01	4619.97	2	2	4.10E-01	5435.77	5	5	1.29E-02
3443.61	3	5	3.46E-01					5436.86	7	5	1.80E-02
3445.22	1	3	4.60E-01	**N VI**				5512.60	3	5	2.69E-03
3454.65	3	3	3.42E-01	24.8980	1	3	5.158E+03	5512.77	5	7	3.58E-03
3461.36	3	1	1.36E+00	28.7870	1	3	1.809E+04	5554.83	3	3	5.83E-03
3463.36	5	5	1.02E+00	*161.220	3	9	2.859E+02	5555.00	5	3	9.71E-03
3474.53	5	3	5.61E-01	173.275	1	3	2.697E+02	5958.39	3	5	6.80E-03
3478.72	3	5	1.06E+00	*173.93	9	15	8.756E+02	5958.58	5	7	9.06E-03
*3480.8	3	9	1.06E+00	185.192	3	5	8.205E+02	6046.23	3	3	1.05E-02
3483.00	3	3	1.06E+00	*1901	3	9	6.780E-01	6046.44	5	3	1.75E-02
3484.93	3	1	1.06E+00	2896.4	1	3	2.079E-01	6046.49	1	3	3.50E-03
3689.94	3	1	9.10E-02	*6991.1	3	9	8.384E-02	6155.99	3	5	2.67E-02
3694.14	3	3	2.27E-02	9622.0	1	3	3.276E-02	6156.78	5	7	5.08E-02
3707.39	5	3	6.73E-02					6158.19	7	9	7.62E-02
3714.43	5	5	1.34E-02	**Oxygen**				6324.84	7	5	3.76E-05
3735.43	7	5	7.37E-02	**O I**				6453.60	3	5	1.65E-02
3747.54	3	5	9.92E-01	791.973	5	5	4.94E+00	6454.44	5	5	2.75E-02
4057.76	3	5	6.62E-01	792.938	1	3	2.19E+00	6455.98	7	5	3.85E-02
4740.26	3	5	1.53E-02	792.967	3	5	1.64E+00	6726.28	5	5	1.18E-05
4747.96	3	3	7.60E-02	877.798	5	3	2.85E+00	6726.54	5	3	6.44E-06
4752.49	5	7	1.13E-02	877.879	5	5	5.12E+00	7001.92	3	5	2.65E-02
4762.09	5	5	6.99E-02	922.008	5	7	1.23E+00	7002.23	5	7	3.53E-02

λ Å	Weights g_i	g_k	A 10^8 s^{-1}	λ Å	Weights g_i	g_k	A 10^8 s^{-1}	λ Å	Weights g_i	g_k	A 10^8 s^{-1}
7254.15	3	3	2.24E-02	2418.46	6	6	2.30E-01	4104.72	4	6	3.14E-01
7254.45	5	3	3.73E-02	2425.57	6	6	1.77E-01	4104.99	4	4	9.14E-01
7254.53	1	3	7.45E-03	2433.54	2	4	4.21E-01	4106.02	8	6	1.70E-02
7771.94	5	7	3.69E-01	2436.06	4	4	1.69E-01	4109.84	6	6	1.21E-02
7774.17	5	5	3.69E-01	2444.25	4	4	7.56E-02	4110.19	6	4	2.54E-01
7775.39	5	3	3.69E-01	2445.53	4	6	4.98E-01	4110.79	4	2	7.70E-01
7981.94	3	3	2.33E-04	2517.96	4	6	7.72E-02	4112.02	6	6	1.81E-01
7982.40	1	3	3.09E-04	2523.21	2	2	9.63E-02	4113.83	8	6	2.41E-01
7986.98	3	5	4.19E-04	2526.87	4	4	1.20E-01	4119.22	6	8	1.33E+00
7987.33	5	5	1.41E-04	2530.28	6	8	8.16E-02	4120.28	6	6	2.15E-01
7995.07	5	7	5.63E-04	2571.46	2	4	1.15E-01	4120.55	6	4	2.60E-01
8221.82	7	7	2.89E-01	2575.28	4	6	1.37E-01	4121.46	2	2	5.60E-01
8227.65	5	3	8.13E-02	3134.73	8	6	1.23E+00	4129.32	4	2	1.79E-01
8230.00	5	5	2.26E-01	3273.43	8	6	9.99E-01	4132.80	2	4	9.13E-01
8233.00	3	3	2.43E-01	3377.15	2	2	1.27E+00	4140.70	4	4	4.09E-02
8235.35	3	5	4.86E-02	3390.21	2	4	1.22E+00	4153.30	4	6	7.91E-01
8446.25	3	1	3.22E-01	3407.28	6	6	1.02E+00	4156.53	6	4	2.11E-01
8446.36	3	5	3.22E-01	3712.74	2	4	2.84E-01	4169.22	6	6	2.71E-01
8446.76	3	3	3.22E-01	3727.32	4	6	5.81E-01	4185.44	5	8	1.91E+00
8820.42	5	7	2.93E-01	3749.48	6	4	9.31E-01	4189.58	8	8	7.06E-02
9260.81	3	1	4.46E-01	3833.07	6	8	1.02E-02	4189.79	8	10	1.98E+00
9260.85	3	3	3.34E-01	3842.81	2	4	7.45E-02	4192.51	6	4	3.21E-01
9260.94	3	5	1.56E-01	3843.58	4	6	3.55E-02	4196.27	4	4	3.56E-02
9262.58	5	3	1.11E-01	3847.89	2	2	1.95E-01	4196.70	4	2	3.56E-01
9262.67	5	5	2.60E-01	3850.80	4	6	6.00E-03	4317.14	2	4	3.70E-01
9262.78	5	7	2.97E-01	3851.03	4	4	1.59E-01	4319.63	4	6	2.55E-01
9265.83	7	5	2.97E-02	3851.47	8	8	2.72E-02	4319.87	2	2	5.62E-01
9265.93	7	7	1.48E-01	3856.13	4	2	2.28E-01	4325.76	2	2	1.47E-01
9266.01	7	9	4.45E-01	3857.16	6	6	6.59E-02	4327.46	6	6	6.76E-01
9482.89	5	3	2.34E-01	3863.50	6	8	6.49E-02	4327.85	6	4	7.24E-02
9622.11	5	3	5.22E-04	3864.13	2	2	9.12E-02	4328.59	4	2	1.12E+00
9622.16	3	3	1.57E-03	3864.43	6	6	2.15E-01	4331.47	4	6	4.82E-02
9625.26	7	5	3.25E-04	3864.67	6	4	1.80E-01	4331.86	4	4	6.50E-01
9625.30	7	7	1.85E-03	3874.09	2	4	3.26E-02	4336.86	4	4	1.57E-01
9694.66	5	7	4.54E-04	3875.80	8	6	3.38E-02	4345.56	4	2	8.31E-01
9694.91	5	5	4.54E-04	3882.19	8	8	5.50E-01	4347.22	6	4	1.19E-01
9695.06	5	3	4.54E-04	3882.45	4	4	8.94E-02	4347.41	4	4	9.32E-01
				3883.14	8	6	1.13E-01	4349.43	6	6	6.91E-01
O II				3893.52	4	4	1.89E-02	4351.26	6	6	9.89E-01
429.918	4	2	4.25E+01	3907.45	6	6	8.64E-02	4351.46	4	6	5.82E-02
430.041	4	4	4.13E+01	3911.96	6	4	1.09E+00	4359.40	4	6	1.44E-02
430.176	4	6	4.36E+01	3912.12	4	4	1.41E-01	4366.89	6	4	3.98E-01
483.760	4	2	2.05E+01	3919.27	4	2	1.22E+00	4369.27	4	4	3.57E-01
483.980	6	4	1.80E+01	3945.04	2	4	2.05E-01	4395.93	6	6	3.91E-01
484.027	4	4	3.22E+00	3954.36	2	2	8.57E-01	4405.98	6	4	4.30E-02
485.087	6	8	2.60E+01	3973.26	4	4	1.04E+00	4414.90	4	6	8.34E-01
485.470	6	6	1.20E+00	3982.71	4	2	4.27E-01	4416.97	2	4	7.13E-01
485.518	4	6	1.93E+01	4069.62	2	4	1.52E+00	4443.01	6	6	5.05E-01
2290.85	2	4	7.41E-02	4069.88	4	6	1.53E+00	4443.52	6	8	1.89E-02
2293.30	2	2	3.25E-01	4072.15	6	8	1.98E+00	4447.68	8	6	2.52E-02
2300.33	4	4	4.17E-01	4075.86	8	10	2.11E+00	4448.19	8	8	5.10E-01
2302.81	4	2	1.67E-01	4078.84	4	4	5.52E-01	4452.38	4	4	1.37E-01
2365.14	4	2	1.52E-01	4084.65	6	8	7.28E-02	4466.24	2	4	9.00E-01
2375.72	6	4	1.35E-01	4085.11	6	6	4.55E-01	4467.46	2	2	9.00E-01
2406.38	6	4	1.85E-01	4092.93	8	8	2.65E-01	4563.18	4	4	7.18E-03
2407.48	4	4	2.25E-01	4094.14	6	4	4.70E-02	4590.97	6	8	8.85E-01
2411.60	4	2	2.05E-01	4096.53	4	6	1.73E-01	4595.96	6	6	4.87E-02
2411.64	2	2	1.10E-01	4097.22	2	4	3.62E-01	4596.18	4	6	8.34E-01
2415.13	4	2	2.20E-01	4103.00	2	2	5.09E-01	4638.86	2	4	3.71E-01

λ Å	g_i	g_k	A 10^8 s^{-1}	λ Å	g_i	g_k	A 10^8 s^{-1}	λ Å	g_i	g_k	A 10^8 s^{-1}
4641.81	4	6	5.96E-01	263.817	5	7	5.97E+01	3017.62	7	7	5.38E-01
4649.13	6	8	7.81E-01	263.861	5	5	1.49E+01	3023.43	3	5	4.79E-01
4650.84	2	2	6.86E-01	277.386	5	7	9.43E+01	3024.36	7	5	9.39E-02
4661.63	4	4	4.10E-01	279.788	5	5	4.25E+01	3024.54	1	3	6.16E-01
4673.73	4	2	1.35E-01	295.942	1	3	5.56E+01	3035.41	3	3	4.59E-01
4676.23	6	6	2.05E-01	303.413	1	3	4.29E+01	3042.07	3	1	1.94E+00
4690.89	2	4	1.86E-01	303.461	3	1	1.29E+02	3047.10	5	5	1.49E+00
4691.42	2	2	7.43E-01	303.517	3	3	3.21E+01	3059.28	5	3	8.72E-01
4696.35	6	4	3.25E-02	303.622	3	5	3.21E+01	3064.98	1	3	2.17E-01
4698.44	6	6	6.59E-02	303.695	5	3	5.34E+01	3068.13	3	1	6.49E-01
4699.01	6	8	9.88E-01	303.800	5	5	9.61E+01	3068.26	3	3	5.41E-02
4699.22	4	6	9.36E-01	305.596	1	3	1.20E+02	3068.67	3	5	2.27E-01
4701.18	4	4	9.23E-01	305.656	3	5	1.62E+02	3074.14	5	7	1.84E-01
4701.71	4	2	3.69E-01	305.702	3	3	9.01E+01	3074.72	5	3	3.76E-01
4703.16	4	6	9.20E-01	305.767	5	7	2.16E+02	3075.13	5	5	1.61E-01
4705.35	6	8	1.10E+00	305.836	5	5	5.40E+01	3075.95	7	9	1.07E-01
4710.01	4	6	2.98E-01	320.978	5	7	2.17E+02	3083.65	7	7	3.20E-01
4741.70	6	6	4.71E-02	328.448	5	5	1.04E+02	3084.64	7	5	2.55E-01
4751.28	6	8	6.39E-02	345.312	1	3	1.35E+02	3088.04	9	9	5.30E-01
4752.69	6	6	1.45E-02	374.073	5	5	2.85E+01	3095.79	9	7	1.35E-01
4844.92	4	6	1.02E-02	395.557	5	3	2.80E+01	3115.67	3	1	1.39E+00
4856.39	4	6	5.58E-02	507.388	1	3	1.61E+01	3121.63	3	3	1.38E+00
4856.76	4	4	1.00E+00	507.680	3	3	4.82E+01	3132.79	3	5	1.37E+00
4860.97	2	4	4.70E-01	508.178	5	3	8.04E+01	3198.18	3	5	9.57E-02
4864.88	4	2	8.07E-02	525.794	5	3	9.60E+01	3201.14	3	3	4.77E-01
4871.52	4	6	5.60E-01	597.814	1	3	1.49E+01	3202.51	5	7	7.08E-02
4872.02	4	4	9.34E-02	599.590	5	5	5.41E+01	3207.61	5	5	4.40E-01
4890.86	4	2	4.80E-01	702.337	1	3	6.06E+00	3210.58	5	3	1.58E-01
4906.83	4	4	4.54E-01	702.838	3	1	1.83E+01	3216.07	7	7	5.58E-01
4924.53	4	6	5.43E-01	832.929	1	3	3.41E+00	3221.21	7	5	9.75E-02
4941.07	2	4	5.87E-01	835.092	5	5	1.44E+00	3260.86	5	7	1.68E+00
4943.01	4	6	7.78E-01	835.289	5	7	5.99E+00	3265.33	7	9	1.88E+00
4955.71	4	4	1.82E-01	1679.03	3	5	6.57E-01	3267.20	3	5	1.58E+00
5159.94	2	2	3.29E-01	1686.73	3	3	6.48E-01	3281.83	5	5	2.89E-01
5175.90	4	2	1.49E-01	1760.41	3	5	8.38E-01	3284.45	7	7	2.06E-01
5190.50	2	4	1.26E-01	1764.46	5	5	2.50E+00	3299.39	1	3	1.64E-01
5206.65	4	4	3.58E-01	1766.63	1	3	1.11E+00	3312.33	3	3	4.60E-01
5583.22	2	4	2.17E-02	1772.28	3	1	3.29E+00	3326.06	3	3	2.65E-01
5611.07	2	2	2.14E-02	1772.97	5	5	1.37E+00	3330.30	3	5	6.81E-01
6627.37	4	4	1.73E-01	2390.43	3	3	1.62E+00	3330.32	3	5	4.76E-01
6641.03	2	2	9.88E-02	2454.97	3	1	3.43E+00	3332.41	5	3	7.92E-01
6666.66	4	2	6.78E-02	2665.68	3	5	6.75E-01	3332.93	5	7	5.04E-01
6677.87	2	4	3.37E-02	2674.58	5	5	1.11E+00	3336.67	3	3	3.76E-01
6717.75	2	2	1.33E-01	2683.66	3	1	1.85E+00	3336.69	5	5	8.77E-02
6721.39	4	2	1.81E-01	2686.15	7	5	1.54E+00	3340.76	5	3	6.57E-01
6810.48	6	8	1.64E-03	2687.55	3	3	1.84E+00	3344.20	5	5	1.25E-01
6844.10	4	6	2.97E-03	2695.48	3	5	1.82E+00	3344.51	5	7	3.48E-01
6846.80	8	8	3.17E-02	2794.14	3	1	1.82E-01	3347.98	7	5	4.86E-01
6869.48	6	6	5.35E-02	2798.93	3	3	4.52E-02	3350.62	5	3	1.12E+00
6884.88	4	4	6.12E-02	2809.66	5	3	1.34E-01	3350.92	7	7	9.91E-01
6895.10	10	8	2.72E-01	2818.70	5	5	2.66E-02	3355.86	7	7	6.89E-01
6906.44	8	6	2.48E-01	2836.31	7	5	1.46E-01	3362.31	7	5	6.87E-01
6907.87	4	2	3.03E-01	2959.69	3	5	1.83E+00	3376.61	3	1	1.49E+00
6910.56	6	4	2.43E-01	2983.78	3	5	2.15E+00	3376.76	3	3	1.12E+00
				2992.08	3	5	9.32E-02	3377.26	3	5	5.20E-01
O III				2996.48	3	3	4.64E-01	3382.61	5	7	9.86E-01
263.694	1	3	3.32E+01	2997.69	5	7	6.88E-02	3383.31	5	3	3.70E-01
263.727	3	5	4.48E+01	3004.34	5	5	4.27E-01	3383.81	5	5	8.62E-01
263.773	3	3	2.49E+01	3008.78	5	3	1.53E-01	3384.90	7	9	1.48E+00

λ Å	Weights g_i	g_k	A 10^8 s^{-1}	λ Å	Weights g_i	g_k	A 10^8 s^{-1}	λ Å	Weights g_i	g_k	A 10^8 s^{-1}
3394.22	7	7	4.88E-01	4440.09	5	3	4.42E-01	2816.53	4	4	5.74E-01
3395.43	7	5	9.75E-02	4447.69	5	5	4.40E-01	2829.17	8	6	1.56E-01
3406.88	1	3	1.93E-01	4461.61	5	7	4.36E-01	2836.27	6	4	8.43E-01
3408.13	3	1	5.79E-01	4524.22	3	1	3.38E-01	2916.31	2	4	1.06E+00
3415.26	3	3	1.44E-01	4532.78	5	3	1.40E-01	2921.46	4	6	1.27E+00
3428.63	3	5	1.42E-01	4535.29	3	3	8.40E-02	2926.18	4	4	2.11E-01
3430.57	5	3	2.37E-01	4555.39	5	5	2.49E-01	3063.43	2	4	1.30E+00
3444.05	5	5	4.21E-01	4557.91	3	5	8.27E-02	3071.60	2	2	1.29E+00
3446.68	3	5	9.71E-01	5268.30	1	3	3.50E-01	3177.89	2	4	7.59E-02
3447.15	1	3	8.09E-01	5508.24	5	5	1.06E-01	3180.77	2	2	1.51E-01
3447.97	5	7	1.19E+00	5592.25	3	3	3.27E-01	3180.99	4	6	7.06E-02
3450.91	7	9	1.44E+00					3185.74	4	4	1.21E-01
3451.30	3	3	8.06E-01	**O IV**				3188.22	6	8	4.28E-02
3454.84	5	5	6.89E-01	238.360	2	4	2.96E+02	3188.64	4	2	1.50E-01
3454.99	9	11	1.72E+00	238.570	4	6	3.54E+02	3194.78	6	6	1.71E-01
3459.48	5	3	1.14E-01	238.579	4	4	5.90E+01	3199.58	6	4	1.04E-01
3459.94	7	7	5.14E-01	279.631	2	2	2.68E+01	3209.65	8	8	2.53E-01
3466.13	9	9	2.84E-01	279.933	4	2	5.34E+01	3216.31	8	6	5.56E-02
3466.85	7	5	6.82E-02	553.329	2	4	1.22E+01	3348.06	2	4	8.51E-01
3475.24	9	7	2.42E-02	554.076	2	2	4.86E+01	3349.11	4	6	1.02E+00
3520.94	1	3	1.50E-01	554.513	4	4	6.06E+01	3354.27	4	2	7.71E-01
3531.22	3	1	4.45E-01	555.263	4	2	2.41E+01	3362.55	4	4	7.65E-01
3533.38	3	3	1.11E-01	608.397	2	2	1.21E+01	3375.40	4	6	7.56E-01
3534.90	3	5	1.11E-01	609.829	4	2	2.40E+01	3378.02	4	4	1.66E-01
3555.24	5	3	1.82E-01	616.952	6	4	2.60E+01	3381.21	4	6	7.19E-01
3556.78	5	5	3.26E-01	617.005	4	4	2.89E+00	3381.30	2	4	4.28E-01
3695.38	3	5	4.01E-01	617.036	4	2	2.89E+01	3385.52	6	8	1.02E+00
3698.72	5	7	7.62E-01	624.619	2	4	1.07E+01	3390.19	2	2	8.49E-01
3703.36	7	9	1.14E+00	625.127	4	4	2.13E+01	3396.80	4	4	5.40E-01
3704.75	3	3	8.53E-01	625.853	6	4	3.19E+01	3405.77	4	2	1.67E-01
3707.27	3	5	7.34E-01	779.736	6	4	1.46E+00	3409.70	6	6	3.00E-01
3709.54	3	1	1.13E+00	779.820	4	4	1.31E+01	3411.30	4	4	1.69E-01
3712.49	5	5	6.59E-01	779.912	6	6	1.36E+01	3411.69	4	6	1.02E+00
3714.03	3	3	4.06E-01	779.997	4	6	9.70E-01	3425.55	6	4	4.94E-02
3715.09	5	7	9.73E-01	787.710	2	4	5.95E+00	3489.89	4	6	7.29E-01
3720.89	7	7	3.74E-01	790.112	4	4	1.18E+00	3492.21	2	4	6.06E-01
3721.95	5	3	2.80E-01	790.199	4	6	7.08E+00	3493.43	4	4	1.21E-01
3725.31	5	5	2.41E-01	921.296	2	4	2.21E+00	3560.39	4	6	1.03E+00
3728.51	5	7	1.29E+00	921.365	2	2	8.83E+00	3563.33	6	8	1.10E+00
3728.84	7	9	1.45E+00	923.367	4	4	1.10E+01	3593.08	6	6	7.15E-02
3729.80	3	5	1.22E+00	923.436	4	2	4.39E+00	3725.89	2	4	5.61E-01
3732.13	5	3	2.67E-02	1338.61	2	4	2.17E+00	3725.94	4	6	6.01E-01
3734.83	7	5	7.40E-02	1342.99	4	4	4.29E-01	3729.03	6	8	6.86E-01
3742.63	5	5	2.24E-01	1343.51	4	6	2.57E+00	3736.68	4	6	2.23E-01
3746.90	7	7	1.59E-01	2120.58	2	4	1.05E+00	3736.85	8	10	7.95E-01
3754.70	3	5	7.53E-01	2132.64	4	4	1.29E+00	3744.89	6	6	1.92E-01
3757.23	1	3	5.56E-01	2493.39	2	4	1.18E+00	3758.39	8	8	1.11E-01
3759.88	5	7	9.79E-01	2493.75	4	6	8.48E-01	3930.68	2	2	3.80E-02
3774.03	3	3	3.91E-01	2493.99	2	2	6.09E-01	3942.06	2	4	9.42E-02
3791.28	5	5	2.24E-01	2499.27	2	2	4.68E-01	3945.31	4	2	1.88E-01
3810.98	5	3	2.37E-02	2501.81	4	4	3.73E-01	3956.77	4	4	2.98E-02
3816.75	5	3	9.63E-02	2507.73	4	2	2.32E+00	3974.58	4	6	6.62E-02
3961.57	5	7	1.25E+00	2509.22	6	6	1.94E+00	3977.09	6	4	9.91E-02
4072.64	1	3	3.37E-01	2510.58	4	2	1.19E+00	3995.08	6	6	1.52E-01
4073.98	3	5	4.54E-01	2517.37	6	4	1.24E+00	4687.03	2	4	2.79E-01
4081.02	5	7	6.02E-01	2781.22	2	2	1.03E-01	4772.60	2	4	1.23E-01
4089.30	3	3	2.49E-01	2803.57	6	4	1.26E-01	4779.10	2	2	2.45E-01
4103.07	5	5	1.48E-01	2805.87	2	4	2.90E-01	4783.42	4	6	2.06E-01
4118.60	5	3	1.63E-02	2812.50	6	6	3.58E-02	4794.18	4	4	1.56E-01

λ Å	Weights g_i	Weights g_k	A 10^8 s^{-1}	λ Å	Weights g_i	Weights g_k	A 10^8 s^{-1}	λ Å	Weights g_i	Weights g_k	A 10^8 s^{-1}
4798.27	6	8	2.91E-01	4213.35	5	3	1.19E-02	2149.1	4	2	3.18E+00
4813.15	6	6	8.65E-02	4522.66	5	3	1.02E-02	2152.9	2	4	4.85E-01
5305.51	4	4	6.10E-02	4554.53	3	5	2.41E-01	2154.1	4	4	1.73E-01
5362.51	6	6	6.12E-02	5114.06	1	3	1.80E-01	2154.1	4	6	5.8E-01
6876.49	2	4	1.88E-02	5339.94	1	3	1.85E-02	2534.0	2	4	2.00E-01
6931.60	2	2	7.35E-02	5349.74	3	1	7.04E-02	2535.6	4	4	9.5E-01
7004.11	4	4	8.90E-02	5372.71	3	3	1.42E-02	2553.3	2	2	7.1E-01
7061.30	4	2	3.48E-02	5414.59	3	5	9.29E-03	2554.9	4	2	3.00E-01
				5428.38	5	3	2.68E-02				
O V				5471.12	5	5	4.86E-02	**P II**			
172.169	1	3	2.94E+02	5571.81	1	3	8.33E-02	1301.9	1	3	5.0E-01
*192.85	9	15	6.90E+02	5580.12	3	5	1.11E-01	1304.5	3	1	1.5E+00
*215.17	9	3	1.83E+02	5583.23	3	3	6.20E-02	1304.7	3	3	3.7E-01
220.353	3	5	4.292E+02	*5589.9	9	15	1.49E-01	1305.5	3	5	3.8E-01
248.460	3	1	5.59E+01	5597.89	5	7	1.48E-01	1309.9	5	3	6.2E-01
629.732	1	3	2.872E+01	5604.27	5	5	3.68E-02	1310.7	5	5	1.1E+00
758.677	3	5	5.547E+00	5607.41	5	3	4.08E-03	4475.3	5	7	1.3E+00
759.442	1	3	7.373E+00	6330.05	5	7	1.21E-01	4499.2	5	7	1.4E+00
760.227	3	3	5.514E+00	6460.12	3	5	9.37E-02	4530.8	3	5	1.0E+00
760.446	5	5	1.652E+01	6466.14	5	7	1.01E-01	4554.8	3	5	9.6E-01
761.128	3	1	2.197E+01	6500.24	7	9	1.11E-01	4588.0	5	7	1.7E+00
762.004	5	3	9.125E+00	6543.77	5	5	1.64E-02	4589.9	3	5	1.6E+00
774.518	3	1	3.804E+01	6601.28	7	7	1.14E-01	4602.1	7	9	1.9E+00
1371.30	3	5	3.336E+00	6764.72	1	3	4.37E-02	4943.5	7	5	6.3E+00
2729.31	3	5	4.52E-01	6789.62	3	5	5.79E-02	5253.5	3	5	1.0E+00
2731.45	1	3	5.90E-01	6817.40	3	3	3.00E-02	5425.9	5	5	6.9E-01
2743.61	3	3	4.38E-01	6828.95	5	7	7.35E-02	6024.2	3	5	5.1E-01
2752.23	3	1	1.82E+00	6878.76	5	5	1.65E-02	6043.1	5	7	6.8E-01
2755.13	5	5	1.37E+00								
2769.69	5	3	7.88E-01	**O VI**				**P III**			
2781.01	3	5	1.40E+00	*150.10	2	6	2.62E+02	1334.8	2	4	5.5E-01
*2784.0	3	9	1.40E+00	*173.03	6	10	8.78E+02	1344.3	4	6	6.4E-01
2786.99	3	3	1.39E+00	1031.91	2	4	4.16E+00	1344.8	4	4	1.1E-01
2789.85	3	1	1.38E+00	1037.61	2	2	4.09E+00	4057.4	4	4	1.0E-01
3058.68	3	5	1.39E+00	3811.35	2	4	5.14E-01	4059.3	6	4	9.0E-01
3144.66	3	5	8.86E-01	3834.24	2	2	5.05E-01	4080.1	4	2	9.9E-01
3219.24	3	1	1.54E-01								
3222.29	1	3	1.16E-01	**O VII**				**Potassium**			
3227.54	3	3	3.38E-02	18.6270	1	3	9.365E+03	**K I**			
3239.21	3	3	3.28E-01	21.6020	1	3	3.309E+04	4044.1	2	4	1.24E-02
3248.28	5	3	1.18E-01	*120.33	3	9	5.334E+02	4047.2	2	2	1.24E-02
3263.54	5	5	1.86E-02	128.411	1	3	8.982E+02	5084.2	2	2	3.50E-03
3275.64	5	3	4.76E-01	*128.46	9	15	1.615E+03	5099.2	4	2	7.0E-03
3297.62	7	5	1.30E-01	135.820	3	5	1.523E+03	5323.3	2	2	6.3E-03
3690.17	3	5	1.97E-02	*1630.3	3	9	7.935E-01	5339.7	4	2	1.26E-02
3698.36	3	3	1.03E-01	2448.98	1	3	2.514E-01	5343.0	2	4	4.0E-03
3702.72	5	7	1.41E-02	*5933.1	3	9	1.002E-01	5359.6	4	6	4.6E-03
3717.31	5	5	9.63E-02	8241.76	1	3	3.864E-02	5782.4	2	2	1.23E-02
3725.63	5	3	2.91E-02					5801.8	4	2	2.46E-02
3746.64	7	7	1.18E-01	**Phosphorus**				5812.2	2	4	2.8E-03
3761.58	7	5	1.61E-02	**P I**				5831.9	4	6	3.2E-03
4119.37	3	5	3.66E-01	1671.7	4	2	3.9E-01	6911.1	2	2	2.72E-02
4120.49	3	1	3.33E-01	1674.6	4	4	4.0E-01	6938.8	4	2	5.4E-02
4123.96	5	7	4.81E-01	1679.7	4	6	3.9E-01	7664.9	2	4	3.87E-01
4125.49	1	3	2.70E-01	1775.0	4	6	2.17E+00	7699.0	2	2	3.82E-01
4134.11	3	3	3.34E-01	1782.9	4	4	2.14E+00				
4153.27	3	3	1.92E-01	1787.7	4	2	2.13E+00	**K II**			
4158.86	3	5	3.39E-01	2135.5	4	4	2.11E-01	607.93	1	3	1.3E-02
4178.46	5	5	1.12E-01	2136.2	6	4	2.83E+00				

λ Å	Weights g_i	g_k	A 10^8 s^{-1}	λ Å	Weights g_i	g_k	A 10^8 s^{-1}	λ Å	Weights g_i	g_k	A 10^8 s^{-1}
K III				3121.76	6	6	1.1E-01	4056.34	6	4	9.5E-03
2550.0	6	4	2.0E+00	3123.70	10	8	4.6E-02	4082.78	6	4	1.4E-01
2635.1	4	4	1.2E+00	3137.71	4	6	3.3E-02	4097.52	2	4	7.0E-02
2992.4	6	8	2.5E+00	3189.05	6	6	3.03E-01	4121.68	6	6	9.8E-02
3052.1	4	6	1.7E+00	3197.13	6	4	4.35E-02	4128.87	6	8	1.73E-01
3202.0	4	4	1.8E+00	3263.14	6	6	1.3E-01	4135.27	8	8	1.0E-01
3289.1	4	6	2.0E+00	3271.61	6	4	2.0E-01	4196.50	6	8	3.9E-02
3322.4	6	6	1.3E+00	3280.55	8	8	2.36E-01	4211.14	8	10	1.62E-01
3421.8	2	4	1.5E+00	3283.57	6	8	4.4E-01	4244.44	4	4	6.5E-03
				3289.14	4	4	1.0E-01	4278.60	4	6	9.2E-03
K XVI				3323.09	8	10	6.3E-01	4288.71	6	8	6.1E-02
206.27	1	3	9.4E+01	3331.09	4	2	5.40E-02	4373.04	2	4	1.8E-02
				3338.54	8	6	3.5E-02	4374.80	8	10	1.64E-01
K XVII				3360.80	4	4	1.2E-01	4379.92	6	6	2.48E-02
22.020	2	4	4.7E+04	3368.38	6	4	1.1E-01	4492.47	6	6	4.5E-03
22.163	4	6	5.6E+04	3396.82	10	10	6.5E-01	4528.72	6	8	1.35E-02
22.18	4	4	9.3E+03	3399.70	6	8	1.2E-01	4548.73	4	6	5.5E-03
22.60	2	2	2.5E+03	3462.04	6	6	6.2E-01	4551.64	4	4	4.00E-02
22.76	4	2	4.7E+03	3470.66	4	4	8.5E-01	4565.19	4	4	1.1E-02
				3478.91	6	6	3.32E-01	4569.00	6	8	1.0E-02
Praseodymium				3484.04	6	8	9.3E-03	4608.12	2	2	2.1E-02
Pr II				3498.73	4	6	2.12E-01	4675.03	8	8	6.4E-03
3997.0	15	15	1.87E-01	3502.52	10	10	4.3E-01	4721.00	6	4	3.43E-03
4062.8	13	15	1.00E+00	3507.32	6	8	3.4E-01	4745.11	6	6	5.2E-03
4100.7	17	19	8.4E-01	3528.02	8	8	8.5E-01	4755.58	4	4	6.0E-03
4143.1	15	17	5.8E-01	3543.95	4	4	4.65E-02	4842.43	6	8	1.6E-03
4179.4	13	15	5.2E-01	3549.54	6	6	2.22E-01	4963.71	2	2	3.0E-02
4222.9	11	13	3.91E-01	3570.18	4	6	1.82E-01	4977.75	4	4	9.8E-03
4241.0	17	15	2.30E-01	3583.10	8	10	2.6E-01	4979.18	4	6	1.0E-02
4359.8	15	15	1.1E-01	3596.19	6	4	5.5E-01	5090.63	6	6	5.0E-03
4405.8	17	17	9.0E-02	3597.15	6	8	5.9E-01	5120.69	6	8	3.1E-03
4429.3	15	15	2.28E-01	3612.47	4	2	8.90E-01	5130.76	4	4	4.35E-03
4449.8	13	13	1.24E-01	3620.46	6	4	8.5E-02	5155.54	2	4	9.8E-03
4468.7	11	13	1.54E-01	3654.87	8	8	6.0E-02	5184.19	6	8	1.6E-03
4510.2	13	15	1.16E-01	3657.99	8	6	8.8E-01	5212.73	4	2	5.95E-03
4534.2	15	17	4.9E-02	3666.22	6	8	8.4E-02	5292.14	10	10	3.7E-03
4734.2	15	13	2.5E-02	3690.70	6	4	3.23E-01	5390.44	4	6	9.5E-03
4879.1	15	15	1.8E-02	3692.36	10	8	9.1E-01	5424.72	4	4	5.0E-03
4886.0	15	15	1.3E-02	3700.91	8	10	3.9E-01	5599.42	6	8	1.3E-02
4912.6	17	15	5.7E-02	3713.02	4	4	8.3E-02	5983.60	10	10	2.1E-02
5034.4	19	19	1.1E-01	3788.47	4	6	1.4E-01				
5110.8	21	19	2.78E-01	3793.22	8	g_k	4.2E-01	**Rubidium**			
5135.1	17	17	1.25E-01	3799.31	8	8	5.5E-01	**Rb I**			
5173.9	19	17	3.18E-01	3806.76	6	6	6.2E-02	3022.5	2	4	4.13E-05
5219.1	15	15	9.5E-02	3818.19	6	4	5.8E-01	3032.0	2	4	4.93E-05
5220.1	17	15	2.35E-01	3822.26	6	6	8.5E-01	3044.2	2	4	8.2E-05
5251.7	15	13	1.1E-02	3828.48	6	6	6.2E-01	3060.2	2	4	1.05E-04
5259.7	15	13	2.24E-01	3833.89	6	4	5.8E-01	3082.0	2	4	1.49E-04
5292.6	13	13	9.3E-02	3856.52	8	10	5.9E-01	3112.6	2	4	2.5E-04
5810.6	17	19	2.3E-02	3872.39	4	6	6.7E-03	3113.1	2	2	1.3E-04
5879.3	15	15	7.6E-02	3877.34	8	6	3.7E-02	3157.5	2	4	3.38E-04
6200.8	15	17	1.8E-02	3913.51	8	8	2.5E-03	3158.3	2	2	2.0E-04
6278.7	13	15	2.6E-02	3922.19	4	2	6.25E-02	3228.0	2	4	6.4E-04
6398.0	11	13	1.9E-02	3934.23	8	8	1.58E-01	3229.2	2	2	3.8E-04
				3942.72	4	2	7.15E-01	3348.7	2	4	1.37E-03
Rhodium				3958.86	6	8	5.5E-01	3350.8	2	2	8.9E-04
Rh I				3984.40	4	4	1.1E-01	3587.1	2	4	3.97E-03
3083.96	8	6	4.8E-02	3995.61	4	6	4.7E-02	3591.6	2	2	2.9E-03
3114.91	6	4	4.45E-02	4053.44	2	2	2.8E-02	4201.8	2	4	1.8E-02

λ (Å)	g_i	g_k	A (10^8 s^{-1})
4215.5	2	2	1.5E-02
7800.3	2	4	3.70E-01
7947.6	2	2	3.40E-01

Scandium
Sc I

λ (Å)	g_i	g_k	A (10^8 s^{-1})
2116.7	4	4	2.0E-01
2120.4	6	6	2.0E-01
2262.3	4	4	5.8E-02
2266.6	4	2	4.8E-01
2270.9	6	4	4.6E-01
2280.8	4	6	2.8E-01
2289.6	6	6	4.1E-02
2311.29	4	6	4.1E-02
2315.69	4	4	2.5E-01
2320.32	6	6	2.4E-01
2324.75	6	4	4.1E-02
2328.19	4	6	4.6E-02
2334.67	4	2	1.7E-01
2346.03	6	4	1.3E-01
2429.19	4	4	2.8E-01
2438.63	6	6	2.1E-01
2468.40	4	2	4.9E-02
2692.78	4	2	1.61E-01
2699.02	4	6	2.4E-02
2706.74	4	4	3.1E-01
2707.93	6	4	1.49E-01
2711.34	6	6	3.2E-01
2965.88	4	6	7.5E-02
2974.01	4	4	5.5E-01
2980.76	6	6	5.4E-01
2988.97	6	4	6.9E-02
3015.37	4	6	7.8E-01
3019.35	6	8	8.7E-01
3030.76	6	6	1.00E-01
3255.68	4	4	3.2E-01
3269.90	4	2	3.13E+00
3273.63	6	4	2.81E+00
3907.48	4	6	1.66E+00
3911.81	6	8	1.79E+00
3933.38	6	6	1.62E-01
3996.60	4	6	1.65E-01
4020.39	4	4	1.63E+00
4023.22	4	4	3.0E-01
4023.68	6	6	1.65E+00
4031.38	6	6	2.9E-01
4036.86	6	4	7.9E-02
4043.80	8	8	3.11E+00
4047.80	6	6	1.54E-01
4051.83	8	6	7.7E-02
4054.54	4	2	1.67E-01
4067.00	6	8	1.91E-01
4067.63	10	8	4.1E-02
4074.96	4	6	3.7E-01
4078.56	2	4	4.3E-01
4080.57	4	4	6.6E-02
4082.39	6	4	2.73E-01
4086.66	6	8	3.7E-01
4087.47	4	6	1.12E-01
4093.12	4	4	1.23E-01
4094.86	6	6	1.44E-01
4098.36	8	8	8.7E-02
4132.98	4	6	1.19E+00
4140.27	6	8	1.17E+00
4147.38	6	6	1.74E-01
4161.85	8	8	1.77E-01
4171.53	6	4	1.36E-01
4186.42	6	8	8.4E-02
4187.61	8	6	1.28E-01
4193.53	4	6	6.1E-02
4204.52	6	8	3.5E-01
4205.20	10	8	1.12E-01
4212.32	4	6	1.58E-01
4212.48	6	6	8.6E-02
4216.08	2	4	2.36E-01
4218.23	4	4	2.26E-01
4225.54	6	8	9.5E-02
4225.69	4	4	7.6E-02
4231.64	4	4	1.31E-01
4233.59	6	6	4.0E-01
4238.05	8	8	7.1E-01
4239.55	6	4	2.27E-01
4246.14	8	8	1.15E-01
4542.55	6	4	1.28E-01
4544.67	8	6	1.33E-01
4706.94	4	6	2.81E-01
4709.31	6	8	4.0E-01
4711.72	2	4	1.81E-01
4714.30	4	4	2.14E-01
4719.31	6	6	1.04E-01
4728.77	8	8	1.16E-01
4729.20	4	4	2.20E-01
4729.24	6	6	1.93E-01
4734.11	4	2	1.10E+00
4737.65	6	4	8.8E-01
4741.02	8	6	9.1E-01
4743.82	10	8	9.8E-01
4973.67	4	2	8.4E-01
4980.36	6	4	5.6E-01
4983.43	4	4	2.58E-01
4991.91	6	6	3.8E-01
4995.00	4	6	5.9E-02
5018.41	6	4	2.09E-01
5021.52	4	6	2.30E-01
5064.31	8	10	7.3E-02
5066.38	6	6	3.6E-02
5070.17	6	8	1.16E-01
5072.71	2	4	2.0E-02
5075.82	4	6	1.15E-01
5080.22	4	4	4.1E-02
5081.56	10	10	7.6E-01
5083.72	8	8	6.2E-01
5085.55	6	6	5.7E-01
5086.94	4	4	6.6E-01
5096.72	6	4	1.69E-01
5099.27	4	6	1.50E-01
5101.12	10	8	8.8E-02
5331.79	4	4	1.11E-01
5339.43	6	6	1.06E-01
5341.07	4	2	3.8E-01
5349.34	6	4	5.9E-01
5350.28	8	8	6.8E-02
5355.79	6	4	3.0E-01
5356.10	8	6	5.7E-01
5375.37	8	6	3.4E-01
5392.06	10	8	4.2E-01
5416.16	4	6	4.4E-02
5416.41	6	6	2.0E-02
5425.55	6	8	4.5E-02
5429.42	2	4	9.0E-02
5432.98	4	4	5.4E-02
5433.25	6	4	9.7E-02
5438.28	4	6	3.4E-02
5439.04	2	2	1.74E-01
5442.62	4	2	2.15E-01
5446.20	8	8	2.8E-01
5451.37	6	6	1.50E-01
5455.24	4	4	6.6E-02
5464.95	4	2	3.2E-02
5468.40	6	4	9.7E-02
5472.19	8	6	9.7E-02
5482.01	8	8	5.2E-01
5484.63	6	6	5.2E-01
5514.23	6	8	4.1E-01
5520.52	8	10	4.3E-01
5526.10	4	4	7.1E-02
5541.07	6	6	5.5E-02
5631.04	2	4	3.0E-02
5671.83	10	12	5.4E-01
5686.86	8	10	4.9E-01
5700.19	6	8	4.6E-01
5708.64	10	10	4.7E-02
5711.79	4	6	4.5E-01
5717.31	8	8	7.5E-02
5724.13	6	6	7.4E-02
5988.43	6	6	6.6E-02
6026.16	4	4	7.2E-02
6146.20	6	8	4.2E-02
6198.43	4	6	3.5E-02
6249.96	6	8	3.2E-02
6262.22	4	6	8.4E-02
6280.16	2	4	4.0E-02
6284.16	6	6	3.9E-02
6284.73	4	4	7.1E-02
6293.02	2	2	1.04E-01
7741.16	10	10	3.8E-02
7800.42	8	8	5.1E-02

Sc II

λ (Å)	g_i	g_k	A (10^8 s^{-1})
1880.6	5	3	5.0E+00
2064.3	7	5	2.2E+00
2068.0	5	3	2.0E+00
2273.1	1	3	7.7E+00
2545.20	5	5	4.0E-01
2552.35	7	5	2.21E+00
2555.79	3	3	6.9E-01
2560.23	5	3	2.01E+00

λ Å	Weights g_i	g_k	A 10^8 s^{-1}	λ Å	Weights g_i	g_k	A 10^8 s^{-1}	λ Å	Weights g_i	g_k	A 10^8 s^{-1}
2563.19	3	1	2.70E+00	5526.79	9	7	3.3E-01	1260.4	2	4	2.0E+01
2611.19	5	5	2.2E+00	5657.91	5	5	1.04E-01	1264.7	4	6	2.3E+01
2667.70	3	5	1.5E+00	5669.06	3	1	1.31E-01	1304.4	2	2	3.6E+00
2746.36	3	1	3.9E+00					1309.3	4	2	7.0E+00
2782.31	5	5	1.3E+00	**Silicon**				1526.7	2	2	3.73E+00
2789.15	7	7	1.3E+00	**Si I**				1533.5	4	2	7.4E+00
2801.31	9	9	1.3E+00	1977.6	1	3	1.8E-01	1808.0	2	4	3.7E-02
2819.49	3	5	2.3E+00	1979.2	3	1	5.1E-01	2904.3	4	6	6.7E-01
2822.12	5	7	2.5E+00	1980.6	3	3	1.3E-01	2905.7	6	8	7.1E-01
2826.64	7	9	2.8E+00	1983.2	3	5	1.4E-01	3210.0	4	6	4.6E-01
2870.85	5	3	1.1E+00	1986.4	5	3	2.1E-01	4128.1	4	6	1.32E+00
2912.98	5	3	1.1E+00	1989.0	5	5	4.1E-01	4130.9	6	8	1.42E+00
2979.68	3	5	1.2E+00	2208.0	1	3	3.11E-01	5041.0	2	4	9.8E-01
2988.92	5	7	2.9E+00	2210.9	3	5	4.16E-01	5056.0	4	6	1.2E+00
3039.92	7	9	3.5E+00	2211.7	3	3	2.32E-01	5957.6	2	2	4.2E-01
3045.73	5	7	3.68E+00	2216.7	5	7	5.5E-01	5978.9	4	2	8.1E-01
3052.92	7	9	3.92E+00	2218.1	5	5	1.38E-01	6347.1	2	4	7.0E-01
3060.54	7	7	3.0E-01	2506.9	3	5	4.66E-01	6371.4	2	2	6.9E-01
3065.12	9	11	4.00E+00	2514.3	1	3	6.1E-01	7848.8	4	6	3.9E-01
3075.36	9	9	2.5E-01	2516.1	5	5	1.21E+00	7849.7	6	8	4.2E-01
3128.27	3	3	1.9E+00	2519.2	3	3	4.56E-01				
3133.07	5	5	1.8E+00	2524.1	3	1	1.81E+00	**Si III**			
3139.72	7	7	2.1E+00	2528.5	5	3	7.7E-01	883.40	5	7	6.3E+01
3190.98	3	3	1.1E+00	2532.4	1	3	2.6E-01	994.79	3	3	7.89E+00
3199.33	5	3	1.9E+00	2631.3	1	3	9.7E-01	997.39	5	3	1.31E+01
3312.72	5	7	1.2E+00	2881.6	5	3	1.89E+00	1141.6	3	5	3.0E+01
3320.40	5	3	1.2E+00	3905.5	1	3	1.18E-01	1144.3	5	7	3.9E+01
3343.23	9	7	1.1E+00	4738.8	3	3	1.0E-02	1161.6	5	5	1.6E+01
3353.72	5	7	1.51E+00	4783.0	5	3	1.7E-02	1206.5	1	3	2.59E+01
3359.67	5	5	2.16E-01	4792.3	5	5	1.7E-02	1206.5	3	5	4.89E+01
3361.26	3	3	3.4E-01	4818.1	5	7	1.1E-02	1207.5	5	5	1.9E+01
3361.93	3	1	1.17E+00	4821.2	3	5	8.0E-03	1294.5	3	5	5.42E+00
3368.94	5	3	8.3E-01	4947.6	3	1	4.2E-02	1296.7	1	3	7.19E+00
3372.15	7	5	9.9E-01	5006.1	3	5	2.8E-02	1298.9	3	3	5.36E+00
3379.16	3	3	2.5E+00	5622.2	3	3	1.6E-02	1299.0	5	5	1.61E+01
3535.71	5	3	6.1E-01	5690.4	3	3	1.2E-02	1301.2	3	1	2.13E+01
3558.53	5	7	3.0E-01	5708.4	5	5	1.4E-02	1303.3	5	3	8.85E+00
3567.70	3	5	3.5E-01	5754.2	5	3	1.5E-02	1328.8	1	3	2.7E+01
3572.53	7	7	1.38E+00	5772.1	3	1	3.6E-02	1417.2	3	1	2.60E+01
3576.34	5	5	1.06E+00	5948.5	3	5	2.2E-02	1435.8	5	7	2.1E+01
3580.93	3	3	1.23E+00	7226.2	3	5	7.9E-03	1589.0	5	3	1.1E+01
3589.63	5	3	4.6E-01	7405.8	3	5	3.7E-02	1778.7	7	9	4.4E+00
3590.47	7	5	2.9E-01	7409.1	5	7	2.3E-02	1783.1	5	7	3.8E+00
3613.83	7	9	1.48E+00	7680.3	3	5	4.6E-02	3241.6	5	3	2.3E+00
3630.74	5	7	1.20E+00	7918.4	3	5	5.2E-02	*3486.9	15	21	1.8E+00
3642.78	3	5	1.13E+00	7932.3	5	7	5.1E-02	3590.5	3	5	3.9E+00
3645.31	7	7	2.74E-01	7944.0	7	9	5.8E-02	4552.6	3	5	1.26E+00
3651.80	5	5	3.0E-01	7970.3	5	5	7.1E-03	4554.0	5	3	7.6E-01
3859.59	7	5	1.1E+00					4567.8	3	3	1.25E+00
4246.82	5	5	1.29E+00	**Si II**				4683.0	5	5	9.5E-01
4314.08	9	7	4.1E-01	989.87	2	4	6.7E+00	4716.7	5	7	2.8E+00
4320.75	7	5	4.0E-01	992.68	4	6	8.0E+00	5451.5	3	5	6.0E-01
4325.00	5	3	4.3E-01	1020.7	2	2	1.3E+00	5473.1	5	7	7.9E-01
4374.46	9	9	1.48E-01	1190.4	2	4	6.9E+00	5716.3	9	7	1.9E-01
4400.39	7	7	1.43E-01	1193.3	2	2	2.8E+01	5739.7	1	3	4.7E-01
4415.54	5	5	1.47E-01	1194.5	4	4	3.6E+01	7462.6	5	3	4.9E-01
4670.41	5	7	1.16E-01	1197.4	4	2	1.4E+01	7466.3	7	5	5.4E-01
5031.01	5	3	3.5E-01	1248.4	4	4	1.3E+01	7612.4	3	5	1.1E+00
5239.81	1	3	1.39E-01	1251.2	6	4	1.9E+01				

λ Å	Weights g_i	g_k	A 10^8 s^{-1}	λ Å	Weights g_i	g_k	A 10^8 s^{-1}	λ Å	Weights g_i	g_k	A 10^8 s^{-1}
Si IV				258.35	4	4	1.4E+02	4747.9	2	2	6.3E-03
457.82	2	4	3.6E+00	261.05	4	2	5.4E+01	4751.8	4	2	1.27E-02
458.16	2	2	3.6E+00	272.00	2	2	3.CE+01	4978.5	2	4	4.1E-02
515.12	2	2	4.1E+00	277.26	4	2	5.7E+01	4982.8	4	4	8.2E-03
516.35	4	2	8.2E+00	287.08	2	4	2.6E+01	4982.8	4	6	4.89E-02
*560.50	6	10	1.0E+00	289.19	4	4	5.0E+01	5148.8	2	2	1.17E-02
*749.94	10	14	1.45E+01	292.22	6	4	7.3E+01	5153.4	4	2	2.33E-02
815.05	2	2	1.23E+01	*347.73	10	10	4.3E+01	5682.6	2	4	1.03E-01
818.13	4	2	2.44E+01	*353.09	6	10	2.1E+01	5688.2	4	6	1.2E-01
*860.74	10	6	1.8E+00					5688.2	4	4	2.1E-02
*1066.6	10	14	3.91E+01	**Si XI**				5890.0	2	4	6.11E-01
1122.5	2	4	2.05E+01	43.763	1	3	6.11E+03	5895.9	2	2	6.10E-01
1128.3	4	4	4.03E+00	*49.116	9	3	2.45E+03	6154.2	2	2	2.6E-02
1128.3	4	6	2.42E+01	49.222	3	5	8.9E+03	6160.8	4	2	5.2E-02
1393.8	2	4	7.73E+00	52.296	3	1	7.6E+02	8183.3	2	4	4.53E-01
1402.8	2	2	7.58E+00	303.30	1	3	6.42E+01	8194.8	4	6	5.4E-01
*1724.1	10	6	5.5E+00	358.29	3	1	1.03E+02	8194.8	4	4	9.0E-02
				358.63	3	5	1.38E+01	11381	2	2	8.9E-02
Si V				361.41	1	3	1.80E+01	11404	4	2	1.76E-01
96.439	1	3	4.8E+02	364.50	3	3	1.32E+01				
97.143	1	3	2.0E+03	365.42	5	5	3.90E+01	**Na II**			
117.86	1	3	3.0E+02	368.28	3	1	5.1E+01	300.15	1	3	3.0E+01
				371.48	5	5	2.07E+01	301.44	1	3	4.9E+01
Si VI				604.14	3	5	1.12E+01	372.08	1	3	3.4E+01
246.00	4	2	1.7E+02	2300.8	1	3	4.34E-01				
249.12	2	2	8.5E+01					**Na III**			
				Si XII				378.14	4	2	7.7E+01
Si VII				*40.924	2	6	4.42E+03	380.10	2	2	3.7E+01
217.83	5	3	4.3E+02	*44.118	6	10	1.4E+04	1991.0	4	6	8.3E+00
272.64	5	3	5.1E+01	499.43	2	4	9.56E+00	2004.2	2	4	4.6E+00
274.18	3	1	1.2E+02	520.72	2	2	8.47E+00	2011.9	6	8	8.4E+00
275.35	5	5	8.9E+01	1862	2	4	1.15E+00	2151.5	2	4	4.4E+00
275.67	3	3	3.0E+01	1949	2	2	1.0E+00	2174.5	4	6	5.3E+00
276.84	1	3	3.9E+01	4620	2	4	4.6E-02	2230.3	6	8	3.7E+00
278.45	3	5	2.9E+01	4942	4	6	4.5E-02	2232.2	4	4	3.3E+00
								2246.7	4	6	2.4E+00
Si VIII				**Silver**				2459.3	4	6	3.0E+00
214.76	4	2	4.1E+02	**Ag I**				2468.9	2	4	2.4E+00
216.92	6	4	3.6E+02	2061.2	2	4	3.1E-02	2497.0	6	6	1.7E+00
232.86	2	2	8.0E+01	2069.9	2	2	1.5E-02				
235.56	4	4	9.7E+01	3280.7	2	4	1.4E+00	**Na V**			
250.45	2	2	7.7E+01	3382.9	2	2	1.3E+00	*307.89	10	6	2.0E+02
250.79	4	2	1.6E+02	5209.1	2	4	7.5E-01	*333.46	6	6	5.6E+01
314.31	4	2	5.2E+01	5465.5	4	6	8.6E-01	*369.01	10	6	1.2E+02
316.20	4	4	5.0E+01	5471.6	4	4	1.4E-01	*400.72	10	10	5.0E+01
319.83	4	6	4.9E+01					*445.14	6	10	7.1E+00
				Sodium				459.90	4	2	2.3E+01
Si IX				**Na I**				461.05	4	4	2.3E+01
223.73	1	3	4.2E+01	3302.4	2	4	2.81E-02	463.26	4	6	2.2E+01
225.03	3	3	1.2E+02	3303.0	2	2	2.81E-02	510.10	2	2	5.6E+01
227.01	5	3	2.0E+02	4390.0	2	4	7.7E-03	511.19	4	4	6.8E+01
227.30	5	3	2.3E+02	4393.3	4	4	1.6E-03				
258.10	5	5	1.04E+02	4393.3	4	6	9.2E-03	**Na VI**			
*294.37	9	9	5.9E+01	4494.2	2	4	1.2E-02	313.75	5	3	1.3E+02
*347.36	9	15	2.2E+01	4497.7	4	6	1.4E-02	361.25	5	5	7.7E+01
				4497.7	4	4	2.4E-03	*416.53	9	9	3.7E+01
Si X				4664.8	2	4	2.33E-02	*492.80	9	15	1.3E+01
253.77	2	4	2.9E+01	4668.6	4	4	4.1E-03	1550.6	5	5	4.35E+00
256.57	2	2	1.1E+02	4668.6	4	6	2.5E-02	1567.8	5	3	2.68E+00

λ Å	g_i	g_k	A 10^8 s^{-1}	λ Å	g_i	g_k	A 10^8 s^{-1}	λ Å	g_i	g_k	A 10^8 s^{-1}
1608.5	3	1	2.6E+00	4607.3	1	3	2.01E+00	4694.1	5	7	6.7E-03
1649.4	5	5	2.05E+00					4695.4	5	5	6.7E-03
1741.5	3	5	2.59E+00	**Sr II**				4696.2	5	3	6.5E-03
1747.5	5	7	3.1E+00	2018.7	2	2	1.2E-01	6403.6	3	5	5.7E-03
				2051.9	4	2	2.4E-01	6408.1	5	5	9.5E-03
Na VII				2282.0	2	4	8.3E-01	6415.5	7	5	1.3E-02
*94.409	6	10	2.7E+03	2322.4	4	6	9.1E-01	*6751.2	15	25	7.9E-02
*105.27	6	2	4.5E+02	2324.5	4	4	1.5E-01	7679.6	3	5	1.2E-02
353.29	4	4	1.0E+02	2423.5	2	2	2.4E-01	7686.1	5	5	2.0E-02
381.30	4	2	4.0E+01	2471.6	4	2	4.8E-01	7696.7	7	5	2.8E-02
397.49	4	4	3.5E+01	3464.5	4	6	3.1E+00				
399.18	6	4	5.2E+01	3474.9	4	4	5.1E-01	**S II**			
*483.28	10	10	2.9E+01	4077.7	2	4	1.42E+00	1124.4	2	4	1.0E+00
486.74	2	4	1.1E+01	4161.8	2	2	6.5E-01	1125.0	4	4	4.6E+00
491.95	4	6	1.3E+01	4215.5	2	2	1.27E+00	1131.0	2	2	3.5E+00
555.80	4	4	2.3E+01	4305.5	4	2	1.4E+00	1131.6	4	2	1.4E+00
777.83	4	6	6.8E+00	4414.8	4	6	1.1E-01	1250.5	4	2	4.6E-01
				4417.5	4	4	1.8E-02	1253.8	4	4	4.2E-01
Na VIII				4585.9	4	2	7.0E-02	1259.5	4	6	3.4E-01
*83.34	9	15	3.94E+03	5303.1	2	4	1.9E-01	4463.6	8	6	5.3E-01
*89.88	9	3	8.09E+02	5379.1	4	6	2.2E-01	4483.4	6	4	3.1E-01
90.536	3	5	2.86E+03	5385.5	4	4	3.7E-02	4486.7	4	2	6.6E-01
411.15	1	3	4.42E+01	5723.7	2	2	7.1E-02	4524.7	4	4	9.3E-02
1239.4	3	3	3.02E+00	5819.0	4	2	1.4E-01	4525.0	6	4	1.2E+00
1802.7	3	1	2.70E+00	8688.9	4	6	5.5E-01	4552.4	4	2	1.2E+00
1867.7	3	5	2.01E+00	8719.6	4	4	9.7E-02	4656.7	2	4	9.0E-02
2059.1	3	5	1.80E+00					4716.2	4	4	2.9E-01
2558.2	5	3	2.26E-02	**Sulfur**				4815.5	6	4	8.8E-01
2772.0	3	5	4.19E-01	**S I**				4885.6	2	4	1.7E-01
3021.0	5	7	4.90E-01	1295.7	5	5	4.9E+00	4917.2	2	2	6.6E-01
3108.9	1	3	2.58E-01	1296.2	5	3	2.7E+00	4924.1	4	6	2.2E-01
3182.3	1	3	2.92E-01	1302.3	3	5	1.8E+00	4925.3	2	4	2.4E-01
				1302.9	3	3	1.6E+00	4942.5	2	2	1.5E-01
Na IX				1303.1	3	1	6.6E+00	4991.9	4	4	1.5E-01
70.615	2	4	1.35E+03	1303.4	5	3	1.9E+00	5009.5	4	2	7.0E-01
70.653	2	2	1.35E+03	1305.9	1	3	2.4E+00	5014.0	4	4	8.4E-01
77.764	2	4	3.6E+03	1401.5	5	3	9.1E-01	5027.2	4	2	2.6E-01
77.911	4	6	4.3E+03	1409.3	3	3	5.0E-01	5032.4	6	6	8.1E-01
681.72	2	4	6.63E+00	1412.9	1	3	1.6E-01	5047.3	4	2	3.6E-01
694.17	2	2	6.30E+00	1425.0	5	7	4.5E+00	5103.3	6	4	5.0E-01
2487.7	2	4	8.32E-01	1425.2	5	5	1.2E+00	5142.3	2	2	1.9E-01
2535.8	2	2	7.89E-01	1433.3	3	5	3.3E+00	5201.0	4	4	7.5E-01
6841.8	2	4	2.59E-02	1433.3	3	3	1.9E+00	5201.3	6	4	6.5E-02
7103.4	4	6	2.78E-02	1437.0	1	3	2.4E+00	5212.6	4	6	9.8E-02
				1448.2	5	3	7.3E+00	5212.6	6	6	8.5E-01
Strontium				1473.0	5	7	4.2E-01	5320.7	6	8	9.2E-01
Sr I				1474.0	5	7	1.6E+00	5345.7	4	6	8.8E-01
2206.2	1	3	6.6E-03	1474.4	5	5	5.0E-01	5345.7	6	6	1.1E-01
2211.3	1	3	8.5E-03	1474.6	5	3	6.2E-02	5428.6	2	4	4.2E-01
2217.8	1	3	1.2E-02	1481.7	3	5	1.7E-01	5432.8	4	6	6.8E-01
2226.3	1	3	1.6E-02	1483.0	3	5	1.2E+00	5453.8	6	8	8.5E-01
2237.7	1	3	2.3E-02	1483.2	3	3	7.5E-01	5473.6	2	2	7.3E-01
2253.3	1	3	3.7E-02	1487.2	1	3	8.7E-01	5509.7	4	4	4.0E-01
2275.3	1	3	6.7E-02	1666.7	5	5	6.3E+00	5526.2	8	8	8.1E-02
2307.3	1	3	1.2E-01	1687.5	1	3	9.4E-01	5536.8	4	6	6.6E-02
2354.3	1	3	1.8E-01	1782.3	1	3	1.9E+00	5556.0	4	2	1.1E-01
2428.1	1	3	1.7E-01	1807.3	5	3	3.8E+00	5564.9	6	6	1.7E-01
2569.5	1	3	5.3E-02	1820.3	3	3	2.2E+00	5578.8	6	6	1.1E-01
2931.8	1	3	1.9E-02	1826.2	1	3	7.2E-01	5606.1	10	8	5.4E-01

λ Å	Weights		A 10⁸ s⁻¹	λ Å	Weights		A 10⁸ s⁻¹	λ Å	Weights		A 10⁸ s⁻¹
	g_i	g_k			g_i	g_k			g_i	g_k	
5616.6	4	4	1.2E-01	249.27	2	2	3.1E+01	4153	4	6	5.7E-02
5640.0	4	6	6.6E-01	388.94	2	2	4.5E+01				
5645.6	6	4	1.8E-02	390.86	4	2	8.8E+01	**Tantalum**			
5647.0	2	4	5.7E-01	706.48	2	4	4.17E+01	**Ta I**			
5659.9	6	4	4.6E-01	712.68	4	6	4.85E+01	3127.9	4	6	5.7E-03
5664.7	4	2	5.8E-01	712.84	4	4	8.1E+00	3168.3	4	4	6.0E-03
5819.2	4	4	8.5E-02	933.38	2	4	1.7E+01	3170.3	8	10	8.5E-02
6305.5	8	6	1.8E-01	944.52	2	2	1.6E+01	3205.5	6	8	5.6E-03
6312.7	6	4	3.0E-01					3260.2	4	4	5.8E-03
				S VII				3337.8	6	6	1.3E-02
S III				60.161	1	3	9.46E+03	3383.9	6	4	5.3E-03
2496.2	7	5	2.5E+00	60.804	1	3	5.1E+02	3406.9	4	6	6.8E-02
2508.2	5	3	2.3E+00	72.029	1	3	8.61E+02	3419.7	8	8	1.91E-02
2636.9	3	5	4.5E-01					3463.8	4	6	2.62E-02
2665.4	5	5	1.4E+00	**S VIII**				3484.6	4	4	8.5E-03
2680.5	1	3	6.2E-01	198.55	4	2	2.5E+02	3488.8	6	4	7.3E-03
2691.8	3	3	4.6E-01	202.61	2	2	1.2E+02	3497.9	6	8	4.9E-02
2702.8	3	1	1.9E+00					3505.0	8	6	2.72E-02
2718.9	3	3	1.2E+00	**S XI**				3553.4	4	6	3.3E-03
2721.4	5	3	7.7E-01	*189.90	9	3	4.3E+02	3607.4	6	8	4.6E-02
2726.8	3	5	6.0E-01	190.37	5	3	2.8E+02	3625.2	10	8	1.0E-02
2731.1	5	5	1.1E+00	215.95	5	5	1.4E+02	3626.6	8	10	7.1E-02
2756.9	7	7	1.4E+00	217.63	1	3	7.2E+01	3642.1	10	12	5.5E-02
2785.5	3	3	6.1E-01	239.81	1	3	2.6E+01	3657.5	6	6	4.3E-03
2856.0	5	7	5.1E+00	242.57	3	5	1.9E+01	3731.0	4	6	5.3E-03
2863.5	7	9	5.7E+00	242.82	3	5	1.9E+01	3754.5	8	8	6.5E-03
2872.0	3	5	4.7E+00	246.90	5	5	5.4E+01	3784.3	4	6	4.3E-02
2950.2	3	5	3.0E+00	247.12	5	3	3.0E+01	3792.1	4	4	9.0E-03
2964.8	5	7	4.0E+00	*288.49	9	15	2.9E+01	3826.9	6	6	5.2E-03
3662.0	3	3	6.4E-01					3836.6	8	10	4.0E-03
3717.8	5	3	1.0E+00	**S XII**				3848.1	10	8	1.30E-02
3778.9	3	5	4.4E-01	212.14	2	4	3.7E+01	3858.6	10	10	2.5E-03
3831.8	1	3	5.6E-01	215.18	2	2	1.4E+02	3918.5	4	2	2.5E-02
3837.8	3	3	4.2E-01	218.20	4	4	1.7E+02	3922.8	4	4	3.98E-02
3838.3	5	5	1.3E+00	221.44	4	2	6.4E+01	3996.2	2	4	3.35E-02
3860.6	3	1	1.6E+00	227.50	2	2	3.7E+01	3999.3	4	4	1.8E-02
3899.1	5	3	6.7E-01	234.48	4	2	6.8E+01	4003.7	10	8	3.1E-03
4253.6	5	7	1.2E+00					4006.8	6	8	7.6E-03
4285.0	3	5	9.0E-01	**S XIII**				4026.9	4	4	3.60E-02
				32.236	1	3	1.09E+04	4029.9	10	10	2.8E-02
S IV				37.600	3	1	1.3E+03	4030.7	8	10	2.3E-03
551.17	2	2	2.06E+01	256.66	1	3	8.7E+01	4040.9	10	12	7.3E-03
554.07	4	2	4.08E+01	299.89	3	5	1.78E+01	4061.4	2	4	6.5E-02
3097.5	2	4	2.6E+00	303.37	1	3	2.28E+01	4064.6	4	4	3.83E-02
3117.7	2	2	2.5E+00	307.36	3	3	1.64E+01	4067.2	6	4	6.8E-03
				308.91	5	5	4.82E+01	4067.9	6	8	8.4E-03
S V				312.68	3	1	6.3E+01	4097.2	10	10	2.1E-03
437.37	1	3	1.12E+01	316.84	5	3	2.50E+01	4105.0	6	4	1.1E-02
438.19	3	3	3.33E+01	500.42	3	5	1.43E+01	4136.2	8	6	1.82E-02
439.65	5	3	5.5E+01					4147.9	10	8	1.79E-02
*661.52	9	15	6.44E+01	**S XIV**				4175.2	6	8	2.8E-02
*679.01	9	15	8.6E+01	*30.434	2	6	8.28E+03	4205.9	8	10	8.9E-03
*690.75	9	9	5.0E+01	*32.517	6	10	2.6E+04	4303.0	6	6	2.08E-02
786.48	1	3	5.25E+01	417.67	2	4	1.2E+01	4378.8	8	6	4.8E-03
*854.85	9	9	4.18E+01	445.71	2	2	1.0E+01	4386.1	4	6	1.0E-02
				1550	2	4	1.4E+00	4402.5	6	6	2.28E-02
S VI				1663	2	2	1.2E+00	4415.7	2	4	2.53E-02
248.99	2	4	3.1E+01	3967	2	4	5.4E-02	4441.0	4	6	7.5E-03

λ Å	Weights g_i	g_k	A 10^8 s^{-1}	λ Å	Weights g_i	g_k	A 10^8 s^{-1}	λ Å	Weights g_i	g_k	A 10^8 s^{-1}
4441.7	10	8	9.0E-03	5518.9	8	10	3.8E-02	2921.5	4	4	7.6E-02
4473.5	6	8	1.36E-02	5620.7	8	10	6.0E-03	3229.8	4	2	1.73E-01
4511.0	10	12	1.56E-02	5640.2	6	8	4.9E-03	3519.2	4	6	1.24E+00
4511.5	10	8	3.6E-03	5645.9	6	8	1.43E-02	3529.4	4	4	2.20E-01
4514.2	10	10	3.1E-03	5699.2	6	6	4.2E-03	3775.7	2	2	6.25E-01
4521.1	10	10	2.3E-03	5767.9	6	8	2.6E-03	5350.5	4	2	7.05E-01
4530.9	4	6	2.42E-02	5780.7	4	6	3.3E-03				
4547.2	4	6	5.3E-03	5811.1	8	6	5.7E-03	**Thulium**			
4553.7	6	8	9.5E-03	5849.7	10	8	2.8E-03	**Tm I**			
4565.9	8	8	2.5E-02	5877.4	10	12	2.3E-02	2513.8	8	10	6.9E-02
4574.3	4	4	1.2E-02	5939.8	2	4	1.6E-02	2527.0	8	8	1.7E-01
4580.7	8	10	2.1E-03	5944.0	4	6	2.13E-02	2596.5	8	10	1.6E-01
4619.5	6	4	5.3E-02	5997.2	10	10	2.4E-02	2601.1	8	6	1.7E-01
4633.1	4	4	1.2E-02	6020.7	2	4	1.0E-02	2622.5	8	10	6.1E-02
4669.1	6	4	2.85E-02	6045.4	6	8	2.6E-02	2841.1	6	6	2.0E-01
4681.9	6	6	1.5E-02	6047.3	8	10	9.0E-03	2854.2	8	6	2.7E-01
4684.9	10	8	2.8E-03	6249.8	6	6	3.5E-03	2914.8	8	8	7.7E-02
4685.3	6	8	3.4E-03	6258.7	6	8	3.3E-03	2933.0	8	6	1.0E-01
4691.9	2	4	4.08E-02	6309.6	4	6	1.83E-02	2973.2	8	8	2.3E-01
4706.1	6	6	1.4E-02	6360.8	6	8	4.6E-03	3046.9	8	8	1.8E-01
4740.2	4	4	5.0E-02	6428.6	6	6	6.0E-03	3081.1	8	8	1.9E-01
4758.0	4	6	7.5E-03	6430.8	8	8	2.9E-02	3122.5	6	6	5.2E-01
4769.0	8	8	2.8E-02	6450.4	8	10	2.2E-02	3142.4	6	6	8.8E-02
4780.9	10	8	2.16E-02	6485.4	10	10	5.8E-02	3172.7	8	8	1.8E-01
4812.8	4	4	1.2E-02	6514.4	6	4	2.2E-02	3233.7	8	10	5.1E-02
4825.4	6	6	2.63E-02	6516.1	6	8	1.25E-02	3247.0	6	8	3.0E-01
4832.2	4	4	1.7E-02	6612.0	6	4	1.9E-02	3251.8	6	4	5.2E-01
4852.2	4	4	1.7E-02	6673.7	2	4	9.0E-03	3380.7	6	8	2.0E-01
4884.0	6	8	1.1E-02	6771.7	4	4	5.8E-03	3406.0	6	8	1.5E-01
4904.6	12	10	1.95E-02	6866.2	8	6	2.58E-02	3410.1	8	10	1.0E-01
4920.9	8	10	2.1E-03	6927.4	10	12	1.01E-02	3416.6	8	8	5.7E-02
4921.3	2	4	1.2E-02	6928.5	10	8	1.69E-02	3418.6	6	6	1.1E-01
4926.0	4	4	1.5E-02	6951.3	10	10	3.7E-03	3563.9	8	6	9.8E-02
4936.4	8	6	4.5E-02	6953.9	6	8	8.3E-03	3567.4	8	10	4.2E-02
4969.7	4	4	1.0E-02	6966.1	8	8	1.2E-02	3744.1	8	8	9.5E-01
5012.5	4	4	1.9E-02	6969.5	10	10	2.9E-03	3751.8	8	10	1.9E-01
5037.4	10	8	4.4E-02	7407.9	6	4	2.0E-02	3798.5	6	4	1.2E+00
5043.3	6	4	2.73E-02					3807.7	6	6	3.9E-01
5067.9	8	6	2.92E-02	**Thallium**				3883.1	8	6	1.0E+00
5069.9	10	12	1.7E-03	**Tl I**				3887.4	8	8	3.8E-01
5082.3	10	12	1.9E-03	2104.6	2	4	4.0E-02	3916.5	6	8	1.5E+00
5087.4	6	4	1.5E-02	2118.9	2	2	2.0E-02	3949.3	6	6	1.0E+00
5090.7	8	6	9.5E-03	2129.3	2	4	5.8E-02	4022.6	6	8	4.0E-02
5095.3	6	6	5.0E-02	2151.9	2	2	3.1E-02	4044.5	6	4	2.9E-01
5136.5	2	2	4.5E-02	2168.6	2	4	9.8E-02	4094.2	8	6	9.0E-01
5141.6	4	2	1.2E-02	2237.8	2	4	1.9E-01	4105.8	8	10	6.0E-01
5143.7	6	4	1.7E-02	2316.0	2	2	7.8E-02	4138.3	6	4	7.0E-01
5147.6	6	4	9.0E-03	2379.7	2	4	4.4E-01	4158.6	6	8	5.5E-02
5161.8	4	6	6.3E-03	2507.9	4	2	1.1E-02	4187.6	8	8	6.1E-01
5218.7	8	6	8.2E-03	2538.2	4	2	1.6E-02	4203.7	8	10	2.5E-01
5235.4	6	6	4.7E-03	2580.1	2	2	1.8E-01	4222.7	6	8	1.5E-01
5295.0	6	6	7.5E-03	2609.0	4	6	1.0E-01	4271.7	6	6	1.1E-01
5336.1	6	8	5.5E-03	2609.8	4	4	1.9E-02	4359.9	8	6	1.3E-01
5349.6	6	4	2.2E-02	2665.6	4	2	5.7E-02	4386.4	8	8	4.2E-02
5354.7	4	4	6.5E-03	2709.2	4	6	1.7E-01	4394.4	6	4	1.1E-01
5396.0	6	8	2.5E-03	2710.7	4	4	3.7E-02	4643.1	6	6	3.4E-02
5402.5	4	2	1.4E-02	2767.9	2	4	1.26E+00	4681.9	6	8	3.9E-02
5435.3	4	6	1.1E-02	2826.2	4	2	8.0E-02	4691.1	6	6	3.9E-02
5499.4	10	10	6.1E-03	2918.3	4	6	4.2E-01	5307.1	8	10	2.3E-02

λ Å	Weights g_i	g_k	A 10^8 s^{-1}	λ Å	Weights g_i	g_k	A 10^8 s^{-1}	λ Å	Weights g_i	g_k	A 10^8 s^{-1}
5658.3	6	8	1.0E-02	6073.5	3	1	6.3E-02	3642.68	7	9	7.74E-01
5675.8	8	10	1.3E-02	6171.5	3	3	4.9E-02	3653.50	9	11	7.54E-01
5760.2	6	6	1.3E-02					3724.57	9	9	9.1E-01
				Sn II				3725.16	5	3	7.3E-01
Tin				2368.3	4	2	4.4E-03	3729.81	5	5	4.27E-01
Sn I				2449.0	4	6	3.7E-01	3741.06	7	7	4.17E-01
2073.1	1	3	3.6E-02	2487.0	6	8	5.5E-01	3752.86	9	9	5.04E-01
2199.3	3	5	2.9E-01	3283.2	4	6	1.0E+00	3786.04	5	3	1.4E+00
2209.7	5	5	5.6E-01	3352.0	6	8	1.0E+00	3948.67	5	3	4.85E-01
2246.1	1	3	1.6E+00	3472.5	2	4	1.6E-01	3956.34	7	5	3.00E-01
2268.9	5	7	1.2E+00	3575.5	4	6	1.3E-01	3958.21	9	7	4.05E-01
2286.7	5	5	3.1E-01	5332.4	2	4	8.6E-01	3981.76	5	5	3.76E-01
2317.2	5	7	2.0E+00	5562.0	4	6	1.2E+00	3989.76	7	7	3.79E-01
2334.8	3	3	6.6E-01	5588.9	4	6	8.5E-01	3998.64	9	9	4.08E-01
2354.8	3	5	1.7E+00	5596.2	4	4	1.5E+00	4013.24	7	5	2.0E-01
2380.7	3	5	3.1E-02	5797.2	6	6	2.8E-01	4055.01	1	3	2.8E-01
2408.2	5	3	1.8E-01	5799.2	6	8	8.1E-01	4060.26	3	5	2.4E-01
2421.7	5	7	2.5E+00	6453.5	2	4	1.2E+00	4064.20	3	3	2.4E-01
2429.5	5	7	1.5E+00	6761.5	2	2	3.2E-01	4065.09	3	1	7.0E-01
2433.5	5	3	8.0E-03	6844.1	2	2	6.6E-01	4186.12	9	9	2.10E-01
2455.2	5	5	1.1E-02					4266.23	5	5	3.1E-01
2476.4	5	3	1.1E-02	**Titanium**				4284.99	5	5	3.2E-01
2483.4	5	5	2.1E-01	**Ti I**				4289.07	5	5	3.0E-01
2491.8	1	3	1.7E-01	2276.75	7	5	1.3E+00	4290.93	3	3	4.5E-01
2495.7	5	5	6.2E-01	2280.00	9	7	9.4E-01	4295.75	3	1	1.3E+00
2523.9	5	5	7.4E-02	2299.86	5	5	6.9E-01	4393.93	9	11	3.3E-01
2546.6	1	3	2.1E-01	2302.75	7	7	5.7E-01	4417.27	11	9	3.6E-01
2558.0	1	3	3.4E-01	2305.69	9	9	5.2E-01	4449.14	11	11	9.7E-01
2571.6	5	7	4.5E-01	2424.26	9	9	1.7E-01	4450.90	9	9	9.6E-01
2594.4	5	5	3.0E-01	2520.54	5	3	3.8E-01	4453.31	5	5	5.98E-01
2636.9	1	3	1.1E-01	2529.87	7	5	3.8E-01	4453.71	7	7	4.7E-01
2661.2	3	3	1.1E-01	2541.92	9	7	4.3E-01	4455.32	7	7	4.8E-01
2706.5	3	5	6.6E-01	2599.91	5	5	6.7E-01	4457.43	9	9	5.6E-01
2761.8	5	5	3.7E-03	2605.16	7	7	6.4E-01	4465.81	5	7	3.28E-01
2779.8	5	7	1.8E-01	2611.29	9	9	6.4E-01	4481.26	7	7	5.7E-01
2785.0	5	3	1.4E-01	2611.47	7	5	3.3E-01	4496.15	7	5	4.4E-01
2788.0	1	3	1.4E-01	2619.94	9	7	2.1E-01	4518.02	7	9	1.72E-01
2812.6	1	3	2.3E-01	2631.55	7	7	1.7E-01	4522.80	5	7	1.9E-01
2813.6	5	5	1.2E-01	2632.42	5	5	2.7E-01	4527.31	3	5	2.2E-01
2840.0	5	5	1.7E+00	2641.12	5	3	1.8E+00	4533.24	11	11	8.83E-01
2850.6	5	5	3.3E-01	2644.28	7	5	1.4E+00	4534.78	9	9	6.87E-01
2863.3	1	3	5.4E-01	2646.65	9	7	1.5E+00	4544.69	5	3	3.3E-01
2913.5	1	3	8.3E-01	2733.27	5	5	1.9E+00	4548.76	7	5	2.85E-01
3009.1	3	3	3.8E-01	2735.30	3	1	4.1E+00	4552.45	9	7	2.1E-01
3032.8	1	3	6.2E-01	2912.07	5	7	1.3E+00	4563.43	9	11	2.1E-01
3034.1	3	1	2.0E+00	2942.00	5	5	1.0E+00	4617.27	7	9	8.51E-01
3141.8	1	3	1.9E-01	2948.26	7	7	9.3E-01	4623.10	5	7	5.74E-01
3175.1	5	3	1.0E+00	2956.13	9	9	9.7E-01	4639.94	3	3	6.64E-01
3218.7	1	3	4.7E-02	2956.80	7	5	1.8E-01	4640.43	3	1	5.0E-01
3223.6	5	5	1.2E-03	3186.45	5	7	8.0E-01	4645.19	3	1	8.57E-01
3262.3	5	3	2.7E+00	3191.99	7	9	8.5E-01	4650.02	5	3	2.6E-01
3330.6	5	5	2.0E-01	3199.92	9	11	9.4E-01	4742.79	9	9	5.3E-01
3655.8	1	3	4.1E-02	3341.88	5	7	6.5E-01	4758.12	11	11	7.13E-01
3801.0	5	3	2.8E-01	3354.63	7	9	6.9E-01	4759.27	13	13	7.40E-01
4524.7	1	3	2.6E-01	3370.44	5	3	7.6E-01	4778.26	9	9	2.0E-01
5631.7	1	3	2.4E-02	3371.45	9	11	7.2E-01	4805.42	5	7	5.8E-01
5970.3	5	3	9.6E-02	3377.58	7	5	6.9E-01	4840.87	5	5	1.76E-01
6037.7	5	5	5.0E-02	3385.94	9	7	5.0E-01	4856.01	13	15	5.2E-01
6069.0	1	3	4.6E-02	3635.46	5	7	8.04E-01	4885.08	11	13	4.90E-01

λ Å	Weights g_i	g_k	A 10^8 s^{-1}	λ Å	Weights g_i	g_k	A 10^8 s^{-1}	λ Å	Weights g_i	g_k	A 10^8 s^{-1}
4913.62	7	9	4.44E-01	2877.47	8	8	5.7E-01	3278.91	6	4	1.0E+00
4928.34	3	5	6.2E-01	2884.13	10	10	5.2E-01	3282.32	2	2	1.6E+00
4981.73	11	13	6.60E-01	2910.65	8	8	4.6E-01	3287.66	8	10	1.4E+00
4989.14	7	5	3.25E-01	2926.64	10	8	8.9E-01	3315.32	2	4	3.8E-01
4991.07	9	11	5.84E-01	2931.10	6	6	3.2E+00	3321.70	4	4	7.2E-01
4999.50	7	9	5.27E-01	2936.02	4	6	2.7E+00	3322.94	10	10	3.96E-01
5000.99	9	7	3.52E-01	2938.57	6	8	2.4E+00	3329.46	8	8	3.25E-01
5007.21	5	7	4.92E-01	2941.90	8	10	1.8E+00	3332.11	6	4	1.1E+00
5014.28	3	5	6.8E-01	2942.97	8	8	1.1E+00	3340.34	4	4	3.6E-01
5036.47	7	9	3.94E-01	2945.30	10	12	2.7E+00	3361.23	8	10	1.1E+00
5038.40	5	7	3.87E-01	2952.00	8	8	3.0E-01	3372.80	6	8	1.11E+00
5062.11	5	3	2.98E-01	2954.59	10	12	4.0E+00	3383.77	4	6	1.09E+00
5210.39	9	9	3.57E-02	2958.80	8	10	4.0E+00	3452.49	2	2	7.7E-01
5222.69	3	3	1.95E-01	2979.06	4	6	1.2E+00	3456.40	4	4	8.2E-01
5224.30	11	11	3.6E-01	2990.06	6	8	5.6E-01	3465.56	4	2	4.1E-01
5259.98	5	7	2.3E-01	3017.17	12	12	3.6E-01	3483.63	10	8	9.7E-01
5351.07	7	7	3.4E-01	3022.64	10	10	1.2E+00	3492.37	8	6	9.8E-01
5503.90	11	9	2.6E-01	3023.67	8	8	1.0E+00	3504.90	10	10	8.2E-01
5774.04	9	11	5.5E-01	3029.76	10	10	3.5E-01	3510.86	8	8	9.3E-01
5785.98	11	13	6.1E-01	3056.75	2	4	3.2E-01	3520.27	2	4	4.8E-01
5804.27	13	15	6.8E-01	3058.08	6	6	5.0E-01	3535.41	4	6	5.5E-01
6098.66	9	7	2.5E-01	3066.34	4	4	3.3E-01	3641.33	4	2	4.9E-01
6220.46	9	7	1.8E-01	3071.25	6	4	3.6E-01	3706.23	4	4	3.1E-01
				3072.99	4	2	1.6E+00	3741.64	6	6	6.2E-01
Ti II				3075.23	6	4	1.13E+00	3757.70	4	4	4.1E-01
2440.91	4	4	5.1E-01	3078.65	8	6	1.09E+00	3759.30	8	8	9.4E-01
2451.18	6	6	4.5E-01	3081.52	10	8	1.1E+00	3761.33	6	6	9.9E-01
2525.59	10	8	5.6E-01	3088.04	10	8	1.25E+00	4911.18	6	4	3.2E-01
2531.28	8	6	4.9E-01	3089.44	8	6	1.3E+00				
2534.63	6	4	5.4E-01	3097.20	4	6	4.4E-01	**Ti III**			
2535.89	4	2	6.8E-01	3103.81	10	8	1.1E+00	865.79	5	3	6.6E+01
2555.99	6	8	3.2E-01	3105.10	2	4	6.3E-01	1002.37	5	5	7.6E+00
2635.44	4	4	1.9E+00	3106.26	6	6	7.8E-01	1004.67	7	5	4.3E+01
2638.56	6	6	1.7E+00	3117.67	4	2	1.1E+00	1005.80	3	3	1.3E+01
2642.02	8	8	1.9E+00	3119.83	6	4	5.9E-01	1007.16	5	3	3.8E+01
2645.86	10	10	2.7E+00	3127.86	6	6	1.6E+00	1008.12	3	1	5.1E+01
2746.54	6	8	2.6E+00	3128.50	8	8	1.1E+00	1286.37	9	9	2.0E+00
2751.59	8	10	3.7E+00	3161.23	4	2	5.9E-01	1289.30	7	7	2.2E+00
2752.68	8	10	1.1E+00	3161.80	6	4	4.6E-01	1291.62	5	5	2.4E+00
2757.62	6	8	7.2E-01	3162.59	8	6	3.9E-01	1293.23	9	7	1.0E+00
2758.35	4	6	9.9E-01	3168.55	10	8	4.1E-01	1298.97	7	5	4.9E+00
2758.79	2	4	4.4E-01	3181.73	6	8	4.6E-01	1327.59	5	3	3.2E+00
2764.28	4	4	7.4E-01	3182.54	4	4	4.3E-01	1420.44	1	3	1.2E+00
2804.82	6	8	4.6E+00	3189.49	4	4	9.2E-01	1421.63	3	1	4.0E+00
2810.30	8	10	5.1E+00	3190.91	6	8	1.3E+00	1422.41	5	5	3.0E+00
2817.83	10	12	3.8E+00	3202.56	4	6	1.1E+00	1424.14	5	3	1.6E+00
2819.87	8	8	6.5E-01	3224.25	12	10	7.0E-01	1455.19	9	7	6.4E+00
2821.26	6	8	7.9E-01	3228.62	4	2	2.0E+00	1498.70	5	5	2.8E+00
2827.12	8	10	1.0E+00	3232.29	8	6	6.0E-01	2007.36	3	3	3.4E+00
2828.06	12	14	4.4E+00	3234.51	10	10	1.38E+00	2007.60	1	3	1.2E+00
2828.64	6	6	1.2E+00	3236.13	4	4	7.0E-01	2010.80	5	3	5.4E+00
2828.83	10	10	9.1E-01	3236.58	8	8	1.11E+00	2097.30	5	7	3.3E+00
2834.02	10	12	7.9E-01	3239.04	6	6	9.87E-01	2099.86	3	5	2.5E+00
2836.47	8	8	1.2E+00	3239.66	6	4	9.4E-01	2104.86	3	3	1.1E+00
2839.64	12	12	8.3E-01	3241.99	4	4	1.16E+00	2105.09	1	3	1.7E+00
2845.93	10	10	1.2E+00	3251.91	6	4	3.38E-01	2199.22	3	3	5.7E+00
2851.11	2	4	4.1E-01	3252.92	8	6	3.9E-01	2237.77	7	7	2.4E+00
2856.10	12	12	1.5E+00	3272.07	2	4	3.2E-01	2331.35	3	1	4.3E+00
2862.33	4	6	4.0E-01	3278.28	4	4	9.6E-01	2331.66	3	3	1.2E+00

λ Å	Weights g_i	Weights g_k	A 10^8 s^{-1}	λ Å	Weights g_i	Weights g_k	A 10^8 s^{-1}	λ Å	Weights g_i	Weights g_k	A 10^8 s^{-1}
2339.00	5	3	3.0E+00	2862.60	4	2	4.1E+00	308.250	3	5	1.3E+02
2346.79	7	5	3.3E+00	3576.44	4	6	4.6E+00	313.229	5	7	1.6E+02
2374.99	5	3	4.0E+00					318	3	1	1.4E+02
2413.99	5	7	3.8E+00	**Ti VIII**				322.75	5	7	1.99E+02
2516.05	7	9	3.4E+00	249	6	4	1.0E+01	323	1	3	1.8E+02
2567.56	3	3	2.3E+00	258.610	6	8	7.5E+02	327.192	3	5	2.9E+02
2984.75	5	5	1.9E+00	269.533	4	6	6.0E+02	332	3	1	3.25E+02
3066.51	3	3	2.5E+00	272.037	4	4	4.3E+02	386.140	1	3	1.48E+02
3228.89	3	3	1.5E+00	272.843	6	4	6.2E+01	408	7	9	1.37E+02
3278.31	7	9	3.4E+00	276.701	2	4	9.3E+01	425.74	3	1	1.2E+02
3320.94	3	5	2.8E+00	277.813	4	4	3.8E+02	446.69	3	1	1.2E+02
3340.20	7	9	3.7E+00	289.375	2	4	3.6E+01	453	5	7	1.3E+02
3346.18	9	11	3.7E+00	478.971	4	4	1.7E+01				
3354.71	11	13	4.4E+00	480.376	6	6	1.5E+01	**Ti XII**			
3397.24	3	1	1.8E+00					52.896	2	4	1.61E+02
3404.46	3	3	1.8E+00	**Ti IX**				53.140	4	6	1.9E+02
3417.62	3	5	1.9E+00	267.941	5	7	5.1E+02	53.433	2	4	2.1E+02
3915.47	9	11	2.1E+00	278.713	5	7	4.7E+02	53.457	2	2	2.1E+02
4119.14	5	5	9.9E-01	281.446	3	1	3.2E+02	55.181	2	4	2.4E+02
4213.26	9	11	2.2E+00	285.128	1	3	4.1E+02	55.443	4	6	2.81E+02
4215.53	9	11	2.2E+00	433.567	1	3	6.9E+00	59.133	2	4	3.72E+02
4247.62	11	13	1.1E+00	439.513	3	3	7.5E+00	59.435	4	6	4.41E+02
4248.54	5	7	2.3E+00	439.745	3	1	2.1E+01	60.701	2	4	3.4E+02
4250.09	3	5	9.5E-01	447.484	5	5	1.6E+01	60.762	2	2	3.5E+02
4259.01	11	13	9.4E-01	447.701	5	3	6.5E+00	61.286	4	2	1.8E+02
4269.84	9	11	1.7E+00	507.174	3	5	6.5E+00	62.433	4	6	2.08E+02
4285.61	13	15	3.0E+00	516.215	5	7	6.9E+00	62.470	6	8	2.22E+02
4288.66	11	13	1.1E+00					65.540	4	6	3.2E+02
4296.70	11	13	1.6E+00	**Ti X**				65.577	6	8	3.5E+02
4319.56	9	11	1.1E+00	253	4	6	2.1E+02	67.171	2	4	6.2E+02
4343.25	3	1	1.0E+00	254	6	8	2.3E+02	67.555	4	6	7.2E+02
4378.94	3	5	1.6E+00	281	2	2	1.1E+02	70.986	4	6	5.7E+02
4433.91	11	13	1.8E+00	289.579	2	4	2.5E+02	71.031	6	8	6.1E+02
4440.66	1	3	1.2E+00	290.294	4	6	1.1E+02	71.545	2	2	1.8E+02
4533.26	3	5	1.5E+00	291	4	2	1.8E+02	71.987	4	2	3.48E+02
4576.53	9	7	1.3E+00	291	2	2	2.3E+02	82.121	2	4	5.9E+02
4628.07	3	1	1.5E+00	292	6	8	1.1E+02	82.307	4	6	1.13E+03
4652.86	7	9	2.6E+00	293.684	6	8	2.97E+02	82.344	2	2	5.8E+02
4874.00	5	7	1.5E+00	293.798	6	8	1.7E+02	82.368	6	8	1.2E+03
4914.32	3	3	1.1E+00	295.584	4	6	2.9E+02	89.844	2	4	9.9E+02
4971.19	9	11	2.1E+00	296	4	6	1.4E+02	90.512	4	6	1.16E+03
5083.80	5	3	9.7E-01	297	4	6	9.9E+01	90.547	4	4	1.9E+02
5278.33	3	3	9.4E-01	298	4	6	4.3E+02	116.497	4	6	3.0E+03
7506.87	11	13	1.1E+00	302	2	2	1.6E+02	116.597	6	8	3.2E+03
				305	2	4	2.5E+02	116.62	6	6	2.1E+02
Ti IV				317	2	2	1.5E+02	139.884	6	4	2.6E+02
423.49	4	6	4.9E+01	355.815	2	2	1.3E+02	140.361	4	2	2.9E+02
424.16	6	8	5.3E+01	360.133	4	4	2.19E+02	141.6	4	6	1.7E+02
433.63	4	2	5.5E+00	363	4	2	2.1E+02	141.7	6	8	1.7E+02
433.76	6	4	5.0E+00	363	6	6	1.3E+02	169.7	4	6	2.8E+02
729.36	4	2	5.7E+00	365.628	4	2	1.2E+02	169.8	6	8	2.9E+02
1183.64	2	2	6.9E+00	382	4	6	1.8E+02	207.2	2	4	1.5E+02
1195.21	4	2	1.4E+01	385	6	8	1.8E+02	208.5	4	6	1.8E+02
1451.74	2	4	1.8E+01	389.99	6	4	1.1E+02	252.8	4	6	4.8E+02
1467.34	4	6	2.1E+01					253.1	6	8	5.2E+02
2067.56	2	4	5.1E+00	**Ti XI**				257.5	4	2	2.4E+02
2103.16	2	2	5.0E+00	65.403	1	3	5.1E+02				
2541.79	4	6	6.9E+00	87.725	1	3	8.5E+02	**Ti XIII**			
2546.88	6	8	7.4E+00	266	5	7	1.8E+02	23.356	1	3	1.02E+05

λ Å	Weights g_i	g_k	A 10^8 s^{-1}	λ Å	Weights g_i	g_k	A 10^8 s^{-1}	λ Å	Weights g_i	g_k	A 10^8 s^{-1}
23.698	1	3	1.2E+04	145.665	6	6	2.3E+02	16.46	3	3	4.4E+04
23.991	1	3	3.4E+02	157.812	4	2	1.32E+02	16.51	5	7	1.0E+05
26.641	1	3	4.06E+03	161.168	4	4	1.2E+02	16.55	5	5	2.7E+04
26.960	1	3	3.06E+03	163.610	4	2	1.92E+02	16.61	3	1	8.0E+04
117.1	3	3	1.3E+02	169.740	4	6	1.0E+02	16.64	3	3	5.3E+04
117.3	3	1	2.8E+02	176.267	2	2	2.45E+02	16.69	1	3	1.02E+05
120.2	5	3	5.4E+02	178.240	4	4	2.52E+02	16.71	3	5	7.3E+04
120.2	7	5	4.4E+02					16.72	5	3	3.3E+04
128.7	3	3	1.2E+02	**Ti XVII**				16.72	5	5	7.3E+04
				18.05	3	3	4.5E+04	16.74	5	7	1.2E+05
Ti XIV				18.13	5	3	2.4E+04	16.77	3	3	2.6E+04
21.341	4	6	9.8E+03	18.13	1	3	8.1E+04	16.80	5	7	1.81E+05
21.522	2	4	4.5E+04	18.176	5	7	9.2E+04	16.85	3	5	4.4E+04
21.657	4	4	1.3E+04	123.654	3	3	2.3E+02	17.08	3	5	8.3E+04
21.733	4	4	8.8E+04	124.553	5	3	5.2E+02	17.36	1	3	9.5E+04
21.82	4	2	6.4E+04	127.782	5	3	4.6E+02				
21.883	2	4	7.0E+04	135.202	3	1	2.93E+02	**Ti XX**			
21.958	2	4	1.2E+04	136.160	5	3	1.95E+02	2.629	2	4	4.9E+04
22.05	2	2	1.4E+04	136.393	3	3	1.14E+02	2.6295	4	4	3.2E+06
24.592	4	2	6.1E+03	141.948	5	5	3.87E+02	2.631	2	2	6.1E+05
24.891	2	2	7.5E+03	142.589	1	3	1.35E+02	2.6319	2	4	1.5E+06
				144.405	5	5	9.4E+01	2.632	2	2	2.7E+06
Ti XV				146.067	7	5	2.6E+02	2.6355	4	6	1.2E+06
20.19	5	7	6.9E+03	154.133	3	1	1.63E+02	8.621	4	2	1.1E+06
20.234	5	7	1.9E+04	156.54	3	1	1.44E+02	9.788	4	6	5.26E+03
20.234	3	3	4.9E+04	158.469	5	5	1.4E+02	10.046	2	4	7.29E+03
20.246	1	3	4.2E+04	159.62	5	3	1.03E+02	10.109	4	6	8.6E+03
20.250	5	3	6.5E+03	163.049	3	1	6.2E+02	*10.278	2	6	8.4E+03
20.29	3	3	1.1E+04	186.863	5	5	2.66E+02	10.620	2	4	1.34E+04
20.30	1	3	3.4E+04	207.73	3	1	1.07E+02	10.690	4	6	1.58E+04
20.30	1	1	5.8E+04					*11.452	2	6	1.7E+04
20.313	5	3	7.5E+04	**Ti XVIII**				11.872	2	4	2.8E+04
20.418	5	7	8.0E+04	17.22	2	4	7.3E+04	11.958	4	6	3.4E+04
20.538	3	3	3.8E+04	17.365	4	6	8.6E+04	11.958	4	4	5.6E+03
20.54	3	1	4.1E+04	17.39	4	4	1.4E+04	15.211	2	4	3.50E+04
20.551	1	3	1.3E+04	133.852	2	4	5.2E+01	15.253	2	2	3.58E+04
20.689	5	7	4.3E+04	144.759	4	4	3.2E+02	15.907	2	4	8.84E+04
20.698	1	3	1.1E+05	150.15	6	4	1.15E+02	16.049	4	6	1.05E+05
20.771	5	3	1.1E+04	153.15	4	2	1.97E+02	16.067	4	4	1.8E+04
20.897	5	7	2.85E+04	153.23	2	4	6.7E+01	31.586	4	6	5.49E+03
20.928	5	5	8.4E+03	159.00	4	4	1.16E+02	45.650	2	4	9.6E+03
21.065	3	3	1.1E+04	166.225	6	4	1.54E+02	45.996	4	6	1.1E+04
21.079	1	3	1.58E+04	179.902	2	4	6.3E+01				
21.102	3	5	1.3E+04	189.663	6	6	9.6E+01	**Ti XXI**			
22.482	5	3	6.4E+03	191.23	4	4	6.6E+01	2.0633	1	3	1.32E+05
22.936	5	5	1.1E+04	197.838	4	6	4.56E+01	2.1108	1	3	2.60E+05
22.966	5	3	1.1E+04	208.07	4	4	1.2E+02	2.2211	1	3	6.35E+05
23.034	1	3	6.3E+03					2.497	3	1	2.4E+06
				Ti XIX				2.505	5	5	3.5E+06
Ti XVI				15.67	3	1	3.3E+04	2.505	1	3	1.4E+06
110.561	4	2	3.36E+02	15.68	5	5	2.7E+04	2.507	3	5	1.4E+06
116.198	4	4	1.45E+02	15.74	5	7	2.7E+04	2.508	3	5	7.9E+05
118.215	6	4	7.4E+02	15.75	3	5	2.4E+04	2.510	3	3	6.9E+05
121.382	4	2	2.4E+02	15.83	1	3	3.2E+04	2.510	1	3	9.6E+05
124.805	4	2	6.1E+02	15.86	1	3	2.9E+04	2.511	3	3	1.4E+06
129.075	4	2	3.81E+02	16.02	3	1	3.1E+04	2.512	5	5	1.8E+06
134.724	2	2	2.6E+02	16.18	3	5	3.8E+04	2.512	3	1	1.4E+06
138.800	6	4	3.5E+02	16.41	1	3	6.1E+04	2.513	3	1	2.7E+06
143.459	4	4	2.8E+02	16.43	3	5	8.2E+04	2.513	3	5	2.4E+06

λ Å	Weights g_i	g_k	A 10^8 s^{-1}	λ Å	Weights g_i	g_k	A 10^8 s^{-1}	λ Å	Weights g_i	g_k	A 10^8 s^{-1}
2.514	5	3	1.2E+06	3780.8	7	5	4.2E-02	4924.6	13	11	1.75E-03
2.520	3	5	2.6E+05	3809.2	7	5	9.0E-03	4931.6	7	5	1.0E-02
2.527	3	1	1.2E+05	3817.5	7	7	3.1E-02	4948.6	9	11	1.36E-03
2.539	3	1	4.1E+05	3829.1	3	3	3.83E-03	4972.6	9	11	3.9E-03
2.6102	1	3	2.40E+06	3835.1	5	5	5.2E-02	4982.6	1	3	4.17E-03
2.6227	1	3	1.12E+05	3846.3	3	5	2.14E-02	4986.9	11	9	6.3E-03
				3847.5	1	3	8.3E-03	5006.2	9	7	1.2E-02
Tungsten				3864.3	5	5	5.6E-03	5015.3	7	9	5.4E-03
W I				3868.0	7	9	4.6E-02	5040.4	3	5	5.2E-03
2879.4	1	3	2.4E-01	3881.4	7	7	3.6E-02	5053.3	3	3	1.9E-02
2911.0	1	3	7.7E-02	3968.5	1	3	5.07E-03	5071.5	13	11	3.4E-03
2923.5	7	9	1.54E-02	3975.5	9	11	4.1E-03	5117.6	11	11	1.61E-03
2935.0	3	5	1.5E-01	4001.4	9	9	5.6E-03	5124.2	5	5	4.0E-03
3013.8	7	9	6.4E-02	4008.8	7	9	1.63E-01	5141.2	7	9	1.12E-03
3016.5	9	11	9.27E-02	4019.3	5	3	6.7E-03	5224.7	7	5	1.2E-02
3017.4	7	9	1.21E-01	4028.8	1	3	2.0E-02	5243.0	9	7	1.1E-02
3024.9	3	3	1.4E-01	4045.6	7	5	2.88E-02	5254.5	7	5	3.86E-03
3046.4	3	5	5.8E-02	4055.2	7	9	1.79E-03	5268.6	9	9	1.4E-03
3049.7	7	5	1.7E-01	4070.0	7	5	3.60E-02	5500.5	11	9	6.9E-03
3064.9	5	7	1.1E-02	4070.6	3	5	5.6E-03	5514.7	5	3	7.3E-03
3084.9	5	5	1.3E-02	4074.4	7	7	1.0E-01	5537.7	9	11	2.2E-03
3093.5	7	9	4.4E-02	4088.3	5	3	4.13E-03	5617.1	7	7	1.47E-03
3107.2	5	7	2.33E-02	4102.7	9	7	4.9E-02	5631.9	9	7	1.43E-03
3108.0	7	9	1.58E-02	4115.6	11	11	4.8E-03	5660.7	13	11	6.8E-03
3145.5	9	9	4.8E-03	4137.5	5	7	8.4E-03	5675.4	5	5	2.20E-03
3170.2	7	5	6.0E-03	4171.2	7	9	8.6E-03	5796.5	9	7	2.21E-03
3176.6	3	5	2.12E-02	4203.8	9	7	4.9E-03	5891.6	7	7	1.47E-03
3183.5	7	7	2.64E-03	4219.4	9	7	6.1E-03	5947.6	5	7	2.40E-03
3184.4	5	3	2.3E-02	4244.4	9	11	1.38E-02	5965.9	7	5	1.0E-02
3191.6	1	3	3.2E-02	4269.4	7	5	3.04E-02	6021.5	5	3	8.7E-03
3198.8	7	9	4.6E-02	4283.8	9	7	1.69E-03	6081.4	5	3	4.7E-03
3207.3	7	9	3.0E-02	4294.6	7	5	1.2E-01	6203.5	7	7	3.0E-03
3208.3	5	5	4.4E-02	4302.1	7	7	3.6E-02	6285.9	7	5	6.6E-03
3215.6	9	11	2.1E-01	4355.2	9	9	5.1E-03	6292.0	3	5	2.26E-03
3221.9	5	7	1.61E-02	4361.8	9	7	1.64E-03	6303.2	9	9	1.84E-03
3223.1	5	3	3.53E-03	4378.5	7	5	3.48E-03	6404.2	5	7	1.50E-03
3232.5	9	9	2.4E-02	4458.1	3	5	4.2E-03	6439.7	9	9	1.29E-03
3235.1	7	5	2.68E-03	4466.3	7	5	1.5E-02	6445.1	7	5	6.4E-03
3259.7	7	7	1.3E-02	4472.5	13	11	1.55E-03	6532.4	3	5	4.6E-03
3300.8	7	9	8.1E-02	4484.2	3	5	5.6E-03	6538.1	11	9	2.7E-03
3311.4	7	5	5.6E-02	4492.3	9	11	3.6E-03	6563.2	5	5	2.04E-03
3363.3	9	7	6.6E-03	4495.3	11	11	3.3E-03	6814.9	9	9	1.46E-03
3371.0	7	5	1.0E-02	4504.8	9	7	7.0E-03	7285.8	13	11	1.47E-03
3371.4	3	3	6.7E-03	4552.5	9	9	1.42E-03	7569.9	5	3	3.73E-03
3386.1	7	7	2.64E-03	4586.8	1	3	4.20E-03	7664.9	5	3	3.80E-03
3413.0	7	9	9.7E-03	4592.6	7	9	3.4E-03	8017.2	5	7	1.6E-03
3459.5	9	9	2.04E-03	4609.9	7	9	1.42E-02	8358.7	5	7	1.89E-03
3510.0	7	9	5.2E-03	4613.3	9	9	2.9E-03	9381.4	9	7	1.53E-03
3545.2	1	3	3.2E-02	4634.8	9	9	8.8E-03				
3570.6	5	3	6.7E-03	4659.9	1	3	1.0E-02	**Uranium**			
3606.1	3	5	9.6E-03	4680.5	7	7	1.4E-02	**U I**			
3617.5	7	7	1.1E-01	4720.4	3	5	3.22E-03	3553.0	13	13	2.0E-02
3631.9	3	5	1.3E-02	4729.6	7	5	7.8E-03	3553.0	9	7	1.4E-02
3675.6	9	11	1.20E-02	4752.6	3	3	5.20E-03	3553.4	15	13	2.2E-02
3682.1	9	11	2.0E-02	4757.5	7	5	2.72E-03	3554.5	11	9	8.4E-03
3707.9	7	7	2.9E-02	4757.8	11	9	4.1E-03	3554.9	15	17	7.9E-03
3757.9	7	9	1.38E-02	4788.4	9	11	2.6E-03	3555.3	13	15	2.7E-02
3760.1	5	7	1.99E-02	4843.8	5	5	1.9E-02	3555.8	13	11	4.1E-03
3768.5	3	3	3.47E-02	4886.9	9	11	8.1E-03	3556.9	13	11	7.5E-03

λ Å	Weights g_i	Weights g_k	A 10^8 s^{-1}	λ Å	Weights g_i	Weights g_k	A 10^8 s^{-1}	λ Å	Weights g_i	Weights g_k	A 10^8 s^{-1}
3557.8	13	13	2.9E-02	3075.93	4	6	2.8E-01	3818.24	4	2	6.73E-01
3558.0	11	13	1.6E-02	3080.33	2	4	2.7E-01	3828.56	6	4	5.33E-01
3558.6	9	7	3.9E-02	3083.54	6	8	2.5E-01	3840.75	8	6	5.48E-01
3559.4	7	9	1.5E-02	3087.06	2	2	9.2E-01	3855.36	4	4	3.30E-01
3560.3	9	7	6.4E-02	3088.11	4	6	4.9E-01	3855.85	10	8	5.78E-01
3561.4	15	13	5.5E-02	3089.13	4	4	5.3E-01	3863.86	8	6	3.1E-01
3561.5	9	9	2.5E-02	3093.79	6	6	4.1E-01	3864.86	6	6	2.70E-01
3561.8	13	11	5.7E-02	3094.69	2	4	4.3E-01	3871.07	10	8	2.8E-01
3563.7	13	13	2.9E-02	3112.92	4	2	5.0E-01	3875.07	8	8	2.36E-01
3563.8	7	7	1.1E-02	3183.41	6	8	2.4E+00	3902.26	10	10	2.68E-01
3565.0	13	11	2.9E-02	3183.96	8	10	2.5E+00	3921.86	4	2	2.7E-01
3566.0	13	15	1.7E-02	3183.98	4	6	2.4E+00	3922.43	6	6	2.6E-01
3566.6	11	11	2.4E-01	3185.38	10	12	2.7E+00	3930.02	10	10	3.3E-01
3568.8	13	13	3.8E-02	3198.01	6	6	3.9E-01	3934.01	8	8	6.2E-01
3569.1	17	15	1.1E-01	3202.39	8	8	4.0E-01	3992.80	12	10	1.2E+00
3569.4	9	9	1.5E-02	3205.58	8	10	1.3E+00	3998.73	14	12	1.0E+00
3570.1	13	11	1.3E-02	3207.41	10	10	2.6E-01	4050.96	10	10	1.4E+00
3570.2	11	9	5.3E-03	3212.43	10	12	1.4E+00	4051.35	12	12	1.3E+00
3570.6	13	15	2.7E-02	3218.87	8	6	3.5E-01	4090.57	8	10	8.5E-01
3570.7	15	15	1.2E-02	3233.19	10	8	3.2E-01	4092.68	8	10	2.30E-01
3571.2	11	11	6.3E-03	3273.03	8	8	2.7E-01	4095.48	6	8	7.2E-01
3571.6	17	15	1.3E-01	3284.36	10	10	2.8E-01	4099.78	6	8	4.10E-01
3572.9	13	15	1.5E-02	3309.18	4	4	3.2E-01	4102.15	4	6	7.1E-01
3573.9	13	11	4.0E-02	3329.85	6	4	7.7E-01	4104.77	10	8	2.1E+00
3574.1	13	15	3.5E-02	3356.35	4	6	3.1E-01	4105.16	4	6	4.9E-01
3574.8	13	15	1.9E-02	3365.55	2	4	4.8E-01	4109.78	2	4	5.00E-01
3577.1	17	15	4.3E-02	3376.05	4	4	3.2E-01	4111.78	10	10	1.01E+00
3577.5	15	13	7.8E-03	3377.39	4	2	9.0E-01	4115.18	8	8	5.80E-01
3577.8	11	11	8.3E-03	3377.62	6	6	6.0E-01	4116.47	6	6	3.2E-01
3577.9	13	13	2.3E-02	3397.58	6	4	2.3E-01	4116.59	2	2	2.90E-01
3578.3	13	11	2.0E-02	3400.39	8	8	2.5E-01	4123.50	4	2	1.00E+00
3580.0	9	9	1.2E-02	3529.73	4	6	4.1E-01	4128.06	6	4	7.70E-01
3580.2	11	9	2.9E-02	3533.68	6	8	5.2E-01	4131.99	8	6	5.5E-01
3580.4	11	13	7.5E-03	3533.76	2	4	3.7E-01	4134.49	10	8	2.90E-01
3580.9	13	13	2.1E-02	3543.49	2	2	6.7E-01	4232.46	10	10	9.8E-01
3582.6	13	13	2.9E-02	3545.33	4	4	3.7E-01	4232.95	8	8	7.7E-01
3584.6	7	5	2.4E-02	3553.27	6	6	2.2E-01	4268.64	14	14	1.2E+00
3584.9	13	15	1.8E-01	3555.14	4	2	2.6E-01	4271.55	12	12	9.6E-01
3585.4	11	11	1.9E-02	3663.60	4	6	3.1E+00	4276.95	10	10	9.4E-01
3585.8	11	9	2.8E-02	3667.74	6	8	2.7E+00	4284.05	8	8	1.2E+00
3587.8	9	11	1.3E-02	3672.41	12	12	9.2E-01	4291.82	12	14	8.8E-01
3588.3	7	9	1.8E-02	3673.41	8	10	2.7E+00	4296.10	10	12	7.7E-01
3589.7	11	13	2.1E-02	3676.70	14	14	1.3E+00	4297.67	8	10	7.0E-01
3589.8	15	13	5.9E-02	3680.12	10	12	2.2E+00	4298.03	6	8	7.8E-01
3590.7	9	7	2.2E-02	3686.26	10	12	2.3E-01	4379.23	10	12	1.1E+00
3591.7	11	9	5.3E-02	3687.50	12	14	2.9E+00	4384.71	8	10	1.1E+00
3593.0	11	11	1.4E-01	3688.07	8	8	3.5E-01	4389.98	6	8	6.9E-01
3593.2	13	15	4.2E-02	3690.28	2	4	4.5E-01	4395.22	4	6	5.5E-01
3593.7	11	11	7.2E-02	3692.22	6	6	5.4E-01	4400.57	2	4	3.4E-01
				3695.34	14	16	2.8E+00	4406.64	10	10	2.2E-01
Vanadium				3695.86	4	4	6.6E-01	4407.63	8	8	4.4E-01
V I				3703.57	10	8	9.2E-01	4408.20	6	6	6.0E-01
3043.12	6	8	2.3E-01	3704.70	8	6	6.6E-01	4416.47	4	2	2.6E-01
3050.39	10	8	5.3E-01	3705.04	6	4	3.6E-01	4452.01	14	16	9.2E-01
3053.65	4	4	1.3E+00	3706.03	10	10	5.2E-01	4457.75	10	12	2.7E-01
3056.33	6	6	1.3E+00	3708.71	12	12	4.4E-01	4460.33	10	8	3.0E-01
3060.46	8	8	1.4E+00	3790.46	10	8	2.3E-01	4462.36	12	14	7.6E-01
3066.37	10	10	2.1E+00	3794.96	10	10	2.3E-01	4468.00	8	10	2.3E-01
3066.53	6	4	3.2E-01	3806.79	10	10	2.5E-01	4469.71	10	12	6.2E-01

λ Å	Weights g_i	g_k	A 10^8 s^{-1}	λ Å	Weights g_i	g_k	A 10^8 s^{-1}	λ Å	Weights g_i	g_k	A 10^8 s^{-1}
4474.04	10	8	4.7E-01	2893.31	9	7	1.2E+00	2371.06	10	12	5.2E+00
4496.06	8	6	4.0E-01	2903.07	3	5	3.4E-01	2373.06	4	6	2.9E+00
4514.18	6	4	3.3E-01	2906.45	7	7	7.8E-01	2382.46	8	10	5.0E+00
4524.21	12	10	3.0E-01	2908.81	11	9	1.6E+00	2393.58	6	8	4.3E+00
4525.17	4	2	4.1E-01	2910.01	5	5	1.1E+00	2404.18	4	6	2.5E+00
4529.58	10	8	2.4E-01	2910.38	3	3	1.2E+00	2516.14	10	10	3.7E+00
4545.40	10	12	7.6E-01	2911.05	7	7	3.7E-01	2521.55	8	8	3.5E+00
4560.72	8	10	7.0E-01	2912.46	11	9	5.0E-01	2548.21	6	4	2.0E+00
4571.79	6	8	6.0E-01	2915.88	9	7	4.9E-01	2554.22	8	6	1.2E+00
4578.73	4	6	6.8E-01	2924.02	11	11	1.7E+00	2593.05	6	6	2.8E+00
4706.16	6	4	2.4E-01	2924.63	9	9	1.2E+00	2595.10	8	8	2.8E+00
4757.47	4	2	7.6E-01	2930.80	7	7	5.8E-01				
4766.62	6	4	5.6E-01	2941.37	11	9	3.5E-01	**V IV**			
4776.36	8	6	5.1E-01	2944.57	9	7	7.6E-01	677.345	9	9	6.7E+00
4786.50	10	8	4.7E-01	2948.08	9	11	4.0E-01	680.632	9	7	1.2E+01
4796.92	12	10	4.8E-01	2952.07	7	5	7.2E-01	681.145	7	5	1.1E+01
4807.52	14	12	5.8E-01	2955.58	7	9	3.3E-01	682.455	7	7	6.5E+00
5193.00	12	12	4.0E-01	2968.37	7	9	7.0E-01	682.923	5	5	6.9E+00
5195.39	8	8	2.3E-01	2972.26	5	7	5.2E-01	684.450	7	5	7.7E+00
5234.08	10	10	4.9E-01	2973.98	9	11	3.5E-01	691.530	5	3	1.1E+01
5240.87	12	12	4.3E-01	2985.18	7	9	4.4E-01	723.537	3	1	1.5E+01
5415.25	12	14	3.1E-01	3001.20	7	7	7.5E-01	724.068	5	5	1.1E+01
5487.91	12	10	2.9E-01	3014.82	5	3	8.9E-01	724.809	5	3	5.6E+00
5507.75	10	8	3.5E-01	3016.78	7	5	5.0E-01	737.854	9	7	2.4E+01
6090.21	8	6	2.60E-01	3020.21	9	7	5.0E-01	750.110	5	5	1.0E+01
				3048.21	11	13	7.0E-01	884.146	1	3	4.7E+00
V II				3063.25	9	11	1.0E+00	1071.05	5	5	6.1E+00
2527.90	13	13	6.1E-01	3100.94	7	7	5.8E-01	1110.72	3	3	5.0E+00
2528.47	9	9	5.2E-01	3113.56	11	11	5.0E-01	1112.20	7	7	6.3E+00
2528.83	11	11	5.3E-01	3122.89	11	13	7.6E-01	1112.44	5	5	5.0E+00
2554.04	9	9	5.4E-01	3134.93	13	13	5.9E-01	1127.84	7	5	8.9E+00
2589.10	9	9	7.7E-01	3136.50	11	11	5.3E-01	1131.26	9	7	9.4E+00
2640.86	5	7	1.2E+00	3139.73	9	9	5.2E-01	1194.46	7	5	1.0E+01
2677.80	3	5	3.4E-01	3151.32	3	5	4.4E-01	1226.52	5	5	1.5E+01
2679.33	7	7	3.4E-01	3190.69	9	9	3.3E-01	1243.72	3	1	9.4E+00
2683.09	1	3	3.4E-01	3250.78	11	9	5.2E-01	1247.07	5	3	4.7E+00
2687.96	9	9	7.6E-01	3251.87	5	7	3.5E-01	1272.97	3	1	2.7E+01
2689.88	3	1	9.2E-01	3271.12	7	9	6.9E-01	1304.17	3	5	1.5E+01
2690.25	7	5	3.4E-01	3276.12	9	11	5.2E-01	1305.42	5	7	7.0E+00
2690.79	5	3	5.2E-01	3279.84	9	11	5.8E-01	1308.06	7	9	7.9E+00
2700.94	9	11	3.5E-01	3287.71	5	7	7.5E-01	1309.50	5	5	8.7E+00
2706.17	7	9	3.4E-01	3337.85	5	7	5.3E-01	1312.72	7	7	8.6E+00
2734.22	9	7	6.2E-01	3517.30	9	7	3.8E-01	1317.57	5	7	8.7E+00
2753.41	13	11	4.2E-01	3530.77	5	3	4.5E-01	1321.92	7	9	9.9E+00
2784.20	9	9	1.3E+00	3545.19	7	5	4.3E-01	1326.81	3	5	4.0E+00
2787.91	7	9	5.0E-01	3556.80	9	7	5.1E-01	1329.29	5	5	1.5E+01
2825.86	9	7	1.2E+00	3592.01	7	5	4.4E-01	1329.97	3	3	4.8E+00
2843.82	7	5	9.9E-01	3618.92	3	5	3.3E-01	1330.36	1	3	6.0E+00
2847.57	9	7	4.6E-01					1331.67	3	1	1.7E+01
2854.34	11	9	5.0E-01	**V III**				1332.46	5	3	7.5E+00
2862.31	11	11	3.6E-01	2318.06	8	10	4.6E+00	1334.49	9	9	8.3E+00
2868.11	5	3	2.1E+00	2323.82	6	8	3.8E+00	1355.13	7	9	2.5E+01
2869.13	13	11	4.8E-01	2330.42	10	10	3.2E+00	1356.53	5	3	4.9E+00
2882.49	5	5	4.2E-01	2331.75	8	8	2.5E+00	1395.00	5	7	1.4E+01
2884.78	3	3	5.6E-01	2334.21	6	6	2.2E+00	1400.42	5	7	7.5E+00
2889.61	3	1	1.9E+00	2337.13	4	4	2.7E+00	1403.62	7	9	8.4E+00
2891.64	5	3	1.4E+00	2343.10	6	8	3.6E+00	1412.69	3	3	1.1E+01
2892.43	9	9	3.6E-01	2358.73	6	8	4.2E+00	1414.41	5	7	1.2E+01
2892.65	7	5	1.3E+00	2366.31	8	10	4.2E+00	1414.84	5	5	4.6E+00

λ Å	g_i	g_k	A 10^8 s^{-1}	λ Å	g_i	g_k	A 10^8 s^{-1}	λ Å	g_i	g_k	A 10^8 s^{-1}
1418.53	7	7	5.2E+00	**Xe II**				4128.30	6	6	1.6E+00
1419.58	7	9	1.3E+01	4180.1	4	4	2.2E+00	4142.84	4	4	1.6E+00
1423.72	3	5	7.1E+00	4330.5	6	8	1.4E+00	4167.51	6	6	2.38E-01
1426.65	9	11	2.2E+01	4414.8	6	6	1.0E+00	4235.93	6	4	3.0E-01
1429.11	5	5	5.0E+00	4603.0	4	4	8.2E-01	4352.40	4	4	6.7E-03
1434.84	7	7	5.4E+00	4844.3	6	8	1.1E+00	4379.33	6	4	7.83E-01
1451.04	3	3	7.0E+00	4876.5	6	8	6.3E-01	4385.47	4	4	6.9E-02
1454.00	5	3	1.1E+01	5260.4	2	4	2.2E-01	4394.01	8	8	1.9E-02
1520.14	5	7	7.2E+00	5262.0	4	4	8.5E-01	4409.70	4	6	2.7E-03
1522.49	3	5	5.5E+00	5292.2	6	6	8.9E-01	4417.43	10	8	3.2E-02
1601.92	3	3	1.2E+01	5372.4	4	2	7.1E-01	4437.34	6	6	8.64E-02
1611.88	7	7	5.2E+00	5419.2	4	6	6.2E-01	4443.65	10	8	1.1E-01
1806.18	5	3	7.3E+00	5439.0	4	2	7.4E-01	4459.01	4	6	1.8E-02
1809.85	3	1	7.2E+00	5472.6	8	8	9.9E-02	4476.95	8	6	2.8E-01
1817.68	5	3	4.8E+00	5531.1	8	6	8.8E-02	4491.74	10	10	2.3E-02
1825.84	7	5	5.3E+00	5719.6	4	6	6.1E-02	4514.01	4	6	3.34E-01
1861.56	5	7	6.6E+00	5976.5	4	4	2.8E-01	4527.78	8	6	8.33E-01
1939.07	7	9	5.8E+00	6036.2	6	6	7.5E-02	4534.09	6	8	4.4E-02
1951.43	5	7	5.0E+00	6051.2	8	6	1.7E-01	4544.31	6	6	4.10E-01
1963.10	3	5	4.8E+00	6097.6	6	4	2.6E-01	4559.36	2	4	4.0E-01
1997.72	7	7	4.7E+00	6270.8	4	6	1.8E-01	4581.33	6	4	1.5E-01
2084.43	5	5	4.0E+00	6277.5	4	6	3.6E-02	4613.00	6	4	1.8E-01
2120.05	7	9	8.1E+00	6805.7	8	6	6.1E-02	4643.70	4	6	1.8E-01
2141.20	3	5	7.0E+00	6990.9	10	8	2.7E-01	4653.78	4	6	1.6E-01
2146.83	7	9	6.6E+00					4674.85	6	8	1.3E-01
2149.85	5	7	5.1E+00	**Ytterbium**				4725.84	4	4	1.5E-01
2151.09	7	9	4.3E+00	**Yb I**				4762.96	6	4	4.2E-02
2155.34	11	13	1.2E+01	2464.5	1	3	9.1E-01	4780.16	2	4	8.9E-02
2446.80	9	11	5.3E+00	2672.0	1	3	1.18E-01	4781.03	8	10	1.0E-01
2570.72	9	11	7.6E+00	3464.4	1	3	6.2E-01	4799.30	6	8	1.6E-01
3284.56	7	9	5.3E+00	3988.0	1	3	1.76E+00	4804.31	6	4	2.6E-01
3496.42	7	9	4.4E+00	5556.5	1	3	1.14E-02	4804.80	4	4	3.84E-01
3514.25	9	11	4.7E+00					4821.63	6	6	1.0E-01
				Yb II				4845.67	8	8	6.8E-01
Xenon				3289.4	2	4	1.8E+00	4852.68	6	6	6.2E-01
Xe I				3694.2	2	2	1.4E+00	4856.71	6	6	2.0E-01
1043.8	1	3	5.9E-01					4859.84	4	4	7.26E-01
1047.1	1	3	1.3E+00	**Yttrium**				4893.44	6	4	2.2E-01
1050.1	1	3	8.5E-02	**Y I**				4900.08	8	6	2.0E-01
1056.1	1	3	2.45E+00	2948.41	4	4	3.5E-01	4906.11	10	8	1.2E-01
1061.2	1	3	1.9E-01	2974.59	4	6	3.5E-01	4950.01	8	6	2.0E-02
1068.2	1	3	3.99E+00	2984.25	6	8	4.8E-01	4963.49	4	4	1.4E-02
1085.4	1	3	4.10E-01	2995.26	6	4	5.1E-02	4981.97	4	6	4.7E-03
1099.7	1	3	4.34E-01	2996.94	4	6	8.4E-02	5004.44	6	4	1.2E-02
1110.7	1	3	1.5E+00	3005.26	4	4	4.8E-02	5205.01	4	4	8.4E-03
1129.3	1	3	4.4E-02	3022.28	6	6	6.6E-02	5258.47	6	6	2.9E-03
1170.4	1	3	1.6E+00	3045.36	6	6	1.07E-01	5271.82	8	6	1.1E-02
1192.0	1	3	6.2E-01	3053.95	6	4	1.9E-03	5380.63	6	4	3.2E-01
1250.2	1	3	1.4E-01	3155.65	4	6	2.7E-02	5381.24	4	4	9.9E-03
1295.6	1	3	2.46E+00	3172.84	4	4	9.9E-03	5388.39	6	8	1.1E-02
1469.6	1	3	2.81E+00	3185.96	6	8	1.2E-03	5390.81	8	8	2.9E-02
4501.0	5	3	6.2E-03	3209.38	6	6	3.0E-03	5401.88	6	8	6.0E-01
4524.7	5	5	2.1E-03	3227.16	6	4	1.10E-03	5424.36	6	4	3.47E-01
4624.3	5	5	7.2E-03	3484.05	4	6	1.2E-02	5466.24	4	4	1.0E-01
4671.2	5	7	1.0E-02	3549.66	6	6	1.0E-03	5466.47	10	12	6.3E-01
4807.0	3	1	2.4E-02	3552.69	4	4	2.3E-01	5469.10	4	6	3.6E-03
7119.6	7	9	6.6E-02	4077.36	4	6	1.1E+00	5513.65	6	6	2.39E-01
7967.3	1	3	3.0E-03	4083.71	4	4	2.5E-01	5519.88	4	6	1.2E-02
8409.2	5	3	1.0E-02	4102.36	6	8	1.3E+00	5526.43	6	4	3.9E-03

λ Å	Weights g_i	g_k	A 10^8 s^{-1}	λ Å	Weights g_i	g_k	A 10^8 s^{-1}	λ Å	Weights g_i	g_k	A 10^8 s^{-1}
5527.56	8	10	5.4E-01	3448.81	5	5	4.1E-02	5119.11	5	7	1.6E-02
5541.63	8	8	5.2E-02	3467.88	5	3	2.7E-02	5200.41	5	5	1.3E-01
5551.00	4	4	6.9E-02	3496.08	1	3	3.49E-01	5205.73	7	7	1.6E-01
5573.03	6	4	1.8E-02	3549.01	5	7	3.97E-01	5289.82	7	5	6.7E-03
5594.12	6	8	5.0E-02	3584.51	3	5	4.02E-01	5320.78	9	7	3.9E-03
5606.34	10	10	5.84E-02	3600.74	7	7	1.4E+00	5473.39	3	5	4.3E-02
5619.96	6	4	2.0E-02	3601.91	3	3	1.13E+00	5480.73	1	3	7.62E-02
5630.14	4	6	4.9E-01	3611.04	5	5	1.04E+00	5497.41	5	5	1.2E-01
5641.78	2	4	1.9E-02	3628.70	5	3	3.3E-01	5509.90	5	5	4.24E-02
5675.27	6	6	9.3E-02	3664.62	7	5	3.7E-01	5544.61	3	1	1.8E-01
5675.64	4	6	4.3E-02	3710.29	7	9	1.5E+00	5546.01	5	3	5.8E-02
5693.63	4	4	1.1E-01	3747.55	3	3	1.9E-01	5728.89	5	5	3.0E-02
5714.94	8	6	2.0E-02	3774.34	5	7	1.1E+00	6613.74	5	7	1.7E-02
5729.25	6	6	2.2E-03	3776.56	5	3	2.42E-01	6832.48	5	5	3.3E-03
5732.09	6	6	7.5E-02	3788.70	3	5	8.1E-01	7264.16	5	3	1.3E-02
5740.22	8	6	4.0E-02	3818.34	5	5	9.70E-02				
5757.59	4	6	7.6E-03	3832.90	7	7	3.0E-01	**Zinc**			
5788.36	4	4	9.4E-03	3878.29	7	5	2.9E-02	**Zn I**			
5844.13	6	4	5.6E-03	3930.66	5	5	2.1E-01	748.29	1	3	6.0E-02
5879.93	4	2	8.5E-02	3950.36	3	5	2.80E-01	765.60	1	3	7.6E-02
5902.91	6	8	4.0E-02	3951.59	5	3	1.5E-02	792.05	1	3	5.7E-02
6087.94	6	4	1.1E-01	3982.60	5	5	2.7E-01	793.85	1	3	1.8E-01
6191.72	4	4	4.7E-02	4124.91	5	7	1.8E-02	809.92	1	3	2.6E-01
6222.58	4	6	5.9E-03	4177.54	5	5	5.27E-01	1109.1	1	3	3.05E-01
6402.01	6	4	2.7E-03	4199.27	3	5	5.36E-03	2138.6	1	3	7.09E+00
6435.02	6	6	4.0E-02	4204.69	1	3	2.20E-02	3075.9	1	3	3.29E-04
6437.17	10	8	4.8E-02	4235.73	5	5	2.3E-02	3282.3	1	3	9.0E-01
6538.57	10	10	1.5E-01	4309.62	7	5	1.29E-01	3302.6	3	5	1.2E+00
6622.48	8	6	4.5E-03	4358.73	3	3	5.55E-02	3302.9	3	3	6.7E-01
6815.15	2	4	7.18E-02	4374.95	5	5	9.97E-01	3345.0	5	7	1.7E+00
7009.89	2	4	4.4E-02	4398.01	5	3	1.16E-01	3345.6	5	5	4.0E-01
7035.15	4	4	6.3E-02	4422.59	3	1	1.83E-01	3345.9	5	3	4.5E-02
				4682.33	5	5	1.9E-02	6362.3	3	5	4.74E-01
Y II				4786.58	7	7	2.1E-02	11054	3	1	2.43E-01
3112.03	1	3	1.3E-02	4823.31	5	5	4.3E-02				
3179.42	3	5	3.8E-02	4854.87	5	3	3.9E-01	**Zn II**			
3195.62	3	3	8.23E-01	4881.44	5	3	1.5E-03	2025.5	2	4	3.3E+00
3200.27	5	5	4.8E-01	4883.69	9	7	4.7E-01	2064.2	2	4	4.6E+00
3203.32	3	1	2.77E+00	4900.11	7	5	4.51E-01	2099.9	4	6	5.6E+00
3216.69	5	3	2.0E+00	4982.13	7	9	1.5E-02	2102.2	4	4	9.3E-01
3242.28	7	5	2.0E+00	5087.42	9	9	2.0E-01	4911.6	4	6	1.6E+00

ELECTRON AFFINITIES
Thomas M. Miller

Electron affinity is defined as the energy difference between the lowest (ground) state of the neutral and the lowest state of the corresponding negative ion. The accuracy of electron affinity measurements has been greatly improved since the advent of laser photodetachment experiments with negative ions. Electron affinities can be determined with optical precision, though a detailed understanding of atomic and molecular states and splittings is required to specify the photodetachment threshold corresponding to the electron affinity.

Atomic and molecular electron affinities are discussed in two excellent articles reviewing photodetachment studies which appear in *Gas Phase Ion Chemistry*, Vol. 3, Bowers, M. T., Ed., Academic Press, Orlando, 1984: Chapter 21 by Drzaic, P. S., Marks, J., and Brauman, J. I., "Electron Photodetachment from Gas Phase Negative Ions", p. 167, and Chapter 22 by Mead, R. D., Stevens, A. E., and Lineberger, W. C., "Photodetachment in Negative Ion Beams", p. 213. Persons interested in photodetachment details should consult these articles and the critical review of Hotop, H., and Lineberger, W. C., *J. Phys. Chem. Ref. Data*, 14, 731, 1985. For simplicity in the tables below, any electron affinity which was discussed in the articles by Drzaic *et al.* or Hotop and Lineberger is referenced to these sources, where original references are given. A great many additional electron affinities have been provided here by G. B. Ellison, W. C. Lineberger, H. Hotop, D. G. Leopold, and K. H. Bowen. Electron affinities for the lanthanides and actinides have not been measured, but theoretical estimates have been made by Bratch, S. G., *Chem. Phys. Lett.*, 98, 113, 1983, and Bratch, S. G., and Lagowski, J. J., *Chem. Phys. Lett.*, 107, 136, 1984. The development of cluster-ion photodetachment apparatuses has brought an explosion of electron affinity estimates for atomic and molecular clusters. [See Arnold, S. T., Eaton, J. G., Patel-Mistra, D., Sarkas, H. W., and Bowen, K. H., in *Ion and Cluster Ion Spectroscopy and Structure*, Maier, J. P., Ed., Elsevier Science, New York, 1989, p. 417.] The policy in this tabulation is to list the electron affinities for the atoms, diatoms, and triatoms, if adiabatic electron affinities have been determined, but to refer the reader to original sources for higher-order clusters. Additional data on molecular electron affinities may be found in Lias, S. G., Bartmess, J. E., Liebman, J. F., Holmes, J. L., Levin, R. D., and Mallard, W. G., *Gas Phase Ion and Neutral Thermochemistry, J. Phys. Chem. Ref. Data*, 17, (Supplement No. 1), 1988.

For the present tabulation the value $e/hc = 8065.5410 \pm 0.0024$ cm^{-1} eV^{-1} (Cohen, E. R., and Taylor, B. N., *Rev. Mod. Phys.*, 59, 1121, 1987) has been used to convert electron affinities from the units used in spectroscopic work, cm^{-1}, into eV for these tables. The uncertainties in the electron affinities given in the tables do not include the 0.3 ppm uncertainty in *e/hc*; only in a few cases, *e.g.*, atomic oxygen, is this additional uncertainty significant.

Abbreviations used in the tables: calc = calculated value; PT = photodetachment threshold using a lamp as a light source; LPT = laser photodetachment threshold; LPES = laser photoelectron spectroscopy; DA = dissociative attachment; e-scat = electron scattering; CT = charge transfer; and CD = collisional detachment.

Table 1
Atomic Electron Affinities

Atomic number	Atom	Electron affinity in eV	Uncertainty in eV	Method	Ref.
1	D	0.754593	0.000074	LPT	89
	H	0.754195	0.000019	LPT	89
		0.754209	0.000003	calc	1
2	He	not stable	—	calc	1
3	Li	0.6180	0.0005	LPT	1
4	Be	not stable	—	calc	1
5	B	0.277	0.010	LPES	1
6	C	1.2629	0.0003	LPT	1
7	N	not stable	—	DA	1
8	O	1.4611103	0.0000007	LPT	4
9	F	3.401190	0.000004	LPT	74
10	Ne	not stable	—	calc	1
11	Na	0.547926	0.000025	LPT	1
12	Mg	not stable	—	e-scat	1
13	Al	0.441	0.010	LPES	1
14	Si	1.385	0.005	LPES	1
15	P	0.7465	0.0003	LPT	1
16	S	2.077104	0.000001	LPT	1
17	Cl	3.61269	0.00006	LPT	52
18	Ar	not stable	—	calc	1
19	K	0.50147	0.00010	LPT	1
20	Ca	0.02455	0.00010	LPT	44
21	Sc	0.188	0.020	LPES	1
22	Ti	0.079	0.014	LPES	1
23	V	0.525	0.012	LPES	1
24	Cr	0.666	0.012	LPES	1
25	Mn	not stable	—	calc	1

Table 1
Atomic Electron Affinities (continued)

Atomic number	Atom	Electron affinity in eV	Uncertainty in eV	Method	Ref.
26	Fe	0.151	0.003	LPES	27
27	Co	0.662	0.003	LPES	27
28	Ni	1.156	0.010	LPES	1
29	Cu	1.235	0.005	LPES	37
30	Zn	not stable	—	e-scat	1
31	Ga	0.3	0.15	PT	1
32	Ge	1.233	0.003	LPES	28
33	As	0.81	0.03	PT	1
34	Se	2.020670	0.000025	LPT	1
35	Br	3.363590	0.000003	LPT	74
36	Kr	not stable	—	calc	1
37	Rb	0.48592	0.00002	LPT	1
38	Sr	0.048	0.006	LPT	122
39	Y	0.307	0.012	LPES	1
40	Zr	0.426	0.014	LPES	1
41	Nb	0.893	0.025	LPES	1
42	Mo	0.748	0.002	LPES	127
43	Tc	0.55	0.20	calc	1
44	Ru	1.05	0.15	calc	1
45	Rh	1.137	0.008	LPES	1
46	Pd	0.562	0.005	LPES	116
47	Ag	1.302	0.007	LPES	1
48	Cd	not stable	—	e-scat	1
49	In	0.3	0.2	PT	1
50	Sn	1.112	0.004	LPES	28
51	Sb	1.046	0.005	LPES	108
52	Te	1.9708	0.0003	LPT	1
53	I	3.059038	0.000010	LPT	92
54	Xe	not stable	—	calc	1
55	Cs	0.471626	0.000025	LPT	1
56	Ba	0.15	—	calc	57
57	La	0.5	0.3	calc	1
72	Hf	≈0	—	calc	1
73	Ta	0.322	0.012	LPES	1
74	W	0.815	0.002	LPES	37
75	Re	0.15	0.15	calc	1
76	Os	1.1	0.2	calc	1
77	Ir	1.565	0.008	LPES	1
78	Pt	2.128	0.002	LPT	1
79	Au	2.30863	0.00003	LPT	1
80	Hg	not stable	—	e-scat	1
81	Tl	0.2	0.2	PT	1
82	Pb	0.364	0.008	LPES	1
83	Bi	0.946	0.010	LPES	1
84	Po	1.9	0.3	calc	1
85	At	2.8	0.2	calc	1
86	Rn	not stable	—	calc	1
87	Fr	0.46	—	calc	82

Table 2
Electron Affinities for Diatomic Molecules

Molecule	Electron affinity in eV	Uncertainty in eV	Method	Ref.	Molecule	Electron affinity in eV	Uncertainty in eV	Method	Ref.
Ag_2	1.023	0.007	LPES	37	MgH	1.05	0.06	PT	2
Al_2	1.10	0.15	LPES	68	MgI	1.899	0.018	LPES	31
AlS	2.60	0.03	LPES	129	MnD	0.866	0.010	LPES	9
As_2	0	—	PT	2	MnH	0.869	0.010	LPES	9
AsH	1.0	0.1	PT	2	MoO	1.290	0.006	LPES	127
Au_2	1.938	0.007	LPES	37	NH	0.370	0.004	LPT	32
BO	3.12	0.09	PT	6	NO	0.026	0.005	LPES	73
BeH	0.7	0.1	PT	2	NS	1.194	0.011	LPES	2
Bi_2	1.271	0.008	LPES	119	Na_2	0.430	0.015	LPES	104
Br_2	2.55	0.10	CT	2	NaBr	0.788	0.010	LPES	30
BrO	2.353	0.006	LPES	88	NaCl	0.727	0.010	LPES	30
C_2	3.269	0.006	LPES	87	NaF	0.520	0.010	LPES	30
CH	1.238	0.008	LPES	2	NaI	0.865	0.010	LPES	30
CN	3.862	0.004	LPES	111	NaK	0.465	0.030	LPES	104
CS	0.205	0.021	LPES	2	Ni_2	0.926	0.010	LPES	112
CaH	0.93	0.05	PT	2	NiCu	0.889	0.010	LPES	128
Cl_2	2.38	0.10	CT	2	NiAg	0.979	0.010	LPES	128
ClO	2.275	0.006	LPES	88	NiD	0.477	0.007	LPES	29
Co_2	1.110	0.008	LPES	27	NiH	0.481	0.007	LPES	29
CoD	0.680	0.010	LPES	29	O_2	0.451	0.007	LPES	73
CoH	0.671	0.010	LPES	29	OD	1.825548	0.000037	LPT	3
Cr_2	0.505	0.005	LPES	114	OH	1.827670	0.000021	LPT	3
CrD	0.568	0.010	LPES	29	P_2	0.589	0.025	LPES	42
CrH	0.563	0.010	LPES	29	PH	1.028	0.010	LPES	2
CrO	1.222	0.010	LPES	5	PO	1.092	0.010	LPES	2
Cs_2	0.469	0.015	LPES	104	Pb_2	1.366	0.010	LPES	117
CsCl	0.455	0.010	LPES	30	PbO	0.722	0.006	LPES	105
CsO	0.273	0.012	LPES	133	Pd_2	1.685	0.008	LPES	112
Cu_2	0.836	0.006	LPES	37	Pt_2	1.898	0.008	LPES	112
CuO	1.777	0.006	LPES	118	PtN	1.240	0.010	LPES	46
F_2	3.08	0.10	CT	2	Rb_2	0.498	0.015	LPES	104
FO	2.272	0.006	LPES	88	RbCl	0.544	0.010	LPES	30
Fe_2	0.902	0.008	LPES	27	RbCs	0.478	0.020	LPES	104
FeD	0.932	0.015	LPES	9	Re_2	1.571	0.008	LPES	33
FeH	0.934	0.011	LPES	9	S_2	1.670	0.015	LPES	53
FeO	1.493	0.005	LPES	45	SD	2.315	0.002	LPES	10
Ge_2	2.035	0.001	LPES	123	SF	2.285	0.006	LPES	93
I_2	2.55	0.05	CT	2	SH	2.314344	0.000004	LPT	47
IBr	2.55	0.10	CT	2	SO	1.125	0.005	LPES	84
IO	2.378	0.006	LPES	88	Sb_2	1.282	0.008	LPES	108
InP	1.95	0.05	LPES	137	Se_2	1.94	0.07	LPES	38
K_2	0.497	0.012	LPES	104	SeH	2.212520	0.000025	LPT	48
KBr	0.642	0.010	LPES	30	SeO	1.456	0.020	LPES	41
KCl	0.582	0.010	LPES	30	Si_2	2.201	0.010	LPES	100
KCs	0.471	0.020	LPES	104	SiH	1.277	0.009	LPES	2
KI	0.728	0.010	LPES	30	Sn_2	1.962	0.010	LPES	117
KRb	0.486	0.020	LPES	104	SnPb	1.569	0.008	LPES	117
LiCl	0.593	0.010	LPES	30	Te_2	1.92	0.07	LPES	38
LiD	0.337	0.012	LPES	102	TeH	2.102	0.015	LPES	39
LiH	0.342	0.012	LPES	102	TeO	1.697	0.022	LPES	40
MgCl	1.589	0.011	LPES	31	ZnH	≤ 0.95	—	PT	2

Table 3
Electron Affinities for Triatomic Molecules

Molecule	Electron affinity in eV	Uncertainty in eV	Method	Ref.	Molecule	Electron affinity in eV	Uncertainty in eV	Method	Ref.
Ag_3	2.32	0.05	LPES	37	GeH_2	1.097	0.015	LPES	28
Al_3	1.4	0.15	LPES	68	HCO	0.313	0.005	LPES	35
Al_2S	0.80	0.12	LPES	129	HCl_2	4.896	0.005	LPES	69
AsH_2	1.27	0.03	PT	2	HNO	0.338	0.015	LPES	14
Au_3	3.7	0.3	LPES	37	HO_2	1.078	0.017	LPES	15
BO_2	3.57	0.13	CT	6	InP_2	1.61	0.05	LPES	137
BO_2	4.3	0.2	CT	98	In_2P	2.36	0.05	LPES	137
Bi_3	1.60	0.03	LPES	119	K_3	0.956	0.050	LPES	18
C_3	1.981	0.020	LPES	11	MnD_2	0.465	0.014	LPES	34
CCl_2	1.591	0.010	LPES	95	MnH_2	0.444	0.016	LPES	34
CD_2	0.645	0.006	LPES	12	N_3	2.70	0.12	PT	2
CDF	0.535	0.005	LPES	95	NCO	3.609	0.005	LPES	111
CF_2	0.165	0.010	LPES	95	NCS	3.537	0.005	LPES	111
CH_2	0.652	0.006	LPES	12	NH_2	0.771	0.005	LPES	58
CHBr	1.454	0.005	LPES	95	N_2O	0.22	0.10	CT	59
CHCl	1.210	0.005	LPES	95	NO_2	2.273	0.005	LPES	63
CHF	0.542	0.005	LPES	95	(NO)R	R=Ar,Kr,Xe	—	LPES	90
CHI	1.42	0.17	LPES	95	Na_3	1.019	0.060	LPES	18
C_2H	2.969	0.006	LPES	87	Ni_3	1.41	0.05	LPES	55
C_2O	1.848	0.027	LPES	13	NiCO	0.804	0.012	LPES	2
COS	0.46	0.20	CD	2	NiD_2	1.926	0.007	LPES	34
CS_2	0.895	0.020	LPES	11	NiH_2	1.934	0.008	LPES	34
CoD_2	1.465	0.013	LPES	34	O_3	2.1028	0.0025	LPT	2
CoH_2	1.450	0.014	LPES	34	O_2Ar	0.52	0.02	LPES	75
CrH_2	>2.5	—	LPES	34	OClO	2.140	0.008	LPES	88
Cr_2D	1.464	0.005	LPES	107	OIO	2.577	0.008	LPES	88
Cr_2H	1.474	0.005	LPES	107	PH_2	1.271	0.010	LPES	2
Cs_3	0.864	0.030	LPES	18	PO_2	3.42	0.01	LPES	124
Cu_3	2.11	0.05	LPES	37	Pt_3	1.87	0.02	LPES	55
DCO	0.301	0.005	LPES	35	Pd_3	<1.5	0.1	LPES	55
DNO	0.330	0.015	LPES	14	Rb_3	0.920	0.030	LPES	18
DO_2	1.089	0.017	LPES	15	S_3	2.093	0.025	LPES	16
DS_2	1.912	0.015	LPES	53	SO_2	1.107	0.008	LPES	16
HS_2	1.907	0.015	LPES	53	S_2O	1.877	0.008	LPES	16
FeCO	1.157	0.005	LPES	103	Sb_3	1.85	0.03	LPES	108
FeD_2	1.038	0.013	LPES	34	SeO_2	1.823	0.050	LPES	38
FeO_2	2.358	0.030	LPES	130	SiH_2	1.124	0.020	LPES	2
Ge_3	2.23	0.01	LPES	123	VO_2	2.3	0.2	CT	101
FeH_2	1.049	0.014	LPES	34					

Table 4
Electron Affinities for Larger Polyatomic Molecules

Molecule	Electron Affinity in eV	Uncertainty in eV	Method	Ref.	Name
Ag_n	n = 1-10	—	LPES	37	
Al_n	n = 3-32	—	LPES	68	
Al_3O	1.00	0.15	LPES	68	
Al_nS_m	n = 1-5	m = 1-3	LPES	129	
$Ar(H_2O)_n$	n = 2,6,7	—	LPES	77	
Au_n	n = 2-5	—	LPES	37	
AuF_6	7.5	estimate	CT	98	
BD_3	0.027	0.014	LPES	62	
BH_3	0.038	0.015	LPES	62	

Table 4
Electron Affinities for Larger Polyatomic Molecules (continued)

Molecule	Electron Affinity in eV	Uncertainty in eV	Method	Ref.	Name
Bi_4	1.05	0.010	LPES	119	
$Br(CO_2)$	3.582	0.017	LPES	131	
C_n	$n = 2\text{-}84$	—	LPES	70	
$(CO_2)_n$	$n = 1,2$	—	LPES	75	
$(CS)_n$	$n = 2$	—	LPES	75	
$(CS_2)_n$	$n = 1,2$	—	LPES	75	
CF_3Br	0.91	0.2	CD	2	
CF_3I	1.57	0.2	CD	2	
$CO_3(H_2O)$	2.1	0.2	PT	2	
CHO_2	3.498	0.005	LPES	109	
CH_2S	0.465	0.023	LPES	53	
CH_3	0.08	0.03	LPES	2	
CH_3I	0.2	0.1	CT	2	
CH_3NO_2	0.48	0.10	CT	61	
CH_3Si	0.852	0.010	LPES	97	$CH_3\text{-}Si$
CH_3Si	2.010	0.010	LPES	97	$CH_2 = SiH$
CD_3O	1.552	0.022	LPES	2	
CH_3O	1.570	0.022	LPES	2	
CD_3S	1.856	0.006	LPT	2	
CH_3S	1.861	0.004	LPT	2	
CD_3S_2	1.748	0.022	LPES	53	
CH_3S_2	1.757	0.022	LPES	53	
CH_3SiH_2	1.19	0.04	LPT	65	
CO_3	2.69	0.14	LPES	2	
C_2F_2	2.255	0.006	LPES	106	difluorovinylidene
C_2DO	2.350	0.020	LPES	13	
C_2HO	2.350	0.020	LPES	13	
C_2D_2	0.492	0.006	LPES	83	vinylidene-d_2
C_2HD	0.489	0.006	LPES	83	vinylidene-d_1
C_2HF	1.718	0.006	LPES	106	monofluorovinylidene
C_2H_2	0.490	0.006	LPES	83	vinylidene
C_2H_2FO	2.22	0.09	PT	2	acetyl fluoride enolate
C_2D_2N	1.538	0.012	LPES	21	cyanomethyl-d_2 radical
C_2D_2N	1.070	0.024	LPES	21	isocyanomethyl-d_2 radical
C_2H_2N	1.543	0.014	LPES	21	cyanomethyl radical
C_2H_2N	1.059	0.024	LPES	21	isocyanomethyl radical
C_2H_3	0.667	0.024	LPES	90	vinyl
C_2D_3O	1.81897	0.00012	LPT	22	acetaldehyde-d_3 enolate
C_2H_3O	1.82476	0.00012	LPT	22	acetaldehyde enolate
C_2H_5N	0.56	0.01	PT	2	ethyl nitrine
C_2D_5O	1.702	0.033	LPES	23	ethoxide-d_3
C_2H_5O	1.726	0.033	LPES	23	ethoxide
C_2H_5S	1.953	0.006	LPT	2	ethyl sulfide
C_2H_5S	0.868	0.051	LPES	53	CH_3SCH_2
$C_2H_7O_2$	2.26	0.08	PT	50	MeOHOMe
C_3Fe	1.69	0.08	LPES	132	
C_3H	1.858	0.023	LPES	11	
C_3HFe	1.58	0.06	LPES	132	
C_3H_2	1.794	0.025	LPES	11	
$C_3H_2F_3O$	2.625	0.010	LPT	113	1,1,1-trifluoroacetone enolate
C_3H_3	0.893	0.025	LPES	24	propargyl radical
C_3H_2D	0.88	0.15	LPES	24	propargyl-d_1 radical
C_3D_2H	0.907	0.023	LPES	24	propargyl-d_2 radical
C_3H_3N	1.247	0.012	LPES	21	$CH_3CH\text{-}CN$
C_3D_5	0.464	0.006	LPES	138	allyl-d_5

Table 4
Electron Affinities for Larger Polyatomic Molecules (continued)

Molecule	Electron Affinity in eV	Uncertainty in eV	Method	Ref.	Name
C_3H_5	0.481	0.008	LPES	138	allyl
C_3H_4D	0.373	0.019	LPES	25	allyl-d_1
C_3H_5O	1.758	0.019	LPT	113	acetone enolate
C_3H_5O	1.621	0.006	LPT	113	propionaldehyde enolate
$C_3H_5O_2$	1.80	0.06	PT	2	methyl acetate enolate
C_3H_7O	1.789	0.033	LPES	23	n-propyl oxide
C_3H_7O	1.839	0.029	LPES	23	isopropyl oxide
C_3H_7S	2.00	0.02	PT	2	n-propyl sulfide
C_3H_7S	2.02	0.02	PT	2	isopropyl sulfide
C_3O	1.34	0.15	LPES	11	
C_3O_2	0.85	0.15	LPES	11	
$C_4F_4O_3$	0.5	0.2	CD	2	tetrafluorosuccinic anhydride
C_4Fe	≤2.2	0.2	LPES	132	
C_4HFe	1.67	0.06	LPES	132	
$C_4H_2O_3$	1.44	0.10	CT	61	maleic anhydride
C_4D_4	0.909	0.015	LPES	125	vinylvinylidene-d_4
C_4H_4	0.914	0.015	LPES	125	vinylvinylidene
C_4H_4N	2.39	0.13	PT	2	pyrrolate
C_4H_5O	1.801	0.008	LPT	113	cyclobutanone enolate
C_4H_6	0.431	0.006	LPES	135	trimethylenemethane
$C_4H_6O_2$	0.69	0.10	CT	61	2,3-butanedione
C_4H_6D	0.493	0.008	LPES	138	2-methylallyl-d_7
C_4H_7	0.505	0.006	LPES	138	2-methylallyl
C_4H_7O	1.67	0.05	PT	2	butyraldehyde enolate
C_4H_5DO	1.67	0.05	PT	2	2-butanone-3-d_1 enolate
$C_4H_5D_2O$	1.75	0.06	PT	2	2-butanone-3,3-d_2 enolate
C_4H_9O	1.912	0.054	LPES	23	t-butoxyl
C_4H_9S	2.03	0.02	PT	2	n-butyl sulfide
C_4H_9S	2.07	0.02	PT	2	t-butyl sulfide
C_4O	2.05	0.15	LPES	11	
C_4O_2	2.0	0.2	LPES	11	
C_5	2.853	0.001	LPT	99	
C_5F_5N	0.68	0.11	CT	67	pentafluoropyridine
$C_5F_6O_3$	1.5	0.2	CD	2	hexafluoroglutaric anhydride
C_5D_5	1.790	0.008	LPES	11	cyclopentadienyl-d_5
C_5H_5	1.804	0.007	LPES	11	cyclopentadienyl
C_5H_7	0.91	0.03	PT	2	pentadienyl
C_5H_7O	1.598	0.007	LPT	113	cyclopentanone enolate
C_5H_9O	1.69	0.05	PT	2	3-penanone enolate
$C_5H_{11}O$	1.93	0.05	LPT	2	neopentoxyl
$C_5H_{11}S$	2.09	0.02	PT	2	n-pentyl sulfide
C_5O_2	1.2	0.2	LPES	11	
C_6	4.180	0.001	LPT	8	
$C_6Br_4O_2$	2.44	0.20	CT	2	tetrabromo-BQ
$C_6Cl_4O_2$	2.78	0.10	CT	61	tetrachloro-BQ
$C_6F_4O_2$	2.70	0.10	CT	61	tetrafluoro-BQ
C_6F_5Br	1.15	0.11	CT	67	pentafluorobromobenzene
C_6F_5Cl	0.82	0.11	CT	67	pentafluorochlorobenzene
C_6F_5I	1.41	0.11	CT	67	pentafluoroiodobenzene
$C_6F_5NO_2$	1.52	0.11	CT	67	pentafluoro-NB
C_6F_6	0.52	0.10	CT	51	hexafluorobenzene
C_6F_{10}	>1.4	0.3	CT	2	perfluorocyclohexane
$C_6H_2Cl_2O_2$	2.48	0.10	CT	61	2,6-dichloro-BQ
$C_6H_3F_2NO_2$	1.17	0.10	CT	61	2,4-difluoro-NB
C_6D_4	0.551	0.010	LPES	36	o-benzyne-d_4

Table 4
Electron Affinities for Larger Polyatomic Molecules (continued)

Molecule	Electron Affinity in eV	Uncertainty in eV	Method	Ref.	Name
C_6H_4	0.560	0.010	LPES	36	o-benzyne
$C_6H_4BrNO_2$	1.16	0.10	CT	61	o-bromo-NB
$C_6H_4BrNO_2$	1.32	0.10	CT	61	m-bromo-NB
$C_6H_4BrNO_2$	1.29	0.10	CT	61	p-bromo-NB
$C_6H_4ClNO_2$	1.14	0.10	CT	61	o-chloro-NB
$C_6H_4ClNO_2$	1.28	0.10	CT	61	m-chloro-NB
$C_6H_4ClNO_2$	1.26	0.10	CT	61	p-chloro-NB
C_6H_4ClO	$\leq$2.58	0.08	PT	2	o-chloroperoxide
$C_6H_4FNO_2$	1.07	0.10	CT	61	o-fluoro-NB
$C_6H_4FNO_2$	1.23	0.10	CT	61	m-fluoro-NB
$C_6H_4FNO_2$	1.12	0.10	CT	61	p-fluoro-NB
$C_6H_4N_2O_4$	1.65	0.10	CT	61	o-diNB
$C_6H_4N_2O_4$	1.65	0.10	CT	61	m-diNB
$C_6H_4N_2O_4$	2.00	0.10	CT	61	p-diNB
$C_6H_4O_2$	1.91	0.10	CT	61	1,4-benzoquinone (BQ)
C_6D_5	1.092	0.020	LPES	26	phenyl-d_5
C_6D_5N	1.44	0.02	LPES	96	phenylnitrene-d_5
C_6H_5	1.096	0.006	LPES	26	phenyl
C_6H_5N	1.429	0.011	LPT	115	phenylnitrene
$C_6H_5NO_2$	1.01	0.10	CT	61	nitrobenzene (NB)
C_6H_5O	2.253	0.006	LPES	26	phenoxyl
C_6H_5S	$\leq$2.47	0.06	PT	2	thiophenoxide
C_6H_5NH	1.70	0.03	PT	2	anilide
C_6H_7	<1.67	0.04	PT	2	methylchylopentadienyl
C_6H_8Si	1.435	0.004	LPT	65	$C_6H_5SiH_3$
C_6H_9O	1.526	0.010	LPT	113	cyclohexanone enolate
C_6H_{10}	0.645	0.015	LPES	126	t-butyl vinylidene
$C_6H_{11}O$	1.755	+0.05/-0.005	LPT	113	pinacolone enolate
$C_6H_{11}O$	1.82	0.06	PT	2	3,3-dimethylbutananl enolate
C_6N_4	2.3	0.3	PT	2	TCNE
C_7F_5N	1.11	0.11	CT	67	pentafluorobenzonitrile
C_7F_8	0.86	0.11	CT	67	octafluorotoluene
C_7F_{14}	1.08	0.10	CT	61	perfluoromethylcyclohexane
C_7HF_5O	1.10	0.11	CT	67	pentafluorobenzaldehyde
$C_7H_3N_3O_4$	2.16	0.10	CT	61	3,5-$(NO_2)_2$-benzonitrile
$C_7H_4F_3NO_2$	1.41	0.10	CT	61	m-trifluoromethyl-NB
$C_7H_4N_2O_2$	1.61	0.10	CT	61	o-cyano-NB
$C_7H_4N_2O_2$	1.56	0.10	CT	61	m-cyno-NB
$C_7H_4N_2O_2$	1.72	0.10	CT	61	p-cyano-NB
C_7H_6FO	2.218	0.010	LPT	2	m-fluoroacetophenone enolate
C_7H_6FO	2.176	0.010	LPT	2	p-fluoroacetophenone enolate
$C_7H_6FeO_3$	0.990	0.10	CT	120	η^4-1,3-butadiene-Fe(CO)$_3$
$C_7H_6N_2O_4$	1.77	0.05	PT	60	3,4-dintrotoluene
$C_7H_6N_2O_4$	1.77	0.05	PT	60	2,3-dinitrotoluene
$C_7H_6N_2O_4$	1.60	0.05	PT	60	2,4-dinitrotoluene
$C_7H_6N_2O_4$	1.55	0.05	PT	60	2,6-dinitrotoluene
C_7H_7	0.912	0.006	LPES	26	benzyl
C_7H_7	0.868	0.006	LPES	136	1-quadricyclanide
C_7H_7	0.962	0.006	LPES	136	2-quadricyclanide
C_7H_7	1.286	0.006	LPES	136	norbornadienide
C_7H_7	0.39	0.04	LPES	136	cycloheptatrienide
C_7H_7	3.046	0.006	LPES	136	1-(1,6-heptadiynide)
C_7H_7	$\geq$1.140	0.006	LPES	136	3-(1,6-heptadiynide)
$C_7H_7NO_2$	0.92	0.10	CT	61	o-methyl-NB
$C_7H_7NO_2$	0.99	0.10	CT	61	m-methyl-NB

Molecule	Electron Affinity in eV	Uncertainty in eV	Method	Ref.	Name
$C_7H_7NO_2$	0.95	0.10	CT	61	p-methyl-NB
$C_7H_7NO_3$	1.04	0.10	CT	61	m-OCH$_3$-NB
$C_7H_7NO_3$	0.91	0.10	CT	61	p-OCH$_3$-NB
C_7H_7O	$\leq$2.36	0.06	PT	2	o-methyl phenoxide
C_7H_7O	2.14	0.02	PT	50	benzyloxide
$C_7H_7O_2$	1.85	0.10	CT	61	o-CH$_3$-BQ
C_7H_8FO	<3.05	0.06	PT	50	PhCH$_2$OHF
C_7H_9	1.27	0.03	PT	2	heptatrienyl
C_7H_9O	1.61	0.05	PT	2	2-norbornanone enolate
C_7H_9Si	1.33	0.04	LPT	65	C$_6$H$_5$(CH$_3$)SiH
$C_7H_{11}O$	1.598	0.007	LPT	113	cycloheptanone enolate
$C_7H_{11}O$	1.49	0.04	PT	2	2,5-dimethylcyclopentanone enolate
$C_7H_{13}O$	1.72	0.06	PT	2	4-heptanone enolate
$C_7H_{13}O$	1.46	0.04	PT	2	di-isopropyl ketone enolate
$C_8F_{14}N_2$	1.89	0.10	CT	51	1,4-(CN)$_2$C$_6$F$_4$
$C_8H_3F_5O$	0.88	0.11	CT	67	pentafluoroacetophenone
$C_8H_3F_6NO_2$	1.79	0.10	CT	61	3,5-(CF$_3$)$_2$-NB
$C_8H_4O_3$	1.21	0.10	CT	61	phthalic anhydride
C_8H_7O	2.057	0.010	PT	2	acetophenone enolate
C_8H_7O	2.10	0.08	LPT	2	phenylacetaldehyde enolate
C_8H_8	0.65	0.04	LPES	134	cyclooctatetraene
C_8H_8	0.919	0.008	LPES	139	m-xylylene
$C_8H_9NO_2$	1.21	0.05	PT	60	3,5-dimethyl-NB
$C_8H_9NO_2$	2.61	0.05	PT	60	2,6-dimethyl-NB
$C_8H_9NO_2$	0.86	0.10	CT	61	2,3-dimethyl-NB
$C_8H_{13}O$	1.63	0.06	PT	2	cyclooctanone enolate
$C_9H_8FeO_3$	0.76	0.10	CT	120	η^4-1,3-cyclohexadiene-Fe(CO)$_3$
C_9H_9O	2.030	0.010	LPT	2	m-methylacetophenone enolate
C_9H_9SiN	1.43	0.10	PT	2	trimethylsilylnitrene
$C_9H_{11}NO_2$	0.70	0.10	CT	61	2,4,6-trimethyl-NB
$C_9H_{15}O$	1.69	0.06	PT	2	cyclononanone enolate
$C_{10}H_4Cl_2O_2$	2.19	0.10	CT	61	2,3-dichloro-1,4-naphthoquinone
$C_{10}H_6N_2O_4$	1.78	0.10	CT	61	1,3-dinitronaphthalene
$C_{10}H_6N_2O_4$	1.77	0.10	CT	61	1,5-dinitronaphthalene
$C_{10}H_6O_2$	1.81	0.10	CT	61	1,4-naphthoquinone
$C_{10}H_7NO_2$	1.23	0.10	CT	61	1-nitronaphthalene
$C_{10}H_7NO_2$	1.18	0.10	CT	61	2-nitronaphthalene
$C_{10}H_8$	0.69	0.10	CT	61	azulene
$C_{10}H_8CrO_3$	0.93	0.10	CT	120	η^4-1,3,5-cycloheptatriene Cr(CO)$_3$
$C_{10}H_8FeO_3$	0.98	0.10	CT	120	η^4-1,3,5-cycloheptatriene-Fe(CO)$_3$
$C_{10}H_{17}O$	1.83	0.06	PT	2	cyclodecanone enolate
$C_{11}H_8FeO_3$	1.29	0.10	CT	120	η^4-1,3-butadiene-Fe(CO)$_3$
$C_{12}F_{10}$	0.82	0.11	CT	67	decafluorobiphenyl
$C_{12}H_4N_4$	2.8	0.3	CD	2	TCNQ
$C_{12}H_9$	1.07	0.10	PT	2	perinaphthenyl
$C_{12}H_{15}O$	2.032	0.010	LPT	2	t-butylacetophenone enolate
$C_{12}H_{21}O$	1.90	0.07	PT	2	cyclododecanone enolate
$C_{13}F_{10}O$	1.52	0.11	CT	67	decafluorobenzophenone
$C_{13}H_9F$	0.64	0.10	CT	61	4-fluorobenzophenone
$C_{13}H_{10}O$	0.62	0.10	CT	61	benzophenone
$C_{14}H_9NO_2$	1.43	0.10	CT	61	9-nitroanthracene
$C_{14}H_{10}$	0.57	0.10	CT	66	anthracene
$C_{18}H_{12}$	1.04	0.10	CT	66	tetracene
$C_{20}H_{12}$	0.79	0.10	CT	66	benz[a]pyrene
$C_{20}H_{12}$	0.97	0.10	CT	66	perylene

Table 4
Electron Affinities for Larger Polyatomic Molecules (continued)

Molecule	Electron Affinity in eV	Uncertainty in eV	Method	Ref.	Name
$C_{22}H_{14}$	1.35	0.10	CT	66	pentacene
CeF_4	3.8	0.4	CT	98	
CoF_4	6.4	0.3	CT	98	
$Cr(CO)_3$	1.349	0.006	LPES	94	
CrO_3	3.6	0.2	CT	98	
Cu_n	$n = 1\text{-}41$	—	LPES	37	
FCO_2	4.277	0.030	LPES	131	
$Fe(CO)_2$	1.22	0.02	LPES	2	
$Fe(CO)_3$	1.8	0.2	LPES	2	
$Fe(CO)_4$	2.4	0.3	LPES	2	
FeF_3	3.6	0.1	CT	98	
FeF_4	6.0	estimate	CT	98	
Ge_n	$n = 3\text{-}15$	—	LPES	71	
Ge_xAs_y	$n = 5\text{-}30$	$n = x + y$	LPES	72	
GeH_3	≤ 1.74	0.04	PT	2	
$(NH_3)_n$	$n = 41\text{-}1100$	—	LPES	77	
$H(NH_3)_n$	$n = 1,2$	—	LPES	76	
HNO_3	0.57	0.15	CD	2	
$(H_2O)_n$	$n = 2\text{-}19$	—	LPES	77	
$I(CO_2)$	3.225	0.001	LPES	131	
In_xP_y	$n = 2\text{-}8$	$n = x + y$	LPES	137	
IrF_4	4.7	0.3	CT	98	
IrF_6	6.5	0.4	CT	98	
K_n	$n = 2\text{-}7$	—	LPES	18	
MnF_4	5.5	0.2	CT	98	
$Mo(CO)_3$	1.337	0.006	LPES	94	
MoF_5	3.5	0.2	CT	98	
MoF_6	3.8	0.2	CT	98	
MoO_3	2.9	0.2	CT	98	
$NH_2(NH_3)_n$	$n = 1,2$	—	LPES	78	
$NO(H_2O)_n$	$n = 1,2$	—	LPES	75	
NO_3	3.937	0.014	LPES	85	
$NO(N_2O)_n$	$n = 1,2$	—	LPES	79	
$(NO)_2$	≥ 2.1	—	LPES	75	
$(N_2O)_n$	$n = 1,2$	—	LPES	81	
Na_n	$n = 2\text{-}5$	—	LPES	18	
$(NaF)_n$	$n = 1\text{-}7,12$	—	LPES	64	
$Na(NaF)_n$	$n = 5,7\text{-}12$	—	LPES	64	
$Ni(CO)_2$	0.643	0.014	LPES	2	
$Ni(CO)_3$	1.077	0.013	LPES	2	
$OH(H_2O)$	<2.95	0.15	PT	2	
OsF_4	3.9	0.3	CT	98	
OsF_6	6.0	0.3	CT	98	
PBr_3	1.59	0.15	CD	2	
PBr_2Cl	1.63	0.20	CD	2	
PCl_2Br	1.52	0.20	CD	2	
PCl_3	0.82	0.10	CD	2	
PF_5	0.75	0.15	CT	121	
PO_3	4.5	0.5	CT	98	
$POCl_2$	3.83	0.25	CD	2	
$POCl_3$	1.41	0.20	CD	2	
PtF_4	5.5	0.3	CT	98	
PtF_6	7.0	0.4	CT	98	
ReF_6	4.7	estimate	CT	98	
RhF_4	5.4	0.3	CT	98	

Table 4
Electron Affinities for Larger Polyatomic Molecules (continued)

Molecule	Electron Affinity in eV	Uncertainty in eV	Method	Ref.	Name
RuF_4	4.8	0.3	CT	98	
RuF_5	5.2	0.4	CT	98	
RuF_6	7.5	0.3	CT	98	
SF_4	1.5	0.2	CT	91	
SF_6	1.05	0.10	CT	56	
SO_3	≥ 1.70	0.15	CD	2	
$(SO_2)_2$	0.6	0.2	LPES	80	
SeF_6	2.9	0.2	CD	2	
Si_4	2.17	0.01	LPES	110	
Si_n	$n = 3\text{-}20$	—	LPES	71	
SiD_3	1.386	0.022	LPES	43	
SiF_3	≤ 2.95	0.10	PT	17	
SiH_3	1.406	0.014	LPES	43	
TeF_6	3.34	0.17	CD	2	
UF_5	3.7	0.2	CT	98	
UF_6	5.1	0.2	CT	98	
UO_3	<2.1	—	CT	98	
VF_4	3.5	0.2	CT	98	
V_4O_{10}	4.2	0.6	CT	101	
$W(CO)_3$	1.859	0.006	LPES	94	
WF_5	1.25	0.3	CD	18	
WF_6	3.36	+0.04/-0.20	CT	19	
WO_3	3.33	+0.04/-0.15	LPT	86	
WO_3	3.9	0.2	CT	98	

REFERENCES

1. Hotop, H., and Lineberger, W. C., *J. Phys. Chem. Ref. Data*, 14, 731, 1985.
2. Drzaic, P. S., Marks, J., and Brauman, J. I., in *Gas Phase Ion Chemistry*, Vol. 3, Bowers, M. T., Ed., Academic Press, Orlando, 1984, p. 167.
3. Schulz, P. A., Mead, R. D., Jones, P. L., and Lineberger, W. C., *J. Chem. Phys.*, 77, 1153, 1982.
4. Neumark, D. M., Lykke, K. R., Anderson, T., and Lineberger, W. C., *Phys. Rev. A*, 32, 1890, 1985.
5. Wenthold, P. G., Gunion, R. F., and Lineberger, W. C., *Chem. Phys. Lett.*, 258, 101, 1996.
6. Srivastava, R. D., Uy, O. M., and Farber, M., *Trans. Faraday Soc.*, 67, 2941, 1971.
7. Klein, R., McGinnis, R. P., and Leone, S. R., *Chem. Phys. Lett.*, 100, 475, 1983.
8. Arnold, C. C., Zhao, Y., Kitsopoulos, T. N., and Neumark, D. M., *J. Chem. Phys.*, 97, 6121, 1992; linear C_6^- to linear C_6.
9. Stevens, A. E., Fiegerle, C. S., and Lineberger, W. C., *J. Chem. Phys.*, 78, 5420, 1983.
10. Breyer, F., Frey, P., and Hotop, H., *Z. Phys.*, A 300, 7, 1981.
11. Oakes, J. M., and Ellison, G. B., *Tetrahedron*, 42, 6263, 1986.
12. Leopold, D. G., Murray, K. K., Miller, A. E. S., and Lineberger, W. C., *J. Chem. Phys.*, 83, 4849, 1985.
13. Oakes, J. M., Jones, M.E., Bierbaum, V. M., and Ellison, G. B., *J. Chem. Phys.*, 87, 4810, 1983.
14. Ellis, H. B., Jr. and Ellison, G. B., *J. Chem. Phys.*, 78, 6541, 1983.
15. Oakes, J. M., Harding, L. B., and Ellison, G. B., *J. Chem. Phys.*, 83, 5400, 1985.
16. Nimlos, M. E., and Ellison, G. B., *J. Chem. Phys.*, 90, 2574, 1986.
17. Richardson, L. M., Stephenson, L. M., and Brauman, J. I., *Chem. Phys. Lett.*, 30, 17, 1975.
18. McHugh, K. M., Eaton, J. G., Lee, G. H., Sarkas, H. W., Kidder, L. H., Snodgrass, J. T., Manaa, M. R., and Bowen, K. H., *J. Chem. Phys.*, 91, 3792, 1989. See also Ref. 104.
19. Viggiano, A. A., Paulson, J. F., Dale, F., Henchman, M., Adams, N. G., and Smith, D., *J. Phys. Chem.*, 89, 2264, 1985.
20. Burnett, S. M., Stevens, A. E., Fiegerle, C. S., and Lineberger, W. C., *Chem. Phys. Lett.*, 100, 124, 1983.
21. Moran, S., Ellis, H. B., DeFrees, D. J., McLean, A. D., and Ellison, G. B., *J. Am. Chem. Soc.*, 109, 5996, 1987; Moran, S., Ellis, H. B., DeFrees, D. J., McLean, A. D., Paulson, S. E., and Ellison, G. B., *J. Am. Chem. Soc.*, 109, 6004, 1987; see also Lykke, K. R., Neumark, D. M., Andersen, T., Trapa, V. J., and Lineberger, W. C., *J. Chem. Phys.*, 87, 6842, 1987.
22. Mead, R. D., Lykke, K. R., Lineberger, W. C., Marks, J., and Brauman, J. I., *J. Chem. Phys.*, 81, 4883, 1984; Lykke, K. R., Mead, R. D., and Lineberger, W. C., *Phys. Rev. Lett.*, 52, 2221, 1984. The EAs are 14717.7 ± 1.0 cm^{-1} for actaldehyde enolate and 14671.0 ± 1.0 cm^{-1} for acetaldehyde-d$_3$ enolate.
23. Ellison, G. B., Engelking, P. C., and Lineberger, W. C., *J. Chem. Phys.*, 86, 4873, 1982.
24. Oakes, J. M., and Ellison, G. B., *J. Am. Chem. Soc.*, 105, 2969, 1983.
25. Ellison, G. B., and Oakes, J. M., *J. Am. Chem. Soc.*, 106, 7734, 1984. EA(allyl) and EA(allyl-d$_5$) are 0.119 and 0.083 eV too low, respectively, in this work, according to Reference 138. Therefore, EA(allyl-d$_1$) is likely too low by a similar amount.
26. Gunion, R. F., Gilles, M. K., Polak, M. L., and Lineberger, W. C., *Int. J. Mass Spectrom. Ion Processes*, 117, 601, 1992; also, Miller, A. E. S., and Lineberger, W. C., unpublished. See also Reference 136.

27. Leopold, D. G., and Lineberger, W. C., *J. Chem. Phys.*, 85, 51, 1986.
28. Miller, T. M., Miller, A. E. S., and Lineberger, W. C., *Phys. Rev. A*, 33, 3558, 1986.
29. Miller, A. E. S., Fiegerle, C. S., and Lineberger, W. C., *J. Chem. Phys.*, 87, 1549, 1987.
30. Miller, T. M., Leopold, D. G., Murray, K. K., and Lineberger, W. C., *J. Chem. Phys.*, 85, 2368, 1986.
31. Miller, T. M., and Lineberger, W. C., *Chem. Phys. Lett.*, 146, 364, 1988.
32. Neumark, D. M., Lykke, K. R., Andersen, T., and Lineberger, W. C., *J. Chem. Phys.*, 83, 4364, 1985.
33. Leopold, D. G., Miller, T. M., and Lineberger, W. C., *J. Am. Chem. Soc.*, 108, 178, 1986.
34. Miller, A. E. S., Fiegerle, C. S., and Lineberger, W. C., *J. Chem. Phys.*, 84, 4127, 1986.
35. Murray, K. K., Miller, T. M., Leopold, D. G., and Lineberger, W. C., *J. Chem. Phys.*, 84, 2520, 1986.
36. Leopold, D. G., Miller, A. E. S., and Lineberger, W. C., *J. Am. Chem. Soc.*, 108, 1379, 1986.
37. Ho, J., Ervin, K. M., and Lineberger, W. C., *J. Chem. Phys.*, 93, 6987, 1990; Leopold, D. G., Ho, J., and Lineberger, W. C., *J. Chem. Phys.*, 86, 1715, 1987; Pettiette, C. L., Yang, S. H., Craycraft, M. J., Conceicao, J., Laaksonen, R. T., Cheshnovsky, O., and Smalley, R. E., *J. Chem. Phys.*, 88, 5377, 1988.
38. Snodgrass, J. T., Coe, J. V., McHugh, K. M., Friedhoff, C. B., and Bowen, K. H., *J. Phys. Chem.*, 93, 1249, 1989.
39. Friedhoff, C. B., Snodgrass, J. T., Coe, J. V., McHugh, K. M., and Bowen, K. H., *J. Chem. Phys.*, 84, 1051, 1986.
40. Friedhoff, C. B., Coe, J. V., Snodgrass, J. T., McHugh, K. M., and Bowen, K. H., *Chem. Phys. Lett.*, 124, 268, 1986.
41. Coe, J. V., Snodgrass, J. T., Friedhoff, C. B., McHugh, K. M., and Bowen, K. H., *J. Chem. Phys.*, 84, 619, 1986.
42. Snodgrass, J. T., Coe, J. V., Friedhoff, C. B., McHugh, K. M., and Bowen, K. H., *Chem. Phys. Lett.*, 122, 352, 1985.
43. Nimlos, M. R., and Ellison, G. B., *J. Am. Chem. Soc.*, 108, 6522, 1986.
44. Peterson, J. R., *Aust. J. Phys.*, 45, 293, 1992. See also Nadeau, M.-J., Zhao, X.-L., Garwan, M. A., and Litherland, A. E., *Phys. Rev. A*, 46, R3588, 1992; also, Fabrikant, I. I., *J. Phys. B: At. Mol. Opt. Phys.*, 26, 2533, 1993.
45. Andersen, T., Lykke, K. R., Neumark, D. M., and Lineberger, W. C., *J. Chem. Phys.*, 86, 1858, 1987.
46. Murray, K. K., Lykke, K. R., and Lineberger, W. C., *Phys. Rev. A*, 36, 699, 1987.
47. Mansour, N. B., and Larson, D. J., *Abstracts of the XV Int. Conf. on the Phys. of Electronic and Atomic Collisions*, p. 70, 1987.
48. Stonemann, R. C., and Larson, D. J., *Phys. Rev. A*, 35, 2928, 1987.
49. Nimlos, M. R., Harding, L. B., and Ellison, G. B., *J. Chem. Phys.*, 87, 5116, 1987.
50. Moylan, C. R., Dodd, J. A., Han, C.-C., and Braumann, J. I., *J. Chem. Phys.*, 86, 5350, 1987.
51. Chowdhury, S., Grimsrud, E. P., Heinis, T., and Kebarle, P., *J. Am. Chem. Soc.*, 108, 3630, 1986.
52. Trainham, R., Fletcher, G. D., and Larson, D. J., *J. Phys. B: At. Mol. Phys.*, 20, L777, 1987.
53. Moran, S., and Ellison, G. B., *J. Phys. Chem.*, 92, 1794, 1988.
54. Murray, K. K., Leopold, D. G., Miller, T. M., and Lineberger, W. C., *J. Chem. Phys.*, 89, 5442, 1988.
55. Ervin, K. M., Ho, J., and Lineberger, W. C., *J. Chem. Phys.*, 89, 4514, 1988.
56. Grimsrud, E. P., Chowdhury, S., and Kebarle, P., *J. Chem. Phys.*, 85, 4989, 1985.
57. Fischer, C. F., *Phys. Rev. A*, 39, 963, 1989.
58. Wickham-Jones, C. T., Ervin, K. M., Ellision, G. B., and Lineberger, W. C., *J. Chem. Phys.*, 91, 2762, 1989.
59. Hopper, D. G., Wahl, A. C., Wu, R. L. C., and Tiernan, T. O., *J. Chem. Phys.*, 65, 5474, 1976.
60. Mock, R. S., and Grimsrud, E. P., *J. Am. Chem. Soc.*, 111, 2861, 1989.
61. Chowdhury, S., Heinis, T., Grimsrud, E. P., and Kebarle, P., *J. Phys. Chem.*, 90, 2747, 1986. The uncertainty and other results are quoted in Ref. 60.
62. Wickham-Jones, C. T., Moran, S., and Ellison, G. B., *J. Chem. Phys.*, 90, 795, 1989.
63. Ervin, K. M., Ho, J., and Lineberger, W. C., *J. Phys. Chem.*, 92, 5405, 1988.
64. Miller, T. M., and Lineberger, W. C., *Int. J. Mass Spectrom. Ion Processes*, 102, 239, 1990.
65. Wetzel, D. M., Salomon, K. E., Berger, S., and Brauman, J. I., *J. Am. Chem. Soc.*, 111, 3835, 1989.
66. Crocker, L., Wang, T., and Kebarle, P., *J. Am. Chem. Soc.*, 115, 7818, 1993.
67. Dillow, G. W., and Kebarle, P., *J. Am. Chem. Soc.*, 111, 5592, 1989.
68. Gantefor, G., Gausa, M., Meiwes-Broer, K. H., and Lutz, H. O., *Z. Phys. D*, 9, 253, 1988; Taylor, K. J., Petteitte, C. L., Craycraft, M. J., Chesnovsky, O., and Smalley, R. E., *Chem. Phys. Lett.*, 152, 347, 1988.
69. Metz, R. B., Kitsopoulos, T., Weaver, A., and Neumark, D. M., *J. Chem. Phys.*, 88, 1463, 1988.
70. Yang, S., Pettiette, C. L., Conceicao, J., Cheshnovsky, O., and Smalley, R. E., *Chem. Phys. Lett.*, 139, 233, 1987; Yang, S., Taylor, K. J., Craycraft, M. J., Conceicao, J., Pettiette, C. L., Cheshnovsky, O., and Smalley, R. E., *Chem. Phys. Lett.*, 144, 431, 1988; Arnold, D. W., Bradforth, S. E., Kitsopoulos, T. N., and Neumark, D. M., *J. Chem. Phys.*, 95, 5479, 1991.
71. Cheshnovsky, O., Yang, S., Pettiette, C. L., Craycraft, M. J., Liu, Y., and Smalley, R. E., *Chem. Phys. Lett.*, 138, 119, 1987.
73. Travers, M. J., Cowles, D. C., and Ellison, G. B., *Chem. Phys. Lett.*, 164, 449, 1989.
74. Blondel, C., Cacciani, P., Delsart, C., and Trainham, R., *Phys. Rev. A*, 40, 3698, 1989.
75. Bowen, K. H., and Eaton, J. G., in *The Structure of Small Molecules and Ions*, Naaman, R., and Vager, Z., Eds., Plenum, New York, 1988, pp. 147-169; Arnold, S. T., Eaton, J. G., Patel-Mistra, D., Sarkas, H. W., and Bowen, K. H., in *Ion and Cluster Ion Spectroscopy and Structure*, Maier, J. P., Ed., Elsevier Science, New York, 1989, p. 417.
76. Snodgrass, J. T., Coe, J. V., Friedhoff, C. B., McHugh, K. M., and Bowen, K. H., *Faraday Disc. Chem. Soc.*, 88, 1988.
77. Lee, G. H., Arnold, S. T., Eaton, J. G., Sarkas, H. W., Bowen, K. H., Ludewigt, C., and Haberland, H., *Z. Phys. D - Atoms, Mol. and Clusters*, 20, 9, 1991; Coe, J. V., Lee, G. H., Eaton, J. G., Arnold, S. T., Sarkas, H. W., Bowen, K. H., Ludewigt, C., Haberland, H., and Worsnop, D. R., *J. Chem. Phys.*, 92, 3980, 1990.
78. Snodgrass, J. T., Coe, J. V., Freidhoff, C. B., McHugh, K. M., and Bowen, K. H., to be published, quoted in Ref. 75.
79. Coe, J. V., Snodgrass, J. T., Friedhoff, C. B., McHugh, K. M., and Bowen, K. H., *J. Chem. Phys.*, 87, 4302, 1987.
80. Friedhoff, C. B., Snodgrass, J. T., and Bowen, K. H., to be published, quoted in Ref. 75.
81. Coe, J. V., Snodgrass, J. T., Friedhoff, C. B., McHugh, K. M., and Bowen, K. H., *Chem. Phys. Lett.*, 124, 274, 1986.
82. Greene, C. H., *Phys. Rev. A*, 42, 1405, 1990.
83. Ervin, K. M., Ho, J., and Lineberger, W. C., *J. Chem. Phys.*, 91, 5974, 1989.
84. Polak, M. L., Fiala, B. L., Ervin, K. M., and Lineberger, W. C., *J. Chem. Phys.*, 94, 6924, 1991.
85. Weaver, A., Arnold, D. W., Bradforth, S. E., Neumark, D. M., *J. Chem. Phys.*, 94, 1740, 1991.
86. Walter, C. W., Devynck, P., Hertzler, C. F., Bae, Y. K., Smith, G. P., and Peterson, J. R., *Bull. Am. Phys. Soc.*, 35, 1163, 1990.
87. Ervin, K. M., and Lineberger, W. C., *J. Phys. Chem.*, 95, 1167, 1991.

88. Gilles, M. K., Polak, M. L., and Lineberger, W. C., *J. Chem. Phys.*, 96, 8012, 1992.
89. Lykke, K. R., Murray, K. K., and Lineberger, W. C., *Phys. Rev. A*, 43, 6104, 1991.
90. Ervin, K. M., Gronert, S., Barlow, S. E., Gilles, M. K., Harrison, A. G., Bierbaum, V. M., DePuy, C. H., Lineberger, W. C., and Ellison, G. B., *J. Am. Chem. Soc.*, 112, 5750, 1990.
91. Viggiano, A. A., Miller, T. M., Miller, A. E. S., Morris, R. A., Van Doren, J. M., and Paulson, J. F., *Int. J. Mass Spectrom. Ion Processes*, 109, 327, 1991.
92. Hanstorp, D., and Gustafsson, M., *J. Phys. B: At. Mol. Opt. Phys.*, 25, 1773, 1992.
93. Polak, M. L., Gilles, M. K., and Lineberger, W. C., *J. Chem. Phys.*, 96, 7191, 1992.
94. Bengali, A. A., Casey, S. M., Cheng, C.-L., Dick, J. P., Fenn, P. T., Villalta, P. W., and Leopold, D. G., *J. Am. Chem. Soc.*, 114, 5257, 1992.
95. Gilles, M. K., Ervin, K. M., Ho, J., and Lineberger, W. C., *J. Phys. Chem.*, 96, 1130, 1992.
96. Travers, M. J., Cowles, D. C., Clifford, E. P., and Ellison, G. B., *J. Am. Chem. Soc.*, 114, 8699, 1992.
97. Bengali, A. A., and Leopold, D. G., *J. Am. Chem. Soc.*, 114, 9192, 1992.
98. Rudnyi, E. B., Kaibicheva, E. A., and Sidorov, L. N., *Rapid Comm. in Mass Spectrom.*, 6, 356, 1992; Sidorov, L. N., *High Temp. Sci.*, 29, 153, 1990.
99. Kitsopoulos, T. N., Chick, C. J., Zhao, Y., and Neumark, D. M., *J. Chem. Phys.*, 95, 5479, 1991.
100. Arnold, C. C., Kitsopoulos, T. N., and Neumark, D. M., *J. Chem. Phys.*, 99, 766, 1993.
101. Rudnyi, E. B., Kaibicheva, E. A., and Sidorov, L. N., *J. Chem. Thermodynamics*, 25, 929, 1993.
102. Sarkas, H. W., Hendricks, J. H., Arnold, S. T., and Bowen, K. H., *J. Chem. Phys.* 100, 1884, 1994.
103. Villalta, P. W., and Leopold, D. G., *J. Chem. Phys.* 98, 7730, 1993.
104. Eaton, J. G., Sarkas, H. W., Arnold, S. T., McHugh, K. M., and Bowen, K. H., *Chem. Phys. Lett.*, 193, 141, 1992. See also Ref. 18.
105. Polak, M. L., Gilles, M. K., Gunion, R. F., and Lineberger, W. C., *Chem. Phys. Lett.*, 210, 55, 1993.
106. Gilles, M. K., Lineberger, W. C., and Ervin, K. M., *J. Am. Chem. Soc.*, 115, 1031, 1993.
107. Casey, S. M., and Leopold, D. G., *Chem. Phys. Lett.*, 201, 205, 1993.
108. Polak, M. L., Gerber, G., Ho, J., and Lineberger, W. C., *J. Chem. Phys.*, 97, 8990, 1992.
109. Neumark, D. M., private communication.
110. Arnold, C. C., and Neumark, D. M., *J. Chem. Phys.*, 99, 3353, 1993.
111. Bradforth, S. E., Kim, E. H., Arnold, D. W., and Neumark, D. M., *J. Chem. Phys.*, 98, 800, 1993.
112. Ho, J., Polak, M. L., Ervin, K. M., and Lineberger, W. C., *J. Chem. Phys.*, 99, 8542, 1993.
113. Brinkman, E. A., Berger, S., Marks, J., and Brauman, J. I., *J. Chem. Phys.*, 99, 7586, 1993.
114. Casey, S. M., and Leopold, D. G., *J. Phys. Chem.*, 97, 816, 1993.
115. McDonald, R. N., and Davidson, S. J., *J. Am. Chem. Soc.*, 115, 10857, 1993.
116. Ho, J., Ervin, K. M., Polak, M. L., Gilles, M. K., and Lineberger, W. C., *J. Chem. Phys.*, 95, 4845, 1991.
117. Ho, J., Polak, M. L., and Lineberger, W. C., *J. Chem. Phys.*, 96, 144, 1992.
118. Polak, M. L., Gilles, M. K., Ho, J., and Lineberger, W. C., *J. Phys. Chem.*, 95, 3460, 1991.
119. Polak, M. L., Ho, J., Gerber, G., and Lineberger, W. C., *J. Chem. Phys.*, 95, 3053, 1991.
120. Sharpe, P., and Kebarle, P., *J. Am. Chem. Soc.*, 115, 782, 1993.
121. Miller, T. M., Miller, A. E. S., Viggiano, A. A., Morris, R. A., and Paulson, J. F., *J. Chem. Phys.*, 100, 7200, 1994.
122. Berkovits, D., Boaretto, E., Gehlberg, S., Heber, O., and Paul, M., *Phys. Rev. Lett.*, 75, 414, 1995.
123. Arnold, C. C., Xu, C., Burton, G. R., and Neumark, D. M., *J. Chem. Phys.*, 102, 6982, 1995. Burton, G. R., Xu, C., Arnold, C. C., and Neumark, D. M., *J. Chem. Phys.*, in press.
124. Xu, C., de Beer, E., and Neumark, D. M., *J. Chem. Phys.*, 104, 2749, 1996.
125. Gunion, R. F., Koppel, H., Leach, G. W., and Lineberger, W. C., *J. Chem. Phys.*, 103, 1250, 1995.
126. Gunion, R. F., and Lineberger, W. C., *J. Phys. Chem.*, 100, 4395, 1996.
127. Gunion, R. F., Dixon-Warren, St. J., Lineberger, W. C., and Morse, M. D., *J. Chem. Phys.*, in press.
128. Dixon-Warren, St. J., Gunion, R. F., and Lineberger, W. C., *J. Chem. Phys.*, in press.
129. Nakajima, A., Zhang, N., Kawamata, H., Hayase, T., Nakao, K., and Kaya, K., *Chem. Phys. Lett.*, 241, 295, 1995; Nakajima, A., Taguwa, T., Nakao, K., Hoshino, K., Iwata, S., and Kaya, K., *J. Chem. Phys.*, 102, 660, 1995.
130. Fan, J., and Wang, L.-S., *J. Chem. Phys.*, 102, 8714, 1995.
131. Arnold, D. W., Bradforth, S. E., Kim, E. H., and Neumark, D. M., *J. Chem. Phys.*, 102, 3493, 1995; Zhao, Y., Arnold, C. C., and Neumark, D. M., *J. Chem. Soc. Faraday Trans. 2*, 89, 1449, 1992.
132. Fan, J., Lou, L., and Wang, L.-S., *J. Chem. Phys.*, 102, 2701, 1995.
133. Sarkas, H. W., Hendricks, J. H., Arnold, S. T., Slager, V. L., and Bowen, K. H., *J. Chem. Phys.*, 100, 3358, 1994.
134. Wenthold, P. G., Hrovat, D. A., Borden, W. T., and Lineberger, W. C., *Science*, 272, 1456, 1996.
135. Wenthold, P. G., Hu, J., Squires, R. R., and Lineberger, W. C., *J. Am. Chem. Soc.*, 118, 475, 1996.
136. Gunion, R. F., Karney, W., Wenthold, P. G., Borden, W. T., and Lineberger, W. C., *J. Am. Chem. Soc.*, 118, 5074, 1996. The numbers in the abstract for 1,6-heptadiyne were misprinted. EA(cycloheptatrienide) quoted here derives from the LPES data combined with other thermochemical data in Reference 136.
137. Xu, C., de Beer, E., Arnold, D. W., Arnold, C. C., and Neumark, D. M., *J. Chem. Phys.*, 101, 5406, 1996.
138. Wenthold, P. G., Polak, M. L., and Lineberger, W. C., *J. Phys. Chem.*, 100, 6920, 1996.
139. Wenthold, P. G., Kim, J. B., and Lineberger, W. C., *J. Am. Chem. Soc.*, in press.

ATOMIC AND MOLECULAR POLARIZABILITIES
Thomas M. Miller

The *polarizability* of an atom or molecule describes the response of the electron cloud to an external field. The atomic or molecular energy shift ΔW due to an external electric field E is proportional to E^2 for external fields which are weak compared to the internal electric fields between the nucleus and electron cloud. The *electric dipole polarizability* α is the constant of proportionality defined by $\Delta W = -\alpha E^2/2$. The induced electric dipole moment is αE. *Hyperpolarizabilities*, coefficients of higher powers of E, are less often required. Technically, the polarizability is a tensor quantity but for spherically symmetric charge distributions reduces to a single number. In any case, an *average polarizability* is usually adequate in calculations. Frequency-dependent or *dynamic polarizabilities* are needed for electric fields which vary in time, except for frequencies which are much lower than electron orbital frequencies, where *static polarizabilities* suffice.

Polarizabilities for atoms and molecules in excited states are found to be larger than for ground states and may be positive or negative. Molecular polarizabilities are very slightly temperature dependent since the size of the molecule depends on its rovibrational state. Only in the case of dihydrogen has this effect been studied enough to warrant consideration in Table 3.

Polarizabilities are normally expressed in cgs units of cm^3. Ground state polarizabilities are in the range of $10^{-24} cm^3 = 1 Å^3$ and hence are often given in $Å^3$ units. Theorists tend to use atomic units of a_o^3 where a_o is the Bohr radius. The conversion is $\alpha(cm^3) = 0.148184 \times 10^{-24} \times \alpha(a_o^3)$. Polarizabilities are only recently encountered in SI units, $C \cdot m^2/V = J/(V/m)^2$. The conversion from cgs units to SI units is $\alpha(C \cdot m^2/V) = 4\pi\varepsilon_o \times 10^{-6} \times \alpha(cm^3)$, where ε_o is the permittivity of free space in SI units and the factor 10^{-6} simply converts cm^3 into m^3. Thus, $\alpha(C \cdot m^2/V) = 1.11265 \times 10^{-16} \times \alpha(cm^3)$. Persons measuring excited state polarizabilities by optical methods tend to use units of $MHz/(V/cm)^2$, where the energy shift, ΔW, is expressed in frequency units with a factor of h understood. The polarizability is $-2\Delta W/E^2$. The conversion into cgs units is $\alpha(cm^3) = 5.95531 \times 10^{-16} \times \alpha[MHz/(V/cm)^2]$.

The polarizability appears in many formulas for low-energy processes involving the valence electrons of atoms or molecules. These formulas are given below in cgs units: the polarizability α is in cm^3; masses m or μ are in grams; energies are in ergs; and electric charges are in esu, where $e = 4.8032 \times 10^{-10}$ esu. The symbol $\alpha(\nu)$ denotes a frequency (ν) dependent polarizability, where $\alpha(\nu)$ reduces to the static polarizability α for $\nu = 0$. For further information and references, see Miller, T. M., and Bederson, B., *Advances in Atomic and Molecular Physics*, 13, 1, 1977. Details on polarizability-related interactions, especially in regard to hyperpolarizabilities and nonlinear optical phenomena, are given by Bogaard, M. P., and Orr, B. J., in *Physical Chemistry*, Series Two, Vol. 2, *Molecular Structure and Properties*, Buckingham, A. D., Ed., Butterworths, London, 1975, pp. 149-194. A tabulation of tensor and hyperpolarizabilities is included. The gas number density, n, in Table 1 is usually taken to be that of 1 atm at 0°C in reporting experimental data.

Table 1
Formulas Involving Polarizability

Description	Formula	Remarks
Lorentz-Lorenz relation	$\alpha(\nu) = \dfrac{3}{4\pi n}\left[\dfrac{\eta^2(\nu)-1}{\eta^2(\nu)+2}\right]$	For a gas of atoms or nonpolar molecules; the index of refraction is $\eta(\nu)$
Refraction by polar molecules	$\alpha(\nu) + \dfrac{d^2}{3kT} = \dfrac{3}{4\pi n}\left[\dfrac{\eta^2(\nu)-1}{\eta^2(\nu)+2}\right]$	The dipole moment is d, in esu·cm ($= 10^{-18}$ D)
Dielectric constant (dimensionless)	$\kappa(\nu) = 1 + 4\pi n\ \alpha(\nu)$	From the Lorentz-Lorenz relation for the usual case of $\kappa(\upsilon) \approx 1$
Index of refraction (dimensionless)	$\eta(\nu) = 1 + 2\pi n\ \alpha(\nu)$	From $\eta^2(\nu) = \kappa(\nu)$
Diamagnetic susceptibility	$\chi_m = e^2\left(a_o N\alpha\right)^{1/2} / 4m_e c^2$	From the approximation that the static polarizability is given by the variational formula $\alpha = (4/9a_o)\Sigma(N_i r_i^2)^2$; N is the number of electrons, m_e is the electron mass; a crude approximation is $\chi_m = (E_i/4m_e c^2)\alpha$, where E_i is the ionization energy
Long-range electron- or ion-molecule interaction energy	$V(r) = -e^2\alpha / 2r^4$	The target molecule polarizability is α
Ion mobility in a gas	$\kappa = 13.87/(\alpha\mu)^{1/2} cm^2 / V \cdot s$	This one formula is not in cgs units. Enter α in $Å^3$ or $10^{-24} cm^3$ units and the reduced mass μ of the ion-molecule pair in amu. Classical limit; pure polarization potential

Table 1
Formulas Involving Polarizability (continued)

Description	Formula	Remarks
Langevin capture cross section	$\sigma(v_o) = (2\pi e/v_o)(\alpha/\mu)^{1/2}$	The relative velocity of approach for an ion-molecule pair is v_o; the target molecular polarizability is α and the reduced mass of the ion-molecule pair is μ
Langevin reaction rate coefficient	$k = 2\pi e(\alpha/\mu)^{1/2}$	Collisional rate coefficient for an ion-molecule reaction
Rate coefficient for polar molecules	$k_d = 2\pi e\left[(\alpha/\mu)^{1/2} + cd(2/\mu\pi kT)^{1/2}\right]$	The dipole moment of the neutral is d in esu·cm; the number c is a "locking factor" that depends on α and d, and is between 0 and 1
Modified effective range cross section for electron-neutral scattering	$\sigma(k) = 4\pi A^2$ $+32\pi^4\mu e^2\alpha Ak/3h^2$ $+\text{L}$	Here, k is the electron momentum divided by $h/2\pi$, where h is Planck's constant; A is called the "scattering length"; the reduced mass is μ
van der Waals constant between two systems A, B	$C_6 = \dfrac{3}{2}\left[\dfrac{\alpha^A\alpha^B E^A E^B}{E^A + E^B}\right]$	For the interaction potential term $V_6(r) = -C_6 r^{-6}$; $E^{A,B}$ represents average dipole transition energies and $\alpha^{A,B}$ the respective polarizabilities of A, B
Dipole-quadrupole constant between two systems A, B	$C_8 = \dfrac{15}{4}\left[\dfrac{\alpha^A\alpha_q^B E^A E_q^B}{E^A + E_q^B}\right]$ $+ \dfrac{15}{4}\left[\dfrac{\alpha_q^A\alpha^B E_q^A E^B}{E_q^A + E^B}\right]$	For the interaction potential term $V_8(r) = -C_8 r^{-8}$; $E_q^{A,B}$ represents average quadrupole transition energies and $\alpha_q^{A,B}$ are the respective quadrupole polarizabilities of A, B
van der Waals constant between an atom and a surface	$C_3 = \dfrac{\alpha g E^A E^S}{8(E^A + E^S)}$	For an interaction potential $V_3(r) = -C_3 r^3$; $E^{A,S}$ are characteristic energies of the atom and surface; $g = 1$ for a free-electron metal and $g = (\varepsilon_\infty - 1)/(\varepsilon_\infty + 1)$ for an ionic crystal
Relation between $\alpha(v)$ and oscillator strengths	$\alpha(v) = \dfrac{e^2 h^2}{4\pi^2 m_e}\Sigma\dfrac{f_k}{E_k^2 - (hv)^2}$	Here, f_k is the oscillator strength from the ground state to an excited state k, with excitation energy E_k. This formula is often used to estimate static polarizabilities ($v = 0$)
Dynamic polarizability	$\alpha(v) = \dfrac{\alpha E_r^2}{E_r^2 - (hv)^2}$	Approximate variation of the frequency-dependent polarizability $\alpha(v)$ from $v = 0$ up to the first dipole-allowed electronic transition, of energy E_r; the static dipole polarizability is $\alpha(0)$; infrared contributions ignored
Rayleigh scattering cross section	$\alpha(v) = \dfrac{8\pi}{9c^4}(2\pi v)^4$ $\times\left[3\alpha^2(v) + 2\gamma^2(v)/3\right]$	The photon frequency is v; the polarizability anisotropy (the difference between polarizabilities parallel and perpendicular to the molecular axis) is $\gamma(v)$
Verdet constant	$V(v) = \dfrac{vn}{2m_e c^2}\left[\dfrac{d\alpha(v)}{dv}\right]$	Defined from $\theta = V(v)B$, where θ is the angle of rotation of linearly polarized light through a medium of number density n, per unit length, for a longitudinal magnetic field strength B (Faraday effect)

Table 2
Static Average Electric Dipole Polarizabiilities for Ground State Atoms (in Units of 10^{-24} cm^3)

Atomic number	Atom	Polarizability	Estimated accuracy (%)	Method	Ref.
1	H	0.666793	"exact"	calc	MB77
2	He	0.204956	"exact"	calc	MB77
		0.2050	0.1	index/diel	NB65/OC67
3	Li	24.3	2	beam	MB77
4	Be	5.60	2	calc	MB77
5	B	3.03	2	calc	MB77
6	C	1.76	2	calc	MB77
7	N	1.10	2	calc/index	MB77
8	O	0.802	2	calc/index	MB77
9	F	0.557	2	calc	MB77
10	Ne	0.3956	0.1	diel	OC67
11	Na	24.08	0.4	interferom	ESCHP94
12	Mg	10.6	2	calc	MB77
13	Al	6.8	4.4	beam	MMD90
14	Si	5.38	2	calc	MB77
15	P	3.63	2	calc	MB77
16	S	2.90	2	calc	MB77
17	Cl	2.18	2	calc	MB77
18	Ar	1.6411	0.05	index/diel	NB65/OC67
19	K	43.4	2	beam	MB77
20	Ca	22.8	2	calc	MB77
		25.0	8	beam	MB77
21	Sc	17.8	25	calc	D84
22	Ti	14.6	25	calc	D84
23	V	12.4	25	calc	D84
24	Cr	11.6	25	calc	D84
25	Mn	9.4	25	calc	D84
26	Fe	8.4	25	calc	D84
27	Co	7.5	25	calc	D84
28	Ni	6.8	25	calc	D84
29	Cu	6.1	25	calc	D84
		7.31	25	calc	G84
30	Zn	7.1	25	calc	MB77
		5.6	25	calc	D84
31	Ga	8.12	2	calc	MB77
32	Ge	6.07	2	calc	MB77
33	As	4.31	2	calc	MB77
34	Se	3.77	2	calc	MB77
35	Br	3.05	2	calc	MB77
36	Kr	2.4844	0.05	diel	OC67
37	Rb	47.3	2	beam	MB77
38	Sr	27.6	8	beam	MB77
39	Y	22.7	25	calc	D84
40	Zr	17.9	25	calc	D84
41	Nb	15.7	25	calc	D84
42	Mo	12.8	25	calc	D84
43	Tc	11.4	25	calc	D84
44	Ru	9.6	25	calc	D84
45	Rh	8.6	25	calc	D84
46	Pd	4.8	25	calc	D84
47	Ag	7.2	25	calc	D84
		8.56	25	calc	G84
48	Cd	7.2	25	calc	D84
49	In	10.2	12	beam	GMBSJ84
		9.1	25	calc	D84

Table 2
Static Average Electric Dipole Polarizabiilities for Ground State Atoms (in Units of 10^{-24} cm^3)
(continued)

Atomic number	Atom	Polarizability	Estimated accuracy (%)	Method	Ref.
50	Sn	7.7	25	calc	D84
51	Sb	6.6	25	calc	D84
52	Te	5.5	25	calc	D84
53	I	5.35	25	index	A56
		4.7	25	calc	D84
54	Xe	4.044	0.5	diel	MB77
55	Cs	59.6	2	beam	MB77
56	Ba	39.7	8	beam	MB77
57	La	31.1	25	calc	D84
58	Ce	29.6	25	calc	D84
59	Pr	28.2	25	calc	D84
60	Nd	31.4	25	calc	D84
61	Pm	30.1	25	calc	D84
62	Sm	28.8	25	calc	D84
63	Eu	27.7	25	calc	D84
64	Gd	23.5	25	calc	D84
65	Tb	25.5	25	calc	D84
66	Dy	24.5	25	calc	D84
67	Ho	23.6	25	calc	D84
68	Er	22.7	25	calc	D84
69	Tm	21.8	25	calc	D84
70	Yb	21.0	25	calc	D84
71	Lu	21.9	25	calc	D84
72	Hf	16.2	25	calc	D84
73	Ta	13.1	25	calc	D84
74	W	11.1	25	calc	D84
75	Re	9.7	25	calc	D84
76	Os	8.5	25	calc	D84
77	Ir	7.6	25	calc	D84
78	Pt	6.5	25	calc	D84
79	Au	5.8	25	calc	D84
		6.48	25	calc	G84
80	Hg	5.7	25	calc	D84
		5.1	15	diel	MB77
81	Tl	7.6	15	beam	NYU84
		7.5	25	calc	D84
82	Pb	6.8	25	calc	D84
83	Bi	7.4	25	calc	D84
84	Po	6.8	25	calc	D84
85	At	6.0	25	calc	D84
86	Rn	5.3	25	calc	D84
87	Fr	48.7	25	calc	D84
88	Ra	38.3	25	calc	D84
89	Ac	32.1	25	calc	D84
90	Th	32.1	25	calc	D84
91	Pa	25.4	25	calc	D84
92	U	24.9	6	beam	KB94
93	Np	24.8	25	calc	D84
94	Pu	24.5	25	calc	D84
95	Am	23.3	25	calc	D84
96	Cm	23.0	25	calc	D84
97	Bk	22.7	25	calc	D84
98	Cf	20.5	25	calc	D84
99	Es	19.7	25	calc	D84

Table 2
Static Average Electric Dipole Polarizabiilities for Ground State Atoms (in Units of 10^{-24} cm^3)
(continued)

Atomic number	Atom	Polarizability	Estimated accuracy (%)	Method	Ref.
100	Fm	23.8	25	calc	D84
101	Md	18.2	25	calc	D84
102	No	17.5	25	calc	D84

Note: calc = calculated value; beam = atomic beam deflection technique; interferom = atomic beam interference; index = determination based on the measured index of refraction; diel = determination based on the measured dielectric constant.

REFERENCES

A56.   Atoji, M., *J. Chem. Phys.*, 25, 174, 1956. Semiempirical method based on molecular polarizabilities and atomic radii.

D84.   Doolen, G. D., Los Alamos National Laboratory, unpublished. A relativistic linear response method was used. The method is that described by Zangwill, A., and Soven, P., *Phys. Rev. A*, 21, 1561, 1980. Adjustments of less than 10% across the periodic table have been made to these results to bring them into agreement with accurate experimental values where avaialble, for the purpose of presenting "recommended" polarizabilities in Table 2.

ESCHP94. Ekstrom, C. R., Schmiedmayer, J., Chapman, M. S., Hammond, T. D., and Pritchard, D. E., *Phys. Rev. Lett.*, in press, 1994.

G84.   Gollisch, H., *J. Phys. B: Atom. Mol. Phys.*, 17, 1463, 1984. Other results and useful references are contained in this paper.

GMBSJ84. Guella, T. P., Miller, T. M., Bederson, B., Stockdale, J. A. D., and Jaduszliwer, B., *Phys. Rev. A*, 29, 2977, 1984.

KB94.   Kadar-Kallen, M. A., and Bonin, K. D., *Phys. Rev. Lett.*, 72, 828, 1994.

MB77.   Miller, T. M., and Bederson, B., *Adv. At. Mol. Phys.*, 13, 1, 1977. For simplicity, any value in Table 2 which has not changed since this 1977 review is referenced as MB77. Persons interested in original references and further details should consult MB77.

MMD90. Milani, P., Moullet, I., and de Heer, W. A., *Phys. Rev. A*, 42, 5150, 1990.

NB65.   Newell, A. C., and Baird, R. D., *J. Appl. Phys.*, 36, 3751, 1965.

NYU84. Preliminary value from the New York University group. See GMBSJ84.

OC67.   Orcutt, R. H., and Cole, R. H., *J. Chem. Phys.*, 46, 697, 1967; see also the later references from this group, given following the tables.

Table 3
Average Electric Dipole Polarizabilities for Ground State Diatomic Molecules (in Units of 10^{-24} cm^3)

Molecule	Polarizability	Ref.	Molecule	Polarizability	Ref.
Al$_2$	19	23	HI	5.44	3
BH	3.32*	1		5.35	2
Br$_2$	7.02	2	HgCl	7.4*	9
CO	1.95	3	ICl	12.3	2
Cl$_2$	4.61	3	K$_2$	77	22
Cs$_2$	104	22		72	21
CsK	89	22	Li$_2$	34	22
D$_2$ ($v = 0, J = 0$)	0.7921*	5	LiCl	3.46*	10
D$_2$ (293 K)	0.7954	6	LiF	10.8*	11
DCl	2.84	2	LiH	3.84*	12
F$_2$	1.38*	7		3.68*	13
H$_2$ ($v = 0, J = 0$)	0.8023*	5		3.88*	14
H$_2$ (293 K)	0.8045*	5	N$_2$	1.7403	6,8
H$_2$ (293 K)	0.8042	6	NO	1.70	2
H$_2$ (322 K)	0.8059	8	Na$_2$	40	22
HBr	3.61	3		38	21
HCl	2.63	3	NaK	51	22
	2.77	2	NaLi	40	4
HD ($v = 0, J = 0$)	0.7976*	5	O$_2$	1.5812	6
HF	0.80	27	Rb$_2$	79	22

Table 4
Average Electric Dipole Polarizabilities for Ground State Triatomic Molecules (in Units of 10^{-24} cm^3)

Molecule	Polarizability	Ref.	Molecule	Polarizability	Ref.	Molecule	Polarizability	Ref.
BeH_2	4.34*	14	HCN	2.59	3	O_3	3.21	2
CO_2	2.911	8		2.46	2	OCS	5.71	2
CS_2	8.74	3	$HgBr_2$	14.5	2		5.2	15
	8.86	2	$HgCl_2$	11.6	2	SO_2	3.72	3
D_2O	1.26	2	HgI_2	19.1	2		4.28	2
H_2O	1.45	2	N_2O	3.03	8			
H_2S	3.78	3	NO_2	3.02	2†			
	3.95	2	Na_3	70	21			

Table 5
Average Electric Dipole Polarizabilities for Ground State Inorganic Polyatomic Molecules (Larger than Triatomic) (in Units of 10^{-24} cm^3)

Molecule	Polarizability	Ref.	Molecule	Polarizability	Ref.
$AsCl_3$	14.9	2	$(NaBr)_2$	26.8	16
AsN_3	5.75	2	$(NaCl)_2$	23.4	16
BCl_3	9.38	20	$(NaF)_2$	20.7	16
BF_3	3.31	2	$(NaI)_2$	26.9	16
$(BN_3)_2$	5.73	2	OsO_4	8.17	2
$(BH_2N)_3$	8.0	2†	PCl_3	12.8	2
ClF_3	6.32	2	PF_5	6.10	2
$(CsBr)_2$	54.5	16	PH_3	4.84	2
$(CsCl)_2$	42.4	16	$(RbBr)_2$	48.2	16
$(CsF)_2$	28.4	16	$(RbCl)_2$	43.2	16
$(CsI)_2$	51.8	16	$(RbF)_2$	40.7	16
$GeCl_4$	15.1	2	$(RbI)_2$	46.3	16
GeH_3Cl	6.7	2†	SF_6	6.54	8
$(HgCl)_2$	14.7	9	$(SF_5)_2$	13.2	2
K_n	n = 2,5,7-9,11,20	21	SO_3	4.84	2
$(KBr)_2$	42.0	16	SO_2Cl_2	10.5	2
$(KCl)_2$	32.1	16	SeF_6	7.33	2
$(KF)_2$	21.0	16	SiF_4	5.45	2
$(KI)_2$	36.3	16	SiH_4	5.44	2
$(LiBr)_2$	18.9	16	$(SiH_3)_2$	11.1	2
$(LiCl)_2$	13.1	16	$SiHCl_3$	10.7	2
$(LiF)_2$	6.9	16	SiH_2Cl_2	8.92	2
$(LiI)_2$	23.4	16	SiH_3Cl	7.02	2
ND_3	1.70	2	$SnBr_4$	22.0	2
NF_3	3.62	2	$SnCl_4$	18.0	2
NH_3	2.81	20		13.8	15
	2.10	2	SnI_4	32.3	2
	2.26	3	TeF_6	9.00	2
$(NO_2)_2$	6.69	2	$TiCl_4$	16.4	2
Na_n	n = 1-40	21	UF_6	12.5	2

Table 6
Average Electric Dipole Polarizabilities for Ground State Hydrocarbon Molecules (in Units of 10^{-24} cm^3)

Molecule	Name	Polarizability	Ref.	Molecule	Name	Polarizability	Ref.
CH_4	methane	2.593	8	C_8H_{10}	ethylbenzene	14.2	2
C_2H_2	acetylene	3.33	3		o-xylene	14.9	2
		3.93	2			14.1	15
C_2H_4	ethylene	4.252	8		p-xylene	13.7	25
C_2H_6	ethane	4.47	3			14.2	15
		4.43	2			14.9	2
C_3H_4	propyne	6.18	2		m-xylene	14.2	15
C_3H_6	propene	6.26	2	C_8H_{16}	ethylcyclohexane	15.9	2
	cyclopropane	5.66	2	C_8H_{18}	n-octane	15.9	2
C_3H_8	propane	6.29	3		3-methylheptane	15.4	27
		6.37	2		2,2,4-trimethylpentane	15.4	27
C_4H_6	1-butyne	7.41	2†	C_9H_{10}	α-methylstyrene	16.05	27
	1,3-butadiene	8.64	2	C_9H_{12}	isopropylbenzene	16.0	2-
C_4H_8	1-butene	7.97	2		mesitylene	15.5	25
		8.52	2			16.1	27
	trans-2-butene	8.49	2	C_9H_{18}	isopropylcyclohexane	17.2	2
	2-methylpropene	8.29	2	C_9H_{20}	nonane	17.4	27
C_4H_{10}	n-butane	8.20	2	$C_{10}H_8$	napthalene	16.5	17
	isobutane	8.14	27			17.5	27
C_5H_6	1,3-cyclopentadiene	8.64	2	$C_{10}H_{14}$	durene	17.3	25
C_5H_8	1-pentyne	9.12	2		t-butylbenzene	17.2	25
	trans-1,3-pentadiene	10.0	2			17 8	2†
	isoprene	9.99	2	$C_{10}H_{20}$	t-butylcyclohexane	19.8	2
C_5H_{10}	cyclopentane	9.15	18	$C_{10}H_{22}$	decane	19.1	27
	1-pentene	9.65	27	$C_{11}H_{10}$	α-methylnaphthalene	19.35	27
	2-pentene	9.84	27		β-methylnaphthalene	19.52	27
C_5H_{12}	pentane	9.99	2	$C_{11}H_{14}$	α,β,β-trimethylstyrene	19.64	27
	neopentane	10.20	18	$C_{11}H_{16}$	pentamethylbenzene	19.1	25
C_6H_6	benzene	10.0	25	$C_{11}H_{24}$	undecane	21.0	27
		10.32	3	$C_{12}H_{10}$	acenaphthene	20.6	27
		10.74	2	$C_{12}H_{12}$	α-ethylnaphthalene	21.19	27
C_6H_{10}	1-hexyne	10.9	2†		β-ethylnaphthalene	21.36	27
	2-ethyl-1,3-butadiene	11.8	2†	$C_{12}H_{18}$	hexamethylbenzene	20.9	25
	3-methyl-1,3-pentadiene	11.8	2†	$C_{12}H_{26}$	dodecane	22.8	27
	2-methyl-1,3-pentadiene	12.1	2†	$C_{13}H_{10}$	fluorene	21.7	27
	2,3-dimethyl-1,3-butadiene	11.8	2†	$C_{14}H_{10}$	anthracene	25.4	17
	cyclohexene	10.7	2†			25.9	27
C_6H_{12}	cyclohexane	11.0	18		phenanthrene	36.8*	17
		10.87	15			24.7	27
	1-hexene	11.7	27	$C_{14}H_{22}$	p-di-t-butylbenzene	24.5	25
C_6H_{14}	n-hexane	11.9	2	$C_{16}H_{10}$	pyrene	28.2	27
C_7H_8	toluene	11.8	25	$C_{17}H_{12}$	2,3-benzfluorene	30.2	27
		12.26	15	$C_{18}H_{12}$	napthacene	32.3	27
		12.3	2		1,2-benzanthracene	32.9	27
C_7H_{12}	1-heptyne	12.8	2†		chrysene	33.1	27
C_7H_{14}	methylcyclohexane	13.1	2		triphenylene	31.1	27
	1-heptene	13.5	27	$C_{18}H_{30}$	1,3,5-tri-t-butylbenzene	31.8	25
C_7H_{16}	n-heptane	13.7	2	$C_{24}H_{12}$	coronene		
C_8H_8	styrene	15.0	2				
		14.4	27				

Table 7
Average Electric Dipole Polarizabilities for Ground State Organic Halides (in Units of 10^{-24} cm^3)

Molecule	Name	Polarizability	Ref.
CBr_2F_2	dibromodifluoromethane	9.0	2[†]
$CClF_3$	chlorotrifluoromethane	5.72	20
		5.59	2
CCl_2F_2	dichlorodifluoromethane	7.93	20
		7.81	2
CCl_2O	phosgene	7.29	2
CCl_2S	thiophosgene	10.2	2
CCl_3F	trichlorofluoromethane	9.47	2
CCl_3NO_2	trichloronitromethane	10.8	2[†]
CCl_4	carbon tetrachloride	11.2	2
		10.5	3
CF_4	carbon tetrafluoride	3.838	8
CF_2O	carbonylfluoride	1.88[*]	17
$CHBr_3$	bromoform	11.8	17
$CHBrF_2$	bromodifluoromethane	5.7	2[†]
$CHClF_2$	chlorodifluoromethane	6.38	20
		5.91	2
$CHCl_2F$	dichlorofluoromethane	6.82	2
$CHCl_3$	chloroform	9.5	8
		8.23	27
CHF_3	fluoroform	3.52	20
		3.57	8
$CHFO$	fluoroformaldehyde	1.76[*]	17
CHI_3	iodoform	18.0	17
CH_2Br_2	dibromomethane	9.32	2
		8.62	27
CH_2ClNO_2	chloronitromethane	6.9	2[†]
CH_2Cl_2	dichloromethane	6.48	3
		7.93	2
CH_2I_2	diiodomethane	12.90	27
CH_3Br	bromomethane	5.87	20
		6.03	2
		5.55	15
CH_3Cl	chloromethane	5.35	20
		4.72	8
CH_3F	fluoromethane	2.97	8
CH_3I	iodomethane	7.97	2
C_2ClF_5	chloropentafluoroethane	6.3	2[†]
$C_2Cl_2F_4$	1,2-dichlorotetrafluoroethane	8.5	2[†]
C_2Cl_3N	trichloroacetonitrile	6.10	18
C_2F_6	hexafluoroethane	6.82	2
C_2HBr	bromoacetylene	7.39	2
C_2HCl	chloroacetylene	6.07	2
C_2HCl_5	pentachloroethane	14.0	2
$C_2H_2Cl_2$	dichloroethylene	10.03	27
	trans-dichloroethylene	9.28	27
	cis-dichloroethylene	9.19	27
$C_2H_2Cl_2F_2$	1,1-dichloro-2,2-difluoroethane	8.4	2[†]
$C_2H_2Cl_2O$	chloroacetyl chloride	8.92	2
$C_2H_2Cl_3F$	1,2,2-trichloro-1-fluoroethane	10.2	2[†]
$C_2H_2Cl_4$	1,1,2,2-tetrachloroethane	12.1	2[†]
C_2H_2ClN	chloroacetonitrile	6.10	18
$C_2H_2F_2$	1,1-difluoroethylene	5.01	20
C_2H_3Br	bromoethylene	7.59	2
C_2H_3Cl	chloroethylene	6.41	2

Table 7
Average Electric Dipole Polarizabilities for Ground State Organic Halides (in Units of 10^{-24} cm^3) (continued)

Molecule	Name	Polarizability	Ref.
$C_2H_3ClF_2$	1-chloro-1,1-difluoroethane	8.05	2
C_2H_3ClO	acetyl chloride	6.62	2
$C_2H_3ClO_2$	ethyl chloroformate	7.1	2[†]
$C_2H_3Cl_3$	1,1,1-trichloroethane	10.7	2
$C_2H_3F_3$	1,1,1-trifluoroethane	4.4	2[†]
C_2H_3I	iodoethylene	9.3	2[†]
C_2H_4BrCl	1-bromo-2-chloroethane	9.5	2[†]
$C_2H_4Br_2$	1,2-dibromomethane	10.7	2[†]
C_2H_4ClF	1-chloro-2-fluoroethane	6.5	2[†]
$C_2H_4ClNO_2$	1-chloro-1-nitroethane	10.9	2
$C_2H_4Cl_2$	1,1-dichloroethane	8.64	2
	1,2-dichloroethane	8.0	2[†]
C_2H_5Br	bromoethane	8.05	2
		7.28	27
C_2H_5Cl	chloroethane	7.27	20
		8.29	2
		6.4	15
C_2H_5ClO	2-chloroethanol	7.1	2[†]
		6.88	27
	chloromethyl methyl ether	7.1	2[†]
C_2H_5F	fluoroethane	4.96	2
C_2H_5I	iodoethane	10.0	2
$C_3H_4Cl_2$	dichloropropene	10.1	2[†]
C_3H_5Cl	chloropropene	8.3	2
C_3H_5ClO	chloroacetone	8.4	2[†]
$C_3H_3ClO_2$	ethyl chloroformate	9.0	2[†]
$C_3H_6ClNO_2$	1-chloro-1-nitropropane	10.4	2[†]
$C_3H_6Cl_2$	dichloropropane	10.9	2[†]
C_3H_7Br	1-bromopropane	9.4	2[†]
		9.07	27
	2-bromopropane	9.6	2[†]
C_3H_7Cl	chloropropane	10.0	2
C_3H_7ClO	β-chloroethyl methyl ether	8.71	27
	2-chloro-1-propanol	8.89	27
	3-chloro-1-propanol	8.84	27
C_3H_7I	1-iodopropane	11.5	2[†]
C_4H_5Cl	4-chloro-1,2-butadiene	10.0	2[†]
C_4H_7Cl	1-chloro-2-methylpropene	10.8	2
$C_4H_7ClO_2$	2-chlorobutyric acid	10.7	27
	3-chlorobutyric acid	10.7	27
	4-chlorobutyric acid	10.6	27
$C_4H_8Cl_2$	1,4-dichlorobutane	12.0	2[†]
C_4H_9Br	bromobutane	13.9	2
		10.86	27
C_4H_9Cl	1-chlorobutane	11.3	2
	1-chloro-2-methylpropane	11.1	2
	2-chloro-2-methylpropane	12.5	2[†]
	2-chlorobutane	12.4	2
C_4H_9ClO	β-chloroethyl ethyl ether	10.56	27
	2-chloro-1-butanol	10.70	27
	3-chloro-1-butanol	10.38	27
C_4H_9I	1-iodobutane	13.3	2[†]
		12.65	27
$C_5H_9ClO_2$	2-chlorobutanoate acid	12.33	27
	3-chlorobutanoate acid	12.31	27

Table 7
Average Electric Dipole Polarizabilities for Ground State Organic Halides (in Units of 10^{-24} cm^3) (continued)

Molecule	Name	Polarizability	Ref.
	4-chlorobutanoate acid	12.27	27
	2-chloropentanoic acid	12.69	27
	3-chloropentanoic acid	12.57	27
	4-chloropentanoic acid	12.53	27
$C_5H_{11}Br$	1-bromopentane	13.1	2[†]
$C_5H_{11}Cl$	1-chloropentane	12.0	2[†]
$C_5H_{11}F$	fluoropentane	9.95	27
C_6F_6	hexafluorobenzene	9.58	27
C_6HF_5	pentafluorobenzene	9.63	27
$C_6H_2Cl_2O_2$	2,5-dichloro-1,4-benzoquinone	18.4	2
$C_6H_2F_4$	1,2,3,4-tetrafluorobenzene	9.69	27
	1,2,4,5-tetrafluorobenzene	9.69	27
$C_6H_3F_3$	1,3,5-trifluorobenzene	9.74	27
C_6H_4BrF	p-bromofluorobenzene	13.4	2[†]
$C_6H_4ClNO_2$	chloronitrobenzene	14.6	2[†]
$C_6H_4Cl_2$	o-dichlorobenzene	14.17	27
	m-dichlorobenzene	14.23	27
	p-dichlorobenzene	14.20	27
C_6H_4FI	p-fluoroiodobenzene	15.5	2[†]
$C_6H_4FNO_2$	p-fluoronitrobenzene	12.8	2[†]
$C_6H_4F_2$	o-difluorobenzene	9.80	27
	m-difluorobenzene	10.3	2[†]
	p-difluorobenzene	9.80	27
C_6H_5Br	bromobenzene	14.7	2
		13.62	27
C_6H_5Cl	chlorobenzene	14.1	2
		12.3	15
C_6H_5ClO	chlorophenol	13.0	2[†]
C_6H_5F	fluorobenzene	10.3	2
C_6H_5I	iodobenzene	15.5	2[†]
$C_6H_{11}ClO_2$	2-chlorobutanoate acid	14.16	27
	3-chlorobutanoate acid	14.13	27
	4-chlorobutanoate acid	14.11	27
$C_6H_{13}Br$	bromohexane	14.44	27
$C_6H_{13}F$	fluorohexane	11.80	27
C_7H_7Br	p-bromotoluene	14.80	27
C_7H_7Cl	p-chlorotoluene	13.70	27
C_7H_7F	p-fluorotoluene	11.70	27
C_7H_7I	p-iodotoluene	17.10	27
$C_7H_{15}Br$	1-bromoheptane	16.8	2[†]
		16.23	27
$C_7H_{15}F$	fluoroheptane	13.66	27
$C_8H_{17}Br$	bromooctane	18.02	27
$C_8H_{17}F$	fluorooctane	15.46	27
$C_9H_{19}Br$	bromononane	19.81	27
$C_9H_{19}F$	fluorononane	17.34	27
$C_{10}F_8$	octafluoronaphthalene	17.64	27
$C_{10}H_7Br$	α-bromonaphthalene	20.34	27
$C_{10}H_7Cl$	α-chloronaphthalene	19.30	27
	β-chloronaphthalene	19.58	27
$C_{10}H_7I$	α-iodonaphthalene	22.41	27
	β-iodonaphthalene	22.95	27
$C_{10}H_{21}Br$	bromodecane	21.60	27
$C_{10}H_{21}F$	fluorodecane	19.18	27

Table 7
Average Electric Dipole Polarizabilities for Ground State Organic Halides (in Units of 10^{-24} cm^3) (continued)

Molecule	Name	Polarizability	Ref.
$C_{11}H_{23}F$	fluoroundecane	21.00	27
$C_{12}H_{25}Br$	bromododecane	25.18	27
$C_{12}H_{25}F$	fluorododecane	22.83	27
$C_{12}H_8Br_2O$	4,4′–dibromodiphenyl ether	27.8	2[†]
$C_{12}H_9BrO$	4-bromodiphenyl ether	24.2	2[†]
$C_{13}H_{11}BrO$	p-bromophenyl-p-tolyl ether	26.6	2[†]
$C_{14}H_9Br$	9-bromoanthracene	28.32	27
$C_{14}H_9Cl$	9-chloroanthracene	27.35	27
$C_{14}H_9F$	fluoranthracene	28.34	27
$C_{14}H_{29}F$	fluorotetradecane	26.57	27
$C_{16}H_{33}Br$	bromohexadecane	32.34	27
$C_{18}H_{37}Br$	bromooctadecane	35.92	27

Table 8
Static Average Electric Dipole Polarizabilities for Other Ground State Organic Molecules (in Units of 10^{-24} cm^3)

Molecule	Name	Polarizability	Ref.
CN_4O_8	tetranitromethane	15.3	2
CH_2O	formaldehyde	2.8	2[†]
		2.45	18
CH_2O_2	formic acid	3.4	2[†]
CH_3NO	formamide	4.2	2[†]
		4.08	18
CH_3NO_2	nitromethane	7.37	2
CH_4O	methanol	3.29	2
		3.23	15
		3.32	18
CH_5N	methyl amine	4.7	2
		4.01	19
C_2N_2	cyanogen	7.99	2
C_2H_2O	ketene	4.4	2[†]
C_2H_3N	acetonitrile	4.40	2[†]
		4.48	18
C_2H_4O	acetaldehyde	4.6	2[†]
		4.59	18
	ethylene oxide	4.43	18
$C_2H_4O_2$	acetic acid	5.1	2[†]
	methyl formate	5.05	27
$C_2H_4O_4$	formic acid dimer	12.7	2
C_2H_5NO	acetamide	5.67	18
	N-methyl formamide	5.91	18
$C_2H_5NO_2$	nitroethane	9.63	2
	ethyl nitrite	7.0	15
C_2H_6O	ethanol	5.41	2
		5.11	18
	methyl ether	5.29	20
		5.84	2
		5.16	15
$C_2H_6O_2$	ethylene glycol	5.7	2[†]
		5.61	27
$C_2H_6O_2S$	dimethyl sulfone	7.3	2[†]

Table 8
Static Average Electric Dipole Polarizabilities for Other Ground State
Organic Molecules (in Units of 10^{-24} cm^3) (continued)

Molecule	Name	Polarizability	Ref.
C_2H_6S	ethanethiol	7.41	2
C_2H_7N	ethyl amine	7.10	2
	dimethyl amine	6.37	2
$C_2H_8N_2$	ethylene diamine	7.2	2†
$C_3H_2N_2$	malononitrile	5.79	18
C_3H_3N	acrylonitrile	8.05	2
$C_3H_4N_2$	pyrazole	7.23	27
C_3H_4O	propenal	6.38	2†
C_3H_5N	propionitrile	6.70	2
		6.24	18
C_3H_6O	acetone	6.33	15
		6.4	2†
		6.39	18
	allyl alcohol	7.65	2
	propionaldehyde	6.50	2
$C_3H_6O_2$	propionic acid	6.9	2†
	ethyl formate	8.01	2
		6.88	27
	methyl acetate	6.94	2
		6.81	27
$C_3H_6O_3$	dimethyl carbonate	7.7	2†
C_3H_7NO	N-methyl acetamide	7.82	18
	N,N-dimethyl formamide	7.81	18
$C_3H_7NO_2$	nitropropane	8.5	2†
C_3H_8O	2-propanol	7.61	2
		6.97	18
	1-propanol	6.74	2
	ethyl methyl ether	7.93	2
$C_3H_8O_2$	dimethoxymethane	7.7	2†
	monomethylether ethylene glycol	7.44	27
C_3H_9N	n-propylamine	7.70	27
		9.20	2
	isopropylamine	7.77	27
	trimethylamine	8.15	2
$C_4H_2N_2$	fumaronitrile	11.8	2
$C_4H_4N_2$	succinonitrile	8.1	2†
	pyrimidene	8.53*	17
	pyridazine	9.27*	17
$C_4H_4O_2$	diketene	8.0	2†
C_4H_4S	thiophene	9.67	2
C_4H_5N	methacrylonitrile	8.0	2†
	trans-crotononitrile	8.2	2†
$C_4H_6N_2$	N-methylpyrazole	8.99	27
C_4H_6O	crotonaldehyde	8.5	2†
	methacrylaldehyde	8.3	2†
$C_4H_6O_2$	biacetyl	8.2	2†
$C_4H_6O_3$	acetic anhydride	8.9	2†
C_4H_6S	divinyl sulfide	10.9	2†
C_4H_7N	butyronitrile	8.4	2†
	isobutyronitrile	8.05	18
C_4H_8O	butanal	8.2	2†
	methyl ethyl ketone	8.13	15
	trans-2,3-epoxy butane	8.22*	17
$C_4H_8O_2$	ethyl acetate	9.7	2
		8.62	27

Table 8
Static Average Electric Dipole Polarizabilities for Other Ground State Organic Molecules (in Units of 10^{-24} cm^3) (continued)

Molecule	Name	Polarizability	Ref.
	1,4-dioxane	10.0	2
	p-dioxane	8.60	18
	2-methyl-1,3-dioxolane	9.44	15
	butyric acid	8.38	27
	methyl propionate	8.97	27
$C_4H_9NO_2$	1-nitrobutane	10.4	2[†]
	2-methyl-2-nitropropane	10.3	2[†]
$C_4H_{10}O$	ethyl ether	10.2	2
		8.73	15
	1-butanol	8.88	2
	2-methylpropanol	8.92	2
	n-propyl methyl ether	8.86	27
$C_4H_{10}O_2$	monoethylether ethylene glycol	9.28	27
$C_4H_{10}S$	ethyl sulfide	10.8	2
$C_4H_{11}N$	n-butylamine	13.5	2
	diethylamine	10.2	2
		9.61	27
C_5H_5N	pyridine	9.5	15
		9.18	27
	4-cyano-1,2-butadiene	10.5	2[†]
$C_5H_8N_2$	1,5-dimethylpyrazole	10.72	27
$C_5H_8O_2$	acetyl acetone	10.5	2[†]
C_5H_9N	valeronitrile	10.4	2
	22-DMPN	9.59	18
$C_5H_{10}O$	diethyl ketone	9.93	15
	methyl propyl ketone	9.93	15
$C_5H_{10}O_2$	ethyl propionate	10.41	27
	methyl butanoate	10.41	27
$C_5H_{10}O_3$	diethyl carbonate	11.3	2
$C_5H_{12}O$	ethyl propyl ether	10.68	27
$C_5H_{12}O_4$	tetramethyl orthocarbonate	13.0	2[†]
$C_6H_4N_2O_4$	p-dinitrobenzene	18.4	2
$C_6H_4O_2$	p-benzoquinone	14.5	2
$C_6H_5NO_2$	nitrobenzene	14.7	2
		12.92	15
C_6H_6O	phenol	11.1	2[†]
		9.94*	17
C_6H_7N	aniline	12.1	2[†]
$C_6H_8N_2$	phenylenediamine	13.8	2[†]
	phenylhydrazine	12.91	27
$C_6H_{10}N_2$	1-ethyl-5-methylpyrazole	12.50	27
$C_6H_{10}O_3$	ethyl acetoacetate	12.9	2[†]
$C_6H_{12}N_2$	dimethylketazine	15.6	2
$C_6H_{12}O$	cyclohexanol	11.56	18
$C_6H_{12}O_2$	amyl formate	14.2	2
$C_6H_{12}O_3$	paraldehyde	17.9	2
$C_6H_{14}O$	propyl ether	12.8	2
		12.5	15
$C_6H_{14}O_2$	1,1-diethoxyethane	13.2	2[†]
	1,2-diethoxyethane	11.3	2[†]
$C_6H_{15}N$	triethylamine	13.1	2
		13.38	27
	di-n-propylamine	13.29	27
$C_7H_4N_2O_2$	p-cyanonitrobenzene	19.0	2
C_7H_5N	benzonitrile	12.5	2[†]

Table 8
Static Average Electric Dipole Polarizabilities for Other Ground State Organic Molecules (in Units of 10^{-24} cm^3) (continued)

Molecule	Name	Polarizability	Ref.
$C_7H_7NO_3$	nitroanisole	15.7	2[†]
C_7H_7O	anisole	13.1	2[†]
C_7H_9NO	o-anisidine	14.2	2[†]
$C_7H_{10}N_2$	1,1-methylphenylhydrazine	14.81	27
$C_7H_{14}O$	cyclohexyl methyl ether	13.4	2[†]
	2,4-dimethyl-3-pentanone	13.5	15
$C_7H_{14}O_2$	pentyl acetate	14.9	2
$C_8H_4N_2$	p-dicyanobenzene	19.2	2
$C_8H_6O_2$	quinoxaline	15.13	27
C_8H_8O	acetophenone	15.0	2
$C_8H_8O_2$	2,5-dimethyl-1,4-benzoquinone	18.8	2
$C_8H_{10}O$	phenetole	14.9	2
$C_8H_{11}N$	N-dimethylaniline	16.2	2[†]
$C_8H_{12}N_2$	1,1-ethylphenylhydrazine	16.62	27
$C_8H_{12}O_2$	ethyl sorbate	17.2	2[†]
	tetramethylcyclobutane-1,3-dione	18.6	2
$C_8H_{14}O_4$	diethyl succinate	16.8	2[†]
$C_8H_{18}O$	n-butyl ether	17.2	2
C_9H_7N	quinoline	15.70	27
	isoquinoline	16.43	27
$C_9H_{10}O_2$	ethyl benzoate	16.9	2[†]
$C_9H_{21}N$	tri-n-propylamine	18.87	27
$C_{10}H_9N$	α-naphthylamine	19.50	27
	β-naphthylamine	19.73	27
	1-methylquinoline	18.65	27
	1-methylisoquinoline	18.28	27
$C_{10}H_{10}Fe$	ferrocene	17.1	26
$C_{10}H_{10}N_2$	2,3-dimethylquinoxaline	18.70	27
$C_{10}H_{14}BeO_4$	beryllium acetylacetonate	34.1	2
$C_{11}H_8O$	1-naphthaldehyde	19.75	27
	2-naphthaldehyde	20.06	27
$C_{14}H_8O_2$	anthraquinone	24.46	27
$C_{12}H_8N_2$	phenazine	23.43	27
$C_{12}H_9NO_3$	4-nitrodiphenyl ether	24.7	2[†]
$C_{14}H_{14}O$	di-p-tolyl ether	24.9	2[†]
$C_{15}H_{21}AlO_6$	aluminum acetylacetonate	51.9	2
$C_{15}H_{21}CrO_6$	chromium acetylacetonate	53.7	2
$C_{15}H_{21}FeO_6$	ferric acetylacetonate	58.1	2
$C_{20}H_{28}O_8Th$	thorium acetylacetonate	79.0	2
C_{60}	buckminsterfullerene	~80	24

Note: All polarizabilities in the tables are experimental values except those values marked by an asterisk (*), which indicates a calculated result. The experimental polarizabilities are mostly determined by measurements of a dielectric constant or refractive index which are quite accurate (0.5% or better). However, one should treat many of the results with several percent of caution because of the age of the data and because some of the results refer to optical frequencies rather than static. Comments given with the references are intended to allow one to judge the degree of caution required. Interested persons should consult these references. In many cases, the reference given is to a theoretical paper in which the experimental results are quoted. These papers, noted in the References, contain valuable information on polarizability calculations and experimental data which often includes the tensor components of the polarizability.

REFERENCES

1. McCullough, E. A., Jr., *J. Chem. Phys.*, 63, 5050, 1975. This calculation is for the parallel component, not the average polarizability.
2. Maryott, A. A., and Buckley, F., U. S. National Bureau of Standards Circular No. 537, 1953. A tabulation of dipole moments, dielectric constants, and molar refractions measured between 1910 and 1952, and used here to determine polarizabilities if no more recent result exists. The polarizability is $3/(4\pi N_o)$ times the molar polarization or molar refraction, where N_o is Avogadro's number. The value $3/(4\pi N_o) = 0.3964308 \times 10^{-24}$ cm^3 was used for this conversion. A dagger (†) following the reference number in the tables indicates that the polarizability was derived from the molar refraction and hence may not include some low-frequency contributions to the static polarizability; these "static" polarizabilities are therefore low by 1 to 30%.
3. Hirschfelder, J. O., Curtis, C. F., and Bird, R. B., *Molecular Theory of Gases and Liquids*, Wiley, New York, 1954, p. 950. Fundamental information on molecular polarizabilities.
4. Miller, T. M., and Bederson, B., *Adv. At. Mol. Phys.*, 13, 1, 1977. Review emphasizing atomic polarizabilities and measurement techniques. The data quoted in Table 3 are accurate to 8 to 12%.
5. Kolos, W., and Wolniewicz, L., *J. Chem. Phys.*, 46, 1426, 1967. Highly accurate molecular hydrogen calculations.
6. Newell, A. C., and Baird, R. C., *J. Appl. Phys.*, 36, 3751, 1965. Highly accurate refractive index measurements at 47.7 GHz (essentially static).
7. Jao, T. C., Beebe, N. H. F., Person, W. B., and Sabin, J. R., *Chem. Phys. Lett.*, 26, 47, 1974. Tensor polarizabilities, derivatives, and other results are reported.
8. Orcutt, R. H., and Cole, R. H., *J. Chem. Phys.*, 46, 697, 1967 (He, Ne, Ar, Kr, H$_2$, N$_2$); Sutter, H., and Cole, R. H., *J. Chem. Phys.*, 52, 132, 1970 (CF$_3$H, CFH$_3$, CClF$_3$, CClH$_3$); Bose, T. K., and Cole, R. H., *J. Chem. Phys.*, 52, 140, 1970 (CO$_2$), and 54, 3829, 1971 (C$_2$H$_2$, C$_2$H$_4$); Nelson, R. D., and Cole, R. H., *J. Chem. Phys.*, 54, 4033, 1971 (SF$_6$, CClF$_3$); Bose, T. K., Sochanski, J. S., and Cole, R. H., *J. Chem. Phys.*, 57, 3592, 1972 (CH$_4$, CF$_4$); Kirouac, S., and Bose, T. K., *J. Chem. Phys.*, 59, 3043, 1973 (N$_2$O), and 64, 1580, 1976 (He). Highly accurate dielectric constant measurements. These modern data give the most accurate polarizabilities available. A criticism of the interpretation of these data in the case of polar molecules is given in Ref. 20, p. 2905.
9. Huestis, D. L., Technical Report #MP 78-25, SRI International (project PYU 6158), Menlo Park, CA 94025. Molar refractions for mercury-chlorine compounds are analyzed.
10. Bounds, D. G., Clarke, J. H. R., and Hinchliffe, A., *Chem. Phys. Lett.*, 45, 367, 1977. Theoretical tensor polarizability for LiCl.
11. Kolker, H. J., and Karplus, M., *J. Chem. Phys.*, 39, 2011, 1963. Theoretical.
12. Cutschick, V. P., and McKoy, V., *J. Chem. Phys.*, 58, 2397, 1973. Theoretical tensor polarizabilities.
13. Gready, J. E., Bacskay, G. B., and Hush, N. S., *Chem. Phys.*, 22, 141, 1977, and 23, 9, 1977. Theoretical.
14. Amos, A. T., and Yoffe, J. A., *J. Chem. Phys.*, 63, 4723, 1975. Theoretical.
15. Stuart, H. A., *Landolt-Börnstein Zahlenwerte and Funktionen*, Vol. 1, Part 3, Eucken, A., and Hellwege, K. H., Eds., Springer-Verlag, Berlin, 1951, p. 511. Tabulation of molecular polarizabilities. Two misprints in the chemical symbols have been corrected.
16. Guella, G., Miller, T. M., Stockdale, J. A. D., Bederson, B., and Vuskovic, L., *J. Chem. Phys.*, 94, 6857, 1991. Beam measurements with accuracies between 12-24%.
17. Marchese, F. T., and Jaffé, *Theoret. Chim. Acta (Berlin)*, 45, 241, 1977. Theoretical and experimental tensor polarizabilities are tabulated in this paper.
18. Applequist, J., Carl, J. R., and Fung, K.-K., *J. Am. Chem. Soc.*, 94, 2952, 1972. Excellent reference on the calculation of molecular polarizabilities, including extensive tables of tensor polarizabilities, both theoretical and experimental, at 589.3 nm wavelength.
19. Bridge, N. J., and Buckingham, A. D., *Proc. Roy. Soc. (London)*, A295, 334, 1966. Measured tensor polarizabilities at 633 nm wavelength.
20. Barnes, A. N. M., Turner, D. J., and Sutton, L. E., *Trans. Faraday Soc.*, 67, 2902, 1971. Dielectric constants yielding polarizabilities accurate from 0.3-8%.
21. Knight, W. D., Clemenger, K., de Heer, W. A., and Saunders, W. A., *Phys. Rev. B*, 31, 2539, 1985. These data probably correspond to a very low internal temperature.
22. Tarnovsky, V., Bunimovicz, M., Vuskovic, L., Stumpf, B., and Bederson, B., *J. Chem. Phys.*, 98, 3894, 1993. These data correspond to internal temperatures 480-948 K.
23. Milani, P., Moullet, I., and de Heer, W. A., *Phys. Rev. A*, 42, 5150, 1990. Beam measurements accurate to 11%.
24. Bonin, K. D., and Kadar-Kallen, M. A., *Int. J. Mod. Phys.*, 8, 3313, 1994. Review article.
25. Aroney, M. J., and Pratten, S. J., *J. Chem. Soc., Faraday Trans. 1*, 80, 1201, 1984. Uncertainties in the range 1-3%.
26. Le Fevre, R. J. W., Murthy, D. S. N., and Saxby, J. D., *Aust. J. Chem.*, 24, 1057, 1971. Kerr effect.
27. No, K. T., Cho, K. H., Jhon, M. S., and Scheraga, H. A., *J. Am. Chem. Soc.*, 115, 2005, 1993. Theoretical; these results are quoted in numerous valuable papers on calculated polarizabilities, e.g., Miller, K. J., and Savchik, J. A., *J. Am. Chem. Soc.*, 101, 7206, 1979.

IONIZATION POTENTIALS OF ATOMS AND ATOMIC IONS

The ionization potentials of neutral and partially ionized atoms are listed in this table. Data were obtained from the compilations cited below, supplemented by results from the recent research literature. All values have been corrected to the currently recommended value of the conversion factor from wave number to energy, namely 1 eV = 8065.541 cm⁻¹ (Reference 5). Values are given in eV.

Following the traditional spectroscopic notation, columns are headed I, II, III, etc. up to XXX, where I indicates the neutral atom, II the singly ionized atom, III the doubly ionized atom, etc. The first section of the table includes spectra I to VIII of all the elements; subsequent sections cover higher spectra (ionization stages) for those elements for which data are available.

REFERENCES

1. Moore, C. E., *Ionization Potentials and Ionization Limits Derived from the Analysis of Optical Spectra*, Natl. Stand. Ref. Data Ser. — Natl. Bur. Stand. (U.S.) No. 34, 1970.
2. Martin, W. C., Zalubas, R., and Hagan, L., *Atomic Energy Levels — The Rare Earth Elements*, Natl. Stand. Ref. Data Ser. — Natl. Bur. Stand. (U.S.), No. 60, 1978.
3. Sugar, J. and Corliss, C., *Atomic Energy Levels of the Iron Period Elements: Potassium through Nickel*, J. Phys. Chem. Ref. Data, Vol.14, Suppl. 2, 1985.
4. References to papers in *J. Phys. Chem. Ref. Data*, in the period 1973—91 covering other elements may be found in the cumulative index to that journal.
5. Cohen, E. R. and Taylor, B. N., *J. Phys. Chem. Ref. Data*, 17, 1795, 1988.
6. Martin, W.C., and Wiese, W.L., in *Atomic, Molecular, and Optical Physics Handbook*, Drake, G.W.F., Ed., AIP Press, New York, 1996.

Neutral Atoms to +7 Ions

Z	Element	I	II	III	IV	V	VI	VII	VIII
1	H	13.59844							
2	He	24.58741	54.41778						
3	Li	5.39172	75.64018	122.45429					
4	Be	9.32263	18.21116	153.89661	217.71865				
5	B	8.29803	25.15484	37.93064	259.37521	340.22580			
6	C	11.26030	24.38332	47.8878	64.4939	392.087	489.99334		
7	N	14.53414	29.6013	47.44924	77.4735	97.8902	552.0718	667.046	
8	O	13.61806	35.11730	54.9355	77.41353	113.8990	138.1197	739.29	871.4101
9	F	17.42282	34.97082	62.7084	87.1398	114.2428	157.1651	185.186	953.9112
10	Ne	21.56454	40.96328	63.45	97.12	126.21	157.93	207.2759	239.0989
11	Na	5.13908	47.2864	71.6200	98.91	138.40	172.18	208.50	264.25
12	Mg	7.64624	15.03528	80.1437	109.2655	141.27	186.76	225.02	265.96
13	Al	5.98577	18.82856	28.44765	119.992	153.825	190.49	241.76	284.66
14	Si	8.15169	16.34585	33.49302	45.14181	166.767	205.27	246.5	303.54
15	P	10.48669	19.7694	30.2027	51.4439	65.0251	220.421	263.57	309.60
16	S	10.36001	23.3379	34.79	47.222	72.5945	88.0530	280.948	328.75
17	Cl	12.96764	23.814	39.61	53.4652	67.8	97.03	114.1958	348.28
18	Ar	15.75962	27.62967	40.74	59.81	75.02	91.009	124.323	143.460
19	K	4.34066	31.63	45.806	60.91	82.66	99.4	117.56	154.88
20	Ca	6.11316	11.87172	50.9131	67.27	84.50	108.78	127.2	147.24
21	Sc	6.56144	12.79967	24.75666	73.4894	91.65	110.68	138.0	158.1
22	Ti	6.8282	13.5755	27.4917	43.2672	99.30	119.53	140.8	170.4
23	V	6.7463	14.66	29.311	46.709	65.2817	128.13	150.6	173.4
24	Cr	6.76664	16.4857	30.96	49.16	69.46	90.6349	160.18	184.7
25	Mn	7.43402	15.63999	33.668	51.2	72.4	95.6	119.203	194.5
26	Fe	7.9024	16.1878	30.652	54.8	75.0	99.1	124.98	151.06
27	Co	7.8810	17.083	33.50	51.3	79.5	102.0	128.9	157.8
28	Ni	7.6398	18.16884	35.19	54.9	76.06	108	133	162
29	Cu	7.72638	20.29240	36.841	57.38	79.8	103	139	166
30	Zn	9.39405	17.96440	39.723	59.4	82.6	108	134	174
31	Ga	5.99930	20.5142	30.71	64				
32	Ge	7.8994	15.93462	34.2241	45.7131	93.5			
33	As	9.8152	18.633	28.351	50.13	62.63	127.6		
34	Se	9.75238	21.19	30.8204	42.9450	68.3	81.7	155.4	
35	Br	11.81381	21.8	36	47.3	59.7	88.6	103.0	192.8
36	Kr	13.99961	24.35985	36.950	52.5	64.7	78.5	111.0	125.802
37	Rb	4.17713	27.285	40	52.6	71.0	84.4	99.2	136
38	Sr	5.69484	11.03013	42.89	57	71.6	90.8	106	122.3
39	Y	6.2171	12.24	20.52	60.597	77.0	93.0	116	129
40	Zr	6.63390	13.13	22.99	34.34	80.348			
41	Nb	6.75885	14.32	25.04	38.3	50.55	102.057	125	

IONIZATION POTENTIALS OF ATOMS AND ATOMIC IONS (continued)

Z	Element	I	II	III	IV	V	VI	VII	VIII
42	Mo	7.09243	16.16	27.13	46.4	54.49	68.8276	125.664	143.6
43	Tc	7.28	15.26	29.54					
44	Ru	7.36050	16.76	28.47					
45	Rh	7.45890	18.08	31.06					
46	Pd	8.3369	19.43	32.93					
47	Ag	7.57624	21.49	34.83					
48	Cd	8.99367	16.90832	37.48					
49	In	5.78636	18.8698	28.03	54				
50	Sn	7.34381	14.63225	30.50260	40.73502	72.28			
51	Sb	8.6084	16.53051	25.3	44.2	56	108		
52	Te	9.0096	18.6	27.96	37.41	58.75	70.7	137	
53	I	10.45126	19.1313	33					
54	Xe	12.12987	21.20979	32.1230					
55	Cs	3.89390	23.15745						
56	Ba	5.21170	10.00390						
57	La	5.5770	11.060	19.1773	49.95	61.6			
58	Ce	5.5387	10.85	20.198	36.758	65.55	77.6		
59	Pr	5.464	10.55	21.624	38.98	57.53			
60	Nd	5.5250	10.73	22.1	40.41				
61	Pm	5.55	10.90	22.3	41.1				
62	Sm	5.6437	11.07	23.4	41.4				
63	Eu	5.6704	11.241	24.92	42.7				
64	Gd	6.1500	12.09	20.63	44.0				
65	Tb	5.8639	11.52	21.91	39.79				
66	Dy	5.9389	11.67	22.8	41.47				
67	Ho	6.0216	11.80	22.84	42.5				
68	Er	6.1078	11.93	22.74	42.7				
69	Tm	6.18431	12.05	23.68	42.7				
70	Yb	6.25416	12.1761	25.05	43.56				
71	Lu	5.42585	13.9	20.9594	45.25	66.8			
72	Hf	6.82507	14.9	23.3	33.33				
73	Ta	7.5496							
74	W	7.8640							
75	Re	7.8335							
76	Os	8.4382							
77	Ir	8.9670							
78	Pt	8.9587	18.563						
79	Au	9.22567	20.5						
80	Hg	10.43750	18.756	34.2					
81	Tl	6.10829	20.428	29.83					
82	Pb	7.41666	15.0322	31.9373	42.32	68.8			
83	Bi	7.2856	16.69	25.56	45.3	56.0	88.3		
84	Po	8.41671							
85	At								
86	Rn	10.74850							
87	Fr	4.0727							
88	Ra	5.27892	10.14716						
89	Ac	5.17	12.1						
90	Th	6.3067	11.5	20.0	28.8				
91	Pa	5.89							
92	U	6.19405							
93	Np	6.2657							
94	Pu	6.0262							
95	Am	5.9738							
96	Cm	6.02							
97	Bk	6.23							
98	Cf	6.30							
99	Es	6.42							
100	Fm	6.50							
101	Md	6.58							
102	No	6.65							

+8 Ions to +15 Ions

Z	Element	IX	X	XI	XII	XIII	XIV	XV	XVI
9	F	1103.1176							
10	Ne	1195.8286	1362.1995						
11	Na	299.864	1465.121	1648.702					
12	Mg	328.06	367.50	1761.805	1962.6650				
13	Al	330.13	398.75	442.00	2085.98	2304.1410			
14	Si	351.12	401.37	476.36	523.42	2437.63	2673.182		
15	P	372.13	424.4	479.46	560.8	611.74	2816.91	3069.842	
16	S	379.55	447.5	504.8	564.44	652.2	707.01	3223.78	3494.1892
17	Cl	400.06	455.63	529.28	591.99	656.71	749.76	809.40	3658.521
18	Ar	422.45	478.69	538.96	618.26	686.10	755.74	854.77	918.03
19	K	175.8174	503.8	564.7	629.4	714.6	786.6	861.1	968
20	Ca	188.54	211.275	591.9	657.2	726.6	817.6	894.5	974
21	Sc	180.03	225.18	249.798	687.36	756.7	830.8	927.5	1009
22	Ti	192.1	215.92	265.07	291.500	787.84	863.1	941.9	1044
23	V	205.8	230.5	255.7	308.1	336.277	896.0	976	1060
24	Cr	209.3	244.4	270.8	298.0	354.8	384.168	1010.6	1097
25	Mn	221.8	248.3	286.0	314.4	343.6	403.0	435.163	1134.7
26	Fe	233.6	262.1	290.2	330.8	361.0	392.2	457	489.256
27	Co	186.13	275.4	305	336	379	411	444	511.96
28	Ni	193	224.6	321.0	352	384	430	464	499
29	Cu	199	232	265.3	369	401	435	484	520
30	Zn	203	238	274	310.8	419.7	454	490	542
36	Kr	230.85	268.2	308	350	391	447	492	541
37	Rb	150	277.1						
38	Sr	162	177	324.1					
39	Y	146.2	191	206	374.0				
42	Mo	164.12	186.4	209.3	230.28	279.1	302.60	544.0	570

+16 Ions to +23 Ions

Z	Element	XVII	XVIII	XIX	XX	XXI	XXII	XXIII	XXIV
17	Cl	3946.2960							
18	Ar	4120.8857	4426.2296						
19	K	1033.4	4610.8	4934.046					
20	Ca	1087	1157.8	5128.8	5469.864				
21	Sc	1094	1213	1287.97	5674.8	6033.712			
22	Ti	1131	1221	1346	1425.4	6249.0	6625.82		
23	V	1168	1260	1355	1486	1569.6	6851.3	7246.12	
24	Cr	1185	1299	1396	1496	1634	1721.4	7481.7	7894.81
25	Mn	1224	1317	1437	1539	1644	1788	1879.9	8140.6
26	Fe	1266	1358	1456	1582	1689	1799	1950	2023
27	Co	546.58	1397.2	1504.6	1603	1735	1846	1962	2119
28	Ni	571.08	607.06	1541	1648	1756	1894	2011	2131
29	Cu	557	633	670.588	1697	1804	1916	2060	2182
30	Zn	579	619	698	738	1856			
36	Kr	592	641	786	833	884	937	998	1051
42	Mo	636	702	767	833	902	968	1020	1082

+24 Ions to +29 Ions

Z	Element	XXV	XXVI	XXVII	XXVIII	XXIX	XXX
25	Mn	8571.94					
26	Fe	8828	9277.69				
27	Co	2219.0	9544.1	10012.12			
28	Ni	2295	2399.2	10288.8	10775.40		
29	Cu	2308	2478	2587.5	11062.38	11567.617	
36	Kr	1151	1205.3	2928	3070	3227	3381
42	Mo	1263	1323	1387	1449	1535	1601

IONIZATION ENERGIES OF GAS-PHASE MOLECULES
Sharon G. Lias

This table presents values for the first ionization energies (IP) of approximately 1000 molecules and atoms. Substances are listed by molecular formula in the modified Hill order (see introduction). Values enclosed in parentheses are considered not to be well established. Data appearing in the 1988 reference, were updated in 1996 for inclusion in the database of ionization energies available at the Internet site of the Standard Reference Data program of the National Institute of Standards and Technology (http://webbook.nist.gov). The list appearing here includes these updates.

The list also includes values for enthalpies of formation of the ions at 298 K, $\Delta_f H_{ion}$, given according to the ion convention used by mass spectrometrists; to convert these values to the electron convention used by thermodynamicists, add 6 kJ/mol. Details on the calculation of $\Delta_f H_{ion}$ as well as data for a much larger number of molecules, may be found in the reference and on the Internet site.

REFERENCE

Lias, S.G., Bartmess, J.E., Liebman, J.F., Holmes, J.L., Levin, R.D., and Mallard, W.G., *Gas-Phase Ion and Neutral Thermochemistry, J. Phys. Chem. Ref. Data*, Vol. 17, Suppl. No. 1, 1988.

Mol. Form.	Name	IP/eV	$\Delta_f H_{ion}$ kJ/mol
	Substances not containing carbon		
Ac	Actinium	5.17	905
Ag	Silver	7.57624	1016
AgCl	Silver(I) chloride	($\leq$ 10.08)	$\leq$ 1065
AgF	Silver(I) fluoride	(11.0 ± 0.3)	1071
Al	Aluminum	5.98577	905
AlBr	Aluminum monobromide	(9.3)	913
AlBr$_3$	Aluminum tribromide	(10.4)	593
AlCl	Aluminum monochloride	9.4	855
AlCl$_3$	Aluminum trichloride	(12.01)	573
AlF	Aluminum monofluoride	9.73 ± 0.01	673
AlF$_3$	Aluminum trifluoride	$\leq$ 15.45	$\leq$ 282
AlI	Aluminum monoiodide	9.3 ± 0.3	965
AlI$_3$	Aluminum triiodide	(9.1)	673
Am	Americium	5.9738 ± 0.0002	860
Ar	Argon	15.75962	1521
As	Arsenic	9.8152	1250
AsCl$_3$	Arsenic(III) chloride	(10.55 ± 0.025)	754
AsF$_3$	Arsenic(III) fluoride	(12.84 ± 0.05)	452
AsH$_3$	Arsine	(9.89)	1021
Au	Gold	9.22567	1254
B	Boron	8.29803	1363
BBr$_3$	Boron tribromide	(10.51)	809
BCl$_3$	Boron trichloride	11.60 ± 0.02	718
BF	Fluoroborane	11.12 ± 0.01	957
	Difluoroborane	(9.4)	317
BF$_3$	Boron trifluoride	15.7 ± 0.3	365
BH	Boron monohydride	(9.77)	1385
BH$_3$	Borane	12.026 ± 0.024	1261
BI$_3$	Boron triiodide	(9.25 ± 0.03)	964
BO$_2$	Boron dioxide	(13.5 ± 0.3)	1001
B$_2$H$_6$	Diborane	11.38 ± 0.05	1134
B$_2$O$_3$	Boron oxide	13.5 ± 0.15	460
B$_4$H$_{10}$	Tetraborane	10.76 ± 0.04	1105
B$_5$H$_9$	Pentaborane(9)	9.90 ± 0.04	1028
B$_6$H$_{10}$	Hexaborane	(9.0)	965
Ba	Barium	5.21170	683
BaO	Barium oxide	6.91 ± 0.06	543
Be	Beryllium	9.32263	1224
BeO	Beryllium oxide	(10.1 ± 0.4)	1111
Bi	Bismuth	7.2855	908
BiCl$_3$	Bismuth trichloride	(10.4)	736
Bk	Berkelium	6.23	911

Mol. Form.	Name	IP/eV	$\Delta_f H_{ion}$ kJ/mol
Br	Bromine (atomic)	11.81381	1252
BrCl	Bromine chloride	11.01	1079
BrF	Bromine fluoride	11.86	1086
BrF_5	Bromine pentafluoride	13.172 ± 0.002	840
BrH	Hydrogen bromide	11.66 ± 0.03	1087
BrH_3Si	Bromosilane	10.6	943
BrI	Iodine bromide	9.790 ± 0.004	986
BrK	Potassium bromide	7.85 ± 0.1	578
BrLi	Lithium bromide	(8.7)	685
BrNO	Nitrosyl bromide	10.17 ± 0.03	1065
BrNa	Sodium bromide	8.31 ± 0.1	660
BrO	Bromine monoxide	10.46 ± 0.02	1135
BrRb	Rubidium bromide	7.94 ± 0.03	583
BrTl	Thallium(I) bromide	9.14 ± 0.02	844
Br_2	Bromine	10.516 ± 0.005	1046
Br_2Hg	Mercury(II) bromide	10.560 ± 0.003	935
Br_2Sn	Tin(II) bromide	9.0	839
Br_3Ga	Gallium(III) bromide	10.40	711
Br_3P	Phosphorus(III) bromide	9.7	798
Br_4Hf	Hafnium(IV) bromide	(10.9)	366
Br_4Sn	Tin(IV) bromide	10.6	709
Br_4Ti	Titanium(IV) bromide	10.3	375
Br_4Zr	Zirconium(IV) bromide	(10.7)	388
Ca	Calcium	6.11316	768
CaCl	Calcium monochloride	5.86 ± 0.07	462
CaO	Calcium oxide	6.66 ± 0.18	668
Cd	Cadmium	8.99367	980
Ce	Cerium	5.5387	957
Cf	Californium	6.30	805
Cl	Chlorine (atomic)	12.96764	1373
ClCs	Cesium chloride	(7.84 ± 0.05)	510
ClF	Chlorine fluoride	12.66 ± 0.01	1171
$ClFO_3$	Perchloryl fluoride	(12.945 ± 0.005)	1224
ClF_2	Chlorine difluoride	(12.77 ± 0.05)	1128
ClF_3	Chlorine trifluoride	(12.65 ± 0.05)	1057
ClF_5S	Sulfur chloride pentafluoride	(12.335 ± 0.005)	144
ClH	Hydrogen chloride	12.749 ± 0.009	1137
ClHO	Hypochlorous acid	(11.12 ± 0.01)	993
ClH_3Si	Chlorosilane	11.4	899
ClI	Iodine chloride	10.088 ± 0.01	991
ClIn	Indium(I) chloride	(9.51)	842
ClK	Potassium chloride	(8.0 ± 0.4)	557
ClLi	Lithium chloride	9.57	727
ClNO	Nitrosyl chloride	10.87 ± 0.01	1099
$ClNO_2$	Nitryl chloride	(11.84)	1155
ClNa	Sodium chloride	8.92 ± 0.06	681
ClO	Chlorine monoxide	10.95	1159
ClO_2	Chlorine dioxide	10.33 ± 0.02	1093
ClRb	Rubidium chloride	(8.50 ± 0.03)	590
ClTl	Thallium(I) chloride	9.70 ± 0.03	869
Cl_2	Chlorine	11.480 ± 0.005	1108
Cl_2CrO_2	Chromyl chloride	11.6	580
Cl_2Ge	Germanium(II) chloride	(10.20 ± 0.05)	813
Cl_2H_2Si	Dichlorosilane	11.4	765
Cl_2Hg	Mercury(II) chloride	11.380 ± 0.003	952
Cl_2O	Chlorine oxide	10.94	1135
Cl_2OS	Thionyl chloride	10.96	844
Cl_2O_2S	Sulfuryl chloride	12.05	807
Cl_2Pb	Lead(II) chloride	(10.2)	791
Cl_2S	Sulfur dichloride	9.45 ± 0.03	895

Mol. Form.	Name	IP/eV	$\Delta_f H_{ion}$ kJ/mol
Cl_2Si	Dichlorosilylene	(10.93 ± 0.10)	887
Cl_2Sn	Tin(II) chloride	(10.0)	760
Cl_3Ga	Gallium(III) chloride	11.52	664
Cl_3HSi	Trichlorosilane	(11.7)	648
Cl_3N	Nitrogen trichloride	(10.12 ± 0.1)	1244
Cl_3OP	Phosphorus(V) oxychloride	11.36 ± 0.02	540
Cl_3OV	Vanadyl trichloride	(11.6)	425
Cl_3P	Phosphorus(III) chloride	9.91	668
Cl_3PS	Phosphorus(V) sulfide trichloride	9.71 ± 0.03	573
Cl_3Sb	Antimony(III) chloride	(≤ 10.7)	s719
Cl_4Ge	Germanium(IV) chloride	11.68 ± 0.05	629
Cl_4Hf	Hafnium(IV) chloride	(11.7)	246
Cl_4Si	Tetrachlorosilane	11.79 ± 0.01	527
Cl_4Sn	Tin(IV) chloride	11.7 ± 0.2	656
Cl_4Ti	Titanium(IV) chloride	(11.5)	349
Cl_4V	Vanadium(IV) chloride	(9.2)	361
Cl_4Zr	Zirconium(IV) chloride	(11.2)	210
Cl_5Mo	Molybdenum(V) chloride	(8.7)	392
Cl_5Nb	Niobium(V) chloride	(10.97)	356
Cl_5P	Phosphorus(V) chloride	(10.2)	608
Cl_5Ta	Tantalum(V) chloride	(11.08)	303
Cl_6W	Tungsten(VI) chloride	(9.5)	348
Cm	Curium	6.02	966
Co	Cobalt	7.8810	1187
Cr	Chromium	6.76664	1050
Cs	Cesium	3.89390	452
CsF	Cesium fluoride	(8.80 ± 0.10)	489
CsNa	Cesium sodium	(4.05 ± 0.04)	535
Cu	Copper	7.72638	1084
CuF	Copper(I) fluoride	10.15 ± 0.02	984
Dy	Dysprosium	5.9389	862
Er	Erbium	6.1078	907
Es	Einsteinium	6.42	753
Eu	Europium	5.6704	723
F	Fluorine (atomic)	17.42282	1761
FGa	Gallium monofluoride	(9.6 ± 0.5)	700
FH	Hydrogen fluoride	16.044 ± 0.003	1276
FHO	Hypofluorous acid	12.71 ± 0.01	1130
FH_3Si	Fluorosilane	11.7	752
FI	Iodine fluoride	10.54 ± 0.01	922
FIn	Indium monofluoride	(9.6 ± 0.5)	740
FNO	Nitrosyl fluoride	12.63 ± 0.03	1152
FNO_2	Nitryl fluoride	(13.09)	1154
FNS	Thionitrosyl fluoride (NSF)	11.51 ± 0.04	1090
FO	Fluorine monoxide	12.78 ± 0.03	1342
FO_2	Fluorine superoxide (FOO)	(12.6 ± 0.2)	1228
FS	Sulfur fluoride	10.09	986
FTl	Thallium(I) fluoride	10.52	835
F_2	Fluorine	15.697 ± 0.003	1515
F_2Ge	Germanium(II) fluoride	(≤ 11.65)	551
F_2HN	Difluoramine	(11.53 ± 0.08)	1046
F_2H_2Si	Difluorosilane	(12.2)	386
F_2Mg	Magnesium fluoride	(13.6 ± 0.3)	588
F_2N	Difluoroamidogen	11.628 ± 0.01	1155
F_2N_2	trans-Difluorodiazine	(12.8)	1315
F_2O	Fluorine monoxide	13.11 ± 0.01	1290
F_2OS	Thionyl fluoride	12.25	688
F_2O_2S	Sulfuryl fluoride	13.04 ± 0.01	501
F_2Pb	Lead(II) fluoride	(11.5)	679
F_2S	Sulfur difluoride	(10.08)	676

Mol. Form.	Name	IP/eV	$\Delta_f H_{ion}$ kJ/mol
F_2Si	Difluorosilylene	10.78 ± 0.05	450
F_2Sn	Tin(II) fluoride	(11.1)	586
F_2Xe	Xenon difluoride	12.35 ± 0.01	1083
F_3HSi	Trifluorosilane	(14.0)	150
F_3N	Nitrogen trifluoride	13.00 ± 0.02	1125
F_3NO	Trifluoramine oxide	13.31 ± 0.06	1121
F_3OP	Phosphorus(V) oxyfluoride	12.76 ± 0.01	-24
F_3P	Phosphorus(III) fluoride	11.60 ± 0.05	161
F_3PS	Phosphorus(V) sulfide trifluoride	$\leq 11.05 \pm 0.035$	≤ 58
F_3Si	Trifluorosilyl	(9.99)	- 32
F_4Ge	Germanium(IV) fluoride	(15.5)	307
F_4N_2	Tetrafluorohydrazine	11.94 ± 0.03	1119
F_4S	Sulfur tetrafluoride	12.0 ± 0.3	399
F_4Si	Tetrafluorosilane	15.24 ± 0.14	-144
F_4Xe	Xenon tetrafluoride	12.65 ± 0.1	1016
F_5I	Iodine pentafluoride	12.943 ± 0.005	408
F_5P	Phosphorus(V) fluoride	(15.1)	-137
F_5S	Sulfur pentafluoride	9.60 ± 0.05	10
F_6Mo	Molybdenum(VI) fluoride	(14.5 ± 0.1)	-159
F_6S	Sulfur hexafluoride	15.32 ± 0.02	258
F_6U	Uranium(VI) fluoride	14.00 ± 0.10	-796
Fe	Iron	7.9024	1177
Fm	Fermium	6.50	627
Ga	Gallium	5.99930	851
GaI_3	Gallium(III) iodide	9.40	765
Gd	Gadolinium	6.1500	991
Ge	Germanium	7.900	1139
GeH_4	Germane	≤ 10.53	≤ 1108
GeI_4	Germanium(IV) iodide	(9.42)	850
GeO	Germanium(II) oxide	11.25 ± 0.01	1044
GeS	Germanium(II) sulfide	(9.98)	1055
H	Hydrogen (atomic)	13.59844	1530
HI	Hydrogen iodide	10.386 ± 0.001	1028
HLi	Lithium hydride	7.7	882
HN	Imidogen	$\leq 13.49 \pm 0.01$	1678
HNO	Nitrosyl hydride	(10.1)	1075
HNO_2	Nitrous acid	≤ 11.3	≤ 1011
HNO_3	Nitric acid	11.95 ± 0.01	1019
HN_3	Hydrazoic acid	10.72 ± 0.025	1328
HO	Hydroxyl	13.0170 ± 0.0002	1294
HO_2	Hydroperoxy	11.35 ± 0.01	1106
HS	Mercapto	10.4219 ± 0.0004	1145
H_2	Hydrogen	15.42593 ± 0.00005	1488
H_2N	Amidogen	11.14 .01	1264
H_2O	Water	12.6206 ± 0.0020	976
H_2O_2	Hydrogen peroxide	10.58 ± 0.04	885
H_2S	Hydrogen sulfide	10.457 ± 0.012	989
H_2Se	Hydrogen selenide	9.892 ± 0.005	984
H_2Si	Silylene	8.244 ± 0.025	1084
H_3N	Ammonia	10.070 ± 0.020	925
H_3NO	Hydroxylamine	(10.00)	923
H_3P	Phosphine	9.869 ± 0.002	958
H_3Sb	Stibine	9.54 ± 0.03	1067
H_4N_2	Hydrazine	8.1 ± 0.15	877
H_4Si	Silane	11.00 ± 0.02	1095
H_4Sn	Stannane	(10.75)	1200
H_6Si_2	Disilane	9.74 ± 0.02	1019
H_8Si_3	Trisilane	(9.2)	1009
He	Helium	24.58741	2372
Hf	Hafnium	6.82507 ± 0.00004	1278

IONIZATION ENERGIES OF GAS-PHASE MOLECULES (continued)

Mol. Form.	Name	IP/eV	$\Delta_f H_{ion}$ kJ/mol
Hg	Mercury	10.43750	1069
HgI$_2$	Mercury(II) iodide	9.5088 ± 0.0022	900
Ho	Holmium	6.0216	882
I	Iodine (atomic)	10.45126	1115
IK	Potassium iodide	(7.21 ± 0.3)	570
ILi	Lithium iodide	(7.5)	633
INa	Sodium iodide	7.64 ± 0.02	659
ITl	Thallium(I) iodide	8.47 ± 0.02	826
I$_2$	Iodine	9.3074 ± 0.0002	960
I$_4$Ti	Titanium(IV) iodide	(9.1)	602
I$_4$Zr	Zirconium(IV) iodide	(9.3)	500
In	Indium	5.78636	802
Ir	Iridium	9.1	1543
K	Potassium	4.34066	508
KLi	Lithium potassium	4.57 ± 0.04	512
KNa	Potassium sodium	4.41636 ± 0.00017	561
K$_2$	Dipotassium	4.0637 ± 0.0002	519
Kr	Krypton	13.99961	1351
La	Lanthanum	5.5770	969
Li	Lithium	5.39172	680
LiNa	Lithium sodium	5.05 ± 0.04	571
LiO	Lithium monoxide	(8.44)	894
LiRb	Lithium rubidium	4.3 ± 0.1	486
Li$_2$	Dilithium	5.1127 ± 0.0003	709
Lu	Lutetium	5.42585	950
Md	Mendelevium	6.58	635
Mg	Magnesium	7.64624	885
MgO	Magnesium oxide	(8.76 ± 0.22)	901
Mn	Manganese	7.43402	998
Mo	Molybdenum	7.09243	1343
N	Nitrogen (atomic)	14.53414	1875
NO	Nitric oxide	9.26438 ± 0.00005	985
NO$_2$	Nitrogen dioxide	9.586 ± 0.002	958
NP	Phosphorus nitride	11.84 ± 0.04	1247
NS	Nitrogen sulfide	8.87 ± 0.01	1119
N$_2$	Nitrogen	15.5808	1503
N$_2$O	Nitrous oxide	12.886	1325
N$_2$O$_4$	Nitrogen tetroxide	(10.8)	1050
N$_2$O$_5$	Nitrogen pentoxide	(11.9)	1161
Na	Sodium	5.13908	603
NaRb	Rubidium sodium	4.32 ± 0.04	480
Na$_2$	Disodium	4.894 ± 0.003	614
Nb	Niobium	6.75885	1384
Nd	Neodymium	5.5250	859
Ne	Neon	21.56454	2081
Ni	Nickel	7.6398	1167
No	Nobelium	6.65	642
Np	Neptunium	6.2657 ± 0.0003	1069
O	Oxygen (atomic)	13.61806	1563
OPb	Lead(II) oxide	9.08 ± 0.10	939
OS	Sulfur monoxide	10.294 ± 0.004	998
OS$_2$	Sulfur oxide (SSO)	10.584 ± 0.005	971
OSi	Silicon monoxide	11.49 ± 0.20	1008
OSn	Tin(II) oxide	9.60 ± 0.02	944
OSr	Strontium oxide	6.6 ± 0.2	623
O$_2$	Oxygen	12.0697 ± 0.0002	1165
O$_2$S	Sulfur dioxide	12.349 ± 0.001	894
O$_2$Th	Thorium(IV) oxide	(8.7 ± 0.15)	342
O$_2$Ti	Titanium(IV) oxide	(9.54 ± 0.1)	623
O$_2$U	Uranium(IV) oxide	(5.4 ± 0.1)	57

Mol. Form.	Name	IP/eV	$\Delta_f H_{ion}$ kJ/mol
O_3	Ozone	12.43	1342
O_3S	Sulfur trioxide	12.82 ± 0.03	841
O_3U	Uranium(VI) oxide	(10.5 ± 0.5)	214
O_4Os	Osmium(VIII) oxide	(12.32)	850
O_4Ru	Ruthenium(VIII) oxide	12.15 ± 0.03	988
O_7Re_2	Rhenium(VII) oxide	(12.7 ± 0.2)	125
Os	Osmium	8.7	1630
P	Phosphorus	10.48669	1328
P_2	Diphosphorus	10.53	1160
Pa	Protactinium	5.89	1133
Pb	Lead	7.41666	911
PbS	Lead(II) sulfide	(8.5 ± 0.5)	954
Pd	Palladium	8.3367	1181
Pm	Promethium	5.55	536
Pr	Praseodymium	5.464	883
Pt	Platinum	9.0	1433
Pu	Plutonium	6.025	926
Ra	Radium	5.27892	668
Rb	Rubidium	4.17713	484
Re	Rhenium	7.88	1530
Rh	Rhodium	7.45890	1276
Rn	Radon	10.74850	1037
Ru	Ruthenium	7.36050	1355
S	Sulfur	10.36001	1277
SSn	Tin(II) sulfide	(8.8)	966
S_2	Disulfur	9.356 ± 0.002	1031
Sb	Antimony	8.64	1096
Sc	Scandium	6.56144	1010
Se	Selenium	9.75238	1168
Si	Silicon	8.15169	1238
Sm	Samarium	5.6437	751
Sn	Tin	7.34381	1011
Sr	Strontium	5.69484	713
Ta	Tantalum	7.89	1544
Tb	Terbium	5.8639	955
Tc	Technetium	7.28	1380
Te	Tellurium	9.0096	1066
Th	Thorium	6.308 ± 0.003	1207
Ti	Titanium	6.8282	1127
Tl	Thallium	6.10829	771
Tm	Thulium	6.18431	827
U	Uranium	6.19405	1129
V	Vanadium	6.746 ± 0.002	1166
W	Tungsten	7.98	1621
Xe	Xenon	12.12987	1170
Y	Yttrium	6.217	1022
Yb	Ytterbium	6.25416	754
Zn	Zinc	9.39405	1037
Zr	Zirconium	6.63390	1251

Substances containing carbon

Mol. Form.	Name	IP/eV	$\Delta_f H_{ion}$ kJ/mol
C	Carbon	11.26030	1803
$CBrClF_2$	Bromochlorodifluoromethane	(11.21)	642
$CBrCl_3$	Bromotrichloromethane	(10.6)	980
$CBrF_3$	Bromotrifluoromethane	(11.40)	451
CBr_2F_2	Dibromodifluoromethane	11.03 ± 0.04	683
CBr_4	Tetrabromomethane	(10.31 ± 0.02)	1079
CCl	Chloromethylidene	(8.9 ± 0.2)	1244
$CClF_3$	Chlorotrifluoromethane	12.6 ± 0.2	505

Mol. Form.	Name	IP/eV	$\Delta_f H_{ion}$ kJ/mol
CClN	Cyanogen chloride	12.34 ± 0.01	1329
CCl$_2$	Dichloromethylene	(9.27)	1058
CCl$_2$F$_2$	Dichlorodifluoromethane	12.05 ± 0.24	685
CCl$_2$O	Carbonyl chloride	(11.5)	888
CCl$_3$F	Trichlorofluoromethane	11.77 ± 0.02	868
CCl$_4$	Tetrachloromethane	11.47 ± 0.01	1010
CF	Fluoromethylidene	9.11 ± 0.01	1134
CFN	Cyanogen fluoride	13.34 ± 0.02	1325
CF$_2$	Difluoromethylene	11.44 ± 0.03	899
CF$_2$O	Carbonyl fluoride	13.035 ± 0.030	617
CF$_3$	Trifluoromethyl	8.7 ± 0.2	379
CF$_3$I	Trifluoroiodomethane	10.23	397
CH	Methylidyne	10.64 ± 0.01	1622
CHBrCl$_2$	Bromodichloromethane	10.6	973
CHBr$_2$Cl	Chlorodibromomethane	10.59 ± 0.01	1030
CHBr$_3$	Tribromomethane	10.48 ± 0.02	1035
CHCl	Chloromethylene	9.84	1247
CHClF$_2$	Chlorodifluoromethane	(12.2)	693
CHCl$_2$F	Dichlorofluoromethane	(11.5)	829
CHCl$_3$	Trichloromethane	11.37 ± 0.02	992
CHF	Fluoromethylene	10.06 ± 0.05	1121
CHF$_3$	Trifluoromethane	(13.86)	643
CHI$_3$	Triiodomethane	9.25 ± 0.02	1010
CHN	Hydrogen cyanide	13.60 ± 0.01	1447
CHN	Hydrogen isocyanide	(12.5 ± 0.1)	1407
CHNO	Isocyanic acid	11.595 ± 0.005	1016
CHNO	Fulminic acid	(10.83)	1263
CHO	Oxomethyl (HCO)	(8.55)	826
CH$_2$	Methylene	10.396 ± .003	1392
CH$_2$BrCl	Bromochloromethane	10.77 ± 0.01	1085
CH$_2$Br$_2$	Dibromomethane	(10.50 ± 0.02)	1013
CH$_2$ClF	Chlorofluoromethane	11.71 ± 0.01	870
CH$_2$Cl$_2$	Dichloromethane	11.32 ± .01	996
CH$_2$F$_2$	Difluoromethane	12.71	774
CH$_2$I$_2$	Diiodomethane	9.46 ± 0.02	1030
CH$_2$N$_2$	Diazomethane	8.999 ± 0.001	1098
CH$_2$N$_2$	Cyanamide	(10.4)	1137
CH$_2$O	Formaldehyde	10.88 ± 0.01	941
CH$_2$O$_2$	Formic acid	11.33 ± 0.01	715
CH$_3$	Methyl	9.843 ± 0.002	1095
CH$_3$BO	Borane carbonyl	11.14 ± 0.02	962
CH$_3$Br	Bromomethane	10.541 ± 0.003	979
CH$_3$Cl	Chloromethane	11.22 ± 0.01	1001
CH$_3$Cl$_3$Si	Methyltrichlorosilane	(11.36 ± 0.03)	548
CH$_3$F	Fluoromethane	12.47 ± 0.02	956
CH$_3$I	Iodomethane	9.538	936
CH$_3$NO	Formamide	10.16 ± 0.06	796
CH$_3$NO$_2$	Nitromethane	11.08 ± 0.07	994
CH$_3$N$_3$	Methyl azide	9.81 ± 0.02	1227
CH$_3$O	Methoxy	(10.72)	1050
CH$_4$	Methane	12.61 ± 0.01	1143
CH$_4$N$_2$O	Urea	9.7	690
CH$_4$O	Methanol	10.85 ± 0.01	845
CH$_4$S	Methanethiol	9.44 ± 0.005	888
CH$_5$N	Methylamine	(8.80)	826
CH$_6$N$_2$	Methylhydrazine	7.7 ± 0.15	835
CH$_6$Si	Methylsilane	(10.7)	1003
CN	Cyanide	13.5984	1748
CNO	Cyanate	11.76 ± 0.01	1290
CO	Carbon monoxide	14.014 ± 0.0003	1242

Mol. Form.	Name	IP/eV	$\Delta_f H_{ion}$ kJ/mol
COS	Carbon oxysulfide	11.18 ± 0.01	936
COSe	Carbon oxyselenide	10.36 ± 0.01	929
CO_2	Carbon dioxide	13.773 ± 0.002	935
CS	Carbon sulfide	11.33 ± 0.01	1361
CS_2	Carbon disulfide	10.0685 ± 0.0020	1089
C_2	Dicarbon	(11.4 ± 0.3)	2000
$C_2Br_2F_4$	1,2-Dibromotetrafluoroethane	(11.1)	280
C_2ClF_3	Chlorotrifluoroethylene	9.81 ± 0.03	373
C_2ClF_5	Chloropentafluoroethane	(12.6)	99
C_2Cl_2	Dichloroacetylene	9.9	1165
$C_2Cl_2F_4$	1,2-Dichlorotetrafluoroethane	12.2	252
$C_2Cl_3F_3$	1,1,1-Trichlorotrifluoroethane	11.5	386
$C_2Cl_3F_3$	1,1,2-Trichlorotrifluoroethane	11.99 ± 0.02	429
C_2Cl_4	Tetrachloroethylene	9.326 ± 0.001	887
$C_2Cl_4F_2$	1,1,2,2-Tetrachloro-1,2-difluoroethane	(11.3)	563
C_2Cl_4O	Trichloroacetyl chloride	(11.0)	827
C_2Cl_6	Hexachloroethane	(11.1)	920
C_2F_3N	Trifluoroacetonitrile	13.93 ± 0.07	845
C_2F_4	Tetrafluoroethylene	10.12 ± 0.02	315
C_2F_6	Hexafluoroethane	(13.6)	-30
C_2H	Ethynyl	(11.61 ± 0.07)	1685
C_2HBr	Bromoacetylene	10.31 ± 0.02	1242
$C_2HBrClF_3$	2-Bromo-2-chloro-1,1,1-trifluoroethane	(11.0)	363
C_2HCl	Chloroacetylene	10.58 ± 0.02	1276
C_2HClF_2	1-Chloro-2,2-difluoroethylene	9.80 ± 0.04	628
C_2HCl_3	Trichloroethylene	9.46 ± 0.02	894
C_2HCl_3O	Dichloroacetyl chloride	(10.9)	809
C_2HCl_5	Pentachloroethane	(11.0)	919
C_2HF	Fluoroacetylene	11.26	1195
C_2HF_3	Trifluoroethylene	10.14	489
$C_2HF_3O_2$	Trifluoroacetic acid	11.46	75
C_2H_2	Acetylene	11.400 ± 0.002	1328
$C_2H_2Cl_2$	1,1-Dichloroethylene	9.81 ± 0.04	949
$C_2H_2Cl_2$	cis-1,2-Dichloroethylene	9.66 ± 0.01	936
$C_2H_2Cl_2$	trans-1,2-Dichloroethylene	9.64 ± 0.02	934
$C_2H_2Cl_2O$	Chloroacetyl chloride	(≤ 10.3)	815
$C_2H_2Cl_4$	1,1,1,2-Tetrachloroethane	(11.1)	920
$C_2H_2Cl_4$	1,1,2,2-Tetrachloroethane	(≤ 11.62)	≤ 971
$C_2H_2F_2$	1,1-Difluoroethylene	10.29 ± 0.01	650
$C_2H_2F_2$	cis-1,2-Difluoroethylene	10.23 ± 0.02	690
C_2H_2O	Ketene	9.617 ± 0.003	880
$C_2H_2O_2$	Glyoxal	10.2	773
$C_2H_2S_2$	Thiirene	8.61	892
C_2H_3Br	Bromoethylene	9.83 ± 0.02	1028
C_2H_3Cl	Chloroethylene	9.99 ± 0.02	985
$C_2H_3ClF_2$	1-Chloro-1,1-difluoroethane	(11.98)	626
C_2H_3ClO	Acetyl chloride	10.82 ± 0.04	801
C_2H_3ClO	Chloroacetaldehyde	(10.48)	815
$C_2H_3ClO_2$	Chloroacetic acid	(10.7)	597
$C_2H_3Cl_3$	1,1,1-Trichloroethane	(11.0)	917
$C_2H_3Cl_3$	1,1,2-Trichloroethane	(11.0)	911
C_2H_3F	Fluoroethylene	10.36 ± 0.01	861
C_2H_3FO	Acetyl fluoride	(11.5)	667
$C_2H_3F_3$	1,1,1-Trifluoroethane	13.3 ± 0.5	536
C_2H_3N	Acetonitrile	12.20 ± 0.01	1253
C_2H_3NO	Methylisocyanate	(10.67)	900
C_2H_4	Ethylene	10.5138 ± 0.0006	1067
$C_2H_4Br_2$	1,2-Dibromoethane	10.35 ± 0.04	961
$C_2H_4Cl_2$	1,1-Dichloroethane	11.04 ± 0.02	935
$C_2H_4Cl_2$	1,2-Dichloroethane	11.04 ± 0.02	931

Mol. Form.	Name	IP/eV	$\Delta_f H_{ion}$ kJ/mol
$C_2H_4F_2$	1,1-Difluoroethane	(11.87)	643
C_2H_4O	Acetaldehyde	10.229 ± 0.0007	821
C_2H_4O	Ethylene oxide	10.56 ± 0.01	966
$C_2H_4O_2$	Acetic acid	10.65 ± 0.02	595
$C_2H_4O_2$	Methyl formate	10.835 ± 0.005	690
C_2H_5Br	Bromoethane	10.29 ± 0.01	931
C_2H_5Cl	Chloroethane	10.98 ± 0.02	947
C_2H_5ClO	2-Chloroethanol	(10.5)	756
C_2H_5F	Fluoroethane	(11.78)	873
C_2H_5I	Iodoethane	9.3492 ± 0.0006	893
C_2H_5N	Ethyleneimine	(9.5 ± 0.3)	1044
C_2H_5NO	Acetamide	9.65 ± 0.03	693
C_2H_5NO	N-Methylformamide	9.83 ± 0.04	760
$C_2H_5NO_2$	Nitroethane	10.88 ± 0.05	948
C_2H_6	Ethane	11.56 ± 0.02	1031
$C_2H_6Cl_2Si$	Dichlorodimethylsilane	(10.7)	576
C_2H_6O	Ethanol	10.43 ± 0.05	772
C_2H_6O	Dimethyl ether	10.025 ± 0.025	783
C_2H_6OS	Dimethyl sulfoxide	9.10 ± 0.03	727
$C_2H_6O_2$	Ethylene glycol	10.16	593
C_2H_6S	Ethanethiol	9.31 ± 0.03	851
C_2H_6S	Dimethyl sulfide	8.69 ± 0.02	801
$C_2H_6S_2$	Dimethyl disulfide	(7.4 ± 0.3)	690
C_2H_7N	Ethylamine	8.86 ± 0.02	808
C_2H_7N	Dimethylamine	8.24 ± 0.08	777
C_2H_7NO	Ethanolamine	8.96	664
$C_2H_8N_2$	1,2-Ethanediamine	(8.6)	812
$C_2H_8N_2$	1,1-Dimethylhydrazine	7.29 ± 0.05	787
C_2N_2	Cyanogen	13.37 ± 0.01	1597
C_3F_6	Perfluoropropene	10.60 ± 0.03	-103
C_3F_6O	Perfluoroacetone	(11.57 ± 0.13)	-282
C_3F_8	Perfluoropropane	(13.38)	-491
C_3HN	Cyanoacetylene	11.64 ± 0.01	1475
C_3H_2O	2-Propynal	(10.7 ± 0.1)	1145
$C_3H_3F_3$	3,3,3-Trifluoropropene	(10.9)	437
C_3H_3N	2-Propenenitrile	10.91 ± 0.01	1237
C_3H_3NO	Oxazole	(9.9)	940
C_3H_3NO	Isoxazole	(9.93)	1038
C_3H_4	Allene	9.692 ± 0.004	1126
C_3H_4	Propyne	10.37 ± 0.01	1187
C_3H_4	Cyclopropene	9.67 ± 0.01	1209
$C_3H_4N_2$	Imidazole	(8.81)	997
C_3H_4O	Propargyl alcohol	10.49 ± 0.02	1060
C_3H_4O	Acrolein	10.103 ± 0.006	900
C_3H_4O	Cyclopropanone	(9.1 ± 0.1)	895
$C_3H_4O_2$	Propenoic acid	10.60	701
$C_3H_4O_2$	2-Oxetanone	(9.70 ± 0.01)	653
C_3H_5Br	3-Bromopropene	(9.96)	1008
C_3H_5Cl	3-Chloropropene	10.04 ± 0.01	965
C_3H_5ClO	Epichlorohydrin	(10.64)	919
$C_3H_5ClO_2$	Methyl chloroacetate	(10.3)	575
C_3H_5F	3-Fluoropropene	(10.11)	821
C_3H_5N	Propanenitrile	11.84 ± 0.02	1194
C_3H_5NO	Acrylamide	(9.5)	720
C_3H_6	Propene	9.73 ± 0.02	959
C_3H_6	Cyclopropane	9.86	1005
$C_3H_6Br_2$	1,2-Dibromopropane	(10.1)	903
$C_3H_6Br_2$	1,3-Dibromopropane	(≤ 10.2)	≤ 919
$C_3H_6Cl_2$	1,2-Dichloropropane	10.8 ± 0.1	886
$C_3H_6Cl_2$	1,3-Dichloropropane	10.89 ± 0.04	892

Mol. Form.	Name	IP/eV	$\Delta_f H_{ion}$ kJ/mol
C_3H_6O	Allyl alcohol	9.67 ± 0.05	808
C_3H_6O	Methyl vinyl ether	8.95 ± 0.01	763
C_3H_6O	Propanal	9.96 ± 0.01	772
C_3H_6O	Acetone	9.703 ± 0.006	719
C_3H_6O	Methyloxirane	(10.22)	892
C_3H_6O	Oxetane	9.65 ± 0.01	851
$C_3H_6O_2$	Propanoic acid	10.525 ± 0.003	568
$C_3H_6O_2$	Ethyl formate	10.61 ± 0.01	639
$C_3H_6O_2$	Methyl acetate	10.25 ± 0.02	579
$C_3H_6O_2$	1,3-Dioxolane	(9.9)	658
$C_3H_6O_3$	1,3,5-Trioxane	(10.3)	528
C_3H_7Br	1-Bromopropane	10.18 ± 0.01	898
C_3H_7Br	2-Bromopropane	10.10 ± 0.03	877
C_3H_7Cl	1-Chloropropane	10.81 ± 0.01	911
C_3H_7Cl	2-Chloropropane	10.79 ± 0.02	896
C_3H_7F	1-Fluoropropane	(11.3)	806
C_3H_7F	2-Fluoropropane	(11.08)	776
C_3H_7I	1-Iodopropane	9.25 ± 0.01	860
C_3H_7I	2-Iodopropane	9.19 ± 0.02	845
C_3H_7N	Allylamine	(8.76)	891
C_3H_7N	Cyclopropylamine	(8.8)	926
C_3H_7N	Propyleneimine	(9.0)	960
C_3H_7NO	N,N-Dimethylformamide	(9.12)	688
$C_3H_7NO_2$	1-Nitropropane	(10.81)	919
$C_3H_7NO_2$	2-Nitropropane	(10.71)	894
C_3H_8	Propane	10.95 ± 0.05	952
C_3H_8O	1-Propanol	10.18 ± 0.06	727
C_3H_8O	2-Propanol	10.17 ± 0.02	709
C_3H_8O	Ethyl methyl ether	9.72 ± 0.07	722
$C_3H_8O_2$	Dimethoxymethane	9.7	588
C_3H_8S	1-Propanethiol	9.20 ± 0.01	819
C_3H_8S	2-Propanethiol	9.145 ± 0.005	806
C_3H_8S	Ethyl methyl sulfide	(8.55)	765
$C_3H_9BO_3$	Trimethyl borate	(10.0)	65
C_3H_9ClSi	Trimethylchlorosilane	(10.15)	624
C_3H_9N	Propylamine	(8.78)	777
C_3H_9N	Isopropylamine	(8.72)	758
C_3H_9N	Trimethylamine	7.82 ± 0.06	731
C_3H_9NO	3-Amino-1-propanol	(9.0)	651
$C_4H_2O_3$	Maleic anhydride	(10.8)	645
C_4H_4	1-Buten-3-yne	9.58 ± 0.02	1230
$C_4H_4N_2$	Succinonitrile	(12.1 ± 0.25)	1377
$C_4H_4N_2$	Pyrimidine	9.23	1087
$C_4H_4N_2$	Pyridazine	8.67 ± 0.03	1112
C_4H_4O	Furan	8.883 ± 0.003	822
$C_4H_4O_2$	Diketene	(9.6 ± 0.02)	736
$C_4H_4O_3$	Succinic anhydride	(10.6)	500
$C_4H_4O_4$	Fumaric acid	(10.7)	355
C_4H_4S	Thiophene	8.86 ± 0.02	970
C_4H_5N	Methylacrylonitrile	10.34	1127
C_4H_5N	Pyrrole	8.207 ± 0.005	900
C_4H_5N	Cyclopropanecarbonitrile	(10.25)	1173
C_4H_6	1,2-Butadiene	(9.03)	1034
C_4H_6	1,3-Butadiene	9.082 ± 0.004	986
C_4H_6	1-Butyne	10.19 ± 0.02	1148
C_4H_6	2-Butyne	9.59 ± 0.03	1071
C_4H_6	Cyclobutene	9.43 ± 0.02	1067
C_4H_6O	Divinyl ether	(8.7)	827
C_4H_6O	trans-2-Butenal	9.73 ± 0.01	835
C_4H_6O	2-Methylpropenal	(9.92)	834

Mol. Form.	Name	IP/eV	$\Delta_f H_{ion}$ kJ/mol
C_4H_6O	Cyclobutanone	(9.35	815
$C_4H_6O_2$	cis-Crotonic acid	(10.08)	625
$C_4H_6O_2$	trans-Crotonic acid	(9.9)	604
$C_4H_6O_2$	Methacrylic acid	(10.15)	611
$C_4H_6O_2$	Vinyl acetate	9.19±0.05	572
$C_4H_6O_2$	Methyl acrylate	(9.9)	641
$C_4H_6O_3$	Acetic anhydride	(10.0)	398
$C_4H_6O_4$	Dimethyl oxalate	(10.0)	287
C_4H_6S	2,5-Dihydrothiophene	(8.4)	898
C_4H_7N	Butanenitrile	(11.2)	1110
C_4H_7N	2-Methylpropanenitrile	(11.3)	1115
C_4H_7NO	2-Pyrrolidone	(9.2)	674
C_4H_8	1-Butene	9.55±0.06	921
C_4H_8	cis-2-Butene	9.11±0.01	871
C_4H_8	trans-2-Butene	9.10±0.01	866
C_4H_8	Isobutene	9.239±0.003	875
C_4H_8	Cyclobutane	(9.82±0.05)	976
C_4H_8	Methylcyclopropane	(9.46)	936
$C_4H_8Br_2$	1,4-Dibromobutane	(10.15)	879
C_4H_8O	Ethyl vinyl ether	(8.98)	709
C_4H_8O	1,2-Epoxybutane	(≤ 10.15)	862
C_4H_8O	Butanal	9.84±0.02	742
C_4H_8O	Isobutanal	9.71±0.01	721
C_4H_8O	2-Butanone	9.52±0.04	678
C_4H_8O	Tetrahydrofuran	9.38±0.05	721
$C_4H_8O_2$	Butanoic acid	10.17±0.05	509
$C_4H_8O_2$	2-Methylpropanoic acid	10.33±0.03	516
$C_4H_8O_2$	Propyl formate	10.52±0.02	555
$C_4H_8O_2$	Ethyl acetate	10.01±0.05	522
$C_4H_8O_2$	Methyl propanoate	10.15±0.03	548
$C_4H_8O_2$	1,3-Dioxane	9.8	607
$C_4H_8O_2$	1,4-Dioxane	9.19±0.01	571
$C_4H_8O_2S$	Sulfolane	(9.8)	577
C_4H_8S	Tetrahydrothiophene	8.38	774
C_4H_9Br	1-Bromobutane	(10.12)	869
C_4H_9Br	2-Bromobutane	10.01±0.02	845
C_4H_9Br	1-Bromo-2-methylpropane	10.09±0.02	861
C_4H_9Br	2-Bromo-2-methylpropane	9.92±0.03	823
C_4H_9Cl	1-Chlorobutane	10.67±0.03	875
C_4H_9Cl	2-Chlorobutane	10.53	857
C_4H_9Cl	1-Chloro-2-methylpropane	10.73±0.07	877
C_4H_9Cl	2-Chloro-2-methylpropane	(10.61)	842
C_4H_9I	1-Iodobutane	9.23±0.01	840
C_4H_9I	2-Iodobutane	9.10±0.02	815
C_4H_9I	1-Iodo-2-methylpropane	9.19±0.01	824
C_4H_9I	2-Iodo-2-methylpropane	(9.02)	798
C_4H_9N	Pyrrolidine	(8.0)	769
C_4H_9NO	N,N-Dimethylacetamide	8.81±0.03	616
C_4H_9NO	Morpholine	(8.2)	841
C_4H_{10}	Butane	10.53±0.10	890
C_4H_{10}	Isobutane	(10.57)	886
$C_4H_{10}O$	1-Butanol	9.99±0.05	689
$C_4H_{10}O$	2-Butanol	9.88±0.03	658
$C_4H_{10}O$	2-Methyl-1-propanol	10.02±0.04	683
$C_4H_{10}O$	2-Methyl-2-propanol	9.90±0.02	642
$C_4H_{10}O$	Diethyl ether	9.51±0.03	666
$C_4H_{10}O$	Methyl propyl ether	9.41±0.07	670
$C_4H_{10}O$	Isopropyl methyl ether	9.45±0.04	661
$C_4H_{10}O_2$	Ethylene glycol monoethyl ether	(9.6)	529
$C_4H_{10}O_2$	Ethylene glycol dimethyl ether	(9.3)	558

Mol. Form.	Name	IP/eV	$\Delta_f H_{ion}$ kJ/mol
$C_4H_{10}S$	1-Butanethiol	9.14 ± 0.01	794
$C_4H_{10}S$	2-Butanethiol	(9.10)	781
$C_4H_{10}S$	2-Methyl-1-propanethiol	(9.12)	783
$C_4H_{10}S$	2-Methyl-2-propanethiol	(9.03)	762
$C_4H_{10}S$	Diethyl sulfide	(8.43)	730
$C_4H_{10}S$	Methyl propyl sulfide	(8.8)	767
$C_4H_{10}S$	Isopropyl methyl sulfide	(8.7)	749
$C_4H_{10}S_2$	Diethyl disulfide	(8.27)	724
$C_4H_{11}N$	Butylamine	8.7 ± 0.1	748
$C_4H_{11}N$	sec-Butylamine	8.46 ± 0.1	711
$C_4H_{11}N$	tert-Butylamine	8.46 ± 0.1	695
$C_4H_{11}N$	Isobutylamine	8.50 ± 0.1	721
$C_4H_{11}N$	Diethylamine	7.85 ± 0.1	684
$C_4H_{12}Si$	Tetramethylsilane	9.80 ± 0.04	713
$C_4H_{12}Sn$	Tetramethylstannane	8.89 ± 0.05	837
C_4NiO_4	Nickel carbonyl	8.27 ± 0.04	200
$C_5H_4O_2$	Furfural	9.22 ± 0.01	739
C_5H_5N	Pyridine	9.25	1031
C_5H_6	1-Penten-3-yne	9.00 ± 0.01	1119
C_5H_6	cis-3-Penten-1-yne	9.14 ± 0.04	1137
C_5H_6	trans-3-Penten-1-yne	9.05 ± 0.01	1128
C_5H_6	2-Methyl-1-buten-3-yne	9.25 ± 0.02	1152
C_5H_6	1,3-Cyclopentadiene	8.55 ± 0.02	955
C_5H_6O	2-Methylfuran	8.38 ± 0.02	729
C_5H_6O	3-Methylfuran	(8.64)	763
C_5H_6S	2-Methylthiophene	(8.14)	867
C_5H_6S	3-Methylthiophene	(8.40)	893
C_5H_8	cis-1,3-Pentadiene	8.63 ± 0.03	914
C_5H_8	trans-1,3-Pentadiene	8.59 ± 0.02	905
C_5H_8	1,4-Pentadiene	9.60 ± 0.02	1032
C_5H_8	2-Methyl-1,3-butadiene	8.84 ± 0.01	928
C_5H_8	1-Pentyne	10.10 ± 0.01	1119
C_5H_8	Cyclopentene	9.01 ± 0.01	905
C_5H_8	Spiropentane	(9.26)	1078
C_5H_8O	Cyclopropyl methyl ketone	(≤ 9.46)	796
C_5H_8O	Cyclopentanone	9.26 ± 0.01	701
C_5H_8O	3,4-Dihydro-2H-pyran	8.35 ± 0.01	681
$C_5H_8O_2$	Ethyl acrylate	(≤ 10.3)	617
$C_5H_8O_2$	Methyl methacrylate	(9.7)	589
$C_5H_8O_2$	2,4-Pentanedione	8.85 ± 0.01	469
C_5H_9NO	N-Methyl-2-pyrrolidone	(≤ 9.17)	≤ 676
C_5H_{10}	1-Pentene	9.51 ± 0.01	896
C_5H_{10}	cis-2-Pentene	9.01 ± 0.03	843
C_5H_{10}	trans-2-Pentene	9.04 ± 0.01	841
C_5H_{10}	2-Methyl-1-butene	9.12 ± 0.01	844
C_5H_{10}	3-Methyl-1-butene	9.52 ± 0.01	891
C_5H_{10}	2-Methyl-2-butene	8.69 ± 0.01	796
C_5H_{10}	Cyclopentane	(10.33 ± 0.15)	918
$C_5H_{10}O$	2,2-Dimethylpropanal	9.51 ± 0.01	675
$C_5H_{10}O$	Cyclopentanol	(9.72)	695
$C_5H_{10}O$	Pentanal	9.74 ± 0.04	709
$C_5H_{10}O$	2-Pentanone	9.38 ± 0.01	646
$C_5H_{10}O$	3-Pentanone	9.31 ± 0.01	640
$C_5H_{10}O$	3-Methyl-2-butanone	9.30 ± 0.01	635
$C_5H_{10}O$	Tetrahydropyran	9.25 ± 0.01	670
$C_5H_{10}O_2$	Pentanoic acid	(≤ 10.53)	≤ 527
$C_5H_{10}O_2$	3-Methylbutanoic acid	(≤ 10.51)	≤ 499
$C_5H_{10}O_2$	Butyl formate	10.52 ± 0.02	584
$C_5H_{10}O_2$	Propyl acetate	(≤ 9.92)	501
$C_5H_{10}O_2$	Isopropyl acetate	9.99 ± 0.03	482

Mol. Form.	Name	IP/eV	$\Delta_f H_{ion}$ kJ/mol
$C_5H_{10}O_2$	Ethyl propanoate	(10.00)	500
$C_5H_{10}O_2$	Methyl butanoate	(10.07)	520
$C_5H_{10}S$	Thiacyclohexane	(8.2)	728
$C_5H_{11}Br$	1-Bromopentane	10.10 ± 0.01	846
$C_5H_{11}I$	1-Iodopentane	9.20 ± 0.01	817
$C_5H_{11}N$	Piperidine	8.03 ± 0.11	726
$C_5H_{11}N$	N-Methylpyrrolidine	$\leq 8.41 \pm 0.02$	≤ 809
C_5H_{12}	Pentane	10.28 ± 0.10	845
C_5H_{12}	Isopentane	10.32 ± 0.05	843
C_5H_{12}	Neopentane	(≤ 10.2)	≤ 818
$C_5H_{12}O$	1-Pentanol	(10.00)	668
$C_5H_{12}O$	2-Pentanol	(9.78)	630
$C_5H_{12}O$	3-Pentanol	9.78	628
$C_5H_{12}O$	2-Methyl-1-butanol	(9.86)	649
$C_5H_{12}O$	2-Methyl-2-butanol	(9.8)	615
$C_5H_{12}O$	3-Methyl-2-butanol	(9.88 ± 0.13)	637
$C_5H_{12}O$	Butyl methyl ether	(9.4 ± 0.1)	648
$C_5H_{12}O$	Methyl $tert$-butyl ether	(9.24)	608
$C_5H_{12}O$	Ethyl propyl ether	(9.45)	640
$C_5H_{12}S$	$tert$-Butyl methyl sulfide	(8.38)	687
$C_5H_{12}S$	Ethyl propyl sulfide	(8.50)	716
$C_5H_{12}S$	Ethyl isopropyl sulfide	(8.35)	689
C_6BrF_5	Bromopentafluorobenzene	(9.67)	222
C_6ClF_5	Chloropentafluorobenzene	(9.72)	126
C_6Cl_6	Hexachlorobenzene	(8.98)	822
C_6F_6	Hexafluorobenzene	9.89 ± 0.04	8
C_6F_{12}	Perfluorocyclohexane	(13.2)	-1095
C_6HF_5	Pentafluorobenzene	(9.63)	122
C_6HF_5O	Pentafluorophenol	(9.20)	-71
$C_6H_2F_4$	1,2,3,4-Tetrafluorobenzene	(9.53)	284
$C_6H_2F_4$	1,2,3,5-Tetrafluorobenzene	(9.53)	263
$C_6H_2F_4$	1,2,4,5-Tetrafluorobenzene	(9.35)	254
$C_6H_3Cl_3$	1,2,4-Trichlorobenzene	(9.04)	880
$C_6H_3Cl_3$	1,3,5-Trichlorobenzene	9.32 ± 0.02	899
$C_6H_4ClNO_2$	1-Chloro-3-nitrobenzene	(9.92 ± 0.1)	995
$C_6H_4ClNO_2$	1-Chloro-4-nitrobenzene	(9.96 ± 0.1)	999
$C_6H_4Cl_2$	o-Dichlorobenzene	9.06 ± 0.02	907
$C_6H_4Cl_2$	m-Dichlorobenzene	9.10 ± 0.02	906
$C_6H_4Cl_2$	p-Dichlorobenzene	8.92 ± 0.02	885
$C_6H_4FNO_2$	1-Fluoro-4-nitrobenzene	(9.90)	826
$C_6H_4F_2$	o-Difluorobenzene	9.29 ± 0.01	602
$C_6H_4F_2$	m-Difluorobenzene	9.33 ± 0.01	591
$C_6H_4F_2$	p-Difluorobenzene	9.1589 ± 0.0003	577
$C_6H_4O_2$	p-Benzoquinone	10.01 ± 0.06	844
C_6H_5Br	Bromobenzene	9.00 ± 0.02	971
C_6H_5Cl	Chlorobenzene	9.07 ± 0.02	930
C_6H_5ClO	m-Chlorophenol	8.655 ± 0.001	680
C_6H_5ClO	p-Chlorophenol	(≤ 8.69)	≤ 692
C_6H_5F	Fluorobenzene	9.20 ± 0.01	772
C_6H_5I	Iodobenzene	8.685	1003
$C_6H_5NO_2$	Nitrobenzene	9.86 ± 0.02	1019
$C_6H_5NO_3$	o-Nitrophenol	(9.1)	782
$C_6H_5NO_3$	m-Nitrophenol	(9.0)	755
$C_6H_5NO_3$	p-Nitrophenol	(9.1)	761
C_6H_6	Benzene	9.24378 ± 0.00007	975
C_6H_6	Fulvene	(8.36)	1031
C_6H_6ClN	o-Chloroaniline	(8.50)	883
C_6H_6ClN	m-Chloroaniline	(8.09)	835
C_6H_6ClN	p-Chloroaniline	(≤ 8.18)	≤ 844
$C_6H_6N_2O_2$	o-Nitroaniline	(8.27)	861

Mol. Form.	Name	IP/eV	$\Delta_f H_{ion}$ kJ/mol
$C_6H_6N_2O_2$	m-Nitroaniline	(8.31)	865
$C_6H_6N_2O_2$	p-Nitroaniline	(8.34)	859
C_6H_6O	Phenol	8.49 ± 0.02	723
$C_6H_6O_2$	p-Hydroquinone	7.94 ± 0.01	503
C_6H_6S	Benzenethiol	(8.32)	915
C_6H_7N	Aniline	7.720 ± 0.002	832
C_6H_7N	2-Methylpyridine	(9.02)	970
C_6H_7N	3-Methylpyridine	(9.04)	979
C_6H_7N	4-Methylpyridine	(9.04)	976
$C_6H_8N_2$	o-Phenylenediamine	(7.2)	787
$C_6H_8N_2$	m-Phenylenediamine	(7.14)	777
$C_6H_8N_2$	p-Phenylenediamine	(6.87 ± 0.05)	759
C_6H_{10}	1,5-Hexadiene	9.27 ± 0.05	978
C_6H_{10}	1-Hexyne	10.03 ± 0.05	1089
C_6H_{10}	3,3-Dimethyl-1-butyne	9.90 ± 0.04	1060
C_6H_{10}	Cyclohexene	8.945 ± 0.01	859
$C_6H_{10}O$	Cyclohexanone	9.14 ± 0.01	656
$C_6H_{10}O$	Mesityl oxide	9.10 ± 0.01	694
$C_6H_{10}O_4$	Diethyl oxalate	(9.8)	205
$C_6H_{11}NO$	Caprolactam	(9.07 ± 0.02)	629
C_6H_{12}	1-Hexene	9.44 ± 0.04	869
C_6H_{12}	cis-2-Hexene	(8.97 ± 0.01)	818
C_6H_{12}	trans-2-Hexene	(8.97 ± 0.01)	814
C_6H_{12}	2-Methyl-1-pentene	(9.08 ± 0.01)	817
C_6H_{12}	4-Methyl-1-pentene	9.45 ± 0.01	862
C_6H_{12}	2-Methyl-2-pentene	(8.58)	761
C_6H_{12}	4-Methyl-cis-2-pentene	8.98 ± 0.01	809
C_6H_{12}	4-Methyl-trans-2-pentene	(8.97 ± 0.01)	804
C_6H_{12}	2-Ethyl-1-butene	(9.06 ± 0.02)	818
C_6H_{12}	2,3-Dimethyl-1-butene	(9.07 ± 0.01)	812
C_6H_{12}	2,3-Dimethyl-2-butene	8.27 ± 0.01	729
C_6H_{12}	Cyclohexane	9.86 ± 0.03	828
C_6H_{12}	Methylcyclopentane	(9.85)	845
$C_6H_{12}O$	Hexanal	9.72 ± 0.05	691
$C_6H_{12}O$	2-Hexanone	9.3 ± 0.1	626
$C_6H_{12}O$	3-Hexanone	9.12 ± 0.02	600
$C_6H_{12}O$	3-Methyl-2-pentanone	9.21 ± 0.01	600
$C_6H_{12}O$	4-Methyl-2-pentanone	9.30 ± 0.01	609
$C_6H_{12}O$	2-Methyl-3-pentanone	9.10 ± 0.01	592
$C_6H_{12}O$	3,3-Dimethyl-2-butanone	9.12 ± 0.02	589
$C_6H_{12}O$	Cyclohexanol	(9.75)	651
$C_6H_{12}O_2$	Hexanoic acid	≤ 10.12	≤ 463
$C_6H_{12}O_2$	Butyl acetate	(9.92 ± .05)	471
$C_6H_{12}O_2$	sec-Butyl acetate	9.90	453
$C_6H_{12}O_2$	Methyl 2,2-dimethylpropanoate	(9.90 ± 0.04)	466
$C_6H_{13}I$	1-Iodohexane	9.179	794
$C_6H_{13}N$	Cyclohexylamine	(8.86)	750
C_6H_{14}	Hexane	10.13	810
C_6H_{14}	2-Methylpentane	(10.12)	802
C_6H_{14}	3-Methylpentane	(10.08)	801
C_6H_{14}	2,2-Dimethylbutane	(10.06)	787
C_6H_{14}	2,3-Dimethylbutane	(10.02)	791
$C_6H_{14}O$	1-Hexanol	(9.89)	639
$C_6H_{14}O$	2-Hexanol	(9.80 ± 0.03)	611
$C_6H_{14}O$	3-Hexanol	(9.63 ± 0.03)	599
$C_6H_{14}O$	Dipropyl ether	(9.27)	602
$C_6H_{14}O$	Diisopropyl ether	9.20 ± 0.05	569
$C_6H_{14}O$	Butyl ethyl ether	(9.36)	610
$C_6H_{14}O$	Methyl pentyl ether	(≤ 9.67)	≤ 657
$C_6H_{14}O_2$	1,1-Diethoxyethane	(9.2)	434

Mol. Form.	Name	IP/eV	$\Delta_f H_{ion}$ kJ/mol
$C_6H_{14}O_3$	Diethylene glycol dimethyl ether	≤ 9.8	≤ 448
$C_6H_{14}S$	Dipropyl sulfide	8.30 ± 0.02	676
$C_6H_{14}S$	Diisopropyl sulfide	(8.2 ± 0.2)	649
$C_6H_{15}N$	Hexylamine	(8.63 ± 0.05)	699
$C_6H_{15}N$	Dipropylamine	(7.84 ± 0.02)	641
$C_6H_{15}N$	Diisopropylamine	(7.73 ± 0.03)	602
$C_6H_{15}N$	Triethylamine	(7.50 ± 0.02)	631
$C_6H_{15}NO_3$	Triethanolamine	(7.9)	206
$C_7H_3F_5$	2,3,4,5,6-Pentafluorotoluene	(9.4)	64
C_7H_5ClO	Benzoyl chloride	(9.53)	815
$C_7H_5Cl_3$	(Trichloromethyl)benzene	(≤ 9.60)	≤ 914
$C_7H_5F_3$	(Trifluoromethyl)benzene	9.685 ± 0.005	335
C_7H_5N	Benzonitrile	9.70 ± 0.01	1154
C_7H_6O	Benzaldehyde	9.49 ± 0.02	878
$C_7H_6O_2$	Benzoic acid	(9.3)	604
C_7H_7Br	p-Bromotoluene	8.67 ± 0.02	908
C_7H_7Cl	o-Chlorotoluene	(8.7 ± 0.1)	856
C_7H_7Cl	m-Chlorotoluene	(8.83)	869
C_7H_7Cl	p-Chlorotoluene	(8.69)	855
C_7H_7Cl	(Chloromethyl)benzene	9.10 ± 0.02	897
C_7H_7F	o-Fluorotoluene	8.91 ± 0.01	709
C_7H_7F	m-Fluorotoluene	8.91 ± 0.01	709
C_7H_7F	p-Fluorotoluene	8.79 ± 0.01	701
C_7H_7NO	Benzamide	(9.25)	792
$C_7H_7NO_2$	o-Nitrotoluene	9.24	946
$C_7H_7NO_2$	m-Nitrotoluene	9.45 ± 0.1	941
$C_7H_7NO_2$	p-Nitrotoluene	9.46 ± 0.05	942
C_7H_8	Toluene	8.8276 ± 0.0006	901
C_7H_8O	o-Cresol	(8.24)	670
C_7H_8O	m-Cresol	8.29 ± 0.07	668
C_7H_8O	p-Cresol	(8.3)	675
C_7H_8O	Benzyl alcohol	(8.3)	701
C_7H_8O	Anisole	8.22 ± 0.03	725
C_7H_9N	Benzylamine	(8.64)	917
C_7H_9N	o-Methylaniline	(7.44 ± 0.02)	772
C_7H_9N	m-Methylaniline	(7.50 ± 0.02)	778
C_7H_9N	p-Methylaniline	(7.24 ± 0.02)	753
C_7H_9N	N-Methylaniline	7.34 ± 0.04	792
C_7H_9N	2,3-Dimethylpyridine	(8.85 ± 0.02)	922
C_7H_9N	2,4-Dimethylpyridine	(8.85 ± 0.03)	918
C_7H_9N	2,5-Dimethylpyridine	(≤ 8.80 ± 0.05)	≤ 916
C_7H_9N	2,6-Dimethylpyridine	8.86 ± 0.03	913
C_7H_9N	3,4-Dimethylpyridine	(≤ 9.15)	≤ 953
C_7H_9N	3,5-Dimethylpyridine	(≤ 9.25)	≤ 965
$C_7H_{10}O$	Dicyclopropyl ketone	(9.1)	1041
C_7H_{14}	1-Heptene	9.34 ± 0.10	839
C_7H_{14}	trans-3-Heptene	(8.92)	790
C_7H_{14}	Cycloheptane	9.97	844
C_7H_{14}	Methylcyclohexane	9.64	775
C_7H_{14}	cis-1,2-Dimethylcyclopentane	(9.92 ± 0.05)	828
C_7H_{14}	trans-1,2-Dimethylcyclopentane	9.7 ± 0.2	799
$C_7H_{14}O$	1-Heptanal	(9.65)	668
$C_7H_{14}O$	2-Heptanone	9.28 ± 0.10	594
$C_7H_{14}O$	3-Heptanone	9.18 ± 0.08	589
$C_7H_{14}O$	4-Heptanone	9.10 ± 0.06	577
$C_7H_{14}O$	5-Methyl-2-hexanone	(9.28)	586
$C_7H_{14}O$	2,4-Dimethyl-3-pentanone	8.95 ± 0.01	552
$C_7H_{14}O$	1-Methylcyclohexanol	(9.8 ± 0.2)	586
C_7H_{16}	Heptane	9.93 ± 0.10	771
$C_7H_{16}O$	1-Heptanol	(9.84)	614

Mol. Form.	Name	IP/eV	$\Delta_f H_{ion}$ kJ/mol
$C_7H_{16}O$	2-Heptanol	(9.70)	580
$C_7H_{16}O$	3-Heptanol	(9.68)	578
$C_7H_{16}O$	4-Heptanol	(9.61)	572
$C_7H_{16}O$	Ethyl pentyl ether	($\leq$ 9.49)	$\leq$ 602
$C_8H_4O_3$	Phthalic anhydride	(10.1)	603
$C_8H_6O_4$	Isophthalic acid	(9.98)	268
$C_8H_6O_4$	Terephthalic acid	(9.86)	232
C_8H_7N	2-Methylbenzonitrile	($\leq$ 9.38)	1085
C_8H_7N	3-Methylbenzonitrile	($\leq$ 9.34)	1085
C_8H_7N	4-Methylbenzonitrile	9.32 ± 0.02	1083
C_8H_7N	Indole	7.7602 ± 0.0006	908
C_8H_8	Styrene	8.464 ± 0.001	964
C_8H_8O	p-Tolualdehyde	(9.33)	825
C_8H_8O	Acetophenone	9.29 ± 0.03	810
$C_8H_8O_2$	o-Toluic acid	(9.1)	558
$C_8H_8O_2$	m-Toluic acid	(9.43)	579
$C_8H_8O_2$	p-Toluic acid	(9.23)	560
$C_8H_8O_2$	Benzeneacetic acid	(8.26)	479
$C_8H_8O_2$	Methyl benzoate	9.32 ± 0.03	611
C_8H_{10}	Ethylbenzene	8.77 ± 0.01	876
C_8H_{10}	o-Xylene	8.56 ± 0.01	844
C_8H_{10}	m-Xylene	8.56 ± 0.01	843
C_8H_{10}	p-Xylene	8.44 ± 0.01	832
$C_8H_{10}O$	p-Ethylphenol	(7.84)	613
$C_8H_{10}O$	2,3-Xylenol	(8.26)	640
$C_8H_{10}O$	2,4-Xylenol	(8.0)	609
$C_8H_{10}O$	2,6-Xylenol	(8.05)	615
$C_8H_{10}O$	3,4-Xylenol	(8.09)	624
$C_8H_{10}O$	Phenetole	(8.13)	683
$C_8H_{11}N$	2,4,6-Trimethylpyridine	($\leq$ 8.9)	$\leq$ 880
$C_8H_{11}N$	N-Ethylaniline	($\leq$ 7.67)	$\leq$ 794
$C_8H_{11}N$	N,N-Dimethylaniline	7.12 ± 0.02	787
C_8H_{14}	1-Octyne	(9.95 ± 0.02)	1040
C_8H_{14}	2-Octyne	9.31 ± 0.01	961
C_8H_{14}	3-Octyne	9.22 ± 0.01	952
C_8H_{14}	4-Octyne	9.20 ± 0.01	946
C_8H_{16}	1-Octene	9.43 ± 0.01	829
C_8H_{16}	Cyclooctane	9.75 ± 0.05	816
C_8H_{16}	Ethylcyclohexane	(9.54)	748
C_8H_{16}	1,1-Dimethylcyclohexane	(9.42)	728
C_8H_{16}	cis-1,2-Dimethylcyclohexane	(<9.78)	772
C_8H_{16}	$trans$-1,2-Dimethylcyclohexane	9.41	728
C_8H_{16}	cis-1,3-Dimethylcyclohexane	(<9.98)	778
C_8H_{16}	$trans$-1,3-Dimethylcyclohexane	9.53	743
C_8H_{16}	cis-1,4-Dimethylcyclohexane	(<9.93)	782
C_8H_{16}	$trans$-1,4-Dimethylcyclohexane	(9.56)	738
C_8H_{16}	Propylcyclopentane	(9.34)	753
$C_8H_{16}O$	2,2,4-Trimethyl-3-pentanone	(8.80)	511
C_8H_{18}	Octane	9.80 ± 0.10	737
C_8H_{18}	2-Methylheptane	(9.84)	734
C_8H_{18}	2,2,4-Trimethylpentane	(9.86)	713
C_8H_{18}	2,2,3,3-Tetramethylbutane	9.8	720
$C_8H_{18}O$	Dibutyl ether	(9.28)	s 560
$C_8H_{18}O$	Di-sec-butyl ether	(9.11)	511
$C_8H_{18}O$	Di-$tert$-butyl ether	8.88 ± 0.07	493
$C_8H_{18}S$	Dibutyl sulfide	(8.2)	624
$C_8H_{18}S$	Di-$tert$-butyl sulfide	(8.0)	583
$C_8H_{18}S$	Diisobutyl sulfide	(8.34)	625
$C_8H_{19}N$	Dibutylamine	(7.69)	586
$C_8H_{19}N$	Diisobutylamine	(7.8)	574

Mol. Form.	Name	IP/eV	$\Delta_f H_{ion}$ kJ/mol
$C_8H_{20}Si$	Tetraethylsilane	(8.9)	595
C_9H_7N	Quinoline	8.62 ± 0.01	1041
C_9H_7N	Isoquinoline	8.53 ± 0.03	1032
C_9H_8	Indene	8.14 ± 0.01	949
C_9H_{10}	o-Methylstyrene	(8.20)	908
C_9H_{10}	m-Methylstyrene	(8.15)	899
C_9H_{10}	p-Methylstyrene	(8.1)	895
C_9H_{10}	Cyclopropylbenzene	(8.35)	956
C_9H_{10}	Indan	(8.3)	864
$C_9H_{10}O_2$	Ethyl benzoate	(8.9)	537
C_9H_{12}	Propylbenzene	8.713 ± 0.010	848
C_9H_{12}	Isopropylbenzene	8.73 ± 0.01	847
C_9H_{12}	1,2,3-Trimethylbenzene	8.42 ± 0.02	803
C_9H_{12}	1,2,4-Trimethylbenzene	8.27 ± 0.01	784
C_9H_{12}	1,3,5-Trimethylbenzene	8.41 ± 0.01	796
$C_9H_{13}N$	N,N-Dimethyl-o-toluidine	7.40 ± 0.02	814
$C_9H_{14}O$	Isophorone	(≤ 9.07)	≤ 670
C_9H_{18}	Butylcyclopentane	(9.95)	793
C_9H_{18}	Propylcyclohexane	(9.46)	720
C_9H_{18}	Isopropylcyclohexane	(9.33)	704
$C_9H_{18}O$	2-Nonanone	(9.16)	545
$C_9H_{18}O$	5-Nonanone	(9.07)	530
$C_9H_{18}O$	2,6-Dimethyl-4-heptanone	9.01 ± 0.06	512
C_9H_{20}	Nonane	9.71 ± 0.10	709
$C_{10}F_8$	Perfluoronaphthalene	8.85	-368
$C_{10}H_7Br$	1-Bromonaphthalene	8.08 ± 0.03	955
$C_{10}H_7Cl$	1-Chloronaphthalene	(8.13)	906
$C_{10}H_8$	Naphthalene	8.1442 ± 0.0009	936
$C_{10}H_8$	Azulene	7.38 ± 0.05	1001
$C_{10}H_8O$	1-Naphthol	7.76 ± 0.03	719
$C_{10}H_8O$	2-Naphthol	7.87 ± 0.06	729
$C_{10}H_{10}O_4$	Dimethyl phthalate	(9.64 ± 0.07)	277
$C_{10}H_{12}$	1,2,3,4-Tetrahydronaphthalene	8.46 ± 0.02	841
$C_{10}H_{14}$	Butylbenzene	8.69 ± 0.02	826
$C_{10}H_{14}$	sec-Butylbenzene	8.68 ± 0.02	820
$C_{10}H_{14}$	tert-Butylbenzene	8.68 ± 0.05	816
$C_{10}H_{14}$	Isobutylbenzene	8.69 ± 0.02	817
$C_{10}H_{14}$	p-Cymene	(8.29)	771
$C_{10}H_{14}$	o-Diethylbenzene	(≤ 8.51)	≤ 804
$C_{10}H_{14}$	m-Diethylbenzene	(8.49)	798
$C_{10}H_{14}$	p-Diethylbenzene	(8.40)	790
$C_{10}H_{14}$	1,2,4,5-Tetramethylbenzene	8.04 ± 0.02	730
$C_{10}H_{14}O$	p-tert-Butylphenol	(7.8)	552
$C_{10}H_{16}$	α-Pinene	(8.07)	808
$C_{10}H_{16}O$	Camphor	(8.76)	577
$C_{10}H_{18}$	cis-Decahydronaphthalene	9.36 ± 0.04	734
$C_{10}H_{18}$	trans-Decahydronaphthalene	9.34 ± 0.04	720
$C_{10}H_{20}$	1-Decene	9.42 ± 0.05	786
$C_{10}H_{20}$	Butylcyclohexane	(9.41)	695
$C_{10}H_{22}$	Decane	(9.65)	682
$C_{11}H_{10}$	1-Methylnaphthalene	7.97 ± 0.03	882
$C_{11}H_{10}$	2-Methylnaphthalene	7.91 ± 0.08	877
$C_{11}H_{16}$	p-tert-Butyltoluene	(8.12)	730
$C_{11}H_{24}$	Undecane	(9.56)	650
$C_{11}H_{24}$	2-Methyldecane	(9.7)	658
$C_{12}H_8$	Acenaphthylene	(8.22)	1053
$C_{12}H_9N$	Carbazole	(7.57)	961
$C_{12}H_{10}$	Acenaphthene	7.75 ± 0.07	903
$C_{12}H_{10}$	Biphenyl	8.23 ± 0.10	977
$C_{12}H_{10}N_2O$	trans-Azoxybenzene	(8.1)	1123

Mol. Form.	Name	IP/eV	$\Delta_f H_{ion}$ kJ/mol
$C_{12}H_{10}O$	Diphenyl ether	(8.09)	766
$C_{12}H_{11}N$	Diphenylamine	7.16 ± 0.04	908
$C_{12}H_{18}$	5,7-Dodecadiyne	(8.67)	1079
$C_{12}H_{18}$	Hexamethylbenzene	7.85 ± 0.01	670
$C_{12}H_{22}$	Cyclohexylcyclohexane	(9.41)	690
$C_{12}H_{27}N$	Tributylamine	(7.4)	492
$C_{13}H_{10}$	9H-Fluorene	7.91 ± 0.02	952
$C_{13}H_{10}O$	Benzophenone	9.08 ± 0.05	926
$C_{13}H_{12}$	Diphenylmethane	(8.55)	963
$C_{14}H_{10}$	Anthracene	7.439 ± 0.006	948
$C_{14}H_{10}$	Phenanthrene	7.8914 ± 0.0006	966
$C_{14}H_{10}$	Diphenylacetylene	7.94 ± 0.03	1168
$C_{14}H_{12}$	cis-Stilbene	(7.80)	1005
$C_{14}H_{12}$	trans-Stilbene	7.656 ± 0.001	973
$C_{14}H_{14}$	1,2-Diphenylethane	8.9 ± 0.1	1002
$C_{16}H_{10}$	Fluoranthene	7.9 ± 0.1	1052
$C_{16}H_{10}$	Pyrene	7.4256 ± 0.0006	935
$C_{18}H_{12}$	Chrysene	7.60 ± 0.01	1017
$C_{18}H_{14}$	o-Terphenyl	(7.99)	1056
$C_{18}H_{14}$	m-Terphenyl	(8.01)	1057
$C_{18}H_{14}$	p-Terphenyl	7.80 ± 0.03	1037
$C_{20}H_{12}$	Perylene	6.960 ± 0.001	981
$C_{24}H_{12}$	Coronene	7.29 ± 0.01	1026

X-RAY ATOMIC ENERGY LEVELS

The energy levels in this tables are the values recommended by Bearden and Burr on the basis of a thorough review of the literature on x-ray wavelengths and related data. All values are in electron volts (eV). Values in parentheses are interpolated, and an asterisk * indicates a level which is not resolved from the level above it. See Reference 1 for uncertainties in the levels and a complete description of how the recommended values were obtained.

REFERENCES

1. Bearden, J. A., and Burr, A. F., *Rev. Mod. Phys.*, 39, 125, 1967; also published as *X-Ray Wavelengths and X-Ray Atomic Energy Levels*, Natl. Stand. Ref. Data Sys.- Natl. Bur. Standards (U.S.), No. 14, 1967.
2. Gray, D. E., Editor, *American Institute of Physics Handbook, Third Edition*, pp. 7-158 to 7-167, McGraw-Hill, New York, 1972.

Level	^{1}H	^{2}He	^{3}Li	^{4}Be	^{5}B	^{6}C	^{7}N	^{8}O
K	13.59811	24.58678	54.75	111.0	188.0	283.8	401.6	532.0
L_I								23.7
$L_{II,III}$					4.7	6.4	9.2	7.1

Level	^{9}F	^{10}Ne	^{11}Na	^{12}Mg	^{13}Al	^{14}Si	^{15}P	^{16}S
K	685.4	866.9	1072.1	1305.0	1559.6	1838.9	2145.5	2472.0
L_I	(31)	(45)	63.3	89.4	117.7	148.7	189.3	229.2
$L_{II,III}$	8.6	18.3	31.1	51.4	73.1	99.2	132.2	164.8

Level	^{17}Cl	^{18}Ar	^{19}K	^{20}Ca	^{21}Sc	^{22}Ti	^{23}V	^{24}Cr
K	2822.4	3202.9	3607.4	4038.1	4492.8	4966.4	5465.1	5989.2
L_I	270.2	320	377.1	437.8	500.4	563.7	628.2	694.6
L_{II}	201.6	247.3	296.3	350.0	406.7	461.5	520.5	583.7
L_{III}	200.0	245.2	293.6	346.4	402.2	455.5	512.9	574.5
M_I	17.5	25.3	33.9	43.7	53.8	60.3	66.5	74.1
$M_{II,III}$	6.8	12.4	17.8	25.4	32.3	34.6	37.8	42.5
$M_{IV,V}$					6.6	3.7	2.2	2.3

Level	^{25}Mn	^{26}Fe	^{27}Co	^{28}Ni	^{29}Cu	^{30}Zn	^{31}Ga	^{32}Ge
K	6539.0	7112.0	7708.9	8332.8	8978.9	9658.6	10367.1	11103.1
L_I	769.0	846.1	925.6	1008.1	1096.1	1193.6	1297.7	1414.3
L_{II}	651.4	721.1	793.6	871.9	951.0	1042.8	1142.3	1247.8
L_{III}	640.3	708.1	778.6	854.7	931.1	1019.7	1115.4	1216.7
M_I	83.9	92.9	100.7	111.8	119.8	135.9	158.1	180.0
M_{II}	48.6	54.0	59.5	68.1	73.6	86.6	106.8	127.9
M_{III}	48.6*	54.0*	59.5*	68.1*	73.6*	86.6*	102.9	120.8
$M_{IV,V}$	3.3	3.6	2.9	3.6	1.6	8.1	17.4	28.7

Level	^{33}As	^{34}Se	^{35}Br	^{36}Kr	^{37}Rb	^{38}Sr	^{39}Y	^{40}Zr
K	11866.7	12657.8	13473.7	14325.6	15199.7	16104.6	17038.4	17997.6
L_I	1526.5	1653.9	1782.0	1921.0	2065.1	2216.3	2372.5	2531.6
L_{II}	1358.6	1476.2	1596.0	1727.2	1863.9	2006.8	2155.5	2306.7
L_{III}	1323.1	1435.8	1549.9	1674.9	1804.4	1939.6	2080.0	2222.3
M_I	203.5	231.5	256.5		322.1	357.5	393.6	430.3
M_{II}	146.4	168.2	189.3	222.7	247.4	279.8	312.4	344.2
M_{III}	140.5	161.9	181.5	213.8	238.5	269.1	300.3	330.5
M_{IV}	41.2	56.7	70.1	88.9	111.8	135.0	159.6	182.4
M_V	41.2*	56.7*	69.0	88.9*	110.3	133.1	157.4	180.0
N_I			27.3	24.0	29.3	37.7	45.4	51.3
N_{II}	2.5	5.6	5.2	10.6	14.8	19.9	25.6	28.7
N_{III}	2.5*	5.6*	4.6	10.6*	14.0	19.9*	25.6*	28.7*

Level	^{41}Nb	^{42}Mo	^{43}Tc	^{44}Ru	^{45}Rh	^{46}Pd	^{47}Ag	^{48}Cd
K	18985.6	19999.5	21044.0	22117.2	23219.9	24350.3	25514.0	26711.2
L$_I$	2697.7	2865.5	3042.5	3224.0	3411.9	3604.3	3805.8	4018.0
L$_{II}$	2464.7	2625.1	2793.2	2966.9	3146.1	3330.3	3523.7	3727.0
L$_{III}$	2370.5	2520.2	2676.9	2837.9	3003.8	3173.3	3351.1	3537.5
M$_I$	468.4	504.6		585.0	627.1	669.9	717.5	770.2
M$_{II}$	378.4	409.7	444.9	482.8	521.0	559.1	602.4	650.7
M$_{III}$	363.0	392.3	425.0	460.6	496.2	531.5	571.4	616.5
M$_{IV}$	207.4	230.3	256.4	283.6	311.7	340.0	372.8	410.5
M$_V$	204.6	227.0	252.9	279.4	307.0	334.7	366.7	403.7
N$_I$	58.1	61.8		74.9	81.0	86.4	95.2	107.6
N$_{II}$	33.9	34.8	38.9	43.1	47.9	51.1	62.6	66.9
N$_{III}$	33.9*	34.8*	38.9*	43.1*	47.9*	51.1*	55.9	66.9*
N$_{IV,V}$	3.2	1.8		2.0	2.5	1.5	3.3	9.3

Level	^{49}In	^{50}Sn	^{51}Sb	^{52}Te	^{53}I	^{54}Xe	^{55}Cs	^{56}Ba
K	27939.9	29200.1	30491.2	31813.8	33169.4	34561.4	35984.6	37440.6
L$_I$	4237.5	4464.7	4698.3	4939.2	5188.1	5452.8	5714.3	5988.8
L$_{II}$	3938.0	4156.1	4380.4	4612.0	4852.1	5103.7	5359.4	5623.6
L$_{III}$	3730.1	3928.8	4132.2	4341.4	4557.1	4782.2	5011.9	5247.0
M$_I$	825.6	883.8	943.7	1006.0	1072.1		1217.1	1292.8
M$_{II}$	702.2	756.4	811.9	869.7	930.5	999.0	1065.0	1136.7
M$_{III}$	664.3	714.4	765.6	818.7	874.6	937.0	997.6	1062.2
M$_{IV}$	450.8	493.3	536.9	582.5	631.3		739.5	796.1
M$_V$	443.1	484.8	527.5	572.1	619.4	672.3	725.5	780.7
N$_I$	121.9	136.5	152.0	168.3	186.4		230.8	253.0
N$_{II}$	77.4	88.6	98.4	110.2	122.7	146.7	172.3	191.8
N$_{III}$	77.4*	88.6*	98.4*	110.2*	122.7*	146.7*	161.6	179.7
N$_{IV}$	16.2	23.9	31.4	39.8	49.6		78.8	92.5
N$_V$	16.2*	23.9*	31.4*	39.8*	49.6*		76.5	89.9
O$_I$	0.1	0.9	6.7	11.6	13.6		22.7	39.1
O$_{II}$	0.8	1.1	2.1	2.3	3.3		13.1	16.6
O$_{III}$	0.8*	1.1*	2.1*	2.3*	3.3*		11.4	14.6

Level	^{57}La	^{58}Ce	^{59}Pr	^{60}Nd	^{61}Pm	^{62}Sm	^{63}Eu	^{64}Gd
K	38924.6	40443.0	41990.6	43568.9	45184.0	46834.2	48519.0	50239.1
L$_I$	6266.3	6548.8	6834.8	7126.0	7427.9	7736.8	8052.0	8375.6
L$_{II}$	5890.6	6164.2	6440.4	6721.5	7012.8	7311.8	7617.1	7930.3
L$_{III}$	5482.7	5723.4	5964.3	6207.9	6459.3	6716.2	6976.9	7242.8
M$_I$	1361.3	1434.6	1511.0	1575.3		1722.8	1800.0	1880.8
M$_{II}$	1204.4	1272.8	1337.4	1402.8	1471.4	1540.7	1613.9	1688.3
M$_{III}$	1123.4	1185.4	1242.2	1297.4	1356.9	1419.8	1480.6	1544.0
M$_{IV}$	848.5	901.3	951.1	999.9	1051.5	1106.0	1160.6	1217.2
M$_V$	831.7	883.3	931.0	977.7	1026.9	1080.2	1130.9	1185.2
N$_I$	270.4	289.6	304.5	315.2		345.7	360.2	375.8
N$_{II}$	205.8	223.3	236.3	243.3	242	265.6	283.9	288.5
N$_{III}$	191.4	207.2	217.6	224.6	242*	247.4	256.6	270.9
N$_{IV,V}$	98.9	110.0	113.2	117.5	120.4	129.0	133.2	140.5
N$_{VI,VII}$		0.1	2.0	1.5		5.5	0.0	0.1
O$_I$	32.3	37.8	37.4	37.5		37.4	31.8	36.1
O$_{II,III}$	14.4	19.8	22.3	21.1		21.3	22.0	20.3

Level	^{65}Tb	^{66}Dy	^{67}Ho	^{68}Er	^{69}Tm	^{70}Yb	^{71}Lu	^{72}Hf
K	51995.7	53788.5	55617.7	57485.5	59389.6	61332.3	63313.8	65350.8
L$_I$	8708.0	9045.8	9394.2	9751.3	10115.7	10486.4	10870.4	11270.7
L$_{II}$	8251.6	8580.6	8917.8	9264.3	9616.9	9978.2	10348.6	10739.4
L$_{III}$	7514.0	7790.1	8071.1	8357.9	8648.0	8943.6	9244.1	9560.7
M$_I$	1967.5	2046.8	2128.3	2206.5	2306.8	2398.1	2491.2	2600.9

Level	^{65}Tb	^{66}Dy	^{67}Ho	^{68}Er	^{69}Tm	^{70}Yb	^{71}Lu	^{72}Hf
M_{II}	1767.7	1841.8	1922.8	2005.8	2089.8	2173.0	2263.5	2365.4
M_{III}	1611.3	1675.6	1741.2	1811.8	1884.5	1949.8	2023.6	2107.6
M_{IV}	1275.0	1332.5	1391.5	1453.3	1514.6	1576.3	1639.4	1716.4
M_V	1241.2	1294.9	1351.4	1409.3	1467.7	1527.8	1588.5	1661.7
N_I	397.9	416.3	435.7	449.1	471.7	487.2	506.2	538.1
N_{II}	310.2	331.8	343.5	366.2	385.9	396.7	410.1	437.0
N_{III}	385.0	292.9	306.6	320.0	336.6	343.5	359.3	380.4
N_{IV}	147.0	154.2	161.0	176.7	179.6	198.1	204.8	223.8
N_V	147.0*	154.2*	161.0*	167.6	179.6*	184.9	195.0	213.7
$N_{VI,VII}$	2.6	4.2	3.7	4.3	5.3	6.3	6.9	17.1
O_I	39.0	62.9	51.2	59.8	53.2	54.1	56.8	64.9
O_{II}	25.4	26.3	20.3	29.4	32.3	23.4	28.0	38.1
O_{III}	25.4*	26.3*	20.3*	29.4*	32.3*	23.4*	28.0*	30.6

Level	^{73}Ta	^{74}W	^{75}Re	^{76}Os	^{77}Ir	^{78}Pt	^{79}Au	^{80}Hg
K	67416.4	69525.0	71676.4	73870.8	76111.0	78394.8	80724.9	83102.3
L_I	11681.5	12099.8	12526.7	12968.0	13418.5	13879.9	14352.8	14839.3
L_{II}	11136.1	11544.0	11958.7	12385.0	12824.1	13272.6	13733.6	14208.7
L_{III}	9881.1	10206.8	10535.3	10870.9	11215.2	11563.7	11918.7	12283.9
M_I	2708.0	2819.6	2931.7	3048.5	3173.7	3296.0	3424.9	3561.6
M_{II}	2468.7	2574.9	2681.6	2792.2	2908.7	3026.5	3147.8	3278.5
M_{III}	2194.0	2281.0	2367.3	2457.2	2550.7	2645.4	2743.0	2847.1
M_{IV}	1793.2	1871.6	1948.9	2030.8	2116.1	2201.9	2291.1	2384.9
M_V	1735.1	1809.2	1882.9	1960.1	2040.4	2121.6	2205.7	2294.9
N_I	565.5	595.0	625.0	654.3	690.1	722.0	758.8	800.3
N_{II}	464.8	491.6	517.9	546.5	577.1	609.2	643.7	676.9
N_{III}	404.5	425.3	444.4	468.2	494.3	519.0	545.4	571.0
N_{IV}	241.3	258.8	273.7	289.4	311.4	330.8	352.0	378.3
N_V	229.3	245.4	260.2	272.8	294.9	313.3	333.9	359.8
N_{VI}	25.0	36.5	40.6	46.3	63.4	74.3	86.4	102.2
N_{VII}	25.0*	33.6	40.6*	46.3*	60.5	71.1	82.8	98.5
O_I	71.1	77.1	82.8	83.7	95.2	101.7	107.8	120.3
O_{II}	44.9	46.8	45.6	58.0	63.0	65.3	71.7	80.5
O_{III}	36.4	35.6	34.6	45.4	50.5	51.7	53.7	57.6
$O_{IV,V}$	5.7	6.1	3.5		3.8	2.2	2.5	6.4

Level	^{81}Tl	^{82}Pb	^{83}Bi	^{84}Po	^{85}At	^{86}Rn	^{87}Fr	^{88}Ra
K	85530.4	88004.5	90525.9	93105.0	95729.9	98404	101137	103921.9
L_I	15346.7	15860.8	16387.5	16939.3	17493	18049	18639	19236.7
L_{II}	14697.9	15200.0	15711.1	16244.3	16784.7	17337.1	17906.5	18484.3
L_{III}	12657.5	13035.2	13418.6	13813.8	14213.5	14619.4	15031.2	15444.4
M_I	3704.1	3850.7	3999.1	4149.4	(4317)	(4482)	(4652)	4822.0
M_{II}	3415.7	3554.2	3696.3	3854.1	4008	4159	4327	4489.5
M_{III}	2956.6	3066.4	3176.9	3301.9	3426	3538	3663	3791.8
M_{IV}	2485.1	2585.6	2687.6	2798.0	2908.7	3021.5	3136.2	3248.4
M_V	2389.3	2484.0	2579.6	2683.0	2786.7	2892.4	2999.9	3104.9
N_I	845.5	893.6	938.2	995.3	(1042)	(1097)	(1153)	1208.4
N_{II}	721.3	763.9	805.3	851	886	929	980	1057.6
N_{III}	609.0	644.5	678.9	705	740	768	810	879.1
N_{IV}	406.6	435.2	463.6	500.2	533.2	566.6	603.3	635.9
N_V	386.2	412.9	440.0	473.4			577	602.7
N_{VI}	122.8	142.9	161.9					298.9
N_{VII}	118.5	138.1	157.4					298.9*
O_I	136.3	147.3	159.3					254.4
O_{II}	99.6	104.8	116.8					200.4
O_{III}	75.4	86.0	92.8					152.8

Level	⁸¹Tl	⁸²Pb	⁸³Bi	⁸⁴Po	⁸⁵At	⁸⁶Rn	⁸⁷Fr	⁸⁸Ra
O_{IV}	15.3	21.8	26.5	31.4				67.2
O_V	13.1	19.2	24.4	31.4*				67.2*
P_I		3.1						43.5
$P_{II,III}$		0.7	2.7					18.8

Level	⁸⁹Ac	⁹⁰Th	⁹¹Pa	⁹²U	⁹³Np	⁹⁴Pu	⁹⁵Am	⁹⁶Cm
K	106755.3	109650.9	112601.4	115606.1	118678	121818	125027	128220
L_I	19840	20472.1	21104.6	21757.4	22426.8	23097.2	23772.9	24460
L_{II}	19083.2	19693.2	20313.7	20947.6	21600.5	22266.2	22944.0	23779
L_{III}	15871.0	16300.3	16733.1	17166.3	17610.0	18056.8	18504.1	18930
M_I	(5002)	5182.3	5366.9	5548.0	5723.2	5932.9	6120.5	6288
M_{II}	4656	4830.4	5000.9	5182.2	5366.2	5541.2	5710.2	5895
M_{III}	3909	4046.1	4173.8	4303.4	4434.7	4556.6	4667.0	4797
M_{IV}	3370.2	3490.8	3611.2	3727.6	3850.3	3972.6	4092.1	4227
M_V	3219.0	3332.0	3441.8	3551.7	3665.8	3778.1	3886.9	3971
N_I	(1269)	1329.5	1387.1	1440.8	1500.7	1558.6	1617.1	1643
N_{II}	1080	1168.2	1224.3	1272.6	1327.7	1372.1	1411.8	1440
N_{III}	890	967.3	1006.7	1044.9	1086.8	1114.8	(1135.7)	1154
N_{IV}	674.9	714.1	743.4	780.4	815.9	848.9	878.7	
N_V		676.4	708.2	737.7	770.3	801.4	827.6	
N_{VI}		344.4	371.2	391.3	415.0	445.8		
N_{VII}		335.2	359.5	380.9	404.4	432.4		
O_I		290.2	309.6	323.7		351.9		385
O_{II}		229.4	222.9	259.3	283.4	274.1		
O_{III}		181.8	222.9*	195.1	206.1	206.5		
O_{IV}		94.3	94.1	105.0	109.3	116.0	115.8	
O_V		87.9	94.1*	96.3	101.3	105.4	103.3	
P_I		59.5		70.7				
P_{II}		49.0		42.3				
P_{III}		43.0		32.3				

Level	⁹⁷Bk	⁹⁸Cf	⁹⁹Es	¹⁰⁰Fm	¹⁰¹Md	¹⁰²No	¹⁰³Lr
K	131590	135960	139490	143090	146780	150540	154380
L_I	25275	26110	26900	27700	28530	29380	30240
L_{II}	24385	25250	26020	26810	27610	28440	29280
L_{III}	19452	19930	20410	20900	21390	21880	22360
M_I	6556	6754	6977	7205	7441	7675	7900
M_{II}	6147	6359	6574	6793	7019	7245	7460
M_{III}	4977	5109	5252	5397	5546	5688	5710
M_{IV}	4366	4497	4630	4766	4903	5037	5150
M_V	4132	4253	4374	4498	4622	4741	4860
N_I	1755	1799	1868	1937	2010	2078	2140
N_{II}	1554	1616	1680	1747	1814	1876	1930
N_{III}	1235	1279	1321	1366	1410	1448	1480
O_I	398	419	435	454	472	484	490

ELECTRON BINDING ENERGIES OF THE ELEMENTS

Gwyn P. Williams

This table gives the binding energies in electron volts (eV) for selected electronic levels of the elements. For metallic elements the binding energy is referred to the Fermi level; for semiconductors, to the valence band maximum; and for gases and insulators, to the vacuum level. The atomic number is listed after the element name.

REFERENCES

1. Fluggle and Martensson, J. *Elect. Spect.*, 21, 275, 1980.
2. Cardona, M. and Ley, L., *Photoemission from Solids*, Springer Verlag, Heidelberg, 1978.
3. Bearden, J. A. and Burr, A. F., *Rev. Mod. Phys.*, 39, 125, 1967.

Actinium (89)

K	1s	106755
L I	2s	19840
L II	$2p_{1/2}$	19083
L III	$2p_{3/2}$	15871
M I	3s	5002
M II	$3p_{1/2}$	4656
M III	$3p_{3/2}$	3909
M IV	$3d_{3/2}$	3370
M V	$3d_{5/2}$	3219
N I	4s	1269[a]
N II	$4p_{1/2}$	1080[a]
N III	$4p_{3/2}$	890[a]
N IV	$4d_{3/2}$	675[a]
N V	$4d_{5/2}$	639[a]
N VI	$4f_{5/2}$	319[a]
N VII	$4f_{7/2}$	319[a]
O I	5s	272[a]
O II	$5p_{1/2}$	215[a]
O III	$5p_{3/2}$	167[a]
O IV	$5d_{3/2}$	80[a]
O V	$5d_{5/2}$	80[a]
P I	6s	—
P II	$6p_{1/2}$	—
P III	$6p_{3/2}$	—

Aluminum (13)

K	1s	1559.0
L I	2s	117.8[a]
L II	$2p_{1/2}$	72.9[a]
L III	$2p_{3/2}$	72.5[a]

Antimony (51)

K	1s	30419
L I	2s	4698
L II	$2p_{1/2}$	4380
L III	$2p_{3/2}$	4132
M I	3s	946[b]
M II	$3p_{1/2}$	812.7[b]
M III	$3p_{3/2}$	766.4[b]
M IV	$3d_{3/2}$	537.5[b]
M V	$3d_{5/2}$	528.2[b]
N I	4s	153.2[b]
N II	$4p_{1/2}$	95.6[b,c]
N III	$4p_{3/2}$	95.6[b]
N IV	$4d_{3/2}$	33.3[b]
N V	$4d_{5/2}$	32.1[b]

Argon (18)

K	1s	3205.9[a]
L I	2s	326.3[a]
L II	$2p_{1/2}$	250.6[a]
L III	$2p_{3/2}$	248.4[a]
M I	3s	29.3[a]
M II	$3p_{1/2}$	15.9[a]
M III	$3p_{3/2}$	15.7[a]

Arsenic (33)

K	1s	11867
L I	2s	1527.0[a,d]
L II	$2p_{1/2}$	1359.1[a,d]
L III	$2p_{3/2}$	1323.6[a,d]
M I	3s	204.7[a]
M II	$3p_{1/2}$	146.2[a]
M III	$3p_{3/2}$	141.2[a]
M IV	$3d_{3/2}$	41.7[a]
M V	$3d_{5/2}$	41.7[a]

Astatine (85)

K	1s	95730
L I	2s	17493
L II	$2p_{1/2}$	16785
L III	$2p_{3/2}$	14214
M I	3s	4317
M II	$3p_{1/2}$	4008
M III	$3p_{3/2}$	3426
M IV	$3d_{3/2}$	2909
M V	$3d_{5/2}$	2787
N I	4s	1042[a]
N II	$4p_{1/2}$	886[a]
N III	$4p_{3/2}$	740[a]
N IV	$4d_{3/2}$	533[a]
N V	$4d_{5/2}$	507[a]
N VI	$4f_{5/2}$	210[a]
N VII	$4f_{7/2}$	210[a]
O I	5s	195[a]
O II	$5p_{1/2}$	148[a]
O III	$5p_{3/2}$	115[a]
O IV	$5d_{3/2}$	40[a]
O V	$5d_{5/2}$	40[a]

Barium (56)

K	1s	37441
L I	2s	5989
L II	$2p_{1/2}$	5624

Bismuth (83)

L III	$2p_{3/2}$	5247
M I	3s	1293[a,d]
M II	$3p_{1/2}$	1137[a,d]
M III	$3p_{3/2}$	1063[a,d]
M IV	$3d_{3/2}$	795.7[a]
M V	$3d_{5/2}$	780.5[a]
N I	4s	253.5[b]
N II	$4p_{1/2}$	192
N III	$4p_{3/2}$	178.6[b]
N IV	$4d_{3/2}$	92.6[b]
N V	$4d_{5/2}$	89.9[b]
N VI	$4f_{5/2}$	—
N VII	$4f_{7/2}$	—
O I	5s	30.3[b]
O II	$5p_{1/2}$	17.0[b]
O III	$5p_{3/2}$	14.8[b]

Beryllium (4)

K	1s	111.5[a]

Bismuth (83)

K	1s	90526
L I	2s	16388
L II	$2p_{1/2}$	15711
L III	$2p_{3/2}$	13419
M I	3s	3999
M II	$3p_{1/2}$	3696
M III	$3p_{3/2}$	3177
M IV	$3d_{3/2}$	2688
M V	$3d_{5/2}$	2580
N I	4s	939[b]
N II	$4p_{1/2}$	805.2[b]
N III	$4p_{3/2}$	678.8[b]
N IV	$4d_{3/2}$	464.0[b]
N V	$4d_{5/2}$	440.1[b]
N VI	$4f_{5/2}$	162.3[b]
N VII	$4f_{7/2}$	157.0[b]
O I	5s	159.3[a,d]
O II	$5p_{1/2}$	119.0[b]
O III	$5p_{3/2}$	92.6[b]
O IV	$5d_{3/2}$	26.9[b]
O V	$5d_{5/2}$	23.8[b]

Boron (5)

K	1s	188[a]

Bromine (35)

K	1s	13474

Cadmium (48)

L I	2s	1782[a]
L II	$2p_{1/2}$	1596[a]
L III	$2p_{3/2}$	1550[a]
M I	3s	257[a]
M II	$3p_{1/2}$	189[a]
M III	$3p_{3/2}$	182[a]
M IV	$3d_{3/2}$	70[a]
M V	$3d_{5/2}$	69[a]

Cadmium (48)

K	1s	26711
L I	2s	4018
L II	$2p_{1/2}$	3727
L III	$2p_{3/2}$	3538
M I	3s	772.0[b]
M II	$3p_{1/2}$	652.6[b]
M III	$3p_{3/2}$	618.4[b]
M IV	$3d_{3/2}$	411.9[b]
M V	$3d_{5/2}$	405.2[b]
N I	4s	109.8[b]
N II	$4p_{1/2}$	63.9[b,c]
N III	$4p_{3/2}$	63.9[b,c]
N IV	$4d_{3/2}$	11.7[b]
N V	$4d_{5/2}$	10.7[b]

Calcium (20)

K	1s	4038.5[a]
L I	2s	438.4[b]
L II	$2p_{1/2}$	349.7[b]
L III	$2p_{3/2}$	346.2[b]
M I	3s	44.3[b]
M II	$3p_{1/2}$	25.4[b]
M III	$3p_{3/2}$	25.4[b]

Carbon (6)

K	1s	284.2[a]

Cerium (58)

K	1s	40443
L I	2s	6548
L II	$2p_{1/2}$	6164
L III	$2p_{3/2}$	5723
M I	3s	1436[a,d]
M II	$3p_{1/2}$	1274[a,d]
M III	$3p_{3/2}$	1187[a,d]
M IV	$3d_{3/2}$	902.4[a]
M V	$3d_{5/2}$	883.8[a]

N I	4s	291.0[a]
N II	4p$_{1/2}$	223.3
N III	4p$_{3/2}$	206.5[a]
N IV	4d$_{3/2}$	109[a]
N V	4d$_{5/2}$	—
N VI	4f$_{5/2}$	0.1
N VII	4f$_{7/2}$	0.1
O I	5s	37.8
O II	5p$_{1/2}$	19.8[a]
O III	5p$_{3/2}$	17.0[a]

Cesium (55)

K	1s	35985
L I	2s	5714
L II	2p$_{1/2}$	5359
L III	2p$_{3/2}$	5012
M I	3s	1211[a,d]
M II	3p$_{1/2}$	1071[a]
M III	3p$_{3/2}$	1003[a]
M IV	3d$_{3/2}$	740.5[a]
M V	3d$_{5/2}$	726.6[a]
N I	4s	232.3[a]
N II	4p$_{1/2}$	172.4[a]
N III	4p$_{3/2}$	161.3[a]
N IV	4d$_{3/2}$	79.8[a]
N V	4d$_{5/2}$	77.5[a]
N VI	4f$_{5/2}$	—
N VII	4f$_{7/2}$	—
O I	5s	22.7
O II	5p$_{1/2}$	14.2[a]
O III	5p$_{3/2}$	12.1[a]

Chlorine (17)

K	1s	2822.0
L I	2s	270[a]
L II	2p$_{1/2}$	202[a]
L III	2p$_{3/2}$	200[a]

Chromium(24)

K	1s	5989
L I	2s	696.0[b]
L II	2p$_{1/2}$	583.8[b]
L III	2p$_{3/2}$	574.1[b]
M I	3s	74.1[b]
M II	3p$_{1/2}$	42.2[b]
M III	3p$_{3/2}$	42.2[b]

Cobalt (27)

K	1s	7709
L I	2s	925.1[b]
L II	2p$_{1/2}$	793.2[b]
L III	2p$_{3/2}$	778.1[b]
M I	3s	101.0[b]
M II	3p$_{1/2}$	58.9[b]
M III	3p$_{3/2}$	58.9[b]

Copper (29)

K	1s	8979
L I	2s	1096.7[b]

L II	2p$_{1/2}$	952.3[b]
L III	2p$_{3/2}$	932.5[b]
M I	3s	122.5[b]
M II	3p$_{1/2}$	77.3[b]
M III	3p$_{3/2}$	75.1[b]

Dysprosium (66)

K	1s	53789
L I	2s	9046
L II	2p$_{1/2}$	8581
L III	2p$_{3/2}$	7790
M I	3s	2047
M II	3p$_{1/2}$	1842
M III	3p$_{3/2}$	1676
M IV	3d$_{3/2}$	1333
M V	3d$_{5/2}$	1292[a]
N I	4s	414.2[a]
N II	4p$_{1/2}$	333.5[a]
N III	4p$_{3/2}$	293.2[a]
N IV	4d$_{3/2}$	153.6[a]
N V	4d$_{5/2}$	153.6[a]
N VI	4f$_{5/2}$	8.0[a]
N VII	4f$_{7/2}$	4.3[a]
O I	5s	49.9[a]
O II	5p$_{1/2}$	26.3
O III	5p$_{3/2}$	26.3

Erbium (68)

K	1s	57486
L I	2s	9751
L II	2p$_{1/2}$	9264
L III	2p$_{3/2}$	8358
M I	3s	2206
M II	3p$_{1/2}$	2006
M III	3p$_{3/2}$	1812
M IV	3d$_{3/2}$	1453
M V	3d$_{5/2}$	1409
N I	4s	449.8[a]
N II	4p$_{1/2}$	366.2
N III	4p$_{3/2}$	320.2[a]
N IV	4d$_{3/2}$	167.6[a]
N V	4d$_{5/2}$	167.6[a]
N VI	4f$_{5/2}$	—
N VII	4f$_{7/2}$	4.7[a]
O I	5s	50.6[a]
O II	5p$_{1/2}$	31.4[a]
O III	5p$_{3/2}$	24.7[a]

Europium (63)

K	1s	48519
L I	2s	8052
L II	2p$_{1/2}$	7617
L III	2p$_{3/2}$	6977
M I	3s	1800
M II	3p$_{1/2}$	1614
M III	3p$_{3/2}$	1481
M IV	3d$_{3/2}$	1158.6[a]
M V	3d$_{5/2}$	1127.5[a]
N I	4s	360

N II	4p$_{1/2}$	284
N III	4p$_{3/2}$	257
N IV	4d$_{3/2}$	133
N V	4d$_{5/2}$	1227[a]
N VI	4f$_{5/2}$	0
N VII	4f$_{7/2}$	0
O I	5s	32
O II	5p$_{1/2}$	22
O III	5p$_{3/2}$	22

Fluorine (9)

K	1s	696.7[a]

Francium (87)

K	1s	101137
L I	2s	18639
L II	2p$_{1/2}$	17907
L III	2p$_{3/2}$	15031
M I	3s	4652
M II	3p$_{1/2}$	4327
M III	3p$_{3/2}$	3663
M IV	3d$_{3/2}$	3136
M V	3d$_{5/2}$	3000
N I	4s	1153[a]
N II	4p$_{1/2}$	980[a]
N III	4p$_{3/2}$	810[a]
N IV	4d$_{3/2}$	603[a]
N V	4d$_{5/2}$	577[a]
N VI	4f$_{5/2}$	268[a]
N VII	4f$_{7/2}$	268[a]
O I	5s	234[a]
O II	5p$_{1/2}$	182[a]
O III	5p$_{3/2}$	140[a]
O IV	5d$_{3/2}$	58[a]
O V	5d$_{5/2}$	58[a]
P I	6s	34
P II	6p$_{1/2}$	15
P III	6p$_{3/2}$	15

Gadolinium (64)

K	1s	50239
L I	2s	8376
L II	2p$_{1/2}$	7930
L III	2p$_{3/2}$	7243
M I	3s	1881
M II	3p$_{1/2}$	1688
M III	3p$_{3/2}$	1544
M IV	3d$_{3/2}$	1221.9[a]
M V	3d$_{5/2}$	1189.6[a]
N I	4s	378.6[a]
N II	4p$_{1/2}$	286
N III	4p$_{3/2}$	271
N IV	4d$_{3/2}$	—
N V	4d$_{5/2}$	142.6[a]
N VI	4f$_{5/2}$	8.6[a]
N VII	4f$_{7/2}$	8.6[a]
O I	5s	36
O II	5p$_{1/2}$	20
O III	5p$_{3/2}$	20

Gallium (31)

K	1s	10367
L I	2s	1299.0[a,d]
L II	2p$_{1/2}$	1143.2[b]
L III	2p$_{3/2}$	1116.4[b]
M I	3s	159.5[b]
M II	3p$_{1/2}$	103.5[b]
M III	3p$_{3/2}$	100.0[b]
M IV	3d$_{3/2}$	18.7[b]
M V	3d$_{5/2}$	18.7[b]

Germanium (32)

K	1s	11103
L I	2s	1414.6[a,d]
L II	2p$_{1/2}$	1248.1[a,d]
L III	2p$_{3/2}$	1217.0[a,d]
M I	3s	180.1[a]
M II	3p$_{1/2}$	124.9[a]
M III	3p$_{3/2}$	120.8[a]
M IV	3d$_{3/2}$	29.8[a]
M V	3d$_{5/2}$	29.2[a]

Gold (79)

K	1s	80725
L I	2s	14353
L II	2p$_{1/2}$	13734
L III	2p$_{3/2}$	11919
M I	3s	3425
M II	3p$_{1/2}$	3148
M III	3p$_{3/2}$	2743
M IV	3d$_{3/2}$	2291
M V	3d$_{5/2}$	2206
N I	4s	762.1[b]
N II	4p$_{1/2}$	642.7[b]
N III	4p$_{3/2}$	546.3[b]
N IV	4d$_{3/2}$	353.2[b]
N V	4d$_{5/2}$	335.1[b]
N VI	4f$_{5/2}$	87.6[b]
N VII	4f$_{7/2}$	83.9[b]
O I	5s	107.2[a,d]
O II	5p$_{1/2}$	74.2[b]
O III	5p$_{3/2}$	57.2[b]

Hafnium (72)

K	1s	65351
L I	2s	11271
L II	2p$_{1/2}$	10739
L III	2p$_{3/2}$	9561
M I	3s	2601
M II	3p$_{1/2}$	2365
M III	3p$_{3/2}$	2107
M IV	3d$_{3/2}$	1176
M V	3d$_{5/2}$	1662
N I	4s	538[a]
N II	4p$_{1/2}$	438.2[a]
N III	4p$_{3/2}$	380.7[a]
N IV	4d$_{3/2}$	220.0[a]
N V	4d$_{5/2}$	211.5[b]
N VI	4f$_{5/2}$	15.9[b]

N VII	$4f_{7/2}$	14.2[b]
O I	5s	64.2[b]
O II	$5p_{1/2}$	38[a]
O III	$5p_{3/2}$	29.9[b]

Helium (2)

K	1s	24.6[a]

Holmium (67)

K	1s	55618
L I	2s	9394
L II	$2p_{1/2}$	8918
L III	$2p_{3/2}$	8071
M I	3s	2128
M II	$3p_{1/2}$	1923
M III	$3p_{3/2}$	1741
M IV	$3d_{3/2}$	1392
M V	$3d_{5/2}$	1351
N I	4s	432.4[a]
N II	$4p_{1/2}$	343.5
N III	$4p_{3/2}$	308.2[a]
N IV	$4d_{3/2}$	160[a]
N V	$4d_{5/2}$	160[a]
N VI	$4f_{5/2}$	8.6[a]
N VII	$4f_{7/2}$	5.2[a]
O I	5s	49.3[a]
O II	$5p_{1/2}$	30.8[a]
O III	$5p_{3/2}$	24.1[a]

Hydrogen (1)

K	1s	13.6

Indium (49)

K	1s	27940
L I	2s	4238
L II	$2p_{1/2}$	3938
L III	$2p_{3/2}$	3730
M I	3s	827.2[b]
M II	$3p_{1/2}$	703.2[b]
M III	$3p_{3/2}$	665.3[b]
M IV	$3d_{3/2}$	451.4[b]
M V	$3d_{5/2}$	443.9[b]
N I	4s	122.9[b]
N II	$4p_{1/2}$	73.5[b,c]
N III	$4p_{3/2}$	73.5[b,c]
N IV	$4d_{3/2}$	17.7[b]
N V	$4d_{5/2}$	16.9[b]

Iodine (53)

K	1s	33169
L I	2s	5188
L II	$2p_{1/2}$	4852
L III	$2p_{3/2}$	4557
M I	3s	1072[a]
M II	$3p_{1/2}$	931[a]
M III	$3p_{3/2}$	875[a]
M IV	$3d_{3/2}$	631[a]
M V	$3d_{5/2}$	620[a]
N I	4s	186[a]
N II	$4p_{1/2}$	123[a]

N III	$4p_{3/2}$	123[a]
N IV	$4d_{3/2}$	50[a]
N V	$4d_{5/2}$	50[a]

Iridium (77)

K	1s	76111
L I	2s	13419
L II	$2p_{1/2}$	12824
L III	$2p_{3/2}$	11215
M I	3s	3174
M II	$3p_{1/2}$	2909
M III	$3p_{3/2}$	2551
M IV	$3d_{3/2}$	2116
M V	$3d_{5/2}$	2040
N I	4s	691.1[b]
N II	$4p_{1/2}$	577.8[b]
N III	$4p_{3/2}$	495.8[b]
N IV	$4d_{3/2}$	311.9[b]
N V	$4d_{5/2}$	296.3[b]
N VI	$4f_{5/2}$	63.8[b]
N VII	$4f_{7/2}$	60.8[b]
O I	52	95.2[a,d]
O II	$5p_{1/2}$	63.0[a,d]
O III	$5p_{3/2}$	48.0[b]

Iron (26)

K	1s	7112
L I	2s	844.6[b]
L II	$2p_{1/2}$	719.9[b]
L III	$2p_{3/2}$	706.8[b]
M I	3s	91.3[b]
M II	$3p_{1/2}$	52.7[b]
M III	$3p_{3/2}$	52.7[b]

Krypton (36)

K	1s	14326
L I	2s	1921
L II	$2p_{1/2}$	1730.9[a]
L III	$2p_{3/2}$	1678.4[a]
M I	3s	292.8[a]
M II	$3p_{1/2}$	222.2[a]
M III	$3p_{3/2}$	214.4[a]
M IV	$3d_{3/2}$	95.0[a]
M V	$3d_{5/2}$	93.8[a]
N I	4s	27.5[a]
N II	$4p_{1/2}$	14.1[a]
N III	$4p_{3/2}$	14.1[a]

Lanthanum (57)

K	1s	38925
L I	2s	6266
L II	$2p_{1/2}$	5891
L III	$2p_{3/2}$	5483
M I	3s	1362[a,d]
M II	$3p_{1/2}$	1209[a,d]
M III	$3p_{3/2}$	1128[a,d]
M IV	$3d_{3/2}$	853[a]
M V	$3d_{5/2}$	836[a]
N I	4s	247.7[a]
N II	$4p_{1/2}$	205.8

N III	$4p_{3/2}$	196.0[a]
N IV	$4d_{3/2}$	105.3[a]
N V	$4d_{5/2}$	102.5[a]
N VI	$4f_{5/2}$	—
N VII	$4f_{7/2}$	—
O I	5s	34.3[a]
O II	$5p_{1/2}$	19.3[a]
O III	$5p_{3/2}$	16.8[a]

Lead (82)

K	1s	88005
L I	2s	15861
L II	$2p_{1/2}$	15200
L III	$2p_{3/2}$	13055
M I	3s	3851
M II	$3p_{1/2}$	3554
M III	$3p_{3/2}$	3066
M IV	$3d_{3/2}$	2586
M V	$3d_{5/2}$	2484
N I	4s	891.8[b]
N II	$4p_{1/2}$	761.9[b]
N III	$4p_{3/2}$	643.5[b]
N IV	$4d_{3/2}$	434.3[b]
N V	$4d_{5/2}$	412.2[b]
N VI	$4f_{5/2}$	141.7[b]
N VII	$4f_{7/2}$	136.9[b]
O I	5s	147[a,d]
O II	$5p_{1/2}$	106.4[b]
O III	$5p_{3/2}$	83.3[b]
O IV	$5d_{3/2}$	20.7[b]
O V	$5d_{5/2}$	18.1[b]

Lithium (3)

K	1s	54.7[a]

Lutetium

K	1s	63314
L I	2s	10870
L II	$2p_{1/2}$	10349
L III	$2p_{3/2}$	9244
M I	3s	2491
M II	$3p_{1/2}$	2264
M III	$3p_{3/2}$	2024
M IV	$3d_{3/2}$	1639
M V	$3d_{5/2}$	1589
N I	4s	506.8[a]
N II	$4p_{1/2}$	412.4[a]
N III	$4p_{3/2}$	359.2[a]
N IV	$4d_{3/2}$	206.1[a]
N V	$4d_{5/2}$	196.3[a]
N VI	$4f_{5/2}$	8.9[a]
N VII	$4f_{7/2}$	7.5[a]
O I	5s	57.3[a]
O II	$5p_{1/2}$	33.6[a]
O III	$5p_{3/2}$	26.7[a]

Magnesium (12)

K	1s	1303.0[b]
L I	2s	88.6[a]
L II	$2p_{1/2}$	49.6[b]
L III	$2p_{3/2}$	49.2[a]

Manganese (25)

K	1s	6539
L I	2s	769.1[b]
L II	$2p_{1/2}$	649.9[b]
L III	$2p_{3/2}$	638.7[h]
M I	3s	82.3[b]
M II	$3p_{1/2}$	47.2[b]
M III	$3p_{3/2}$	47.2[b]

Mercury (80)

K	1s	83102
L I	2s	14839
L II	$2p_{1/2}$	14209
L III	$2p_{3/2}$	12284
M I	3s	3562
M II	$3p_{1/2}$	3279
M III	$3p_{3/2}$	2847
M IV	$3d_{3/2}$	2385
M V	$3d_{5/2}$	2295
N I	4s	802.2[b]
N II	$4p_{1/2}$	680.2[b]
N III	$4p_{3/2}$	576.6[b]
N IV	$4d_{3/2}$	378.2[b]
N V	$4d_{5/2}$	358.8[b]
N VI	$4f_{5/2}$	104.0[b]
N VII	$4f_{7/2}$	99.9[b]
O I	5s	127[b]
O II	$5p_{1/2}$	83.1[b]
O III	$5p_{3/2}$	64.5[b]
O IV	$5d_{3/2}$	9.6[b]
O V	$5d_{5/2}$	7.8[b]

Molybdenum (42)

K	1s	20000
L I	2s	2866
L II	$2p_{1/2}$	2625
L III	$2p_{3/2}$	2520
M I	3s	506.3[b]
M II	$3p_{1/2}$	411.6[b]
M III	$3p_{3/2}$	394.0[b]
M IV	$3d_{3/2}$	231.1[b]
M V	$3d_{5/2}$	227.9[b]
N I	4s	63.2[b]
N II	$4p_{1/2}$	37.6[b]
N III	$4p_{3/2}$	35.5[b]

Neodymium (60)

K	1s	43569
L I	2s	7126
L II	$2p_{1/2}$	6722
L III	$2p_{3/2}$	6208
M I	3s	1575
M II	$3p_{1/2}$	1403
M III	$3p_{3/2}$	1297
M IV	$3d_{3/2}$	1003.3[a]
M V	$3d_{5/2}$	980.4[a]
N I	4s	319.2[a]
N II	$4p_{1/2}$	243.3
N III	$4p_{3/2}$	224.6
N IV	$4d_{3/2}$	120.5[a]

Level	Orbital	Energy
N V	$4d_{5/2}$	120.5[a]
N VI	$4f_{5/2}$	1.5
N VII	$4f_{7/2}$	1.5
O I	5s	37.5
O II	$5p_{1/2}$	21.1
O III	$5p_{3/2}$	21.1

Neon (10)

Level	Orbital	Energy
K	1s	870.2[a]
L I	2s	48.5[a]
L II	$2p_{1/2}$	21.7[a]
L III	$2p_{3/2}$	21.6[a]

Nickel (28)

Level	Orbital	Energy
K	1s	8333
L I	2s	1008.6[b]
L II	$2p_{1/2}$	870.0[b]
L III	$2p_{3/2}$	852.7[b]
M I	3s	110.8[b]
M II	$3p_{1/2}$	68.0[b]
M III	$3p_{3/2}$	66.2[b]

Niobium (41)

Level	Orbital	Energy
K	1s	18986
L I	2s	2698
L II	$2p_{1/2}$	2465
L III	$2p_{3/2}$	2371
M I	3s	466.6[b]
M II	$3p_{1/2}$	376.1[b]
M III	$3p_{3/2}$	360.6[b]
M IV	$3d_{3/2}$	205.0[b]
M V	$3d_{5/2}$	202.3[b]
N I	4s	56.4[b]
N II	$4p_{1/2}$	32.6[b]
N III	$4p_{3/2}$	30.8[b]

Nitrogen (7)

Level	Orbital	Energy
K	1s	409.9[a]
L I	2s	37.3[a]

Osmium (76)

Level	Orbital	Energy
K	1s	73871
L I	2s	12968
L II	$2p_{1/2}$	12385
L III	$2p_{3/2}$	10871
M I	3s	3049
M II	$3p_{1/2}$	2792
M III	$3p_{3/2}$	2457
M IV	$3d_{3/2}$	2031
M V	$3d_{5/2}$	1960
N I	4s	658.2[b]
N II	$4p_{1/2}$	549.1[b]
N III	$4p_{3/2}$	470.7[b]
N IV	$4d_{3/2}$	293.1[b]
N V	$4d_{5/2}$	278.5[b]
N VI	$4f_{5/2}$	53.4[b]
N VII	$4f_{7/2}$	50.7[b]
O I	5s	84[a]
O II	$5p_{1/2}$	58[a]
O III	$5p_{3/2}$	44.5[b]

Oxygen (8)

Level	Orbital	Energy
K	1s	543.1[a]
L I	2s	41.6[a]

Palladium (46)

Level	Orbital	Energy
K	1s	24350
L I	2s	3604
L II	$2p_{1/2}$	3330
L III	$2p_{3/2}$	3173
M I	3s	671.6[b]
M II	$3p_{1/2}$	559.9[b]
M III	$3p_{3/2}$	532.3[b]
M IV	$3d_{3/2}$	340.5[b]
M V	$3d_{5/2}$	335.2[b]
N I	4s	87.1[a,d]
N II	$4p_{1/2}$	55.7[b,c]
N III	$4p_{3/2}$	50.9[b,c]

Phosphorus (15)

Level	Orbital	Energy
K	1s	2145.5
L I	2s	189[a]
L II	$2p_{1/2}$	136[a]
L III	$2p_{3/2}$	135[a]

Platinum (78)

Level	Orbital	Energy
K	1s	78395
L I	2s	13880
L II	$2p_{1/2}$	13273
L III	$2p_{3/2}$	11564
M I	3s	3296
M II	$3p_{1/2}$	3027
M III	$3p_{3/2}$	2645
M IV	$3d_{3/2}$	2202
M V	$3d_{5/2}$	2122
N I	4s	725.4[b]
N II	$4p_{1/2}$	609.1[b]
N III	$4p_{3/2}$	519.4[b]
N IV	$4d_{3/2}$	331.6[b]
N V	$4d_{5/2}$	314.6[b]
N VI	$4f_{5/2}$	74.5[b]
N VII	$4f_{7/2}$	71.2[b]
O I	5s	101.7[a,d]
O II	$5p_{1/2}$	65.3[a,b]
O III	$5p_{3/2}$	51.7[b]

Polonium (84)

Level	Orbital	Energy
K	1s	93105
L I	2s	16939
L II	$2p_{1/2}$	16244
L III	$2p_{3/2}$	13814
M I	3s	4149
M II	$3p_{1/2}$	3854
M III	$3p_{3/2}$	3302
M IV	$3d_{3/2}$	2798
M V	$3d_{5/2}$	2683
N I	4s	995[a]
N II	$4p_{1/2}$	851[a]
N III	$4p_{3/2}$	705[a]
N IV	$4d_{3/2}$	500[a]
N V	$4d_{5/2}$	473[a]
N VI	$4f_{5/2}$	184[a]
N VII	$4f_{7/2}$	184[a]
O I	5s	177[a]
O II	$5p_{1/2}$	132[a]
O III	$5p_{3/2}$	104[a]
O IV	$5d_{3/2}$	31[a]
O V	$5d_{5/2}$	31[a]

Potassium (19)

Level	Orbital	Energy
K	1s	3608.4[a]
L I	2s	378.6[a]
L II	$2p_{1/2}$	297.3[a]
L III	$2p_{3/2}$	294.6[a]
M I	3s	34.8[a]
M II	$3p_{1/2}$	18.3[a]
M III	$3p_{3/2}$	18.3[a]

Praseodymium (59)

Level	Orbital	Energy
K	1s	41991
L I	2s	6835
L II	$2p_{1/2}$	6440
L III	$2p_{3/2}$	5964
M I	3s	1511
M II	$3p_{1/2}$	1337
M III	$3p_{3/2}$	1242
M IV	$3d_{3/2}$	948.3[a]
M V	$3d_{5/2}$	928.8[a]
N I	4s	304.5
N II	$4p_{1/2}$	236.3
N III	$4p_{3/2}$	217.6
N IV	$4d_{3/2}$	115.1[a]
N V	$4d_{5/2}$	115.1[a]
N VI	$4f_{5/2}$	2.0
N VII	$4f_{7/2}$	2.0
O I	5s	37.4
O II	$5p_{1/2}$	22.3
O III	$5p_{3/2}$	22.3

Promethium (61)

Level	Orbital	Energy
K	1s	45184
L I	2s	7428
L II	$2p_{1/2}$	7013
L III	$2p_{3/2}$	6459
M I	3s	—
M II	$3p_{1/2}$	1471.4
M III	$3p_{3/2}$	1357
M IV	$3d_{3/2}$	1052
M V	$3d_{5/2}$	1027
N I	4s	—
N II	$4p_{1/2}$	242
N III	$4p_{3/2}$	242
N IV	$4d_{3/2}$	120
N V	$4d_{5/2}$	120

Protactinium (91)

Level	Orbital	Energy
K	1s	112601
L I	2s	21105
L II	$2p_{1/2}$	20314
L III	$2p_{3/2}$	16733
M I	3s	5367
M II	$3p_{1/2}$	5001
M III	$3p_{3/2}$	4174
M IV	$3d_{3/2}$	3611
M V	$3d_{5/2}$	3442
N I	4s	1387[a]
N II	$4p_{1/2}$	1224[a]
N III	$4p_{3/2}$	1007[a]
N IV	$4d_{3/2}$	743[a]
N V	$4d_{5/2}$	708[a]
N VI	$4f_{5/2}$	371[a]
N VII	$4f_{7/2}$	360[a]
O I	5s	310[a]
O II	$5p_{1/2}$	232[a]
O III	$5p_{3/2}$	232[a]
O IV	$5d_{3/2}$	94[a]
O V	$5d_{5/2}$	94[a]
P I	6s	—
P II	$6p_{1/2}$	—
P III	$6p_{3/2}$	—

Radium (88)

Level	Orbital	Energy
K	1s	103922
L I	2s	19237
L II	$2p_{1/2}$	18484
L III	$2p_{3/2}$	15444
M I	3s	4822
M II	$3p_{1/2}$	4490
M III	$3p_{3/2}$	3792
M IV	$3d_{3/2}$	3248
M V	$3d_{5/2}$	3105
N I	4s	1208[a]
N II	$4p_{1/2}$	1058
N III	$4p_{3/2}$	879[a]
N IV	$4d_{3/2}$	636[a]
N V	$4d_{5/2}$	603[a]
N VI	$4f_{5/2}$	299[a]
N VII	$4f_{7/2}$	299[a]
O I	5s	254[a]
O II	$5p_{1/2}$	200[a]
O III	$5p_{3/2}$	153[a]
O IV	$5d_{3/2}$	68[a]
O V	$5d_{5/2}$	68[a]
P I	6s	44
P II	$6p_{1/2}$	19
P III	$6p_{3/2}$	19

Radon (86)

Level	Orbital	Energy
K	1s	98404
L I	2s	18049
L II	$2p_{1/2}$	17337
L III	$2p_{3/2}$	14619
M I	3s	4482
M II	$3p_{1/2}$	4159
M III	$3p_{3/2}$	3538
M IV	$3d_{3/2}$	3022
M V	$3d_{5/2}$	2892

N I	4s	1097[a]
N II	4p1/2	929[a]
N III	4p3/2	768[a]
N IV	4d3/2	567[a]
N V	4d5/2	541[a]
N VI	4f5/2	238[a]
N VII	4f7/2	238[a]
O I	5s	214[a]
O II	5p1/2	164[a]
O III	5p3/2	127[a]
O IV	5d3/2	48[a]
O V	5d5/2	48[a]
P I	6s	26

Rhenium (75)

K	1s	71676
L I	2s	12527
L II	2p1/2	11959
L III	2p3/2	10535
M I	3s	2932
M II	3p1/2	2682
M III	3p3/2	2367
M IV	3d3/2	1949
M V	3d5/2	1883
N I	4s	625.4[b]
N II	4p1/2	518.7[b]
N III	4p3/2	446.8[b]
N IV	4d3/2	273.9[b]
N V	4d5/2	260.5[b]
N VI	4f5/2	42.9[a]
N VII	4f7/2	40.5[a]
O I	5s	83[b]
O II	5p1/2	45.6[b]
O III	5p3/2	34.6[a,d]

Rhodium (45)

K	1s	23220
L I	2s	3412
L II	2p1/2	3146
L III	2p3/2	3004
M I	3s	628.1[b]
M II	3p1/2	521.3[b]
M III	3p3/2	496.5[b]
M IV	3d3/2	311.9[b]
M V	3d5/2	307.2[b]
N I	4s	81.4[a,d]
N II	4p1/2	50.5[b]
N III	4p3/2	47.3[b]

Rubidium (37)

K	1s	15200
L I	2s	2065
L II	2p1/2	1864
L III	2p3/2	1804
M I	3s	326.7[a]
M II	3p1/2	248.7[a]
M III	3p3/2	239.1[a]
M IV	3d3/2	113.0[a]
M V	3d5/2	112[a]

N I	4s	30.5[a]
N II	4p1/2	16.3[a]
N III	4p3/2	15.3[a]

Ruthenium (44)

K	1s	22117
L I	2s	3224
L II	2p1/2	2967
L III	2p3/2	2838
M I	3s	586.2[b]
M II	3p1/2	483.3[b]
M III	3p3/2	461.5[b]
M IV	3d3/2	284.2[b]
M V	3d5/2	280.0[b]
N I	4s	75.0[b]
N II	4p1/2	46.5[b]
N III	4p3/2	43.2[b]

Samarium (62)

K	1s	46834
L I	2s	7737
L II	2p1/2	7312
L III	2p3/2	6716
M I	3s	1723
M II	3p1/2	1541
M III	3p3/2	1419.8
M IV	3d3/2	1110.9[a]
M V	3d5/2	1083.4[a]
N I	4s	347.2[a]
N II	4p1/2	265.6
N III	4p3/2	247.4
N IV	4d3/2	129.0
N V	4d5/2	129.0
N VI	4f5/2	5.2
N VII	4f7/2	5.2
O I	5s	37.4
O II	5p1/2	21.3
O III	5p3/2	21.3

Scandium (21)

K	1s	4492
L I	2s	498.0[a]
L II	2p1/2	403.6[a]
L III	2p3/2	389.7[a]
M I	3s	51.1[a]
M II	3p1/2	28.3[a]
M III	3p3/2	28.3[a]

Selenium (34)

K	1s	12658
L I	2s	1652.0[a,d]
L II	2p1/2	1474.3[a,d]
L III	2p3/2	1433.9[a,d]
M I	3s	229.6[a]
M II	3p1/2	166.5[a]
M III	3p3/2	160.7[a]
M IV	3d3/2	55.5[a]
M V	3d5/2	54.6[a]

Silicon (14)

K	1s	1839
L I	2s	149.7[a,d]
L II	2p1/2	99.8[a]
L III	2p3/2	99.2[a]

Silver (47)

K	1s	25514
L I	2s	3806
L II	2p1/2	3524
L III	2p3/2	3351
M I	3s	719.0[b]
M II	3p1/2	603.8[b]
M III	3p3/2	573.0[b]
M IV	3d3/2	374.0[b]
M V	3d5/2	368.0[b]
N I	4s	97.0[b]
N II	4p1/2	63.7[b]
N III	4p3/2	58.3[b]

Sodium (11)

K	1s	1070.8[b]
L I	2s	63.5[b]
L II	2p1/2	30.4[b]
L III	2p3/2	30.5[a]

Strontium (38)

K	1s	16105
L I	2s	2216
L II	2p1/2	2007
L III	2p3/2	1940
M I	3s	358.7[b]
M II	3p1/2	280.3[b]
M III	3p3/2	270.0[b]
M IV	3d3/2	136.0[b]
M V	3d5/2	134.2[b]
N I	4s	38.9[b]
N II	4p1/2	21.6[b]
N III	4p3/2	20.1[b]

Sulfur (16)

K	1s	2472
L I	2s	230.9[a,d]
L II	2p1/2	163.6[a]
L III	2p3/2	162.5[a]

Tantalum (73)

K	1s	67416
L I	2s	11682
L II	2p1/2	11136
L III	2p3/2	9881
M I	3s	2708
M II	3p1/2	2469
M III	3p3/2	2194
M IV	3d3/2	1793
M V	3d5/2	1735
N I	4s	563.4[b]
N II	4p1/2	463.4[b]

N III	4p3/2	400.9[b]
N IV	4d3/2	237.9[b]
N V	4d5/2	226.4[b]
N VI	4f5/2	23.5[b]
N VII	4f7/2	21.6[b]
O I	5s	69.7[b]
O II	5p1/2	42.2[a]
O III	5p3/2	32.7[b]

Technetium (43)

K	1s	21044
L I	2s	3043
L II	2p1/2	2793
L III	2p3/2	2677
M I	3s	586.1[a]
M II	3p1/2	447.6[a]
M III	3p3/2	417.7[a]
M IV	3d3/2	257.6[a]
M V	3d5/2	253.9[a]
N I	4s	69.5[a]
N II	4p1/2	42.3[a]
N III	4p3/2	39.9[a]

Tellurium (52)

K	1s	31814
L I	2s	4939
L II	2p1/2	4612
L III	2p3/2	4341
M I	3s	1006[b]
M II	3p1/2	870.8[b]
M III	3p3/2	820.0[b]
M IV	3d3/2	583.4[b]
M V	3d5/2	573.0[b]
N I	4s	169.4[b]
N II	4p1/2	103.3[b,c]
N III	4p3/2	103.3[b,c]
N IV	4d3/2	41.9[b]
N V	4d5/2	40.4[b]

Terbium (65)

K	1s	51996
L I	2s	8708
L II	2p1/2	8252
L III	2p3/2	7514
M I	3s	1968
M II	3p1/2	1768
M III	3p3/2	1611
M IV	3d3/2	1267.9[a]
M V	3d5/2	1241.1[a]
N I	4s	396.0[a]
N II	4p1/2	322.4[a]
N III	4p3/2	284.1[a]
N IV	4d3/2	150.5[a]
N V	4d5/2	150.5[a]
N VI	4f5/2	7.7[a]
N VII	4f7/2	2.4[a]
O I	5s	45.6[a]
O II	5p1/2	28.7[a]
O III	5p3/2	22.6[a]

Thallium (81)

K	1s	85530
L I	2s	15347
L II	2p$_{1/2}$	14698
L III	2p$_{3/2}$	12658
M I	3s	3704
M II	3p$_{1/2}$	3416
M III	3p$_{3/2}$	2957
M IV	3d$_{3/2}$	2485
M V	3d$_{5/2}$	2389
N I	4s	846.2[b]
N II	4p$_{1/2}$	720.5[b]
N III	4p$_{3/2}$	609.5[b]
N IV	4d$_{3/2}$	405.7[b]
N V	4d$_{5/2}$	385.0[b]
N VI	4f$_{5/2}$	122.2[b]
N VII	4f$_{7/2}$	117.8[b]
O I	5s	136[a,d]
O II	5p$_{1/2}$	94.6[b]
O III	5p$_{3/2}$	73.5[b]
O IV	5d$_{3/2}$	14.7[b]
O V	5d$_{5/2}$	12.5[b]

Thorium (90)

K	1s	109651
L I	2s	20472
L II	2p$_{1/2}$	19693
L III	2p$_{3/2}$	16300
M I	3s	5182
M II	3p$_{1/2}$	4830
M III	3p$_{3/2}$	4046
M IV	3d$_{3/2}$	3491
M V	3d$_{5/2}$	3332
N I	4s	1330[a]
N II	4p$_{1/2}$	1168[a]
N III	4p$_{3/2}$	966.4[b]
N IV	4d$_{3/2}$	712.1[b]
N V	4d$_{5/2}$	675.2[b]
N VI	4f$_{5/2}$	342.4[b]
N VII	4f$_{7/2}$	333.1[b]
O I	5s	290[a,c]
O II	5p$_{1/2}$	229[a,c]
O III	5p$_{3/2}$	182[a,c]
O IV	5d$_{3/2}$	92.5[b]
O V	5d$_{5/2}$	85.4[b]
P I	6s	41.4[b]
P II	6p$_{1/2}$	24.5[b]
P III	6p$_{3/2}$	16.6[b]

Thulium (69)

K	1s	59390
L I	2s	10116
L II	2p$_{1/2}$	9617
L III	2p$_{3/2}$	8648
M I	3s	2307
M II	3p$_{1/2}$	2090
M III	3p$_{3/2}$	1885
M IV	3d$_{3/2}$	1515

M V	3d$_{5/2}$	1468
N I	4s	470.9[a]
N II	4p$_{1/2}$	385.9[a]
N III	4p$_{3/2}$	332.6[a]
N IV	4d$_{3/2}$	175.5[a]
N V	4d$_{5/2}$	175.5[a]
N VI	4f$_{5/2}$	—
N VII	4f$_{7/2}$	4.6
O I	5s	54.7[a]
O II	5p$_{1/2}$	31.8[a]
O III	5p$_{3/2}$	25.0[a]

Tin (50)

K	1s	29200
L I	2s	4465
L II	2p$_{1/2}$	4156
L III	2p$_{3/2}$	3929
M I	3s	884.7[b]
M II	3p$_{1/2}$	756.5[b]
M III	3p$_{3/2}$	714.6[b]
M IV	3d$_{3/2}$	493.2[b]
M V	3d$_{5/2}$	484.9[b]
N I	4s	137.1[b]
N II	4p$_{1/2}$	83.6[b,c]
N III	4p$_{3/2}$	83.6[b,c]
N IV	4d$_{3/2}$	24.9[b]
N V	4d$_{5/2}$	23.9[b]

Titanium (22)

K	1s	4966
L I	2s	560.9[b]
L II	2p$_{1/2}$	460.2[b]
L III	2p$_{3/2}$	453.8[b]
M I	3s	58.7[b]
M II	3p$_{1/2}$	32.6[b]
M III	3p$_{3/2}$	32.6[b]

Tungsten (74)

K	1s	69525
L I	2s	12100
L II	2p$_{1/2}$	11544
L III	2p$_{3/2}$	10207
M I	3s	2820
M II	3p$_{1/2}$	2575
M III	3p$_{3/2}$	2281
M IV	3d$_{3/2}$	1949
M V	3d$_{5/2}$	1809
N I	4s	594.1[b]
N II	4p$_{1/2}$	490.4[b]
N III	4p$_{3/2}$	423.6[b]
N IV	4d$_{3/2}$	255.9[b]
N V	4d$_{5/2}$	243.5[b]
N VI	4f$_{5/2}$	33.6[a]
N VII	4f$_{7/2}$	31.4[b]
O I	5s	75.6[b]
O II	5p$_{1/2}$	453[a,d]
O III	5p$_{3/2}$	36.8[b]

Uranium (92)

K	1s	115606
L I	2s	21757
L II	2p$_{1/2}$	20948
L III	2p$_{3/2}$	17166
M I	3s	5548
M II	3p$_{1/2}$	5182
M III	3p$_{3/2}$	4303
M IV	3d$_{3/2}$	3728
M V	3d$_{5/2}$	3552
N I	4s	1439[a,d]
N II	4p$_{1/2}$	1271[a,d]
N III	4p$_{3/2}$	1043[b]
N IV	4d$_{3/2}$	778.3[b]
N V	4d$_{5/2}$	736.2[b]
N VI	4f$_{5/2}$	388.2[a]
N VII	4f$_{7/2}$	377.4[b]
O I	5s	321[a,c,d]
O II	5p$_{1/2}$	257[a,c,d]
O III	5p$_{3/2}$	192[a,c,d]
O IV	5d$_{3/2}$	102.8[b]
O V	5d$_{5/2}$	94.2[b]
P I	6s	43.9[b]
P II	6p$_{1/2}$	26.8[b]
P III	6p$_{3/2}$	16.8[b]

Vanadium (23)

K	1s	5465
L I	2s	626.7[b]
L II	2p$_{1/2}$	519.8[b]
L III	2p$_{3/2}$	521.1[b]
M I	3s	66.3[b]
M II	3p$_{1/2}$	37.2[b]
M III	3p$_{3/2}$	37.2[b]

Xenon (54)

K	1s	34561
L I	2s	5453
L II	2p$_{1/2}$	5107
L III	2p$_{3/2}$	4786
M I	3s	1148.7[a]
M II	3p$_{1/2}$	1002.1[a]
M III	3p$_{3/2}$	940.6[a]
M IV	3d$_{3/2}$	689.0[a]
M V	3d$_{5/2}$	676.4[a]
N I	4s	213.2[a]
N II	4p$_{1/2}$	146.7
N III	4p$_{3/2}$	145.5[a]
N IV	4d$_{3/2}$	69.5[a]
N V	4d$_{5/2}$	67.5[a]
N VI	4f$_{5/2}$	—
N VII	4f$_{7/2}$	—
O I	5s	23.3[a]
O II	5p$_{1/2}$	13.4[a]
O III	5p$_{3/2}$	12.1[a]

Ytterbium (70)

K	1s	61332

L I	2s	10486
L II	2p$_{1/2}$	9978
L III	2p$_{3/2}$	8944
M I	3s	2398
M II	3p$_{1/2}$	2173
M III	3p$_{3/2}$	1950
M IV	3d$_{3/2}$	1576
M V	3d$_{5/2}$	1528
N I	4s	480.5[a]
N II	4p$_{1/2}$	388.7[a]
N III	4p$_{3/2}$	339.7[a]
N IV	4d$_{3/2}$	191.2[a]
N V	4d$_{5/2}$	182.4[a]
N VI	4f$_{5/2}$	2.5[a]
N VII	4f$_{7/2}$	1.3[a]
O I	5s	52.0[a]
O II	5p$_{1/2}$	30.3[a]
O III	5p$_{3/2}$	24.1[a]

Yttrium (39)

K	1s	17038
L I	2s	2373
L II	2p$_{1/2}$	2156
L III	2p$_{3/2}$	2080
M I	3s	392.0[a,d]
M II	3p$_{1/2}$	310.6[a]
M III	3p$_{3/2}$	298.8[a]
M IV	3d$_{3/2}$	157.7[b]
M V	3d$_{5/2}$	155.8[b]
N I	4s	43.8[a]
N II	4p$_{1/2}$	24.4[a]
N III	4p$_{3/2}$	23.1[a]

Zinc (30)

K	1s	9659
L I	2s	1196.2[a]
L II	2p$_{1/2}$	1044.9[a]
L III	2p$_{3/2}$	1021.8[a]
M I	3s	139.8[a]
M II	3p$_{1/2}$	91.4[a]
M III	3p$_{3/2}$	88.6[a]
M IV	3d$_{3/2}$	10.2[a]
M V	3d$_{5/2}$	10.1[a]

Zirconium (40)

K	1s	17998
L I	2s	2532
L II	2p$_{1/2}$	2307
L III	2p$_{3/2}$	2223
M I	3s	430.3[b]
M II	3p$_{1/2}$	343.5[b]
M III	3p$_{3/2}$	329.8[b]
M IV	3d$_{3/2}$	181.1[b]
M V	3d$_{5/2}$	178.8[b]
N I	4s	50.6
N II	4p$_{1/2}$	28.5[b]
N III	4p$_{3/2}$	27.1[b]

[a] Reference 1.
[b] Reference 2 (remaining values from Reference 3).
[c] One-particle approximation not valid.
[d] Derived using energy differences from Reference 3.

NATURAL WIDTH OF X-RAY LINES

Natural widths of K X-ray lines in eV:

Element	Kα₁	Kα₂	Kβ₁	Kβ₃	Element	Kα₁	Kα₂	Kβ₁	Kβ₃
Ca	1.00	0.98			Ce	18.60	19.50	20.60	18.60
Ti	1.45	2.13			Nd	21.50	21.50	23.25	21.33
Cr	2.05	2.64			Sm	26.00	24.70	25.65	24.65
Fe	2.45	3.20			Gd	29.50	28.00	29.37	28.00
Ni	3.00	3.70			Dy	33.90	32.20	32.73	32.00
Zn	3.40	3.96			Er	35.00	35.50	36.20	35.70
Ge	3.75	4.18			Yb	38.80	40.60	41.43	41.15
Se	4.10	4.43			Hf	42.70	44.30	46.00	46.10
Kr	4.23	4.62			W	46.80	48.00	51.83	51.50
Sr	5.17	4.97			Os	49.00	49.40	55.90	55.95
Zr	5.70	5.25			Pt	54.10	54.30	59.98	62.13
Mo	6.82	6.80			Hg	64.75	68.20	65.75	68.95
Ru	7.41	7.96			Pb	67.10	72.30	72.20	73.80
Pd	8.80	9.20			Po	73.20	75.10	78.60	80.10
Cd	9.80	10.40			Rn	80.00	81.50	85.50	86.50
Sn	11.20	12.40	11.80	11.00	Ra	87.00	88.20	94.20	95.50
Te	12.80	14.20	13.30	13.10	Th	94.70	95.00	99.70	101.00
Xe	14.20	15.10	15.30	14.50	U	103.00	104.30	105.00	107.30
Ba	16.10	16.80	18.15	16.70					

From Salem, S. I. and Lee, P. L., *At. Data Nucl. Data Tables,* 18, 233, 1976.

Natural widths of L X-ray lines in eV:

Element	Lα₁	Lα₂	Lβ₁	Lβ₂	Lβ₃	Lβ₄	Lγ₁
Zr	1.68	1.52	1.87	5.13	5.50	5.60	3.34
Mo	1.86	1.80	2.03	5.30	5.90	5.78	3.76
Ru	2.03	1.98	2.18	5.45	6.35	5.96	4.15
Pd	2.21	2.16	2.36	5.63	6.80	6.18	4.50
Gd	2.43	2.40	2.54	5.82	7.23	6.28	4.83
Sn	2.62	2.62	2.75	6.10	7.70	6.60	5.23
Tc	2.88	2.88	2.96	6.25	8.22	6.82	5.60
Xe	3.15	3.15	3.20	6.43	8.70	7.15	5.95
Ba	3.39	3.45	3.45	6.70	9.20	7.42	6.35
Ce	3.70	3.78	3.73	6.86	9.70	7.82	6.75
Nd	3.93	4.08	4.00	7.18	10.30	8.15	7.16
Sm	4.13	4.50	4.33	7.42	10.80	8.60	7.50
Cd	4.46	4.90	4.63	7.70	11.20	9.08	7.83
Dy	4.81	5.35	5.03	7.90	11.50	9.60	8.30
Er	5.17	5.73	5.45	8.28	11.85	10.03	8.75
Yb	5.40	6.22	5.90	8.58	12.20	11.00	9.20
Hf	5.83	6.70	6.36	8.92	12.40	12.80	9.63
W	6.50	7.20	6.90	9.06	13.10	14.60	10.20
Os	7.04	7.70	7.42	9.60	14.60	16.50	10.65
Pt	7.60	8.28	8.00	9.95	16.10	18.00	11.20
Hg	8.10	8.80	8.70	10.40	17.40	19.70	11.80
Pb	8.82	9.35	9.35	10.75	18.65	21.30	12.30
Po	9.50	9.95	10.10	11.25	19.90	22.70	13.05
Rn	10.03	10.50	10.65	11.65	21.00	24.00	13.55
Ra	11.00	11.20	11.60	12.20	22.00	25.20	14.30
Th	11.90	11.80	12.40	12.80	22.85	26.35	15.00
U	12.40	12.40	13.50	13.30	23.70	27.50	15.70
Pu	13.20	13.00	14.10	13.90	24.10	28.30	16.40
Cm	14.80	13.60	15.70	14.60	25.00	29.40	17.10

PHOTON ATTENUATION COEFFICIENTS

Martin J. Berger and John H. Hubbell

This table gives mass attenuation coefficients for photons for all elements at energies between 1 keV (soft x-rays) and 1 GeV (hard gamma rays). The mass attenuation coefficient μ describes the attenuation of radiation as it passes through matter by the relation

$$I(x)/I_o = e^{-\mu \rho x}$$

where I_o is the initial intensity, $I(x)$ the intensity after path length x, and ρ is the mass density of the element in question. To a high approximation the mass attenuation coefficient is additive for the elements present, independent of the way in which they are bound in chemical compounds.

The power of ten is indicated beside each number in the table; i.e., $7.41 + 03$ means 7.41×10^3. A vertical line between two columns indicates that an absorption edge lies between those energy values. The various edges are labeled at the bottom of the table.

The attenuation coefficients were calculated with the computer program XCOM (Reference 1), which uses a cross-section database compiled at the Photon and Charged Particle Data Center at the National Institute of Standards and Technology. Their accuracy has been confirmed at all energies by extensive comparisons with experimental attenuation coefficients. Such comparisons for X-ray energies up to 100 keV can be found in Reference 2.

REFERENCES

1. Berger, M. J. and Hubbell, J. H., National Bureau of Standards Report NBSIR-87-3597, 1987.
2. Saloman, E. B., Hubbell, J. H., and Scofield, J. H., *Atomic Data and Nuclear Data Tables*, 38, 1, 1988.

Mass attenuation coefficient, cm²/g

	Atomic no.	Photon energy, MeV								
		0.001	0.002	0.005	0.01	0.02	0.05	0.1	0.2	0.5
H	1	7.21 + 00	1.06 + 00	4.19-01	3.85-01	3.69-01	3.36-01	2.94-01	2.43-01	1.73-01
He	2	6.08 + 01	6.86 + 00	5.77-01	2.48-01	1.96-01	1.70-01	1.49-01	1.22-01	8.71-02
Li	3	2.34 + 02	2.71 + 01	1.62 + 00	3.40-01	1.86-01	1.49-01	1.29-01	1.06-01	7.53-02
Be	4	6.04 + 02	7.47 + 01	4.37 + 00	6.47-01	2.25-01	1.55-01	1.33-01	1.09-01	7.74-02
B	5	1.23 + 03	1.60 + 02	9.68 + 00	1.25 + 00	3.01-01	1.66-01	1.39-01	1.14-01	8.07-02
C	6	2.21 + 03	3.03 + 02	1.91 + 01	2.37 + 00	4.42-01	1.87-01	1.51-01	1.23-01	8.72-02
N	7	3.31 + 03	4.77 + 02	3.14 + 01	3.88 + 00	6.18-01	1.98-01	1.53-01	1.23-01	8.72-02
O	8	4.59 + 03	6.95 + 02	4.79 + 01	5.95 + 00	8.65-01	2.13-01	1.55-01	1.24-01	8.73-02
F	9	5.65 + 03	9.05 + 02	6.51 + 01	8.21 + 00	1.13 + 00	2.21-01	1.50-01	1.18-01	8.27-02
Ne	10	7.41 + 03	1.24 + 03	9.34 + 01	1.20 + 01	1.61 + 00	2.58-01	1.60-01	1.24-01	8.66-02
Na	11	6.54 + 02	1.52 + 03	1.19 + 02	1.56 + 01	2.06 + 00	2.80-01	1.59-01	1.20-01	8.37-02
Mg	12	9.22 + 02	1.93 + 03	1.58 + 02	2.11 + 01	2.76 + 00	3.29-01	1.69-01	1.24-01	8.65-02
Al	13	1.19 + 03	2.26 + 03	1.93 + 02	2.62 + 01	3.44 + 00	3.68-01	1.70-01	1.22-01	8.44-02
Si	14	1.57 + 03	2.78 + 03	2.45 + 02	3.39 + 01	4.46 + 00	4.38-01	1.84-01	1.28-01	8.75-02
P	15	1.91 + 03	3.02 + 02	2.86 + 02	4.04 + 01	5.35 + 00	4.92-01	1.87-01	1.25-01	8.51-02
S	16	2.43 + 03	3.85 + 02	3.49 + 02	5.01 + 01	6.71 + 00	5.85-01	2.02-01	1.30-01	8.78-02
Cl	17	2.83 + 03	4.52 + 02	3.90 + 02	5.73 + 01	7.74 + 00	6.48-01	2.05-01	1.27-01	8.45-02
Ar	18	3.18 + 03	5.12 + 02	4.23 + 02	6.32 + 01	8.63 + 00	7.01-01	2.04-01	1.20-01	7.96-02
K	19	4.06 + 03	6.59 + 02	5.19 + 02	7.91 + 01	1.09 + 01	8.68-01	2.34-01	1.32-01	8.60-02
Ca	20	4.87 + 03	8.00 + 02	6.03 + 02	9.34 + 01	1.31 + 01	1.02 + 00	2.57-01	1.38-01	8.85-02
Sc	21	5.24 + 03	8.70 + 02	6.31 + 02	9.95 + 01	1.41 + 01	1.09 + 00	2.58-01	1.31-01	8.31-02
Ti	22	5.87 + 03	9.86 + 02	6.84 + 02	1.11 + 02	1.59 + 01	1.21 + 00	2.72-01	1.31-01	8.19-02
V	23	6.50 + 03	1.11 + 03	9.29 + 01	1.22 + 02	1.77 + 01	1.35 + 00	2.88-01	1.32-01	8.07-02
Cr	24	7.40 + 03	1.28 + 03	1.08 + 02	1.39 + 02	2.04 + 01	1.55 + 00	3.17-01	1.38-01	8.28-02
Mn	25	8.09 + 03	1.42 + 03	1.21 + 02	1.51 + 02	2.25 + 01	1.71 + 00	3.37-01	1.39-01	8.19-02
Fe	26	9.09 + 03	1.63 + 03	1.40 + 02	1.71 + 02	2.57 + 01	1.96 + 00	3.72-01	1.46-01	8.41-02
Co	27	9.80 + 03	1.78 + 03	1.54 + 02	1.84 + 02	2.80 + 01	2.14 + 00	3.95-01	1.48-01	8.32-02
Ni	28	9.86 + 03	2.05 + 03	1.79 + 02	2.09 + 02	3.22 + 01	2.47 + 00	4.44-01	1.58-01	8.70-02
Cu	29	1.06 + 04	2.15 + 03	1.90 + 02	2.16 + 02	3.38 + 01	2.61 + 00	4.58-01	1.56-01	8.36-02
Zn	30	1.55 + 03	2.37 + 03	2.12 + 02	2.33 + 02	3.72 + 01	2.89 + 00	4.97-01	1.62-01	8.45-02
Ga	31	1.70 + 03	2.52 + 03	2.27 + 02	3.42 + 01	3.93 + 01	3.08 + 00	5.20-01	1.62-01	8.24-02
Ge	32	1.89 + 03	2.71 + 03	2.47 + 02	3.74 + 01	4.22 + 01	3.34 + 00	5.55-01	1.66-01	8.21-02
As	33	2.12 + 03	2.93 + 03	2.71 + 02	4.12 + 01	4.56 + 01	3.63 + 00	5.97-01	1.72-01	8.26-02
Se	34	2.32 + 03	3.10 + 03	2.90 + 02	4.41 + 01	4.82 + 01	3.86 + 00	6.28-01	1.74-01	8.13-02
Br	35	2.62 + 03	3.41 + 03	3.21 + 02	4.91 + 01	5.27 + 01	4.26 + 00	6.86-01	1.84-01	8.33-02
Kr	36	2.85 + 03	3.60 + 03	3.43 + 02	5.26 + 01	5.55 + 01	4.52 + 00	7.22-01	1.87-01	8.23-02
Rb	37	3.17 + 03	3.41 + 03	3.74 + 02	5.77 + 01	5.98 + 01	4.92 + 00	7.80-01	1.96-01	8.36-02
Sr	38	3.49 + 03	2.59 + 03	4.06 + 02	6.27 + 01	6.39 + 01	5.31 + 00	8.37-01	2.04-01	8.44-02
Y	39	3.86 + 03	7.42 + 02	4.42 + 02	6.87 + 01	6.86 + 01	5.76 + 00	9.05-01	2.15-01	8.61-02
Zr	40	4.21 + 03	8.12 + 02	4.76 + 02	7.42 + 01	7.24 + 01	6.17 + 00	9.66-01	2.24-01	8.69-02
Nb	41	4.60 + 03	8.89 + 02	5.13 + 02	8.04 + 01	7.71 + 01	6.64 + 00	1.04 + 00	2.34-01	8.83-02
Mo	42	4.94 + 03	9.60 + 02	5.45 + 02	8.58 + 01	1.31 + 01	7.04 + 00	1.10 + 00	2.42-01	8.85-02
Tc	43	5.36 + 03	1.04 + 03	5.84 + 02	9.23 + 01	1.41 + 01	7.52 + 00	1.17 + 00	2.53-01	8.97-02
Ru	44	5.72 + 03	1.12 + 03	6.17 + 02	9.80 + 01	1.50 + 01	7.92 + 00	1.23 + 00	2.62-01	8.99-02
Rh	45	6.17 + 03	1.21 + 03	6.59 + 02	1.05 + 02	1.61 + 01	8.45 + 00	1.31 + 00	2.74-01	9.13-02
Pd	46	6.54 + 03	1.29 + 03	6.91 + 02	1.11 + 02	1.70 + 01	8.85 + 00	1.38 + 00	2.83-01	9.13-02
Ag	47	7.04 + 03	1.40 + 03	7.39 + 02	1.19 + 02	1.84 + 01	9.45 + 00	1.47 + 00	2.97-01	9.32-02
Cd	48	7.35 + 03	1.47 + 03	7.69 + 02	1.24 + 02	1.92 + 01	9.78 + 00	1.52 + 00	3.04-01	9.25-02
In	49	7.81 + 03	1.58 + 03	8.13 + 02	1.32 + 02	2.04 + 01	1.03 + 01	1.61 + 00	3.17-01	9.37-02
Sn	50	8.16 + 03	1.66 + 03	8.47 + 02	1.38 + 02	2.15 + 01	1.07 + 01	1.68 + 00	3.26-01	9.37-02

L₃ L₁ L₂   K EDGE

Mass attenuation coefficient, cm²/g

	Atomic no.	Photon energy, MeV								
		1.0	2.0	5.0	10.0	20.0	50.0	100.0	500.0	1000.0
H	1	1.26-01	8.77-02	5.05-02	3.25-02	2.15-02	1.42-02	1.19-02	1.14-02	1.16-02
He	2	6.36-02	4.42-02	2.58-02	1.70-02	1.18-02	8.61-03	7.78-03	7.79-03	7.95-03
Li	3	5.50-02	3.83-02	2.26-02	1.53-02	1.11-02	8.68-03	8.21-03	8.61-03	8.87-03
Be	4	5.65-02	3.94-02	2.35-02	1.63-02	1.23-02	1.02-02	9.94-03	1.08-02	1.12-02
B	5	5.89-02	4.11-02	2.48-02	1.76-02	1.37-02	1.19-02	1.19-02	1.32-02	1.37-02
C	6	6.36-02	4.44-02	2.71-02	1.96-02	1.58-02	1.43-02	1.46-02	1.64-02	1.70-02
N	7	6.36-02	4.45-02	2.74-02	2.02-02	1.67-02	1.57-02	1.63-02	1.85-02	1.92-02
O	8	6.37-02	4.46-02	2.78-02	2.09-02	1.77-02	1.71-02	1.79-02	2.06-02	2.13-02
F	9	6.04-02	4.23-02	2.66-02	2.04-02	1.77-02	1.75-02	1.86-02	2.14-02	2.21-02
Ne	10	6.32-02	4.43-02	2.82-02	2.20-02	1.95-02	1.96-02	2.11-02	2.43-02	2.51-02
Na	11	6.10-02	4.28-02	2.75-02	2.18-02	1.97-02	2.03-02	2.19-02	2.53-02	2.62-02
Mg	12	6.30-02	4.43-02	2.87-02	2.31-02	2.13-02	2.23-02	2.42-02	2.81-02	2.90-02
Al	13	6.15-02	4.32-02	2.84-02	2.32-02	2.17-02	2.31-02	2.52-02	2.93-02	3.03-02
Si	14	6.36-02	4.48-02	2.97-02	2.46-02	2.34-02	2.52-02	2.76-02	3.23-02	3.34-02
P	15	6.18-02	4.36-02	2.91-02	2.45-02	2.36-02	2.58-02	2.84-02	3.33-02	3.45-02
S	16	6.37-02	4.50-02	3.04-02	2.59-02	2.53-02	2.79-02	3.08-02	3.62-02	3.75-02
Cl	17	6.13-02	4.33-02	2.95-02	2.55-02	2.52-02	2.81-02	3.11-02	3.67-02	3.80-02
Ar	18	5.76-02	4.07-02	2.80-02	2.45-02	2.45-02	2.76-02	3.07-02	3.62-02	3.75-02
K	19	6.22-02	4.40-02	3.05-02	2.70-02	2.74-02	3.11-02	3.46-02	4.09-02	4.24-02
Ca	20	6.39-02	4.52-02	3.17-02	2.84-02	2.90-02	3.32-02	3.71-02	4.40-02	4.56-02
Sc	21	5.98-02	4.24-02	3.00-02	2.72-02	2.80-02	3.23-02	3.62-02	4.30-02	4.45-02
Ti	22	5.89-02	4.18-02	2.98-02	2.73-02	2.84-02	3.30-02	3.71-02	4.40-02	4.56-02
V	23	5.79-02	4.11-02	2.96-02	2.74-02	2.88-02	3.36-02	3.78-02	4.49-02	4.65-02
Cr	24	5.93-02	4.21-02	3.06-02	2.86-02	3.03-02	3.56-02	4.01-02	4.76-02	4.93-02
Mn	25	5.85-02	4.16-02	3.04-02	2.87-02	3.07-02	3.63-02	4.09-02	4.86-02	5.04-02
Fe	26	5.99-02	4.26-02	3.15-02	2.99-02	3.22-02	3.83-02	4.33-02	5.15-02	5.33-02
Co	27	5.91-02	4.20-02	3.13-02	3.00-02	3.26-02	3.88-02	4.40-02	5.23-02	5.41-02
Ni	28	6.16-02	4.39-02	3.29-02	3.18-02	3.48-02	4.17-02	4.73-02	5.61-02	5.81-02
Cu	29	5.90-02	4.20-02	3.18-02	3.10-02	3.41-02	4.10-02	4.66-02	5.53-02	5.72-02
Zn	30	5.94-02	4.24-02	3.22-02	3.18-02	3.51-02	4.24-02	4.82-02	5.72-02	5.91-02
Ga	31	5.77-02	4.11-02	3.16-02	3.13-02	3.48-02	4.22-02	4.80-02	5.70-02	5.89-02
Ge	32	5.73-02	4.09-02	3.16-02	3.16-02	3.53-02	4.30-02	4.89-02	5.80-02	6.00-02
As	33	5.73-02	4.09-02	3.19-02	3.21-02	3.60-02	4.40-02	5.01-02	5.95-02	6.15-02
Se	34	5.62-02	4.01-02	3.14-02	3.19-02	3.60-02	4.41-02	5.03-02	5.97-02	6.17-02
Br	35	5.73-02	4.09-02	3.23-02	3.29-02	3.74-02	4.60-02	5.24-02	6.22-02	6.43-02
Kr	36	5.63-02	4.02-02	3.20-02	3.28-02	3.74-02	4.61-02	5.26-02	6.25-02	6.46-02
Rb	37	5.69-02	4.06-02	3.25-02	3.36-02	3.85-02	4.75-02	5.43-02	6.45-02	6.67-02
Sr	38	5.71-02	4.08-02	3.29-02	3.41-02	3.93-02	4.87-02	5.56-02	6.61-02	6.83-02
Y	39	5.80-02	4.14-02	3.35-02	3.50-02	4.05-02	5.03-02	5.75-02	6.83-02	7.06-02
Zr	40	5.81-02	4.15-02	3.38-02	3.55-02	4.12-02	5.13-02	5.87-02	6.98-02	7.22-02
Nb	41	5.87-02	4.18-02	3.44-02	3.63-02	4.22-02	5.27-02	6.03-02	7.17-02	7.42-02
Mo	42	5.84-02	4.16-02	3.44-02	3.65-02	4.26-02	5.33-02	6.10-02	7.26-02	7.51-02
Tc	43	5.88-02	4.19-02	3.48-02	3.71-02	4.35-02	5.45-02	6.24-02	7.43-02	7.68-02
Ru	44	5.85-02	4.16-02	3.48-02	3.73-02	4.39-02	5.50-02	6.30-02	7.51-02	7.77-02
Rh	45	5.89-02	4.20-02	3.53-02	3.80-02	4.48-02	5.63-02	6.45-02	7.69-02	7.94-02
Pd	46	5.85-02	4.16-02	3.52-02	3.80-02	4.50-02	5.66-02	6.49-02	7.73-02	8.00-02
Ag	47	5.92-02	4.21-02	3.58-02	3.88-02	4.61-02	5.81-02	6.67-02	7.93-02	8.20-02
Cd	48	5.83-02	4.14-02	3.54-02	3.85-02	4.59-02	5.79-02	6.64-02	7.91-02	8.18-02
In	49	5.85-02	4.15-02	3.56-02	3.90-02	4.65-02	5.88-02	6.75-02	8.04-02	8.32-02
Sn	50	5.80-02	4.11-02	3.55-02	3.90-02	4.66-02	5.90-02	6.78-02	8.07-02	8.35-02

Mass attenuation coefficients, cm²/g

Photon energy, MeV

	Atomic no.	0.001	0.002	0.005	0.01	0.02	0.05	0.1	0.2	0.5
Sb	51	8.58 + 03	1.77 + 03	8.85 + 02	1.46 + 02	2.27 + 01	1.12 + 01	1.76 + 00	3.38-01	9.45-02
Te	52	8.43 + 03	1.83 + 03	9.01 + 02	1.50 + 02	2.34 + 01	1.14 + 01	1.80 + 00	3.43-01	9.33-02
I	53	9.10 + 03	2.00 + 03	8.43 + 02	1.63 + 02	2.54 + 01	1.23 + 01	1.94 + 00	3.66-01	9.70-02
Xe	54	9.41 + 03	2.09 + 03	6.39 + 02	1.69 + 02	2.65 + 01	1.27 + 01	2.01 + 00	3.76-01	9.70-02
Cs	55	9.37 + 03	2.23 + 03	2.30 + 02	1.79 + 02	2.82 + 01	1.34 + 01	2.12 + 00	3.94-01	9.91-02
ZB	56	8.54 + 03	2.32 + 03	2.41 + 02	1.86 + 02	2.94 + 01	1.38 + 01	2.20 + 00	4.05-01	9.92-02
La	57	9.09 + 03	2.46 + 03	2.58 + 02	1.97 + 02	3.12 + 01	1.45 + 01	2.32 + 00	4.24-01	1.01-01
Ce	58	9.71 + 03	2.61 + 03	2.74 + 02	2.08 + 02	3.31 + 01	1.52 + 01	2.45 + 00	4.45-01	1.04-01
Pr	59	1.06 + 04	2.77 + 03	2.92 + 02	2.21 + 02	3.53 + 01	1.60 + 01	2.59 + 00	4.69-01	1.07-01
Nd	60	6.63 + 03	2.88 + 03	3.06 + 02	2.30 + 02	3.68 + 01	1.65 + 01	2.69 + 00	4.84-01	1.08-01
Pm	61	2.06 + 03	3.05 + 03	3.26 + 02	2.44 + 02	3.92 + 01	1.73 + 01	2.84 + 00	5.10-01	1.12-01
Sm	62	2.11 + 03	3.12 + 03	3.36 + 02	2.50 + 02	4.03 + 01	1.77 + 01	2.90 + 00	5.19-01	1.11-01
Eu	63	2.22 + 03	3.28 + 03	3.54 + 02	2.63 + 02	4.24 + 01	1.85 + 01	3.04 + 00	5.43-01	1.14-01
Gd	64	2.29 + 03	3.36 + 03	3.65 + 02	2.69 + 02	4.36 + 01	3.86 + 00	3.11 + 00	5.54-01	1.14-01
Tb	65	2.40 + 03	3.51 + 03	3.84 + 02	2.82 + 02	4.59 + 01	4.06 + 00	3.25 + 00	5.77-01	1.17-01
Dy	66	2.49 + 03	3.47 + 03	3.99 + 02	2.90 + 02	4.76 + 01	4.23 + 00	3.36 + 00	5.95-01	1.18-01
Ho	67	2.62 + 03	3.59 + 03	4.17 + 02	3.01 + 02	4.98 + 01	4.43 + 00	3.49 + 00	6.18-01	1.20-01
Er	68	2.75 + 03	3.52 + 03	4.36 + 02	3.13 + 02	5.20 + 01	4.63 + 00	3.63 + 00	6.41-01	1.23-01
Tm	69	2.90 + 03	3.69 + 03	4.57 + 02	2.83 + 02	5.45 + 01	4.87 + 00	3.78 + 00	6.68-01	1.26-01
Yb	70	3.02 + 03	3.80 + 03	4.72 + 02	2.94 + 02	5.63 + 01	5.04 + 00	3.88 + 00	6.86-01	1.27-01
Lu	71	3.19 + 03	3.45 + 03	4.94 + 02	2.21 + 02	5.88 + 01	5.28 + 00	4.03 + 00	7.13-01	1.30-01
Hf	72	3.34 + 03	3.60 + 03	5.11 + 02	2.30 + 02	6.09 + 01	5.48 + 00	4.15 + 00	7.34-01	1.32-01
Ta	73	3.51 + 03	3.77 + 03	5.33 + 02	2.38 + 02	6.33 + 01	5.72 + 00	4.30 + 00	7.60-01	1.35-01
W	74	3.68 + 03	3.92 + 03	5.53 + 02	9.69 + 01	6.57 + 01	5.95 + 00	4.44 + 00	7.84-01	1.38-01
Re	75	3.87 + 03	3.77 + 03	5.76 + 02	1.01 + 02	6.84 + 01	6.21 + 00	4.59 + 00	8.12-01	1.41-01
Os	76	4.03 + 03	2.22 + 03	5.93 + 02	1.04 + 02	7.04 + 01	6.41 + 00	4.70 + 00	8.33-01	1.43-01
Ir	77	4.24 + 03	1.03 + 03	6.18 + 02	1.09 + 02	7.32 + 01	6.69 + 00	4.86 + 00	8.63-01	1.46-01
Pt	78	4.43 + 03	1.08 + 03	6.40 + 02	1.13 + 02	7.57 + 01	6.95 + 00	4.99 + 00	8.90-01	1.49-01
Au	79	4.65 + 03	1.14 + 03	6.66 + 02	1.18 + 02	7.88 + 01	7.26 + 00	5.16 + 00	9.22-01	1.53-01
Hg	80	4.83 + 03	1.18 + 03	6.87 + 02	1.22 + 02	8.12 + 01	7.50 + 00	5.28 + 00	9.46-01	1.56-01
Tl	81	5.01 + 03	1.23 + 03	7.07 + 02	1.26 + 02	8.36 + 01	7.75 + 00	5.40 + 00	9.69-01	1.58-01
Pb	82	5.21 + 03	1.29 + 03	7.30 + 02	1.31 + 02	8.64 + 01	8.04 + 00	5.55 + 00	9.99-01	1.61-01
Bi	83	5.44 + 03	1.35 + 03	7.58 + 02	1.36 + 02	8.95 + 01	8.38 + 00	5.74 + 00	1.03 + 00	1.66-01
Po	84	5.72 + 03	1.42 + 03	7.93 + 02	1.43 + 02	9.35 + 01	8.80 + 00	5.99 + 00	1.08 + 00	1.71-01
At	85	5.87 + 03	1.49 + 03	8.25 + 02	1.49 + 02	9.70 + 01	9.19 + 00	6.17 + 00	1.12 + 00	1.77-01
Rn	86	5.83 + 03	1.49 + 03	8.16 + 02	1.48 + 02	9.56 + 01	9.12 + 00	6.09 + 00	1.10 + 00	1.73-01
Fr	87	6.08 + 03	1.56 + 03	8.49 + 02	1.54 + 02	9.93 + 01	9.52 + 00	1.66 + 00	1.14 + 00	1.78-01
Ra	88	6.20 + 03	1.62 + 03	8.74 + 02	1.59 + 02	1.02 + 02	9.85 + 00	1.71 + 00	1.17 + 00	1.82-01
Ac	89	6.47 + 03	1.70 + 03	8.69 + 02	1.65 + 02	1.06 + 02	1.03 + 01	1.79 + 00	1.21 + 00	1.87-01
Th	90	6.61 + 03	1.74 + 03	8.88 + 02	1.69 + 02	9.37 + 01	1.05 + 01	1.83 + 00	1.23 + 00	1.90-01
Pa	91	6.53 + 03	1.83 + 03	8.76 + 02	1.77 + 02	7.03 + 01	1.10 + 01	1.92 + 00	1.29 + 00	1.97-01
U	92	6.63 + 03	1.86 + 03	8.89 + 02	1.79 + 02	7.11 + 01	1.12 + 01	1.95 + 00	1.30 + 00	1.98-01
Np	93	6.95 + 03	1.96 + 03	9.32 + 02	1.87 + 02	7.45 + 01	1.18 + 01	2.05 + 00	1.35 + 00	2.05-01
Pu	94	7.19 + 03	2.04 + 03	9.65 + 02	1.94 + 02	7.71 + 01	1.22 + 01	2.13 + 00	1.39 + 00	2.10-01
Am	95	7.37 + 03	2.10 + 03	9.90 + 02	1.98 + 02	7.93 + 01	1.25 + 01	2.19 + 00	1.42 + 00	2.14-01
Cm	96	7.54 + 03	2.15 + 03	1.02 + 03	2.03 + 02	8.14 + 01	1.28 + 01	2.25 + 00	1.44 + 00	2.18-01
Bk	97	7.84 + 03	2.25 + 03	1.06 + 03	2.10 + 02	8.39 + 01	1.34 + 01	2.35 + 00	1.50 + 00	2.25-01
Cf	98	7.89 + 03	2.31 + 03	9.27 + 02	2.15 + 02	8.58 + 01	1.37 + 01	2.41 + 00	1.52 + 00	2.29-01
Es	99	7.79 + 03	2.40 + 03	9.59 + 02	2.22 + 02	4.01 + 01	1.42 + 01	2.51 + 00	1.57 + 00	2.36-01
Fm	100	7.13 + 03	2.46 + 03	9.77 + 02	2.26 + 02	4.09 + 01	1.45 + 01	2.57 + 00	1.59 + 00	2.39-01

N_5 N_4 N_3 N_2 N_1 M_5 M_4 M_3 M_2 M_1 L_3 L_2 L_1 K EDGE

Mass attenuation coefficients, cm^2/g

	Atomic no.	Photon Energy, MeV								
		1.0	2.0	5.0	10.0	20.0	50.0	100.0	500.0	1000.0
Sb	51	5.80-02	4.10-02	3.56-02	3.92-02	4.70-02	5.96-02	6.85-02	8.16-02	8.44-02
Te	52	5.67-02	4.01-02	3.49-02	3.86-02	4.64-02	5.89-02	6.77-02	8.07-02	8.35-02
I	53	5.84-02	4.12-02	3.61-02	4.00-02	4.82-02	6.13-02	7.04-02	8.40-02	8.69-02
Xe	54	5.78-02	4.08-02	3.58-02	3.99-02	4.82-02	6.12-02	7.04-02	8.40-02	8.69-02
Cs	55	5.85-02	4.12-02	3.64-02	4.06-02	4.91-02	6.25-02	7.19-02	8.58-02	8.88-02
ZB	56	5.80-02	4.08-02	3.61-02	4.04-02	4.90-02	6.25-02	7.19-02	8.58-02	8.88-02
La	57	5.88-02	4.12-02	3.66-02	4.11-02	5.00-02	6.37-02	7.34-02	8.76-02	9.06-02
Ce	58	5.96-02	4.18-02	3.73-02	4.19-02	5.10-02	6.52-02	7.50-02	8.96-02	9.27-02
Pr	59	6.07-02	4.24-02	3.80-02	4.29-02	5.23-02	6.68-02	7.69-02	9.19-02	9.50-02
Nd	60	6.07-02	4.24-02	3.81-02	4.30-02	5.26-02	6.72-02	7.74-02	9.25-02	9.56-02
Pm	61	6.19-02	4.31-02	3.88-02	4.40-02	5.38-02	6.89-02	7.94-02	9.48-02	9.81-02
Sm	62	6.11-02	4.24-02	3.83-02	4.35-02	5.34-02	6.84-02	7.88-02	9.41-02	9.73-02
Eu	63	6.19-02	4.28-02	3.88-02	4.42-02	5.42-02	6.96-02	8.02-02	9.57-02	9.90-02
Gd	64	6.12-02	4.23-02	3.84-02	4.38-02	5.38-02	6.91-02	7.97-02	9.51-02	9.83-02
Tb	65	6.20-02	4.27-02	3.89-02	4.45-02	5.47-02	7.03-02	8.11-02	9.67-02	1.00-01
Dy	66	6.20-02	4.26-02	3.90-02	4.46-02	5.49-02	7.06-02	8.15-02	9.72-02	1.00-01
Ho	67	6.26-02	4.29-02	3.93-02	4.50-02	5.55-02	7.14-02	8.24-02	9.83-02	1.02-01
Er	68	6.32-02	4.32-02	3.96-02	4.55-02	5.61-02	7.23-02	8.34-02	9.95-02	1.03-01
Tm	69	6.40-02	4.36-02	4.01-02	4.61-02	5.70-02	7.35-02	8.48-02	1.01-01	1.04-01
Yb	70	6.40-02	4.35-02	4.00-02	4.61-02	5.70-02	7.35-02	8.49-02	1.01-01	1.04-01
Lu	71	6.48-02	4.39-02	4.05-02	4.66-02	5.77-02	7.45-02	8.60-02	1.02-01	1.06-01
Hf	72	6.50-02	4.39-02	4.05-02	4.68-02	5.80-02	7.48-02	8.64-02	1.03-01	1.06-01
Ta	73	6.57-02	4.41-02	4.08-02	4.72-02	5.85-02	7.56-02	8.73-02	1.04-01	1.07-01
W	74	6.62-02	4.43-02	4.10-02	4.75-02	5.89-02	7.62-02	8.80-02	1.05-01	1.08-01
Re	75	6.69-02	4.46-02	4.14-02	4.79-02	5.95-02	7.70-02	8.89-02	1.06-01	1.09-01
Os	76	6.71-02	4.46-02	4.13-02	4.79-02	5.96-02	7.71-02	8.90-02	1.06-01	1.10-01
Ir	77	6.79-02	4.50-02	4.17-02	4.84-02	6.02-02	7.80-02	9.01-02	1.07-01	1.11-01
Pt	78	6.86-02	4.52-02	4.20-02	4.87-02	6.06-02	7.86-02	9.08-02	1.08-01	1.12-01
Au	79	6.95-02	4.57-02	4.24-02	4.93-02	6.14-02	7.95-02	9.19-02	1.09-01	1.13-01
Hg	80	6.99-02	4.57-02	4.25-02	4.94-02	6.15-02	7.98-02	9.22-02	1.10-01	1.13-01
Tl	81	7.03-02	4.58-02	4.25-02	4.94-02	6.16-02	8.00-02	9.24-02	1.10-01	1.14-01
Pb	82	7.10-02	4.61-02	4.27-02	4.97-02	6.21-02	8.06-02	9.31-02	1.11-01	1.15-01
Bi	83	7.21-02	4.66-02	4.32-02	5.03-02	6.28-02	8.15-02	9.42-02	1.12-01	1.16-01
Po	84	7.39-02	4.75-02	4.40-02	5.12-02	6.40-02	8.32-02	9.61-02	1.15-01	1.18-01
At	85	7.54-02	4.82-02	4.46-02	5.20-02	6.49-02	8.44-02	9.76-02	1.16-01	1.20-01
Rn	86	7.30-02	4.65-02	4.30-02	5.01-02	6.26-02	8.14-02	9.42-02	1.12-01	1.16-01
Fr	87	7.45-02	4.72-02	4.36-02	5.08-02	6.35-02	8.26-02	9.56-02	1.14-01	1.18-01
Ra	88	7.53-02	4.75-02	4.38-02	5.10-02	6.38-02	8.31-02	9.61-02	1.15-01	1.19-01
Ac	89	7.69-02	4.82-02	4.44-02	5.17-02	6.47-02	8.43-02	9.75-02	1.16-01	1.20-01
Th	90	7.71-02	4.81-02	4.42-02	5.15-02	6.45-02	8.40-02	9.72-02	1.16-01	1.20-01
Pa	91	7.94-02	4.93-02	4.52-02	5.26-02	6.59-02	8.60-02	9.95-02	1.19-01	1.23-01
U	92	7.90-02	4.88-02	4.46-02	5.19-02	6.51-02	8.49-02	9.83-02	1.17-01	1.21-01
Np	93	8.13-02	4.99-02	4.56-02	5.30-02	6.65-02	8.68-02	1.01-01	1.20-01	1.24-01
Pu	94	8.26-02	5.05-02	4.60-02	5.34-02	6.71-02	8.76-02	1.01-01	1.21-01	1.25-01
Am	95	8.33-02	5.06-02	4.60-02	5.34-02	6.70-02	8.77-02	1.02-01	1.21-01	1.25-01
Cm	96	8.41-02	5.08-02	4.60-02	5.34-02	6.70-02	8.77-02	1.02-01	1.21-01	1.26-01
Bk	97	8.62-02	5.18-02	4.68-02	5.42-02	6.81-02	8.92-02	1.03-01	1.24-01	1.28-01
Cf	98	8.70-02	5.20-02	4.68-02	5.42-02	6.81-02	8.92-02	1.04-01	1.24-01	1.28-01
Es	99	8.89-02	5.28-02	4.74-02	5.48-02	6.89-02	9.04-02	1.05-01	1.25-01	1.29-01
Fm	100	8.94-02	5.28-02	4.72-02	5.45-02	6.86-02	9.00-02	1.05-01	1.25-01	1.29-01

CLASSIFICATION OF ELECTROMAGNETIC RADIATION

Hans Dolezalek

Basic Conversions: $c = \lambda\nu = \nu/k$; $\nu = c/\lambda = ck$; $\lambda = c/\nu = 1/k$; $k = \nu/c = 1/\lambda$

c = speed of light = 2.99792458×10^8 m/s

Frequency (ν)	Wavelength (λ)	Wave number (k)	Names of bands	Approximate photon energies
$3 \times 10^0 - 3 \times 10^1$ Hz 3 — 30 Hz	$10^8 - 10^7$ m 100 — 10 m	$10^{-8} - 10^{-7}$ m^{-1} 10 — 100 Gm^{-1}	ELF-(ELF 1), ITU band no. 1	
$3 \times 10^1 - 3 \times 10^2$ Hz 30 — 300 Hz	$10^7 - 10^6$ m 10 — 1 Mm	$10^{-7} - 10^{-6}$ m^{-1} 100 Gm^{-1} — 1 Mm^{-1}	SLF-(ELF 2), ITU band no. 2, mega-meter waves	
$3 \times 10^2 - 3 \times 10^3$ Hz 300 Hz — 3 kHz	$10^6 - 10^5$ m 1 Mm — 100 km	$10^{-6} - 10^{-5}$ m^{-1} 1 — 10 Mm^{-1}	ULF-(ELF 3), ITU band no. 3	
$3 \times 10^3 - 3 \times 10^4$ Hz 3 — 30 kHz	$10^5 - 10^4$ m 100 — 10 km	$10^{-5} - 10^{-4}$ m^{-1} 10 — 100 Mm^{-1}	VLF, ITU band no. 4, myriameter waves	
$3 \times 10^4 - 3 \times 10^5$ Hz 30 — 300 kHz	$10^4 - 10^3$ m 10 — 1 km	$10^{-4} - 10^{-3}$ m^{-1} 100 Mm^{-1} — 1 km^{-1}	LF, ITU band no. 5, kilometer waves	
$3 \times 10^5 - 3 \times 10^6$ Hz 300 kHz — 3 MHz	$10^3 - 10^2$ m 1 km — 100 m	$10^{-3} - 10^{-2}$ m^{-1} 1 — 10 km^{-1}	MF, ITU band no. 6, hectometer waves	
$3 \times 10^6 - 3 \times 10^7$ Hz 3 — 30 MHz	$10^2 - 10^1$ m 100 — 10 m	$10^{-2} - 10^{-1}$ m^{-1} 10 — 100 km^{-1}	HF, ITU band no. 7, decameter waves	
$3 \times 10^7 - 3 \times 10^8$ Hz 30 — 300 MHz	$10^1 - 10^0$ m 10 — 1 m	$10^{-1} - 10^0$ m^{-1} 100 km^{-1} — 1 m^{-1}	VHF, ITU band no. 8, meter waves	
$3 \times 10^8 - 3 \times 10^9$ Hz 300 MHz — 3 GHz	$10^0 - 10^{-1}$ m 1 m — 100 mm	$10^0 - 10^1$ m^{-1} 1 — 10 m^{-1}	UHF, ITU band no. 9, decimeter waves[a]	
$3 \times 10^9 - 3 \times 10^{10}$ Hz 3 — 30 GHz	$10^{-1} - 10^{-2}$ m 100 — 10 mm	$10^1 - 10^2$ m^{-1} 1 — 100 m^{-1}	SHF, ITU band no. 10, centimeter waves[a]	
$3 \times 10^{10} - 3 \times 10^{11}$ Hz 30 — 300 GHz	$10^{-2} - 10^{-3}$ m 10 1 mm	$10^2 - 10^3$ m^{-1} 100 m^{-1} — 1 mm^{-1} (1 — 10 cm^{-1})	EHF, ITU band no. 11, millimeter waves	
$3 \times 10^{11} - 3 \times 10^{12}$ Hz 300 GHz — 3 THz	$10^{-3} - 10^{-4}$ m 1 mm — 100 μm	$10^3 - 10^4$ m^{-1} 1 — 10 mm^{-1} (10 — 100 cm^{-1})	Part of micrometer waves, includes part of far or thermal infrared; ITU band no. 12	
$3 \times 10^{12} - 3 \times 10^{13}$ Hz 3 — 30 THz	$10^{-4} - 10^{-5}$ m 100 — 10 μm	$10^4 - 10^5$ m^{-1} 10 — 100 mm^{-1} (100 — 1000 cm^{-1})	Part of micrometer waves includes part of far (thermal) infrared	
$3 \times 10^{13} - 3 \times 10^{14}$ Hz 30 — 300 THz	$10^{-5} - 10^{-6}$ m 10 — 1 μm (100,000 — 10,000 Å)	$10^5 - 10^6$ m^{-1} 100 mm^{-1} — 1 μm^{-1}	Part of μm waves, part of infrared	$(1.6-16) \times 10^{-20}$ joule {0.1 — 1 eV}
$3 \times 10^{14} - 3 \times 10^{15}$ Hz 300 THz — 3 PHz	$10^{-6} - 10^{-7}$ m 1 μm — 100 nm (10,000 — 1000 Å)	$10^6 - 10^7$ m^{-1} 1 — 10 μm^{-1}	Near infrared, visible, near ultraviolet	$(1.6-16) \times 10^{-19}$ joule {1 — 10 eV}
$3 \times 10^{15} - 3 \times 10^{16}$ Hz 3 — 30 PHz	$10^{-7} - 10^{-8}$ m 100 — 10 nm (1000 — 100 Å)	$10^7 - 10^8$ m^{-1} 10 — 100 μm^{-1}	Part of "vacuum" - ultraviolet	$(1.6-16) \times 10^{-18}$ joule {10 — 100 eV}
$3 \times 10^{16} - 3 \times 10^{17}$ Hz 30 — 300 PHz	$10^{-8} - 10^{-9}$ m 10 — 1 nm (100 — 10 Å)	$10^8 - 10^9$ m^{-1} 100 μm^{-1} — 1 nm^{-1}	Part of soft X-rays	$(1.6-16) \times 10^{-17}$ joule {100 — 1000 eV)}
$3 \times 10^{17} - 3 \times 10^{18}$ Hz 300 PHz — 3 EHz	$10^{-9} - 10^{-10}$ m 1 nm — 100 pm (10 — 1 Å)	$10^9 - 10^{10}$ m^{-1} 1 — 10 nm^{-1}	Part of soft X-rays	$(1.6-16) \times 10^{-16}$ joule {1 — 10 keV}
$3 \times 10^{18} - 3 \times 10^{19}$ Hz 3 — 30 EHz	$10^{-10} - 10^{-11}$ m 100 — 10 pm (1 — 0.1 Å)	$10^{10} - 10^{11}$ m^{-1} 10 — 100 nm^{-1}	Hard X-rays and part of soft γ-rays	$(1.6-16) \times 10^{-15}$ joule {10 — 100 keV}
$3 \times 10^{19} - 3 \times 10^{20}$ Hz 30 — 300 EHz	$10^{-11} - 10^{-12}$ m 10 — 1 pm (0.1 — 0.01 Å)	$10^{11} - 10^{12}$ m^{-1} 100 nm^{-1} — 1 pm^{-1}	Part of soft and part of hard γ-rays (limit at 510 keV)	$(1.6-16) \times 10^{-14}$ joule {100 keV — 1 MeV}
$3 \times 10^{20} - 3 \times 10^{21}$ Hz 300 — 3000 EHz	$10^{-12} - 10^{-13}$ m 1 pm — 100 fm (0.01 — 0.001 Å)	$10^{12} - 10^{13}$ m^{-1} 1 — 10 pm^{-1}	Part of hard γ-rays and part of "cosmic" γ-rays	$(1.6-16) \times 10^{-13}$ joule {1 — 10 MeV}
$3 \times 10^{21} - 3 \times 10^{22}$ Hz 3000 — 30,000 EHz	$10^{-13} - 10^{-14}$ m 100 — 10 fm (0.001 — 0.0001 Å)	$10^{13} - 10^{14}$ m^{-1} 10 — 100 pm^{-1}	γ-rays produced by cosmic rays	$(1.6-16) \times 10^{-12}$ joule {10 — 100 MeV}

CLASSIFICATION OF ELECTROMAGNETIC RADIATION (continued)

Note: Abbreviations used in this table: Å—ångstrom (1 Å = 10^{-10} m); EHz—exahertz (10^{18} hertz); EHF—extremely high frequency; ELF—extremely low frequency; eV—electron volt (1 eV = 1.60219×10^{-19} joule); PHz—petahertz (10^{15} hertz); fm—femtometer (10^{-15} m); GHz—gigahertz (10^9 hertz); Gm—gigameter (10^9 m); HF—high frequency; Hz—hertz (s^{-1}); ITU—International Telecommunications Union; keV—kiloelectron volt (10^3 eV); km—kilometer (10^3 m); LF—low frequency; m—meter; MeV—megaelectron volt (10^6 eV); MF—medium frequency; MHz—megahertz (10^6 hertz); Mm—megameter (10^6 meter); mm—millimeter (10^{-3} meter); μm—micrometer (10^{-6} meter); nm—nanometer (10^{-9} meter); pm—picometer (10^{-12} meter); SHF—super high frequency; SLF—super low frequency; THz—terahertz; UHF—ultra high frequency; ULF—ultra low frequency; VHF—very high frequency; VLF—very low frequency.

[a] Also called "microwaves"; not to be confused with "micrometer waves".

LETTER DESIGNATIONS OF MICROWAVE BANDS

Frequency (GHz)	Wavelength (cm)	Wavenumber (cm^{-1})	Band
1—2	30—15	0.033—0.067	L-Band
1—4	15—7.5	0.067—0.133	S-Band
4—8	7.5—3.7	0.133—0.267	C-Band
8—12	3.7—2.5	0.267—0.4	X-Band
12—18	2.5—1.7	0.4—0.6	Ku-Band
18—27	1.7—1.1	0.6—0.9	K-Band
27—40	1.1—0.75	0.9—1.33	Ka-Band

BLACK BODY RADIATION

The total power radiated from an ideal black body and the wavelength corresponding to maximum power are given here as a function of absolute temperature. Constants used in the calculation are taken from the table "Fundamental Physical Constants" in Section 1. The radiated power in a band $\Delta\lambda$ at λ_{max} may be calculated from:

$$P_{max} = 0.657548\,(\Delta\lambda/\lambda_{max})\,P_{tot}$$

T/K	P_{tot}	$\lambda_{max}/\mu m$	T/K	P_{tot}	$\lambda_{max}/\mu m$	T/K	P_{tot}	$\lambda_{max}/\mu m$
50	0.354 W/m²	57.955	740	17.004	3.916	1520	302.689	1.906
100	5.671	28.978	750	17.942	3.864	1540	318.937	1.882
150	28.707	19.318	760	18.918	3.813	1560	335.831	1.858
200	90.728	14.489	770	19.934	3.763	1580	353.387	1.834
250	221.504	11.591	780	20.989	3.715	1600	371.623	1.811
273	314.973	10.614	790	22.087	3.668	1620	390.555	1.789
280	348.541	10.349	800	23.226	3.622	1640	410.202	1.767
290	401.064	9.992	810	24.410	3.577	1660	430.581	1.746
300	459.311	9.659	820	25.638	3.534	1680	451.710	1.725
310	523.684	9.348	830	26.911	3.491	1700	473.607	1.705
320	594.596	9.055	840	28.232	3.450	1720	496.290	1.685
330	672.478	8.781	850	29.600	3.409	1740	519.779	1.665
340	757.771	8.523	860	31.018	3.369	1760	544.093	1.646
350	850.931	8.279	870	32.486	3.331	1780	569.249	1.628
360	952.428	8.049	880	34.006	3.293	1800	595.267	1.610
370	1.063 kW/m²	7.832	890	35.578	3.256	1820	622.168	1.592
380	1.182	7.626	900	37.204	3.220	1840	649.970	1.575
390	1.312	7.430	910	38.886	3.184	1860	678.694	1.558
400	1.452	7.244	920	40.623	3.150	1880	708.359	1.541
410	1.602	7.068	930	42.418	3.116	1900	738.987	1.525
420	1.764	6.899	940	44.272	3.083	1920	770.597	1.509
430	1.939	6.739	950	46.187	3.050	1940	803.210	1.494
440	2.125	6.586	960	48.162	3.018	1960	836.848	1.478
450	2.325	6.439	970	50.201	2.987	1980	871.531	1.464
460	2.539	6.299	980	52.303	2.957	2000	907.282	1.449
470	2.767	6.165	990	54.471	2.927	2020	944.121	1.435
480	3.010	6.037	1000	56.705	2.898	2040	982.071	1.420
490	3.269	5.914	1020	61.379	2.841	2060	1.021 MW/m²	1.407
500	3.544	5.796	1040	66.337	2.786	2080	1.061	1.393
510	3.836	5.682	1060	71.589	2.734	2100	1.103	1.380
520	4.146	5.573	1080	77.147	2.683	2120	1.145	1.367
530	4.474	5.467	1100	83.022	2.634	2140	1.189	1.354
540	4.822	5.366	1120	89.227	2.587	2160	1.234	1.342
550	5.189	5.269	1140	95.773	2.542	2180	1.281	1.329
560	5.577	5.175	1160	102.672	2.498	2200	1.328	1.317
570	5.986	5.084	1180	109.939	2.456	2220	1.377	1.305
580	6.417	4.996	1200	117.584	2.415	2240	1.428	1.294
590	6.871	4.911	1220	125.621	2.375	2260	1.479	1.282
600	7.349	4.830	1240	134.063	2.337	2280	1.532	1.271
610	7.851	4.750	1260	142.924	2.300	2300	1.587	1.260
620	8.379	4.674	1280	152.217	2.264	2320	1.643	1.249
630	8.933	4.600	1300	161.955	2.229	2340	1.700	1.238
640	9.514	4.528	1320	172.154	2.195	2360	1.759	1.228
650	10.122	4.458	1340	182.827	2.163	2380	1.819	1.218
660	10.760	4.391	1360	193.989	2.131	2400	1.881	1.207
670	11.427	4.325	1380	205.655	2.100	2420	1.945	1.197
680	12.124	4.261	1400	217.838	2.070	2440	2.010	1.188
690	12.853	4.200	1420	230.556	2.041	2460	2.077	1.178
700	13.615	4.140	1440	243.822	2.012	2480	2.145	1.168
710	14.410	4.081	1460	257.652	1.985	2500	2.215	1.159
720	15.239	4.025	1480	272.063	1.958	2550	2.398	1.136
730	16.103	3.970	1500	287.070	1.932	2600	2.591	1.115

T/K	P_{tot}	$\lambda_{max}/\mu m$	T/K	P_{tot}	$\lambda_{max}/\mu m$	T/K	P_{tot}	$\lambda_{max}/\mu m$
2650	2.796	1.093	3600	9.524	0.805	5100	38.362	0.568
2700	3.014	1.073	3650	10.065	0.794	5200	41.461	0.557
2750	3.243	1.054	3700	10.627	0.783	5300	44.743	0.547
2800	3.485	1.035	3750	11.214	0.773	5400	48.217	0.537
2850	3.741	1.017	3800	11.824	0.763	5500	51.889	0.527
2900	4.011	0.999	3850	12.458	0.753	5600	55.767	0.517
2950	4.294	0.982	3900	13.118	0.743	5700	59.858	0.508
3000	4.593	0.966	3950	13.804	0.734	5800	64.170	0.500
3050	4.907	0.950	4000	14.517	0.724	5900	68.712	0.491
3100	5.237	0.935	4100	16.024	0.707	6000	73.490	0.483
3150	5.583	0.920	4200	17.645	0.690	6500	101.222	0.446
3200	5.946	0.906	4300	19.386	0.674	7000	136.149	0.414
3250	6.326	0.892	4400	21.254	0.659	7500	179.418	0.386
3300	6.725	0.878	4500	23.253	0.644	8000	232.264	0.362
3350	7.142	0.865	4600	25.389	0.630	8500	296.004	0.341
3400	7.578	0.852	4700	27.670	0.617	9000	372.042	0.322
3450	8.033	0.840	4800	30.101	0.604	9500	461.867	0.305
3500	8.509	0.828	4900	32.689	0.591	10000	567.051	0.290
3550	9.006	0.816	5000	35.441	0.580			

The curves below show, for various temperatures, the fraction of radiant power as a function of wavelength. The function plotted is $P_\lambda/\Delta\lambda\, P_{tot}$, where P_λ is the power at wavelength λ in a small interval $\Delta\lambda$ (in μm), and P_{tot} is the total power.

CHARACTERISTICS OF INFRARED DETECTORS

This graph summarizes the wavelength response of some semiconductors used as detectors for infrared radiation. The quantity $D^*(\lambda)$ is the signal to noise ratio for an incident radiant power density of 1 W/cm^2 and a bandwidth of 1 Hz (60° field of view). The Ge, InAs, and InSb detectors are photovoltaics, while the HgCdTe series are photoconductive devices. The cutoff wavelength of the latter can be varied by adjusting the relative amounts of Hg, Cd, and Te (three examples are shown at 77 K). The graph also shows the theoretical background limited sensitivity for ideal detectors which introduce no intrinsic noise.

REFERENCE

Infrared Detectors 1995, EG&G Judson, Montgomeryville, PA.

REFRACTIVE INDEX AND TRANSMITTANCE OF REPRESENTATIVE GLASSES

Typical values of the index of refraction and internal transmittance (fraction of light transmitted through a one centimeter thickness) are tabulated here for selected types of glasses, as well as for synthetic fused (vitreous) silica. Nominal compositions are given in the first part of the table. The second part gives the index of refraction, relative to air, and the internal transmittance for representative samples of each glass at wavelengths in the infrared, visible, and near-ultraviolet regions. It should be emphasized that wide variation of these parameters may be found among subtypes of each glass. More detailed data may be found in Reference 3.

Assuming that the Lambert-Beer Law is followed, the transmittance of a glass plate of thickness d (in centimeters) can be obtained by raising the transmittance value in the table to the power d.

REFERENCES

1. Weber, M.J., *CRC Handbook of Laser Science and Technology*, Vol. IV, Part 2, CRC Press, Boca Raton, FL ,1988.
2. Gray, D.E., Ed., *American Institute of Physics Handbook, Third Edition*, McGraw Hill, New York, 1972.
3. *Schott Optical Glass*, Schott Glass Technologies, Inc., 400 York Ave., Duryea, PA 18642.
4. Kaye, G.W.C., and Laby, T.H., *Tables of Physical and Chemical Constants, 15th Edition*, Longman, London, 1986.

Type	Name	Composition in percent by mass									
		SiO_2	B_2O_3	Al_2O_3	Na_2O	K_2O	CaO	BaO	ZnO	PbO	P_2O_5
PK	Phosphate crown		3	10		12	5				70
PSK	Dense phosphate crown		3	5			4	28			60
BK	Borosilicate crown	70	10		8	8	1	3			
K	Crown	74			9	11	6				
ZK	Zinc crown	71			17				12		
BaK	Barium crown	60	3		3	10		19	5		
SK	Dense crown	39	15	5				41			
KF	Crown flint	67		2	16				3	12	
BaLF	Barium light flint	51			6	5		20	14	4	
SSK	Extra dense crown	35	10	5				42	8		
LLF	Extra light flint	63			5	8				24	
BaF	Barium flint	46				8		16	8	22	
LF	Light flint	53			5	8				34	
F	Flint	47			2	7				44	
BaSF	Dense barium flint	43			1	7		11	5	33	
SF	Dense flint	33				5				62	
KzFS	Short flint										
SiO_2	Fused silica	100									

Type	Index of refraction				Transmittance of 1 cm plate			
	1.060 µm	546.1 nm	365.0 nm	312.6 nm	1.060 µm	546.1 nm	365.0 nm	310 nm
PK	1.51519	1.52736	1.54503	1.5574	0.997	0.998	0.987	0.46
PSK	1.54154	1.55440	1.57342	1.5868	0.996	0.998	0.984	0.46
BK	1.50669	1.51872	1.53627	1.5486	0.999	0.998	0.987	0.35
K	1.50091	1.51314	1.53189	1.5454	0.998	0.998	0.988	0.40
ZK	1.52220	1.53534	1.55588	1.5708	0.996	0.998	0.976	0.27
BaK	1.55695	1.57124	1.59407	1.6108	0.998	0.997	0.986	0.28
SK	1.59490	1.60994	1.63398		0.998	0.998	0.959	0.28
KF	1.50586	1.51978	1.54251	1.5600	0.998	0.996	0.989	0.49
BaLF	1.57579	1.59166	1.61804		0.996	0.998	0.933	0.010
SSK	1.60402	1.61993	1.64595		0.999	0.998	0.915	0.010
LaK	1.69710	1.71616	1.74573		0.999	0.998	0.882	0.17
LLF	1.52775	1.54344	1.57038		0.998	0.997	0.990	0.32
BaF	1.56873	1.58565	1.61524		0.999	0.997	0.992	0.004
LF	1.56594	1.58482	1.61926		0.999	0.998	0.981	0.008
F	1.58636	1.60718	1.64606		0.997	0.998	0.959	
BaSF	1.60889	1.62987	1.66926		0.999	0.998	0.857	
SF	1.71350	1.74620	1.8145		0.998	0.997	0.650	
KzFS	1.59680	1.61639	1.64849	1.6739	0.998	0.998	0.672	0.012
SiO_2	1.44968	1.46008	1.47435[a]	1.53430[b]				

[a] At 366.3 nm.

[b] At 213.9 nm.

INDEX OF REFRACTION OF WATER

This table gives the index of refraction of liquid water at atmospheric pressure, relative to a vacuum, at several temperatures and wavelengths. It is generated from the formulation in Reference 1, which covers a wide range of temperature, pressure, and wavelength. The wavelengths listed here correspond to prominent lines of cadmium (226.50 and 361.05 nm), potassium (404.41 nm), sodium (589.00 nm), Ne (632.80 nm, from a helium - neon laser), and mercury (1.01398 μm).

REFERENCES

1. Schiebener, P., Straub, J., Levelt Sengers, J.M.H., and Gallagher, J.S., *J. Phys. Chem. Ref. Data*, 19, 677 (1990); 19, 1617, 1990.
2. Marsh, K.N., Editor, *Recommended Reference Materials for the Realization of Physicochemical Properties*, Blackwell Scientific Publications, Oxford, 1987.

$T/°C$	226.50 nm	361.05 nm	404.41 nm	589.00 nm	632.80 nm	1.01398 μm
0	1.39450	1.34896	1.34415	1.33432	1.33306	1.32612
10	1.39422	1.34870	1.34389	1.33408	1.33282	1.32591
20	1.39336	1.34795	1.34315	1.33336	1.33211	1.32524
30	1.39208	1.34682	1.34205	1.33230	1.33105	1.32424
40	1.39046	1.34540	1.34065	1.33095	1.32972	1.32296
50	1.38854	1.34373	1.33901	1.32937	1.32814	1.32145
60	1.38636	1.34184	1.33714	1.32757	1.32636	1.31974
70	1.38395	1.33974	1.33508	1.32559	1.32438	1.31784
80	1.38132	1.33746	1.33284	1.32342	1.32223	1.31576
90	1.37849	1.33501	1.33042	1.32109	1.31991	1.31353
100	1.37547	1.33239	1.32784	1.31861	1.31744	1.31114

INDEX OF REFRACTION OF LIQUIDS FOR CALIBRATION PURPOSES

This table gives the index of refraction of six liquids which are available in highly pure form and whose index of refraction has been accurately measured as a function of wavelength and temperature. They are therefore useful for calibration of refractometers. The estimated uncertainty in the values is:

2,2,4-Trimethylpentane	±0.00003
Hexadecane	±0.00008
trans-Bicyclo[4.0.0]decane	±0.00008
1-Methylnaphthalene	±0.00008
Toluene	±0.00003
Methylcyclohexane	±0.00003

Full details are given in the references. This table is reprinted from Reference 1 by permission of the International Union of Pure and Applied Chemistry.

REFERENCES

1. Marsh, K. N., Editor, *Recommended Reference Materials for the Realization of Physicochemical Properties,* Blackwell Scientific Publications, Oxford, 1987.
2. Tilton, L. W., J. *Opt. Soc. Am.*, 32, 71, 1941.

λ	2,2,4-Trimethylpentane			Hexadecane		
nm	20°C	25°C	30°C	20°C	25°C	30°C
667.81	1.38916	1.38670	1.38424	1.43204	1.43001	1.42798
656.28	1.38945	1.38698	1.38452	1.43235	1.43032	1.42829
589.26	1.39145	1.38898	1.38650	1.43453	1.43250	1.43047
546.07	1.39316	1.39068	1.38820	1.43640	1.43436	1.43232
501.57	1.39544	1.39294	1.39044	1.43888	1.43684	1.43480
486.13	1.39639	1.39389	1.39138	1.43993	1.43788	1.43583
435.83	1.40029	1.39776	1.39523	1.44419	1.44213	1.44007

λ	*trans*-Bicyclo[4.4.0]decane			1-Methylnaphthalene		
nm	20°C	25°C	30°C	20°C	25°C	30°C
667.81	1.46654	1.46438	1.46222	1.60828	1.60592	1.60360
656.28	1.46688	1.46472	1.46256	1.60940	1.60703	1.60471
589.26	1.46932	1.46715	1.46498	1.61755	1.61512	1.61278
546.07	1.47141	1.46923	1.46705	1.62488	1.62240	1.62005
501.57	1.47420	1.47200	1.46980	1.63513	1.63259	1.63022
486.13	1.47535	1.47315	1.47095	1.63958	1.63701	1.63463
435.83	1.48011	1.47789	1.47567		1.65627	1.65386

λ	Toluene			Methylcyclohexane		
nm	20°C	25°C	30°C	20°C	25°C	30°C
667.81	1.49180	1.48903	1.48619	1.42064	1.41812	1.41560
656.28	1.49243	1.48966	1.48682	1.42094	1.41842	1.41591
589.26	1.49693	1.49413	1.49126	1.42312	1.42058	1.41806
546.07	1.50086	1.49803	1.49514	1.42497	1.42243	1.41989
501.57	1.50620	1.50334	1.50041	1.42744	1.42488	1.42233
486.13	1.50847	1.50559	1.50265	1.42847	1.42590	1.42334
435.83	1.51800	1.51506	1.51206	1.43269	1.43010	1.42752

INDEX OF REFRACTION OF AIR

This is a table of the index of refraction n of dry air at 15°C and a pressure of 101.325 kPa and containing 0.03% by volume of carbon dioxide ("standard air"). The index of refraction is defined by $n = \lambda_{vac}/\lambda_{air}$, where λ is the wavelength of the radiation. The index is calculated from the expression:

$$(n-1) \times 10^8 = 8342.13 + 2406030(130 - \sigma^2)^{-1} + 15997(38.9 - \sigma^2)^{-1}$$

where $\sigma = 1/\lambda_{vac}$ and λ_{vac} has units of μm. The equation is valid for λ_{vac} from 200 nm to 2 μm.

The table also gives the correction $(n-1)\lambda_{air}$ which must be added to the wavelength in air to obtain λ_{vac}.

If the air is at a temperature t in °C and a pressure p in pascals, a value of $(n-1)$ from this table should be multiplied by

$$\frac{p\left[1 + p(61.3 - t) \times 10^{-10}\right]}{96095.4(1 + 0.003661t)}$$

REFERENCE

Edlen, B., *Metrologia*, 2, 71, 1966.

λ_{vac}	$(n-1) \times 10^8$	$\lambda_{vac} - \lambda_{air}$	λ_{vac}	$(n-1) \times 10^8$	$\lambda_{vac} - \lambda_{air}$	λ_{vac}	$(n-1) \times 10^8$	$\lambda_{vac} - \lambda_{air}$
200 nm	32408	0.0648 nm	540	27803	0.1501	880	27461	0.2416
210	31746	0.0666	550	27782	0.1528	890	27457	0.2443
220	31224	0.0687	560	27763	0.1554	900	27452	0.2470
230	30799	0.0708	570	27745	0.1581	910	27448	0.2497
240	30445	0.0730	580	27728	0.1608	920	27444	0.2524
250	30146	0.0753	590	27712	0.1635	930	27440	0.2551
260	29890	0.0777	600	27697	0.1661	940	27436	0.2578
270	29669	0.0801	610	27682	0.1688	950	27432	0.2605
280	29475	0.0825	620	27669	0.1715	960	27429	0.2632
290	29306	0.0850	630	27656	0.1742	970	27425	0.2660
300	29155	0.0874	640	27643	0.1769	980	27422	0.2687
310	29022	0.0899	650	27631	0.1796	990 nm	27419	0.2714 nm
320	28902	0.0925	660	27620	0.1822			
330	28795	0.0950	670	27609	0.1849	1.00 μm	27416	0.000274 μm
340	28698	0.0975	680	27599	0.1876	1.05	27401	0.000288
350	28611	0.1001	690	27589	0.1903	1.10	27389	0.000301
360	28531	0.1027	700	27579	0.1930	1.15	27378	0.000315
370	28458	0.1053	710	27570	0.1957	1.20	27368	0.000328
380	28392	0.1079	720	27562	0.1984	1.25	27360	0.000342
390	28331	0.1105	730	27553	0.2011	1.30	27352	0.000355
400	28275	0.1131	740	27545	0.2038	1.35	27346	0.000369
410	28223	0.1157	750	27538	0.2065	1.40	27340	0.000383
420	28175	0.1183	760	27530	0.2092	1.45	27334	0.000396
430	28131	0.1209	770	27523	0.2119	1.50	27330	0.000410
440	28090	0.1236	780	27516	0.2146	1.55	27325	0.000423
450	28052	0.1262	790	27510	0.2173	1.60	27321	0.000437
460	28016	0.1288	800	27504	0.2200	1.65	27318	0.000451
470	27983	0.1315	810	27498	0.2227	1.70	27314	0.000464
480	27952	0.1341	820	27492	0.2254	1.75	27311	0.000478
490	27923	0.1368	830	27486	0.2281	1.80	27309	0.000491
500	27896	0.1394	840	27481	0.2308	1.85	27306	0.000505
510	27870	0.1421	850	27476	0.2335	1.90	27304	0.000519
520	27846	0.1448	860	27471	0.2362	1.95	27302	0.000532
530	27824	0.1474	870	27466	0.2389	2.00 μm	27300	0.000546 μm

CHARACTERISTICS OF LASER SOURCES
William F. Krupke

Light Amplification by Stimulated Emission of Radiation was first demonstrated by Maiman in 1960, the result of a population inversion produced between energy levels of chromium ions in a ruby crystal when irradiated with a xenon flashlamp. Since then population inversions and coherent emission have been generated in literally thousands of substances (neutral and ionized gases, liquids, and solids) using a variety of incoherent excitation techniques (optical pumping, electrical discharges, gas-dynamic flow, electron-beams, chemical reactions, nuclear decay).

The extrema of laser output parameters which have been demonstrated to date and the laser media used are summarized in Table 1. Note that the extreme power and energy parameters listed in this table were attained with laser *systems* rather than with simple laser oscillators.

Laser sources are commonly classified in terms of the state-of-matter of the active medium: gas, liquid, and solid. Each of these classes is further subdivided into one or more types as shown in Table 2. A well-known representative example of each type of laser is also given in Table 2 together with its nominal operation wavelength and the methods by which it is pumped.

The various lasers together cover a wide spectral range from the far ultraviolet to the far infrared. The particular wavelength of emission (usually a narrow line) is presented for some six dozen lasers in Figures 1A and 1B.

By suitably designing the excitation source and/or by controlling the laser resonator structure, laser systems can provide continuous or pulsed radiation as shown in Table 3.

Besides the method of excitation and the temporal behavior of a laser, there are many other parameters that characterize its operation and efficiency, as shown in Tables 4 and 5.

Although many lasers only emit in one or more narrow spectral "lines", an increasing number of lasers can be tuned by changing the composition or the pressure of the medium, or by varying the wavelength of the pump bands. The spectral regions in which these tunable lasers operate are presented in Figure 2.

REFERENCE

Krupke, W. F., in *Handbook of Laser Science and Technology,* Vol. I, Weber, M. J., Ed., CRC Press, Boca Raton, FL, 1986.

TABLE 1
Extrema of Output Parameters of Laser Devices or Systems

Parameter	Value	Laser medium
Peak power	1×10^{14} W (collimated)	Nd:glass
Peak power density	10^{18} W/cm^2 (focused)	Nd:glass
Pulse energy	$>10^5$ J	CO_2, Nd:glass
Average power	10^5 W	CO_2
Pulse duration	3×10^{-15} s continuous wave (cw)	Rh6G dye; various gases, liquids, solids
Wavelength	60 nm $\leftrightarrow$ 385 μm	Many required
Efficiency (nonlaser pumped)	70%	CO
Beam quality	Diffraction limited	Various gases, liquids, solids
Spectral linewidth	20 Hz (for 10^{-1} s)	Neon-helium
Spatial coherence	10 m	Ruby

TABLE 2
Classes, Types, and Representative Examples of Laser Sources

Class	Type (characteristic)	Representative example	Nominal operating wavelength (nm)	Method(s) of excitation
Gas	Atom, neutral (electronic transition)	Neon-Helium (Ne-He)	633	Glow discharge
	Atom, ionic (electronic transition)	Argon (Ar⁺)	488	Arc discharge
	Molecule, neutral (electronic transition)	Krypton fluoride (KrF)	248	Glow discharge; e-beam
	Molecule, neutral (vibrational transition)	Carbon dioxide (CO_2)	10600	Glow discharge; gas-dynamic flow
	Molecule, neutral (rotational transition)	Methyl fluoride (CH_3F)	496000	Laser pumping
	Molecule, ionic (electronic transition)	Nitrogen ion (N_2^+)	420	E-beam
Liquid	Organic solvent (dye-chromophore)	Rhodamine dye (Rh6G)	580–610	Flashlamp; laser pumping
	Organic solvent (rare earth chelate)	Europium:TTF	612	Flashlamp
	Inorganic solvent (trivalent rare earth ion)	Neodymium:POCl₄	1060	Flashlamp
Solid	Insulator, crystal (impurity)	Neodymium:YAG	1064	Flashlamp, arc lamp
	Insulator, crystal (stoichiometric)	Neodymium:UP(NdP₅O₁₄)	1052	Flashlamp
	Insulator, crystal (color center)	F_2^-:LiF	1120	Laser pumping
	Insulator, amorphous (impurity)	Neodymium:glass	1061	Flashlamp
	Semiconductor (p-n junction)	GaAs	820	Injection current
	Semiconductor (electron-hole plasma)	GaAs	890	E-beam, laser pumping

Table 3
Temporal Characteristics of Lasers and Laser Systems

Form	Technique	Pulse width range (s)
Continuous wave	Excitation is continuous; resonator Q is held constant at some moderate value	∞
Pulsed	Excitation is pulsed; resonator Q is held constant at some moderate value	10^{-8}–10^{-3}
Q-Switched	Excitation is continuous or pulsed; resonator Q is switched from a very low value to a moderate value	10^{-8}–10^{-6}
Cavity dumped	Excitation is continuous or pulsed; resonator Q is switched from a very high value to a low value	10^{-7}–10^{-5}
Mode locked	Excitation is continuous or pulsed; phase or loss of the resonator modes is modulated at a rate related to the resonator transit time	10^{-12}–10^{-9}

Table 4
Properties and Performance of Some Continuous Wave (CW) Lasers

Parameter	Unit	Gas			Liquid	Solid	
		Neon helium	Argon ion	Carbon dioxide	Rhodamine 6G dye	Nd:YAG	GaAs
Excitation method		DC discharge	DC discharge	DC discharge	Ar^+ laser pump	Krypton arc lamp	DC injection
Gain medium composition		Neon:helium	Argon	CO_2:N_2:He	Rh 6G:H_2O	Nd:YAG	p:n:GaAs
Gain medium density	Torr, ions/cm^3	0.1:1.0	0.4	0.4:0.8:5.0			
Wavelength	nm	633	488	10600	590	1064	810
Laser cross-section	cm^{-2}	3(-13)	1.6(-12)	1.5(-16)	1.8(-16)	7(-19)	~6(-15)
Radiative lifetime (upper level)	s	~1(-7)	7.5(-9)	4(-3)	6.5(-9)	2.6(-4)	~1(-9)
Decay lifetime (upper level)	s	~1(-7)	~5.0(-9)	~4(-3)	6.0(-9)	2.3(-4)	~1(-9)
Gain bandwidth	nm	2(-3)	5(-3)	1.6(-2)	80	0.5	10
Type, gain saturation		Inhomogeneous	Inhomogeneous	Homogeneous	Homogeneous	Homogeneous	Homogeneous
Homogeneous saturation flux	W cm^{-2}	~1(-8)	~4(-10)	~20	3(5)	2.3(3)	~2(4)
Decay lifetime (lower level)	s	~1(9)	2(10)	~5(-6)[b]	<1(-12)	<1(-7)	<1(-12)
Inversion density	cm^{-3}	~1(-3)	~3(-2)	2(15)	2(16)	6(16)	1(16)
Small signal gain coefficient	cm^{-1}	3	900	1(-2)	4	5(-2)	40
Pump power density	W cm^{-3}	3	900	0.15	1(6)	150	7(7)
Output power density	W cm^{-3}	2.6(-3)	~1	2(-2)	3(5)	95	5(6)
Laser size (diameter:length)	cm:cm	0.5:100	0.3:100	5.0:600	1(-3):0.3	0.6:10	5(-4):7(-3):2(-2)[a]
Excitation current/voltage	A/V	3(-2):2(3)	30:300	0.1:1.5(4)		90:125	1.0/1.7
Excitation current density	A cm^{-2}	0.15	600	6(-3)		140	4.5(3)
Excitation power	W	60	9(3)	1.5(3)	4	1.1(4)	1.7
Output power	W	0.06	10	240	0.3	300	0.12
Efficiency	%	0.1	0.1	13	7	2.6	7

[a] Junction thickness:width:length.
[b] Pressure dependent.

Table 5

Properties and Performance of Some Pulsed Lasers

Parameter	Unit	Gas				Liquid	Solid	
		Carbon dioxide		Krypton fluoride		Rhodamine 6G	Nd:YAG	Nd:glass
Excitation method		TEA-discharge	E-beam/sust.	Glow discharge	E-beam	Xenon flashlamp	Xenon flashlamp	Xenon flashlamp
Gain medium composition		CO_2:N_2:He	CO_2:N_2:He	He:Kr:F_2	Ar:Kr:F_2	Rh6G:alcohol	Nd:YAG	Nd:Glass
Gain medium density	torr / ions/cm^3	100:50:600	240:240:320	1070:70:3	1235:52:3	1(18):1.5(22)	1.5(20):1(22)	3(20):2(22)
Wavelength	nm	10600	10600	249	249	590	1064	1061
Laser cross-section	cm^{-2}	2(-18)	2(-18)	2(-16)	2(-16)	1.8(-16)	7(-19)	2.8(-20)
Radiative lifetime (upper level)	s	4(-3)	4(-3)	7(-9)	7(-9)	6.5(-9)	2.6(-4)	4.1(-4)
Decay lifetime (upper level)	s	~1(-4)	5(-5)	2(-9)	3(-9)	6.0(-9)	2.3(-4)	3.7(-4)
Gain bandwidth	nm	1	1	2	2	80	0.5	26
Homogeneous saturation fluence	J/cm^2	0.2	0.2	4(-3)	4(-3)	2(-3)	0.6	~5
Decay lifetime (lower level)	s	5(-8)[a]	1(-8)[a]	<1(-12)	<1(-12)	<1(-12)	<1(-7)	<1(-8)
Inversion density	cm^{-3}	3(17)	6(17)	4(14)	2(14)	2(16)	4(17)	3(18)
Small signal gain coefficient	cm^{-1}	2(-2)	4(-2)	8–92	4(-2)	4	0.3	8(-2)
Medium excitation energy density	J/cm^3	0.1	0.36	0.15	0.13	2.8	0.15	0.6
Output energy density	J/cm^3	2(-2)	1.8(-2)	1.5(-3)	1.2(-2)	0.85	5(-2)	2(-2)
Laser dimensions	cm:cm:cm	4.5:4.5:87	10:10:100	1.5:4.5:100	8.5:10:100	1.2φ25	0.6φ7.5	0.6φ8.3
Excitation current/voltage	A/V	6(4)/3.3(3)	2.4(4)/4(4)	2.5(4)/1.5(5)	1.24(4)/2.5(5)	2(5)/2.5(4)		
Excitation current density	A cm^2	8.5	22	170	11.5	2.6(3)		
Excitation peak power	W	2(8)	9(8)	4(9)	3(9)	5.4(9)	4(4)	9(4)
Output pulse energy	J	35	180	1	102	32	0.1	1.0
Output pulse length	s	1(-6)	4(-6)	2.5(-8)	6(-7)	3.2(-6)	2(-8)	1(-4)
Output pulse power	W	3.5(7)	4(7)	4(7)	2(8)	1(7)	5(6)	1(4)
Efficiency	%	17	5	1	10[b]	0.2	1.5	3.7

[a] Pressure dependent.
[b] Intrinsic efficiency ≡ energy output/energy deposited in gas.

FIGURE 1A. Wavelengths of lasers operating in the 120 to 1200 nm spectral region.

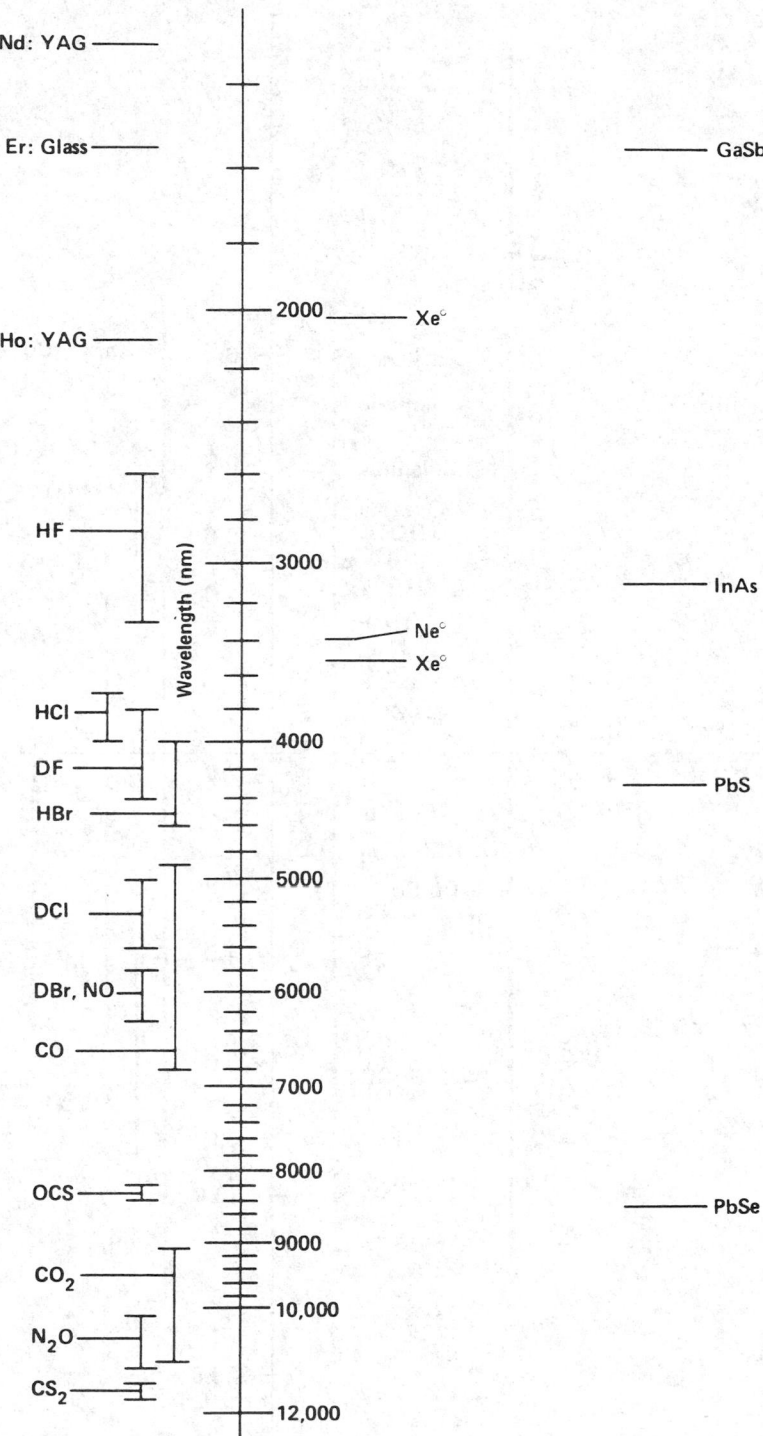

FIGURE 1B. Wavelength of lasers operating in the 1300 to 12,000 nm spectral region.

FIGURE 2. Spectral tuning ranges of various types of tunable lasers.

INFRARED LASER FREQUENCIES
Arthur Maki

The CO_2 laser has been the subject of a number of very accurate frequency measurements. Most of the earlier measurements are given by Bradley et al.[1] That analysis was based on a single absolute frequency measurement and many laser frequency differences. New measurements of the methane frequency[2-4] have made it necessary to slightly revise that single absolute frequency measurement. In addition, there have been several other absolute frequency measurements[5-7] that have been used here to improve the accuracy of the present tables. New frequency difference measurements have also been added to the database used for the present tables.[8]

REFERENCES

1. Bradley, L. C., Soohoo, K. L., and Freed, C., *IEEE J. Quantum Electron.*, QE-22, 234-267, 1986.
2. Clairon, A., Dahmani, B., Filimon, A., and Rutman, J., *IEEE Trans. Inst. Meas.*, IM-34, 265-268, 1985.
3. Weiss, C. O., Kramer, G., Lipphardt, B., and Garcia, E., *IEEE J. Quantum Electron.*, QE-24, 1970-1972, 1988.
4. Bagayev, S. N., Baklanov, A. E., Chebotayev, V. P., and Dychkov, A. S., *Appl. Phys.*, B-48, 31-35, 1989.
5. Blaney, T. G., Bradley, C. C., Edwards, G. J., Jolliffe, B. W., Knight, D. J. E., Rowley, W. R. C., Shotten, K. C., and Woods, P. T., *Proc. R. Soc. Lond.*, A-355, 61-88, 1977.
6. Chardonnet, Ch., Van Lerberghe, A., and Bordé, Ch. J., *Opt. Comm.*, 58, 333-337, 1986.
7. Clairon, A., Acef, O., Chardonnet, Ch., and Bordé, Ch. J., *Frequency Standards and Metrology*, De Marchi, A., Ed., Springer-Verlag, Berlin, Heidelberg, 1989, p. 212.
8. Evenson, K., private communication.

Frequencies for the $00°1$-$(10°0,02°0)_I$ and $00°1$-$(10°0,02°0)_{II}$ Bands of $^{12}C^{16}O_2$ with the Estimated 2-σ Uncertainties

Line	Band I Frequency (MHz)	Uncertainty (MHz)	Line	Band II Frequency (MHz)	Uncertainty (MHz)
P(70)	26721305.4647	0.1680	P(70)	29789856.3783	0.0308
P(68)	26794232.6712	0.1217	P(68)	29861850.7690	0.0192
P(66)	26866318.8073	0.0867	P(66)	29933216.1760	0.0122
P(64)	26937571.7234	0.0606	P(64)	30003944.2861	0.0086
P(62)	27007998.9216	0.0415	P(62)	30074026.9127	0.0072
P(60)	27077607.5643	0.0279	P(60)	30143456.0039	0.0066
P(58)	27146404.4834	0.0185	P(58)	30212223.6504	0.0061
P(56)	27214396.1873	0.0121	P(56)	30280322.0930	0.0055
P(54)	27281588.8696	0.0081	P(54)	30347743.7306	0.0049
P(52)	27347988.4161	0.0057	P(52)	30414481.1273	0.0044
P(50)	27413600.4119	0.0043	P(50)	30480527.0196	0.0041
P(48)	27478430.1487	0.0036	P(48)	30545874.3239	0.0039
P(46)	27542482.6310	0.0032	P(46)	30610516.1429	0.0039
P(44)	27605762.5826	0.0030	P(44)	30674445.7724	0.0039
P(42)	27668274.4525	0.0028	P(42)	30737656.7080	0.0039
P(40)	27730022.4206	0.0027	P(40)	30800142.6511	0.0039
P(38)	27791010.4036	0.0026	P(38)	30861897.5150	0.0038
P(36)	27851242.0594	0.0025	P(36)	30922915.4310	0.0037
P(34)	27910720.7927	0.0024	P(34)	30983190.7534	0.0037
P(32)	27969449.7593	0.0023	P(32)	31042718.0652	0.0037
P(30)	28027431.8708	0.0022	P(30)	31101492.1833	0.0036
P(28)	28084669.7981	0.0021	P(28)	31159508.1631	0.0037
P(26)	28141165.9762	0.0020	P(26)	31216761.3029	0.0037
P(24)	28196922.6067	0.0019	P(24)	31273247.1487	0.0037
P(22)	28251941.6622	0.0017	P(22)	31328961.4978	0.0037
P(20)	28306224.8888	0.0016	P(20)	31383900.4028	0.0037
P(18)	28359773.8090	0.0014	P(18)	31438060.1749	0.0037
P(16)	28412589.7245	0.0012	P(16)	31491437.3872	0.0036
P(14)	28464673.7184	0.0011	P(14)	31544028.8776	0.0036
P(12)	28516026.6574	0.0009	P(12)	31595831.7516	0.0036
P(10)	28566649.1935	0.0008	P(10)	31646843.3843	0.0035
P(8)	28616541.7661	0.0008	P(8)	31697061.4225	0.0035
P(6)	28665704.6027	0.0008	P(6)	31746483.7868	0.0035

INFRARED LASER FREQUENCIES (continued)

Line	Band I Frequency (MHz)	Uncertainty (MHz)	Line	Band II Frequency (MHz)	Uncertainty (MHz)
P(4)	28714137.7205	0.0008	P(4)	31795108.6724	0.0035
P(2)	28761840.9272	0.0008	P(2)	31842934.5511	0.0035
R(0)	28832026.2198	0.0008	R(0)	31913172.5691	0.0035
R(2)	28877902.4382	0.0007	R(2)	31958996.0621	0.0034
R(4)	28923046.4303	0.0006	R(4)	32004017.3822	0.0034
R(6)	28967457.0657	0.0005	R(6)	32048236.2498	0.0034
R(8)	29011133.0054	0.0003	R(8)	32091652.6619	0.0034
R(10)	29054072.7010	0.0001	R(10)	32134266.8917	0.0034
R(12)	29096274.3935	0.0003	R(12)	32176079.4878	0.0034
R(14)	29137736.1129	0.0005	R(14)	32217091.2721	0.0035
R(16)	29178455.6759	0.0007	R(16)	32257303.3386	0.0036
R(18)	29218430.6852	0.0009	R(18)	32296717.0510	0.0037
R(20)	29257658.5269	0.0010	R(20)	32335334.0408	0.0038
R(22)	29296136.3689	0.0011	R(22)	32373156.2044	0.0039
R(24)	29333861.1583	0.0012	R(24)	32410185.7003	0.0041
R(26)	29370829.6191	0.0011	R(26)	32446424.9459	0.0042
R(28)	29407038.2491	0.0011	R(28)	32481876.6140	0.0042
R(30)	29442483.3168	0.0011	R(30)	32516543.6293	0.0042
R(32)	29477160.8582	0.0012	R(32)	32550429.1641	0.0042
R(34)	29511066.6733	0.0013	R(34)	32583536.6340	0.0042
R(36)	29544196.3221	0.0015	R(36)	32615869.6937	0.0041
R(38)	29576545.1205	0.0017	R(38)	32647432.2320	0.0040
R(40)	29608108.1360	0.0019	R(40)	32678228.3665	0.0039
R(42)	29638880.1831	0.0022	R(42)	32708262.4386	0.0038
R(44)	29668855.8183	0.0024	R(44)	32737539.0081	0.0039
R(46)	29698029.3350	0.0027	R(46)	32766062.8469	0.0041
R(48)	29726394.7582	0.0032	R(48)	32793838.9334	0.0045
R(50)	29753945.8385	0.0037	R(50)	32820872.4463	0.0055
R(52)	29780676.0464	0.0042	R(52)	32847168.7576	0.0071
R(54)	29806578.5659	0.0047	R(54)	32872733.4269	0.0099
R(56)	29831646.2878	0.0052	R(56)	32897572.1935	0.0141
R(58)	29855871.8032	0.0058	R(58)	32921690.9701	0.0202
R(60)	29879247.3960	0.0074	R(60)	32945095.8355	0.0288
R(62)	29901765.0357	0.0113	R(62)	32967793.0268	0.0407
R(64)	29923416.3695	0.0186	R(64)	32989788.9322	0.0567
R(66)	29944192.7145	0.0302	R(66)	33011090.0831	0.0780
R(68)	29964085.0488	0.0475	R(68)	33031703.1467	0.1060
R(70)	29983084.0036	0.0720	R(70)	33051634.9172	0.1423

Frequencies for the $00^\circ1$-$(10^\circ0,02^\circ0)_I$ and $00^\circ1$-$(10^\circ0,02^\circ0)_{II}$ Bands of $^{13}C^{16}O_2$ with the Estimated 2-σ Uncertainties

Line	Band I Frequency (MHz)	Uncertainty (MHz)	Line	Band II Frequency (MHz)	Uncertainty (MHz)
P(66)	25523832.1808	0.7836	P(66)	28512082.5283	1.2894
P(64)	25590013.4703	0.5415	P(64)	28585121.9396	0.9194
P(62)	25655543.6502	0.3629	P(62)	28657449.4180	0.6420
P(60)	25720428.2487	0.2339	P(60)	28729056.6374	0.4375
P(58)	25784672.4840	0.1430	P(58)	28799935.4147	0.2897
P(56)	25848281.2771	0.0810	P(56)	28870077.7187	0.1853
P(54)	25911259.2627	0.0405	P(54)	28939475.6771	0.1135
P(52)	25973610.8005	0.0157	P(52)	29008121.5846	0.0659
P(50)	26035339.9857	0.0045	P(50)	29076007.9109	0.0357
P(48)	26096450.6582	0.0079	P(48)	29143127.3077	0.0180
P(46)	26156946.4123	0.0101	P(46)	29209472.6164	0.0090

INFRARED LASER FREQUENCIES (continued)

Line	Band I Frequency (MHz)	Uncertainty (MHz)	Line	Band II Frequency (MHz)	Uncertainty (MHz)
P(44)	26216830.6053	0.0101	P(44)	29275036.8754	0.0058
P(42)	26276106.3655	0.0090	P(42)	29339813.3270	0.0050
P(40)	26334776.6003	0.0077	P(40)	29403795.4243	0.0044
P(38)	26392844.0030	0.0068	P(38)	29466976.8383	0.0037
P(36)	26450311.0599	0.0063	P(36)	29529351.4635	0.0032
P(34)	26507180.0565	0.0061	P(34)	29590913.4252	0.0029
P(32)	26563453.0836	0.0060	P(32)	29651657.0844	0.0028
P(30)	26619132.0428	0.0058	P(30)	29711577.0447	0.0028
P(28)	26674218.6515	0.0055	P(28)	29770668.1566	0.0031
P(26)	26728714.4479	0.0054	P(26)	29828925.5239	0.0035
P(24)	26782620.7952	0.0054	P(24)	29886344.5074	0.0041
P(22)	26835938.8858	0.0054	P(22)	29942920.7308	0.0046
P(20)	26888669.7451	0.0055	P(20)	29998650.0838	0.0051
P(18)	26940814.2347	0.0055	P(18)	30053528.7271	0.0054
P(16)	26992373.0555	0.0055	P(16)	30107553.0955	0.0055
P(14)	27043346.7508	0.0054	P(14)	30160719.9016	0.0055
P(12)	27093735.7083	0.0052	P(12)	30213026.1388	0.0054
P(10)	27143540.1624	0.0051	P(10)	30264469.0839	0.0054
P(8)	27192760.1962	0.0049	P(8)	30315046.2994	0.0054
P(6)	27241395.7431	0.0048	P(6)	30364755.6359	0.0055
P(4)	27289446.5880	0.0047	P(4)	30413595.2335	0.0056
P(2)	27336912.3682	0.0046	P(2)	30461563.5231	0.0057
R(0)	27407012.8882	0.0045	P(0)	30531879.5415	0.0057
R(2)	27453013.4589	0.0043	P(2)	30577664.6138	0.0056
R(4)	27498426.5430	0.0040	P(4)	30622575.1885	0.0054
R(6)	27543251.1200	0.0037	P(6)	30666611.0128	0.0051
R(8)	27587486.0225	0.0034	P(8)	30709772.1257	0.0047
R(10)	27631129.9356	0.0031	P(10)	30752058.8571	0.0045
R(12)	27674181.3963	0.0029	P(12)	30793471.8269	0.0044
R(14)	27716638.7917	0.0029	P(14)	30834011.9425	0.0043
R(16)	27758500.3577	0.0029	P(16)	30873680.3976	0.0044
R(18)	27799764.1770	0.0029	P(18)	30912478.6694	0.0044
R(20)	27840428.1773	0.0030	P(20)	30950408.5159	0.0044
R(22)	27880490.1283	0.0029	P(22)	30987471.9732	0.0043
R(24)	27919947.6395	0.0029	P(24)	31023671.3517	0.0042
R(26)	27958798.1567	0.0028	P(26)	31059009.2327	0.0042
R(28)	27997038.9591	0.0028	P(28)	31093488.4642	0.0042
R(30)	28034667.1551	0.0027	P(30)	31127112.1569	0.0043
R(32)	28071679.6785	0.0027	P(32)	31159883.6793	0.0045
R(34)	28108073.2842	0.0026	P(34)	31191806.6529	0.0046
R(36)	28143844.5432	0.0026	P(36)	31222884.9469	0.0048
R(38)	28178989.8377	0.0026	P(38)	31253122.6730	0.0053
R(40)	28213505.3554	0.0028	P(40)	31282524.1795	0.0061
R(42)	28247387.0838	0.0033	P(42)	31311094.0452	0.0077
R(44)	28280630.8035	0.0046	P(44)	31338837.0736	0.0108
R(46)	28313232.0818	0.0083	P(46)	31365758.2858	0.0173
R(48)	28345186.2652	0.0161	P(48)	31391862.9147	0.0295
R(50)	28376488.4720	0.0301	P(50)	31417156.3972	0.0505
R(52)	28407133.5839	0.0531	P(52)	31441644.3679	0.0845
R(54)	28437116.2372	0.0887	P(54)	31465332.6516	0.1366
R(56)	28466430.8141	0.1419	P(56)	31488227.2557	0.2138
R(58)	28495071.4324	0.2188	P(58)	31510334.3631	0.3247
R(60)	28523031.9357	0.3271	P(60)	31531660.3243	0.4800
R(62)	28550305.8819	0.4763	P(62)	31552211.6497	0.6932
R(64)	28576886.5323	0.6781	P(64)	31571995.0017	0.9805
R(66)	28602766.8393	0.9467	P(66)	31591017.1868	1.3619

Frequencies for the $00°1$-$(10°0,02°0)_I$ and $00°1$-$(10°0,02°0)_{II}$ Bands of $^{12}C^{18}O_2$ with the Estimated 2-σ Uncertainties

Line	Band I Frequency (MHz)	Uncertainty (MHz)	Line	Band II Frequency (MHz)	Uncertainty (MHz)
P(70)	27045326.3119	0.4540	P(70)	30695237.5856	0.0858
P(68)	27114914.0922	0.3324	P(68)	30755520.2231	0.0570
P(66)	27183635.7945	0.2392	P(66)	30815311.4928	0.0364
P(64)	27251496.4118	0.1688	P(64)	30874607.2084	0.0223
P(62)	27318500.7361	0.1165	P(62)	30933403.2309	0.0131
P(60)	27384653.3618	0.0783	P(60)	30991695.4724	0.0075
P(58)	27449958.6881	0.0510	P(58)	31049479.9009	0.0049
P(56)	27514420.9224	0.0319	P(56)	31106752.5446	0.0041
P(54)	27578044.0828	0.0191	P(54)	31163509.4964	0.0040
P(52)	27640832.0010	0.0108	P(52)	31219746.9183	0.0040
P(50)	27702788.3248	0.0059	P(50)	31275461.0455	0.0039
P(48)	27763916.5206	0.0035	P(48)	31330648.1908	0.0039
P(46)	27824219.8762	0.0028	P(46)	31385304.7490	0.0039
P(44)	27883701.5029	0.0026	P(44)	31439427.2006	0.0039
P(42)	27942364.3379	0.0025	P(42)	31493012.1163	0.0038
P(40)	28000211.1464	0.0024	P(40)	31546056.1605	0.0038
P(38)	28057244.5242	0.0022	P(38)	31598556.0954	0.0037
P(36)	28113466.8992	0.0021	P(36)	31650508.7847	0.0037
P(34)	28168880.5335	0.0020	P(34)	31701911.1970	0.0037
P(32)	28223487.5256	0.0019	P(32)	31752760.4093	0.0037
P(30)	28277289.8118	0.0017	P(30)	31803053.6105	0.0037
P(28)	28330289.1679	0.0016	P(28)	31852788.1043	0.0038
P(26)	28382487.2111	0.0015	P(26)	31901961.3125	0.0038
P(24)	28433885.4012	0.0013	P(24)	31950570.7773	0.0038
P(22)	28484485.0420	0.0012	P(22)	31998614.1649	0.0038
P(20)	28534287.2828	0.0011	P(20)	32046089.2669	0.0037
P(18)	28583293.1193	0.0010	P(18)	32092994.0036	0.0037
P(16)	28631503.3952	0.0010	P(16)	32139326.4254	0.0036
P(14)	28678918.8025	0.0009	P(14)	32185084.7154	0.0036
P(12)	28725539.8830	0.0010	P(12)	32230267.1907	0.0036
P(10)	28771367.0288	0.0010	P(10)	32274872.3041	0.0037
P(8)	28816400.4829	0.0010	P(8)	32318898.6455	0.0038
P(6)	28860640.3403	0.0011	P(6)	32362344.9434	0.0039
P(4)	28904086.5477	0.0011	P(4)	32405210.0652	0.0041
P(2)	28946738.9048	0.0011	P(2)	32447493.0185	0.0041
R(0)	29009228.1702	0.0010	P(0)	32509824.0580	0.0042
R(2)	29049894.0586	0.0010	P(2)	32550648.1723	0.0042
R(4)	29089764.2368	0.0009	P(4)	32590887.7542	0.0042
R(6)	29128837.8426	0.0008	P(6)	32630542.4457	0.0041
R(8)	29167113.8668	0.0008	P(8)	32669612.0295	0.0041
R(10)	29204591.1529	0.0009	P(10)	32708096.4282	0.0040
R(12)	29241268.3964	0.0010	P(12)	32745995.7040	0.0040
R(14)	29277144.1444	0.0011	P(14)	32783310.0573	0.0040
R(16)	29312216.7955	0.0012	P(16)	32820039.8258	0.0040
R(18)	29346484.5984	0.0012	P(18)	32856185.4827	0.0040
R(20)	29379945.6517	0.0013	P(20)	32891747.6358	0.0040
R(22)	29412597.9024	0.0013	P(22)	32926727.0254	0.0040
R(24)	29444439.1458	0.0013	P(24)	32961124.5220	0.0040
R(26)	29475467.0236	0.0014	P(26)	32994941.1249	0.0040
R(28)	29505679.0230	0.0015	P(28)	33028177.9594	0.0040
R(30)	29535072.4755	0.0016	P(30)	33060836.2743	0.0040
R(32)	29563644.5557	0.0018	P(32)	33092917.4394	0.0041
R(34)	29591392.2794	0.0020	P(34)	33124422.9429	0.0043
R(36)	29618312.5023	0.0023	P(36)	33155354.3878	0.0046
R(38)	29644401.9182	0.0028	P(38)	33185713.4894	0.0049

Line	Band I Frequency (MHz)	Uncertainty (MHz)	Line	Band II Frequency (MHz)	Uncertainty (MHz)
R(40)	29669657.0575	0.0036	P(40)	33215502.0716	0.0056
R(42)	29694074.2853	0.0053	P(42)	33244722.0637	0.0068
R(44)	29717649.7992	0.0082	P(44)	33273375.4969	0.0092
R(46)	29740379.6276	0.0128	P(46)	33301464.5003	0.0134
R(48)	29762259.6274	0.0200	P(48)	33328991.2976	0.0199
R(50)	29783285.4820	0.0307	P(50)	33355958.2027	0.0294
R(52)	29803452.6988	0.0461	P(52)	33382367.6161	0.0427
R(54)	29822756.6072	0.0681	P(54)	33408222.0209	0.0607
R(56)	29841192.3558	0.0985	P(56)	33433523.9780	0.0848
R(58)	29858754.9100	0.1401	P(58)	33458276.1228	0.1165
R(60)	29875439.0495	0.1960	P(60)	33482481.1601	0.1576
R(62)	29891239.3658	0.2702	P(62)	33506141.8605	0.2104
R(64)	29906150.2589	0.3673	P(64)	33529261.0556	0.2775
R(66)	29920165.9352	0.4930	P(66)	33551841.6335	0.3621
R(68)	29933280.4042	0.6540	P(68)	33573886.5352	0.4679
R(70)	29945487.4756	0.8581	P(70)	33595398.7493	0.5992

Frequencies for the $00°1$-$(10°0,02°0)_I$ and $00°1$-$(10°0,02°0)_{II}$ Bands of $^{13}C^{18}O_2$ with the Estimated 2-σ Uncertainties

Line	Band I Frequency (MHz)	Uncertainty (MHz)	Line	Band II Frequency (MHz)	Uncertainty (MHz)
P(70)	25967863.7652	1.1146	P(70)	28960476.2278	0.4069
P(68)	26033448.2798	0.8152	P(68)	29022326.9578	0.2861
P(66)	26098273.9159	0.5860	P(66)	29083661.3546	0.1961
P(64)	26162346.4813	0.4129	P(64)	29144473.5795	0.1303
P(62)	26225671.5466	0.2844	P(62)	29204757.8761	0.0833
P(60)	26288254.4494	0.1906	P(60)	29264508.5768	0.0507
P(58)	26350100.2984	0.1237	P(58)	29323720.1086	0.0290
P(56)	26411213.9778	0.0772	P(56)	29382386.9988	0.0152
P(54)	26471600.1504	0.0459	P(54)	29440503.8809	0.0073
P(52)	26531263.2618	0.0258	P(52)	29498065.4997	0.0038
P(50)	26590207.5442	0.0138	P(50)	29555066.7172	0.0032
P(48)	26648437.0195	0.0077	P(48)	29611502.5178	0.0031
P(46)	26705955.5026	0.0057	P(46)	29667368.0132	0.0031
P(44)	26762766.6051	0.0055	P(44)	29722658.4475	0.0034
P(42)	26818873.7378	0.0055	P(42)	29777369.2022	0.0039
P(40)	26874280.1143	0.0056	P(40)	29831495.8006	0.0044
P(38)	26928988.7531	0.0056	P(38)	29885033.9125	0.0049
P(36)	26983002.4809	0.0056	P(36)	29937979.3584	0.0053
P(34)	27036323.9351	0.0055	P(34)	29990328.1139	0.0054
P(32)	27088955.5657	0.0054	P(32)	30042076.3132	0.0055
P(30)	27140899.6384	0.0051	P(30)	30093220.2534	0.0055
P(28)	27192158.2363	0.0049	P(28)	30143756.3978	0.0054
P(26)	27242733.2620	0.0047	P(26)	30193681.3793	0.0053
P(24)	27292626.4396	0.0044	P(24)	30242992.0038	0.0052
P(22)	27341839.3165	0.0042	P(22)	30291685.2529	0.0051
P(20)	27390373.2651	0.0040	P(20)	30339758.2870	0.0049
P(18)	27438229.4843	0.0037	P(18)	30387208.4477	0.0048
P(16)	27485409.0008	0.0035	P(16)	30434033.2603	0.0046
P(14)	27531912.6704	0.0033	P(14)	30480230.4356	0.0045
P(12)	27577741.1795	0.0031	P(12)	30525797.8725	0.0044
P(10)	27622895.0455	0.0031	P(10)	30570733.6593	0.0043
P(8)	27667374.6182	0.0031	P(8)	30615036.0750	0.0043
P(6)	27711180.0803	0.0033	P(6)	30658703.5912	0.0044
P(4)	27754311.4480	0.0034	P(4)	30701734.8727	0.0045

INFRARED LASER FREQUENCIES (continued)

Line	Band I Frequency (MHz)	Uncertainty (MHz)	Line	Band II Frequency (MHz)	Uncertainty (MHz)
P(2)	27796768.5718	0.0036	P(2)	30744128.7785	0.0045
R(0)	27859189.3155	0.0036	P(0)	30806522.5414	0.0045
R(2)	27899959.0889	0.0035	P(2)	30847319.2956	0.0044
R(4)	27940052.7921	0.0033	P(4)	30887476.2168	0.0043
R(6)	27979469.5315	0.0031	P(6)	30926993.0424	0.0042
R(8)	28018208.2478	0.0028	P(8)	30965869.7046	0.0041
R(10)	28056267.7161	0.0026	P(10)	31004106.3298	0.0040
R(12)	28093646.5448	0.0025	P(12)	31041703.2379	0.0040
R(14)	28130343.1757	0.0025	P(14)	31078660.9408	0.0040
R(16)	28166355.8825	0.0025	P(16)	31114980.1420	0.0040
R(18)	28201682.7706	0.0025	P(18)	31150661.7340	0.0041
R(20)	28236321.7757	0.0025	P(20)	31185706.7976	0.0042
R(22)	28270270.6628	0.0024	P(22)	31220116.5992	0.0043
R(24)	28303527.0249	0.0024	P(24)	31253892.5891	0.0043
R(26)	28336088.2817	0.0023	P(26)	31287036.3991	0.0044
R(28)	28367951.6781	0.0024	P(28)	31319549.8396	0.0043
R(30)	28399114.2823	0.0025	P(30)	31351434.8973	0.0043
R(32)	28429572.9843	0.0026	P(32)	31382693.7318	0.0042
R(34)	28459324.4940	0.0028	P(34)	31413328.6728	0.0042
R(36)	28488365.3390	0.0029	P(36)	31443342.2165	0.0041
R(38)	28516691.8625	0.0029	P(38)	31472737.0219	0.0040
R(40)	28544300.2211	0.0031	P(40)	31501515.9074	0.0039
R(42)	28571186.3823	0.0032	P(42)	31529681.8467	0.0040
R(44)	28597346.1222	0.0032	P(44)	31557237.9646	0.0042
R(46)	28622775.0223	0.0038	P(46)	31584187.5329	0.0046
R(48)	28647468.4672	0.0071	P(48)	31610533.9656	0.0057
R(50)	28671421.6417	0.0148	P(50)	31636280.8146	0.0088
R(52)	28694629.5272	0.0286	P(52)	31661431.7650	0.0151
R(54)	28717086.8993	0.0510	P(54)	31685990.6298	0.0261
R(56)	28738788.3239	0.0852	P(56)	31709961.3449	0.0434
R(58)	28759728.1540	0.1355	P(58)	31733347.9642	0.0693
R(60)	28779900.5263	0.2075	P(60)	31756154.6537	0.1068
R(62)	28799299.3572	0.3078	P(62)	31778385.6867	0.1594
R(64)	28817918.3393	0.4447	P(64)	31800045.4375	0.2317
R(66)	28835750.9374	0.6283	P(66)	31821138.3761	0.3291
R(68)	28852790.3843	0.8707	P(68)	31841669.0622	0.4581
R(70)	28869029.6768	1.1863	P(70)	31861642.1394	0.6268

Frequencies for the $01^{1e}1$-$(11^{1e}0,03^{1e}0)_I$ and $01^{1e}1$-$(11^{1e}0,03^{1e}0)_{II}$ Bands of $^{12}C^{16}O_2$ with the Estimated 2-σ Uncertainties

Line	Band I Frequency (MHz)	Uncertainty (MHz)	Line	Band II Frequency (MHz)	Uncertainty (MHz)
P(59)	26125213.2723	1.6633	P(59)	30427055.2899	0.1962
P(57)	26191576.6703	1.0880	P(57)	30494640.3229	0.1332
P(55)	26257240.7898	0.6844	P(55)	30561557.5929	0.0865
P(53)	26322208.2302	0.4094	P(53)	30627802.0344	0.0530
P(51)	26386481.4313	0.2286	P(51)	30693368.7014	0.0306
P(49)	26450062.6783	0.1155	P(49)	30758252.7710	0.0175
P(47)	26512954.1076	0.0498	P(47)	30822449.5469	0.0123
P(45)	26575157.7109	0.0191	P(45)	30885954.4624	0.0114
P(43)	26636675.3402	0.0160	P(43)	30948763.0834	0.0109
P(41)	26697508.7115	0.0182	P(41)	31010871.1119	0.0100
P(39)	26757659.4084	0.0177	P(39)	31072274.3882	0.0091
P(37)	26817128.8857	0.0160	P(37)	31132968.8940	0.0091
P(35)	26875918.4726	0.0144	P(35)	31192950.7549	0.0102

INFRARED LASER FREQUENCIES (continued)

Line	Band I Frequency (MHz)	Uncertainty (MHz)	Line	Band II Frequency (MHz)	Uncertainty (MHz)
P(33)	26934029.3751	0.0131	P(33)	31252216.2430	0.0118
P(31)	26991462.6787	0.0119	P(31)	31310761.7788	0.0134
P(29)	27048219.3509	0.0106	P(29)	31368583.9339	0.0147
P(27)	27104300.2431	0.0096	P(27)	31425679.4328	0.0155
P(25)	27159706.0925	0.0093	P(25)	31482045.1550	0.0157
P(23)	27214437.5237	0.0097	P(23)	31537678.1367	0.0154
P(21)	27268495.0505	0.0104	P(21)	31592575.5725	0.0147
P(19)	27321879.0769	0.0108	P(19)	31646734.8172	0.0137
P(17)	27374589.8987	0.0108	P(17)	31700153.3868	0.0127
P(15)	27426627.7040	0.0104	P(15)	31752828.9602	0.0119
P(13)	27477992.5747	0.0098	P(13)	31804759.3803	0.0113
P(11)	27528684.4867	0.0096	P(11)	31855942.6551	0.0113
P(9)	27578703.3113	0.0101	P(9)	31906376.9582	0.0116
P(7)	27628048.8151	0.0113	P(7)	31956060.6304	0.0122
P(5)	27676720.6609	0.0127	P(5)	32004992.1796	0.0129
P(3)	27724718.4080	0.0141	P(3)	32053170.2819	0.0136
R(1)	27841759.7696	0.0152	P(1)	32170312.0391	0.0149
R(3)	27887393.2105	0.0146	P(3)	32215845.0845	0.0151
R(5)	27932349.2934	0.0135	P(5)	32260620.8121	0.0152
R(7)	27976627.0108	0.0124	P(7)	32304638.8261	0.0152
R(9)	28020225.2521	0.0115	P(9)	32347898.8990	0.0150
R(11)	28063142.8031	0.0110	P(11)	32390400.9714	0.0148
R(13)	28105378.3457	0.0109	P(13)	32432145.1513	0.0145
R(15)	28146930.4576	0.0109	P(15)	32473131.7137	0.0142
R(17)	28187797.6116	0.0107	P(17)	32513361.0997	0.0140
R(19)	28227978.1750	0.0103	P(19)	32552833.9153	0.0140
R(21)	28267470.4088	0.0099	P(21)	32591550.9309	0.0141
R(23)	28306272.4666	0.0099	P(23)	32629513.0796	0.0143
R(25)	28344382.3939	0.0107	P(25)	32666721.4564	0.0144
R(27)	28381798.1267	0.0122	P(27)	32703177.3164	0.0142
R(29)	28418517.4902	0.0141	P(29)	32738882.0732	0.0136
R(31)	28454538.1976	0.0165	P(31)	32773837.2976	0.0136
R(33)	28489857.8477	0.0213	P(33)	32808044.7156	0.0174
R(35)	28524473.9240	0.0312	P(35)	32841506.2063	0.0279
R(37)	28558383.7917	0.0486	P(37)	32874223.8000	0.0462
R(39)	28591584.6963	0.0754	P(39)	32906199.6761	0.0735
R(41)	28624073.7602	0.1131	P(41)	32937436.1606	0.1114
R(43)	28655847.9806	0.1644	P(43)	32967935.7238	0.1624
R(45)	28686904.2261	0.2328	P(45)	32997700.9775	0.2292
R(47)	28717239.2334	0.3239	P(47)	33026734.6728	0.3151
R(49)	28746849.6038	0.4465	P(49)	33055039.6965	0.4238
R(51)	28775731.7988	0.6142	P(51)	33082619.0689	0.5595
R(53)	28803882.1361	0.8465	P(53)	33109475.9403	0.7272

Frequencies for the $01^{1f}1$-$(11^{1f}0,03^{1f}0)_I$ and $01^{1f}1$-$(11^{1f}0,03^{1f}0)_{II}$ Bands of $^{12}C^{16}O_2$ with the Estimated 2-σ Uncertainties

Line	Band I Frequency (MHz)	Uncertainty (MHz)	Line	Band II Frequency (MHz)	Uncertainty (MHz)
P(60)	26051570.0104	4.4521	P(60)	30355115.0204	0.2752
P(58)	26120964.4932	3.0629	P(58)	30425283.5969	0.1926
P(56)	26189552.8496	2.0516	P(56)	30494732.8293	0.1301
P(54)	26257339.6006	1.3305	P(54)	30563455.6325	0.0840
P(52)	26324329.0344	0.8289	P(52)	30631445.1076	0.0512
P(50)	26390525.2136	0.4901	P(50)	30698694.5456	0.0292
P(48)	26455931.9824	0.2698	P(48)	30765197.4310	0.0163

INFRARED LASER FREQUENCIES (continued)

Line	Band I Frequency (MHz)	Uncertainty (MHz)	Line	Band II Frequency (MHz)	Uncertainty (MHz)
P(46)	26520552.9722	0.1334	P(46)	30830947.4444	0.0111
P(44)	26584391.6075	0.0551	P(44)	30895938.4662	0.0104
P(42)	26647451.1105	0.0181	P(42)	30960164.5794	0.0105
P(40)	26709734.5057	0.0151	P(40)	31023620.0723	0.0105
P(38)	26771244.6242	0.0174	P(38)	31086299.4415	0.0107
P(36)	26831984.1067	0.0157	P(36)	31148197.3941	0.0114
P(34)	26891955.4069	0.0126	P(34)	31209308.8510	0.0126
P(32)	26951160.7945	0.0105	P(32)	31269628.9481	0.0138
P(30)	27009602.3576	0.0096	P(30)	31329153.0395	0.0147
P(28)	27067282.0045	0.0092	P(28)	31387876.6994	0.0151
P(26)	27124201.4662	0.0090	P(26)	31445795.7236	0.0149
P(24)	27180362.2977	0.0089	P(24)	31502906.1318	0.0141
P(22)	27235765.8792	0.0090	P(22)	31559204.1695	0.0128
P(20)	27290413.4182	0.0093	P(20)	31614686.3091	0.0113
P(18)	27344305.9494	0.0096	P(18)	31669349.2515	0.0098
P(16)	27397444.3368	0.0097	P(16)	31723189.9280	0.0086
P(14)	27449829.2733	0.0096	P(14)	31776205.5007	0.0081
P(12)	27501461.2824	0.0096	P(12)	31828393.3642	0.0085
P(10)	27552340.7179	0.0101	P(10)	31879751.1463	0.0095
P(8)	27602467.7649	0.0111	P(8)	31930276.7092	0.0107
P(6)	27651842.4399	0.0125	P(6)	31979968.1497	0.0120
P(4)	27700464.5912	0.0139	P(4)	32028823.8002	0.0131
P(2)	27748333.8988	0.0148	P(2)	32076842.2290	0.0139
R(2)	27864709.8633	0.0146	P(2)	32193218.1935	0.0150
R(4)	27909939.2762	0.0135	P(4)	32238298.4853	0.0151
R(6)	27954412.3294	0.0122	P(6)	32282538.0393	0.0153
R(8)	27998127.7801	0.0112	P(8)	32325936.7244	0.0153
R(10)	28041084.2173	0.0107	P(10)	32368494.6458	0.0154
R(12)	28083280.0620	0.0108	P(12)	32410212.1438	0.0155
R(14)	28124713.5668	0.0110	P(14)	32451089.7941	0.0155
R(16)	28165382.8151	0.0112	P(16)	32491128.4063	0.0154
R(18)	28205285.7213	0.0111	P(18)	32530329.0234	0.0152
R(20)	28244420.0302	0.0110	P(20)	32568692.9211	0.0149
R(22)	28282783.3158	0.0114	P(22)	32606221.6061	0.0147
R(24)	28320372.9812	0.0129	P(24)	32642916.8154	0.0144
R(26)	28357186.2574	0.0149	P(26)	32678780.5147	0.0140
R(28)	28393220.2023	0.0168	P(28)	32713814.8971	0.0133
R(30)	28428471.6994	0.0175	P(30)	32748022.3813	0.0120
R(32)	28462937.4565	0.0165	P(32)	32781405.6101	0.0106
R(34)	28496614.0042	0.0142	P(34)	32813967.4482	0.0122
R(36)	28529497.6934	0.0163	P(36)	32845710.9809	0.0212
R(38)	28561584.6939	0.0309	P(38)	32876639.5111	0.0385
R(40)	28592870.9914	0.0593	P(40)	32906756.5580	0.0646
R(42)	28623352.3850	0.1054	P(42)	32936065.8540	0.1015
R(44)	28653024.4839	0.1779	P(44)	32964571.3426	0.1518
R(46)	28681882.7038	0.2907	P(46)	32992277.1760	0.2184
R(48)	28709922.2632	0.4635	P(48)	33019187.7118	0.3048
R(50)	28737138.1785	0.7231	P(50)	33045307.5105	0.4152

INFRARED AND FAR-INFRARED ABSORPTION FREQUENCY STANDARDS
Arthur Maki

Aside from the CO_2 laser transitions, the absorption spectrum of CO has been more accurately and thoroughly measured than any other spectrum. A bibliography of earlier measurements on CO is given by Maki and Wells,[1] and the present tables were calculated from the measurements referred to in that work. In addition, some new and very accurate frequency measurements[2,3] have been made and were incorporated in the present tables. The frequencies of the rotational transitions of HF and HCl were calculated from constants obtained from fitting the measurements of Evenson et al.[4,5] and Jennings and Wells.[6]

A new report on infrared wavenumber standards from the International Union of Pure and Applied Chemistry, Commission on Molecular Structure and Spectroscopy, may be found in Reference 7.

REFERENCES

1. Maki, A. G. and Wells, J. S., *Wavenumber Calibration Tables from Heterodyne Frequency Measurements*, NIST Special Publication 821, U.S. Dept. of Commerce, Washington, D.C., 1991.
2. Evenson, K. and Stroh, F., private communication.
3. George, T., Urban, W., and co-workers, private communication.
4. Jennings, D. A., Evenson, K. M., Zink, L. R., Demuynck, C., Destombes, J. L., Lemoine, B., and Johns, J. W. C., *J. Mol. Spectrosc.*, 122, 477-480, 1987.
5. Nolt, I. G., Radostitz, J. V., DiLonardo, G., Evenson, K. M., Jennings, D. A., Leopold, K. R., Vanek, M. D., Zink, L. R., Hinz, A., and Chance, K. V., *J. Mol. Spectrosc.*, 125, 274-287, 1987.
6. Jennings, D. A. and Wells, J. S., *J. Mol. Spectrosc.*, 130, 267-268, 1988.
7. *High Resolution Wavenumber Standards for the Infrared, Pure Appl. Chemistry*, 68, 193, 1996.

Wavenumbers for the ν = 1-0 Band of CO

Wavenumber (unc)* cm^{-1}	Transition	Wavenumber (unc) cm^{-1}	Transition
		2147.081132(01)	R(0)
2139.426071(01)	P(1)	2150.856006(01)	R(1)
2135.546178(01)	P(2)	2154.595581(01)	R(2)
2131.631574(01)	P(3)	2158.299710(01)	R(3)
2127.682404(01)	P(4)	2161.968245(01)	R(4)
2123.698816(01)	P(5)	2165.601041(01)	R(5)
2119.680957(01)	P(6)	2169.197949(01)	R(6)
2115.628973(01)	P(7)	2172.758824(01)	R(7)
2111.543012(01)	P(8)	2176.283519(01)	R(8)
2107.423221(01)	P(9)	2179.771887(01)	R(9)
2103.269746(01)	P(10)	2183.223782(01)	R(10)
2099.082734(01)	P(11)	2186.639057(01)	R(11)
2094.862333(01)	P(12)	2190.017565(01)	R(12)
2090.608688(01)	P(13)	2193.359161(01)	R(13)
2086.321947(01)	P(14)	2196.663698(01)	R(14)
2082.002256(01)	P(15)	2199.931030(01)	R(15)
2077.649762(01)	P(16)	2203.161010(01)	R(16)
2073.264612(01)	P(17)	2206.353492(01)	R(17)
2068.846952(01)	P(18)	2209.508331(02)	R(18)
2064.396929(01)	P(19)	2212.625379(02)	R(19)
2059.914688(02)	P(20)	2215.704492(02)	R(20)
2055.400377(02)	P(21)	2218.745522(02)	R(21)
2050.854140(02)	P(22)	2221.748326(03)	R(22)
2046.276126(03)	P(23)	2224.712755(03)	R(23)
2041.666479(03)	P(24)	2227.638666(03)	R(24)
2037.025345(03)	P(25)	2230.525912(04)	R(25)
2032.352870(04)	P(26)	2233.374349(04)	R(26)
2027.649200(04)	P(27)	2236.183829(04)	R(27)
2022.914480(04)	P(28)	2238.954210(05)	R(28)
2018.148857(05)	P(29)	2241.685344(05)	R(29)
2013.352474(05)	P(30)	2244.377088(06)	R(30)
2008.525477(06)	P(31)	2247.029296(07)	R(31)
2003.668012(06)	P(32)	2249.641824(08)	R(32)
1998.780224(07)	P(33)	2252.214527(10)	R(33)
1993.862257(09)	P(34)	2254.747262(14)	R(34)

Wavenumber (unc)* cm⁻¹	Transition	Wavenumber (unc) cm⁻¹	Transition
1988.914257(11)	P(35)	2257.239883(18)	R(35)
1983.936367(14)	P(36)	2259.692248(24)	R(36)
1978.928733(18)	P(37)	2262.104213(33)	R(37)
1973.891500(25)	P(38)	2264.475634(45)	R(38)
1968.824811(34)	P(39)	2266.806368(61)	R(39)
1963.728813(46)	P(40)	2269.096273(81)	R(40)
1958.603648(61)	P(41)	2271.345206(106)	R(41)
1953.449462(82)	P(42)	2273.553027(139)	R(42)

* The uncertainty in the last digits (twice the standard error) is given in parentheses.

Wavenumbers for the $v = 2$-0 Band of CO

Wavenumber (unc)* cm⁻¹	Transition	Wavenumber (unc) cm⁻¹	Transition
		4263.837198(02)	R(0)
4256.217140(02)	P(1)	4267.542066(02)	R(1)
4252.302244(02)	P(2)	4271.176630(02)	R(2)
4248.317633(02)	P(3)	4274.740746(02)	R(3)
4244.263453(02)	P(4)	4278.234264(02)	R(4)
4240.139852(02)	P(5)	4281.657039(02)	R(5)
4235.946975(02)	P(6)	4285.008924(02)	R(6)
4231.684972(02)	P(7)	4288.289772(02)	R(7)
4227.353987(02)	P(8)	4291.499437(02)	R(8)
4222.954169(02)	P(9)	4294.637773(02)	R(9)
4218.485665(02)	P(10)	4297.704631(02)	R(10)
4213.948620(02)	P(11)	4300.699868(02)	R(11)
4209.343182(02)	P(12)	4303.623334(02)	R(12)
4204.669499(02)	P(13)	4306.474886(02)	R(13)
4199.927716(02)	P(14)	4309.254375(02)	R(14)
4195.117980(02)	P(15)	4311.961657(02)	R(15)
4190.240439(02)	P(16)	4314.596584(02)	R(16)
4185.295239(02)	P(17)	4317.159011(02)	R(17)
4180.282526(02)	P(18)	4319.648791(02)	R(18)
4175.202447(02)	P(19)	4322.065779(03)	R(19)
4170.055149(03)	P(20)	4324.409829(03)	R(20)
4164.840777(03)	P(21)	4326.680794(03)	R(21)
4159.559478(03)	P(22)	4328.878530(03)	R(22)
4154.211398(03)	P(23)	4331.002889(04)	R(23)
4148.796683(04)	P(24)	4333.053728(04)	R(24)
4143.315479(04)	P(25)	4335.030899(05)	R(25)
4137.767932(04)	P(26)	4336.934259(06)	R(26)
4132.154187(05)	P(27)	4338.763661(07)	R(27)
4126.474391(06)	P(28)	4340.518961(09)	R(28)
4120.728689(07)	P(29)	4342.200014(11)	R(29)
4114.917226(09)	P(30)	4343.806675(16)	R(30)
4109.040148(12)	P(31)	4345.338799(21)	R(31)
4103.097600(16)	P(32)	4346.796243(29)	R(32)
4097.089728(21)	P(33)	4348.178862(40)	R(33)
4091.016676(29)	P(34)	4349.486513(54)	R(34)
4084.878591(40)	P(35)	4350.719052(73)	R(35)
4078.675618(54)	P(36)	4351.876336(96)	R(36)
4072.407901(73)	P(37)	4352.958224(127)	R(37)
4066.075588(97)	P(38)	4353.964572(166)	R(38)
4059.678822(127)	P(39)	4354.895240(214)	R(39)

* The uncertainty in the last digits (twice the standard error) is given in parentheses.

Wavenumbers for the v = 3-0 Band of CO

Wavenumber (unc)* cm⁻¹	Transition	Wavenumber (unc) cm⁻¹	Transition
		6354.179057(13)	R(0)
6346.594000(13)	P(1)	6357.813923(13)	R(1)
6342.644103(13)	P(2)	6361.343487(13)	R(2)
6338.589491(13)	P(3)	6364.767599(13)	R(3)
6334.430309(13)	P(4)	6368.086115(13)	R(4)
6330.166705(13)	P(5)	6371.298887(13)	R(5)
6325.798826(13)	P(6)	6374.405768(12)	R(6)
6321.326819(13)	P(7)	6377.406611(12)	R(7)
6316.750831(12)	P(8)	6380.301271(12)	R(8)
6312.071008(12)	P(9)	6383.089600(12)	R(9)
6307.287498(12)	P(10)	6385.771452(12)	R(10)
6302.400447(12)	P(11)	6388.346680(13)	R(11)
6297.410003(12)	P(12)	6390.815139(13)	R(12)
6292.316311(13)	P(13)	6393.176681(13)	R(13)
6287.119520(13)	P(14)	6395.431160(13)	R(14)
6281.819775(13)	P(15)	6397.578430(13)	R(15)
6276.417224(13)	P(16)	6399.618344(13)	R(16)
6270.912012(13)	P(17)	6401.550757(13)	R(17)
6265.304287(13)	P(18)	6403.375523(13)	R(18)
6259.594194(13)	P(19)	6405.092495(14)	R(19)
6253.781880(13)	P(20)	6406.701527(14)	R(20)
6247.867492(14)	P(21)	6408.202474(14)	R(21)
6241.851176(14)	P(22)	6409.595189(15)	R(22)
6235.733077(14)	P(23)	6410.879527(15)	R(23)
6229.513342(15)	P(24)	6412.055343(16)	R(24)
6223.192117(15)	P(25)	6413.122491(17)	R(25)
6216.769547(16)	P(26)	6414.080825(19)	R(26)
6210.245778(17)	P(27)	6414.930201(23)	R(27)
6203.620957(19)	P(28)	6415.670474(28)	R(28)
6196.895229(23)	P(29)	6416.301500(37)	R(29)
6190.068739(28)	P(30)	6416.823133(50)	R(30)
6183.141633(37)	P(31)	6417.235231(67)	R(31)
6176.114058(50)	P(32)	6417.537649(90)	R(32)
6168.986159(67)	P(33)		
6161.758082(90)	P(34)		

* The uncertainty in the last digits (twice the standard error) is given in parentheses.

Frequencies and Wavenumbers for the Rotational Lines of CO

Frequency MHz	Uncertainty* MHz	J'	J''	Wavenumber cm⁻¹	Uncertainty* cm⁻¹
115271.2029	0.0004	1	0	3.84503345	0.00000001
230538.0016	0.0008	2	1	7.68991999	0.00000003
345795.9923	0.0012	3	2	11.53451273	0.00000004
461040.7712	0.0016	4	3	15.37866477	0.00000005
576267.9350	0.0019	5	4	19.22222923	0.00000006
691473.0809	0.0021	6	5	23.06505926	0.00000007
806651.8065	0.0023	7	6	26.90700800	0.00000008
921799.7104	0.0025	8	7	30.74792863	0.00000008
1036912.3919	0.0027	9	8	34.58767438	0.00000009
1151985.4515	0.0029	10	9	38.42609848	0.00000010
1267014.4906	0.0031	11	10	42.26305422	0.00000010
1381995.1119	0.0034	12	11	46.09839491	0.00000011
1496922.9195	0.0038	13	12	49.93197392	0.00000013

INFRARED AND FAR-INFRARED ABSORPTION FREQUENCY STANDARDS (continued)

Frequency MHz	Uncertainty* MHz	J'	J"	Wavenumber cm⁻¹	Uncertainty* cm⁻¹
1611793.5189	0.0042	14	13	53.76364468	0.00000014
1726602.5173	0.0047	15	14	57.59326065	0.00000016
1841345.5237	0.0052	16	15	61.42067535	0.00000017
1956018.1486	0.0057	17	16	65.24574239	0.00000019
2070616.0050	0.0061	18	17	69.06831542	0.00000020
2185134.7075	0.0065	19	18	72.88824816	0.00000022
2299569.8733	0.0069	20	19	76.70539441	0.00000023
2413917.1217	0.0071	21	20	80.51960806	0.00000024
2528172.0747	0.0073	22	21	84.33074306	0.00000024
2642330.3567	0.0074	23	22	88.13865346	0.00000025
2756387.5949	0.0075	24	23	91.94319341	0.00000025
2870339.4194	0.0077	25	24	95.74421713	0.00000026
2984181.4631	0.0080	26	25	99.54157896	0.00000027
3097909.3621	0.0085	27	26	103.33513334	0.00000028
3211518.7558	0.0090	28	27	107.12473480	0.00000030
3325005.2869	0.0096	29	28	110.91023800	0.00000032
3438364.6013	0.0102	30	29	114.69149772	0.00000034
3551592.3489	0.0107	31	30	118.46836884	0.00000036
3664684.1829	0.0111	32	31	122.24070637	0.00000037
3777635.7608	0.0118	33	32	126.00836545	0.00000039
3890442.7435	0.0137	34	33	129.77120137	0.00000046
4003100.7965	0.0179	35	34	133.52906952	0.00000060
4115605.5892	0.0254	36	35	137.28182546	0.00000085
4227952.7954	0.0370	37	36	141.02932487	0.00000123
4340138.0932	0.0531	38	37	144.77142361	0.00000177
4452157.1657	0.0746	39	38	148.50797766	0.00000249
4564005.7001	0.1025	40	39	152.23884318	0.00000342

* The uncertainty given is twice the standard error.

Frequencies and Wavenumbers for the Rotational Lines of HF

Frequency MHz	Uncertainty* MHz	J'	J"	Wavenumber cm⁻¹	Uncertainty* cm⁻¹
1232476.21	0.12	1	0	41.110981	0.000004
2463428.09	0.19	2	1	82.171116	0.000006
3691334.81	0.25	3	2	123.129676	0.000008
4914682.58	0.51	4	3	163.936165	0.000017
6131968.11	1.10	5	4	204.540439	0.000037
7341702.00	2.00	6	5	244.892818	0.000067
8542412.1	3.21	7	6	284.944197	0.000107
9732646.8	4.72	8	7	324.646153	0.000157
10910978.2	6.51	9	8	363.951056	0.000217
12076004.8	8.55	10	9	402.81216	0.000285
13226355.2	10.81	11	10	441.18372	0.000361
14360689.8	13.25	12	11	479.02105	0.00044
15477704.4	15.86	13	12	516.28065	0.00053
16576131.8	18.61	14	13	552.92024	0.00062
17654744.4	21.48	15	14	588.89888	0.00072
18712356.5	24.44	16	15	624.17703	0.00082
19747825.6	27.43	17	16	658.71656	0.00092
20760054.3	30.32	18	17	692.4809	0.00101
21747991.7	32.91	19	18	725.4349	0.00110
22710634.7	34.94	20	19	757.5452	0.00117
23647028.7	36.08	21	20	788.7800	0.00120
24556268.8	35.93	22	21	819.1090	0.00120
25437499.9	34.12	23	22	848.5037	0.00114

INFRARED AND FAR-INFRARED ABSORPTION FREQUENCY STANDARDS (continued)

Frequency MHz	Uncertainty* MHz	J'	J''	Wavenumber cm^{-1}	Uncertainty* cm^{-1}
26289917.4	30.32	24	23	876.9373	0.00101
27112767.2	24.41	25	24	904.38457	0.00081
27905345.6	16.88	26	25	930.82214	0.00056
28666999.3	10.80	27	26	956.22817	0.00036
29397124.8	14.65	28	27	980.58253	0.00049
30095168.2	24.62	29	28	1003.86676	0.00082
30760624.2	33.36	30	29	1026.0640	0.00111
31393035.7	36.17	31	30	1047.1590	0.00121

* The uncertainty given is twice the standard error.

Frequencies and Wavenumbers for the Rotational Lines of H^{35}Cl

Frequency MHz	Uncertainty* MHz	J'	J''	Wavenumber cm^{-1}	Uncertainty* cm^{-1}
1876226.517	0.065	3	2	62.584180	0.000002
2499864.439	0.066	4	3	83.386502	0.000002
3121986.563	0.064	5	4	104.138262	0.000002
3742216.601	0.076	6	5	124.826909	0.000003
4360180.042	0.098	7	6	145.439951	0.000003
4975504.51	0.11	8	7	165.964966	0.000004
5587820.10	0.12	9	8	186.389615	0.000004
6196759.76	0.22	10	9	206.701656	0.000007
6801959.63	0.50	11	10	226.888951	0.000017
7403059.41	1.02	12	11	246.939481	0.000034
7999702.7	1.8	13	12	266.841359	0.000062
8591537.3	3.1	14	13	286.582837	0.000103
9178215.8	4.8	15	14	306.152324	0.000161

* The uncertainty given is twice the standard error.

Frequencies and Wavenumbers for the Rotational Lines of H^{37}Cl

Frequency MHz	Uncertainty* MHz	J'	J''	Wavenumber cm^{-1}	Uncertainty* cm^{-1}
1873410.72	0.05	3	2	62.490255	0.000002
2496115.33	0.05	4	3	83.261445	0.000002
3117308.69	0.05	5	4	103.982225	0.000002
3736615.64	0.06	6	5	124.640082	0.000002
4353662.84	0.08	7	6	145.222561	0.000003
4968079.04	0.09	8	7	165.717279	0.000003
5579495.53	0.10	9	8	186.111938	0.000003
6187546.42	0.19	10	9	206.394332	0.000006
6791869.04	0.45	11	10	226.552365	0.000015
7392104.3	0.9	12	11	246.574057	0.000030
7987896.9	1.6	13	12	266.447561	0.000054
8578896.1	2.7	14	13	286.161170	0.000089

* The uncertainty given is twice the standard error.

Section 11
Nuclear and Particle Physics

SUMMARY TABLES OF PARTICLE PROPERTIES

July 1996

Particle Data Group

R.M. Barnett, C.D. Carone, D.E. Groom, T.G. Trippe, C.G. Wohl,
B. Armstrong*, P.S. Gee*, G.S. Wagman*[†], F. James, M. Mangano,
K. Mönig, L. Montanet, J.L. Feng, H. Murayama, J.J. Hernández,
A. Manohar, M. Aguilar-Benitez, C. Caso, R.L. Crawford, M. Roos,
N.A. Törnqvist, K.G. Hayes, K. Hagiwara, K. Nakamura, M. Tanabashi,
K. Olive, K. Honscheid, P.R. Burchat, R.E. Shrock, S. Eidelman,
R.H. Schindler, A. Gurtu, K. Hikasa, G. Conforto, R.L. Workman,
C. Grab, and C. Amsler
*Technical Associate
[†]Deceased

Reprinted from Physical Review **D54** 1 (1996)

(Approximate closing date for data: January 1, 1996)

GAUGE AND HIGGS BOSONS

γ

$$I(J^{PC}) = 0,1(1^{--})$$

Mass $m < 6 \times 10^{-16}$ eV, CL = 99.7%
Charge $q < 5 \times 10^{-30}$ e
Mean life τ = Stable

g or gluon

$$I(J^{P}) = 0(1^{-})$$

Mass $m = 0$ [a]
SU(3) color octet

W

$$J = 1$$

Charge $= \pm 1$ e
Mass $m = 80.33 \pm 0.15$ GeV
$m_Z - m_W = 10.85 \pm 0.15$ GeV
$m_{W^+} - m_{W^-} = -0.2 \pm 0.6$ GeV
Full width $\Gamma = 2.07 \pm 0.06$ GeV

W^- modes are charge conjugates of the modes below.

W^+ DECAY MODES		Fraction (Γ_i/Γ)	Confidence level	p (MeV/c)
$\ell^+\nu$	[b]	(10.8 ± 0.4) %		40110
$e^+\nu$		(10.8 ± 0.4) %		40110
$\mu^+\nu$		(10.4 ± 0.6) %		40110
$\tau^+\nu$		(10.9 ± 1.0) %		40110
hadrons		(67.9 ± 1.5) %		–
$\pi^+\gamma$		$< 5 \times 10^{-4}$	95%	40110

Z

$$J = 1$$

Charge $= 0$
Mass $m = 91.187 \pm 0.007$ GeV [c]
Full width $\Gamma = 2.490 \pm 0.007$ GeV
$\Gamma(\ell^+\ell^-) = 83.83 \pm 0.27$ MeV [b]
$\Gamma(\text{invisible}) = 498.3 \pm 4.2$ MeV [d]
$\Gamma(\text{hadrons}) = 1740.7 \pm 5.9$ MeV
$\Gamma(\mu^+\mu^-)/\Gamma(e^+e^-) = 1.000 \pm 0.005$
$\Gamma(\tau^+\tau^-)/\Gamma(e^+e^-) = 0.998 \pm 0.005$ [e]

Average charged multiplicity

$\langle N_{charged} \rangle = 20.99 \pm 0.14$

Couplings to leptons

$g_V^\ell = -0.0376 \pm 0.0012$
$g_A^\ell = -0.5008 \pm 0.0008$
$g^{\nu_e} = 0.53 \pm 0.09$
$g^{\nu_\mu} = 0.502 \pm 0.017$

Asymmetry parameters [f]

$A_e = 0.156 \pm 0.008$ (S = 1.2)
$A_\tau = 0.145 \pm 0.009$
$A_c = 0.59 \pm 0.19$
$A_b = 0.89 \pm 0.11$

Charge asymmetry (%) at Z pole

$A_{FB}^{(0\ell)} = 1.59 \pm 0.18$
$A_{FB}^{(0s)} = 13 \pm 4$
$A_{FB}^{(0c)} = 7.22 \pm 0.67$
$A_{FB}^{(0b)} = 9.92 \pm 0.35$

Z DECAY MODES			Fraction (Γ_i/Γ)	Confidence level	p (MeV/c)
e^+e^-			(3.366 ± 0.008) %		45600
$\mu^+\mu^-$			(3.367 ± 0.013) %		45600
$\tau^+\tau^-$			(3.360 ± 0.015) %		45600
$\ell^+\ell^-$		[b]	(3.366 ± 0.006) %		45600
invisible			(20.01 ± 0.16) %		–
hadrons			(69.90 ± 0.15) %		–
$(u\bar{u}+c\bar{c})/2$			(9.6 ± 1.3) %		–
$(d\bar{d}+s\bar{s}+b\bar{b})/3$			(16.9 ± 0.9) %		–
$c\bar{c}$			(11.0 ± 0.7) %		–
$b\bar{b}$			(15.46 ± 0.14) %		–
$\pi^0\gamma$			$< 5.2 \times 10^{-5}$	95%	45600
$\eta\gamma$			$< 5.1 \times 10^{-5}$	95%	45600
$\omega\gamma$			$< 6.5 \times 10^{-4}$	95%	45600
$\eta'(958)\gamma$			$< 4.2 \times 10^{-5}$	95%	45600
$\gamma\gamma$			$< 5.2 \times 10^{-5}$	95%	45600
$\gamma\gamma\gamma$			$< 1.0 \times 10^{-5}$	95%	45600
$\pi^\pm W^\mp$		[g]	$< 7 \times 10^{-5}$	95%	10300
$\rho^\pm W^\mp$		[g]	$< 8.3 \times 10^{-5}$	95%	10300
$J/\psi(1S)X$			$(3.80 \pm 0.27) \times 10^{-3}$		–
$\psi(2S)X$			$(1.60 \pm 0.33) \times 10^{-3}$		–
$\chi_{c1}(1P)X$			$(6.0 \pm 1.9) \times 10^{-3}$		–
ΥX			$(1.0 \pm 0.5) \times 10^{-4}$		–
$(D^0/\overline{D}^0)X$			(20.7 ± 2.0) %		–
$D^\pm X$			(12.2 ± 1.7) %		–
$D^*(2010)^\pm X$		[g]	(11.4 ± 1.3) %		–
$B_s^0 X$			seen		–
anomalous γ+ hadrons		[h]	$< 3.2 \times 10^{-3}$	95%	–
$e^+e^-\gamma$		[h]	$< 5.2 \times 10^{-4}$	95%	45600
$\mu^+\mu^-\gamma$		[h]	$< 5.6 \times 10^{-4}$	95%	45600
$\tau^+\tau^-\gamma$		[h]	$< 7.3 \times 10^{-4}$	95%	45600
$\ell^+\ell^-\gamma\gamma$		[i]	$< 6.8 \times 10^{-6}$	95%	45600
$q\bar{q}\gamma\gamma$		[i]	$< 5.5 \times 10^{-6}$	95%	–
$\nu\bar{\nu}\gamma\gamma$		[i]	$< 3.1 \times 10^{-6}$	95%	45600
$e^\pm\mu^\mp$	LF	[g]	$< 1.7 \times 10^{-6}$	95%	45600
$e^\pm\tau^\mp$	LF	[g]	$< 9.8 \times 10^{-6}$	95%	45600
$\mu^\pm\tau^\mp$	LF	[g]	$< 1.7 \times 10^{-5}$	95%	45600

Gauge & Higgs Boson Summary Table

Higgs Bosons — H^0 and $H^\pm$, Searches for

H^0 Mass $m > 58.4$ GeV, CL = 95%

H_1^0 in Supersymmetric Models ($m_{H_1^0} < m_{H_2^0}$) [j]

Mass $m > 44$ GeV, CL = 95%

A^0 Pseudoscalar Higgs Boson in Supersymmetric Models [j]

Mass $m > 24.3$ GeV, CL = 95% $\tan\beta > 1$, $m_t < 200$ GeV

$H^\pm$ Mass $m > 43.5$ GeV, CL = 95%

See the Particle Listings for a Note giving details of Higgs Bosons.

Heavy Bosons Other Than Higgs Bosons, Searches for

Additional W Bosons

W_R — right-handed W

Mass $m > 406$ GeV, CL = 90%
(assuming light right-handed neutrino)

W' with standard couplings decaying to $e\nu$, $\mu\nu$

Mass $m > 652$ GeV, CL = 95%

Additional Z Bosons

Z'_{SM} with standard couplings

Mass $m > 505$ GeV, CL = 95% ($p\overline{p}$ direct search)
Mass $m > 779$ GeV, CL = 95% (electroweak fit)

Z_{LR} of SU(2)$_L$×SU(2)$_R$×U(1)
(with $g_L = g_R$)

Mass $m > 445$ GeV, CL = 95% ($p\overline{p}$ direct search)
Mass $m > 389$ GeV, CL = 95% (electroweak fit)

Z_χ of SO(10) → SU(5)×U(1)$_\chi$
(coupling constant derived from G.U.T.)

Mass $m > 425$ GeV, CL = 95% ($p\overline{p}$ direct search)
Mass $m > 321$ GeV, CL = 95% (electroweak fit)

Z_ψ of E_6 → SO(10)×U(1)$_\psi$
(coupling constant derived from G.U.T.)

Mass $m > 415$ GeV, CL = 95% ($p\overline{p}$ direct search)
Mass $m > 160$ GeV, CL = 95% (electroweak fit)

Z_η of E_6 → SU(3)×SU(2)×U(1)×U(1)$_\eta$
(coupling constant derived from G.U.T.);
charges are $Q_\eta = \sqrt{3/8}Q_\chi - \sqrt{5/8}Q_\psi$

Mass $m > 440$ GeV, CL = 95% ($p\overline{p}$ direct search)
Mass $m > 182$ GeV, CL = 95% (electroweak fit)

Scalar Leptoquarks

Mass $m > 116$ GeV, CL = 95% (1st generation, pair prod.)
Mass $m > 230$ GeV, CL = 95% (1st gener., single prod.)
Mass $m > 97$ GeV, CL = 95% (2nd gener., pair prod.)
Mass $m > 73$ GeV, CL = 95% (2nd gener., single prod.)
Mass $m > 45$ GeV, CL = 95% (3rd gener., pair prod.)
(The second, fourth, and fifth limits above are for charge
$-1/3$, weak isoscalar.)

Axions (A^0) and Other Very Light Bosons, Searches for

The standard Peccei-Quinn axion is ruled out. Variants with reduced couplings or much smaller masses are constrained by various data. The Particle Listings in the full *Review* contain a Note discussing axion searches.

The best limit for the half-life of neutrinoless double beta decay with Majoron emission is $> 7.2 \times 10^{24}$ years (CL = 90%).

NOTES

In this Summary Table:

When a quantity has "(S = ...)" to its right, the error on the quantity has been enlarged by the "scale factor" S, defined as S = $\sqrt{\chi^2/(N-1)}$, where N is the number of measurements used in calculating the quantity. We do this when S > 1, which often indicates that the measurements are inconsistent. When S > 1.25, we also show in the Particle Listings an ideogram of the measurements. For more about S, see the Introduction.

A decay momentum p is given for each decay mode. For a 2-body decay, p is the momentum of each decay product in the rest frame of the decaying particle. For a 3-or-more-body decay, p is the largest momentum any of the products can have in this frame.

[a] Theoretical value. A mass as large as a few MeV may not be precluded.
[b] ℓ indicates each type of lepton (e, μ, and τ), not sum over them.
[c] The Z-boson mass listed here corresponds to a Breit-Wigner resonance parameter. It lies approximately 34 MeV above the real part of the position of the pole (in the energy-squared plane) in the Z-boson propagator.
[d] This partial width takes into account Z decays into $\nu\overline{\nu}$ and any other possible undetected modes.
[e] This ratio has not been corrected for the τ mass.
[f] Here $A \equiv 2g_V g_A/(g_V^2 + g_A^2)$.
[g] The value is for the sum of the charge states of particle/antiparticle states indicated.
[h] See the Z Particle Listings for the γ energy range used in this measurement.
[i] For $m_{\gamma\gamma} = (60 \pm 5)$ GeV.
[j] The limits assume no invisible decays.

LEPTONS

e

$J = \frac{1}{2}$

Mass $m = 0.51099907 \pm 0.00000015$ MeV [a]
$= (5.48579903 \pm 0.00000013) \times 10^{-4}$ u
$(m_{e^+} - m_{e^-})/m < 4 \times 10^{-8}$, CL = 90%
$|q_{e^+} + q_{e^-}|/e < 4 \times 10^{-8}$
Magnetic moment $\mu = 1.001159652193 \pm 0.000000000010\ \mu_B$
$(g_{e^+} - g_{e^-})\ /\ g_{average} = (-0.5 \pm 2.1) \times 10^{-12}$
Electric dipole moment $d = (-0.3 \pm 0.8) \times 10^{-26}$ ecm
Mean life $\tau > 4.3 \times 10^{23}$ yr, CL = 68% [b]

μ

$J = \frac{1}{2}$

Mass $m = 105.658389 \pm 0.000034$ MeV [c]
$= 0.113428913 \pm 0.000000017$ u
Mean life $\tau = (2.19703 \pm 0.00004) \times 10^{-6}$ s
$\tau_{\mu^+}/\tau_{\mu^-} = 1.00002 \pm 0.00008$
$c\tau = 658.654$ m
Magnetic moment $\mu = 1.001165923 \pm 0.000000008\ e\hbar/2m_\mu$
$(g_{\mu^+} - g_{\mu^-})\ /\ g_{average} = (-2.6 \pm 1.6) \times 10^{-8}$
Electric dipole moment $d = (3.7 \pm 3.4) \times 10^{-19}$ ecm

Decay parameters [d]

$\rho = 0.7518 \pm 0.0026$
$\eta = -0.007 \pm 0.013$
$\delta = 0.749 \pm 0.004$
$\xi P_\mu = 1.003 \pm 0.008$ [e]
$\xi P_\mu \delta/\rho > 0.99682$, CL = 90% [e]
$\xi' = 1.00 \pm 0.04$
$\xi'' = 0.7 \pm 0.4$
$\alpha/A = (0 \pm 4) \times 10^{-3}$
$\alpha'/A = (0 \pm 4) \times 10^{-3}$
$\beta/A = (4 \pm 6) \times 10^{-3}$
$\beta'/A = (2 \pm 6) \times 10^{-3}$
$\bar{\eta} = 0.02 \pm 0.08$

μ^+ modes are charge conjugates of the modes below.

μ^- DECAY MODES		Fraction (Γ_i/Γ)	Confidence level	p (MeV/c)
$e^- \bar{\nu}_e \nu_\mu$		$\approx 100\%$		53
$e^- \bar{\nu}_e \nu_\mu \gamma$	[f]	$(1.4 \pm 0.4)\%$		53
$e^- \bar{\nu}_e \nu_\mu e^+ e^-$	[g]	$(3.4 \pm 0.4) \times 10^{-5}$		53
Lepton Family number (LF) violating modes				
$e^- \nu_e \bar{\nu}_\mu$	LF	[h] < 1.2 %	90%	53
$e^- \gamma$	LF	$< 4.9 \times 10^{-11}$	90%	53
$e^- e^+ e^-$	LF	$< 1.0 \times 10^{-12}$	90%	53
$e^- 2\gamma$	LF	$< 7.2 \times 10^{-11}$	90%	53

τ

$J = \frac{1}{2}$

Mass $m = 1777.00^{+0.30}_{-0.27}$ MeV
Mean life $\tau = (291.0 \pm 1.5) \times 10^{-15}$ s
$c\tau = 87.2\ \mu$m
Electric dipole moment $d < 5 \times 10^{-17}$ ecm, CL = 95%

Weak dipole moment

$\mathrm{Re}(d_\tau^w) < 7.8 \times 10^{-18}$ ecm, CL = 95%
$\mathrm{Im}(d_\tau^w) < 4.5 \times 10^{-17}$ ecm, CL = 95%

Decay parameters

See the τ Particle Listings for a note concerning τ-decay parameters.

$\rho^\tau(e \text{ or } \mu) = 0.742 \pm 0.027$
$\rho^\tau(e) = 0.736 \pm 0.028$
$\rho^\tau(\mu) = 0.74 \pm 0.04$
$\xi^\tau(e \text{ or } \mu) = 1.03 \pm 0.12$
$\zeta^\tau(e) = 1.03 \pm 0.25$
$\xi^\tau(\mu) = 1.23 \pm 0.24$
$\eta^\tau(e \text{ or } \mu) = -0.01 \pm 0.14$
$\eta^\tau(\mu) = -0.24 \pm 0.29$
$(\delta\xi)^\tau(e \text{ or } \mu) = 0.76 \pm 0.11$ (S = 1.3)
$(\delta\xi)^\tau(e) = 1.11 \pm 0.18$
$(\delta\xi)^\tau(\mu) = 0.71 \pm 0.15$
$\xi^\tau(\pi) = 0.99 \pm 0.06$
$\xi^\tau(\rho) = 1.04 \pm 0.07$
$\xi^\tau(a_1) = 1.01 \pm 0.04$
$\xi^\tau(\text{all hadronic modes}) = 1.011 \pm 0.027$

τ^+ modes are charge conjugates of the modes below. "$h^\pm$" stands for $\pi^\pm$ or $K^\pm$. "ℓ" stands for e or μ. "Neutral" means neutral hadron whose decay products include γ's and/or π^0's.

τ^- DECAY MODES		Fraction (Γ_i/Γ)	Scale factor/ Confidence level	p (MeV/c)
Modes with one charged particle				
particle$^- \geq 0$ neutrals $\geq 0K_L^0\ \nu_\tau$ ("1-prong")		$(84.96 \pm 0.14)\%$	S=1.3	–
particle$^- \geq 0$ neutrals $\geq 0K^0\ \nu_\tau$		$(85.53 \pm 0.14)\%$	S=1.3	–
$\mu^- \bar{\nu}_\mu \nu_\tau$	[i]	$(17.35 \pm 0.10)\%$		885
$\mu^- \bar{\nu}_\mu \nu_\tau \gamma$ ($E_\gamma > 37$ MeV)		$(2.3 \pm 1.0) \times 10^{-3}$		–
$e^- \bar{\nu}_e \nu_\tau$	[i]	$(17.83 \pm 0.08)\%$		888
$h^- \geq 0$ neutrals $\geq 0K_L^0\ \nu_\tau$		$(49.78 \pm 0.17)\%$	S=1.2	–
$h^- \geq 0K_L^0\ \nu_\tau$		$(12.51 \pm 0.13)\%$	S=1.1	–
$h^- \nu_\tau$		$(12.03 \pm 0.14)\%$	S=1.1	–
$\pi^- \nu_\tau$	[i]	$(11.31 \pm 0.15)\%$	S=1.1	883
$K^- \nu_\tau$	[i]	$(7.1 \pm 0.5) \times 10^{-3}$		820
$h^- \geq 1\pi^0 \nu_\tau$		$(36.97 \pm 0.18)\%$	S=1.1	–
$h^- \pi^0 \nu_\tau$		$(25.76 \pm 0.15)\%$	S=1.1	–
$\pi^- \pi^0 \nu_\tau$	[i]	$(25.24 \pm 0.16)\%$	S=1.1	878
$\pi^- \pi^0$ non-$\rho(770) \nu_\tau$		$(3.0 \pm 3.2) \times 10^{-3}$		878
$K^- \pi^0 \nu_\tau$	[i]	$(5.2 \pm 0.5) \times 10^{-3}$		814
$h^- \geq 2\pi^0 \nu_\tau$		$(10.95 \pm 0.16)\%$	S=1.1	–
$h^- 2\pi^0 \nu_\tau$		$(9.50 \pm 0.14)\%$	S=1.1	–
$h^- 2\pi^0 \nu_\tau$ (ex.K^0)		$(9.35 \pm 0.14)\%$	S=1.1	–
$\pi^- 2\pi^0 \nu_\tau$ (ex.K^0)	[i]	$(9.27 \pm 0.14)\%$	S=1.1	862
$K^- 2\pi^0 \nu_\tau$ (ex.K^0)	[i]	$(8.1 \pm 2.7) \times 10^{-4}$		796
$h^- \geq 3\pi^0 \nu_\tau$		$(1.46 \pm 0.11)\%$	S=1.1	–
$h^- 3\pi^0 \nu_\tau$		$(1.28 \pm 0.10)\%$		–
$\pi^- 3\pi^0 \nu_\tau$ (ex.K^0)	[i]	$(1.14 \pm 0.14)\%$		836
$K^- 3\pi^0 \nu_\tau$ (ex.K^0)	[i]	$(5.0^{+10.0}_{-3.3}) \times 10^{-4}$		766
$h^- 4\pi^0 \nu_\tau$ (ex.K^0)		$(1.8 \pm 0.6) \times 10^{-3}$		–
$h^- 4\pi^0 \nu_\tau$ (ex.K^0, η)		$(1.2 \pm 0.6) \times 10^{-3}$		–
$K^- \geq 1 (\pi^0 \text{ or } K^0)\ \nu_\tau$		$(9.4 \pm 1.0) \times 10^{-3}$		–
Modes with K^0's				
$h^- \overline{K}^0 \geq 0$ neutrals $\geq 0K_L^0 \nu_\tau$		$(1.54 \pm 0.10)\%$	S=1.3	–
$h^- \overline{K}^0 \nu_\tau$		$(9.2 \pm 0.8) \times 10^{-3}$	S=1.3	–
$\pi^- \overline{K}^0 \nu_\tau$	[i]	$(7.7 \pm 0.8) \times 10^{-3}$	S=1.3	812
$\pi^- \overline{K}^0$ (non-$K^*(892)^-) \nu_\tau$		$< 1.7 \times 10^{-3}$	CL=95%	812
$K^- K^0 \nu_\tau$	[i]	$(1.55 \pm 0.28) \times 10^{-3}$		737
$h^- \overline{K}^0 \pi^0 \nu_\tau$		$(5.5 \pm 0.5) \times 10^{-3}$		–
$\pi^- \overline{K}^0 \pi^0 \nu_\tau$	[i]	$(4.1 \pm 0.6) \times 10^{-3}$		794
$K^- K^0 \pi^0 \nu_\tau$	[i]	$(1.38 \pm 0.32) \times 10^{-3}$		685
$h^- K_S^0 K_S^0 \nu_\tau$		$(2.5 \pm 0.6) \times 10^{-4}$		–
$\pi^- K^0 \overline{K}^0 \nu_\tau$	[i]	$(1.01 \pm 0.23) \times 10^{-3}$		682
$K^- K^0 \geq 0$ neutrals ν_τ		$(2.9 \pm 0.4) \times 10^{-3}$		–
$K^- \geq 0\pi^0 \geq 0K^0\ \nu_\tau$		$(1.65 \pm 0.10)\%$		–
K^0 (particles)$^- \nu_\tau$		$(1.58 \pm 0.10)\%$	S=1.2	–
$K^0 h^+ h^- h^- \geq 0$ neut. ν_τ		$< 1.7 \times 10^{-3}$	CL=95%	–

Lepton Summary Table

Modes with three charged particles

Mode		Fraction	S	#
$h^-h^-h^+ \geq 0$ neut. ν_τ ("3-prong")		$(14.91\pm 0.14)\,\%$	S=1.3	–
$h^-h^-h^+ \geq 0$ neutrals ν_τ (ex. $K^0_S \to \pi^-\pi^+$)		$(14.36\pm 0.14)\,\%$	S=1.3	–
$\pi^-\pi^+\pi^- \geq 0$ neutrals ν_τ		$(14.09\pm 0.31)\,\%$		–
$h^-h^-h^+\nu_\tau$		$(9.80\pm 0.10)\,\%$	S=1.1	–
$h^-h^-h^+\nu_\tau$ (ex.K^0)		$(9.48\pm 0.10)\,\%$	S=1.1	–
$h^-h^-h^+\nu_\tau$ (ex.K^0,ω)	[l]	$(9.44\pm 0.10)\,\%$	S=1.1	–
$h^-h^-h^+ \geq 1$ neutrals ν_τ		$(5.08\pm 0.11)\,\%$	S=1.2	–
$h^-h^-h^+ \geq 1$ neutrals ν_τ (ex. $K^0_S \to \pi^+\pi^-$)		$(4.88\pm 0.11)\,\%$	S=1.2	–
$h^-h^-h^+\pi^0\nu_\tau$		$(4.44\pm 0.09)\,\%$	S=1.1	–
$h^-h^-h^+\pi^0\nu_\tau$ (ex.K^0)		$(4.25\pm 0.09)\,\%$	S=1.1	–
$h^-h^-h^+\pi^0\nu_\tau$ (ex. K^0,ω)	[l]	$(2.55\pm 0.09)\,\%$		–
$h^-(\rho\pi)^0\nu_\tau$		$(2.84\pm 0.34)\,\%$		–
$(a_1(1260)h)^-\nu_\tau$		$< 2.0\ \%$	CL=95%	–
$h^-\rho\pi^0\nu_\tau$		$(1.33\pm 0.20)\,\%$		–
$h^-\rho^+h^-\nu_\tau$		$(4.4 \pm 2.2) \times 10^{-3}$		–
$h^-\rho^-h^+\nu_\tau$		$(1.15\pm 0.23)\,\%$		–
$h^-h^-h^+2\pi^0\nu_\tau$		$(5.2 \pm 0.5) \times 10^{-3}$		–
$h^-h^-h^+2\pi^0\nu_\tau$ (ex.K^0)		$(5.1 \pm 0.5) \times 10^{-3}$		–
$h^-h^-h^+2\pi^0\nu_\tau$ (ex.K^0,ω,η)	[l]	$(1.0 \pm 0.4) \times 10^{-3}$		–
$h^-h^-h^+ \geq 3\pi^0\nu_\tau$	[l]	$(1.1 \pm 0.6) \times 10^{-3}$		–
$K^-h^+h^- \geq 0$ neutrals ν_τ		$< 6\ \times 10^{-3}$	CL=90%	–
$K^-\pi^+\pi^- \geq 0$ neut. ν_τ		$(3.9\,^{+1.9}_{-1.6}) \times 10^{-3}$	S=1.5	–
$K^-\pi^+K^- \geq 0$ neut. ν_τ		$< 9\ \times 10^{-4}$	CL=95%	–
$K^-K^+\pi^- \geq 0$ neut. ν_τ		$(1.5\,^{+0.9}_{-0.8}) \times 10^{-3}$		–
$K^-K^+\pi^-\nu_\tau$		$(2.2\,^{+1.8}_{-1.2}) \times 10^{-3}$		685
$\phi\pi^-\nu_\tau$		$< 3.5\ \times 10^{-4}$	CL=90%	585
$K^-K^+K^- \geq 0$ neut. ν_τ		$< 2.1\ \times 10^{-3}$	CL=95%	–
$\pi^-K^+\pi^- \geq 0$ neut. ν_τ		$< 2.5\ \times 10^{-3}$	CL=95%	–
$e^-e^-e^+\overline{\nu}_e\nu_\tau$		$(2.8 \pm 1.5) \times 10^{-5}$		888
$\mu^-e^-e^+\overline{\nu}_\mu\nu_\tau$		$< 3.6\ \times 10^{-5}$	CL=90%	885

Modes with five charged particles

Mode		Fraction	CL	#
$3h^-2h^+ \geq 0$ neutrals ν_τ (ex. $K^0_S \to \pi^-\pi^+$) ("5-prong")		$(9.7 \pm 0.7) \times 10^{-4}$		–
$3h^-2h^+\nu_\tau$ (ex.K^0)	[l]	$(7.5 \pm 0.7) \times 10^{-4}$		–
$3h^-2h^+\pi^0\nu_\tau$ (ex.K^0)	[l]	$(2.2 \pm 0.5) \times 10^{-4}$		–
$3h^-2h^+2\pi^0\nu_\tau$		$< 1.1\ \times 10^{-4}$	CL=90%	–

Miscellaneous other allowed modes

Mode		Fraction	CL	#
$(5\pi)^-\nu_\tau$		$(3.3 \pm 0.7) \times 10^{-3}$		–
$4h^-3h^+ \geq 0$ neutrals ν_τ ("7-prong")		$< 1.9\ \times 10^{-4}$	CL=90%	–
$K^*(892)^- \geq 0(h^0 \neq K^0_S)\nu_\tau$		$(1.94\pm 0.31)\,\%$		–
$K^*(892)^- \geq 0$ neutrals ν_τ		$(1.33\pm 0.13)\,\%$		–
$K^*(892)^-\nu_\tau$		$(1.28\pm 0.08)\,\%$		665
$K^*(892)^0 K^- \geq 0$ neutrals ν_τ		$(3.2 \pm 1.4) \times 10^{-3}$		–
$K^*(892)^0 K^-\nu_\tau$		$(2.0 \pm 0.6) \times 10^{-3}$		539
$\overline{K}^*(892)^0 \pi^- \geq 0$ neutrals ν_τ		$(3.8 \pm 1.7) \times 10^{-3}$		–
$\overline{K}^*(892)^0 \pi^-\nu_\tau$		$(2.5 \pm 1.1) \times 10^{-3}$		653
$K_1(1270)^-\nu_\tau$		$(4 \pm 4) \times 10^{-3}$		433
$K_1(1400)^-\nu_\tau$		$(8 \pm 4) \times 10^{-3}$		335
$K^*_2(1430)^-\nu_\tau$		$< 3\ \times 10^{-3}$	CL=95%	317
$\eta\pi^-\nu_\tau$		$< 1.4\ \times 10^{-3}$	CL=95%	798
$\eta\pi^-\pi^0\nu_\tau$	[l]	$(1.71\pm 0.28) \times 10^{-3}$		778
$\eta\pi^-\pi^0\pi^0\nu_\tau$		$< 4.3\ \times 10^{-4}$	CL=95%	746
$\eta K^-\nu_\tau$		$(2.6 \pm 0.7) \times 10^{-4}$		720
$\eta\pi^+\pi^-\pi^- \geq 0$ neutrals ν_τ		$< 3\ \times 10^{-4}$	CL=90%	–
$\eta\eta\pi^-\nu_\tau$		$< 1.1\ \times 10^{-4}$	CL=95%	637
$\eta\eta\pi^-\pi^0\nu_\tau$		$< 2.0\ \times 10^{-4}$	CL=95%	559
$h^-\omega \geq 0$ neutrals ν_τ		$(2.32\pm 0.11)\,\%$		–
$h^-\omega\nu_\tau$	[l]	$(1.91\pm 0.09)\,\%$		–
$h^-\omega\pi^0\nu_\tau$	[l]	$(4.1 \pm 0.6) \times 10^{-3}$		–

Lepton Family number (LF), Lepton number (L), or Baryon number (B) violating modes
(In the modes below, ℓ means a sum over e and μ modes)

L means lepton number violation (e.g. $\tau^- \to e^+\pi^-\pi^-$). Following common usage, LF means lepton family violation *and not* lepton number violation (e.g. $\tau^- \to e^-\pi^+\pi^-$).

Mode	type	limit	CL	#
$e^-\gamma$	LF	$< 1.1\ \times 10^{-4}$	CL=90%	888
$\mu^-\gamma$	LF	$< 4.2\ \times 10^{-6}$	CL=90%	885
$e^-\pi^0$	LF	$< 1.4\ \times 10^{-4}$	CL=90%	883
$\mu^-\pi^0$	LF	$< 4.4\ \times 10^{-5}$	CL=90%	880
e^-K^0	LF	$< 1.3\ \times 10^{-3}$	CL=90%	819
μ^-K^0	LF	$< 1.0\ \times 10^{-3}$	CL=90%	815
$e^-\eta$	LF	$< 6.3\ \times 10^{-5}$	CL=90%	804
$\mu^-\eta$	LF	$< 7.3\ \times 10^{-5}$	CL=90%	800
$e^-\rho^0$	LF	$< 4.2\ \times 10^{-6}$	CL=90%	722
$\mu^-\rho^0$	LF	$< 5.7\ \times 10^{-6}$	CL=90%	718
$e^-K^*(892)^0$	LF	$< 6.3\ \times 10^{-6}$	CL=90%	663
$\mu^-K^*(892)^0$	LF	$< 9.4\ \times 10^{-6}$	CL=90%	657
$\pi^-\gamma$	L	$< 2.8\ \times 10^{-4}$	CL=90%	883
$\pi^-\pi^0$	L	$< 3.7\ \times 10^{-4}$	CL=90%	878
$e^-e^+e^-$	LF	$< 3.3\ \times 10^{-6}$	CL=90%	888
$e^-\mu^+\mu^-$	LF	$< 3.6\ \times 10^{-6}$	CL=90%	882
$e^+\mu^-\mu^-$	L	$< 3.5\ \times 10^{-6}$	CL=90%	882
$\mu^-e^+e^-$	LF	$< 3.4\ \times 10^{-6}$	CL=90%	885
$\mu^+e^-e^-$	L	$< 3.4\ \times 10^{-6}$	CL=90%	885
$\mu^-\mu^+\mu^-$	LF	$< 1.9\ \times 10^{-6}$	CL=90%	873
$e^-\pi^+\pi^-$	LF	$< 4.4\ \times 10^{-6}$	CL=90%	877
$e^+\pi^-\pi^-$	L	$< 4.4\ \times 10^{-6}$	CL=90%	877
$\mu^-\pi^+\pi^-$	LF	$< 7.4\ \times 10^{-6}$	CL=90%	866
$\mu^+\pi^-\pi^-$	L	$< 6.9\ \times 10^{-6}$	CL=90%	866
$e^-\pi^+K^-$	LF	$< 7.7\ \times 10^{-6}$	CL=90%	813
$e^-\pi^-K^+$	LF	$< 4.6\ \times 10^{-6}$	CL=90%	813
$e^+\pi^-K^-$	L	$< 4.5\ \times 10^{-6}$	CL=90%	813
$\mu^-\pi^+K^-$	LF	$< 8.7\ \times 10^{-6}$	CL=90%	800
$\mu^-\pi^-K^+$	LF	$< 1.5\ \times 10^{-5}$	CL=90%	800
$\mu^+\pi^-K^-$	L	$< 2.0\ \times 10^{-5}$	CL=90%	800
$\overline{p}\gamma$	L,B	$< 2.9\ \times 10^{-4}$	CL=90%	641
$\overline{p}\pi^0$	L,B	$< 6.6\ \times 10^{-4}$	CL=90%	632
$\overline{p}\eta$	L,B	$< 1.30\ \times 10^{-3}$	CL=90%	475
$e^-\overline{K}^*(892)^0$	LF	$< 1.1\ \times 10^{-5}$	CL=90%	663
$\mu^-\overline{K}^*(892)^0$	LF	$< 8.7\ \times 10^{-6}$	CL=90%	657
e^- light boson	LF	$< 2.7\ \times 10^{-3}$	CL=95%	–
μ^- light boson	LF	$< 5\ \times 10^{-3}$	CL=95%	–

Heavy Charged Lepton Searches

$L^\pm$ – charged lepton
Mass $m > 42.7$ GeV, CL = 95% $m_\nu \approx 0$

$L^\pm$ – stable charged heavy lepton
Mass $m > 42.8$ GeV, CL = 95%

Neutrinos

See the Particle Listings for a Note giving details of neutrinos, masses, mixing, and the status of experimental searches.

ν_e $J = \frac{1}{2}$

Mass m: Unexplained effects have resulted in significantly negative m^2 in the new, precise tritium beta decay experiments. It is felt that a real neutrino mass as large as 10–15 eV would cause observable spectral distortions even in the presence of the end-point count excesses.
Mean life/mass, $\tau/m_{\nu_e} > 300$ s/eV, CL = 90%
Magnetic moment $\mu < 1.8 \times 10^{-10}\,\mu_B$, CL = 90%

ν_μ $J = \frac{1}{2}$

Mass $m < 0.17$ MeV, CL = 90%
Mean life/mass, $\tau/m_{\nu_\mu} > 15.4$ s/eV, CL = 90%
Magnetic moment $\mu < 7.4 \times 10^{-10}\,\mu_B$, CL = 90%

ν_τ $J = \frac{1}{2}$

     Mass $m <$ 24 MeV, CL = 95%
     Magnetic moment $\mu < 5.4 \times 10^{-7}$ μ_B, CL = 90%

Number of Light Neutrino Types

(including ν_e, ν_μ, and ν_τ)
     Number $N = 2.991 \pm 0.016$ (Standard Model fits to LEP data)
     Number $N = 3.09 \pm 0.13$ (Direct measurement of invisible Z
     width)

Massive Neutrinos and Lepton Mixing, Searches for

For excited leptons, see Compositeness Limits below.

See the Particle Listings for a Note giving details of neutrinos, masses, mixing, and the status of experimental searches.

No direct, uncontested evidence for massive neutrinos or lepton mixing has been obtained. Sample limits are:
     Mass $m >$ 45.0, CL = 95%   (Dirac)
     Mass $m >$ 39.5, CL = 95%   (Majorana)

ν oscillation: $\nu_\mu \to \nu_e$ (θ = mixing angle)
     Mass $m >$ 19.6 GeV, CL = 95%   (all $|U_{\ell j}|^2$) (Dirac)
     Mass $m >$ 45.7 GeV   or $m < 25$, CL = 95% ($|U_{\ell j}|^2 > 10^{-13}$
     (Dirac)

ν oscillation: $\overline{\nu}_e \not\to \overline{\nu}_e$
     $\Delta(m^2) <$ 0.0075 eV^2, CL = 90%   (if $\sin^2 2\theta = 1$)
     $\sin^2 2\theta <$ 0.02, CL = 90%   (if $\Delta(m^2)$ is large)

ν oscillation: $\nu_\mu \to \nu_e$ (θ = mixing angle)
     $\Delta(m^2) <$ 0.09 eV^2, CL = 90%   (if $\sin^2 2\theta = 1$)
     $\sin^2 2\theta <$ 2.5 $\times 10^{-3}$, CL = 90%   (if $\Delta(m^2)$ is large)

NOTES

In this Summary Table:

When a quantity has "(S = ...)" to its right, the error on the quantity has been enlarged by the "scale factor" S, defined as $S = \sqrt{\chi^2 / (N-1)}$, where N is the number of measurements used in calculating the quantity. We do this when $S > 1$, which often indicates that the measurements are inconsistent. When $S > 1.25$, we also show in the Particle Listings an ideogram of the measurements. For more about S, see the Introduction.

A decay momentum p is given for each decay mode. For a 2-body decay, p is the momentum of each decay product in the rest frame of the decaying particle. For a 3-or-more-body decay, p is the largest momentum any of the products can have in this frame.

[a] The uncertainty in the electron mass in unified atomic mass units (u) is ten times smaller than that given by the 1986 CODATA adjustment, quoted in the Table of Physical Contants (Section 1). The conversion to MeV via the factor 931.49432(28) MeV/u is more uncertain because of the electron charge uncertainty. Our value in MeV differs slightly from the 1986 CODATA result.

[b] This is the best "electron disappearance" limit. The best limit for the mode $e^- \to \nu\gamma$ is $> 2.35 \times 10^{25}$ yr (CL=68%).

[c] The muon mass is most precisely known in u (unified atomic mass units). The conversion factor to MeV via the factor 931.49432(28) MeV/u is more uncertain because of the electron charge uncertainty.

[d] See the "Note on Muon Decay Parameters" in the μ Particle Listings for definitions and details.

[e] P_μ is the longitudinal polarization of the muon from pion decay. In standard $V-A$ theory, $P_\mu = 1$ and $\rho = \delta = 3/4$.

[f] This only includes events with the γ energy > 10 MeV. Since the $e^- \overline{\nu}_e \nu_\mu$ and $e^- \overline{\nu}_e \nu_\mu \gamma$ modes cannot be clearly separated, we regard the latter mode as a subset of the former.

[g] See the μ Particle Listings for the energy limits used in this measurement.

[h] A test of additive vs. multiplicative lepton family number conservation.

[i] Basis mode for the τ.

QUARKS

The u-, d-, and s-quark masses are estimates of so-called "current-quark masses," in a mass-independent subtraction scheme such as $\overline{MS}$ at a scale $\mu \approx 1$ GeV. The c- and b-quark masses are estimated from charmonium, bottomonium, D, and B masses. They are the "running" masses in the $\overline{MS}$ scheme. These can be different from the heavy quark masses obtained in potential models.

u $I(J^P) = \frac{1}{2}(\frac{1}{2}^+)$

Mass $m = 2$ to 8 MeV [a]   Charge $= \frac{2}{3} e$ $I_z = +\frac{1}{2}$
$m_u/m_d = 0.25$ to 0.70

d $I(J^P) = \frac{1}{2}(\frac{1}{2}^+)$

Mass $m = 5$ to 15 MeV [a]   Charge $= -\frac{1}{3} e$ $I_z = -\frac{1}{2}$
$m_s/m_d = 17$ to 25

s $I(J^P) = 0(\frac{1}{2}^+)$

Mass $m = 100$ to 300 MeV [a]   Charge $= -\frac{1}{3} e$ Strangeness $= -1$
$(m_s - (m_u + m_d)/2)/(m_d - m_u) = 34$ to 51

c $I(J^P) = 0(\frac{1}{2}^+)$

Mass $m = 1.0$ to 1.6 GeV   Charge $= \frac{2}{3} e$ Charm $= +1$

b $I(J^P) = 0(\frac{1}{2}^+)$

Mass $m = 4.1$ to 4.5 GeV   Charge $= -\frac{1}{3} e$ Bottom $= -1$

t $I(J^P) = 0(\frac{1}{2}^+)$

Charge $= \frac{2}{3} e$ Top $= +1$

Mass $m = 180 \pm 12$ GeV   (direct observation of top events)
Mass $m = 179 \pm 8^{+17}_{-20}$ GeV   (Standard Model electroweak fit)

b' (4th Generation) Quark, Searches for

Mass $m > 85$ GeV, CL $= 95\%$ ($p\bar{p}$, charged current decays)
Mass $m > 46.0$ GeV, CL $= 95\%$ ($e^+ e^-$, all decays)

Free Quark Searches

All searches since 1977 have had negative results.

NOTES

[a] The ratios m_u/m_d and m_s/m_d are extracted from pion and kaon masses using chiral symmetry. The estimates of u and d masses are not without controversy and remain under active investigation. Within the literature there are even suggestions that the u quark could be essentially massless. The s-quark mass is estimated from SU(3) splittings in hadron masses.

LIGHT UNFLAVORED MESONS
$(S = C = B = 0)$

For $I = 1$ (π, b, ρ, a): $u\bar{d}$, $(u\bar{u}-d\bar{d})/\sqrt{2}$, $d\bar{u}$;
for $I = 0$ $(\eta, \eta', h, h', \omega, \phi, f, f')$: $c_1(u\bar{u} + d\bar{d}) + c_2(s\bar{s})$

$\pi^\pm$ $I^G(J^P) = 1^-(0^-)$

Mass $m = 139.56995 \pm 0.00035$ MeV
Mean life $\tau = (2.6033 \pm 0.0005) \times 10^{-8}$ s $(S = 1.2)$
$c\tau = 7.8045$ m

$\pi^\pm \to \ell^\pm \nu \gamma$ form factors [a]

$F_V = 0.017 \pm 0.008$
$F_A = 0.0116 \pm 0.0016$ $(S = 1.3)$
$R = 0.059^{+0.009}_{-0.008}$

π^- modes are charge conjugates of the modes below.

π^+ DECAY MODES		Fraction (Γ_i/Γ)		p (MeV/c)
$\mu^+\nu_\mu$	[b]	(99.98770 ± 0.00004) %		30
$\mu^+\nu_\mu\gamma$	[c]	$(1.24 \pm 0.25) \times 10^{-4}$		30
$e^+\nu_e$	[b]	$(1.230 \pm 0.004) \times 10^{-4}$		70
$e^+\nu_e\gamma$	[c]	$(1.61 \pm 0.23) \times 10^{-7}$		70
$e^+\nu_e\pi^0$		$(1.025 \pm 0.034) \times 10^{-8}$		4
$e^+\nu_e e^+ e^-$		$(3.2 \pm 0.5) \times 10^{-9}$		70
$e^+\nu_e\nu\bar{\nu}$		$< 5 \times 10^{-6}$	90%	70

Lepton Family number (LF) or Lepton number (L) violating modes

$\mu^+\bar{\nu}_e$	L	[d] $< 1.5 \times 10^{-3}$	90%	30
$\mu^+\nu_e$	LF	[d] $< 8.0 \times 10^{-3}$	90%	30
$\mu^-e^+e^+\nu$	LF	$< 1.6 \times 10^{-6}$	90%	30

π^0 $I^G(J^{PC}) = 1^-(0^{-+})$

Mass $m = 134.9764 \pm 0.0006$ MeV
$m_{\pi^\pm} - m_{\pi^0} = 4.5936 \pm 0.0005$ MeV
Mean life $\tau = (8.4 \pm 0.6) \times 10^{-17}$ s $(S = 3.0)$
$c\tau = 25.1$ nm

π^0 DECAY MODES		Fraction (Γ_i/Γ)	Scale factor/ Confidence level	p (MeV/c)
2γ		(98.798 ± 0.032) %	$S=1.1$	67
$e^+e^-\gamma$		(1.198 ± 0.032) %	$S=1.1$	67
γ positronium		$(1.82 \pm 0.29) \times 10^{-9}$		67
$e^+e^+e^-e^-$		$(3.14 \pm 0.30) \times 10^{-5}$		67
e^+e^-		$(7.5 \pm 2.0) \times 10^{-8}$		67
4γ		$< 2 \times 10^{-8}$	CL=90%	67
$\nu\bar{\nu}$	[e]	$< 8.3 \times 10^{-7}$	CL=90%	67
$\nu_e\bar{\nu}_e$		$< 1.7 \times 10^{-6}$	CL=90%	67
$\nu_\mu\bar{\nu}_\mu$		$< 3.1 \times 10^{-6}$	CL=90%	67
$\nu_\tau\bar{\nu}_\tau$		$< 2.1 \times 10^{-6}$	CL=90%	67

Charge conjugation (C) or Lepton Family number (LF) violating modes

3γ	C	$< 3.1 \times 10^{-8}$	CL=90%	67
$\mu^+e^- + e^-\mu^+$	LF	$< 1.72 \times 10^{-8}$	CL=90%	26

η $I^G(J^{PC}) = 0^+(0^{-+})$

Mass $m = 547.45 \pm 0.19$ MeV $(S = 1.6)$
Full width $\Gamma = 1.18 \pm 0.11$ keV [f] $(S = 1.8)$

C-nonconserving decay parameters [g]

$\pi^+\pi^-\pi^0$	Left-right asymmetry $= (0.09 \pm 0.17) \times 10^{-2}$
$\pi^+\pi^-\pi^0$	Sextant asymmetry $= (0.18 \pm 0.16) \times 10^{-2}$
$\pi^+\pi^-\pi^0$	Quadrant asymmetry $= (-0.17 \pm 0.17) \times 10^{-2}$
$\pi^+\pi^-\gamma$	Left-right asymmetry $= (0.9 \pm 0.4) \times 10^{-2}$
$\pi^+\pi^-\gamma$	β (D-wave) $= 0.05 \pm 0.06$ $(S = 1.5)$

η DECAY MODES		Fraction (Γ_i/Γ)	Scale factor/ Confidence level	p (MeV/c)
neutral modes		(71.4 ± 0.6) %	$S=1.3$	–
2γ	[f]	(39.25 ± 0.31) %	$S=1.3$	274
$3\pi^0$		(32.1 ± 0.4) %	$S=1.2$	180
$\pi^0 2\gamma$		$(7.1 \pm 1.4) \times 10^{-4}$		258
other neutral modes		< 2.8 %	CL=90%	–

charged modes		(28.6 ± 0.6) %	$S=1.3$	–
$\pi^+\pi^-\pi^0$		(23.2 ± 0.5) %	$S=1.3$	175
$\pi^+\pi^-\gamma$		(4.78 ± 0.12) %	$S=1.2$	236
$e^+e^-\gamma$		$(4.9 \pm 1.1) \times 10^{-3}$		274
$\mu^+\mu^-\gamma$		$(3.1 \pm 0.4) \times 10^{-4}$		253
e^+e^-		$< 3 \times 10^{-4}$	CL=90%	274
$\mu^+\mu^-$		$(5.8 \pm 0.8) \times 10^{-6}$		253
$\pi^+\pi^-e^+e^-$		$(1.3 ^{+1.2}_{-0.8}) \times 10^{-3}$		236
$\pi^+\pi^- 2\gamma$		$< 2.1 \times 10^{-3}$		236
$\pi^+\pi^-\pi^0\gamma$		$< 6 \times 10^{-4}$	CL=90%	175
$\pi^0\mu^+\mu^-\gamma$		$< 3 \times 10^{-6}$	CL=90%	211

Charge conjugation (C), Parity (P), or Charge conjugation $\times$ Parity (CP) violating modes

$\pi^+\pi^-$	P,CP	$< 1.5 \times 10^{-3}$		236
3γ	C	$< 5 \times 10^{-4}$	CL=95%	274
$\pi^0 e^+e^-$	C	[h] $< 4 \times 10^{-5}$	CL=90%	258
$\pi^0\mu^+\mu^-$	C	[h] $< 5 \times 10^{-6}$	CL=90%	211

$f_0(400-1200)$ [i] or σ $I^G(J^{PC}) = 0^+(0^{++})$

The interpretation of this entry as a particle is controversial. See the "Note on scalar mesons" in the Particle Listings under the $f_0(1370)$.

Mass $m = (400-1200)$ MeV
Full width $\Gamma = (600-1000)$ MeV

$f_0(400-1200)$ DECAY MODES	Fraction (Γ_i/Γ)	p (MeV/c)
$\pi\pi$	dominant	–
$\gamma\gamma$	seen	–

$\rho(770)$ [j] $I^G(J^{PC}) = 1^+(1^{--})$

Mass $m = 768.5 \pm 0.6$ MeV $(S = 1.2)$
Full width $\Gamma = 150.7 \pm 1.2$ MeV
$\Gamma_{ee} = 6.77 \pm 0.32$ keV

$\rho(770)$ DECAY MODES		Fraction (Γ_i/Γ)	Scale factor/ Confidence level	p (MeV/c)
$\pi\pi$		~ 100 %		358
$\rho(770)^\pm$ decays				
$\pi^\pm\gamma$		$(4.5 \pm 0.5) \times 10^{-4}$	$S=2.2$	372
$\pi^\pm\eta$		$< 6 \times 10^{-3}$	CL=84%	146
$\pi^\pm\pi^+\pi^-$		$< 2.0 \times 10^{-3}$	CL=84%	249
$\rho(770)^0$ decays				
$\pi^+\pi^-\gamma$		$(9.9 \pm 1.6) \times 10^{-3}$		358
$\pi^0\gamma$		$(7.9 \pm 2.0) \times 10^{-4}$		372
$\eta\gamma$		$(3.8 \pm 0.7) \times 10^{-4}$		189
$\mu^+\mu^-$	[k]	$(4.60 \pm 0.28) \times 10^{-5}$		369
e^+e^-	[k]	$(4.48 \pm 0.22) \times 10^{-5}$		384
$\pi^+\pi^-\pi^0$		$< 1.2 \times 10^{-4}$	CL=90%	319
$\pi^+\pi^-\pi^+\pi^-$		$< 2 \times 10^{-4}$	CL=90%	246
$\pi^+\pi^-\pi^0\pi^0$		$< 4 \times 10^{-5}$	CL=90%	252

$\omega(782)$ $I^G(J^{PC}) = 0^-(1^{--})$

Mass $m = 781.94 \pm 0.12$ MeV $(S = 1.5)$
Full width $\Gamma = 8.43 \pm 0.10$ MeV
$\Gamma_{ee} = 0.60 \pm 0.02$ keV

$\omega(782)$ DECAY MODES	Fraction (Γ_i/Γ)	Confidence level	p (MeV/c)
$\pi^+\pi^-\pi^0$	(88.8 ± 0.7) %		327
$\pi^0\gamma$	(8.5 ± 0.5) %		379
$\pi^+\pi^-$	(2.21 ± 0.30) %		365
neutrals (excluding $\pi^0\gamma$)	$(5.3 ^{+8.7}_{-3.5}) \times 10^{-3}$		–
$\eta\gamma$	$(8.3 \pm 2.1) \times 10^{-4}$		199
$\pi^0 e^+e^-$	$(5.9 \pm 1.9) \times 10^{-4}$		379
$\pi^0\mu^+\mu^-$	$(9.6 \pm 2.3) \times 10^{-5}$		349

$e^+ e^-$	$(7.15 \pm 0.19) \times 10^{-5}$			391
$\pi^+ \pi^- \pi^0 \pi^0$	< 2	%	90%	261
$\pi^+ \pi^- \gamma$	< 3.6	$\times 10^{-3}$	95%	365
$\pi^+ \pi^- \pi^+ \pi^-$	< 1	$\times 10^{-3}$	90%	256
$\pi^0 \pi^0 \gamma$	$(7.2 \pm 2.5) \times 10^{-5}$			367
$\mu^+ \mu^-$	< 1.8	$\times 10^{-4}$	90%	376
3γ	< 2	$\times 10^{-4}$	90%	391

Charge conjugation (C)

$\eta \pi^0$	C	< 1	$\times 10^{-3}$	90%	162
$3\pi^0$	C	< 3	$\times 10^{-4}$	90%	329

$\eta'(958)$ $\qquad I^G(J^{PC}) = 0^+(0^{-+})$

Mass $m = 957.77 \pm 0.14$ MeV
Full width $\Gamma = 0.201 \pm 0.016$ MeV (S = 1.3)

$\eta'(958)$ DECAY MODES	Fraction (Γ_i/Γ)	Scale factor/ Confidence level	p (MeV/c)
$\pi^+ \pi^- \eta$	(43.7 ± 1.5) %	S=1.1	232
$\rho^0 \gamma$	(30.2 ± 1.3) %	S=1.1	169
$\pi^0 \pi^0 \eta$	(20.8 ± 1.3) %	S=1.2	239
$\omega \gamma$	(3.02 ± 0.30) %		160
$\gamma \gamma$	(2.12 ± 0.13) %	S=1.2	479
$3\pi^0$	$(1.55 \pm 0.26) \times 10^{-3}$		430
$\mu^+ \mu^- \gamma$	$(1.04 \pm 0.26) \times 10^{-4}$		467
$\pi^+ \pi^- \pi^0$	< 5 %	CL=90%	427
$\pi^0 \rho^0$	< 4 %	CL=90%	118
$\pi^+ \pi^-$	< 2 %	CL=90%	458
$\pi^0 e^+ e^-$	< 1.3 %	CL=90%	469
$\eta e^+ e^-$	< 1.1 %	CL=90%	322
$\pi^+ \pi^- \pi^+ \pi^-$	< 1 %	CL=90%	372
$\pi^+ \pi^+ \pi^- \pi^-$ neutrals	< 1 %	CL=95%	–
$\pi^+ \pi^+ \pi^- \pi^- \pi^0$	< 1 %	CL=90%	298
6π	< 1 %	CL=90%	189
$\pi^+ \pi^- e^+ e^-$	< 6 $\times 10^{-3}$	CL=90%	458
$\pi^0 \pi^0$	< 9 $\times 10^{-4}$	CL=90%	459
$\pi^0 \gamma \gamma$	< 8 $\times 10^{-4}$	CL=90%	469
$4\pi^0$	< 5 $\times 10^{-4}$	CL=90%	379
3γ	< 1.0 $\times 10^{-4}$	CL=90%	479
$\mu^+ \mu^- \pi^0$	< 6.0 $\times 10^{-5}$	CL=90%	445
$\mu^+ \mu^- \eta$	< 1.5 $\times 10^{-5}$	CL=90%	274
$e^+ e^-$	< 2.1 $\times 10^{-7}$	CL=90%	479

$f_0(980)$ [l] $\qquad I^G(J^{PC}) = 0^+(0^{++})$

Mass $m = 980 \pm 10$ MeV
Full width $\Gamma = 40$ to 100 MeV

$f_0(980)$ DECAY MODES	Fraction (Γ_i/Γ)	Confidence level	p (MeV/c)
$\pi \pi$	(78.1 ± 2.4) %		470
$K \overline{K}$	(21.9 ± 2.4) %		–
$\gamma \gamma$	$(1.19 \pm 0.33) \times 10^{-5}$		490
$e^+ e^-$	< 3 $\times 10^{-7}$	90%	490

$a_0(980)$ [l] $\qquad I^G(J^{PC}) = 1^-(0^{++})$

Mass $m = 983.5 \pm 0.9$ MeV
Full width $\Gamma = 50$ to 100 MeV

$a_0(980)$ DECAY MODES	Fraction (Γ_i/Γ)		p (MeV/c)
$\eta \pi$	dominant		321
$K \overline{K}$	seen		–
$\gamma \gamma$	seen		492

$\phi(1020)$ $\qquad I^G(J^{PC}) = 0^-(1^{--})$

Mass $m = 1019.413 \pm 0.008$ MeV
Full width $\Gamma = 4.43 \pm 0.05$ MeV
$\Gamma_{ee} = 1.37 \pm 0.05$ keV

$\phi(1020)$ DECAY MODES	Fraction (Γ_i/Γ)	Scale factor/ Confidence level	p (MeV/c)
$K^+ K^-$	(49.1 ± 0.6) %	S=1.2	127
$K_L^0 K_S^0$	(34.1 ± 0.5) %	S=1.1	110
$\rho \pi$	(12.9 ± 0.7) %		181

$\pi^+ \pi^- \pi^0$	(2.7 ± 0.9) %	S=1.1	462
$\eta \gamma$	(1.26 ± 0.06) %	S=1.1	363
$\pi^0 \gamma$	$(1.31 \pm 0.13) \times 10^{-3}$		501
$e^+ e^-$	$(3.00 \pm 0.06) \times 10^{-4}$	S=1.1	510
$\mu^+ \mu^-$	$(2.48 \pm 0.34) \times 10^{-4}$		499
$\eta e^+ e^-$	$(1.3 ^{+0.8}_{-0.6}) \times 10^{-4}$		363
$\pi^+ \pi^-$	$(8 ^{+5}_{-4}) \times 10^{-5}$	S=1.5	490
$\omega \gamma$	< 5 %	CL=84%	210
$\rho \gamma$	< 2 %	CL=84%	219
$\pi^+ \pi^- \gamma$	< 7 $\times 10^{-3}$	CL=90%	490
$\pi^0 \pi^0 \gamma$	< 1 $\times 10^{-3}$	CL=90%	492
$\pi^+ \pi^- \pi^+ \pi^-$	< 8.7 $\times 10^{-4}$	CL=90%	410
$\eta'(958) \gamma$	< 4.1 $\times 10^{-4}$	CL=90%	60
$\pi^+ \pi^+ \pi^- \pi^- \pi^0$	< 1.5 $\times 10^{-4}$	CL=95%	341
$\pi^0 e^+ e^-$	< 1.2 $\times 10^{-4}$	CL=90%	501
$a_0(980) \gamma$	< 5 $\times 10^{-3}$	CL=90%	36

$h_1(1170)$ $\qquad I^G(J^{PC}) = 0^-(1^{+-})$

Mass $m = 1170 \pm 20$ MeV
Full width $\Gamma = 360 \pm 40$ MeV

$h_1(1170)$ DECAY MODES	Fraction (Γ_i/Γ)	p (MeV/c)
$\rho \pi$	seen	310

$b_1(1235)$ $\qquad I^G(J^{PC}) = 1^+(1^{+-})$

Mass $m = 1231 \pm 10$ MeV [l]
Full width $\Gamma = 142 \pm 8$ MeV (S = 1.1)

$b_1(1235)$ DECAY MODES	Fraction (Γ_i/Γ)	Confidence level	p (MeV/c)
$\omega \pi$	dominant		348
$\quad$ [D/S amplitude ratio = 0.26 ± 0.04]			
$\pi^\pm \gamma$	$(1.6 \pm 0.4) \times 10^{-3}$		608
$\eta \rho$	seen		–
$\pi^+ \pi^+ \pi^- \pi^0$	< 50 %	84%	536
$(K \overline{K})^\pm \pi^0$	< 8 %	90%	248
$K_S^0 K_S^0 \pi^\pm$	< 6 %	90%	238
$K_S^0 K_S^0 \pi^\pm$	< 2 %	90%	238
$\pi \phi$	< 1.5 %	84%	146

$a_1(1260)$ [m] $\qquad I^G(J^{PC}) = 1^-(1^{++})$

Mass $m = 1230 \pm 40$ MeV [l]
Full width $\Gamma \sim 400$ MeV

$a_1(1260)$ DECAY MODES	Fraction (Γ_i/Γ)	p (MeV/c)
$\rho \pi$	dominant	356
$\pi \gamma$	seen	607
$K \overline{K}^*(892)$	possibly seen	–

$f_2(1270)$ $\qquad I^G(J^{PC}) = 0^+(2^{++})$

Mass $m = 1275 \pm 5$ MeV [l]
Full width $\Gamma = 185 \pm 20$ MeV [l]

$f_2(1270)$ DECAY MODES	Fraction (Γ_i/Γ)	Scale factor/ Confidence level	p (MeV/c)
$\pi \pi$	$(84.7 ^{+2.6}_{-1.2})$ %	S=1.3	622
$\pi^+ \pi^- 2\pi^0$	$(7.2 ^{+1.4}_{-2.9})$ %	S=1.3	562
$K \overline{K}$	(4.6 ± 0.5) %	S=2.8	403
$2\pi^+ 2\pi^-$	(2.8 ± 0.4) %	S=1.2	559
$\eta \eta$	$(4.5 \pm 1.0) \times 10^{-3}$	S=2.4	327
$4\pi^0$	$(3.0 \pm 1.0) \times 10^{-3}$		564
$\gamma \gamma$	$(1.32 ^{+0.18}_{-0.16}) \times 10^{-5}$		637
$\eta \pi \pi$	< 8 $\times 10^{-3}$	CL=95%	475
$K^0 K^- \pi^+ +$ c.c.	< 3.4 $\times 10^{-3}$	CL=95%	293
$e^+ e^-$	< 9 $\times 10^{-9}$	CL=90%	637

$f_1(1285)$ — $I^G(J^{PC}) = 0^+(1^{++})$

Mass $m = 1282.2 \pm 0.7$ MeV [l] (S = 1.7)
Full width $\Gamma = 24.8 \pm 1.3$ MeV [l] (S = 1.3)

$f_1(1285)$ DECAY MODES	Fraction (Γ_i/Γ)	Scale factor/ Confidence level	p (MeV/c)
4π	(29 ± 6) %		563
$\pi^0\pi^0\pi^+\pi^-$	$(15 \, {}^{+\,9}_{-\,8})$ %	S=1.1	566
$2\pi^+2\pi^-$	(15 ± 6) %		563
$\rho^0\pi^+\pi^-$	dominates $2\pi^+2\pi^-$		340
$4\pi^0$	$< 7 \quad \times 10^{-4}$	CL=90%	568
$\eta\pi\pi$	(54 ± 15) %		479
$a_0(980)\pi$ [ignoring $a_0(980) \to$ $K\overline{K}$]	(44 ± 7) %	S=1.1	234
$\eta\pi\pi$ [excluding $a_0(980)\pi$]	$(10 \, {}^{+\,7}_{-\,6})$ %	S=1.1	–
$K\overline{K}\pi$	(9.7 ± 1.6) %	S=1.2	308
$K\overline{K}^*(892)$	not seen		–
$\gamma\rho^0$	(6.6 ± 1.3) %	S=1.5	410
$\phi\gamma$	$(8.0 \pm 3.1) \times 10^{-4}$		236

$\eta(1295)$ — $I^G(J^{PC}) = 0^+(0^{-+})$

Mass $m = 1295 \pm 4$ MeV
Full width $\Gamma = 53 \pm 6$ MeV

$\eta(1295)$ DECAY MODES	Fraction (Γ_i/Γ)	p (MeV/c)
$\eta\pi^+\pi^-$	seen	488
$a_0(980)\pi$	seen	245

$\pi(1300)$ — $I^G(J^{PC}) = 1^-(0^{-+})$

Mass $m = 1300 \pm 100$ MeV [l]
Full width $\Gamma = 200$ to 600 MeV

$\pi(1300)$ DECAY MODES	Fraction (Γ_i/Γ)	p (MeV/c)
$\rho\pi$	seen	406
$\pi(\pi\pi)_{S\text{-wave}}$	seen	612

$a_2(1320)$ — $I^G(J^{PC}) = 1^-(2^{++})$

Mass $m = 1318.1 \pm 0.7$ MeV (S = 1.2)
Full width $\Gamma = 107 \pm 5$ MeV [l] ($K^\pm K^0_S$ and $\eta\pi$ modes)

$a_2(1320)$ DECAY MODES	Fraction (Γ_i/Γ)	Scale factor/ Confidence level	p (MeV/c)
$\rho\pi$	(70.1 ± 2.7) %	S=1.2	419
$\eta\pi$	(14.5 ± 1.2) %		535
$\omega\pi\pi$	(10.6 ± 3.2) %	S=1.3	362
$K\overline{K}$	(4.9 ± 0.8) %		437
$\eta'(958)\pi$	$(5.7 \pm 1.1) \times 10^{-3}$		287
$\pi^\pm\gamma$	$(2.8 \pm 0.6) \times 10^{-3}$		652
$\gamma\gamma$	$(9.7 \pm 1.0) \times 10^{-6}$		659
$\pi^+\pi^-\pi^-$	< 8 %	CL=90%	621
e^+e^-	$< 2.3 \times 10^{-7}$	CL=90%	659

$f_0(1370)$ [l] was $f_0(1300)$ — $I^G(J^{PC}) = 0^+(0^{++})$

Mass $m = 1200$ to 1500 MeV
Full width $\Gamma = 300$ to 500 MeV

In two-particle decay modes the $\pi\pi$ decay is dominant. We include here the resonance observed in 4π under the same entry as the one decaying to 2 pseudoscalars. See also the minireview under non-$q\bar{q}$ candidates.

$f_0(1370)$ DECAY MODES	Fraction (Γ_i/Γ)	p (MeV/c)
$\pi\pi$	seen	–
4π	seen	–
$2\pi^+2\pi^-$	seen	–
$\pi^+\pi^-2\pi^0$	seen	–

$\eta\eta$	seen	–
$K\overline{K}$	seen	–
$\gamma\gamma$	seen	–
e^+e^-	not seen	–

$f_1(1420)$ [n] — $I^G(J^{PC}) = 0^+(1^{++})$

Mass $m = 1426.8 \pm 2.3$ MeV (S = 1.3)
Full width $\Gamma = 53 \pm 5$ MeV

$f_1(1420)$ DECAY MODES	Fraction (Γ_i/Γ)	p (MeV/c)
$K\overline{K}\pi$	dominant	439
$\eta\pi\pi$	possibly seen	571

$\omega(1420)$ [o] — $I^G(J^{PC}) = 0^-(1^{--})$

Mass $m = 1419 \pm 31$ MeV
Full width $\Gamma = 174 \pm 60$ MeV

$\omega(1420)$ DECAY MODES	Fraction (Γ_i/Γ)	p (MeV/c)
$\rho\pi$	dominant	488

$\eta(1440)$ [p] — $I^G(J^{PC}) = 0^+(0^{-+})$

Mass $m = 1415 \pm 10$ MeV [l]
Full width $\Gamma = 60 \pm 20$ MeV [l]

$\eta(1440)$ DECAY MODES	Fraction (Γ_i/Γ)	p (MeV/c)
$K\overline{K}\pi$	seen	429
$\eta\pi\pi$	seen	564
$a_0(980)\pi$	seen	347
4π	seen	637

$\rho(1450)$ [q] — $I^G(J^{PC}) = 1^+(1^{--})$

Mass $m = 1465 \pm 25$ MeV [l]
Full width $\Gamma = 310 \pm 60$ MeV [l]

$\rho(1450)$ DECAY MODES	Fraction (Γ_i/Γ)	Confidence level	p (MeV/c)
$\pi\pi$	seen		719
4π	seen		665
e^+e^-	seen		732
$\eta\rho$	<4 %		317
$\omega\pi$	<2.0 %	95%	512
$\phi\pi$	<1 %		358
$K\overline{K}$	$<1.6 \times 10^{-3}$	95%	541

$f_0(1500)$ [r] was $f_0(1525)$ and $f_0(1590)$ — $I^G(J^{PC}) = 0^+(0^{++})$

Mass $m = 1503 \pm 11$ MeV
Full width $\Gamma = 120 \pm 19$ MeV

$f_0(1500)$ DECAY MODES	Fraction (Γ_i/Γ)	p (MeV/c)
$\eta\eta'(958)$	seen	–
$\eta\eta$	seen	515
$4\pi^0$	seen	690
$\pi^0\pi^0$	seen	739
$2\pi^+2\pi^-$	seen	686

$f_1(1510)$ — $I^G(J^{PC}) = 0^+(1^{++})$

Mass $m = 1512 \pm 4$ MeV
Full width $\Gamma = 35 \pm 15$ MeV

$f_1(1510)$ DECAY MODES	Fraction (Γ_i/Γ)	p (MeV/c)
$K\overline{K}^*(892)+$ c.c.	seen	292

$f'_2(1525)$ $I^G(J^{PC}) = 0^+(2^{++})$

Mass $m = 1525 \pm 5$ MeV [l]
Full width $\Gamma = 76 \pm 10$ MeV [l]

$f'_2(1525)$ DECAY MODES	Fraction (Γ_i/Γ)	p (MeV/c)
$K\overline{K}$	(88.8 ± 3.1) %	581
$\eta\eta$	(10.3 ± 3.1) %	531
$\pi\pi$	$(8.2 \pm 1.5) \times 10^{-3}$	750
$\gamma\gamma$	$(1.32\pm0.21) \times 10^{-6}$	763

$\omega(1600)$ [s] $I^G(J^{PC}) = 0^-(1^{--})$

Mass $m = 1649 \pm 24$ MeV   (S = 2.3)
Full width $\Gamma = 220 \pm 35$ MeV   (S = 1.6)

$\omega(1600)$ DECAY MODES	Fraction (Γ_i/Γ)	p (MeV/c)
$\rho\pi$	seen	637
$\omega\pi\pi$	seen	601
e^+e^-	seen	824

$\omega_3(1670)$ $I^G(J^{PC}) = 0^-(3^{--})$

Mass $m = 1667 \pm 4$ MeV
Full width $\Gamma = 168 \pm 10$ MeV [l]

$\omega_3(1670)$ DECAY MODES	Fraction (Γ_i/Γ)	p (MeV/c)
$\rho\pi$	seen	647
$\omega\pi\pi$	seen	614
$b_1(1235)\pi$	possibly seen	359

$\pi_2(1670)$ $I^G(J^{PC}) = 1^-(2^{-+})$

Mass $m = 1670 \pm 20$ MeV [l]
Full width $\Gamma = 258 \pm 18$ MeV [l]   (S = 1.7)
$\Gamma_{ee} = 1.35 \pm 0.26$ keV

$\pi_2(1670)$ DECAY MODES	Fraction (Γ_i/Γ)	p (MeV/c)
3π	(95.8 ± 1.4) %	806
$f_2(1270)\pi$	(56.2 ± 3.2) %	325
$\rho\pi$	(31 ± 4) %	649
$f_0(1370)\pi$	(8.7 ± 3.4) %	–
$K\overline{K}^*(892)+$ c.c.	(4.2 ± 1.4) %	453
$\gamma\gamma$	$(5.2\pm1.1) \times 10^{-6}$	835

$\phi(1680)$ $I^G(J^{PC}) = 0^-(1^{--})$

Mass $m = 1680 \pm 20$ MeV [l]
Full width $\Gamma = 150 \pm 50$ MeV [l]

$\phi(1680)$ DECAY MODES	Fraction (Γ_i/Γ)	p (MeV/c)
$K\overline{K}^*(892)+$ c.c.	dominant	463
$K^0_S K\pi$	seen	620
$K\overline{K}$	seen	681
e^+e^-	seen	840
$\omega\pi\pi$	not seen	622

$\rho_3(1690)$ $I^G(J^{PC}) = 1^+(3^{--})$

J^P from the 2π and $K\overline{K}$ modes.
Mass $m = 1691 \pm 5$ MeV [l]
Full width $\Gamma = 160 \pm 10$ MeV [l]   (S = 1.5)

$\rho_3(1690)$ DECAY MODES	Fraction (Γ_i/Γ)	Scale factor	p (MeV/c)
4π	(71.1 ± 1.9) %		788
$\pi^{\pm}\pi^+\pi^-\pi^0$	(67 ± 22) %		788
$\pi\pi$	(23.6 ± 1.3) %		834
$\omega\pi$	(16 ± 6) %		656
$K\overline{K}\pi$	(3.8 ± 1.2) %		628
$K\overline{K}$	(1.58 ± 0.26) %	1.2	686
$\eta\pi^+\pi^-$	seen		728

$\rho(1700)$ [q] $I^G(J^{PC}) = 1^+(1^{--})$

Mass $m = 1700 \pm 20$ MeV [l]   ($\eta\rho^0$ and $\pi^+\pi^-$ modes)
Full width $\Gamma = 235 \pm 50$ MeV [l]   ($\eta\rho^0$ and $\pi^+\pi^-$ modes)

$\rho(1700)$ DECAY MODES	Fraction (Γ_i/Γ)	p (MeV/c)
$\rho\pi\pi$	dominant	640
$\rho^0\pi^+\pi^-$	large	640
$\rho^{\pm}\pi^{\mp}\pi^0$	large	642
$2(\pi^+\pi^-)$	large	792
$\pi^+\pi^-$	seen	838
$K\overline{K}^*(892)+$ c.c.	seen	479
$\eta\rho$	seen	533
$K\overline{K}$	seen	692
e^+e^-	seen	850

$f_J(1710)$ [t] $I^G(J^{PC}) = 0^+(\text{even}^{++})$

Mass $m = 1697 \pm 4$ MeV   (S = 1.4)
Full width $\Gamma = 175 \pm 9$ MeV   (S = 1.7)

$f_J(1710)$ DECAY MODES	Fraction (Γ_i/Γ)	p (MeV/c)
$K\overline{K}$	seen	690
$\eta\eta$	seen	648
$\pi\pi$	seen	837

$\phi_3(1850)$ $I^G(J^{PC}) = 0^-(3^{--})$

Mass $m = 1854 \pm 7$ MeV
Full width $\Gamma = 87^{+28}_{-23}$ MeV   (S = 1.2)

$\phi_3(1850)$ DECAY MODES	Fraction (Γ_i/Γ)	p (MeV/c)
$K\overline{K}$	seen	785
$K\overline{K}^*(892)+$ c.c.	seen	602

$f_2(2010)$ $I^G(J^{PC}) = 0^+(2^{++})$

Seen by one group only.
Mass $m = 2011^{+60}_{-80}$ MeV
Full width $\Gamma = 202 \pm 60$ MeV

$f_2(2010)$ DECAY MODES	Fraction (Γ_i/Γ)	p (MeV/c)
$\phi\phi$	seen	–

$f_4(2050)$ $I^G(J^{PC}) = 0^+(4^{++})$

Mass $m = 2044 \pm 11$ MeV   (S = 1.4)
Full width $\Gamma = 208 \pm 13$ MeV   (S = 1.2)

$f_4(2050)$ DECAY MODES	Fraction (Γ_i/Γ)	p (MeV/c)
$\omega\omega$	(26 ± 6) %	658
$\pi\pi$	(17.0 ± 1.5) %	1012
$K\overline{K}$	$(6.8^{+3.4}_{-1.8}) \times 10^{-3}$	895
$\eta\eta$	$(2.1\pm0.8) \times 10^{-3}$	863
$4\pi^0$	< 1.2 %	977

$f_2(2300)$ $I^G(J^{PC}) = 0^+(2^{++})$

Mass $m = 2297 \pm 28$ MeV
Full width $\Gamma = 149 \pm 40$ MeV

$f_2(2300)$ DECAY MODES	Fraction (Γ_i/Γ)	p (MeV/c)
$\phi\phi$	seen	529

$f_2(2340)$

$$I^G(J^{PC}) = 0^+(2^{++})$$

Mass $m = 2339 \pm 60$ MeV
Full width $\Gamma = 319^{+80}_{-70}$ MeV

$f_2(2340)$ DECAY MODES	Fraction (Γ_i/Γ)	p (MeV/c)
$\phi\phi$	seen	573

STRANGE MESONS
$(S = \pm 1, C = B = 0)$
$K^+ = u\bar{s}$, $K^0 = d\bar{s}$, $\overline{K^0} = \bar{d}s$, $K^- = \bar{u}s$, similarly for K^*'s

---

$K^\pm$

$$I(J^P) = \tfrac{1}{2}(0^-)$$

Mass $m = 493.677 \pm 0.016$ MeV [u] (S = 2.8)
Mean life $\tau = (1.2386 \pm 0.0024) \times 10^{-8}$ s (S = 2.0)
$c\tau = 3.713$ m

Slope parameter g [v]

(See Particle Listings for quadratic coefficients)
$K^+ \to \pi^+\pi^+\pi^- = -0.2154 \pm 0.0035$ (S = 1.4)
$K^- \to \pi^-\pi^-\pi^+ = -0.217 \pm 0.007$ (S = 2.5)
$K^\pm \to \pi^\pm\pi^0\pi^0 = 0.594 \pm 0.019$ (S = 1.3)

$K^\pm$ decay form factors [a,w]

K^+_{e3} $\lambda_+ = 0.0286 \pm 0.0022$
$K^+_{\mu3}$ $\lambda_+ = 0.033 \pm 0.008$ (S = 1.6)
$K^+_{\mu3}$ $\lambda_0 = 0.004 \pm 0.007$ (S = 1.6)
K^+_{e3} $|f_S/f_+| = 0.084 \pm 0.023$ (S = 1.2)
K^+_{e3} $|f_T/f_+| = 0.38 \pm 0.11$ (S = 1.1)
$K^+_{\mu3}$ $|f_T/f_+| = 0.02 \pm 0.12$
$K^+ \to e^+\nu_e\gamma$ $|F_A + F_V| = 0.148 \pm 0.010$
$K^+ \to \mu^+\nu_\mu\gamma$ $|F_A + F_V| < 0.23$, CL = 90%
$K^+ \to e^+\nu_e\gamma$ $|F_A - F_V| < 0.49$
$K^+ \to \mu^+\nu_\mu\gamma$ $|F_A - F_V| = -2.2$ to 0.3

K^- modes are charge conjugates of the modes below.

K^+ DECAY MODES		Fraction (Γ_i/Γ)	Scale factor/ Confidence level	p (MeV/c)
$\mu^+\nu_\mu$		(63.51 ± 0.18) %	S=1.3	236
$e^+\nu_e$		$(1.55 \pm 0.07) \times 10^{-5}$		247
$\pi^+\pi^0$		(21.16 ± 0.14) %	S=1.1	205
$\pi^+\pi^+\pi^-$		(5.59 ± 0.05) %	S=1.8	125
$\pi^+\pi^0\pi^0$		(1.73 ± 0.04) %	S=1.2	133
$\pi^0\mu^+\nu_\mu$		(3.18 ± 0.08) %	S=1.5	215
Called $K^+_{\mu3}$.				
$\pi^0 e^+\nu_e$		(4.82 ± 0.06) %	S=1.3	228
Called K^+_{e3}.				
$\pi^0\pi^0 e^+\nu_e$		$(2.1 \pm 0.4) \times 10^{-5}$		206
$\pi^+\pi^- e^+\nu_e$		$(3.91 \pm 0.17) \times 10^{-5}$		203
$\pi^+\pi^-\mu^+\nu_\mu$		$(1.4 \pm 0.9) \times 10^{-5}$		151
$\pi^0\pi^0\pi^0 e^+\nu_e$		$< 3.5 \times 10^{-6}$	CL=90%	135
$\pi^+\gamma\gamma$	[x]	$< 1 \times 10^{-6}$	CL=90%	227
$\pi^+3\gamma$	[x]	$< 1.0 \times 10^{-4}$	CL=90%	227
$\mu^+\nu_\mu\nu\overline{\nu}$		$< 6.0 \times 10^{-6}$	CL=90%	236
$e^+\nu_e\nu\overline{\nu}$		$< 6 \times 10^{-5}$	CL=90%	247
$\mu^+\nu_\mu e^+ e^-$		$(1.06 \pm 0.32) \times 10^{-6}$		236
$e^+\nu_e e^+ e^-$		$(2.1^{+2.1}_{-1.1}) \times 10^{-7}$		247
$\mu^+\nu_\mu\mu^+\mu^-$		$< 4.1 \times 10^{-7}$	CL=90%	185
$\mu^+\nu_\mu\gamma$	[x,y]	$(5.50 \pm 0.28) \times 10^{-3}$		236
$\pi^+\pi^0\gamma$	[x,y]	$(2.75 \pm 0.15) \times 10^{-4}$		205
$\pi^+\pi^0\gamma$(DE)	[x,z]	$(1.8 \pm 0.4) \times 10^{-5}$		205
$\pi^+\pi^+\pi^-\gamma$	[x,y]	$(1.04 \pm 0.31) \times 10^{-4}$		125
$\pi^+\pi^0\pi^0\gamma$	[x,y]	$(7.5^{+5.5}_{-3.0}) \times 10^{-6}$		133
$\pi^0\mu^+\nu_\mu\gamma$	[x,y]	$< 6.1 \times 10^{-5}$	CL=90%	215
$\pi^0 e^+\nu_e\gamma$	[x,y]	$(2.62 \pm 0.20) \times 10^{-4}$		228
$\pi^0 e^+\nu_e\gamma$(SD)	[aa]	$< 5.3 \times 10^{-5}$	CL=90%	228
$\pi^0\pi^0 e^+\nu_e\gamma$		$< 5 \times 10^{-6}$	CL=90%	206

Lepton Family number (LF), Lepton number (L), $\Delta S = \Delta Q$ (SQ) violating modes, or $\Delta S = 1$ weak neutral current ($S1$) modes

$\pi^+\pi^+ e^-\overline{\nu}_e$	SQ	$< 1.2 \times 10^{-8}$	CL=90%	203
$\pi^+\pi^+\mu^-\overline{\nu}_\mu$	SQ	$< 3.0 \times 10^{-6}$	CL=95%	151
$\pi^+ e^+ e^-$	$S1$	$(2.74 \pm 0.23) \times 10^{-7}$		227
$\pi^+\mu^+\mu^-$	$S1$	$< 2.3 \times 10^{-7}$	CL=90%	172
$\pi^+\nu\overline{\nu}$	$S1$	$< 2.4 \times 10^{-9}$	CL=90%	227
$\mu^-\nu e^+ e^+$	LF	$< 2.0 \times 10^{-8}$	CL=90%	236
$\mu^+\nu_e$	LF	[d] $< 4 \times 10^{-3}$		236
$\pi^+\mu^+ e^-$	LF	$< 2.1 \times 10^{-10}$	CL=90%	214
$\pi^+\mu^- e^+$	LF	$< 7 \times 10^{-9}$	CL=90%	214
$\pi^-\mu^+ e^+$	L	$< 7 \times 10^{-9}$	CL=90%	214
$\pi^- e^+ e^+$	L	$< 1.0 \times 10^{-8}$	CL=90%	227
$\pi^-\mu^+\mu^+$	L	$< 1.5 \times 10^{-4}$	CL=90%	172
$\mu^+\overline{\nu}_e$	L	[d] $< 3.3 \times 10^{-3}$		236
$\pi^0 e^+\overline{\nu}_e$	L	[d] $< 3 \times 10^{-3}$	CL=90%	228

K^0

$$I(J^P) = \tfrac{1}{2}(0^-)$$

50% K_S, 50% K_L
Mass $m = 497.672 \pm 0.031$ MeV
$m_{K^0} - m_{K^\pm} = 3.995 \pm 0.034$ MeV (S = 1.1)
$|m_{K^0} - m_{\overline{K^0}}| / m_{\text{average}} < 9 \times 10^{-19}$ [bb]

K^0_S

$$I(J^P) = \tfrac{1}{2}(0^-)$$

Mean life $\tau = (0.8927 \pm 0.0009) \times 10^{-10}$ s
$c\tau = 2.6762$ cm

CP-violation parameters [cc]

$\text{Im}(\eta_{+-0}) = -0.015 \pm 0.030$
$\text{Im}(\eta_{000})^2 < 0.1$, CL = 90%

K^0_S DECAY MODES		Fraction (Γ_i/Γ)	Scale factor/ Confidence level	p (MeV/c)
$\pi^+\pi^-$		(68.61 ± 0.28) %	S=1.2	206
$\pi^0\pi^0$		(31.39 ± 0.28) %	S=1.2	209
$\pi^+\pi^-\gamma$	[y,dd]	$(1.78 \pm 0.05) \times 10^{-3}$		206
$\gamma\gamma$		$(2.4 \pm 0.9) \times 10^{-6}$		249
$\pi^+\pi^-\pi^0$		$(3.9^{+5.5}_{-1.9}) \times 10^{-7}$		133
$3\pi^0$		$< 3.7 \times 10^{-5}$	CL=90%	139
$\pi^\pm e^\mp\nu$	[ee]	$(6.70 \pm 0.07) \times 10^{-4}$	S=1.3	229
$\pi^\pm\mu^\mp\nu$	[ee]	$(4.69 \pm 0.06) \times 10^{-4}$	S=1.2	216

$\Delta S = 1$ weak neutral current ($S1$) modes

$\mu^+\mu^-$	$S1$	$< 3.2 \times 10^{-7}$	CL=90%	225
$e^+ e^-$	$S1$	$< 2.8 \times 10^{-6}$	CL=90%	249
$\pi^0 e^+ e^-$	$S1$	$< 1.1 \times 10^{-6}$	CL=90%	231

K^0_L

$$I(J^P) = \tfrac{1}{2}(0^-)$$

$m_{K_L} - m_{K_S} = (0.5304 \pm 0.0014) \times 10^{10} \, \hbar \, \text{s}^{-1}$
$\qquad = (3.491 \pm 0.009) \times 10^{-12}$ MeV
Mean life $\tau = (5.17 \pm 0.04) \times 10^{-8}$ s (S = 1.1)
$c\tau = 15.51$ m

Slope parameter g [v]

(See Particle Listings for quadratic coefficients)
$K^0_L \to \pi^+\pi^-\pi^0 = 0.670 \pm 0.014$ (S = 1.6)

K_L decay form factors [w]

K^0_{e3} $\lambda_+ = 0.0300 \pm 0.0016$ (S = 1.2)
$K^0_{\mu3}$ $\lambda_+ = 0.034 \pm 0.005$ (S = 2.3)
$K^0_{\mu3}$ $\lambda_0 = 0.025 \pm 0.006$ (S = 2.3)
K^0_{e3} $|f_S/f_+| < 0.04$, CL = 68%
K^0_{e3} $|f_T/f_+| < 0.23$, CL = 68%
$K^0_{\mu3}$ $|f_T/f_+| = 0.12 \pm 0.12$
$K_L \to e^+ e^-\gamma$: $\alpha_{K^*} = -0.28 \pm 0.08$

CP-violation parameters [cc]

$\delta = (0.327 \pm 0.012)\%$

$|\eta_{00}| = (2.275 \pm 0.019) \times 10^{-3}$ $(S = 1.1)$

$|\eta_{+-}| = (2.285 \pm 0.019) \times 10^{-3}$

$|\eta_{00}/\eta_{+-}| = 0.9956 \pm 0.0023$ [ff] $(S = 1.8)$

$\epsilon'/\epsilon = (1.5 \pm 0.8) \times 10^{-3}$ [ff] $(S = 1.8)$

$\phi_{+-} = (43.7 \pm 0.6)°$

$\phi_{00} = (43.5 \pm 1.0)°$

$\phi_{00} - \phi_{+-} = (-0.2 \pm 0.8)°$

j for $K_L^0 \to \pi^+\pi^-\pi^0 = 0.0011 \pm 0.0008$

$|\eta_{+-\gamma}| = (2.35 \pm 0.07) \times 10^{-3}$

$\phi_{+-\gamma} = (44 \pm 4)°$

$|\epsilon'_{+-\gamma}|/\epsilon < 0.3,\ CL = 90\%$

$\Delta S = -\Delta Q$ in $K_{\ell 3}^0$ decay

Re $x = 0.006 \pm 0.018$ $(S = 1.3)$

Im $x = -0.003 \pm 0.026$ $(S = 1.2)$

CPT-violation parameters

Re $\Delta = 0.018 \pm 0.020$

Im $\Delta = 0.02 \pm 0.04$

K_L^0 DECAY MODES		Fraction (Γ_i/Γ)	Scale factor/ Confidence level	p (MeV/c)
$3\pi^0$		$(21.12 \pm 0.27)\%$	$S=1.1$	139
$\pi^+\pi^-\pi^0$		$(12.56 \pm 0.20)\%$	$S=1.7$	133
$\pi^\pm\mu^\mp\nu$	[gg]	$(27.17 \pm 0.25)\%$	$S=1.1$	216
Called $K_{\mu 3}^0$.				
$\pi^\pm e^\mp \nu_e$	[gg]	$(38.78 \pm 0.27)\%$	$S=1.1$	229
Called K_{e3}^0.				
2γ		$(5.92 \pm 0.15) \times 10^{-4}$		249
3γ		$< 2.4 \times 10^{-7}$	CL=90%	249
$\pi^0 2\gamma$	[hh]	$(1.70 \pm 0.28) \times 10^{-6}$		231
$\pi^0\pi^\pm e^\mp\nu$	[gg]	$(5.18 \pm 0.29) \times 10^{-5}$		207
$(\pi\mu\text{atom})\nu$		$(1.06 \pm 0.11) \times 10^{-7}$		–
$\pi^\pm e^\mp \nu_e \gamma$	[y,gg,hh]	$(1.3 \pm 0.8)\%$		229
$\pi^+\pi^-\gamma$	[y,hh]	$(4.61 \pm 0.14) \times 10^{-5}$		206
$\pi^0\pi^0\gamma$		$< 5.6 \times 10^{-6}$		209

Charge conjugation × Parity (CP, CPV) or Lepton Family number (LF) violating modes, or $\Delta S = 1$ weak neutral current (S1) modes

			Fraction (Γ_i/Γ)		p (MeV/c)
$\pi^+\pi^-$	CPV		$(2.067 \pm 0.035) \times 10^{-3}$	$S=1.1$	206
$\pi^0\pi^0$	CPV		$(9.36 \pm 0.20) \times 10^{-4}$		209
$\mu^+\mu^-$	S1		$(7.2 \pm 0.5) \times 10^{-9}$	$S=1.4$	225
$\mu^+\mu^-\gamma$	S1		$(3.23 \pm 0.30) \times 10^{-7}$		225
e^+e^-	S1		$< 4.1 \times 10^{-11}$	CL=90%	249
$e^+e^-\gamma$	S1		$(9.1 \pm 0.5) \times 10^{-6}$		249
$e^+e^-\gamma\gamma$	S1	[hh]	$(6.5 \pm 1.2) \times 10^{-7}$		249
$\pi^+\pi^- e^+e^-$	S1		$< 2.5 \times 10^{-6}$	CL=90%	206
$\mu^+\mu^- e^+e^-$	S1		$< 4.9 \times 10^{-6}$	CL=90%	225
$e^+e^- e^+e^-$	S1	[ii]	$(4.1 \pm 0.8) \times 10^{-8}$	$S=1.2$	249
$\pi^0\mu^+\mu^-$	CP,S1	[jj]	$< 5.1 \times 10^{-9}$	CL=90%	177
$\pi^0 e^+e^-$	CP,S1	[jj]	$< 4.3 \times 10^{-9}$	CL=90%	231
$\pi^0\nu\overline{\nu}$	CP,S1	[kk]	$< 5.8 \times 10^{-5}$	CL=90%	231
$e^\pm\mu^\mp$	LF	[gg]	$< 3.3 \times 10^{-11}$	CL=90%	238

$K^*(892)$ $I(J^P) = \frac{1}{2}(1^-)$

$K^*(892)^\pm$ mass $m = 891.59 \pm 0.24$ MeV $(S = 1.1)$
$K^*(892)^0$ mass $m = 896.10 \pm 0.28$ MeV $(S = 1.4)$
$K^*(892)^\pm$ full width $\Gamma = 49.8 \pm 0.8$ MeV
$K^*(892)^0$ full width $\Gamma = 50.5 \pm 0.6$ MeV $(S = 1.1)$

$K^*(892)$ DECAY MODES	Fraction (Γ_i/Γ)	Confidence level	p (MeV/c)
$K\pi$	$\sim 100\%$		291
$K^0\gamma$	$(2.30 \pm 0.20) \times 10^{-3}$		310
$K^\pm\gamma$	$(1.01 \pm 0.09) \times 10^{-3}$		309
$K\pi\pi$	$< 7 \times 10^{-4}$	95%	224

$K_1(1270)$ $I(J^P) = \frac{1}{2}(1^+)$

Mass $m = 1273 \pm 7$ MeV [l]
Full width $\Gamma = 90 \pm 20$ MeV [l]

$K_1(1270)$ DECAY MODES	Fraction (Γ_i/Γ)	p (MeV/c)
$K\rho$	$(42 \pm 6)\%$	76
$K_0^*(1430)\pi$	$(28 \pm 4)\%$	–
$K^*(892)\pi$	$(16 \pm 5)\%$	301
$K\omega$	$(11.0 \pm 2.0)\%$	–
$K f_0(1370)$	$(3.0 \pm 2.0)\%$	–

$K_1(1400)$ $I(J^P) = \frac{1}{2}(1^+)$

Mass $m = 1402 \pm 7$ MeV
Full width $\Gamma = 174 \pm 13$ MeV $(S = 1.6)$

$K_1(1400)$ DECAY MODES	Fraction (Γ_i/Γ)	p (MeV/c)
$K^*(892)\pi$	$(94 \pm 6)\%$	401
$K\rho$	$(3.0 \pm 3.0)\%$	298
$K f_0(1370)$	$(2.0 \pm 2.0)\%$	–
$K\omega$	$(1.0 \pm 1.0)\%$	285

$K^*(1410)$ $I(J^P) = \frac{1}{2}(1^-)$

Mass $m = 1412 \pm 12$ MeV $(S = 1.1)$
Full width $\Gamma = 227 \pm 22$ MeV $(S = 1.1)$

$K^*(1410)$ DECAY MODES	Fraction (Γ_i/Γ)	Confidence level	p (MeV/c)
$K^*(892)\pi$	$> 40\%$	95%	408
$K\pi$	$(6.6 \pm 1.3)\%$		611
$K\rho$	$< 7\%$	95%	309

$K_0^*(1430)$ [ll] $I(J^P) = \frac{1}{2}(0^+)$

Mass $m = 1429 \pm 6$ MeV
Full width $\Gamma = 287 \pm 23$ MeV

$K_0^*(1430)$ DECAY MODES	Fraction (Γ_i/Γ)	p (MeV/c)
$K\pi$	$(93 \pm 10)\%$	621

$K_2^*(1430)$ $I(J^P) = \frac{1}{2}(2^+)$

$K_2^*(1430)^\pm$ mass $m = 1425.4 \pm 1.3$ MeV $(S = 1.1)$
$K_2^*(1430)^0$ mass $m = 1432.4 \pm 1.3$ MeV
$K_2^*(1430)^\pm$ full width $\Gamma = 98.4 \pm 2.3$ MeV
$K_2^*(1430)^0$ full width $\Gamma = 109 \pm 5$ MeV $(S = 1.9)$

$K_2^*(1430)$ DECAY MODES	Fraction (Γ_i/Γ)	Scale factor/ Confidence level	p (MeV/c)
$K\pi$	$(49.7 \pm 1.2)\%$		622
$K^*(892)\pi$	$(25.2 \pm 1.7)\%$		423
$K^*(892)\pi\pi$	$(13.0 \pm 2.3)\%$		375
$K\rho$	$(8.8 \pm 0.8)\%$	$S=1.2$	331
$K\omega$	$(2.9 \pm 0.8)\%$		319
$K^+\gamma$	$(2.4 \pm 0.5) \times 10^{-3}$		627
$K\eta$	$(1.4 ^{+2.8}_{-0.9}) \times 10^{-3}$	$S=1.1$	492
$K\omega\pi$	$< 7.2 \times 10^{-4}$	CL=95%	110
$K^0\gamma$	$< 9 \times 10^{-4}$	CL=90%	631

$K^*(1680)$ $I(J^P) = \frac{1}{2}(1^-)$

Mass $m = 1714 \pm 20$ MeV (S = 1.1)
Full width $\Gamma = 323 \pm 110$ MeV (S = 4.2)

$K^*(1680)$ DECAY MODES	Fraction (Γ_i/Γ)	p (MeV/c)
$K\pi$	(38.7 ± 2.5) %	779
$K\rho$	$(31.4^{+4.7}_{-2.1})$ %	571
$K^*(892)\pi$	$(29.9^{+2.2}_{-4.7})$ %	615

$K_2(1770)$ [mm] $I(J^P) = \frac{1}{2}(2^-)$

Mass $m = 1773 \pm 8$ MeV
Full width $\Gamma = 186 \pm 14$ MeV

$K_2(1770)$ DECAY MODES	Fraction (Γ_i/Γ)	p (MeV/c)
$K\pi\pi$		–
$K_2^*(1430)\pi$	dominant	287
$K^*(892)\pi$	seen	653
$K f_2(1270)$	seen	–
$K\phi$	seen	441
$K\omega$	seen	608

$K_3^*(1780)$ $I(J^P) = \frac{1}{2}(3^-)$

Mass $m = 1770 \pm 10$ MeV (S = 1.7)
Full width $\Gamma = 164 \pm 17$ MeV (S = 1.1)

$K_3^*(1780)$ DECAY MODES	Fraction (Γ_i/Γ)	Scale factor/ Confidence level	p (MeV/c)
$K\rho$	(45 ± 4) %	S=1.4	612
$K^*(892)\pi$	(27.3 ± 3.2) %	S=1.5	651
$K\pi$	(19.3 ± 1.0) %		810
$K\eta$	(8.0 ± 1.5) %	S=1.4	715
$K_2^*(1430)\pi$	< 21 %	CL=95%	284

$K_2(1820)$ [nn] $I(J^P) = \frac{1}{2}(2^-)$

Mass $m = 1816 \pm 13$ MeV
Full width $\Gamma = 276 \pm 35$ MeV

$K_2(1820)$ DECAY MODES	Fraction (Γ_i/Γ)	p (MeV/c)
$K\phi$	possibly seen	481
$K_2^*(1430)\pi$	seen	325
$K^*(892)\pi$	seen	680
$K f_2(1270)$	seen	186
$K\omega$	seen	638

$K_4^*(2045)$ $I(J^P) = \frac{1}{2}(4^+)$

Mass $m = 2045 \pm 9$ MeV (S = 1.1)
Full width $\Gamma = 198 \pm 30$ MeV

$K_4^*(2045)$ DECAY MODES	Fraction (Γ_i/Γ)	p (MeV/c)
$K\pi$	(9.9 ± 1.2) %	958
$K^*(892)\pi\pi$	(9 ± 5) %	800
$K^*(892)\pi\pi\pi$	(7 ± 5) %	764
$\rho K\pi$	(5.7 ± 3.2) %	742
$\omega K\pi$	(5.0 ± 3.0) %	736
$\phi K\pi$	(2.8 ± 1.4) %	591
$\phi K^*(892)$	(1.4 ± 0.7) %	363

CHARMED MESONS ($C = \pm 1$)

$D^+ = c\overline{d}$, $D^0 = c\overline{u}$, $\overline{D}^0 = \overline{c}u$, $D^- = \overline{c}d$,   similarly for D^*'s

$D^\pm$ $I(J^P) = \frac{1}{2}(0^-)$

Mass $m = 1869.3 \pm 0.5$ MeV (S = 1.1)
Mean life $\tau = (1.057 \pm 0.015) \times 10^{-12}$ s
$c\tau = 317$ μm

CP-violation decay-rate asymmetries
$A_{CP}(K^+ K^- \pi^\pm) = -0.03 \pm 0.07$
$A_{CP}(K^\pm K^{*0}) = -0.12 \pm 0.13$
$A_{CP}(\phi\pi^\pm) = 0.07 \pm 0.09$

$D^+ \to \overline{K}^*(892)^0 \ell^+ \nu_\ell$ form factors
$r_2 = 0.73 \pm 0.15$
$r_v = 1.90 \pm 0.25$
$\Gamma_L/\Gamma_T = 1.23 \pm 0.13$
$\Gamma_+/\Gamma_- = 0.16 \pm 0.04$

D^- modes are charge conjugates of the modes below.

D^+ DECAY MODES	Fraction (Γ_i/Γ)	Scale factor/ Confidence level	p (MeV/c)
Inclusive modes			
e^+ anything	(17.2 ± 1.9) %		–
K^- anything	(24.2 ± 2.8) %	S=1.4	–
$\overline{K}^0$ anything + K^0 anything	(59 ± 7) %		–
K^+ anything	(5.8 ± 1.4) %		–
η anything	[oo] < 13 %	CL=90%	–
Leptonic and semileptonic modes			
$\mu^+ \nu_\mu$	< 7.2 $\times 10^{-4}$	CL=90%	932
$\overline{K}^0 \ell^+ \nu_\ell$	[pp] (6.7 ± 0.8) %		868
$\overline{K}^0 e^+ \nu_e$	(6.6 ± 0.9) %		868
$\overline{K}^0 \mu^+ \nu_\mu$	$(7.0^{+3.0}_{-2.0})$ %		865
$K^- \pi^+ e^+ \nu_e$	$(4.2^{+0.9}_{-0.7})$ %		863
$\overline{K}^*(892)^0 e^+ \nu_e$	(3.2 ± 0.33) %		720
$\quad \times B(\overline{K}^{*0} \to K^- \pi^+)$			
$K^- \pi^+ e^+ \nu_e$ nonresonant	< 7 $\times 10^{-3}$	CL=90%	863
$K^- \pi^+ \mu^+ \nu_\mu$	(3.2 ± 0.4) %	S=1.1	851
$\overline{K}^*(892)^0 \mu^+ \nu_\mu$	(3.0 ± 0.4) %		715
$\quad \times B(\overline{K}^{*0} \to K^- \pi^+)$			
$K^- \pi^+ \mu^+ \nu_\mu$ nonresonant	$(2.7 \pm 1.1) \times 10^{-3}$		851
$(\overline{K}^*(892)\pi)^0 e^+ \nu_e$	< 1.2 %	CL=90%	714
$(\overline{K}\pi\pi)^0 e^+ \nu_e$ non-$\overline{K}^*(892)$	< 9 $\times 10^{-3}$	CL=90%	846
$K^- \pi^+ \pi^0 \mu^+ \nu_\mu$	< 1.4 $\times 10^{-3}$	CL=90%	825
$\pi^0 \ell^+ \nu_\ell$	[qq] $(5.7 \pm 2.2) \times 10^{-3}$		930

Fractions of some of the following modes with resonances have already appeared above as submodes of particular charged-particle modes.

$\overline{K}^*(892)^0 \ell^+ \nu_\ell$	[pp] (4.8 ± 0.4) %		720
$\overline{K}^*(892)^0 e^+ \nu_e$	(4.8 ± 0.5) %		720
$\overline{K}^*(892)^0 \mu^+ \nu_\mu$	(4.5 ± 0.6) %	S=1.1	715
$\rho^0 e^+ \nu_e$	< 3.7 $\times 10^{-3}$	CL=90%	776
$\rho^0 \mu^+ \nu_\mu$	$(2.0^{+1.5}_{-1.3}) \times 10^{-3}$		772
$\phi e^+ \nu_e$	< 2.09 %	CL=90%	657
$\phi \mu^+ \nu_\mu$	< 3.72 %	CL=90%	651
$\eta'(958) \mu^+ \nu_\mu$	< 9 $\times 10^{-3}$	CL=90%	684
Hadronic modes with a $\overline{K}$ or $\overline{K}K\overline{K}$			
$\overline{K}^0 \pi^+$	(2.74 ± 0.29) %		862
$K^- \pi^+ \pi^+$	[rr] (9.1 ± 0.6) %		845
$\overline{K}^*(892)^0 \pi^+$	(1.28 ± 0.13) %		712
$\quad \times B(\overline{K}^{*0} \to K^- \pi^+)$			
$\overline{K}_0^*(1430)^0 \pi^+$	(2.3 ± 0.3) %		368
$\quad \times B(\overline{K}_0^*(1430)^0 \to K^- \pi^+)$			
$\overline{K}^*(1680)^0 \pi^+$	$(3.7 \pm 0.8) \times 10^{-3}$		65
$\quad \times B(\overline{K}^*(1680)^0 \to K^- \pi^+)$			
$K^- \pi^+ \pi^+$ nonresonant	(8.6 ± 0.9) %		845

$\overline{K}^0 \pi^+ \pi^0$	[rr]	(9.7 ±3.0) %	S=1.1	845
$\overline{K}^0 \rho^+$		(6.6 ±2.5) %		680
$\overline{K}^*(892)^0 \pi^+$		(6.4 ±0.6) × 10^{-3}		712
$\times$ B($\overline{K}^{*0} \to \overline{K}^0 \pi^0$)				
$\overline{K}^0 \pi^+ \pi^0$ nonresonant		(1.3 ±1.1) %		845
$K^- \pi^+ \pi^+ \pi^0$	[rr]	(6.4 ±1.1) %		816
$\overline{K}^*(892)^0 \rho^+$ total		(1.4 ±0.9) %		423
$\times$ B($\overline{K}^{*0} \to K^- \pi^+$)				
$\overline{K}_1(1400)^0 \pi^+$		(2.2 ±0.6) %		390
$\times$ B($\overline{K}_1(1400)^0 \to K^- \pi^+ \pi^0$)				
$K^- \rho^+ \pi^+$ total		(3.1 ±1.1) %		616
$K^- \rho^+ \pi^+$ 3-body		(1.1 ±0.4) %		616
$\overline{K}^*(892)^0 \pi^+ \pi^0$ total		(4.5 ±0.9) %		687
$\times$ B($\overline{K}^{*0} \to K^- \pi^+$)				
$\overline{K}^*(892)^0 \pi^+ \pi^0$ 3-body		(2.8 ±0.9) %		687
$\times$ B($\overline{K}^{*0} \to K^- \pi^+$)				
$K^*(892)^- \pi^+ \pi^+$ 3-body		(7 ±3) × 10^{-3}		688
$\times$ B($K^{*-} \to K^- \pi^0$)				
$K^- \pi^+ \pi^+ \pi^0$ nonresonant	[ss]	(1.2 ±0.6) %		816
$\overline{K}^0 \pi^+ \pi^+ \pi^-$	[rr]	(7.0 ±1.0) %		814
$\overline{K}^0 a_1(1260)^+$		(4.0 ±0.9) %		328
$\times$ B($a_1(1260)^+ \to \pi^+ \pi^+ \pi^-$)				
$\overline{K}_1(1400)^+ \pi^+$		(2.2 ±0.6) %		390
$\times$ B($\overline{K}_1(1400)^0 \to \overline{K}^0 \pi^+ \pi^-$)				
$K^*(892)^- \pi^+ \pi^+$ 3-body		(1.4 ±0.6) %		688
$\times$ B($K^{*-} \to \overline{K}^0 \pi^-$)				
$\overline{K}^0 \rho^0 \pi^+$ total		(4.2 ±0.9) %		614
$\overline{K}^0 \rho^0 \pi^+$ 3-body		(5 ±5) × 10^{-3}		614
$\overline{K}^0 \pi^+ \pi^+ \pi^-$ nonresonant		(8 ±4) × 10^{-3}		814
$K^- \pi^+ \pi^+ \pi^+ \pi^-$		(8.2 ±1.4) × 10^{-3}		772
$\overline{K}^*(892)^0 \pi^+ \pi^+ \pi^-$		(6.8 ±1.8) × 10^{-3}		642
$\times$ B($\overline{K}^{*0} \to K^- \pi^+$)				
$\overline{K}^*(892)^0 \rho^0 \pi^+$		(5.1 ±2.2) × 10^{-3}		242
$\times$ B($\overline{K}^{*0} \to K^- \pi^+$)				
$K^- \pi^+ \pi^+ \pi^0 \pi^0$		(2.2 $^{+5.0}_{-0.9}$) %		775
$\overline{K}^0 \pi^+ \pi^+ \pi^- \pi^0$		(5.4 $^{+3.0}_{-1.4}$) %		773
$\overline{K}^0 \pi^+ \pi^+ \pi^+ \pi^- \pi^-$		(8 ±7) × 10^{-4}		714
$K^- \pi^+ \pi^+ \pi^+ \pi^- \pi^0$		(2.0 ±1.8) × 10^{-3}		718
$\overline{K}^0 \overline{K}^0 K^+$		(1.8 ±0.8) %		545

Fractions of some of the following modes with resonances have already appeared above as submodes of particular charged-particle modes.

$\overline{K}^0 \rho^+$		(6.6 ±2.5) %		680
$\overline{K}^0 a_1(1260)^+$		(8.1 ±1.7) %		328
$\overline{K}^0 a_2(1320)^+$		< 3 × 10^{-3}	CL=90%	199
$\overline{K}^*(892)^0 \pi^+$		(1.92±0.19) %		712
$\overline{K}^*(892)^0 \rho^+$ total		(2.1 ±1.4) %		423
$\overline{K}^*(892)^0 \rho^+$ S-wave	[ss]	(1.7 ±1.6) %		423
$\overline{K}^*(892)^0 \rho^+$ P-wave		< 1 × 10^{-3}	CL=90%	423
$\overline{K}^*(892)^0 \rho^+$ D-wave		(10 ±7) × 10^{-3}		423
$\overline{K}^*(892)^0 \rho^+$ D-wave longitudinal		< 7 × 10^{-3}	CL=90%	423
$\overline{K}_1(1270)^0 \pi^+$		< 7 × 10^{-3}	CL=90%	487
$\overline{K}_1(1400)^0 \pi^+$		(5.0 ±1.3) %		390
$\overline{K}^*(1410)^0 \pi^+$		< 7 × 10^{-3}	CL=90%	382
$\overline{K}_0^*(1430)^0 \pi^+$		(3.7 ±0.4) %		368
$\overline{K}^*(1680)^0 \pi^+$		(1.45±0.31) %		65
$\overline{K}^*(892)^0 \pi^+ \pi^0$ total		(6.7 ±1.4) %		687
$\overline{K}^*(892)^0 \pi^+ \pi^0$ 3-body		(4.2 ±1.4) %		687
$K^*(892)^- \pi^+ \pi^+$ 3-body		(2.1 ±0.9) %		688
$K^- \rho^+ \pi^+$ total		(3.1 ±1.1) %		616
$K^- \rho^+ \pi^+$ 3-body		(1.1 ±0.4) %		616
$\overline{K}^0 \rho^0 \pi^+$ total		(4.2 ±0.9) %	CL=90%	614
$\overline{K}^0 \rho^0 \pi^+$ 3-body		(5 ±5) × 10^{-3}		614
$\overline{K}^0 f_0(980) \pi^+$		< 5 × 10^{-3}	CL=90%	461
$\overline{K}^*(892)^0 \pi^+ \pi^+ \pi^-$		(1.02±0.27) %		642
$\overline{K}^*(892)^0 \rho^0 \pi^+$		(7.7 ±3.3) × 10^{-3}		242

Pionic modes

$\pi^+ \pi^0$		(2.5 ±0.7) × 10^{-3}		925
$\pi^+ \pi^+ \pi^-$		(3.2 ±0.6) × 10^{-3}		908
$\rho^0 \pi^+$		< 1.4 × 10^{-3}	CL=90%	769
$\pi^+ \pi^+ \pi^-$ nonresonant		(2.5 ±0.7) × 10^{-3}		908

$\pi^+ \pi^+ \pi^- \pi^0$		(1.9 $^{+1.5}_{-1.2}$) %		882
$\eta \pi^+ \times$ B($\eta \to \pi^+ \pi^- \pi^0$)		(1.8 ±0.6) × 10^{-3}		848
$\omega \pi^+ \times$ B($\omega \to \pi^+ \pi^- \pi^0$)		< 6 × 10^{-3}	CL=90%	764
$\pi^+ \pi^+ \pi^+ \pi^- \pi^-$		(1.0 $^{+0.8}_{-0.7}$) × 10^{-3}		845
$\pi^+ \pi^+ \pi^+ \pi^- \pi^- \pi^0$		(2.9 $^{+2.9}_{-2.0}$) × 10^{-3}		799

Fractions of some of the following modes with resonances have already appeared above as submodes of particular charged-particle modes.

$\eta \pi^+$		(7.5 ±2.5) × 10^{-3}		848
$\rho^0 \pi^+$		< 1.4 × 10^{-3}	CL=90%	769
$\omega \pi^+$		< 7 × 10^{-3}	CL=90%	764
$\eta \rho^+$		< 1.2 %	CL=90%	658
$\eta'(958) \pi^+$		< 9 × 10^{-3}	CL=90%	680
$\eta'(958) \rho^+$		< 1.5 %	CL=90%	355

Hadronic modes with a $K\overline{K}$ pair

$K^+ \overline{K}^0$		(7.2 ±1.2) × 10^{-3}		792
$K^+ K^- \pi^+$	[rr]	(8.9 ±0.8) × 10^{-3}		744
$\phi \pi^+ \times$ B($\phi \to K^+ K^-$)		(3.0 ±0.3) × 10^{-3}		647
$K^+ \overline{K}^*(892)^0$		(2.8 ±0.4) × 10^{-3}		610
$\times$ B($\overline{K}^{*0} \to K^- \pi^+$)				
$K^+ K^- \pi^+$ nonresonant		(4.6 ±0.9) × 10^{-3}		744
$K^0 \overline{K}^0 \pi^+$				741
$K^*(892)^+ \overline{K}^0$		(2.0 ±0.9) %		611
$\times$ B($K^{*+} \to K^0 \pi^+$)				
$K^+ K^- \pi^+ \pi^0$				682
$\phi \pi^+ \pi^0 \times$ B($\phi \to K^+ K^-$)		(1.1 ±0.5) %		619
$\phi \rho^+ \times$ B($\phi \to K^+ K^-$)		< 7 × 10^{-3}	CL=90%	268
$K^+ K^- \pi^+ \pi^0$ non-ϕ		(1.5 $^{+0.7}_{-0.6}$) %		682
$K^+ \overline{K}^0 \pi^+ \pi^-$		< 2 %	CL=90%	678
$K^0 K^- \pi^+ \pi^+$		(1.0 ±0.6) %		678
$K^*(892)^+ \overline{K}^*(892)^0$		(1.2 ±0.5) %		273
$\times$ B^2($K^* \to K\pi^+$)				
$K^0 K^- \pi^+ \pi^+$ non-$K^{*+}\overline{K}^{*0}$		< 7.9 × 10^{-3}	CL=90%	678
$K^+ K^- \pi^+ \pi^+ \pi^-$				600
$\phi \pi^+ \pi^+ \pi^-$		< 1 × 10^{-3}	CL=90%	565
$\times$ B($\phi \to K^+ K^-$)				
$K^+ K^- \pi^+ \pi^+ \pi^-$ nonresonant		< 3 %	CL=90%	600

Fractions of the following modes with resonances have already appeared above as submodes of particular charged-particle modes.

$\phi \pi^+$		(6.1 ±0.6) × 10^{-3}		647
$\phi \pi^+ \pi^0$		(2.3 ±1.0) %		619
$\phi \rho^+$		< 1.5 %	CL=90%	268
$\phi \pi^+ \pi^+ \pi^-$		< 2 × 10^{-3}	CL=90%	565
$K^+ \overline{K}^*(892)^0$		(4.2 ±0.5) × 10^{-3}		610
$K^*(892)^+ \overline{K}^0$		(3.0 ±1.4) %		611
$K^*(892)^+ \overline{K}^*(892)^0$		(2.6 ±1.1) %		273

Doubly Cabibbo suppressed (DC) modes,
$\Delta C = 1$ weak neutral current (C1) modes, or
Lepton Family number (LF) or Lepton number (L) violating modes

$K^+ \pi^+ \pi^-$	DC	(6.5 ±2.6) × 10^{-4}		845
$K^+ \rho^0$	DC	< 6 × 10^{-4}	CL=90%	681
$K^*(892)^0 \pi^+$	DC	< 1.9 × 10^{-4}	CL=90%	712
$K^+ K^+ K^-$	DC	< 1.5 × 10^{-4}	CL=90%	550
ϕK^+	DC	< 1.3 × 10^{-4}	CL=90%	527
$\pi^+ e^+ e^-$	C1	< 6.6 × 10^{-5}	CL=90%	929
$\pi^+ \mu^+ \mu^-$	C1	< 1.8 × 10^{-5}	CL=90%	917
$\rho^+ \mu^+ \mu^-$	C1	< 5.6 × 10^{-4}	CL=90%	759
$K^+ e^+ e^-$	[tt]	< 4.8 × 10^{-3}	CL=90%	869
$K^+ \mu^+ \mu^-$	[tt]	< 3.2 × 10^{-4}	CL=90%	856
$\pi^+ e^\pm \mu^\mp$	LF [gg]	< 3.8 × 10^{-3}	CL=90%	926
$\pi^+ e^+ \mu^-$	LF	< 3.3 × 10^{-3}	CL=90%	926
$\pi^+ e^- \mu^+$	LF	< 3.3 × 10^{-3}	CL=90%	926
$K^+ e^+ \mu^-$	LF	< 3.4 × 10^{-3}	CL=90%	866
$K^+ e^- \mu^+$	LF	< 3.4 × 10^{-3}	CL=90%	866
$\pi^- e^+ e^+$	L	< 4.8 × 10^{-3}	CL=90%	929
$\pi^- \mu^+ \mu^+$	L	< 2.2 × 10^{-4}	CL=90%	917
$\pi^- e^+ \mu^+$	L	< 3.7 × 10^{-3}	CL=90%	926
$\rho^- \mu^+ \mu^+$	L	< 5.6 × 10^{-4}	CL=90%	759
$K^- e^+ e^+$	L	< 9.1 × 10^{-3}	CL=90%	869
$K^- \mu^+ \mu^+$	L	< 3.2 × 10^{-4}	CL=90%	856
$K^- e^+ \mu^+$	L	< 4.0 × 10^{-3}	CL=90%	866
$K^*(892)^- \mu^+ \mu^+$	L	< 8.5 × 10^{-4}	CL=90%	703

D^0		$I(J^P) = \frac{1}{2}(0^-)$

Mass $m = 1864.5 \pm 0.5$ MeV (S = 1.1)

$\left| m_{D_1^0} - m_{D_2^0} \right| < 21 \times 10^{10}\ \hbar\ \text{s}^{-1}$, CL = 90% [uu]

$m_{D^\pm} - m_{D^0} = 4.78 \pm 0.10$ MeV

Mean life $\tau = (0.415 \pm 0.004) \times 10^{-12}$ s

$\quad c\tau = 124.4\ \mu$m

$\left| \Gamma_{D_1^0} - \Gamma_{D_2^0} \right| / \Gamma_{D^0} < 0.17$, CL = 90% [uu]

$\dfrac{\Gamma(K^+\pi^-\ \text{or}\ K^+\pi^-\pi^+\pi^-(\text{via}\overline{D^0}))}{\Gamma(K^-\pi^+\ \text{or}\ K^-\pi^+\pi^+\pi^-)} < 0.0037$, CL = 90%

$\Gamma(\mu^- X\ (\text{via}\ \overline{D^0}))/\Gamma(\mu^+ X) < 0.0056$, CL = 90%

CP-violation decay-rate asymmetries

$A_{CP}(K^+K^-) = 0.06 \pm 0.05$

$A_{CP}(K_S^0 \phi) = -0.03 \pm 0.09$

$A_{CP}(K_S^0 \pi^0) = -0.018 \pm 0.030$

$\overline{D^0}$ modes are charge conjugates of the modes below.

D^0 DECAY MODES		Fraction (Γ_i/Γ)	Scale factor/ Confidence level	p (MeV/c)
Inclusive modes				
e^+ anything		$(7.7\ \pm 1.2\ )$ %	S=1.1	–
μ^+ anything	[vv]	$(6.8\ \pm 1.0\ )$ %		–
K^- anything		$(53\ \pm 4\ )$ %	S=1.3	–
$\overline{K^0}$ anything + K^0 anything		$(42\ \pm 5\ )$ %		–
K^+ anything		$(3.4\ ^{+0.6}_{-0.4})$ %		–
η anything	[oo]	< 13 %	CL=90%	–
Semileptonic modes				
$K^- \ell^+ \nu_\ell$	[pp]	(3.48 ± 0.16) %	S=1.1	867
$K^- e^+ \nu_e$		(3.64 ± 0.20) %	S=1.1	867
$K^- \mu^+ \nu_\mu$		(3.23 ± 0.19) %		863
$K^- \pi^0 e^+ \nu_e$		$(1.6\ ^{+1.3}_{-0.5})$ %		861
$\overline{K^0} \pi^- e^+ \nu_e$		$(2.8\ ^{+1.7}_{-0.9})$ %		860
$\overline{K}{}^*(892)^- e^+ \nu_e$ $\quad \times B(K^{*-} \to \overline{K^0}\pi^-)$		(1.34 ± 0.22) %		719
$K^- \pi^+ \pi^- \mu^+ \nu_\mu$		$< 1.2\quad \times 10^{-3}$	CL=90%	821
$(\overline{K}{}^*(892)\pi)^- \mu^+ \nu_\mu$		$< 1.4\quad \times 10^{-3}$	CL=90%	693
$\pi^- e^+ \nu_e$		$(3.8\ ^{+1.2}_{-1.0}) \times 10^{-3}$		927

A fraction of the following resonance mode has already appeared above as a submode of a charged-particle mode.

$K^*(892)^- e^+ \nu_e$		(2.01 ± 0.33) %		719
Hadronic modes with a $\overline{K}$ or $\overline{K}K\overline{K}$				
$K^- \pi^+$		(3.83 ± 0.12) %		861
$\overline{K^0} \pi^0$		(2.11 ± 0.21) %	S=1.1	860
$\overline{K^0} \pi^+ \pi^-$	[rr]	$(5.4\ \pm 0.4\ )$ %	S=1.2	842
$\overline{K^0} \rho^0$		(1.20 ± 0.17) %		676
$\overline{K^0} f_0(980)$ $\quad \times B(f_0 \to \pi^+\pi^-)$		$(3.0\ \pm 0.8\ ) \times 10^{-3}$		549
$\overline{K^0} f_2(1270)$ $\quad \times B(f_2 \to \pi^+\pi^-)$		$(2.3\ \pm 0.9\ ) \times 10^{-3}$		263
$\overline{K^0} f_0(1370)$ $\quad \times B(f_0 \to \pi^+\pi^-)$		$(4.3\ \pm 1.3\ ) \times 10^{-3}$		–
$K^*(892)^- \pi^+$ $\quad \times B(K^{*-} \to \overline{K^0}\pi^-)$		$(3.3\ \pm 0.3\ )$ %		711
$K_0^*(1430)^- \pi^+$ $\quad \times B(K_0^*(1430)^- \to \overline{K^0}\pi^-)$		$(6.4\ \pm 1.6\ ) \times 10^{-3}$		364
$\overline{K^0} \pi^+ \pi^-$ nonresonant		(1.46 ± 0.24) %		842
$K^- \pi^+ \pi^0$	[rr]	$(13.9\ \pm 0.9\ )$ %	S=1.3	844
$K^- \rho^+$		$(10.8\ \pm 1.0\ )$ %		678
$K^*(892)^- \pi^+$ $\quad \times B(K^{*-} \to K^-\pi^0)$		$(1.7\ \pm 0.2\ )$ %		711
$\overline{K}{}^*(892)^0 \pi^0$ $\quad \times B(\overline{K}{}^{*0} \to K^-\pi^+)$		$(2.1\ \pm 0.3\ )$ %		709
$K^- \pi^+ \pi^0$ nonresonant		$(6.9\ \pm 2.5\ ) \times 10^{-3}$		844
$\overline{K^0} \pi^0 \pi^0$				843
$\overline{K}{}^*(892)^0 \pi^0$ $\quad \times B(\overline{K}{}^{*0} \to \overline{K^0}\pi^0)$		$(1.0\ \pm 0.2\ )$ %		709
$\overline{K^0} \pi^0 \pi^0$ nonresonant		$(7.8\ \pm 2.0\ ) \times 10^{-3}$		843

$K^- \pi^+ \pi^+ \pi^-$	[rr]	$(7.5\ \pm 0.4\ )$ %	S=1.1	812
$K^- \pi^+ \rho^0$ total		$(6.3\ \pm 0.4\ )$ %		612
$K^- \pi^+ \rho^0$ 3-body		$(4.7\ \pm 2.1\ ) \times 10^{-3}$		612
$\overline{K}{}^*(892)^0 \rho^0$ $\quad \times B(\overline{K}{}^{*0} \to K^-\pi^+)$		$(9.8\ \pm 2.2\ ) \times 10^{-3}$		418
$K^- a_1(1260)^+$ $\quad \times B(a_1(1260)^+ \to \pi^+\pi^+\pi^-)$		$(3.6\ \pm 0.6\ )$ %		327
$\overline{K}{}^*(892)^0 \pi^+ \pi^-$ total $\quad \times B(\overline{K}{}^{*0} \to K^-\pi^+)$		$(1.5\ \pm 0.4\ )$ %		683
$\overline{K}{}^*(892)^0 \pi^+ \pi^-$ 3-body $\quad \times B(\overline{K}{}^{*0} \to K^-\pi^+)$		$(9.5\ \pm 2.1\ ) \times 10^{-3}$		683
$K_1(1270)^- \pi^+$ $\quad \times B(K_1(1270)^- \to K^-\pi^+\pi^-)$	[ss]	$(3.6\ \pm 1.0\ ) \times 10^{-3}$		483
$K^- \pi^+ \pi^+ \pi^-$ nonresonant		(1.75 ± 0.25) %		812
$\overline{K^0} \pi^+ \pi^- \pi^0$	[rr]	$(10.0\ \pm 1.2\ )$ %		812
$\overline{K^0} \eta \times B(\eta \to \pi^+\pi^-\pi^0)$		$(1.6\ \pm 0.3\ ) \times 10^{-3}$		772
$\overline{K^0} \omega \times B(\omega \to \pi^+\pi^-\pi^0)$		$(1.9\ \pm 0.4\ ) \times 10^{-3}$		670
$K^*(892)^- \rho^+$ $\quad \times B(K^{*-} \to \overline{K^0}\pi^-)$		$(4.0\ \pm 1.6\ ) \times 10^{-3}$		422
$\overline{K}{}^*(892)^0 \rho^0$ $\quad \times B(\overline{K}{}^{*0} \to \overline{K^0}\pi^0)$		$(4.9\ \pm 1.1\ ) \times 10^{-3}$		418
$K_1(1270)^- \pi^+$ $\quad \times B(K_1(1270)^- \to \overline{K^0}\pi^-\pi^0)$	[ss]	$(5.1\ \pm 1.4\ ) \times 10^{-3}$		483
$\overline{K}{}^*(892)^0 \pi^+ \pi^-$ 3-body $\quad \times B(\overline{K}{}^{*0} \to K^-\pi^+)$		$(4.7\ \pm 1.1\ ) \times 10^{-3}$		683
$\overline{K^0} \pi^+ \pi^- \pi^0$ nonresonant		$(2.1\ \pm 2.1\ )$ %		812
$K^- \pi^+ \pi^0 \pi^0$		$(15\ \pm 5\ )$ %		815
$K^- \pi^+ \pi^+ \pi^-$		$(4.0\ \pm 0.4\ )$ %		771
$\overline{K}{}^*(892)^0 \pi^+ \pi^- \pi^0$ $\quad \times B(\overline{K}{}^{*0} \to K^-\pi^+)$		$(1.2\ \pm 0.6\ )$ %		641
$\overline{K}{}^*(892)^0 \eta$ $\quad \times B(\overline{K}{}^{*0} \to K^-\pi^+)$ $\quad \times B(\eta \to \pi^+\pi^-\pi^0)$		$(3.0\ \pm 0.8\ ) \times 10^{-3}$		580
$K^- \pi^+ \omega \times B(\omega \to \pi^+\pi^-\pi^0)$		$(2.7\ \pm 0.5\ )$ %		605
$\overline{K}{}^*(892)^0 \omega$ $\quad \times B(\overline{K}{}^{*0} \to K^-\pi^+)$ $\quad \times B(\omega \to \pi^+\pi^-\pi^0)$		$(7\ \pm 3\ ) \times 10^{-3}$		406
$\overline{K^0} \pi^+ \pi^+ \pi^- \pi^-$		$(5.8\ \pm 1.6\ ) \times 10^{-3}$		768
$\overline{K^0} \pi^+ \pi^- \pi^0 \pi^0 (\pi^0)$		$(10.6\ ^{+7.3}_{-3.0})$ %		771
$\overline{K^0} K^+ K^-$		$(9.3\ \pm 1.0\ ) \times 10^{-3}$		544
$\overline{K^0} \phi \times B(\phi \to K^+K^-)$		$(4.2\ \pm 0.5\ ) \times 10^{-3}$		520
$\overline{K^0} K^+ K^-$ non-ϕ		$(5.0\ \pm 0.8\ ) \times 10^{-3}$		544
$K_S^0 K_S^0 K_S^0$		$(9.7\ \pm 2.3\ ) \times 10^{-4}$		538
$K^+ K^- K^- \pi^+$		$(2.1\ \pm 0.5\ ) \times 10^{-4}$		434
$K^+ K^- \overline{K^0} \pi^0$		$(7.2\ ^{+4.8}_{-3.5}) \times 10^{-3}$		435

Fractions of many of the following modes with resonances have already appeared above as submodes of particular charged-particle modes. (Modes for which there are only upper limits and $\overline{K}{}^*(892)\rho$ submodes only appear below.)

$\overline{K^0} \eta$		$(7.0\ \pm 1.0\ ) \times 10^{-3}$		772
$\overline{K^0} \rho^0$		(1.20 ± 0.17) %		676
$K^- \rho^+$		$(10.8\ \pm 1.0\ )$ %	S=1.2	678
$\overline{K^0} \omega$		$(2.1\ \pm 0.4\ )$ %		670
$\overline{K^0} \eta'(958)$		(1.70 ± 0.26) %		565
$\overline{K^0} f_0(980)$		$(5.7\ \pm 1.6\ ) \times 10^{-3}$		549
$\overline{K^0} \phi$		$(8.5\ \pm 1.0\ ) \times 10^{-3}$		520
$K^- a_1(1260)^+$		$(7.3\ \pm 1.1\ )$ %		327
$\overline{K^0} a_1(1260)^0$		< 1.9 %	CL=90%	322
$\overline{K^0} f_2(1270)$		$(4.1\ \pm 1.5\ ) \times 10^{-3}$		263
$\overline{K^0} f_0(1370)$		$(6.9\ \pm 2.1\ ) \times 10^{-3}$		–
$K^- a_2(1320)^+$		$< 2\quad \times 10^{-3}$	CL=90%	197
$K^*(892)^- \pi^+$		$(5.0\ \pm 0.4\ )$ %	S=1.2	711
$\overline{K}{}^*(892)^0 \pi^0$		$(3.1\ \pm 0.4\ )$ %		709
$\overline{K}{}^*(892)^0 \pi^+ \pi^-$ total		$(2.3\ \pm 0.5\ )$ %		683
$\overline{K}{}^*(892)^0 \pi^+ \pi^-$ 3-body		(1.42 ± 0.32) %		683
$K^- \pi^+ \rho^0$ total		$(6.3\ \pm 0.4\ )$ %		612
$K^- \pi^+ \rho^0$ 3-body		$(4.7\ \pm 2.1\ ) \times 10^{-3}$		612
$\overline{K}{}^*(892)^0 \rho^0$		(1.47 ± 0.33) %		418
$\overline{K}{}^*(892)^0 \rho^0$ transverse		$(1.5\ \pm 0.5\ )$ %		418
$\overline{K}{}^*(892)^0 \rho^0$ S-wave		$(2.8\ \pm 0.6\ )$ %		418
$\overline{K}{}^*(892)^0 \rho^0$ S-wave long.		$< 3\quad \times 10^{-3}$	CL=90%	418
$\overline{K}{}^*(892)^0 \rho^0$ P-wave		$< 3\quad \times 10^{-3}$	CL=90%	418
$\overline{K}{}^*(892)^0 \rho^0$ D-wave		$(1.9\ \pm 0.6\ )$ %		418

$K^*(892)^- \rho^+$	(6.0 ±2.4) %		422
$K^*(892)^- \rho^+$ longitudinal	(2.9 ±1.2) %		422
$K^*(892)^- \rho^+$ transverse	(3.2 ±1.8) %		422
$K^*(892)^- \rho^+$ P-wave	< 1.5 %	CL=90%	422
$K^- \pi^+ f_0(980)$	< 1.1 %	CL=90%	459
$\overline{K}^*(892)^0 f_0(980)$	< 7 × 10^{-3}	CL=90%	–
$K_1(1270)^- \pi^+$	[ss] (1.06±0.29) %		483
$K_1(1400)^- \pi^+$	< 1.2 %	CL=90%	386
$\overline{K}_1(1400)^0 \pi^0$	< 3.7 %	CL=90%	387
$K^*(1410)^- \pi^+$	< 1.2 %	CL=90%	378
$K_0^*(1430)^- \pi^+$	(1.04±0.26) %		364
$K_2^*(1430)^- \pi^+$	< 8 × 10^{-3}	CL=90%	367
$\overline{K}_2^*(1430)^0 \pi^0$	< 4 × 10^{-3}	CL=90%	363
$\overline{K}^*(892)^0 \pi^+ \pi^- \pi^0$	(1.8 ±0.9) %		641
$\overline{K}^*(892)^0 \eta$	(1.9 ±0.5) %		580
$K^- \pi^+ \omega$	(3.0 ±0.6) %		605
$\overline{K}^*(892)^0 \omega$	(1.1 ±0.4) %		406
$K^- \pi^+ \eta'(958)$	(7.0 ±1.8) × 10^{-3}		479
$\overline{K}^*(892)^0 \eta'(958)$	< 1.1 × 10^{-3}	CL=90%	99

Pionic modes

$\pi^+ \pi^-$	(1.52±0.11) × 10^{-3}		922
$\pi^0 \pi^0$	(8.4 ±2.2) × 10^{-4}		922
$\pi^+ \pi^- \pi^0$	(1.6 ±1.1) %	S=2.7	907
$\pi^+ \pi^+ \pi^- \pi^-$	(7.4 ±0.6) × 10^{-3}		879
$\pi^+ \pi^+ \pi^- \pi^- \pi^0$	(1.9 ±0.4) %		844
$\pi^+ \pi^+ \pi^+ \pi^- \pi^- \pi^-$	(4.0 ±3.0) × 10^{-4}		795

Hadronic modes with a $K\overline{K}$ pair

$K^+ K^-$	(4.33±0.27) × 10^{-3}		791
$K^0 \overline{K}^0$	(1.3 ±0.4) × 10^{-3}		788
$K^0 K^- \pi^+$	(6.4 ±1.0) × 10^{-3}	S=1.1	739
$\overline{K}^*(892)^0 K^0$ $\times$ B($\overline{K}^{*0} \to K^- \pi^+$)	< 1.1 × 10^{-3}	CL=90%	605
$K^*(892)^+ K^-$ $\times$ B($K^{*+} \to K^0 \pi^+$)	(2.3 ±0.5) × 10^{-3}		610
$K^0 K^- \pi^+$ nonresonant	(2.3 ±2.3) × 10^{-3}		739
$\overline{K}^0 K^+ \pi^-$	(4.9 ±1.0) × 10^{-3}		739
$K^*(892)^0 \overline{K}^0$ $\times$ B($K^{*0} \to K^+ \pi^-$)	< 5 × 10^{-4}	CL=90%	605
$K^*(892)^- K^+$ $\times$ B($K^{*-} \to \overline{K}^0 \pi^-$)	(1.2 ±0.7) × 10^{-3}		610
$\overline{K}^0 K^+ \pi^-$ nonresonant	(3.8 $^{+2.3}_{-1.9}$) × 10^{-3}		739
$K^+ K^- \pi^+ \pi^-$	[ww] (2.58±0.28) × 10^{-3}		676
$\phi \pi^+ \pi^- \times$ B($\phi \to K^+ K^-$)	(5.3 ±1.4) × 10^{-4}		614
$\phi \rho^0 \times$ B($\phi \to K^+ K^-$)	(5.3 ±1.4) × 10^{-4}		260
$K^+ K^- \rho^0$ 3-body	(9.0 ±2.3) × 10^{-4}		309
$K^*(892)^0 K^- \pi^+$ $\times$ B($K^{*0} \to K^+ \pi^-$)	(2.1 ±0.9) × 10^{-3}		528
$\overline{K}^*(892)^0 K^+ \pi^-$ $\times$ B($\overline{K}^{*0} \to K^- \pi^+$)	(1.1 ±0.8) × 10^{-3}		528
$K^*(892)^0 \overline{K}^*(892)^0$ $\times$ B²($K^{*0} \to K^+ \pi^-$)	(6 ±2) × 10^{-4}		257
$K^+ K^- \pi^+ \pi^-$ non-ϕ	(1.7 ±0.5) × 10^{-3}		676
$K^+ K^- \pi^+ \pi^-$ nonresonant	< 8 × 10^{-4}	CL=90%	676
$K^0 \overline{K}^0 \pi^+ \pi^-$	(6.8 ±2.7) × 10^{-3}		673
$K^+ K^- \pi^+ \pi^- \pi^0$	(3.1 ±2.0) × 10^{-3}		600

Fractions of most of the following modes with resonances have already appeared above as submodes of particular charged-particle modes.

$\overline{K}^*(892)^0 K^0$	< 1.6 × 10^{-3}	CL=90%	605
$K^*(892)^+ K^-$	(3.5 ±0.8) × 10^{-3}		610
$K^*(892)^0 \overline{K}^0$	< 8 × 10^{-4}	CL=90%	605
$K^*(892)^- K^+$	(1.8 ±1.0) × 10^{-3}		610
$\phi \pi^0$	< 1.4 × 10^{-3}	CL=90%	644
$\phi \eta$	< 2.8 × 10^{-3}	CL=90%	489
$\phi \omega$	< 2.1 × 10^{-3}	CL=90%	239
$\phi \pi^+ \pi^-$	(1.07±0.29) × 10^{-3}		614
$\phi \rho^0$	(1.07±0.29) × 10^{-3}		260
$\phi \pi^+ \pi^-$ 3-body	< 5 × 10^{-4}	CL=90%	614
$K^*(892)^0 K^- \pi^+$	(3.2 ±1.3) × 10^{-3}		528
$\overline{K}^*(892)^0 K^+ \pi^-$	(1.7 ±1.2) × 10^{-3}		528
$K^*(892)^0 \overline{K}^*(892)^0$	(1.4 ±0.5) × 10^{-3}		257

Doubly Cabibbo suppressed (DC) modes, $\Delta C = 2$ forbidden via mixing (C2M) modes, $\Delta C = 1$ weak neutral current (C1) modes, or Lepton Family number (LF) violating modes

$K^+ \pi^-$	DC	(2.9 ±1.4) × 10^{-4}		861
$K^+ \pi^-$ (via $\overline{D}^0$)	C2M	< 1.9 × 10^{-4}	CL=90%	861
$K^+ \pi^- \pi^+ \pi^-$	DC	< 1.4 × 10^{-3}	CL=90%	812
$K^+ \pi^- \pi^+ \pi^-$ (via $\overline{D}^0$)	C2M	< 4 × 10^{-4}	CL=90%	812
μ^- anything (via $\overline{D}^0$)	C2M	< 4 × 10^{-4}	CL=90%	–
$e^+ e^-$	C1	< 1.3 × 10^{-5}	CL=90%	932
$\mu^+ \mu^-$	C1	< 7.6 × 10^{-6}	CL=90%	926
$\pi^0 e^+ e^-$	C1	< 4.5 × 10^{-5}	CL=90%	927
$\pi^0 \mu^+ \mu^-$	C1	< 1.8 × 10^{-4}	CL=90%	915
$\eta e^+ e^-$	C1	< 1.1 × 10^{-4}	CL=90%	852
$\eta \mu^+ \mu^-$	C1	< 5.3 × 10^{-4}	CL=90%	838
$\rho^0 e^+ e^-$	C1	< 1.0 × 10^{-4}	CL=90%	773
$\rho^0 \mu^+ \mu^-$	C1	< 2.3 × 10^{-4}	CL=90%	756
$\omega e^+ e^-$	C1	< 1.8 × 10^{-4}	CL=90%	768
$\omega \mu^+ \mu^-$	C1	< 8.3 × 10^{-4}	CL=90%	751
$\phi e^+ e^-$	C1	< 5.2 × 10^{-5}	CL=90%	654
$\phi \mu^+ \mu^-$	C1	< 4.1 × 10^{-4}	CL=90%	631
$\overline{K}^0 e^+ e^-$	[tt] < 1.1 × 10^{-4}	CL=90%	866
$\overline{K}^0 \mu^+ \mu^-$	[tt] < 2.6 × 10^{-4}	CL=90%	852
$\overline{K}^*(892)^0 e^+ e^-$	[tt] < 1.4 × 10^{-4}	CL=90%	717
$\overline{K}^*(892)^0 \mu^+ \mu^-$	[tt] < 1.18 × 10^{-3}	CL=90%	698
$\pi^+ \pi^- \pi^0 \mu^+ \mu^-$	C1 < 8.1 × 10^{-4}	CL=90%	863
$\mu^\pm e^\mp$	LF [gg] < 1.9 × 10^{-5}	CL=90%	929
$\pi^0 e^\pm \mu^\mp$	LF [gg] < 8.6 × 10^{-5}	CL=90%	924
$\eta e^\pm \mu^\mp$	LF [gg] < 1.0 × 10^{-4}	CL=90%	848
$\rho^0 e^\pm \mu^\mp$	LF [gg] < 4.9 × 10^{-5}	CL=90%	769
$\omega e^\pm \mu^\mp$	LF [gg] < 1.2 × 10^{-4}	CL=90%	764
$\phi e^\pm \mu^\mp$	LF [gg] < 3.4 × 10^{-5}	CL=90%	648
$\overline{K}^0 e^\pm \mu^\mp$	LF [gg] < 1.0 × 10^{-4}	CL=90%	862
$\overline{K}^*(892)^0 e^\pm \mu^\mp$	LF [gg] < 1.0 × 10^{-4}	CL=90%	712

$D^*(2007)^0$ $I(J^P) = \frac{1}{2}(1^-)$
I, J, P need confirmation.

Mass $m = 2006.7 \pm 0.5$ MeV (S = 1.1)
$m_{D^{*0}} - m_{D^0} = 142.12 \pm 0.07$ MeV
Full width $\Gamma < 2.1$ MeV, CL = 90%

$\overline{D}^*(2007)^0$ modes are charge conjugates of modes below.

$D^*(2007)^0$ DECAY MODES	Fraction (Γ_i/Γ)	p (MeV/c)
$D^0 \pi^0$	(61.9±2.9) %	43
$D^0 \gamma$	(38.1±2.9) %	137

$D^*(2010)^\pm$ $I(J^P) = \frac{1}{2}(1^-)$
I, J, P need confirmation.

Mass $m = 2010.0 \pm 0.5$ MeV (S = 1.1)
$m_{D^*(2010)^+} - m_{D^+} = 140.64 \pm 0.09$ MeV
$m_{D^*(2010)^+} - m_{D^0} = 145.42 \pm 0.05$ MeV
Full width $\Gamma < 0.131$ MeV, CL = 90%

$D^*(2010)^-$ modes are charge conjugates of the modes below.

$D^*(2010)^\pm$ DECAY MODES	Fraction (Γ_i/Γ)	p (MeV/c)
$D^0 \pi^+$	(68.3±1.4) %	39
$D^+ \pi^0$	(30.6±2.5) %	38
$D^+ \gamma$	(1.1 $^{+2.1}_{-0.7}$) %	136

$D_1(2420)^0$ $I(J^P) = \frac{1}{2}(1^+)$
I, J, P need confirmation.

Mass $m = 2422.2 \pm 1.8$ MeV (S = 1.2)
Full width $\Gamma = 18.9 ^{+4.6}_{-3.5}$ MeV

$\overline{D}_1(2420)^0$ modes are charge conjugates of modes below.

$D_1(2420)^0$ DECAY MODES	Fraction (Γ_i/Γ)	p (MeV/c)
$D^*(2010)^+ \pi^-$	seen	355
$D^+ \pi^-$	not seen	474

$\boxed{D_2^*(2460)^0}$ $\qquad I(J^P) = \frac{1}{2}(2^+)$

$J^P = 2^+$ assignment strongly favored (ALBRECHT 89B).

Mass $m = 2458.9 \pm 2.0$ MeV $\quad$ (S = 1.2)
Full width $\Gamma = 23 \pm 5$ MeV

$\overline{D}_2^*(2460)^0$ modes are charge conjugates of modes below.

$D_2^*(2460)^0$ DECAY MODES	Fraction (Γ_i/Γ)	p (MeV/c)
$D^+\pi^-$	seen	503
$D^*(2010)^+\pi^-$	seen	387

$\boxed{D_2^*(2460)^\pm}$ $\qquad I(J^P) = \frac{1}{2}(2^+)$

$J^P = 2^+$ assignment strongly favored (ALBRECHT 89B).
Mass $m = 2459 \pm 4$ MeV $\quad$ (S = 1.7)
$m_{D_2^*(2460)^\pm} - m_{D_2^*(2460)^0} = 0.9 \pm 3.3$ MeV $\quad$ (S = 1.1)
Full width $\Gamma = 25^{+8}_{-7}$ MeV

$D_2^*(2460)^-$ modes are charge conjugates of modes below.

$D_2^*(2460)^\pm$ DECAY MODES	Fraction (Γ_i/Γ)	p (MeV/c)
$D^0\pi^+$	seen	508
$D^{*0}\pi^+$	seen	390

CHARMED, STRANGE MESONS
$(C = S = \pm 1)$

$D_s^+ = c\bar{s}$, $D_s^- = \bar{c}s$, $\quad$ similarly for D_s^*'s

$\boxed{\begin{array}{c} D_s^\pm \\ \text{was } F^\pm \end{array}}$ $\qquad I(J^P) = 0(0^-)$

Mass $m = 1968.5 \pm 0.6$ MeV $\quad$ (S = 1.1)
$m_{D_s^\pm} - m_{D^\pm} = 99.2 \pm 0.5$ MeV $\quad$ (S = 1.1)
Mean life $\tau = (0.467 + 0.017) \times 10^{-12}$ s
$c\tau = 140$ μm

D_s^+ form factors
$r_2 = 1.6 \pm 0.4$
$r_v = 1.5 \pm 0.5$
$\Gamma_L/\Gamma_T = 0.72 \pm 0.18$

Branching fractions for modes with a resonance in the final state include all the decay modes of the resonance. D_s^- modes are charge conjugates of the modes below.

D_s^+ DECAY MODES		Fraction (Γ_i/Γ)	Confidence level	p (MeV/c)
Inclusive modes				
K^- anything		$(13 \, ^{+14}_{-12})$ %		–
$\overline{K}^0$ anything $+ K^0$ anything		(39 ± 28) %		–
K^+ anything		$(20 \, ^{+18}_{-14})$ %		–
non-$K\overline{K}$ anything		(64 ± 17) %		–
e^+ anything		< 20 %	90%	–
Leptonic and semileptonic modes				
$\mu^+\nu_\mu$		$(9 \pm 4) \times 10^{-3}$		981
$\phi\ell^+\nu_\ell$	[xx]	(1.9 ± 0.5) %		–
$\eta\ell^+\nu_\ell + \eta'(958)\ell^+\nu_\ell$		(3.3 ± 1.0) %		–
$\eta\ell^+\nu_\ell$		(2.5 ± 0.7) %		–
$\eta'(958)\ell^+\nu_\ell$		$(8.7 \pm 3.4) \times 10^{-3}$		–

Hadronic modes with a $K\overline{K}$ pair (including from a ϕ)				
$K^+\overline{K}^0$		(3.6 ± 1.1) %		850
$K^+K^-\pi^+$	[rr,yy]	(4.6 ± 1.2) %		805
$\phi\pi^+$		(3.6 ± 0.9) %		712
$K^+\overline{K}^*(892)^0$		(3.4 ± 0.9) %		682
$f_0(980)\pi^+$		(1.1 ± 0.4) %		732
$K^+\overline{K}_0^*(1430)^0$		$(7 \pm 4) \times 10^{-3}$		186
$f_J(1710)\pi^+ \to K^+K^-\pi^+$	[zz]	$(1.5 \pm 2.0) \times 10^{-3}$		204
$K^+K^-\pi^+$ nonresonant		$(9 \pm 4) \times 10^{-3}$		805
$K^0\overline{K}^0\pi^+$				802
$K^*(892)^+\overline{K}^0$		(4.3 ± 1.4) %		683
$K^+K^-\pi^+\pi^0$				748
$\phi\pi^+\pi^0$		(9 ± 5) %		687
$\phi\rho^+$		(6.7 ± 2.3) %		407
$\phi\pi^+\pi^0$ 3-body		< 2.6 %	90%	687
$K^+K^-\pi^+\pi^0$ non-ϕ		< 9 %	90%	748
$K^+\overline{K}^0\pi^+\pi^-$		< 2.8 %	90%	744
$K^0K^-\pi^+\pi^+$		(4.3 ± 1.5) %		744
$K^*(892)^+\overline{K}^*(892)^0$		(5.8 ± 2.5) %		412
$K^0K^-\pi^+\pi^+$ non-$K^{*+}\overline{K}^{*0}$		< 2.9 %	90%	744
$K^+K^-\pi^+\pi^+\pi^-$				673
$\phi\pi^+\pi^+\pi^-$		(1.8 ± 0.6) %		640
$K^+K^-\pi^+\pi^+\pi^-$ non-ϕ		$(3.0 \, ^{+3.0}_{-2.0}) \times 10^{-3}$		673
Other hadronic modes (0, 1, or 3 K's)				
$\pi^+\pi^+\pi^-$		(1.4 ± 0.4) %		959
$\rho^0\pi^+$		$< 2.9 \times 10^{-3}$	90%	827
$f_0(980)\pi^+$		(1.2 ± 0.5) %		732
$\pi^+\pi^+\pi^-$ nonresonant		(1.0 ± 0.4) %		959
$\pi^+\pi^+\pi^-\pi^0$		< 12 %	90%	935
$\eta\pi^+$		(2.0 ± 0.6) %		902
$\omega\pi^+$		< 1.8 %	90%	822
$\pi^+\pi^+\pi^+\pi^-\pi^-$		$(3.0 \, ^{+4.0}_{-3.0}) \times 10^{-3}$		899
$\pi^+\pi^+\pi^-\pi^0\pi^0$				902
$\eta\rho^+$		(10.3 ± 3.2) %		727
$\eta\pi^+\pi^0$ 3-body		< 3.0 %	90%	886
$\pi^+\pi^+\pi^+\pi^-\pi^-\pi^0$		(4.9 ± 3.2) %		856
$\eta'(958)\pi^+$		(4.9 ± 1.8) %		743
$\pi^+\pi^+\pi^+\pi^-\pi^-\pi^0\pi^0$				803
$\eta'(958)\rho^+$		(12 ± 4) %		470
$\eta'(958)\pi^+\pi^0$ 3-body		< 3.1 %	90%	720
$K^0\pi^+$		$< 8 \times 10^{-3}$	90%	916
$K^+\pi^+\pi^-$		(1.0 ± 0.4) %		900
$K^+\rho^0$		$< 2.9 \times 10^{-3}$	90%	747
$K^*(892)^0\pi^+$		$(6.5 \pm 2.8) \times 10^{-3}$		773
$K^+K^+K^-$		$< 6 \times 10^{-4}$	90%	628
ϕK^+		$< 5 \times 10^{-4}$	90%	607
$\Delta C = 1$ weak neutral current ($C1$) modes, or Lepton number (L) violating modes				
$\pi^+\mu^+\mu^-$	[aaa]	$< 4.3 \times 10^{-4}$	90%	968
$K^+\mu^+\mu^-$	C1	$< 5.9 \times 10^{-4}$	90%	909
$K^*(892)^+\mu^+\mu^-$	C1	$< 1.4 \times 10^{-3}$	90%	765
$\pi^-\mu^+\mu^+$	L	$< 4.3 \times 10^{-4}$	90%	968
$K^-\mu^+\mu^+$	L	$< 5.9 \times 10^{-4}$	90%	909
$K^*(892)^-\mu^+\mu^+$	L	$< 1.4 \times 10^{-3}$	90%	765

$I(J^P) = ?(?^?)$

Mass $m = 2112.4 \pm 0.7$ MeV   (S = 1.1)

$m_{D_s^{*\pm}} - m_{D_s^{\pm}} = 143.8 \pm 0.4$ MeV

Full width $\Gamma < 1.9$ MeV, CL = 90%

D_s^{*-} modes are charge conjugates of the modes below.

D_s^{*+} DECAY MODES	Fraction (Γ_i/Γ)	p (MeV/c)
$D_s^+ \gamma$	seen	139
$D_s^+ \pi^0$	seen	48

$I(J^P) = 0(1^+)$

I, J, P need confirmation.

Mass $m = 2535.35 \pm 0.34$ MeV

Full width $\Gamma < 2.3$ MeV, CL = 90%

$D_{s1}(2536)^-$ modes are charge conjugates of the modes below.

$D_{s1}(2536)^+$ DECAY MODES	Fraction (Γ_i/Γ)	p (MeV/c)
$D^*(2010)^+ K^0$	seen	150
$D^*(2007)^0 K^+$	seen	169
$D^+ K^0$	not seen	382
$D^0 K^+$	not seen	392
$D_s^{*+} \gamma$	possibly seen	389

$D_{sJ}(2573)^\pm$

$I(J^P) = ?(?^?)$

J^P is natural, width and decay modes consistent with 2^{+2}.

Mass $m = 2573.5 \pm 1.7$ MeV

Full width $\Gamma = 15^{+5}_{-4}$ MeV

$D_{sJ}(2573)^-$ modes are charge conjugates of the modes below.

$D_{sJ}(2573)^+$ DECAY MODES	Fraction (Γ_i/Γ)	p (MeV/c)
$D^0 K^+$	seen	436
$D^*(2007)^0 K^+$	seen	245

BOTTOM MESONS
($B = \pm 1$)

$B^+ = u\overline{b}$, $B^0 = d\overline{b}$, $\overline{B^0} = \overline{d}b$, $B^- = \overline{u}b$,   similarly for B^*'s

B-particle organization

Many measurements of B decays involve admixtures of B hadrons. Previously we arbitrarily included such admixtures in the $B^\pm$ section, but because of their importance we have created two new sections: "$B^\pm/B^0$ Admixture" for $\Upsilon(4S)$ results and "$B^\pm/B^0/B_s^0/b$-baryon Admixture" for results at higher energies. Most inclusive decay branching fractions are found in the Admixture sections. B^0-$\overline{B^0}$ mixing data are found in the B^0 section, while B_s^0-$\overline{B_s^0}$ mixing data and B-$\overline{B}$ mixing data for a B^0/B_s^0 admixture are found in the B_s^0 section. CP-violation data are found in the B^0 section. b-baryons are found near the end of the Baryon section.

The organization of the B sections is now as follows, where bullets indicate particle sections and brackets indicate reviews.

[Production and Decay of b-flavored Hadrons]

[Semileptonic Decays of B Mesons]

- $B^\pm$
  - mass
  - mean life
  - branching fractions
- B^0
  - mass
  - mean life
  - branching fractions
  - polarization in B^0 decay
  - B^0-$\overline{B^0}$ mixing
  - [B^0-$\overline{B^0}$ Mixing and CP Violation in B Decay]
  - CP violation
- $B^\pm B^0$ Admixtures
  - branching fractions
- $B^\pm/B^0/B_s^0/b$-baryon Admixtures
  - mean life
  - production fractions
  - branching fractions
- B^*
  - mass
- B_s^0
  - mass
  - mean life
  - branching fractions
  - polarization in B_s^0 decay
  - B_s^0-$\overline{B_s^0}$ mixing
  - B-$\overline{B}$ mixing (admixture of B^0, B_s^0)

At end of Baryon Listings:

- Λ_b
  - mass
  - mean life
  - branching fractions

$B^{\pm}$		$I(J^P) = \frac{1}{2}(0^-)$

I, J, P need confirmation. Quantum numbers shown are quark-model predictions.

Mass $m_{B^{\pm}} = 5278.9 \pm 1.8$ MeV
Mean life $\tau_{B^{\pm}} = (1.62 \pm 0.06) \times 10^{-12}$ s
$c\tau = 462$ μm

B^- modes are charge conjugates of the modes below. Modes which do not identify the charge state of the B are listed in the $B^{\pm}/B^0$ ADMIXTURE section.

The branching fractions listed below assume 50% $B^0 \overline{B}{}^0$ and 50% $B^+ B^-$ production at the $\Upsilon(4S)$. We have attempted to bring older measurements up to date by rescaling their assumed $\Upsilon(4S)$ production ratio to 50:50 and their assumed D, D_s, D^*, and ψ branching ratios to current values whenever this would affect our averages and best limits significantly.

Indentation is used to indicate a subchannel of a previous reaction. All resonant subchannels have been corrected for resonance branching fractions to the final state so the sum of the subchannel branching fractions can exceed that of the final state.

B^+ DECAY MODES	Fraction (Γ_i/Γ)	Scale factor/ Confidence level	p (MeV/c)
Semileptonic and leptonic modes			
$\ell^+ \nu_\ell$ anything	[qq] (10.1 ± 2.3) %		–
$\overline{D}{}^0 \ell^+ \nu_\ell$	[qq] (1.6 ± 0.7) %		–
$\overline{D}{}^*(2007)^0 \ell^+ \nu_\ell$	[qq] (5.3 ± 0.8) %		–
$\pi^0 e^+ \nu_e$	$< 2.2 \times 10^{-3}$	CL=90%	2638
$\omega \ell^+ \nu_\ell$	[qq] $< 2.1 \times 10^{-4}$	CL=90%	–
$\rho^0 \ell^+ \nu_\ell$	[qq] $< 2.1 \times 10^{-4}$	CL=90%	–
$e^+ \nu_e$	$< 1.5 \times 10^{-5}$	CL=90%	2639
$\mu^+ \nu_\mu$	$< 2.1 \times 10^{-5}$	CL=90%	2638
$\tau^+ \nu_\tau$	$< 1.8 \times 10^{-3}$	CL=90%	2340
D, D^*, or D_s modes			
$\overline{D}{}^0 \pi^+$	$(5.3 \pm 0.5) \times 10^{-3}$		2308
$\overline{D}{}^0 \rho^+$	(1.34 ± 0.18) %		2238
$\overline{D}{}^0 \pi^+ \pi^+ \pi^-$	(1.1 ± 0.4) %		2289
$\overline{D}{}^0 \pi^+ \pi^+ \pi^-$ nonresonant	$(5 \pm 4) \times 10^{-3}$		2289
$\overline{D}{}^0 \pi^+ \rho^0$	$(4.2 \pm 3.0) \times 10^{-3}$		2209
$\overline{D}{}^0 a_1(1260)^+$	$(5 \pm 4) \times 10^{-3}$		2123
$D^*(2010)^- \pi^+ \pi^+$	$(2.1 \pm 0.6) \times 10^{-3}$		2247
$D^- \pi^+ \pi^+$	$< 1.4 \times 10^{-3}$	CL=90%	2299
$\overline{D}{}^*(2007)^0 \pi^+$	$(5.2 \pm 0.8) \times 10^{-3}$		2256
$\overline{D}{}^*(2007)^0 \rho^+$	(1.55 ± 0.31) %		2183
$\overline{D}{}^*(2007)^0 \pi^+ \pi^+ \pi^-$	$(9.4 \pm 2.6) \times 10^{-3}$		2236
$\overline{D}{}^*(2007)^0 a_1(1260)^+$	(1.9 ± 0.5) %		2062
$D^*(2010)^- \pi^+ \pi^+ \pi^0$	(1.5 ± 0.7) %		2235
$D^*(2010)^- \pi^+ \pi^+ \pi^+ \pi^-$	< 1 %	CL=90%	2217
$\overline{D}_1(2420)^0 \pi^+$	$(1.5 \pm 0.6) \times 10^{-3}$	S=1.3	2081
$\overline{D}_1(2420)^0 \rho^+$	$< 1.4 \times 10^{-3}$	CL=90%	1997
$\overline{D}_2^*(2460)^0 \pi^+$	$< 1.3 \times 10^{-3}$	CL=90%	2064
$\overline{D}_2^*(2460)^0 \rho^+$	$< 4.7 \times 10^{-3}$	CL=90%	1979
$\overline{D}{}^0 D_s^+$	(1.7 ± 0.6) %		1815
$\overline{D}{}^0 D_s^{*+}$	(1.2 ± 1.0) %		1734
$\overline{D}{}^*(2007)^0 D_s^+$	$(10 \pm 7) \times 10^{-3}$		1737
$\overline{D}{}^*(2007)^0 D_s^{*+}$	(2.3 ± 1.4) %		1650
$D_s^+ \pi^0$	$< 2.0 \times 10^{-4}$	CL=90%	2270
$D_s^{*+} \pi^0$	$< 3.3 \times 10^{-4}$	CL=90%	2214
$D_s^+ \eta$	$< 5 \times 10^{-4}$	CL=90%	2235
$D_s^{*+} \eta$	$< 8 \times 10^{-4}$	CL=90%	2177
$D_s^+ \rho^0$	$< 4 \times 10^{-4}$	CL=90%	2198
$D_s^{*+} \rho^0$	$< 5 \times 10^{-4}$	CL=90%	2139
$D_s^+ \omega$	$< 5 \times 10^{-4}$	CL=90%	2195
$D_s^{*+} \omega$	$< 7 \times 10^{-4}$	CL=90%	2136
$D_s^+ a_1(1260)^0$	$< 2.2 \times 10^{-3}$	CL=90%	2079
$D_s^{*+} a_1(1260)^0$	$< 1.6 \times 10^{-3}$	CL=90%	2014
$D_s^+ \phi$	$< 3.2 \times 10^{-4}$	CL=90%	2141
$D_s^{*+} \phi$	$< 4 \times 10^{-4}$	CL=90%	2079
$D_s^+ \overline{K}{}^0$	$< 1.1 \times 10^{-3}$	CL=90%	2241
$D_s^{*+} \overline{K}{}^0$	$< 1.1 \times 10^{-3}$	CL=90%	2184
$D_s^+ \overline{K}{}^*(892)^0$	$< 5 \times 10^{-4}$	CL=90%	2171
$D_s^{*+} \overline{K}{}^*(892)^0$	$< 4 \times 10^{-4}$	CL=90%	2110
$D_s^- \pi^+ K^+$	$< 8 \times 10^{-4}$	CL=90%	2222
$D_s^{*-} \pi^+ K^+$	$< 1.2 \times 10^{-3}$	CL=90%	2164
$D_s^- \pi^+ K^*(892)^+$	$< 6 \times 10^{-3}$	CL=90%	2137
$D_s^{*-} \pi^+ K^*(892)^+$	$< 8 \times 10^{-3}$	CL=90%	2075
Charmonium modes			
$J/\psi(1S) K^+$	$(1.01 \pm 0.14) \times 10^{-3}$		1683
$J/\psi(1S) K^+ \pi^+ \pi^-$	$(1.4 \pm 0.6) \times 10^{-3}$		1612
$J/\psi(1S) K^*(892)^+$	$(1.7 \pm 0.5) \times 10^{-3}$		1571
$J/\psi(1S) \pi^+$	$(4.4 \pm 2.4) \times 10^{-5}$		1727
$\psi(2S) K^+$	$(6.9 \pm 3.1) \times 10^{-4}$	S=1.3	1284
$\psi(2S) K^*(892)^+$	$< 3.0 \times 10^{-3}$	CL=90%	1115
$\psi(2S) K^*(892)^+ \pi^+ \pi^-$	$(1.9 \pm 1.2) \times 10^{-3}$		909
$\chi_{c1}(1P) K^+$	$(1.0 \pm 0.4) \times 10^{-3}$		1411
$\chi_{c1}(1P) K^*(892)^+$	$< 2.1 \times 10^{-3}$	CL=90%	1265
K or K^* modes			
$K^0 \pi^+$	$< 4.8 \times 10^{-5}$	CL=90%	2614
$K^+ \pi^0$	$< 1.4 \times 10^{-5}$	CL=90%	2615
$K^*(892)^0 \pi^+$	$< 4.1 \times 10^{-5}$	CL=90%	2561
$K^*(892)^+ \pi^0$	$< 9.9 \times 10^{-5}$	CL=90%	2562
$K^+ \pi^- \pi^+$ (no charm)	$< 1.9 \times 10^{-4}$	CL=90%	2609
$K_1(1400)^0 \pi^+$	$< 2.6 \times 10^{-3}$	CL=90%	2451
$K_2^*(1430)^0 \pi^+$	$< 6.8 \times 10^{-3}$	CL=90%	2443
$K^+ \rho^0$	$< 1.9 \times 10^{-5}$	CL=90%	2559
$K^0 \rho^+$	$< 4.8 \times 10^{-5}$	CL=90%	2559
$K^*(892)^+ \pi^+ \pi^-$	$< 1.1 \times 10^{-3}$	CL=90%	2556
$K^*(892)^+ \rho^0$	$< 9.0 \times 10^{-4}$	CL=90%	2505
$K_1(1400)^+ \rho^0$	$< 7.8 \times 10^{-4}$	CL=90%	2389
$K_2^*(1430)^+ \rho^0$	$< 1.5 \times 10^{-3}$	CL=90%	2382
$K^+ K^- K^+$	$< 3.1 \times 10^{-4}$	CL=90%	2522
$K^+ \phi$	$< 1.2 \times 10^{-5}$	CL=90%	2516
$K^*(892)^+ K^+ K^-$	$< 1.6 \times 10^{-3}$	CL=90%	2466
$K^*(892)^+ \phi$	$< 7.0 \times 10^{-5}$	CL=90%	2460
$K_1(1400)^+ \phi$	$< 1.1 \times 10^{-3}$	CL=90%	2339
$K_2^*(1430)^+ \phi$	$< 3.4 \times 10^{-3}$	CL=90%	2332
$K^+ f_0(980)$	$< 8 \times 10^{-5}$	CL=90%	2524
$K^*(892)^+ \gamma$	$(5.7 \pm 3.3) \times 10^{-5}$		2564
$K_1(1270)^+ \gamma$	$< 7.3 \times 10^{-3}$	CL=90%	2486
$K_1(1400)^+ \gamma$	$< 2.2 \times 10^{-3}$	CL=90%	2453
$K_2^*(1430)^+ \gamma$	$< 1.4 \times 10^{-3}$	CL=90%	2447
$K^*(1680)^+ \gamma$	$< 1.9 \times 10^{-3}$	CL=90%	2361
$K_3^*(1780)^+ \gamma$	$< 5.5 \times 10^{-3}$	CL=90%	2343
$K_4^*(2045)^+ \gamma$	$< 9.9 \times 10^{-3}$	CL=90%	2243
Light unflavored meson modes			
$\pi^+ \pi^0$	$< 1.7 \times 10^{-5}$	CL=90%	2636
$\pi^+ \pi^+ \pi^-$	$< 1.9 \times 10^{-4}$	CL=90%	2630
$\rho^0 \pi^+$	$< 4.3 \times 10^{-5}$	CL=90%	2582
$\pi^+ f_0(980)$	$< 1.4 \times 10^{-4}$	CL=90%	2547
$\pi^+ f_2(1270)$	$< 2.4 \times 10^{-4}$	CL=90%	2483
$\pi^+ \pi^0 \pi^0$	$< 8.9 \times 10^{-4}$	CL=90%	2631
$\rho^+ \pi^0$	$< 7.7 \times 10^{-5}$	CL=90%	2582
$\pi^+ \pi^- \pi^+ \pi^0$	$< 4.0 \times 10^{-3}$	CL=90%	2621
$\rho^+ \rho^0$	$< 1.0 \times 10^{-3}$	CL=90%	2525
$a_1(1260)^+ \pi^0$	$< 1.7 \times 10^{-3}$	CL=90%	2494
$a_1(1260)^0 \pi^+$	$< 9.0 \times 10^{-4}$	CL=90%	2494
$\omega \pi^+$	$< 4.0 \times 10^{-4}$	CL=90%	2580
$\eta \pi^+$	$< 7.0 \times 10^{-4}$	CL=90%	2609
$\pi^+ \pi^+ \pi^+ \pi^- \pi^-$	$< 8.6 \times 10^{-4}$	CL=90%	2608
$\rho^0 a_1(1260)^+$	$< 6.2 \times 10^{-4}$	CL=90%	2434
$\rho^0 a_2(1320)^+$	$< 7.2 \times 10^{-4}$	CL=90%	2411
$\pi^+ \pi^+ \pi^+ \pi^- \pi^- \pi^0$	$< 6.3 \times 10^{-3}$	CL=90%	2592
$a_1(1260)^+ a_1(1260)^0$	< 1.3 %	CL=90%	2335
Baryon modes			
$p \overline{p} \pi^+$	$< 1.6 \times 10^{-4}$	CL=90%	2439
$p \overline{p} \pi^+ \pi^+ \pi^-$	$< 5.2 \times 10^{-4}$	CL=90%	2369
$p \overline{\Lambda}$	$< 6 \times 10^{-5}$	CL=90%	2430
$p \overline{\Lambda} \pi^+ \pi^-$	$< 2.0 \times 10^{-4}$	CL=90%	2367
$\overline{\Delta}{}^0 p$	$< 3.8 \times 10^{-4}$	CL=90%	2402
$\Delta^{++} \overline{p}$	$< 1.5 \times 10^{-4}$	CL=90%	2402

Lepton Family number (*LF*) or Lepton number (*L*) violating modes, or
$\Delta B = 1$ weak neutral current (*B1*) modes

$\pi^+ e^+ e^-$	B1	< 3.9	$\times 10^{-3}$	CL=90%	2638
$\pi^+ \mu^+ \mu^-$	B1	< 9.1	$\times 10^{-3}$	CL=90%	2633
$K^+ \mu^+ \mu^-$	B1	< 1.7	$\times 10^{-4}$	CL=90%	2612
$K^*(892)^+ e^+ e^-$	B1	< 6.9	$\times 10^{-4}$	CL=90%	2564
$K^*(892)^+ \mu^+ \mu^-$	B1	< 1.2	$\times 10^{-3}$	CL=90%	2560
$\pi^+ e^+ \mu^-$	LF	< 6.4	$\times 10^{-3}$	CL=90%	2637
$\pi^+ e^- \mu^+$	LF	< 6.4	$\times 10^{-3}$	CL=90%	2637
$K^+ e^+ \mu^-$	LF	< 6.4	$\times 10^{-3}$	CL=90%	2615
$K^+ e^- \mu^+$	LF	< 6.4	$\times 10^{-3}$	CL=90%	2615
$\pi^- e^+ e^+$	L	< 3.9	$\times 10^{-3}$	CL=90%	2638
$\pi^- \mu^+ \mu^+$	L	< 9.1	$\times 10^{-3}$	CL=90%	2633
$\pi^- e^+ \mu^+$	L	< 6.4	$\times 10^{-3}$	CL=90%	2637
$K^- e^+ e^+$	L	< 3.9	$\times 10^{-3}$	CL=90%	2616
$K^- \mu^+ \mu^+$	L	< 9.1	$\times 10^{-3}$	CL=90%	2612
$K^- e^+ \mu^+$	L	< 6.4	$\times 10^{-3}$	CL=90%	2615

$\boxed{B^0}$ $I(J^P) = \tfrac{1}{2}(0^-)$

I, J, P need confirmation. Quantum numbers shown are quark-model predictions.

Mass $m_{B^0} = 5279.2 \pm 1.8$ MeV
$m_{B^0} - m_{B^\pm} = 0.35 \pm 0.29$ MeV   (S = 1.1)
Mean life $\tau_{B^0} = (1.56 \pm 0.06) \times 10^{-12}$ s
     $c\tau = 468$ μm
$\tau_{B^+}/\tau_{B^0} = 1.02 \pm 0.05$ (average of direct and inferred)
$\tau_{B^+}/\tau_{B^0} = 1.03 \pm 0.06$ (direct measurements)
$\tau_{B^+}/\tau_{B^0} = 0.93 \pm 0.22$ (inferred from branching fractions)

B^0-$\overline{B}^0$ mixing parameters

$\chi_d = 0.175 \pm 0.016$
$\Delta m_{B^0} = m_{B^0_H} - m_{B^0_L} = (0.474 \pm 0.031) \times 10^{12}$ $\hbar$ s^{-1}
$x_d = \Delta m_{B^0}/\Gamma_{B^0} = 0.73 \pm 0.05$

CP violation parameters

$|\text{Re}(\epsilon_{B^0})| < 0.045$

$\overline{B}^0$ modes are charge conjugates of the modes below. Reactions indicate the weak decay vertex and do not include mixing. Modes which do not identify the charge state of the *B* are listed in the $B^\pm/B^0$ ADMIXTURE section.

The branching fractions listed below assume 50% $B^0 \overline{B}^0$ and 50% $B^+ B^-$ production at the $\Upsilon(4S)$. We have attempted to bring older measurements up to date by rescaling their assumed $\Upsilon(4S)$ production ratio to 50:50 and their assumed D, D_s, D^*, and ψ branching ratios to current values whenever this would affect our averages and best limits significantly.

Indentation is used to indicate a subchannel of a previous reaction. All resonant subchannels have been corrected for resonance branching fractions to the final state so the sum of the subchannel branching fractions can exceed that of the final state.

B^0 DECAY MODES	Fraction (Γ_i/Γ)	Scale factor/ Confidence level	*p* (MeV/c)
Semileptonic and leptonic modes			
$\ell^+ \nu_\ell$ anything	[qq] (10.3 ±1.0) %		–
$D^- \ell^+ \nu_\ell$	[qq] (1.9 ±0.5) %		–
$D^*(2010)^- \ell^+ \nu_\ell$	[qq] (4.56±0.27) %		–
$\rho^- \ell^+ \nu_\ell$	[qq] < 4.1 $\times 10^{-4}$	CL=90%	–
D, D^*, or D_s modes			
$D^- \pi^+$	(3.0 ±0.4) $\times 10^{-3}$		2306
$D^- \rho^+$	(7.8 ±1.4) $\times 10^{-3}$		2236
$\overline{D}^0 \pi^+ \pi^-$	< 1.6 $\times 10^{-3}$	CL=90%	2301
$D^*(2010)^- \pi^+$	(2.6 ±0.4) $\times 10^{-3}$		2254
$D^- \pi^+ \pi^+ \pi^-$	(8.0 ±2.5) $\times 10^{-3}$		2287
$\quad (D^- \pi^+ \pi^+ \pi^-)$ nonresonant	(3.9 ±1.9) $\times 10^{-3}$		2287
$\quad D^- \pi^+ \rho^0$	(1.1 ±1.0) $\times 10^{-3}$		2207
$\quad D^- a_1(1260)^+$	(6.0 ±3.3) $\times 10^{-3}$		2121
$D^*(2010)^- \pi^+ \pi^0$	(1.5 ±0.5) %		2247
$\quad D^*(2010)^- \rho^+$	(7.3 ±1.5) $\times 10^{-3}$		2181
$D^*(2010)^- \pi^+ \pi^+ \pi^-$	(7.6 ±1.7) $\times 10^{-3}$	S=1.3	2235
$\quad (D^*(2010)^- \pi^+ \pi^+ \pi^-)$ non-resonant	(0.0 ±2.5) $\times 10^{-3}$		2235
$\quad D^*(2010)^- \pi^+ \rho^0$	(5.7 ±3.1) $\times 10^{-3}$		2151
$\quad D^*(2010)^- a_1(1260)^+$	(1.30±0.27) %		2061
$D^*(2010)^- \pi^+ \pi^+ \pi^- \pi^0$	(3.4 ±1.8) %		2218
$\overline{D}_2^*(2460)^- \pi^+$	< 2.2 $\times 10^{-3}$	CL=90%	2064
$\overline{D}_2^*(2460)^- \rho^+$	< 4.9 $\times 10^{-3}$	CL=90%	1979
$D^- D_s^+$	(7 ±4) $\times 10^{-3}$		1812
$D^*(2010)^- D_s^+$	(1.2 ±0.6) %		1735
$D^- D_s^{*+}$	(2.0 ±1.5) %		1731
$D^*(2010)^- D_s^{*+}$	(1.9 ±1.2) %		1649
$D_s^+ \pi^-$	< 2.8 $\times 10^{-4}$	CL=90%	2270
$D_s^{*+} \pi^-$	< 5 $\times 10^{-4}$	CL=90%	2214
$D_s^+ \rho^-$	< 7 $\times 10^{-4}$	CL=90%	2198
$D_s^{*+} \rho^-$	< 8 $\times 10^{-4}$	CL=90%	2139
$D_s^+ a_1(1260)^-$	< 2.6 $\times 10^{-3}$	CL=90%	2079
$D_s^{*+} a_1(1260)^-$	< 2.2 $\times 10^{-3}$	CL=90%	2014
$D_s^- K^+$	< 2.4 $\times 10^{-4}$	CL=90%	2242
$D_s^{*-} K^+$	< 1.7 $\times 10^{-4}$	CL=90%	2185
$D_s^- K^*(892)^+$	< 9.9 $\times 10^{-4}$	CL=90%	2172
$D_s^{*-} K^*(892)^+$	< 1.1 $\times 10^{-4}$	CL=90%	2112
$D_s^- \pi^+ K^0$	< 5 $\times 10^{-3}$	CL=90%	2221
$D_s^{*-} \pi^+ K^0$	< 3.1 $\times 10^{-3}$	CL=90%	2164
$D_s^- \pi^+ K^*(892)^0$	< 4 $\times 10^{-3}$	CL=90%	2136
$D_s^{*-} \pi^+ K^*(892)^0$	< 2.0 $\times 10^{-3}$	CL=90%	2074
$\overline{D}^0 \pi^0$	< 4.8 $\times 10^{-4}$	CL=90%	2308
$\overline{D}^0 \rho^0$	< 5.5 $\times 10^{-4}$	CL=90%	2238
$\overline{D}^0 \eta$	< 6.8 $\times 10^{-4}$	CL=90%	2274
$\overline{D}^0 \eta'$	< 8.6 $\times 10^{-4}$	CL=90%	2198
$\overline{D}^0 \omega$	< 6.3 $\times 10^{-4}$	CL=90%	2235
$\overline{D}^*(2007)^0 \pi^0$	< 9.7 $\times 10^{-4}$	CL=90%	2256
$\overline{D}^*(2007)^0 \rho^0$	< 1.17 $\times 10^{-3}$	CL=90%	2183
$\overline{D}^*(2007)^0 \eta$	< 6.9 $\times 10^{-4}$	CL=90%	2220
$\overline{D}^*(2007)^0 \eta'$	< 2.7 $\times 10^{-3}$	CL=90%	2141
$\overline{D}^*(2007)^0 \omega$	< 2.1 $\times 10^{-3}$	CL=90%	2180
Charmonium modes			
$J/\psi(1S) K^0$	(7.5 ±2.1) $\times 10^{-4}$		1683
$J/\psi(1S) K^+ \pi^-$	(1.1 ±0.6) $\times 10^{-3}$		1652
$J/\psi(1S) K^*(892)^0$	(1.58±0.27) $\times 10^{-3}$		1570
$J/\psi(1S) \pi^0$	< 6.9 $\times 10^{-3}$	CL=90%	1728
$\psi(2S) K^0$	< 8 $\times 10^{-4}$	CL=90%	1283
$\psi(2S) K^+ \pi^-$	< 1 $\times 10^{-3}$	CL=90%	1238
$\psi(2S) K^*(892)^0$	(1.4 ±0.9) $\times 10^{-3}$		1113
$\chi_{c1}(1P) K^0$	< 2.7 $\times 10^{-3}$	CL=90%	1411
$\chi_{c1}(1P) K^*(892)^0$	< 2.1 $\times 10^{-3}$	CL=90%	1263
K or K^* modes			
$K^+ \pi^-$	< 1.7 $\times 10^{-5}$	CL=90%	2615
$K^0 \pi^0$	< 4.0 $\times 10^{-5}$	CL=90%	2614
$K^+ K^-$	< 4 $\times 10^{-6}$	CL=90%	2593
$K^+ \rho^-$	< 3.5 $\times 10^{-5}$	CL=90%	2559
$K^0 \rho^0$	< 3.9 $\times 10^{-5}$	CL=90%	2559
$K^0 f_0(980)$	< 3.6 $\times 10^{-4}$	CL=90%	2523
$K^*(892)^+ \pi^-$	< 7.2 $\times 10^{-5}$	CL=90%	2562
$K^*(892)^0 \pi^0$	< 2.8 $\times 10^{-5}$	CL=90%	2562
$K_2^*(1430)^+ \pi^-$	< 2.6 $\times 10^{-3}$	CL=90%	2445
$K^0 \overline{K}^+ K^-$	< 1.3 $\times 10^{-3}$	CL=90%	2522
$K^0 \phi$	< 8.8 $\times 10^{-5}$	CL=90%	2516
$K^- \pi^+ \pi^+ \pi^-$	[bbb] < 2.1 $\times 10^{-4}$	CL=90%	2600
$K^*(892)^0 \pi^+ \pi^-$	< 1.4 $\times 10^{-3}$	CL=90%	2556
$\quad K^*(892)^0 \rho^0$	< 4.6 $\times 10^{-4}$	CL=90%	2504
$\quad K^*(892)^0 f_0(980)$	< 1.7 $\times 10^{-4}$	CL=90%	2467
$K_1(1400)^+ \pi^-$	< 1.1 $\times 10^{-3}$	CL=90%	2451
$K^- a_1(1260)^+$	[bbb] < 3.9 $\times 10^{-4}$	CL=90%	2471
$K^*(892)^0 K^+ K^-$	< 6.1 $\times 10^{-4}$	CL=90%	2466
$K^*(892)^0 \phi$	< 4.3 $\times 10^{-5}$	CL=90%	2459
$K_1(1400)^0 \rho^0$	< 3.0 $\times 10^{-3}$	CL=90%	2389
$K_1(1400)^0 \phi$	< 5.0 $\times 10^{-3}$	CL=90%	2339
$K_2^*(1430)^0 \rho^0$	< 1.1 $\times 10^{-3}$	CL=90%	2380
$K_2^*(1430)^0 \phi$	< 1.4 $\times 10^{-3}$	CL=90%	2330
$K^*(892)^0 \gamma$	(4.0 ±1.9) $\times 10^{-5}$		2564
$K_1(1270)^0 \gamma$	< 7.0 $\times 10^{-3}$	CL=90%	2486
$K_1(1400)^0 \gamma$	< 4.3 $\times 10^{-3}$	CL=90%	2453
$K_2^*(1430)^0 \gamma$	< 4.0 $\times 10^{-4}$	CL=90%	2445
$K^*(1680)^0 \gamma$	< 2.0 $\times 10^{-3}$	CL=90%	2361
$K_3^*(1780)^0 \gamma$	< 1.0 %	CL=90%	2343
$K_4^*(2045)^0 \gamma$	< 4.3 $\times 10^{-3}$	CL=90%	2244
$\phi\phi$	< 3.9 $\times 10^{-5}$	CL=90%	2435

Light unflavored meson modes

Mode		Fraction		CL	p
$\pi^+\pi^-$		< 2.0	$\times 10^{-5}$	CL=90%	2636
$\pi^0\pi^0$		< 9.1	$\times 10^{-6}$	CL=90%	2636
$\eta\pi^0$		< 2.5	$\times 10^{-4}$	CL=90%	2609
$\eta\eta$		< 4.1	$\times 10^{-4}$	CL=90%	2582
$\pi^+\pi^-\pi^0$		< 7.2	$\times 10^{-4}$	CL=90%	2631
$\rho^0\pi^0$		< 2.4	$\times 10^{-5}$	CL=90%	2582
$\rho^\mp\pi^\pm$	[gg]	< 8.8	$\times 10^{-5}$	CL=90%	2582
$\pi^+\pi^-\pi^+\pi^-$		< 2.8	$\times 10^{-4}$	CL=90%	2621
$\rho^0\rho^0$		< 2.8	$\times 10^{-4}$	CL=90%	2525
$a_1(1260)^\mp\pi^\pm$	[gg]	< 4.9	$\times 10^{-4}$	CL=90%	2494
$a_2(1320)^\mp\pi^\pm$	[gg]	< 3.0	$\times 10^{-4}$	CL=90%	2473
$\pi^+\pi^-\pi^0\pi^0$		< 3.1	$\times 10^{-3}$	CL=90%	2622
$\rho^+\rho^-$		< 2.2	$\times 10^{-3}$	CL=90%	2525
$a_1(1260)^0\pi^0$		< 1.1	$\times 10^{-3}$	CL=90%	2494
$\omega\pi^0$		< 4.6	$\times 10^{-4}$	CL=90%	2580
$\pi^+\pi^+\pi^-\pi^-\pi^0$		< 9.0	$\times 10^{-3}$	CL=90%	2609
$a_1(1260)^+\rho^-$		< 3.4	$\times 10^{-3}$	CL=90%	2434
$a_1(1260)^0\rho^0$		< 2.4	$\times 10^{-3}$	CL=90%	2434
$\pi^+\pi^+\pi^+\pi^-\pi^-\pi^-$		< 3.0	$\times 10^{-3}$	CL=90%	2592
$a_1(1260)^+ a_1(1260)^-$		< 2.8	$\times 10^{-3}$	CL=90%	2336
$\pi^+\pi^+\pi^+\pi^-\pi^-\pi^-\pi^0$		< 1.1	%	CL=90%	2572

Baryon modes

Mode		Fraction		CL	p
$p\overline{p}$		< 3.4	$\times 10^{-5}$	CL=90%	2467
$p\overline{p}\pi^+\pi^-$		< 2.5	$\times 10^{-4}$	CL=90%	2406
$p\overline{\Lambda}\pi^-$		< 1.8	$\times 10^{-4}$	CL=90%	2401
$\Delta^0\overline{\Delta}{}^0$		< 1.5	$\times 10^{-3}$	CL=90%	2334
$\Delta^{++}\Delta^{--}$		< 1.1	$\times 10^{-4}$	CL=90%	2334
$\overline{\Sigma}_c^{--}\Delta^{++}$		< 1.2	$\times 10^{-3}$	CL=90%	1839

Lepton Family number (LF) violating modes, or $\Delta B = 1$ weak neutral current (B1) modes

Mode			Fraction		CL	p
$\gamma\gamma$	B1		< 3.9	$\times 10^{-5}$	CL=90%	2640
e^+e^-	B1		< 5.9	$\times 10^{-6}$	CL=90%	2640
$\mu^+\mu^-$	B1		< 5.9	$\times 10^{-6}$	CL=90%	2637
$K^0 e^+e^-$	B1		< 3.0	$\times 10^{-4}$	CL=90%	2616
$K^0\mu^+\mu^-$	B1		< 3.6	$\times 10^{-4}$	CL=90%	2612
$K^*(892)^0 e^+e^-$	B1		< 2.9	$\times 10^{-4}$	CL=90%	2564
$K^*(892)^0\mu^+\mu^-$	B1		< 2.3	$\times 10^{-5}$	CL=90%	2559
$e^\pm\mu^\mp$	LF	[gg]	< 5.9	$\times 10^{-6}$	CL=90%	2639
$e^\pm\tau^\mp$	LF	[gg]	< 5.3	$\times 10^{-4}$	CL=90%	2341
$\mu^\pm\tau^\mp$	LF	[gg]	< 8.3	$\times 10^{-4}$	CL=90%	2339

$B^\pm/B^0$ ADMIXTURE

The branching fraction measurements are for an admixture of B mesons at the $\Upsilon(4S)$. The values quoted assume that $B(\Upsilon(4S) \to B\overline{B}) = 100\%$.

For inclusive branching fractions, e.g., $B \to D^\pm$ anything, the treatment of multiple D's in the final state must be defined. One possibility would be to count the number of events with one-or-more D's and divide by the total number of B's. Another possibility would be to count the total number of D's and divide by the total number of B's, which is the definition of average multiplicity. The two definitions are identical when only one of the specified particles is allowed in the final state. Even though the "one-or-more" definition seems sensible, for practical reasons inclusive branching fractions are almost always measured using the multiplicity definition. For heavy final state particles, authors call their results inclusive branching fractions while for light particles some authors call their results multiplicities. In the B sections, we list all results as inclusive branching fractions, adopting a multiplicity definition. This means that inclusive branching fractions can exceed 100% and that inclusive partial widths can exceed total widths, just as inclusive cross sections can exceed total cross sections.

$\overline{B}$ modes are charge conjugates of the modes below. Reactions indicate the weak decay vertex and do not include mixing.

B DECAY MODES

B DECAY MODES		Fraction (Γ_i/Γ)	Scale factor/ Confidence level	p (MeV/c)
Semileptonic and leptonic modes				
$e^+\nu_e$ anything	[ccc]	(10.4 ±0.4) %	S=1.3	–
$\overline{p}e^+\nu_e$ anything		< 1.6	$\times 10^{-3}$	CL=90%	–
$\mu^+\nu_\mu$ anything	[ccc]	(10.3 ±0.5) %		–
$\ell^+\nu_\ell$ anything	[qq,ccc]	(10.43±0.24) %		–
$D^-\ell^+\nu_\ell$ anything	[qq]	(2.7 ±0.8) %		–
$\overline{D}{}^0\ell^+\nu_\ell$ anything	[qq]	(7.0 ±1.4) %		–
$\overline{D}{}^{**}\ell^+\nu_\ell$	[qq,ddd]	(2.7 ±0.7) %		–
$\overline{D}(1)(2420)^0\ell^+\nu_\ell$ anything		seen		–
$\overline{D}(2)^*(2460)^0\ell^+\nu_\ell$ anything		not seen		–
$D^{*-}\pi^+\ell^+\nu_\ell$ anything		(1.00±0.34) %		–
$D_s^-\ell^+\nu_\ell$ anything	[qq]	< 9	$\times 10^{-3}$	CL=90%	–
$D_s^-\ell^+\nu_\ell K^+$ anything	[qq]	< 6	$\times 10^{-3}$	CL=90%	–
$D_s^-\ell^+\nu_\ell K^0$ anything	[qq]	< 9	$\times 10^{-3}$	CL=90%	–
$K^+\ell^+\nu_\ell$ anything	[qq]	(6.0 ±0.5) %		–
$K^-\ell^+\nu_\ell$ anything	[qq]	(10 ±4) $\times 10^{-3}$		–
$K^0/\overline{K}{}^0\ell^+\nu_\ell$ anything	[qq]	(4.4 ±0.5) %		–
D, D*, or D_s modes				
$D^\pm$ anything		(24.2 ±3.3) %		–
$D^0/\overline{D}{}^0$ anything		(58 ±5) %	S=1.1	–
$D^*(2010)^\pm$ anything		(23.1 ±3.3) %	S=1.1	–
$D_s^\pm$ anything	[gg]	(8.6 ±1.6) %		–
$D_s D, D_s^* D, D_s D^*,$ or $D_s^* D^*$	[gg]	(4.9 ±1.1) %		–
$D^*(2010)\gamma$		< 1.1	$\times 10^{-3}$	CL=90%	–
$D_s^+\pi^-, D_s^{*+}\pi^-, D_s^+\rho^-,$ $D_s^{*+}\rho^-, D_s^+\pi^0, D_s^{*+}\pi^0,$ $D_s^+\eta, D_s^{*+}\eta, D_s^+\rho^0,$ $D_s^{*+}\rho^0, D_s^+\omega, D_s^{*+}\omega$	[gg]	< 5	$\times 10^{-4}$	CL=90%	–
Charmonium modes				
$J/\psi(1S)$ anything		(1.14±0.06) %		–
$J/\psi(1S)$(direct) anything		(8.0 ±0.8) $\times 10^{-3}$		–
$\psi(2S)$ anything		(3.5 ±0.5) $\times 10^{-3}$		–
$\chi_{c1}(1P)$ anything		(4.2 ±0.7) $\times 10^{-3}$		–
$\chi_{c1}(1P)$(direct) anything		(3.7 ±0.7) $\times 10^{-3}$		–
$\chi_{c2}(1P)$ anything		< 3.8	$\times 10^{-3}$	CL=90%	–
$\eta_c(1S)$ anything		< 9	$\times 10^{-3}$	CL=90%	–
K or K* modes				
$K^\pm$ anything	[gg]	(78.9 ±2.5) %		–
K^+ anything		(66 ±5) %		–
K^- anything		(13 ±4) %		–
$K^0/\overline{K}{}^0$ anything	[gg]	(64 ±4) %		–
$K^*(892)^\pm$ anything		(18 ±6) %		–
$K^*(892)^0/\overline{K}{}^*(892)^0$ anything	[gg]	(14.6 ±2.6) %		–
$K_1(1400)\gamma$		< 4.1	$\times 10^{-4}$	CL=90%	–
$K_2^*(1430)\gamma$		< 8.3	$\times 10^{-4}$	CL=90%	–
$K_2(1770)\gamma$		< 1.2	$\times 10^{-3}$	CL=90%	–
$K_3^*(1780)\gamma$		< 3.0	$\times 10^{-3}$	CL=90%	–
$K_4^*(2045)\gamma$		< 1.0	$\times 10^{-3}$	CL=90%	–
$b \to s\gamma$		(2.3 ±0.7) $\times 10^{-4}$		–
Light unflavored meson modes				
$\pi^\pm$ anything	[gg,eee]	(359 ±7) %		–
ρ^0 anything		(21 ±5) %		–
ω anything		< 81	%	CL=90%	–
ϕ anything		(3.5 ±0.7) %	S=1.8	–
Baryon modes				
charmed-baryon anything		(6.4 ±1.1) %		–
$\overline{\Sigma}_c^{--}$ anything		(4.8 ±2.5) $\times 10^{-3}$		–
$\overline{\Sigma}_c^-$ anything		< 1.1	%	CL=90%	–
$\overline{\Sigma}_c^0$ anything		(5.2 ±2.5) $\times 10^{-3}$		–
$\overline{\Sigma}_c^0 N (N = p$ or $n)$		< 1.7	$\times 10^{-3}$	CL=90%	–
$p/\overline{p}$ anything	[gg]	(8.0 ±0.4) %		–
$p/\overline{p}$(direct) anything	[gg]	(5.5 ±0.5) %		–
$\Lambda/\overline{\Lambda}$ anything	[gg]	(4.0 ±0.5) %		–
$\Xi^-/\overline{\Xi}{}^+$ anything	[gg]	(2.7 ±0.6) $\times 10^{-3}$		–
baryons anything		(6.8 ±0.6) %		–
$p\overline{p}$ anything		(2.47±0.23) %		–
$\Lambda\overline{p}/\overline{\Lambda}p$ anything	[gg]	(2.5 ±0.4) %		–
$\Lambda\overline{\Lambda}$ anything		< 5	$\times 10^{-3}$	CL=90%	–

$\Delta B = 1$ weak neutral current ($B1$) modes				
$e^+ e^-$ anything	$B1$	< 2.4	$\times 10^{-3}$ CL=90%	–
$\mu^+ \mu^-$ anything	$B1$	< 2.4	$\times 10^{-3}$ CL=90%	–

$B^\pm / B^0 / B_s^0 / b$-baryon ADMIXTURE

These measurements are for an admixture of bottom particles at high energy (LEP, Tevatron, $Sp\bar{p}S$).

Mean life $\tau = (1.549 \pm 0.020) \times 10^{-12}$ s

Mean life $\tau = (1.72 \pm 0.10) \times 10^{-12}$ s   Charged b-hadron admixture

Mean life $\tau = (1.58 \pm 0.14) \times 10^{-12}$ s   Neutral b-hadron admixture

$\tau_{\text{charged } b\text{-hadron}} / \tau_{\text{neutral } b\text{-hadron}} = 1.09 \pm 0.13$

The branching fraction measurements are for an admixture of B mesons and baryons at energies above the $\Upsilon(4S)$. Only the highest energy results (LEP, Tevatron, $Sp\bar{p}S$) are used in the branching fraction averages. The production fractions give our best current estimate of the admixture at LEP.

For inclusive branching fractions, *e.g.*, $B \to D^\pm$ anything, the treatment of multiple D's in the final state must be defined. One possibility would be to count the number of events with one-or-more D's and divide by the total number of B's. Another possibility would be to count the total number of D's and divide by the total number of B's, which is the definition of average multiplicity. The two definitions are identical when only one of the specified particles is allowed in the final state. Even though the "one-or-more" definition seems sensible, for practical reasons inclusive branching fractions are almost always measured using the multiplicity definition. For heavy final state particles, authors call their results inclusive branching fractions while for light particles some authors call their results multiplicities. In the B sections, we list all results as inclusive branching fractions, adopting a multiplicity definition. This means that inclusive branching fractions can exceed 100% and that inclusive partial widths can exceed total widths, just as inclusive cross sections can exceed total cross sections.

The modes below are listed for a $\bar{b}$ initial state. b modes are their charge conjugates. Reactions indicate the weak decay vertex and do not include mixing.

$\bar{b}$ DECAY MODES		Fraction (Γ_i/Γ)	Confidence level	p (MeV/c)

PRODUCTION FRACTIONS

The production fractions for weakly decaying b-hadrons at the Z have been calculated from the best values of mean lives, mixing parameters, and branching fractions in this edition by O. Hayes (CERN) and M. Jimack (U. Birmingham) as described in the note "Production and Decay of b-Flavored Hadrons" in the $B^\pm$ Particle Listings. Values assume

$\text{B}(\bar{b} \to B^+) = \text{B}(\bar{b} \to B^0)$
$\text{B}(\bar{b} \to B^+) + \text{B}(\bar{b} \to B^0) + \text{B}(\bar{b} \to B_s^0) + \text{B}(b \to \Lambda_b) = 100\%$.

The notation for production fractions varies in the literature (f_{B^0}, $f(b \to \bar{B}^0)$, $\text{Br}(b \to \bar{B}^0)$). We use our own branching fraction notation here, $\text{B}(\bar{b} \to B^0)$.

$\bar{b} \to B^+$		(37.8 $\pm$ 2.2) %	–
$\bar{b} \to B^0$		(37.8 $\pm$ 2.2) %	–
$\bar{b} \to B_s^0$		(11.2 $^{+1.8}_{-1.9}$) %	–
$b \to \Lambda_b$		(13.2 $\pm$ 4.1) %	–

DECAY MODES

Semileptonic and leptonic modes

$\bar{b} \to e^+ \nu_e$ anything	[ccc]	(11.1 $\pm$ 1.0) %		–
$\bar{b} \to \mu^+ \nu_\mu$ anything	[ccc]	(10.7 $\pm$ 0.7) %		–
$\bar{b} \to \ell^+ \nu_\ell$ anything	[qq,ccc]	(11.13 $\pm$ 0.29) %		–
$\bar{b} \to D^- \ell^+ \nu_\ell$ anything	[qq]	(2.01 $\pm$ 0.29) %		–
$\bar{b} \to \overline{D}^0 \ell^+ \nu_\ell$ anything	[qq]	(6.6 $\pm$ 0.6) %		–
$\bar{b} \to D^{*-} \ell^+ \nu_\ell$ anything	[qq]	(2.76 $\pm$ 0.29) %		–
$\bar{b} \to \overline{D}_j^0 \ell^+ \nu_\ell$ anything	[qq,fff]	seen		–
$\bar{b} \to D_j^- \ell^+ \nu_\ell$ anything	[qq,fff]	seen		–
$\bar{b} \to \overline{D}_2^*(2460)^0 \ell^+ \nu_\ell$ anything		seen		–
$\bar{b} \to D_2^*(2460)^- \ell^+ \nu_\ell$ anything		seen		–
$\bar{b} \to \tau^+ \nu_\tau$ anything		(2.7 $\pm$ 0.4) %		–
$\bar{b} \to \bar{b} \to \bar{c} \to \ell^- \bar{\nu}_\ell$ anything	[qq]	(7.9 $\pm$ 0.8) %		–

Charmonium modes

$\bar{b} \to J/\psi(1S)$ anything	(1.16 $\pm$ 0.10) %		–
$\bar{b} \to \psi(2S)$ anything	(4.8 $\pm$ 2.4) $\times 10^{-3}$		–
$\bar{b} \to \chi_{c1}(1P)$ anything	(1.8 $\pm$ 0.5) %		–

K or K^* modes

$\bar{b} \to \bar{s}\gamma$	< 1.2 $\times 10^{-3}$	90%	–
$\bar{b} \to K^\pm$ anything	(88 $\pm$ 19) %		–
$\bar{b} \to K_S^0$ anything	(29.0 $\pm$ 2.9) %		–

Baryon modes

$\bar{b} \to p/\bar{p}$ anything	(14 $\pm$ 6) %		–
$\bar{b} \to \Lambda/\overline{\Lambda}$ anything	(5.9 $\pm$ 1.1) %		–

Other modes

$\bar{b} \to$ charged anything	[eee] (584 $\pm$ 40) %		–

$\Delta B = 1$ weak neutral current ($B1$) modes

$\bar{b} \to \mu^+ \mu^-$ anything	$B1$	< 5.0	$\times 10^{-5}$	90%	–
$\bar{b} \to \nu \bar{\nu}$ anything	$B1$	< 3.9	$\times 10^{-4}$		–

B^* $I(J^P) = \frac{1}{2}(1^-)$

I, J, P need confirmation. Quantum numbers shown are quark-model predictions.

Mass $m_{B^*} = 5324.8 \pm 1.8$ MeV
$m_{B^*} - m_B = 45.7 \pm 0.4$ MeV

B^* DECAY MODES	Fraction (Γ_i/Γ)	p (MeV/c)
$B\gamma$	dominant	46

BOTTOM, STRANGE MESONS ($B = \pm 1$, $S = \mp 1$)

$B_s^0 = s\bar{b}$, $\overline{B}_s^0 = \bar{s}b$,   similarly for B_s^*'s

B_s^0 $I(J^P) = 0(0^-)$

I, J, P need confirmation. Quantum numbers shown are quark-model predictions.

Mass $m_{B_s^0} = 5369.3 \pm 2.0$ MeV
Mean life $\tau = (1.61 ^{+0.10}_{-0.09}) \times 10^{-12}$ s
$c\tau = 483$ μm

B_s^0-$\overline{B}_s^0$ mixing parameters

$\chi_s > 0.49$, CL = 95%
χ_B at high energy $= f_d\chi_d + f_s\chi_s = 0.126 \pm 0.008$
$\Delta m_{B_s^0} = m_{B_{sH}^0} - m_{B_{sL}^0} > 5.9 \times 10^{12}$ $\hbar$ s^{-1}, CL = 95%
$x_s = \Delta m_{B_s^0} / \Gamma_{B_s^0} > 9.5$, CL = 95%

These branching fractions all scale with $\text{B}(\bar{b} \to B_s^0)$, the LEP B_s^0 production fraction. The first four were evaluated using $\text{B}(\bar{b} \to B_s^0) = (11.2^{+1.8}_{-1.9})\%$ and the rest assume $\text{B}(\bar{b} \to B_s^0) = 12\%$.

The branching fraction $\text{B}(B_s^0 \to D_s^- \ell^+ \nu_\ell$ anything) is not a pure measurement since the measured product branching fraction $\text{B}(\bar{b} \to B_s^0) \times \text{B}(B_s^0 \to D_s^- \ell^+ \nu_\ell$ anything) was used to determine $\text{B}(\bar{b} \to B_s^0)$, as described in the note on "Production and Decay of b-Flavored Hadrons."

B_s^0 DECAY MODES		Fraction (Γ_i/Γ)	Confidence level	p (MeV/c)
D_s^- anything		(87 $\pm$ 31) %		–
$D_s^- \ell^+ \nu_\ell$ anything	[ggg]	(7.6 $\pm$ 2.4) %		–
$D_s^- \pi^+$		< 12	%	2321
$J/\psi(1S)\phi$		< 6	$\times 10^{-3}$	1590
$\psi(2S)\phi$		seen		1122
$\pi^0 \pi^0$		< 2.1	$\times 10^{-4}$	90%	2861
$\eta \pi^0$		< 1.0	$\times 10^{-3}$	90%	2655
$\eta \eta$		< 1.5	$\times 10^{-3}$	90%	2628
$\pi^+ K^-$		< 2.6	$\times 10^{-4}$	90%	2660
$K^+ K^-$		< 1.4	$\times 10^{-4}$	90%	2639

$\Delta B = 1$ weak neutral current ($B1$) modes

$\gamma\gamma$	$B1$	< 1.48	$\times 10^{-4}$	90%	2685

$c\bar{c}$ MESONS

$\eta_c(1S)$ $\quad I^G(J^{PC}) = 0^+(0^{-+})$

Mass $m = 2979.8 \pm 2.1$ MeV   (S = 2.1)
Full width $\Gamma = 13.2^{+3.8}_{-3.2}$ MeV

$\eta_c(1S)$ DECAY MODES	Fraction (Γ_i/Γ)	Confidence level	p (MeV/c)
Decays involving hadronic resonances			
$\eta'(958)\pi\pi$	(4.1 ± 1.7) %		1319
$\rho\rho$	(2.6 ± 0.9) %		1275
$K^*(892)^0 K^-\pi^+ +$ c.c.	(2.0 ± 0.7) %		1273
$K^*(892)\overline{K}^*(892)$	$(8.5 \pm 3.1) \times 10^{-3}$		1193
$\phi\phi$	$(7.1 \pm 2.8) \times 10^{-3}$		1086
$a_0(980)\pi$	< 2 %	90%	1323
$a_2(1320)\pi$	< 2 %	90%	1193
$K^*(892)\overline{K} +$ c.c.	< 1.28 %	90%	1307
$f_2(1270)\eta$	< 1.1 %	90%	1142
$\omega\omega$	$< 3.1 \times 10^{-3}$	90%	1268
Decays into stable hadrons			
$K\overline{K}\pi$	(5.5 ± 1.7) %		1378
$\eta\pi\pi$	(4.9 ± 1.8) %		1425
$\pi^+\pi^- K^+ K^-$	$(2.0^{+0.7}_{-0.6})$ %		1342
$2(K^+ K^-)$	(2.1 ± 1.2) %		1053
$2(\pi^+\pi^-)$	(1.2 ± 0.4) %		1457
$p\overline{p}$	$(1.2 \pm 0.4) \times 10^{-3}$		1157
$K\overline{K}\eta$	< 3.1 %	90%	1262
$\pi^+\pi^- p\overline{p}$	< 1.2 %	90%	1023
$\Lambda\overline{\Lambda}$	$< 2 \times 10^{-3}$	90%	987
Radiative decays			
$\gamma\gamma$	$(3.0 \pm 1.2) \times 10^{-4}$		1489

$J/\psi(1S)$ $\quad I^G(J^{PC}) = 0^-(1^{--})$

Mass $m = 3096.88 \pm 0.04$ MeV
Full width $\Gamma = 87 \pm 5$ keV
$\Gamma_{ee} = 5.26 + 0.37$ keV   (Assuming $\Gamma_{ee} = \Gamma_{\mu\mu}$)

$J/\psi(1S)$ DECAY MODES	Fraction (Γ_i/Γ)	Scale factor/ Confidence level	p (MeV/c)
hadrons	(87.7 ± 0.5) %		–
virtual $\gamma \to$ hadrons	(17.0 ± 2.0) %		–
$e^+ e^-$	(6.02 ± 0.19) %		1548
$\mu^+\mu^-$	(6.01 ± 0.19) %		1545
Decays involving hadronic resonances			
$\rho\pi$	(1.28 ± 0.10) %		1449
$\rho^0\pi^0$	$(4.2 \pm 0.5) \times 10^{-3}$		1449
$a_2(1320)\rho$	(1.09 ± 0.22) %		1125
$\omega\pi^+\pi^+\pi^-\pi^-$	$(8.5 \pm 3.4) \times 10^{-3}$		1392
$\omega\pi^+\pi^-$	$(7.2 \pm 1.0) \times 10^{-3}$		1435
$K^*(892)^0\overline{K}^*_2(1430)^0 +$ c.c.	$(6.7 \pm 2.6) \times 10^{-3}$		1005
$\omega K^*(892)\overline{K} +$ c.c.	$(5.3 \pm 2.0) \times 10^{-3}$		1098
$\omega f_2(1270)$	$(4.3 \pm 0.6) \times 10^{-3}$		1143
$K^+\overline{K}^*(892)^- +$ c.c.	$(5.0 \pm 0.4) \times 10^{-3}$		1373
$K^0\overline{K}^*(892)^0 +$ c.c.	$(4.2 \pm 0.4) \times 10^{-3}$		1371
$\omega\pi^0\pi^0$	$(3.4 \pm 0.8) \times 10^{-3}$		1436
$b_1(1235)^\pm\pi^\mp$	$[gg]$ $(3.0 \pm 0.5) \times 10^{-3}$		1299
$\omega K^\pm K^0_S \pi^\mp$	$[gg]$ $(3.0 \pm 0.7) \times 10^{-3}$		1210
$b_1(1235)^0\pi^0$	$(2.3 \pm 0.6) \times 10^{-3}$		1299
$\phi K^*(892)\overline{K} +$ c.c.	$(2.04 \pm 0.28) \times 10^{-3}$		969
$\omega K\overline{K}$	$(1.9 \pm 0.4) \times 10^{-3}$		1268
$\omega f_J(1710) \to \omega K\overline{K}$	$(4.8 \pm 1.1) \times 10^{-4}$		878
$\phi 2(\pi^+\pi^-)$	$(1.60 \pm 0.32) \times 10^{-3}$		1318
$\Delta(1232)^{++}\overline{p}\pi^-$	$(1.6 \pm 0.5) \times 10^{-3}$		1030
$\omega\eta$	$(1.58 \pm 0.16) \times 10^{-3}$		1394
$\phi K\overline{K}$	$(1.48 \pm 0.22) \times 10^{-3}$		1179
$\phi f_J(1710) \to \phi K\overline{K}$	$(3.6 \pm 0.6) \times 10^{-4}$		875
$p\overline{p}\omega$	$(1.30 \pm 0.25) \times 10^{-3}$	S=1.3	769
$\Delta(1232)^{++}\overline{\Delta}(1232)^{--}$	$(1.10 \pm 0.29) \times 10^{-3}$		938
$\Sigma(1385)^-\overline{\Sigma}(1385)^+$ (or c.c.)	$[gg]$ $(1.03 \pm 0.13) \times 10^{-3}$		692
$p\overline{p}\eta'(958)$	$(9 \pm 4) \times 10^{-4}$	S=1.7	596
$\phi f'_2(1525)$	$(8 \pm 4) \times 10^{-4}$	S=2.7	871

$\phi\pi^+\pi^-$	$(8.0 \pm 1.2) \times 10^{-4}$		1365
$\phi K^\pm K^0_S \pi^\mp$	$[gg]$ $(7.2 \pm 0.9) \times 10^{-4}$		1114
$\omega f_1(1420)$	$(6.8 \pm 2.4) \times 10^{-4}$		1062
$\phi\eta$	$(6.5 \pm 0.7) \times 10^{-4}$		1320
$\Xi(1530)^-\overline{\Xi}^+$	$(5.9 \pm 1.5) \times 10^{-4}$		597
$pK^-\overline{\Sigma}(1385)^0$	$(5.1 \pm 3.2) \times 10^{-4}$		645
$\omega\pi^0$	$(4.2 \pm 0.6) \times 10^{-4}$	S=1.4	1447
$\phi\eta'(958)$	$(3.3 \pm 0.4) \times 10^{-4}$		1192
$\phi f_0(980)$	$(3.2 \pm 0.9) \times 10^{-4}$	S=1.9	1182
$\Xi(1530)^0\overline{\Xi}^0$	$(3.2 \pm 1.4) \times 10^{-4}$		608
$\Sigma(1385)^-\overline{\Sigma}^+$ (or c.c.)	$[gg]$ $(3.1 \pm 0.5) \times 10^{-4}$		857
$\phi f_1(1285)$	$(2.6 \pm 0.5) \times 10^{-4}$	S=1.1	1032
$\rho\eta$	$(1.93 \pm 0.23) \times 10^{-4}$		1398
$\omega\eta'(958)$	$(1.67 \pm 0.25) \times 10^{-4}$		1279
$\omega f_0(980)$	$(1.4 \pm 0.5) \times 10^{-4}$		1271
$\rho\eta'(958)$	$(1.05 \pm 0.18) \times 10^{-4}$		1283
$p\overline{p}\phi$	$(4.5 \pm 1.5) \times 10^{-5}$		527
$a_2(1320)^\pm\pi^\mp$	$[gg]$ $< 4.3 \times 10^{-3}$	CL=90%	1263
$K\overline{K}^*_2(1430) +$ c.c.	$< 4.0 \times 10^{-3}$	CL=90%	1159
$K^*_2(1430)^0\overline{K}^*_2(1430)^0$	$< 2.9 \times 10^{-3}$	CL=90%	588
$K^*(892)^0\overline{K}^*(892)^0$	$< 5 \times 10^{-4}$	CL=90%	1263
$\phi f_2(1270)$	$< 3.7 \times 10^{-4}$	CL=90%	1036
$p\overline{p}\rho$	$< 3.1 \times 10^{-4}$	CL=90%	779
$\phi\eta(1440) \to \phi\eta\pi\pi$	$< 2.5 \times 10^{-4}$	CL=90%	946
$\omega f'_2(1525)$	$< 2.2 \times 10^{-4}$	CL=90%	1003
$\Sigma(1385)^0\overline{\Lambda}$	$< 2 \times 10^{-4}$	CL=90%	911
$\Delta(1232)^+\overline{p}$	$< 1 \times 10^{-4}$	CL=90%	1100
$\Sigma^0\overline{\Lambda}$	$< 9 \times 10^{-5}$	CL=90%	1032
$\phi\pi^0$	$< 6.8 \times 10^{-6}$	CL=90%	1377
Decays into stable hadrons			
$2(\pi^+\pi^-)\pi^0$	(3.37 ± 0.26) %		1496
$3(\pi^+\pi^-)\pi^0$	(2.9 ± 0.6) %		1433
$\pi^+\pi^-\pi^0$	(1.50 ± 0.20) %		1533
$\pi^+\pi^-\pi^0 K^+ K^-$	(1.20 ± 0.30) %		1368
$4(\pi^+\pi^-)\pi^0$	$(9.0 \pm 3.0) \times 10^{-3}$		1345
$\pi^+\pi^- K^+ K^-$	$(7.2 \pm 2.3) \times 10^{-3}$		1407
$K\overline{K}\pi$	$(6.1 \pm 1.0) \times 10^{-3}$		1440
$p\overline{p}\pi^+\pi^-$	$(6.0 \pm 0.5) \times 10^{-3}$	S=1.3	1107
$2(\pi^+\pi^-)$	$(4.0 \pm 1.0) \times 10^{-3}$		1517
$3(\pi^+\pi^-)$	$(4.0 \pm 2.0) \times 10^{-3}$		1466
$n\overline{n}\pi^+\pi^-$	$(4 \pm 4) \times 10^{-3}$		1106
$\Sigma\overline{\Sigma}$	$(3.8 \pm 0.5) \times 10^{-3}$		992
$2(\pi^+\pi^-)K^+ K^-$	$(3.1 \pm 1.3) \times 10^{-3}$		1320
$p\overline{p}\pi^+\pi^-\pi^0$	$[hhh]$ $(2.3 \pm 0.9) \times 10^{-3}$	S=1.9	1033
$p\overline{p}$	$(2.14 \pm 0.10) \times 10^{-3}$		1232
$p\overline{p}\eta$	$(2.09 \pm 0.18) \times 10^{-3}$		948
$p\overline{n}\pi^-$	$(2.00 \pm 0.10) \times 10^{-3}$		1174
$n\overline{n}$	$(1.9 \pm 0.5) \times 10^{-3}$		1231
$\Xi\overline{\Xi}$	$(1.8 \pm 0.4) \times 10^{-3}$	S=1.8	818
$\Lambda\overline{\Lambda}$	$(1.35 \pm 0.14) \times 10^{-3}$	S=1.2	1074
$p\overline{p}\pi^0$	$(1.09 \pm 0.09) \times 10^{-3}$		1176
$\Lambda\overline{\Sigma}^-\pi^+$ (or c.c.)	$[gg]$ $(1.06 \pm 0.12) \times 10^{-3}$		945
$pK^-\overline{\Lambda}$	$(8.9 \pm 1.6) \times 10^{-4}$		876
$2(K^+ K^-)$	$(7.0 \pm 3.0) \times 10^{-4}$		1131
$pK^-\overline{\Sigma}^0$	$(2.9 \pm 0.8) \times 10^{-4}$		820
$K^+ K^-$	$(2.37 \pm 0.31) \times 10^{-4}$		1468
$\Lambda\overline{\Lambda}\pi^0$	$(2.2 \pm 0.7) \times 10^{-4}$		998
$\pi^+\pi^-$	$(1.47 \pm 0.23) \times 10^{-4}$		1542
$K^0_S K^0_L$	$(1.08 \pm 0.14) \times 10^{-4}$		1466
$\Lambda\overline{\Sigma} +$ c.c.	$< 1.5 \times 10^{-4}$	CL=90%	1032
$K^0_S K^0_S$	$< 5.2 \times 10^{-6}$	CL=90%	1466
Radiative decays			
$\gamma\eta_c(1S)$	(1.3 ± 0.4) %		116
$\gamma\pi^+\pi^- 2\pi^0$	$(8.3 \pm 3.1) \times 10^{-3}$		1518
$\gamma\eta\pi\pi$	$(6.1 \pm 1.0) \times 10^{-3}$		1487
$\gamma\eta(1440) \to \gamma K\overline{K}\pi$	$[p]$ $(9.1 \pm 1.5) \times 10^{-3}$		1223
$\gamma\eta(1440) \to \gamma\gamma\rho^0$	$(6.4 \pm 1.4) \times 10^{-5}$		1223
$\gamma\rho\rho$	$(4.5 \pm 0.8) \times 10^{-3}$		1343
$\gamma\eta'(958)$	$(4.31 \pm 0.30) \times 10^{-3}$		1400
$\gamma 2\pi^+ 2\pi^-$	$(2.8 \pm 0.5) \times 10^{-3}$	S=1.9	1517
$\gamma f_4(2050)$	$(2.7 \pm 0.7) \times 10^{-3}$		874
$\gamma\omega\omega$	$(1.59 \pm 0.33) \times 10^{-3}$		1337
$\gamma\eta(1440) \to \gamma\rho^0\rho^0$	$(1.7 \pm 0.4) \times 10^{-3}$	S=1.3	1223
$\gamma f_2(1270)$	$(1.38 \pm 0.14) \times 10^{-3}$		1286
$\gamma f_J(1710) \to \gamma K\overline{K}$	$(9.7 \pm 1.2) \times 10^{-4}$		1075
$\gamma\eta$	$(8.6 \pm 0.8) \times 10^{-4}$		1500

$\gamma f_1(1420) \to \gamma K \overline{K} \pi$	$(8.3 \pm 1.5) \times 10^{-4}$		1220
$\gamma f_1(1285)$	$(6.5 \pm 1.0) \times 10^{-4}$		1283
$\gamma f_2'(1525)$	$(6.3 \pm 1.0) \times 10^{-4}$		1173
$\gamma \phi \phi$	$(4.0 \pm 1.2) \times 10^{-4}$	S=2.1	1166
$\gamma \rho \overline{p}$	$(3.8 \pm 1.0) \times 10^{-4}$		1232
$\gamma \eta(2225)$	$(2.9 \pm 0.6) \times 10^{-4}$		834
$\gamma \eta(1760) \to \gamma \rho^0 \rho^0$	$(1.3 \pm 0.9) \times 10^{-4}$		1048
$\gamma \pi^0$	$(3.9 \pm 1.3) \times 10^{-5}$		1546
$\gamma p \overline{p} \pi^+ \pi^-$	$< 7.9 \times 10^{-4}$	CL=90%	1107
$\gamma \gamma$	$< 5 \times 10^{-4}$	CL=90%	1548
$\gamma \Lambda \overline{\Lambda}$	$< 1.3 \times 10^{-4}$	CL=90%	1074
3γ	$< 5.5 \times 10^{-5}$	CL=90%	1548
$\gamma f_0(1370)$	$(3.4 \pm 0.7) \times 10^{-4}$		–
$\gamma f_0(1500)$	$(8.2 \pm 1.5) \times 10^{-4}$		1184

$\chi_{c0}(1P)$ $I^G(J^{PC}) = 0^+(0^{++})$

Mass $m = 3415.1 \pm 1.0$ MeV
Full width $\Gamma = 14 \pm 5$ MeV

$\chi_{c0}(1P)$ DECAY MODES	Fraction (Γ_i/Γ)	Confidence level	p (MeV/c)
Hadronic decays			
$2(\pi^+ \pi^-)$	(3.7 ± 0.7) %		1679
$\pi^+ \pi^- K^+ K^-$	(3.0 ± 0.7) %		1580
$\rho^0 \pi^+ \pi^-$	(1.6 ± 0.5) %		1608
$3(\pi^+ \pi^-)$	(1.5 ± 0.5) %		1633
$K^+ \overline{K}^*(892)^0 \pi^- +$ c.c.	(1.2 ± 0.4) %		1522
$\pi^+ \pi^-$	$(7.5 \pm 2.1) \times 10^{-3}$		1702
$K^+ K^-$	$(7.1 \pm 2.4) \times 10^{-3}$		1635
$\pi^+ \pi^- p \overline{p}$	$(5.0 \pm 2.0) \times 10^{-3}$		1320
$\pi^0 \pi^0$	$(3.1 \pm 0.6) \times 10^{-3}$		1702
$\eta \eta$	$(2.5 \pm 1.1) \times 10^{-3}$		1617
$p \overline{p}$	$< 9.0 \times 10^{-4}$	90%	1427
Radiative decays			
$\gamma J/\psi(1S)$	$(6.6 \pm 1.8) \times 10^{-3}$		303
$\gamma \gamma$	$(4.0 \pm 2.3) \times 10^{-4}$		1708

$\chi_{c1}(1P)$ $I^G(J^{PC}) = 0^+(1^{++})$

Mass $m = 3510.53 \pm 0.12$ MeV
Full width $\Gamma = 0.88 \pm 0.14$ MeV

$\chi_{c1}(1P)$ DECAY MODES	Fraction (Γ_i/Γ)	p (MeV/c)
Hadronic decays		
$3(\pi^+ \pi^-)$	(2.2 ± 0.8) %	1683
$2(\pi^+ \pi^-)$	(1.6 ± 0.5) %	1727
$\pi^+ \pi^- K^+ K^-$	$(9 \pm 4) \times 10^{-3}$	1632
$\rho^0 \pi^+ \pi^-$	$(3.9 \pm 3.5) \times 10^{-3}$	1659
$K^+ \overline{K}^*(892)^0 \pi^- +$ c.c.	$(3.2 \pm 2.1) \times 10^{-3}$	1576
$\pi^+ \pi^- p \overline{p}$	$(1.4 \pm 0.9) \times 10^{-3}$	1381
$p \overline{p}$	$(8.6 \pm 1.2) \times 10^{-5}$	1483
$\pi^+ \pi^- + K^+ K^-$	$< 2.1 \times 10^{-3}$	–
Radiative decays		
$\gamma J/\psi(1S)$	(27.3 ± 1.6) %	389

$\chi_{c2}(1P)$ $I^G(J^{PC}) = 0^+(2^{++})$

Mass $m = 3556.17 \pm 0.13$ MeV
Full width $\Gamma = 2.00 \pm 0.18$ MeV

$\chi_{c2}(1P)$ DECAY MODES	Fraction (Γ_i/Γ)	Confidence level	p (MeV/c)
Hadronic decays			
$2(\pi^+ \pi^-)$	(2.2 ± 0.5) %		1751
$\pi^+ \pi^- K^+ K^-$	(1.9 ± 0.5) %		1656
$3(\pi^+ \pi^-)$	(1.2 ± 0.8) %		1707
$\rho^0 \pi^+ \pi^-$	$(7 \pm 4) \times 10^{-3}$		1683
$K^+ \overline{K}^*(892)^0 \pi^- +$ c.c.	$(4.8 \pm 2.8) \times 10^{-3}$		1601
$\pi^+ \pi^- p \overline{p}$	$(3.3 \pm 1.3) \times 10^{-3}$		1410
$\pi^+ \pi^-$	$(1.9 \pm 1.0) \times 10^{-3}$		1773
$K^+ K^-$	$(1.5 \pm 1.1) \times 10^{-3}$		1708
$p \overline{p}$	$(10.0 \pm 1.0) \times 10^{-5}$		1510
$\pi^0 \pi^0$	$(1.10 \pm 0.28) \times 10^{-3}$		1773
$\eta \eta$	$(8 \pm 5) \times 10^{-4}$		1692
$J/\psi(1S) \pi^+ \pi^- \pi^0$	< 1.5 %	90%	185

Radiative decays

$\gamma J/\psi(1S)$	(13.5 ± 1.1) %	430
$\gamma \gamma$	$(1.6 \pm 0.5) \times 10^{-4}$	1778

$\psi(2S)$ $I^G(J^{PC}) = 0^-(1^{--})$

Mass $m = 3686.00 \pm 0.09$ MeV
Full width $\Gamma = 277 \pm 31$ keV (S = 1.1)
$\Gamma_{ee} = 2.14 \pm 0.21$ keV (Assuming $\Gamma_{ee} = \Gamma_{\mu\mu}$)

$\psi(2S)$ DECAY MODES	Fraction (Γ_i/Γ)	Scale factor/ Confidence level	p (MeV/c)
hadrons	(98.10 ± 0.30) %		–
virtual $\gamma \to$ hadrons	(2.9 ± 0.4) %		–
$e^+ e^-$	$(8.8 \pm 1.3) \times 10^{-3}$		1843
$\mu^+ \mu^-$	$(7.7 \pm 1.7) \times 10^{-3}$		1840
Decays into $J/\psi(1S)$ and anything			
$J/\psi(1S)$ anything	(57 ± 4) %		–
$J/\psi(1S)$ neutrals	(23.2 ± 2.6) %		–
$J/\psi(1S) \pi^+ \pi^-$	(32.4 ± 2.6) %		477
$J/\psi(1S) \pi^0 \pi^0$	(18.4 ± 2.7) %		481
$J/\psi(1S) \eta$	(2.7 ± 0.4) %	S=1.7	200
$J/\psi(1S) \pi^0$	$(9.7 \pm 2.1) \times 10^{-4}$		527
Hadronic decays			
$3(\pi^+ \pi^-) \pi^0$	$(3.5 \pm 1.6) \times 10^{-3}$		1746
$2(\pi^+ \pi^-) \pi^0$	$(3.1 \pm 0.7) \times 10^{-3}$		1799
$\pi^+ \pi^- K^+ K^-$	$(1.6 \pm 0.4) \times 10^{-3}$		1726
$\pi^+ \pi^- p \overline{p}$	$(8.0 \pm 2.0) \times 10^{-4}$		1491
$K^+ \overline{K}^*(892)^0 \pi^- +$ c.c.	$(6.7 \pm 2.5) \times 10^{-4}$		1673
$2(\pi^+ \pi^-)$	$(4.5 \pm 1.0) \times 10^{-4}$		1817
$\rho^0 \pi^+ \pi^-$	$(4.2 \pm 1.5) \times 10^{-4}$		1751
$\overline{p} p$	$(1.9 \pm 0.5) \times 10^{-4}$		1586
$3(\pi^+ \pi^-)$	$(1.5 \pm 1.0) \times 10^{-4}$		1774
$\overline{p} p \pi^0$	$(1.4 \pm 0.5) \times 10^{-4}$		1543
$K^+ K^-$	$(1.0 \pm 0.7) \times 10^{-4}$		1776
$\pi^+ \pi^- \pi^0$	$(9 \pm 5) \times 10^{-5}$		1830
$\pi^+ \pi^-$	$(8 \pm 5) \times 10^{-5}$		1838
$\Lambda \overline{\Lambda}$	$< 4 \times 10^{-4}$	CL=90%	1467
$\Xi^- \overline{\Xi}^+$	$< 2 \times 10^{-4}$	CL=90%	1285
$\rho \pi$	$< 8.3 \times 10^{-5}$	CL=90%	1760
$K^+ K^- \pi^0$	$< 2.96 \times 10^{-5}$	CL=90%	1754
$K^+ \overline{K}^*(892)^- +$ c.c.	$< 5.4 \times 10^{-5}$	CL=90%	1698
Radiative decays			
$\gamma \chi_{c0}(1P)$	(9.3 ± 0.8) %		261
$\gamma \chi_{c1}(1P)$	(8.7 ± 0.8) %		171
$\gamma \chi_{c2}(1P)$	(7.8 ± 0.8) %		127
$\gamma \eta_c(1S)$	$(2.8 \pm 0.6) \times 10^{-3}$		639
$\gamma \pi^0$	$< 5.4 \times 10^{-4}$	CL=95%	1841
$\gamma \eta'(958)$	$< 1.1 \times 10^{-3}$	CL=90%	1719
$\gamma \gamma$	$< 1.6 \times 10^{-4}$	CL=90%	1843
$\gamma \gamma \eta(1440) \to \gamma K \overline{K} \pi$	$< 1.2 \times 10^{-4}$	CL=90%	1569

$\psi(3770)$ $I^G(J^{PC}) = ?^?(1^{--})$

Mass $m = 3769.9 \pm 2.5$ MeV (S = 1.8)
Full width $\Gamma = 23.6 \pm 2.7$ MeV (S = 1.1)
$\Gamma_{ee} = 0.26 \pm 0.04$ keV (S = 1.2)

$\psi(3770)$ DECAY MODES	Fraction (Γ_i/Γ)	Scale factor	p (MeV/c)
$D \overline{D}$	dominant		242
$e^+ e^-$	$(1.12 \pm 0.17) \times 10^{-5}$	1.2	1885

$\psi(4040)$ [lll] $I^G(J^{PC}) = ?^?(1^{--})$

Mass $m = 4040 \pm 10$ MeV
Full width $\Gamma = 52 \pm 10$ MeV
$\Gamma_{ee} = 0.75 \pm 0.15$ keV

$\psi(4040)$ DECAY MODES	Fraction (Γ_i/Γ)	p (MeV/c)
$e^+ e^-$	$(1.4 \pm 0.4) \times 10^{-5}$	2020
$D^0 \overline{D}^0$	seen	777
$D^*(2007)^0 \overline{D}^0 +$ c.c.	seen	578
$D^*(2007)^0 \overline{D}^*(2007)^0$	seen	232

$\psi(4160)$ [iii]

$I^G(J^{PC}) = ?^?(1^{--})$

Mass $m = 4159 \pm 20$ MeV
Full width $\Gamma = 78 \pm 20$ MeV
$\Gamma_{ee} = 0.77 \pm 0.23$ keV

$\psi(4160)$ DECAY MODES	Fraction (Γ_i/Γ)	p (MeV/c)
$e^+ e^-$	$(10 \pm 4) \times 10^{-6}$	2079

$\psi(4415)$ [iii]

$I^G(J^{PC}) = ?^?(1^{--})$

Mass $m = 4415 \pm 6$ MeV
Full width $\Gamma = 43 \pm 15$ MeV (S = 1.8)
$\Gamma_{ee} = 0.47 \pm 0.10$ keV

$\psi(4415)$ DECAY MODES	Fraction (Γ_i/Γ)	p (MeV/c)
hadrons	dominant	–
$e^+ e^-$	$(1.1 \pm 0.4) \times 10^{-5}$	2207

$b\bar{b}$ MESONS

$\Upsilon(1S)$

$I^G(J^{PC}) = 0^-(1^{--})$

Mass $m = 9460.37 \pm 0.21$ MeV (S = 2.7)
Full width $\Gamma = 52.5 \pm 1.8$ keV
$\Gamma_{ee} = 1.32 \pm 0.05$ keV

$\Upsilon(1S)$ DECAY MODES	Fraction (Γ_i/Γ)	Scale factor/ Confidence level	p (MeV/c)
$\tau^+ \tau^-$	$(2.67^{+0.14}_{-0.16})$ %		4384
$e^+ e^-$	(2.52 ± 0.17) %		4730
$\mu^+ \mu^-$	(2.48 ± 0.07) %	S=1.1	4729
Hadronic decays			
$J/\psi(1S)$ anything	$(1.1 \pm 0.4) \times 10^{-3}$		4223
$\rho \pi$	$< 2 \quad \times 10^{-4}$	CL=90%	4698
$\pi^+ \pi^-$	$< 5 \quad \times 10^{-4}$	CL=90%	4728
$K^+ K^-$	$< 5 \quad \times 10^{-4}$	CL=90%	4704
$p \bar{p}$	$< 5 \quad \times 10^{-4}$	CL=90%	4636
Radiative decays			
$\gamma 2h^+ 2h^-$	$(7.0 \pm 1.5) \times 10^{-4}$		4720
$\gamma 3h^+ 3h^-$	$(5.4 \pm 2.0) \times 10^{-4}$		4703
$\gamma 4h^+ 4h^-$	$(7.4 \pm 3.5) \times 10^{-4}$		4679
$\gamma \pi^+ \pi^- K^+ K^-$	$(2.9 \pm 0.9) \times 10^{-4}$		4686
$\gamma 2\pi^+ 2\pi^-$	$(2.5 \pm 0.9) \times 10^{-4}$		4720
$\gamma 3\pi^+ 3\pi^-$	$(2.5 \pm 1.2) \times 10^{-4}$		4703
$\gamma 2\pi^+ 2\pi^- K^+ K^-$	$(2.4 \pm 1.2) \times 10^{-4}$		4658
$\gamma \pi^+ \pi^- p \bar{p}$	$(1.5 \pm 0.6) \times 10^{-4}$		4604
$\gamma 2\pi^+ 2\pi^- p \bar{p}$	$(4 \pm 6) \times 10^{-5}$		4563
$\gamma 2K^+ 2K^-$	$(2.0 \pm 2.0) \times 10^{-5}$		4601
$\gamma \eta'(958)$	$< 1.3 \quad \times 10^{-3}$	CL=90%	4682
$\gamma \eta$	$< 3.5 \quad \times 10^{-4}$	CL=90%	4714
$\gamma f_2'(1525)$	$< 1.4 \quad \times 10^{-4}$	CL=90%	4607
$\gamma f_2(1270)$	$< 1.3 \quad \times 10^{-4}$	CL=90%	4644
$\gamma \eta(1440)$	$< 8.2 \quad \times 10^{-5}$	CL=90%	4624
$\gamma f_J(1710) \to \gamma K \overline{K}$	$< 2.6 \quad \times 10^{-4}$	CL=90%	4576
$\gamma f_0(2200) \to \gamma K^+ K^-$	$< 2 \quad \times 10^{-4}$	CL=90%	4475
$\gamma f_J(2220) \to \gamma K^+ K^-$	$< 1.5 \quad \times 10^{-4}$	CL=90%	4469
$\gamma \eta(2225) \to \gamma \phi\phi$	$< 3 \quad \times 10^{-3}$	CL=90%	4469
γX	$< 3 \quad \times 10^{-5}$	CL=90%	–
X = pseudoscalar with $m < 7.2$ GeV			
$\gamma X \overline{X}$	$< 1 \quad \times 10^{-3}$	CL=90%	–
$X \overline{X}$ = vectors with $m < 3.1$ GeV			

$\chi_{b0}(1P)$ [iii]

$I^G(J^{PC}) = 0^+(0^{++})$
J needs confirmation.

Mass $m = 9859.8 \pm 1.3$ MeV

$\chi_{b0}(1P)$ DECAY MODES	Fraction (Γ_i/Γ)	Confidence level	p (MeV/c)
$\gamma \Upsilon(1S)$	< 6 %	90%	391

$\chi_{b1}(1P)$ [iii]

$I^G(J^{PC}) = 0^+(1^{++})$
J needs confirmation.

Mass $m = 9891.9 \pm 0.7$ MeV

$\chi_{b1}(1P)$ DECAY MODES	Fraction (Γ_i/Γ)	p (MeV/c)
$\gamma \Upsilon(1S)$	(35 ± 8) %	422

$\chi_{b2}(1P)$ [iii]

$I^G(J^{PC}) = 0^+(2^{++})$
J needs confirmation.

Mass $m = 9913.2 \pm 0.6$ MeV

$\chi_{b2}(1P)$ DECAY MODES	Fraction (Γ_i/Γ)	p (MeV/c)
$\gamma \Upsilon(1S)$	(22 ± 4) %	443

$\Upsilon(2S)$

$I^G(J^{PC}) = 0^-(1^{--})$

Mass $m = 10.02330 \pm 0.00031$ GeV
Full width $\Gamma = 44 \pm 7$ keV
$\Gamma_{ee} = 0.52 \pm 0.03$ keV

$\Upsilon(2S)$ DECAY MODES	Fraction (Γ_i/Γ)	Confidence level	p (MeV/c)
$\Upsilon(1S)\pi^+ \pi^-$	(18.5 ± 0.8) %		475
$\Upsilon(1S)\pi^0 \pi^0$	(8.8 ± 1.1) %		480
$\tau^+ \tau^-$	(1.7 ± 1.6) %		4686
$\mu^+ \mu^-$	(1.31 ± 0.21) %		5011
$e^+ e^-$	seen		5012
$\Upsilon(1S)\pi^0$	$< 8 \quad \times 10^{-3}$	90%	531
$\Upsilon(1S)\eta$	$< 2 \quad \times 10^{-3}$	90%	127
$J/\psi(1S)$ anything	$< 6 \quad \times 10^{-3}$	90%	4533
Radiative decays			
$\gamma \chi_{b1}(1P)$	(6.7 ± 0.9) %		131
$\gamma \chi_{b2}(1P)$	(6.6 ± 0.9) %		110
$\gamma \chi_{b0}(1P)$	(4.3 ± 1.0) %		162
$\gamma f_J(1710)$	$< 5.9 \quad \times 10^{-4}$	90%	4866
$\gamma f_2'(1525)$	$< 5.3 \quad \times 10^{-4}$	90%	4896
$\gamma f_2(1270)$	$< 2.41 \quad \times 10^{-4}$	90%	4931

$\chi_{b0}(2P)$ [iii]

$I^G(J^{PC}) - 0^+(0^{++})$
J needs confirmation.

Mass $m = 10.2321 \pm 0.0006$ GeV

$\chi_{b0}(2P)$ DECAY MODES	Fraction (Γ_i/Γ)	p (MeV/c)
$\gamma \Upsilon(2S)$	(4.6 ± 2.1) %	210
$\gamma \Upsilon(1S)$	$(9 \pm 6) \times 10^{-3}$	746

$\chi_{b1}(2P)$ [iii]

$I^G(J^{PC}) = 0^+(1^{++})$
J needs confirmation.

Mass $m = 10.2552 \pm 0.0005$ GeV
$m_{\chi_{b1}(2P)} - m_{\chi_{b0}(2P)} = 23.5 \pm 1.0$ MeV

$\chi_{b1}(2P)$ DECAY MODES	Fraction (Γ_i/Γ)	Scale factor	p (MeV/c)
$\gamma \Upsilon(2S)$	(21 ± 4) %	1.5	229
$\gamma \Upsilon(1S)$	(8.5 ± 1.3) %	1.3	764

$\chi_{b2}(2P)$ [iii]

$I^G(J^{PC}) = 0^+(2^{++})$
J needs confirmation.

Mass $m = 10.2685 \pm 0.0004$ GeV
$m_{\chi_{b2}(2P)} - m_{\chi_{b1}(2P)} = 13.5 \pm 0.6$ MeV

$\chi_{b2}(2P)$ DECAY MODES	Fraction (Γ_i/Γ)	p (MeV/c)
$\gamma \Upsilon(2S)$	(16.2 ± 2.4) %	242
$\gamma \Upsilon(1S)$	(7.1 ± 1.0) %	776

Meson Summary Table

$\Upsilon(3S)$		$I^G(J^{PC}) = 0^-(1^{--})$

Mass $m = 10.3553 \pm 0.0005$ GeV
Full width $\Gamma = 26.3 \pm 3.5$ keV

$\Upsilon(3S)$ DECAY MODES	Fraction (Γ_i/Γ)	Scale factor	p (MeV/c)
$\Upsilon(2S)$ anything	(10.6 ± 0.8) %		296
$\Upsilon(2S)\pi^+\pi^-$	(2.8 ± 0.6) %	2.2	177
$\Upsilon(2S)\pi^0\pi^0$	(2.00 ± 0.32) %		190
$\Upsilon(2S)\gamma\gamma$	(5.0 ± 0.7) %		327
$\Upsilon(1S)\pi^+\pi^-$	(4.48 ± 0.21) %		814
$\Upsilon(1S)\pi^0\pi^0$	(2.06 ± 0.28) %		816
$\mu^+\mu^-$	(1.81 ± 0.17) %		5177
e^+e^-	seen		5177
Radiative decays			
$\gamma\chi_{b2}(2P)$	(11.4 ± 0.8) %	1.3	87
$\gamma\chi_{b1}(2P)$	(11.3 ± 0.6) %		100
$\gamma\chi_{b0}(2P)$	(5.4 ± 0.6) %	1.1	123

$\Upsilon(4S)$ or $\Upsilon(10580)$		$I^G(J^{PC}) = ?^?(1^{--})$

Mass $m = 10.5800 \pm 0.0035$ GeV
Full width $\Gamma = 21 \pm 4$ MeV $\quad$ (S = 2.3)
$\Gamma_{ee} = 0.248 \pm 0.031$ keV $\quad$ (S = 1.3)

$\Upsilon(4S)$ DECAY MODES	Fraction (Γ_i/Γ)	Confidence level	p (MeV/c)
$B\bar{B}$	dominant		–
e^+e^-	$(2.8 \pm 0.7) \times 10^{-5}$		5290
$J/\psi(3097)$ anything	$(2.2 \pm 0.7) \times 10^{-3}$		–
D^{*+} anything + c.c.	< 7.4 %	90%	5099
ϕ anything	< 2.3 $\times 10^{-3}$	90%	5240
$\Upsilon(1S)$ anything	< 4 $\times 10^{-3}$	90%	1053
non-$B\bar{B}$	< 4 %	95%	–

$\Upsilon(10860)$		$I^G(J^{PC}) = ?^?(1^{--})$

Mass $m = 10.865 \pm 0.008$ GeV $\quad$ (S = 1.1)
Full width $\Gamma = 110 \pm 13$ MeV
$\Gamma_{ee} = 0.31 \pm 0.07$ keV $\quad$ (S = 1.3)

$\Upsilon(10860)$ DECAY MODES	Fraction (Γ_i/Γ)	p (MeV/c)
e^+e^-	$(2.8 \pm 0.7) \times 10^{-6}$	5432

$\Upsilon(11020)$		$I^G(J^{PC}) = ?^?(1^{--})$

Mass $m = 11.019 \pm 0.008$ GeV
Full width $\Gamma = 79 \pm 16$ MeV
$\Gamma_{ee} = 0.130 \pm 0.030$ keV

$\Upsilon(11020)$ DECAY MODES	Fraction (Γ_i/Γ)	p (MeV/c)
e^+e^-	$(1.6 \pm 0.5) \times 10^{-6}$	5509

NOTES

In this Summary Table:

When a quantity has "(S = ...)" to its right, the error on the quantity has been enlarged by the "scale factor" S, defined as S = $\sqrt{\chi^2/(N-1)}$, where N is the number of measurements used in calculating the quantity. We do this when S > 1, which often indicates that the measurements are inconsistent. When S > 1.25, we also show in the Particle Listings an ideogram of the measurements. For more about S, see the Introduction.

A decay momentum p is given for each decay mode. For a 2-body decay, p is the momentum of each decay product in the rest frame of the decaying particle. For a 3-or-more-body decay, p is the largest momentum any of the products can have in this frame.

[a] See the "Note on $\pi^\pm \to \ell^\pm \nu \gamma$ and $K^\pm \to \ell^\pm \nu \gamma$ Form Factors" in the $\pi^\pm$ Particle Listings for definitions and details.

[b] Measurements of $\Gamma(e^+\nu_e)/\Gamma(\mu^+\nu_\mu)$ always include decays with γ's, and measurements of $\Gamma(e^+\nu_e\gamma)$ and $\Gamma(\mu^+\nu_\mu\gamma)$ never include low-energy γ's. Therefore, since no clean separation is possible, we consider the modes with γ's to be subreactions of the modes without them, and let $[\Gamma(e^+\nu_e) + \Gamma(\mu^+\nu_\mu)]/\Gamma_{total} = 100\%$.

[c] See the $\pi^\pm$ Particle Listings for the energy limits used in this measurement; low-energy γ's are not included.

[d] Derived from an analysis of neutrino-oscillation experiments.

[e] Astrophysical and cosmological arguments give limits of order 10^{-13}; see the π^0 Particle Listings.

[f] See the "Note on the Decay Width $\Gamma(\eta \to \gamma\gamma)$" in our 1994 edition, Phys. Rev. **D50**, 1 August 1994, Part I, p. 1451.

[g] See the "Note on η Decay Parameters" in the η Particle Listings.

[h] C parity forbids this to occur as a single-photon process.

[i] See the "Note on scalar mesons" in the $f_0(1370)$ Particle Listings.

[j] See the "Note on $\rho(770)$" in the $\rho(770)$ Particle Listings.

[k] The e^+e^- branching fraction is from $e^+e^- \to \pi^+\pi^-$ experiments only. The $\omega\rho$ interference is then due to $\omega\rho$ mixing only, and is expected to be small. If $e\mu$ universality holds, $\Gamma(\rho^0 \to \mu^+\mu^-) = \Gamma(\rho^0 \to e^+e^-) \times 0.99785$.

[l] This is only an educated guess; the error given is larger than the error on the average of the published values. See the Particle Listings for details.

[m] See the "Note on $a_1(1260)$" in the $a_1(1260)$ Particle Listings.

[n] See the "Note on the $f_1(1420)$" in the $f_1(1420)$ Particle Listings.

[o] See also the $\omega(1600)$ Particle Listings.

[p] See the "Note on the $\eta(1440)$" in the $\eta(1440)$ Particle Listings.

[q] See the "Note on the $\rho(1450)$ and the $\rho(1700)$" in the $\rho(1700)$ Particle Listings.

[r] See the "Note on non-$q\bar{q}$ mesons" in the Particle Listings (see the index for the page number).

[s] See also the $\omega(1420)$ Particle Listings.

[t] See the "Note on $f_J(1710)$" in the $f_J(1710)$ Particle Listings.

[u] See the note in the $K^\pm$ Particle Listings.

[v] The definition of the slope parameter g of the $K \to 3\pi$ Dalitz plot is as follows (see also "Note on Dalitz Plot Parameters for $K \to 3\pi$ Decays" in the $K^\pm$ Particle Listings):
$$|M|^2 = 1 + g(s_3 - s_0)/m_{\pi^+}^2 + \cdots .$$

[w] For more details and definitions of parameters see the Particle Listings.

[x] See the $K^\pm$ Particle Listings for the energy limits used in this measurement.

[y] Most of this radiative mode, the low-momentum γ part, is also included in the parent mode listed without γ's.

[z] Direct-emission branching fraction.

[aa] Structure-dependent part.

[bb] Derived from measured values of ϕ_{+-}, ϕ_{00}, $|\eta|$, $\tau_{K_S^0}$, and $|m_{K_L^0} - m_{K_S^0}|$, as described in the introduction to "Tests of Conservation Laws."

[cc] The *CP*-violation parameters are defined as follows (see also "Note on *CP* Violation in $K_S \to 3\pi$" and "Note on *CP* Violation in K_L^0 Decay" in the Particle Listings):

$$\eta_{+-} = |\eta_{+-}|e^{i\phi_{+-}} = \frac{A(K_L^0 \to \pi^+\pi^-)}{A(K_S^0 \to \pi^+\pi^-)} = \epsilon + \epsilon'$$

$$\eta_{00} = |\eta_{00}|e^{i\phi_{00}} = \frac{A(K_L^0 \to \pi^0\pi^0)}{A(K_S^0 \to \pi^0\pi^0)} = \epsilon - 2\epsilon'$$

$$\delta = \frac{\Gamma(K_L^0 \to \pi^-\ell^+\nu) - \Gamma(K_L^0 \to \pi^+\ell^-\nu)}{\Gamma(K_L^0 \to \pi^-\ell^+\nu) + \Gamma(K_L^0 \to \pi^+\ell^-\nu)},$$

$$\text{Im}(\eta_{+-0})^2 = \frac{\Gamma(K_S^0 \to \pi^+\pi^-\pi^0)^{CP\ \text{viol.}}}{\Gamma(K_L^0 \to \pi^+\pi^-\pi^0)},$$

$$\text{Im}(\eta_{000})^2 = \frac{\Gamma(K_S^0 \to \pi^0\pi^0\pi^0)}{\Gamma(K_L^0 \to \pi^0\pi^0\pi^0)}.$$

where for the last two relations *CPT* is assumed valid, *i.e.*, $\text{Re}(\eta_{+-0}) \simeq 0$ and $\text{Re}(\eta_{000}) \simeq 0$.

[dd] See the K_S^0 Particle Listings for the energy limits used in this measurement.

[ee] Calculated from K_L^0 semileptonic rates and the K_S^0 lifetime assuming $\Delta S = \Delta Q$.

[ff] ϵ'/ϵ is derived from $|\eta_{00}/\eta_{+-}|$ measurements using theoretical input on phases.

[gg] The value is for the sum of the charge states of particle/antiparticle states indicated.

[hh] See the K_L^0 Particle Listings for the energy limits used in this measurement.

[ii] $m_{e^+e^-} > 470$ MeV.

[jj] Allowed by higher-order electroweak interactions.

[kk] Violates *CP* in leading order. Test of direct *CP* violation since the indirect *CP*-violating and *CP*-conserving contributions are expected to be suppressed.

[ll] See the "Note on $f_0(1370)$" in the $f_0(1370)$ Particle Listings and in the 1994 edition.

[mm] See the note in the $L(1770)$ Particle Listings in Reviews of Modern Physics **56** No. 2 Pt. II (1984), p. S200. See also the "Note on $K_2(1770)$ and the $K_2(1820)$" in the $K_2(1770)$ Particle Listings.

[nn] See the "Note on $K_2(1770)$ and the $K_2(1820)$" in the $K_2(1770)$ Particle Listings.

[oo] This is a weighted average of $D^\pm$ (44%) and D^0 (56%) branching fractions. See "D^+ and $D^0 \to (\eta \text{anything})\ /\ (\text{total } D^+ \text{ and } D^0)$" under "$D^+$ Branching Ratios" in the Particle Listings.

[pp] This value averages the e^+ and μ^+ branching fractions, after making a small phase-space adjustment to the μ^+ fraction to be able to use it as an e^+ fraction; hence our ℓ^+ is really an e^+.

[qq] ℓ indicates e or μ mode, not sum over modes.

[rr] The branching fractions for this mode may differ from the sum of the submodes that contribute to it, due to interference effects. See the relevant papers in the Particle Listings.

[ss] The two experiments determining this ratio are in serious disagreement. See the Particle Listings.

[tt] This mode is not a useful test for a $\Delta C=1$ weak neutral current because both quarks must change flavor in this decay.

[uu] The D_1^0-D_2^0 limits are inferred from the D^0-$\overline{D}^0$ mixing ratio $\Gamma(K^+\pi^-$ or $K^+\pi^-\pi^+\pi^-$ via $\overline{D}^0)\ /\ \Gamma(K^-\pi^+$ or $K^-\pi^+\pi^+\pi^-)$.

[vv] This value is calculated from the ratio $\Gamma(K^-\mu^+\nu_\mu)/\Gamma(\mu^+ \text{anything})$ in the D^0 Particle Listings.

[ww] The experiments on the division of this charge mode amongst its submodes disagree, and the submode branching fractions here add up to considerably more than the charged-mode fraction.

[xx] For now, we average together measurements of the $\phi e^+\nu_e$ and $\phi\mu^+\nu_\mu$ branching fractions. This is the *average*, not the *sum*.

[yy] This branching fraction is calculated from appropriate fractions of the next three branching fractions.

[zz] This value includes only K^+K^- decays of the $f_J(1710)$, because branching fractions of this resonance are not known.

[aaa] This mode is not a useful test for a $\Delta C=1$ weak neutral current because both quarks must change flavor in this decay.

[bbb] B^0 and B_s^0 contributions not separated. Limit is on weighted average of the two decay rates.

[ccc] These values are model dependent. See 'Note on Semileptonic Decays' in the B^+ Particle Listings.

[ddd] D^{**} stands for the sum of the $D(1\,^1P_1)$, $D(1\,^3P_0)$, $D(1\,^3P_1)$, $D(1\,^3P_2)$, $D(2\,^1S_0)$, and $D(2\,^1S_1)$ resonances.

[eee] Inclusive branching fractions have a multiplicity definition and can be greater than 100%.

[fff] D_J represents an unresolved mixture of pseudoscalar and tensor D^{**} (*P*-wave) states.

[ggg] Not a pure measurement. See note at head of B_s^0 Decay Modes.

[hhh] Includes $p\overline{p}\pi^+\pi^-\gamma$ and excludes $p\overline{p}\eta$, $p\overline{p}\omega$, $p\overline{p}\eta'$.

[iii] J^{PC} known by production in e^+e^- via single photon annihilation. I^G is not known; interpretation of this state as a single resonance is unclear because of the expectation of substantial threshold effects in this energy region.

[jjj] Spectroscopic labeling for these states is theoretical, pending experimental information.

Meson Summary Table

See also the table of suggested $q\bar{q}$ quark-model assignments in the Quark Model section.

- Indicates particles that appear in the preceding Meson Summary Table. We do not regard the other entries as being established.

† Indicates that the value of J given is preferred, but needs confirmation.

LIGHT UNFLAVORED ($S = C = B = 0$)

Particle	$I^G(J^{PC})$	Particle	$I^G(J^{PC})$
• $\pi^\pm$	$1^-(0^-)$	$f_2(1640)$	$0^+(2^{++})$
• π^0	$1^-(0^{-+})$	• $\omega_3(1670)$	$0^-(3^{--})$
• η	$0^+(0^{-+})$	• $\pi_2(1670)$	$1^-(2^{-+})$
• $f_0(400-1200)$	$0^+(0^{++})$	• $\phi(1680)$	$0^-(1^{--})$
• $\rho(770)$	$1^+(1^{--})$	• $\rho_3(1690)$	$1^+(3^{--})$
• $\omega(782)$	$0^-(1^{--})$	• $\rho(1700)$	$1^+(1^{--})$
• $\eta'(958)$	$0^+(0^{-+})$	• $f_J(1710)$	$0^+(\text{even}^{++})$
• $f_0(980)$	$0^+(0^{++})$	$X(1740)$	$0^+(\text{even}^{++})$
• $a_0(980)$	$1^-(0^{++})$	$\eta(1760)$	$0^+(0^{-+})$
• $\phi(1020)$	$0^-(1^{--})$	$\pi(1800)$	$1^-(0^{-+})$
• $h_1(1170)$	$0^-(1^{+-})$	$X(1775)$	$1^-(?^{-+})$
• $b_1(1235)$	$1^+(1^{+-})$	$f_2(1810)$	$0^+(2^{++})$
• $a_1(1260)$	$1^-(1^{++})$	• $\phi_3(1850)$	$0^-(3^{--})$
• $f_2(1270)$	$0^+(2^{++})$	$\eta_2(1870)$	$0^+(2^{-+})$
• $f_1(1285)$	$0^+(1^{++})$	$X(1910)$	$0^+(?^{?+})$
• $\eta(1295)$	$0^+(0^{-+})$	$f_2(1950)$	$0^+(2^{++})$
• $\pi(1300)$	$1^-(0^{-+})$	$X(2000)$	$1^-(?^{?+})$
• $a_2(1320)$	$1^-(2^{++})$	• $f_2(2010)$	$0^+(2^{++})$
• $f_0(1370)$	$0^+(0^{++})$	$a_4(2040)$	$1^-(4^{++})$
$h_1(1380)$	$?^-(1^{+?})$	• $f_4(2050)$	$0^+(4^{++})$
$\hat{\rho}(1405)$	$1^-(1^{-+})$	$\pi_2(2100)$	$1^-(2^{-+})$
• $f_1(1420)$	$0^+(1^{++})$	$f_2(2150)$	$0^+(2^{++})$
• $\omega(1420)$	$0^-(1^{--})$	$\rho(2150)$	$1^+(1^{--})$
$f_2(1430)$	$0^+(2^{++})$	$f_0(2200)$	$0^+(0^{++})$
• $\eta(1440)$	$0^+(0^{-+})$	$f_J(2220)$	$0^+(2^{++} \text{ or }$
$a_0(1450)$	$1^-(0^{++})$		$4^{++})$
• $\rho(1450)$	$1^+(1^{--})$	$\eta(2225)$	$0^+(0^{-+})$
• $f_0(1500)$	$0^+(0^{++})$	$\rho_3(2250)$	$1^+(3^{--})$
• $f_1(1510)$	$0^+(1^{++})$	• $f_2(2300)$	$0^+(2^{++})$
• $f_2'(1525)$	$0^+(2^{++})$	$f_4(2300)$	$0^+(4^{++})$
$f_2(1565)$	$0^+(2^{++})$	• $f_2(2340)$	$0^+(2^{++})$
• $\omega(1600)$	$0^-(1^{--})$	$\rho_5(2350)$	$1^+(5^{--})$
$X(1600)$	$2^+(2^{++})$	$a_6(2450)$	$1^-(6^{++})$
		$f_6(2510)$	$0^+(6^{++})$
		$X(3250)$	$?^?(?^{??})$

OTHER LIGHT UNFLAVORED ($S = C = B = 0$)

$e^+e^-(1100-2200)$	$?^?(1^{--})$
$\overline{N}N(1100-3600)$	
$X(1900-3600)$	

STRANGE ($S = \pm 1$, $C = B = 0$)

Particle	$I(J^P)$
• $K^\pm$	$1/2(0^-)$
• K^0	$1/2(0^-)$
• K_S^0	$1/2(0^-)$
• K_L^0	$1/2(0^-)$
• $K^*(892)$	$1/2(1^-)$
• $K_1(1270)$	$1/2(1^+)$
• $K_1(1400)$	$1/2(1^+)$
• $K^*(1410)$	$1/2(1^-)$
• $K_0^*(1430)$	$1/2(0^+)$
• $K_2^*(1430)$	$1/2(2^+)$
$K(1460)$	$1/2(0^-)$
$K_2(1580)$	$1/2(2^-)$
$K_1(1650)$	$1/2(1^+)$
• $K^*(1680)$	$1/2(1^-)$
• $K_2(1770)$	$1/2(2^-)$
• $K_3^*(1780)$	$1/2(3^-)$
• $K_2(1820)$	$1/2(2^-)$
$K(1830)$	$1/2(0^-)$
$K_0^*(1950)$	$1/2(0^+)$
$K_2^*(1980)$	$1/2(2^+)$
• $K_4^*(2045)$	$1/2(4^+)$
$K_2(2250)$	$1/2(2^-)$
$K_3(2320)$	$1/2(3^+)$
$K_5^*(2380)$	$1/2(5^-)$
$K_4(2500)$	$1/2(4^-)$
$K(3100)$	$?^?(?^{??})$

CHARMED ($C = \pm 1$)

Particle	$I(J^P)$
• $D^\pm$	$1/2(0^-)$
• D^0	$1/2(0^-)$
• $D^*(2007)^0$	$1/2(1^-)$
• $D^*(2010)^\pm$	$1/2(1^-)$
• $D_1(2420)^0$	$1/2(1^+)$
$D_1(2420)^\pm$	$1/2(?^?)$
• $D_2^*(2460)^0$	$1/2(2^+)$
• $D_2^*(2460)^+$	$1/2(2^+)$

CHARMED, STRANGE ($C = S = \pm 1$)

Particle	$I(J^P)$
• $D_s^\pm$	$0(0^-)$
• $D_s^{*\pm}$	$?(?^?)$
• $D_{s1}(2536)^\pm$	$0(1^+)$
• $D_{sJ}(2573)^\pm$	$?(?^?)$

BOTTOM ($B = \pm 1$)

Particle	$I(J^P)$
• $B^\pm$	$1/2(0^-)$
• B^0	$1/2(0^-)$
• B^*	$1/2(1^-)$
$B_J^*(5732)$	$?(?^?)$

BOTTOM, STRANGE ($B = \pm 1$, $S = \mp 1$)

Particle	$I^G(J^{PC})$
• B_s^0	$0(0^-)$
B_s^*	$?(?^?)$
$B_{sJ}^*(5850)$	$?(?^?)$

$c\bar{c}$

Particle	$I^G(J^{PC})$
• $\eta_c(1S)$	$0^+(0^{-+})$
• $J/\psi(1S)$	$0^-(1^{--})$
• $\chi_{c0}(1P)$	$0^+(0^{++})$
• $\chi_{c1}(1P)$	$0^+(1^{++})$
$h_c(1P)$	$?^?(?^{??})$
• $\chi_{c2}(1P)$	$0^+(2^{++})$
$\eta_c(2S)$	$?^?(?^{?+})$
• $\psi(2S)$	$0^-(1^{--})$
• $\psi(3770)$	$?^?(1^{--})$
• $\psi(4040)$	$?^?(1^{--})$
• $\psi(4160)$	$?^?(1^{--})$
• $\psi(4415)$	$?^?(1^{--})$

$b\bar{b}$

Particle	$I^G(J^{PC})$
• $\Upsilon(1S)$	$0^-(1^{--})$
• $\chi_{b0}(1P)$	$0^+(0^{++})$
• $\chi_{b1}(1P)$	$0^+(1^{++})$
• $\chi_{b2}(1P)$	$0^+(2^{++})$
• $\Upsilon(2S)$	$0^-(1^{--})$
• $\chi_{b0}(2P)$	$0^+(0^{++})$
• $\chi_{b1}(2P)$	$0^+(1^{++})$
• $\chi_{b2}(2P)$	$0^+(2^{++})$
• $\Upsilon(3S)$	$0^-(1^{--})$
• $\Upsilon(4S)$	$?^?(1^{--})$
• $\Upsilon(10860)$	$?^?(1^{--})$
• $\Upsilon(11020)$	$?^?(1^{--})$

NON-$q\bar{q}$ CANDIDATES

Non-$q\bar{q}$ Candidates

This short table gives the name, the quantum numbers (where known), and the status of baryons in the Review. Only the baryons with 3- or 4-star status are included in the main Baryon Summary Table. Due to insufficient data or uncertain interpretation, the other entries in the short table are not established as baryons. The names with masses are of baryons that decay strongly. See our 1986 edition (Physics Letters **170B**) for listings of evidence for Z baryons (KN resonances).

p	P_{11}	****	$\Delta(1232)$	P_{33}	****	Λ	P_{01}	****	Σ^+	P_{11}	****	Ξ^0	P_{11}	****
n	P_{11}	****	$\Delta(1600)$	P_{33}	***	$\Lambda(1405)$	S_{01}	****	Σ^0	P_{11}	****	Ξ^-	P_{11}	****
$N(1440)$	P_{11}	****	$\Delta(1620)$	S_{31}	****	$\Lambda(1520)$	D_{03}	****	Σ^-	P_{11}	****	$\Xi(1530)$	P_{13}	****
$N(1520)$	D_{13}	****	$\Delta(1700)$	D_{33}	****	$\Lambda(1600)$	P_{01}	***	$\Sigma(1385)$	P_{13}	****	$\Xi(1620)$		*
$N(1535)$	S_{11}	****	$\Delta(1750)$	P_{31}	*	$\Lambda(1670)$	S_{01}	****	$\Sigma(1480)$		*	$\Xi(1690)$		***
$N(1650)$	S_{11}	****	$\Delta(1900)$	S_{31}	***	$\Lambda(1690)$	D_{03}	****	$\Sigma(1560)$		**	$\Xi(1820)$	D_{13}	***
$N(1675)$	D_{15}	****	$\Delta(1905)$	F_{35}	****	$\Lambda(1800)$	S_{01}	***	$\Sigma(1580)$	D_{13}	**	$\Xi(1950)$		***
$N(1680)$	F_{15}	****	$\Delta(1910)$	P_{31}	****	$\Lambda(1810)$	P_{01}	***	$\Sigma(1620)$	S_{11}	**	$\Xi(2030)$		***
$N(1700)$	D_{13}	***	$\Delta(1920)$	P_{33}	***	$\Lambda(1820)$	F_{05}	****	$\Sigma(1660)$	P_{11}	***	$\Xi(2120)$		*
$N(1710)$	P_{11}	***	$\Delta(1930)$	D_{35}	***	$\Lambda(1830)$	D_{05}	****	$\Sigma(1670)$	D_{13}	****	$\Xi(2250)$		**
$N(1720)$	P_{13}	****	$\Delta(1940)$	D_{33}	*	$\Lambda(1890)$	P_{03}	****	$\Sigma(1690)$		**	$\Xi(2370)$		**
$N(1900)$	P_{13}	**	$\Delta(1950)$	F_{37}	****	$\Lambda(2000)$		*	$\Sigma(1750)$	S_{11}	***	$\Xi(2500)$		*
$N(1990)$	F_{17}	**	$\Delta(2000)$	F_{35}	**	$\Lambda(2020)$	F_{07}	*	$\Sigma(1770)$	P_{11}	*			
$N(2000)$	F_{15}	**	$\Delta(2150)$	S_{31}	*	$\Lambda(2100)$	G_{07}	****	$\Sigma(1775)$	D_{15}	****	Ω^-		****
$N(2080)$	D_{13}	**	$\Delta(2200)$	G_{37}	*	$\Lambda(2110)$	F_{05}	***	$\Sigma(1840)$	P_{13}	*	$\Omega(2250)^-$		***
$N(2090)$	S_{11}	*	$\Delta(2300)$	H_{39}	**	$\Lambda(2325)$	D_{03}	*	$\Sigma(1880)$	P_{11}	**	$\Omega(2380)^-$		**
$N(2100)$	P_{11}	*	$\Delta(2350)$	D_{35}	*	$\Lambda(2350)$	H_{09}	***	$\Sigma(1915)$	F_{15}	****	$\Omega(2470)^-$		**
$N(2190)$	G_{17}	****	$\Delta(2390)$	F_{37}	*	$\Lambda(2585)$		**	$\Sigma(1940)$	D_{13}	***			
$N(2200)$	D_{15}	**	$\Delta(2400)$	G_{39}	**				$\Sigma(2000)$	S_{11}	*	Λ_c^+		****
$N(2220)$	H_{19}	****	$\Delta(2420)$	$H_{3,11}$	****				$\Sigma(2030)$	F_{17}	****	$\Lambda_c(2593)^+$		***
$N(2250)$	G_{19}	****	$\Delta(2750)$	$I_{3,13}$	**				$\Sigma(2070)$	F_{15}	*	$\Lambda_c(2625)^+$		***
$N(2600)$	$I_{1,11}$	***	$\Delta(2950)$	$K_{3,15}$	**				$\Sigma(2080)$	P_{13}	**	$\Sigma_c(2455)$		****
$N(2700)$	$K_{1,13}$	**							$\Sigma(2100)$	G_{17}	*	$\Sigma_c(2530)$		*
									$\Sigma(2250)$		***	Ξ_c^+		***
									$\Sigma(2455)$		**	Ξ_c^0		***
									$\Sigma(2620)$		**	$\Xi_c(2645)$		***
									$\Sigma(3000)$		*	Ω_c^0		***
									$\Sigma(3170)$		*			
												Λ_b^0		***
												Ξ_b^0, Ξ_b^-		*

****   Existence is certain, and properties are at least fairly well explored.

***   Existence ranges from very likely to certain, but further confirmation is desirable and/or quantum numbers, branching fractions, etc. are not well determined.

**   Evidence of existence is only fair.

*   Evidence of existence is poor.

N BARYONS
($S = 0$, $I = 1/2$)

p, $N^+ = uud$; n, $N^0 = udd$

$\boxed{p}$ $I(J^P) = \frac{1}{2}(\frac{1}{2}^+)$

Mass $m = 938.27231 \pm 0.00028$ MeV [a]
$= 1.007276470 \pm 0.000000012$ u
$|\frac{q_p}{m_{\bar{p}}}|/(\frac{q_p}{m_p}) = 1.0000000015 \pm 0.0000000011$
$|q_p + q_{\bar{p}}|/e < 2 \times 10^{-5}$
$|q_p + q_e|/e < 1.0 \times 10^{-21}$ [b]
Magnetic moment $\mu = 2.79284739 \pm 0.00000006$ μ_N
Electric dipole moment $d = (-4 \pm 6) \times 10^{-23}$ $e$cm
Electric polarizability $\overline{\alpha} = (12.1 \pm 0.9) \times 10^{-4}$ fm^3
Magnetic polarizability $\overline{\beta} = (2.1 \pm 0.9) \times 10^{-4}$ fm^3
Mean life $\tau > 1.6 \times 10^{25}$ years (independent of mode)
$= > 10^{31} - 5 \times 10^{32}$ years [c] (mode dependent)

Below, for N decays, p and n distinguish proton and neutron partial life-times. See also the "Note on Nucleon Decay" in our 1994 edition (Phys. Rev. **D50**, 1673) for a short review.

The "partial mean life" limits tabulated here are the limits on τ/B_i, where τ is the total mean life and B_i is the branching fraction for the mode in question.

p DECAY MODES	Partial mean life (10^{30} years)	Confidence level	p (MeV/c)
Antilepton + meson			
$N \to e^+ \pi$	> 130 (n), > 550 (p)	90%	459
$N \to \mu^+ \pi$	> 100 (n), > 270 (p)	90%	453
$N \to \nu \pi$	> 100 (n), > 25 (p)	90%	459
$p \to e^+ \eta$	> 140	90%	309
$p \to \mu^+ \eta$	> 69	90%	296
$n \to \nu \eta$	> 54	90%	310
$N \to e^+ \rho$	> 58 (n), > 75 (p)	90%	153
$N \to \mu^+ \rho$	> 23 (n), > 110 (p)	90%	119
$N \to \nu \rho$	> 19 (n), > 27 (p)	90%	153
$p \to e^+ \omega$	> 45	90%	142
$p \to \mu^+ \omega$	> 57	90%	104
$n \to \nu \omega$	> 43	90%	144
$N \to e^+ K$	> 1.3 (n), > 150 (p)	90%	337
$p \to e^+ K^0_S$	> 76	90%	337
$p \to e^+ K^0_L$	> 44	90%	337
$N \to \mu^+ K$	> 1.1 (n), > 120 (p)	90%	326
$p \to \mu^+ K^0_S$	> 64	90%	326
$p \to \mu^+ K^0_L$	> 44	90%	326
$N \to \nu K$	> 86 (n), > 100 (p)	90%	339
$p \to e^+ K^*(892)^0$	> 52	90%	45
$N \to \nu K^*(892)$	> 22 (n), > 20 (p)	90%	45
Antilepton + mesons			
$p \to e^+ \pi^+ \pi^-$	> 21	90%	448
$p \to e^+ \pi^0 \pi^0$	> 38	90%	449
$n \to e^+ \pi^- \pi^0$	> 32	90%	449
$p \to \mu^+ \pi^+ \pi^-$	> 17	90%	425
$p \to \mu^+ \pi^0 \pi^0$	> 33	90%	427
$n \to \mu^+ \pi^- \pi^0$	> 33	90%	427
$n \to e^+ K^0 \pi^-$	> 18	90%	319
Lepton + meson			
$n \to e^- \pi^+$	> 65	90%	459
$n \to \mu^- \pi^+$	> 49	90%	453
$n \to e^- \rho^+$	> 62	90%	154
$n \to \mu^- \rho^+$	> 7	90%	120
$n \to e^- K^+$	> 32	90%	340
$n \to \mu^- K^+$	> 57	90%	330
Lepton + mesons			
$p \to e^- \pi^+ \pi^+$	> 30	90%	448
$n \to e^- \pi^+ \pi^0$	> 29	90%	449
$p \to \mu^- \pi^+ \pi^+$	> 17	90%	425
$n \to \mu^- \pi^+ \pi^0$	> 34	90%	427
$p \to e^- \pi^+ K^+$	> 20	90%	320
$p \to \mu^- \pi^+ K^+$	> 5	90%	279

	Partial mean life	Confidence level	p (MeV/c)
Antilepton + photon(s)			
$p \to e^+ \gamma$	> 460	90%	469
$p \to \mu^+ \gamma$	> 380	90%	463
$n \to \nu \gamma$	> 24	90%	470
$p \to e^+ \gamma \gamma$	> 100	90%	469
Three leptons			
$p \to e^+ e^+ e^-$	> 510	90%	469
$p \to e^+ \mu^+ \mu^-$	> 81	90%	457
$p \to e^+ \nu \nu$	> 11	90%	469
$n \to e^+ e^- \nu$	> 74	90%	470
$n \to \mu^+ e^- \nu$	> 47	90%	464
$n \to \mu^+ \mu^- \nu$	> 42	90%	458
$p \to \mu^+ e^+ e^-$	> 91	90%	464
$p \to \mu^+ \mu^+ \mu^-$	> 190	90%	439
$p \to \mu^+ \nu \nu$	> 21	90%	463
$p \to e^- \mu^+ \mu^+$	> 6	90%	457
$n \to 3\nu$	> 0.0005	90%	470
Inclusive modes			
$N \to e^+$ anything	> 0.6 (n, p)	90%	–
$N \to \mu^+$ anything	> 12 (n, p)	90%	–
$N \to e^+ \pi^0$ anything	> 0.6 (n, p)	90%	–

$\Delta B = 2$ dinucleon modes

The following are lifetime limits per iron nucleus.

$pp \to \pi^+ \pi^+$	> 0.7	90%	–
$pn \to \pi^+ \pi^0$	> 2	90%	–
$nn \to \pi^+ \pi^-$	> 0.7	90%	–
$nn \to \pi^0 \pi^0$	> 3.4	90%	–
$pp \to e^+ e^+$	> 5.8	90%	–
$pp \to e^+ \mu^+$	> 3.6	90%	–
$pp \to \mu^+ \mu^+$	> 1.7	90%	–
$pn \to e^+ \overline{\nu}$	> 2.8	90%	–
$pn \to \mu^+ \overline{\nu}$	> 1.6	90%	–
$nn \to \nu_e \overline{\nu}_e$	> 0.000012	90%	–
$nn \to \nu_\mu \overline{\nu}_\mu$	> 0.000006	90%	–

$\overline{p}$ DECAY MODES

$\overline{p}$ DECAY MODES	Partial mean life (years)	Confidence level	p (MeV/c)
$\overline{p} \to e^- \gamma$	> 1848	95%	469
$\overline{p} \to e^- \pi^0$	> 554	95%	459
$\overline{p} \to e^- \eta$	> 171	95%	309
$\overline{p} \to e^- K^0_S$	> 29	95%	337
$\overline{p} \to e^- K^0_L$	> 9	95%	337

$\boxed{n}$ $I(J^P) = \frac{1}{2}(\frac{1}{2}^+)$

Mass $m = 939.56563 \pm 0.00028$ MeV [a]
$= 1.008664904 \pm 0.000000014$ u
$m_n - m_p = 1.293318 \pm 0.000009$ MeV
$= 0.001388434 \pm 0.000000009$ u
Mean life $\tau = 887.0 \pm 2.0$ s (S = 1.3)
$c\tau = 2.659 \times 10^8$ km
Magnetic moment $\mu = -1.9130428 \pm 0.0000005$ μ_N
Electric dipole moment $d < 1.1 \times 10^{-25}$ $e$cm, CL = 95%
Electric polarizability $\alpha = (0.98^{+0.19}_{-0.23}) \times 10^{-3}$ fm^3 (S = 1.1)
Charge $q = (-0.4 \pm 1.1) \times 10^{-21}$ e
Mean $n\overline{n}$-oscillation time $> 1.2 \times 10^8$ s, CL = 90% [d] (bound n)
$> 0.86 \times 10^8$ s, CL = 90% (free n)

Decay parameters [e]

$pe^- \overline{\nu}_e$	$g_A/g_V = -1.2601 \pm 0.0025$ (S = 1.1)
"	$A = -0.1139 \pm 0.0011$ (S = 1.3)
"	$B = 0.990 \pm 0.008$
"	$a = -0.102 \pm 0.005$
"	$\phi_{AV} = (180.07 \pm 0.18)°$ [f]
"	$D = (-0.5 \pm 1.4) \times 10^{-3}$

n DECAY MODES	Fraction (Γ_i/Γ)	Confidence level	p (MeV/c)
$pe^- \overline{\nu}_e$	100 %		1.19
Charge conservation (Q) violating mode			
$p\nu_e \overline{\nu}_e$	Q $< 9 \times 10^{-24}$	90%	1.29

$N(1440) \, P_{11}$ $\qquad I(J^P) = \frac{1}{2}(\frac{1}{2}^+)$

Mass $m = 1430$ to 1470 (≈ 1440) MeV
Full width $\Gamma = 250$ to 450 (≈ 350) MeV
$p_{beam} = 0.61$ GeV/c $\qquad 4\pi\lambda^2 = 31.0$ mb

N(1440) DECAY MODES	Fraction (Γ_i/Γ)	p (MeV/c)
$N\pi$	60–70 %	397
$N\pi\pi$	30–40 %	342
$\Delta\pi$	20–30 %	143
$N\rho$	<8 %	†
$N(\pi\pi)_{S\text{-wave}}^{I=0}$	5–10 %	–
$p\gamma$	0.035–0.048 %	414
$\quad p\gamma$, helicity=1/2	0.035–0.048 %	414
$n\gamma$	0.009–0.032 %	413
$\quad n\gamma$, helicity=1/2	0.009–0.032 %	413

$N(1520) \, D_{13}$ $\qquad I(J^P) = \frac{1}{2}(\frac{3}{2}^-)$

Mass $m = 1515$ to 1530 (≈ 1520) MeV
Full width $\Gamma = 110$ to 135 (≈ 120) MeV
$p_{beam} = 0.74$ GeV/c $\qquad 4\pi\lambda^2 = 23.5$ mb

N(1520) DECAY MODES	Fraction (Γ_i/Γ)	p (MeV/c)
$N\pi$	50–60 %	456
$N\pi\pi$	40–50 %	410
$\Delta\pi$	15–25 %	228
$N\rho$	15–25 %	†
$N(\pi\pi)_{S\text{-wave}}^{I=0}$	<8 %	–
$p\gamma$	0.46–0.56 %	470
$\quad p\gamma$, helicity=1/2	0.001–0.034 %	470
$\quad p\gamma$, helicity=3/2	0.44–0.53 %	470
$n\gamma$	0.30–0.53 %	470
$\quad n\gamma$, helicity=1/2	0.04–0.10 %	470
$\quad n\gamma$, helicity=3/2	0.25–0.45 %	470

$N(1535) \, S_{11}$ $\qquad I(J^P) = \frac{1}{2}(\frac{1}{2}^-)$

Mass $m = 1520$ to 1555 (≈ 1535) MeV
Full width $\Gamma = 100$ to 250 (≈ 150) MeV
$p_{beam} = 0.76$ GeV/c $\qquad 4\pi\lambda^2 = 22.5$ mb

N(1535) DECAY MODES	Fraction (Γ_i/Γ)	p (MeV/c)
$N\pi$	35–55 %	467
$N\eta$	30–55 %	182
$N\pi\pi$	1–10 %	422
$\Delta\pi$	<1 %	242
$N\rho$	<4 %	†
$N(\pi\pi)_{S\text{-wave}}^{I=0}$	<3 %	–
$N(1440)\pi$	<7 %	†
$p\gamma$	0.08–0.27 %	481
$\quad p\gamma$, helicity=1/2	0.08–0.27 %	481
$n\gamma$	0.004–0.29 %	480
$\quad n\gamma$, helicity=1/2	0.004–0.29 %	480

$N(1650) \, S_{11}$ $\qquad I(J^P) = \frac{1}{2}(\frac{1}{2}^-)$

Mass $m = 1640$ to 1680 (≈ 1650) MeV
Full width $\Gamma = 145$ to 190 (≈ 150) MeV
$p_{beam} = 0.96$ GeV/c $\qquad 4\pi\lambda^2 = 16.4$ mb

N(1650) DECAY MODES	Fraction (Γ_i/Γ)	p (MeV/c)
$N\pi$	55–90 %	547
$N\eta$	3–10 %	346
ΛK	3–11 %	161
$N\pi\pi$	10–20 %	511
$\Delta\pi$	1–7 %	344
$N\rho$	4–12 %	†

$N(\pi\pi)_{S\text{-wave}}^{I=0}$	<4 %	–
$N(1440)\pi$	<5 %	147
$p\gamma$	0.04–0.18 %	558
$\quad p\gamma$, helicity=1/2	0.04–0.18 %	558
$n\gamma$	0.003–0.17 %	557
$\quad n\gamma$, helicity=1/2	0.003–0.17 %	557

$N(1675) \, D_{15}$ $\qquad I(J^P) = \frac{1}{2}(\frac{5}{2}^-)$

Mass $m = 1670$ to 1685 (≈ 1675) MeV
Full width $\Gamma = 140$ to 180 (≈ 150) MeV
$p_{beam} = 1.01$ GeV/c $\qquad 4\pi\lambda^2 = 15.4$ mb

N(1675) DECAY MODES	Fraction (Γ_i/Γ)	p (MeV/c)
$N\pi$	40–50 %	563
ΛK	<1 %	209
$N\pi\pi$	50–60 %	529
$\Delta\pi$	50–60 %	364
$N\rho$	< 1–3 %	†
$p\gamma$	0.004–0.023 %	575
$\quad p\gamma$, helicity=1/2	0.0–0.015 %	575
$\quad p\gamma$, helicity=3/2	0.0–0.011 %	575
$n\gamma$	0.02–0.12 %	574
$\quad n\gamma$, helicity=1/2	0.006–0.046 %	574
$\quad n\gamma$, helicity=3/2	0.01–0.08 %	574

$N(1680) \, F_{15}$ $\qquad I(J^P) = \frac{1}{2}(\frac{5}{2}^+)$

Mass $m = 1675$ to 1690 (≈ 1680) MeV
Full width $\Gamma = 120$ to 140 (≈ 130) MeV
$p_{beam} = 1.01$ GeV/c $\qquad 4\pi\lambda^2 = 15.2$ mb

N(1680) DECAY MODES	Fraction (Γ_i/Γ)	p (MeV/c)
$N\pi$	60–70 %	567
$N\pi\pi$	30–40 %	532
$\Delta\pi$	5–15 %	369
$N\rho$	3–15 %	†
$N(\pi\pi)_{S\text{-wave}}^{I=0}$	5–20 %	–
$p\gamma$	0.21–0.32 %	578
$\quad p\gamma$, helicity=1/2	0.001–0.011 %	578
$\quad p\gamma$, helicity=3/?	0.20–0.32 %	578
$n\gamma$	0.021–0.046 %	577
$\quad n\gamma$, helicity=1/2	0.004–0.029 %	577
$\quad n\gamma$, helicity=3/2	0.01–0.024 %	577

$N(1700) \, D_{13}$ $\qquad I(J^P) = \frac{1}{2}(\frac{3}{2}^-)$

Mass $m = 1650$ to 1750 (≈ 1700) MeV
Full width $\Gamma = 50$ to 150 (≈ 100) MeV
$p_{beam} = 1.05$ GeV/c $\qquad 4\pi\lambda^2 = 14.5$ mb

N(1700) DECAY MODES	Fraction (Γ_i/Γ)	p (MeV/c)
$N\pi$	5–15 %	580
ΛK	<3 %	250
$N\pi\pi$	85–95 %	547
$N\rho$	<35 %	†
$p\gamma$	0.01–0.05 %	591
$\quad p\gamma$, helicity=1/2	0.0–0.024 %	591
$\quad p\gamma$, helicity=3/2	0.002–0.026 %	591
$n\gamma$	0.01–0.13 %	590
$\quad n\gamma$, helicity=1/2	0.0–0.09 %	590
$\quad n\gamma$, helicity=3/2	0.01–0.05 %	590

Baryon Summary Table

$N(1710)\ P_{11}$ $\qquad I(J^P) = \frac{1}{2}(\frac{1}{2}^+)$

Mass $m = 1680$ to $1740\ (\approx 1710)$ MeV
Full width $\Gamma = 50$ to $250\ (\approx 100)$ MeV
$p_{beam} = 1.07$ GeV/c $\qquad 4\pi\lambda^2 = 14.2$ mb

$N(1710)$ DECAY MODES	Fraction (Γ_i/Γ)	p (MeV/c)
$N\pi$	10–20 %	587
ΛK	5–25 %	264
$N\pi\pi$	40–90 %	554
$\quad \Delta\pi$	15–40 %	393
$\quad N\rho$	5–25 %	48
$\quad N(\pi\pi)^{I=0}_{S\text{-wave}}$	10–40 %	–
$p\gamma$	0.002–0.05%	598
$\quad p\gamma$, helicity=1/2	0.002–0.05%	598
$n\gamma$	0.0–0.02%	597
$\quad n\gamma$, helicity=1/2	0.0–0.02%	597

$N(1720)\ P_{13}$ $\qquad I(J^P) = \frac{1}{2}(\frac{3}{2}^+)$

Mass $m = 1650$ to $1750\ (\approx 1720)$ MeV
Full width $\Gamma = 100$ to $200\ (\approx 150)$ MeV
$p_{beam} = 1.09$ GeV/c $\qquad 4\pi\lambda^2 = 13.9$ mb

$N(1720)$ DECAY MODES	Fraction (Γ_i/Γ)	p (MeV/c)
$N\pi$	10–20 %	594
ΛK	1–15 %	278
$N\pi\pi$	>70 %	561
$\quad N\rho$	70–85 %	104
$p\gamma$	0.003–0.10 %	604
$\quad p\gamma$, helicity=1/2	0.003–0.08 %	604
$\quad p\gamma$, helicity=3/2	0.001–0.03 %	604
$n\gamma$	0.002–0.39 %	603
$\quad n\gamma$, helicity=1/2	0.0–0.002 %	603
$\quad n\gamma$, helicity=3/2	0.001–0.39 %	603

$N(2190)\ G_{17}$ $\qquad I(J^P) = \frac{1}{2}(\frac{7}{2}^-)$

Mass $m = 2100$ to $2200\ (\approx 2190)$ MeV
Full width $\Gamma = 350$ to $550\ (\approx 450)$ MeV
$p_{beam} = 2.07$ GeV/c $\qquad 4\pi\lambda^2 = 6.21$ mb

$N(2190)$ DECAY MODES	Fraction (Γ_i/Γ)	p (MeV/c)
$N\pi$	10–20 %	888

$N(2220)\ H_{19}$ $\qquad I(J^P) = \frac{1}{2}(\frac{9}{2}^+)$

Mass $m = 2180$ to $2310\ (\approx 2220)$ MeV
Full width $\Gamma = 320$ to $550\ (\approx 400)$ MeV
$p_{beam} = 2.14$ GeV/c $\qquad 4\pi\lambda^2 = 5.97$ mb

$N(2220)$ DECAY MODES	Fraction (Γ_i/Γ)	p (MeV/c)
$N\pi$	10–20 %	905

$N(2250)\ G_{19}$ $\qquad I(J^P) = \frac{1}{2}(\frac{9}{2}^-)$

Mass $m = 2170$ to $2310\ (\approx 2250)$ MeV
Full width $\Gamma = 290$ to $470\ (\approx 400)$ MeV
$p_{beam} = 2.21$ GeV/c $\qquad 4\pi\lambda^2 = 5.74$ mb

$N(2250)$ DECAY MODES	Fraction (Γ_i/Γ)	p (MeV/c)
$N\pi$	5–15 %	923

$N(2600)\ I_{1,11}$ $\qquad I(J^P) = \frac{1}{2}(\frac{11}{2}^-)$

Mass $m = 2550$ to $2750\ (\approx 2600)$ MeV
Full width $\Gamma = 500$ to $800\ (\approx 650)$ MeV
$p_{beam} = 3.12$ GeV/c $\qquad 4\pi\lambda^2 = 3.86$ mb

$N(2600)$ DECAY MODES	Fraction (Γ_i/Γ)	p (MeV/c)
$N\pi$	5–10 %	1126

Δ BARYONS ($S = 0,\ I = 3/2$)

$\Delta^{++} = uuu,\quad \Delta^+ = uud,\quad \Delta^0 = udd,\quad \Delta^- = ddd$

$\Delta(1232)\ P_{33}$ $\qquad I(J^P) = \frac{3}{2}(\frac{3}{2}^+)$

Mass $m = 1230$ to $1234\ (\approx 1232)$ MeV
Full width $\Gamma = 115$ to $125\ (\approx 120)$ MeV
$p_{beam} = 0.30$ GeV/c $\qquad 4\pi\lambda^2 = 94.8$ mb

$\Delta(1232)$ DECAY MODES	Fraction (Γ_i/Γ)	p (MeV/c)
$N\pi$	>99 %	227
$N\gamma$	0.54–0.61 %	259
$\quad N\gamma$, helicity=1/2	0.12–0.14 %	259
$\quad N\gamma$, helicity=3/2	0.41–0.47 %	259

$\Delta(1600)\ P_{33}$ $\qquad I(J^P) = \frac{3}{2}(\frac{3}{2}^+)$

Mass $m = 1550$ to $1700\ (\approx 1600)$ MeV
Full width $\Gamma = 250$ to $450\ (\approx 350)$ MeV
$p_{beam} = 0.87$ GeV/c $\qquad 4\pi\lambda^2 = 18.6$ mb

$\Delta(1600)$ DECAY MODES	Fraction (Γ_i/Γ)	p (MeV/c)
$N\pi$	10–25 %	512
$N\pi\pi$	75–90 %	473
$\quad \Delta\pi$	40–70 %	301
$\quad N\rho$	<25 %	†
$\quad N(1440)\pi$	10–35 %	74
$N\gamma$	0.001–0.02 %	525
$\quad N\gamma$, helicity=1/2	0.0–0.02 %	525
$\quad N\gamma$, helicity=3/2	0.001–0.005 %	525

$\Delta(1620)\ S_{31}$ $\qquad I(J^P) = \frac{3}{2}(\frac{1}{2}^-)$

Mass $m = 1615$ to $1675\ (\approx 1620)$ MeV
Full width $\Gamma = 120$ to $180\ (\approx 150)$ MeV
$p_{beam} = 0.91$ GeV/c $\qquad 4\pi\lambda^2 = 17.7$ mb

$\Delta(1620)$ DECAY MODES	Fraction (Γ_i/Γ)	p (MeV/c)
$N\pi$	20–30 %	526
$N\pi\pi$	70–80 %	488
$\quad \Delta\pi$	30–60 %	318
$\quad N\rho$	7–25 %	†
$N\gamma$	0.004–0.044 %	538
$\quad N\gamma$, helicity=1/2	0.004–0.044 %	538

$\Delta(1700)\ D_{33}$ $\qquad I(J^P) = \frac{3}{2}(\frac{3}{2}^-)$

Mass $m = 1670$ to $1770\ (\approx 1700)$ MeV
Full width $\Gamma = 200$ to $400\ (\approx 300)$ MeV
$p_{beam} = 1.05$ GeV/c $\qquad 4\pi\lambda^2 = 14.5$ mb

$\Delta(1700)$ DECAY MODES	Fraction (Γ_i/Γ)	p (MeV/c)
$N\pi$	10–20 %	580
$N\pi\pi$	80–90 %	547
$\quad \Delta\pi$	30–60 %	385
$\quad N\rho$	30–55 %	†
$N\gamma$	0.12–0.26 %	591
$\quad N\gamma$, helicity=1/2	0.08–0.16 %	591
$\quad N\gamma$, helicity=3/2	0.025–0.12 %	591

$\Delta(1900)\ S_{31}$ $\qquad I(J^P) = \frac{3}{2}(\frac{1}{2}^-)$

Mass $m = 1850$ to $1950\ (\approx 1900)$ MeV
Full width $\Gamma = 140$ to $240\ (\approx 200)$ MeV
$p_{beam} = 1.44$ GeV/c $\qquad 4\pi\lambda^2 = 9.71$ mb

$\Delta(1900)$ DECAY MODES	Fraction (Γ_i/Γ)	p (MeV/c)
$N\pi$	10–30 %	710

$\Delta(1905)$ F_{35}	$I(J^P) = \frac{3}{2}(\frac{5}{2}^+)$

Mass $m = 1870$ to 1920 (≈ 1905) MeV
Full width $\Gamma = 280$ to 440 (≈ 350) MeV
$p_{beam} = 1.45$ GeV/c $4\pi\lambda^2 = 9.62$ mb

$\Delta(1905)$ DECAY MODES	Fraction (Γ_i/Γ)	p (MeV/c)
$N\pi$	5–15 %	713
$N\pi\pi$	85–95 %	687
$\Delta\pi$	<25 %	542
$N\rho$	>60 %	421
$N\gamma$	0.01–0.03 %	721
$N\gamma$, helicity=1/2	0.0–0.1 %	721
$N\gamma$, helicity=3/2	0.004–0.03 %	721

$\Delta(1910)$ P_{31}	$I(J^P) = \frac{3}{2}(\frac{1}{2}^+)$

Mass $m = 1870$ to 1920 (≈ 1910) MeV
Full width $\Gamma = 190$ to 270 (≈ 250) MeV
$p_{beam} = 1.46$ GeV/c $4\pi\lambda^2 = 9.54$ mb

$\Delta(1910)$ DECAY MODES	Fraction (Γ_i/Γ)	p (MeV/c)
$N\pi$	15–30 %	716
$N\gamma$	0.0–0.2 %	725
$N\gamma$, helicity=1/2	0.0–0.2 %	725

$\Delta(1920)$ P_{33}	$I(J^P) = \frac{3}{2}(\frac{3}{2}^+)$

Mass $m = 1900$ to 1970 (≈ 1920) MeV
Full width $\Gamma = 150$ to 300 (≈ 200) MeV
$p_{beam} = 1.48$ GeV/c $4\pi\lambda^2 = 9.37$ mb

$\Delta(1920)$ DECAY MODES	Fraction (Γ_i/Γ)	p (MeV/c)
$N\pi$	5–20 %	722

$\Delta(1930)$ D_{35}	$I(J^P) = \frac{3}{2}(\frac{5}{2}^-)$

Mass $m = 1920$ to 1970 (≈ 1930) MeV
Full width $\Gamma = 250$ to 450 (≈ 350) MeV
$p_{beam} = 1.50$ GeV/c $4\pi\lambda^2 = 9.21$ mb

$\Delta(1930)$ DECAY MODES	Fraction (Γ_i/Γ)	p (MeV/c)
$N\pi$	10–20 %	729
$N\gamma$	0.0–0.02 %	737
$N\gamma$, helicity=1/2	0.0–0.01 %	737
$N\gamma$, helicity=3/2	0.0–0.01 %	737

$\Delta(1950)$ F_{37}	$I(J^P) = \frac{3}{2}(\frac{7}{2}^+)$

Mass $m = 1940$ to 1960 (≈ 1950) MeV
Full width $\Gamma = 290$ to 350 (≈ 300) MeV
$p_{beam} = 1.54$ GeV/c $4\pi\lambda^2 = 8.91$ mb

$\Delta(1950)$ DECAY MODES	Fraction (Γ_i/Γ)	p (MeV/c)
$N\pi$	35–40 %	741
$N\pi\pi$		716
$\Delta\pi$	20–30 %	574
$N\rho$	<10 %	469
$N\gamma$	0.08–0.13 %	749
$N\gamma$, helicity=1/2	0.03–0.055 %	749
$N\gamma$, helicity=3/2	0.05–0.075 %	749

$\Delta(2420)$ $H_{3,11}$	$I(J^P) = \frac{3}{2}(\frac{11}{2}^+)$

Mass $m = 2300$ to 2500 (≈ 2420) MeV
Full width $\Gamma = 300$ to 500 (≈ 400) MeV
$p_{beam} = 2.64$ GeV/c $4\pi\lambda^2 = 4.68$ mb

$\Delta(2420)$ DECAY MODES	Fraction (Γ_i/Γ)	p (MeV/c)
$N\pi$	5–15 %	1023

Λ BARYONS
$(S = -1, I = 0)$
$\Lambda^0 = uds$

Λ	$I(J^P) = 0(\frac{1}{2}^+)$

Mass $m = 1115.684 \pm 0.006$ MeV
Mean life $\tau = (2.632 \pm 0.020) \times 10^{-10}$ s   (S = 1.6)
$c\tau = 7.89$ cm
Magnetic moment $\mu = -0.613 \pm 0.004$ μ_N
Electric dipole moment $d < 1.5 \times 10^{-16}$ $e$cm, CL = 95%

Decay parameters

$p\pi^-$	$\alpha_- = 0.642 \pm 0.013$
"	$\phi_- = (-6.5 \pm 3.5)°$
"	$\gamma_- = 0.76$ [g]
"	$\Delta_- = (8 \pm 4)°$ [g]
$n\pi^0$	$\alpha_0 = +0.65 \pm 0.05$
$pe^-\bar{\nu}_e$	$g_A/g_V = -0.718 \pm 0.015$ [e]

Λ DECAY MODES	Fraction (Γ_i/Γ)	p (MeV/c)
$p\pi^-$	(63.9 ± 0.5) %	101
$n\pi^0$	(35.8 ± 0.5) %	104
$n\gamma$	$(1.75 \pm 0.15) \times 10^{-3}$	162
$p\pi^-\gamma$	[h] $(8.4 \pm 1.4) \times 10^{-4}$	101
$pe^-\bar{\nu}_e$	$(8.32 \pm 0.14) \times 10^{-4}$	163
$p\mu^-\bar{\nu}_\mu$	$(1.57 \pm 0.35) \times 10^{-4}$	131

$\Lambda(1405)$ S_{01}	$I(J^P) = 0(\frac{1}{2}^-)$

Mass $m = 1407 \pm 4$ MeV
Full width $\Gamma = 50.0 \pm 2.0$ MeV
Below $\overline{K}N$ threshold

$\Lambda(1405)$ DECAY MODES	Fraction (Γ_i/Γ)	p (MeV/c)
$\Sigma\pi$	100 %	152

$\Lambda(1520)$ D_{03}	$I(J^P) = 0(\frac{3}{2}^-)$

Mass $m = 1519.5 \pm 1.0$ MeV [i]
Full width $\Gamma = 15.6 \pm 1.0$ MeV [i]
$p_{beam} = 0.39$ GeV/c $4\pi\lambda^2 = 82.8$ mb

$\Lambda(1520)$ DECAY MODES	Fraction (Γ_i/Γ)	p (MeV/c)
$N\overline{K}$	45 ± 1%	244
$\Sigma\pi$	42 ± 1%	267
$\Lambda\pi\pi$	10 ± 1%	252
$\Sigma\pi\pi$	0.9 ± 0.1%	152
$\Lambda\gamma$	0.8 ± 0.2%	351

$\Lambda(1600)$ P_{01}	$I(J^P) = 0(\frac{1}{2}^+)$

Mass $m = 1560$ to 1700 (≈ 1600) MeV
Full width $\Gamma = 50$ to 250 (≈ 150) MeV
$p_{beam} = 0.58$ GeV/c $4\pi\lambda^2 = 41.6$ mb

$\Lambda(1600)$ DECAY MODES	Fraction (Γ_i/Γ)	p (MeV/c)
$N\overline{K}$	15–30 %	343
$\Sigma\pi$	10–60 %	336

Baryon Summary Table

$\Lambda(1670)$ S_{01} $\qquad I(J^P) = 0(\frac{1}{2}^-)$

Mass $m = 1660$ to 1680 (≈ 1670) MeV
Full width $\Gamma = 25$ to 50 (≈ 35) MeV
$p_{beam} = 0.74$ GeV/c $\qquad 4\pi\lambda^2 = 28.5$ mb

$\Lambda(1670)$ DECAY MODES	Fraction (Γ_i/Γ)	p (MeV/c)
$N\overline{K}$	15–25 %	414
$\Sigma\pi$	20–60 %	393
$\Lambda\eta$	15–35 %	64

$\Lambda(1690)$ D_{03} $\qquad I(J^P) = 0(\frac{3}{2}^-)$

Mass $m = 1685$ to 1695 (≈ 1690) MeV
Full width $\Gamma = 50$ to 70 (≈ 60) MeV
$p_{beam} = 0.78$ GeV/c $\qquad 4\pi\lambda^2 = 26.1$ mb

$\Lambda(1690)$ DECAY MODES	Fraction (Γ_i/Γ)	p (MeV/c)
$N\overline{K}$	20–30 %	433
$\Sigma\pi$	20–40 %	409
$\Lambda\pi\pi$	~ 25 %	415
$\Sigma\pi\pi$	~ 20 %	350

$\Lambda(1800)$ S_{01} $\qquad I(J^P) = 0(\frac{1}{2}^-)$

Mass $m = 1720$ to 1850 (≈ 1800) MeV
Full width $\Gamma = 200$ to 400 (≈ 300) MeV
$p_{beam} = 1.01$ GeV/c $\qquad 4\pi\lambda^2 = 17.5$ mb

$\Lambda(1800)$ DECAY MODES	Fraction (Γ_i/Γ)	p (MeV/c)
$N\overline{K}$	25–40 %	528
$\Sigma\pi$	seen	493
$\Sigma(1385)\pi$	seen	345
$N\overline{K}^*(892)$	seen	†

$\Lambda(1810)$ P_{01} $\qquad I(J^P) = 0(\frac{1}{2}^+)$

Mass $m = 1750$ to 1850 (≈ 1810) MeV
Full width $\Gamma = 50$ to 250 (≈ 150) MeV
$p_{beam} = 1.04$ GeV/c $\qquad 4\pi\lambda^2 = 17.0$ mb

$\Lambda(1810)$ DECAY MODES	Fraction (Γ_i/Γ)	p (MeV/c)
$N\overline{K}$	20–50 %	537
$\Sigma\pi$	10–40 %	501
$\Sigma(1385)\pi$	seen	356
$N\overline{K}^*(892)$	30–60 %	†

$\Lambda(1820)$ F_{05} $\qquad I(J^P) = 0(\frac{5}{2}^+)$

Mass $m = 1815$ to 1825 (≈ 1820) MeV
Full width $\Gamma = 70$ to 90 (≈ 80) MeV
$p_{beam} = 1.06$ GeV/c $\qquad 4\pi\lambda^2 = 16.5$ mb

$\Lambda(1820)$ DECAY MODES	Fraction (Γ_i/Γ)	p (MeV/c)
$N\overline{K}$	55–65 %	545
$\Sigma\pi$	8–14 %	508
$\Sigma(1385)\pi$	5–10 %	362

$\Lambda(1830)$ D_{05} $\qquad I(J^P) = 0(\frac{5}{2}^-)$

Mass $m = 1810$ to 1830 (≈ 1830) MeV
Full width $\Gamma = 60$ to 110 (≈ 95) MeV
$p_{beam} = 1.08$ GeV/c $\qquad 4\pi\lambda^2 = 16.0$ mb

$\Lambda(1830)$ DECAY MODES	Fraction (Γ_i/Γ)	p (MeV/c)
$N\overline{K}$	3–10 %	553
$\Sigma\pi$	35–75 %	515
$\Sigma(1385)\pi$	>15 %	371

$\Lambda(1890)$ P_{03} $\qquad I(J^P) = 0(\frac{3}{2}^+)$

Mass $m = 1850$ to 1910 (≈ 1890) MeV
Full width $\Gamma = 60$ to 200 (≈ 100) MeV
$p_{beam} = 1.21$ GeV/c $\qquad 4\pi\lambda^2 = 13.6$ mb

$\Lambda(1890)$ DECAY MODES	Fraction (Γ_i/Γ)	p (MeV/c)
$N\overline{K}$	20–35 %	599
$\Sigma\pi$	3–10 %	559
$\Sigma(1385)\pi$	seen	420
$N\overline{K}^*(892)$	seen	233

$\Lambda(2100)$ G_{07} $\qquad I(J^P) = 0(\frac{7}{2}^-)$

Mass $m = 2090$ to 2110 (≈ 2100) MeV
Full width $\Gamma = 100$ to 250 (≈ 200) MeV
$p_{beam} = 1.68$ GeV/c $\qquad 4\pi\lambda^2 = 8.68$ mb

$\Lambda(2100)$ DECAY MODES	Fraction (Γ_i/Γ)	p (MeV/c)
$N\overline{K}$	25–35 %	751
$\Sigma\pi$	~ 5 %	704
$\Lambda\eta$	<3 %	617
ΞK	<3 %	483
$\Lambda\omega$	<8 %	443
$N\overline{K}^*(892)$	10–20 %	514

$\Lambda(2110)$ F_{05} $\qquad I(J^P) = 0(\frac{5}{2}^+)$

Mass $m = 2090$ to 2140 (≈ 2110) MeV
Full width $\Gamma = 150$ to 250 (≈ 200) MeV
$p_{beam} = 1.70$ GeV/c $\qquad 4\pi\lambda^2 = 8.53$ mb

$\Lambda(2110)$ DECAY MODES	Fraction (Γ_i/Γ)	p (MeV/c)
$N\overline{K}$	5–25 %	757
$\Sigma\pi$	10–40 %	711
$\Lambda\omega$	seen	455
$\Sigma(1385)\pi$	seen	589
$N\overline{K}^*(892)$	10–60 %	524

$\Lambda(2350)$ H_{09} $\qquad I(J^P) = 0(\frac{9}{2}^+)$

Mass $m = 2340$ to 2370 (≈ 2350) MeV
Full width $\Gamma = 100$ to 250 (≈ 150) MeV
$p_{beam} = 2.29$ GeV/c $\qquad 4\pi\lambda^2 = 5.85$ mb

$\Lambda(2350)$ DECAY MODES	Fraction (Γ_i/Γ)	p (MeV/c)
$N\overline{K}$	~ 12 %	915
$\Sigma\pi$	~ 10 %	867

Σ BARYONS
(S = −1, I = 1)

$\Sigma^+ = uus$, $\Sigma^0 = uds$, $\Sigma^- = dds$

 Σ⁺ $I(J^P) = 1(\frac{1}{2}^+)$

Mass $m = 1189.37 \pm 0.07$ MeV   (S = 2.2)
Mean life $\tau = (0.799 \pm 0.004) \times 10^{-10}$ s
    $c\tau = 2.396$ cm
Magnetic moment $\mu = 2.458 \pm 0.010$ μ_N (S = 2.1)
$\Gamma(\Sigma^+ \to n\ell^+\nu)/\Gamma(\Sigma^- \to n\ell^-\bar{\nu}) < 0.043$

Decay parameters

$p\pi^0$	$\alpha_0 = -0.980^{+0.017}_{-0.015}$
"	$\phi_0 = (36 \pm 34)°$
"	$\gamma_0 = 0.16$ [g]
"	$\Delta_0 = (187 \pm 6)°$ [g]
$n\pi^+$	$\alpha_+ = 0.068 \pm 0.013$
"	$\phi_+ = (167 \pm 20)°$ (S = 1.1)
"	$\gamma_+ = -0.97$ [g]
"	$\Delta_+ = (-73^{+133}_{-10})°$ [g]
$p\gamma$	$\alpha_\gamma = -0.76 \pm 0.08$

Σ⁺ DECAY MODES		Fraction (Γ_i/Γ)	Confidence level	p (MeV/c)
$p\pi^0$		(51.57 ± 0.30) %		189
$n\pi^+$		(48.31 ± 0.30) %		185
$p\gamma$		$(1.23 \pm 0.05) \times 10^{-3}$		225
$n\pi^+\gamma$	[h]	$(4.5 \pm 0.5) \times 10^{-4}$		185
$\Lambda e^+ \nu_e$		$(2.0 \pm 0.5) \times 10^{-5}$		71

$\Delta S = \Delta Q$ (SQ) violating modes or
$\Delta S = 1$ weak neutral current (S1) modes

$ne^+\nu_e$	SQ	< 5	$\times 10^{-6}$ 90%	224
$n\mu^+\nu_\mu$	SQ	< 3.0	$\times 10^{-5}$ 90%	202
pe^+e^-	S1	< 7	$\times 10^{-6}$	225

Σ⁰ $I(J^P) = 1(\frac{1}{2}^+)$

J^P not measured; assumed to be the same as for the Σ^+ and Σ^-.

Mass $m = 1192.55 \pm 0.08$ MeV   (S = 1.2)
$m_{\Sigma^-} - m_{\Sigma^0} = 4.88 \pm 0.08$ MeV   (S − 1.2)
$m_{\Sigma^0} - m_\Lambda = 76.87 \pm 0.08$ MeV   (S = 1.2)
Mean life $\tau = (7.4 \pm 0.7) \times 10^{-20}$ s
    $c\tau = 2.22 \times 10^{-11}$ m
Transition magnetic moment $|\mu_{\Sigma\Lambda}| = 1.61 \pm 0.08$ μ_N

Σ⁰ DECAY MODES		Fraction (Γ_i/Γ)	Confidence level	p (MeV/c)
$\Lambda\gamma$		100 %		74
$\Lambda\gamma\gamma$		< 3 %	90%	74
$\Lambda e^+ e^-$	[J]	5×10^{-3}		74

Σ⁻ $I(J^P) = 1(\frac{1}{2}^+)$

Mass $m = 1197.436 \pm 0.033$ MeV   (S = 1.2)
$m_{\Sigma^-} - m_{\Sigma^+} = 8.07 \pm 0.08$ MeV   (S = 1.9)
$m_{\Sigma^-} - m_\Lambda = 81.752 \pm 0.034$ MeV   (S = 1.2)
Mean life $\tau = (1.479 \pm 0.011) \times 10^{-10}$ s   (S = 1.3)
    $c\tau = 4.434$ cm
Magnetic moment $\mu = -1.160 \pm 0.025$ μ_N (S = 1.7)

Decay parameters

$n\pi^-$	$\alpha_- = -0.068 \pm 0.008$
"	$\phi_- = (10 \pm 15)°$
"	$\gamma_- = 0.98$ [g]
"	$\Delta_- = (249^{+12}_{-120})°$ [g]
$ne^-\bar{\nu}_e$	$g_A/g_V = 0.340 \pm 0.017$ [e]
"	$f_2(0)/f_1(0) = 0.97 \pm 0.14$
"	$D = 0.11 \pm 0.10$
$\Lambda e^-\bar{\nu}_e$	$g_V/g_A = 0.01 \pm 0.10$ [e] (S = 1.5)
"	$g_{WM}/g_A = 2.4 \pm 1.7$ [e]

Σ⁻ DECAY MODES		Fraction (Γ_i/Γ)	p (MeV/c)
$n\pi^-$		(99.848 ± 0.005) %	193
$n\pi^-\gamma$	[h]	$(4.6 \pm 0.6) \times 10^{-4}$	193
$ne^-\bar{\nu}_e$		$(1.017 \pm 0.034) \times 10^{-3}$	230
$n\mu^-\bar{\nu}_\mu$		$(4.5 \pm 0.4) \times 10^{-4}$	210
$\Lambda e^-\bar{\nu}_e$		$(5.73 \pm 0.27) \times 10^{-5}$	79

Σ(1385) P_{13} $I(J^P) = 1(\frac{3}{2}^+)$

$\Sigma(1385)^+$mass $m = 1382.8 \pm 0.4$ MeV   (S = 2.0)
$\Sigma(1385)^0$ mass $m = 1383.7 \pm 1.0$ MeV   (S = 1.4)
$\Sigma(1385)^-$ mass $m = 1387.2 \pm 0.5$ MeV   (S = 2.2)
$\Sigma(1385)^+$full width $\Gamma = 35.8 \pm 0.8$ MeV
$\Sigma(1385)^0$ full width $\Gamma = 36 \pm 5$ MeV
$\Sigma(1385)^-$full width $\Gamma = 39.4 \pm 2.1$ MeV   (S = 1.7)
    Below $\overline{K}N$ threshold

Σ(1385) DECAY MODES	Fraction (Γ_i/Γ)	p (MeV/c)
$\Lambda\pi$	88 ± 2 %	208
$\Sigma\pi$	12 ± 2 %	127

Σ(1660) P_{11} $I(J^P) = 1(\frac{1}{2}^+)$

Mass $m = 1630$ to 1690 (≈ 1660) MeV
Full width $\Gamma = 40$ to 200 (≈ 100) MeV
$p_{beam} = 0.72$ GeV/c $4\pi\lambda^2 = 29.9$ mb

Σ(1660) DECAY MODES	Fraction (Γ_i/Γ)	p (MeV/c)
$N\overline{K}$	10–30 %	405
$\Lambda\pi$	seen	439
$\Sigma\pi$	seen	385

Σ(1670) D_{13} $I(J^P) = 1(\frac{3}{2}^-)$

Mass $m = 1665$ to 1685 (≈ 1670) MeV
Full width $\Gamma = 40$ to 80 (≈ 60) MeV
$p_{beam} = 0.74$ GeV/c $4\pi\lambda^2 = 28.5$ mb

Σ(1670) DECAY MODES	Fraction (Γ_i/Γ)	p (MeV/c)
$N\overline{K}$	7–13 %	414
$\Lambda\pi$	5–15 %	447
$\Sigma\pi$	30–60 %	393

Σ(1750) S_{11} $I(J^P) = 1(\frac{1}{2}^-)$

Mass $m = 1730$ to 1800 (≈ 1750) MeV
Full width $\Gamma = 60$ to 160 (≈ 90) MeV
$p_{beam} = 0.91$ GeV/c $4\pi\lambda^2 = 20.7$ mb

Σ(1750) DECAY MODES	Fraction (Γ_i/Γ)	p (MeV/c)
$N\overline{K}$	10–40 %	486
$\Lambda\pi$	seen	507
$\Sigma\pi$	<8 %	455
$\Sigma\eta$	15–55 %	81

11-35

$\Sigma(1775)\ D_{15}$ $\qquad I(J^P) = 1(\frac{5}{2}^-)$

Mass $m = 1770$ to 1780 (≈ 1775) MeV
Full width $\Gamma = 105$ to 135 (≈ 120) MeV
$p_{\text{beam}} = 0.96$ GeV/c $\qquad 4\pi\lambda^2 = 19.0$ mb

$\Sigma(1775)$ DECAY MODES	Fraction (Γ_i/Γ)	p (MeV/c)
$N\overline{K}$	37–43%	508
$\Lambda\pi$	14–20%	525
$\Sigma\pi$	2–5%	474
$\Sigma(1385)\pi$	8–12%	324
$\Lambda(1520)\pi$	17–23%	198

$\Sigma(1915)\ F_{15}$ $\qquad I(J^P) = 1(\frac{5}{2}^+)$

Mass $m = 1900$ to 1935 (≈ 1915) MeV
Full width $\Gamma = 80$ to 160 (≈ 120) MeV
$p_{\text{beam}} = 1.26$ GeV/c $\qquad 4\pi\lambda^2 = 12.8$ mb

$\Sigma(1915)$ DECAY MODES	Fraction (Γ_i/Γ)	p (MeV/c)
$N\overline{K}$	5–15 %	618
$\Lambda\pi$	seen	622
$\Sigma\pi$	seen	577
$\Sigma(1385)\pi$	<5 %	440

$\Sigma(1940)\ D_{13}$ $\qquad I(J^P) = 1(\frac{3}{2}^-)$

Mass $m = 1900$ to 1950 (≈ 1940) MeV
Full width $\Gamma = 150$ to 300 (≈ 220) MeV
$p_{\text{beam}} = 1.32$ GeV/c $\qquad 4\pi\lambda^2 = 12.1$ mb

$\Sigma(1940)$ DECAY MODES	Fraction (Γ_i/Γ)	p (MeV/c)
$N\overline{K}$	<20 %	637
$\Lambda\pi$	seen	639
$\Sigma\pi$	seen	594
$\Sigma(1385)\pi$	seen	460
$\Lambda(1520)\pi$	seen	354
$\Delta(1232)\overline{K}$	seen	410
$N\overline{K}^*(892)$	seen	320

$\Sigma(2030)\ F_{17}$ $\qquad I(J^P) = 1(\frac{7}{2}^+)$

Mass $m = 2025$ to 2040 (≈ 2030) MeV
Full width $\Gamma = 150$ to 200 (≈ 180) MeV
$p_{\text{beam}} = 1.52$ GeV/c $\qquad 4\pi\lambda^2 = 9.93$ mb

$\Sigma(2030)$ DECAY MODES	Fraction (Γ_i/Γ)	p (MeV/c)
$N\overline{K}$	17–23 %	702
$\Lambda\pi$	17–23 %	700
$\Sigma\pi$	5–10 %	657
ΞK	<2 %	412
$\Sigma(1385)\pi$	5–15 %	529
$\Lambda(1520)\pi$	10–20 %	430
$\Delta(1232)\overline{K}$	10–20 %	498
$N\overline{K}^*(892)$	<5 %	438

$\Sigma(2250)$ $\qquad I(J^P) = 1(?^?)$

Mass $m = 2210$ to 2280 (≈ 2250) MeV
Full width $\Gamma = 60$ to 150 (≈ 100) MeV
$p_{\text{beam}} = 2.04$ GeV/c $\qquad 4\pi\lambda^2 = 6.76$ mb

$\Sigma(2250)$ DECAY MODES	Fraction (Γ_i/Γ)	p (MeV/c)
$N\overline{K}$	<10 %	851
$\Lambda\pi$	seen	842
$\Sigma\pi$	seen	803

Ξ BARYONS
$(S = -2,\ I = 1/2)$

$\Xi^0 = uss,\quad \Xi^- = dss$

Ξ^0 $\qquad I(J^P) = \frac{1}{2}(\frac{1}{2}^+)$

P is not yet measured; $+$ is the quark model prediction.
Mass $m = 1314.9 \pm 0.6$ MeV
$m_{\Xi^-} - m_{\Xi^0} = 6.4 \pm 0.6$ MeV
Mean life $\tau = (2.90 \pm 0.09) \times 10^{-10}$ s
$c\tau = 8.71$ cm
Magnetic moment $\mu = -1.250 \pm 0.014\ \mu_N$

Decay parameters

$\Lambda\pi^0$	$\alpha = -0.411 \pm 0.022$	$(S = 2.1)$
"	$\phi = (21 \pm 12)^\circ$	
"	$\gamma = 0.85$ [g]	
"	$\Delta = (218^{+12}_{-19})^\circ$ [g]	
$\Lambda\gamma$	$\alpha = 0.4 \pm 0.4$	
$\Sigma^0\gamma$	$\alpha = 0.20 \pm 0.32$	

Ξ^0 DECAY MODES	Fraction (Γ_i/Γ)	Confidence level	p (MeV/c)
$\Lambda\pi^0$	(99.54 ± 0.05) %		135
$\Lambda\gamma$	$(1.06 \pm 0.16) \times 10^{-3}$		184
$\Sigma^0\gamma$	$(3.5 \pm 0.4) \times 10^{-3}$		117
$\Sigma^+ e^- \overline{\nu}_e$	$< 1.1 \times 10^{-3}$	90%	120
$\Sigma^+ \mu^- \overline{\nu}_\mu$	$< 1.1 \times 10^{-3}$	90%	64

$\Delta S = \Delta Q$ (SQ) violating modes or
$\Delta S = 2$ forbidden (S2) modes

$\Sigma^- e^+ \nu_e$	SQ	$< 9 \times 10^{-4}$	90%	112
$\Sigma^- \mu^+ \nu_\mu$	SQ	$< 9 \times 10^{-4}$	90%	49
$p\pi^-$	S2	$< 4 \times 10^{-5}$	90%	299
$pe^- \overline{\nu}_e$	S2	$< 1.3 \times 10^{-3}$		323
$p\mu^- \overline{\nu}_\mu$	S2	$< 1.3 \times 10^{-3}$		309

Ξ^- $\qquad I(J^P) = \frac{1}{2}(\frac{1}{2}^+)$

P is not yet measured; $+$ is the quark model prediction.
Mass $m = 1321.32 \pm 0.13$ MeV
Mean life $\tau = (1.639 \pm 0.015) \times 10^{-10}$ s
$c\tau = 4.91$ cm
Magnetic moment $\mu = -0.6507 \pm 0.0025\ \mu_N$

Decay parameters

$\Lambda\pi^-$	$\alpha = -0.456 \pm 0.014$	$(S = 1.8)$
"	$\phi = (4 \pm 4)^\circ$	
"	$\gamma = 0.89$ [g]	
"	$\Delta = (188 \pm 8)^\circ$ [g]	
$\Lambda e^- \overline{\nu}_e$	$g_A/g_V = -0.25 \pm 0.05$ [e]	

Ξ^- DECAY MODES	Fraction (Γ_i/Γ)	Confidence level	p (MeV/c)
$\Lambda\pi^-$	(99.887 ± 0.035) %		139
$\Sigma^- \gamma$	$(1.27 \pm 0.23) \times 10^{-4}$		118
$\Lambda e^- \overline{\nu}_e$	$(5.63 \pm 0.31) \times 10^{-4}$		190
$\Lambda \mu^- \overline{\nu}_\mu$	$(3.5^{+3.5}_{-2.2}) \times 10^{-4}$		163
$\Sigma^0 e^- \overline{\nu}_e$	$(8.7 \pm 1.7) \times 10^{-5}$		122
$\Sigma^0 \mu^- \overline{\nu}_\mu$	$< 8 \times 10^{-4}$	90%	70
$\Xi^0 e^- \overline{\nu}_e$	$< 2.3 \times 10^{-3}$	90%	6

$\Delta S = 2$ forbidden (S2) modes

$n\pi^-$	S2	$< 1.9 \times 10^{-5}$	90%	303
$ne^- \overline{\nu}_e$	S2	$< 3.2 \times 10^{-3}$	90%	327
$n\mu^- \overline{\nu}_\mu$	S2	< 1.5 %	90%	314
$p\pi^- \pi^-$	S2	$< 4 \times 10^{-4}$	90%	223
$p\pi^- e^- \overline{\nu}_e$	S2	$< 4 \times 10^{-4}$	90%	304
$p\pi^- \mu^- \overline{\nu}_\mu$	S2	$< 4 \times 10^{-4}$	90%	250
$p\mu^- \mu^-$	L	$< 4 \times 10^{-4}$	90%	272

$\Xi(1530)\ P_{13}$		$I(J^P) = \frac{1}{2}(\frac{3}{2}^+)$		

$\Xi(1530)^0$ mass $m = 1531.80 \pm 0.32$ MeV (S = 1.3)
$\Xi(1530)^-$ mass $m = 1535.0 \pm 0.6$ MeV
$\Xi(1530)^0$ full width $\Gamma = 9.1 \pm 0.5$ MeV
$\Xi(1530)^-$ full width $\Gamma = 9.9^{+1.7}_{-1.9}$ MeV

$\Xi(1530)$ DECAY MODES	Fraction (Γ_i/Γ)	Confidence level	p (MeV/c)
$\Xi\pi$	100 %		152
$\Xi\gamma$	<4 %	90%	200

$\Xi(1690)$		$I(J^P) = \frac{1}{2}(?^?)$

Mass $m = 1690 \pm 10$ MeV [l]
Full width $\Gamma < 50$ MeV

$\Xi(1690)$ DECAY MODES	Fraction (Γ_i/Γ)	p (MeV/c)
$\Lambda \overline{K}$	seen	240
$\Sigma \overline{K}$	seen	51
$\Xi^-\pi^+\pi^-$	possibly seen	214

$\Xi(1820)\ D_{13}$		$I(J^P) = \frac{1}{2}(\frac{3}{2}^-)$

Mass $m = 1823 \pm 5$ MeV [l]
Full width $\Gamma = 24^{+15}_{-10}$ MeV [l]

$\Xi(1820)$ DECAY MODES	Fraction (Γ_i/Γ)	p (MeV/c)
$\Lambda \overline{K}$	large	400
$\Sigma \overline{K}$	small	320
$\Xi\pi$	small	413
$\Xi(1530)\pi$	small	234

$\Xi(1950)$		$I(J^P) - \frac{1}{2}(?^?)$

Mass $m = 1950 \pm 15$ MeV [l]
Full width $\Gamma = 60 \pm 20$ MeV [l]

$\Xi(1950)$ DECAY MODES	Fraction (Γ_i/Γ)	p (MeV/c)
$\Lambda \overline{K}$	seen	522
$\Sigma \overline{K}$	possibly seen	460
$\Xi\pi$	seen	518

$\Xi(2030)$		$I(J^P) = \frac{1}{2}(\geq \frac{5}{2}^?)$

Mass $m = 2025 \pm 5$ MeV [l]
Full width $\Gamma = 20^{+15}_{-5}$ MeV [l]

$\Xi(2030)$ DECAY MODES	Fraction (Γ_i/Γ)	p (MeV/c)
$\Lambda \overline{K}$	$\sim$ 20 %	589
$\Sigma \overline{K}$	$\sim$ 80 %	533
$\Xi\pi$	small	573
$\Xi(1530)\pi$	small	421
$\Lambda \overline{K}\pi$	small	501
$\Sigma \overline{K}\pi$	small	430

Ω BARYONS
$(S = -3, I = 0)$
$\Omega^- = sss$

Ω^-		$I(J^P) = 0(\frac{3}{2}^+)$

J^P is not yet measured; $\frac{3}{2}^+$ is the quark model prediction.

Mass $m = 1672.45 \pm 0.29$ MeV
Mean life $\tau = (0.822 \pm 0.012) \times 10^{-10}$ s
  $c\tau = 2.46$ cm
Magnetic moment $\mu = -2.02 \pm 0.05\ \mu_N$

Decay parameters

ΛK^-	$\alpha = -0.026 \pm 0.026$
$\Xi^0\pi^-$	$\alpha = 0.09 \pm 0.14$
$\Xi^-\pi^0$	$\alpha = 0.05 \pm 0.21$

Ω^- DECAY MODES	Fraction (Γ_i/Γ)	Confidence level	p (MeV/c)
ΛK^-	(67.8±0.7) %		211
$\Xi^0\pi^-$	(23.6±0.7) %		294
$\Xi^-\pi^0$	(8.6±0.4) %		290
$\Xi^-\pi^+\pi^-$	($4.3^{+3.4}_{-1.3}$) $\times 10^{-4}$		190
$\Xi(1530)^0\pi^-$	($6.4^{+5.1}_{-2.0}$) $\times 10^{-4}$		17
$\Xi^0 e^-\overline{\nu}_e$	(5.6±2.8) $\times 10^{-3}$		319
$\Xi^-\gamma$	< 4.6 $\times 10^{-4}$	90%	314

$\Delta S = 2$ forbidden (S2) modes

$\Lambda\pi^-$	S2	< 1.9 $\times 10^{-4}$	90%	449

$\Omega(2250)^-$		$I(J^P) = 0(?^?)$

Mass $m = 2252 \pm 9$ MeV
Full width $\Gamma - 55 \pm 18$ MeV

$\Omega(2250)^-$ DECAY MODES	Fraction (Γ_i/Γ)	p (MeV/c)
$\Xi^-\pi^+ K^-$	seen	531
$\Xi(1530)^0 K^-$	seen	437

CHARMED BARYONS
$(C = +1)$
$\Lambda_c^+ = udc,\ \Sigma_c^{++} = uuc,\ \Sigma_c^+ = udc,\ \Sigma_c^0 = ddc,$
$\Xi_c^+ = usc,\ \Xi_c^0 = dsc,\ \Omega_c^0 = ssc$

Λ_c^+		$I(J^P) = 0(\frac{1}{2}^+)$

J not confirmed; $\frac{1}{2}$ is the quark model prediction.

Mass $m = 2284.9 \pm 0.6$ MeV
Mean life $\tau = (0.206 \pm 0.012) \times 10^{-12}$ s
  $c\tau = 61.8\ \mu$m

Decay asymmetry parameters

$\Lambda\pi^+$	$\alpha = -0.98 \pm 0.19$
$\Sigma^+\pi^0$	$\alpha = -0.45 \pm 0.32$
$\Lambda\ell^+\nu_\ell$	$\alpha = -0.82^{+0.11}_{-0.07}$

Λ_c^+ DECAY MODES	Fraction (Γ_i/Γ)	Scale factor/ Confidence level	p (MeV/c)
Hadronic modes with a p and one $\overline{K}$			
$p\overline{K}^0$	(2.2 ± 0.4) %		872
$pK^-\pi^+$	(4.4 ± 0.6) %		822
$p\overline{K}^*(892)^0$	(1.6 ± 0.4) % [k]		681
$\Delta(1232)^{++}K^-$	(7 ± 4) $\times 10^{-3}$		709
$\Lambda(1520)\pi^+$	($4.0^{+2.0}_{-1.7}$) $\times 10^{-3}$ [k]		626
$pK^-\pi^+$ nonresonant	($2.5^{+0.5}_{-0.6}$) %		822
$p\overline{K}^0\eta$	(1.10± 0.29) %		567

			p (MeV/c)
$p \overline{K}^0 \pi^+ \pi^-$		$(2.1 \pm 0.8)\%$	753
$p K^- \pi^+ \pi^0$		seen	758
$p K^*(892)^- \pi^+$	[k]	$(9 \pm 5) \times 10^{-3}$	579
$p (K^- \pi^+)_{\text{nonresonant}} \pi^0$		$(3.2 \pm 0.7)\%$	758
$\Delta(1232) \overline{K}^*(892)$		seen	416
$p K^- \pi^+ \pi^+ \pi^-$		$(10 \pm 7) \times 10^{-4}$	670
$p K^- \pi^+ \pi^0 \pi^0$		$(7.0 \pm 3.5) \times 10^{-3}$	676
$p K^- \pi^+ \pi^0 \pi^0 \pi^0$		$(4.4 \pm 2.8) \times 10^{-3}$	573

Hadronic modes with a p and zero or two K's

			p (MeV/c)
$p \pi^+ \pi^-$		$(3.0 \pm 1.6) \times 10^{-3}$	926
$p f_0(980)$	[k]	$(2.4 \pm 1.6) \times 10^{-3}$	621
$p \pi^+ \pi^+ \pi^- \pi^-$		$(1.6 \pm 1.0) \times 10^{-3}$	851
$p K^+ K^-$		$(2.0 \pm 0.6) \times 10^{-3}$	615
$p \phi$	[k]	$(1.06 \pm 0.33) \times 10^{-3}$	589

Hadronic modes with a hyperon

			p (MeV/c)
$\Lambda \pi^+$		$(7.9 \pm 1.8) \times 10^{-3}$	863
$\Lambda \pi^+ \pi^0$		$(3.2 \pm 0.9)\%$	843
$\Lambda \rho^0$		$< 4 \%$ CL=95%	638
$\Lambda \pi^+ \pi^+ \pi^-$		$(2.9 \pm 0.6)\%$	806
$\Lambda \pi^+ \eta$		$(1.5 \pm 0.4)\%$	690
$\Sigma(1385)^+ \eta$	[k]	$(7.5 \pm 2.4) \times 10^{-3}$	569
$\Lambda K^+ \overline{K}^0$		$(5.3 \pm 1.4) \times 10^{-3}$	441
$\Sigma^0 \pi^+$		$(8.8 \pm 2.0) \times 10^{-3}$	824
$\Sigma^+ \pi^0$		$(8.8 \pm 2.2) \times 10^{-3}$	826
$\Sigma^+ \eta$		$(4.8 \pm 1.7) \times 10^{-3}$	712
$\Sigma^+ \pi^+ \pi^-$		$(3.0 \pm 0.6)\%$	803
$\Sigma^+ \rho^0$		$< 1.2 \%$ CL=95%	578
$\Sigma^- \pi^+ \pi^+$		$(1.6 \pm 0.6)\%$	798
$\Sigma^0 \pi^+ \pi^0$		$(1.6 \pm 0.6)\%$	802
$\Sigma^0 \pi^+ \pi^+ \pi^-$		$(9.2 \pm 3.4) \times 10^{-3}$	762
$\Sigma^+ \pi^+ \pi^- \pi^0$			766
$\Sigma^+ \omega$	[k]	$(2.4 \pm 0.7)\%$	568
$\Sigma^+ \pi^+ \pi^+ \pi^- \pi^-$		$(2.6 ^{+3.5}_{-1.8}) \times 10^{-3}$	707
$\Sigma^+ K^+ K^-$		$(3.1 \pm 0.8) \times 10^{-3}$	346
$\Sigma^+ \phi$	[k]	$(3.0 \pm 1.3) \times 10^{-3}$	292
$\Sigma^+ K^+ \pi^-$		$(5.7 ^{+5.3}_{-3.2}) \times 10^{-3}$	668
$\Xi^0 K^+$		$(3.4 \pm 0.9) \times 10^{-3}$	652
$\Xi^- K^+ \pi^+$		$(4.3 \pm 1.1) \times 10^{-3}$	564
$\Xi(1530)^0 K^+$	[k]	$(2.3 \pm 0.7) \times 10^{-3}$	471

Semileptonic modes

			p (MeV/c)
$\Lambda \ell^+ \nu_\ell$	[l]	$(2.3 \pm 0.5)\%$	–
e^+ anything		$(4.5 \pm 1.7)\%$	–
$p e^+$ anything		$(1.8 \pm 0.9)\%$	–
Λe^+ anything		$(1.6 \pm 0.6)\%$	–
$\Lambda \mu^+$ anything		$(1.5 \pm 0.9)\%$	–

Inclusive modes

			p (MeV/c)
p anything		$(50 \pm 16)\%$	–
p anything (no Λ)		$(12 \pm 19)\%$	–
n anything		$(50 \pm 16)\%$	–
n anything (no Λ)		$(29 \pm 17)\%$	–
Λ anything		$(35 \pm 11)\%$ S=1.4	–
$\Sigma^\pm$ anything	[m]	$(10 \pm 5)\%$	–

$\Delta C = 1$ weak neutral current ($C1$) modes, or Lepton number (L) violating modes

					p (MeV/c)
$p \mu^+ \mu^-$	$C1$	< 3.4	$\times 10^{-4}$	CL=90%	936
$\Sigma^- \mu^+ \mu^+$	L	< 7.0	$\times 10^{-4}$	CL=90%	811

$\Lambda_c(2593)^+$ $\qquad I(J^P) = 0(\frac{1}{2}^-)$

The spin-parity follows from the fact that $\Sigma_c(2455)\pi$ decays, with little available phase space, dominate.

Mass $m = 2593.6 \pm 1.0$ MeV (S = 1.2)
$m - m_{\Lambda_c^+} = 308.6 \pm 0.8$ MeV (S = 1.3)
Full width $\Gamma = 3.9 ^{+2.4}_{-1.6}$ MeV

$\Lambda_c^+ \pi \pi$ and $\Sigma_c(2455)\pi$ — the latter just barely — are the only strong decays allowed to an excited Λ_c^+ having this mass; and the $\Lambda_c^+ \pi^+ \pi^-$ mode seems to be largely via $\Sigma_c^{++} \pi^-$ or $\Sigma_c^0 \pi^+$.

$\Lambda_c(2593)^+$ DECAY MODES	Fraction (Γ_i/Γ)	p (MeV/c)
$\Lambda_c^+ \pi^+ \pi^-$	seen	124
$\Sigma_c(2455)^{++} \pi^-$	large	17
$\Sigma_c(2455)^0 \pi^+$	large	23
$\Lambda_c^+ \pi^+ \pi^-$ 3-body	small	124
$\Lambda_c^+ \pi^0$	not seen	261
$\Lambda_c^+ \gamma$	not seen	290

$\Lambda_c(2625)^+$ $\qquad I(J^P) = 0(?^?)$

J^P is expected to be $3/2^-$.

Mass $m = 2626.4 \pm 0.9$ MeV (S = 1.3)
$m - m_{\Lambda_c^+} = 341.5 \pm 0.8$ MeV (S = 1.9)
Full width $\Gamma < 1.9$ MeV, CL = 90%

$\Lambda_c^+ \pi \pi$ and $\Sigma(2455)\pi$ are the only strong decays allowed to an excited Λ_c^+ having this mass.

$\Lambda_c(2625)^+$ DECAY MODES	Fraction (Γ_i/Γ)	p (MeV/c)
$\Lambda_c^+ \pi^+ \pi^-$	seen	184
$\Sigma_c(2455)^{++} \pi^-$	small	100
$\Sigma_c(2455)^0 \pi^+$	small	101
$\Lambda_c^+ \pi^+ \pi^-$ 3-body	large	184
$\Lambda_c^+ \pi^0$	not seen	293
$\Lambda_c^+ \gamma$	not seen	319

$\Sigma_c(2455)$ $\qquad I(J^P) = 1(\frac{1}{2}^+)$

J^P not confirmed; $\frac{1}{2}^+$ is the quark model prediction.

$\Sigma_c(2455)^{++}$ mass $m = 2452.9 \pm 0.6$ MeV
$\Sigma_c(2455)^+$ mass $m = 2453.5 \pm 0.9$ MeV
$\Sigma_c(2455)^0$ mass $m = 2452.1 \pm 0.7$ MeV
$m_{\Sigma_c^{++}} - m_{\Lambda_c^+} = 167.95 \pm 0.25$ MeV
$m_{\Sigma_c^+} - m_{\Lambda_c^+} = 168.5 \pm 0.7$ MeV (S = 1.1)
$m_{\Sigma_c^0} - m_{\Lambda_c^+} = 167.2 \pm 0.4$ MeV (S = 1.1)
$m_{\Sigma_c^{++}} - m_{\Sigma_c^0} = 0.79 \pm 0.33$ MeV (S = 1.2)
$m_{\Sigma_c^+} - m_{\Sigma_c^0} = 1.4 \pm 0.6$ MeV

$\Lambda_c^+ \pi$ is the only strong decay allowed to a Σ_c having this mass.

$\Sigma_c(2455)$ DECAY MODES	Fraction (Γ_i/Γ)	p (MeV/c)
$\Lambda_c^+ \pi$	$\approx 100\%$	90

Ξ_c^+ $\qquad I(J^P) = \frac{1}{2}(\frac{1}{2}^+)$

$I(J^P)$ not confirmed; $\frac{1}{2}(\frac{1}{2}^+)$ is the quark model prediction.

Mass $m = 2465.6 \pm 1.4$ MeV
Mean life $\tau = (0.35 ^{+0.07}_{-0.04}) \times 10^{-12}$ s
$c\tau = 106$ μm

Ξ_c^+ DECAY MODES	Fraction (Γ_i/Γ)	p (MeV/c)
$\Lambda K^- \pi^+ \pi^+$	seen	784
$\Lambda \overline{K}^*(892)^0 \pi^+$	not seen	601
$\Sigma(1385)^+ K^- \pi^+$	not seen	676
$\Sigma^+ K^- \pi^+$	seen	808
$\Sigma^+ \overline{K}^*(892)^0$	seen	653
$\Sigma^0 K^- \pi^+ \pi^+$	seen	733
$\Xi^0 \pi^+$	seen	875
$\Xi^- \pi^+ \pi^+$	seen	850
$\Xi(1530)^0 \pi^+$	not seen	748
$\Xi^0 \pi^+ \pi^0$	seen	854
$\Xi^0 \pi^+ \pi^+ \pi^-$	seen	817
$\Xi^0 e^+ \nu_e$	seen	882

$I(J^P) = \frac{1}{2}(\frac{1}{2}^+)$

$I(J^P)$ not confirmed; $\frac{1}{2}(\frac{1}{2}^+)$ is the quark model prediction.

Mass $m = 2470.3 \pm 1.8$ MeV (S = 1.3)
$m_{\Xi_c^0} - m_{\Xi_c^+} = 4.7 \pm 2.1$ MeV (S = 1.2)
Mean life $\tau = (0.098^{+0.023}_{-0.015}) \times 10^{-12}$ s
$c\tau = 29$ μm

Ξ_c^0 DECAY MODES	Fraction (Γ_i/Γ)	p (MeV/c)
$\Lambda \overline{K}^0$	seen	864
$\Xi^- \pi^+$	seen	875
$\Xi^- \pi^+ \pi^+ \pi^-$	seen	816
$p K^- \overline{K}^*(892)^0$	seen	406
$\Omega^- K^+$	seen	522
$\Xi^- e^+ \nu_e$	seen	882
$\Xi^- \ell^+$ anything	seen	–

| $\Xi_c(2645)$ | | $I(J^P) = ?(?^?)$ |

Mass $m = 2643.8 \pm 1.8$ MeV
$m_{\Xi_c(2645)^0} - m_{\Xi_c^+} = 178.2 \pm 1.1$ MeV
Full width Γ < 5.5 MeV, CL = 90%

$\Xi_c \pi$ is the only strong decay allowed to a Ξ_c resonance having this mass.

$\Xi_c(2645)$ DECAY MODES	Fraction (Γ_i/Γ)	p (MeV/c)
$\Xi_c^+ \pi^-$	seen	107

$I(J^P) = 0(\frac{1}{2}^+)$

$I(J^P)$ not confirmed; $0(\frac{1}{2}^+)$ is the quark model prediction.

Mass $m = 2704 \pm 4$ MeV (S = 1.8)
Mean life $\tau = (0.064 \pm 0.020) \times 10^{-12}$ s
$c\tau = 19$ μm

Ω_c^0 DECAY MODES	Fraction (Γ_i/Γ)	p (MeV/c)
$\Sigma^+ K^- K^- \pi^+$	seen	697
$\Xi^- K^- \pi^+ \pi^+$	seen	838
$\Omega^- \pi^+$	seen	827
$\Omega^- \pi^- \pi^+ \pi^+$	seen	759

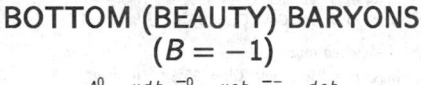

BOTTOM (BEAUTY) BARYONS
($B = -1$)

$\Lambda_b^0 = u d b$, $\Xi_b^0 = u s b$, $\Xi_b^- = d s b$

$I(J^P) = 0(\frac{1}{2}^+)$

$I(J^P)$ not yet measured; $0(\frac{1}{2}^+)$ is the quark model prediction.

Mass $m = 5641 \pm 50$ MeV
Mean life $\tau = (1.14 \pm 0.08) \times 10^{-12}$ s
$c\tau = 342$ μm

These branching fractions are actually an average over weakly decaying b-baryons weighted by their production rates in Z decay (or high-energy $p\overline{p}$), branching ratios, and detection efficiencies. They scale with the LEP Λ_b production fraction $B(b \to \Lambda_b)$ and are evaluated for our value $B(b \to \Lambda_b) = (13.2 \pm 4.1)\%$.

The branching fractions $B(\Lambda_b^0 \to \Lambda \ell^- \overline{\nu}_\ell \text{anything})$ and $B(\Lambda_b^0 \to \Lambda_c^+ \ell^- \overline{\nu}_\ell \text{anything})$ are not pure measurements because the underlying measured products of these with $B(b \to \Lambda_b)$ were used to determine $B(b \to \Lambda_b)$, as described in the note "Production and Decay of b-Flavored Hadrons."

Λ_b^0 DECAY MODES	Fraction (Γ_i/Γ)	p (MeV/c)
$J/\psi(1S) \Lambda$	(1.4 $\pm$ 0.9) %	1756
$p D^0 \pi^-$	seen	2383
$\Lambda_c^+ \pi^+ \pi^- \pi^-$	seen	2336
$p \mu^- \overline{\nu}$ anything	(3.7 $\pm$ 1.7) %	–
$\Lambda \ell^- \overline{\nu}_\ell$ anything	[n] (2.5 $\pm$ 0.5) %	–
$\Lambda_c^+ \ell^- \overline{\nu}_\ell$ anything	[n] (10.0 $\pm$ 3.0) %	–
$\Lambda/\overline{\Lambda}$ anything	(17 $^{+11}_{-8}$) %	–

NOTES

This Summary Table only includes established baryons. The Particle Listings include evidence for other baryons. The masses, widths, and branching fractions for the resonances in this Table are Breit-Wigner parameters. The Particle Listings also give, where available, pole parameters. See, in particular, the *Note on N and Δ Resonances.*

For most of the resonances, the parameters come from various partial-wave analyses of more or less the same sets of data, and it is not appropriate to treat the results of the analyses as independent or to average them together. Furthermore, the systematic errors on the results are not well understood. Thus, we usually only give ranges for the parameters. We then also give a best guess for the mass (as part of the name of the resonance) and for the width. The *Note on N and Δ Resonances* and the *Note on Λ and Σ Resonances* in the Particle Listings review the partial-wave analyses.

When a quantity has "(S = ...)" to its right, the error on the quantity has been enlarged by the "scale factor" S, defined as $S = \sqrt{\chi^2/(N-1)}$, where N is the number of measurements used in calculating the quantity. We do this when S > 1, which often indicates that the measurements are inconsistent. When S > 1.25, we also show in the Particle Listings an ideogram of the measurements. For more about S, see the Introduction.

A decay momentum p is given for each decay mode. For a 2-body decay, p is the momentum of each decay product in the rest frame of the decaying particle. For a 3-or-more-body decay, p is the largest momentum any of the products can have in this frame. For any resonance, the *nominal* mass is used in calculating p. A dagger ("†") in this column indicates that the mode is forbidden when the nominal masses of resonances are used, but is in fact allowed due to the nonzero widths of the resonances.

[a] The masses of the p and n are most precisely known in u (unified atomic mass units). The conversion factor to MeV, 1 u = 931.49432 $\pm$ 0.00028 MeV, is less well known than are the masses in u.

[b] The limit is from neutrality-of-matter experiments; it assumes $q_n = q_p + q_e$. See also the charge of the neutron.

[c] The first limit is geochemical and independent of decay mode. The second entry, a range of limits, assumes the dominant decay modes are among those investigated. For antiprotons the best limit, inferred from the observation of cosmic ray $\overline{p}$'s is $\tau_{\overline{p}} > 10^7$ yr, the cosmic-ray storage time, but this limit depends on a number of assumptions. The best direct observation of stored antiprotons gives $\tau_{\overline{p}}/B(\overline{p} \to e^- \gamma) > 1848$ yr.

[d] There is some controversy about whether nuclear physics and model dependence complicate the analysis for bound neutrons (from which the best limit comes). The second limit here is from reactor experiments with free neutrons.

[e] The parameters g_A, g_V, and g_{WM} for semileptonic modes are defined by $\overline{B}_f[\gamma_\lambda(g_V + g_A\gamma_5) + i(g_{WM}/m_{B_i}) \sigma_{\lambda\nu} q^\nu]B_i$, and ϕ_{AV} is defined by $g_A/g_V = |g_A/g_V|e^{i\phi_{AV}}$. See the "Note on Baryon Decay Parameters" in the neutron Particle Listings.

[f] Time-reversal invariance requires this to be 0° or 180°.

[g] The decay parameters γ and Δ are calculated from α and ϕ using

$$\gamma = \sqrt{1-\alpha^2}\cos\phi, \qquad \tan\Delta = -\frac{1}{\alpha}\sqrt{1-\alpha^2}\sin\phi.$$

See the "Note on Baryon Decay Parameters" in the neutron Particle Listings.

[h] See the Particle Listings for the pion momentum range used in this measurement.

[i] The error given here is only an educated guess. It is larger than the error on the weighted average of the published values.

[j] A theoretical value using QED.

[k] This branching fraction includes all the decay modes of the final-state resonance.

[l] ℓ indicates e or μ mode, not sum over modes.

[m] The value is for the sum of the charge states of particle/antiparticle states indicated.

[n] Not a pure measurement. See note at head of Λ_b^0 Decay Modes.

MONOPOLES, SUPERSYMMETRY, COMPOSITENESS, etc., SEARCHES FOR

Magnetic Monopole Searches

Isolated candidate events have not been confirmed. Most experiments obtain negative results.

Supersymmetric Particle Searches

Limits are based on the Minimal Supersymmetric Standard Model. Assumptions include: 1) $\tilde{\chi}_1^0$ (or $\tilde{\gamma}$) is lightest supersymmetric particle; 2) R-parity is conserved; 3) $m_{\tilde{t}_L} = m_{\tilde{t}_R}$, and all scalar quarks (except $\tilde{t}_L$ and $\tilde{t}_R$) are degenerate in mass.

See the Particle Listings for a Note giving details of supersymmetry.

$\tilde{\chi}_i^0$ — neutralinos (mixtures of $\tilde{\gamma}$, $\tilde{Z}^0$, and $\tilde{H}_i^0$)

Mass $m_{\tilde{\gamma}} > 15$ GeV, CL = 90%   [if $m_{\tilde{t}} = 100$ GeV (from cosmology)]

Mass $m_{\tilde{\chi}_1^0} > 23$ GeV, CL = 95%   [$\tan\beta > 3$]

Mass $m_{\tilde{\chi}_2^0} > 52$ GeV, CL = 95%   [$\tan\beta > 3$]

Mass $m_{\tilde{\chi}_3^0} > 84$ GeV, CL = 95%   [$\tan\beta > 3$]

Mass $m_{\tilde{\chi}_4^0} > 127$ GeV, CL = 95%   [$\tan\beta > 3$]

$\tilde{\chi}_i^\pm$ — charginos (mixtures of $\tilde{W}^\pm$ and $\tilde{H}_i^\pm$)

Mass $m_{\tilde{\chi}_1^\pm} > 45$ GeV, CL = 95%   [all $m_{\tilde{\chi}_1^0}$]

Mass $m_{\tilde{\chi}_2^\pm} > 99$ GeV, CL = 95%   [GUT relations assumed]

$\tilde{\nu}$ — scalar neutrino (sneutrino)

Mass $m > 37.1$ GeV, CL = 95%   [one flavor]

Mass $m > 41.8$ GeV, CL = 95%   [three degenerate flavors]

$\tilde{e}$ — scalar electron (selectron)

Mass $m > 65$ GeV, CL = 95%   [if $m_{\tilde{\gamma}} = 0$]

Mass $m > 50$ GeV, CL = 95%   [if $m_{\tilde{\gamma}} < 5$ GeV]

Mass $m > 45$ GeV, CL = 95%   [if $m_{\tilde{\chi}_1^0} < 41$ GeV]

$\tilde{\mu}$ — scalar muon (smuon)

Mass $m > 45$ GeV, CL = 95%   [if $m_{\tilde{\chi}_1^0} < 41$ GeV]

$\tilde{\tau}$ — scalar tau (stau)

Mass $m > 45$ GeV, CL = 95%   [if $m_{\tilde{\chi}_1^0} < 38$ GeV]

$\tilde{q}$ — scalar quark (squark)

These limits include the effects of cascade decays, evaluated assuming a fixed value of the parameters μ and $\tan\beta$. The limits are weakly sensitive to these parameters over much of parameter space. Limits assume GUT relations between gaugino masses and the gauge coupling; in particular that for $|\mu|$ not small, $m_{\tilde{\chi}_1^0} \approx m_{\tilde{g}}/6$.

Mass $m > 176$ GeV, CL = 95%   [any $m_{\tilde{g}} < 300$ GeV, $\mu = -250$ GeV, $\tan\beta = 2$]

Mass $m > 224$ GeV, CL = 95%   [$m_{\tilde{g}} \le m_{\tilde{q}}$, $\mu = -400$ GeV, $\tan\beta = 4$]

$\tilde{g}$ — gluino

There is some controversy about a low-mass window ($1 \lesssim m_{\tilde{g}} \lesssim 4$ GeV). Several experiments cast doubt on the existence of this window.

These limits include the effects of cascade decays, evaluated assuming a fixed value of the parameters μ and $\tan\beta$. The limits are weakly sensitive to these parameters over much of parameter space. Limits assume GUT relations between gaugino masses and the gauge coupling; in particular that for $|\mu|$ not small, $m_{\tilde{\chi}_1^0} \approx m_{\tilde{g}}/6$.

Mass $m > 154$ GeV, CL = 95%   [$m_{\tilde{g}} \le m_{\tilde{q}}$, $\mu = -400$ GeV, $\tan\beta = 4$]

Mass $m > 212$ GeV, CL = 95%   [$m_{\tilde{g}} \ge m_{\tilde{q}}$, $\mu = -250$ GeV, $\tan\beta = 2$]

Quark and Lepton Compositeness, Searches for

Scale Limits Λ for Contact Interactions (the lowest dimensional interactions with four fermions)

If the Lagrangian has the form

$$\pm \frac{g^2}{2\Lambda^2} \overline{\psi}_L \gamma_\mu \psi_L \overline{\psi}_L \gamma^\mu \psi_L$$

(with $g^2/4\pi$ set equal to 1), then we define $\Lambda \equiv \Lambda_{LL}^\pm$. For the full definitions and for other forms, see the Note in the Listings on Searches for Quark and Lepton Compositeness in the full *Review* and the original literature.

$\Lambda_{LL}^+(eeee)$ > 1.6 TeV, CL = 95%

$\Lambda_{LL}^-(eeee)$ > 3.6 TeV, CL = 95%

$\Lambda_{LL}^+(ee\mu\mu)$ > 2.6 TeV, CL = 95%

$\Lambda_{LL}^-(ee\mu\mu)$ > 1.9 TeV, CL = 95%

$\Lambda_{LL}^+(ee\tau\tau)$ > 1.9 TeV, CL = 95%

$\Lambda_{LL}^-(ee\tau\tau)$ > 2.9 TeV, CL = 95%

$\Lambda_{LL}^+(\ell\ell\ell\ell)$ > 3.5 TeV, CL = 95%

$\Lambda_{LL}^-(\ell\ell\ell\ell)$ > 2.8 TeV, CL = 95%

$\Lambda_{LL}^+(eeqq)$ > 2.3 TeV, CL = 95%

$\Lambda_{LL}^-(eeqq)$ > 2.2 TeV, CL = 95%

$\Lambda_{LL}^+(\mu\mu qq)$ > 1.4 TeV, CL = 95%

$\Lambda_{LL}^-(\mu\mu qq)$ > 1.6 TeV, CL = 95%

$\Lambda_{LR}^\pm(\nu_\mu \nu_e \mu e)$ > 3.1 TeV, CL = 90%

$\Lambda_{LL}^\pm(qqqq)$ > 1.4 TeV, CL = 95%

Recent CDF measurements of the inclusive jet cross section in $p\overline{p}$ collisions could be interpreted as tentative evidence for a four-quark contact interaction with $\Lambda_{LL}^\pm(qqqq) \sim 1.6$ TeV. However, CDF notes that uncertainty in the parton distribution functions, higher-order QCD corrections, and detector calibration may possibly account for the effect.

Excited Leptons

The limits from $\ell^{*+}\ell^{*-}$ do not depend on λ (where λ is the $\ell\ell^*$ transition coupling). The λ-dependent limits assume chiral coupling, except for the third limit for e^* which is for nonchiral coupling. For chiral coupling, this limit corresponds to $\lambda_\gamma = \sqrt{2}$.

$e^{*\pm}$ — excited electron

Mass $m > 46.1$ GeV, CL = 95%   (from $e^{*+}e^{*-}$)

Mass $m > 91$ GeV, CL = 95%   (if $\lambda_Z > 1$)

Mass $m > 146$ GeV, CL = 95%   (if $\lambda_\gamma = 1$)

$\mu^{*\pm}$ — excited muon

Mass $m > 46.1$ GeV, CL = 95%   (from $\mu^{*+}\mu^{*-}$)

Mass $m > 91$ GeV, CL = 95%   (if $\lambda_Z > 1$)

$\tau^{*\pm}$ — excited tau

Mass $m > 46.0$ GeV, CL = 95%   (from $\tau^{*+}\tau^{*-}$)

Mass $m > 90$ GeV, CL = 95%   (if $\lambda_Z > 0.18$)

ν^* — excited neutrino

Mass $m > 47$ GeV, CL = 95%   (from $\nu^*\overline{\nu}^*$)

Mass $m > 91$ GeV, CL = 95%   (if $\lambda_Z > 1$)

q^* — excited quark

Mass $m > 45.6$ GeV, CL = 95%   (from $q^*\overline{q}^*$)

Mass $m > 88$ GeV, CL = 95%   (if $\lambda_Z > 1$)

Mass $m > 570$ GeV, CL = 95%   ($p\overline{p} \to q^* X$)

Color Sextet and Octet Particles

Color Sextet Quarks (q_6)

Mass $m > 84$ GeV, CL = 95%   (Stable q_6)

Color Octet Charged Leptons (ℓ_8)

Mass $m > 86$ GeV, CL = 95%   (Stable ℓ_8)

Color Octet Neutrinos (ν_8)

Mass $m > 110$ GeV, CL = 90%   ($\nu_8 \to \nu g$)

TABLE OF THE ISOTOPES
(Revised 1995)
Norman E. Holden

This table presents an evaluated set of values for the experimental quantities which characterize the decay of radioactive nuclides. A list of the major references used in this evaluation is given below. When uncertainties are not listed, they are assumed to be five or less in the last digit quoted. If they exceed five in the last digit, the value is prefaced by an approximate sign. The effective literature cutoff date for data in this edition of the Table is January, 1995. The assistance of Felicia Kramer in assembling this file is gratefully acknowledged.

TABLE LAYOUT

Column No.	Column Title	Description
1	Isotope or Element	For elements, the atomic number and chemical symbol are listed. For nuclides, the mass number and chemical symbol are listed. Isomers are indicated by the addition of m, m1, or m2.
2	Isotopic Abundance	in atom percent.
3	Atomic Mass or Atomic Weight	Atomic mass relative to $^{12}C = 12$. Atomic weight is given on the same scale.
4	Half-life	Half-life in decimal notation. μs = microseconds; ms = milliseconds; s = seconds; m = minutes; h = hours; d = days; and y = years.
5	Decay Mode/Energy	Decay modes are α = alpha particle emission; β^- = negative beta emission; β^+ = positron emission; EC = orbital electron capture; IT = isomeric transition from upper to lower isomeric state; n = neutron emission; SF = spontaneous fission. Total disintegration energy in MeV units.
6	Particle Energy/Intensity	End point energies of beta transitions and discrete energies of alpha particles in MeV and their intensities in percent.
7	Spin and Parity	Nuclear spin or angular momentum of the nuclides in units of $h/2\pi$; parity is positive or negative.
8	Magnetic Dipole Moment	Magnetic dipole moments in nuclear magneton units.
9	Electric Quadrupole Moment	Electric quadrupole moments in barn units (10^{-24} cm^2).
10	Gamma Ray Energy/Intensity	Gamma ray energies in MeV and intensities in percent. Ann. rad. refers to the 511.006 keV photons emitted in the annihilation of positrons in matter.

GENERAL NUCLEAR DATA REFERENCES

The following references represent the major sources of the nuclear data presented, along with subsequent published journals and reports:

1. G. Audi, A.H. Wapstra, *The 1993 Atomic Mass Evaluation,* Nuclear Physics *A565*, 1 (1993).
2. International Commission on Atomic Weights, *Atomic Weights of the Elements - 1993,* Pure & Applied Chemistry *66*, 2423 (1994).
3. J.R. Parrington, S. Breneman, F. Feiner, H. Knox, *Chart of the Nuclides, 15th Edition,* Knolls Atomic Power Lab. (To be published).
4. N.E. Holden, *Total and Spontaneous Fission Half-lives for Uranium, Plutonium, Americium and Curium Nuclides,* Pure & Applied Chemistry *61*, 1483 (1989).
5. N.E. Holden, *Half-lives of Selected Nuclides,* Pure & Applied Chemistry *62*, 941 (1990).
6. N.E. Holden, *Review of Thermal Neutron Cross Sections and Isotopic Composition of the Elements,* BNL-NCS-42224 (March 1989).
7. P. Ragahavan, *Table of Nuclear Moments,* Atomic Data Nuclear Data Tables *42*, 189 (1989).
8. E. Brown, R. Firestone, *Radioactivity Handbook,* Wiley Interscience Press (1986).
9. J.K. Tuli, *Nuclear Wallet Cards,* Brookhaven National Laboratory (July 1990).

* This research was carried out under the auspices of the U.S. Department of Energy Contract No. DE-AC02–76CH00016.

Elem. or Isot.	Natural Abundance (%)	Atomic Mass or Weight	Half-Life	Decay Mode/Energy (/MeV)	Particle Energy /Intensity (MeV / %)	Spin (h/2π)	Nuclear Magnetic Mom.(nm)	Elect. Quadr. Mom. (b)	γ-ray Energy /Intensity (MeV / %)
₀n		1.00866492	10.3 m	β⁻/0.78235	0.782/100.	1/2+	-1.913043		
₁H		1.00794(7)							
¹H	99.985(1)	1.007825032				1/2+	+2.79285		
²H	0.015(1)	2.014101778				1+	+0.85744	+2.86 mb	
³H		3.01604927	12.32 y	β⁻/0.01859	0.01860/100.	1/2+	+2.97896		
₂He		4.002602(2)							
³He	1.37x10⁻⁴	3.01602931				1/2+	-2.12762		
⁴He	≈100.	4.00260325				0+			
⁵He		5.01222	7.6x10⁻²² s	n,α		3/2-			
⁶He		6.018888	0.807 s	β⁻/3.508,d	3.510/100.	0+			
⁷He		7.02803	3.x10⁻²¹ s	n		(3/2)-			
⁸He		8.03392	0.119 s	β⁻/10.65, t n/	13/88. /12.	0+			0.9807/84. 0.4776/5.
⁹He		9.0438		n					
¹⁰He				2n					
₃Li		6.941(2)							
⁵Li		5.01254	≈3.x10⁻²² s			3/2-			
⁶Li	7.5(2)	6.015122				1+	+0.82205	-0.8 mb	
⁷Li	92.5(2)	7.016004				3/2-	+3.25644	-0.041	
⁸Li		8.022486	0.84 s	β⁻/16.004 α/	12.5/100. α(1.6)	2+	+1.6536	+0.032	
⁹Li		9.026789	0.178 s	β⁻/13.606 β⁻/	13.5/75. 11/25.	3/2-	3.439	0.028	
¹⁰Li		10.03590	4. x 10⁻²² s	β⁻/20.84					
¹¹Li		11.04379	8.7 ms	β⁻/20.6		(1/2-)	3.668	0.031	3.367/35.
₄Be		9.012182(3)							
⁶Be		6.01973	5.0x10⁻²¹ s	2p,α		0+			
⁷Be		7.016929	53.28 d	EC/0.862		3/2-			0.4776/10.4
⁸Be		8.0053051	≈ 7.x10⁻¹⁷ s	2α/0.046		0+			
⁹Be	100.	9.012182		β⁻/0.5559		3/2-	-1.1776	+0.053	
¹⁰Be		10.013534	1.52x10⁶ y	β⁻/0.5559	0.555/100.	0+			
¹¹Be		11.02166	13.8 s	β⁻,β⁻α/11.51	11.48/61.	1/2+			2.125/35.5 (0.478-7.97)
¹²Be		12.02692	24. ms	β⁻/11.71		0+			(0.95 - 4.4)
¹⁴Be		14.0428	4. ms	β⁻/16.2		0+			
₅B		10.811(5)							
⁷B		7.0299	4.x10⁻²² s	p					
⁸B		8.024607	0.770 s	β⁺, 2α/17.979	13.7(β⁺)/93.	2+	1.0355	0.068	ann.rad.
⁹B		9.013329	8x10⁻¹⁹ s	p2α/		3/2-			
¹⁰B	19.9(2)	10.012937				3+	+1.8006	+0.085	
¹¹B	80.1(2)	11.009306				3/2-	+2.6886	+0.0406	
¹²B		12.014352	0.0202 s	β⁻/13.369 β⁻α/1.6/		1+	+1.0031	0.0132	4.438/1.3 3.215/0.00065
¹³B		13.017780	0.0174 s	β⁻/13.437 β⁻n/0.25/	13.4 2.43(n)/0.09 3.55(n)/0.16	3/2-	+3.17778	0.037	3.68/7.6
¹⁴B		14.02540	14. ms	β⁻/20.64		2-			6.094/90.
¹⁵B		15.03110	10.4 ms	β⁻,(n)/19.09					
¹⁷B		17.0469	5.1 ms	β⁻/22.7					
¹⁹B		19.0637		β⁻/26.5					
₆C		12.011(1)							
⁸C		8.03768	2.0x10⁻²¹ s	p		(3/2-)			ann.rad.
⁹C		9.031040	127. ms	β⁺,p, 2α/16.498		(3/2-)			ann.rad. 0.71829/100.
¹⁰C		10.016853	19.3 s	β⁺ /3.648	1.865	0+			
¹¹C		11.011433	20.3 m	β⁺, EC/1.982	0.9608/99.	3/2-	-0.964	0.0333	ann.rad.
¹²C	98.89(1)	12.000000				0+			
¹³C	1.11(1)	13.00335484				1/2-	+0.70241		
¹⁴C		14.00324199	5715. y	β⁻ /0.15648	0.1565/100.	0+			
¹⁵C		15.010599	2.45 s	β⁻ /9.772	4.51/68. 9.82/32.	1/2+	1.32		5.298/68. (7.30-9.05)
¹⁶C		16.014701	0.75 s	β⁻,n/8.012					
¹⁷C		17.02258	0.19 s	β⁻,n/13.17					1.375 1.849 1.906

Elem. or Isot.	Natural Abundance (%)	Atomic Mass or Weight	Half-Life	Decay Mode/Energy (/MeV)	Particle Energy /Intensity (MeV / %)	Spin (h/2π)	Nuclear Magnetic Mom.(nm)	Elect. Quadr. Mom. (b)	γ-ray Energy /Intensity (MeV / %)
^{18}C		18.02676	0.09 s	β^- ,n/11.81					
^{19}C		19.0353	0.05 s	n					
^{20}C		20.0403	0.01 s						
$_7$N		**14.00674(7)**							
^{12}N		12.018613	11.00 ms	β^+ ,β^+ α/17.338	16.38/95.	1+	+0.457	+0.026	ann.rad. 4.438/2.
^{13}N		13.0057386	9.97 m	β^+ /2.2204	1.190/100.	1/2-	0.3222		
^{14}N	99.634(9)	14.00307401				1+	+0.40376	+0.020	
^{15}N	0.366(9)	15.0001090				1/2-	-0.28319		
^{16}N		16.006100	7.13 s	β^- /10.419 β^-, α	4.27/68. 10.44/26. 1.85/.0012	2-			6.129/68.8 7.115/4.7 (0.99-8.87)
^{17}N		17.00845	4.17 s	β^- ,β^- n/8.68 0.4-1.7n/95. β^- α/	3.7/100. 8.0, 8.2	1/2-			0.871/3. 2.1842/0.3
^{18}N		18.01408	0.62 s	β^- /13.90	9.4/100.	1-			0.822/61. 1.65/60.5 1.982/98. (0.535-7.13)
^{19}N		19.01703	0.32 s	β^-/12.53					(0.096-3.14)
^{20}N		20.02337	0.14 s	β^- /17.97					
^{21}N		21.0271	0.08 s						
^{22}N		22.0344	0.02s						
^{23}N		23.0405							
$_8$O		**15.9994(3)**							
^{12}O		12.03442	≈ 1.x10^{-21} s	2p					
^{13}O		13.02481	8.9 ms	β^+ ,p/17.77	1.56 (p)/	(3/2-)			ann.rad. 4.438/0.56
^{14}O		14.0085953	70.60 s	β^+ /5.1430	1.81/99.	0+			ann.rad. 2.312/99.4
^{15}O		15.003065	122.2 s	β^+ /2.754	1.723/100.	1/2-	0.719		ann.rad.
^{16}O	99.762(15)	15.99491462				0+			
^{17}O	0.038(3)	16.9991315				5/2+	-1.8938	-0.026	
^{18}O	0.200(12)	17.999160				0+			
^{19}O		19.003577	26.9 s	β^- /4.820	3.25/60. 4.60/40.	5/2+			0.197/95.9 1.3569/50.4 (0.11-4.18)
^{20}U		20.004076	13.5 s	β^- /3.814		0+			1.057/100. (0.28-4.6)
^{21}O		21.00866	3.4 s	β^- /8.11					(0.64-1.86)
^{22}O		22.00997	2.2 s	β^- /6.5					
^{23}O		23.0157	0.08 s						
^{24}O		24.0204	0.06 s						
$_9$F		**18.9984032(5)**							
^{15}F		15.0180	5.x10^{-22} s	p		(1/2+)			
^{16}F		16.01147	≈ 1.x10^{-20} s	p		0-			
^{17}F		17.0020952	64.5 s	β^+ /2.761	1.75/	5/2+	+4.722	0.058	ann.rad.
^{18}F		18.000938	1.830 h	β^+ ,EC/1.656	0.635/97.	1+			
^{19}F	100.	18.9984032				1/2+	+2.62887	0.072	
^{20}F		19.9999813	11.00 s	β^- /7.0245	5.398/100.	2+	+2.094	0.042	1.634/100. 3.33/0.009
^{21}F		20.999949	4.16 s	β^- /5.684	3.7/8. 5.0/63. 5.4/29.	5/2+			0.3507/90. 1.395/15. (1.746-4.684)
^{22}F		22.00300	4.23 s	β^- /10.82	3.48/15. 4.67/7. 5.50/62.	4+			1.2746/100. 2.0826/82. (0.82-4.37)
^{23}F		23.00357	2.2 s	β^- /8.5		5/2+			1.701/48. 2.129/34. (0.493-3.83)
^{24}F		24.0081	0.3 s	β^- /13.5					1.9816/
^{25}F		25.0121	0.06 s						
^{26}F		26.0196	0.03 s						
^{27}F		27.0269							
$_{10}$Ne		**20.1797(6)**							
^{16}Ne		16.02575	4.x10^{-21} s	2p					
^{17}Ne		17.01770	109. ms	β^+ .p/14.53	1.4-10.6/	1/2-			ann.rad./
^{18}Ne		18.005710	1.67 s	β^+ /4.446	3.416/92.	0+			ann.rad./ 1.0413/7.8 (0.658-1.70)

Elem. or Isot.	Natural Abundance (%)	Atomic Mass or Weight	Half-Life	Decay Mode/Energy (/MeV)	Particle Energy /Intensity (MeV / %)	Spin (h/2π)	Nuclear Magnetic Mom.(nm)	Elect. Quadr. Mom. (b)	γ-ray Energy /Intensity (MeV / %)
^{19}Ne		19.001880	17.22 s	β^+ /3.238	2.24/99.	1/2+	-1.885		ann.rad./ (0.11-1.55)
^{20}Ne	90.48(3)	19.99244018				0+			
^{21}Ne	0.27(1)	20.9938467				3/2+	-0.66180	+0.103	
^{22}Ne	9.25(3)	21.9913855				0+		-0.19	
^{23}Ne		22.9944673	37.2 s	β^- /4.376	3.95/32. 4.39/67.	5/2+	-1.08		0.440/33. (1.64-2.98)
^{24}Ne		23.99362	3.38 m	β^- /2.47	1.10/8. 1.98/92.	0+			0.4723/100. 0.874/7.9
^{25}Ne		24.99779	0.61 s	β^- /7.30	6.3/ 7.3/	1/2+			0.0895/96. (0.98-3.69)
^{26}Ne		26.00046	197 ms	β^- /7.3					0.233/
^{27}Ne		27.0076	32 ms	β^-, n/12.7					
^{28}Ne		28.0121	17 ms	β^-, n/12.3					
^{29}Ne		29.0194	0.2 s	β^- /15.4					
$_{11}$**Na**		**22.989770(2)**							
^{19}Na		19.01388	0.03 s	β^+, p/11.18					
^{20}Na		20.00735	0.446 s	β^+ /13.89 α	2.15/	2+	+0.3694		ann.rad./ 1.634/79.
^{21}Na		20.997655	22.48 s	β^+ /3.547	2.50/95.	3/2+	+2.3863	+0.05	ann.rad./ 0.351/5.
^{22}Na		21.994437	2.605 y	β^+ /90/2.842 EC/10/	0.545/90.	3+	+1.746		ann.rad./ 1.2745/99.9
^{23}Na	100.	22.9897697				3/2+	+2.21752	+0.101	
^{24m}Na			20.2 ms	I.T., β^-		1+			0.4723/100.
^{24}Na		23.9909633	14.96 h	β^- /5.5158	1.389/ > 99.	4+	+1.690		1.3686/100. 2.754/100. (0.997-4.238)
^{25}Na		24.989954	59.3 s	β^- /3.835	2.6/7. 3.15/25. 4.0/65.	5/2+	+3.683	-0.10	0.3897/12.7 0.5850/13. 0.9747/14.9 (0.836-2.80)
^{26}Na		25.99259	1.07 s	β^- /9.31		3+	+2.851	-0.08	1.809/98.9
^{27}Na		26.99401	0.290 s	β^- /9.01 β^-, n/	7.95/	5/2+	+3.90	-0.06	0.9847/87.4 1.698/11.9
^{28}Na		27.9989	31. ms	β^- /14.0 β^-, n/	12.3/	1+	+2.43	-0.02	1.473/37. 2.389/18.6
^{29}Na		29.0028	44. ms	β^-, n/13.3	11.5/	3/2+	+2.45	-0.03	2.560/36. (1.04-3.99)
^{30}Na		30.0092	50. ms	β^- /17.5		2	+2.08		1.483/46.
^{31}Na		31.0136	17.2 ms	/15.9 β^-, n		(3/2-)	+2.31		1.483/14. (0.05-3.54)
^{32}Na		32.0197	13.5 ms	β^- /19.1					0.886/60.
^{33}Na		33.027	8.1 ms	β^- /20.					0.886/16.
^{34}Na		34.035	5. ms	β^- /24.					0.886/60.
^{35}Na		35.044	1.5 ms	β^- /24					
$_{12}$**Mg**		**24.3050(6)**							
^{20}Mg		20.01886	0.1 s	β^+, p/10.73					
^{21}Mg		21.01171	122. ms	β^+, p/13.10		5/2+			0.332/51.
^{22}Mg		21.999574	3.86 s	β^+ /4.786	3.05/	0+			0.0729/60. 0.5820/100. (1.28-1.93)
^{23}Mg		22.994125	11.32 s	β^+ /4.057	3.09/92.	3/2+	0.536		0.440/8.2
^{24}Mg	78.99(3)	23.9850419				0+			
^{25}Mg	10.00(1)	24.9858370				5/2+	-0.85545	+0.199	
^{26}Mg	11.01(2)	25.9825930				0+			
^{27}Mg		26.9843407	9.45 m	β^- /2.6103	1.59/41. 1.75/58. 2.65/0.3	1/2+			0.17068/0.8 0.84376/73. 1.01443/29.
^{28}Mg		27.983877	20.9 h	β^- /1.832	0.459/95.	0+			0.0306/95. 0.4006/36. 0.9418/36. 1.342/54.
^{29}Mg		28.98855	1.3 s	β^- /7.55	5.4/	3/2+			0.960/15. 1.398/16. 2.224/36.
^{30}Mg		29.99046	0.32 s	β^- /7.0		0+			0.224/85.
^{31}Mg		30.99655	0.24 s	β^- /11.7		(3/2+)			1.61/26.
^{32}Mg		31.9992	0.12 s	β^- /10.3					2.765/25.
^{33}Mg		33.0056	0.09 s	β^- /13.7					1.848/
^{34}Mg		34.0091	0.02 s	β^- /11.3					

11-44

Elem. or Isot.	Natural Abundance (%)	Atomic Mass or Weight	Half-Life	Decay Mode/Energy (/MeV)	Particle Energy /Intensity (MeV / %)	Spin (h/2π)	Nuclear Magnetic Mom.(nm)	Elect. Quadr. Mom. (b)	γ-ray Energy /Intensity (MeV / %)
$_{18}$Ar		39.948(1)							
^{31}Ar		31.0121	15. ms	β^+ /18.4 β^+, 2p β^+, 3p					
^{32}Ar		31.99766	98. ms	β^+ ,p/11.2					ann.rad./
^{33}Ar		32.98993	174. ms	β^+ /11.62 β^+ ,p/	3.12/	1/2+			ann.rad./ 0.810(2)/48.
^{34}Ar		33.980270	0.844 s	β^+/6.061	5.0/95.	0+			ann.rad./ 0.6658(1)/2.5 3.1290(1)/1.3
^{35}Ar		34.975257	1.77 s	β^+/5.965	4.94/93.	3/2+	+0.633		ann.rad./ 1.2185(5)/1.22 1.763(1)/0.25 2.964(1)/0.2
^{36}Ar	0.3365(30)	35.9675463				0+			
^{37}Ar		36.9667759	35.0 d	EC/.813		3/2+	+1.15		
^{38}Ar	0.0632(5)	37.9627322				0+			
^{39}Ar		38.964313	268. y	β^-/0.565	0.565/100.	7/2-	-1.3		
^{40}Ar	99.6003(30)	39.962383124				0+			
^{41}Ar		40.964501	1.82 h	β^-/2.492	1.198/	7/2-			1.29364(5)/99. 1.6770(3)/0.05
^{42}Ar		41.96305	33. y	β^-/0.60	0.60/100.	0+			
^{43}Ar		42.9657	5.4 m	β^-/4.6					0.4791(2)/10. 0.7380(1)/43. 0.9752(1)/100. 1.4400(3)/39.
^{44}Ar		43.96537	11.87 m	β^-/3.55		0+			0.182-1.866
^{45}Ar		44.96809	21.5 s	β^-/6.9		7/2-			0.0610/25. 1.020/35. 3.707/34.
^{46}Ar		45.96809	8.4 s	β^-/5.70		0+			1.944/
^{47}Ar		46.9722							
$_{19}$K		39.0983(1)							
^{35}K		34.98801	0.19 s	β^+ /11.88 β^+ ,p/		3/2+			ann.rad./ 1.751/14. 2.5698/26. 2.9827/51.
^{36}K		35.98129	0.342 s	β^+ /12.81	5.3/42. 9.9/44.	2+	+0.548		ann.rad./ 1.97044(5)/82. 2.20783(5)/30. 2.43343(2)/32.
^{37}K		36.9733769	1.23 s	β^+ /6.149	5.13/	3/2+	+0.2032		ann.rad./ 2.7944(8)/2. 3.602(2)/0.05
^{38m}K			0.924 s	β^+ /6.742	5.02/100.	0+			ann.rad./
^{38}K		37.969080	7.63 m	β^+ /5.913	2.60/99.8	3+	+1.37		ann.rad./ 2.1675(3)/99.8 3.9356(5)/0.2
^{39}K	93.2581(44)	38.9637069				3/2+	+0.39146	+0.049	
^{40}K	0.0117(1)	39.9639987	1.26x10^9 y	β^- /1.3111 β^+ ,EC/1.505	1.312/89. 1.50/10.7	4-	-1.298	-0.061	ann.rad./ 1.46081(5)/10.5
^{41}K	6.7302(44)	40.9618260				3/2+	+0.21487	+0.060	
^{42}K		41.9624031	12.36 h	β^- /3.525	1.97/19. 3.523/81.	2-	-1.1425		0.31260(2)/0.3 1.5246(3)/18.
^{43}K		42.96072	22.3 h	β^- /1.82	0.465/8. 0.825/87. 1.24/3.5 1.814/1.3	3/2+	+0.163		0.2211(2)/4. 0.3729(2)/88. 0.3971(2)/11. 0.6178(2)/81.
^{44}K		43.96156	22.1 m	β^- /5.66	5.66/34.	2-	-0.856		0.368207(1)/2.2 1.15700(1)/58. 2.15079(2)/22.
^{45}K		44.96070	17.8 m	β^- /4.20	1.1/23. 2.1/69. 4.0/8.	3/2+	+0.1734		0.1743(5)/80. 1.2607(8)/7. 1.7056(6)/69. 2.3542(5)/14.
^{46}K		45.96198	1.8 m	β^- /7.72	6.3/	2-	-1.05		1.347(1)/91. 3.700(5)/28.
^{47}K		46.96168	17.5 s	β^- /6.64	4.1/99. 6.0/1.	1/2+	+1.93		0.56474(3)/15. 0.58575(3)/85. 2.01313(3)/100.
^{48}K		47.96551	6.8 s	β^- /12.09	5.0/	(2-)			0.67122(1)/4. 0.6723(5)/20. 0.78016(1)/32. 3.83153(7)/80.
^{49}K		48.96745	1.26 s	β^- /11.0					2.025/ 2.252/

Elem. or Isot.	Natural Abundance (%)	Atomic Mass or Weight	Half-Life	Decay Mode/Energy (/MeV)	Particle Energy /Intensity (MeV / %)	Spin (h/2π)	Nuclear Magnetic Mom.(nm)	Elect. Quadr. Mom. (b)	γ-ray Energy /Intensity (MeV / %)
^{22}Al		22.0195	70. ms	β^+ /18.6 β^+ ,p/		4+			ann.rad./
^{23}Al		23.00727	0.47 s	β^+ /12.24 β^+ ,p/					ann.rad./
^{24m}Al			0.129 s	I.T./0.4259 β^+ /	13.3/	1+			1.3686/5.3
^{24}Al		23.999941	2.07 s	β^+ /13.878,p	3.40/48. 4.42/41. 6.80/3. 8.74/8.	4+			1.078(2)/16. 1.368(2)/96. 2.753(2)/43. 4.315(3)/15. 5.392(3)/20. 7.0662(2)/41.
^{25}Al		24.990429	7.17 s	β^+ /4.277	3.27/	5/2+	3.646		ann.rad./ 1.6115(2)/100. 0.975(2)/5.
^{26m}Al			6.345 s	β^+ /	3.2/	0+			ann.rad./
^{26}Al		25.9868917	7.1×10^5 y	β^+ /82/4.0042 EC/18/	1.16/	5+			ann.rad./ 1.80865(7)/99.8
^{27}Al	100.	26.9815384				5/2+	+3.64151	+0.15	
^{28}Al		27.9819102	2.25 m	β^- /4.6422	2.865/100.	3+	3.24	0.18	1.7778(6)/100.
^{29}Al		28.980445	6.5 m	β^- /3.680	1.4/30. 2.5/70.	5/2+			1.2732(8)/89. 2.0282(8)/4. 2.4262(8)/7.
^{30}Al		29.98296	3.68 s	β^- /8.56	5.05/	3+			1.26313(3)/35. 2.23525(5)/65.
^{31}Al		30.98395	0.64 s	β^- /8.00	6.25/				0.75223(3)/18. 1.69473(3)/59. 2.31664(4)/73.
^{32}Al		31.9881	33. ms	β^- /13.0		1+			
^{33}Al		32.9909							
^{34}Al		33.9969	0.05 s	β^- /17.1					
^{35}Al		34.9999							
^{36}Al		36.0064							
$_{14}$Si		28.0855(3)							
^{22}Si		22.0345							
^{24}Si		24.01155	0.10 s	β^+ ,p/10.81					ann.rad./
^{25}Si		25.00411	221 ms	β^+ ,p/12.74		5/2+			ann.rad./
^{26}Si		25.992330	2.23 s	β^+ /5.066	3.282/	0+			ann.rad./ 0.8294(8)/22.
^{27}Si		26.9867041	4.14 s	β^+ /4.8118	3.85/100.	5/2+	-0.8554		ann.rad./ 2.211(5)/0.2
^{28}Si	92.23(2)	27.9769265				0+			
^{29}Si	4.67(2)	28.9764947				1/2+	-0.5553		
^{30}Si	3.10(1)	29.9737702				0+			
^{31}Si		30.9753632	2.62 h	β^- /1.4920	1.471/99.9	3/2+			1.2662(5)/0.05
^{32}Si		31.974148	1.6×10^2 y	β^- /0.224	0.213/100.	0+			
^{33}Si		32.97800	6.1 s	β^- /5.85	3.92	(3/2+)			1.4313(5)/13. 1.84769(5)/100. 2.538(2)/10.
^{34}Si		33.97858	2.8 s	β^- /4.60	3.09/	0+			0.42907(5)/60. 1.17852(2)/64. 1.60756(5)/36.
^{35}Si		34.98458	0.9 s	β^- /10.50					
^{36}Si		35.9867	0.5 s	β^- /7.9					
^{37}Si		36.9930							
^{38}Si		37.9960							
^{39}Si									
$_{15}$P		30.973761(2)							
^{26}P		26.0118	≈20. ms	β^+ ,p/18.1		3+			
^{27}P		26.99919	0.3 s	β^- /11.63		1/2+			
^{28}P		27.992312	270. ms	β^+ /14.332	3.94/13. 5.25/13. 6.96/16. 8.8/7. 11.49/52.	3+			ann.rad./ 1.779(2)/98. 2.839(2)/2.8 3.040(2)/3.2 4.498(2)/12. 7.537(2)/9.
^{29}P		28.981801	4.14 s	β^+ /4.9431	3.945/98.	1/2+	1.2349		ann.rad./ 1.273/1.32 2.426/0.39
^{30}P		29.978314	2.50 m	β^+ /4.2323	3.245/99.9	1+			ann.rad./ 2.230(3)/0.07
^{31}P	100.	30.9737615				1/2+	+1.13160		
^{32}P		31.9739071	14.28 d	β^- /1.7106	1.710/100.	1+	-0.2524		
^{33}P		32.971725	25.3 d	β^- /0.249	0.249/100.	1/2+			

Elem. or Isot.	Natural Abundance (%)	Atomic Mass or Weight	Half-Life	Decay Mode/Energy (/MeV)	Particle Energy /Intensity (MeV / %)	Spin (h/2π)	Nuclear Magnetic Mom.(nm)	Elect. Quadr. Mom. (b)	γ-ray Energy /Intensity (MeV / %)
³⁴P		33.973636	12.4 s	β⁻ /5.374	3.2/15. 5.1/85.	1+			1.78-4.1/ 2.127(5)/15.
³⁵P		34.973314	47. s	β⁻ /3.989	2.34/100.	1/2+			1.572(1)/100.
³⁶P		35.97826	5.7 s	β⁻ /10.41					0.902/77. 3.291/100.
³⁷P		36.97961	2.3 s	β⁻ /7.90					0.6462/ 1.5829/
³⁸P		37.9845	0.6 s	β⁻ /12.4					1.2923/ 2.224/
³⁹P		38.9864	≈ 0.16 s						
⁴⁰P		39.9911							
⁴¹P		40.9948	0.12 s						
⁴²P		42.0001	0.11 s						
₁₆S	**32.066(6)**								
²⁷S		27.0188	21. ms	β⁺, 2p /18.3					
²⁹S		28.99661	0.188 s	β⁺ /13.79 β⁺,p/		5/2+			ann.rad./
³⁰S		29.984903	1.18 s	β⁺ /6.138	4.42/78. 5.08/20.	0+			ann.rad./ 0.678/79.
³¹S		30.979555	2.56 s	β⁺ /5.396	4.39/99.	1/2+	0.48793		ann.rad./ 1.2662(5)/1.2
³²S	95.02(9)	31.9720707				0+			
³³S	0.75(1)	32.9714585				3/2+	+0.64382	-0.07	
³⁴S	4.21(8)	33.9678669				0+			
³⁵S		34.9690322	87.2 d	β⁻ /0.1672	0.1674/100.	3/2+	+1.00	+0.047	
³⁶S	0.02(1)	35.9670809				0+			
³⁷S		36.9711257	5.05 m	β⁻ /4.8653	1.64/94. 4.75/5.6				0.9083(4)/0.06 3.1033(2)/94.2
³⁸S		37.97116	2.84 h	β⁻ /2.94	1.00/	0+			0.1962(4)/0.2 1.9421(3)/84.
³⁹S		38.97514	11.5 s	β⁻ /6.64					1.301/52. 1.697/44.
⁴⁰S		39.9755	9. s	β⁻ /4.7					
⁴¹S		40.9800							
⁴²S		41.9815							
⁴³S		42.987							
⁴⁴S		43.988	0.12 s	β⁻, n/9.					
₁₇Cl	**35.4527(9)**								
³¹Cl		30.99242	0.15 s	β⁺ ,p/11.98		3/2+			ann.rad./
³²Cl		31.98569	297. ms	β⁺ /12.69	4.75/25. 6.18/10. 7.48/14. 9.47/50. 11.6/1.	1+			ann.rad./ 1.548(2)/3.5 2.2305(1)/92. 2.4638(1)/4. 2.885(1)/1. 4.770(1)/20.
³³Cl		32.977452	2.511 s	β⁺ /5.583	4.51/98.	3/2+	+0.752		ann.rad./ 0.8409/0.52 1.966/0.45 2.866/0.44
³⁴ᵐCl			32.2 m	β⁺ / I.T./	1.35/24. 2.47/28.	3+			ann.rad./ 0.1457(8)/42. 2.1276(5)/42.
³⁴Cl		33.9737630	1.528 s	β⁺ /5.4922	4.50/100.	0+			ann.rad./
³⁵Cl	75.77(7)	34.96885271				3/2+	+0.82187	-0.08	
³⁶Cl		35.9683069	3.01x10⁵ y	β⁻ /0.7086 β⁺,EC/1.1421	0.7093/98. 0.115/0.002	0+	+1.28547	-0.018	ann.rad./
³⁷Cl	24.23(7)	36.9659026				3/2+	+0.68412	-0.065	
³⁸ᵐCl			0.715 s	I.T./		5-			0.67138(2)/100.
³⁸Cl		37.9680106	37.2 m	β⁻ /4.9168	1.11/31. 2.77/11. 4.91/58.	2-	2.05		1.64216(1)/31. 2.16760(2)/42.
³⁹Cl		38.968009	55.6 m	β⁻ /3.442	1.91/85. 2.18/8. 3.45/7.	3/2+			0.25026(1)/47. 1.26720(5)/54. 0.986-1.517
⁴⁰Cl		39.97042	1.38 m	β⁻ /7.48		2-			0.6431(3)/6. 1.4608(1)/77. 2.8402(2)/17.
⁴¹Cl		40.9707	34. s	β⁻ /5.7	3.8/				(0.167-1.359)
⁴²Cl		41.9732	6.8 s	β⁻ /9.4					
⁴³Cl		42.9742	3.3 s	β⁻ /8.0					
⁴⁴Cl		43.9785		β⁻ /12.3					
⁴⁵Cl		44.980	0.41	β⁻, n/11.					
⁴⁶Cl		45.984	0.20	β⁻, n/14.9					
⁴⁷Cl		46.988		β⁻, n/15.					

Elem. or Isot.	Natural Abundance (%)	Atomic Mass or Weight	Half-Life	Decay Mode/Energy (/MeV)	Particle Energy /Intensity (MeV / %)	Spin (h/2π)	Nuclear Magnetic Mom.(nm)	Elect. Quadr. Mom. (b)	γ-ray Energy /Intensity (MeV / %)
^{50}K		49.9728	0.472 s	β^- /14.2					
^{51}K			0.365 s	β^- /					
^{52}K			0.105 s	β^-					
^{53}K			30. ms	β^-	3/2+				
^{54}K			10. ms	β^-					
$_{20}$Ca		**40.078(4)**							
^{35}Ca		35.0048	0.05 s	β^+ /15.6					
^{36}Ca		35.99309	0.1 s	β^+ /10.99 β^+ ,n/					ann.rad./
^{37}Ca		36.98587	0.173 s	β^+ /11.64 β^+ ,n/		3/2+			ann.rad./
^{38}Ca		37.976319	0.44 s	β^+ /6.74		0+			ann.rad./ 1.5677(5)/25. 3.210(2)/1.
^{39}Ca		38.970718	0.861 s	β^+ /6.531	5.49/100.	3/2+	1.02168		ann.rad./
^{40}Ca	96.941(18)	39.9625912				0+			
^{41}Ca		40.9622783	1.02x10^5 y	EC/0.4214		7/2-	-1.595	-0.08	
^{42}Ca	0.647(9)	41.9586183				0+			
^{43}Ca	0.135(6)	42.9587668				7/2-	-1.3173	-0.05	
^{44}Ca	2.086(12)	43.955481				0+			
^{45}Ca		44.956186	162.7 d	β^- /0.257	0.257/100.	7/2-	-1.327	+0.05	
^{46}Ca	0.004(3)	45.953693				0+			
^{47}Ca		46.954546	4.536 d	β^- /1.992	0.684/84. 1.98/16.	7/2-	-1.38	+0.02	1.297/75 (0.041-1.88)
^{48}Ca	0.187(4)	47.952533				0+			
^{49}Ca		48.955673	8.72 m	β^- /5.262	0.89/7. 1.95/92.	3/2-			3.0844(1)/92. 4.0719(1)/7.
^{50}Ca		49.95752	14. s	β^- /4.97	3.12/	0+			0.2569/98. (0.0715 - 1.59)
^{51}Ca		50.9615	10. s	β^- /7.3		(3/2-)			
^{52}Ca		51.9651	4.6 s	β^- /8.0					
^{53}Ca		52.9700	0.09 s	β^- /10.9					
$_{21}$Sc		**44.955910(8)**							
^{40}Sc		39.977964	0.182 s	β^+ /14.320	5.73/50. 7.53/15. 8.76/15. 9.58/20.	4-			ann.rad./ 0.752/41. 3.732/99.5 (1.12-3.92)
^{41}Sc		40.9692513	0.596 s	β^+ /6.4953	5.61/100.	7/2-	5.431	0.166	ann.rad./
^{42m}Sc			61.6 s	β^+ /	2.82/	7+			ann.rad./ 0.4375(5)/100. 1.2270(5)/100. 1.5245(5)/100.
^{42}Sc		41.9655168	0.682 s	β^+ /6.4259	5.32/100.	0+			ann.rad./
^{43}Sc		42.961151	3.89 h	β^+ ,EC/2.221	0.82/22. 1.22/78.	7/2-	+4.62	-0.26	ann.rad./ 0.3729(1)/22.
^{44m}Sc			2.44 d	I.T./0.27 EC/3.926		6+	+3.88		0.27124(1)/87. (1.00-1.16)
^{44}Sc		43.959403	3.93 h	β^+, EC/3.653	1.47/	2+	+2.56	+0.10	ann.rad./ 1.15700(1)/100.
^{45}Sc	100.	44.955910				7/2-	+4.75649	-0.22	
^{46m}Sc			18.7 s	I.T./0.14253		1-			0.14253(2)/62.
^{46}Sc		45.955170	83.81 d	β^- /2.367	0.357/100.	4+	+3.03	+0.12	0.88925(3)/100. 1.12051(1)/100.
^{47}Sc		46.952408	3.349 d	β^- /0.600	0.439/69. 0.601/31.	7/2-	+5.34	-0.22	0.15938(1)/68.
^{48}Sc		47.952235	43.7 h	β^- /3.99	0.655/	6+			0.98350(1)/100. 1.03750(1)/97. 1.31209(3)/100.
^{49}Sc		48.950024	57.3 m	β^- /2.006	2.00/99.9.	7/2-			1.7619(3)/0.05
^{50}Sc		49.95219	1.71 m	β^- /6.89	3.05/76. 3.60/24.	(5+)			0.5235(1)/88. 1.1210(1)/100. 1.5537(2)/100.
^{51}Sc		50.95360	12.4 s	β^- /6.51	4.4/ 5.0/	7/2-			1.4373(4)/52. 0.718-2.144
^{52}Sc		51.9566	8.2 s	β^- /9.0		(3+)			
^{53}Sc		52.9584		β^- /8.1					
^{54}Sc		53.9635		β^- /11.6					
^{55}Sc		54.969		β^- /13					
$_{22}$Ti		**47.867(1)**							
^{39}Ti		39.0013	28. ms	β^+ /15.4					
^{40}Ti		39.9905	0.06 s	β^+ /11.7					
^{41}Ti		40.98313	80. ms	β^+ ,p/12.93		3/2+			ann.rad./

Elem. or Isot.	Natural Abundance (%)	Atomic Mass or Weight	Half-Life	Decay Mode/Energy (/MeV)	Particle Energy /Intensity (MeV / %)	Spin (h/2π)	Nuclear Magnetic Mom.(nm)	Elect. Quadr. Mom. (b)	γ-ray Energy /Intensity (MeV / %)
^{42}Ti		41.97303	0.20 s	β^+ /7.000	6.0/				ann.rad./ 0.6107(5)/56.
^{43}Ti		42.96852	0.50 s	β^+ /6.87	5.80/	7/2-	0.85		ann.rad./
^{44}Ti		43.959690	67. y	EC/0.268		0+			0.067868(1)/91. 0.078322(1)/97.
^{45}Ti		44.958124	3.078 h	β^+/86/2.062 EC/14/	1.04	7/2-	0.095	0.015	ann.rad./ (0.36-1.66)/weak
^{46}Ti	8.25(3)	45.952630				0+			
^{47}Ti	7.44(2)	46.951764				5/2-	-0.78848	+0.30	
^{48}Ti	73.72(3)	47.947947				0+			
^{49}Ti	5.41(2)	48.947871				7/2-	-1.10417	+0.24	
^{50}Ti	5.18(2)	49.944792				0+			
^{51}Ti		50.946616	5.76 m	β^- /2.471	1.50/92. 2.13/	3/2-			0.3197(2)/93. 0.6094-0.9291
^{52}Ti		51.94690	1.7 m	β^- /1.97	1.8/100.	0+			0.0170(5)/100. 0.12445(5)/100.
^{53}Ti		52.9497	33. s	β^- /5.0	(2.2-3)/	3/2-			0.1008(1)/20. 0.1276(1)/45. 0.2284(1)/39. 1.6755(5)/45. (1.72-2.8)/
^{54}Ti		53.9510		β^- /4.3					
^{55}Ti		54.9552		β^- /13					
^{56}Ti		55.9580		β^- /7.0					
^{57}Ti		56.964		β^- /11.					
^{58}Ti									
$_{23}$V		50.9415(1)							
^{43}V		42.9807	>0.8 s	β^+ /11.3					
^{44}V		43.9744	0.09 s	β^+ ,α/13.7					
^{45}V		44.96578	0.54 s	β^+ /7.13		7/2-			
^{46}V		45.96020	0.4223 s	β^+ /7.051	6.03/100.	0+			ann.rad./
^{47}V		46.954907	32.6 m	β^+ ,EC/2.928	1.90/99.+	3/2-			ann.rad./ 1.7949(8)/0.19 (0.2-2.16)
^{48}V		47.952254	15.98 d	β^+ /4.012	0.698/50.	4+	2.01		ann.rad./ 0.98350(2)/100. (1.3-2.4)
^{49}V		48.948517	337. d	EC/0.602		7/2-	4.47		
^{50}V	0.250(2)	49.947163	>1.4x10^{17}y	EC, β		6+	+3.34569	+0.21	
^{51}V	99.750(2)	50.943964				7/2-	+5.148706	-0.04	
^{52}V		51.944780	3.76 m	β^- /3.976	2.47/	3+			1.4341(1)/100.
^{53}V		52.944342	1.61 m	β^- /3.436	2.52/	7/2-			1.0060(5)/90. 1.2891(3)/10.
^{54}V		53.94644	49.8 s	β^- /7.04	1.00/5. 2.00/12. 2.95/45. 5.20/11.	3+			0.8348/97. 0.9887/80. 2.259/46. (0.56-3.38)
^{55}V		54.9472	6.5 s	β^- /6.0	6.0/	(7/2-)			0.5177/73. (0.224-1.21)
^{56}V		55.9505		β^- /9.1					
^{57}V		56.9525		β^- /8.1					
^{58}V		57.9567		β^- /11.6					
^{59}V		58.9593		β^- /9.9					
^{60}V		59.965		β^- /14.					
^{61}V									
$_{24}$Cr		51.9961(6)							
^{43}Cr		42.9977	21. ms						
^{44}Cr		43.9855	53. ms	β^+ ,(p) /10.3					
^{45}Cr		44.9792	0.05 s	β^+ ,p/12.5		7/2-			ann.rad./
^{46}Cr		45.96836	0.3 s	β^+ /7.60		0+			ann.rad./
^{47}Cr		46.96291	0.51 s	β^+ /7.45		3/2-			ann.rad./
^{48}Cr		47.95404	21.6 h	EC/1.66					ann.rad./ 0.116(2)/95. 0.305(10)/100.
^{49}Cr		48.951341	42.3 m	β^+ ,EC/2.631	1.39/ 1.45/ 1.54/	5/2-	0.476		ann.rad./ 0.09064(1)/51. 0.15293(1)/27. (0.062-1.6)
^{50}Cr	4.345(13)	49.946050				0+			
^{51}Cr		50.944772	27.70 d	EC/0.7527		7/2-	-0.934		0.320076(1)/10.2
^{52}Cr	83.789(18)	51.940512				0+			
^{53}Cr	9.501(17)	52.940653				3/2-	-0.47454	-0.15	
^{54}Cr	2.365(7)	53.938885				0+		-0.2	

Elem. or Isot.	Natural Abundance (%)	Atomic Mass or Weight	Half-Life	Decay Mode/Energy (/MeV)	Particle Energy /Intensity (MeV / %)	Spin (h/2π)	Nuclear Magnetic Mom.(nm)	Elect. Quadr. Mom. (b)	γ-ray Energy /Intensity (MeV / %)
^{55}Cr		54.940844	3.497 m	β^- /2.603	2.5/	3/2-			1.5282(2)/0.04 (0.13-2.37)
^{56}Cr		55.94065	5.9 m	β^- /1.62	1.50/100.	0+			0.026(2)/100. 0.083(3)/100.
^{57}Cr		56.9438	21. s	β^- /5.1	3.3/ 3.5/	3/2-	0.0834		0.850/8. (0.083-2.62)
^{58}Cr		57.9443	7.0 s	β^- /4.0					
^{59}Cr		58.9487	1.0 s	β^- /7.7					
^{60}Cr		59.9497	0.6 s	β^- /6.0					
^{61}Cr		60.9541		β^- /8.8					
^{62}Cr		61.9558		β^- /7.3					
^{63}Cr									
$_{25}$Mn		54.938049(9)							
^{46}Mn		45.9867	≈41. ms	β^+ /17.1					
^{47}Mn		46.9761		β^+ /12.3					
^{48}Mn		47.9686	0.15 s	β^+ /13.5	5.79/58. 4.43/10.	4+			
^{49}Mn		48.95962	0.38 s	β^+ /7.72	6.69/	5/2-			ann.rad./
^{50m}Mn			1.74 m	β^+ /7.887	3.54/	5+			ann.rad./ 1.0980/94. 0.783/91. (0.66-3.11)
^{50}Mn		49.954244	0.283 s	β^+ /7.6330	6.61/	0+			ann.rad./
^{51}Mn		50.948215	46.2 m	β^+ ,EC/3.208	2.2/	5/2-	3.568	0.4	ann.rad./ 0.7491(1)/0.26 (1.148-1.164)
^{52m}Mn			21.1 m	β^+ /98/5.09 I.T. /2/0.378	2.631/	2+	0.0076		ann.rad./ 0.3778 (I.T.) 1.43406(1)/98. (0.7-4.8)
^{52}Mn		51.945570	5.591 d	β^+ /4.712 EC/	0.575/	6+	+3.063	+0.5	ann.rad./ 0.74421(1)/90. 1.43406(1)/100.
^{53}Mn		52.941294	3.7x10^6 y	EC/0.5970		7/2-	5.024		
^{54}Mn		53.940363	312.2 d	EC/1.377		3+	+3.282	+0.33	0.83403(5)/100.
^{55}Mn	100.	54.938049				5/2-	+3.4687	+0.32	
^{56}Mn		55.938909	2.579 h	β^- /3.6954	0.718/18. 1.028/34.	3+	+3.2266		0.84675(2)/98.9 1.81072(4)/27. 2.11305(5)/14.5
^{57}Mn		56.938287	1.45 m	β^- /2.691		5/2-			
^{58}Mn		57.93999	65 s	β^- /6.25	3.8/ 5.1/	3+			0.45916(2)/20. 0.81076(1)/82. 1.32309(5)/53.
^{59}Mn		58.94045	4.6 s	β^- /5.19	4.5/				0.471/ 0.531-0.726
^{60m}Mn			1.77 s	β^- /IT	5.7/	3+			0.824/ 1.969/
^{60}Mn		59.9433	50. s	β^- /8.6		0+			
^{61}Mn		60.9446	0.71 s	β^- /7.4		(5/2)-			
^{62}Mn		61.9480	0.9 s	β^- /10.4		(3+)			0.877/ 0.942-1.299
^{63}Mn		62.9498		β^- /8.8					
^{64}Mn		63.9537		β^- /11.8					
^{65}Mn		64.956		β^- /10.					
^{66}Mn									
$_{26}$Fe		55.845(2)							
^{46}Fe		46.0008	≈0.02 s	β^+ /13.1					
^{47}Fe		46.9929	≈0.03 s	β^+ /15.6					
^{48}Fe		47.9806		β^+ /11.2					
^{49}Fe		48.9763	0.08 s	β^+ /13.0		(7/2-)			ann.rad./
^{50}Fe		49.9630	0.15 s	β^+ /8.2					0.651
^{51}Fe		50.95683	0.31 s	β^+ /8.02		(5/2-)			ann.rad./
^{52m}Fe			46. s	β^+ /4.4		(12+)			ann.rad./ (0.622-2.286)/
^{52}Fe		51.94812	8.28 h	β^+ /57/2.37 EC/43/ I.T./	0.804/	0+			ann.rad./ 0.16868(1)/99. 0.377 (I.T.)/
^{53m}Fe			2.6 m	I.T./3.0407		19/2-			0.7011(1)/99. 1.0115(1)/87. 1.3281(1)/87. 2.3396(1)/13.
^{53}Fe		52.945312	8.51 m	β^+ /3.743	2.40/42. 2.80/57.	7/2-			ann.rad./ 0.3779(1)/42. (1.2 - 3.2)
^{54}Fe	5.845(35)	53.939615				0+			
^{55}Fe		54.938298	2.73 y	EC/0.2316		3/2-			

Elem. or Isot.	Natural Abundance (%)	Atomic Mass or Weight	Half-Life	Decay Mode/Energy (/MeV)	Particle Energy /Intensity (MeV / %)	Spin (h/2π)	Nuclear Magnetic Mom.(nm)	Elect. Quadr. Mom. (b)	γ-ray Energy /Intensity (MeV / %)
^{56}Fe	91.754(36)	55.934942				0+			
^{57}Fe	2.119(10)	56.935398				1/2-	+0.0906		
^{58}Fe	0.282(4)	57.933280				0+			
^{59}Fe		58.934880	44.51 d	β^- /1.565	0.273/48. 0.475/51.	3/2-	0.29		1.099/57 1.292/43. (0.14-1.48)
^{60}Fe		59.934077	1.5x10^6 y	β^- /0.237	0.184/100.	0+			0.0586/100.(IT)
^{61}Fe		60.93675	6.0 m	β^- /3.98	2.5/13. 2.63/54. 2.80/31.				1.205/44. 1.028/43. (0.12-3.37)
^{62}Fe		61.93677	68. s	β^- /2.53	2.5/100.	0+			0.5061(1)/100.
^{63}Fe		62.9404	6. s	β^- /6.3		5/2-			0.995/ (1.365-1.427)
^{64}Fe		63.9411	2.0 s	β^- /4.9					
^{65}Fe		64.9449	≈0.45 s	β^- /7.9					
^{66}Fe		65.9460		β^- /5.7					
^{67}Fe		66.9500		β^- /8.8					
^{68}Fe		67.953	≈0.10 s	β^- / ≈7.6					
^{69}Fe									
$_{27}$Co		58.933200(9)							
^{50}Co		49.9812		β^+ /17.0					
^{51}Co		50.9705		β^+ /12.8					
^{52}Co		51.9632		β^+ /14.0					
^{53m}Co			0.25 s	β^+ ,p/		19/2-			ann.rad./
^{53}Co		52.95423	0.26 s	β^+ /8.30		7/2-			ann.rad./
^{54m}Co			1.46 m	β^+ /8.44	4.25/100.	7+			ann.rad./ 0.411(1)/99. 1.130(1)/100. 1.408(1)/100.
^{54}Co		53.948464	0.1932 s	β^+ /8.2430	7.34/100.	0+			ann.rad./
^{55}Co		54.942003	17.53 h	β^+ /3.4513 EC/	0.53/ 1.03/ 1.50/	7/2-	+4.822		ann.rad./ 0.9312/75. 0.4772/20. (0.092-3.11)
^{56}Co		55.939844	77.3 d	β^+ /4.566 EC/	1.459/18.	4+	3.85	+0.25	ann.rad./ 0.84678(6)/99.9 1.2383/68. (0.26-3.61)
^{57}Co		56.936296	271.8 d	EC/0.8361		7/2-	+4.72	+0.5	0.12206(2)/85.6 (0.014-0.706)
^{58m}Co			9.1 h	I.T./		5+			0.02489/0.035
^{58}Co		57.935757	70.88 d	β^+ /2.307 EC/		2+	+4.04	+0.22	ann.rad./ 0.810755(3)/99.
^{59}Co	100.	58.933200				7/2-	+4.63	+0.41	
^{60m}Co			10.47 m	I.T./99.8/0.059 β^- /0.2/1.56		2+	+4.40	+0.3	0.058603(1)/2.0
^{60}Co		59.933822	5.271 y	β^- /2.824	0.315/99.7	5+	+3.799	+0.44	1.173210(2)/100. 1.332470(2)/100.
^{61}Co		60.932479	1.650 h	β^- /1.322	1.22/95.	7/2-			0.067415(1)/86. (0.8417-0.9092)
^{62m}Co			13.9 m	β^- /	0.88/25. 2.88/75.	5+			1.1635(3)/70. 1.1730(3)/98. 2.0039(3)/19.
^{62}Co		61.93405	1.50 m	β^- /5.32	1.03/10. 1.76/5. 2.9/20. 4.05/60.	2+			1.1292(3)/13. 1.1730(3)/83. 1.9851(3)/3. 2.3020(1)/19.
^{63}Co		62.93362	27.5 s	β^- /3.67	3.6/	7/2-			0.08713(1)/49. 0.9817(3)/2.6 0.156-2.17
^{64}Co		63.93581	0.30 s	β^- /7.31	7.0/	1+			
^{65}Co		64.93648	1.14 s	β^- /5.96		(7/2)-			
^{66}Co		65.9398	≈ 0.23 s	β^- /10.0					
^{67}Co		66.9406		β^- /8.4					
^{68}Co		67.9444	≈ 0.18 s	β^- /11.7					
^{69}Co		68.9452	0.27 s	β^- /9.3					
^{70}Co		69.9498		β^- 13.					
^{71}Co									
$_{28}$Ni		58.6934(2)							
^{51}Ni		50.9877		β^+ /16.0					
^{52}Ni		51.9757		β^+ /11.7					
^{53}Ni		52.9685	0.05 s	β^+ ,p/13.3		7/2-			ann.rad./
^{54}Ni		53.95791		β^+ /8.80					
^{55}Ni		54.95134	0.20 s	β^+ /8.70	7.66/	7/2-			ann.rad./

Elem. or Isot.	Natural Abundance (%)	Atomic Mass or Weight	Half-Life	Decay Mode/Energy (/MeV)	Particle Energy /Intensity (MeV / %)	Spin (h/2π)	Nuclear Magnetic Mom.(nm)	Elect. Quadr. Mom. (b)	γ-ray Energy /Intensity (MeV / %)
^{56}Ni		55.94214	6.08 d	EC/2.14		0+			0.15838(3)/98.8 0.81185(3)/87. 0.2695-0.7500
^{57}Ni		56.939800	35.6 h	β^+ /3.264 EC/	0.712/10. 0.849/76.	3/2-	0.88		ann.rad./ 1.3776/78. (0.127-3.177)
^{58}Ni	68.077(9)	57.935348				0+			
^{59}Ni		58.934351	≈7.6x10⁴ y	EC/		3/2-			
^{60}Ni	26.223(8)	59.930790				0+			
^{61}Ni	1.140(1)	60.931060				3/2-	-0.75002	+0.16	
^{62}Ni	3.634(2)	61.928348				0+			
^{63}Ni		62.929673	100. y	β^- /0.066945	0.065/	1/2-			
^{64}Ni	0.926(1)	63.927969				0+			
^{65}Ni		64.930088	2.517 h	β^- /2.137	0.65/30. 1.020/11. 2.140/58.	5/2-	0.69		0.36627(3)/5. 1.11553(4)/16. 1.48184(5)/23.
^{66}Ni		65.92912	54.6 h	β^- /0.23		0+			
^{67}Ni		66.93157	21. s	β^- /3.56	3.8/	1/2-			1.0722/100. 1.6539/100. (0.10-1.98)
^{68}Ni		67.93185	19. s	β^- /2.06					0.6807(3)/100.
^{69}Ni		68.9352	10. s	β^- /5.4					
^{70}Ni		69.9361		β^- /3.5					
^{71}Ni		70.9400	1.9 s	β^- /6.9					
^{72}Ni		71.9413	2.1 s	β^- /5.2					
^{73}Ni		72.946	0.9 s	β^- /9.					
^{74}Ni		73.948	1.1 s	β^- /7.					
$_{29}$Cu		63.546(3)							
^{55}Cu		54.9655		β^+ /13.2					
^{56}Cu		55.9586		β^+ /15.3					
^{57}Cu		56.94922	0.23 s	β^+ /8.77		3/2-			
^{58}Cu		57.944541	3.21 s	β^+ /8.563 EC/	4.5/15. 7.439/83.	1+			ann.rad./ 0.0403(4)/5. 1.4483(2)/11. 1.4546(2)/16.
^{59}Cu		58.939504	1.36 m	β^+ /4.800	1.9/ 3.75/	3/2-			ann.rad./ 0.3393(1)/8. 0.8780(1)/12. 1.3015(1)/15. (0.4 - 2.6)
^{60}Cu		59.937368	23.7 m	β^+ /6.127 EC/	2.00/69. 3.00/18. 3.92/6.	2+	+1.219		ann.rad./ 1.3325/88. 1.7915/45. (0.12-5.048)
^{61}Cu		60.933462	3.35 h	β^+ /2.237	0.56/3. 0.94/5. 1.15/2. 1.220/51.	3/2-	+2.14		ann.rad./ 0.2830/13. 0.6560/11. (0.067-2.123)
^{62}Cu		61.932587	9.74 m	β^+ /98/3.948 EC/	2.93/98.	1+	-0.380		ann.rad./ 1.17302(1)/0.6 (0.87-3.37)
^{63}Cu	69.17(3)	62.929601				3/2-	+2.2233	-0.211	
^{64}Cu		63.929768	12.701 h	β^- /39/0.579 β^+ /19/1.6751 EC/41/	0.578/ 0.65/	1+	-0.217		ann.rad./ 1.3459(3)/0.6
^{65}Cu	30.83(3)	64.927794				3/2-	+2.3817	-0.195	
^{66}Cu		65.928873	5.09 m	β^- /2.642	1.65/6. 2.7/94.	1+	-0.282		0.8330(1)/0.22 1.0392(2)/9.2
^{67}Cu		66.92775	2.580 d	β^- /0.58	0.395/56. 0.484/23. 0.577/20.	3/2-			0.09125(1)/7. 0.09325(1)/17. 0.18453(1)/47.
^{68m}Cu			3.79 m	I.T./86/ β^- /14/1.8		6-			0.0843(5)/70. 0.1112(5)/18. 0.5259(5)/74. (0.64-1.34)
^{68}Cu		67.92964	31. s	β^- /4.46	3.5/40. 4.6/31.	1+			1.0774(5)/58. 1.2613(5)/17. (0.15-2.34)
^{69}Cu		68.92943	2.8 m	β^- /2.68	2.48/80.	3/2-			0.5307(3)/3. 0.8340(5)/6. 1.0065(8)/10.
^{70m}Cu			47. s	β^- /	2.52/10.	5-			0.8848(2)/100. 0.9017(2)/90. 1.2517(5)/60. (0.39-3.06)
^{70}Cu		69.93241	5. s	β^- /6.60	5.42/54. 6.09/46.	1+			0.8848(2)/54.

Elem. or Isot.	Natural Abundance (%)	Atomic Mass or Weight	Half-Life	Decay Mode/Energy (/MeV)	Particle Energy /Intensity (MeV / %)	Spin (h/2π)	Nuclear Magnetic Mom.(nm)	Elect. Quadr. Mom. (b)	γ-ray Energy /Intensity (MeV / %)
^{71}Cu		70.93262	20. s	β^- /4.56		3/2-			0.490/
^{72}Cu		71.9357	6.6 s	β^- /8.2		(1+)			0.652/
^{73}Cu		72.9365	3.9 s	β^- /6.3					0.450/
^{74}Cu		73.9401	1.6 s	β^- /9.9					
^{75}Cu		74.9414	1.2 s	β^- /7.9					
^{76}Cu		75.9455	0.64 s	β^- /11.					
^{77}Cu		76.947	0.47 s	β^- /$\approx$10.					
^{78}Cu		77.952	0.34 s	β^- /12.					
^{79}Cu		78.954	0.19 s	β^- /11.					
$_{30}$**Zn**		**65.39(2)**							
^{57}Zn		56.9649	0.04 s	β^+,p/14.6		(7/2-)			ann.rad./
^{59}Zn		58.94927	183. ms	β^+,p/9.09	8.1/	3/2-			ann.rad./ (0.491-0.914)
^{60}Zn		59.94183	2.40 m	β^+ /97/4.16 EC/3/		0+			ann.rad./ 0.669/47. (0.062-0.947)
^{61}Zn		60.93951	1.485 m	β^+ /5.64	4.38/68.	3/2-			ann.rad./ 0.4748/17. (0.15-3.52)
^{62}Zn		61.93433	9.22 h	β^+ /3/1.63 EC/93/	0.66/7.	0+			ann.rad./ 0.0408/25 0.5967/26. (0.20-1.526)/
^{63}Zn		62.933215	38.5 m	β^+ /93/3.367 EC/7/	1.02/ 1.40/ 1.71/ 2.36/84.	3/2-	-0.28164	+0.29	ann.rad./ 0.66962(5)/8.4 0.96206(5)/6.6 (0.24-3.1)
^{64}Zn	48.6(3)	63.929146				0+			
^{65}Zn		64.929245	243.8 d	β^+ /98/1.3514 EC/1.5/	0.325/	5/2-	+0.7690	-0.023	ann.rad./ 1.11552(2)/50.8
^{66}Zn	27.9(2)	65.926036				0+			
^{67}Zn	4.1(1)	66.927131				5/2-	+0.8755	+0.15	
^{68}Zn	18.8(4)	67.924847				0+			
^{69m}Zn			13.76 h	I.T./99+/0.439		9/2+			0.4390(2)/95.
^{69}Zn		68.926553	56. m	β^- /0.906	0.905/99.9	1/2-			0.318/
^{70}Zn	0.6(1)	69.925325				0+			
^{71m}Zn			3.97 h	β^- /	1.45/	9/2+			0.3864/93. 0.4874/62. 0.6203/57. (0.099-2.489)
^{71}Zn		70.92773	2.4 m	β^- /2.81		1/2-			0.5116(1)/30. 0.9103(1)/7.5 (0.12-2.29)
^{72}Zn		71.92686	46.5 h	β^- /0.46	0.25/14. 0.30/86.	0+			0.0164(3)/8. 0.1447(1)/83. 0.1915(2)/9.4
^{73m}Zn			6. s	I.T./0.196		(7/2+)			0.042
^{73}Zn		72.92978	24. s	β^- /4.29	4.7/	(1/2-)			0.216(1)/100. 0.496-0.911
^{74}Zn		73.92946	1.60 m	β^- /2.3	2.1/				0.0565/ 0.1401/ (0.05-0.35)
^{75}Zn		74.9329	10.2 s	β^- /6.0					0.229/
^{76}Zn		75.9334	5.7 s	β^- /4.2	3.6/				0.119/
^{77m}Zn			1.0 s	β^- /		(1/2-)			0.772
^{77}Zn		76.9371	2.1 s	β^- /7.3	4.8/				0.189/
^{78}Zn		77.9386	1.5 s	β^- /6.4					0.225/
^{79}Zn		78.9421	1.0 s	β^- /8.6					0.702/
^{80}Zn		79.9444	0.54 s	β^- /7.3					0.713/ 0.2248/
^{81}Zn		80.9505	0.29 s	β^- /11.9					
$_{31}$**Ga**		**69.723(1)**							
^{61}Ga		60.9492	0.15 s	β^+ /9.0					
^{62}Ga		61.94418	0.116 s	β^+ /9.17 EC/	8.3/	0+			ann.rad./
^{63}Ga		62.9391	32. s	β^+ /5.5 EC/	4.5/				ann.rad./ 0.6271(2)/10. 0.6370(2)/11. 1.0652(4)/45.
^{64}Ga		63.936838	2.63 m	β^+ /7.165	2.79/ 6.05/	0+			ann.rad./ 0.80785(1)/14. 0.99152(1)/43. 1.38727(1)/12. 3.3659(1)/13.

Elem. or Isot.	Natural Abundance (%)	Atomic Mass or Weight	Half-Life	Decay Mode/Energy (/MeV)	Particle Energy /Intensity (MeV / %)	Spin (h/2π)	Nuclear Magnetic Mom.(nm)	Elect. Quadr. Mom. (b)	γ-ray Energy /Intensity (MeV / %)
⁶⁵Ga		64.9394	15.2 m	β⁺ /86/3.255 EC/	0.82/10. 1.39/19. 2.113/56. 2.237/15.	3/2-			ann.rad./ 0.1151(2)/55. 0.1530(2)/96. 0.2069(2)/39. (0.06-2.4)
⁶⁶Ga		65.931592	9.5 h	β⁺ /56/5.175 EC/43/	0.74/1. 1.84/54. 4.153/51.	0+			ann.rad./ 1.03935(8)/38. 2.7523(1)/23. (0.28-5.01)
⁶⁷Ga		66.928205	3.260 d	EC/1.001		3/2-	+1.8507	0.20	0.09332/37. 0.18459/20. 0.30024/17. (0.091-0.89)
⁶⁸Ga		67.927983	1.130 h	β⁺ /90/2.921 EC/10/	1.83/	1+	0.01175	0.028	ann.rad./ 1.0774(1)/3. (0.57-2.33)/
⁶⁹Ga	60.108(6)	68.925581				3/2-	+2.01659	+0.17	
⁷⁰Ga		69.926027	21.1 m	EC/0.2/0.655 β⁻ /99.8/1.656	1.65/99.	1+			0.1755(5)/0.15 1.042(5)/0.48
⁷¹Ga	39.892(6)	70.924707				3/2-	+2.56227	+0.11	
⁷²Ga		71.926372	14.10 h	β⁻ /4.001	0.64/40. 1.51/9. 2.52/8. 3.15/11.	3-	-0.13224	+0.5	0.62986(5)/24. 2.2016(2)/26. 2.5077(2)/12.8 (0.11-3.3)/
⁷³Ga		72.92517	74.87 h	β⁻ /1.59		3/2-			0.05344(5)/10. 0.29732(5)/47. (0.01-1.00)/
⁷⁴ᵐGa			10. s	I.T./		1+			0.0565(1)/75.
⁷⁴Ga		73.92694	8.1 m	β⁻ /5.4	2.6/	3-			0.5959/92. 2.354/45. (0.23-3.99)
⁷⁵Ga		74.92650	2.10 m	β⁻ /3.39	3.3/	3/2-			0.2529/ 0.5746/ (0.12-2.10)
⁷⁶Ga		75.9289	29. s	β⁻ /7.0		3-			0.5629/66. 0.5455/26. (0.34-4.25)
⁷⁷Ga		76.9293	13.0 s	β⁻ /5.3	5.2/				0.469/ 0.459/
⁷⁸Ga		77.9317	5.09 s	β⁻ /8.2		3+			0.619/77. 1.187/20.
⁷⁹Ga		78.9329	2.85 s	β⁻ /7.0	4.6/				0.465/
⁸⁰Ga		79.9366	1.68 s	β⁻ /10.4	10./				0.659/
⁸¹Ga		80.9377	1.22 s	β⁻ /8.3	5.1/				0.217/
⁸²Ga		81.9432	0.599 s	β⁻ /12.6					1.348/
⁸³Ga		82.9469	0.308 s	β⁻ / ≈ 11.5					
⁸⁴Ga		83.952	≈0.085 s	β⁻ /14					
₃₂Ge		**72.61(2)**							
⁶¹Ge		60.9638	0.04 s	β⁺ /13.6					
⁶³Ge		62.9496	0.10 s	β⁻ /9.8					
⁶⁴Ge		63.9416	1.06 m	β⁺ /4.4 EC/	3.0/	0+			ann.rad./ 0.1282(2)/11. 0.4270(3)/37. 0.6671(3)/17.
⁶⁵Ge		64.9394	31. s	β⁺ /6.2 EC/	0.82/10. 1.39/19. 2.113/56. 2.237/15.				ann.rad./ 0.0620/27. 0.6497/33. 0.8091/21. (0.19-3.28)
⁶⁶Ge		65.93385	2.26 h	β⁺ /27/2.10 EC/73/		0+			ann.rad./ 0.0438/29. 0.3819/28. (0.022-1.77)
⁶⁷Ge		66.932738	19.0 m	β⁺ /96/4.225 EC/4/	1.6/ 2.3/ 3.15/	1/2-			ann.rad./ 0.1670/84. (0.25-3.73)
⁶⁸Ge		67.92810	270.8 d	EC/0.11		0+			Ga k x-ray/39.
⁶⁹Ge		68.927973	1.63 d	β⁺ /36/2.2273 EC/64/	0.70/ 1.2/	5/2-	0.735	0.02	ann.rad./ 0.574/13. 1.1068/36. (0.2-2.04)
⁷⁰Ge	21.23(4)	69.924250				0+			
⁷¹ᵐGe			20.4 ms	I.T./0.0234		9/2+			0.1749
⁷¹Ge		70.924954	11.2 d	EC/0.229		1/2-	+0.547		
⁷²Ge	27.66(3)	71.922076				0+			
⁷³Ge	7.73(1)	72.923460				9/2+	-0.879467	-0.17	
⁷⁴Ge	35.94(2)	73.921178				0+			

11-54

Elem. or Isot.	Natural Abundance (%)	Atomic Mass or Weight	Half-Life	Decay Mode/Energy (/MeV)	Particle Energy /Intensity (MeV / %)	Spin (h/2π)	Nuclear Magnetic Mom.(nm)	Elect. Quadr. Mom. (b)	γ-ray Energy /Intensity (MeV / %)
^{75m}Ge			48. s	I.T./		7/2+			0.13968(3)/39.
^{75}Ge		74.922860	1.380 h	β⁻ /1.177	1.19/	1/2-	+0.510		0.26461(5)/11.
									0.41931(5)/0.2
^{76}Ge	7.44(2)	75.921403				0+			
^{77m}Ge			53. s	I.T./20/		1/2-			0.1597(1)/11.
				β⁻ /80/2.861	2.9/				0.21551/21.
									(0.19-0.61)
^{77}Ge		76.923549	11.30 h	β⁻ /2.702	0.71/23.	7/2+			0.2110/29.
					1.38/35.				0.2155/27.
					2.19/42.				0.2644/51.
									(0.15-2.35)
^{78}Ge		77.922853	1.45 h	β⁻ /0.95	0.70/	0+			0.2773(5)/96.
									0.2939(5)/4.
^{79m}Ge			39. s	β⁻ /IT		7/2+			
^{79}Ge		78.9254	19.1 s	β⁻ /4.2	4.0/20.	1/2-			0.1096/21.
					4.3/80.				(0.10-2.59)
									0.5427(4)/15.
^{80}Ge		79.92545	29.5 s	β⁻ /2.67	2.4/	0+			0.1104(4)/6.
									0.2656(4)/25.
^{81m}Ge			≈7.6 s	β⁻ /	3.75/	1/2+			0.3362(4)/
									0.7935(4)/
^{81}Ge		80.9288	≈7.6 s	β⁻ /6.2	3.44/	9/2+			0.1976(4)/21.
									0.3362(4)/100.
^{82}Ge		81.9296	4.6 s	β⁻ /4.7		0+			1.093/
^{83}Ge		82.9345	1.9 s	β⁻ /8.9					
^{84}Ge		83.9373	0.98 s	β⁻ /7.7					
^{85}Ge		84.943	0.54 s	β⁻ /10.					
$_{33}$As		74.92160(2)							
^{65}As		64.9495	≈0.19 s	β⁺ /9.4					
^{66}As		65.94410	95.8 ms	β⁺ /9.55					
^{67}As		66.9392	42. s	β⁺ /6.0	5.0/	5/2-			0.121/
				EC/					0.123/
									0.244/
^{68}As		67.9368	2.53 m	β⁺ /8.1		3+			ann. rad./
									0.652/32.
									0.762/33.
									1.016/77.
									(0.61-3.55)
^{69}As		68.93228	15.2 m	β⁺ /98/4.01	2.95/	5/2-	1.2		ann.rad./
				EC/2/					0.0868(5)/1.5
									0.1458(3)/2.4
^{70}As		69.93093	52.6 m	β⁺ /84/6.22	1.44/	4+	+2.1061	+0.09	ann.rad./
				EC/16/2.14					1.0395(7)/82.
				/2.89					(0.17-4.4)/
^{71}As		70.927114	2.72 d	β⁺ /32/2.013		5/2-	+1.6735	-0.02	ann.rad./
				EC/68/					0.1749(2)/84.
									1.0957(2)/4.2
^{72}As		71.926753	26.0 h	β⁺ /77/4.356	0.669/5.	2-	-2.1566	-0.08	ann.rad./
					1.884/12.				0.83395(5)/80.
					2.498/62.				1.0507(1)/9.6
					3.339/19.				(0.1-4.0)
^{73}As		72.923825	80.3 d	EC/0.341		3/2-			0.013263(1)/0.1
									0.053437(1)/10.5
									Se k x-ray/90.
^{74}As		73.923829	17.78 d	β⁺ /31/2.562	0.94/26.	2-	-1.597		ann.rad./
				EC/37/	1.53/3.				0.59588(1)/60.
				β⁻ /1.353	0.71/16.				0.6084(1)/0.6
					1.35/16.				0.6348(1)/15.
^{75}As	100.	74.921597				3/2-	+1.43947	+0.31	
^{76}As		75.922394	26.3 h	β⁻ /2.962	0.54/3.	2-	-0.906		0.5591(1)/45.
					1.785/8.				0.65703(5)/6.2
					2.410/36.				1.21602(1)/3.4
					2.97/51.				(0.3-2.67)
^{77}As		76.920648	38.8 h	β⁻ /0.683	0.70/98.	3/2-			0.2391(2)/1.6
									0.2500(3)/0.4
									0.52078(1)/0.43
^{78}As		77.92183	1.512 h	β⁻ /4.21	3.00/12.	2-			0.6136(3)/54.
					3.70/17.				0.6954(3)/18.
					4.42/37.				1.3088(3)/10.
^{79}As		78.92095	9.0 m	β⁻ /2.28	1.80/95.	3/2-			0.0955(5)/16.
									0.3645(5)/1.9
^{80}As		79.92258	16. s	β⁻ /5.64	3.38/	1+			0.6662(2)/42.
									(2.5-3.0)
^{81}As		80.92213	33. s	β⁻ /3.856		3/2-			0.4676(2)/20.
									0.4911(2)/8.

Elem. or Isot.	Natural Abundance (%)	Atomic Mass or Weight	Half-Life	Decay Mode/Energy (/MeV)	Particle Energy /Intensity (MeV / %)	Spin (h/2π)	Nuclear Magnetic Mom.(nm)	Elect. Quadr. Mom. (b)	γ-ray Energy /Intensity (MeV / %)
^{82m}As			13.7 s	β^- /	3.6/	5-			0.6544(1)/72. 0.8186(4)/27. 1.7313(2)/27. 1.8954(2)/38.
^{82}As		81.9246	19. s	β^- /7.4	7.2/80.	1+			0.6544(1)/15.
^{83}As		82.9250	13.4 s	β^- /5.5					0.7345/100. 1.1131/34. 2.0767/28.
^{84m}As			0.6 s	β^-					
^{84}As		83.9291	4. s	β^-, n/7.2		1-			0.6671(2)/21. 1.4439(5)/49. (0.325-5.150)
^{85}As		84.9318	2.03 s	β^-, n/8.9		3/2-			0.667(1)/42. 1.4551(2)/100.
^{86}As		85.9362	0.95 s	β^-, n/11.4					0.704/
^{87}As		86.9396	0.49 s	β^-, n/10.					0.704/
$_{34}$Se		**78.96(3)**							
^{65}Se		64.965	0.011 s	β^+ /60/14. β^+, p	3.55/				
^{67}Se		66.9501	0.11 s	β^+ /10.2					ann.rad./ 0.352
^{68}Se		67.9419	36. s	β^+ /4.7					ann.rad./ (0.050-0.426)
^{69}Se		68.93956	27.4 s	β^+ /6.78 EC/	5.006/				ann.rad./ 0.0664(4)/27. 0.0982(4)/63.
^{70}Se		69.9335	41.1 m	β^+ /2.4		0+			ann.rad 0.04951(5)/35. 0.4262(2)/29.
^{71}Se		70.9319	4.7 m	β^+ /4.4 EC/	3.4/36.	5/2-			ann.rad 0.1472(3)/47. 0.8309(3)/13. 1.0960(3)/10.
^{72}Se		71.92711	8.5 d	EC/0.34		0+			0.0460(2)/57.
^{73m}Se			40. m	I.T./73/0.0257 β^+ /27/2.77	0.85 1.45/ 1.70/	3/2-			ann.rad. 0.0257(2)/27. 0.2538(1)/2.5
^{73}Se		72.92678	7.1 h	β^+ /65/2.74 EC/35/	0.80/ 1.32/95. 1.68/1.	9/2+			ann.rad 0.0670(1)/72. 0.3609(1)/97. (0.6-1.5)
^{74}Se	0.89(2)	73.922477				0+			
^{75}Se		74.922524	119.78 d	EC/0.864		5/2+	0.67	1.0	0.136000(1)/55. 0.264651(1)/58.
^{76}Se	9.36(11)	75.919214				0+			
^{77m}Se			17.4 s	I.T./		7/2+			0.1619(2)/52.
^{77}Se	7.63(6)	76.919915				1/2-	+0.53506		
^{78}Se	23.78(9)	77.917310				0+			
^{79m}Se			3.92 m	I.T./					0.09573(3)/9.5
^{79}Se		78.918500	6.5x10^4 y	β^- /0.151		7/2+	-1.018	+0.8	
^{80}Se	49.61(10)	79.916522				0+			
^{81m}Se			57.3 m	I.T./99/0.1031		7/2+			0.1031(3)/9.7 0.2602(2)/0.06 0.27599(1)/0.06
^{81}Se		80.917993	18.5 m	β^- /1.585	1.6/98.	1/2-			0.27594(5)/0.85 0.29008(5)/0.75 0.82827(5)/0.32
^{82}Se	8.73(6)	81.916700				0+			
^{83m}Se			1.17 m	β^- /3.96	2.88/ 3.92/	1/2-			0.35666(6)/17. 0.9879(1)/15. 1.0305(1)/21. 2.0514(2)/11. (0.19-3.1)
^{83}Se		82.919119	22.3 m	β^- /3.668	0.93/ 1.51/	9/2+			0.22516(6)/33. 0.35666(6)/69. 0.51004(8)/45. (0.21-2.42)
^{84}Se		83.91847	3.3 m	β^- /1.83	1.41/100.	0+			0.4088(5)/100.
^{85}Se		84.92225	32. s	β^- /6.18	5.9/	5/2+			0.3450(1)/22. 0.6094(1)/41. 2.0124(1)/24. 2.4433(8)/45.
^{86}Se		85.92428	15. s	β^- /5.10		5/2+			2.6619(1)/49. 0.468(1)/100. 1.4979(1)/23.
^{87}Se		86.92853	5.4 s	β^- /7.28 n/					0.5346/
^{88}Se		87.93143	1.5 s	β^-, n/6.85					
^{89}Se		88.9360	0.41s	β^-, n/9.0					

Elem. or Isot.	Natural Abundance (%)	Atomic Mass or Weight	Half-Life	Decay Mode/Energy (/MeV)	Particle Energy /Intensity (MeV / %)	Spin (h/2π)	Nuclear Magnetic Mom.(nm)	Elect. Quadr. Mom. (b)	γ-ray Energy /Intensity (MeV / %)
^{91}Se		90.945	0.27 s	β^-,n/8.					
$_{35}$**Br**		**79.904(1)**							
^{69}Br		68.9499		β^+/9.6					
^{70}Br		69.9442	0.08 s	β^+/10.0					
^{71}Br		70.9393	21. s	β^+/6.9					
^{72}Br		71.9365	1.31 m	β^+/8.7		3	≈0.55		0.4547-1.3167
^{73}Br		72.9318	3.4 m	β^+/4.7	3.7/	3/2-			ann.rad (0.0649-0.6995)
^{74m}Br			46. m	β^+/	4.5/	4-	1.8		ann.rad 0.6348 0.7285 (0.2 - 4.38)
^{74}Br		73.92989	25.4 m	β^+/6.91					ann.rad 0.6341 0.6348 (0.2-4.7)
^{75}Br		74.92578	1.62	β^+/76/3.03		3/2-	0.75(11)		ann.rad 0.28650 (0.1-1.56)
^{76m}Br			1.4 s	I.T./5.05		4+			0.104548 0.05711
^{76}Br		75.92454	16.0 h	β^+/57 /4.96	1.9/ 3.68/	1-	0.54821	0.270	ann.rad 0.55911 1.85368 (0.4-4.6)
^{77m}Br			4.3 m	I.T./0.1059		9/2+	0.97		0.1059
^{77}Br		76.921380	2.376 d	EC/99 /1.365		3/2-			ann.rad. 0.23898 0.52069 (0.08-1.2)
^{78}Br		77.921146	6.45 m	β^+/92/3.574 EC/8 /	1.2/ 2.5/	1+	0.1		ann.rad. 0.61363 (0.7-3.0)
^{79m}Br			4.86 s	I.T./0.207		9/2+	+2.106		0.2072
^{79}Br	50.69(7)	78.918338				3/2-	+2.106400	+0.331	Br k x-ray 0.03705/39.1 0.04885/0.3
^{80m}Br			4.42 h	I.T./0.04885		5-	+1.3177		
^{80}Br		79.918530	17.66 m	β^-/92 /2.004 EC/5.7/1.8706 β^+ /2.6/	1.38 β^- /7.6 1.99 β^- /82 0.85 β^+ /2.8	1+	0.5140	0.196	ann.rad. 0.6169/6.7 (0.64-1.45)
^{81}Br	49.31(7)	80.916291				3/2-	+2.270562	+0.276	
^{82m}Br			6.1 m	I.T./98/0.046 β^-/2 /3.139		2-			0.046/0.24 (0.62-2.66)
^{82}Br		81.916805	1.471 d	β^- /3.093	0.444/	5-	+1.6270	0.751	0.5544/71 0.61905/43 0.77649/84 (0.013-1.96)
^{83}Br		82.915181	2.40 h	β^- /0.972	0.395/1 0.925/99	3/2-			0.52964 (0.12-0.68)
^{84m}Br			6.0 m	β^- /4.97	2.2/100	(6-)			0.4240/100 0.8817/98 1.4637/101
^{84}Br		83.91651	31.8 m	β^- /4.65	2.70/11 3.81/20 4.63/34	2-	2.		0.8816/41 1.8976/13 (0.23-4.12)
^{85}Br		84.91561	2.87 m	β^- /2.87	2.57	3/2-			0.80241/2.56 0.92463/1.6 (0.09-2.4)
^{86}Br		85.91880	55.5 s	β^- /7.63	3.3 7.4	(2-)			1.56460/64 2.75106/21 (0.5-6.8)
^{87}Br		86.92072	55.6 s	β^- /6.85 n/	6.1/	3/2-			1.41983 1.4762 (0.2-6.1)
^{88}Br		87.92407	16.3 s	β^- /8.96 n/		1-			0.7649 0.7753 0.8021 (0.1-6.99)
^{89}Br		88.92640	4.35 s	β^- /8.16 n/		3/2-			0.7753 1.0978
^{90}Br		89.9306	1.91 s	β^- /10.4 n/	8.3/ 9.8/	2-			0.6555 0.7071 1.3626
^{91}Br		90.9339	0.54 s	β^- /90 /9.80 β^- n/10 /					0.263 0.803

Elem. or Isot.	Natural Abundance (%)	Atomic Mass or Weight	Half-Life	Decay Mode/Energy (/MeV)	Particle Energy /Intensity (MeV / %)	Spin (h/2π)	Nuclear Magnetic Mom.(nm)	Elect. Quadr. Mom. (b)	γ-ray Energy /Intensity (MeV / %)
92Br		91.9392	0.31 s	β⁻ /12.20 β⁻ n/					0.740
93Br		92.9431	0.10 s	β⁻ n/11.1					
94Br			0.07 s	β⁻ n/					
36Kr		83.80(1)							
71Kr		70.9505	0.10 s	β⁺ ,EC/10.5					
72Kr		71.9419	17. s	β⁺ /5.0 EC/		0+			ann.rad. 0.3100/29 0.4150/36 (0.12-0.58)
73Kr		72.9389	27. s	β⁺ /6.7 EC/ β⁺ ,p/					ann.rad. 0.1781/66 (0.06-0.86)
74Kr		73.9333	11.5 m	β⁺ /3.1 EC/		0+			ann.rad. 0.08970/31 0.2030/20 (0.010-1.06)
75Kr		74.93104	4.3 m	β⁺ /4.90 EC/	3.2/				ann.rad. 0.1325/68 0.1547/21 (0.02-1.7)
76Kr		75.92595	14.8 h	EC/1.31		0+			Br k x-ray 0.270/21 0.3158/39 (0.03-1.07)
77Kr		76.92467	1.24 h	β⁺ /80 /3.06 EC/20 /	1.55/ 1.70/ 1.87/	5/2+			ann.rad. 0.1297/80 0.1465/38 (0.02-2.3)
78Kr	0.35(2)	77.92039				0+			
79mKr			53. s	I.T./0.1299		7/2+			Kr x-ray
79Kr		78.920083	1.455 d	β⁺ /7 /1.626 EC/93 /		1/2-			ann.rad. 0.2613/13 0.39756/19 0.6061/8 (0.04-1.3)
80Kr	2.25(2)	79.916379				0+			
81mKr			13.1 s	I.T./0.1904		1/2-			0.1904
81Kr		80.916593	2.1x10⁵ y	EC/0.2807		7/2+			Br k x-ray 0.2760
82Kr	11.6(1)	81.913485				0+			
83mKr			1.86 h	I.T./0.0416		1/2-			Kr k x-ray 0.00940 0.03216
83Kr	11.5(1)	82.914137				9/2+	-0.970699	+0.253	
84Kr	57.0(3)	83.911508				0+			
85mKr			4.48 h	β⁻ /79 / I.T./21 /0.305	0.83/79	1/2-			0.30487 0.15118
85Kr		84.912530	10.73 y	β⁻ /0.687	0.15/0.4	9/2+	1.005	+0.43	0.51399
86Kr	17.3(2)	85.910615				0+			
87Kr		86.913359	1.27 h	β⁻ /3.887	1.33/8 3.49/43 3.89/30	5/2+	-1.018		0.40258/49.6 2.5548/9.2 (0.13-3.31)
88Kr		87.91445	2.84 h	β⁻ /2.91		0+			0.19632/26. 2.392/34.6 (0.03-2.8)
89Kr		88.91764	3.15 m	β⁻ /4.99	3.8/ 4.6/ 4.9/	5/2+			0.19746 0.2209/19.9 0.5858/16.4 1.4728/6.8 (0.2-4.7)
90Kr		89.91953	32.3 s	β⁻ /4.39	2.6/77 2.8/6	0+			0.12182/32.9 0.5395/28.6 1.1187/36.2 (0.1 - 4.2)
91Kr		90.9234	8.6 s	β⁻ /6.4	4.33/ 4.59/	5/2+			0.10878/43.5 0.50658/19. (0.2-4.4)
92Kr		91.92611	1.84 s	β⁻ /5.99 n/					0.1424/66. (0.14 - 3.7)
93Kr		92.9312	1.29 s	β⁻ /8.6 n/	7.1/				0.1820 0.2534/42. 0.32309/24.6 (0.057-4.03)

Elem. or Isot.	Natural Abundance (%)	Atomic Mass or Weight	Half-Life	Decay Mode/Energy (/MeV)	Particle Energy /Intensity (MeV / %)	Spin (h/2π)	Nuclear Magnetic Mom.(nm)	Elect. Quadr. Mom. (b)	γ-ray Energy /Intensity (MeV / %)
⁹⁴Kr		93.9343	0.21 s	β⁻ /7.3					0.2196/67 0.6293/100. (0.098-0.985)
⁹⁵Kr		94.9397	0.78 s	β⁻ /9.7					
⁹⁷Kr			< 0.1 s	β⁻					
₃₇Rb		**85.4678(3)**							
⁷⁴Rb		73.9445	65. ms	β⁺ /10.4					
⁷⁵Rb		74.93857	19. s	β⁺ /7.02	2.31/				ann. rad. 0.179
⁷⁶Rb		75.93508	39. s	β⁺ /8.50	4.7/	1-	-0.372623	+0.4	ann. rad. 0.4240/92. (0.064-1.68)
⁷⁷Rb		76.93041	3.8 m	β⁺ /5.34	3.86/	3/2-	+0.654468	+0.70	ann. rad. 0.0665/59 (0.04 - 2.82)
⁷⁸ᵐRb			5.7 m	I.T./0.1034 β⁺ / EC/	3.4	4-			ann. rad. 0.4553/81. (0.103-4.01)
⁷⁸Rb		77.92814	17.7 m	β⁺ /7.22 EC/		0+			ann. rad. 0.4553/63. (0.42-5.57)
⁷⁹Rb		78.92400	23. m	β⁺ /84 /3.65 EC/16 /		5/2+	+0.3358	-0.10	ann. rad. 0.68812/23. (0.017-3.02)
⁸⁰Rb		79.92252	34. s	β⁺ /5.72	4.1/22 4.7/74	1+	-0.0836	+0.35	ann. rad. 0.6167/25.
⁸¹ᵐRb			30.5 m	I.T./0.85 β⁺,EC/	1.4	9/2+			ann.rad. (0.085-1.9)
⁸¹Rb		80.91900	4.57 h	β⁺/27 /2.24 EC/73	1.05/	3/2-	+2.060	+0.40	ann.rad./ 0.19030/64. (0.05 - 1.9)
⁸²ᵐRb			6.47 h	β⁺/26 / EC/74 /	0.80/	5-	+1.5		ann.rad./ 0.5544/63. 0.7765/85. (0.092 - 2.3)
⁸²Rb		81.91821	1.258 m	β⁺/96 /4.40 EC/4 /	3.3/	1+	+0.554508	+0.19	ann.rad./ 0.7665/13. (0.47 - 3.96)
⁸³Rb		82.91511	86.2 d	EC/0.91		5/2-	+1.425	+0.20	Kr x-ray 0.5205/46. (0.03-0.80)
⁸⁴ᵐRb			20.3 m	I.T./0.216		6-			0.2163/34. 0.2482/63. 0.4645/32.
⁸⁴Rb		83.914387	32.9 d	β⁺/22 /2.681 EC/75 / β⁻/3 /0.894	0.780/11 1.658/11 0.893/	2-	-1.32412	-0.015	ann.rad./ 0.8817/68. (1.02-1.9)
⁸⁵Rb	72.165(20)	84.911792				5/2-	+1.353	+0.23	
⁸⁶ᵐRb			1.018 m	I.T./0.5560		6-			0.556/98.
⁸⁶Rb		85.911170	18.65 d	β⁻/1.775	1.774/8.8	2-	-1.6920		1.0768/8.8
⁸⁷Rb	27.835(20)	86.909186	4.88x10¹⁰y	β⁻/0.283	0.273/100	3/2-	+2.7512		
⁸⁸Rb		87.911323	17.7 m	β⁻/5.316	5.31	2-	0.508		0.8980/14. 1.8360/21. (0.34-4.85)
⁸⁹Rb		88.91229	15.4 m	β⁻/4.50	1.26/38 1.9/5 2.2/34 4.49/18	3/2-	+2.304	+0.14	1.032/58. 1.248/42. 2.1960/13 (0.12-4.09)
⁹⁰ᵐRb			4.3 m	β⁻/4.50	1.7/ 6.5/	4-			0.1069(IT) 0.8317/94 (0.20-5.00)
⁹⁰Rb		89.91481	2.6 m	β⁻/6.59	6.6	1-			0.8317/28. (0.31-5.60)
⁹¹Rb		90.91649	58.0 s	β⁻/5.861	5.9	3/2-	+2.182	+0.15	0.0936/34. (0.35-4.70)
⁹²Rb		91.91968	4.48 s	β⁻/8.11	8.1/94	1-			0.8148/8. (0.1-6.1)
⁹³Rb		92.92195	5.85 s	β⁻/7.46 n/1	7.4/	5/2	+1.410	+0.18	0.2134/4.8 0.4326/12.5 0.9861/4.9 (0.16-5.41)
⁹⁴Rb		93.92643	2.71 s	β⁻/10.31 n/10	9.5/	3	+1.498	+0.16	0.8369/87. 1.5775/32. (0.12-6.35)
⁹⁵Rb		94.92929	0.377 s	β⁻/9.30 n/8	8.6/	5/2	+1.334	+0.21	0.352/65. 0.680/22. (0.20-2.27)

Elem. or Isot.	Natural Abundance (%)	Atomic Mass or Weight	Half-Life	Decay Mode/Energy (/MeV)	Particle Energy /Intensity (MeV / %)	Spin (h/2π)	Nuclear Magnetic Mom.(nm)	Elect. Quadr. Mom. (b)	γ-ray Energy /Intensity (MeV / %)
⁹⁶Rb		95.93427	0.199 s	β⁻/11.76 n/13/	10.8/	2+	+1.466	+0.25	0.815/76. (0.20-5.42)
⁹⁷Rb		96.93733	0.169 s	β⁻/10.42 n/27/	10.0	3/2	+1.841	+0.58	0.167/100. 0.585/79. 0.599/56. 1.258/52. (0.14-2.08)
⁹⁸Rb		97.94174	0.107 s	β⁻/12.34 n/13	0.144/				(0.07-3.68)
⁹⁹Rb		98.9453	59. ms	β⁻/11.3					
¹⁰⁰Rb		99.9499	53. ms	β⁻ /13.5					
¹⁰¹Rb		100.9532	0.03 s	β⁻ /11.8					
¹⁰²Rb			0.09 s	β⁻					
₃₈Sr		87.62(1)							
⁷⁶Sr		75.9416	8.9 s	β⁺ /6.1					
⁷⁷Sr		76.9378	9.0 s	β⁺ /6.9	5.6		-0.35	1.4	0.147 (0.047-0.793)
⁷⁸Sr		77.93218	2.7 m	β⁺ /3.76					
⁷⁹Sr		78.92971	2.1 m	β⁺ /5.32	4.1	3/2-	-0.474	+0.74	ann.rad./ 0.039/28. 0.105/22. (0.135-0.612)
⁸⁰Sr		79.92453	1.77 h	β⁺ /1.87		0+			ann.rad./ 0.174/10. 0.589/39. (0.24-0.55)
⁸¹Sr		80.92322	22.3 m	β⁺ /87 /3.93 EC/13 /	2.43/ 2.68/	1/2-	+0.544		ann.rad./ 0.148/31. 0.1534/35 (0.06-1.7)
⁸²Sr		81.91840	25.36 d	EC/0.18					Rb x-ray
⁸³ᵐSr			5.0 s	I.T./0.2591		1/2-			0.2591/87.5
⁸³Sr		82.91756	1.350 d	β⁺/24 /2.28 EC/76/	0.465/ 0.803/ 1.227/	7/2+	-0.898	+0.8	ann.rad./ 0.3816/12. 0.3816 0.7627/30. (0.094-2.15)
⁸⁴Sr	0.56(1)	83.913426				0+			
⁸⁵ᵐSr			1.127 h	I.T./87 /0.2387 EC/13		1/2-			0.2318/84. (0.15-0.24)
⁸⁵Sr		84.912936	64.85 d	EC/1.065		9/2+	-1.001	+0.32	0.51399/99.3
⁸⁶Sr	9.86(1)	85.909265				0+			
⁸⁷ᵐSr			2.81 h	I.T./0.3884		1/2-			0.3884(IT)
⁸⁷Sr	7.00(1)	86.908882				9/2+	-1.09360	+0.34	
⁸⁸Sr	82.58(1)	87.905617				0+			
⁸⁹Sr		88.907455	50.52 d	β⁻/1.497	1.492/100	5/2+	-1.149	-0.3	0.9092
⁹⁰Sr		89.907738	29.1 y	β⁻/0.546	0.546/100	0+			
⁹¹Sr		90.91020	9.5 h	β⁻/2.70	0.61/7 1.09/33 1.36/29 2.66/26	5/2+	-0.887	+0.044	0.5556/61. 0.7498/24. 1.0243/33. (0.12-2.4)
⁹²Sr		91.91098	2.71 h	β⁻/1.91	0.55/96 1.5/3	0+			1.3831/90. (0.24-1.1)
⁹³Sr		92.91394	7.4 m	β⁻/4.08	2.2/10 2.6/25 3.2/65	5/2+	-0.794	+0.27	0.5903/ 0.7104 0.87573 0.8883/ (0.17-3.97)
⁹⁴Sr		93.91537	1.25 m	β⁻/3.511	2.1/ 3.3/	0+			0.6219 0.7043 0.7241 0.8064 1.4283
⁹⁵Sr		94.91931	25.1 s	β⁻/6.08	6.1/50	1/2+	-0.5379		0.6859 0.8269 2.7173 2.9332
⁹⁶Sr		95.92165	1.06 s	β⁻/5.37	4.2/	0+			0.1222 0.5305 0.8094 0.9318
⁹⁷Sr		96.92615	0.42 s	β⁻/7.47	5.3	(1/2+)	-0.500		0.2164 0.3071 0.6522 0.9538 1.2580 1.9050

Elem. or Isot.	Natural Abundance (%)	Atomic Mass or Weight	Half-Life	Decay Mode/Energy (/MeV)	Particle Energy /Intensity (MeV / %)	Spin (h/2π)	Nuclear Magnetic Mom.(nm)	Elect. Quadr. Mom. (b)	γ-ray Energy /Intensity (MeV / %)
^{98}Sr		97.92845	0.65 s	β^-/5.83	5.1				0.0365 0.1190 0.4286 0.4447 0.5636
^{99}Sr		98.9333	0.27 s	β^-/8.0					
^{100}Sr		99.9354	0.201 s	β^-/7.1					
^{101}Sr		100.9405	0.115 s	β^-/9.5					
^{102}Sr		101.9430	68. ms	β^-/8.8					
$_{39}$Y		**88.90585(2)**							
^{78}Y		77.9435		β^+/10.5					
^{79}Y		78.9374	15. s	β^+/7.1					(0.152-1.106)
^{80}Y		79.9320	36. s	β^+/7.0	5.5 5.0/	(4)			ann.rad./ 0.3858 0.5951 1.1852
^{81}Y		80.9291	1.21 m	β^+/5.5	3.7/ 4.2/				ann.rad./ 0.428 0.469
^{82}Y		81.9268	9.5 s	β^+/7.8	6.3/	1+			ann.rad./ 0.5736 0.6017 0.7375
^{83m}Y			2.85 m	β^+/95/4.6 EC/5 /	2.9	1/2-			ann.rad./ 0.2591 0.4218 0.4945
^{83}Y		82.92235	7.1 m	β^+/4.47 EC/	3.3	9/2+			ann.rad./ 0.0355 0.4899 0.8821 (0.03 - 3.4)
^{84m}Y			4.6 s	β^+/ EC/		1+			ann.rad./ 0.7930
^{84}Y		83.9203	40. m	β^+/6.4 EC/	1.64/47 2.24/25 2.64/21 3.15/7	5-			ann.rad./ 0.4628 0.6606 0.7931 0 9744 1.0398 (0.2 - 3.3)
^{85m}Y			4.9 h	β^+/70 / EC/30 /		9/2+			ann.rad./ 0.2317 0.5356 0.7673 2.1238 (0.1 - 3.1)
^{85}Y		84.91643	2.6 h	β^+/55 /3.26 EC/45 /	1.54/	1/2-			ann.rad./ 0.2317 0.5045 0.9140 (0.07 - 1.4)
^{86m}Y			48. m	I.T./99 / β^+/ EC/		8+			ann.rad./ 0.0102(IT) 0.2080 (0.09 - 1.1)
^{86}Y		85.91489	14.74 h	β^+/5.24 EC/		4-	<0.6		ann.rad./ 0.3070 0.6277 1.0766 1.1531 1.9207 (0.1 - 3.8)
^{87m}Y			13. h	I.T./98 / β^+/0.7 / EC/	1.15/0.7	9/2+			0.3807
^{87}Y		86.910880	3.35 d	EC/99+/1.862	0.78/	1/2-			0.3880 0.4870
^{88}Y		87.909506	106.6 d	EC/99+ /3.623 β^+/0.2 /	0.76/	4-			ann.rad./ 0.89802 1.83601 2.73404 3.2190
^{89m}Y			15.7 s	I.T./0.909		9/2+			0.9092(IT)
^{89}Y	100.	88.905849				1/2-	-0.13742		

Elem. or Isot.	Natural Abundance (%)	Atomic Mass or Weight	Half-Life	Decay Mode/Energy (/MeV)	Particle Energy /Intensity (MeV / %)	Spin (h/2π)	Nuclear Magnetic Mom.(nm)	Elect. Quadr. Mom. (b)	γ-ray Energy /Intensity (MeV / %)
^{90m}Y			3.24 h	I.T./99+ /0.68204 β⁻/0.002/		7+			0.2025 0.4794 0.6820
^{90}Y		89.907152	2.67 d	β⁻/2.282	2.28/	2-	-1.630	-0.155	
^{91m}Y			49.7 m	I.T./0.555		9/2+	6.0		0.5556(IT)
^{91}Y		90.907301	58.5 d	β⁻/1.544	1.545/	1/2-	0.1641		1.208
^{92}Y		91.90893	3.54 h	β⁻/3.63	3.64/	2-			0.4485 0.5611 0.9345 1.4054 (0.4 - 3.3)
^{93m}Y			0.82 s	I.T./0.759		9/2+			0.1686(IT) 0.5902
^{93}Y		92.90956	10.2 h	β⁻/2.87	2.88/90	1/2-			0.2669 0.9471 1.9178
^{94}Y		93.91160	18.7 m	β⁻/4.919	4.92/	2-			0.3816 0.9188 1.1389 (0.3 - 4.1)
^{95}Y		94.91279	10.3 m	β⁻/4.42		1/2-			0.4324 0.9542 2.1760 3.5770
^{96m}Y			9.6 s	β⁻/		(3+)			0.1467 0.6174 0.9150 1.1071 1.7507 1.594
^{96}Y		95.91588	6.2 s	β⁻ /7.09	7.12/	0-			0.1614 0.9700 1.1030
^{97m}Y			1.21 s	β⁻ /7.4	4.8/ 6.0/	9/2+			0.2969 1.9960 3.2876 3.4013
^{97}Y		96.91813	3.76 s	β⁻ /6.69	6.7	1/2-			0.2415 0.6205 0.6473 1.2228 1.8016
^{98m}Y			2.1 s	β⁻ /9.8	5.5/	(4-)			0.2131 1.2228 1.5907 2.9413 4.4501
^{98}Y		97.92224	0.59 s	β⁻ /8.83	8.7/	1+			0.1218/43.8 0.5362 0.7242 1.0130
^{99}Y		98.92463	1.47 s	β⁻ /7.57		1/2-			
^{100m}Y			0.94 s	β⁻ /		3+			
^{100}Y		99.9278	0.73 s	β⁻ /9.3		1+			
^{101}Y		100.9303	0.43 s	β⁻ /8.6		(5/2)			
^{102}Y		101.9336	0.36 s	β⁻ /9.9					
$_{40}$**Zr**		**91.224(2)**							
^{80}Zr		79.9406		β⁺ /8.0					0.290 0.538
^{81}Zr		80.9368	15. s	β⁺ /7.2	6.1				
^{82}Zr		81.9311	32. s	β⁺ /4.0	3.				ann.rad./
^{83m}Zr			7. s	β⁺ /7.0		(7/2+)			ann.rad./
^{83}Zr		82.9287	44. s	β⁺ /5.9 EC	4.8	(1/2-)			ann.rad./ 0.0556 0.1050 0.2560 0.474 1.525
^{84}Zr		83.9233	26. m	β⁺ /2.7 EC/		0+			ann.rad./ 0.0449 0.1125 0.3729 0.667
^{85m}Zr			10.9 s	I.T./0.2922 β⁺ ,EC/		1/2-			ann.rad./ 0.2922(IT) 0.4165

Elem. or Isot.	Natural Abundance (%)	Atomic Mass or Weight	Half-Life	Decay Mode/Energy (/MeV)	Particle Energy /Intensity (MeV / %)	Spin (h/2π)	Nuclear Magnetic Mom.(nm)	Elect. Quadr. Mom. (b)	γ-ray Energy /Intensity (MeV / %)
90mNb			18.8 s	I.T./0.1246		4-			0.002 0.1225
90Nb		89.911263	14.6 h	β+ /53 /6.111 EC/47 /	0.86/5 1.5/92	8+	4.961		ann.rad./ 0.1412/ 1.1292 2.1862 2.3189 (0.1 - 3.3)
91mNb			62. d	I.T./97 / EC/3 /		1/2-			0.1045(IT) 1.2050
91Nb		90.906989	7x10²y	EC/1.253		9/2+			Mo k x-ray
92mNb			10.13 d	EC/99+ /		2+	6.114		0.9126 0.9345 1.8475
92Nb		91.907192	3.7x10⁷y	EC/2.006		7+			0.5611 0.9345
93mNb			16.1 y	I.T./0.0304		1/2-			Nb x-ray 0.0304
93Nb	100.	92.906376				9/2+	+6.1705	-0.32	
94mNb			6.26 m	I.T./99+ /2.086 β⁻ /0.5 /		3+			Nb k x-ray 0.0409 0.87109
94Nb		93.907282	2.4x10⁴y	β⁻ /2.045	0.47/	6+			0.70263 0.87109
95mNb			3.61 d	I.T./97.5 /0.2357 β⁻ /2.5 /		1/2-			0.2040 0.2356
95Nb		94.906834	34.97 d	β⁻ /0.926	0.160/	9/2+	6.141		0.76578
96Nb		95.908099	23.4 h	β⁻ /3.187	0.5/10 0.75/90	6+	4.976		0.7782 0.2191-1.498
97mNb			58.1 s	I.T./0.7434	0.734/98	1/2-			0.7434
97Nb		96.908096	1.23 h	β⁻ /1.934	1.27/98	9/2+	6.15		0.4809 0.6579 0.7874
98mNb			51. m	β⁻ /4.67		5+			0.1726-1.89
98Nb		97.91033	2.9 s	β⁻ /4.59	4.6/	1+			0.6451 0.7874 1.0243
99mNb			2.6 m	β⁻ /	3.2/	1/2-			0.0978/100 (0.138-3.010)
99Nb		98.91162	15.0 s	β⁻ /3.64	3.5/100	9/2+			0.0977 0.1378/3.1
100mNb			3.0 s	β⁻ /6.74	5.8				Nb k x-ray 0.159 0.6364 1.0637
100Nb		99.91418	1.5 s	β⁻ /6.25	6.2/ 5.3/				0.5354 0.6001-1.566
101Nb		100.91525	7.1 s	β⁻ /4.57	4.3/				0.1105-0.810
102mNb			4.3 s	β⁻ /					
102Nb		101.91804	1.3 s	β⁻ /7.21	7.2/				0.2960-2.184
103Nb		102.91914	1.5 s	β⁻ /5.53	5.3/	5/2+			
104mNb			0.9 s	β⁻ /					
104Nb		103.9225	4.8 s	β⁻ /8.1					
105Nb		104.9239	3.0 s	β⁻ /6.5					
106Nb		105.9281	1.0 s	β⁻ /9.3					
107Nb			0.3 s	β⁻ /7.9					
₄₂Mo		95.94(1)							
84Mo		83.9401		β+ /6.					
85Mo		84.9366		β+/8.1					
86Mo		85.9302	20. s	β+ /4.8					
87Mo		86.9273	14. s	EC, β+/6.5					
88Mo		87.92195	8.0 m	β+ /3.4 EC		0+			ann.rad./ 0.0800 0.1399 0.1707
89mMo			0.19 s	I.T./0.118		1/2-			0.118(IT) 0.268
89Mo		88.91948	2.2 m	β+ /5.58 EC/		9/2+			ann.rad./ 0.659 0.803 1.155 1.272
90Mo		89.91394	5.7 h	β+ /25 /2.489 EC/75 /	1.085/	0+			ann.rad./ 0.04274 0.12237 0.25734

Elem. or Isot.	Natural Abundance (%)	Atomic Mass or Weight	Half-Life	Decay Mode/Energy (/MeV)	Particle Energy /Intensity (MeV / %)	Spin (h/2π)	Nuclear Magnetic Mom.(nm)	Elect. Quadr. Mom. (b)	γ-ray Energy /Intensity (MeV / %)
^{85}Zr		84.9215	7.9 m	β^+ /4.7 EC/	3.1	7/2+			ann.rad./ 0.2663 0.4163 0.4543
^{86}Zr		85.91647	16.5 h	EC/1.47		0+			0.0280 0.243 0.612
^{87m}Zr			14.0 s	I.T./0.3362		1/2-			0.1352(IT) 0.2010
^{87}Zr		86.91482	1.73 h	β^+ /3.67 EC/	2.26	9/2+			ann.rad./ 0.3811 1.228
^{88}Zr		87.91023	83.4 d	EC/0.67		0+			0.3929
^{89m}Zr			4.18 m	I.T./94 /0.5877 β^+ /1.5 / EC/4.7 /		1/2-			ann.rad./ 0.5877(IT) 1.507
^{89}Zr		88.908889	3.27 d	β^+ /23 /2.832 EC/77 /	0.9/	9/2+			ann.rad./ 0.9092
^{90m}Zr			0.809 s	I.T./		5-			0.1326 2.1862 2.3189(IT)
^{90}Zr	51.45(3)	89.904702				0+			
^{91}Zr	11.22(4)	90.905643				5/2+	-1.30362	-0.21	
^{92}Zr	17.15(2)	91.905039				0+			
^{93}Zr		92.906474	1.5×10^6y	β^- /0.091		5/2+			0.0304
^{94}Zr	17.38(4)	93.906314				0+			
^{95}Zr		94.908041	64.02 d	β^- /1.125	0.366/55 0.400/44	5/2+	1.10	+0.29	0.7242 0.7567
^{96}Zr	2.80(2)	95.908275				0+			
^{97}Zr		96.910950	16.8 h	β^- /2.658	1.91/	1/2-			0.7434
^{98}Zr		97.91276	30.7 s	β^- /2.26	2.2/100	0+			
^{99}Zr		98.91651	2.2 s	β^- /4.56	3.9/ 3.5/	1/2+			0.4692/55.2 0.5459 0.5940
^{100}Zr		99.91776	7.1 s	β^- /3.34		0+			0.4006 0.5043
^{101}Zr		100.92114	2.1 s	β^- /5.49	6.2/	3/2-			0.1194 0.2057 0.2089
^{102}Zr		101.92298	2.9 s	β^- /4.61					
^{103}Zr		102.9266	1.3 s	β^- /7.0					
^{104}Zr		103.9288	1.2 s	β^- /5.9					
^{105}Zr		104.9331	$\approx 1.$ s	β^- /8.5					
$_{41}$Nb		92.90638(2)							
^{82}Nb		81.9431		β^+ /11.					
^{83}Nb		82.9367		β^+ /7.5					
^{84}Nb		83.9336	12. s	β^+ ,EC/9.6		(3+)			
^{85}Nb		84.9279	2.3 m	β^+ /6.0					
^{86m}Nb			56. s	β^+					
^{86}Nb		85.9250	1.46 m	β^+ /8.0					ann.rad./ 0.751 1.003
^{87m}Nb			3.7 m	β^+ / EC/		1/2-			ann.rad./ 0.1352 0.2010
^{87}Nb		86.92036	2.6 m	β^+ 5.2/ EC/		(9/2+)			ann.rad./ 0.2010 0.4706 0.6165 1.0665 1.8842
^{88m}Nb			7.7 m	β^+ / EC/		4-			ann.rad./ 0.2625 0.3996 1.0569 1.0825
^{88}Nb		87.9183	14.3 m	β^+ /7.6 EC/	3.2/	8+			ann.rad./ 1.0570 1.0828 (0.07 - 2.5)
^{89m}Nb			2.0 h	β^+ / EC/	3.3/	9/2+			0.5880/10(D) (0.17 - 4.0)
^{89}Nb		88.91349	1.10 h	β^+ /74 /4.29 EC/26 /	2.8/	1/2-			ann.rad./ 0.5074 0.5880 0.7696 1.2775

Elem. or Isot.	Natural Abundance (%)	Atomic Mass or Weight	Half-Life	Decay Mode/Energy (/MeV)	Particle Energy /Intensity (MeV / %)	Spin (h/2π)	Nuclear Magnetic Mom.(nm)	Elect. Quadr. Mom. (b)	γ-ray Energy /Intensity (MeV / %)
^{91m}Mo			1.08 m	I.T./50 /0.653 $β^+$,EC/50 /	2.5/ 2.8/ 4.0/	1/2-			ann.rad./ 0.6529 1.2081 1.5080 2.2407
^{91}Mo		90.91175	15.5 m	$β^+$ /94 /4.43 EC/6 /	3.44/94	9/2-			ann.rad./ 1.6373 2.6321 3.0286 (0.1 - 4.2)
^{92}Mo	14.84(4)	91.906810				0+			
^{93m}Mo			6.9 h	I.T./99+ /2.425	21/2+	+9.21			0.26306(IT) 0.68461 1.47711
^{93}Mo		92.906811	3.5x10³y	EC/0.405		5/2+			0.0304
^{94}Mo	9.25(3)	93.905087				0+			
^{95}Mo	15.92(5)	94.905841				5/2+	-0.9142	-0.02	
^{96}Mo	16.68(5)	95.904678				0+			
^{97}Mo	9.55(3)	96.906020				5/2+	-0.9335	+0.26	
^{98}Mo	24.13(7)	97.905407				0+			
^{99}Mo		98.907711	2.7476 d	$β^-$ /1.357	0.45/14 0.84/2 1.21/84	1/2+	0.375		0.144048 0.18109 0.36644 0.73947
^{100}Mo	9.63(3)	99.90748				0+			
^{101}Mo		100.91035	14.6 m	$β^-$ /2.82	2.23/ 0.7/	1/2+			0.0063 0.19193 0.5909 (0.0809-2.405)
^{102}Mo		101.91030	11.3 m	$β^-$ /1.01	1.2/	0+			0.1493/89. 0.2116/100. 0.2243/32.
^{103}Mo		102.91320	1.13 m	$β^-$ /3.8		3/2+			0.1028(2)/ 0.1440(2) 0.2511(2)
^{104}Mo		103.91376	1.00 m	$β^-$ /2.16		0+			0.0686(1)/100. 0.4239(4)/21.
^{105}Mo		104.9170	36. s	$β^-$ /4.95		3/2+			0.0642/ 0.0856/ 0.2495/
^{106}Mo		105.91814	8.4 s	$β^-$ /3.52		0+			0.1894(2)/22. 0.3644(2)/6. 0.3723(2)/12.
^{107}Mo		106.9217	3.5 s	$β^-$ /6.2					
^{108}Mo		107.9240	1.5 s	$β^-$ /5.1					
^{109}Mo		108.9277	0.5 s	$β^-$ /7.2					
^{110}Mo		109.9295	0.30 s	$β^-$ /5.7					Tc k x-ray 0.142 (0.121-0.599)
$_{43}$Tc									
^{86}Tc		85.9430		$β^+$ /11.9					
^{87}Tc		86.9365		$β^+$ /8.6					
^{88}Tc		87.9328		$β^+$ /10.1					
^{89m}Tc			13. s						
^{89}Tc		88.9275	13. s	$β^+$ /7.5					
^{90m}Tc			49.2 s	$β^+$	5.3/	6+			ann.rad./ 0.9479/ 1.0542/
^{90}Tc		89.9235	8.3 s	$β^+$ /8.9	7.0/15 7.9/95.	1+			ann.rad./ 0.9479/
^{91m}Tc			3.3 m	$β^+$ EC		1/2+			ann.rad./170. 0.3375(2)/1.1 0.8110(5)/5. 1.6052(1)/7.8 1.6339(1)/9.1 1.9023(1)/6. 2.4509(1)/13.5
^{91}Tc		90.9184	3.14 m	$β^+$ /6.2	5.2	9/2+			ann.rad./200.
^{92}Tc		91.91526	4.4 m	$β^+$ /7.87 EC	4.1	8+			ann.rad./200. 0.0850/ 0.1475 0.3293 0.7731 1.5096
^{93m}Tc			43. m	I.T./13 EC/20		1/2-			0.3924(IT) 0.9437 2.6445

Elem. or Isot.	Natural Abundance (%)	Atomic Mass or Weight	Half-Life	Decay Mode/Energy (/MeV)	Particle Energy /Intensity (MeV / %)	Spin (h/2π)	Nuclear Magnetic Mom.(nm)	Elect. Quadr. Mom. (b)	γ-ray Energy /Intensity (MeV / %)
^{93}Tc		92.910248	2.73 h	β$^+$ /13/3.201 EC/87/	0.81	9/2+	6.26		ann.rad./ 1.3629 1.4771 1.5203 (0.1 - 3.0)
^{94m}Tc			52. m	β$^+$ /72/4.33 EC/28/		2+			ann.rad./ 0.8710 1.8686
^{94}Tc		93.909655	4.88 h	β$^+$ /11/4.256 EC/89/		7+	5.08		ann.rad./ 0.4491 0.7026 0.8496 0.8710
^{95m}Tc			61. d	I.T./4/ β$^+$ /0.3 EC/96	0.5/ 0.7/	1/2-			ann.rad./ 0.0389(IT) 0.2041 0.5821 0.5821 0.8351
^{95}Tc		94.90766	20.0 h	EC/100/1.691		9/2+	5.89		0.7657 1.0738
^{96m}Tc			52. m	I.T./90/ β$^+$,EC/2/		4+			0.0342(IT) 0.7782 1.2002
^{96}Tc		95.90787	4.3 d	EC/2.973		7+	+5.04		Mo k x-ray 0.7782 0.8125 0.8498 1.12168
^{97m}Tc			90. d	I.T./0.0965		1/2-			Tc k x-ray 0.0965
^{97}Tc		96.906364	2.6x10^6 y	EC/100/0.320		9/2+			Mo k x-ray
^{98}Tc		97.907215	4.2x10^6 y	β$^-$ /1.80	0.40/100	6+			0.65241 0.74535
^{99m}Tc			6.01 h	I.T./100/0.142		1/2-			Tc k x-ray 0.14049 0.14261
^{99}Tc		98.906254	2.13x10^5 y	β$^-$ /0.294	0.293/100	9/2+	+5.6847	-0.129	
^{100}Tc		99.907657	15.8 s	β$^-$ /3.202 EC /1.8(10)$^{-3}$/0.17	2.2/ 2.9/ 3.3	1+			0.5396 0.5908 1.5122 (0.3 - 2.6)
^{101}Tc		100.90731	14.2 m	β$^-$ /1.61	1.32/	9/2+			0.1272 0.1841 0.3068 0.5451 (0.073-0.969)
^{102m}Tc			4.4 m	I.T./2/4.8 β$^-$ /98/	1.8/				0.4184 0.4752 0.6281 0.6302 1.0464 1.1033 1.6163 2.2447
^{102}Tc		101.90921	5.3 s	β$^-$ /4.53	3.4/ 4.2 2.2/	1+			0.4686 0.4751 1.1055
^{103}Tc		102.90918	54. s	β$^-$ /2.66	2.0/ 2.2/	5/2+			0.1361 0.1743 0.2104 0.3464 0.5629 (0.13 - 1.0)
^{104}Tc		103.91144	18.2 m	β$^-$ /5.60	5.3/	(3+)			0.3483 0.3580 0.5305 0.5351 0.8844 0.8931 1.6768 (0.3 - 3.7)
^{105}Tc		104.91166	7.6 m	β$^-$ /3.6	3.4/	5/2+			0.1079 0.1432 0.3215

Elem. or Isot.	Natural Abundance (%)	Atomic Mass or Weight	Half-Life	Decay Mode/Energy (/MeV)	Particle Energy /Intensity (MeV / %)	Spin (h/2π)	Nuclear Magnetic Mom.(nm)	Elect. Quadr. Mom. (b)	γ-ray Energy /Intensity (MeV / %)
¹⁰⁶Tc		105.91436	36. s	β⁻ /6.55		2+			0.2703 0.5222 1.9694 2.2393 2.7893
¹⁰⁷Tc		106.9151	21.2 s	β⁻ /4.8					0.1027 0.1063 0.1770 0.4587
¹⁰⁸Tc		107.9185	5.1 s	β⁻ /7.72		(3)			0.2422 0.4656 0.7078 0.7326 1.5835
¹⁰⁹Tc		108.9200	1.4 s	β⁻ /6.3					
¹¹⁰Tc		109.9234	0.83 s	β⁻ /8.8					0.2407
¹¹¹Tc		110.9251	0.30 s	β⁻ /7.0					
¹¹³Tc		112.931	0.13 s	β⁻ /8.					
₄₄**Ru**		**101.07(2)**							
⁸⁸Ru									
⁸⁹Ru		88.936		β⁺ /8.					
⁹⁰Ru		89.9298	11. s	β⁺ /5.9					ann.rad./ 0.155 - 1.551
⁹¹Ru		90.9264	9. s	β⁺ ,EC/7.4		9/2+			ann.rad./
⁹²Ru		91.9201	3.7 m	β⁺ /53/4.5 EC/47/		0+			ann.rad./ 0.1346 0.2138 0.2593
⁹³ᵐRu			10.8 s	I.T./21/ β⁺ ,EC/79/	5.3/	1/2-			ann.rad./ 0.7344 1.1112 1.3962 2.0931
⁹³Ru		92.9171	1.0 m	β⁺ /6.3 EC/		9/2+			ann.rad./ 0.6807 1.4349 (0.5- 4.2)weak
⁹⁴Ru		93.91137	52. m	EC/100/1.59					0.3672 0.5247 0.8922
⁹⁵Ru		94.91042	1.64 h	EC/85/2.57 β⁺ /15/	1.20/ 0.91/	5/2+	0.86		ann.rad./ 0.2904 0.3364 0.6268
⁹⁶Ru	5.52(6)	95.90760				0+			
⁹⁷Ru		96.90756	2.89 d	EC/1.12		5/2+	-0.78		Tc k x-ray 0.2157 0.3245 0.4606
⁹⁸Ru	1.88(6)	97.90529				0+			
⁹⁹Ru	12.7(1)	98.905939				5/2+	-0.6413	+0.079	
¹⁰⁰Ru	12.6(1)	99.904219				0+			
¹⁰¹Ru	17.0(1)	100.905582				5/2+	-0.7188	+0.46	
¹⁰²Ru	31.6(2)	101.904349				0+			
¹⁰³Ru		102.906323	39.27 d	β⁻ /0.763	0.223	3/2+	0.20	+0.62	0.05329 0.29498 0.4438 0.49708 0.55704 0.61033 (0.04 - 1.6)
¹⁰⁴Ru	18.7(2)	103.905430				0+			
¹⁰⁵Ru		104.907750	4.44 h	β⁻ /1.917	1.11/22 1.134/13 1.187/49	3/2+	-0.3		0.12968 / 0.0 0.1491 0.2629 0.31664 0.46943 0.67634 0.72420 (0.1 - 1.8)
¹⁰⁶Ru		105.90733	1.020 y	β⁻ /0.0394	0.0394/100	0+			
¹⁰⁷Ru		106.9099	3.8 m	β⁻ /2.9	2.1/ 3.2/				0.1939 0.3741 0.4625 0.8488

Elem. or Isot.	Natural Abundance (%)	Atomic Mass or Weight	Half-Life	Decay Mode/Energy (/MeV)	Particle Energy /Intensity (MeV / %)	Spin (h/2π)	Nuclear Magnetic Mom.(nm)	Elect. Quadr. Mom. (b)	γ-ray Energy /Intensity (MeV / %)
^{108}Ru		107.9102	4.5 m	β^- /1.4	1.2/				0.0923
									0.1651
									0.4339
									0.4975
^{109}Ru		108.91320	34.5 s	β^- /4.2					0.6189
									0.1164
									0.3584
^{110}Ru		109.9140	15. s	β^- /2.81					0.1121
									0.3737
									0.4397
									0.7967
^{111}Ru		110.9176	1.5 s	β^- /5.5					
^{112}Ru		111.9188	4.5 s	β^- /4.5					
^{113}Ru		112.9225	2.7 s	β^- /7.					
^{114}Ru		113.9239	0.57 s	β^- /6.1					0.127/24
									(0.053-0.180)
^{115}Ru		114.928	≈ 0.74 s	β^- /8.					
$_{45}$Rh		102.90550(2)							
^{90}Rh									
^{91}Rh									
^{92}Rh		91.9320		β^+ /11.1					
^{93}Rh		92.9257		β^+ /8.1					
^{94m}Rh			25.8 s	β^+ /		8+			ann.rad./
									0.1264
									0.3117
									0.7562
									1.0752
									1.4307
^{94}Rh		93.9217	1.18 m	β^+ /9.6	6.4/	3+			ann.rad./
									0.1461
									0.3117
									0.7562
									1.4307
^{95m}Rh			1.96 m	I.T./88/		1/2+			ann.rad./
				β^+ ,EC/12/					0.5433(IT)
									0.7837
^{95}Rh		94.9159	5.0 m	β^+ /5.1	3.2	9/2+			ann.rad./
									0.2293
									0.4103
									0.6610
									0.9416
									1.3520
									(0.2 - 3.8)
^{96m}Rh			1.51 m	I.T./60 /0.052		2+			ann.rad./
				β^+ ,EC/40/	4.70/				Tc,Ru x-rays
									0.8326
									1.0985
									1.6921
									(0.4 - 3.3)
^{96}Rh		95.91452	9.6 m	β^+/6.45	3.3/	5+			ann.rad./
				EC/					0.4299
									0.6315
									0.6853
									0.7418
									0.8326
									(0.2 - 3.4)
^{97m}Rh			46.m	I.T./5 /	2.6/	1/2-			ann.rad./
				β^+,EC/95 /					0.1886
									0.4215
									2.2452
^{97}Rh		96.91134	31.0m	β^+ /3.52	2.1/	9/2+			ann.rad./
									0.1886
									0.3892
									0.4515
									0.8398
									0.8788
									(0.2 - 3.5)
^{98m}Rh			3.5 m	β^+ /		5+			ann.rad./
									0.6154
									0.6524
									0.7452
^{98}Rh		97.91072	8.7 m	β^+ /90 /5.06	3.4/	2+			ann.rad./
									0.6524
									0.7623

Elem. or Isot.	Natural Abundance (%)	Atomic Mass or Weight	Half-Life	Decay Mode/Energy (/MeV)	Particle Energy /Intensity (MeV / %)	Spin (h/2π)	Nuclear Magnetic Mom.(nm)	Elect. Quadr. Mom. (b)	γ-ray Energy /Intensity (MeV / %)
99mRh			4.7 h	β+ /8 / EC/92 /	.74/	9/2+			ann.rad./ 0.2766/ 0.3408 0.6178 1.2612
99Rh		98.90820	16. d	β+/4 /2.10 EC/97 /	0.54/ 0.68/	1/2-			ann.rad./ 0.0894/ 0.3530 0.5277 (0.1 - 2.0)
100mRh			4.7 m	I.T./99 / β+ /0.4 /		5+			ann.rad./ 0.0748/ 0.2647(IT)
100Rh		99.90812	20.8 h	β+ /3.63 EC/	2.62/ 2.07/	1-			0.4462 0.5396 0.5882 0.8225 1.5534 2.3761
101mRh			4.35 d	EC/92 / I.T./8 /0.1573		9/2+	+5.51		Rh k x-ray 0.1272/ 0.3069 0.5451
101Rh		100.90616	3.3 y	EC/0.54		1/2-			Ru k x-ray 0.1272 0.1980 0.3252
102mRh			207. d	I.T./5 / β− /19/ β+ /14/ EC/62/					ann.rad./ 0.4686 0.4751 0.5566 0.6280 1.1032 (0.4 - 1.6)
102Rh		101.906842	2.9 y	EC/2.323		6+	4.04		Ru k x-ray 0.4751 0.6313 0.6975 0.7668 1.0466 1.1032
103mRh			56.12 m	IT		7/2+			
103Rh	100.	102.905504				1/2-	-0.0884		
104mRh			4.36 m	I.T./99+ / β−	1.3/	5+			Rh k x-ray 0.0514 0.0971 0.5558
104Rh		103.906655	42.3 s	β− /99+ /2.441 EC/0.4/1.141	1.88/2 2.44/98	1+			0.3581 0.5558 1.2370 (0.35 - 1.8)
105mRh			40. s	I.T./1.296		1/2-			Rh k x-ray 0.1296
105Rh		104.905692	35.4 h	β− /0.567	0.247/30 0.567/70	7/2+	+4.45		0.2801 0.3061 0.3189
106mRh			2.18 h	β− /	0.92/	6+			0.2217 0.4510 0.5119 0.6162 0.7173 0.7484 1.0458 1.5277
106Rh		105.90729	29.9 s	β− /3.54	2.4/2 3.0/12 3.54/79	1+	2.58		0.51186/ 0.61612 0.62187 (0.05 - 3.04)
107Rh		106.90675	21.7 m	β− /1.51	1.20/65 1.5/17	7/2+			0.2776 0.3028 0.3925
108mRh			6.0 m	β− /	1.57/				0.4339 0.4973 0.6189

Elem. or Isot.	Natural Abundance (%)	Atomic Mass or Weight	Half-Life	Decay Mode/Energy (/MeV)	Particle Energy /Intensity (MeV / %)	Spin (h/2π)	Nuclear Magnetic Mom.(nm)	Elect. Quadr. Mom. (b)	γ-ray Energy /Intensity (MeV / %)
¹⁰⁸Rh		107.9087	17. s	β⁻ /4.5		1+			0.4046
									0.4339
									0.4973
									0.5811
									0.6146
									0.9014
									0.9471
¹⁰⁹Rh		108.90874	1.34 m	β⁻ /2.59	2.25/	7/2 +			0.1134
									0.1780
									0.2914
									0.3254
									0.3268
									0.4261
									(0.1 - 1.6)
¹¹⁰ᵐRh			29. s	β⁻ /	[.6/				0.3737
									0.4397
									0.7967
¹¹⁰Rh		109.9110	3.1 s	β⁻ /5.4	5.5/	1+			0.3737
									0.4400
									0.5463
									0.6877
									0.8381
									0.9045
¹¹¹Rh		110.9117	11. s	β⁻ /3.7					0.275
¹¹²ᵐRh			6.8 s	β⁻ /					
¹¹²Rh		111.9140	4. s	β⁻ /6.2		1+			0.3489
¹¹³Rh		112.9154	0.9 s	β⁻ /4.9					0.1285
¹¹⁴ᵐRh			1.8 s	β⁻ /					
¹¹⁴Rh		113.9173	1.8 s	β⁻ /6.5		1+			
¹¹⁵Rh		114.9201	0.99 s	β⁻ /6.0					
¹¹⁶ᵐRh			0.9 s	β⁻ /					0.3405
¹¹⁶Rh		115.9228	0.7 s	β⁻ /8.0		1+			
¹¹⁷Rh		116.925	0.44 s	β⁻ /7.					0.0346
									0.1317
₄₆Pd		106.42(1)							
⁹²Pd									
⁹³Pd									
⁹⁴Pd		93.9288	9. s	EC,β⁺ / ≈ 6.6					0.5582
									(0.0546-0.798)
⁹⁵ᵐPd		94.92684	13.4 s	EC,β⁺ /10.2		21/2 +			
⁹⁵Pd									
⁹⁶Pd		95.9182	2.03 m	EC,β⁺ /3.5	1.15/				0.1248
									0.4995
⁹⁷Pd		96.9165	3.1 m	β⁺ ,EC/4.8	3.5/	5/2 +			ann.rad./
									0.2653
									0.4752
									0.7927
									(0.2 - 3.4)
⁹⁸Pd		97.91273	17.7 m	β⁺ /1.87 EC/		0+			ann.rad./
									0.0677
									0.1125
									0.6630
									0.8379
⁹⁹Pd		98.91181	21.4 m	β⁺ /49 /3.37 EC/51 /	2.18/	5/2 +			ann.rad./
									0.1360
									0.2636
									0.6734
									(0.2 - 2.85)
¹⁰⁰Pd		99.90851	3.7 d	EC/0.36		0+			0.03271
									0.0748
									0.0840
¹⁰¹Pd		100.90829	8.4 h	β⁺ /5 /1.980 EC/95 /	0.776/	5/2 +	-0.66		ann.rad./
									0.0244
									0.2963
									0.5904
¹⁰²Pd	1.02(1)	101.905607				0+			
¹⁰³Pd		102.906087	16.99 d	EC/0.543		5/2 +			Rh k x-ray
									0.03975
									0.3575
									0.4971
¹⁰⁴Pd	11.14(8)	103.904034				0+			
¹⁰⁵Pd	22.33(8)	104.905083				5/2 +	-0.642	+0.66	
¹⁰⁶Pd	27.33(3)	105.903484				0+			
¹⁰⁷ᵐPd			20.9 s	I.T./0.2149		11/2-			Pd k x-ray
									0.2149(IT)
¹⁰⁷Pd		106.90513	6.5x10⁶y	β⁻ /0.033	0.03/	5/2 +			
¹⁰⁸Pd	26.46(9)	107.903895				0+			

Elem. or Isot.	Natural Abundance (%)	Atomic Mass or Weight	Half-Life	Decay Mode/Energy (/MeV)	Particle Energy /Intensity (MeV / %)	Spin (h/2π)	Nuclear Magnetic Mom.(nm)	Elect. Quadr. Mom. (b)	γ-ray Energy /Intensity (MeV / %)
^{109m}Pd			4.75 m	I.T./0.1889		11/2-			Pd x-ray 0.1889(IT)
^{109}Pd		108.905954	13.5 h	β⁻ /1.116	1.028	5/2+			0.0880 (0.08 - 1.0)
^{110}Pd	11.72(9)	109.905153				0+			
^{111m}Pd			5.5 h	I.T. /73 /0.172 β⁻ /27 /	0.35 0.77	11/2-			0.0704 0.1722 0.3912 (0.1 - 1.97)
^{111}Pd		110.90764	23.4 m	β⁻ /2.19	2.2/95	5/2+			0.0598 0.2454 0.5800 0.6504 1.3885 1.4590
^{112}Pd		111.90731	21.04 h	β⁻ /0.29	0.28/	0+			0.018
^{113m}Pd			1.48 m	β⁻ /		5/2+			0.0959
^{113}Pd		112.91015	1.64 m	β⁻ /3.34					0.0958 0.4824 0.6436 0.7394
^{114}Pd		113.91037	2.48 m	β⁻ /1.45		0+			0.1266 0.2320 0.5582 0.5760
^{115}Pd		114.9137	47. s	β⁻ /4.58					0.1255 0.2554 0.3428
^{116}Pd		115.9142	12.7 s	β⁻ /2.61					0.1015 0.1147 0.1778
^{117}Pd		116.9178	4.4 s	β⁻ /5.7					0.2473 (0.0766-0.4025)
^{118}Pd		117.9189	2.4 s	β⁻ /4.1					0.1254 (0.0284-0.5957)
^{119}Pd		118.9227	0.9 s	β⁻ /6.5					0.2566 (0.0699-0.3261)
^{120}Pd		119.9240	0.5 s	β⁻ /5.0					0.1581 (0.0525-0.5952)
$_{47}$Ag		107.8682(2)							
^{94}Ag			0.42 s	β⁺, p/					
^{95}Ag			2.0 s	β⁺, p/					
^{96}Ag		95.9307	5.1 s	β⁺ /11.6 EC/					ann.rad./ 0.1248 0.4995
^{97}Ag		96.9240	19. s	β⁺ /7.0 EC/					ann.rad./ 0.6862 1.2941
^{98}Ag		97.9218	47. s	β⁺ /8.4 EC/		5+			ann.rad./ 0.5711 0.6786 0.8631
^{99m}Ag			11. s	I.T./100/		1/2-			Ag k x-ray 0.1636(IT) 0.3426
^{99}Ag		98.9176	2.07 m	β⁺ /87 /5.4 EC/13 /		9/2+			ann.rad./ 0.2199 0.2645 0.8056 0.8323 (0.2 - 3.5)
^{100m}Ag			2.3 m	β⁺ / EC/		2+			ann.rad./ 0.6657 1.5941
^{100}Ag		99.9161	2.0m	β⁺/7.1 EC/	4.7/	5+			ann.rad./ 0.2807 0.4503 0.6657 0.7508 0.7732
^{101m}Ag			3.1 s	I.T./0.23		1/2-			Ag k x-ray 0.0981 0.176(IT)

Elem. or Isot.	Natural Abundance (%)	Atomic Mass or Weight	Half-Life	Decay Mode/Energy (/MeV)	Particle Energy /Intensity (MeV / %)	Spin (h/2π)	Nuclear Magnetic Mom.(nm)	Elect. Quadr. Mom. (b)	γ-ray Energy /Intensity (MeV / %)
^{101}Ag		100.9128	11.1 m	β^+ /69 /4.2 EC/31 /	2.7/ 2.18/ 2.73/ 3.38/	9/2+	5.7		ann.rad./ 0.2610 0.2747 0.3269 0.4392 0.6673 1.1739 (0.2 - 3.1)
^{102m}Ag			7.8 m	β^+ /38 / EC/13 / I.T./49 /	3.4	2+	+4.14		ann.rad./ 0.5567 0.9777 1.8347 2.0545 2.1594 3.2386
^{102}Ag		101.91197	13.0 m	β^+ /78 /5.92 EC/22 /	2.26/	5+	4.6		ann.rad./ 0.5567 0.7194 0.8354 1.2571 1.5816 1.7446
^{103m}Ag			5.7 s	I.T./0.134		1/2-			Ag k x-ray 0.1344
^{103}Ag		102.90897	1.10 h	β^+ /28 /2.69 EC/72 /	1.7 1.3	7/2+	+4.47		ann.rad./ 0.1187 0.1482
^{104m}Ag			33. m	β^+ /64/ EC/36/ I.T./0.07/	2.71/	2+	+3.7		ann.rad./ 0.5558 0.7657 (0.5 - 3.4)
^{104}Ag		103.90863	69. m	β^+ /16 /4.28 EC/84 /	0.99/	5+	3.92		ann.rad./ 0.5558 0.9259 0.9416 (0.18 - 2.27)
^{105m}Ag			7.2 m	I.T./98 /0.0255 EC/2 /		7/2+			Ag x-ray 0.3063 0.3192 (0.1 - 1.0)
^{105}Ag		104.90653	41.3 d	EC/1.35		1/2-	0.1014		0.0640 0.2804 0.3445 0.4434
^{106m}Ag			8.4 d	EC/		6+	3.71		Pd k x-ray 0.4510 0.5118 0.7173 1.0458
^{106}Ag		105.90667	24.0 m	β^+ /59 /2.965 EC/41 /	/1.96	1+	+2.85		ann.rad./ 0.5119
^{107m}Ag			44.2 s	I.T./0.093		7/2+			Ag x-ray 0.0931
^{107}Ag	51.839(7)	106.905093				1/2-	-0.11357		
^{108m}Ag			1.3×10^2y	EC/92 / I.T./8 /0.079		6+	3.580		Ag k x-ray Pd k x-ray 0.43392 0.61427 0.72290
^{108}Ag		107.905954	2.39 m	β^- /97/1.65 EC/2/ β^- /1/1.92	1.02/1.7 1.65/96 0.88/0.3	1+	+2.6884		ann.rad./ 0.43392 0.61885 0.63298
^{109m}Ag			39.8 s	I.T./0.088		7/2+			Ag k x-ray 0.0880
^{109}Ag	48.161(7)	108.904756				1/2-	-0.13056		
^{110m}Ag			249.8 d	β^- /99 / I.T./1 /0.1164	0.087 0.530	6+	+3.607		0.65774 0.76393 0.88467 0.93748 1.38427 (0.447-1.56)
^{110}Ag		109.906111	24.6 s	β^- /2.892	2.22/5 2.89/95	1+	+2.7271		0.65774 0.8154 1.1257
^{111m}Ag			1.08 m	I.T./99 /0.0598 β^- /1/		7/2+			Ag k x-ray 0.0598 0.2454

Elem. or Isot.	Natural Abundance (%)	Atomic Mass or Weight	Half-Life	Decay Mode/Energy (/MeV)	Particle Energy /Intensity (MeV / %)	Spin (h/2π)	Nuclear Magnetic Mom.(nm)	Elect. Quadr. Mom. (b)	γ-ray Energy /Intensity (MeV / %)
^{111}Ag		110.905295	7.47 d	β⁻ /1.037	1.035/	1/2-	-0.146		0.2454
									0.3421
^{112}Ag		111.90701	3.13 h	β⁻ /3.96	3.94/	2-	0.0547		0.6067
					3.4				0.6174
									1.3877
									(0.4 - 2.9)
^{113m}Ag			1.14 m	I.T./80 /0.043		7/2+			0.1422
				β⁻ /20 /	1.5				0.2983
									0.3161
									0.3923
^{113}Ag		112.90657	5.3 h	β⁻ /2.02	2.01/	1/2-	0.159		0.2588
									0.2986
^{114}Ag		113.90881	4.6 s	β⁻ /5.08	4.9/	1+			0.5582
									0.5760
									1.9946
^{115m}Ag			18.7 s	β⁻ /		7/2+			0.1134
									0.1315
									0.2288
									0.3887
^{115}Ag		114.90876	20. m	β⁻ /3.10		1/2-			0.1316
									0.2128
									0.2291
									0.4727
									(0.13 - 2.49)
^{116m}Ag			10.5 s	I.T./2 /	3.2/	5+			0.1027
				β⁻ /98 /	2.9				0.2549
									0.5134
									0.7055
									1.0289
^{116}Ag		115.91137	2.68 m	β⁻ /6.16	5.3	2-			0.5134
									0.6993
									2.4779
^{117m}Ag			5.3 s	β⁻ /	3.2/	7/2+			0.1354
									0.2981
									0.3868
^{117}Ag		116.91171	1.22 m	β⁻ /4.18	2.3	1/2-			0.1354
									0.1571
									0.3377
^{118m}Ag			2.8 s	β⁻ /59 /					0.1277
				I.T./41 /0.1277					0.4878
									0.6771
									0.7709
									1.0586
^{118}Ag		117.9145	4.0 s	β⁻ /7.1					0.4878
									0.6771
									3.2259
^{119}Ag		118.9157	2.1 s	β⁻ /5.35		7/2+			0.0674
									0.3662
									0.3991
									0.6264
^{120m}Ag			0.32 s	β⁻ /					0.2030
				I.T./					0.5059
									0.6978
									0.8300
									0.9258
^{120}Ag		119.9187	1.23 s	β⁻ /8.2					0.5059
									0.6978
									0.8171
									1.3231
^{121}Ag		120.9200	0.78 s	β⁻ /6.4					0.1150
									0.3148
									0.3537
									0.3696
									0.5007
									1.5105
									(0.11 - 2.5)
^{122m}Ag			1. s	β⁻ /					
^{122}Ag		121.9233	0.44 s	β⁻ /9.2					
^{123}Ag		122.9249	0.31 s	β⁻ /7.4					
^{124}Ag		123.9285	0.22 s	β⁻ /10.1					
$_{48}$Cd		112.411(8)							
^{97}Cd			3. s	β⁺,(p)					
^{98}Cd		97.9276	9.2 s	β⁺ /5.4					ann.rad./
^{99}Cd		98.9250	16. s	β⁺ ,EC/6.9					ann.rad./
^{100}Cd		99.9203	1.1 m	β⁺ ,EC/3.9					(0.0935-0.935)

Elem. or Isot.	Natural Abundance (%)	Atomic Mass or Weight	Half-Life	Decay Mode/Energy (/MeV)	Particle Energy /Intensity (MeV / %)	Spin (h/2π)	Nuclear Magnetic Mom.(nm)	Elect. Quadr. Mom. (b)	γ-ray Energy /Intensity (MeV / %)
^{101}Cd		100.9187	1.2 m	β⁺ /83 /5.5 EC/17 /	4.5	5/2+			In k x-ray 0.0985 1.7225 0.31 - 2.84)
^{102}Cd		101.91474	5.8 m	β⁺ /27 /2.59 EC/73		0+			ann.rad./ 0.0974 0.4810 1.0366 1.3598
^{103}Cd		102.91342	7.5 m	β⁺ /33 /4.14 EC/67 /		5/2+	-0.81	-0.8	ann.rad./ Ag k x-ray 1.0799 1.4487 1.4618 (0.1 - 2.8)
^{104}Cd		103.90985	58. m	EC/1.14		0+			Ag k x-ray 0.0835 0.7093
^{105}Cd		104.90947	55.5 m	β⁺ /26 /2.739 EC/74 /	1.69/	5/2+	-0.7393	+0.43	Ag k x-ray 0.3469 0.6072 0.9618 1.3025 (0.25 - 2.4)
^{106}Cd	1.25(4)	105.90646				0+			
^{107}Cd		106.90661	6.52 h	EC/99+ /1.417 β⁺ /		5/2+	-0.615055	+0.68	Ag k x-ray 0.0931 0.8289
^{108}Cd	0.89(2)	107.90418				0+			
^{109}Cd		108.904985	462.0 d	EC/0.214		5/2+	-0.827846	+0.69	Ag k x-ray 0.08804
^{110}Cd	12.49(12)	109.903006				0+			
^{111m}Cd			48.5 m	I.T./		11/2-			Cd k x-ray 0.1508(IT) 0.2454
^{111}Cd	12.80(8)	110.904182				1/2+	-0.594886		
^{112}Cd	24.13(14)	111.902758				0+			
^{113m}Cd			14.1 y	β⁻ /99.9 /0.59	0.59/99.9	11/2-	-1.087		0.2637
^{113}Cd	12.22(8)	112.904401				1/2+	-0.622301		
^{114}Cd	28.73(28)	113.903359				0+			
^{115m}Cd			44.6 d	β⁻ /1.629	0.68/1.6 1.62/97	11/2-	-1.042		0.48450 0.93381 1.29064
^{115}Cd		114.905431	2.228 d	β⁻ /1.446	0.593/42 1.11/58	1/2+	-0.648426		0.23141 0.26085 0.33624 0.49227 0.52780
^{116}Cd	7.49(12)	115.904756				0+			
^{117m}Cd			3.4 h	β⁻ /2.66	0.72/	11/2-			0.1586 0.5529 0.37 - 2.42
^{117}Cd		116.907219	2.49 h	β⁻ /2.52	0.67/51 2.2/10	1/2+			0.2209 0.2733 0.3445 1.3033
^{118}Cd		117.90692	50.3 m	β⁻ /0.52		0+			
^{119m}Cd			2.20 m	β⁻ /		11/2-			0.1056 0.7208 1.0250 2.0213
^{119}Cd		118.90992	2.69 m	β⁻ /3.8	≈ 3.5/	1/2+			0.1340 0.2929 0.3429
^{120}Cd		119.90985	50.8 s	β⁻ /1.76	1.5/	0+			
^{121m}Cd			8. s	β⁻ /		11/2-			0.1008 0.9878 1.0209 1.1815 2.0594
^{121}Cd		120.9131	13.5 s	β⁻ /4.9		(3/2+)			0.2102 0.3242 0.3492 1.0403
^{122}Cd		121.9135	5.3 s	β⁻ /3.0		0+			
^{123m}Cd			1.9 s	β⁻ /					
^{123}Cd		122.91770	2.09 s	β⁻ /6.12		3+			

Elem. or Isot.	Natural Abundance (%)	Atomic Mass or Weight	Half-Life	Decay Mode/Energy (/MeV)	Particle Energy /Intensity (MeV / %)	Spin (h/2π)	Nuclear Magnetic Mom.(nm)	Elect. Quadr. Mom. (b)	γ-ray Energy /Intensity (MeV / %)
¹²⁴Cd		123.9177	1.24 s	β⁻ /4.17		0+			0.0365 0.0628 0.1799
¹²⁵ᵐCd			0.66 s	β⁻ /					
¹²⁵Cd		124.92129	0.68 s	β⁻ /7.16		3/+			
¹²⁶Cd		125.9224	0.52 s	β⁻ /5.49		0+			0.2601
¹²⁷Cd		126.9264	0.4 s	β⁻ /8.5		3/+			
¹²⁸Cd		127.9278	0.28 s	β⁻ /7.1		0+			0.247
¹²⁹Cd			0.27 s	β⁻ /5.9					0.281
¹³⁰Cd			0.20 s	β⁻ /		0+			
₄₉In		114.818(3)							
⁹⁹In		98.9346		β⁺ /8.9					
¹⁰⁰In		99.9316	5. s	β⁺,(p)/10.5					
¹⁰¹In		100.9266		β⁺ /7.3					
¹⁰²In		101.9243	23. s	EC/8.9		(5)			0.1566 0.7767 0.8612
¹⁰³In		102.91991	1.1 m	β⁺,EC/6.05	4.2	9/2+			ann.rad./ 0.1879 0.7200
¹⁰⁴ᵐIn			16. s	IT/0.0935					
¹⁰⁴In		103.9183	1.84 m	β⁺,EC/7.9	4.8	5+	+4.44	+0.7	ann.rad./ 0.6580 0.8341 0.8781
¹⁰⁵ᵐIn			43. s	I.T.		1/2-			In k x-ray 0.6740
¹⁰⁵In		104.91467	5.1 m	β⁺,EC/4.85	3.7	9/2+	+5.675	+0.83	0.1310 0.2600 0.6038
¹⁰⁶ᵐIn			5.3 m	β⁺ /85 / EC/15 /	4.90	3+			ann.rad./ 0.6326 0.8611 1.7164
¹⁰⁶In		105.91346	6.2 m	β⁺ /65 /6.52 EC/35 /	2.6	7+	+4.92	+0.97	ann.rad./ 0.2259 0.6327 0.8611 0.9978 1.0091
¹⁰⁷ᵐIn			51. s	I.T./0.6786		1/2-			In k x-ray 0.6785
¹⁰⁷In		106.91029	32.4 m	β⁺ /35 /3.43 E.C/65 /	2.20/	9/2+	+5.59	+0.81	ann.rad./ Cd k x-ray 0.2050 0.3209 0.5055 (0.2 - 2.99)
¹⁰⁸ᵐIn			57. m	β⁺ /53 / EC/47 /	1.3	6+			ann.rad./ Cd k x-ray 0.6329 1.9863 3.4522
¹⁰⁸In		107.90971	40. m	β⁺ /33 /5.15 EC/67 /	3.49/	3+	+4.56	+1.01	ann.rad./ Cd k x-ray 0.2429 0.6331 0.8756
¹⁰⁹ᵐIn			1.3 m	I.T./0.650		1/2-			In k x-ray 0.6498
¹⁰⁹In		108.90715	4.2 h	β⁺ /8 /2.02 EC/92 /	0.79/	9/2+	+5.54	+0.84	ann.rad./ Cd k x-ray 0.2035 0.6235
¹¹⁰ᵐIn			4.9 h	EC/		7+	+4.72	+1.00	Cd k x-ray 0.6577 0.8847 0.9375 (0.1 - 1.98)
¹¹⁰In		109.90717	1.15 h	β⁺ /62 /3.88 EC/38 /	2.22/	2+	+4.37	+0.35	ann.rad./ Cd k x-ray 0.6577 (0.6 - 3.6)
¹¹¹ᵐIn			7.7 m	I.T./0.537		1/2-	+5.53		In k x-ray 0.537

Elem. or Isot.	Natural Abundance (%)	Atomic Mass or Weight	Half-Life	Decay Mode/Energy (/MeV)	Particle Energy /Intensity (MeV / %)	Spin (h/2π)	Nuclear Magnetic Mom.(nm)	Elect. Quadr. Mom. (b)	γ-ray Energy /Intensity (MeV / %)
111In		110.90511	2.8049 d	EC/0.866		9/2+	+5.50	+0.80	Cd k x-ray 0.1712 0.2453
112mIn			20.8 m	I.T./0.155		4+			In k x-ray 0.1555
112In		111.90553	14.4 m	β+ /22 /2.586 EC/34 / β- /0.663		1+	+2.82	+0.09	ann.rad./ Cd k x-ray 0.6171
113mIn			1.658 h	I.T./0.3917		1/2-	-0.210		In k x-ray 0.3917
113In	4.29(2)	112.904062				9/2+	+5.529	+0.80	
114mIn			49.51 d	I.T./97 /0.190 EC/3 /		5+	+4.7		In k x-ray 0.19027
114In		113.904918	1.198 m	β- /97 /1.989 EC/3 /1.453	1.984/	1+	+2.82		Cd k x-ray 0.5584 0.5727 1.2998
115mIn			4.486 h	I.T./95 /0.336 β- /5 /0.83		1/2-	-0.255		In k x-ray 0.3362 0.4974
115In	95.71(2)	114.903879	4.4x10^{14} y	β- /0.495		9/2+	+5.541	+0.81	
116m2In			2.16 s	I.T./0.162		8-			In k x-ray 0.1624
116m1In			54.1 m	β- /	1.0	5+	+4.3		0.13792 0.41688 1.09723 1.29349
116In		115.905261	14.1 s	β- /3.274	3.3/99	1+	2.788	0.11	0.46313 1.2526 1.29349
117mIn			1.94 h	β- /53 /1.769 I.T./47 /	1.77/	1/2-	0.25		In k x-ray 0.15855 0.31531 0.55294
117In		116.90452	44. m	β- /1.455	0.74/	9/2+	+5.52	+0.83	0.15855 0.3966 0.55294
118m2In			8.5 s	I.T./98 / β- /2 /		(8-)			In k x-ray 0.1382
118m1In			4.40 m	β- /	1.3 2.0	5+			0.2086 0.6833 1.2295
118In		117.90636	5.0 s	β- /4.42	4.2/	1+			0.5282 1.1734 1.2295 2.0432
119mIn			17.9 m	β- /97 / I.T./3 /0.311	2.7/	1/2-			0.3114 0.7631
119In		118.90585	2.3 m	β- /2.36	1.6/	9/2+	+5.52	+0.85	0.0239 0.6495 0.7631 1.2149
120mIn			47. s	β- /	2.2/	8-			0.7042 1.1725 2.0398 2.3902
120In		119.90796	3.1 s	β- /5.37	5.6/ 3.1/	(1+)	+4.30	+0.81	0.4146 0.5924 0.8637 1.0232 1.1714 (0.4 - 2.7)
121mIn			3.8 m	β- /99 / I.T./1 /0.313	3.7/	1/2-			0.0601 0.3136 0.9256 1.0412 1.1022 1.1204
121In		120.90785	23. s	β- /3.36	2.5	9/2+	+5.50	+0.81	0.2620 0.6573 0.9256 1.0014 1.1403
122mIn			10. s	β- /	4.4/	8-			0.2391 1.0014 1.1403 1.164 1.1903
122In		121.91028	1.5 s	β- /6.37	5.3/	(1+)			

Elem. or Isot.	Natural Abundance (%)	Atomic Mass or Weight	Half-Life	Decay Mode/Energy (/MeV)	Particle Energy /Intensity (MeV / %)	Spin (h/2π)	Nuclear Magnetic Mom.(nm)	Elect. Quadr. Mom. (b)	γ-ray Energy /Intensity (MeV / %)
^{123m}In			47. s	β^- /	4.6/	(1/2-)	-0.40		0.1258 1.170 3.234
^{123}In		122.91044	6.0 s	β^- /4.39	3.3/	(9/2+)	+5.49	+0.76	0.6188 1.0197 1.1305
^{124m}In			3.4 s	β^-		8-	+3.89	+0.66	0.1029 0.9699 1.0729 1.1316
^{124}In		123.91318	3.18 s	β^- /7.36	5/	3+	+4.04	+0.61	0.7070 0.9978 1.1316 3.2142 (0.3 - 4.6)
^{125m}In			12.2 s	β^- /	5.5/	1/2-	-0.43		0.1876
^{125}In		124.91360	2.33 s	β^- /5.42	4.1/	9/2+	+5.50	+0.71	0.4260 1.0318 1.3350
^{126m}In			1.53 s		4.9/	3+	+4.03	+0.49	0.9086 0.9696 1.1411
^{126}In		125.91646	1.63 s	β^- /8.21	4.2/	8-	+4.06		0.1118 0.9086 1.1411
^{127m}In			3.73 s	β^- /	6.4/	(1/2-)			0.2523 3.074
^{127}In		126.91734	1.14 s	β^- /6.51	4.9/	(9/2+)	+5.52	+0.59	0.4680 0.6461 0.8051 1.5977
^{128m}In			0.7 s	β^- /	5.4/	(8-)			1.8670 1.9739 (0.1205-2.12)
^{128}In		127.92017	0.80 s	β^- /8.98	5.0/	3+			0.9352 1.1688 3.5198 4.2970
^{129m}In			1.23 s	β^- /98 / n/2 /	≈ 7.5/	1/2-			0.3153 0.9067 1.2220
^{129}In		128.9217	0.63 s	β^- /7.66	5.5/	9/2+			0.2853 0.7693 1.8650 2.1180
^{130m2}In			0.53 s	β^- /	8.8/	5+			0.0892 0.7744 1.2212
^{130m1}In			0.51 s	β^- /	6.1/	10-			0.0892 0.1298 0.7744 1.2212 1.9052
^{130}In		129.92486	0.29 s	β^- /10.25	10.0/	1-			
^{131m2}In			0.3 s	β^- /		(21/2+)			
^{131m1}In			0.35 s	β^- /		(1/2-)			
^{131}In		130.9268	0.28 s	β^- /9.18	6.4/	(9/2+)			0.3328 2.433
^{132}In		131.9323	0.20 s	β^- /13.6	6.0/ 8.8/	(7-)			0.1320 0.2992 0.3747 4.0406
^{133}In			0.18 s	β^- ,(n)					
$_{50}$Sn		118.710(7)							
^{100}Sn		99.9394		β^+ /7.3					
^{101}Sn		100.9361		β^+ /9.					
^{102}Sn		101.9243		β^+ /5.8					
^{103}Sn		102.9281	7. s	β^+ /7.7					
^{104}Sn		103.9232	21. s	β^+, EC/4.5					
^{105}Sn		104.9214	28. s	β^+ /6.3					
^{106}Sn		105.91688	2.0 m	β^+ /20 /3.18 EC/80 /					ann.rad./ In k x-ray 0.3865 0.4772

Elem. or Isot.	Natural Abundance (%)	Atomic Mass or Weight	Half-Life	Decay Mode/Energy (/MeV)	Particle Energy /Intensity (MeV / %)	Spin (h/2π)	Nuclear Magnetic Mom.(nm)	Elect. Quadr. Mom. (b)	γ-ray Energy /Intensity (MeV / %)
¹⁰⁷Sn		106.9157	2.92 m	EC/5.0 β⁺ /	1.2/				0.4218 0.6105 0.6785 1.0013 1.1290 1.542
¹⁰⁸Sn		107.91196	10.3 m	β⁺ /1 /2.09 EC/99 /	0.36/	0+			In k x-ray 0.2724 0.3965 (0.105-1.68)
¹⁰⁹Sn		108.91129	18.0 m	β⁺ /9 /3.85 EC/91 /	1.52/	7/2+	-1.08	+0.3	ann.rad./ In k x-ray 0.6498 1.0992
¹¹⁰Sn		109.90785	4.1 h	EC/0.64		0+			In k x-ray 0.283
¹¹¹Sn		110.90774	35. m	β⁺ /31 /2.45 EC/69 /	1.5/	7/2+	+0.61	+0.2	In k x-ray 0.7620 1.1530 1.9147
¹¹²Sn	0.97(1)	111.904822				0+			
¹¹³ᵐSn			21.4 m	I.T./92 /0.077 EC/8 /		7/2+			Sn k x-ray In x-ray 0.0774
¹¹³Sn		112.905174	115.1 d	EC/1.036		1/2+	-0.879		In k x-ray 0.25511 0.39169
¹¹⁴Sn	0.65(1)	113.902783				0+			
¹¹⁵Sn	0.34(1)	114.903347				1/2+	-0.9188		
¹¹⁶Sn	14.54(11)	115.901745				0+			
¹¹⁷ᵐSn			13.60 d	I.T./0.3146		11/2-	-1.40	-0.4	Sn k x-ray 0.15856
¹¹⁷Sn	7.68(7)	116.902955				1/2+	-1.0010		
¹¹⁸Sn	24.22(11)	117.901608				0+			
¹¹⁹ᵐSn			293. d	I.T./0.0896		11/2-	-1.4	0.21	Sn k x-ray 0.02387
¹¹⁹Sn	8.59(4)	118.903311				1/2+	-1.0473		
¹²⁰Sn	32.59(10)	119.902199				0+			
¹²¹ᵐSn			≈ 55. y	I.T./78 /0.006 β⁻ /22 /	0.354/	11/2-			Sn k x-ray 0.03715
¹²¹Sn		120.904239	1.128 d	β⁻ /0.388	0.383/100	3/2+	0.698	-0.02	
¹²²Sn	4.63(3)	121.903441				0+			
¹²³ᵐSn			40.1 m	β⁻ /1.428	1.26/99	3/2+			0.1603 0.3814
¹²³Sn		122.905723	129.2 d	β⁻ /1.404	1.42/99.4	11/2-	-1.370	+0.03	0.1603 1.0302 1.0886
¹²⁴Sn	5.79(5)	123.905275				0+			
¹²⁵ᵐSn			9.51 m	β⁻ /2.387	2.03/98	3/2+			0.3321 1.4040
¹²⁵Sn		124.907785	9.63 d	β⁻ /2.364	2.35/82	11/2-	-1.35	+0.1	1.0671 (0.2-2.3)
¹²⁶Sn		125.90765	1.x10⁵y	β⁻ /0.38	0.25/100	0+			0.0643 0.0876 0.4148 0.6663 0.6950
¹²⁷ᵐSn			4.15 m	β⁻ /3.21	2.72/	3/2+			0.4909 1.3480 1.5640
¹²⁷Sn		126.91035	2.12 h	β⁻ /3.20	2.42/ 3.2/	11/2-			0.8231 1.0956 (0.120-2.84)
¹²⁸ᵐSn			6.5 s	IT/0.091		(7-)			
¹²⁸Sn		127.91054	59.1 m	β⁻ /1.27	0.48/ 0.63/	0+			0.4823 0.5573 0.6805 1.1611
¹²⁹ᵐSn			6.9 m	β⁻ /		11/2-			0.6456
¹²⁹Sn		128.9134	2.4 m	β⁻ /4.0		3/2+			0.1449
¹³⁰ᵐSn			1.7 m	β⁻ /		(7-)			0.8992
¹³⁰Sn		129.91386	3.7 m	β⁻ /2.15	1.10/	0+			0.0700 0.1925 0.7798
¹³¹ᵐSn			1.02m	β⁻/	3.4/	11/2-			0.3043 0.4500 0.7985 1.2260 (0.08 - 3.21)

Elem. or Isot.	Natural Abundance (%)	Atomic Mass or Weight	Half-Life	Decay Mode/Energy (/MeV)	Particle Energy /Intensity (MeV / %)	Spin (h/2π)	Nuclear Magnetic Mom.(nm)	Elect. Quadr. Mom. (b)	γ-ray Energy /Intensity (MeV / %)
^{131}Sn		130.9169	39. s	β^- /4.64	3.8/	3/2+			see ^{131m}Sn
^{132}Sn		131.91775	40. s	β^- /3.3	1.8/				0.0855
									0.2467
									0.3402
									0.8985
^{133}Sn		132.9236	1.44 s	β^- /7.8	7.5/	7/2-			
^{134}Sn		133.9278	1.04 s	β^- /6.8					
$_{51}$Sb		**121.760(1)**							
^{106}Sb		105.9282		β^+ /10.5					
^{107}Sb		106.9242		β^+ /7.9					
^{108}Sb		107.9222	7.0 s	β^+ /9.5					
^{109}Sb		108.91814	17. s	β^+ /6.38 EC/	4.42/ 4.67/ 4.33/	5/2+			ann.rad./ 0.2467 0.6645 0.9254 1.0617 1.4958
^{110}Sb		109.9175	24. s	β^+ /9.0 EC/	6.8/	3+			ann.rad./ 0.6365 0.9847 1.2117 1.2433
^{111}Sb		110.91254	1.25 m	β^+ /87 /4.47 EC/13 /	3.3/	5/2+			ann.rad./ 0.1002 0.1545 0.4891 1.0326
^{112}Sb		111.91240	51.4 s	β^+ /90 /7.06 EC/10 /	4.75/	3+			ann.rad./ 0.6700 0.9909 1.2571 (0.3 - 3.6)
^{113}Sb		112.90937	6.7 m	β^+ /65 /3.91 EC/35 /	2.42/	5/2+			ann.rad./ Sn k x-ray 0.3324 0.4980
^{114}Sb		113.9091	3.49 m	β^+ /78 /5.9 EC/22 /	3.4/	3+	1.7		ann.rad./ Sn k x-ray 0.8876 1.2999
^{115}Sb		114.90660	32.1 m	β^+ /67 /3.03 EC/33 /	1.51/	5/2+	+3.46	-0.4	ann.rad./ Sn k x-ray 0.4973
^{116m}Sb			1.00 h	β^+ /78 / EC/22 /	1.16/	8-			ann.rad./ Sn k x-ray 0.4073 0.5429 0.9725 1.2935 (0.0998-1.501)
^{116}Sb		115.90680	16. m	β^+ /50 /4.707 EC/50 /	1.3/ 2.3/	3+	2.72		ann.rad./ Sn k x-ray 0.93180 1.29354 (0.138-3.903)
^{117}Sb		116.90484	2.80 h	β^+ /2 /1.76 EC/98 /	0.57/	5/2+	+3.4		Sn k x-ray 0.1586
^{118m}Sb			5.00 h	EC/99 /		8-	2.3		Sn k x-ray 0.25368 1.05069 1.22964
^{118}Sb		117.905533	3.6 m	β^+ /74 /3.657 EC/26 /	2.65/	1+	2.5		ann.rad./ Sn k x-ray 1.22964
^{119}Sb		118.90395	38.1 h	EC/0.59		5/2+	+3.45	-0.4	Sn k x-ray 0.0239
^{120m}Sb			5.76 d	EC/		8-	2.34		Sn k x-ray 0.0898 0.19730 1.02301 1.17121
^{120}Sb		119.90508	15.89 m	β^+ /41 /2.68 EC/59 /	1.72/	1+	+2.3		ann.rad./ Sn k x-ray 0.7038 1.17121
^{121}Sb	57.21(5)	120.903822				5/2+	+3.363	-0.4	

Elem. or Isot.	Natural Abundance (%)	Atomic Mass or Weight	Half-Life	Decay Mode/Energy (/MeV)	Particle Energy /Intensity (MeV / %)	Spin (h/2π)	Nuclear Magnetic Mom.(nm)	Elect. Quadr. Mom. (b)	γ-ray Energy /Intensity (MeV / %)
^{122m}Sb			4.19 m	I.T./0.162		8-			Sb x-ray 0.0614 0.0761
^{122}Sb		121.90518	2.72 d	β^- /98 /1.979 β^+ /2 /1.620	1.414/65 1.980/26	2-	-1.90	+0.9	0.56409 0.69277 1.14050 1.2569
^{123}Sb	42.79(5)	122.904216				7/2 +	+2.550	-0.5	
^{124m2}Sb			20.3 m	I.T./0.035		8-			
^{124m1}Sb			1.6 m	I.T./80 / β^- /20 /	1.2/ 1.7/	5+			0.4984 0.6027 0.6458 1.1010
^{124}Sb		123.905938	60.20 d	β^- /2.905	0.61/52 2.301/23	3-	1.2	+1.9	0.60271/97.8 0.64583/7.4 0.72277/10.5 1.69094/48.2 (0.0274-2.808)
^{125}Sb		124.905247	2.758 y	β^- /0.767	0.13/30 0.302/45 0.62/13	7/2 +	+2.63		0.0355 0.17632 0.38044 0.42786 0.46336 0.60060 0.63595
^{126m2}Sb			11. s	I.T./		3-			L x-ray 0.0227
^{126m1}Sb			19.0 m	β^- /86 / I.T./14 /	1.9	5+			0.4148 0.6663 0.6950
^{126}Sb		125.90725	12.4 d	β^- /3.67	1.9	8-	1.3		0.2786 0.4148 0.6663 0.6950 0.7205
^{127}Sb		126.906914	3.84 d	β^- /1.581	0.89/ 1.10/ 1.50/	7/2 +	2.70		0.2524 0.2908 0.4121 0.4370 0.6857 0.7837
^{128m}Sb			10.1 m	β^- /96 / I.T./4 /	2.6/	5+			0.3140 0.5941 0.7432 0.7539
^{128}Sb		127.90917	9.1 h	β^- /4.38	2.3/	8-	1.3		0.2148 0.3141 0.5265 0.7433 0.7540
^{129m}Sb			17.7 m	β^- /					0.4338 0.6578 0.7598
^{129}Sb		128.90915	4.40 h	β^- /2.38	0.65/	7/2-			0.0278 0.1808 0.3594 0.4596 0.5447 0.8128 0.9146 1.0301
^{130m}Sb			6.5 m	β^- /2.6	2.12/				0.1023 0.7934 0.8394
^{130}Sb		129.91155	38.4 m	β^- /4.96	2.9/	8-			0.1823 0.3309 0.4680 0.7394 0.8394
^{131}Sb		130.9120	23.0 m	β^- /3.2	1.31/ 3.0/	7/2 +			0.6423 0.6579 0.9331 0.9434
^{132m}Sb			2.8 m	β^- /	3.9/	4+			0.1034 0.3538 0.6968 0.9739 0.9896

Elem. or Isot.	Natural Abundance (%)	Atomic Mass or Weight	Half-Life	Decay Mode/Energy (/MeV)	Particle Energy /Intensity (MeV / %)	Spin (h/2π)	Nuclear Magnetic Mom.(nm)	Elect. Quadr. Mom. (b)	γ-ray Energy /Intensity (MeV / %)
^{132}Sb		131.91420	4.2 m	β^- /5.29		8-			0.1034 0.1506 0.6968 0.9739
^{133}Sb		132.9152	2.5 m	β^- /4.00	1.20/	7/2+			0.4235 0.6318 0.8165 1.0764
^{134m}Sb			10.4 s	β^- /	6.1	7-			
^{134}Sb		133.9206	0.8 s	β^- /8.4	8.4	0-			0.1152 0.2970 0.7063 1.2791
^{135}Sb		134.9252	1.71 s	β^- /8.12		7/2+			1.127 1.279
^{136}Sb		135.9301	0.82 s	β^- /9.3					
$_{52}$Te		127.60(3)							
^{106}Te		105.9377	0.06 ms	α/4.32					
^{107}Te		106.9350	3.1 ms	α/ 70/ β^+ ,EC/10.1	3.86(1)/				
^{108}Te		107.9295	2.1 s	α/68 / β^+ ,EC/32 /6.8	3.314(4)/	0+			
^{109}Te		108.9275	4.6 s	β^+ EC/96 /8.7 α/4 /	3.107(4)/				
^{110}Te		109.9224	19. s	β^+ ,EC/4.5		0+			ann.rad./ 0.2191 0.6059
^{111}Te		110.9211	19.3 s	β^+ ,EC/8.0		(7/2+)			ann.rad./ 0.267 0.322 0.341
^{112}Te		111.9171	2.0 m	β^+ ,EC/4.3		0+			ann.rad./ 0.2962 0.3727 0.4187
^{113}Te		112.9154	1.7 s	β^+ /85 /5.7 EC/15 /	4.5/	(7/2+)			ann.rad./ Sb k x-ray 0.8144 1.0181 1.1812
^{114}Te		113.9125	15. m	β^+ /40 /3.2 EC/60 /		0+			ann.rad./ Sb k x-ray 0.0838 0.0903
^{115m}Te			6.7 m	β^+ /45 / EC/55 /		(1/2+)			ann.rad./ Sb k x-ray 0.7236 0.7704
^{115}Te		114.9116	5.8 m	β^+ /45 /4.6 EC/55 /	2.7/	7/2+			ann.rad./ Sb k x-ray 0.7236 1.3268 1.3806 (0.22 - 2.7)
^{116}Te		115.9084	2.49 h	EC/1.5		0+			Sb k x-ray 0.0937
^{117}Te		116.90864	1.03 h	EC/75 /3.54 β^+ /25 /	1.78/	1/2+			ann.rad./ Sb k x-ray 0.9197 1.7164 2.3000
^{118}Te		117.90583	6.00 d	EC/0.28		0+			Sb k x-ray
^{119m}Te			4.69 d	EC/		11/2-	0.89		Sb k x-ray 0.15360 0.2705 1.21271
^{119}Te		118.90641	16.0 h	β^+ /2 /2.293 EC/98 /	0.627/	1/2+	0.25		ann.rad./ Sb k x-ray 0.6440 0.6998
^{120}Te	0.096(2)	119.90403				0+			
^{121m}Te			≈ 154. d	I.T.(89%) EC(11%)		11/2-			Te k x-ray 0.2122
^{121}Te		120.90494	16.8 d	EC/1.04		1/2+			Sb k x-ray 0.5076 0.5731

Elem. or Isot.	Natural Abundance (%)	Atomic Mass or Weight	Half-Life	Decay Mode/Energy (/MeV)	Particle Energy /Intensity (MeV / %)	Spin (h/2π)	Nuclear Magnetic Mom.(nm)	Elect. Quadr. Mom. (b)	γ-ray Energy /Intensity (MeV / %)
^{122}Te	2.603(4)	121.903056				0+			
^{123m}Te			119.7 d	I.T./0.247		11/2-	-0.93		Te k x-ray 0.1590/84.1
^{123}Te	0.908(2)	122.904271	1.3x10^{13} y	EC/0.051		1/2+	-0.73695		
^{124}Te	4.816(6)	123.902819				0+			
^{125m}Te			58. d	I.T./0.145		11/2-	-0.99	-0.06	Te k x-ray 0.0355
^{125}Te	7.139(6)	124.904424				1/2+	-0.8885		
^{126}Te	18.952(11)	125.903305				0+			
^{127m}Te			109. d	I.T./98 /0.088 β$^-$ /2 /0.77		11/2-	-1.04		Te k x-ray 0.0883 0.3603
^{127}Te		126.905217	9.4 h	β$^-$ /0.698	0.696/	3/2+	0.64		
^{128}Te	31.687(11)	127.904462				0+			
^{129m}Te			33.6 d	I.T./63 /0.105 β$^-$ /37 /	1.60/	11/2-			Te k x-ray 0.45984 0.6959
^{129}Te		128.906596	1.16 h	β$^-$ /1.498	0.99/9 1.45/89	3/2+	0.70	0.06	0.0278 0.45984 0.48728
^{130}Te	33.799(10)	129.906223	2.5x10^{21} y			0+			0.0811
^{131m}Te			1.35 d	β$^-$ /78 /2.4 I.T./22 /0.18	0.42/	11/2-	-1.04		0.1021 0.14973 0.77369 0.79375 0.85225
^{131}Te		130.908522	25.0 m	β$^-$ /2.233	1.35/12 1.69/22 2.14/60	3/2+	0.70		0.14973 0.45327 0.49269
^{132}Te		131.90852	3.26 d	β$^-$ /0.493	0.215	0+			0.049725 0.11198 0.22830
^{133m}Te			55.4 m	β$^-$ /82 / I.T./18 /0.334	2.4/30	11/2-			Te k x-ray 0.0949 0.1689 0.3121 0.3341
^{133}Te		132.9109	12.4 m	β$^-$ /2.9	2.25/25 2.65	3/2+			0.3121 0.4079 1.3334
^{134}Te		133.9116	42. m	β$^-$ /1.6	0.6/ 0.7/	0+			0.7672/29 0.0794-0.9255
^{135}Te		134.9165	19.0 s	β$^-$ /6.0	5.4/ 6.0				0.267 0.603 0.870
^{136}Te		135.92010	17.5 s	β$^-$ /5.1	2.5/	0+			2.0779/25 0.0873-3.235
^{137}Te		136.9253	2.5 s	β$^-$ /98 /6.9 n/2 /	6.8	7/2-			0.2436
^{138}Te		137.9292	1.4 s	β$^-$ /6.4					
$_{53}$I		**126.90447(3)**							
^{108}I		107.9436	0.04 s	α/91 /4. p	3.95				
^{109}I		108.9382	0.11 ms						
^{110}I		109.9346	0.65 s	β$^+$,EC/83 /11.4 α/17 / ≈3.6 p /11 /	3.457(10)/				ann.rad./
^{111}I		110.9303	2.5 s	β$^+$,E../8.5					ann.rad./ 0.2665 0.3215 0.3412
^{112}I		111.9280	3.4 s	β$^+$,EC/10.2					ann.rad./ 0.6889 0.7869
^{113}I		112.9237	5.9 s	β$^+$,EC/7.6					ann.rad./ 0.4625/100 0.6224/74 0.0550-1.422
^{114}I		113.9219	2.1 s	β$^+$,EC/8.7					ann.rad./ 0.6826 0.7088
^{115}I		114.9188	1.3 m	β$^+$,EC/6.7		5/2+			ann.rad./ 0.275 0.284 0.460 0.709

Elem. or Isot.	Natural Abundance (%)	Atomic Mass or Weight	Half-Life	Decay Mode/Energy (/MeV)	Particle Energy /Intensity (MeV / %)	Spin (h/2π)	Nuclear Magnetic Mom.(nm)	Elect. Quadr. Mom. (b)	γ-ray Energy /Intensity (MeV / %)
^{116}I		115.9167	2.9 s	β^+ /97 /7.8 EC/3 /	6.7/	1+			ann.rad./ 0.5402 0.6789
^{117}I		116.9136	2.22 m	β^+ ,EC/4.7	3.2/	(5/2+)	3.1		ann.rad./ 0.2744 0.3259
^{118m}I			8.5 m	β^+ ,EC/ I.T.	4.9/	7-	4.2		ann.rad./ 0.104 0.5998 0.6052 0.6138
^{118}I		117.9134	14. m	β^+ ,EC/7.0		2-	2.0		ann.rad./ 0.5448 0.6052 1.3384
^{119}I		118.9102	19. m	β^+ /54 /3.5 EC/46 /	2.4/	(5/2+)	+2.9		ann.rad./ Te k x-ray 0.2575
^{120m}I			53. m	β^+ /80 / EC/20 /	3.8		4.2		ann.rad. Te k x-ray 0.4257 0.5604 0.6147 1.3459
^{120}I		119.91005	1.35 h	β^+ /81 /5.62 EC/19 /	4.03 4.60	2-	1.23		ann.rad./ Tek x-ray 0.5604 0.6411 1.5230 (0.43 - 3.1)
^{121}I		120.90737	2.12 h	β^+ /13 /2.27 EC/87 /	1.2/	5/2+	2.3		ann.rad./ Te k x-ray 0.2122 (0.14 - 1.1)
^{122}I		121.90760	3.6 m	β^+ /4.234 EC/	3.1/	1+	0.94		ann.rad./ Te k x-ray 0.5641
^{123}I		122.905605	13.2 h	EC/1.242		5/2+	2.82		Te k x-ray 0.1590
^{124}I		123.906211	4.18 d	β^+ /23 /3.160 EC/77 /	1.54/ 2.14/ 0.75/	2-	1.44		ann.rad./ Te k x-ray 0.6027/62.9 0.7228/10.3 1.6910/11.2 (0.31-1.73)
^{125}I		124.904624	59.4 d	EC/0.1861		5/2+	2.82	-0.89	Te k x-ray 0.0355
^{126}I		125.905619	13.0 d	EC/ β^+ /2.155 β^- /1.258	1.13/ 0.87/ 1.25/	2-	1.44		ann.rad./ Te k x-ray 0.3887 0.6622
^{127}I	100.	126.904468				5/2+	+2.8133	-0.79	
^{128}I		127.905805	25.00 m	β^- /2.118 EC/1.251	2.13/	1+			Te k x-ray 0.44287 0.52658
^{129}I		128.904988	1.7x10^7 y	β^- /0.194	0.15/	7/2+	+2.621	-0.55	Xe k x-ray 0.0396
^{130m}I			9.0 m	I.T./83 /0.048 β^- /17 /		2+			I k x-ray 0.5361
^{130}I		129.906674	12.36 h	β^- /2.949	1.04/ 0.62	5+	3.35		0.4180 0.5361 0.6685 0.7395
^{131}I		130.906125	8.040 d	β^- /0.971	0.606/	7/2+	+2.742	-0.40	0.08017 0.28431 0.36446 0.63699
^{132m}I			1.39 h	IT		8-			
^{132}I		131.90800	2.28 h	β^- /14 /3.58 I.T./86 /	0.80/ 1.03/ 1.2/ 1.6/ 2.16/	4+	3.09	0.09	I k x-ray 0.0980 0.5059 0.52264 0.63019 0.6506 0.66768 0.77260 0.95457

Elem. or Isot.	Natural Abundance (%)	Atomic Mass or Weight	Half-Life	Decay Mode/Energy (/MeV)	Particle Energy /Intensity (MeV / %)	Spin (h/2π)	Nuclear Magnetic Mom.(nm)	Elect. Quadr. Mom. (b)	γ-ray Energy /Intensity (MeV / %)
^{133m}I			9. s	I.T./1.63		19/2-			I kx-ray 0.0730 0.6474 0.9126
^{133}I		132.90781	20.8 h	β⁻ /1.77	1.24/85	7/2+	+2.86	-0.27	0.51056 0.52989 0.87537
^{134m}I			3.7 m	I.T./98 /0.316 β⁻ /2 /		8-			I k x-ray 0.0444 0.2719
^{134}I		133.9099	52.6 m	β⁻ /4.2	1.2/	4+			0.1354 0.84702 0.88409
^{135}I		134.91005	6.57 h	β⁻ /2.65	0.9/ 1.3/	7/2+			0.2884 0.41768 0.52658 1.13156 1.26046
^{136m}I			47. s	β⁻ /	4.7/ 5.2/	6-			0.1973 0.3468 0.3701 0.3814 1.3130 (0.16 - 2.36)
^{136}I		135.91466	1.39 m	β⁻ /6.93	4.3/ 5.6/	2-			0.3447 1.3130 1.3211 2.2896 (0.3 - 6.1)
^{137}I		136.91787	24.5 s	β⁻ /5.88	5.0/	(7/2+)			0.6010 1.2180 1.2201 1.3026 1.5343 (0.25 - 4.4)
^{138}I		137.9224	6.5 s	β⁻ /7.8	6.9/ 7.4/	2-			0.4836 0.5888 0.8752 (0.4 - 5.3)
^{139}I		138.92609	2.30 s	β⁻ /6.81 n/					0.192 0.198 0.273 0.382 0.386 0.468 0.683 1.313
^{140}I		139.9310	0.86 s	β⁻ /8.8 n/		(3)			0.372 0.377 0.457
^{141}I		140.9351	0.45 s	β⁻ /7.8					
^{142}I			≈ 0.2 s	β⁻					
$_{54}$Xe		131.29(2)							
^{110}Xe		109.9445	0.2 s	β⁺ /9.2					
^{111m}Xe			0.9 s	EC,β⁺					
^{111}Xe		110.9416	0.7 s	EC,β⁺ /10.6 α/	3.58(1)/				
^{112}Xe		111.9357	3. s	EC,β⁺ /7.2 α/0.8 /					
^{113}Xe		112.9334	2.8 s	EC,β⁺ /9.1					
^{114}Xe		113.9281	10.0 s	β⁺ ,EC/5.9		0+			ann.rad./ 0.1031 0.1616 0.3085 0.6826 0.7088
^{115}Xe		114.9270	18. s	β⁺ ,EC/7.6		(5/2+)			ann.rad./
^{116}Xe		115.9214	56. s	β⁺ ,EC/4.3	3.3/	0+			ann.rad./ 0.1042 0.1916 0.2477 0.3107 0.4127
^{117}Xe		116.9206	1.02 m	β⁺ ,EC/6.5		(5/2+)			ann.rad./ 0.2214 0.5190 0.6389 0.6613

Elem. or Isot.	Natural Abundance (%)	Atomic Mass or Weight	Half-Life	Decay Mode/Energy (/MeV)	Particle Energy /Intensity (MeV / %)	Spin (h/2π)	Nuclear Magnetic Mom.(nm)	Elect. Quadr. Mom. (b)	γ-ray Energy /Intensity (MeV / %)
^{118}Xe		117.917	≈ 4. m	β^+ ,EC/3.	2.7/	0+			ann.rad./ 0.0535 0.0600 0.1199
^{119}Xe		118.9156	5.8 m	β^+ ,EC/5.0	3.5/	7/2+	0.6		0.0873 0.1000 0.2318 0.4615
^{120}Xe		119.91216	40. m	β^+ ,EC/97 /1.96 β^+ /3 /		0+			I k x-ray 0.0251 0.0726 0.1781 (0.1 - 1.03)
^{121}Xe		120.91138	39. m	β^+ /44 /3.73 EC/56 /	2.8/	5/2+	0.65		ann.rad./ I k x-ray 0.1328 0.2527 0.4452 (0.1 - 3.1)
^{122}Xe		121.9086	20.1 h	EC/0.9		0+			I k x-ray 0.3501
^{123}Xe		122.90848	2.00 h	β^+ /23 /2.68 EC/77 /	1.51/	1/2+			ann.rad./ I k x-ray 0.1489 0.1781 (0.1 - 2.1)
^{124}Xe	0.10(1)	123.905895							
^{125m}Xe			57. s	I.T./0.252		(9/2-)			Xe k x-ray 0.1111 0.141
^{125}Xe		124.906398	17.1 h	EC/1.653	0.47/	1/2+			I k x-ray 0.1884 0.2434
^{126}Xe	0.09(1)	125.90427				0+			
^{127m}Xe			1.15 m	I.T./0.297		(9/2-)			Xe k x-ray 0.1246 0.1725
^{127}Xe		126.905179	36.4 d	EC/0.662		1/2+	-0.504		I k x-ray 0.1721 0.2029 0.3750
^{128}Xe	1.91(3)	127.903531				0+			
^{129m}Xe			8.89 d	I.T./0.236		11/2-			Xe k x-ray 0.0396 0.1966
^{129}Xe	26.4(6)	128.904780				1/2+	-0.7780		
^{130}Xe	4.1(1)	129.903509				0+			
^{131m}Xe			11.9 d	I.T./0.164		11/2-	+0.6908		Xe k x-ray 0.16398
^{131}Xe	21.2(4)	130.905083				3/2+	+0.69186	-0.12	
^{132}Xe	26.9(5)	131.904155				0+			
^{133m}Xe			2.19 d	I.T./0.233		11/2-	-1.082		Xe k x-ray 0.23325
^{133}Xe		132.905906	5.243 d	β^- /0.427	0.346/99	3/2+	+0.813	+0.14	Cs k x-ray 0.080998 0.1606
^{134}Xe	10.4(2)	133.905395				0+			
^{135m}Xe			15.3 m	I.T./		11/2-	1.103		Xe k x-ray 0.52658
^{135}Xe		134.90721	9.10 h	β^- /1.15	0.91/	3/2+	0.903		0.24975 0.60807
^{136}Xe	8.9(1)	135.90722				0+			
^{137}Xe		136.91156	3.82 m	β^- /4.17	4.1/ 3.6/	7/2-			0.45549 0.8489 0.9822 1.2732 1.7834 2.8498
^{138}Xe		137.91399	14.1 m	β^- /2.77	0.8/ 2.4/	0+			0.1538 0.2426 0.2583 0.4345 1.76826 2.0158
^{139}Xe		138.91879	39.7 s	β^- /5.06	4.5/ 5.0/				0.1750 0.2186 0.2965 (0.1 - 3.37)

Elem. or Isot.	Natural Abundance (%)	Atomic Mass or Weight	Half-Life	Decay Mode/Energy (/MeV)	Particle Energy /Intensity (MeV / %)	Spin (h/2π)	Nuclear Magnetic Mom.(nm)	Elect. Quadr. Mom. (b)	γ-ray Energy /Intensity (MeV / %)
¹⁴⁰Xe		139.9216	13.6 s	β⁻ /4.1	2.6	0+			0.0801 0.6220 0.8055 1.4137 (0.04 - 2.3)
¹⁴¹Xe		140.9267	1.72 s	β⁻ /6.2	6.2/	5/2+			0.1187 0.9095 (0.05 - 2.55)
¹⁴²Xe		141.9297	1.22 s	β⁻ /5.0	3.7/ 4.2/	0+			0.0338 0.0729 0.2038 0.3091 0.4145 0.5382 0.5718 0.6181 0.6448
¹⁴³ᵐXe			0.96 s	β⁻					
¹⁴³Xe		142.9352	0.30 s	β⁻ /7.3					
¹⁴⁴Xe		143.9385	1.2 s	β⁻ /6.1					
¹⁴⁵Xe			0.9 s	β⁻ ,(n)					
₅₅Cs		132.90545(2)							
¹¹²Cs		111.9503	0.5 ms	p	0.81				
¹¹³Cs		112.9445	17. μs	p	0.96				
¹¹⁴Cs		113.9408	0.58 s	β⁺ ,EC/11.8		1+			ann. rad./ 0.6826 0.7088
¹¹⁵Cs		114.9359	≈ 1.4 s	β⁺ ,EC/8.4					ann. rad./
¹¹⁶ᵐCs			0.7 s	β ,EC/					ann. rad./ 0.3935
¹¹⁶Cs		115.9330	3.8 s	β⁺ ,EC/10.8					ann. rad./ 0.3935 0.5243 0.6151 0.6223
¹¹⁷ᵐCs			6.5 s	β⁺ ,EC/					
¹¹⁷Cs		116.9286	≈ 8.4 s	β⁺ ,EC/7.5					ann. rad./
¹¹⁸ᵐCs			17. s	β⁺ ,EC/					
¹¹⁸Cs		117.92654	14. s	β⁺ ,EC/9.		2			ann.rad./ 0.3372 0.4727 0.5865 0.5906
¹¹⁹ᵐCs			28. s			3/2			
¹¹⁹Cs		118.92234	38. s	β⁺ ,EC/6.3		9/2+	+5.5	+2.8	ann. rad./ 0.169 0.176 0.224 0.257
¹²⁰ᵐCs			60. s	β⁺ ,EC/					
¹²⁰Cs		119.92066	64. s	β⁺ ,EC/7.92		2+	+3.87	+1.45	ann.rad./ 0.3224 0.4735 0.5534 (0.3 - 3.28)
¹²¹ᵐCs			2.0 m	I.T./60 / β⁺ /40 /	4.4	(9/2+)	+5.41	+2.7	ann.rad./ 0.1794 0.1961
¹²¹Cs		120.91718	2.3 m	β⁺ ,EC/5.40	4.38/	3/2+	+0.77	+0.84	ann.rad./ 0.1537 (0.08 - 0.56)
¹²²ᵐ²Cs			4.4 m	β⁺ ,EC		8-	+4.77	+3.3	ann.rad./
¹²²ᵐ¹Cs			0.36 s	IT					0.3311 0.4971 0.6385 (0.27 - 2.22)
¹²²Cs		121.91614	21. s	β⁺ ,EC/7.1	5.8/	(1+)	-0.133	-0.19	ann.rad./ 0.3311 0.5120 0.8179
¹²³ᵐCs			1.6 s	I.T./		11/2-			Cs k x-ray 0.0946

Elem. or Isot.	Natural Abundance (%)	Atomic Mass or Weight	Half-Life	Decay Mode/Energy (/MeV)	Particle Energy /Intensity (MeV / %)	Spin (h/2π)	Nuclear Magnetic Mom.(nm)	Elect. Quadr. Mom. (b)	γ-ray Energy /Intensity (MeV / %)
^{123}Cs		122.91299	5.87 m	β⁺ /75 /4.20 EC/25 /	3.0/	1/2+	+1.38		ann.rad./ Xe k x-ray 0.0974 0.5964
^{124m}Cs			6.3 s	IT		7+			
^{124}Cs		123.91225	30. s	β⁺ /92 /5.92 EC/8 /	≈ 5.	1+	+0.673	-0.74	ann.rad./ Xe k x-ray 0.3539 0.4925 0.9418
^{125}Cs		124.90972	45. m	β⁺ /40 /3.09 EC/60 /	2.06/	1/2+	+1.41		ann.rad./ Xe k x-ray 0.112 0.526
^{126}Cs		125.90945	1.64 m.	β⁺ /81 /4.83 EC/19 /	3.4 3.7/	1+	+0.78	-0.68	ann.rad./ Xe k x-ray 0.3886 0.4912 0.9252
^{127}Cs		126.90741	6.2 h	β⁺ /96 /2.08 EC/4 /	0.65/ 1.06	1/2+	+1.46		Xe k x-ray 0.1247 0.4119
^{128}Cs		127.90775	3.62 m	β⁺ /68 /3.930 EC/32 /	2.44/ 2.88/	1+	+0.97	-0.57	ann.rad./ Xe k x-ray 0.4429
^{129}Cs		128.90606	1.336 d	EC/1.195		1/2+	+1.49		Xe k x-ray 0.3719 0.4115
^{130m}Cs			3.5 m	IT,β⁺ ,EC		5-			
^{130}Cs		129.90671	29.21 m	β⁺ /55 /2.98 EC/43 / β⁻ /1.6 /0.37	1.98/ 0.44/1.6	1+	+1.46	-0.06	ann.rad./ Xe k x-ray 0.5361
^{131}Cs		130.90546	9.69 d	EC/0.352		5/2+	+3.54	-0.58	Xe k x-ray
^{132}Cs		131.906430	6.48 d	EC/98 / β⁺ /0.3 /2.120 β⁻ / /1.280		2-	+2.22	+0.51	Xe k x-ray 0.4646 0.6302 0.66769
^{133}Cs	100.	132.905447				7/2+	+2.582	-0.0037	
^{134m}Cs			2.91 h	I.T./0.139		8-	+1.098	+1.0	Cs k x-ray 0.12749
^{134}Cs		133.906714	2.065 y	β⁻ /2.059 EC/1.22	0.089/27 0.658/70	4+	+2.994	+0.39	0.56327 0.56935 0.60473 0.79584
^{135m}Cs			53. m	I.T./1.627		19/2-	+2.18	+0.9	0.7869 0.8402
^{135}Cs		134.905972	2.3x10⁶ y	β⁻ /0.269	0.205/100	7/2+	+2.732	+0.05	
^{136m}Cs			19. s	I.T./		8			
^{136}Cs		135.907307	13.16 d	β⁻ /2.548	0.341/	5+	+3.71	+0.2	0.06691 0.34057 0.81850 1.04807
^{137}Cs		136.907085	30.2 y	β⁻ /1.176	0.514/95	7/2+	+2.84	+0.05	Ba k x-ray 0.66164
^{138m}Cs			2.9 m	I.T./75 /0.080 β⁻ /25 /	3.3	6-	+1.71	-0.40	Cs k x-ray 0.0799 0.1917 0.4628 1.43579
^{138}Cs		137.91101	32.2 m	β⁻ /5.37	2.9/	3-	+0.700	+0.12	0.1381 0.46269 1.00969 1.43579 2.21788
^{139}Cs		138.913359	9.3 m	β⁻ /4.213	4.21	7/2+	+2.70	-0.07	0.6272 1.2832 (0.4 - 3.66)
^{140}Cs		139.91727	1.06 m	β⁻ /6.22	5.7/ 6.21/	1-	+0.13390	-0.11	0.5283 0.6023 0.9084 (0.41 - 3.94)

Elem. or Isot.	Natural Abundance (%)	Atomic Mass or Weight	Half-Life	Decay Mode/Energy (/MeV)	Particle Energy /Intensity (MeV / %)	Spin (h/2π)	Nuclear Magnetic Mom.(nm)	Elect. Quadr. Mom. (b)	γ-ray Energy /Intensity (MeV / %)
^{141}Cs		140.92005	24.9 s	β^- /5.26	5.20/	7/2+	+2.44	-0.4	Ba k x-ray 0.0485 0.5616 0.5887 1.1940 (0.05 - 3.33)
^{142}Cs		141.92430	1.8 s	β^- /7.31	6.9/ 7.28				0.3596 0.9668 1.1759 1.3265
^{143}Cs		142.92732	1.78 s	β^- /6.24	6.1	(3/2+)	+0.87	+0.47	0.1955 0.2324 0.3064 (0.17 - 1.98)
^{144}Cs		143.93203	1.01 s	β^- /8.47	8.46/ 7.9/	1	-0.546	+0.30	0.1993 0.5598 0.6392 0.7587
^{145}Cs		144.93541	0.59 s	β^- /7.89	7.4/ 7.9/	3/2+	+0.784	+0.6	0.1126 0.1755 0.1990
^{146}Cs		145.94024	0.322 s	β^- ,(n)/9.38	≈ 9.0	2-			
^{147}Cs		146.9439	0.227 s	β^- ,(n)/9.3					
^{148}Cs		147.9490	0.15 s	β^- ,(n)/10.5					
$_{56}$**Ba**		**137.327(7)**							
^{117}Ba		116.9377	1.8 s	β^+ ,EC/8.4		(3/2-)			
^{119}Ba		118.931	5.4 s	β^+ ,EC/8.					
^{120}Ba		119.9260	24. s	β^+ ,EC/5.0		0+			ann.rad./ 0.140 (0.075-0.146)
^{121}Ba		120.9245	30. s	β^+ ,EC/6.8		5/2	+0.660	+1.8	ann.rad./
^{122}Ba		121.9203	2.0 m	β^+ ,EC/3.8		0+			ann.rad./
^{123}Ba		122.9189	2.7 m	β^+ ,EC/5.5			-0.68	+1.5	ann.rad./ 0.0306 0.0927 0.1161 0.1235
^{124}Ba		123.91509	12. m	β^+ ,EC/2.65					ann.rad./ 0.1695 0.1888 1.2160
^{125m}Ba			8. m	β^+ ,EC/	4.5				ann.rad./
^{125}Ba		124.9146	3.5 m	β^+ ,EC/4.6	3.4	1/2+	+0.18		0.0550 0.0776 0.0854 0.1409
^{126}Ba		125.91124	1.65 h	β^+ /2 /1.67 EC/98 /		0+			Cs k x-ray 0.2179 0.2336 0.2576
^{127}Ba		126.9111	12.9 m	β^+ /54 /3.5 EC/46 /		1/2+	+0.09		ann.rad./ Cs k x-ray 0.1148 0.1808 (0.07 - 2.5)
^{128}Ba		127.90831	2.43 d	EC/0.52		0+			Cs k x-ray 0.27344
^{129m}Ba			2.17 h	EC/98 / β^+ /2 /		7/2+			Cs k x-ray 0.1769 0.1823 0.2023 1.4593
^{129}Ba		128.90868	2.2 h	β^+ /20 /2.43 EC/80 /	1.42/	1/2+	-0.40		ann.rad./ Cs k x-ray 0.1291 0.2143 0.2208
^{130}Ba	0.106(2)	129.90631				0+			Ba k x-ray
^{131m}Ba			14.6 m	I.T./0.187		9/2-	-0.87	+1.5	0.1085

Elem. or Isot.	Natural Abundance (%)	Atomic Mass or Weight	Half-Life	Decay Mode/Energy (/MeV)	Particle Energy /Intensity (MeV / %)	Spin (h/2π)	Nuclear Magnetic Mom.(nm)	Elect. Quadr. Mom. (b)	γ-ray Energy /Intensity (MeV / %)
¹³¹Ba		130.90693	11.7 d	EC/1.37		1/2+	0.7081		Cs k x-ray 0.12381 0.21608 0.49636
¹³²Ba	0.101(3)	131.905056				0+			
^{133m}Ba			1.621 d	I.T./0.288		11/2-	-0.91	+0.9	Ba k x-ray 0.2761
¹³³Ba		132.906003	10.53 y	EC/0.517		1/2+	0.7717		Cs k x-ray 0.08099 0.35600
¹³⁴Ba	2.417(27)	133.904504				0+			
^{135m}Ba			1.20 d	I.T./0.2682		11/2-	-1.00	+1.0	Ba k x-ray 0.2682
¹³⁵Ba	6.592(18)	134.905684				3/2+	+0.838	+0.16	
^{136m}Ba			0.308 s	I.T./2.0305		7-			Ba k x-ray 0.8185 1.0481
¹³⁶Ba	7.854(36)	135.904571				0+			
^{137m}Ba			2.552 m	I.T./0.6617		11/2-	-0.99	+0.8	Ba k x-ray 0.66164
¹³⁷Ba	11.23(4)	136.905822				3/2+	+0.9374	+0.245	
¹³⁸Ba	71.70(7)	137.905242				0+			
¹³⁹Ba		138.908836	1.396 h	β⁻/2.317	2.14/27 2.27/72	7/2-	-0.97	-0.57	0.16585 1.2544 1.42033
¹⁴⁰Ba		139.91060	12.75 d	β⁻/1.05	0.48 1.0/ 1.02/	0+			0.16268 0.30485 0.53727
¹⁴¹Ba		140.91441	18.3 m	β⁻/3.22	2.59/ 2.73/	3/2-	-0.34	+0.45	0.1903 0.2770 0.3042 (0.1 - 2.5)
¹⁴²Ba		141.91645	10.7 m	β⁻/2.212	1.0/ 1.10/	0+			0.23152 0.25512 0.3090 1.2040
¹⁴³Ba		142.92061	14.3 s	β⁻/4.24	4.2/	5/2+	+0.44	-0.88	0.1786 0.21148 0.7988 (0.17 - 2.4)
¹⁴⁴Ba		143.92294	11.4 s	β⁻/3.1	2.4/ 2.9/	0+			La k x-ray 0.10386 0.1566 0.1728 0.3882 0.43048
¹⁴⁵Ba		144.9269	4.0 s	β⁻/4.9	4.9/	(5/2-)	-0.28	+1.22	La k x-ray 0.0918 0.09709
¹⁴⁶Ba		145.9302	2.20 s	β⁻/4.12	3.9/	0+			0.0644 0.2513 0.3270 0.3329 0.3622
¹⁴⁷Ba		146.9340	0.892 s	β⁻/5.75	5.5/				
¹⁴⁸Ba		147.9377	0.64 s	β⁻,n/5.11					
¹⁴⁹Ba		148.9421	0.36 s	β⁻,(n)/7.3					
₅₇La		138.9055(2)							
¹²⁰La		119.938	2.8 s	EC,β⁺/11.					
¹²²La		121.931	9. s	EC,β⁺/ ≈ 9.7					
¹²³La		122.9262	17. s	EC /7.					
¹²⁴La		123.9245	30. s	EC/ ≈ 8.8		(7+)			
¹²⁵La		124.9207	1.2 m	β⁺,EC/5.6		11/2-			ann.rad./ 0.0436 0.0676
¹²⁶La		125.9194	1.0 m	β⁺,EC/7.6					ann.rad./ 0.2561 0.340 0.4555 0.6214
¹²⁷La		126.9162	3.8 m	β⁺,EC/4.7		3/2+			ann.rad./ 0.025 0.0562

Elem. or Isot.	Natural Abundance (%)	Atomic Mass or Weight	Half-Life	Decay Mode/Energy (/MeV)	Particle Energy /Intensity (MeV / %)	Spin (h/2π)	Nuclear Magnetic Mom.(nm)	Elect. Quadr. Mom. (b)	γ-ray Energy /Intensity (MeV / %)
^{128}La		127.9155	5.0 m	β^+ /80 /6.7 EC/20 /		(5-)			ann.rad./ Ba k x-ray 0.2841 0.4793 0.6436
^{129m}La			0.56 s	IT		(11/2-)			
^{129}La		128.91267	11.6 m	β^+ /58 /3.72 EC/42 /	2.42/	3/2+			ann.rad./ Ba k x-ray 0.1105 0.2786 (0.1 - 1.8)
^{130}La		129.9123	8.7 m	β^+ /78 /5.6 EC/22 /		3+			ann.rad./ Ba k x-ray 0.3573 0.5444 0.5506 0.9079
^{131}La		130.9101	59. m	β^+ /76 /3.0 EC/24 /	1.42/ 1.94/	3/2+			ann.rad./ Ba k x-ray 0.1085 0.3658 0.5263
^{132m}La			24. m	I.T./76 / β^+ ,EC/24 /		6-			La k x-ray / 0.1352 0.4645
^{132}La		131.91011	4.8 h	β^+ /40 /4.71 EC/60 /	2.6/ 3.2 3.7/	2-			ann.rad./ Ba k x-ray 0.4645 0.5671
^{133}La		132.9084	3.91 h	β^+ /4 /2.2 EC/96 /	1.2/	5/2+			Ba k x-ray 0.2788 0.2901 0.3024
^{134}La		133.90849	6.5 m	β^+ /63 /3.71 EC/37/	2.67/	1+			ann.rad./ Ba k x-ray 0.6047 (0.5 - 1.9)
^{135}La		134.90697	19.5 h	EC/1.20		5/2+			Ba k x-ray 0.4805
^{136}La		135.9077	9.87 m	β^+ /36 /2.9 EC/64 /	1.8/	1+			ann.rad./ Ba k x-ray 0.8185
^{137}La		136.90647	6x10^4 y	EC/0.60		7/2+			
^{138}La	0.0902(2)	137.907107	1.06x10^{11} y			5+	+2.70	+0.2	
^{139}La	99.9098(2)	138.906349				7/2+			
^{140}La		139.909473	1.678 d	β^- /3.762	1.35 1.24/ 1.67/	3-			
^{141}La		140.910958	3.90 h	β^- /2.502	2.43/	7/2+			
^{142}La		141.91408	1.54 h	β^- /4.505	2.11/ 2.98/ 4.52/	2-			
^{143}La		142.91606	14.1 m	β^- /3.43	3.3/	7/2-			
^{144}La		143.9196	40.7 s	β^- /5.5	4.1/				
^{145}La		144.9217	24. s	β^- /4.1	4.1/	3/2+			
^{146m}La			10.0 s	β^- /6.7	5.5/	(6)			
^{146}La		145.9258	6.3 s	β^- /6.6	6.2/	(2-)			
^{147}La		146.9278	4.02 s	β^- /5.0	4.6/				
^{148}La		147.9322	1.1 s	β^- /7.26		2-			
^{149}La		148.9342	1.10 s	β^- /5.5					
$_{58}$Ce		**140.115(4)**							
^{123}Ce		122.936	3.8 s	β^+ ,EC/ $\approx$ 8.6					ann.rad./
^{124}Ce		123.931	6. s	EC / $\approx$ 5.6					
^{125}Ce		124.929	10. s	β^+ ,EC/7.		(5/2+)			ann.rad./
^{126}Ce		125.9241	50. s	EC/4.					
^{127}Ce		126.9228	32. s	β^+ ,EC/6.1					ann.rad./ 0.06-0.25
^{128}Ce		127.9189	6. m	β^+ ,EC/3.2					ann.rad./
^{129}Ce		128.9187	3.5 m	β^+ ,EC/5.6					ann.rad./ 0.0675
^{130}Ce		129.9147	26. m	β^+ ,EC/2.2		0+			ann.rad./ La k x-ray

Elem. or Isot.	Natural Abundance (%)	Atomic Mass or Weight	Half-Life	Decay Mode/Energy (/MeV)	Particle Energy /Intensity (MeV / %)	Spin (h/2π)	Nuclear Magnetic Mom.(nm)	Elect. Quadr. Mom. (b)	γ-ray Energy /Intensity (MeV / %)
^{131m}Ce			5. m	β^+ ,EC/					ann.rad./ 0.2304 0.3955 0.4213
^{131}Ce		130.9144	10. m	β^+ ,EC/4.0	2.8/				ann.rad./ 0.119 0.169 0.414
^{132}Ce		131.9115	3.5 h	EC/1.3		0+			La k x-ray 0.1554 0.1821
^{133m}Ce			1.6 h	β^+ ,EC/		1/2+			ann.rad./ 0.0769 0.0973 0.5577
^{133}Ce		132.9116	5.4 h	β^+ /8 /2.9 EC/92 /	1.3/	9/2-			ann.rad./ La k x-ray 0.0584 0.1308 0.4722 0.5104
^{134}Ce		133.9090	3.16 d	EC/0.5		0+			La k x-ray 0.1304 0.1623 0.6047
^{135m}Ce			20. s	I.T./0.446		11/2-			Ce k x-ray 0.0826 0.1497 0.2134
^{135}Ce		134.90915	17.7 h	β^+ /1 /2.026 EC/99 /	0.8/	1/2+			La k x-ray 0.0345 0.2656 0.3001 0.6068
^{136}Ce	0.19(1)	135.90714				0+			
^{137m}Ce			1.43 d	I.T./99 /0.254 EC/0.8 /		11/2-	0.70		Ce k x-ray 0.1693 0.2543
^{137}Ce		136.90778	9.0 h	β^+ /1.222		3/2+	0.9		La k x-ray 0.4472
^{138}Ce	0.25(1)	137.90599				0+			
^{139m}Ce			56.4 s	I.T./0.7542		11/2-			Ce k x-ray 0.7542
^{139}Ce		138.90665	137.6 d	EC/0.28		3/2+	0.9		La k x-ray 0.16585
^{140}Ce	88.48(10)	139.905435				0+			
^{141}Ce		140.908272	32.50 d	β^- /0.581	0.436/69 0.581/31	7/2-	1.1		Pr k x-ray 0.14544/48.0
^{142}Ce	11.08(10)	141.909241				0+			
^{143}Ce		142.912382	1.38 d	β^- /1.462	1.404/ 1.110/47	3/2-	≈ 1.		Pr k x-ray 0.0574 0.2933
^{144}Ce		143.913643	284.6 d	β^- /0.319	0.185/20 0.318/	0+			Pr k x-ray 0.0801 0.1335
^{145}Ce		144.91723	3.00 m	β^- /2.54	1.7/24 1.3	3/2-			Pr k x-ray 0.0627 0.7245
^{146}Ce		145.9187	13.5 m	β^- /1.04	0.7/90	0+			Pr k x-ray 0.0986 0.2182 0.3167
^{147}Ce		146.9225	56. s	β^- /3.29	3.3/				0.0930 0.2687
^{148}Ce		147.9244	56. s	β^- /2.1	1.66/	0+			0.0904 0.0985 0.1212 0.2918
^{149}Ce		148.9283	5.2 s	β^- /4.2					0.0577 0.0864 0.3800
^{150}Ce		149.9302	4.4 s	β^- /3.0					0.1099
^{151}Ce		150.9340	1.0 s	β^- /5.3					0.0526
^{152}Ce		151.9366	1.4 s	β^- /4.4					Pr k x-ray 0.098 0.115
$_{59}$Pr		140.90765(2)							

Elem. or Isot.	Natural Abundance (%)	Atomic Mass or Weight	Half-Life	Decay Mode/Energy (/MeV)	Particle Energy /Intensity (MeV / %)	Spin (h/2π)	Nuclear Magnetic Mom.(nm)	Elect. Quadr. Mom. (b)	γ-ray Energy /Intensity (MeV / %)
^{124}Pr		123.943	1.2 s	β⁺ ,EC/12.					ann.rad./
^{126}Pr		125.935	3. s	β⁺ ,EC/ ≈10.4					ann.rad./
^{127}Pr		126.931	≈6. s	β⁺ / ≈7.5					ann.rad./ (0.12-0.16)
^{128}Pr		127.9288	3.2 s	β⁺ ,EC/ ≈9.3					ann.rad./
^{129}Pr		128.9249	0.5 m	β⁺ ,EC/5.8					ann.rad./ (0.04-0.20)
^{130}Pr		129.9234	40. s	β⁺ ,EC/8.1					ann.rad./
^{131m}Pr			5.7 s						(0.06 - 0.16)
^{131}Pr		130.9201	1.7 m	β⁺ ,EC/5.3				≈5.5	ann.rad./ (0.07-0.34)
^{132}Pr		131.9191	1.6 m	β⁺ ,EC/7.1					ann.rad./ 0.325 0.496 0.533
^{133}Pr		132.9162	6.5 m	β⁺ ,EC/4.3		5/2+			ann.rad./ 0.074 0.1343 0.2419 0.3156 0.3308 0.4650
^{134m}Pr			≈ 11. m	β⁺ ,EC/					ann.rad./ 0.294 0.460 0.495 0.632
^{134}Pr		133.9157	17. m	β⁺ ,EC/6.2		2+			ann.rad./ 0.294 0.495
^{135}Pr		134.9131	24. m	β⁺ ,EC/3.7	2.5/	3/2+			ann.rad./ 0.0826 0.2135 0.2961 0.5832
^{136}Pr		135.91265	13.1 m	β⁺ /57 /5.13 EC/43	2.98/	2+			ann.rad./ Ce k x-ray 0.5398 0.5522
^{137}Pr		136.91068	1.28 h	β⁺ /26 /2.70 EC/74 /	1.68/	5/2+			ann.rad./ Ce k x-ray 0.4339 0.5140 0.8367 (0.16 - 1.8)
^{138m}Pr			2.1 h	β⁺ /24 / EC/76 /	1.65/	7-			ann.rad./ Ce k x-ray 0.3027 0.7887 1.0378 (0.07 - 2.0)
^{138}Pr		137.91075	1.45 m	β⁺ /75 /4.44 EC/25 /	3.42/	1+			ann.rad./ Ce k x-ray 0.7887
^{139}Pr		138.90893	4.41 h	β⁺ /8 /2.129 EC/92 /	1.09/	5/2+			ann.rad./ Ce k x-ray 0.2551 1.3473 1.6307
^{140}Pr		139.90907	3.39 m	β⁺ /51 /3.39 EC/49 /	2.37/	1+			ann.rad./ Ce k x-ray 0.3069 1.5965
^{141}Pr	100.	140.907648				5/2+	+4.275	-0.059	
^{142m}Pr			14.6 m	I.T./0.004	c.e./	5-	2.2		
^{142}Pr		141.910041	19.12 h	β⁻ /2.162 EC/0.744	0.58/4 2.16/96	2-	+0.234	+0.030	0.5088 1.57580
^{143}Pr		142.910813	13.57 d	β⁻ /0.934	0.933/	7/2+	+2.7	+0.8	0.7420
^{144m}Pr			7.2 m	I.T./99+ /0.059 β⁻ /		3-			Pr k x-ray 0.0590 0.6965 0.8142
^{144}Pr		143.913301	17.28 m	β⁻ /2.998	0.807/1 2.30/ 2.996/98	0-			0.69649 1.48912 2.18562
^{145}Pr		144.91451	5.98 h	β⁻ /1.81	1.80/97	7/2+			0.0725 0.6758 0.7483

Elem. or Isot.	Natural Abundance (%)	Atomic Mass or Weight	Half-Life	Decay Mode/Energy (/MeV)	Particle Energy /Intensity (MeV / %)	Spin (h/2π)	Nuclear Magnetic Mom.(nm)	Elect. Quadr. Mom. (b)	γ-ray Energy /Intensity (MeV / %)
[146]Pr		145.9176	24.2 m	β⁻ /4.2	2.2/30 3.7/10 4.2/40	2-			0.4539/48 1.5247
[147]Pr		146.91898	13.4 m	β⁻ /2.69	1.5/ 2.1/	3/2+			0.3146/24. 0.5779/16 0.6413/19.
[148m]Pr			2.0 m	β⁻ /	4.0/ 3.8/	(4)			0.3016 0.4506 0.6975
[148]Pr		147.9222	2.27 m	β⁻ /4.9	4.8/ 4.5/	1-			0.3017
[149]Pr		148.92379	2.3 m	β⁻ /3.40	3.0	(5/2+)			0.1085 0.1385 0.1651
[150]Pr		149.9270	6.2 s	β⁻ /5.7	≈ 5.5	1-			0.1302 0.8044 0.8527
[151]Pr		150.9283	22.4 s	β⁻ /4.2					
[152]Pr		151.9319	3.2 s	β⁻ /6.7		(3)			0.0726 0.164 0.285
[153]Pr		152.9339	4.3 s	β⁻ /5.5					
[154]Pr		153.9381	2.3 s	β⁻ /7.9					
₆₀Nd		144.24(3)							
[127]Nd		126.941	1.8 s	β⁺ ,EC/9.		(5/2)			ann.rad./
[128]Nd		127.935	4. s	β⁺ ,EC/6.					ann.rad./
[129]Nd		128.933	4.9 s	β⁺ ,EC/8.		5/2(-)			ann.rad./ (0.091-0.33)
[130]Nd		129.929	28. s	β⁺ ,EC/5.					ann.rad./
[131]Nd		130.9271	0.5 m	β⁺ ,EC/6.6					ann.rad./ (0.09-0.36)
[132]Nd		131.9231	1.5 m	β⁺ ,EC/3.7					ann.rad./ (0.099-0.567)
[133]Nd		132.9222	1.2 m	β⁺ ,EC/5.6					ann.rad./ (0.06-0.37)
[134]Nd		133.9187	≈ 8.5 m	β⁺ /17 /2.8 EC/83 /		0+			ann.rad./ Pr k x-ray 0.1631/58 (0.09-1.00)
[135m]Nd			5.5 m	β⁺ /					
[135]Nd		134.9182	12. m	β⁺ /65 /4.8 EC/35 /		9/2-	0.78	+2.1	ann.rad./ Pr k x-ray 0.0415/23. 0.204/51. (0.11-1.8)
[136]Nd		135.9150	50.6 m	EC/94 /2.21 β⁺ /6 /	1.04/	0+			Pr kx-ray 0.0401/21. 0.1091/35. (0.10-0.97)
[137m]Nd			1.6 s	I.T./0.5196		11/2-			Nd k x-ray 0.1084 0.1775 0.2337
[137]Nd		136.9146	38. m	β⁺ /40 /3.69 EC/60 /	1.7/20 2.40/20	1/2+	-0.63		ann.rad./ Pr k x-ray 0.0755 0.5806
[138]Nd		137.9119	5.1 h	EC/1.1		0+			Pr k x-ray 0.1995 0.3258
[139m]Nd			5.5 h	I.T./12 /0.231 β⁺ /88 /	1.17/	11/2-			Nd k x-ray Pr k x-ray 0.1139/34. 0.7382/30.
[139]Nd		138.91192	30. m	β⁺ /25 /2.79 EC/75 /	1.77/	3/2+	0.91	+0.3	ann.rad./ Pr k x-ray 0.4050
[140]Nd		139.90931	3.37 d	EC /0.22		0+			Pr k x-ray
[141m]Nd			1.04 m	I.T./99+ /0.756		11/2-			Nd k x-ray 0.7565
[141]Nd		140.909605	2.49 h	EC/98 /1.823 β⁺ /2 /	0.802/	3/2+	1.01	+0.3	Pr k x-ray (0.15-1.7)
[142]Nd	27.13(12)	141.907719				0+			
[143]Nd	12.18(6)	142.909810				7/2-	-1.07	-0.6	
[144]Nd	23.80(12)	143.910083	2.1x10¹⁵ y			0+			
[145]Nd	8.30(6)	144.912569				7/2-	-0.66	-0.33	
[146]Nd	17.19(9)	145.913113				0+			

Elem. or Isot.	Natural Abundance (%)	Atomic Mass or Weight	Half-Life	Decay Mode/Energy (/MeV)	Particle Energy /Intensity (MeV / %)	Spin (h/2π)	Nuclear Magnetic Mom.(nm)	Elect. Quadr. Mom. (b)	γ-ray Energy /Intensity (MeV / %)
¹⁴⁷Nd		146.916096	10.98 d	β⁻ /0.896	0.805/	5/2-	0.58	0.9	Pr k x-ray 0.09111 0.53102
¹⁴⁸Nd	5.76(3)	147.916889				0+			
¹⁴⁹Nd		148.920145	1.73 h	β⁻ /1.691	1.03/25 1.13/26 1.42/	5/2-	0.35	1.3	Pr k x-ray 0.1143/19. 0.2113/27. (0.06 - 1.6)
¹⁵⁰Nd	5.64(3)	149.920887				0+			
¹⁵¹Nd		150.923825	12.4 m	β⁻ /2.442	1.2/	(3/2+)			Pm k x-ray 0.1168 0.2557 1.1806 (0.10 - 1.9)m
¹⁵²Nd		151.92468	11.4 m	β⁻ /1.1		0+			0.2785/29. 0.2501/18. (0.016 - 0.66)
¹⁵³Nd		152.9280	28.9 s	β⁻ /3.6					0.418
¹⁵⁴Nd		153.9296	25.9 s	β⁻ /2.8					0.1519 0.7998
¹⁵⁵Nd		154.9334	8.9 s	β⁻ /5.0					0.1807
¹⁵⁶Nd		155.9355	5.5 s	β⁻ /4.1					0.0848
₆₁Pm									
¹³⁰Pm		129.940	2.2 s	β⁺ ,EC/11.					ann.rad./
¹³²Pm		131.934	6. s	β⁺ ,EC/10.					ann.rad./
¹³³Pm		132.930	12. s	β⁺ ,EC/ ≈ 7.0					ann.rad./
¹³⁴Pm		133.9282	24. s	β⁺ ,EC/ ≈ 8.9		(5+)			ann.rad./ 0.294 0.495 (0.13-0.47)
¹³⁵Pm		134.9247	0.8 m	β⁺ ,EC/6.0		11/2-			ann.rad./
¹³⁶Pm		135.9235	1.8 m	β⁺ /89 /7.9 EC/11 /		(3+)			Nd k x-ray 0.3735 0.6027
¹³⁷Pm		136.9206	2.4 m	β⁺ ,EC/5.6		(11/2-)			ann.rad./ 0.1086 0.1775
¹³⁸ᵐPm			3.2 m	β⁺ /50 / ≈ 7.0 EC/50 /	3.9/	3+			ann.rad./ Nd k x-ray 0.5209 0.7290
¹³⁸Pm		137.9193	10. s	β⁺ /6.9	6.1/	1+			ann.rad./ 0.1887
¹³⁹ᵐPm			0.18 s	IT/		(11/2-)			
¹³⁹Pm		138.91678	4.14 m	β⁺ /68 /4.52 EC/32 /	3.52/	(5/2+)			ann.rad./ Nd k x-ray 0.4028 (0.27 - 2.4)
¹⁴⁰ᵐPm			5.87 m	β⁺ /70 / EC/30 /	3.2	7/2-			ann.rad./ Nd k x-ray 0.4199 0.7738 1.0283
¹⁴⁰Pm		139.91585	9.2 s	β⁺ /89 /6.09 EC/11 /	5.07/74	1+			ann.rad./ Nd k x-ray 0.7738 1.4898
¹⁴¹Pm		140.91359	20.9 m	β⁺ /52 /3.72 EC/48 /	2.71	5/2+			ann.rad./ Nd k x-ray 0.8862 1.2233
¹⁴²Pm		141.91295	40.5 s	β⁺ /86 /4.87 EC/20 /	3.8/	1+			ann.rad./ Nd k x-ray 0.6414 1.5758
¹⁴³Pm		142.910928	265. d	EC/1.041 β⁺ /<6 x 10⁻⁶/		5/2+	3.8		Nd k x-ray 0.7420
¹⁴⁴Pm		143.912586	360. d	EC/2.332 β⁺ /<8 x 10⁻⁵/		5-	1.7		Nd k x-ray 0.6180 0.6965
¹⁴⁵Pm		144.912745	17.7 y	EC/0.163		5/2+			Nd k x-ray 0.0723
¹⁴⁶Pm		145.914693	5.53 y	EC/63 /1.472 β⁻ /37 /1.542	0.795/	3-			Nd k x-ray 0.4538 0.7362 0.7474

Elem. or Isot.	Natural Abundance (%)	Atomic Mass or Weight	Half-Life	Decay Mode/Energy (/MeV)	Particle Energy /Intensity (MeV / %)	Spin (h/2π)	Nuclear Magnetic Mom.(nm)	Elect. Quadr. Mom. (b)	γ-ray Energy /Intensity (MeV / %)
[147]Pm		146.915134	2.6234 y	β⁻ /0.224	0.224/	7/2+	+2.6	+0.7	0.1213
[148m]Pm			41.3 d	β⁻ /95 /2.6 I.T./5 /0.137	0.4/60 0.5/17 0.7/21	6-	1.8		0.1974 0.5503/94. 0.6300/89. 0.7257/33
[148]Pm		147.91747	5.37 d	β⁻ /2.47	1.02/ 2.47/	1-	+2.0	+0.2	0.5503 0.9149 1.4651
[149]Pm		148.918330	2.212 d	β⁻ /1.071	0.78/9 1.072/90	7/2+	3.3		0.2859 0.5909 0.8594
[150]Pm		149.92098	2.68 h	β⁻ /3.45	1.6/ 2.3/ 1.8/	(1-)			0.3339/69. 1.1658/16. 1.3245/17. (0.25 - 2.9)
[151]Pm		150.92120	1.183 d	β⁻ /1.187	0.84/	5/2+	+1.8	1.9	0.1677/8 0.2751/7 0.3401/22 (0.14-1.4)
[152m2]Pm			15. m	β⁻ ,I.T./		(>6)			
[152m1]Pm			7.5 m	β⁻ /		(4-)			0.1218 0.2447 0.3404 1.0971 1.4375
[152]Pm		151.9235	4.1 m	β⁻ /3.5	3.5/20 3.50/60	1+			0.1218 (0.12 - 2.1)
[153]Pm		152.92414	5.4 m	β⁻ /1.90	1.7/	(5/2-)			0.0910 0.1198 0.1273
[154m]Pm			2.7 m	β⁻ /	2.0/				0.0820 0.1848 1.4403
[154]Pm		153.9266	1.7 m	β⁻ /4.1	1.9/				0.0820 0.8396 1.3940 2.0589 (0.08 - 2.8)
[155]Pm		154.9280	48. s	β⁻ /3.2		(5/2-)			(0.05-0.78)
[156]Pm		155.93106	26.7 s	β⁻ /5.16					
[157]Pm		156.9332	10.9 s	β⁻ /4.6					
[158]Pm		157.9367	5. s	β⁻ /6.3					
₆₂Sm		150.36(3)							
[131]Sm			1.2 s	β⁺ ,EC/					ann.rad./
[133]Sm		132.939	2.9 s	β⁺ ,EC/ ≈8.4		5/2+			ann.rad./
[134]Sm		133.934	11. s	β⁺ ,EC/5.		0+			ann.rad./
[135]Sm		134.932	10. s	β⁺ ,EC/7.		7/2+			ann.rad./
[136]Sm		135.9283	42. s	β⁺ ,EC/4.5		0+			ann.rad./
[137]Sm		136.9271	45. s	β⁺ ,EC/6.1					ann.rad./
[138]Sm		137.9235	3.0 m	β⁺ ,EC/3.9		0+			ann.rad./ 0.0536 0.0747
[139m]Sm			10. s	I.T./94 /0.457 β⁺ /6 /	4.7	(11/2-)			Sm k x-ray 0.1118 0.1553 0.1901 0.2673
[139]Sm		138.9226	2.6 m	β⁺ /75 /5.5 EC/25 /	4.1/	1/2+	-0.53		Pm k x-ray 0.3678 0.4028 (0.27 - 2.4)
[140]Sm		139.9195	14.8 m	β⁺ ,EC/3.4	1.9/	0+			ann.rad./ Pm k x-ray 0.1396 0.2255 (0.07 - 1.7)
[141m]Sm			22.6 m	β⁺ /32 / EC/68 / I.T./0.3 /0.1758	1.6/ 2.19/	11/2-	-0.83	+1.6	ann.rad./ Pm k x-ray 0.1966 0.4318 0.7774
[141]Sm		140.91847	10.2 m	β⁺ /52 /4.54 EC/48 /	3.2/	1/2+	-0.74		ann.rad./ Pm k x-ray 0.4382
[142]Sm		141.91520	1.208 h	β⁺ /6 /2.10 EC/94 /	1.0/	0+			ann.rad./ Pm k x-ray
[143m]Sm			1.10 m	I.T./99 /0.7540		11/2-			Sm k x-ray 0.7540

Elem. or Isot.	Natural Abundance (%)	Atomic Mass or Weight	Half-Life	Decay Mode/Energy (/MeV)	Particle Energy /Intensity (MeV / %)	Spin (h/2π)	Nuclear Magnetic Mom.(nm)	Elect. Quadr. Mom. (b)	γ-ray Energy /Intensity (MeV / %)
^{143}Sm		142.914624	8.83 m	β^+ /46 /3.443 EC/54 /	2.47/	3/2+	+1.01	+0.4	ann.rad./ Pm k x-ray 1.0565
^{144}Sm	3.1(1)	143.911996				0+			
^{145}Sm		144.913407	340. d	EC/0.617		7/2-	-1.1	-0.6	Pm k x-ray 0.0613 0.4924
^{146}Sm		145.913038	1.03×10^8y	α/	2.50/	0+			
^{147}Sm	15.0(2)	146.914894	1.06×10^{11}y	α/	2.23/	7/2-	-0.815	-0.26	
^{148}Sm	11.3(1)	147.914818	7×10^{15}y	α/	1.96/	0+			
^{149}Sm	13.8(1)	148.917180	10^{16}y	α/		7/2-	-0.672	+0.075	
^{150}Sm	7.4(1)	149.917272				0+			
^{151}Sm		150.919929	90. y	β^- /0.0768	0.076/	5/2-	-0.363	+0.7	0.02154
^{152}Sm	26.7(2)	151.919729				0+			
^{153}Sm		152.922094	1.929 d	β^- /0.808	0.64/ 0.69/	3/2+	-0.0216	+1.3	Eu k x-ray 0.069676 0.10318
^{154}Sm	22.7(2)	153.922206				0+			
^{155}Sm		154.924636	22.2 m	β^- /1.627	1.52	3/2-		1.1	Eu k x-ray 0.1043/75.
^{156}Sm		155.92553	9.4 h	β^- /0.72	0.43/ 0.71/	0+			0.0872 0.1657 0.2038
^{157}Sm		156.9283	8.0 m	β^- /2.7	2.4/	3/2-			Eu k x-ray 0.1964 0.1978 0.3942
^{158}Sm		157.9299	5.5 m	β^- /2.0		0+			0.1894/100. 0.3636/82.
^{159}Sm		158.9332	11.3 s	β^- /3.8					0.1898
^{160}Sm		159.9353	9.6 s	β^- /3.6		0+			0.110
$_{63}$Eu		**151.965(9)**							
^{134}Eu			0.5 s	EC,β^+					ann.rad./
^{135}Eu		134.942	1.5 s	EC,β^+ / ≈ 8.7					ann.rad./ 0.255
^{136m}Eu			≈ 3.2 s			7+			
^{136}Eu		135.940	≈ 3.9 s	EC,β^+ /10.		1+			ann.rad./
^{137}Eu		136.935	11. s	EC/ ≈7.5		11/2-			ann.rad./
^{138}Eu		137.9335	12. s	EC,β^+ / ≈ 9.2		7+			ann.rad./
^{139}Eu		138.9298	18. s	EC,β^+ /6.7					ann.rad./
^{140m}Eu			0.125 s	EC,β^+					ann.rad./
^{140}Eu		139.9285	1.51 s	EC,β^+ /8.4		1-			ann.rad./
^{141m}Eu			3.0 s	β^+ /58 / EC/9 / I.T./33 /0.0964		11/2-			Eu k x-ray (0.09 - 1.6)
^{141}Eu		140.9244	40. s	β^+/81 /5.6 EC/15 /		5/2+	+3.49	+0.85	ann.rad./ Sm k x-ray 0.3845 0.3940
^{142m}Eu			1.22 m	β^+/83 / EC/17 /	4.8/	8-	+2.98	+1.4	ann.rad./ Sm k x-ray 0.5566 0.7680 1.0233
^{142}Eu		141.9231	2.4 s	β^- /94/7.4 EC/6 /	7.0/	1+	+1.54	+0.12	ann.rad./ 0.7680
^{143}Eu		142.92017	2.62 m	β^+ /72/5.17 EC/28/	4.1/ 5.1/	5/2+	+3.67	+0.51	ann.rad./ Sm k x-ray 0.1107/7 1.5368/3. 1.9127/2.
^{144}Eu		143.91879	10.2 s	β^+ /86 /6.33 EC/13 /	5.31/	1+	+1.89	+0.10	ann.rad./ Sm k x-ray 1.6601
^{145}Eu		144.916263	5.93 d	β^+ /2 /2.660 EC/98 /1.71	0.79/	5/2+	+3.99	+0.29	ann.rad./ Sm k x-ray 0.6535 0.8937 1.6587
^{146}Eu		145.91720	4.57 d	β^+ /5 /3.88 EC/95 /	1.47/	4-	+1.43	-0.18	ann.rad./ Sm k x-ray 0.6336 0.6341 0.7470 (0.27 - 2.64)

Elem. or Isot.	Natural Abundance (%)	Atomic Mass or Weight	Half-Life	Decay Mode/Energy (/MeV)	Particle Energy /Intensity (MeV / %)	Spin (h/2π)	Nuclear Magnetic Mom.(nm)	Elect. Quadr. Mom. (b)	γ-ray Energy /Intensity (MeV / %)
[147]Eu		146.916742	24.4 d	EC/99. /1.722 β⁺ /0.4 /		5/2+	+3.72	+0.55	Sm k x-ray 0.12113 0.19725 0.6776
[148]Eu		147.91815	54.5 d	EC/3.11	0.92	5-	+2.34	+0.35	Sm k x-ray 0.5503/99. 0.6299/71. (0.067-2.17)
[149]Eu		148.91792	93.1 d	EC/0.692		5/2+	+3.57	+0.75	Sm k x-ray 0.2770 0.3275
[150]Eu		149.91970	36. y	EC/2.26		5-	+2.71	+1.13	Sm k x-ray 0.3340 0.4394 0.5843 (0.25 - 1.8)
[150m]Eu			12.8 h	β⁻ /92 / β⁺ /0.4 / EC/8 /	1.013/ 1.24/	0-			Sm k x-ray 0.3339 0.4065
[151]Eu	47.8(1.5)	150.919846				5/2+	+3.472	+0.90	
[152m2]Eu			1.60 h	I.T./0.1478		8-			Eu k x-ray 0.0898
[152m1]Eu			9.30 h	β⁻ /72 / EC/28 /	1.85/ 0.89/	0-			Sm k x-ray 0.12178 0.84153 0.96334
[152]Eu		151.921741	13.5 y	EC/72 /1.874 β⁻ /28 /1.818	0.69/ 1.47/	3-	-1.941	+2.71	Sm k x-ray Gd k x-ray 0.12178 0.34427 1.40802 (0.252-1.528)
[153]Eu	52.2(15)	152.921227				5/2+	+1.533	+2.41	
[154m]Eu			46.1 m	I.T./≈ 0.16		8-			Eu k x-ray 0.0682 0.1009
[154]Eu		153.922976	8.59 y	β⁻ /99.9 /1.969 EC/0.02 /0.717	0.27/29 0.58/38 0.84/17 0.98/4 1.87/11	3-	-2.01	+2.8	Gd k x-ray 0.12299/40. 0.72331/20. 1.2745/36 (0.059-1.90)
[155]Eu		154.922890	4.76 y	β⁻ /0.252	0.15/	5/2+	+1.6	+2.3	Gd k x-ray 0.0865/30 0.1053/20
[156]Eu		155.92475	15.2 d	β⁻ /2.451	0.30/11 0.49/30 1.2/12 2.45/31	1+	≈ 1.1		0.08899/9. 0.64623/7. 0.723441/6. 0.8118/10.
[157]Eu		156.92542	15.13 h	β⁻ /1.36	0.98/ 1.30/41	(5/2+)			Gd k x-ray 0.0639/100. 0.3705/48. 0.4107/76.
[158]Eu		157.9278	45.9 m	β⁻ /3.5	2.5/	(1-)			0.0795 0.8976 0.9442 0.9771
[159]Eu		158.92909	18.1 m	β⁻ /2.51	2.4/ 2.57/	(5/2+)			0.0678 0.0786 0.0957
[160]Eu		159.9315	38. s	β⁻ /4.1	2.7/ 4.1/	(0-)			0.0753 0.1735 0.4131 0.5155 0.8217 0.9110 0.9246
[161]Eu		160.9337	27. s	β⁻ /3.7					0.0719
[162]Eu		161.9370	11. s	β⁻ /5.6					
₆₄Gd		157.25(3)							
[137]Gd		136.945	7. s	EC,β⁺ / ≈ 8.8					ann.rad./
[139]Gd		138.9381	5. s	EC,β⁺ / ≈ 7.7				ann.rad./	
[140]Gd		139.934	16. s	EC/4.8		0+			0.1748
[141m]Gd			25. s	EC,β⁺ /		11/2-			ann.rad./
[141]Gd		140.9322	21. s	β⁺ /7.3		0+			ann.rad./
[142]Gd		141.9276	1.17 m	EC,β⁺ /4.2		1/2+			ann.rad./

Elem. or Isot.	Natural Abundance (%)	Atomic Mass or Weight	Half-Life	Decay Mode/Energy (/MeV)	Particle Energy /Intensity (MeV / %)	Spin (h/2π)	Nuclear Magnetic Mom.(nm)	Elect. Quadr. Mom. (b)	γ-ray Energy /Intensity (MeV / %)
¹⁴³ᵐGd			1.84 m	β⁺ /67 / EC/33 / I.T./		11/2-			ann.rad./ Eu k x-ray 0.1176 0.2719 0.5880 0.6681 0.7999
¹⁴³Gd		142.9266	39. s	β⁺ /82 /6.0 EC/18 /		1/2+			ann.rad./ Eu k x-ray 0.2048 0.2588
¹⁴⁴Gd		143.9234	4.5 m	β⁺ /45 /4.3 EC/55 /	3.3/	0+			ann.rad./ Eu k x-ray 0.3332
¹⁴⁵ᵐGd			1.44 m	I.T./95 /0.749 β⁺ /4 /5.7		11/2-			0.0273 0.3295 0.3866 0.7214
¹⁴⁵Gd		144.92169	23.4 m	β⁺ /33 /5.05 EC/67 /	2.5/	1/2+			ann.rad./ Eu k x-ray 1.7579 1.8806 (0.32 - 3.69)
¹⁴⁶Gd		145.91831	48.3 d	EC/99.9 /1.03 β⁺ /0.2	0.35/	0+			Eu k x-ray 0.1147 0.1155 0.1546
¹⁴⁷Gd		146.919090	1.588 d	EC/99.8 /2.188 EC/0.2 /	0.93/	7/2-	1.0		Eu k x-ray 0.2293 0.3699 0.3960 0.9289 (0.1 - 1.8)
¹⁴⁸Gd		147.918111	75. y	α/3.27	3.1828/	0+	0+		
¹⁴⁹Gd		148.919339	9.3 d	EC/1.32		7/2-	0.9		Eu k x-ray 0.1496 0.2985 0.3465
¹⁵⁰Gd		149.91866	1.8x10⁶ y	α/2.80	2.73/	0+			
¹⁵¹Gd		150.920345	124. d	EC/0.464		7/2-	0.8		Eu k x-ray 0.1536 0.2432
¹⁵²Gd	0.20(1)	151.919789				0+			
¹⁵³Gd		152.921747	241.6 d	EC/0.485		3/2-	0.4		Eu k x-ray 0.09743 0.10318
¹⁵⁴Gd	2.18(3)	153.920862				0+			
¹⁵⁵Gd	14.80(5)	154.922619				3/2-	-2.59	+1.30	
¹⁵⁶Gd	20.47(4)	155.922120				0+			
¹⁵⁷Gd	15.65(3)	156.923957				3/2-	-3.40	+1.36	
¹⁵⁸Gd	24.84(12)	157.924101				0+			
¹⁵⁹Gd		158.926385	18.6 h	β⁻ 0.971	0.60/11 0.89/26 0.96/63	3/2-	-0.44		Tb k x-ray 0.05845 0.36351
¹⁶⁰Gd	21.86(4)	159.927051				0+			
¹⁶¹Gd		160.929666	3.66 m	β⁻ /1.956	1.56/85	5/2-			Tb k x-ray 0.1023 0.3149 0.3609 0.4030 0.4421
¹⁶²Gd		161.930981	8.4 m	β⁻ /1.39	1.0/	0+			0.2868 0.214 1.685
¹⁶³Gd		162.9340	1.13 m	β⁻ /3.1					
¹⁶⁴Gd		163.9359	45. s	β⁻ /2.3					
₆₅Tb		158.92534(2)							
¹⁴⁰Tb		139.946	2.4 s	β⁺ ,EC/11					
¹⁴¹Tb		140.941	3.5 s	β⁺ ,EC/ ≈ 8.3					
¹⁴²ᵐTb			0.30 s	β⁺ ,EC/		4-			
¹⁴²Tb		141.939	0.60 s	β⁺ ,EC/10.		0+			
¹⁴³Tb		142.9346	12. s	β⁺ ,EC/7.4		11/2-			
¹⁴⁴ᵐTb			4.1 s	IT		5-			
¹⁴⁴Tb		143.9324	< 1.5 s	β⁺ ,EC/8.4		1+			
¹⁴⁵ᵐTb			30. s	β⁺ ,EC/ ≈ 6.6		11/2-			ann.rad./ 0.2577 0.5370 0.9876

Elem. or Isot.	Natural Abundance (%)	Atomic Mass or Weight	Half-Life	Decay Mode/Energy (/MeV)	Particle Energy /Intensity (MeV / %)	Spin (h/2π)	Nuclear Magnetic Mom.(nm)	Elect. Quadr. Mom. (b)	γ-ray Energy /Intensity (MeV / %)
^{145}Tb		144.9287		β^+ ,EC/6.5		1/2+			
^{146m}Tb			23. s	β^+ /76 / EC/24 /		(5-)			ann.rad./ Gd k x-ray 1.0789 1.5795
^{146}Tb		145.9270	≈ 8. s	β^+ /8.1		1+			
^{147m}Tb			1.8 m	β^+ /35 / EC/65 /		11/2-			ann.rad./ Gd k x-ray 1.3977 1.7978
^{147}Tb		146.92404	1.6 h	β^+ /42 /4.61 EC/58 /		5/2+			ann.rad./ Gd k x-ray 0.1398 0.6944 1.1522
^{148m}Tb			2.3 m	β^+ /25 / EC/75 /		9+			ann.rad./ Gd k x-ray 0.3945 0.6319 0.7845 0.8824
^{148}Tb		147.92422	1.00 h	β^+ ,EC/5.69		2-			ann.rad./ Gd k x-ray 0.4888 0.7845 (0.14 - 3.8)
^{149m}Tb			4.16 m	EC/88 / β^+ /12 /		11/2-			ann.rad./ Gd k x-ray 0.1650 0.7960
^{149}Tb		148.923243	4.13 h	β^+ /4 /3.636 α/16/	1.8/ 3.97/	1/2+	<0.9		Gd k x-ray 0.1650 0.3522 0.3886 (0.1 - 3.2)
^{150m}Tb			6.0 m	β^+ /17 / EC/83 /					ann.rad./ Gd k x-ray 0.4384 0.6380 0.6504 0.8275
^{150}Tb		149.92366	3.3 h	β^+ ,EC/4.66		2-			ann.rad./ 0.4963 0.6380 (0.3 - 4.29)
^{151m}Tb			25. s	I.T./95 / β^+ ,EC/7 /		11/2-			0.0229 0.0495 0.3797 0.8305
^{151}Tb		150.923099	17.61 h	β^+ /1 /2.565 EC/99 /	0.70/	1/2+			Gd k x-ray 0.1083 0.2517 0.2870 (0.1 - 1.8)
^{152m}Tb			4.3 m	I.T./79 /0.5018 EC/21 /4.35		(8+)			Tb k x-ray Gd k x-ray 0.2833 0.3443 0.4111
^{152}Tb		151.92407	17.5 h	β^+ /20 /3.99 EC/80 /	2.5/ 2.8/	2-	0.9		ann.rad./ Gd k x-ray 0.3443 (0.2 - 2.88)
^{153}Tb		152.923433	2.34 d	EC/1.570		5/2+	3.5		Gd k x-ray 0.2119 (0.05 - 1.1)
^{154m2}Tb			23.1 h	EC/98 / I.T./2 /		(7-)			Gd k x-ray 0.1231 0.2479 0.3467 1.4199
^{154m1}Tb			9. h	β^+ /78 / I.T./22 /		(3-)	1.8	+3.	Gd k x-ray 0.1231 0.2479 0.5401 (0.12 - 2.57)

Elem. or Isot.	Natural Abundance (%)	Atomic Mass or Weight	Half-Life	Decay Mode/Energy (/MeV)	Particle Energy /Intensity (MeV / %)	Spin (h/2π)	Nuclear Magnetic Mom.(nm)	Elect. Quadr. Mom. (b)	γ-ray Energy /Intensity (MeV / %)
^{154}Tb		153.92469	21.5 h	EC/99 /3.56 β^+ /1 /	1.86/ 2.45	0-	0.9		Gd k x-ray 0.1231 1.2744 2.1872 (0.12 - 3.14)
^{155}Tb		154.92350	5.3 d	EC/0.82		3/2+	2.0		Gd k x-ray 0.08654 0.10530
^{156m2}Tb			1.02 d	I.T./		(7-)			Tb k x-ray 0.0496
^{156m1}Tb			5.3 h	I.T./0.0884		(0+)			Tb k x-ray 0.0884
^{156}Tb		155.924744	5.3 d	EC/2.444		3-	1.4	+2.3	Gd k x-ray 0.08896 0.19921 0.53435 1.22245
^{157}Tb		156.924021	1.1×10^2 y	EC/0.0601		3/2+	2.0		Gd k x-ray 0.0545
^{158m}Tb			10.5 s	I.T./0.11		0-			Gd k x-ray 0.0110
^{158}Tb		157.925410	1.8×10^2 y	EC/80 /1.220 β^- /20 /0.937		3-	+1.76	+2.7	Gd k x-ray 0.0795 0.9442 0.9621
^{159}Tb	100.	158.925343				3/2+	+2.014	+1.43	
^{160}Tb		159.927164	72.3 d	β^- /1.835	0.57/47 0.86/27	3-	+1.79	3.8	Dy k x-ray 0.08678 0.29857 0.87936 0.96615
^{161}Tb		160.927566	6.91 d	β^- /0.593	0.46/23 0.52/66 0.6/10	3/2+	2.2	+1.2	Dy k x-ray 0.02565 0.04892 0.07458
^{162}Tb		161.92948	7.6 m	β^- /2.51	1.4	(1/2-)			Dy k x-ray 0.2600 0.8075 0.8882
^{163}Tb		162.930644	19.5 m	β^- /1.785	0.80/	3/2+			Dy k x-ray 0.3511 0.3897 0.4945
^{164}Tb		163.9334	3.0 m	β^- /3.9	1.7/	(5+)			Dy k x-ray 0.1689 0.2157 0.6110 0.6885 0.7548
^{165}Tb		164.9349	2.1 m	β^- /3.0		3/2+			0.5389 1.1785 1.2920 1.6648
$_{66}$Dy		162.50(3)							
^{141}Dy		140.951	0.9 s	EC,β^+ /9.					
^{142}Dy		141.946	2.3 s	EC,β^+ /7.1					
^{143}Dy		142.9440	3.9 s	EC,β^+ / $\approx$ 8.8					
^{144}Dy		143.9391	9.1 s	EC,β^+ / $\approx$ 6.2					
^{145m}Dy		144.9365	14. s	EC,β^+		11/2-			
^{146m}Dy			0.15 s	I.T.		10+			
^{146}Dy		145.9325	30. s	EC,β^+ /5.2					
^{147m}Dy			56. s	I.T./40 / β^+ ,EC/60 /		(11/2-)	-0.66	+0.7	Dy k x-ray 0.072 0.6787
^{147}Dy		146.9309	75. s	EC,β^+ /6.37		1/2+	-0.92	0.1007	ann.rad./ 0.2534 0.3653
^{148}Dy		147.92710	3.1 m	β^+ /4 /2.68 EC/96 /	1.2/	0+			ann.rad./ Tb k x-ray 0.6202

Elem. or Isot.	Natural Abundance (%)	Atomic Mass or Weight	Half-Life	Decay Mode/Energy (/MeV)	Particle Energy /Intensity (MeV / %)	Spin (h/2π)	Nuclear Magnetic Mom.(nm)	Elect. Quadr. Mom. (b)	γ-ray Energy /Intensity (MeV / %)
^{149}Dy		148.92734	4.2 m	β^+ ,EC/3.81		(7/2-)	-0.12	-0.62	ann. rad./ 0.1008 0.1063 0.2534 0.6536 0.7894 1.7765 1.8062
^{150}Dy		149.92558	7.18 m	β^+ ,EC/67 /1.79 α /33 /	4.233/	0+			Tb k x-ray 0.3967
^{151}Dy		150.926181	17. m	β^+ /5 /2.871 EC/89 / α /6 /	4.067/	7/2-	-0.95	-0.30	Tb k x-ray 0.1764 0.3030 0.3861 0.5463 (0.16 - 2.09)
^{152}Dy		151.92472	2.37 h	EC/0.60 α /	3.63/	0+			Tb k x-ray 0.2569
^{153}Dy		152.925763	6.3 h	β^+ /1 /2.171 EC/99 / α /0.01 /	0.89/ 3.46/	(7/2-)	-0.78	-0.02	Tb k x-ray 0.0807 0.0997 0.2137 (0.08 - 1.66)
^{154}Dy		153.92442	3.x10^6 y	α/2.95	2.87/	0+			
^{155}Dy		154.92575	9.9 h	β^+ /2 /2.095 EC/98 /	0.845/	3/2-	-0.385	+1.04	Tb k x-ray 0.0655 0.2269
^{156}Dy	0.06(1)	155.92428				0+			
^{157}Dy		156.92546	8.1 h	EC/1.34		3/2-	-0.301	+1.30	Tb k x-ray 0.3262
^{158}Dy	0.10(1)	157.924405				0+			
^{159}Dy		158.925736	144. d	EC/0.366		3/2-	-0.354	+1.37	Tb k x-ray 0.3262
^{160}Dy	2.34(6)	159.925194				0+			
^{161}Dy	18.9(2)	160.926930				5/2+	-0.480	+2.51	
^{162}Dy	25.5(2)	161.926795				0+			
^{163}Dy	24.9(2)	162.928728				5/2-	+0.673	+2.65	
^{164}Dy	28.2(2)	163.929171				0+			
^{165m}Dy			1.26 m	I.T./98 /0.108 β^- /2 /		1/2-			Dy k x-ray 0.1082 0.5155
^{165}Dy		164.931700	2.33 h	β /1.286	1.29/	7/2+	-0.52	-3.5	Ho k x-ray 0.09468
^{166}Dy		165.932803	3.400 d	β^- /0.486	0.40/	0+			Ho k x-ray 0.0282 0.0825
^{167}Dy		166.9357	6.2 m	β^- / $\approx$ 2.35	1.78	(1/2-)			Ho k x-ray 0.2593 0.3103 0.5697 (0.06 - 1.4)
^{168}Dy		167.9372	8.5 m	β^- /1.6		0+			Ho k x-ray 0.1925 0.4867
^{169}Dy		168.9403	$\approx$ 39. s	β^- /3.2					
$_{67}$Ho		164.93032(2)							
^{144}Ho		143.952	0.7 s	β^+ ,EC/12					ann. rad./
^{146}Ho		145.9440	3.3 s	β^+ ,EC/10.7		(10+)			ann. rad./
^{147}Ho		146.9396	5.8 s	β^+ ,EC/8.2		11/2-			ann. rad./
^{148m}Ho			9. s	β^+ ,EC/		4-			ann. rad./
^{148}Ho		147.9372	2. s	β^+ ,EC/9.4		1+			ann. rad./ 0.6615 1.6883
^{149m}Ho			21. s	β^+ ,EC/		11/2-			ann. rad./ 1.0733 1.0911
^{149}Ho		148.93379	> 30. s	β^+ ,EC/6.01		1/2+			
^{150m}Ho			25. s	β^+ ,EC/		(9+)			ann. rad./ 0.3939 0.5511 0.6534 0.8034
^{150}Ho		149.9326	1.3 m	β^+ ,EC/6.6					ann. rad./ 0.5913 0.6534 0.8034

Elem. or Isot.	Natural Abundance (%)	Atomic Mass or Weight	Half-Life	Decay Mode/Energy (/MeV)	Particle Energy /Intensity (MeV / %)	Spin (h/2π)	Nuclear Magnetic Mom.(nm)	Elect. Quadr. Mom. (b)	γ-ray Energy /Intensity (MeV / %)
¹⁵¹ᵐHo			47. s	β⁺ ,EC/87 / α/13	4.605/				ann.rad./ 0.2102 0.4889 0.6948 0.7762
¹⁵¹Ho		150.93169	35.2 s	β⁺ ,EC/80/5.13 α/20 /	4.519/				ann.rad./ 0.3522 0.5274 0.9676 1.0471
¹⁵²ᵐHo			50. s	β⁺ ,EC/90 / α/10 /	4.453/	(9+)	+5.9	-1.	ann.rad./ 0.4929 0.6138 0.6474 0.6835
¹⁵²Ho		151.93166	2.4 m	β⁺ ,EC/88 /6.47 α/12 /	4.387/	(3+)	-1.02	+0.1	ann.rad./ 0.6140 0.6476
¹⁵³ᵐHo			9.3 m	β⁺ ,EC/99+ /4.12 α/	4.01/	5/2	+1.19		ann.rad./ 0.0905 0.1089 0.1618 0.2302 0.2707 0.3659 0.4565
¹⁵³Ho		152.93020	2.0 m	β⁺ ,EC/99+ /4.13 α/	3.91/	11/2-	+6.8	-1.	ann.rad./ 0.2958 0.3346 0.4381 0.6383
¹⁵⁴ᵐHo			3.3 m	β⁺ ,EC/		(8+)	5.7	-.1.	ann.rad./ 0.3346 0.4124 0.4771
¹⁵⁴Ho		153.93060	12. m	β⁺ ,EC/5.75		1-	-0.64	+0.2	ann.rad./ Dy k x-ray 0.3346 0.5700 0.8734
¹⁵⁵Ho		154.92908	48. m	β⁺ /6 /3.10 EC/94 /		(5/2+)	+3.51	+1.5	ann.rad./ Dy k x-ray 0.0474 0.1363 0.3254 (0.06 - 2.24)
¹⁵⁶ᵐHo			5.8 m	I.T./0.0352 β⁺ /25 / EC/75 /	1.8/ 2.9/		+2.99	+2.3	ann.rad./ Dy k x-ray 0.1378 0.2666 (0.28 - 2.9)
¹⁵⁶Ho		155.9290	56. m	β⁺ ,EC/4.4		(5+)			ann.rad./ 0.1378 0.2665
¹⁵⁷Ho		156.92819	12.6 m	β⁺ /5 /2.54 EC/95 /	1.18/	7/2-	+4.35	+3.0	ann.rad./ Dy k x-ray 0.2800 0.3411
¹⁵⁸ᵐ²Ho			28. m	I.T./44 / EC/56 /		2-	+2.44	+1.6	ann.rad./ Dy k x-ray 0.0989 0.2182
¹⁵⁸ᵐ¹Ho			21. m	β⁺ ,EC/		(9+)			ann.rad./ 0.0981 0.1664 0.2182 0.3205 0.4062 0.9774 1.0532 0.4846
¹⁵⁸Ho		157.92895	11.3 m	β⁺ /8 /4.24 EC/92 /	1.30/	5+	+3.77	+4.1	ann.rad./ Dy k x-ray 0.0989 0.2182 0.9488
¹⁵⁹ᵐHo			8.3 s	I.T./0.206		1/2+			Ho k x-ray 0.1660 0.2059

Elem. or Isot.	Natural Abundance (%)	Atomic Mass or Weight	Half-Life	Decay Mode/Energy (/MeV)	Particle Energy /Intensity (MeV / %)	Spin (h/2π)	Nuclear Magnetic Mom.(nm)	Elect. Quadr. Mom. (b)	γ-ray Energy /Intensity (MeV / %)
159Ho		158.927708	33.0 m	EC/1.838		7/2-	+4.28	+3.2	Dy k x-ray 0.1210 0.1320 0.2529 0.3096 (0.06 - 1.2)
160m2Ho			3. s			1+			
160Ho			5.0 h	I.T./67 /0.060 EC/33 /3.35		2-	+2.52	+1.8	0.0868 0.1970 0.6464 0.7281 0.8791 0.9619 0.9658
160Ho		159.92873	25.6 m	β+ ,EC/3.29	0.57/	5+	+3.71	+4.0	See Ho[166m] 0.7282 0.8794
161mHo			6.8 s	I.T./0.211					Ho k x-ray 0.2112
161Ho		160.927852	2.48 h	EC/0.859		7/2-	+4.25	+3.2	Dy k x-ray 0.0256 0.0592 0.0774 0.1031
162mHo			1.12 h	I.T./61 / EC/39 /		6-	+3.60	+4.	Dy k x-ray Ho k x-ray 0.0807 0.1850 0.2828 0.9372 1.2200
162Ho		161.929092	15. m	EC/96 /0.295 β+ /4 /		1+			Dy k x-ray 0.0807 1.3196 1.3728
163mHo			1.09 s	I.T./0.298		(1/2+)			Ho k x-ray 0.2798
163Ho		162.928730	4.57x10³ y	EC/0.00258		7/2-	+4.23	+3.6	Dy M x-rays
164mHo			38. m	I.T./0.140		(6-)			Ho k x-ray 0.0373 0.0566 0.0940
164Ho		163.930231	29. m	EC/58 /0.987 β- /42 /0.963		1+			Dy k x-ray 0.0734 0.0914
165Ho	100.	164.930319				7/2-	+4.17	+3.49	
166mHo			1.2x10³ y	β- /		7-	3.6	-3.	Er k x-ray 0.18407 0.71169 0.81031
166Ho		165.932281	1.117 d	β- /1.855	1.776/48 1.855/51	0-			Er k x-ray 0.08057 1.37943
167Ho		166.933127	3.1 h	β- /1.007	0.31/43 0.61/21 0.96/15 0.97/15	(7/2-)			Er k x-ray 0.0793 0.0835 0.2379 0.3213 0.3465
168mHo			2.2 m	I.T./					
168Ho		167.93550	3.0 m	β- /2.91	2.0/	3+			Er k x-ray 0.7413 0.8159 0.8211 (0.08 - 2.34)
169Ho		168.93687	4.7 m	β- /2.12	1.2/ 2.0/	(7/2-)			0.1496 0.7610 0.7784 0.7884 0.8529
170mHo			43. s	β- /		1+			0.0787 0.8123 1.8940 1.9726

Elem. or Isot.	Natural Abundance (%)	Atomic Mass or Weight	Half-Life	Decay Mode/Energy (/MeV)	Particle Energy /Intensity (MeV / %)	Spin (h/2π)	Nuclear Magnetic Mom.(nm)	Elect. Quadr. Mom. (b)	γ-ray Energy /Intensity (MeV / %)
¹⁷⁰Ho		169.93962	2.8 m	β⁻ /3.87		6+			Er k x-ray 0.1816 0.2582 0.8902 0.9321 0.9414 1.1387
¹⁷²Ho			25. s	β⁻ /					Er k x-ray (0.077-1.186)
₆₈Er		**167.26(3)**							
¹⁴⁷Er		146.9494	2.5 s	E.C.,β⁺ / ≈ 9.1					
¹⁴⁸Er		147.9444	4.5 s	β⁺ ,EC/6.8					
¹⁴⁹ᵐEr			10. s	IT		11/2-			
¹⁴⁹Er		148.9425	10.7 s	ECβ⁺ /8.1		1/2+			
¹⁵⁰Er		149.9370	18. s	β⁺ /36 /4.11 EC/64 /		0+			ann.rad./ Ho k x-ray 0.4758
¹⁵¹Er		150.9373	23. s	β⁺ ,EC/5.2		7/2-			ann.rad./
¹⁵²Er		151.93500	10.2 s	β⁺ ,EC/10 /3.11 α/90 /	4.804/	0+			ann.rad./
¹⁵³Er		152.93509	37.1 s	α/ β⁺ ,EC/47 /4.56	4.674 4.35/				0.351 (0.0945-1.700)
¹⁵⁴Er		153.93278	3.7 m	β⁺ ,EC/99+ /2.03 α/0.5 /	4.166/	0+			ann.rad./
¹⁵⁵Er		154.93321	5.3 m	β⁺ ,EC/47 /3.84 EC/53 /		(7/2-)	-0.669	-0.27	ann.rad./ Ho k x-ray 0.1101 0.2415
¹⁵⁶Er		155.9308	20. m	β⁺ ,EC/1.7		0+			ann.rad./ 0.0298 0.0352 0.0522 0.1336
¹⁵⁷Er		156.9319	25. m	β⁺ ,EC/3.5		3/2-	-0.412	+0.92	ann.rad./ 0.117 0.385 1.320 1.660 1.820 2.000
¹⁵⁸Er		157.93087	2.2 h	EC/99.5 /1.78 β⁺ /0.5 /	0.74/	0+			Ho k x-ray 0.0719 0.2486 0.3868
¹⁵⁹Er		158.930681	36. m	β⁺ /7 /2.769 EC/93 /		3/2-	-0.304	+1.17	ann.rad./ Ho k x-ray 0.6245 0.6493 (0.07 - 2.5)
¹⁶⁰Er		159.92908	1.191 d	EC/0.33		0+			Ho k x-ray (0.05 - 0.96)
¹⁶¹Er		160.93000	3.21 h	EC/2.00		3/2-	-0.37	+1.36	Ho k x-ray 0.8265 (0.07 - 1.74)
¹⁶²Er	0.14(1)	161.928775				0+			
¹⁶³Er		162.93003	1.25 h	EC/1.210		5/2-	+0.557	+2.55	Ho k x-ray 0.4361 0.4399 1.1135
¹⁶⁴Er	1.61(2)	163.929197				0+			
¹⁶⁵Er		164.930723	10.36 h	EC/0.376		5/2-	+0.643	+2.71	Ho k x-ray
¹⁶⁶Er	33.6(2)	165.930290				0+			
¹⁶⁷ᵐEr			2.27 s	I.T./0.208		1/2-			Er k x-ray 0.2078
¹⁶⁷Er	22.95(15)	166.932046				7/2+	-0.5639	+3.57	
¹⁶⁸Er	26.8(2)	167.932368				0+			
¹⁶⁹Er		168.934588	9.40 d	β⁻ /0.351	0.35/ ≈ 100	1/2-	+0.515		Tm k x-ray 0.1098 0.1182
¹⁷⁰Er	14.9(2)	169.935461				0+			
¹⁷¹Er		170.938026	7.52 h	β⁻ /1.491		5/2-	0.66	2.9	Tm k x-ray 0.11160 0.29591 0.30832 (0.08 - 1.4)

Elem. or Isot.	Natural Abundance (%)	Atomic Mass or Weight	Half-Life	Decay Mode/Energy (/MeV)	Particle Energy /Intensity (MeV / %)	Spin (h/2π)	Nuclear Magnetic Mom.(nm)	Elect. Quadr. Mom. (b)	γ-ray Energy /Intensity (MeV / %)
^{172}Er		171.939352	2.05 d	β^- /0.891	0.28/48 0.36/46				Tm k x-ray 0.0597 0.4073 0.6101
^{173}Er		172.9424	1.4 m	β^- /2.6		(7/2-)			Tm k x-ray 0.1928 0.1992 0.8952
^{174}Er		173.9441	3.1 m	β^- /1.8					Tm k x-ray (0.100-0.152)
$_{69}$Tm		168.93421(2)							
^{146}Tm		145.967		β^+ /14. p	1.119/ 1.189/				
^{147}Tm		146.961	0.6 s	EC, β^+ /85/ ≈10.7 p/15/	1.052/				
^{148m}Tm		147.9573	0.7 s	β^+ ,EC/12.					ann.rad./
^{148}Tm									
^{149}Tm		148.9524	0.9 s	β^+ ,EC/ ≈9.2		11/2-			
^{150}Tm		149.9494	2.3 s	β^+ ,EC/ ≈11.5		6-			
^{151}Tm		150.9454	4. s	β^+ ,EC/7.5					ann.rad./
^{152m}Tm			8. s	β^+ ,EC/		9+			
^{152}Tm		151.9443	5. s	β^+ ,EC/8.8					ann.rad./
^{153}Tm		152.94203	1.6 s	β^+ ,EC/10 /6.46 α/90 /	5.109/				ann.rad./
^{154m}Tm			3.3 s	β^+ ,EC/15 / α /	5.031/				ann.rad./
^{154}Tm		153.9407	8.1 s	β^+ ,EC/56 /7.4 α /44 /	4.45/	9+			ann.rad./
^{155}Tm		154.93919	30. s	β^+ ,EC/5.58 α/	4.46/				0.0315 0.0638 0.0881 0.2268 0.5320 0.6067
^{156m}Tm			19. s	α/	4.46/				
^{156}Tm		155.9389	1.40 m	β^+ ,EC/7.6 α/	4.23/	2-	+0.40	-0.5	ann.rad./ 0.3446 0.4529 0.5860
^{157}Tm		156.9367	3.6 m	β^+ ,EC/4.5 α/	2.6 3.97/	1/2	+0.48		ann.rad./ 0.1104 0.3484 0.3855 0.4550 (0.1 - 1.58)
^{158}Tm		157.9379	4.0 m	β^+ ,EC/74 /6.5 EC/26 /		(2-)	+0.04	+0.7	ann.rad./ Er k x-ray 0.1921 0.3351 0.6280 1.1498 (0.18 - 2.81)
^{159}Tm		158.9348	9.1 m	β^+ /23 /3.9 EC/77 /		5/2+	+3.42	+1.9	ann.rad./ Er k x-ray 0.0591 0.0848 0.2713 (0.05 - 1.27)
^{160m}Tm			1.24 m	IT		(5)			
^{160}Tm		159.9354	9.4 m	β^+ /15 /5.9 EC/85 /		1-	+0.16	+0.58	ann.rad./ Er k x-ray 0.1264 0.2642 0.7285 0.8544 0.8614 1.3685
^{161}Tm		160.9334	31. m	β^+ ,EC/3.2		7/2+	+2.40	+2.9	ann.rad./ Er k x-ray 0.0595 0.0844 1.6481 (0.04 - 2.15)

Elem. or Isot.	Natural Abundance (%)	Atomic Mass or Weight	Half-Life	Decay Mode/Energy (/MeV)	Particle Energy /Intensity (MeV / %)	Spin (h/2π)	Nuclear Magnetic Mom.(nm)	Elect. Quadr. Mom. (b)	γ-ray Energy /Intensity (MeV / %)
^{162m}Tm			24. s	I.T./90 / β^+ ,EC/10 /		5+			Tm k x-ray Er k x-ray 0.0669 0.8115 0.9003
^{162}Tm		161.93394	21.7 m	β^+ /8 /4.81 EC/92 /		1-	+0.07	+0.69	ann.rad./ Er k x-ray 0.1020 0.7987 (0.1 - 3.75)m
^{163}Tm		162.93265	1.81 h	EC/98 /2.439 β^+ /1 /		1/2+	0.082		Er k x-ray 0.0692 0.1043 0.2414
^{164m}Tm			5.1 m	I.T./80 / β^+ ,EC/20 /		6-			0.0914 0.1394 0.2081 0.2405 0.3149
^{164}Tm		163.93345	2.0 m	β^+ /36 /3.96 EC/64 /	2.94/	1+	+2.38	+0.71	ann.rad./ Er k x-ray 0.0914
^{165}Tm		164.932433	1.253 d	EC/1.593		1/2+	0.139		Er k x-ray 0.0472 0.0544 0.29728 0.80636
^{166}Tm		165.93355	7.70 h	EC/98 /3.04 β^+ /2 /		2+	+0.092	+2.14	Er k x-ray 0.0806 0.1844 0.7789 1.2734 2.0524
^{167}Tm		166.932849	9.24 d	EC/0.748		1/2+	-0.197		Er k x-ray 0.0571 0.20778
^{168}Tm		167.934171	93.1 d	EC/1.679		3+	0.23	+3.2	Er k x-ray 0.19825 0.4475 0.81595
^{169}Tm	100	168.934211				1/2+	-0.2316	-1.2	
^{170}Tm		169.935798	128.6 d	β^- /99.8 /0.968 EC/0.2 /0.314	0.883/24 0.968/76	1-	0.2476	+0.74	Yb k x-ray 0.08425
^{171}Tm		170.936426	1.92 y	β^- /0.096	0.03/2 0.096/98	1/2+	-0.2303		0.06674
^{172}Tm		171.93840	2.65 d	β^- /1.88	1.79/36 1.88/29	2-			Yb k x-ray 0.07879 1.38722 1.46601 1.52982 1.60861
^{173}Tm		172.93960	8.2 h	β^- /1.298	0.80/21 0.86/71	1/2+			Yb k x-ray 0.3988 0.4613
^{174}Tm		173.94216	5.4 m	β^- /3.08	0.70/14 1.20/83	(4-)			Yb k x-ray 0.07664 0.17669 0.27332 0.3666 0.99205 (0.08 - 1.6)
^{175}Tm		174.94383	15.2 m	β^- /2.39	0.9/36 1.9/23	(1/2+)			Yb k x-ray 0.36396 0.51487 0.94125 0.98247
^{176}Tm		175.9471	1.9 m	β^- /4.2	2.0/ 1.2/	(4+)			Yb k x-ray 0.1898 0.3819 1.0691
$_{70}$Yb		173.04(3)							
^{151}Yb		150.9545	1.6 s	β^+ /8.5					
^{152}Yb		151.9502	3.2 s	β^+ EC/5.5					
^{153}Yb		152.9492	4. s	β^+ EC/6.7					
^{154}Yb		153.9455	0.40 s	β^+ EC/7 /4.49 α/93 /	5.32/				ann.rad./

Elem. or Isot.	Natural Abundance (%)	Atomic Mass or Weight	Half-Life	Decay Mode/Energy (/MeV)	Particle Energy /Intensity (MeV / %)	Spin (h/2π)	Nuclear Magnetic Mom.(nm)	Elect. Quadr. Mom. (b)	γ-ray Energy /Intensity (MeV / %)
¹⁵⁵Yb		154.9456	1.7 s	β⁺ ,EC/16 /6.0 α/84 /	5.19/		-0.8		ann.rad./
¹⁵⁶Yb		155.94277	26. s	β⁺ ,EC/21 /3.57 α/79 /	4.69/	0+			ann.rad./
¹⁵⁷Yb		156.9427	39. s	β⁺ ,EC/99+ /5.5 α/0.5 /	4.69/				ann.rad./ 0.231 (0.035-0.670)
¹⁵⁸Yb		157.93986	1.5 m	β⁺ ,EC/1.9		0+			ann.rad./ 0.0741 0.2526
¹⁵⁹Yb		158.9402	1.4 m	EC,β⁺ /5.1					Tm k x-ray 0.1661 0.1772 0.3297 0.3903
¹⁶⁰Yb		159.9376	4.8 m	β⁺ ,EC/2.0		0+			ann.rad./ 0.1404 0.1737 0.2158
¹⁶¹Yb		160.9375	4.2 m	β⁺ ,EC/3.9		3/2-	-0.33	+1.03	ann.rad./ Tm k x-ray 0.0782 0.5999 0.6315
¹⁶²Yb		161.9358	18.9 m	β⁺ ,EC/1.7		0+			ann.rad./ Tm k x-ray 0.1188 0.1635
¹⁶³Yb		162.9363	11.1 m	β⁺ /26 /3.4	1.4/	3/2-	-0.37	+1.2	ann.rad./ Tm k x-ray 0.0636 0.8603 (0.06 - 1.9)
¹⁶⁴Yb		163.9345	1.26 h	EC/1.0		0+			Tm k x-ray 0.0914 0.6752
¹⁶⁵Yb		164.93540	9.9 m	β⁺ /10 /2.76 EC/90 /	1.58/	(5/2-)	+0.48	+2.48	ann.rad./ Tm k x-ray 0.0801 1.0903
¹⁶⁶Yb		165.93388	2.363 d	EC/0.30		0+			Tm k x-ray 0.0828 0.1844 0.7789 1.2734 2.0524
¹⁶⁷Yb		166.934947	17.5 m	β⁺ /0.5 /1.954 EC/99.5 /	0.639/	5/2-	+0.62	+2.70	Tm k x-ray 0.06296 0.10616 0.11337 0.17633
¹⁶⁸Yb	0.13(1)	167.933895				0+			
¹⁶⁹ᵐYb			46. s	I.T./0.0242		1/2-			Yb L x-ray 0.0242
¹⁶⁹Yb		168.935187	32.03 d	EC/0.909		7/2+	-0.63	+3.5	Tm k x-ray 0.06306 0.10977 0.17718 0.19795
¹⁷⁰Yb	3.05(6)	169.934759				0+			
¹⁷¹Yb	14.3(2)	170.936323				1/2-	+0.49367		
¹⁷²Yb	21.9(3)	171.936378				0+			
¹⁷³Yb	16.12(21)	172.938207				5/2-	-0.67989	+2.80	
¹⁷⁴Yb	31.8(4)	173.938858				0+			
¹⁷⁵Yb		174.941273	4.19 d	β⁻ /0.470	0.466/73 0.071/21 0.353/6.2	7/2-	0.6		Lu k x-ray 0.3963/13 (0.114 - 0.28)
¹⁷⁶ᵐYb			11.4 s	I.T./1.051		(8-)			Yb k x-ray 0.0961 0.1901 0.2929 0.3897
¹⁷⁶Yb	12.7(2)	175.942569				0+			
¹⁷⁷ᵐYb			6.41 s	I.T./0.3315		1/2-			Yb k x-ray 0.1131 0.2084
¹⁷⁷Yb		176.945257	1.9 h	β⁻ /1.399	1.40	9/2+			Lu k x-ray 0.1504

Elem. or Isot.	Natural Abundance (%)	Atomic Mass or Weight	Half-Life	Decay Mode/Energy (/MeV)	Particle Energy /Intensity (MeV / %)	Spin (h/2π)	Nuclear Magnetic Mom.(nm)	Elect. Quadr. Mom. (b)	γ-ray Energy /Intensity (MeV / %)
^{178}Yb		177.94664	1.23 h	β^- /0.65	0.25/	0+			0.1415
									0.3246
									0.3516
									0.3815
									0.6125
^{179}Yb		178.9499	8. m	β^- /2.4					
^{180}Yb			2. m	β^-					
$_{70}$Lu		174.967(1)							
^{150}Lu		149.973	≈35. ms	p					
^{151}Lu		150.967	0.08 s	p/1.231					
^{152}Lu		151.963	0.7 s						
^{153}Lu		152.959							
^{154}Lu		153.9571	1.0 s	β^+ ,EC/10.8					
^{155m}Lu			2.6 ms	α/7.41					
^{155}Lu		154.9542	0.07 s	EC/8.0					
				α/	5.66/				
					5.57/				
^{156m}Lu			0.20 s	α/					
^{156}Lu		155.9529	≈ 0.5 s	β^+ ,EC/9.5					ann.rad./
				α/	5.45/				
^{157m}Lu			≈9.6 s	α	4.925/				
^{157}Lu		156.95010	4.8 s	β^+ ,EC/94 /6.93					ann.rad./
				α/	5.00/				
^{158}Lu		157.94984	10.4 s	β^+ ,EC/99 /8.0					ann.rad./
				α/	4.67/				0.3682
									0.4770
^{159}Lu		158.9467	12.3 s	β^+ ,EC/6.0					ann.rad./
									0.1505
									0.1875
									0.3693
^{160}Lu		159.94654	36.1 s	β^+ ,EC/7.3					ann.rad./
									0.2434
									0.3957
									0.5773
^{161}Lu		160.9432	1.2 m	β^+ ,EC/5.3					ann.rad./
									0.0437
									0.0671
									0.1003
									0.1108
									0.1562
									0.2562
^{162m}Lu			≈ 1.5 m	EC/		4-			
^{162}Lu		161.9432	1.37 m	β^+ ,EC/6.9		1-			ann.rad./
									0.1666
									0.6314
^{163}Lu		162.9412	4.1 m	β^+ ,EC/4.6					ann.rad./
									0.0539
									0.0581
									0.1504
									0.1631
									0.3717
^{164}Lu		163.9412	3.14 m	β^+ ,EC/6.3	1.6/				0.1238
					3.8/				0.2621
									0.7404
									0.8639
									0.8804
^{165}Lu		164.9396	10.7 m	β^+ ,EC/3.9	2.06/	1/2 +			ann.rad./
									0.1206
									0.1324
									0.1742
									0.2036
									(0.04 - 2.0)
^{166m2}Lu			2.1 m	β^+ /35 / EC/65 /		(0-)			ann.rad./
									Yb k x-ray
									1.0673
									1.2566
									2.0986
^{166m1}Lu			1.4 m	β^+ ,EC/58 / I.T./42 /0.0344		(3-)			ann.rad./
									0.1024
									0.2281
									0.2861
									0.8119
									0.8301

Elem. or Isot.	Natural Abundance (%)	Atomic Mass or Weight	Half-Life	Decay Mode/Energy (/MeV)	Particle Energy /Intensity (MeV / %)	Spin (h/2π)	Nuclear Magnetic Mom.(nm)	Elect. Quadr. Mom. (b)	γ-ray Energy /Intensity (MeV / %)
^{166}Lu		165.9398	2.8 m	β^+ /25 /5.5 EC/75 /		(6-)			ann.rad./ Yb k x-ray 0.1024 0.2281 0.3375 0.3679
^{167}Lu		166.9383	52. m	β^+ /2 /3.1 EC/98 /	2.1/	7/2+			Yb k x-ray 0.0297 0.2392 (0.03 - 2.0)
^{168m}Lu			6.7 m	β^+ /12 / EC/88 /		3+			ann.rad./ Yb k x-ray 0.1988 0.8960 0.9792
^{168}Lu		167.9387	5.5 m	β^+ /6 /4.5 EC/94 /	1.2/	(6-)			ann.rad./ Yb k x-ray 0.1114 0.1124 0.2286 0.3483 1.4836
^{169m}Lu			2.7 m	I.T./0.0290		1/2-			Lu L x-ray 0.0290
^{169}Lu		168.93765	1.419 d	EC/2.293	1.271/	7/2+			Yb k x-ray 0.19121 0.9606 (0.08 - 2.1)
^{170m}Lu			0.7 s	I.T./0.0929		4-			Lu L x-ray 0.04449 0.0484
^{170}Lu		169.93847	2.01 d	EC/3.46	2.44/	0+			Yb k x-ray 0.58711 0.5908 1.28029 (0.1 - 3.38)
^{171m}Lu			1.31 m	I.T./0.0711		1/2-			Lu k x-ray 0.07119
^{171}Lu		170.937910	8.24 d	EC/1.479	0.362/	7/2+	2.0		Yb k x-ray 0.01939 0.66744 (0.02 - 1.3)
^{172m}Lu			3.7 m	I.T./0.0419		1-			Lu L x-rays 0.04186
^{172}Lu		171.939082	6.70 d	EC/2.519		4-	2.25		Yb k x-ray 0.18156 1.09367 (0.07 - 2.2)
^{173}Lu		172.938927	1.37 y	EC/0.671		7/2+	2.3		Yb k x-ray 0.07860 0.27198
^{174m}Lu			142. d	I.T./99.3 /0.17086 EC/0.7 /		6-	1.5		Lu k x-ray 0.067055
^{174}Lu		173.940334	3.3 y	EC/1.374		1-	1.9		Yb k x-ray 0.07664 1.2419
^{175}Lu	97.41(2)	174.940768				7/2+	+2.2327	+3.49	
^{176m}Lu			3.66 h	β^- /1.315	1.229/ 1.317/	1-	+0.318	-1.47	Hf k x-ray 0.088372
^{176}Lu	2.59(2)	175.942683	3.8x10^{10} y	β^- /1.192		7-	+3.169	+4.92	Hf k x-ray 0.20187 0.30691
^{177m}Lu			160.7 d	I.T./22 /0.9702 β^- /78		23/2-	2.9	4.2	Lu k x-ray Hf k x-ray 0.11295 0.20836 0.37850 0.41853
^{177}Lu		176.943755	6.75 d	β^- /0.498	0.497/	7/2+	+2.239	+3.39	0.11295 0.20836
^{178m}Lu			23.1 m	β^- /		(9-)			0.2166 0.3317
^{178}Lu		177.945952	28.5 m	β^- /2.099	2.03/	1+			Hf k x-ray 0.0932 1.3099 1.3408 (0.09 - 1.7)
^{179}Lu		178.94732	4.6 h	β^- /1.405	1.35/	7/2+			0.2143 0.3377

Elem. or Isot.	Natural Abundance (%)	Atomic Mass or Weight	Half-Life	Decay Mode/Energy (/MeV)	Particle Energy /Intensity (MeV / %)	Spin (h/2π)	Nuclear Magnetic Mom.(nm)	Elect. Quadr. Mom. (b)	γ-ray Energy /Intensity (MeV / %)
^{180}Lu		179.9499	5.7 m	β^- /3.1	1.49/				0.40795/50. (0.07-1.9)
^{181}Lu		180.9518	3.5 m	β^- /2.5		(7/2+)			0.0458 0.2059 0.5749
^{182}Lu			2.0 m	β^- / ≈ 4.1					0.0978 0.7208 0.8182
^{183}Lu			58. s	β^- /		7/2+			
$_{72}$Hf		**178.49(2)**							
^{154}Hf		153.964	2. s	EC,β^+ / ≈ 6.7					
^{155}Hf		154.963	0.9 s	EC,β^+ /8.					
^{156}Hf		155.9593	25. ms	α/					
^{157}Hf		156.9581	0.11 s	α/					
^{158}Hf		157.9539	2.9 s	EC/54 /5.1 α/46 /	5.27/	0+			
^{159}Hf		158.9538	5.6 s	β^+ ,EC/88 /6.9 α/12 /	5.09/				ann.rad./
^{160}Hf		159.95063	≈ 12. s	β^+ ,EC/97 /4.9 α /4.78		0+			ann.rad./
^{161}Hf		160.9503	17. s	α/	4.60/				ann.rad./
^{162}Hf		161.94720	38. s	β^+ ,EC/3.7		0+			0.1739 0.1963 0.4101
^{163}Hf		162.9471	40. s	β^+ ,EC/5.5					ann.rad./ 0.0454 0.0621 0.0710 0.6882
^{164}Hf		163.9536	2.8 m	EC,β^+ /3.0					ann.rad./
^{165}Hf		164.9445	1.7 m	EC/4.6		11/2-			Lu k x-ray
^{166}Hf		165.9423	6.8 m	EC/93 /2.3 β^+ /7 /					0.0788
^{167}Hf		166.9426	2.0 m	β^+ /40 /4.0 EC/60 /		(5/2-)			ann.rad./ Lu k x-ray 0.1754 0.3152
^{168}Hf		167.9406	25.9 m	β^+ ,EC/1.8		0+			ann.rad./ 0.1572 0.1838 0.1988
^{169}Hf		168.9412	3.25 m	EC/85 /3.3 β^+ /15 /		(5/2-)			ann.rad./ Lu k x-ray 0.3695 0.4929
^{170}Hf		169.9397	16.0 h	EC/1.1		0+			Lu k x-ray 0.0985 0.1202 0.1647 0.5729 0.6207
^{171}Hf		170.9405	12.2 h	EC,β^+ /2.4		7/2+			ann.rad./ Lu k x-ray 0.1221 0.6620 1.0714
^{172}Hf		171.93946	1.87 y	EC/0.35		0+			Lu k x-ray 0.02399 0.12582 (0.0818-0.123)
^{173}Hf		172.9407	23.6 h	EC/1.6		1/2-			Lu k x-ray 0.12367 0.13963 0.29697 0.31124 (0.1 - 2.1)
^{174}Hf	0.162(3)	173.940042	2.0x10^{15} y			0+			
^{175}Hf		174.941504	70. d	EC/0.686		5/2-	0.54	+2.8	Lu k x-ray 0.08936 0.34340
^{176}Hf	5.206(5)	175.941403				0+			

Elem. or Isot.	Natural Abundance (%)	Atomic Mass or Weight	Half-Life	Decay Mode/Energy (/MeV)	Particle Energy /Intensity (MeV / %)	Spin (h/2π)	Nuclear Magnetic Mom.(nm)	Elect. Quadr. Mom. (b)	γ-ray Energy /Intensity (MeV / %)
177m2Hf			51.4 m	I.T./2.740		37/2-			Hf k x-ray 0.2140 0.2951 0.3115 0.3267
177m1Hf			1.1 s	I.T./		23/2+			Hf k x-ray 0.20836 0.22847 0.37851
177Hf	18.606(4)	176.943220				7/2-	+0.7935	+0.337	
178m2Hf			31. y	I.T./		16+	+8.2	+6.0	Hf k x-ray 0.21342 0.25761 0.32555 0.42635 0.57418
178m1Hf			4.0 s	I.T./		8-			Hf k x-ray 0.21342 0.32555 0.42635
178Hf	27.297(4)	177.943698				0+			
179m2Hf			25.1 d	I.T./1.1057		25/2-	7.4		Hf k x-ray 0.1227 0.1461 0.3626 0.4537
179m1Hf			18.7 s	I.T./0.375		1/2-			Hf k x-ray 0.1607 0.2141
179Hf	13.629(6)	178.945815				9/2+	-0.6409	+3.79	
180mHf			5.52 h	I.T./1.1416		8-	+8.7	+4.6	Hf k x-ray 0.2152 0.3323 0.4432
180Hf	35.100(7)	179.946549				0+			
181Hf		180.949099	42.4 d	β⁻ /1.027	0.408/	1/2-			Ta k x-ray 0.13294 0.48200
182mHf			62. m	β⁻ /54 /1.60 I.T./46 /1.1729	0.49/43 0.95/10	8-			Hf k x-ray 0.0509 0.2244 0.3441 0.4558 0.5066 0.9428
182Hf		181.95055	9.x10⁶ y	β⁻ /0.37		0+			Ta k x-ray 0.2704
183Hf		182.95353	1.07 h	β⁻ /2.01	1.18/68 1.54/25	3/2-			Ta k x-ray 0.0732 0.4591 0.7837
184Hf		183.95545	4.1 h	β⁻ /1.34	0.74/38 0.85/16 1.10/46	0+			Ta k x-ray 0.0414 0.1391 0.3449
185Hf			≈3.5m	β⁻ /					0.165
73Ta		180.9479(1)							
156Ta		155.972	> 0.01 s	β⁺ / ≈11.6 p/	1.02/ ≈100				
157Ta		156.968	5. ms	α/6.22					
158Ta		157.9664	37. ms	α/6.02					
159Ta		158.9629	0.6 s	β⁺ ,EC/20 /8.5 α/80 /	5.60				ann.rad./
160Ta		159.9615	1.4 s	β⁺ ,EC/10.1 α	5.41/				ann.rad./
161Ta		160.9584	2.9 s	β⁺ ,EC/7.5 α/	5.15				ann.rad./
162Ta		161.9564	4. s	EC/8.6					
163Ta		162.9544	10.6 s	EC/6.8					
164Ta		163.9536	14.2 s	β⁺ /8.5 α/	4.62/	3+			ann.rad./ 0.2110 0.3768
165Ta		164.9508	31. s	ECβ⁺ /5.9					

Elem. or Isot.	Natural Abundance (%)	Atomic Mass or Weight	Half-Life	Decay Mode/Energy (/MeV)	Particle Energy /Intensity (MeV / %)	Spin (h/2π)	Nuclear Magnetic Mom.(nm)	Elect. Quadr. Mom. (b)	γ-ray Energy /Intensity (MeV / %)
^{166}Ta		165.9505	34. s	β^+ /82 /7.7 EC/18 /					ann. rad./ Hf k x-ray 0.1587 0.3117 0.8101
^{167}Ta		166.9486	1.4 m	β^+ ,EC/5.6					ann.rad./
^{168}Ta		167.9478	2.4 m	β^+ /77 /6.7 EC/23 /		3+			ann.rad./ Hf k x-ray 0.1239 0.2615 0.7502
^{169}Ta		168.9459	4.9 m	β^+ ,EC/4.4					ann.rad./ 0.0288 0.1535 0.1924
^{170}Ta		169.9461	6.8 m	β^+ /70 /6.0 EC/35 /		(3+)			ann.rad./ Hf k x-ray 0.1008 0.2212 0.0496 0.5018 0.5064 (0.05 - 1.02)
^{171}Ta		170.9445	23.3 m	β^+ ,EC/3.7		(5/2-)			
^{172}Ta		171.9447	36.8 m	β^+ /25 /4.9 EC/75 /		(3-)			ann.rad./ Hf k x-ray 0.21396 1.10923 (0.09 - 3.8)
^{173}Ta		172.9446	3.6 h	β^+ /24 /3.7 EC/76 /		(5/2-)	1.7	-1.9	ann.rad./ Hf k x-ray 0.06972 0.17219 (0.06 - 2.7)
^{174}Ta		173.9442	1.12 h	β^+ /27 /3.8 EC/73 /		(3+)			ann.rad./ Hf k x-ray 0.09089 0.20638 (0.09 - 3.64)
^{175}Ta		174.9437	10.5 h	EC/2.0		7/2+	2.27	+3.7	Hf k x-ray 0.2077 0.2671 0.3487
^{176}Ta		175.9447	8.1 h	EC/3.1		1-			Hf k x-ray 0.08837 1.15735
^{177}Ta		176.944472	2.356 d	EC/1.166		7/2+	2.25		Hf k x-ray 0.11295 (0.07 - 1.06)
^{178m}Ta			2.4 h	EC/		(7-)			Hf k x-ray 0.08886 0.21342 0.32555 0.42635
^{178}Ta		177.9458	9.29 m	EC/99 /1.9 β^+ /1 /		1+	+2.74	+0.65	ann.rad./ Hf k x-ray 0.09316
^{179}Ta		178.94593	1.8 y	EC/0.110		7/2+			Hf k x-ray
^{180m}Ta	0.012(2)		> 1.2x10^{15} y			(9-)	4.77		
^{180}Ta		179.947466	8.15 h	EC/87 /0.854 β^- /13 /0.708	0.61/3 0.71/10	1+			Hf k x-ray W k x-ray 0.09333 0.10340
^{181}Ta	99.988(2)	180.947996				7/2+	+2.370	+3.3	
^{182m}Ta			15.8 m	I.T./0.5198		10-			Ta k x-ray 0.14678 0.17157
^{182}Ta		181.950152	114.43 d	β^- /1.814	0.25/30 0.44/20 0.52/40	3-	+3.02		W k x-ray 0.06775 0.10010 1.12127 1.22138
^{183}Ta		182.951373	5.1 d	β^- /1.070	0.45/5 0.62/91	7/2+	+2.36		W k x-ray 0.0847 0.0991 0.1079 0.2461 0.3540

Elem. or Isot.	Natural Abundance (%)	Atomic Mass or Weight	Half-Life	Decay Mode/Energy (/MeV)	Particle Energy /Intensity (MeV / %)	Spin (h/2π)	Nuclear Magnetic Mom.(nm)	Elect. Quadr. Mom. (b)	γ-ray Energy /Intensity (MeV / %)
^{184}Ta		183.95401	8.7 h	β^- /2.87	1.11/15 1.17/81	(5-)			W k x-ray 0.2528/44. 0.4140/74. (0.09-1.4)
^{185}Ta		184.95556	49. m	β^- /1.99	1.21/5 1.77/81	(7/2+)			W k x-ray 0.0697 0.1739 0.1776
^{186}Ta		185.9586	10.5 m	β^- /3.9	2.2/	(3-)			W k x-ray 0.1979 0.2149 0.5106 (0.09 - 1.5)
$_{74}$W		183.84(1)							
^{158m}W			<1. ms	α	8.28(3)/				
^{158}W		157.974	0.9 ms	α/	6.433				
^{159}W		158.972	7. ms	α/					
^{160}W		159.9684	0.08 s	α/	5.92/	0+			
^{161}W		160.9671	0.41 s	β^+ ,EC/18 /8.1 α/82 /	5.78/				
^{162}W		161.9626	1.39 s	β^+ ,EC/54 /5.8 α/46 /	5.54/	0+			
^{163}W		162.9624	2.8 s	β^+ ,EC/59 /7.5 α/41 /	5.38/				
^{164}W		163.95890	6. s	β^+ ,EC/97 /5.0 α/3 /	5.15/	0+			ann.rad./
^{165}W		164.9583	5.1 s	β^+ ,EC/99 /7.0 α/1 /	4.91/				ann.rad./
^{166}W		165.95502	16. s	β^+ ,EC/99 /4.2 α/1 /	4.74/	0+			ann.rad./
^{167}W		166.9547	20. s	EC/5.6					
^{168}W		167.9519	53. s	EC/3.8 α/10^{-5}/	4.40(1)				ann.rad./ Ta k x-ray 0.1755 (0.037-0.573)
^{169}W		168.9518	1.3 m	EC/5.4					ann.rad./ Ta k x-ray 0.123 (0.097-0.699)
^{170}W		169.9485	2.4 m	EC/2.2					ann.rad./ Ta k x-ray 0.3162 (0.060-0.144)
^{171}W		170.9494	2.4 m	EC/4.6					ann.rad./ Ta k x-ray 0.1842 (0.052-0.479)
^{172}W		171.9474	6.6 m	β^+ ,EC/2.5					ann.rad./ Ta k x-ray 0.0389 (0.034-0.674)
^{173}W		172.9489	6.3 m	EC/4.0					ann.rad./ Ta k x-ray 0.4576 (0.035-0.623)
^{174}W		173.9462	35. m	EC/1.9		0+			ann.rad./ Ta k x-ray 0.3287 0.4288 (0.056-0.429)
^{175}W		174.9468	35. m	EC/2.9		1/2-			(0.015-0.27)
^{176}W		175.9456	2.5 h	β^+ ,EC/0.8		0+			0.03358 0.06129 0.09487 0.10020
^{177}W		176.9466	2.21 h	EC/2.0		(1/2-)			Ta k x-ray 0.15505 0.18569 0.42694
^{178}W		177.9459	21.6 d	EC/0.091		0+			Ta k x-ray
^{179m}W			6.4 m	I.T./99.7 /0.222 EC/0.3 /		(1/2-)			W k x-ray 0.2220
^{179}W		178.94707	38. m	EC/1.06		(7/2-)			Ta k x-ray 0.0307
^{180}W	0.120(1)	179.946706				0+			

Elem. or Isot.	Natural Abundance (%)	Atomic Mass or Weight	Half-Life	Decay Mode/Energy (/MeV)	Particle Energy /Intensity (MeV / %)	Spin (h/2π)	Nuclear Magnetic Mom.(nm)	Elect. Quadr. Mom. (b)	γ-ray Energy /Intensity (MeV / %)
¹⁸¹W		180.94820	121.2 d	EC/0.188		9/2+			Ta k x-ray 0.13617 0.15221
¹⁸²W ¹⁸³ᵐW	26.498(29)	181.948205	5.15 s	I.T./		0+ (11/2+)			W k x-ray 0.0465 0.0526 0.0991 0.1605
¹⁸³W ¹⁸⁴W ¹⁸⁵ᵐW	14.314(4) 30.642(8)	182.950224 183.950932	1.6 m	I.T./0.1974		1/2- 0+ 11/2+	+0.1177848		W k x-ray 0.0659 0.1315 0.1737
¹⁸⁵W ¹⁸⁶W ¹⁸⁷W	28.426(37)	184.953420 185.954362 186.957158	74.8 d 23.9 h	β⁻/0.433 β⁻/1.311	0.433/99.9 0.624/ 1.315/	3/2- 0+ 3/2-	0.62		0.12536 Re k x-ray 0.0725 0.47951 0.68572
¹⁸⁸W		187.958487	69.4 d	β⁻/0.349	0.349/99	0+			0.0636 0.2271 0.2907
¹⁸⁹W		188.9619	11.5 m	β⁻/2.5	1.4/ 2.5/	(3/2-)			0.258 0.417 0.550
¹⁹⁰W		189.9632	30. m	β⁻/1.3	0.95/	0+			Re k x-ray 0.1576 0.1621
₇₅Re		**186.207(1)**							
¹⁶⁰Re		159.981	0.8 ms	p/ α/	1.261(6)/91 6.54/				
¹⁶¹Re		160.978	0.01 s	α/					
¹⁶²Re		161.9757	0.10 s	α/	6.12/				
¹⁶³Re		162.9721	0.26 s	β⁺,EC/9.0 α/	5.92				
¹⁶⁴Re		163.9704	0.9 s	β⁺,EC/10.7 α/	5.78/				
¹⁶⁵Re		164.9671	2. s	β⁺,EC/87 /8.1 α/	5.51/				
¹⁶⁶Re		165.9651	2.5 s	β⁺,EC/9.4 α/	5.50/				
¹⁶⁷ᵐRe ¹⁶⁷Re		166.9626	6.2 s 3.4 s	α, EC/ β⁺,EC/7.4 α/	5.015/				
¹⁶⁸Re		167.9616	4.4 s	β⁺,EC/9.1 α/	4.833/				0.1117
¹⁶⁹ᵐRe			8.1 s	α	4.70/ 4.87/				
¹⁶⁹Re ¹⁷⁰Re		168.9588 169.9582	16. s 9.2 s	β⁺, EC/9.0					0.1560 0.3055 0.4125
¹⁷¹Re ¹⁷²ᵐRe		170.9555	15.2 s 55. s	EC/ ≈ 5.7 β⁺,EC/		(2)			ann.rad./ 0.1234 0.2537 0.3504
¹⁷²Re		171.9553	15. s	β⁺,EC/7.3					ann.rad./ 0.1234 0.2537
¹⁷³Re ¹⁷⁴Re		172.9531 173.9521	2.0 m 2.4 m	EC/ ≈ 3.9 β⁺,EC/5.6					ann.rad./ ann.rad./ 0.1119 0.2430
¹⁷⁵Re ¹⁷⁶Re		174.9514 175.9516	5.8 m 5.3 m	β⁺,EC/4.3 β⁺,EC/5.6		(3+)			ann.rad./ ann.rad./ 0.1089 0.2406
¹⁷⁷Re		176.9503	14. m	EC/78 /3.4 β⁺ /22 /		(5/2-)			ann.rad./ W k x-ray 0.0797 0.0843 0.1968

Elem. or Isot.	Natural Abundance (%)	Atomic Mass or Weight	Half-Life	Decay Mode/Energy (/MeV)	Particle Energy /Intensity (MeV / %)	Spin (h/2π)	Nuclear Magnetic Mom.(nm)	Elect. Quadr. Mom. (b)	γ-ray Energy /Intensity (MeV / %)
178Re		177.9509	13.2 m	β⁺ /11 /4.7 EC/89 /	3.3/	(3)			ann.rad./ W k x-ray 0.1059 0.2373 0.9391
179Re		178.9500	19.7 m	EC/99 /2.71 β⁺ /1 /	0.95/	(5/2+)			W k x-ray 0.1199 0.2900 0.4154 0.4302 1.6803
180Re		179.95079	2.45 m	EC/92 /3.80 β⁺ /8 /	1.76/	1-			ann.rad./ W k x-ray 0.1036 0.9028 (0.07 - 2.2)
181Re		180.95006	20. h	EC /1.74		5/2+	3.19		W k x-ray 0.3607 0.3655 0.6390
182mRe			12.7 h	EC/	0.55/ 1.74/	2+	3.3	+1.8	W k x-ray 0.0677 1.1214 1.2215 (0.06 - 2.2)
182Re		181.9512	2.67 d	EC/2.8		(7+)	2.8	+4.1	W k x-ray 0.0678 0.2293 1.1213 1.2214
183Re		182.95082	70. d	EC/0.56		(5/2+)	+3.17	+2.3	W k x-ray 0.16232
184mRe			165. d	I.T./75 /0.188 EC/25 /		8+	+2.9		Re k x-ray 0.1047 0.2165 0.92093 (0.10 - 1.1)
184Re		183.95252	38. d	EC/1.48		3-	+2.53	+2.8	W k x-ray 0.79207 0.90328 (0.1 - 1.4)
185Re	37.40(2)	184.952955				5/2+	+3.1871	+2.18	
186mRe			2.0x10⁵ y	I.T./0.150		8+			Re k x-ray 0.0590
186Re		185.954986	3.718 d	β⁻ /92 /1.070 EC/8 /0.582	0.973/21 1.07/71	1-	+1.739	+0.62	W k x-ray 0.1227/0.6 0.1372/9.5 (0.63-0.77)
187Re	62.60(2)	186.955751	4.4x10¹⁰ y	β⁻ /0.00266	0.0025/	5/2+	+3.2197	+2.07	
188mRe			18.6 m	I.T./0.172		(6-)			Re k x-ray 0.0925 0.1059
188Re		187.958112	16.94 h	β⁻ /2.120	1.962/20 2.118/79	1-	+1.788	+0.57	Os k x-ray 0.15502
189Re		188.959228	24. h	β⁻ /1.01	1.01/	(5/2+)			0.1471 0.2167 0.2194 0.2451
190mRe			3.0 h	β⁻ /51 / I.T./49 /		(6-)			Re k x-ray 0.1191 0.2238 0.6731 (0.1 - 1.79)
190Re		189.9618	3.0 m	β⁻ /3.2	1.8/	(2-)			Os k x-ray 0.1867 0.5580 0.6051
191Re		190.96312	9.7 m	β⁻ /2.05	1.8/				
192Re		191.9660	16. s	β⁻ /4.2	≈ 2.5/				(0.2-0.75)
76Os		190.23(3)							
162Os		161.984	≈ 1.9 ms	α/					
163Os		162.982		α					
164Os		163.9779	0.04 s	α					
165Os		164.9765	0.07 s	α					
166Os		165.9718	0.18 s	β⁺ ,EC/28 /6.3 α/72 /	6.27/ 5.98/	0+			ann. rad./

Elem. or Isot.	Natural Abundance (%)	Atomic Mass or Weight	Half-Life	Decay Mode/Energy (/MeV)	Particle Energy /Intensity (MeV / %)	Spin (h/2π)	Nuclear Magnetic Mom.(nm)	Elect. Quadr. Mom. (b)	γ-ray Energy /Intensity (MeV / %)
[167]Os		166.9714	0.7 s	β⁺,EC/76 /8.2 α/24 /	5.84/				ann.rad./
[168]Os		167.96775	2.2 s	β⁺,EC/51 /5.7 α/49 /		0+			ann. rad./
[169]Os		168.9671	3.3 s	β⁺,EC/89 /7.7 α/17 /	5.57/				ann.rad./
[170]Os		169.96357	7.1 s	β⁺,EC/5.0 α/	5.40/	0+			ann.rad./
[171]Os		170.9630	8. s	β⁺,EC/98 /7.1 α/2 /	5.24/				ann.rad./
[172]Os		171.9601	19. s	β⁺,EC/99 /4.5 α/1 /	5.10/	0+			ann.rad./ 0.177 0.187
[173]Os		172.9598	16. s	β⁺,EC/6.3 α/0.02 /	4.94/				ann.rad./
[174]Os		173.9563	44. s	β⁺,EC/3.9 α/0.02 /	4.76/	0+			ann.rad./ 0.118 0.138 / 0.001 0.158 0.325
[175]Os		174.9570	1.4 m	β⁺,EC/5.3					0.125 0.181 0.248
[176]Os		175.9550	3.6 m	β⁺,EC/3.2		0+			0.8155 0.7758 0.8573 1.2093 1.2909
[177]Os		176.9551	2.8 m	β⁺,EC/4.5		(1/2-)			0.0848 0.1958 0.3002 1.2686
[178]Os		177.9534	5.0 m	β⁺,EC/2.3		0+			ann.rad./ 0.5946 0.6850 0.9687 1.3311
[179]Os		178.9539	7. m	β⁺,EC/3.7					ann.rad./ 0.0654 0.2186 0.5938
[180]Os		179.9524	21.5 m	β⁺,EC/1.5		0+			Re k x-ray 0.0202-0.7174
[181m]Os			1.75 h	EC/		(1/2-)			ann.rad./ 0.14493
[181]Os		180.9532	2.7 m	EC/2.9		(7/2-)			ann.rad./ 0.11794 0.23868 0.8267 (0.07 - 2.64)
[182]Os		181.95219	21.5 h	EC/0.9		0+			Re k x-ray 0.1802 0.5100
[183m]Os			9.9 h	EC/84 / I.T./16 /		1/2-			Os k x-ray Re k x-ray 1.1020 1.1080
[183]Os		182.9531	13. h	EC/2.1		9/2+	-0.79	+3.1	Re k x-ray 0.1144 0.3818
[184]Os	0.020(3)	183.952491				0+			
[185]Os		184.954043	93.6 d	EC/1.013		1/2-			Re k x-ray 0.6461 0.8748 0.8805
[186]Os	1.58(10)	185.953838	2.x10¹⁵ y	α/	≈ 2.75/	0+			
[187]Os	1.6(1)	186.955748				1/2-	+0.0646519		
[188]Os	13.3(2)	187.955836				0+			
[189m]Os			5.8 h	I.T./0.0308		9/2-			Os L x-ray 0.0308
[189]Os	16.1(3)	188.958145				3/2+	+0.65993	+0.86	
[190m]Os			9.9 m	I.T./1.705		10-	-0.6		Os k x-ray 0.1867 0.3611 0.5026 0.6161
[190]Os	26.4(4)	189.958445				0+			
[191m]Os			13.1 h	I.T./0.0744		3/2-			Os k x-ray 0.0744

Elem. or Isot.	Natural Abundance (%)	Atomic Mass or Weight	Half-Life	Decay Mode/Energy (/MeV)	Particle Energy /Intensity (MeV / %)	Spin (h/2π)	Nuclear Magnetic Mom.(nm)	Elect. Quadr. Mom. (b)	γ-ray Energy /Intensity (MeV / %)
^{191}Os		190.960928	15.4 d	β⁻ /0.314	0.140/100	9/2-		+2.5	Ir k x-ray 0.1294
^{192m}Os			6.0 s	I.T./2.0154		(10-)			Os k x-ray 0.2058 0.3024 0.4531 0.5692
^{192}Os	41.0(3)	191.961479				0+			
^{193}Os		192.964148	30.5 h	β⁻ /1.141	1.04/20	3/2-	+0.730	+0.47	Ir k x-ray 0.1389 0.4605
^{194}Os		193.965179	6.0 y	β⁻ /0.097	0.054/33 0.096/67	0+			Ir L x-ray 0.0429
^{195}Os		194.9681	6.5 m	β⁻ /2.0	2.0/				
^{196}Os		195.96962	34.9 m	β⁻ /1.16	0.84/	0+			0.1262/5 0.4079/5.9
$_{77}$Ir		**192.217(3)**							
^{166}Ir		165.9855	≥ 5. ms	α/					
^{167}Ir		166.9817	≥ 5. ms	α/					
^{168}Ir		167.9799		α/					
^{169}Ir		168.9764	0.4 s	α/					
^{170}Ir		169.9743	1.0 s	α/	6.03/				
^{171}Ir		170.9718	1.5 s	α/	5.91/				
^{172}Ir		171.9706	2.1 s	α/	5.811/				
^{173}Ir		172.9677	3.0 s	α/	5.665/				0.228 (0.379-0.475)
^{174}Ir		173.9668	4. s	α/	5.478/				0.0493 (0.092-0.296)
^{175}Ir		174.9641	≈ 4.5 s	α/	5.393/				0.1587 (0.276-1.33)
^{176}Ir		175.9635	8. s	EC, β⁺/80 α/3.2/	5.118/				0.1056 0.260 (0.135-0.415)
^{177}Ir		176.9612	30. s	EC, β⁺/5.7 α/0.06/	5.011/				0.184 (0.062-0.194)
^{178}Ir		177.9601	12. s	β⁺ ,EC/6.3					0.1320 0.2667 0.3633
^{179}Ir		178.9592	4. m	EC/4.9					0.0975 (0.045-0.220)
^{180}Ir		179.9593	1.5 m	EC/6.4					0.2765 ((0.132-1.106)
^{181}Ir		180.9576	4.9 m	β⁺ ,EC/4.1		(7/2+)			ann.rad./ 0.1076 0.2270 0.3189
^{182}Ir		181.9582	15. m	β⁺ /44 /5.6 EC/56 /					1.6396 ann.rad./ Os k x-ray 0.1273
^{183}Ir		182.9568	57. m	β⁺ ,EC/3.5					0.2370 ann.rad./ 0.0877 0.2285
^{184}Ir		183.9574	3.0 h	β⁺ /12 /4.6 EC/88 /	2.3/ 2.9/	5-	0.70	+2.0	0.2824 ann.rad./ Os k x-ray 0.11968 0.2640
^{185}Ir		184.9566	14. h	β⁺ /3 /2.4 EC/97 /		(5/2-)	2.60	-2.1	0.3904 ann.rad./ Os k x-ray 0.2543
^{186m}Ir			1.7 h	EC /		(2-)			1.8288 Os k x-ray 0.1371
^{186}Ir		185.95795	15.7 h	EC/98 /3.83 β⁺ /2 /		(5+)	3.9	-2.5	0.7675 Os k x-ray 0.1372 0.2968 0.4348 (0.13 - 3.0)

Elem. or Isot.	Natural Abundance (%)	Atomic Mass or Weight	Half-Life	Decay Mode/Energy (/MeV)	Particle Energy /Intensity (MeV / %)	Spin (h/2π)	Nuclear Magnetic Mom.(nm)	Elect. Quadr. Mom. (b)	γ-ray Energy /Intensity (MeV / %)
[187]Ir		186.95736	10.5 h	EC/1.50		3/2+			Os k x-ray 0.0743 0.4009 0.4271 0.6109 0.9128
[188]Ir		187.95885	1.72 d	β⁺ /2.81 EC/99+ /	1.13/ 1.64/	(2-)	0.30	+5.4	Os k x-ray 0.1550 0.4780 0.6330 2.2146
[189]Ir		188.95872	13.2 d	EC/0.53		3/2+	0.13	+1.0	Os k x-ray 0.2449
[190m2]Ir			3.2 h	β⁺,EC/95 / I.T./5 /		(11-)			
[190m1]Ir			1.2 h	I.T. /0.0263		7+			Ir L x-ray
[190]Ir		189.9606	11.8 d	EC/2.0		(4+)	0.04	+2.8	Os k x-ray 0.1867 0.4072 0.5186 0.5580 0.6051 (0.2 - 1.4)
[191m]Ir			4.93 s	I.T./0.1714		11/2-	0.603		Ir k x-ray 0.1294
[191]Ir	37.3(5)	190.960591				3/2+	+0.151	+0.82	Ir k x-ray
[192m2]Ir			241. y	I.T./0.161		(9+)			Ir L x-ray
[192m1]Ir			1.44 m	I.T./0.0580		(1+)			0.0580 0.3165
[192]Ir		191.962602	73.83 d	β⁻ /1.460		(4-)	+1.92	+2.3	Pt k x-ray 0.31649/83. 0.46806/48.
[193m]Ir			10.53 d	I.T./0.0802		11/2-			Ir L x-ray 0.0803
[193]Ir	62.7(5)	192.962923				3/2+	+0.164	+0.75	
[194m]Ir			170. d	β⁻ /		11			Pt k x-ray 0.3284 0.4829 0.5624
[194]Ir		193.965075	19.3 h	β⁻ /2.247	1.92/9 2.25/86	1-	0.39	+0.34	0.2935 0.3284 0.6451 (0.1 - 2.2)
[195m]Ir			3.9 h	β⁻ /	0.41/ 0.97/	(11/2-)			Pt k x-ray 0.3199/9.6 0.3649/9.5 0.4329/9.6 0.6849/9.6
[195]Ir		194.965976	2.8 h	β⁻ /1.120	1.0/80 1.11/13	(3/2+)			Pt k x-ray 0.0989/9.7
[196m]Ir			1.40 h	β⁻ /	1.16/				Pt k x-ray 0.3557 0.3935 0.4471 0.5214 0.6473
[196]Ir		195.96838	52.s	β⁻ /3.21	2.1/15 3.2/80	0-			0.3329 0.3557 0.7796
[197m]Ir			8.9 m	β⁻ / I.T./		(11/2-)			0.3465 See Ir[197]
[197]Ir		196.96964	5.8 m	β⁻ /2.16	1.5/ 2.0/	(3/2+)			0.0531 0.1351 0.4306 0.4697
[198]Ir		197.9723	8. s	β⁻ /4.1					0.4074 0.5070
₇₈Pt		195.08(3)							
[168]Pt		167.9880		α					
[169]Pt		168.9864	3. ms	α					
[170]Pt		169.9816	6. ms	α	6.55				
[171]Pt		170.9811	0.03 s	α					
[172]Pt		171.97730	0.10 s	α /	6.31/94	0+			
[173]Pt		172.9765	0.34 s	β⁺,EC/8.2 α/	6.20/				

Elem. or Isot.	Natural Abundance (%)	Atomic Mass or Weight	Half-Life	Decay Mode/Energy (/MeV)	Particle Energy /Intensity (MeV / %)	Spin (h/2π)	Nuclear Magnetic Mom.(nm)	Elect. Quadr. Mom. (b)	γ-ray Energy /Intensity (MeV / %)
[174]Pt		173.97281	0.89 s	β+ ,EC/17 /5.6 α/83 /	6.040/	0+			
[175]Pt		174.9723	2.5 s	β+ ,EC/65 /7.6 α/35 /	5.831/5 5.96/54 6.038/				0.0774 0.1354 0.2128
[176]Pt		175.9690	6.3 s	β+ ,EC/60 /5.1 α/40 /	5.528/0.6 5.750/41	0+			ann.rad./ 0.2277
[177]Pt		176.9685	11. s	EC/91 /6.8 α/9 /	5.53/ 5.485/3 5.525/6				0.0908
[178]Pt		177.9649	21. s	EC/93 /4.5 α/7 /	5.286/0.2 5.442/7	0+			
[179]Pt		178.9653	33. s	β+ ,EC/5.7 α /	5.16/				
[180]Pt		179.9632	52. s	β+ ,EC/99.7 /3.7 α/0.3 /	5.140/	0+			
[181]Pt		180.9632	51. s	β+ ,EC/5.2					
[182]Pt		181.9613	2.7 m	β+ ,EC/2.9		0+			ann.rad./ 0.1360 0.1460 0.2100
[183m]Pt			43. s	β+ ,EC/ I.T./		(7/2-)			ann.rad./ 0.3132 0.3164 0.6296
[183]Pt		182.9617	7. m	β+ ,EC/4.6					ann.rad./
[184]Pt		183.9599	17.3 m	β+ ,EC/2.3					ann.rad./ 0.1549 0.1919 0.5484
[185m]Pt			33. m	β+ ,EC/		1/2-			
[185]Pt		184.9607	1.18 h	β+ ,EC/3.8		(9/2+)			ann.rad./ 0.1353 0.1974 0.2296 0.2551
[186]Pt		185.95943	2.0 h	β+ ,EC/1.38		0+			ann.rad./ 0.6115 0.6892
[187]Pt		186.9607	2.35 h	β+ ,EC/3.1		3/2			ann.rad./ Ir k x-ray 0.1064 0.1100 0.2015 0.2849 0.7092
[188]Pt		187.95940	10.2 d	EC/0.51		0+			Ir k x-ray 0.1876 0.1951
[189]Pt		188.96083	10.9 h	β+ ,EC/1.97		3/2-	0.43	-0.6	Ir k x-ray 0.0943 0.6076 0.7214 (0.09 - 1.47)
[190]Pt	0.01(1)	189.95993	6.5x10[11] y			0+			
[191]Pt		190.961684	2.96 d	EC/1.02		(3/2-)	0.50	-0.6	Ir k x-ray 0.3599 0.4094 0.5389
[192]Pt	0.79(6)	191.961035				0+			
[193m]Pt			4.33 d	I.T./0.1498		13/2+	-0.75		Pt k x-ray 0.1355
[193]Pt		192.962984	60. y	EC/0.0566		(1/2-)			Ir k x-rays
[194]Pt	32.9(6)	193.962663				0+			
[195m]Pt			4.02 d	I.T./0.2952		13/2+	0.61	+1.4	Pt k x-ray 0.0989
[195]Pt	33.8(6)	194.964774				1/2-	+0.6095		
[196]Pt	25.3(6)	195.964934				0+			
[197m]Pt			1.590 h	I.T./97 / β- /3 /		13/2+			Pt k x-ray 0.0530 0.3465
[197]Pt		196.967323	18.3 h	β- /0.719		1/2-	0.51		Au k x-ray 0.1914 0.2688
[198]Pt	7.2(2)	197.967875				0+			

Elem. or Isot.	Natural Abundance (%)	Atomic Mass or Weight	Half-Life	Decay Mode/Energy (/MeV)	Particle Energy /Intensity (MeV / %)	Spin (h/2π)	Nuclear Magnetic Mom.(nm)	Elect. Quadr. Mom. (b)	γ-ray Energy /Intensity (MeV / %)
^{199m}Pt			13.6 s	I.T./0.424		13/2+			Pt k x-ray 0.3919
^{199}Pt		198.970576	30.8 m	β^- /1.70	0.90/18 1.14/14	(5/2-)			0.0772 0.18579 0.31703 0.49375 0.54298
^{200}Pt		199.97142	12.5 h	β^- /≈0.66		0+			Au k x-ray 0.13590 0.22747 0.24371
^{201}Pt		200.9745	2.5 m	β^- /2.66		(5/2-)			0.070 0.152 0.222 1.760
^{202}Pt			1.8 d						0.440
$_{79}$Au		196.96655(2)							
^{172}Au			4 ms	α/7.02	6.86				
^{173}Au		172.9864	0.06 s	α					
^{174}Au		173.9842	0.12 s	α					
^{175}Au		174.9817	0.20 s	α					
^{176}Au		175.9803	1.2 s	β^+ ,EC/10.5 α /	6.260/80 6.290/20				
^{177}Au		176.9772	1.2 s	α /	6.115/ 6.150/				
^{178}Au		177.9760	2.6 s	α /	5.920/				
^{179}Au		178.9732	7.5 s	α /	5.85/				
^{180}Au		179.9724	8.1 s	EC/8.6 α/	5.65 5.61 5.50				0.1522 0.2564 0.5242 0.6765 0.8084 0.8597
^{181}Au		180.9700	11.4 s	EC/97.5/6.3 α/2.7/	5.482/				
^{182}Au		181.9686	21. s	β^+ ,EC/6.9 α/0.13/					ann.rad./ 0.1549 0.2649 (0.13 - 1.4)
^{183}Au		182.9676	42. s	EC/5.5 α/0.8/			+1.97		0.1630 0.2730 0.3625
^{184}Au		183.9675	53. s	EC,β^+ /7.1 α/0.013/		(5+)	+1.65		
^{185m}Au			6.8 m	β^+ ,EC/ I.T./0.145					
^{185}Au		184.9657	4.3 m	β^+ ,EC/4.71 α/0.26/		(5/2-)	+2.17	-1.1	ann.rad./ 0.1915
^{186m}Au			< 2. m	β^+ ,EC/					
^{186}Au		185.9659	10.7 m	β^+ ,EC/6.0 α/8(10)$^{-4}$/		3-	-1.26	+3.1	ann.rad./ 0.1915 0.2988
^{187m}Au			2.3 s	IT		9/2-			
^{187}Au		186.9646	8.3 m	β^+ ,EC/3.60		1/2+	+0.54		ann.rad./ 0.9152 1.2668 1.3321 1.4081
^{188}Au		187.9651	8.8 m	β^+ ,EC/5.3		(1-)	-0.07		ann.rad./ 0.2660 0.3404 0.6061
^{189m}Au			4.6 m	β^+ ,EC/		11/2-	+6.19		0.1667
^{189}Au		188.9642	28.7 m	EC/96 /3.2 β^+ /4 /		1/2+	+0.49		ann.rad./ Pt k x-ray 0.4478 0.7133 0.8128
^{190}Au		189.96470	43. m	β^+ /2 /4.44 EC/98 /		1-	-0.07		ann.rad./ Pt k x-ray 0.2958 0.3018 0.5977
^{191m}Au			0.9 s	I.T./0.2663		(11/2-	6.6		Au k x-ray 0.2414 0.2526

Elem. or Isot.	Natural Abundance (%)	Atomic Mass or Weight	Half-Life	Decay Mode/Energy (/MeV)	Particle Energy /Intensity (MeV / %)	Spin (h/2π)	Nuclear Magnetic Mom.(nm)	Elect. Quadr. Mom. (b)	γ-ray Energy /Intensity (MeV / %)
^{191}Au		190.96365	3.2 h	EC/1.83		3/2+	+0.137	+0.72	Pt k x-ray 0.2779 0.2839 0.5864 0.6742 (0.08 - 1.3)
^{192}Au		191.96481	4.9 h	β$^+$ /5 /3.52 EC/95 /	2.19/ 2.49/	1-	-0.011	-0.23	ann.rad./ Pt k x-ray 0.2959 0.3165
^{193m}Au			3.9 s	I.T./0.2901		11/2-	6.2		Au k x-ray 0.2580
^{193}Au		192.96413	17.6 h	EC/1.07		3/2+	+0.140	+0.66	Pt k x-ray 0.1862 0.2556
^{194}Au		193.96534	1.64 d	β$^+$ /3 /2.49 EC/97 /	1.49/	1-	+0.075	-0.24	ann.rad./ Pt k x-ray 0.2935 0.3284/61
^{195m}Au			30.5 s	I.T./0.3186		11/2-	6.2	+1.4	Au k x-ray 0.2617
^{195}Au		194.965017	186.12 d	EC/0.227		3/2+	+0.149	+0.61	Pt k x-ray
^{196m2}Au			9.7 h	I.T./0.5954		12-	5.7		Au k x-ray 0.1478 0.1883
^{196m1}Au			8.1 s	I.T./0.0846		8+			0.0847
^{196}Au		195.966551	6.18 d	EC/92 /1.506		2-	+0.591	0.81	Pt k x-ray
^{197m}Au			7.8 s	I.T./0.4094 β$^-$ /8 /0.686		11/2-	+6.0	+1.4	Au k x-ray 0.1302 0.2790
^{197}Au	100.	196.966551				3/2+	+0.14575	+0.55	
^{198m}Au			2.30 d	I.T./0.812		(12-)			Au k x-ray 0.0972 0.1803 0.2419
^{198}Au		197.968225	2.694 d	β$^-$ /1.372	0.290/1 0.961/99	2-	+0.5934	+0.68	Hg k x-ray 0.411794
^{199}Au		198.968748	3.14 d	β$^-$ /0.453	0.25/22 0.292/72 0.462/6	3/2+	+0.2715	0.55	Hg k x-ray 0.15837 0.20820
^{200m}Au			18.7 h	β$^-$ /84 /1.0 I.T./16 /	0.56/	12-	5.9		Au k x-ray 0.1111 0.2559 0.36797 0.4978 0.5793 0.7595
^{200}Au		199.97072	48.4 m	β$^-$ /2.24	0.7/15 2.2/77	1-			0.3679 1.2254 (0.3 - 1.6)
^{201}Au		200.97165	26. m	β$^-$ /1.28	1.27/82	3/2+			0.1674 0.5170 0.5426 0.6132
^{202}Au		201.9738	29. s	β$^-$ /3.0		(1-)			0.4396 (0.04-0.37)
^{203}Au		202.97515	1.0 m	β$^-$ /2.14	≈ 1.9/	3/2+			0.4366
^{204}Au		203.9783	40. s	β$^-$ /4.5		(2-)			1.5113 (0.38 - 1.33)
^{205}Au			31. s	β$^-$ /					
$_{80}$Hg		200.59(2)							
^{175}Hg		174.9912	0.02 s	α					
^{176}Hg		175.98733	0.03 s	α	6.77				
^{177}Hg		176.9863	0.13 s	α					
^{178}Hg		177.98248	0.26 s	EC/50 /6.1 α/50 /	6.43/	0+			
^{179}Hg		178.9818	1.09 s	EC/8.0 α /	6.29/				
^{180}Hg		179.9783	2.6 s	EC/5.5 α /	6.12/33 5.69/.03	0+			0.1250 0.3005 0.3812
^{181}Hg		180.9778	3.6 s	β$^+$ EC/74 / ≈7.3 α/26 /		(1/2-)	+0.5071		0.0663 0.0811 0.0924 0.1474 0.1587 0.2142 0.2398

Elem. or Isot.	Natural Abundance (%)	Atomic Mass or Weight	Half-Life	Decay Mode/Energy (/MeV)	Particle Energy /Intensity (MeV / %)	Spin (h/2π)	Nuclear Magnetic Mom.(nm)	Elect. Quadr. Mom. (b)	γ-ray Energy /Intensity (MeV / %)
[182]Hg		181.9739	10.8 s	β^+ ,EC/85 /5.0 α/15 /	5.87/8.6 5.45/0.03	0+			0.1289 0.2168 0.4126
[183]Hg		182.9744	9. s	β^+ ,EC/77 /6.3 α /	5.83/ 5.91/	1/2-	+0.524		0.0714 0.0874 0.1538
[184]Hg		183.9719	30.9 s	β^+ ,EC/99 /4.1 α/1 /	5.54/1.3 5.07/2 x 10^{-3}	0+			0.0915 0.1265 0.1560 0.2362
[185m]Hg			21. s	β^+ ,EC,IT,α /	5.37/	13/2+	-1.02	+0.2	0.211 0.292
[185]Hg		184.9720	51. s	β^+ ,EC/95 /5.8		1/2-	+0.509		(0.02 - 0.55)
[186]Hg		185.9695	1.4 m	β^+ ,EC/3.3 α	5.09/0.02	0+			0.1119 0.2518
[187m]Hg			1.7 m	β^+ ,EC/		13/2+	-1.04	+0.5	See Hg[187]
[187]Hg		186.9698	2.4 m	β^+ ,EC/4.9		3/2-	-0.594		0.1034/32. 0.2334/100. 0.2403/33. 0.27151/31. 0.3763/38. 0.5254/30. (0.10-2.18)
[188]Hg		187.9676	3.2 m	β^+ ,EC/2.3 α	4.61	0+			0.0988 0.1148 0.1424 0.1900
[189m]Hg			8.6 m	EC/		13/2+	-1.06	+0.7	0.0780 0.3210 0.4345 0.5655 (0.08 - 2.10)
[189]Hg		188.9687	7.6 m	EC/4.2		3/2-	-0.6086	-0.8	0.2005 0.2038 0.2386 0.2485
[190]Hg		189.9663	20.0 m	EC/1.5		0+			0.1296 0.1426
[191m]Hg			51. m	β^+ /6 / EC/94 /		13/2+	-1.07	+0.6	ann.rad./ Au k x-ray 0.2741 0.4203 0.5787 (0.07 - 1.9)
[191]Hg		190.9671	50. m	β^+ ,EC/3.2		(3/2-)	-0.62	-0.8	0.1963 0.2247 0.2524
[192]Hg		191.9653	5.0 h	EC/ $\approx$0.5		0+			Au k x-ray 0.1572 0.2748 0.3065
[193m]Hg			11.8 h	β^+ ,EC/91 / I.T./9 /0.2901		13/2+	-1.05843	+0.92	Hg k x-ray 0.1866 0.2580 0.4076 0.5733 0.9324 (0.1 - 1.96)
[193]Hg		192.96664	3.8 h	EC,B$^+$/2.34		3/2-	-0.6276		0.1866 0.2580 0.8611
[194]Hg		193.96538	520. y	EC/0.04		0+			Au L x-rays
[195m]Hg			1.67 d	I.T./(54)/0.3186 EC/(46)/		13/2+	-1.04465	+1.1	Hg k x-ray Au k x-ray 0.2617 0.5603 0.7798
[195]Hg		194.96664	9.5 h	EC/1.51		1/2-	+0.541475		Au k x-ray 0.0614 0.7798
[196]Hg	0.15(1)	195.965814				0+			
[197m]Hg			23.8 h	I.T./(93)/0.2989		13/2+	-1.02768	+1.2	Hg k x-ray Au k x-ray 0.13398
[197]Hg		196.967195	2.672 d	EC/0.600		1/2-	+0.527374		Au k x-ray 0.07735
[198]Hg	9.97(8)	197.966752				0+			
[199m]Hg			42.6 m	I.T./0.532		13/2+	-1.014703	+1.2	Hg k x-ray 0.15841

Elem. or Isot.	Natural Abundance (%)	Atomic Mass or Weight	Half-Life	Decay Mode/Energy (/MeV)	Particle Energy /Intensity (MeV / %)	Spin (h/2π)	Nuclear Magnetic Mom.(nm)	Elect. Quadr. Mom. (b)	γ-ray Energy /Intensity (MeV / %)
^{199}Hg	16.87(10)	198.968262				1/2-	+0.505885		
^{200}Hg	23.10(16)	199.968309				0+			
^{201}Hg	13.18(8)	200.970285				3/2-	-0.560226	+0.39	
^{202}Hg	29.86(20)	201.970625				0+			
^{203}Hg		202.972857	46.61 d	β^-/0.492	0.213/100	5/2-	+0.8489	+0.34	Tl k x-ray 0.279188
^{204}Hg	6.87(4)	203.973475				0+			
^{205}Hg		204.976056	5.2 m	β^-/1.531	1.33/4	1/2-	+0.6010		0.20378 (0.2 - 1.4)
^{206}Hg		205.97750	8.2 m	β^-/1.31	0.935/34 1.3/63	0+			Tl k x-ray 0.3052 0.6502
^{207}Hg		206.9825	2.9 m	β^-/4.8		(9/2+)			
^{208}Hg			0.7 h	β^-					0.474
$_{81}$**Tl**		**204.3833(2)**							
^{179}Tl		178.9917	0.2 s	α					
^{182}Tl		181.9856	3. s	β^+, EC/10.9					0.351 (0.26 - 0.41)
^{183m}Tl			0.06 s	α		9/2-			
^{183}Tl		182.9826	5. s	β^+, EC/7.7		1/2+			0.208
^{184}Tl		183.9818	11. s	β^+, EC/(98)/9.2 α/(2)/	6.16/				0.2868 0.3399 0.3667
^{185m}Tl			1.8 s	I.T./0.453 α/5.97	6.01/	(9/2-)			0.1688 0.2840
^{185}Tl		184.9791	20. s	EC/β^+/6.6					
^{186m}Tl			4. s	I.T./0.374					0.3738
^{186}Tl		185.9776	28. s	β^+, EC/7.5					0.3567 0.4026 0.4053
^{187m}Tl			15.6 s	I.T./$\approx$ 0.33		(9/2+)	3.8	-2.4	0.2995
^{187}Tl		186.9762	50. s	β^+, EC/6.0		1/2+	1.6		Hg k x-ray
^{188m}Tl			1.18 m	β^+, EC/		(7+)			0.4129 0.5043 0.5921
^{188}Tl		187.9759	1.2 m	β^+, EC/7.8		(2-)	0.48	0.13	See Tl[188m] 0.4129
^{189}Tl			1.4 m	β^+, EC/		(9/2-)	+3.878	-2.29	0.2156 0.2284 0.3175 0.4452
^{189}Tl		188.9743	2.3 m	β^+, EC/5.2		(1/2+)			0.3337 0.4510 0.5223 0.9422
^{190m}Tl			3.7 m	β^+, EC/	4.2/	(7+)	+0.495	0.29	0.1968 0.4164 0.7311
^{190}Tl		189.9738	2.6 m	β^+, EC/7.0	5.7/	(2-)	+0.25	-0.33	0.4164 0.6254 0.6838 1.0999
^{191m}Tl			5.2 m	β^+, EC/(98)/		(9/2+)	+3.903	-2.3	0.2157 0.2647 0.3256 0.3359
^{191}Tl		190.9723				(1/2)	1.59		
^{192m}Tl			10.8 m	β^+, EC/		(7+)	+0.518	0.46	0.1740 0.4228 0.6348 0.7863 0.7455
^{192}Tl		191.972	9.6 m	β^+, EC/6.4		(2-)	+0.20	-0.33	0.3975 0.4228 0.6908
^{193m}Tl			2.1 m	I.T./(75)/		(9/2-)	+3.948	-2.2	0.3650
^{193}Tl		192.9706	22. m	β^+, EC/3.6		(1/2+)	+1.591		0.2077 0.3244 0.3440 0.6761 1.0447 1.5793

Elem. or Isot.	Natural Abundance (%)	Atomic Mass or Weight	Half-Life	Decay Mode/Energy (/MeV)	Particle Energy /Intensity (MeV / %)	Spin (h/2π)	Nuclear Magnetic Mom.(nm)	Elect. Quadr. Mom. (b)	γ-ray Energy /Intensity (MeV / %)
¹⁹⁴ᵐTl			32.8 m	β⁺ /(20)/ ≈ 0.30 EC/(80)/		(7+)	+0.540	0.62	ann.rad./ Hg k x-ray 0.4282 0.6363 0.7490
¹⁹⁴Tl		193.9711	34. m	β⁺ ,EC/5.3		2-	0.14	-0.28	0.3955 0.4282 0.6363
¹⁹⁵ᵐTl			3.6 s	I.T./0.483		9/2-			Tl k x-ray 0.0990 0.3836
¹⁹⁵Tl		194.9697	1.16 h	EC/(97)/2.8 β⁺ /(3)/		1/2+	+1.58		ann.rad./ Hg k x-ray 0.2422 0.5635 0.8845 1.3639 (0.13 - 2.5)
¹⁹⁶ᵐTl			1.41 h	β⁺ ,EC/(95)/4.9		(7+)	0.55	+0.76	0.0840 0.4261 0.6353 0.6954 (0.08 - 1.0)
¹⁹⁶Tl		195.9705	1.84 h	β⁺ /(15)/4.4 EC/(85)/		2-	0.07	-0.18	ann.rad./ Hg k x-ray 0.4257 0.6105 (0.03 - 2.4)
¹⁹⁷ᵐTl			0.54 s	I.T./(53)/0.608 β⁺ ,EC/(47)/		9/2-			Tl k x-ray 0.2262 0.4118 0.5872 0.6367
¹⁹⁷Tl		196.96954	2.83 h	β⁺ /(1)/2.18 EC/(99)/		1/2+	+1.58		Hg k x-ray 0.1522/8.2 0.4258
¹⁹⁸ᵐTl			1.87 h	β⁺ ,EC/(53)/ I.T./(47)/0.5347		7+	+0.64		Hg k x-ray Tl k x-ray 0.4118 0.5872 0.6367
¹⁹⁸Tl		197.9405	5.3 h	EC,β⁺ /(1)/3.5	1.4/ 2.1/ 2.4/	2-	0.00(1)		Hg k x-ray 0.4118 0.6367 0.6759 (0.23 - 2.8)
¹⁹⁹Tl		198.9698	7.4 h	EC/1.4		1/2-	+1.60		Hg k x-ray 0.2082 0.2473 0.4555
²⁰⁰Tl		199.97095	1.087 d	EC/2.46	1.07/ 1.44/	2-	0.04		Hg k x-ray 0.36799 1.2057 (0.11 - 2.3)
²⁰¹Tl		200.97080	3.040 d	EC/0.48		1/2+	+1.605		Hg k x-ray 0.13528 0.16740/10.0
²⁰²Tl		201.97209	12.23 d	EC/1.36		2-	0.06		Hg k x-ray 0.43957
²⁰³Tl	29.524(14)	202.972329				1/2+	+1.622258		
²⁰⁴Tl		203.973848	3.78 y	β⁻ /(97)/0.7637 EC/(3)/0.347	0.763/97	2-	0.09		Hg k x-ray
²⁰⁵Tl	70.476(14)	204.974412				1/2+	+1.638215		
²⁰⁶ᵐTl			3.76 m	I.T./2.644		12-			Tl k x-ray 0.2166 0.2661 0.4534 0.6866 1.0219
²⁰⁶Tl		205.976095	4.20 m	β⁻ /1.533	1.53/99.9	0-			Pb k x-ray 0.80313
²⁰⁷ᵐTl			1.3 s	I.T./1.350		11/2-			Tl k x-ray 0.3501 1.0000
²⁰⁷Tl		206.97741	4.77 m	β⁻ /1.423	1.43/99.8	1/2+	+1.88		0.89723

Elem. or Isot.	Natural Abundance (%)	Atomic Mass or Weight	Half-Life	Decay Mode/Energy (/MeV)	Particle Energy /Intensity (MeV / %)	Spin (h/2π)	Nuclear Magnetic Mom.(nm)	Elect. Quadr. Mom. (b)	γ-ray Energy /Intensity (MeV / %)
²⁰⁸Tl		207.982004	3.053 m	β⁻ /5.001	1.28/23 1.52/22 1.796/51	(5+)	+0.29		Pb k x-ray 0.27728 0.51061 0.58302 2.61448
²⁰⁹Tl		208.98535	2.16 m	β⁻ /3.98	1.8 /100	(1/2+)			Pb k x-ray 1.5670/100 0.4651/95 (0.12 - 1.33)
²¹⁰Tl		209.99006	1.30 m	β⁻ /5.48	1.3/25 1.9/56	(5+)			Pb k x-ray 0.081 0.2981 0.79788
₈₂Pb		**207.2(1)**							
¹⁸²Pb		181.99268	0.06 s	α	6.92				
¹⁸³Pb		182.9919	0.3 s	α/		1/2+			
¹⁸⁴Pb		183.9882	0.6 s	α/	6.63/	0+			
¹⁸⁵Pb		184.9876	4.1 s	α/	6.34/ 6.40/ 6.48/				
¹⁸⁶Pb		185.9835	5. s	β⁺ ,EC/(95)/5.5 α/(5)/	6.32/ 6.34/<100 6.01/<0.2	0+			
¹⁸⁷ᵐPb			15.2 s	β⁺ ,EC/	5.99/ 6.19/	(1/2-)			0.0674 0.2080 0.2755 0.2995 0.4487 0.7477
¹⁸⁷Pb		186.9839	18.3 s	EC/7.2 α/	6.08/	13/2 +			0.1930 0.3314 0.3435 0.3934
¹⁸⁸Pb		187.9811	23. s	EC/(78)/4.8 α/(22)/	5.98/<10 5.61/<0.1	0+			0.1850 0.7582
¹⁸⁹Pb		188.9809	51. s	EC/6.1 α/	5.58/				
¹⁹⁰Pb		189.9782	1.2 m	β⁺ (13)/4.1 EC/(86)/ α/(0.9)/	5.58/	0+			ann.rad./ Tl k x-ray 0.1415 0.1512 0.9422
¹⁹¹ᵐPb			2.2 m	β⁺ ,EC/		13/2 +			ann.rad./ 0.3871 0.6135 0.7122
¹⁹¹Pb		190.9782	1.3 m	β⁺ ,EC/5.5					ann.rad./ 0.9368
¹⁹²Pb		191.9758	3.5 m	β⁺ ,EC/≈3.4 α/.006/	5.11	0+			ann.rad./ 0.1675 0.6082 1.1954
¹⁹³ᵐPb			5.8 m	β⁺ ,EC/		13/2 +	-1.16		ann.rad./ 0.3650 0.3922
¹⁹³Pb		192.9761	≈ 2. m	EC/5.2		3/2			
¹⁹⁴Pb		193.9740	11. m	β⁺ ,EC/2.7 α	4.64	0+			ann.rad./ 0.2036
¹⁹⁵ᵐPb			15. m	β⁺ /(8)/ EC/(92)/		13/2 +	-1.132	+0.29	ann.rad./ Tl k x-ray 0.3836 0.3942 0.8784
¹⁹⁵Pb		194.976	≈ 15. m	β⁺,EC/5.8					ann.rad./ 0.3836 0.3937 0.7776
¹⁹⁶Pb		195.9727	37. m	β⁺ ,EC/2.1		0+			Tl k x-ray 0.2531 0.5021

Elem. or Isot.	Natural Abundance (%)	Atomic Mass or Weight	Half-Life	Decay Mode/Energy (/MeV)	Particle Energy /Intensity (MeV / %)	Spin (h/2π)	Nuclear Magnetic Mom.(nm)	Elect. Quadr. Mom. (b)	γ-ray Energy /Intensity (MeV / %)
^{197m}Pb			43. m	EC/(79)/ β^+/(2)/ I.T./(19)/0.3193		13/2+	-1.105	+0.5	Tl k x-ray 0.3079 0.3877 0.7743 (0.2 - 2.2)
^{197}Pb		196.9734	≈ 8. m	EC/(97)/3.6 β^+/(3)/		(3/2-)	-1.075	-0.08	Tl k x-ray 0.3755 0.3858 0.7611
^{198}Pb		197.9720	2.4 h	EC/1.4		0+			Tl k x-ray 0.1734 0.2903 0.3654
^{199m}Pb			12.2 m	I.T./(93)/0.4248 β^+,EC/(7)/		13/2+	-1.074	+0.08	Pb k x-ray 0.4255
^{199}Pb		198.9729	1.5 h	EC/(99)/2.9 β^+/(1)/		5/2-			Tl k x-ray 0.3534 0.7202 1.1350 (0.22 - 2.4)
^{200}Pb		199.97182	21.5 h	EC/0.81		0+			Tl k x-ray 0.14763
^{201m}Pb			1.02 m	I.T./0.6291		13/2+			Pb k x-ray 0.6288
^{201}Pb		200.97285	9.33 h	EC/1.90		5/2-	+0.675	-0.009	Tl k x-ray 0.33120 0.36131 (0.11 - 1.8)
^{202m}Pb			3.53 h	I.T./(90)/2.170 β^+/(10)/		9-	-0.228	+0.58	Pb k x-ray Tl k x-ray 0.42219 0.78700 0.96271
^{202}Pb		201.97214	5.3x10^4 y	EC/0.05		0+			Tl L x-ray
^{203m}Pb			6.2 s	I.T./0.8252		13/2+			Pb k x-ray 0.8203 0.8252
^{203}Pb		202.97338	2.1615 d	EC/0.98		5/2-	+0.686	+0.10	Tl k x-ray 0.279188
^{204m}Pb			1.12 h	I.T./2.185		9-			Pb k x-ray 0.37481 0.89922 0.91175
^{204}Pb	1.4(1)	203.973028				0+			
^{205}Pb		204.974467	1.51x10^7 y	EC/0.0512		5/2-	+0.712	+0.23	Tl L x-ray
^{206}Pb	24.1(1)	205.974449				0+			
^{207m}Pb			0.80 s	I.T./1.632		13/2+			Pb k x-ray 0.56915 1.06310
^{207}Pb	22.1(1)	206.975880				1/2-	+0.59258		
^{208}Pb	52.4(1)	207.976636				0+			
^{209}Pb		208.981075	3.25 h	β^-/0.644	0.645/100	9/2+	-1.474	-0.3	
^{210}Pb		209.984174	22.6 y	β^-/0.0635	0.017/81 0.061/19 α 3.72	0+			
^{211}Pb		210.988732	36.1 m	β^-/1.37	0.57/5 1.36/92	(9/2+)	-1.404	+0.09	0.40486 0.42700 0.83186 (0.09 - 1.27)
^{212}Pb		211.991887	10.64 h	β^-/0.574	0.28/83 0.57/12	0+			Bi k x-ray 0.23858
^{213}Pb		212.9966	10.2 m	β^-/2.1					
^{214}Pb		213.999797	26.9 m	β^-/1.0	0.67/48 0.73/42	0+			Bi k x-ray 0.24192 0.29509 0.35187
$_{83}$Bi		208.98038(2)							
^{187m}Bi			≈ 8. ms	α					
^{187}Bi		186.9935	35. ms	α					
^{188}Bi		187.9922		α					
^{189m}Bi			≈ 5. ms	α					
^{189}Bi		188.9895	0.68 s	α					
^{190}Bi		189.9875	5. s	β^+,EC/(10)/8.7 α/(90)/	6.45/				
^{191}Bi		190.9861	12. s	β^+,EC/(60)/7.3 α/(40)/	6.32/				

Elem. or Isot.	Natural Abundance (%)	Atomic Mass or Weight	Half-Life	Decay Mode/Energy (/MeV)	Particle Energy /Intensity (MeV / %)	Spin (h/2π)	Nuclear Magnetic Mom.(nm)	Elect. Quadr. Mom. (b)	γ-ray Energy /Intensity (MeV / %)
^{192}Bi		191.9854	40. s	β⁺,EC/(80)/9.0 α/(20)/	6.06/				
^{193m}Bi			3.2 s	β⁺,EC/ α/	6.48/	1/2+			
^{193}Bi		192.9837	1.11 m	β⁺,EC/(40)/7.1 α/(60)/	5.91/	9/2+			
^{194}Bi		193.9828	1.8 m	β⁺,EC/(99.9)/8.2 α/0.1/		(10-)			0.1661 0.1740 0.2802 0.421 0.5754 0.9650
^{195m}Bi			1.45 m	β⁺,EC/(94) α/(6)/	6.11/				
^{195}Bi		194.9811	2.9 m	β⁺,EC/(99.8)/5.8 α/(0.2)	5.45/	3/2-			
^{196}Bi		195.9806	5. m	EC/≈7.4					0.1376 0.3720 0.6880 1.0486
^{197}Bi		196.9789	5. m	β⁺,EC/5.2		1/2+			
^{198m}Bi			7.7 s	I.T./0.2485		(10-)			0.2485
^{198}Bi		197.9790	11.8 m	β⁺,EC/6.6		(7+)			0.0900 0.1976 0.5624 1.0635 ann.rad./
^{199m}Bi			24.7 m	β⁺,EC/					
^{199}Bi		198.9776	27. m	β⁺,EC/4.3		9/2-			0.7203 0.8374 0.8417 0.9460 1.0528 1.3056 (0.12 - 3.2)
^{200m}Bi			31. m	β⁺,EC/		(2+)			0.2453 0.4198 0.4624 1.0265 ann.rad./
^{200}Bi		199.9781	36. m	EC/(90)/5.9 β⁺/(10)/		7+			Pb k x-ray 0.4198 0.4623 1.0265
^{201m}Bi			59.1 m	I.T./0.846 β⁺,EC/		(1/2+)			Bi k x-ray 0.8464
^{201}Bi		200.97697	1.8 h	EC/3.84		9/2-			Pb k x-ray 0.6288 0.9357 1.0138 (0.13 - 2.4) ann.rad./
^{202}Bi		201.97768	1.72 h	β⁺/(3)/5.16 EC/(97)/		5+			Pb k x-ray 0.57860 0.92734 (0.08 - 3.5)
^{203}Bi		202.97687	11.8 h	EC/(99.8)/3.25 β⁺/(0.2)/	1.35/	9/2-	+4.62	-0.7	Pb k x-ray 0.1865 0.8203 0.8969 1.8475 (0.1 - 2.9)
^{204}Bi		203.97779	11.2 h	EC/4.44		6+	+4.28	-0.43	Pb k x-ray 0.37481 0.89922 0.98409
^{205}Bi		204.97737	15.31 d	EC/2.71		9/2-	+4.16		Pb k x-ray 0.70347 1.76435
^{206}Bi		205.97848	6.243 d	EC/3.76		6+	+4.60	-0.20	Pb k x-ray 0.51619 0.80313 0.88100
^{207}Bi		206.978456	35. y	EC/2.399		9/2-	4.08	-0.6	Pb k x-ray 0.56915 1.06310
^{208}Bi		207.979727	3.68x10⁵ y	EC/2.880		5+			Pb k x-ray 2.61435
^{209}Bi	100.	208.980384				9/2-	+4.111	-0.37	

Elem. or Isot.	Natural Abundance (%)	Atomic Mass or Weight	Half-Life	Decay Mode/Energy (/MeV)	Particle Energy /Intensity (MeV / %)	Spin (h/2π)	Nuclear Magnetic Mom.(nm)	Elect. Quadr. Mom. (b)	γ-ray Energy /Intensity (MeV / %)
^{210m}Bi			3.0x10⁶ y	α/	4.420(3)/0.29 4.569(3)/3.9 4.584(3)/1.4 4.908(4)/39 4.946(3)/55	9-			Tl k x-ray 0.2661 0.3052 0.6502
^{210}Bi		209.984105	5.01 d	β⁻ /1.163	1.16/99	1-	-0.0445	+0.136	0.2661 0.3.52
^{211}Bi		210.98726	2.14 m	α/(99.7)/ β⁻ /(0.3)/0.58	6.279/16 6.623/84	9/2-			Tl k x-ray 0.3501
^{212m2}Bi			7. m	β⁻ /		(15-)			
^{212m1}Bi			25.0 m	α/(93)/ β⁻ /(7)/	6.300/40 6.340/53	(9-)			0.120 0.233 0.275 0.404 0.727
^{212}Bi		211.991271	1.009 h	β⁻ /(64)/2.254 α/(36)/	6.051/25 6.090/9.6	(1-)	0.41		Tl k x-ray Po k x-ray 0.2881 0.72725 0.78551 1.62066
^{213}Bi		212.99437	45.6 m	β⁻ /(98)/1.43 α/(2)/	1.02/31 1.42/66 5.549/0.16 5.869/2.0	9/2-	+3.9		Po k x-ray 0.4404 (0.15 - 1.328) 1.10006
^{214}Bi		213.99870	19.7 m	β⁻ /3.27					0.60931 1.12027 1.76449 (0.19 - 3.2)
^{215}Bi		215.0018	7.7 m	β⁻ /2.3					0.2937 (0.27 - 0.835)
^{216}Bi		216.0062	3.6 m	β⁻/4.0					0.5498 0.4192
₈₄Po									
^{192}Po		191.9915	34. ms	α/	7.17				
^{193m}Po			0.24 s	α/	7.00				
^{193}Po		192.9911	0.45 s	α/	6.95				
^{194}Po		193.9883	0.39 s	α/	6.84/93 6.19/0.22	0+			
^{195m}Po			1.9 s	α/	6.70/				
^{195}Po		194.9881	4.6 s	α/	6.61/				
^{196}Po		195.9855	5.8 s	α/(95)/ β⁺ ,EC/(5)/ ≈4.6	6.52/94 5.77/0.02	0+			
^{197m}Po			25.8 s	α/(84)/ β⁺ ,EC/(16)/	6.385(3)/55	13/2+			
^{197}Po		196.9856	53. s	α/(44)/ β⁺ ,EC/(56)/6.2	6.282(4)/76	(3/2-)			
^{198}Po		197.9834	1.76 m	α/(70)/ β⁺ ,EC/(30)/4.0	6.18/57 5.27/7.6 x 10⁻⁴	0+			
^{199m}Po			4.2 m	β⁺ ,EC/(51)/ α/(39)/	6.059/24	13/2+	0.99		ann.rad./ 0.2745 0.4998 1.0020
^{199}Po		198.985	5.2 m	β⁺ ,EC/(88)/7. α/(12)/	5.952/7.5	(3/2-)			Bi k x-ray 0.1877 0.3616 1.0214 1.0344
^{200}Po		199.9817	11.5 m	β⁺ ,EC/(85)/3.4 α/(15)/	5.863/11.1	0+			0.14748 0.32792 0.6176 0.6709
^{201m}Po			8.9 m	β⁺ ,EC/(57)/ I.T./(40)/0.418 α/(3)/	5.786/ ≈3.	13/2+	1.0		Bi k x-ray Po k x-ray 0.2726 0.4123 0.4179 0.9670
^{201}Po		200.9822	15.3 m	β⁺ ,EC/(98)/4.9 α/(2)/	5.683(3)/1.1	3/2-	0.94		Bi k x-ray 0.2056 0.2250 0.8483 0.9048

Elem. or Isot.	Natural Abundance (%)	Atomic Mass or Weight	Half-Life	Decay Mode/Energy (/MeV)	Particle Energy /Intensity (MeV / %)	Spin (h/2π)	Nuclear Magnetic Mom.(nm)	Elect. Quadr. Mom. (b)	γ-ray Energy /Intensity (MeV / %)
²⁰²Po		201.9807	45. m	β⁺,EC/(98)/2.8 α/(2)/	5.588/1.9	0+			0.0410 0.1656 0.3158 0.6884 Bi k x-ray Po k x-ray
²⁰³ᵐPo			1.2 m	I.T./(96)/0.6414 β⁻ EC/(4)/		13/2+			0.6414
²⁰³Po		202.9814	35. m	β⁺,EC/4.2		5/2-	+0.74		0.17516 0.21477 0.89350 0.90863 1.09095 Bi k x-ray
²⁰⁴Po		203.98031	3.53 h	EC/2.34 α	5.377/0.66	0+			0.2702 0.8844 1.0162 (0.11 - 1.9) Bi k x-ray
²⁰⁵Po		204.98117	1.7 h	β⁺,EC/3.53		5/2-	+0.76	+0.17	0.83681 0.84983 0.87241 1.00124 (0.12 - 2.7) Bi k x-ray
²⁰⁶Po		205.98047	8.8 d	EC/(95)/1.85 α/(5)/	5.223/5.5	0+			0.28644 0.31156 0.51134 0.80737 1.03228 (0.11 - 1.5) Po k x-ray
²⁰⁷ᵐPo			2.8 s	I.T./1.383		19/2-			0.2682 0.30074 0.81448 Bi k x-ray
²⁰⁷Po		206.98158	5.80 h	EC,β⁺ /2.91		5/2-	+0.79	+0.28	0.74263 0.91176 0.99225
²⁰⁸Po		207.981231	2.898 y	α/5.213	4.233/0.0002 5.1158/100	0+			
²⁰⁹Po		208.982415	102. y	α/4.9/6	4.624/0.56 4.879/99.2	1/2-	+0.77		0.26049 0.8964
²¹⁰Po		209.982857	138.38 d	α/5.407	4.516/0.001 5.304/100	0+			0.80313
²¹¹ᵐPo			25.2 s	α/	7.273/91 7.994/1.7 8.316/0.25 8.875/7.0	25/2+			Pb k x-ray 0.32808 0.56915 0.89723 1.06310
²¹¹Po		210.986637	0.516 s	α/7.594	6.570/0.54 6.892/0.55 7.450/98.9	9/2+			0.56915 0.89723
²¹²ᵐPo			45. s	α/	8.514/2.0 9.086/1.0 11.650/97	16+			
²¹²Po		211.988852	0.298 μs	α/8.953	8.784/100	0+			
²¹³Po		212.992843	4. μs	α/8.537	7.614/0.003 8.375/100	9/2+			
²¹⁴Po		213.995186	163.7 μs	α/7.833	6.904/0.01 7.686/99.99	0+			
²¹⁵Po		214.999415	1.780 ms	α/7.526	6.950/0.02 6.957/0.03 7.386/100	(9/2+)			
²¹⁶Po		216.001905	0.145 s	α/6.906	5.895/0.002 6.778/99.99	0+			
²¹⁷Po		217.0064	< 10. s	α/6.662	6.539/				
²¹⁸Po		218.008965	3.04 m	α/6.114	5.181/1.00	0+			
₈₅At									
¹⁹⁶At		195.9957	0.3 s	α/	7.06/				
¹⁹⁷ᵐAt			4. s	α		(1/2+)			
¹⁹⁷At		196.9939	0.35 s	β⁺,EC/7.8 α/	6.96/	(9/2-)			
¹⁹⁸ᵐAt			1.5 s	β⁺,EC/(75)/ α/(25)/	6.85/				
¹⁹⁸At		197.9928	5. s	α/	6.75/				
¹⁹⁹At		198.9910	7.1 s	β⁺,EC/(8)/5.6 α/(92)/	6.64/	9/2-			

Elem. or Isot.	Natural Abundance (%)	Atomic Mass or Weight	Half-Life	Decay Mode/Energy (/MeV)	Particle Energy /Intensity (MeV / %)	Spin (h/2π)	Nuclear Magnetic Mom.(nm)	Elect. Quadr. Mom. (b)	γ-ray Energy /Intensity (MeV / %)
^{200m}At			4.3 s	β^+,EC/(80) α/(20)/	6.536/	10-			
^{200}At		199.990	43. s	β^+,EC/(65)/ ≈8.0 α/(35)/	6.412/21 6.465/14	5+			
^{201}At		200.9885	1.48 s	β^+,EC/(29)/5.9 α/(71)/6.474	6.344/	9/2-			
^{202m}At			≤ 1.5 s	I.T./0.391					
^{202}At		201.9885	3.02 m	β^+,EC/(88)/7.2 α/(12)/	6.135/7.7 6.225/4.3	5+			ann.rad./ 0.4413 0.5697 0.6753
^{203}At		202.9868	7.4 m	β^+,EC/(69)/5.1 α/(31)/6.210	6.088/	9/2-			0.1458 0.2459 0.6414 1.0020 1.0340
^{204}At		203.9873	9.1 m	β^+,EC/(95)/6.5 α/(5)/	5.951/	(5+)			Po k x-ray 0.3271 0.4254 0.5156 0.6837
^{205}At		204.98604	26. m	β^+,EC/(90)/4.54 α/(10)/6.020	5.902/	(9/2-)			Po k x-ray 0.1543 0.6696 0.7194
^{206}At		205.98660	29.4 m	β^+,EC/(99)/5.72 α/(1)/5.881	5.703/	5+			Po k x-ray 0.20186 0.39561 0.47716 0.70071
^{207}At		206.98578	1.81 h	β^+,EC/(90)/3.91 α/(10)/5.873	5.758/	9/2-			Po k x-ray 0.16801 0.58842 0.81448
^{208}At		207.98657	1.63 h	β^+,EC/(99)/4.97 α/(1)/5.752	5.626/0.01 5.641/0.53	(6+)			Po k x-ray 0.1770 0.2060 0.6601 0.6852 0.8450 1.0281
^{209}At		208.98616	5.4 h	β^+,EC/(96)/3.49 α/(4)/5.757	5.647/4.1	(6+)			Po k x-ray 0.10422 0.54503 0.78189 0.79020 (0.1 - 2.6)
^{210}At		209.98713	8.1 h	EC/99.8/3.98 α/(0.2)/5.632	5.361/0.05 5.442/0.05	5+			Po k x-ray 0.24535 0.52758 1.18143 1.43678 1.48335 (0.04 - 2.4)
^{211}At		210.987481	7.21 h	EC/(58)/0.787 α/(42)/5.980	5.211/0.004 5.868/42	9/2-			Po k x-ray 0.66956 0.6870 0.74263
^{212m}At			0.119 s	α/	7.837/65 7.897/33	(9-)			
^{212}At		211.990735	0.314 s	α/7.828	7.058/0.4 7.088/0.6 7.618/15 7.681/84	(1-)			
^{213}At		212.992922	0.11 µs	α/9.254	9.080/	9/2-			
^{214m}At			0.76 µs	α/8.762		(9-)			
^{214}At		213.996357	0.56 µs	α/8.987	8.819/100	(1-)			
^{215}At		214.99864	0.10 ms	α/8.178	7.626/0.045 8.023/99.9	(9/2-)			0.40486
^{216}At		216.002408	0.30 ms	α/7.947	7.595/0.2 7.697/2.1 7.800/97	(1-)			
^{217}At		217.00471	32. ms	α/7.202	6.812/0.06 7.067/99.9	(9/2-)			0.2595 0.3345 0.5940
^{218}At		218.00868	1.6 s	α/6.883	6.654/6 6.695/90 6.748/4				

Elem. or Isot.	Natural Abundance (%)	Atomic Mass or Weight	Half-Life	Decay Mode/Energy (/MeV)	Particle Energy /Intensity (MeV / %)	Spin (h/2π)	Nuclear Magnetic Mom.(nm)	Elect. Quadr. Mom. (b)	γ-ray Energy /Intensity (MeV / %)
²¹⁹At		219.0113	50. s	α/6.390	6.275/				
²²⁰At		220.0153	3.71 m	β⁻/3.7					(0.24-0.70)
²²¹At			2.3 m						
²²²At			5.4 s						
²²³At			50. s						
₈₆Rn									
¹⁹⁸Rn		197.9988	0.05 s	α					
¹⁹⁹ᵐRn			0.3 s	α		(13/2+)			
¹⁹⁹Rn		198.9983	0.62 s	α/		3/2-			
²⁰⁰Rn		199.9957	1.06 s	α/(98)/ EC/(2)/5.	6.909/	0+			
²⁰¹ᵐRn			3.8 s	EC/(10)/ α/(90)/	6.770/	13/2+			
²⁰¹Rn		200.9955	7.0 s	α/(80)/6.636 EC/(20)/6.6	6.721/	(3/2-)			
²⁰²Rn		201.9932	9.9 s	α/(12)/6.771 EC/(88)/4.5	6.636(3)/	0+			
²⁰³ᵐRn			28. s	α/ 6.548(3)/		13/2+	-0.96	+1.3	
²⁰³Rn		202.9948	45. s	α/(66)/6.629 EC/(34)/≈7.4	6.498/	0			
²⁰⁴Rn		203.9914	1.24 m	α/(68)/6.546 EC/(32)/3.8	6.417(3)/	0+			
²⁰⁵Rn		204.9917	2.8 m	α/(23)/6.390 EC/(77)/5.2	6.123(3)/0.02 6.262(3)/23	(5/2-)	+0.80	+0.06	0.2652 0.3553 0.4648 0.6205 0.6753 0.7300
²⁰⁶Rn		205.9902	5.7 m	α/(68)/6.384 EC/(32)/3.3	6.258(3)/	0+			0.06170 0.0968 0.3245 0.3862 0.4822 0.4973 0.7728
²⁰⁷Rn		206.9907	9.3 m	β⁺,EC/(77)/4.6 α/(23)/6.252	5.995(4)/0.02 6.068(3)/0.15 6.126(3)/22.8	5/2-	+0.82	+0.22	At k x-ray 0.32947 0.34455 0.36767 0.40267 0.74723 (0.18 - 1.4)
²⁰⁸Rn		207.98963	24.3 m	α/(60)/6.260 EC/(40)/2.85	5.469(2)/0.003 6.140(2)/60	0+			
²⁰⁹Rn		208.99038	29. m	β⁺ /(83)/3.93 α/(17)/	2.16/2.3 5.887(3)/0.04 5.898(3)/0.02 6.039(2)/16.9	5/2-	+0.8388	+0.31	At k x-ray 0.27933 0.33753 0.40841 0.68942 0.74594 (0.18 - 3.2)
²¹⁰Rn		209.98968	2.4 h	α/(96)/6.157 EC/(4)/2.37	5.351(2)/0.005 6.039(2)/96	0+			At k x-ray 0.19625 0.45824 0.57104 0.64868 (0.14 - 1.7)
²¹¹Rn		210.99059	14.6 h	β⁺,EC/(74)/2.89 α/(26)/5.964	5.619(1)/0.7 5.784(1)/16.4 5.851(1)/8.8	1/2-	+0.60		At k x-ray 0.16877 0.25022 0.37049 0.67412 0.67839 1.36298 (0.11 - 2.7)
²¹²Rn		211.990689	24. m	α/6.385	5.587(4)/0.05 6.260(4)/99.95	0+			
²¹³Rn		212.99387	25.0 ms	α/8.243	7.552(8)/1.0 8.087(8)/99	9/2+			
²¹⁴Rn		213.99535	0.27 μs	α/9.209	9.037(9)/	0+			
²¹⁵Rn		214.99873	2.3 μs	α/8.840	8.674(8)/	(9/2+)			
²¹⁶Rn		216.00026	45. μs	α					
²¹⁷Rn		217.003915	0.6 ms	α/7.885	7.500/0.1 7.742(4)/100	9/2+			
²¹⁸Rn		218.005586	35. ms	α/7.267	6.534(1)/0.16 7.133(1)/99.8	0+			

Elem. or Isot.	Natural Abundance (%)	Atomic Mass or Weight	Half-Life	Decay Mode/Energy (/MeV)	Particle Energy /Intensity (MeV / %)	Spin (h/2π)	Nuclear Magnetic Mom.(nm)	Elect. Quadr. Mom. (b)	γ-ray Energy /Intensity (MeV / %)
²¹⁹Rn		219.009475	3.96 s	α/6.946(1)	6.3130(5)/0.05 6.425(3)/7.5 6.5309(4)/0.12 6.5531(3)/12.2 6.8193(3)/81	(5/2+)	-0.44	+0.93	Po k x-ray 0.13057 0.27113 0.40170 (0.1 - 1.05)
²²⁰Rn		220.011384	55.6 s	α/6.404	5.7486(5)/0.07 6.2883(1)/99.9	0+			
²²¹Rn		221.0156	25. m	α/(22)/6.148 β⁻/(78)/1.2	5.778(3)/1.8 5.788(3)/2.2 6.037(3)/18	7/2+	-0.020	-0.38	Fr L x-ray 0.07384 0.08323 0.0610 0.18639
²²²Rn		222.017570	3.8235 d	α/5.590	4.987(1)/0.08 5.4897(3)/99.9	0+			0.510
²²³Rn			23. m	β⁻/					
²²⁴Rn			1.8 h	β⁻/		0+			0.1085 0.2601 0.2655
²²⁵Rn			4.5 m	β⁻/		7/2	-0.70	+0.84	
²²⁶Rn			7.4 m	β⁻/					
²²⁷Rn			2. s	β⁻/					
²²⁸Rn			65. s	β⁻/					

₈₇Fr

Elem. or Isot.	Natural Abundance (%)	Atomic Mass or Weight	Half-Life	Decay Mode/Energy (/MeV)	Particle Energy /Intensity (MeV / %)	Spin (h/2π)	Nuclear Magnetic Mom.(nm)	Elect. Quadr. Mom. (b)	γ-ray Energy /Intensity (MeV / %)
²⁰¹Fr		201.0046	0.05 s	α/7.54	7.388(15)/	(9/2-)			
²⁰²Fr		202.0033	0.34 s	α/7.590	7.237(8)/100				
²⁰³Fr		203.0014	0.55 s	α/7.280	7.132(5)/	(9/2-)			
²⁰⁴Fr		204.001	2.1 s	α/7.170	6.967(5)/30 7.027(5)/70				
²⁰⁵Fr		204.9987	3.9 s	α/7.050	6.914(5)/	(9/2-)			
²⁰⁶ᵐFr			0.7 s	α/	6.93				0.531(IT)
²⁰⁶Fr		205.9985	16.0 s	α/7.416	6.792(5)/84				
²⁰⁷Fr		206.9969	14.8 s	α/6.900	6.766(5)/	9/2-	+3.9	-0.16	
²⁰⁸Fr		207.99713	59.1 s	α/(77)/6.770 EC/(23)/6.99	6.636(5)/	7+	-4.8	+0.004	
²⁰⁹Fr		208.99592	50.0 s	α/(89)/5.1 EC/(11)/5.16	6.646(3)/	9/2-	+3.9	-0.24	
²¹⁰Fr		209.99640	3.2 m	α/6.670 EC/6.26	6.543(5)/	6+	+4.4	+0.19	0.2030 0.6438 0.8175 0.9008
²¹¹Fr		210.99553	3.10 m	α/6.660 EC/4.61	6.534(5)/	9/2-	+4.0	-0.19	0.220 0.2799 0.5389 0.9169
²¹²Fr		211.99618	20. m	EC/(57)/5.12 α/(43)/6.529	6.261(1)/16 6.335(1)/4 6.335(1)/4 6.343(1)/1.3 6.383(1)/10 6.406(1)/9.5 6.08-6.18	(5+)	+4.6	-0.10	Rn x-ray 0.08107 0.08378 0.2277 1.1856 1.2748 0.014-1.178
²¹³Fr		212.99617	34.6 s	α/6.905	8.476(4)/51	9/2-	+4.0	-0.14	
²¹⁴ᵐFr			3.4 ms	α/	8.547(4)/46 6.775-8.046	9-			
²¹⁴Fr		213.99895	5.1 ms	α/8.587	7.409(3)/0.3 7.605(8)/1.0 7.940(3)/1.0 8.355(3)/4.7 8.427(3)/93	(1-)			
²¹⁵Fr		215.00033	0.12 μs	α/9.537	9.360(8)/	(9/2-)			
²¹⁶Fr		216.00319	0.70 μs	α/9.175	9.005(10)/				
²¹⁷Fr		217.00462	0.016 ms	α/8.471	8.315(8)/	(9/2-)			
²¹⁸ᵐFr			22. ms	α					
²¹⁸Fr		218.00756	1. ms	α/8.014	7.384(10)/0.5 7.542(15)/1.0 7.572(10)/5 7.732(10)/0.5 7.867(2)/93	(1-)			
²¹⁹Fr		219.00924	21. ms	α/8.132	6.802(2)/0.25 6.967(2)/0.6 7.146(2)/0.25 7.313(2)/99	(9/2-)			
²²⁰Fr		220.012313	27.4 s	α/6.800	6.582(1)/10 6.630(2)/6 6.641(1)/12 6.686(1)/61 6.39-6.58	1+	-0.67	+0.47	0.0450 0.061 0.1060 0.1539 0.1617

Elem. or Isot.	Natural Abundance (%)	Atomic Mass or Weight	Half-Life	Decay Mode/Energy (/MeV)	Particle Energy /Intensity (MeV / %)	Spin (h/2π)	Nuclear Magnetic Mom.(nm)	Elect. Quadr. Mom. (b)	γ-ray Energy /Intensity (MeV / %)
^{221}Fr		221.01425	4.8 m	α/6.457	5.9393(7)/0.17 5.9797(7)/0.49 6.0751(7)/0.15 6.1270(7)/ 6.2433(3)/1.3 6.3410(7)/83.4	(5/2-)	+1.58	-1.0	At k x-ray 0.0995 0.21798 0.4091
^{222}Fr		222.01754	14.3 m	β⁻/2.03 α/5.850	1.78/	2-	+0.63	+0.51	
^{223}Fr		223.019731	22.0 m	β⁻/1.149	1.17/65	(3/2+)	+1.17	+1.17	0.05014 0.07972 (0.13 - 0.9)
^{224}Fr		224.02323	3.0 m	β⁻/2.82		1-	+0.40	+0.517	0.13150 0.21575 0.8367 (0.1 - 2.21)
^{225}Fr		225.02561	3.9 m	β⁻/1.87		3/2	+1.07	+1.3	
^{226}Fr		226.0293	49. s	β⁻/3.6		1	+0.071	-1.35	0.18606 0.25373
^{227}Fr		227.0318	2.48 m	β⁻/2.5		1/2	+1.50		
^{228}Fr		228.035	39. s	β⁻/≈3.5		2-	-0.76	+2.4	
^{229}Fr			50. s	β⁻/					
^{230}Fr			19. s	β⁻/		(3)			
^{231}Fr			17. s	β⁻/					
^{232}Fr			5. s	β⁻/					

$_{88}$Ra

Elem. or Isot.	Natural Abundance (%)	Atomic Mass or Weight	Half-Life	Decay Mode/Energy (/MeV)	Particle Energy /Intensity (MeV / %)	Spin (h/2π)	Nuclear Magnetic Mom.(nm)	Elect. Quadr. Mom. (b)	γ-ray Energy /Intensity (MeV / %)
^{206}Ra		206.0038	0.4 s	α/7.416	7.272(5)/	0+			
^{207}Ra		207.005	1.3 s	α/7.270	7.133(5)/				
^{208}Ra		208.0018	1.4 s	α/7.273	7.133(5)/	0+			
^{209}Ra		209.0019	4.6 s	α/7.150	7.008(5)/	5/2	+0.87	+0.38	
^{210}Ra		210.0005	3.7 s	α/7.610	7.020(5)/	0+			
^{211}Ra		211.0009	13. s	α/7.046 EC/5.0	6.912(5)/	(5/2-)	+0.878	+0.46	
^{212}Ra		211.99978	13.0 s	α/7.033	6.901(2)/	0+			
^{213m}Ra			2.1 ms	IT					
^{213}Ra		213.00034	2.7 m	EC/(20)/3.88 α/(80)/6.860	6.521(3)/4.8 6.622(3)/39 6.730(3)/36	(1/2-)	+0.613		0.1024 0.11010 0.2125
^{214}Ra		214.00009	2.46 s	α/7.272	7.136(4)/	0+			
^{215}Ra		215.00270	1.6 ms	α/8.864	7.883(6)/2.8 8.171(3)/1.4 8.700(3)/95.9	(9/2+)			
^{216}Ra		216.00352	0.18 μs	α/9.526	9.349(8)/	0+			
^{217}Ra		217.00631	1.6 μs	α/9.161	8.992(8)/	9/2-			
^{218}Ra		218.00712	26. μs	α/8.547	8.390(8)/	0+			
^{219}Ra		219.01006	0.010 s	α/8.132	7.680(10)/65 7.982(9)/35				
^{220}Ra		220.01101	25. ms	α/7.593	6.998(7)/1.0 7.455(7)/99	0+			0.465
^{221}Ra		221.01391	29. s	α/6.879	6.254(10)/0.7 6.578(5)/3 6.585(3)/8 6.608(3)/35 6.669(3)/21 6.758(3)/31	5/2	-0.180	+1.9	
^{222}Ra		222.015361	38. s	α/5.590	6.237(2)/3.0 6.556(2)/97	0+			
^{223}Ra		223.018497	11.435 d	α/5.979	5.287(1)/0.15 5.338(1)/0.13 5.365(1)/0.13 5.433(5)/2.3 5.502(1)/1.0 5.540(1)/9.2 5.607(3)/24 5.716(3)/52 5.747(1)/9 5.857(1)/0.32 5.872(1)/0.85	(3/2+)	+0.271	+1.2	Rn k x-ray 0.12231 0.14418 0.15418 0.15859 0.26939 0.32388 0.33328 0.44494 (0.10 - 0.7)
^{224}Ra		224.020202	3.66 d	α/5.789	5.034(10)/0.003 5.047(1)/0.007 5.164(5)/0.007 5.449(2)/4.9 5.685(2)/95	0+			Rn k x-ray 0.2407 0.4093 0.6501
^{225}Ra		225.023603	14.9 d	β⁻/0.36	0.32/100	(3/2+)	-0.734		Ac k x-ray 0.0434

Elem. or Isot.	Natural Abundance (%)	Atomic Mass or Weight	Half-Life	Decay Mode/Energy (/MeV)	Particle Energy /Intensity (MeV / %)	Spin (h/2π)	Nuclear Magnetic Mom.(nm)	Elect. Quadr. Mom. (b)	γ-ray Energy /Intensity (MeV / %)
²²⁶Ra		226.025402	1599. y	α/4.870	4.194(1)/0.001 4.343(1)/0.006 4.601(1)/5.5 4.784(1)/94	0+			Rn k x-ray 0.1861 0.2624
²²⁷Ra		227.029170	42. m	β⁻ /1.325	1.03/ 1.30/	(3/2+)	−0.404	+1.5	Ac L x-ray Ac k x-ray 0.02739
²²⁸Ra		228.031063	5.76 y	β⁻ /0.046		0+			0.0135 0.016
²²⁹Ra		229.0348	4.0 m	β⁻ /1.76	1.76/	(3/2+)	+0.503	+3.0	
²³⁰Ra		230.03708	1.5 h	β⁻ /1.0	0.7/	0+			0.0631 0.0720 0.2028 0.4698 0.4787
²³¹Ra			1.7 m	β⁻					
²³²Ra			4. m	β⁻					
²³³Ra			30. s	β⁻					
²³⁴Ra			≈ 30. s	β⁻/					
₈₉Ac									
²⁰⁷Ac			≈0.02 s	α/7.71					
²⁰⁸ᵐAc			≈25. ms	α/7.76					
²⁰⁸Ac			≈0.1 s	α/7.57					
²⁰⁹Ac		209.0096	≈0.09 s	α/7.58					
²¹⁰Ac		210.0093	0.4 s	α/7.610	7.462(8)/				
²¹¹Ac		211.0076	0.3 s	α/7.620	7.480(8)/				
²¹²Ac		212.0078	0.9 s	α/7.520	7.379(8)/				
²¹³Ac		213.0066	0.8 s	α/7.500	7.364(8)/	(9/2-)			
²¹⁴Ac		214.0069	8.2 s	α/(86)/7.350 EC/(14)/6.34	7.007(8)/3 7.082(5)/38 7.214(5)/45	(5+)			
²¹⁵Ac		215.0065	0.17 s	α/7.750	7.604(5)/	(9/2-)			
²¹⁶ᵐAc			0.33 ms	α/	8.198(8)/1.7 8.283(8)/2.5 9.028(5)/49 9.106(5)/46	(9-)			
²¹⁶Ac		216.00871	≈ 0.3 ms	α/9.241	8.990(2)/10 9.070(8)/90	(1)			
²¹⁷ᵐAc			0.7 μs	α/	10.540/100				
²¹⁷Ac		217.00933	0.07 μs	α/9.832	9.650(10)/100	9/2-			
²¹⁸Ac		218.01162	1.1 μs	α/9.380	9.205(15)/				
²¹⁹Ac		219.01241	0.012 ms	α/8.830	8.664(10)/	(9/2-)			
²²⁰Ac		220.0148	26. ms	α/8.350	7.610(20)/23 4.680(20)/21 7.790(10)/13 7.850(10)/24 7.985(10)/4 8.005(10)/5 8.060(10)/6 8.195(10)/3				
²²¹Ac		221.01558	52. ms	α/7.790	7.170(10)/2 7.375(10)/10 7.440(15)/20 7.645(10)/70				
²²²ᵐAc			63. s	α/(>89)/ EC/(1)/ I.T./(<10)/	6.710(20)/7 6.750(20)/13 6.810(20)/24 6.840(20)/9 6.890(20)/13 6.970(20)/7 7.000(20)/13				
²²²Ac		222.01782	5. s	α/7.141	6.967(10)/6 7.013(2)/94	1-			
²²³Ac		223.01913	2.1 m	α/(99)/6.783 EC/(1)/0.59	6.131(2)/0.12 6.177(2)/0.94 6.293(1)/0.47 6.326(1)/0.3 6.332(2)/0.14 6.360(1)/0.22 6.397(1)/0.13 6.448(1)/0.2 6.473(1)/3.1 6.523(2)/0.6 6.528(1)/3.1 6.563(1)/13.6 6.582(3)/0.3 6.646(1)/44 6.661(1)/31	(5/2-)			0.0725 0.0839 0.0927 0.0990 0.1917 0.2158 0.3588 0.4768

Elem. or Isot.	Natural Abundance (%)	Atomic Mass or Weight	Half-Life	Decay Mode/Energy (/MeV)	Particle Energy /Intensity (MeV / %)	Spin (h/2π)	Nuclear Magnetic Mom.(nm)	Elect. Quadr. Mom. (b)	γ-ray Energy /Intensity (MeV / %)
^{224}Ac		224.021708	2.7 h	EC/(90)/1.403 α/(10)/6.323	5.841(1)/0.5 5.860(1)/0.75 5.875(1)/1.7 5.941(1)/4.4 6.000(1)/6.7 6.013(1)/1.4 6.056(1)/22 6.138(1)/26 6.154(1)/1.0 6.204(1)/12 6.210(1)/20	0-			Ra L kx-ray Ra k x-ray 0.08426 0.13150 0.1571 0.21575 0.2619 (0.03 - 0.3)
^{225}Ac		225.02322	10.0 d	α/5.935	5.286(1)/0.2 5.444(3)/0.1 5.554(1)/0.1 5.608(1)/1.1 5.636(1)/4.5 5.681(1)/1.4 5.722(1)/2.9 5.731(1)/10 5.791(1)/9 5.793(1)/18	3/2			Fr k x-ray 0.9958 0.9982 0.1084 0.1116 0.1451 0.1539 0.15724 0.18799 0.19575 0.2162 0.21686 (0.025 - 0.52)
^{226}Ac		226.026089	1.224 d	EC/(17)/0.640 β^-/(83)/1.116 α/(0.006)/5.510	5.399(5)/0.006	(1-)			Ra k x-ray Th k x-ray 0.07218 0.15816 0.23034
^{227}Ac		227.027747	21.77 y	β^-/(98.6)/0.045 α/(1.4)/5.043	0.045/54 4.869(1)/0.09 4.938(1)/0.52 4.951(1)/0.65	(3/2-)	+1.1	+1.7	0.0838/23. 0.0811/14. 0.2696/13. (0.044 - 1.27)
^{228}Ac		228.031014	6.15 h	β^-/2.127	1.11/32 1.85/12 2.18/11	(3+)			Th L x-ray Th k x-ray 0.12903 0.33842 0.91116 0.96897 (0.2 - 1.96)
^{229}Ac		229.03293	1.04 h	β^-/1.10	1.1/	(3/2+)			0.07450 0.16451 0.26188 0.5085 0.56916
^{230}Ac		230.0360	2.03 m	β^-/2.7	1.4/	1+			Th k x-ray 0.45497 0.50820 (0.12 - 2.5)
^{231}Ac		231.0386	7.5 m	β^-/2.1	2.1/100	(1/2+)			0.14379 0.18574 0.22140 0.28250 0.3070
^{232}Ac		232.0420	2.0 m	β^-/3.7		(2-)			
^{233}Ac			2.4 m	β^-/		(1/2+)			
^{234}Ac			40. s	β^-/		(1+)			
$_{90}$Th		232.03805(2)							
^{212}Th		212.0129	≈ 30. ms	α/	7.80/	0+			
^{213}Th		213.0130	0.14 s	α/7.840	7.692(10)/				
^{214}Th		214.0115	0.09 s	α/7.825	7.677(10)/	0+			
^{215}Th		215.0117	1.2 s	α/7.660	7.33(10)/8 7.395(8)/52	(1/2-)			7.524(8)/40
^{216m}Th			0.18 ms	α					
^{216}Th		216.01105	28. ms	α/8.071	7.921(8)/	0+			
^{217}Th		217.01306	0.25 ms	α/9.424	9.250(10)/				
^{218}Th		218.01327	0.11 μs	α/9.847	9.665(10)/	0+			
^{219}Th		219.01552	1.05 μs	α/9.510	9.340(20)/				
^{220}Th		220.01573	10. μs	α/8.953	8.790(20)/	0+			
^{221}Th		221.01817	1.7 ms	α/8.628	7.743(8)/6 8.146(5)/56 8.4272(5)/39				
^{222}Th		222.01845	2.8 ms	α/8.129	7.982(8)/9.7 7.600(15)/3	0+			

Elem. or Isot.	Natural Abundance (%)	Atomic Mass or Weight	Half-Life	Decay Mode/Energy (/MeV)	Particle Energy /Intensity (MeV / %)	Spin (h/2π)	Nuclear Magnetic Mom.(nm)	Elect. Quadr. Mom. (b)	γ-ray Energy /Intensity (MeV / %)
²²³Th		223.02079	0.65 s	α/7.454	7.29(1)/41(5) 7.32(1)/29(5) 7.350(15)/20(5) 7.390(15)/10(4)				
²²⁴Th		224.02146	1.05 s	α/7.305	6.768(5)/1.2 6.997(5)/19 7.170(5)/79				
²²⁵Th		225.02394	8.72 m	EC/(10)/0.68 α/(90)/6.920	6.441(2)/15 6.479(2)/43 6.501(3)/14 6.627(3)/3 6.650(5)/3 6.700(5)/2 6.743(3)/7 6.796(2)/9	(3/2+)			
²²⁶Th		226.024891	30.6 m	α/6.454	6.026(1)/0.2 6.041(1)/0.19 6.098(1)/1.3 6.2283(4)/23 6.3375(4)/75	0+			Ra k x-ray 0.11110 0.13100 0.19028 0.20621 0.24210 (0.1 - 0.8)
²²⁷Th		227.027699	18.72 d	α/6.146		(3/2+)			Ra L x-ray Ra k x-ray 0.05014 0.23597 0.25624 (0.02 - 1.0)
²²⁸Th		228.028731	1.913 y	α/5.520	5.1770(2)/0.18 5.2114(1)/0.4 5.3405(1)/26.7 5.4233(1)/73	0+			
²²⁹Th		229.031754	7.9x10³ y	α/5.168	4.814/9.3 4.845(5)/56 4.9008(5)/10.2 4.689-5.077	5/2+	+0.46	+4.	
²³⁰Th		230.033126	7.54x10⁴ y	α/4.771	4.4383(6)/0.03 4.4798(6)/0.12 4.6211(6)/23.4 4.6876(6)/76.3	0+			0.0677/0.46 0.1439/0.078
²³¹Th		231.036296	1.063 d	β⁻/0.390	0.138/22 0.218/20 0.305/52	5/2+			Pa L x-ray Pa k x-ray 0.02564 0.084203/ (0.02 - 0.3)
²³²Th	100.	232.038050	1.4x10¹⁰ y	α/4.081	3.830(10)/0.2 3.952(5)/23 4.010(5)/77	0+			0.0590 0.124
²³³Th		233.041576	22.3 m	β⁻/1.245	1.245/	1/2+			Pa L x-ray Pa k x-ray 0.02938 0.08653 0.45930 (0.02 - 1.2)
²³⁴Th		234.043596	24.10 d	β⁻/0.273	0.102/20 0.198/72	0+			Pa L x-ray 0.06329/4.1 0.09235/2.4 0.09278/2.4
²³⁵Th		235.04751	7.2 m	β⁻/1.9					0.4162 0.6594 0.7272 0.747 0.9318
²³⁶Th			37.5 m	β⁻/≈ 1.0					Pa k x-ray 0.1107
²³⁷Th			5.0 m	β⁻					
₉₁Pa									
²¹⁵Pa		215.0190	≈ 14. ms	α					
²¹⁶Pa		216.0190	0.19 s	α/8.010	7.720/ 7.820/ 7.920/				
²¹⁷ᵐPa			2. ms	α/	10.16/				
²¹⁷Pa		217.0183	5. ms	α/8.490	8.340(10)/				
²¹⁸Pa		218.0200	0.12 ms	α/	9.54/ 9.61/				
²²¹Pa		221.0219	6. µs	α	9.08(3)				

Elem. or Isot.	Natural Abundance (%)	Atomic Mass or Weight	Half-Life	Decay Mode/Energy (/MeV)	Particle Energy /Intensity (MeV / %)	Spin (h/2π)	Nuclear Magnetic Mom.(nm)	Elect. Quadr. Mom. (b)	γ-ray Energy /Intensity (MeV / %)
^{222}Pa		222.0237	≈ 4.3 ms	α/8.700	8.180/50 8.330/20 8.540/30				
^{223}Pa		223.0240	≈ 6.5 ms	α/8.340	8.006(10)/55 8.196(10)/45				
^{224}Pa		224.0256	≈ 0.95 s	α/7.630	7.555(10)/75(3) 7.46(1)/25(3)				
^{225}Pa		225.0261	1.8 s	α/7.380	7.195(10)/30 7.245(10)/70				
^{226}Pa		226.02792	1.8 m	α/(74)/6.987 EC/(26)/2.83	6.728(10)/0.7 6.823(10)/35 6.863(10)/39				
^{227}Pa		227.02879	38.3 m	α/(85)/6.582 EC/(15)/1.02	6.357(4)/7 6.376(10)/2.2 6.401(4)/8 6.416(4)/13 6.423(10)/10 6.465(4)/43	(5/2-)			0.0649 0.0669 0.1100
^{228}Pa		228.03100	22. h	EC/(98)/2.111 α/(2)	5.779/0.23 5.805/0.15 6.078/0.4 6.105/0.25 6.118/0.22	(3+)	+3.5		Th k x-ray 0.20939 0.27026 0.28202 0.32767 0.32807 0.46310 0.91116 0.96464 0.96897 (0.1 - 1.96)
^{229}Pa		229.03209	1.5 d	EC/(99.8)/0.32 α/(0.2)/5.836	5.536(2)/0.02 5.579(2)/0.09 5.668(2)/0.05	(5/2)			0.04244 (0.024 - 0.18)
^{230}Pa		230.034532	17.4 d	EC/(90)/1.310 β$^-$/(10)/0.563	0.51/	(2-)	2.0		Th L x-ray Th k x-ray 0.4437 0.45477 0.89876 0.91856 0.95199 (0.053-1.07)
^{231}Pa		231.035878	3.25x10^4 y	α/5.148	4.6781(5)/1.5 4.7102(5)/1.0 4.7343(5)/8.4 4.8513(5)/1.4 4.9339(5)/3 4.9505(5)/22.8 4.9858(5)/1.4 5.0131(5)/25.4 5.0292(5)/20 5.0318(5)/2.5 5.0587(5)/11	3/2-	2.01	-1.	Ac L x-ray Ac k x-ray 0.01899 0.027396 0.03823 0.04639 0.25586 0.26029 0.28367 0.30007 0.30264 0.33007 (0.02 - 0.61)
^{232}Pa		232.03858	1.31 d	β$^-$/1.34		(2-)			U k x-ray 0.10900 0.15009 0.89439 0.96934 (0.10 - 1.17)
^{233}Pa		233.040239	27.0 d	β$^-$/0.571	0.15/40 0.256/60	3/2-	+4.0	-3.0	U L x-ray U k x-ray 0.30017 0.31201 0.34059
^{234m}Pa			1.17 m	β$^-$/(99.9)/2.29 I.T./(0.13)/		(0-)			U k x-ray 0.25818/0.07 0.76641/0.32 1.0009/0.85 (0.06 - 1.96)
^{234}Pa		234.043303	6.69 h	β$^-$/2.197	0.51/	(4+)			U L x-ray U k x-ray 0.1312/0.03 0.5695/0.02 0.9256/0.02 (0.02 - 1.99)
^{235}Pa		235.04544	24.4 m	β$^-$/1.41	1.4/97	(3/2-)			0.0308-0.65893

Elem. or Isot.	Natural Abundance (%)	Atomic Mass or Weight	Half-Life	Decay Mode/Energy (/MeV)	Particle Energy /Intensity (MeV / %)	Spin (h/2π)	Nuclear Magnetic Mom.(nm)	Elect. Quadr. Mom. (b)	γ-ray Energy /Intensity (MeV / %)
²³⁶Pa		236.0487	9.1 m	β⁻ /2.9	1.1/40 2.0/50 3.1/10	(1-)			U k x-ray 0.64235 0.68759 1.7630 (0.04 - 2.18)
²³⁷Pa		237.0511	8.7 m	β⁻ /2.3	1.1/60 1.6/30 2.3/10	(1/2+)			0.4986 0.5293 0.5407 0.8536 0.8650 (0.04 - 1.4)
²³⁸Pa		238.0545	2.3 m	β⁻ /3.5	1.2/ 1.7/	(3-)			0.10350 0.1785 0.4484 0.6350 0.6800 1.01446 (0.04 - 2.5)
₉₂U		**238.0289(1)**							
²¹⁸U			≈0.002 s	α	8.63(3)/				
²¹⁹U			0.04 ms	α	9.68(4)/				
²²²U		222.0261	≈ 1.μs	α					
²²³U		223.0277	0.02 s	α/	8.78(4)/				
²²⁴U		224.02759	≈ 1. ms	α/	8.46/100				
²²⁵U		225.02938	0.09 s	α/	7.88/85 7.82/15				
²²⁶U		226.02933	0.5 s	α/7.560	7.430/	0+			
²²⁷U		227.03113	1.1 m	α/7.200	6.870/				
²²⁸U		228.03137	9.1 m	α/6.803	6.404(6)/0.6 6.440(5)/0.7 6.589(5)/29 6.681(6)/70	0+			0.095 0.152 0.187 0.246
²²⁹U		229.03350	58. m	EC/(80)/1.31 α/(20)/6.473	6.223/3 6.297(3)/11 6.332(3)/20 6.360(3)/64	(3/2+)			
²³⁰U		230.033927	20.8 d	α/5.992	5.5866(3)/0.01 5.6624(3)/0.26 5.6663(3)/0.38 5.8178(3)/32 5.8887(3)/67	0+			Th L x-ray 0.07218 0.15421 0.23034
²³¹U		231.03626	4.2 d	EC/0.36 α/(10⁻³)	5.46/1.6 x 10⁻³ 5.47/1.4 x 10⁻³ 5.40/1. x 10⁻³	(5/2-)			Pa L x-ray Pa k x-ray 0.02564 0.08420
²³²U		232.037146	68.9 y	α/5.414	4.9979(1)/0.003 5.1367(1)/0.3 5.2635(1)/31 5.3203(1)/69	0+			
²³³U		233.039627	1.59x10⁵ y	α/4.909	4.7830(8)/13.2 4.8247(8)/84.4 4.510-4.804	5/2+	+0.59	3.66	Th L x-ray 0.04244 0.09714 (0.0252-1.119)
²³⁴U	0.0055(5)	234.040945	2.45x10⁵ y	α/4.856	4.604(1)/0.24 4.7231(1)/27.5 4.776(1)/72.5	0+			0.05323/0.156 0.12091
²³⁵ᵐU			26. m	I.T./0.0007		1/2+			
²³⁵U	0.720(1)	235.043922	7.04x10⁸ y	α/4.6793	4.1525(9)/0.9 4.2157(9)/5.7 4.3237(9)/4.6 4.3641(9)/11 4.370(4)/6 4.3952(9)/55 4.4144(9)/2.1 4.5025(9)/1.7 4.5558(9)/4.2 4.5970(9)/5.0	7/2-	-0.38	4.6	Th L x-ray Th k x-ray 0.10917 0.14378 0.16338 0.18574 0.20213 0.20533 0.22140 (0.03 - 0.79)
²³⁶U		236.045561	2.34x10⁷ y	α/4.569	4.332(8)/0.26 4.445(5)/26 4.494(3)/74	0+			Th L x-ray 0.04937 0.11275
²³⁷U		237.048723	6.75 d	β⁻ /0.519	0.24/ 0.25/	1/2+			Np L x-ray Np k x-ray 0.05953 0.20801

Elem. or Isot.	Natural Abundance (%)	Atomic Mass or Weight	Half-Life	Decay Mode/Energy (/MeV)	Particle Energy /Intensity (MeV / %)	Spin (h/2π)	Nuclear Magnetic Mom.(nm)	Elect. Quadr. Mom. (b)	γ-ray Energy /Intensity (MeV / %)
^{238}U	99.2745(15)	238.050784	4.46x10⁹ y	α	4.039(5)/0.23 4.147(5)/23 4.196(5)/77	0+			Th L x-ray 0.04955/.06 0.1135/.01
^{239}U		239.054289	23.5 m	β⁻ /1.265	1.2/ 1.3/	5/2 +			0.04354 0.07467
^{240}U		240.056585	14.1 h	β⁻ /0.39	0.36/	0+			Np L x-ray 0.04410 0.05558 0.06760
^{242}U			16.8 m	β⁻ /≈ 1.2					
$_{93}$**Np**									
^{226}Np		226.0351	0.03 s	α/	8.04(2)/				
^{227}Np		227.0350	0.51 s	α/	7.65(2)/ 7.68(1)/				
^{228}Np		228.0362	61. s	EC/60(7)/ α/40(7)/,SF					
^{229}Np		229.0363	4.0 m	α/7.010	6.890(20)				
^{230}Np		230.0378	4.6 m	EC/97 /3.6 α/3	6.660(20)				
^{231}Np		231.03823	48.8 m	EC/98 /1.8 α/2 /6.368	6.280/2	5/2			0.2629 0.3475 0.3703
^{232}Np		232.0400	14.7 m	EC/99 /2.7		(4-)			U L x-ray U k x-ray 0.3268 0.81925 0.86683
^{233}Np		233.0410	36.2 m	EC/1.2		(5/2 +)			U L x-ray U k x-ray 0.29887 0.31201
^{234}Np		234.04289	4.4 d	β⁺ ,EC/1.81	0.79/	(0+)			U L x-ray U k x-ray 1.5272 1.5587 1.6022
^{235}Np		235.044055	1.085 y	EC/99.9 /0.124 α/0.001/5.191		5/2 +			U k x-ray
^{236m}Np			22.5 h	EC/52 / β⁻ /48 /		(1-)			U L x-ray Pu L x-ray U k x-ray 0.64235 0.68759
^{236}Np		236.04657	1.55x10⁵ y	EC/91 /0.94 β⁻ /9 /0.49		(6-)			U L x-ray U k x-ray 0.10423 0.16031
^{237}Np		237.048166	2.14x10⁶ y	α/4.957	4.6395(5)/6.5 4.766(5)/9.7 4.7715(5)/22.7 4.7884(5)/47.8 4.558-4.873	5/2 +	+3.14	+3.89	Pa L x-ray Pa k x-ray 0.029378/15 0.08653/12 (0.03-0.28)
^{238}Np		238.050940	2.117 d	β⁻ /1.292	1.2/	2+			Pu L x-ray Pu k x-ray 0.98447/25.2 1.02855/18.3 (.044-1.026)
^{239}Np		239.052931	2.355 d	β⁻ /0.722	0.341/30 0.438/48	5/2 +			Pu L x-ray Pu k x-ray 0.10613 0.228186/11 0.27760/15 (0.04-0.50)
^{240m}Np			7.22 m	β⁻ /99.9 / I.T./0.1 /	2.18/	(1+)			0.25143 0.26333 0.55454 0.59735
^{240}Np		240.05617	1.032 h	β⁻ /2.20	0.89/	5+			0.1471/ 0.5664 0.6008
^{241}Np		241.0583	13.9 m	β⁻ /1.3	1.3/	5/2 +			0.1330/ 0.1740 0.280
^{242m}Np			2.2 m	β⁻ /		(1+)			0.15910 0.2651/ 0.78570 0.9448/

Elem. or Isot.	Natural Abundance (%)	Atomic Mass or Weight	Half-Life	Decay Mode/Energy (/MeV)	Particle Energy /Intensity (MeV / %)	Spin (h/2π)	Nuclear Magnetic Mom.(nm)	Elect. Quadr. Mom. (b)	γ-ray Energy /Intensity (MeV / %)
^{242}Np		242.0616	5.5 m	β⁻ /2.7	2.7/	6+			0.6209 0.73620 0.78074 1.47340 (0.04-2.37)
$_{94}$Pu									
^{228}Pu				α/	7.81(2)/				
^{229}Pu				α/	7.46(3)/				
^{230}Pu		230.03964		α/	7.05/				
^{232}Pu		232.04118	34. m	EC/ > 80/1.1 α/ < 20/6.716	6.542(10)/38 6.600(10)/62	0+			
^{233}Pu		233.04299	20.9 m	EC(99.9)/1.9 α/0.1 /6.416	6.300(20)/0.1				0.1503 0.1804 0.2353 0.5002 0.5346/ 1.0352/
^{234}Pu		234.04331	8.8 h	EC/94 /0.39 α/6 /6.310	6.035(3)/0.024 6.149(3)/1.9 6.200(3)/4.0	0+			
^{235}Pu		235.0453	25.3 m	EC/99+ /1.2 α/0.003/5.957	5.850(20)/0.003	(5/2 +)			
^{236}Pu		236.046048	2.87 y	α/5.867	5.611/0.21 5.7210/30.5 5.7677(1)/69.3	0+			0.0476/0.07 0.109/0.02 (0.17 - 0.97)
^{237}Pu		237.048403	45.7 d	EC/99.9 /0.220 α/0.003 /5.747	5.334(4)/0.0015 5.356(4)/0.0006 5.650(4)/0.0007	7/2-			Np L x-ray Np k x-ray 0.026344 0.03319 0.05954 (0.03-0.5)
^{238}Pu		238.049553	87.74 y	α/5.593	5.3583(1)/0.10 5.465(1)/28.3 5.4992(1)/71.6	0+			U k x-ray 0.04347 (0.04-1.1)
^{239}Pu		239.052156	2.411 x 10⁴ y	α/5.244	5.055/0.047 5.076/0.078 5.106/11.9 5.144/17.1 5.157/70.8 (4.74 -5.03)	1/2 +	+0.203		U k x-ray 0.05162 0.05682 0.12928 0.37502 0.41369
^{240}Pu		240.053807	6537. y	α/5.255	5.0212(1)/0.07 5.1237(1)/26.4 5.1681(1)/73.5	0+			U L x-ray 0.04524 0.10423 (0.04-0.97)
^{241}Pu		241.056844	14.4 y	β⁻ /99+ /0.0208 α/0.002 /5.139	4.853(7)/3x10⁻⁴ 4.8966(7)/0.002	5/2 +	-0.683	+6.	0.14854
^{242}Pu		242.058736	3.76 x 10⁵ y	α/4.983	4.7546(7)/0.098 4.8564(7)/22.4 4.9006(7)/78	0+			U L x-ray 0.04491 0.10350
^{243}Pu		243.061996	4.956 h	β⁻ /0.582	0.49/21 0.58/60	7/2 +			Am L x-ray 0.0417 0.0839
^{244}Pu		244.064197	8.2x10⁷ y	α/99.9/4.665 S.F./0.1 /	4.546(1)/19.4 4.589(1)/80.5	0+		U L x-ray	0.0439
^{245}Pu		245.06774	10.5 h	β⁻ /1.21	0.93/57 1.21/11	(9/2-)			Am L x-ray Am k x-ray 0.2804 / 0.30832 0.32752 0.56014 (0.03-1.2)
^{246}Pu		246.07020	10.85 d	β⁻ /0.40	0.150/85 0.35/10	0+			Am L x-ray Am k x-ray 0.04379 0.22371
$_{95}$Am									
^{232}Am			0.9 m	EC/ ≈ 5.0					
^{234}Am		234.0478	2.3 m	EC/4.2					
^{237}Am		237.0503	1.22 h	EC/99.98 /1.7 α/0.02 /6.20	6.042(5)/0.02	(5/2-)			Pu k x-ray 0.14559 0.28026 0.43845

Elem. or Isot.	Natural Abundance (%)	Atomic Mass or Weight	Half-Life	Decay Mode/Energy (/MeV)	Particle Energy /Intensity (MeV / %)	Spin (h/2π)	Nuclear Magnetic Mom.(nm)	Elect. Quadr. Mom. (b)	γ-ray Energy /Intensity (MeV / %)
^{238}Am		238.05198	1.63 h	EC/2.26 α/0.0001 /6.04	5.940/0.0001	1+			Pu L x-ray Pu k x-ray 0.91870 0.96278
^{239}Am		239.053018	11.9 h	EC/99.99 /0.803 a/0.01 /5.924	5.734(2)/0.001 5.776(2)/0.008	5/2-			Pu L x-ray Pu k x-ray 0.18172 0.22818 0.27760
^{240}Am		240.05529	2.12 d	EC/1.38 α/5.592	5.378(1)/16x10^{-4}	(3-)			Pu L x-ray Pu k x-ray 0.88878 0.98764 (0.1-1.3)
^{241}Am		241.056822	432.2 y	α/5.637	5.2443(1)/0.002 5.3221(1)/0.015 5.3884(1)/1.4 5.4431(1)/12.8 5.4857(1)/85.2 5.5116(1)/0.20 5.5442(1)/0.34	5/2-	+1.61	+4.	Np L x-ray 0.02634 0.033192 0.059536 (0.03-1.128)
^{242m}Am			141. y	I.T./99.5 /0.048 α/0.5 /5.62	5.141(4)/0.026 5.2070(2)/0.4	5-	+1.0	+6.5	Am L x-ray 0.04863 0.08648 0.10944 0.16304
^{242}Am		242.059542	16.02 h	β⁻ /83 /0.665 EC/17 /0.750	0.63/46 0.67/37	1-	+0.388	-2.4	Pu L x-ray Cm L x-ray Pu k x-ray 0.0422 0.04453
^{243}Am		243.061372	7.37x10^3 y	α/5.438	5.1798(5)/1.1 5.2343(5)/11 5.2766(5)/88 5.394(5)/0.12 5.3500(5)/0.16	5/2-	+1.61	+4.	0.04354 0.07467 0.08657 0.11770 0.14197
^{244m}Am			≈ 26. m	β⁻ /1.498		(1-)			0.0429
^{244}Am		244.064279	10.1 h	β⁻ /1.428					Am L x-ray Cm k x-ray 0.7460 0.9000
^{245}Am		245.066444	2.05 h	β⁻ /0.894	0.65/19 0.90/77	(5/2+)			Cm L x-ray Cm k x-ray 0.25299
^{246m}Am			25.0 m	β⁻ /	1.3/79. 1.60/14 2.1/7	2-			Cm L x-ray Cm k x-ray 0.27002 0.79881 1.06201 1.07885 (0.04-2.29)
^{246}Am		246.06977	39. m	β⁻ /2.38	1.2/	(7-)			Cm L x-ray Cm k x-ray 0.1529 0.2046 0.6786
^{247}Am		247.0722	22. m	β⁻ /1.7					Cm L x-ray Cm k x-ray 0.2267 / 0.2853 /
$_{96}$Cm									
^{236}Cm		236.0514		EC/1.7					
^{237}Cm		237.0529		EC/2.5					
^{238}Cm		238.05302	2.4 h	EC/>90 /0.97 α/<10 /6.632	6.520(50)/<10	0+			
^{239}Cm		239.0548	≈ 3. h	EC/1.7					0.0407 0.1466 0.1874
^{240}Cm		240.055519	27. d	α/6.397	5.989/0.014 6.147/0.05 6.2478(6) /28.8 6.2906(6) /70.6	0+			

Elem. or Isot.	Natural Abundance (%)	Atomic Mass or Weight	Half-Life	Decay Mode/Energy (/MeV)	Particle Energy /Intensity (MeV / %)	Spin (h/2π)	Nuclear Magnetic Mom.(nm)	Elect. Quadr. Mom. (b)	γ-ray Energy /Intensity (MeV / %)
²⁴¹Cm		241.057646	32.8 d	EC/99 /0.768 α/1 /6.184	5.8842(4)/0.12 5.9291(4)/0.18 5.9389(4)/0.69	1/2+			Am k x-ray 0.13241 0.16505 0.18028 0.43063 0.47181
²⁴²Cm		242.058828	162.8 d	α/6.216	5.9694(1)/0.035 6.069(1)/25 6.1129(1)/74	0+			Pu L x-ray 0.04408 0.10189 (0.04-1.2)
²⁴³Cm		243.061381	28.5 y	α/6.167	5.6815(5) /0.2 5.6856(5)/1.6 5.7420(5)/10.6 5.7859(5)/73.3 5.9922(5)/6.5 6.0103(5)/1.0 6.0589(5)/5 6.0666(5)/1.5	5/2+		0.41	Pu L x-ray Pu k x-ray 0.10612 0.20975 0.22819 0.27760 0.28546 0.33431 (0.04-0.7)
²⁴⁴Cm		244.062745	18.11 y	α/5.902	5.6656(1)/0.02 5.7528(1)/24 5.8050(1)/76	0+			Pu L x-ray 0.04282 0.09885 0.15262
²⁴⁵Cm		245.065485	8.5x10³ y	α /5.623	5.235(10)/0.3 5.3038(10)/5.0 5.3620(7)/93 5.4927(11)/0.8 5.5331(11)/0.6	7/2+		0.5	Pu L x-ray Pu k x-ray 0.04195 0.13299 0.13606 0.17494
²⁴⁶Cm		246.067217	4.78x10³ y	α/5.476	5.343(3)/21 5.386(3)/79	0+			Pu L x-ray 0.04453
²⁴⁷Cm		247.070346	1.56x10⁷ y	α/5.352	4.818(4)/4.7 4.8690(20)/71 4.941(4)/1.6 4.9820(20)/2.0 5.1436(20)/1.2 5.2104(20)/5.7 5.2659(20)/13.8	9/2- 9/2-		0.37	Pu k x-ray 0.2792 0.2886 0.3471 0.4035
²⁴⁸Cm		248.072341	3.4x10⁵ y	α/92 /5.162 S.F./8/	4.931(5)/0.07 5.0349(2)/16.5 5.0784(2)/(75)/1	0+			
²⁴⁹Cm		249.075946	64.15 m	β⁻ /0.900	0.9/	1/2+			Bk k x-ray 0.56039 0.63431
²⁵⁰Cm		250.07835	≈ 9.7x10³ y	S.F./ α/5.27		0+			
²⁵¹Cm		251.08228	16.8 m	β⁻ /1.42	0.90/16	(1/2+)			0.3896 / 0.5299 0.5425
₉₇**Bk**									
²³⁸Bk		238.0583	2.4 m	EC/5.0					
²⁴²Bk		242.0621	7.0 m	EC/3.0					
²⁴³Bk		243.063001	4.5 h	EC/99.8 /1.508 α/0.15 /6.871	6.542(4)/0.03 6.5738(2)/0.04 6.7180(22)/0.02 6.7581(20)/0.02	(3/2-)			0.1466 0.1874 0.755 0.840 0.946
²⁴⁴Bk		244.0652	4.4 h	EC/99.99 /2.26 α/0.01 /6.778	6.625(4)/0.003 6.667(4)/0.003	(4-)			0.1445 0.1876 0.2176 0.9815 0.9215/
²⁴⁵Bk		245.066355	4.94 d	EC/99.9 /0.810 α/0.1 /6.453	5.8851(5)/0.03 6.1176(9)/0.01 6.1467(5)/0.02 6.3087(5)/0.014 6.3492(5)/0.018	3/2-			Cm L x-ray Cm k x-ray 0.25299 0.3809 0.3851
²⁴⁶Bk		246.0687	1.80 d	EC/1.35		(2-)			Cm L x-ray Cm k x-ray 0.79881 1.08142

Elem. or Isot.	Natural Abundance (%)	Atomic Mass or Weight	Half-Life	Decay Mode/Energy (/MeV)	Particle Energy /Intensity (MeV / %)	Spin (h/2π)	Nuclear Magnetic Mom.(nm)	Elect. Quadr. Mom. (b)	γ-ray Energy /Intensity (McV / %)
^{247}Bk		247.07030	1.4x10³ y	α/5.889	5.465(5)/1.5 5.501(5)/7 5.532(5)/45 5.6535(20)/5.5 5.678(2)/13 5.712(2)/17 5.753(2)/4.3 5.794(2)/5.5	(3/2-)			0.04175 0.0839 0.268
^{248}Bk		248.07311	23.7 h	β⁻ /70 /0.87 EC/30 /0.72	0.86/	(1-)			Cm L x-ray Cf L x-ray Cm k x-ray Cf k x-ray 0.5507
^{249}Bk		249.074980	320. d	β⁻ /0.125 α/0.001 /5.525	0.125/100 5.390(1)/0.0002 5.4174(6)/0.001	7/2+	2.0		0.327/10⁻⁵ 0.308/10⁻⁶
^{250}Bk		250.078309	3.217 h	β⁻ /1.780	0.74/	2-			Cf L x-ray Cf k x-ray 0.98912 1.03184 (0.04-1.6)
^{251}Bk		251.08075	56. m	β⁻ /1.09		(3/2-)			0.02481 0.1528 0.1776
$_{98}$Cf									
^{239}Cf		239.0626	≈ 0.7 m	α					
^{240}Cf		240.0623	1.1 m	α/7.719	7.590(10)/	0+			
^{241}Cf		241.0637	4. m	EC/3.3 α/7.60	7.335(5)/				
^{242}Cf		242.06369	3.5 m	α/7.509	7.351(6)/20 7.385(4)/80	0+			
^{243}Cf		243.0654	11. m	EC/86 /2.2 α/14 /7.40	7.060(6)/20 7.170/4	(1/2+)			
^{244}Cf		244.065990	20. m	α/7.328	7.168(5)/25 7.210(5)/75	0+			
^{245}Cf		245.068038	44. m	α/30 /7.255 EC/70 /1.569	6.886/ 6.983/ 7.036/ 7.084/ 7.137(2)/				
^{246}Cf		246.068798	1.49 h	α/6.869	6.6156(10)/0.18 6.7086(7)/21.8 6.7501(7)/78.0	0+			Cm L x-ray 0.04221 0.0945 0.147
^{247}Cf		247.07099	3.11 h	EC/99.96 /0.65 α/0.04 /6.55	6.301(5)/	7/2+			Bk k x-ray 0.2941 0.4778
^{248}Cf		248.07218	334. d	α/6.369	6.220(5)/17 6.262(5)/83	0+			
^{249}Cf		249.074846	351. y	α/6.295	5.7582(2)/3.7 5.8119(2)/84 5.8488(2)/1.0 5.9029(2)/2.8 5.9451(2)/4.0 6.1401(2)/1.1 6.1940(2)/2.2	9/2-			Cm L x-ray Cm k x-ray 0.25299 0.33351 0.38832
^{250}Cf		250.076399	13.1 y	α/6.129	5.8913(4)/0.3 5.9889(4)/15 6.0310(4)/84.5	0+			Cm L x-ray 0.04285
^{251}Cf		251.079579	9.0x10² y	α/6.172	5.56448(7)/1.5 5.632(1)/4.5 5.648(1)/3.5 5.6773(6)/35 5.762(3)/3.8 5.7937(7)/2.0 5.8124(8)/4.2 5.8514(6)/27 6.0140(7)/11.6 6.0744(7)/2.7	1/2+			
^{252}Cf		252.081619	2.65 y	α/96.9 /6.217 S.F./3.1/	5.7977(1)/0.23 6.0756(4)/15.2 6.1184(4)/81.6	0+			Cm L x-ray 0.04339 0.1002
^{253}Cf		253.08512	17.8 d	β⁻ /99.7 /0.29 α/0.3 /6.126	0.27/100 5.921(5)/0.02	(7/2+)			
^{254}Cf		254.08732	60.5 d	S.F./99.7/ α/0.3/5.930	5.792(5)/0.05 5.834(5)/0.26	0+			

Elem. or Isot.	Natural Abundance (%)	Atomic Mass or Weight	Half-Life	Decay Mode/Energy (/MeV)	Particle Energy /Intensity (MeV / %)	Spin (h/2π)	Nuclear Magnetic Mom.(nm)	Elect. Quadr. Mom. (b)	γ-ray Energy /Intensity (MeV / %)
^{255}Cf		255.0910	1.4 h	β^- /0.7					
^{255}Cf			12. m	SF					
$_{99}$Es									
^{243}Es		243.0696	21. s	α/ > 30 / EC/ < 70 /4.0	7.89/ > 30				
^{244}Es		244.0709	37. s	EC/76 /4.6 α/4 /	7.57/4				
^{245}Es		245.0713	1.3 m	α/40 /7.858 EC/60 /3.1	7.74				
^{246}Es		246.0730	7.7 m	EC/90 /3.9 α/10 /	7.35				
^{247}Es		247.07365	4.8 m	EC/93 /2.48 α/7 /	7.32				
^{248}Es		248.0755	26. m	EC/99.7 /3.1 α/0.3 /	6.87				
^{249}Es		249.07640	1.70 h	EC/99.4 /1.45 α/0.6 /	6.77	(7/2+)			0.3795 0.8132
^{250m}Es			2.2 h	EC/ β^+		(1-)			Cf L x-ray Cf k x-ray 0.9891 1.0319
^{250}Es		250.0787	8.6 h	EC/2.1		(6+)			Cf L x-ray Cf k x-ray 0.30339 0.34948 0.82883
^{251}Es		251.07998	1.38 d	EC/99.5 /0.38 α/0.5 /	6.462/0.05 6.492/0.4	(3/2-)			
^{252}Es		252.08297	1.29 y	α/76 / EC/24 /1.26	6.632/61.0 6.562/10.3	(5-)			
^{253}Es		253.084818	20.47 d	α/	6.633/89.8 6.5916/6.6	7/2+	+4.10	7.	0.04180 0.3892
^{254m}Es			1.64 d	β^- /99.6 / α/0.3 /6.67	0.475 6.382	2+ 2+	2.9	3.7	Fm L x-ray Fm k x-ray 0.6488 0.6938
^{254}Es		254.088017	276. d	α/	6.429	(7+)			0.064
^{255}Es		255.09027	40. d	β^- /92 /0.29 α/8 / S.F./0.004/	6.26 6.300	(7/2+)			
^{256m}Es			7.6 h	β^- /		(8+)			0.218 0.232 0.862
^{256}Es		256.0936	25. m	β^- /1.7		(1+)			
$_{100}$Fm									
^{242}Fm			0.8 ms	SF					
^{243}Fm		243.0745	0.2 s	α/	8.55				
^{244}Fm		244.0741	3.7 ms	S.F./		0+			
^{245}Fm		245.0754	4. s	α/	8.15/				
^{246}Fm		246.07528	1.2 s	α/92 / S.F./8/	8.24/	0+			
^{247m}Fm			9. s	α/	8.18/				
^{247}Fm		247.0768	35. s	α/8.20 EC/2.9	7.87/70 7.93/30				
^{248}Fm		248.07718	36. s	α/99.9 /8.001 S.F./0.1/	7.83/20 7.87/80	0+			
^{249}Fm		249.0790	3. m	EC/2.4 α/	7.53	(7/2+)			
^{250m}Fm			1.8 s	I.T./					
^{250}Fm		250.07951	30. m	α/ EC/0.8	7.43/	0+			
^{251}Fm		251.08157	5.3 h	EC/98 /1.47 α/2 /	6.833	(9/2-)			
^{252}Fm		252.08246	1.058 d	α/7.154 SF	6.998/15 7.039/85	0+			
^{253}Fm		253.085175	3.0 d	EC(88%)/0.333 α/12 /	6.676/ 6.943/	1/2+			Es k x-ray 0.2719
^{254}Fm		254.086847	3.240 h	α/ S.F./0.06/	7.150 7.192	0+			
^{255}Fm		255.089955	20.1 h	α/	6.9635(5)/5.0 7.0225(5)/93.4	7/2+			
^{256}Fm		256.09177	2.63 h	S.F./92/ α/18 /	6.92/	0+			

Elem. or Isot.	Natural Abundance (%)	Atomic Mass or Weight	Half-Life	Decay Mode/Energy (/MeV)	Particle Energy /Intensity (MeV / %)	Spin (h/2π)	Nuclear Magnetic Mom.(nm)	Elect. Quadr. Mom. (b)	γ-ray Energy /Intensity (MeV / %)
²⁵⁷Fm		257.09510	100.5 d	α/99.8 / S.F./0.2	6.519	(9/2+)			0.1794 0.2410
²⁵⁸Fm		258.0971	0.37 ms	S.F./					
²⁵⁹Fm		259.1006	1.5 s	S.F./					
₁₀₁Md									
²⁴⁷Md		247.0817	3. s	α	8.43				
²⁴⁸Md		248.0828	7. s	EC/80 /5.3 α/20 /	8.32/15 8.36/5				
²⁴⁹Md		249.0830	24. s	E.C>/<80 /3.7 α/>20 /8.46	8.030(20)/				
²⁵⁰Md		250.0845	50. s	EC/94 /4.6 α/6 /8.25	7.75/4 7.83/2				
²⁵¹Md		251.0849	4.0 m	EC/>94 /3.1 α/<6 /	7.55/				
²⁵²Md		252.0866	2. m	EC/>50 /3.9 α/<50 /	7.73/				
²⁵³Md		253.0873	≈6 m	EC/2.0					
²⁵⁴ᵐMd			30. m	EC/					
²⁵⁴Md		254.0897	10. m	EC/2.7					
²⁵⁵Md		255.09108	27. m	EC/92 /1.04 α/8 /	7.33	(7/2-)			0.4531 0.4056
²⁵⁶Md		256.0941	1.30 h	EC/89 /2.13 α/11 /	7.22/47 7.155/21				Fm k x-ray (0.112-0.834)
²⁵⁷Md		257.095535	5.5 h	EC/85 /0.41 α/15 /, S.F.	7.074 7.014	(7/2-)			Fm k x-ray (0.181-0.389)
²⁵⁸ᵐMd			57. m	EC/		(1-)			Fm k x-ray
²⁵⁸Md		258.098427	51.5 d	α/7.40	6.718(2)/ 6.763(4)/	(8-)			0.3678 (0.057 - 0.448)
²⁵⁹Md		259.1005	1.64 h	S.F./ α/<1.3		7/2+			
²⁶⁰Md		260.104	32. d	S.F./					
₁₀₂No									
²⁵⁰No			0.25 ms	S.F./		0+			
²⁵¹No		251.0889	0.8 s	α/	8.60/80 8.68/20				
²⁵²No		252.08897	2.3 s	α/73 /8.551 S.F./27/	8.42 8.37	0+			
²⁵³No		253.0907	1.7 m	α/ EC/3.2	8.010(20)	(9/2-)			
²⁵⁴ᵐNo			0.28 s	I.T./					
²⁵⁴No		254.09095	55. s	α/ EC/1.1	8.09	0+			
²⁵⁵No		255.09323	3.1 m	α/62 / EC/38/2.01	8.12/ 7.93 8.08	1/2+			0.187
²⁵⁶No		256.09428	2.9 s	α/	8.43	0+			
²⁵⁷No		257.09685	25. s	α/	8.22 8.27 8.32	(7/2+)			
²⁵⁸No		258.0983	≈ 1.2 ms	S.F./		0+			
²⁵⁹No		259.1011	58. m	α/78 /7.794 EC/22/0.5	7.52 7.55	(9/2+)			
²⁶²No		262.108	≈ 8. ms	S.F./					
₁₀₃Lr									
²⁵³Lr		253.0953	1.3 s	α/	8.80 8.72				
²⁵⁴Lr		254.0965	13. s	α/ EC/5.2	8.45				
²⁵⁵Lr		255.0967	22. s	α/ EC/3.2	8.37/60 8.43/40				
²⁵⁶Lr		256.0988	28. s	α/99.7 /8.554 EC/4.2	8.43/ 8.39				
²⁵⁷Lr		257.0996	0.65 s	α/ EC/2.5	8.80 8.80	7/2+			
²⁵⁸Lr		258.1019	3.9 s	α/ EC/3.4	8.60 8.62 8.56				
²⁵⁹Lr		259.1030	6.1 s	α/80 SF/20	8.44(1)				
²⁶⁰Lr		260.1056	3. m	α/	8.03				
²⁶¹Lr		261.1069	40. m	SF					
²⁶²Lr		262.110	3.6 h	EC/2.					

Elem. or Isot.	Natural Abundance (%)	Atomic Mass or Weight	Half-Life	Decay Mode/Energy (/MeV)	Particle Energy /Intensity (MeV / %)	Spin (h/2π)	Nuclear Magnetic Mom.(nm)	Elect. Quadr. Mom. (b)	γ-ray Energy /Intensity (MeV / %)
$_{104}$104									
^{253}104			≈ 1.8 s	SF					
^{254}104			0.5 ms	SF					
^{255}104		255.1015	1.7 s	SF					
^{256}104		256.10118	7. ms	SF,α/8.92	8.81				
^{257}104		257.1032	4.7 s	α/9.22	8.77				0.117
					9.01				
					8.95				
					8.62				
^{258}104		258.1035	12. ms	SF/					
^{259}104		259.1056	3.4 s.	α/9.09	8.77(2)/				
				SF/	8.86/				
^{260}104		260.1065	20. ms	SF/					
^{261}104		261.1089	1.1 m	α/8.78	8.28/				
^{262}104		262.1101	≈1.2 s	SF/					
^{263}104			<20. m	SF					
$_{105}$105									
^{255}105			≈ 1.5 s	SF					
^{257}105		257.1079	1.5 s	α/	8.97/				
					9.07/				
					9.16/				
^{258}105		258.1093	4.2 s	α/	9.17/				
				E.C/5.3	9.08/				
^{259}105		259.1097	≈ 1.2 s	S.F./					
^{260}105		260.1114	1.5 s	α/	9.05/				
				S.F./	9.08/				
					9.13/				
^{261}105		261.1121	1.8 s	α/	8.93/				
				S.F./					
^{262}105		262.1144	34. s	S.F./					
				α/	8.45/				
					8.53/				
					8.67/				
^{263}105		263.1153	≈0.45 m	S.F./57/, α/43/	8.35/43				
$_{106}$106									
^{259}106		259.1147	0.5 s	α,S.F./	9.62				
^{260}106		260.11444	4. ms	α,S.F./	9.77				
^{261}106		261.1164	0.3 s	α,S.F./	9.56				
^{263}106		263.1186	0.8 s	S.F.,α	9.06				
					9.25/				
^{265}106		265.121	<30. s	α/ >50					
				SF/ <50					
^{266}106		266.122	<30. s	α/ >50	8.63(5)				
				SF/ <50					
$_{107}$107									
^{261}107		261.1218	12. ms	α/,SF	10.40				
					10.10				
					10.03				
^{262m}107			8. ms	α/	10.37				
					10.24				
^{262}107		262.1231	0.10 s	α/	10.06				
					9.91				
					9.74				
^{264}107									
$_{108}$108									
^{263}108				α/					
^{264}108		264.1286	≈ 0.08 ms	α/,SF	11.0				
^{265}108		265.1306	2. ms	α/	10.36				
$_{109}$109									
^{266}109		266.1378	≈ 3.4 ms	α	11.10				
^{268}109									
$_{110}$110									
^{269}110			0.17 ms	α					
$_{111}$111									
^{272}111			1.5 ms	α					

NEUTRON SCATTERING AND ABSORPTION PROPERTIES*
(Revised 1996)

Norman E. Holden

Reactor Division
Brookhaven National Laboratory
Upton, New York 11973

This Table presents an evaluated set of values for the experimental quantities, which characterize the properties for scattering and absorption of neutrons. The neutron cross section is given for room temperature neutrons, 20.43° C, corresponds to a thermal neutron energy of 0.0253 electron volts (eV) or a neutron velocity of 2200 meters/second. The neutron resonance integral is defined over the energy range from 0.5 eV to 0.1×10^6 eV, or 0.1 MeV. Bound neutron scattering lengths and neutron cross sections averaged over a maxwellian spectrum at 30 keV for astrophysical applications are also presented. A list of the major references used is given below. The literature cutoff date is December 1996. Uncertainties are given in parentheses. Parentheses with two or more numbers indicate values to the excited state(s) and to the ground state of the product nucleus. Discussions with Frank Feiner and Harold Knox of KAPL on thermal cross sections and resonance integrals and the assistance of Felicia Kramer in assembling this file are gratefully acknowledged.

TABLE LAYOUT

Column No.	Column Title	Description
1	Isotope or Element	For elements, the atomic number and chemical symbol are listed. For nuclides, the mass number and chemical symbol are listed. Isomers are indicated by the addition of m, m1, or m2.
2	Isotopic Abundance	in atom percent.
3	Half-life	Half-life in decimal notation. μs = microseconds; ms = milliseconds; s = seconds; m = minute; h = hours; d = days; and y = years.
4	Thermal Neutron Cross Section	Cross sections for neutron capture reactions in barns (10^{24} cm^2) or in millibarns (mb). Proton, alpha production and fission reactions are designated by σ_p, σ_α, σ_f, respectively. Separate values are listed for isomeric production.
5	Neutron Resonance Integral	Resonance integrals for neutron capture reactions in barns (10^{24} cm^2) or millibarns (mb). Proton, alpha production and fission reactions are designated by RI_p, RI_α, RI_f, respectively. Separate values are listed for isomeric production.
6	Neutron Scattering Length	Bound Coherent scattering length for neutron scattering reactions in units of femtometer (fm), which is equal to fermis (10^{-13} cm).
7	Maxwellian Averaged Cross Section	Astrophysical Cross Sections, averaged over a stellar neutron maxwellian spectrum characterized by a thermal energy of 30 keV, expressed in barns (10^{24} cm^2), millibarns (mb) or microbarns (μb).

GENERAL NUCLEAR DATA REFERENCES

The following references represent the major sources of the neutron data presented:

1. Mughabghab, S.F.; Divadeenam, M.; Holden, N.E.; Neutron Cross Sections, Vol. I *Neutron Resonance Parameters and Thermal Cross Sections* Part A Z=1-60. Academic Press Inc., New York, New York (1981); Mughabghab, S.F.; Part B, Z-61-100. Academic Press Inc., Orlando, Florida (1984).

2. Parrington, J.R.; Knox, H.D.; Breneman, S.L.; Baum, E.M.; Feiner, F.: *Chart of the Nuclides, 15th Edition*, Knolls Atomic Power Laboratory (Oct. 1996).

3. Holden, N.E.; *Fifty Years with Nuclear Fission* Conference, Wash., D.C., Gaithersburg, Md. April 26-29, 1989, p. 946. American Nuclear Society, LaGrange Park, Illinois (1989).

4. Holden, N.E.; *Review of Thermal Neutron Cross Sections and Isotopic Composition of the Elements*, BNL-NCS-42224 (March 1989) and Holden, N.E.; *Thermal Neutron Cross Sections for the 1991 Table of the Isotopes*, BNL-45255 (Feb. 1991).

5. Tuli, J.K.; *Nuclear Wallet Cards*, Brookhaven National Laboratory (July 1995).

6. Holden, N.E.; *Half-lives of Selected Nuclides*, Pure & Applied Chemistry 62, 941 (1990).

7. Koester, L; Rauch, H.; Seymann, E.; *Neutron Scattering Lengths: A Survey of Experimental Data and Methods*, Atomic Data Nuclear Data Tables 49, 65 (1991).

8. Sears, V.F.; *Neutron Scattering Lengths and Cross Sections,* Neutron News 3, (3), 26 (1992).

*Research carried out under auspices of the US Department of Energy, Contract No. DE-AC02-76CH00636

Elem. or Isot.	Natural Abundance (%)	Half-life	Thermal Neut. Cross Section (barns)	Resonance Integral (barns)	Coh. Scat. Length (fm)	σ(30KEV) Maxw. Avg. (barns)
$_1$H			0.332(2)	0.149(1)	- 3.739(1)	
^{1}H	99.985(1)		0.332(2)	0.149(1)	- 3.741(1)	0.25(2) mb*
^{2}H	0.015(1)		0.51(1)mb	0.23(2) mb	6.671(4)	
^{3}H		12.32 y	< 6. μb		4.79(3)	
$_2$He			< 0.05		3.26(3)	
^{3}He	1.37x10^{-6}		σ_p=5.33(1)x10^3	RI$_p$=2.39(1)x10^3	5.74(7)	
			0.05(1) mb			8.(1) μb*
^{4}He	$\approx$ 100.				3.26(3)	
$_3$Li			71.(2)	32.(1)	- 1.90(2)	
^{6}Li	7.5(2)		σ_α=9.4(1)x10^2	RI$_\alpha$=422.(4)	2.0(1)	
			39.(5) mb	17.(2) mb		
^{7}Li	92.5(2)		45.(5) mb	20.(2) mb	- 2.22(2)	39(6) μb
$_4$Be			8.8(4) mb	3.9(2) mb	7.79(1)	
^{7}Be		53.28 d	σ_p=3.9(1)x10^4	RI$_p$=1.75(5)x10^4		σ_p 16(4)
			$\sigma_\alpha \approx$ 0.1			
^{9}Be	100.		8.8(4) mb	3.9(2) mb	7.79(1)	
^{10}Be		1.52x10^6 y	<1. mb			
$_5$B			7.6(1)x10^2	343.(5)	5.30(4)	
^{10}B	19.9(2)		σ_α=38.4(1)x10^2	RI$_\alpha$=17.3(1)x10^2	- 0.1(3)	
			0.3(1)	0.13(4)		
			σ_p=7.(1) mb			
			σ_t=8.(2) mb			
^{11}B	80.1(2)		5.(3) mb	2(1) mb	6.65(4)	
$_6$C			3.5(1) mb	1.6(1) mb	6.646(1)	
^{12}C	98.89(1)		3.5(1) mb	1.6(1) mb	6.651(2)	16(1) μb*
^{13}C	1.11(1)		1.4(1) mb	1.7(2) mb	6.19(9)	0.2(1) mb
^{14}C		5715. y	<1. μb			1.5(4) μb*
$_7$N			1.9(1)	0.85(5)	9.36(2)	
^{14}N	99.634(9)		σ_p=1.83(7)	RI$_p$=0.82(3)	9.37(2)	σ_p 1.8(2) mb*
			0.080(1)	0.034(1)		0.04(7) mb
^{15}N	0.366(9)		0.04(1) mb	0.11(3) mb	6.44(3)	4.9(5) μb*
$_8$O			0.29(1) mb	0.40(4) mb	5.805(4)	
^{16}O	99.76(1)		0.19(1) mb	0.36(4) mb	5.805(5)	34(4) μb
^{17}O	0.04		σ_α=0.24(1)	0.11(1)	5.8(2)	σ_α 3.9(5) mb*
			0.54(7) mb	0.39(5) mb		
^{18}O	0.20(1)		0.16(1) mb	0.81(4) mb	5.84(7)	7.2(6) μb*

Elem. or Isot.	Natural Abundance (%)	Half-life	Thermal Neut. Cross Section (barns)	Resonance Integral (barns)	Coh. Scat. Length (fm)	σ(30KEV) Maxw. Avg. (barns)
$_9$F			9.5(7) mb	21.(3) mb	5.65(1)	
^{19}F	100.		9.5(7) mb	21.(3) mb	5.65(1)	6.(1) mb
$_{10}$Ne			42(5) mb	19(3) mb	4.566(6)	
^{20}Ne	90.48(3)		39(5) mb	18(3) mb	4.631(6)	0.12(1) mb
^{21}Ne	0.27(1)		0.7(1),σ_α<1.5	0.31(5)	6.7(2)	$\approx$ 1.6 mb
^{22}Ne	9.25(3)		51(5) mb	23(3) mb	3.87(1)	60.(5) μb*
$_{11}$Na			0.53(2)	0.32(2)	3.63(2)	2.3(2) mb
^{22}Na		2.605 y	σ_p=2.8(3)x10^4 σ_α=2.6(4)x10^2	< 2x10^5 1.2(2) x 10^2		
^{23}Na	100.		σ_m=0.43(3)	RI$_m$=0.30(6)	3.63(2)	
$_{12}$Mg			66(6) mb	38.(5) mb	5.375(4)	
^{24}Mg	78.99(3)		0.053(6)	32.(4) mb	5.7(2)	4.2(2) mb
^{25}Mg	10.00(1)		.20(1)	98.(15) mb	3.6(2)	6.5(3) mb
^{26}Mg	11.01(2)		.038(1)	25.(2) mb	4.9(2)	66.(3) μb
^{27}Mg		9.45 m	0.07(2)	0.03(1)		
$_{13}$Al			0.230(2)	0.17(1)	3.45(1)	
^{26}Al			$\sigma_p \approx$ 2.1			0.14(2)
^{27}Al	100.		0.230(2)	0.17(1)	3.45(1)	3.8(3) mb
$_{14}$Si			0.166(9)	0.12(2)	4.15(1)	
^{28}Si	92.21(2)		0.17(1)	0.11(2)	4.11(1)	1.8(2) mb
^{29}Si	4.69(2)		0.12(1)	0.08(2)	4.7(1)	7.9(9) mb
^{30}Si	3.10(1)		0.107(4)	0.62(6)	4.61(1)	6.4(6) mb
^{31}Si		2.62 h	73.(6) mb	33.(3) mb		
$_{15}$P			0.17(1)	0.08(1)	5.13(1)	
^{31}P	100.		0.17(1)	0.08(1)	5.13(1)	1.7(1) mb
$_{16}$S			0.54(2)	0.24(2)	2.847(1)	
^{32}S	95.02(9)		0.55(5) σ_α < 0.5 mb	0.25(2)	2.804(2)	4.7(2) mb
^{33}S	0.75(4)		0.46(3) σ_α = 0.12(1) σ_p = 2 mb	0.21(2) RI$_\alpha$ = 0.05(1)	4.7(2)	7.4(15) mb σ_α = 0.18(1)
^{34}S	4.21(8)		0.30(2)	0.13	3.48(3)	2.9(9) mb
^{36}S	0.02(1)		0.23(2)	0.26(3)		0.17(1) mb
$_{17}$Cl			33.6(3)	15.(2)	9.58(1)	
^{35}Cl	75.77(7)		43.7(4);σ_p=0.44(1) $\sigma_\alpha \approx$ 0.08 mb	20.(2);RI$_p$=0.2	11.7(1)	10.(3) mb σ_p 1.7(2) mb

Elem. or Isot.	Natural Abundance (%)	Half-life	Thermal Neut. Cross Section (barns)	Resonance Integral (barns)	Coh. Scat. Length (fm)	σ(30KEV) Maxw. Avg. (barns)
^{36}Cl		3.01×10^5 y	σ_p=46.(2) mb <10. σ_α = 0.59(7) mb	RI$_p$=0.02		σ_p 91.(8) mb σ_α 0.9(2) mb
^{37}Cl	24.23(7)		(0.05 + 0.38)	(0.04+0.26)	3.1(1)	2.15(8) mb
$_{18}$Ar			0.66(3)	0.42(5)	1.91(1)	
^{36}Ar	0.337(3)		5(1);σ_p<1.5 mb σ_α 6(1) mb	2.(1)	24.9(1)	
^{37}Ar		35.0 d	σ_α=2.0(3)x10^3 σ_p = 69(14)	RI$_\alpha$=900. RI$_p$=31.		
^{38}Ar	0.063(1)		0.8(2)	0.4(1)	3.5(35)	
^{39}Ar		268. y	6.(2)x10^2			
^{40}Ar	99.600(3)		0.64(3)	0.41(5)	1.83(1)	2.3(2)mb
^{41}Ar		1.82 h	0.5(1)	0.2(1)		
$_{19}$K			2.1(1)	1.0(1)	3.67(2)	
^{39}K	93.2581(44)		2.1(2) σ_α = 4.3(5) mb σ_p < 0.05 mb	0.9(1)	3.74(2)	11.8(4)mb
^{40}K	0.0117(1)	1.26×10^9 y	30.(8) σ_p=4.4(4) σ_α = 0.42(8)	13.(4) 2.0(2)		σ_p .7(1)mb σ_α 40(6)mb
^{41}K	6.7302(44)		1.46(3)	1.4(2)	2.69(8)	22.0(7)mb
$_{20}$Ca			0.43(2)	0.23(2)	4.70(2)	
^{40}Ca	96.941(18)		0.41(3) σ_α 0.13(4) mb	0.22(4)	4.80(2)	6.7(7)mb
^{41}Ca		1.02×10^5 y	≈ 4. σ_α = 0.19			
^{42}Ca	0.647(9)		0.65(10)	0.39(4)	3.4(1)	16.(2)mb
^{43}Ca	0.135(6)		6.(1)	3.9(2)	- 1.56(9)	51.(6)mb
^{44}Ca	2.086(12)		0.8(2)	0.56(1)	1.42(6)	9.(1)mb
^{45}Ca		162.7 d	≈ 15.			
^{46}Ca	0.004(3)		0.70(3)	0.9(1)	3.6(2)	5.7(5)mb
^{48}Ca	0.187(4)		1.1(1)	0.5(1)	0.39(9)	1.0(2)mb
$_{21}$Sc			27.2(2)	12.(1)	12.3(1)	
^{45}Sc	100.		(10.+17.)	(5.6+6.4)	12.3(1)	90.(6)mb
^{46}Sc		83.81 d	8.(1)	3.6(5)		
$_{22}$Ti			6.1(1)	2.8(2)	- 3.438(2)	
^{46}Ti	8.25(3)		0.6(2)	0.4(1)	4.93(6)	27.(3)mb
^{47}Ti	7.44(2)		1.6(2)	1.6(2)	3.63(1)	64.(8)mb
^{48}Ti	73.72(3)		7.9(9)	3.6(2)	- 6.09(2)	31.(5)mb
^{49}Ti	5.41(2)		1.9(5)	1.2(2)	1.04(5)	22.(2)mb
^{50}Ti	5.18(2)		0.179(3)	0.12(2)	6.18(8)	4.0(5)mb

Elem. or Isot.	Natural Abundance (%)	Half-life	Thermal Neut. Cross Section (barns)	Resonance Integral (barns)	Coh. Scat. Length (fm)	σ(30KEV) Maxw. Avg. (barns)
$_{23}$V			5.0(2)	2.8(1)	- 0.382(1)	
^{50}V	0.250(4)	>1.4x10^{17} y	21(4) $\sigma_p = 0.7(4)$ mb	50.(20)	7.6(6)	
^{51}V	99.750(4)		4.9(1)	2.7(2)	- 0.402(2)	42.(3)mb
$_{24}$Cr			3.0(2)	1.7(1)	3.635(7)	
^{50}Cr	4.345(13)		15.(1)	8.(1)	- 4.5(1)	52.(1)mb
^{51}Cr		27.70 d	< 10.			
^{52}Cr	83.79(2)		0.8(1)	0.6(2)	4.91(2)	8.8(4)mb
^{53}Cr	9.50(2)		18.(2)	9(1)	- 4.2(1)	0.06(1)
^{54}Cr	2.365(7)		0.36(4)	0.25(5)	4.6(1)	7.(2)mb
$_{25}$Mn			13.3(1)	14.0(3)	- 3.75(2)	
^{53}Mn		3.7x10^6 y	70.(10)	32.(5)		
^{54}Mn		312.2 d	< 10.			
^{55}Mn	100.		13.3(1)	14.0(3)	- 3.75(2)	40.(2)mb
$_{26}$Fe			2.6(1)	1.4(2)	9.45(2)	
^{54}Fe	5.85(4)		2.7(5) $\sigma_\alpha = 10$ μb	1.3(2) $RI_\alpha = 1.1(1)$ mb	4.2(1)	29.(2)mb
^{55}Fe		2.73 y	13.(2) $\sigma_\alpha = 0.03$	6.(1)		
^{56}Fe	91.75(4)		2.6(2)	1.4(2)	9.93(3)	13.5(5)mb
^{57}Fe	2.12(1)		2.5(5)	1.6(2)	2.3(1)	35.(4)mb
^{58}Fe	0.28(1)		1.3(1)	1.3(2)	15.(7)	13.(2)mb
^{59}Fe		44.51 d	13.(3)	6.(1)		
$_{27}$Co			37.19(8)	74.(2)	2.49(2)	
^{58m}Co		9.1 h	1.4(1)x10^5	2.5(10)x10^5		
^{58}Co		70.88 d	1.9(2)x10^3	7.(1)x10^3		
^{59}Co	100.		(20.7+16.5)	(39.+35.)	2.49(2)	38.(4)mb
^{60m}Co		10.47 m	58.(3)	230.(50)		
^{60}Co		5.271 y	2.0(2)	4.3(10)		
^{28}Ni			4.5(2)	2.3(2)	10.3(1)	
^{58}Ni	68.077(9)		4.6(4) $\sigma_\alpha < 0.03$ mb	2.3(2)	14.4(1)	42.(3)mb
^{59}Ni		$\approx$7.6x10^4 y	σabs=92.(4) $\sigma_\alpha = 14(2)$ $\sigma_p = 2(1)$	RIabs=1.4(1)x10^2		
^{60}Ni	26.223(8)		2.9(3)	1.5(2)	2.8(1)	27.(2)mb
^{61}Ni	1.140(1)		2.5(5) $\sigma_\alpha = 0.03$ mb	1.5(4)	7.60(6)	82.(1)mb
^{62}Ni	3.634(2)		15.(1)	6.8(3)	- 8.7(2)	36.(4)mb
^{63}Ni		100. y	20(5)	9.(2)	26.(8)mb	
^{64}Ni	0.926(1)		1.6(1)	1.(2)	- 0.37(7)	10.(1)mb
^{65}Ni		2.517 h	22.(2)	10.(1)		

Elem. or Isot.	Natural Abundance (%)	Half-life	Thermal Neut. Cross Section (barns)	Resonance Integral (barns)	Coh. Scat. Length (fm)	σ(30KEV) Maxw. Avg. (barns)
$_{29}$Cu			3.8(1)	4.1(4)	7.718(4)	
^{63}Cu	69.17(3)		4.5(2)	5.(1)	6.43(15)	0.09(1)
^{64}Cu		12.701 h	$\approx$ 270.			
^{65}Cu	30.83(3)		2.17(3)	2.2(1)	10.61(19)	53(5)mb
^{66}Cu		5.09 m	$1.4(1) \times 10^2$	60.(20)		
$_{30}$Zn			1.1(2)	2.8(4)	5.680(5)	
^{64}Zn	48.6(6)		0.74(5) $\sigma_p < 12 \ \mu b$ $\sigma_\alpha = 11(3) \ \mu b$	1.4(3)	5.23(4)	59(5)mb
^{65}Zn		243.8 d	66.(8) $\sigma_\alpha = 2.0(2)$	30.(4)		
^{66}Zn	27.9(3)		0.9(3) $\sigma_\alpha < 0.02$ mb	1.8(2)	5.98(5)	35(3)mb
^{67}Zn	4.1(1)		6.9(1.4) $\sigma_\alpha = 0.4$ mb	25.(5)	7.58(8)	0.15(2)
^{68}Zn	18.8(5)		(0.072 + 0.8) $\sigma_\alpha < 0.02$ mb	(0.2 + 2.9)	6.04(3)	19(2)mb
^{70}Zn	0.6(3)		(8.1+83.) mb	0.9(2)		
$_{31}$Ga			2.9(1)	22.(3)	7.288(2)	
^{69}Ga	60.11(3)		1.68(7)	16.(2)	7.88(4)	0.15(1)
^{71}Ga	39.89(3)		4.7(2);σ_m=0.15(5)	31.(3)	6.40(3)	0.12(1)
$_{32}$Ge			2.2(1)	6.(2)	8.19(2)	
^{70}Ge	21.23(4)		(0.3 + 2.7)	2.3(1)	10.0(1)	86(5)mb
^{72}Ge	27.66(3)		0.9(2)	0.8(3)	8.5(1)	0.06(2)
^{73}Ge	7.73(1)		15.(1)	66.(20)	5.02(4)	0.28(8)
^{74}Ge	35.94(2)		(0.14 + 0.28)	(0.4+0.5)	7.6(1)	54(6)mb
^{76}Ge	7.44(2)		(0.09 + 0.06)	(1.3+0.6)	8.2(15)	16(1)mb
$_{33}$As			4.0(4)	61.(5)	6.58(1)	
^{75}As	100.		4.0(4)	61.(5)	6.58(1)	0.46(2)
$_{34}$Se			12.(1)	14.(3)	7.970(9)	
^{74}Se	0.89(4)		50.(2)	520(50)	0.8(3)	
^{75}Se		119.78 d	$3.3(10) \times 10^2$			
^{76}Se	9.37(29)		(22. + 63.)	(9.+31.)	12.2(1)	164(8)mb
^{77}Se	7.63(16)		42.(4),σ_α=0.97(3)μb	30.(5)	8.25(8)	
^{78}Se	23.77(28)		σ_m=0.38(2)	RI$_m$=4.3(4)	8.24(9)	0.08
^{80}Se	49.61(41)		(0.06+0.35)	(0.36+1.0)	7.48(3)	44.(3)mb
^{82}Se	8.73(22)		(39.+ 5.2) mb	39.(4)mb	6.34(8)	
$_{35}$Br			6.8(2)	92.(8)	6.79(2)	
^{76}Br		16.0 h	224.(42)			
^{79}Br	50.69(7)		(2.5+8.3)	(36.+96)	6.79(7)	0.74(3)
^{81}Br	49.31(7)		(2.4+0.24)	51.(5)	6.78(7)	0.24(1)

Elem. or Isot.	Natural Abundance (%)	Half-life	Thermal Neut. Cross Section (barns)	Resonance Integral (barns)	Coh. Scat. Length (fm)	σ(30KEV) Maxw. Avg. (barns)
$_{36}$Kr			24.(1)	39.(6)	7.81(2)	
^{78}Kr	0.35(1)		(0.17+6.)	20.(1)		0.34(4)
^{80}Kr	2.28(6)		(4.6+7.)	57.(6)		0.24(1)
^{82}Kr	11.58(14)		(14.+7.)	130.(13)		79(6)mb
^{83}Kr	11.50(6)		183.(30)	183.(20)	0.24(2)	
^{84}Kr	56.99(4)		σ_m=0.09;σ_m+g=0.11	2.4(3)		(17+17)mb
^{85}Kr		10.73 y	1.7(2)	1.8(10)		0.07(2)
^{86}Kr	17.30(22)		3.(2)mb	$\approx$1. mb	8.1(3)	3.5(3)mb
$_{37}$Rb			0.39(4)	6.(3)	7.08(2)	
^{84}Rb		32.9 d	σ_p=12.(2)			
^{85}Rb	72.17(2)		(0.06+0.38)	(0.7+7.)	7.0(1)	240(9)mb
^{86}Rb		18.65 d	<20.			
^{87}Rb	27.83(2)	4.88x10^{10} y	0.10(1)	2.3(4)	7.3(1)	18.0(5)mb
^{88}Rb		17.7 m	1.2(3)	0.5(1)		
$_{38}$Sr			1.2(1)	10.(1)	7.02(2)	
^{84}Sr	0.56(1)		(0.6+0.2)	(9.+1.)		
^{86}Sr	9.86(1)		σ_m=0.81(4)	RI$_m$=4.(1)	5.68(5)	(48+22)mb
^{87}Sr	7.00(1)		16.(3)	118.(30)	7.41(7)	97.(5)mb
^{88}Sr	82.58(1)		5.8(4)mb	0.07(3)	7.16(6)	6.1(2)mb
^{89}Sr		50.52 d	0.42(4)	0.2		
^{90}Sr		29.1 y	9.7(7)mb	4. mb		
$_{39}$Y			1.25(5)	1.0(1)	7.75(2)	
^{89}Y	100.		(0.001+1.25)	(0.006+1.0)	7.75(2)	19.(1)mb
^{90}Y		2.67 d	<6.5			
^{91}Y		58.5 d	1.4(3)	0.6(1)		
$_{40}$Zr			0.19(1);σ_α<0.1mb	0.95(9)	7.16(3)	
^{90}Zr	51.45(40)		$\approx$0.014	0.2(1)	6.4(1)	21.(2)mb
^{91}Zr	11.22(5)		1.2(3)	5.(2)	8.8(1)	60.(8)mb
^{92}Zr	17.15(8)		0.2(1)	0.6(2)	7.5(2)	33.(4)mb
^{93}Zr		1.5x10^6 y	<4.	16.(5)		0.10(1)
^{94}Zr	17.38(28)		0.049(6)	0.25(3)	8.3(2)	26.(1)mb
^{96}Zr	2.80(9)		0.020(3)	5.0(5)	5.5(1)	10.0(3)mb
$_{41}$Nb			1.11(1);σ_α<0.1mb	8.5(6)	7.14(3)	
^{93}Nb	100.		1.1;σ_m=0.86	(6.3+2.2)	7.14(3)	266.(5)mb
^{94}Nb		2.4x10^4 y	σ_m+g=15.(1) σ_m=0.6(1)	126.(13)		
^{95}Nb		34.97 d	<7.	<200.		
$_{42}$Mo			2.5(1),σ_α<0.1mb	26.(5)	6.72(2)	
^{92}Mo	14.84(35)		0.06,σ_m=0.2μb	$\approx$0.8	6.93(8)	
^{94}Mo	9.25(12)		0.02	$\approx$0.8	6.82(7)	
^{95}Mo	15.92(13)		13.4(3),σ_α=30.(4)μb	109.(5)	6.93(6)	0.29(1)

Elem. or Isot.	Natural Abundance (%)	Half-life	Thermal Neut. Cross Section (barns)	Resonance Integral (barns)	Coh. Scat. Length (fm)	σ(30KEV) Maxw. Avg. (barns)
[96]Mo	16.68(2)		0.5	17.(3)	6.22(6)	112.(5)mb
[97]Mo	9.55(8)		2.5(2), σ_α=0.4(2)μb	14.(3)	7.26(8)	0.34(1)
[98]Mo	24.13(31)		0.14(1)	7.2(7)	6.60(7)	99.(5)mb
[100]Mo	9.63(23)		0.19(1)	3.6(3)	6.75(7)	
$_{43}$Tc						
[98]Tc		4.2x10[6] y	σ_m = 0.9(2)			
[99]Tc		2.13x10[5] y	23.(2)	4.0(4)x10[2]	6.8(3)	0.78(4)
$_{44}$Ru			2.6 (1)	48.(5)	7.03(3)	
[96]Ru	5.52(20)		0.23(4)	7.(2)		
[98]Ru	1.86(9)		< 8.			
[99]Ru	12.74(26)		4.(1)	195.(20)		
[100]Ru	12.60(19)		5.8(6)	11.(2)		0.21(1)
[101]Ru	17.05(7)		5.(1), σ_α<0.15μb	1.1(3)x10[2]		1.00(4)
[102]Ru	31.57(31)		1.2(1)	4.3(5)		0.19(1)
[103]Ru		39.27 d	<20.	$\approx$30.		
[104]Ru	18.66(44)		0.49(2)	6.(2)		0.16(1)
[105]Ru		4.44 h	0.29(3)	0.13(1)		
[106]Ru		1.020 y	0.15(4)	2.0(6)		
$_{45}$Rh			145.(2)	1.2(1)x10[3]	5.88(4)	
[103]Rh	100.		(11.+ 134.)	(0.08+1.1)x10[3]	5.88(4)	0.81(1)
[104m]Rh		4.36 m	800.(100)			
[104]Rh		42.3 s	40.(30)			
[105]Rh		35.4 h	1.1(3)x10[4]	1.7(4)x10[4]		
$_{46}$Pd			7.(1)	82.(8)	5.91(6)	
[102]Pd	1.02(1)		3.2(10)	10.(2)		
[104]Pd	11.14(11)			16.(2)		
[105]Pd	22.33(13)		22.(2), σ_α=0.5(2)μb	60.(20)	5.5(3)	1.20(6)
[106]Pd	27.33(6)		(0.013+0.28)	(0.2+5.5)	6.4(4)	0.25(3)
[107]Pd		6.5x10[6] y	1.8(2)	108.(4)		1.34(6)
[108]Pd	26.46(15)		(0.19+8.5)	(2.+240.)	4.1(3)	0.20(2)
[110]Pd	11.72(16)		(0.033+0.7)	(0.7+8.)		0.15(2)
$_{47}$Ag			62.(1)	767.(60)	5.922(7)	
[107]Ag	51.839(8)		(0.37+35.)	(1.2+105.)	7.56(1)	0.80(2)
[109]Ag	48.161(8)		(4.2 + 87.)	(0.7+14.1)x10[2]	4.17(1)	0.78(2)
[110m]Ag		249.8 d	82.(11)	20.(4))		
[111]Ag		7.47 d	3.(2)	105.(20)		
$_{48}$Cd			2.52(5)x10[3]	73.(8)	4.87(5)	
[106]Cd	1.25(6)		0.20(3)	4.(1)		0.56(6)
[108]Cd	0.89(3)		1.	14.(3)	5.4(1)	0.41(7)
[109]Cd		462.0 d	$\approx$180., σ_α<0.05	6.7(12)x10[3]		
[110]Cd	12.49(18)		(0.06+11.)	(6.+34.)	5.9(1)	0.25(3)
[111]Cd	12.80(12)		3.5(20)	51.(6)	6.5(1)	1.06(13)

Elem. or Isot.	Natural Abundance (%)	Half-life	Thermal Neut. Cross Section (barns)	Resonance Integral (barns)	Coh. Scat. Length (fm)	σ(30KEV) Maxw. Avg. (barns)
^{112}Cd	24.13(21)		(0.012+2.2)	15.	6.4(1)	0.23(3)
^{113}Cd	12.22(12)	9x10^{15} y	2.06(4)x10^4,σ_α<1.μb	390.(40)	- 8.0(2)	0.73(8)
^{114}Cd	28.73(42)		(0.04+0.29)	16.(7)	7.5(1)	0.16(3)
^{116}Cd	7.49(18)		(26.+52.)mb	1.2	6.3(1)	0.09(1)
$_{49}$In			197.(4)	3.3(2)x10^3	4.07(2)	
^{113}In	4.29(5)		(3.1+5.0+3.9)	(220.+90)	5.39(6)	0.79(7)
^{115}In	95.71(5)	4.4x10^{14} y	(88.+73.+44.)	(1.5+1.2+0.7)x10^3	4.01(2)	0.74(7)
$_{50}$Sn			0.61(3)	8.(2)	6.225(2)	
^{112}Sn	0.97(1)		(0.15+0.40)	(8.+19.)		0.20(5)
^{113}Sn		115.1 d	≈ 9.	210.(50)		
^{114}Sn	0.66(1)		≈ 0.12	5.(1)	6.2(3)	134.(3) mb
^{115}Sn	0.34(1)		σ_α=0.06 mb	29.(6)		0.34(1)
^{116}Sn	14.54(9)		(0.006+0.14)	(0.5+11.)	5.93(5)	91.(2)mb
^{117}Sn	7.68(7)		1.1(1)	16.(5)	6.48(5)	319.(7)mb
^{118}Sn	24.22(9)		σ_m=4. mb	4.7(5)	6.07(5)	62.(1)mb
^{119}Sn	8.58(4)		2.(1)	2.9(5)	6.12(5)	0.25(5)
^{120}Sn	32.59(9)		(0.001+0.13)	1.2(3)	6.49(5)	(0.5+36)mb
^{122}Sn	4.63(3)		(0.15+0.001)	0.81(4)	5.74(5)	
^{124}Sn	5.79(5)		(0.13+0.005)	RI$_m$=8.0(2)	5.97(5)	
$_{51}$Sb			5.2(2)	169.(20)	5.57(3)	
^{121}Sb	57.21(5)		(0.4+5.6)	(13.+192.)	5.71(6)	0.53(2)
^{123}Sb	42.79(5)		(0.02+0.04+4.0)	(1.+119.)	5.38(7)	0.30(1)
^{124}Sb		60.20 d	17.(3)	≈8.		
$_{52}$Te			4.2(1)	47.(3)	5.80(3)	
^{120}Te	0.096(2)		(0.25+2.0)	≈1.	5.3(5)	
^{122}Te	2.603(4)		2.4(5)	80.(20)	3.8(2)	295.(3)mb
^{123}Te	0.908(2)	1.3x10^{13} y	370.(40),σ_α=0.05mb	4.5(3)x10^3	- 0.05	0.83(1)
^{124}Te	4.816(6)		(0.05+7.)	5.2(7)	8.0(1)	155.(2)mb
^{125}Te	7.139(6)		1.6(2)	21.(4)	5.02(8)	431.(4)mb
^{126}Te	18.952(11)		(0.12+0.8)	8.2(6)	5.56(7)	81.(1)mb
^{128}Te	31.687(11)		(0.016+0.20)	(0.08+1.5)	5.89(7)	
^{130}Te	33.799(10)	2.5x10^{21} y	(0.03+0.20)	(0.08+0.34)	6.02(7)	
$_{53}$I			6.15(10)	148.(7)	5.28(2)	
^{125}I		59.4 d	900.(100)	1.4(2)x10^4		
^{127}I	100.		6.15(10)	148.(7)	5.28(2)	0.64(3)
^{128}I		25.00 m	22.(4)	≈10.		
^{129}I		1.7x10^7 y	(20.7+10.3)	36.(4)		0.44(2)
^{130}I		12.36 h	18.(3)	≈8.		
^{131}I		8.040 d	≈ 0.7	8.(4)		
$_{54}$Xe			25.(1)	263.(50)	4.92(3)	
^{124}Xe	0.09(1)		(28.+137.)	(0.6+3.0)x10^3		
^{125}Xe		17.1 h	σ_α≤ <0.03			

Elem. or Isot.	Natural Abundance (%)	Half-life	Thermal Neut. Cross Section (barns)	Resonance Integral (barns)	Coh. Scat. Length (fm)	σ(30KEV) Maxw. Avg. (barns)
^{126}Xe	0.09(1)		(0.45+3.)	(8.+52.)		
^{127}Xe		36.4 d	$\sigma_\alpha \leq 0.01$			
^{128}Xe	1.92(3)		σ_m=0.48	RI_m=38.(10)		0.35(11)
^{129}Xe	26.44(24)		22.(5)	250.(50)		0.76(7)
^{130}Xe	4.08(2)		σ_m=0.45	RI_m=16.(4)		0.18(5)
^{131}Xe	21.18(3)		90.(10)	$9.(1) \times 10^2$		0.45(8)
^{132}Xe	26.89(6)		(0.05+0.4)	(0.9+3.7)		65.(5)mb
^{133}Xe		5.243 d	190.(90)			
^{134}Xe	10.44(10)		(0.003 + 0.26)	0.40(4)		20.(2)mb
^{135}Xe		9.10 h	$2.65(11) \times 10^6$	$7.6(5) \times 10^3$		
^{136}Xe	8.87(16)		0.26(2)	0.7(2)		0.9(1)mb
$_{55}$Cs			30.4(8)	422.(50)	5.42(2)	
^{132}Cs		6.48 d	$\sigma_\alpha < 0.15$			
^{133}Cs	100.		(2.7+27.7)	(32.+390.)	5.42(2)	0.51(2)
^{134}Cs		2.065 y	140.(10)	54.(9)		
^{135}Cs		2.3×10^6 y	8.9(5)	90.(20)		
^{137}Cs		30.2 y	0.25(1)	0.36(7)		
$_{56}$Ba			1.3(2)	10.(2)	5.07(3)	
^{130}Ba	0.106(1)		(1.+8.)	(25.+200.)	- 3.6(6)	
^{132}Ba	0.101(1)		(0.84+9.7)	(4.7+24.)	7.8(3)	
^{133}Ba		10.53 y	4.(1)	85.(30)		
^{134}Ba	2.417(18)		(0.1+1.3)	(5.6+18.)	5.7(1)	0.18(6)
^{135}Ba	6.592(12)		(0.014+5.8)	(0.47+131.)	4.7(1)	0.30(6)
^{136}Ba	7.854(24)		(0.010+0.44)	(0.1+1.5)	4.91(8)	62.(2)mb
^{137}Ba	11.232(24)		5.(1)	4.(1)	6.8(1)	0.05(1)
^{138}Ba	71.698(42)		0.41(2)	0.4(1)	4.84(8)	4.1(2)mb
^{139}Ba		1.396 h	5.(1)	2.2(5)		
^{140}Ba		12.75 d	1.6(3)	14.(1)		
$_{57}$La			9.2(2)	12.(1)	8.24(4)	
^{138}La	0.090(1)	1.06×10^{11} y	57.(6)	$4.1(9) \times 10^2$		
^{139}La	99.910(1)		9.2(2)	12.(1)	8.24(4)	38.(3)mb
^{140}La		1.678 d	2.7(3)	69.(4)		
$_{58}$Ce			0.64(4)	0.71(6)	4.84(2)	
^{136}Ce	0.185(2)		(1.0+6.5)	58.(12)	5.80(9)	(28+272)mb
^{138}Ce	0.251(2)		(0.018+1.)	(1.5+5.2)	6.70(9)	179.(5)mb
^{140}Ce	88.450(18)		0.58(4)	0.50(5)	4.84(9)	11.0(4)mb
^{141}Ce		32.50 d	29.(3)	13.(2)		
^{142}Ce	11.114(17)		0.97(3)	1.3(3)	4.75(9)	28.(1)mb
^{143}Ce		1.38 d	6.1(7)	2.7(3)		
^{144}Ce		284.6 d	1.0(1)	2.6(3)		
$_{59}$Pr			11.5(4)	14.(3)	4.58(5)	
^{141}Pr	100.		(4.+7.5)	14.(3)	4.58(5)	0.12(2)
^{142}Pr		19.12 h	20.(3)	9.(1)		
^{143}Pr		13.57 d	90.(10)	190.(25)		

Elem. or Isot.	Natural Abundance (%)	Half-life	Thermal Neut. Cross Section (barns)	Resonance Integral (barns)	Coh. Scat. Length (fm)	σ(30KEV) Maxw. Avg. (barns)
$_{60}$Nd			51.(2)	49.(5)	7.69(5)	
^{142}Nd	27.2(5)		19.(1)	34.(11)	7.7(3)	46.(4)mb
^{143}Nd	12.2(2)		330.(10),σ_α=17.mb	128.(30)		0.24(1)
^{144}Nd	23.8(3)	2.1x10^{15} y	3.6(3)	3.9(5)	2.8(3)	0.11(1)
^{145}Nd	8.30(1)		47.(6),σ_α=12.μb	260.(40)		0.49(10)
^{146}Nd	17.2(3)		1.5(2)	3.0(4)	8.7(2)	0.16(4)
^{147}Nd		10.98 d	440.(150)	200.		
^{148}Nd	5.7(1)		2.4(1)	13.(2)	5.7(3)	0.19(4)
^{150}Nd	5.6(2)		1.0(1)	14.(2)	5.3(2)	
$_{61}$Pm						
^{146}Pm		5.53 y	8.4(1.7)x10^3			
^{147}Pm		2.6234 y	(84.+96.)	(1000.+1280.)	12.6(4)	
^{148m}Pm		41.3 d	10600.(800)			
^{148}Pm		5.37 d	$\approx 10^3$	2.6(2.4)x10^3		
^{149}Pm		2.212 d	1400.(200)			
^{151}Pm		1.183 d	$\approx$ 150.			
$_{62}$Sm			5.6(1)x10^3	1.4(2)x10^3		
^{144}Sm	3.08(7)		1.6(1)	2.4(3)		
^{145}Sm		340. d	280.(20)	600.(90)		
^{147}Sm	14.99(17)	1.06x10^{11} y	56.(4),σ_α=0.6mb	710.(50)	14.(3)	0.97(1)
^{148}Sm	11.24(10)	7x10^{15} y	2.4(6)	27.(14)		241.(2)mb
^{149}Sm	13.82(7)	10^{16} y	40100.(600),σ_α=31.mb	3100.(500)		1.82(2)
^{150}Sm	7.38(1)		102.(5)	290.(30)	14.(3)	422.(4)mb
^{151}Sm		90. y	15200.(300)	3520.(60)		
^{152}Sm	26.74(14)		206.(15)	3000.(300)	- 5.0(6)	473.(4)mb
^{153}Sm		1.929 d	420.(180)			
^{154}Sm	22.75(27)		7.5(3)	32.(6)	9.(1)	
$_{63}$Eu			4570.(100)	3.8(5)x10^3	5.3(3)	
^{151}Eu	47.81(6)		(4.+3150.+6000.) σ_α=8.7(3)μb	(2000.+4000.)		4.4(2)
^{152m1}Eu		9.30 h	68000.(15000)	< 100000.		
^{152}Eu		13.5 y	11000.(2000)	1600.(200)		
^{153}Eu	52.19(6)		300.(20),σ_α<1.μb	1800.(400)	8.2(1)	3.2(3)
^{154}Eu		8.59 y	1500.(300)	1600.(200)		4.4(7)
^{155}Eu		4.76 y	3900.(200)	16000.(2000)		
$_{64}$Gd			48.8(6)x10^3	400.(10)	9.5(2)	
^{152}Gd	0.20(1)		700.(200),σ_α<7.mb	700.(200)		1.05(7)
^{153}Gd		241.6 d	20000.(10000),σ_α=0.03			
^{154}Gd	2.18(3)		(0.06+60.)	230.(50)		0.88(3)
^{155}Gd	14.80(12)		61.(1)x10^3,σ_α=.08mb	1540.(100)		2.72(9)
^{156}Gd	20.47(9)		$\approx$2.0	104.(15)	6.3(4)	0.64(6)
^{157}Gd	15.65(2)		2.54(3)x10^5,σ_α<0.05	800.(100)		1.36(4)
^{158}Gd	24.84(7)		2.3(3)	73.(7)	9.(2)	0.22(2)
^{160}Gd	21.86(19)		1.5(7)	6.(1)	9.15(5)	
^{161}Gd		3.66 m	2.0(6)x10^4			

Elem. or Isot.	Natural Abundance (%)	Half-life	Thermal Neut. Cross Section (barns)	Resonance Integral (barns)	Coh. Scat. Length (fm)	σ(30KEV) Maxw. Avg. (barns)
$_{65}$Tb			23.2(5)	420.(50)	7.38(3)	
^{159}Tb	100.		23.2(5)	420.(50)	7.38(3)	1.75(10)
^{160}Tb		72.3 d	570.(110)			
$_{66}$Dy			9.5(2)x10^2	1.5(2)x10^3	16.9(3)	
^{156}Dy	0.06(1)		33.(3),σ_α<9mb	1000.(100)		
^{158}Dy	0.10(1)		43.(6),σ_α<6mb	120.(10)	6.1(5)	
^{159}Dy		144. d	8.(2)x10^3			
^{160}Dy	2.34(8)		60.(10),σ_α<0.3mb	1100.(200)	6.7(4)	0.77(4)
^{161}Dy	18.91(24)		600.(50),σ_α<1μb	1100.(100)	10.3(4)	2.01(6)
^{162}Dy	25.51(26)		170.(20)	2755.(300)	1.4(5)	0.47(5)
^{163}Dy	24.90(16)		120.(10),σ_α<20μb	1600.(400)	5.0(4)	1.14(4)
^{164}Dy	28.18(36)		(1.7+1.0)x10^3	(420.+100.)	49.4(2)	0.27(3)
^{165m}Dy		1.26 m	2.0(6)x10^3			
^{165}Dy		2.33 h	3.5(3)x10^3	2.2(3)x10^4		
$_{67}$Ho			61.(2)	670.(40)	8.01(8)	
^{165}Ho	100.		(3.1+58.),σ_α<20μb	(?+670.)	8.01(8)	1.32(7)
^{166}Ho		1.2x10^3 y	9140.(650)	1140.(90)		
$_{68}$Er			169.(20)	730.(10)	7.79(2)	
^{162}Er	0.14(1)		19.(3),σ_α<11mb	480.(50)	8.8(2)	
^{164}Er	1.61(3)		13.(3),σ_α<1.2mb	105.(10)	8.2(2)	0.71(6)
^{166}Er	33.61(35)		(15.+5.),σ_α<70μb	96.(12)	10.6(2)	0.61(6)
^{167}Er	22.93(17)		700.(200),σ_α=3μb	2970.(70)	3.0(3)	1.5(2)
^{168}Er	26.78(26)		2.0(6),σ_α=0.09mb	37.(5)	7.4(4)	0.37(4)
^{170}Er	14.93(27)		6.(1)	26.(4)	9.6(5)	0.22(3)
^{171}Er		7.52 h	370.(40)	170.(20)		
$_{69}$Tm			106.(5)	1.5(2)x10^3	7.07(3)	
^{169}Tm	100		(8.+98.)	1.5(2)x10^3	7.07(3)	1.13(6)
^{170}Tm		128.6 d	100.(20)	460.(50)		
^{171}Tm		1.92 y	$\approx$ 160.	118.(6)		
$_{70}$Yb				52.(10)	1.7(2)x10^2	12.43(3)
^{168}Yb	0.13(1)		2.4(2)x103,σ_α<0.1mb	2.0(5)x10^4	-4.07(2)	
^{169}Yb		32.03 d	3.6(3)x10^3	5200.(500)		
^{170}Yb	3.04(15)		12.(2),σ_α<10μb	320.(30)	6.8(1)	0.74(3)
^{171}Yb	14.3(6)		53.(5),σ_α<1.5μb	315.(30)	9.7(1)	1.41(5)
^{172}Yb	21.8(7)		$\approx$ 1.3,σ_α<1μb	25.(3)	9.4(1)	0.41(3)
^{173}Yb	16.13(3)		16.(2),σ_α<1μb	380.(30)	9.56(7)	0.87(7)
^{174}Yb	31.8(9)		(46.+74.),σ_α<0.02mb	(13.+47.)	19.3(1)	0.18(2)
^{176}Yb	12.8(4)		3.1(2),σ_α<1μb	8.(2)	8.7(1)	
$_{71}$Lu			78.(7)	8.3(7)x10^2	7.21(3)	
^{175}Lu	97.41(2)		(16.+8.)	(550.+270.)	7.24(3)	1.18(4)
^{176}Lu	2.59(2)	3.8x10^{10} y	(2.+2100.)	(3.+1100.)	6.1(2)	1.54(6)

Elem. or Isot.	Natural Abundance (%)	Half-life	Thermal Neut. Cross Section (barns)	Resonance Integral (barns)	Coh. Scat. Length (fm)	σ(30KEV) Maxw. Avg. (barns)
^{177m}Lu		160.7 d	3.2(3)	1.4(2)		
^{177}Lu		6.75 d	1000.(300)			
$_{72}$Hf			106.(3)	19.7(5)x10^2	7.8(1)	
^{174}Hf	0.162(3)	2.0x10^{15} y	600.(50)	400.(50)	11.(1)	
^{176}Hf	5.206(5)		23.(4)	700.(100)	6.6(2)	0.46(2)
^{177}Hf	18.606(13)		(1.+375.),σ_α<20μb	7170.(200)		1.37(6)
^{178}Hf	27.297(4)		(53.+32.)	(340.+1500.)	5.9(2)	0.31(1)
^{179}Hf	13.629(6)		(0.43+46.)	(6.8+620.)	7.5(2)	0.99(3)
^{180}Hf	35.100(7)		13.(1),σ_α<13μb	32.(2)	13.2(3)	175.(5)mb
^{181}Hf		42.4 d	30.(25)			
$_{73}$Ta			20.(1)	650(20.)	6.91(7)	
^{180m}Ta	0.012(2)	> 1.2x10^{15} y	≈ 560.	1350.(100)		
^{181}Ta	99.998(2)		(0.012 + 20),σ_α<1μb	(0.4+650.)	6.91(7)	0.77(2)
^{182}Ta		114.43 d	8200.(600)	900.(90)		
$_{74}$W			18.(1)	3.6(3)x10^2	4.86(2)	
^{180}W	0.12(1)		≈ 4.	210.(30)		
^{182}W	26.50(16)		20.(1)	600.(90)	6.97(4)	274.(8)mb
^{183}W	14.31(4)		10.5(3)	340.(50)	6.53(4)	0.52(2)
^{184}W	30.64(2)		(0.002 + 2.0)	15.(2)	7.48(6)	0.23(1)
^{185}W		74.8 d	≈ 3.3	300.(50)		
^{186}W	28.43(19)		37.(2)	510.(50)	- 0.72(4)	176.(5)mb
^{187}W		23.9 h	70.(10)	2760.(550)		
$_{75}$Re			90.(4)	8.4(2)x10^2	9.2(3)	
^{185}Re	37.40(2)		(0.33+110.)	1700.(50)	9.0(3)	1.54(6)
^{187}Re	62.60(2)	4.4x10^{10} y	(2.+72.)	(9.+310.)	9.3(3)	1.16(6)
$_{76}$Os			17.(1)	1.5(1)x10^2	10.7(2)	
^{184}Os	0.020(3)		3.3(3)x10^3,σ_α<10mb	1.4(1)x10^3		
^{186}Os	1.58(2)	2.x10^{15} y	≈ 80.,σ_α<0.1mb	280.(40)	12(2)	0.42(2)
^{187}Os	1.6(4)		2.(1)x10^2,σ_α<0.1mb	500.(70)		0.87(3)
^{188}Os	13.3(1)		≈ 5.,σ_α<30μb	150.(20)	7.6(3)	0.40(2)
^{189}Os	16.1(1)		(0.00026+40),σ_α<10μb	(0.013+670.)	10.7(3)	1.17(5)
^{190}Os	26.4(2)		(9.+4.),σ_α<20μb	(22.+8.)	11.0(3)	0.30(5)
^{191}Os		15.4 d	3.8(6)x10^2			
^{192}Os	41.0(3)		3.(1),σ_α<10μb	6.(1)	11.5(4)	0.31(5)
^{193}Os		30.5 h	40.(10)			
$_{77}$Ir			4.2(1)x10^2	2.8(4)x10^3	10.6(3)	
^{191}Ir	37.3(2)		(660.+260.)	(1000.+4200.)		1.35(7)
^{192}Ir		73.83 d	1450.(250)	4800.(700)		
^{193}Ir	62.7(2)		(6.+110.)	1400.(200)		0.86(4)
^{194}Ir		19.3 h	1.5(3)x10^3			
$_{78}$Pt			10.(1)	1.3(1)x10^2	9.60(1)	

Elem. or Isot.	Natural Abundance (%)	Half-life	Thermal Neut. Cross Section (barns)	Resonance Integral (barns)	Coh. Scat. Length (fm)	σ(30KEV) Maxw. Avg. (barns)
^{190}Pt	0.014(1)	6.5×10^{11} y	$1.5(1) \times 10^2, \sigma_\alpha < 8$mb	70.(10)	9(1)	
^{192}Pt	0.782(7)		$(2.0+6.), \sigma_\alpha < 0.2$mb	115.(20)	9.9(5)	0.20(6)
^{194}Pt	32.97(10)		$(0.1+1.1), \sigma_\alpha < 5\mu$b	(4. + ?)	10.55(8)	0.37(8)
^{195}Pt	33.832(10)		$28.(1), \sigma_\alpha < 5\mu$b	365.(50)	8.8(1)	0.9(2)
^{196}Pt	25.242(41)		(0.045+0.55)	7.(2)	9.89(8)	0.20(2)
^{198}Pt	7.16(6)		(0.027+3.6)	(0.5+56.)	7.8(1)	
^{199}Pt		30.8 m	$\approx$ 16.			
$_{79}$Au			98.7(1)	$1.55(3) \times 10^3$	7.63(6)	
^{197}Au	100.		98.7(1)	$1.55(3) \times 10^3$	7.63(6)	582.(9)mb
^{198}Au		2.694 d	$26.5(15) \times 10^3$	$\approx 4. \times 10^4$		
^{199}Au		3.14 d	$\approx$ 30.			
$_{80}$Hg			$3.7(1) \times 10^2$	87.(5)	12.69(2)	
^{196}Hg	0.15(1)		(105.+3000.)	(53.+410.)	30.(1)	
^{198}Hg	9.97(20)		(0.017+2.)	(1.7+70.)		0.17(2)
^{199}Hg	16.87(22)		$2.1(2) \times 10^3$	435(20)	16.9(4)	0.37(2)
^{200}Hg	23.10(19)		< 60.	2.1(5)		0.12(1)
^{201}Hg	13.18(9)		< 60.	30.(3)		0.26(1)
^{202}Hg	29.86(26)		4.9(5)	4.5(2)		74.(6)mb
^{204}Hg	6.87(15)		0.4(1)	0.8(2)		42.(4)mb
$_{81}$Tl			3.3(1)	12.5(8)	8.776(5)	
^{203}Tl	29.52(1)		$11.(1), \sigma_\alpha < 0.3$mb	41.(2)	7.0(2)	124.(8)mb
^{204}Tl		3.78 y	22.(2)	90.(20)		
^{205}Tl	70.48(1)		0.11(2)	0.6(2)	9.52(7)	54.(4)mb
$_{82}$Pb			0.172(2)	0.14(4)	9.402(2)	
^{204}Pb	1.4(1)		0.68(7)	2.0(2)	10.6(20)	90.(6)mb
^{205}Pb		1.51×10^7 y	$\approx$ 5.			
^{206}Pb	24.1(1)		0.030(1)	0.10(1)	9.23(5)	14.(1)mb
^{207}Pb	22.1(1)		0.70(1)	0.38(1)	9.28(4)	8.4(4)mb
^{208}Pb	52.4(1)		$0.49(3)$mb, $\sigma_\alpha < 8\mu$b	2.0(2) mb	9.50(6)	0.36(2)mb
^{210}Pb		22.6 y	< 0.5			
$_{83}$Bi			0.034(1)	0.19(2)	8.532(2)	
^{209}Bi	100.		$(11+23)$mb, $\sigma_\alpha < 0.3\mu$b	0.19(2)	8.532(2)	3.1(3)mb
^{210m}Bi		3.0×10^6 y	54(4)mb	0.20(3)		
$_{84}$Po						
^{210}Po		138.38 d	$\sigma_m < 0.5$mb, $\sigma_\alpha < $ 2.mb $\sigma_g < 30.$mb, $\sigma_f < $ 0.1			
$_{85}$At						
$_{86}$Rn						
^{220}Rn		55.6 s	<0.2			

Elem. or Isot.	Natural Abundance (%)	Half-life	Thermal Neut. Cross Section (barns)	Resonance Integral (barns)	Coh. Scat. Length (fm)	σ(30KEV) Maxw. Avg. (barns)
^{222}Rn		3.8235 d	0.74(5)			
$_{88}$Ra						
^{223}Ra		11.435d	$1.3(2) \times 10^2, \sigma_f < 0.7$			
^{224}Ra		3.66 d	12.0(5)			
^{226}Ra		1599. y	$\approx 13., \sigma_f < 7. \mu b$	280.(50)	10.(1)	
^{228}Ra		5.76 y	$36.(5), \sigma_f < 2.$			
$_{89}$Ac						
^{227}Ac		21.77 y	$8.8(7) \times 10^2, \sigma_f < .35$ mb	$1.5(4) \times 10^3$		
$_{90}$Th			7.4	85.(3)	10.31(3)	
^{227}Th		18.72 d	$\sigma_f = 2.0(2) \times 10^2$			
^{228}Th		1.913 y	$1.2(2) \times 10^2, \sigma_f < 0.3$	1014.(400)		
^{229}Th		7.9×10^3 y	$\approx 60.$	$1.0(2) \times 10^3$		
			$\sigma_f = 30(3)$	$RI_f = 466.(75)$		
^{230}Th		7.54×10^4 y	23.4(5)	$1.0(1) \times 10^3$		
			$\sigma_f < 0.5$mb			
^{232}Th	100.	1.4×10^{10}y	7.37(4)	85.(3)	10.31(3)	
			$\sigma_f = 3.(1) \mu b$			
			$\sigma_\alpha < 1. \mu b$			
^{233}Th		22.3 m	$1.5(1) \times 10^3$	$4.(1) \times 10^2$		
			$\sigma_f = 15.(2)$			
^{234}Th		24.10 d	1.8(5)			
			$\sigma_f < 0.01$			
$_{91}$Pa						
^{230}Pa		17.4 d	$1.5(3) \times 10^3$			
^{231}Pa		3.25×10^4 y	$2.0(1) \times 10^2$	750.(80)	9.1(3)	
			$\sigma_f = 20.(1)$ mb	$RI_f = 0.05(1)$		
^{232}Pa		1.31 d	$4.6(10) \times 10^2$	300.(70)		
			$\sigma_f = 7.(1) \times 10^2$			
^{233}Pa		27.0 d	39.(2)	$(460. + 440)$		
			$\sigma_m = 20.(4)$			
			$\sigma_g = 19.(3)$			
			$\sigma_f < 0.1$			
$_{92}$U			$3.4(3); \sigma_f = 4.2(1)$	$280.(20), RI_f = 2.0$	8.417(5)	
^{230}U		20.8 d	$\sigma_f \approx 25.$			
^{231}U		4.2 d	$\sigma_f \approx 250.$			
^{232}U		68.9 y	73.(2)	280.(15)		
			$\sigma_f = 74.(8)$	$RI_f = 350.(30)$		
^{233}U		1.59×10^5 y	47.(2)	137.(6)	10.1(2)	
			$\sigma_f = 5.3(1) \times 10^2$	$RI_f = 760.(17)$		
			$\sigma_\alpha < 0.2$ mb			
^{234}U	0.0055(5)	2.45×10^5 y	96.(2)	660.(70)	12.(4)	
			$\sigma_f < 5.$ mb	$RI_f = 6.5$		

Elem. or Isot.	Natural Abundance (%)	Half-life	Thermal Neut. Cross Section (barns)	Resonance Integral (barns)	Coh. Scat. Length (fm)	σ(30KEV) Maxw. Avg. (barns)
^{235}U	0.720(1)	7.04×10^8 y	95.(5) σ_f=586.(2) σ_α< 0.1 mb	144.(6) RI_f=275(5)	10.47(4)	
^{236}U		2.34×10^7 y	5.1(3) σ_f=0.04	360.(15) RI_f=4.38(50)		
^{237}U		6.75 d	$\approx 10^2$ σ_f< 0.35	1200.(200)		
^{238}U	99.2745(15)	4.46×10^9 y	2.7(1) $\sigma_f \approx 3.\mu b$ σ_α=1.4(5) μb	277.(3) 1.54(15) mb	8.402(5)	
^{239}U		23.5 m	22.(2) σ_f=15.(3)			
$_{93}$Np						
^{234}Np		4.4 d	σ_f=9.(3)$\times 10^2$			
^{235}Np		1.085 y	1.6(1)$\times 10^2$			
^{236m}Np		22.5 h	σ_f=2.7(2)$\times 10^3$	700.(400)		
^{236}Np		1.55×10^5 y	σ_f=2.6(5)$\times 10^3$	1030.(100)		
^{237}Np		2.14×10^6 y	160.(10) σ_f=20.(1)mb	8.(2)$\times 10^2$ RI_f=0.32	10.6(1)	
^{238}Np		2.117 d	σ_f=2.1(1)$\times 10^3$	9.(1)$\times 10^2$		
^{239}Np		2.355 d	(32.+19.) σ_f< 1.			
$_{94}$Pu						
^{236}Pu		2.87 y	σ_f=1.6(3)$\times 10^2$	1000.(60)		
^{237}Pu		45.7 d	σ_f=2.3(3)$\times 10^3$			
^{238}Pu		87.74 y	5.1(2)$\times 10^2$ σ_f=17.(1)	1.6(2)$\times 10^2$ RI_f=33.(5)	14.1(5)	
^{239}Pu		2.411 x 10^4y	2.7(1)$\times 10^2$ σ_f=752.(3) $\sigma_\alpha \leq 0.3$ mb	2.0(2)$\times 10^2$ 3.0(1)$\times 10^2$	7.7(1)	
^{240}Pu		6537. y	2.9(1)$\times 10^2$ $\sigma_f \approx 44$.mb	8.4(3)$\times 10^3$ RI_f=2.4	3.5(1)	
^{241}Pu		14.4 y	3.7(1)$\times 10^2$,σ_α<0.2mb σ_f=1.01(1)$\times 10^3$	1.6(1)$\times 10^2$ 5.7(4)$\times 10^2$		
^{242}Pu		3.76 x 10^5 y	19.(1) σ_f< 0.2	1.1(1)$\times 10^3$ RI_f=0.23	8.1(1)	
^{243}Pu		4.956 h	<100. σ_f=2.0(2)$\times 10^2$			
^{244}Pu		8.2×10^7 y	1.7(1)	41.(3)		
^{245}Pu		10.5 h	1.5(3)$\times 10^2$	220(40)		
$_{95}$Am						
^{241}Am		432.2 y	(0.5+5.7)$\times 10^2$ σ_f=3.1(2)	(2.0+12.3)$\times 10^2$ 14.(1)		
^{242m}Am		141. y	1.7(4)$\times 10^3$ σ_f=7.0(3)$\times 10^3$	$\approx 200.$ RI_f=1.8(1)$\times 10^3$		
^{242}Am		16.02 h	σ_f=2.1(2)$\times 10^3$	RI_f= < 300.		
^{243}Am		7.37×10^3y	(75.+5.) σ_f=74.(4) mb	(17.1+1.0)$\times 10^2$ RI_f=0.056	8.3(2)	

Elem. or Isot.	Natural Abundance (%)	Half-life	Thermal Neut. Cross Section (barns)	Resonance Integral (barns)	Coh. Scat. Length (fm)	σ(30KEV) Maxw. Avg. (barns)
^{244m}Am		$\approx$ 26. m	σ_f=1.6(3)x10^3			
^{244}Am		10.1 h	σ_f=2.2(3)x10^3			
$_{96}$Cm						
^{242}Cm		162.8 d	$\approx$ 20. $\sigma_f \approx$ 5.	120.(50)		
^{243}Cm		28.5 y	1.3(1)x10^2 σ_f=6.2(2)x10^2	214.(20) RI$_f$=1.6(1)x10^3		
^{244}Cm		18.11 y	15.(1) σ_f=1.1(2)	640.(50) RI$_f$=10.8(8)	9.5(3)	
^{245}Cm		8.5x10^3 y	3.5(2)x10^2 σ_f=2.1(1)x10^3	110(10) RI$_f$=8.(1)x10^2		
^{246}Cm		4.78x10^{3}y	1.2(2) σ_f=0.16(7)	120.(10) 13.(2)	9.3(2)	
^{247}Cm		1.56x10^{7}y	60.(30) σ_f=82.(5)	5.(1)x10^2 7.3(7)x10^2		
^{248}Cm		3.4x10^5 y	2.6(3) σ_f=0.36(7)	270.(30) 13.(2)	7.7(2)	
^{249}Cm		64.15 m	$\approx$1.6			
^{250}Cm		$\approx$ 9.7x10^3 y	$\approx$80.			
$_{97}$Bk						
^{249}Bk		320. d	7.(1)x10^2 $\sigma_f \approx$0.1	9.(1)x10^2		
^{250}Bk		3.217 h	σ_f=1.0(2)x10^3			
$_{98}$Cf						
^{249}Cf		351. y	5.0(3)x10^2 σ_f=1.7(1)x10^3	7.7(4)x10^2 RI$_f$=2.1(3)x10^3		
^{250}Cf		13.1 y	2.0(2)x10^3 σ_f=110.(90)	12.(2)x10^3 RI$_f$=160.(40)		
^{251}Cf		9.0x10^2 y	2.9(2)x10^3 σ_f=4.5(5)x10^3	1.6(1)x10^3 RI$_f$=5.5(3)x10^3		
^{252}Cf		2.65 y	20.(2) σ_f=32.(4)	43.(3) RI$_f$=1.1(3)x10^2		
^{253}Cf		17.8 d	18.(2) σ_f=1.3(2)x10^3	8.(1)		
^{254}Cf		60.5 d	4.5(10)	2.		
$_{99}$Es						
^{253}Es		20.47 d	(180.+5.8)	(37.5+1.1)x10^2		
^{254m}Es		1.64 d	σ_f=1.8(1)x10^3			
^{254}Es		276. d	28.(3) σ_f=1.8(2)x10^3	18.(2) RI$_f$=1.2(3)x10^3		
^{255}Es		40. d	$\approx$ 55.			
$_{100}$Fm						
^{255}Fm		20.1 h	26.(3) σ_f=3.3(2)x10^3	14.(2)		
^{257}Fm		100.5 d	σ_f=3.0(2)x10^3			

COSMIC RADIATION

A.G. Gregory and R.W. Clay

The Nature of Cosmic Rays

Primary cosmic radiation, in the form of high energy nuclear particles, electrons and photons from outside the solar system and from the Sun, continually bombards our atmosphere. Secondary radiation, resulting from the interaction of the primary cosmic rays with atmospheric gas, is present at sea-level and throughout the atmosphere.

The secondary radiation is collimated by absorption and scattering in the atmosphere and consists of a number of components associated with different particle species. High energy primary particles can produce large numbers of secondary particles forming an extensive air shower. Thus, a number of particles may then be detected simultaneously at sea-level.

Primary particle energies accessible in the vicinity of the earth range from ~10^8 eV to ~10^{20} eV. At the lower energies, the limit is determined by the inability of charged particles to traverse the heliosphere to us through the outward-moving solar wind. The upper energy limit is set by the practicality of building detectors to record particles with the extremely low fluxes found at those energies (J.G. Wilson, 1976; O.C. Allkofer, 1975a).

Primary Cosmic Rays

Primary Particle Energy Spectrum

Figure 1 shows the spectrum of primary particle energies. This includes all particle species. In differential form it is roughly a power law of intensity versus energy with an index of ~ –3. There appears to be a knee (a steepening) at a little above 10^{15} eV and an ankle (a flattening) above ~10^{18} eV. Figure 2 emphasizes the features in the spectrum at the highest energies through multiplying the flux with a strongly rising power law of energy. This figure should be used with caution as errors for the two axes are not now independent.

Data on the high energy cosmic ray spectrum are uncertain largely because of limited event statistics due to the very low flux which might best be measured in particles per square kilometer per century. The highest energy event recorded to 1995 had an energy of 3×10^{20} eV (D.J. Bird et al., 1993).

It is expected that the highest energy cosmic rays will interact with the 2.7 K cosmic microwave background through photoproduction or photodisintegration. These interactions will appreciably reduce the observed flux of cosmic rays with energies above 5×10^{19} eV if they travel further than ~150 million light years. This process is known as the Greisen-Zatsepin-Kuz'min (GZK) cut off (P. Sokolsky, 1989).

At energies below ~10^{13} eV, solar system magnetic fields and plasma can modulate the primary component and Figure 3 shows the extent of this modulation between solar maximum and minimum (E. Juliusson, 1975; J. Linsley, 1981).

Primary Particle Energy Density

If the above spectrum is corrected for solar effects, the energy density above a particle energy of 10^9 eV outside the solar system is found to be ~5×10^5 eV m^{-3}. As the threshold energy is increased, the energy density decreases rapidly, being 2×10^4 eV m^{-3} above 10^{12} eV and 10^2 eV m^{-3} above 10^{15} eV. The energy density at lower energies outside the heliosphere is unknown but may be substantially greater if the particle rest mass energy is included together with the kinetic energy (A. W. Wolfendale, 1979).

Primary Particle Isotropy

This is measured as an anisotropy $(I_{max} - I_{min})/(I_{max} + I_{min}) \times 100\%$, where I, the intensity (m^{-2}s^{-1}sr^{-1}), is usually measured with an angular resolution of a few degrees.

The measured anisotropy is small and energy dependent. It is roughly constant in amplitude at between 0.05 and 0.1% (with a phase of 0 to 6 hours in right ascension) for energies between 10^{11} eV and 10^{14} eV and appears to increase at higher energies roughly as $0.4 \times (\text{Energy(eV)}/10^{16})^{0.5}\%$ up to ~10^{18} eV. The latter rise may well be an artifact of the progressively more limited statistics as the flux drops rapidly with energy. It appears possible that a real anisotropy has been observed at the highest energies (above a few times 10^{19} eV) with a directional preference for the supergalactic plane (this plane reflects the directions of galaxies within about 100 million light years) (A.W. Wolfendale, 1979; R.W. Clay, 1987; T. Stanev et al., 1995).

Primary Particle Composition

The composition of low energy cosmic rays is close to universal nuclear abundances except where propagation effects are present. For example, Li, Be, and B which are spallation products, are over-abundant by about six orders of magnitude.

Composition at 10^{11} eV per nucleus

Charge	1	2	(3–5)	(6–8)	(10–14)	(16–24)	(26–28)	≥30
% Composition	50	25	1	12	7	4	4	0.1
(10% uncertainty)								

Measurements at higher energies indicate that there is an increase in the relative abundances of nuclei with charge greater than 6 at energies above 50 TeV/nucleus (K. Asakimori et al., 1993) (1 TeV = 10^{12} eV).

Cosmic ray composition at low energies is often quoted at a fixed energy per nucleon. When presented in this way, protons constitute roughly 90% of the flux, helium nuclei about 10% and the remainder sum to a total of about 1%.

Certain radioactive isotopic ratios show lifetime effects. The ratio of Be^{10}/B^9 abundances is used to measure an "age" of cosmic rays since Be^{10} is unstable with a half life of about 1.6×10^6 years. A ratio of 0.6 is expected in the absence of Be^{10} decay and a ratio of about 0.2 is found experimentally (E. Juliusson, 1975; P. Meyer, 1981).

FIGURE 1. The energy spectrum of cosmic ray particles. This spectrum is of a differential form and can be converted to an integral spectrum by integration over all energies above a required threshold (E). Insofar as the spectrum approximates a power law of index −3, a simple conversion to the integral at an energy $E/1.8$ is obtained by multiplying the differential flux by the energy and dividing by 0.62.

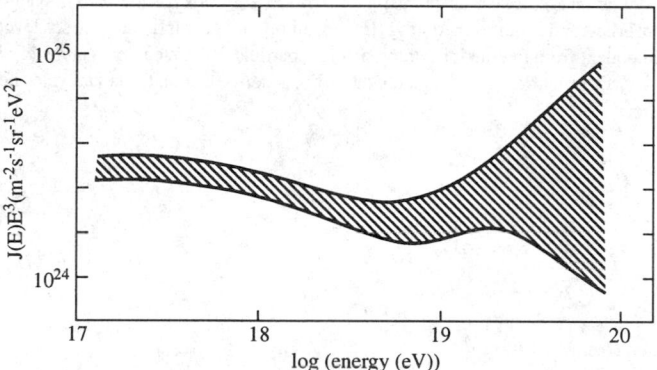

FIGURE 2. Energy spectrum at the highest energies. This spectrum (after Yoshida et al., 1995) has the differential spectrum multiplied by energy cubed. It is from a compilation of a number of measurements and indicates the good general agreement at the lower energies and a spread due to inadequate statistics at the highest energies.

FIGURE 3. Energy spectrum of particles at lower energies. (a) Solar minimum proton energy spectrum. (b) Solar maximum proton energy spectrum. (c) Gamma-ray energy spectrum. (d) Local interstellar electron spectrum.

At higher energies, composition determinations are indirect and are rather contradictory and controversial. Experiments aim to differentiate between broad composition models. The measurement technique is based on studies of cosmic ray shower development. A rather direct technique for such studies is to use fluorescence observations of the shower development to determine the atmospheric depth of maximum development of the shower. Such observations suggest a heavy composition (large atomic number) at energies ~10^{17} eV which changes with increasing energy to a light composition (perhaps protonic) above ~10^{19} eV (T. K. Gaisser et al., 1993).

Primary Electrons

Primary electrons constitute about 1% of the cosmic ray beam. The positron to negative electron ratio is about 10% (J. M. Clem et al., 1995).

Antimatter in the Primary eam

The ratio of antiprotons to protons in the primary cosmic ray beam (at about 400 MeV) is about 10^{-5}. At about 10 GeV the ratio is about 10^{-3}. At the highest measured energies (10 TeV), the upper limit to the ratio is about 20% (S. Orito et al., 1995; M. Amenomori et al., 1995).

Primary Gamma-Rays

The flux of primary gamma-rays is low at high energies. At 1 GeV the ratio of gamma-rays to protons is about 10^{-6}. The arrival directions of these gamma-rays are strongly concentrated in the plane of the Milky Way although there is a diffuse, near isotropic background flux and some point sources have been detected.

Since the absorption cross section for gamma-rays above 100 MeV is approximately 20 mbarn/electron, less than 10% of gamma-rays reach mountain altitudes (A. W. Wolfendale, 1979; P. F. Michelson, 1994).

Sea Level Cosmic Radiation

The sea level cosmic ray dose is 300 millirad·yr^{-1} and the sea level ionization is 2.2×10^6 ion pairs m^{-3}s^{-1}. The sea level flux has a soft component, which can be absorbed in about 100 mm of lead (about 100 g·cm^{-2} of absorber) and a more penetrating (largely muon) hard component. The sea level radiation is largely produced in the atmosphere and is a secondary component from interactions of the primary particles. The steep primary energy spectrum means that most secondaries at sea level are from rather low energy primaries. Thus the secondary flux is dependent on the solar cycle and the geomagnetic latitude of the observer.

Absolute Flux of the Hard Component

Vertical Integral Intensity I(0) ~100 m^{-2}s^{-1}sr^{-1}
Angular dependence I(θ) ~ I(0) cos^2(θ)
Integrated Intensity ~200 m^{-2}s^{-1}
(O.C. Allkofer, 1975b).

Flux of the Soft Component

In free air, the soft component comprises about one third of the total cosmic ray flux.

Latitude Effect

The geomagnetic field influences the trajectories of lower energy cosmic rays approaching the Earth. As a result, the background flux is reduced by about 7% at the geomagnetic equator. The effect decreases towards the poles and is negligible at latitudes above about 40°.

Flux of Protons

The proton component is strongly attenuated by the atmosphere with an attenuation length (reduction by a factor of e) of about 120 g·cm^{-2}. It constitutes about 1% of the total vertical sea level flux.

Absorption

The soft component is absorbed in about 100 g·cm^{-2} of matter. The hard component is absorbed much more slowly:

Absorption in lead, 6% per 100 g·cm^{-2}
Absorption in rock, 8.5% per 100 g·cm^{-2}
Absorption in water, 10% per 100 g·cm^{-2}
(Absorption for depths less than 100 g·pd cm^{-2} is given by K. Greisen, 1943.)

Altitude Dependence

The cosmic ray background in the atmosphere has a maximum intensity of about 15 times that at sea level at a depth of about 150 g·cm^{-2} (15 km altitude). At maximum intensity, the soft and hard components contribute roughly equally but the hard component is then attenuated more slowly (S. Hayakawa, 1969).

Cosmic Ray Showers

High energy cosmic rays produce particle cascades in the atmosphere which can be detected at sea level provided that their energy exceeds about 100 GeV (such low energy cascades may be detected by using the most sensitive atmospheric Cerenkov detectors). The primary particle progressively loses energy which is transferred through the production of successive generations of secondary particles to a cascade of hadrons, an electromagnetic shower component (both positively and negatively charged electrons and gamma-rays) and muons. The secondary particles are relativistic and all travel effectively at the speed of light. As a result, they reach sea level at approximately the same time but, due to Coulomb scattering (for the electrons) and production angles (for the pions producing the muons), are spread laterally into a disk-like shower front with a characteristic lateral width of several tens of meters and thickness (near the central shower core) of 2 to 3 m. The number of particles at sea level is roughly proportional to the primary particle energy:

Number of particles at sea level $\sim 10^{-10} \times$ energy (eV).

At altitudes below a few kilometers, the number of particles in a shower attenuates with an *attenuation length* of about 200 g·cm^{-2}.

i.e., particle number = original number $\times \exp(-(\text{depth increase})/200)$

The above applies to an individual shower. The rate of observation of showers of a given size (particle number at the detector) at different depths of absorber attenuates with an *absorption length* of about 100 g·cm^{-2} (J.G. Wilson, 1976).

Atmospheric Background Light from Cosmic Rays

Cosmic ray particles produce Cerenkov light in the atmosphere and produce fluorescent light through the excitation of atmospheric molecules.

Cerenkov Light

High energy charged particles will cause the emission of Cerenkov light in air if their energies are above about 30 MeV (electrons). This threshold is pressure (and hence altitude) dependent. A typical Cerenkov light pulse (at sea level, 100 m from the central shower core) has a time spread of a few nanoseconds. Over this time, the photon flux between 430 and 530 nm would be $\sim 10^{14}$ m^{-2}s^{-1} for a primary particle energy of 10^{16} eV. For comparison, the night sky background flux is $\sim 6 \times 10^{11}$ photons m^{-2}s^{-1}sr^{-1} in the same wavelength band (J.V. Jelley, 1967).

Fluorescence Light

Cosmic ray particles in the atmosphere excite atmospheric molecules which then emit fluorescence light. This is weak compared to the highly collimated Cerenkov component when viewed in the direction of the incident cosmic ray particle but is emitted isotropically. Typical pulse widths are longer than 50 ns and may be up to several microseconds for the total pulse from distant large showers (R.M. Baltrusaitis et al., 1985).

Effects of Cosmic Rays

Cerenkov Effects in Transparent Media

Background cosmic ray particles will produce Cerenkov light in transparent material with a photon yield between wavelengths λ_1 and λ_2

$$\sim (2\pi/137)\sin^2\left(\theta_c\right)\int_{\lambda_1}^{\lambda_2} d\lambda/\lambda^2 \text{ photons (unit length)}^{-1}$$

where θ_c (the Cerenkov angle) = $\cos^{-1}$ (1/refractive index).

This background light is known to affect light detectors, e.g., photomultipliers, and can be a major source of background noise (R.W. Clay and A.G. Gregory, 1977).

Effects on Electronic Components

If background cosmic ray particles pass through electronic components, they may deposit sufficient energy to affect the state of, e.g., a transistor flip-flop. This effect may be significant where reliability is of great importance or the background flux is high. For instance, it has been estimated that, in communication satellite operation, an error rate of about 2×10^{-3} per transistor per year may be found. Permanent damage may also result. A significant error rate may be found even at sea level in large electronic memories. This error rate is dependent on the sensitivity of the component devices to the deposition of electrons in their sensitive volumes (J.F. Ziegler, 1981).

Biophysical Significance

When cosmic rays interact with living tissue, they produce radiation damage. The amount of the damage depends on the total dose of radiation. At sea level, this dose is small compared with doses from other sources but both the quantity and quality of the radiation change rapidly with altitude. Approximate dose rates under various conditions are:

Dose rates (mrem·yr^{-1})
Sea level cosmic rays, 30
Cosmic rays at 10 km (subsonic jets), 2000
Cosmic rays at 18 km (supersonic transports), 10,000
(c.f., mean total sea level dose, 300)

Astronauts would be subject to radiation from galactic (0.05 rads per day) and solar (a few hundred rads per solar flare) cosmic rays as well as large fluxes of low energy radiation when passing through the Van Allen belts (about 0.3 rads per traverse).
Both astronauts and SST travellers would be subject to a small flux of low energy heavy nuclei stopping in the body. Such particles are capable of destroying cell nuclei and could be particularly harmful in the early stages of the development of an embryo. The rates of heavy nuclei stopping in tissue in supersonic transports and spacecraft are approximately as follows:

Stopping nuclei ((cm^3 tissue)$^{-1}$ hr^{-1})
Supersonic transport (16 km), 0.0005
Supersonic transport (20 km), 0.005
Spacecraft, 0.15
(O. C. Allkofer, 1975a; O. C. Allkofer et al., 1974).

Carbon Dating

Radiocarbon is produced in the atmosphere due to the action of cosmic ray slow neutrons. Solar cycle modulation of the very low energy cosmic rays causes an anticorrelation of the atmospheric ^{14}C activity with sunspot number with a mean amplitude of about 0.5%. In the long term, modulation of cosmic rays by a varying magnetic field may be important (A.A. Burchuladze et al., 1979).

Practical Uses of Cosmic Rays

There are few direct practical uses of cosmic rays. Their attenuation in water and snow have, however, enabled automatic monitors of water and snow depth to be constructed. A search for hidden cavities in pyramids has been carried out using a muon "telescope".

Other Effects

Stellar X-rays have been observed to affect the transmission times of radio signals between distant stations by altering the depth of the ionospheric reflecting layer. It has also been suggested that variations in ionization of the atmosphere due to solar modulation may have observable effects on climatic conditions.

REFERENCES

O.C. Allkofer, (1975a) *Introduction to Cosmic Radiation,* Verlag Karl Thiemig, Munchen, Germany.
O.O. Allkofer, (1975b) *J. Phys. G: Nucl. Phys.,* 1, L51.
O.C. Allkofer and W. Heinrich, (1974) *Health Phys.,* 27, 543.
M. Amenomori et al., (1995) Proc. 24th Int. Cosmic Ray Conf. Rome, 3, 85. Universita La Sapienza, Roma.
K. Asakimori et al., (1993) Proc. 23rd Int. Cosmic Ray Conf. Calgary, 2, 25, University of Calgary, Canada.
R.M. Baltrusaitis et al., (1985) *Nucl. Inst. Meth.,* A420, 410.
D.J. Bird et al., (1993) *Phys. Rev. Lett.,* 71, 3401.
A.A. Burchuladze, S.V. Pagava, P. Povinec, G. I. Togonidze, S. Usacev, (1979) Proc. 16th Int. Cosmic Ray Conf. Kyoto, 3, 201, Univ. of Tokyo, Japan.
R.W. Clay, (1987) *Aust. J. Phys.,* 40, 423.
R.W. Clay and A.G. Gregory, (1977) *J. Phys. A: Math. Gen.,* 10, 135.
J.M. Clem et al., (1995) Proc. 24th Int. Cosmic Ray Conf. Rome, 3, 5, Universita La Sapienza, Roma.
T.K. Gaisser et al., (1993) *Phys. Rev. D,* 47, 1919.
K. Greisen, (1943) *Phys. Rev.,* 63, 323.
S. Hayakawa, (1969) *Cosmic Ray Physics,* Wiley-Interscience, New York.
J.V. Jelley, (1967) *Prog. in Elementary Particle and Cosmic Ray Physics,* 9, 41.
E. Juliusson, (1975) Proc. 14th Int. Cosmic Ray Conf. Munich, 8, 2689, Max Planck Institute fur Extraterrestriche Physik, Munchen, Germany.
J. Linsley, (1981) *Origin of Cosmic Rays,* I.A.U. Symposium 94, 53, D. Reidel Publishing Co Dordrecht, Holland.
P. Meyer, (1981) *Origin of Cosmic Rays,* I.A.U. Symposium 94, 7, D. Reidel Publishing Co. Dordrecht, Holland.
P.F. Michelson (1994) in *Towards a Major Atmospheric Cerenkov Detector III,* 257, Ed. T. Kifune, Universal Academy Press Inc., Tokyo, Japan.
P. Sokolsky, (1989) *Introduction to Ultrahigh Energy Cosmic Ray Physics,* Addison Wesley Publishing Company.
T. Stanev et al., (1995) *Phys. Rev. Lett.,* 75, 3056.
S. Orito et al., (1995) Proc. 24th Int. Cosmic Ray Conf. Rome, 3, 76. Universita La Sapienza, Roma.
J.G. Wilson, (1976) *Cosmic Rays,* Wykeham Pub. (London) Lt., U.K.
A.W. Wolfendale, (1979) *Pramana,* 12, 631.
S.Yoshida et al., (1995) *Astroparticle Phys.,* 3, 105.
J.F. Ziegler, (1981) IEEE Trans. Electron Devices, ED-28, 560.

Section 12
Properties of Solids

TECHNIQUES FOR MATERIALS CHARACTERIZATION

EXPERIMENTAL TECHNIQUES USED TO DETERMINE THE COMPOSITION, STRUCTURE, AND ENERGY STATES OF SOLIDS AND LIQUIDS

H.P.R.Frederikse

The many experimental methods, originally designed to study the chemical and physical behavior of solids and liquids, have grown into a new field known as Materials Characterization (or Materials Analysis). During the past 30 years a host of techniques aimed at the study of surfaces and thin films has been added to the many tools for the analysis of bulk samples. The field has benefitted particularly from the development of computers and microprocessors, which have vastly increased the speed and accuracy of the measuring devices and the recording of their output. Materials characterization was and is a very important tool in the search for new physical and chemical phenomena. It plays an essential role in new applications of solids and liquids in industry, communications, and medicine. Many of its techniques are used in quality control, in safety regulations, and in the fight against pollution.

In most Materials Characterization experiments the sample is subjected to some kind of radiation: electromagnetic, acoustic, thermal, or particles (electrons, ions, neutrons, etc.). The surface analysis techniques usually require a high vacuum. As a result of interactions between the solid (or liquid) and the incoming radiation a beam of a similar (or a different) nature will emerge from the sample. Measurement of the physical and/or chemical attributes of this emerging radiation will yield qualitative, and often quantitative, information about the composition and the properties of the material being probed.

The modern tendency of describing practically everything in this world by a combination of a few letters (acronyms) has also penetrated the field of Materials Characterization. The table below gives the meaning of the acronym for every technique listed, the form and size of the required sample (bulk, surface, film, liquid, powder, etc.), the nature of the incoming radiation, the depth and the lateral spatial resolution that can be probed, and the information obtained from the experiment. The last column lists one or two major references to the technique described.

OPTICAL AND MASS SPECTROSCOPIES FOR CHEMICAL ANALYSIS

Technique	Sample	In	Out	Depth	Lateral resolution	Information obtained	Ref.
1. AAS — Atomic Absorption Spectroscopy	Atomize (flame, electro, thermal, etc.)	Light e.g., glow discharge	Absorption spectrum	—	—	Concentration of atomic species (quantitative, using standards)	1,2
2. ICP-AES — Induct. Coupled Plasma — Atomic Emission Spectroscopy	Atomize (flame, electro, thermal, ICP, etc.)	—	Emission spectrum	—	—	Concentration of atomic species (quantitative, using standards)	3
3. Dynamic SIMS — Dynamic Secondary Ion Mass Spectroscopy	Surface	Ion beam (1–20 keV)	Secondary ions; analysis with mass spectrometer	2 nm–1 µm (or deeper: ion milling)	0.50 nm	Elemental and isotopic analysis; depth profile (all elements); detection limits: ppb-ppm	4
4. Static SIMS — Static Secondary Ion Mass Spectroscopy	Surface	Ion beam (0.5–20 keV)	Secondary ions, analysis with mass spectrometer	0.1–0.5 nm	10 µm	Elemental analysis of surface layers; molecular analysis; detection limits: ppb-ppm	4
5. SNMS — Sputtered Neutral Mass Spectroscopy	Surface, bulk	Plasma discharge; noble gases: 0.5–20 keV	Sputtered atoms ionized by atoms or electrons; then mass analyzed	0.1–0.5 nm (or deeper: ion milling)	1 cm	Elemental analysis $Z \geq 3$; depth profile; detection limit: ppm	4,6
6. SALI — Surface Analysis by Laser Ionization	Surface	e-beam, ion-beam, or laser for sputtering	Sputtered atoms ionized by laser; then mass analyzed	0.1–0.5 nm up to 3 µm in milling mode	60 nm	Surface analysis; depth profiling	7
7. LIMS — Laser Ionization Mass Spectroscopy	Surface, bulk	u.v. laser (ns pulses)	Ionized species; analyzed with mass spectrometer	50–150 nm	5 µm–1 mm	Elemental (micro)analysis; detection limits: 1–100 ppm	8
8. SSMS — Spark Source Mass Spectroscopy	Sample in the form of two electrodes	High voltage R.F. spark produces ions	Ions — analyzed in mass spectrometer	1–5 µm	—	Survey of trace elements; detection limit: 0.01–0.05 ppm	9
9. GDMS — Glow Discharge Mass Spectroscopy	Sample forms the cathode for a D.C. glow discharge	Sputtered atoms ionized in plasma	Ions — analyzed in mass spectrometer	0.1–100 µm	3–4 mm	(Bulk) trace element analysis; detection limit: sub-ppb	9,10
10. ICPMS — Induct. Coupled Plasma Mass Spectroscopy	Liquid-dissolved sample carried by gas stream into R.F. induction coil	Ions produced in argon plasma	Ions — analyzed in quadrupole mass spectrometer	—	—	High sensitivity analysis of trace elements	11

TECHNIQUES FOR MATERIALS CHARACTERIZATION (continued)

PHOTONS — ABSORPTION, REFLECTION AND ELECTRON EMISSION

Technique	Sample	In	Out	Depth	Lateral resolution	Information obtained	Ref.
11. IRS Infrared Spectroscopy	Thin crystal, glass, liquid	I.R. light (W-filament, globar, Hg-arc)	I.R. spectrum	—	—	Electronic transitions (mainly in semiconductors and superconductors); vibrational modes (in crystals and molecules)	12,13, 14
12. FTIR Fourier Transform I.R. Spectroscopy	Solid, liquid; transmission or reflection	White light (all frequencies)	Fourier Transform of spectrum (interferometer)	—	—	Spectra obtained at higher speed and resolution	15
13. ATR Attenuated Total Reflection	Surface or thin crystal	—	—	μm's	—	Atomic or molecular spectra of surfaces and films	16
14. (μ)-RS (Micro-) Raman Spectroscopy	Solid, liquid (1 μm–1 cm)	Laser beam, e.g., Ar-line, YAG-line	Raman spectra	0.5 μm	0.5 μm	Molecular and crystal vibrations	12,14, 17
15. CARS Coherent Anti-Stokes Raman Spectroscopy	Solid, liquid (50 μm–3 cm)	Pump beam (ω_0)+ probe beam (ω_s)	Anti-Stokes spectrum	—	—	High resolution Raman spectra	14
16. Ellipsometry	Transparent films, crystals, adsorbed layers	Polarized light	Change in polarization	0.05 nm–5 μm	25 μm (or sample thickness)	Refractive index and absorption	18,19
17. UPS Ultraviolet Photo-electron Spectroscopy	Surfaces, adsorbed layers	u.v. light, 10–100 eV; 200 eV (synchrotron)	Electrons	0.2–10 nm	0.1–10 nm	Energies of electronic states of surfaces and free molecules	20,21
18. PSD Photon Stimulated Desorption	Surfaces with adsorbed species	Far u.v. light $E > 10$ eV	Ions — analyzed with mass spectrometer	0.1–2 nm	—	Structure and desorption kinetics of adsorbed atoms and molecules	22

X-RAYS

Technique	Sample	In	Out	Depth	Lateral resolution	Information obtained	Ref.
19. XRD X-Ray Diffraction	Single crystals, powders films	X-rays: λ = 0.05–0.2 nm (6–17 keV)	Diffracted X-ray beam	1–1000 μm	0.1–10 mm	Identification of crystallographic structures; all elements (low Z difficult)	23,24
20. XRF/EDS X-Ray Fluorescence/Energy Dispersive Spectroscopy	Thin films, single layer	Prim. X-ray beam λ = 0.02–0.1 nm 12–80 keV	Fluorescent X-rays	1–100 μm	10 mm	Elemental analysis; all elements except H, He, Li — (EDS also used in XRD, SEM, TEM and EPMA)	25,26
21. EXAFS Extended X-Ray Absorption Fine Structure	Films, foils	High intensity X-rays (synchrotron)	Spectrum near absorption edge	nm–μm	—	Local atomic structure: order/disorder in vicinity of absorbing atom	27
22. XPS/ESCA X-Ray Photo-electron Spectroscopy/Electron Spect. for Chemical Analysis	Surfaces, thin films (≈20 atomic layers)	Soft X-rays (1–20 keV)	Core electrons; valence electrons	0.5–10 nm	5 nm–50 μm	(Quantitative) identification of all elements in surface layer or film	28,29

ELECTRONS

Technique	Sample	In	Out	Depth	Lateral resolution	Information obtained	Ref.
23. CL Cathode Luminescence	Insulators, semiconductors	Electrons 5–50 keV	Photons 0.1–5 eV	1 nm–2 μn	1 or 2 μm	Energy levels of impurities and point defects	30
24. APS Appearance Potential Spectroscopy	Surface (≈20 atomic layers)	Electrons (energy scan) 50–2000 eV	X-rays to pinpoint electron energy threshold	—	—	Identification of surface species	21, see also C

TECHNIQUES FOR MATERIALS CHARACTERIZATION (continued)

	Technique	Sample	In	Out	Depth	Lateral resolution	Information obtained	Ref.
25.	AES Auger Electron Spectroscopy	Thin films, surfaces	Electrons 3–10 keV	Auger electrons 20–2000 eV	0.3–3 nm	≈30 nm	Elemental composition of surface (except H, He); detection limit 0.1–1%	28,29
26.	EELS Electron Energy Loss Spectroscopy	Very thin samples (<200 nm)	Electrons (100–400 keV)	(Retarded) electrons; minus 1–1000 eV	<200 nm	1–100 nm	Local elemental concentration; electronic structure, chem. bonding; interatomic distances	31
27.	EXELFS Extended Electron Energy Loss Fine Structure	Thin films	Electrons (100–400 keV)	Electrons energies 0–30 eV above edge	<200 nm	1–100 nm	Density of states of valence electrons (above Fermi level)	27,32
28.	ESD Electron Stimulated Desorption	Adsorbed species	Electrons E > 10 eV	Ions — analyzed with mass spectrometer	—	—	Structure and desorption properties of adsorbed atoms and molecules	22
29.	ESDIAD ESD-Ion Angular Distribution	(See ESD)	(See ESD)	Directional dependence of emitted ions	—	—	Geometries of adsorbed species (atoms or molecules)	22
30.	EPMA Electron Probe (X-Ray) Micro Analysis	Solid conductors and insulators <1 cm thick	Electrons 5–30 keV	Characteristic X-ray 0.1–15 keV	100 nm–5 μm	1 μm	Elemental analysis, Z ≤ 4, major, minor and trace amounts	33,34
31.	LEED Low Energy Electron Diffraction	Surface	Mono-energetic electron beam 10–1000 eV	Diffracted electrons	0.4–2 nm	<5 μm	Crystallographic structure of surface; resolution: 0.01 nm	35
32.	RHEED Reflection High Energy Electron Diffraction	Surface	Electron beam at grazing angle 5–50 keV	Reflected electrons	0.2–10 nm	<5 μm	Surface symmetry	36,37
33.	SEM Scanning Electron Microscopy	Bulk, films (conducting)	High energy electrons usually ≈30 keV	Secondary and backscattered electrons	1 nm–5 μm	1–20 nm	Surface image, defect structure; resolution 5–15 nm; magnification 300,000×	33,34
34.	(S)TEM (Scanning) Transmission Electron Microscopy	Thin specimen — <200 nm	High energy electrons typically 300 keV	Transmitted and diffracted electrons	(Sample thickness)	2–20 nm	(Defect) structure of cryst. solids; microchemistry; high resol.: 0.2 nm	33
35.	FEM Field Emission Microscopy	Metals, alloys (sharp point)	—	Electron emission (with appl. electric field — 50 kV)	≈0.5 nm	10–100 nm	Surface image, crystallographic structure	34
36.	STM Scanning Tunneling Microscopy	Polished or cleaved surface (conducting)	Tunneling current controls distance between sample and very sharp tip		1–5 nm	2–10 nm	Atomic-scale relief map of surface; resolution: vert. 0.002 nm, hor. 0.2 nm	39
37.	SPM Scanned Probe Microscopy	Very flat surface	Any field: e.g. mechan. vibration recorded with laser probe; same with magnetic, electric or thermal field		1–100 nm	1–100 nm	Surface-magnetic field, surface-thermal conductivity, etc.	39a
38.	AFM Atomic Force Microscopy	Very flat surface	Similar to STM; force measured with cantilever spring		0.5–5 nm	0.2–130 nm	Surface topography with atomic resolution; interatomic force	40
	IONS AND NEUTRONS							
39.	ISS (or LEIS) Ion Scattering Spectroscopy (Low Energy Ion Scattering)	Surface	Ion beam He+ or Ne+ <3 keV	Sputtered ions (energy analysis)	0.1–0.5 nm	1–100 μm	Elemental analysis (better for low Z) detection limits: 0.01–1%	41
40.	FIM Field Ion Microscopy	Surface: metals, alloys; very sharp tip	(He gas above sample)	He ions + high electric field produce image	≈0.1 nm	0.1–2 nm	Atomic structure of surface	34,42
41.	RBS Rutherford Back Scattering	Solids, thin films	Mono-energetic ions (H+ or He++) 0.5–3 MeV	Backscattered ions	10 nm–1 μm	1 mm	Element identification (Li to U) detection limit: 0.01–1%	46

TECHNIQUES FOR MATERIALS CHARACTERIZATION (continued)

Technique	Sample	In	Out	Depth	Lateral resolution	Information obtained	Ref.
42. NRA Nuclear Reaction Analysis	Solids, thin films	Mono-energetic ions (Li, Be, B, etc.) 200 keV–6 MeV	Protons, deuterons ^{3}He, α-particles, γ-rays	0.1–5 μm	10 μm– 10 mm	Element identification (all) detection limit: 10^{-12}–10^{-2}	47
43. PIXE Particle Induced X-ray Emission	Thin films, surface layers	High energy ions (H^+ or He^{++})	Characteristic X-rays	<10 μm	1 μm–2 mm	Trace impurities: Z >3 detection limit: 0.1–100 ppm (depending on sample thickness)	48
44. INS Ion Neutralization Spectroscopy	Surface	He-ions (≈5 eV)	Electrons	—	—	Energies of valence electrons	49
45. NAA Neutron Activation Analysis	Bulk, >0.5 g	Thermal neutrons	Characteristic γ-rays, (≈1 MeV)	Bulk	—	Trace concentrations (of isotopes) of elements: trans. metals, Pt-group; detection limit: 10^8–10^{14} atoms/cm^3	43
46. N(P)D Neutron (Powder) Diffraction	Crystalline solids	Thermal neutrons E =0.0025 eV	Diffracted neutrons	Bulk	—	Crystallographic structure; porosity, particle size	44
47. SANS Small Angle Neutron Scattering	Inhomogeneous solids; powders; porous samples	Thermal neutrons $2\theta = 10^{-2}$–10^{-4}	Scattered neutrons	1–25 mm	—	Average size of inhomogeneities; range: 1 nm–1 mm	45
ACOUSTIC							
48. SLAM Scanning Laser Acoustic Microscopy	Bulk, film	Acoustic wave produced by laser 1 MHz–1 GHz	Reflected acoustic wave	μm–cm	0.1–20 mm	Defect structure; thickness measurement	50
THERMAL							
49. DTA Differential Thermal Analysis	Specimen and reference sample	Uniform heating	Temperature difference	Bulk	—	Phase transitions, crystallization	51
50. DSC Differential Scanning Calorimetry	Specimen and ref. sample	Controlled heating	Measure heat required for equal temperature	Bulk	—	Phase transitions, crystallization; activation energies	51
51. TGA Thermo Gravimetric Analysis	Bulk, 1–100 g	Controlled heating	Weight as function of temperature (and time)	Bulk	—	Decomposition, non-stoichiometry, kinetics of reaction	52
RESONANCE							
52. EPR (ESR) Electron Paramagnetic (Spin) Resonance	Paramagnetic solids or liquids	Microwave radiation in magnetic field 3–300 GHz; 1–100 kG	Microwave absorption (at resonance)	Bulk	—	Local environment of paramagnetic ion; concentration of paramagnetic species; detection limit: 10^{11} spins/cm^3	53,54
53. ECR Electron Cyclotron Resonance	Semiconductors; metals; free electrons (low temperature)	Microwave radiation in magnetic field 10–30 GHz; 5–10 kG	Microwave absorption (at resonance)	Bulk	—	Electronic energy bands, effective masses	55
54. Mössbauer Effect	Source and absorber	Mono-energetic γ-rays: 5–100 keV	Mössbauer spectrum (Doppler shifted (lines)	50 m	1 cm	Interaction between nucleus and its environment (local electric, magnetic fields; bonds; valency; diffusion, etc.)	56
55. NMR (MRI) Nuclear Magnetic Resonance (Magnetic Resonance Imaging)	Solids, liquids	R.F. radiation + magnetic field; e.g. for protons: 60 MHz, 14 kG	R.F. absorption	<1 cm	1 cm	Quant. analysis; local magnetic environment; diffusion; imaging	58
56. ENDOR Electron Nuclear Double Resonance	Solids, liquids	R.F. + microwave radiation in magn. field.	Microwave absorption	—	—	Hyperfine interaction → local atomic structure	54

12-4

TECHNIQUES FOR MATERIALS CHARACTERIZATION (continued)

Technique	Sample	In	Out	Depth	Lateral resolution	Information obtained	Ref.
57. NQR Nuclear Quadrupole Resonance	Solids	R.F. radiation 0.5—1000 MHz	R.F. absorption	—	—	Asymmetry of the charge distribution at the nucleus	55,59

OTHER

Technique	Sample	In	Out	Depth	Lateral resolution	Information obtained	Ref.
58. BET Brunauer-Emmett-Teller	(Large) surface area 1–20 m²/g	Adsorbed gas (e.g. N_2 at low temp.) as function of pressure (monolayer coverage)		—	—	Surface area measurement	60

REFERENCES

General References

A. Wachtman, J. B., *Characterization of Materials*, Butterworth-Heinemann, Boston, 1993.

B. Brundle, C. R., Evans, C. A., and Wilson, S., Eds., *Encyclopedia of Materials*, Butterworth-Heinemann, Boston, 1992.

C. Woodruff, D. P. and Delchar, T. A., *Modern Techniques of Surface Science*, Cambridge University Press, Cambridge, 1986.

D. *Metals Handbook*, 9th Edition, Vol. 10, Materials Characterization, Whan, R. E., Coordinator, American Society for Metals, Metals Park, OH, 1986.

Specific References

1. Slavin, M., *Atomic Absorption Spectroscopy*, 2nd Edition, John Wiley & Sons, New York, 1978.
2. Schrenk, W. G., *Analytical Atomic Spectroscopy*, Plenum Press, New York, 1975.
3. Dean, J. A. and Rains, T. E., *Flame Emission and Atomic Absorption Spectroscopy*, Vols. 1—3, Marcel Dekker, New York, 1969.
4. Benninghoven, A., Rudenauer, F. G., and Werner, H. W., *Secondary Ion Mass Spectroscopy*, John Wiley & Sons, New York, 1987.
5. Bird, J. R. and Williams, J. S., Eds., in *Ion Beams for Materials Analysis*, Academic Press, New York, 1989, pp. 515—537.
6. Smith, G. C., *Quantitative Surface Analysis for Materials Science*, The Institute of Metals, London, 1991.
7. Becker, E. H., in *Ion Spectroscopies for Surface Analysis*, Czanderna, A. W. and Hercules, D. M., Eds., Plenum Press, New York, 1991, p. 273.
8. Simons, D. S., *Int. J. Mass Spectrometry and Ion Processes*, 55, 15, 1983.
9. White, F. A. and Wood, G. M., *Mass Spectrometry: Applications in Science and Engineering*, John Wiley & Sons, New York, 1986.
10. Harrison, W. W. and Bentz, B. L., *Prog. Anal. Spectrometry*, 11, 53, 1988.
11. Bowmans, P. W. J. M., *Inductively Coupled Plasma Emission Spectroscopy*, Parts I and II, John Wiley & Sons, New York, 1987.
12. Brame, Jr., E. G. and Grasselli, J., *Infrared and Raman Spectroscopy*, Practical Spectroscopy Series, Vol. I, Marcel Dekker, New York, 1976.
13. Hollas, J. M., *Modern Spectroscopy*, John Wiley & Sons, New York, 1987.
14. Turrell, G., *Infrared and Raman Spectroscopy of Crystals*, Academic Press, New York and London, 1972.
15. Griffith, P. R. and Haseth, J. A., *Fourier Transform Infrared Spectroscopy*, John Wiley & Sons, New York, 1986.
16. Barnowski, M. K., *Fundamentals of Optical Fiber Communications*, Academic Press, New York, 1976.
17. Long, D. A., *Raman Spectroscopy*, McGraw-Hill, New York, 1977.
18. Azzam, R. M. A., *Ellipsometry and Polarized Light*, Elsevier-North Holland, Amsterdam, 1977.
19. Hecht, E., *Optics*, 2nd Edition, Addison-Wesley, Reading MA, 1987.
20. Brundle, C. R., in *Molecular Spectroscopy*, West, A. R., Ed., Heyden, London, 1976.
21. Park, R. L., in *Experimental Methods in Catalytic Research*, Vol. III, Anderson, R. B. and Dawson, P. T., Academic Press, New York, 1976, pp. 1—39.
22. Madey, T. E. and Stockbauer, R., in *Solid State Physics: Surfaces*, Vol. 22 of Methods of Experimental Physics, Park, R.L. and Lagally, M. G., Eds., Academic Press, New York, 1985.
23. Cullity, B. D., *Elements of X-Ray Diffraction*, 2nd Edition, Addison-Wesley, Reading, MA, 1978.
24. Schwartz, L. H. and Cohen, J. B., *Diffraction from Materials*, Springer Verlag, Berlin, 1987.

12-5

TECHNIQUES FOR MATERIALS CHARACTERIZATION (continued)

25. deBoer, D. K. G., in *Advances in X-Ray Analysis*, Vol. 34, Barrett, C. S. et. al., Eds., Plenum Press, New York, 1991.

26. Birks, L. S., *X-Ray Spectrochemical Analysis*, 2nd Edition, John Wiley & Sons, New York, 1969.

27. Bonnelle, C. and Mande, C., *Advances in X-Ray Spectroscopy*, Pergamon Press, Oxford, 1982.

28. *Practical Surface Analysis by Auger and X-Ray Photo-Electric Spectroscopy*, Briggs, D. and Seah, M. P., Eds., John Wiley & Sons, New York, 1983.

29. Powell, C. J. and Seah, M. P., *J. Vac. Sci. Technol. A*, Vol. 8, 735, 1990.

30. Yacobi, G. G. and Holt, D. B., *Cathodeluminescence Microscopy of Inorganic Solids*, Plenum Press, New York, 1990.

31. Egerton, R. F., *Electron Energy Loss Spectroscopy in the Electron Microscope*, Plenum Press, New York, 1986.

32. Disko, M. M., Krivanek, O. L., and Rez, P., *Phys. Rev.*, B25, 4252, 1982.

33. Goldstein, J. I., et. al., *Scanning Electron Microscopy and X-Ray Microanalysis*, 2nd Edition, Plenum Press, New York, 1986.

34. Murr, L. E., *Electron and Ion Microscopy and Microanalysis*, Marcel Dekker, New York, 1982.

35. Armstrong, R. A., in *Experimental Methods in Catalytic Research*, Vol. III, Anderson, R. B., and Dawson, P. T.,Eds., Academic Press, New York, 1976.

36. Dobson, P. J. et. al., *Vacuum*, 33, 593, 1983.

37. Ryner, T. B., *Electron Diffraction*, Methuen, London, 1970.

38. Reimer, L., *Transmission Election Microscopy*, Springer-Verlag, Berlin, 1984.

39. *Scanning Tunneling Microscopy and Related Methods*, Behm, R. J., Garcia, N., and Rohrer, H., Kluwer, Eds., Academic Publishers, Norwell, MA, 1990.

39a. Wikramasinghe, H.K., *Scientific American*, Vol. 261, No. 4, pp. 98—105, Oct. 1989.

40. Rugar, D. and Hansma, P., *Physics Today*, 43(10), pp. 23—30, 1990.

41. Feldman, C. C. and Mayer, J. W., *Fundamentals of Surface and Thin Film Analysis*, North-Holland, Amsterdam, 1986.

42. Muller, E. W. and Tsong, T. T., *Field Ion Microscopy*, Elsevier, Amsterdam, 1969.

43. Amiel, S., *Nondestructive Activation Analysis*, Elsevier, Amsterdam, 1981.

44. Bacon, G. E., *Neutron Diffraction*, 3rd Edition, Clarendon Press, Oxford, 1975.

45. Neutron Scattering, Part A., in *Methods of Experimental Physics*, Vol. 23, Skold, K. and Price, D. L., Eds., Academic Press, New York, 1986.

46. Chu, W. K., Mayer, J. W., and Nicolet, M. A., *Backscattering Spectroscopy*, Academic Press, New York, 1987.

47. Rickey, F. A., in *High Energy and Heavy Ion Beams in Materials Analysis*, Tesmer, J. R., et. al., Eds., MRS, 1990, pp. 3—26.

48. Johansson, S. A. E. and Campbell, J. L., *PIXE: A Novel Technique for Elemental Analysis*, John Wiley & Sons, New York, 1988.

49. Hagstrum, H. D., in *Inelastic Ion-Surface Collisions*, Tolk, N. H. et. al., Eds., Academic Press, New York, 1977, pp. 1—46.

50. Nikoonahad, M., in *Research Techniques in Nondestructive Testing*, Vol. VI, Sharpe, R.S., Ed., Academic Press, New York, 1984, pp. 217—257.

51. Gallagher, P. K., *Characterization of Materials by Thermoanalytical Techniques*, MRS - Bulletin, Vol. 13, No. 7, pp. 23—27, 1988.

52. Earnest, C. M., *Compositional Analysis by Thermogravimetry*, ASTM Special Technical Publication 997, 1988.

53. Poole, C. P., *Electron Spin Resonance — A Comprehensive Treatise on Experimental Techniques*, 2nd Edition, John Wiley & Sons, New York, 1983.

54. Atherton, N. M., *Principles of Electron Spin Resonance*, Ellis Horwood Ltd., Chichester, U.K., 1993.

55. Kittel, C., *Introduction to Solid State Physics*, 6th Edition, John Wiley & Sons, New York, 1986, p. 196.

56. Gibb, T. C., *Principles of Mössbauer Spectroscopy*, Chapman & Hall, London, 1976.

57. Slichter, C. P., *Principles of Magnetic Resonance*, 3rd Edition, Springer-Verlag, Berlin, 1990.

58. *NMR Spectroscopy Techniques*, Dybrowski, C. and Lichter, R. L., Eds., Marcel Dekker, New York, 1987.

59. Das, T. P. and Hahn, E. L., *Nuclear Quadrupole Resonance Spectroscopy*, Academic Press, New York, 1958.

60. Somorjai, G. A., *Principles of Surface Chemistry*, Prentice-Hall, Englewood Cliffs, NJ, 1972, p. 216.

SYMMETRY OF CRYSTALS

L. I. Berger

The ability of a body to coincide with itself in its different positions regarding a coordinate system is called its symmetry. This property reveals itself in iteration of the parts of the body in space. The iteration may be done by reflection in mirror planes, rotation about certain axes, inversions and translations. These actions are called the symmetry operations. The planes, axes, points, etc., are known as symmetry elements. Essentially, mirror reflection is the only truly primitive symmetry operation. All other operations may be done by a sequence of reflections in certain mirror planes. Hence, the mirror plane is the only true basic symmetry element. But for clarity, it is convenient to use the other symmetry operations, and accordingly, the other aforementioned symmetry elements. The symmetry elements and operations are presented in Table 1.

The entire set of symmetry elements of a body is called its symmetry class. There are thirty-two symmetry classes that describe all crystals which have ever been noted in mineralogy or been synthesized (more than 150,000). The denominations and symbols of the symmetry classes are presented in Table 2.

There are several known approaches to classification of individual crystals in accordance with their symmetry and crystallochemistry. The particles which form a crystal are distributed in certain points in space. These points are separated by certain distances (translations) equal to each other in any chosen direction in the crystal. Crystal lattice is a diagram that describes the location of particles (individual or groups) in a crystal. The lattice parameters are three non-coplanar translations that form the crystal lattice. Three basic translations form the unit cell of a crystal. August Bravais (1848) has shown that all possible crystal lattice structures belong to one or another of fourteen lattice types (Bravais lattices). The Bravais lattices, both primitive and non-primitive, are the contents of Table 3.

Among the three-dimensional figures, there is a group of polyhedrons that are called regular, which have all faces of the same shape and all edges of the same size (regular polygons). It has been shown that there are only five regular polyhedrons. Because of their importance in crystallography and solid state physics, a brief description of these polyhedrons is included in Table 4.

The systematic description of crystal structures is presented primarily in the well known *Structurbericht*. The classification of crystals by the Structurbericht does not reflect their crystal class, the Bravais lattice, but is based on the crystallochemical type. This makes it inconvenient to use the Structurbericht categories for comparison of some individual crystals. Thus, there have been several attempts to provide a more convenient classification of crystals. Table 5 presents a compilation of different classifications which allows the reader to correlate the Structurbericht type with the international and Schoenflies point and space groups and with Pearson's symbols, based on the Bravais lattice and chemical composition of the class prototype. The information included in Table 5 has been chosen as an introduction to a more detailed crystallophysical and crystallochemical description of solids.

TABLE 1
Symmetry Operations and Elements

Symmetry operation	Symmetry element Name	Symbol International (Hermann-Mauguin)	Schoenflies	Presentation on the stereographic projection Parallel	Perpendicular
Reflection in a plane	Plane	m	C_s		
Rotation by angle $\alpha = 360°/n$ about an axis	Axis	$n = 1, 2, 3, 4$ or 6	C_n		
		$n = 2$	C_2		
		$n = 3$	C_3		
		$n = 4$	C_4		
		$n = 6$	C_6		
Rotation about an axis and inversion in a symmetry center lying on the axis	Inversion (improper) axis	$\bar{n} = \bar{3}, \bar{4}, \bar{6}$	C_{ni}		
		$\bar{n} = \bar{3}$	C_{3i}		
		$\bar{n} = \bar{4}$	C_{4i}		

TABLE 1
Symmetry Operations and Elements (continued)

Symmetry operation	Name	Symmetry element — Symbol		Presentation on the stereographic projection	
		International (Hermann-Mauguin)	Schoenflies	Parallel	Perpendicular
		$\bar{n} = \bar{6}$	C_{6i}		
Inversion in a point	Center	$\bar{1}$	C_i		
Parallel translation	Translation vector a, b, c				
Reflection in a plane and translation parallel to the plane	Glide–plane	a, b, c, n, d			
Rotation about an axis and translation parallel to the axis	Screw axis	n_m $(m = 1, 2, .., n-1)$			
Rotation about an axis and reflection in a plane perpendicular to the axis	Rotatory-reflection axis	$\tilde{n}$ $\tilde{n} = \tilde{1}, \tilde{2}, \tilde{3}, \tilde{4}, \tilde{6}$	S_n		

TABLE 2
The Thirty-Two Symmetry Classes

Crystal symbol	Class name[a] and its symbol — International (Int) and Schoenflies (Sch)													
	Primitive		Central		Planal		Axial		Plane-axial		Inversion primitive		Inversion-planal	
	Int	Sch	Int	Sch	Int	Sch	Int	Sch	Int	Sch	Int	Sch	Int	Sch
Triclinic	1	C_1	$\bar{1}$	C_i										
Monoclinic					m	C_s	2	C_2	2/m	C_{2h}				
Orthorhombic					mm2	C_{2v}	222	D_2	mmm	D_{2h}				
Trigonal	3	C_3	$\bar{3}$	C_{3i}	3m	C_{3v}	32	D_3	$\bar{3}m$	C_{3d}				
Tetragonal	4	C_4	4/m	C_{4h}	4mm	C_{4v}	422	D_4	4/mmm	D_{4h}	$\bar{4}$	S_4	$\bar{4}2m$	D_{2d}
Hexagonal	6	C_6	6/m	C_{6h}	6mm	C_{6v}	622	D_6	6/mmm	D_{6h}	$\bar{6}$	C_{3h}	$\bar{6}m2$	D_{3h}
Cubic	23	T	m3	T_h	$\bar{4}3m$	T_d	432	O	m3m	O_h				

[a] Per Fedorov Institute of Crystallography, USSR Academy of Sciences, nomenclature.

SYMMETRY OF CRYSTALS (continued)

TABLE 3

The Fourteen Possible Space Lattices (Bravais Lattices)

Crystal system	Metric category of the system	No. of different lattices in the system	Lattice type[a] (marked by +)					No. of identi-points per unit cell	Characteristic parameters (marked by +)						Description of characteristic parameters $a \subset X$, $b \subset Y$, $c \subset Z$; $\alpha \equiv (b,c)$, $\beta \equiv (a,c)$, $\gamma \equiv (a,b)$	Symmetry of the lattice	
			P	C	I	F	R		a	b	c	α	β	γ		Int	Sch
Triclinic	Trimetric	1	+					1	+	+	+	+	+	+	$a \neq b \neq c$, $\alpha \neq \beta \neq \gamma$	1	C
Monoclinic	Trimetric	2	+	+				1 or 2	+	+	+		+		$a \neq b \neq c$, $\alpha = \gamma = 90° \neq \beta$	2/m	C_{2h}
Orthorhombic	Trimetric	4	+	+	+	+		1, 2 or 4	+	+	+				$a \neq b \neq c$, $\alpha = \beta = \gamma = 90°$	mmm	D_{2h}
Trigonal (rhombohedral)	Dimetric	1					+	1	+			+			$a = b = c$, $120° > \alpha = \beta = \gamma \neq 90°$	3m	D_{3d}
Tetragonal	Dimetric	2	+		+			1 or 2	+		+				$a = b \neq c$, $\alpha = \beta = \gamma = 90°$	4/mmm	D_{4h}
Hexagonal	Dimetric	1	+					1	+		+				$a = b \neq c$, $\alpha = \beta = 90°$, $\gamma = 120°$	6/mmm	D_{6h}
Isometric (cubic)	Monometric	3	+		+	+		1, 2 or 4	+						$a = b = c$, $\alpha = \beta = \gamma = 90°$	m3m	O_h

[a] Designations of the space-lattice types: P — primitive, C — side-centered (base-centered), I — body-centered, F — face-centered, R — rhombohedral.

TABLE 4
The Five Possible Regular Polyhedrons

| Polyhedron | Symmetry (Schoenflies) | | Form of faces | Number of[a] | | |
	Class	Elements		Faces (F)	Edges (E)	Vertices (V)
Tetrahedron	T	$4C_3 3C_2$	Equilateral triangle	4	6	4
Cube (hexahedron)	O	$3C_4 4C_3 6C_2$	Square	6	12	8
Octahedron	O	$3C_4 4C_3 6C_2$	Equilateral triangle	8	12	6
Pentagonal dodecahedron	J	$6C_5 10C_3 15C_2$	Regular pentagon	12	30	20
Icosahedron	J	$6C_5 10C_3 15C_2$	Equilateral triangle	20	30	12

[a] Per formula by Leonhard Euler: $F + V - E = 2$

TABLE 5
Classification of Crystals

| Strukturbericht symbol 1 | Structure name 2 | Symmetry group | | Pearson symbol[a] 5 | Standard ASTM E157-82a symbol[b] 6 |
		International 3	Schoenflies 4		
A1	Cu	Fm3m	O^4_h	cF4	F
A2	W	Im3m	O^9_h	cI2	B
A3	Mg	P6$_3$/mmc	D^4_{6h}	hP2	H
A4	C	Fd3m	O^7_h	cF8	F
A5	Sn	If$_1$/amd	D^{19}_{4h}	tI4	U
A6	In	I4/mmm	D^{17}_{4h}	tI2	U
A7	As	R$\bar{3}$m	D^5_{3d}	hR2	R
A8	Se	P3$_1$21 or P3$_2$21	D^4_3 (D^6_3)	hP3	H
A10	Hg	R$\bar{3}$m	D^5_{3d}	hR1	R
A11	Ga	Cmca	D^{18}_{2h}	oC8	Q
A12	α-Mn	I$\bar{4}$3m	T^3_d	cI58	B
A13	β-Mn	P4$_1$32	O^7	cP20	C
A15	OW$_3$	Pm3n	O^3_h	cP8	C
A20	α-U	Cmcm	D^{17}_{2h}	oC4	Q
B1	ClNa	Fm3m	O^5_h	cF8	F
B2	ClCs	Pm3m	O^1_h	cP2	C
B3	SZn	F$\bar{4}$3m	T^2_d	cF8	F
B4	SZn	P6$_3$mc	C^4_{6v}	hP4	H
B8$_1$	AsNi	P6$_3$/mmc	D^4_{6h}	hP4	H
B8$_2$	InNi$_2$	P6$_3$/mmc	D^4_{6h}	hP6	H
B9	HgS	P3$_1$21 or P3$_2$21	D^4_3 or D^6_3	hP6	H
B10	OPb	P4/nmm	D^7_{4h}	tP4	T
B11	γ-CuTi	P4/nmm	D^7_{4h}	tP4	T
B13	NiS	R$\bar{3}$m	D^5_{3d}	hR6	R
B16	GeS	Pnma	D^{16}_{2h}	oP8	O
B17	PtS	P4$_2$/mmc	D^9_{4h}	tP4	T
B18	CuS	P6$_3$/mmc	D^4_{6h}	hP12	H
B19	AuCd	Pmma	D^5_{2h}	oP4	O
B20	FeSi	P2$_1$3	T^4	cP8	C

TABLE 5
Classification of Crystals (continued)

Strukturbericht symbol	Structure name	Symmetry group International	Symmetry group Schoenflies	Pearson symbol[a]	Standard ASTM E157-82a symbol[b]
1	2	3	4	5	6
B27	BFe	Pnma	D_{2h}^{16}	oP8	O
B31	MnP	Pnma	D_{2h}^{16}	oP8	O
B32	NaTl	Fd3m	O_h^7	cF16	F
B34	Pds	$P4_2/m$	C_{4h}^2	tP16	T
B35	CoSn	P6/mmm	D_{6h}^1	hP6	H
B37	SeTl	I4/mcm	D_{4h}^{18}	tI16	U
B_e	CdSb	Pbca	D_{2h}^{15}	oP16	O
B_f (B33)	ξ-BCr	Cmcm	D_{2h}^{17}	oC8	Q
B_g	BMo	$I4_1/amd$	D_{4h}^{19}	tI4	U
B_h	CW	P6m2	D_{3h}^1	hP2	H
B_i	γ-CMo (AsTi)	$P6_3/mmc$	D_{6h}^4	hP8	H
C1	CaF_2	Fm3m	O_h^5	cF12	F
$C1_b$	AgAsMg	$\bar{F}43m$	T_d^2	cF12	F
C2	FeS_2	Pa3	T_h^6	cP12	C
C3	Cu_2O	Pn3m	O_h^4	cP6	C
C4	O_2Ti	$P4_2/mnm$	D_{4h}^{14}	tP6	T
C6	CdI_2	$P\bar{3}m1$	D_{3d}^3	hP3	H
C7	MoS_2	$P6_3/mmc$	D_{6h}^4	hP6	H
$C11_a$	C_2Ca	I4/mmm	D_{4h}^{17}	tI6	U
$C11_b$	$MoSi_2$	I4/mmm	D_{4h}^{17}	tI6	U
C12	$CaSi_2$	$R\bar{3}m$	D_{3d}^5	hR6	R
C14	$MgZn_2$	$P6_3/mmc$	D_{6h}^4	hP12	H
C15	Cu_2Mg	Fd3m	O_h^7	cF24	F
$C15_b$	$AuBe_5$	$\bar{F}43m$ or F23	T_d^2 or T^2	cF24	F
C16	Al_2Cu	I4/mcm	D_{4h}^{18}	tI12	U
C18	FeS_2	Pnnm	D_{2h}^{12}	oP6	O
C19	$CdCl_2$	$R\bar{3}m$	D_{3d}^5	hR3	R
C22	Fe_2P	$P\bar{2}6m$	D_{3h}^1	hP9	H
C23	Cl_2Pb	Pnma	D_{2h}^{16}	oP12	O
C32	AlB_2	P6/mmm	D_{6h}^1	hP3	H
C33	Bi_2STe_2	$R\bar{3}m$	D_{3d}^5	hR5	R
C34	$AuTe_2$	C2/m (P2/m)	C_{2h}^3 (C_{2h}^1)	mC6	N
C36	$MgNi_2$	$P6_3/mmc$	D_{6h}^4	hP24	H
C38	Cu_2Sb	P4/nmm	D_{4h}^7	tP6	T
C40	$CrSi_2$	$P6_222$	D_6^4	hP9	H
C42	SiS_2	Ibam	D_{2h}^{26}	oI12	P
C44	GeS_2	Fdd2	C_{2v}^{19}	oF72	S
C46	$AuTe_2$	Pma2	C_{2v}^4	oP24	O
C49	Si_2Zr	Cmcm	D_{2h}^{17}	oC12	Q
C54	Si_2Ti	Fddd	D_{2h}^{24}	oF24	S
C_c	Si_2Th	$I4_1/amd$	D_{4h}^{19}	tI12	U
C_e	$CoGe_2$	Aba2	C_{2v}^{17}	oC23	Q
DO_2	As_3Co	Im3	T_h^5	cI32	B
DO_3	BiF_3	Fm3m	O_h^5	cF16	F
DO_9	O_3Re	Pm3m	O_h^1	cP4	C
DO_{11}	CFe_3	Pnma	D_{2h}^{16}	oP16	O
DO_{18}	$AsNa_3$	$P6_3/mmc$	D_{6h}^4	hP8	H
DO_{19}	Ni_3Sn	$P6_3/mmc$	D_{6h}^4	hP8	H
DO_{20}	Al_3Ni	Pnma	D_{2h}^{16}	oP16	O

TABLE 5
Classification of Crystals (continued)

Strukturbericht symbol 1	Structure name 2	Symmetry group International 3	Symmetry group Schoenflies 4	Pearson symbol[a] 5	Standard ASTM E157-82a symbol[b] 6
DO_{21}	Cu_3P	$P\bar{3}c1$	D^4_{3d}	hP24	H
DO_{22}	Cu_3P	I4/mmm	D^{17}_{4h}	tI8	U
DO_{23}	Al_3Zr	I4/mmm	D^{17}_{4h}	tI16	U
DO_{24}	Ni_3Ti	$P6_3/mmc$	D^4_{6h}	hP16	H
DO_c	SiU_3	I4/mcm	D^{18}_{4h}	tI16	U
DO_e	Ni_3P	$I\bar{4}$	S^2_4	tI32	U
$D1_3$	Al_4Ba	I4/mmm	D^{17}_{4h}	tI10	U
$D1_a$	$MoNi_4$	I4/m	C^5_{4h}	tI10	U
$D1_b$	Al_4U	Imma	D^{28}_{2h}	oI20	P
$D1_c$	$PtSn_4$	Aba2	C^{17}_{2v}	oC20	Q
$D1_e$	B_4Th	P4/mbm	D^5_{4h}	tP20	T
$D1_f$	BMn_4	Fddd	D^{24}_{2h}	oF40	S
$D2_1$	B_6Ca	Pm3m	O^1_h	cP7	C
$D2_3$	$NaZn_{13}$	Fm3m	O^5_h	cF112	F
$D2_b$	$Mn_{12}Th$	I4/mmm	D^{17}_{4h}	tI26	U
$D2_c$	MnU_6	I4/mcm	D^{18}_{4h}	tI28	U
$D2_d$	$CaCu_5$	P6/mmm	D^1_{6h}	hP6	H
$D2_f$	$B_{12}U$	Fm3m	O^5_h	cF52	F
$D2_h$	Al_6Mn	Cmcm	D^{17}_{2h}	oC28	Q
$D5_1$	$\alpha\text{-}Al_2O_3$	$R\bar{3}c$	D^6_{3d}	hR10	R
$D5_2$	La_2O_3	$P\bar{3}m1$	D^3_{3d}	hP5	H
$D5_3$	Mn_2O_3	Ia3	T^7_h	cI80	B
$D5_8$	S_3Sb_2	Pnma	D^{16}_{2h}	oP20	O
$D5_9$	P_2Zn_3	$P4_2/mmc$	D^9_{4h}	tP40	T
$D5_{10}$	C_2C_3	Pnma	D^{16}_{2h}	oP20	O
$D5_{13}$	Al_3Ni_2	$P\bar{3}m1$	D^3_{3d}	hP5	H
$D5_a$	Si_2U_3	P4/mbm	D^5_{4h}	tP10	T
$D5_c$	C_3Pu_2	$I\bar{4}3d$	T^6_d	cI40	B
$D7_1$	Al_4C_3	$R\bar{3}m$	D^5_{3d}	hR7	R
$D7_3$	P_4Th_3	$I\bar{4}3d$	T^6_d	cI28	B
$D7_b$	B_4Ta_3	Immm	D^{25}_{2h}	oI14	P
$D8_1$	Fe_3Zn_{10}	Im3m	O^9_h	cI52	B
$D8_2$	Cu_5Zn_8	$I\bar{4}3m$	T^3_d	cI52	B
$D8_3$	Al_4Cu_9	$P\bar{4}3m$	T^1_d	cP52	C
$D8_4$	C_6Cr23	Fm3m	O^5_h	cF116	F
$D8_5$	Fe_7W_6	$R\bar{3}m$	D^5_{3d}	hR13	R
$D8_6$	$Cu_{15}Si_4$	$I\bar{4}3m$	T^3_d	cI76	B
$D8_8$	Mn_5Si_3	$P6_3/mcm$	D^3_{6h}	hP16	H
$D8_9$	Co_9S_8	Fm3m	O^5_h	cF68	F
$D8_{10}$	Al_8Cr_5	R3m	C^5_{3v}	hR26	R
$D8_{11}$	Al_5Co_2	$P6_3/mcm$	D^3_{6h}	hP28	H
$D8_a$	$Mn_{23}Th_6$	Fm3m	O^5_h	cF116	F
$D8_b$	σ-phase of Cr-Fe	$p4_2/mnm$	D^{14}_{4h}	tP30	T
$D8_e$	$(Al,Zn)_{49}Mg_{32}$	Im3	T^5_h	cI162	B
$D8_f$	Ge_7Ir_3	Im3m	O^9_h	cI40	B
$D8_h$	B_5W_2	$P6_3/mmc$	D^4_{6h}	hP14	H
$D8_i$	B_5Mo_2	$R\bar{3}m$	D^5_{3d}	hR7	R
$D8_l$	B_3Cr_5	I4/mcm	D^{18}_{4h}	tI32	U
$D8_m$	Si_3W_5	I4/mcm	D^{18}_{4h}	tI32	U

TABLE 5
Classification of Crystals (continued)

Strukturbericht symbol 1	Structure name 2	Symmetry group		Pearson symbol[a] 5	Standard ASTM E157-82a symbol[b] 6
		International 3	Schoenflies 4		
$D10_1$	C_3Cr_7	P31c	C^4_{3v}	hP80	H
$D10_2$	Fe_3Th_7	$P6_3mc$	C^4_{6v}	hP20	H
$E0_1$	ClFPb	P4/nmm	D^7_{4h}	tP6	T
$E1_1$	$CuFeS_2$	$\bar{I}42d$	D^{12}_{2d}	tI16	U
$E2_1$	CaO_3Ti	Pm3m	O^1_h	cP5	C
$E2_4$	S_3Sn_2	Pnma	D^{16}_{2h}	oP20	O
E3	Al_2CdS_4	$\bar{I}4$	S^2_4	tI14	U
$E9_3$	$SiFe_3W_3$	Fd3m	O^7_h	cF112	F
$E9_a$	Al_7Cu_2Fe	P4/mnc	D^6_{4h}	tP40	T
$E9_b$	$AlLi_3N_2$	Ia3	T^7_h	cI96	B
$F0_1$	NiSSb	$P2_13$	T^4	cP12	C
$F5_1$	$CrNaS_2$	$R\bar{3}m$ or R32	D^5_{3d} or D^7_3	hR4	R
$F5_6$	CuS_2Sb	Pnma	D^{16}_{2h}	oP16	O
$H1_1$	Al_2MgO_4	Fd3m	O^7_h	cF56	F
$H2_4$	Cu_3S_4V	$P\bar{4}3m$	T^1_d	cP8	C
$H2_5$	$AsCu_3S_4$	$Pmn2_1$	C^7_{2v}	oP16	O
$L1_0$	AuCu	P4/mmm	D^1_{4h}	tP4	T
$L1_2$	$AlCu_3$	Pm3m	O^1_h	cP4	C
$L2_1$	$AlCu_2Mn$	Fm3m	O^5_h	cF16	F
$L2_2$	Sb_2Tl_7	Im3m	O^9_h	cI54	B
$L'2_b$	H_2Th	I4/mmm	D^{17}_{4h}	tI6	U
$L'3$	Fe_2N	$P6_3/mmc$	D^4_{6h}	hP3	H
$L6_0$	$CuTi_3$	P4/mmm	D^1_{4h}	tP4	T

[a] The first letter denotes the crystal system: triclinic (a), monoclinic (m), orthorhombic (o), tetragonal (t), hexagonal (h) and cubic (c). Trigonal (rhombohedral) system is presented by combination hR. The second letter of Pearson's symbol denotes lattice type: primitive (P), edge- (base-) centered (C), body-centered (I) or face-centered (F). The following number denotes amount of atoms in the crystal unit cell.

[b] Standard ASTM E157-82a has the Bravais lattices designations as following: C — primitive cubic; B — body-centered cubic; F — face-centered cubic; T — primitive tetragonal; U — body-centered tetragonal; R — rhombohedral; H — hexagonal; O — primitive orthorhombic; P — body-centered orthorhombic; Q — base-centered orthorhombic; S — face-centered orthorhombic; M — primitive monoclinic; N — centered monoclinic; A — triclinic.

REFERENCES

1. A. Schoenflies, *Kristallsysteme und Kristallstructur*, Leipzig, 1891.
2. E. S. Fedorow, *Zusammenstellung der kristallographischen Resultate*, *Zs. Krist.*, 20, 1892.
3. P. Groth, *Elemente der physikalischen und chemischen Krystallographie*, R. Oldenbourg, München/Berlin, 1921.
4. N. V. Belov, *Class Method of Deriving Space Groups of Symmetry*, Trudy Instituta Kristallodraffi imeni Fedorova (Transactions of the Fedorov Inst. of Crystallography), 5, 25, 1951, in Russian.
5. W. B. Pearson, *Handbook of Lattice Spacings and Structures of Metals and Alloys*, Vol. 1, Pergamon Press, 1958; Vol. 2, 1967.
6. Ch. Kittel, *Introduction to Solid State Physics*, John Wiley & Sons, 1956.
7. G. S. Zhdanov, *Fizika Tverdogo Tela (Solid State Physics)*, Moscow University Press, 1962, in Russian.
8. M. J. Buerger, *Elementary Crystallography*, John Wiley & Sons, 1963.
9. F. D. Bloss, *Crystallography & Crystal Chemistry*, Holt, Rinehart & Winston, 1971.
10. T. Janssen, *Crystallographic Groups*, North-Holland/American Elsevier, 1973.
11. M. P. Shaskolskaya, *Kristallografiya (Crystallography)*, Vysshaya Shkola, Moscow, 1976, in Russian.
12. T. Hahn, Ed., Internat. *Tables for Crystallography*, Vol. A, D. Reidel Publishing, Boston, 1983.
13. Crystal Data. Determinative Tables, Volumes 1—6, 1966—1983, JCPDS-Intern Centre for Diffraction Data and U.S. Dept. of Commerce.

IONIC RADII IN CRYSTALS

Ionic radii are a useful tool for predicting and visualizing crystal structures. This table lists a set of ionic radii R_i in Å units for the most common coordination numbers CN of positive and negative ions. The values are based on experimental crystal structure determinations, supplemented by empirical relationships, and theoretical calculations. The notation sq after the coordination number indicates a square configuration, while py indicates pyramidal.

The advice of Howard T. Evans and Marvin J. Weber in preparing this table is appreciated.

REFERENCES

1. Shannon, R. D., *Acta Crystallogr.* A32, 751, 1976.
2. Jia, Y.Q., *J. Solid State Chem.* 95, 184, 1991.

Ion	CN	R_i/Å	Ion	CN	R_i/Å	Ion	CN	R_i/Å
Anions			C^{+4}	4	0.15	Er^{+3}	6	0.89
F^{-1}	6	1.33		6	0.16		8	1.00
Cl^{-1}	6	1.81	Ca^{+2}	6	1.00	Eu^{+2}	6	1.17
Br^{-1}	6	1.96		8	1.12		8	1.25
I^{-1}	6	2.20		10	1.23		10	1.35
OH^{-1}	4	1.35		12	1.34	Eu^{+3}	6	0.95
	6	1.37	Cd^{+2}	4	0.78		8	1.07
O^{-2}	2	1.21		6	0.95	F^{+7}	6	0.08
	6	1.40		8	1.10	Fe^{+2}	4	0.63
	8	1.42		12	1.31		6	0.61
S^{-2}	6	1.84	Ce^{+3}	6	1.01		8	0.92
Se^{-2}	6	1.98		8	1.14	Fe^{+3}	4	0.49
Te^{-2}	6	2.21		10	1.25		6	0.55
				12	1.34		8	0.78
Cations			Ce^{+4}	6	0.87	Fr^{+1}	6	1.80
Ac^{+3}	6	1.12		8	0.97	Ga^{+3}	4	0.47
Ag^{+1}	4	1.00		10	1.07		6	0.62
	6	1.15		12	1.14	Gd^{+3}	6	0.94
	8	1.28	Cf^{+3}	6	0.95		8	1.05
Ag^{+2}	4sq	0.79	Cf^{+4}	6	0.82	Ge^{+2}	6	0.73
	6	0.94		8	0.92	Ge^{+4}	4	0.39
Al^{+3}	4	0.39	Cl^{+5}	3py	0.12		6	0.53
	5	0.48	Cl^{+7}	4	0.08	Hf^{+4}	4	0.58
	6	0.54	Cm^{+3}	6	0.97		6	0.71
Am^{+3}	6	0.98	Cm^{+4}	6	0.85		8	0.83
	8	1.09		8	0.95	Hg^{+1}	6	1.19
Am^{+4}	6	0.85	Co^{+2}	4	0.56	Hg^{+2}	2	0.69
	8	0.95		6	0.65		4	0.96
As^{+3}	6	0.58		8	0.90		6	1.02
As^{+5}	4	0.34	Co^{+3}	6	0.55		8	1.14
	6	0.46	Cr^{+2}	6	0.73	I^{+5}	3py	0.44
Au^{+1}	6	1.37	Cr^{+3}	6	0.62		6	0.95
Au^{+3}	4sq	0.64	Cr^{+4}	4	0.41	I^{+7}	4	0.42
	6	0.85		6	0.55		6	0.53
Ba^{+2}	6	1.35	Cr^{+6}	4	0.26	In^{+3}	4	0.62
	8	1.42		6	0.44		6	0.80
	12	1.61	Cs^{+1}	6	1.67	Ir^{+3}	6	0.68
Be^{+2}	4	0.27		8	1.74	Ir^{+4}	6	0.63
	6	0.45		10	1.81	Ir^{+5}	6	0.57
Bi^{+3}	5	0.96		12	1.88	K^{+1}	4	1.37
	6	1.03	Cu^{+1}	2	0.46		6	1.38
	8	1.17		4	0.60		8	1.51
Bi^{+5}	6	0.76		6	0.77		12	1.64
Bk^{+3}	6	0.96	Cu^{+2}	4sq	0.57	La^{+3}	6	1.03
Bk^{+4}	6	0.83		6	0.73		8	1.16
	8	0.93	Dy^{+2}	6	1.07		10	1.27
Br^{+5}	3py	0.31		8	1.19		12	1.36
Br^{+7}	4	0.25	Dy^{+3}	6	0.91	Li^{+1}	4	0.59
	6	0.39		8	1.03		6	0.76

Ion	CN	R_i/Å
	8	0.92
Lu+3	6	0.86
	8	0.97
Mg+2	4	0.57
	6	0.72
	8	0.89
Mn+2	4	0.66
	6	0.83
	8	0.96
Mn+3	6	0.58
Mn+4	4	0.39
	6	0.53
Mn+5	4	0.33
Mn+6	4	0.26
Mn+7	4	0.25
Mo+3	6	0.69
Mo+4	6	0.65
Mo+5	4	0.46
	6	0.61
Mo+6	4	0.41
	6	0.59
	7	0.73
N+3	6	0.16
N+5	6	0.13
Na+1	4	0.99
	6	1.02
	8	1.18
	9	1.24
	12	1.39
Nb+3	6	0.72
	8	0.79
Nb+4	6	0.68
Nb+5	4	0.48
	6	0.64
	8	0.74
Nd+3	6	0.98
	8	1.12
	9	1.16
	12	1.27
Ni+2	4sq	0.49
	6	0.69
Ni+3	6	0.56
Np+3	6	1.01
Np+4	6	0.87
Np+5	6	0.75
Np+6	6	0.72
Os+4	6	0.63
Os+5	6	0.58
Os+6	6	0.55
Os+8	4	0.39
P+5	4	0.17
	6	0.38
Pa+3	6	1.04
Pa+4	6	0.90
Pa+5	6	0.78
Pb+2	6	1.19
	8	1.29
	10	1.40
	12	1.49
Pb+4	4	0.65

Ion	CN	R_i/Å
	6	0.78
	8	0.94
Pd+2	4sq	0.64
	6	0.86
Pd+3	6	0.76
Pd+4	6	0.62
Pm+3	6	0.97
	8	1.09
Po+4	6	0.97
Pr+3	6	0.99
	8	1.13
Pr+4	6	0.85
	8	0.96
Pt+2	4sq	0.60
	6	0.80
Pt+4	6	0.63
Pu+3	6	1.00
Pu+4	6	0.86
Pu+5	6	0.74
Pu+6	6	0.71
Ra+2	8	1.48
	12	1.70
Rb+1	6	1.52
	8	1.61
	10	1.66
	12	1.72
Re+4	6	0.63
Re+5	6	0.58
Re+6	6	0.55
Re+7	4	0.38
	6	0.53
Rh+3	6	0.67
Rh+4	6	0.60
Rh+5	6	0.55
Ru+3	6	0.68
Ru+4	6	0.62
Ru+5	6	0.57
Ru+7	4	0.38
Ru+8	4	0.36
S+4	6	0.37
S+6	4	0.12
	6	0.29
Sb+3	4py	0.76
	6	0.76
Sb+5	6	0.60
Sc+3	6	0.75
	8	0.87
Se+4	6	0.50
Se+6	4	0.28
	6	0.42
Si+4	4	0.26
	6	0.40
Sm+2	6	1.19
	8	1.27
Sm+3	6	0.96
	8	1.08
	12	1.24
Sn+4	4	0.55
	6	0.69
	8	0.81

Ion	CN	R_i/Å
Sr+2	6	1.18
	8	1.26
	10	1.36
	12	1.44
Ta+3	6	0.72
Ta+4	6	0.68
Ta+5	6	0.64
Tb+3	6	0.92
	8	1.04
Tb+4	6	0.76
	8	0.88
Tc+4	6	0.65
Te+4	4	0.66
	6	0.97
Te+6	4	0.43
	6	0.56
Th+4	6	0.94
	8	1.05
	10	1.13
	12	1.21
Ti+2	6	0.86
Ti+3	6	0.67
Ti+4	4	0.42
	6	0.61
	8	0.74
Tl+1	6	1.50
	8	1.59
	12	1.70
Tl+3	4	0.75
	6	0.89
	8	0.98
Tm+2	6	1.01
	7	1.09
Tm+3	6	0.88
	8	0.99
U+3	6	1.03
U+4	6	0.89
	8	1.00
	12	1.17
U+5	6	0.76
U+6	2	0.45
	4	0.52
	6	0.73
	8	0.86
V+2	6	0.79
V+3	6	0.64
V+4	5	0.53
	6	0.58
	8	0.72
V+5	4	0.36
	5	0.46
	6	0.54
W+4	6	0.66
W+5	6	0.62
W+6	4	0.42
	5	0.51
	6	0.60
Y+3	6	0.90
	8	1.02
	9	1.08

Ion	CN	R_i/Å	Ion	CN	R_i/Å	Ion	CN	R_i/Å
Yb^{+2}	6	1.02	Zn^{+2}	4	0.60		6	0.72
	8	1.14		6	0.74		8	0.84
Yb^{+3}	8	0.99		8	0.90		9	0.89
	9	1.04	Zr^{+4}	4	0.59			

POLARIZABILITIES OF ATOMS AND IONS IN SOLIDS
H. P. R. Frederikse

The polarization of a solid dielectric medium, P, is defined as the dipole moment per unit volume averaged over the volume of a crystal cell. A component of P can be expanded as a function of the electric field E:

$$P_i = \sum_j a_j E_j + \sum_{jk} b_{jk} E_j E_k$$

For relatively small electric fields in isotropic substances $P = \chi_e E$, where χ_e is the electric susceptibility. If the medium is made up of N atoms (or ions) per unit volume, the polarization is $P = N p_m$ where p_m is the average dipole moment per atom. The polarizability α can be defined as $p_m = \alpha E_0$, where E_0 is the local field at the position of the atom. Using the Lorentz method to calculate the local field one finds:

$$P = N\alpha(E + 4\pi P) = \chi_e E$$

Together with the definition of the dielectric constant (relative permittivity), $\varepsilon = 1 + 4\pi\chi_e$, this leads to:

$$\alpha = \frac{3}{4\pi N}\left(\frac{\varepsilon - 1}{\varepsilon + 2}\right)$$

This expression is known as the Clausius-Mossotti equation.

The total polarization associated with atoms, ions, or molecules is due to three different sources:

1. Electronic polarization arises because the center of the local electronic charge cloud around the nucleus is displaced under the action of the field: $P_e = N\alpha_e E_0$ where α_e is the *electronic polarizability*.
2. Ionic polarization occurs in ionic materials because the electric field displaces cations and anions in opposite directions: $P_i = N\alpha_i E_0$, where α_i is the *ionic polarizability*.
3. Orientational polarization can occur in substances composed of molecules that have permanent electric dipoles. The alignment of these dipoles depends on temperature and leads to an *orientational polarizability* per molecule: $\alpha_{or} = p^2/3kT$, where p is the permanent dipole moment per molecule, k is the Boltzmann constant, and T is the temperature.

Because of the different nature of these three polarization processes the response of a dielectric solid to an applied electric field will strongly depend on the frequency of the field. The resonance of the electronic excitation in insulators (dielectrics) takes place in the ultraviolet part of the spectrum; the characteristic frequency of the lattice vibrations is located in the infrared, while the orientation of dipoles requires fields of much lower frequencies (below 10^{10} Hz). This response to electric fields of different frequencies is shown in Figure 1. Values of the electronic polarizabilities for selected atoms and ions are given in Table 1.

REFERENCES

1. Kittel, C., *Introduction to Solid State Physics*, Fourth Edition, John Wiley & Sons, New York, 1971.
2. Lerner, R.G., and Trigg, G.L., Editors, *Encyclopedia of Physics, Second Edition*, VCH Publishers, New York, 1990.
3. Ralls, K.M., Courtney, T.H., and Wulff, J., *An Introduction to Materials Science and Engineering*, John Wiley & Sons, New York, 1976.

Figure 1. Frequency dependence of the three contributions to the polarizability. (Reprinted with permission from Kittel, C., *Introduction to Solid State Physics*, Fourth Edition, John Wiley & Sons, New York, 1971.)

TABLE 1
Electronic Polarizabilities of Ions, in 10^{-24} cm³

		He	Li⁺	Be²⁺	B³⁺	C⁴⁺
Pauling		0.201	0.029	0.008	0.003	0.0013
JS			0.029			

	O²⁻	F⁻	Ne	Na⁺	Mg²⁺	Al³⁺	Si⁴⁺
Pauling	3.88	1.04	0.390	0.179	0.094	0.052	0.0165
JS-(TKS)	(2.4)	0.858		0.290			

	S²⁻	Cl⁻	Ar	K⁺	Ca²⁺	Sc³⁺	Ti⁴⁺
Pauling	10.2	3.66	1.62	0.83	0.47	0.286	0.185
JS-(TKS)	(5.5)	2.947		1.133	(1.1)		(0.19)

	Se²⁻	Br⁻	Kr	Rb⁺	Sr²⁺	Y³⁺	Zr⁴⁺
Pauling	10.5	4.77	2.46	1.40	0.86	0.55	0.37
JS-(TKS)	(7)	4.091		1.679	(1.6)		

	Te²⁻	I⁻	Xe	Cs⁺	Ba²⁺	La³⁺	Ce⁴⁺
Pauling	14.0	7.10	3.99	2.42	1.55	1.04	0.73
JS-(TKS)	(9)	6.116		2.743	(2.5)		

Values from L. Pauling, *Proc. R. Soc. London*, A114, 181, 1927; S.S. Jaswal and T.P. Sharma, *J. Phys. Chem. Solids*, 34, 509, 1973; and J. Tessman, A. Kahn, and W. Shockley, *Phys. Rev.*, 92, 890, 1953. The TKS polarizabilities are at the frequency of the D lines of sodium. The values are in CGS; to convert to SI, multiply by $1/9 \times 10^{-15}$. (Reprinted with permission from Kittel, C., *Introduction to Solid State Physics*, Fourth Edition, John Wiley & Sons, New York, 1971.

CRYSTAL STRUCTURES AND LATTICE PARAMETERS OF ALLOTROPES OF THE ELEMENTS

H. W. King

The crystal structures of the allotropic forms of the elements are presented in terms of the Pearson symbol, the Strukturbericht designation, and the prototype of the structure. The temperatures of the phase transformations are listed in degrees Celsius and the pressures are in GPa. A consistent nomenclature is used, whereby all allotropes are labeled by Greek letters. The lattice parameters of the unit cells are given in nanometers (nm) and are considered to be accurate to ±2 in the last reported digit.

This compilation is restricted to changes in crystal structure that occur as a result of a change in temperature or pressure. Low-temperature structures are included for the diatomic and rare gases, which show many similarities with respect to the metallic elements. The elements identified with an asterisk (*) have polymorphic structures based on different molecular configurations. The crystal data given for these elements refer to the most stable structure at room temperature.

Reprinted with the permission of ASM International from T. B. Massalski, Ed., Binary Alloy Phase Diagrams, ASM International, Metals Park, Ohio, 1986; certain data on rare earth elements were provided by K. A. Gschneidner.

Element	Temperature, °C	Pressure, GPa	Pearson symbol	Space group	Strukturbericht designation	Prototype	Lattice parameters, nm — a	b	c	Comment, c/a or α or β
Ac	25	atm	cF4	Fm3m	A1	Cu	0.5311	...	...	...
Ag	25	atm	cF4	Fm3m	A1	Cu	0.40857	...	...	...
αAl	25	atm	cF4	Fm3m	A1	Cu	0.40496	...	...	...
βAl	25	>20.5	hP2	P6₃/mmc	A3	Mg	0.2693	...	0.4398	1.6331
α'Am	25	atm	hP4	P6₃/mmc	A3'	αLa	0.34681	...	1.1241	2*1.621
αAm	>769	atm	cF4	Fm3m	A1	Cu	0.4894	...	...	...
βAm	>1074	atm	cI2	Im3m	A2	W	?	...	...	...
γAm	25	>15	oC4	Cmcm	A20	αU	0.3063	0.5968	0.5169	...
αAr	<−189.35	atm	cF4	Fm3m	A1	Cu	0.5316	...	...	...
(βAr)	<−189.40	atm	hP2	P6₃/mmc	A3	Mg	0.3760	...	0.6141	1.633
αAs	25	atm	hR2	R3m	A7	αAs	0.41319	...	...	α = 54.12°
εAs	>448	atm	oC8	Cmca	...	P (black)	0.362	1.085	0.448	...
Au	25	atm	cF4	Fm3m	A1	Cu	0.40782	...	...	...
βB	25	atm	hR105	R3m	...	βB	1.017	...	...	α = 65.12°
αBa	25	atm	cI2	Im3m	A2	W	0.50227	...	...	...
βBa	25	>5.33	hP2	P6₃/mmc	A3	Mg	0.3901	...	0.6154	1.5775
γBa	25	>23	?	?	...	...	...	...	...	...
αBe	25	atm	hP2	P6₃/mmc	A3	Mg	0.22859	...	0.35845	1.5681
βBe	>1270	atm	cI2	Im3m	A2	W	0.25515	...	...	...
γBe	25	>9.3	?	...	...	...	...	...	...	...
αBi	25	atm	hR2	R3m	A7	αAs	0.47460	...	...	α = 57.23°
βBi	25	>2.6	mC4	C2/m	...	βBi	0.6674	0.6117	0.3304	β − 110.33°
γBi	25	>3.0	mP3	?	...	...	0.605	0.42	0.465	β = 85.33°
σBi	25	>4.3	?	?	...	...	...	...	...	...
εBi	25	>6.5	?	?	...	...	...	...	...	...
ζBi	25	>9.0	cI2	Im3m	A2	W	0.3800	...	...	...
αBk	25	atm	hP4	P6₃/mmc	A3'	αLa	0.3416	...	1.1069	2*1.620
βBk	>977	atm	cF4	Fm3m	A1	Cu	0.4997	...	...	...
Br	<7.25	atm	oC8	Cmca	...	Cl	0.668	0.449	0.874	...
C (graphite)	25	atm	hP4	P6₃/mmc	A9	C (graphite)	0.24612	...	0.6709	2.7258
C (diamond)	25	>60	cF8	Fd3m	A4	C (diamond)	0.35669	...	...	...
C (hd)	25	HP	hP4	P6₃/mmc	...	C (hd)	0.2522	...	0.4119	1.633
αCa	25	atm	cF4	Fm3m	A1	Cu	0.55884	...	...	...
βCa	>443	atm	cI2	Im3m	A2	W	0.4480	...	...	...
γCa	25	>1.5	?	...	...	...	...	...	...	...
Cd	25	atm	hP2	P6₃/mmc	A3	Mg	0.29793	...	0.56196	1.8862
αCe	<−177	atm	cF4	Fm3m	A1	Cu	0.485	...	...	...
βCe	25	atm	hP4	P6₃/mmc	A3'	αLa	0.36810	...	1.1857	2*1.611
γCe	25	atm	cF4	Fm3m	A1	Cu	0.51610	...	...	...
δ-Ce	>726	atm	cI2	Im3m	A2	W	0.412	...	...	...
α'Ce	25	>5.4	oC4	Cmcm	A20	αU	0.3049	0.5998	0.5215	...
αCf	25	atm	hP4	P6₃/mmc	A3'	αLa	0.339	...	1.1015	2*1.625
βCf	>590	atm	cF4	Fm3m	A1	Cu	?	...	...	...
Cl	25	atm	oC8	Cmca	...	Cl	0.624	0.448	0.826	...
αCm	25	atm	hP4	P6₃/mmc	A3'	αLa	0.3496	...	1.1331	2*1.621
βCm	>1277	atm	cF4	Fm3m	A1	Cu	0.4382	...	...	...
εCo	25	atm	hP2	P6₃/mmc	A3	Mg	0.25071	...	0.40686	1.6228
αCo	>422	utm	cF4	Fm3m	A1	Cu	0.35447	...	...	...
αCr	25	atm	cI2	Im3m	A2	W	0.38848	...	...	...
α'Cr	25	HP	tI2	I4/mmm	...	α'Cr	0.2882	...	0.2887	1.002
αCs	25	atm	cI2	Im3m	A2	W	0.6141	...	...	...
βCs	25	>2.37	cF4	Fm3m	A1	Cu	0.6465	...	...	...
β'Cs	25	>4.22	cF4	Fm3m	A1	Cu	0.5800	...	...	...
γCs	25	>4.27	?	...	...	...	...	...	...	...
Cu	25	atm	cF4	Fm3m	A1	Cu	0.36146	...	...	...
α'Dy	<−187	atm	oC4	Cmcm	...	α'Dy	0.3595	0.6184	0.5678	...
αDy	25	atm	hP2	P6₃/mmc	A3	Mg	0.35915	...	0.56501	1.5732

Element	Temperature, °C	Pressure, GPa	Pearson symbol	Space group	Strukturbericht designation	Prototype	Lattice parameters, nm			Comment, c/a or α or β
							a	b	c	
βDy	>1381	atm	cI2	Im3m	A2	W	0.403	···	···	···
γDy	25	>7.5	hR3	R3m	···	αSm	0.3436	···	2.483	4.5*1.606
Er	25	atm	hP2	P6₃/mmc	A3	Mg	0.35592	···	0.55850	1.5692
αEs	25	atm	hP4	P6₃/mmc	A3′	αLa	?	···	···	···
βEs	?	atm	cF4	Fm3m	A1	Cu	?	···	···	···
Eu	25	atm	cI2	Im3m	A2	W	0.45827	···	···	···
αF	<−227.6	atm	mC8	C2/c	···	αF	0.550	0.338	0.728	β = 102.17°
βF	<−219.67	atm	cP16	Pm3n	···	γO	0.667	···	···	···
αFe	25	atm	cI2	Im3m	A2	W	0.28665	···	···	···
γFe	>912	atm	cF4	Fm3m	A1	Cu	0.36467	···	···	···
σFe	>1394	atm	cI2	Im3m	A2	W	0.29315	···	···	···
εFe	25	>13	hP2	P6₃/mmc	A3	Mg	0.2468	···	0.396	1.603
αGa	25	atm	oC8	Cmca	A11	αGa	0.45186	0.76570	0.45258	···
βGa	25	>1.2	tI2	I4/mmm	A6	In	0.2808	···	0.4458	1.588
γGa	−53	>3.0	oC40	Cmcm	···	γGa	1.0593	1.3523	0.5203	···
αGd	25	atm	hP2	P6₃/mmc	A3	Mg	0.36336	···	0.57810	1.5910
βGd	>1235	atm	cI2	Im3m	A2	W	0.406	···	···	···
γGd	25	>3.0	hR3	R3m	···	αSm	0.361	···	2.603	4*1.60
αGe	25	atm	cF8	Fd3m	A4	C (diamond)	0.56574		···	···
βGe	25	>12	tI4	I4₁/amd	A5	βSn	0.4884		0.2692	0.551
γGe	25	>12 → atm	tP12	P4₁2₁2	···	σGe	0.593		0.698	1.18
σGe	LT	>12	cI16	Im3m	···	γSi	0.692		···	···
αH	<−271.9	atm	cF4	Fm3m	A1	Cu	0.5338		···	···
βH	<−259.34	atm	hP2	P6₃/mmc	A3	Mg	0.3776		0.6162	1.632
αHe	<−268.94	atm	hP2	P6₃/mmc	A3	Mg	0.3555		0.5798	1.631
βHe	>−258	0.125	cF4	Fm3m	A1	Cu	0.4240		···	···
γHe	<−271.47	0.03	cI2	Im3m	A2	W	0.4110		···	···
αHf	25	atm	hP2	P6₃/mmc	A3	Mg	0.31946		0.50510	1.5811
βHf	>1995	atm	cI2	Im3m	A2	W	0.3610		···	···
αHg	<−38.84	atm	hR1	R3m	A10	αHg	0.3005	···	···	α = 70.53°
βHg	<−194	HP	tI2	I4/mmm	···	βHg	0.3995	···	0.2825	0.707
γHg	<−194	c.w.	hR1	?	···	···	···	···	···	···
αHo	25	atm	hP2	P6₃/mmc	A3	Mg	0.35778	···	0.56178	1.5702
βHo	25	>7.5	hR3	R3m	···	αSm	0.334	···	2.45	4.5*1.63
I	25	atm	oC8	Cmca	···	Cl	0.72697	0.47903	0.97942	···
In	25	atm	tI2	I4/mmm	A6	In	0.3253	···	0.49470	1.5210
Ir	25	atm	cF4	Fm3m	A1	Cu	0.38392	···	···	···
K	25	atm	cI2	Im3m	A2	W	0.5321	···	···	···
Kr	<−157.39	atm	cF4	Fm3m	A1	Cu	0.5810	···	···	···
αLa	25	atm	hP4	P6₃/mmc	A3′	αLa	0.37740	···	1.2171	2*1.6125
βLa	>310	atm	cF4	Fm3m	A1	Cu	0.5303	···	···	···
γLa	>865	atm	cI2	Im3m	A2	W	0.426	···	···	···
β′La	25	>2.0	cF4	Fm3m	A1	Cu	0.517	···	···	···
αLi	<−193	atm	hP2	P6₃/mmc	A3	Mg	0.3111	···	0.5093	1.637
βLi	25	atm	cI2	Im3m	A2	W	0.35093	···	···	···
γLi	<−201	c.w.	cF4	Fm3m	A1	Cu	0.4388	···	···	···
Lu	25	atm	hP2	P6₃/mmc	A3	Mg	0.35052	···	0.55494	1.5832
Mg	25	atm	hP2	P6₃/mmc	A3	Mg	0.32094	···	0.52107	1.6236
αMn	25	atm	cI58	I43m	A12	αMn	0.89126	···	···	···
βMn	>710	atm	cP20	P4₁32	A13	βMn	0.63152	···	···	···
γMn	>1079	atm	cF4	Fm3m	A1	Cu	0.3860	···	···	···
σMn	>1143	atm	cI2	Im3m	A2	W	0.3080	···	···	···
Mo	25	atm	cI2	Im3m	A2	W	0.31470	···	···	···
αN	<−237.6	atm	cP8	Pa3	···	αN	0.5661	···	···	···
βN	<−210.00	atm	hP4	P6₃/mmc	···	βN	0.4050	···	0.6604	1.631
γN	<−253	>3.3	tP4	P4₂/mnm	···	γN	0.3957	···	0.5109	1.291
αNa	<−233	atm	hP2	P6₃/mmc	A3	Mg	0.3767	···	0.6154	1.634
βNa	25	atm	cI2	Im3m	A2	W	0.42906	···	···	···
Nb	25	atm	cI2	Im3m	A2	W	0.33004	···	···	···
αNd	25	atm	hP4	P6₃/mmc	A3′	αLa	0.36582	···	1.17966	2*1.6124
βNd	>863	atm	cI2	Im3m	A2	W	0.413	···	···	···
γNd	25	>5.0	cF4	Fm3m	A1	Cu	0.480	···	···	···
Ne	<−243.59	atm	cF4	Fm3m	A1	Cu	0.4462	···	···	···
Ni	25	atm	cF4	Fm3m	A1	Cu	0.35240	···	···	···
αNp	25	atm	oP8	Pnma	A_c	αNp	0.6663	0.4723	0.4887	···
βNp	>280	atm	tP4	P42₁2	A_d	βNp	0.4883	···	0.3389	0.694
γNp	>576	atm	cI2	Im3m	A2	W	0.352	···	···	···
αO	<−243.3	atm	mC4	C2m	···	αO	0.5403	0.3429	0.5086	β = 132.53°
βO	<−229.6	atm	hR2	R3m	···	βO	0.4210	···	···	α = 46.27°
γO	<−218.79	atm	cP16	Pm3n	···	γO	0.683	···	···	···
Os	25	atm	hP2	P6₃/mmc	A3	Mg	0.27341	···	0.43918	1.6063
P (black)	25	atm	oC8	Cmca	···	P (black)	0.33136	1.0478	0.43763	···
αPa	25	atm	tI2	I4/mmm	A_a	αPa	0.3921	···	0.3235	0.825
βPa	>1170	atm	cI2	Im3m	A2	W	0.381	···	···	···
αPb	25	atm	cF4	Fm3m	A1	Cu	0.49502	···	···	···
βPb	25	>10.3	hP2	P6₃/mmc	A3	Mg	0.3265	···	0.5387	1.650

CRYSTAL STRUCTURES AND LATTICE PARAMETERS OF ALLOTROPES OF THE ELEMENTS
(continued)

Element	Temperature, °C	Pressure, GPa	Pearson symbol	Space group	Strukturbericht designation	Prototype	Lattice parameters, nm			Comment, c/a or α or β
							a	b	c	
Pd	25	atm	cF4	$Fm3m$	A1	Cu	0.38903	...	...	...
αPm	25	atm	hP4	$P6_3/mmc$	A3'	αLa	0.365	...	1.165	2*1.60
βPm	>890	atm	cI2	$Im3m$	A2	W	(0.410)	...	...	...
αPo	25	atm	cP1	$Pm3m$	A_h	αPo	0.3366	...	...	...
βPo	>54	atm	hR1	$R3m$	...	βPo	0.3373	...	...	α = 98.08°
αPr	25	atm	hP4	$P6_3/mmc$	A3'	αLa	0.36721	...	1.18326	2*1.6111
βPr	>795	atm	cI2	$Im3m$	A2	W	0.413	...	...	...
γPr	25	>4.0	cF4	$Fm3m$	A1	Cu	0.488	...	...	...
Pt	25	atm	cF4	$Fm3m$	A1	Cu	0.39236	...	...	...
αPu	25	atm	mP16	$P2_1/m$	...	αPu	0.6183	0.4822	1.0963	β = 101.97°
βPu	>125	atm	mI34	$I2/m$	...	βPu	0.9284	1.0463	0.7859	β = 92.13°
γPu	>215	atm	oF8	$Fddd$	...	γPu	0.31587	0.57682	1.0162	...
σPu	>320	atm	cF4	$Fm3m$	A1	Cu	0.46371	...	...	...
σ'Pu	>463	atm	tI2	$I4/mmm$	A6	In	0.33261	...	0.44630	1.3418
εPu	>483	atm	cI2	$Im3m$	A2	W	0.36343	...	...	...
Ra	25	atm	cI2	$Im3m$	A2	W	0.5148	...	...	...
αRb	25	atm	cI2	$Im3m$	A2	W	0.5705	...	...	...
βRb	25	>1.08	?	...	...	...	...	...	...	...
γRb	25	>2.05	?	...	...	...	...	...	...	...
Re	25	atm	hP2	$P6_3/mmc$	A3	Mg	0.27609	...	0.4458	1.6145
Rh	25	atm	cF4	$Fm3m$	A1	Cu	0.38032	...	...	...
Ru	25	atm	hP2	$P6_3/mmc$	A3	Mg	0.27058	...	0.42816	1.5824
αS	25	atm	oF128	$Fddd$	A16	αS	1.0464	1.28660	2.44860	...
αSb	25	atm	hR2	$R3m$	A7	αAs	0.45067	...	...	α = 57.11°
βSb	25	>5.0	cP1	$Pm3m$	A_h	αPo	0.2992	...	...	...
γSb	25	>7.5	hP2	$P6_3/mmc$	A3	Mg	0.3376	...	0.5341	1.582
σSb	25	>14.0	mP3	?	...	...	0.556	0.404	0.422	β = 86.0°
αSc	25	atm	hP2	$P6_3/mmc$	A3	Mg	0.33088	...	0.52680	1.5921
βSc	>1337	atm	cI2	$Im3m$	A2	W	(0.373)	...	...	...
γSe	25	atm	hP3	$P3_121$	A8	γSe	0.43659	...	0.49537	1.1346
αSi	25	atm	cF8	$Fd3m$	A4	C (diamond)	0.54306	...	...	...
βSi	25	>9.5	tI4	$I4_1/amd$	A5	βSn	0.4686	...	0.2585	0.552
γSi	25	>16.0	cI16	$Im3m$	...	γSi	0.6636	...	...	...
σSi	25	>16 → atm	hP4	$P6_3/mmc$	A3'	αLa	0.380	...	0.628	1.653
αSm	25	atm	hR3	$R3m$	...	αSm	0.36290	...	2.6207	4*1.6048
βSm	>734	atm	hP2	$P6_3/mmc$	A3	Mg	0.36630	...	0.58448	1.5956
γ'Sm	>922	atm	cI2	$Im3m$	A2	W	(0.410)	...	...	...
σSm	25	>4.0	hP4	$P6_3/mmc$	A3'	αLa	0.3618	...	1.166	2*1.611
αSn	<13	atm	cF8	$Fd3m$	A4	C (diamond)	0.64892	...	...	...
βSn	25	atm	tI4	$I4_1/amd$	A5	βSn	0.58318	...	0.31818	0.5456
γSn	25	>9.0	tI2	?	...	γSn	0.370	...	0.337	0.91
αSr	25	atm	cF4	$Fm3m$	A1	Cu	0.6084	...	...	...
βSr	>547	atm	cI2	$Im3m$	A2	W	0.487	...	...	...
β'Sr	25	>3.5	cI2	$Im3m$	A2	W	0.4437	...	...	...
Ta	25	atm	cI2	$Im3m$	A2	W	0.33030	...	...	...
α'Tb	<-53	atm	oC4	$Cmcm$	...	α'Dy	0.3605	0.6244	0.5706	...
αTb	25	atm	hP2	$P6_3/mmc$	A3	Mg	0.36055	...	0.56966	1.5800
βTb	>1289	atm	cI2	$Im3m$	A2	W	(0.407)	...	...	...
γTb	25	>6.0	hR3	$R3m$	...	αSm	0.341	...	2.45	4*1.60
Tc	25	atm	hP2	$P6_3/mmc$	A3	Mg	0.2738	...	0.4393	1.604
αTe	25	atm	hP3	$P3_121$	A8	γSe	0.44566	...	0.59264	1.3298
βTe	25	>2.0	hR2	$R3m$	A7	αAs	0.469	...	...	α = 53.30°
γTe	25	>7.0	hR1	$R3m$	...	βPo	0.3002	...	...	α = 103.3°
αTh	25	atm	cF4	$Fm3m$	A1	Cu	0.50842	...	...	...
βTh	>1360	atm	cI2	$Im3m$	A2	W	0.411	...	...	...
αTi	25	atm	hP2	$P6_3/mmc$	A3	Mg	0.29506	...	0.46835	1.5873
βTi	>882	atm	cI2	$Im3m$	A2	W	0.33065	...	...	...
ωTi	25	HP → atm	hP3	$P6/mmm$	...	ωTi	0.4625	...	0.2813	0.6082
αTl	25	atm	hP2	$P6_3/mmc$	A3	Mg	0.34566	...	0.55248	1.5983
βTl	>230	atm	cI2	$Im3m$	A2	W	0.3879	...	...	...
γTl	25	HP	cF4	$Fm3m$	A1	Cu	?	...	...	...
Tm	25	atm	hP2	$P6_3/mmc$	A3	Mg	0.35375	...	0.55540	1.5700
αU	25	atm	oC4	$Cmcm$	A20	αU	0.28537	0.58695	0.49548	...
βU	>668	atm	tP30	$P4_2/mnm$	A_b	βU	1.0759	...	0.5656	0.526
γU	>776	atm	cI2	$Im3m$	A2	W	0.3524	...	...	...
V	25	atm	cI2	$Im3m$	A2	W	0.30240	...	...	...
W	25	atm	cI2	$Im3m$	A2	W	0.31652	...	...	...
Xe	<-111.76	atm	cF4	$Fm3m$	A1	Cu	0.6350	...	...	...
αY	25	atm	hP2	$P6_3/mmc$	A3	Mg	0.36482	...	0.57318	1.5711
βY	>1478	atm	cI2	$Im3m$	A2	W	(0.410)	...	...	...
αYb	<-3	atm	hP2	$P6_3/mmc$	A3	Mg	0.38799	...	0.63859	1.6459
βYb	25	atm	cF4	$Fm3m$	A1	Cu	0.54848	...	...	...
γYb	>795	atm	cI2	$Im3m$	A2	W	0.444	...	...	...
Zn	25	atm	hP2	$P6_3/mmc$	A3	Mg	0.26650	...	0.49470	1.8563
αZr	25	atm	hP2	$P6_3/mmc$	A3	Mg	0.32316	...	0.51475	1.5929
βZr	>863	atm	cI2	$Im3m$	A2	W	0.36090	...	...	...
ωZr	25	HP → atm	hP2	$P6/mmm$	...	ωTi	0.5036	...	0.3109	0.617

LATTICE ENERGIES

H. D. B. Jenkins

Table 1 contains calculated values of the lattice energies (total lattice potential energies), U_{POT}, of crystalline salts, M_aX_b.
U_{POT} is expressed in units of kilojoules per mole, kJ mol^{-1}. M and X can be either simple or complex ions.

Also listed in the table is the lattice energy, U_{POT}^{BFHC}, obtained from the application of the Born-Fajans-Haber cycle (BFHC) described below, using thermochemical data from Reference 1 plus certain other data which are given in Table 2 of this section (Reference 2).

The lattice enthalpy, ΔH_L, is given by the cycle:

where (ss) is the standard state of the element concerned.

The lattice enthalpy, ΔH_L, is obtained using the equation:

$$\Delta H_L = a\Delta H_f^{\circ}\left(M^{Za+}\right)(g) + b\Delta H_f^{\circ}\left(X^{Zb-}\right)(g) - \Delta H_f^{\circ}\left(M_aX_b\right)(c)$$

and is further related to the total lattice potential energy, U_{POT}^{BFHC}, by the relationship:

$$U_{POT}^{BFHC}\left(M_aX_b\right) = \Delta H_L - \left[a\left(\frac{n_{M^{Za+}}}{2} - 2\right) + b\left(\frac{n_{X^{Zb-}}}{2} - 2\right)\right]RT$$

where $n_{M^{Za+}}$ and $n_{X^{Zb-}}$ is equal to 3 for monatomic ions, 5 for linear polyatomic ions and 6 for polyatomic non-linear ions.

Substances are arranged by chemical class.

REFERENCES

1. Wagman, D. D., et al., *The NBS Tables of Chemical Thermodynamic Properties, J. Phys. Chem. Ref. Data*, 11, Suppl. 2, 1982.
2. *Adv. Inorg. Chem. Radiochem.*, 22, 1, 1978.

Table 1
LATTICE ENERGIES

Substance	Calc. U_{POT} (kJ mol^{-1})	U_{POT}^{BFHC} (kJ mol^{-1})	Substance	Calc. U_{POT} (kJ mol^{-1})	U_{POT}^{BFHC} (kJ mol^{-1})
Acetates			Mn(CH$_3$COO)$_2$	2548	2616
Li(CH$_3$COO)	—	881	Cu(CH$_3$COO)$_2$	—	2835
Na(CH$_3$COO)	761	763	Zn(CH$_3$COO)$_2$	2615	2750
K(CH$_3$COO)	686	682	Hg(CH$_3$COO)$_2$	2368	2595
Rb(CH$_3$COO)	715	656	Pb(CH$_3$COO)$_2$	2247	2225
Cs(CH$_3$COO)	682	—	Acetylides		
NH$_4$ (CH$_3$COO)	725	695	CaC$_2$	2911	2904
Ag(CH$_3$COO)	—	863	SrC$_2$	2788	2784
Tl(CH$_3$COO)	—	750	BaC$_2$	2647	2654
Ca(CH$_3$COO)$_2$	2431	2294	Ammonium salts		
Sr(CH$_3$COO)$_2$	2280	2166	NH$_4$HF$_2$	705	705
Ba(CH$_3$COO)$_2$	2180	2033	NH$_4$HCO$_3$	—	741

Substance	Calc. U_{POT} (kJ mol^{-1})	U_{POT}^{BFHC} (kJ mol^{-1})	Substance	Calc. U_{POT} (kJ mol^{-1})	U_{POT}^{BFHC} (kJ mol^{-1})
NH_4HSO_4	640	(645)	$Cd(N_3)_2$	2446	2454
NH_4BF_4	582	—	$Pb(N_3)_2$	—	2173
NH_4NCO	724	—	Bihalide salts		
NH_4CN	617	670	$LiHF_2$	893	866
NH_4HCO_2	715	—	$NaHF_2$	788	788
$NH_4(NH_2CH_2CO_2)$	650	—	KHF_2	703	698
NH_4IO_3	—	808	$RbHF_2$	674	676
$NH_4IO_2F_2$	678	685	$CsHF_2$	646	628
NH_4HS	661	666	NH_4HF_2	705	705
NH_4NO_3	661	676	$CsHCl_2$	601	—
NH_4ClO_4	583	580	Me_4NHCl_2	427	—
$(NH_4)_2S$	2008	(2026)	Et_4NHCl_2	346	—
$(NH_4)_2SO_4$	1766	1777	Bu_4NHCl_2	290	—
$(NH_4)_2GeF_6$	1657	—	Bicarbonates		
$(NH_4)_2IrCl_6$	1442	1440	$NaHCO_3$	820	818
$(NH_4)_2OsCl_6$	1433	—	$KHCO_3$	741	736
$(NH_4)_2PdCl_6$	1481	—	$RbHCO_3$	707	714
$(NH_4)_2PbCl_6$	1355	—	$CsHCO_3$	678	709
$(NH_4)_2PbCl_6$	1468	—	NH_4HCO_3	—	741
$(NH_4)_2ReCl_6$	1402	1390	$Ca(HCO_3)_2$	2402	(2403)
$(NH_4)_2SiF_6$	1657	1727	$Sr(HCO_3)_2$	2255	(2272)
$(NH_4)_2SnCl_6$	1370	1334	$Ba(HCO_3)_2$	2155	(2159)
$(NH_4)_2SnBr_6$	1319	—	Bisulphates		
$(NH_4)_2TeCl_6$	1318	—	NH_4HSO_4	640	(645)
$(NH_4)_2TeBr_6$	1294	—	Borides		
$(NH_4)_2TiCl_6$	1413	—	CaB_6	5146	—
NH_4CNS	605	611	SrB_6	5104	—
$(NH_4)_2SeCl_6$	1420	—	BaB_6	5021	—
$(NH_4)_2SeBr_6$	1380	—	YB_6	7447	—
Arsenates			LaB_6	7406	—
$Mg_3(AsO_4)_2$	10669	10716	CeB_6	10083	—
$Ca_3(AsO_4)_2$	9749	9653	PrB_6	7447	—
$Sr_3(AsO_4)_2$	9330	9266	NdB_6	7447	—
$Ba_3(AsO_4)_2$	8870	8985	PmB_6	7406	—
$AlAsO_4$	7255	—	SmB_6	7447	—
$GaAsO_4$	7243	—	EuB_6	5104	—
Astatides			GdB_6	7489	—
$LiAt$	720	—	TbB_6	7489	—
$NaAt$	657	—	DyB_6	7489	—
KAt	615	—	HoB_6	7489	—
$RbAt$	594	—	ErB_6	7489	—
$CsAt$	586	—	TmB_6	7489	—
$FrAt$	573	—	YbB_6	5146	—
Azides			LuB_6	7489	—
LiN_3	812	818	ThB_6	10167	—
NaN_3	732	731	Borohydrides		
KN_3	659	658	$LiBH_4$	778	—
RbN_3	637	632	$NaBH_4$	703	—
CsN_3	612	604	KBH_4	665	—
AgN_3	854	—	$RbBH_4$	648	—
TlN_3	689	686	$CsBH_4$	628	—
$Ca(N_3)_2$	2186	2162	Borohalides		
$Sr(N_3)_2$	2056	2066	$LiBF_4$	699	—
$Ba(N_3)_2$	2021	1965	$NaBF_4$	657	619
$Mn(N_3)_2$	2408	2416	KBF_4	611	631
$Cu(N_3)_2$	2730	2738	$RbBF_4$	577	605
$Zn(N_3)_2$	2840	2848	$CsBF_4$	556	—

Table 1 (continued)

Substance	Calc. U_{POT} (kJ mol^{-1})	U_{POT}^{BFHC} (kJ mol^{-1})	Substance	Calc. U_{POT} (kJ mol^{-1})	U_{POT}^{BFHC} (kJ mol^{-1})
NH$_4$BF$_4$	582	—	Sr(HCO$_2$)$_2$	2221	2261
Co(BF$_4$)$_2$	2127	—	Ba(HCO$_2$)$_2$	2092	2134
Ni(BF$_4$)$_2$	2136	—	Mn(HCO$_2$)$_2$	2598	2701
Zn(BF$_4$)$_2$	2063	—	Co(HCO$_2$)$_2$·	—	2792
Cd(BF$_4$)$_2$	1937	—	Ni(HCO$_2$)$_2$	—	2880
KBCl$_4$	506	—	Cu(HCO$_2$)$_2$	2870	2913
RbBCl$_4$	489	—	Zn(HCO$_2$)$_2$	2791	2847
CsBCl$_4$	473	—	Cd(HCO$_2$)$_2$	2556	—
Carbonates			Pb(HCO$_2$)$_2$	2276	2330
Li$_2$CO$_3$	2523	2269	Germanates		
Na$_2$CO$_3$	2301	2030	Mg$_2$GeO$_4$	7991	—
K$_2$CO$_3$	2084	1858	Ca$_2$GeO$_4$	7301	7306
Rb$_2$CO$_3$	2000	1795	Sr$_2$GeO$_4$	6987	—
Cs$_2$CO$_3$	1920	1702	Ba$_2$GeO$_4$	6653	6643
MgCO$_3$	3180	3122	Glycinates		
CaCO$_3$	2804	2810	Na(NH$_2$CH$_2$COO$_2$)	739	—
SrCO$_3$	2720	2688	K(NH$_2$CH$_2$COO$_2$)	668	—
BaCO$_3$	2615	2554	Rb(NH$_2$CH$_2$COO$_2$)	648	—
MnCO$_3$	3046	3151	NH$_4$(NH$_2$CH$_2$COO$_2$)	650	—
FeCO$_3$	3121	3171	Cu(NH$_2$CH$_2$CO$_2$)$_2$	2694	—
CoCO$_3$	3443	3232	Halates		
NiCO$_3$	—	3297	LiBrO$_3$	883	897
CuCO$_3$	3494	3327	NaBrO$_3$	803	814
ZnCO$_3$	3121	3273	KBrO$_3$	740	745
CdCO$_3$	2929	3052	RbBrO$_3$	720	742
SnCO$_3$	2904	(2853)	CsBrO$_3$	694	681
PbCO$_3$	2728	2750	NaClO$_3$	770	770
Cyanates			KClO$_3$	711	706
LiNCO	770	—	RbClO$_3$	690	687
NaNCO	736	—	CsClO$_3$	—	647
KNCO	653	—	LiIO$_3$	975	—
RbNCO	615	—	NaIO$_3$	883	882
CsNCO	586	—	KIO$_3$	820	806
NH$_4$NCO	724	—	RbIO$_3$	791	—
Cyanides			CsIO$_3$	761	769
LiCN	—	849	NH$_4$IO$_3$	—	808
NaCN	738	739	Mg(ClO$_3$)$_2$	2535	(2475)
KCN	674	669	Ca(ClO$_3$)$_2$	2259	2286
RbCN	646	(640)	Sr(ClO$_3$)$_2$	2138	2155
CsCN	612	602	Ba(ClO$_3$)$_2$	2021	2027
NH$_4$CN	617	670	Halides		
Ca(CN)$_2$	2268	2191	LiF	1030	1036
Sr(CN)$_2$	2138	(2076)	LiCl	834	853
Ba(CN)$_2$	2046	1960	LiBr	788	807
CuCN	—	1035	LiI	730	757
AgCN	741	914	NaF	910	923
Zn(CN)$_2$	2431	2768	NaCl	769	786
Cd(CN)$_2$	2284	2542	NaBr	732	747
Formates			NaI	682	704
Li(HCO$_2$)	865	872	KF	808	821
Na(HCO$_2$)	791	811	KCl	701	715
K(HCO$_2$)	713	729	KBr	671	682
Rb(HCO$_2$)	685	682	KI	632	649
Cs(HCO$_2$)	651	644	RbF	774	785
NH$_4$(HCO$_2$)	715	—	RbCl	680	689
Mg(HCO$_2$)$_2$	2674	—	RbBr	651	660
Ca(HCO$_2$)$_2$	2360	2390	RbI	617	630

Table 1 (continued)

Substance	Calc. U_{POT} (kJ mol^{-1})	U_{POT}^{BFHC} (kJ mol^{-1})	Substance	Calc. U_{POT} (kJ mol^{-1})	U_{POT}^{BFHC} (kJ mol^{-1})
CsF	744	740	VBr$_2$	—	2523
CsCl	657	659	VI$_2$	—	2456
CsBr	632	631	CrF$_2$	2778	2917
CsI	600	604	CrCl$_2$	2455	2586
FrF	715	—	CrBr$_2$	2377	2523
FrCl	632	—	CrI$_2$	2269	2425
FrBr	611	—	MoCl$_2$	2485	2733
FrI	582	—	MoBr$_2$	2448	2742
CuCl	921	996	MoI$_2$	2422	2630
CuBr	879	979	MnF$_2$	2644	—
CuI	835	966	MnCl$_2$	2368	2537
AgF	953	967	MnBr$_2$	2304	2471
AgCl	864	915	MnI$_2$	2212	—
AgBr	830	904	FeF$_2$	2769	—
AgI	808	889	FeCl$_2$	2525	2631
AuCl	1013	1066	FeBr$_2$	2464	2569
AuBr	1015	1061	FeI$_2$	2382	2480
AuI	1015	1070	CoF$_2$	2878	3018
InCl	—	763	CoCl$_2$	2709	2691
InBr	—	767	CoBr$_2$	2648	2629
InI	—	732	CoI$_2$	2569	2545
TlF	—	845	NiF$_2$	2845	3066
TlCl	782	751	NiCl$_2$	2753	2772
TlBr	713	735	NiBr$_2$	2699	2709
TlI	687	709	NiI$_2$	2607	2623
BeF$_2$	3150	3505	PdCl$_2$	2766	2778
BeCl$_2$	3004	3020	PdBr$_2$	—	2741
BeBr$_2$	2950	2914	PdI$_2$	—	2748
BeI$_2$	2653	2800	CuF$_2$	3046	3082
MgF$_2$	2913	2957	CuCl$_2$	2774	2811
MgCl$_2$	2326	2526	CuBr$_2$	2711	2763
MgBr$_2$	2097	2440	CuI$_2$	2640	—
MgI$_2$	1944	2327	AgF$_2$	2919	2942
CaF$_2$	2609	2630	ZnF$_2$	2930	3032
CaCl$_2$	2223	2258	ZnCl$_2$	2690	2734
CaBr$_2$	2132	2176	ZnBr$_2$	2632	2678
CaI$_2$	1905	2074	ZnI$_2$	2549	2605
SrF$_2$	2476	2492	CdF$_2$	2740	2809
SrCl$_2$	2127	2156	CdCl$_2$	2226	2552
SrI$_2$	2008	2075	CdBr$_2$	2468	2507
SrI$_2$	1937	1963	CdI$_2$	2406	2441
BaF$_2$	2341	2352	HgF$_2$	2757	(2798)
BaCl$_2$	2033	2056	HgCl$_2$	2569	2651
BaBr$_2$	1950	1985	HgBr$_2$	2598	2628
BaI$_2$	1831	1877	HgI$_2$	2569	2610
RaF$_2$	2284	—	SnF$_2$	2551	—
RaCl$_2$	2004	—	SnCl$_2$	2276	2297
RaBr$_2$	1929	—	SnBr$_2$	2211	2245
RaI$_2$	1803	—	SnI$_2$	2123	2193
ScCl$_2$	2380	—	PbF$_2$	2460	2522
ScBr$_2$	2291	—	PbCl$_2$	2229	2269
ScI$_2$	2201	—	PbBr$_2$	2169	2219
TiF$_2$	2724	—	PbI$_2$	2086	2163
TiCl$_2$	2431	2501	ScF$_3$	5096	5492
TiBr$_2$	2360	2419	ScCl$_3$	4874	4866
TiI$_2$	2259	2329	ScBr$_3$	4711	4729
VCl$_2$	2607	2579	ScI$_3$	4640	—

Substance	Calc. U_{POT} (kJ mol^{-1})	U_{POT}^{BFHC} (kJ mol^{-1})	Substance	Calc. U_{POT} (kJ mol^{-1})	U_{POT}^{BFHC} (kJ mol^{-1})
YF$_3$	4983	—	GaBr$_3$	4966	5552
YCl$_3$	4447	4506	GaI$_3$	4611	5476
YBr$_3$	4410	—	InCl$_3$	4736	5187
YI$_3$	4125	4240	InBr$_3$	4535	5124
TiF$_3$	5644	—	InI$_3$	4234	5005
TiCl$_3$	5134	5134	TlF$_3$	5493	—
TiBr$_3$	5012	5007	TlCl$_3$	5252	5258
TiI$_3$	4845	—	TlBr$_3$	5171	—
ZrF$_3$	—	(5400)	TlI$_3$	5088	—
ZrCl$_3$	—	4791	AsBr$_3$	—	5497
ZrBr$_3$	—	4758	AsI$_3$	(3758)	4824
ZrI$_3$	—	(4591)	SbF$_3$	—	5295
VF$_3$	5895		SbCl$_3$	—	5032
VCl$_3$	5322	5315	SbBr$_3$	—	4954
BVr$_3$	5192	5214	SbI$_3$	—	4867
VI$_3$	5058	5121	BiCl$_3$	—	4689
NbCl$_3$	5062	—	BiBr$_3$	—	—
NbBr$_3$	4980	—	BiI$_3$	(3774)	—
NbI$_3$	4860	—	LaF$_3$	4682	—
CrF$_3$	5958	6033	LaCl$_3$	4343	4242
CrCl$_3$	5473	5509	LaBr$_3$	4209	—
CrBr$_3$	5355	—	LaI$_3$	3916	3986
CrI$_3$	5201	5274	CeCl$_3$	4297	4284
MoF$_3$	(6459)	5230	CeI$_3$	—	4029
MoCl$_3$	—	5230	PrCl$_3$	4322	4326
MoBr$_3$	—	5156	PrI$_3$	—	4071
MoI$_3$	—	5073	NdCl$_3$	4343	—
MnF$_3$	6017	—	SmCl$_3$	4376	—
MnCl$_3$	5544	—	EuCl$_3$	4393	—
MnBr$_3$	5448	—	GdCl$_3$	4406	—
MnI$_3$	5330	—	DyCl$_3$	4481	—
TcCl$_3$	5270	—	HoCl$_3$	4501	—
TcBr$_3$	5215	—	ErCl$_3$	4527	—
TcI$_3$	5188	—	TmCl$_3$	4548	4550
FeF$_3$	5870	—	TmI$_3$	—	4314
FeCl$_3$	5364	5359	YbCl$_3$	—	4546
FeBr$_3$	5268	5333	AcCl$_3$	4096	—
FeI$_3$	5117	—	UCl$_3$	4243	—
RuCl$_3$	—	5245	NpCl$_3$	4268	—
RuBr$_3$	—	5223	PuCl$_3$	4289	—
RuI$_3$	—	5222	PuBr$_3$	(3959)	—
CoF$_3$	5991	6118	AmCl$_3$	4293	—
RhCl$_3$	—	5641	TiF$_4$	10012	9908
IrF$_3$	(6112)	—	TiCl$_4$	9431	—
IrBr$_3$	(4794)	—	TiBr$_4$	9288	9039
NiF3	(6111)	—	TiI$_4$	9108	8893
AuF$_3$	(5777)	—	ZrF$_4$	8853	8971
AuCl$_3$	(4605)	—	ZrCl$_4$	8096	8144
ZnCl$_3$	5832	—	ZrBr$_4$	7916	7984
ZnBr$_3$	5732	—	ZrI$_4$	7661	7801
ZnI$_3$	5636	—	MoF$_4$	8795	—
AlF$_3$	5924	5215	MoCl$_4$	8556	9573
AlCl$_3$	5376	5492	MoBr$_4$	8510	9475
AlBr$_3$	5247	5361	MoI$_4$	8427	—
AlI$_3$	5070	5218	SnCl$_4$	8355	—
GaF$_3$	—	6205	SnBr$_4$	7970	8833
GaCl$_3$	5217	5645	PbF$_4$	—	9461

Substance	Calc. U_{POT} (kJ mol^{-1})	U_{POT}^{BFHC} (kJ mol^{-1})	Substance	Calc. U_{POT} (kJ mol^{-1})	U_{POT}^{BFHC} (kJ mol^{-1})
CrF_2Cl	5795	—	SrH_2	2250	2253
CrF_2Br	5753	—	BaH_2	2121	2121
CrF_2I	5669	—	ScH_2	2711	2659
$CrCl_2Br$	5448	—	YH_2	(2598)	2670
$CrCl_2I$	5381	—	LaH_2	2380	2522
$CrBr_2I$	5330	—	CeH_2	2414	2484
$CuFCl$	2891	—	PrH_2	2448	2398
$CuFBr$	2853	—	NdH_2	2464	2367
$CuFI$	2803	—	PmH_2	2519	—
$CuClBr$	2753	—	SmH_2	2510	2389
$CuClI$	2694	—	GdH_2	2494	2706
$CuBrI$	2669	—	AcH_2	2372	—
FeF_2Cl	5711	—	ThH_2	2711	—
FeF_2Br	5653	—	PuH_2	2519	—
FeF_2I	5569	—	AmH_2	2544	—
$FeCl_2Br$	5339	—	TiH_2	2866	2845
$FeCl_2I$	5272	—	ZrH_2	2711	2999
FBr_2I	5209	—	CuH_2	2941	—
$LiIO_2F_2$	845	—	ZnH_2	2870	—
$NaIO_2F_2$	766	764	HgH_2	2707	—
KIO_2F_2	699	697	AlH_3	5924	—
$RbIO_2F_2$	674	671	FeH_3	5724	—
$CsIO_2F_2$	636	—	ScH_3	5439	—
$NH_4IO_2F_2$	678	—	YH_3	5063	—
$AgIO_2F_2$	736	685	LaH_3	4895	4493
Hydrides			FeH_3	5724	—
LiH	858	920	GaH_3	5690	—
NaH	782	808	InH_3	5092	—
KH	699	714	TlH_3	5092	—
RbH	674	685	**Hydroselenides**		
CsH	648	644	$NaHse$	703	—
TiH	—	1407	$KHSe$	644	—
ZrH	—	1590	$RbHSe$	623	—
VH	—	(1344)	$CsHSe$	598	—
NbH	—	(1633)	**Hydrosulphides**		
PdH	—	(1368)	$LiHS$	759	821
CuH	—	1254	$NaHS$	704	747
TiH	996	1407	KHS	650	659
ZrH	916	1590	$RbHS$	623	637
HfH	904	—	$CsHS$	582	595
LaH	828	—	NH_4HS	661	666
VH	1184	(1344)	$Ca(HS)_2$	2184	(2171)
NbH	1163	(1633)	$Sr(HS)_2$	2063	—
TaH	1021	—	$Ba(HS)_2$	1979	(1956)
CrH	1050	—	**Hydroxides**		
NiH	929	—	$LiOH$	1021	1039
PdH	979	1368	$NaOH$	887	900
PtH	937	—	KOH	789	804
CuH	828	1254	$RbOH$	766	773
AgH	941	—	$CsOH$	721	724
AuH	1033	—	$Be(OH)_2$	3477	3629
TlH	745	—	$Mg(OH)_2$	2870	3006
GeH	950	—	$Ca(OH)_2$	2506	2645
PbH	778	—	$Sr(OH)_2$	2330	2483
BeH_2	3205	3295	$Ba(OH)_2$	2141	2339
MgH_2	2791	2706	$Ti(OH)_2$	—	2962
CaH_2	2410	2394	$Mn(OH)_2$	2909	3008

Table 1 (continued)

Substance	Calc. U_{POT} (kJ mol^{-1})	U_{POT}^{BFHC} (kJ mol^{-1})	Substance	Calc. U_{POT} (kJ mol^{-1})	U_{POT}^{BFHC} (kJ mol^{-1})
Fe(OH)$_2$	2653	3055	Ba(NO$_3$)$_2$	1975	2016
Co(OH)$_2$	2786	3115	Mn(NO$_3$)$_2$	2318	2519
Ni(OH)$_2$	2832	3193	Fe(NO$_3$)$_2$	—	(2563)
Pd(OH)$_2$	—	3175	CO(NO$_3$)$_2$	2560	2626
Cu(OH)$_2$	2870	3237	Ni(NO$_3$)$_2$	—	2709
CuOH	1006	—	Cu(NO$_3$)$_2$	—	2720
AgOH	918	—	Zn(NO$_3$)$_2$	2376	2628
AuOH	1033	—	Cd(NO$_3$)$_2$	2238	2443
TlOH	705	—	Ha(NO$_3$)$_2$	2255	—
Zn(OH)$_2$	2795	3158	Sn(NO$_3$)$_2$	2155	2254
Cd(OH)$_2$	2607	2918	Pb(NO$_3$)$_2$	2067	2189
Hg(OH)$_2$	2669	—	Nitrides		
Sn(OH)$_2$	2489	2729	ScN	7547	7506
Pb(OH)$_2$	2376	2623	LaN	6876	6793
Sc(OH)$_3$	5063	—	TiN	8130	8033
Y(OH)$_3$	4707	—	ZrN	7633	7723
La(OH)$_3$	4443	—	VN	8283	8233
Cr(OH)$_3$	5556	—	NbN	7939	8022
Mn(OH)$_3$	6213	—	CrN	8269	8358
Al(OH)$_3$	5627	—	Nitrites		
Ga(OH)$_3$	5732	—	NaNO$_2$	774	748
In(OH)$_3$	5280	—	KNO$_2$	660	664
Tl(OH)$_3$	5314	—	RbNO$_2$	638	765
Ti(OH)$_4$	9456	—	CsNO$_2$	598	—
Zr(OH)$_4$	8619	—	Ca(NO$_2$)$_2$	2460	2225
Mn(OH)$_4$	10933	—	Sr(NO$_2$)$_2$	2305	2111
Sn(OH)$_4$	9188	—	Ba(NO$_2$)$_2$	2205	1987
Imides			Oxides		
CaNH	3293	—	Li$_2$O	2799	—
SrNH	3146	—	Na$_2$O	2481	—
BaNH	2975	—	K$_2$O	2238	—
Metaniobates			Rb$_2$O	2163	—
NaNbO$_3$	789	—	Cu$_2$O	3273	—
Ca(NbO$_3$)$_2$	2315	—	Ag$_2$O	3002	—
Fe(NBO$_3$)$_2$	2502	—	Tl$_2$O	2659	—
Metatantalates			LiO$_2$	(878)	(872)
NaTaO$_3$	789	—	NaO$_2$	799	796
Ca(TaO$_3$)$_2$	2315	—	KO$_2$	741	725
Fe(TaO$_3$)$_2$	2502	—	RbO$_2$	706	695
Metavanadates			CsO$_2$	679	668
Li$_3$VO$_4$	3945	—	Li$_2$O$_2$	2592	256
Na$_3$VO$_4$	3766	—	Na$_2$O$_2$	2309	2305
K$_3$VO$_4$	3376	—	K$_2$O$_2$	2114	2078
Rb$_3$VO$_4$	3243	—	Rb$_2$O$_2$	2025	2006
Cs$_3$VO$_4$	3137	—	Cs$_2$O$_2$	1948	1861
Nitrates			MgO$_2$	3356	3526
LiNO$_3$	848	848	CaO$_2$	3144	3133
NaNO$_3$	755	756	SrO$_2$	3037	2849
KNO$_3$	685	687	KO$_3$	697	—
RbNO$_3$	662	658	BeO	4293	4443
CsNO$_3$	648	625	MgO	3795	3791
NH$_4$NO$_3$	661	676	CaO	3414	3401
AgNO$_3$	820	822	SrO	3217	3223
TlNO$_3$	690	700	BaO	3029	3054
Mg(NO$_3$)$_2$	2468	2503	TiO	3832	3811
Ca(NO$_3$)$_2$	2209	2228	VO	3932	3863
Sr(NO$_3$)$_2$	2092	2132	MnO	3724	3745

Substance	Calc. U_{POT} (kJ mol^{-1})	U_{POT}^{BFHC} (kJ mol^{-1})	Substance	Calc. U_{POT} (kJ mol^{-1})	U_{POT}^{BFHC} (kJ mol^{-1})
FeO	3795	3865	NaClO$_4$	643	648
CoO	3837	3910	KClO$_4$	599	602
NiO	3908	4010	RbClO$_4$	564	582
PdO	3736	—	CsClO$_4$	636	(542)
CuO	4135	4050	NH$_4$ClO$_4$	583	580
ZnO	4142	3971	Ca(ClO$_4$)$_2$	1958	1971
CdO	3806	—	Sr(ClO$_4$)$_2$	1862	1862
HgO	3907	—	Ba(ClO$_4$)$_2$	1795	1769
GeO	3919	—	NaMnO$_4$	661	—
SnO	3652	—	KMnO$_4$	607	—
PbO	3520	—	RbMnO$_4$	586	—
Sc$_2$O$_3$	13557	13708	CsMnO$_4$	565	—
Y$_2$O$_3$	12705	—	Ca(MnO$_4$)$_2$	1937	—
La$_2$O$_3$	12452	—	Sr(MnO$_4$)$_2$	1845	—
Ce$_2$O$_3$	12661	—	Ba(MnO$_4$)$_2$	1778	—
Pr$_2$O$_3$	12703	—	Phosphates		
Nd$_2$O$_3$	12736	—	Mg$_4$(PO$_4$)$_2$	11632	11407
Pm$_2$O$_3$	12811	—	Ca$_3$(PO$_4$)$_2$	10602	10479
Sm$_2$O$_3$	12878	—	Sr$_3$(PO$_4$)$_2$	10125	10075
Eu$_2$O$_3$	12945	—	Ba$_3$(PO$_4$)$_3$	9652	9654
Gd$_2$O$_3$	12996	—	MnPO$_4$	7397	—
Tb$_2$O$_3$	13071	—	FePO$_4$	7251	7303
Dy$_2$O$_3$	13138	—	BPO$_4$	8201	—
Ho$_2$O$_3$	13180	—	AlPO$_4$	7427	7509
Er$_2$O$_3$	13263	—	GaPO$_4$	7381	
Tm$_2$O$_3$	13322	—	Phosphonium salts		
Yb$_2$O$_3$	13380	—	PH$_4$Br	616	—
Lu$_2$O$_3$	13665	—	PH$_4$I	590	—
Ac$_2$O$_3$	12573	—	Selenides		
Ti$_2$O$_3$	—	14149	Li$_2$Se	2364	—
V$_2$O$_3$	15096	14520	Na$_2$Se	2130	—
Cr$_2$O$_3$	15276	14957	K$_2$Se	1933	—
Mn$_2$O$_3$	15146	15035	Rb$_2$Se	1837	—
Fe$_2$O$_3$	14309	14774	CS$_2$Se	1745	—
Al$_2$O$_3$	15916	—	Ag$_2$Se	2686	—
Ga$_2$O$_3$	15590	15220	Tl$_2$Se	2209	—
In$_2$O$_3$	13928	—	BeSe	3431	—
Ti$_2$O$_3$	14702	—	MgSe	3071	—
Pb$_2$O$_3$	(14841)	—	CaSe	2858	2862
CeO$_2$	9627	—	SrSe	2736	—
ThO$_2$	10397	—	BaSe	2611	—
PaO$_2$	10573	—	MnSe	3176	3194
VO$_2$	10644	—	FeSe	3499	3396
NpO$_2$	10707	—	CoSe	3554	3471
PuO$_2$	10786	—	NiSe	3658	3558
AmO$_2$	10799	—	CuSe	3736	3662
CmO$_2$	10832	—	ZnSe	3502	3514
TiO$_2$	12150	—	CdSe	3330	—
ZrO$_2$	11188	—	HgSe	3501	—
MoO$_2$	11648	—	SnSe	3058	—
MnO$_2$	12970	—	PbSe	3050	—
SiO$_2$	13125	—	Selenites		
GeO$_2$	12828	—	Li$_2$SeO$_3$	2171	—
SnO$_2$	11807	—	Na$_2$SeO$_3$	1950	1931
PbO$_2$	11217		K$_2$SeO$_3$	1774	—
Perchlorates			Rb$_2$SeO$_3$	1715	1675
LiClO$_4$	709	723	Cs$_2$SeO$_3$	1640	—

Table 1 (continued)

Substance	Calc. U_{POT} (kJ mol^{-1})	U_{POT}^{BFHC} (kJ mol^{-1})	Substance	Calc. U_{POT} (kJ mol^{-1})	U_{POT}^{BFHC} (kJ mol^{-1})
Tl_2SeO_3	1879	—	Rb_2SO_4	1636	1748
Ag_2SeO_3	2113	—	Cs_2SO_4	1596	1658
$BeSeO_3$	3322	—	$(NH_4)_2SO_4$	1766	1777
$MgSeO_3$	3012	2996	Cu_2SO_4	2276	2166
$CaSeO_3$	2732	—	Ag_2SO_4	2104	1989
$SrSeO_3$	2586	2586	Tl_2SO_4	1828	1722
$BaSeO_3$	2460	2448	Hg_2SO_4	—	2127
$RaSeO_3$	2456	—	$CaSO_4$	2489	2480
$MnSeO_3$	2975	—	$SrSO_4$	2577	2484
$FeSeO_3$	2895	—	$BaSO_4$	2469	2374
$CoSeO_3$	3155	—	$MnSO_4$	2920	2825
$NiSeO_3$	2945	—	$FeSO_4$	2983	2921
$CuSeO_3$	3209	—	$CoSO_4$	3088	2917
$ZnSeO_3$	3167	—	$NiSO_4$	3167	3044
$CdSeO_3$	2962	—	$CuSO_4$	3167	3066
$PbSeO_3$	2669	—	$ZnSO_4$	3100	3006
Selenates			$CdSO_4$	2891	—
Li_2SeO_4	2054	—	$PbSO_4$	2635	2534
Na_2SeO_4	1879	—	$Sn(SO_4)_2$	—	9616
K_2SeO_4	1732	—	Ternary salts		
Rb_2SeO_4	1686	—	Cs_2CuCl_4	1393	—
Cs_2SeO_4	1615	—	Rb_2ZnCl_4	1529	—
Cu_2SeO_4	2201	—	Cs_2ZnCl_4	1492	—
Ag_2SeO_4	2033	—	Rb_2ZnBr_4	1498	—
Tl_2SeO_4	1766	—	Cs_2ZnBr_4	1454	—
Hg_2SeO_4	2163	—	$(Me_4N)_2ZnBr_4$	1364	—
$BeSeO_4$	3448	—	Cs_2ZnI_4	1386	—
$MgSeO_4$	2895	—	$CsGaCl_4$	494	—
$CaSeO_4$	2632	—	$NaAlCl_4$	556	—
$SrSeO_4$	2489	—	$CsAlCl_4$	486	—
$BaSeO_4$	2385	—	$NaFeCl_4$	492	—
$RaSeO_4$	2364	—	Rb_2CoCl_4	1447	—
$GdSeO_4$	2753	—	Cs_2CoCl_4	1391	—
$MnSeO_4$	2837	—	K_2PtCl_4	1574	1550
$FeSeO_4$	3008	—	Cs_2GeF_6	1573	—
$CoSeO_4$	2912	—	$(NH_4)_2GeF_6$	1657	—
$NiSeO_4$	3079	—	Cs_2GeCl_6	1375	1375
$CuSeO_4$	3104	—	K_2HfCl_6	1345	1461
$ZnSeO_4$	3021	—	K_2IrCl_6	1442	1440
$CdSeO_4$	2833	—	$(NH_4)_2IrCl_6$	1442	—
$HgSeO_4$	2845	—	Na_2MoCl_6	1526	—
$PbSeO_4$	2561	—	K_2MoCl_6	1418	1433
Sulphides			Rb_2MoCl_6	1399	1399
Li_2S	2464	2472	Cs_2MoCl_6	1347	1347
Na_2S	2192	2203	K_2NbCl_6	1375	1398
K_2S	1979	(2052)	Rb_2NbCl_6	1371	1385
Rb_2S	1929	1949	Cs_2NbCl_6	1381	1344
Cs_2S	1892	1850	K_2OsCl_6	1447	1447
$(NH_4)_2S$	2008	(2026)	Cs_2OsCl_6	1409	—
Cu_2S	2786	2865	$(NH_4)_2OsCl_6$	1433	—
Ag_2S	2606	2677	K_2OsBr_6	1396	—
Au_2S	2908	—	K_2PdCl_6	1450	1466
Tl_2S	2298	2258	Rb_2PdCl_6	1449	—
Sulphates			Cs_2PdCl_6	1426	—
Li_2SO_4	2229	2142	$(NH_4)_2PdCl_6$	1481	—
Na_2SO_4	1827	1938	Rb_2PbCl_6	1343	1343
K_2SO_4	1700	1796	Cs_2PbCl_6	1344	—

Substance	Calc. U_{POT} (kJ mol^{-1})	U_{POT}^{BFHC} (kJ mol^{-1})	Substance	Calc. U_{POT} (kJ mol^{-1})	U_{POT}^{BFHC} (kJ mol^{-1})
$(NH_4)_2PbCl_6$	1355	—	Cs_2TeI_6	1246	—
K_2PtCl_6	1468	—	K_2TiCl_6	1412	1443
Rb_2PtCl_6	1464	—	Rb_2TiCl_6	1415	1425
Cs_2PtCl_6	1444	—	Cs_2TiCl_6	1402	1370
$(NH_4)_2PtCl_6$	1468	—	Tl_2TiCl_6	1560	1553
Tl_2PtCl_6	1546	—	$(NH_4)_2TiCl_6$	1413	—
Ag_2PtCl_6	1773	—	K_2TiBr_6	1379	1379
$BaPtCl_6$	2047	—	Rb_2TiBr_6	1341	1365
K_2PtBr_6	1423	—	Cs_2TiBr_6	1339	1316
Ag_2PtBr_6	1791	—	Na_2UBr_6	1504	—
K_2PtI_6	1421	—	K_2UBr_6	1484	—
K_2ReCl_6	1416	1442	Rb_2UBr_6	1473	—
Rb_2ReCl_6	1414	—	Cs_2UBr_6	1459	—
Cs_2ReCl_6	1398	—	K_2WCl_6	1398	1423
$(NH_4)_2ReCl_6$	1402	1390	Rb_2WCl_6	1397	1434
K_2ReBr_6	1375	1375	Cs_2WCl_6	1392	1366
K_2SiF_6	1670	1628	K_2WBr_6	1408	1408
Rb_2SiF_6	1639	1621	Rb_2WBr_6	1361	1391
Cs_2SiF_6	1604	1498	Cs_2WBr_6	1362	1332
Tl_2SiF_6	1675	—	K_2ZrCl_6	1339	1371
$(NH_4)_2SiF_6$	1657	1727	Rb_2ZrCl_6	1341	—
K_2SnCl_6	1363	1390	Cs_2ZrCl_6	1339	1308
Rb_2SnCl_6	1361	1363	**Tetraalkyl ammonium salts**		
Cs_2SnCl_6	1358	—	Me_4NCl	566	—
Tl_2SnCl_6	1437	—	Me_4NBr	553	—
$(NH_4)_2SnCl_6$	1370	1334	Me_4NI	544	—
Rb_2SnBr_6	1309	—	**Tellurides**		
Cl_2SnBr_6	1306	—	Li_2Te	2212	—
$(NH_4)_2SnBr_6$	1319	—	Na_2Te	1997	2095
Rb_2SnI_6	1226	—	K_2Te	1830	—
Cs_2SnI_6	1243	—	Rb_2Te	1837	—
K_2TeCl_6	1318	1320	Cs_2Te	1745	
Rb_2TeCl_6	1321	—	Cu_2Te	2706	2683
Cs_2TeCl_6	1323	—	Ag_2Te	2607	2600
Tl_2TeCl_6	1392	—	Tl_2Te	2084	2172
$(NH_4)_2TeCl_6$	1318	—	$BeTe$	3319	—
K_2RuCl_6	1451	—	$MgTe$	2878	3081
Rb_2CoF_6	1688	—	$CaTe$	2721	—
Cs_2CoF_6	1632	—	$SrTe$	2599	—
K_2NiF_6	1721	—	$BaTe$	2473	—
Rb_2NiF_6	1688	—	$RaTe$	2481	—
Rb_2SbCl_6	1357	—	$MnTe$	3041	—
Rb_2SeCl_6	1409	—	$FeTe$	3399	—
Cs_2SeCl_6	1397	—	$CoTe$	3429	—
$(NH_4)_2SeCl_6$	1420	—	$NiTe$	3534	—
K_2SeBr_6	1379	—	$CuTe$	3639	—
$(NH_4)_2SeBr_6$	1380	—	$ZnTe$	3416	—
$(NH_4)_2PoCl_6$	1338	—	$CdTe$	3212	—
Cs_2PoBr_6	1286	—	$PbTe$	2930	—
$(NH_4)_2PoBr_6$	1292	—	**Thiocyanates**		
Cs_2CrF_6	1603	—	$LiCNS$	764	(765)
Rb_2MnF_6	1688	—	$NaCNS$	682	682
Cs_2MnF_6	1620	—	$KCNS$	623	616
K_2MnCl_6	1462	—	$RbCNS$	623	619
Rb_2MnCl_6	1451	—	$CsCNS$	623	568
$(NH_4)_2MnCl_6$	1464	—	NH_4CNS	605	611
Cs_2TeBr_6	1306	—	$Ca(CNS)_2$	2184	2118
$(NH_4)_2TeBr_6$	1294	—	$Sr(CNS)_2$	2063	1957

Substance	Calc. U_{POT} (kJ mol^{-1})	U_{POT}^{BFHC} (kJ mol^{-1})	Substance	Calc. U_{POT} (kJ mol^{-1})	U_{POT}^{BFHC} (kJ mol^{-1})
Ba(CNS)$_2$	1979	1852	Vanadates		
Mn(CNS)$_2$	2280	2351	LiVO$_3$	810	—
Zn(CNS)$_2$	2335	2560	NaVO$_3$	761	—
Cd(CNS)$_2$	2201	2374	KVO$_3$	686	—
Hg(CNS)$_2$	2146	2492	RbVO$_3$	657	—
Sn(CNS)$_2$	2117	2142	CsVO$_3$	628	—
Pb(CNS)$_2$	2058	—			

Table 2
ANCILLARY THERMODYNAMIC DATA

Substance	State	ΔH_f° (kJ mol^{-1})	Substance	State	ΔH_f° (kJ mol^{-1})
Li$^+$	g	687.16	OsCl$_6^{-2}$	g	−752
Na$^+$	g	609.84	IrCl$_6^{-2}$	g	−785
K$^+$	g	514.19	PdCl$_6^{-2}$	g	−749
Cs$^+$	g	452.3	PtCl$_6^{-2}$	g	−774
O$_2^-$	g	−74	PtBr$_6^{-2}$	g	−645
CN$^-$	g	41	SiF$_6^{-2}$	g	−2207
NH$_4^+$	g	630.2	GeCl$_6^{-2}$	g	−981
C$_2^{-2}$	g	918	SuCl$_6^{-2}$	g	−1156
O$_2^{-2}$	g	553	PbCl$_6^{-2}$	g	−940
NO$_3^-$	g	−320.1	TeCl$_6^{-2}$	g	−902
IO$_2$F$_2^-$	g	−693	NO$_2^-$	g	−202
IO$_3^-$	g	−208	LiO$_2$	c	−259.4
ClO$_3^-$	g	−200	NaO$_2$	c	−260.7
BrO$_3^-$	g	−145	KO$_2$	c	−284.5
HSO$_4^-$	g	(−1012)	CsO$_2$	c	−289.5
HCO$_3^-$	g	−738	Li$_2$O$_2$	c	−632.6
NH$_2$CH$_2$CO$_2^-$	g	−564	Na$_2$O$_2$	c	−531.2
N^{-3}	g	(2588)	K$_2$O$_2$	c	−495.8
CO$_3^{-2}$	g	−321	Cs$_2$O$_2$	c	−402.5
PO$_4^{-3}$	g	(291)	ThO$_2$	c	−1230.5
AsO$_4^{-3}$	g	(289)	MgO$_2$	c	−623.0
GeO$_4^{-4}$	g	(1460)	NaIO$_2$F$_2$	c	−848.7
CH$_3$COO$^-$	g	−554	KIO$_2$F$_2$	c	−876.6
ClO$_4^-$	g	−344	RbIO$_2$F$_2$	c	−875.3
NO$_2^-$	g	−219	NH$_4$IO$_2$F$_2$	c	−747.7
SeO$_3^{-2}$	g	(−249)	LiBrO$_3$	c	−356.4
HCO$_2^-$	g	−463	CsBrO$_3$	c	−374.5
BH$_4^-$	g	−96	CsHCO$_3$	c	−995.6
HS$^-$	g	−120	Ca(HCO$_3$)$_2$	c	(−1950)
TiCl$_6^{-2}$	g	−1330	Sr(HCO$_3$)$_2$	c	(−1954)
TiBr$_6^{-2}$	g	−1142	Ba(HCO$_3$)$_2$	c	(−1971)
ZrCl$_6^{-2}$	g	−1526	CH$_3$COONH$_4$	c	−618.4
HfCl$_6^{-2}$	g	−1640	CH$_3$COOLi	c	−748.9
NbCl$_6^{-2}$	g	−1224	CH$_3$COORb	c	−720.9
TaCl$_6^{-2}$	g	−1275	LiClO$_4$	c	−380.7
MoCl$_6^{-2}$	g	−1070	CsClO$_4$	c	(−435.1)
WCl$_6^{-2}$	g	−985	LiH	c	−90.62
WBr$_6^{-2}$	g	−705	NaH	c	−56.4
ReCl$_6^{-2}$	g	−919	KH	c	−57.8
ReBr$_6^{-2}$	g	−689	CsH	c	−49.9

Table 2
ANCILLARY THERMODYNAMIC DATA (continued)

Substance	State	ΔH_f° (kJ mol^{-1})	Substance	State	ΔH_f° (kJ mol^{-1})
LiN$_3$	c	10.8	Ca(ClO$_3$)$_2$	c	−757
NaN$_3$	c	21.25	Sr(ClO$_3$)$_2$	c	−766
KN$_3$	c	−1.38	ScN	c	−284.5
RbN$_3$	c	−0.29	Li$_2$CO$_3$	c	−1216.0
CsN$_3$	c	−9.92	Cs$_2$CO$_3$	c	−1118.8
TlN$_3$	c	233.4	NiCO$_3$	c	−689.1
AgN$_3$	c	310.3	CuCO$_3$	c	−595.0
CsHS	c	−263.2	SnCO$_3$	c	(−740.6)
Ca(HS)$_2$	c	(−481)	HCO$_2$Li	c	−649.3
Ba(HS)$_2$	c	(−531.4)	HCO$_2$Rb	c	−656.0
LiHS	c	−254.4	TiH	c	−130.5
K$_2$TiCl$_6$	c	−1747	ZrH	c	−173.5
Rb$_2$TiCl$_6$	c	−1767	VH	c	(−32)
Cs$_2$TiCl$_6$	c	−1797	NbH	c	(−100)
Tl$_2$TiCl$_6$	c	−1330	PdH	c	(−37)
K$_2$TiBr$_6$	c	−1493	CuH	c	(−21.4)
Rb$_2$TiBr$_6$	c	−1517	LaH$_2$	c	−230.1
Cs$_2$TiBr$_6$	c	−1553	CeH$_2$	c	−141.8
K$_2$ZrCl$_6$	c	−1932	PrH$_2$	c	−200.0
Cs$_2$ZrCl$_6$	c	−1992	NdH$_2$	c	−187.4
K$_2$HfCl$_6$	c	−1957	SmH$_2$	c	−187.4
K$_2$NbCl$_6$	c	−1594	GdH$_2$	c	−196.2
Rb$_2$NbCl$_6$	c	−1619	CsOH	c	−406.7
Cs$_2$NbCl$_6$	c	−1663	Ti(OH)$_2$	c	(−778.0)
K$_2$ReCl$_6$	c	−1333	LiOH	c	−487.2
(NH$_4$)$_2$ReCl$_6$	c	−1056	K$_2$TaCl$_6$	c	−1648
K$_2$ReBr$_6$	c	−1036	Rb$_2$TaCl$_6$	c	−1669
K$_2$OsCl$_6$	c	−1171	Cs$_2$TaCl$_6$	c	1711
K$_2$IrCl$_6$	c	−1197	Na$_2$MoCl$_6$	c	−1376
K$_2$SiF$_6$	c	−2807	K$_2$MoCl$_6$	c	−1475
Rb$_2$SiF$_6$	c	−2838	Rb$_2$MoCl$_6$	c	−1479
Cs$_2$SiF$_6$	c	−2801	Cs$_2$MoCl$_6$	c	−1512
(NH$_4$)$_2$SiF$_6$	c	−2681	K$_2$WCl$_6$	c	−1380
Rb$_2$PbCl$_6$	c	−1293	Rb$_2$WCl$_6$	c	−1429
Mn(CN)$_2$	c	60.7	Cs$_2$WCl$_6$	c	−1446
RbCN	c	(−108.8)	K$_2$WBr$_6$	c	−1085
CsCN	c	(−108.7)	Rb$_2$WBr$_6$	c	−1106
LiCN	c	−121	Cs$_2$WBr$_6$	c	−1132
Sr(CN)$_2$	c	(−205)	K$_2$PdCl$_6$	c	−1187
LiNO$_3$	c	−482.3	Ag$_2$PtCl$_6$	c	−527
RbNO$_3$	c	−489.7	BaPtCl$_6$	c	−1180
CsNO$_3$	c	−494.2	K$_2$PtBr$_6$	c	−1040
Fe(NO$_3$)$_2$	c	(−447.7)	Ag$_2$PtBr$_6$	c	−398
Sn(NO$_3$)$_2$	c	−456.1	Cs$_2$GeCl$_6$	c	−1451
NaBF$_4$	c	−1774	K$_2$SnCl$_6$	c	−1518
CsIO$_3$	c	−526.3	Rb$_2$SnCl$_6$	c	−1529
CsClO$_3$	c	−395.8	(NH$_4$)$_2$SnCl$_6$	c	−1237
Mg(ClO$_3$)$_2$	c	(−523)	Rb$_2$TeCl$_6$	c	−1237

THE MADELUNG CONSTANT AND CRYSTAL LATTICE ENERGY

If U is the crystal lattice energy and M is the Madelung constant, then[a]

$$U = \frac{N M z_i z_j e^2}{r}(1 - 1/n)$$

Substance	Ion type	Crystal form[b]	M
Sodium chloride, NaCl	M^+, X^-	FCC	1.74756
Cesium chloride, CsCl	M^+, X^-	BCC	1.76267
Calcium chloride, $CaCl_2$	M^{++}, $2X^-$	Cubic	2.365
Calcium fluoride (fluorite), CaF_2	M^{++}, $2X^-$	Cubic	2.51939
Cadmium chloride, $CdCl_2$	M^{++}, $2X^-$	Hexagonal	2.244[c]
Cadmium iodide (α), CdI_2	M^{++}, $2X^-$	Hexagonal	2.355[c]
Magnesium fluoride, MgF_2	M^{++}, $2X^-$	Tetragonal	2.381[c]
Cuprous oxide (cuprite), Cu_2O	$2M^+$, X^{--}	Cubic	2.22124
Zinc oxide, ZnO	M^{++}, X^{--}	Hexagonal	1.4985[c]
Sphalerite (zinc blende), ZnS	M^{++}, X^{--}	FCC	1.63806
Wurtzite, ZnS	M^{++}, X^{--}	Hexagonal	1.64132[c]
Titanium dioxide (anatase), TiO_2	M^{4+}, $2X^{--}$	Tetragonal	2.400[c]
Titanium dioxide (rutile), TiO_2	M^{4+}, $2X^{--}$	Tetragonal	2.408[c]
β-Quartz, SiO_2	M^{4+}, $2X^{--}$	Hexagonal	2.2197[c]
Corundum, Al_2O_3	$2M^{3+}$, $3X^{--}$	Rhombohedral	4.1719

[a] N is Avogadro's number, z_i and z_j are the integral charges on the ions (in units of e), and e is the charge on the electron in electrostatic units ($e = 4.803 \times 10^{-10}$ esu). r is the shortest distance between cation-anion pairs in centimeters. Then U is in ergs (1 erg = 10^{-7} J).

[b] FCC = face centered cubic; BCC = body centered cubic.

[c] For tetragonal and hexagonal crystals the value of M depends on the details of the lattice parameters.

The Born Exponent, n is:

Ion type	n
He, Li^+	5
Ne, Na^+, F^-	7
Ar, K^+, Cu^+, Cl^-	9
Kr, Rb^+, Ag^+, Br^-	10
Xe, Cs^+, Au^+, I^-	12

For a crystal with a mixed-ion type, an average of the values of n in this table is to be used (6 for LiF, for example).

ELASTIC CONSTANTS OF SINGLE CRYSTALS

H. P. R. Frederikse

This table gives selected values of elastic constants for single crystals. The values believed most reliable were selected from the original literature. The substances are arranged by crystal system and, within each system, alphabetically by name. A reference to the original literature is given for each value; a useful compilation of published values from many sources may be found in Reference 1 below.

Data are given for the single-crystal density and for the elastic constants c_{ij}, in units of 10^{11} N/m^2, which is equivalent to 10^{12} dyn/cm^2.

GENERAL REFERENCES

1. Simmons, G., and Wang, H., *Single Crystal Elastic Constants and Calculated Aggregate Properties: A Handbook, Second Edition,* The MIT Press, Cambridge, MA, 1971.
2. Gray, D.E., Ed., *American Institute of Physics Handbook, Third Edition,* McGraw-Hill, New York, 1972.

CUBIC CRYSTALS

Name	Formula	ρ/g cm^{-3}	T/K	Ref.	C_{11}	C_{12}	C_{44}
Aluminum	Al	2.6970	298	1	1.0675	0.6041	0.2834
Aluminum antimonide	AlSb	4.3600	300	2	0.8939	0.4427	0.4155
Ammonium bromide	NH$_4$Br	2.4314	300	3	0.3414	0.0782	0.0722
Ammonium chloride	NH$_4$Cl	1.5279	290	4	0.3814	0.0866	0.0903
Argon	Ar	1.7710	4.2	5	0.0529	0.0135	0.0159
Barium fluoride	BaF$_2$	4.8860	298	6	0.9199	0.4157	0.2568
Barium nitrate	Ba(NO$_3$)$_2$	3.2560	293	7	0.2925	0.2065	0.1277
Calcium fluoride	CaF$_2$	3.810	298	8	1.6420	0.4398	0.8406
Calcium telluride	CaTe	5.8544	298	9	0.5351	0.3681	0.1994
Cesium	Cs	1.9800	78	10	0.0247	0.0206	0.0148
Cesium bromide	CsBr	4.4560	298	11	0.3063	0.0807	0.0750
Cesium chloride	CsCl	3.9880	298	11	0.3644	0.0882	0.0804
Cesium iodide	CsI	4.5250	298	11	0.2446	0.0661	0.0629
Chromite	FeCr$_2$O$_4$	4.4500	RT	12	3.2250	1.4370	1.1670
Chromium	Cr	7.20	298	13	3.398	0.586	0.990
Cobalt oxide	CoO	6.44	298	14	2.6123	1.4699	0.8300
Cobalt zinc ferrite	CoZnFeO$_2$	5.43	303	12	2.660	1.530	0.780
Copper	Cu	8.932	298	15	1.683	1.221	0.757
Gallium antimonide	GaSb	5.6137	298	16	0.8839	0.4033	0.4316
Gallium arsenide	GaAs	5.3169	298	17	1.1877	0.5372	0.5944
Gallium phosphide	GaP	4.1297	300	18	1.4120	0.6253	0.7047
Garnet (yttrium-iron)	Y$_3$Fe$_2$(FeO$_4$)$_3$	5.17	298	19	2.680	1.106	0.766
Germanium	Ge	5.313	298	20	1.2835	0.4823	0.6666
Gold	Au	19.283	296.5	21	1.9244	1.6298	0.4200
Indium antimonide	InSb	5.7890	298	22	0.6720	0.3670	0.3020
Indium arsenide	InAs	5.6720	293	23	0.8329	0.4526	0.3959
Indium phosphide	InP	4.78	RT	24	1.0220	0.5760	0.4600
Iridium	Ir	22.52	300	25	5.80	2.42	2.56
Iron	Fe	7.8672	298	26	2.26	1.40	1.16
Lead	Pb	11.34	296	27	0.4966	0.4231	0.1498
Lead fluoride	PbF$_2$	7.79	300	28	0.8880	0.4720	0.2454
Lead nitrate	Pb(NO$_3$)$_2$	4.547	293	29	0.3729	0.2765	0.1347
Lead telluride	PbTe	8.2379	303.2	30	1.0795	0.0764	0.1343
Lithium	Li	0.5326	298	31	0.1350	0.1144	0.0878
Lithium bromide	LiBr	3.47	RT	32	0.3940	0.1880	0.1910
Lithium chloride	LiCl	2.068	295	33	0.4927	0.2310	0.2495
Lithium fluoride	LiF	2.638	RT	34	1.1397	0.4767	0.6364
Lithium iodide	LiI	4.061	RT	32	0.2850	0.1400	0.1350
Magnesium oxide	MgO	3.579	298	20	2.9708	0.9536	1.5613
Magnetite	Fe$_3$O$_4$	5.18	RT	32	2.730	1.060	0.971
Manganese oxide	MnO	5.39	298	35	2.23	1.20	0.79
Mercury telluride	HgTe	8.079	290	36	0.548	0.381	0.204
Molybdenum	Mo	10.2284	273	37	4.637	1.578	1.092

Name	Formula	ρ/g cm^{-3}	T/K	Ref.	C_{11}	C_{12}	C_{44}
Nickel	Ni	8.91	298	15	2.481	1.549	1.242
Niobium	Nb	8.578	300	38	2.4650	1.3450	0.2873
Palladium	Pd	12.038	300	39	2.2710	1.7604	0.7173
Platinum	Pt	21.50	300	40	3.4670	2.5070	0.7650
Potassium	K	0.851	295	41	0.0370	0.0314	0.0188
Potassium bromide	KBr	2.740	298	11	0.3468	0.0580	0.0507
Potassium chloride	KCl	1.984	298	11	0.4069	0.0711	0.0631
Potassium cyanide	KCN	1.553	RT	32	0.1940	0.1180	0.0150
Potassium fluoride	KF	2.480	295	33	0.6490	0.1520	0.1232
Potassium iodide	KI	3.128	300	42	0.2710	0.0450	0.0364
Pyrite	FeS$_2$	5.016	RT	43	3.818	0.310	1.094
Rubidium	Rb	1.58	170	44	0.0296	0.0250	0.0171
Rubidium bromide	RbBr	3.350	300	45	0.3152	0.0500	0.0380
Rubidium chloride	RbCl	2.797	300	45	0.3624	0.0612	0.0468
Rubidium iodide	RbI	3.551	300	45	0.2556	0.0382	0.0278
Silicon	Si	2.331	298	46	1.6578	0.6394	0.7962
Silver	Ag	10.50	300	47	1.2399	0.9367	0.4612
Silver bromide	AgBr	5.585	300	48	0.5920	0.3640	0.0616
Sodium	Na	0.971	299	49	0.0739	0.0622	0.0419
Sodium bromate	NaBrO$_3$	3.339	RT	32	0.5450	0.1910	0.1500
Sodium bromide	NaBr	3.202	300	33	0.3970	0.1001	0.0998
Sodium chlorate	NaClO$_3$	2.485	RT	50	0.4920	0.1420	0.1160
Sodium chloride	NaCl	2.163	298	11	0.4947	0.1288	0.1287
Sodium fluoride	NaF	2.804	300	51	0.9700	0.2380	0.2822
Sodium iodide	NaI	3.6689	300	52	0.3007	0.0912	0.0733
Spinel	MgAl$_2$O$_4$	3.6193	298	53	2.9857	1.5372	1.5758
Strontium fluoride	SrF$_2$	4.277	300	54	1.2350	0.4305	0.3128
Strontium nitrate	Sr(NO$_3$)$_2$	2.989	293	29	0.4255	0.2921	0.1590
Strontium oxide	SrO	4.99	300	55	1.601	0.435	0.590
Strontium titanate	SrTiO$_3$	5.123	RT	56	3.4817	1.0064	4.5455
Tantalum	Ta	16.626	298	57	2.6023	1.5446	0.8255
Tantalum carbide	TaC	14.65	RT	58	5.05	0.73	0.79
Thallium bromide	TlBr	7.4529	298	59	0.3760	0.1458	0.0757
Thorium	Th	11.694	300	60	0.7530	0.4890	0.4780
Thorium oxide	ThO$_2$	9.991	298	61	3.670	1.060	0.797
Tin telluride	SnTe	6.445	300	62	1.1250	0.0750	0.1172
Titanium carbide	TiC	4.940	RT	107	5.00	1.13	1.75
Tungsten	W	19.257	297	64	5.2239	2.0437	1.6083
Uranium carbide	UC	13.63	300	65	3.200	0.850	0.647
Uranium dioxide	UO$_2$	10.97	298	66	3.960	1.210	0.641
Vanadium	V	6.022	300	67	2.287	1.190	0.432
Zinc selenide	ZnSe	5.262	298	68	0.8096	0.4881	0.4405
Zinc sulfide	ZnS	4.088	298	68	1.0462	0.6534	0.4613
Zinc telluride	ZnTe	5.636	298	68	0.7134	0.4078	0.3115
Zirconium carbide	ZrC	6.606	298	63	4.720	0.987	1.593

TETRAGONAL CRYSTALS

Name	Formula	ρ/g cm^{-3}	T/K	Ref.	C_{11}	C_{12}	C_{13}	C_{16}	C_{33}	C_{44}	C_{66}
Ammonium dihydrogen arsenate (ADA)	$NH_4H_2AsO_4$	2.3110	298	69	0.6747	-0.106	0.1652		0.3022	0.0685	0.0639
Ammonium dihydrogen phosphate (ADP)	$NH_4H_2PO_4$	1.8030	293	69	0.6200	-0.050	0.1400		0.3000	0.0910	0.0610
Barium titanate	$BaTiO_3$	5.9988	298	70	2.7512	1.7897	1.5156		1.6486	0.5435	1.1312
Calcium molybdate	$CaMoO_4$	4.255	298	79	1.447	0.664	0.466	0.134	1.265	0.369	0.451
Indium	In	7.300	RT	71	0.4450	0.3950	0.4050		0.4440	0.0655	0.1220
Magnesium fluoride	MgF_2	3.177	RT	72	1.237	0.732	0.536		1.770	0.552	0.978
Nickel sulfate hexahydrate	$NiSO_4 \cdot 6H_2O$	2.070	RT	73	0.3209	0.2315	0.0209		0.2931	0.1156	0.1779
Potassium dihydrogen arsenate (KDA)	KH_2AsO_4	2.867	RT	12	0.530	-0.060	-0.020		0.370	0.120	0.070
Potassium dihydrogen phosphate (KDP)	KH_2PO_4	2.388	RT	71	0.7140	-0.049	0.1290		0.5620	0.1270	0.0628
Rubidium dihydrogen phosphate (RDP)	RbH_2PO_4	2.800	298	74	0.5562	-0.064	0.0279		0.4398	0.1142	0.0350
Rutile	TiO_2	4.260	298	75	2.7143	1.7796	1.4957		4.8395	1.2443	1.9477
Tellurium oxide	TeO_2	5.99	RT	76	0.5320	0.4860	0.2120		1.0850	0.2440	0.5520
Tin (white)	Sn	7.29	288	77	0.7529	0.6156	0.4400		0.9552	0.2193	0.2336
Zircon	$ZrSiO_4$	4.70	RT	78	2.585	1.791	1.542		3.805	0.733	1.113

ORTHORHOMBIC CRYSTALS

Name	Formula	ρ/g cm^{-3}	T/K	Ref.	C_{11}	C_{12}	C_{13}	C_{22}	C_{23}	C_{33}	C_{44}	C_{55}	C_{66}
Acenaphthene	$C_{12}H_{10}$	1.220	293	80	0.1380	0.0210	0.0410	0.1262	0.0460	0.1117	0.0265	0.0290	0.0185
Ammonium sulfate	$(NH_4)_2SO_4$	1.774	293	81	0.3607	0.1651	0.1580	0.2981	0.1456	0.3534	0.1025	0.0717	0.0974
Aragonite	$CaCO_3$	2.93	RT	82	1.5958	0.3663	0.0197	0.8697	0.1597	0.8503	0.4132	0.2564	0.4274
Barite	$BaSO_4$	4.40	RT	82	0.8941	0.4614	0.2691	0.7842	0.2676	1.0548	0.1190	0.2874	0.2778
Benzene	C_6H_6	1.061	250	83	0.0614	0.0352	0.0401	0.0656	0.0390	0.0583	0.0197	0.0378	0.0153
Benzophenone	$(C_6H_5)_2CO$	1.219	RT	32	0.1070	0.0550	0.0169	0.1000	0.0321	0.0710	0.0203	0.0155	0.0353
Bronzite	$(MgFe)SiO_3$	3.38	RT	78	1.876	0.686	0.605	1.578	0.561	2.085	0.700	0.592	0.544
Calcium sulfate	$CaSO_4$	2.962	RT	84	0.9382	0.1650	0.1520	1.845	0.3173	1.1180	0.3247	0.2653	0.0926
Celestite	$SrSO_3$	3.96	RT	12	1.044	0.773	0.605	1.061	0.619	1.286	0.135	0.279	0.266
Cesium sulfate	Cs_2SO_4	4.243	293	81	0.4490	0.1958	0.1815	0.4283	0.1800	0.3785	0.1326	0.1319	0.1323
Fosterite	Mg_2SiO_4	3.224	298	85	3.2848	0.6390	0.6880	1.9980	0.7380	2.3530	0.6515	0.8120	0.8088
Iodic acid	HIO_3	4.630	RT	73	0.3030	0.1194	0.1169	0.5448	0.0548	0.4359	0.1835	0.2193	0.1736
Lithium ammonium tartrate	$LiNH_4C_4H_4O_6 \cdot 4H_2O$	1.71	RT	12	0.3864	0.1655	0.0875	0.5393	0.2007	0.3624	0.1190	0.0667	0.2326
Magnesium sulfate heptahydrate	$MgSO_4 \cdot 7H_2O$	1.68	RT	86	0.325	0.174	0.182	0.288	0.182	0.315	0.078	0.156	0.090
Natrolite	$(Na,Al)SiO_3$	2.25	RT	78	0.716	0.261	0.297	0.632	0.297	1.378	0.196	0.248	0.423
Nickel sulfate heptahydrate	$NiSO_4 \cdot 7H_2O$	1.948	RT	86	0.353	0.198	0.201	0.311	0.201	0.335	0.091	0.172	0.099
Olivine	$(MgFe)SiO_4$	3.324	RT	87	3.240	0.590	0.790	1.980	0.780	2.490	0.667	0.810	0.793
Potassium pentaborate	$KB_5O_8 \cdot 4H_2O$	1.74	RT	71	0.582	0.229	0.174	0.359	0.231	0.255	0.164	0.046	0.057
Potassium sulfate	K_2SO_4	2.665	293	81	0.5357	0.1999	0.2095	0.5653	0.1990	0.5523	0.195	0.1879	0.1424
Rochelle salt	$NaK(C_4H_4O_6) \cdot 4H_2O$	1.79	RT	71	0.255	0.141	0.116	0.381	0.146	0.371	0.134	0.032	0.098
Rubidium sulfate	Rb_2SO_4	3.621	293	81	0.5029	0.1965	0.1999	0.5098	0.1925	0.4761	0.1626	0.1589	0.1407
Sodium ammonium tartrate	$NaNH_4C_4H_4O_6 \cdot 4H_2O$	1.587	RT	12	0.3685	0.2725	0.3083	0.5092	0.3472	0.5541	0.1058	0.0303	0.0870
Sodium tartrate	$Na_2C_4H_4O_6 \cdot 2H_2O$	1.794	RT	12	0.461	0.286	0.320	0.547	0.352	0.665	0.124	0.031	0.098
Strontium formate dihydrate	$Sr(CHO_2)_2 \cdot 2H_2O$	2.25	RT	12	0.4391	0.1037	-0.149	0.3484	-0.014	0.3746	0.1538	0.1075	0.1724
Sulfur	S	2.07	RT	12	0.240	0.133	0.171	0.205	0.159	0.483	0.043	0.087	0.076
Thallium sulfate	$TlSO_4$	6.776	293	81	0.4106	0.2573	0.2288	0.3885	0.2174	0.4268	0.1125	0.1068	0.0751
Topaz	$Al_2SiO_3(OH,F)_2$	3.52	RT	82	2.8136	1.2582	0.8464	3.8495	0.8815	2.9452	1.0811	1.3298	1.3089
Uranium (alpha)	U	19.0453	293	88	2.1486	0.4622	0.2176	1.9983	1.0764	2.6763	1.2479	0.7379	0.7454
Zinc sulfate heptahydrate	$ZnSO_4 \cdot 7H_2O$	1.970	RT	86	0.3320	0.1720	0.2000	0.2930	0.1980	0.3200	0.0780	0.1530	0.0830

ELASTIC CONSTANTS OF SINGLE CRYSTALS (continued)

MONOCLINIC CRYSTALS

Name	Formula	ρ/g cm^{-3}	T/K	Ref.	C_{11}	C_{12}	C_{13}	C_{15}	C_{22}
Aegirine	(NaFe)Si$_2$O$_6$	3.50	RT	89	1.858	0.685	0.707	0.098	1.813
Anthracene	C$_{14}$H$_{10}$	1.258	RT	90	0.0852	0.0672	0.0590	-0.0192	0.1170
Cobalt sulfate heptahydrate	CoSO$_4$·7H$_2$O	1.948	RT	86	0.335	0.205	0.158	0.016	0.378
Diopside	(CaMg)Si$_2$O$_6$	3.31	RT	91	2.040	0.884	0.0883	-0.193	1.750
Dipotassium tartrate	KHC$_4$H$_4$O$_6$	1.97	RT	12	0.4294	0.1399	0.3129	-0.0105	0.3460
Feldspar (microceine)	KAlSi$_3$O$_8$	2.56	RT	92	0.664	0.438	0.259	-0.033	1.710
Ferrous sulfate heptahydrate	FeSO$_4$·7H$_2$O	1.898	RT	86	0.349	0.208	0.174	-0.020	0.376
Lithium sulfate monohydrate	Li$_2$SO$_4$·H$_2$O	2.221	RT	32	0.5250	0.1715	0.1730	-0.0196	0.5060
Naphthalene	C$_{10}$H$_8$	1.127	RT	93	0.0780	0.0445	0.0340	-0.006	0.0990
Potassium tartrate	K$_2$C$_4$H$_4$O$_6$	1.987	RT	32	0.3110	0.1720	0.1690	0.0287	0.3900
Sodium thiosulfate	Na$_2$S$_2$O$_3$	1.7499	RT	12	0.3323	0.1814	0.1875	0.0225	0.2953
Stilbene	(C$_6$H$_5$CH)$_2$	1.60	RT	94	0.0930	0.0570	0.0670	-0.003	0.0920
Triglycine sulfate (TGS)	(NH$_2$CH$_2$COOH)$_3$·H$_2$SO$_4$	1.68	RT	32	0.4550	0.1720	0.1980	-0.030	0.3210

Name	C_{23}	C_{25}	C_{33}	C_{35}	C_{44}	C_{46}	C_{55}	C_{66}
Aegirine	0.626	0.094	2.344	0.214	0.692	0.077	0.510	0.474
Anthracene	0.0375	-0.0170	0.1522	-0.0187	0.0272	0.0138	0.0242	0.0399
Cobalt sulfate heptahydrate	0.158	-0.018	0.371	-0.047	0.060	0.016	0.058	0.101
Diopside	0.482	-0.196	2.380	-0.336	0.675	-0.113	0.588	0.705
Dipotassium tartrate	0.1173	0.0176	0.6816	0.0294	0.0961	-0.0044	0.1270	0.0841
Feldspar (microceine)	0.192	-0.148	1.215	-0.131	0.143	-0.015	0.238	0.361
Ferrous sulfate heptahydrate	0.172	-0.019	0.360	-0.014	0.064	0.001	0.056	0.096
Lithium sulfate monohydrate	0.0368	0.0571	0.5400	-0.0254	0.1400	-0.0054	0.1565	0.2770
Naphthalene	0.0230	-0.0270	0.1190	0.0290	0.0330	-0.0050	0.0210	0.0415
Potassium tartrate	0.1330	0.0182	0.5540	0.0710	0.0870	0.0072	0.1040	0.0826
Sodium thiosulfate	0.1713	0.0983	0.4590	-0.0678	0.0569	-0.0268	0.1070	0.0598
Stilbene	0.0485	-0.005	0.0790	-0.005	0.0325	0.0050	0.0640	0.0245
Triglycine sulfate (TGS)	0.2080	-0.0036	0.2630	-0.0500	0.0950	-0.0026	0.1110	0.0620

ELASTIC CONSTANTS OF SINGLE CRYSTALS (continued)

HEXAGONAL CRYSTALS

Name	Formula	ρ/g cm^{-3}	T/K	Ref.	C_{11}	C_{12}	C_{13}	C_{33}	C_{55}
Apatite	Ca$_5$(PO$_4$)$_3$(OH,F,Cl)	3.218	RT	12	1.667	0.131	0.655	1.396	0.663
Beryl	Be$_3$Al$_2$Si$_6$O$_{18}$	2.68	RT	12	2.800	0.990	0.670	2.480	0.658
Beryllium	Be	1.8477	300	95	2.923	0.267	0.140	3.364	1.625
Beryllium oxide	BeO	3.01	RT	96	4.70	1.68	1.19	4.94	1.53
Cadmium	Cd	8.652	300	97	1.1450	0.3950	0.3990	0.5085	0.1985
Cadmium selenide	CdSe	5.655	298	68	0.7046	0.4516	0.3930	0.8355	0.1317
Cadmium sulfide	CdS	4.824	298	98	0.8431	0.5208	0.4567	0.9183	0.1458
Cobalt	Co	8.836	298	99	3.071	1.650	1.027	3.581	0.755
Dysprosium	Dy	8.560	298	100	0.7466	0.2616	0.2233	0.7871	0.2427
Erbium	Er	9.064	298	100	0.8634	0.3050	0.2270	0.8554	0.2809
Gadolinium	Gd	7.888	298	101	0.6667	0.2499	0.2132	0.7191	0.2089
Hafnium	Hf	12.727	298	102	1.881	0.772	0.661	1.969	0.557
Ice	H$_2$O(solid)	0.920	250	103	0.1410	0.0660	0.0624	0.1515	0.0288
Indium	In	7.2788	300	104	0.4535	0.4006	0.4151	0.4515	0.0651
Magnesium	Mg	1.7364	298	105	0.5950	0.2612	0.2180	0.6155	0.1635
Rhenium	Re	21.024	298	100	6.1820	2.7530	2.0780	6.8350	1.6060
Ruthenium	Ru	12.3615	298	100	5.6260	1.8780	1.6820	6.2420	1.8060
Thallium	Tl	11.560	300	106	0.4080	0.3540	0.2900	0.5280	0.0726
Titanium	Ti	4.5063	298	102	1.6240	0.9200	0.6900	1.8070	0.4670
Titanium diboride	TiB$_2$	4.95	RT	107	6.90	4.10	3.20	4.40	2.50
Yttrium	Y	4.472	300	108	0.7790	0.2850	0.2100	0.7690	0.2431
Zinc	Zn	7.134	295	109	1.6368	0.3640	0.5300	0.6347	0.3879
Zinc oxide	ZnO	5.6760	298	110	2.0970	1.2110	1.0510	2.1090	0.4247
Zinc sulfide	ZnS	4.089	298	96	1.2420	0.6015	0.4554	1.4000	0.2864
Zirconium	Zr	6.505	298	102	1.434	0.728	0.653	1.648	0.320

TRIGONAL CRYSTALS

Name	Formula	ρ/g cm^{-3}	T/K	Ref.	C_{11}	C_{12}	C_{13}	C_{14}	C_{33}	C_{44}
Aluminum oxide	Al$_2$O$_3$	3.986	300	111	4.9735	1.6397	1.1220	-0.2358	4.9911	1.4739
Aluminum phosphate	AlPO$_4$	2.556	RT	73	1.0503	0.2934	0.6927	-0.1271	1.3353	0.2314
Antimony	Sb	6.70	295	112	1.0130	0.3450	0.2920	0.2090	0.4500	0.3930
Bismuth	Bi	9.80	295	112	0.6370	0.2490	0.2470	0.0717	0.3820	0.1123
Calcite	CaCO$_3$	2.712	300	113	1.4806	0.5578	0.5464	-0.2058	0.8557	0.3269
Hematite	Fe$_2$O$_3$	5.240	RT	82	2.4243	0.5464	0.1542	-0.1247	2.2734	0.8569
Lithium niobate	LiNbO$_3$	4.70	RT	114	2.030	0.530	0.750	0.090	2.450	0.600
Lithium tantalate	LiTaO$_3$	7.45	RT	114	2.330	0.470	0.800	-0.110	2.750	0.940
Quartz	SiO$_2$	2.6485	298	115	0.8680	0.0704	0.1191	-0.1804	1.0575	0.5820
Selenium	Se	4.838	300	116	0.1870	0.0710	0.2620	0.0620	0.7410	0.1490
Sodium nitrate	NaNO$_3$	2.27	RT	12	0.8670	0.1630	0.1600	0.0820	0.3740	0.2130
Tourmaline		3.05	RT	82	2.7066	0.6927	0.0872	-0.0774	1.6070	0.6682

REFERENCES

1. Thomas, J. F., *Phys. Rev.*, 175, 955-962, 1968.
2. Bolef, D. I. and M. Menes, *J. Appl. Phys.*, 31, 1426-1427, 1960.
3. Garland, C. W. and C. F. Yarnell, *J. Chem. Phys.*, 44, 1112-1120, 1966.
4. Garland, C. W. and R. Renard, *J. Chem. Phys.*, 44, 1130-1139, 1966.

5. Gsänger, M., H. Egger and E. Lüscher, *Phys. Letters*, 27A, 695-696, 1968.
6. Wong, C. and D. E. Schuele, *J. Phys. Chem. Solids*, 29, 1309-1330, 1968.
7. Haussühl, S., *Phys. Stat. Sol.*, 3, 1072-1076, 1963.
8. Wong, C. and D. E. Schuele, *J. Phys. Chem. Solids*, 28, 1225-1231, 1967.
9. McSkimin, H. J. and D. G. Thomas, *J. Appl. Phys.*, 33, 56-59, 1962.
10. Kollarits, F. J. and J. Trivisonno, *J. Phys. Chem. Solids*, 29, 2133-2139, 1968.
11. Slagle, D. D. and H. A. McKinstry, *J. Appl. Phys.*, 38, 446-458, 1967.
12. Hearmon, R. F. S., *Adv. Phys.*, 5, 323-382, 1956.
13. Sumer, A. and J. F. Smith, *J. Appl. Phys.*, 34, 2691-2694, 1963.
14. Alexandrov, K. S. et. al., *Sov. Phys. Sol. State*, 10, 1316-1321, 1968.
15. Epstein, S. G. and O. N. Carlson, *Acta Metal.*, 13, 487-491, 1965.
16. McSkimin, H. J., et. al., *J. Appl. Phys.*, 39, 4127-4128, 1968.
17. McSkimin, H. J., et. al., *J. Appl. Phys.*, 38, 2362-2364, 1967.
18. Weil, R. and W. O. Groves, *J. Appl. Phys.*, 39, 4049-4051, 1968.
19. Bateman, T. B., *J. Appl. Phys.*, 37, 2194-2195, 1966.
20. Bogardus, E. H., *J. Appl. Phys.*, 36, 2504-2513, 1965.
21. Golding, B., S. C. Moss and B. L. Averbach, *Phys. Rev.*, 158, 637-645, 1967.
22. Bateman, T. B., H. J. McSkimin and J. M. Whelan, *J. Appl. Phys.*, 30, 544-545, 1959.
23. Gerlich, D., *J. Appl. Phys.*, 35, 3062, 1964.
24. Hickernell, F. S. and W. R. Gayton, *J. Appl. Phys.*, 37, 462, 1966.
25. MacFarlane, R. E., et. al., *Phys. Letters*, 20, 234-235, 1966.
26. Leese, J. and A. E. Lord Jr., *J. Appl. Phys.*, 39, 3986-3988, 1968.
27. Miller, R. A. and D. E. Schuele, *J. Phys. Chem. Solids*, 30, 589-600, 1969.
28. Wasilik, J. H. and M. L. Wheat, *J. Appl. Phys.*, 36, 791-793, 1965.
29. Haussühl, S., *Phys. Stat. Sol.*, 3, 1072-1076, 1963.
30. Houston, B., et. al., *J. Appl. Phys.*, 39, 3913-3916, 1968.
31. Trivisonno, J. and C. S. Smith, *Acta Metal.*, 9, 1064-1071, 1961.
32. Alexandrov, K. S. and T. V. Ryzhova, *Sov. Phys. Cryst.*, 6, 228-252, 1961.
33. Lewis, J. T., A. Lehoczky and C. V. Briscoe, *Phys. Rev.*, 161, 877-887, 1967.
34. Drabble, J. R. and R. E. B. Strathen, *Proc. Phys. Soc.*, 92, 1090-1995, 1967.
35. Oliver, D. W., *J. Appl. Phys.*, 40, 893, 1969.
36. Alper, T., and G. A. Saunders, *J. Phys. Chem. Solids*, 28, 1637-1642, 1967.
37. Dickinson, J. M. and P. E. Armstrong, *J. Appl. Phys.*, 38, 602-606, 1967.
38. Bolef, D. I., *J. Appl. Phys.*, 32, 100-105, 1961.
39. Rayne, J. A., *Phys. Rev.*, 112, 1125-1130, 1958.
40. MacFarlane, R. E., et. al., *Phys. Letters*, 18, 91-92, 1965.
41. Smith, P. A. and C. S. Smith, *J. Phys. Chem. Solids*, 26, 279-289, 1965.
42. Norwood, M. H. and C. V. Briscoe, *Phys. Rev.*, 112, 45-48, 1958.
43. Simmons, G. and F. Birch, *J. Appl. Phys.*, 34, 2736-2738, 1963.
44. Gutman, E. J. and J. Trivisonno, *J. Phys. Chem. Sol.*, 28, 805-809, 1967.
45. Ghafelehbashi, M., et. al., *J. Appl. Phys.*, 41, 652-666, 1970.
46. McSkimin, H. J. and P. Andreatch Jr., *J. Appl. Phys.*, 35, 2161-2165, 1964.
47. Neighbours, J. R. and G. A. Alers, *Phys. Rev.*, 111, 707-712, 1958.
48. Hidshaw, W., J. T. Lewis, and C. V. Briscoe, *Phys. Rev.*, 163, 876-881, 1967.
49. Daniels, W. B., *Phys. Rev.*, 119, 1246-1252, 1960.
50. Viswanathan, R., *J. Appl. Phys.*, 37, 884-886, 1966.
51. Miller, R. A. and C. S. Smith, *J. Phys. Chem. Sol.*, 25, 1279-1292, 1964.
52. Claytor, R. N. and B. J. Marshall, *Phys. Rev.*, 120, 332-334, 1960.
53. Schreiber, E., *J. Appl. Phys.*, 38, 2508-2511, 1967.
54. Gerlich, D., *Phys. Rev.*, 136, A1366-A1368, 1964.
55. Johnston, D. L., P. H. Thrasher and R. J. Kearney, *J. Appl. Phys.*, 41, 427-428, 1970.
56. Poindexter, E. and A. A. Giardini, *Phys. Rev.*, 110, 1069, 1958.
57. Soga, N., *J. Appl. Phys.*, 37, 3416-3420, 1966.
58. Bartlett, R. W. and C. W. Smith, *J. Appl. Phys.*, 38, 5428-5429, 1967.
59. Morse, G. E. and A. W. Lawson, *J. Phys. Chem. Sol.*, 28, 939-950, 1967.
60. Armstrong, P. E., O. N. Carlson and J. F. Smith, *J. Appl. Phys.*, 30, 36-41, 1959.
61. Macedo, P. M., W. Capps and J. B. Wachtman, *J. Am. Cer. Soc.*, 47, 651, 1964.
62. Beattie, A. G., *J. Appl. Phys.*, 40, 4818-4821, 1969.
63. Chang, R. and L. J. Graham, *J. Appl Phys.*, 37, 3778-3783, 1966.
64. Lowrie, R. and A. M. Gonas, *J. Appl. Phys.*, 38, 4505-4509. 1967.
65. Graham, L. J., H. Nadler and R. Chang, *J. Appl. Phys.*, 34, 1572-1573, 1963.

66. Wachtman, J. B. Jr., et. al., *J. Nucl. Mat.*, 16, 39-41, 1965.
67. Bolef, D. I., *J. Appl. Phys.*, 32, 100-105, 1961.
68. Berlincourt, D., H. Jaffe and L. R. Shiozawa, *Phys. Rev.*, 129, 1009-1017, 1963.
69. Adhav. R. S. J. *Acoust. Soc. Am.*, 43, 835-838, 1968.
70. Berlincourt, D. and H. Jaffe, *Phys. Rev.*, 111, 143-148, 1958.
71. Huntington, H. B., in *Solid State Pysics, Vol. 7*, Seitz, F., and Turnbull, D., Ed., pp. 213-285; Academic Press, New York 1958.
72. Cutler, H. R., J. J. Gibson and K. A. McCarthy, *Sol. State Comm.*, 6, 431-433, 1968.
73. Mason, W. P., *Piezoelectric Crystals and Their Application to Ultrasonics*, D. Van Nostrand Co., Inc., New York, 1950.
74. Adhav, R. S., *J. Appl. Phys.*, 40, 2725-2727, 1969.
75. Manghnani, M. H., *J. Geophys. Res.*, 74, 4317-4328, 1969.
76. Uchida, N. and Y. Ohmachi, *J. Appl Phys.*, 40, 4692-4695, 1969.
77. House, D. G. and E. Y. Vernon, *Br. J. Appl. Phys.*, 11, 254-259, 1960.
78. Ryzhova, T. V., et. al., *Bull. Acad. Sci. USSR, Earth Phys. Ser.*, English Transl., no. 2, 111-113, 1966.
79. Alton, W. J. and A. J. Barlow, *J. Appl. Phys.*, 38, 3817-3820, 1967.
80. Michard, F., et. al., *C. R. Acad. Sci., Paris*, 265, 565-567, 1967.
81. Haussühl, S., *Acta Cryst.*, 18, 839-842, 1965.
82. Hearmon, R. F. S., *Rev. Mod. Phys.*, 18, 409-440, 1946.
83. Heseltine, J. C. W., D. W. Elliott and O. B. Wilson, *J. Chem. Phys.*, 40, 2584-2587, 1964.
84. Schwerdtner, W. M., et. al., *Canad. J. Earth Sci.*, 2, 673-683, 1965.
85. Kumazawa, M. and O. L. Anderson, *J. Geophys. Res.*, 74, 5961-5972, 1969 .
86. Alexandrov, K. S., et. al., *Sov. Phys. Cryst.*, 7, 753-755, 1963.
87. Verma, R. K., *J. Geophys. Soc.*, 65, 757-766, 1960.
88. McSkimin, H. J. and E. S. Fisher, *J. Appl. Phys.*, 31, 1627-1639, 1960.
89. Alexandrov, K. S. and T.V. Ryzhova, *Bull. Acad. Sci. USSR, Geophys. Ser.*, English Transl., no.8, 871-875, 1961.
90. Afanaseva, G. K., et. al, *Phys. Stat. Sol.*, 24, K61-K63, 1967.
91. Alexandrov, K. S., et. al., *Sov. Phys. Cryst.*, 8, 589-591, 1964.
92. Alexandrov, K. S. and T. V Ryzhova, *Bull Acad. Sci. USSR, Geophys. Ser.*, English Transl., no.2, 129-131, 1962.
93. Alexandrov, K. S., et. al., *Sov. Phys. Cryst.*, 8, 164-166, 1963.
94. Teslenko, V. F., et. al., *Sov. Phys. Cryst.*, 10, 744-747, 1966.
95. Smith, J. F. and C. L. Arbogast, *J. Appl. Phys.*, 31, 99-102, 1960.
96. Cline, C. F., H. L. Dunegan and G. M. Henderson, *J. Appl. Phys.*, 38, 1944-1948, 1967.
97. Chang, Y. A. and L. Himmel, *J. Appl. Phys.*, 37, 3787-3790, 1966.
98. Gerlich, D., *J. Phys. Chem. Solids*, 28, 2575-2579, 1967.
99. McSkimin, H. J., *J. Appl. Phys.*, 26, 406-409, 1955.
100. Fisher, E. S. and D. Dever, *Trans. Met. Soc. AIME*, 239, 48-57, 1967.
101. Fisher, E. S. and D. Dever, *Proc. Conf. Rare Earth Res.*, 6th, Gatlinburg, Tenn., 522-533, 1967.
102. Fisher, E. S. and C. J. Renken, Phys. Rev., 135, A482-A494, 1964.
103. Proctor, T. M. Jr., *J. Acoust. Soc. Am.*, 39, 972-977, 1966.
104. Chandrasekhar, B. S. and J. A. Rayne, *Phys. Rev.*, 124, 1011-1041, 1961.
105. Wazzan, A. R. and L. B. Robinson, *Phys. Rev.*, 155, 586-594, 1967.
106. Ferris, R. W., et. al., *J. Appl. Phys.*, 34, 768-770, 1963.
107. Gilman, J. J. and B. W. Roberts, *J. Appl. Phys.*, 32, 1405, 1961.
108. Smith, J. F. and J. A. Gjevre, *J. Appl. Phys.*, 31, 645-647, 1960.
109. Alers, G. A. and J. R. Neighbours, *J. Phys. Chem. Solids*, 7, 58-64, 1908.
110. Bateman, T. B., *J. Appl. Phys.*, 33, 3309-3312, 1962.
111. Tefft, W. E., *J. Res. Natl. Bur. Stand.*, 70A, 277-280, 1966.
112. DeBretteville, Jr., A. et. al., *Phys. Rev.*, 148, 575-579, 1966.
113. Dandekar, D. P. and A. L. Ruoff, *J. Appl. Phys.*, 39, 6004-6009, 1968.
114. Warner, A. W., M. Onoe and G. A. Coquin, *J. Acoust. Soc. Am.*, 42, 1223-1231, 1967.
115. McSkimin, H. J., P. Andreatch and R. N. Thurston, *J. Appl. Phys.*, 36, 1624-1632, 1965.
116. Mort, J., *J. Appl. Phys.*, 38, 3414-3415, 1967.

ELECTRICAL RESISTIVITY OF PURE METALS

The first part of this table gives the electrical resistivity, in units of 10^{-8} Ω m, for 28 common metallic elements as a function of temperature. The data refer to polycrystalline samples. The number of significant figures indicates the accuracy of the values. However, at low temperatures (especially below 50 K) the electrical resistivity is extremely sensitive to sample purity. Thus the low-temperature values refer to samples of specified purity and treatment. The references should be consulted for further information on this point, as well as for values at additional temperatures.

The second part of the table gives resistivity values in the neighborhood of room temperature for other metallic elements that have not been studied over an extended temperature range.

REFERENCES

1. C. Y. Ho, et al., *J. Phys. Chem. Ref. Data*, 12, 183—322, 1983; 13, 1069—1096, 1984; 13, 1097—1130, 1984, 13, 1131—1172, 1984.
2. R. A. Matula, *J. Phys Chem. Ref. Data*, 8, 1147—1298, 1979.
3. T. C. Chi, *J. Phys. Chem. Ref. Data*, 8, 339—438, 1979; 8, 439—498, 1979.
4. K. H. Hellwege, Ed., *Landolt-Börnstein Numerical Data and Functional Relationships in Science and Technology*, Group III, Vol. 15, Subvolume a, Springer-Verlag, Heidelberg, 1982.
5. L. A. Hall, *Survey of Electrical Resistivity Measurements on 16 Pure Metals in the Temperature Range 0 to 273 K*, NBS Technical Note 365, U.S. Superintendent of Documents, 1968.

ELECTRICAL RESISTIVITY IN 10^{-8} Ω m

T/K	Aluminum	Barium	Beryllium	Calcium	Cesium	Chromium	Copper
1	0.000100	0.081	0.0332	0.045	0.0026		0.00200
10	0.000193	0.189	0.0332	0.047	0.243		0.00202
20	0.000755	0.94	0.0336	0.060	0.86		0.00280
40	0.0181	2.91	0.0367	0.175	1.99		0.0239
60	0.0959	4.86	0.067	0.40	3.07		0.0971
80	0.245	6.83	0.075	0.65	4.16		0.215
100	0.442	8.85	0.133	0.91	5.28	1.6	0.348
150	1.006	14.3	0.510	1.56	8.43	4.5	0.699
200	1.587	20.2	1.29	2.19	12.2	7.7	1.046
273	2.417	30.2	3.02	3.11	18.7	11.8	1.543
293	2.650	33.2	3.56	3.36	20.5	12.5	1.678
298	2.709	34.0	3.70	3.42	20.8	12.6	1.712
300	2.733	34.3	3.76	3.45	21.0	12.7	1.725
400	3.87	51.4	6.76	4.7		15.8	2.402
500	4.99	72.4	9.9	6.0		20.1	3.090
600	6.13	98.2	13.2	7.3		24.7	3.792
700	7.35	130	16.5	8.7		29.5	4.514
800	8.70	168	20.0	10.0		34.6	5.262
900	10.18	216	23.7	11.4		39.9	6.041

T/K	Gold	Hafnium	Iron	Lead	Lithium	Magnesium	Manganese
1	0.0220	1.00	0.0225		0.007	0.0062	7.02
10	0.0226	1.00	0.0238		0.008	0.0069	18.9
20	0.035	1.11	0.0287		0.012	0.0123	54
40	0.141	2.52	0.0758		0.074	0.074	116
60	0.308	4.53	0.271		0.345	0.261	131
80	0.481	6.75	0.693	4.9	1.00	0.557	132
100	0.650	9.12	..28	6.4	1.73	0.91	132
150	1.061	15.0	3.15	9.9	3.72	1.84	136
200	1.462	21.0	5.20	13.6	5.71	2.75	139
273	2.051	30.4	8.57	19.2	8.53	4.05	143
293	2.214	33.1	9.61	20.8	9.28	4.39	144
298	2.255	33.7	9.87	21.1	9.47	4.48	144
300	2.271	34.0	9.98	21.3	9.55	4.51	144
400	3.107	48.1	16.1	29.6	13.4	6.19	147
500	3.97	63.1	23.7	38.3		7.86	149

T/K	Gold	Hafnium	Iron	Lead	Lithium	Magnesium	Manganese
600	4.87	78.5	32.9			9.52	151
700	5.82		44.0			11.2	152
800	6.81		57.1			12.8	
900	7.86					14.4	

T/K	Molybdenum	Nickel	Palladium	Platinum	Potassium	Rubidium	Silver
1	0.00070	0.0032	0.0200	0.002	0.0008	0.0131	0.00100
10	0.00089	0.0057	0.0242	0.0154	0.0160	0.109	0.00115
20	0.00261	0.0140	0.0563	0.0484	0.117	0.444	0.0042
40	0.0457	0.068	0.334	0.409	0.480	1.21	0.0539
60	0.206	0.242	0.938	1.107	0.90	1.94	0.162
80	0.482	0.545	1.75	1.922	1.34	2.65	0.289
100	0.858	0.96	2.62	2.755	1.79	3.36	0.418
150	1.99	2.21	4.80	4.76	2.99	5.27	0.726
200	3.13	3.67	6.88	6.77	4.26	7.49	1.029
273	4.85	6.16	9.78	9.6	6.49	11.5	1.467
293	5.34	6.93	10.54	10.5	7.20	12.8	1.587
298	5.47	7.12	10.73	10.7	7.39	13.1	1.617
300	5.52	7.20	10.80	10.8	7.47	13.3	1.629
400	8.02	11.8	14.48	14.6			2.241
500	10.6	17.7	17.94	18.3			2.87
600	13.1	25.5	21.2	21.9			3.53
700	15.8	32.1	24.2	25.4			4.21
800	18.4	35.5	27.1	28.7			4.91
900	21.2	38.6	29.4	32.0			5.64

T/K	Sodium	Strontium	Tantalum	Tungsten	Vanadium	Zinc	Zirconium
1	0.0009	0.80	0.10	0.000016		0.0100	0.250
10	0.0015	0.80	0.102	0.000137	0.0145	0.0112	0.253
20	0.016	0.92	0.146	0.00196	0.039	0.0387	0.357
40	0.172	1.70	0.751	0.0544	0.304	0.306	1.44
60	0.447	2.68	1.65	0.266	1.11	0.715	3.75
80	0.80	3.64	2.62	0.606	2.41	1.15	6.64
100	1.16	4.58	3.64	1.02	4.01	1.60	9.79
150	2.03	6.84	6.19	2.09	8.2	2.71	17.8
200	2.89	9.04	8.66	3.18	12.4	3.83	26.3
273	4.33	12.3	12.2	4.82	18.1	5.46	38.8
293	4.77	13.2	13.1	5.28	19.7	5.90	42.1
298	4.88	13.4	13.4	5.39	20.1	6.01	42.9
300	4.93	13.5	13.5	5.44	20.2	6.06	43.3
400		17.8	18.2	7.83	28.0	8.37	60.3
500		22.2	22.9	10.3	34.8	10.82	76.5
600		26.7	27.4	13.0	41.1	13.49	91.5
700		31.2	31.8	15.7	47.2		104.2
800		35.6	35.9	18.6	53.1		114.9
900			40.1	21.5	58.7		123.1

Element	T/K	Electrical resistivity 10^{-8} Ω m
Antimony	273	39
Bismuth	273	107
Cadmium	273	6.8
Cerium (β, hex)	290—300	82.8
Cerium (γ, cub)	298	74.4
Cobalt	273	5.6
Dysprosium	290—300	92.6
Erbium	290—300	86.0
Europium	290—300	90.0
Gadolinium	290—300	131
Gallium	273	13.6
Holmium	290—300	81.4
Indium	273	8.0
Iridium	273	4.7
Lanthanum	290—300	61.5
Lutetium	290—300	58.2
Mercury	273	94.1
Neodymium	290—300	64.3
Niobium	273	15.2
Osmium	273	8.1
Polonium	273	40
Praseodymium	290—300	70.0
Promethium	290—300	75 est.
Protactinium	273	17.7
Rhenium	273	17.2
Rhodium	273	4.3
Ruthenium	273	7.1
Samarium	290—300	94.0
Scandium	290—300	56.2
Terbium	290—300	115
Thallium	273	15
Thorium	273	14.7
Thulium	290—300	67.6
Tin	273	11.5
Titanium	273	39
Uranium	273	28
Ytterbium	290—300	25.0
Yttrium	290—300	59.6

Values of the resistivity are given in units of 10^{-8} Ω m. General comments in the preceeding table for pure metals also apply here.

REFERENCE

C. Y. Ho, et al., *J. Phys. Chem. Ref. Data*, 12, 183—322, 1983.

Alloy Aluminum-Copper

Wt % Al	273 K	293 K	300 K	350 K	400 K
99[a]	2.51	2.74	2.82	3.38	3.95
95[a]	2.88	3.10	3.18	3.75	4.33
90[b]	3.36	3.59	3.67	4.25	4.86
85[b]	3.87	4.10	4.19	4.79	5.42
80[b]	4.33	4.58	4.67	5.31	5.99
70[b]	5.03	5.31	5.41	6.16	6.94
60[b]	5.56	5.88	5.99	6.77	7.63
50[b]	6.22	6.55	6.67	7.55	8.52
40[c]	7.57	7.96	8.10	9.12	10.2
30[c]	11.2	11.8	12.0	13.5	15.2
25[f]	16.3[aa]	17.2	17.6	19.8	22.2
15[h]	—	12.3	—	—	—
19[g]	10.8[aa]	11.0	11.1	11.7	12.3
5[e]	9.43	9.61	9.68	10.2	10.7
1[b]	4.46	4.60	4.65	5.00	5.37

	273 K	293 K	300 K	350 K	400 K
60[j]	45.43	45.38	45.35	45.20	45.01
50[j]	50.19	50.05	50.01	49.73	49.50
40[c]	47.42	47.73	47.82	48.28	48.49
30[j]	40.19	41.79	42.34	44.51	45.40
25[c]	33.46	35.11	35.69	39.67[aa]	42.81[aa]
15[c]	22.00	23.35	23.85	27.60	31.38
10[c]	16.65	17.82	18.26	21.51	25.19
5[c]	11.49	12.50	12.90	15.69	18.78
1[c]	7.23	8.08	8.37	10.63[aa]	13.18[aa]

Alloy—Aluminum-Magnesium

Wt % Al	273 K	293 K	300 K	350 K	400 K
99[c]	2.96	3.18	3.26	3.82	4.39
95[c]	5.05	5.28	5.36	5.93	6.51
90[c]	7.52	7.76	7.85	8.43	9.02
85	—	—	—	—	—
80	—	—	—	—	—
70	—	—	—	—	—
60	—	—	—	—	—
50	—	—	—	—	—
40	—	—	—	—	—
30	—	—	—	—	—
25	—	—	—	—	—
15	—	—	—	—	—
10[b]	17.1	17.4	17.6	18.4	19.2
5[b]	13.1	13.4	13.5	14.3	15.2
1[a]	5.92	6.25	6.37	7.20	8.03

Alloy—Copper-Gold

Wt % Cu	273 K	293 K	300 K	350 K	400 K
99[c]	1.73	1.86[aa]	1.91[aa]	2.24[aa]	2.58[aa]
95[c]	2.41	2.54[aa]	2.59[aa]	2.92[aa]	3.26[aa]
90[c]	3.29	4.42[aa]	3.46[aa]	3.79[aa]	4.12[aa]
85[c]	4.20	4.33	4.38[aa]	4.71[aa]	5.05[aa]
80[c]	5.15	5.28	5.32	5.65	5.99
70[c]	7.12	7.25	7.30	7.64	7.99
60[c]	9.18	9.13	9.36	9.70	10.05
50[c]	11.07	11.20	11.25	11.60	11.94
40[c]	12.70	12.85	12.90[aa]	13.27[aa]	13.65[aa]
30[c]	13.77	13.93	13.99[aa]	14.38[aa]	14.78[aa]
25[c]	13.93	14.09	14.14	14.54	14.94
15[c]	12.75	12.91	12.96[aa]	13.36[aa]	13.77
10[c]	10.70	10.86	10.91	11.31	11.72
5[c]	7.25	7.41[aa]	7.46	7.87	8.28
1[c]	3.40	3.57	3.62	4.03	4.45

Alloy—Copper-Nickel

Wt % Cu	273 K	293 K	300 K	350 K	400 K
99[c]	2.71	2.85	2.91	3.27	3.62
95[c]	7.60	7.71	7.82	8.22	8.62
90[c]	13.69	13.89	13.96	14.40	14.81
85[c]	19.63	19.83	19.90	2032	20.70
80[c]	25.46	25.66	25.72	26.12[aa]	26.44[aa]
70[j]	36.67	36.72	36.76	36.85	36.89

Alloy—Copper-Palladium

Wt % Cu	273 K	293 K	300 K	350 K	400 K
99[c]	2.10	2.23	2.27	2.59	2.92
95[c]	4.21	4.35	4.40	4.74	5.08
90[c]	6.89	7.03	7.08	7.41	7.74
85[c]	9.48	9.61	9.66	10.01	10.36
80[c]	11.99	12.12	12.16	12.51[aa]	12.87
70[c]	16.87	17.01	17.06	17.41	17.78
60[c]	21.73	21.87	21.92	22.30	22.69
50[c]	27.62	27.79	27.86	28.25	28.64
40[c]	35.31	35.51	35.57	36.03	36.47
30[c]	46.50	46.66	46.71	47.11	47.47
25[c]	46.25	46.45	46.52	46.99[aa]	47.43[aa]
15[c]	36.52	36.99	37.16	38.28	39.35
10[c]	28.90	29.51	29.73	31.19[aa]	32.56[aa]
5[c]	20.00	20.75	21.02	22.84[aa]	24.54[aa]
1[c]	11.90	12.67	12.93[aa]	14.82[aa]	16.68[aa]

Alloy—Copper-Zinc

Wt % Cu	273 K	293 K	300 K	350 K	400 K
99[b]	1.84	1.97	2.02	2.36	2.71
95[b]	2.78	2.92	2.97	3.33	3.69
90[b]	3.66	3.81	3.86	4.25	4.63
85[b]	4.37	4.54	4.60	5.02	5.44
80[b]	5.01	5.19	5.26	5.71	6.17
70[b]	5.87	6.08	6.15	6.67	7.19
60	—	—	—	—	—
50	—	—	—	—	—
40	—	—	—	—	—
30	—	—	—	—	—
25	—	—	—	—	—
15	—	—	—	—	—
10	—	—	—	—	—
5	—	—	—	—	—
1	—	—	—	—	—

Alloy—Gold-Palladium

Wt % Au	273 K	293 K	300 K	350 K	400 K
99[c]	2.69	2.86	2.91	3.32	3.73
95[c]	5.21	5.35	5.41	5.79	6.17
90[j]	8.01	8.17	8.22	8.56	8.93
85[b]	10.50[aa]	10.66	10.72[aa]	11.100[aa]	11.48[aa]
80[b]	12.75	12.93	12.99	13.45	13.93
70[c]	18.23	18.46	18.54	19.10	19.67
60[b]	26.70	26.94	27.02	27.63[aa]	28.23[aa]
50[a]	27.23	27.63	27.76	28.64[aa]	29.42[aa]
40[a]	24.65	25.23	25.42	26.74	27.95
30[b]	20.82	21.49	21.72	23.35	24.92
25[b]	18.86	19.53	19.77	21.51	23.19
15[a]	15.08	15.77	16.01	17.80	19.61
10[a]	13.25	13.95	14.20[aa]	16.00[aa]	17.81[aa]
5[a]	11.49[aa]	12.21	12.46[aa]	14.26[aa]	16.07[aa]
1[a]	10.07	10.85[aa]	11.12[aa]	12.99[aa]	14.80[aa]

Alloy—Gold-Silver

Wt % Au	273 K	293 K	300 K	350 K	400 K
99[b]	2.58	2.75	2.80[aa]	3.22[aa]	3.63[aa]
95[a]	4.58	4.74	4.79	5.19	5.59
90[j]	6.57	6.73	6.78	7.19	7.58
85[j]	8.14	8.30	8.36[aa]	8.75	9.15
80[j]	9.34	9.50	9.55	9.94	10.33
70[j]	10.70	10.86	10.91	11.29	11.68[aa]
60[j]	10.92	11.07	11.12	11.50	11.87
50[j]	10.23	10.37	10.42	10.78	11.14
40[j]	8.92	9.06	9.11	9.46[aa]	9.81
30[a]	7.34	7.47	7.52	7.85	8.19
25[a]	6.46	6.59	6.63	6.96	7.30[aa]
15[a]	4.55	4.67	4.72	5.03	5.34
10[a]	3.54	3.66	3.71	4.00	4.31
5[i]	2.52	2.64[aa]	2.68[aa]	2.96[aa]	3.25[aa]
1[b]	1.69	1.80	1.84[aa]	2.12[aa]	2.42[aa]

(continued)

	273 K	293 K	300 K	350 K	400 K
60[c]	53.8	57.1	58.2	—	73.9
50[d]	28.4	30.6	31.4	—	43.7
40[d]	19.6	21.6	22.5	—	34.0
30[c]	15.3	17.1	17.7	—	27.4
25[b]	14.3	15.9	16.4	—	25.1
15[c]	12.6	13.8	14.2	—	21.1
10[c]	11.4	12.5	12.9	—	18.9
5[c]	9.66	10.6	10.9	—	16.1[aa]
1[b]	7.17	7.94	8.12	—	12.8

Alloy—Iron-Nickel

Wt % Fe	273 K	293 K	300 K	350 K	400 K
99[a]	10.9	12.0	12.4	—	18.7
95[c]	18.7	19.9	20.2	—	26.8
90[c]	24.2	25.5	25.9	—	33.2
85[c]	27.8	29.2	29.7	—	37.3
80[c]	30.1	31.6	32.2	—	40.0
70[b]	32.3	33.9	34.4	—	42.4

Alloy—Silver-Palladium

Wt % Ag	273 K	293 K	300 K	350 K	400 K
99[b]	1.891	2.007	2.049	2.35	2.66
95[b]	3.58	3.70	3.74	4.04	4.34
90[b]	5.82	5.94	5.98	6.28	6.59
85[k]	7.92[aa]	8.04[aa]	8.08	8.38[aa]	8.68[aa]
80[k]	10.01	10.13	10.17	10.47	10.78
70[k]	14.53	14.65	14.69	14.99	15.30
60[j]	20.9	21.1	21.2	21.6	22.0
50[k]	31.2	31.4	31.5	32.0	32.4
40[m]	42.2	42.2	42.2	42.3	42.3
30[b]	40.4	40.6	40.7	41.3	41.7
25[k]	36.67[aa]	37.06	37.19	38.1[aa]	38.8[aa]
15[i]	27.08[aa]	26.68[aa]	27.89[aa]	29.3[aa]	30.6[aa]
10[i]	21.69	22.39	22.63	24.3	25.9
5[b]	15.98	16.72	16.98	18.8[aa]	20.5[aa]
1[a]	11.06	11.82	12.08[aa]	13.92[aa]	15.70[aa]

[a] Uncertainty in resistivity is ± 2%.
[b] Uncertainty in resistivity is ± 3%.
[c] Uncertainty in resistivity is ± 5%.
[d] Uncertainty in resistivity is ± 7% below 300 K and ± 5% at 300 and 400 K.
[e] Uncertainty in resistivity is ± 7%.
[f] Uncertainty in resistivity is ± 8%.
[g] Uncertainty in resistivity is ± 10%.
[h] Uncertainty in resistivity is ± 12%.
[i] Uncertainty in resistivity is ± 4%.
[j] Uncertainty in resistivity is ± 1%.
[k] Uncertainty in resistivity is ± 3% up to 300 K and ± 4% above 300 K.
[m] Uncertainty in resistivity is ± 2% up to 300 K and + 4% above 300 K.
[n] Crystal usually a mixture of α-hcp and fcc lattice.
[aa] In temperature range where no experimental data are available.

PERMITTIVITY (DIELECTRIC CONSTANT) OF INORGANIC SOLIDS

H. P. R. Frederikse

This table lists the permittivity ϵ, frequently called the dielectric constant, of a number of inorganic solids. When the material is not isotropic, the individual components of the permittivity are given. A superscript S indicates a measurement made under constant strain ("clamped" dielectric constant). If the constraint is removed, the measurement yields ϵ^T, the "unclamped" or free dielectric constant.

The temperature of the measurement is given when available; the symbol r.t. indicates a value at nominal room temperature. The frequency of the measurement is given in the last column (i.r. indicates a measurement in the infrared).

Substances are listed in alphabetical order by chemical formula.

REFERENCE

Young, K. F. and Frederikse, H. P. R., *J. Phys. Chem. Ref. Data,* 2, 313, 1973.

Formula	Name	ϵ_{ijk}	T/K	ν/Hz
Ag_3AsS_3	Silver thioarsenate (Proustite)	$\epsilon_{11}^T = 16.5,\ \epsilon_{11}^S = 14.5$	r.t.	2×10^7
		$\epsilon_{33}^T = 20.0,\ \epsilon_{33}^S = 18.0$	r.t.	2×10^7
$AgBr$	Silver bromide	12.50	r.t.	
$AgCN$	Silver cyanide	5.6	r.t.	10^6
$AgCl$	Silver chloride	11.15	r.t.	
$AgNO_3$	Silver nitrate	9.0	293	5×10^5
$AgNa(NO_2)_2$	Silver sodium nitrite	4.5 ± 0.5	r.t.	9.4×10^9
Ag_2O	Silver oxide	8.8	r.t.	
$(AlF)_2SiO_4$	Aluminum fluosilicate (topaz)	$\epsilon_{11} = 6.62$	297	7×10^3
		$\epsilon_{22} = 6.58$	297	7×10^3
		$\epsilon_{33} = 6.95$	297	7×10^3
Al_2O_3	Aluminum oxide (alumina)	$\epsilon_{11} = \epsilon_{22} = 9.34$	298	10^2—8×10^9
		$\epsilon_{33} = 11.54$	298	10^2—8×10^9
$AlPO_4$	Aluminum phosphate	$\epsilon_{11}^T = 6.05$	r.t.	10^3
$AlSb$	Aluminum antimonide	11.21	300	i.r.
AsF_3	Arsenic trifluoride	5.7	r.t.	
BN	Boron nitride	7.1	r.t.	i.r.
$BaCO_3$	Barium carbonate	8.53	291	2×10^5
$Ba(COOH)_2$	Barium formate	$\epsilon_{11} = 7.9$	r.t.	10^3
		$\epsilon_{22} = 5.9$	r.t.	10^3
		$\epsilon_{33} = 7.5$	r.t.	10^3
$BaCl_2$	Barium chloride	9.81	r.t.	
$BaCl_2 \cdot 2H_2O$	Barium chloride dihydrate	9.00	r.t.	10^3
BaF_2	Barium fluoride	7.32	292	5×10^2—10^{11}
$Ba(NO_3)_2$	Barium nitrate	4.95	292	2×10^5
$Ba_2NaNb_5O_{15}$	Barium sodium niobate ("Bananas")	$\epsilon_{11}^S = 222,\ \epsilon_{11}^T = 235$	296	10^4
		$\epsilon_{22}^S = 227,\ \epsilon_{22}^T = 247$	296	
		$\epsilon_{33}^S = 32,\ \epsilon_{33}^T = 51$	296	
BaO	Barium oxide (baria)	34 ± 1	248, 333	60×10^7
BaO_2	Barium peroxide	10.7	r.t.	2×10^6
BaS	Barium sulfide	19.23	r.t.	7.25×10^6
$BaSO_4$	Barium sulfate	11.4	288	10^8
$BaSnO_3$	Barium stannate	18	298	25×10^5
$BaTiO_3$	Barium titanate	$\epsilon_{11}^T = 3600$	298	10^5
		$\epsilon_{11}^S = 2300$	298	2.5×10^8
		$\epsilon_{33}^T = 150$	298	10^5
		$\epsilon_{33}^S = 80$	298	2.5×10^8
$Ba_6Ti_2Nb_8O_{30}$	Barium titanium niobate	$\epsilon_{11} = \epsilon_{22} \approx 190$	298	
		$\epsilon_{33} \approx 220$	298	
$BaWO_4$	Barium tungstate	$\epsilon_{11} = \epsilon_{22} = 35.5 \pm 0.2$	297.5	1.6×10^3
		$\epsilon_{33} = 37.2 \pm 0.2$	297.5	1.6×10^3
$BaZrO_3$	Barium zirconate	43	r.t.	

Formula	Name	ϵ_{ijk}	T/K	ν/Hz
$Be_3Al_2Si_6O_{18}$	Beryllium aluminum silicate (Beryl)	$\epsilon_{33} = 5.95$	297	7×10^3
		$\epsilon_{11} = \epsilon_{22} = 6.86$	297	7×10^3
$BeCO_3$	Beryllium carbonate	9.7	291	2×10^5
BeO	Beryllium oxide (beryllia)	7.35 ± 0.2	293	2×10^6
$BiFeO_3$	Bismuth iron oxide	40 ± 3	300	9.4×10^9
$Bi_{12}GeO_{20}$	Bismuth germanite	$\epsilon_{11}^S = 38$	r.t.	
$Bi(GeO_4)_3$	Bismuth germanate	16	293	
Bi_2O_3	Bismuth sesquioxide	18.2	r.t.	2×10^6
$Bi_4Ti_3O_{12}$	Bismuth titanate	112	r.t.	10^3
C	Diamond			
	Type I	5.87 ± 0.19	300	10^3
	Type IIa	5.66 ± 0.04	300	10^3
$C_4H_4O_6$	Tartaric acid	$\epsilon_{11} = \epsilon_{22} = 4.3$	298	
		$\epsilon_{33} = 4.5$	298	
		$\epsilon_{13} = 0.55$	298	
$C_6H_{14}N_2O_6$	Ethylene diamine tartrate (EDT)	$\epsilon_{11}^T = 5.0$	293	
		$\epsilon_{22}^T = 8.3$	293	
		$\epsilon_{33}^T = 6.0$	293	
		$\epsilon_{13}^T = 0.7$	293	
$C_6H_{12}O_6NaBr$	Dextrose sodium bromide	$\epsilon_{11}^T = 4.0$	r.t.	10^3
$(CH_3NH_3)Al(SO_4)_2 \cdot 2H_2O$	Methyl ammonium alum (MASD)	19	197	
$Ca_2B_6O_{11} \cdot 5H_2O$	Colemanite	$\epsilon_{11} = 20$	293	10^3
		$\epsilon_{33} = 25$	293	10^3
$CaCO_3$	Calcium carbonate	$\epsilon_{11} = 8.67$	r.t.	9.4×10^{10}
		$\epsilon_{22} = 8.69$	r.t.	9.4×10^{10}
		$\epsilon_{33} = 8.31$	r.t.	9.4×10^{10}
$CaCeO_3$	Calcium cerate	21	r.t.	
CaF_2	Calcium fluoride	6.81	300	$5 \times 10^2 - 10^{11}$
$CaMoO_4$	Calcium molybdate	$\epsilon_{11} = \epsilon_{22} = 24.0 \pm 0.2$	297.5	<10
		$\epsilon_{33} = 20.0 \pm 0.2$	297.5	<10
$Ca(NO_3)_2$	Calcium nitrate	6.54	292	2×10^5
$CaNb_2O_6$	Calcium niobate	$\epsilon_{11} = 22.8 \pm 1.9$	r.t.	$(5—500) \times 10^3$
$Ca_2Nb_2O_7$	Calcium pyroniobate	~45	r.t.	5×10^7
CaO	Calcium oxide	11.8 ± 0.3	283	2×10^6
CaS	Calcium sulfide	6.699	r.t.	7.25×10^6
$CaSO_4 \cdot 2H_2O$	Calcium sulfate dihydrate	$\epsilon_{11} = 5.10$	r.t.	
		$\epsilon_{22} = 5.24$	r.t.	
		$\epsilon_{33} = 10.30$	r.t.	
$CaTiO_3$	Calcium titanate	165	r.t.	
$CaWO_4$	Calcium tungstate	$\epsilon_{11} = \epsilon_{22} = 11.7 \pm 0.1$	297.5	1.59×10^3
		$\epsilon_{33} = 9.5 \pm 0.2$	297.5	1.59×10^3
Cd_3As_2	Cadmium arsenide	$\epsilon_{33} = 18.5$	4	
$CdBr_2$	Cadmium bromide	8.6	293	5×10^5
CdF_2	Cadmium fluoride	8.33 ± 0.08	300	$10^5 - 10^7$
CdS	Cadmium sulfide	$\epsilon_{11} = \epsilon_{22} = 8.7$	300	i.r.
		$\epsilon_{33} = 9.25$	300	i.r.
		$\epsilon_{11} = \epsilon_{22} = 8.37$	8	i.r.
		$\epsilon_{33} = 9.00$	8	i.r.
		$\epsilon_{11}^T = 8.48$	77	10^4
		$\epsilon_{33}^T = 9.48$	77	10^4
		$\epsilon_{11}^S = 9.02, \epsilon_{11}^T = 9.35$	298	10^4
		$\epsilon_{33}^S = 9.53, \epsilon_{33}^T = 10.33$	298	10^4
$CdSe$	Cadmium selenide	$\epsilon_{11}^S = 9.53, \epsilon_{11}^T = 9.70$	298	10^4
		$\epsilon_{33}^S = 10.2, \epsilon_{33}^T = 10.65$	298	10^4
$CdTe$	Cadmium telluride	$\epsilon_{11} = \epsilon_{22} = 10.60 \pm 0.15$	297	i.r.
		$\epsilon_{33} = 7.05 \pm 0.05$	297	i.r.
$Cd_2Nb_2O_7$	Cadmium pyroniobate	500—580	293	10^3
CeO_2	Cerium oxide	7.0	r.t.	2×10^6

Formula	Name	ϵ_{ijk}	T/K	ν/Hz
$CoNb_2O_6$	Cobalt niobate	$\epsilon_{11} = 18.4 \pm 0.6$	r.t.	$(5-500) \times 10^3$
		$\epsilon_{22} = 21.4 \pm 1.1$	r.t.	$(5-500) \times 10^3$
		$\epsilon_{33} = 33.0 \pm 0.7$	r.t.	$(5-500) \times 10^3$
CoO	Cobalt oxide	12.9	298	10^2-10^{10}
Cr_2O_3	Chromic sesquioxide	$\epsilon_{11} = \epsilon_{22} = 13.3$	298.5	10^3
		$\epsilon_{33} = 11.9$	298.5	10^3
		8	315 (T_N)	6×10^{10}
$CsAl(SO_4)_2 \cdot 12H_2O$	Cesium alum	5.0	r.t.	$20-20 \times 10^3$
$CsBr$	Cesium bromide	6.38	298	1.6×10^3
Cs_2CO_3	Cesium carbonate	6.53	291	2×10^5
$CsCl$	Cesium chloride	7.2	298	
$Cs_2H_2AsO_4$	Cesium dihydrogen arsenate (CDA)	4.8	273	9.5×10^9
$Cs_2H_2PO_4$	Cesium dihydrogen phosphate (CDP)	6.15	285	9.5×10^9
$CsH_3(SeO_3)_2$	Cesium trihydrogen selenite	$\epsilon_{11} = 80$	273	10^5
		$\epsilon_{22} = 63$	273	10^5
		$\epsilon_{33} = 12$	273	10^5
CsI	Cesium iodide	6.31	298	1.6×10^3
$CsNO_3$	Cesium nitrate	$\epsilon_{11} = \epsilon_{22} = 9.4$	r.t.	5×10^5
		$\epsilon_{33} = 8.3$	r.t.	5×10^5
$CsPbCl_3$	Cesium lead chloride	14.37	300	10^5-10^6
$CuBr$	Cuprous bromide	8.0	293	5×10^5
$CuCl$	Cuprous chloride	9.8 ± 0.5	r.t.	10^3
CuO	Cupric oxide	18.1	r.t.	2×10^6
Cu_2O	Cuprous oxide (Cuprite)	7.60 ± 0.06	r.t.	10^5
$CuSO_4 \cdot 5H_2O$	Cupric sulfate pentahydrate	6.60	r.t.	
EuF_2	Europium fluoride	7.7 ± 0.2	298	$(1-300) \times 10^3$
$Eu_2(MoO_4)_3$	Europium molybdate	9.5	298	
EuS	Europium sulfide	13.10 ± 0.04	80	$5 \times 10^2-10^5$
FeO	Ferrous oxide	14.2	r.t.	2×10^6
Fe_2O_3	Ferric sesquioxide	4.5	r.t.	10^5-10^7
$Fe_2O_3-\alpha$	Ferric sesquioxide (hematite)	12		6×10^{10}
Fe_3O_4	Ferrosoferric oxide (magnetite)	20	r.t.	10^5-10^7
$GaAs$	Gallium arsenide	13.13	300	
		12.90	4	i.r.
GaP	Gallium phosphide	11.1	r.t.	
		10.75 ± 0.1	1.6	i.r.
$GaSb$	Gallium antimonide	15.69	r.t.	
		15.7	4	i.r.
$Gd_2(MoO_4)_3$	Gadolinium molybdate	$\epsilon^T = 10$	298	
		$\epsilon^S = 9.5$	298	10^3
Ge	Germanium	16.0 ± 0.3	4	9.2×10^9
		15.8 ± 0.2	r.t.	$500-3 \times 10^{10}$
GeO_2	Germanium dioxide	$\epsilon_{11} = \epsilon_{22} = 7.44$	r.t.	i.r.
HIO_3	Iodic acid	$\epsilon_{11} = 7.5$	r.t.	10^3
		$\epsilon_{22} = 12.4$	r.t.	10^3
		$\epsilon_{33} = 8.1$	r.t.	10^3
$HNH_4(ClCH_2COO)_2$	Hydrogen ammonium dichloroacetate	$\epsilon_{[102]} = 5.9$	r.t.	10^5
H_2O	Ice I (P = 0 kbar)	99	243	
	Ice III (P = 3 kbar)	117	243	
	Ice V(P = 5 kbar)	114	243	
	Ice VI (P = 8 kbar)	193	243	
$HgCl$	Mercurous chloride (Calumel)	$\epsilon_{11} = \epsilon_{22} = 14.0$	r.t.	10^{12}
$HgCl_2$	Mercuric chloride	6.5	r.t.	10^{12}
HgS	Mercurous sulfide (Cinnabar)	$\epsilon_{11} = \epsilon_{22} = 18.0$	r.t.	i.r.
		$\epsilon_{33} = 32.5$	r.t.	i.r.
$HgSe$	Mercurous selenide	25.6	r.t.	10^4-10^6
I_2	Iodine	$\epsilon_{11} = 6$	r.t.	$5 \times 10^4-10^7$
		$\epsilon_{22} = 3$	r.t.	$5 \times 10^4-10^7$
		$\epsilon_{33} = 40$	r.t.	$5 \times 10^4-10^7$

Formula	Name	ϵ_{ijk}	T/K	ν/Hz
InAs	Indium arsenide	14.55 ± 0.3	r.t.	i.r.
		15.15	4	i.r.
InP	Indium phosphide	12.61	r.t.	i.r.
InSb	Indium antimonide	17.88	4	i.r.
$KAl(SO_4)_2 \cdot 12H_2O$	Potassium alum	6.5	r.t.	$20—20 \times 10^3$
KBr	Potassium bromide	4.88	300	
		4.53	4.2	
$KBrO_3$	Potassium bromate	7.3	r.t.	2×10^6
KCN	Potassium cyanide	6.15	r.t.	2×10^6
K_2CO_3	Potassium carbonate	4.96	291	2×10^5
$K_2C_4H_4O_6 \cdot {}^1/_2\,H_2O$	Dipotassium tartrate (DKT)	$\epsilon_{11} = 6.44$	r.t.	
		$\epsilon_{22} = 5.80$	r.t.	
		$\epsilon_{33} = 6.49$	r.t.	
		$\epsilon_{13} = 0.005$	r.t.	
KCl	Potassium chloride	4.86 ± 0.02	r.t.	5×10^3
		4.50	4.2	
$KClO_3$	Potassium chlorate	5.1	r.t.	2×10^6
$KClO_4$	Potassium perchlorate	5.9	r.t.	2×10^6
K_2CrO_4	Potassium chromate	7.3	r.t.	6×10^7
$KCr(SO_4)_2 \cdot 12H_2O$	Potassium chrome alum	6.5	100—240	175×10^3
KD_2AsO_4	Potassium dideuterium arsenate (KDDA)	$\epsilon_{11} = 70$	298	
		$\epsilon_{33} = 31$	298	
KD_2PO_4	Potassium dideuterium phosphate (KDDP)	50 ± 2	297	10^3
KF	Potassium fluoride	6.05		2×10^6
KH_2AsO_4	Potassium dihydrogen arsenate (KDA)	$\epsilon_{11} = 60$	298	
		$\epsilon_{33} = 24$	298	
KH_2PO_4	Potassium dihydrogen phosphate (KDP)	46	298	10^3
		$\epsilon_{11} = 42$	r.t.	
		$\epsilon_{33} = 21$	r.t.	
K_2HPO_4	Dipotassium monohydrogen orthophosphate	9.05	r.t.	2×10^6
KI	Potassium iodide	5.00	r.t.	9.4×10^{10}
KIO_3	Potassium iodate	170	255	10^5
		10	293	10^5
		$\epsilon_{[101]} \approx 40,70$	r.t.	10^5
		16.85	r.t.	2×10^6
$(K,H)Al_3(SiO_4)_3$	Mica (muscovite)	5.4	299	$10^2—3 \times 10^9$
$(K,H)Mg_3Al(SiO_4)_3$	Mica (Canadian)	$\epsilon_{11} = \epsilon_{22} = 6.9$	298	$10^2—10^4$
		$\epsilon_{33} = 7.3$	298	10^4
KNO_2	Potassium nitrite	25	305	
KNO_3	Potassium nitrate	4.37	293	2×10^5
$KNbO_3$	Potassium niobate	700	r.t.	
K_3PO_4	Potassium orthophosphate	7.75	r.t.	2×10^6
KSCN	Potassium thiocyanate	7.9	r.t.	2×10^6
K_2SO_4	Potassium sulfate	6.4	r.t.	2×10^6
$K_2S_3O_6$	Potassium trithionate	5.7	293	1.8×10^6
$K_2S_4O_6$	Potassium tetrathionate	5.5	293	1.8×10^6
$K_2S_5O_6 \cdot H_2O$	Potassium pentathionate	7.8	293	1.8×10^6
$K_2S_6O_6$	Potassium hexathionate	7.8	293	1.8×10^6
K_2SeO_4	Potassium selenate	$\epsilon_{11} = 5.9$	r.t.	10^3
		$\epsilon_{22} = 7.7$	r.t.	10^3
$KSr_2Nb_5O_{15}$	Potassium strontium niobate	$\epsilon_{11} = \epsilon_{11} \approx 1200$	298	
		$\epsilon_{33} \approx 800$	298	
$KTaNbO_3$	Potassium tantalate niobate (KTN)	$34,000$	273	10^4
		$6,000$	293	10^4
$KTaO_3$	Potassium tantalate	242	298	2×10^5
$LaScO_3$	Lanthanum scandate	30	r.t.	
LiBr	Lithium bromide	12.1	r.t.	2×10^6
Li_2CO_3	Lithium carbonate	4.9	291	2×10^5

Formula	Name	ϵ_{ijk}	T/K	ν/Hz
LiCl	Lithium chloride	11.05	r.t.	2×10^6
LiD	Lithium deuteride	14.0 ± 0.5	r.t.	i.r.
LiF	Lithium fluoride	9.00	298	10^2—10^7
		9.11	353	10^2—10^7
$LiGaO_2$	Lithium metagallate	$\epsilon_{11}^T = 7.0$, $\epsilon_{22}^T = 6.0$	r.t.	
		$\epsilon_{33}^T = 9.5$	r.t.	
		$\epsilon_{11}^S = 6.8$, $\epsilon_{22}^S = 5.8$	r.t.	
Li^6H	Lithium-6 hydride	13.2 ± 0.5	r.t.	
Li^7H	Lithium-7 hydride	12.9 ± 0.5	r.t.	
$LiH_3(SeO_3)_2$	Lithium trihydrogen selenite	29	298	10^4
		$\epsilon_{11} = 13.0$	r.t.	
		$\epsilon_{22} = 12.9$	r.t.	
		$\epsilon_{33} = 46$	r.t.	
LiI	Lithium iodide	11.03	r.t.	2×10^6
$LiIO_3$	Lithium iodate	$\epsilon_{11} = \epsilon_{22} = 65$	294.5	10^3
		$\epsilon_{33} = 554$	298	
$LiNH_4C_4H_4O_6 \cdot H_2O$	Lithium ammonium tartrate (LAT)	$\epsilon_{11}^T = 7.2$	298	
		$\epsilon_{22}^T = 8.0$	298	
		$\epsilon_{33}^T = 6.9$	298	
$LiNa_3CrO_4 \cdot 6H_2O$	Lithium trisodium chromate	8.0	r.t.	10^3
$LiNa_3MoO_4 \cdot 6H_2O$	Lithium trisodium molybdate	$\epsilon_{11} = 6.7$	r.t.	10^3
		$\epsilon_{33} = 5.3$	r.t.	10^3
$LiNbO_3$	Lithium niobate	$\epsilon_{11} = \epsilon_{22} = 82$	298	10^5
		$\epsilon_{33} = 30$	298	10^5
$Li_2SO_4 \cdot H_2O$	Lithium sulfate monohydrate	$\epsilon_{11} = 5.6$	298	
		$\epsilon_{22} = 10.3$	298	
		$\epsilon_{33} = 6.5$	298	
		$\epsilon_{13} = 0.07$	298	
$LiTaO_3$	Lithium tantalate	$\epsilon_{11} = \epsilon_{22} = 53$	r.t.	10^5
		$\epsilon_{33} = 46$	r.t.	10^5
		$\epsilon_{11}^S = \epsilon_{22}^S = 41$	r.t.	
		$\epsilon_{33}^S = 43$	r.t.	
		$\epsilon_{11}^T = \epsilon_{22}^T = 51$	r.t.	
		$\epsilon_{33}^T = 45$	r.t.	
$LiTlC_4H_4O_6 \cdot H_2O$	Lithium thallium tartrate (LTT)	$\epsilon_{11} \approx 20$	80	
$Mg_3B_7O_{13}Cl$	Magnesium borate monochloride (boracite)	$\epsilon_{11} = 14.1$	r.t.	5×10^5
$MgCO_3$	Magnesium carbonate	8.1	291	2×10^5
$MgNb_2O_6$	Magnesium niobate	$\epsilon_{11} = 16.4 \pm 0.5$	r.t.	$(5$—$500) \times 10^3$
		$\epsilon_{22} = 20.9 \pm 0.5$	r.t.	$(5$—$500) \times 10^3$
		$\epsilon_{33} = 32.4 \pm 0.5$	r.t.	$(5$—$500) \times 10^3$
MgO	Magnesium oxide (Periclase)	9.65	298	10^2—10^8
$(MgO)_xAl_2O_3$	Spinel	8.6	r.t.	
$MgSO_4$	Magnesium sulfate	8.2	r.t.	
$MgSO_4 \cdot 7H_2O$	Magnesium sulfate septa hydrate	5.46	r.t.	
$MgTiO_3$	Magnesium titanate	13.5	r.t.	
$MgWO_4$	Magnesium tungstate	$\epsilon_{11} = 18.0 \pm 1$	r.t.	$(5$—$500) \times 10^3$
		$\epsilon_{22} = 18.0 \pm 1$	r.t.	$(5$—$500) \times 10^3$
$MnNb_2O_6$	Manganese niobate	$\epsilon_{11} = 17.4 \pm 2$	r.t.	$(5$—$500) \times 10^3$
		$\epsilon_{22} = 16.1 \pm 0.5$	r.t.	$(5$—$500) \times 10^3$
		$\epsilon_{33} = 30.7 \pm 1$	r.t.	$(5$—$500) \times 10^3$
MnO	Manganese oxide (Pyrolusite)	12.8	r.t.	6×10^{10}
MnO_2	Manganese dioxide	$\sim 10^4$	298	10^4
Mn_2O_3	Manganese sesquioxide	8	r.t.	6×10^{10}
$MnWO_4$	Manganese tungstate	$\epsilon_{11} = 19.3 \pm 1.3$	r.t.	$(5$—$500) \times 10^3$
		$\epsilon_{22} = 14.3 \pm 0.5$	r.t.	$(5$—$500) \times 10^3$
		$\epsilon_{33} = 16.5 \pm 1.1$	r.t.	$(5$—$500) \times 10^3$
$N(CH_3)_4HgBr_3$	Tetramethylammonium tribromo mercurate (TTM)	~ 10	233—373	

Formula	Name	ϵ_{ijk}	T/K	ν/Hz
$N(CH_3)_4HgI_3$	Tetramethylammonium triiodo mercurate (TTM)	~10	233—373	
$N_4(CH_2)_6$	Hexamethylene tetramine (HMTA)	2.6 ± 0.2	r.t.	10^9—10^{10}
$(ND_4)_2BeF_4$	Deuteroammonium fluoberyllate	$\epsilon_{11} = 10$	r.t.	
		$\epsilon_{22} = 9$	r.t.	
		$\epsilon_{33} = 9$	r.t.	
$(ND_4)_2SO_4$	Deuteroammonium sulfate	$\epsilon_{11} = 9$	r.t.	
		$\epsilon_{22} = 10$	r.t.	
		$\epsilon_{33} = 9$	r.t.	
$(NH_2 \cdot CH_2COOH)_3 \cdot H_2SO_4$	Triglycine sulfate (TGS)	$\epsilon_{11} = 9$	273	10^4
		$\epsilon_{22} = 30$	273	10^4
		$\epsilon_{33} = 6.5$	273	10^4
$(NH_2 \cdot CH_2COOH)_3 \cdot H_2SeO_4$	Triglycine selenate (TGSe)	200	293	1.6×10^3
$(NH_2 \cdot CH_2 COOH)_3 \cdot H_2BeF_4$	Triglycine fluorberyllate (TGFB)	$\epsilon_{22} = 12$	273	10^4
$NH_4Al(SO_4)_2 \cdot 12H_2O$	Ammonium alum	6	r.t.	10^{12}
$(NH_4)_2BeF_4$	Ammonium fluorberyllate	$\epsilon_{11} = \epsilon_{22} = 7.8$	123	10^5
		$\epsilon_{33} = 7.1$	123	10^5
		$\epsilon_{11} = \epsilon_{22} = 8.8$	293	10^5
		$\epsilon_{33} = 9.2$	293	10^5
NH_4Br	Ammonium bromide	7.1	r.t.	7×10^5
NH_4I	Ammonium iodide	9.8	r.t.	
$(NH_4)_2C_2H_6O_6$	Ammonium tartrate	$\epsilon_{11} = 6.45$	r.t.	10^3
		$\epsilon_{22} = 6.8$	r.t.	10^3
		$\epsilon_{33} = 6.0$	r.t.	10^3
$(NH_4)_2Cd_2(SO_4)_3$	Ammonium cadmium sulfate	10.0	r.t.	10^4
NH_4Cl	Ammonium chloride	6.9	r.t.	7×10^5
$NH_4(ClCH_2COO)$	Ammonium monochloroacetate	5	r.t.	2×10^6
$NH_4Cr(SO_4)_2 \cdot 12H_2O$	Ammonium chrome alum	6.5	r.t.	175×10^3
NH_4HSO_4	Ammonium bisulfate	165	273	5×10^4
$NH_4H_2AsO_4$	Ammonium dihydrogen arsenate (ADA)	5.1	265	9.5×10^9
		$\epsilon_{11} = \epsilon_{22} = 85$	298	10^3
		$\epsilon_{33} = 22$	298	
$NH_4H_2PO_4$	Ammonium dihydrogen phosphate (ADP)	$\epsilon_{11} = \epsilon_{22} = 57.1 \pm 0.6$	294.5	10^5—35×10^9
		$\epsilon_{33} = 14.0 \pm 0.3$	294	10^5—36×10^9
$ND_4D_2PO_4$	Ammonium dideuterium phosphate (ADDP)	$\epsilon_{11} = \epsilon_{22} = 74, \epsilon_{33} = 24$	300	
NH_4NO_3	Ammonium nitrate	10.7	322	$(5$—$50) \times 10^3$
$(NH_4)_2SO_4$	Ammonium sulfate	$\epsilon_{11} = \epsilon_{22} = 8.0$	123	10^5
		$\epsilon_{33} = 6.3$	123	10^5
		$\epsilon_{11} = \epsilon_{22} = 10.0$	293	10^5
		$\epsilon_{33} = 9.3$	293	10^5
$(NH_4)_2UO_2(C_2O_4)_2$	Ammonium uranyl oxalate	8.03	r.t.	10^4—3.3×10^9
$(NH_4)_2UO_2(C_2O_4)_2 \cdot 3H_2O$	Ammonium uranyl oxalate trihydrate	6.06	r.t.	10^4—3.3×10^9
$NaBr$	Sodium bromide	6.44	298	1.6×10^3
$NaBrO_3$	Sodium bromate	$\epsilon_{11}^T = 5.70$	298	10^3
$NaCN$	Sodium cyanide	7.55	293	10^5
$NaCO_3$	Sodium carbonate	8.75	291	2×10^5
$NaCO_3 \cdot 10H_2O$	Sodium carbonate decahydrate	5.3	r.t.	6×10^7
$NaCl$	Sodium chloride	5.9	298	10^2—10^7
		5.45	4.2	
$NaClO_3$	Sodium chlorate	$\epsilon_{11}^T = 5.76$	301	10^3
		5.28	r.t.	10^3
$NaClO_4$	Sodium perchlorate	5.76	r.t.	10^3
NaF	Sodium fluoride	5.08 ± 0.02	r.t.	5×10^3
$NaH_3(SeO_3)_2$	Sodium trihydrogen selenite	$\epsilon_{11} \approx 75$	273	2×10^5
$NaD_3(SeO_3)_2$	Sodium trideuterium selenite	$\epsilon_{11} \approx 220$	273	2×10^5
NaI	Sodium iodide	7.28 ± 0.03	r.t.	
$NaK(C_4H_2D_2O_6) \cdot 4D_2O$	Sodium potassium tartrate tetradeutrate (double deuterated Rochelle salt)	$\epsilon_{11} = 70$	273	10^3
		$\epsilon_{22} = 8.9$	273	10^3

Formula	Name	ϵ_{ijk}	T/K	ν/Hz
$NaK(C_4H_4O_6) \cdot 4H_2O$	Sodium potassium tartrate tetrahydrate (Rochelle salt)	$\epsilon_{11} = 170$	273	10^3
		$\epsilon_{22} = 9.1$	273	10^3
$NaNH_4(C_4H_4O_6) \cdot 4H_2O$	Sodium ammonium tartrate (Ammonium Rochelle salt)	$\epsilon_{11} = 8.4$	298	
		$\epsilon_{22} = 9.2$	298	
		$\epsilon_{33} = 9.5$	298	
$NaNbO_3$	Sodium niobate	$\epsilon_{33} = 670 \pm 13$	r.t.	
		$\epsilon_{11} = \epsilon_{22} = 76 \pm 2$	r.t.	
$NaNO_2$	Sodium nitrite	$\epsilon_{11} = 7.4$	r.t.	5×10^5
		$\epsilon_{22} = 5.5$	r.t.	5×10^5
		$\epsilon_{33} = 5.0$	r.t.	5×10^5
$NaNO_3$	Sodium nitrate	6.85	292	2×10^5
$NaSO_4$	Sodium sulfate	7.90	r.t.	
$NaSO_4 \cdot 10H_2O$	Sodium sulfate decahydrate	5.0	r.t.	
$Na_2S_2O_3 \cdot 5H_2O$	Sodium sulfate pentahydrate	7	250—290	$300—10^4$
$Na_2UO_2(C_2O_4)_2$	Sodium uranyl oxalate	5.18	r.t.	
$NdAlO_3$	Neodymium aluminate	17.5	r.t.	
$NdScO_3$	Neodymium scandate	27	r.t.	
$Ni_3B_7O_{13}I$	Nickel iodine boracite	$\epsilon_{11} = 14$	260	
$NiNb_2O_6$	Nickel niobate	$\epsilon_{11} = 16.0 \pm 0.5$	r.t.	$(5—500) \times 10^3$
		$\epsilon_{22} = 23.8 \pm 1.8$	r.t.	$(5—500) \times 10^3$
		$\epsilon_{33} = 31.3 \pm 2.5$	r.t.	$(5—500) \times 10^3$
NiO	Nickel oxide	11.9	298	10^5
$NiSO_4 \cdot 6H_2O$	Nickel sulfate hexahydrate	$\epsilon_{11} = 6.2$	r.t.	
		$\epsilon_{33} = 6.8$	r.t.	
$NiWO_4$	Nickel tungstate	$\epsilon_{11} = 17.4 \pm 2.4$	r.t.	$(5—500) \times 10^3$
		$\epsilon_{22} = 13.6 \pm 1.0$	r.t.	$(5—500) \times 10^3$
		$\epsilon_{33} = 19.7 \pm 0.6$	r.t.	$(5—500) \times 10^3$
P	Phosphorous (red)	4.1	r.t.	10^8
	Phosphorous (yellow)	3.6	r.t.	10^8
$[P(CH_3)_4]HgBr_3$	Tetramethylphosphonium tribromo mercurate (TTM)	~ 10	233—373	
$PbBr_2$	Lead bromide	>30	293	$(0.5—3) \times 10^6$
$PbCO_3$	Lead carbonate	18.6	288	10.8
$Pb(C_2H_3O_2)_2$	Lead acetate	2.6	290—295	10^6
$PbCl_2$	Lead chloride	33.5	273	$(0.5—3) \times 10^6$
Pb_2CoWO_6	Lead cobalt tungstate	~ 250	r.t.	
PbF_2	Lead fluoride	26.3	r.t.	
$PbHfO_3$	Lead hafnate	390	300	10^5
		185	400	
PbI_2	Lead iodide	20.8	293	$(0.5—3) \times 10^6$
$Pb_3MgNb_2O_9$	Lead magnesium niobate	10,000	297	
$PbMoO_4$	Lead molybdate	$\epsilon_{11} = 34.0 \pm 0.4$	297.5	1.6×10^3
		$\epsilon_{33} = 40.6 \pm 0.2$	297.5	1.6×10^3
$Pb(NO_3)_2$	Lead nitrate	16.8	r.t.	$(0.5—3) \times 10^6$
$PbNb_2O_6$	Lead niobate	$\epsilon_{33}^T = 180$	298	
PbO	Lead oxide	25.9	r.t.	2×10^6
PbS	Lead sulfide (Galena)	190	77	i.r.
		200 ± 35	r.t.	i.r.
$PbSO_4$	Lead sulfate	14.3	290—295	10^6
$PbSe$	Lead selenide	280	r.t.	i.r.
$PbTa_2O_6$	Lead metatantalate	$\epsilon_{11} = \epsilon_{22} \approx 300$	r.t.	10^4
		$\epsilon_{33} = 150$	r.t.	10^4
$PbTe$	Lead telluride	450	r.t.	i.r.
		40	77	$10^4—15 \times 10^4$
		430	4.2	$10^4—15 \times 10^4$
$PbTiO_3$	Lead titanate	~ 200	r.t.	10^3
$PbWO_4$	Lead tungstate	$\epsilon_{11} = \epsilon_{22} = 23.6 \pm 0.3$	297.5	1.59×10^3
		$\epsilon_{33} = 31.0 \pm 0.4$	297.5	1.59×10^3
$Pb(Zn_{1/3}Nb_{2/3})O_3$	Lead zinc niobate	7	300	$10^3, 300 \times 10^3$
$PbZrO_3$	Lead zirconate	200	400	

Formula	Name	ϵ_{ijk}	T/K	ν/Hz
$RbAl(SO_4)_2 \cdot 12H_2O$	Rubidium alum	5.1	r.t.	10^{12}
RbBr	Rubidium bromide	4.83	300	
Rb_2CO_3	Rubidium carbonate	4.87 ± 0.02	r.t.	5×10^3
RbCl	Rubidium chloride	4.91 ± 0.02	r.t.	5×10^3
$RbCr(SO_4)_2 \cdot 12H_2O$	Rubidium chrome alum	5.0	r.t.	10^{12}
RbF	Rubidium fluoride	5.91	r.t.	2×10^6
$RbHSO_4$	Rubidium bisulfate	$\epsilon_{11} = 7$	r.t.	10^5
		$\epsilon_{22} = 8$	r.t.	10^5
		$\epsilon_{33} = 10$	r.t.	10^5
RbH_2AsO_4	Rubidium dihydrogen arsenate (RDA)	3.90	273	9.5×10^9
RbH_2PO_4	Rubidium dihydrogen phosphate (RDP)	6.15	285	9.5×10^9
RbI	Rubidium iodide	4.94 ± 0.02	r.t.	5×10^3
$RbInSO_4$	Rubidium indium sulfate	6.85	r.t.	
$RbNO_3$	Rubidium nitrate	20—380	433—488	10^6
		30	488—538	10^6
S	Sulfur	$\epsilon_{11} = 3.75$	298	10^2—10^3
		$\epsilon_{22} = 3.95$	298	10^2—10^3
		$\epsilon_{33} = 4.44$	298	10^2—10^3
	sublimed	3.69	298	10^2—10^3
$SC(NH_2)_2$	Thiourea	$\epsilon_{11} = \epsilon_{33} \approx 3$	77—300	10^3
		$\epsilon_{22} = 35$	300	10^3
Sb_2O_3	Antimonous sesquioxide	12.8	r.t.	$(1.5—2) \times 10^3$
Sb_2S_3	Antimonous sulfide (stibnite)	$\epsilon_{11} = \epsilon_{22} = 15$	r.t.	10^3
		$\epsilon_{33} = 180$	r.t.	10^3
Sb_2Se_3	Antimonous selenide	~110	r.t.	$(10—16.5) \times 10^9$
SbSI	Antimonous sulfide iodide	2000	273	10^5
		$\epsilon_{11} = \epsilon_{22} \approx 25$	r.t.	10^3—10^5
		$\epsilon_{33} \approx 5 \times 10^4$	295	10^3—10^5
Se	Selenium	$\epsilon_{11} = \epsilon_{22} = 11$	300	24×10^9
	(monocrystal)			
		$\epsilon_{33} = 21$	300	24×10^9
	(amorphous)	6.0	298	10^2—10^{10}
Si	Silicon	12.1	4.2	10^7—10^9
SiC	Silicon carbide			
	cubic	9.72	r.t.	i.r.
	6H	$\epsilon_{11} = \epsilon_{22} = 9.66$	r.t.	i.r.
		$\epsilon_{33} = 10.03$	r.t.	i.r.
		9.7 ± 0.1	1.8	i.r.
Si_3N_4	Silicon nitride	4.2 (film)	r.t.	10^3
SiO	Silicon monoxide	5.8	r.t.	10^3
SiO_2	Silicon dioxide	$\epsilon_{11} = 4.42$	r.t.	9.4×10^{10}
		$\epsilon_{22} = 4.41$	r.t.	9.4×10^{10}
		$\epsilon_{33} = 4.60$	r.t.	9.4×10^{10}
$Sm_2(MoO_4)_3$	Samarium molybdate	12	298	
SnO_2	Stannic dioxide	$\epsilon_{11} = \epsilon_{22} = 14 \pm 2$	r.t.	10^4—10^{10}
		$\epsilon_{33} = 9.0 \pm 0.5$	r.t.	10^4—10^{10}
SnSb	Tin antimonide	147	r.t.	10^4—10^6
SnTe	Tin telluride	1770 ± 300	r.t.	i.r.
$Sr(COOH)_2 \cdot 2H_2O$	Strontium formate dihydrate	6.1	r.t.	10^3
$SrCO_3$	Strontium carbonate	8.85	298	2×10^5
$SrCl_2$	Strontium chloride	9.19	r.t.	
$Sr_4Cl_2 \cdot 6H_2O$	Strontium chloride hexahydrate	8.52	r.t.	
SrF_2	Strontium fluoride	6.50	300	5×10^2—10^{11}
$SrMoO_4$	Strontium molybdate	$\epsilon_{11} = \epsilon_{22} = 31.7 \pm 0.2$	297.5	1.59×10^3
		$\epsilon_{33} = 41.7 \pm 0.2$	297.5	1.59×10^3
$Sr(NO_3)_2$	Strontium nitrate	5.33	292	2×10^5
$Sr_2Nb_2O_7$	Strontium niobate	$\epsilon_{11} = 75$	r.t.	10^3
		$\epsilon_{22} = 46$	r.t.	10^3
		$\epsilon_{33} = 43$	r.t.	10^3

Formula	Name	ϵ_{ijk}	T/K	ν/Hz
SrO	Strontium oxide	13.3 ± 0.3	273	2×10^6
SrS	Strontium sulfide	11.3	r.t.	7.25×10^6
SrSO$_4$	Strontium sulfate	11.5	r.t.	
SrTiO$_3$	Strontium titanate	332	298	10^3
		2080	78	10^3
SrWO$_4$	Strontium tungstate	$\epsilon_{11} = \epsilon_{22} = 25.7 \pm 0.2$	297.5	1.6×10^3
		$\epsilon_{33} = 34.1 \pm 0.2$	297.5	1.6×10^3
Ta$_2$O$_5$	Tantalum pentoxide (tantala)			
	α phase	$\epsilon_{11} = \epsilon_{22} = 30$	77	10^3
		$\epsilon_{33} = 65$	77	10^3
	β phase	24	292	10^3
Tb(MoO$_4$)$_3$	Terbium molybdate	11	298	
		$\epsilon_{11} = \epsilon_{22} = 33$	100—200	9.4×10^9
		$\epsilon_{33} = 53$	100—200	9.4×10^9
Te	Tellurium	$\epsilon_{11} = \epsilon_{22} = 33$	r.t.	
		$\epsilon_{33} = 54$	r.t.	
	polycrystalline	27.5	r.t.	i.r.
	monocrystalline	28.0	r.t.	i.r.
ThO$_2$	Thorium dioxide	18.9 ± 0.4	r.t.	3×10^5
TiO$_2$	Titanium dioxide (rutile)	$\epsilon_{11} = \epsilon_{22} = 86$	300	10^4—10^6
		$\epsilon_{33} = 170$	300	10^4—10^6
Ti$_2$O$_3$	Titanium sesquioxide	30	77	6×10^{10}
TlBr	Thallium bromide	30	293	10^3—10^7
TlCl	Thallous chloride	32.2 ± 0.2	293	10^3—10^5
TlI	Thallous iodide (orthorhombic)	20.7 ± 0.2	293	10^4
		37.3	193	10^7
TlNO$_3$	Thallous nitrate	16.5	293	5×10^5
TlSO$_4$	Thallous sulfate	25.5	293	5×10^5
UO$_2$	Uranium dioxide	24	r.t.	3×10^5
WO$_3$	Tungsten trioxide	300		
YMnO$_3$	Yttrium manganate	20	r.t.	2×10^7
Y$_2$O$_3$	Yttrium sesquioxide	10	r.t.	10^6
YbMnO$_3$	Ytterbium manganate	20	r.t.	2×10^7
Yb$_2$O$_3$	Ytterbium sesquioxide	5.0 (film)	r.t.	10^3
ZnO	Zinc monoxide	$\epsilon_{11}^S = 8.33$	r.t.	
		$\epsilon_{33}^S = 8.84$	r.t.	
		$\epsilon_{11}^T = 9.26$	r.t.	
		$\epsilon_{33}^T = 11.0$	r.t.	
		$\epsilon_{11} = 9.26$	r.t.	
		$\epsilon_{33} = 8.2$	r.t.	
		8.15	r.t.	i.r.
ZnS	Zinc sulfide	$\epsilon_{11}^S = 8.08 \pm 2\%$	77	10^4
		$\epsilon_{11}^S = 8.32 \pm 2\%$	298	10^4
		$\epsilon_{11}^T = 8.14 \pm 2\%$	77	10^4
		$\epsilon_{11}^T = 8.37 \pm 2\%$	298	10^4
ZnSe	Zinc selenide	$\epsilon_{11}^T = \epsilon_{11}^S = 9.12 \pm 2\%$	298	10^4
ZnTe	Zinc telluride	$\epsilon_{11}^T = \epsilon_{11}^S = 10.10 \pm 2\%$	r.t.	
ZnWO$_4$	Zinc tungstate	$\epsilon_{22} = 16.1 \pm 0.5$	r.t.	$(5—500) \times 10^3$
ZrO$_2$	Zirconium dioxide (zirconia)	12.5	r.t.	2×10^6

CURIE TEMPERATURE OF SELECTED FERROELECTRIC CRYSTALS

H. P. R. Frederikse

The following table lists the major ferroelectric crystals and their Curie temperatures, T_C.

REFERENCE

Young, K. F. and Frederikse, H. P. R., *J. Phys. Chem. Ref. Data*, 2, 313, 1973.

Name or acronym	Formula	T_C/K
Potassium dihydrogen phosphate group		
KDP	KH_2PO_4	123
KDA	KH_2AsO_4	97
KDDP	KD_2PO_4	213
KDDA	KD_2AsO_4	162
RDP	RbH_2PO_4	146
RDA	RbH_2AsO_4	111
RDDP	RbD_2PO_4	218
RDDA	RbD_2AsO_4	178
CDP	CsH_2PO_4	159
CDA	CsH_2AsO_4	143
CDDA	CsD_2AsO_4	212
Rochelle salt group		
Rochelle salt	$NaKC_4H_4O_6 \cdot 4H_2O$	255—297
Deuterated Rochelle salt	$NaKC_4H_2D_2O_6 \cdot 4H_2O$	251—308
Ammonium Rochelle salt	$NaNH_4C_4H_4O_6 \cdot 4H_2O$	109
LAT	$LiNH_4C_4H_4O_6 \cdot H_2O$	106
Triglycine sulfate group		
TGS	$(NH_2CH_2COOH)_3 \cdot H_2SO_4$	322
TGSe	$(NH_2CH_2COOH)_3 \cdot H_2SeO_4$	295
TGFB	$(NH_2CH_2COOH)_3 \cdot H_2BeF_4$	346
AFB	$(NH_4)_2BeF_4$	176
HADA	$HNH_4(ClCH_2COO)_2$	128
Perovskites and related compounds		
Barium titanate	$BaTiO_3$	406, 278, 193
Lead titanate	$PbTiO_3$	765
Potassium niobate	$KNbO_3$	712
Potassium tantalate niobate	$KTa_{2/3}Nb_{1/3}O_3$	271, 220, 170
Lithium niobate	$LiNbO_3$	1483
Lithium tantalate	$LiTaO_3$	891
Barium titanium niobate	$Ba_6Ti_2Nb_8O_{30}$	521
Ba-Na niobate ("Bananas")	$Ba_2NaNb_5O_{15}$	833
Potassium iodate	KIO_3	485, 343, 257—263, 83
Lithium iodate	$LiIO_3$	529
Potassium nitrate	KNO_3	397
Sodium nitrate	$NaNO_3$	548
Rubidium nitrate	$RbNO_3$	437—487
Miscellaneous compounds		
Cesium trihydrogen selenite	$CsH_3(SeO_3)_2$	143
Lithium trihydrogen selenite	$LiH_3(SeO_3)_2$	$T_C > T_{mp}$

Name or acronym	Formula	T_C/K
Potassium selenate	K_2SeO4	93
Methyl ammonium alum (MASD)	$CH_3NH_3Al(SO_4)_2 \cdot 12H_2O$	177
Ammonium cadmium sulfate	$(NH_4)_2Cd_2(SO_4)_3$	95
Ammonium bisulfate	$(NH_4)HSO_4$	271
Ammonium sulfate	$(NH_4)_2SO_4$	224
Ammonium nitrate	NH_4NO_3	398, 357, 305, 255
Colemanite	$CaB_3O_4(OH)_3 \cdot H_2O$	266
Cadmium pyroniobite	$Cd_2Nb_2O_7$	185
Gadolinium molybdate	$Gd_2(MoO_4)_3$	432

PROPERTIES OF ANTIFERROELECTRIC CRYSTALS

H. P. R. Frederikse

Some important antiferroelectric crystals are listed here with their Curie Temperatures T_C. The last column gives the constant T_O which appears in the Curie-Weiss law describing the dielectric constant of these materials above the Curie Temperature:

$$\varepsilon = \text{const.}/(T - T_O)$$

Name or acronym	Formula	T_C/K	T_O/K
ADP	$NH_4H_2PO_4$	148	
ADA	$NH_4H_2AsO_4$	216	
ADDP	$NH_4D_2PO_4$	242, 245	
ADDA	$NH_4D_2AsO_4$	299	
A_dDDP	$ND_4D_2PO_4$	243	
A_dDDA	$ND_4D_2AsO_4$	304	
Sodium niobate	$NaNbO_3$	911, 793	
Lead hafnate	$PbHfO_3$	476	378
Lead zirconate	$PbZrO_3$	503	475
Lead metaniobate	$PbNb_2O_6$	843	530
Lead metatantalate	$PbTa_2O_6$	543	533
Tungsten trioxide	WO_3	1010	
Potassium strontium niobate	$KSr_2Nb_5O_{15}$	427	413
Sodium nitrite	$NaNO_2$	437	437
Sodium trihydrogen selenite	$NaH_3(SeO_3)_2$	193	192
Sodium trideuterium selenite	$NaD_3(SeO_3)_2$	271	245
Ammonium trihydrogen iodate	$(NH_4)_2H_3IO_3$	245	

DIELECTRIC CONSTANTS OF GLASSES

Type	Dielectric constant at 100 MHz (20°C)	Volume resistivity (350°C megohm-cm)	Loss factor[a]
Corning 0010	6.32	10	0.015
Corning 0080	6.75	0.13	0.058
Corning 0120	6.65	100	0.012
Pyrex 1710	6.00	2,500	0.025
Pyrex 3320	4.71	—	0.019
Pyrex 7040	4.65	80	0.013
Pyrex 7050	4.77	16	0.017
Pyrex 7052	5.07	25	0.019
Pyrex 7060	4.70	13	0.018
Pyrex 7070	4.00	1,300	0.0048
Vycor 7230	3.83	—	0.0061
Pyrex 7720	4.50	16	0.014
Pyrex 7740	5.00	4	0.040
Pyrex 7750	4.28	50	0.011
Pyrex 7760	4.50	50	0.0081
Vycor 7900	3.9	130	0.0023
Vycor 7910	3.8	1,600	0.00091
Vycor 7911	3.8	4,000	0.00072
Corning 8870	9.5	5,000	0.0085
G. E. Clear (silica glass)	3.81	4,000—30,000	0.00038
Quartz (fused)	3.75 4.1 (1 MHz)	—	0.0002 (1 MHz)

[a] Power factor × dielectric constant equals loss factor.

PROPERTIES OF SUPERCONDUCTORS

L. I. Berger and B. W. Roberts

The following tables include superconductive properties of selected elements, compounds, and alloys. Individual tables are given for thin films, elements at high pressures, superconductors with high critical magnetic fields, and high critical temperature superconductors.

The historically first observed and most distinctive property of a superconductive body is the near total loss of resistance at a critical temperature (T_c) that is characteristic of each material. Figure 1(a) below illustrates schematically two types of possible transitions. The sharp vertical discontinuity in resistance is indicative of that found for a single crystal of a very pure element or one of a few well annealed alloy compositions. The broad transition, illustrated by broken lines, suggests the transition shape seen for materials that are not homogeneous and contain unusual strain distributions. Careful testing of the resistivity limit for superconductors shows that it is less than 4×10^{-23} ohm cm, while the lowest resistivity observed in metals is of the order of 10^{-13} ohm cm. If one compares the resistivity of a superconductive body to that of copper at room temperature, the superconductive body is at least 10^{17} times less resistive.

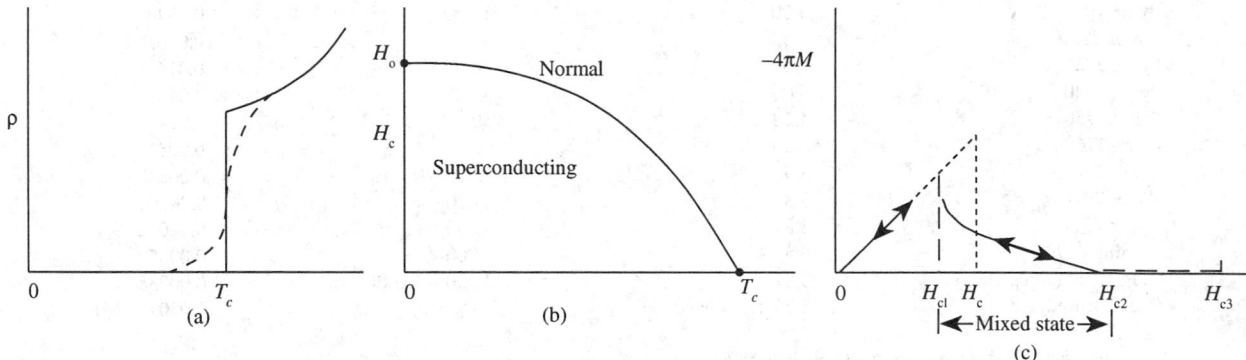

FIGURE 1. Physical properties of superconductors. (a) Resistivity vs. temperature for a pure and perfect lattice (solid line); impure and/or imperfect lattice (broken line). (b) Magnetic-field temperature dependence for Type-I or "soft" superconductors. (c) Schematic magnetization curve for "hard" or Type-II superconductors.

The temperature interval ΔT_c, over which the transition between the normal and superconductive states takes place, may be of the order of as little as 2×10^{-5} K or several K in width, depending on the material state. The narrow transition width was attained in 99.9999% pure gallium single crystals.

A Type-I superconductor below T_c, as exemplified by a pure metal, exhibits perfect diamagnetism and excludes a magnetic field up to some critical field H_c, whereupon it reverts to the normal state as shown in the H-T diagram of Figure 1(b).

The magnetization of a typical high-field superconductor is shown in Figure 1(c). The discovery of the large current-carrying capability of Nb_3Sn and other similar alloys has led to an extensive study of the physical properties of these alloys. In brief, a high-field superconductor, or Type-II superconductor, passes from the perfect diamagnetic state at low magnetic fields to a mixed state and finally to a sheathed state before attaining the normal resistive state of the metal. The magnetic field values separating the four stages are given as H_{c1}, H_{c2}, and H_{c3}. The superconductive state below H_{c1} is perfectly diamagnetic, identical to the state of most pure metals of the "soft" or Type-I superconductor. Between H_{c1} and H_{c2} a "mixed superconductive state" is found in which fluxons (a minimal unit of magnetic flux) create lines of normal flux in a superconductive matrix. The volume of the normal state is proportional to $-4\pi M$ in the "mixed state" region. Thus at H_{c2} the fluxon density has become so great as to drive the interior volume of the superconductive body completely normal. Between H_{c2} and H_{c3} the superconductor has a sheath of current-carrying superconductive material at the body surface, and above H_{c3} the normal state exists. With several types of careful measurement, it is possible to determine H_{c1}, H_{c2}, and H_{c3}. Table 6 contains some of the available data on high-field superconductive materials.

High-field superconductive phenomena are also related to specimen dimension and configuration. For example, the Type-I superconductor, Hg, has entirely different magnetization behavior in high magnetic fields when contained in the very fine sets of filamentary tunnels found in an unprocessed Vycor glass. The great majority of superconductive materials are Type-II. The elements in very pure form and a very few precisely stoichiometric and well annealed compounds are Type I with the possible exceptions of vanadium and niobium.

Metallurgical Aspects. The sensitivity of superconductive properties to the material state is most pronounced and has been used in a reverse sense to study and specify the detailed state of alloys. The mechanical state, the homogeneity, and the presence of impurity atoms and other electron-scattering centers are all capable of controlling the critical temperature and the current-carrying capabilities in high-magnetic fields. Well annealed specimens tend to show sharper transitions than those that are strained or inhomogeneous. This sensitivity to mechanical state underlines a general problem in the tabulation of properties for superconductive materials. The occasional divergent values of the critical temperature and of the critical fields quoted for a Type-II superconductor may lie in the variation in sample preparation. Critical temperatures of materials studied early in the history of superconductivity must be evaluated in light of the probable metallurgical state of the material, as well as the availability of less pure starting elements. It has been noted that recent work has given extended consideration to the metallurgical aspects of sample preparation.

Symbols in tables: T_c: Critical temperature; H_o: Critical magnetic field in the $T = 0$ limit; θ_D: Debye temperature; and γ: Electronic specific heat.

TABLE 1
Selective Properties of Superconductive Elements

Element	$T_c(K)$	H_0(oersted)	$\theta_D(K)$	γ(mJ mol^{-1}K^{-1})
Al	1.175 ± 0.002	104.9 ± 0.3	420	1.35
Am* (α,?)	0.6			
Am* (β,?)	1.0			
Be	0.026			0.21
Cd	0.517 ± 0.002	28 ± 1	209	0.69
Gd	1.083 ± 0.001	58.3 ± 0.2	325	0.60
Gd (β)	5.9, 6.2	560		
Gd (γ)	7	950, HF[a]		
Gd (Δ)	7.85	815, HF		
Hf	0.128	12.7		2.21
Hg (α)	4.154 ± 0.001	411 ± 2	87, 71.9	1.81
Hg (β)	3.949	339	93	1.37
In	3.408 ± 0.001	281.5 ± 2	109	1.672
Ir	0.1125 ± 0.001	16 ± 0.05	425	3.19
La (α)	4.88 ± 0.02	800 ± 10	151	9.8
La (β)	6.00 ± 0.1	1096, 1600	139	11.3
Lu	0.1 ± 0.03	350 ± 50		
Mo	0.915 ± 0.005	96 ± 3	460	1.83
Nb	9.25 ± 0.02	2060 ± 50, HF	276	7.80
Os	0.66 ± 0.03	70	500	2.35
Pa	1.4			
Pb	7.196 ± 0.006	803 ± 1	96	3.1
Re	1.697 ± 0.006	200 ± 5	4.5	2.35
Ru	0.49 ± 0.015	69 ± 2	580	2.8
Sn	3.722 ± 0.001	305 ± 2	195	1.78
Ta	4.47 ± 0.04	829 ± 6	258	6.15
Tc	7.8 ± 0.1	1410, HF	411	6.28
Th	1.38 ± 0.02	1.60 ± 3	165	4.32
Ti	0.40 ± 0.04	56	415	3.3
Tl	2.38 ± 0.02	178 ± 2	78.5	1.47
U	0.2			
V	5.40 ± 0.05	1408	383	9.82
W	0.0154 ± 0.0005	1.15 ± 0.03	383	0.90
Zn	0.85 ± 0.01	54 ± 0.3	310	0.66
Zr	0.61 ± 0.15	47	290	2.77
Zr (ω)	0.65, 0.95			

TABLE 2
Range of Critical Temperatures Observed for Superconductive Elements in Thin Films Condensed Usually at Low Temperatures

Element	T_c Range (K)	Comments	Element	T_c Range (K)	Comments
Al	1.15—5.7	HF[a]	Nb	2.0—10.1	
Be	5—9.75	HF	Pb	1.8—7.5	
Bi	6.17—6.6		Re	1.7—7	
Cd			Sn	3.5—6	
(Disordered)	0.79—0.91		Ta	<1.7—4.51	HF[a]
(Ordered)	0.53—0.59		Tc	4.6—7.7	
Ga	2.5—8.5	HF	Ti	1.3 Max	
Hg	3.87—4.5		Tl	2.33—2.96	
In	2.2—5.6	HF	V	1.8—6.02	
La	3.55—6.74		W	<1.0—4.1	
Mo	3.3—8.0		Zn	0.77—1.9	

[a]   HF denotes high magnetic field superconductive properties.

TABLE 3
Elements Exhibiting Superconductivity Under or After Application of High Pressure

Element	T_c Range (K)	Pressure (kbar)	Element	T_c Range (K)	Pressure (kbar)
Al	1.98—0.075	0—62	Pb II	3.55	160
As	0.31—0.5	220—140	Re II	2.3 Max.	"Plastic"
	0.2—0.25	140—100			compression
Ba II	1—1.8	55—85	Sb (prepared 120	2.6—2.7	
III	1.8—5	85—144	kbar, held below		
IV	4.5—5.4	144—190	77K)		
Bi II	3.9	25—27	Sb II	3.55—3.40	85—150
III	6.55—7.25	28—38	Se II	6.75, 6.95	130
IV	7.0, 8.7—6.0	43, 43—62	Si	6.7—7.1	120—130
V	6.7, 8.3	48—80	Sn II	5.2—4.85	125—160
VI	8.55	90, 92—101	III	5.30	113
VII(?)	8.2	30	Te II	2.4—5.1	38—55
Ce (α)	0.020—0.045	20—35		4.1—4.2	53—62
Ce (α')	1.9—1.3	45—125	IV	4.72—4	63—80
Cs V	1.5	>125	( )	3.3—2.8	100—260
Ga II	6.38	≥35	Tl (cubic form)	1.45	35
II′	7.5	≥35 then P	(hexagonal form)	1.95	35
		removed	U	2.4—0.4	10—85
Ge	5.35	115	Y	1.7—2.5	110—160
La	5.5—12.9	0—210	Zr (omega form,	1—1.7	60—130
Lu	0.022—1.0	45—190	metastable)		
P	5.8	170			

TABLE 4
Superconductive Compounds and Alloys

All compositions are denoted on an atomic basis, i.e., AB, AB_2, or AB_3 for compounds, unless noted. Solid solutions or odd compositions may be denoted as A_zB_{1-z} or A_zB. A series of three or more alloys is indicated as A_xB_{1-x} or by actual indication of the atomic fraction range, such as $A_{0—0.6}B_{1—0.4}$. The critical temperature of such a series of alloys is denoted by a range of values or possibly the maximum value.

The selection of the critical temperature from a transition in the effective permeability, or the change in resistance, or possibly the incremental changes in frequency observed by certain techniques is not often obvious from the literature. Most authors choose the mid-point of such curves as the probable critical temperature of the idealized material, while others will choose the highest temperature at which a deviation from the normal state property is observed. In view of the previous discussion concerning the variability of the superconductive properties as a function of purity and other metallurgical aspects, it is recommended that appropriate literature be checked to determine the most probable critical temperature or critical field of a given alloy.

A very limited amount of data on critical fields, H_o, is available for these compounds and alloys; these values are given at the end of the table.

A. SUPERCONDUCTORS WITH $T_c < 10$ K

Substance	T_c, K	Crystal structure type
$Ag_{3.3}Al$	0.34	A12-cI58 (Mn)
$Ag_xAl_yZn_{1-x-y}$	0.15	Cubic
$AgBi_2$	2.87—3.0	
$Ag_7F_{0.25}N_{0.75}O_{10.25}$	0.85—0.90	
Ag_2F	0.0066	
Ag_7FO_8	0.3	Cubic
$Ag_{0.8—0.3}Ga_{0.2—0.7}$	6.5—8	
Ag_4Ge	0.85	Hex., c.p.
$Ag_{0.438}Hg_{0.562}$	0.64	$D8_2$
$AgIn_2$	~2.4	C16

TABLE 4
Superconductive Compounds and Alloys (continued)

Substance	T_c, K	Crystal structure type
$Ag_{0.1}In_{0.9}Te$ ($n = 1.4 \times 10^{22}$)*	1.2—1.89	B1
$Ag_{0.2}In_{0.8}Te$ ($n = 1.07 \times 10^{22}$)	0.77—1.00	B1
AgLa	0.94	B2-cP2 (CsCl)
AgLa (9.5 kbar)	1.2	B2
AgLu	0.33	B2-cP2
$AgMo_4S_5$	9.1	hR15 (Mo_6PbS_8)
$Ag_{1.2}Mo_6Se_8$	5.9	Same
Ag_7NO_{11}	1.04	Cubic
Ag_xPb_{1-x}	7.2 max.	
Ag_4Sn	0.1	h**
Ag_xSn_{1-x}	1.5—3.7	
Ag_xSn_{1-x} (film)	2.0—3.8	
$AgTe_3$	2.6	Cubic
AgTh	2.2	C16-tI12 (Al_2Cu)
$AgTh_2$	2.26	C16
$Ag_{0.03}Tl_{0.97}$	2.67	
$Ag_{0.94}Tl_{0.06}$	2.32	
AgY	0.33	B2-cP2 (CsCl)
Ag_xZn_{1-x}	0.5—0.845	
$AlAu_4$	0.4—0.7	Like A13
Al_2Au	0.1	C1-cF12 (CaF_2)
Al_2CMo_3	9.8—10.2	A13+trace 2nd. phase)
Al_2CaSi	5.8	
$Al_{0.131}Cr_{0.088}V_{0.781}$	1.46	Cubic
$AlGe_2$	1.75	
Al_2Ge_2U	1.6	LI_2-cP4 (Cu_3Au)
$AlLa_3$	5.57	DO_{19}
Al_2La	3.23	C15
Al_2Lu	1.02	C15-cF24 (Cu_2Mg)
Al_3Mg_2	0.84	F.C.C.
$AlMo_3$	0.58	A15
$AlMo_6Pd$	2.1	
AlN	1.55	B4
Al_2NNb_3	1.3	A13
Al_3Nb	0.64	tI8 (Al_3Ti)
AlOs	0.39	B2
Al_3Os	5.90	
AlPb (film)	1.2—7	
Al_2Pt	0.48—0.55	C1
Al_5Re_{24}	3.35	A12
AlSb	2.8	B4-tI4 (Sn)
Al_2Sc	1.02	C15-cF24 (Cu_2Mg)
Al_2Si_2U	1.34	LI_2-cP4 (Cu_3Au)
$AlTh_2$	0.1	C16-tI12 (Al_2Cu)
Al_3Th	0.75	DO_{19}
$Al_xTi_yV_{1-x-y}$	2.05—3.62	Cubic
$Al_{0.108}V_{0.892}$	1.82	Cubic
Al_2Y	0.35	C15-cF24 (Cu_2Mg)
Al_3Yb	0.94	LI_2-cP4 (Cu_3Au)
Al_xZn_{1-x}	0.5—0.845	
$AlZr_3$	0.73	LI_2
AsBiPb	9.0	
AsBiPbSb	9.0	
AsHfOs	3.2	C22-hP9 (Fe_2P)
AsHfRu	4.9	same

TABLE 4
Superconductive Compounds and Alloys (continued)

Substance	T_c, K	Crystal structure type
$As_{0.33}InTe_{0.67}$ ($n = 1.24 \times 10^{22}$)	0.85—1.15	B1
$As_{0.5}InTe_{0.5}$ ($n = 0.97 \times 10^{22}$)	0.44—0.62	B1
As_4La_3	0.6	cI28 (Th_3P_4)
$AsNb_3$	0.3	$L1_2$-tP32
$As_{0.50}Ni_{0.06}Pd_{0.44}$	1.39	C2
$AsNi_{0.25}Pd_{0.75}$	1.6	$B8_1$-hP4 (NiAs)
$AsOsZr$	8.0	C22-hP9 (Fe_2P)
$AsPb$	8.4	
$AsPd_2$ (low-temp. phase)	0.60	Hexagonal
$AsPd_2$ (high-temp. phase)	1.70	C22
$AsPd_5$	0.46	Complex
As_3Pd_5	1.9	
$AsRh$	0.58	B31
$AsRh_{1.4—1.6}$	< 0.03—0.56	Hexagonal
$AsSn$	4.10	
$AsSn$ ($n = 2.14 \times 10^{22}$)	3.41—3.65	B1
$As_{-2}Sn_{-3}$	3.5—3.6; 1.21—1.17	
As_3Sn_4 ($n = 0.56 \times 10^{22}$)	1.16—1.19	Rhombohedral
AsV_3	0.20	A15-cP8 (Cr_3Si)
Au_5Ba	0.4—0.7	$D2_d$
$AuBe$	2.64	B20
Au_2Bi	1.80	C15
Au_5Ca	0.34—0.38	$C15_b$
$AuGa_2$	1.6	C1-cF12 (CaF_2)
$AuGa$	1.2	B31
$Au_{0.40—0.92}Ge_{0.60—0.08}$	<0.32—1.63	Complex
$AuIn_2$	0.2	C1-cF12
$AuIn$	0.4—0.6	Complex
$AuLu$	<0.35	B2
$AuNb_3$	1.2	A2
$AuPb_2$	3.15	
$AuPb_2$ (film)	4.3	
$AuPb_3$	4.40	
$AuPb_3$ (film)	4.25	
Au_2Pb	1.18; 6—7	C15
$AuSb_2$	0.58	C2
$AuSn$	1.25	$B8_1$
Au_xSn_{1-x} (film)	2.0—3.8	
Au_5Sn	0.7—1.1	A3
$AuTa_{4.3}$	0.55	A15-cP8 (Cr_3Si)
Au_3Te_5	1.62	Cubic
$AuTh_2$	3.08	C16
$AuTl$	1.92	
AuV_3	0.74	A15
Au_xZn_{1-x}	0.50—0.845	
$AuZn_3$	1.21	Cubic
Au_xZr_y	1.7—2.8	A3
$AuZr_3$	0.92	A15
$B_2Ba_{0.67}Pt_3$	5.60	hP12 (B_2BaPt_3)
$BCMo_2$	5.4	Orthorhombic
$BCMo_2$	5.3—7.0	Same
$B_2Ca_{0.67}Pt_3$	1.57	hP12
B_4ErIr_4	2.1	tP18 (B_4CeCo_4)
B_4ErRh_4	4.3	oC108 (B_4LuRh_4)
B_4ErRh_4	8.7	tP18 (B_4CeCo_4)

TABLE 4
Superconductive Compounds and Alloys (continued)

Substance	T_c, K	Crystal structure type
BHf	3.1	Cubic
B_4HoIr_4	2.0	tP18
B_4HoRh_4	1.4	oC108
B_2Ir_3La	1.65	hP6 ($CaCu_5$)
B_2Ir_3Th	2.09	Same
B_4Ir_4Tm	1.6	tP18
B_6La	5.7	
B_2LaRh_3	2.82	hP6
$B_{12}Lu$	0.48	
B_2LuOs	2.66	oP16 (B_2LuRu)
B_2LuOs_3	4.62	hP6
B_4LuRh_4	6.2	oC108
B_2LuRu	9.86	oP16
B_4LuRu_4	2.0	tI72 (B_4LuRu_4)
BMo	0.5 (extrapol.)	
BMo_2	4.74	C16
BNb	8.25	B_f
B_4NdRh_4	5.3	tP18
B_2OsSc	1.34	oP16
B_2OsY	2.22	oP16
$B_2Pt_3Sr_{0.67}$	2.78	hP12 (B_2BaPt_3)
BRe_2	2.80; 4.6	
$B_4Rh_{3.4}Ru_{0.6}$	8.38	tI72
B_4Rh_4Sm	2.7	tP18
B_4Rh_4Th	4.3	Same
B_4Rh_4Tm	9.8	Same
B_4Rh_4Tm	5.4	oC108
$B_{0.3}Ru_{0.7}$	2.58	$D10_2$
B_4Ru_4Sc	7.2	tI72
B_2Ru_3Th	1.79	hP6
B_2Ru_3Y	2.85	Same
$B_2Ru\,Y$	7.80	oP16
B_4Ru_4Y	1.4	tI72
$B_{12}Sc$	0.39	
BTa	4.0	B_f
BTa_2	3.12	C16-tI12 (Al_2Cu)
B_6Th	0.74	
BW_2	3.1	C16
B_6Y	6.5—7.1	
$B_{12}Y$	4.7	
BZr	3.4	Cubic
$B_{12}Zr$	5.82	
$BaBi_3$	5.69	Tetragonal
$Ba_2Mo_{15}Se_{19}$	2.75	hP15 (Mo_6PbS_8)
$Ba_xO_3Sr_{1-x}Ti$ ($n = 4.2 \times 10^{19}$)	<0.1—0.55	
$Ba_{0.13}O_3W$	1.9	Tetragonal
$Ba_{0.14}O_3W$	<1.25—2.2	Hexagonal
$BaRh_2$	6.0	C15
$Be_{22}Mo$	2.51	Cubic ($Be_{22}Re$)
$Be_8Nb_5Zr_2$	5.2	
$Be_{0.98—0.92}Re_{0.02—0.08}$ (quenched)	9.5—9.75	Cubic
$Be_{0.957}Re_{0.043}$	9.62	Cubic ($Be_{22}Re$)
BeTc	5.21	Cubic
$Be_{22}W$	4.12	Cubic ($Be_{22}Re$)

TABLE 4
Superconductive Compounds and Alloys (continued)

Substance	T_c, K	Crystal structure type
$Be_{13}W$	4.1	Tetragonal
Bi_3Ca	2.0	
$Bi_{0.5}Cd_{0.13}Pb_{0.25}Sn_{0.12}$ (weight fractions)	8.2	
BiCo	0.42—0.49	
Bi_2Cs	4.75	C15
Bi_xCu_{1-x} (electrodeposited)	2.2	
BiCu	1.33—1.40	
Bi_3Fe	1.0	m**
$Bi_{0.019}In_{0.981}$	3.86	
$Bi_{0.05}In_{0.95}$	4.65	α-phase
$Bi_{0.10}In_{0.90}$	5.05	Same
$Bi_{0.15-0.30}In_{0.85-0.70}$	5.3—5.4	α- and β-phases
$Bi_{0.34-0.48}In_{0.66-0.52}$	4.0—4.1	
Bi_3In_5	4.1	
$BiIn_2$	5.65	β-phase
Bi_2Ir	1.7—2.3	
Bi_2Ir (quenched)	3.0—3.96	
BiK	3.6	
Bi_2K	3.58	C15
BiLi	2.47	$L1_o$, α-phase
$Bi_{4-9}Mg$	0.7—~1.0	
Bi_3Mo	3—3.7	
BiNa	2.25	$L1_o$
$BiNb_3$	4.5	A15-cP8 (Cr_3Si)
$BiNb_3$ (high pressure and temperature)	3.05	A15
BiNi	4.25	$B8_1$
Bi_3Ni	4.06	Orthorhombic
$BiNi_{0.5}Rh_{0.5}$	3.0	$B8_1$-hP4 (AsNi)
$Bi_{0.5}NiSb_{0.5}$	2.0	Same
$Bi_{1-0}Pb_{0-1}$	7.26—9.14	
$Bi_{1-0}Pb_{0-1}$ (film)	7.25—8.67	
$Bi_{0.05-0.40}Pb_{0.95-0.60}$	7.35—8.4	H.C.P. to ε-phase
Bi_2Pb	4.25	t**
BiPbSb	8.9	
$Bi_{0.5}Pb_{0.31}Sn_{0.19}$ (weight fractions)	8.5	
$Bi_{0.5}Pb_{0.25}Sn_{0.25}$	8.5	
$BiPd_2$	4.0	
$Bi_{0.4}Pd_{0.6}$	3.7—4	Hexagonal, ordered
BiPd	3.7	Orthorhombic
Bi_2Pd	1.70	Monoclinic, α-phase
Bi_2Pd	4.25	Tetragonal, β-phase
$BiPd_{0.45}Pt_{0.55}$	3.7	$B8_1$-hP4 (NiAs)
BiPdSe	1.0	C2
BiPdTe	1.2	C2
BiPt	1.21	$B8_1$
$Bi_{0.1}PtSb_{0.9}$	2.05; 1.5	$B8_1$-hP4 (NiAs)
BiPtSe	1.45	C2
BiPtTe	1.15	C2
Bi_2Pt	0.155	Hexagonal
Bi_2Rb	4.25	C15
$BiRe_2$	1.9—2.2	
BiRh	2.06	$B8_1$

TABLE 4
Superconductive Compounds and Alloys (continued)

Substance	T_c, K	Crystal structure type
Bi_3Rh	3.2	Orthorhombic (NiB_3)
Bi_4Rh	2.7	Hexagonal
$BiRu$	5.7	m**
Bi_3Sn	3.6—3.8	
$BiSn$	3.8	
Bi_xSn_y	3.85—4.18	
Bi_3Sr	5.62	$L1_2$
Bi_3Te	0.75—1.0	
Bi_5Tl_3	6.4	
$Bi_{0.26}Tl_{0.74}$	4.4	Cubic, disordered
$Bi_{0.26}Tl_{0.74}$	4.15	$L1_2$, ordered (?)
Bi_2Y_3	2.25	
Bi_3Zn	0.8—0.9	
$Bi_{0.3}Zr_{0.7}$	1.51	
$BiZr_3$	2.4—2.8	
$BrMo_6Se_7$	7.1	hP15 (Mo_6PbS_8)
$Br_3Mo_6Se_5$	7.1	Same
CCs_x	0.020—0.135	Hexagonal
CFe_3	1.30	DO_{11}-oP16 (Fe_3C)
$CGaMo_2$	3.7—4.1	Hexagonal
$CHf_{0.5}Mo_{0.5}$	3.4	B1
$CHf_{0.3}Mo_{0.7}$	5.5	B1
$CHf_{0.25}Mo_{0.75}$	6.6	B1
$CHf_{0.7}Nb_{0.3}$	6.1	B1
$CHf_{0.6}Nb_{0.4}$	4.5	B1
$CHf_{0.5}Nb_{0.5}$	4.8	B1
$CHf_{0.4}Nb_{0.6}$	5.6	B1
$CHf_{0.25}Nb_{0.75}$	7.0	B1
$CHf_{0.2}Nb_{0.8}$	7.8	B1
$CHf_{0.9—0.1}Ta_{0.1—0.9}$	5.0—9.0	B1
CK (excess K)	0.55	Hexagonal
C_8K	0.39	Hexagonal
C_2La	1.66	tI6 (CaC_2)
C_2Lu	3.33	Same
$C_{0.40—0.44}Mo_{0.60—0.56}$	9—13	
C_3MoRe	3.8	B1-cF8
$C_{0.6}Mo_{4.8}Si_3$	7.6	$D8_8$
$CMo_{0.2}Ta_{0.8}$	7.5	B1
$CMo_{0.5}Ta_{0.5}$	7.7	B1
$CMo_{0.75}Ta_{0.25}$	8.5	B1
$CMo_{0.8}Ta_{0.2}$	8.7	B1
$CMo_{0.85}Ta_{0.15}$	8.9	B1
CMo_xV_{1-x}	2.9—9.3	B1
CMo_xZr_{1-x}	9.8	B1
$C_{0.984}Nb$	9.8	B1
CNb_2	9.1	
CNb_xTi_{1-x}	<4.2—8.8	B1
$CNb_{0.1—0.9}Zr_{0.9—0.1}$	4.2—8.4	B1
CRb_x (Au)	0.023—0.151	Hexagonal
$CRe_{0.06}W$	5.0	
CRu	2.00	hP2 (CW)
$C_{0.98}7Ta$	9.7	
$C_{0.848—0.987}$	2.04—9.7	
CTa (film)	5.09	B1
CTa_2	3.26	L'_3

TABLE 4
Superconductive Compounds and Alloys (continued)

Substance	T_c, K	Crystal structure type
$CTa_{0.4}Ti_{0.6}$	4.8	B1
$Cta_{1-0.4}W_{0-0.6}$	8.5—10.5	B1
$CTa_{0.2-0.9}Zr_{0.8-0.1}$	4.6—8.3	B1
CTc (excess C)	3.85	Cubic
$CTi_{0.5-0.7}W_{0.5-0.3}$	6.7—2.1	B1
CW	1.0	
CW_2	2.74	L'_3
CW_2	5.2	F.C.C.
C_2Y	3.88	tI6 (CaC_2)
$Ca_3Co_4Sn_{13}$	5.9	cP40 ($Pr_3Rh_2Sn_{13}$)
$Ca_3Ge_{13}Rh_4$	2.1	Same
CaHg	1.6	B2-cP2 (CsCl)
$CaHg_3$	1.6	hP8 (Ni_3Sn)
$CaIr_2$	6.15	C15
$Ca_3Ir_4Sn_{13}$	7.1	cP40
$Ca_xO_3Sr_{1-x}Ti$ (n = 3.7—11 × 10^{19})	<0.1—0.55	
$Ca_{0.1}O_3W$	1.4—3.4	Hexagonal
CaPb	7.0	
$CaRh_2$	6.40	C15
$CaRh_{1.2}Sn_{4.5}$	8.7	cP40
$CaTl_3$	2.0	B2-cP2
$Cd_{0.3-0.5}Hg_{0.7-0.5}$	1.70—1.92	
CdHg	1.77; 2.15	Tetragonal
$Cd_{0.0075-0.05}In_{0.9925-0.95}$	3.24—3.36	Tetragonal
$Cd_{0.97}Pb_{0.03}$	4.2	
CdSn	3.65	
$Cd_{0.17}Tl_{0.83}$	2.3	
$Cd_{0.18}Tl_{0.82}$	2.54	
$CeCo_2$	0.84	C15
$CeCo_{1.67}Ni_{0.33}$	0.46	C15
$CeCo_{1.67}Rh_{0.33}$	0.47	C15
$Ce_xGd_{1-x}Ru_2$	3.2—5.2	C15
$CeIr_3$	3.34	
$CeIr_5$	1.82	
$Ce_{0.005}La_{0.995}$	4.6	
Ce_xLa_{1-x}	1.3—6.3	
$Ce_xPr_{1-x}Ru_2$	1.4—5.3	C15
Ce_xPt_{1-x}	0.7—1.55	
$CeRu_2$	6.0	C15
$Ce_3Mo_6Se_5$	5.7	hR15 (Mo_6PbS_8)
$Ce_2Mo_6Te_6$	1.7	Same
$Co_xFe_{1-x}Si_2$	1.4 (max.)	C1
$CoHf_2$	0.56	$E9_3$
$CoLa_3$	4.28	
$Co_4La_3Sn_{13}$	2.8	cP40
$CoLu_3$	~0.35	
Co_xLuSn_y	1.5	cP40
$Co_{0-0.01}Mo_{0.8}Re_{0.2}$	2—10	
$Co_{0.02-0.10}Nb_3Rh_{0.98-0.90}$	2.28—1.90	A15
$Co_xNi_{1-x}Si_2$	1.4 (max.)	C1
$Co_{0.5}Rh_{0.5}Si_2$	2.5	
$Co_xRh_{1-x}Si_2$	3.65 (max.)	
$Co_{0.3}So_{0.7}$	~0.35	
$Co_4Sc_5Si_{10}$	5.0	tP38 ($Co_4Sc_5Si_{10}$)
$CoSi_2$	1.40; 1.22	C1

TABLE 4
Superconductive Compounds and Alloys (continued)

Substance	T_c, K	Crystal structure type
Co_xSn_yYb	2.5	cP40
Co_3Th_7	1.83	$D10_2$
Co_xTi_{1-x}	2.8 (max.)	Co in α-Ti
Co_xTi_{1-x}	3.8 (max.)	Co in β-Ti
$CoTi_2$	3.44	$E9_3$
CoTi	0.71	A2
CoU	1.7	B2, distorted
CoU_6	2.29	$D2_c$
$Co_{0.28}Y_{0.72}$	0.34	
CoY_3	<0.34	
$CoZr_2$	6.3	C16
$Co_{0.1}Zr_{0.9}$	3.9	A3
$Cr_{0.6}Ir_{0.4}$	0.4	H.C.P.
$Cr_{0.65}Ir_{0.35}$	0.59	H.C.P.
$Cr_{0.7}Ir_{0.3}$	0.76	H.C.P.
$Cr_{0.72}Ir_{0.28}$	0.83	
Cr_3Ir	0.45	A15
$Cr_{0—0.1}Nb_{1—0.9}$	4.6—9.2	A2
$Cr_{0.80}Os_{0.20}$	2.5	Cubic
Cr_3Os	4.68	A15-cP8 (Cr_3Si)
Cr_xRe_{1-x}	1.2—5.2	
$Cr_{0.4}Re_{0.6}$	2.15	$D8_b$
$Cr_{0.8—0.6}Rh_{0.2—0.4}$	0.5—1.10	A3
Cr_3Rh	0.3	A15-cP8
Cr_3Ru (annealed)	3.3	A15
Cr_2Ru	2.02	$D8_b$
Cr_3Ru_2	2.10	$D8_b$-tP30 (CrFe)
$Cr_{0.1—0.5}Ru_{0.9—0.5}$	0.34—1.65	A3
Cr_xTi_{1-x}	3.6 (max.)	Cr in α-Ti
Cr_xTi_{1-x}	4.2 (max.)	Cr in β-Ti
$Cr_{0.1}Ti_{0.3}V_{0.6}$	5.6	
$Cr_{0.0175}U_{0.9825}$	0.75	β-phase
$Cs_{0.32}O_3W$	1.12	Hexagonal
$Cu_{0.15}In_{0.85}$ (film)	3.75	
$Cu_{0.04—0.08}In_{0.94—0.92}$	4.4	
CuLa	5.85	
$Cu_2Mo_6O_2S_6$	9	hR15 (Mo_6PbS_8)
$Cu_2Mo_6Se_8$	5.9	Same
Cu_xPb_{1-x}	5.7—7.7	
CuS	1.62	B18
CuS_2	1.48—1.53	C18
CuSSe	1.5—2.0	C18
$CuSe_2$	2.3—2.43	C18
CuSeTe	1.6—2.0	C18
Cu_xSn_{1-x}	3.2—3.7	
Cu_xSn_{1-x} (film, made at 10K)	3.6—7	
Cu_xSn_{1-x} (film, made at 300K)	2.8—3.7	
$CuTe_2$	<1.25—1.3	C18
$CuTh_2$	3.49	C16
$Cu_{0—0.027}V$	3.9—5.3	A2
CuY	0.33	B2-cP2 (CsCl)
Cu_xZn_{1-x}	0.5—0.845	
$DyMo_6S_8$	2.1	hR15
Er_xLa_{1-x}	1.4—6.3	
$ErMo_6S_8$	2.2	hR15

TABLE 4
Superconductive Compounds and Alloys (continued)

Substance	T_c, K	Crystal structure type
$ErMo_6Se_8$	6.2	hR15
$Fe_3Lu_2Si_5$	6.1	tP40 ($Fe_3Sc_2Si_5$)
$Fe_{0-0.04}Mo_{0.8}Re_{0.2}$	1—10	
$Fe_{0.05}Ni_{0.05}Zr_{0.90}$	~3.9	
Fe_3Re_2	6.55	$D8_b$-tP30 (FeCr)
$Fe_3Sc_2Si_5$	4.52	tP40
Fe_3Si_5Tm	1.3	Same
$Fe_3Si_5Y_2$	2.4	Same
Fe_3Th_7	1.86	D10
Fe_xTi_{1-x}	3.2 (max.)	Fe in α-Ti
Fe_xTi_{1-x}	3.7 (max.)	Fe in β-Ti
$Fe_xTi_{0.6}V_{1-x}$	6.8 (max.)	
FeU_6	3.86	$D2_c$
$Fe_{0.1}Zr_{0.9}$	1.0	A3
$Ga_{0.5}Ge_{0.5}Nb_3$	7.3	A15
Ga_2Ge_2U	0.87	B2-cP2
$GaHf_2$	0.21	C16-tI12 (Al_2Cu)
$GaLa_3$	5.84	
Ga_3Lu	2.3	B2-cP2
Ga_2Mo	9.5	
$GaMo_3$	0.76	A15
GaN (black)	5.85	B4
$Ga_{0.7}Pt_{0.3}$	2.9	C1
GaPt	1.74	B20
GaSb (120kbar, 77K, annealed)	4.24	A5
GaSb (unannealed)	~5.9	
$Ga_{0-1}Sn_{1-0}$ (quenched)	3.47—4.18	
$Ga_{0-1}Sn_{1-0}$ (annealed)	2.6—3.85	
GaTe	0.17	mC24 (GaTe)
Ga_5V_2	3.55	Tetragonal (Mn_2Hg_5)
$GaV_{4.5}$	9.15	
Ga_3Zr	1.38	
Ga_3Zr_5	3.8	$D8_b$-hP16 (Mn_5Si_3)
Gd_xLa_{1-x}	< 1.0—5.5	
$GdMo_6S_8$	3.5	hR15
$GdMo_6Se_8$	5.6	hR15
$Gd_xOs_2Y_{1-x}$	1.4—4.7	
$Gd_xRu_2Th_{1-x}$	3.6 (max.)	C15
$Ge_{10}As_4Y_5$	9.06	tP38 ($C0_4Sc_5Si_{10}$)
GeIr	4.7	B31
GeIrLa	1.64	tI12 (LaPtSi)
$Ge_{10}Ir_4Lu_5$	2.60	tP38
$Ge_{10}Ir_4Y_5$	2.62	tP38
Ge_2La	1.49; 2.2	Orthorhombic, distorted (Mn_2Hg_5)
GeLaPt	3.53	tI12
$Ge_{13}Lu_3Os_4$	3.6	cP40 ($Pr_3Rh_2Sn_{13}$)
$Ge_{10}Lu_5Rh_4$	2.79	tP38
$Ge_{13}Lu_3Ru_4$	2.3	cP40
$GeMo_3$	1.43	A15
$GeNb_2$	1.9	
$Ge_{0.29}Nb_{0.71}$	6	A15
GePt	0.40	B31
Ge_3Rh_5	2.12	Orthorhombic, related to $InNi_2$

TABLE 4
Superconductive Compounds and Alloys (continued)

Substance	T_c, K	Crystal structure type
GeRh	0.96	B31-oP8 (MnP)
$Ge_{13}Rh_4Sc_3$	1.9	c P40
$Ge_{10}Rh_4Y_5$	1.35	tP38
$Ge_{13}Ru_4Y_3$	1.7	cP40
Ge_2So	1.3	
$GeTa_3$	8.0	A15-cP8 (Cr_3Si)
Ge_3Te_4 ($n = 1.06 \times 10^{22}$)	1.55—1.80	Rhombohedral
Ge_xTe_{1-x} ($n = 8.5$—64×10^{20})	0.07—0.41	R1
GeV_3	6.01	A15
Ge_2Y	3.80	C_c
$Ge_{1.62}Y$	2.4	
Ge_2Zr	0.30	oC12 ($ZrSi_2$)
$GeZr_3$	0.4	$L1_2$-tP32 (Ti_3P)
$H_{0.33}Nb_{0.67}$	7.28	B.C.C.
$H_{0.1}Nb_{0.9}$	7.38	Same
$H_{0.05}Nb_{0.95}$	7.83	Same
$H_{0.12}Ta_{0.88}$	2.81	B.C.C.
$H_{0.08}Ta_{0.92}$	3.26	Same
$H_{0.04}Ta_{0.96}$	3.62	Same
HfIrSi	3.50	C37-cP12 (Co_2Si)
$HfMo_2$	0.05	hP24 (Ni_2Mn)
$HfN_{0.989}$	6.6	B1
$Hf_{0—0.5}Nb_{1—0.5}$	8.3—9.5	A2
$Hf_{0.75}Nb_{0.25}$	>4.2	
$HfOs_2$	2.69	C14
HfOsP	6.1	C22-hP9 (Fe_2P)
HfPRu	9.9	Same
$HfRe_2$	4.80	C14
$Hf_{0.14}Re_{0.86}$	5.86	A12
$Hf_{0.99—0.96}Rh_{0.01—0.04}$	0.85—1.51	
$Hf_{0—0.55}Ta_{1—0.45}$	4.4—6.5	A2
HfV_2	8.9—9.6	C15
Hg_xIn_{1-x}	3.14—4.55	
HgIn	3.81	
Hg_2K	1.20	Orthorhombic
Hg_3K	3.18	
Hg_4K	3.27	
Hg_8K	3.42	
Hg_3Li	1.7	Hexagonal
$HgMg_3$	0.17	hP8 (Na_3As)
Hg_2Mg	4.0	tI6 ($MoSi_2$)
Hg_3Mg_5	0.48	$D8_b$-hP16 (Mn_5Si_3)
Hg_2Na	1.62	Hexagonal
Hg_4Na	3.05	
Hg_xPb_{1-x}	4.14—7.26	
HgSn	4.2	
Hg_xTl_{1-x}	2.30—4.19	
Hg_5Tl_2	3.86	
Ho_xLa_{1-x}	1.3—6.3	
$Ho_{1.2}Mo_6Se_8$	6.1	$D10_2$-hR12 (Be_3Nb)
$In_{1—0.86}Mg_{0—0.14}$	3.395—3.363	
$In_2Mo_6Te_6$	2.6	hR15 (Mo_6PbS_8)
$InNb_3$ (high pressure and temp.)	4—8; 9.2	A15
$In_{0.5}Nb_3Zr_{0.5}$	6.4	
$In_{0.11}O_3W$	< 1.25—2.8	Hexagonal

TABLE 4
Superconductive Compounds and Alloys (continued)

Substance	T_c, K	Crystal structure type
$In_{0.95—0.85}Pb_{0.05—0.15}$	3.6—5.05	
$In_{0.98—0.91}Pb_{0.02—0.09}$	3.45—4.2	
InPb	6.65	
InPd	0.7	B2
InSb (quenched from 170 kbar into liquid N_2)	4.8	Like A5
InSb	2.1	B4
$(InSb)_{0.95—0.10}Sn_{0.05—0.90}$ (various heat treatments)	3.8—5.1	
$(InSb)_{0—0.07}Sn_{1—0.93}$	3.67—3.74	
In_3Sn	~5.5	
In_xSn_{1-x}	3.4—7.3	
$In_{0.82—1}Te$ ($n = 0.83—1.71 \times 10^{22}$)	1.02—3.45	B1
$In_{1.000}Te_{1.002}$	3.5—3.7	B1
In_3Te_4 ($n = 4.7 \times 10^{21}$)	1.15—1.25	Rhombohedral
In_xTl_{1-x}	2.7—3.374	
$In_{0.8}Tl_{0.2}$	3.223	
$In_{0.62}Tl_{0.38}$	2.760	
$In_{0.78—0.69}Tl_{0.22—0.31}$	3.18—3.32	Tetragonal
$In_{0.69—0.62}Tl_{0.31—0.38}$	2.98—3.3	F.C.C.
Ir_2La	0.48	C15
Ir_3La	2.32	$D10_2$
Ir_3La_7	2.24	$D10_2$
Ir_5La	2.13	
$IrLaSi_2$	2.03	oC16 ($CeNiSi_2$)
$IrLaSi_3$	2.7	tI10 ($BaNiSn_3$)
Ir_2Lu	2.47	C15
Ir_3Lu	2.89	C15
$Ir_4Lu_5Si_{10}$	3.9	tP38 ($Co_4Sc_5Si_{10}$)
IrMo	< 1.0	A3
$IrMo_3$	9.6	A15
$IrMo_3$	6.8	$D8_b$
$IrNb_3$	1.9	A15
$Ir_{0.4}Nb_{0.6}$	9.8	$D8_b$
$Ir_{0.37}Nb_{0.63}$	2.32	$D8_b$
IrNb	7.9	$D8_b$
$Ir_{1.15}Nb_{0.85}$	4.6	oP12 (IrTa)
$Ir_{0.02}Nb_3Rh_{0.98}$	2.43	A15
$Ir_{0.05}Nb_3Rh_{0.95}$	2.38	A15
$Ir_{0.287}O_{0.14}Ti_{0.573}$	5.5	$E9_3$
$Ir_{0.265}O_{0.035}Ti_{0.65}$	2.30	$E9_3$
Ir_xOs_{1-x}	0.3—0.98	
$Ir_{1.5}Os_{0.5}$	2.4	C14
IrOsY	2.6	C15
IrSiY	2.70	C37-oP12 (Co_2Si)
IrSiZr	2.04	Same
Ir_2Sc	2.07	C15
$Ir_{2.5}Sc$	2.46	C15
$Ir_4Sc_5Si_{10}$	8.46	tP38
Ir_2Si_2Th	2.14	tI10
$IrSi_3Th$	1.75	tI10
IrSiTh	6.50	tI12 (LaPtSi)
Ir_2Si_2Y	2.60	tI10 (Al4Ba)
$Ir_4Si_{10}Y_5$	3.10	tP38
$Ir_3Si_5Y_2$	2.83	oI40

TABLE 4
Superconductive Compounds and Alloys (continued)

Substance	T_c, K	Crystal structure type
$IrSn_2$	0.65—0.78	C1
Ir_2Sr	5.70	C15
Ir_7Ta_{13}	1.2	$D8_b$-tP30 (FeCr)
$Ir_{0.5}Te_{0.5}$	~3	
$IrTe_3$	1.18	C2
IrTh	< 0.37	B_f
Ir_2Th	6.50	C15
Ir_3Th	4.71	
Ir_3Th_7	1.52	$D10_2$
Ir_5Th	3.93	$D2_d$
$IrTi_3$	5.40	A15
IrV_2	1.39	A15
IrW_3	3.82	
$Ir_{0.28}W_{0.72}$	4.49	
Ir_2Y	2.18; 1.38	C15
$Ir_{0.69}Y_{0.31}$	1.98; 1.44	C15
$Ir_{0.70}Y_{0.30}$	2.16	C15
Ir_2Y_3	1.61	
Ir_3Y	3.50	$D10_2$-hR13 (Be_3Nb)
Ir_xY_{1-x}	0.3—3.7	
Ir_2Zr	4.10	C15
$Ir_{0.1}Zr_{0.9}$	5.5	A3
$K_2Mo_{15}S_{19}$	3.32	hR15
$K_{0.27—0.31}O_3W$	0.50	Hexagonal
$K_{0.40—0.57}O_3W$	1.5	Tetragonal
$La_{0.55}Lu_{0.45}$	2.2	Hexagonal, La type
$La_{0.8}Lu_{0.2}$	3.4	Same
$LaMg_2$	1.05	C15
$LaMo_6S_8$	7.1	hR15
LaN	1.35	
$LaOs_2$	6.5	C15
$LaPt_2$	0.46	C15
$La_{0.28}Pt_{0.72}$	0.54	C15
LaPtSi	3.48	tI12
$LaRh_3$	2.60	
$LaRh_5$	1.62	
La_7Rh_3	2.58	$D10_2$
$LaRhSi_2$	3.42	oC16 ($CeNiSi_2$)
$La_2Rh_3Si_5$	4.45	oI40 ($Co_3Si_5U_2$)
$LaRhSi_3$	2.7	tI10 ($BaNiSn_3$)
$LaRh_2Si_2$	3.90	tI10 (Al_4Ba)
$LaRu_2$	1.63	C15
La_3S_4	6.5	$D7_3$
La_3Se_4	8.6	$D7_3$
$LaSi_2$	2.3	C_c
La_xY_{1-x}	1.7—5.4	
LaZn	1.04	B2
$Li_2Mo_6S_8$	4.2	hR15
LiPb	7.2	
$LuOs_2$	3.49	C14
$Lu_{0.275}Rh_{0.725}$	1.27	C15
$LuRh_5$	0.49	
$Lu_5Rh_4Si_{10}$	3.95	tP38 ($Co_4So_5Si_{10}$)
$LuRu_2$	0.86	C14
$Mg_{1.14}Mo_{6.6}S_8$	3.5	hR15

TABLE 4
Superconductive Compounds and Alloys (continued)

Substance	T_c, K	Crystal structure type
Mg_2Nb	5.6	
$Mg_{-0.47}Tl_{-0.53}$	2.75	B2
$MgZn$	0.9	A3-oP4 (AuCd)
Mn_xTi_{1-x}	2.3 (max.)	Mn in -Ti
Mn_xTi_{1-x}	1.1—3.0	Mn in -Ti
MnU_6	2.32	$D2_c$
Mo_2N	5.0	F.C.C.
$Mo_6Na_2S_8$	8.6	hR15
Mo_xNb_{1-x}	0.016—9.2	
$Mo_{5.25}Nb_{0.75}Se_8$	6.2	hR15
Mo_6NdSa_8	8.2	hR15
Mo_3Os	7.2	A15
$Mo_{0.62}Cs_{0.38}$	5.65	$D8_b$
Mo_3P	5.31	DO_e
$Mo_6Pb_{1.2}Se_8$	6.75	hR15
$Mo_{0.5}Pd_{0.5}$	3.52	A3
Mo_6PrSe_8	9.2	hR15
$MoRe$	7.8	$D8_b$-tP30
$MoRe_3$	9.25; 9.89	A12
Mo_xRe_{1-x}	1.2—12.2	
$Mo_{0.42}Re_{0.58}$	6.35	$D8_b$
$MoRh$	1.97	A3
Mo_xRh_{1-x}	1.5—8.2	B.C.C.
$MoRu$	9.5—10.5	A3
$Mo_{0.61}Ru_{0.39}$	7.18	$D8_b$
$Mo_{0.2}Ru_{0.8}$	1.66	A3
Mo_3Ru_2	7.0	$D8_b$-tP30
$Mo_4Ru_2Te_8$	1.7	hR15
Mo_6S_8	1.85	hR15
Mo_6S_8Sc	3.6	hR15
$Mo_6S_8Sm_{1.2}$	2.9	hR15
Mo_6S_8Tb	2.0	hR15
Mo_6S_8Tl	8.7	hR15
$Mo_6S_8Tm_{1.2}$	2.1	hR15
$Mo_6S_8Y_{1.2}$	3.0	hR15
Mo_6S_8Yb	9.2	hR15
$Mo_{6.6}S_8Zn_{11}$	3.6	hR15
Mo_3Sb_4	2.1	
Mo_6Se_8	6.3	hR15
$Mo_6Se_8Sm_{1.2}$	6.8	hR15
$Mo_6Se_8Sn_{1.2}$	6.8	hR15
Mo_6Se_8Tb	5.7	hR15
Mo_3Se_3Tl	4.0	hP14
$Mo_6Se_8Tm_{1.2}$	6.3	hR15
Mo_6Se_8Yb	6.2	hR15
Mo_3Si	1.30	A15
$MoSi_{0.7}$	1.34	
Mo_xSiV_{3-x}	4.54—16.0	A15
$Mo_{5.25}Ta_{0.75}Te_8$	1.7	hR15
Mo_6Te_8	1.7	hR15
$Mo_{0.16}Ti_{0.84}$	4.18; 4.25	
$Mo_{0.913}Ti_{0.087}$	2.95	
$Mo_{0.04}Ti_{0.96}$	2.0	Cubic
$Mo_{0.025}Ti_{0.975}$	1.8	
Mo_xU_{1-x}	0.7—2.1	

TABLE 4
Superconductive Compounds and Alloys (continued)

Substance	T_c, K	Crystal structure type
Mo_xV_{1-x}	0—~5.3	
Mo_2Zr	4.25—4.75	C15
NNb (film)	6—9	B1
$N_xO_yTi_z$	2.9—5.6	Cubic
$N_xO_yV_z$	5.8—8.2	Cubic
$N_{0.34}Re$	4—5	F.C.C.
NTa (film)	4.84	B1
$N_{0.6—0.987}Ti$	<1.17—5.8	B1
$N_{0.82—0.99}V$	2.9—7.9	B1
NZr	9.8	B1
$N_{0.906—0.984}Zr$	3.0—9.5	B1
$Na_{0.28—0.35}O_3W$	0.56	Tetragonal
$Na_{0.28}Pb_{0.72}$	7.2	
NbO	1.25	
$NbOs_2$	2.52	A12
Nb_3Os	1.05	A15
$Nb_{0.6}Os_{0.4}$	1.89; 1.78	$D8_b$
$Nb_3Os_{0.02—0.10}Rh_{0.98—0.90}$	2.42—2.30	A15
Nb_3P	1.8	$L1_2$tP32 (Ti_3P)
NbPRh	4.08	C37-oP12 (Co_2Si)
$Nb_{0.6}Pd_{0.4}$	1.60	$D8_f$ plus cubic
$Nb_3Pd_{0.02—0.10}Rh_{0.92—0.90}$	2.49—2.55	A15
$Nb_{0.62}Pt_{0.38}$	4.21	$D8_b$
Nb_5Pt_3	3.73	$D8_b$
$Nb_3Pt_{0.02—0.98}Rh_{0.98—0.02}$	2.52—9.6	A15
$NbRe_3$	5.27	$D8_b$-tP30 (FeCr)
$Nb_{0.38—0.18}Re_{0.62—0.82}$	2.43—9.70	A15
NbRe	3.8	$D8_b$-tP30
NbReSi	5.1	oI36 (FeTiSi)
Nb_3Rh	2.64	A15
$Nb_{0.6}Rh_{0.40}$	4.21	$D8_b$ plus other
$Nb_{0.9}Rh_{1.1}$	3.07	A3-oP4 (AuCd)
$Nb_3Rh_{0.98—0.90}Ru_{0.02—0.10}$	2.42—2.44	A15
Nb_xRu_{1-x}	1.2—4.8	
NbRuSi	2.65	oI36
NbS_2	6.1—6.3	Hexagonal, $NbSe_2$ type
NbS_2	5.0—5.5	Hexagonal, three-layer type
Nb_3Sb	0.2	$L1_2$-tP32 (Ti_3P)
$Nb_3Sb_{0—0.7}Sn_{1—0.3}$	6.8—18	A15
$NbSe_2$	5.15—5.62	Hexagonal
$Nb_{1—1.05}Se_2$	2.2—7.0	Same
Nb_3Se_4	2.0	hP14
Nb_3Si	1.5	$L1_2$
Nb_3SiSnV_3	4.0	
$NbSn_2$	2.60	Orthorhombic
Nb_6Sn_5	2.8	oI44 (Sn_5Ti_6)
NbSnTaV	6.2	A15
$NbSnV_2$	5.5	A15
Nb_2SnV	9.8	A15
Nb_xTa_{1-x}	4.4—9.2	A2
Nb_3Te_4	1.8	hP14
Nb_xTi_{1-x}	0.6—9.8	
$Nb_{0.6}Ti_{0.4}$	9.8	
Nb_xU_{1-x}	1.95 (max.)	
$Nb_{0.88}V_{0.12}$	5.7	A2

TABLE 4
Superconductive Compounds and Alloys (continued)

Substance	T_c, K	Crystal structure type
$Nb_{0.5}V_{1.5}Zr$	4.3	C15-hP12 ($MgZn_2$)
$Ni_{0.3}Th_{0.7}$	1.98	$D10_2$
$NiZr_2$	1.52	
$Ni_{0.1}Zr_{0.9}$	1.5	A3
$O_3Rb_{0.27—0.29}W$	1.98	Hexagonal
OSn	3.81	tP4 (PbO)
O_3SrTi (n = $1.7—12.0 \times 10^{19}$)	0.12—0.37	
O_3SrTi (n = $10^{18}—10^{21}$)	0.05—0.47	
O_3SrTi (n = 10^{20})	0.47	
$O_3Sr_{0.08}W$	2—4	Hexagonal
OTi	0.58	
$O_3Tl_{0.30}W$	2.0—2.14	Hexagonal
OV_3Zr_3	7.5	$E9_3$
OW_3 (film)	3.35; 1.1	A15
OsPti	1.2	C22-hP9 (Fe_2P)
OsPZr	7.4	Same
OsReY	2.0	C14
Os_2Sc	4.6	C14
OsTa	1.95	A12
Os_3Th_7	1.51	$D10_2$
Os_xW_{1-x}	0.9—4.1	
OsW_3	~3	
Os_2Y	4.7	C14
Os_2Zr	3.0	C14
Os_xZr_{1-x}	1.5—5.6	
PPb	7.8	
OsW_2	3.81	$D8_b$-tP30 (FeCr)
$PPd_{3.0—3.2}$	<0.35—0.7	DO_{11}
P_3Pd_7 (high temperature)	1.0	Rhombohedral
P_3Pd_7 (low temperature)	0.70	Complex
PRh	1.22	
PRh_2	1.3	C1
P_4Rh_5	1.22	oP28 ($CaFe_2O_4$)
PRhTa	4.41	C37-oP12 (Co_2Si)
PRhZr	1.55	Same
PRuTi	1.3	C22-hP9 (Fe_2P)
PRuZr	3.46	C37-oP12
PW_3	2.26	DO_e
Pb_2Pd	2.95	C16
Pb_4Pt	2.80	Related to C16
Pb_2Rh	2.66	C16
PbSb	6.6	
PbTe (plus 0.1 w/o Pb)	5.19	
PbTe (plus 0.1 w/o Te)	5.24—5.27	
$PbTl_{0.27}$	6.43	
$PbTl_{0.17}$	6.73	
$PbTl_{0.12}$	6.88	
$PbTl_{0.075}$	6.98	
$PbTl_{0.04}$	7.06	
$Pb_{1—0.26}Tl_{0—0.74}$	7.20—3.68	
$PbTl_2$	3.75—4.1	
Pb_3Zr_5	4.60	$D8_8$
$PbZr_3$	0.76	A15
$Pd_{0.9}Pt_{0.1}Te_2$	1.65	C6
$Pd_{0.05}Ru_{0.05}Zr_{0.9}$	~9	

TABLE 4
Superconductive Compounds and Alloys (continued)

Substance	T_c, K	Crystal structure type
$Pd_{2.2}S$ (quenched)	1.63	Cubic
$PdSb_2$	1.25	C2
PdSb	1.5	$B8_1$
PdSbSe	1.0	C2
PdSbTe	1.2	C2
Pd_4Se	0.42	Tetragonal
$Pd_{6-7}Se$	0.66	Like Pd_4Te
$Pd_{2.8}Se$	2.3	
Pd_xSe_{1-x}	2.5 (max.)	
PdSi	0.93	B31
PdSn	0.41	B31
$PdSn_2$	3.34	
Pd_2Sn	0.41	C37
Pd_3Sn	0.47—0.64	$B8_2$
Pd_2SnTm	1.77	DO_3-cF16 (BiF_3)
Pd_2SnY	4.92	Same
Pd_2SnYb	1.79	Same
PdTe	2.3; 3.85	$B8_1$
$PdTe_{1.02-1.08}$	2.56—1.88	$B8_1$
$PdTe_2$	1.69	C6
$PdTe_{2.1}$	1.89	C6
$PdTe_{2.3}$	1.85	C6
$Pd_{1.1}Te$	4.07	$B8_1$
Pd_3Te	0.76	cI2 (W)
$PdTh_2$	0.85	C16
$Pd_{0.1}Zr_{0.9}$	7.5	A3
PtSb	2.1	$B8_1$
PtSi	0.88	B31
PtSn	0.37	$B8_1$
$PtSn_4$	2.38	C16-oC20 ($PdSn_4$)
Pt_3Ta_7	1.5	$D8_b$-tP30
$PtTa_3$	0.4	A15-cP8 (Cr_3Si)
PtTe	0.59	Orthorhombic
PtTh	0.44	B_f
Pt_3Th_7	0.98	$D10_2$
Pt_5Th	3.13	
$PtTi_3$	0.58	A15
$Pt_{0.02}U_{0.98}$	0.87	β-phase
$PtV_{2.5}$	1.36	A15
PtV_3	2.87—3.20	A15
$PtV_{3.5}$	1.26	A15
$Pt_{0.5}W_{0.5}$	1.45	A1
Pt_xW_{1-x}	0.4—2.7	
Pt_2Y_3	0.90	
Pt_2Y	1.57; 1.70	C15
Pt_3Y_7	0.82	$D10_2$
PtZr	3.0	A3
Re_2Sc	4.2	C15-hP12 ($MgZn_2$)
$Re_{24}Sc_5$	2.2	A12-cI58 (Mg)
ReSiTa	4.4	oI36 (FeTiSi)
$Re_3Si_5Y_2$	1.76	tP40 ($Fe_3Sc_2Si_5$)
Re_3Ta_2	1.4	$D8_b$-tP30 (FeCr)
$Re_{0.64}Ta_{0.36}$	1.46	A12
Re_3Ta	6.78	A12-cI58 (Mn)
$Re_{24}Ti_5$	6.60	A12

TABLE 4
Superconductive Compounds and Alloys (continued)

Substance	T_c, K	Crystal structue type
Re_xTi_{1-x}	6.6 (max.)	
$Re_{0.76}V_{0.24}$	4.52	$D8_b$
Re_3V	6.26	$D8_b$-tP30
$Re_{0.92}V_{0.08}$	6.8	A3
$Re_{0.6}W_{0.4}$	6.0	
$Re_{0.5}W_{0.5}$	5.12	$D8_b$
$Re_{13}W_{12}$	5.2	$D8_b$-tP30
Re_3W	9.0	A12-cI58
Re_2Y	1.83	C14
Re_2Zr	5.9	C14
Re_3Zr	7.40	A12-cI58
Re_6Zr	7.40	Same
$Rh_{17}S_{15}$	5.8	Cubic
$Rh_{-0.24}Sc_{-0.76}$	0.88; 0.92	
$Rh_4Sc_5Si_{10}$	8.54	tP38
$Rh_4Sc_3Sn_{13}$	4.5	cP40
Rh_xSe_{1-x}	6.0 (max.)	
$RhSi_3Th$	1.76	tI10
$Rh_{0.86}Sc_{1.04}Th$	6.45	tI12
Rh_2Si_2Y	3.11	tI10
$Rh_3Si_5Y_2$	2.70	oI40
$Rh_4Sn_{13}Sr_3$	4.3	cP40
Rh_xSn_yTh	1.9	cI2 (W)
Rh_xSn_yTm	2.3	cP40
$Rh_4Sn_{13}Y_3$	3.2	cP40
Rh_2Sr	6.2	C15
$Rh_{0.4}Ta_{0.6}$	2.35	$D8_b$
$RhTe_2$	1.51	C2
$Rh_{0.67}Te_{0.33}$	0.49	
Rh_xTe_{1-x}	1.51 (max.)	
$RhTh$	0.36	B_f
Rh_3Th_7	2.15	$D10_2$
Rh_5Th	1.07	
Rh_xTi_{1-x}	2.25—3.95	
$Rh_{0.02}U_{0.98}$	0.96	
RhV_3	0.38	A15
RhW	~3.4	A3
RhY_3	0.65	
Rh_2Y_3	1.48	
Rh_3Y	1.07	C15
Rh_5Y	0.56	
Rh_3Y_7	0.32	hP20 (Fe_3Th_7)
$Rh_{0.005}Zr$ (annealed)	5.8	
$Rh_{0—0.45}Zr_{1—0.55}$	2.1—10.8	
$Rh_{0.1}Zr_{0.9}$	9.0	H.C.P.
Ru_2Sc	1.67	C14
$RuSiTa$	3.15	oI36
Ru_3Si_2Th	3.98	hP12
Ru_3Si_2Y	3.51	hP12
$Ru_{1.1}Sn_{3.1}Y$	1.3	cP40
Ru_2Th	3.56	C15
$RuTi$	1.07	B2
$Ru_{0.05}Ti_{0.95}$	2.5	
$Ru_{0.1}Ti_{0.9}$	3.5	
$Ru_xTi_{0.6}V_y$	6.6 (max.)	

TABLE 4
Superconductive Compounds and Alloys (continued)

Substance	T_c, K	Crystal structue type
Ru_3U	0.15	$L1_2$-cP4
$Ru_{0.45}V_{0.55}$	4.0	B2
RuW	7.5	A3
Ru_2Y	1.52	C14
Ru_2Zr	1.84	C14
$Ru_{0.1}Zr_{0.9}$	5.7	A3
STh	0.5	B1-cF8 (NaCl)
SbSn	1.30—1.42	B1 or distorted
$SbTa_3$	0.72	A15-cP8 (Cr_3Si)
$SbTi_3$	5.8	Same
Sb_2Ti_7	5.2	
$Sb_{0.01—0.03}V_{0.99—0.97}$	3.76—2.63	A2
SbV_3	0.80	A15
SeTh	1.7	B1-cF8
$SiMo_3$	1.4	A15-cP8
Si_2Th	3.2	C_c, α-phase
Si_2Th	2.4	C32, β-phase
$SiV_{2.7}Ru_{0.3}$	2.9	A15
Si_2W_3	2.8; 2.84	
$SiZr_3$	0.5	$L1_2$-tP32 (Ti_3P)
$Sn_{0.174—0.104}Ta_{0.826—0.896}$	6.5—< 4.2	A15
$SnTa_3$	8.35	A15, highly ordered
$SnTa_3$	6.2	A15, partially ordered
$SnTaV_2$	2.8	A15
$SnTa_2V$	3.7	A15
Sn_xTe_{1-x} (n = 10.5—20 × 10^{20})	0.07—0.22	B1
Sn_3Th	3.33	$L1_2$-cP4
$SnTi_3$	5.80	A15-cP8
Sn_xTl_{1-x}	2.37—5.2	
SnV_3	3.8	A15
$Sn_{0.02—0.057}V_{0.98—0.943}$	2.87—~1.6	A2
$SnZr_3$	0.92	A15-cP8
$Ta_{0.025}Ti_{0.975}$	1.3	Hexagonal
$Ta_{0.05}Ti_{0.95}$	2.9	Hexagonal
$Ta_{0.05—0.75}V_{0.95—0.25}$	4.30—2.65	A2
$Ta_{0.8—1}W_{0.2—0}$	1.2—4.4	A2
$Tc_{0.1—0.4}W_{0.9—0.6}$	1.25—7.18	Cubic
$Tc_{0.50}W_{0.50}$	7.52	α plus
$Tc_{0.60}W_{0.40}$	7.88	plus α
Tc_6Zr	9.7	A12
TeY	1.02	B1-cF8
$ThTl_3$	0.87	$L1_2$-cP4
$Th_{0—0.55}Y_{1—0.45}$	1.2—1.8	
$Ti_{0.70}V_{0.30}$	6.14	Cubic
Ti_xV_{1-x}	0.2—7.5	
$Ti_{0.5}Zr_{0.5}$ (annealed)	1.23	
$Ti_{0.5}Zr_{0.5}$ (quenched)	2.0	
Tl_3Y	1.52	$L1_2$-cP4
V_2Zr	8.80	C15
$V_{0.26}Zr_{0.74}$	5.9	
W_2Zr	2.16	C15
YZn	0.33	B2-cP2 (CsCl)

* n denotes current carriers concentration in cm^{-3}.

TABLE 4
Superconductive Compounds and Alloys (continued)

B. SUPERCONDUCTORS WITH $T_c > 10K$

Substance	T_c,K	Crystal structure type	
Al_2CMo_3	10.0	A13	
$Al_{0.5}Ge_{0.5}Nb$	12.6	A15	
$Al_{-0.8}Ge_{-0.2}Nb_3$	20.7	A15	
$AlNb_3$	18.0	A15	(Cr_3Si)
$AlNb_3$	12.0		$(FeCr)$
Al_xNb_{1-x}	<4.2—13.5	$D8_b$	
Al_xNb_{1-x}	12—17.5	A15	
$Al_{0.27}Nb_{0.73-0.48}V_{0-0.25}$	14.5—17.5	A15	
$Al\ Nb_xV_{1-x}$	4.4—13.5		
$Al_{0.1}Si_{0.9}V_3$	14.05		
AlV_3	11.8	A15	(Cr_3Si)
$AuNb_3$	11.5	A15	
$Au_{0-0.3}Nb_{1-0.7}$	1.1—11.0		
$Au_{0.02-0.98}Nb_3Rh_{0.98-0.02}$	2.53—10.9	A15	
$AuNb_{3(1-x)}V_{3x}$	1.5—11.0	A15	
$B_{0.03}C_{0.51}Mo_{0.47}$	12.5		
B_4LuRh_4	11.7		(B_4CeCo_4)
B_2LuRu	10		
B_4Rh_4Y	11.3		(B_4CeCo_4)
$B_{0.1}Si_{0.9}V_3$	15.8	A15	
$BaBi_{0.2}O_3Pb_{0.8}$	13.2		
$Ba_2CaCu_2O_8Tl_2$	120		
$Ba_2Cu_3LaO_6$	80		
$Ba_2Cu_3O_7Tm$	101		
$Ba_2Cu_3O_7Y$	90		
$(Ba,La)_2CuO_4$	36	A15	(K_2NiF_4)
$Bi_2CaCu_2O_8Sr_2$	110		
$Br_2Mo_6S_6$	13.8		(Mo_6PbS_8)
C_3La	11.0		(C_3Pu_2)
CMo	14.3	B1	$(NaCl)$
CMo_2	12.2	o**	
$C_{0.5}Mo_xNb_{1-x}$	10.8—12.5	B1	
CMo_xTi_{1-x}	10.2(max)	B1	
$CMo_{0.83}Ti_{0.17}$	10.2	B1	
$C_{0-0.38}N_{1-0.62}Ta$	10.0—11.3		
CNb (whiskers)	7.5—10.5		
CNb	11.5	B1	
$C_{0.7-1.0}Nb_{0.3-0}$	6—11	B1	
CNb_xTa_{1-x}	8.2—13.9		
$CNb_{0.6-0.9}W_{0.4-0.1}$	12.5—11.6	B1	
$C_{0.1}Si_{0.9}V_3$	16.4	A15	
CTa	10.3	B1	
$CTa_{1-0.4}W_{0-0.6}$	8.5—10.5	B1	
$C_{0.66}Th_{0.13}Y_{0.21}$	17		(C_3Pu_2)
C_3Y_2	11.5		(C_3Pu_2)
CW	10	B1	
$(Ca,La)_2CuO_4$	18		(K_2NiF_4)
$Cu(La,Sr)_2O_4$	39		
$Cu_{1.8}Mo_6S_8$	10.8		(Mo_6PbS_8)
$Cr_{0.3}SiV_{2.7}$	11.3	A15	
$GaNb_3$	14.5	A15	(Cr_3Si)
$Ga_xNb_3Sn_{1-x}$	14—18.37	A15	

TABLE 4
Superconductive Compounds and Alloys (continued)

Substance	T_cK	Crystal structure type	
GaV_3	16.8	A15	
$GaV_{2.1-3.5}$	6.3—14.45	A15	
$GeNb_3$	23.2	A15	
$GeNb_3$ (quenched)	6—17	A15	
$Ge_xNb_3Sn_{1-x}$	17.6—18.0	A15	
$Ge_{0.5}Nb_3Sn_{0.5}$	11.3		
$Ge_{0.1}Si_{0.9}V_3$	14.0	A15	
GeV_3	11	A15	
$InLa_3$	9.83; 10.4	LI_2	$(AuCu_3)$
$InLa_3$ (0—35 kbar)	9.75—10.55		
$In_{0-0.3}Nb_3Sn_{1-0.7}$	18.0—18.19	A15	
InV_3	13.9	A15	
$Ir_{0.4}Nb_{0.6}$	10		(FeCr)
$LaMo_6Se_8$	11.4		(Mo_6PbS_8)
LiO_4Ti_2	13.7		(Al_2MgO_4)
MoN	12; 14.8	h*	
Mo_3Os	12.7	A15	
$Mo_6Pb_{0.9}S_{7.5}$	15.2		(Mo_6PbS_8)
Mo_3Re	10.0; 15	A15	
Mo_xRe_{1-x}	1.2—12.2		
$Mo_{0.52}Re_{0.48}$	11.1		
$Mo_{0.57}Re_{0.43}$	14.0		
$Mo_{-0.60}Re_{0.395}$	10.6		
MoRu	9.5—10.5	A3	
Mo_3Ru	10.6	A15	
Mo_6Se_8Tl	12.2		(Mo_6PbS_8)
$Mo_{0.3}SiV_{2.7}$	11.7	A15	
Mn_3Si	12.5	A15	
Mo_3Tc	15	A15	
$Mo_{0.3}Tc_{0.7}$	12.0	A15	
Mo_xTc_{1-x}	10.8—15.8		
$MoTc_3$	15.8		
NNb (whiskers)	10—14.5		
NNb (diffusion wires)	16.10		
$N_{0.988}Nb$	14.9; 17.3	B1	
$N_{0.824-0.988}Nb$	14.4—15.3	B1	
$N_{0.7-0.795}Nb$	11.3—12.9		
NNb_xO_y	13.5—17.0	B1	
NNb_xO_y	6.0—11		
$N_{100-42w/o}Nb_{0-58w/o}Ti$	15—16.8		
$N_{100-75w/o}Nb_{0-25w/o}Zr$	12.5—16.35		
NNb_xZr_{1-x}	9.8—13.8	B1	
$N_{0.93}Nb_{0.85}Zr_{0.15}$	13.8	B1	
NTa	12—14	B1	
NZr	10.7	B1	
Nb_3Pt	10.9	A15	
$Nb_{0.18}Re_{0.82}$	10		(Mn)
Nb_3Si	19	A15	
$Nb_{0.3}SiV_{2.7}$	12.8	A15	
Nb_3Sn	18.05	A15	
$Nb_{0.8}Sn_{0.2}$	18.18; 18.5	A15	
Nb_xSn_{1-x} (film)	2.6—18.5	o*	
Nb_3Sn_2	16.6	t*	
$NbSnTa_2$	10.8	A15	
Nb_2SnTa	16.4	A15	

TABLE 4
Superconductive Compounds and Alloys (continued)

Substance	T_c,K	Crystal structure type
$Nb_{2.5}SnTa_{0.5}$	17.6	A15
$Nb_{2.75}SnTa_{0.25}$	17.8	A15
$Nb_{3x}SnTa_{3(1-x)}$	6.0—18.0	
$Nb_2SnTa_{0.5}V_{0.5}$	12.2	A15
$NbTc_3$	10.5	A12
$Nb_{0.75}Zr_{0.25}$	10.8	
$Nb_{0.66}Zr_{0.33}$	10.8	
$PbTa_3$	17	A15
$RhTa_3$	10	A15
$RhZr_2$	10.8; 11.3	C16 (Al$_2$Cu)
$Rh_{0—0.45}Zr_{1—0.55}$	2.1—10.8	
$SiTi_{0.3}V_{2.7}$	10.9	A15
SiV_3	17.1	A15
$SiV_{2.7}Zr_{0.3}$	13.2	A15

TABLE 5
Critical Field Data

Substance	H_o oersteds	Substance	H_o oersteds
Ag_2F	2.5	InSb	1100
Ag_7NO_{11}	57	In_xTl_{1-x}	252—284
Al_2CMo_3	1700	$In_{0.8}Tl_{0.2}$	252
$BaBi_3$	740	$Mg_{0.47}Tl_{0.53}$	220
Bi_2Pt	10	$Mo_{0.16}Ti_{0.84}$	<985
Bi_3Sr	530	$NbSn_2$	620
Bi_5Tl_3	>400	$PbTl_{0.27}$	756
CdSn	>266	$PbTl_{0.17}$	796
$CoSi_2$	105	$PbTl_{0.12}$	849
$Cr_{0.1}Ti_{0.3}V_{0.6}$	1360	$PbTl_{0.075}$	880
$In_{1-0.86}Mg_{0-0.14}$	272.4—259.2	$PbTl_{0.04}$	864

TABLE 6
High Critical Magnetic-Field Superconductive Compounds and Alloys

Substance	T_c, K	H_{c1}, kOe	H_{c2}, kOe	H_{c3}, kOe	T_{obs}, K[a]
Al_2CMo_3	9.8—10.2	0.091	156		1.2
$AlNb_3$		0.375			
$Ba_xO_3Sr_{1-x}Ti$	<0.1—0.55	0.0039 max.			
$Bi_{0.5}Cd_{0.1}Pb_{0.27}Sn_{0.13}$			>24		3.06
Bi_xPb_{1-x}	7.35—8.4	0.122 max.	30 max.		4.2
$Bi_{0.56}Pb_{0.44}$	8.8		15		4.2
$Bi_{7.5w/o}Pb_{92.5w/o}$[b]			2.32		
$Bi_{0.099}Pb_{0.901}$		0.29	2.8		
$Bi_{0.02}Pb_{0.98}$		0.46	0.73		
$Bi_{0.53}Pb_{0.32}Sn_{0.16}$			>25		3.06
$Bi_{1-0.93}Sn_{0-0.07}$			0—0.032		3.7
Bi_5Tl_3	6.4		>5.6		3.35
C_8K (excess K)	0.55		0.160 (H⊥c)		0.32
			0.730 (H∥c)		0.32

TABLE 6
High Critical Magnetic-Field Superconductive Compounds and Alloys (continued)

Substance	T_c, K	H_{c1}, kOe	H_{c2}, kOe	H_{c3}, kOe	T_{obs}, K[a]
C_8K	0.39		0.025 (H⊥c)		0.32
			0.250 (H∥c)		0.32
$C_{0.44}Mo_{0.56}$	12.5—13.5	0.087	98.5		1.2
CNb	8—10	0.12	16.9		4.2
$CNb_{0.4}Ta_{0.6}$	10—13.6	0.19	14.1		1.2
CTa	9—11.4	0.22	4.6		1.2
$Ca_xO_3Sr_{1-x}Ti$	<0.1—0.55	0.002—0.004			
$Cd_{0.1}Hg_{0.9}$ (by weight)		0.23	0.34		2.04
$Cd_{0.05}Hg_{0.95}$		0.28	0.31		2.16
$Cr_{0.10}Ti_{0.30}V_{0.60}$	5.6	0.071	84.4		0
GaN	5.85	0.725			4.2
Ga_xNb_{1-x}			>28		4.2
$GaSb$ (annealed)	4.24		2.64		3.5
$GaV_{1.95}$	5.3		73e		
$GaV_{2.1-3.5}$	6.3—14.45		230—300d		0
GaV_3		0.4	350e		0
			500d		
$GaV_{4.5}$	9.15		121c		0
Hf_xNb_y			>52—>102		1.2
Hf_xTa_y			>28—>86		1.2
$Hg_{0.05}Pb_{0.95}$		0.235	2.3		
$Hg_{0.101}Pb_{0.899}$		0.23	4.3		4.2
$Hg_{0.15}Pb_{0.85}$	6.75		>13		2.93
$In_{0.98}Pb_{0.02}$	3.45	0.1		0.12	2.76
$In_{0.96}Pb_{0.04}$	3.68	0.1	0.12	0.25	2.94
$In_{0.94}Pb_{0.06}$	3.90	0.095	0.18	0.35	3.12
$In_{0.913}Pb_{0.087}$	4.2	~10.17	0.55	2.65	
$In_{0.316}Pb_{0.684}$		0.155	3.7		4.2
$In_{0.17}Pb_{0.83}$			2.8	5.5	4.2
$In_{1.000}Te_{1.002}$	3.5—3.7		1.2c		0
$In_{0.95}Tl_{0.05}$		0.263	0.263		3.3
$In_{0.90}Tl_{0.10}$		0.257	0.257		3.25
$In_{0.83}Tl_{0.17}$		0.242	0.39		3.21
$In_{0.75}Tl_{0.25}$		0.216	0.50		3.16
LaN	1.35	0.45			0.76
La_3S_4	6.5	≈0.15	>25		1.3
La_3Se_4	8.6	≈0.2	>25		1.25
$Mo_{0.52}Re_{0.48}$	11.1		14—21	22—33	4.2
			18—28	37—43	1.3
$Mo_{0.6}Re_{0.395}$	10.6		14—20	20—37	4.2
			19—26	26—37	1.3
$Mo_{0.5}Ti_{0.5}$			75c		0
$Mo_{0.16}Ti_{0.84}$	4.18	0.028	98.7c		0
			36—38		3.0
$Mo_{0.913}Ti_{0.087}$	2.95	0.060	15		4.2
$Mo_{0.1-0.3}U_{0.9-0.7}$	1.85—2.06		>25		
$Mo_{0.17}Zr_{0.83}$			30		
$N_{(12.8 w/o)}Nb$	15.2		>9.5		13.2
NNb (wires)	16.1		153c		0
			132		4.2
			95		8
			53		12
NNb_xO_{1-x}	13.5—17.0		38		

TABLE 6
High Critical Magnetic-Field Superconductive Compounds and Alloys (continued)

Substance	T_c, K	H_{c1}, kOe	H_{c2}, kOe	H_{c3}, kOe	T_{obs}, K[a]
NNb_xZr_{1-x}	9.8—13.8		4- >130		4.2
$N_{0.93}Nb_{0.85}Zr_{0.15}$	13.8		>130		4.2
$Na_{0.086}Pb_{0.914}$		0.19	6.0		
$Na_{0.016}Pb_{0.984}$		0.28	2.05		
Nb	9.15		2.020		1.4
			1.710		4.2
Nb		0.4—1.1	3—5.5		4.2
Nb (unstrained)		1.1—1.8	3.40	6—9.1	4.2
Nb (strained)		1.25—1.92	3.44	6.0—8.7	4.2
Nb (cold-drawn wire)		2.48	4.10	≈10	4.2
Nb (film)			>25		4.2
NbSc			>30		
Nb_3Sn		0.170	221		4.2
			70		14.15
			54		15
			34		16
			17		17
$Nb_{0.1}Ta_{0.9}$		0.084	0.154		4.195
$Nb_{0.2}Ta_{0.8}$			10		4.2
$Nb_{0.65-0.73}Ta_{0.02-0.10}Zr_{0.25}$			>70—>90		4.2
Nb_xTi_{1-x}			148 max.		1.2
			120 max.		4.2
$Nb_{0.222}U_{0.778}$		1.98	23		1.2
Nb_xZr_{1-x}			127 max.		1.2
			94 max.		4.2
O_3SrTi	0.43	0.0049[c]	0.504[c]		0
O_3SrTi	0.33	0.00195[c]	0.420[c]		0
$PbSb_{1\,w/o}$(quenched)			>1.5		4.2
$PbSb_{1\,w/o}$(annealed)			>0.7		4.2
$PbSb_{2.8\,w/o}$(quenched)			>2.3		4.2
$PbSb_{2.8\,w/o}$(annealed)			>0.7		4.2
$Pb_{0.871}Sn_{0.129}$		0.45	1.1		
$Pb_{0.965}Sn_{0.035}$		0.53	0.56		
$Pb_{1-0.26}Tl_{0-0.74}$	7.20—3.68		2—6.9[c]		0
$PbTl_{0.17}$	6.73		4.5[c]		0
$Re_{0.26}W_{0.74}$			>30		
$Sb_{0.93}Sn_{0.07}$			0.12		3.7
SiV_3	17.0	0.55	156[e]		
Sn_xTe_{1-x}		0.00043—0.00236	0.005—0.0775		0.012—0.079
Ta (99.95%)		0.425	1.850		1.3
		0.325	1.425		2.27
		0.275	1.175		2.66
		0.090	0.375		3.72
$Ta_{0.5}Nb_{0.5}$			3.55		4.2
$Ta_{0.65-0}Ti_{0.35-1}$	4.4—7.8		>14—138		1.2
$Ta_{0.5}Ti_{0.5}$			138		1.2
Te	3.3	0.25[c]			0
Tc_xW_{1-x}	5.75—7.88		8—44		4.2
Ti				2.7	4.2
$Ti_{0.75}V_{0.25}$	5.3	0.029[c]	199[c]		0
$Ti_{0.775}V_{0.225}$	4.7	0.024[c]	172[c]		0
$Ti_{0.615}V_{0.385}$	7.07	0.050	34		4.2
$Ti_{0.516}V_{0.484}$	7.20	0.062	28		4.2
$Ti_{0.415}V_{0.585}$	7.49	0.078	25		4.2

TABLE 6
High Critical Magnetic-Field Superconductive Compounds and Alloys (continued)

Substance	T_c, K	H_{c1}, kOe	H_{c2}, kOe	H_{c3}, kOe	T_{obs}, K[a]
$Ti_{0.12}V_{0.88}$			17.3	28.1	4.2
$Ti_{0.09}V_{0.91}$			14.3	16.4	4.2
$Ti_{0.06}V_{0.94}$			8.2	12.7	4.2
$Ti_{0.03}V_{0.97}$			3.8	6.8	4.2
Ti_xV_{1-x}			108 max.		1.2
V	5.31	0.8	3.4		1.79
		0.75	3.15		2
		0.45	2.2		3
		0.30	1.2		4
$V_{0.26}Zr_{0.74}$	≈5.9	0.238			1.05
		0.227			1.78
		0.185			3.04
		0.165			3.5
W (film)	1.7—4.1		>34		1

[a] Temperature of critical field measurement.
[b] w/o denotes weight percent.
[c] Extrapolated.
[d] Linear extrapolation.
[e] Parabolic extrapolation.

REFERENCES

1. B. W. Roberts, in *Superconductive Materials and Some of Their Properties. Progress in Cryogenics*, Vol. IV, 1964, pp. 160—231.
2. B. W. Roberts, Superconductive Materials and Some of Their Properties, NBS Technical Notes 408 and 482, U.S. Government Printing Office, 1966 and 1969; B. W. Roberts, *J. Phys. Chem. Ref. Data*, 5, 581, 1976.
3. B. W. Roberts, Properties of Selected Superconductive Materials, 1978 Supplement, NBS Technical Note 983, 1978.
4. T. Claeson, *Phys. Rev.*, 147, 340, 1966.
5. C. J. Raub, W. H. Zachariasen, T. H. Geballe, and B. T. Matthias, *J. Phys. Chem. Solids*, 24, 1093, 1963.
6. T. H. Geballe, B. T. Matthias, V. B. Compton, E. Corenzwit, G. W. Hull, Jr., and L. D. Longinotti, *Phys. Rev.*, 1A, 119, 1965.
7. C. J. Raub, V. B. Compton, T. H. Geballe, B. T. Matthias, J. P. Maita, and G. W. Hull, Jr., *J. Phys. Chem. Solids*, 26, 2051, 1965.
8. R. D. Blaugher, J. K. Hulm, and P. N. Yocom, *J. Phys. Chem. Solids*, 26, 2037, 1965.
9. T. Claeson and H. L. Luo, *J. Phys. Chem. Solids*, 27, 1081, 1966.
10. S. C. Ng and B. N. Brockhouse, *Solid State Comm.*, 5, 79, 1967.
11. O. I. Shulishova and I. A. Shcherbak, *Izv. AN SSSR, Neorg. Materials*, 3, 1495, 1967.
12. T. F. Smith and H. L. Luo, *J. Phys. Chem. Solids*, 28, 569, 1967.
13. A. C. Lawson, *J. Less–Common Metals*, 23, 103, 1971.
14. R. Chevrel, M. Sergent, and J. Prigent, *J. Solid State Chem.*, 3, 515, 1971.
15. M. Marezio, P. D. Dernier, J. P. Remeika, and B. T. Matthias, *Mat. Res. Bull.*, 8, 657, 1973.
16. J. K. Hulm and R. D., *Blaugher in Superconductivity in d– and f–Band Metals*, D. H. Douglass, Ed., American Institute of Physics, 4, 1, 1972.
17. R. N. Shelton, A. C. Lawson, and D. C. Johnston, *Mat. Res. Bull.*, 10, 297, 1975.
18. H. D. Wiesinger, *Phys. Status Sol.*, 41A, 465, 1977.
19. O. Fisher, *Applied Phys.*, 16, 1, 1978.
20. D. C. Johnston, *Solid State Comm.*, 24, 699, 1977.
21. H. C. Ku and R. H. Shelton, *Mat. Res. Bull.*, 15, 1441, 1980.
22. H. Barz, *Mat. Res. Bull.*, 15, 1489, 1980.
23. G. P. Espinosa, A. S. Cooper, H. Barz, and J. P. Remeika, *Mat. Res. Bull.*, 15, 1635, 1980.
24. E. M. Savitskii, V. V. Baron, Yu. V. Efimov, M. I. Bychkova, and L. F. Myzenkova, in *Superconducting Materials*, Plenum Press, 1981, p. 107.
25. R. Fluckiger and R. Baillif, in Topics in *Current Physics*, O. Fischer and M. B. Maple, Eds., Springer Verlag, 34, 113, 1982.
26. R. N. Shelton, in *Superconductivity in d– and f–Band Metals*, W. Buckel and W. Weber, Eds., Kernforschungszentrum, Karlsruhe, 1982, p. 123.
27. D. C. Johnston and H. F. Braun, *Topics in Current Phys.*, 32, 11, 1982.
28. R. Chevrel and M. Sergent, *Topics in Current Phys.*, 32, 25, 1982.

29. G. P. Espinosa, A. S. Cooper, and H. Barz, *Mat. Res. Bull.*, 17, 963, 1982.
30. R. Muller, R. N. Shelton, J. W. Richardson, Jr., and R. A. Jacobson, *J. Less–Comm. Met.*, 92, 177, 1983.
31. You–Xian Zhao and Shou–An He, in *High Pressure in Science and Technology*, North Holland, 22, 51, 1983.
32. You–Xian Zhao and Shou–An He, *Solid State Comm.*, 24, 699, 1983.
33. G. P. Meisner and H. C. Ku, *Appl. Phys.*, A31, 201, 1983.
34. R. J. Cava, D. W. Murphy, and S. M. Zahurak, *J. Electrochem. Soc.*, 130, 2345, 1983.
35. R. N. Shelton, *J. Less–Comm. Met.*, 94, 69, 1983.
36. B. Chevalier, P. Lejay, B. Lloret, Wang Xian–Zhong, J. Etourneau, and P. Hagenmuller, *Annales de Chemie*, 9, 191, 1984.
37. G. Venturini, M. Meot–Meyer, E. McRae, J. F. Mareche, and B. Rogues, *Mat. Res. Bull.*, 19, 1647, 1984.
38. J. M. Tarascon, F. G. DiSalvo, D. W. Murphy, G. Hull, and J. V. Waszczak, *Phys. Rev.*, 29B, 172, 1984.
39. G. V. Subba and G. Balakrishnan, *Bull. Mat. Sci.*, 6, 283, 1984.
40. B. Batlog, *Physica*, 126B, 275, 1984.
41. M. J. Johnson, Ames Lab (USA) Report IS-T-1140, 1984.
42. I. M. Chapnik, *J. Mat. Sci. Lett.*, 4, 370, 1985.
43. W. Rong–Yao, L. Qi–Guang, and Z. Xiao, *Phys. Status Sol.*, 90A, 763, 1985.
44. W. Xian–Zhong, B. Chevalier, J. Etourneau, and P. Hagenmuller, *Mat. Res. Bull.*, 20, 517, 1985.
45. H. R. Ott, F. Hulliger, H. Rudigier, and Z. Fisk, *Phys. Rev.*, 31B, 1329, 1985.
46. P. Villars and L. D. Calver, *Pearson's Handbook of Crystallographic Data for Intermetallic Phases*, Vol. 1—3, ASM, 1985.
47. G. V. Subba Rao, K. Wagner, G. Balakhrishnan, J. Jakani, W. Paulus, and R. Scollhorn, *Bull. Mat. Sci.*, 7, 215, 1985.
48. J. G. Bednorz and K. A. Muller, *Zs. Physik*, B64, 189, 1986.
49. W. Rong–Yao, *Phys. Status Sol.*, 94A, 445, 1986.
50. H. D. Yang, R. N. Shelton, and H. F. Braun, *Phys. Rev.*, 33B, 5062, 1986.
51. G. Venturini, M. Kanta, E. McRae, J. F. Mareche, B. Malaman, and B. Roques, *Mat. Res. Bull.*, 21, 1203, 1986.
52. W. Rong–Yao, *J. Mat. Sci. Lett.*, 5, 87, 1986.
53. M. K. Wu, J. R. Ashburn, C. J. Torng, P. H. Hor, R. L. Meng, L. Gao, Z. J. Huang, Y. Q. Wang, and C. W. Chu, *Phys. Rev. Lett.*, 58, 908, 1987.
54. R. J. Cava, R. B. Van Dover, B. Batlog, and E. A. Rietman, *Phys. Rev. Lett.*, 58, 408, 1987.
55. L. C. Porter, T. J. Thorn, U. Geiser, A. Umezawa, H. H. Wang, W. K. Kwok, H–C. I. Kao, M. R. Monaghan, G. W. Crabtree, K. D. Carlson, and J. M. Williams, *Inorg. Chem.*, 26, 1645, 1987.
56. A. M. Kini, U. Geiser, H–C. I. Kao, K. D. Carlson, H. H. Wang, M. R. Monaghan, and K. M. Williams, *Inorg. Chem.*, 26, 1834, 1987.
57. T. Penney, S. von Molnar, D. Kaiser, F. Holtzberg, and A. W. Kleinsasser, *Phys. Rev.*, B38, 2918, 1988.
58. Y. K. Tao, J. S. Swinnea, A. Manthiram, J. S. Kim, J. B. Goodenoug, and H. Steinfink, *J. Mat. Res.*, 3, 248, 1988.
59. G. G. Peterson, B. R. Weinberger, L. Lynds, and H. A. Krasinski, *J. Mat. Res.*, 3, 605, 1988.
60. J. B. Torrance, Y. Tokura, A. Nazzai, and S. S. P. Parkin, *Phys. Rev. Lett.*, 60, 542, 1988.
61. K. Kourtakis, M. Robbins, P. K. Gallagher, and T. Teifel, *J. Mat. Res.*, 4, 1289, 1989.
62. J. C. Phillips, *Physics of High-T$_c$ Superconductors*, Academic Press, 1989, p. 336.
63. Shui Wai Lin and L. I. Berger, *Rev. Sci. Instrum.*, 60, 507, 1989.

HIGH TEMPERATURE SUPERCONDUCTORS
C. N. R. Rao and A. K. Raychaudhuri

The following tables give properties of a number of high temperature superconductors. Table 1 lists the crystal structure (space group and lattice constants) and the critical transition temperature T_c for the more important high temperature superconductors so far studied. Table 2 gives energy gap, critical current density, and penetration depth in the superconducting state. Table 3 gives electrical and thermal properties of some of these materials in the normal state. The tables were prepared in November 1992 and updated in November 1994.

REFERENCES

1. Ginsburg, D.M., Ed., *Physical Properties of High-Temperature Superconductors*, Vols. I—III, World Scientific, Singapore, 1989—1992.
2. Rao, C.N.R., Ed., *Chemistry of High-Temperature Superconductors*, World Scientific, Singapore, 1991.
3. Shackelford, J.F., *The CRC Materials Science and Engineering Handbook*, CRC Press, Boca Raton, 1992, 98—99 and 122—123.
4. Kaldis, E., Ed., *Materials and Crystallographic Aspects of HT_c-Superconductivity*, Kluwer Academic Publ., Dordrecht, The Netherlands, 1992.
5. Malik, S.K. and Shah, S.S., Ed., *Physical and Material Properties of High Temperature Superconductors*, Nova Science Publ., Commack, N.Y., 1994.
6. Chmaissem, O. et. al., *Physica*, C230, 231—238, 1994.
7. Antipov, E.V. et. al., *Physica*, C215, 1—10, 1993.

Table 1
Structural Parameters and Approximate T_c Values of High-Temperature Superconductors

Material	Structure	T_c/K (maximum value)
$La_2CuO_{4+\delta}$	Bmab; $a = 5.355$, $b = 5.401$, $c = 13.15$ Å	39
$La_{2-x}Sr_x(Ba_x)CuO_4$	I4/mmm; $a = 3.779$, $c = 13.23$ Å	35
$La_2Ca_{1-x}Sr_xCu_2O_6$	I4/mmm; $a = 3.825$, $c = 19.42$ Å	60
$YBa_2Cu_3O_7$	Pmmm; $a = 3.821$, $b = 3.885$, $c = 11.676$ Å	93
$YBa_2Cu_4O_8$	Ammm; $a = 3.84$, $b = 3.87$, $c = 27.24$ Å	80
$Y_2Ba_4Cu_7O_{15}$	Ammm; $a = 3.851$, $b = 3.869$, $c = 50.29$ Å	93
$Bi_2Sr_2CuO_6$	Amaa; $a = 5.362$, $b = 5.374$, $c = 24.622$ Å	10
$Bi_2CaSr_2Cu_2O_8$	A_2aa; $a = 5.409$, $b = 5.420$, $c = 30.93$ Å	92
$Bi_2Ca_2Sr_2Cu_3O_{10}$	A_2aa; $a = 5.39$, $b = 5.40$, $c = 37$ Å	110
$Bi_2Sr_2(Ln_{1-x}Ce_x)_2Cu_2O_{10}$	P4/mmm; $a = 3.888$, $c = 17.28$ Å	25
$Tl_2Ba_2CuO_6$	A_2aa; $a = 5.468$, $b = 5.472$, $c = 23.238$ Å;	
	I4/mmm; $a = 3.866$, $c = 23.239$ Å	92
$Tl_2CaBa_2Cu_2O_8$	I4/mmm; $a = 3.855$, $c = 29.318$ Å	119
$Tl_2Ca_2Ba_2Cu_3O_{10}$	I4/mmm; $a = 3.85$, $c = 35.9$ Å	128
$Tl(BaLa)CuO_5$	P4/mmm; $a = 3.83$, $c = 9.55$ Å	40
$Tl(SrLa)CuO_5$	P4/mmm; $a = 3.7$, $c = 9$ Å	40
$(Tl_{0.5}Pb_{0.5})Sr_2CuO_5$	P4/mmm; $a = 3.738$, $c = 9.01$ Å	40
$TlCaBa_2Cu_2O_7$	P4/mmm; $a = 3.856$, $c = 12.754$ Å	103
$(Tl_{0.5}Pb_{0.5})CaSr_2Cu_2O_7$	P4/mmm; $a = 3.80$, $c = 12.05$ Å	90
$TlSr_2Y_{0.5}Ca_{0.5}Cu_2O_7$	P4/mmm; $a = 3.80$, $c = 12.10$ Å	90
$TlCa_2Ba_2Cu_3O_8$	P4/mmm; $a = 3.853$, $c = 15.913$ Å	110
$(Tl_{0.5}Pb_{0.5})Sr_2Ca_2Cu_3O_9$	P4/mmm; $a = 3.81$, $c = 15.23$ Å	120
$TlBa_2(La_{1-x}Ce_x)_2Cu_2O_9$	I4/mmm; $a = 3.8$, $c = 29.5$ Å	40
$Pb_2Sr_2La_{0.5}Ca_{0.5}Cu_3O_8$	Cmmm; $a = 5.435$, $b = 5.463$, $c = 15.817$ Å	70
$Pb_2(Sr,La)_2Cu_2O_6$	$P22_12$; $a = 5.333$, $b = 5.421$, $c = 12.609$ Å	32
$(Pb,Cu)Sr_2(La,Ca)Cu_2O_7$	P4/mmm; $a = 3.820$, $c = 11.826$ Å	50
$(Pb,Cu)(Sr,Eu)(Eu,Ce)Cu_2O_x$	I4/mmm; $a = 3.837$, $c = 29.01$ Å	25
$Nd_{2-x}Ce_xCuO_4$	I4/mmm; $a = 3.95$, $c = 12.07$ Å	30
$Ca_{1-x}Sr_xCuO_2$	P4/mmm; $a = 3.902$, $c = 3.35$ Å	110
$Sr_{1-x}Nd_xCuO_2$	P4/mmm; $a = 3.942$, $c = 3.393$ Å	40
$Ba_{0.6}K_{0.4}BiO_3$	Pm3m; $a = 4.287$ Å	31
Rb_2CsC_{60}	$a = 14.493$ Å	31
$NdBa_2Cu_3O_7$	Pmmm; $a = 3.878$, $b = 3.913$, $c = 11.753$	58

Table 1
Structural Parameters and Approximate T_c Values of High-Temperature Superconductors
(continued)

Material	Structure	T_c/K (maximum value)
$SmBaSrCu_3O_7$	I4/mmm; $a = 3.854$, $c = 11.62$	84
$EuBaSrCu_3O_7$	I4/mmm; $a = 3.845$, $c = 11.59$	88
$GdBaSrCu_3O_7$	I4/mmm; $a = 3.849$, $c = 11.53$	86
$DyBaSrCu_3O_7$	Pmmm; $a = 3.802$, $b = 3.850$, $c = 11.56$	90
$HoBaSrCu_3O_7$	Pmmm; $a = 3.794$, $b = 3.849$, $c = 11.55$	87
$ErBaSrCu_3O_7$ (multiphase)	Pmmm; $a = 3.787$, $b = 3.846$, $c = 11.54$	82
$TmBaSrCu_3O_7$ (multiphase)	Pmmm; $a = 3.784$, $b = 3.849$, $c = 11.55$	88
$YBaSrCu_3O_7$	Pmmm; $a = 3.803$, $b = 3.842$, $c = 11.54$	84
$HgBa_2CuO_4$	I4/mmm; $a = 3.878$, $c = 9.507$	94
$HgBa_2CaCu_2O_6$ (annealed in O_2)	I4/mmm; $a = 3.862$, $c = 12.705$	127
$HgBa_2Ca_2Cu_3O_8$	Pmmm; $a = 3.85$, $c = 15.85$	133
$HgBa_2Ca_3Cu_4O_{10}$	Pmmm; $a = 3.854$, $c = 19.008$	126

Table 2
Superconducting Properties

$J_c (0)$: Critical current density extrapolated to 0 K
λ_{ab}: Penetration depth in a-b plane
k_B: Boltzmann constant

Material	Form	Energy gap (Δ)		$10^{-6} \times J_c (0)$/A cm^{-2}	λ_{ab}/Å
		$2\Delta_{pp}/k_B T_c$*	$2\Delta_{fit}/k_B T_c$†		
$Y Ba_2Cu_3O_7$	Single Crystal	5–6	4–5	30 (film)	1400
$Bi_2Sr_2CaCu_2O_8$	Single Crystal	8–9	5.5–6.5	2	2700
$Tl_2Ba_3CaCu_2O_8$	Ceramic	6–7	4–6	10 (film, 80 K)	2000
$La_{2-x}Sr_xCuO_4$, $x = 0.15$	Ceramic	7–9	4–6		
$Nd_{2-x}Ce_xCuO_4$	Ceramic	8	4–5	0.2 (film)	

* Obtained from peak to peak value.
† Obtained from fit to BCS-type relation.

Table 3
Normal State Properties

ρ_{ab}:   Resistivity in the a-b plane
ρ_c: Resistivity along the c axis
+ve: ρ_c has positive temperature coefficient of resistivity
−ve: ρ_c has negative temperature coefficient of resistivity
n_H: Hall density
k: Thermal conductivity
in plane: Along a-b plane
out of plane: Perpendicular to a-b plane

Material	Form	$\rho_{ab}/\mu\Omega\,cm$ 300 K	$\rho_{ab}/\mu\Omega\,cm$ 100 K	$\rho_c/m\Omega\,cm$ 300 K	$d\rho_c/dT$	$10^{-21} \times n_H/cm^{-3}$ 300 K	$10^{-21} \times n_H/cm^{-3}$ 100 K	$k/(mW/cm\,K)$ at 300 K in plane	$k/(mW/cm\,K)$ at 300 K out of plane
$YBa_2Cu_3O_7$	Single Crystal	110	35	5	+ve	11–16	4–6	120	3
	film	200–300	60–100			5–9	2–3		
$YBa_2Cu_4O_8$	Single Crystal	75	20	10	−ve	14			
	film	100–200	20–50						
$Bi_2Sr_2CuO_6$	Single Crystal	300	150	5000	−ve	22	17		
$Bi_2Sr_2CaCu_2O_8$	Single Crystal	150	50	>1000	−ve	6	5	60	8
$Tl_2Ba_2CuO_6$	Single Crystal	300–400	50–75	200–300	+ve	4	3		
$Tl_2Ba_2Ca_2Cu_3O_{10}$	Ceramic	***	**		+ve for T >225 K	3.1	2.5		
$La_{2-x}Sr_xCuO_4$, $x = 0.12$	Single Crystal	900	350	200	+ve for T >150 K	2.5	≈ 2*		
$La_{2-x}Sr_xCuO_4$, $x = 0.20$	Single Crystal	400	200	80		10	6.3	50 (for x = 0.04)	20
	film	400	160			8.4	17		
$Nd_{2-x}Ce_xCuO_4$, $x = 0.17$	Single Crystal	500	275			53	11	250 (for x = 0.15)	
$x = 0.15$	film	140–180	35			32			

*   At 200 K
** ρ ~0.4 mΩ cm at 120 K
*** ρ ~1.5 mΩ cm at 300 K

ORGANIC SUPERCONDUCTORS
H.P.R. Frederikse

Although the vast majority of organic compounds are insulating gases, liquids or solids, a small number of organic solids show a considerable amount of electrical conductivity. Some of these appear to be superconductors. The superconducting organics fall primarily into two groups: those containing fulvalenes (pentagonal rings containing sulfur or selenium) and fullerenes based on the nearly spherical cluster C_{60}.

The transition temperatures T_c of the fulvalene derivatives are shown in Table 1. The abbreviations of the various molecular groups are listed in Table 2 and their chemical structures are depicted in Figure 1. Most of the T_c's are between 1 and 12 K. Several of the compounds only show superconductivity under pressure.

The fullerenes are A_3C_{60} compounds, where A represents a single or a combination of alkali atoms. The C_{60} cluster is shown in Figure 2a, while Figure 2b illustrates how the alkali atoms fit into the A_3C_{60} molecule to form the A15 crystallographic structure. Their superconducting transition temperatures range from 8 to 31.3 K (see Table 3).

REFERENCES

1. Ishigura, T. and Yamaji, K., *Organic Superconductors*, Springer-Verlag, Berlin, 1990.
2. Williams, Jack M. et al., *Organic Superconductors (Including Fullerenes)*, Prentice Hall, Englewood Cliffs, N.J., 1992.
3. *The Fullerenes*, Ed.: Krato, H.W., Fisher, J.E., and Cox, D.E., Pergamon Press, Oxford, 1993.
4. Schluter, M. et al., in *The Fullerenes* (Ref. 3), p. 303.

Table 1
Critical Pressure and Maximum Critical Temperature of Organic Superconductors

Material	P_c/kbar	T_c/K	Material	P_c/kbar	T_c/K
$(TMTSF)_2PF_6$	6.5	1.2	β-$(ET)_2IBr_2$	0	2.8
$(TMTSF)_2AsF_6$	9	1.3	β-$(ET)_2AuI_2$	0	4.8
$(TMTSF)_2SbF_6$	11	0.4	$(ET)_4Hg_{2.89}Cl_8$	0	4.2
$(TMTSF)_2TaF_6$	12	1.4	$(ET)_4Hg_{2.89}Br_8$	12	1.8
$(TMTSF)_2ClO_4$	0	1.4	$(ET)_3Cl_2(H_2O)_2$	16	2
$(TMTSF)_2ReO_4$	9.5	1.3	κ-$(ET)_2Cu(NCS)_2$	0	10.4
$(TMTSF)_2FSO_3$	5	3	κ-$(d$-$ET)_2Cu(NCS)_2$	0	11.4
$(ET)_4(ReO_4)_2$	4.5	2	$(DMET)_2Au(CN)_2$	1.5	0.9
β_L-$(ET)_2I_3$	0	1.4	$(DMET)_2AuI_2$	5	0.6
β_H-$(ET)_2I_3$	0	8.1	$(DMET)_2AuBr_2$	0	1.9
γ-$(ET)_3I_{2.5}$	0	2.5	$(DMET)_2AuCl_2$	0	0.9
ϵ-$(ET)_2I_3(I_8)_{0.5}$	0	2.5	$(DMET)_2I_3$	0	0.6
α-$(ET)_2I_3I_2$-doped	0	3.3	$(DMET)_2IBr_2$	0	0.7
α_t-$(ET)_2I_3$	0	8	$(MDT$-$TTF)_2AuI_2$	0	3.5
$\epsilon \rightarrow \beta$-$(ET)_2I_3$[a]	0	6	$TTF[Ni(dmit)_2]_2$	2	1.6[b]
θ-$(ET)_2I_3$	0	3.6	$TTF[Pd(dmit)_2]_2$	20	6.5
κ-$(ET)_2I_3$	0	3.6	$(CH_3)_4N[Ni(dmit)_2]_2$	7	5

[a] Converted form ϵ-type to β-type by thermal treatment.

[b] For 7 kbar.

From Ishigura, T. and Yamaji, K., *Organic Superconductors*, Springer-Verlag, Berlin, 1990. With permission.

Table 2
List of Symbols and Abbreviations

TTF	tetrathiafulvalene
TMTSF	tetramethyltetraselenafulvalene
BEDT-TTF or "ET"	bis(ethylenedithio)tetrathiafulvalene
MDT-TTF	methylenedithiotetrathiafulvalene
DMET	[dimethyl(ethylenedithio)diselenadithiafulvalene]
dmit	4,5-dimercapto-1,3-dithiole-2-thione
T_c	transition temperature to superconducting state
P_c	minimum pressure required for superconducting transition

TMTSF

Tetramethyltetraselenafulvalene

TTF

Tetrathiafulvalene

BEDT – TTF or ET

Bis(ethylenedithio)tetrathiafulvalene

DMET

Dimethyl(ethylenedithio)diselenadithiafulvalene

MDT – TTF

Methylenedithiotetrathiafulvalene

M=Ni, Pd, Pt

$M(dmit)_2^{2-}$

Ligand is 4,5-dimercapto-1.3-dithiole-2-thione

FIGURE 1. Structures of various donor molecules and acceptor species.

a

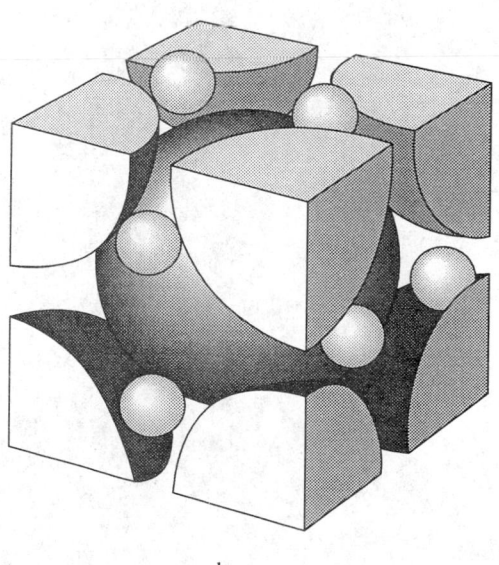

b

FIGURE 2. (a) C_{60} cluster placed in a fcc lattice. Each crystal axis crosses a double bond shared by two hexagons. (b) A hypothetical A_3C_{60} with the A15 structure. The structure can be seen to be an ordered defect structure of A_6C_{60}.

Table 3
Unit Cell and T_c for FCC-$A_3 C_{60}$

	Lattice parameter(s) (Å)	T_c/K
$Na_2Rb_{0.5}Cs_{0.5}C_{60}$	14.148(3)	8.0
Na_2CsC_{60} No. 1[a]	14.132(2)	10.5
Na_2CsC_{60} No. 2[a]	14.176(9)	14.0
K_3C_{60}	14.253(3)	19.3
K_2RbC_{60}	14.299(2)	21.8
Rb_2KC_{60} No. 1[a]	14.336(1)	24.4
Rb_2KC_{60} No. 2[a]	14.364(5)	26.4
Rb_3C_{60}	14.436(2)	29.4
Rb_2CsC_{60}	14.493(2)	31.3

[a] Samples labeled No. 1 and No. 2 have the same nominal composition.

From Schluter, M et. al., *The Fullerenes*, Ed.: Krato, H.W., Fisher, J.E., and Cox, D.E., Pergamon Press, Oxford, 1993. With permission.

PROPERTIES OF SEMICONDUCTORS

L. I. Berger and B. R. Pamplin

The term "semiconductor" is applied to a material in which electric current is carried by electrons or holes and whose electrical conductivity, when extremely pure, rises exponentially with temperature and may be increased from its low "intrinsic" value by many orders of magnitude by "doping" with electrically active impurities.

Semiconductors are characterized by an energy gap in the allowed energies of electrons in the material which separates the normally filled energy levels of the *valence band* (where "missing" electrons behave like positively charged current carriers "holes") and the *conduction band* (where electrons behave rather like a gas of free negatively charged carriers with an effective mass dependent on the material and the direction of the electrons' motion). This energy gap depends on the nature of the material and varies with direction in anisotropic crystals. It is slightly dependent on temperature and pressure, and this dependence is usually almost linear at normal temperatures and pressures.

Data are presented in three tables. Table I "General Properties of Semiconductors" lists the main crystallographic and semiconducting properties of a large number of semiconducting materials in three main categories: "Tetrahedral Semiconductors" in which every atom is tetrahedrally co-ordinated to four nearest neighbor atoms (or atomic sites) as for example in the diamond structure; "Octahedral Semiconductors" in which every atom is octahedrally co-ordinated to six nearest neighbor atoms — as for examples the halite structure; and "Other Semiconductors."

Table II gives more detailed information about some better known semiconductors, while Table III gives some information about the electronic energy band structure parameters of the best known materials.

Table I
PHYSICO-CHEMICAL PROPERTIES OF SEMICONDUCTORS
(LISTED BY CRYSTAL STRUCTURE)

Substance	Molecular mass	Average atomic mass	Lattice parameters (Å room temp.)	Density (g/cm³)	Melting point (K)	Microhardness, N/mm² (M-Mohs Scale)	Specific heat, J/kg·K (300K)	Debye temp. (K)	Coefficient of thermal linear expansion [10⁻⁶ K⁻¹ (300K)]	Thermal conductivity [mW/cm·K (300K)]

PART A. ADAMANTINE SEMICONDUCTORS

§A1. Diamond Structure Elements (Strukturbericht symbol A4, Space Group Fd3m-O_h^7)

Substance	Molecular mass	Average atomic mass	Lattice parameters	Density	Melting point	Microhardness	Specific heat	Debye temp.	Coeff. thermal linear exp.	Thermal conductivity
C	12.01		3.56683	3.51	≈3850 Transition to graphite > 980	10(M)	471.5	2340	1.18	9900(I) 23200(IIA) 13600(IIB)
Si	28.09		5.43072	2.3283	1685 ± 2	11270	702	645	2.49	1240
Ge	72.59		5.65754	5.3234	1231	7644	321.9	374	6.1	640
α-Sn	118.69		6.4912	5.765 (281 K)	505.2 (Tr. 286.4)		213	230	5.4 (220 K)	

§A2. Sphalerite (Zinc Blende) Structure Compounds (Strukturbericht symbol B3 Space Group F$\overline{4}$3m-T_d^2)

Substance	Molecular mass	Average atomic mass	Lattice parameters	Density	Melting point	Microhardness	Specific heat	Debye temp.	Coeff. thermal linear exp.	Thermal conductivity
I VII Compounds										
CuF	82.54	41.27	4.255		1181					
CuCl	98.99	49.49	5.4057	3.53	695	2.3 (M)	490	240	12.1	8.4
CuBr	143.36	71.73	5.6905	4.98	770	2.5(M)	381	207	15.4	12.5
CuI	190.46	95.23	6.60427	5.63	878	192	276	181	19.2	16.8
AgBr	187.78	93.89		6.473	>1570 (Tr. 410)	2.5(M)	270			
AgI	234.77	117.39	6.502	5.67	831	2.5(M)	232	134	−2.5	4.2
II VI Compounds										
BeS	41.08	20.54	4.865	2.36						
BeSe	87.97	43.99	5.139	4.315						
BeTe	136.61	68.31	5.626	5.090						
BePo	(218)	(109)	5.838	7.3						
ZnO	81.37	40.69	4.63	5.675	2248	5.0 (M)	494	416	2.9	234
ZnS	97.43	48.72	5.4093	4.079	2100 (Tr. 1295)	1780	472	530	6.36	251
ZnSe	144.34	72.17	5.6676	5.42	1790	1350	339	400	7.2	140
ZnTe	192.99	96.5	6.101	6.34	1568	900	264	223	8.19	108
ZnPo	(274)	(137)	6.309							
CdS	144.46	72.23	5.832	4.826	1750	1250	330	219	4.7	200
CdSe	191.36	95.68	6.05	5.674	1512	1300	255	181	3.8	90
CdTe	240.00	120.00	6.477	5.86	1365	600	205	200	4.9	58.5
CdPo	(321)	(161)	6.665							
HgS	232.65	116.33	5.8517	7.73	1820	3(M)	210			
HgSe	279.55	139.78	6.084	8.25	1070	2.5(M)	178	151	5.46	10
HgTe	328.19	164.10	6.4623	8.17	943	300	164	242	4.6	20
III V Compounds										
BN	24.82	12.41	3.615	3.49	≈3300	10(M)	793	≈1900		200
BP(L.T.)	41.78	20.87	4.538	2.9	≈2800	37000		≈980		
BAs	85.73	42.87	4.777		≈2300	19000		≈625		

Substance	Molecular mass	Average atomic mass	Lattice parameters (Å room temp.)		Density (g/cm³)	Melting point (K)	Microhardness, N/mm² (M-Mohs Scale)	Specific heat, J/kg·K (300 K)	Debye temp. (K)	Coefficient of thermal linear expansion [10^{-6} K^{-1}] (300K)	Thermal conductivity [mW/cm·K] (300K)]
AlP	57.95	28.98	5.451		2.42	≈2100	5.5(M)		588		920
AlAs	101.90	50.95	5.6622		3.81	2013	5000		417	3.5	840
AlSb	148.73	74.37	6.1355		4.218	1330	4000		292	4.2	600
GaP	100.69	50.35	5.4505		4.13	1750	9450		446	5.3	752
GaAs	144.64	72.32	5.65315		5.316	1510	7500		344	5.4	560
GaSb	191.47	95.74	6.0954		5.619	980	4480	320	265	6.1	270
InP	145.79	72.90	5.86875		4.787	1330	4100		321	4.6	800
InAs	189.74	94.87	6.05838		5.66	1215	3300	268	249	4.7	290
InSb	236.57	118.29	6.47877		5.775	798	2200	144	202	4.7	160
Other sphalerite structure compounds											
MnS	87.0	43.5	5.011								
MnSe	133.9	66.95	5.82								
β-SiC	40.1	20.1	4.348		3.21	3070					
Ga₂Se₃	376.32	75.26	5.429		4.92	1020	3160			8.9	50
Ga₂Te₃	522.24	104.45	5.899		5.75	1063	2370				47
In₂Te₃(H.T.)	608.44	121.7	6.150		5.8	940	1660				69
MgGeP₂	158.84	39.71	5.652								
ZnSnP₂	246.00	61.5	5.65			1200					
ZnSnAs₂(H.T.)	333.90	82.38	5.851		5.53	1050					76
ZnSnSb₂	427.56	106.89	6.281		5.67	870	2500				76

§A3. Wurtzite (Zincite) Structure Compounds (Strukturbericht symbol B4, Space Group P 6_3mc-C$_{6v}^4$)

Substance	Molecular mass	Average atomic mass	Lattice parameters (Å room temp.)		Density (g/cm³)	Melting point (K)	Microhardness, N/mm² (M-Mohs Scale)	Specific heat, J/kg·K (300 K)	Debye temp. (K)	Coefficient of thermal linear expansion [10^{-6} K^{-1}] (300K)	Thermal conductivity [mW/cm·K] (300K)]
I VII Compounds											
CuCl	99.0	49.5	3.91	6.42		T_c680K					
CuBr	143.46	71.73	4.06	6.66		T_c658K					
CuI	190.46	95.23	4.31	7.09							
AgI	234.80	117.40	4.580	7.494							
II VI Compounds											
BeO	25.01	12.51	2.698	4.380			2800				
MgTe	151.9	76.0	4.54	7.39	3.85		≈2800				
ZnO	81.37	40.69	3.24950	5.2069	5.66		2250				600
ZnS	97.43	48.72	3.8140	6.2576	4.1		2100				460
ZnTe	192.99	46.50	4.27	6.99			1568				
CdS	144.46	72.23	4.1348	6.7490	4.82		1748				401
CdSe	191.36	95.68	4.299	7.010	5.66		1512				316
CdTe	240.00	120.00	4.57	7.47							
III V Compounds											
BP(H.T.)	41.79	20.90	3.562	5.900							
AlN	40.99	20.50	3.111	4.978	3.26		≈2500				823
GaN	83.73	41.87	3.190	5.189	6.10		1500				656
InN	128.83	64.42	3.533	5.693	6.88		1200				556
Other wurtzite structure compounds											
MnS	87.0	43.5	3.985	6.45	3.248						
MnSe	133.9	66.95	4.12	6.72							
SiC	40.1	20.1	3.076	5.048							
MnTe	182.54	91.27	4.078	6.701							
Al₂S₃	150.14	30.03	3.579	5.829	2.55		1400				
Al₂Se₃	290.84	58.17	3.890	6.30	3.91		1250				

§A4. Chalcopyrite Structure Compounds (Strukturbericht symbol E1₁, Space Group I $\bar{4}$ 2d-D$_{2d}^{12}$)

Substance	Molecular mass	Average atomic mass	Lattice parameters (Å room temp.)		Density (g/cm³)	Melting point (K)	Microhardness, N/mm² (M-Mohs Scale)	Specific heat, J/kg·K (300 K)	Debye temp. (K)	Coefficient of thermal linear expansion [10^{-6} K^{-1}] (300K)	Thermal conductivity [mW/cm·K] (300K)]
I III VI₂ Compounds											
CuAlS₂	154.65	38.66	5.323	10.44	3.47	2500					
CuAlSe₂	248.45	62.11	5.617	10.92	4.70	2260					
CuAlTe₂	345.73	86.43	5.976	11.80	5.50	2550					
CuGaS₂	197.39	49.35	5.360	10.49	4.35	2300					
CuGaSe₂	291.19	72.80	5.618	11.01	5.56	1970	4200		275	5.4	42
CuGaTe₂	388.47	97.12	6.013	11.93	5.99	2400	3500			6.9	27
CuInS₂	242.49	60.62	5.528	11.08	4.75	1400	2550				
CuInSe₂	336.29	84.07	5.785	11.56	5.77	1600	2050			6.6	37
CuInTe₂	433.57	108.39	6.179	12.365	6.10	1660	400		195	7.1	49
CuTlS₂	332.05	83.01	5.580	11.17	6.32						
CuTlSe₂(L.T.)	425.85	106.46	5.844	11.65	7.11	900					
CuFeS₂	183.51	45.88	5.25	10.32	4.088						
CuFeSe₂	277.31	69.33				850					

Substance	Molecular mass	Average atomic mass	Lattice parameters (Å room temp.)		Density (g/cm³)	Melting point (K)	Microhardness, N/mm² (M-Mohs Scale)	Specific heat, J/kg·K (300 K)	Debye temp. (K)	Coefficient of thermal linear expansion [10^{-6} K^{-1} (300K)]	Thermal conductivity [mW/cm·K (300K)]
$CuLaS_2$	266.58	66.65	5.65	10.86							
$AgAlS_2$	198.97	49.74	5.707	10.28	3.94						
$AgAlSe_2$	292.77	73.19	5.968	10.77	5.07	1220					
$AgAlTe_2$	390.05	97.51	6.309	11.85	6.18	1000					
$AgGaS_2$	241.71	60.43	5.755	10.28	4.72						
$AgGaSe_2$	335.51	83.88	5.985	10.90	5.84	1120	4400				
$AgGaTe_2$	432.79	108.2	6.301	11.96	6.05	990	1800		212		10
$AgInS_2$(L.T.)	286.87	71.70	5.828	11.19	5.00		2250				
$AgInSe_2$	380.61	95.15	6.102	11.69	5.81	1053	1850				30
$AgInTe_2$	477.89	119.47	6.42	12.59	6.12	965				9.49, 0.69	
$AgFeS_2$	227.83	56.96	5.66	10.30	4.53						
II IV V_2 Compounds											
$ZnSiP_2$	155.40	38.85	5.400	10.441	3.39	1640	11000				
$ZnGeP_2$	199.90	49.98	5.465	10.771	4.17	1295	8100				180
$ZnSnP_2$	246.00	61.5					6500				
$CdSiP_2$	202.43	50.61	5.678	10.431	4.00	≈1470	10500		282		
$CdGeP_2$	246.94	61.74	5.741	10.775	4.48	1049	5650				110
$CdSnP_2$	243.03	73.26	5.900	11.518			5000		195		140
$ZnSiAs_2$	242.20	60.55	5.61	10.88	4.70	1311	9200				
$ZnGeAs_2$	287.80	71.95	5.672	11.153	5.32	1150	6800		263		110
$ZnSnAs_2$	333.90	83.48	5.8515	11.704	5.53	1048	4550		271		150
$CdSiAs_2$	290.34	72.58	5.884	10.882			6850				
$CdGeAs_2$	334.83	83.71	5.9427	11.2172	5.60	938	4700				48
$CdSnAs_2$	380.93	95.23	6.0944	11.9182	5.72	880	3450				40

§A5. Other Ternary Semiconductors with Tetrahedral Coordination

Substance	Molecular mass	Average atomic mass	Lattice parameters (Å room temp.)		Density (g/cm³)	Melting point (K)	Microhardness, N/mm² (M-Mohs Scale)	Specific heat, J/kg·K (300 K)	Debye temp. (K)	Coefficient of thermal linear expansion [10^{-6} K^{-1} (300K)]	Thermal conductivity [mW/cm·K (300K)]
I_2 IV VI_3 Compounds											
Cu_2SiS_3(H.T.)	251.36	41.89	3.684	6.004	3.81	1200					23
Cu_2SiS_3(L.T.)			5.290	10.156	3.63						
Cu_2SiTe_3	537.98	89.66	5.93		5.47						
Cu_2GeS_3(H.T.)	295.88	49.31	5.317		4.45	1210	4550	510	254	7.2	12
Cu_7GeS_3(L.T.)			5.327	5.215	4.46						
Cu_2GeSe_3	436.56	72.76	5.589	5.485	5.57	1030	3840	340	168	8.4	24
Cu_2GeTe_3	582.51	97.09	5.958	5.935	5.92		2890				130
Cu_2SnS_3	341.98	57.00	5.436		5.02	1110	2770	440	214	7.8	28
$CuSnSe_3$	482.66	80.44	5.687		5.94	960	2510	310	148	8.9	35
Cu_2SnTe_3	628.61	104.77	6.048		6.51	680	1970				144
Ag_2GeSe_3	525.21	87.54									
Ag_2SnSe_3	571.31	95.22									
Ag_2GeTe_3	671.13	111.86									
Ag_2SnTe_3	717.23	119.54									
I_3 V VI_4 Compounds											
Cu_3PS_4	349.85	40.73	7.44		6.19						
Cu_3AsS_4	393.79	49.22	6.43		6.14	4.37				3.2	30.2
Cu_3AsSe_4	581.37	72.67	5.570	10.957	5.61				169	9.5	19
Cu_3SbS_4	440.64	55.08	5.38	16.76	4.90						
Cu_3SbSe_4	628.22	78.53	5.654	11.256	6.0				131	12.4	14.6
I IV_2 V_3 Compounds											
$CuSi_2P_3$	212.64	35.44	5.25								
$CuGe_2P_3$	301.65	50.28	5.375		4.318	1113	8500	429		8.21	37.6
$AgGe_2P_3$	345.97	57.66				1015	6150				

§A6. "Defect Chalcopyrite" Structure Compounds (Strukturbericht symbol E3, Space Group I $\overline{4}$-S_4^2)

Substance	Molecular mass	Average atomic mass	Lattice parameters (Å room temp.)		Density (g/cm³)	Melting point (K)	Microhardness, N/mm² (M-Mohs Scale)	Specific heat, J/kg·K (300 K)	Debye temp. (K)	Coefficient of thermal linear expansion [10^{-6} K^{-1} (300K)]	Thermal conductivity [mW/cm·K (300K)]
$ZnAl_2Se_4$	435.18	62.17	5.503	10.90	4.37						
$ZnAl_2Te_4$(?)	629.74	84.96	5.904	12.05	4.95						
$ZnGa_2S_4$(?)	333.06	47.58	5.274	10.44	3.80						
$ZnGa_2Se_4$(?)	520.66	74.38	5.496	10.99	5.21						
$ZnGa_2Te_4$(?)	715.22	102.17	5.937	11.87	5.67						
$ZnIn_2Se_4$	610.86	87.27	5.711	11.42	5.44	1250					
$ZnIn_2Te_4$	805.42	115.06	6.122	12.24	5.83	1075					
$CdAl_2S_4$	294.61	42.09	5.564	10.32	3.06						
$CdAl_2Se_4$	482.21	68.89	5.747	10.68	4.54						
$CdAl_2Te_4$(?)	676.77	97.68	6.011	12.21	5.10						

Substance	Molecular mass	Average atomic mass	Lattice parameters (Å room temp.)		Density (g/cm³)	Melting point (K)	Microhardness, N/mm² (M-Mohs Scale)	Specific heat, J/kg·K (300 K)	Debye temp. (K)	Coefficient of thermal linear expansion [10⁻⁶ K⁻¹ (300K)]	Thermal conductivity [mW/cm·K (300K)]
$CdGa_2S_4$	380.09	54.30	5.577	10.08	4.03						
$CdGa_2Se_4$	567.69	81.10	5.743	10.73	5.32						
$CdGa_2Te_4$	762.25	108.89	6.093	11.81	5.77						
$CdIn_2Te_4$	852.45	121.78	6.205	12.41	5.9	1060					
$HgAl_2S_4$	382.79	54.68	5.488	10.26	4.11						
$HgAl_2Se_4$	570.39	82.48	5.708	10.74	5.05						
$HgAl_2Te_4(?)$	764.48	109.28	6.004	12.11	5.81						
$HgGa_2S_4$	468.27	66.90	5.507	10.23	5.00						
$HgGa_2Se_4$	655.87	93.70	5.715	10.78	6.18						
$HgIn_2Se_4$	746.07	106.58	5.764	11.80	6.3	1100					
$HgIn_2Te_4(?)$	940.63	134.38	6.186	12.37	6.3	980					

§A7. Other Adamantine Compounds

Substance	Molecular mass	Average atomic mass	Lattice parameters (Å room temp.)	Density (g/cm³)	Melting point (K)
αSiC	40.1	20.1	3.0817 15.1183	3.21	3070
$Hg_5Ga_2Te_8$	2163.19	144.21	6.235		
$Hg_5In_2Te_8$	2253.39	150.23	6.328		
$CdIn_2Se_4$	657.89	93.98	$a = c = 5.823$		

PART B. OCTAHEDRAL SEMICONDUCTORS

§B1. HALITE STRUCTURE SEMICONDUCTORS (Strukturbericht symbol B1, Space Group Fm3m-O_h^5)

Substance	Molecular mass	Average atomic mass	Lattice parameters (Å room temp.)	Density (g/cm³)	Melting point (K)	Microhardness, N/mm² (M-Mohs Scale)	Thermal conductivity [mW/cm·K (300K)]
GeTe	200.19	100.1	5.98	6.14			
SnSe	197.65	98.83	6.020		1133		
SnTe	246.29	123.15	6.313	6.45	1080 (max)		91
PbS	239.26	119.63	5.9362	7.61	1390		23
PbSe	286.16	143.08	6.1243	8.15	1340		17
PbTe	334.8	167.4	6.454	8.16	1180		23
Selected other binary halites							
BiSe	287.94	143.97	5.99	7.98	880		
BiTe	336.58	168.29	6.47				
EuSe	230.92	115.46	6.191		2300		2.4
GdSe	236.21	118.11	5.771		2400		
NiD	60.71	30.35	4.1684	6.6	2260		
CdO	128.41	64.21	4.6953		1700		7
SrSW	119.68	59.84	6.0199	3.643	3000		

PART C. OTHER SEMICONDUCTORS

§C1. Antifluorite Structure Compounds Fm3m (225)

Substance	Molecular mass	Average atomic mass	Lattice parameters (Å room temp.)	Density (g/cm³)	Melting point (K)	Thermal conductivity [mW/cm·K (300K)]
Mg_2Si	76.70	25.57	6.338	1.88	1375	
Mg_2Ge	121.20	40.4	6.380	3.08	1388	
Mg_2Sn	167.3	55.77	6.765	3.53	1051	92
Mg_2Pb	225.81	85.27	6.836	5.1	823	

§C2. Tetradymite Structure Compounds ($\overline{R}3m - D_{3d}^5$)

Substance	Molecular mass	Average atomic mass	Lattice parameters (Å room temp.)		Density (g/cm³)	Melting point (K)	Specific heat, J/kg·K (300 K)	Coefficient of thermal linear expansion [10⁻⁶ K⁻¹ (300K)]	Thermal conductivity [mW/cm·K (300K)]
Sb_2Te_3	626.3	125.26	4.25	30.3	6.44	895			24
Bi_2Se_3	654.84	130.97	4.14	28.7	7.51	979	167		24
Bi_2Te_3	800.76	160.15	4.38	30.45	7.73	858	155	16	30

§C3. Selected Multinary Compounds

Substance	Molecular mass	Average atomic mass	Lattice parameters (Å room temp.)	Density (g/cm³)	Melting point (K)	Thermal conductivity [mW/cm·K (300K)]
$AgSbSe_2$	387.54	96.88	5.786	6.60	910	10.5
$AgSbTe_2$ (or $Ag_{19}Sb_{29}Te_{52}$)	484.82	121.2	6.078	7.12	830	86, 0.3
$AgBiS_2$(H.T.)	380.97	95.24	5.648			
$AgBiSe_2$(H.T.)	474.77	118.69	5.82			
$AgBiTe_2$(H.T.)	572.05	143.01	6.155			
Cu_2CdSnS_4	486.43	60.80	5.586	10.83		

Table I
PHYSICO-CHEMICAL PROPERTIES OF SEMICONDUCTORS
(LISTED BY CRYSTAL STRUCTURE) (continued)

Substance	Molecular mass	Average atomic mass	Lattice parameters (Å room temp.)	Density (g/cm³)	Melting point (K)	Microhardness, N/mm² (M-Mohs Scale)	Specific heat, J/kg·K (300 K)	Debye temp. (K)	Coefficient of thermal linear expansion [10⁻⁶ K⁻¹ (300K)]	Thermal conductivity [mW/cm·K (300K)]
§C4. Some Elemental Semiconductors										
B		10.81	4.91 12.6	2.34	2348	9.5(M)	1277	1370	8.3	600
Se(gray)		78.96	4.36 4.95	4.81	493	350	292.6		(‖C) 17.89 (⊥C) 74.09	(‖C) 45.2 (⊥C) 13.1
Te		127.6	4.45 5.91	6.23	723		196.5		16.8	(‖C) 33.8 (⊥C) 19.7

Table II
BASIC THERMODYNAMIC, ELECTRICAL, AND MAGNETIC PROPERTIES OF SEMICONDUCTORS (LISTED BY CRYSTAL STRUCTURE)

Substance	Heat of formation [kJ/mole (300K)]	Volume compressibility (10⁻¹⁰ m²/N)	Static dielectric constant	Atomic magnetic susceptibility (10⁻⁶ CGS)	Index of refraction	Minimum room temperature energy gap (eV)	Mobility (Room temp.) (cm²/V·s) Electrons	Holes	Optical transition	Remarks
PART A. ADAMANTINE SEMICONDUCTORS										
§A1. Diamond Structure Elements (Strukturbericht symbol A4, Space Group Fd3m-O$_h^7$)										
C	714.4	18	5.7	−5.88	2.419 (589 nm)	5.4	1800	1400	i*	
Si	324	0.306	11.8	−3.9	3.49 (589 nm)	1.107	1900	500	i	
Ge	291	0.768	16	−0.12	3.99 (589 nm)	0.67	3800	1820	i	
α-Sn	267.5		24		2.75 (589 nm)	0.0; 0.8	2500	2400		
§A2. Sphalerite (Zinc Blende) Structure Compounds (Strukturbericht symbol B3 Space Group F $\bar{4}$ 3m-T$_d^2$)										
I VII Compounds										
CuF										
CuCl	481	0.26	7.9		1.93	3.17			d	Nantokite
CuBr	481	0.26	7.9		2.12	2.91			d	
CuI	439	0.27	6.5		2.346	2.95			d	Marshite
AgBr	486		12.4		2.253	2.50	4000		i	Bromirite
AgI	389	0.41	10		2.22	2.22	30		d	Miersite
II VI Compounds										
BeS						4.17			i	
BeSe						3.61			i	
BeTe						1.45		20	d	
BePo										
ZnO										See A3
ZnS	477		8.9	−9.9	2.356	3.54	180	5(400°C)	d	See also A3
ZnSe	422		9.2		2.89	2.58	540	28	d	
ZnTe	376		10.4		3.56	2.26	340	100	d	
ZnPo										
CdS										See A3
CdSe										See A3
CdTe	339		7.2		2.50	1.44	1200	50	d	
CdPo										
HgS					2.85		250		d	Metacinnabarite
HgSe	247					2.10 (α)	20000	≈1.5	s	Tiemannite
HgTe	242					−0.06	25000	350	s	Coloradoite
III V Compounds										
BN	815					4.6				Borazone
BP(L.T.)						≈2.1	500	70		Ignites 470K
BAs						≈1.5				
AlP						2.45	80		i	
AlAs	627		10.9			2.16	1200	420	i	
AlSb	585	0.571	11		3.2	1.60	200—400	550	i	
GaP	635	0.110	11.1	−13.8	3.2	2.24	300	150	i	
GaAs	535	0.771	13.2	−16.2	3.30	1.35	8800	400	d	
GaSb	493	0.457	15.7	−14.2	3.8	0.67	4000	1400	d	
InP	560	0.735	12.4	−22.8	3.1	1.27	4600	150	d	
InAs	477	0.549	14.6	−27.7	3.5	0.36	33000	460	d	
InSb	447	0.442	17.7	−32.9	3.96	0.163	78000	750	d	

* i = indirect, d = direct, s = semimetal.

Substance	Heat of formation [kJ/mole (300K)]	Volume compressibility $(10^{-10}\ m^2/N)$	Static dielectric constant	Atomic magnetic susceptibility $(10^{-6}\ CGS)$	Index of refraction	Minimum room temperature energy gap (eV)	Mobility (Room temp.) $(cm^2/V \cdot s)$		Optical transition	Remarks
							Electrons	Holes		
Other sphalerite structure compounds										See also §A3
MnS										See also §A3
MnSe										
β-SiC					2.697	2.3	4000			
Ga$_2$Te$_3$	271			−13.5		1.35	50			
In$_2$Te$_3$(H.T.)	198			−13.6		1.04	50			
MgGeP$_2$										E1—T^{d12}
ZnSnP$_2$						2.1				Same
ZnSnAs$_2$(H.T.)						≈0.7				Same
ZnSnSb$_2$						0.4				Same

§A3. Wurtzite (Zincite) Structure Compounds (Strukturbericht symbol B4, Space Group P 6$_3$mc-C$_{6v}^4$)

Substance	Heat of formation	Volume compressibility	Static dielectric constant	Atomic magnetic susceptibility	Index of refraction	Minimum room temperature energy gap (eV)	Electrons	Holes	Optical transition	Remarks
I VII Compounds										
CuCl										
CuBr										
CuI										
AgI						2.63				Iodargirite
II VI Compounds										
BeO										
MgTe										
ZnO	−350					3.2	180			
ZnS	−206					3.67				
ZnTe	−163									
CdS			8.45; 9.12		2.32	2.42	350	40	d	Greenockide
CdSe						1.74	900	50	d	Cadmoselite
CdTe						1.50	650			
III V Compounds										
BP(H.T.)						6.02				
AlN						3.34				
GaN						2.0				
InN										
Other wurtzite structure compounds										
MnS										
MnSe										
SiC					2.654					
MnTe						≈1.0				
Al$_2$S$_3$	426					4.1				
Al$_2$Se$_3$	367					3.1				

§A4. Chalcopyrite Structure Compounds (Strukturbericht symbol E1$_1$, Space Group I $\bar{4}$ 2d-D$_{2d}^{12}$)

Substance	Heat of formation	Volume compressibility	Static dielectric constant	Atomic magnetic susceptibility	Index of refraction	Minimum room temperature energy gap (eV)	Electrons	Holes	Optical transition	Remarks
I III VI$_2$ Compounds										
CuAlS$_2$	0.106					2.5				
CuAlSe$_2$						1.1				
CuAlTe$_2$						0.88				
CuGaS$_2$	0.106					2.38				
CuGaSe$_2$	0.141					0.96, 1.63				
CuGaTe$_2$	0.227					0.82, 1.0				
CuInS$_2$	0.141					1.2				
CuInSe$_2$	0.187					0.86, 0.92				
CuInTe$_2$	0.278					0.95				
CuTlS$_2$										
CuTlSe$_2$(L.T.)						1.07				
CuFeS$_2$						0.53				Chalcopyrite
CuFeSe$_2$						0.16				
CuLaS$_2$										
AgAlS$_2$										
AgAlSe$_2$						0.7				
AgAlTe$_2$						0.56				
AgGaS$_2$	0.150					1.66				
AgGaSe$_2$	0.182					1.1				
AgGaTe$_2$	0.280					1.9				
AgInS$_2$(L.T.)	0.185					1.18				
AgInSe$_2$	0.238					0.96, 0.52				
AgInTe$_2$	0.338									
AgFeS$_2$										

Substance	Heat of formation [kJ/mole] (300K)	Volume compressibility (10^{-10} m²/N)	Static dielectric constant	Atomic magnetic susceptibility (10^{-6} CGS)	Index of refraction	Minimum room temperature energy gap (eV)	Mobility (Room temp.) (cm²/V·s) Electrons	Holes	Optical transition	Remarks
II IV V$_2$ Compounds										
ZnSiP$_2$	312					2.3	1000			
ZnGeP$_2$	293					2.2				
ZnSnP$_2$	275					1.45				
CdSiP$_2$		0.103				2.2	1000			
CdGeP$_2$	289					1.8				
CdSnP$_2$	270					1.5				
ZnSiAs$_2$	290					1.7		50		
ZnGeAs$_2$	271			-14.4		0.85				
ZnSnAs$_2$	252			-18.4		0.65		300		Disorders at 910K
CdSiAs$_2$		0.143				1.6				
CdGeAs$_2$	266			-23.4		0.53	70	25		Disorders at 903
CdSnAs$_2$	247		13.7	-21.5		0.26	22000	250		

§A5. Other Ternary Semiconductors with Tetrahedral Coordination

Substance	Heat of formation [kJ/mole] (300K)	Volume compressibility (10^{-10} m²/N)	Static dielectric constant	Atomic magnetic susceptibility (10^{-6} CGS)	Index of refraction	Minimum room temperature energy gap (eV)	Mobility Electrons	Holes	Optical transition	Remarks
I$_2$ IV VI$_3$ Compounds										
Cu$_2$SiS$_3$(H.T.)										Wurtzite
Cu$_2$SiS$_3$(L.T.)										Tetragonal
Cu$_2$SiTe$_3$										Cubic
Cu$_2$GeS$_3$(H.T.)				-18.7						Cubic
Cu$_2$GeS$_3$(L.T.)							360			Tetragonal
Cu$_2$GeSe$_3$	211.5			-21.3		0.94	238			Same
Cu$_2$GeTe$_3$	190.2			-23.4						Same
Cu$_2$SnS$_3$				-18.2		0.91	405			Cubic
CuSnSe$_3$				-21.0		0.66	870			Cubic
Cu$_2$SnTe$_3$				-28.4						Cubic
Ag$_2$GeSe$_3$				-29.6		0.91 (77K)				
Ag$_2$SnSe$_3$				-29.5		0.81				
Ag$_2$GeTe$_3$				-31.4		0.25				
Ag$_2$SnTe$_3$				-31.0		0.08				
I$_3$ V VI$_4$ Compounds										
Cu$_3$PS$_4$										Enargite
Cu$_3$AsS$_4$	269.6			-15.8		1.24				Famatinite
Cu$_3$AsSe$_4$	161.3			-13.1		0.88				Famatinite
Cu$_3$SbS$_4$				-8.3		0.74				Famatinite
Cu$_3$SbSe$_4$	127.1			-20.5		0.31				
I IV$_2$ V$_3$ Compounds										
CuSi$_2$P$_3$										E1
CuGe$_2$P$_3$		0.12				0.9				E1
AgGe$_2$P$_3$										

§A6. "Defect Chalcopyrite" Structure Compounds (Strukturbericht symbol E3, Space Group I $\bar{4}$-S$_4^2$)

Substance	Heat of formation [kJ/mole] (300K)	Volume compressibility (10^{-10} m²/N)	Static dielectric constant	Atomic magnetic susceptibility (10^{-6} CGS)	Index of refraction	Minimum room temperature energy gap (eV)	Mobility Electrons	Holes	Optical transition	Remarks
ZnAl$_2$Se$_4$										
ZnAl$_2$Te$_4$(?)										
ZnGa$_2$S$_4$(?)						≈3.4				
ZnGa$_2$Se$_4$(?)						≈2.2				
ZnGa$_2$Te$_4$(?)						1.35				
ZnIn$_2$Se$_4$	206					1.82	35			
ZnIn$_2$Te$_4$	198					1.2				
CdAl$_2$S$_4$										
CdAl$_2$Se$_4$										
CdAl$_2$Te$_4$(?)										
CdGa$_2$S$_4$	256					3.44	60			
CdGa$_2$Se$_4$	216					2.43	33			
CdGa$_2$Te$_4$										
CdIn$_2$Te$_4$	195					(1.26 or 0.9)	4000			
HgAl$_2$S$_4$										
HgAl$_2$Se$_4$										
HgAl$_2$Te$_4$(?)										
HgGa$_2$S$_4$	249					2.84				
HgGa$_2$Se$_4$	204					1.95	400			
HgIn$_2$Se$_4$	196					0.6	290			
HgIn$_2$Te$_4$(?)	188					0.86	200			

Substance	Heat of formation [kJ/mole] (300K)]	Volume compressibility (10^{-10} m²/N)	Static dielectric constant	Atomic magnetic susceptibility (10^{-6} CGS)	Index of refraction	Minimum room temperature energy gap (eV)	Mobility (Room temp.) (cm²/V·s)		Optical transition	Remarks
							Electrons	Holes		
§A7. Other Adamantine Compounds										
αSiC		10.2		−6.4	2.67	2.86	400			6H structure
Hg$_5$Ga$_2$Te$_8$										B3 with superlattice
Hg$_5$In$_2$Te$_8$						0.7	2000			B3 with superlattice
CdIn$_2$Se$_4$						1.55				

PART B. OCTAHEDRAL SEMICONDUCTORS

§B1. HALITE STRUCTURE SEMICONDUCTORS (Strukturbericht symbol B1, Space Group Fm3m-O$_h^5$)

Substance	Heat of formation [kJ/mole] (300K)]	Volume compressibility (10^{-10} m²/N)	Static dielectric constant	Atomic magnetic susceptibility (10^{-6} CGS)	Index of refraction	Minimum room temperature energy gap (eV)	Electrons	Holes	Optical transition	Remarks
GeTe										
SnSe										
SnTe										
PbS	435					0.5	600	600		
PbSe	393		161			0.37	1000	900		
PbTe	393		280			0.26	1600	600		Altaite
			360			0.25				
Selected other binary halites										
BiSe										
BiTe						0.4				
EuSe										
GdSe						1.8	4			
NiD						2.0 or 3.7	100			
CdO	531					2.5				
SrSW						4.1				

PART C. OTHER SEMICONDUCTORS

§C1. Antifluorite Structure Compounds Fm3m (225)

Substance	Heat of formation [kJ/mole] (300K)]	Volume compressibility (10^{-10} m²/N)	Static dielectric constant	Atomic magnetic susceptibility (10^{-6} CGS)	Index of refraction	Minimum room temperature energy gap (eV)	Electrons	Holes	Optical transition	Remarks
Mg$_2$Si						0.77	405	70		
Mg$_2$Ge						0.74	520	110		
Mg$_2$Sn						0.36	320	260		
Mg$_2$Pb						0.1				

§C2. Tetradymite Structure Compounds ($R\overline{3}m − D_{3d}^5$)

Substance	Heat of formation [kJ/mole] (300K)]	Volume compressibility (10^{-10} m²/N)	Static dielectric constant	Atomic magnetic susceptibility (10^{-6} CGS)	Index of refraction	Minimum room temperature energy gap (eV)	Electrons	Holes	Optical transition	Remarks
Sb$_2$Te$_3$						0.3		360		
Bi$_2$Se$_3$						0.35	600			
Bi$_2$Te$_3$						0.21	1140	680		R3m (166)

§C3. Selected Multinary Compounds

Substance	Heat of formation [kJ/mole] (300K)]	Volume compressibility (10^{-10} m²/N)	Static dielectric constant	Atomic magnetic susceptibility (10^{-6} CGS)	Index of refraction	Minimum room temperature energy gap (eV)	Electrons	Holes	Optical transition	Remarks
AgSbSe$_2$						0.58				
AgSbTe$_2$ (or Ag$_{19}$Sb$_{29}$Te$_{52}$)						0.7, 0.27				
AgBiS$_2$(H.T.)										
AgBiSe$_2$(H.T.)										
AgBiTe$_2$(H.T.)										
Cu$_2$CdSnS$_4$						1.16	<2			

§C4. Some Elemental Semiconductors

Substance	Heat of formation [kJ/mole] (300K)]	Volume compressibility (10^{-10} m²/N)	Static dielectric constant	Atomic magnetic susceptibility (10^{-6} CGS)	Index of refraction	Minimum room temperature energy gap (eV)	Electrons	Holes	Optical transition	Remarks
B	397.1			−6.7	3.4	1.55	10			
Se(gray)		6.6 (0.1 GHz)		−22.1	2.5	1.5		5		P3$_1$21(152)
Te				−39.5	3.3	0.33	1700	1200		Same

Table III
SEMICONDUCTING PROPERTIES OF SELECTED MATERIALS

Substance	Minimum Energy Gap (eV)		$\frac{dE_g}{dT}$ x 10^4 eV/°C	$\frac{dE_g}{dP}$ x 10^6 eV·cm²/kg	Density of States Electron Effective Mass m_{d_n} (m_o)	Electron Mobility and Temperature Dependence		Density of States Hole Effective Mass m_{d_p} (m_o)	Hole Mobility and Temperature Dependence	
	R.T.	0 K				μ_n cm²/V·s	$-x$		μ_p cm²/V·s	$-x$
Si	1.107	1.153	−2.3	−2.0	1.1	1,900	2.6	0.56	500	2.3
Ge	0.67	0.744	−3.7	+7.3	0.55	3,800	1.66	0.3	1,820	2.33
αSn	0.08	0.094	−0.5		0.02	2,500	1.65	0.3	2,400	2.0
Te	0.33				0.68	1,100		0.19	560	
III–V Compounds										
AlAs	2.2	2.3				1,200			420	
AlSb	1.6	1.7	−3.5	−1.6	0.09	200	1.5	0.4	500	1.8
GaP	2.24	2.40	−5.4	−1.7	0.35	300	1.5	0.5	150	1.5
GaAs	1.35	1.53	−5.0	+9.4	0.068	9,000	1.0	0.5	500	2.1
GaSb	0.67	0.78	−3.5	+12	0.050	5,000	2.0	0.23	1,400	0.9
InP	1.27	1.41	−4.6	+4.6	0.067	5,000	2.0		200	2.4
InAs	0.36	0.43	−2.8	+8	0.022	33,000	1.2	0.41	460	2.3
InSb	0.165	0.23	−2.8	+15	0.014	78,000	1.6	0.4	750	2.1
II–VI Compounds										
ZnO	3.2		−9.5	+0.6	0.38	180	1.5			
ZnS	3.54		−5.3	+5.7		180			5(400°C)	
ZnSe	2.58	2.80	−7.2	+6		540			28	
ZnTe	2.26			+6		340			100	
CdO	2.5 ± .1		−6		0.1	120				
CdS	2.42		−5	+3.3	0.165	400		0.8		
CdSe	1.74	1.85	−4.6		0.13	650	1.0	0.6		
CdTe	1.44	1.56	−4.1	+8	0.14	1,200		0.35	50	
HgSe	0.30				0.030	20,000	2.0			
HgTe	0.15		−1		0.017	25,000		0.5	350	
Halite Structure Compounds										
PbS	0.37	0.28	+4		0.16	800		0.1	1,000	2.2
PbSe	0.26	0.16	+4		0.3	1,500		0.34	1,500	2.2
PbTe	0.25	0.19	+4	−7	0.21	1,600		0.14	750	2.2
Others										
ZnSb	0.50	0.56			0.15	10				1.5
CdSb	0.45	0.57	−5.4		0.15	300			2,000	1.5
Bi₂S₃	1.3					200			1,100	
Bi₂Se₃	0.27					600			675	
Bi₂Te₃	0.13		−0.95		0.58	1,200	1.68	1.07	510	1.95
Mg₂Si		0.77	−6.4		0.46	400	2.5		70	
Mg₂Ge		0.74	−9			280	2		110	
Mg₂Sn	0.21	0.33	−3.5		0.37	320			260	
Mg₃Sb₂		0.32				20			82	
Zn₃As₂	0.93					10	1.1		10	
Cd₃As₂	0.55				0.046	100,000	0.88			
GaSe	2.05		3.8						20	
GaTe	1.66	1.80	−3.6			14	−5			
InSe	1.8					900				
TlSe	0.57		−3.9		0.3	30		0.6	20	1.5
CdSnAs₂	0.23				0.05	25,000	1.7			
Ga₂Te₃	1.1	1.55	−4.8							
α-In₂Te₃	1.1	1.2			0.7				50	1.1
β-In₂Te₃	1.0								5	
Hg₅In₂Te₈	0.5								11,000	
SnO₂									78	

Table IV
BAND PROPERTIES OF SEMICONDUCTORS

PART A. DATA ON VALENCE BANDS OF SEMICONDUCTORS (ROOM TEMPERATURES)

Substance	Band Curvature Effective Mass			Energy Separation of "Split-off" Band (eV)	Measured (Light) Hole Mobility cm²/V·s
	Heavy Holes	Light Holes	"Split-off" Band Holes		
	(Expressed as fraction of free electron mass)				
Semiconductors with Valence Band Maximum at the Center of the Brillouin Zone ('Γ'')					
Si	0.52	0.16	0.25	0.044	500
Ge	0.34	0.043	0.08	0.3	1,820
Sn	0.3				2,400
AlAs					
AlSb	0.4			0.7	550
GaP				0.13	100
GaAs	0.8	0.12	0.20	0.34	400
GaSb	0.23	0.06		0.7	1,400
InP				0.21	150
InAs	0.41	0.025	0.083	0.43	460
InSb	0.4	0.015		0.85	750
CdTe	0.35				50
HgTe	0.5				350

Table IV
BAND PROPERTIES OF SEMICONDUCTORS (continued)

Semiconductors with Multiple Valence Band Maxima

Substance	Number of Equivalent Valleys and Direction	Band Curvature Effective Masses Longitudinal m_L	Transverse m_T	Anisotropy $K = m_L/m_T$	Measured (Light) Hole Mobility $cm^2/V\cdot s$
PbSe	4 "L" [111]	0.095	0.047	2.0	1,500
PbTe	4 "L" [111]	0.27	0.02	10	750
Bi_2Te_3	6	0.207	~0.045	4.5	515

PART B. DATA ON CONDUCTION BANDS OF SEMICONDUCTORS (Room Temperature Data)

Single Valley Semiconductors

Substance	Energy Gap (eV)	Effective Mass (m_o)	Mobility ($cm^2/V.s$)	Comments
GaAs	1.35	0.067	8,500	3(or 6?) equivalent [100] valleys 0.36 eV above this maximum with a mobility of ~50
InP	1.27	0.067	5,000	3(or 6?) equivalent [100] valleys 0.4 eV above this minimum.
InAs	0.36	0.022	33,000	equivalent valleys ~1.0 eV above this minimum.
InSb	0.165	0.014	78,000	
CdTe	1.44	0.11	1,000	4(or 8?) equivalent [111] valleys 0.51 eV above this minimum.

Multivalley Semiconductors

Substance	Energy Gap	Number of Equivalent Valleys and Direction	Band Curvature Effective Mass Longitudinal m_L	Transverse m_T	Anisotropy $K = m_L/m_T$	Comments
Si	1.107	6 in [100] "△"	0.90	0.192	4.7	
Ge	0.67	4 in [111] at "L"	1.588	0.0815	19.5	
GaSb	0.67	as Ge (?)	~1.0	~0.2	~5	
PbSe	0.26	4 in [111] at "L"	0.085	0.05	1.7	
PbTe	0.25	4 in [111] at "L"	0.21	0.029	5.5	
Bi_2Te_3	0.13	6			~0.05	

Table V
RESISTIVITY OF SEMICONDUCTING MINERALS

Mineral	ρ (ohm $\cdot$ m)	Mineral	ρ (ohm $\cdot$ m)
Diamond (C)	2.7	Gersdorffite, NiAsS	1 to 160×10^{-6}
Sulfides		Glaucodote, (Co, Fe)AsS	5 to 100×10^{-6}
Argentite, Ag_2S	1.5 to 2.0×10^{-3}	Antimonide	
Bismuthinite, Bi_2S_3	3 to 570	Dyscrasite, Ag_3Sb	0.12 to 1.2×10^{-6}
Bornite, $Fe_2S_3 \cdot nCu_2S$	1.6 to 6000×10^{-6}	Arsenides	
Chalcocite, Cu_2S	80 to 100×10^{-6}	Allemonite, $SbAs_3$	70 to 60,000
Chalcopyrite, $Fe_2S_3 \cdot Cu_2S$	150 to 9000×10^{-6}	Lollingite, $FeAs_2$	2 to 270×10^{-6}
Covellite, CuS	0.30 to 83×10^{-6}	Nicollite, NiAs	0.1 to 2×10^{-6}
Galena, PbS	6.8×10^{-6} to 9.0×10^{-2}	Skutterudite, $CoAs_3$	1 to 400×10^{-6}
Haverite, MnS_2	10 to 20	Smaltite, $CoAs_2$	1 to 12×10^{-6}
Marcasite, FeS_2	1 to 150×10^{-3}	Tellurides	
Metacinnabarite, 4HgS	2×10^{-6} to 1×10^{-3}	Altaite, PbTe	20 to 200×10^{-6}
Millerite, NiS	2 to 4×10^{-7}	Calavarite, $AuTe_2$	6 to 12×10^{-6}
Molybdenite, MoS_2	0.12 to 7.5	Coloradoite, HgTe	4 to 100×10^{-6}
Pentlandite, $(Fe, Ni)_9S_8$	1 to 11×10^{-6}	Hessite, Ag_2Te	4 to 100×10^{-6}
Pyrrhotite, Fe_7S_8	2 to 160×10^{-6}	Nagyagite, $Pb_6Au(S, Te)_{14}$	20 to 80×10^{-6}
Pyrite, FeS_2	1.2 to 600×10^{-3}	Sylvanite, $AgAuTe_4$	4 to 20×10^{-6}
Sphalerite, ZnS	2.7×10^{-3} to 1.2×10^4	Oxides	
Antimony-sulfur compounds		Braunite, Mn_2O_3	0.16 to 1.0
Berthierite, $FeSb_2S_4$	0.0083 to 2.0	Cassiterite, SnO_2	4.5×10^{-4} to 10,000
Boulangerite, $Pb_5Sb_4S_{11}$	2×10^3 to 4×10^4	Cuprite, Cu_2O	10 to 50
Cylindrite, $Pb_3Sn_4Sb_2S_{14}$	2.5 to 60	Hollandite, $(Ba, Na, K)Mn_8O_{16}$	2 to 100×10^{-3}
Franckeite, $Pb_5Sn_3Sb_2S_{14}$	1.2 to 4	Ilmenite, $FeTiO_3$	0.001 to 4
Hauchecornite, $Ni_9(Bi, Sb)_2S_8$	1 to 83×10^{-6}	Magnetite, Fe_3O_4	52×10^{-6}
Jamesonite, $Pb_4FeSb_6S_{14}$	0.020 to 0.15	Manganite, $MnO \cdot OH$	0.018 to 0.5
Tetrahedrite, Cu_3SbS_3	0.30 to 30,000	Melaconite, CuO	6000
Arsenic-sulfur compounds		Psilomelane, $KMnO \cdot MnO_2 \cdot nH_2O$	0.04 to 6000
Arsenopyrite, FeAsS	20 to 300×10^{-6}	Pyrolusite, MnO_2	0.007 to 30
Cobaltite, CoAsS	6.5 to 130×10^{-3}	Rutile, TiO_2	29 to 910
Enargite, Cu_3AsS_4	0.2 to 40×10^{-3}	Uraninite, UO	1.5 to 200

From Carmichael, R. S., ed., *Handbook of Physical Properties of Rocks*, Vol. I, CRC Press, 1982.

REFERENCES

1. Beer, A. C., *Galvanomagnetic Effects in Semiconductors*, Academic Press, 1963.
2. Goryunova, N. A., *The Chemistry of Diamond-Like Semiconductors*, The MIT Press, 1965.
3. Abrikosov, N. Kh., Bankina, V. F., Poretskaya, L. E., Shelimova, L. E., and Skudnova, E. V., *Semiconducting II-VI, IV-VI, and V-VI Compounds*, Plenum Press, 1969.
4. Berger, L. I. and Prochukhan, V. D., *Ternary Diamond-Like Semiconductors*, Cons. Bureau/Plenum Press, 1969.
5. Shay, J. L. and Wernick, J. H., *Ternary Chalcopyrite Semiconductors: Growth, Electronic Properties, and Applications*, Pergamon Press, 1975.
6. Bergman, R., *Thermal Conductivity in Solids*, Clarendon, Oxford, 1976.
7. Handbook of Semiconductors, Vol. 1, Moss, T. S. and Paul, W., Eds., *Band Theory and Transport Properties*; Vol. 2, Moss, T. S. and Balkanski, M., Eds., *Optical Properties of Solids*; Vol. 3, Moss, T. S. and Keller, S. P., Eds., *Materials Properties and Preparation*, North Holland Publ. Co., 1980.
8. Böer, K. W., *Survey of Semiconductor Physics*, Van Nostrand Reinhold, 1990.
9. Rowe, D. M., Ed., *CRC Handbook of Thermoelectrics*, CRC Press, Boca Raton, FL, 1995.
10. Berger, L. I., *Semiconductor Materials*, CRC Press, Boca Raton, FL, 1997.
11. Glazov, V. M., Chizhevskaya, S. N., and Glagoleva, N. N., *Liquid Semiconductors*, Plenum Press, New York, 1969.
12. Phillips, J. C., *Bonds and Bands in Semiconductors*, Academic Press, New York, 1973.
13. Harrlson, W. A., *Electronic Structure and the Properties of Solids*, Freeman Publ. House, San Francisco, 1980.
14. Balkanski, M., Ed., *Optical Properties of Solids*, North-Holland, Amsterdam, 1980.
15. *Landolt-Börnstein. Numerical Data and Functional Relationships in Science and Technology, New Series, Group III: Crystal and Solid State Physics*, Hellwege, K.-H., and Madelung, O., Eds., Volumes 17 and 22, Springer Verlag, Berlin, 1984 (and further).
16. Shklovskii, B. L., and Efros, A. L., *Electronic Processes in Doped Semiconductors*, Springer Verlag, Berlin, 1984.
17. Cohen, M. L., and Chelikowsky, J. R., *Electronic Structure and Optical Properties of Semiconductors*, Springer Verlag, New York, 1988.
18. Glass, J. T., Messier, R. F., and Fujimori, N., Eds., *Diamond, Silicon Carbide, and Related Wide Bandgap Semiconductors*, MRS Symposia Proc. 1652, Mater. Res. Soc., Pittsburgh, 1990.
19. Palik, E., Ed., *Handbook of Optical Constants of Solids II*, Academic Press, New York, 1991.
20. Reed, M., Ed., *Semiconductors and Semimetals*, Volume 35, Academic Press, Boston, 1992.
21. Haug, H., and Koch, S. W., *Quantum Theory of the Optical and Electronic Properties of Semiconductors*, 2nd Edition, World Scientific, Singapore, 1993.
22. Lockwood, D. J., Ed., *Proc. 22nd Intl. Conf. on the Physics of Semiconductors, Vancouver, 1994*, World Scientific, Singapore, 1994.
23. Berger, L. I., *Semiconductor Materials*, CRC Press, Boca Raton, 1997.

DIFFUSION DATA FOR SEMICONDUCTORS

B. L. Sharma

The diffusion coefficient D in many semiconductors may be expressed by an Arrhenius-type relation

$$D = D_o \exp(-Q/kT)$$

where D_o is a frequency factor, Q is the activation energy for diffusion, k is the Boltzmann constant, and T is the absolute temperature. This table lists D_o and Q for various diffusants in common semiconductors.

Abbreviations used in the table are

AES — Auger Electron Spectroscopy
DLTS — Deep Level Transient Spectroscopy
SEM — Scanning Electron Microscopy
SIMS — Secondary Ion Mass Spectrometry
D(c) — Concentration Dependent Diffusion Coefficient
D_{max} — Maximum Diffusion Coefficient

(f) — Fast Diffusion Component
(i) — Interstitial Diffusion Component
(s) — Slow Diffusion Component
($\parallel$) — Parallel to c Direction
($\perp$) — Perpendicular to c Direction

Semiconductor	Diffusant	Frequency factor, D_o (cm²/s)	Activation energy, Q (eV)	Temperature range (°C)	Method of measurement	Ref.
Si	Li	2.5×10^{-3}	0.65	25—1350	Electrical	1
	Na	1.65×10^{-3}	0.77	800—1100	Electrical and flame photometry	2
	K	1.1×10^{-3}	0.76	800—1100	Electrical and flame photometry	2
	Cu	4×10^{-2}	1.0	800—1100	Radioactive	3
		4.7×10^{-3}	0.43 (i)	300—700	Radioactive	4
	Ag	2×10^{-3}	1.6	1100—1350	Radioactive	5
	Au	2.4×10^{-4}	0.39 (i)	700—1300	Radioactive	6
		2.75×10^{-3}	2.05 (s)			
	Be	$(D \sim 10^{-7})$	—	1050	Electrical	7
	Zn	1×10^{-1}	1.4	980—1270	Electrical	8
	B	2.46	3.59	1100—1250	Electrical	9
		2.4×10^{1}	3.87	840—1250	Electrical	10
	Al	1.38	3.41	1119—1390	Electrical	11
		1.8	3.2	1025—1175	Electrical	12
	Ga	3.74×10^{-1}	3.39	1143—1393	Electrical	11
		6×10^{1}	3.89	900—1050	Radioactive	13
	In	7.85×10^{-1}	3.63	1180—1389	Electrical	11
		1.94×10^{1}	3.86	1150—1242	Radioactive	14
	Tl	1.37	3.7	1244—1338	Electrical	11
		1.65×10^{1}	3.9	1105—1360	Electrical	15
	Ti	1.45×10^{-2}	1.79	950—1200	DLTS	16
	C	3.3×10^{-1}	2.92	1070—1400	Radioactive	17
	Si (self)	1.54×10^{2}	4.65	855—1175	SIMS	18
		1.6×10^{3}	4.77	1200—1400	Radioactive	19
	Ge	3.5×10^{-1}	3.92	855—1000	Radioactive	20
		2.5×10^{3}	4.97	1030—1302	Radioactive	20
		7.55×10^{3}	5.08	1100—1300	SIMS	21
	Sn	3.2×10^{1}	4.25	1050—1200	Neutron activation	22
	P	2.02×10^{1}	3.87	1100—1250	Electrical	9
		1.1	3.4	900—1200	Radioactive	23
		7.4×10^{-2}	3.3	1130—1405	Electrical	24
	As	6.0×10^{1}	4.2	950—1350	Radioactive	25
		6.55×10^{-2}	3.44	1167—1394	Electrical	26
		2.29×10^{1}	4.1	900—1250	Electrical	27
	Sb	1.29×10^{1}	3.98	1190—1398	Radioactive	28
		2.14×10^{-1}	3.65	1190—1405	Electrical	26
	Bi	1.03×10^{3}	4.64	1220—1380	Electrical	15
		1.08	3.85	1190—1394	Electrical	26

Semiconductor	Diffusant	Frequency factor, D_o (cm²/s)	Activation energy, Q (eV)	Temperature range (°C)	Method of measurement	Ref.
	Cr	1×10^{-2}	1	1100—1250	Radioactive	29
	O	7×10^{-2}	2.44	700—1250	SIMS	30
		1.4×10^{-1}	2.53	700—1160	SIMS	31
	S	5.95×10^{-3}	1.83	975—1200	Radioactive	32
	Se	9.5×10^{-1}	2.6	1050—1250	Electrical	33
	Te	$(D < 10^{-11})$	—	1050—1250	Electrical	33
	Mn	6.9×10^{-4}	0.63	900—1200	Radioactive	34
	Fe	1.3×10^{-3}	0.68	30—1250	Radioactive	35
	Co	2×10^{-3}	0.69	700—1300	Radioactive	36
	Ni	2×10^{-3}	0.47	800—1300	Radioactive	37
	Rh	$(D \sim 10^{-6}$—$10^{-4})$	—	1000—1200	Electrical	38
	Os	$(D \sim 2 \times 10^{-6})$	—	1280	Electrical	39
	Ir	4.2×10^{-2}	1.3	950—1250	Electrical	40
	Pt	$(D \sim 9.8 \times 10^{-16}$— $3.6 \times 10^{-14})$	—	700—800	DLTS	41
Ga	Li	1.3×10^{-3}	0.46	350—800	Electrical	42
		9.1×10^{-3}	0.57	800—500	Electrical	43
	Na	3.95×10^{-1}	2.03	700—850	Radioactive	44
	Cu	1.9×10^{-4}	0.18 (i)	750—900	Radioactive	45
		4×10^{-2}	0.99 (s)	600—700		
		4×10^{-3}	0.33 (i)	350—750	Radioactive	4
	Ag	4.4×10^{-2}	1.0 (i)	700—900	Radioactive	46, 47
		4×10^{-2}	2.23 (s)	800—900	Radioactive	48
	Au	2.25×10^{2}	2.5	600—900	Radioactive	49
	Be	5×10^{-1}	2.5	720—900	Electrical	50
	Zn	5	2.7	600—900	Radioactive and electrical	51
	Cd	1.75×10^{9}	4.4	760—915	Radioactive	52
	B	1.8×10^{9}	4.55	600—900	Electrical	51
	Al	1.0×10^{3}	3.45	554—905	SIMS	53
		$\sim 1.6 \times 10^{2}$	~ 3.24	750—850	Electrical	54
	Ga	1.4×10^{2}	3.35	554—916	SIMS	55
		3.4×10^{1}	3.1	600—900	Electrical	51
	In	1.8×10^{4}	3.67	554—919	SIMS	56
		3.3×10^{1}	3.02	700—855	Radioactive	57
	Tl	1.7×10^{3}	3.4	800—930	Radioactive	58
	Si	2.4×10^{-1}	2.9	650—900	(γ) resonance	59
	Ge (self)	2.48×10^{1}	3.14	549—891	Radioactive	60
		7.8	2.95	766—928	Radioactive	61
	Sn	1.7×10^{-2}	1.9	—	Radioactive	45
	P	3.3	2.5	600—900	Electrical	51
	As	2.1	2.39	700—900	Electrical	62
	Sb	3.2	2.41	700—855	Radioactive	57
		1.0×10^{1}	2.5	600—900	Radioactive and electrical	51
	Bi	3.3	2.57	650—850	—	63
	O	4×10^{-1}	2.08	—	Optical	64
	S	$(D \sim 10^{-9})$	—	920	—	65
	Se	$(D \sim 10^{-10})$	—	920	—	65
	Te	5.6	2.43	750—900	Radioactive	66
	Fe	1.3×10^{-1}	1.08	750—900	Radioactive	67
	Co	1.6×10^{-1}	1.12	750—850	Radioactive	47
	Ni	8×10^{-1}	0.9	670—900	Electrical	68
GaAs	Li	5.3×10^{-1}	1.0	250—500	Electrical and chemical	69
	Cu	3×10^{-2}	0.53	100—500	Radioactive	70

Semiconductor	Diffusant	Frequency factor, D_0 (cm²/s)	Activation energy, Q (eV)	Temperature range (°C)	Method of measurement	Ref.
		6×10^{-2}	0.98	450—750	Ultrasonic	71
		1.5×10^{-3}	0.6	800—1000	Radioactive	72
	Ag	4×10^{-4}	0.8	500—1150	Radioactive	73
	Au	1×10^{-3}	1.0	740—1025	Radioactive	74
	Be	7.3×10^{-6}	1.2	800—990	Electrical	75
	Mg	4×10^{-5}	1.22	800—1200	Electrical	76
	Zn	1.5×10^{1}	2.49	600—980	Radioactive	77
		2.5×10^{-1}	3.0	750—1000	Radioactive	78
	Cd	1.3×10^{-3}	2.2	800—1100	Radioactive	79
		5×10^{-2}	2.43	868—1149	Radioactive	77
	Hg	$(D \sim 5 \times 10^{-14})$	—	1100	Radioactive	80
	Al	$(D \sim 4 \times 10^{-18}$— $10^{-14})$	4.3	850—1100	AES	81
	Ga (self)	4×10^{-5}	2.6	1025—1100	Radioactive	82
		1×10^{7}	5.6	1125—1230	Radioactive	83
	Si	1.1×10^{-1}	2.5	850—1050	SIMS	84
	Ge	1.6×10^{-5}	2.06	650—850	SIMS	85
	Sn	6×10^{-4}	2.5	1060—1200	Radioactive	86
		1×10^{-5}	2	800—1000	Radioactive	87
	P	$(D \sim 10^{-12}$—$10^{-10})$	2.9	800—1150	Reflectance measurements	88
	As (self)	7×10^{-1}	3.2	—	Radioactive	78
	Cr	2.04×10^{-6}	0.83 (f)	750—1000	SIMS	89
			1.7 (s)	700—900		
		7.9×10^{-3}	2.2	800—1100	Chemical analysis	90
	O	2×10^{-3}	1.1	700—900	Mass spectroscopy	91
	S	1.85×10^{-2}	2.6	1000—1300	Radioactive	92
		1.1×10^{1}	2.95	750—900	Electrical	93
	Se	3×10^{3}	4.16	1025—1200	Radioactive	83
	Te	1.5×10^{-1}	3.5	1000—1150	Radioactive	94
	Mn	6.5×10^{-1}	2.49	850—1100	Radioactive	95
	Fe	4.2×10^{-2}	1.8	850—1150	Radioactive	96
		2.2×10^{-3}	2.32	750—1050	Radioactive	97
	Co	5×10^{2}	2.5	800—1000	Radioactive	98
		1.2×10^{-1}	2.64	750—1050	Radioactive	97
	Tm	2.3×10^{-16}	1.0	800—1000	Radioactive	99
GaSb	Li	2.3×10^{-4}	1.9 (s)	527—657	Electrical and flame photometry	100
		1.2×10^{-1}	0.7 (f)	277—657		
	Cu	4.7×10^{-3}	0.9	470—650	Radioactive	101
	Zn	$(D \sim 2 \times 10^{-13}$— $1 \times 10^{-11})$	2	510—600	Radioactive	102
	Cd	1.5×10^{-6}	0.72	640—800	Electrical	103
	Ga (self)	3.2×10^{3}	3.15	658—700	Radioactive	104
	In	1.2×10^{-7}	0.53	320—650	Radioactive	105
	Sn	2.4×10^{-5}	0.8	320—650	Radioactive	106
		1.3×10^{-5}	1.1	500—650	Radioactive	107
	Sb (self)	3.4×10^{4}	3.45	658—700	Radioactive	104
	Se	$(D \sim 2.4 \times 10^{-13}$— $1.37 \times 10^{-11})$	—	400—500	Radioactive	106
	Te	3.8×10^{-4}	1.20	320—650	Radioactive	106
	Fe	5×10^{-2}	1.9 (I)	500—650	Radioactive	108
		5×10^{2}	2.3 (II)	500—650		
GaP	Ag	—	—	1000—1300	Radioactive	109
	Au	8	2.5 (I)	1050—1250	Radioactive	110
		20	2.4 (II)	1100—1250	Diffusion (I) A face and (II) B face	

Semiconductor	Diffusant	Frequency factor, D_o (cm^2/s)	Activation energy, Q (eV)	Temperature range (°C)	Method of measurement	Ref.
	Be	$(D_{max} \sim 2.4 \times 10^{-9}$—$8.5 \times 10^{-8})$	—	900—1000	Atomic absorption analysis	111
	Mg	5×10^{-5}	1.4	700—1050	Electrical	112
	Zn	1.0	2.1	700—1300	Radioactive	113
	Ge	—	—	900—1000	Radioactive	114
	Cr	6.2×10^{-4}	1.2	900—1130	Radioactive; ESR	115
	S	3.2×10^3	4.7	1120—1305	Radioactive	92
	Mn	2.1×10^9	4.7	$T < 950$	Radioactive; ESR	116
		1.1×10^{-6}	0.9	950—1130		
	Fe	1.6×10^{-1}	2.3	980—1180	Radioactive	117
	Co	2.8×10^{-3}	2.9	850—1100	Radioactive	118
InP	Cu	3.8×10^{-3}	0.69	600—900	Radioactive	119
	Ag	3.6×10^{-4}	0.59	500—900	Radioactive	120
	Au	1.32×10^{-5}	0.48	600—820	Radioactive	121
		1.37×10^{-4}	0.73	600—900	Radioactive	122
	Zn	1.6×10^{-8}	0.3	750—900	Electrical	123
		$(D \sim 2 \times 10^{-9}$—$4 \times 10^{-8})$	—	700—900	Radioactive	124
	Cd	1.8	1.9	700—900	Radioactive	125
		1.1×10^{-7}	0.72	700—900	Electrical	123
		$(D \sim 7 \times 10^{-13}$—$2 \times 10^{-10})$	—	450—650	Electrical	126
	In (self)	1×10^5	3.85	830—990	Radioactive	83
	Sn	$(D \sim 3 \times 10^{-8})$	—	550	Etching and cathodoluminescence	127
	P (self)	7×10^{10}	5.65	900—1000	Radioactive	83
	Cr	—	—	600—900	Radioactive	128
	S	—	—	607—710	Electrical	129
	Mn	—	2.9	650—750	SIMS	130
	Fe	3	2	600—950	Radioactive	131
		6.8×10^5	3.4	600—700	SIMS	132
	Co	9×10^{-1}	1.8	600—950	Radioactive	131
InAs	Cu	3.6×10^{-3}	0.52	342—875	Radioactive	133
		2.2×10^{-2}	0.54	525—890	Radioactive	134
	Ag	7.3×10^{-4}	0.26	450—900	Radioactive	135
	Au	5.8×10^{-3}	0.65	600—900	Radioactive	136
	Mg	1.98×10^{-6}	1.17	600—900	Electrical	137
	Zn	4.2×10^{-3}	0.96	600—900	Radioactive	138
		3.11×10^{-3}	1.17	600—900	Electrical	137
	Cd	7.4×10^{-4}	1.15	650—900	Radioactive	139
	Hg	1.45×10^{-5}	1.32	650—850	Radioactive	140
	In (self)	6×10^5	4.0	740—900	Radioactive	141
	Ge	3.74×10^{-6}	1.17	600—900	Electrical	137
	Sn	1.49×10^{-6}	1.17	600—900	Electrical	137
	As (self)	3×10^7	4.45	740—900	Radioactive	141
	S	6.78	2.2	600—900	Electrical	137
	Se	12.6	2.2	600—900	Electrical	137
	Te	3.43×10^{-5}	1.28	600—900	Electrical	137
InSb	Li	7×10^{-4}	0.28	0—210	Electrical	142
	Cu	9×10^{-4}	1.08	200—500	Radioactive	143
		3×10^{-5}	0.37	230—490	Radioactive	144
	Ag	1×10^{-7}	0.25	440—510	Radioactive	145
	Au	7×10^{-4}	0.32	140—510	Radioactive	144
	Zn	5×10^{-1}	1.35	362—508	Radioactive	146
		—	1.5	355—455	SIMS	147
	Cd	1×10^{-5}	1.1	250—500	Radioactive	148

Semiconductor	Diffusant	Frequency factor, D_o (cm²/s)	Activation energy, Q (eV)	Temperature range (°C)	Method of measurement	Ref.
		1.3×10^{-4}	1.2	360—500	Electrical	149
	Hg	4×10^{-6}	1.17	425—500	Radioactive	150
	In (self)	5×10^{-2}	1.82	478—520	Radioactive	104
		1.8×10^{13}	4.3	475—517	Radioactive	151
	Sn	5.5×10^{-8}	0.75	390—512	Radioactive	152
	Pb	$(D \sim 2.7 \times 10^{-15})$	—	500	Radioactive	153
	Sb (self)	5×10^{-2}	1.94	478—520	Radioactive	104
		3.1×10^{13}	4.3	475—517	Radioactive	151
	S	9×10^{-2}	1.4	360—500	Electrical	154
	Se	1.6	1.87	380—500	Electrical	155
	Te	1.7×10^{-7}	0.57	300—500	Radioactive	156
	Fe	1×10^{-7}	0.25	440—510	Radioactive	145
	Co	2.7×10^{-11}	0.39	420—500	Radioactive	157
AlAs	Ga	$(D \sim 2 \times 10^{-18}$— $10^{-15})$	3.6	850—1100	AES	81
	Zn	$(D \sim 9 \times 10^{-11})$	—	557	SEM	158
AlSb	Cu	3.5×10^{-3}	0.36	150—500	Radioactive	159
	Zn	3.3×10^{-1}	1.93	660—860	Radioactive	160
	Cd	$D(c) \sim 4 \times 10^{-12}$— 3×10^{-10}	—	900	Radioactive	161
	Al (self)	2	1.88	570—620	X-ray	162
	Sb (self)	1	1.7	570—620	X-ray	162
ZnS	Cu	2.6×10^{-3}	0.79	470—750	Radioactive	163
		4.3×10^{-4}	0.64	250—1200	Electroluminescence	164
		9.75×10^{-3}	1.04	400—800	Luminescence	165
	Au	1.75×10^{-4}	1.16	500—800	Radioactive	166
	Zn (self)	3×10^{-4}	1.5	925<T<940	Radioactive	167
		1.5×10^4	3.26	940<T<1030		
		1×10^{16}	6.5	1030<T<1075		
	Cd	$(D \sim 10^{-10})$	—	1100	Luminescence	168
	Al	5.69×10^{-4}	1.28	800—1000	Luminescence	167
	In	3×10^1	2.2	750—1000	Radioactive	163
	S (self)	2.16×10^4	3.15	600—800	Radioactive	166
		8×10^{-5}	2.2	740—1100	Radioactive	169
	Se	$(D \sim 5 \times 10^{-13})$	—	1070	X-ray microprobe	170
	Mn	2.3×10^3	2.46	500—800	Radioactive	166
ZnSe	Cu	1×10^{-4}	0.66	400—800	Luminescence	171
		1.7×10^{-5}	0.56	200—570	Radioactive	172
	Ag	2.2×10^{-2}	1.18	400—800	Luminescence	171
	Zn (self)	9.8	3.0	760—1150	Radioactive	173
	Cd	6.39×10^{-4}	1.87	700—950	Photoluminescence	174
	Al	2.3×10^{-2}	1.8	800—1100	Luminescence	171
	Ga	1.81×10^2	3.0	900—1100	Luminescence	165
		—	1.3	700—850	Electron probe	175
	In	$(D \sim 2 \times 10^{-12})$	—	940	—	176
	S	$(D \sim 8 \times 10^{-12})$	—	1060	X-ray microprobe	170
	Se (self)	1.3×10^1	2.5	860—1020	Radioactive	173
		2.3×10^{-1}	2.7	1000—1050	Radioactive	177
	Ni	$(D \sim 1.5 \times 10^{-8}$— $1.7 \times 10^{-7})$	—	740—910	Luminescence	165
ZnTe	Li	2.9×10^{-2}	1.22 (s)	400—700	Nuclear and chemical analysis	178
		1.7×10^{-4}	0.78 (f)			
	Zn (self)	2.34	2.56	760—860	Radioactive	179
		1.4×10^1	2.69	667—1077	Radioactive	180
	Al	—	2.0	700—1000	Electrical and optical	181

Semiconductor	Diffusant	Frequency factor, D_o (cm²/s)	Activation energy, Q (eV)	Temperature range (°C)	Method of measurement	Ref.
CdS	In	4	1.96	1100—1300	Radioactive	182
	Te (self)	2×10^4	3.8	727—977	Radioactive	180
	Li	3×10^{-6}	0.68	610—960	Microhardness	183
	Na	$(D \sim 3 \times 10^{-7})$	—	800	Radioactive	184
	Cu	1.5×10^{-3}	0.76	400—700	Radioactive	185
		1.2×10^{-2}	1.05	300—700	Ultrasonic	186
		8×10^{-5}	0.72	20—200	Electrical	187
	Ag	2.5×10^1	1.2 (s)	300—500	Radioactive	188
		2.4×10^{-1}	0.8 (f)			
	Au	2×10^2	1.8	500—800	Radioactive	189
	Zn	1.27×10^{-9}	0.86 (s)	720—1000	Radioactive	190
		1.22×10^{-8}	0.66 (f)			
	Cd (self)	3.4	2.0	700—1100	Radioactive	191
	Ga	—	—	667—967	Optical and microprobe	192
	In	6×10^1	2.3 ($\parallel$)	650—930	Radioactive, optical and microprobe	193
		1×10^1	2.03 ($\perp$)			
	P	6.5×10^{-4}	1.6	800—1100	Radioactive	194
	S (self)	1.6×10^{-2}	2.05	800—900	Radioactive	195
		—	2.4	750—1050	Radioactive	196
	Se	$(D \sim 1.2 \times 10^{-9})$	—	900	Radioactive	177
	Te	1.3×10^{-7}	10.4	700—1000	Radioactive	197
	Cl	$(D \sim 3 \times 10^{-10})$	—	800	Electrical	184
	I	$(D \sim 5 \times 10^{-12})$	—	1000	Radioactive	184
	Ni	6.75×10^{-3}	10.9	570—900	Luminescence	165
	Yb	$(D \sim 1.3 \times 10^{-9})$	—	960	Photoluminescence	198
CdSe	Ag	2×10^{-4}	0.53	22—400	Ultrasonic	199
	Cd (self)	1.6×10^{-3}	1.5	700—1000	Radioactive	200
		6.3×10^{-2}	1.25 (I)	600—900	Radioactive;	201
		4.12×10^{-2}	2.18 (II)	600—900	(I) saturated Cd and (II) saturated Se pressure	
	P	$(D \sim 5.3 \times 10^{-12}$— $6 \times 10^{-11})$	—	900—1000	Radioactive	202
	Se (self)	2.6×10^3	1.55	700—1000	Radioactive; saturated Se pressure	177
CdTe	Li	$(D \sim 1.5 \times 10^{-10})$	—	300	Ion microprobe	203
	Cu	3.7×10^{-4}	0.67	97—300	Radioactive	204
		8.2×10^{-8}	0.64	290—350	Ion backscattering	205
	Ag	—	—	700—800	Electrical and photoluminescence	206
	Au	6.7×10^1	2.0	600—1000	Radioactive	207
	Cd (self)	1.26	2.07	700—1000	Radioactive	208
		3.26×10^2	2.67 (I)	650—900	Radioactive;	209
		1.58×10^1	2.44 (II)		(I) saturated Cd and (II) saturated Te pressure	
	In	8×10^{-2}	1.61	650—1000	Radioactive	210
		$1.17 \times^2$	2.21 (I)	500—850	Radioactive; (I) saturated Cd and (II) saturated Te pressure	211
		6.48×10^{-4}	1.15 (II)			
	Sn	8.3×10^{-2}	2.2	700—925	Radioactive	212
	P	$(D \sim 1.2 \times 10^{-10})$	—	900	Radioactive	202
	As	—	—	850	—	213
	O	5.6×10^{-9}	1.22	200—650	Mass spectrometry	214
		6.0×10^{-10}	0.29	650—900		
	Se	1.7×10^{-4}	1.35	700—1000	Radioactive	215

Semiconductor	Diffusant	Frequency factor, D_o (cm²/s)	Activation energy, Q (eV)	Temperature range (°C)	Method of measurement	Ref.
	Te (self)	8.54×10^{-7}	1.42 (I)	600—900	Radioactive; (I) saturated Cd and (II) saturated Te pressure	209
		1.66×10^{-4}	1.38 (II)	500—800		
	Cl	7.1×10^{-2}	1.6	520—800	Radioactive	216
	Fe	$(D \sim 4 \times 10^{-8})$	0.77	900	Radioactive	217
HgSe	Sb	6.3×10^{-5}	0.85	540—630	Radioactive	218
	Se (self)	—	—	200—400	Radioactive	219
HgTe	Ag	6×10^{-4}	0.8	250—350	Radioactive	220
	Zn	5×10^{-8}	0.6	250—350	Radioactive	220
	Cd	3.1×10^{-4}	0.66	250—350	Radioactive	221
	Hg (self)	2×10^{-8}	0.6	200—350	Radioactive	220
	In	6×10^{-6}	0.9	200—300	Radioactive	220
	Sn	1.72×10^{-6}	0.66 (s)	200—300	Radioactive	222
		1.8×10^{-3}	0.80 (f)			
	Te (self)	10^{-6}	1.4	200—400	Radioactive	220
	Mn	1.5×10^{-4}	1.3	250—350	Radioactive	220
PbS	Cu	4.6×10^{-4}	0.36	150—450	Electrical	223
		5×10^{-3}	0.31	100—400	Electrical	224
	Pb (self)	8.6×10^{-5}	1.52	500—800	Radioactive	225
	S (self)	6.8×10^{-5}	1.38	500—750	Radioactive	226
	Ni	1.78×10^{1}	0.95	200—500	Electrical	227
PbSe	Na	1.5×10^{1}	1.74 (s)	400—850	Radioactive	228
		5.6×10^{-6}	0.4 (f)			
	Cu	2×10^{-5}	0.31	93—520	Radioactive	229
	Ag	7.4×10^{-4}	0.35	400—850	Radioactive	228
	Pb (self)	4.98×10^{-6}	0.83	400—800	Radioactive	230
	Sb	3.4×10^{-1}	2.0	650—850	Radioactive	231
	Se (self)	2.1×10^{-5}	1.2	650—850	Radioactive	231
	Cl	1.6×10^{-8}	0.45	400—850	Radioactive	232
	Ni	$(D \sim 1 \times 10^{-10})$	—	700	Radioactive	233
PbTe	Na	1.7×10^{-1}	1.91	600—850	Radioactive	234
	Sn	3.1×10^{-2}	1.56	500—800	Radioactive	231
	Pb (self)	2.9×10^{-5}	0.6	250—500	Radioactive	231
	Sb	4.9×10^{-2}	1.54	500—800	Radioactive	231
	Te	2.7×10^{-6}	0.75	500—800	Radioactive	231
	Cl	$(D > 2.3 \times 10^{-10})$	—	700	Radioactive	235
	Ni	$(D > 1 \times 10^{-6})$	—	700	Radioactive	235

REFERENCES

1. E. M. Pell, *Phys. Rev.,* 119, 1960; 119, 1014, 1960.
2. L. Svob, *Solid State Electron,* 10, 991, 1967.
3. B. I. Boltaks and I. I. Sosinov, *Zh. Tekh. Fiz.,* 28, 3, 1958.
4. R. N. Hall and J. N. Racette, *J. Appl. Phys.,* 35, 379, 1964.
5. B. I. Boltaks and Hsueh Shih-Yin, *Sov. Phys. Solid State,* 2, 2383, 1961.
6. W. R. Wilcox and T. J. LaChapelle, *J. Appl. Phys.,* 35, 240, 1964.
7. E. A. Taft and R. O. Carlson, *J. Electrochem. Soc.,* 117, 711, 1970.
8. R. Sh. Malkovich and N. A. Alimbarashvili, *Sov. Phys. Solid State,* 4, 1725, 1963.
9. R. N. Ghoshtagore, *Solid State Electron,* 15, 1113, 1972.
10. C. Hill, *Semiconductor Silicon 1981,* H. R. Huff, R. J. Kreiger, and Y. Takeishi, Eds., p. 988, *Electrochem. Soc.,* 1981.
11. R. N. Ghoshtagore, *Phys. Rev. B,* 3, 2507, 1971.
12. W. Rosnowski, *J. Electrochem. Soc.,* 125, 957, 1978.

13. J. S. Makris and B. J. Masters, *J. Appl. Phys.,* 42, 3750, 1971.
14. M. F. Millea, *J. Phys. Chem. Solids,* 27, 315, 1965 (refer Reference 2).
15. C. S. Fuller and J. A. Ditzenberger, *J. Appl. Phys.,* 27, 544, 1956.
16. S. Hocine and D. Mathiot, *Appl. Phys. Lett.,* 53, 1269, 1988.
17. R. C. Newman and J. Wakefield, *J. Phys. Chem. Solids,* 19, 230, 1961.
18. L. Kalinowski and R. Seguin, *Appl. Phys. Lett.,* 35, 211, 1979; *Appl. Phys. Lett.,* 36, 171, 1980.
19. R. F. Peart, *Phys. Stat. Sol.,* 15, K 119, 1966.
20. G. Hettich, H. Mehrer and K. Maler, *Inst. Phys. Conf. Ser.,* 46, 500, 1979.
21. M. Ogina, Y. Oana and M. Watanabe, *Phys. Stat. Sol. (a),* 72, 535, 1982.
22. T. H. Yeh, S. M. Hu, and R. H. Kastl, *Appl. Phys.,* 39, 4266, 1968.
23. I. Franz and W. Langheinrich, *Solid State Electron,* 14, 835, 1971.
24. R. N. Ghoshtagore, *Hys. Rev. B,* 3, 389, 1971.
25. B. J. Masters and J. M. Fairfield, *J. Appl. Phys.,* 40, 2390, 1969.
26. R. N. Goshtagore, *Phys. Rev. B,* 3, 397, 1971.
27. R. S. Fair and J. C. C. Tsai, *J. Electrochem. Soc.,* 122, 1689, 1975.
28. J. J. Rohan, N. E. Pickering and J. Kennedy, *J. Electrochem. Soc.,* 106, 705, 1969.
29. W. Wuerker, K. Roy, and J. Hesse, *Matsr. Res. Bull.,* 9, 971, 1974.
30. J. C. Mikkelsen, Jr., *Appl. Phys. Lett.,* 40, 336, 1982.
31. S. Tang Lee and D. Nicols, *Appl. Phys. Lett.,* 47, 1001, 1985.
32. P. L. Gruzin, S. V. Zemskii, A. D. Bullkin, and N. M. Makarov, *Sov. Phys. Sem.,* 7, 1241, 1974.
33. N. S. Zhdanovich and Yu. I. Kozlov, *Svoistva Legir, Poluprovodn.,* V. S. Zemskov, Ed., Nauka, Moscow, 1977, 115-120; *Fiz Tekh. Poluprovod.,* 9, 1594, 1975.
34. D. Gilles, W. Bergholze, and W. Schroeter, *J. Appl. Phys.,* 59, 3590, 1986.
35. E. R. Weber, *Appl. Phys. A,* 30, 1, 1983.
36. E. R. Weber, Properties of Silicon, EMIS Datareviews Ser. No. 4, INSPEC Publications, 1988, 409-451.
37. M. K. Bakhadyrkhanov, S. Zainabidinov, and A. Khamidov, *Sov. Phys. Sem.,* 14, 243, 1980.
38. S. A. Azimov, M. S. Yunosov, F. K. Khatamkulov, and G. Nasyrov, *Poluprovod.,* N. Kh. Abrikosov and V. S. Zemskov, Eds., Nauka, Moscow, 1975, 21-23.
39. S. A. Azimov, M. S. Yunosov, G. Nurkuziev, and F. R. Karimov, *Sov. Phys. Sem.,* 12, 981, 1978.
40. S. A. Azimov, B. V. Umarov, and M. S. Yunusov, *Sov. Phys. Sem.,* 10, 842, 1976.
41. S. Mantovani, F. Nava, C. Nobili, and G. Ottaviani, *Phys. Rev. B,* 33, 5536, 1986.
42. C. S. Fuller and J. A. Ditzenberger, *Phys. Rev.,* 91, 193, 1953.
43. B. Pratt and F. Friedman, *J. Appl. Phys.,* 37, 1893, 1966.
44. M. Stojic, V. Spiric, and D. Kostoski, *Inst. Phys. Conf. Ser.,* 31, 304, 1976.
45. B. I. Boltaks, *Diffusion in Semiconductors,* Inforsearch, London, 1963, 162.
46. A. A. Bugai, V. E. Kosenko, and E. G. Miselyuk, *Zh. Tekh. Fiz.,* 27, 67, 1957.
47. L. Y. Wei, *J. Phys. Chem. Solids,* 18, 162, 1961.
48. V. E. Kosenko, *Sov. Phys. Solid State,* 4, 42, 1962.
49. W. C. Dunlap, Jr., *Phys. Rev.,* 97, 614, 1955
50. Yu. I. Belyaev and V. A. Zhidkov, *Sov. Phys. Solid State,* 3, 133, 1961.
51. W. C. Dunlap, Jr. *Phys. Rev.,* 94, 1531, 1954.
52. V. E. Kosenko, *Sov. Phys. Solid State,* 1, 1481, 1960.
53. P. Dorner, W. Gust, A. Lodding, H. Odelius, B. Predel, and U. Roll, *Acta Metall.,* 30, 941, 1982.
54. W. Meer and D. Pommerrening, *Z. Agnew. Phys.,* 23, 369, 1967.
55. U. Sodervall, H. Odelius, A. Lodding, U. Roll, B. Predel, W. Gust, and P. Dorner, *Phil. Mag. A,* 54, 539, 1986.
56. P. Dorner, W. Gust, A. Lodding, H. Odelius, B. Predel, and U. Roll, *Z. Metalkd.,* 73, 325, 1982.
57. P. V. Pavlov, *Sov. Phys. Solid State,* 8, 2377, 1967.
58. V. I. Tagirov and A. A. Kuliev, *Sov. Phys. Solid State,* 4, 196, 1962.
59. J. Raisanen, J. Hirvonen, and A. Anttila, *Solid State Electron.,* 24, 333, 1981.
60. C. Vogel, G. Hettich, and H. Mehrer, *J. Phys. C.,* 16, 6197, 1983.
61. H. Letaw, Jr., W. M. Portnoy, and L. Slifkin, *Phys. Rev.,* 102, 363, 1956.
62. W. Bosenberg, *Z. Naturforsch.,* 10a, 285, 1955.
63. V. M. Glazov and V. S. Zemskov, Physicochemical Principles of Semiconductor Doping, Israel Program for Scientific Translation, Jerusalem, 1968.
64. J. W. Corbett, R. S. McDonald, and G. D. Watkins, *J. Phys. Chem. Solids,* 25, 873, 1964.
65. W. W. Tyler, *J. Phys. Chem. Solids,* 8, 59, 1959.
66. V. D. Ignatkov and V. E. Kosenko, *Sov. Phys. Solid State,* 4, 1193, 1962.
67. A. A. Bugal, V. E. Kosenko, and E. G. Miseluk, *Zh. Tekh. Fiz.,* 27, 210, 1957.
68. F. van der Maesen and J. A. Brenkman, *Phillips Res. Rep.,* 9, 255, 1954.
69. C. S. Fuller and K. B. Wolfstrin, *J. Appl. Phys.,* 33, 2507, 1962.

70. R. N. Hall and J. H. Racette, *J. Appl. Phys.,* 25, 379, 1964.

72. V. S. Vasilev, I. N. Kamevoskii, and V. B. Osvenskii, *Sov. Phys. Sem.,* 2, 1495, 1969.

73. B. I. Boltaks and F. S. Shishiyanu, *Sov. Phys. Solid State,* 5, 1680, 1964.

74. V. I. Sokolov and F. S. Shishiyanu, *Sov. Phys. Solid State,* 6, 265, 1964.

75. E. A. Poltoratskii and V. M. Stuchebnikov, *Sov. Phys. Solid State,* 8, 770, 1966.

76. E. I. Andreivckii, S. V. Mashkin, and S. S. Zludkov, *Diffusion in Semiconductor,* (Russian), Gorkii, 1969.

77. B. Goldstein, *Phys. Rev.,* 118, 1024, 1960.

78. D. L. Kendall, *Semiconductors and Semimetals,* Vol. 4, R. K. Willardson and A. C. Beer, Eds., Academic, 163-259.

79. M. Fujimoto, K. Kudo, and N. Hishinuma, *Jpn. J. Appl. Phys.,* 8, 725, 1969.

80. J. A. Kanz, (p. 227 Reference 78).

81. L. L. Chang and A. Koma, *Appl. Physics Lett.,* 29, 138, 1976.

82. H. D. Palfrey, M. Brown, and A. F. W. Willoughby, *J. Electrochem. Soc.,* 128, 2224, 1981.

83. B. Goldstein, *Phys. Rev.,* 121, 1305, 1961.

84. M. E. Greiner and J. B. Gibbons, *Appl. Phys. Lett.,* 44, 750, 1984.

85. K. Sarma, R. Dalby, K. Rose, O. Aina, W. Katz, and N. Lewis, *J. Appl. Phys.,* 56, 2703, 1984.

86. S. Goldstein and H. Keller, *J. Appl. Phys.,* 32, 1180, 1961.

87. H. Yamazaki, Y. Kawaski, M. Futjimoto, and K. Kudo, *Jpn. J. Appl. Phys.,* 14, 717, 1975.

88. G. C. Jain, D. K. Sadana, and B. K. Das, *Solid State Electron.,* 19, 731, 1976.

89. M. D. Deal and D. A. Stevenson, *J. Appl. Phys.,* 59, 2398, 1986.

90. S. S. Khludkov, G. L., Prikhodko, and T. A. Karchina, *Izv. Akad. Nauk SSR, Heorg. Mater.,* 8, 1044, 1972.

91. J. Rachmann and R. Biermann, *Solid State Commun.,* 7, 1771, 1969.

92. A. S. Young and G. L. Pearson, *J. Phys. Chem. Solids,* 31, 517, 1970.

93. H. Matino, *Solid State Electron.,* 17, 35, 1974.

94. T. A. Karelina, T. T. Lavrischev, G. L. Prokhodko, and S. S. Khludkov, *Izv. Akad. Nauk SSSR, Neorg. Mater.,* 10, 228, 1974.

95. M. S. Seltzer, *J. Phys. Chem. Solids,* 26, 243, 1965.

96. B. I. Boltaks, G. S. Kulikov, I. N. Nikulista, and F. S. Shishiyanu, *Inorg. Mater.,* 11, 292, 1975.

97. V. A. Uskov and V. P. Sorvina, *Izv. Akad. Nauk SSSR, Neorg. Mater.,* 8, 758, 1972.

98. T. D. Dzhafarov, E. A. Skoryatina, E. S. Guds, and I. E. Moronchuk, *Phys. Stat. Sol. (a),* 51, K221, 1979.

99. H. C. Casey, Jr. and G. L. Pearson, *J. Appl. Phys.,* 35, 3401, 1965.

100. M. H. van Maaren, *Phys. Stat. Sol.,* 24, K125, 1967.

101. A. I. Blashku and T. D. Dzhafarov, *Sov. Phys. Solid State,* 15, 536, 1973.

102. A. I. Blashku, B. I. Boltaks, I. I. Burdiyan, T. D. Dzhafarov, and M. A. Rzaev, *Sov. Phys. Sem.,* 6, 402, 1972.

103. J. Bougnot, L. Szepessy, and S. F. DaCunka, *Phys. Stat. Sol.,* 26, K127, 1968.

104. F. H. Eissen and C. E. Birchenall, *Acta Met.,* 5, 265, 1957.

105. R. K. Willardson, *Ultrapurification of Materials,* (Conf. Report, Boston), Macmillan, 1962.

106. B. I. Boltaks and Yu. A. Gutorov, *Sov. Phys. Solid State,* 1, 930, 1960.

107. V. A. Uskov, *Fiz. i Tekhn. Poluprov.,* 8, 2414, 1974.

108. F. S. Shishiyanu, *Diffusion and Degradation in Semiconductor Materials,* (Russian), Kishinev, 1978, 55.

109. S. V. Shudyakov, *Sov. Phys. Sem.,* 15, 4, 1981.

110. T. D. Dzhafarov, A. A. Litvin, and S. V. Khudyakov, *Sov. Phys. Solid State,* 20, 152, 1978.

111. M. Ilegems and W. C. O'Mara, *J. Appl. Phys.,* 43, 1190, 1972.

112. F. S. Shishiyanu, *Diffusion and Degradation in Semiconductor Materials,* (Russian), Kishinev, 1978, 78.

113. H. W. Allison, *J. Appl. Phys.,* 34, 231, 1963.

114. M. Schneider and E. Nebaur, *Phys. Stat. Sol. (a),* 32, 333, 1975.

115. V. I. Kirillov, N. N. Pribylov, S. I. Rembeza, and A. I. Spirin, *Sov. Phys. Sem.,* 12, 1342, 1978.

116. V. I. Kirillov, N. N. Pribylov, S. I. Rembeza, and A. I. Spirin, *Sov. Phys. Solid State,* 22, 1945, 1980.

117. F. S. Shishiyanu and V. G. Georgiu, *Sov. Phys. Sem.,* 10, 1301, 1976.

118. T. D. Dzhafarov, Yu. P. Demakov, and N. N. Pribylov, *Fiz. Tverd. Tela,* 17, 3110, 1975.

119. K. A. Arseni, *Sov. Phys. Solids State,* 10, 2263, 1969.

120. K. A. Arseni and B. I. Boltaks, *Sov. Phys. Solid State,* 10, 2190, 1969.

121. S. I. Rembeza, *Sov. Phys. Sem.,* 3, 519, 1969.

122. K. A. Arseni, *Sov. Phys. Sem.,* 2, 1454, 1969.

123. R. M. Kundukhov, S. G. Metreveli, and N. V. Siukaev, *Sov. Phys. Sem.,* 1, 765, 1967.

124. L. L. Chang and H. Y. Casey, Jr., *Solid State Electron,* 7, 481, 1964.

125. K. A. Arseni, B. I. Boltaks, V. I. Gordin, and Ya. A. Ugai, *Izv. Akad. Nauk SSSR, Neorg. Mater.,* 3, 1679, 1967.

126. K. Ohtsuka, T. Nishino, and Y. Hamakawa, *Jpn. J. Appl. Phys.,* 21, 1170, 1982.

127. A. K. Chin, I. Camlibel, B. V. Dutt, V. Swaminathan, W. A. Bonner, and A. A. Ballman, *Appl. Phys. Lett.,* 42, 901, 1983.

128. M. R. Brozel, E. J. Foulkes, and B. Tuck, *Phys. Stat. Sol. (a),* 72, K159, 1982; *Electron. Lett.,* 17, 532, 1981.

129. T. Kawakami and M. Okamura, *Electron. Lett.,* 15, 503, 1979.

130. R. Chaplain, M. Gaunear, H. H. L'Haridon, and R. Rupert, *J. Appl. Phys.,* 58, 1803, 1985.

131. F. S. Shishiyanu, V. Gh. Gheorgiu, and S. K. Palazov, *Phys. Stat. Sol. (a)*, 40, 29, 1977.
132. R. E. Holmes and R. G. Wilson, *J. Appl. Phys.*, 52, 3396, 1981.
133. C. S. Fuller and K. B. Wolfstirn, *J. Electrochem. Soc.*, 114, 856, 1967.
134. B. I. Boltaks, S. I. Rembeza, and M. K. Bakhadyrkhanov, *Sov. Phys. Solid State*, 10, 432, 1968.
135. B. I. Boltaks, S. I. Rembeza, and B. L. Sharma, *Sov. Phys. Sem.*, 1, 196, 1967.
136. S. I. Rembeza, *Sov. Phys. Sem.*, 1, 516, 1967.
137. E. Schillman, *Z. Naturforsch.*, 11a, 472, 1956.
138. B. I. Boltaks and S. I. Rembeza, *Sov. Phys. Solid State*, 8, 2177, 1967.
139. K. A. Arseni, B. I. Boltaks and S. I. Rembeza, *Sov. Phys. Solid State*, 8, 2248, 1967.
140. B. L. Sharma, R. K. Purohit, and S. N. Mukerjee, *J. Phys. Chem. Solids*, 32, 1397, 1971.
141. H. Kato, M. Yokozawa, R. Kohara, Y. Okabayashi, and S. Takayanagi, *Solid State Electron.*, 12, 137, 1969.
142. T. T. Takabatake, H. Ikari, and Y. Uyeda, *Jpn. J. Appl. Phys.*, 5, 839, 966.
143. H. J. Stocker, *Phys. Rev.*, 130, 2160, 1963.
144. B. I. Boltaks and V. I. Sokolov, *Sov. Phys. Solid State*, 6, 600, 1964.
145. L. A. K. Watt and W. S. Chen, *Bull. Am. Phys. Soc.*, 7, 89, 1962.
146. B. Goldstein, *Properties of Elemental and Compound Semiconductors*, Interscience, 1960, 155.
147. K. Nishitani, K. Hagahama, and T. Murotani, *Jpn. J. Appl. Phys.*, 22, 836, 1983.
148. B. I. Boltaks and V. I. Sokolov, *Sov. Phys. Solid State*, 5, 785, 1963.
149. R. B. Wilson and E. L. Heasell, *Proc. Soc. (London)*, 79, 403, 1962.
150. I. A. Gusev and A. N. Murin, *Sov. Phys. Solid State*, 6, 1229, 1964.
151. D. L. Kendall and R. A. Huggins, *J. Appl. Phys.*, 40, 2750, 1969.
152. S. M. Sze and L. Y. Wei, *Phys. Rev.*, 124, 84, 1961.
153. D. L. Kendall, *Semiconductors and Semimetals*, Vol. 4, R. K. Willardson and A. C. Beer, Eds., Academic, 1968, 255.
154. G. I. Rekalova, U. Kebe, and L. A. Mezrinea, *Sov. Phys. Sem.*, 5, 685, 1971.
155. G. I. Rekalova, A. A. Shakov, and V. V. Gaurushko, *Sov. Phys. Sem.*, 2, 1452, 1969.
156. B. I. Boltaks and G. S. Kulikov, *Sov. Phys. Tech. Phys.*, 2, 67, 1957.
157. I. A. Gusev and A. N. Nurin, *Sov. Phys. Solid State*, 6, 2274, 1965.
158. S. W. Kirchoefer, N. Holonyak, J. J. Coleman, and P. D. Dapkus, *J. Appl. Phys.*, 53, 766, 1982.
159. R. H. Wieber, H. C. Gorton, and C. S. Peet, *J. Appl. Phys.*, 31, 608, 1960.
160. D. Shaw, P. Joans, and D. Hazelby, *Proc. Phys. Soc. (London)*, 80, 167, 1962.
161. D. Shaw and S. R. Showan, *Phys. Stat. Sol.*, 34, 475, 1969.
162. B. Y. Pines and E. F. Shaikovskii, *Sov. Phys. Solid State*, 1, 864, 1959.
163. H. Nelkowski and G. Bollman, *Z. Naturforsch.*, A24, 1302, 1969.
164. V. M. Korsun and A. M. Nemchenko, *Sov. Phys. Solid State*, 8, 2988, 1967.
165. T. Lukaszewicz, *Phys. Stat. Sol. (a)*, 73, 611, 1982.
166. V. A. Williams, *J. Mat. Sci.*, 7, 807, 1972.
167. E. A. Secco, *J. Chem. Phys.*, 29, 406, 1958.
168. H. J. Biter and F. Williams, *J. Luminescence*, 3, 395, 1971.
169. H. Gobrecht, H. Nelkowski, J. W. Bears, and M. Weight, *Solid State Comm.*, 5, 777, 1967.
170. S. Asami, A. Ebina, and T. Takahashi, *Jpn. J. Appl. Phys.*, 17, 779, 1978.
171. T. Lukaszewicz and J. Zmija, *Phys. Stat. Sol. (A)*, 62, 695, 1980.
172. M. Aven and R. E. Halsted, *Phys. Rev.*, 137, A228, 1965.
173. M. M. Henneberg and D. A. Stevenson, *Phys. Stat. Sol. (b)*, 48, 255, 1971.
174. W. E. Martin, *J. Appl. Phys.*, 44, 5639, 1973.
175. T. Muranoi and M. Furukoshi, *Thin Solid Films*, 86, 307, 1981.
176. A. K. Kun and R. J. Robinson, *J. Electron. Mat.*, 5, 23, 1976.
177. H. H. Woodbury and R. B. Hall, *Phys. Rev.*, 157, 641, 1967.
178. P. Martin and A. Bontemps, *J. Phys. Chem. Solids*, 41, 1171, 1980.
179. E. A. Secco and S. C. Yeo, *Can. J. Chem.*, 49, 1953, 1971.
180. R. A. Reynold and D. A. Stevenson, *J. Phys. Chem. Solids*, 30, 139, 1969.
181. M. Aven and E. L. Kreiger, *J. Appl. Phys.*, 41, 1930, 1970.
182. M. Yokozawa, H. Kato, and S. Takayanagi, *Denki Kagaku*, 36, 282, 1968.
183. E. R. Dobrovinska, N. I. Krainyukov, Ya. A. Obukhouski, and L. A. Sysoev, *Ukr. Fiz. Zh.*, 13, 861, 1968.
184. H. H. Woodbury, *II-VI Semiconducting Compounds*, D. G. Thomas, Ed., Benjamin, NY, 1967, 244-276.
185. R. L. Clarke, *J. Appl. Phys.*, 30, 957, 1959.
186. J. L. Sullivan, *J. Phys. D: Appl. Phys.*, 6, 552, 1973.
187. B. Lepley, P. H. Nguyen, C. Boutrit, and S. Revelet, *J. Phys. D: Appl. Phys.*, 12, 145, 1979.
188. H. H. Woodbury, *J. Appl. Phys.*, 36, 2287, 1965.
189. E. Nebauer, *Phys. Stat. Sol.*, 29, 269, 1968.
190. J. Zimija and L. Sados, *Biul. Wojek, Akad. Tech.*, 20, No. 4, 105, 1971.

191. H. H. Woodbury, *Phys. Rev.,* 134, A492, 1964.
192. E. D. Jones and H. Mykura, *J. Phys. Chem. Solids,* 41, 1261, 1980.
193. E. D. Jones and H. Mykura, *J. Phys. Chem. Solids,* 39, 11, 1978.
194. E. Nebauer, *Phys. Stat. Sol. (b),* 60, K57, 1973.
195. V. Kumar and F. A. Kroger, *J. Solid State Chem.,* 3, 387, 1971.
196. L. A. Sysoev, A. Ya, Gelfman, A. D. Kovaleva, and N. G. Kravchencko, *Izv. Akad. Nauk. SSSR., Neorg. Mater.,* 5, 2208, 1969.
197. E. Nebauer and J. Lautenbach, *Phys. Stat. Sol. (b),* 48, 657, 1971.
198. D. G. Girton and W. E. Anderson, *Trans. Met. Soc. AIME,* 245, 465, 1969.
199. J. L. Sullivan, *Thin Solid Films,* 25, 245, 1975.
200. N. V. Dmitrieva, A. V. Vanyukov, and S. G. Yakovlev, *Elektron. Tekh. Nauk Tekh. Sb. Mater.,* No. 5, 150, 1970.
201. J. Zmija, *Acta Phys. Pol.,* A43, 345, 1973.
202. R. B. Hall and H. H. Woodbury, *J. Appl. Phys.,* 39, 5361, 1968.
203. L. Svob and C. Grattepain, *Calloq. Metall.,* 19, 725, 1976.
204. H. H. Woodbury and M. Aven, *J. Appl. Phys.,* 39, 5485, 1968.
205. H. Mann, G. Linker, and O. Meyer, *Solid State Comm.,* 11, 475, 1972.
206. J. P. Chamonal, E. Molva, J. L. Pautrat, and L. Revoil, *J. Cryst. Growth,* 59, 297, 1982.
207. I. Teramoto and S. Takayanagi, *J. Phys. Soc. Jpn.,* 17, 1137, 1982.
208. R. C. Whelan and D. Shaw, *II-IV Semiconducting Compounds,* D. C. Thomas, Ed., Benjamin, NY, 1967, 451.
209. P. M. Borsenberger and D. A. Stevenson, *J. Phys. Chem. Solids,* 29, 1277, 1968.
210. M. Yokozawa, H. Kato, and S. Takayanagi, *Denki Kagaku,* 34, 828, 1966.
211. E. Watson and D. Shaw, *J. Phys. C: Solid State Phys.,* 16, 515, 1983.
212. O. E. Panchuk, L. P. Shcherbak, P. I. Felchuk, and A. V. Savitskii, *Izv. Akad. Nauk SSSR, Neorg. Mater.,* 14, 54, 1978.
213. F. F. Morehead and G. Mandel, *Phys. Lett.,* 10, 5, 1964.
214. F. F. Vodovatov, G. V. Indenbaum, and A. V. Vanyukov, *Sov. Phys. Solid State,* 12, 17, 1970.
215. H. Kato, M. Yokozawa, and S. Takayanagi, *Jpn. J. Appl. Phys.,* 4, 1019, 1965.
216. D. Shaw and E. Watson, *J. Phys. C: Solid State Phys.,* 17, 4945, 1985.
217. O. E. Panchuk, R. N. Fesh, A. V. Savitskii, and L. P. Shcherbak, *Izv. Akad. Nauk SSSR, Neorg. Mater.,* 17, 1354, 1981.
218. B. I. Boltaks, *Diffusion in Semiconductors,* Inforsearch, 1963, 300.
219. F. F. Kharakhorin, *Izv. Akad. Nauk SSSR, Neorg. Mater.,* 5, 2212, 1969.
220. F. A. Zaitov, V. I. Stafeev, G. S., Khodakov, *Sov. Phys. Solid State,* 14, 2628, 1973.
221. V. V. Belov, F. A. Zaitov, and G. E. Popovyan, *Sov. Phys. Solid State,* 11, 1627, 1970.
222. F. A. Zaitov, *Sov. Phys. Solid State,* 13, 219, 1971.
223. V. S. Andramonov, N. S. Baryshev, and I. S. Aver'yanov, *Soviet Phys. Solid State,* 4, 1626, 1963.
224. J. Bloem and F. A. Kroger, *Philips Res. Rep.,* 12, 281, 1957.
225. G. Simkovich and J. B. Wagner, Jr., *J. Chem. Phys.,* 38, 1368, 1963.
226. M. S. Seltzer and J. B. Wagner, Jr., *J. Phys. Chem. Solids,* 26, 233, 1965.
227. J. Bloem and F. A. Kroger, Philips Res. Rep., 12, 303, 1957.
228. N. A. Fedorovich, *Soviet Phys. Solid State,* 7, 1289, 1965.
229. F. F. Kharakhorin, D. A. Gambarova, and V. V. Aksenov, *Soviet Phys. Solid State,* 7, 2813, 1966.
230. M. S. Seltzer and J. B. Wagner, Jr., *J. Chem. Phys.,* 36, 130, 1962.
231. B. I. Boltaks and Yu. N. Mokhov, *Zh. Tekh. Fiz.,* 28, 1046, 1958; *Zh. Tekh. Fiz.,* 26, 2448, 1956.
232. N. A. Fedorovich, *Soviet Phys. Solid State,* 7, 1291, 1965.
233. B. Swaroop and J. B. Wagner, Jr., *Appl. Phys. Lett.,* 12, 267, 1968.
234. A. J. Crocker and B. F. Dorning, *J. Phys. Chem. Solids,* 29, 155, 1968.
235. T. D. George and J. B. Wagner, Jr., *J. Electrochem. Soc.,* 115, 956, 1968; 116, 848, 1969.

PROPERTIES OF MAGNETIC MATERIALS

H. P. R. Frederikse

Glossary of Symbols

Quantity	Symbol	Units SI	emu
Magnetic field	H	A m^{-1}	Oe (oersted)
Magnetic induction	B	T (tesla)	G (gauss)
Magnetization	M	A m^{-1}	emu cm^{-3}
Spontaneous magnetization	M_s	A m^{-1}	emu cm^{-3}
Saturation magnetization	M_0	A m^{-1}	emu cm^{-3}
Magnetic flux	Φ	Wb (weber)	maxwell
Magnetic moment	m	A m^2	erg/G
Coercive field	H_c	A m^{-1}	Oe
Remanence	B_r	T	G
Saturation magnetic polarization	J_s	T	G
Magnetic susceptibility	χ		
Magnetic permeability	μ	H m^{-1} (henry/meter)	
Magnetic permeability of free space	μ_0	H m^{-1}	
Saturation magnetostriction	$\lambda\ (\Delta l/l)$		
Curie temperature	T_C	K	K
Néel temperature	T_N	K	K

Magnetic moment $\mu = \gamma \hbar J = g\, \mu_B\, J$
where
   γ = gyromagnetic ratio; J = angular momentum; g = spectroscopic splitting factor (~2)
   μ_B = bohr magneton = $9.2741 \cdot 10^{-24}$ J/T = $9.2741 \cdot 10^{-21}$ erg/G

Earth's magnetic field $H = 56$ A m^{-1} = 0.7 Oe
For iron: $M_0 = 1.7 \cdot 10^6$ A m^{-1}; $B_r = 0.8 \cdot 10^6$ A m^{-1}
1 Oe − $(1000/4\pi)$ A m^{-1}; 1 G = 10^{-4} T; 1 emu cm^{-3} = 10^3 A m^{-1}
1 maxwell = 10^{-8} Wb
$\mu_0 = 4\pi \cdot 10^{-7}$ H m^{-1}

Relation Between Magnetic Induction and Magnetic Field

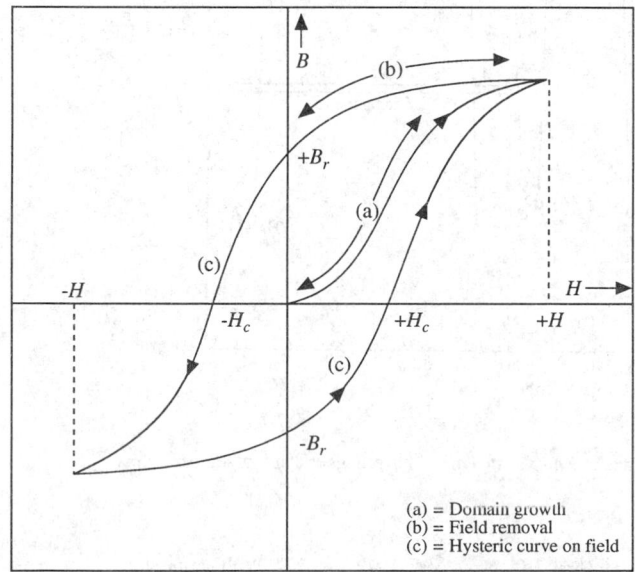

(a) = Domain growth
(b) = Field removal
(c) = Hysteric curve on field

Figure 1. Magnetic induction B vs. magnetic field H relationship for a ferromagnetic material. Along (a), starting from the demagnetized state, the more favorably oriented domains grow at the expense of less favored domains. At high fields, nearing saturation, the induction increases only by domain rotation. Upon removal of the applied field [curve (b)], the material maintains a remnant induction, B, and can only be demagnetized by imposing a negative field equal in magnitude to the coercivity, H_c. Cyclic applications of fields [such as in curve (c)] lead to a hysteresis loop for ferromagnetic materials. The area within the B-H loop represents an energy loss per unit volume of material per cycle.

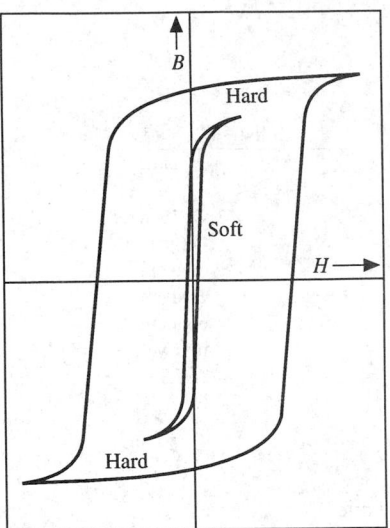

Figure 2. Schematic induction-field curves for a hard magnetic material and for a soft material. Hard magnetic materials are characterized by high remnant inductions, high coercivities, and large hysteresis losses per cycle.

REFERENCE

Ralls, K.M., Courtney, T.H., and Wulff, J., *Introduction to Materials Science and Engineering*, J. Wiley & Sons, New York, 1976, p. 577, 582. With permission.

Magnetic Susceptibility of the Elements

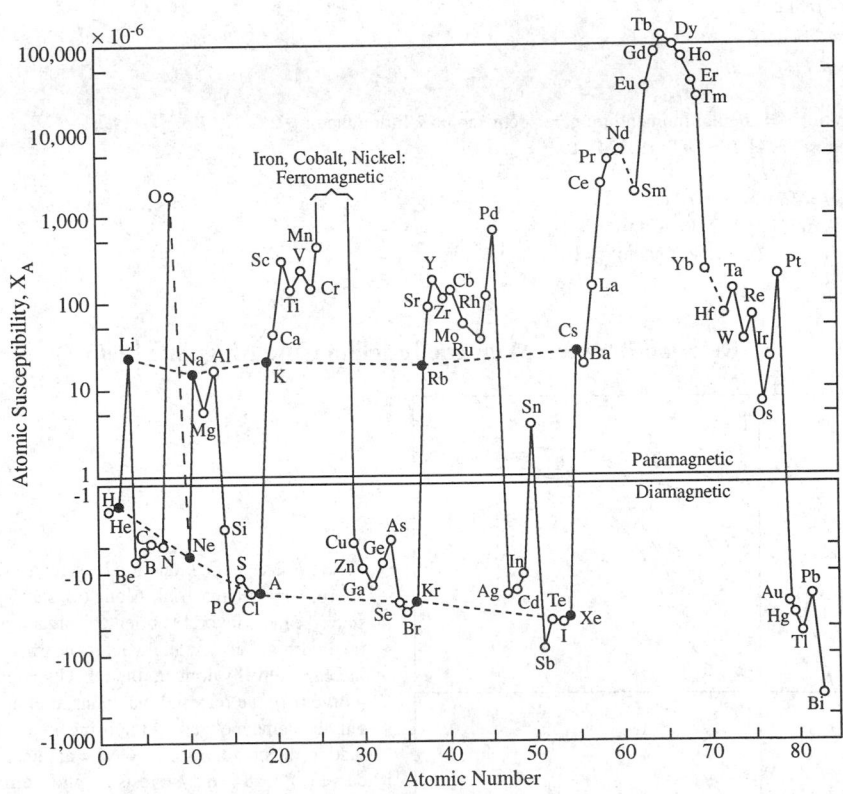

Figure 3. Atomic susceptibility of the elements at room temperature. The elements Gd to Tm are ferromagnetic at low temperatures.

REFERENCE

Gray, D.E., Ed., *American Institute of Physics Handbook, Third Edition*, McGraw Hill, New York, 1972, p. 5-224. With permission.

Ground State of Ions with Partly Filled d or f Shells

Z	Element	n	S	L	J	Gr. state	p_{calc}[a]	p_{calc}[b]	p_{meas}
22	Ti^{3+}	1	1/2	2	3/2	$^2D_{3/2}$	1.73	1.55	1.8
23	V^{4+}	1	1/2	2	3/2	$^2D_{3/2}$	1.73	1.55	1.8
23	V^{3+}	2	1	3	2	$^3F_2$	2.83	1.63	2.8
23	V^{2+}	3	3/2	3	3/2	$^4F_{3/2}$	3.87	0.77	3.8
24	Cr^{3+}	3	3/2	3	3/2	$^4F_{3/2}$	3.87	0.77	3.7
25	Mn^{4+}	3	3/2	3	3/2	$^4F_{3/2}$	3.87	0.77	4.0
24	Cr^{2+}	4	2	2	0	$^5D_0$	4.90	0	4.9
25	Mn^{3+}	4	2	2	0	$^5D_0$	4.90	0	5.0
25	Mn^{2+}	5	5/2	0	5/2	$^6S_{5/2}$	5.92	5.92	5.9
26	Fe^{3+}	5	5/2	0	5/2	$^6S_{5/2}$	5.92	5.92	5.9
26	Fe^{2+}	6	2	2	4	$^5D_4$	4.90	6.70	5.4
27	Co^{2+}	7	3/2	3	9/2	$^4F_{9/2}$	3.87	6.54	4.8
28	Ni^{2+}	8	1	3	4	$^3F_4$	2.83	5.59	3.2
29	Cu^{2+}	9	1/2	2	5/2	$^2D_{5/2}$	1.73	3.55	1.9

Z	Element	n	S	L	J	Gr. state	p_{calc}[c]		p_{meas}
58	Ce^{3+}	1	1/2	3	5/2	$^2F_{5/2}$	2.54		2.4
59	Pr^{3+}	2	1	5	4	$^3H_4$	3.58		3.5
60	Nd^{3+}	3	3/2	6	9/2	$^4I_{9/2}$	3.62		3.5
61	Pm^{3+}	4	2	6	4	$^5I_4$	2.68		
62	Sm^{3+}	5	5/2	5	5/2	$^6H_{5/2}$	0.84		1.5
63	Eu^{3+}	6	3	3	0	$^7F_0$	0.0		3.4
64	Gd^{3+}	7	7/2	0	7/2	$^8S_{7/2}$	7.94		8.0
65	Tb^{3+}	8	3	3	6	$^7F_6$	9.72		9.5
66	Dy^{3+}	9	5/2	5	15/2	$^6H_{15/2}$	10.63		10.6
67	Ho^{3+}	10	2	6	8	$^5I_8$	10.60		10.4
68	Er^{3+}	11	3/2	6	15/2	$^4I_{15/2}$	9.59		9.5
69	Tm^{3+}	12	1	5	6	$^3H_6$	7.57		7.3
70	Yb^{3+}	13	1/2	3	7/2	$^2\Gamma_{7/2}$	4.54		4.5

[a] $p_{calc} = 2[S(S + 1)]^{1/2}$
[b] $p_{calc} = 2[J(J + 1)]^{1/2}$
[c] $p_{calc} = g[J(J + 1)]^{1/2}$

REFERENCES

1. Jiles, D., *Magnetism and Magnetic Materials*, Chapman & Hall, London, 1991, p. 243.
2. Kittel, C., *Introduction to Solid State Physics, 6th Edition*, J. Wiley & Sons, New York, 1986, pp. 405—406.
3. Ashcroft, N.W. and Mermin, N.D., *Solid State Physics*, Holt, Rinehart, and Winston, New York, 1976, p. 652.

Ferro- and Antiferromagnetic Elements

M_0 is the saturation magnetization at $T = 0$ K
n_B is the number of Bohr magnetons per atom
T_C is the Curie temperature
T_N is the Néel temperature

	M_0/gauss	n_B	T_C/K	T_N/K	Comments
Fe	1752	2.22	1043		
Co	1446	1.72	1388		
Ni	510	0.62	627		
Cr				311	
Mn				100	
Ce				12.5	c-Axis antiferromagnetic
Nd				19.2	Basal plane modulation on hexagonal sites

Ferro- and Antiferromagnetic Elements (continued)

	M_0/gauss	n_B	T_C/K	T_N/K	Comments
				7.8	Cubic sites order (periodicity different from high-T phase)
Sm				106	Ordering on hexagonal sites
				13.8	Cubic site order
Eu				90.5	Spiral along cube axis
Gd	1980	7	293		
Tb		9	220		Basal plane ferromagnet
				230.2	Basal plane spiral
Dy		10	87		Basal plane ferromagnet
				176	Basal plane spiral
Ho		10	20		Bunched cone structure
				133	Basal plane spiral
Er		9	32		c-Axis ferrimagnetic cone structure
				80	c-Axis modulated structure
Tm		7	32		c-Axis ferrimagnetic cone structure
				56	c-Axis modulated structure

REFERENCES

1. Ashcroft, N.W., and Mermin, N.D., *Solid State Physics,* Holt, Rinehart, and Winston, New York, 1976, p.652.
2. Gschneidner, K.A., and Eyring, L., *Handbook on the Physics and Chemistry of Rare Earths*, North Holland Publishing Co., Amsterdam, 1978.

Selected Ferromagnetic Compounds

M_0 is the saturation magnetization at $T = 293$ K
T_C is the Curie temperature

Compound	M_0/gauss	T_C/K	Crystal system
MnB	152	578	orthorh(FeB)
MnAs	670	318	hex(FeB)
MnBi	620	630	hex(FeB)
MnSb	710	587	hex(FeB)
M_4N	183	743	
MnSi		34	cub(FeSi)
CrTe	247	339	hex(NiAs)
$CrBr_3$	270	37	hex(BiI_3)
CrI_3		68	hex(BiI_3)
CrO_2	515	386	tetr(TiO_2)
EuO	1910*	77	cub
EuS	1184*	16.5	cub
$GdCl_3$	550*	2.2	orthorh
FeB		598	orthorh
Fe_2B		1043	tetr ($CuAl_2$)
$FeBe_5$		75	cub($MgCu_2$)
Fe_3C		483	orthorh
FeP		215	orthorh (MnP)

* At $T = 0$ K

REFERENCES

1. Kittel, C., *Introduction to Solid State Physics, 6th Edition*, J. Wiley & Sons, New York, 1986.
2. Ashcroft, N.W., and Mermin, N.D., *Solid State Physics*, Holt, Rinehart, and Winston, New York, 1976.

PROPERTIES OF MAGNETIC MATERIALS (continued)

Magnetic Properties of High-Permeability Metals and Alloys (Soft)

μ_i is the initial permeability
μ_m is the maximum permeability
H_c is the coercive force
J_s is the saturation polarization
W_H is the hysteresis loss per cycle
T_C is the Curie temperature

Material	Composition (mass %)	μ_i/μ_0	μ_m/μ_0	H_c/A m^{-1}	J_s/T	W_H/J m^{-3}	T_C/K
Iron	Commercial 99Fe	200	6000	70	2.16	500	1043
Iron	Pure 99.9Fe	25000	350000	0.8	2.16	60	1043
Silicon-iron	96Fe-4Si	500	7000	40	1.95	50-150	1008
Silicon-iron (110) [001]	97Fe-3Si	9000	40000	12	2.01	35-140	1015
Silicon-iron {100} <100>	97Fe-3Si		100000	6	2.01		1015
Mild steel	Fe-0.1C-0.1Si-0.4Mn	800	1100	200			
Hypernik	50Fe-50Ni	4000	70000	4	1.60	22	753
Deltamax {100} <100>	50Fe-50Ni	500	200000	16	1.55		773
Isoperm {100} <100>	50Fe-50Ni	90	100	480	1.60		
78 Permalloy	78Ni-22Fe	4000	100000	4	1.05	50	651
Supermalloy	79Ni-16Fe-5Mo	100000	1000000	0.15	0.79	2	673
Mumetal	77Ni-16Fe-5Cu-2Cr	20000	100000	4	0.75	20	673
Hyperco	64Fe-35Co-0.5Cr	650	10000	80	2.42	300	1243
Permendur	50Fe-50Co	500	6000	160	2.46	1200	1253
2V-Permendur	49Fe-49Co-2V	800	4000	160	2.45	600	1253
Supermendur	49Fe-49Co-2V		60000	16	2.40	1150	1253
25Perminvar	45Ni-30Fe-25Co	400	2000	100	1.55		
7Perminvar	70Ni-23Fe-7Co	850	4000	50	1.25		
Perminvar (magnet. annealed)	43Ni-34Fe-23Co		400000	2.4	1.50		
Alfenol (or Alperm)	84Fe-16Al	3000	55000	3.2	0.8		723
Alfer	87Fe-13Al	700	3700	53	1.20		673
Aluminum-Iron	96.5Fe-3.5Al	500	19000	24	1.90		
Sendust	85Fe-10Si-5Al	36000	120000	1.6	0.89		753

REFERENCES

1. McCurrie, R.A., *Ferromagnetic Materials*, Academic Press, London, 1994, p. 42.
2. Gray, D.E., Ed., *American Institute of Physics Handbook, Third Edition*, McGraw Hill, New York, 1972, p. 5-224.

Applications of High-Permeability Materials

Applications	Requirements

POWER APPLICATIONS

Distribution and power transformers	Low core losses, high permeability, high saturation magnetic polarization
High-quality motors and generators, stators and armatures, switched-mode power supplies	

INSTRUMENT TRANSFORMERS

Audiofrequency transformers	Low core losses, high permeability, high magnetic polarization
Pulse transformers	High permeability

Applications of High-Permeability Materials (continued)

Applications	Requirements

CORES FOR INDUCTOR COILS

Applications	Requirements
Audiofrequency	Low hysteresis, high permeability
Carrier frequency	Very low hysteresis and eddy current loss
Radiofrequency	High permeability at low fields

MISCELLANEOUS

Applications	Requirements
Relays, switches } Earth leakage circuit	High permeability, low remanence, low coercivity
Magnetic shielding	Low core loss for AC applications
Magnetic recording heads	High initial permeability, low or zero remanence
Magnetic amplifiers Saturable reactors Saturable transformers Transformer cores }	Rectangular hysteresis loops, low hysteresis loss
Magnetic shunts for temperature compensation in magnetic circuits	Low Curie temperature, appropriate decrease in permeability with increase in temperature
Electromagnets in indicating instruments, fire detection, quartz watches, electromechanical devices	High permeability, high saturation magnetic polarization
Magnetic yokes in permanent magnet devices, such as lifting and holding magnets, loudspeakers	High permeability, high saturation magnetic polarization

REFERENCE

McCurrie, R.A., *Ferromagnetic Materials*, Academic Press, London, 1994. With permission.

Saturation Magnetostriction of Selected Materials

The tabulated parameter λ_s is related to the fractional change in length $\Delta l/l$ by $\Delta l/l = (3/2)\lambda_s(\cos^2\theta - 1/3)$, where θ is the angle of rotation.

Material	$\lambda_s \times 10^6$
Iron	-7
Fe - 3.2% Si	+9
Nickel	-33
Cobalt	-62
45 Permalloy, 45% Ni - 55% Fe	+27
Permalloy, 82% Ni - 18% Fe	0
Permendur, 49% Co - 49% Fe - 2% V	+70
Alfer, 87% Fe - 13% Al	+30
Magnetite, Fe_3O_4	+40
Cobalt ferrite, $CoFe_2O_4$	-110
$SmFe_2$	-1560
$TbFe_2$	+1753
$Tb_{0.3}Dy_{0.7}Fe_{1.93}$ (Terfenol D)	+2000
$Fe_{66}Co_{18}B_{15}Si$ (amorphous)	+35
$Co_{72}Fe_3B_6A_{13}$ (amorphous)	0

REFERENCE

McCurrie, R.A., *Ferromagnetic Materials*, Academic Press, London, 1994, p. 91; additional data provided by A.E. Clark, Adelphi, MD.

PROPERTIES OF MAGNETIC MATERIALS (continued)

Properties of Various Permanent Magnetic Materials (Hard)

B_r is the remanence
$_BH_c$ is the flux coercivity
$_iH_c$ is the intrinsic coercivity
$(BH)_{max}$ is the maximum energy product
T_C is the Curie temperature
T_{max} is the maximum operating temperature

Composition	B_r/T	$_BH_c/10^3$ A m^{-1}	$_iH_c/10^3$ A m^{-1}	$(BH)_{max}$/ kJ m^{-3}	T_C/°C	T_{max}/°C
Alnico1 20Ni;12Al;5Co	0.72		35	25		
Alnico2 17Ni;10Al;12.5Co;6Cu	0.72		40-50	13-14		
Alnico3 24-30Ni;12-14Al;0-3Cu	0.5-0.6		40-54	10		
Alnico4 21-28Ni;11-13Al;3-5Co;2-4Cu	0.55-0.75		36-56	11-12		
Alnico5 14Ni;8Al;24Co;3Cu	1.25	53	54	40	850	520
Alnico6 16Ni;8Al;24Co;3Cu;2Ti	1.05		75	52		
Alnico8 15Ni;7Al;35Co;4Cu;5Ti	0.83	1.6	160	45		
Alnico9 15Ni;7Al;35Co;4Cu;5Ti	1.10	1.45	1.45	75	850	520
Alnico12 13.5Ni;8Al;24.5Co;2Nb	1.20		64	76.8		
$BaFe_{12}O_{19}$ (Ferroxdur)	0.4	1.6	192	29	450	400
$SrFe_{12}O_{19}$	0.4	2.95	3.3	30	450	400
$LaCo_5$	0.91			164	567	
$CeCo_5$	0.77			117	380	
$PrCo_5$	1.20			286	620	
$NdCo_5$	1.22			295	637	
$SmCo_5$	1.00	7.9	696	196	700	250
$Sm(Co_{0.76}Fe_{0.10}Cu_{0.14})_{6.8}$	1.04	4.8	5	212	800	300
$Sm(Co_{0.65}Fe_{0.28}Cu_{0.05}Zr_{0.02})_{7.7}$	1.2	10	16	264	800	300
$Nd_2Fe_{14}B$ sintered	1.22	8.4	1120	280	300	100
Fe;52Co;14V (Vicalloy II)	1.0	42		28	700	500
Fe;24Cr;15Co;3Mo (anisotropic)	1.54	67		76	630	500
Fe;28Cr;10.5Co (Chromindur II)	0.98	32		16	630	500
Fe;23Cr;15Co;3V;2Ti	1.35	4		44	630	500
Cu;20Ni;20Fe (Cunife)	0.55	4		12	410	350
Cu;21Ni;29Fe (Cunico)	0.34	0.5		8		
Pt;23Co	0.64	4		76	480	350
Mn;29.5Al;0.5C (anisotropic)	0.61	2.16	2.4	56	300	120

REFERENCES

1. McCurrie, R.A., *Ferromagnetic Materials*, Academic Press, London, 1994, p. 204.
2. Gray, D.E., Ed., *American Institute of Physics Handbook, Third Edition*, McGraw Hill, New York, 1972, p. 5-165.
3. Jiles, D., *Magnetism and Magnetic Materials,* Chapman & Hall, London, 1991.

Selected Ferrites

J_s is the saturation magnetic polarization
T_C is the Curie temperature
ΔH is the line width

Material	J_s/T	T_C/°C	ΔH/ kA m^{-1}	Applications
Spinels				
γ-Fe_2O_3	0.52	575		
Fe_3O_4	0.60	585		
$NiFe_2O_4$	0.34	575	350	Microwave devices
$MgFe_2O_4$	0.14	440	70	
$NiZnFe_2O_4$	0.50	375	120	Transformer cores
$MnFe_2O_4$	0.50	300	50	Microwave devices

Selected Ferrites (continued)

Material	J_s/T	$T_C/°C$	$\Delta H/$ kA m^{-1}	Applications
$NiCoFe_2O_4$	0.31	590	140	Microwave devices
$NiCoAlFe_2O_4$	0.15	450	330	Microwave devices
$NiAl_{0.35}Fe_{1.65}O_4$	0.12	430	67	Microwave devices
$NiAlFe_2O_4$	0.05	1860	32	Microwave devices
$Mg_{0.9}Mn_{0.1}Fe_2O_4$	0.25	290	56	Microwave devices
$Ni_{0.5}Zn_{0.5}Al_{0.8}Fe_{1.2}O_4$	0.14		17	Microwave devices
$CuFe_2O_4$	0.17	455		Electromechanical transducers
$CoFe_2O_4$	0.53	520		
$LiFe_5O_8$	0.39	670		Microwave devices
Garnets				
$Y_3Fe_5O_{12}$	0.178	280	55	Microwave devices
$Y_3Fe_5O_{12}$ (single crys.)	0.178	292	0.5	Microwave devices
$(Y,Al)_3Fe_5O_{12}$	0.12	250	80	Microwave devices
$(Y,Gd)_3Fe_5O_{12}$	0.06	250	150	Microwave devices
$Sm_3Fe_5O_{12}$	0.170	305		Microwave devices
$Eu_3Fe_5O_{12}$	0.116	293		Microwave devices
$GdFe_5O_{12}$	0.017	291		Microwave devices
Hexagonal crystals				
$BaFe_{12}O_{19}$	0.45	430	1.5	Permanent magnets
$Ba_3Co_2Fe_{24}O_{41}$	0.34	470	12	Microwave devices
$Ba_2Zn_2Fe_{12}O_{22}$	0.28	130	25	Microwave devices
$Ba_3Co_{1.35}Zn_{0.65}Fe_{24}O_{41}$		390	16	Microwave devices
$Ba_2Ni_2Fe_{12}O_{22}$	0.16	500	8	Microwave devices
$SrFe_{12}O_{19}$	0.4	450		Permanent magnets

REFERENCE

McCurrie, R.A., *Ferromagnetic Materials*, Academic Press, London, 1994.

Spinel Structure (AB$_2$O$_4$)

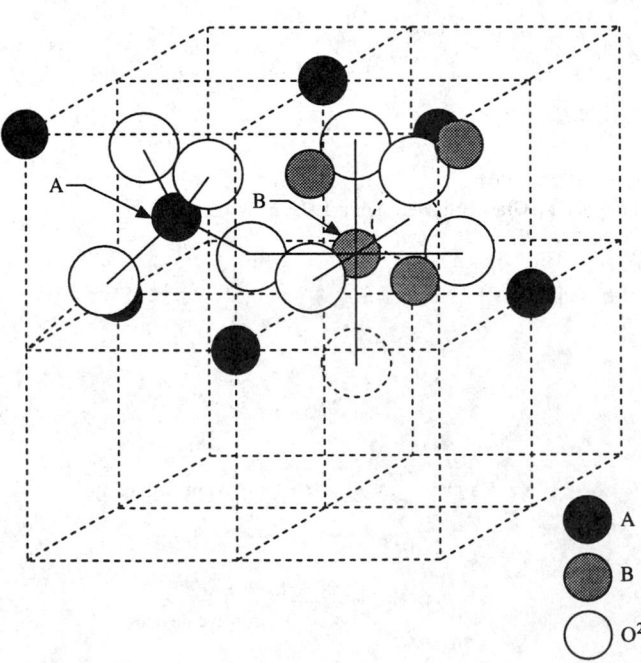

Figure 4. Arrangement of metal ions in the two octants A and B, showing tetrahedrally (A) and octahedrally (B) coordinated sites. (Reprinted from McCurrie, R.A., *Ferromagnetic Materials*, Academic Press, London, 1994. With permission.)

Selected Antiferromagnetic Solids

T_N is the Néel temperature

Material	Structure	T_N/K	Material	Structure	T_N/K
Binary oxides			**NiAs and related structures**		
MnO	cub(fcc)	122	CrAs	orth	300
FeO	cub(fcc)	198	CrSb	hex	705-723
CoO	cub(fcc)	291	CrSe	hex	300
NiO	cub(fcc)	525	MnTe	hex	320-323
α-Mn_2O_3	cub	90	NiS	hex	263
CuO	monocl	230	CrS	monocl	460
UO_2	cub	30.8	**Rutile and related structures**		
Er_2O_3	cub	3.4	CoF_2	tetr	38
Gd_2O_3	cub	1.6	CrF_2	monocl	53
Perovskites			FeF_2	tetr	79
$LaCrO_3$	orth	282	MnF_2	tetr	67
$LaMnO_3$	orth	100	NiF_2	tetr	83
$LaFeO_3$	orth	750	$CrCl_2$	orth	20
$NdCrO_3$	orth	224	MnO_2	tetr	84
$NdFeO_3$	orth	760	FeOF	tetr	315
$YbCrO_3$	orth	118	**Corundum and related structures**		
$CaMnO_3$	cub	110	Cr_2O_3	rhomb	318
$EuTiO_3$	cub	5.3	α-Fe_2O_3	rhomb	948
$YCrO_3$	orth	141	$FeTiO_3$	rhomb	68
$BiFeO_3$	cub*	673	$MnTiO_3$	rhomb	41
$KCoF_3$	cub	125	$CoTiO_3$	rhomb	38
$KMnF_3$	cub*	88.3	**VF_3 and related structures**		
$KFeF_3$	cub	115	CoF_3	rhomb	460
$KNiF_3$	cub	275	CrF_3	rhomb	80
$NaMnF_3$	cub*	60	FeF_3	rhomb	394
$NaNiF_3$	orth	149	MnF_3	monocl	43
$RbMnF_3$	cub	82	MoF_3	rhomb	185
Spinels			**Miscellaneous**		
Co_3O_4	cub	40	K_2NiF_4	tetr	97
$NiCr_2O_4$	tetr	65	MnI_2	hex	3.4
$ZnCr_2O_4$	cub	15	$CoUO_4$	orth	12
$ZnFe_2O_4$	cub	9	$CaMn_2O_4$	orth	225
$GeFe_2O_4$	cub	10	CrN	cub*	273
MgV_2O_4	cub	45	CeC_2	tetr	33
$MnGa_2O_4$	cub	33	FeSn	hex	373
			Mn_2P	hex	103

* Distorted.

REFERENCES

1. Gray, D.E., Ed., *American Institute of Physics Handbook, Third Edition*, McGraw Hill, New York, 1972, p. 5-168 to 183.
2. Kittel, C., *Introduction to Solid State Physics, 6th Edition*, J. Wiley & Sons, New York, 1986.
3. Ashcroft, N.W., and Mermin, N.D., *Solid State Physics*, Holt, Rinehart, and Winston, New York, 1976, p. 697.

ELECTRON WORK FUNCTION OF THE ELEMENTS

The electron work function Φ is a measure of the minimum energy required to extract an electron from the surface of a solid. It is defined more precisely as the energy difference between the state in which an electron has been removed to a distance from the surface of a single crystal face that is large enough that the image force is negligible but small compared to the distance to any other face (typically about 10^{-4} cm) and the state in which the electron is in the bulk solid. In general, Φ differs for each face of a monocrystalline sample.

Since Φ is dependent on the cleanliness of the surface, measurements reported in the literature often cover a considerable range. This table contains selected values for the electron work function of the elements which may be regarded as typical values for a reasonably clean surface. The method of measurement is indicated for each value. The following abbreviations appear:

TE — Thermionic emission
PE — Photoelectric effect
FE — Field emission
CPD — Contact potential difference
polycr — Polycrystalline sample
amorp — Amorphous sample

Values in parentheses are only approximate.

REFERENCES

1. Hölzl, J., and Schulte, F. K., Work Functions of Metals, in *Solid Surface Physics*, Höhler, G., Editor, Springer-Verlag, Berlin, 1979.
2. Riviere, J. C., Work Function: Measurements and Results, in *Solid State Surface Science, Vol.1*, Green, M., Editor, Decker, New York, 1969.
3. Michaelson, H. B., *J. Appl. Phys.*, 48, 4729, 1977.

Element	Plane	Φ/eV	Method	Element	Plane	Φ/eV	Method	Element	Plane	Φ/eV	Method
Ag	100	4.64	PE		210	5.00	PE	Ru	polycr	4.71	PE
	110	4.52	PE	K	polycr	2.29	PE	Sb	amorp	4.55	
	111	4.74	PE	La	polycr	3.5	PE		100	4.7	
Al	100	4.20	PE	Li	polycr	2.93	FE	Sc	polycr	3.5	PE
	110	4.06	PE	Lu	polycr	(3.3)	CPD	Se	polycr	5.9	PE
	111	4.26	PE	Mg	polycr	3.66	PE	Si	n	4.85	CPD
As	polycr	(3.75)	PE	Mn	polycr	4.1	PE		p 100	(4.91)	CPD
Au	100	5.47	PE	Mo	100	4.53	PE		p 111	4.60	PE
	110	5.37	PE		110	4.95	PE	Sm	polycr	2.7	PE
	111	5.31	PE		111	4.55	PE	Sn	polycr	4.42	CPD
B	polycr	(4.45)	TH		112	4.36	PE	Sr	polycr	(2.59)	TH
Ba	polycr	2.52	TH		114	4.50	PE	Ta	polycr	4.25	TH
Be	polycr	4.98	PE		332	4.55	PE		100	4.15	TH
Bi	polycr	4.34	PE	Na	polycr	2.36	PE		110	4.80	TH
C	polycr	(5.0)	CPD	Nb	001	4.02	TH		111	4.00	TH
Ca	polycr	2.87	PE		110	4.87	TH	Tb	polycr	3.0	PE
Cd	polycr	4.08	CPD		111	4.36	TH	Te	polycr	4.95	PE
Ce	polycr	2.9	PE		112	4.63	TH	Th	polycr	3.4	TH
Co	polycr	5.0	PE		113	4.29	TH	Ti	polycr	4.33	PE
Cr	polycr	4.5	PE		116	3.95	TH	Tl	polycr	(3.84)	CPD
Cs	polycr	1.95	PE		310	4.18	TH	U	polycr	3.63	PE
Cu	100	5.10	FE	Nd	polycr	3.2	PE		100	3.73	PE
	110	4.48	PE	Ni	100	5.22	PE		110	3.90	PE
	111	4.94	PE		110	5.04	PE		113	3.67	PE
	112	4.53	PE		111	5.35	PE	V	polycr	4.3	PE
Eu	polycr	2.5	PE	Os	polycr	5.93	PE	W	polycr	4.55	CPD
Fe	100	4.67	PE	Pb	polycr	4.25	PE		100	4.63	FE
	111	4.81	PE	Pd	polycr	5.22	PE		110	5.22	FE
Ga	polycr	4.32	PE		111	5.6	PE		111	4.45	FE
Gd	polycr	2.90	CPD	Pt	polycr	5.64	PE		113	4.46	FE
Ge	polycr	5.0	CPD		110	5.84	FE		116	4.32	TH
Hf	polycr	3.9	PE		111	5.93	FE	Y	polycr	3.1	PE
Hg	liquid	4.475	PE		320	5.22	FE	Zn	polycr	3.63	PE
In	polycr	4.09	PE		331	5.12	FE		polycr	(4.9)	CPD
Ir	100	5.67	PE	Rb	polycr	2.261	PE	Zr	polycr	4.05	PE
	110	5.42	PE	Re	polycr	4.72	TE				
	111	5.76	PE	Rh	polycr	4.98	PE				

SECONDARY ELECTRON EMISSION

The secondary emission yield, or secondary emission ratio, δ, is the average number of secondary electrons emitted from a bombarded material for every incident primary electron. It is a function of the primary electron energy E_p. The maximum yield δ_{max} corresponds to a primary electron energy E_{pmax} (see figure). The two primary electron energies corresponding to a yield of unity are denoted the first and second crossovers (E_I and E_{II}). An insulating target, or a conducting target that is electrically floating, will charge positively or negatively depending on the primary electron energy. For $E_I < E_p < E_{II}$, $\delta > 1$ and the surface charges positively provided there is a collector present that is positive with respect to the target. For $E_p < E_I$ or $E_p > E_{II}$, $\delta < 1$, and the surface charges negatively with respect to the potential of the source of primary electrons.

Primary Electron Energy (Ep)

Element	δ_{max}	E_{pmax} (eV)	E_I (eV)	E_{II} (eV)	Element	δ_{max}	E_{pmax} (eV)	E_I (eV)	E_{II} (eV)
Ag	1.5	800	200	>2000	Li	0.5	85	None	None
Al	1.0	300	300	300	Mg	0.95	300	None	None
Au	1.4	800	150	>2000	Mo	1.25	375	150	1200
B	1.2	150	50	600	Na	0.82	300	None	None
Ba	0.8	400	None	None	Nb	1.2	375	150	1050
Bi	1.2	550	None	None	Ni	1.3	550	150	>1500
Be	0.5	200	None	None	Pb	1.1	500	250	1000
C (diamond)	2.8	750	None	>5000	Pd	>1.3	>250	120	None
C (graphitc)	1.0	300	300	300	Pt	1.8	700	350	3000
C (soot)	0.45	500	None	None	Rb	0.9	350	None	None
Cd	1.1	450	300	700	Sb	1.3	600	250	2000
Co	1.2	600	200	None	Si	1.1	250	125	500
Cs	0.7	400	None	None	Sn	1.35	500	None	None
Cu	1.3	600	200	1500	Ta	1.3	600	250	>2000
Fe	1.3	400	120	1400	Th	1.1	800	None	None
Ga	1.55	500	75	None	Ti	0.9	280	None	None
Ge	1.15	500	150	900	Tl	1.7	650	70	>1500
Hg	1.3	600	350	>1200	W	1.4	650	250	>1500
K	0.7	200	None	None	Zr	1.1	350	None	None

Compound	δ_{max}	E_{pmax} (eV)	Compound	δ_{max}	E_{pmax} (eV)
Alkali halides			NaCl (layer)	6.8	600
CsCl	6.5		NaF (crystal)	14	1200
KBr (crystal)	14	1800	NaF (layer)	5.7	
KCl (crystal)	12	1600	NaI (crystal)	19	1300
KCl (layer)	7.5	1200	NaI (layer)	5.5	
KI (crystal)	10	1600	RbCl (layer)	5.8	
KI (layer)	5.6		Oxides		
LiF (crystal)	8.5		Ag_2O	1.0	
LiF (layer)	5.6	700	Al_2O_3 (layer)	2—9	
NaBr (crystal)	24	1800	BaO (layer)	2.3—4.8	400
NaBr (layer)	6.3		BeO	3.4	2000
NaCl (crystal)	14	1200	CaO	2.2	500

Compound	δ_{max}	E_{pmax} (eV)	Compound	δ_{max}	E_{pmax} (eV)
Cu_2O	1.2	400	Others		
MgO (crystal)	20—25	1500	BaF_2 (layer)	4.5	
MgO (layer)	3—15	400—1500	CaF_2 (layer)	3.2	
MoO_2	1.2		$BiCs_3$	6	1000
SiO_2 (quartz)	2.1—4	400	BiCs	1.9	1000
SnO_2	3.2	640	GeCs	7	700
Sulfides			Rb_3Sb	7.1	450
MoS_2	1.1		$SbCs_3$	6	700
PbS	1.2	500	Mica	2.4	350
WS_2	1.0		Glasses	2—3	300—450
ZnS	1.8	350			

OPTICAL PROPERTIES OF METALS AND SEMICONDUCTORS
J. H. Weaver and H. P. R. Frederikse

These tables list the index of refraction n, the extinction coefficient k, and the normal incidence reflection R ($\phi = 0$) as a function of photon energy E, which is expressed in electron volts (eV). To convert the energy in eV to wavelength in μm, use $\lambda = 1.2398/E$. To compute the dielectric function $\tilde{\varepsilon} = \varepsilon_1 + i\varepsilon_2$ from the complex index of refraction $\tilde{N} = n + ik$, use $\varepsilon_1 = n^2 - k^2$ and $\varepsilon_2 = 2nk$.

The optical constants in these tables are abridged from three more extensive tabulations:

- *Optical Properties of Metals* (OPM), Volumes I and II, *Physics Data, Nr.* 18-1 and 18-2, J. H. Weaver, C. Krafka, D. W. Lynch, and E. E. Koch, Fachinformationzentrum, Karlsruhe, Germany.
- *Handbook of Optical Constants* (HOC), Vol. I, 1985, and Vol. II, 1991. Edited by E. D. Palik, published by Academic Press, Inc.
- *American Institute of Physics Handbook* (AIPH), 3rd Edition, Coord. Editor D. E. Gray, published by McGraw-Hill Book Co., New York, 1972.

The first two of these major sources provide detailed comparisons of all optical data available in the literature at the time of the compilation. For critical applications the reader should refer to the original work. References for individual metals and semiconductors are listed at the end of the tables. Generally, tabulated values for the optical properties are accurate to better than 10%. Data in parentheses are extrapolated or interpolated values. For most elements the spectral range covered is from the far infrared (0.010 or 0.10 eV) to the far ultraviolet (10, 30 or 300 eV). The intervals between successive energies in the tables are chosen in such a way that the major spectral features are preserved.

Very small values of k are expressed in exponential notation, e.g., 1.23E-5 means 1.23×10^{-5}.

The following table is convenient for identifying the energy entries in these tables with the corresponding wavelengths:

λ	E/eV	λ	E/eV
1 mm	0.00124	6000 Å	2.066
500 μm	0.00248	5000 Å	2.480
100 μm	0.01240	4000 Å	3.100
50 μm	0.02480	3000 Å	4.133
10 μm	0.12398	2000 Å	6.199
5 μm	0.24797	1000 Å	12.398
1 μm	1.240	400 Å	30.996

Energy (eV)	n	k	R($\phi = 0$)	Energy (eV)	n	k	R($\phi = 0$)	Energy (eV)	n	k	R($\phi = 0$)
Aluminium[1]				2.200	1.018	6.846	0.9200	14.200	0.053	0.373	0.8312
0.040	98.595	203.701	0.9923	2.400	0.826	6.283	0.9228	14.400	0.058	0.327	0.8102
0.050	74.997	172.199	0.9915	2.600	0.695	5.800	0.9238	14.600	0.067	0.273	0.7802
0.060	62.852	150.799	0.9906	2.800	0.598	5.385	0.9242	14.800	0.086	0.211	0.7202
0.070	53.790	135.500	0.9899	3.000	0.523	5.024	0.9241	15.000	0.125	0.153	0.6119
0.080	45.784	123.734	0.9895	3.200	0.460	4.708	0.9243	15.200	0.178	0.108	0.4903
0.090	39.651	114.102	0.9892	3.400	0.407	4.426	0.9245	15.400	0.234	0.184	0.3881
0.100	34.464	105.600	0.9889	3.600	0.363	4.174	0.9246	15.600	0.280	0.073	0.3182
0.125	24.965	89.250	0.9884	3.800	0.326	3.946	0.9247	15.800	0.318	0.065	0.2694
0.150	18.572	76.960	0.9882	4.000	0.294	3.740	0.9248	16.000	0.351	0.060	0.2326
0.175	14.274	66.930	0.9879	4.200	0.267	3.552	0.9248	16.200	0.380	0.055	0.2031
0.200	11.733	59.370	0.9873	4.400	0.244	3.380	0.9249	16.400	0.407	0.050	0.1789
0.250	8.586	48.235	0.9858	4.600	0.223	3.222	0.9249	16.750	0.448	0.045	0.1460
0.300	6.759	40.960	0.9844	4.800	0.205	3.076	0.9249	17.000	0.474	0.042	0.1278
0.350	5.438	35.599	0.9834	5.000	0.190	2.942	0.9244	17.250	0.498	0.040	0.1129
0.400	4.454	31.485	0.9826	6.000	0.130	2.391	0.9257	17.500	0.520	0.038	0.1005
0.500	3.072	25.581	0.9817	6.500	0.110	2.173	0.9260	17.750	0.540	0.036	0.0899
0.600	2.273	21.403	0.9806	7.000	0.095	1.983	0.9262	18.000	0.558	0.035	0.0809
0.700	1.770	18.328	0.9794	7.500	0.082	1.814	0.9265	18.500	0.591	0.032	0.0664
0.800	1.444	15.955	0.9778	8.000	0.072	1.663	0.9269	19.000	0.620	0.030	0.0554
0.900	1.264	14.021	0.9749	8.500	0.063	1.527	0.9272	19.500	0.646	0.028	0.0467
1.000	1.212	12.464	0.9697	9.000	0.056	1.402	0.9277	20.000	0.668	0.027	0.0398
1.100	1.201	11.181	0.9630	9.500	0.049	1.286	0.9282	20.500	0.689	0.025	0.0342
1.200	1.260	10.010	0.9521	10.000	0.044	1.178	0.9286	21.000	0.707	0.024	0.0296
1.300	1.468	8.949	0.9318	10.500	0.040	1.076	0.9293	21.500	0.724	0.023	0.0258
1.400	2.237	8.212	0.8852	11.000	0.036	0.979	0.9298	22.000	0.739	0.022	0.0226
1.500	2.745	8.309	0.8678	11.500	0.033	0.883	0.9283	22.500	0.753	0.021	0.0199
1.600	2.625	8.597	0.8794	12.000	0.033	0.791	0.9224	23.000	0.766	0.021	0.0177
1.700	2.143	8.573	0.8972	12.500	0.034	0.700	0.9118	23.500	0.778	0.020	0.0157
1.800	1.741	8.205	0.9069	13.000	0.038	0.609	0.8960	24.000	0.789	0.019	0.0140
1.900	1.488	7.821	0.9116	13.500	0.041	0.517	0.8789	24.500	0.799	0.018	0.0126
2.000	1.304	7.479	0.9148	14.000	0.048	0.417	0.8486	25.000	0.809	0.018	0.0113

Energy (eV)	n	k	R(φ = 0)	Energy (eV)	n	k	R(φ = 0)	Energy (eV)	n	k	R(φ = 0)
25.500	0.817	0.017	0.0102	0.39		3.67E-05		8.75	3.247	0.855	0.308
26.000	0.826	0.016	0.0092	0.40		3.58E-05		9.00	3.272	0.910	0.314
27.000	0.840	0.015	0.0076	0.41		3.25E-05		9.25	3.308	0.978	0.322
28.000	0.854	0.014	0.0063	0.4133	2.3795		0.167	9.50	3.348	1.055	0.331
29.000	0.865	0.014	0.0053	0.42		2.94E-05		9.75	3.398	1.147	0.342
30.000	0.876	0.013	0.0044	0.43		2.87E-05		10.00	3.453	1.258	0.355
35.000	0.915	0.010	0.0020	0.44		3.14E-05		10.25	3.514	1.403	0.371
40.000	0.940	0.008	0.0010	0.45		3.62E-05		10.50	3.565	1.581	0.389
45.000	0.957	0.007	0.0005	0.46		3.22E-05		10.75	3.600	1.813	0.411
50.000	0.969	0.006	0.0003	0.47		1.57E-05		11.00	3.582	2.078	0.434
55.000	0.979	0.005	0.0001	0.48		6.17E-06		11.25	3.507	2.380	0.460
60.000	0.987	0.004	0.0000	0.4959	2.3801		0.167	11.50	3.346	2.693	0.488
65.000	0.995	0.004	0.0000	0.6199	2.3813		0.167	11.75	3.090	2.986	0.518
70.000	1.006	0.004	0.0000	0.8266	2.3837		0.167	12.00	2.736	3.228	0.551
72.500	1.025	0.004	0.0002	1.240	2.3905		0.168	12.20	2.383	3.354	0.580
75.000	1.011	0.024	0.0002	1.378	2.3934		0.169	12.40	1.983	3.382	0.610
77.500	1.008	0.025	0.0002	1.459	2.3953		0.169	12.60	1.532	3.265	0.641
80.000	1.007	0.024	0.0002	1.550	2.3975		0.169	12.80	1.312	2.953	0.627
85.000	1.007	0.028	0.0002	1.653	2.4003		0.170	13.00	1.223	2.722	0.604
90.000	1.005	0.031	0.0002	1.771	2.4036		0.170	13.50	1.129	2.379	0.557
95.000	0.999	0.036	0.0003	1.889	2.4073		0.171	14.00	1.070	2.178	0.526
100.000	0.991	0.030	0.0002	1.926	2.4084		0.171	14.50	1.018	2.034	0.504
110.000	0.994	0.025	0.0002	2.066	2.4133		0.171	15.00	0.972	1.929	0.489
120.000	0.991	0.024	0.0002	2.105	2.4147		0.172	15.50	0.917	1.845	0.482
130.000	0.987	0.021	0.0001	2.271	2.4210		0.173	16.00	0.861	1.767	0.477
140.000	0.989	0.016	0.0001	2.480	2.4299		0.174	16.50	0.805	1.692	0.474
150.000	0.990	0.015	0.0001	2.650	2.4380		0.175	17.00	0.753	1.619	0.471
160.000	0.989	0.014	0.0001	2.845		3.82E-07		17.50	0.707	1.546	0.467
170.000	0.989	0.011	0.0001	3.100	2.4627		0.178	18.00	0.665	1.476	0.463
180.000	0.990	0.010	0.0000	3.434	2.4849		0.182	18.50	0.626	1.408	0.459
190.000	0.990	0.009	0.0000	3.576	2.4955		0.183	19.00	0.589	1.341	0.455
200.000	0.991	0.007	0.0000	3.961		8.97E-07		19.50	0.557	1.273	0.449
220.000	0.992	0.006	0.0000	4.160	2.5465		0.190	20.00	0.527	1.203	0.442
240.000	0.993	0.005	0.0000	4.511		1.29E-06		21.00	0.487	1.052	0.413
260.000	0.993	0.004	0.0000	4.8187	2.6205	1.47E-06	0.200	22.00	0.518	0.888	0.330
280.000	0.994	0.003	0.0000	5.00	2.6383		0.203	23.00	0.597	0.850	0.270
300.000	0.995	0.002	0.0000	5.30		2.98E-06		24.00	0.586	0.829	0.268
				5.35		6.45E-06		25.00	0.562	0.787	0.265
Carbon (diamond)[2]				5.40		1.04E-05		26.00	0.538	0.736	0.260
0.06199	2.3741		0.166	5.50		3.41E-05		27.00	0.516	0.679	0.252
0.06888	2.3741		0.166	5.55		5.48E-04		28.00	0.501	0.616	0.239
0.07749	2.3745		0.166	5.60	2.740	1.48E-03	0.216	29.00	0.494	0.552	0.221
0.08856	2.3750		0.166	5.80	2.780	5.02E-03	0.222	30.00	0.493	0.490	0.201
0.1033	2.3757		0.166	6.00	2.826	7.99E-03	0.228				
0.1240	2.3765		0.166	6.10	2.852	8.62E-03	0.231	**Cesium (evaporated)**[3]			
0.1550	2.3772		0.166	6.20	2.879	9.30E-03	0.235	2.145	0.264	1.123	0.631
0.1907		3.1 E-05		6.30	2.910	9.74E-03	0.239	2.271	0.278	0.950	0.561
0.2066	2.3779	5.7 E-05	0.166	6.40	2.944	9.87E-03	0.243	2.845	0.425	0.438	0.235
0.22		1.21E-04		6.50	2.985	1.10E-02	0.248	3.064	0.540	0.320	0.127
0.23		2.36E-04		6.60	3.031	1.47E-02	0.254	3.397	0.671	0.233	0.057
0.24		3.82E-04		6.70	3.085	2.20E-02	0.261	3.966	0.827	0.174	0.018
0.25		5.21E-04		6.80	3.146	3.44E-02	0.268	4.889	0.916	0.143	0.007
0.26		2.96E-04		6.90	3.220	5.24E-02	0.277				
0.27		4.39E-04		7.00	3.322	9.35E-02	0.289	**Chromium**[4]			
0.28		2.75E-04		7.10	3.444	0.210	0.304	0.06	21.19	42.00	0.962
0.29		7.82E-05		7.15	3.464	0.307	0.308	0.10	11.81	29.76	0.955
0.30		1.32E-04		7.20	3.437	0.388	0.307	0.14	15.31	26.36	0.936
0.31	2.3787	1.30E-04	0.167	7.30	3.376	0.473	0.303	0.18	8.73	25.37	0.53
0.32		1.11E-04		7.40	3.335	0.515	0.300	0.22	5.30	20.62	0.954
0.33		2.99E-05		7.50	3.321	0.533	0.299	0.26	3.91	17.12	0.951
0.34		1.89E-05		7.60	3.306	0.592	0.300	0.30	3.15	14.28	0.943
0.35		2.11E-05		7.80	3.276	0.659	0.300	0.42	3.47	8.97	0.862
0.36		2.47E-05		8.00	3.251	0.712	0.300	0.54	3.92	7.06	0.788
0.37		2.80E-05		8.25	3.232	0.765	0.301	0.66	3.96	5.95	0.736
0.38		3.11E-05		8.50	3.228	0.806	0.303	0.78	4.13	5.03	0.680

Energy (eV)	n	k	$R(\phi = 0)$	Energy (eV)	n	k	$R(\phi = 0)$	Energy (eV)	n	k	$R(\phi = 0)$
0.90	4.43	4.60	0.650	20.00	0.77	0.64	0.130	5.60	1.16	1.75	0.400
1.00	4.47	4.43	0.639	20.5	0.76	0.63	0.129	5.80	1.10	1.73	0.406
1.12	4.53	4.31	0.631	21.0	0.74	0.58	0.121	6.00	1.03	1.68	0.407
1.24	4.50	4.28	0.629	21.5	0.72	0.55	0.116	6.20	0.97	1.62	0.401
1.36	4.42	4.30	0.631	22.0	0.71	0.52	0.112	6.40	0.94	1.53	0.386
1.46	4.31	4.32	0.632	22.5	0.70	0.50	0.109	6.60	0.91	1.46	0.368
1.77	3.84	4.37	0.639	23.0	0.69	0.48	0.105	6.80	0.91	1.38	0.345
2.00	3.48	4.36	0.644	23.5	0.68	0.45	0.101	7.00	0.91	1.32	0.326
2.20	3.18	4.41	0.656	24.0	0.68	0.43	0.096	7.00	0.91	1.26	0.305
2.40	2.75	4.46	0.677	24.5	0.67	0.39	0.089	7.40	0.92	1.21	0.286
2.60	2.22	4.36	0.698	25.0	0.68	0.36	0.080	7.60	0.93	1.17	0.269
2.80	1.80	4.06	0.703	25.5	0.68	0.33	0.072	7.80	0.94	1.13	0.253
3.00	1.54	3.71	0.695	26.0	0.70	0.31	0.063	8.00	0.95	1.09	0.239
3.20	1.44	3.40	0.670	26.5	0.71	0.28	0.055				
3.40	1.39	3.24	0.657	27.0	0.72	0.26	0.048	**Cobalt, single crystal,** $\vec{E} \perp \hat{c}$ [5]			
3.60	1.26	3.12	0.661	27.5	0.73	0.25	0.043	0.10	5.83	32.36	0.979
3.80	1.12	2.95	0.660	28.0	0.75	0.23	0.037	0.15	4.24	21.37	0.965
4.00	1.02	2.76	0.651	29.0	0.77	0.22	0.032	0.20	3.87	15.53	0.042
4.20	0.94	2.58	0.639	30.0	0.78	0.21	0.030	0.30	4.34	10.01	0.865
4.40	0.90	2.42	0.620					0.40	4.66	7.39	0.785
4.50	0.89	2.35	0.607	**Cobalt, single crystal,** $\vec{E} \parallel \hat{c}$ [5]				0.50	5.17	5.75	0.709
4.60	0.88	2.28	0.598	0.10	6.71	37.87	0.982	0.60	5.77	5.17	0.682
4.70	0.86	2.21	0.586	0.15	4.66	25.47	0.973	0.70	6.15	5.20	0.685
4.80	0.86	2.13	0.572	0.20	3.55	18.78	0.962	0.80	6.08	5.61	0.702
4.90	0.86	2.07	0.557	0.25	3.98	14.59	0.933	0.90	5.57	5.93	0.715
5.00	0.85	2.01	0.542	0.30	4.04	12.16	0.907	1.00	4.83	5.94	0.721
5.10	0.86	1.94	0.523	0.40	4.24	9.13	0.847	1.10	4.31	5.60	0.711
5.20	0.87	1.87	0.503	0.50	4.41	7.19	0.782	1.20	4.02	5.34	0.701
5.40	0.93	1.80	0.466	0.60	4.91	6.13	0.729	1.30	3.78	5.16	0.694
5.60	0.95	1.74	0.443	0.70	5.24	5.85	0.713	1.40	3.55	5.05	0.692
5.80	0.97	1.74	0.437	0.80	5.17	5.89	0.716	1.50	3.26	4.93	0.692
6.00	0.94	1.73	0.444	0.90	4.94	5.95	0.720	1.60	3.03	4.74	0.687
6.20	0.89	1.69	0.446	1.00	4.46	5.86	0.722	1.70	2.83	4.60	0.684
6.40	0.85	1.66	0.447	1.10	4.07	5.61	0.715	1.80	2.61	4.45	0.683
6.60	0.80	1.59	0.444	1.20	3.81	5.36	0.706	1.90	2.41	4.27	0.677
6.80	0.75	1.51	0.439	1.30	3.60	5.20	0.701	2.00	2.25	4.09	0.670
7.00	0.74	1.45	0.425	1.40	3.37	5.09	0.701	2.10	2.13	3.89	0.659
7.20	0.71	1.39	0.414	1.50	3.10	4.96	0.701	2.20	2.04	3.72	0.646
7.40	0.69	1.33	0.404	1.60	2.84	4.77	0.697	2.30	1.99	3.56	0.632
7.60	0.66	1.23	0.378	1.70	2.66	4.57	0.690	2.40	1.95	3.44	0.620
7.80	0.67	1.15	0.347	1.80	2.45	4.41	0.687	2.50	1.90	3.34	0.611
8.00	0.68	1.07	0.315	1.90	2.31	4.18	0.675	2.60	1.86	3.26	0.605
8.20	0.71	1.00	0.278	2.00	2.21	4.00	0.664	2.70	1.79	3.19	0.602
8.50	0.74	0.92	0.235	2.10	2.13	3.85	0.654	2.80	1.72	3.11	0.596
9.0	0.83	0.81	0.170	2.20	2.07	3.70	0.642	2.90	1.66	3.03	0.591
9.50	0.92	0.74	0.132	2.30	2.01	3.59	0.634	3.00	1.60	2.94	0.586
10.00	0.98	0.73	0.120	2.40	1.95	3.49	0.627	3.20	1.50	2.78	0.571
10.50	1.01	0.72	0.112	2.50	1.88	3.40	0.622	3.40	1.42	2.62	0.553
11.00	1.05	0.69	0.103	2.60	1.81	3.32	0.618	3.60	1.36	2.47	0.533
11.50	1.09	0.69	0.100	2.70	1.73	3.24	0.615	3.80	1.33	2.33	0.511
12.00	1.13	0.70	0.101	2.80	1.66	3.13	0.607	4.00	1.31	2.21	0.488
12.50	1.15	0.73	0.108	2.90	1.61	3.05	0.600	4.20	1.28	2.12	0.471
13.00	1.15	0.77	0.119	3.00	1.55	2.96	0.594	4.40	1.27	2.03	0.452
13.50	1.12	0.80	0.128	3.20	1.46	2.80	0.579	4.60	1.26	1.95	0.435
14.00	1.09	0.82	0.135	3.40	1.38	2.64	0.563	4.80	1.25	1.90	0.423
14.50	1.03	0.82	0.142	3.60	1.31	2.48	0.544	5.00	1.24	1.84	0.411
15.00	1.00	0.82	0.143	3.80	1.28	2.33	0.519	5.20	1.22	1.80	0.403
15.50	0.96	0.80	0.141	4.00	1.26	2.20	0.495	5.40	1.21	1.78	0.399
16.00	0.92	0.77	0.139	4.20	1.25	2.10	0.471	5.60	1.17	1.76	0.400
16.50	0.31	0.75	0.134	4.40	1.24	2.01	0.452	5.80	1.11	1.74	0.406
17.00	0.90	0.73	0.132	4.60	1.24	1.94	0.435	6.00	1.04	1.69	0.407
17.50	0.88	0.72	0.130	4.80	1.23	1.88	0.423	6.20	0.98	1.62	0.401
18.00	0.87	0.70	0.129	5.00	1.22	1.83	0.411	6.40	0.94	1.54	0.386
18.50	0.84	0.69	0.130	5.20	1.21	1.79	0.403	6.60	0.92	1.46	0.368
19.00	0.82	0.68	0.131	5.40	1.19	1.77	0.399	6.80	0.91	1.38	0.345

Energy (eV)	n	k	R(φ = 0)	Energy (eV)	n	k	R(φ = 0)	Energy (eV)	n	k	R(φ = 0)
7.00	0.91	1.32	0.326	27.00	0.88	0.38	0.043	0.02108		1.60E-03	
7.20	0.91	1.26	0.305	28.00	0.86	0.35	0.039	0.02232		1.55E-03	
7.40	0.92	1.21	0.285	29.00	0.85	0.30	0.032	0.02356		1.53E-03	
7.60	0.93	1.17	0.269	30.00	0.86	0.26	0.025	0.02480		1.50E-03	
7.80	0.94	1.13	0.253	31.00	0.88	0.24	0.020	0.02604		1.25E-03	
				32.00	0.89	0.22	0.017	0.02728		8.50E-04	
Copper[6]				33.00	0.90	0.21	0.015	0.02852		6.50E-04	
0.10	29.69	71.57	0.980	34.00	0.91	0.20	0.014	0.02976		7.00E-04	
0.50	1.71	17.63	0.979	35.00	0.92	0.20	0.013	0.03100	3.9827	8.50E-04	0.358
1.00	0.44	8.48	0.976	36.00	0.92	0.19	0.012	0.03224		1.55E-03	
1.50	0.26	5.26	0.965	37.00	0.92	0.19	0.011	0.03348		2.75E-03	
1.70	0.22	4.43	0.958	38.00	0.93	0.18	0.010	0.03472		3.55E-03	
1.75	0.21	4.25	0.956	39.00	0.93	0.17	0.009	0.03596	(3.9900)	3.05E-03	0.359
1.80	0.21	4.04	0.952	40.00	0.93	0.17	0.009	0.03720		2.75E-03	
1.85	0.22	3.85	0.947	41.00	0.94	0.16	0.008	0.03844		2.70E-03	
1.90	0.21	3.67	0.943	42.00	0.94	0.16	0.007	0.03968	(3.9930)	2.90E-03	0.359
2.00	0.27	3.24	0.910	43.00	0.94	0.15	0.007	0.04092		2.95E-03	
2.10	0.47	2.81	0.814	44.00	0.95	0.15	0.007	0.04215		3.20E-03	
2.20	0.83	2.60	0.673	45.00	0.95	0.15	0.006	0.04339		6.30E-03	
2.30	1.04	2.59	0.618	46.00	0.95	0.15	0.006	0.04463		3.40E-03	
2.40	1.12	2.60	0.602	47.00	0.95	0.14	0.006	0.04587	(3.9955)	2.50E-03	0.360
2.60	1.15	2.50	0.577	48.00	0.95	0.14	0.006	0.04711		2.10E-03	
2.80	1.17	2.36	0.545	49.00	0.95	0.14	0.005	0.04835		2.00E-03	
3.00	1.18	2.21	0.509	50.00	0.95	0.13	0.005	0.04959		8.00E-04	
3.20	1.23	2.07	0.468	51.00	0.95	0.13	0.005	0.05083		1.40E-03	
3.40	1.27	1.95	0.434	52.00	0.95	0.13	0.005	0.05207		1.35E-03	
3.60	1.31	1.87	0.407	53.00	0.96	0.12	0.004	0.05331		1.10E-03	
3.80	1.34	1.81	0.387	54.00	0.96	0.12	0.004	0.05455		8.00E-04	
4.00	1.34	1.72	0.364	55.00	0.96	0.12	0.004	0.05579		6.00E-04	
4.20	1.42	1.64	0.336	56.00	0.96	0.11	0.004	0.05703		9.0 E-04	
4.40	1.49	1.64	0.329	57.00	0.96	0.11	0.004	0.05827		6.5 E-04	
4.60	1.52	1.67	0.334	58.00	0.96	0.11	0.004	0.05951		4.6 E-04	
4.80	1.53	1.71	0.345	59.00	0.97	0.11	0.003	0.06075		4.0 E-04	
5.00	1.47	1.78	0.366	60.00	0.97	0.11	0.003	0.06199	3.9992	3.98E-04	0.360
5.20	1.38	1.80	0.380	61.00	0.97	0.11	0.003	0.06323		4.0 E-04	
5.40	1.28	1.78	0.389	62.00	0.97	0.11	0.003	0.06447		4.3 E-04	
5.60	1.18	1.74	0.391	63.00	0.96	0.10	0.003	0.06571		4.4 E-04	
5.80	1.10	1.67	0.389	64.00	0.96	0.10	0.003	0.06695	(4.0000)	4.3 E-04	0.360
6.00	1.04	1.59	0.380	65.00	0.97	0.10	0.003	0.06819		3.1 E-04	
6.50	0.96	1.37	0.329	66.00	0.97	0.10	0.003	0.06943		3.3 E-04	
7.00	0.97	1.20	0.271	67.00	0.97	0.09	0.003	0.07067		3.8 E-04	
7.50	1.00	1.09	0.230	68.00	0.97	0.09	0.002	0.07191		3.3 E-04	
8.00	1.03	1.03	0.206	69.00	0.97	0.09	0.002	0.07315		2.5 E-04	
8.50	1.03	0.98	0.189	70.00	0.97	0.09	0.002	0.07439		1.9 E-04	
9.00	1.03	0.92	0.171	75.00	0.98	0.09	0.002	0.07514		1.58E-04	
9.50	1.03	0.87	0.154	80.00	0.98	0.09	0.002	0.07749	4.0009	9.55E-05	0.360
10.00	1.04	0.82	0.139	85.00	0.97	0.09	0.002	0.07999	4.0011	1.71E-04	0.360
11.00	1.07	0.75	0.118	90.00	0.96	0.08	0.002	0.08266	4.0013	9.78E-05	0.360
12.00	1.09	0.73	0.111					0.08551	4.0015	5.77E-05	0.360
13.00	1.08	0.72	0.109	**Gallium (liquid)[7]**				0.08920		3.98E-05	
14.00	1.06	0.72	0.111	1.425	2.40	9.20	0.900	0.09460		4.59E-05	
14.50	1.03	0.72	0.111	1.550	2.09	8.50	0.898	0.09840		3.51E-05	
15.00	1.01	0.71	0.111	1.771	1.65	7.60	0.898	0.1	4.0063	3.70E-05	0.361
15.50	0.98	0.69	0.109	2.066	1.25	6.60	0.897	0.2	4.0108		0.361
16.00	0.95	0.67	0.106	2.480	0.89	5.60	0.898	0.3	4.0246		0.362
17.00	0.91	0.62	0.097	3.100	0.59	4.50	0.896	0.4	4.0429		0.364
18.00	0.89	0.56	0.084					0.5	(4.074)		0.367
19.00	0.88	0.51	0.071	**Germanium, single crystal[8]**				0.6	(4.104)	6.58E-07	0.370
20.00	0.88	0.45	0.059	0.01240	(4.0065)	3.00E-03	0.361	0.7	4.180	1.27E-04	0.377
21.00	0.90	0.41	0.048	0.01364	4.0063	2.40E-03	0.361	0.8	4.275	5.67E-03	0.385
22.00	0.92	0.38	0.040	0.01488	(4.0060)	1.70E-03	0.361	0.9	4.285	7.45E-02	0.386
23.00	0.94	0.37	0.035	0.01612	(4.0060)	1.55E-03	0.361	1.0	4.325	8.09E-02	0.390
24.00	0.96	0.37	0.035	0.01736	(4.0060)	1.50E-03	0.361	1.1	4.385	0.103	0.395
25.00	0.96	0.40	0.040	0.01860		1.50E-03		1.2	4.420	0.123	0.398
26.00	0.92	0.40	0.044	0.01984		1.60E-03		1.3	4.495	0.167	0.405

Energy (eV)	n	k	R(φ = 0)	Energy (eV)	n	k	R(φ = 0)	Energy (eV)	n	k	R(φ = 0)
1.4	4.560	0.190	0.411	40.0		0.0604		8.80	1.31	0.86	0.140
1.5	4.635	0.298	0.418					9.00	1.30	0.83	0.133
1.6	4.763	0.345	0.428	**Gold, electropolished, Au (110)[9]**				9.20	1.31	0.81	0.126
1.7	4.897	0.401	0.439	0.10	8.17	82.83	0.995	9.40	1.33	0.78	0.122
1.8	5.067	0.500	0.453	0.20	2.13	41.73	0.995	9.60	1.36	0.78	0.121
1.9	5.380	0.540	0.475	0.30	0.99	27.82	0.995	9.80	1.37	0.79	0.124
2.0	5.588	0.933	0.495	0.40	0.59	20.83	0.995	10.00	1.37	0.80	0.126
2.1	5.748	1.634	0.523	0.50	0.39	16.61	0.994	10.20	1.36	0.80	0.127
2.2	5.283	2.049	0.516	0.60	0.28	13.78	0.994	10.40	1.35	0.80	0.125
2.3	5.062	2.318	0.519	0.70	0.22	11.75	0.994	10.60	1.34	0.79	0.123
2.4	4.610	2.455	0.508	0.80	0.18	10.21	0.993	10.80	1.34	0.77	0.120
2.5	4.340	2.384	0.492	0.90	0.15	9.01	0.993	11.00	1.34	0.76	0.116
2.6	4.180	2.309	0.480	1.00	0.13	8.03	0.992	11.20	1.34	0.74	0.113
2.7	4.082	2.240	0.471	1.20	0.10	6.54	0.991	11.40	1.35	0.73	0.111
2.8	4.035	2.181	0.464	1.40	0.08	5.44	0.989	11.60	1.36	0.72	0.109
2.9	4.037	2.140	0.461	1.60	0.08	4.56	0.986	11.80	1.38	0.71	0.108
3.0	4.082	2.145	0.463	1.80	0.09	3.82	0.979	12.00	1.39	0.71	0.109
3.1	4.141	2.215	0.471	2.00	0.13	3.16	0.953	12.40	1.44	0.73	0.115
3.2	4.157	2.340	0.482	2.10	0.18	2.84	0.925	12.80	1.45	0.79	0.127
3.3	4.128	2.469	0.490	2.20	0.24	2.54	0.880	13.20	1.42	0.84	0.137
3.4	4.070	2.579	0.497	2.40	0.50	1.86	0.647	13.60	1.37	0.86	0.140
3.5	4.020	2.667	0.502	2.50	0.82	1.59	0.438	14.00	1.33	0.86	0.140
3.6	3.985	2.759	0.509	2.60	1.24	1.54	0.331	14.40	1.29	0.86	0.139
3.7	3.958	2.863	0.517	2.70	1.43	1.72	0.356	14.80	1.26	0.84	0.135
3.8	3.936	2.986	0.527	2.80	1.46	1.77	0.368	15.20	1.24	0.83	0.132
3.9	3.920	3.137	0.539	2.90	1.50	1.79	0.368	15.60	1.22	0.81	0.127
4.0	3.905	3.336	0.556	3.00	1.54	1.80	0.369	16.00	1.21	0.79	0.123
4.1	3.869	3.614	0.579	3.10	1.54	1.81	0.371	16.40	1.20	0.78	0.119
4.2	3.745	4.009	0.612	3.20	1.54	1.80	0.368	16.80	1.19	0.76	0.116
4.3	3.338	4.507	0.659	3.30	1.55	1.78	0.362	17.20	1.19	0.75	0.114
4.4	2.516	4.669	0.705	3.40	1.56	1.76	0.356	17.60	1.19	0.74	0.111
4.5	1.953	4.297	0.713	3.50	1.58	1.73	0.349	18.00	1.19	0.74	0.109
4.6	1.720	3.960	0.702	3.60	1.62	1.73	0.346	18.40	1.19	0.73	0.109
4.7	1.586	3.709	0.690	3.70	1.64	1.75	0.351	18.80	1.20	0.74	0.110
4.8	1.498	3.509	0.677	3.80	1.63	1.79	0.360	19.20	1.21	0.76	0.116
4.9	1.435	3.342	0.664	3.90	1.59	1.81	0.366	19.60	1.21	0.80	0.125
5.0	1.394	3.197	0.650	4.00	1.55	1.81	0.369	20.00	1.18	0.83	0.133
5.1	1.370	3.073	0.636	4.10	1.51	1.79	0.368	20.40	1.14	0.85	0.141
5.2	1.364	2.973	0.622	4.20	1.48	1.78	0.367	20.80	1.10	0.87	0.149
5.3	1.371	2.897	0.609	4.30	1.45	1.77	0.368	21.20	1.05	0.88	0.156
5.4	1.383	2.854	0.600	4.40	1.41	1.76	0.370	21.60	1.00	0.88	0.162
5.5	1.380	2.842	0.598	4.50	1.35	1.74	0.370	22.00	0.94	0.86	0.164
5.6	1.360	2.846	0.602	4.60	1.30	1.69	0.364	22.40	0.89	0.83	0.163
5.7	1.293	2.163	0.479	4.70	1.27	1.64	0.354	22.80	0.85	0.79	0.157
5.8	1.209	2.873	0.632	4.80	1.25	1.59	0.344	23.20	0.82	0.75	0.149
5.9	1.108	2.813	0.641	4.90	1.23	1.54	0.332	23.60	0.80	0.70	0.138
6.0	1.30	2.34	0.517	5.00	1.22	1.49	0.319	24.00	0.80	0.66	0.125
6.5	1.10	2.05	0.489	5.20	1.21	1.40	0.295	24.40	0.80	0.62	0.113
7.0	1.00	1.80	0.448	5.40	1.21	1.33	0.275	24.80	0.80	0.58	0.101
7.5		1.60		5.60	1.21	1.27	0.256	25.20	0.82	0.56	0.090
8.0	0.92	1.40	0.348	5.80	1.21	1.20	0.236	25.60	0.83	0.54	0.084
8.5	0.92	1.20	0.282	6.00	1.22	1.14	0.218	26.00	0.84	0.52	0.079
9.0	0.92	1.14	0.262	6.20	1.24	1.09	0.203	26.40	0.85	0.51	0.074
9.5		1.00		6.40	1.25	1.05	0.190	26.80	0.85	0.50	0.071
10.0	0.93	0.86	0.167	6.60	1.27	1.01	0.177	27.20	0.86	0.49	0.068
20.0		0.237		6.80	1.30	0.97	0.167	27.60	0.86	0.49	0.065
22.0		0.179		7.00	1.34	0.95	0.162	28.00	0.87	0.48	0.063
24.0		0.144		7.20	1.36	0.95	0.161	28.40	0.88	0.48	0.062
26.0		0.110		7.40	1.38	0.96	0.164	28.80	0.88	0.48	0.062
28.0		0.0747		7.60	1.38	0.98	0.169	29.20	0.88	0.48	0.062
30.0		0.1020		7.80	1.35	0.99	0.171	29.60	0.87	0.48	0.064
32.0		0.0999		8.00	1.31	0.96	0.165	30.00	0.86	0.48	0.064
34.0		0.0856		8.20	1.30	0.92	0.155				
36.0		0.0740		8.40	1.30	0.89	0.147	**Hafnium, single crystal, $\vec{E} \parallel \hat{c}$[10]**			
38.0		0.0651		8.60	1.31	0.88	0.144	0.52	1.48	4.11	0.747

Energy (eV)	n	k	$R(\phi=0)$
0.56	1.84	3.29	0.615
0.60	2.34	2.62	0.486
0.66	3.21	2.13	0.428
0.70	3.70	2.03	0.441
0.76	4.31	2.10	0.476
0.80	4.61	2.31	0.504
0.86	4.71	2.70	0.533
0.90	4.64	2.85	0.541
0.95	4.54	2.96	0.545
1.00	4.45	3.00	0.545
1.10	4.28	3.08	0.547
1.20	4.08	3.10	0.544
1.30	3.87	3.04	0.536
1.40	3.72	2.95	0.525
1.50	3.60	2.85	0.514
1.60	3.52	2.73	0.500
1.70	3.52	2.61	0.488
1.80	3.57	2.56	0.485
1.90	3.63	2.59	0.489
2.00	3.65	2.67	0.498
2.10	3.64	2.81	0.511
2.20	3.53	2.99	0.526
2.30	3.34	3.09	0.534
2.40	3.15	3.11	0.537
2.50	2.99	3.13	0.540
2.60	2.83	3.12	0.542
2.70	2.68	3.10	0.542
2.80	2.54	3.08	0.543
2.90	2.40	3.04	0.544
3.00	2.27	3.00	0.544
3.10	2.14	2.95	0.544
3.20	2.00	2.89	0.544
3.30	1.87	2.79	0.538
3.40	1.78	2.68	0.528
3.50	1.71	2.58	0.517
3.60	1.66	2.48	0.503
3.70	1.63	2.40	0.491
3.80	1.60	2.33	0.481
3.90	1.56	2.27	0.473
4.00	1.52	2.21	0.466
4.10	1.48	2.14	0.455
4.20	1.45	2.07	0.442
4.30	1.43	2.01	0.431
4.40	1.41	1.95	0.420
4.50	1.39	1.89	0.407
4.60	1.39	1.83	0.394
4.70	1.39	1.79	0.382
4.80	1.38	1.75	0.373
4.90	1.38	1.71	0.364
5.00	1.37	1.68	0.356
5.20	1.36	1.61	0.341
5.40	1.35	1.55	0.324
5.60	1.35	1.51	0.314
5.80	1.32	1.48	0.308
6.00	1.28	1.41	0.295
6.20	1.26	1.35	0.278
6.40	1.26	1.28	0.258
6.60	1.27	1.22	0.240
6.80	1.28	1.16	0.224
7.00	1.31	1.13	0.212
7.20	1.33	1.10	0.204
7.40	1.34	1.07	0.197
7.60	1.36	1.05	0.191
7.80	1.37	1.02	0.183
8.00	1.40	1.01	0.179

Energy (eV)	n	k	$R(\phi=0)$
8.20	1.43	1.01	0.178
8.40	1.45	1.01	0.180
8.60	1.47	1.02	0.183
8.80	1.48	1.04	0.186
9.00	1.49	1.07	0.193
9.20	1.50	1.10	0.201
9.40	1.48	1.14	0.211
9.60	1.46	1.18	0.222
9.80	1.41	1.21	0.230
10.00	1.36	1.22	0.235
10.20	1.32	1.22	0.238
10.40	1.28	1.22	0.240
10.60	1.24	1.21	0.241
10.80	1.20	1.20	0.242
11.00	1.16	1.19	0.242
11.20	1.13	1.17	0.241
11.40	1.10	1.16	0.241
11.60	1.07	1.14	0.239
11.80	1.04	1.12	0.238
12.00	1.02	1.10	0.236
12.40	0.96	1.06	0.232
12.80	0.92	1.01	0.225
13.20	0.88	0.96	0.218
13.60	0.84	0.90	0.205
14.00	0.83	0.83	0.186
14.40	0.83	0.80	0.172
14.80	0.81	0.76	0.167
15.20	0.79	0.70	0.153
15.60	0.79	0.64	0.132
16.00	0.83	0.60	0.111
16.40	0.81	0.60	0.114
16.80	0.79	0.55	0.105
17.20	0.79	0.50	0.089
17.60	0.80	0.46	0.077
18.00	0.81	0.42	0.064
18.40	0.84	0.38	0.051
18.80	0.87	0.34	0.040
19.00	0.89	0.33	0.036
19.60	0.93	0.32	0.030
20.00	0.94	0.31	0.027
20.60	0.97	0.30	0.023
21.00	0.99	0.29	0.022
21.60	1.01	0.28	0.020
22.00	1.03	0.28	0.020
22.60	1.06	0.28	0.020
23.00	1.07	0.28	0.021
23.60	1.09	0.29	0.022
24.00	1.09	0.30	0.023
24.60	1.10	0.31	0.024

Hafnium, single crystal, $\vec{E} \perp \hat{c}$ [10]

Energy (eV)	n	k	$R(\phi=0)$
0.52	2.25	4.65	0.723
0.56	2.34	3.66	0.623
0.60	2.84	2.89	0.512
0.66	3.71	2.35	0.469
0.70	4.26	2.21	0.482
0.76	4.97	2.33	0.521
0.80	5.41	2.62	0.554
0.86	5.46	3.36	0.593
0.90	5.22	3.62	0.601
0.95	4.95	3.72	0.602
1.00	4.76	3.76	0.602
1.10	4.43	3.80	0.601
1.20	4.07	3.74	0.594
1.30	3.79	3.55	0.578

Energy (eV)	n	k	$R(\phi=0)$
1.40	3.61	3.36	0.561
1.50	3.55	3.13	0.540
1.60	3.58	3.01	0.529
1.70	3.63	2.98	0.526
1.80	3.66	3.02	0.530
1.90	3.63	3.14	0.541
2.00	3.51	3.26	0.551
2.10	3.35	3.33	0.558
2.20	3.18	3.36	0.563
2.30	2.99	3.39	0.568
2.40	2.78	3.35	0.569
2.50	2.65	3.26	0.562
2.60	2.54	3.22	0.560
2.70	2.42	3.17	0.559
2.80	2.31	3.13	0.558
2.90	2.20	3.08	0.558
3.00	2.08	3.05	0.561
3.10	1.94	2.98	0.560
3.20	1.83	2.88	0.555
3.30	1.74	2.78	0.547
3.40	1.68	2.69	0.538
3.50	1.62	2.61	0.529
3.60	1.57	2.52	0.519
3.70	1.53	2.45	0.510
3.80	1.49	2.38	0.501
3.90	1.45	2.32	0.493
4.00	1.41	2.25	0.484
4.10	1.38	2.18	0.474
4.20	1.35	2.11	0.462
4.30	1.33	2.05	0.451
4.40	1.31	1.99	0.438
4.50	1.30	1.93	0.427
4.60	1.29	1.88	0.415
4.70	1.28	1.82	0.402
4.80	1.28	1.77	0.389
4.90	1.27	1.73	0.379
5.00	1.27	1.69	0.367
5.20	1.27	1.62	0.349
5.40	1.27	1.57	0.335
5.60	1.26	1.52	0.322
5.80	1.24	1.48	0.313
6.00	1.21	1.42	0.302
6.20	1.19	1.36	0.285
6.40	1.18	1.29	0.265
6.60	1.19	1.22	0.244
6.80	1.21	1.18	0.230
7.00	1.22	1.14	0.217
7.20	1.23	1.10	0.206
7.40	1.26	1.06	0.194
7.60	1.28	1.04	0.187
7.80	1.30	1.02	0.180
8.00	1.33	1.00	0.174
8.20	1.35	0.99	0.173
8.40	1.38	0.99	0.173
8.60	1.40	1.00	0.174
8.80	1.42	1.02	0.178
9.00	1.43	1.04	0.184
9.20	1.45	1.08	0.193
9.40	1.43	1.12	0.204
9.60	1.40	1.16	0.214
9.80	1.37	1.19	0.223
10.00	1.32	1.21	0.230
10.20	1.27	1.21	0.234
10.40	1.23	1.20	0.235
10.60	1.19	1.20	0.237

OPTICAL PROPERTIES OF METALS AND SEMICONDUCTORS (continued)

Energy (eV)	n	k	$R(\phi = 0)$	Energy (eV)	n	k	$R(\phi = 0)$	Energy (eV)	n	k	$R(\phi = 0)$
10.80	1.15	1.19	0.237	2.40	2.07	4.14	0.689	17.60	1.30	0.87	0.140
11.00	1.12	1.17	0.237	2.50	1.98	4.00	0.682	18.00	1.30	0.93	0.154
11.20	1.08	1.16	0.237	2.60	1.91	3.86	0.673	18.40	1.27	0.97	0.166
11.40	1.05	1.14	0.236	2.70	1.85	3.73	0.665	18.80	1.24	1.00	0.176
11.60	1.03	1.12	0.235	2.80	1.81	3.61	0.655	19.20	1.20	1.03	0.187
11.80	1.00	1.10	0.233	2.90	1.77	3.51	0.646	19.60	1.15	1.05	0.197
12.00	0.97	1.08	0.231	3.00	1.73	3.43	0.640	20.00	1.10	1.06	0.205
12.40	0.92	1.04	0.226	3.20	1.62	3.26	0.629	20.50	1.04	1.05	0.210
12.80	0.88	0.99	0.219	3.40	1.53	3.05	0.610	21.00	0.99	1.04	0.215
13.20	0.83	0.94	0.211	3.60	1.52	2.81	0.573	21.50	0.94	1.02	0.220
13.60	0.80	0.88	0.196	3.80	1.61	2.69	0.541	22.00	0.89	1.00	0.222
14.00	0.79	0.81	0.177	4.00	1.64	2.68	0.535	22.50	0.84	0.99	0.228
14.40	0.80	0.77	0.160	4.20	1.58	2.71	0.549	23.00	0.79	0.96	0.232
14.80	0.77	0.73	0.154	4.40	1.45	2.68	0.561	23.50	0.76	0.92	0.228
15.20	0.76	0.68	0.140	4.60	1.31	2.60	0.567	24.00	0.73	0.87	0.223
15.60	0.76	0.61	0.119	4.80	1.18	2.49	0.570	24.50	0.70	0.83	0.218
16.00	0.81	0.58	0.099	5.00	1.10	2.35	0.559	25.00	0.69	0.79	0.209
16.40	0.78	0.57	0.102	5.20	1.04	2.22	0.543	25.50	0.68	0.76	0.200
16.80	0.77	0.53	0.092	5.40	1.00	2.09	0.522	26.00	0.67	0.72	0.192
17.20	0.77	0.48	0.077	5.60	0.98	1.98	0.499	26.50	0.67	0.69	0.181
17.60	0.79	0.44	0.065	5.80	0.96	1.86	0.474	27.00	0.66	0.66	0.174
18.00	0.80	0.39	0.053	6.00	0.95	1.78	0.454	27.50	0.66	0.63	0.166
18.40	0.82	0.36	0.041	6.20	0.94	1.68	0.427	28.00	0.66	0.61	0.158
18.80	0.86	0.33	0.032	6.40	0.94	1.59	0.401	28.50	0.66	0.59	0.151
19.00	0.88	0.32	0.030	6.60	0.94	1.50	0.375	29.00	0.65	0.57	0.148
19.60	0.91	0.31	0.025	6.80	0.95	1.42	0.345	29.50	0.64	0.55	0.145
20.00	0.93	0.30	0.023	7.00	0.97	1.34	0.318	30.00	0.64	0.53	0.140
20.60	0.96	0.29	0.021	7.20	0.99	1.27	0.290	32.00	0.62	0.44	0.119
21.00	0.97	0.29	0.020	7.40	1.02	1.20	0.262	34.00	0.64	0.35	0.091
21.60	1.00	0.28	0.019	7.60	1.03	1.14	0.241	36.00	0.69	0.27	0.059
22.00	1.01	0.28	0.019	7.80	1.08	1.06	0.208	38.00	0.73	0.24	0.044
22.60	1.03	0.27	0.018	8.00	1.13	1.03	0.191	40.00	0.76	0.22	0.034
23.00	1.05	0.28	0.019	8.20	1.18	1.00	0.179				
23.60	1.06	0.28	0.020	8.40	1.22	0.98	0.171	**Iron[5]**			
24.00	1.07	0.29	0.021	8.60	1.26	0.96	0.164	0.10	6.41	33.07	0.978
24.60	1.09	0.30	0.022	8.80	1.29	0.95	0.160	0.15	6.26	22.82	0.956
				9.00	1.33	0.94	0.157	0.20	3.68	18.23	0.958
Iridium[11]				9.20	1.36	0.95	0.159	0.26	4.98	13.68	0.911
0.10	28.49	60.62	0.975	9.40	1.39	0.95	0.161	0.30	4.87	12.05	0.892
0.15	15.32	45.15	0.973	9.60	1.42	0.97	0.163	0.36	4.68	10.44	0.867
0.20	9.69	35.34	0.972	9.80	1.44	0.99	0.169	0.40	4.42	9.75	0.858
0.25	6.86	28.84	0.969	10.00	1.45	1.01	0.175	0.50	4.14	8.02	0.817
0.30	5.16	24.25	0.967	10.20	1.45	1.04	0.182	0.60	3.93	6.95	0.783
0.35	4.11	20.79	0.964	10.40	1.44	1.07	0.187	0.70	3.78	6.17	0.752
0.40	3.42	18.06	0.960	10.60	1.43	1.09	0.193	0.80	3.65	5.60	0.725
0.45	3.05	15.82	0.954	10.80	1.41	1.12	0.200	0.90	3.52	5.16	0.700
0.50	2.98	14.06	0.944	11.00	1.38	1.13	0.206	1.00	3.43	4.79	0.678
0.60	2.79	11.58	0.925	11.20	1.34	1.14	0.208	1.10	3.33	4.52	0.660
0.70	2.93	9.78	0.895	11.40	1.31	1.13	0.208	1.20	3.24	4.26	0.641
0.80	3.14	8.61	0.862	11.60	1.28	1.12	0.206	1.30	3.16	4.07	0.626
0.90	3.19	7.88	0.840	11.80	1.25	1.10	0.203	1.40	3.12	3.87	0.609
1.00	3.15	7.31	0.822	12.00	1.24	1.08	0.199	1.50	3.05	3.77	0.601
1.10	3.04	6.84	0.808	12.40	1.21	1.05	0.191	1.60	3.00	3.60	0.585
1.20	2.96	6.41	0.791	12.80	1.19	1.01	0.181	1.70	2.98	3.52	0.577
1.30	2.85	6.07	0.779	13.20	1.18	0.98	0.173	1.80	2.92	3.46	0.573
1.40	2.72	5.74	0.767	13.60	1.17	0.95	0.165	1.90	2.89	3.37	0.563
1.50	2.65	5.39	0.750	14.00	1.16	0.91	0.155	2.00	2.85	3.36	0.563
1.60	2.68	5.08	0.728	14.40	1.17	0.88	0.147	2.10	2.80	3.34	0.562
1.70	2.69	4.92	0.716	14.80	1.18	0.87	0.142	2.20	2.74	3.33	0.563
1.80	2.64	4.81	0.710	15.20	1.19	0.84	0.136	2.30	2.65	3.34	0.567
1.90	2.57	4.68	0.704	15.60	1.20	0.83	0.133	2.40	2.56	3.31	0.567
2.00	2.50	4.57	0.699	16.00	1.21	0.83	0.131	2.50	2.46	3.31	0.570
2.10	2.40	4.48	0.697	16.40	1.23	0.82	0.129	2.60	2.34	3.30	0.576
2.20	2.29	4.38	0.695	16.80	1.25	0.82	0.127	2.70	2.23	3.25	0.575
2.30	2.18	4.26	0.692	17.20	1.28	0.83	0.131	2.80	2.12	3.23	0.580

Energy (eV)	n	k	$R(\phi = 0)$	Energy (eV)	n	k	$R(\phi = 0)$	Energy (eV)	n	k	$R(\phi = 0)$
2.90	2.01	3.17	0.580	13.17	0.84	0.79	0.161	24.00	0.74	0.27	0.045
3.00	1.88	3.12	0.583	13.33	0.84	0.78	0.160	24.17	0.74	0.26	0.044
3.10	1.78	3.04	0.580	13.50	0.83	0.77	0.159	24.33	0.74	0.26	0.043
3.20	1.70	2.96	0.576	13.67	0.82	0.76	0.157	24.50	0.74	0.25	0.042
3.30	1.62	2.87	0.572	13.83	0.81	0.75	0.154	24.67	0.75	0.25	0.040
3.40	1.55	2.79	0.565	14.00	0.81	0.73	0.151	24.83	0.75	0.24	0.039
3.50	1.50	2.70	0.556	14.17	0.80	0.72	0.149	25.00	0.75	0.24	0.038
3.60	1.47	2.63	0.548	14.33	0.80	0.71	0.146	26.00	0.76	0.21	0.031
3.70	1.43	2.56	0.542	14.50	0.79	0.79	0.144	27.00	0.78	0.18	0.026
3.83	1.38	2.49	0.534	14.67	0.79	0.69	0.141	28.00	0.79	0.16	0.021
4.00	1.30	2.39	0.527	14.83	0.78	0.67	0.138	29.00	0.81	0.14	0.017
4.17	1.26	2.27	0.510	15.00	0.78	0.66	0.135	30.00	0.82	0.13	0.014
4.33	1.23	2.18	0.494	15.17	0.78	0.65	0.131				
4.50	1.20	2.10	0.482	15.33	0.78	0.64	0.238	**Lithium**[12]			
4.67	1.16	2.02	0.470	15.50	0.77	0.63	0.126	0.14	0.659	38.0	0.998
4.83	1.14	1.93	0.451	15.67	0.77	0.62	0.123	0.54	0.661	12.6	0.984
5.00	1.14	1.87	0.435	15.83	0.77	0.61	0.119	0.75	0.561	7.68	0.963
5.17	1.12	1.81	0.425	16.00	0.77	0.60	0.116	1.05	0.448	5.58	0.946
5.33	1.11	1.75	0.408	16.17	0.78	0.58	0.112	1.35	0.338	4.36	0.935
5.50	1.09	1.17	0.401	16.33	0.78	0.58	0.110	1.65	0.265	3.55	0.925
5.67	1.09	1.65	0.383	16.50	0.78	0.57	0.107	1.95	0.221	2.94	0.913
5.83	1.10	1.61	0.373	16.67	0.77	0.56	0.106	2.25	0.206	2.48	0.892
6.00	1.09	1.59	0.366	16.83	0.78	0.55	0.103	2.55	0.217	2.11	0.854
6.17	1.08	1.57	0.365	17.00	0.78	0.55	0.102	2.85	0.247	1.82	0.797
6.33	1.04	1.55	0.365	17.17	0.78	0.54	0.100	3.15	0.304	1.60	0.715
6.50	1.02	1.51	0.358	17.33	0.78	0.54	0.098	3.45	0.334	1.45	0.656
6.67	1.00	1.47	0.351	17.50	0.77	0.53	0.097	3.75	0.345	1.32	0.611
6.83	0.97	1.43	0.346	17.67	0.77	0.52	0.095	4.05	0.346	1.21	0.578
7.00	0.96	1.39	0.333	17.83	0.78	0.51	0.092	4.35	0.333	1.11	0.557
7.17	0.94	1.35	0.327	18.00	0.78	0.51	0.091	4.65	0.317	1.01	0.540
7.33	0.94	1.30	0.311	18.17	0.78	0.51	0.090	4.95	0.302	0.906	0.520
7.50	0.94	1.26	0.298	18.33	0.78	0.50	0.089	5.25	0.299	0.795	0.484
7.67	0.94	1.23	0.288	18.50	0.77	0.50	0.089	5.55	0.310	0.688	0.434
7.83	0.94	1.21	0.279	18.67	0.77	0.50	0.088	5.85	0.342	0.594	0.365
8.00	0.94	1.18	0.272	18.83	0.77	0.49	0.087	6.15	0.376	0.522	0.306
8.17	0.94	1.16	0.265	19.00	0.77	0.49	0.087	6.45	0.408	0.460	0.256
8.33	0.94	1.14	0.258	19.17	0.76	0.49	0.088	6.75	0.440	0.407	0.214
8.50	0.94	1.12	0.251	19.33	0.76	0.48	0.087	7.05	0.466	0.364	0.183
8.67	0.94	1.10	0.246	19.50	0.75	0.47	0.086	7.35	0.492	0.320	0.155
8.83	0.92	1.08	0.240	19.67	0.75	0.47	0.085	7.65	0.517	0.282	0.131
9.00	0.93	1.07	0.236	19.83	0.75	0.46	0.084	7.95	0.545	0.246	0.109
9.17	0.92	1.06	0.233	20.00	0.74	0.45	0.083	8.25	0.572	0.214	0.091
9.33	0.91	1.04	0.231	20.17	0.74	0.44	0.081	8.55	0.601	0.189	0.075
9.50	0.90	1.02	0.226	20.33	0.74	0.44	0.081	8.85	0.624	0.163	0.063
9.67	0.90	1.00	0.221	20.50	0.74	0.42	0.080	9.15	0.657	0.144	0.050
9.83	0.89	0.99	0.218	20.67	0.73	0.43	0.079	9.45	0.680	0.130	0.042
10.00	0.88	0.97	0.213	20.83	0.73	0.42	0.078	9.75	0.708	0.119	0.034
10.17	0.87	0.94	0.203	21.00	0.73	0.41	0.077	10.1	0.726	0.108	0.029
10.33	0.87	0.91	0.196	21.17	0.72	0.40	0.076	10.4	0.743	0.102	0.025
10.50	0.87	0.89	0.189	21.33	0.72	0.39	0.074	10.6	0.753	0.080	0.022
10.67	0.88	0.87	0.179	21.50	0.72	0.38	0.073				
10.83	0.89	0.85	0.170	21.67	0.72	0.38	0.071	**Magnesium (evaporated)**[13]			
11.00	0.91	0.83	0.162	21.83	0.72	0.37	0.070	2.145	0.48	3.71	0.880
11.17	0.92	0.83	0.159	22.00	0.72	0.36	0.068	2.270	0.57	3.47	0.843
11.33	0.93	0.84	0.159	22.17	0.71	0.35	0.067	2.522	0.53	2.92	0.805
11.50	0.93	0.84	0.160	22.33	0.72	0.34	0.064	2.845	0.52	2.65	0.777
11.67	0.93	0.84	0.162	22.50	0.72	0.34	0.063	3.064	0.52	2.05	0.681
11.83	0.92	0.84	0.163	22.67	0.72	0.33	0.062	5.167	0.10	1.60	0.894
12.00	0.91	0.84	0.163	22.83	0.72	0.32	0.059	5.636	0.15	1.50	0.832
12.17	0.90	0.84	0.165	23.00	0.72	0.31	0.058	6.200	0.20	1.40	0.765
12.33	0.89	0.83	0.164	23.17	0.72	0.30	0.056	6.889	0.25	1.30	0.693
12.50	0.98	0.83	0.165	23.33	0.72	0.29	0.054	7.750	0.20	1.20	0.722
12.67	0.87	0.82	0.166	23.50	0.73	0.28	0.050	8.857	0.15	0.95	0.730
12.83	0.86	0.81	0.166	23.67	0.73	0.28	0.049	10.335	0.25	0.40	0.419
13.00	0.85	0.80	0.162	23.83	0.74	0.27	0.047				

Energy (eV)	n	k	$R(\phi = 0)$	Energy (eV)	n	k	$R(\phi = 0)$	Energy (eV)	n	k	$R(\phi = 0)$
Manganese[14]				2.4	1.384	4.407	0.779	17.0	1.177	0.367	0.034
0.64	3.89	5.95	0.738	2.6	1.186	4.090	0.779	17.5	1.184	0.366	0.034
0.77	3.78	5.41	0.710	2.8	1.027	3.802	0.779	18.0	1.191	0.367	0.035
0.89	3.65	5.02	0.688	3.0	0.898	3.538	0.777	18.5	1.195	0.367	0.035
1.02	3.48	4.74	0.673	3.2	0.798	3.294	0.773	19.0	1.200	0.366	0.035
1.14	3.30	4.53	0.662	3.4	0.713	3.074	0.770	19.5	1.208	0.364	0.035
1.26	3.10	4.35	0.653	3.6	0.644	2.860	0.763				
1.39	2.97	4.18	0.643	3.8	0.589	2.665	0.755	**Molybdenum[16]**			
1.51	2.83	4.03	0.634	4.0	0.542	2.502	0.749	0.10	18.53	68.51	0.985
1.64	2.70	3.91	0.627	4.2	0.507	2.341	0.738	0.15	8.78	47.54	0.985
1.76	2.62	3.78	0.617	4.4	0.477	2.195	0.727	0.20	5.10	35.99	0.985
1.88	2.56	3.65	0.606	4.6	0.452	2.058	0.715	0.25	3.36	28.75	0.984
2.01	2.51	3.54	0.596	4.8	0.431	1.929	0.701	0.30	2.44	23.80	0.983
2.13	2.47	3.43	0.585	5.0	0.414	1.806	0.685	0.34	2.00	20.84	0.982
2.26	2.39	3.33	0.577	5.2	0.401	1.687	0.666	0.38	1.70	18.44	0.980
2.38	2.32	3.23	0.567	5.4	0.394	1.569	0.642	0.42	1.57	16.50	0.978
2.50	2.25	3.14	0.559	5.6	0.386	1.454	0.617	0.46	1.46	14.91	0.975
2.63	2.19	3.06	0.552	5.7	0.386	1.396	0.601	0.50	1.37	13.55	0.971
2.75	2.11	2.98	0.545	5.8	0.386	1.341	0.585	0.54	1.35	12.36	0.966
2.88	2.06	2.90	0.536	5.9	0.385	1.287	0.569	0.58	1.34	11.34	0.960
3.00	2.00	2.82	0.528	6.0	0.386	1.232	0.551	0.62	1.38	10.44	0.952
3.12	1.96	2.74	0.518	6.1	0.388	1.176	0.531	0.66	1.43	9.67	0.942
3.25	1.92	2.67	0.509	6.2	0.390	1.118	0.510	0.70	1.48	8.99	0.932
3.37	1.89	2.59	0.498	6.3	0.399	1.058	0.481	0.74	1.51	8.38	0.921
3.50	1.89	2.51	0.484	6.4	0.412	1.002	0.450	0.78	1.60	7.83	0.906
3.62	1.87	2.45	0.475	6.5	0.428	0.949	0.418	0.82	1.64	7.35	0.892
3.74	1.86	2.38	0.463	6.6	0.436	0.898	0.392	0.86	1.70	6.89	0.876
3.87	1.86	2.32	0.451	6.7	0.438	0.836	0.367	9.90	1.74	6.48	0.859
3.99	1.86	2.25	0.438	6.8	0.459	0.756	0.320	1.00	1.94	5.58	0.805
4.12	1.86	2.19	0.427	6.9	0.510	0.676	0.255	1.10	2.15	4.85	0.743
4.24	1.85	2.14	0.417	7.0	0.585	0.617	0.191	1.20	2.44	4.22	0.671
4.36	1.85	2.08	0.406	7.1	0.663	0.589	0.148	1.30	2.77	3.74	0.608
4.49	1.86	2.03	0.395	7.2	0.717	0.584	0.128	1.40	3.15	3.40	0.562
4.61	1.85	1.99	0.388	7.3	0.769	0.575	0.111	1.50	3.53	3.30	0.550
4 74	1.84	1.94	0.378	7.4	0.817	0.574	0.100	1.60	3.77	3.41	0.562
4.86	1.83	1.91	0.372	7.5	0.860	0.580	0.094	1.70	3.84	3.51	0.570
4.98	1.82	1.86	0.362	7.6	0.893	0.597	0.093	1.80	3.81	3.58	0.576
5.11	1.82	1.82	0.354	7.8	0.929	0.623	0.096	1.90	3.74	3.58	0.576
5.23	1.81	1.79	0.348	8.0	0.946	0.639	0.098	2.00	3.68	3.52	0.571
5.36	1.78	1.76	0.342	8.2	0.952	0.645	0.099	2.10	3.68	3.45	0.565
5.48	1.74	1.73	0.337	8.4	0.953	0.638	0.097	2.20	3.76	3.41	0.562
5.60	1.73	1.70	0.331	8.6	0.956	0.624	0.093	2.30	3.79	3.61	0.578
5.73	1.72	1.67	0.325	8.8	0.965	0.607	0.087	2.40	3.59	3.78	0.594
5.85	1.70	1.64	0.319	9.0	0.975	0.588	0.082	2.50	3.36	3.73	0.591
5.98	1.67	1.61	0.313	9.2	0.988	0.568	0.076	2.60	3.22	3.61	0.582
6.10	1.63	1.58	0.307	9.4	1.009	0.548	0.069	2.70	3.13	3.51	0.573
6.22	1.62	1.55	0.301	9.6	1.044	0.541	0.066	2.80	3.08	3.42	0.565
6.35	1.59	1.52	0.295	9.8	1.061	0.557	0.069	2.90	3.05	3.33	0.566
6.47	1.55	1.50	0.292	10.0	1.062	0.567	0.071	3.00	3.04	3.27	0.550
6.60	1.48	1.47	0.288	10.2	1.054	0.569	0.072	3.10	3.03	3.21	0.544
				10.4	1.045	0.561	0.070	3.20	3.05	3.18	0.540
				10.6	1.041	0.550	0.068	3.30	3.06	3.18	0.540
Mercury (liquid)[15]				10.8	1.039	0.537	0.065	3.40	3.06	3.19	0.541
0.2	13.99	14.27	0.869	11.0	1.039	0.523	0.062	3.50	3.06	3.21	0.543
0.3	11.37	11.95	0.846	11.5	1.050	0.491	0.055	3.60	3.05	3.23	0.546
0.4	9.741	10.65	0.830	12.0	1.064	0.467	0.050	3.70	3.04	3.27	0.550
0.5	8.528	9.805	0.818	12.5	1.078	0.445	0.045	3.80	3.04	3.31	0.554
0.6	7.574	9.195	0.808	13.0	1.092	0.430	0.042	3.90	3.04	3.40	0.564
0.8	6.086	8.312	0.796	13.5	1.104	0.416	0.040	4.00	3.01	3.51	0.576
1.0	4.962	7.643	0.789	14.0	1.115	0.404	0.038	4.20	2.77	3.77	0.610
1.2	4.050	7.082	0.786	14.5	1.125	0.394	0.037	4.40	2.39	3.88	0.640
1.4	3.324	6.558	0.785	15.0	1.135	0.383	0.035	4.60	2.06	3.84	0.658
1.6	2.746	6.054	0.783	15.5	1.146	0.374	0.034	4.80	1.75	3.76	0.678
1.8	2.284	5.582	0.782	16.0	1.159	0.368	0.034	5.00	1.46	3.62	0.695
2.0	1.910	5.150	0.782	16.5	1.170	0.367	0.034	5.20	1.22	3.42	0.706
2.2	1.620	4.751	0.780								

Energy (eV)	n	k	$R(\phi = 0)$	Energy (eV)	n	k	$R(\phi = 0)$	Energy (eV)	n	k	$R(\phi = 0)$
5.40	1.07	3.20	0.706	27.00	0.73	0.29	0.050	4.80	1.53	2.11	0.435
5.60	0.96	2.99	0.700	27.50	0.76	0.28	0.041	5.00	1.40	2.10	0.449
5.80	0.89	2.80	0.688	28.00	0.79	0.27	0.036	5.20	1.27	2.04	0.454
6.00	0.85	2.64	0.674	28.50	0.81	0.26	0.031	5.40	1.16	1.94	0.449
6.20	0.81	2.50	0.660	29.00	0.83	0.26	0.028	5.60	1.09	1.83	0.435
6.40	0.79	2.36	0.641	29.50	0.86	0.26	0.025	5.80	1.04	1.73	0.417
6.60	0.78	2.24	0.619	30.00	0.88	0.26	0.023	6.20	1.00	1.54	0.371
6.80	0.78	2.13	0.592	31.00	0.92	0.29	0.024	6.40	1.01	1.46	0.345
7.00	0.80	2.04	0.568	32.00	0.92	0.32	0.030	6.60	1.01	1.40	0.325
7.20	0.81	1.98	0.548	33.00	0.90	0.33	0.032	6.80	1.02	1.35	0.308
7.40	0.81	1.95	0.542	34.00	0.91	0.34	0.034	7.00	1.03	1.30	0.291
7.60	0.75	1.90	0.552	35.00	0.87	0.37	0.043	7.20	1.03	1.27	0.282
7.80	0.71	1.81	0.542	36.00	0.82	0.34	0.043	7.40	1.03	1.24	0.273
8.00	0.69	1.73	0.530	37.00	0.81	0.30	0.038	7.60	1.02	1.22	0.265
8.20	0.67	1.65	0.512	38.00	0.81	0.27	0.033	7.80	1.01	1.18	0.256
8.40	0.66	1.57	0.495	39.00	0.82	0.25	0.029	8.00	1.01	1.15	0.248
8.60	0.65	1.49	0.475	40.00	0.83	0.23	0.025	8.20	1.00	1.13	0.242
8.80	0.65	1.41	0.450					8.40	0.99	1.11	0.235
9.00	0.65	1.33	0.420	**Nickel**[17]				8.60	0.98	1.08	0.228
9.20	0.67	1.25	0.385	0.10	9.54	45.82	0.983	8.80	0.97	1.05	0.220
9.40	0.69	1.19	0.355	0.15	5.45	30.56	0.978	9.00	0.97	1.01	0.211
9.60	0.71	1.12	0.320	0.20	4.12	22.48	0.969	9.20	0.96	0.99	0.203
9.80	0.74	1.05	0.285	0.25	4.25	17.68	0.950	9.40	0.95	0.96	0.194
10.00	0.77	0.99	0.250	0.30	4.19	15.05	0.934	9.60	0.95	0.93	0.185
10.20	0.81	0.93	0.217	0.35	4.03	13.05	0.918	9.80	0.95	0.89	0.175
10.40	0.86	0.88	0.188	0.40	3.84	11.43	0.900	10.00	0.95	0.87	0.166
10.60	0.91	0.83	0.162	0.50	4.03	9.64	0.864	10.20	0.95	0.83	0.155
10.80	0.98	0.79	0.138	0.60	3.84	8.35	0.835	10.40	0.95	0.80	0.145
11.00	1.05	0.77	0.125	0.70	3.59	7.48	0.813	10.60	0.97	0.76	0.129
11.20	1.12	0.78	0.123	0.80	3.38	6.82	0.794	10.80	0.99	0.75	0.123
11.40	1.18	0.80	0.125	0.90	3.18	6.23	0.774	11.00	1.01	0.73	0.115
11.60	1.23	0.85	0.135	1.00	3.06	5.74	0.753	11.25	1.04	0.72	0.111
11.80	1.25	0.89	0.145	1.10	2.97	5.38	0.734	11.50	1.05	0.71	0.109
12.00	1.26	0.92	0.154	1.20	2.85	5.10	0.721	11.75	1.07	0.71	0.108
12.40	1.25	0.98	0.168	1.30	2.74	4.85	0.708	12.00	1.07	0.71	0.108
12.80	1.23	1.00	0.178	1.40	2.65	4.63	0.695	12.25	1.07	0.71	0.107
13.20	1.20	1.02	0.185	1.50	2.53	4.47	0.688	12.50	1.08	0.71	0.106
13.60	1.17	1.02	0.187	1.60	2.43	4.31	0.679	12.75	1.08	0.71	0.106
14.00	1.15	1.01	0.185	1.70	2.28	4.18	0.677	13.00	1.08	0.71	0.105
14.40	1.13	1.00	0.182	1.80	2.14	4.01	0.670	13.25	1.08	0.71	0.105
14.80	1.13	0.99	0.179	1.90	2.02	3.82	0.659	13.50	1.07	0.70	0.105
15.00	1.14	0.99	0.179	2.00	1.92	3.65	0.649	13.75	1.07	0.70	0.105
15.60	1.15	1.01	0.184	2.10	1.85	3.48	0.634	14.00	1.07	0.71	0.106
16.00	1.14	1.04	0.194	2.20	1.80	3.33	0.620	14.25	1.06	0.70	0.106
16.60	1.10	1.10	0.216	2.30	1.75	3.19	0.605	14.50	1.05	0.70	0.106
17.00	1.04	1.12	0.233	2.40	1.71	3.06	0.590	14.75	1.04	0.70	0.107
17.60	0.94	1.14	0.257	2.50	1.67	2.93	0.575	15.00	1.03	0.70	0.107
18.00	0.87	1.12	0.270	2.60	1.65	2.81	0.557	15.25	1.02	0.69	0.106
18.60	0.77	1.08	0.283	2.70	1.64	2.71	0.542	15.50	1.01	0.69	0.105
19.00	0.71	1.02	0.284	2.80	1.63	2.61	0.525	15.75	1.00	0.68	0.104
19.60	0.66	0.94	0.275	2.90	1.62	2.52	0.509	16.00	0.99	0.67	0.103
20.00	0.64	0.89	0.264	3.00	1.61	2.44	0.495	16.50	0.98	0.66	0.101
20.60	0.62	0.81	0.245	3.10	1.61	2.36	0.480	17.00	0.96	0.64	0.098
21.00	0.61	0.77	0.234	3.20	1.61	2.30	0.467	17.50	0.94	0.63	0.096
21.60	0.61	0.71	0.215	3.30	1.61	2.23	0.454	18.00	0.92	0.61	0.092
22.00	0.60	0.69	0.207	3.40	1.62	2.17	0.441	18.50	0.91	0.58	0.087
22.60	0.59	0.63	0.195	3.50	1.63	2.11	0.428	19.00	0.90	0.56	0.082
23.00	0.58	0.60	0.185	3.60	1.64	2.07	0.416	19.50	0.90	0.54	0.077
23.60	0.58	0.53	0.166	3.70	1.66	2.02	0.405	20.00	0.89	0.51	0.071
24.00	0.58	0.49	0.151	3.80	1.69	1.99	0.397	20.50	0.89	0.49	0.066
24.60	0.60	0.43	0.124	3.90	1.72	1.98	0.393	21.00	0.90	0.47	0.061
25.00	0.62	0.39	0.106	4.00	1.73	1.98	0.392	21.50	0.91	0.46	0.057
25.60	0.66	0.35	0.085	4.20	1.74	2.01	0.396	22.00	0.91	0.45	0.055
26.00	0.68	0.33	0.072	4.40	1.71	2.06	0.409	22.50	0.91	0.44	0.053
26.50	0.71	0.31	0.060	4.60	1.63	2.09	0.421	23.00	0.92	0.44	0.051

Energy (eV)	n	k	R(φ = 0)	Energy (eV)	n	k	R(φ = 0)	Energy (eV)	n	k	R(φ = 0)
23.50	0.91	0.44	0.052	4.60	2.39	2.56	0.470	23.60	0.88	0.30	0.029
24.00	0.90	0.43	0.051	4.80	2.32	2.52	0.465	24.00	0.91	0.29	0.025
24.50	0.90	0.43	0.051	5.00	2.26	2.57	0.475	24.60	0.94	0.28	0.022
25.00	0.89	0.42	0.050	5.20	2.16	2.62	0.487	25.00	0.96	0.27	0.020
26.00	0.88	0.39	0.046	5.40	2.00	2.68	0.505	25.60	0.99	0.26	0.018
27.00	0.87	0.37	0.042	5.60	1.81	2.67	0.518	26.00	1.00	0.26	0.017
28.00	0.87	0.35	0.040	5.80	1.63	2.60	0.522	26.60	1.03	0.25	0.016
29.00	0.86	0.34	0.037	6.00	1.49	2.49	0.520	27.00	1.04	0.25	0.015
30.00	0.86	0.32	0.034	6.20	1.38	2.38	0.512	27.60	1.06	0.25	0.015
35.00	0.86	0.24	0.022	6.40	1.31	2.25	0.496	28.00	1.08	0.24	0.015
40.00	0.87	0.18	0.014	6.60	1.26	2.14	0.480	28.60	1.11	0.24	0.016
45.00	0.88	0.13	0.008	6.80	1.24	2.04	0.460	29.00	1.13	0.25	0.017
50.00	0.92	0.10	0.004	7.00	1.23	1.96	0.441	29.60	1.16	0.26	0.020
60.00	0.96	0.08	0.002	7.20	1.22	1.91	0.430	30.00	1.18	0.28	0.023
65.00	0.98	0.09	0.002	7.40	1.20	1.88	0.427	31.00	1.18	0.31	0.026
68.00	0.96	0.12	0.004	7.60	1.14	1.85	0.430	32.00	1.20	0.34	0.031
70.00	0.94	0.11	0.004	7.80	1.07	1.78	0.428	33.00	1.21	0.38	0.038
75.00	0.94	0.09	0.003	8.00	1.02	1.69	0.412	34.00	1.20	0.42	0.044
80.00	0.94	0.07	0.002	8.20	1.00	1.60	0.390	35.20	1.17	0.47	0.051
90.00	0.94	0.06	0.002	8.40	0.99	1.51	0.365	36.00	1.15	0.50	0.056
				8.60	0.99	1.43	0.340	37.50	1.07	0.53	0.064
Niobium[18]				8.70	0.99	1.39	0.328	39.50	0.95	0.50	0.063
0.12	15.99	53.20	0.979	8.80	1.00	1.36	0.315	40.50	0.92	0.47	0.059
0.20	7.25	34.14	0.976	9.00	1.01	1.29	0.290				
0.24	5.47	28.88	0.975	9.20	1.04	1.22	0.265	**Osmium (Polycrystalline)[9]**			
0.28	4.26	24.95	0.974	9.40	1.07	1.18	0.245	0.10	4.08	50.23	0.994
0.35	3.11	20.03	0.970	9.60	1.10	1.13	0.227	0.15	2.90	33.60	0.990
0.45	2.28	15.58	0.964	9.80	1.13	1.09	0.209	0.20	2.44	25.11	0.985
0.55	1.83	12.67	0.956	10.00	1.18	1.05	0.194	0.25	2.35	19.99	0.977
0.65	1.57	10.59	0.947	10.20	1.23	1.04	0.187	0.30	2.23	16.54	0.969
0.75	1.41	9.00	0.935	10.40	1.27	1.04	0.185	0.35	2.33	14.06	0.955
0.85	1.35	7.74	0.918	10.60	1.30	1.06	0.190	0.40	2.45	12.32	0.940
0.95	1.35	6.70	0.893	10.80	1.32	1.08	0.195	0.45	2.43	11.02	0.927
1.05	1.44	5.86	0.857	11.00	1.32	1.10	0.200	0.50	2.41	9.97	0.913
1.15	1.55	5.18	0.814	11.20	1.31	1.12	0.204	0.55	2.33	9.12	0.901
1.25	1.65	4.63	0.768	11.40	1.30	1.13	0.207	0.60	2.21	8.37	0.890
1.35	1.76	4.13	0.715	11.60	1.28	1.13	0.209	0.65	2.11	7.68	0.877
1.45	1.95	3.68	0.650	11.80	1.27	1.13	0.210	0.70	2.02	7.04	0.862
1.55	2.15	3.37	0.595	12.00	1.25	1.12	0.209	0.75	2.00	6.46	0.842
1.65	2.36	3.13	0.552	12.40	1.24	1.10	0.204	0.80	2.00	5.95	0.820
1.75	2.54	2.99	0.527	12.80	1.24	1.09	0.200	0.85	2.01	5.51	0.796
1.85	2.69	2.89	0.510	13.20	1.24	1.09	0.201	0.90	2.03	5.10	0.769
1.95	2.82	2.86	0.505	13.60	1.23	1.12	0.208	0.95	2.05	4.74	0.742
2.05	2.89	2.87	0.505	14.00	1.20	1.13	0.216	1.00	2.09	4.41	0.712
2.15	2.92	2.87	0.505	14.40	1.16	1.15	0.225	1.10	2.15	3.84	0.651
2.25	2.93	2.87	0.505	14.80	1.11	1.16	0.234	1.20	2.16	3.35	0.592
2.35	2.92	2.88	0.506	15.00	1.08	1.16	0.238	1.30	2.25	2.77	0.506
2.45	2.89	2.90	0.509	15.60	0.99	1.14	0.247	1.40	2.49	2.23	0.419
2.55	2.83	2.92	0.512	16.00	0.92	1.11	0.250	1.50	2.84	1.80	0.369
2.65	2.74	2.90	0.511	16.60	0.85	1.04	0.245	1.60	3.36	1.62	0.379
2.75	2.66	2.86	0.507	17.00	0.80	0.99	0.240	1.70	3.70	1.75	0.411
2.85	2.58	2.80	0.500	17.20	0.79	0.96	0.236	1.80	3.78	1.83	0.423
3.00	2.51	2.68	0.485	17.40	0.77	0.93	0.230	1.90	3.81	1.75	0.418
3.10	2.48	2.60	0.475	17.80	0.75	0.87	0.217	2.00	3.98	1.60	0.418
3.20	2.45	2.53	0.465	18.00	0.74	0.85	0.209	2.10	4.26	1.54	0.432
3.30	2.44	2.45	0.453	18.60	0.73	0.77	0.185	2.20	4.58	1.62	0.457
3.40	2.46	2.38	0.442	19.00	0.72	0.72	0.170	2.30	4.84	1.76	0.479
3.50	2.48	2.33	0.435	19.60	0.72	0.66	0.150	2.40	5.10	2.01	0.506
3.60	2.52	2.29	0.428	20.00	0.72	0.62	0.137	2.50	5.28	2.38	0.532
3.70	2.56	2.27	0.426	20.60	0.71	0.55	0.119	2.60	5.36	2.82	0.557
3.80	2.59	2.28	0.427	21.00	0.72	0.50	0.100	2.70	5.30	3.29	0.580
3.90	2.62	2.29	0.429	21.60	0.75	0.43	0.075	2.80	5.07	3.78	0.603
4.00	2.64	2.33	0.434	22.00	0.78	0.40	0.063	2.90	4.65	4.18	0.624
4.20	2.64	2.42	0.447	22.60	0.82	0.35	0.045	3.00	4.05	4.40	0.639
4.40	2.53	2.56	0.467	23.00	0.85	0.33	0.038	3.20	3.29	3.96	0.614

Energy (eV)	n	k	$R(\phi = 0)$	Energy (eV)	n	k	$R(\phi = 0)$	Energy (eV)	n	k	$R(\phi = 0)$
3.40	2.93	3.79	0.607	20.00	0.96	1.10	0.239	2.50	1.41	3.48	0.685
3.60	2.75	3.45	0.577	20.40	0.93	1.09	0.240	2.60	1.37	3.36	0.676
3.80	2.73	3.32	0.562	20.80	0.89	1.05	0.240	2.70	1.32	3.25	0.668
4.00	2.71	3.34	0.565	21.20	0.86	1.02	0.237	2.80	1.29	3.13	0.658
4.20	2.53	3.44	0.584	21.60	0.83	0.99	0.235	2.90	1.26	3.03	0.648
4.40	2.24	3.44	0.599	22.00	0.80	0.96	0.230	3.00	1.23	2.94	0.639
4.60	2.01	3.31	0.598	22.40	0.78	0.93	0.226	3.10	1.20	2.85	0.630
4.80	1.88	3.19	0.592	22.80	0.77	0.90	0.220	3.20	1.17	2.77	0.622
5.00	1.74	3.12	0.596	23.20	0.75	0.88	0.217	3.30	1.14	2.68	0.613
5.20	1.58	3.00	0.597	23.60	0.75	0.86	0.211	3.40	1.12	2.60	0.602
5.40	1.46	2.88	0.593	24.00	0.73	0.84	0.209	3.50	1.10	2.52	0.591
5.60	1.36	2.77	0.589	24.40	0.72	0.82	0.207	3.60	1.08	2.45	0.581
5.80	1.27	2.65	0.582	24.80	0.70	0.80	0.205	3.70	1.07	2.38	0.570
6.00	1.20	2.54	0.575	25.20	0.69	0.77	0.202	3.80	1.06	2.31	0.558
6.20	1.13	2.44	0.571	25.60	0.67	0.75	0.199	3.90	1.05	2.25	0.547
6.40	1.06	2.33	0.562	26.00	0.66	0.72	0.195	4.00	1.03	2.19	0.537
6.60	1.01	2.21	0.548	26.40	0.65	0.69	0.189	4.20	1.04	2.09	0.510
6.80	0.97	2.11	0.532	26.80	0.63	0.66	0.183	4.40	1.03	2.01	0.493
7.00	0.95	2.00	0.514	27.20	0.65	0.62	0.165	4.60	1.03	1.94	0.476
7.20	0.92	1.91	0.497	28.00	0.64	0.59	0.156	4.80	1.01	1.90	0.470
7.40	0.91	1.81	0.476	28.40	0.64	0.57	0.148	5.00	0.96	1.86	0.472
7.60	0.90	1.72	0.451	28.80	0.65	0.55	0.140	5.20	0.90	1.79	0.474
7.80	0.90	1.63	0.426	29.20	0.65	0.53	0.134	5.40	0.85	1.70	0.463
8.00	0.91	1.55	0.400	29.60	0.65	0.51	0.128	5.60	0.81	1.62	0.449
8.20	0.91	1.48	0.375	30.00	0.65	0.49	0.121	5.80	0.78	1.54	0.437
8.40	0.94	1.40	0.344	31.00	0.65	0.45	0.111	6.00	0.76	1.45	0.418
8.60	0.96	1.34	0.319	32.00	0.66	0.41	0.095	6.20	0.74	1.37	0.397
8.80	0.98	1.29	0.296	33.00	0.68	0.37	0.079	6.40	0.73	1.29	0.375
9.00	1.01	1.24	0.274	34.00	0.70	0.34	0.068	6.60	0.72	1.21	0.350
9.20	1.04	1.19	0.255	35.00	0.72	0.31	0.057	6.80	0.73	1.13	0.316
9.40	1.08	1.16	0.238	36.00	0.74	0.29	0.048	7.00	0.73	1.05	0.287
9.60	1.10	1.14	0.229	37.00	0.77	0.27	0.040	7.20	0.75	0.98	0.255
9.80	1.13	1.11	0.217	38.00	0.79	0.26	0.035	7.40	0.77	0.91	0.223
10.00	1.16	1.10	0.209	39.00	0.81	0.26	0.031	7.60	0.79	0.85	0.195
10.20	1.19	1.08	0.203	40.00	0.84	0.26	0.026	7.80	0.83	0.78	0.163
10.30	1.20	1.08	0.201					8.00	0.88	0.73	0.133
10.40	1.22	1.08	0.200	**Palladium[19]**				8.20	0.94	0.70	0.117
10.50	1.23	1.09	0.201	0.10	4.13	54.15	0.994	8.40	0.96	0.70	0.114
10.60	1.24	1.10	0.203	0.15	3.13	35.82	0.990	8.60	1.00	0.65	0.097
10.80	1.25	1.11	0.206	0.20	3.07	26.59	0.983	8.80	1.04	0.65	0.094
11.00	1.24	1.13	0.213	0.26	3.11	20.15	0.971	9.00	1.07	0.64	0.090
11.20	1.23	1.14	0.217	0.30	3.56	17.27	0.955	9.50	1.12	0.65	0.089
11.40	1.19	1.15	0.223	0.36	3.98	14.41	0.932	10.00	1.14	0.65	0.088
11.60	1.17	1.12	0.216	0.40	4.27	13.27	0.916	10.50	1.16	0.65	0.087
11.80	1.16	1.10	0.211	0.46	4.27	12.11	0.902	11.00	1.18	0.64	0.086
12.00	1.15	1.08	0.205	0.50	4.10	11.44	0.896	11.50	1.19	0.65	0.087
12.40	1.14	1.03	0.191	0.56	3.92	10.49	0.883	12.00	1.20	0.66	0.089
12.80	1.15	1.01	0.183	0.60	3.80	9.96	0.876	12.50	1.19	0.67	0.091
13.20	1.16	0.98	0.174	0.72	3.51	8.70	0.854	13.00	1.18	0.67	0.091
13.60	1.17	0.97	0.170	0.80	3.35	8.06	0.840	13.50	1.18	0.67	0.092
14.00	1.17	0.96	0.169	1.00	2.99	6.89	0.811	14.00	1.17	0.67	0.093
14.40	1.16	0.94	0.165	1.10	2.81	6.46	0.800	14.50	1.15	0.68	0.095
14.80	1.16	0.91	0.156	1.20	2.65	6.10	0.790	15.00	1.13	0.69	0.098
15.20	1.17	0.89	0.148	1.30	2.50	5.78	0.781	15.50	1.10	0.68	0.096
15.60	1.20	0.86	0.140	1.40	2.34	5.50	0.774	16.00	1.08	0.66	0.092
16.00	1.25	0.87	0.140	1.50	2.17	5.22	0.767	16.50	1.06	0.63	0.086
16.40	1.28	0.90	0.147	1.60	2.08	4.95	0.755	17.00	1.07	0.61	0.081
16.80	1.28	0.94	0.157	1.70	2.00	4.72	0.745	17.50	1.06	0.61	0.080
17.20	1.27	0.97	0.167	1.80	1.92	4.54	0.737	18.00	1.07	0.59	0.077
17.60	1.26	1.01	0.178	1.90	1.82	4.35	0.729	18.50	1.07	0.59	0.077
18.00	1.23	1.04	0.189	2.00	1.75	4.18	0.721	19.00	1.08	0.59	0.077
18.40	1.19	1.08	0.200	2.10	1.67	4.03	0.714	19.50	1.08	0.61	0.080
18.80	1.14	1.10	0.210	2.20	1.60	3.88	0.707	20.00	1.07	0.65	0.090
19.20	1.10	1.10	0.219	2.30	1.53	3.75	0.700	20.50	1.03	0.67	0.098
19.60	1.05	1.11	0.227	2.40	1.47	3.61	0.693	21.00	0.99	0.67	0.103

Energy (eV)	n	k	$R(\phi = 0)$	Energy (eV)	n	k	$R(\phi = 0)$	Energy (eV)	n	k	$R(\phi = 0)$
21.50	0.95	0.66	0.103	6.00	1.38	1.40	0.276	28.50	0.75	0.59	0.121
22.00	0.91	0.64	0.103	6.20	1.39	1.35	0.261	29.00	0.75	0.58	0.118
22.50	0.88	0.62	0.101	6.40	1.42	1.29	0.246	29.50	0.74	0.58	0.120
23.00	0.86	0.59	0.097	6.60	1.45	1.26	0.236	30.00	0.73	0.58	0.124
23.50	0.85	0.56	0.091	6.80	1.48	1.24	0.231				
24.00	0.84	0.54	0.086	7.00	1.50	1.24	0.230	**Potassium**[21]			
25.00	0.81	0.51	0.084	7.20	1.50	1.25	0.231	0.55	0.139	7.10	0.989
26.40	0.80	0.43	0.066	7.40	1.49	1.23	0.228	0.58	0.119	6.72	0.990
27.80	0.81	0.38	0.052	7.60	1.48	1.22	0.225	0.63	0.106	6.32	0.990
29.20	0.82	0.35	0.046	7.80	1.48	1.20	0.221	0.67	0.091	5.79	0.990
				8.00	1.47	1.18	0.216	0.73	0.079	5.30	0.989
Platinum[20]				8.20	1.47	1.17	0.212	0.81	0.066	4.75	0.989
0.10	13.21	44.72	0.976	8.40	1.47	1.15	0.209	0.92	0.056	4.19	0.988
0.15	8.18	31.16	0.969	8.60	1.47	1.14	0.205	1.05	0.044	3.58	0.987
0.20	5.90	23.95	0.962	8.80	1.47	1.13	0.202	1.23	0.040	3.04	0.985
0.25	4.70	19.40	0.954	9.00	1.48	1.12	0.200	1.44	0.040	2.56	0.979
0.30	3.92	16.16	0.945	9.20	1.49	1.11	0.198	1.65	0.044	2.19	0.970
0.35	3.28	13.66	0.936	9.40	1.49	1.12	0.200	1.87	0.050	1.84	0.955
0.40	2.81	11.38	0.922	9.60	1.49	1.13	0.203	2.07	0.053	1.62	0.943
0.45	3.03	9.31	0.882	9.80	1.48	1.15	0.207	2.27	0.049	1.43	0.938
0.50	3.91	7.71	0.813	10.00	1.46	1.15	0.209	2.45	0.046	1.28	0.933
0.55	4.58	7.14	0.777	10.20	1.43	1.16	0.211	2.64	0.043	1.14	0.928
0.60	5.13	6.75	0.753	10.40	1.40	1.15	0.210	2.82	0.043	1.02	0.919
0.65	5.52	6.66	0.746	10.60	1.37	1.14	0.207	2.95	0.041	0.898	0.913
0.70	5.71	6.83	0.751	10.80	1.35	1.12	0.203	3.06	0.041	0.799	0.905
0.75	5.57	7.02	0.759	11.00	1.33	1.10	0.199	3.40	0.052	0.549	0.852
0.80	5.31	7.04	0.762	11.20	1.31	1.08	0.194	3.71	0.089	0.288	0.719
0.85	5.05	6.98	0.763	11.40	1.30	1.06	0.188	3.97	0.287	0.091	0.310
0.90	4.77	6.91	0.765	11.60	1.29	1.04	0.183	4.00	0.34	0.08	0.245
0.95	4.50	6.77	0.763	11.80	1.29	1.01	0.177	4.065	0.38	0.07	0.204
1.00	4.25	6.62	0.762	12.00	1.29	1.00	0.173	4.133	0.41	0.07	0.177
1.10	3.86	6.24	0.753	12.40	1.29	0.97	0.165	4.203	0.45	0.06	0.145
1.20	3.55	5.92	0.746	12.80	1.29	0.94	0.158	4.275	0.48	0.06	0.125
1.30	3.29	5.61	0.736	13.20	1.31	0.93	0.155	4.350	0.52	0.05	0.101
1.40	3.10	5.32	0.725	13.60	1.31	0.93	0.155	4.428	0.55	0.05	0.085
1.50	2.92	5.07	0.716	14.00	1.31	0.93	0.155	4.509	0.58	0.05	0.072
1.60	2.76	4.84	0.706	14.40	1.30	0.93	0.156	4.592	0.61	0.05	0.060
1.70	2.63	4.64	0.697	14.80	1.27	0.93	0.157	4.679	0.64	0.04	0.049
1.80	2.51	4.43	0.686	15.20	1.27	0.93	0.155	4.769	0.66	0.04	0.043
1.90	2.38	4.26	0.678	15.60	1.25	0.92	0.151	4.862	0.68	0.04	0.037
2.00	2.30	4.07	0.664	16.00	1.24	0.89	0.146	4.959	0.70	0.04	0.032
2.10	2.23	3.92	0.654	16.50	1.24	0.87	0.142	5.061	0.72	0.04	0.027
2.20	2.17	3.77	0.642	17.00	1.25	0.86	0.138	5.166	0.74	0.04	0.023
2.30	2.10	3.67	0.636	17.50	1.27	0.85	0.135	5.276	0.76	0.04	0.019
2.40	2.03	3.54	0.626	18.00	1.31	0.88	0.142	5.391	0.78	0.04	0.016
2.50	1.96	3.42	0.616	18.50	1.30	0.94	0.157	5.510	0.79	0.05	0.015
2.60	1.91	3.30	0.605	19.00	1.28	0.99	0.171	5.637	0.81	0.05	0.012
2.70	1.87	3.20	0.595	19.50	1.23	1.03	0.184	5.767	0.83	0.05	0.009
2.80	1.83	3.10	0.585	20.00	1.18	1.06	0.197	6.048	0.85	0.05	0.007
2.90	1.79	3.01	0.575	20.50	1.11	1.09	0.212	6.199	0.87	0.05	0.006
3.00	1.75	2.92	0.565	21.00	1.03	1.10	0.226	6.358	0.88	0.05	0.005
3.20	1.68	2.76	0.546	21.50	0.94	1.08	0.238	6.526	0.90	0.06	0.004
3.40	1.63	2.62	0.527	22.00	0.87	1.04	0.240	6.702	0.91	0.06	0.003
3.60	1.58	2.48	0.507	22.50	0.81	0.98	0.235	6.888	0.92	0.06	0.003
3.80	1.53	2.37	0.491	23.00	0.77	0.92	0.226	7.085	0.92	0.06	0.003
4.00	1.49	2.25	0.472	23.50	0.75	0.87	0.213	7.293	0.93	0.06	0.002
4.20	1.45	2.14	0.452	24.00	0.74	0.82	0.201	7.514	0.93	0.06	0.002
4.40	1.43	2.04	0.432	24.50	0.73	0.77	0.187	7.749	0.94	0.06	0.002
4.60	1.39	1.95	0.415	25.00	0.73	0.73	0.174	7.999	0.94	0.06	0.002
4.80	1.38	1.85	0.392	25.50	0.73	0.70	0.162	8.260	0.94	0.06	0.002
5.00	1.36	1.76	0.372	26.00	0.74	0.67	0.150	8.551	0.94	0.06	0.002
5.20	1.36	1.67	0.350	26.50	0.74	0.65	0.142	8.856	0.94	0.05	0.002
5.40	1.36	1.61	0.332	27.00	0.74	0.63	0.136	9.184	0.94	0.05	0.002
5.60	1.36	1.54	0.315	27.50	0.74	0.62	0.130	9.537	0.94	0.04	0.001
5.80	1.36	1.47	0.295	28.00	0.75	0.60	0.125	9.919	0.94	0.04	0.001

Energy (eV)	n	k	$R(\phi = 0)$	Energy (eV)	n	k	$R(\phi = 0)$	Energy (eV)	n	k	$R(\phi = 0)$
10.33	0.94	0.03	0.001	7.60	1.12	1.99	0.470	31.00	0.62	0.29	0.086
11.0		0.03		7.80	1.08	1.89	0.454	32.00	0.66	0.26	0.065
12.0		0.028		8.00	1.05	1.80	0.435	33.00	0.68	0.24	0.054
				8.20	1.05	1.71	0.411	34.00	0.72	0.21	0.041
Rhenium, single crystal, $\vec{E} \parallel \hat{c}$ [9]				8.40	1.05	1.62	0.386	35.00	0.76	0.20	0.031
0.10	6.06	51.03	0.991	8.60	1.06	1.55	0.360	36.00	0.79	0.20	0.025
0.15	4.66	33.96	0.984	8.80	1.09	1.48	0.336	37.00	0.82	0.19	0.021
0.20	4.16	25.36	0.975	9.00	1.11	1.43	0.317	38.00	0.85	0.20	0.018
0.25	4.03	20.10	0.962	9.20	1.13	1.39	0.301	39.00	0.89	0.21	0.016
0.30	4.37	16.69	0.943	9.40	1.16	1.34	0.281	40.00	0.88	0.26	0.022
0.35	4.50	14.53	0.925	9.60	1.18	1.32	0.274	42.00	0.88	0.26	0.022
0.40	4.53	12.96	0.909	9.80	1.20	1.29	0.264	44.00	0.89	0.29	0.026
0.45	4.53	11.78	0.893	10.00	1.23	1.26	0.252	46.00	0.85	0.32	0.035
0.50	4.53	10.88	0.878	10.20	1.25	1.25	0.246	48.00	0.82	0.30	0.036
0.55	4.50	10.26	0.867	10.40	1.28	1.25	0.242	50.00	0.80	0.30	0.038
0.60	4.29	9.75	0.861	10.60	1.29	1.25	0.242	52.00	0.78	0.30	0.044
0.65	4.07	9.35	0.856	10.80	1.30	1.26	0.244	54.00	0.72	0.30	0.055
0.70	3.80	8.94	0.853	11.00	1.30	1.27	0.247	56.00	0.66	0.24	0.061
0.75	3.48	8.55	0.850	11.20	1.29	1.28	0.249	58.00	0.65	0.16	0.055
0.80	3.21	8.10	0.846	11.40	1.28	1.28	0.252				
0.85	2.96	7.68	0.841	11.60	1.26	1.28	0.252	**Rhenium, single crystal, $\vec{E} \perp \hat{c}$ [9]**			
0.90	2.73	7.24	0.835	11.80	1.24	1.26	0.249	0.10	4.25	42.83	0.991
0.95	2.56	6.79	0.826	12.00	1.23	1.24	0.244	0.15	3.28	28.08	0.984
1.00	2.45	6.36	0.813	12.40	1.22	1.21	0.237	0.20	3.28	20.66	0.971
1.10	2.38	5.61	0.778	12.80	1.21	1.18	0.230	0.25	3.47	16.27	0.951
1.20	2.35	5.02	0.742	13.20	1.22	1.16	0.222	0.30	3.73	13.44	0.926
1.30	2.39	4.54	0.702	13.60	1.22	1.13	0.215	0.35	3.93	11.54	0.900
1.40	2.44	4.13	0.662	14.00	1.24	1.12	0.209	0.40	3.99	10.15	0.875
1.50	2.50	3.79	0.624	14.40	1.27	1.11	0.204	0.45	4.17	9.03	0.846
1.60	2.59	3.49	0.587	14.80	1.29	1.15	0.213	0.50	4.34	8.26	0.821
1.70	2.70	3.27	0.557	15.20	1.29	1.19	0.225	0.55	4.45	7.73	0.801
1.80	2.82	3.10	0.535	15.60	1.26	1.22	0.236	0.60	4.53	7.40	0.788
1.90	2.90	3.00	0.520	16.00	1.23	1.25	0.248	0.65	4.44	7.26	0.784
2.00	2.97	2.91	0.510	16.40	1.19	1.27	0.259	0.70	4.13	7.09	0.784
2.10	3.03	2.86	0.504	16.80	1.14	1.29	0.269	0.75	3.77	6.75	0.779
2.20	3.06	2.84	0.501	17.00	1.12	1.30	0.275	0.80	3.55	6.32	0.766
2.30	3.07	2.82	0.499	17.40	1.07	1.30	0.286	0.85	3.39	5.95	0.752
2.40	3.06	2.81	0.498	18.00	0.99	1.30	0.300	0.90	3.26	5.61	0.737
2.50	3.02	2.80	0.497	18.40	0.93	1.29	0.311	0.95	3.17	5.27	0.719
2.60	2.96	2.77	0.493	18.80	0.87	1.28	0.321	1.00	3.09	4.96	0.701
2.70	2.89	2.68	0.482	19.20	0.81	1.25	0.330	1.10	3.05	4.39	0.658
2.80	2.89	2.57	0.468	19.60	0.77	1.21	0.332	1.20	3.08	3.89	0.613
2.90	2.99	2.47	0.457	20.00	0.73	1.18	0.333	1.30	3.20	3.56	0.578
3.00	3.11	2.57	0.470	20.40	0.70	1.14	0.332	1.40	3.23	3.38	0.559
3.20	2.90	2.68	0.482	20.80	0.67	1.11	0.332	1.50	3.23	3.12	0.532
3.40	2.83	2.50	0.459	21.20	0.64	1.08	0.334	1.60	3.29	2.88	0.507
3.60	2.93	2.48	0.457	21.60	0.61	1.04	0.335	1.70	3.38	2.72	0.491
3.80	2.86	2.56	0.467	22.00	0.58	1.01	0.340	1.80	3.47	2.59	0.480
4.00	2.81	2.51	0.460	22.40	0.55	0.97	0.341	1.90	3.54	2.50	0.473
4.20	2.86	2.55	0.466	22.80	0.53	0.93	0.338	2.00	3.63	2.43	0.469
4.40	2.81	2.74	0.489	23.20	0.51	0.89	0.334	2.10	3.74	2.40	0.470
4.60	2.56	2.83	0.504	23.60	0.50	0.85	0.329	2.20	3.83	2.38	0.472
4.80	2.41	2.71	0.493	24.00	0.48	0.80	0.319	2.30	3.93	2.44	0.481
5.00	2.39	2.68	0.488	24.40	0.48	0.76	0.207	2.40	4.00	2.55	0.492
5.20	2.34	2.75	0.500	24.80	0.47	0.72	0.296	2.50	4.01	2.70	0.505
5.40	2.20	2.81	0.515	25.20	0.47	0.68	0.282	2.60	3.90	2.84	0.514
5.60	2.02	2.84	0.530	25.60	0.47	0.65	0.270	2.70	3.74	2.92	0.517
5.80	1.83	2.80	0.538	26.00	0.47	0.61	0.255	2.80	3.57	2.88	0.511
6.00	1.65	2.71	0.541	26.40	0.48	0.57	0.240	2.90	3.49	2.75	0.497
6.20	1.54	2.59	0.532	26.80	0.48	0.54	0.225	3.00	3.53	2.71	0.493
6.40	1.45	2.50	0.526	27.20	0.49	0.51	0.208	3.20	3.55	2.84	0.506
6.80	1.32	2.31	0.508	27.60	0.50	0.48	0.193	3.40	3.34	2.88	0.508
7.00	1.26	2.23	0.500	28.00	0.51	0.45	0.176	3.60	3.25	2.83	0.501
7.20	1.20	2.15	0.493	29.00	0.54	0.39	0.145	3.80	3.24	2.84	0.502
7.40	1.16	2.06	0.480	30.00	0.57	0.33	0.114	4.00	3.19	2.94	0.513

Energy (eV)	n	k	$R(\phi = 0)$	Energy (eV)	n	k	$R(\phi = 0)$	Energy (eV)	n	k	$R(\phi = 0)$
4.20	3.05	3.06	0.526	22.80	0.55	0.92	0.325	3.00	1.53	4.29	0.753
4.40	2.88	3.15	0.539	23.20	0.53	0.89	0.322	3.10	1.41	4.20	0.760
4.60	2.67	3.18	0.548	23.60	0.52	0.85	0.317	3.20	1.30	4.09	0.764
4.80	2.44	3.17	0.554	24.00	0.50	0.82	0.314	3.30	1.20	3.97	0.767
5.00	2.25	3.12	0.556	24.40	0.49	0.79	0.309	3.40	1.11	3.84	0.769
5.20	2.10	3.04	0.555	24.80	0.48	0.75	0.303	3.50	1.04	3.71	0.768
5.40	1.96	2.96	0.553	25.20	0.47	0.72	0.295	3.60	0.99	3.58	0.764
5.60	1.84	2.88	0.551	25.60	0.47	0.68	0.286	3.70	0.95	3.45	0.759
5.80	1.73	2.81	0.549	26.00	0.46	0.64	0.276	3.80	0.91	3.34	0.753
6.00	1.61	2.74	0.549	26.40	0.46	0.61	0.263	3.90	0.88	3.23	0.747
6.20	1.51	2.64	0.545	26.80	0.46	0.57	0.249	4.00	0.86	3.12	0.739
6.40	1.42	2.56	0.541	27.20	0.47	0.53	0.231	4.20	0.83	2.94	0.722
6.80	1.28	2.37	0.526	27.60	0.48	0.50	0.216	4.40	0.80	2.76	0.706
7.00	1.22	2.28	0.517	28.00	0.49	0.47	0.198	4.60	0.78	2.60	0.684
7.20	1.16	2.19	0.508	29.00	0.51	0.41	0.164	4.80	0.79	2.46	0.659
7.40	1.12	2.08	0.493	30.00	0.55	0.34	0.129	5.00	0.79	2.34	0.635
7.60	1.12	1.98	0.468	31.00	0.59	0.29	0.097	5.20	0.79	2.23	0.613
7.80	1.08	1.93	0.463	32.00	0.64	0.26	0.072	5.40	0.80	2.14	0.591
8.00	1.05	1.83	0.443	33.00	0.67	0.24	0.060	5.60	0.80	2.06	0.573
8.20	1.05	1.74	0.418	34.00	0.70	0.22	0.047	5.80	0.79	2.00	0.561
8.40	1.05	1.66	0.397	35.00	0.74	0.20	0.036	6.00	0.76	1.93	0.556
8.60	1.06	1.58	0.372	36.00	0.77	0.19	0.029	6.20	0.73	1.85	0.544
8.80	1.07	1.52	0.351	37.00	0.80	0.19	0.023	6.40	0.70	1.77	0.534
9.00	1.09	1.46	0.327	38.00	0.84	0.19	0.018	6.60	0.68	1.69	0.518
9.20	1.11	1.41	0.309	39.00	0.88	0.21	0.016	6.80	0.67	1.60	0.498
9.40	1.14	1.36	0.290	40.00	0.87	0.25	0.023	7.00	0.66	1.52	0.476
9.60	1.17	1.31	0.273	42.00	0.87	0.25	0.023	7.20	0.66	1.43	0.452
9.80	1.20	1.27	0.258	44.00	0.88	0.28	0.026	7.40	0.66	1.35	0.423
10.00	1.24	1.24	0.244	46.00	0.84	0.31	0.035	7.60	0.67	1.27	0.394
10.20	1.29	1.22	0.234	48.00	0.82	0.30	0.036	7.80	0.68	1.20	0.363
10.40	1.33	1.23	0.233	50.00	0.80	0.30	0.039	8.00	0.69	1.12	0.329
10.60	1.36	1.25	0.238	52.00	0.77	0.30	0.044	8.20	0.71	1.04	0.288
10.80	1.38	1.28	0.245	54.00	0.71	0.29	0.055	8.40	0.74	0.97	0.252
11.00	1.37	1.31	0.253	56.00	0.66	0.23	0.061	8.60	0.78	0.89	0.212
11.20	1.36	1.33	0.259	58.00	0.64	0.16	0.055	8.80	0.83	0.83	0.179
11.40	1.33	1.34	0.264					9.00	0.88	0.77	0.148
11.60	1.31	1.34	0.266	**Rhodium**[11]				9.20	0.95	0.73	0.125
11.80	1.28	1.33	0.266	0.10	18.48	69.43	0.986	9.40	1.01	0.71	0.110
12.00	1.26	1.32	0.264	0.20	8.66	37.46	0.977	9.60	1.07	0.69	0.102
12.40	1.23	1.29	0.257	0.30	5.85	25.94	0.967	9.80	1.12	0.69	0.098
12.80	1.22	1.26	0.251	0.40	4.74	19.80	0.955	10.00	1.17	0.69	0.098
13.20	1.20	1.23	0.245	0.50	4.20	16.07	0.941	10.60	1.26	0.73	0.106
13.60	1.19	1.20	0.236	0.60	3.87	13.51	0.925	11.00	1.29	0.76	0.113
14.00	1.20	1.16	0.225	0.70	3.67	11.72	0.908	11.60	1.32	0.80	0.124
14.40	1.22	1.13	0.214	0.80	3.63	10.34	0.887	12.00	1.32	0.82	0.127
14.80	1.27	1.12	0.207	0.90	3.62	9.36	0.867	12.60	1.32	0.82	0.129
15.20	1.31	1.17	0.218	1.00	3.71	8.67	0.848	13.00	1.32	0.83	0.131
15.60	1.31	1.23	0.234	1.10	3.67	8.26	0.837	13.60	1.32	0.85	0.134
16.00	1.28	1.28	0.251	1.20	3.51	7.94	0.832	14.00	1.32	0.86	0.138
16.40	1.24	1.33	0.270	1.30	3.26	7.63	0.829	14.60	1.30	0.89	0.144
16.80	1.17	1.37	0.288	1.40	3.01	7.31	0.827	15.00	1.28	0.90	0.147
17.00	1.14	1.38	0.297	1.50	2.78	6.97	0.823	15.60	1.25	0.90	0.147
17.40	1.06	1.39	0.314	1.60	2.60	6.64	0.818	16.00	1.24	0.89	0.147
18.00	0.95	1.38	0.334	1.70	2.42	6.33	0.813	16.50	1.23	0.88	0.145
18.40	0.88	1.36	0.346	1.80	2.30	6.02	0.805	17.00	1.22	0.88	0.144
18.80	0.82	1.33	0.355	1.90	2.20	5.76	0.798	17.50	1.22	0.87	0.143
19.20	0.76	1.29	0.360	2.00	2.12	5.51	0.789	18.00	1.23	0.88	0.145
19.60	0.72	1.25	0.363	2.10	2.05	5.30	0.780	18.50	1.25	0.92	0.155
20.00	0.67	1.21	0.369	2.20	2.00	5.11	0.772	19.00	1.24	0.98	0.172
20.40	0.64	1.15	0.364	2.30	1.94	4.94	0.765	19.50	1.18	1.05	0.193
20.80	0.61	1.10	0.357	2.40	1.90	4.78	0.756	20.00	1.10	1.09	0.213
21.20	0.60	1.06	0.349	2.50	1.88	4.65	0.748	20.50	1.00	1.09	0.230
21.60	0.58	1.02	0.342	2.60	1.85	4.55	0.743	21.00	0.91	1.05	0.234
22.00	0.57	0.98	0.336	2.70	1.80	4.49	0.742	21.50	0.86	1.00	0.228
22.40	0.56	0.95	0.328	2.90	1.63	4.36	0.748	22.00	0.83	0.95	0.219

Energy (eV)	n	k	$R(\phi = 0)$	Energy (eV)	n	k	$R(\phi = 0)$	Energy (eV)	n	k	$R(\phi = 0)$
22.50	0.81	0.92	0.214	4.20	1.29	3.08	0.649	26.00	0.68	0.63	0.154
23.00	0.79	0.90	0.213	4.40	1.22	2.93	0.639	27.00	0.67	0.57	0.140
23.50	0.75	0.87	0.214	4.60	1.16	2.79	0.628	28.00	0.66	0.51	0.124
24.00	0.73	0.84	0.210	4.80	1.11	2.67	0.617	29.00	0.67	0.46	0.107
24.50	0.70	0.81	0.208	5.00	1.06	2.56	0.607	30.00	0.67	0.43	0.097
25.00	0.69	0.77	0.202	5.20	1.01	2.46	0.600	31.00	0.67	0.37	0.084
25.50	0.67	0.74	0.195	5.40	0.95	2.35	0.593	32.00	0.69	0.33	0.070
26.00	0.66	0.70	0.188	5.60	0.92	2.23	0.576	33.00	0.71	0.30	0.058
26.50	0.65	0.66	0.176	5.80	0.90	2.14	0.559	34.00	0.73	0.27	0.048
27.00	0.65	0.64	0.168	6.00	0.88	2.05	0.545	35.00	0.75	0.25	0.039
27.50	0.65	0.61	0.159	6.20	0.87	1.98	0.531	36.00	0.77	0.24	0.035
28.00	0.65	0.59	0.152	6.40	0.84	1.91	0.521	37.00	0.79	0.23	0.039
29.00	0.65	0.54	0.137	6.60	0.82	1.84	0.510	38.00	0.80	0.22	-0.027
30.00	0.66	0.51	0.127	6.80	0.79	1.77	0.500	39.00	0.82	0.22	0.024
31.00	0.64	0.49	0.127	7.00	0.76	1.69	0.489	40.00	0.83	0.22	0.022
32.00	0.61	0.44	0.126	7.20	0.75	1.61	0.472				
33.00	0.60	0.37	0.110	7.40	0.73	1.54	0.455	**Ruthenium, single crystal, $\vec{E} \perp \hat{c}$[5]**			
34.00	0.65	0.30	0.074	7.60	0.73	1.46	0.433	0.10	11.85	50.81	0.983
35.00	0.69	0.28	0.058	7.80	0.73	1.39	0.411	0.20	6.68	27.18	0.966
36.00	0.73	0.27	0.049	8.00	0.72	1.33	0.391	0.30	4.94	18.92	0.950
37.00	0.74	0.28	0.047	8.20	0.72	1.26	0.366	0.40	3.90	14.51	0.933
38.00	0.74	0.27	0.045	8.40	0.73	1.20	0.342	0.50	3.27	11.63	0.915
39.00	0.75	0.25	0.041	8.60	0.74	1.14	0.318	0.60	2.98	9.54	0.888
				8.80	0.74	1.08	0.295	0.70	2.82	7.99	0.856
Ruthenium, single crystal, $\vec{E} \parallel \hat{c}$[9]				9.00	0.75	1.02	0.267	0.80	2.73	6.71	0.815
0.10	11.50	51.38	0.984	9.20	0.77	0.97	0.243	0.90	2.82	5.54	0.751
0.20	5.93	27.14	0.970	9.40	0.79	0.91	0.217	1.00	3.17	4.59	0.670
0.30	4.33	18.50	0.953	9.60	0.82	0.86	0.190	1.10	3.69	3.91	0.604
0.40	3.60	13.97	0.933	9.80	0.85	0.81	0.167	1.20	4.28	3.66	0.585
0.50	3.18	11.04	0.909	10.00	0.88	0.76	0.144	1.30	4.66	3.72	0.593
0.60	3.28	8.89	0.865	10.20	0.92	0.72	0.125	1.40	4.86	3.79	0.601
0.70	3.62	7.73	0.822	10.40	0.96	0.69	0.110	1.50	4.99	3.89	0.609
0.80	3.42	7.02	0.801	10.60	1.01	0.67	0.100	1.60	5.08	4.03	0.618
0.90	3.25	6.12	0.766	10.80	1.05	0.66	0.094	1.70	5.12	4.22	0.629
1.00	3.39	5.33	0.715	11.00	1.09	0.65	0.090	1.80	5.10	4.45	0.642
1.10	3.66	4.83	0.675	11.20	1.12	0.65	0.088	1.90	4.96	4.78	0.660
1.20	3.84	4.57	0.654	11.40	1.15	0.65	0.087	2.00	4.61	5.06	0.677
1.30	3.94	4.38	0.638	11.60	1.18	0.65	0.088	2.10	4.21	5.09	0.682
1.40	4.02	4.19	0.624	11.80	1.21	0.66	0.090	2.20	3.94	5.00	0.681
1.50	4.16	4.07	0.614	12.00	1.23	0.67	0.092	2.30	3.69	4.97	0.684
1.60	4.33	4.08	0.615	12.40	1.26	0.69	0.098	2.40	3.44	4.88	0.684
1.70	4.42	4.21	0.624	12.80	1.27	0.72	0.104	2.50	3.27	4.77	0.681
1.80	4.40	4.38	0.636	13.20	1.28	0.74	0.108	2.60	3.14	4.66	0.677
1.90	4.29	4.61	0.651	13.60	1.28	0.75	0.111	2.70	3.06	4.59	0.674
2.00	4.04	4.81	0.667	14.00	1.28	0.76	0.114	2.80	2.99	4.59	0.676
2.10	3.69	4.90	0.679	14.40	1.27	0.76	0.114	2.90	2.87	4.64	0.686
2.20	3.35	4.82	0.683	14.80	1.27	0.76	0.114	3.00	2.64	4.69	0.701
2.30	3.09	4.70	0.681	15.00	1.27	0.76	0.114	3.10	2.40	4.64	0.710
2.40	2.89	4.55	0.677	15.60	1.28	0.77	0.115	3.20	2.18	4.55	0.717
2.50	2.74	4.40	0.671	16.00	1.30	0.78	0.118	3.30	2.00	4.43	0.721
2.60	2.64	4.25	0.663	16.50	1.32	0.80	0.123	3.40	1.84	4.30	0.723
2.70	2.58	4.14	0.656	17.00	1.34	0.85	0.136	3.50	1.71	4.16	0.723
2.80	2.54	4.05	0.650	17.50	1.32	0.93	0.155	3.60	1.60	4.03	0.722
2.90	2.48	4.03	0.650	18.00	1.26	0.99	0.173	3.70	1.50	3.90	0.721
3.00	2.38	4.03	0.656	18.50	1.18	1.02	0.185	3.80	1.41	3.77	0.718
3.10	2.26	4.00	0.661	19.00	1.11	1.02	0.192	3.90	1.35	3.64	0.713
3.20	2.13	3.96	0.666	19.50	1.05	1.02	0.199	4.00	1.29	3.53	0.707
3.30	2.00	3.91	0.671	20.00	0.99	1.02	0.208	4.20	1.21	3.31	0.694
3.40	1.87	3.83	0.673	20.50	0.92	0.99	0.212	4.40	1.16	3.13	0.679
3.50	1.76	3.74	0.674	21.00	0.86	0.94	0.209	4.60	1.13	2.97	0.662
3.60	1.66	3.65	0.675	21.50	0.83	0.90	0.203	4.80	1.09	2.86	0.652
3.70	1.57	3.55	0.673	22.00	0.81	0.86	0.193	5.00	1.03	2.75	0.648
3.80	1.49	3.45	0.672	23.00	0.77	0.79	0.182	5.20	0.97	2.64	0.643
3.90	1.42	3.35	0.668	24.00	0.74	0.74	0.171	5.40	0.91	2.52	0.635
4.00	1.37	3.24	0.661	25.00	0.71	0.69	0.163	5.60	0.88	2.40	0.622

Energy (eV)	n	k	$R(\phi=0)$	Energy (eV)	n	k	$R(\phi=0)$	Energy (eV)	n	k	$R(\phi=0)$
5.80	0.86	2.29	0.605	34.00	0.67	0.28	0.065	7.0	1.84	1.45	0.276
6.00	0.84	2.20	0.591	35.00	0.70	0.26	0.054	8.0	1.35	1.68	0.353
6.20	0.82	2.11	0.576	36.00	0.72	0.25	0.047	9.0	1.35	1.64	0.342
6.40	0.81	2.04	0.564	37.00	0.73	0.23	0.041	10.0	0.92	1.07	0.238
6.60	0.78	1.97	0.556	38.00	0.75	0.22	0.035	12.0	1.00	1.10	0.232
6.80	0.76	1.89	0.545	39.00	0.77	0.22	0.031	14.0	0.81	0.91	0.211
7.00	0.73	1.82	0.538	40.00	0.79	0.22	0.028	16.0	0.65	0.61	0.160
7.20	0.70	1.75	0.527					18.0	0.65	0.48	0.120
7.40	0.68	1.67	0.513	**Selenium, single crystal, $\vec{E} \parallel \hat{c}$[22]**				20.0	0.69	0.36	0.076
7.60	0.67	1.59	0.496	0.01364	2.914	0.248	0.242	22.0	0.81	0.25	0.030
7.80	0.66	1.51	0.476	0.01488	3.175	9.95E-02	0.272	24.0	0.91	0.18	0.011
8.00	0.66	1.44	0.454	0.01612	3.263	2.13E-03	0.282	26.0	0.86	0.15	0.012
8.20	0.65	1.36	0.430	0.01736	3.306	3.81E-02	0.287	28.0	0.85	0.13	0.011
8.40	0.66	1.29	0.403	0.01860	3.330	7.04E-03	0.290	30.0	0.87	0.11	0.008
8.60	0.66	1.22	0.378	0.01984	3.346	4.23E-02	0.291				
8.80	0.68	1.15	0.346	0.02108	3.358	3.40E-03	0.293	**Selenium, single crystal, $\vec{E} \perp \hat{c}$[22]**			
9.00	0.69	1.09	0.317	0.02232	3.366	5.31E-02	0.294	0.01364	2.854	0.0239	0.231
9.20	0.70	1.02	0.286	0.02356	3.372	1.96E-03	0.294	0.01488	2.932	0.0325	0.241
9.40	0.73	0.95	0.251	0.02480	3.377	2.39E-02	0.295	0.01612	3.140	0.1750	0.269
9.60	0.77	0.89	0.216	0.02604	3.380		0.295	0.01736	2.959	1.3300	0.321
9.80	0.82	0.84	0.185	0.02728		1.16E-02		0.01860	2.111	0.2550	0.133
10.00	0.86	0.81	0.163	0.02976		7.96E-03		0.01984	2.356	0.0746	0.164
10.20	0.90	0.77	0.143	0.03224		8.57E-03		0.02108	2.462	0.0276	0.178
10.40	0.94	0.74	0.127	0.03472		2.70E-02		0.02232	2.502	0.0442	0.184
10.60	0.99	0.72	0.115	0.03720	3.397	1.72E-02	0.297	0.02356	2.543	0.0097	0.190
10.80	1.04	0.71	0.108	0.04463		1.13E-02		0.02480	2.550	0.0239	0.191
11.00	1.08	0.70	0.104	0.04959	3.403	2.79E-03	0.298	0.02604	2.582		0.195
11.20	1.11	0.70	0.102	0.05703		1.56E-03		0.02728	2.600	0.0101	0.198
11.40	1.14	0.70	0.101	0.06199	3.405	1.35E-03	0.298	0.02976	2.576	9.95E-03	0.194
11.60	1.17	0.71	0.102	0.06819		5.79E-04		0.03224	2.598	1.16E-02	0.197
11.80	1.20	0.72	0.104	0.07439	3.407	4.44E-04	0.298	0.03472	2.607	1.68E-02	0.199
12.00	1.22	0.73	0.107	0.08059		4.41E-04		0.03720	2.613	1.54E-02	0.199
12.40	1.25	0.76	0.113	0.08679	3.408	4.32E-04	0.298	0.04463		1.17E-02	
12.80	1.26	0.78	0.118	0.09299		2.44E-04		0.04959	2.627	3.58E-03	0.201
13.20	1.27	0.81	0.124	0.09919	3.409	3.23E-04	0.299	0.05703		8.65E-04	
13.60	1.27	0.83	0.129	0.1116	3.409	2.87E-04	0.299	0.06199	2.632	2.07E-03	0.202
14.00	1.26	0.84	0.132	0.1240	3.410	2.71E-04	0.299	0.06819		2.89E-04	
14.40	1.25	0.84	0.132	0.2480	3.417	2.67E-04	0.299	0.07439	2.635	1.59E-04	0.202
14.80	1.25	0.84	0.133	0.3720	3.427	1.90E-04	0.301	0.08059		1.35E-04	
15.00	1.25	0.84	0.133	0.4959	3.442	1.41E-04	0.302	0.08679	2.636	1.42E-04	0.202
15.60	1.25	0.85	0.134	0.6199	3.462	1.12E-04	0.304	0.09299		1.04E-04	
16.00	1.27	0.85	0.134	0.7439	3.486	9.42E-05	0.307	0.09919	2.637	8.95E-05	0.203
16.50	1.28	0.89	0.145	0.8679	3.516	8.07E-05	0.310	0.1116	2.638	8.84E-05	0.203
17.00	1.28	0.94	0.158	0.9919	3.551	7.11E-05	0.314	0.1240	2.639	8.51E-05	0.203
17.50	1.25	1.00	0.175	1.116	3.592	6.37E-05	0.319	0.2480	2.645	5.97E-05	0.204
18.00	1.19	1.04	0.190	1.240	3.640	5.81E-05	0.324	0.3720	2.652	5.44E-05	0.205
18.50	1.12	1.05	0.200	1.50		1.33E-04		0.4959	2.654	4.58E-05	0.205
19.00	1.07	1.05	0.205	1.60		1.59E-04		0.6199	2.675	3.82E-05	0.208
19.50	1.02	1.04	0.212	1.70		6.27E-04		0.7439	2.692	3.32E-05	0.210
20.00	0.97	1.04	0.219	1.80	4.46	2.20E-02	0.402	0.8679	2.713	2.96E-05	0.213
20.50	0.91	1.03	0.228	2.0	4.79	0.76	0.438	0.9919	2.739	2.69E-05	0.216
21.00	0.85	1.01	0.234	2.2	4.49	1.19	0.431	1.116	2.772	2.48E-05	0.221
21.50	0.80	0.97	0.234	2.4	4.28	1.21	0.417	1.240	2.816	2.31E-05	0.226
22.00	0.77	0.94	0.233	2.6	4.40	1.32	0.430	1.50		7.37E-05	
23.00	0.71	0.87	0.229	2.8	4.59	1.70	0.462	1.60		8.63E-05	
24.00	0.67	0.79	0.218	3.0	4.44	2.29	0.490	1.70		3.60E-04	
25.00	0.64	0.73	0.205	3.2	3.92	2.59	0.493	1.80	3.32	0.11	0.289
26.00	0.61	0.66	0.194	3.4	3.69	2.76	0.502	2.00	3.38	0.65	0.310
27.00	0.60	0.59	0.177	3.6	3.39	3.01	0.521	2.20	3.07	0.73	0.282
28.00	0.60	0.53	0.155	3.8	(3.00)			2.40	2.93	0.61	0.259
29.00	0.61	0.48	0.134	4.0	(2.65)			2.60	3.00	0.53	0.263
30.00	0.62	0.45	0.123	4.2	(2.30)			2.80	3.12	0.58	0.279
31.00	0.61	0.40	0.114	4.5	1.92	2.78	0.528	3.00	3.30	0.70	0.305
32.00	0.63	0.34	0.093	5.0	1.50	2.31	0.482	3.20	3.35	1.01	0.328
33.00	0.65	0.31	0.077	6.0	1.57	1.49	0.288	3.40	3.22	1.24	0.334

Energy (eV)	n	k	R(φ = 0)
3.60	3.06	1.47	0.344
3.80	2.84	1.66	0.351
4.00	2.51	1.81	0.356
4.20	2.18	1.83	0.352
4.50	1.75	1.94	0.382
5.00	1.25	1.50	0.316
6.00	1.32	0.73	0.107
7.00	1.62	0.61	0.105
8.00	1.81	0.69	0.135
9.00	1.66	1.02	0.182
10.00	1.72	0.95	0.171
12.00	1.25	1.02	0.181
14.00	0.98	0.92	0.178
16.00	0.68	0.96	0.274
18.00	0.61	0.65	0.191
20.00	0.73	0.48	0.094
22.00	0.78	0.39	0.060
24.00	0.78	0.32	0.046
26.00	0.78	0.26	0.036
28.00	0.80	0.19	0.023
30.00	0.79	0.14	0.020

Silicon, single crystal[23]

Energy (eV)	n	k	R(φ = 0)
0.01240	3.4185	2.90E-04	0.300
0.01488	3.4190	2.30E-04	0.300
0.01736	3.4192	1.90E-04	0.300
0.01984	3.4195	1.70E-04	0.300
0.02480	3.4197		0.300
0.03100	3.4199		0.300
0.04092	3.4200		0.300
0.04463		1.08E-04	
0.04959	3.4201	9.15E-05	0.300
0.05703		1.56E-04	
0.06199	3.4204	2.86E-04	0.300
0.06943		3.84E-04	
0.07439		7.16E-04	
0.08059	(3.4207)	1.52E-04	0.300
0.08679		1.02E-04	
0.09299		2.59E-04	
0.09919		1.77E-04	
0.1054		1.53E-04	
0.1116		2.02E-04	
0.1178		1.22E-04	
0.1240	3.4215	6.76E-05	0.300
0.1364		5.49E-05	
0.1488		2.41E-05	
0.1612		2.49E-05	
0.1736	(3.4230)	1.68E-05	0.300
0.1798		2.45E-05	
0.1860		2.66E-06	
0.1922		1.74E-06	
0.1984		8.46E-07	
0.2046		5.64E-07	
0.2108	(3.4244)	4.17E-07	0.300
0.2170		4.05E-07	
0.2232		3.94E-07	
0.2294		3.26E-07	
0.2356		2.97E-07	
0.2418		2.82E-07	
0.2480	3.4261	1.99E-07	0.300
0.3100	3.4294		0.301
0.3626	3.4327		0.301
0.4568	3.4393	2.50E-09	0.302
0.6199	3.4490		0.303
0.8093	3.4784		0.306

Energy (eV)	n	k	R(φ = 0)
1.033	3.5193		0.311
1.1	(3.5341)	1.30E-05	0.312
1.2		1.80E-04	
1.3		2.26E-03	
1.4		7.75E-03	
1.5	3.673	5.00E-03	0.327
1.6	3.714	8.00E-03	0.331
1.7	3.752	1.00E-02	0.335
1.8	3.796	0.013	0.340
1.9	3.847	0.016	0.345
2.0	3.906	0.022	0.351
2.1	3.969	0.030	0.357
2.2	4.042	0.032	0.364
2.3	4.123	0.048	0.372
2.4	4.215	0.060	0.380
2.5	4.320	0.073	0.390
2.6	4.442	0.090	0.400
2.7	4.583	0.130	0.412
2.8	4.753	0.163	0.426
2.9	4.961	0.203	0.442
3.0	5.222	0.269	0.461
3.1	5.570	0.387	0.486
3.2	6.062	0.630	0.518
3.3	6.709	1.321	0.561
3.4	6.522	2.705	0.592
3.5	5.610	3.014	0.575
3.6	5.296	2.987	0.564
3.7	5.156	3.058	0.563
3.8	5.065	3.182	0.568
3.9	5.016	3.346	0.577
4.0	5.010	3.587	0.591
4.1	5.020	3.979	0.614
4.2	4.888	4.639	0.652
4.3	4.086	5.395	0.703
4.4	3.120	5.344	0.726
4.5	2.451	5.082	0.740
4.6	1.988	4.678	0.742
4.7	1.764	4.278	0.728
4.8	1.658	3.979	0.710
4.9	1.597	3.749	0.693
5.0	1.570	3.565	0.675
5.1	1.571	3.429	0.658
5.2	1.589	3.354	0.646
5.3	1.579	3.353	0.647
5.4	1.471	3.366	0.663
5.5	1.340	3.302	0.673
5.6	1.247	3.206	0.675
5.7	1.180	3.112	0.673
5.8	1.133	3.045	0.672
5.9	1.083	2.982	0.673
6.0	1.010	2.909	0.677
6.5	0.847	2.73	0.688
7.0	0.682	2.45	0.691
7.5	0.563	2.21	0.693
8.0	0.478	2.00	0.691
8.5	0.414	1.82	0.688
9.0	0.367	1.66	0.683
9.5	0.332	1.51	0.672
10.0	0.306	1.38	0.661
12.0	0.257	0.963	0.590
14.0	0.275	0.641	0.460
16.0	0.345	0.394	0.297
18.0	0.455	0.219	0.159
20.0	0.567	0.0835	0.079
22.14	0.675	0.0405	0.038

Energy (eV)	n	k	R(φ = 0)
24.31	0.752	0.0243	0.020
26.38	0.803	0.0178	0.012
28.18	0.834	0.0152	0.008
30.24	0.860	0.0138	0.006
31.79	0.877	0.0132	0.004
34.44	0.899	0.0121	0.003
36.47	0.913	0.0113	0.002
38.75	0.925	0.0104	0.002
40.00	0.930	0.0100	0.001

Silver[6]

Energy (eV)	n	k	R(φ = 0)
0.10	9.91	90.27	0.995
0.20	2.84	45.70	0.995
0.30	1.41	30.51	0.994
0.40	0.91	22.89	0.993
0.50	0.67	18.32	0.992
1.00	0.28	9.03	0.987
1.50	0.27	5.79	0.969
2.00	0.27	4.18	0.944
2.50	0.24	3.09	0.914
3.00	0.23	2.27	0.864
3.25	0.23	1.86	0.816
3.50	0.21	1.42	0.756
3.60	0.23	1.13	0.671
3.70	0.30	0.77	0.475
3.77	0.53	0.40	0.154
3.80	0.73	0.30	0.053
3.90	1.30	0.36	0.040
4.00	1.61	0.60	0.103
4.10	1.73	0.85	0.153
4.20	1.75	1.06	0.194
4.30	1.73	1.13	0.208
4.50	1.69	1.28	0.238
4.75	1.61	1.34	0.252
5.00	1.55	1.36	0.257
5.50	1.45	1.34	0.257
6.00	1.34	1.28	0.246
6.50	1.25	1.18	0.225
7.00	1.18	1.06	0.196
7.50	1.14	0.91	0.157
8.00	1.16	0.75	0.114
9.00	1.33	0.56	0.074
10.00	1.46	0.56	0.082
11.00	1.52	0.56	0.088
12.00	1.61	0.59	0.100
13.00	1.66	0.64	0.112
14.00	1.72	0.78	0.141
14.50	1.64	0.88	0.152
15.00	1.56	0.92	0.156
16.00	1.42	0.91	0.151
17.00	1.33	0.86	0.139
18.00	1.28	0.80	0.124
19.00	1.27	0.75	0.111
20.00	1.29	0.71	0.103
21.00	1.35	0.75	0.112
21.50	1.37	0.80	0.124
22.00	1.34	0.87	0.141
22.50	1.26	0.93	0.157
23.00	1.17	0.94	0.163
23.50	1.10	0.93	0.165
24.00	1.04	0.90	0.165
24.50	0.99	0.87	0.160
25.00	0.95	0.83	0.154
25.50	0.91	0.78	0.144
26.00	0.90	0.74	0.133

OPTICAL PROPERTIES OF METALS AND SEMICONDUCTORS (continued)

Energy (eV)	n	k	$R(\phi=0)$	Energy (eV)	n	k	$R(\phi=0)$	Energy (eV)	n	k	$R(\phi=0)$
26.50	0.89	0.69	0.121	6.358	0.454		0.141	3.20	2.73	2.31	0.432
27.00	0.89	0.65	0.109	6.526	0.485		0.120	3.40	2.61	2.33	0.435
27.50	0.89	0.62	0.099	6.702	0.533		0.093	3.60	2.49	2.30	0.430
28.00	0.90	0.59	0.090	6.888	0.574		0.073	3.80	2.40	2.22	0.418
28.50	0.91	0.57	0.084	7.130	0.616		0.056	4.00	2.36	2.14	0.406
29.00	0.92	0.56	0.079	7.328	0.641		0.048	4.20	2.35	2.06	0.392
30.00	0.93	0.54	0.074	7.583	0.674		0.038	4.40	2.39	2.01	0.384
31.00	0.93	0.53	0.072	7.847	0.700		0.031	4.60	2.45	2.00	0.384
32.00	0.92	0.53	0.072	8.015	0.710		0.029	4.80	2.53	2.06	0.394
33.00	0.90	0.51	0.071	8.634	0.762		0.018	5.00	2.58	2.20	0.416
34.00	0.88	0.49	0.067	9.143	0.800		0.012	5.20	2.52	2.44	0.450
35.00	0.86	0.45	0.061	9.709	0.819		0.010	5.40	2.31	2.61	0.480
36.00	0.89	0.44	0.055	10.20	0.843		0.007	5.60	2.06	2.67	0.501
38.00	0.89	0.39	0.043	11.08	0.870		0.005	5.80	1.83	2.63	0.510
40.00	0.90	0.37	0.039	11.83	0.887		0.004	6.00	1.63	2.56	0.515
42.00	0.90	0.35	0.036	12.73	0.907		0.002	6.20	1.48	2.45	0.512
44.00	0.90	0.33	0.033	13.05	0.913		0.002	6.40	1.37	2.33	0.504
46.00	0.90	0.32	0.031	13.42	0.914		0.002	6.60	1.29	2.22	0.492
48.00	0.89	0.31	0.030	13.73	0.917		0.002	6.80	1.23	2.11	0.478
50.00	0.88	0.29	0.027	14.07	0.922		0.002	7.00	1.18	2.01	0.462
52.00	0.89	0.28	0.024	14.83	0.934		0.001	7.20	1.15	1.91	0.445
54.00	0.88	0.17	0.024	15.05	0.936		0.001	7.40	1.13	1.82	0.425
56.00	0.87	0.26	0.024	15.46	0.942		0.001	7.60	1.12	1.75	0.406
58.00	0.87	0.24	0.021	16.21	0.948		0.001	7.80	1.11	1.68	0.390
60.00	0.87	0.22	0.018	18.10	0.964		0.000	8.00	1.11	1.61	0.370
62.00	0.88	0.21	0.016	21.12	0.979		0.000	8.20	1.12	1.55	0.350
64.00	0.88	0.21	0.016	25.51	0.993		0.000	8.40	1.13	1.50	0.332
66.00	0.88	0.21	0.016	26.95	1.00		0.000	8.60	1.14	1.45	0.317
68.00	0.87	0.21	0.017	27.68	1.01		0.000	8.80	1.17	1.41	0.301
70.00	0.83	0.20	0.021	28.37	1.01		0.000	9.00	1.19	1.40	0.294
72.00	0.85	0.18	0.016	29.52	1.02		0.000	9.20	1.21	1.38	0.289
74.00	0.85	0.17	0.014					9.40	1.21	1.38	0.287
76.00	0.85	0.16	0.013	**Tantalum[16]**				9.60	1.21	1.38	0.285
78.00	0.85	0.15	0.013	0.10	10.14	66.39	0.984	9.80	1.21	1.37	0.285
80.00	0.85	0.14	0.012	0.15	9.45	46.41	0.9834	10.00	1.20	1.37	0.286
85.00	0.85	0.11	0.011	0.20	5.77	35.46	0.982	10.20	1.19	1.37	0.286
90.00	0.85	0.08	0.009	0.26	3.67	27.53	0.981	10.40	1.18	1.37	0.287
95.00	0.86	0.06	0.007	0.30	2.87	23.90	0.980	10.60	1.16	1.36	0.288
100.00	0.87	0.04	0.005	0.38	2.03	18.87	0.978	10.80	1.15	1.36	0.289
				0.50	1.37	14.26	0.974	11.00	1.13	1.35	0.290
Sodium[24]				0.58	1.15	12.19	0.970	11.20	1.11	1.35	0.292
0.55	0.262	9.97	0.990	0.70	0.96	9.92	0.962	11.40	1.09	1.34	0.293
0.58	0.241	9.45	0.989	0.78	0.89	8.77	0.956	11.60	1.07	1.33	0.294
0.63	0.207	8.80	0.990	0.90	0.84	7.38	0.942	11.80	1.05	1.32	0.295
0.67	0.175	8.09	0.990	1.00	0.89	6.47	0.992	12.00	1.02	1.31	0.296
0.73	0.147	7.42	0.990	1.10	0.93	5.75	0.899	12.20	1.00	1.29	0.295
0.81	0.123	6.67	0.989	1.20	0.98	5.14	0.872	12.40	0.98	1.28	0.294
0.92	0.099	5.82	0.989	1.30	1.00	4.62	0.842	12.60	0.96	1.26	0.292
1.05	0.078	5.11	0.989	1.40	1.04	4.15	0.805	12.80	0.94	1.24	0.289
1.23	0.064	4.35	0.987	1.50	1.09	3.73	0.762	13.00	0.93	1.22	0.286
1.44	0.053	3.72	0.986	1.60	1.15	3.33	0.707	13.60	0.91	1.16	0.272
1.65	0.050	3.22	0.983	1.70	1.24	2.95	0.640	14.00	0.90	1.15	0.272
1.87	0.049	2.76	0.978	1.80	1.35	2.60	0.560	14.60	0.85	1.15	0.285
2.07	0.053	2.48	0.971	1.90	1.57	2.24	0.460	15.00	0.80	1.13	0.293
2.27	0.059	2.23	0.961	2.00	1.83	1.99	0.388	15.60	0.72	1.08	0.301
2.45	0.063	2.07	0.953	2.10	2.10	1.84	0.354	16.00	0.68	1.04	0.304
2.64	0.066	1.88	0.943	2.20	2.36	1.81	0.351	16.60	0.63	0.97	0.301
2.82	0.068	1.76	0.936	2.30	2.56	1.86	0.365	17.00	0.60	0.92	0.296
2.95	0.068	1.63	0.928	2.40	2.68	1.92	0.378	17.60	0.60	0.92	0.296
3.06	0.069	1.54	0.921	2.50	2.75	1.98	0.388	18.00	0.55	0.79	0.274
3.20	0.065	1.47	0.921	2.60	2.80	2.02	0.395	18.60	0.53	0.71	0.254
3.40	0.061	1.33	0.916	2.70	2.84	2.08	0.405	19.00	0.53	0.65	0.236
3.71	0.055	1.13	0.908	2.80	2.85	2.14	0.412	19.60	0.53	0.57	0.207
3.97	0.049	1.01	0.908	2.90	2.84	2.20	0.420	20.00	0.54	0.52	0.185
6.199	0.390		0.193	3.00	2.81	2.24	0.425	20.60	0.55	0.44	0.153

Energy (eV)	n	k	$R(\phi = 0)$	Energy (eV)	n	k	$R(\phi = 0)$	Energy (eV)	n	k	$R(\phi = 0)$
21.00	0.57	0.39	0.127	0.7	7.00	0.24	0.563	0.09919		4.28E-05	
21.60	0.64	0.34	0.089	0.8	7.23	0.48	0.574	0.1240	4.796	3.18E-05	0.429
22.00	0.64	0.32	0.081	0.9	7.48	0.94	0.589	0.1550	4.809		0.430
22.60	0.69	0.27	0.058	1.0	7.70	1.56	0.606	0.2066	4.838		0.432
23.00	0.73	0.24	0.043	1.2	6.99	2.22	0.593	0.2480	4.864	2.19E-05	0.434
23.60	0.80	0.26	0.033	1.4	7.11	2.46	0.604	0.31	4.929	3.18E-05	0.439
24.00	0.80	0.26	0.034	1.6	6.75	2.91	0.606	0.35		7.89E-02	
24.60	0.82	0.25	0.029	1.8	6.89	3.70	0.637	0.41		0.149	
25.00	0.83	0.25	0.026	2.0	4.67	4.67	0.654	0.5	4.90		0.437
25.60	0.86	0.24	0.022	2.2	4.94	5.16	0.681	0.6	4.93		0.439
26.00	0.88	0.25	0.022	2.4	3.94	5.08	0.686	0.7	4.95	0.11	0.441
26.60	0.87	0.26	0.023	2.6	3.25	4.77	0.681	0.8	5.10	0.13	0.452
27.00	0.87	0.25	0.022	2.8	2.73	4.42	0.674	0.9	5.22	0.22	0.461
27.60	0.89	0.23	0.019	3.0	2.30	4.16	0.674	1.0	5.35	0.45	0.472
28.00	0.90	0.23	0.017	3.5	1.69	3.44	0.646	1.2	5.17	0.63	0.462
28.60	0.91	0.22	0.015	4.0	1.33	2.64	0.571	1.4	5.56	0.63	0.488
29.00	0.92	0.22	0.014	4.5	1.32	1.96	0.428	1.6	5.88	1.15	0.517
29.60	0.94	0.22	0.014	5.0	1.63	1.60	0.312	1.8	6.10	1.80	0.545
30.00	0.95	0.22	0.014	5.5	1.72	1.57	0.302	2.0	5.94	2.69	0.571
31.00	0.97	0.23	0.014	6.0	1.73	1.45	0.276	2.2	5.10	3.61	0.594
32.00	0.98	0.24	0.015	6.5	1.78	1.36	0.257	2.4	4.24	3.77	0.593
33.00	0.98	0.25	0.015	7.0	1.83	1.36	0.257	2.6	3.57	3.75	0.591
34.00	0.99	0.25	0.016	7.5	1.72	1.51	0.289	2.8	3.03	3.63	0.588
35.00	0.99	0.26	0.017	8.0	1.54	1.37	0.260	3.0	2.51	3.39	0.578
36.00	0.99	0.27	0.018	8.5	1.55	1.23	0.226	3.5	1.72	2.70	0.532
37.00	0.99	0.28	0.019	9.0	0.99	0.93	0.179	4.0	1.32	2.01	0.440
38.00	0.98	0.28	0.021	9.5	1.47	1.25	0.233	4.5	1.28	1.28	0.251
39.00	0.97	0.29	0.022	10.0	0.86	0.86	0.181	5.0	1.47	0.82	0.132
40.00	0.95	0.29	0.023	11.0	0.80	0.77	0.165	5.5	1.74	0.51	0.104
				12.0	0.79	0.76	0.164	6.0	1.94	0.39	0.118
Tellurium, $\vec{E} \parallel \hat{c}$ [25]				14.0	0.67	0.59	0.146	6.5	2.19	0.32	0.148
0.01364	4.82	0.118	0.431	16.0	0.59	0.49	0.147	7.0	2.48	0.40	0.192
0.01488	5.26	0.0505	0.463	18.0	0.48	0.31	0.160	7.5	2.60	0.69	0.226
0.01612	5.47	0.0278	0.477	20.0	0.74	0.20	0.035	8.0	2.59	0.91	0.245
0.01736	5.59	0.0174	0.485	22.0	0.83	0.18	0.018	8.5	2.39	1.00	0.235
0.01860		0.0796		24.0	0.85	0.15	0.013	9.0	1.11	1.24	0.259
0.01984		0.0696		26.0	0.87	0.12	0.009	9.5	2.08	1.11	0.224
0.02108		0.0749		28.0	0.89	0.090	0.006	10.0	0.99	1.04	0.215
0.02232		0.1900		30.0	0.90	0.045	0.003	11.0	0.84	1.01	0.237
0.02356		0.2220						12.0	0.87	0.87	0.182
0.02480		0.0716		**Tellurium, $\vec{E} \perp \hat{c}$ [25]**				14.0	0.59	0.87	0.282
0.02604		0.0682		0.01364	2.61	0.2980	0.204	16.0	0.64	0.55	0.144
0.02728		0.0832		0.01488	3.65	0.0894	0.325	18.0	0.52	0.41	0.161
0.02976		0.0149		0.01612	4.10	0.0535	0.370	20.0	0.50	0.38	0.165
0.03224		2.14E-03		0.01736	4.63	0.4990	0.420	22.0	0.56	0.29	0.110
0.03472		1.71E-02		0.01860		0.1170		24.0	0.54	0.25	0.113
0.03720	5.94	3.71E-03	0.507	0.01984		0.0343		26.0	0.50	0.20	0.127
0.03968		2.44E-03		0.02108	(4.42)	0.0421	0.398	28.0	0.48	0.17	0.135
0.04339	5.96	1.59E-03	0.508	0.02232		0.1060		30.0	0.46	0.088	0.140
0.04711		7.85E-04		0.02356		0.0880					
0.05083		7.38E-04		0.02480		0.0458		**Titanium (Polycrystalline) [14]**			
0.05579		3.89E-04		0.02604		0.0928		0.10	5.03	23.38	0.965
0.06199	5.98	3.09E-04	0.509	0.02728		0.0886		0.15	3.00	15.72	0.954
0.07439		2.52E-04		0.02976		0.0232		0.20	2.12	11.34	0.939
0.08679		2.96E-04		0.03224		3.06E-03		0.25	2.05	8.10	0.890
0.09919		3.68E-04		0.03472		1.25E-02		0.30	6.39	9.94	0.833
0.12400	6.246	3.34E-04	0.524	0.03720	4.71	2.65E-03	0.422	0.35	2.74	6.21	0.792
0.15500	6.253		0.525	0.03968		1.89E-03		0.40	2.49	4.68	0.708
0.20660	6.286		0.526	0.04339	4.74	1.41E-03	0.425	0.45	3.35	3.25	0.545
0.24800	6.316	7.48E-05	0.528	0.04711		8.38E-04		0.50	4.43	3.22	0.555
0.31	6.372	1.18E-05	0.531	0.05083		6.79E-04		0.60	4.71	3.77	0.597
0.35		4.93E-04		0.05579		1.59E-04		0.70	4.38	3.89	0.603
0.41		6.74E-03		0.06199	4.77	1.16E-04	0.427	0.80	4.04	3.82	0.596
0.5	6.53	2.30E-02	0.539	0.07439		7.23E-05		0.90	3.80	3.65	0.582
0.6	6.71	7.50E-02	0.549	0.08679		5.34E-05		1.00	3.62	3.52	0.570

Energy (eV)	n	k	$R(\phi=0)$	Energy (eV)	n	k	$R(\phi=0)$	Energy (eV)	n	k	$R(\phi=0)$
1.10	3.47	3.40	0.560	11.00	0.79	0.72	0.152	0.86	2.92	4.37	0.661
1.20	3.35	3.30	0.550	11.20	0.81	0.69	0.139	0.90	3.11	4.44	0.660
1.30	3.28	3.25	0.546	11.40	0.81	0.69	0.139	0.94	3.15	4.43	0.658
1.40	3.17	3.28	0.549	11.60	0.79	0.68	0.139	0.98	3.15	4.36	0.653
1.50	2.98	3.32	0.557	11.80	0.78	0.67	0.137	1.00	3.14	4.32	0.649
1.60	2.74	3.30	0.559	12.00	0.77	0.65	0.132	1.10	3.05	4.04	0.627
1.70	2.54	3.23	0.557	12.80	0.76	0.55	0.106	1.20	3.00	3.64	0.590
1.80	2.36	3.11	0.550	13.20	0.76	0.52	0.097	1.30	3.12	3.24	0.545
1.90	2.22	2.99	0.540	13.60	0.76	0.48	0.087	1.40	3.29	2.96	0.515
2.00	2.11	2.88	0.530	14.00	0.77	0.45	0.077	1.50	3.48	2.79	0.500
2.10	2.01	2.77	0.520	14.40	0.77	0.42	0.069	1.60	3.67	2.68	0.494
2.20	1.92	2.67	0.509	14.80	0.79	0.38	0.058	1.70	3.84	2.79	0.507
2.30	1.86	2.56	0.495	15.20	0.79	0.36	0.052	1.80	3.82	2.91	0.518
2.40	1.81	2.47	0.483	15.60	0.79	0.32	0.045	1.90	3.70	2.94	0.518
2.50	1.78	2.39	0.471	16.00	0.83	0.31	0.037	2.00	3.60	2.89	0.512
2.60	1.75	2.34	0.462	16.40	0.84	0.28	0.030	2.10	3.54	2.84	0.506
2.70	1.71	2.29	0.456	16.80	0.87	0.27	0.025	2.20	3.49	2.76	0.497
2.80	1.68	2.25	0.451	17.20	0.90	0.25	0.020	2.30	3.49	2.72	0.494
2.90	1.63	2.21	0.447	17.60	0.93	0.25	0.017	2.40	3.45	2.72	0.493
3.00	1.59	2.17	0.444	18.00	0.94	0.24	0.165	2.50	3.38	2.68	0.487
3.10	1.55	2.15	0.442	18.40	0.94	0.23	0.017	2.60	3.34	2.62	0.480
3.20	1.50	2.12	0.442	18.80	0.95	0.24	0.016	2.70	3.31	2.55	0.472
3.30	1.44	2.09	0.442	19.20	0.96	0.25	0.016	2.80	3.31	2.49	0.466
3.40	1.37	2.06	0.443	19.60	0.97	0.25	0.017	2.90	3.32	2.45	0.461
3.50	1.30	2.01	0.443	20.00	0.98	0.27	0.018	3.00	3.35	2.42	0.459
3.60	1.24	1.96	0.441	20.40	0.98	0.27	0.019	3.10	3.39	2.41	0.460
3.70	1.17	1.90	0.436	20.60	1.00	0.29	0.020	3.20	3.43	2.45	0.465
3.80	1.11	1.83	0.430	21.20	0.99	0.31	0.023	3.30	3.45	2.55	0.476
3.85	1.08	1.78	0.423	21.60	0.99	0.31	0.024	3.40	3.39	2.66	0.485
3.90	1.06	1.73	0.413	22.00	0.98	0.32	0.025	3.50	3.24	2.70	0.488
4.00	1.04	1.62	0.389	22.40	0.98	0.33	0.027	3.60	3.13	2.67	0.482
4.20	1.05	1.45	0.333	22.80	0.97	0.33	0.028	3.70	3.05	2.62	0.476
4.40	1.13	1.33	0.284	23.20	0.96	0.34	0.030	3.80	2.99	2.56	0.468
4.60	1.17	1.29	0.265	23.60	0.95	0.35	0.031	3.90	2.96	2.50	0.460
4.80	1.21	1.23	0.244	24.00	0.92	0.35	0.033	4.00	2.95	2.43	0.451
5.00	1.24	1.21	0.236	24.5	0.91	0.34	0.032	4.20	3.02	2.33	0.440
5.20	1.27	1.20	0.228	25.0	0.91	0.33	0.032	4.40	3.13	2.32	0.442
5.40	1.17	1.16	0.228	25.5	0.89	0.33	0.032	4.60	3.24	2.41	0.455
5.60	1.24	1.21	0.234	26.0	0.89	0.33	0.032	4.80	3.33	2.57	0.475
5.80	1.21	1.22	0.241	26.5	0.88	0.32	0.032	5.00	3.40	2.85	0.505
6.00	1.15	1.21	0.244	27.0	0.86	0.31	0.032	5.20	3.27	3.27	0.548
6.20	1.11	1.18	0.240	27.5	0.85	0.30	0.033	5.40	2.92	3.58	0.586
6.40	1.08	1.14	0.232	28.0	0.84	0.29	0.033	5.60	2.43	3.70	0.618
6.60	1.04	1.06	0.212	28.5	0.82	0.26	0.029	5.80	2.00	3.61	0.637
6.80	1.05	1.02	0.198	29.0	0.83	0.25	0.027	6.00	1.70	3.42	0.643
7.00	1.06	0.97	0.182	30.0	0.84	0.22	0.022	6.20	1.47	3.24	0.646
7.20	1.07	0.95	0.175					6.40	1.32	3.04	0.640
7.40	1.11	0.94	0.167	**Tungsten[27]**				6.60	1.21	2.87	0.631
7.60	1.09	0.92	0.165	0.10	14.06	54.71	0.983	6.80	1.12	2.70	0.619
7.80	1.11	0.93	0.165	0.20	3.87	28.30	0.981	7.00	1.06	2.56	0.607
8.00	1.10	0.94	0.169	0.25	2.56	22.44	0.980	7.20	1.01	2.43	0.593
8.20	1.10	0.95	0.171	0.30	1.83	18.32	0.979	7.40	0.98	2.30	0.573
8.40	1.08	0.95	0.175	0.34	1.71	15.71	0.973	7.60	0.95	2.18	0.556
8.60	1.04	0.96	0.181	0.38	1.86	13.88	0.963	7.80	0.93	2.06	0.533
8.80	1.02	0.95	0.181	0.42	1.92	12.63	0.954	8.00	0.94	1.95	0.505
9.00	1.00	0.94	0.182	0.46	1.69	11.59	0.952	8.20	0.94	1.86	0.481
9.20	0.97	0.93	0.182	0.50	1.40	10.52	0.952	8.40	0.96	1.76	0.449
9.40	0.95	0.91	0.181	0.54	1.23	9.45	0.948	8.60	0.99	1.70	0.422
9.60	0.94	0.90	0.179	0.58	1.17	8.44	0.938	8.80	1.01	1.65	0.401
9.80	0.91	0.88	0.179	0.62	1.28	7.52	0.917	9.00	1.01	1.60	0.388
10.00	0.89	0.88	0.180	0.66	1.45	6.78	0.888	9.20	1.02	1.55	0.369
10.20	0.86	0.85	0.178	0.70	1.59	6.13	0.856	9.40	1.03	1.50	0.352
10.40	0.85	0.83	0.175	0.74	1.83	5.52	0.810	9.60	1.05	1.44	0.329
10.60	0.81	0.79	0.167	0.78	2.12	5.00	0.759	9.80	1.09	1.38	0.307
10.80	0.80	0.76	0.162	0.82	2.36	4.61	0.710	10.00	1.13	1.34	0.287

Energy (eV)	n	k	$R(\phi=0)$	Energy (eV)	n	k	$R(\phi=0)$	Energy (eV)	n	k	$R(\phi=0)$
10.20	1.19	1.33	0.274	**Vanadium**[9]				13.50	0.87	0.53	0.079
10.40	1.24	1.34	0.270	0.10	12.83	45.89	0.978	14.00	0.86	0.51	0.075
10.60	1.27	1.36	0.274	0.20	3.90	24.30	0.975	14.50	0.86	0.49	0.070
10.80	1.29	1.39	0.282	0.28	2.13	17.35	0.973	15.00	0.86	0.47	0.065
11.00	1.28	1.42	0.290	0.36	1.54	13.32	0.966	15.50	0.86	0.46	0.062
11.20	1.27	1.44	0.297	0.44	1.28	10.74	0.957	16.00	0.85	0.45	0.061
11.40	1.25	1.46	0.305	0.52	1.16	8.93	0.945	16.50	0.84	0.43	0.059
11.60	1.22	1.48	0.313	0.60	1.10	7.59	0.929	17.00	0.84	0.41	0.056
11.80	1.20	1.48	0.318	0.68	1.07	6.54	0.909	17.50	0.83	0.40	0.054
12.00	1.16	1.48	0.323	0.76	1.08	5.67	0.882	18.00	0.82	0.38	0.051
12.40	1.10	1.47	0.329	0.80	1.10	5.30	0.864	18.50	0.82	0.37	0.048
12.80	1.04	1.44	0.333	0.90	1.18	4.50	0.811	19.00	0.82	0.35	0.045
13.20	0.98	1.40	0.332	1.00	1.34	3.80	0.730	19.50	0.82	0.34	0.043
13.60	0.94	1.35	0.325	1.10	1.60	3.26	0.632	20.00	0.81	0.32	0.041
14.00	0.91	1.28	0.312	1.20	1.93	2.88	0.543	20.50	0.81	0.31	0.038
14.40	0.90	1.23	0.296	1.30	2.25	2.71	0.498	21.00	0.81	0.29	0.036
14.80	0.90	1.17	0.276	1.40	2.48	2.72	0.491	21.50	0.81	0.28	0.033
15.20	0.93	1.13	0.255	1.50	2.57	2.79	0.499	22.00	0.81	0.27	0.032
15.60	0.97	1.12	0.246	1.60	2.57	2.84	0.507	22.50	0.81	0.25	0.029
16.00	0.98	1.14	0.249	1.70	2.52	2.88	0.512	23.00	0.82	0.24	0.027
16.40	0.97	1.17	0.260	1.80	2.45	2.88	0.515	23.50	0.82	0.23	0.025
16.80	0.94	1.19	0.273	1.90	2.36	2.85	0.514	24.00	0.82	0.22	0.024
17.20	0.90	1.21	0.289	2.00	2.34	2.81	0.509	24.50	0.83	0.21	0.022
17.60	0.85	1.21	0.304	2.10	2.31	2.78	0.506	25.00	0.83	0.20	0.020
18.00	0.80	1.20	0.317	2.20	2.28	2.80	0.510	25.50	0.83	0.19	0.019
18.40	0.74	1.18	0.330	2.30	2.23	2.83	0.516	26.00	0.83	0.18	0.018
18.80	0.69	1.15	0.340	2.40	2.15	2.88	0.528	26.50	0.84	0.17	0.016
19.20	0.64	1.11	0.347	2.50	2.02	2.91	0.540	27.00	0.84	0.16	0.015
19.60	0.60	1.07	0.353	2.60	1.89	2.92	0.552	27.50	0.85	0.16	0.014
20.00	0.56	1.02	0.354	2.70	1.74	2.89	0.561	28.00	0.85	0.15	0.013
20.40	0.54	0.97	0.350	2.80	1.61	2.85	0.569	28.50	0.86	0.14	0.012
20.80	0.52	0.92	0.342	2.90	1.48	2.80	0.577	29.00	0.86	0.14	0.011
21.20	0.50	0.87	0.331	3.00	1.36	2.73	0.582	29.50	0.86	0.13	0.010
21.60	0.50	0.82	0.318	3.20	1.16	2.55	0.585	30.00	0.87	0.13	0.009
22.00	0.49	0.77	0.303	3.40	0.99	2.37	0.586	31.00	0.88	0.12	0.008
22.40	0.49	0.73	0.287	3.60	0.87	2.17	0.575	32.00	0.90	0.11	0.007
22.80	0.49	0.69	0.272	3.80	0.80	1.96	0.547	33.00	0.90	0.10	0.005
23.20	0.49	0.66	0.263	4.00	0.78	1.76	0.503	34.00	0.91	0.10	0.005
23.60	0.48	0.62	0.252	4.20	0.80	1.60	0.449	35.00	0.92	0.09	0.004
24.00	0.49	0.57	0.234	4.40	0.83	1.47	0.400	36.00	0.94	0.10	0.004
24.40	0.50	0.53	0.213	4.60	0.87	1.38	0.355	37.00	0.94	0.10	0.004
24.80	0.51	0.49	0.191	4.80	0.90	1.31	0.326	38.00	0.95	0.11	0.004
25.20	0.53	0.46	0.171	5.00	0.91	1.26	0.304	39.00	0.95	0.12	0.004
25.60	0.55	0.43	0.150	5.25	0.93	1.18	0.271	40.00	0.95	0.13	0.005
26.00	0.57	0.40	0.132	5.50	0.94	1.14	0.258				
26.40	0.59	0.38	0.117	5.75	0.96	1.09	0.235	**Zinc, $\vec{E} \parallel \hat{c}$**[28]			
26.80	0.61	0.37	0.105	6.00	0.98	1.06	0.223	0.7514	1.9241	7.5619	0.883
27.00	0.62	0.36	0.099	6.25	0.97	1.02	0.212	0.827	1.7921	6.9973	0.874
27.50	0.64	0.34	0.085	6.50	0.97	0.98	0.199	0.866	1.5571	6.7753	0.881
28.00	0.67	0.32	0.073	6.75	0.97	0.94	0.185	0.952	1.4824	6.2296	0.868
28.50	0.69	0.31	0.065	7.00	0.98	0.91	0.175	0.992	1.5762	5.8843	0.847
29.00	0.71	0.30	0.057	7.33	0.97	0.89	0.170	1.033	1.5407	5.3192	0.823
29.50	0.73	0.30	0.052	7.66	0.98	0.87	0.162	1.078	1.5853	4.9013	0.793
30.00	0.75	0.29	0.047	8.00	0.98	0.85	0.155	1.127	1.7768	4.5307	0.748
31.00	0.78	0.29	0.042	8.33	0.98	0.81	0.146	1.181	1.9808	4.2004	0.701
32.00	0.79	0.29	0.040	8.66	0.98	0.81	0.145	1.240	2.8821	3.4766	0.575
33.00	0.82	0.28	0.033	9.00	0.96	0.79	0.142	1.305	3.2039	3.0042	0.520
34.00	0.84	0.29	0.032	9.50	0.94	0.77	0.136	1.377	2.9459	3.5761	0.584
35.00	0.85	0.31	0.033	10.00	0.91	0.74	0.133	1.459	3.2523	4.2447	0.640
36.00	0.85	0.32	0.036	10.50	0.89	0.71	0.126	1.550	3.8086	4.6212	0.657
37.00	0.84	0.33	0.039	11.00	0.87	0.65	0.112	1.653	3.7577	4.6239	0.659
38.00	0.83	0.33	0.040	11.50	0.88	0.58	0.091	1.722	3.5908	4.4614	0.650
39.00	0.81	0.33	0.042	12.00	0.90	0.58	0.089	1.823	3.4234	4.3232	0.642
40.00	0.80	0.33	0.045	12.50	0.89	0.57	0.086	1.937	3.0132	3.9974	0.624
				13.00	0.88	0.55	0.082	1.984	1.8562	3.9706	0.690

Energy (eV)	n	k	R(φ = 0)	Energy (eV)	n	k	R(φ = 0)	Energy (eV)	n	k	R(φ = 0)
2.066	1.4856	4.0555	0.737	0.80	4.03	1.42	0.168	9.20	1.63	0.90	0.025
2.094	1.2525	3.9961	0.762	0.90	3.74	1.37	0.149	9.40	1.60	0.89	0.024
2.119	1.0017	3.8683	0.789	0.96	3.69	1.36	0.145	9.60	1.57	0.89	0.023
2.275	0.7737	3.9129	0.832	1.00	3.66	1.35	0.143	9.80	1.52	0.87	0.021
2.445	0.6395	3.4013	0.821	1.10	3.65	1.35	0.142	10.00	1.47	0.86	0.020
2.666	0.4430	3.1379	0.851	1.20	3.53	1.33	0.134	10.20	1.42	0.84	0.018
2.917	0.3589	2.8140	0.853	1.30	3.25	1.27	0.116	10.40	1.35	0.82	0.016
3.220	0.3069	2.5088	0.847	1.40	3.10	1.25	0.106	10.50	1.32	0.81	0.016
3.594	0.2737	2.1737	0.828	1.50	3.02	1.23	0.100	10.60	1.28	0.80	0.015
4.065	0.2510	1.8528	0.799	1.60	2.88	1.20	0.091	10.80	1.23	0.78	0.014
4.678	0.2354	1.6357	0.776	1.70	2.68	1.16	0.078	11.00	1.19	0.77	0.014
				1.80	2.49	1.12	0.067	11.20	1.16	0.76	0.013
Zinc, $\vec{E} \perp \hat{c}$[28]				2.00	2.14	1.03	0.047	11.40	1.13	0.75	0.013
0.751	1.4469	7.4158	0.905	2.10	1.99	1.00	0.040	11.60	1.11	0.74	0.013
0.827	1.4744	6.9688	0.892	2.20	1.87	0.97	0.034	11.80	1.09	0.74	0.013
0.866	1.3628	6.6886	0.892	2.30	1.78	0.94	0.030	12.00	1.08	0.73	0.013
0.952	1.3165	6.2212	0.881	2.40	1.71	0.92	0.027	12.40	1.05	0.72	0.012
0.992	1.3835	5.8910	0.863	2.50	1.62	0.90	0.024	12.80	1.01	0.71	0.012
1.033	1.2889	5.4001	0.850	2.60	1.54	0.88	0.022	13.20	0.98	0.70	0.012
1.078	1.3095	4.9025	0.822	2.70	1.46	0.86	0.019	13.60	0.95	0.69	0.013
1.127	1.6897	4.4062	0.746	2.80	1.40	0.84	0.018	14.00	0.92	0.68	0.013
1.181	1.9701	4.0176	0.684	2.90	1.34	0.82	0.016	14.40	0.89	0.67	0.013
1.240	2.8717	3.2873	0.555	3.00	0.30	0.81	0.016	14.80	0.90	0.67	0.013
1.305	3.3991	2.7684	0.497	3.10	1.26	0.80	0.015	15.20	0.92	0.68	0.013
1.377	3.1807	3.4709	0.569	3.30	1.19	0.77	0.014	15.60	0.95	0.69	0.013
1.459	3.5064	4.1994	0.630	3.40	1.16	0.76	0.013	16.00	0.98	0.70	0.012
1.550	4.1241	4.7768	0.664	3.50	1.13	0.75	0.013	16.40	1.01	0.71	0.012
1.653	4.0269	4.8027	0.667	3.60	1.10	0.74	0.013	16.80	1.04	0.72	0.012
1.722	3.9369	4.6356	0.657	3.70	1.07	0.73	0.013	17.20	1.09	0.74	0.013
1.823	3.7549	4.3042	0.635	3.80	1.04	0.72	0.012	17.60	1.13	0.75	0.013
1.937	3.4512	4.1942	0.631	3.90	1.01	0.71	0.012	18.00	1.17	0.76	0.014
1.984	3.2515	4.2980	0.644	4.00	0.98	0.70	0.012	18.40	1.21	0.78	0.014
2.066	2.0802	4.7231	0.738	4.20	0.94	0.68	0.013	18.80	1.24	0.79	0.014
2.094	1.7084	4.7923	0.774	4.40	0.89	0.67	0.013	19.20	1.27	0.80	0.015
2.119	1.3329	4.4751	0.791	4.60	0.85	0.65	0.014	19.60	1.29	0.80	0.015
2.275	0.9725	4.2879	0.825	4.80	0.81	0.64	0.014	20.00	1.30	0.81	0.015
2.455	0.7568	3.7627	0.824	5.00	0.78	0.63	0.015	20.60	1.29	0.80	0.015
2.666	0.5470	3.4277	0.845	5.20	0.77	0.62	0.016	21.00	1.27	0.80	0.015
2.917	0.4774	3.0476	0.834	5.40	0.77	0.62	0.016	21.60	1.23	0.78	0.014
3.220	0.3911	2.7463	0.835	5.60	0.80	0.63	0.014	22.00	1.20	0.77	0.014
3.594	0.3147	2.3041	0.821	5.80	0.87	0.66	0.013	22.60	1.15	0.76	0.013
4.065	0.3013	2.0077	0.789	6.00	1.00	0.71	0.012	23.00	1.12	0.75	0.013
4.678	0.2806	1.7997	0.770	6.20	1.11	0.75	0.013	23.60	1.08	0.73	0.013
				6.40	1.23	0.78	0.015	24.00	1.05	0.73	0.013
Zirconium (Polycrystalline)[28]				6.60	1.33	0.81	0.016	24.60	1.02	0.71	0.012
0.10	6.18	1.76	0.300	6.80	1.42	0.84	0.018	25.00	1.00	0.71	0.012
0.15	3.37	1.30	0.123	7.00	1.49	0.86	0.020	25.60	0.97	0.69	0.012
0.20	2.34	1.08	0.058	7.20	1.54	0.88	0.022	26.00	0.95	0.69	0.013
0.26	2.24	1.06	0.052	7.40	1.58	0.89	0.023	26.60	0.91	0.67	0.013
0.30	2.59	1.14	0.073	7.60	1.61	0.90	0.024	27.00	0.88	0.66	0.013
0.36	3.17	1.26	0.110	7.80	1.63	0.90	0.025	27.60	0.84	0.65	0.014
0.40	3.09	1.24	0.105	8.00	1.66	0.91	0.026	28.00	0.83	0.64	0.014
0.46	3.36	1.30	0.123	8.20	0.67	0.91	0.026	28.60	0.82	0.64	0.014
0.50	4.13	1.44	0.175	8.40	1.68	0.92	0.026	29.00	0.81	0.64	0.014
0.56	5.01	1.58	0.231	8.60	1.68	0.92	0.026	29.60	0.82	0.64	0.014
0.60	5.18	1.61	0.242	8.80	1.66	0.91	0.026	30.00	0.82	0.64	0.014
0.70	4.54	1.51	0.202	9.00	1.65	0.91	0.025				

REFERENCES

1. Shiles, E., Sasaki, T., Inokuti, M., and Smith, D. Y., *Phys. Rev. Sect. B*, 22, 1612, 1980.
2. Edwards, D. F., and Philipp, H. R., in *HOC-I*, p.665.
3. Ives, H. E., and Briggs, N. B., *J. Opt. Soc. Am.*, 27, 395, 1937.
4. Bos, L. W., and Lynch, D. W., *Phys. Rev. Sect. B*, 2, 4567, 1970.
5. Weaver, J. H., Colavita, E., Lynch, D. W., and Rosei, R., *Phys. Rev. Sect. B*, 19, 3850, 1979.
6. Hagemann, H. J., Gudat, W., and Kunz, C., *J. Opt. Soc. Am.*, 65, 742, 1975.
7. Schulz, L. G., *J. Opt. Soc. Am.*, 47, 64, 1957.
8. Potter, R. F., in *HOC-I*, p.465.
9. Olson, C. G., Lynch, D. W., and Weaver, J. H., unpublished.
10. Lynch, D. W., Olson, C. G., and Weaver, J. H., unpublished.
11. Weaver, J. H., Olson, C. G., and Lynch, D. W., *Phys. Rev. Sect. B*, 15, 4115, 1977.
12. Lynch, D. W., and Hunter, W. R., in *HOC-II*, p.345.
13. Priol, M. A., Daudé, A., and Robin, S., *Compt. Rend.*, 264, 935, 1967.
14. Johnson, P. B., and Christy, R. W., *Phys. Rev. Sect. B*, 9, 5056, 1974.
15. Arakawn, E. T., and Inagaki, T., in *HOC-II*, p.461.
16. Weaver, J. H., Lynch, D. W., and Olson, D. G., *Phys. Rev. Sect. B*, 10, 501, 1973.
17. Lynch, D. W., Rosei, R., and Weaver, J. H., *Solid State Commun.*, 9, 2195, 1971.
18. Weaver, J. H., Lynch, D. W., and Olson, C. G., *Phys. Rev. Sect. B*, 7, 4311, 1973.
19. Weaver, J. H., and Benbow, R. L., *Phys. Rev. Sect. B*, 12, 3509, 1975.
20. Weaver, J. H., *Phys. Rev.,Sect. B*, 11, 1416, 1975.
21. Lynch, D. W., and Hunter, W. R., in *HOC-II*, p.364.
22. Palik, E. D., in *HOC-II*, p. 691.
23. Edwards, D. F., in *HOC-I*, p. 547.
24. Lynch, D. W., and Hunter, W. R., in *HOC-II*, p.354.
25. Palik, E. D., in *HOC-II*, p. 709.
26. Lynch, D. W., Olson, C. G., and Weaver, J. H., *Phys. Rev. Sect. B*, 11, 3671, 1975.
27. Weaver, J. H., Lynch, D. W., and Olson, C. G., *Phys. Rev. Sect. B*, 12, 1293, 1975.
28. Lanham, A. P., and Terherne, D. M., *Proc. Phys. Soc.*, 83, 1059, 1964.

ELASTO-OPTIC, ELECTRO-OPTIC, AND MAGNETO-OPTIC CONSTANTS

When a crystal is subjected to a stress field, an electric field, or a magnetic field, the resulting optical effects are in general dependent on the orientation of these fields with respect to the crystal axes. It is useful, therefore, to express the optical properties in terms of the refractive index ellipsoid (or indicatrix):

$$\frac{x^2}{n_x^2} + \frac{y^2}{n_y^2} + \frac{z^2}{n_z^2} = 1$$

or

$$\sum_{ij} B_{ij} x_i y_j = 1 \quad (i, j = 1, 2, 3)$$

where

$$B_{ij} = \left[\frac{1}{\varepsilon}\right]_{ij} = \left[\frac{1}{n^2}\right]_{ij}$$

ε is the dielectric constant or permeability; the quantity B_{ij} has the name impermeability.

A crystal exposed to a *stress* S will show a change of its impermeability. The photo-elastic (or elasto-optic) constants, P_{ijkl}, are defined by

$$\Delta\left[\frac{1}{\varepsilon}\right]_{ij} = \Delta\left[\frac{1}{n^2}\right]_{ij} = \sum_{kl} P_{ijkl} S_{kl}$$

where n is the refractive index and S_{kl} are the strain tensor elements; the P_{ijkl} are the elements of a 4th rank tensor.

When a crystal is subjected to an *electric field E* two possible changes of the refractive index may occur depending on the symmetry of the crystal.

1. All materials, including isotropic solids and polar liquids, show an electro-optic birefringence (Kerr effect) which is proportional to the square of the electric field, E:

$$\left[\frac{1}{n^2}\right]_{ij} = \sum_{k,l=1,2,3} K_{ijkl} E_k E_l = \sum_{k,l=1,2,3} g_{ijkl} P_k P_l$$

where E_k and E_l are the components of the electric field and P_k and P_l the electric polarizations. The coefficients, K_{ijkl}, are the quadratic electro-optic coefficients, while the constants g_{ijkl} are known as the Kerr constants.

2. The other electro-optic effect only occurs in the 20 piezo-electric crystal classes (no center of symmetry). This effect is known as the Pockels effect. The optical impermeability changes linearly with the static field

$$\Delta\left[\frac{1}{n^2}\right]_{ij} = \sum_{k} r_{ij,k} E_k$$

The coefficients $r_{ij,k}$ have the name (linear) electro-optic coefficients.

The values of the electro-optic coefficients depend on the boundary conditions. If the superscripts T and S denote respectively the conditions of zero stress (free) and zero strain (clamped) one finds:

$$r_{ij}^T = r_{ij}^S + q_{ik}^E e_{jk} = r_{ij}^S + P_{ik}^E d_{jk}$$

where $e_{jk} = (\partial T_k/\partial E_j)_S$ and $d_{jk} = (\partial S_k/\partial E_j)_T$ are the appropriate piezo-electric coefficients.

The interaction between a *magnetic field* and a light wave propagating in a solid or in a liquid gives rise to a rotation of the plane of polarization. This effect is known as *Faraday rotation*. It results from a difference in propagation velocity for left and right circular polarized light.

The Faraday rotation, θ_F, is linearly proportional to the magnetic field H:

$$\theta_F = VlH$$

where l is the light path length and V is the *Verdet* constant (minutes/oersted·cm).

For ferromagnetic, ferrimagnetic, and antiferromagnetic materials the magnetic field in the above expression is replaced by the magnetization M and the magneto-optic coefficient in this case is known as the Kund constant K:

$$\text{Specific Faraday rotation } F = KM$$

In the tables below the *Faraday rotation* is listed at the saturation magnetization per unit length, together with the absorption coefficient α, the temperature T, the critical temperature T_C (or T_N), and the wavelength of the measurement.

ELASTO-OPTIC, ELECTRO-OPTIC, AND MAGNETO-OPTIC CONSTANTS (continued)

In the tables which follow, the properties are presented in groups:
- Elasto-optic coefficients (photoelastic constants)
- Linear electro-optic coefficients (Pockels constants)
- Quadratic electro-optic coefficients (Kerr constants)
- Magneto-optic coefficients:
  - Verdet constants
  - Faraday rotation parameters

Within each group, materials are classified by crystal system or physical state. References are given at the end of each group of tables.

ELASTO-OPTIC COEFFICIENTS (PHOTOELASTIC CONSTANTS)

Name

Cubic (43m, 432, m3m)	Formula	$\lambda/\mu m$	p_{11}	p_{12}	p_{44}	p_{11}-p_{12}	Ref.
Sodium fluoride	NaF	0.633	0.08	0.20	-0.03	-0.12	1
Sodium chloride	NaCl	0.589	0.115	0.159	-0.011	-0.042	2
Sodium bromide	NaBr	0.589	0.148	0.184	-0.0036	-0.035	1
Sodium iodide	NaI	0.589	—	—	0.0048	-0.0141	3
Potassium fluoride	KF	0.546	0.26	0.20	-0.029	0.06	1
Potassium chloride	KCl	0.633	0.22	0.16	-0.025	0.06	4
Potassium bromide	KBr	0.589	0.212	0.165	-0.022	0.047	5
Potassium iodide	KI	0.590	0.212	0.171	—	0.041	6
Rubidium chloride	RbCl	0.589	0.288	0.172	-0.041	0.116	7,8
Rubidium bromide	RbBr	0.589	0.293	0.185	-0.034	0.108	7,8
Rubidium iodide	RbI	0.589	0.262	0.167	-0.023	0.095	7,8
Lithium fluoride	LiF	0.589	0.02	0.13	-0.045	-0.11	5
Lithium chloride	LiCl	0.589	—	—	-0.0177	-0.0407	3
Ammonium chloride	NH$_4$Cl	0.589	0.142	0.245	0.042	-0.103	9
Cadmium telluride	CdTe	1.06	-0.152	-0.017	-0.057	-0.135	10
Calcium fluoride	CaF$_2$	0.55-0.65	0.038	0.226	0.0254	-0.183	11
Copper chloride	CuCl	0.633	0.120	0.250	-0.082	-0.130	12
Copper bromide	CuBr	0.633	0.072	0.195	-0.083	-0.123	12
Copper iodide	CuI	0.633	0.032	0.151	-0.068	-0.119	12
Diamond	C	0.540-0.589	-0.278	0.123	-0.161	-0.385	13
Germanium	Ge	3.39	-0.151	-0.128	-0.072	-0.023	14
Gallium arsenide	GaAs	1.15	-0.165	-0.140	-0.072	-0.025	15
Gallium phosphide	GaP	0.633	-0.151	-0.082	-0.074	-0.069	15
Strontium fluoride	SrF$_2$	0.633	0.080	0.269	0.0185	-0.189	16
Strontium titanite	SrTiO$_3$	0.633	0.15	0.095	0.072	—	17
KRS-5	Tl(Br,I)	0.633	-0.140	0.149	-0.0725	-0.289	18,20
KRS-6	Tl(Br,Cl)	0.633	-0.451	-0.337	-0.164	-0.114	19,20
Zinc sulfide	Zn	0.633	0.091	-0.01	0.075	0.101	15

Rare Gases	Formula	$\lambda/\mu m$	p_{11}	p_{12}	p_{44}	p_{11}-p_{12}	Ref.
Neon (T = 24.3 K)	Ne	0.488	0.157	0.168	0.004	-0.011	21
Argon (T = 82.3 K)	Ar	0.488	0.256	0.302	0.015	-0.046	22
Krypton (T = 115.6 K)	Kr	0.488	0.34	0.34	0.037	0	21
Xenon (T = 160.5 K)	Xe	0.488	0.284	0.370	0.029	-0.086	22

Garnets	Formula	$\lambda/\mu m$	p_{11}	p_{12}	p_{44}	p_{11}-p_{12}	Ref.
GGG	Gd$_3$Ga$_5$O$_{12}$	0.514	-0.086	-0.027	-0.078	-0.059	23
YIG	Y$_3$Fe$_5$O$_{12}$	1.15	0.025	0.073	0.041	—	15
YGG	Y$_3$Ga$_5$O$_{12}$	0.633	0.091	0.019	0.079	—	17
YAG	Y$_3$Al$_5$O$_{12}$	0.633	-0.029	0.0091	-0.0615	-0.038	15

ELASTO-OPTIC COEFFICIENTS (PHOTOELASTIC CONSTANTS) continued

Name

Cubic (23, m3)	Formula	$\lambda/\mu m$	p_{11}	p_{12}	p_{44}	p_{13}	Ref.
Barium nitrate	Ba(NO3)2	0.589	—	$p_{11}-p_{22}=0.992$	-0.0205	$p_{11}-p_{13}=0.713$	13
Lead nitrate	$Pb(NO_3)_2$	0.589	0.162	0.24	-0.0198	0.20	24,25
Sodium bromate	$NaBrO_3$	0.589	0.185	0.218	-0.0139	0.213	26
Sodium chlorate	$NaClO_3$	0.589	0.162	0.24	-0.0198	0.20	26
Strontium nitrate	$Sr(NO_3)_2$	0.41	0.178	0.362	-0.014	0.316	27

Hexagonal (mmc,6mm)	Formula	$\lambda/\mu m$	p_{11}	p_{12}	p_{13}	p_{31}	p_{33}	p_{44}	Ref.
Beryl	$Be_3Al_2Si_6O_{18}$	0.589	0.0099	0.175	0.191	0.313	0.023	-0.152	28
Cadmium sulfide	CdS	0.633	-0.142	-0.066	-0.057	-0.041	-0.20	-0.099	15,2
Zinc oxide	ZnO	0.633	±0.222	±0.099	-0.111	±0.088	-0.235	0.0585	30
Zinc sulfide	ZnS	0.633	-0.115	0.017	0.025	0.0271	-0.13	-0.0627	31

Trigonal (3m,32,$\overline{3}$m)	Formula	$\lambda/\mu m$	p_{11}	p_{12}	p_{13}	p_{14}	p_{31}
Sapphire	Al_2O_3	0.644	-0.23	-0.03	0.02	0.00	-0.04
Calcite	$CaCO_3$	0.514	0.062	0.147	0.186	-0.011	0.241
Lithium niobate	$LiNbO_3$	0.633	±0.034	±0.072	±0.139	±0.066	±0.178
Lithium tantalate	$LiTaO_3$	0.633	-0.081	0.081	0.093	-0.026	0.089
Cinnabar	HgS	0.633			±0.445		
Quartz	SiO_2	0.589	0.16	0.27	0.27	-0.030	0.29
Proustite	Ag_3AsS_3	0.633	±0.10	±0.19	±0.22		±0.24
Sodium nitrite	$NaNO_3$	0.633		±0.21	±0.215	±0.027	±0.25
Tellurium	Te	10.6	0.155	0.130	—	—	—

Trigonal (3m,32,$\overline{3}$m) (continued)	p_{33}	p_{41}	p_{44}	Ref.
Sapphire	-0.20	0.01	-0.10	15,32
Calcite	0.139	-0.036	-0.058	33
Lithium niobate	+-0.060	±0.154	±0.300	15,34
Lithium tantalate	-0.044	-0.085	0.028	15,35
Cinnabar	+-0.115	—	—	36
Quartz	0.10	-0.047	-0.079	37
Proustite	+-0.20	—	—	38
Sodium nitrite		0.055	-0.06	39
Tellurium		—	—	15

Tetragonal (4/mmm,$\overline{4}$2m,422)	Formula	$\lambda/\mu m$	p_{11}	p_{12}	p_{13}	p_{31}
Ammonium dihydrogen phosphate	ADP	0.589	0.319	0.277	0.169	0.197
Barium titanate	$BaTiO_3$	0.633	0.425	—	—	—
Cesium dihydrogen arsenate	CDA	0.633	0.267	0.225	0.200	0.195
Magnesium fluoride	MgF_2	0.546	—	—	—	—
Calomel	Hg_2Cl_2	0.633	±0.551	±0.440	±0.256	±0.137
Potassium dihydrogen phosphate	KDP	0.589	0.287	0.282	0.174	0.241
Rubidium dihydrogen arsenate	RDA	0.633	0.227	0.239	0.200	0.205
Rubidium dihydrogen phosphate	RDP	0.633	0.273	0.240	0.218	0.210
Strontium barium niobate	$Sr_{0.75}Ba_{0.25}Nb_2O_6$	0.633	0.16	0.10	0.08	0.11
Strontium barium niobate	$Sr_{0.5}Ba_{0.5}Nb_2O_6$	0.633	0.06	0.08	0.17	0.09
Tellurium oxide	TeO_2	0.633	0.0074	0.187	0.340	0.090
Rutile	TiO_2	0.633	0.017	0.143	-0.139	-0.080

ELASTO-OPTIC COEFFICIENTS (PHOTOELASTIC CONSTANTS) (continued)

Name Tetragonal $(4/mmm, \bar{4}2m, 422)$ (continued)	p_{33}	p_{44}	p_{66}	Ref.
Ammonium dihydrogen phosphate	0.167	-0.058	-0.091	40
Barium titanate	—	—	—	41
Cesium dihydrogen arsenate	0.227	—	—	42
Magnesium fluoride	—	±0.0776	±0.0488	43
Calomel	±0.010	—	±0.047	44
Potassium dihydrogen phosphate	0.122	-0.019	-0.064	45
Rubidium dihydrogen arsenate	0.182	—	—	41
Rubidium dihydrogen phosphate	0.208	—	—	41
Strontium barium niobate	0.47	—	—	46
Strontium barium niobate	0.23	—	—	46
Tellurium oxide	0.240	-0.17	-0.046	47
Rutile	-0.057	-0.009	-0.060	48

Tetragonal $(4,\bar{4},4/m)$	Formula	$\lambda/\mu m$	p_{11}	p_{12}	p_{13}	p_{16}	p_{31}
Cadmium molybdate	$CdMoO_4$	0.633	0.12	0.10	0.13	—	0.11
Lead molybdate	$PbMoO_4$	0.633	0.24	0.24	0.255	0.017	0.175
Sodium bismuth molybdate	$NaBi(MoO_4)_2$	0.633	0.243	0.205	0.25	—	0.21

Tetragonal $(4,\bar{4},4/m)$ (continued)	p_{33}	p_{44}	p_{45}	p_{61}	p_{66}	Ref.
Cadmium molybdate	0.18	—	—	—	—	49
Lead molybdate	0.300	0.067	-0.01	0.013	0.05	52
Sodium bismuth molybdate	0.29	—	—	—	—	

Orthorhombic (222,m22,mmm)	Formula	$\lambda/\mu m$	p_{11}	p_{12}	p_{13}	p_{21}	p_{22}	p_{23}
Ammonium chlorate	NH_4ClO_4	0.633	—	0.24	0.18	0.23	—	0.20
Ammonium sulfate	$(NH_4)_2SO_4$	0.633	0.26	0.19	±0.260	±0.230	±0.27	±0.254
Rochelle salt	$NaKC_4H_4O_6$	0.589	0.35	0.41	0.42	0.37	0.28	0.34
Iodic acid (α)	HIO_3	0.633	0.302	0.496	0.339	0.263	0.412	0.304
Sulfur (α)	S	0.633	0.324	0.307	0.268	0.272	0.301	0.310
Barite	$BaSO_4$	0.589	0.21	0.25	0.16	0.34	0.24	0.19
Topaz	$Al_2SiO_4(OH,F)_2$	—	-0.085	0.069	0.052	0.095	-0.120	0.065

Orthorhombic (222,m22,mmm) (continued)	p_{31}	p_{32}	p_{33}	p_{44}	p_{55}	p_{66}	Ref.
Ammonium chlorate	0.19	0.18	±0.02	<±0.02	—	±0.04	51
Ammonium sulfate	0.20	±0.26	0.26	0.015	±0.0015	0.012	52
Rochelle salt	0.36	0.35	0.36	-0.030	0.0046	-0.025	53
Iodic Acid (α)	0.251	0.345	0.336	0.084	-0.030	0.098	54
Sulfur (α)	0.203	0.232	0.270	0.143	0.019	0.118	54
Barite	0.28	0.22	0.31	0.002	-0.012	0.037	55
Topaz	0.095	0.085	-0.083	-0.095	-0.031	0.098	28

Monoclinic (2,m,2/m)	Formula	$\lambda/\mu m$			
Taurine	$C_2H_7NO_3S$	0.589	$p_{11} = 0.313$	$p_{25} = -0.0025$	$p_{51} = -0.014$
			$p_{12} = 0.251$	$p_{31} = 0.362$	$p_{52} = 0.006$
			$p_{13} = 0.270$	$p_{32} = 0.275$	$p_{53} = 0.0048$
			$p_{15} = -0.10$	$p_{33} = 0.308$	$p_{55} = 0.047$

Name

Monoclinic (2,m,2/m)	Formula	$\lambda/\mu m$			
Taurine (continued)			$p_{21} = 0.281$	$p_{35} = -0.003$	$p_{64} = 0.0024$
			$p_{22} = 0.252$	$p_{44} = 0.0025$	$p_{66} = 0.0028$
			$p_{23} = 0.272$	$p_{46} = -0.0056$	

Isotropic	Formula	$\lambda/\mu m$	p_{11}	p_{12}	p_{44}	Ref.
Fused silica	SiO_2	0.633	0.121	0.270	-0.075	15
Water	H_2O	0.633	±0.31	±0.31		15
Polystyrene		0.633	±0.30	±0.31		25
Lucite		0.633	±0.30	0.28		25
Orpiment	As_2S_3-glass	1.15	0.308	0.299	0.0045	15
Tellurium oxide	TeO_2-glass	0.633	0.257	0.241	0.0079	56
Laser glasses	LGS-247-2	0.488	±0.168	±0.230		57
	LGS-250-3		±0.135	±0.198		
	LGS-1		±0.214	±0.250		
	KGSS-1621		±0.205	±0.239		
Dense flint glasses	LaSF	0.633	0.088	0.147	-0.030	58
(examples)	SF_4		0.215	0.243	-0.014	
	U10502		0.172	0.179	-0.004	
	$TaFd_7$		0.099	0.138	-0.020	

REFERENCES

A. Narasimhamurty, T. S., *Photoelastic and Electro-Optic Properties of Crystals*, Plenum Press, New York, 1981; pp. 290-293.

B. Weber, M. J., Ed., *CRC Handbook of Laser Science and Technology*, Volume IV, Part 2, CRC Press, Boca Raton, FL, 1986; pp. 324-331.

1. Petterson, H. E., *J. Opt. Soc. Am.*, 63, 1243, 1973.
2. Burstein, E. and Smith, P. L., *Phys. Rev.*, 74, 229, 1948.
3. Pakhnev, A. V., et al., *Sov. Phys. J. (transl.)*, 18, 1662, 1975.
4. Feldman, A., Horovitz, D., and Waxler, R. M., *Appl. Opt.*, 16, 2925, 1977.
5. Iyengar, K. S., *Nature (London)*, 176, 1119, 1955.
6. Bansigir, K. G. and Iyengar, K. S., *Acta Crystallogr.*, 14, 727, 1961.
7. Pakhev, A. V., et al., *Sov. Phys. J. (transl.)*, 20, 648, 1975.
8. Bansigir, K. G., *Acta Crystallogr.*, 23, 505, 1967.
9. Krishna Rao, K. V. and Krishna Murty, V. G., *Ind. J. Phys.*, 41, 150, 1967.
10. Weil, R. and Sun, M. J., *Proc. Int. Symp. CdTe (Detectors)*, Strasbourg Centre de Rech. Nucl., 1971, XIX-1 to 6, 1972.
11. Schmidt, E. D. D. and Vedam, K., *J. Phys. Chem. Solids*, 27, 1563, 1966.
12. Biegelsen, D. K., et al., *Phys. Rev. B*, 14, 3578, 1976.
13. Helwege, K. H., *Landolt-Börnstein, New Series Group III*, Vol. II, Springer-Verlag Berlin, 1979.
14. Feldman, A., Waxler, R. M., and Horovitz, D., *J. Appl. Phys.*, 49, 2589, 1978.
15. Dixon, R. W., *J. Appl. Phys.*, 38, 5149, 1967.
16. Shabin, O. V., et al., *Sov. Phys. Sol. State (transl.)*, 13, 3141, 1972.
17. Reintjes, J. and Schultz, M. B., *J. Appl. Phys.*, 39, 5254, 1968.
18. Rivoallan, L. and Favre, F., *Opt. Commun.*, 8, 404, 1973.
19. Rivoallan, L. and Favre, F., *Opt. Commun.*, 11, 296, 1974.
20. Afanasev, I. I., et al., *Sov. J. Opt. Technol.*, 46, 663, 1979.
21. Rand, S. C., et al., *Phys. Rev. B*, 19, 4205, 1979.
22. Sipe, J. E., *Can J. Phys.*, 56, 199, 1978.
23. Christyi, I. L., et al., *Sov. Phys. Sol. State (transl.)*, 17, 922, 1975.
24. Narasimhamurty, T. S., *Curr. Sci. (India)*, 23, 149, 1954.
25. Smith, T. M. and Korpel, A., *IEEE J. Quant. Electron.*, QE-1, 283, 1965.
26. Narasimhamurty, T. S., *Proc. Ind. Acad. Sci.*, A40, 164, 1954.
27. Rabman, A., *Bhagarantam Commem. Vol.*, Bangalore Print. and Publ., 173, 1969.
28. Eppendahl, R., *Ann. Phys. (IV)*, 61, 591, 1920.
29. Laurenti, J. P. and Rouzeyre, M., *J. Appl. Phys.*, 52, 6484, 1981.
30. Sasaki, H., et al., *J. Appl. Phys.*, 47, 2046, 1976.
31. Uchida, N. and Saito, S., *J. Appl. Phys.*, 43, 971, 1972.

ELASTO-OPTIC COEFFICIENTS (PHOTOELASTIC CONSTANTS) (continued)

32. Waxler, R. M. and Farabaugh, E. M., *J. Res. Natl. Bur. Stand.*, A74, 215, 1970.
33. Nelson, D. F., Lazay, P. D., and Lax, M., *Phys. Rev.*, B6, 3109, 1972.
34. O'Brien, R. J., Rosasco, G. J., and Weber, A., *J. Opt. Soc. Am.*, 60, 716, 1970.
35. Avakyants, L. P., et al., *Sov. Phys.*, 18, 1242, 1976.
36. Sapriel, J., *Appl. Phys. Litt.*, 19, 533, 1971.
37. Narasimhamurty, T. S., *J. Opt. Soc. Am.*, 59, 682, 1969.
38. Zubrinov, I. I., et al., *Sov. Phys. Sol. State (transl.)*, 15, 1921, 1974.
39. Kachalov, O. V. and Shpilko, I. O., *Sov. Phys. JETP (transl.)*, 35, 957, 1972.
40. Narasimhamurty, T. S., et al., *J. Mater. Sci.*, 8, 577, 1973.
41. Tada, K. and Kikuchi, K., *Jpn. J. Appl. Phys.*, 19, 1311, 1980.
42. Aleksandrov, K. S., et al., *Sov. Phys. Sol. State (transl.)*, 19, 1090, 1977.
43. Afanasev, I. I., et al., *Sov. Phys. Sol. State (transl.)*, 17, 2006, 1975.
44. Silvestrova, I. M., et al., *Sov. Phys. Cryst. (transl.)*, 20, 649, 1975.
45. Veerabhadra Rao, K. and Narasimhamurty, T. S., *J. Mater. Sci.*, 10, 1019, 1975.
46. Venturini, E. L., et al., *J. Appl. Phys.*, 40, 1622, 1969.
47. Vehida, N. and Ohmachi, Y., *J. Appl. Phys.*, 40, 4692, 1969.
48. Grimsditch, M. H. and Ramdus, A. K., *Phys. Rev. B*, 22, 4094, 1980.
49. Schinke, D. P. and Viehman, W., unpublished Data.
50. Coquin, G. A., et al., *J. Appl. Phys.*, 42, 2162, 1971.
51. Vasquez, F., et al., *J. Phys. Chem. Solids*, 37, 451, 1976.
52. Luspin, Y. and Hauret, G., *C.R.Ac. Sci. Paris*, B274, 995 1972.
53. Narasimhamurty, T. S., *Phys. Rev.*, 186, 945, 1969.
54. Haussühl, S. and Weber, H. J., *Z. Kristall.*, 132, 266, 1970.
55. Vedam, K., *Proc. Ind. Ac. Sci.*, A34, 161, 1951.
56. Yano, T., Fukumoto, A., and Watanabe, A., *J. Appl. Phys.*, 42, 3674, 1971.
57. Manenkov, A. A. and Ritus, A. I., *Sov. J. Quant. Electr.*, 8, 78, 1978.
58. Eschler, H. and Weidinger, F., *J. Appl. Phys.*, 46, 65, 1975.

LINEAR ELECTRO-OPTIC COEFFICIENTS

Name

Cubic ($\bar{4}$3m)	Formula	$\lambda/\mu m$	r_{41} pm/V
Cuprous bromide	CuBr	0.525	0.85
Cuprous chloride	CuCl	0.633	3.6
Cuprous iodide	CuI	0.55	-5.0
Eulytite (BSO)	$Bi_4Si_3O_{12}$	0.63	0.54
Germanium eulytite (BGO)	$Bi_4Ge_3O_{12}$	0.63	1.0
Gallium arsenide	GaAs	10.6	1.6
Gallium phosphide	GaP	0.56	-1.07
Hexamethylenetetramine	$C_6H_{12}N_4$	0.633	0.78
Sphalerite	ZnS	0.65	2.1
Zinc selenide	ZnSe	0.546	2.0
Zinc telluride	ZnTe	3.41	4.2
Cadmium telluride	CdTe	3.39	6.8

Cubic (23)	Formula	$\lambda/\mu m$	r_{41} pm/V
Ammonium chloride (77 K)	NH_4Cl	—	1.5
Ammonium cadmium langbeinite	$(NH_4)_2Cd_2(SO_4)_3$	0.546	0.70
Ammonium manganese langbeinite	$(NH_4)_2Mn_2(SO_4)_3$	0.546	0.53
Thallium cadmium langbeinite	$Tl_2Cd_2(SO_4)_3$	0.546	0.37
Potassium magnesium langbeinite	$K_2Mg_2(SO_4)_3$	0.546	0.40
Bismuth monogermanate	$Bi_{12}GeO_{20}$	—	3.3
Bismuth monosilicate	$Bi_{12}SiO_{20}$	—	3.3
Sodium chlorate	$NaClO_3$	0.589	0.4
Sodium uranyl acetate	$NaUO_2(CH_3COO)_3$	0.546	0.87

Name

Cubic (23)	Formula	$\lambda/\mu m$	r_{41} pm/V
Trenhydrobromide	$N(CH_2CH_2NH_2)_3$ 3HBr	—	1.5
Trenhydrochloride	$N(CH_2CH_2NH_2)_3$ 3HCl	—	1.7

Tetragonal ($\bar{4}$2m)	Formula	T_{tran} K	r_{41} pm/V	r_{63} pm/V
Ammonium dihydrogen phosphate (ADP)	$NH_4H_2PO_4$	148	24.5	-8.5
Ammonium dideuterium phosphate (AD*P)	$NH_4D_2PO_4$	242	—	11.9
Ammonium dihydrogen arsenate (ADA)	$NH_4H_2AsO_4$	—	—	9.2
Cesium dihydrogen arsenate (CsDA)	CsH_2AsO_4	143	—	18.6
Cesium dideuterium arsenate (CsD*A)	CsD_2AsO_4	212	—	36.6
Potassium dihydrogen phosphate (KDP)	KH_2PO_4	123	8.6	-10.5
Potassium dideuterium phosphate (KD*P)	KD_2PO_4	222	8.8	23.8
Potassium dihydrogen arsenate (KDA)	KH_2AsO_4	97	12.5	10.9
Potassium dideuterium arsenate (KD*A)	KD_2AsO_4	162	—	18.2
Rubidium dihydrogen phosphate (RDP)	RbH_2PO_4	147	—	15.5
Rubidium dihydrogen arsenate (RDA)	RbH_2AsO_4	110	—	13.0
Rubidium dideuterium arsenate (RD*A)	RbD_2AsO_4	178	—	21.4

Tetragonal (4mm)	Formula	T_{tran} K	r_{13} pm/V	r_{33} pm/V	r_{51} pm/V
Barium titanate	$BaTiO_3$	406	8	28	—
Potassium lithium niobate	$K_3Li_2Nb_5O_{15}$	693	8.9	5.9	—
Lead titanate	$PbTiO_3$	765	13.8	5.9	—
Strontium barium niobate (SBN75)	$Sr_{0.75}Ba_{0.25}Nb_2O_6$	330	6.7	1340	42
Strontium barium niobate (SBN46)	$Sr_{0.46}Ba_{0.54}Nb_2O_6$	602	~180	35	—

Hexagonal (6mm)	Formula	r_{13} pm/V	r_{33} pm/V	r_{42} pm/V	r_{51} pm/V
Greenockite	CdS	3.1	2.9	2.0	3.7
Greenockite (const. strain)	CdS	1.1	2.4	—	—
Wurzite	ZnS	0.9	1.8	—	—
Zincite	ZnO	-1.4	+2.6	—	—

Hexagonal (6)	Formula	r_{13} pm/V	r_{33} pm/V	r_{42} pm/V	r_{51} pm/V
Lithium iodate	$LiIO_3$	4.1	6.4	1.4	3.3
Lithium potassium sulfate	$LiKSO_4$	$r_{13}-r_{33} = 1.6$	—	—	—

Trigonal (3m)	Formula	T_{tran} K	r_{13} pm/V	r_{22} pm/V	r_{33} pm/V	r_{42} pm/V
Cesium nitrate	$CsNO_3$	425	—	0.43	—	—
Lithium niobate	$LiNbO_3$	1483	8.6	7.0	30.8	28
Lithium tantalate	$LiTaO_3$	890	8.4	—	30.5	—
Lithium sodium sulfate	$LiNaSO_4$	—	—	<0.02	—	—
Tourmaline	—	—	—	0.3	—	—

Trigonal (32)	Formula	T_{tran} K	r_{11} pm/V	r_{41} pm/V
Cesium tartrate	$Cs_2C_4H_4O_6$	—	1.0	—
Cinnabar	HgS	659	3.1	1.5

Name Trigonal (32)	Formula	T_{tran} K	r_{11} pm/V	r_{41} pm/V
Potassium dithionate	$K_2S_2O_6$	—	0.26	—
Strontium dithionate	$SrS_2O_6 \cdot 4H_2O$	—	0.1	—
Quartz	SiO_2	1140	-0.47	0.2
Selenium	Se	398	2.5	

Name Orthorhombic (222)	Formula	T_{tran} K	r_{41} pm/V	r_{52} pm/V	r_{63} pm/V
Ammonium oxalate	$(NH_4)_2C_2O_4 \cdot 4H_2O$	—	230	330	250
Rochelle salt	$KNaC_4H_4O_6 \cdot 4H_2O$	$T_u = 297$ $T_l = 255$	-2.0	-1.7	+0.32

Name Orthorhombic (mm2)	Formula	T_{trans} K	r_{13} pm/V	r_{23} pm/V	r_{33} pm/V	r_{42} pm/V	r_{51} pm/V
Barium sodium niobate (BSN)	Ba_2NaNbO_{15}	833	15	13	48	92	90
Potassium niobate	$KNbO_3$	476	28	1.3	64	380	105

Name Monoclinic (2)	Formula	T_{trans} K	r_{22} pm/V	r_{32} pm/V
Calcium pyroniobate	$Ca_2Nb_2O_7$	—	0.33	13.7
Triglycine sulfate (TGS)	$(NH_2CH_2COOH)_3 \cdot H_2SO_4$	322	7.2	13.6

REFERENCES

1. Narasimhamurty, T. S., *Photoelastic and Electro-Optic Properties of Crystals*, Plenum Press, New York, 1981, pp. 405-407.
2. Weber, M. J., Ed., *CRC Handbook of Laser Science and Technology*, Vol. IV, CRC Press, Boca Raton, FL, 1986, pp. 258-278.

QUADRATIC ELECTRO-OPTIC COEFFICIENTS
Kerr Constants of Ferroelectric Crystals[1,2]

Name	Formula	T_{tran} K	λ μm	g_{11} 10^{10} esu	g_{12} 10^{10} esu	g_{11}-g_{12} 10^{10} esu	g_{44} 10^{10} esu
Barium titanate	$BaTiO_3$	406	0.633	1.33	-0.11	1.44	
Strontium titanate	$SrTiO_3$	—	0.633	—	—	1.56	—
Potassium tantalate niobate	$KTa_{0.65}Nb_{0.35}O_3$	330	0.633	1.50	-0.42	1.92	1.63
Potassium tantalate	$KTaO_3$	13	0.633	—	—	1.77	1.33
Lithium niobate	$LiNbO_3$	1483	—	0.94	0.25	0.7	0.6
Lithium tantalate	$LiTaO_3$	938	—	1.0	0.17	0.8	0.7
Barium sodium niobate (BSN)	$Ba_{0.8}Na_{0.4}Nb_2O_6$	833	—	1.55	0.44	1.11	

Kerr Constants of Selected Liquids[2]
K is the Kerr constant at a wavelength of 589 nm and at room temperature; ε is the static dielectric constant; T_m is the melting point; and T_b is the normal boiling point

Name	Molecular formula	K 10^{-7} esu	ε	T_m °C	T_b °C
Carbon disulfide	CS_2	+3.23	2.63	-111.5	+46.3
Acetone	C_3H_6O	+16.3	21.0	-94.8	+56.1
Methyl ethyl ketone	C_4H_8O	+13.6	18.56	-86.67	+79.6

QUADRATIC ELECTRO-OPTIC COEFFICIENTS (continued)
Kerr Constants of Selected Liquids (continued)[2]

Name	Molecular formula	K 10^{-7} esu	ε	T_m °C	T_b °C
Pyridine	C_5H_5N	+20.4	13.26	-42	+115.23
Ethyl cyanoacetate	$C_5H_7NO_2$	+38.8	31.6	-22.5	205
o-Dichlorobenzene	$C_6H_4Cl_2$	+42.6	10.12	-16.7	180
Benzenesulfonyl chloride	$C_6H_5ClO_2S$	+89.9	28.90	+14.5	247
Nitrobenzene	$C_6H_5NO_2$	+326	35.6	+5.7	210.8
Ethyl 3-aminocrotonate	$C_6H_{11}NO_2$	+31.0	—	+33.9	210
Paraldehyde	$C_6H_{12}O_3$	-23.0	14.7 12.0[a]	+12.6	124
Benzaldehyde	C_7H_6O	+80.8	17.85 14.1[a]	-26	179.05
p-Chlorotoluene	C_7H_7Cl	+23.0	6.25	+7.5	162.4
o-Nitrotoluene	$C_7H_7NO_2$	+174	26.26	-10	222.3
m-Nitrotoluene	$C_7H_7NO_2$	+177	24.95	+15.5	232
p-Nitrotoluene	$C_7H_7NO_2$	+222	22.2	+51.6	238.3
Benzyl alcohol	C_7H_8O	-15.4	11.92 10.8[a]	-15.3	205.8
m-Cresol	C_7H_8O	+21.2	12.44 5.0[a]	+11.8	202.27
m-Chloroacetophenone	C_8H_7ClO	+69.1			
Acetophenone	C_8H_8O	+66.6	17.44 15.8[a]	+19.7	202.3
Quinoline	C_9H_7N	+15.0	9.16	-14.78	237.16
Ethyl salicylate	$C_9H_{10}O_3$	+19.6	8.48	+1.3	231.5
Carvone	$C_{10}H_{14}O$	+23.6	11.2	<0	230
Ethyl benzoylacetate	$C_{11}H_{12}O_3$	+16.0	13.50	<0	270
Water	H_2O	+4.0	80.10	0.00	100.0

[a] Dielectric constant at radiofrequencies (10^8-10^9 Hz).

REFERENCES

1. Narasimhamurty, T. S., *Photoelastic and Electro-Optic Properties of Crystals*, Plenum Press, New York, 1981, p. 408.
2. Gray, D. E., Ed., *AIP Handbook of Physics*, McGraw Hill, New York, 1972, p. 6-241.

MAGNETO-OPTIC CONSTANTS
Verdet Constants of Non-Magnetic Crystals[1]
V is the Verdet constant; n is the refractive index; and λ is the wavelength

Material	T K	λ nm	n	V min/Oe cm
Al_2O_3	300	546.1	1.771	0.0240
	300	589.3	1.768	0.0210
$BaTaO_3$	403	427		0.95
	403	496		0.38
	403	620		0.18
	403	826		0.072
$Bi_4Ge_3O_{12}$	300	442	2.077	0.289
	300	632.8	2.048	0.099
	300	1064	2.031	0.026
C (diamond)	300	589.3	2.417	0.0233
$CaCO_3$	300	589.3	1.658	0.019
CaF_2	300	589.3	1.434	0.0088
$Cd_{0.55}Mn_{0.45}Te$	300	632.8		6.87

MAGNETO-OPTIC CONSTANTS (continued)
Verdet Constants of Non-Magnetic Crystals (continued)[1]

Material	T K	λ nm	n	V min/Oe cm
CuCl	300	546.1	1.93	0.20
GaSe	298	632.8		0.80
KAl(SO$_4$)$_2$·12H$_2$O	300	589.3	1.456	0.0124
KBr	300	546.1	1.564	0.0500
	300	589.3	1.560	0.0425
KCl	300	589.3	1.490	0.0275
KI	300	546.1	1.673	0.083
	300	589.3	1.666	0.070
KTaO$_3$	296	352		0.44
	296	413		0.19
	296	496		0.096
	296	620		0.051
	296	826		0.022
LaF$_3$	300	325	1.639	0.054
(H‖c)	300	442	1.615	0.028
	300	632.8	1.601	0.012
	300	1064	1.592	0.006
MgAl$_2$O$_4$	300	589.3	1.718	0.021
NH$_4$AlSO$_4$·12H$_2$O	300	589.3	1.459	0.0128
NH$_4$Br	300	589.3	1.711	0.0504
NH$_4$Cl	300	546.1		0.0410
	300	589.3	1.643	0.0362
NaBr	300	546.1		0.0621
NaCl	300	546.1		0.0410
	300	589.3	1.544	0.0345
NaClO$_3$	300	546.1		0.0105
	300	589.3	1.515	0.0081
NiSO$_4$·6H$_2$O	297	546.1		0.0256
	297	589.3	1.511	0.0221
SiO$_2$	300	546.1	1.546	0.0195
	300	589.3	1.544	0.0166
SrTiO$_3$	298	413	2.627	0.78
	298	496		0.31
	298	620		0.14
	298	826		0.066
ZnS	300	546.1		0.287
	300	589.3	2.368	0.226
ZnSe	300	476	2.826	1.50
	300	496	2.759	1.04
	300	514	2.721	0.839
	300	587	2.627	0.529
	300	632.8	2.592	0.406

Verdet Constants of Rare-Earth Aluminum Garnets at Various Wavelengths[1]
The absorption coefficient α for these materials ranges from 0.2 to 0.6 cm^{-1} at 300 K

Material	T/K	V in min/Oe cm							
		λ = 405 nm	450 nm	480 nm	520 nm	546 nm	578 nm	635 nm	670 nm
Tb$_2$Al$_5$O$_{12}$	300	-2.266	-1.565	-1.290	-1.039	-0.912	-0.787	-0.620	-0.542
	77		-102.16	-83.45	-3.425	-3.051	-2.603	-2.008	-1.815
	4.2				-64.80	-58.35	-53.77	48.39	-45.15
	1.45		-200.95	-172.52	-139.28	-125.07	-111.27	97.47	-93.42
Dy$_3$Al$_5$O$_{12}$	300	-1.241	-0.942	-0.803	-0.667	-0.592	-0.518	-0.411	-0.359
Ho$_3$Al$_5$O$_{12}$	300	-0.709	-0.320	-0.260	-0.335	-0.304	-0.299		-0.206

MAGNETO-OPTIC CONSTANTS (continued)
Verdet Constants of Rare-Earth Aluminum Garnets at Various Wavelengths (continued)[1]

Material	T/K	λ = 405 nm	450 nm	480 nm	520 nm	546 nm	578 nm	635 nm	670 nm
					V in min/Oe cm				
$Er_3Al_5O_{12}$	300	-0.189	-0.240	-0.154	-0.162	-0.157	-0.145	-0.105	-0.089
$Tm_3Al_5O_{12}$	300	+0.151	+0.103	+0.093	0.076	0.069	+0.059	+0.048	
$Yb_3Al_5O_{12}$	298	0.287	0.215	0.186	0.140	0.133	0.116	0.094	
	77	0.718	0.540	0.481	0.393	0.342	0.302	0.239	

Verdet Constants for KDP-Type Crystals[1]
Measurements refer to T = 298 K and
λ = 632.8 nm, with $k \parallel$ [001]

Material	V min/Oe cm
KH_2PO_4 (KDP)	0.0124
$KH_{0.3}D_{1.7}PO_4$ (KD*P)	0.145
$NH_4H_2PO_4$ (ADP)	0.138
KH_2AsO_4 (KDA)	0.238
$KH_{0.1}D_{1.9}AsO_4$ (KD*A)	0.245
$NH_4H_2AsO_4$ (ADH)	0.244

Verdet Constants of Gases[2]
Values refer to T = 0°C and P = 101.325 kPa (760 mmHg); n_D is the refractive index at a wavelength of 589 nm

Gas	$(n_D - 1) \times 10^3$	$10^6 \times V$ min/Oe cm
He	0.036	+0.40
Ar	2.81	+9.36
H_2		+6.29
N_2	0.297	+6.46
O_2	0.272	+5.69
Air	0.293	+6.27
Cl_2	0.773	+31.9
HCl	0.447	+21.5
H_2S	0.63	+41.5
NH_3	0.376	+19.0
CO	0.34	+11.0
CO_2	0.45	+9.39
NO	0.297	-58
CH4	0.444	+17.4
n-C4H10		+44.0

Verdet Constants of Liquids[2]
n_D is the refractive index at a wavelength of 589 nm and a temperature of 20°C, unless otherwise indicated. V is the Verdet constant

Liquid	λ/nm	T/°C	$10^2 \times V$ min/Oe cm	n_D
P	589	33	+13.3	
S	589	114	+8.1	1.929 (110°C)
H_2O	589	20	+1.309	1.3328
D_2O	589	19.7	+1.257	1.3384
H_3PO_4	578	97.4	+1.35	
CS_2	589	20	+4.255	1.6255
CCl_4	578-589	25.1	+1.60	1.463 (15°C)
$SbCl_5$	578	18	+7.45	1.601 (14°C)
$TiCl_4$	578	17	-1.65	1.61
$TiBr_4$	578	46	-5.3	
Methanol	589	18.7	+0.958	1.3289
Acetone	578-589	20.0	+1.116	1.3585
Toluene	578-589	15.0	+2.71	1.4950
Benzene	578-589	15.0	+3.00	1.5005
Chlorobenzene	589	15	+2.92	1.5246
Nitrobenzene	589	15	+2.17	1.5523
Bromoform	589	17.9	+3.13	1.5960

MAGNETO-OPTIC CONSTANTS (continued)

Verdet Constants of Rare Earth Paramagnetic Crystals[1]

n is the refractive index, and V is the Verdet constant at the wavelength and temperature indicated

Rare Earth	Host	T/K	λ/nm	n	V min/Oe cm
Ce^{3+}(30%)	CaF_2	300	325	1.516	-0.956
		300	442	1.502	-0.297
		300	633	1.494	-0.111
		300	1064	1.489	-0.035
Ce^{3+}	CeF_3	300	442	1.613	-1.05
		300	633	1.598	-0.406
		77	633		-1.418
		300	1064	1.589	-0.113
Pr^{3+}(5%)	CaF_2	300	266	1.471	-0.172
		300	325	1.461	-0.0818
		300	442	1.451	-0.0089
		300	633	1.445	-0.0168
		300	1064	1.441	-0.0045
Nd^{3+}(2.9%)	CaF_2	4.2	426		-0.19
Nd^{3+}	NdF_3	300	442	1.60	-0.553
		290	633	1.59	-0.209
		77	633		-0.755
		300	1064	1.58	-0.097
Eu^{3+}(3%)	CaF_2	4.2	430		29
		4.2	440		22
Eu^{2+}	EuF_2	300	450		-4.5
		300	500		-2.6
		300	550		-1.6
		300	600		-1.1
		300	650		-0.8
		300	1064		-0.19
Tb^{3+}	KTb_3F_{10}	300	325	1.531	-2.174
		300	442	1.518	-0.933
		300	633	1.510	-0.386
		77	633		-1.94
		300	1064	1.505	-0.114
Tb^{3+}	$LiTbF_4$	300	325	1.493	-1.9
		300	442	1.481	-0.98
		300	633	1.473	-0.44
		300	1064	1.469	-0.13
Tb^{3+}	$Tb_3Ga_5O_{12}$	300	500	1.989	-0.749
		300	570	1.981	-0.581
		300	633	1.976	-0.461
		300	830	1.967	-0.21
		300	1060	1.954	-0.12

MAGNETO-OPTIC CONSTANTS (continued)

Verdet Constants of Paramagnetic Glasses[1]

The Verdet constant V is given at room temperature for the wavelengths indicated

Rare earth phosphate glasses of composition $R_2O_3 \cdot xP_2O_5$, where x is given in the second column

		Verdet constant V in min/Oe cm									
R	x	λ=405 nm	λ=436 nm	λ=480 nm	λ=500 nm	λ=520 nm	λ=546 nm	λ=578 nm	λ=600 nm	λ=635 nm	λ=670 nm
La		0.037	0.030	0.024	0.022	0.020	0.018	0.015	-0.014	0.013	—
Ce	2.67	-0.672	0.510	-0.366	-0.326	-0.287	-0.253	-0.217	-0.197	-0.173	-0.150
Pr	3.09	-0.447	-0.332	-0.283	-0.261	-0.236	-0.208	-0.182	-0.170	-0.150	-0.132
Nd	2.92	-0.250	-0.209	-0.167	-0.155	-0.136	-0.134	-0.094	-0.080	-0.080	-0.071
Sm	2.87	0.026	0.024	0.020	0.020	0.017	0.015	0.014	0.012	0.011	0.010
Eu	2.93	-0.025	-0.017	-0.010	-0.006	-0.006	-0.005	-0.004	-0.003	-0.002	-0.002
Gd	3.01	0.018	0.015	0.014	0.012	0.012	0.011	0.011	0.010	0.009	0.009
Tb	2.94	-0.560	-0.458	-0.357	-0.323	-0.295	-0.261	-0.226	-0.206	-0.190	-0.164
Dy	2.51	-0.540	-0.453	-0.359	-0.331	-0.301	0.268	-0.237	-0.217	-0.197	-0.173
Ho	2.94	-0.299	-0.313	-0.156	-0.153	-0.138	-0.138	-0.119	-0.110	-0.098	-0.084
Er	3.01	-0.139	-0.121	-0.100	-0.111	-0.095	-0.062	-0.060	-0.057	-0.051	-0.044
Tm	2.79	0.019	0.013	0.012	0.009	0.008	0.006	0.005	0.004	0.004	0.007
Yb	3.01	0.087	0.072	0.056	0.050	0.045	0.041	0.036	0.032	0.029	0.024

The following are rare earth borate glasses with composition:
for La and Pr: $R_2O_3 \cdot xP_2O_5$; for Tb-Pr and Dy-Pr: $R_2O_3 \cdot xB_2O_3$; and for other elements: $R_2O_3 \cdot 0.85La_3O_3 \cdot xB_2O_3$.

R	x	λ=405 nm	λ=436 nm	λ=480 nm	λ=500 nm	λ=520 nm	λ=546 nm	λ=578 nm	λ=600 nm	λ=635 nm	λ=670 nm
La	3.04	0.043	0.036	0.029	0.026	0.023	0.022	0.019	0.018	0.016	0.014
Pr-La	5.44	-0.380	-0.307	-0.230	-0.220	-0.201	-0.178	-0.153	-0.146	-0.128	-0.110
Nd-La	5.41	-0.180	-0.147	-0.120	-0.111	-0.096	-0.094	-0.100	-0.059	-0.056	-0.046
Sm-La	4.97	0.032	0.030	0.025	0.024	0.022	0.019	0.017	0.016	0.014	0.012
Eu-La	4.69	-0.081	-0.060	-0.038	-0.033	-0.029	-0.024	-0.019	-0.016	-0.014	-0.012
Gd-La	4.71	0.032	0.026	0.024	0.022	0.021	0.020	0.018	0.017	0.015	0.013
Tb-La	4.73	-0.512	-0.419	-0.319	-0.288	-0.262	-0.234	-0.205	-0.186	-0.167	-0.142
Dy-La	4.88	-0.436	-0.361	-0.299	-0.273	-0.246	-0.220	-0.193	-0.177	-0.159	-0.138
Ho-La	4.36	-0.269	-0.252	-0.123	-0.131	-0.112	-0.128	-0.104	-0.096	—	-0.074
Er-La	4.50	-0.093	-0.078	-0.068	-0.082	—	-0.045	-0.042	-0.040	-0.035	-0.034
Tm-La	4.75	0.060	0.046	0.039	0.034	0.031	0.026	0.023	0.021	0.018	0.016
Yb-La	8.58	0.115	0.094	0.073	0.066	0.060	0.054	0.046	0.043	0.037	0.033
Tb-Pr	4.99	-0.940	-0.786	-0.560	-0.536	-0.489	-0.436	-0.380	-0.348	-0.306	-0.265
Dy-Pr	4.63	-0.850	—	—	-0.497	-0.465	-0.413	-0.358	-0.332	-0.290	-0.252
Pr	2.56	-0.843	-0.646	-0.471	-0.480	-0.432	-0.390	-0.334	-0.317	-0.271	-0.243

MAGNETO-OPTIC CONSTANTS (continued)
Verdet Constants of Diamagnetic Glasses[1]
The Verdet constant V is given at room temperature for the wavelengths indicated

Glass type	Composition (wt. %)	Verdet constant V in min/Oe cm			
		$\lambda = 325$ nm	$\lambda = 442$ nm	$\lambda = 633$ nm	$\lambda = 1064$ nm
SiO_2	100% SiO_2			0.013	
B_2O_3	100% B_2O_3			0.010	
CdO	47.5% CdO, 52.5% P_2O_5	0.079	0.033	0.022	
ZnO	36.4% ZnO, 63.6% P_2O_5	0.072	0.044	0.020	
TeO_2	88.9% TeO_2, 11.1% P_2O_5		0.196	0.076	0.022
ZrF_4	63.1% ZrF_4, 14.9% BaF_2, 7.2% LaF_3, 1.9% AlF_3, 9.1% PbF_2, 3.8% LiF			0.011	

Glass type	Composition (wt. %)	$\lambda = 700$ nm	$\lambda = 853$ nm	$\lambda = 1060$ nm
Bi_2O_3	95% Bi_2O_3, 5% B_2O_3	0.086	0.051	0.033
PbO	95% PbO, 5% B_2O_3	0.093	0.061	0.031
	82% PbO, 18% SiO_2	0.077	0.045	0.027
	50% PbO, 15% K_2O, 35% SiO_2	0.032	0.020	0.011
Tl_2O	95% Tl_2O, 5% B_2O_3	0.092	0.061	0.032
	82% Tl_2O, 18% SiO_2	0.100	0.067	0.043
	50% Tl_2O, 15% K_2O, 35% SiO_2	0.036	0.022	0.012
SnO	76% SnO, 13% B_2O_3, 11% SiO_2	0.071	0.046	0.026
TeO_3	75% TeO_2, 25% Sb_2O_3	0.076	0.052	0.032
	80% TeO_2, 20% $ZnCl_2$	0.073	0.046	0.025
	84% TeO_2, 16% BaO	0.056	0.041	0.029
	70% TeO_2, 30% WO_3	0.052	0.035	0.022
	20% TeO_2, 80% PbO	0.128	0.075	0.048
Sb_2O_3	25% Sb_2O_3, 75% TeO_2	0.076	0.050	0.032
	75% Sb_2O_3, 75% Cs_2O, 5% Al_2O_3	0.074	0.044	0.025
	75% Sb_2O_3, 10% Cs_2O, 10% Rb_2O, 5% Al_2O_3	0.078	0.052	0.030

MAGNETO-OPTIC CONSTANTS (continued)
Verdet Constants of Commercial Glasses[1]

This table gives the density, ρ, refractive index at 589 nm, n_D, and Verdet constant, V, for the wavelengths indicated; the data refer to room temperature

Glass type	ρ g/cm³	n_D	V in min/Oe cm $\lambda = 365.0$ nm	$\lambda = 404.7$ nm	$\lambda = 435.8$ nm	$\lambda = 546.1$ nm	$\lambda = 578.0$ nm
BSC	2.49	1.5096	0.0499	0.0392	0.0333	0.02034	0.01798
HC	2.53	1.5189	0.0561	0.0440	0.0372	0.0225	0.01995
LBC	2.87	1.5406	0.0609	0.0477	0.0403	0.0245	0.0216
LF	3.23	1.5785	0.1143	0.0850	0.0693	0.0394	0.0344
BLF	3.48	1.6047	0.1112	0.0832	0.0685	0.0393	0.0344
DBC	3.56	1.6122	0.0662	0.0517	0.0435	0.0261	0.0231
DF	3.63	1.6203	0.1473	0.1076	0.0872	0.0485	0.0423
EDF	3.9	1.6533	0.1725	0.1248	0.1007	0.0556	0.0483

The composition of the glasses in weight percent is:

Glass type	SiO_2	B_2O_3	K_2O	CaO	Al_2O_3	As_2O_3	Na_2O	BaO	ZnO	PbO
BSC	69.6	6.7	20.5	2.9	0.3	0.1	—	—	—	—
HC	72.0	—	10.1	11.4	0.3	0.2	6.1	—	—	—
LBC	57.1	1.8	13.7	0.3	0.2	0.1	—	26.9	—	—
LF	52.5	—	9.5	0.3	0.2	0.1	—	—	—	37.6
BLF	45.2	—	7.8	—	—	0.4	—	16.0	8.3	22.2
DBC	36.2	7.7	0.2	0.2	3.5	0.7	—	44.6	6.7	—
DF	46.3	—	1.1	0.3	0.2	0.1	5.0	—	—	47.0
EDF	40.6	—	7.5	0.2	0.2	0.2	0.1	—	—	51.5

REFERENCES

1. Weber, M. J., *CRC Handbook of Laser Science and Technology*, Vol. IV, Part 2, CRC Press, Boca Raton, FL, 1988, p. 299-310.
2. Gray, D. E., Ed., *American Institute of Physics Handbook*, Third Edition, McGraw Hill, New York, 1972, p. 6-230.

FARADAY ROTATION
Ferro-, Ferri-, and Antiferromagnetic Solids

Material	T_c K	$4\pi M_s$ gauss	F deg/cm	α cm⁻¹	$2F/\alpha$	T K	λ nm
Fe	1043	21,800	4.4×10^5	6.5×10^5	1.4	300	500
			6.5×10^5	5.0×10^5	2.6	300	1000
			7×10^5	4.2×10^5	3.3	300	1500
			7×10^5	3.5×10^5	4.0	300	2000
Co	1390	18,200	2.9×10^5	—	—	300	500
			5.5×10^5	6.1×10^5	1.8	300	1000
			5.5×10^5	4.5×10^5	2.4	300	1500
			5.5×10^5	3.6×10^5	2.7	300	2000
Ni	633	6,400	0.8×10^5	—	—	300	500
			2.6×10^5	5.8×10^5	0.9	300	1000
			1.5×10^5	4.8×10^5	0.6	300	1500
			1×10^5	4.1×10^5	0.25	300	2000
Permalloy (Ni/Fe = 82/18)	803	10,700	1.2×10^5	6×10^5	0.4	300	500

Material	T_c K	$4\pi M_s$ gauss	F deg/cm	α cm^{-1}	$2F/\alpha$	T K	λ nm
Ni/Fe = 100/0		6,000	1.2×10^5	7.05×10^5	0.34	300	632.8
Ni/Fe = 80/20		10,800	2.2×10^5	7.10×10^5	0.62	300	632.8
Ni/Fe = 60/40		14,900	2.9×10^5	7.54×10^5	0.77	300	632.8
Ni/Fe = 40/60		14,400	2.2×10^5	8.17×10^5	0.54	300	632.8
Ni/Fe = 20/80		19,400	3.3×10^5	8.10×10^5	0.81	300	632.8
Ni/Fe = 0/100	639	21,600	3.5×10^5	8.13×10^5	0.86	300	632.8
MnBi		7,700	4.2×10^5	6.1×10^5	1.4	300	450
			7.5×10^5	4.2×10^5	3.6	300	900
MnAs	313	—	0.44×10^5	5.0×10^5	0.174	300	500
			0.62×10^5	4.4×10^5	0.28	300	900
CrTe	334	1015	0.5×10^5	2.0×10^5	0.5	300	550
			0.4×10^5	1.2×10^5	0.7	300	900
FeRh	333	—	0.9×10^5	3.3×10^5	0.56	348	700
$Y_3Fe_5O_{12}$ (YIG)	560	2500	2400	1500	3.2	300	555
			1250	1400	1.8	300	625
			750	450	3.3	300	770
			175	<0.06	$>3 \times 10^3$	300	5000 to 1500
$Gd_3Fe_5O_{12}$ (GdIG)	Tn = 564	7300	-2000	6000	0.6	300	500
	T = 286		-1050	900	2.3	300	600
			-300	100	6.0	300	800
			-80	70	2.3	300	1000
$NiFe_2O_4$	858	3350	2.0×10^4	5.9×10^4	0.7	300	286
			-1.0×10^4	10×10^4	0.2	300	500
			-120	38	6	300	1500
			+75	15	10	300	3000
			+110	32	7	300	5000
$CoFe_2O_4$	793	4930	2.75×10^4	12×10^4	0.5	300	286
			3.6×10^4	17×10^4	0.4	300	400
			-2.5×10^4	6×10^4	0.8	300	660
$MgFe_2O_4$	593-713[e]	1450[e]	-60	100	1	300	2500
			0	12	0	300	4000
			+35	6	11	300	6000
$Li_{0.5}Fe_{2.5}O_4$	863-953[e]	3240[e] to 3900	-440	150	6	300	1500
			+10	85	0.2	300	3000
			+110	44	5	300	5000
			+135	80	3	300	7000
$BaFe_{12}O_{19}$	723	—	-50	-38	3	300	2000
			+75	20	7.5	300	3000
			+150	20	15	300	5000
			+165	22	15	300	7000
$Ba_2Zn_2Fe_{12}O_{19}$	—	—	90	120	1.5	300	5000
			75	65	2.0	300	7000
$RbNiF_3$	220	1250	360	35	20	77	450[a]
			70	10	14	77	600[a]
			310	70	9	77	800[a]
			75	25	6	77	1000[a]
$RbNi_{0.75}Co_{0.25}F_3$	109	—	180	9	40	77	600[b]
$RbFeF_3$	102	—	3400	7	900	82	300[c]
			1600	3	1100	82	400[c]
			620	1.5	830	82	600[c]
			300	2.5	240	82	800[c]
FeF_3	365	40 at 300 K	670	14	95	300	349[d]
			180	4.4	82	300	522.5[d]
$CrCl_3$	16.8	3880	2000	200	20	1.5	410

FARADAY ROTATION (continued)
Ferro-, Ferri-, and Antiferromagnetic Solids (continued)

Material	T_c K	$4\pi M_s$ gauss	F deg/cm	α cm^{-1}	$2F/\alpha$	T K	λ nm
			-500	300	3	1.5	450
			-1000	70	30	1.5	590
CrBr$_3$	32.5	3390	3×10^5	3×10^3	200	1.5	478
			1.6×10^5	1.4×10^4	23	1.5	500
CrI$_3$	68	2690	1.1×10^5	6.3×10^3	35	1.5	970
			0.8×10^5	3×10^3	53	1.5	1000
FeBO$_3$	348	115	3200	140	45	300	500
		at 300 K	450	38	24	300	700
EuO	69	23700	-1.0×10^5	0.5×10^4	40	5	1100
			5×10^5	9.7×10^4	10	5	700
			0.5×10^5	7.8×10^4	1.3	5	500
			3×10^4	>0.5	~105	20	2500
			660	>1.0	1300	20	10600
EuS	16.3	—	-1.6×10^5	0	—	6	825
			-9.6×10^5	3.3×10^4	58	6	690
			$+5.5 \times 10^5$	1.2×10^5	9.2	6	563
EuSe	7.0	13,200	1.45×10^5	80	3600	4.2	750
			0.95×10^5	60	3170	4.2	800

[a] Measured along the C-axis (magnetic hard axis).

[b] Measured along the C-axis (magnetic easy axis).

[c] Measured along the C-axis ([100]-direction at room temperature).

[d] Strong natural birefringence interferes with the Faraday effect.

[e] Depends on heat treatment.

REFERENCE

1. Weber, M. J., Ed., *CRC Handbook of Laser Science and Technology,* Vol. IV, Part 2, CRC Press, Boca Raton, FL, 1988, pp. 288-296.

NONLINEAR OPTICAL CONSTANTS
H. P. R. Frederikse

The relation between the polarization density P of a dielectric medium and the electric field E is linear when E is small, but becomes nonlinear as E acquires values comparable with interatomic electric fields (10^5 to 10^8 V/cm). Under these conditions the relation between P and E can be expanded in a Taylor's series

$$P = \varepsilon_0 \chi^{(1)} E + 2\chi^{(2)} E^2 + 4\chi^{(3)} E^3 + \cdots \tag{1}$$

where ε_0 is the permittivity of free space, while $\chi^{(1)}$ is the linear and $\chi^{(2)}$, $\chi^{(3)}$ etc. the nonlinear optical susceptibilities.

If we consider two optical fields, the first $E_j^{\omega_1}$ (along the j-direction at frequency ω_1) and the second $E_k^{\omega_2}$ (along the k-direction at frequency ω_2) one can write the second term of the Taylor's series as follows

$$P_i(\omega_1\omega_2) = 2\chi_{ijk}^{\omega_3 = \omega_1 \pm \omega_2} E_j^{\omega_1} E_k^{\omega_2}$$

When $\omega_1 \neq \omega_2$ the (parametric) mixing of the two fields gives rise to two new polarizations at the frequencies $\omega_3 = \omega_1 + \omega_2$ and $\omega_3' = \omega_1 - \omega_2$. When the two frequencies are equal, $\omega_1 = \omega_2 = \omega$, the result is Second Harmonic Generation (SHG) $\chi_{ijk}(2\omega, \omega, \omega)$, while equal and opposite frequencies, $\omega_1 = \omega$ and $\omega_2 = -\omega$ leads to Optical Rectification (OR): $\chi_{ijk}(0,\omega,-\omega)$. In the SHG case the following convention is adopted: the second order nonlinear coefficient d is equal to one half of the second order nonlinear susceptibility

$$d_{ijk} = 1/2 \chi^{(2)}$$

Because of the symmetry of the indices j and k one can replace these two by a single index (subscript) m. Consequently the notation for the SHG nonlinear coefficient in reduced form is d_{im} where m takes the values 1 to 6. Only noncentrosymmetric crystals can possess a nonvanishing d_{ijk} tensor (third rank). The unit of the SHG coefficients is m/V (in the MKSQ/SI system).

In centrosymmetric media the dominant nonlinearity is of the third order. This effect is represented by the third term in the Taylor's series (Equation 1); it is the result of the interaction of a number of optical fields (one to three) producing a new frequency $\omega_4 = \omega_1 + \omega_2 + \omega_3$. The third order polarization is given by

$$P_j(\omega_1\omega_2\omega_3) = g_4 \chi_{jklm} E_k^{\omega_1} E_l^{\omega_2} E_m^{\omega_3}$$

Third Harmonic Generation (THG) is achieved when $\omega_1 = \omega_2 = \omega_3 = \omega$. In this case the constant $g_4 = 1/4$. The third order nonlinear coefficient C is related to the third order susceptibility as follows

$$C_{jklm} = 1/4 \chi_{jklm}$$

This coefficient is a fourth rank tensor. In the THG case the matrices must be invariant under permutation of the indices k, l, and m; as a result the notation for the third order nonlinear coefficient can be simplified to C_{jn}. The unit of C_{jn} is $m^2 \cdot V^{-2}$ (in the MKSQ/SI system).

Applications of second order nonlinear optical materials include the generation of higher (up to sixth) optical harmonics, the mixing of monochromatic waves to generate sum or difference frequencies (frequency conversion), the use of two monochromatic waves to amplify a third wave (parametric amplification) and the addition of feedback to such an amplifier to create an oscillation (parametric oscillation).

Third order nonlinear optical materials are used for THG, self-focusing, four wave mixing, optical amplification, and optical conjugation. Many of these effects — as well as the variation and modulation of optical propagation caused by mechanical, electric, and magnetic fields (see the preceeding table on "Elasto-Optic, Electro-Optic, and Magneto-Optic Constants") — are used in the areas of optical communication, optical computing, and optical imaging.

REFERENCES (NONLINEAR OPTICS)

1. *Handbook of Laser Science and Technology*, Vol. 111, Part 1; Ed.: Marvin J. Weber, Publ.: CRC Press, Inc., Boca Raton, FL, 1986.
2. Dmitriev, V.G., Gurzadyan, G.G., and Nikogosyan, D., *Handbook of Nonlinear Optical Crystals*, Springer-Verlag, Berlin, 1991.
3. Shen, Y.R., *The Principles of Nonlinear Optics*, John Wiley, New York, 1984.
4. Yariv, A., *Quantum Electronics*, 3rd edition, John Wiley, New York, 1988.
5. Bloembergen, N., *Nonlinear Optics,* W.A. Benjamin, New York, 1965.
6. Zernike F. and Midwinter, J.E., *Applied Nonlinear Optics*, John Wiley, New York, 1973.
7. Hopf, F.A. and Stegeman, G.I., *Applied Classical Electrodynamics*, Volume 2: Nonlinear Optics, John Wiley, New York, 1986.
8. *Nonlinear Optical Properties of Organic Molecules and Crystals,* Eds.: D.S. Chemla and J. Zyss, Publ.: Academic Press, Orlando, FL, 1987.
9. *Optical Phase Conjugation*, Ed.: R.A. Fisher, Publ.: Academic Press, New York, 1983.
10. Zyss, J., *Molecular Nonlinear Optics: Materials, Devices and Physics*, Academic Press, Boston, 1994.
11. Nonlinear Optics, 5 articles in *Physics Today, (Am. Inst. of Phys.)*, Vol. 47, No. 5, May, 1994.

Selected SHG Coefficients of NLO Crystals*

Material	Symmetry class	$d_{im} \times 10^{12}$ m/V	λ µm
GaAs	$\bar{4}3$ m	$d_{14} = 134.1 \pm 42$	10.6
GaP	$\bar{4}3$ m	$d_{14} = 71.8 \pm 12.3$	1.058
InAs	$\bar{4}3$ m	$d_{14} = 364 \pm 47$	1.058
		$d_{14} = 210$	10.6
ZnSe	$\bar{4}3$ m	$d_{14} = 78.4 \pm 29.3$	10.6
		$d_{36} = 26.6 \pm 1.7$	1.058
β-ZnS	$\bar{4}3$ m	$d_{14} = 30.6 \pm 8.4$	10.6
		$d_{36} = 20.7 \pm 1.3$	1.058
ZnTe	$\bar{4}3$ m	$d_{14} = 92.2 \pm 33.5$	10.6
		$d_{14} = 83.2 \pm 8.4$	1.058
		$d_{36} = 89.6 \pm 5.7$	1.058
CdTe	$\bar{4}3$ m	$d_{14} = 167.6 \pm 63$	10.6
Bi_4GeO_{12}	$\bar{4}3$ m	$d_{14} = 1.28$	1.064
$N_4(CH_2)_6$ (hexamine)	$\bar{4}3$ m	$d_{14} = 4.1$	1.06
$LiIO_3$	6	$d_{33} = -7.02$	1.06
		$d_{31} = -5.53 \pm 0.3$	1.064
ZnO	6 mm	$d_{33} = -5.86 \pm 0.16$	1.058
		$d_{31} = 1.76 \pm 0.16$	1.058
		$d_{15} = 1.93 \pm 0.16$	1.058
α-ZnS	6 mm	$d_{33} = 11.37 \pm 0.07$	1.058
		$d_{33} = 37.3 \pm 12.6$	10.6
		$d_{31} = -18.9 \pm 6.3$	10.6
		$d_{15} = 21.37 \pm 8.4$	10.6
CdS	6 mm	$d_{33} = 25.8 \pm 1.6$	1.058
		$d_{31} = -13.1 \pm 0.8$	1.058
		$d_{15} = 14.4 \pm 0.8$	1.058
CdSe	6 mm	$d_{33} = 54.5 \pm 12.6$	10.6
		$d_{31} = -26.8 \pm 2.7$	10.6
$BaTiO_3$	4 mm	$d_{33} = 6.8 \pm 1.0$	1.064
		$d_{31} = 15.7 \pm 1.8$	1.064
		$d_{15} = 17.0 \pm 1.8$	1.064
$PbTiO_3$	4 mm	$d_{33} = 7.5 \pm 1.2$	1.064
		$d_{31} = 37.6 \pm 5.6$	1.064
		$d_{15} = 33.3 \pm 5$	1.064
$K_3Li_2Nb_5O_{15}$	4 mm	$d_{33} = 11.2 \pm 1.6$	1.064
		$d_{31} = 6.18 \pm 1.28$	1.064
		$d_{15} = 5.45 \pm 0.54$	1.064
$K_{0.8}Na_{0.2}Ba_2Nb_5O_{15}$	4 mm	$d_{31} = 13.6 \pm 1.6$	1.064
$SrBaNb_5O_{15}$	4 mm	$d_{33} = 11.3 \pm 3.3$	1.064
		$d_{31} = 4.31 \pm 1.32$	1.064
		$d_{15} = 5.98 \pm 2$	1.064
$NH_4H_2PO_4$ (ADP)	$\bar{4}2$ m	$d_{36} = 0.53$	1.064
		$d_{36} = 0.85$	0.694
KH_2PO_4 (KDP)	$\bar{4}2$ m	$d_{36} = 0.44$	1.064
		$d_{36} = 0.47 \pm 0.07$	0.694
KD_2PO_4 (KD*P)	$\bar{4}2$ m	$d_{36} = 0.38 \pm 0.016$	1.058
		$d_{36} = 0.34 \pm 0.06$	0.694
		$d_{14} = 0.37$	1.058
KH_2AsO_4 (KDA)	$\bar{4}2$ m	$d_{36} = 0.43 \pm 0.025$	1.06
		$d_{36} = 0.39 \pm 0.4$	0.694
$CdGeAs_2$	$\bar{4}2$ m	$d_{36} = 351 \pm 105$	10.6
$AgGaS_2$	$\bar{4}2$ m	$d_{36} = 18 \pm 2.7$	10.6
$AgGaSe_2$	$\bar{4}2$ m	$d_{36} = 37.4 \pm 6.0$	10.6
$(NH_2)_2CO$ (urea)	$\bar{4}2$ m	$d_{36} = 1.3$	1.06
$AlPO_4$	32	$d_{11} = 0.35 \pm 0.03$	1.058
Se	32	$d_{11} = 97 \pm 25$	10.6

Material	Symmetry class	$d_{im} \times 10^{12}$ m/V	λ μm
Te	32	$d_{11} = 650 \pm 30$	10.6
SiO_2 (quartz)	32	$d_{11} = 0.335$	1.064
HgS	32	$d_{11} = 50.3 \pm 17$	10.6
$(C_6H_5CO)_2$ [benzil]	32	$d_{11} = 3.6 \pm 0.5$	1.064
β-BaB_2O_4 [BBO]	3 m	$d_{22} = 2.22 \pm 0.09$	1.06
		$d_{31} = 0.16 \pm 0.08$	1.06
$LiNbO_3$	3 m	$d_{33} = 34.4$	1.06
		$d_{31} = -5.95$	1.06
		$d_{22} = 2.76$	1.06
$LiTaO_3$	3 m	$d_{33} = -16.4 \pm 2$	1.058
		$d_{31} = -1.07 \pm 0.2$	1.058
		$d_{22} = +1.76 \pm 0.2$	1.058
Ag_3AsS_3 [proustite]	3 m	$d_{31} = 11.3 \pm 2.5$	10.6
		$d_{22} = 18.0 \pm 2.5$	10.6
Ag_3SbS_3 [pyrargerite]	3m	$d_{31} = 12.6 \pm 4$	10.6
		$d_{22} = 13.4 \pm 4$	10.6
α-HIO_3	222	$d_{36} = 5.15 \pm 0.16$	1.064
$NO_2 \cdot CH_3NOC_5H_4 \cdot$ (POM)	222	$d_{36} = 6.4 \pm 1.0$	1.064
$Ba_2NaNb_5O_{15}$ [Banana]	mm 2	$d_{33} = -17.6 \pm 1.28$	1.064
		$d_{31} = -12.8 \pm 1.28$	1.064
$C_6H_4(NO_2)_2$ [MDB]	mm 2	$d_{33} = 0.74$	1.064
		$d_{32} = 2.7$	1.064
		$d_{31} = 1.78$	1.064
$Gd_2(MoO_4)_3$	mm 2	$d_{33} = -0.044 \pm 0.008$	1.064
		$d_{32} = +2.42 \pm 0.36$	1.064
		$d_{31} = -2.49 \pm 0.37$	1.064
$KNbO_3$	mm 2	$d_{33} = -19.58 \pm 1.03$	1.064
		$d_{32} = +11.34 \pm 1.03$	1.064
		$d_{31} = -12.88 \pm 1.03$	1.064
$KTiOPO_4$ [KTP]	mm 2	$d_{33} = 13.7$	1.06
		$d_{32} = \pm 5.0$	1.06
		$d_{31} = \pm 6.5$	1.06
$NO_2C_6H_4 \cdot NH_2$ [mNA]	mm 2	$d_{33} = 13.12 \pm 1.28$	1.064
		$d_{32} = 1.02 \pm 0.22$	1.064
		$d_{31} = 12.48 \pm 1.28$	1.064
$C_{10}H_{12}N_3O_6$ [MAP]	2	$d_{23} = 10.67 \pm 1.3$	1.064
		$d_{22} = 11.7 \pm 1.3$	1.064
		$d_{21} = 2.35 \pm 0.5$	1.064
		$d_{25} = -0.35 \pm 0.3$	1.064
$(NH_2CH_2COOH)_3H_2SO_4$ [TGS]	2	$d_{23} = 0.32$	0.694

* These data are taken from References 1 and 2.

Selected THG Coefficients of Some NLO Materials*

Material	NLO process	$C_{jn} \times 10^{20}$ m²/V⁻²	λ μm
$NH_4H_2PO_4$ [ADP]	$(-3\omega,\omega,\omega,\omega)$	$C_{11} = 0.0104$	1.06
		$C_{18} = 0.0098$	1.06
C_6H_6 [benzene]	$(-3\omega,\omega,\omega,\omega)$	$C_{11} = 0.0184 \pm 0.0042$	1.89
$CdGeAs_2$	$(-3\omega,\omega,\omega,\omega)$	$C_{11} = 182 \pm 84$	10.6
p-type: 5×10^{16} cm⁻³		$C_{16} = 175$	10.6
		$C_{18} = -35$	10.6
$C_{40}H_{56}$ [β-carotene]	$(-3\omega,\omega,\omega,\omega)$	$C_{11}\ 0.263 \pm 0.08$	1.89
GaAs	$(-3\omega,\omega,\omega,-\omega)$	$C_{11} = 62 \pm 31$	1.06
high-resistivity			
Ge	$(-3\omega,\omega,\omega,-\omega)$	$C_{11} = 23.5 \pm 12$	1.06
$LiIO_3$	$(-3\omega,\omega,\omega,-\omega)$	$C_{12} = 0.2285$	1.06
		$C_{35} = 6.66 \pm 1$	1.06
KBr	$(-3\omega,\omega,\omega,-\omega)$	$C_{11} = 0.0392$	1.06
		$C_{18}/C_{11} = 0.3667$	1.06
KCl	$(-3\omega,\omega,\omega,-\omega)$	$C_{11} = 0.0168$	1.06
		$C_{18}/C_{11} = 0.28$	1.06
KH_2PO_4 [KDP]	$(-3\omega,\omega,\omega,-\omega)$	$C_{11}-3C_{18} = 0.04$	1.06
Si	$(-3\omega,\omega,\omega,-\omega)$	$C_{11} = 82.8 \pm 25$	1.06
p-type: 10^{14} cm⁻³			
NaCl	$(-3,\omega,\omega,\omega,-\omega)$	$C_{11} = 0.0168$	1.06
		$C_{18}/C_{11} = 0.4133$	1.06
NaF	$(-3\omega,\omega,\omega,-\omega)$	$C_{11} = 0.0035$	1.06

* These data are taken from Reference 1.

PHASE DIAGRAMS
H. P. R. Frederikse

A phase is a structurally homogeneous portion of matter. Regardless of the number of chemical constituents of a gas, there is only one vapor phase. This is true also for the liquid form of a pure substance, although a mixture of several liquid substances may exist as one or several phases, depending on the interactions among the substances. On the other hand a pure solid may exist in several phases at different temperatures and pressures because of differences in crystal structure (Reference 1). At the phase transition temperature, T_{tr}, the chemical composition of the solid remains the same, but often a change in the physical properties will take place. Such changes are found in ferroelectric crystals (example $BaTiO_3$) which develop a spontaneous polarization below T_{tr}, in superconductors (example Pb) which loose all electrical resistance below the transition point, and in many other classes of solids.

In quite a few cases it is difficult to bring about the phase transition, and the high- (or low-) temperature phase persists in its metastable form. Many liquids remain in the liquid state for shorter or longer periods of time when cooled below the melting point (supercooling). However, often the slightest disturbance will cause solidification. Persistence of the high temperature phase in solid-solid transitions is usually of much longer duration. An example of this behavior is found in white tin; although gray tin is the thermodynamically stable form below T_{tr} (286.4 K), the metal remains in its undercooled, white tin state all the way to $T = 0$ K, and crystals of gray tin are very difficult to produce.

A *phase diagram* is a map which indicates the areas of stability of the various phases as a function of external conditions (temperature and pressure). Pure materials, such as mercury, helium, water, and methyl alcohol are considered one-component systems and they have *unary* phase diagrams. The equilibrium phases in two-component systems are presented in *binary* phase diagrams. Because many important materials consist of three, four, and more components, many attempts have been made to deduce their multicomponent phase diagrams. However, the vast majority of systems with three or more components are very complex, and no overall maps of the phase relationships have been worked out.

It has been shown during the last 20 to 25 years that very useful partial phase diagrams of complex systems can be obtained by means of thermodynamic modeling (References 2, 3). Especially for complicated, multicomponent alloy systems the CALPHAD method has proved to be a successful approach for producing valuable portions of very intricate phase diagrams (Reference 4). With this method thermodynamic descriptions of the free energy functions of various phases are obtained which are consistent with existing (binary) phase diagram information and other thermodynamic data. Extrapolation methods are then used to extend the thermodynamic functions into a ternary system. Comparison of the results of this procedure with available experimental data is then used to fine-tune the phase diagram and add ternary interaction functions if necessary. In principle this approximation strategy can be extended to four, five, and more component systems.

The nearly two dozen phase diagrams shown below present the reader with examples of some important types of single and multicomponent systems, especially for ceramics and metal alloys. This makes it possible to draw attention to certain features like the kinetic aspects of phase transitions (see Figure 22, which presents a time-temperature-transformation, or TTT, diagram for the precipitation of α-phase particles from the β-phase in a Ti-Mo alloy; Reference 1, pp.358-360). The general references listed below and the references to individual figures contain phase diagrams for many additional systems.

GENERAL REFERENCES

1. Ralls, K.M., Courtney, T.H., and Wulff, J., *Introduction to Materials Science and Engineering*, Chapters 16 and 17, John Wiley & Sons, New York, 1976.
2. Kaufman, L., and Bernstein, H., *Computer Calculation of Phase Diagrams*, Academic Press, New York, 1970.
3. Kattner, U.R., Boettinger, W.J.B., and Coriell, S.R., *Z. Metallkd.*, 87, 9, 1996.
4. Dinsdale, A.T., Editor, *CALPHAD*, Vol. 1–20, Pergamon Press, Oxford, 1977–1996 and continuing.
5. Baker, H., Editor, *ASM Handbook, Volume 3: Alloy Phase Diagrams*, ASM International, Materials Park, OH, 1992.
6. Massalski, T.B., Editor, *Binary Alloy Phase Diagrams, Second Edition*, ASM International, Materials Park, OH, 1990.
7. Roth. R.S., Editor, *Phase Diagrams for Ceramists*, Vol. I (1964) to Volume XI (1995), American Ceramic Society, Waterville, OH.

REFERENCES TO INDIVIDUAL PHASE DIAGRAMS

Figure 1. Carbon: Reference 7, Vol. X (1994), Figure 8930. Reprinted with permission.
Figure 2. Si-Ge : Ref.5, p. 2.231. Reprinted with permission.
Figure 3. H_2O (ice): Lerner, R.G., and Trigg, G.L., Editors, *Encyclopedia of Physics*, VCH Publishers, New York, 1990. Reprinted with permission.
Figure 4. SiO_2: Reference 7, Vol. XI (1995), Figure 9174. Reprinted with permission.
Figure 5. Fe-O: Darken, L.S., and Gurry, R.W., *J. Am. Chem. Soc.*, 68, 798, 1946. Reprinted with permission.
Figure 6. Ti-O: Reference 5, p. 2.324. Reprinted with permission.
Figure 7. $BaO-TiO_2$: Reference 7, Vol. III (1975), Figure 4302. Reprinted with permission.
Figure 8. $MgO-Al_2O_3$: Reference 7, Vol. XI (1995), Figure 9239. Reprinted with permission.
Figure 9. $Y_2O_3-ZrO_2$: Reference 7, Vol. XI (1995), Figure 9348. Reprinted with permission.
Figure 10. Si-N-Al-O (Sialon): Reference 7, Vol. X (1994), Figure 8759. Reprinted with permission.
Figure 11. $PbO-ZrO_2-TiO_2$ (PZT): Reference 7, Vol. III (1975), Figure 4587. Reprinted with permission.
Figure 12. Al-Si-Ca-O: Reference 7 (1964), Vol. I, Figure 630. Reprinted with permission.
Figure 13. Y-Ba-Cu-O: Whitler, J.D., and Roth, R.S., *Phase Diagrams for High T_c Superconductors*, Figure S-082, American Ceramic Society, Waterville, OH, 1990. Reprinted with permission.
Figure 14. Al-Cu: Reference 5, p. 2.44. Reprinted with permission.
Figure 15. Fe-C: Ralls, K.M., Courtney, T.H., and Wulff, J., *Introduction to Materials Science and Engineering*, Figure 16.13 , John Wiley & Sons, New York, 1976. Reprinted with permission.
Figure 16. Fe-Cr: Reference 5, p. 2.152. Reprinted with permission.

Figure 17. Cu-Sn: Reference 5, p. 2.178. Reprinted with permission.
Figure 18. Cu-Ni: Reference 5, p. 2.173. Reprinted with permission.
Figure 19. Pb-Sn (solder): Reference 5, p. 2.335. Reprinted with permission.
Figure 20. Cu-Zn (brass): Subramanian, P.R., Chakrabarti, D.J., and Laughlin, D.E., Editors, *Phase Diagrams of Binary Copper Alloys*, p. 487, ASM
    International, Materials Park, OH, 1994. Reprinted with permission.
Figure 21. Co-Sm: Reference 5, p. 2.148. Reprinted with permission.
Figure 22. Ti-Mo: Reference 5, p. 2.296; Reference 1, p. 359. Reprinted with permission.
Figure 23: Fe-Cr-Ni: Reference 5, Figure 48. Reprinted with permission.

Figure 1

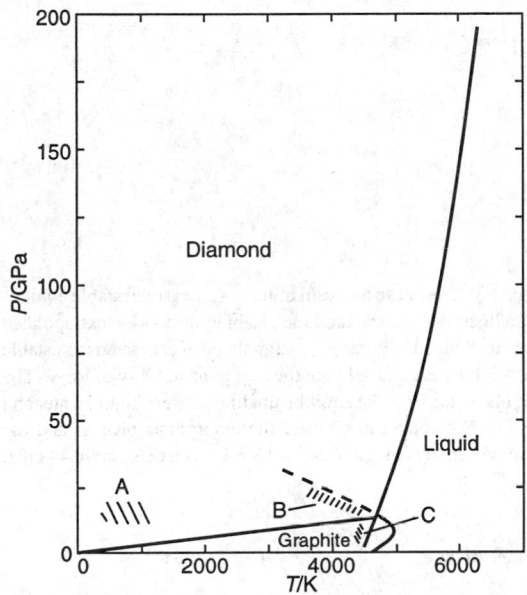

Figure 1. Phase diagram of carbon. (A) Martensitic transition: hex graphite →
hex diamond. (B) Fast graphite-to-diamond transition. (C) Fast diamond-to-
graphite transition.

Figure 2

Figure 2. Si-Ge system.

Phase	Composition, mass % Si	Pearson symbol	Space group
(Ge,Si)	0 to 100	*cF*8	*Fd$\bar{3}$m*
High-pressure phases			
GeII	—	*tI*4	*I*4$_1$/*amd*
SiII	—	*tI*4	*I*4$_1$/*amd*

FIGURE 3

Figure 3. The phase diagram of ice: —, measured stable equilibrium lines; ·—·—·, measured metastable lines; - - - -, extrapolated or estimated stable lines; ····, extrapolated or estimated metastable lines. Ice IV is metastable in the region of stability of ice V. The line above ice IV is the equilibrium line between liquid water and ice IV. Not shown are ice Ic, the amorphous phases, and the ordered versions of ice V and ice VI. Ice X exists above 44 GPa.

Structural Data on the Phases of Ice

Ice phase	Crystal system	Space group	Cell dimensions[a] (pm, deg)	No. of molecules in unit cell	No. of nearest neighbors	Density[a] (Mg m^{-3})
Ih	Hexagonal	$P6_3/mmc$	$a = 450; c = 732$	4	4	0.93
		$Fd3m$				0.93
Ic	Cubic	—	$a = 635$	8	4	0.94
a_I	Amorphous	—	—	—	4	1.17
a_{II}	Amorphous	—	—	—	4	
II	Rhombohedral	$R\bar{3}$	$a = 778; \alpha = 113.1$	12	4	1.18
III	Tetragonal	$P4_12_12$	$a = 673; c = 683$	12	4	1.15
IV	Rhombohedral	$R\bar{3}c$	$a = 760; \alpha = 70.1$	16	4	1.27
V	Monoclinic	$A2/a$	$a = 922; b = 754$ $c = 1035; \beta = 109.2$	28	4	1.24
VI	Tetragonal	$P4_2/nmc$	$a = 627; c = 579$	10	4	1.33
VII	Cubic	$Pn3m$	$a = 341$	2	8	1.56
VIII	Tetragonal	$I4_1/amd$	$a = 480; c = 699$	8	8	1.56
IX	Tetragonal	$P4_12_12_1$	$a = 673; c = 683$	12	4	1.16
X	Cubic	$Pn3m$	$a = 283$	2	8	2.51

[a] Cell dimensions and density are at 110 K and atmospheric pressure. Ice X at 44 GPa and 25°C.

Figure 4

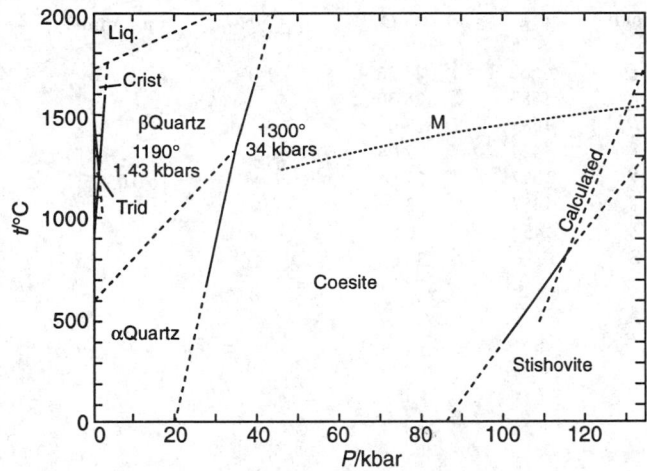

Figure 4. SiO₂ system. Crist = cristobalite; Trid = tridymite.

Figure 5

Figure 5. Fe-O system.

Figure 5 (continued)

Point	$t/°C$	% O	p_{CO_2}/p_{CO}	Point	$t/°C$	% O	p_{CO_2}/p_{CO}	p_{O_2}/atm
A	1539			Q	560	23.26	1.05	
B	1528	0.16	0.209	R	1583	28.30		1
C	1528	22.60	0.209	R'	1583	28.07		1
G	1400[a]	22.84	0.263	S	1424	27.64	16.2	
H	1424	25.60	16.2	V	1597	27.64		0.0575
I	1424	25.31	16.2	Y	1457	28.36		1
J	1371	23.16	0.282	Z	1457	30.04		1
L	911[a]	23.10	0.447	Z'		30.6		
N	1371	22.91	0.282					

[a] Values for pure iron.

Figure 6

Figure 6. Ti-O system.

Phase	Composition, mass % O	Pearson symbol	Space group
(βTi)	0 to 3	cI2	$Im\bar{3}m$
(αTi)	0 to 13.5	hP2	$P6_3/mmc$
Ti_3O	~8 to ~13	hP~16	$P\bar{3}c$
Ti_2O	~10 to 14.4	hP3	$P\bar{3}m1$
γTiO	15.2 to 29.4	cF8	$Fm\bar{3}m$
Ti_3O_2	~18	hP~5	$P6/mmm$
βTiO	~24 to ~29.4	c**	—
αTiO	~25.0	mC16	$A2/m$ or $B*/*$
$\beta Ti_{1-x}O$	~29.5	oI12	$I222$
$\alpha Ti_{1-x}O$	~29.5	tI18	$I4/m$
βTi_2O_3	33.2 to 33.6	hR30	$R\bar{3}c$
αTi_2O_3	33.2 to 33.6	hR30	$R\bar{3}c$
βTi_3O_5	35.8	m**	—
αTi_3O_5	35.8	mC32	$C2/m$
$\alpha' Ti_3O_5$	35.8	mC32	Cc

Figure 6 (continued)

Phase	Composition, mass % O	Pearson symbol	Space group
γTi_4O_7	36.9	$aP44$	$P\bar{1}$
βTi_4O_7	36.9	$aP44$	$P\bar{1}$
$\alpha Ti4O_7$	36.9	$aP44$	$P\bar{1}$
γTi_5O_9	37.6	$aP28$	$P\bar{1}$
βTi_6O_{11}	38.0	$aC68$	$A\bar{1}$
Ti_7O_{13}	38.3	$aP40$	$P\bar{1}$
Ti_8O_{15}	38.5	$aC92$	$A\bar{1}$
Ti_9O_{17}	38.7	$aP52$	$P\bar{1}$
Rutile TiO_2	40.1	$tP6$	$P4_2/mnm$
Metastable phases			
Anatase	—	$tI12$	$I4_1/amd$
Brookite	—	$oP24$	$Pbca$
High-pressure phases			
TiO_2-II	—	$oP12$	$Pbcn$
TiO_2-III	—	$hP{\sim}48$	—

FIGURE 7

Figure 7. BaO-TiO_2 system.

Figure 8

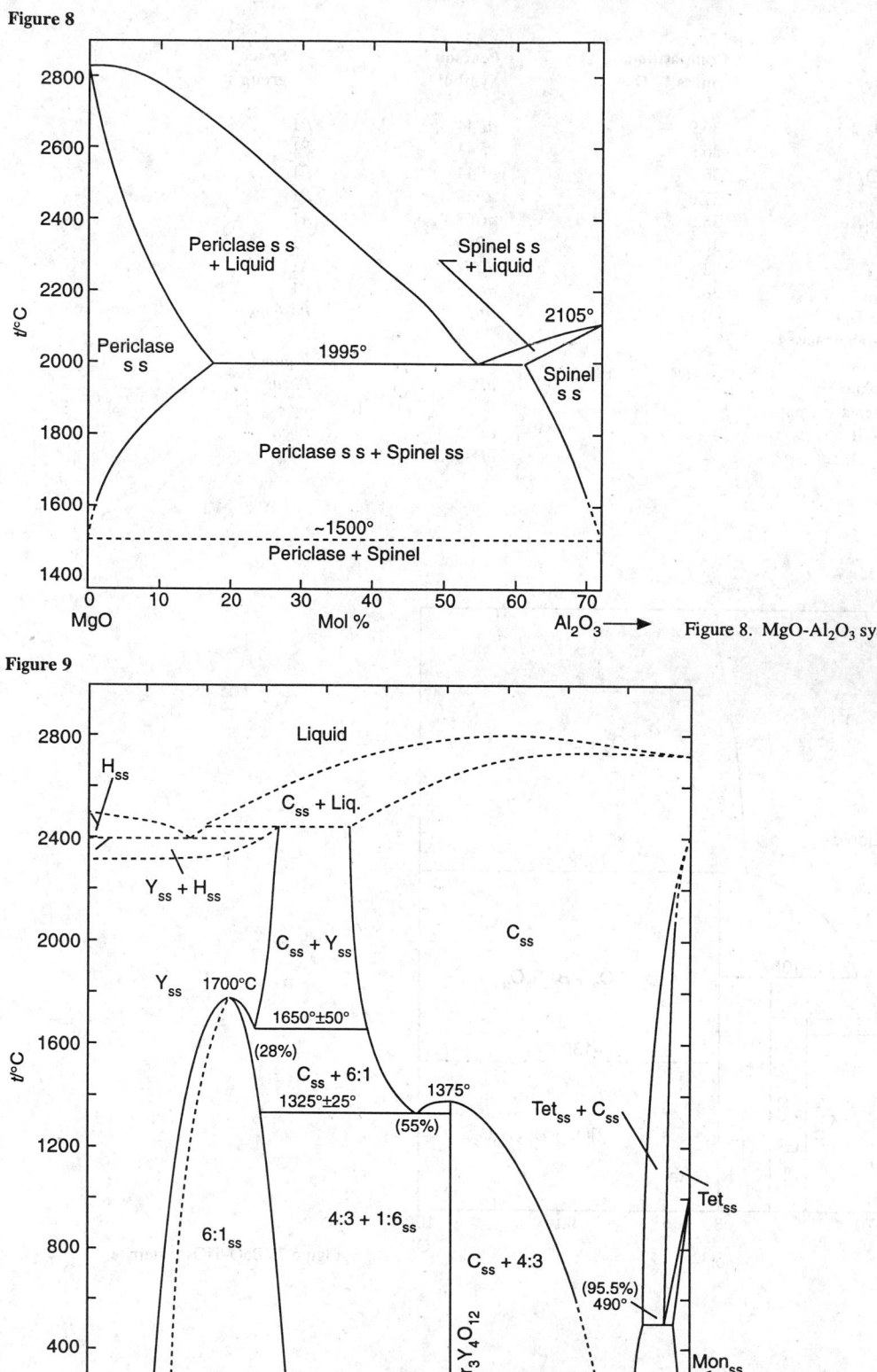

Figure 8. MgO-Al₂O₃ system.

Figure 9

Figure 9. Y₂O₃-ZrO₂ system. C$_{ss}$ = cubic ZrO₂ ss (fluorite-type ss); Y$_{ss}$ = cubic Y₂O₃ ss; Tet$_{ss}$ = tetragonal ZrO₂ ss; Mon$_{ss}$ = monoclinic ZrO₂ ss; H$_{ss}$ = hexagonal Y₂O₃ ss; 3:4 = Zr₃Y₄O₁₂; 1:6 = ZrY₆O₁₁ ss.

Figure 10

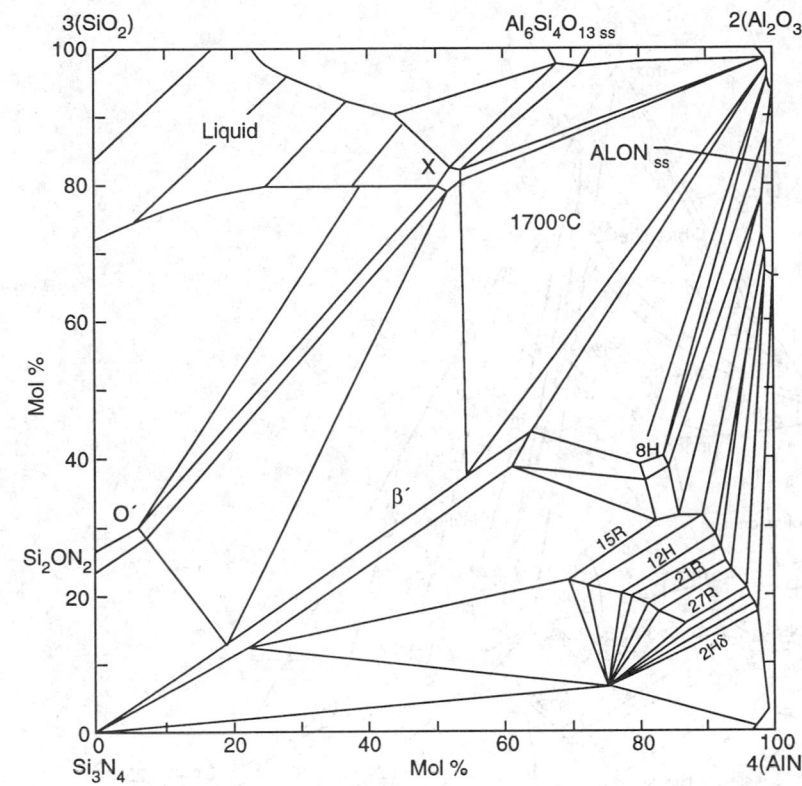

Figure 10. $3(SiO_2)$-Si_3N_4-$4(AlN)$-$2(Al_2O_3)$ system. "Behavior" diagram at 1700°C. The labels 8H, 15R, 12H, 21R, 27R, $2H^\delta$ indicate defect AlN polytypes. β' = 3-sialon ($Si_{6-x}Al_xO_xN_{8-x}$); O' = sialon of Si_2ON_2 type; X = $SiAlO_2N$ ("nitrogen mullite"). ALON ss = aluminum oxynitride ss extending from approximately Al_7O_9N to Al_3O_3N.

Figure 11

Figure 11. PbO-ZrO_2-TiO_2 (PZT) system, subsolidus at 1100°C. $P = PbO$; $T = TiO_2$; $Z = ZrO_2$.

Figure 12

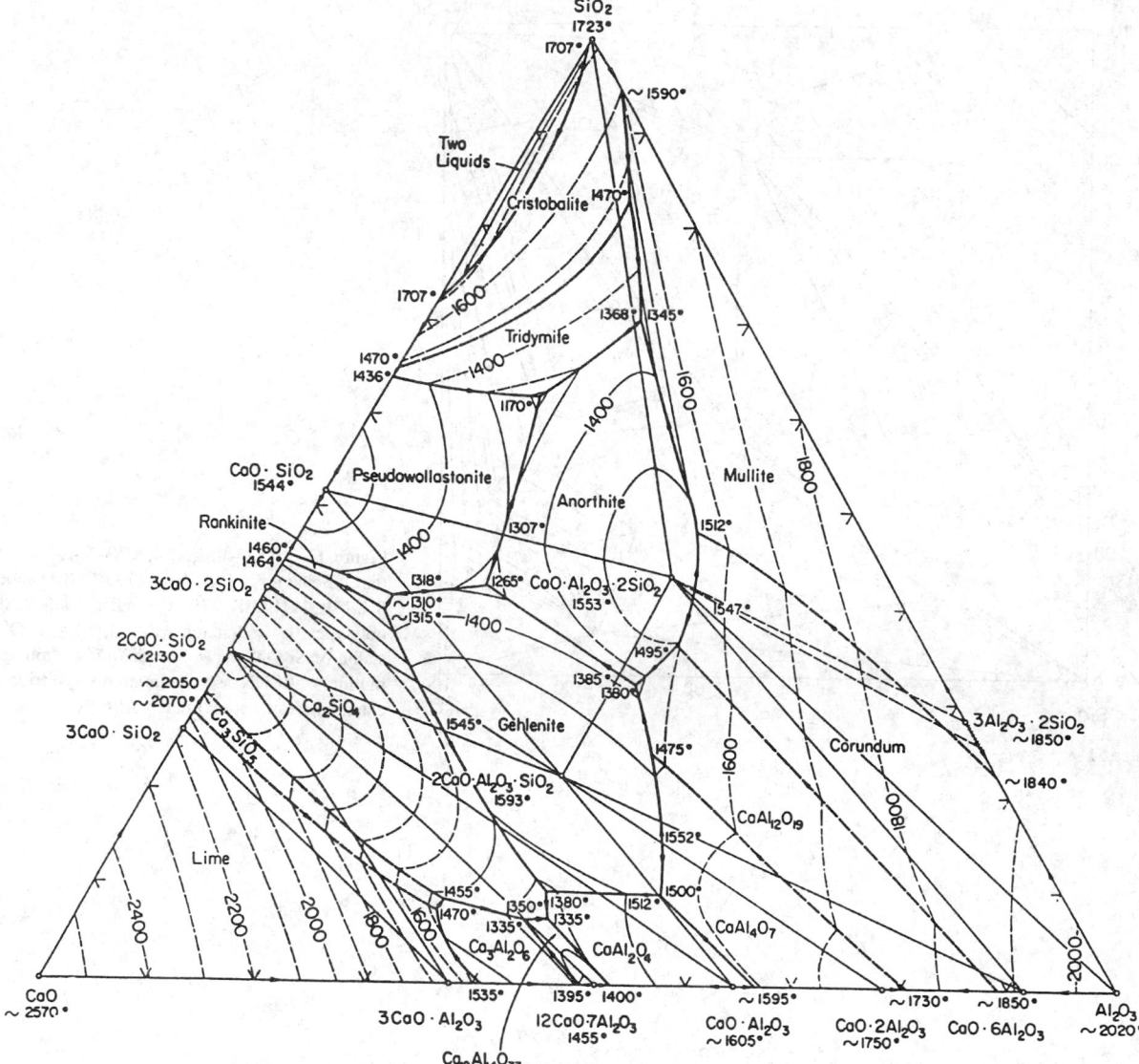

Figure 12. CaO-Al₂O₃-SiO₂ system (temperatures in °C).

Crystalline Phases

Notation	Oxide formula
Cristobalite } Tridymite	SiO_2
Pseudowollastonite	$CaO \cdot SiO_2$
Rankinite	$3CaO \cdot 2SiO_2$
Lime	CaO
Corundum	Al_2O_3
Mullite	$3Al_2O_3 \cdot 2SiO_2$
Anorthite	$CaO \cdot Al_2O_3 \cdot 2SiO_2$
Gehlenite	$2CaO \cdot Al_2O_3 \cdot SiO_2$

Temperatures up to approximately 1550°C are on the Geophysical Laboratory Scale; those above 1550°C are on the 1948 International Scale.

Figure 13

Figure 13. BaO-Y$_2$O$_3$-CuO system. 2:1:3 = Ba$_2$YCu$_3$O$_{7-x}$; 1:2:1 = BaY$_2$CuO$_5$; 4:1:2 = Ba$_4$YCu$_2$O$_{7.5+x}$; and 5:1:3 = Ba$_5$YCu$_3$O$_{9.5+x}$. The superconducting 2:1:3 phase was prepared using barium peroxide.

Figure 14

Figure 14. Al-Cu system.

Figure 14 (continued)

Phase	Composition, wt % Cu	Pearson symbol	Space group
(Al)	0 to 5.65	cF4	$Fm\bar{3}m$
θ	52.5 to 53.7	tI12	I4/mcm
η_1	70.0 to 72.2	oP16 or oC16	Pban or Cmmm
η_2	70.0 to 72.1	mC20	C2/m
ζ_1	74.4 to 77.8	hP42	P6/mmm
ζ_2	74.4 to 75.2	(a)	—
ε_1	77.5 to 79.4	(b)	—
ε_2	72.2 to 78.7	hP4	P63/mmc
δ	77.4 to 78.3	(c)	$R\bar{3}m$
γ_0	77.8 to 84	(d)	—
γ_1	79.7 to 84	cP52	$P\bar{4}3m$
β_0	83.1 to 84.7	(d)	—
β	85.0 to 91.5	cI2	$Im\bar{3}m$
α_2	88.5 to 89	(e)	—
(Cu)	90.6 to 100	cF4	$Fm\bar{3}m$
Metastable phases			
θ′	—	tP6	—
β′	—	cF16	$Fm\bar{3}m$
Al_3Cu_2	61 to 70	hp5	$P\bar{3}m1$

(a) Monoclinic? (b) Cubic? (c) Rhombohedral. (d) Unknown. (e) DO_{22}-type long-period superlattice.

Figure 15

Figure 15. Fe-C system.

Phase	Composition, mass % C	Pearson symbol	Space group
(δFe)	0 to 0.09	cI2	$Im\bar{3}m$
(γFe)	0 to 2.1	cF4	$Fm\bar{3}m$
(αFe)	0 to 0.021	cI2	$Im\bar{3}m$
(C)	100	hP4	P63/mmc
Metastable/high-pressure phases			
(εFe)	0	hP2	P63/mmc
Martensite	< 2.1	tI4	I4/mmm

Figure 15 (continued)

Phase	Composition, mass % C	Pearson symbol	Space group
Fe$_4$C	5.1	$cP5$	$P\bar{4}3m$
Fe$_3$C (θ)	6.7	$oP16$	$Pnma$
Fe$_5$C$_2$ (χ)	7.9	$mC28$	$C2/c$
Fe$_7$C$_3$	8.4	$hP20$	$P6_3mc$
Fe$_7$C$_3$	8.4	$oP40$	$Pnma$
Fe$_2$C (η)	9.7	$oP6$	$Pnnm$
Fe$_2$C (ε)	9.7	$hP*$	$P6_322$
Fe$_2$C	9.7	$hP*$	$P\bar{3}m1$
(C)	100	$cF8$	$Fd\bar{3}m$

Figure 16

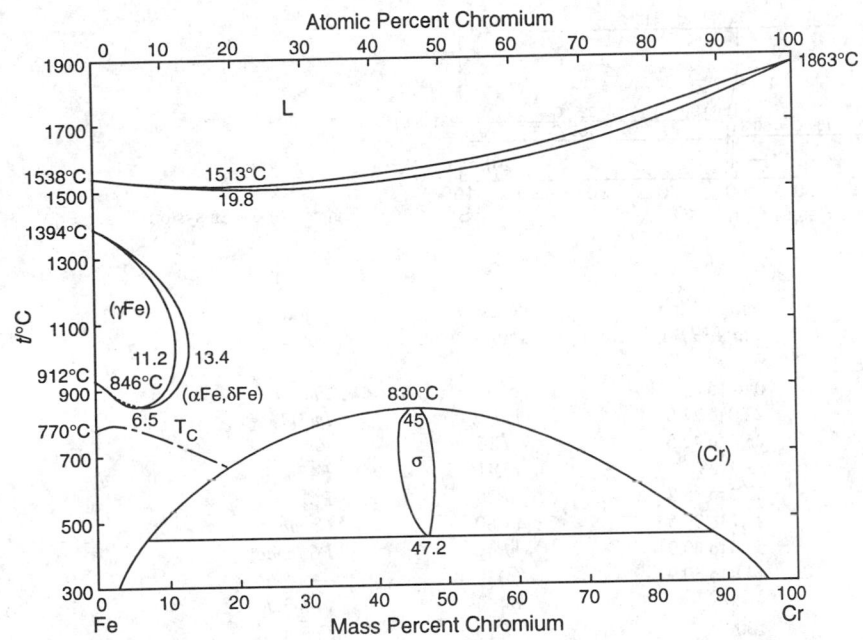

Figure 16. Fe-Cr system.

Phase	Composition, mass % Cr	Pearson symbol	Space group
(aFe, Cr)	0 to 100	$cI2$	$Im\bar{3}m$
(γFe)	0 to 11.2	$cF4$	$Fm\bar{3}m$
σ	42.7 to 48.2	$tP30$	$P4_2/mnm$

Figure 17

Figure 17. Cu-Sn system.

Phase	Composition, mass % Sn	Pearson symbol	Space group
α	0 to 15.8	$cF4$	$Fm\bar{3}m$
β	22.0 to 27.0	$cI2$	$Im\bar{3}m$
γ	25.5 to 41.5	$cF16$	$Fm\bar{3}m$
δ	32 to 33	$cF416$	$F\bar{4}3m$
ζ	32.2 to 35.2	$hP26$	$P6_3$
ε	27.7 to 39.5	$oC80$	$Cmcm$
η	59.0 to 60.9	$hP4$	$P6_3/mmc$
η′	44.8 to 60.9	(a)	—
(βSn)	~100	$tI4$	$I4_1/amd$
(αSn)	100	$cF8$	$Fd\bar{3}m$

(a) Hexagonal; superlattice based on NiAs-type structure.

Figure 18

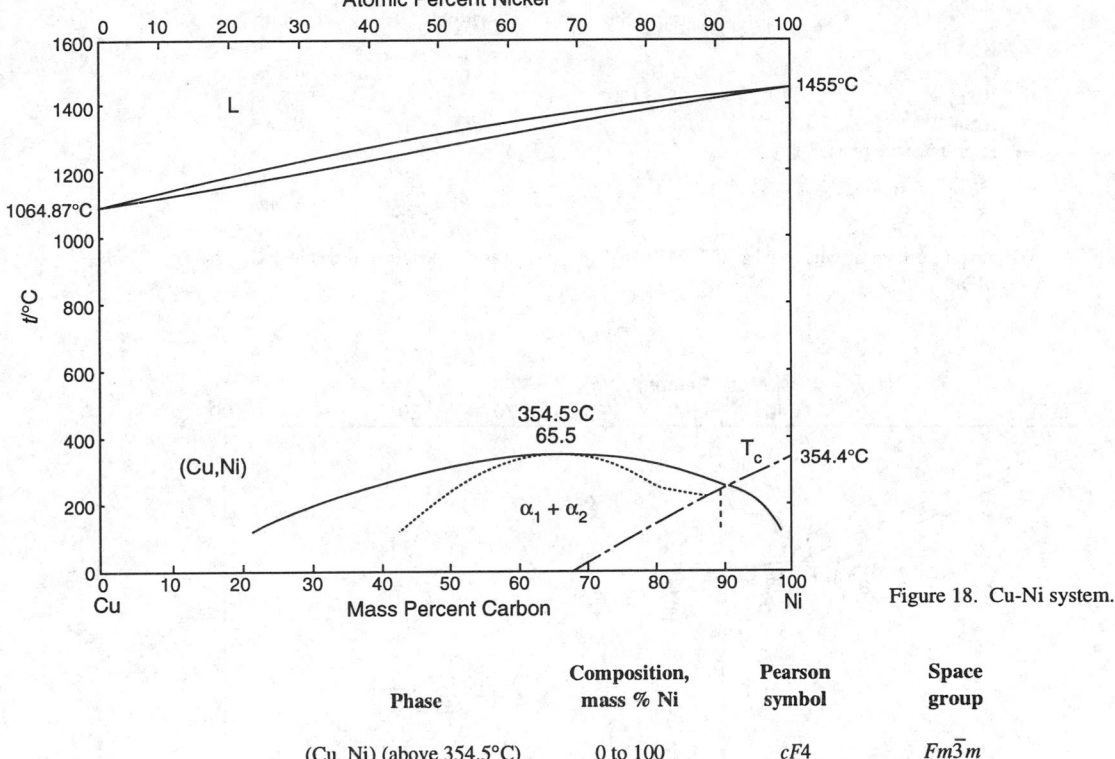

Figure 18. Cu-Ni system.

Phase	Composition, mass % Ni	Pearson symbol	Space group
(Cu, Ni) (above 354.5°C)	0 to 100	cF4	$Fm\bar{3}m$

Figure 19

Figure 19. Pb-Sn system.

Figure 19 (continued)

Phase	Composition, mass % Sn	Pearson symbol	Space group
(Pb)	0 to 18.3	$cF4$	$Fm\bar{3}m$
(βSn)	97.8 to 100	$tI4$	$I4_1/amd$
(αSn)	100	$cF8$	$Fd\bar{3}m$
High-pressure phases			
ε(a)	52 to 74	$hP1$	$P6/mmm$
ε'(b)	52	$hP2$	$P6_3/mmc$

(a) From phase diagram calculated at 2500 MPa. (b) This phase was claimed for alloys at 350°C and 5500 MPa.

Figure 20

Figure 20. Cu-Zn system.

Phase	Composition, mass % Zn	Pearson symbol	Space group
α or (Cu)	0 to 38.95	$cF4$	$Fm\bar{3}m$
β	36.8 to 56.5	$cI2$	$Im\bar{3}m$
β'	45.5 to 50.7	$cP2$	$Pm\bar{3}m$
γ	57.7 to 70.6	$cI52$	$I\bar{4}3m$
δ	73.02 to 76.5	$hP3$	$P\bar{6}$
ε	78.5 to 88.3	$hP2$	$P6_3/mmc$
η or (Zn)	97.25 to 100	$hP2$	$P6_3/mmc$

Figure 21

Figure 21. Co-Sm system.

Phase	Composition, mass % Sm	Pearson symbol	Space group
(αCo)	0 to ~3.7	cF4	$Fm\bar{3}m$
(εCo)	~0	hP2	$P6_3/mmc$
βCo₁₇Sm₂	~23.0	hP38	$P6_3/mmc$
αCo₁₇Sm₂	~23.0	hR19	$R\bar{3}m$
		hP8	$P6/mmm$
Co₅₊ₓSm	~33 to 34	—	—
Co₅₋ₓSm	~34 to 35	—	—
Co₁₉Sm₅	~40.1	hR24	$R\bar{3}m$
		hP48	$P6_3/mmc$
αCo₇Sm₂	~42.1	hR18	$R\bar{3}m$
βCo₇Sm₂	~42.1	hP36	$P6_3/mmc$
Co₃Sm	46	hR12	$R\bar{3}m$
Co₂Sm	56.0	hR4	$R\bar{3}m$
		cF24	$Fd\bar{3}m$
Co₄Sm₉	~85.1	o**	—
CoSm₃	88	oP16	$Pnma$
(γSm)	~100	cI2	$Im\bar{3}m$
(βSm)	~100	hP2	$P6_3/mmc$
(αSm)	~100	hR3	$R\bar{3}m$
Other reported phases			
Co₅Sm	~33.8	hP6	$P6/mmm$
Co₂Sm₅	~86.4	mC28	$C2/c$

Figure 22

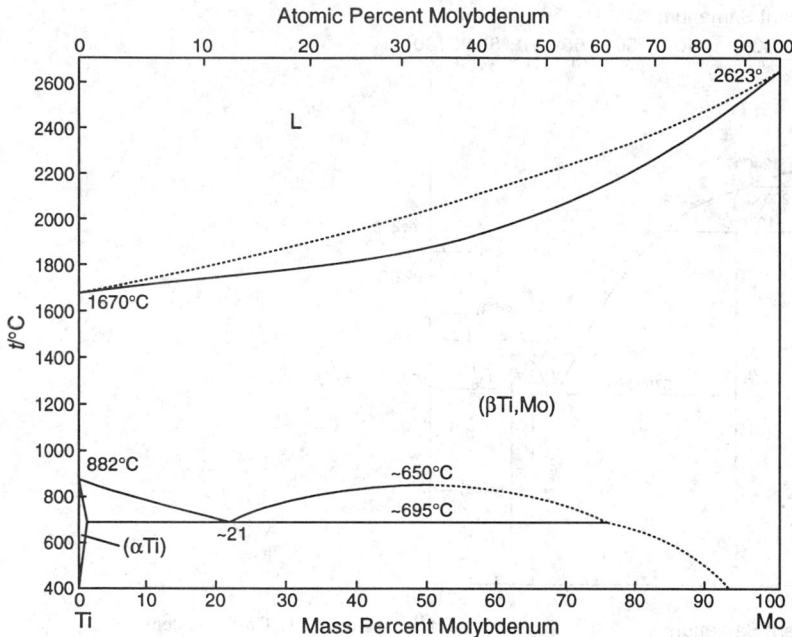

Figure 22. Ti-Mo system.

Phase	Composition, mass % Mo	Pearson symbol	Space group
(βTi, Mo)	0 to 100	cI2	Im3̄m
(αTi)	0 to 0.8	hP2	P6₃/mmc
α′	(a)	hP2	P6₃/mmc
α″	(a)	oC4	Cmcm
ω	(a)	hP3	P6/mmm

(a) Metastable.

Experimental time-temperature-transformation (TTT) diagram for Ti-Mo. The start and finish times of the isothermal precipitation reaction vary with temperature as a result of the temperature dependence of the nucleation and growth processes. Precipitation is complete, at any temperature, when the equilibrium fraction of α is established in accordance with the lever rule. The solid horizontal line represents the athermal (or nonthermally activated) martensitic transformation that occurs when the β phase is quenched.

Figure 23

Figure 23. The isothermal section at 900°C (1652°F) of the iron-chromium-nickel ternary phase diagram, showing the nominal composition of 18-8 stainless steel.

HEAT CAPACITY OF SELECTED SOLIDS

This table gives the molar heat capacity at constant pressure of representative metals, semiconductors, and other crystalline solids as a function of temperature in the range 200 to 600 K.

REFERENCES

1. Chase, M. W., et al., *JANAF Thermochemical Tables, 3rd ed., J. Phys. Chem. Ref. Data,* 14, Suppl. 1, 1985.
2. Garvin, D., Parker, V. B., and White, H. J., *CODATA Thermodynamic Tables,* Hemisphere Press, New York, 1987.
3. DIPPR Database of Pure Compound Properties, Design Institute for Physical Properties Data, American Institute of Chemical Engineers, New York, 1987.

Name	C_p in J/mol K						
	200 K	250 K	300 K	350 K	400 K	500 K	600 K
Aluminum	21.33	23.08	24.25	25.11	25.78	26.84	27.89
Aluminum oxide	51.12	67.05	79.45	88.91	96.14	106.17	112.55
Anthracene	138.6	173.9	210.7	248.8	288.4		
Benzoic acid	102.7	123.5	147.4	172.0			
Beryllium	9.98	13.58	16.46	18.53	19.95	21.94	23.34
Biphenyl	131.0	162.5	197.2				
Boron	5.99	8.82	11.40	13.65	15.69	18.72	20.78
Calcium	24.54	25.41	25.94	26.32	26.87	28.49	30.38
Calcium carbonate	66.50	75.66	83.82	91.51	96.97	104.52	109.86
Calcium oxide	33.64	38.59	42.18	45.07	46.98	49.33	50.72
Cesium chloride	50.13	51.34	52.48	53.58	54.68	56.90	59.10
Chromium	19.86	22.30	23.47	24.39	25.23	26.63	27.72
Cobalt	22.23	23.98	24.83	25.68	26.53	28.20	29.66
Copper	22.63	23.77	24.48	24.95	25.33	25.91	26.48
Copper oxide	34.80		42.41	44.95	46.78	49.19	50.83
Copper sulfate	77.01	89.25	99.25	107.65	114.93	127.19	136.31
Germanium			23.25	23.85	24.31	24.96	25.45
Gold			25.41	25.37	25.51	26.06	26.65
Graphite	5.01	6.82	8.58	10.24	11.81	14.62	16.84
Hexachlorobenzene	162.7	183.6	202.4				
Iodine	51.57	53.24	54.51	58.60			
Iron	21.59	23.74	25.15	26.28	27.39	29.70	32.05
Lead	25.87	26.36	26.85	27.30	27.72	28.55	29.40
Lithium	21.57	23.42	24.64	25.96	27.60	29.28	
Lithium chloride	43.35	46.08	48.10	49.66	50.97	53.34	55.59
Magnesium	22.72	24.02	24.90	25.57	26.14	27.17	28.18
Magnesium oxide			37.38	40.59	42.77	45.56	47.30
Manganese	23.05	24.95	26.35	27.52	28.53	30.29	31.90
Naphthalene	105.8	134.1	167.8	204.1			
Potassium	27.00	28.01	29.60				
Potassium chloride	48.44	50.10	51.37	52.31	53.08	54.71	56.35
Silicon	15.64	18.22	20.04	21.28	22.14	23.33	24.15
Silicon dioxide	32.64	39.21	44.77	49.47	53.43	59.64	64.42
Silver			25.36	25.55	25.79	26.36	26.99
Sodium	22.45	27.01	28.20	30.14			
Sodium chloride	46.89	48.85	50.21	51.25	52.14	53.96	55.81
Tantalum	24.08	24.86	25.31	25.60	25.84	26.35	26.84
Titanium	22.37	24.07	25.28	26.17	26.86	27.88	28.60
Tungsten	22.49	23.69	24.30	24.65	24.92	25.36	25.79
Vanadium	21.88	23.70	24.93	25.68	26.23	26.94	27.49
Zinc	24.05	25.02	25.45	25.88	26.35	27.39	28.59
Zirconium	23.87	24.69	25.22	25.61	25.93	26.56	27.28

THERMAL AND PHYSICAL PROPERTIES OF PURE METALS

This table gives the following properties for the metallic elements:

t_m: Melting point in °C
t_b: Normal boiling point in °C, at a pressure of 101.325 kPa (760 Torr)
$\Delta_{fus} H$:   Enthalpy of fusion at the melting point in J/g
ρ_{25}:   Density at 25°C in g/cm³
α: Coefficient of linear expansion at 25°C in K⁻¹ (the quantity listed is $10^6 \times \alpha$)
c_p: Specific heat capacity at constant pressure at 25°C in J/g K
λ: Thermal conductivity at 27°C in W/cm K

REFERENCES

1. Dinsdale, A. T., *CALPHAD*, 15, 317, 1991 (melting points, enthalpy of fusion).
2. Touloukian, Y. S., *Thermophysical Properties of Matter*, Vol. 12, Thermal Expansion, IFI/Plenum, New York, 1975 (coefficient of expansion, density).
3. Ho, C. Y., Powell, R. W., and Liley, P. E., *J. Phys. Chem. Ref. Data*, 3, Suppl. 1, 1974 (thermal conductivity).
4. Cox, J. D., Wagman, D. D., and Medvedev, V. A., *CODATA Key Values for Thermodynamics*, Hemisphere Publishing Corp., New York, 1989 (heat capacity).
5. Glushko, V. P., Ed., *Thermal Constants of Substances*, VINITI, Moscow, (enthalpy of fusion, heat capacity).
6. Wagman, D. D., et. al., *The NBS Tables of Chemical Thermodynamic Properties, J. Phys. Chem. Ref. Data*, 11, Suppl. 2, 1982 (heat capacity).
7. Chase, M. W., et. al., *JANAF Thermochemical Tables*, 3rd ed., J. Phys. Chem. Ref. Data, 14, Suppl. 1, 1985 (heat capacity, enthalpy of fusion).
8. Gschneidner, K. A., *Bull. Alloy Phase Diagrams*, 11, 216—224, 1990 (various properties of the rare earth metals).
9. Hellwege, K. H., Ed., *Landolt Börnstein, Numerical Values and Functions in Physics, Chemistry, Astronomy, Geophysics, and Technology*, Vol. 2, Part 1, Mechanical-Thermal Properties of State, 1971 (density).

Metal (symbol)	Atomic weight	t_m °C	t_b °C	$\Delta_{fus} H$ J/g	ρ_{25} g/cm³	$\alpha \times 10^6$ K⁻¹	c_p J/g K	λ W/cm K
Actinium (Ac)		1051	3198		10			
Aluminum (Al)	26.98	660.32	2519	397	2.70	23.1	0.897	2.37
Antimony (Sb)	121.76	630.63	1587	163.2	6.68	11.0	0.207	0.243
Barium (Ba)	137.33	727	1897	52	3.62	20.6	0.204	0.184
Beryllium (Be)	9.01	1287	2471	877	1.85	11.3	1.825	2.00
Bismuth (Bi)	208.98	271.40	1564	54.1	9.79	13.4	0.122	0.0787
Cadmium (Cd)	112.41	321.07	767	55.1	8.69	30.8	0.232	0.968
Calcium (Ca)	40.08	842	1484	213	1.54	22.3	0.647	2.00
Cerium (Ce)	140.11	798	3443	39.0	6.77	6.3	0.192	0.113
Cesium (Cs)	132.91	28.44	671	15.8	1.93		0.242	0.359
Chromium (Cr)	52.00	1907	2671	404	7.15	4.9	0.449	0.937
Cobalt (Co)	58.93	1495	2927	275	8.86	13.0	0.421	1.00
Copper (Cu)	63.55	1084.62	2562	208.7	8.96	16.5	0.385	4.01
Dysprosium (Dy)	162.50	1412	2567	68.1	8.55	9.9	0.170	0.107
Erbium (Er)	167.26	1529	2868	119	9.07	12.2	0.168	0.145
Europium (Eu)	151.96	822	1529	60.6	5.24	35.0	0.182	0.139[a]
Gadolinium (Gd)	157.25	1313	3273	63.6	7.90	9.4[b]	0.236	0.105
Gallium (Ga)	69.72	29.76	2204	80.2	5.91		0.371	0.406
Gold (Au)	196.97	1064.18	2856	63.7	19.3	14.2	0.129	3.17
Hafnium (Hf)	178.49	2233	4603	152.4	13.3	5.9	0.144	0.230
Holmium (Ho)	164.93	1474	2700	103[a]	8.80	11.2	0.165	0.162
Indium (In)	114.82	156.60	2072	28.6	7.31	32.1	0.233	0.816
Iridium (Ir)	192.22	2446	4428	213.9	22.5	6.4	0.131	1.47
Iron (Fe)	55.85	1538	2861	247.3	7.87	11.8	0.449	0.802
Lanthanum (La)	138.91	918	3464	44.6	6.15	12.1	0.195	0.134
Lead (Pb)	207.20	327.46	1749	23.0	11.3	28.9	0.129	0.353
Lithium (Li)	6.94	180.5	1342	432	0.534	46	3.582	0.847
Lutetium (Lu)	174.97	1663	3402	126[a]	9.84	9.9	0.154	0.164
Magnesium (Mg)	24.30	650	1090	349	1.74	24.8	1.023	1.56

Metal (symbol)	Atomic weight	t_m °C	t_b °C	$\Delta_{fus} H$ J/g	ρ_{25} g/cm^3	$\alpha \times 10^6$ K^{-1}	c_p J/g K	λ W/cm K
Manganese (Mn)	54.94	1246	2061	235.0	7.3	21.7	0.479	0.0782
Mercury (Hg)	200.59	-38.83	356.73	11.4	13.5336		0.140	0.0834
Molybdenum (Mo)	95.94	2623	4639	390.7	10.2	4.8	0.251	1.38
Neodymium (Nd)	144.24	1021	3074	49.5	7.01	9.6	0.190	0.165
Neptunium (Np)		644			20.2			0.063
Nickel (Ni)	58.69	1455	2913	298	8.90	13.4	0.444	0.907
Niobium (Nb)	92.91	2477	4744	323	8.57	7.3	0.265	0.537
Osmium (Os)	190.23	3033	5012	304.2	22.59	5.1	0.130	0.876
Palladium (Pd)	106.42	1554.9	2963	157.3	12.0	11.8	0.246	0.718
Platinum (Pt)	195.08	1768.4	3825	113.6	21.5	8.8	0.133	0.716
Plutonium (Pu)		640	3228		19.7	46.7		0.0674
Polonium (Po)		254	962		9.20			0.20
Potassium (K)	39.10	63.38	759	59.3	0.89		0.757	1.024
Praseodymium (Pr)	140.91	931	3520	48.9	6.77	6.7	0.193	0.125
Promethium (Pm)		1042	3000[a]	53[a]	7.26	11[a]	0.19[a]	0.15[a]
Protactinium (Pa)	231.04	1572		53.4	15.4			
Radium (Ra)		700			5			
Rhenium (Re)	186.21	3186	5596	324.5	20.8	6.2	0.137	0.479
Rhodium (Rh)	102.91	1964	3695	258.4	12.4	8.2	0.243	1.50
Rubidium (Rb)	85.47	39.31	688	25.6	1.53		0.363	0.582
Ruthenium (Ru)	101.07	2334	4150	381.8	12.1	6.4	0.238	1.17
Samarium (Sm)	150.36	1074	1794	57.3	7.52	12.7	0.197	0.133
Scandium (Sc)	44.96	1541	2836	314	2.99	10.2	0.568	0.158
Silver (Ag)	107.87	961.78	2162	104.8	10.5	18.9	0.235	4.29
Sodium (Na)	22.99	97.72	883	113	0.97	71	1.228	1.41
Strontium (Sr)	87.62	777	1382	84.8	2.64	22.5	0.301	0.353
Tantalum (Ta)	180.95	3017	5458	202.1	16.4	6.3	0.140	0.575
Technetium (Tc)		2157	4265		11			0.506
Terbium (Tb)	158.93	1356	3230	67.9	8.23	10.3	0.182	0.111
Thallium (Tl)	204.38	304	1473	20.3	11.8	29.9	0.129	0.461
Thorium (Th)	232.04	1750	4788	59.5	11.7	11.0	0.113	0.540
Thulium (Tm)	168.93	1545	1950	99.4	9.32	13.3	0.160	0.169
Tin (Sn)	118.71	231.93	2602	59.2	7.26	22.0	0.228	0.666
Titanium (Ti)	47.88	1668	3287	295	4.51	8.6	0.523	0.219
Tungsten (W)	183.84	3422	5555	284.5	19.3	4.5	0.132	1.74
Uranium (U)	238.03	1135	4131	38.4	19.1	13.9	0.116	0.276
Vanadium (V)	50.94	1910	3407	422	6.0	8.4	0.489	0.307
Ytterbium (Yb)	173.04	819	1196	44.3	6.90	26.3	0.155	0.385
Yttrium (Y)	88.91	1522	3345	128	4.47	10.6	0.298	0.172
Zinc (Zn)	65.39	419.53	907	112	7.14	30.2	0.388	1.16
Zirconium (Zr)	91.22	1855	4409	230	6.52	5.7	0.278	0.227

[a] Estimated.
[b] At 100°C.

THERMAL CONDUCTIVITY OF METALS AND SEMICONDUCTORS AS A FUNCTION OF TEMPERATURE

This table gives the temperature dependence of the thermal conductivity of several metals and of carbon, germanium, and silicon. For graphite, separate entries are given for the thermal conductivity parallel (∥) and perpendicular (⊥) to the layer planes. The thermal conductivity of all these materials is very sensitive to impurities at low temperatures, especially below 100 K. Therefore, the values given here should be regarded as typical values for a highly purified specimen; the thermal conductivity of different specimens can vary by more than an order of magnitude in the low-temperature range. See Reference 2 for details.

REFERENCES

1. Ho, C. Y., Powell, R. W., and Liley, P. E., *J. Phys. Chem. Ref. Data*, 1, 279, 1972.
2. White, G. K., and Minges, M. L., *Thermophysical Properties of Some Key Solids*, CODATA Bulletin No. 59, 1985.

Thermal Conductivity in W/cm K

T/K	Ag	Al	Au	Diamond (type) I	IIa	IIb	Pyrolytic graphite ∥	⊥	Cr	Cu
1	39.4	41.1	5.46						0.402*	42.2
2	78.3	81.8	10.9	0.0138*	0.033*	0.0200*			0.803	84.0
3	115	121	16.1	0.0461	0.111	0.0676			1.20	125
4	147	157	20.9	0.108	0.261	0.160			1.60	162
5	172	188	25.2	0.206	0.494	0.307			2.00	195
6	187	213	28.5	0.344	0.820	0.510			2.39	222
7	193	229	30.9	0.523	1.24	0.778			2.27	239
8	190	237	32.3	0.762	1.77	1.12			3.14	248
9	181	239	32.7	1.05	2.41	1.53			3.50	249
10	168	235	32.4	1.40	3.17	2.03	0.811	0.0116	3.85	243
15	96.0	176	24.6	3.96	8.65	5.66			5.24	171
20	51.0	117	15.8	7.87	16.8	11.2	4.20	0.0397	5.93	108
30	19.3	49.5	7.55	18.8	38.9	26.5	9.86	0.0786	5.49	44.5
40	10.5	24.0	5.15	29.4	65.9	44.0	16.4	0.120	4.25	21.7
50	7.0	13.5	4.21	35.3	92.1	59.1	23.1	0.152	3.17	12.5
60	5.5	8.5	3.74	37.4	112	67.5	29.8	0.173	2.48	8.29
70	4.97	5.85	3.48	36.9	119	69.1	36.6	0.181	2.07	6.47
80	4.71	4.32	3.32	35.1	117	65.7	42.8	0.181	1.84	5.57
90	4.60	3.42	3.28	32.7	109	60.0	47.5	0.176	1.69	5.08
100	4.50	3.02	3.27	30.0	100	54.2	49.7	0.168	1.59	4.82
150	4.32	2.48	3.25	19.5	60.2	32.5	45.1	0.125	1.29	4.29
200	4.30	2.37	3.23	14.1	40.3	22.6	32.3	0.0923	1.11	4.13
250	4.29	2.35	3.21	11.0	29.7	17.0	24.4	0.0711	1.00	4.06
300	4.29	2.37	3.17	8.95	23.0	13.5	19.5	0.0570	0.937	4.01
350	4.27	2.40	3.14	7.55*	18.5*	11.1*	16.2	0.0477	0.929	3.96
400	4.25	2.40	3.11	6.5*	15.4*	9.32*	13.9	0.0409	0.909	3.93
500	4.19	2.36	3.04				10.8	0.0322	0.860	3.86
600	4.12	2.31	2.98				8.92	0.0268	0.807	3.79
800	3.96	2.18	2.84				6.67	0.0201	0.713	3.66
1000	3.79		2.70				5.34	0.0160	0.654	3.52
1200	3.61*		2.55				4.48	0.0134	0.619	3.39
1400							3.84	0.0116	0.588	
1600							3.33	0.0100	0.556	
1800							2.93	0.00895	0.526*	
2000							2.62	0.00807	0.494*	

T/K	Fe	Ge[a]	Mg	Ni	Pb	Pt	Si[a]	Sn	Ti	W
1	1.71	0.274	9.86	2.17	27.9	2.31	0.0693*	183	0.0144*	14.4
2	3.42	2.06	19.6	4.34	44.6	4.60	0.454	323	0.0288*	28.7
3	5.11	5.35	29.0	6.49	35.8	6.79	1.38	297	0.0432	42.8
4	6.77	8.77	37.6	8.59	22.2	8.8	2.97	181	0.0575	56.3
5	8.39	11.6	45.0	10.6	13.8	10.5	5.27	117	0.0719	68.7
6	9.93	13.9	50.8	12.5	8.10	11.8	8.23	76	0.0863	79.5
7	11.4	15.5	54.7	14.2	4.86	12.6	11.7	52	0.101	88.0
8	12.7	16.6	56.7	15.8	3.20	12.9	15.5	36	0.115	93.8
9	13.9	17.3	57.0	17.1	2.30	12.8	19.5	26	0.129	96.8
10	14.8	17.7	55.8	18.1	1.78	12.3	23.3	19.3	0.143	97.1
15	17.0	17.3	41.1	19.5	0.845	8.41	41.6	6.3	0.212	72.0
20	15.4	14.9	27.2	16.5	0.591	4.95	49.8	3.2	0.275	40.5
30	10.0	10.8	12.9	9.56	0.477	2.15	48.1	1.79	0.365	14.4
40	6.23	7.98	7.19	5.82	0.451	1.39	35.3	1.33	0.390	6.92
50	4.05	6.15	4.65	4.00	0.436	1.09	26.8	1.15	0.374	4.27
60	2.85	4.87	3.27	3.08	0.425	0.947	21.1	1.04	0.355	3.14
70	2.16	3.93	2.49	2.50	0.416	0.862	16.8	0.96	0.340	2.58
80	1.75	3.25	2.02	2.10	0.409	0.815	13.4	0.915	0.326	2.29
90	1.50	2.70	1.78	1.83	0.403	0.789	10.8	0.880	0.315	2.17
100	1.34	2.32	1.69	1.64	0.397	0.775	8.84	0.853	0.305	2.08
150	1.04	1.32	1.61	1.22	0.379	0.740	4.09	0.779	0.270	1.92
200	0.94	0.968	1.59	1.07	0.367	0.726	2.64	0.733	0.245	1.85
250	0.865	0.749	1.57	0.975	0.360	0.718	1.91	0.696	0.229	1.80
300	0.802	0.599	1.56	0.907	0.353	0.716	1.48	0.666	0.219	1.74
350	0.744	0.495	1.55	0.850	0.347	0.717	1.19	0.642	0.210	1.67
400	0.695	0.432	1.53	0.802	0.340	0.718	0.989	0.622	0.204	1.59
500	0.613	0.338	1.51	0.722	0.328	0.723	0.762	0.596	0.197	1.46
600	0.547	0.273	1.49	0.656	0.314	0.732	0.619		0.194	1.37
800	0.433	0.198	1.46*	0.676		0.756	0.422		0.197	1.25
1000	0.323	0.174		0.718		0.787	0.312		0.207	1.18
1200	0.283	0.174		0.762		0.826	0.257		0.220	1.12
1400	0.312			0.804		0.871	0.235		0.236	1.08
1600	0.330					0.919	0.221		0.253	1.04
1800	0.345*					0.961			0.270*	1.01
2000						0.994*				0.98

[a] Values below 300 K are typical values.

* Extrapolated.

THERMAL CONDUCTIVITY OF ALLOYS AS A FUNCTION OF TEMPERATURE

This table lists the thermal conductivity of selected alloys at various temperatures. The indicated compositions refer to weight percent. Since the thermal conductivity is sensitive to exact composition and processing history, especially at low temperatures, these values should be considered approximate.

REFERENCES

1. Powell, R. L., and Childs, G. E., in *American Institute of Physics Handbook, 3rd Edition*, Gray, D. E., Ed., McGraw-Hill, New York, 1972.
2. Ho, C. Y., et al., *J. Phys. Chem. Ref. Data*, 7, 959, 1978.

Thermal conductivity in W/m K

	Alloy	4 K	20 K	77 K	194 K	273 K	373 K	573 K	973 K
Aluminum:	1100	50	240	270	220	220			
	2024	3.2	17	56	95	130			
	3003	11	58	140	150	160			
	5052	4.8	25	77	120	140			
	5083, 5086	3	17	55	95	120			
	Duralumin	5.5	30	91	140	160	180		
Bismuth:	Rose metal		5.5	8.3	14	16			
	Wood's metal	4	17	23					
Copper:	electrolytic tough pitch	330	1300	550	400	390	380	370	350
	free cutting, leaded	200	800	460	380	380			
	phosphorus, deoxidized	7.5	42	120	190	220			
	brass, leaded	2.3	12	39	70	120			
	bronze, 68% Cu; 32% Zn	2.3	16	48	92	110			
	beryllium	2	17	36	70	90	113	172	
	german silver	0.75	7.5	17	20	23	25	30	40
	silicon bronze A		3.4	11	23	30			
	manganin	0.48	3.2	14	17	22			
	constantan	0.9	8.6	17	19	22			
Ferrous:	commercial pure iron	15	72	106	82	76	66	54	34
	plain carbon steel(AISI 1020)	13	20	58	65	65			
	plain carbon steel(AISI 1095)		8.5	31	41	45			
	3% Ni; 0.7% Cr; 0.6% Mo		6	22		33	35	36	30
	4% Si					20	24	28	26
	stainless steel	0.3	2	8	13	14	16	19	25
	27% Ni; 15% Cr		1.7	55		11	12	16	21
Gold:	colbalt thermocouple	1.2	8.6	20					
	65% Au; 35% Ag		12	24		61	89		
Indium:	85.5% In; 14.5% Pb	1.9	7.8	24	41				
Lead:	60% Pb; 40% Sn (soft solder)		28	44					
	64.35% Pb; 35.65% In	0.8	3.26	9.1		20.2			
Nickel:	80% Ni; 20% Cr					12	14	17	23
	contracid	0.2	2	7.3	9.5	13			
	inconel	0.5	4.2	12.5	13	15	16	19	26
	monel	0.9	7.1	15	20	21	24	30	43
Platinum:	90% Pt; 10% Ir					31	31.4		
	90% Pt; 10% Rh					30.1	30.5		
Silver:	silver solder		12	34	58				
	normal Ag thermocouple	48	230	310					
Tin:	60% Sn; 40% Pb	16	55	51					
Titanium:	5.5% Al; 2.5% Sn; 0.2% Fe		1.8	4.3	6.4	7.8	8.4	10.8	
	4.7% Mn; 3.99% Al; 0.14% C		1.7	4.5	6.5	8.5			

THERMAL CONDUCTIVITY OF CRYSTALLINE DIELECTRICS

This table lists the thermal conductivity of a number of crystalline dielectrics, including some which find use as optical materials. Values are given at temperatures for which data are available.

REFERENCE

Powell, R. L., and Childs, G. E., in *American Institute of Physics Handbook, 3rd Edition*, Gray, D. E., Ed., McGraw-Hill, New York, 1972.

Material	T/K	Ther. cond. W/m K	Material	T/K	Ther. cond. W/m K
AgCl	223	1.3	BeO	4.2	0.3
	273	1.2		20	16
	323	1.1		77	270
	373	1.1		373	210
Al,B silicate (tourmaline)	398	2.9		573	120
‖ to c axis	540	3.2		1273	29
	723	3.5	Bi_2Te_3	80	6.4
Al,Be silicate (beryl)	315	6.4		204	2.8
Al,F silicate (topaz)	315	17.7		303	3.6
‖ to c axis	358	15.6		370	4.6
	417	13.3	C (diamond)	4.2	13
Al,Fe silicate (garnet)	315	35.8	type I	20	800
	358	35.4		77	3550
	377	35.6		194	1450
Al_2O_3 (sapphire):				273	1000
36° to c axis	4.2	110	$CaCO_3$		
	20	3500	‖ to c axis	83	25
	35	6000		273	5.5
	77	1100	⊥ to c axis	83	17
⊥ to c axis	373	2.6		194	6.5
	523	3.9		273	4.6
	773	5.8		373	3.6
Al_2O_3 (sintered)	4.2	0.5	CaF_2	83	39
	20	23		223	18
	77	150		273	10
	194	48		323	9.2
	273	35		373	9
	373	26	$CaWO_4$ (scheelite)	422	11.3
	973	8	CdTe	160	7.0
Ar	8	6.0		297	3.6
	10	3.7		422	2.9
	20	1.4	CsBr	223	1.2
	77	0.31		273	0.94
As_2S_3 (glass)	283	0.16		323	0.81
	323	0.21		373	0.77
	373	0.27	CsI	223	1.4
BN	1047	36.2		273	1.2
	1475	22.7		323	1
	1928	21.9		373	0.95
	2111	18.5	Cu_2O (cuprite)	102	3.74
BaF_2	225	20		163	7.76
	260	13.4		299	5.58
	305	10.9		360	4.86
	370	10.5	Fe_3O_4 (magnetite)	4.5	27.4
$BaTiO_3$	5	4.2		20.5	293.0
	30	24.0		126.5	7.4
	40	25.0		304	7.0
	100	12.0	Glass:		
	250	4.8	phoenix	4.2	0.095
	300	6.2		20	0.13
				77	0.37

Material	T/K	Ther. cond. W/m K	Material	T/K	Ther. cond. W/m K
plastic perspex	4.2	0.058	NaCl	4.2	440
	20	0.074		20	300
pyrex	77	0.44		77	30
	194	0.88		273	6.4
	273	1		323	5.6
H$_2$ (para + 0.5% ortho)	2.5	100		373	5.4
	3	150	NaF	5	1100
	4	200		50	250
	6	30		100	90
	10	3	Ne	2	3.0
H$_2$O (ice)	173	3.5		3	4.6
	223	2.8		4.2	4.2
	273	2.2		10	0.8
He^3 (high pressure)	0.6	25		20	0.3
	1	2	NH$_4$Cl	77	17
	1.5	0.57		194	23
	2	0.21		230	38
He^4 (high pressure)	0.5	42		273	27
	0.8	120	NH$_4$H$_2$PO$_4$		
	1	24	‖ to optic axis	315	0.71
	2	0.18		339	0.71
I$_2$	300	0.45	⊥ to optic axis	313	1.26
	325	0.42		342	1.34
	350	0.4	NiO	4.2	5.9
KBr	2	150		40	400
	4.2	360		194	82
	100	12	SiO$_2$ (quartz)		
	273	5	‖ to c axis	20	720
	323	4.8		194	20
	373	4.8		273	12
KCl	4.2	500	⊥ to c axis	20	370
	25	140		194	10
	80	35		273	6.8
	194	10	SiO$_2$ (fused silica)	4.2	0.25
	273	7.0		20	0.7
	323	6.5		77	0.8
	373	6.3		194	1.2
KI	4.2	700		273	1.4
	80	13		373	1.6
	194	4.6		673	1.8
	273	3.1	SrTiO$_3$	5	2.4
Kr	4.2	0.48		30	21.0
	10	1.7		40	19.2
	20	1.2		100	18.5
	77	0.36		250	12.5
LaF$_3$	78	7.8		300	11.2
	197	5.0	TlBr	316	0.59
	274	5.4	TlCl	311	0.75
LiF	4.2	620	TiO$_2$ (rutile)		
	20	1800	‖ to optic axis	4.2	200
	77	150		20	1000
MgO·Al$_2$O$_3$ (spinel)	373	13		273	13
	773	8.5	⊥ to optic axis	4.2	160
MnO	4.2	0.25		20	690
	40	55		273	9
	120	8			
	573	3.5			

THERMAL CONDUCTIVITY OF CERAMICS AND OTHER INSULATING MATERIALS

Thermal conductivity values for ceramics, refractory oxides, and miscellaneous insulating materials are given here. The thermal conductivity refers to samples with density indicated in the second column. Since most of these materials are highly variable, the values should only be considered as a rough guide.

REFERENCES

1. Powell, R. L., and Childs, G. E., in *American Institute of Physics Handbook, 3rd Edition*, Gray, D. E., Ed., McGraw-Hill, New York, 1972.
2. Perry, R. H., and Green, D., *Perry's Chemical Engineers' Handbook, Sixth Edition*, McGraw-Hill, New York, 1984.

Material	Dens. g/cm^3	t °C	Ther. cond. W/m K
Alumina (Al$_2$O$_3$)	3.8	100	30
		400	13
		1300	6
		1800	7.4
	3.5	100	17
		800	7.6
Al$_2$O$_3$ + MgO		100	15
		400	10
		1000	5.6
Asbestos	0.4	-100	0.07
		100	1
		0	0.09
Asbestos + 85% MgO	0.3	30	0.08
Asphalt	2.1	20	0.06
Beryllia (BeO)	2.8	100	210
		400	90
		1000	20
		1800	15
	1.85	50	64
		200	40
		600	23
Brick, dry	1.54	0	0.04
Brick, refractory:			
alosite		1000	1.3
aluminous	1.99	400	1.2
		1000	1.3
diatomaceous	0.77	100	0.2
		500	0.24
	0.4	100	0.08
		500	0.1
fireclay	2	400	1
		1000	1.2
silicon carbide	2	200	2
		600	2.4
vermiculite	0.77	200	0.26
		600	0.31
Calcium oxide		100	16
		400	9
		1000	7.5
Cement mortar	2	90	0.55
Charcoal	0.2	20	0.055
Coal	1.35	20	0.26
Concrete	1.6	0	0.8
Cork	0.05	0	0.03
		100	0.04
	0.35	0	0.06
		100	0.08
Cotton wool	0.08	30	0.04
Diatomite	0.2	0	0.05
		400	0.09
	0.5	0	0.09
		400	0.16
Ebonite	1.2	0	0.16
Felt, flax	0.2	30	0.05
	0.3	30	0.04
Fuller's earth	0.53	30	0.1
Glass wool	0.2	-200 to 20	0.005
		50	0.04
		100	0.05
		300	0.08
Graphite			
100 mesh	0.48	40	0.18
20-40 mesh	0.7	40	1.29
Linoleum cork	0.54	20	0.08
Magnesia (MgO)		100	36
		400	18
		1200	5.8
		1700	9.2
MgO + SiO$_2$		100	5.3
		400	3.5
		1500	2.3
Mica:			
muscovite		100	0.72
		300	0.65
		600	0.69
phlogopite		100	0.66
Canadian		300	0.19
		600	0.2
Micanite		30	0.3
Mineral wool	0.15	30	0.04
Perlite, expanded	0.1	-200 to 20	0.002
Plastics:			
bakelite	1.3	20	1.4
celluloid	1.4	30	0.02
polystyrene foam	0.05	-200 to 20	0.033
mylar foil	0.05	-200 to 20	0.0001
nylon		-253	0.10
		-193	0.23
		25	0.30
teflon		-253	1.43
		-193	2.15
		25	2.25
		230	2.5
urethane foam	0.07	20	0.06
Porcelain		90	1

Material	Dens. g/cm^3	t °C	Ther. cond. W/m K	Material	Dens. g/cm^3	t °C	Ther. cond. W/m K
Rock:				Uranium dioxide		100	9.8
basalt		20	2			400	5.5
chalk		20	0.92			1000	3.4
granite	2.8	20	2.2	Wood:			
limestone	2	20	1	balsa, ⊥	0.11	30	0.04
sandstone	2.2	20	1.3	fir, ⊥	0.54	20	0.14
slate, ⊥		95	1.4	fir, ‖	0.54	20	0.35
slate, ‖		95	2.5	oak		20	0.16
Rubber:				plywood		20	0.11
sponge	0.2	20	0.05	pine, ⊥	0.45	60	0.11
92 percent		25	0.16	pine, ‖	0.45	60	0.26
Sand, dry	1.5	20	0.33	walnut, ⊥	0.65	20	0.14
Sawdust	0.2	30	0.06	Wool	0.09	30	0.04
Shellac		20	0.23	Zinc oxide		200	17
Silica aerogel	0.1	-200 to 20	0.003			800	5.3
Snow	0.25	0	0.16	Zirconia (ZrO$_2$)		100	2
Steel wool	0.1	55	0.09			400	2
Thoria (ThO$_2$)		100	10			1500	2.5
		400	5.8	Zirconia + silica		200	5.6
		1500	2.4			600	4.6
Titanium dioxide		100	6.5			1500	3.7
		400	3.8				
		1200	3.3				

THERMAL CONDUCTIVITY OF GLASSES

This table gives the composition of various types of glasses and the thermal conductivity k as a function of temperature. Because of the variability of glasses, the data should be regarded as only approximate.

Type of glass	SiO₂ (wt%)	Other oxides (wt%)		t °C	k W/m K
Vitreous silica	100			−150	0.85
				−100	1.05
				−50	1.20
				0	1.30
				50	1.40
				100	1.50
Vycor glass	96	B_2O_3	3	−100	1.00
				0	1.25
				100	1.40
Pyrex type chemically-resistant borosilicate glasses	80–81	B_2O_3	12–13	−100	0.90
		Na_2O	4	0	1.10
		Al	2	100	1.25
Borosilicate crown glasses	60–65	B_2O_3	15–20	−100	0.65–0.75
				0	0.90–0.95
				100	1.00–1.05
	65–70	B_2O_3	10–15	−100	0.75–0.80
				0	0.95–1.00
				100	1.05–1.15
	70–75	B_3O_3	5–10	−100	0.80–0.85
				0	1.05–1.10
				100	1.15–1.20
Zinc crown glasses (i)	55–65	ZnO Remainder: B_2O_3, Al_2O_3	5–15	−100	0.88–0.92
				0	1.10–1.15
				100	1.15–1.25
		ZnO Remainder: Na_2O, K_2O	5–15	−100	0.60–0.70
				0	0.70–0.90
				100	0.85–0.95
		ZnO Remainder: B_2O_3, Al_2O_3	15–25	−100	0.88–0.92
				0	1.10–1.15
				100	1.15–1.20
		ZnO Remainder: Na_2O, K_2O	15–25	−100	0.65–0.80
				0	0.85–0.95
				100	0.90–1.05
Zinc crown glasses (ii)	65–75	ZnO Remainder: B_2O_3, Al_2O_3	5–15	−100	0.88–0.92
				0	1.15–1.15
				100	1.20–1.30
		ZnO Remainder: Na_2O, K_2O	5–15	−100	0.70–0.85
				0	0.90–1.05
				100	1.00–1.15

Type of glass	SiO$_2$ (wt%)	Other oxides (wt%)		t °C	k W/m K
		ZnO	15–25	−100	0.90–0.95
		Remainder:		0	1.15–1.15
		B$_2$O$_3$, Al$_2$O$_3$		100	1.20–1.25
		ZnO	15–25	−100	0.65–0.85
		Remainder:		0	0.85–1.00
		Na$_2$O, K$_2$O		100	1.05–1.20
Barium crown glasses	31	B$_2$O$_3$	12	−100	0.55
		Al$_2$O$_3$	8	0	0.70
		BaO	48	100	0.80
	41	B$_2$O$_3$	6	−100	0.60
		Al$_2$O$_3$	2	0	0.75
		ZnO	8	100	0.85
		BaO	43		
	47	B$_2$O$_3$	4	−100	0.65
		Na$_2$O	1	0	0.75
		K$_2$O	7	100	0.90
		ZnO	8		
		BaO	32		
	65	B$_2$O$_3$	2	−100	0.70
		Na$_2$O	5	0	0.90
		K$_2$O	15	100	1.00
		ZnO	2		
		BaO	10		
Borate glasses					
Borate flint glass	9	B$_2$O$_3$	36	−100	0.55
		Na$_2$O	1	0	0.65
		K$_2$O	2	100	0.80
		PbO	36		
		Al$_2$O$_3$	10		
		ZnO	6		
Borate flint glass	0	B$_2$O$_3$	56	−100	0.50
		Al$_2$O$_3$	12	0	0.65
		PbO	32	100	0.85
Borate flint glass	0	B$_2$O$_3$	43	−100	0.40
		Al$_2$O$_3$	5	0	0.55
		PbO	52	100	0.70
Borate glass	4	B$_2$O$_3$	55	−100	0.65
		Al$_2$O$_3$	14	0	0.80
		PbO	11	100	0.90
		K$_2$O	4		
		ZnO	12		
Borate crown glass	0	B$_2$O$_3$	64	−100	0.50
		Na$_2$O	8	0	0.65
		K$_2$O	3	100	0.85
		BaO	4		
		PbO	3		
		Al$_2$O$_3$	18		

Type of glass	SiO$_2$ (wt%)	Other oxides (wt%)		t °C	k W/m K
Light borate crown glass	0	B$_2$O$_3$	69	−100	0.55
		Na$_2$O	8	0	0.70
		BaO	5	100	0.90
		Al$_2$O$_3$	18		
Zinc borate glass	0	B$_2$O$_3$	40	−100	0.65
		ZnO	60	0	0.75
				100	0.85
Phosphate crown glasses					
Potash phosphate glass	0	P$_2$O$_5$	70	0	0.75
		B$_2$O$_3$	3	100	0.85
		K$_2$O	12		
		Al$_2$O$_3$	10		
		MgO	4		
Baryta phosphate glass	0	P$_2$O$_5$	60	45	0.75
		B$_2$O$_3$	3		
		Al$_2$O$_3$	8		
		BaO	28		
Soda-lime glasses	75	Na$_2$O	17	−100	0.75
		CaO	8	0	0.95
				100	1.10
	75	Na$_2$O	12	−100	0.90
		CaO	13	0	1.10
				100	1.15
	72	Na$_2$O	15	−100	0.80
		CaO	11	0	1.00
		Al$_2$O$_3$	2	100	1.15
	65	Na$_2$O	25	−100	0.65
		CaO	10	0	0.85
				100	0.95
	65	Na$_2$O	15	−100	0.85
		CaO	20	0	1.00
				100	1.10
	60	Na$_2$O	20	−100	0.75
		CaO	20	0	0.90
				100	1.00
Other crown glasses					
Crown glass	75	Na$_2$O	9	−100	0.80
		K$_2$O	11	0	1.00
		CaO	5	100	1.10
High dispersion crown glass	68	Na$_2$O	16	−100	0.65
		ZnO	3	0	0.85
		PbO	13	100	1.00

Type of glass	SiO$_2$ (wt%)	Other oxides (wt%)		t °C	k W/m K
Miscellaneous flint glasses					
(i) Silicate flint glasses					
Light flint glasses	65	PbO	25	−100	0.65–0.70
		Others	10	0	0.88–0.92
				100	1.00–1.05
	55	PbO	35	−100	0.60–0.65
		Others	10	0	0.75–0.85
				100	0.88–0.92
Ordinary flint glass	45	PbO	45	−100	0.50–0.60
		Others	10	0	0.65–0.75
				100	0.80–0.85
Heavy flint glass	35	PbO	60	−100	0.45–0.50
		Others	5	0	0.60–0.65
				100	0.70–0.75
Very heavy flint glasses	25	PbO	73	−100	0.40–0.45
		Others	2	0	0.55–0.60
				100	0.63–0.67
	20	PbO	80	−100	0.40
				0	0.50
				100	0.60
(ii) Borosilicate flint glass	33	B$_2$O$_3$	31	−100	0.65
		PbO	25	0	0.85
		Al$_2$O$_3$	7	100	0.95
		K$_2$O	3		
		Na$_2$O	1		
(iii) Barium flint glass	50	BaO	24	−100	0.60
		PbO	6	0	0.70
		K$_2$O	8	100	0.85
		Na$_2$O	3		
		ZnO	8		
		Sb$_2$O$_3$	1		
Other glasses					
Potassium glass	59	K$_2$O	33	50	0.88–0.92
		CaO	8		
Iron glasses	63	Fe$_2$O$_3$	10	−100	0.80
		Na$_2$O	17	0	0.95
		MgO	4	100	1.05
		CaO	3		
		Al$_2$O$_3$	2		
	67	Fe$_2$O$_3$	15	0	0.88—0.92
		Na$_2$O$_3$	18	100	1.00—1.05
	62	Fe$_2$O$_3$	20	0	0.85—0.90
		Na$_2$O	18	100	0.95—1.00
Rock glasses					
Obsidian				0	1.35
Artificial diabase				100	1.25

COMMERCIAL METALS AND ALLOYS

This table gives typical values of mechanical, thermal, and electrical properties of several common commercial metals and alloys. Values refer to ambient temperature (0 to 25°C). All values should be regarded as typical, since these properties are dependent on the particular type of alloy, heat treatment, and other factors. Values for individual specimens can vary widely.

REFERENCES

1. *ASM Metals Reference Book, Second Edition*, American Society for Metals, Metals Park, OH, 1983.
2. Lynch, C. T., *CRC Practical Handbook of Materials Science*, CRC Press, Boca Raton, FL, 1989.
3. Shackelford, J. F., and Alexander, W., *CRC Materials Science and Engineering Handbook*, CRC Press, Boca Raton, FL, 1991.

Common name	Thermal conductivity W/cm K	Density g/cm³	Coeff. of linear expansion $10^{-16}/°C$	Electrical resistivity μΩ cm	Modulus of elasticity GPa	Tensile strength MPa	Approx. melting point °C
Ingot iron	0.7	7.86	11.7	9.7	205	-	1540
Plain carbon steel AISI-SAE 1020	0.52	7.86	11.7	18	205	450	1515
Stainless steel type 304	0.15	7.9	17.3	72	195	550	1425
Cast gray iron	0.47	7.2	10.5	67	90	180	1175
Malleable iron		7.3	12	30	170	345	1230
Hastelloy C	0.12	8.94	11.3	125·	200	780	1350
Inconel	0.15	8.25	11.5	103	200	800	1370
Aluminum alloy 3003, rolled	1.9	2.73	23.2	3.7	70	110	650
Aluminum alloy 2014, annealed	1.9	2.8	23.0	3.4	70	185	650
Aluminum alloy 360	1.5	2.64	21.0	7.5	70	325	565
Copper, electrolytic (ETP)	3.9	8.94	16.5	1.7	120	300	1080
Yellow brass (high brass)	1.2	8.47	20.3	6.4	100	300-800	930
Aluminum bronze	0.7	7.8	16.4	12	120	400-600	1050
Beryllium copper 25	0.8	8.23	17.8	7	130	500-1400	925
Cupronickel 30%	0.3	8.94	16.2		150	400-600	1200
Red brass, 85%	1.6	8.75	18.7	11	90	300-700	1000
Chemical lead	0.35	11.34	29.3	21	13	17	327
Antimonial lead (hard lead)	0.3	10.9	26.5	23	20	47	290
Solder 50-50	0.5	8.89	23.4	15	-	42	215
Magnesium alloy AZ31B	1.0	1.77	26	9	45	260	620
Monel	0.3	8.84	14.0	58	180	545	1330
Nickel (commercial)	0.9	8.89	13.3	10	200	460	1440
Cupronickel 55-45 (constantan)	0.2	8.9	18.8	49	160	-	1260
Titanium (commercial)	1.8	4.5	8.5	43	110	330-500	1670
Zinc (commercial)	1.1	7.14	32.5	6	-	130	419
Zirconium (commercial)	0.2	6.5	5.85	41	95	450	1855

HARDNESS OF MINERALS AND CERAMICS

There are several hardness scales for describing the resistance of a material to indentation or scratching. This table lists a number of common materials in order of increasing hardness. Values are given, when available, on three different hardness scales: the original Mohs Scale (range 1 to 10); the modified Mohs Scale (range 1 to 15), and the Knoop Hardness Scale. In the last case, a load of 100 g is assumed.

REFERENCE

Shackelford, J. F. and Alexander, W., *CRC Materials Science and Engineering Handbook*, CRC Press, Boca Raton, FL, 1991.

Material	Formula	Mohs	Modified mohs	Knoop
Graphite	C	0.5		
Talc	$3MgO \cdot 4SiO_2 \cdot H_2O$	1	1	
Alabaster	$CaSO_4 \cdot 2H_2O$	1.7		
Gypsum	$CaSO_4 \cdot 2H_2O$	2	2	32
Halite (rock salt)	NaCl	2		
Stibnite (antimonite)	Sb_2S_3	2.0		
Galena	PbS	2.5		
Mica		2.8		
Calcite	$CaCO_3$	3	3	135
Barite	$BaSO_4$	3.3		
Marble		3.5		
Aragonite	$CaCO_3$	3.5		
Dolomite	$CaMg(CO_3)_2$	3.5		
Fluorite	CaF_2	4	4	163
Magnesia	MgO	5		370
Apatite	$CaF_2 \cdot 3Ca_3(PO_4)_2$	5	5	430
Opal		5		
Feldspar (orthoclase)	$K_2O \cdot Al_2O \cdot 6SiO_2$	6	6	560
Augite		6		
Hematite	Fe_2O_3	6		750
Magnetite	Fe_3O_4	6		
Rutile	TiO_2	6.2		
Pyrite	FeS_2	6.3		
Agate	SiO_2	6.5		
Uranium dioxide	UO_2	6.7		600
Silica (fused)	SiO_2		7	
Quartz	SiO_2	7	8	820
Flint		7		
Silicon	Si	7		
Andalusite	Al_2OSiO_4	7.5		
Zircon	$ZrSiO_4$	7.5		
Zirconia	ZrO_2			1200
Aluminum nitride	AlN			1225
Beryl	$Be_3Al_2Si_6O_{18}$	7.8		
Beryllia	BeO			1300
Topaz	$Al_2SiO_4(OH,F)_2$	8	9	1340
Garnet	$Al_2O_3 \cdot 3FeO \cdot 3SiO_2$		10	1360
Emery	Al_2O_3 (impure)	8		
Zirconium nitride	ZrN	8+		1510
Zirconium boride	ZrB_2			1560
Titanium nitride	ZrN	9		1770
Zirconia (fused)	ZrO_2		11	
Tantalum carbide	TaC			1800
Tungsten carbide	WC			1880
Corundum (alumina)	Al_2O_3	9		2025
Zirconium carbide	ZrC			2150
Alumina (fused)	Al_2O_3		12	
Beryllium carbide	Be_2C			2400

Material	Formula	Mohs	Modified mohs	Knoop
Titanium carbide	TiC			2470
Carborundum (silicon carbide)	SiC	9.3	13	2500
Aluminum boride	AlB			2500
Tantalum boride	TaB_2			2600
Boron carbide	B_4C		14	2800
Boron	B	9.5		—
Titanium boride	TiB_2			2850
Diamond	C	10	15	7000

Section 13
Polymer Properties

NAMING ORGANIC POLYMERS

Robert B. Fox

Organic polymers have been traditionally named on the basis of the monomer used, a hypothetical monomer, or a semisystematic structure. Alternatively, they may be named in the same way as organic compounds, i.e., on the basis of a structure as written. While the former methods, often called "source-based nomenclature", frequently result in ambiguity and multiple names for a single material, the latter method, termed "structure-based nomenclature", generates a sometimes cumbersome unique name for a given polymer, independent of its source. Within their limitations, both types of names are acceptable and well-documented.[1] The use of stereochemical descriptors with both types of polymer nomenclature has been published.[1]

TRADITIONAL POLYMER NAMES

Monomer-Based Names

"Polystyrene" is the name of a polymer made from the monomer styrene. Generally, the name of the monomer should be enclosed in parentheses, as in "poly(methyl methacrylate)" or "poly(4-bromostyrene)" to more clearly identify the monomer. Many names are often possible for a given polymer with this method: "poly(ethylene glycol)", "poly(ethylene oxide)", and "poly(oxirane)" describe the same polymer. Sometimes, the name of a hypothetical monomer is used, as in "poly(vinyl alcohol)". Even though a name like "polyethylene" covers a multitude of materials, the system does provide understandable names when a single monomer is involved in the synthesis of a single polymer. When one monomer can yield more than one polymer, as with 1,3-butadiene, some sort of structural notation must be used to identify the product, and one is not far from a formal structure-based name.

Copolymers, Block Polymers, and Graft Polymers. Monomer-based names are more complex when more than one monomer is involved. Some common polymers have been given names based on an apparent structure, as with "poly(ethylene terephthalate)". A better system has been approved by the IUPAC.[1] With this method, the arrangement of the monomeric units is introduced through use of an italicized connective placed between the names of the monomers. For monomer names represented by A, B, and C, the various types of arrangements are

Type	Connective	Example
unspecified	*-co-*	poly(A-*co*-B)
statistical	*-stat-*	poly(A-*stat*-B)
random	*-ran-*	poly(A-*ran*-B)
alternating	*-alt-*	poly(A-*alt*-B)
periodic	*-per-*	poly(A-*per*-B-*per*-C)
block	*-block-*	polyA-*block*-polyB
graft	*-graft-*	polyA-*graft*-polyB

A long dash may be used in place of the word "block". These connectives can be used in combination and with small joining units. Names such as "poly[ethylene-*ran*-(vinyl chloride)]" and "polystyrene-*block*-dimethylsilylene-*block*-polybutadiene" result.

STRUCTURE-BASED POLYMER NOMENCLATURE

Regular Single-Strand Polymers

Structure-based nomenclature has been approved by the IUPAC[4] and is currently being updated; it is used by *Chemical Abstracts*.[5] Monomer names are not involved. To the extent that a polymer chain can be described by a repeating segment of the chain, it can be named "poly(repeating unit)", where "repeating unit" is, in most cases, a bivalent group.

Since there are usually many possible repeating units in a given chain, it is necessary to select systematically one, called the "constitutional repeating unit" (CRU), to provide a unique and unambiguous name, "poly(CRU)", where "CRU" is a recitation of the names of successive units as one proceeds through the CRU. For this purpose, a portion of the main chain structure that includes at least two repeating sequences is written out. These sequences will be composed of bivalent subunits such as -CH_2-, -O-, and groups from ring systems, each of which can be named by the usual rules of nomenclature.[6,7]

Where a chain is simply one long sequence of a single subunit, that subunit is itself the CRU, as in "poly(methylene)" or "poly(1,4-phenylene)". In chains having more than one kind of subunit, a seniority system is used to determine the beginning of the CRU and the direction to move along the main chain atoms (following the shortest path in rings) to complete the CRU. For citation of the first, most senior, subunit, the descending order of seniority is (1) heterocyclic rings, (2) hetero atoms, (3) carbocyclic rings, and, lowest, (4) acyclic groups containing only carbon atoms.

Within each of these classes, there is a further order of seniority that follows the usual rules of nomenclature. Thus, among hetero atoms, the descending order of seniority is O, S, Se, N, P, As, Si, Ge, B, with other atoms fitted in according to their place in the Periodic Table. The descending order of seniority among heterocylic rings is (a) a ring system containing N; (b) a ring system containing a hetero atom (other than N) as high as possible in the order of hetero atoms given above; (c) a ring system having the greatest number of rings; (d) a ring system with the largest individual ring; (e) a ring system with the largest number of hetero atoms. Among equivalent ring systems, preference is given to the one having lowest locants for the free valences in the subunit, and among otherwise identical ring systems, the one having least hydrogenation is senior. Lowest locants in unsaturated chains are also given preference. Lowest locants for substituents, independent of kind except for alphabetical order, are the final determinant of seniority.

Direction within the repeating unit depends upon the shortest path, in main chain atoms, from the most senior subunit to another subunit of the same kind or to a subunit next lower in seniority. Once identification of the CRU is established, the name of the CRU is formed from the names of largest possible subunits within the CRU. For example, the main chain of the polymer traditionally named "poly(ethylene terephthalate)" has the structure

$$...CH_2CH_2OC(O) - \bigcirc - C(O) [OCH_2CH_2OC(O) - \bigcirc - C(O)]OCH_2CH_2OC(O) - \bigcirc - C(O) O...$$

The CRU is enclosed in brackets and read from left to right. It is selected because -O- is the most senior subunit, and two carbon atoms is the shortest path to the next -O-. The structure-based name of this polymer is then "poly(oxyethyleneoxyterephthaloyl)", not much longer than the traditional name and much more adaptable to the complexities of substitution. As organic nomenclature evolves, more systematic names can be applied to the subunits, such as "ethane-1,2-diyl" instead of "ethylene".

Structure-based nomenclature can also be used when the CRU has no carbon atoms. An example is the polymer traditionally named "poly(dimethylsiloxane)", which on the basis of structure would be named "poly(oxydimethylsilylene)" or "poly(oxydimethylsilanediyl)". This nomenclature method has also been applied to inorganic and coordination polymers[8] and to double-strand (ladder and spiro) organic polymers.[9]

Additional examples of both structure-based and traditional names will be found in the tables of glass transition temperatures (13-4 to 11) in this *Handbook* as well as in the *Chemical Abstracts* indexes and in the IUPAC documents.

Irregular Single-Strand Polymers

Polymers that cannot be described by the repetition of a single CRU or comprise units not all connected identically in a directional sense can also be named on a structure basis.[10] These include copolymers, block and graft polymers, star polymers, and chemically modified polymers. They are given names of the type "poly(A/B/C...)", where A, B, C, etc. are the names of the component constitutional units, the number of which are minimized. The constitutional units may include regular or irregular blocks as well as atoms or atomic groupings, and each is named by the method described above or by the rules of organic nomenclature.

The oblique stroke denotes an unspecified arrangement of the units within the main chain. For example, a statistical copolymer derived from styrene and vinyl chloride with the monomeric units joined head-to-tail is named "poly(1-chloroethylene/1-phenylethylene)". A polymer obtained by 1,4 polymerization and both head-to-head and head-to-tail 1,2 polymerization of 1,3-butadiene would be named "poly(but-2-ene-1,4-diyl/1-vinylethylene/ 2-vinyl-ethylene)". A long dash is used separate components of block polymers, as in "poly(A)—poly(B)—poly(C)" or "poly(A)—X—poly(B)". Graft polymers are named in the same way as a substituted polymer but without the ending "yl" for the grafted chain: "poly[Z/poly(A)Z]" is the name of a regular polymer consisting of Z units in which some have grafts of "poly(A)". Star polymers are treated as a central unit with substituent blocks, as in "tetrakis(polymethylene)silane".

REFERENCES

1. International Union of Pure and Applied Chemistry, *Compendium of Macromolecular Nomenclature*, Blackwell Scientific Publications, Oxford, 1991.
2. International Union of Pure and Applied Chemistry, Stereochemical Definitions and Notations Relating to Polymers (Recommendations 1980), *Pure & Appl. Chem.*, 53, 733—752, 1981.
3. International Union of Pure and Applied Chemistry, Source-Based Nomenclature for Copolymers (Recommendations 1985), *Pure & Appl. Chem.*, 57, 1427—1440, 1985)
4. International Union of Pure and Applied Chemistry, Nomenclature of Regular Single-Strand Organic Polymers (1975), *Pure & Appl. Chem.*, 48, 373—385, 1976.
5. Chemical Abstracts Service, Naming and Indexing of Chemical Substances for Chemical Abstracts, Appendix IV, *Chemical Abstracts 1994 Index Guide*.
6. International Union of Pure and Applied Chemistry, *A Guide to IUPAC Nomenclature of Organic Compounds (1993)*, Blackwell Scientific Publications, Oxford, 1993.
7. International Union of Pure and Applied Chemistry, *Nomenclature of Organic Chemistry, Sections A, B, C, D, E, F, and H*, Pergamon Press, Oxford, 1979.
8. International Union of Pure and Applied Chemistry, Nomenclature of Regular Double-Strand and Quasi-Single-Strand Inorganic and Coordination Polymers (Recommendations 1984), *Pure & Appl. Chem.*, 57, 149—168, 1985.
9. International Union of Pure and Applied Chemistry, Nomenclature of Regular Double-Strand (Ladder and Spiro) Organic Polymers (Recommendations 1993), *Pure & Appl. Chem.*, 65, 1561—1580, 1993.
10. International Union of Pure and Applied Chemistry, Structure-Based Nomenclature for Irregular Single-Strand Organic Polymers (Recommendations 1994), *Pure & Appl. Chem.*, 66, 873—889, 1994.

SOLVENTS FOR COMMON POLYMERS

Abbreviations: HC: hydrocarbons; MEK: methyl ethyl ketone; THF: tetrahydrofuran; DMF: dimethylformamide; DMSO: dimethylsulfoxide

Polyethylene (HDPE)	HC and halogenated HC
Polypropylene (atactic)	HC and halogenated HC
Polybutadiene	HC, THF, ketones
Polystyrene	ethylbenzene, $CHCl_3$, CCl_4, THF, MEK
Polyacrylates	aromatic HC, chlorinated HC, THF, esters, ketones
Polymethacrylates	aromatic HC, chlorinated HC, THF, esters, MEK
Polyacrylamide	water
Poly(vinyl ethers)	halogenated HC, MEK, butanol
Poly(vinyl alcohol)	glycols (hot), DMF
Poly(vinyl acetate)	aromatic HC, chlorinated HC, THF, esters, DMF
Poly(vinyl chloride)	THF, DMF, DMSO
Poly(vinylidene chloride)	THF (hot), dioxane, DMF
Poly(vinyl fluoride)	DMF, DMSO (hot)
Polyacrylonitrile	DMF, DMSO
Poly(oxyethylene)	aromatic HC, $CHCl_3$, alcohols, esters, DMF
Poly(2,6-dimethylphenylene oxide)	aromatic HC, halogenated HC
Poly(ethylene terephthalate)	phenol, DMSO (hot)
Polyurethanes (linear)	aromatic HC, THF, DMF
Polyureas	phenol, formic acid
Polysiloxanes	HC, THF, DMF
Poly[bis(2,2,2-trifluoroethoxy)-phosphazene]	THF, ketones, ethyl acetate

GLASS TRANSITION TEMPERATURE FOR SELECTED POLYMERS

Robert B. Fox

Polymer names are based on the IUPAC structure-based nomenclature system described in the table "Naming Organic Polymers". Within each category, names are listed in alphabetical order. Source-based and trivial names are also given (in italics) for the most common polymers. The table does not include polymers for which T_g is not clearly defined because of variability of structure or because of reactions taking place near the glass transition.

All values of T_g cited in this table have been determined by differential scanning calorimetry (DSC) except those values indicated by:

(D)   dynamic method
(Dil)   dilatometry
(M)   mechanical method

Polymer name	Glass transition temperature (T_g/K)
ACYCLIC CARBON CHAINS	
Polyalkadienes	
Poly(alkenylene) *Polyalkadiene* $-[CH=CHCH_2CH_2]-$	
Poly(*cis*-1-butenylene)	171
cis-1,3-polybutadiene [PBD]	
Poly(*trans*-1-butenylene)	215
trans-1,3-polybutadiene [PBD]	
Poly(1-chloro-*cis*-1-butenylene)	253
cis-1,3-polychloroprene	
Poly(1-chloro-*trans*-1-butenylene)	233
trans-1,3-polychloroprene	
Poly(1-methyl-*cis*-1-butenylene)	200
cis-1,3-polyisoprene	
Poly(1-methyl-*trans*-1-butenylene)	207
trans-1,3-polyisoprene	
Poly(1,4,4-trifluoro-1-butenylene)	238
Polyalkenes	
Poly(alkylethylene) *Poly(alkylethylene)* $-[RCHCH_2]-$	
Poly(1-benzylethylene)	333
Poly(1-butylethylene)	223
Poly(1-cyclohexylethylene) (atactic)	393
Poly(1-cyclohexylethylene) (isotactic)	406 (D)
Poly(1,1-dimethylethylene)	200
Polyisobutylene [PIB]	
Poly(ethylene)	148
Poly(methylene)	155
Poly(1-phenethylethylene)	283
Poly(propylene) (isotactic)	272
Poly(propylene) (syndiotactic)	ca. 265
Poly[1-(2-pyridyl)ethylene]	377
Poly[1-(4-pyridyl)ethylene]	415
Poly(1-vinylethylene)	273
Polyacrylics	
Poly[1-(alkoxycarbonyl)ethylene] *Poly(alkyl acrylate)* $-[(ROCO)CHCH_2]-$	
Poly[1-(benzyloxycarbonyl)ethylene]	279
Poly[1-(butoxycarbonyl)ethylene]	219 (M)
Poly(butyl acrylate) [PBA]	

Polymer name	Glass transition temperature (T_g/K)
Poly[1-(*sec*-butoxycarbonyl)ethylene]	251
Poly[1-(butoxycarbonyl)-1-cyanoethylene]	358
Poly[1-(butylcarbamoyl)ethylene]	319 (M)
Poly(1-carbamoylethylene)	438
Polyacrylamide [PAM]	
Poly(1-carboxyethylene)	379
Poly(acrylic acid) [PAA]	
Poly[1-(2-chlorophenoxycarbonyl)ethylene]	326
Poly[1-(4-chlorophenoxycarbonyl)ethylene]	331
Poly[1-(4-cyanobenzyloxycarbonyl)ethylene]	317
Poly[1-(2-cyanoethoxycarbonyl)ethylene]	277
Poly[1-(cyanomethoxycarbonyl)ethylene)]	433 Dil
Poly[1-(4-cyanophenoxycarbonyl)ethylene]	363
Poly[1-(cyclohexyloxycarbonyl)ethylene]	292
Poly[1-(2,4-dichlorophenoxycarbonyl)ethylene]	333
Poly[1-(dimethylcarbamoyl)ethylene]	362
Poly[1-(ethoxycarbonyl)ethylene]	249
Poly(ethyl acrylate) [PEA]	
Poly[1-(ethoxycarbonyl)-1-fluoroethylene]	316
Poly[1-(2-ethoxycarbonylphenoxycarbonyl)ethylene]	303
Poly[1-(3-ethoxycarbonylphenoxycarbonyl)ethylene]	297
Poly[1-(4-ethoxycarbonylphenoxycarbonyl)ethylene]	310
Poly[1-(2-ethoxyethoxycarbonyl)ethylene]	223
Poly[1-(3-ethoxypropoxycarbonyl)ethylene]	218
Poly[1-(isopropoxycarbonyl)ethylene]	267-270
Poly[1-(methoxycarbonyl)ethylene]	283
Poly(methyl acrylate) [PMA]	
Poly[1-(2-methoxycarbonylphenoxycarbonyl)ethylene]	319
Poly[1-(3-methoxycarbonylphenoxycarbonyl)ethylene]	311
Poly[1-(4-methoxycarbonylphenoxycarbonyl)ethylene]	340
Poly[1-(2-methoxyethoxycarbonyl)ethylene]	223
Poly[1-(4-methoxyphenoxycarbonyl)ethylene]	324
Poly[1-(3-methoxypropoxycarbonyl)ethylene]	198
Poly[1-(2-naphthyloxycarbonyl)ethylene]	358
Poly[1-(pentachlorophenoxycarbonyl)ethylene]	420
Poly[1-(phenethoxycarbonyl)ethylene]	270
Poly[1-(phenoxycarbonyl)ethylene]	330
Poly[1-(*m*-tolyloxycarbonyl)ethylene]	298
Poly[1-(*o*-tolyloxycarbonyl)ethylene]	325
Poly[1-(*p*-tolyloxycarbonyl)ethylene]	316
Poly[1-(2,2,2-trifluorethoxycarbonyl)ethylene]	263

Polymethacrylics
    Poly[1-(alkoxycarbonyl)-1-methylethylene] *Poly(alkyl methacrylate)* –[(ROCO)(Me)CCH$_2$]–

Poly[1-(benzyloxycarbonyl)-1-methylethylene]	327
Poly[1-(2-bromoethoxycarbonyl)-1-methylethylene]	325
Poly[(1-(butoxycarbonyl)-1-methylethylene]	293
Poly(butyl methacrylate) [PBMA]	
Poly[1-(*sec*-butoxycarbonyl)-1-methylethylene]	333
Poly[1-(*tert*-butoxycarbonyl)-1-methylethylene)]	391
Poly[1-(2-chloroethoxycarbonyl)-1-methylethylene]	ca 315
Poly[1-(2-cyanoethoxycarbonyl)-1-methylethylene]	364
Poly[1-(4-cyanophenoxycarbonyl)-1-methylethylene]	428
Poly[1-(cyclohexyloxycarbonyl)-1-methylethylene] (atactic)	356
Poly[1-(cyclohexyloxycarbonyl)-1-methylethylene)] (isotactic)	324

Polymer name	Glass transition temperature (T_g/K)
Poly[1-(dimethylaminoethoxycarbonyl)-1-methylethylene]	292
Poly[1-(ethoxycarbonyl)-1-ethylethylene]	300
Poly[1-(ethoxycarbonyl)-1-methylethylene] (atactic) *Poly(ethyl methacrylate)* [PEMA]	338
Poly[1-(ethoxycarbonyl)-1-methylethylene] (isotactic)	285
Poly[1-(ethoxycarbonyl)-1-methylethylene]] (syndiotactic)	339
Poly[1-(hexyloxycarbonyl)-1-methylethylene]	268
Poly[1-(isobutoxycarbonyl)-1-methylethylene]	326
Poly[1-(isopropoxycarbonyl)-1-methylethylene]	354
Poly[1-(mcthoxycarbonyl)-1-methylethylene] (atactic) *Poly(methyl methacrylate)* [PMMA]	378
Poly[1-(methoxycarbonyl)-1-methylethylene]] (isotactic)	311
Poly[1-(methoxycarbonyl)-1-methylethylene]] (syndiotactic)	378
Poly[1-(4-methoxycarbonylphenoxy)-1-methylethylene]	379
Poly[1-(methoxycarbonyl)-1-phenylethylene]] (atactic)	391
Poly[1-(methoxycarbonyl)-1-phenylethylene]] (isotactic)	397
Poly[1-methyl-1-(phenethoxycarbonyl)ethylene]	299
Poly[1-methyl-1-(phenoxycarbonyl)ethylene]	383

Polyvinyl ethers, alcohols, and ketones
    Poly(1-alkoxyethylene) *Poly(alkyl vinyl ether)* –[ROCHCH$_2$]–
    Poly(1-hydroxyethylene) *Poly(vinyl alcohol)* –[HOCHCH$_2$]–
    Poly(1-alkanoylethylene) *Poly(alkyl vinyl ketone)* –[RCOCHCH$_2$]–

Poly(1-butoxyethylene)	218
Poly(1-*sec*-butoxyethylene)	253
Poly(1-*tert*-butoxyethylene)	361
Poly[1-(butylthio)ethylene]	253
Poly(1-ethoxyethylene)	230
Poly[1-(4-ethylbenzoyl)ethylene]	325
Poly(1-hydroxyethylene)	358 (D)
Poly(vinyl alcohol) [PVA]	
Poly(hydroxymethylene)	407
Poly(1-isopropoxyethylene)	270
Poly[1-(4-methoxybenzoyl)ethylene]	319 (M)
Poly(1-methoxyethylene)	242
Poly(methyl vinyl ether) [PMVE]	
Poly[1-(methylthio)ethylene]	272
Poly(1-propoxyethylene)	224
Poly[1-(trifluoromethoxy)trifluoroethylene]	268

Polyvinyl halides and nitriles
    Poly(1-haloethylene) *Poly(vinyl halide)* –[XCHCH$_2$]–
    Poly(1-cyanoethylene) *Poly(acrylonitrile)* –[NCCHCH$_2$]–

Poly(1-chloroethylene)	354
Poly(vinyl chloride) [PVC]	
Poly(chlorotrifluoroethylene)	373
Poly(1-cyanoethylene)	370
Polyacrylonitrile [PAN]	
Poly(1-cyano-1-methylethylene)	393
Polymethacrylonitrile	
Poly(1,1-dichloroethylene)	255
Poly(vinylidene chloride)	
Poly(1,1-difluoroethylene)	ca 233
Poly(vinylidene fluoride)	
Poly(1-fluoroethylene)	314 (M)
Poly(vinyl fluoride)	

Polymer name	Glass transition temperature (T_g/K)
Poly(1-hexafluoropropylene)	425
Poly[1-(2-iodoethyl)ethylene]	343
Poly(tetrafluoroethylene)	(160)
Poly[1-(trifluoromethyl)ethylene]	300

Polyvinyl esters
      Poly[1-(alkanoyloxy)ethylene] *Poly(vinyl alkanoate)* –[RCOOCHCH$_2$]–

Polymer name	Glass transition temperature (T_g/K)
Poly(1-acetoxyethylene)	305
Poly(vinyl acetate) [PVAc]	
Poly[1-(benzoyloxy)ethylene]	344
Poly[1-(4-bromobenzoyloxy)ethylene]	365
Poly[1-(2-chlorobenzoyloxy)ethylene]	335
Poly[1-(3-chlorobenzoyloxy)ethylene]	338
Poly[1-(4-chlorobenzoyloxy)ethylene]	357
Poly[1-(cyclohexanoyloxy)ethylene]	349 (M)
Poly[1-(4-ethoxybenzoyloxy)ethylene]	343
Poly[1-(4-ethylbenzoyloxy)ethylene]	326
Poly[1-(4-isopropylbenzoyloxy)ethylene]	342
Poly[1-(2-methoxybenzoyloxy)ethylene]	338
Poly[1-(3-methoxybenzoyloxy)ethylene]	ca 317
Poly[1-(4-methoxybenzoyloxy)ethylene]	360
Poly[1-(4-methylbenzoyloxy)ethylene]	343
Poly[1-(4-nitrobenzoyloxy)ethylene]	395
Poly[1-(propionoyloxy)ethylene]	283 (M)

Polystyrenes
      Poly(1-phenylethylene) *Polystyrene* –[C$_6$H$_5$CHCH$_2$]–

Polymer name	Glass transition temperature (T_g/K)
Poly[1-(4-acetylphenyl)ethylene]	389 (M)
Poly[1-(4-benzoylphenyl)ethylene]	371 (M)
Poly[1-(4-bromophenyl)ethylene]	391
Poly[1-(4-butoxyphenyl)ethylene]	ca 320 (M)
Poly[1-(4-butoxycarbonylphenyl)ethylene]	349 (M)
Pol[(1-(4-butylphenyl)ethylene]	279
Poly[1-(4-carboxyphenyl)ethylene]	386 (M)
Poly[1-(2-chlorophenyl)ethylene]	392
Poly[1-(3-chlorophenyl)ethylene]	363
Poly[1-(4-chlorophenyl)ethylene]	383
Poly[1-(2,4-dichlorophenyl)ethylene]	406
Poly[1-(2,5-dichlorophenyl)ethylene]	379
Poly[1-(2,6-dichlorophenyl)ethylene]	440
Poly[1-(3,4-dichlorophenyl)ethylene]	401
Poly[1-(2,4-dimethylphenyl)ethylene]	385
Poly[1-(4-(dimethylamino)phenyl)ethylene]	398 (M)
Poly[1-(4-ethoxyphenyl)ethylene]	ca 359 (M)
Poly[1-(4-ethoxycarbonylphenyl)ethylene]	367 (M)
Poly[1-(4-fluorophenyl)ethylene]	368
Poly[1-(4-iodophenyl)ethylene]	429
Poly[1-(4-methoxyphenyl)ethylene]	386
Poly[1-(4-methoxycarbonylphenyl)ethylene]	386 (M)
Poly(1-methyl-1-phenylethylene)	373
Poly(α-methylstyrene)	
Poly[1-(2-(methylamino)phenyl)ethylene]	462 (M)
Poly(1-phenylethylene)	373
Polystyrene [PS]	

Polymer name	Glass transition temperature (T_g/K)
Poly[1-(4-propoxyphenyl)ethylene]	343 (M)
Poly[1-(4-propoxycarbonylphenyl)ethylene]	365 (M)
Poly(1-*o*-tolylethylene)	409

CHAINS WITH CARBOCYCLIC UNITS

Poly(arylenealkylene) –[–Ar–$(CH_2)_n$]–

Poly[1-(2-bromo-1,4-phenylene)ethylene]	353 (M)
Poly[1-(2-chloro-1,4-phenylene)ethylene]	343 (M)
Poly[1-(2-cyano-1,4-phenylene)ethylene]	363 (M)
Poly[1-(2,5-dimethyl-1,4-phenylene)ethylene]	373 (M)
Poly[1-(2-ethyl-1,4-phenylene)ethylene]	298 (M)
Poly[1-(1,4-naphthylene)ethylene]	433 (M)
Poly[1-(1,4-phenylene)ethylene]	ca 353 (M)

CHAINS WITH HETEROATOM UNITS

Main chain oxide units
Poly(oxyalkylene) *Poly(alkylene oxide)* –[$O(CH_2)_n$]–

Poly[oxy(1,1-bis(chloromethyl)trimethylene)]	265
Poly[oxy(1-(bromomethyl)ethylene)]	259
Poly[oxy(1-(butoxymethyl)ethylene)]	194
Poly[oxy(1-butylethylene)]	203
Poly[oxy(1-*tert*-butylethylene)]	308
Poly[oxy(1-(chloromethyl)ethylene)]	251
Poly(epichlorohydrin)	
Poly[oxy(2,6-dimethoxy-1,4-phenylene)]	440
Poly[oxy(1,1-dimethylethylene)]	264
Poly[oxy(2,6-dimethyl-1,4-phenylene)]	482
Poly[oxy(2,6-diphenyl-1,4-phenylene)]	493
Poly[oxy(1-ethylethylene)]	203
Poly(oxyethylidene)	243
Polyacetaldehyde	
Poly[oxy(1-(methoxymethyl)ethylene)]	211
Poly[oxy(2-methyl-6-phenyl-1,4-phenylene)]	428
Poly[oxy(1-methyltrimethylene)]	223 (D)
Poly[oxy(2-methyltrimethylene)]	218
Poly(oxy-1,4-phenylene)	358
Poly(phenylene oxide) [PPO]	
Poly[oxy(1-phenylethylene)]	313
Poly(oxytetramethylene)	189
Poly(tetrahydrofuran) [PTMO]	
Poly(oxytrimethylene)	195

Main-chain ester or anhydride units
Poly(oxyalkyleneoxyalkanedioyl) *Poly(alkylene alkanedioate)*--[$O(CH_2)_mOCO(CH_2)_nCO$]–

Poly(oxyadipoyloxydecamethylene)	217
Poly(oxyadipoyloxy-1,4-phenyleneisopropylidene-1,4-phenylene)	341
Poly(oxycarbonyloxy-1,4-phenylene-isopropylidene-1,4-phenylene)	422
Bisphenol A polycarbonate	
Poly(oxycarbonylpentamethylene)	213
Poly(oxycarbonyl-1,4-phenylenemethylene-1,4-phenylene)	395
Poly(oxycarbonyl-1,4-phenyleneisopropylidene-1,4-phenylene)	333

Polymer name	Glass transition temperature (T_g/K)
Poly[oxy(2,6-dimethyl-1,4-phenyleneisopropylidene-3,5-dimethyl-1,4-phenylene)oxysebacoyl]	318
Poly(oxyethylenecarbonyl-1,4-cyclohexylenecarbonyl) (trans)	291
Poly(oxyethyleneoxycarbonyl-1,4-naphthylenecarbonyl)	337
Poly(oxyethyleneoxycarbonyl-1,5-naphthylenecarbonyl)	344
Poly(oxyethyleneoxycarbonyl-2,6-naphthylenecarbonyl)	386
Poly(oxyethyleneoxycarbonyl-2,7-naphthylenecarbonyl)	392
Poly(oxyethyleneoxyterephthaloyl)	342
Poly(ethylene terephthalate) [PET]	
Poly(oxyisophthaloyl)	403 (D)
Poly(oxy(1-oxo-2,2-dimethyltrimethylene))	263
Poly(pivalolactone)	
Poly(oxy-1,4-phenyleneisopropylidene-1,4-phenyleneoxysebacoyl)	280
Poly(oxy-1,4-phenyleneoxy-1,4-phenyleneoxy-carbonyl-1,4-phenylene) [PEEK]	416
Poly(oxypropyleneoxyterephthaloyl)	341
Poly[oxyterephthaloyloxy(2,6-dimethyl-1,4-phenyleneisopropylidene-3,5-dimethyl-1,4-(D)phenylene)]	498
Poly(oxyterephthaloyloxyoctamethylene)	318 (D)
Poly(oxyterephthaloyloxy-1,4-phenyleneisopropylidene-1,4-phenylene)	478
Poly(bisphenol A terephthalate)	
Poly(oxytetramethyleneoxyterephthaloyl)	323
Poly(butylene terephthalate) [PBT]	

Main-chain amide units
    Poly(iminoalkyleneiminoalkanedioyl) *Poly(alkylene alkanediamide)*–[NH(CH$_2$)$_m$NHCO(CH$_2$)$_n$CO]–

Polymer name	Glass transition temperature (T_g/K)
Poly(iminoadipoyliminodecamethylene)	313
Nylon 10,6	
Poly(iminoadipoyliminohexamethylene)	ca 323
Nylon 6,6	
Poly(iminoadipoyliminooctamethylene)	318
Nylon 8,6	
Poly[iminoadipoyliminotrimethylene(methylimino)trimethylene]	278
Poly(iminocarbonyl-1,4-cyclohexylenemethylene)	466
Poly[iminocarbonyl-1,4-phenylene(2-oxoethylene)iminohexamethylene]	377
Poly(iminoethylene-1,4-phenyleneethyleneiminosebacoyl)	378 (D)
Poly(iminohexamethyleneiminoazelaoyl)	331
Nylon 6,9	
Poly(iminohexamethyleneiminododecanedioyl)	319
Nylon 6, 12	
Poly(iminohexamethyleneiminopimeloyl)	331
Nylon 6,7	
Poly(iminohexamethyleneiminosebacoyl)	323
Nylon 6,10	
Poly(iminohexamethyleneiminosuberoyl)	330
Nylon 6,8	
Poly(iminoisophthaloylimino-4,4′-biphenylylene)	558
Poly(iminoisophthaloyliminohexamethylene)	390
Poly(iminoisophthaloyliminomethylene-1,4-cyclohexylenemethylene)	481
Poly(iminoisophthaloyliminomethylene-1,3-phenylenemethylene)	438 (M)
Poly[iminomethylene(2,5-dimethyl-1,4-phenylene)methyleneiminosuberoyl]	351
Poly(imino-1,5-naphthyleneiminoisophthaloyl)	598
Poly(imino-1,5-naphthyleneiminoterephthaloyl)	578
Poly(iminooctamethyleneiminodecanedioyl)	333
Nylon 8,10	
Poly(iminooxalyliminohexamethylene)	430
Nylon 6,2	
Poly[imino(1-oxohexamethylene)]	326
Nylon 6	

Polymer name	Glass transition temperature (T_g/K)
Poly[imino(1-oxodecamethylene)]	315
Nylon 10	
Poly[imino(1-oxoheptamethylene)]	325
Nylon 7	
Poly[imino(1-oxo-3-methyltrimethylene]	369
Poly[imino(1-oxononamethylene)]	319
Nylon 9	
Poly[imino(1-oxooctamethylene)]	323
Nylon 8	
Poly[imino(1-oxotrimethylene)]	384
Nylon 3	
Poly(iminopentamethyleneiminoadipoyl)	318
Nylon 5,6	
Poly[iminopentamethyleneiminocarbonyl-1,4-phenylene(2-oxoethylene)]	376
Poly(imino-1,3-phenyleneiminoisophthaloyl)	553 (M)
Poly(imino-1,4-phenyleneiminoterephthaloyl)	618
Poly(iminopimeloyliminoheptamethylene)	328
Nylon 7,7	
Poly(iminoterephthaloylimino-4,4′-biphenylylene)	613
Poly(iminotetramethyleneiminoadipoyl)	316
Nylon 4,6	
Poly[iminotetramethyleneiminocarbonyl-1,4-phenylene(2-oxoethylene)]	357
Poly(iminotrimethyleneiminoadipoyliminotrimethylene)	307
Poly[iminotrimethyleneiminocarbonyl-1,4-phenylene(2-oxoethylene)]	382
Poly(oxy-1,4-phenyleneiminoterephthaloyl-imino-1,4-phenylene)	613
Poly(sulfonylimino-1,4-phenyleneiminoadipoylimino-1,4-phenylene)	467

Main-chain urethane units
   Poly(oxyalkyleneoxycarbonyliminoalkyleneiminocarbonyl)–[O(CH$_2$)$_m$OCONH(CH$_2$)$_n$NHCO]–

Poly(oxyethyleneoxycarbonyliminohexamethyleneiminocarbonyl)	329
Poly[oxyethyleneoxycarbonylimino(6-methyl-1,3-phenylene)iminocarbonyl]	325
Poly(oxyethyleneoxycarbonylimino-1,4-phenylenemethylene-1,4-phenyleneiminocarbonyl)	412
Poly(oxyhexamethyleneoxycarbonyliminohexamethyleneiminocarbonyl)	332
Poly[oxyhexamethyleneoxycarbonylimino(6-methyl-1,3-phenylene)iminocarbonyl]	305
Poly(oxyhexamethyleneoxycarbonylimino-1,4-phenylenemethylene-1,4-phenyleneiminocarbonyl)	364
Poly(oxyoctamethyleneoxycarbonyliminohexamethyleneiminocarbonyl)	331
Poly[oxyoctamethyleneoxycarbonylimino(6-methyl-1,3-phenylene)iminocarbonyl]	337
Poly(oxyoctamethyleneoxycarbonylimino-1,4-phenylenemethylene-1,4-phenyleneiminocarbonyl)	352
Poly(oxytetramethyleneoxycarbonyliminohexamethyleneiminocarbonyl)	332
Poly[oxytetramethyleneoxycarbonylimino(6-methyl-1,3-phenylene)iminocarbonyl]	315
Poly(oxytetramethyleneoxycarbonylimino-1,4-phenylenemethylene-1,4-phenyleneiminocarbonyl)	382

Main-chain siloxanes
   Poly[oxy(dialkylsilylene)] *Poly(dialkylsiloxane)* –[O(R$_2$Si)]–

Poly[oxy(dimethylsilylene)]	148
Poly(dimethylsiloxane) [PDMS]	
Poly[oxy(dimethylsilylene)oxy-1,4-phenylene]	363 (M)
Poly[oxy(dimethylsilylene)oxy-1,4-phenyleneisopropylidene-1,4-phenylene]	318 (M)
Poly[oxy(diphenylsilylene)]	238
Poly(diphenylsiloxane)	
Poly[oxy(diphenylsilylene)-1,3-phenylene]	ca 331
Poly[oxy((methyl)phenylsilylene)]	187
Poly[oxy((methyl)-3,3,3-trifluoropropylsilylene]	<193

Polymer name	Glass transition temperature (T_g/K)
Main-chain sulfur-containing units	
Poly(dithioethylene)	223
Poly(dithiomethylene-1,4-phenylenemethylene)	296
Poly(oxy-4,4′-biphenylylene-1,4-phenylenesulfonyl-1,4-phenylene)	503 (M)
Poly(oxycarbonyloxy-1,4-phenylenethio-1,4-phenylene)	ca 383
Poly(oxyethylenedithioethylene)	220 (M)
Poly[oxy(2-hydroxytrimethylene)oxy-1,4-phenylenesulfonyl-1,4-phenylene]	428
Poly(oxymethyleneoxyethylenedithioethylene)	214
Poly(oxy-1,4-phenylenesulfinyl-1,4-phenyleneoxy-1,4-phenylenecarbonyl-1,4-phenylene)	478 (M)
Poly(oxy-1,4-phenylenesulfinyl-1,4-phenyleneoxy-1,4-phenyleneisopropylidene-1,4-phenylene)	438 (M)
Poly(oxy-1,4-phenylenesulfonyl-1,4-phenylene)	487
Poly(oxy-1,4-phenylenesulfonyl-4,4′-biphenylylenesulfonyl-1,4-phenylene)	533
Poly[oxy-1,4-phenylenesulfonyl-1,4-phenyleneoxy(2,6-dimethyl-1,4-phenylene)isopropylidene (3,5-dimethyl-1,4-phenylene)]	508 (M)
Poly(oxy-1,4-phenylenesulfonyl-1,4-phenyleneoxy-1,4-phenylenecarbonyl-1,4-phenylene)	478 (M)
Poly[oxy-1,4-phenylenesulfonyl-1,4-phenyleneoxy-1,4-phenylene(hexafluoroisopropylidene)1,4-phenylene]	478 (M)
Poly(oxy-1,4-phenylenesulfonyl-1,4-phenyleneoxy-1,4-phenyleneisopropylidene-1,4-phenylene)	449
Poly(oxy-1,4-phenylenesulfonyl-1,4-phenyleneoxy-1.4-phenylenemethylene-1,4-phenylene)	453 (M)
Poly(oxy-1,4-phenylenesulfonyl-1,4-phenyleneoxy-1.4-phenylenethio-1,4-phenylene)	448 (M)
Poly(oxy-1,4-phenylenesulfonyl-1,4-phenyleneoxyterephthaloyl)	522
Poly(oxytetramethylenedithiotetramethylene)	197
Poly(sulfonyl-1,2-cyclohexylene)	401
Poly(sulfonyl-1,3-cyclohexylene)	381
Poly(sulfonyl-1,4-phenylenemethylene-1,4-phenylene)	497
Poly(thio-1,3-cyclohexylene)	221
Poly[thio(difluoromethylene)]	155
Poly(thioethylene)	223
Poly[thio(1-ethylethylene]	218
Poly[thio(1-methyl-3-oxotrimethylene)]	285
Poly[thio(1-methyltrimethylene)]	214
Pol[(thio(1-oxohexamethylene)]	292
Poly(thio-1,4-phenylene)	370
Poly(thiopropylene)	226
Main-chain heterocyclic units	
Poly(1,3-dioxa-4,6-cyclohexylenemethylene)	378
Poly(vinyl formal)	
Poly[(2,6-dioxopiperidine-1,4-diyl)trimethylene]	363
Poly[(2-methyl-1,3-dioxa-4,6-cyclohexylene)methylene]	355
Poly(vinyl acetal)	
Poly(1,4-piperazinediylcarbonyloxyethyleneoxycarbonyl)	333
Poly(1,4-piperazinediylisophthaloyl)	465 (M)
Poly[(2-propyl-1,3-dioxa-4,6-cyclohexylene)methylene]	322
Poly(vinyl butyral)	
Poly(3,6-pyridazinediyloxy-1,4-phenyleneisopropylidene-1,4-phenyleneoxy)	453 (M)
Poly(2,5-pyridinediylcarbonyliminohexamethyleneiminocarbonyl)	322

DIELECTRIC CONSTANT OF SELECTED POLYMERS

This table lists typical values of the dielectric constant (more properly called relative permittivity) of some important polymers. Values are given for frequencies of 1 kHz, 1 MHz, and 1 GHz; in most cases the dielectric constant at frequencies below 1 kHz does not differ significantly from the value at 1 kHz. Since the dielectric constant of a polymeric material can vary with density, degree of crystallinity, and other details of a particular sample, the values given here should be regarded only as typical or average values.

REFERENCES

1. Gray, D.E., Ed., *American Institute of Physics Handbook, Third Edition*, p. 5-132, McGraw Hill, New York, 1972.
2. Anderson, H.L., Editor, *A Physicist's Desk Reference*, American Institute of Physics, New York, 1989.
3. Brandrup, J., and Immergut, E.H., *Polymer Handbook, Third Edition*, John Wiley & Sons, New York, 1989.

Name	$t/°C$	1 kHz	1 MHz	1 GHz
Polyacrylonitrile	25	5.5	4.2	
Polyamides (nylons)	25	3.50	3.14	2.8
	84	11	4.4	2.8
Polybutadiene	25	2.5		
Polycarbonate	23	2.92	2.8	
Polychloroprene (neoprene)	25	6.6	6.3	4.2
Polychlorotrifluoroethylene	23	2.65	2.46	2.39
Polyethylene	23	2.3		
Poly(ethylene terephthalate) (Mylar)	23	3.25	3.0	2.8
Polyisoprene (natural rubber)	27	2.6	2.5	2.4
Poly(methyl methacrylate)	27	3.12	2.76	2.6
	80	3.80	2.7	2.6
Polyoxymethylene (polyformaldehyde)	25	3.8		
Poly(phenylene oxide)	23	2.59	2.59	
Polypropylene	25	2.3	2.3	2.3
Polystyrene	25	2.6	2.6	2.6
Polysulfones	25	3.13	2.10	
Polytetrafluoroethylene (teflon)	25	2.1	2.1	2.1
Poly(vinyl acetate)	50		3.5	
	150		8.3	
Poly(vinyl chloride)	25	3.39	2.9	2.8
	100	7.8	3.3	2.7
Poly(vinylidene chloride)	23	4.6	3.2	2.7
Poly(vinylidene fluoride)	23	12.2	8.9	4.7

Section 14
Geophysics, Astronomy, and Acoustics

ASTRONOMICAL CONSTANTS
Victor Abalakin

The constants in this table are based primarilarly on the set of constants adopted by the International Astronomical Union (IAU) in 1976. Updates have been made when new data were available. All values are given in SI Units; thus masses are expressed in kilograms and distances in meters. The astronomical unit of time is a time interval of one day (1 d) equal to 86400 s. An interval of 36525 d is one Julian century (1 cy).

REFERENCES

1. Seidelmann, P. K., *Explanatory Supplement to the Astronomical Almanac*, University Science Books, Mill Valley, CA, 1990.
2. Lang, K. R., *Astrophysical Data: Planets and Stars*, Springer-Verlag, New York, 1992.

Defining constants

Gaussian gravitational constant	$k = 0.01720209895$ m^3 kg^{-1} s^{-2}
Speed of light	$c = 299792458$ m s^{-1}

Primary constants

Light-time for unit distance (1 AU)	$\tau_A = 499.004782$ s
Equatorial radius of earth	$a_e = 6378140$ m
Equatorial radius of earth (IUGG value)	$a_e = 6378136$ m
Dynamical form-factor for earth	$J_2 = 0.001082626$
Geocentric gravitational constant	$GE = 3.986005 \times 10^{14}$ m^{3}s^{-2}
Constant of gravitation	$G = 6.672 \times 10^{-11}$ m^{3}kg^{-1}s^{-2}
Ratio of mass of moon to that of earth	$\mu = 0.01230002$
	$1/\mu = 81.300587$
General precession in longitude, per Julian century, at standard epoch J2000	$\rho = 5029''.0966$
Obliquity of the ecliptic at standard epoch J2000	$\varepsilon = 23°26'21''.448$

Derived constants

Constant of nutation at standard epoch J2000	$N = 9''.2025$
Unit distance (AU $= c\tau_A$)	AU $= 1.49597870 \times 10^{11}$ m
Solar parallax ($\pi_0 = \arcsin(a_e/\text{AU})$)	$\pi_0 = 8''.794148$
Constant of aberration for standard epoch J2000	$\kappa = 20''.49552$
Flattening factor for the earth	$f = 1/298.257 = 0.00335281$
Heliocentric gravitational constant ($GS = A^3k^2/D^2$)	$GS = 1.32712438 \times 10^{20}$ m^3 s^{-2}
Ratio of mass of sun to that of the earth (S/E) $= (GS)/(GE)$)	$S/E = 332946.0$
Ratio of mass of sun to that of earth + moon	$(S/E)/(1 + \mu) = 328900.5$
Mass of the sun ($S = (GS)/G$)	$S = 1.9891 \times 10^{30}$ kg

Ratios of mass of sun to masses of the planets

Mercury	6023600
Venus	408523.5
Earth + moon	328900.5
Mars	3098710
Jupiter	1047.355
Saturn	3498.5
Uranus	22869
Neptune	19314
Pluto	3000000

PROPERTIES OF THE SOLAR SYSTEM

The following tables give various properties of the planets and characteristics of their orbits in the solar system. Certain properties of the sun and of the earth's moon are also included.

Explanations of the column headings:

- *Den.*: mean density in g/cm^3
- *Radius*: radius at the equator in km
- *Flattening*: degree of oblateness, defined as $(r_e - r_p)/r_e$, where r_e and r_p are the equatorial and polar radii, respectively
- *Potential coefficients*: coefficients in the spherical harmonic representation of the gravitational potential U by the equation

$$U(r,\phi) = (GM/r) \left[1 - \sum J_n (a/r)^n P_n(\sin \phi)\right]$$

  where G is the gravitational constant, r the distance from the center of the planet, a the radius of the planet, M the mass, ϕ the latitude, and P_n the Legendre polynomial of degree n.
- *Gravity*: acceleration due to gravity at the surface
- *Escape velocity*: velocity needed at the surface of the planet to escape the gravitational pull
- *Dist. to sun*: semi-major axis of the elliptical orbit (1 AU $= 1.496 \times 10^8$ km)
- ε: eccentricity of the orbit
- *Ecliptic angle*: angle between the planetary orbit and the plane of the earth's orbit around the sun
- *Inclin.*: angle between the equatorial plane and the plane of the planetary orbit
- *Rot. period*: period of rotation of the planet measured in earth days
- *Albedo*: ratio of the light reflected from the planet to the light incident on it
- T_{sur}: mean temperature at the surface
- P_{sur}: pressure of the atmosphere at the surface

The following general information on the solar system is of interest:

Mass of the earth $= M_e = 5.9742 \times 10^{24}$ kg
Total mass of planetary system $= 2.669 \times 10^{27}$ kg $= 447\ M_e$
Total angular momentum of planetary system $= 3.148 \times 10^{43}$ kg m^2/s
Total kinetic energy of the planets $= 1.99 \times 10^{35}$ J
Total rotational energy of planets $= 0.7 \times 10^{35}$ J

Properties of the sun:

Mass $= 1.9891 \times 10^{30}$ kg $= 332946.0\ M_e$
Radius $= 6.9599 \times 10^8$ m
Surface area $= 6.087 \times 10^{18}$ m^2
Volume $= 1.412 \times 10^{27}$ m^3
Mean density $= 1.409$ g/cm^3
Gravity at surface $= 27398$ cm/s^2
Escape velocity at surface $= 6.177 \times 10^5$ m/s
Effective temperature $= 5780$ K
Total radiant power emitted (luminosity) $= 3.86 \times 10^{26}$ W
Surface flux of radiant energy $= 6.340 \times 10^7$ W/m^2
Flux of radiant energy at the earth (Solar Constant) $= 1373$ W/m^2

REFERENCES

1. Seidelmann, P. K., Editor, *Explanatory Supplement to the Astronomical Almanac*, University Science Books, Mill Valley, CA, 1992.
2. Lang, K. R., *Astrophysical Data: Planets and Stars*, Springer-Verlag, New York, 1992.
3. Allen, C. W., *Astrophysical Quantities, Third Edition*, Athlone Press, London, 1977.

Planet	Mass 10^{24} kg	Den. g/cm³	Radius km	Flattening	$10^3 J_2$	$10^6 J_3$	$10^6 J_4$	Gravity cm/s²	Escape vel. km/s
					Potential coeffients				
Mercury	0.33022	5.43	2439.7	0				370	4.25
Venus	4.8690	5.24	6051.9	0	0.027			887	10.4
Earth	5.9742	5.515	6378.140	0.00335364	1.08263	-2.54	-1.61	980	11.2
(Moon)	0.073483	3.34	1738	0	0.2027			162	2.37
Mars	0.64191	3.94	3397	0.00647630	1.964	36		371	5.02
Jupiter	1898.8	1.33	71492	0.0648744	14.75	-580		2312	59.6
Saturn	568.50	0.70	60268	0.0979624	16.45	-1000		896	35.5
Uranus	86.625	1.30	25559	0.0229273	12			777	21.3
Neptune	102.78	1.76	24764	0.0171	4			1100	23.3
Pluto	0.015	1.1	1151	0				72	1.1

Planet	Dist. to sun AU	ε	Ecliptic angle	Inclin.	Rot. period d	Albedo	No. of satellites
Mercury	0.38710	0.2056	7.00°	0°	58.6462	0.106	0
Venus	0.72333	0.0068	3.39°	177.3°	-243.01	0.65	0
Earth	1.00000	0.0167		23.45°	0.99726968	0.367	1
(Moon)				6.68°	27.321661	0.12	
Mars	1.52369	0.0933	1.85°	25.19°	1.02595675	0.150	2
Jupiter	5.20283	0.048	1.31°	3.12°	0.41354	0.52	16
Saturn	9.53876	0.056	2.49°	26.73°	0.4375	0.47	18
Uranus	19.19139	0.046	0.77°	97.86°	-0.65	0.51	15
Neptune	30.06107	0.010	1.77°	29.56°	0.768	0.41	8
Pluto	39.52940	0.248	17.15°	118°	-6.3867	0.3	1

Planet	T_{sur} K	P_{sur} bar	CO_2	N_2	O_2	H_2O	H_2	He	Ar	Ne	CO
			Atmospheric composition								
Mercury	440	2×10^{-15}					2%	98%			
Venus	730	90	96.4%	3.4%	69 ppm	0.1%			4 ppm		20 ppm
Earth	288	1	0.03%	78.08%	20.95%	0 to 3%			0.93%	18 ppm	1 ppm
Mars	218	0.007	95.32%	2.7%	0.13%	0.03%			1.6%	3 ppm	0.07%
Jupiter	129						86.1%	13.8%			
Saturn	97						92.4%	7.4%			
Uranus	58						89%	11%			
Neptune	56						89%	11%			
Pluto	50	1×10^{-5}									

SATELLITES OF THE PLANETS

This table gives characteristics of the known satellites of the planets. The parameters covered are:

- Orbital period in units of earth days. An R following the value indicates a retrograde motion.
- Distance from the planet, as measured by the semi-major axis of the orbit.
- Eccentricity of the orbit.
- Inclination of the satellite orbit with respect to the equator of the planet.
- Mass of the satellite relative to the planet.
- Radius of the satellite in km.
- Mean density of the satellite.
- Geometric albedo, which is a measure of the fraction of incident sunlight reflected by the satellite.

REFERENCES

1. Seidelmann, P. K., Editor, *Explanatory Supplement to the Astronomical Almanac*, University Science Books, Mill Valley, CA, 1992.
2. Lang, K. R., *Astrophysical Data: Planets and Stars*, Springer-Verlag, New York, 1992.
3. Burns, J. A., and Matthews, M. S., Eds., *Satellites*, University of Arizona Press, Tucson, 1986.

Planet		Satellite	Orb. Period d	Distance 10^3 km	Eccentricity	Inclination	Rel. mass	Radius km	Den. g/cm³	Albedo
Earth		Moon	27.321661	384.400	0.054900489	18.28-28.58°	0.0123002	1738	3.34	0.12
Mars	I	Phobos	0.31891023	9.378	0.015	1.0°	1.5×10^{-8}	$13.5 \times 10.8 \times 9.4$	<2	0.06
	II	Deimos	1.2624407	23.459	0.0005	0.9-2.7°	3×10^{-9}	$7.5 \times 6.1 \times 5.5$	<2	0.07
Jupiter	I	Io	1.769137786	422	0.004	0.04°	4.68×10^{-5}	1815	3.55	0.61
	II	Europa	3.551181041	671	0.009	0.47°	2.52×10^{-5}	1569	3.04	0.64
	III	Ganymede	7.15455296	1070	0.002	0.21°	7.80×10^{-5}	2631	1.93	0.42
	IV	Callisto	16.6890184	1883	0.007	0.51°	5.66×10^{-5}	2400	1.83	0.20
	V	Amalthea	0.49817905	181	0.003	0.40°	3.8×10^{-9}	$135 \times 83 \times 75$		0.05
	VI	Himalia	250.5662	11480	0.15798	27.63°	5.0×10^{-9}	93		0.03
	VII	Elara	259.6528	11737	0.20719	24.77°	4×10^{-10}	38		0.03
	VIII	Pasiphae	735 R	23500	0.378	145°	1×10^{-10}	25		
	IX	Sinope	758 R	23700	0.275	153°	0.4×10^{-10}	18		
	X	Lysithea	259.22	11720	0.107	29.02°	0.4×10^{-10}	18		
	XI	Carme	692 R	22600	0.20678	164°	0.5×10^{-10}	20		
	XII	Ananke	631 R	21200	0.16870	147°	0.2×10^{-10}	15		
	XIII	Leda	238.72	11094	0.14762	26.07°	0.03×10^{-10}	8		
	XIV	Thebe	0.6745	222	0.015	0.8°	4×10^{-10}	55×45		0.05
	XV	Adrastea	0.29826	129			0.1×10^{-10}	$12.5 \times 10 \times 7.5$		0.05
	XVI	Metis	0.294780	128			0.5×10^{-10}	20		0.05
Saturn	I	Mimas	0.942421813	185.52	0.0202	1.53°	8.0×10^{-8}	196	1.44	0.5
	II	Enceladus	1.370217855	238.02	0.00452	1.86°	1.3×10^{-7}	250	1.13	1.0
	III	Tethys	1.887802160	294.66	0.00000	1.86°	1.3×10^{-6}	530	1.20	0.9
	IV	Dione	2.736914742	377.40	0.002230	0.02°	1.85×10^{-6}	560	1.41	0.7
	V	Rhea	4.517500436	527.04	0.00100	0.35°	4.4×10^{-6}	765	1.33	0.7
	VI	Titan	15.94542068	1221.83	0.029192	0.33°	2.38×10^{-4}	2575	1.88	0.21
	VII	Hyperion	21.2766088	1481.1	0.104	0.43°	3×10^{-8}	$205 \times 130 \times 110$		0.3
	VIII	Iapetus	79.3301825	3561.3	0.02828	14.72°	3.3×10^{-6}	730	1.15	0.2
	IX	Phoebe	550.48 R	12952	0.16326	177°	7×10^{-10}	110		0.06

14-4

SATELLITES OF THE PLANETS (continued)

Planet		Satellite	Orb. Period d	Distance 10³ km	Eccentricity	Inclination	Rel. mass	Radius km	Den. g/cm³	Albedo
	X	Janus	0.6945	151.472	0.007	0.14°		$110 \times 100 \times 80$		0.8
	XI	Epimetheus	0.6942	151.422	0.009	0.34°		$70 \times 60 \times 50$		0.8
	XII	Helene	2.7369	377.40	0.005	0.0°		$18 \times 16 \times 15$		0.7
	XIII	Telesto	1.8878	294.66				$17 \times 14 \times 13$		0.5
	XIV	Calypso	1.8878	294.66				$17 \times 11 \times 11$		0.6
	XV	Atlas	0.6019	137.670	0.000	0.3°		20×10		0.9
	XVI	Prometheus	0.6130	139.353	0.003	0.0°		$70 \times 50 \times 40$		0.6
	XVII	Pandora	0.6285	141.700	0.004	0.0°		$55 \times 45 \times 35$		0.9
	XVIII	Pan	0.5750	133.583				10		0.5
Uranus	I	Ariel	2.52037935	191.02	0.0034	0.3°	1.56×10^{-5}	579	1.55	0.34
	II	Umbriel	4.1441772	266.30	0.0050	0.36°	1.35×10^{-5}	586	1.58	0.18
	III	Titania	8.7058717	435.91	0.0022	0.14°	4.06×10^{-5}	790	1.69	0.27
	IV	Oberon	13.4632389	583.52	0.0008	0.10°	3.47×10^{-5}	762	1.64	0.24
	V	Miranda	1.41347925	129.39	0.0027	4.2°	0.08×10^{-5}	240	1.25	0.27
	VI	Cordelia	0.335033	49.77	<0.001	0.1°		13		0.07
	VII	Ophelia	0.376409	53.79	0.010	0.1°		15		0.07
	VIII	Bianca	0.434577	59.17	<0.001	0.2°		21		0.07
	IX	Cressida	0.463570	61.78	<0.001	0.0°		31		0.07
	X	Desdemona	0.473651	62.68	<0.001	0.2°		27		0.07
	XI	Juliet	0.493066	64.35	<0.001	0.1°		42		0.07
	XII	Portia	0.513196	66.09	<0.001	0.1°		54		0.07
	XIII	Rosalind	0.558459	69.94	<0.001	0.3°		27		0.07
	XIV	Belinda	0.623525	75.26	<0.001	0.0°		33		0.07
	XV	Puck	0.761832	86.01	<0.001	0.31°		77		0.07
Neptune	I	Triton	5.8768541 R	354.76	0.000016	157.345°	2.09×10^{-4}	1353	2.05	0.7
	II	Nereid	360.13619	5513.4	0.7512	27.6°	2×10^{-7}	170		0.4
	III	Naiad	0.294396	117.6	<0.001	4.74°		29		0.06
	IV	Thalassa	0.311485	73.6	<0.001	0.21°		40		0.06
	V	Despina	0.334655	52.6	<0.001	0.07°		74		0.06
	VI	Galatea	0.428745	62.0	<0.001	0.05°		79		0.06
	VII	Larissa	0.554654	50.0	<0.0014	0.20°		104×89		0.06
	VIII	Proteus	1.122315	48.2	<0.001	0.55°		$218 \times 208 \times 201$		0.06
Pluto	I	Charon	6.38725	19.6	<0.001	99°	0.22	593		0.5

MASS, DIMENSIONS, AND OTHER PARAMETERS OF THE EARTH

This table is a collection of data on various properties of the Earth. Most of the values are given in SI units. Note that 1 AU (astronomical unit) = 149,597,870 km.

REFERENCES

1. Seidelmann, P. K., Editor, *Explanatory Supplement to the Astronomical Almanac*, University Science Books, Mill Valley, CA, 1992.
2. Lang, K. R., *Astrophysical Data: Planets and Stars*, Springer-Verlag, New York, 1992.

Quantity	Symbol	Value	Unit
Mass	M	$5.9742 \cdot 10^{27}$	g
Major orbital semi-axis	a_{orb}	1.000000	AU
		$1.4959787 \cdot 10^8$	km
Distance from sun at perihelion	r_π	0.9833	AU
Distance from sun at aphelion	r_α	1.0167	AU
Moment of perihelion passage	T_π	Jan. 2, 4 h 52 min	
Moment of aphelion passage	T_α	July 4, 5 h 05 min	
Siderial rotation period around sun	P_{orb}	$31.5581 \cdot 10^6$	s
		365.25636	d
Mean rotational velocity	U_{orb}	29.78	km/s
Mean equatorial radius	$\overline{a}$	6378.140	km
Mean polar compression (flattening factor)	α	1/298.257	
Difference in equatorial and polar semi-axes	$a - c$	21.385	km
Compression of meridian of major equatorial axis	α_a	1/295.2	
Compression of meridian of minor equatorial axis	α_b	1/298.0	
Equatorial compression	ε	1/30 000	
Difference in equatorial semi-axes	$a - b$	213	m
Difference in polar semi-axes	$c_N - c_S$	~70	m
Polar asymmetry	η	$\sim 1 \cdot 10^{-5}$	
Mean acceleration of gravity at equator	g_e	9.78036	m/s^2
Mean acceleration of gravity at poles	g_p	9.83208	m/s^2
Difference in acceleration of gravity at pole and at equator	$g_p - g_e$	5.172	cm/s^2
Mean acceleration of gravity for entire surface of terrestrial ellipsoid	g	9.7978	m/s^2
Mean radius	R	6371.0	km
Area of surface	S	$5.10 \cdot 10^8$	km^2
Volume	V	$1.0832 \cdot 10^{12}$	km^3
Mean density	ρ	5.515	g/cm^3
Siderial rotational period	P	86,164.09	s
Rotational angular velocity	ω	$7.292116 \cdot 10^{-5}$	rad/s
Mean equatorial rotational velocity	v	0.46512	km/s
Rotational angular momentum	L	$5.861 \cdot 10^{33}$	J s
Rotational energy	E	$2.137 \cdot 10^{29}$	J
Ratio of centrifugal force to force of gravity at equator	q_c	0.0034677 = 1/288	
Moment of inertia	I	$8.070 \cdot 10^{37}$	kg m^2
Relative braking of earth's rotation due to tidal friction	$\Delta\omega_e/\omega$	$-4.2 \cdot 10^{-8}$	century^{-1}
Relative secular acceleration of earth's rotation	$\Delta\omega_i/\omega$	$+1.4 \cdot 10^{-8}$	century^{-1}
Not secular braking of earth's rotation	$\Delta\omega/\omega$	$-2.8 \cdot 10^{-8}$	century^{-1}
Probable value of total energy of tectonic deformation of earth	E_t	$\sim 1 \cdot 10^{23}$	J/century
Secular loss of heat of earth through radiation into space	$\Delta'E_k$	$1 \cdot 10^{23}$	J/century
Portion of earth's kinetic energy transformed into heat as a result of lunar and solar tides in the hydrosphere	$\Delta''E_k$	$1.3 \cdot 10^{23}$	J/century

MASS, DIMENSIONS, AND OTHER PARAMETERS OF THE EARTH (continued)

Quantity	Symbol	Value	Unit
Differences in duration of days in March and August	ΔP	0.0025 (March-August)	s
Corresponding relative annual variation in earth's rotational velocity	$\Delta^*\omega/\omega$	$2.9 \cdot 10^{-8}$ (Aug.-March)	
Presumed variation in earth's radius between August and March	Δ^*R	-9.2 (Aug.-March)	cm
Annual variation in level of world ocean	Δh_o	~10 (Sept.-March)	cm
Area of continents	S_C	$1.49 \cdot 10^8$	km^2
		29.2	% of surface
Area of world ocean	S_o	$3.61 \cdot 10^8$	km^2
		70.8	% of surface
Mean height of continents above sea level	h_C	875	m
Mean depth of world ocean	h_o	3794	m
Mean thickness of lithosphere within the limits of the continents	$h_{c.l.}$	35	km
Mean thickness of lithosphere within the limits of the ocean	$h_{o.l.}$	4.7	km
Mean rate of thickening of continental lithosphere	$\Delta h/\Delta t$	10 - 40	$m/10^6$ y
Mean rate of horizontal extension of continental lithosphere	$\Delta l/\Delta t$	0.75 - 20	$km/10^6$ y
Mass of crust	m_l	$2.36 \cdot 10^{22}$	kg
Mass of mantle		$4.05 \cdot 10^{24}$	kg
Amount of water released from the mantle and core in the course of geological time		$3.40 \cdot 10^{21}$	kg
Total reserve of water in the mantle		$2 \cdot 10^{23}$	kg
Present content of free and bound water in the earth's lithosphere		$2.4 \cdot 10^{21}$	kg
Mass of hydrosphere	m_h	$1.664 \cdot 10^{21}$	kg
Amount of oxygen bound in the earth's crust		$1.300 \cdot 10^{21}$	kg
Amount of free oxygen		$1.5 \cdot 10^{18}$	kg
Mass of atmosphere	m_a	$5.136 \cdot 10^{18}$	kg
Mass of biosphere	m_b	$1.148 \cdot 10^{16}$	kg
Mass of living matter in the biosphere		$3.6 \cdot 10^{14}$	kg
Density of living matter on dry land		0.1	g/cm^2
Density of living matter in ocean		$15 \cdot 10^{-8}$	g/cm^3
Age of the earth		$4.55 \cdot 10^9$	y
Age of oldest rocks		$4.0 \cdot 10^9$	y
Age of most ancient fossils		$3.4 \cdot 10^9$	y

GEOLOGICAL TIME SCALE

Period or Epoch	Beginning and end, in 10^6 years	Key events
Cenozoic era		
Quaternian		
Contemporary	0–10,000 y ± 2,000 y	
Pleistocene	10,000–1,000,000 y ± 50,000 y	Homo Erectus breakout
Tertiary		
Pliocene	1.8–5.3	Ape man fossils
Miocene	5–25	Origin of grass
Oligocene	25–37	Rise of cats, dogs, pigs
Eocene	37–55	Debut of hoofed mammals
Paleocene	55–67	Earliest primates
Mesozoic era		
Cretaceous	67–138	Demise of dinosaurs
Jurassic	138–208	First birds
Triassic	208–245	Appearance of dinosaurs
Paleozoic era		
Permian	245–290	Flowers, insect pollination
Carboniferous	290–360	First conifers
Devonian	360–410	First vertebrates ashore
Silurian	410–435	Spore-bearing plants
Ordovician	435–520	First animals ashore
Cambrian	520–570	Vertebrates appear
Pre-Cambrian		
Pre-Cambrian III (Proterozoic)	570–2500	First plants, jellyfish
Pre-Cambrian II (Archean)	2500–3800	Photosynthetic bacteria
Pre-Cambrian I (Hadean)	3800–4450	Earth formed 4600 million years ago

Reference: Calder, N., *Timescale - An Atlas of the Fourth Dimension,* Viking Press, New York, 1983.

ACCELERATION DUE TO GRAVITY

The acceleration due to gravity is tabulated here as a function of latitude and height above the earth's surface. Values were calculated from the expression

$$g/(\text{m/s}^2) = 9.780356 \, (1 + 0.0052885 \sin^2 \phi - 0.0000059 \sin^2 2\phi)$$
$$- 0.003086 \, H$$

where ϕ is the latitude and H is the height in kilometers.

REFERENCE

Jursa, A. S., Ed., *Handbook of Geophysics and the Space Environment,* 4th ed., Air Force Geophysics Laboratory, 1985, p. 14-17.

ϕ	$H = 0$	$H = 1$ km	$H = 5$ km	$H = 10$ km
0	9.78036	9.77727	9.76493	9.74950
5	9.78075	9.77766	9.76532	9.74989
10	9.78191	9.77882	9.76648	9.75105
15	9.78381	9.78072	9.76838	9.75295
20	9.78638	9.78330	9.77095	9.75552
25	9.78956	9.78647	9.77413	9.75870
30	9.79324	9.79016	9.77781	9.76238
35	9.79732	9.79424	9.78189	9.76646
40	9.80167	9.79858	9.78624	9.77081
45	9.80616	9.80307	9.79073	9.77530
50	9.81065	9.80757	9.79522	9.77979
55	9.81501	9.81193	9.79958	9.78415
60	9.81911	9.81602	9.80368	9.78825
65	9.82281	9.81972	9.80738	9.79195
70	9.82601	9.82292	9.81058	9.79515
75	9.82860	9.82551	9.81317	9.79774
80	9.83051	9.82743	9.81508	9.79965
85	9.83168	9.82860	9.81625	9.80082
90	9.83208	9.82899	9.81665	9.80122

DENSITY, PRESSURE, AND GRAVITY AS A FUNCTION OF DEPTH WITHIN THE EARTH

This table gives the density ρ, pressure p, and acceleration due to gravity g as a function of depth below the earth's surface, as calculated from the model of the structure of the earth in Reference 1. The model assumes a radius of 6371 km for the earth. The boundary between the crust and mantle (the Mohorovicic discontinuity) is taken as 21 km, while in reality it varies considerable with location.

REFERENCES

1. Anderson, D. L., and Hart, R. S., *J. Geophys. Res.*, 81, 1461, 1976.
2. Carmichael, R. S., *CRC Practical Handbook of Physical Properties of Rocks and Minerals*, p.467, CRC Press, Boca Raton, FL, 1989.

Depth km	ρ g/cm^3	p kbar	g cm/s^2	Depth km	ρ g/cm^3	p kbar	g cm/s^2
Crust				1771	4.96	752	994
				2071	5.12	903	1002
0	1.02	0	981	2371	5.31	1061	1017
3	1.02	3	982	2671	5.45	1227	1042
3	2.80	3	982	2886	5.53	1352	1069
21	2.80	5	983				
Mantle (solid)				**Outer core (liquid)**			
				2886	9.96	1352	1069
21	3.49	5	983	2971	10.09	1442	1050
41	3.51	12	983	3371	10.63	1858	953
61	3.52	19	984	3671	11.00	2154	874
81	3.48	26	984	4071	11.36	2520	760
101	3.44	33	984	4471	11.69	2844	641
121	3.40	39	985	4871	11.99	3116	517
171	3.37	56	987	5156	12.12	3281	427
221	3.34	73	989				
271	3.37	89	991	**Inner core (solid)**			
321	3.47	106	993				
371	3.59	124	994	5156	12.30	3281	427
571	3.95	199	999	5371	12.48	3385	355
871	4.54	328	997	5771	12.52	3529	218
1171	4.67	466	992	6071	12.53	3592	122
1471	4.81	607	991	6371	12.58	3617	0

OCEAN PRESSURE AS A FUNCTION OF DEPTH AND LATITUDE

The following table is based upon an ocean model which takes into account the equation of state of standard seawater and the dependence on latitude of the acceleration of gravity. The tabulated pressure value is the excess pressure over the ambient atmospheric pressure at the surface.

REFERENCES

1. *International Oceanographic Tables, Volume 4*, Unesco Technical Papers in Marine Science No. 40, Unesco, Paris, 1987.
2. Saunders, P.M., and Fofonoff, N.P., *Deep-Sea Res.* 23, 109-111, 1976.

Pressure in MPa at the Specified Latitude

Depth (meters)	0°	15°	30°	45°	60°	75°	90°
0	0.0000	0.0000	0.0000	0.0000	0.0000	0.0000	0.0000
500	5.0338	5.0355	5.0404	5.0471	5.0537	5.0586	5.0605
1000	10.0796	10.0832	10.0930	10.1064	10.1198	10.1296	10.1333
1500	15.1376	15.1431	15.1577	15.1778	15.1980	15.2127	15.2182
2000	20.2076	20.2148	20.2344	20.2613	20.2882	20.3080	20.3153
2500	25.2895	25.2985	25.3231	25.3568	25.3905	25.4153	25.4244
3000	30.3831	30.3940	30.4236	30.4641	30.5047	30.5345	30.5453
3500	35.4886	35.5012	35.5358	35.5832	35.6307	35.6654	35.6782
4000	40.6056	40.6201	40.6598	40.7140	40.7683	40.8082	40.8229
4500	45.7342	45.7505	45.7952	45.8564	45.9176	45.9626	45.9791
5000	50.8742	50.8924	50.9421	51.0102	51.0785	51.1285	51.1469
5500	56.0255	56.0456	56.1004	56.1755	56.2508	56.3059	56.3262
6000	61.1882	61.2100	61.2700	61.3521	61.4344	61.4947	61.5168
6500	66.3619	66.3857	66.4508	66.5399	66.6292	66.6947	66.7187
7000	71.5467	71.5724	71.6427	71.7388	71.8352	71.9059	71.9318
7500	76.7426	76.7701	76.8456	76.9488	77.0523	77.1282	77.1560
8000	81.9493	81.9788	82.0594	82.1697	82.2804	82.3614	82.3911
8500	87.1669	87.1983	87.2841	87.4016	87.5193	87.6057	87.6373
9000	92.3950	92.4284	92.5194	92.6440	92.7689	92.8606	92.8941
9500	97.6346	97.6698	97.7661	97.8978	98.0300	98.1269	98.1624
10000	102.8800	102.9170	103.0185	103.1572	103.2961	103.3981	103.4355

PROPERTIES OF SEAWATER

In addition to the dependence on temperature and pressure, the physical properties of seawater vary with the concentration of the dissolved constituents. A convenient parameter for describing the composition is the salinity, S, which is defined in terms of the electrical conductivity of the seawater sample. The defining equation for the practical salinity is:

$$S = a_0 + a_1 K^{1/2} + a_2 K + a_3 K^{3/2} + a_4 K^2 + a_5 K^{5/2},$$

where K is the ratio of the conductivity of the seawater sample at 15°C and atmospheric pressure to the conductivity of a potassium chloride solution in which the mass fraction of KCl is 0.0324356, at the same temperature and pressure. The values of the coefficients are:

$$a_0 = 0.080 \qquad a_3 = 14.0941$$
$$a_1 = -0.1692 \qquad a_4 = -7.0261$$
$$a_2 = 25.3851 \qquad a_5 = 2.7081$$
$$\Sigma\, a_i = 35.0000$$

Thus when $K = 1$, $S = 35$ exactly (S is normally quoted in units of ‰, i.e., parts per thousand). The value of S can be roughly equated with the mass of dissolved material in grams per kilogram of seawater. Salinity values in the oceans at mid latitudes typically fall between 34 and 36.

The freezing point of seawater at normal atmospheric pressure varies with salinity as follows:

S	0	5	10	15	20	25	30	35	40
t_f/°C	0.000	-0.274	-0.542	-0.812	-1.083	-1.358	-1.636	-1.922	-2.212

The first table below gives several properties of seawater as a function of temperature for a salinity of 35. The second gives electrical conductivity as a function of salinity at several temperatures, and the third lists typical concentrations of the main constituents of seawater as a function of salinity.

REFERENCES

1. *The Practical Salinity Scale 1978 and the International Equation of State of Seawater 1980*, Unesco Technical Papers in Marine Science No. 36, Unesco, Paris, 1981; sections No. 37, 38, 39, and 40 in this series give background papers and detailed tables.
2. Kennish, M.J., *CRC Practical Handbook of Marine Science*, CRC Press, Boca Raton, FL, 1989.
3. Poisson, A. *IEEE J. Ocean. Eng.* OE-5, 50, 1981.

Properties of Seawater as a Function of Temperature at Salinity S=35 and Normal Atmospheric Pressure (100 kPa)

ρ = density in g/cm^3
$\beta = (1/\rho)\,(d\rho/dS)$ = fractional change in density per unit change in salinity
$\alpha = (1/\rho)\,(d\rho/dt)$ = fractional change in density per unit change in temperature (°C^{-1})
κ = electrical conductivity in S/cm
η = viscosity in mPa s (equal to cP)
c_p = specific heat in J/kg
v = speed of sound in m/s

t/°C	ρ/g cm^{-3}	$10^7\cdot\beta$	$10^7\cdot\alpha$/°C^{-1}	κ/S cm^{-1}	η/mPa s	c_p/J kg^{-1} °C^{-1}	v/m s^{-1}
0	1.028106	7854	526	0.029048	1.892	3986.5	1449.1
5	1.027675	7717	1136	0.033468	1.610		
10	1.026952	7606	1668	0.038103	1.388	3986.3	1489.8
15	1.025973	7516	2141	0.042933	1.221		
20	1.024763	7444	2572	0.047934	1.085	3993.9	1521.5
25	1.023343	7385	2970	0.053088	0.966		
30	1.021729	7338	3341	0.058373	0.871	4000.7	1545.6
35	1.019934	7300	3687				
40		7270	4004			4003.5	1563.2

PROPERTIES OF SEAWATER (continued)

Electrical Conductivity of Seawater in S/cm as a Function of Temperature and Salinity

$t/°C$	$S = 5$	$S = 10$	$S = 15$	$S = 20$	$S = 25$	$S = 30$	$S = 35$	$S = 40$
0	0.004808	0.009171	0.013357	0.017421	0.021385	0.025257	0.029048	0.032775
5	0.005570	0.010616	0.015441	0.020118	0.024674	0.029120	0.033468	0.037734
10	0.006370	0.012131	0.017627	0.022947	0.028123	0.033171	0.038103	0.042935
15	0.007204	0.013709	0.019905	0.025894	0.031716	0.037391	0.042933	0.048355
20	0.008068	0.015346	0.022267	0.028948	0.035438	0.041762	0.047934	0.053968
25	0.008960	0.017035	0.024703	0.032097	0.039276	0.046267	0.053088	0.059751
30	0.009877	0.018771	0.027204	0.035330	0.043213	0.050888	0.058373	0.065683

Composition of Seawater and Ionic Strength at Various Salinities
Concentration of major constituents expressed as molality (moles per kilogram of H_2O)

Constituent	$S = 30$	$S = 35$	$S = 40$
Cl^-	0.482	0.562	0.650
Br^-	0.00074	0.00087	0.00100
F^-		0.00007	
SO_4^{2-}	0.0104	0.0114	0.0122
HCO_3^-	0.00131	0.00143	0.00100
$NaSO_4^-$	0.0085	0.0108	0.0139
KSO_4^-	0.00010	0.00012	0.00015
Na^+	0.405	0.472	0.544
K^+	0.00892	0.01039	0.01200
Mg^{2+}	0.0413	0.0483	0.0561
Ca^{2+}	0.00131	0.00143	0.00154
Sr^+	0.00008	0.00009	0.00011
$MgHCO_3^+$	0.00028	0.00036	0.00045
$MgSO_4$	0.00498	0.00561	0.00614
$CaSO_4$	0.00102	0.00115	0.00126
$NaHCO_3$	0.00015	0.00020	0.00024
H_3BO_3	0.00032	0.00037	0.00042
Ionic strength (mol/kg)	0.5736	0.6675	0.7701

ABUNDANCE OF ELEMENTS IN THE EARTH'S CRUST AND IN THE SEA

This table gives the estimated abundance of the elements in the continental crust (in mg/kg, equivalent to parts per million by mass) and in seawater near the surface (in mg/L). Values represent the median of reported measurements. The concentrations of the less abundant elements may vary with location by several orders of magnitude.

REFERENCES

1. Carmichael, R. S., Ed., *CRC Practical Handbook of Physical Properties of Rocks and Minerals,* CRC Press, Boca Raton, FL, 1989.
2. Bodek, I., et al., *Environmental Inorganic Chemistry*, Pergamon Press, New York, 1988.
3. Ronov, A. B., and Yaroshevsky, A. A., "Earth's Crust Geochemistry", in *Encyclopedia of Geochemistry and Environmental Sciences,* Fairbridge, R. W., Ed., Van Nostrand, New York, 1969.

Element	Abundance Crust mg/kg	Abundance Sea mg/L	Element	Abundance Crust mg/kg	Abundance Sea mg/L
Ac	5.5×10^{-10}		N	1.9×10^1	5×10^{-1}
Ag	7.5×10^{-2}	4×10^{-5}	Na	2.36×10^4	1.08×10^4
Al	8.23×10^4	2×10^{-3}	Nb	2.0×10^1	1×10^{-5}
Ar	3.5	4.5×10^{-1}	Nd	4.15×10^1	2.8×10^{-6}
As	1.8	3.7×10^{-3}	Ne	5×10^{-3}	1.2×10^{-4}
Au	4×10^{-3}	4×10^{-6}	Ni	8.4×10^1	5.6×10^{-4}
B	1.0×10^1	4.44	O	4.61×10^5	8.57×10^5
Ba	4.25×10^2	1.3×10^{-2}	Os	1.5×10^{-3}	
Be	2.8	5.6×10^{-6}	P	1.05×10^3	6×10^{-2}
Bi	8.5×10^{-3}	2×10^{-5}	Pa	1.4×10^{-6}	5×10^{-11}
Br	2.4	6.73×10^1	Pb	1.4×10^1	3×10^{-5}
C	2.00×10^2	2.8×10^1	Pd	1.5×10^{-2}	
Ca	4.15×10^4	4.12×10^2	Po	2×10^{-10}	1.5×10^{-14}
Cd	1.5×10^{-1}	1.1×10^{-4}	Pr	9.2	6.4×10^{-7}
Ce	6.65×10^1	1.2×10^{-6}	Pt	5×10^{-3}	
Cl	1.45×10^2	1.94×10^4	Ra	9×10^{-7}	8.9×10^{-11}
Co	2.5×10^1	2×10^{-5}	Rb	9.0×10^1	1.2×10^{-1}
Cr	1.02×10^2	3×10^{-4}	Re	7×10^{-4}	4×10^{-6}
Cs	3	3×10^{-4}	Rh	1×10^{-3}	
Cu	6.0×10^1	2.5×10^{-4}	Rn	4×10^{-13}	6×10^{-16}
Dy	5.2	9.1×10^{-7}	Ru	1×10^{-3}	7×10^{-7}
Er	3.5	8.7×10^{-7}	S	3.50×10^2	9.05×10^2
Eu	2.0	1.3×10^{-7}	Sb	2×10^{-1}	2.4×10^{-4}
F	5.85×10^2	1.3	Sc	2.2×10^1	6×10^{-7}
Fe	5.63×10^4	2×10^{-3}	Se	5×10^{-2}	2×10^{-4}
Ga	1.9×10^1	3×10^{-5}	Si	2.82×10^5	2.2
Gd	6.2	7×10^{-7}	Sm	7.05	4.5×10^{-7}
Ge	1.5	5×10^{-5}	Sn	2.3	4×10^{-6}
H	1.40×10^3	1.08×10^5	Sr	3.70×10^2	7.9
He	8×10^{-3}	7×10^{-6}	Ta	2.0	2×10^{-6}
Hf	3.0	7×10^{-6}	Tb	1.2	1.4×10^{-7}
Hg	8.5×10^{-2}	3×10^{-5}	Te	1×10^{-3}	
Ho	1.3	2.2×10^{-7}	Th	9.6	1×10^{-6}
I	4.5×10^{-1}	6×10^{-2}	Ti	5.65×10^3	1×10^{-3}
In	2.5×10^{-1}	2×10^{-2}	Tl	8.5×10^{-1}	1.9×10^{-5}
Ir	1×10^{-3}		Tm	5.2×10^{-1}	1.7×10^{-7}
K	2.09×10^4	3.99×10^2	U	2.7	3.2×10^{-3}
Kr	1×10^{-4}	2.1×10^{-4}	V	1.20×10^2	2.5×10^{-3}
La	3.9×10^1	3.4×10^{-6}	W	1.25	1×10^{-4}
Li	2.0×10^1	1.8×10^{-1}	Xe	3×10^{-5}	5×10^{-5}
Lu	8×10^{-1}	1.5×10^{-7}	Y	3.3×10^1	1.3×10^{-5}
Mg	2.33×10^4	1.29×10^3	Yb	3.2	8.2×10^{-7}
Mn	9.50×10^2	2×10^{-4}	Zn	7.0×10^1	4.9×10^{-3}
Mo	1.2	1×10^{-2}	Zr	1.65×10^2	3×10^{-5}

SOLAR SPECTRAL IRRADIANCE

The solar luminosity (total radiant power emitted) is $3.86 \cdot 10^{26}$ W, of which 1373 W/m^2 reaches the top of the earth's atmosphere. To a zeroth approximation the sun can be considered a black body with an effective temperature of 5780 K, which implies a peak in the radiation at around 0.520 μm (5200 Å). The actual solar spectral emission is more complex, especially at ultraviolet and shorter wavelengths. The graph below, which was taken from Reference 1, summarizes the solar irradiance at the top of the atmosphere in the range 0.3 to 10 μm.

REFERENCES

1. Jursa, A.S., ed., *Handbook of Geophysics and the Space Environment*, Air Force Geophysics Laboratory, 1985.
2. Pierce, A.K., and Allen, R.G., "The Solar Spectrum between 0.3 and 10 μm", in *The Solar Output and its Variation*, White, O.R., Ed., Colorado Associated University Press, Boulder, CO, 1977.
3. Lang, K.R., *Astrophysical Data. Planets and Stars*, Springer-Verlag, New York, 1992.

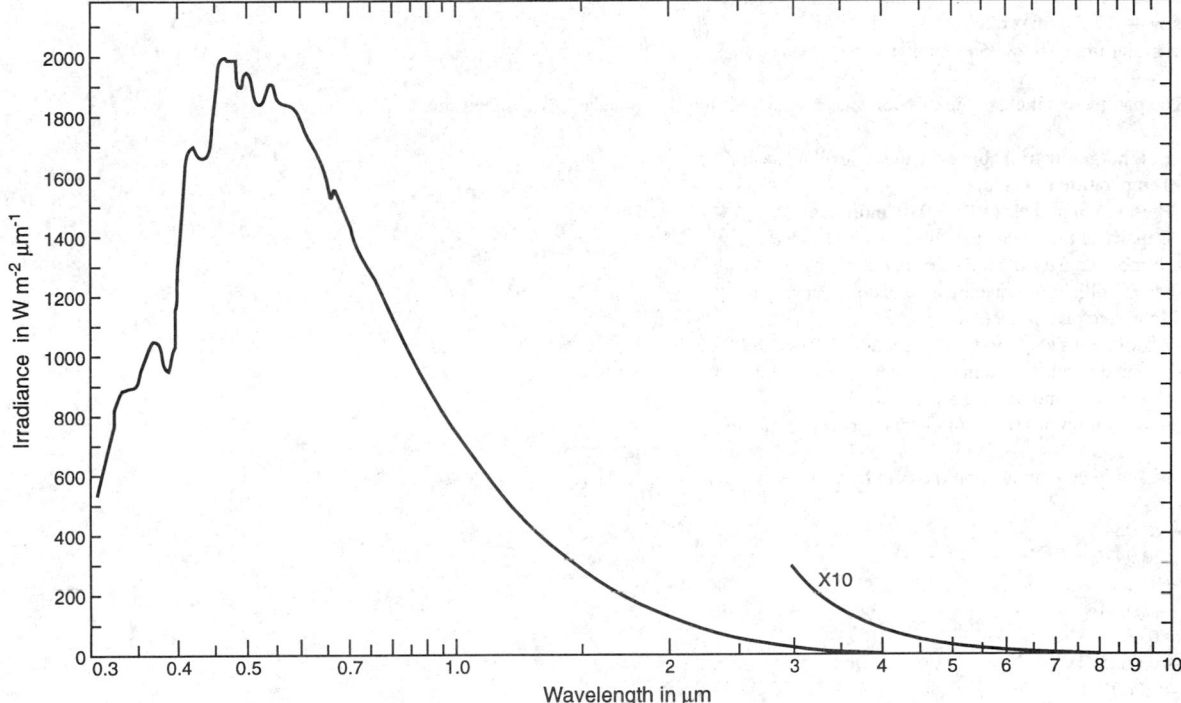

U.S. STANDARD ATMOSPHERE (1976)

A Standard Atmosphere is a hypothetical vertical distribution of atmospheric temperature, pressure, and density which is roughly representative of year-round, midlatitude conditions. Typical uses are to serve as a basis for pressure altimeter calibrations, aircraft performance calculations, aircraft and rocket design, ballistic tables, meteorological diagrams, and various types of atmospheric modeling. The air is assumed to be dry and to obey the perfect gas law and the hydrostatic equation which, taken together, relate temperature, pressure, and density with vertical position. The atmosphere is considered to rotate with the earth and to be an average over the diurnal cycle, the semiannual variation, and the range from active to quiet geomagnetic and sunspot conditions.

The U.S. Standard Atmosphere, 1976 is an idealized, steady-state representation of mean annual conditions of the earth's atmosphere from the surface to 1000 km at latitude 45°N, as it is assumed to exist during a period with moderate solar activity. The defining meteorological elements are sea-level temperature and pressure and a temperature-height profile to 1000 km. The 1976 Standard Atmosphere uses the following sea-level values which have been standard for many decades:

Temperature — 288.15 K (15°C)
Pressure — 101325 Pa (1013.25 mbar, 760 mm of Hg, or 29.92 in. of Hg)
Density — 1225 g/m^3 (1.225 g/L)
Mean molar mass — 28.964 g/mol

The parameters included in this condensed version of the U.S. Standard Atmosphere are:

Z — Height (geometric) above mean sea level in meters
T — Temperature in kelvins
P — Pressure in pascals (1 Pa = 0.01 millibars)
ρ — Density in kilograms per cubic meter (1 kg/m^3 = 1 g/L)
n — Number density in molecules per cubic meter
ν — Mean collision frequency in collisions per second
l — Mean free path in meters
η — Absolute viscosity in pascal seconds (1 Pa s = 1000 cP)
k — Thermal conductivity in joules per meter second kilogram (W/m K)
v_s — Speed of sound in meters per second
g — Acceleration of gravity in meters per second square

The sea-level composition (percent by volume) is taken to be:

N_2 — 78.084%
O_2 — 20.9476
Ar — 0.934
CO_2 — 0.0314
Ne — 0.001818
He — 0.000524
Kr — 0.000114
Xe — 0.0000087
CH_4 — 0.0002
H_2 — 0.00005

Details of the procedure for generating the U.S. Standard Atmosphere and information on other reference atmospheres may be found in Reference 2.

REFERENCES

1. COESA, U.S. Standard Atmosphere, 1976, U.S. Government Printing Office, Washington, D.C., 1976.
2. Jursa, A.S., ed., *Handbook of Geophysics and the Space Environment*, Air Force Geophysics Laboratory, 1985.

Z/m	T/K	P/Pa	ρ/kg m⁻³	n/m⁻³	ν/s⁻¹	l/m	η/Pa s	k/J m⁻¹s⁻¹K⁻¹	v_s/m s⁻¹	g/m s⁻²
-5000	320.68	1.778E+05	1.931	4.015E+25	1.151E+10	4.208E-08	1.942E-05	0.02788	359.0	9.822
-4500	317.42	1.685E+05	1.849	3.845E+25	1.096E+10	4.395E-08	1.927E-05	0.02763	357.2	9.830
-4000	314.17	1.596E+05	1.770	3.680E+25	1.044E+10	4.592E-08	1.912E-05	0.02738	355.3	9.819
-3500	310.91	1.511E+05	1.693	3.520E+25	9.933E+09	4.800E-08	1.897E-05	0.02713	353.5	9.818
-3000	307.66	1.430E+05	1.619	3.366E+25	9.448E+09	5.019E-08	1.882E-05	0.02688	351.6	9.816
-2500	304.41	1.352E+05	1.547	3.217E+25	8.982E+09	5.252E-08	1.867E-05	0.02663	349.8	9.814
-2000	301.15	1.278E+05	1.478	3.102E+25	8.623E+09	5.447E-08	1.852E-05	0.02638	347.9	9.813
-1500	297.90	1.207E+05	1.411	2.935E+25	8.106E+09	5.757E-08	1.836E-05	0.02613	346.0	9.811
-1000	294.65	1.139E+05	1.347	2.801E+25	7.693E+09	6.032E-08	1.821E-05	0.02587	344.1	9.810
-500	291.40	1.075E+05	1.285	2.672E+25	7.298E+09	6.324E-08	1.805E-05	0.02562	342.2	9.808
0	288.15	1.013E+05	1.225	2.547E+25	6.919E+09	6.633E-08	1.789E-05	0.02533	340.3	9.807
500	284.90	9.546E+04	1.167	2.427E+25	6.556E+09	6.961E-08	1.774E-05	0.02511	338.4	9.805
1000	281.65	8.988E+04	1.112	2.311E+25	6.208E+09	7.310E-08	1.758E-05	0.02485	336.4	9.804
1500	278.40	8.456E+04	1.058	2.200E+25	5.874E+09	7.680E-08	1.742E-05	0.02459	334.5	9.802
2000	275.15	7.950E+04	1.007	2.093E+25	5.555E+09	8.073E-08	1.726E-05	0.02433	332.5	9.801
2500	271.91	7.469E+04	0.957	1.990E+25	5.250E+09	8.491E-08	1.710E-05	0.02407	330.6	9.799
3000	268.66	7.012E+04	0.909	1.891E+25	4.959E+09	8.937E-08	1.694E-05	0.02381	328.6	9.797
3500	265.41	6.579E+04	0.863	1.795E+25	4.680E+09	9.411E-08	1.678E-05	0.02355	326.6	9.796
4000	262.17	6.166E+04	0.819	1.704E+25	4.414E+09	9.917E-08	1.661E-05	0.02329	324.6	9.794
4500	258.92	5.775E+04	0.777	1.616E+25	4.160E+09	1.046E-07	1.645E-05	0.02303	322.6	9.793
5000	255.68	5.405E+04	0.736	1.531E+25	3.918E+09	1.103E-07	1.628E-05	0.02277	320.6	9.791
5500	252.43	5.054E+04	0.697	1.450E+25	3.687E+09	1.165E-07	1.612E-05	0.02250	318.5	9.790
6000	249.19	4.722E+04	0.660	1.373E+25	3.467E+09	1.231E-07	1.595E-05	0.02224	316.5	9.788
6500	245.94	4.408E+04	0.664	1.299E+25	3.258E+09	1.302E-07	1.578E-05	0.02197	314.4	9.787
7000	242.70	4.111E+04	0.590	1.227E+25	3.058E+09	1.377E-07	1.561E-05	0.02170	312.3	9.785
7500	239.46	3.830E+04	0.557	1.159E+25	2.869E+09	1.458E-07	1.544E-05	0.02144	310.2	9.784
8000	236.22	3.565E+04	0.526	1.093E+25	2.689E+09	1.545E-07	1.527E-05	0.02117	308.1	9.782
8500	232.97	3.315E+04	0.496	1.031E+25	2.518E+09	1.639E-07	1.510E-05	0.02090	306.0	9.781
9000	229.73	3.080E+04	0.467	9.711E+24	2.356E+09	1.740E-07	1.493E-05	0.02063	303.9	9.779
9500	226.49	2.858E+04	0.440	9.141E+24	2.202E+09	1.848E-07	1.475E-05	0.02036	301.7	9.777
10000	223.25	2.650E+04	0.414	8.598E+24	2.056E+09	1.965E-07	1.458E-05	0.02009	299.5	9.776
10500	220.01	2.454E+04	0.389	8.079E+24	1.918E+09	2.091E-07	1.440E-05	0.01982	297.4	9.774
11000	216.77	2.270E+04	0.365	7.585E+24	1.787E+09	2.227E-07	1.422E-05	0.01954	295.2	9.773
11500	216.65	2.098E+04	0.337	7.016E+24	1.653E+09	2.408E-07	1.422E-05	0.01953	295.1	9.771
12000	216.65	1.940E+04	0.312	6.486E+24	1.528E+09	2.605E-07	1.422E-05	0.01953	295.1	9.770
12500	216.65	1.793E+04	0.288	5.996E+24	1.412E+09	2.818E-07	1.422E-05	0.01953	295.1	9.768
13000	216.65	1.658E+04	0.267	5.543E+24	1.306E+09	3.048E-07	1.422E-05	0.01953	295.1	9.767
13500	216.65	1.533E+04	0.246	5.124E+24	1.207E+09	3.297E-07	1.422E-05	0.01953	295.1	9.765
14000	216.65	1.417E+04	0.228	4.738E+24	1.116E+09	3.566E-07	1.422E-05	0.01953	295.1	9.764
14500	216.65	1.310E+04	0.211	4.380E+24	1.032E+09	3.857E-07	1.422E-05	0.01953	295.1	9.762
15000	216.65	1.211E+04	0.195	4.049E+24	9.538E+08	4.172E-07	1.422E-05	0.01953	295.1	9.761
16000	216.65	1.035E+04	0.166	3.461E+24	8.153E+08	4.881E-07	1.422E-05	0.01953	295.1	9.758
17000	216.65	8.850E+03	0.142	2.959E+24	6.969E+08	5.710E-07	1.422E-05	0.01953	295.1	9.754
18000	216.65	7.565E+03	0.122	2.529E+24	5.958E+08	6.680E-07	1.422E-05	0.01953	295.1	9.751
19000	216.65	6.467E+03	0.104	2.162E+24	5.093E+08	7.814E-07	1.422E-05	0.01953	295.1	9.748

Z/m	T/K	P/Pa	ρ/kg m⁻³	n/m⁻³	ν/s⁻¹	l/m	η/Pa s	k/J m⁻¹s⁻¹K⁻¹	v_s/m s⁻¹	g/m s⁻²
20000	216.65	5.529E+03	8.891E-02	1.849E+24	4.354E+08	9.139E-07	1.422E-05	0.01953	295.1	9.745
21000	217.58	4.729E+03	7.572E-02	1.574E+24	3.716E+08	1.073E-06	1.427E-05	0.01961	295.1	9.742
22000	218.57	4.048E+03	6.451E-02	1.341E+24	3.173E+08	1.260E-06	1.432E-05	0.01970	296.4	9.739
23000	219.57	3.467E+03	5.501E-02	1.144E+24	2.712E+08	1.477E-06	1.438E-05	0.01978	297.1	9.736
24000	220.56	2.972E+03	4.694E-02	9.759E+23	2.319E+08	1.731E-06	1.443E-05	0.01986	297.7	9.733
25000	221.55	2.549E+03	4.008E-02	8.334E+23	1.985E+08	2.027E-06	1.448E-05	0.01995	298.4	9.730
26000	222.54	2.188E+03	3.426E-02	7.123E+23	1.700E+08	2.372E-06	1.454E-05	0.02003	299.1	9.727
27000	223.54	1.880E+03	2.930E-02	6.092E+23	1.458E+08	2.773E-06	1.459E-05	0.02011	299.7	9.724
28000	224.53	1.610E+03	2.508E-02	5.214E+23	1.250E+08	3.240E-06	1.465E-05	0.02020	300.4	9.721
29000	225.52	1.390E+03	2.148E-02	4.466E+23	1.073E+08	3.783E-06	1.470E-05	0.02028	301.1	9.718
30000	226.51	1.197E+03	1.841E-02	3.828E+23	9.219E+07	4.414E-06	1.475E-05	0.02036	301.7	9.715
31000	227.50	1.031E+03	1.579E-02	3.283E+23	7.925E+07	5.146E-06	1.481E-05	0.02044	302.4	9.712
32000	228.49	8.891E+02	1.356E-02	2.813E+23	6.818E+07	5.995E-06	1.486E-05	0.02053	303.0	9.709
33000	230.97	7.673E+02	1.157E-02	2.406E+23	5.852E+07	7.021E-06	1.499E-05	0.02073	304.7	9.706
34000	233.74	6.634E+02	9.887E-03	2.056E+23	5.030E+07	8.218E-06	1.514E-05	0.02096	306.5	9.703
35000	236.51	5.746E+02	8.463E-03	1.760E+23	4.331E+07	9.601E-06	1.529E-05	0.02119	308.3	9.700
36000	239.28	4.985E+02	7.258E-03	1.509E+23	3.736E+07	1.120E-05	1.543E-05	0.02142	310.1	9.697
38000	244.82	3.771E+02	5.367E-03	1.116E+23	2.794E+07	1.514E-05	1.572E-05	0.02188	313.7	9.690
40000	250.35	2.871E+02	3.996E-03	8.308E+22	2.104E+07	2.034E-05	1.601E-05	0.02233	317.2	9.684
42000	255.88	2.200E+02	2.995E-03	6.227E+22	1.594E+07	2.713E-05	1.629E-05	0.02278	320.7	9.678
44000	261.40	1.695E+02	2.259E-03	4.697E+22	1.215E+07	3.597E-05	1.657E-05	0.02323	324.1	9.672
46000	266.93	1.313E+02	1.714E-03	3.564E+22	9.318E+06	4.740E-05	1.685E-05	0.02376	327.5	9.666
48000	270.65	1.023E+02	1.317E-03	2.738E+22	7.208E+06	6.171E-05	1.704E-05	0.02397	329.8	9.660
50000	270.65	7.978E+01	1.027E-03	2.135E+22	5.620E+06	7.913E-05	1.703E-05	0.02397	329.8	9.654
52000	269.03	6.221E+01	8.056E-04	1.675E+22	4.397E+06	1.009E-04	1.696E-05	0.02384	328.8	9.648
54000	263.52	4.834E+01	6.390E-04	1.329E+22	3.452E+06	1.272E-04	1.660E-05	0.02340	325.4	9.642
56000	258.02	3.736E+01	5.045E-04	1.049E+22	2.696E+06	1.611E-04	1.640E-05	0.02296	322.0	9.636
58000	252.52	2.872E+01	3.963E-04	8.239E+21	2.095E+06	2.051E-04	1.612E-05	0.02251	318.6	9.632
60000	247.02	2.196E+01	3.097E-04	6.439E+21	1.620E+06	2.624E-04	1.584E-05	0.02206	315.1	9.624
65000	233.29	1.093E+01	1.632E-04	3.393E+21	8.294E+05	4.979E-04	1.512E-05	0.02093	306.2	9.609
70000	219.59	5.221	8.283E-05	1.722E+21	4.084E+05	9.810E-04	1.438E-05	0.01978	297.1	9.594
75000	208.40	2.388	3.992E-05	8.300E+20	1.918E+05	2.035E-03	1.376E-05	0.01883	289.4	9.579
80000	198.64	1.052	1.846E-05	3.838E+20	8.656E+04	4.402E-03	1.321E-05	0.01800	282.5	9.564
85000	188.89	4.457E-01	8.220E-06	1.709E+20	3.766E+04	9.886E-03	1.265E-05	0.01716	275.5	9.550
90000	186.87	1.836E-01	3.416E-06	7.116E+19	1.560E+04	2.370E-02				9.535
95000	188.42	7.597E-02	1.393E-06	2.920E+19	6.440E+03	5.790E-02				9.520
100000	195.08	3.201E-02	5.604E-07	1.189E+19	2.680E+03	1.420E-01				9.505
110000	240.00	7.104E-03	9.708E-08	2.144E+18	5.480E+02	7.880E-01				9.476
120000	360.00	2.538E-03	2.222E-08	5.107E+17	1.630E+02	3.310				9.447
130000	469.27	1.251E-03	8.152E-09	1.930E+17	7.100E+01	8.800				9.418
140000	559.63	7.203E-04	3.831E-09	9.322E+16	3.800E+01	1.800E+01				9.389
150000	634.39	4.542E-04	2.076E-09	5.186E+16	2.300E+01	3.300E+01				9.360
160000	696.29	3.040E-04	1.233E-09	3.162E+16	1.500E+01	5.300E+01				9.331
170000	747.57	2.121E-04	7.815E-10	2.055E+16	1.000E+01	8.200E+01				9.302
180000	790.07	1.527E-04	5.194E-10	1.400E+16	7.200	1.200E+02				9.274

Z/m	T/K	P/Pa	$\rho/kg\ m^{-3}$	n/m^{-3}	v/s^{-1}	l/m	$\eta/Pa\ s$	$k/J\ m^{-1}s^{-1}K^{-1}$	$v_s/m\ s^{-1}$	$g/m\ s^{-2}$
190000	825.16	1.127E-04	3.581E-10	9.887E+15	5.200	1.700E+02				9.246
200000	854.56	8.474E-05	2.541E-10	7.182E+15	3.900	2.400E+02				9.218
220000	899.01	5.015E-05	1.367E-10	4.040E+15	2.300	4.200E+02				9.162
240000	929.73	3.106E-05	7.858E-11	2.420E+15	1.400	7.000E+02				9.106
260000	950.99	1.989E-05	4.742E-11	1.515E+15	9.300E-01	1.100E+03				9.051
280000	965.75	1.308E-05	2.971E-11	9.807E+14	6.100E-01	1.700E+03				8.997
300000	976.01	8.770E-06	1.916E-11	6.509E+14	4.200E-01	2.600E+03				8.943
320000	983.16	5.980E-06	1.264E-11	4.405E+14	2.900E-01	3.800E+03				8.889
340000	988.15	4.132E-06	8.503E-12	3.029E+14	2.000E-01	5.600E+03				8.836
360000	991.65	2.888E-06	5.805E-12	2.109E+14	1.400E-01	8.000E+03				8.784
380000	994.10	2.038E-06	4.013E-12	1.485E+14	1.000E-01	1.100E+04				8.732
400000	995.83	1.452E-06	2.803E-12	1.056E+14	7.200E-02	1.600E+04				8.680
450000	998.22	6.447E-07	1.184E-12	4.678E+13	3.300E-02	3.600E+04				8.553
500000	999.24	3.024E-07	5.215E-13	2.192E+13	1.600E-02	7.700E+04				8.429
550000	999.67	1.514E-07	2.384E-13	1.097E+13	8.400E-03	1.500E+05				8.307
600000	999.85	8.213E-08	1.137E-13	5.950E+12	4.800E-03	2.800E+05				8.188
650000	999.93	4.887E-08	5.712E-14	3.540E+12	3.100E-03	4.800E+05				8.072
700000	999.97	3.191E-08	3.070E-14	2.311E+12	2.200E-03	7.300E+05				7.958
750000	999.98	2.260E-08	1.788E-14	1.637E+12	1.700E-03	1.000E+06				7.846
800000	999.99	1.704E-08	1.136E-14	1.234E+12	1.400E-03	1.400E+06				7.737
850000	1000.00	1.342E-08	7.824E-15	9.717E+11	1.200E-03	1.700E+06				7.630
900000	1000.00	1.087E-08	5.759E-15	7.876E+11	1.000E-03	2.100E+06				7.525
950000	1000.00	8.982E-09	4.453E-15	6.505E+11	8.700E-04	2.600E+06				7.422
1000000	1000.00	7.514E-09	3.561E-15	5.442E+11	7.500E-04	3.100E+06				7.322

IGURE 1. Temperature-height profile for U.S. Standard Atmosphere –86 km.

FIGURE 3. Mean molecular weight as a function of geometric altitude.

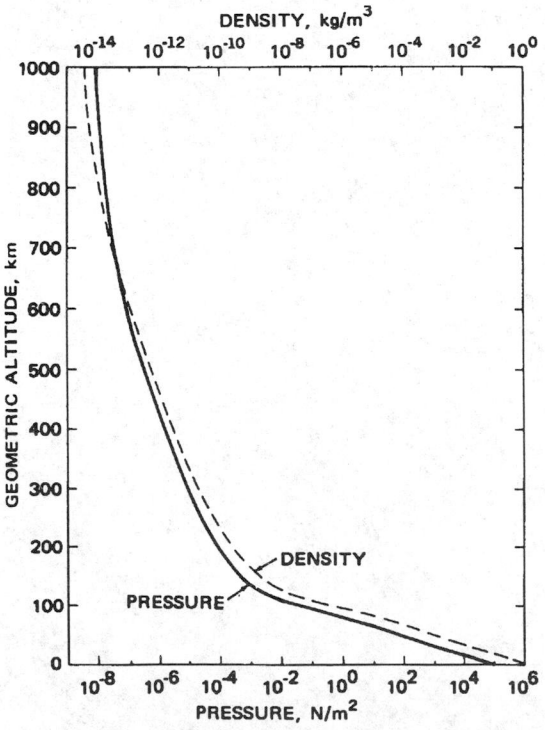

FIGURE 2. Total pressure and mass density as a function of geometric altitude.

FIGURE 4. Number density of individual species and total number density as a function of geometric altitude.

FIGURE 5. Collision frequency as a function of geometric altitude.

FIGURE 7. Mean air-particle speed as a function of geometric altitude.

FIGURE 6. Mean free path as a function of geometric altitude.

FIGURE 8. Dynamic viscosity as a function of geometric altitude.

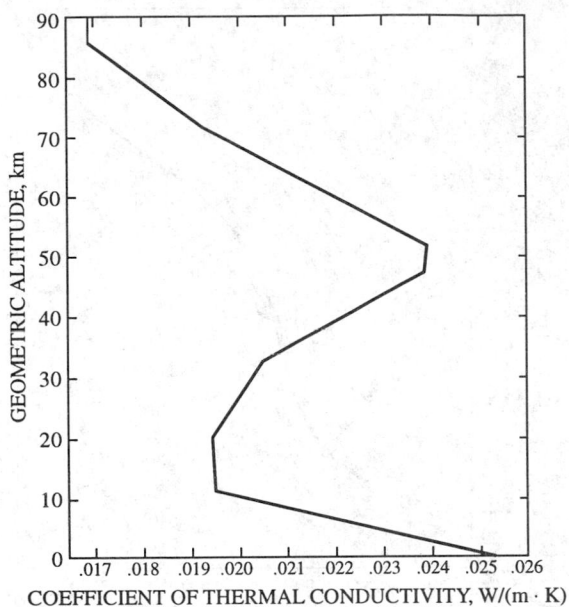

FIGURE 9. Coefficient of thermal conductivity as a function of geometric altitude.

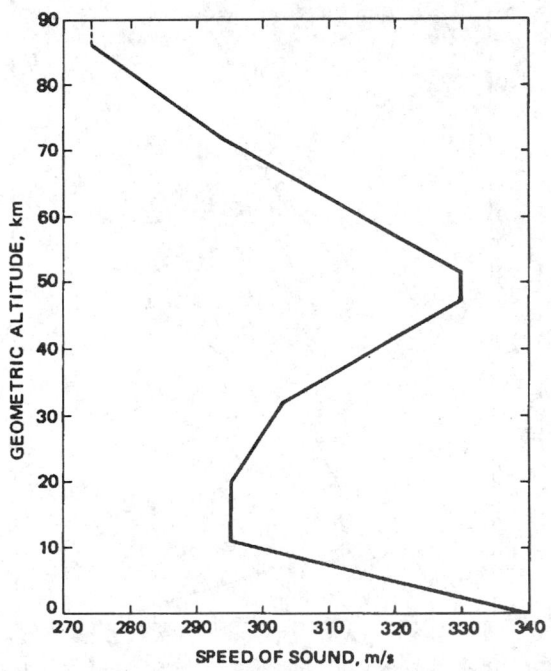

FIGURE 10. Speed of sound as a function of geometric altitude.

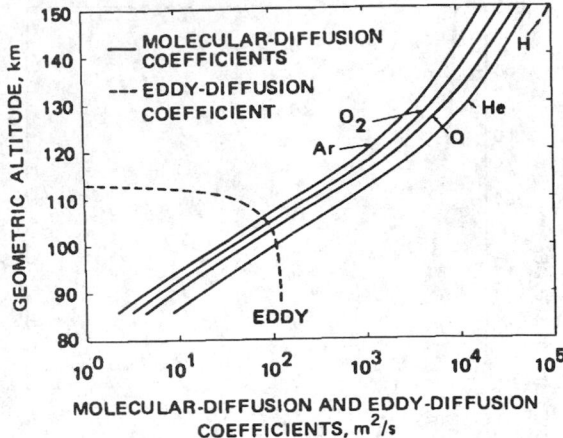

FIGURE 11. Molecular-diffusion and eddy-diffusion coefficients as a function of geometric altitude.

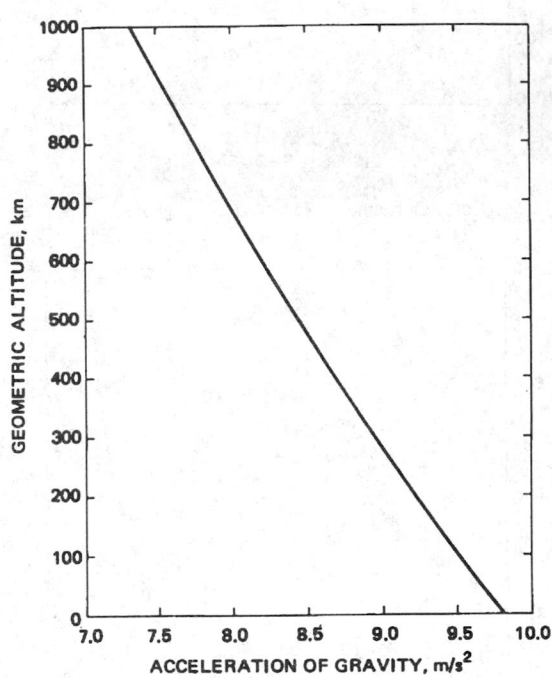

FIGURE 12. Acceleration of gravity as a function of geometric altitude.

GEOGRAPHICAL AND SEASONAL VARIATION IN SOLAR RADIATION

This table gives the amount of solar radiation reaching a unit area at the top of the earth's atmosphere per day as a function of latitude and approximate date. It is based upon a solar constant (total energy per unit area at the earth's average orbital distance) of 1373 W/m^2. Absorption of radiation by the atmosphere is not taken into consideration.

REFERENCE

List, R.J., *Smithsonian Meteorological Tables, Seventh Edition*, Smithsonian Institution Press, Washington, D.C., 1962.

Daily Solar Radiation in MJ/m^2

Lat.	Mar. 21	Apr. 13	May 6	May 29	Jun. 2	Jul. 15	Aug. 8	Aug. 31
90°		18.0	32.8	42.4	45.7	42.2	32.5	17.7
80	6.6	18.0	32.3	41.8	45.0	41.6	32.0	17.7
70	13.0	22.3	31.8	39.9	43.0	39.7	31.5	22.0
60	19.0	27.0	34.4	39.7	41.6	39.4	34.0	26.7
50	24.4	31.1	36.8	40.7	42.0	40.5	36.5	30.8
40	29.1	34.3	38.6	41.3	42.1	41.1	38.3	33.9
30	32.9	36.7	39.4	41.1	41.4	40.8	39.1	36.3
20	35.7	38.0	39.2	39.7	39.7	39.5	38.9	37.5
10	37.4	38.1	37.9	37.4	37.1	37.2	37.6	37.7
0	38.0	37.1	35.5	34.1	33.5	34.0	35.2	36.6
−10	37.4	35.0	32.3	30.0	29.2	29.9	32.0	34.6
−20	35.7	31.8	28.0	25.2	24.1	25.1	27.8	31.5
−30	32.9	27.8	23.1	19.7	18.5	19.7	22.8	27.4
−40	29.1	22.8	17.5	14.0	12.6	13.9	17.4	22.6
−50	24.4	17.3	11.7	8.2	7.0	8.2	11.6	17.2
−60	19.0	11.4	5.9	2.9	2.0	2.9	5.9	11.3
−70	13.0	5.4	1.0				1.0	5.3
−80	6.6	0.3						0.3
−90								

Lat.	Sep. 23	Oct. 16	Nov. 8	Nov. 30	Dec. 22	Jan. 13	Feb. 4	Feb. 26
90°								
80	6.5	0.3						0.3
70	12.9	5.5	1.0				1.0	5.6
60	18.8	11.6	6.2	3.1	2.1	3.1	6.2	11.7
50	24.1	17.6	12.1	8.7	7.5	8.7	12.3	17.8
40	28.7	23.1	18.2	14.8	13.5	14.9	18.4	23.5
30	32.5	28.2	23.9	20.9	19.8	21.0	24.1	28.4
20	35.3	32.3	29.1	26.6	25.7	26.7	29.3	32.7
10	37.0	35.5	33.5	31.8	31.1	31.9	33.8	35.9
0	37.6	37.6	36.9	36.1	35.8	36.3	37.3	38.0
−10	37.0	38.6	39.4	39.5	39.6	39.7	39.7	39.1
−20	35.3	38.5	40.7	42.0	42.4	42.2	41.1	39.0
−30	32.5	37.2	40.9	43.3	44.2	43.5	41.3	37.7
−40	28.7	34.8	40.1	43.6	45.0	43.8	40.5	35.2
−50	24.1	31.5	38.3	43.1	44.8	43.2	38.6	31.9
−60	18.8	27.3	35.7	41.9	44.4	42.1	36.0	27.7
−70	12.9	22.6	33.0	42.2	45.9	42.4	33.3	22.9
−80	6.5	18.2	33.5	44.2	48.1	44.4	33.8	18.4
−90		18.2	34.0	44.8	48.8	45.1	34.4	18.4

INFRARED ABSORPTION BY THE EARTH'S ATMOSPHERE

This graph summarizes the absorption by various atmospheric constituents in the wavelength range from 1 to 13 μm (wavenumber range 10,000 to 770 cm⁻¹). The vertical scale is in arbitrary units and does not take into account the wavelength variation of either the solar background radiation or the infrared detector response. Thus the intensities of the absorption bands have only qualitative significance.

WAVELENGTH IN MICROMETERS

ATMOSPHERIC CONCENTRATION OF CARBON DIOXIDE, 1958—1990

The data were taken at Mauna Loa Observatory in Hawaii and represent averages adjusted to the 15th of each month. Concentration of CO_2 is given in parts per million by volume.

REFERENCE

TRENDS 90, A Compendium of Data on Global Change, Carbon Dioxide Information Analysis Center, Oak Ridge National Laboratory, P.O. Box 2008, Oak Ridge, TN 37831-6335.

CO₂ Concentrations (ppmv)

Year	Jan	Feb	Mar	April	May	June	July	Aug	Sept	Oct	Nov	Dec	Ann
1958			315.55	317.29	317.34		315.69	314.78	313.03		313.18	314.50	
1959	315.42	316.32	316.49	317.56	318.13	318.00	316.39	314.66	313.68	313.18	314.66	315.43	315.83
1960	316.27	316.81	317.42	318.87	319.87	319.43	318.01	315.75	314.00	313.68	314.84	316.03	316.75
1961	316.73	317.54	318.38	319.31	320.42	319.61	318.42	316.64	314.83	315.15	315.95	316.85	317.49
1962	317.78	318.40	319.53	320.41	320.85	320.45	319.44	317.25	316.12	315.27	316.53	317.53	318.30
1963	318.58	318.92	319.70	321.22	322.08	321.31	319.58	317.61	316.05	315.83	316.91	318.20	318.83
1964	319.41				322.06	321.73	320.27	318.54	316.54	316.71	317.53	318.55	
1965	319.27	320.28	320.73	321.97	322.00	321.71	321.05	318.71	317.65	317.14	318.71	319.25	319.87
1966	320.46	321.43	322.22	323.54	323.91	323.59	322.26	320.21	318.48	317.94	319.63	320.87	321.21
1967	322.17	322.34	322.88	324.25	324.83	323.93	322.39	320.76	319.10	319.23	320.56	321.80	322.02
1968	322.40	322.99	323.73	324.86	325.41	325.19	323.97	321.92	320.10	319.96	320.97	322.48	322.83
1969	323.52	323.89	325.04	326.01	326.67	325.96	325.13	322.90	321.61	321.01	322.08	323.37	323.93
1970	324.34	325.30	326.29	327.54	327.54	327.21	325.98	324.42	322.91	322.90	323.85	324.96	325.27
1971	326.01	326.51	327.01	327.62	328.76	328.40	327.20	325.28	323.20	323.40	324.64	325.85	326.16
1972	326.60	327.47	327.58	329.56	329.90	328.92	327.89	326.17	324.68	325.04	326.34	327.39	327.29
1973	328.37	329.40	330.14	331.33	332.31	331.90	330.70	329.15	327.34	327.02	327.99	328.48	329.51
1974	329.18	330.55	331.32	332.48	332.92	332.08	331.02	329.24	327.28	327.21	328.29	329.41	330.08
1975	330.23	331.24	331.87	333.14	333.80	333.42	331.73	329.90	328.40	328.17	329.32	330.59	330.99
1976	331.58	332.39	333.33	334.41	334.71	334.17	332.88	330.77	329.14	328.77	330.14	331.52	331.98
1977	332.75	333.25	334.53	335.90	336.57	336.10	334.76	332.59	331.41	330.98	332.24	333.68	333.73
1978	334.80	335.22	336.47	337.59	337.84	337.72	336.37	334.51	332.60	332.37	333.75	334.79	335.34
1979	336.05	336.59	337.79	338.71	339.30	339.12	337.56	335.92	333.74	333.70	335.13	336.56	336.68
1980	337.84	338.19	339.90	340.60	341.29	341.00	339.39	337.43	335.72	335.84	336.93	338.04	338.52
1981	339.06	340.30	341.21	342.33	342.74	342.07	340.32	338.27	336.52	336.68	338.19	339.44	339.76
1982	340.57	341.44	342.53	343.39	343.96	343.18	341.88	339.65	337.80	337.69	339.09	340.32	340.96
1983	341.20	342.35	342.93	344.77	345.58	345.14	343.81	342.22	339.69	339.82	340.98	342.82	342.61
1984	343.52	344.33	345.11	346.88	347.25	346.61	345.22	343.11	340.90	341.17	342.80	344.04	344.25
1985	344.79	345.82	347.25	348.17	348.75	348.07	346.38	344.52	342.92	342.63	344.06	345.38	345.73
1986	346.12	346.79	347.69	349.38	350.04	349.38	347.78	345.75	344.70	344.01	345.50	346.75	346.99
1987	347.86	348.32	349.26	350.84	351.70	351.11	349.37	347.97	346.31	346.22	347.68	348.82	348.79
1988	350.29	351.58	352.08	353.45	354.08	353.66	352.25	350.30	348.58	348.74	349.93	351.21	351.35
1989	352.62	352.93	353.54	355.27	355.52	354.97	353.74	351.51	349.63	349.82	351.12	352.35	352.75
1990	353.47	354.51	355.18	355.98	356.94	355.99	354.58	352.68	350.72	350.92	352.55	353.91	353.95

MEAN TEMPERATURES IN THE UNITED STATES,
1901—1987

Historical records of atmospheric temperatures in the U.S. have been analyzed to obtain mean temperatures in °C for 23 climatically-distinct regions for the period 1901—1987. This table gives the average of the values for these 23 regions covering the entire country. Data for the individual regions and for other parts of the world may be found in the reference.

REFERENCE

TRENDS 90, A Compendium on Global Change, Carbon Dioxide Information Analysis Center, ORNL, P.O. Box 2008, Oak Ridge, Tenn. 37831-6335.

Year	Ann*	Win	Spr	Sum	Fall	Ann†
1901	11.46	0.55	10.59	22.92	12.36	11.61
1902	11.33	−0.21	11.39	21.75	12.46	11.35
1903	10.79	−0.57	10.94	21.23	11.70	10.82
1904	11.03	−0.82	10.74	21.31	12.56	10.95
1905	11.03	−1.37	11.39	21.97	12.25	11.06
1906	11.44	0.88	10.27	21.77	12.15	11.27
1907	11.31	1.51	10.43	21.32	12.07	11.33
1908	11.59	1.20	11.56	21.62	12.21	11.65
1909	11.21	1.35	9.93	22.31	12.68	11.57
1910	11.78	−1.60	12.74	22.10	12.75	11.50
1911	11.65	1.02	11.53	22.35	11.60	11.63
1912	10.61	−1.09	10.02	21.40	11.96	10.57
1913	11.39	0.04	10.62	22.42	12.41	11.37
1914	11.55	0.83	11.17	22.51	12.91	11.85
1915	11.27	−0.39	10.47	20.98	12.88	10.98
1916	11.01	0.30	10.89	21.99	11.59	11.19
1917	10.43	−1.07	9.10	21.85	11.79	10.42
1918	11.55	−1.17	11.66	22.66	12.06	11.30
1919	11.38	1.14	10.99	22.51	12.00	11.66
1920	11.05	−0.28	9.88	21.58	12.26	10.86
1921	12.63	2.04	12.09	22.83	13.15	12.52
1922	11.64	0.22	10.93	22.58	13.12	11.71
1923	11.41	0.88	10.14	22.02	12.08	11.28
1924	10.80	0.83	9.69	21.98	12.32	11.21
1925	11.86	0.36	12.21	22.49	11.57	11.66
1926	11.48	1.16	10.61	22.14	12.17	11.52
1927	11.65	1.39	11.12	21.28	13.30	11.77
1928	11.51	0.27	10.97	21.70	12.28	11.31
1929	10.97	−1.24	11.26	22.14	11.52	10.92
1930	11.59	0.96	11.26	22.64	12.03	11.72
1931	12.52	1.63	10.58	23.12	13.95	12.32
1932	11.51	2.06	10.45	22.65	11.95	11.78
1933	12.24	0.33	10.91	23.14	13.36	11.93
1934	12.75	2.22	12.59	23.61	13.29	12.93
1935	11.56	1.30	10.70	22.61	11.99	11.65
1936	11.69	−1.78	11.87	23.72	12.22	11.51
1937	11.32	−0.26	10.66	22.98	12.30	11.42
1938	12.25	1.43	11.78	22.64	12.85	12.18
1939	12.29	0.96	11.74	22.68	13.15	12.13
1940	11.49	0.34	10.89	22.59	12.42	11.56
1941	12.06	1.61	11.28	22.33	12.96	12.04
1942	11.57	0.76	11.29	22.30	12.58	11.73
1943	11.65	1.08	10.78	22.87	11.98	11.68
1944	11.54	1.12	10.41	22.21	12.75	11.62

MEAN TEMPERATURES IN THE UNITED STATES,
1901—1987 (continued)

Year	Ann*	Win	Spr	Sum	Fall	Ann†
1945	11.46	0.53	11.30	21.69	12.60	11.53
1946	12.16	0.48	12.41	22.10	12.32	11.83
1947	11.55	1.21	10.46	22.21	12.98	11.72
1948	11.44	−0.12	11.01	22.31	12.52	11.43
1949	11.74	−0.13	11.41	22.62	12.92	11.71
1950	11.32	1.40	10.05	21.42	12.63	11.38
1951	11.14	0.88	10.38	21.90	11.63	11.19
1952	11.78	1.21	10.52	22.98	12.07	11.70
1953	12.33	2.28	10.88	22.71	13.42	12.32
1954	12.33	2.32	10.95	22.78	13.24	12.32
1955	11.43	0.16	11.41	22.39	12.17	11.53
1956	11.78	0.32	10.88	22.46	12.57	11.56
1957	11.65	1.64	10.82	22.43	11.58	11.62
1958	11.55	1.27	10.62	22.23	12.98	11.77
1959	11.69	0.21	11.32	22.81	11.71	11.51
1960	11.26	0.49	10.10	22.42	13.05	11.51
1961	11.47	0.71	10.65	22.42	11.95	11.43
1962	11.49	−0.11	10.91	21.88	12.91	11.40
1963	11.64	−0.43	11.74	22.28	14.05	11.91
1964	11.39	−0.57	10.83	22.09	12.25	11.15
1965	11.41	0.55	10.24	21.61	12.60	11.25
1966	11.29	0.05	11.08	22.24	12.37	11.44
1967	11.45	0.93	11.05	21.73	12.11	11.45
1968	11.24	0.22	11.00	21.99	12.24	11.36
1969	11.28	−0.21	10.60	22.24	11.96	11.14
1970	11.38	0.32	10.50	22.60	12.00	11.36
1971	11.42	0.54	10.21	22.26	12.56	11.39
1972	11.25	0.81	11.51	21.95	11.54	11.45
1973	11.75	−0.05	11.02	22.44	12.98	11.60
1974	11.72	1.02	11.77	22.07	12.01	11.72
1975	11.29	0.84	9.82	22.06	12.35	11.27
1976	11.23	1.72	11.15	21.82	10.84	11.38
1977	11.88	−0.83	12.29	22.79	12.76	11.76
1978	11.02	−1.29	10.91	22.42	12.64	11.16
1979	10.96	−2.47	10.86	21.92	12.48	10.69
1980	11.71	1.06	10.60	22.90	12.46	11.75
1981	12.12	1.77	11.81	22.65	12.60	12.20
1982	11.28	−0.25	10.96	21.87	12.06	11.16
1983	11.50	2.11	10.02	22.79	12.97	11.98
1984	11.57	−0.49	10.41	22.47	12.14	11.14
1985	11.21	−0.34	12.48	22.14	11.84	11.53
1986	12.30	0.76	12.40	22.72	12.21	12.03
1987	12.28	1.62	12.24	22.66	12.58	12.27

*Calendar year mean (Jan–Dec).
†Season year mean (Win = Dec–Feb; Spr = Mar–May; Sum = Jun–Aug; Fall = Sep–Nov).

ATMOSPHERIC ELECTRICITY
Hans Dolezalek, Hannes Tammet, John Latham, and Martin A. Uman

I. SURVEY AND GLOBAL CIRCUIT
Hans Dolezalek

The science of atmospheric electricity originated in 1752 by an experimental proof of a related earlier hypothesis (that lightning is an electrical event). In spite of a large effort, in part by such eminent physicists as Coulomb, Lord Kelvin, and many others, an overall, proven theory able to generate models with sufficient resolution is not yet available. Generally accepted and encompassing text books are now more than 20 years old. The voluminous proceedings of the, so far, nine international atmospheric electricity conferences (1954 to 1992) give much valuable detail and demonstrate impressive progress, as do a number of less comprehensive textbooks published in the last 20 years, but a general theory as indicated above is not yet created. Only now, certain related measuring techniques and mathematical possibilities are emerging.

Applications to practical purposes do exist in the field of lightning research (including the electromagnetic radiation emanating from lightning) by the establishment of lightning-location networks and by the now developing possibility to detect electrified clouds which pose hazards to aircraft. Application of atmospheric electricity to other parts of meteorology seems to be promising but so far has seldom been instituted. Because some atmospheric electric signals propagate around the earth and because of the existence of a global circuit, applications for the monitoring of global change processes and conditions are now being proposed. Significant secular changes in the global circuit would indicate a change in the global climate; the availability of many old data (about a span of 100 years) could help detect a long term trend.

The concept of the "global circuit" is based on the theory of the global spherical capacitor: both, the solid (and liquid) earth as one electrode, and the high atmospheric layers (about the ionosphere) as the other, are by orders of magnitude more electrically conductive than the atmosphere between them. According to the "classical picture of atmospheric electricity", this capacitor is continuously charged by the common action of all thunderstorms to a d.c. voltage difference of several hundred kilovolts, the earth being negative. The much smaller but still existing conductivity of the atmosphere allows a current flowing from the ionosphere to the ground, integrated for all sink areas of the whole earth, of the order of 1.5 kA. In this way, a global circuit is created with many generators and sink-areas both interspaced and distributed over the whole globe, all connected to two nodes: ionosphere and ground. Within the scope of the global circuit, for each location, the current density (order of several pA/m^2) is determined by the voltage difference between ionosphere and ground (which is the same for all locations but varying in time) and the columnar resistance reaching from the ground up to the ionosphere (in the order of $10^{17} \Omega\, m^2$).

Natural processes, especially meteorological processes and some human activity, which produce or move electric charges ("space charges") or affect the ion distribution, constitute local generators and thereby "local circuits", horizontally and/or in parallel or antiparallel to the local part of the global circuit. In many cases, the local currents are much stronger than the global ones, making the measurement of the global current at a given location and/or during a period of time very difficult or, often, impossible. The strongest local circuits usually occur with certain weather conditions (precipitation, fog, high wind, blown-up dust or snow, heavy cloudiness) which make measurement of the global circuit impossible everywhere; but even in their absence local generators exist in varying magnitudes and of different characters. The separation of the local and global shares in the measured values of current density is a central problem of the science of atmospheric electricity. Aerological measurements are of high value in this regard.

The above description is within the "classical picture" of atmospheric electricity, a group of hypotheses to explain the electricifation of the atmosphere. It is probably fundamentally correct but certainly not complete; it has not yet been confirmed by systems of measurements resulting in no inner contradictions. In particular, extraterrestrial influences must be permitted; their general significance is still under debate.

Within this "classical picture" a kind of electric standard atmosphere may be constructed as shown in Table 1.

Values with a star, *, are rough average values from measurement. A star in parentheses, (*), points to a typical value from one or a few measurements. All other values have been calculated from starred values, under the assumption that at 2 km 50% and at 12 km 90% of the columnar resistance is reached. Voltage drop along one of the partial columns can be calculated by subtracting the value for the lower column from that of the upper one. Columnar resistances, conductances, and capacitances are valid for that particular part of the column which is indicated at the left. Capacitances are calculated with the formula for plate capacitors, and this fact must be considered also for the time constants for columns.

According to measurements, U, the potential difference between 0 m and 65 km may vary by a factor of approximately 2. The total columnar resistance, R_c, is estimated to vary up to a factor of 3, the variation being due to either reduction of conductivity in the exchange layer (about lowest 2 km of this table) or to the presence of high mountains; in both cases the variation is caused in the troposphere. Smaller variations in the stratosphere and mesosphere are being discussed because of aerosols there. The air-earth current density in fair weather varies by a factor of 3 to 6 accordingly. Conductivity near the ground varies by a factor of about 3 but only decreasing; increase of conductivity due to extraordinary radioactivity is a singular event. The field strength near the ground varies as a consequence of variations of air-earth current density and conductivity from about 1/3 to about 10 times of the value quoted in the table. Conductivity near the ground shows a diurnal and an annual variation which depends strongly on the locality: air-earth current density shows a diurnal and annual variation because the earth-ionosphere potential difference undergoes such variations, and also because the columnar resistance is supposed to have a diurnal and probably an annual variation.

Conductivities and air-earth current densities on high mountains are greater than at sea level by factors of up to 10. Conductivity decreases when atmospheric humidity increases. Values for space charges are not quoted because measurements are too few to allow calculation of average values. Values of parameters over the oceans are still rather uncertain.

Theoretically, in fair-weather conditions, Ohm's law must be fulfilled for the electric field, the conduction current density, and the electrical conductivity of the atmosphere. Deviations point to shortcomings in the applied measuring techniques. Data which are representative for a large area (in the extreme, "globally representative data", i.e. data on the global circuit), can on the ground be obtained only by stations on an open plane and only if local generators are either small or constant or are independently measured. Certain measurements with instrumented aircraft provide globally representative information valid for the period of the actual measurement.

TABLE 1

Electrical Parameters of the Clear (Fair-Weather) Atmosphere, Pertinent to the Classical Picture of Atm. Electricity (Electric Standard Atmosphere)

Part of atmosphere for which the values are calculated (elements are in free, cloudless atmosphere)	Currents, I, in A; and current densities, i, in A/m²	Potential differences, U, in V; field strength E in V/m; $U = 0$ at sea level	Resistances, R, in Ω; columnar resistances, R_c, in Ω m² and resistivities, ρ, in Ω m	Conductances, G, in Ω⁻¹; columnar conductances G_c, in Ω⁻¹ m²; total conductivities, γ, in Ω⁻¹ m⁻¹	Capacitances, C, in F; columnar capacitances, C_c, in F m⁻² and capacitivities, ε, in F m⁻¹	Time constants τ, in seconds
Volume element at about sea level, 1 m³	$i = 3 \times 10^{-12}$*	$E_0 = 1.2 \times 10^2$*	$\rho_0 = 4 \times 10^{13}$	$\gamma_0 = 2.5 \times 10^{-14}$	$\varepsilon_0 = 8.9 \times 10^{-12}$*	$\tau_0 = 3.6 \times 10^2$
Lower column of 1 m² cross section from sea level to 2 km height	$i = 3 \times 10^{-12}$	At upper end: $U_1 = 1.8 \times 10^5$	$R_{c1} = 6 \times 10^{15}$	$G_{c1} = 1.7 \times 10^{-17}$	$C_{c1} = 4.4 \times 10^{-15}$	$\tau_{c1} = 2.6 \times 10^2$
Volume element at about 2 km height, 1 m³	$i = 3 \times 10^{-12}$	$E_2 = 6.6 \times 10^1$	$\rho_2 = 2.2 \times 10^{13(*)}$	$\gamma_2 = 4.5 \times 10^{-14}$	$\varepsilon_2 = 8.9 \times 10^{-12}$	$\tau_2 = 2 \times 10^2$
Center column of 1 m² cross section from 2 to 12 km	$i = 3 \times 10^{-12}$	At upper end: $U_m = 3.15 \times 10^5$	$R_{cm} = 4.5 \times 10^{16}$	$G_{cm} = 5 \times 10^{-17}$	$C_{cm} = 8.8 \times 10^{-16}$	$\tau_{cm} = 1.8 \times 10^1$
Volume element at about 12 km height, 1 m³	$i = 3 \times 10^{-12}$	$E_{12} = 3.9 \times 10^0$	$\rho_{12} = 1.3 \times 10^{12(*)}$	$\gamma_{12} = 7.7 \times 10^{-13}$	$\varepsilon_{12} = 8.9 \times 10^{-12}$	$\tau_{12} = 1.2 \times 10^1$
Upper column of 1 m² cross section from 12 to 65 km height	$i = 3 \times 10^{-12}$	At upper end: $U_u = 3.5 \times 10^5$	$R_{cu} = 1.5 \times 10^{16}$	$G_{cu} = 2.5 \times 10^{-17}$	$C_{cu} = 1.67 \times 10^{-16}$	$\tau_{cm} = 6.7 \times 10^0$
Whole column of 1 m² cross section from 0 to 65 km height	$i = 3 \times 10^{-12}$	At upper end: $U = 3.5 \times 10^5$	$R_c = 1.2 \times 10^{17}$	$G_c = 8.3 \times 10^{-18}$	$C_c = 1.36 \times 10^{-16}$	$\tau_c = 1.64 \times 10^1$
Total spherical capacitor area: 5 × 10¹⁴ m²	$1 = 1.5 \times 10^3$	$U = 3.5 \times 10^5$*	$R = 2.4 \times 10^2$	$G = 4.2 \times 10^{-3}$	$C = 6.8 \times 10^{-2}$	$\tau = 1.64 \times 10^1$

Note: All currents and fields listed are part of the global circuit, i.e., circuits of local generators are not included. Values are subject to variations due to latitude and altitude of the point of observation above sea level, locality with respect to sources of disturbances, meteorological and climatological factors, and man-made changes. For more explanations, see text.

II. AIR IONS
Hannes Tammet

The term " air ions" signifies all airborne particles which are the carriers of the electrical current in the air and have drift velocities determined by the electric field.

The probability of electrical dissociation of molecules in the atmospheric air under thermodynamic equilibrium is near to zero. The average ionization at the ground level over the ocean is $2 \cdot 10^6$ ion pairs $m^{-3}s^{-1}$. This ionization is produced mainly by cosmic rays. Over the continents the ionizing radiation from soil and from radioactive substances in the air each add about $4 \cdot 10^6$ $m^{-3}s^{-1}$. The total average ionization rate of 10^7 $m^{-3}s^{-1}$ is equivalent to 17 $\mu R/h$ which is a customary expression of the background level of the ionizing radiations. The ionization rate over the ground varies in space due to the radioactivity of soil, and in time depending on the exchange of air between the atmosphere and radon-containing soil. Radioactive pollution increases the ionization rate. A temporary increase of about 10 times was registered in Sweden after the Chernobyl accident in 1986. The emission of Kr^{85} from nuclear power plants can noticeably increase the global ionization rate in the next century. The ionization rate decreases with altitude near the ground and increases at higher altitudes up to 15 km, where it has a maximum about $5 \cdot 10^7$ $m^{-3}s^{-1}$. Solar X-ray and extreme UV radiation cause a new increase at altitudes over 60 km.

Local sources of air ions are point discharges in strong electric fields, fluidization of charged drops from waves, etc.

The enhanced chemical activity of an ion results in a chain of ion-molecule reactions with the colliding neutrals, and, in the first microsecond of the life of an air ion, a charged molecular cluster called the *cluster ion* is formed. According to theoretical calculations in the air free from exotic trace gases the following cluster ions should be dominant:

$$NO_3^- \cdot (HNO_3) \cdot H_2O, \; NO_2^- \cdot (H_2O)_2, \; NO_3^- \cdot H_2O, \; O_2^- \cdot (H_2O)_4, \; O_2^- \cdot (H_2O)_5,$$
$$H_3O^+ \cdot (H_2O)_6, \; NH_4^+ \cdot (H_2O)_2, \; NH_4^+ \cdot (H_2O), \; H_3O^+ \cdot (H_2O)_5, \; NH_4^+ \cdot NH_3$$

A measurable parameter of air ions is the electrical mobility k, characterizing the drift velocity in the unit electric field. The mobility is inversely proportional to the density of air, and the results of measurements are as a rule reduced to normal conditions. According to mobility the air ions are called: fast or small or light ions with mobility $k > 5 \cdot 10^{-5}$ $m^2V^{-1}s^{-1}$, intermediate ions, and slow or large or heavy ions with mobility $k < 10^{-6}$ m^2V^{-1} s^{-1}. The boundary between intermediate and slow ions is conventional.

Cluster ions are fast ions. The masses of cluster ions may be measured with mass spectrometers, but the possible ion-molecule reactions during the passage of the air through nozzles to the vacuum chamber complicate the measurement. Mass and mobility of cluster ions are highly correlated. The experimental results[5] can be expressed by the empirical formula

$$m \approx \frac{850 \text{ u}}{\left[0.3 + k / \left(10^{-4} \, m^2 V^{-1} s^{-1} \right) \right]^3}$$

where u is the unified atomic mass unit.

The value of the transport cross-section of a cluster ion is needed to calculate its mobility according to the kinetic theory of Chapman and Enskog. The theoretical estimation of transport cross-sections is rough and cannot be used to identify the chemical structures of cluster ions. Mass spectrometry is the main technique of identification of cluster ions.[2]

Märk and Castleman (1986) presented an overview of over 1000 publications on the experimental studies of cluster ions. Most of them present information about ions of millisecond age range. The low concentration makes it difficult to get detailed information about masses and mobilities of the natural atmospheric ions at ground level. The results of a 1-year continuous measurement[6] are as follows:

	+ ions	– ions	unit
Average mobility	1.36	1.56	10^{-4} $m^2V^{-1}s^{-1}$
The corresponding mass	190	130	u
The corresponding diameter	0.69	0.61	nm
The average concentration	400	360	10^6 m^{-3}
The corresponding conductivity	8.7	9.0	fS

The distribution of tropospheric cluster ions according to the mobility and estimated mass is depicted in Figure 1.

The problems and results of direct mass spectrometry of natural cluster ions are analysed by Eisele (1986) for ground level and by Meyerott, Reagan and Joiner (1980) for stratospheric measurements. Air ions in the high atmosphere are a subject of ionospheric physics.

During its lifetime (about 1 min), a cluster ion at ground level collides with nearly 10^{12} molecules. Thus the cluster ions are able to concentrate trace gases of very low concentration if they have an extra high electron or proton affinity. For example, Eisele (1986) demonstrated that a considerable fraction of positive atmospheric cluster ions in the unpolluted atmosphere at ground level probably consist of a molecule derived from pyridine. The concentration of these constituents is estimated to be about 10^{-12}. Therefore, air-ion mass and mobility spectrometry is considered as a promising technique for trace analysis in the air. Mass and mobility spectrometry of millisecond-age air ions has been developed as a technique of chemical analysis known as "plasma chromatography" (Carr, 1984). The sensitivity of the detection grows with the age of the cluster ions measured.

The mechanisms of annihilation of cluster ions are ion-ion recombination (on the average 3%) and sedimentation on aerosol particles (on the average 97% of cluster ions at ground level). The result of the combination of a cluster ion and neutral particle is a charged particle called an *aerosol ion*. In conditions of detailed thermodynamic equilibrium the probability that a spherical particle of diameter d carries q elementary charges is calculated from the Boltzmann distribution:

$$p_q(d) = (2 \, \pi d/d_0)^{1/2} \exp(-q^2 \, d_0/2d)$$

Figure 1. Average mobility and mass spectra of natural tropospheric cluster ions. Concentrations of the mobility fractions were measured in a rural site every 5 min over 1 year.[6] Ion mass is estimated according to the above empirical formula.

where $d_0 = 115$ nm (at 18°C). The supposition about the detailed equilibrium is an approximation and the formula is not valid for particles less than d_0. On the basis of numerical calculations by Hoppel and Frick (1990) the following charge probabilities can be derived:

d	3	10	30	100	300	1000	3000	nm
p_0	98	90	70	42	24	14	8	%
$p_{-1} + p_1$	2	10	30	48	41	25	15	%
$p_{-2} + p_2$	0	0	0	10	23	21	14	%
$p_{q > 2}$	0	0	0	0	12	40	63	%
k_1	15000	1900	250	28	5.1	1.11	0.33	$10^{-9} m^2 V^{-1} s^{-1}$

The last line of the table presents the mobility of a particle carrying one elementary charge. The distribution of the atmospheric aerosol ions over mobility is demonstrated in Figure 2.

Although the concentration of aerosol in continental air at ground level is an order of magnitude higher than the concentration of cluster ions, the mobilities of aerosol ions are so small that their percentage in air conductivity is less than 1%.

A specific class of aerosol ions are condensed aerosol ions produced as a result of the condensation of gaseous matter on the cluster ions. In aerosol physics the process is called ion-induced nucleation; it is considered as one among the processes of gas-to-particle conversion. The condensed aerosol ions have an inherent charge. Their sizes and mobilities are between the sizes and mobilities of cluster ions and of ordinary aerosol ions. Water and standard constituents of atmospheric air are not able to condense on the cluster ions in the real atmosphere. Thus the concentration of condensed aerosol ions depends on the trace constituents in the air and is very low in unpolluted air. Knowledge about condensed aerosol ions is poor because of measurement difficulties.

Figure 2. Mobility and size spectra of tropospheric aerosol ions.[6] The wide bars mark the fraction concentrations theoretically estimated on the basis of the standard size distribution of tropospheric aerosol. The pin bars with head + and - mark average values of positive and negative aerosol ion fraction concentrations measured in a rural site every 5 min during 4 months.

REFERENCES

1. Carr, T. W., Ed., *Plasma Chromatography*, Plenum Press, New York and London, XII + 259 pp., 1984.
2. Eisele, F. L., Identification of Tropospheric Ions, *J. Geophys. Res.*, vol. 91, no. D7, pp. 7897-7906, 1986.

3. Hoppel, W. A., and Frick, G. M., The Nonequilibrium Character of the Aerosol Charge Distributions Produced by Neutralizers, *Aerosol Sci. Technol.*, vol. 12, no. 3, pp. 471-496, 1990.
4. Mark, T. D., and Castleman, A. W., Experimental Studies on Cluster Ions, in *Advances in Atomic and Molecular Physics*, vol. 20, pp. 65.-172, Academic Press, 1985.
5. Meyerott, R. E., Reagan, J. B., and Joiner, R. G., The Mobility and Concentration of Ions and the Ionic Conductivity in the Lower Stratosphere, *J. Geophys. Res.*, vol. 85, no. A3, pp. 1273-1278, 1980.
6. Salm, J., Tammet, H., Iher, H., and Hörrak, U., Atmospheric Electrical Measurements in Tahkuse, Estonia (in Russian), in *Voprosy Atmosfernogo Elektrichestva*, pp. 168-175, Gidrometeoizdat, Leningrad, 1990.

III. THUNDERSTORM ELECTRICITY
John Latham

The development of improved radar techniques and instruments for in-cloud electrical and physical measurements, coupled with a much clearer recognition by the research community that establishment of the mechanism or mechanisms responsible for electric field development in thunderclouds, culminating in lightning, is inextricably linked to the concomitant dynamical and microphysical evolution of the clouds, has led to significant progress over the past decade.

Field studies indicate that in most thunderclouds the electrical development is associated with the process of glaciation, which can occur in a variety of incompletely understood ways. In the absence of ice, field growth is slow, individual hydrometeor charges are low, and lightning is produced only rarely. Precipitation — in the solid form, as graupel — also appears to be a necessary ingredient for significant electrification, as does significant convective activity and mixing between the clouds and their environments, via entrainment.

Increasingly, the view is being accepted that charge transfer leading to field-growth is largely a consequence of rebounding collisions between graupel pellets and smaller vapor-grown ice crystals, followed by the separation under gravity of these two types of hydrometeor. These collisions occur predominantly within the temperature range -15 to -30°C, and for significant charge transfer need to occur in the presence of supercooled cloud droplets.

The field evidence is inconsistent with an inductive mechanism, and extensive laboratory studies indicate that the principal charging mechanism is non-inductive and associated — in ways yet to be identified — with differences in surface characteristics of the interacting hydrometeors.

Laboratory studies indicate that the two most favored sites for corona emission leading to the lightning discharge are the tips of ephemeral liquid filaments, produced during the glancing collisions of supercooled raindrops, and protuberances on large ice crystals or graupel pellets. The relative importance of these alternatives will depend on the hydrometeor characteristics and the temperature in the regions of strongest fields; these features are themselves dependent on air-mass characteristics and climatological considerations.

A recently identified but unresolved question is why, in continental Northern Hemisphere thunderclouds at least, the sign of the charge brought to ground by lightning is predominantly negative in summer but more evenly balanced in winter.

IV. LIGHTNING
Martin A. Uman

From both ground-based weather-station data and satellite measurements, it has been estimated that there are about 100 lightning discharges, both cloud and ground flashes, over the whole earth each second; representing an average global lightning flash density of about $6 \text{ km}^{-2}\text{yr}^{-1}$. Most of this lightning occurs over the earth's land masses. For example, in central Florida, where thunderstorms occur about 90 days/yr, the flash density for discharges to earth is about $15 \text{ km}^{-2}\text{yr}^{-1}$. Some tropical areas of the earth have thunderstorms up to 300 days/yr.

Lightning can be defined as a transient, high-current electric discharge whose path length is measured in kilometers and whose most common source is the electric charge separated in the ordinary thunderstorm or cumulonimbus cloud. Well over half of all lightning discharges occur totally within individual thunderstorm clouds and are referred to as intracloud discharges. Cloud-to-ground lightning, however, has been studied more extensively than any other lightning form because of its visibility and its more practical interest. Cloud-to-cloud and cloud-to-air discharges are less common than intracloud or cloud-to-ground lightning.

Lightning between the cloud and earth can be categorized in terms of the direction of motion, upward or downward, and the sign of the charge, positive or negative, of the developing discharge (called a *leader*) which initiates the overall event. Over 90% of the worldwide cloud-to-ground discharges is initiated in the thundercloud by downward-moving negatively-charged leaders and subsequently results in the lowering of negative charge to earth. Cloud-to-ground lightning can also be initiated by downward-moving positive leaders, less than 10% of the worldwide cloud-to-ground lightning being of this type although the exact percentage is a function of season and latitude. Lightning between cloud and ground can also be initiated by leaders which develop upward from the earth. These upward-initiated discharges are relatively rare, may be of either polarity, and generally occur from mountaintops and tall man-made structures.

We discuss next the most common type of cloud-to-ground lightning. A negative cloud-to-ground discharge or *flash* has an overall duration of some tenths of a second and is made up of various components, among which are typically three or four high-current pulses called *strokes*. Each stroke lasts about a millisecond, the separation time between strokes being typically several tens of milliseconds. Such lightning often appears to "flicker" because the human eye can just resolve the individual light pulse associated with each stroke. A drawing of the components of a negative cloud-to-ground flash is found in Figure 3. Some values for salient parameters are found in Table 1. The negatively-charged *stepped leader* initiates the first stroke in a flash by propagating from cloud to ground through virgin air in a series of discrete steps. Photographically observed leader steps in clear

Figure 3. Sequence of steps in cloud-to-ground lightning.

air are typically 1 µs in duration and tens of meters in length, with a pause time between steps of about 50 µs. A fully developed stepped leader lowers up to 10 or more coulombs of negative cloud charge toward ground in tens of milliseconds with an average downward speed of about 2×10^5 m/s. The average leader current is in the 100 to 1000 A range. The steps have pulse currents of at least 1 kA. Associated with these currents are electric- and magnetic-field pulses with widths of about 1 µs or less and risetimes of about 0.1 µs or less. The stepped leader, during its trip toward ground, branches in a downward direction, resulting in the characteristic downward-branched geometrical structure commonly observed. The electric potential of the bottom of the negatively-charged leader channel with respect to ground has a magnitude in excess of 10^7 V. As the leader tip nears ground, the electric field at sharp objects on the ground or at irregularities of the ground itself exceeds the breakdown value of air, and one or more upward-moving discharges (often called upward leaders) are initiated from those points, thus beginning the *attachment process*. An understanding of the physics of the attachment process is central to an understanding of the operation of lightning protection of ground-based objects and the effects of lightning on humans and animals, since it is the attachment process that determines where the lightning connects to objects on the ground and the value of the early currents which flow. When one of the upward-moving discharges from the ground (or from a lightning rod or an individual) contacts the tip of the downward-moving stepped leader, typically some tens of meters above the ground, the leader tip is effectively connected to ground potential. The negatively-charged leader channel is then discharged to earth when a ground potential wave, referred to as the first *return stroke*, propagates continuously up the leader path. The upward speed of a return stroke near the ground is typically near one third the speed of light, and the speed decreases with height. The first return stroke produces a peak current near ground of typically 30 kA, with a time from zero to peak of a few microseconds. Currents measured at the ground fall to half of the peak value in about 50 µs, and currents of the order of hundreds of amperes may flow for times of a few milliseconds up to several hundred milliseconds. The longer-lasting currents are known as *continuing currents*. The rapid release of return stroke energy heats the leader channel to a temperature near 30,000 K and creates a high-pressure channel which expands and generates the shock waves that eventually become thunder, as further discussed later. The return stroke effectively lowers to ground the charge originally deposited onto the stepped-leader channel and additionally initiates the lowering of other charge which may be available to the top of its channel. First return-stroke electric fields exhibit a microsecond scale rise to peak with a typical peak value of 5 V/m, normalized to a distance of 100 km by an inverse distance relationship. Roughly half of the field rise to peak, the so-called "fast transition", takes place in tenths of a microsecond, an observation that can only be made if the field propagation is over a highly conducting surface such as salt water.

After the first return-stroke current has ceased to flow, the flash, including charge motion in the cloud, may end. The lightning is then called a single-stroke flash. On the other hand, if additional charge is made available to the top of the channel, a continuous or *dart leader* may propagate down the residual first-stroke channel at a typical speed of about 1×10^7 m/s. The dart leader lowers a charge of the order of 1 C by virtue of a current of about 1 kA. The dart leader then initiates the second (or any subsequent) return stroke. Subsequent return-stroke currents generally have faster zero-to-peak rise times than do first-stroke currents, but similar maximum rates of change, about 100 kA/μs. Some leaders begin as dart leaders, but toward the end of their trip toward ground become stepped leaders. These leaders are known as *dart-stepped leaders* and may have different ground termination points (and separate upward leaders) from the first stroke. Most often the dart-stepped leaders are associated with the second stroke of the flash. Nearly half of all flashes exhibit more than one termination point on ground with the distance between separate terminations being up to several kilometers. Subsequent return-stroke radiated electric and magnetic fields are similar to, but usually a factor of two or so smaller, than first return-stroke fields. About one third of all multiple-stroke flashes has at least one subsequent stroke which is larger than the first stroke.

Cloud-to-ground flashes that lower positive charge, though not common, are of considerable practical interest because their peak currents and total charge transfer can be much larger than for the more common negative ground flash. The largest recorded peak currents, those in the 200- to 300-kA range, are due to the return strokes of positive lightning. Such positive flashes to ground are initiated by downward-moving leaders which do not exhibit the distinct steps of their negative counterparts. Rather, they show a luminosity which is more or less continuous but modulated in intensity. Positive flashes are generally composed of a single stroke followed by a period of continuing current. Positive flashes are probably initiated from the upper positive charge in the thundercloud charge dipole when that cloud charge is horizontally separated from the negative charge beneath it, the source of the usual negative cloud-to-ground lightning. Positive flashes are relatively common in winter thunderstorms (snow storms), which produce few flashes overall, and are relatively uncommon in summer thunderstorms. The fraction of positive lightning in summer thunderstorms apparently increases with increasing latitude and with increasing height of the ground above sea level.

Distant lightning return stroke fields are often referred to as sferics (called "atmospherics" in the older literature). The peak in the sferics frequency spectrum is near 5 kHz due to the bipolar or ringing nature of the distant return-stroke electromagnetic signal and to the effects of propagation.

Thunder, the acoustic radiation associated with lightning, is sometimes divided into the categories "audible", sounds that one can hear, and "infrasonic", below a few tens of hertz, a frequency range that is inaudible. This division is made because it is thought that the mechanisms that produce audible and infrasonic thunder are different. Audible thunder is thought to be due to the expansion of a rapidly heated return stroke channel, as noted earlier, whereas infrasonic thunder is thought to be associated with the conversion to sound of the energy stored in the electrostatic field of the thundercloud when lightning rapidly reduces that cloud field.

The technology of artificially initiating lightning by firing upward small rocket trailing grounded wire of a few hundred meters length has been well-developed during the past decade. Such "triggered" flashes are similar to natural upward-initiated discharges from tall structure. They often contain subsequent strokes which, when they occur, are similar to the subsequent strokes in natural lightning. These triggered subsequent strokes have been the subject of considerable recent research.

Also in the past 10 years or so sophisticated lightning locating equipment has been installed throughout the world. For example, all ground flashes in the U.S. are now centrally monitored for research, for better overall weather prediction, and for hazard warning for aviation, electric utilities and other lightning-sensitive facilities.

Information on lightning physics can be found in M. A. Uman, *The Lightning Discharge*, Academic Press, San Diego, 1987; on lightning death and injury in *Medical Aspects of Lightning Injury*, editors C. Andrews, M. A. Cooper, M. Darveniza, and D. Mackerras, CRC Press, 1992. Ground flash location information for the U.S., in real time or archived, is available from Geomet Data Service of Tucson, AZ, which is also a source of the names of providers of those data in other countries.

Table 2 has data for cloud-to-ground lightning discharges bringing negative charge to earth. The values listed are intended to convey a rough feeling for the various physical parameters of lightning. No great accuracy is claimed since the results of different investigators are often not in good agreement. These values may, in fact, depend on the particular environment in which the lightning discharge is generated. The choice of some of the entries in the table is arbitrary.

TABLE 2
Data for Cloud-to-Ground Lightning Discharges

	Minimum[a]	Representative values	Maximum[a]
Stepped leader			
Length of step, m	3	50	200
Time interval between steps, μs	30	50	125
Average speed of propagation of stepped leader, m/s[b]	1.0×10^5	2.0×10^5	3.0×10^6
Charged deposited on stepped-leader channel, coulombs	3	5	20
Dart leader			
Speed of propagation, m/s[b]	1.0×10^6	1.0×10^7	2.4×10^7
Charged deposited on dart-leader channel, coulombs	0.2	1	6
Return stroke[c]			
Speed of propagation, m/s[b]	2.0×10^7	1.0×10^8	2.0×10^8
Maximum current rate of increase, kA/μs	<1	100	400
Time to peak current, μs	<1	2	30
Peak current, kA	2	30	200
Time to half of peak current, μs	10	50	250
Charge transferred excluding continuing current, coulombs	0.02	3	20
Channel length, km	2	5	15
Lightning flash			
Number of strokes per flash	1	4	26
Time interval between strokes in absence of continuing current, ms	3	60	100
Time duration of flash, s	10^{-2}	0.5	2
Charge transferred including continuing current, coulombs	3	30	200

[a] The words maximum and minimum are used in the sense that most measured values fall between these limits.

[b] Speeds of propagation are generally determined from photographic data and are "two-dimensional". Since many lightning flashes are not vertical, values stated are probably slight underestimates of actual values.

[c] First return strokes have longer times to current peak and generally larger charge transfer than do subsequent return strokes.

Adapted from Uman, M. A., *Lightning*, Dover Paperbook, New York, 1986, and Uman, M. A., *The Lightning Discharge*, Academic Press, San Diego, 1987.

SPEED OF SOUND IN VARIOUS MEDIA

The speed of sound in various solids, liquids, and gases is given in these tables. While only a single parameter v is needed for liquids and gases, sound propagation in isotropic solids is characterized by three velocity parameters. For a solid of infinite extent (or of finite extent if all dimensions are much larger than a wavelength), there are two relevant quantities,

v_l: velocity of longitudinal waves
v_s: velocity of shear waves.

For a cylindrical rod with diameter much smaller than a wavelength,

v_{ext}: velocity of extensional waves along the rod. (Torsional waves in the rod are propagated at the same speed as sheer waves in an infinite solid.)

Table 1 lists values for a variety of solid materials. Table 2 covers gases liquids and gases; values for cryogenic liquids are given at the normal boiling point. Table 3 gives the speed of sound in pure water and in seawater of salinity $S = 3.5\%$ as a function of temperature. All values are in meters per second and are given for normal atmospheric pressure.

REFERENCES

1. Gray, D.E., Ed., *American Institute of Physics Handbook, Third Edition*, McGraw Hill, New York, 1972.
2. Anderson, H.L., Ed., *A Physicist's Desk Reference*, American Institute of Physics, New York, 1989.
3. Younglove, B.A., *Thermophysical Proeprties of Fluids. Part I, J. Phys. Chem. Ref. Data,* 11, Suppl. 1, 1982.
4. Younglove, B.A., and Ely, J.F., *Thermophysical Properties of Fluids. Part II, J. Phys. Chem. Ref. Data,* 16, 577, 1987.
5. Haar, L., Gallagher, J.S., and Kell, G.S., *NBS/NRC Steam Tables,* Hemisphere Publishing Corp., New York, 1984.
6. Mason, W.P., *Physical Acoustics and the Properties of Solids*, D. Van Nostrand Co., Princeton, N.J., 1958.
7. *Landolt-Börnstein, Numerical Data and Functional Relationships in Science and Technology, New Series, II/5, Molecular Acoustics*, Springer-Verlag, Heidelberg, 1967.

TABLE 1
Speed of Sound in Solids at Room Temperature

Name	v_l/m s^{-1}	v_s/m s^{-1}	v_{ext}/m s^{-1}	Name	v_l/m s^{-1}	v_s/m s^{-1}	v_{ext}/m s^{-1}
Metals				Steel, 347 Stainless	5790	3100	5000
Aluminum, rolled	6420	3040	5000	Steel, K9	5940	3250	5250
Berylium	12890	8880	12870	Tin, rolled	3320	1670	2730
Brass (70 Cu, 30 Zn)	4700	2110	3480	Titanium	6070	3125	5090
Constantan	5177	2625	4270	Tungsten, annealed	5220	2890	4620
Copper, annealed	4760	2325	3810	Tungsten, drawn	5410	2640	4320
Copper, rolled	5010	2270	3750	Zinc, rolled	4210	2440	3850
Duralumin 17S	6320	3130	5150				
Gold, hard-drawn	3240	1200	2030	**Other materials**			
Iron, cast	4994	2809	4480	Fused silica	5968	3764	5760
Iron, electrolytic	5950	3240	5120	Glass, heavy silicate flint	3980	2380	3720
Iron, Armco	5960	3240	5200	Glass, light borate crown	5100	2840	4540
Lead, annealed	2160	700	1190	Glass, pyrex	5640	3280	5170
Lead, rolled	1960	690	1210	Lucite	2680	1100	1840
Magnesium, annealed	5770	3050	4940	Nylon 6-6	2620	1070	1800
Molybdenum	6250	3350	5400	Polyethylene	1950	540	920
Monel metal	5350	2720	4400	Polystyrene	2350	1120	1840
Nickel	6040	3000	4900	Rubber, butyl	1830		
Platinum	3260	1730	2800	Rubber, gum	1550		
Silver	3650	1610	2680	Rubber, neoprene	1600		
Steel (1% C)	5940	3220	5180	Tungsten carbide	6655	3980	6220

TABLE 2
Speed of Sound in Liquids and Gases

Name	$t/^\circ C$	$v/\text{m s}^{-1}$	Name	$t/^\circ C$	$v/\text{m s}^{-1}$
Liquids			Pentane	20	1008
Acetone	20	1203	Propane	-42.1	1158
Argon	-185.9	813	1-Propanol	20	1223
Benzene	25	1310	Tetrachloromethane	25	930
Bromobenzene	20	1169	Trichloromethane	25	987
Butane	-0.5	1034	1-Undecene	20	1275
1-Butanol	20	1258	Water	25	1497
Carbon disulphide	25	1140	Water (sea, $S = 3.5\%$)	25	1535
Chlorobenzene	20	1311			
Cyclohexane	19	1280	**Gases at 1 atm**		
1-Decene	20	1250	Air, dry	25	346
Diethyl ether	25	976	Ammonia	0	415
Ethane	-88.6	1326	Argon	27	323
Ethanol	20	1162	Carbon monoxide	0	338
Ethylene	-103.8	1309	Carbon dioxide	0	259
Ethylene glycol	25	1658	Chlorine	0	206
Fluorobenzene	20	1183	Deuterium	0	890
Glycerol	25	1904	Ethane	27	312
Helium	-268.9	180	Ethylene	27	331
Heptane	20	1162	Helium	0	965
1-Heptene	20	1128	Hydrogen	27	1310
Hexane	20	1083	Hydrogen bromide	0	200
Hydrogen	-252.9	1101	Hydrogen chloride	0	296
Iodobenzene	20	1114	Hydrogen iodide	0	157
Mercury	25	1450	Hydrogen sulfide	0	289
Methane	-161.5	1337	Methane	27	450
Methanol	20	1121	Neon	0	435
Nitrobenzene	25	1463	Nitric oxide	10	325
Nitrogen	-195.8	939	Nitrogen	27	353
1-Nonene	20	1218	Nitrous oxide	0	263
Octane	20	1197	Oxygen	27	330
1-Octene	20	1184	Sulfur dioxide	0	213
Oxygen	-183.0	906	Water (steam)	100	473
1-Pentadecene	20	1351			

TABLE 3
Speed of Sound in Water and Seawater (S = 3.5%) at Different Temperatures

	$v/\text{m s}^{-1}$	
$t/^\circ C$	Water	Seawater
0	1401.0	1449.4
10	1447.8	1490.4
20	1483.2	1522.2
25	1497.4	1535.1
30	1509.5	1546.2
40	1528.4	
50	1541.4	
60	1549.5	
70	1553.2	
80	1552.8	

ATTENUATION AND SPEED OF SOUND IN AIR AS A FUNCTION OF HUMIDITY AND FREQUENCY

This table gives the attenuation and speed of sound as a function of frequency at various values of relative humidity. All values refer to still air at 20°C.

REFERENCES

1. Tables of Absorption and Velocity of Sound in Still Air at 68°F (20°C), AD-738576, National Technical Information Service, Springfield, VA.
2. Evans, L. B., Bass, H. E., and Sutherland, L. C., *J. Acoust. Soc. Am.,* 51, 1565, 1972.

Frequency (Hz)	Attenuation (dB/km)	Speed (m/s)	Frequency (Hz)	Attenuation (dB/km)	Speed (m/s)
Relative humidity 0%			**Relative humidity 60%**		
20	0.51	343.477	20	0.02	344.182
40	1.07	343.514	40	0.06	344.183
50	1.26	343.525	50	0.09	344.183
63	1.43	343.536	63	0.15	344.184
100	1.67	343.550	100	0.34	344.185
200	1.84	343.559	200	0.99	344.190
400	1.96	343.561	400	1.94	344.197
630	2.11	343.562	630	2.57	344.200
800	2.27	343.562	800	2.94	344.201
1250	2.82	343.562	1250	4.01	344.202
2000	4.14	343.562	2000	6.55	344.203
4000	8.84	343.564	4000	18.73	344.204
6300	14.89	343.565	6300	42.51	344.204
10000	26.28	343.566	10000	101.84	344.206
12500	35.81	343.566	12500	155.67	344.208
16000	52.15	343.567	16000	247.78	344.211
20000	75.37	343.567	20000	373.78	344.215
40000	267.01	343.567	40000	1195.37	344.238
63000	644.66	343.567	63000	2220.64	344.262
80000	1032.14	343.567	80000	2951.71	344.274
Relative humidity 30%			**Relative humidity 100%**		
20	0.03	343.807	20	0.01	344.685
40	0.11	343.808	40	0.04	344.685
50	0.17	343.810	50	0.06	344.685
63	0.25	343.810	63	0.09	344.685
100	0.50	343.814	100	0.22	344.686
200	1.01	343.821	200	0.77	344.689
400	1.59	343.826	400	2.02	344.695
630	2.24	343.827	630	3.05	344.699
800	2.85	343.828	800	3.57	344.701
1250	5.09	343.828	1250	4.59	344.704
2000	10.93	343.829	2000	6.29	344.705
4000	38.89	343.831	4000	13.58	344.706
6300	90.61	343.836	6300	27.72	344.706
10000	204.98	343.846	10000	63.49	344.706
12500	294.08	343.854	12500	96.63	344.707
16000	422.51	343.865	16000	154.90	344.708
20000	563.66	343.877	20000	237.93	344.709
40000	1110.97	343.911	40000	884.28	344.718
63000	1639.47	343.924	63000	1973.62	344.731
80000	2083.08	343.929	80000	2913.01	344.742

SPEED OF SOUND IN DRY AIR

The values in this table were calculated from the equation of state for dry air (average molecular weight 28.96) treated as a real gas. Values refer to standard atmospheric pressure. The speed of sound varies only slightly with pressure; at two atmospheres and -100°C the value decreases by 0.13%, while at two atmospheres and 80°C the speed increases by 0.04%.

REFERENCE

Sytchev, V.V., Vasserman, A.A., Kozlov, A.D., Spiridonov, G.A., and Tsymarny, V.A., *Thermodynamic Properties of Air*, Hemisphere Publishing Corp., New York, 1987.

$t/°C$	$v_s/\text{m s}^{-1}$	$t/°C$	$v_s/\text{m s}^{-1}$	$t/°C$	$v_s/\text{m s}^{-1}$
-100	263.5	-35	309.5	30	349.1
-95	267.3	-30	312.7	35	352.0
-90	271.1	-25	315.9	40	354.8
-85	274.8	-20	319.1	45	357.6
-80	278.5	-15	322.3	50	360.4
-75	282.1	-10	325.4	55	363.2
-70	285.7	-5	328.4	60	365.9
-65	289.2	0	331.5	65	368.6
-60	292.7	5	334.5	70	371.3
-55	296.1	10	337.5	75	374.0
-50	299.5	15	340.4	80	376.7
-45	302.9	20	343.4		
-40	306.2	25	346.3		

MUSICAL SCALES

EQUAL TEMPERED CHROMATIC SCALE
$A_4 = 440$

American Standard pitch. Adopted by the American Standards Association in 1936

Note	Frequency	Note	Frequency	Note	Frequency	Note	Frequency
C_0	16.35	C_2	65.41	C_4	261.63	C_6	1046.50
$C\#_0$	17.32	$C\#_2$	69.30	$C\#_4$	277.18	$C\#_6$	1108.73
D_0	18.35	D_2	73.42	D_4	293.66	D_6	1174.66
$D\#_0$	19.45	$D\#_2$	77.78	$D\#_4$	311.13	$D\#_6$	1244.51
E_0	20.60	E_2	82.41	E_4	329.63	E_6	1318.51
F_0	21.83	F_2	87.31	F_4	349.23	F_6	1396.91
$F\#_0$	23.12	$F\#_2$	92.50	$F\#_4$	369.99	$F\#_6$	1479.98
G_0	24.50	G_2	98.00	G_4	392.00	G_6	1567.98
$G\#_0$	25.96	$G\#_2$	103.83	$G\#_4$	415.30	$G\#_6$	1661.22
A_0	27.50	A_2	110.00	A_4	440.00	A_6	1760.00
$A\#_0$	29.14	$A\#_2$	116.54	$A\#_4$	466.16	$A\#_6$	1864.66
B_0	30.87	B_2	123.47	B_4	493.88	B_6	1975.53
C_1	32.70	C_3	130.81	C_5	523.25	C_7	2093.00
$C\#_1$	34.65	$C\#_3$	138.59	$C\#_5$	554.37	$C\#_7$	2217.46
D_1	36.71	D_3	146.83	D_5	587.33	D_7	2349.32
$D\#_1$	38.89	$D\#_3$	155.56	$D\#_5$	622.25	$D\#_7$	2489.02
E_1	41.20	E_3	164.81	E_5	659.26	E_7	2637.02
F_1	43.65	F_3	174.61	F_5	698.46	F_7	2793.83
$F\#1$	46.25	$F\#_3$	185.00	$F\#_5$	739.99	$F\#_7$	2959.96
G_1	49.00	G_3	196.00	G_5	783.99	G_7	3135.96
$G\#_1$	51.91	$G\#_3$	207.65	$G\#_5$	830.61	$G\#_7$	3322.44
A_1	55.00	A_3	220.00	A_5	880.00	A_7	3520.00
$A\#_1$	58.27	$A\#_3$	233.08	$A\#_5$	932.33	$A\#_7$	3729.31
B_1	61.74	B_3	246.94	B_5	987.77	B_7	3951.07
						C_8	4186.01

EQUAL TEMPERED CHROMATIC SCALE
$A_4 = 435$

International Pitch, adopted 1891

Note	Frequency	Note	Frequency	Note	Frequency	Note	Frequency
C_0	16.17	C_2	64.66	C_4	258.65	C_6	1034.61
$C\#_0$	17.13	$C\#_2$	68.51	$C\#_4$	274.03	$C\#_6$	1096.13
D_0	18.15	D_2	72.58	D_4	290.33	D_6	1161.31
$D\#_0$	19.22	$D\#_2$	76.90	$D\#_4$	307.59	$D\#_6$	1230.37
E_0	20.37	E_2	81.47	E_4	325.88	E_6	1303.53
F_0	21.58	F_2	86.31	F_4	345.26	F_6	1381.04
$F\#_0$	22.86	$F\#_2$	91.45	$F\#_4$	365.79	$F\#_6$	1463.16
G_0	24.22	G_2	96.89	G_4	387.54	G_6	1550.16
$G\#_0$	25.66	$G\#_2$	102.65	$G\#_4$	410.59	$G\#_6$	1642.34
A_0	27.19	A_2	108.75	A_4	435.00	A_6	1740.00
$A\#_0$	28.80	$A\#_2$	115.22	$A\#_4$	460.87	$A\#_6$	1843.47
B_0	30.52	B_2	122.07	B_4	488.27	B_6	1953.08
C_1	32.33	C_3	129.33	C_5	517.31	C_7	2069.22
$C\#_1$	34.25	$C\#_3$	137.02	$C\#_5$	548.07	$C\#_7$	2192.26
D_1	36.29	D_3	145.16	D_5	580.66	D_7	2322.62
$D\#_1$	38.45	$D\#_3$	153.80	$D\#_5$	615.18	$D\#_7$	2460.73
E_1	40.74	E_3	162.94	E_5	651.76	E_7	2607.05
F_1	43.16	F_3	172.63	F_5	690.52	F_7	2762.08
$F\#_1$	45.72	$F\#_3$	182.89	$F\#_5$	731.58	$F\#_7$	2926.32
G_1	48.44	G_3	193.77	G_5	775.08	G_7	3100.33
$G\#_1$	51.32	$G\#_3$	205.29	$G\#_5$	821.17	$G\#_7$	3284.68
A_1	54.38	A_3	217.50	A_5	870.00	A_7	3480.00
$A\#_1$	57.61	$A\#_3$	230.43	$A\#_5$	921.73	$A\#_7$	3686.93
B_1	61.03	B_3	244.14	B_5	976.54	B_7	3906.17
						C_8	4138.44

SCIENTIFIC OR JUST SCALE
$C_4 = 256$

Note	Frequency	Note	Frequency	Note	Frequency	Note	Frequency
C_0	16	C_2	64	C_4	256	C_6	1024
D_0	18	D_2	72	D_4	288	D_6	1152
E_0	20	E_2	80	E_4	320	E_6	1280
F_0	21.33	F_2	85.33	F_4	341.33	F_6	1365.33
G_0	24	G_2	96	G_4	384	G_6	1536
A_0	26.67	A_2	106.67	A_4	426.67	A_6	1706.67
B_0	30	B_2	120	B_4	480	B_6	1920
C_1	32	C_3	128	C_5	512	C_7	2048
D_1	36	D_3	144	D_5	576	D_7	2304
E_1	40	E_3	160	E_5	640	E_7	2560
F_1	42.67	F_3	170.67	F_5	682.67	F_7	2730.67
G_1	48	G_3	192	G_5	768	G_7	3072
A_1	53.33	A_3	213.33	A_5	853.33	A_7	3413.33
B_1	60	B_3	240	B_5	960	B_7	3840
						C_8	4096

Section 15
Practical Laboratory Data

STANDARD ITS-90 THERMOCOUPLE TABLES

The Instrument Society of America (ISA) has assigned standard letter designations to a number of thermocouple types having specified emf-temperature relations. These designations and the approximate metal compositions which meet the required relations, as well as the useful temperature ranges, are given below:

Type B	(Pt + 30% Rh) vs. (Pt + 6% Rh)	0 to 1820°C
Type E	(Ni + 10% Cr) vs. (Cu + 43% Ni)	-270 to 1000°C
Type J	Fe vs. (Cu + 43%Ni)	-210 to 1200°C
Type K	(Ni + 10% Cr) vs. (Ni + 2% Al + 2% Mn + 1% Si)	-270 to 1372°C
Type N	(Ni + 14% Cr + 1.5% Si) vs. (Ni + 4.5% Si + 0.1% Mg)	-270 to 1300°C
Type R	(Pt + 13% Rh) vs. Pt	-50 to 1768°C
Type S	(Pt + 10% Rh) vs. Pt	-50 to 1768°C
Type T	Cu vs. (Cu + 43% Ni)	-270 to 400°C

The compositions are given in weight percent, and the positive leg is listed first. It should be emphasized that the standard letter designations do not imply a precise composition but rather that the specified emf-temperature relation is satisfied.

The first set of tables below lists, for each thermocouple type, the emf as a function of temperature on the International Temperature Scale of 1990 (ITS-90). The coefficients in the equation used to generate the table are also given. The second set of tables gives the inverse relationships, i.e., the coefficients in the polynomial equation which expresses the temperature as a function of thermocouple emf. The accuracy of these equations is also stated.

Further details and tables at closer intervals may be found in Reference 1.

REFERENCES

1. Burns, G. W., Seroger, M. G., Strouse, G. F., Croarkin, M. C., and Guthrie, W. F., *Temperature-Electromotive Force Reference Functions and Tables for the Letter-Designated Thermocouple Types Based on the ITS-90*, Natl. Inst. Stand. Tech. (U.S.) Monogr. 175, 1993.
2. Schooley, J. F., *Thermometry*, CRC Press, Boca Raton, FL, 1986.

Type B thermocouples: emf-temperature (°C) reference table and equations.

Thermocouple emf as a Function of Temperature in Degrees Celsius (ITS-90)

emf in Millivolts   Reference Junctions at 0 °C

°C	0	10	20	30	40	50	60	70	80	90	100
0	0.000	−0.002	−0.003	−0.002	−0.000	0.002	0.006	0.011	0.017	0.025	0.033
100	0.033	0.043	0.053	0.065	0.078	0.092	0.107	0.123	0.141	0.159	0.178
200	0.178	0.199	0.220	0.243	0.267	0.291	0.317	0.344	0.372	0.401	0.431
300	0.431	0.462	0.494	0.527	0.561	0.596	0.632	0.669	0.707	0.746	0.787
400	0.787	0.828	0.870	0.913	0.957	1.002	1.048	1.095	1.143	1.192	1.242
500	1.242	1.293	1.344	1.397	1.451	1.505	1.561	1.617	1.675	1.733	1.792
600	1.792	1.852	1.913	1.975	2.037	2.101	2.165	2.230	2.296	2.363	2.431
700	2.431	2.499	2.569	2.639	2.710	2.782	2.854	2.928	3.002	3.078	3.154
800	3.154	3.230	3.308	3.386	3.466	3.546	3.626	3.708	3.790	3.873	3.957
900	3.957	4.041	4.127	4.213	4.299	4.387	4.475	4.564	4.653	4.743	4.834
1000	4.834	4.926	5.018	5.111	5.205	5.299	5.394	5.489	5.585	5.682	5.780
1100	5.780	5.878	5.976	6.075	6.175	6.276	6.377	6.478	6.580	6.683	6.786
1200	6.786	6.890	6.995	7.100	7.205	7.311	7.417	7.524	7.632	7.740	7.848
1300	7.848	7.957	8.066	8.176	8.286	8.397	8.508	8.620	8.731	8.844	8.956
1400	8.956	9.069	9.182	9.296	9.410	9.524	9.639	9.753	9.868	9.984	10.099
1500	10.099	10.215	10.331	10.447	10.563	10.679	10.796	10.913	11.029	11.146	11.263
1600	11.263	11.380	11.497	11.614	11.731	11.848	11.965	12.082	12.199	12.316	12.433
1700	12.433	12.549	12.666	12.782	12.898	13.014	13.130	13.246	13.361	13.476	13.591
1800	13.591	13.706	13.820								

Temperature Ranges and Coefficients of Equations Used to Compute the Above Table

The equations are of the form: $E = c_0 + c_1 t + c_2 t^2 + c_3 t^3 + \ldots c_n t^n$, where E is the emf in millivolts, t is the temperature in degrees Celsius (ITS-90), and c_0, c_1, c_2, c_3, etc. are the coefficients. These coefficients are extracted from NIST Monograph 175.

		0 °C to 630.615 °C	630.615 °C to 1820 °C
c_0	=	0.000 000 000 0 …	−3.893 816 862 1 …
c_1	=	−2.465 081 834 6 X 10^{-4}	2.857 174 747 0 X 10^{-2}
c_2	=	5.904 042 117 1 X 10^{-6}	−8.488 510 478 5 X 10^{-5}
c_3	=	−1.325 793 163 6 X 10^{-9}	1.578 528 016 4 X 10^{-7}
c_4	=	1.566 829 190 1 X 10^{-12}	−1.683 534 486 4 X 10^{-10}
c_5	=	−1.694 452 924 0 X 10^{-15}	1.110 979 401 3 X 10^{-13}
c_6	=	6.299 034 709 4 X 10^{-19}	−4.451 543 103 3 X 10^{-17}
c_7	=	. . . .	9.897 564 082 1 X 10^{-21}
c_8	=	. . . .	−9.379 133 028 9 X 10^{-25}

Type E thermocouples: emf-temperature (°C) reference table and equations.

Thermocouple emf as a Function of Temperature in Degrees Celsius (ITS-90)

emf in Millivolts   Reference Junctions at 0 °C

°C	0	−10	−20	−30	−40	−50	−60	−70	−80	−90	−100
−200	−8.825	−9.063	−9.274	−9.455	−9.604	−9.718	−9.797	−9.835			
−100	−5.237	−5.681	−6.107	−6.516	−6.907	−7.279	−7.632	−7.963	−8.273	−8.561	−8.825
0	0.000	−0.582	−1.152	−1.709	−2.255	−2.787	−3.306	−3.811	−4.302	−4.777	−5.237

°C	0	10	20	30	40	50	60	70	80	90	100
0	0.000	0.591	1.192	1.801	2.420	3.048	3.685	4.330	4.985	5.648	6.319
100	6.319	6.998	7.685	8.379	9.081	9.789	10.503	11.224	11.951	12.684	13.421
200	13.421	14.164	14.912	15.664	16.420	17.181	17.945	18.713	19.484	20.259	21.036
300	21.036	21.817	22.600	23.386	24.174	24.964	25.757	26.552	27.348	28.146	28.946
400	28.946	29.747	30.550	31.354	32.159	32.965	33.772	34.579	35.387	36.196	37.005
500	37.005	37.815	38.624	39.434	40.243	41.053	41.862	42.671	43.479	44.286	45.093
600	45.093	45.900	46.705	47.509	48.313	49.116	49.917	50.718	51.517	52.315	53.112
700	53.112	53.908	54.703	55.497	56.289	57.080	57.870	58.659	59.446	60.232	61.017
800	61.017	61.801	62.583	63.364	64.144	64.922	65.698	66.473	67.246	68.017	68.787
900	68.787	69.554	70.319	71.082	71.844	72.603	73.360	74.115	74.869	75.621	76.373
1000	76.373										

Temperature Ranges and Coefficients of Equations Used to Compute the Above Table
The equations are of the form: $E = c_0 + c_1 t + c_2 t^2 + c_3 t^3 + \ldots c_n t^n$, where E is the emf in millivolts, t is the temperature in degrees Celsius (ITS-90), and c_0, c_1, c_2, c_3, *etc.* are the coefficients. These coefficients are extracted from NIST Monograph 175.

		−270 °C to 0 °C	0 °C to 1000 °C
c_0	=	0.000 000 000 0 . . .	0.000 000 000 0 . . .
c_1	=	$5.866\ 550\ 870\ 8 \times 10^{-2}$	$5.866\ 550\ 871\ 0 \times 10^{-2}$
c_2	=	$4.541\ 097\ 712\ 4 \times 10^{-5}$	$4.503\ 227\ 558\ 2 \times 10^{-5}$
c_3	=	$-7.799\ 804\ 868\ 6 \times 10^{-7}$	$2.890\ 840\ 721\ 2 \times 10^{-8}$
c_4	=	$-2.580\ 016\ 084\ 3 \times 10^{-8}$	$-3.305\ 689\ 665\ 2 \times 10^{-10}$
c_5	=	$-5.945\ 258\ 305\ 7 \times 10^{-10}$	$6.502\ 440\ 327\ 0 \times 10^{-13}$
c_6	=	$-9.321\ 405\ 866\ 7 \times 10^{-12}$	$-1.919\ 749\ 550\ 4 \times 10^{-16}$
c_7	=	$-1.028\ 760\ 553\ 4 \times 10^{-13}$	$-1.253\ 660\ 049\ 7 \times 10^{-18}$
c_8	=	$-8.037\ 012\ 362\ 1 \times 10^{-16}$	$2.148\ 921\ 756\ 9 \times 10^{-21}$
c_9	=	$-4.397\ 949\ 739\ 1 \times 10^{-18}$	$-1.438\ 804\ 178\ 2 \times 10^{-24}$
c_{10}	=	$-1.641\ 477\ 635\ 5 \times 10^{-20}$	$3.596\ 089\ 948\ 1 \times 10^{-28}$
c_{11}	=	$-3.967\ 361\ 951\ 6 \times 10^{-23}$	. . . .
c_{12}	=	$-5.582\ 732\ 872\ 1 \times 10^{-26}$	. . . .
c_{13}	=	$-3.465\ 784\ 201\ 3 \times 10^{-29}$	. . . .

Type J thermocouples: emf-temperature (°C) reference table and equations.

Thermocouple emf as a Function of Temperature in Degrees Celsius (ITS-90)

emf in Millivolts   Reference Junctions at 0 °C

°C	0	−10	−20	−30	−40	−50	−60	−70	−80	−90	−100
−200	−7.890	−8.095									
−100	−4.633	−5.037	−5.426	−5.801	−6.159	−6.500	−6.821	−7.123	−7.403	−7.659	−7.890
0	0.000	−0.501	−0.995	−1.482	−1.961	−2.431	−2.893	−3.344	−3.786	−4.215	−4.633

°C	0	10	20	30	40	50	60	70	80	90	100
0	0.000	0.507	1.019	1.537	2.059	2.585	3.116	3.650	4.187	4.726	5.269
100	5.269	5.814	6.360	6.909	7.459	8.010	8.562	9.115	9.669	10.224	10.779
200	10.779	11.334	11.889	12.445	13.000	13.555	14.110	14.665	15.219	15.773	16.327
300	16.327	16.881	17.434	17.986	18.538	19.090	19.642	20.194	20.745	21.297	21.848
400	21.848	22.400	22.952	23.504	24.057	24.610	25.164	25.720	26.276	26.834	27.393
500	27.393	27.953	28.516	29.080	29.647	30.216	30.788	31.362	31.939	32.519	33.102
600	33.102	33.689	34.279	34.873	35.470	36.071	36.675	37.284	37.896	38.512	39.132
700	39.132	39.755	40.382	41.012	41.645	42.281	42.919	43.559	44.203	44.848	45.494
800	45.494	46.141	46.786	47.431	48.074	48.715	49.353	49.989	50.622	51.251	51.877
900	51.877	52.500	53.119	53.735	54.347	54.956	55.561	56.164	56.763	57.360	57.953
1000	57.953	58.545	59.134	59.721	60.307	60.890	61.473	62.054	62.634	63.214	63.792
1100	63.792	64.370	64.948	65.525	66.102	66.679	67.255	67.831	68.406	68.980	69.553
1200	69.553										

Temperature Ranges and Coefficients of Equations Used to Compute the Above Table

The equations are of the form: $E = c_0 + c_1 t + c_2 t^2 + c_3 t^3 + \ldots c_n t^n$, where E is the emf in millivolts, t is the temperature in degrees Celsius (ITS-90), and c_0, c_1, c_2, c_3, *etc.* are the coefficients. These coefficients are extracted from NIST Monograph 175.

		−210 °C to 760 °C	760 °C to 1200 °C
c_0	=	0.000 000 000 0 . . .	$2.964\ 562\ 568\ 1 \times 10^2$
c_1	=	$5.038\ 118\ 781\ 5 \times 10^{-2}$	$-1.497\ 612\ 778\ 6$ . . .
c_2	=	$3.047\ 583\ 693\ 0 \times 10^{-5}$	$3.178\ 710\ 392\ 4 \times 10^{-3}$
c_3	=	$-8.568\ 106\ 572\ 0 \times 10^{-8}$	$-3.184\ 768\ 670\ 1 \times 10^{-6}$
c_4	=	$1.322\ 819\ 529\ 5 \times 10^{-10}$	$1.572\ 081\ 900\ 4 \times 10^{-9}$
c_5	=	$-1.705\ 295\ 833\ 7 \times 10^{-13}$	$-3.069\ 136\ 905\ 6 \times 10^{-13}$
c_6	=	$2.094\ 809\ 069\ 7 \times 10^{-16}$	. . . .
c_7	=	$-1.253\ 839\ 533\ 6 \times 10^{-19}$	. . . .
c_8	=	$1.563\ 172\ 569\ 7 \times 10^{-23}$	. . . .

STANDARD ITS-90 THERMOCOUPLE TABLES (continued)

Type K thermocouples: emf-temperature (°C) reference table and equations.

Thermocouple emf as a Function of Temperature in Degrees Celsius (ITS-90)

emf in Millivolts   Reference Junctions at 0 °C

°C	0	−10	−20	−30	−40	−50	−60	−70	−80	−90	−100
−200	−5.891	−6.035	−6.158	−6.262	−6.344	−6.404	−6.441	−6.458			
−100	−3.554	−3.852	−4.138	−4.411	−4.669	−4.913	−5.141	−5.354	−5.550	−5.730	−5.891
0	0.000	−0.392	−0.778	−1.156	−1.527	−1.889	−2.243	−2.587	−2.920	−3.243	−3.554

°C	0	10	20	30	40	50	60	70	80	90	100
0	0.000	0.397	0.798	1.203	1.612	2.023	2.436	2.851	3.267	3.682	4.096
100	4.096	4.509	4.920	5.328	5.735	6.138	6.540	6.941	7.340	7.739	8.138
200	8.138	8.539	8.940	9.343	9.747	10.153	10.561	10.971	11.382	11.795	12.209
300	12.209	12.624	13.040	13.457	13.874	14.293	14.713	15.133	15.554	15.975	16.397
400	16.397	16.820	17.243	17.667	18.091	18.516	18.941	19.366	19.792	20.218	20.644
500	20.644	21.071	21.497	21.924	22.350	22.776	23.203	23.629	24.055	24.480	24.905
600	24.905	25.330	25.755	26.179	26.602	27.025	27.447	27.869	28.289	28.710	29.129
700	29.129	29.548	29.965	30.382	30.798	31.213	31.628	32.041	32.453	32.865	33.275
800	33.275	33.685	34.093	34.501	34.908	35.313	35.718	36.121	36.524	36.925	37.326
900	37.326	37.725	38.124	38.522	38.918	39.314	39.708	40.101	40.494	40.885	41.276
1000	41.276	41.665	42.053	42.440	42.826	43.211	43.595	43.978	44.359	44.740	45.119
1100	45.119	45.497	45.873	46.249	46.623	46.995	47.367	47.737	48.105	48.473	48.838
1200	48.838	49.202	49.565	49.926	50.286	50.644	51.000	51.355	51.708	52.060	52.410
1300	52.410	52.759	53.106	53.451	53.795	54.138	54.479	54.819			

Temperature Ranges and Coefficients of Equations Used to Compute the Above Table

The equations are of the form: $E = c_0 + c_1 t + c_2 t^2 + c_3 t^3 + \ldots c_n t^n$, where E is the emf in millivolts, t is the temperature in degrees Celsius (ITS-90), and c_0, c_1, c_2, c_3, *etc.* are the coefficients. In the 0 °C to 1372 °C range there is also an exponential term that must be evaluated and added to the equation. The exponential term is of the form: $c_0 e^{c_1 (t - 126.9686)^2}$, where t is the temperature in °C, e is the natural logarithm base, and c_0 and c_i are the coefficients. These coefficients are extracted from NIST Monograph 175.

		−270 °C to 0 °C	0 °C to 1372 °C	0 °C to 1372 °C (exponential term)
c_0	=	0.000 000 000 0 . . .	−1.760 041 368 6 X 10^{-2}	1.185 976 X 10^{-1}
c_1	=	3.945 012 802 5 X 10^{-2}	3.892 120 497 5 X 10^{-2}	−1.183 432 X 10^{-4}
c_2	=	2.362 237 359 8 X 10^{-5}	1.855 877 003 2 X 10^{-5}	. . . .
c_3	=	−3.285 890 678 4 X 10^{-7}	−9.945 759 287 4 X 10^{-6}	. . . .
c_4	=	−4.990 482 877 7 X 10^{-9}	3.184 094 571 9 X 10^{-10}	. . . .
c_5	=	−6.750 905 917 3 X 10^{-11}	−5.607 284 488 9 X 10^{-13}	. . . .
c_6	=	−5.741 032 742 8 X 10^{-13}	5.607 505 905 9 X 10^{-16}	. . . .
c_7	=	−3.108 887 289 4 X 10^{-15}	−3.202 072 000 3 X 10^{-19}	. . . .
c_8	=	−1.045 160 936 5 X 10^{-17}	9.715 114 715 2 X 10^{-23}	. . . .
c_9	=	−1.988 926 687 8 X 10^{-20}	−1.210 472 127 5 X 10^{-26}	. . . .
c_{10}	=	−1.632 269 748 6 X 10^{-23}	. . . .	. . . .

Type N thermocouples: emf-temperature (°C) reference table and equations.

Thermocouple emf as a Function of Temperature in Degrees Celsius (ITS-90)

emf in Millivolts   Reference Junctions at 0 °C

°C	0	−10	−20	−30	−40	−50	−60	−70	−80	−90	−100
−200	−3.990	−4.083	−4.162	−4.226	−4.277	−4.313	−4.336	−4.345			
−100	−2.407	−2.612	−2.808	−2.994	−3.171	−3.336	−3.491	−3.634	−3.766	−3.884	−3.990
0	0.000	−0.260	−0.518	−0.772	−1.023	−1.269	−1.509	−1.744	−1.972	−2.193	−2.407

°C	0	10	20	30	40	50	60	70	80	90	100
0	0.000	0.261	0.525	0.793	1.065	1.340	1.619	1.902	2.189	2.480	2.774
100	2.774	3.072	3.374	3.680	3.989	4.302	4.618	4.937	5.259	5.585	5.913
200	5.913	6.245	6.579	6.916	7.255	7.597	7.941	8.288	8.637	8.988	9.341
300	9.341	9.696	10.054	10.413	10.774	11.136	11.501	11.867	12.234	12.603	12.974
400	12.974	13.346	13.719	14.094	14.469	14.846	15.225	15.604	15.984	16.366	16.748
500	16.748	17.131	17.515	17.900	18.286	18.672	19.059	19.447	19.835	20.224	20.613
600	20.613	21.003	21.393	21.784	22.175	22.566	22.958	23.350	23.742	24.134	24.527
700	24.527	24.919	25.312	25.705	26.098	26.491	26.883	27.276	27.669	28.062	28.455
800	28.455	28.847	29.239	29.632	30.024	30.416	30.807	31.199	31.590	31.981	32.371
900	32.371	32.761	33.151	33.541	33.930	34.319	34.707	35.095	35.482	35.869	36.256
1000	36.256	36.641	37.027	37.411	37.795	38.179	38.562	38.944	39.326	39.706	40.087
1100	40.087	40.466	40.845	41.223	41.600	41.976	42.352	42.727	43.101	43.474	43.846
1200	43.846	44.218	44.588	44.958	45.326	45.694	46.060	46.425	46.789	47.152	47.513
1300	47.513										

Temperature Ranges and Coefficients of Equations Used to Compute the Above Table

The equations are of the form: $E = c_0 + c_1 t + c_2 t^2 + c_3 t^3 + \ldots c_n t^n$, where E is the emf in millivolts, t is the temperature in degrees Celsius (ITS-90), and c_0, c_1, c_2, c_3, *etc.* are the coefficients. These coefficients are extracted from NIST Monograph 175.

	−270 °C to 0 °C	0 °C to 1300 °C
c_0 =	0.000 000 000 0 . . .	0.000 000 000 0 . . .
c_1 =	$2.615\ 910\ 596\ 2 \times 10^{-2}$	$2.592\ 939\ 460\ 1 \times 10^{-2}$
c_2 =	$1.095\ 748\ 422\ 8 \times 10^{-5}$	$1.571\ 014\ 188\ 0 \times 10^{-5}$
c_3 =	$-9.384\ 111\ 155\ 4 \times 10^{-6}$	$4.382\ 562\ 723\ 7 \times 10^{-8}$
c_4 =	$-4.641\ 203\ 975\ 9 \times 10^{-11}$	$-2.526\ 116\ 979\ 4 \times 10^{-10}$
c_5 =	$-2.630\ 335\ 771\ 6 \times 10^{-12}$	$6.431\ 181\ 933\ 9 \times 10^{-13}$
c_6 =	$-2.265\ 343\ 800\ 3 \times 10^{-14}$	$-1.006\ 347\ 151\ 9 \times 10^{-15}$
c_7 =	$-7.608\ 930\ 079\ 1 \times 10^{-17}$	$9.974\ 533\ 899\ 2 \times 10^{-19}$
c_8 =	$-9.341\ 966\ 783\ 5 \times 10^{-20}$	$-6.086\ 324\ 560\ 7 \times 10^{-22}$
c_9 =	. . . .	$2.084\ 922\ 933\ 9 \times 10^{-26}$
c_{10} =	. . . .	$-3.068\ 219\ 615\ 1 \times 10^{-29}$

Type R thermocouples: emf-temperature (°C) reference table and equations.

Thermocouple emf as a Function of Temperature in Degrees Celsius (ITS-90)

emf in Millivolts   Reference Junctions at 0 °C

°C	0	−10	−20	−30	−40	−50	−60	−70	−80	−90	−100
0	0.000	−0.051	−0.100	−0.145	−0.188	−0.226					

°C	0	10	20	30	40	50	60	70	80	90	100
0	0.000	0.054	0.111	0.171	0.232	0.296	0.363	0.431	0.501	0.573	0.647
100	0.647	0.723	0.800	0.879	0.959	1.041	1.124	1.208	1.294	1.381	1.469
200	1.469	1.558	1.648	1.739	1.831	1.923	2.017	2.112	2.207	2.304	2.401
300	2.401	2.498	2.597	2.696	2.796	2.896	2.997	3.099	3.201	3.304	3.408
400	3.408	3.512	3.616	3.721	3.827	3.933	4.040	4.147	4.255	4.363	4.471
500	4.471	4.580	4.690	4.800	4.910	5.021	5.133	5.245	5.357	5.470	5.583
600	5.583	5.697	5.812	5.926	6.041	6.157	6.273	6.390	6.507	6.625	6.743
700	6.743	6.861	6.980	7.100	7.220	7.340	7.461	7.583	7.705	7.827	7.950
800	7.950	8.073	8.197	8.321	8.446	8.571	8.697	8.823	8.950	9.077	9.205
900	9.205	9.333	9.461	9.590	9.720	9.850	9.980	10.111	10.242	10.374	10.506
1000	10.506	10.638	10.771	10.905	11.039	11.173	11.307	11.442	11.578	11.714	11.850
1100	11.850	11.986	12.123	12.260	12.397	12.535	12.673	12.812	12.950	13.089	13.228
1200	13.228	13.367	13.507	13.646	13.786	13.926	14.066	14.207	14.347	14.488	14.629
1300	14.629	14.770	14.911	15.052	15.193	15.334	15.475	15.616	15.758	15.899	16.040
1400	16.040	16.181	16.323	16.464	16.605	16.746	16.887	17.028	17.169	17.310	17.451
1500	17.451	17.591	17.732	17.872	18.012	18.152	18.292	18.431	18.571	18.710	18.849
1600	18.849	18.988	19.126	19.264	19.402	19.540	19.677	19.814	19.951	20.087	20.222
1700	20.222	20.356	20.488	20.620	20.749	20.877	21.003				

Temperature Ranges and Coefficients of Equations Used to Compute the Above Table
The equations are of the form: $E = c_0 + c_1 t + c_2 t^2 + c_3 t^3 + \ldots c_n t^n$, where E is the emf in millivolts, t is the temperature in degrees Celsius (ITS-90), and c_0, c_1, c_2, c_3, *etc.* are the coefficients. These coefficients are extracted from NIST Monograph 175.

	−50 °C to 1064.18 °C	1064.18 °C to 1664.5 °C	1664.5 °C to 1768.1 °C
$c_0 =$	0.000 000 000 00 . . .	2.951 579 253 16 . . .	$1.522\ 321\ 182\ 09 \times 10^2$
$c_1 =$	$5.289\ 617\ 297\ 65 \times 10^{-3}$	$-2.520\ 612\ 513\ 32 \times 10^{-3}$	$-2.688\ 198\ 885\ 45 \times 10^{-1}$
$c_2 =$	$1.391\ 665\ 897\ 82 \times 10^{-5}$	$1.595\ 645\ 018\ 65 \times 10^{-5}$	$1.712\ 802\ 804\ 71 \times 10^{-4}$
$c_3 =$	$-2.388\ 556\ 930\ 17 \times 10^{-8}$	$-7.640\ 859\ 475\ 76 \times 10^{-9}$	$-3.458\ 957\ 064\ 53 \times 10^{-8}$
$c_4 =$	$3.569\ 160\ 010\ 63 \times 10^{-11}$	$2.053\ 052\ 910\ 24 \times 10^{-12}$	$-9.346\ 339\ 710\ 46 \times 10^{-15}$
$c_5 =$	$-4.623\ 476\ 662\ 98 \times 10^{-14}$	$-2.933\ 596\ 681\ 73 \times 10^{-16}$	. . . .
$c_6 =$	$5.007\ 774\ 410\ 34 \times 10^{-17}$	. . . .	. . . .
$c_7 =$	$-3.731\ 058\ 861\ 91 \times 10^{-20}$	. . . .	. . . .
$c_8 =$	$1.577\ 164\ 823\ 67 \times 10^{-23}$	. . . .	. . . .
$c_9 =$	$-2.810\ 386\ 252\ 51 \times 10^{-27}$	. . . .	. . . .

Type S thermocouples: emf-temperature (°C) reference table and equations.

Thermocouple emf as a Function of Temperature in Degrees Celsius (ITS-90)

emf in Millivolts   Reference Junctions at 0 °C

°C	0	−10	−20	−30	−40	−50	−60	−70	−80	−90	−100
0	0.000	−0.053	−0.103	−0.150	−0.194	−0.236					

°C	0	10	20	30	40	50	60	70	80	90	100
0	0.000	0.055	0.113	0.173	0.235	0.299	0.365	0.433	0.502	0.573	0.646
100	0.646	0.720	0.795	0.872	0.950	1.029	1.110	1.191	1.273	1.357	1.441
200	1.441	1.526	1.612	1.698	1.786	1.874	1.962	2.052	2.141	2.232	2.323
300	2.323	2.415	2.507	2.599	2.692	2.786	2.880	2.974	3.069	3.164	3.259
400	3.259	3.355	3.451	3.548	3.645	3.742	3.840	3.938	4.036	4.134	4.233
500	4.233	4.332	4.432	4.532	4.632	4.732	4.833	4.934	5.035	5.137	5.239
600	5.239	5.341	5.443	5.546	5.649	5.753	5.857	5.961	6.065	6.170	6.275
700	6.275	6.381	6.486	6.593	6.699	6.806	6.913	7.020	7.128	7.236	7.345
800	7.345	7.454	7.563	7.673	7.783	7.893	8.003	8.114	8.226	8.337	8.449
900	8.449	8.562	8.674	8.787	8.900	9.014	9.128	9.242	9.357	9.472	9.587
1000	9.587	9.703	9.819	9.935	10.051	10.168	10.285	10.403	10.520	10.638	10.757
1100	10.757	10.875	10.994	11.113	11.232	11.351	11.471	11.590	11.710	11.830	11.951
1200	11.951	12.071	12.191	12.312	12.433	12.554	12.675	12.796	12.917	13.038	13.159
1300	13.159	13.280	13.402	13.523	13.644	13.766	13.887	14.009	14.130	14.251	14.373
1400	14.373	14.494	14.615	14.736	14.857	14.978	15.099	15.220	15.341	15.461	15.582
1500	15.582	15.702	15.822	15.942	16.062	16.182	16.301	16.420	16.539	16.658	16.777
1600	16.777	16.895	17.013	17.131	17.249	17.366	17.483	17.600	17.717	17.832	17.947
1700	17.947	18.061	18.174	18.285	18.395	18.503	18.609				

Temperature Ranges and Coefficients of Equations Used to Compute the Above Table

The equations are of the form: $E = c_0 + c_1t + c_2t^2 + c_3t^3 + \ldots c_nt^n$, where E is the emf in millivolts, t is the temperature in degrees Celsius (ITS-90), and c_0, c_1, c_2, c_3, etc. are the coefficients. These coefficients are extracted from NIST Monograph 175.

	−50 °C to 1064.18 °C	1064.18 °C to 1664.5 °C	1664.5 °C to 1768.1 °C
c_0 =	0.000 000 000 00 …	1.329 004 440 85 …	1.466 282 326 36 $\times 10^2$
c_1 =	5.403 133 086 31 $\times 10^{-3}$	3.345 093 113 44 $\times 10^{-3}$	−2.584 305 167 52 $\times 10^{-1}$
c_2 =	1.259 342 897 40 $\times 10^{-5}$	6.548 051 928 18 $\times 10^{-6}$	1.636 935 746 41 $\times 10^{-4}$
c_3 =	−2.324 779 686 89 $\times 10^{-8}$	−1.648 562 592 09 $\times 10^{-9}$	−3.304 390 469 87 $\times 10^{-8}$
c_4 =	3.220 288 230 36 $\times 10^{-11}$	1.299 896 051 74 $\times 10^{-14}$	−9.432 236 906 12 $\times 10^{-15}$
c_5 =	−3.314 651 963 89 $\times 10^{-14}$	. . . .	. . . .
c_6 =	2.557 442 517 86 $\times 10^{-17}$	. . . .	. . . .
c_7 =	−1.250 688 713 93 $\times 10^{-20}$	. . . .	. . . .
c_8 =	2.714 431 761 45 $\times 10^{-24}$	. . . .	. . . .

Type T thermocouples: emf-temperature (°C) reference table and equations.

Thermocouple emf as a Function of Temperature in Degrees Celsius (ITS-90)

emf in Millivolts   Reference Junctions at 0 °C

°C	0	−10	−20	−30	−40	−50	−60	−70	−80	−90	−100
−200	−5.603	−5.753	−5.888	−6.007	−6.105	−6.180	−6.232	−6.258			
−100	−3.379	−3.657	−3.923	−4.177	−4.419	−4.648	−4.865	−5.070	−5.261	−5.439	−5.603
0	0.000	−0.383	−0.757	−1.121	−1.475	−1.819	−2.153	−2.476	−2.788	−3.089	−3.379

°C	0	10	20	30	40	50	60	70	80	90	100
0	0.000	0.391	0.790	1.196	1.612	2.036	2.468	2.909	3.358	3.814	4.279
100	4.279	4.750	5.228	5.714	6.206	6.704	7.209	7.720	8.237	8.759	9.288
200	9.288	9.822	10.362	10.907	11.458	12.013	12.574	13.139	13.709	14.283	14.862
300	14.862	15.445	16.032	16.624	17.219	17.819	18.422	19.030	19.641	20.255	20.872
400	20.872										

Temperature Ranges and Coefficients of Equations Used to Compute the Above Table

The equations are of the form: $E = c_0 + c_1 t + c_2 t^2 + c_3 t^3 + \dots c_n t^n$, where E is the emf in millivolts, t is the temperature in degrees Celsius (ITS-90), and c_0, c_1, c_2, c_3, *etc.* are the coefficients. These coefficients are extracted from NIST Monograph 175.

		−270 °C to 0 °C	0 °C to 400 °C
c_0	=	0.000 000 000 0 . . .	0.000 000 000 0 . . .
c_1	=	3.874 810 636 4 X 10^{-2}	3.874 810 636 4 X 10^{-2}
c_2	=	4.419 443 434 7 X 10^{-5}	3.329 222 788 0 X 10^{-5}
c_3	=	1.184 432 310 5 X 10^{-7}	2.061 824 340 4 X 10^{-7}
c_4	=	2.003 297 355 4 X 10^{-8}	−2.188 225 684 6 X 10^{-9}
c_5	=	9.013 801 955 9 X 10^{-10}	1.099 688 092 8 X 10^{-11}
c_6	=	2.265 115 659 3 X 10^{-11}	−3.081 575 877 2 X 10^{-14}
c_7	=	3.607 115 420 5 X 10^{-13}	4.547 913 529 0 X 10^{-17}
c_8	=	3.849 393 988 3 X 10^{-15}	−2.751 290 167 3 X 10^{-20}
c_9	=	2.821 352 192 5 X 10^{-17}	. . . .
c_{10}	=	1.425 159 477 9 X 10^{-19}	. . . .
c_{11}	=	4.876 866 228 6 X 10^{-22}	. . . .
c_{12}	=	1.079 553 927 0 X 10^{-24}	. . . .
c_{13}	=	1.394 502 706 2 X 10^{-27}	. . . .
c_{14}	=	7.979 515 392 7 X 10^{-31}	. . . .

Type B thermocouples: coefficients (c_i) of polynomials for the computation of temperatures in °C as a function of the thermocouple emf in various temperature and emf ranges.

Temperature Range:	250 °C to 700 °C	700 °C to 1820 °C
emf Range:	0.291 mV to 2.431 mV	2.431 mV to 13.820 mV
$c_0 =$	$9.842\ 332\ 1 \times 10^1$	$2.131\ 507\ 1 \times 10^2$
$c_1 =$	$6.997\ 150\ 0 \times 10^2$	$2.851\ 050\ 4 \times 10^2$
$c_2 =$	$-8.476\ 530\ 4 \times 10^2$	$-5.274\ 288\ 7 \times 10^1$
$c_3 =$	$1.005\ 264\ 4 \times 10^3.$	$9.916\ 080\ 4\ \dots$
$c_4 =$	$-8.334\ 595\ 2 \times 10^2$	$-1.296\ 530\ 3\ \dots$
$c_5 =$	$4.550\ 854\ 2 \times 10^2$	$1.119\ 587\ 0 \times 10^{-1}$
$c_6 =$	$-1.552\ 303\ 7 \times 10^2$	$-6.062\ 519\ 9 \times 10^{-3}$
$c_7 =$	$2.988\ 675\ 0 \times 10^1$	$1.866\ 169\ 6 \times 10^{-4}$
$c_8 =$	$-2.474\ 286\ 0\ \dots$	$-2.487\ 858\ 5 \times 10^{-6}$

NOTE—The above coefficients are extracted from NIST Monograph 175 and are for an expression of the form shown in Section 10.3.2. They yield approximate values of temperature that agree within ± 0.03 °C with the values given in Table 10.2.

Type E thermocouples: coefficients (c_i) of polynomials for the computation of temperatures in °C as a function of the thermocouple emf in various temperature and emf ranges.

Temperature Range:	−200 °C to 0 °C	0 °C to 1000 °C
emf Range:	−8.825 mV to 0.0 mV	0.0 mV to 76.373 mV
$c_0 =$	$0.000\ 000\ 0\ \dots$	$0.000\ 000\ 0\ \dots$
$c_1 =$	$1.697\ 728\ 8 \times 10^1$	$1.705\ 703\ 5 \times 10^1$
$c_2 =$	$-4.351\ 497\ 0 \times 10^{-1}$	$-2.330\ 175\ 9 \times 10^{-1}$
$c_3 =$	$-1.585\ 969\ 7 \times 10^{-1}$	$6.543\ 558\ 5 \times 10^{-3}$
$c_4 =$	$-9.250\ 287\ 1 \times 10^{-2}$	$-7.356\ 274\ 9 \times 10^{-5}$
$c_5 =$	$-2.608\ 431\ 4 \times 10^{-2}$	$-1.789\ 600\ 1 \times 10^{-6}$
$c_6 =$	$-4.136\ 019\ 9 \times 10^{-3}$	$8.403\ 616\ 5 \times 10^{-8}$
$c_7 =$	$-3.403\ 403\ 0 \times 10^{-4}$	$-1.373\ 587\ 9 \times 10^{-9}$
$c_8 =$	$-1.156\ 489\ 0 \times 10^{-5}$	$1.062\ 982\ 3 \times 10^{-11}$
$c_9 =$	$\dots\dots$	$-3.244\ 708\ 7 \times 10^{-14}$

NOTE—The above coefficients are extracted from NIST Monograph 175 and are for an expression of the form shown in Section 10.3.2. They yield approximate values of temperature that agree within ± 0.02 °C with the values given in Table 10.4.

Type J thermocouples: coefficients (c_i) of polynomials for the computation of temperatures in °C as a function of the thermocouple emf in various temperature and emf ranges.

Temperature Range:	−210 °C to 0 °C	0 °C to 760 °C	760 °C to 1200 °C
emf Range:	−8.095 mV to 0.0 mV	0.0 mV to 42.919 mV	42.919 mV to 69.553 mV
c_0 =	0.000 000 0 . . .	0.000 000 . . .	−3.113 581 87 X 10^3
c_1 =	1.952 826 8 X 10^1	1.978 425 X 10^1	3.005 436 84 X 10^2
c_2 =	−1.228 618 5 . . .	−2.001 204 X 10^{-1}	−9.947 732 30 . . .
c_3 =	−1.075 217 8 . . .	1.036 969 X 10^{-2}	1.702 766 30 X 10^{-1}
c_4 =	−5.908 693 3 X 10^{-1}	−2.549 687 X 10^{-4}	−1.430 334 68 X 10^{-3}
c_5 =	−1.725 671 3 X 10^{-1}	3.585 153 X 10^{-6}	4.738 860 84 X 10^{-6}
c_6 =	−2.813 151 3 X 10^{-2}	−5.344 285 X 10^{-8}	. . . .
c_7 =	−2.396 337 0 X 10^{-3}	5.099 890 X 10^{-10}	. . . .
c_8 =	−8.382 332 1 X 10^{-5}	. . . .	. . . .

NOTE—The above coefficients are extracted from NIST Monograph 175 and are for an expression of the form shown in Section 10.3.2. They yield approximate values of temperature that agree within ± 0.05 °C with the values given in Table 10.6.

Type K thermocouples: coefficients (c_i) of polynomials for the computation of temperatures in °C as a function of the thermocouple emf in various temperature and emf ranges.

Temperature Range:	−200 °C to 0 °C	0 °C to 500 °C	500 °C to 1372 °C
emf Range:	−5.891 mV to 0.0 mV	0.0 mV to 20.644 mV	20.644 mV to 54.886 mV
c_0 =	0.000 000 0 . . .	0.000 000 0 . . .	−1.318 058 X 10^2
c_1 =	2.517 346 2 X 10^1	2.508 355 X 10^1	4.830 222 X 10^1
c_2 =	−1.166 287 8 . . .	7.860 106 X 10^{-2}	−1.646 031 . . .
c_3 =	−1.083 363 8 . . .	−2.503 131 X 10^{-1}	5.464 731 X 10^{-2}
c_4 =	−8.977 354 0 X 10^{-1}	8.315 270 X 10^{-2}	−9.650 715 X 10^{-4}
c_5 =	−3.734 237 7 X 10^{-1}	−1.228 034 X 10^{-2}	8.802 193 X 10^{-6}
c_6 =	−8.663 264 3 X 10^{-2}	9.804 036 X 10^{-4}	−3.110 810 X 10^{-8}
c_7 =	−1.045 059 8 X 10^{-2}	−4.413 030 X 10^{-5}	. . . .
c_8 =	−5.192 057 7 X 10^{-4}	1.057 734 X 10^{-6}	. . . .
c_9 =	. . . .	−1.052 755 X 10^{-8}	. . . .

NOTE—The above coefficients are extracted from NIST Monograph 175 and are for an expression of the form shown in Section 10.3.2. They yield approximate values of temperature that agree within ± 0.05 °C with the values given in Table 10.8.

Type N thermocouples: coefficients (c_i) of polynomials for the computation of temperatures in °C as a function of the thermocouple emf in various temperature and emf ranges.

	-200 °C to 0 °C	0 °C to 600 °C	600 °C to 1300 °C
Temperature Range:			
emf Range:	-3.990 mV to 0.0 mV	0.0 mV to 20.613 mV	20.613 mV to 47.513 mV
$c_0 =$	$0.000\,000\,0\ldots$	$0.000\,00\ldots$	$1.972\,485 \times 10^1$
$c_1 =$	$3.843\,684\,7 \times 10^1$	$3.868\,96 \times 10^1$	$3.300\,943 \times 10^1$
$c_2 =$	$1.101\,048\,5\ldots$	$-1.082\,67\ldots$	$-3.915\,159 \times 10^{-1}$
$c_3 =$	$5.222\,931\,2\ldots$	$4.702\,05 \times 10^{-2}$	$9.855\,391 \times 10^{-3}$
$c_4 =$	$7.206\,052\,5\ldots$	$-2.121\,69 \times 10^{-6}$	$-1.274\,371 \times 10^{-4}$
$c_5 =$	$5.848\,858\,6\ldots$	$-1.172\,72 \times 10^{-4}$	$7.767\,022 \times 10^{-7}$
$c_6 =$	$2.775\,491\,6\ldots$	$5.392\,80 \times 10^{-6}$	$\ldots\ldots\ldots$
$c_7 =$	$7.707\,516\,6 \times 10^{-1}$	$-7.981\,56 \times 10^{-8}$	$\ldots\ldots\ldots$
$c_8 =$	$1.158\,266\,5 \times 10^{-1}$	$\ldots\ldots\ldots$	$\ldots\ldots\ldots$
$c_9 =$	$7.313\,886\,8 \times 10^{-3}$	$\ldots\ldots\ldots$	$\ldots\ldots\ldots$

NOTE—The above coefficients are extracted from NIST Monograph 175 and are for an expression of the form shown in Section 10.3.2. They yield approximate values of temperature that agree within ± 0.04 °C with the values given in Table 10.10.

Type R thermocouples: coefficients (c_i) of polynomials for the computation of temperatures in °C as a function of the thermocouple emf in various temperature and emf ranges.

	-50 °C to 250 °C	250 °C to 1200 °C	1064 °C to 1664.5 °C	1664.5 °C to 1768.1 °C
Temperature Range:				
emf Range:	-0.226 mV to 1.923 mV	1.923 mV to 13.228 mV	11.361 mV to 19.739 mV	19.739 mV to 21.103 mV
$c_0 =$	$0.000\,000\,0\ldots$	$1.334\,584\,505 \times 10^1$	$-8.199\,599\,416 \times 10^1$	$3.406\,177\,836 \times 10^4$
$c_1 =$	$1.889\,138\,0 \times 10^2$	$1.472\,644\,573 \times 10^2$	$1.553\,962\,042 \times 10^2$	$-7.023\,729\,171 \times 10^3$
$c_2 =$	$-9.383\,529\,0 \times 10^1$	$-1.844\,024\,844 \times 10^1$	$-8.342\,197\,663\ldots$	$5.582\,903\,813 \times 10^2$
$c_3 =$	$1.306\,861\,9 \times 10^2$	$4.031\,129\,726\ldots$	$4.279\,433\,549 \times 10^{-1}$	$-1.952\,394\,635 \times 10^1$
$c_4 =$	$-2.270\,358\,0 \times 10^2$	$-6.249\,428\,360 \times 10^{-1}$	$-1.191\,577\,910 \times 10^{-2}$	$2.560\,740\,231 \times 10^{-1}$
$c_5 =$	$3.514\,565\,9 \times 10^2$	$6.468\,412\,046 \times 10^{-2}$	$1.492\,290\,091 \times 10^{-4}$	$\ldots\ldots\ldots$
$c_6 =$	$-3.895\,390\,0 \times 10^2$	$-4.458\,750\,426 \times 10^{-3}$	$\ldots\ldots\ldots$	$\ldots\ldots\ldots$
$c_7 =$	$2.823\,947\,1 \times 10^2$	$1.994\,710\,149 \times 10^{-4}$	$\ldots\ldots\ldots$	$\ldots\ldots\ldots$
$c_8 =$	$-1.260\,728\,1 \times 10^2$	$-5.313\,401\,790 \times 10^{-6}$	$\ldots\ldots\ldots$	$\ldots\ldots\ldots$
$c_9 =$	$3.135\,361\,1 \times 10^1$	$6.481\,976\,217 \times 10^{-8}$	$\ldots\ldots\ldots$	$\ldots\ldots\ldots$
$c_{10} =$	$-3.318\,776\,9\ldots$	$\ldots\ldots\ldots$	$\ldots\ldots\ldots$	$\ldots\ldots\ldots$

NOTE—The above coefficients are extracted from NIST Monograph 175 and are for an expression of the form shown in Section 10.3.2. They yield approximate values of temperature that agree within ± 0.02 °C with the values given in Table 10.12.

STANDARD ITS-90 THERMOCOUPLE TABLES (continued)

Type S thermocouples: coefficients (c_i) of polynomials for the computation of temperatures in °C as a function of the thermocouple emf in various temperature and emf ranges.

Temperature Range:	−50 °C to 250 °C	250 °C to 1200 °C	1064 °C to 1664.5 °C	1664.5 °C to 1768.1 °C
emf Range:	−0.235 mV to 1.874 mV	1.874 mV to 11.950 mV	10.332 mV to 17.536 mV	17.536 mV to 18.693 mV
c_0 =	0.000 000 00 . . .	$1.291\ 507\ 177 \times 10^1$	$-8.087\ 801\ 117 \times 10^1$	$5.333\ 875\ 126 \times 10^4$
c_1 =	$1.849\ 494\ 60 \times 10^2$	$1.466\ 298\ 863 \times 10^2$	$1.621\ 573\ 104 \times 10^2$	$-1.235\ 892\ 298 \times 10^4$
c_2 =	$-8.005\ 040\ 62 \times 10^1$	$-1.534\ 713\ 402 \times 10^1$	$-8.536\ 869\ 453$ . . .	$1.092\ 657\ 613 \times 10^3$
c_3 =	$1.022\ 374\ 30 \times 10^2$	$3.145\ 945\ 973$ . . .	$4.719\ 686\ 976 \times 10^{-1}$	$-4.265\ 693\ 686 \times 10^1$
c_4 =	$-1.522\ 485\ 92 \times 10^2$	$-4.163\ 257\ 839 \times 10^{-1}$	$-1.441\ 693\ 666 \times 10^{-2}$	$6.247\ 205\ 420 \times 10^{-1}$
c_5 =	$1.888\ 213\ 43 \times 10^2$	$3.187\ 963\ 771 \times 10^{-2}$	$2.081\ 618\ 890 \times 10^{-4}$	. . . .
c_6 =	$-1.590\ 859\ 41 \times 10^2$	$-1.291\ 637\ 500 \times 10^{-3}$	. . . .	. . . .
c_7 =	$8.230\ 278\ 80 \times 10^1$	$2.183\ 475\ 087 \times 10^{-5}$	. . . .	. . . .
c_8 =	$-2.341\ 819\ 44 \times 10^1$	$-1.447\ 379\ 511 \times 10^{-7}$	. . . .	. . . .
c_9 =	$2.797\ 862\ 60$ . . .	$8.211\ 272\ 125 \times 10^{-9}$	. . . .	. . . .

NOTE—The above coefficients are extracted from NIST Monograph 175 and are for an expression of the form shown in Section 10.3.2. They yield approximate values of temperature that agree within ± 0.02 °C with the values given in Table 10.14.

Type T thermocouples: coefficients (c_i) of polynomials for the computation of temperatures in °C as a function of the thermocouple emf in various temperature and emf ranges.

Temperature Range:	−200 °C to 0 °C	0 °C to 400 °C
emf Range:	−5.603 mV to 0.0 mV	0.0 mV to 20.872 mV
c_0 =	0.000 000 0 . . .	0.000 000 . . .
c_1 =	$2.594\ 919\ 2 \times 10^1$	$2.592\ 800 \times 10^1$
c_2 =	$-2.131\ 696\ 7 \times 10^{-1}$	$-7.602\ 961 \times 10^{-1}$
c_3 =	$7.901\ 869\ 2 \times 10^{-1}$	$4.637\ 791 \times 10^{-2}$
c_4 =	$4.252\ 777\ 7 \times 10^{-1}$	$-2.165\ 394 \times 10^{-3}$
c_5 =	$1.330\ 447\ 3 \times 10^{-1}$	$6.048\ 144 \times 10^{-5}$
c_6 =	$2.024\ 144\ 6 \times 10^{-2}$	$-7.293\ 422 \times 10^{-7}$
c_7 =	$1.266\ 817\ 1 \times 10^{-3}$	. . . .

NOTE—The above coefficients are extracted from NIST Monograph 175 and are for an expression of the form shown in Section 10.3.2. They yield approximate values of temperature that agree within ± 0.04 °C with the values given in Table 10.16.

PROPERTIES OF COMMON LABORATORY SOLVENTS

This table give properties of 200 organic solvents which are frequently used in laboratory and industrial applications. Compounds are listed in alphabetical order by the most common name; synonyms are given in some cases. The properties tabulated are:

MF: Molecular formula
CAS RN: Chemical Abstracts Service Registry Number
M_r: Molecular weight
t_m: Melting point in °C
t_b: Normal boiling point in °C
ρ: Density in g/cm^3 at the temperature in °C indicated by the superscript
c_p: Specific heat capacity of the liquid at constant pressure at 25°C in J/g K
vp: Vapor pressure at 25°C in kPa (1 kPa = 7.50 mmHg)
μ: Electric dipole moment in debye units. Values in parentheses are measurements on the pure liquid or in solution; these are less reliable than the other values, which were obtained in the gas phase.
FP: Flash point temperature in °C. The fact that no flash point is listed does not necessarily mean that the substance is nonflammable, because some liquids will burn if the quantity is large or impurities are present.
Fl. Lim.: Flammable (explosive) range in air in percent by volume
Ign. Temp.: Autoignition temperature in °C
TLV: Threshold limit for allowable airborne concentration, given in parts per million by volume at 25°C and atmospheric pressure (see table "Threshold Limit Values for Airborne Contaminants" in Section 16)

REFERENCES

1. Lide, D.R., *Handbook of Organic Solvents*, CRC Press, Boca Raton, FL, 1994.
2. Lide, D.R., and Kehiaian, H.V., *CRC Handbook of Thermophysical and Thermochemical Data*, CRC Press, Boca Raton, FL, 1994.
3. Riddick, J.A., Bunger, W.B., and Sakano, T.K., *Organic Solvents*, Fourth Edition, John Wiley & Sons, New York, 1986.
4. *Fire Protection Guide to Hazardous Materials*, 10th Edition , National Fire Protection Association, Quincy, MA, 1991.
5. Urben, P.G., Ed., *Bretherick's Handbook of Reactive Chemical Hazards, 5th Edition*, Butterworth-Heinemann, Oxford, 1995.

Name	MF	CAS RN	M_r	t_m/°C	t_b/°C	ρ/g cm^{-3}	c_p/J g^{-1}K^{-1}	vp/kPa	μ/D	FP/°C	Fl. Lim.	Ign. Temp./°C	TLV
Acetal (1,1-Diethoxyethane)	C$_6$H$_{14}$O$_2$	105-57-7	118.18	-100	102.2	0.8254^{20}	2.01	3.68	(1.4)	-21	2-10%	230	10
Acetic acid	C$_2$H$_4$O$_2$	64-19-7	60.05	17	118	1.0492^{20}	2.06	2.07	1.70	39	4-20%	463	10
Acetone	C$_3$H$_6$O	67-64-1	58.08	-95	56	0.7899^{20}	2.18	30.8	2.88	-20	3-13%	465	750
Acetonitrile	C$_2$H$_3$N	75-05-8	41.05	-44	82	0.7857^{20}	2.23	11.8	3.92	6	3-16%	524	40
Acetylacetone	C$_5$H$_8$O$_2$	123-54-6	100.12	-23	138	0.9721^{25}	2.08	1.02	(2.8)	34		340	
Acrylonitrile	C$_3$H$_3$N	107-13-1	53.06	-83.5	77.3	0.8060^{20}	2.05	14.1	3.87	0	3-17%	481	2
Adiponitrile	C$_6$H$_8$N$_2$	111-69-3	108.14	1	295	0.9676^{20}	1.19	<0.01		93	2-5%	550	2
Allyl alcohol	C$_3$H$_6$O	107-18-6	58.08	-129	97.0	0.8540^{20}	2.39	3.14	1.60	21	3-18%	378	2
Allylamine	C$_3$H$_7$N	107-11-9	57.10	-88.2	53.3	0.758^{20}		33.1	1.2	-29	2-22%	374	2
2-Aminoisobutanol	C$_4$H$_{11}$NO	124-68-5	89.14	25.5	165.5	0.934^{20}				67			
Benzal chloride	C$_7$H$_6$Cl$_2$	98-87-3	161.03	-17	205	1.26^{25}		0.06	(2.1)				
Benzaldehyde	C$_7$H$_6$O	100-52-7	106.12	-26	179.0	1.0415^{10}	1.62	0.17	(3.0)	63		192	
Benzene	C$_6$H$_6$	71-43-2	78.11	6	80	0.8765^{20}	1.74	12.7	0	-11	1-8%	498	10
Benzonitrile	C$_7$H$_5$N	100-47-0	103.12	-12.7	191.1	1.0093^{15}	1.60	0.11	4.18				
Benzyl chloride	C$_7$H$_7$Cl	100-44-7	126.59	-45	179	1.1004^{20}	1.44	0.16	(1.8)	67	1%-	585	1
Bromochloromethane	CH$_2$BrCl	74-97-5	129.38	-87.9	68.0	1.9344^{20}	0.41	19.5	(1.7)				200
Bromoform (Tribromomethane)	CHBr$_3$	75-25-2	252.73	8.0	149	2.899^{15}	0.52	0.73	0.99	83			0.5

PROPERTIES OF COMMON LABORATORY SOLVENTS (continued)

Name	MF	CAS RN	M_r	t_m/°C	t_b/°C	ρ/g cm^{-3}	c_p/J g^{-1} K^{-1}	vp/kPa	μ/D	FP/°C	Fl. Lim.	Ign. Temp./°C	TLV
Butyl acetate	$C_6H_{12}O_2$	123-86-4	116.16	-78	126	0.8825^{20}	1.96	1.66	(1.9)	22	2-8%	425	150
Butyl alcohol	$C_4H_{10}O$	71-36-3	74.12	-90	118	0.8098^{20}	2.39	0.86	1.66	37	1-11%	343	50
sec-Butyl alcohol	$C_4H_{10}O$	78-92-2	74.12	-114.7	99.5	0.8063^{20}	2.66	2.32	(1.8)	24	2-10%	405	100
tert-Butyl alcohol	$C_4H_{10}O$	75-65-0	74.12	26	82	0.7887^{20}	2.97	5.52	(1.7)	11	2-8%	478	100
Butylamine	$C_4H_{11}N$	109-73-9	73.14	-49	77	0.7414^{20}	2.45	12.2	1.0	-12	2-10%	312	5
tert-Butylamine	$C_4H_{11}N$	75-64-9	73.14	-67	44	0.6958^{20}	2.63	48.4	(1.3)	-9	2-9%	380	
Butyl methyl ketone	$C_6H_{12}O$	591-78-6	100.16	-56	128	0.8113^{20}	2.13	1.54	(2.7)	25	1-8%	423	5
p-tert-Butyltoluene	$C_{11}H_{16}$	98-51-1	148.25	-52	190	0.8612^{20}		0.09	=0	68			10
γ-Butyrolactone	$C_4H_6O_2$	96-48-0	86.09	-43.3	204	1.1284^{16}	1.64	0.43	4.27	98			
Caprolactam	$C_6H_{11}NO$	105-60-2	113.16	69	270		1.38	<0.01	(3.9)	125			5
Carbon disulfide	CS_2	75-15-0	76.14	-112	46	1.2632^{20}	1.00	48.2	0	-30	1-50%	90	10
Carbon tetrachloride	CCl_4	56-23-5	153.82	-23	77	1.5940^{20}	0.85	15.2	0				5
1-Chloro-1,1-difluoroethane	$C_2H_3ClF_2$	75-68-3	100.50	-131	-10	1.107^{25}	1.30	351	2.14		1-10%		
Chlorobenzene	C_6H_5Cl	108-90-7	112.56	-45	132	1.1058^{20}	1.33	1.6	1.69	28		593	10
Chloroform	$CHCl_3$	67-66-3	119.38	-64	61	1.4832^{20}	0.96	26.2	1.04				10
Chloropentafluoroethane	C_2ClF_5	76-15-3	154.47	-99	-38	1.5678^{-42}	1.19	912	0.52				1000
Cumene (Isopropylbenzene)	C_9H_{12}	98-82-8	120.19	-95.0	152	0.8618^{20}	1.75	0.61	0.79	36	1-7%	424	50
Cyclohexane	C_6H_{12}	110-82-7	84.16	7	81	0.7785^{20}	1.84	13.0	=0	-20	1-8%	245	300
Cyclohexanol	$C_6H_{12}O$	108-93-0	100.16	25	161	0.9624^{20}	2.08	0.10		68	1-9%	300	50
Cyclohexanone	$C_6H_{10}O$	108-94-1	98.14	-31	155	0.9478^{20}	1.86	0.53	2.87	44	1-9%	420	25
Cyclohexylamine	$C_6H_{13}N$	108-91-8	99.18	-18	134	0.8191^{20}		1.20	(1.3)	31	1-9%	293	10
Cyclopentane	C_5H_{10}	287-92-3	70.13	-93.8	49.3	0.7457^{20}	1.84	42.3	=0	<-7	2%-	361	600
Cyclopentanone	C_5H_8O	120-92-3	84.12	-51.3	130.5	0.9487^{20}	1.84	1.55	3.3	26			
p-Cymene	$C_{10}H_{14}$	99-87-6	134.22	-69	177	0.8573^{20}	1.76	0.19	=0	47	1-6%	436	
cis-Decalin	$C_{10}H_{18}$	493-01-6	138.25	-42.9	195.8	0.8965^{20}	1.68	0.10	=0	54	1-5%	255	
trans-Decalin	$C_{10}H_{18}$	493-02-7	138.25	-30.3	187.3	0.8699^{20}	1.65	0.16	=0	58		643	
Diacetone alcohol	$C_6H_{12}O_2$	123-42-2	116.16	-44	168	0.9387^{20}	1.91	0.22			2-7%		50
1,2-Dibromoethane	$C_2H_4Br_2$	106-93-4	187.86	9.9	131.6	2.1791^{20}	0.72	1.55	(3.2)				
Dibromofluoromethane	$CHBr_2F$	1868-53-7	191.83	-78	64.9	2.421^{20}			(1.2)				
Dibromomethane	CH_2Br_2	74-95-3	173.83	-52.5	97	2.4969^{20}	0.61	6.12	1.43				
1,2-Dibromotetrafluoroethane	$C_2Br_2F_4$	124-73-2	259.82	-110.4	47.3	2.149^{25}	0.69	43.4					
Dibutylamine	$C_8H_{19}N$	111-92-2	129.25	-62	160	0.7670^{20}	2.27	0.34	(1.0)	47	1-6%		
o-Dichlorobenzene	$C_6H_4Cl_2$	95-50-1	147.00	-17	180	1.3059^{20}	1.10	0.18	2.50	66	2-9%	648	25
1,1-Dichloroethane	$C_2H_4Cl_2$	75-34-3	98.96	-97	57	1.1757^{20}	1.28	30.5	2.06	-17	5-11%	458	100
1,2-Dichloroethane	$C_2H_4Cl_2$	107-06-2	98.96	-36	84	1.2351^{20}	1.30	10.6	(1.8)	13	6-16%	413	10
1,1-Dichloroethylene	$C_2H_2Cl_2$	75-35-4	96.94	-122.5	31.6	1.213^{20}	1.15	80.0	1.34	-15	7-16%	570	5
cis-1,2-Dichloroethylene	$C_2H_2Cl_2$	156-59-2	96.94	-80	60	1.2837^{20}	1.20	26.8	1.90	6	3-15%	460	200
trans-1,2-Dichloroethylene	$C_2H_2Cl_2$	156-60-5	96.94	-50	49	1.2565^{20}	1.20	44.2	0	2	6-13%	460	200
Dichloroethyl ether	$C_4H_8Cl_2O$	111-44-4	143.01	-52	179	1.22^{20}	1.54	0.14	(2.6)	55	3%-	369	5
Dichloromethane	CH_2Cl_2	75-09-2	84.93	-95	40	1.3266^{20}	1.19	58.2	1.60		13-23%	556	50
1,2-Dichloropropane	$C_3H_6Cl_2$	78-87-5	112.99	-100	96	1.1560^{20}	1.32	6.62	(1.8)	16	3-15%	557	75
1,2-Dichlorotetrafluoroethane	$C_2Cl_2F_4$	76-14-2	170.92	-94	4	1.518^4	0.96	215	0.5				1000
Diethanolamine	$C_4H_{11}NO_2$	111-42-2	105.14	28	269	1.0966^{20}	2.22	<0.01	(2.8)	172	2-13%	662	0.46
Diethylamine	$C_4H_{11}N$	109-89-7	73.14	-50	55	0.7056^{20}	2.31	30.1	0.92	-23	2-10%	312	5
Diethyl carbonate	$C_5H_{10}O_3$	105-58-8	118.13	-43	126	0.9752^{20}	1.80	1.63	1.10	25			
Diethylene glycol	$C_4H_{10}O_3$	111-46-6	106.12	-10	246	1.1197^{15}	2.31	<0.01	(2.3)	124		224	
Diethylene glycol dimethyl ether	$C_6H_{14}O_3$	111-96-6	134.18	-68	162	0.9434^{20}	2.04	0.31	(2.0)	67			
Diethylene glycol monoethyl ether	$C_6H_{14}O_3$	111-90-0	134.18		196	0.9885^{20}	2.24	0.02	(1.6)	96	2-17%		

PROPERTIES OF COMMON LABORATORY SOLVENTS (continued)

Name	MF	CAS RN	M_r	t_m/°C	t_b/°C	ρ/g cm⁻³	c_p/J g⁻¹K⁻¹	vp/kPa	μ/D	FP/°C	Fl. Lim.	Ign. Temp./°C	TLV
Diethylene glycol monoethyl ether acetate	$C_8H_{16}O_4$	112-15-2	176.21	-25	218.5	1.0096^{20}		0.03	(1.8)	110	1-23%	425	
Diethylene glycol monomethyl ether	$C_5H_{12}O_3$	111-77-3	120.15		193	1.035^{20}	2.26	0.02	(1.6)	96		240	1
Diethylenetriamine	$C_4H_{13}N_3$	111-40-0	103.17	-39	207	0.9569^{20}	2.46	0.03	(1.9)	98	2-7%	358	
Diethyl ether	$C_4H_{10}O$	60-29-7	74.12	-116	34	0.7138^{20}	2.33	71.7	1.15	-45	2-36%	180	400
Diisobutyl ketone	$C_9H_{18}O$	108-83-8	142.24	-42	169	0.8062^{20}	2.09	0.23	(2.7)	49	1-7%	396	25
Diisopropyl ether	$C_6H_{14}O$	108-20-3	102.18	-87	69	0.7241^{20}	2.12	19.9	1.13	-28	1-8%	443	250
N,N-Dimethylacetamide	C_4H_9NO	127-19-5	87.12	-20	165	0.9366^{25}	2.02	0.07	(3.7)	70	2-12%	490	10
Dimethylamine	C_2H_7N	124-40-3	45.08	-92	7	0.6804^{0}	3.05	203	1.01	20	3-14%	400	5
Dimethyl disulfide	$C_2H_6S_2$	624-92-0	94.20	-85	109.8	1.0625^{20}	1.55	3.82	(1.8)	24			
N,N-Dimethylformamide	C_3H_7NO	68-12-2	73.09	-60	153	0.944^{25}	2.06	0.44	3.82	58	2-15%	445	10
Dimethyl sulfoxide	C_2H_6OS	67-68-5	78.14	19	189	1.1014^{20}	1.96	0.08	3.96	95	3-42%	215	
1,4-Dioxane	$C_4H_8O_2$	123-91-1	88.11	12	101	1.0337^{20}	1.74	4.95	0	12	2-22%	180	25
1,3-Dioxolane	$C_3H_6O_2$	646-06-0	74.08	-95	78	1.060^{20}	1.59	14.6	1.19	2			
Dipentene	$C_{10}H_{16}$	7705-14-8	136.24	-95.5	178	0.8402^{21}	1.83	0.26		45		237	
Epichlorohydrin	C_3H_5ClO	106-89-8	92.52	-26	116	1.1812^{20}	1.42	2.2	(1.8)	31	4-21%	411	2
Ethanolamine (Glycinol)	C_2H_7NO	141-43-5	61.08	11	171	1.0180^{20}	3.20	0.05	(2.3)	86	3-24%	410	3
Ethyl acetate	$C_4H_8O_2$	141-78-6	88.11	-84	77	0.9003^{20}	1.94	12.6	1.78	-4	2-12%	426	400
Ethyl acetoacetate	$C_6H_{10}O_3$	141-97-9	130.14	-45	180.8	1.0368^{10}	1.91	0.09		57	1-10%	295	
Ethyl alcohol	C_2H_6O	64-17-5	46.07	-114	78	0.7893^{20}	2.44	7.87	1.69	13	3-19%	363	1000
Ethylamine	C_2H_7N	75-04-7	45.08	-81	17	0.686^{17}	2.88	142	1.22	<-18	4-14%	385	5
Ethylbenzene	C_8H_{10}	100-41-4	106.17	-95	136	0.8670^{20}	1.73	1.28	0.59	21	1-7%	432	100
Ethyl bromide	C_2H_5Br	74-96-4	108.97	-118.6	38.5	1.4604^{20}	0.93	62.5	2.03		7-8%	511	5
Ethyl chloride	C_2H_5Cl	75-00-3	64.51	-139	12	0.909^{12}	1.62	160	2.05	-50	4-15%	519	1000
Ethylene carbonate	$C_3H_4O_3$	96-49-1	88.06	36.4	248	1.3214^{39}	1.52	<0.01	(4.9)	143			
Ethylenediamine	$C_2H_8N_2$	107-15-3	60.10	11	117	0.8979^{20}	2.87	1.62	1.99	40	3-12%	385	10
Ethylene glycol	$C_2H_6O_2$	107-21-1	62.07	-13	197	1.1088^{20}	2.41	0.01	2.28	111	3-22%	398	50
Ethylene glycol diethyl ether	$C_6H_{14}O_2$	629-14-1	118.18	-74	119.4	0.8484^{20}	2.19	4.33		35			
Ethylene glycol dimethyl ether	$C_4H_{10}O_2$	110-71-4	90.12	-58	85	0.8691^{20}	2.14	9.93		-2	4-13%	202	
Ethylene glycol monobutyl ether	$C_6H_{14}O_2$	111-76-2	118.18	-75	168	0.9015^{20}	2.38	0.15	(2.1)	69	3-18%	238	25
Ethylene glycol monoethyl ether	$C_4H_{10}O_2$	110-80-5	90.12	-70	135	0.9297^{20}	2.34	0.71	(2.1)	43	2-8%	235	5
Ethylene glycol ethyl ether acetate	$C_6H_{12}O_3$	111-15-9	132.16	-62	156	0.9740^{20}	2.85	0.24	(2.2)	56	2-14%	379	5
Ethylene glycol monomethyl ether	$C_3H_8O_2$	109-86-4	76.10	-85	124	0.9647^{20}	2.25	1.31	2.36	39	2-12%	285	5
Ethylene glycol monomomethyl ether acetate	$C_5H_{10}O_3$	110-49-6	118.13	-70	143	1.0074^{19}	2.62	0.67	(2.1)	49		392	5
Ethyl formate	$C_3H_6O_2$	109-94-4	74.08	-80	54	0.9168^{20}	2.02	32.3	1.9	-20	3-16%	455	100
Furan	C_4H_4O	110-00-9	68.08	-86	31	0.9514^{20}	1.69	80.0	0.66	<0	2-14%		2
Furfural	$C_5H_4O_2$	98-01-1	96.09	-37	162	1.1594^{20}	1.70	0.29	(3.5)	60	2-19%	316	2
Furfuryl alcohol	$C_5H_6O_2$	98-00-0	98.10	-31	171	1.1296^{20}	2.08	0.10	(1.9)	75	2-16%	491	10
Glycerol	$C_3H_8O_3$	56-81-5	92.09	18	290	1.2613^{20}	2.38	<0.01	(2.6)	199		370	
Heptane	C_7H_{16}	142-82-5	100.20	-91	98	0.6837^{20}	2.24	6.09	=0	-4	1-7%	204	400
1-Heptanol	$C_7H_{16}O$	111-70-6	116.20	-34	176.4	0.8219^{20}	2.34						
Hexane	C_6H_{14}	110-54-3	86.18	-95	69	0.6548^{25}	2.27	20.2	=0	-22	1-8%	225	50
1-Hexanol (Caproyl alcohol)	$C_6H_{14}O$	111-27-3	102.18	-44.6	157.6	0.8136^{20}	2.35	0.11		63	1-9%		
Hexylene glycol	$C_6H_{14}O_2$	107-41-5	118.18	-50	197	0.923^{15}	2.84	<0.01	(2.9)	102		306	25
Hexyl methyl ketone	$C_8H_{16}O$	111-13-7	128.21	-16	172.5	0.820^{20}	2.13		(2.7)	52			
Isobutyl acetate	$C_6H_{12}O_2$	110-19-0	116.16	-99	117	0.8712^{20}	2.01	2.39	(1.9)	18	1-11%	421	150
Isobutyl alcohol	$C_4H_{10}O$	78-83-1	74.12	-108	108	0.8018^{20}	2.44	1.39	1.64	28	2-11%	415	50

Name	MF	CAS RN	M_r	t_m/°C	t_b/°C	ρ/g cm⁻³	c_p/J g⁻¹K⁻¹	vp/kPa	μ/D	FP/°C	Fl. Lim.	Ign. Temp./°C	TLV
Isobutylamine	C₄H₁₁N	78-81-9	73.14	-87	68	0.724 [25]	2.50	19.0	(1.3)	-9	2-12%	378	100
Isopentyl acetate	C₇H₁₄O₂	123-92-2	130.19	-79	143	0.876 [15]	1.91	0.73	(1.9)	25	1-8%	360	5
Isophorone	C₉H₁₄O	78-59-1	138.21	-8	215	0.9255 [20]	1.83	0.06		84	1-4%	460	250
Isopropyl acetate	C₅H₁₀O₂	108-21-4	102.13	-73	89	0.8718 [20]	1.95	8.1		2	2-8%	460	400
Isopropyl alcohol	C₃H₈O	67-63-0	60.10	-90	82	0.7855 [20]	2.58	6.02	1.56	12	2-13%	399	
Isoquinoline	C₉H₇N	119-65-3	129.16	26.47	243.2	1.0910 [30]	1.52	0.28	2.73	49			
d-Limonene (Citrene)	C₁₀H₁₆	5989-27-5	136.24	-97	178	0.8411 [20]	1.83	0.75	(1.7)				25
2,6-Lutidine	C₇H₉N	108-48-5	107.16	-6.1	144.1	0.9226 [20]	1.73	0.75					
Mesitylene	C₉H₁₂	108-67-8	120.19	-45	165	0.8652 [20]	1.74	0.33	0	50	1-5%	559	25
Mesityl oxide	C₆H₁₀O	141-79-7	98.14	-59	130	0.8653 [20]	2.17	1.47	(2.8)	31	1-7%	344	15
Methyl acetate	C₃H₆O₂	79-20-9	74.08	-98	57	0.9342 [20]	1.92	28.8	1.72	-10	3-16%	454	200
Methylal	C₃H₈O₂	109-87-5	76.10	-105	42	0.8593 [20]	2.12	53.1	(0.7)	-32	2-14%	237	1000
Methyl alcohol	CH₄O	67-56-1	32.04	-98	65	0.7914 [20]	2.53	16.9	1.70	11	6-36%	464	200
Methylamine	CH₅N	74-89-5	31.06	-93	-6	0.656 [25]	3.29	353	1.31	0	5-21%	430	5
Methyl benzoate	C₈H₈O₂	93-58-3	136.15	-15	199	1.0933 [15]	1.63	0.05	(1.9)	83			
Methylcyclohexane	C₇H₁₄	108-87-2	98.19	-127	101	0.7694 [20]	1.88	6.18	=0	-4	1-7%	250	400
Methyl ethyl ketone	C₄H₈O	78-93-3	72.11	-87	80	0.8054 [20]	2.20	12.6	2.78	-9	1-11%	404	200
N-Methylformamide	C₂H₅NO	123-39-7	59.07	-3.8	199.5	1.011 [19]	2.10		3.83				
Methyl formate	C₂H₄O₂	107-31-3	60.05	-99	32	0.9742 [20]	1.98	78.1	1.77	-19	5-23%	449	100
Methyl iodide	CH₃I	74-88-4	141.94	-66.4	42.5	2.279 [20]	0.89	53.9	1.62			448	2
Methyl isobutyl ketone	C₆H₁₂O	108-10-1	100.16	-84	116	0.7978 [20]	2.13	2.64		18	1-8%	448	50
Methyl isopentyl ketone	C₇H₁₄O	110-12-3	114.19		144	0.888 [20]		0.69		36	1-8%	191	50
2-Methylpentane	C₆H₁₄	107-83-5	86.18	-153.7	60.2	0.650 [25]	2.25	28.2	=0	-29	1-7%	264	
4-Methyl-2-pentanol	C₆H₁₄O	108-11-2	102.18	-90	132	0.8075 [20]	2.67	0.70	(2.6)	41	1-6%	393	25
Methyl pentyl ketone	C₇H₁₄O	110-43-0	114.19	-35	151	0.8111 [20]	2.04	0.49	(2.7)	39	1-8%	452	50
Methyl propyl ketone	C₅H₁₀O	107-87-9	86.13	-77	102	0.809 [20]	2.14	4.97	(2.7)	7	2-8%	452	200
N-Methyl-2-pyrrolidone	C₅H₉NO	872-50-4	99.13	-24	202	1.0230 [25]	3.11	0.04	(4.1)	96	1-10%	346	
Morpholine	C₄H₉NO	110-91-8	87.12	-5	128	1.0005 [20]	1.89	1.34	1.55	37	1-11%	290	20
Nitrobenzene	C₆H₅NO₂	98-95-3	123.11	6	211	1.2037 [20]	1.51	0.03	4.22	88	2-9%	482	1
Nitroethane	C₂H₅NO₂	79-24-3	75.07	-90	114	1.0448 [25]	1.79	2.79	3.23	28	3-17%	414	100
Nitromethane	CH₃NO₂	75-52-5	61.04	-29	101	1.1371 [20]	1.75	4.79	3.46	35	7-22%	418	20
1-Nitropropane	C₃H₇NO₂	108-03-2	89.09	-108	131.1	0.9961 [25]	1.97	1.36	3.66	36	2%-	421	25
2-Nitropropane	C₃H₇NO₂	79-46-9	89.09	-91	120	0.9821 [25]	1.91	2.3	3.73	24	3-11%	428	10
Octane	C₈H₁₈	111-65-9	114.23	-57	126	0.6986 [25]	2.23	1.86	=0	13	1-7%	206	300
1-Octanol	C₈H₁₈O	111-87-5	130.23	-15.5	195.1	0.8262 [25]	2.34	0.01	(1.8)	81			
Pentachloroethane	C₂HCl₅	76-01-7	202.29	-29	160	1.6796 [20]	0.86	0.48	0.92				
Pentamethylene glycol	C₅H₁₂O₂	111-29-5	104.15	-18	239	0.9914 [20]	3.08		(2.5)	129		335	
Pentane	C₅H₁₂	109-66-0	72.15	-130	36	0.6262 [20]	2.32	68.3	=0	<-40	2-8%	260	600
1-Pentanol	C₅H₁₂O	71-41-0	88.15	-79	138	0.8144 [20]	2.36	0.26	(1.7)	33	1-10%	300	
Pentyl acetate	C₇H₁₄O₂	628-63-7	130.19	-71	149	0.8756 [20]	2.00	0.60	1.75	16	1-8%	360	100
2-Picoline	C₆H₇N	109-06-8	93.13	-67	129	0.9443 [20]	1.70	1.5	1.85	39		538	
α-Pinene	C₁₀H₁₆	80-56-8	136.24	-64	156	0.8539 [25]		0.64		35		275	
β-Pinene	C₁₀H₁₆	127-91-3	136.24	-61.5	166	0.860 [25]		0.61		38		275	
Piperidine	C₅H₁₁N	110-89-4	85.15	-11	106	0.8606 [20]	2.11	4.28	(1.2)	16	1-10%	512	
Propanenitrile	C₃H₅N	107-12-0	55.08	-93	97	0.7818 [20]	2.17	6.14	4.05	2	3-14%	450	
Propyl acetate	C₅H₁₀O₂	109-60-4	102.13	-93	102	0.8878 [20]	1.92	4.49	(1.8)	13	2-8%	412	200
Propyl alcohol	C₃H₈O	71-23-8	60.10	-126	97	0.8035 [20]	2.39	2.76	1.55	23	2-14%	318	200
Propylamine	C₃H₉N	107-10-8	59.11	-83	47	0.7173 [20]	2.75	42.1	1.17	-37	2-10%		
Propylbenzene	C₉H₁₂	103-65-1	120.19	-99.5	159.2	0.8620 [20]	1.79		=0	30	1-6%	450	

PROPERTIES OF COMMON LABORATORY SOLVENTS (continued)

Name	MF	CAS RN	M_r	t_m/°C	t_b/°C	ρ/g cm^{-3}	c_p/J g^{-1} K^{-1}	vp/kPa	μ/D	FP/°C	Fl. Lim.	Ign. Temp./°C	TLV
Propylene glycol	C3H8O2	57-55-6	76.10	-60	188	1.0361^{20}	2.51	0.02	(2.2)	99	3-13%	371	
Pseudocumene	C9H12	95-63-6	120.19	-44	169	0.8758^{20}	1.79	0.30	≈0	44	1-6%	500	25
Pyridine	C5H5N	110-86-1	79.10	-42	115	0.9819^{20}	1.68	2.76	2.21	20	2-12%	482	5
Pyrrole	C4H5N	109-97-7	67.09	-23.4	129.7	0.9698^{20}	1.90	1.10	1.74	39			
Pyrrolidine	C4H9N	123-75-1	71.12	-57.8	86.5	0.8586^{20}	2.20	8.40	(1.6)	3			
2-Pyrrolidone	C4H7NO	616-45-5	85.11	25	251	1.120^{20}	1.99		(3.5)	129			
Quinoline	C9H7N	91-22-5	129.16	-14.78	237.1	1.0977^{15}	1.51		2.29			480	
Styrene	C8H8	100-42-5	104.15	-31	145	0.9060^{20}	1.75	0.81		31	1-7%	490	50
Sulfolane	C4H8O2S	126-33-0	120.17	28	287	1.2723^{18}	1.50	<0.01	(4.8)	177			
α-Terpinene	C10H16	99-86-5	136.24		174	0.8375^{19}							
1,1,1-Tetrachloro-2,2-difluoroethane	C2Cl4F2	76-11-9	203.83	40.6	91.5	1.649^{25}		7.36					500
1,1,2,2-Tetrachloro-1,2-difluoroethane	C2Cl4F2	76-12-0	203.83	26	93	1.6447^{25}	0.85	7.51			5-12%		500
1,1,1,2-Tetrachloroethane	C2H2Cl4	630-20-6	167.85	-70	131	1.5406^{20}	0.92	1.6	1.32		20-54%		1
1,1,2,2-Tetrachloroethane	C2H2Cl4	79-34-5	167.85	-44	146	1.5953^{20}	0.97	0.62					50
Tetrachloroethylene	C2Cl4	127-18-4	165.83	-22	121	1.6227^{20}	0.86	2.42	0				200
Tetraethylene glycol	C8H18O5	112-60-7	194.23	-6.2	328	1.1285^{15}	2.21			182			
Tetrahydrofuran	C4H8O	109-99-9	72.11	-108	65	0.8892^{20}	1.72	21.6	1.75	-14	2-12%	321	200
1,2,3,4-Tetrahydronaphthalene	C10H12	119-64-2	132.21	-36	208	0.9660^{25}	1.65	0.05	≈0	71	1-5%	385	
Tetrahydropyran	C5H10O	142-68-7	86.13	-45	88	0.8814^{20}	1.82	9.54	1.74	-20	1-7%		
Tetramethylsilane	C4H12Si	75-76-3	88.22	-99.0	26.6	0.648^{19}	2.31	94.2	0		1%-		
Toluene	C7H8	108-88-3	92.14	-95	111	0.8669^{20}	1.70	3.79	0.37	4	1-7%	480	50
o-Toluidine	C7H9N	95-53-4	107.16	-16.3	200.3	0.9984^{20}	1.96	0.04	(1.6)	85		482	2
Triacetin	C9H14O6	102-76-1	218.21	-78	259	1.1583^{20}	1.76	<0.01	(0.8)	138		433	
Tributylamine	C12H27N	102-82-9	185.35	-70	217	0.7770^{20}		0.01	(0.8)	86			
1,1,1-Trichloroethane	C2H3Cl3	71-55-6	133.40	-30	74	1.3390^{20}	1.08	16.5	1.76		8-13%	537	350
1,1,2-Trichloroethane	C2H3Cl3	79-00-5	133.40	-37	114	1.4397^{20}	1.13	3.1	(1.4)	32	6-28%	460	10
Trichloroethylene	C2HCl3	79-01-6	131.39	-85	87	1.4642^{20}	0.95	9.91	(0.8)	32	8-11%	420	50
Trichlorofluoromethane	CCl3F	75-69-4	137.37	-111	24	1.478^{24}	0.89	106	0.46				1000
1,1,2-Trichlorotrifluoroethane	C2Cl3F3	76-13-1	187.38	-35	48	1.5635^{25}	0.91	44.8					1000
Triethanolamine	C6H15NO3	102-71-6	149.19	21	335	1.1242^{20}	2.61	<0.01	(3.6)	179	1-10%		0.5
Triethylamine	C6H15N	121-44-8	101.19	-115	89	0.7275^{20}	2.17	7.70	0.66	-7	1-8%	249	0.1
Triethylene glycol	C6H14O4	112-27-6	150.17	-7	285	1.1274^{15}	2.18			177	1-9%	371	
Triethyl phosphate	C6H15O4P	78-40-0	182.16	-56.4	215.5	1.0695^{20}			(3.1)	115		454	
Trimethylamine	C3H9N	75-50-3	59.11	-117	3	0.627^{25}	2.33	215	0.61	-7	2-12%	190	5
Trimethylene glycol	C3H8O2	504-63-2	76.10	-26.7	214.4	1.0538^{20}			(2.5)	107		400	
Trimethyl phosphate	C3H9O4P	512-56-1	140.08	-46	197.2	1.2144^{20}		0.11	(3.2)				
Veratrole	C8H10O2	91-16-7	138.17	22.5	206	1.0810^{25}			(1.3)				
o-Xylene	C8H10	95-47-6	106.17	-25	144	0.8802^{10}	1.75	0.88	0.64	32	1-7%	463	100
m-Xylene	C8H10	108-38-3	106.17	-48	139	0.8642^{20}	1.72	1.13	≈0	27	1-7%	527	100
p-Xylene	C8H10	106-42-3	106.17	13	138	0.8611^{20}	1.71	1.19	0	27	1-7%	528	100

DEPENDENCE OF BOILING POINT ON PRESSURE

The normal boiling point of a liquid is defined as the temperature at which the vapor pressure reaches standard atmospheric pressure, 101.325 kPa. The change in boiling point with pressure may be calculated from the representation of the vapor pressure by the Antoine Equation,

$$\ln p = A_1 - A_2/(T + A_3)$$

where p is the vapor pressure, T the absolute temperature, and A_1, A_2, and A_3 are constants. This table, which has been calculated using the Antoine constants in Reference 1, gives values of $\Delta t/\Delta p$ for a number of liquids, in units of both °C/kPa and °C/mmHg. The correction to the boiling point is generally accurate to 0.1 to 0.2 °C as long as the pressure is within 10% of standard atmospheric pressure.

A slightly less accurate estimate of $\Delta t/\Delta p$ may be obtained from the Claussius-Clapeyron equation, with the assumption that the change in volume upon vaporization equals the ideal-gas volume of the vapor. This leads to the equation

$$\Delta t/\Delta p = RT_b^2/p_0 \Delta_{vap}H(T_b)$$

where R is the molar gas constant, p_0 is 101.325 kPa, T_b is the normal boiling point temperature (absolute), and $\Delta_{vap}H(T_b)$ is the molar enthalpy of vaporization at the normal boiling point. Values of the last quantity may be obtained from the table "Enthalpy of Vaporization" in Section 6.

REFERENCE

1. Lide, D.R., and Kehiaian, H.V., *CRC Handbook of Thermophysical and Thermochemical Data*, CRC Press, Boca Raton, FL, 1994, pp. 49-59.

Compound	t_b °C	$\Delta t/\Delta p$ °C/kPa	°C/mmHg	Compound	t_b °C	$\Delta t/\Delta p$ °C/kPa	°C/mmHg
Acetaldehyde	20.1	0.261	0.0348	1-Hexanol	157.6	0.318	0.0424
Acetic acid	117.9	0.324	0.0432	Hydrogen fluoride	20.1	0.276	0.0368
Acetone	56.0	0.289	0.0385	Iodomethane	42.5	0.291	0.0388
Acetonitrile	81.6	0.316	0.0421	Isobutane	-11.7	0.254	0.0339
Ammonia	-33.33	0.198	0.0264	Methanol	64.6	0.251	0.0335
Aniline	184.1	0.378	0.0504	Methyl acetate	56.8	0.282	0.0376
Anisole	153.7	0.367	0.0489	Methyl formate	31.7	0.582	0.0776
Benzaldehyde	179.0	0.392	0.0523	N-Methylaniline	196.2	0.396	0.0528
Benzene	80.0	0.321	0.0428	N-Methylformamide	199.5	0.371	0.0495
Bromine	58.8	0.300	0.0400	Nitrobenzene	210.8	0.418	0.0557
Butane	-0.5	0.267	0.0356	Nitromethane	101.1	0.320	0.0427
1-Butanol	117.7	0.278	0.0371	1-Octanol	195.1	0.360	0.0480
Carbon disulfide	46.2	0.304	0.0405	Pentane	36.0	0.289	0.0385
Chlorine	-34.04	0.224	0.0299	1-Pentanol	137.9	0.296	0.0395
Chlorobenzene	131.7	0.365	0.0487	Phenol	181.8	0.349	0.0465
1-Chlorobutane	78.6	0.321	0.0428	Propane	-42.1	0.224	0.0299
Chloroethane	12.3	0.262	0.0349	1-Propanol	97.2	0.261	0.0348
Chloroethylene	-13.3	0.241	0.0321	2-Propanol	82.3	0.247	0.0329
Cyclohexane	80.7	0.328	0.0437	Pyridine	115.2	0.340	0.0453
Cyclohexanol	160.8	0.344	0.0459	Pyrrole	129.7	0.330	0.0440
Cyclohexanone	155.4	0.382	0.0509	Pyrrolidine	86.5	0.309	0.0412
Decane	174.1	0.388	0.0517	Styrene	145.1	0.369	0.0492
Dibutyl ether	140.2	0.363	0.0484	Sulfur dioxide	-10.05	0.221	0.0295
Dichloromethane	39.6	0.276	0.0368	Tetrachloroethylene	121.3	0.354	0.0472
Diethyl ether	34.5	0.278	0.0371	Tetrachloromethane	76.8	0.325	0.0433
Dimethyl sulfoxide	189.0	0.379	0.0505	Toluene	110.6	0.353	0.0471
1,4-Dioxane	101.5	0.321	0.0428	Trichloroethylene	87.2	0.330	0.0440
Dipropyl ether	90.0	0.326	0.0435	Trichloromethane	61.1	0.302	0.0403
Ethanol	78.2	0.249	0.0332	Trimethylamine	2.8	0.248	0.0331
Ethyl acetate	77.1	0.300	0.0400	Water	100.0	0.276	0.0368
Ethylene glycol	197.3	0.331	0.0441	o-Xylene	144.5	0.373	0.0497
Heptane	98.5	0.336	0.0448	m-Xylene	139.1	0.368	0.0491
Hexafluorobenzene	80.2	0.305	0.0407	p-Xylene	138.3	0.369	0.0492
Hexane	68.7	0.314	0.0419				

EBULLIOSCOPIC CONSTANTS FOR CALCULATION OF BOILING POINT ELEVATION

The boiling point T_b of a dilute solution of a non-volatile, non-dissociating solute is elevated relative to that of the pure solvent. If the solution is ideal (i.e., follows Raoult's Law), the amount of elevation depends only on the number of particles of solute present. Hence the change in boiling point ΔT_b can be expressed as

$$\Delta T_b = E_b\, m_2$$

where m_2 is the molality (moles of solute per kilogram of solvent) and E_b is the Ebullioscopic Constant, a characteristic property of the solvent. The Ebullioscopic Constant may be calculated from the relation

$$E_b = R\, T_b^2\, M/\Delta_{vap}H$$

where R is the molar gas constant, T_b is the normal boiling point temperature (absolute) of the solvent, M the molar mass of the solvent, and $\Delta_{vap}H$ the molar enthalpy (heat) of vaporization of the solvent at its normal boiling point.

This table lists E_b values for some common solvents, as calculated from data in the table "Enthalpy of Vaporization" in Section 6.

Compound	E_b/K kg mol^{-1}	Compound	E_b/K kg mol^{-1}
Acetic acid	3.22	Hexane	2.90
Acetone	1.80	Iodomethane	4.31
Acetonitrile	1.44	Methanol	0.86
Aniline	3.82	Methyl acetate	2.21
Anisole	4.20	N-Methylaniline	4.3
Benzaldehyde	4.24	N-Methylformamide	2.2
Benzene	2.64	Nitrobenzene	5.2
1-Butanol	2.17	Nitromethane	2.09
Carbon disulfide	2.42	1-Octanol	5.06
Chlorobenzene	4.36	Phenol	3.54
1-Chlorobutane	3.13	1-Propanol	1.66
Cyclohexane	2.92	2-Propanol	1.58
Cyclohexanol	3.5	Pyridine	2.83
Decane	6.10	Pyrrole	2.33
Dichloromethane	2.42	Pyrrolidine	2.32
Diethyl ether	2.20	Tetrachloroethylene	6.18
Dimethyl sulfoxide	3.22	Tetrachloromethane	5.26
1,4-Dioxane	3.01	Toluene	3.40
Ethanol	1.23	Trichloroethylene	4.52
Ethyl acetate	2.82	Trichloromethane	3.80
Ethylene glycol	2.26	Water	0.513
Heptane	3.62	o-Xylene	4.25

CRYOSCOPIC CONSTANTS FOR CALCULATION OF FREEZING POINT DEPRESSION

The freezing point T_f of a dilute solution of a non-volatile, non-dissociating solute is depressed relative to that of the pure solvent. If the solution is ideal (i.e., follows Raoult's Law), this lowering is a function only of the number of particles of solute present. Thus the absolute value of the lowering of freezing point ΔT_f can be expressed as

$$\Delta T_f = E_f m_2$$

where m_2 is the molality (moles of solute per kilogram of solvent) and E_f is the Cryoscopic Constant, a characteristic property of the solvent. The Cryoscopic Constant may be calculated from the relation

$$E_f = R\, T_f^2\, M/\Delta_{fus}H$$

where R is the molar gas constant, T_b is the freezing point temperature (absolute) of the solvent, M the molar mass of the solvent, and $\Delta_{fus}H$ the molar enthalpy (heat) of fusion of the solvent.

This table lists cryscopic constants for selected substances, as calculated from data in the table "Enthalpy of Fusion" in Section 6.

Compound	E_f/K kg mol^{-1}	Compound	E_f/K kg mol^{-1}
Acetamide	3.92	1,4-Dioxane	4.63
Acetic acid	3.63	Diphenylamine	8.38
Acetophenone	5.16	Ethylene glycol	3.11
Aniline	5.23	Formamide	4.25
Benzene	5.07	Formic acid	2.38
Benzonitrile	5.35	Glycerol	3.56
Benzophenone	8.58	Methylcyclohexane	2.60
(+)-Camphor	37.8	Naphthalene	7.45
1-Chloronaphthalene	7.68	Nitrobenzene	6.87
o-Cresol	5.92	Phenol	6.84
m-Cresol	7.76	Pyridine	4.26
p-Cresol	7.20	Quinoline	6.73
Cyclohexane	20.8	Succinonitrile	19.3
Cyclohexanol	42.2	1,1,2,2-Tetrabromoethane	21.4
cis-Decahydronaphthalene	6.42	1,1,2,2-Tetrachloro-1,2-difluoroethane	41.0
trans-Decahydronaphthalene	4.70	Toluene	3.55
Dibenzyl ether	6.17	p-Toluidine	4.91
p-Dichlorobenzene	7.57	Tribromomethane	15.0
Diethanolamine	3.16	Water	1.86
Dimethyl sulfoxide	3.85	p-Xylene	4.31

DETERMINATION OF RELATIVE HUMIDITY FROM DEW POINT

The relative humidity of a water vapor-air mixture is defined as 100 times the partial pressure of water divided by the saturation vapor pressure of water at the same temperature. The relative humidity may be determined from the dew point t_{dew}, which is the temperature at which liquid water first condenses when the mixture is cooled from an initial temperature t. This table gives relative humidity as a function of the dew point depression $t - t_{dew}$ for several values of the dew point. Values are calculated from the vapor pressure table in Section 6.

$t - t_{dew}$	t_{dew}/°C -10	0	10	20	30	$t - t_{dew}$	t_{dew}/°C -10	0	10	20	30
0.0	100	100	100	100	100	8.2	54	56	59	61	63
0.2	99	99	99	99	99	8.4	53	56	58	60	63
0.4	97	97	97	98	98	8.6	52	55	57	60	62
0.6	95	96	96	96	97	8.8	51	54	57	59	61
0.9	94	94	95	95	96	9.0	51	53	56	58	61
1.0	92	93	94	94	94	9.2	50	53	55	58	60
1.2	91	92	92	93	93	9.4	49	52	55	57	59
1.4	90	90	91	92	92	9.6	48	51	54	56	59
1.6	88	89	90	91	91	9.8	48	51	53	56	58
1.8	87	88	89	90	90	10.0	47	50	53	55	57
2.0	86	87	88	88	89	10.5	45	48	51	54	56
2.2	84	85	86	87	89	11.0	44	47	49	52	55
2.4	83	84	85	86	87	11.5	42	45	48	51	53
2.6	82	83	84	85	86	12.0	41	44	47	49	52
2.8	80	82	83	84	85	12.5	39	42	45	48	50
3.0	79	81	82	83	84	13.0	38	41	44	46	49
3.2	78	80	81	82	83	13.5	37	40	43	45	48
3.4	77	79	80	81	82	14.0	35	38	41	44	47
3.6	76	77	79	80	82	14.5	34	37	40	43	45
3.8	75	76	78	79	81	15.0	33	36	39	42	44
4.0	73	75	77	78	80	15.5	32	35	38	40	
4.2	72	74	76	77	79	16.0	31	34	37	39	
4.4	71	73	75	77	78	16.5	30	33	36	38	
4.6	70	72	74	76	77	17.0	29	32	35	37	
4.8	69	71	73	75	76	17.5	28	31	34	36	
5.0	69	70	72	74	75	18.0	27	30	33	35	
5.2	67	69	71	73	75	18.5	26	29	32	34	
5.4	66	68	70	72	74	19.0	25	28	31	33	
5.6	65	67	69	71	73	19.5	24	27	30	33	
5.9	64	66	69	70	72	20.0	24	26	29	32	
6.0	63	66	68	70	71	21.0	22	25	27	30	
6.2	62	65	67	69	71	22.0	21	23	26	29	
6.4	61	64	66	68	70	23.0	19	22	24	27	
6.6	60	63	65	67	69	24.0	18	21	23	26	
6.8	60	62	64	66	68	25.0	17	19	22	24	
7.0	59	61	63	66	68	26.0	16	18	21	23	
7.2	58	60	63	65	67	27.0	15	17	20	22	
7.4	57	60	62	64	66	28.0	14	16	19	21	
7.6	56	59	61	63	65	29.0	13	15	18	20	
7.8	55	58	60	63	65	30.0	12	14	17	19	
8.0	54	57	60	62	64						

DETERMINATION OF RELATIVE HUMIDITY FROM WET AND DRY BULB TEMPERATURES

Relative humidity may be determined by comparing temperature readings of wet and dry bulb thermometers. The following table, extracted from more extensive U.S. National Weather Service tables, gives the relative humidity as a function of air temperature t_d (dry bulb) and the difference $t_d - t_w$ between dry and wet bulb temperatures. The data assume a pressure near normal atmospheric pressure and an instrumental configuration with forced ventilation.

$t_d/°C$	$(t_d - t_w)/°C$											
	0.5	1.0	1.5	2.0	2.5	3.0	3.5	4.0	4.5	5.0	5.5	6.0
-10	83	67	51	35	19							
-8	86	71	57	43	29	15						
-6	88	74	61	49	37	25	8					
-4	89	77	66	55	44	33	23	12				
-2	90	79	69	60	50	40	31	22	12			
0	91	81	72	64	55	46	38	29	21	13	5	
2	91	84	76	68	60	52	44	37	29	22	14	7
4	92	85	78	71	63	57	49	43	36	29	22	16
6	93	86	79	73	66	60	54	48	41	35	29	24
8	93	87	81	75	69	63	57	51	46	40	35	29
10	94	88	82	77	71	66	60	55	50	44	39	34
12	94	89	83	78	73	68	63	58	53	48	43	39
14	95	90	85	79	75	70	65	60	56	51	47	42
16	95	90	85	81	76	71	67	63	58	54	50	46
18	95	91	86	82	77	73	69	65	61	57	53	49
20	96	91	87	83	78	74	70	66	63	59	55	51
22	96	92	87	83	80	76	72	68	64	61	57	54
24	96	92	88	84	80	77	73	69	66	62	59	56
26	96	92	88	85	81	78	74	71	67	64	61	58
28	96	93	89	85	82	78	75	72	69	65	62	59
30	96	93	89	86	83	79	76	73	70	67	64	61
35	97	94	90	87	84	81	78	75	72	69	67	64
40	97	94	91	88	85	82	80	77	74	72	69	67

$t_d/°C$	$(t_d - t_w)/°C$											
	6.5	7.0	7.5	8.0	8.5	9.0	10.0	11.0	12.0	13.0	14.0	15.0
4	9											
6	17	11	5									
8	24	19	14	8								
10	29	24	20	15	10	6						
12	34	29	25	21	16	12	5					
14	38	34	30	26	22	18	10					
16	42	38	34	30	26	23	15	8				
18	45	41	38	34	30	27	20	14	7			
20	48	44	41	37	34	31	24	18	12	6		
22	50	47	44	40	37	34	28	22	17	11	6	
24	53	49	46	43	40	37	31	26	20	15	10	5
26	54	51	49	46	43	40	34	29	24	19	14	10
28	56	53	51	48	45	42	37	32	27	22	18	13
30	58	55	52	50	47	44	39	35	30	25	21	17
32	60	57	54	51	49	46	41	37	32	28	24	20
34	61	58	56	53	51	48	43	39	35	30	26	23
36	62	59	57	54	52	50	45	41	37	33	29	25
38	63	61	58	56	54	51	47	43	39	35	31	27
40	64	62	59	57	54	53	48	44	40	36	33	29

CONSTANT HUMIDITY SOLUTIONS

Anthony Wexler

An excess of a water soluble salt in contact with its saturated solution and contained within an enclosed space produces a constant relative humidity and water vapor pressure according to

$$RH = A \exp(B/T)$$

where RH is the percent relative humidity (generally accurate to ± 2 %), T is the temperature in kelvin, and the constants A and B and the range of valid temperatures are given in the table below. The vapor pressure, p, can be calculated from

$$p = (RH/100) \times p_0$$

where p_0 is the vapor pressure of pure water at temperature T as given in the table in Section 6 titled "Vapor Pressure of Water from 0 to 370°C".

REFERENCES

1. Wexler, A. S. and Seinfeld, J. H., *Atmospheric Environment*, 25A, 2731, 1991.
2. Greenspan, L., *J. Res. National Bureau of Standards*, 81A, 89, 1977.
3. Broul, et al., *Solubility of Inorganic Two-Component Systems*, Elsevier, New York, 1981.
4. Wagman, D. D. et al., *J. Phys. Chem. Ref. Data*, Vol. 11, Suppl. 2, 1982.

Compound	Temperature range (°C)	RH 25°C	A	B
NaOH · H$_2$O	15—60	6	5.48	27
LiBr · 2H$_2$O	10—30	6	0.23	996
ZnBr$_2$ · 2H$_2$O	5—30	8	1.69	455
KOH · 2H$_2$O	5—30	9	0.014	1924
LiCl · H$_2$O	20—65	11	14.53	−75
CaBr$_2$ · 6H$_2$O	11—22	16	0.17	1360
LiI · 3H$_2$O	15—65	18	0.15	1424
CaCl$_2$· 6H$_2$O	15—25	29	0.11	1653
MgCl$_2$· 6H$_2$O	5—45	33	29.26	34
NaI· 2H$_2$O	5—45	38	3.62	702
Ca(NO$_3$)$_2$ · 4H$_2$O	10—30	51	1.89	981
Mg(NO$_3$)$_2$ · 6H$_2$O	5—35	53	25.28	220
NaBr · 2H$_2$O	0—35	58	20.49	308
NH$_4$NO$_3$	10—40	62	3.54	853
KI	5—30	69	29.35	254
SrCl$_2$ · 6H$_2$O	5—30	71	31.58	241
NaNO$_3$	10—40	74	26.94	302
NaCl	10—40	75	69.20	25
NH$_4$Cl	10—40	79	35.67	235
KBr	5—25	81	40.98	203
(NH$_4$)$_2$SO$_4$	10—40	81	62.06	79
KCl	5—25	84	49.38	159
Sr(NO$_3$)$_2$ · 4H$_2$O	5—25	85	28.34	328
BaCl$_2$ · 2H$_2$O	5—25	90	69.99	75
CsI	5—25	91	70.77	75
KNO$_3$	0—50	92	43.22	225
K$_2$SO$_4$	10—50	97	86.75	34

STANDARD SALT SOLUTIONS FOR HUMIDITY CALIBRATION

Saturated aqueous solutions of inorganic salts are convenient secondary standards for calibration of instruments for measurement of relative humidity. The International Union of Pure and Applied Chemistry has recommended the salt solutions in this table for calibrations in the range of 10 to 90% relative humidity. Details on the preparation and use of these standards may be found in Reference 1. Data for other salts are given in Reference 2.

REFERENCES

1. Marsh, K.N., Editor, *Recommended Reference Materials for the Realization of Physicochemical Properties*, Blackwell Scientific Publications, Oxford, 1987, pp.157-162.
2. Greenspan, L., *J. Res. Natl. Bur. Stand.*, 81A, 89, 1977.

$t/°C$	LiCl	$MgCl_2$	$Mg(NO_3)_2$	NaCl	KCl
0		33.66 ± 0.33	60.35 ± 0.55	75.51 ± 0.34	88.61 ± 0.53
5		33.60 ± 0.28	58.86 ± 0.43	75.65 ± 0.27	87.67 ± 0.45
10		33.47 ± 0.24	57.36 ± 0.33	75.67 ± 0.22	86.77 ± 0.39
15		33.30 ± 0.21	55.87 ± 0.27	75.61 ± 0.18	85.92 ± 0.33
20	11.31 ± 0.31	33.07 ± 0.18	54.38 ± 0.23	75.47 ± 0.14	85.11 ± 0.29
25	11.30 ± 0.27	32.78 ± 0.16	52.89 ± 0.22	75.29 ± 0.12	84.34 ± 0.26
30	11.28 ± 0.24	32.44 ± 0.14	51.40 ± 0.24	75.09 ± 0.11	83.62 ± 0.25
35	11.25 ± 0.22	32.05 ± 0.13	49.91 ± 0.29	74.87 ± 0.12	82.95 ± 0.25
40	11.21 ± 0.21	31.60 ± 0.13	48.42 ± 0.37		82.32 ± 0.25
45	11.16 ± 0.21	31.10 ± 0.13	46.93 ± 0.47		81.74 ± 0.28
50	11.10 ± 0.22	30.54 ± 0.14	45.44 ± 0.60		81.20 ± 0.31
55	11.03 ± 0.23	29.93 ± 0.16			80.70 ± 0.35
60	10.95 ± 0.26	29.26 ± 0.18			80.25 ± 0.41
65	10.86 ± 0.29	28.54 ± 0.21			79.85 ± 0.48
70	10.75 ± 0.33	27.77 ± 0.25			79.49 ± 0.57
75	10.64 ± 0.38	26.94 ± 0.29			79.17 ± 0.66
80	10.51 ± 0.44	26.05 ± 0.34			78.90 ± 0.77

LOW TEMPERATURE BATHS FOR MAINTAINING CONSTANT TEMPERATURE

A liquid-solid slurry is a convenient means of maintaining a constant temperature environment below room temperature. The following is a list of readily available organic liquids suitable for this purpose, arranged in order of their melting (freezing) points t_m. The normal boiling points t_b are also given.

Compound	$t_m/°C$	$t_b/°C$
Isopentane (2-Methylbutane)	-159.9	27.8
Methylcyclopentane	-142.5	71.8
3-Chloropropene (Allyl chloride)	-134.5	45.1
Pentane	-129.7	36.0
Allyl alcohol	-129	97.0
Ethanol	-114.1	78.2
Carbon disulfide	-111.5	46
Isobutyl alcohol	-108	107.8
Toluene	-94.9	110.6
Acetone	-94.8	56.0
Ethyl acetate	-83.6	77.1
Dry ice + acetone	-78	
p-Cymene	-68.9	177.1
Trichloromethane (Chloroform)	-63.6	61.1
N-Methylaniline	-57	196.2
Chlorobenzene	-45.2	131.7
Anisole	-37.5	153.7
Bromobenzene	-30.6	156.0
Tetrachloromethane (Carbon tetrachloride)	-23	76.8
Benzonitrile	-12.7	191.1

WIRE TABLES

The resistance per unit length of wires of various metals is tabulated here. Values were calculated from resistivity values in the tables "Electrical Resistivity of Pure Metals" and "Electrical Resistivity of Selected Alloys", which appear in Section 12. In practice, resistance may vary because of differing heat treatments and metal composition. The values in the table refer to 20°C, but values at other temperatures may be calculated from the following resistivity data:

Metal	Resistivity in 10^{-8} Ω m at temperature			
	0°C	20°C	25°C	100°C
Aluminum	2.417	2.650	2.709	3.56
Brass (70% Cu, 30% Zn)	5.87	6.08	6.13	6.91
Constantan (60% Cu, 40% Ni)	45.43	45.38	45.35	45.11
Copper	1.543	1.678	1.712	2.22
Nichrome (79% Ni, 21% Cr)	107.3	107.5	107.6	108.3
Platinum	9.6	10.5	10.7	13.6
Silver	1.467	1.587	1.617	2.07
Tungsten	4.82	5.28	5.39	7.18

Resistance per unit length at 20°C in Ω/m

B & S Gauge	Diameter (mm)	Aluminum	Brass	Constantan	Copper	Nichrome	Platinum	Silver	Tungsten
0	8.252	0.000495	0.00114	0.00848	0.000314	0.0201	0.00196	0.000297	0.00099
2	6.543	0.000788	0.00181	0.0135	0.000499	0.0320	0.00312	0.000472	0.00157
4	5.189	0.00125	0.00287	0.0214	0.000793	0.0508	0.00496	0.000750	0.00250
6	4.115	0.00199	0.00457	0.0341	0.00126	0.0808	0.00789	0.00119	0.00397
8	3.264	0.00317	0.00727	0.0542	0.00200	0.128	0.0125	0.00190	0.00631
10	2.588	0.00504	0.0115	0.0863	0.00319	0.204	0.0200	0.00302	0.0100
12	2.053	0.00800	0.0184	0.137	0.00507	0.325	0.0317	0.00479	0.0159
14	1.628	0.0127	0.0292	0.218	0.00806	0.516	0.0504	0.00762	0.0254
16	1.291	0.0202	0.0464	0.347	0.0128	0.821	0.0802	0.0121	0.0403
18	1.024	0.0322	0.0738	0.551	0.0204	1.30	0.127	0.0193	0.0641
20	0.8118	0.0512	0.117	0.877	0.0324	2.08	0.203	0.0307	0.102
22	0.6439	0.0814	0.187	1.39	0.0515	3.30	0.322	0.0487	0.162
24	0.5105	0.129	0.297	2.22	0.0820	5.25	0.513	0.0775	0.258
26	0.4049	0.206	0.472	3.52	0.130	8.35	0.815	0.123	0.410
28	0.3211	0.327	0.751	5.60	0.207	13.3	1.30	0.196	0.652
30	0.2548	0.520	1.19	8.90	0.329	21.1	2.06	0.311	1.03
32	0.2019	0.828	1.90	14.2	0.524	33.6	3.28	0.496	1.65
34	0.1601	1.32	3.02	22.5	0.833	53.4	5.22	0.788	2.62
36	0.1270	2.09	4.80	35.8	1.32	84.9	8.29	1.25	4.17
38	0.1007	3.33	7.63	57.0	2.11	135	13.2	1.99	6.63
40	0.07988	5.29	12.1	90.5	3.35	214	20.9	3.17	10.5

CHARACTERISTICS OF PARTICLES AND PARTICLE DISPERSOIDS

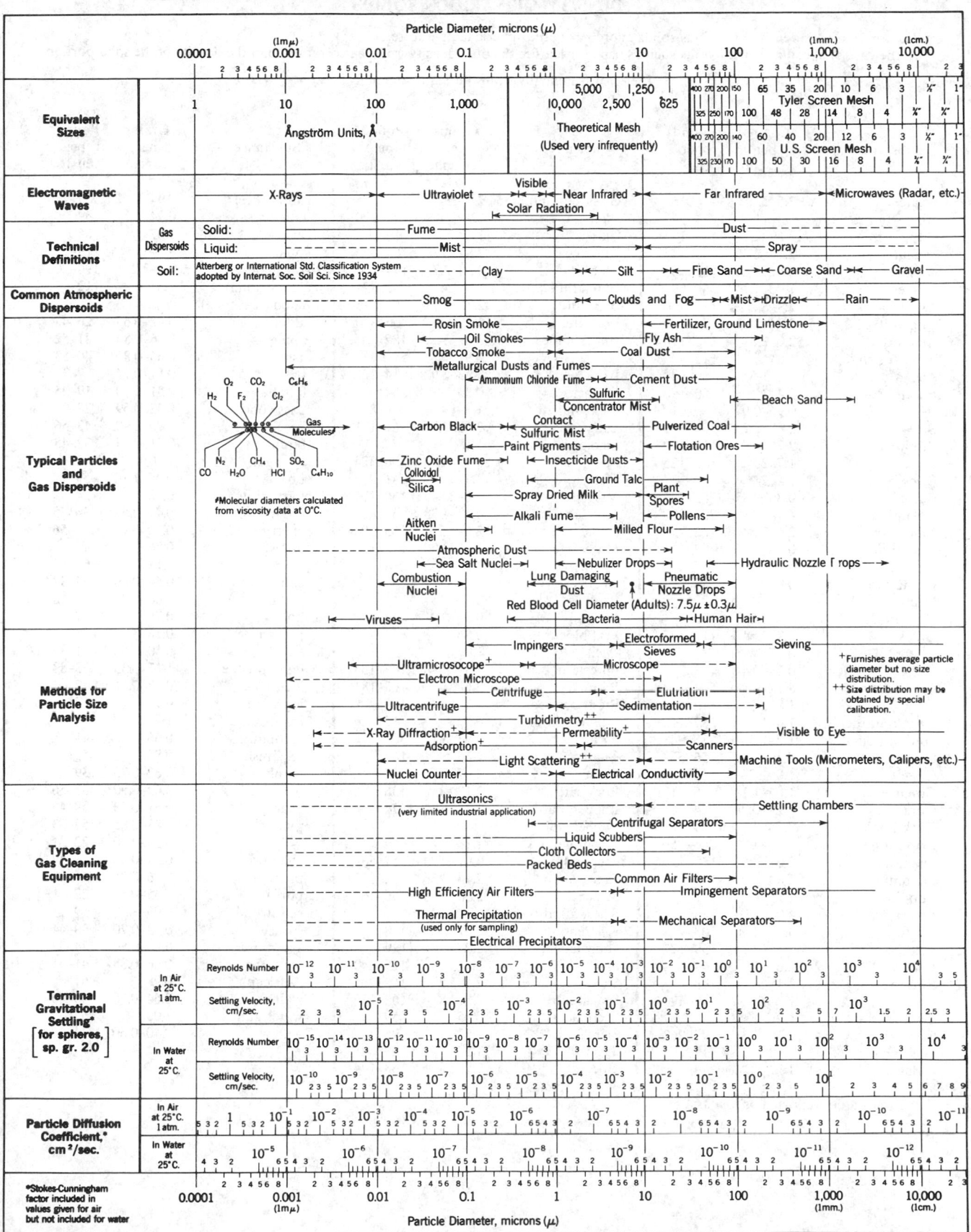

C. E. Lapple, Stanford Research
Institute Journal, Vol. 5, p.95
(Third Quarter, 1961)

DENSITY OF VARIOUS SOLIDS

The approximate density of various solids at ordinary atmospheric temperature.
In the case of substances with voids such as paper or leather the bulk density is indicated rather than the density of the solid portion.

(Selected principally from the Smithsonian Tables.)

Substance	Grams per cu cm	Pounds per cu ft	Substance	Grams per cu cm	Pound per cu ft	Substance	Grams per cu cm	Pounds per cu ft
Agate	2.5–2.7	156–168	Glass, common	2.4–2.8	150–175	Tallow, beef	0.94	59
Alabaster, carbon-			flint	2.9–5.9	180–370	mutton	0.94	59
ate	2.69–2.78	168–173	Glue	1.27	79	Tar	1.02	66
sulfate	2.26–2.32	141–145	Granite	2.64–2.76	165–172	Topaz	3.5–3.6	219–223
Albite	2.62–2.65	163–165	Graphite*	2.30–2.72	144–170	Tourmaline	3.0–3.2	190–200
Amber	1.06–1.11	66–69	Gum arabic	1.3–1.4	81–87	Wax, sealing	1.8	112
Amphiboles	2.9–3.2	180–200	Gypsum	2.31–2.33	144–145	Wood (seasoned)		
Anorthite	2.74–2.76	171–172	Hematite	4.9–5.3	306–330	alder	0.42–0.68	26–42
Asbestos	2.0–2.8	125–175	Hornblende	3.0	187	apple	0.66–0.84	41–52
Asbestos slate	1.8	112	Ice	0.917	57.2	ash	0.65–0.85	40–53
Asphalt	1.1–1.5	69–94	Ivory	1.83–1.92	114–120	balsa	0.11–0.14	7–9
Basalt	2.4–3.1	150–190	Leather, dry	0.86	54	bamboo	0.31–0.40	19–25
Beeswax	0.96–0.97	60–61	Lime, slaked	1.3–1.4	81–87	basswood	0.32–0.59	20–37
Beryl	2.69–2.7	168–169	Limestone	2.68–2.76	167–171	beech	0.70–0.90	43–56
Biotite	2.7–3.1	170–190	Linoleum	1.18	74	birch	0.51–0.77	32–48
Bone	1.7–2.0	106–125	Magnetite	4.9–5.2	306–324	blue gum	1.00	62
Brick	1.4–2.2	87–137	Malachite	3.7–4.1	231–256	box	0.95–1.16	59–72
Butter	0.86–0.87	53–54	Marble	2.6–2.84	160–177	butternut	0.38	24
Calamine	4.1–4.5	255–280	Meerschaum	0.99–1.28	62–80	cedar	0.49–0.57	30–35
Calcspar	2.6–2.8	162–175	Mica	2.6–3.2	165–200	cherry	0.70–0.90	43–56
Camphor	0.99	62	Muscovite	2.76–3.00	172–187	dogwood	0.76	47
Caoutchouc	0.92–0.99	57–62	Ochre	3.5	218	ebony	1.11–1.33	69–83
Cardboard	0.69	43	Opal	2.2	137	elm	0.54–0.60	34–37
Celluloid	1.4	87	Paper	0.7–1.15	44–72	hickory	0.60–0.93	37–58
Cement, set	2.7–3.0	170.190	Paraffin	0.87–0.91	54–57	holly	0.76	47
Chalk	1.9–2.8	118–175	Peat blocks	0.84	52	juniper	0.56	35
Charcoal, oak	0.57	35	Pitch	1.07	67	larch	0.50–0.56	31–35
pine	0.28–0.44	18–28	Porcelain	2.3–2.5	143–156	lignum vitae	1.17–1.33	73–83
Cinnabar	8.12	507	Porphyry	2.6–2.9	162–181	locust	0.67–0.71	42–44
Clay	1.8–2.6	112–162	Pressed wood			logwood	0.91	57
Coal, anthracite	1.4–1.8	87–112	pulp board	0.19	12	mahogany		
bituminous	1.2–1.5	75–94	Pyrite	4.95–5.1	309–318	Honduras	0.66	41
Cocoa butter	0.89–0.91	56–57	Quartz	2.65	165	Spanish	0.85	53
Coke	1.0–1.7	62–105	Resin	1.07	67	maple	0.62–0.75	39–47
Copal	1.04–1.14	65–71	Rock salt	2.18	136	oak	0.60–0.90	37–56
Cork	0.22–0.26	14–16	Rubber, hard	1.19	74	pear	0.61–0.73	38–45
Cork linoleum	0.54	34	Rubber, soft			pine, pitch	0.83–0.85	52–53
Corundum	3.9–4.0	245–250	commercial	1.1	69	white	0.35–0.50	22–31
Diamond	3.01–3.52	188–220	pure gum	0.91–0.93	57–58	yellow	0.37–0.60	23–37
Dolomite	2.84	177	Sandstone	2.14–2.36	134–147	plum	0.66–0.78	41–49
Ebonite	1.15	72	Serpentine	2.50–2.65	156–165	poplar	0.35–0.5	22–31
Emery	4.0	250	Silica, fused trans-			satinwood	0.95	59
Epidote	3.25–3.50	203–218	parent	2.21	138	spruce	0.48–0.70	30–44
Feldspar	2.55–2.75	159–172	translucent	2.07	129	sycamore	0.40–0.60	24–37
Flint	2.63	164	Slag	2.0–3.9	125–240	teak, Indian	0.66–0.88	41–55
Fluorite	3.18	198	Slate	2.6–3.3	162–205	African	0.98	61
Galena	7.3–7.6	460–470	Soapstone	2.6–2.8	162–175	walnut	0.64–0.70	40–43
Gamboge	1.2	75	Spermacéti	0.95	59	water gum	1.00	62
Garnet	3.15–4.3	197–268	Starch	1.53	95	willow	0.40–0.60	24–37
Gas carbon	1.88	117	Sugar	1.59	99			
Gelatin	1.27	79	Talc	2.7–2.8	168–174			

* Some values reported as low as 1.6.

15-28

Section 16
Health and Safety Information

HANDLING AND DISPOSAL OF CHEMICALS IN LABORATORIES

Robert Joyce and Blaine C. McKusick

The following material has been extracted from two books prepared under the auspices of the Committee on Hazardous Substances in the Laboratory of the National Academy of Sciences — National Research Council. Readers are referred to these books for full details:

Prudent Practices for Handling Hazardous Chemicals in Laboratories, National Academy Press, Washington, 1981.
Prudent Practices for Disposal of Chemicals from Laboratories, National Academy Press, Washington, 1983.

The permission of the National Academy Press to use these extracts is gratefully acknowledged.

INCOMPATIBLE CHEMICALS

The term "incompatible chemicals" refers to chemicals that can react with each other

- Violently
- With evolution of substantial heat
- To produce flammable products
- To produce toxic products

Good laboratory safety practice requires that incompatible chemicals be stored, transported, and disposed of in ways that will prevent their coming together in the event of an accident. Tables 1 and 2 give some basic guidelines for the safe handling of acids, bases, reactive metals, and other chemicals. Neither of these tables is exhaustive, and additional information on incompatible chemicals can be found in the following references.

1. L. Bretherick, *Handbook of Reactive Chemical Hazards*, 3rd ed., Butterworths, London–Boston, 1985.
2. L. Bretherick, Ed., *Hazards in the Chemical Laboratory*, 3rd ed., Royal Society of Chemistry, London, 1981.
3. *Manual of Hazardous Chemical Reactions, A Compilation of Chemical Reactions Reported to be Potentially Hazardous*, National Fire Protection Association, NFPA 491M, 1975, NFPA, 470 Atlantic Avenue, Boston, MA 02210.

TABLE 1
General Classes of Incompatible Chemicals

A	B
Acids	Bases, reactive metals
Oxidizing agents[a]	Reducing agents[a]
Chlorates	Ammonia, anhydrous and aqueous
Chromates	Carbon
Chromium trioxide	Metals
Dichromates	Metal hydrides
Halogens	Nitrites
Halogenating agents	Organic compounds
Hydrogen peroxide	Phosphorus
Nitric acid	Silicon
Nitrates	Sulfur
Perchlorates	
Peroxides	
Permanganates	
Persulfates	

[a] The examples of oxidizing and reducing agents are illustrative of common laboratory chemicals; they are not intended to be exhaustive.

TABLE 2
Examples of Incompatible Chemicals

Chemical	Is incompatible with
Acetic acid	Chromic acid, nitric acid, hydroxyl compounds, ethylene glycol, perchloric acid, peroxides, permanaganates
Acetylene	Chlorine, bromine, copper, fluorine, silver, mercury
Acetone	Concentrated nitric and sulfuric acid mixtures
Alkali and alkaline earth metals (such as powdered aluminum or magnesium, calcium, lithium, sodium, potassium)	Water, carbon tetrachloride or other chlorinated hydrocarbons, carbon dioxide, halogens
Ammonia (anhydrous)	Mercury (in manometers, for example), chlorine, calcium hypochlorite, iodine, bromine, hydrofluoric acid (anhydrous)
Ammonium nitrate	Acids, powdered metals, flammable liquids, chlorates, nitrites, sulfur, finely divided organic or combustible materials
Aniline	Nitric acid, hydrogen peroxide
Arsenical materials	Any reducing agent
Azides	Acids
Bromine	See Chlorine
Calcium oxide	Water
Carbon (activated)	Calcium hypochlorite, all oxidizing agents
Carbon tetrachloride	Sodium
Chlorates	Ammonium salts, acids, powdered metals, sulfur, finely divided organic or combustible materials
Chromic acid and chromium troioxide	Acetic acid, naphthalene, camphor, glycerol, alcohol, flammable liquids in general
Chlorine	Ammonia, acetylene, butadiene, butane, methane, propane (or other petroleum gases), hydrogen, sodium carbide, benzene, finely divided metals, turpentine
Chlorine dioxide	Ammonia, methane, phosphine, hydrogen sulfide
Copper	Acetylene, hydrogen peroxide
Cumene hydroperoxide	Acids (organic or inorganic)
Cyanides	Acids
Flammable liquids	Ammonium nitrate, chromic acid, hydrogen peroxide, nitric acid, sodium peroxide, halogens
Fluorine	Everything
Hydrocarbons (such as butane, propane, benzene)	Fluorine, chlorine, bromine, chromic acid, sodium peroxide
Hydrocyanic acid	Nitric acid, alkali
Hydrofluoric acid (anhydrous)	Ammonia (aqueous or anhydrous)
Hydrogen peroxide	Copper, chromium, iron, most metals or their salts, alcohols, acetone, organic materials, aniline, nitromethane, combustible matierals
Hydrogen sulfide	Fuming nitric acid, oxidizing gases
Hypochlorites	Acids, activated carbon
Iodine	Acetylene, ammonia (aqueous or anhydrous), hydrogen
Mercury	Acetylene, fulminic acid, ammonia
Nitrates	Sulfuric acid
Nitric acid (concentrated)	Acetic acid, aniline, chromic acid, hydrocyanic acid, hydrogen sulfide, flammable liquids, flammable gases, copper, brass, any heavy metals
Nitrites	Acids
Nitroparaffins	Inorganic bases, amines
Oxalic acid	Silver, mercury

TABLE 2
Examples of Incompatible Chemicals (continued)

Chemical	Is incompatible with
Oxygen	Oils, grease, hydrogen, flammable liquids, solids, or gases
Perchloric acid	Acetic anhydride, bismuth and its alloys, alcohol, paper, wood, grease, oils
Peroxides, organic	Acids (organic or mineral), avoid friction, store cold
Phosphorus (white)	Air, oxygen, alkalis, reducing agents
Potassium	Carbon tetrachloride, carbon dioxide, water
Potassium chlorate	Sulfuric and other acids
Potassium perchlorate (see also chlorates)	Sulfuric and other acids
Potassium permanganate	Glycerol, ethylene glycol, benzaldehyde, surfuric acid
Selenides	Reducing agents
Silver	Acetylene, oxalic acid, tartartic acid, ammonium compounds, fulminic acid
Sodium	Carbon tetrachloride, carbon dioxide, water
Sodium nitrite	Ammonium nitrate and other ammonium salts
Sodium peroxide	Ethyl or methyl alcohol, glacial acetic acid, acetic anhydride, benzaldehyde, carbon disulfide, glycerin, ethylene glycol, ethyl acetate, methyl acetate, furfural
Sulfides	Acids
Sulfuric acid	Potassium chlorate, potassium perchlorate, potassium permanganate (similar compounds of light metals, such as sodium, lithium)
Tellurides	Reducing agents

EXPLOSION HAZARDS

Table 3 lists some common classes of laboratory chemicals that have potential for producing a violent explosion when subjected to shock or friction. These chemicals should never be disposed of as such, but should be handled by procedures given in *Prudent Practices for Disposal of Chemicals from Laboratories*, National Academy Press, 1983, chapters 6 and 7. Additional information on these, as well as on some less common classes of explosives, can be found in L. Bretherick, *Handbook of Reactive Chemical Hazards,* 3rd ed., Butterworths, London–Boston, 1985.

Table 4 lists some illustrative combinations of common laboratory reagents that can produce explosions when they are brought together or that form reaction products that can explode without any apparent external initiating action. This list is not exhaustive, and additional information on potentially explosive reagent combinations can be found in Manual of Hazardous Chemical Reactions, A Compilation of Chemical Reactions Reported to be Potentially Hazardous, National Fire Protection Association, NFPA 491M, 1975, NFPA, 470 Atlantic Avenue, Boston, MA 02210.

WATER-REACTIVE CHEMICALS

Table 5 lists some common laboratory chemicals that react violently with water and that should always be stored and handled so that they do not come into contact with liquid water or water vapor. Procedures for decomposing laboratory quantities are given in Prudent Practices for Disposal of Chemicals from Laboratories, chapter 6; the pertinent section of that chapter is given in parentheses.

PYROPHORIC CHEMICALS

Many members of the classes of readily oxidized, common laboratory chemicals listed in Table 6 ignite spontaneously in air. A more extensive list can be found in L. Bretherick, *Handbook of Reactive Chemical Hazards*, 3rd ed., Butterworths, London-Boston, 1985. Pyrophoric chemicals should be stored in tightly closed containers under an inert atmosphere (or, for some, an inert liquid), and all transfers and manipulations of them must be carried out under an inert atmosphere or liquid. Suggested procedures for decomposing them are given in Prudent Practices for Disposal of Chemicals from Laboratories, chapter 6; the pertinent section of that chapter is given in parentheses.

TABLE 3
Shock–Sensitive Compounds

Acetylenic compounds, especially polyacetylenes, haloacetylenes, and heavy metal salts of acetylenes (copper, silver, and mercury salts are particularly sensitive)

Acyl nitrates

Alkyl nitrates, particularly polyol nitrates such as nitrocellulose and nitroglycerine

Alkyl and acyl nitrites

Alkyl perchlorates

Amminemetal oxosalts: metal compounds with coordinated ammonia, hydrazine, or similar nitrogenous donors and ionic perchlorate, nitrate, permanganate, or other oxidizing group

Azides, including metal, nonmetal, and organic azides

Chlorite salts of metals, such as $AgClO_2$ and $Hg(ClO_2)_2$

Diazo compounds such as CH_2N_2

Diazonium slats, when dry

Fulminates (silver fulminate, AgCNO, can form in the reaction mixture from the Tollens' test for aldehydes if it is allowed to stand for some time; this can be prevented by adding dilute nitric acid to the test mixture as soon as the test has been completed)

Hydrogen peroxide becomes increasingly treacherous as the concentration rises above 30%, forming explosive mixtures with organic materials and decomposing violently in the presence of traces of transition metals

N–Halogen compounds such as difluoroamino compounds and halogen azides

N–Nitro compounds such as N–nitromethylamine, nitrourea, nitroguanidine, and nitric amide

Oxo salts of nitrogenous bases: perchlorates, dichromates, nitrates, iodates, chlorites, chlorates, and permanganates of ammonia, amines, hydroxylamine, guanidine, etc.

Perchlorate salts. Most metal, nonmetal, and amine perchlorates can be detonated and may undergo violent reaction in contact with combustible materials

Peroxides and hydroperoxides, organic (see Chapter 6, Section II.P)

Peroxides (solid) that crystallize from or are left from evaporation of peroxidizable solvents (see Chapter 6 and Appendix I)

Peroxides, transition–metal salts

Picrates, especially salts of transition and heavy metals, such as Ni, Pb, Hg, Cu, and Zn; picric acid is explosive but is less sensitive to shock or friction than its metal salts and is relatively safe as a water–wet paste (see Chapter 7)

Polynitroalkyl compounds such as tetranitromethane and dinitroacetonitrile

Polynitroaromatic compounds, especially polynitro hydrocarbons, phenols, and amines

TABLE 4
Potentially Explosive Combinations of Some Common Reagents

Acetone + chloroform in the presence of base

Acetylene + copper, silver, mercury, or their salts

Ammonia (including aqueous solutions) + Cl_2, Br_2, or I_2

Carbon disulfide + sodium azide

Chlorine + an alcohol

Chloroform or carbon tetrachloride + powdered Al or Mg

Decolorizing carbon + an oxidizing agent

Diethyl ether + chlorine (including a chlorine atmosphere)

Dimethyl sulfoxide + an acyl halide, $SOCl_2$ or $POCl_3$

Dimethyl sulfoxide + CrO_3

Ethanol + calcium hypochlorite

Ethanol + silver nitrate

Nitric acid + acetic anhydride or acetic acid

Picric acid + a heavy–metal salt, such as of Pb, Hg, or Ag

Silver oxide + ammonia + ethanol

Sodium + a cholrinated hydrocarbon

Sodium hypochlorite + an amine

TABLE 5
Water–Reactive Chemicals

Alkali metals (III.D)
Alkali metal hydrides (III.C.2)
Alkali metal amides (III.C.7)
Metal alkyls, such as lithium alkyls and aluminum alkyls (IV.A)
Grignard reagents (IV.A)
Halides of nonmetals, such as BCl_3, BF_3, PCl_3, PCl_5, $SiCl_4$, S_2Cl_2 (III.F)
Inorganic acid halides, such as $POCl_3$, $SOCl_2$, SO_2Cl_2 (III.F)
Anhydrous metal halides, such as $AlCl_3$, $TiCl_4$, $ZrCl_4$, $SnCl_4$ (III.E)
Phosphorus pentoxide (III.I)
Calcium carbide (IV.E)
Organic acid halides and anhydrides of low molecular weight (II.J)

TABLE 6
Classes of Pyrophoric Chemicals

Grignard reagents, RMgX (IV.A)
Metal alkyls and aryls, such as RLi, RNa, R_3Al, R_2Zn (IV.A)
Metal carbonyls, such as Ni $(CO)_4$, $Fe(CO)_5$, $Co_2(CO)_8$ (IV.B)
Alkali metals such as Na, K (III.D.l)
Metal powders, such as Al, Co, Fe, Mg, Mn, Pd, Pt, Ti, Sn, Zn, Zr (III.D.2)
Metal hydrides, such as NaH, $LiAlH_4$ (IV.C.2)
Nonmetal hydrides, such as B_2H_6 and other boranes, PH_3, AsH_3 (III.G)
Nonmetal alkyls, such as R_3B, R_3P, R_3As (IV.C)
Phosphorus (white) (III.H)

HAZARDS FROM PEROXIDE FORMATION

Many common laboratory chemicals can form peroxides when allowed access to air over a period of time. A single opening of a container to remove some of the contents can introduce enough air for peroxide formation to occur. Some types of compounds form peroxides that are treacherously and violently explosive in concentrated solution or as solids. Accordingly, peroxide–containing liquids should never be evaporated near to or to dryness. Peroxide formation can also occur in many polymerizable unsaturated compounds, and these peroxides can initiate a runaway, sometimes explosive, polymerization reaction. Procedures for testing for peroxides and for removing small amounts from laboratory chemicals are given in Prudent Practices for Disposal of Chemicals from Laboratories, chapter 6, Section II.P.

Table 7 provides a list of structural characteristics in organic compounds that can peroxidize. These structures are listed in approximate order of decreasing hazard. Reports of serious incidents involving the last five structural types are extremely rare, but these structures are listed because laboratory workers should be aware that they can form peroxides that can influence the course of experiments in which they are used.

Table 8 gives examples of common laboratory chemicals that are prone to form peroxides on exposure to air. The lists are not exhaustive, and analogous organic compounds that have any of the structural features given in Table 7 should be tested for peroxides before being used as solvents or reagents, or before being distilled. The recommended retention times begin with the date of synthesis or of opening the original container.

DISPOSAL OF TOXIC CHEMICALS

It is often desirable to precipitate toxic cations or hazardous anions from solution to facilitate recovery or disposal. Table 9 lists precipitants for many common cations, and Table 10 gives precipitants for some hazardous anions. Many cations can be precipitated as sulfides by adding sodium sulfide solution (preferable to the highly toxic hydrogen sulfide) to a neutral solution of the cation (Table 11). Control of pH is important because some sulfides will redissolve in excess sulfide ion. After precipitation, excess sulfide can be destroyed by addition of hypochlorite.

Most metal cations are precipitated as hydroxides or oxides at high pH. Since many of these precipitates will redissolve in excess base, it is often necessary to control pH. Table 12 shows the recommended pH range for precipitating many cations in their most common oxidation state. The notation "1 N" in the right–hand column indicates that the precipitate will not dissolve in 1 N sodium hydroxide (pH 14).

The distinctions between high and low toxicity or hazard are based on toxicological and other data, and are relative. There is no implication of a sharp distinction between high and low, or that any cations or anions are totally without hazard.

TABLE 7
Types of Chemicals That Are Prone to Form Peroxides

A. Organic structures (in approximate order of decreasing hazard)

1.
$$\begin{array}{c} H \\ | \\ \diagdown C-O- \\ \diagup \end{array}$$
Ethers and acetals with a hydrogen atoms

2.
$$\begin{array}{c} \qquad H \\ \qquad | \quad | \\ \diagdown C=C-C \diagup \\ \diagup \qquad \diagdown \end{array}$$
Olefins with allylic hydrogen atoms

3.
$$\begin{array}{c} X \\ | \\ \diagdown C=C- \\ \diagup \end{array}$$
Chloroolefins and fluoroolefins

4. $CH_2=C\diagdown$ Vinyl halides, esters, and ethers

5.
$$\begin{array}{c} \diagdown \quad | \quad | \quad \diagup \\ C=C-C=C \\ \diagup \qquad \diagdown \end{array}$$
Dienes

6.
$$\begin{array}{c} H \\ | \\ \diagdown C=C-C\equiv CH \\ \diagup \end{array}$$
Vinylacetylenes with α hydrogen atoms

7.
$$\begin{array}{c} H \\ \diagdown | \\ C-C\equiv CH \\ \diagup \end{array}$$
Alkylacetylenes with α hydrogen atoms

8.
$$\begin{array}{c} H \\ \diagdown | \\ C-Ar \\ \diagup \end{array}$$
Alkylarenes that contain tertiary hydrogen atoms

9.
$$\begin{array}{c} | \\ -C-H \\ | \end{array}$$
Alkanes and cycloalkanes that contain tertiary hydrogen atoms

10.
$$\begin{array}{c} \diagdown \\ C=C-CO_2R \\ \diagup \end{array}$$
Acrylates and methacrylates

11.
$$\begin{array}{c} H \\ \diagdown | \\ C-OH \\ \diagup \end{array}$$
Secondary alcohols

12.
$$\begin{array}{c} O \quad H \\ \| \quad | \diagup \\ -C-C \\ \diagdown \end{array}$$
Ketones that contain a hydrogen atoms

13.
$$\begin{array}{c} H \\ | \\ -C=O \end{array}$$
Aldehydes

14.
$$\begin{array}{c} O \quad H \quad H \\ \| \quad | \quad | \diagup \\ -C-N-C \\ \diagdown \end{array}$$
Ureas, amides, and lactams that have a hydrogen atom on a carbon atom attached to nitrogen

TABLE 7
Types of Chemicals That Are Prone to Form Peroxides (continued)

B. Inorganic substances

1. Alkali metals, especially potassium, rubidium, and cesium (see Chapter 6, Section III.D)
2. Metal amides (see Chapter 6, Section III.C.7)
3. Organometallic compounds with a metal atom bonded to carbon (see Chapter 6, Section IV)
4. Metal alkoxides

TABLE 8
Common Peroxide-Forming Chemicals

LIST A
Severe Peroxide Hazard on Storage with Exposure to Air
Discard within 3 months

- Diisopropyl ether (isopropyl ether)
- Divinylacetylene (DVA)[a]
- Potassium metal
- Potassium amide
- Sodium amide (sodamide)
- Vinylidene chloride (1,1–dichloro-ethylene)[a]

LIST B
Peroxide Hazard on Concentration; Do Not Distill or Evaporate Without First Testing for the Presence of Peroxides

Discard or test for peroxides after 6 months

- Acetaldehyde diethyl acetal (acetal)
- Cumene (isopropylbenzene)
- Cyclohexene
- Cyclopentene
- Decalin (decahydronaphthalene)
- Diacetylene
- Dicyclopentadiene
- Diethyl ether (ether)
- Diethylene glycol dimethyl ether (diglyme)
- Dioxane
- Ethylene glycol dimethyl ether (glyme)
- Ethylene glycol ether acetates
- Ethylene glycol monoethers (cello–solves)
- Furan
- Methylacetylene
- Methylcyclopentane
- Methyl isobutyl ketone
- Tetrahydrofuran (THF)
- Tetralin (tetrahydronaphthalene)
- Vinyl ethers[a]

LIST C
Hazard of Rapid Polymerization Initiated by Internally Formed Peroxides[a]
a. Normal Liquids; Discard or test for peroxides after 6 months[b]

- Chloroprene (2–chloro–1,3–buta–diene)[c]
- Styrene
- Vinyl acetate
- Vinylpyridine

b. Normal Gases; Discard after 12 months[d]

- Butadiene[c]
- Tetrafluoroethylene (TFE)[c]
- Vinylacetylene (MVA)[c]
- Vinyl chloride

[a] Polymerizable monomers should be stored with a polymerization inhibitor from which the monomer can be separated by distillation just before use.
[b] Although common acrylic monomers such as acrylonitrile, acrylic acid, ethyl acrylate, and methyl methacrylate can form peroxides, they have not been reported to develop hazardous levels in normal use and storage.
[c] The hazard from peroxides in these compounds is substantially greater when they are stored in the liquid phase, and if so stored without an inhibitor they should be considered as in LIST A.

TABLE 8
Common Peroxide-Forming Chemicals (continued)

d   Although air will not enter a gas cylinder in which gases are stored under pressure, these gases are sometimes transferred from the original cylinder to another in the laboratory, and it is difficult to be sure that there is no residual air in the receiving cylinder. An inhibitor should be put into any such secondary cylinder before one of these gases is transferred into it; the supplier can suggest inhibitors to be used. The hazard posed by these gases is much greater if there is a liquid phase in such a secondary container, and even inhibited gases that have been put into a secondary container under conditions that create a liquid phase should be discarded within 12 months.

Note: Laboratory workers should label all containers of peroxidizable solvents or reagents with one of the following:

[LIST A]

Peroxidizable compound
Received   Opened
Date   _____   _____
Discard 3 months after opening

[LISTS B AND C]

Peroxidizable compound
Received   Opened
Date   _____   _____
Discard or test for peroxides
6 months after opening

TABLE 9
Relative Toxicity of Cations

High toxic hazard	Precipitant[a]	Low toxic hazard	Precipitant[a]
Antimony	OH^-, S^{2-}	Aluminum	OH^-
Arsenic	S^{2-}	Bismuth	OH^-, S^{2-}
Barium	SO_4^{2-}, CO_3^{2-}	Calcium	SO_4^{2-}, CO_3^{2-}
Beryllium	OH^-	Cerium	OH^-
Cadmium	OH^-, S^{2-}	Cesium	
Chromium (III)[b]	OH^-	Copper[c]	OH^-, S^{2-}
Cobalt (II)[b]	OH^-, S^{2-}	Gold	OH^-, S^{2-}
Gallium	OH^-	Iron[c]	OH^-, S^{2-}
Germanium	OH^-, S^{2-}	Lanthanides	OH^-
Hafnium	OH^-	Lithium	
Indium	OH^-, S^{2-}	Magnesium	OH^-
Iridium	OH^-, S^{2-}	Molybdenum (VI)[b,d]	
Lead	OH^-, S^{2-}	Niobium (V)	OH^-
Manganese (II)[b]	OH^-, S^{2-}	Palladium	OH^-, S^{2-}
Mercury	OH^-, S^{2-}	Potassium	
Nickel	OH^-, S^{2-}	Rubidium	
Osmium (IV)[b,e]	OH^-, S^{2-}	Scandium	OH^-
Platinum (II)[b]	OH^-, S^{2-}	Sodium	
Rhenium (VII)[b]	S^{2-}	Strontium	SO_4^{2-}, CO_3^{2-}
Rhodium (III)[b]	OH^-, S^{2-}	Tantalum	OH^-
Ruthenium (III)[b]	OH^-, S^{2-}	Tin	OH^-, S^{2-}
Selenium	S^{2-}	Titanium	OH^-
Silver	Cl^-, OH^-, S^{2-}	Yttrium	OH^-
Tellurium	S^{2-}	Zinc[c]	OH^-, S^{2-}
Thallium	OH^-, S^{2-}	Zirconium	OH^-
Tungsten (VI)[b,d]			
Vanadium	OH^-, S^{2-}		

TABLE 9
Relative Toxicity of Cations (continued)

[a] Precipitants are listed in order of preference:
  OH^- = base (sodium hydroxide or sodium carbonate)
  S^{2-} = sulfide
  Cl^- = chloride
  SO_4^{2-} = sulfate
  CO_3^{2-} = carbonate
[b] The precipitant is for the indicated valence state.
[c] Maximum tolerance levels have been set for these low–toxicity ions by the U.S. Public Health Service, and large amounts should not be put into public sewer systems. The small amounts typically used in laboratories will not normally affect water supplies.
[d] These ions are best precipitated as calcium molybdate or calcium tungstate.
[e] CAUTION: OsO_4, a volatile, extremely poisonous substance, is formed from almost any osmium compound under acid conditions in the presence of air.

TABLE 10
Relative Hazard of Anions

High–hazard anions			
Ion	Hazard type[a]	Precipitant	Low-hazard anions
Aluminum hydride, AlH_4	F	—	Bisulfite, HSO_3^-
Amide, NH_2	F, E[b]	—	Borate, BO_3^{3-}, $B_4O_7^{2-}$
Arsenate, AsO_3^-, AsO_4^{3-}	T	Cu^{2+}, Fe^{2+}	Bromide, Br^-
Arsenite, AsO_2^-, AsO_3^{3-}	T	Pb^{2+}	Carbonate, CO_3^{2-}
Azide, N_3^-	E, T	—	Chloride, Cl^-
Borohydride, BH_4^-	F	—	Cyanate, OCN^-
Bromate, BrO_3	O, E	—	Hydroxide, OH^-
Chlorate, ClO_3^-	O, E	—	Iodide, I^-
Chromate, CrO_4^{2-}, $Cr_2O_7^{2-}$	T, O	[c]	Oxide, O^{2-}
Cyanide, CN^-	T	—	Phosphate, PO_4^{3-}
Ferricyanide, $Fe(CN)_6^{3-}$	T	Fe^{2+}	Sulfate, SO_4^{2-}
Ferrocyanide, $Fe(CN)_6^{4-}$	T	Fe^{3+}	Sulfite, SO_3^{2-}
Fluoride, F^-	T	Ca^{2+}	Thiocyanate, SCN^-
Hydride, H^-	F	—	
Hydroperoxide, O_2H^-	O, E	—	
Hydrosulfide, SH^-	T	—	
Hypochlorite, OCl^-	O	—	
Iodate, IO_3^-	O, E	—	
Nitrate, NO_3^-	O	—	
Nitrite, NO_2^-	T, O	—	
Perchlorate, ClO_4^-	O, E	—	
Permanganate, MnO_4^-	T, O	[d]	
Peroxide, O_2^{2-}	O, E	—	
Persulfate, $S_2O_8^{2-}$	O	—	
Selenate, SeO_4^{2-}	T	Pb^{2+}	
Selenide, Se^{2-}	T	Cu^{2+}	
Sulfide, S^{2-}	T	[e]	

[a] Toxic, T: oxidant, O; flammable, F; explosive, E.
[b] Metal amides readily form explosive peroxides on exposure to air.
[c] Reduce and precipitate as Cr(III); see Table 9.
[d] Reduce and precipitate as Mn(II); see Table 9.
[e] See Table 11.

TABLE 11
Precipitation of Sulfides

Precipitated at pH 7	Not precipitated at low pH	Forms a soluble complex at high pH
Ag^+		
As^{3+a}		X
Au^{+a}		X
Bi^{3+}		
Cd^{2+}		
Co^{2+}	X	
Cr^{3+a}		
Cu^{2+}		
Fe^{2+a}	X	
Ge^{2+}		X
Hg^{2+}		X
In^{3+}	X	
Ir^{4+}		X
Mn^{2+a}	X	
Mo^{3+}		X
Ni^{2+}	X	
Os^{4+}		
Pb^{2+}		
Pd^{2+a}		
Pt^{2+a}		X
Re^{4+}		
Rh^{2+a}		
Ru^{4+}		
Sb^{3+a}		X
Se^{2+}		X
Sn^{2+}		X
Te^{4+}		X
Tl^{+a}	X	
V^{4+a}		
Zn^{2+}	X	

[a] Higher oxidation states of this ion are reduced by sulfide ion and precipitated as this sulfide.

TABLE 12
pH Range for Precipitation of Metal Hydroxides and Oxides

	1	2	3	4	5	6	7	8	9	10	
Ag^{1+}									⊢──→		1 N
Al^{3+}							⊢──┤				
As^{3+}	Not precipitated (precipitate as sulfide)										
As^{5+}	Not precipitated (precipitate as sulfide)										
Au^{3+}							⊢──┤				
Be^{2+}							⊢──┤				
Bi^{3+}							⊢──────→				1 N
Cd^{2+}							⊢──────→				1 N
Co^{2+}								⊢─────→			1 N
Cr^{3+}							⊢──────→				1 N
Cu^{1+}								⊢────→			1 N

TABLE 12
pH Range for Precipitation of Metal
Hydroxides and Oxides (continued)

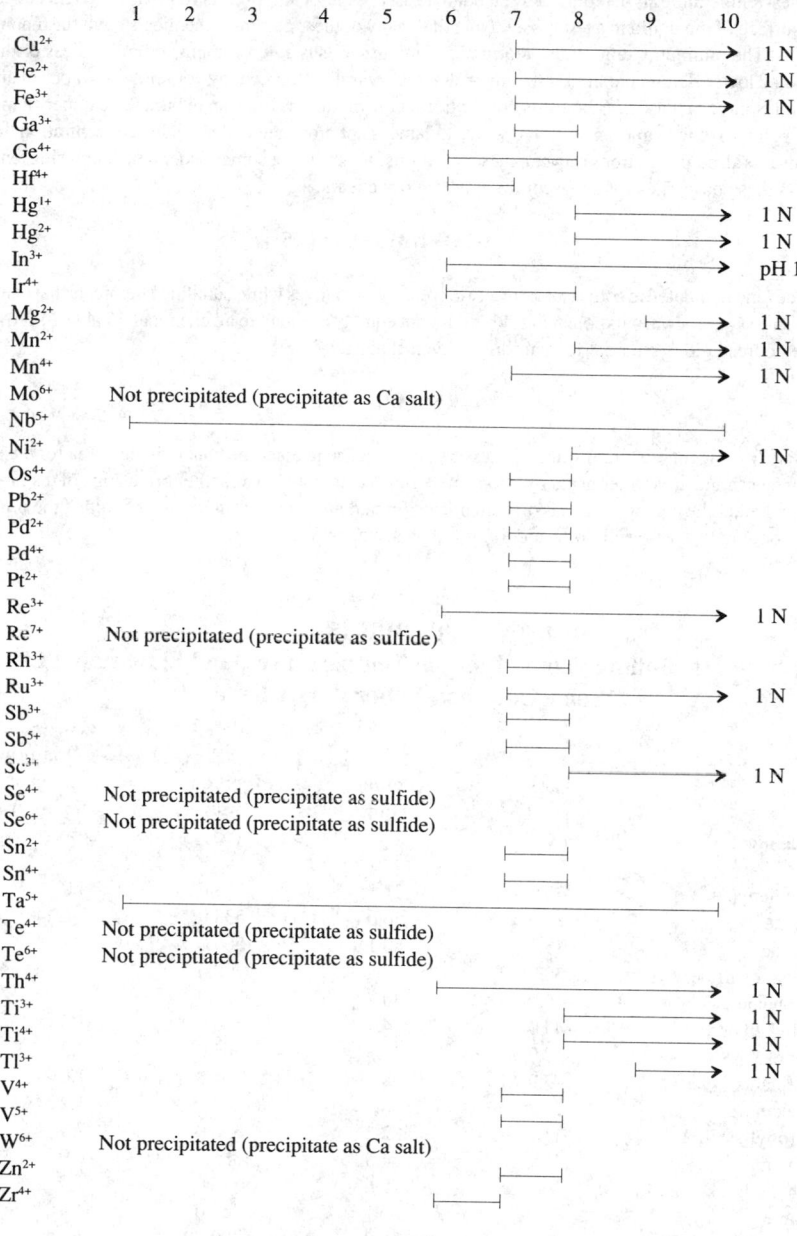

	1	2	3	4	5	6	7	8	9	10	
Cu^{2+}											1 N
Fe^{2+}											1 N
Fe^{3+}											1 N
Ga^{3+}											
Ge^{4+}											
Hf^{4+}											
Hg^{1+}											1 N
Hg^{2+}											1 N
In^{3+}											pH 13
Ir^{4+}											
Mg^{2+}											1 N
Mn^{2+}											1 N
Mn^{4+}											1 N
Mo^{6+}	Not precipitated (precipitate as Ca salt)										
Nb^{5+}											
Ni^{2+}											1 N
Os^{4+}											
Pb^{2+}											
Pd^{2+}											
Pd^{4+}											
Pt^{2+}											
Re^{3+}											1 N
Re^{7+}	Not precipitated (precipitate as sulfide)										
Rh^{3+}											
Ru^{3+}											1 N
Sb^{3+}											
Sb^{5+}											
Sc^{3+}											1 N
Se^{4+}	Not precipitated (precipitate as sulfide)										
Se^{6+}	Not precipitated (precipitate as sulfide)										
Sn^{2+}											
Sn^{4+}											
Ta^{5+}											
Te^{4+}	Not precipitated (precipitate as sulfide)										
Te^{6+}	Not preciptiated (precipitate as sulfide)										
Th^{4+}											1 N
Ti^{3+}											1 N
Ti^{4+}											1 N
Tl^{3+}											1 N
V^{4+}											
V^{5+}											
W^{6+}	Not precipitated (precipitate as Ca salt)										
Zn^{2+}											
Zr^{4+}											

REFERENCES

L. Erdey, *Gravimetric Analysis,* Part II, Pergamon Press, New York, 1965.

D. T. Burns, A. Towsend, and A. H. Carter, *Inorganic Reaction Chemistry,* Vol. 2, Ellis Horwood, New York, 1981.

HANDLING AND DISPOSAL OF CHEMICALS IN LABORATORIES (continued)

FIRE HAZARDS

Flammable solvents are a common source of laboratory fires. The relative ease with which some common laboratory solvents can be ignited is indicated by the following properties.

Flash Point — The lowest temperature, as determined by standard tests, at which a liquid emits vapor in sufficient concentration to form an ignitable mixture with air near the surface of the liquid in a test vessel. Note that many of these common chemicals have flash points below room temperature.

Ignition Temperature — The minimum temperature required to initiate self–sustained combustion, regardless of the heat source.

Flammable Limits — The lower flammable limit is the minimum concentration (percent by volume) of a vapor in air below which a flame is not propagated when an ignition source is present. Below this concentration the mixture is too lean to burn. The upper flammable limit is the maximum concentration (percent by volume) of the vapor in air above which a flame is not propagated. Above this concentration the mixture is too rich to burn. The flammable range comprises all concentrations between these two limits. This range becomes wider with increasing temperature and in oxygen–rich atmospheres. Table 13 lists these properties for a few common laboratory chemicals.

GLOVE MATERIALS

It is good safety practice (and mandated in some laboratories) to wear rubber gloves while handling chemicals that can cause injury when in contact with, or absorbed through, the skin. The various common rubbers are not equally resistant to all chemicals. Table 14 provides guidelines for selecting the best, and avoiding the poorest, glove material for handling a given chemical.

RESPIRATORS

In the event of a laboratory accident or spill, it will be necessary for someone to enter the contaminated area for cleanup. If significant quantities of a chemical are spilled, or even minor quantities of a known toxic material, it is essential to wear the correct kind of respirator equipment when entering the area. If it is not known whether the contamination is of a chemical "Immediately dangerous to life or health", the prudent course is to assume that it is, and to use the corresponding type of respirator. Guidelines are presented in Table 15.

TABLE 13
Flash Points, Boiling Points, Ignition Temperatures, and Flammable Limits of Some Common Laboratory Chemicals

Chemical	Flash point (°C)	Boiling point (°C)	Ignition temp. (°C)	Flammable limit (percent by volume in air)	
				Lower	Upper
Acetaldehyde	−37.8	21.1	175.0	4.0	60.0
Acetone	−19.0	56.0	538.0	2.6	12.8
Benzene	−11.1	80.1	560.0	1.4	8.0
Carbon disulfide	−30.0	45.8	90.0	1.0	44.0
Cyclohexane	−18.0	80.7	260.0	1.3	8.0
Diethyl ether	−45.0	34.4	160.0	1.8	48.0
Ethanol	12.0	78.3	363.0	3.3	19.0
n–Heptane	−3.9	98.4	204.0	1.0	6.7
n–Hexane	−21.7	68.7	223.0	1.2	7.5
Isopropyl alcohol	11.7	82.2	398.9	2.0	12.0
Methanol	11.1	64.5	385.0	6.0	36.5
Methyl ethyl ketone	−6.1	79.6	515.6	1.9	11.0
Pentane	−40.0	36.1	260.0	1.4	7.8
Styrene	31.0	145.0	490.0	1.1	6.1
Toluene	4.4	110.6	530.0	1.3	7.0
p–Xylene	25.0	132.4	529.0	1.1	7.0

Note: For a more extensive listing, see the table "Properties of Common Solvents" in Section 15.

TABLE 14
Resistance to Chemicals of Common Glove Materials
(E = Excellent, G = Good, F = Fair, P = Poor)

Chemical	Natural rubber	Neoprene	Nitrile	Vinyl
Acetaldehyde	G	G	E	G
Acetic acid	E	E	E	E
Acetone	G	G	G	F
Acrylonitrile	P	G	—	F
Ammonium hydroxide (sat)	G	E	E	E
Aniline	F	G	E	G
Benzaldehyde	F	F	E	G
Benzene[a]	P	F	G	F
Benzyl chloride[a]	F	P	G	P
Bromine	G	G	—	G
Butane	P	E	—	P
Butyraldehyde	P	G	—	G
Calcium hypochlorite	P	G	G	G
Carbon disulfide	P	P	G	F
Carbon tetrachloride[a]	P	F	G	F
Chlorine	G	G	—	G
Chloroacetone	F	E	—	P
Chloroform[a]	P	F	G	P
Chromic acid	P	F	F	E
Cyclohexane	F	E	—	P
Dibenzyl ether	F	G	—	P
Dibutyl phthalate	F	G	—	P
Diethanolamine	F	E	—	E
Diethyl ether	F	G	E	P
Dimethyl sulfoxide[b]	—	—	—	—
Ethyl acetate	F	G	G	F
Ethylene dichloride[a]	P	F	G	
Ethylene glycol	G	G	E	E
Ethylene trichloride[a]	P	P	—	P
Fluorine	G	G	—	G
Formaldehyde	G	E	E	E
Formic acid	G	E	E	E
Glycerol	G	G	E	E
Hexane	P	E	—	P
Hydrobromic acid (40%)	G	E	—	E
Hydrochloric acid (conc)	G	G	G	E
Hydrofluoric acid (30%)	G	G	G	E
Hydrogen peroxide	G	G	G	E
Iodine	G	G	—	G
Methylamine	G	G	E	E
Methyl cellosolve	F	E	—	E
Methyl chloride[a]	P	E	—	P
Methyl ethyl ketone	F	G	G	P
Methylene chloride[a]	F	F	G	F
Monoethanolamine	F	E	—	E
Morpholine	F	E	—	E
Naphthalene[a]	G	G	E	G
Nitric acid (conc)	P	P	P	G
Perchloric acid	F	G	F	E
Phenol	G	E	—	E
Phosphoric acid	G	E	—	E
Potassium hydroxide (sat)	G	G	G	E

TABLE 14
Resistance to Chemicals of Common Glove Materials
(E = Excellent, G = Good, F = Fair, P = Poor) (continued)

Chemical	Natural rubber	Neoprene	Nitrile	Vinyl
Propylene dichloride[a]	P	F	—	P
Sodium hydroxide	G	G	G	E
Sodium hypochlorite	G	P	F	G
Sulfuric acid (conc)	G	G	F	G
Toluene[a]	P	F	G	F
Trichloroethylene[a]	P	F	G	F
Tricresyl phosphate	P	F	—	F
Triethanolamine	F	E	E	E
Trinitrotoluene	P	E	—	P

[a] Aromatic and halogenated hydrocarbons will attack all types of natural and synthetic glove materials. Should swelling occur, the user should change to fresh gloves and allow the swollen gloves to dry and return to normal.

[b] No data on the resistance to dimethyl sulfoxide of natural rubber, neoprene, nitrile rubber, or vinyl materials are available; the manufacturer of the substance recommends the use of butyl rubber gloves.

TABLE 15
Guide for Selection of Respirators

Type of hazard	Type of respirator
Oxygen deficiency	Self–contained breathing apparatus Hose mask with blower Combination of air–line respirator and auxiliary self–contained air supply or air–storage receiver with alarm
Gas and vapor contaminants 　Immediately dangerous to 　　life or health	Self–contained breathing apparatus Hose mask with blower Air–purifying full–facepiece respirator with chemical canister (gas mask) Self–rescue mouthpiece respirator (for escape only) Combination of air–line respirator and auxiliary self–contained air supply or air–storage receiver with alarm
Not immediately dangerous 　　to life or health	Air–line respirator Hose mask with blower Air–purifying half–mask or mouthpiece respirator with chemical cartridge
Particulate Contaminants 　Immediately dangerous to 　　life or health	Self–contained breathing apparatus Hose mask with blower Air–purifying full–facepiece respirator with appropriate filter Self-rescue mouthpiece respirator (for escape only) Combination of air–line respirator and auxiliary self–contained air supply or air–storage receiver with alarm

TABLE 15
Guide for Selection of Respirators (continued)

Type of hazard	Type of respirator
Not immediately dangerous to life or health	Air–purifying half–mask or mouthpiece respirator with filter pad or cartridge
	Air–line respirator
	Air–line abrasive–blasting respirator
	Hose mask with blower
Combination of gas, vapor, and particulate contaminants	Self–contained breathing apparatus
Immediately dangerous to life or health	Hose mask with blower
	Air–purifying full–facepiece respirator with chem–ical canister and appropriate filter (gas mask with filter)
	Self–rescue mouthpiece respirator (for escape only)
	Combination of air–line respirator and auxiliary self–contained air supply or air–storage receiver with alarm
Not immediately dangerous to life or health	Air–line respirator
	Hose mask without blower
	Air–purifying half–mask or mouthpiece respirator with chemical cartridge and appropriate filter

Source: ANSI Standard Z88.2 (1969).

FLAMMABILITY OF CHEMICAL SUBSTANCES

This table gives properties related to the flammability of about 900 chemical substances. The properties listed are:

t_B : Normal boiling point in °C (at 101.325 kPa pressure).

FP: Flash point, which is the minimum temperature at which the vapor pressure of a liquid is sufficient to form an ignitable mixture with air near the surface of the liquid. Flash point is not an intrinsic physical property but depends on the conditions of measurement (see Reference 1).

Fl. Limits: Flammable limits (often called explosive limits), which specify the range of concentration of the vapor in air (in percent by volume) for which a flame can propagate. Below the lower flammable limit, the gas mixture is too lean to burn; above the upper flammable limit, the mixture is too rich. Values refer to ambient temperature and pressure and are dependent on the precise test conditions. A ? indicates that one of the limits is not known.

IT: Ignition temperature (sometimes called autoignition temperature), which is the minimum temperature required for self-sustained combustion in the absence of an external ignition source. As in the case of flash point, the value depends on specified test conditions.

Even in cases where very careful measurements of flash point have been replicated in several laboratories, observed values can differ by 3 to 6°C (Reference 4). For more typical measurements, larger uncertainties should be assumed in both flash points and autoignition temperatures. The absence of a flash point entry in this table does not mean that the substance is nonflammable, but only that no reliable value is available.

Compounds are listed by molecular formula following the Hill convention. Substances not containing carbon are listed first, followed by those that contain carbon. To locate an organic compound by name or CAS Registry Number when the molecular formula is not known, use the table "Physical Constants of Organic Compounds" in Section 3 and its indexes to determine the molecular formula.

REFERENCES

1. *Fire Protection Guide to Hazardous Materials, 11th Edition*, National Fire Protection Association, Quincy, MA, 1994.
2. Urben, P.G., Editor, *Bretherick's Handbook of Reactive Chemical Hazards, 5th Edition*, Butterworth-Heinemann, Oxford, 1995.
3. Daubert, T.E., Danner, R.P., Sibul, H.M., and Stebbins, C.C., *Physical and Thermodynamic Properties of Pure Compounds: Data Compilation*, extant 1994 (core with 4 supplements), Taylor & Francis, Bristol, PA.
4. *Report of Investigation: Flash Point Reference Materials*, National Institute of Standards and Technology, Standard Reference Materials Program, Gaithersburg, MD, 1995.

Mol. Form.	Name	t_B/°C	FP/°C	Fl. Limits	IT/°C
Compounds not containing carbon					
B_2H_6	Diborane	-92.4	-90	1-98%	≈40
B_5H_9	Pentaborane(9)	60	30	0.4-?	35
BrH_3Si	Bromosilane	1.9	<0		≈20
Br_3HSi	Tribromosilane	109			≈20
Cl_2H_2Si	Dichlorosilane	8.3		4.1-99%	36
Cl_3HSi	Trichlorosilane	33	-50		104
GeH_4	Germane	-88.1			≈20
Ge_2H_6	Digermane	30.8			≈50
H_2	Hydrogen	-252.8		4-74%	
H_2S	Hydrogen sulfide	-59.55		4-44%	260
H_2S_2	Hydrogen disulfide	70.7	<22		
H_2Te	Hydrogen telluride	-2			-50
H_3N	Ammonia	-33.33		16-25%	
H_3P	Phosphine	-87.75		1.8-?	
H_4N_2	Hydrazine	113.55	38	5-100%	
H_4P_2	Diphosphine	63.5			≈20
H_4Si	Silane	-111.9	-112	1.4-?	≈20
H_6Si_2	Disilane	-14.3	-14		≈20
H_8Si_3	Trisilane	52.9	<0		≈20
P	Phosphorus (white)	280.5			38
Compounds containing carbon					
CHN	Hydrogen cyanide	26	-18	6-40%	538
CH_2Cl_2	Dichloromethane	40		13-23%	556
CH_2N_2	Cyanamide		141		

Mol. Form	Name	t_B/°C	FP/°C	Fl Limits	AIT/°C
CH_2O	Formaldehyde	-19.1	85	7.0-73%	424
$(CH_2O)_x$	Paraformaldehyde		70	7.0-73%	300
CH_2O_2	Formic acid	101	50	18-57%	434
CH_3Br	Bromomethane	3.5		10-16%	537
CH_3Cl	Chloromethane	-24.0		8.1-17.4%	632
CH_3Cl_3Si	Methyltrichlorosilane	65.6	-9	7.6->20%	>404
CH_3NO	Formamide	220	154		
CH_3NO_2	Nitromethane	101.1	35	7.3-?	418
CH_4	Methane	-161.5		5.0-15.0%	537
CH_4Cl_2Si	Dichloromethylsilane	41	-9	6.0-55%	316
CH_4O	Methanol	64.6	11	6.0-36%	464
CH_4S	Methanethiol	5.9	-18	3.9-21.8%	
CH_5N	Methylamine	-6.3	0	4.9-20.7%	430
CH_6N_2	Methylhydrazine	87.5	-8	2.5-92%	194
CO	Carbon monoxide	-191.5		12.5-74%	609
COS	Carbon oxysulfide	-50		12-29%	
CS_2	Carbon disulfide	46	-30	1.3-50.0%	90
C_2ClF_3	Chlorotrifluoroethylene	-27.8		8.4-16.0%	
C_2F_4	Tetrafluoroethylene	-75.9		10.0-50.0%	200
C_2HCl_3	Trichloroethylene	87.2		8-10.5%	420
C_2HCl_3O	Dichloroacetyl chloride	108	66		
C_2H_2	Acetylene	-84.7		2.5-100%	305
$C_2H_2Cl_2$	1,1-Dichloroethylene	31.6	-28	6.5-15.5%	570
$C_2H_2Cl_2$	cis-1,2-Dichloroethylene	60.1	6	3-15%	460
$C_2H_2Cl_2$	trans-1,2-Dichloroethylene	48.7	2	6-13%	460
$C_2H_2F_2$	1,1-Difluoroethylene	-85.7		5.5-21.3%	
C_2H_3Br	Bromoethylene	15.8		9-15%	530
C_2H_3Cl	Chloroethylene	-13.3	-78	3.6-33.0%	472
$C_2H_3ClF_2$	1-Chloro-1,1-difluoroethane	-9.7		6-18%	632
C_2H_3ClO	Acetyl chloride	50.7	4		390
$C_2H_3Cl_2NO_2$	1,1-Dichloro-1-nitroethane	123.5	76		
$C_2H_3Cl_3$	1,1,1-Trichloroethane	74.0		8-10.5%	500
$C_2H_3Cl_3$	1,1,2-Trichloroethane	113.8	32	6-28%	460
$C_2H_3Cl_3Si$	Trichlorovinylsilane	91.5	21		
C_2H_3F	Fluoroethylene	-72		2.6-21.7%	
C_2H_3N	Acetonitrile	81.6	6	3.0-16.0%	524
C_2H_3NO	Methyl isocyanate	39.5	-7	5.3-26%	534
C_2H_4	Ethylene	-103.7		2.7-36%	450
$C_2H_4ClNO_2$	1-Chloro-1-nitroethane	124.5	56		
$C_2H_4Cl_2$	1,1-Dichloroethane	57.4	-17	5.4-11.4%	458
$C_2H_4Cl_2$	1,2-Dichloroethane	83.5	13	6.2-16%	413
C_2H_4O	Acetaldehyde	20.1	-39	4.0-60%	175
C_2H_4O	Ethylene oxide	10.6	-20	3.0-100%	429
$C_2H_4O_2$	Acetic acid	117.9	39	4.0-19.9%	463
$C_2H_4O_2$	Methyl formate	31.7	-19	4.5-23%	449
$C_2H_4O_3$	Ethaneperoxoic acid	110	41		
C_2H_5Br	Bromoethane	38.5		6.8-8.0%	511
C_2H_5Cl	Chloroethane	12.3	-50	3.8-15.4%	519
C_2H_5ClO	Ethylene chlorohydrin	128.6	60	4.9-15.9%	425
$C_2H_5Cl_3Si$	Trichloroethylsilane	100.5	22		
C_2H_5N	Ethyleneimine	56	-11	3.3-54.8%	320
$C_2H_5NO_2$	Nitroethane	114.0	28	3.4-17%	414
$C_2H_5NO_2$	Ethyl nitrite	18	-35	4.0-50%	90
$C_2H_5NO_3$	Ethyl nitrate	87.2	10	4-?	
C_2H_6	Ethane	-88.6		3.0-12.5%	472
$C_2H_6Cl_2Si$	Dichlorodimethylsilane	70.3	<21	3.4-9.5%	
C_2H_6O	Ethanol	78.2	13	3.3-19%	363
C_2H_6O	Dimethyl ether	-24.8	-41	3.4-27.0%	350
C_2H_6OS	2-Mercaptoethanol	158	74		
C_2H_6OS	Dimethyl sulfoxide	189	95	2.6-42%	215
$C_2H_6O_2$	Ethylene glycol	197.3	111	3.2-22%	398

Mol. Form.	Name	$t_B/°C$	FP/°C	Fl. Limits	IT/°C
$C_2H_6O_4S$	Dimethyl sulfate		83		188
C_2H_6S	Ethanethiol	35.1	-17	2.8-18.0%	300
C_2H_6S	Dimethyl sulfide	37.3	-37	2.2-19.7%	206
$C_2H_6S_2$	Dimethyl disulfide	109.8	24		
C_2H_7N	Ethylamine	16.5	-16	3.5-14%	385
C_2H_7N	Dimethylamine	6.8	20	2.8-14.4%	400
C_2H_7NO	Ethanolamine	171	86	3.0-23.5%	410
$C_2H_8N_2$	1,2-Ethanediamine	117	40	2.5-12.0%	385
$C_2H_8N_2$	1,1-Dimethylhydrazine	63.9	-15	2-95%	249
C_2N_2	Cyanogen	-21.1		6.6-32%	
C_3H_3Br	3-Bromo-1-propyne	89	10	3.0-?	324
C_3H_3N	2-Propenenitrile	77.3	0	3.0-17.0%	481
C_3H_4	Propyne	-23.2		2.1-12.5%	
C_3H_4ClN	3-Chloropropanenitrile	175.5	76		
$C_3H_4Cl_2$	2,3-Dichloropropene	94	15	2.6-7.8%	
C_3H_4O	Propargyl alcohol	113.6	36		
C_3H_4O	Acrolein	52.6	-26	2.8-31%	220
$C_3H_4O_2$	Propenoic acid	141	50	2.4-8.0%	438
$C_3H_4O_2$	2-Oxetanone	162	74	2.9-?	
$C_3H_4O_3$	Ethylene carbonate	248	143		
C_3H_5Br	3-Bromopropene	70.1	-1	4.4-7.3%	295
C_3H_5Cl	2-Chloropropene	22.6	-37	4.5-16%	
C_3H_5Cl	3-Chloropropene	45.1	-32	2.9-11.1%	485
C_3H_5ClO	Epichlorohydrin	118	31	3.8-21.0%	411
C_3H_5ClO	Propanoyl chloride	80	12		
$C_3H_5ClO_2$	2-Chloropropanoic acid	185	107		500
$C_3H_5ClO_2$	Ethyl chloroformate	95	16		500
$C_3H_5ClO_2$	Methyl chloroacetate	129.5	57	7.5-18.5%	
$C_3H_5Cl_2NO_2$	1,1-Dichloro-1-nitropropane	145	66		
$C_3H_5Cl_3$	1,2,3-Trichloropropane	157	71	3.2-12.6%	
$C_3H_5Cl_3Si$	Trichloro-2-propenylsilane	117.5	35		
C_3H_5N	Propanenitrile	97.1	2	3.1-14%	512
C_3H_5NO	3-Hydroxypropanenitrile	221	129		
$C_3H_5N_3O_9$	Trinitroglycerol				270
C_3H_6	Propene	-47.6		2.0-11.1%	455
C_3H_6	Cyclopropane	-32.8		2.4-10.4%	498
$C_3H_6ClNO_2$	1-Chloro-1-nitropropane	142	62		
$C_3H_6ClNO_2$	2-Chloro-2-nitropropane		57		
$C_3H_6Cl_2$	1,2-Dichloropropane	96.4	21	3.4-14.5%	557
$C_3H_6Cl_2O$	1,3-Dichloro-2-propanol	176	74		
$C_3H_6N_2$	Dimethylcyanamide	163.5	71		
C_3H_6O	Allyl alcohol	97.0	21	2.5-18.0%	378
C_3H_6O	Methyl vinyl ether	5.5			287
C_3H_6O	Propanal	48	-30	2.6-17%	207
C_3H_6O	Acetone	56.0	-20	2.5-12.8%	465
C_3H_6O	Methyloxirane	35	-37	3.1-27.5%	449
$C_3H_6O_2$	Propanoic acid	141.1	52	2.9-12.1%	465
$C_3H_6O_2$	Ethyl formate	54.4	-20	2.8-16.0%	455
$C_3H_6O_2$	Methyl acetate	56.8	-10	3.1-16%	454
$C_3H_6O_2$	1,3-Dioxolane	78	2		
$C_3H_6O_3$	Dimethyl carbonate	90.5	19		
$C_3H_6O_3$	1,3,5-Trioxane	114.5	45	3.6-29%	414
C_3H_7Br	1-Bromopropane	71.1			490
C_3H_7Cl	1-Chloropropane	46.5	<-18	2.6-11.1%	520
C_3H_7Cl	2-Chloropropane	35.7	-32	2.8-10.7%	593
C_3H_7ClO	2-Chloro-1-propanol	133.5	52		
C_3H_7ClO	1-Chloro-2-propanol	127	52		
$C_3H_7Cl_3Si$	Trichloropropylsilane	123.5	37		
C_3H_7N	Allylamine	53.3	-29	2.2-22%	374
C_3H_7NO	N,N-Dimethylformamide	153	58	2.2-15.2%	445
$C_3H_7NO_2$	1-Nitropropane	131.1	36	2.2-?	421

Mol. Form.	Name	t_B/°C	FP/°C	Fl. Limits	IT/°C
$C_3H_7NO_2$	2-Nitropropane	120.2	24	2.6-11.0%	428
$C_3H_7NO_3$	Propyl nitrate	110	20	2-100%	175
C_3H_8	Propane	-42.1	-104	2.1-9.5%	450
C_3H_8O	1-Propanol	97.2	23	2.2-13.7%	412
C_3H_8O	2-Propanol	82.3	12	2.0-12.7%	399
C_3H_8O	Ethyl methyl ether	7.4	-37	2.0-10.1%	190
$C_3H_8O_2$	1,2-Propylene glycol	187.6	99	2.6-12.5%	371
$C_3H_8O_2$	1,3-Propylene glycol	214.4			400
$C_3H_8O_2$	Ethylene glycol monomethyl ether	124.1	39	1.8-14%	285
$C_3H_8O_2$	Dimethoxymethane	42	-32	2.2-13.8%	237
$C_3H_8O_3$	Glycerol	290	199	3-19%	370
$C_3H_9BO_3$	Trimethyl borate	67.5	-8		
C_3H_9ClSi	Trimethylchlorosilane	60	-28		395
C_3H_9N	Propylamine	47.2	-37	2.0-10.4%	318
C_3H_9N	Isopropylamine	31.7	-37		402
C_3H_9N	Trimethylamine	2.8	-5	2.0-11.6%	190
C_3H_9NO	3-Amino-1-propanol	187.5	80		
C_3H_9NO	1-Amino-2-propanol	159.4	77		374
C_3H_9NO	N-Methyl-2-ethanolamine	158	74		
$C_3H_9O_3P$	Trimethyl phosphite	111.5	54		
$C_3H_9O_4P$	Trimethyl phosphate	197.2	107		
$C_3H_{10}N_2$	1,3-Propanediamine	139.8	24		
C_4Cl_6	Hexachloro-1,3-butadiene	215			610
$C_4H_2O_3$	Maleic anhydride	202	102	1.4-7.1%	477
C_4H_4	1-Buten-3-yne	5.1		21-100%	
$C_4H_4N_2$	Succinonitrile	266	132		
C_4H_4O	Furan	31.5	-36	2.3-14.3%	
$C_4H_4O_2$	Diketene	126.1	34		
C_4H_4S	Thiophene	84.0	-1		
C_4H_5Cl	2-Chloro-1,3-butadiene	59.4	-20	4.0-20.0%	
C_4H_5N	2-Butenenitrile	120.5	16		
C_4H_5N	Methylacrylonitrile	90.3	1	2-6.8%	
C_4H_5N	Pyrrole	129.7	39		
C_4H_6	1,3-Butadiene	-4.4		2.0-12.0%	420
C_4H_6	2-Butyne	26.9	-31	1.4-?	
C_4H_6O	Divinyl ether	28.3	<-30	1.7-27%	360
C_4H_6O	Ethoxyacetylene	50	<-7		
C_4H_6O	trans-2-Butenal	102.2	13	2.1-15.5%	232
C_4H_6O	3-Buten-2-one	81.4	-7	2.1-15.6%	491
C_4H_6O	Vinyloxirane	68	<-50		
$C_4H_6O_2$	Methacrylic acid	162.5	77	1.6-8.8%	68
$C_4H_6O_2$	Vinyl acetate	72.5	-8	2.6-13.4%	402
$C_4H_6O_2$	Methyl acrylate	80.7	-3	2.8-25%	468
$C_4H_6O_2$	2,3-Butanedione	88	27		
$C_4H_6O_2$	gamma-Butyrolactone	204	98		
$C_4H_6O_3$	Acetic anhydride	139.5	49	2.7-10.3%	316
$C_4H_6O_3$	Propylene carbonate	242	135		
$C_4H_6O_6$	L-Tartaric acid		210		425
C_4H_7Br	1-Bromo-2-butene	104.5		4.6-12.0%	
$C_4H_7BrO_2$	Ethyl bromoacetate	168.5	48		
C_4H_7Cl	2-Chloro-1-butene	58.5	-19	2.3-9.3%	
C_4H_7Cl	3-Chloro-2-methylpropene	71.5	-12	3.2-8.1%	
C_4H_7ClO	2-Chloroethyl vinyl ether	108	27		
$C_4H_7ClO_2$	Ethyl chloroacetate	144.3	64		
C_4H_7N	Butanenitrile	117.6	24	1.6-?	501
C_4H_7N	2-Methylpropanenitrile	103.9	8		482
C_4H_7NO	Acetone cyanohydrin		74	2.2-12.0%	688
C_4H_7NO	2-Pyrrolidone	251	129		
C_4H_8	1-Butene	-6.2		1.6-10.0%	385
C_4H_8	cis-2-Butene	3.7		1.7-9.0%	325
C_4H_8	trans-2-Butene	0.8		1.8-9.7%	324

Mol. Form.	Name	$t_b/°C$	FP/°C	Fl. Limits	IT/°C
C_4H_8	Isobutene	-6.9		1.8-9.6%	465
C_4H_8	Cyclobutane	12.6	<10	1.8-?	
$C_4H_8Cl_2$	1,2-Dichlorobutane	124.1			275
$C_4H_8Cl_2$	1,4-Dichlorobutane	161	52		
$C_4H_8Cl_2O$	Bis(2-chloroethyl) ether	178.5	55	2.7-?	369
C_4H_8O	2-Buten-1-ol	121.5	27	4.2-35.3%	349
C_4H_8O	2-Methyl-2-propenol	114.5	33		
C_4H_8O	Ethyl vinyl ether	35.5	<-46	1.7-28%	202
C_4H_8O	1,2-Epoxybutane	63.4	-22	1.7-19%	439
C_4H_8O	Butanal	74.8	-22	1.9-12.5%	218
C_4H_8O	Isobutanal	64.5	-18	1.6-10.6%	196
C_4H_8O	2-Butanone	79.5	-9	1.4-11.4%	404
C_4H_8O	Tetrahydrofuran	65	-14	2-11.8%	321
C_4H_8OS	1,4-Oxathiane	147	42		
$C_4H_8O_2$	Butanoic acid	163.7	72	2.0-10.0%	443
$C_4H_8O_2$	2-Methylpropanoic acid	154.4	56	2.0-9.2%	481
$C_4H_8O_2$	Propyl formate	80.9	-3		455
$C_4H_8O_2$	Isopropyl formate	68.2	-6		485
$C_4H_8O_2$	Ethyl acetate	77.1	-4	2.0-11.5%	426
$C_4H_8O_2$	Methyl propanoate	79.8	-2	2.5-13%	469
$C_4H_8O_2$	3-Hydroxybutanal		66		250
$C_4H_8O_2$	1,4-Dioxane	101.5	12	2.0-22%	180
$C_4H_8O_2S$	Sulfolane	287.3	177		
$C_4H_8O_3$	Methyl lactate	144.8	49	2.2-?	385
$C_4H_8O_3$	Ethylene glycol monoacetate	188	102		
C_4H_9Br	1-Bromobutane	101.6	18	2.6-6.6%	265
C_4H_9Br	2-Bromobutane	91.2	21		
C_4H_9Cl	1-Chlorobutane	78.6	-12	1.9-10.1%	240
C_4H_9Cl	2-Chlorobutane	68.2	-10		
C_4H_9Cl	1-Chloro-2-methylpropane	68.5	-6	2.0-8.7%	
C_4H_9Cl	2-Chloro-2-methylpropane	50.9	0		
$C_4H_9Cl_3Si$	Butyltrichlorosilane	148.5	54		
C_4H_9N	Pyrrolidine	86.5	3		
C_4H_9NO	N-Ethylacetamide	205	110		
C_4H_9NO	N,N-Dimethylacetamide	165	70	1.8-11.5%	490
C_4H_9NO	Butanal oxime	154	58		
C_4H_9NO	2-Butanone oxime	152.5	≈70		
C_4H_9NO	Morpholine	128	37	1.4-11.2%	290
$C_4H_9NO_2$	N-Acetylethanolamine		179		460
$C_4H_9NO_3$	Butyl nitrate	133	36		
C_4H_{10}	Butane	-0.5	-60	1.9-8.5%	287
C_4H_{10}	Isobutane	-11.7	-87	1.8-8.4%	460
$C_4H_{10}N_2$	Piperazine	146	81		
$C_4H_{10}O$	1-Butanol	117.7	37	1.4-11.2%	343
$C_4H_{10}O$	2-Butanol	99.5	24	1.7-9.8%	405
$C_4H_{10}O$	2-Methyl-1-propanol	107.8	28	1.7-10.6%	415
$C_4H_{10}O$	2-Methyl-2-propanol	82.4	11	2.4-8.0%	478
$C_4H_{10}O$	Diethyl ether	34.5	-45	1.9-36.0%	180
$C_4H_{10}O$	Methyl propyl ether	39.1	-20	2.0-14.8%	
$C_4H_{10}O_2$	1,2-Butanediol	190.5	40		
$C_4H_{10}O_2$	1,3-Butanediol	207.5	121		395
$C_4H_{10}O_2$	1,4-Butanediol	235	121		
$C_4H_{10}O_2$	2,3-Butanediol	182.5			402
$C_4H_{10}O_2$	Ethylene glycol monoethyl ether	135	43	3-18%	235
$C_4H_{10}O_2$	Ethylene glycol dimethyl ether	85	-2		202
$C_4H_{10}O_2$	tert-Butyl hydroperoxide		27		
$C_4H_{10}O_2S$	2,2'-Thiodiethanol	282	160		298
$C_4H_{10}O_3$	Diethylene glycol	245.8	124	2-17%	224
$C_4H_{10}O_4S$	Diethyl sulfate	208	104		436
$C_4H_{10}S$	1-Butanethiol	98.5	2		
$C_4H_{10}S$	2-Butanethiol	85	-23		

Mol. Form.	Name	t_b/°C	FP/°C	Fl. Limits	IT/°C
$C_4H_{10}S$	2-Methyl-1-propanethiol	88.5	2		
$C_4H_{10}S$	2-Methyl-2-propanethiol	64.3	<-29		
$C_4H_{10}Se$	Diethyl selenide	108		2.5-?	
$C_4H_{11}N$	Butylamine	77.0	-12	1.7-9.8%	312
$C_4H_{11}N$	sec-Butylamine	63.5	-9		
$C_4H_{11}N$	tert-Butylamine	44.0	-9	1.7-8.9%	380
$C_4H_{11}N$	Isobutylamine	67.7	-9	2-12%	378
$C_4H_{11}N$	Diethylamine	55.5	-23	1.8-10.1%	312
$C_4H_{11}NO$	2-Amino-1-butanol	178	74		
$C_4H_{11}NO$	2-Amino-2-methyl-1-propanol	165.5	67		
$C_4H_{11}NO_2$	Diethanolamine	268.8	172	2-13%	662
$C_4H_{12}Sn$	Tetramethylstannane	78	-12	1.9-?	
$C_4H_{13}N_3$	Diethylenetriamine	207	98	2-6.7%	358
$C_5H_4O_2$	Furfural	161.7	60	2.1-19.3%	316
C_5H_5N	Pyridine	115.2	20	1.8-12.4%	482
C_5H_6	2-Methyl-1-buten-3-yne	32	<-7		
$C_5H_6N_2$	2-Methylpyrazine	137	50		
C_5H_6O	3-Methylfuran	66	-30		
$C_5H_6O_2$	Furfuryl alcohol	171	75	1.8-16.3%	491
C_5H_7N	1-Methylpyrrole	115	16		
C_5H_7NO	2-Furanmethanamine	145.5	37		
$C_5H_7NO_2$	Ethyl cyanoacetate	205	110		
C_5H_8	2-Methyl-1,3-butadiene	34.0	-54	1.5-8.9%	395
C_5H_8	1-Pentyne	40.1	<-20		
C_5H_8	Cyclopentene	44.2	-29		395
C_5H_8O	3-Methyl-3-buten-2-one	98		1.8-9.0%	
C_5H_8O	Cyclopentanone	130.5	26		
C_5H_8O	3,4-Dihydro-2H-pyran	86	-18		
$C_5H_8O_2$	Allyl acetate	103.5	22		374
$C_5H_8O_2$	Isopropenyl acetate	94	26		432
$C_5H_8O_2$	Vinyl propanoate	91.2	1		
$C_5H_8O_2$	Ethyl acrylate	99.4	10	1.4-14%	372
$C_5H_8O_2$	Methyl methacrylate	100.5	10	1.7-8.2%	
$C_5H_8O_2$	2,4-Pentanedione	138	34		340
$C_5H_8O_3$	Methyl acetoacetate	171.7	77		280
C_5H_9NO	N-Methyl-2-pyrrolidone	202	96	1-10%	346
C_5H_{10}	1-Pentene	29.9	-18	1.5-8.7%	275
C_5H_{10}	cis-2-Pentene	36.9	<-20		
C_5H_{10}	trans-2-Pentene	36.3	<-20		
C_5H_{10}	2-Methyl-1-butene	31.2	-20		
C_5H_{10}	3-Methyl-1-butene	20.1	-7	1.5-9.1%	365
C_5H_{10}	2-Methyl-2-butene	38.5	-20		
C_5H_{10}	Cyclopentane	49.3	-25	1.5-?	361
$C_5H_{10}Cl_2$	1,5-Dichloropentane	179	>27		
$C_5H_{10}N_2$	3-(Dimethylamino)propanenitrile	173	65		
$C_5H_{10}O$	Cyclopentanol	140.4	51		
$C_5H_{10}O$	Pentanal	103	12		222
$C_5H_{10}O$	2-Pentanone	102.2	7	1.5-8.2%	452
$C_5H_{10}O$	3-Pentanone	101.9	13	1.6-?	450
$C_5H_{10}O$	Tetrahydropyran	88	-20		
$C_5H_{10}O$	2-Methyltetrahydrofuran	78	-11		
$C_5H_{10}O_2$	Pentanoic acid	186.1	96		400
$C_5H_{10}O_2$	3-Methylbutanoic acid	176.5			416
$C_5H_{10}O_2$	Butyl formate	106.1	18	1.7-8.2%	322
$C_5H_{10}O_2$	Isobutyl formate	98.2	5	2-9%	320
$C_5H_{10}O_2$	Propyl acetate	101.5	13	1.7-8%	450
$C_5H_{10}O_2$	Isopropyl acetate	88.6	2	1.8-8%	460
$C_5H_{10}O_2$	Ethyl propanoate	99.1	12	1.9-11%	440
$C_5H_{10}O_2$	Methyl butanoate	102.8	14		
$C_5H_{10}O_2$	3-Ethoxypropanal	135.2	38		
$C_5H_{10}O_2$	Tetrahydrofurfuryl alcohol	178	75	1.5-9.7%	282

Mol. Form.	Name	t_B/°C	FP/°C	Fl. Limits	IT/°C
$C_5H_{10}O_3$	Diethyl carbonate	126	25		
$C_5H_{10}O_3$	Ethylene glycol monomethyl ether acetate	143	49		
$C_5H_{10}O_3$	Ethyl lactate	154.5	46	1.5-12.3%	392
$C_5H_{11}Br$	1-Bromopentane	129.8	32	1.5-?	400
$C_5H_{11}Cl$	1-Chloropentane	107.8	13		
$C_5H_{11}Cl$	2-Chloro-2-methylbutane	85.6		1.6-8.6%	260
$C_5H_{11}Cl$	1-Chloro-3-methylbutane	98.9	<21	1.5-7.4%	345
$C_5H_{11}Cl_3Si$	Trichloropentylsilane	172	63	1.5-7.4%	
$C_5H_{11}N$	Piperidine	106.2	16		
$C_5H_{11}N$	N-Methylpyrrolidine	81	-14		
$C_5H_{11}NO$	4-Methylmorpholine	116	24		
$C_5H_{11}NO_2$	Isopentyl nitrite	99.2			210
C_5H_{12}	Pentane	36.0	-40	1.4-8.0%	260
C_5H_{12}	Isopentane	27.8	-51	1.4-7.6%	420
C_5H_{12}	Neopentane	9.4	-65	1.4-7.5%	450
$C_5H_{12}N_2$	1-Methylpiperazine	138	42		
$C_5H_{12}N_2O$	Tetramethylurea	176.5	77		
$C_5H_{12}O$	1-Pentanol	137.9	33	1.2-10.0%	300
$C_5H_{12}O$	2-Pentanol	119.3	34	1.2-9.0%	343
$C_5H_{12}O$	3-Pentanol	116.2	41	1.2-9.0%	435
$C_5H_{12}O$	2-Methyl-1-butanol	128	50		385
$C_5H_{12}O$	3-Methyl-1-butanol	131.1	43	1.2-9.0%	350
$C_5H_{12}O$	2-Methyl-2-butanol	102.4	19	1.2-9.0%	437
$C_5H_{12}O$	3-Methyl-2-butanol	112.9	38		
$C_5H_{12}O$	2,2-Dimethyl-1-propanol	113.5	37		
$C_5H_{12}O$	Ethyl propyl ether	63.2	<-20	1.7-9.0%	
$C_5H_{12}O_2$	1,5-Pentanediol	239	129		335
$C_5H_{12}O_2$	2-Isopropoxyethanol	145	33		
$C_5H_{12}O_2$	2,2-Dimethyl-1,3-propanediol	208	129		399
$C_5H_{12}O_3$	Diethylene glycol monomethyl ether	193	96	1.38-22.7%	240
$C_5H_{12}S$	1-Pentanethiol	126.6	18		
$C_5H_{12}S$	3-Methyl-2-butanethiol		3		
$C_5H_{13}N$	Pentylamine	104.3	-1	2.2-22%	
$C_5H_{13}N$	Butylmethylamine	91	13		
$C_6H_2Cl_4$	1,2,4,5-Tetrachlorobenzene	244.5	155		
$C_6H_3ClN_2O_4$	1-Chloro-2,4-dinitrobenzene	315	194	2.0-22%	
$C_6H_3Cl_3$	1,2,4-Trichlorobenzene	213.5	105	2.5-6.6%	571
$C_6H_4ClNO_2$	1-Chloro-4-nitrobenzene	242	127		
$C_6H_4Cl_2$	o-Dichlorobenzene	180	66	2.2-9.2%	648
$C_6H_4Cl_2$	m-Dichlorobenzene	173	72		
$C_6H_4Cl_2$	p-Dichlorobenzene	174	66		
$C_6H_4Cl_2O$	2,4-Dichlorophenol	210	114		
C_6H_5Br	Bromobenzene	156.0	51		565
C_6H_5Cl	Chlorobenzene	131.7	28	1.3-9.6%	593
C_6H_5ClO	o-Chlorophenol	174.9	64		
C_6H_5ClO	p-Chlorophenol	220	121		
$C_6H_5Cl_2N$	3,4-Dichloroaniline	272	166		
$C_6H_5Cl_3Si$	Trichlorophenylsilane	201	91		
C_6H_5F	Fluorobenzene	84.7	-15		
$C_6H_5NO_2$	Nitrobenzene	210.8	88	1.8-?	482
$C_6H_5N_3O_4$	2,4-Dinitroaniline		224		
C_6H_6	1,5-Hexadien-3-yne	85	<-20	1.5-?	
C_6H_6	Benzene	80.0	-11	1.2-7.8%	498
$C_6H_6N_2O_2$	p-Nitroaniline	332	199		
C_6H_6O	Phenol	181.8	79	1.8-8.6%	715
$C_6H_6O_2$	1,2-Benzenediol	245	127		
$C_6H_6O_2$	Resorcinol		127	1.4-?	608
$C_6H_6O_2$	p-Hydroquinone	287	165		516
C_6H_7N	Aniline	184.1	70	1.3-11%	615
C_6H_7N	2-Methylpyridine	129.3	39		538
C_6H_7N	4-Methylpyridine	145.3	57		

Mol. Form.	Name	t_b/°C	FP/°C	Fl. Limits	IT/°C
C_6H_8ClN	Aniline, hydrochloride		193		
$C_6H_8Cl_2O_2$	Hexanedioyl dichloride		72		
$C_6H_8N_2$	Adiponitrile	295	93	1.0-?	550
$C_6H_8N_2$	o-Phenylenediamine	257	156	1.5-?	
$C_6H_8N_2$	Phenylhydrazine	243.5	88		
$C_6H_8N_2$	2,5-Dimethylpyrazine	155	64		
C_6H_8O	2,5-Dimethylfuran	93.5	7		
$C_6H_8O_4$	Dimethyl maleate	202	113		
C_6H_{10}	1,4-Hexadiene	65	-21	2.0-6.1%	
C_6H_{10}	2-Methyl-1,3-pentadiene	75.8	-12		
C_6H_{10}	4-Methyl-1,3-pentadiene	76.5	-34		
C_6H_{10}	2-Hexyne	84.5	-10		
C_6H_{10}	Cyclohexene	82.9	-12	1.2-?	310
$C_6H_{10}O$	Diallyl ether	94	-7		
$C_6H_{10}O$	Cyclohexanone	155.4	44	1.1-9.4%	420
$C_6H_{10}O$	Mesityl oxide	130	31	1.4-7.2%	344
$C_6H_{10}O_2$	Vinyl butanoate	116.7	20	1.4-8.8%	
$C_6H_{10}O_2$	Ethyl 2-butenoate	136.5	2		
$C_6H_{10}O_2$	Ethyl methacrylate	117	20		
$C_6H_{10}O_2$	2,5-Hexanedione	194	79		499
$C_6H_{10}O_3$	Ethyl acetoacetate	180.8	57	1.4-9.5%	295
$C_6H_{10}O_3$	Propanoic anhydride	170	63	1.3-9.5%	285
$C_6H_{10}O_4$	Adipic acid	337.5	196		420
$C_6H_{10}O_4$	Diethyl oxalate	185.7	76		
$C_6H_{10}O_4$	Ethylene glycol diacetate	190	88	1.6-8.4%	482
$C_6H_{11}Cl$	Chlorocyclohexane	142	32		
$C_6H_{11}NO$	Caprolactam	270	125		
$C_6H_{11}NO_2$	Nitrocyclohexane	205	88		
$C_6H_{11}NO_2$	4-Acetylmorpholine		113		
C_6H_{12}	1-Hexene	63.4	-26	1.2-6.9%	253
C_6H_{12}	cis-2-Hexene	68.8	-21		
C_6H_{12}	2-Methyl-1-pentene	62.1	-28		300
C_6H_{12}	4-Methyl-1-pentene	53.9	-7		300
C_6H_{12}	4-Methyl-cis-2-pentene	56.3	-32		
C_6H_{12}	4-Methyl-trans-2-pentene	58.6	-29		
C_6H_{12}	2-Ethyl-1-butene	64.7	<-20		315
C_6H_{12}	2,3-Dimethyl-1-butene	55.6	<-20		360
C_6H_{12}	2,3-Dimethyl-2-butene	73.3	<-20		401
C_6H_{12}	Cyclohexane	80.7	-20	1.3-8%	245
C_6H_{12}	Methylcyclopentane	71.8	-29	1.0-8.35%	258
C_6H_{12}	Ethylcyclobutane	70.8	-15	1.2-7.7%	210
C_6H_{12}	2-Methyl-2-pentene	67.3	<-7		
$C_6H_{12}Cl_2O_2$	1,2-Bis(2-chloroethoxy)ethane	232	121		
$C_6H_{12}O$	cis-3-Hexen-1-ol	156.5	54		
$C_6H_{12}O$	Butyl vinyl ether	94	-9		255
$C_6H_{12}O$	Isobutyl vinyl ether	83	-9		
$C_6H_{12}O$	Hexanal	131	32		
$C_6H_{12}O$	2-Ethylbutanal		21	1.2-7.7%	
$C_6H_{12}O$	2-Methylpentanal	117	17		199
$C_6H_{12}O$	2-Hexanone	127.6	25	1-8%	423
$C_6H_{12}O$	3-Hexanone	123.5	35	1-8%	
$C_6H_{12}O$	4-Methyl-2-pentanone	116.5	18	1.2-8.0%	448
$C_6H_{12}O$	Cyclohexanol	160.8	68	1-9%	300
$C_6H_{12}O_2$	Hexanoic acid	205.2	102		380
$C_6H_{12}O_2$	2-Methylpentanoic acid	195.6	107		378
$C_6H_{12}O_2$	Diethylacetic acid	194	99		400
$C_6H_{12}O_2$	Pentyl formate	130.4	26		
$C_6H_{12}O_2$	Butyl acetate	126.1	22	1.7-7.6%	425
$C_6H_{12}O_2$	sec-Butyl acetate	112	31	1.7-9.8%	
$C_6H_{12}O_2$	Isobutyl acetate	116.5	18	1.3-10.5%	421
$C_6H_{12}O_2$	Propyl propanoate	122.5	79		

Mol. Form.	Name	t_B/°C	FP/°C	Fl. Limits	IT/°C
$C_6H_{12}O_2$	Ethyl butanoate	121.5	24		463
$C_6H_{12}O_2$	Ethyl 2-methylpropanoate	110.1	13		
$C_6H_{12}O_2$	Diacetone alcohol	167.9	58	1.8-6.9%	643
$C_6H_{12}O_3$	Ethylene glycol monoethyl ether acetate	156.4	56	2-8%	379
$C_6H_{12}O_3$	Paraldehyde	124.3	36	1.3-?	238
$C_6H_{12}S$	Cyclohexanethiol	158.9	43		
$C_6H_{13}Cl$	1-Chlorohexane	135	35		
$C_6H_{13}N$	Cyclohexylamine	134	31	1.9-9.4%	293
$C_6H_{13}NO$	N-Butylacetamide	229	116		
$C_6H_{13}NO$	2,6-Dimethylmorpholine	146.6	44		
$C_6H_{13}NO$	N-Ethylmorpholine	138.5	32		
$C_6H_{13}NO_2$	4-Morpholineethanol	227	99		
C_6H_{14}	Hexane	68.7	-22	1.1-7.5%	225
C_6H_{14}	2-Methylpentane	60.2	<-29	1.0-7.0%	264
C_6H_{14}	3-Methylpentane	63.2	-7	1.2-7.0%	278
C_6H_{14}	2,2-Dimethylbutane	49.7	-48	1.2-7.0%	405
C_6H_{14}	2,3-Dimethylbutane	57.9	-29	1.2-7.0%	405
$C_6H_{14}N_2O$	1-Piperazineethanol	246	124		
$C_6H_{14}O$	1-Hexanol	157.6	63		
$C_6H_{14}O$	2-Methyl-1-pentanol	149	54	1.1-9.65%	310
$C_6H_{14}O$	4-Methyl-2-pentanol	131.6	41	1.0-5.5%	
$C_6H_{14}O$	2-Ethyl-1-butanol	147	57		
$C_6H_{14}O$	Dipropyl ether	90.0	21	1.3-7.0%	188
$C_6H_{14}O$	Diisopropyl ether	68.5	-28	1.4-7.9%	443
$C_6H_{14}O$	Butyl ethyl ether	92.3	4		
$C_6H_{14}O_2$	2,5-Hexanediol	218	110		
$C_6H_{14}O_2$	2-Methyl-2,4-pentanediol	197.1	102	1-9%	306
$C_6H_{14}O_2$	Ethylene glycol monobutyl ether	168.4	69	4-13%	238
$C_6H_{14}O_2$	1,1-Diethoxyethane	102.2	-21	1.6-10.4%	230
$C_6H_{14}O_2$	Ethylene glycol diethyl ether	119.4	27		205
$C_6H_{14}O_3$	1,2,6-Hexanetriol		191		
$C_6H_{14}O_3$	Diethylene glycol monoethyl ether	196	96		
$C_6H_{14}O_3$	Diethylene glycol dimethyl ether	162	67		
$C_6H_{14}O_3$	Trimethylolpropane		149		
$C_6H_{14}O_4$	Triethylene glycol	285	177	0.9-9.2%	371
$C_6H_{15}N$	Hexylamine	132.8	29		
$C_6H_{15}N$	Butylethylamine	107.5	18		
$C_6H_{15}N$	Dipropylamine	109.3	17		299
$C_6H_{15}N$	Diisopropylamine	83.9	-1	1.1-7.1%	316
$C_6H_{15}N$	Triethylamine	89	-7	1.2-8.0%	249
$C_6H_{15}NO_2$	Diisopropanolamine	250	127		374
$C_6H_{15}NO_3$	Triethanolamine	335.4	179	1-10%	
$C_6H_{15}N_3$	1-Piperazineethanamine	220	93		
$C_6H_{15}O_4P$	Triethyl phosphate	215.5	115		454
$C_6H_{16}N_2$	N,N-Diethylethylenediamine	144	46		
$C_7H_3ClF_3NO_2$	1-Chloro-4-nitro-2-(trifluoromethyl)benzene	232	135		
$C_7H_4ClF_3$	1-Chloro-2-(trifluoromethyl)benzene	152.2	59		
$C_7H_4F_3NO_2$	1-Nitro-3-(trifluoromethyl)benzene	202.8	103		
C_7H_5ClO	Benzoyl chloride	197.2	72		
C_7H_5ClO	4-Chlorobenzaldehyde	213.5	88		
$C_7H_5Cl_3$	(Trichloromethyl)benzene	221	127		211
$C_7H_5F_3$	(Trifluoromethyl)benzene	102.1	12		
$C_7H_6N_2O_4$	1-Methyl-2,4-dinitrobenzene		207		
C_7H_6O	Benzaldehyde	179.0	63		192
$C_7H_6O_2$	Benzoic acid	249.2	121		570
$C_7H_6O_2$	Salicylaldehyde	197	78		
$C_7H_6O_3$	Salicylic acid		157	1.1-?	540
C_7H_7Br	o-Bromotoluene	181.7	79		
C_7H_7Br	p-Bromotoluene	184.3	85		
C_7H_7Cl	(Chloromethyl)benzene	179	67	1.1-?	585
$C_7H_7NO_2$	o-Nitrotoluene	222	106		

Mol. Form.	Name	t_B/°C	FP/°C	Fl. Limits	IT/°C
$C_7H_7NO_2$	m-Nitrotoluene	232	106		
$C_7H_7NO_2$	p-Nitrotoluene	238.3	106		
C_7H_8	Toluene	110.6	4	1.1-7.1%	480
C_7H_8	Bicyclo[2.2.1]hepta-2,5-diene	89.5	-21		
C_7H_8O	o-Cresol	191.0	81	1.4-?	599
C_7H_8O	m-Cresol	202.2	86	1.1-?	558
C_7H_8O	p-Cresol	201.9	86	1.1-?	558
C_7H_8O	Benzyl alcohol	205.3	93		436
C_7H_8O	Anisole	153.7	52		475
$C_7H_8O_2$	4-Methoxyphenol	243	132		421
$C_7H_8O_3S$	p-Toluenesulfonic acid		184		
C_7H_9N	o-Methylaniline	200.3	85		482
C_7H_9N	p-Methylaniline	200.4	87		482
C_7H_9NO	o-Anisidine	224	118		
$C_7H_{10}O$	3-Cyclohexene-1-carboxaldehyde	105	57		
$C_7H_{10}O_4$	3,3-Diacetoxy-1-propene	180	82		
C_7H_{12}	4-Methylcyclohexene	102.7	-1		
$C_7H_{12}O_2$	Butyl acrylate	145	29	1.7-9.9%	292
$C_7H_{12}O_2$	Isobutyl acrylate	132	30		427
$C_7H_{12}O_2$	Cyclohexyl formate	162	51		
$C_7H_{12}O_4$	Diethyl malonate	200	93		
C_7H_{14}	1-Heptene	93.6	-1		260
C_7H_{14}	trans-2-Heptene	98	<0		
C_7H_{14}	Cycloheptane	118.4	<21	1.1-6.7%	
C_7H_{14}	Methylcyclohexane	100.9	-4	1.2-6.7%	250
C_7H_{14}	Ethylcyclopentane	103.5	<21	1.1-6.7%	260
$C_7II_{14}O$	2-Heptanone	151.0	39	1.1-7.9%	393
$C_7H_{14}O$	3-Heptanone	147	46		
$C_7H_{14}O$	4-Heptanone	144	49		
$C_7H_{14}O$	5-Methyl-2-hexanone	144	36	1.0-8.2%	191
$C_7H_{14}O$	cis-2-Methylcyclohexanol	165	65		296
$C_7H_{14}O$	trans-2-Methylcyclohexanol	167.5	65		296
$C_7H_{14}O$	cis-3-Methylcyclohexanol	174.5	70		295
$C_7H_{14}O$	trans-3-Methylcyclohexanol	174.5	70		295
$C_7H_{14}O$	cis-4-Methylcyclohexanol	173	70		295
$C_7H_{14}O$	trans-4-Methylcyclohexanol	174	70		295
$C_7H_{14}O_2$	Pentyl acetate	149.2	16	1.1-7.5%	360
$C_7H_{14}O_2$	Isopentyl acetate	142.5	25	1.0-7.5%	360
$C_7H_{14}O_2$	sec-Pentyl acetate	130.5	32		
$C_7H_{14}O_2$	Butyl propanoate	146.8	32		426
$C_7H_{14}O_2$	Propyl butanoate	143.0	37		
$C_7H_{15}NO_2$	Ethyl N-butylcarbamate	202	92		
C_7H_{16}	Heptane	98.5	-4	1.05-6.7%	204
C_7H_{16}	2-Methylhexane	90.0	-1	1.0-6.0%	280
C_7H_{16}	3-Methylhexane	92	-4		280
C_7H_{16}	2,3-Dimethylpentane	89.7	-56	1.1-6.7%	335
C_7H_{16}	2,4-Dimethylpentane	80.4	-12		
C_7H_{16}	2,2,3-Trimethylbutane	80.8	<0		412
$C_7H_{16}N_2O$	4-Morpholinepropanamine	220	104		
$C_7H_{16}O$	2-Heptanol	159	71		
$C_7H_{16}O$	3-Heptanol	157	60		
$C_7H_{16}O$	2,4-Dimethyl-3-pentanol	138.7	49		
$C_7H_{16}O$	2,3,3-Trimethyl-2-butanol	131	<0		375
$C_7H_{17}N$	Heptylamine	156	54		
$C_7H_{18}N_2$	N,N-Diethyl-1,3-propanediamine	168.5	59		
$C_8H_4O_3$	Phthalic anhydride	295	152	1.7-10.5%	570
$C_8H_6O_4$	Phthalic acid		168		
$C_8H_6O_4$	Terephthalic acid		260		496
C_8H_7ClO	α-Chloroacetophenone	247	118		
C_8H_7N	Benzeneacetonitrile	233.5	113		
C_8H_8	Styrene	145	31	0.9-6.8%	490

Mol. Form.	Name	t_B/°C	FP/°C	Fl. Limits	IT/°C
C_8H_8O	Phenyloxirane	194.1	74		498
C_8H_8O	Benzeneacetaldehyde	195	71		
C_8H_8O	Acetophenone	202	77		570
$C_8H_8O_2$	Benzeneacetic acid	265.5	>100		
$C_8H_8O_2$	Phenyl acetate	196	80		
$C_8H_8O_2$	Methyl benzoate	199	83		
$C_8H_8O_2$	2-Methoxybenzaldehyde	243.5	118		
$C_8H_8O_3$	Methyl salicylate	222.9	96		454
C_8H_9Cl	1-Chloro-4-ethylbenzene	184.4	64		
C_8H_9NO	Acetanilide	304	169		530
$C_8H_9NO_2$	Methyl 2-aminobenzoate	256	>100		
C_8H_{10}	Ethylbenzene	136.1	21	0.8-6.7%	432
C_8H_{10}	o-Xylene	144.5	32	0.9-6.7%	463
C_8H_{10}	m-Xylene	139.1	27	1.1-7.0%	527
C_8H_{10}	p-Xylene	138.3	27	1.1-7.0%	528
$C_8H_{10}O$	p-Ethylphenol	217.9	104		
$C_8H_{10}O$	Benzeneethanol	218.2	96		
$C_8H_{10}O$	α-Methylbenzyl alcohol	205	93		
$C_8H_{10}O$	Phenetole	169.8	63		
$C_8H_{10}O$	Benzyl methyl ether	170	135		
$C_8H_{10}O$	4-Methylanisole	175.5	60		
$C_8H_{10}O_2$	2-Phenoxyethanol	245	121		
$C_8H_{11}N$	N-Ethylaniline	203.0	85		
$C_8H_{11}N$	N,N-Dimethylaniline	194.1	63		371
$C_8H_{11}N$	2,3-Xylidine	221.5	97	1.0-?	
$C_8H_{11}N$	2,6-Xylidine	215	96		
$C_8H_{11}N$	α-Methylbenzylamine	187	79		
$C_8H_{11}N$	5-Ethyl-2-picoline	178.3	68	1.1-6.6%	
$C_8H_{11}NO$	N-Phenylethanolamine	279.5	152		
$C_8H_{11}NO$	o-Phenetidine	232.5	115		
$C_8H_{11}NO$	p-Phenetidine	254	116		
C_8H_{12}	1,5-Cyclooctadiene	150.8	35		
C_8H_{12}	4-Vinylcyclohexene	128	16		269
$C_8H_{12}O_4$	Diethyl maleate	223	121		350
$C_8H_{12}O_4$	Diethyl fumarate	214	104		
$C_8H_{14}O_2$	Cyclohexyl acetate	173	58		335
$C_8H_{14}O_2$	Butyl methacrylate	160	52		
$C_8H_{14}O_3$	Butanoic anhydride	200	54	0.9-5.8%	279
$C_8H_{14}O_3$	2-Methylpropanoic anhydride	183	59	1.0-6.2%	329
$C_8H_{14}O_3$	Butyl acetoacetate		85		
$C_8H_{14}O_4$	Ethyl succinate	217.7	90		
$C_8H_{14}O_5$	Diethylene glycol diacetate	200	135		
$C_8H_{14}O_6$	Diethyl tartrate	281	93		
$C_8H_{15}ClO$	Octanoyl chloride	195.6	82		
C_8H_{16}	1-Octene	121.2	21		230
C_8H_{16}	2,4,4-Trimethyl-1-pentene	101.4	-5	0.8-4.8%	391
C_8H_{16}	2,4,4-Trimethylpentene	101	-5	0.8-4.8%	391
C_8H_{16}	2,4,4-Trimethyl-2-pentene	104.9	2		305
C_8H_{16}	Ethylcyclohexane	131.9	35	0.9-6.6%	238
C_8H_{16}	cis-1,2-Dimethylcyclohexane	129.8	16		304
C_8H_{16}	trans-1,2-Dimethylcyclohexane	123.5	11		304
C_8H_{16}	cis-1,4-Dimethylcyclohexane	124.4	16		
C_8H_{16}	Propylcyclopentane	131			269
$C_8H_{16}O$	Octanal	171	52		
$C_8H_{16}O$	2-Ethylhexanal	163	44	0.85-7.2%	190
$C_8H_{16}O$	2-Octanone	172.5	52		
$C_8H_{16}O_2$	Hexyl acetate	171.5	45		
$C_8H_{16}O_2$	sec-Hexyl acetate	147.5	45		
$C_8H_{16}O_2$	2-Ethylbutyl acetate	162.5	54		
$C_8H_{16}O_2$	Pentyl propanoate	168.6	41		378
$C_8H_{16}O_2$	Butyl butanoate	166	53		

Mol. Form.	Name	$t_B/°C$	FP/°C	Fl. Limits	IT/°C
$C_8H_{16}O_2$	Isobutyl butanoate	156.9	50		
$C_8H_{16}O_2$	Isobutyl isobutanoate	148.6	38	0.96-7.59%	432
$C_8H_{16}O_2$	Ethyl hexanoate	167	49		
$C_8H_{16}O_2$	1,4-Cyclohexanedimethanol	283	167		316
$C_8H_{16}O_3$	Pentyl lactate		79		
$C_8H_{16}O_4$	Diethylene glycol monoethyl ether acetate	218.5	110		425
$C_8H_{17}Cl$	1-Chlorooctane	181.5	70		
$C_8H_{17}Cl$	3-(Chloromethyl)heptane	172	60		
C_8H_{18}	Octane	125.6	13	1.0-6.5%	206
C_8H_{18}	2,3-Dimethylhexane	115.6	7		438
C_8H_{18}	2,4-Dimethylhexane	109.5	10		
C_8H_{18}	3-Ethyl-2-methylpentane	115.6	<21		460
C_8H_{18}	2,2,3-Trimethylpentane	110	<21		346
C_8H_{18}	2,2,4-Trimethylpentane	99.2	-12		418
C_8H_{18}	2,3,3-Trimethylpentane	114.8	<21		425
$C_8H_{18}O$	1-Octanol	195.1	81		
$C_8H_{18}O$	2-Octanol	180	88		
$C_8H_{18}O$	2-Ethyl-1-hexanol	184.6	73	0.88-9.7%	231
$C_8H_{18}O$	Dibutyl ether	140.2	25	1.5-7.6%	194
$C_8H_{18}O_2$	2-Ethyl-1,3-hexanediol	244	127		360
$C_8H_{18}O_2$	2,2,4-Trimethyl-1,3-pentanediol	235	113		346
$C_8H_{18}O_2$	Di-*tert*-butyl peroxide	111	18		
$C_8H_{18}O_3$	Diethylene glycol diethyl ether	188	82		
$C_8H_{18}O_4$	2,5,8,11-Tetraoxadodecane	216	111		
$C_8H_{18}O_5$	Tetraethylene glycol	328	182		
$C_8H_{18}S$	1-Octanethiol	199.1	69		
$C_8H_{18}S$	Dibutyl sulfide	185	76		
$C_8H_{19}N$	Octylamine	179.6	60		
$C_8H_{19}N$	Dibutylamine	159.6	47	1.1-6%	
$C_8H_{19}N$	Diisobutylamine	139.6	29		
$C_8H_{19}N$	2-Ethylhexylamine	169.2	60		
$C_8H_{20}O_4Si$	Ethyl silicate	168.8	52		
$C_8H_{23}N_5$	Tetraethylenepentamine	341.5	163		321
$C_9H_6N_2O_2$	Toluene-2,4-diisocyanate	251	127	0.9-9.5%	
C_9H_7N	Quinoline	237.1			480
C_9H_{10}	*o*-Methylstyrene	169.8	53	0.8-11.0%	538
C_9H_{10}	*m*-Methylstyrene	164	53	0.8-11.0%	538
C_9H_{10}	*p*-Methylstyrene	172.8	53	0.8-11.0%	538
C_9H_{10}	Isopropenylbenzene	165.4	54	1.9-6.1%	574
$C_9H_{10}O$	1-Phenyl-1-propanone	217.5	99		
$C_9H_{10}O$	4-Methylacetophenone	226	96		
$C_9H_{10}O_2$	Ethyl benzoate	212	88		490
$C_9H_{10}O_2$	Benzyl acetate	213	90		460
$C_9H_{10}O_2$	Methyl 2-phenylacetate	216.5	91		
$C_9H_{11}NO$	4-Methylacetanilide	307	168		
C_9H_{12}	Propylbenzene	159.2	30	0.8-6.0%	450
C_9H_{12}	Isopropylbenzene	152.4	36	0.9-6.5%	424
C_9H_{12}	*o*-Ethyltoluene	165.2			440
C_9H_{12}	*m*-Ethyltoluene	161.3			480
C_9H_{12}	*p*-Ethyltoluene	162			475
C_9H_{12}	1,2,3-Trimethylbenzene	176.1	44	0.8-6.6%	470
C_9H_{12}	1,2,4-Trimethylbenzene	169.3	44	0.9-6.4%	500
C_9H_{12}	1,3,5-Trimethylbenzene	164.7	50	1-5%	559
$C_9H_{12}O$	α–Ethylbenzyl alcohol	219	100		
$C_9H_{12}O_2$	Ethylene glycol monobenzyl ether	256	129		352
$C_9H_{12}O_3S$	Ethyl *p*-toluenesulfonate		158		
$C_9H_{13}N$	Amphetamine	203	<100		
$C_9H_{14}O$	Phorone	197.5	85		
$C_9H_{14}O$	Isophorone	215.2	84	0.8-3.8%	460
$C_9H_{14}O_6$	Triacetin	259	138	1.0-?	433
C_9H_{16}	Octahydroindene	167			296

Mol. Form.	Name	t_B/°C	FP/°C	Fl. Limits	IT/°C
C₉H₁₆O₂	Allyl hexanoate	186	66		
C₉H₁₈	1-Nonene	146.9	26		
C₉H₁₈	Propylcyclohexane	156.7			248
C₉H₁₈	Isopropylcyclohexane	154.8			283
C₉H₁₈	Butylcyclopentane	156.6			250
C₉H₁₈O	2-Nonanone	195.3	60	0.9-5.9%	360
C₉H₁₈O	Diisobutyl ketone	169.4	49	0.8-7.1%	396
C₉H₁₈O₂	Pentyl butanoate	186.4	57		
C₉H₁₈O₂	Isopentyl butanoate	179	59		
C₉H₁₈O₂	Butyl 3-methylbutanoate		53		
C₉H₂₀	Nonane	150.8	31	0.8-2.9%	205
C₉H₂₀	3-Ethyl-4-methylhexane	140	24		
C₉H₂₀	4-Ethyl-2-methylhexane	133.8	<21	0.7-?	280
C₉H₂₀	2,2,5-Trimethylhexane	124.0	13		
C₉H₂₀	3,3-Diethylpentane	146.3		0.7-5.7%	290
C₉H₂₀	3-Ethyl-2,4-dimethylpentane	136.7	390		
C₉H₂₀	2,2,3,3-Tetramethylpentane	140.2	<21	0.8-4.9%	430
C₉H₂₀	2,2,3,4-Tetramethylpentane	133.0	<21		
C₉H₂₁BO₃	Triisopropyl borate	140	28		
C₉H₂₁N	Tripropylamine	156	41		
C₉H₂₁NO₃	Triisopropanolamine	160			320
C₁₀H₇Cl	1-Chloronaphthalene	259	121		>558
C₁₀H₈	Naphthalene	217.9	79	0.9-5.9%	526
C₁₀H₈O	2-Naphthol	285	153		
C₁₀H₉N	1-Naphthalenamine	300.8	157		
C₁₀H₁₀O₂	Safrole	234.5	100		
C₁₀H₁₀O₄	Dimethyl phthalate	283.7	146	0.9-?	490
C₁₀H₁₀O₄	Dimethyl isophthalate	282	138		
C₁₀H₁₀O₄	Dimethyl terephthalate	288	153		518
C₁₀H₁₁NO₂	Acetoacetanilide		185		
C₁₀H₁₂	1,2,3,4-Tetrahydronaphthalene	207.6	71	0.8-5.0%	385
C₁₀H₁₂O₂	Isopropyl benzoate	216	99		
C₁₀H₁₂O₂	Ethyl phenylacetate	227	99		
C₁₀H₁₄	Butylbenzene	183.3	71	0.8-5.8%	410
C₁₀H₁₄	sec-Butylbenzene	173.3	52	0.8-6.9%	418
C₁₀H₁₄	tert-Butylbenzene	169.1	60	0.7-5.7%	450
C₁₀H₁₄	Isobutylbenzene	172.7	55	0.8-6.0%	427
C₁₀H₁₄	p-Cymene	177.1	47	0.7-5.6%	436
C₁₀H₁₄	1,2,3,4-Tetramethylbenzene	205	74		427
C₁₀H₁₄	1,2,3,5-Tetramethylbenzene	198	71		427
C₁₀H₁₄	1,2,4,5-Tetramethylbenzene	196.8	54		
C₁₀H₁₄	o-Diethylbenzene	184	57		395
C₁₀H₁₄	m-Diethylbenzene	181.1	56		450
C₁₀H₁₄	p-Diethylbenzene	183.7	55	0.7-6.0%	430
C₁₀H₁₄O	Butyl phenyl ether	210	82		
C₁₀H₁₄O₂	4-tert-Butyl-1,2-benzenediol	285	130		
C₁₀H₁₅N	N-Butylaniline	243.5	107		
C₁₀H₁₅N	N,N-Diethylaniline	216.3	85		630
C₁₀H₁₅NO₂	N-Phenyl-N,N-diethanolamine		196	0.7-?	387
C₁₀H₁₆	Dipentene	178	45		237
C₁₀H₁₆	d-Limonene	178	45	0.7-6.1%	237
C₁₀H₁₆	α-Pinene	156.2	33		255
C₁₀H₁₆	β-Pinene	166	38		275
C₁₀H₁₆	β-Phellandrene	171.5	49		
C₁₀H₁₆O	Camphor	207.4	66	0.6-3.5%	466
C₁₀H₁₈	trans-Decahydronaphthalene	187.3	54	0.7-5.4%	255
C₁₀H₁₈	Decahydronaphthalene	155.5	58	0.7-4.9%	250
C₁₀H₁₈O	Borneol		66		
C₁₀H₁₈O	Linalol	198	71		
C₁₀H₁₈O	α-Terpineol	220	90		
C₁₀H₁₈O	Cineole	176.4	48		

Mol. Form.	Name	$t_B/°C$	FP/°C	Fl. Limits	IT/°C
$C_{10}H_{18}O$	trans-Geraniol	230	>100		
$C_{10}H_{18}O_4$	Dibutyl oxalate	241	104		
$C_{10}H_{19}NO_2$	N-tert-Butylaminoethyl methacrylate		96		
$C_{10}H_{20}$	1-Decene	170.5	<55		235
$C_{10}H_{20}$	Butylcyclohexane	180.9			246
$C_{10}H_{20}$	Isobutylcyclohexane	171.3			274
$C_{10}H_{20}$	tert-Butylcyclohexane	171.5			342
$C_{10}H_{20}O$	Citronellol	224	96		
$C_{10}H_{20}O_2$	2-Ethylhexyl acetate	199	71	0.76-8.14%	268
$C_{10}H_{20}O_2$	Ethyl octanoate	208.5	79		
$C_{10}H_{21}N$	N-Butylcyclohexanamine		93		
$C_{10}H_{22}$	Decane	174.1	51	0.8-5.4%	210
$C_{10}H_{22}$	2-Methylnonane	167.1			210
$C_{10}H_{22}$	3-Ethyloctane	166.5			230
$C_{10}H_{22}$	4-Ethyloctane	163.7			229
$C_{10}H_{22}O$	1-Decanol	231.1	82		288
$C_{10}H_{22}O$	Dipentyl ether	190	57		170
$C_{10}H_{22}O_2$	Ethylene glycol dibutyl ether	203.3	85		
$C_{10}H_{22}O_5$	Tetraethylene glycol dimethyl ether	275.3	141		
$C_{10}H_{22}S$	Dipentyl sulfide		85		
$C_{10}H_{23}N$	Decylamine	220.5	99		
$C_{10}H_{23}N$	Dipentylamine	202.5	51		
$C_{11}H_{10}$	1-Methylnaphthalene	244.7			529
$C_{11}H_{12}O_3$	Ethyl benzoylacetate		141		
$C_{11}H_{14}O_2$	Butyl benzoate	250.3	107		
$C_{11}H_{16}$	p-tert-Butyltoluene	190	68		
$C_{11}H_{16}$	Pentylbenzene	205.4	66		
$C_{11}H_{16}$	1,3-Diethyl-5-methylbenzene	205			455
$C_{11}H_{16}$	Pentamethylbenzene	232	93		427
$C_{11}H_{16}O$	4-tert-Butyl-2-methylphenol	237	118		
$C_{11}H_{17}N$	p-tert-Pentylaniline	260.5	102		
$C_{11}H_{20}O_2$	2-Ethylhexyl acrylate		82		252
$C_{11}H_{22}$	Pentylcyclohexane	203.7			239
$C_{11}H_{22}O$	2-Undecanone	231.5	89		
$C_{11}H_{22}O_2$	Nonyl acetate	210	68		
$C_{11}H_{24}$	Undecane	195.9	69		
$C_{11}H_{24}$	2-Methyldecane	189.3			225
$C_{11}H_{24}O$	2-Undecanol	228	113		
$C_{12}H_9Br$	4-Bromo-1,1'-Biphenyl	310	144		
$C_{12}H_{10}$	Biphenyl	256.1	113	0.6-5.8%	540
$C_{12}H_{10}Cl_2Si$	Dichlorodiphenylsilane	305	142		
$C_{12}H_{10}O$	o-Phenylphenol	286	124		530
$C_{12}H_{10}O$	Diphenyl ether	258.0	112	0.8-1.5%	618
$C_{12}H_{11}N$	2-Aminobiphenyl	299			450
$C_{12}H_{11}N$	Diphenylamine	302	153		634
$C_{12}H_{12}$	1-Ethylnaphthalene	258.6			480
$C_{12}H_{14}O_4$	Diethyl phthalate	295	161	0.7-?	457
$C_{12}H_{14}O_4$	Diethyl terephthalate	302	117		
$C_{12}H_{16}$	Cyclohexylbenzene	240.1	99		
$C_{12}H_{16}O_3$	Pentyl salicylate	270	132		
$C_{12}H_{17}NO$	N-Butyl-N-phenylacetamide	281	141		
$C_{12}H_{18}$	1,5,9-Cyclododecatriene	240	71		
$C_{12}H_{20}O_4$	Dibutyl maleate	280	141		
$C_{12}H_{22}O_4$	Dimethyl sebacate		145		
$C_{12}H_{22}O_6$	Dibutyl tartrate	320	91		284
$C_{12}H_{23}N$	Dicyclohexylamine		>99		
$C_{12}H_{24}$	1-Dodecene	213.8	79		
$C_{12}H_{24}O_2$	Ethyl decanoate	241.5	>100		
$C_{12}H_{25}Br$	1-Bromododecane	276	144		
$C_{12}H_{26}$	Dodecane	216.3	74	0.6-?	203
$C_{12}H_{26}O$	1-Dodecanol	259	127		275

Mol. Form.	Name	t_B/°C	FP/°C	Fl. Limits	IT/°C
C$_{12}$H$_{26}$O	2-Butyl-1-octanol	246.5	110		
C$_{12}$H$_{26}$O$_3$	Diethylene glycol dibutyl ether	256	118		310
C$_{12}$H$_{26}$S	1-Dodecanethiol	277	128		
C$_{12}$H$_{27}$BO$_3$	Tributyl borate	234	93		
C$_{12}$H$_{27}$N	Tributylamine	216.5	63		
C$_{12}$H$_{27}$O$_4$P	Tributyl phosphate	289	146		
C$_{13}$H$_{12}$	2-Methylbiphenyl	255.5	137		502
C$_{13}$H$_{12}$	Diphenylmethane	265.0	130		485
C$_{13}$H$_{14}$N$_2$	p,p'-Diaminodiphenylmethane	398	220		
C$_{13}$H$_{26}$	1-Tridecene	232.8	79		
C$_{13}$H$_{26}$O	2-Tridecanone	263	107		
C$_{13}$H$_{28}$	Tridecane	235.4	79		
C$_{13}$H$_{28}$O	1-Tridecanol		121		
C$_{14}$H$_8$O$_2$	9,10-Anthracenedione	377	185		
C$_{14}$H$_{10}$	Anthracene	339.9	121	0.6-?	540
C$_{14}$H$_{10}$	Phenanthrene	340	171		
C$_{14}$H$_{12}$O$_2$	Benzyl benzoate	323.5	148		480
C$_{14}$H$_{12}$O$_3$	Benzyl salicylate	320	>100		
C$_{14}$H$_{14}$	1,1-Diphenylethane	272.6	>100		440
C$_{14}$H$_{14}$O	Dibenzyl ether	298	135		
C$_{14}$H$_{16}$	1-Butylnaphthalene	289.3	360		
C$_{14}$H$_{16}$N$_2$O$_2$	o-Dianisidine		206		
C$_{14}$H$_{23}$N	N,N-Dibutylaniline	274.8	110		
C$_{14}$H$_{28}$	1-Tetradecene	233	110		235
C$_{14}$H$_{30}$	Tetradecane	253.5	112	0.5-?	200
C$_{14}$H$_{30}$O	1-Tetradecanol	289	141		
C$_{15}$H$_{18}$	1-Pentylnaphthalene	307	124		
C$_{15}$H$_{24}$	Nonylbenzene	280.5	99		
C$_{15}$H$_{24}$O	2,6-Di-tert-butyl-4-methylphenol	265	127		
C$_{15}$H$_{26}$O$_6$	Tributyrin	307.5	180	0.5-?	407
C$_{15}$H$_{33}$N	Tripentylamine	242.5	102		
C$_{16}$H$_{14}$O	1,3-Diphenyl-2-buten-1-one	342.5	177		
C$_{16}$H$_{18}$	2-Butyl-1,1'-biphenyl		>100		430
C$_{16}$H$_{22}$O$_4$	Dibutyl phthalate	340	157	0.5-?	402
C$_{16}$H$_{26}$	Decylbenzene	298	107		
C$_{16}$H$_{34}$	Hexadecane	286.8	136		202
C$_{16}$H$_{34}$O	Dioctyl ether	283	>100		205
C$_{16}$H$_{35}$N	Bis(2-ethylhexyl)amine		132		
C$_{17}$H$_{20}$N$_2$O	N,N'-Diethylcarbanilide		150		
C$_{17}$H$_{34}$O	2-Heptadecanone	320	120		
C$_{17}$H$_{36}$O	1-Heptadecanol	333	154		
C$_{18}$H$_{14}$	o-Terphenyl	332	163		
C$_{18}$H$_{14}$	m-Terphenyl	363	191		
C$_{18}$H$_{15}$O$_3$P	Triphenyl phosphite	360	218		
C$_{18}$H$_{15}$O$_4$P	Triphenyl phosphate		220		
C$_{18}$H$_{15}$P	Triphenylphosphine		180		
C$_{18}$H$_{30}$	Dodecylbenzene	328	140		
C$_{18}$H$_{32}$O$_7$	Butyl citrate		157		368
C$_{18}$H$_{34}$O$_2$	Oleic acid	360	189		363
C$_{18}$H$_{34}$O$_4$	Dibutyl sebacate	344.5	178	0.4-?	365
C$_{18}$H$_{36}$O$_2$	Stearic acid		196		395
C$_{18}$H$_{37}$Cl$_3$Si	Trichlorooctadecylsilane		89		
C$_{18}$H$_{38}$	Octadecane	316.3	>100		227
C$_{18}$H$_{38}$O	1-Octadecanol				450
C$_{19}$H$_{16}$	Triphenylmethane	359	>100		
C$_{19}$H$_{38}$O	2-Nonadecanone		124		
C$_{19}$H$_{38}$O$_2$	Methyl stearate	443	153		
C$_{19}$H$_{40}$	Nonadecane	329.9	>100		230
C$_{20}$H$_{14}$O$_4$	Diphenyl phthalate		224		
C$_{20}$H$_{28}$	1-Decylnaphthalene	379	177		
C$_{20}$H$_{42}$	Eicosane	343	>100		232

Mol. Form.	Name	$t_B/°C$	FP/°C	Fl. Limits	IT/°C
$C_{21}H_{21}O_4P$	Tri-o-cresyl phosphate	410	225		385
$C_{21}H_{26}O_3$	4-Octylphenyl salicylate		216		416
$C_{21}H_{32}O_2$	Methyl abietate		180		
$C_{22}H_{42}O_2$	Butyl oleate		180		
$C_{22}H_{42}O_4$	Bis(2-ethylhexyl) adipate		206	0.4-?	377
$C_{22}H_{44}O_2$	Butyl stearate	343	160		355
$C_{23}H_{46}O_2$	Pentyl stearate		185		
$C_{24}H_{20}Sn$	Tetraphenylstannane	420	232		
$C_{24}H_{38}O_4$	Bis(2-ethylhexyl) phthalate	384	218		
$C_{25}H_{48}O_4$	Bis(2-ethylhexyl) azelate		227	0.3-?	374

THRESHOLD LIMIT VALUES FOR AIRBORNE CONTAMINANTS

This table gives threshold limits for selected airborne contaminants in the workplace. The data are derived from a more extensive list in the publication:

> Threshold Limit Values and Biological Exposure Indices for 1995-1996
> American Conference of Governmental Industrial Hygienists
> 1330 Kemper Meadow Drive
> Cincinnati, OH 45240

Threshold limit values are airborne concentrations under which it is believed that nearly all workers may be repeatedly exposed day after day without adverse effect. However, it should be noted that a small percentage of individuals may be affected at concentrations at or below these limits because of unusual susceptibility or pre-existing conditions. The publication cited above should be consulted for full details.

In this table the threshold limit valve (TLV) is given in parts per million (ppm) by volume in air at 25°C and 101.325 kPa (1 atmosphere). For most substances the quantity tabulated is the time weighted average (TWA) for a normal 8-hr workday and 40-hr workweek. However, a C following a TLV entry indicates a ceiling valve, which should not be exceeded even for short periods.

Explanation of abbreviations and notes:

> Carcinogen: indicates that the compound is a suspected or confirmed human carcinogen
> Asphyxiant: indicates a compound that may not have significant physiological effects but is an asphyxiant when present in high concentrations. Fire and explosion hazards may also exist for these compounds.
> CAS RN: Chemical Abstracts Serviced Registry Number.
> C: indicates a ceiling value, not to be exceeded even for short periods.

Compounds are listed by systematic name. A synonym index follows the table

Substance	Molecular formula	CAS RN	TLV-TWA ppm		Note
Acetaldehyde	C_2H_4O	75-07-0	100		carcinogen
Acetic acid	$C_2H_4O_2$	64-19-7	10		
Acetic anhydride	$C_4H_6O_3$	108-24-7	5		
Acetone	C_3H_6O	67-64-1	750		
Acetonitrile	C_2H_3N	75-05-8	40		
Acetylene	C_2H_2	74-86-2			asphyxiant
Allyl alcohol	C_3H_6O	107-18-6	2		
Ammonia	H_3N	7664-41-7	25		
Anisidine (o, p)	C_7H_9NO	29191-52-4	0.1		carcinogen
Aniline	C_6H_7N	62-53-3	2		carcinogen
Argon	Ar	7440-37-1			asphyxiant
Arsine	AsH_3	7784-42-1	0.05		
Benzene*	C_6H_6	71-43-2	10		carcinogen
Benzenethiol	C_6H_6S	108-98-5	0.5		
p-Benzoquinone	$C_6H_4O_2$	106-51-4	0.1		
Benzoyl chloride	C_7H_5ClO	98-88-4	0.5	C	
Benzyl acetate	$C_9H_{10}O_2$	140-11-4	10		
Benzyl chloride	C_7H_7Cl	100-44-7	1		carcinogen
Biphenyl	$C_{12}H_{10}$	92-52-4	0.2		
Bis(2-amimoethyl)amine	$C_4H_{13}N_3$	111-40-0	1		
Bis(2-chloroethyl) ether	$C_4H_8Cl_2O$	111-44-4	5		
Bis(chloromethyl) ether	$C_2H_4Cl_2O$	542-88-1	0.001		carcinogen
Boron tribromide	BBr_3	10294-33-4	1	C	
Boron trifluoride	BF_3	7637-07-2	1	C	
Bromine	Br_2	7726-95-6	0.1		
Bromine pentafluoride	BrF_5	7789-30-2	0.1		
Bromochloromethane	CH_2BrCl	74-97-5	200		
Bromoethane	C_2H_5Br	74-96-4	5		carcinogen
Bromomethane	CH_3Br	74-83-9	5		carcinogen
1,3-Butadiene	C_4H_6	106-99-0	2		carcinogen
Butane	C_4H_{10}	106-97-8	800		
1-Butanethiol	$C_4H_{10}S$	109-79-5	0.5		
1-Butanol*	$C_4H_{10}O$	71-36-3	50	C	
2-Butanol	$C_4H_{10}O$	78-92-2	100		

* Changes have been proposed for these substances, but not yet adopted.

Substance	Molecular formula	CAS RN	TLV-TWA ppm	Note
2-Butanone	C_4H_8O	78-93-3	200	
2-Butoxyethanol	$C_6H_{14}O_2$	111-76-2	25	
Butyl acetate*	$C_6H_{12}O_2$	123-86-4	150	
sec-Butyl acetate	$C_6H_{12}O_2$	105-46-4	200	
tert-Butyl acetate	$C_6H_{12}O_2$	540-88-5	200	
Butyl acrylate	$C_7H_{12}O_2$	141-32-2	10	
Butylamine	$C_4H_{11}N$	109-73-9	5	C
p-tert-Butyltoluene	$C_{11}H_{16}$	98-51-1	10	
Camphor	$C_{10}H_{16}O$	76-22-2	2	
Caprolactam	$C_6H_{11}NO$	105-60-2	5	
Carbon dioxide	CO_2	124-38-9	5000	
Carbon disulfide	CS_2	75-15-0	10	
Carbon monoxide	CO	630-08-0	25	
Carbonyl chloride	CCl_2O	75-44-5	0.1	
Carbonyl fluoride	CF_2O	353-50-4	2	
Chlorine	Cl_2	7782-50-5	0.5	
Chlorine dioxide	ClO_2	10049-04-4	0.1	
Chlorine trifluoride	ClF_3	7790-91-2	0.1	C
Chloroacetaldehyde	C_2H_3ClO	107-20-0	1	C
Chloroacetyl chloride	$C_2H_2Cl_2O$	79-04-9	0.05	
Chlorobenzene	C_6H_5Cl	108-90-7	10	
2-Chloro-1,3-butadiene	C_4H_5Cl	126-99-8	10	carcinogen
Chlorodifluoromethane	$CHClF_2$	75-45-6	1000	
Chloroethane	C_2H_5Cl	75-00-3	100	
2-Chloroethanol	C_2H_5ClO	107-07-3	1	C
Chloroethylene	C_2H_3Cl	75-01-4	5	carcinogen
Chloromethane	CH_3Cl	74-87-3	50	carcinogen
Chloropentafluoroethane	C_2ClF_5	76-15-3	1000	
3-Chloropropene	C_3H_5Cl	107-05-1	1	carcinogen
o-Chlorotoluene	C_7H_7Cl	95-49-8	50	
Chromyl chloride (CrO_2Cl_2)	Cl_2CrO_2	14977-61-8	0.025	carcinogen
Chrysene	$C_{18}H_{12}$	218-01-9		carcinogen
o-Cresol	C_7H_8O	95-48-7	5	
m-Cresol	C_7H_8O	108-39-4	5	
p-Cresol	C_7H_8O	106-44-5	5	
Cumene	C_9H_{12}	98-82-8	50	
Cyanogen	C_2N_2	460-19-5	10	
Cyanogen chloride	CClN	506-77-4	0.3	C
Cyclohexane	C_6H_{12}	110-82-7	300	
Cyclohexanol	$C_6H_{12}O$	108-93-0	50	
Cyclohexanone	$C_6H_{10}O$	108-94-1	25	
Cyclohexene	C_6H_{10}	110-83-8	300	
Cyclohexylamine	$C_6H_{13}N$	108-91-8	10	
1,3-Cyclopentadiene	C_5H_6	542-92-7	75	
Cyclopentane	C_5H_{10}	287-92-3	600	
Diacetone alcohol	$C_6H_{12}O_2$	123-42-2	50	
Diazomethane	CH_2N_2	334-88-3	0.2	carcinogen
Diborane	B_2H_6	19287-45-7	0.1	
Dibromodifluoromethane	CBr_2F_2	75-61-6	100	
1,2-Dibromoethane	$C_2H_4Br_2$	106-93-4		carcinogen
Dichloroacetylene	C_2Cl_2	7572-29-4	0.1	C carcinogen
o-Dichlorobenzene	$C_6H_4Cl_2$	95-50-1	25	C
p-Dichlorobenzene	$C_6H_4Cl_2$	106-46-7	10	carcinogen
Dichlorodifluoromethane	CCl_2F_2	75-71-8	1000	
1,1-Dichloroethane	$C_2H_4Cl_2$	75-34-3	100	
1,2-Dichloroethane	$C_2H_4Cl_2$	107-06-2	10	carcinogen
1,1-Dichloroethylene	$C_2H_2Cl_2$	75-35-4	5	carcinogen
cis-1,2-Dichloroethylene	$C_2H_2Cl_2$	156-59-2	200	
trans-1,2-Dichloroethylene	$C_2H_2Cl_2$	156-60-5	200	

Substance	Molecular formula	CAS RN	TLV-TWA ppm		Note
Dichlorofluoromethane	$CHCl_2F$	75-43-4	10		
Dichloromethane	CH_2Cl_2	75-09-2	50		carcinogen
1,2-Dichloropropane	$C_3H_6Cl_2$	78-87-5	75		carcinogen
1,3-Dichloropropene	$C_3H_4Cl_2$	542-75-6	1		carcinogen
1,2-Dichlorotetrafluoroethane	$C_2Cl_2F_4$	76-14-2	1000		
Diethanolamine	$C_4H_{11}NO_2$	111-42-2	0.46		
Diethylamine	$C_4H_{11}N$	109-89-7	5		
Diethyl ether	$C_4H_{10}O$	60-29-7	400		
Diisopropylamine	$C_6H_{15}N$	108-18-9	5		
Dimethoxymethane	$C_3H_8O_2$	109-87-5	1000		
N,N-Dimethylacetamide	C_4H_9NO	127-19-5	10		
Dimethylamine	C_2H_7N	124-40-3	5		
N,N-Dimethylaniline	$C_8H_{11}N$	121-69-7	5		
N,N-Dimethylformamide	C_3H_7NO	68-12-2	10		
2,6-Dimethyl-4-heptanone	$C_9H_{18}O$	108-83-8	25		
1,1-Dimethylhydrazine	$C_2H_8N_2$	57-14-7	0.5		carcinogen
1,4-Dioxane*	$C_4H_8O_2$	123-91-1	25		carcinogen
Epichlorohydrin*	C_3H_5ClO	106-89-8	2		carcinogen
Ethane	C_2H_6	74-84-0			asphyxiant
1,2-Ethanediamine	$C_2H_8N_2$	107-15-3	10		
Ethanethiol	C_2H_6S	75-08-1	0.5		
Ethanol	C_2H_6O	64-17-5	1000		
Ethanolamine	C_2H_7NO	141-43-5	3		
2-Ethoxyethanol	$C_4H_{10}O_2$	110-80-5	5		
Ethyl acetate	$C_4H_8O_2$	141-78-6	400		
Ethyl acrylate	$C_5H_8O_2$	140-88-5	5		carcinogen
Ethylamine	C_2H_7N	75-04-7	5		
Ethylbenzene	C_8H_{10}	100-41-4	100		
Ethylene	C_2H_4	74-85-1			asphyxiant
Ethylene glycol	$C_2H_6O_2$	107-21-1	39.4	C	
Ethyleneimine	C_2H_5N	151-56-4	0.05		carcinogen
Ethylene oxide	C_2H_4O	75-21-8	1		carcinogen
Ethyl formate	$C_3H_6O_2$	109-94-4	100		
Fluorine	F_2	7782-41-4	1		
Formaldehyde	CH_2O	50-00-0	0.3	C	carcinogen
Formamide	CH_3NO	75-12-7	10		
Formic acid	CH_2O_2	64-18-6	5		
Furfural	$C_5H_4O_2$	98-01-1	2		
Furfuryl alcohol	$C_5H_6O_2$	98-00-0	10		
Heptane	C_7H_{16}	142-82-5	400		
2-Heptanone	$C_7H_{14}O$	110-43-0	50		
Hexachloro-1,3-butadiene	C_4Cl_6	87-68-3	0.02		carcinogen
Hexachloroethane	C_2Cl_6	67-72-1	1		carcinogen
Hexane	C_6H_{14}	110-54-3	50		
2-Hexanone	$C_6H_{12}O$	591-78-6	5		
Hexylene glycol	$C_6H_{14}O_2$	107-41-5	25		
Hydrazine	H_4N_2	302-01-2	0.01		carcinogen
Hydrogen bromide	BrH	10035-10-6	3	C	
Hydrogen chloride	ClH	7647-01-0	5	C	
Hydrogen cyanide	CHN	74-90-8	4.7	C	
Hydrogen fluoride	FH	7664-39-3	3	C	
Hydrogen peroxide	H_2O_2	7722-84-1	1		
Hydrogen selenide	H_2Se	7783-07-5	0.05		
Hydrogen sulfide	H_2S	7783-06-4	10		
Indene	C_9H_8	95-13-6	10		
Iodine	I_2	7553-56-2	0.1	C	
Iodomethane	CH_3I	74-88-4	2		carcinogen
Isobutyl acetate	$C_6H_{12}O_2$	110-19-0	150		
Isopentyl acetate	$C_7H_{14}O_2$	123-92-2	100		

Substance	Molecular formula	CAS RN	TLV-TWA ppm	Note
Isophorone	$C_9H_{14}O$	78-59-1	5	C
Isopropyl acetate	$C_5H_{10}O_2$	108-21-4	250	
Isopropylamine	C_3H_9N	75-31-0	5	
Ketene	C_2H_2O	463-51-4	0.5	
Maleic anhydride	$C_4H_2O_3$	108-31-6	0.25	
Mesitylene	C_9H_{12}	108-67-8	25	
Mesityl oxide	$C_6H_{10}O$	141-79-7	15	
Methacrylic acid	$C_4H_6O_2$	79-41-4	20	
Methane	CH_4	74-82-8		asphyxiant
Methanethiol	CH_4S	74-93-1	0.5	
Methanol	CH_4O	67-56-1	200	
2-Methoxyethanol	$C_3H_8O_2$	109-86-4	5	
2-Methoxyethyl acetate	$C_5H_{10}O_3$	110-49-6	5	
Methyl acetate	$C_3H_6O_2$	79-20-9	200	
Methyl acrylate*	$C_4H_6O_2$	96-33-3	10	
Methylacrylonitrile	C_4H_5N	126-98-7	1	
Methylamine	CH_5N	74-89-5	5	
o-Methylaniline	C_7H_9N	95-53-4	2	carcinogen
m-Methylaniline	C_7H_9N	108-44-1	2	
p-Methylaniline	C_7H_9N	106-49-0	2	carcinogen
N-Methylaniline	C_7H_9N	100-61-8	0.5	
Methyl tert-butyl ether	$C_5H_{12}O$	1634-04-4	40	
Methylcyclohexane	C_7H_{14}	108-87-2	400	
Methyl formate	$C_2H_4O_2$	107-31-3	100	
Methylhydrazine	CH_6N_2	60-34-4	0.01	
Methylisocyanate	C_2H_3NO	624-83-9	0.02	
Methyl methacrylate	$C_5H_8O_2$	80-62-6	100	
Methyloxirane	C_3H_6O	75-56-9	20	carcinogen
4-Methyl-2-pentanone	$C_6H_{12}O$	108-10-1	50	
2-Methyl-2-propanol	$C_4H_{10}O$	75-65-0	100	
2-Methyl-1-propanol	$C_4H_{10}O$	78-83-1	50	
o-Methylstyrene	C_9H_{10}	611-15-4	50	
m-Methylstyrene	C_9H_{10}	100-80-1	50	
p-Methylstyrene	C_9H_{10}	622-97-9	50	
Morpholine	C_4H_9NO	110-91-8	20	
Naphthalene	$C_{10}H_8$	91-20-3	10	
Nickel carbonyl (Ni(CO)₄)*	C_4NiO_4	13463-39-3	0.05	carcinogen
Nitric oxide	NO	10102-43-9	25	
Nitric acid	HNO_3	7697-37-2	2	
Nitrobenzene	$C_6H_5NO_2$	98-95-3	1	
Nitroethane	$C_2H_5NO_2$	79-24-3	100	
Nitrogen dioxide	NO_2	10102-44-0	3	
Nitrogen trifluoride	F_3N	7783-54-2	10	
Nitromethane	CH_3NO_2	75-52-5	20	
2-Nitropropane	$C_3H_7NO_2$	79-46-9	10	carcinogen
o-Nitrotoluene	$C_7H_7NO_2$	88-72-2	2	
m-Nitrotoluene	$C_7H_7NO_2$	99-08-1	2	
p-Nitrotoluene	$C_7H_7NO_2$	99-99-0	2	
Nonane	C_9H_{20}	111-84-2	200	
Octane	C_8H_{18}	111-65-9	300	
2-Oxetanone	$C_3H_4O_2$	57-57-8	0.05	carcinogen
Oxygen difluoride	F_2O	7783-41-7	0.05	C
Ozone*	O_3	10028-15-6	0.1	C
Pentaborane	B_5H_9	19624-22-7	0.005	
Pentanal	$C_5H_{10}O$	110-62-3	50	
Pentane	C_5H_{12}	109-66-0	600	
2-Pentanone	$C_5H_{10}O$	107-87-9	200	
3-Pentanone	$C_5H_{10}O$	96-22-0	200	
Pentyl acetate	$C_7H_{14}O_2$	628-63-7	100	

Substance	Molecular formula	CAS RN	TLV-TWA ppm	Note
Perchloryl fluoride	$ClFO_3$	7616-94-6	3	
Perfluoroacetone	C_3F_6O	684-16-2	0.1	
Phenol	C_6H_6O	108-95-2	5	
Phenylhydrazine	$C_6H_8N_2$	100-63-0	0.1	carcinogen
Phosphine	H_3P	7803-51-2	0.3	
Phosphorus pentachloride	Cl_5P	10026-13-8	0.1	
Phosphorus trichloride	Cl_3P	7719-12-2	0.2	
Phosphoryl chloride	Cl_3OP	10025-87-3	0.1	
Phthalic anhydride	$C_8H_4O_3$	85-44-9	1	
Propane	C_3H_8	74-98-6		asphyxiant
Propanoic acid	$C_3H_6O_2$	79-09-4	10	
1-Propanol	C_3H_8O	71-23-8	200	
2-Propanol	C_3H_8O	67-63-0	400	
Propargyl alcohol	C_3H_4O	107-19-7	1	
Propenal	C_3H_4O	107-02-8	0.1	
Propenamide	C_3H_5NO	79-06-1		carcinogen
Propenenitrile	C_3H_3N	107-13-1	2	carcinogen
Propenoic acid	$C_3H_4O_2$	79-10-7	2	
Propyl acetate	$C_5H_{10}O_2$	109-60-4	200	
Propyleneimine	C_3H_7N	75-55-8	2	carcinogen
Propyne	C_3H_4	74-99-7	1000	
Pyridine	C_5H_5N	110-86-1	5	
Selenium fluoride (SeF_6)	F_6Se	7783-79-1	0.05	
Silane	H_4Si	7803-62-5	5	
Stibine	H_3Sb	7803-52-3	0.1	
Styrene*	C_8H_8	100-42-5	50	carcinogen
Sulfur dioxide	O_2S	7446-09-5	2	
Sulfur fluoride (SF_4)	F_4S	7783-60-0	0.1	C
Sulfur fluoride (SF_5)	F_5S	10546-01-7	0.01	C
Sulfur fluoride (SF_6)	F_6S	2551-62-4	1000	
Sulfuryl fluoride	F_2O_2S	2699-79-8	5	
Tetrabromomethane	CBr_4	558-13-4	0.1	
Tetrachloro-1,2-difluoroethane	$C_2Cl_4F_2$	76-12-0	500	
1,1,2,2-Tetrachloroethane	$C_2H_2Cl_4$	79-34-5	1	carcinogen
Tetrachloroethylene	C_2Cl_4	127-18-4	50	carcinogen
Tetrachloromethane	CCl_4	56-23-5	5	carcinogen
Tetrahydrofuran	C_4H_8O	109-99-9	200	
Thionyl chloride	Cl_2OS	7719-09-7	1	C
Toluene	C_7H_8	108-88-3	50	
Toluene-2,4-diisocyanate	$C_9H_6N_2O_2$	584-84-9	0.005	carcinogen
Tribromomethane	$CHBr_3$	75-25-2	0.5	
1,2,4-Trichlorobenzene	$C_6H_3Cl_3$	120-82-1	5	C
1,1,1-Trichloroethane	$C_2H_3Cl_3$	71-55-6	350	
1,1,2-Trichloroethane	$C_2H_3Cl_3$	79-00-5	10	carcinogen
Trichloroethylene	C_2HCl_3	79-01-6	50	carcinogen
Trichlorofluoromethane	CCl_3F	75-69-4	1000	C
Trichloromethane	$CHCl_3$	67-66-3	10	carcinogen
1,1,2-Trichlorotrifluoroethane	$C_2Cl_3F_3$	76-13-1	1000	
Triethanolamine	$C_6H_{15}NO_3$	102-71-6	0.5	
Triethylamine	$C_6H_{15}N$	121-44-8	1	
Triiodomethane	CHI_3	75-47-8	0.6	
Trimethylamine	C_3H_9N	75-50-3	5	
1,2,4-Trimethylbenzene	C_9H_{12}	95-63-6	25	
Vinyl acetate	$C_4H_6O_2$	108-05-4	10	carcinogen
o-Xylene	C_8H_{10}	95-47-6	100	
m-Xylene	C_8H_{10}	108-38-3	100	
p-Xylene	C_8H_{10}	106-42-3	100	

SYNONYM INDEX

Acetyl acetate: see Acetic anhydride
Acetylene tetrachloride: see 1,1,2,2-Tetrachloroethane
Acrolein: see Propenal
Acrylamide: see Propenamide
Acrylic acid: see Propenoic acid
Acrylonitrile: see Propenenitrile
Allyl chloride: see 3-Chloropropene
2-Aminoethanol: see Ethanolamine
2-Aminopropane: see Isopropylamine
Amyl acetate: see Pentyl acetate
Azine: see Pyridine
Aziridine: see Ethyleneimine
Benzenamine: see Aniline
1,2-Benzenedicarboxylic anhydride: see Phthalic anhydride
1,2-Benzophenanthrene: see Chrysene
Bis(2-hydroxyethyl)amine: see Diethanolamine
Bromoform: see Tribromomethane
cis-Butenedioic anhydride: see Maleic anhydride
Butyl alcohol: see 1-Butanol
sec-Butyl alcohol: see 2-Butanol
tert-Butyl alcohol: see 2-Methyl-2-propanol
Butyl mercaptan: see 1-Butanethiol
Butyl methyl ketone: see 2-Hexanone
Butyl-2-propenoate: see Butyl acrylate
Carbon tetrabromide: see Tetrabromomethane
Carbon tetrachloride: see Tetrachloromethane
Chloroform: see Trichloromethane
(Chloromethyl)benzene: see Benzyl chloride
(Chloromethyl)oxirane: see Epichlorohydrin
Cyclohexyl alcohol: see Cyclohexanol
Dichloroethyl ether: see Bis(2-chloroethyl) ether
Diethylenetriamine: see Bis(2-aminoethyl)amine
Diethyl ketone: see 3-Pentanone
Diisobutyl ketone: see 2,6-Dimethyl-4-heptanone
2,4-Diisocyanatotoluene: see Toluene-2,4-diisocyanate
1,2-Dimethylbenzene: see o-Xylene
1,3-Dimethylbenzene: see m-Xylene
1,4-Dimethylbenzene: see p-Xylene
Diphenyl: see Biphenyl
Ethanal: see Acetaldehyde
Ethanedinitrile: see Cyanogen
1,2-Ethanediol: see Ethylene glycol
Ethanenitrile: see Acetonitrile
Ethene: see Ethylene
Ethenone: see Ketene
Ethyl alcohol: see Ethanol
Ethyl bromide: see Bromoethane
Ethyl chloride: see Chloroethane
Ethylene chlorohydrin: see 2-Chloroethanol
Ethylenediamine: see 1,2-Ethanediamine
Ethylene dibromide: see 1,2-Dibromoethane
Ethylene dichloride: see 1,2-Dichloroethane
Ethylene glycol monobutyl ether: see 2-Butoxyethanol
Ethylene glycol monoethyl ether: see 2-Ethoxyethanol
Ethylene glycol monomethyl ether: see 2-Methoxyethanol
Ethyl ether: see Diethyl ether
Ethylidene dichloride: see 1,1-Dichloroethane
Ethyl mercaptan: see Ethanethiol

Ethyl propenoate: see Ethyl acrylate
Ethyne: see Acetylene
Fluorine oxide (F_2O): see Oxygen difluoride
2-Furaldehyde: see Furfural
2-Furanmethanol: see Furfuryl alcohol
Hexafluoro-2-propanone: see Perfluoroacetone
Hexahydro-2-azepinone: see Caprolactam
Hydrobromic acid: see Hydrogen bromide
Hydrochloric acid: see Hydrogen chloride
Hydrocyanic acid: see Hydrogen cyanide
Hydrofluoric acid: see Hydrogen fluoride
Hydroxybenzene: see Phenol
4-Hydroxy-4-methyl-2-pentanone: see Diacetone alcohol
Iodoform: see Triiodomethane
Isoamyl acetate: see Isopentyl acetate
Isobutyl alcohol: see 2-Methyl-1-propanol
Isobutyl methyl ketone: see 4-Methyl-2-pentanone
Isopropyl alcohol: see 2-Propanol
Isopropylbenzene: see Cumene
Methanal: see Formaldehyde
Methanamide: see Formamide
2-Methoxyethyl ethanoate: see 2-Methoxyethyl acetate
Methylacetylene: see Propyne
Methylal: see Dimethoxymethane
Methyl alcohol: see Methanol
2-Methylaziridine: see Propyleneimine
Methylbenzene: see Toluene
Methyl bromide: see Bromomethane
Methyl chloride: see Chloromethane
Methyl chloroform: see 1,1,1-Trichloroethane
Methylene chloride: see Dichloromethane
Methyl ethyl ketone: see 2-Butanone
Methyl iodide: see Iodomethane
Methyl mercaptan: see Methanethiol
Methyl 2-methylpropenoate: see Methyl methacrylate
2-Methyl-2,4-pentanediol: see Hexylene glycol
4-Methyl-3-penten-2-one: see Mesityl oxide
Methyl pentyl ketone: see 2-Heptanone
2-Methylphenol: see o-Cresol
3-Methylphenol: see m-Cresol
4-Methylphenol: see p-Cresol
2-Methylpropenenitrile: see Methylacrylonitrile
Methyl propenoate: see Methyl acrylate
2-Methylpropenoic acid: see Methacrylic acid
Methyl propyl ketone: see 2-Pentanone
Nitrogen monoxide: see Nitric oxide
Oxirane: see Ethylene oxide
Oxolane: see Tetrahydrofuran
Phenyl chloride: see Chlorobenzene
Phenyl mercaptan: see Benzenethiol
Phosgene: see Carbonyl chloride
Phosphorus oxychloride: see Phosphoryl chloride
2-Propanone: see Acetone
1-Propen-3-ol: see Allyl alcohol
β-Propiolactone: see 2-Oxetanone
Propionic acid: see Propanoic acid
Propyl alcohol: see 1-Propanol
Propylene dichloride: see 1,2-Dichloropropane

SYNONYM INDEX (continued)

1,2-Propylene oxide: see Methyloxirane
1-Propyne-3-ol: see Propargyl alcohol
Pseudocumene: see 1,2,4-Trimethylbenzene
Sulfinyl dichloride: see Thionyl chloride
Sulfonyl difluoride: see Sulfuryl fluoride
Sulfur hexafluoride: see Sulfur fluoride (SF$_6$)
Tetrahydro-1,4-oxazine: see Morpholine
o-Toluidine: see o-Methyl aniline
m-Toluidine: see m-Methyl aniline
p-Toluidine: see p-Methylaniline
o-Tolyl chloride: see o-Chlorotoluene

Trichloroethene: see Trichloroethylene
1,3,5-Trimethylbenzene: see Mesitylene
3,3,5-Trimethyl-2-cyclohexen-1-one: see Isophorone
Tris(2-hydroxyethyl)amine: see Triethanolamine
Valeraldehyde: see Pentanal
Vinylbenzene: see Styrene
Vinyl chloride: see Chloroethylene
Vinylidene chloride: see 1,1-Dichloroethylene
2-Vinyltoluene: see o-Methylstyrene
3-Vinyltoluene: see m-Methylstyrene
4-Vinyltoluene: see p-Methylstyrene

OCTANOL-WATER PARTITION COEFFICIENTS

The octanol-water partition coefficient, P, is a widely used parameter for correlating biological effects of organic substances. It is a property of the two-phase system in which water and 1-octanol are in equilibrium at a fixed temperature and the substance is distributed between the water-rich and octanol-rich phases. P is defined as the ratio of the equilibrium concentration of the substance in the octanol-rich phase to that in the water-rich phase, in the limit of zero concentration. In general, P tends to be large for compounds with extended non-polar structures (such as long chain or multi-ring hydrocarbons) and small for compounds with highly polar groups. Thus P (or, in its more common form of expression, $\log P$) provides a measure of the lipophilic vs. hydrophilic nature of a compound, which is an important consideration in assessing the potential toxicity. A discussion of methods of measurement and accuracy considerations for $\log P$ may be found in Reference 1.

This table gives selected values of $\log P$ for about 450 organic compounds, including many of environmental importance. All values refer to a nominal temperature of 25°C. The source of each value is indicated in the last column. These references contain data on many more compounds than are included here.

Compounds are listed by molecular formula following the Hill convention. To locate a compound by name or CAS Registry Number when the molecular formula is not known, use the table "Physical Constants of Organic Compounds" in Section 3 and its indexes to determine the molecular formula.

REFERENCES

1. Sangster, J., *J. Phys. Chem. Ref. Data*, 18, 1111, 1989.
2. Mackay, D., Shiu, W.Y., and Ma, K.C., *Illustrated Handbook of Physical-Chemical Properties and Environmental Fate for Organic Chemicals*, Lewis Publishers/CRC Press, Boca Raton, FL, 1992.
3. Shiu, W.Y., and Mackay, D., *J. Phys. Chem. Ref. Data*, 15, 911, 1986.
4. Pinsuwan, S., Li, L., and Yalkowsky, S.H., *J. Chem. Eng. Data*, 40, 623, 1995.
5. *Solubility Data Series, International Union of Pure and Applied Chemistry, Vol. 20*, Pergamon Press, Oxford, 1985.
6. *Solubility Data Series, International Union of Pure and Applied Chemistry, Vol. 38*, Pergamon Press, Oxford, 1985.
7. Miller, M.M., Ghodbane, S., Wasik, S.P., Tewari, Y.B., and Martire, D.E., *J. Chem. Eng. Data*, 29, 184, 1984.

Mol. Form.	Name	$\log P$	Ref.	Mol. Form.	Name	$\log P$	Ref.
CCl_2F_2	Dichlorodifluoromethane	2.16	2	C_2H_4O	Acetaldehyde	0.45	1
CCl_3F	Trichlorofluoromethane	2.53	2	C_2H_4O	Ethylene oxide	-0.30	1
CCl_4	Tetrachloromethane	2.64	2	$C_2H_4O_2$	Acetic acid	-0.17	1
$CHBr_3$	Tribromomethane	2.38	2	C_2H_5Br	Bromoethane	1.6	2
$CHCl_3$	Trichloromethane	1.97	2	C_2H_5Cl	Chloroethane	1.43	2
CH_2BrCl	Bromochloromethane	1.41	2	C_2H_5I	Iodoethane	2	2
CH_2Br_2	Dibromomethane	2.3	2	C_2H_5NO	Acetamide	-1.26	1
CH_2Cl_2	Dichloromethane	1.25	2	$C_2H_5NO_2$	Nitroethane	0.18	1
CH_2F_2	Difluoromethane	0.20	1	C_2H_6O	Ethanol	-0.30	1
CH_2I_2	Diiodomethane	2.5	2	C_2H_6O	Dimethyl ether	0.10	1
CH_2O	Formaldehyde	0.35	1	C_2H_6OS	Dimethyl sulfoxide	-1.35	1
CH_2O_2	Formic acid	-0.54	1	$C_2H_6O_2S$	Dimethyl sulfone	-1.41	1
CH_3Br	Bromomethane	1.19	2	C_2H_7N	Ethylamine	-0.13	1
CH_3Cl	Chloromethane	0.91	2	C_2H_7N	Dimethylamine	-0.38	1
CH_3F	Fluoromethane	0.51	1	C_3H_3N	2-Propenenitrile	0.25	1
CH_3I	Iodomethane	1.5	2	$C_3H_4Cl_2$	*cis*-1,3-Dichloropropene	2.03	2
CH_3NO	Formamide	-1.51	1	C_3H_4O	Propargyl alcohol	-0.38	1
CH_3NO_2	Nitromethane	-0.33	1	C_3H_4O	Acrolein	-0.01	1
CH_4O	Methanol	-0.74	1	C_3H_5Br	3-Bromopropene	1.79	1
CH_5N	Methylamine	-0.57	1	C_3H_5ClO	Epichlorohydrin	0.30	2
$C_2Cl_3F_3$	1,1,2-Trichlorotrifluoroethane	3.16	2	$C_3H_5Cl_3$	1,2,3-Trichloropropane	2.63	2
C_2Cl_4	Tetrachloroethylene	2.88	2	C_3H_5N	Propanenitrile	0.16	1
C_2Cl_6	Hexachloroethane	4.00	4	C_3H_5NO	Acrylamide	-0.78	1
C_2HCl_3	Trichloroethylene	2.53	2	$C_3H_6Cl_2$	1,2-Dichloropropane	2.0	2
C_2HCl_5	Pentachloroethane	2.89	2	C_3H_6O	Allyl alcohol	0.17	1
$C_2H_2Cl_2$	1,1-Dichloroethylene	2.13	2	C_3H_6O	Propanal	0.59	1
$C_2H_2Cl_2$	*cis*-1,2-Dichloroethylene	1.86	2	C_3H_6O	Acetone	-0.24	1
$C_2H_2Cl_2$	*trans*-1,2-Dichloroethylene	1.93	2	C_3H_6O	Methyloxirane	0.03	1
$C_2H_2Cl_4$	1,1,2,2-Tetrachloroethane	2.39	2	$C_3H_6O_2$	Propanoic acid	0.33	1
C_2H_3Cl	Chloroethylene	1.38	2	$C_3H_6O_2$	Methyl acetate	0.18	1
$C_2H_3Cl_3$	1,1,1-Trichloroethane	2.49	2	C_3H_7Br	1-Bromopropane	2.1	2
$C_2H_3Cl_3$	1,1,2-Trichloroethane	2.38	2	C_3H_7Br	2-Bromopropane	1.9	2
C_2H_3N	Acetonitrile	-0.34	1	C_3H_7Cl	1-Chloropropane	2.04	1
$C_2H_4Cl_2$	1,1-Dichloroethane	1.79	2	C_3H_7Cl	2-Chloropropane	1.90	1
$C_2H_4Cl_2$	1,2-Dichloroethane	1.48	2	C_3H_7I	1-Iodopropane	2.5	2

Mol. Form.	Name	log P	Ref.	Mol. Form.	Name	log P	Ref.
C_3H_7N	Allylamine	0.03	1	C_5H_{10}	Cyclopentane	3.00	1
C_3H_7NO	N,N-Dimethylformamide	-1.01	1	$C_5H_{10}O$	2-Pentanone	0.84	1
C_3H_7NO	N-Methylacetamide	-1.05	1	$C_5H_{10}O$	3-Pentanone	0.82	1
$C_3H_7NO_2$	1-Nitropropane	0.87	1	$C_5H_{10}O$	3-Methyl-2-butanone	0.56	1
C_3H_8O	1-Propanol	0.25	1	$C_5H_{10}O$	Tetrahydropyran	0.82	1
C_3H_8O	2-Propanol	0.05	1	$C_5H_{10}O$	2-Methyltetrahydrofuran	1.85	2
C_3H_8S	1-Propanethiol	1.81	1	$C_5H_{10}O_2$	Pentanoic acid	1.39	1
C_3H_9N	Propylamine	0.48	1	$C_5H_{10}O_2$	Propyl acetate	1.24	1
C_3H_9N	Isopropylamine	0.26	1	$C_5H_{10}O_2$	Ethyl propanoate	1.21	1
C_3H_9N	Ethylmethylamine	0.15	1	$C_5H_{10}O_3$	Diethyl carbonate	1.21	1
C_3H_9N	Trimethylamine	0.16	1	$C_5H_{11}Br$	1-Bromopentane	3.37	1
C_4H_4O	Furan	1.34	1	$C_5H_{11}F$	1-Fluoropentane	2.33	1
C_4H_4S	Thiophene	1.81	1	$C_5H_{11}N$	Piperidine	0.84	1
C_4H_5N	Pyrrole	0.75	1	$C_5H_{11}NO_2$	1-Nitropentane	2.01	1
C_4H_6	1,3-Butadiene	1.99	1	C_5H_{12}	Pentane	3.45	1
C_4H_6	2-Butyne	1.46	1	C_5H_{12}	Neopentane	3.11	1
C_4H_6O	2,5-Dihydrofuran	0.46	1	$C_5H_{12}O$	1-Pentanol	1.51	1
$C_4H_6O_2$	Methacrylic acid	0.93	1	$C_5H_{12}O$	2-Pentanol	1.25	1
$C_4H_6O_2$	Vinyl acetate	0.73	1	$C_5H_{12}O$	3-Pentanol	1.21	1
$C_4H_6O_2$	Methyl acrylate	0.80	1	$C_5H_{12}O$	3-Methyl-1-butanol	1.28	1
C_4H_7N	Butanenitrile	0.60	1	$C_5H_{12}O$	2-Methyl-2-butanol	0.89	1
C_4H_8	cis-2-Butene	2.33	1	$C_5H_{12}O$	3-Methyl-2-butanol	1.28	1
C_4H_8	$trans$-2-Butene	2.31	1	$C_5H_{12}O$	2,2-Dimethyl-1-propanol	1.31	1
C_4H_8	Isobutene	2.35	1	$C_5H_{12}O$	Methyl $tert$-butyl ether	0.94	1
$C_4H_8Cl_2O$	Bis(2-chloroethyl) ether	1.12	2	$C_5H_{13}N$	Pentylamine	1.49	1
C_4H_8O	Ethyl vinyl ether	1.04	1	C_6Cl_6	Hexachlorobenzene	5.47	5
C_4H_8O	Butanal	0.88	1	C_6HCl_5	Pentachlorobenzene	5.03	5
C_4H_8O	2-Butanone	0.29	1	C_6HCl_5O	Pentachlorophenol	5.07	4
C_4H_8O	Tetrahydrofuran	0.46	1	$C_6H_2Cl_4$	1,2,3,4-Tetrachlorobenzene	4.55	5
$C_4H_8O_2$	Butanoic acid	0.79	1	$C_6H_2Cl_4$	1,2,3,5-Tetrachlorobenzene	4.65	5
$C_4H_8O_2$	Propyl formate	0.83	1	$C_6H_2Cl_4$	1,2,4,5-Tetrachlorobenzene	4.51	5
$C_4H_8O_2$	Ethyl acetate	0.73	1	$C_6H_3Cl_3$	1,2,3-Trichlorobenzene	4.04	5
C_4H_9Br	1-Bromobutane	2.75	1	$C_6H_3Cl_3$	1,2,4-Trichlorobenzene	3.98	5
C_4H_9Cl	1-Chlorobutane	2.64	2	$C_6H_3Cl_3$	1,3,5-Trichlorobenzene	4.02	5
C_4H_9F	1-Fluorobutane	2.58	1	$C_6H_4Cl_2$	o-Dichlorobenzene	3.38	5
C_4H_9I	1-Iodobutane	3	2	$C_6H_4Cl_2$	m-Dichlorobenzene	3.48	5
C_4H_9N	Pyrrolidine	0.46	1	$C_6H_4Cl_2$	p-Dichlorobenzene	3.38	5
C_4H_9NO	Butanamide	-0.21	1	$C_6H_4Cl_2O$	2,4-Dichlorophenol	3.23	4
C_4H_9NO	N,N-Dimethylacetamide	-0.77	1	C_6H_5Br	Bromobenzene	2.99	2
$C_4H_9NO_2$	1-Nitrobutane	1.47	1	C_6H_5Cl	Chlorobenzene	2.84	1
C_4H_{10}	Isobutane	2.8	2	C_6H_5F	Fluorobenzene	2.27	2
$C_4H_{10}O$	1-Butanol	0.84	1	C_6H_5I	Iodobenzene	3.28	2
$C_4H_{10}O$	2-Butanol	0.65	1	$C_6H_5NO_2$	Nitrobenzene	1.85	1
$C_4H_{10}O$	2-Methyl-1-propanol	0.76	1	C_6H_6	Benzene	2.13	1
$C_4H_{10}O$	2-Methyl-2-propanol	0.35	1	C_6H_6O	Phenol	1.48	4
$C_4H_{10}O$	Diethyl ether	0.89	1	C_6H_6S	Benzenethiol	2.52	1
$C_4H_{10}S$	1-Butanethiol	2.28	1	C_6H_7N	Aniline	0.90	1
$C_4H_{10}S$	Diethyl sulfide	1.95	1	C_6H_7N	2-Methylpyridine	1.11	1
$C_4H_{11}N$	Butylamine	0.86	1	C_6H_7N	3-Methylpyridine	1.20	1
$C_4H_{11}N$	$tert$-Butylamine	0.40	1	C_6H_7N	4-Methylpyridine	1.22	1
$C_4H_{11}N$	Diethylamine	0.58	1	C_6H_8	1,4-Cyclohexadiene	2.3	2
C_5H_5N	Pyridine	0.65	1	C_6H_8O	5-Hexyn-2-one	0.58	1
C_5H_6O	2-Methylfuran	1.85	1	C_6H_8O	2-Cyclohexen-1-one	0.61	1
C_5H_7N	1-Methylpyrrole	1.21	1	C_6H_8O	2-Ethylfuran	2.40	1
C_5H_8	1,4-Pentadiene	2.48	1	C_6H_{10}	1,5-Hexadiene	2.8	2
C_5H_8	1-Pentyne	1.98	1	C_6H_{10}	1-Hexyne	2.73	2
$C_5H_8O_2$	Methyl methacrylate	1.38	1	C_6H_{10}	Cyclohexene	2.86	1
$C_5H_8O_2$	Ethyl acrylate	1.32	1	$C_6H_{10}O$	5-Hexen-2-one	1.02	1
C_5H_9N	Pentanenitrile	0.94	1	$C_6H_{10}O$	Cyclohexanone	0.81	1
C_5H_{10}	1-Pentene	2.2	2	$C_6H_{10}O_2$	Ethyl methacrylate	1.94	1

Mol. Form.	Name	log P	Ref.	Mol. Form.	Name	log P	Ref.
$C_6H_{11}Br$	Bromocyclohexane	3.20	1	C_7H_{16}	Heptane	4.50	1
$C_6H_{11}N$	Hexanenitrile	1.66	1	$C_7H_{16}O$	1-Heptanol	2.62	1
C_6H_{12}	1-Hexene	3.40	1	$C_7H_{16}O$	2-Heptanol	2.31	1
C_6H_{12}	4-Methyl-1-pentene	2.5	2	$C_7H_{16}O$	3-Heptanol	2.24	1
C_6H_{12}	Cyclohexane	3.44	1	$C_7H_{16}O$	4-Heptanol	2.22	1
C_6H_{12}	Methylcyclopentane	3.37	2	$C_7H_{17}N$	Heptylamine	2.57	1
$C_6H_{12}O$	Cyclohexanol	1.23	1	C_8H_6	Phenylacetylene	2.40	1
$C_6H_{12}O$	Hexanal	1.78	1	C_8H_6O	Benzofuran	2.67	1
$C_6H_{12}O$	2-Hexanone	1.38	1	C_8H_6S	Benzo[b]thiophene	3.12	1
$C_6H_{12}O$	4-Methyl-2-pentanone	1.31	1	C_8H_7N	Benzeneacetonitrile	1.56	1
$C_6H_{12}O_2$	Hexanoic acid	1.92	1	C_8H_7N	Indole	2.14	1
$C_6H_{12}O_2$	Butyl acetate	1.82	1	C_8H_8	Styrene	3.05	1
$C_6H_{13}Br$	1-Bromohexane	3.80	1	C_8H_8O	Acetophenone	1.63	1
$C_6H_{13}N$	Cyclohexylamine	1.49	1	C_8H_8O	2-Methylbenzaldehyde	2.26	1
C_6H_{14}	Hexane	4.00	1	C_8H_8O	Benzeneacetaldehyde	1.78	1
C_6H_{14}	3-Methylpentane	3.60	2	C_8H_8O	2,3-Dihydrobenzofuran	2.14	1
C_6H_{14}	2,2-Dimethylbutane	3.82	1	C_8H_8O	Phenyloxirane	1.61	1
C_6H_{14}	2,3-Dimethylbutane	3.85	2	$C_8H_8O_2$	o-Toluic acid	2.32	4
$C_6H_{14}O$	1-Hexanol	2.03	1	$C_8H_8O_2$	m-Toluic acid	2.37	1
$C_6H_{14}O$	2-Hexanol	1.76	1	$C_8H_8O_2$	p-Toluic acid	2.34	1
$C_6H_{14}O$	3-Hexanol	1.65	1	$C_8H_8O_2$	Benzeneacetic acid	1.41	1
$C_6H_{14}O$	3,3-Dimethyl-2-butanol	1.48	1	$C_8H_8O_2$	Phenyl acetate	1.49	1
$C_6H_{14}O$	Dipropyl ether	2.03	1	$C_8H_8O_2$	Methyl benzoate	2.20	1
$C_6H_{14}O$	Diisopropyl ether	1.52	1	C_8H_{10}	Ethylbenzene	3.15	1
$C_6H_{15}N$	Hexylamine	2.06	1	C_8H_{10}	o-Xylene	3.12	1
$C_6H_{15}N$	Dipropylamine	1.67	1	C_8H_{10}	m-Xylene	3.20	1
$C_6H_{15}N$	Triethylamine	1.45	1	C_8H_{10}	p-Xylene	3.15	1
$C_7H_5BrO_2$	2-Bromobenzoic acid	2.20	4	$C_8H_{10}O$	o-Ethylphenol	2.47	1
$C_7H_5BrO_2$	3-Bromobenzoic acid	2.87	4	$C_8H_{10}O$	m-Ethylphenol	2.50	1
$C_7H_5BrO_2$	4-Bromobenzoic acid	2.86	4	$C_8H_{10}O$	p-Ethylphenol	2.50	1
C_7H_5N	Benzonitrile	1.56	1	$C_8H_{10}O$	2,4-Xylenol	2.35	1
C_7H_6O	Benzaldehyde	1.48	1	$C_8H_{10}O$	2,5-Xylenol	2.34	1
$C_7H_6O_2$	Benzoic acid	1.88	4	$C_8H_{10}O$	2,6-Xylenol	2.36	1
$C_7H_6O_2$	Phenyl formate	1.26	1	$C_8H_{10}O$	3,4-Xylenol	3.23	1
$C_7H_6O_3$	Salicylic acid	2.20	4	$C_8H_{10}O$	3,5-Xylenol	2.35	1
C_7H_7Br	(Bromomethyl)benzene	2.92	1	$C_8H_{10}O$	Benzeneethanol	1.36	1
C_7H_7Cl	o-Chlorotoluene	3.42	1	$C_8H_{10}O$	α-Methylbenzyl alcohol	1.42	1
C_7H_7Cl	m-Chlorotoluene	3.28	1	$C_8H_{10}O$	3-Methylbenzenemethanol	1.60	1
C_7H_7Cl	p-Chlorotoluene	3.33	1	$C_8H_{10}O$	4-Methylbenzenemethanol	1.58	1
C_7H_7Cl	(Chloromethyl)benzene	2.30	1	$C_8H_{10}O$	Phenetole	2.51	1
$C_7H_7NO_2$	p-Nitrotoluene	2.42	1	$C_8H_{10}O$	Benzyl methyl ether	1.35	1
C_7H_8	Toluene	2.73	1	$C_8H_{10}O$	2-Methylanisole	2.74	1
C_7H_8	1,3,5-Cycloheptatriene	2.63	2	$C_8H_{10}O$	3-Methylanisole	2.66	1
C_7H_8O	o-Cresol	1.98	1	$C_8H_{10}O$	4-Methylanisole	2.81	1
C_7H_8O	m-Cresol	1.98	1	$C_8H_{11}N$	p-Ethylaniline	1.96	1
C_7H_8O	p-Cresol	1.97	1	$C_8H_{11}N$	N,N-Dimethylaniline	2.31	1
C_7H_8O	Benzyl alcohol	1.05	1	$C_8H_{11}N$	Benzeneethanamine	1.41	1
C_7H_8O	Anisole	2.11	1	$C_8H_{14}O_2$	Butyl methacrylate	2.88	1
C_7H_9N	Benzylamine	1.09	1	$C_8H_{15}N$	Octanenitrile	2.75	1
C_7H_9N	o-Methylaniline	1.32	1	C_8H_{16}	1-Octene	4.57	1
C_7H_9N	m-Methylaniline	1.40	1	C_8H_{16}	Cyclooctane	4.45	2
C_7H_9N	p-Methylaniline	1.39	1	$C_8H_{16}O$	2-Octanone	2.37	1
C_7H_9N	N-Methylaniline	1.66	1	$C_8H_{16}O_2$	Octanoic acid	3.05	1
C_7H_{14}	1-Heptene	3.99	1	$C_8H_{17}Br$	1-Bromooctane	4.89	1
C_7H_{14}	Methylcyclohexane	3.88	1	C_8H_{18}	Octane	5.15	1
$C_7H_{14}O$	2-Heptanone	1.98	1	$C_8H_{18}O$	1-Octanol	3.07	1
$C_7H_{14}O$	5-Methyl-2-hexanone	1.88	1	$C_8H_{18}O$	2-Octanol	2.90	1
$C_7H_{15}Br$	1-Bromoheptane	4.36	1	$C_8H_{18}O$	4-Octanol	2.68	1
$C_7H_{15}Cl$	1-Chloroheptane	4.15	1	$C_8H_{18}O$	Dibutyl ether	3.21	1
$C_7H_{15}I$	1-Iodoheptane	4.70	1	C_9H_7N	Quinoline	2.03	1

Mol. Form.	Name	log P	Ref.	Mol. Form.	Name	log P	Ref.
C_9H_7N	Isoquinoline	2.08	1	$C_{12}H_2Cl_8$	2,2′,3,3′,5,5′,6,6′-Octachlorobiphenyl	7.10	3
C_9H_8	Indene	2.92	1	$C_{12}H_3Cl_7$	2,2′,3,3′,4,4′,6-Heptachlorobiphenyl	6.70	3
$C_9H_8O_2$	trans-Cinnamic acid	2.13	1	$C_{12}H_4Cl_6$	2,2′,3,3′,4,4′-Hexachlorobiphenyl	7.00	3
C_9H_9N	Benzenepropanenitrile	1.72	1	$C_{12}H_4Cl_6$	2,2′,4,4′,6,6′-Hexachlorobiphenyl	7.00	3
C_9H_{10}	Indan	3.33	1	$C_{12}H_4Cl_6$	2,2′,3,3′,6,6′-Hexachlorobiphenyl	6.70	3
$C_9H_{10}O$	1-Phenyl-1-propanone	2.19	1	$C_{12}H_5Cl_5$	2,3,4,5,6-Pentachlorobiphenyl	6.30	3
$C_9H_{10}O$	1-Phenyl-2-propanone	1.44	1	$C_{12}H_5Cl_5$	2,2′,4,5,5′-Pentachlorobiphenyl	6.40	3
$C_9H_{10}O$	4-Methylacetophenone	2.19	1	$C_{12}H_6Cl_4$	2,3,4,5-Tetrachlorobiphenyl	5.72	3
$C_9H_{10}O_2$	2-Phenylpropanoic acid	1.80	1	$C_{12}H_6Cl_4$	2,2′,4′,5-Tetrachlorobiphenyl	5.73	7
$C_9H_{10}O_2$	Benzyl acetate	1.96	1	$C_{12}H_7Cl_3$	2,4,5-Trichlorobiphenyl	5.60	3
$C_9H_{10}O_2$	4-Methylphenyl acetate	2.11	1	$C_{12}H_7Cl_3$	2,4,6-Trichlorobiphenyl	5.47	3
$C_9H_{10}O_2$	Ethyl benzoate	2.64	1	$C_{12}H_8Cl_2$	2,5-Dichlorobiphenyl	5.10	3
C_9H_{12}	Propylbenzene	3.69	1	$C_{12}H_8Cl_2$	2,6-Dichlorobiphenyl	5.00	3
C_9H_{12}	Isopropylbenzene	3.66	1	$C_{12}H_8O$	Dibenzofuran	4.12	1
C_9H_{12}	o-Ethyltoluene	3.53	1	$C_{12}H_9Cl$	2-Chlorobiphenyl	4.52	1
C_9H_{12}	p-Ethyltoluene	3.63	2	$C_{12}H_9Cl$	3-Chlorobiphenyl	4.58	1
C_9H_{12}	1,2,3-Trimethylbenzene	3.60	1	$C_{12}H_9Cl$	4-Chlorobiphenyl	4.61	1
C_9H_{12}	1,2,4-Trimethylbenzene	3.63	1	$C_{12}H_9N$	Carbazole	3.72	1
C_9H_{12}	1,3,5-Trimethylbenzene	3.42	1	$C_{12}H_{10}$	Acenaphthene	3.96	4
$C_9H_{12}O$	2-Propylphenol	2.93	1	$C_{12}H_{10}$	Biphenyl	3.76	6
$C_9H_{12}O$	4-Propylphenol	3.20	1	$C_{12}H_{10}N_2$	Azobenzene	3.82	1
$C_9H_{12}O$	2,3,6-Trimethylphenol	2.67	1	$C_{12}H_{10}O$	Diphenyl ether	4.21	1
$C_9H_{12}O$	2,4,6-Trimethylphenol	2.46	1	$C_{12}H_{10}S$	Diphenyl sulfide	4.45	1
$C_9H_{12}O$	Benzenepropanol	1.88	1	$C_{12}H_{11}N$	Diphenylamine	3.44	4
$C_9H_{13}N$	N,N-Dimethylbenzylamine	1.98	1	$C_{12}H_{12}$	1-Ethylnaphthalene	4.40	1
$C_9H_{13}N$	Amphetamine	1.76	1	$C_{12}H_{12}$	1,2-Dimethylnaphthalene	4.31	1
C_9H_{18}	1-Nonene	5.15	1	$C_{12}H_{12}$	1,4-Dimethylnaphthalene	4.37	1
$C_9H_{18}O$	2-Nonanone	3.16	1	$C_{12}H_{14}O$	4-Phenylcyclohexanone	2.45	1
$C_9H_{18}O$	5-Methyl-2-octanone	2.92	1	$C_{12}H_{18}$	Hexylbenzene	5.52	1
C_9H_{20}	Nonane	5.65	1	$C_{12}H_{18}$	Hexamethylbenzene	4.69	4
$C_9H_{20}O$	1-Nonanol	4.02	1	$C_{12}H_{22}O$	Cyclododecanone	4.10	1
$C_9H_{21}N$	Tripropylamine	2.79	1	$C_{12}H_{24}O_2$	Dodecanoic acid	4.6	1
$C_{10}H_7Cl$	1-Chloronaphthalene	3.90	1	$C_{12}H_{26}O$	1-Dodecanol	5.13	1
$C_{10}H_7Cl$	2-Chloronaphthalene	3.98	1	$C_{13}H_8O$	9H-Fluoren-9-one	3.58	1
$C_{10}H_8$	Naphthalene	3.34	4	$C_{13}H_9N$	Acridine	3.40	1
$C_{10}H_8$	Azulene	3.22	1	$C_{13}H_{10}$	9H-Fluorene	4.20	4
$C_{10}H_8O$	1-Naphthol	2.84	1	$C_{13}H_{10}O$	Benzophenone	3.18	1
$C_{10}H_8O$	2-Naphthol	2.70	1	$C_{13}H_{10}O_2$	Phenyl benzoate	3.59	1
$C_{10}H_{12}O_2$	Isopropyl benzoate	3.18	1	$C_{13}H_{11}NO$	N-Phenylbenzamide	2.62	1
$C_{10}H_{14}$	Butylbenzene	4.26	1	$C_{13}H_{12}$	Diphenylmethane	4.14	1
$C_{10}H_{14}$	tert-Butylbenzene	4.11	1	$C_{13}H_{12}$	4-Methylbiphenyl	4.63	1
$C_{10}H_{14}$	Isobutylbenzene	4.01	2	$C_{13}H_{12}O$	Diphenylmethanol	2.67	1
$C_{10}H_{14}$	p-Cymene	4.10	1	$C_{13}H_{12}O$	Benzyl phenyl ether	3.79	1
$C_{10}H_{14}$	1,2,4,5-Tetramethylbenzene	4.10	2	$C_{14}H_{10}$	Anthracene	4.56	4
$C_{10}H_{14}$	1,2,3,4-Tetramethylbenzene	4.00	1	$C_{14}H_{10}$	Phenanthrene	4.52	4
$C_{10}H_{14}$	1,2,3,5-Tetramethylbenzene	4.10	1	$C_{14}H_{12}$	trans-Stilbene	4.81	1
$C_{10}H_{14}O$	4-Butylphenol	3.65	1	$C_{14}H_{12}$	1-Methylfluorene	4.97	1
$C_{10}H_{20}O$	2-Decanone	3.77	1	$C_{14}H_{12}O$	2-Phenylacetophenone	3.18	1
$C_{10}H_{20}O_2$	Decanoic acid	4.09	1	$C_{14}H_{12}O_2$	Benzyl benzoate	3.97	1
$C_{10}H_{22}$	Decane	6.25	1	$C_{14}H_{14}$	1,2-Diphenylethane	4.70	1
$C_{10}H_{22}O$	1-Decanol	4.57	1	$C_{14}H_{14}$	4,4′-Dimethylbiphenyl	5.09	1
$C_{11}H_9N$	4-Phenylpyridine	2.59	1	$C_{14}H_{22}$	Octylbenzene	6.30	1
$C_{11}H_{10}$	1-Methylnaphthalene	3.87	1	$C_{14}H_{28}O_2$	Tetradecanoic acid	6.1	1
$C_{11}H_{10}$	2-Methylnaphthalene	4.00	1	$C_{15}H_{12}$	2-Methylanthracene	5.15	2
$C_{11}H_{16}$	Pentylbenzene	4.90	1	$C_{15}H_{12}$	9-Methylanthracene	5.07	1
$C_{11}H_{16}$	Pentamethylenzene	4.56	1	$C_{15}H_{12}$	1-Methylphenanthrene	5.14	2
$C_{11}H_{22}O$	2-Undecanone	4.09	1	$C_{16}H_{10}$	Fluoranthene	5.07	4
$C_{11}H_{22}O_2$	Methyl decanoate	4.41	1	$C_{16}H_{10}$	Pyrene	5.08	4
$C_{12}Cl_{10}$	Decachlorobiphenyl	8.26	3	$C_{16}H_{14}$	9,10-Dimethylanthracene	5.69	1
$C_{12}HCl_9$	2,2′,3,3′,4,5,5′,6,6′-Nonachlorobiphenyl	8.16	3	$C_{16}H_{32}O_2$	Hexadecanoic acid	7.17	1

Mol. Form.	Name	log P	Ref.	Mol. Form.	Name	log P	Ref.
$C_{17}H_{12}$	11H-Benzo[a]fluorene	5.40	1	$C_{18}H_{36}O_2$	Stearic acid	8.23	1
$C_{17}H_{12}$	11H-Benzo[b]fluorene	5.75	1	$C_{19}H_{16}O$	Triphenylmethanol	3.68	1
$C_{18}H_{12}$	Benz[a]anthracene	5.91	1	$C_{20}H_{12}$	Perylene	6.25	1
$C_{18}H_{12}$	Chrysene	5.73	4	$C_{20}H_{12}$	Benzo[a]pyrene	6.20	4
$C_{18}H_{12}$	Naphthacene	5.76	1	$C_{20}H_{32}O_2$	Arachidonic acid	6.98	1
$C_{18}H_{12}$	Triphenylene	5.49	4	$C_{20}H_{40}O_2$	Arachidic acid	9.29	1
$C_{18}H_{15}N$	Triphenylamine	5.74	1	$C_{21}H_{16}$	1,2-Dihydro-3-methylbenz[j]aceanthrylene	6.75	1
$C_{18}H_{30}O_2$	Linolenic acid	6.46	1				
$C_{18}H_{32}O_2$	Linoleic acid	7.05	1	$C_{22}H_{12}$	Benzo[ghi]perylene	6.90	1
$C_{18}H_{34}O_2$	Oleic acid	7.64	1	$C_{24}H_{12}$	Coronene	6.05	4

PROTECTION AGAINST IONIZING RADIATION

The following data and rules of thumb are helpful in estimating the penetrating capability of and danger of exposure to various types of ionizing radiation. More precise data should be used for critical applications.

Alpha Particles

Alpha particles of at least 7.5 MeV are required to penetrate the epidermis, the protective layer of skin, 0.07 mm thick.

Electrons

Electrons of at least 70 keV are required to penetrate the epidermis, the protective layer of skin, 0.07 mm thick.

The range of electrons in g/cm^2 is approximately equal to the maximum energy (E) in MeV divided by 2.

The range of electrons in air is about 3.65 m per MeV; for example, a 3 MeV electron has a range of about 11 m in air.

A chamber wall thickness of 30 mg/cm^2 will transmit 70% of the initial fluence of 1 MeV electrons and 20% of that of 0.4 MeV electrons.

When electrons of 1 to 2 MeV pass through light materials such as water, aluminum, or glass, less than 1% of their energy is dissipated as bremsstrahlung.

The bremsstrahlung from 1 Ci of ^{32}P aqueous solution in a glass bottle is about 1 mR/h at 1 meter distance.

When electrons from a 1 Ci source of ^{90}Sr - ^{90}Y are absorbed, the bremsstrahlung hazard is approximately equal to that presented by the gamma radiation from 12 mg of radium. The average energy of the bremsstrahlung is about 300 keV.

Gamma Rays

The air-scattered radiation (sky-shine) from a 100 Ci ^{60}Co source placed 1 ft behind a 4 ft high shield is about 100 mrad/h at 6 ft from the outside of the shield.

Within ±20% for point source gamma emitters with energies between 0.07 and 4 MeV, the exposure rate (R/h) at 1 ft is $6C \cdot E \cdot n$ where C is the activity in curies, E is the energy in MeV, and n is the number of gammas per disintegration.

Neutrons

An approximate HVL (thickness of absorber for which the neutron flux falls to half its initial value) for 1 MeV neutrons is 3.2 cm of paraffin; that for 5 MeV neutrons is 6.9 cm of paraffin).

Miscellaneous

The activity of any radionuclide is reduced to less than 1% after 7 half-lives (i.e., $2^{-7} = 0.8\%$).

For nuclides with a half-life greater than 6 days, the change in activity in 24 hours will be less than 10%.

10 HVL (half-value layers) attenuates approximately by 10^{-3}.

There is 0.64 mm^3 of radon gas at STP in transient equilibrium with 1 Ci of radium.

The natural background from all sources in most parts of the world leads to an equivalent dose rate of about 0.04 to 4 mSv per year for the average person. About 84% of this comes from terrestrial sources, the remainder from cosmic rays. The U. S. average is about 3.6 mSv/yr but can range up to 50 mSv/yr in some areas. A passenger in a plane flying at 12,000 meters receives 5 μSv/hr from cosmic rays (as compared to about 0.03 μSv/hr at sea level).

The ICRP recommended exposure limit to man-made sources of ionizing radiation (Reference 2) is 20 mSv/yr averaged over 5 years, with the dose in any one year not to exceed 50 mSv.

A whole-body dose of about 3 Gy over a short time interval will typically lead to 50% mortality in 30 days assuming no medical treatment.

Units

The gray (Gy) is the SI unit of absorbed dose; it is a measure of the mean energy imparted to a sample of irradiated matter, divided by the mass of the sample. Gy is a special name for the SI unit J/kg.

The sievert (Sv) is the SI unit of equivalent dose, which is defined as the absorbed dose multiplied by a weighting factor that expresses the long-term biological risk from low-level chronic exposure to a specified type of radiation. The Sv is another special name for J/kg.

1 curie (Ci) = $3.7 \cdot 10^{10}$ becquerel (Bq); i.e., $3.7 \cdot 10^{10}$ disintegrations per second.

1 roentgen (R) = $2.58 \cdot 10^{-4}$ coulomb per kilogram (C/kg); a measure of the charge (positive or negative) liberated by x-ray or gamma radiation in air, divided by the mass of air.

1 rad = 0.01 Gy

1 rem = 0.01 Sv

REFERENCES

1. Padikal, T.N., and Fivozinsky, S.P., *Medical Physics Data Book, National Bureau of Standards Handbook 138*, U. S. Government Printing Office, Washington, D.C., 1981.
2. *1990 Recommendations of the International Commission on Radiological Protection*, ICRP Publication 60, *Annals of the ICRP*, Pergamon Press, Oxford, 1991.
3. *Radiation: Doses, Effects, Risks*, United Nations Sales No. E.86.III.D.4, 1985.
4. *Review of Particle Properties, Phys. Rev. D*, 50, 1173, 1994 (p. 1268).

ANNUAL LIMITS ON INTAKES OF RADIONUCLIDES

K. F. Eckerman

The following table lists, for workers, the annual limits on oral and inhalation intakes (ALI) for selected radionuclides based on the occupational radiation protection guidance of the International Commission on Radiological Protection (References 1 and 2). An intake of one ALI corresponds to an annual whole body dose of 0.02 Sv (2 rem).

The ALI is expressed in the SI unit of activity, the becquerel (Bq), and in the conventional unit, the microcurie (μCi); 1 μCi = $3.7 \cdot 10^4$ Bq. The chemical form of inhaled radionuclides is, in most instances, stated in terms of the rate of absorption to blood from the lungs and the fractional absorption from the small intestine. Type F, M, and S denote chemical forms which are absorbed from the lungs at rates characterized as fast, moderate, and slow, respectively. The time to absorb 90% of the deposited radionuclide, in the absence of radioactive decay, corresponds to about 10 minutes, 150 days, and 7000 days for Type F, M, and S compounds, respectively. Type F compounds can be considered to be more soluble than M or S, S being the most insoluble. Chemical form consideration for ingestion is specified by the fractional absorption from the small intestine, denoted as f_1. The f_1 values range from 10^{-5} to 1. Higher fractional absorption is associated with greater solubility of the compound.

REFERENCES

1. *1990 Recommendations of the International Commission on Radiological Protection, ICRP Publication 60, Annals of the ICRP 21, (1—3)*, Pergamon Press, Oxford, 1991.
2. *Dose Coefficients for Intakes of Radionuclides by Workers, ICRP Publication 68, Annals of the ICRP*, 24(4), Pergamon Press, Oxford, 1995.

	Physical half-life	Inhalation intakes			Oral intakes		
		Chemical form Type/f_1	ALI Bq	ALI μCi	Chemical form f_1	ALI Bq	ALI μCi
^{3}H	12.3 y	HT gas	1.1E+13	3.0E+08	1.000	1.1E+13	3.0E+08
		HTO vapor	1.1E+09	3.0E+04			
^{11}C	0.340 h	CO	1.7E+10	4.5E+05	1.000	8.3E+08	2.3E+04
		CO_2	9.1E+09	2.5E+05			
		Organic compounds	6.2E+09	1.7E+05			
^{14}C	5730 y	CO	2.5E+10	6.8E+05	1.000	3.4E+07	9.3E+02
		CO_2	3.1E+09	8.3E+04			
		Organic compounds	3.4E+07	9.3E+02			
^{18}F	1.83 h	F 1.000	3.7E+08	1.0E+04	1.000	4.1E+08	1.1E+04
		M 1.000	2.2E+08	6.1E+03			
		S 1.000	2.2E+08	5.8E+03			
^{22}Na	2.60 y	F 1.000	1.0E+07	2.7E+02	1.000	6.3E+06	1.7E+02
^{24}Na	15.0 h	F 1.000	3.8E+07	1.0E+03	1.000	4.7E+07	1.3E+03
^{32}P	14.3 d	F 0.800	1.8E+07	4.9E+02	0.800	8.3E+06	2.3E+02
		M 0.800	6.9E+06	1.9E+02			
^{35}S	87.4 d	Inorganic compounds					
		F 0.800	2.5E+08	6.8E+03	0.800	1.4E+08	3.9E+03
		M 0.800	1.8E+07	4.9E+02	0.100	1.1E+08	2.8E+03
		Vapor	1.7E+08	4.5E+03			
		Organic compounds			1.000	2.6E+07	7.0E+02
^{42}K	12.4 h	F 1.000	1.0E+08	2.7E+03	1.000	4.7E+07	1.3E+03
^{43}K	22.6 h	F 1.000	7.7E+07	2.1E+03	1.000	8.0E+07	2.2E+03
^{45}Ca	163 d	M 0.300	8.7E+06	2.4E+02	0.300	2.6E+07	7.1E+02
^{47}Ca	4.53 d	M 0.300	9.5E+06	2.6E+02	0.300	1.3E+07	3.4E+02
^{51}Cr	27.7 d	F 0.100	6.7E+08	1.8E+04	0.100	5.3E+08	1.4E+04
		M 0.100	5.9E+08	1.6E+04	0.010	5.4E+08	1.5E+04
		S 0.100	5.6E+08	1.5E+04			
^{54}Mn	312 d	F 0.100	1.8E+07	4.9E+02	0.100	2.8E+07	7.6E+02
		M 0.100	1.7E+07	4.5E+02			
^{52}Fe	8.28 h	F 0.100	2.9E+07	7.8E+02	0.100	1.4E+07	3.9E+02
		M 0.100	2.1E+07	5.7E+02			
^{55}Fe	2.70 y	F 0.100	2.2E+07	5.9E+02	0.100	6.1E+07	1.6E+03
		M 0.100	6.1E+07	1.6E+03			

	Physical half-life	Inhalation intakes			Oral intakes		
		Chemical form Type/f_1	ALI Bq	ALI μCi	Chemical form f_1	ALI Bq	ALI μCi
^{59}Fe	44.5 d	F 0.100	6.7E+06	1.8E+02	0.100	1.1E+07	3.0E+02
		M 0.100	6.3E+06	1.7E+02			
^{57}Co	271 d	M 0.100	5.1E+07	1.4E+03	0.100	9.5E+07	2.6E+03
		S 0.050	3.3E+07	9.0E+02	0.050	1.1E+08	2.8E+03
^{58}Co	70.8 d	M 0.100	1.4E+07	3.9E+02	0.100	2.7E+07	7.3E+02
		S 0.050	1.2E+07	3.2E+02	0.050	2.9E+07	7.7E+02
^{60}Co	5.27 y	M 0.100	2.8E+06	7.6E+01	0.100	5.9E+06	1.6E+02
		S 0.050	1.2E+06	3.2E+01	0.050	8.0E+06	2.2E+02
^{64}Cu	12.7 h	F 0.500	2.9E+08	7.9E+03	0.500	1.7E+08	4.5E+03
		M 0.500	1.3E+08	3.6E+03			
		S 0.500	1.3E+08	3.6E+03			
^{59}Ni	75000 y	F 0.050	9.1E+07	2.5E+03	0.050	3.2E+08	8.6E+03
		M 0.050	2.1E+08	5.8E+03			
		Vapor	2.4E+07	6.5E+02			
^{63}Ni	96.0 y	F 0.050	3.8E+07	1.0E+03	0.050	1.3E+08	3.6E+03
		M 0.050	6.5E+07	1.7E+03			
		Vapor	1.0E+07	2.7E+02			
^{65}Zn	244 d	S 0.500	7.1E+06	1.9E+02	0.500	5.1E+06	1.4E+02
^{67}Ga	3.26 d	F 0.001	1.8E+08	4.9E+03	0.001	1.1E+08	2.8E+03
		M 0.001	7.1E+07	1.9E+03			
^{68}Ga	1.13 h	F 0.001	4.1E+08	1.1E+04	0.001	2.0E+08	5.4E+03
		M 0.001	2.5E+08	6.7E+03			
^{68}Ge	288 d	F 1.000	2.4E+07	6.5E+02	1.000	1.5E+07	4.2E+02
		M 1.000	2.5E+06	6.8E+01			
^{75}Se	120 d	F 0.800	1.4E+07	3.9E+02	0.800	7.7E+06	2.1E+02
		M 0.800	1.2E+07	3.2E+02	0.050	4.9E+07	1.3E+03
^{79}Se	65000 y	F 0.800	1.3E+07	3.4E+02	0.800	6.9E+06	1.9E+02
		M 0.800	6.5E+06	1.7E+02	0.050	5.1E+07	1.4E+03
^{86}Rb	18.6 d	F 1.000	1.5E+07	4.2E+02	1.000	7.1E+06	1.9E+02
^{85}Sr	64.8 d	F 0.300	3.6E+07	9.7E+02	0.300	3.6E+07	9.7E+02
		S 0.010	3.1E+07	8.4E+02	0.010	6.1E+07	1.6E+03
^{87m}Sr	2.80 h	F 0.300	9.1E+08	2.5E+04	0.300	6.7E+08	1.8E+04
		S 0.010	5.7E+08	1.5E+04	0.010	6.1E+08	1.6E+04
^{89}Sr	50.5 d	F 0.300	1.4E+07	3.9E+02	0.300	7.7E+06	2.1E+02
		S 0.010	3.6E+06	9.7E+01	0.010	8.7E+06	2.4E+02
^{90}Sr	29.1 y	F 0.300	6.7E+05	1.8E+01	0.300	7.1E+05	1.9E+01
		S 0.010	2.6E+05	7.0E+00	0.010	7.4E+06	2.0E+02
^{99}Mo	2.75 d	F 0.800	5.6E+07	1.5E+03	0.800	2.7E+07	7.3E+02
		S 0.050	1.8E+07	4.9E+02	0.050	1.7E+07	4.5E+02
^{99m}Tc	6.02 h	F 0.800	1.0E+09	2.7E+04	0.800	9.1E+08	2.5E+04
		M 0.800	6.9E+08	1.9E+04			
^{99}Tc	213000 y	F 0.800	5.0E+07	1.4E+03	0.800	2.6E+07	6.9E+02
		M 0.800	6.3E+06	1.7E+02			
^{106}Ru	1.01 y	F 0.050	2.0E+06	5.5E+01	0.050	2.9E+06	7.7E+01
		M 0.050	1.2E+06	3.2E+01			
		S 0.050	5.7E+05	1.5E+01			
^{111}In	2.83 d	F 0.020	9.1E+07	2.5E+03	0.020	6.9E+07	1.9E+03
		M 0.020	6.5E+07	1.7E+03			
^{113m}In	1.66 h	F 0.020	1.1E+09	2.8E+04	0.020	7.1E+08	1.9E+04
		M 0.020	6.3E+08	1.7E+04			
^{113}Sn	115 d	F 0.020	2.5E+07	6.8E+02	0.020	2.7E+07	7.4E+02
		M 0.020	1.1E+07	2.8E+02			
^{123}I	13.2 h	F 1.000	1.8E+08	4.9E+03	1.000	9.5E+07	2.6E+03
		Vapor	9.5E+07	2.6E+03			
^{125}I	60.1 d	F 1.000	2.7E+06	7.4E+01	1.000	1.3E+06	3.6E+01
		Vapor	1.4E+06	3.9E+01			
^{129}I	$1.57 \cdot 10^7$ y	F 1.000	3.9E+05	1.1E+01	1.000	1.8E+05	4.9E+00

	Physical half-life	Inhalation intakes			Oral intakes		
		Chemical form Type/f_1	ALI Bq	ALI µCi	Chemical form f_1	ALI Bq	ALI µCi
^{131}I	8.04 d	Vapor	2.1E+05	5.6E+00			
		F 1.000	1.8E+06	4.9E+01	1.000	9.1E+05	2.5E+01
^{129}Cs	1.34 d	Vapor	1.0E+06	2.7E+01			
		F 1.000	2.5E+08	6.7E+03	1.000	3.3E+08	9.0E+03
^{134}Cs	2.06 y	F 1.000	2.1E+06	5.6E+01	1.000	1.1E+06	2.8E+01
^{136}Cs	13.1 d	F 1.000	1.1E+07	2.8E+02	1.000	6.7E+06	1.8E+02
^{137}Cs	30.0 y	F 1.000	3.0E+06	8.1E+01	1.000	1.5E+06	4.2E+01
^{141}Ce	32.5 d	M 5.0E-04	7.4E+06	2.0E+02	5.0E-04	2.8E+07	7.6E+02
		S 5.0E-04	6.5E+06	1.7E+02			
^{144}Ce	284 d	M 5.0E-04	8.7E+05	2.4E+01	5.0E-04	3.8E+06	1.0E+02
		S 5.0E-04	6.9E+05	1.9E+01			
^{133}Ba	10.7 y	F 0.100	1.1E+07	3.0E+02	0.100	2.0E+07	5.4E+02
^{140}Ba	12.7 d	F 0.100	1.3E+07	3.4E+02	0.100	8.0E+06	2.2E+02
^{169}Yb	32.0 d	M 5.0E-04	9.5E+06	2.6E+02	5.0E-04	2.8E+07	7.6E+02
		S 5.0E-04	8.3E+06	2.3E+02			
^{198}Au	2.69 d	F 0.100	5.1E+07	1.4E+03	0.100	2.0E+07	5.4E+02
		M 0.100	2.0E+07	5.5E+02			
		S 0.100	1.8E+07	4.9E+02			
^{198m}Au	2.30 d	F 0.100	3.4E+07	9.2E+02	0.100	1.5E+07	4.2E+02
		M 0.100	1.0E+07	2.7E+02			
		S 0.100	1.1E+07	2.8E+02			
^{197}Hg	2.67 d	Inorganic compounds					
		F 0.400	2.4E+08	6.4E+03	1.000	2.0E+08	5.5E+03
					0.400	1.2E+08	3.2E+03
		Vapor	4.5E+06	1.2E+02			
		Organic compounds					
		F 0.020	2.0E+08	5.4E+03	0.020	8.7E+07	2.4E+03
		M 0.020	7.1E+07	1.9E+03			
^{203}Hg	46.6 d	Inorganic compounds					
		F 0.400	2.7E+07	7.2E+02	1.000	1.1E+07	2.8E+02
					0.400	1.8E+07	4.9E+02
		Vapor	2.9E+06	7.7E+01			
		Organic compounds					
		F 0.020	3.4E+07	9.2E+02	0.020	3.7E+07	1.0E+03
		M 0.020	1.1E+07	2.8E+02			
^{201}Tl	3.04 d	F 1.000	2.6E+08	7.1E+03	1.000	2.1E+08	5.7E+03
^{210}Pb	22.3 y	F 0.200	1.8E+04	4.9E-01	0.200	2.9E+04	7.9E-01
^{207}Bi	38.0 y	F 0.050	2.4E+07	6.4E+02	0.050	1.5E+07	4.2E+02
		M 0.050	6.3E+06	1.7E+02			
^{210}Po	138 d	F 0.100	2.8E+04	7.6E-01	0.100	8.3E+04	2.3E+00
		M 0.100	9.1E+03	2.5E-01			
^{224}Ra	3.66 d	M 0.200	8.3E+03	2.3E-01	0.200	3.1E+05	8.3E+00
^{226}Ra	1600 y	M 0.200	1.7E+03	4.5E-02	0.200	7.1E+04	1.9E+00
^{228}Ra	5.75 y	M 0.200	1.2E+04	3.2E-01	0.200	3.0E+04	8.1E-01
^{228}Th	1.91 y	M 5.0E-04	8.7E+02	2.4E-02	5.0E-04	2.9E+05	7.7E+00
		S 2.0E-04	6.3E+02	1.7E-02	2.0E-04	5.7E+05	1.5E+01
^{230}Th	77000 y	M 5.0E-04	7.1E+02	1.9E-02	5.0E-04	9.5E+04	2.6E+00
		S 2.0E-04	2.8E+03	7.5E-02	2.0E-04	2.3E+05	6.2E+00
^{232}Th	1.40·10^{10} y	M 5.0E-04	6.9E+02	1.9E-02	5.0E-04	9.1E+04	2.5E+00
		S 2.0E-04	1.7E+03	4.5E-02	2.0E-04	2.2E+05	5.9E+00
^{234}U	2.44·10^5 y	F 0.020	3.1E+04	8.4E-01	0.020	4.1E+05	1.1E+01
		M 0.020	9.5E+03	2.6E-01	0.002	2.4E+06	6.5E+01
		S 0.002	2.9E+03	7.9E-02			

ANNUAL LIMITS ON INTAKES OF RADIONUCLIDES (continued)

	Physical half-life	Inhalation intakes			Oral intakes		
		Chemical form Type/f_1	ALI		Chemical form f_1	ALI	
			Bq	μCi		Bq	μCi
^{235}U	$7.04 \cdot 10^8$ y	F 0.020	3.3E+04	9.0E-01	0.020	4.3E+05	1.2E+01
		M 0.020	1.1E+04	3.0E-01	0.002	2.4E+06	6.5E+01
		S 0.002	3.3E+03	8.9E-02			
^{238}U	$4.47 \cdot 10^9$ y	F 0.020	3.4E+04	9.3E-01	0.020	4.5E+05	1.2E+01
		M 0.020	1.3E+04	3.4E-01	0.002	2.6E+06	7.1E+01
		S 0.002	3.5E+03	9.5E-02			
^{237}Np	$2.14 \cdot 10^6$ y	M 5.0E-04	1.3E+03	3.6E-02	5.0E-04	1.8E+05	4.9E+00
^{239}Np	2.36 d	M 5.0E-04	1.8E+07	4.9E+02	5.0E-04	2.5E+07	6.8E+02
^{238}Pu	87.7 y	M 5.0E-04	6.7E+02	1.8E-02	5.0E-04	8.7E+04	2.4E+00
		S 1.0E-05	1.8E+03	4.9E-02	1.0E-05	2.3E+06	6.1E+01
					1.0E-04	4.1E+05	1.1E+01
^{239}Pu	24100 y	M 5.0E-04	6.3E+02	1.7E-02	5.0E-04	8.0E+04	2.2E+00
		S 1.0E-05	2.4E+03	6.5E-02	1.0E-05	2.2E+06	6.0E+01
					1.0E-04	3.8E+05	1.0E+01
^{241}Pu	14.4 y	M 5.0E-04	3.4E+04	9.3E-01	5.0E-04	4.3E+06	1.2E+02
		S 1.0E-05	2.4E+05	6.4E+00	1.0E-05	1.8E+08	4.9E+03
					1.0E-04	2.1E+07	5.6E+02
^{241}Am	432 y	M 5.0E-04	7.4E+02	2.0E-02	5.0E-04	1.0E+05	2.7E+00
^{244}Cm	18.1 y	M 5.0E-04	1.2E+03	3.2E-02	5.0E-04	1.7E+05	4.5E+00
^{252}Cf	2.64 y	M 5.0E-04	1.5E+03	4.2E-02	5.0E-04	2.2E+05	6.0E+00

CHEMICAL CARCINOGENS

The following substances are listed in the *Seventh Annual Report on Carcinogens: Summary 1994*, released by the National Institute of Environmental Health Sciences (NIEHS) under the National Toxicology Program (NTP). Substances are grouped into three classes:

1. Known to be carcinogens from studies in humans
2. Reasonably anticipated to be carcinogens, based on limited human studies or studies on experimental animals
3. Associated with technological processes that are known to be carcinogenic

The table includes the name used by the NTP, the Chemical Abstracts Service Registry Number (CAS RN) as given in the NTP report, and, where possible, the systematic name of the substance (generally the CAS index name).

REFERENCE

Seventh Annual Report on Carcinogens: Summary 1994, Public Health Service, National Toxicology Program, Central Data Management, P.O. Box 12233, MD A0-01, Research Triangle Park, NC 27709.

NTP name	CAS RN	Systematic name
SUBSTANCES KNOWN TO BE CARCINOGENIC		
Aflatoxins	1402-68-2	
Aminobiphenyl	92-67-1	[1,1′-Biphenyl]-4-amine
Analgesic mixtures containing phenacetin		Acetamide, N-(4-ethoxyphenyl)- (in analgesic mixtures)
Arsenic	7440-38-2	Arsenic
Certain arsenic compounds		
Asbestos	1332-21-4	Asbestos
Azathioprine	446-86-6	Purine, 6-((1-methyl-4-nitroimidazol-5-yl)thio)
Benzene	71-43-2	Benzene
Benzidine	92-87-5	[1,1′-Biphenyl]-4,4′-diamine
Bis(chloromethyl) ether	542-88-1	Methane, oxybis[chloro-
Chloromethyl methyl ether (techical grade)	107-30-2	Methane, chloromethoxy-
1,4-Butanediol dimethyl-sulfonate (Myleran)	55-98-1	1,4-Butanediol, dimethanesulfonate
Chlorambucil	305-03-3	Benzenebutanoic acid, 4-[bis(2-chloroethyl)amino]-
(2-Chloroethyl)-3-(4-methylcyclohexyl)-1-nitrosourea (MeCCNU)	13909-09-6	Urea, 1-(2-chloroethyl)-3-(4-methylcyclohexyl)-1-nitroso-, (*E*)
Chromium	7440-47-3	Chromium
Certain chromium compounds		
Conjugated estrogens		
Cyclophosphamide	50-18-0	2H-1,3,2-Oxazaphosphorine, 2(bis(2-chloroethyl)amino)-tetrahydro, 2-oxide
Diethylstilbestrol	56-53-1	Phenol, 4,4′-(1,2-diethyl-1,2-ethenediyl)bis-, (*E*)-
Erionite	66733-21-9	Erionite (Calcium sodium aluminum silicate)
Melphalan	148-82-3	Alanine, 3-(*p*-(bis(2-chloroethyl)amino)phenyl)-, *L*
Methoxsalen with ultraviolet A therapy (PUVA)		
Mustard gas	505-60-2	Ethane, 1,1′-thiobis[2-chloro-
2-Naphthylamine	91-59-8	2-Naphthalenamine
Radon	10043-92-2	Radon
Thorium dioxide	1314-20-1	Thorium(IV) oxide
Vinyl chloride	75-01-4	Ethene, chloro-
SUBSTANCES ANTICIPATED TO BE CARCINOGENIC		
Acetaldehyde	75-07-0	Acetaldehyde
2-Acetylaminofluorene	53-96-3	Acetamide, *N*-9H-fluoren-2-yl-
Acrylamide	79-06-1	2-Propenamide
Acrylonitrile	107-13-1	2-Propenenitrile
2-Aminoanthraquinone	117-79-3	9,10-Anthracenedione, 2-amino-
o-Aminoazotoluene	97-56-3	Benzenamine, 2-methyl-4-[(2-methylphenyl)azo]-
1-Amino-2-methylanthraquinone	82-28-0	9,10-Anthracenedione, 1-amino-2-methyl-
Amitrole	61-82-5	1H-1,2,4-Triazol-3-amine
o-Anisidine hydrochloride	134-29-2	Benzenamine, 2-methoxy-, hydrochloride
Benzotrichloride	98-07-7	Benzene, (trichloromethyl)-

NTP name	CAS RN	Systematic name
Benzotrichloride	98-07-7	Benzene, (trichloromethyl)-
Beryllium	7440-41-7	Beryllium
Certain beryllium compounds		
Bischloroethyl nitrosourea	154-93-8	Urea, N,N'-bis(2-chloroethyl)-N-nitroso-
Bromodichloromethane	75-27-4	Methane, bromodichloro-
1,3-Butadiene	106-99-0	1,3-Butadiene
Butylated hydroxyanisole	25013-16-5	Phenol, (1,1-dimethylethyl)-4-methoxy-
Cadmium	7440-43-9	Cadmium
Certain cadmium compounds		
Carbon tetrachloride	56-23-5	Methane, tetrachloro-
Ceramic fibers (respirable size)		
Chlorendic acid	115-28-6	5-Norbornene-2,3-dicarboxylic acid, 1,4,5,6,7,7-hexachloro-
Chlorinated paraffins (C_{12}, 60% chlorine)	108171-26-2	
1-(2-Chloroethyl)-3-cyclohexyl-1-nitrosourea (CCNU)	13010-47-4	Urea, 1-(2-chloroethyl)-3-cyclohexyl-1-nitroso-
Chloroform	67-66-3	Methane, trichloro-
3-Chloro-2-methylpropene	563-47-3	1-Propene, 3-chloro-2-methyl-
4-Chloro-o-phenylenediamine	95-83-0	1,2-Benzenediamine, 4-chloro-
C.I. Basic Red 9 Monohydrochloride	569-61-9	Benzenamine, 4-[(4-aminophenyl)(4-imino-2,5-cyclohexadien-1- ylidene)methyl]-, monohydrochloride
Cisplatin	15663-27-1	Platinum(II), diamminedichloro-, cis
p-Cresidine	120-71-8	Benzenamine, 2-methoxy-5-methyl-
Cupferron	135-20-6	Benzenamine, N-hydroxy-N-nitroso-, ammonium salt
DDT	50-29-3	Benzene, 1,1'-(2,2,2-trichloroethylidene)bis[4-chloro-
2,4-Diaminoanisole sulfate	39156-41-7	1,3-Benzenediamine, 4-methoxy-, sulfate
2,4-Diaminotoluene	95-80-7	1,3-Benzenediamine, 4-methyl-
1,2-Dibromo-3-chloropropane	96-12-8	Propane, 1,2-dibromo-3-chloro-
1,2-Dibromoethane (Ethylene dibromide)	106-93-4	Ethane, 1,2-dibromo-
1,4-Dichlorobenzene	106-46-7	Benzene, 1,4-dichloro-
3,3'-Dichlorobenzidine	91-94-1	[1,1'-Biphenyl]-4,4'-diamine, 3,3'-dichloro-
3,3'-Dichlorobenzidine 2HCl	612-83-9	[1,1'-Biphenyl]-4,4'-diamine, 3,3'-dichloro-, dihydrochloride
1,2-Dichloroethane	107-06-2	Ethane, 1,2-dichloro-
Dichloromethane (Methylene chloride)	75-09-2	Methane, dichloro-
1,3-Dichloropropene (technical grade)	542-75-6	1-Propene, 1,3-dichloro- (mixed isomers)
Diepoxybutane	1464-53-5	2,2'-Bioxirane
Di(2-ethylhexyl) phthalate	117-81-7	1,2-Benzenedicarboxylic acid, bis(2-ethylhexyl) ester
Diethyl sulfate	64-67-5	Sulfuric acid, diethyl ester
Diglycidyl resorcinol ether	101-90-6	Oxirane, 2,2'-[1,3-phenylenebis(oxymethylene)]bis-
3,3'-Dimethoxybenzidine	119-90-4	[1,1'-Biphenyl]-4,4'-diamine, 3,3'-dimethoxy-
3,3'-Dimethoxybenzidine 2HCl	20325-40-0	[1,1'-Biphenyl]-4,4'-diamine, 3,3'-dimethoxy-, dihydrochloride
4-Dimethylamino-azobenzene	60-11-7	Benzenamine, N,N-dimethyl-4-(phenylazo)-
3,3'-Dimethylbenzidine	119-93-7	[1,1'-Biphenyl]-4,4'-diamine, 3,3'-dimethyl-
Dimethylcarbamoyl chloride	79-44-7	Carbamic chloride, dimethyl-
1,1-Dimethylhydrazine	57-14-7	Hydrazine, 1,1-dimethyl-
Dimethyl sulfate	77-78-1	Sulfuric acid, dimethyl ester
Dimethylvinyl chloride	513-37-1	1-Propene, 1-chloro-2-methyl-
1,4-Dioxane	123-91-1	1,4-Dioxane
Direct Black 38	1937-37-7	2,7-Naphthalenedisulfonic acid, 4-amino-3-((4'-((2,4-diaminophenyl)azo)(1,1'-biphenyl)-4-yl)azo)-5-hydroxy-6-(phenylazo)-, disodium salt
Direct Blue 6	2602-46-2	2,7-Naphthalenedisulfonic acid, 3,3'-((4,4'-biphenylylene)bis(azo))bis(5-amino-4-hydroxy-, tetrasodium salt
Epichlorohydrin	106-89-8	Propane, 1-chloro-2,3-epoxy-
Estrogens (not conjugated): Estradiol-17beta	50-28-2	Estra-1,3,5(10)-triene-3,17-diol (17ß)-
Estrogens (not conjugated): Estrone	53-16-7	Estra-1,3,5(10)-trien-17-one, 3-hydroxy-
Estrogens (not conjugated): Ethinylestradiol	57-63-6	19-Norpregna-1,3,5(10)-trien-20-yne-3,17-diol, (17α)-
Estrogens (not conjugated): Mestranol	72-33-3	17-α-19-Norpregna-1,3,5(10)-trien-20-yn-17-ol, 3-methoxy
Ethyl acrylate	140-88-5	2-Propenoic acid, ethyl ester
Ethylene oxide	75-21-8	Oxirane

NTP name	CAS RN	Systematic name
Ethylene thiourea	96-45-7	2-Imidazolidinethione
Ethyl methanesulfonate	62-50-0	Methanesulfonic acid, ethyl ester
Formaldehyde (gas)	50-00-0	Formaldehyde
Glasswool (respirable size)		
Glycidol	556-52-5	1-Propanol, 2,3-epoxy
Hexachlorobenzene	118-74-1	Benzene, hexachloro-
Hexachloroethane	67-72-1	Ethane, hexachloro-
Hexamethyl-phosphoramide	680-31-9	Phosphoric triamide, hexamethyl-
Hydrazine	302-01-2	Hydrazine
Hydrazine sulfate	10034-93-2	Hydrazine sulfate
Hydrazobenzene	122-66-7	Hydrazine, 1,2-diphenyl-
Iron dextran complex	9004-66-4	Imferon
Kepone (chlordecone)	143-50-0	1,3,4-Metheno-2H-cyclobuta[cd]pentalen-2-one, 1,1a,3,3a,4,5,5,5a,5b,6-decachlorooctahydro-
Lead acetate	301-04-2	Lead(II) acetate
Lead phosphate	7446-27-7	Lead(II) phosphate
Lindane and other hexachlorocyclohexane isomers		Cyclohexane, 1,2,3,4,5,6-hexachloro- (mixed isomers)
2-Methylaziridine (Propyleneimine)	75-55-8	Aziridine, 2-methyl-
4,4´-Methylenebis(2-chloroaniline) (MBOCA)	101-14-4	Benzenamine, 4,4´-methylenebis[2-chloro-
4,4´-Methylenebis(N,N-dimethylbenzenamine)	101-61-1	Benzenamine, 4,4´-methylenebis[N,N-dimethyl-
4,4´-Methylenedianiline	101-77-9	Benzenamine, 4,4´-methylenebis-
4,4´-Methylenedianiline 2HCl	13552-44-8	Aniline, 4,4´-methylenedi-, dihydrochloride
Methyl methanesulfonate	66-27-3	Methanesulfonic acid, methyl ester
N-Methyl-N-nitro-N-nitrosoguanidine	70-25-7	Guanidine, N-methyl-N´-nitro-N-nitroso-
Metronidazole	443-48-1	1H-Imidazole-1-ethanol, 2-methyl-5-nitro-
Michler´s ketone	90-94-8	Methanone, bis[4-(dimethylamino)phenyl]-
Mirex	2385-85-5	1,3,4-Metheno-1H-cyclobuta[cd]pentalene, 1,1a,2,2,3,3a,4,5,5,5a,5b,6-dodecachlorooctahydro-
Nickel	7440-02-0	Nickel
Certain nickel compounds		
Nitrilotriacetic acid	139-13-9	Glycine, N,N-bis(carboxymethyl)-
Nitrofen	1836-75-5	Benzene, 2,4-dichloro-1-(4-nitrophenoxy)-
Nitrogen mustard hydrochloride	55-86-7	Ethanamine, 2-chloro-N-(2-chloroethyl)-N-methyl-, hydrochloride
2-Nitropropane	79-46-9	Propane, 2-nitro-
N-Nitrosodi-n-butylamine	924-16-3	1-Butanamine, N-butyl-N-nitroso-
N-Nitrosodiethanolamine	1116-54-7	Ethanol, N-nitrosoiminodi-
N-Nitrosodiethylamine	55-18-5	Ethanamine, N-ethyl-N-nitroso-
N-Nitrosodimethylamine	62-75-9	Methanamine, N-methyl-N-nitroso-
N-Nitrosodi-n-propylamine	621-64-7	1-Propanamine, N-nitroso-N-propyl-
N-Nitroso-n-ethylurea	759-73-9	Urea, N-ethyl-N-nitroso-
4-(N-Nitrosomethyl-amino)-1-(3-pyridyl)-1-butanone (NNK)	64091-91-4	Ketone, 3-pyridyl-3-(N-methyl-N-nitrosamino)propyl-
N-Nitroso-n-methylurea	684-93-5	Urea, N-methyl-N-nitroso-
N-Nitrosomethyl-vinylamine	4549-40-0	Ethenylamine, N-methyl-N-nitroso-
N-Nitrosomorpholine	59-89-2	Morpholine, 4-nitroso-
N-Nitrosonornicotine	16543-55-8	Pyridine, 3-(1-nitroso-2-pyrrolidinyl)-, (S)-
N-Nitrosopiperidine	100-75-4	Piperidine, 1-nitroso-
N-Nitrosopyrrolidine	930-55-2	Pyrrolidine, 1-nitroso-
N-Nitrososarcosine	13256-22-9	Glycine, N-methyl-N-nitroso-
Norethisterone	68-22-4	19-Nor-17-α-pregn-4-en-20-yn-3-one, 7-hydroxy
Ochratoxin A	303-47-9	Alanine, N-((5-chloro-8-hydroxy-3-methyl-1-oxo-7-isochromanyl)carbonyl)-3-phenyl-, (-)
4,4´-Oxydianiline	101-80-4	Benzenamine, 4,4´-oxybis-
Oxymetholone	434-07-1	5-α,17-ß-Androstan-3-one, 17-hydroxy-2-(hydroxymethylene)-17-methyl-
Phenacetin	62-44-2	Acetamide, N-(4-ethoxyphenyl)-
Phenazopyridine hydrochloride	136-40-3	Pyridine, 2,6-diamino-3-(phenylazo), monohydrochloride
Phenoxybenzamine hydrochloride	63-92-3	Benzenemethanamine, N-(2-chloroethyl)-N-(1-methyl-2-phenoxyethyl)-, hydrochloride

CHEMICAL CARCINOGENS (continued)

NTP name	CAS RN	Systematic name
Phenytoin	57-41-0	2,4-Imidazolidinedione, 5,5-diphenyl-
Polybrominated biphenyls		
Polychlorinated biphenyls	1336-36-3	
Polycyclic aromatic hydrocarbons, 15 listings		
Procarbazine hydrochloride	366-70-1	*p*-Toluamide, *N*-isopropyl-α-(2-methylhydrazino)-, monohydrochloride
Progesterone	57-83-0	Pregn-4-ene-3,20-dione
1,3-Propane sultone	1120-71-4	1,2-Oxathiolane, 2,2-dioxide
beta-Propiolactone	57-57-8	2-Oxetanone
Propylene oxide	75-56-9	Propane, 1,2-epoxy
Propylthiouracil	51-52-5	4(1H)-Pyrimidinone, 2,3-dihydro-6-propyl-2-thioxo-
Reserpine	50-55-5	Yohimban-16-carboxylic acid, 11,17-dimethoxy-18-[(3,4,5-trimethoxybenzoyl)oxy]-, methyl ester, (3ß,16ß,17α18ß,20α)-
Saccharin	128-44-9	1,2-Benzisothiazol-3(2H)-one, 1,1-dioxide, sodium salt
Safrole	94-59-7	1,3-Benzodioxole, 5-(2-propenyl)-
Selenium sulfide	7446-34-6	Selenium sulfide (SSe)
Silica, crystalline (respirable size)		Silicon dioxide
Streptozotocin	18883-66-4	Glucopyranose, 2-deoxy-2-(3-methyl-3-nitrosoureido)-, *D*
Sulfallate	95-06-7	Carbamodithioic acid, diethyl-, 2-chloro-2-propenyl ester
2,3,7,8-Tetrachlorodibenzo-*p*-dioxin (TCDD)	1746-01-6	Dibenzo-*p*-dioxin, 2,3,7,8-tetrachloro-
Tetrachloroethylene (perchloroethylene)	127-18-4	Ethene, tetrachloro-
Tetranitromethane	509-14-8	Methane, tetranitro-
Thioacetamide	62-55-5	Ethanethioamide
Thiourea	62-56-6	Thiourea
Toluene diisocyanate	26471-62-5	Benzene, 1,3-diisocyanato-methyl- (mixed isomers)
o-Toluidine	95-53-4	Benzenamine, 2-methyl-
o-Toluidine hydrochloride	636-21-5	Benzenamine, 2-methyl-, hydrochloride
Toxaphene	8001-35-2	Camphene, polychlorinated
2,4,6-Trichlorophenol	88-06-2	Phenol, 2,4,6-trichloro-
Tris(1-aziridinyl)phosphine sulfide (ThioTEPA)	52-24-4	Phosphine sulfide, tris(1-aziridinyl)
Tris(2,3-dibromopropyl) phosphate	126-72-7	1-Propanol, 2,3-dibromo-, phosphate (3:1)
Urethane	51-79-6	Carbamic acid, ethyl ester
4-Vinyl-1-cyclohexene diepoxide	106-87-6	7-Oxabicyclo[4.1.0]heptane, 3-oxiranyl-

SUBSTANCES ASSOCIATED WITH TECHNOLOGICAL PROCESSES KNOWN TO BE CARCINOGENIC

Coke oven emissions
Soots, tars, and mineral oils

Appendix A
Mathematical Tables

MISCELLANEOUS MATHEMATICAL CONSTANTS

π CONSTANTS

$$\pi = 3.14159\ 26535\ 89793\ 23846\ 26433\ 83279\ 50288\ 41971\ 69399\ 37511$$
$$1/\pi = 0.31830\ 98861\ 83790\ 67153\ 77675\ 26745\ 02872\ 40689\ 19291\ 48091$$
$$\pi^2 = 9.86960\ 44010\ 89358\ 61883\ 44909\ 99876\ 15113\ 53136\ 99407\ 24079$$
$$\log_e \pi = 1.14472\ 98858\ 49400\ 17414\ 34273\ 51353\ 05871\ 16472\ 94812\ 91531$$
$$\log_{10} \pi = 0.49714\ 98726\ 94133\ 85435\ 12682\ 88290\ 89887\ 36516\ 78324\ 38044$$
$$\log_{10} \sqrt{2\pi} = 0.39908\ 99341\ 79057\ 52478\ 25035\ 91507\ 69595\ 02099\ 34102\ 92128$$

CONSTANTS INVOLVING e

$$e = 2.71828\ 18284\ 59045\ 23536\ 02874\ 71352\ 66249\ 77572\ 47093\ 69996$$
$$1/e = 0.36787\ 94411\ 71442\ 32159\ 55237\ 70161\ 46086\ 74458\ 11131\ 03177$$
$$e^2 = 7.38905\ 60989\ 30650\ 22723\ 04274\ 60575\ 00781\ 31803\ 15570\ 55185$$
$$M = \log_{10} e = 0.43429\ 44819\ 03251\ 82765\ 11289\ 18916\ 60508\ 22943\ 97005\ 80367$$
$$1/M = \log_e 10 = 2.30258\ 50929\ 94045\ 68401\ 79914\ 54684\ 36420\ 76011\ 01488\ 62877$$
$$\log_{10} M = 9.63778\ 43113\ 00536\ 78912\ 29674\ 98645\ -10$$

π^e AND e^π CONSTANTS

$$\pi^e = 22.45915\ 77183\ 61045\ 47342\ 71522$$
$$e^\pi = 23.14069\ 26327\ 79269\ 00572\ 90864$$
$$e^{-\pi} = 0.04321\ 39182\ 63772\ 24977\ 44177$$
$$e^{1/2\pi} = 4.81047\ 73809\ 65351\ 65547\ 30357$$
$$i^i = e^{-1/2\pi} = 0.20787\ 95763\ 50761\ 90854\ 69556$$

NUMERICAL CONSTANTS

$$\sqrt{2} = 1.41421\ 35623\ 73095\ 04880\ 16887\ 24209\ 69807\ 85696\ 71875\ 37695$$
$$\sqrt[3]{2} = 1.25992\ 10498\ 94873\ 16476\ 72106\ 07278\ 22835\ 05702\ 51464\ 70151$$
$$\log_e 2 = 0.69314\ 71805\ 59945\ 30941\ 72321\ 21458\ 17656\ 80755\ 00134\ 36026$$
$$\log_{10} 2 = 0.30102\ 99956\ 63981\ 19521\ 37388\ 94724\ 49302\ 67881\ 89881\ 46211$$
$$\sqrt{3} = 1.73205\ 08075\ 68877\ 29352\ 74463\ 41505\ 87236\ 69428\ 05253\ 81039$$
$$\sqrt[3]{3} = 1.44224\ 95703\ 07408\ 38232\ 16383\ 10780\ 10958\ 83918\ 69253\ 49935$$
$$\log_e 3 = 1.09861\ 22886\ 68109\ 69139\ 52452\ 36922\ 52570\ 46474\ 90557\ 82275$$
$$\log_{10} 3 = 0.47712\ 12547\ 19662\ 43729\ 50279\ 03255\ 11530\ 92001\ 28864\ 19070$$

OTHER CONSTANTS

$$\text{Euler's Constant } \gamma = 0.57721\ 56649\ 01532\ 86061$$
$$\log_e \gamma = -0.54953\ 93129\ 81644\ 82234$$
$$\text{Golden Ratio } \phi = 1.61803\ 39887\ 49894\ 84820\ 45868\ 34365\ 63811\ 77203\ 09180$$

EXPONENTIAL AND HYPERBOLIC FUNCTIONS AND THEIR COMMON LOGARITHMS

x	e^x Value	$\log^{10}$	e^{-x} (value)	sinh x Value	$\log^{10}$	cosh x Value	$\log^{10}$	tanh x (value)
0.00	1.0000	0.00000	1.00000	0.0000	— ∞	1.0000	0.00000	0.00000
0.01	1.0101	.00434	0.99005	.0100	$\overline{2}$.00001	1.0001	.00002	.01000
0.02	1.0202	.00869	.98020	.0200	$\overline{2}$.30106	1.0002	.00009	.02000
0.03	1.0305	.01303	.97045	.0300	$\overline{2}$.47719	1.0005	.00020	.02999
0.04	1.0408	.01737	.96079	.0400	$\overline{2}$.60218	1.0008	.00035	.03998
0.05	1.0513	.02171	.95123	.0500	$\overline{2}$.69915	1.0013	.00054	.04996
0.06	1.0618	.02606	.94176	.0600	$\overline{2}$.77841	1.0018	.00078	.05993
0.07	1.0725	.03040	.93239	.0701	$\overline{2}$.84545	1.0025	.00106	.06989
0.08	1.0833	.03474	.92312	.0801	$\overline{2}$.90355	1.0032	.00139	.07983
0.09	1.0942	.03909	.91393	.0901	$\overline{2}$.95483	1.0041	.00176	.08976
0.10	1.1052	.04343	.90484	.1002	$\overline{1}$.00072	1.0050	.00217	.09967
0.11	1.1163	.04777	.89583	1102	$\overline{1}$.04227	1.0061	.00262	.10956
0.12	1.1275	.05212	.88692	.1203	$\overline{1}$.08022	1.0072	.00312	.11943
0.13	1.1388	.05646	.87809	.1304	$\overline{1}$.11517	1.0085	.00366	.12927
0.14	1.1503	.06080	.86936	.1405	$\overline{1}$.14755	1.0098	.00424	.13909
0.15	1.1618	.06514	.86071	.1506	$\overline{1}$.17772	1.0113	.00487	.14889
0.16	1.1735	.06949	.85214	.1607	$\overline{1}$.20597	1.0128	.00554	.15865
0.17	1.1853	.07383	.84366	.1708	$\overline{1}$.23254	1.0145	.00625	.16838
0.18	1.1972	.07817	.83527	.1810	$\overline{1}$.25762	1.0162	.00700	.17808
0.19	1.2092	.08252	.82696	.1911	$\overline{1}$.28136	1.0181	.00779	.18775
0.20	1.2214	.08686	.81873	.2013	$\overline{1}$.30392	1.0201	.00863	.19738
0.21	1.2337	.09120	.81058	.2115	$\overline{1}$.32541	1.0221	.00951	.20697
0.22	1.2461	.09554	.80252	.2218	$\overline{1}$.34592	1.0243	.01043	.21652
0.23	1.2586	.09989	.79453	.2320	$\overline{1}$.36555	1.0266	.01139	.22603
0.24	1.2712	.10423	.78663	.2423	$\overline{1}$.38437	1.0289	.01239	.23550
0.25	1.2840	.10857	.77880	.2526	$\overline{1}$.40245	1.0314	.01343	.24492
0.26	1.2969	.11292	.77105	.2629	$\overline{1}$.41986	1.0340	.01452	.25430
0.27	1.3100	.11726	.76338	.2733	$\overline{1}$.43663	1.0367	.01564	.26362
0.28	1.3231	12160	.75578	.2837	$\overline{1}$.45282	1.0395	.01681	.27291
0.29	1.3364	.12595	.74826	.2941	$\overline{1}$.46847	1.0423	.01801	.28213
0.30	1.3499	.13029	.74082	.3045	$\overline{1}$.48362	1.0453	.01926	.29131
0.31	1.3634	.13463	.73345	.3150	$\overline{1}$.49830	1.0484	.02054	.30044
0.32	1.3771	.13897	.72615	.3255	$\overline{1}$.51254	1.0516	.02187	.30951
0.33	1.3910	.14332	.71892	.3360	$\overline{1}$.52637	1.0549	.02323	.31852
0.34	1.4049	.14766	.71177	.3466	$\overline{1}$.53981	1.0584	.02463	.32748
0.35	1.4191	.15200	.70469	.3572	$\overline{1}$.55290	1.0619	.02607	.33638
0.36	1.4333	.15635	.69768	.3678	$\overline{1}$.56564	1.0655	.02755	.34521
0.37	1.4477	.16069	.69073	.3785	$\overline{1}$.57807	1.0692	.02907	.35399
0.38	1.4623	.16503	.68386	.3892	$\overline{1}$.59019	1.0731	.03063	.36271
0.39	1.4770	.16937	.67706	.4000	$\overline{1}$.60202	1.0770	.03222	.37136
0.40	1.4918	.17372	.67032	.4108	$\overline{1}$.61358	1.0811	.03385	.37995
0.41	1.5063	.17806	.66365	.4216	$\overline{1}$.62488	1.0852	.03552	.33847
0.42	1.5220	.18240	.65705	.4325	$\overline{1}$.63594	1.0895	.03723	.39693
0.43	1.5373	.18675	.65051	.4434	$\overline{1}$.64677	1.0939	.03897	.40532
0.44	1.5527	.19109	.64404	.4543	$\overline{1}$.65738	1.0984	.04075	.41364
0.45	1.5683	.19543	.63763	.4653	$\overline{1}$.66777	1.1030	.04256	.42190
0.46	1.5841	.19978	.63128	.4764	$\overline{1}$.67797	1.1077	.04441	.43008
0.47	1.6000	.20412	.62500	.4875	$\overline{1}$.68797	1.1125	.04630	.43820
0.48	1.6161	.20846	.61878	.4986	$\overline{1}$.69779	1.1174	.04822	.44624
0.49	1.6323	.21280	.61263	.5098	$\overline{1}$.70744	1.1225	.05018	.45422
0.50	1.6487	.21715	.60653	.5211	$\overline{1}$.71692	1.1276	.05217	.46212
0.51	1.6653	.22149	.60050	.5324	$\overline{1}$.72624	1.1329	.05419	.46995
0.52	1.6820	.22583	.59452	.5438	$\overline{1}$.73540	1.1383	.05625	.47770
0.53	1.6989	.23018	.58860	.5552	$\overline{1}$.74442	1.1438	.05834	.48538
0.54	1.7160	.23452	.58275	.5666	$\overline{1}$.75330	1.1494	.06046	.49299
0.55	1.7333	.23886	.57695	.5782	$\overline{1}$.76204	1.1551	.06262	.50052
0.56	1.7507	.24320	.57121	.5897	$\overline{1}$.77065	1.1609	.06481	.50798
0.57	1.7683	.24755	.56553	.6014	$\overline{1}$.77914	1.1669	.06703	.51536
0.58	1.7860	.25189	.55990	.6131	$\overline{1}$.78751	1.1730	.06929	.52267
0.59	1.8040	.25623	.55433	.6248	$\overline{1}$.79576	1.1792	.07157	.52990
0.60	1.8221	.26058	.54881	.6367	$\overline{1}$.80390	1.1855	.07389	.53705
0.61	1.8404	.26492	.54335	.6485	$\overline{1}$.81194	1.1919	.07624	.54413
0.62	1.8589	.26926	.53794	.6605	$\overline{1}$.81987	1.1984	.07861	.55113
0.63	1.8776	.27361	.53259	.6725	$\overline{1}$.82770	1.2051	.08102	.55805
0.64	1.8965	.27795	.52729	.6846	$\overline{1}$.83543	1.2119	.08346	.56490
0.65	1.9155	.28229	.52205	.6967	$\overline{1}$.84308	1.2188	.08593	.57167
0.66	1.9348	.28664	.51685	.7090	$\overline{1}$.85063	1.2258	.08843	.57836
0.67	1.9542	.29098	.51171	.7213	$\overline{1}$.85809	1.2330	.09095	.58498
0.68	1.9739	.29532	.50662	.7336	$\overline{1}$.86548	1.2402	.09351	.59152
0.69	1.9937	.29966	.50158	.7461	$\overline{1}$.87278	1.2476	.09609	.59798
0.70	2.0138	.30401	.49659	.7586	$\overline{1}$.88000	1.2552	.09870	.60437
0.71	2.0340	.30835	.49164	.7712	$\overline{1}$.88715	1.2628	.10134	.61068
0.72	2.0544	.31269	.48675	.7838	$\overline{1}$.89423	1.2706	.10401	.61691

x	e^x Value	e^x log^{10}	e^{-x} (value)	sinh x Value	sinh x log^{10}	cosh x Value	cosh x log^{10}	tanh x (value)
0.73	2.0751	.31703	.48191	.7966	$\overline{1}$.90123	1.2785	.10670	.62307
0.74	2.0959	.32138	.47711	.8094	$\overline{1}$.90817	1.2865	.10942	.62915
0.75	2.1170	.32572	.47237	.8223	$\overline{1}$.91504	1.2947	.11216	.63515
0.76	2.1383	.33006	.46767	.8353	$\overline{1}$.92185	1.3030	.11493	.64108
0.77	2.1598	.33441	.46301	.8484	$\overline{1}$.92859	1.3114	.11773	.64693
0.78	2.1815	.33875	.45841	.8615	$\overline{1}$.93527	1.3199	.12055	.65721
0.79	2.2034	.34309	.45384	.8748	$\overline{1}$.94190	1.3286	.12340	.65841
0.80	2.2255	.34744	.44933	.8881	$\overline{1}$.94846	1.3374	.12627	.66404
0.81	2.2479	.35178	.44486	.9015	$\overline{1}$.95498	1.3464	.12917	.66959
0.82	2.2705	.35612	.44043	.9150	$\overline{1}$.96144	1.3555	.13209	.67507
0.83	2.2933	.36046	.43605	.9286	$\overline{1}$.96784	1.3647	.13503	.68048
0.84	2.3164	.36481	.43171	.9423	$\overline{1}$.97420	1.3740	.13800	.68581
0.85	2.3396	.36915	.42741	.9561	$\overline{1}$.98051	1.3835	.14099	.69107
0.86	2.3632	.37349	.42316	.9700	$\overline{1}$.98677	1.3932	.14400	.69626
0.87	2.3869	.37784	.41895	.9840	$\overline{1}$.99299	1.4029	.14704	.70137
0.88	2.4100	.38218	.41478	.9981	1.99916	1.4128	.15009	.70642
0.89	2.4351	.38652	.41066	1.0122	0.00528	1.4229	.15317	.71139
0.90	2.4596	.39087	.40657	1.0265	.01137	1.4331	.15627	.21630
0.91	2.4843	39521	.40242	1.0409	.01741	1.4434	.15939	.72113
0.92	2.5093	.39955	.39852	1.0554	.02341	1.4539	.16254	.72590
0.93	2.5345	.40389	.39455	1.0700	.02937	1.4645	.16570	.73059
0.94	2.5600	.40824	.39063	1.0847	.03530	1.4753	.16888	.73522
0.95	2.5857	.41258	.38674	1.0995	.04119	1.4862	.17208	.73978
0.96	2.6117	.41692	.38289	1.1144	.04704	1.4973	.17531	.74428
0.97	2.6379	.42127	.37908	1.1294	.05286	1.5085	.17855	.74870
0.98	2.6645	.42561	.37531	1.1446	.05864	1.5199	.18181	.75307
0.99	2.6912	.42995	.37158	1.1598	.06439	1.5314	.18509	.75736
1.00	2.7183	.43429	.36788	1.1752	.07011	1.5431	.18839	.76159
1.01	2.7456	.43864	.36422	1.1907	.07580	1.5549	.19171	.76576
1.02	2.7732	.44298	.36060	1.2063	.06146	1.5669	.19504	.76987
1.03	2.8011	.44732	.35701	1.2220	.08708	1.5790	.19839	.77391
1.04	2.8292	.45167	.35345	1.2379	.09268	1.5913	.20176	.77789
1.05	2.8577	.45601	.34994	1.2539	.09825	1.6038	.20515	.78181
1.06	2.8864	.46035	.34646	1.2700	.10379	1.6164	.20855	.78566
1.07	2.9154	.46470	.34301	1.2862	.10930	1.6292	.21197	.78946
1.08	2.9447	.46904	.33960	1.3025	.11479	1.6421	.21541	.79320
1.09	2.9743	.47338	33622	1.3190	.12025	1.6552	.21886	.79688
1.10	3.0042	.47772	.33287	1.3356	.12569	1.6685	.22233	.80050
1.11	3.0344	.48207	.32956	1.3524	.13111	1.6820	.22582	.80406
1.12	3.0659	.48641	.32628	1.3693	.13649	1.6956	.22931	.80757
1.13	3.0957	.49075	.32303	1.3863	.14186	1.7083	.23283	.81102
1.14	3.1268	.49510	.31982	1.4035	.14720	1.7233	.23636	.81441
1.15	3.1582	.49944	.31644	1.4208	.15253	1.7374	.23990	.81775
1.16	3.1899	.50378	.31349	1.4382	.15783	1.7517	.24346	.82104
1.17	3.2220	.50812	.31037	1.4558	.16311	1.7662	.24703	.82427
1.18	3.2544	.51247	.30728	1.4735	.16836	1.7808	.25062	.82745
1.19	3.2871	.51681	.30422	1.4914	.17360	1.7957	.25422	.83058
1.20	3.3201	.52115	.30119	1.5095	.17882	1.8107	.25784	.83365
1.21	3.3535	.52550	.29820	1.5276	.18402	1.8258	.26146	.83668
1.22	3.3872	.52984	.29523	1.5460	.18920	1.8412	.26510	.83965
1.23	3.4212	.53418	.29229	1.5645	.19437	1.8568	.26876	.84258
1.24	3.4556	.53853	.28938	1.5831	.19951	1.8725	.27242	.83546
1.25	3.4903	.54287	.28650	1.6019	.20464	1.8884	.27610	.84828
1.26	3.5254	.54721	.28365	1.6209	.20975	1.9045	.27979	.85106
1.27	3.5609	.55155	.28083	1.6400	.21485	1.9208	.28349	.85380
1.28	3.5996	.55590	.27804	1.6593	.21993	1.9373	.28721	.85648
1.29	3.6328	.56024	.27527	1.6788	.22499	1.9540	.29093	.85913
1.30	3.6693	.56458	.27253	1.6984	.23004	1.9709	.29467	.86172
1.31	3.7062	.56893	.26982	1.7182	.23507	1.9880	.29842	.86428
1.32	3.7434	.57327	.26714	1.7381	.24009	2.0053	.30217	.86678
1.33	3.7810	.57761	.26448	1.7583	.24509	2.0228	.30594	.86925
1.34	3.8190	.58195	.26185	1.7786	.25008	2.0404	.30972	.87167
1.35	3.8574	.58630	.25924	1.7991	.25505	2.0583	.31352	.87405
1.36	3.8962	.59064	.25666	1.8198	.26002	2.0764	.31732	.87639
1.37	3.9354	.59498	.25411	1.8406	.26496	2.0947	.32113	.87869
1.38	3.9749	.59933	.25158	1.8617	.26990	2.1132	.32495	.88095
1.39	4.0149	60367	.24908	1.8829	.27482	2.1320	.32878	.88317
1.40	4.0552	.60801	.24660	1.9043	.27974	2.1509	.33262	.88535
1.41	4.0960	.61236	.24414	1.9259	.28464	2.1700	.33647	.88749
1.42	4.1371	.61670	.24171	1.9477	.28952	2.1894	.34033	.88960
1.43	4.1787	.62104	.23931	1.9697	.29440	2.2090	.34420	.89167

EXPONENTIAL AND HYPERBOLIC FUNCTIONS AND THEIR COMMON LOGARITHMS
(continued)

x	e^x Value	$\log^{10}$	e^{-x} (value)	sinh x Value	$\log^{10}$	cosh x Value	$\log^{10}$	tanh x (value)
1.44	4.2207	.62538	.23693	1.9919	.29926	2.2288	.34807	.89370
1.45	4.2631	.62973	.23457	2.0143	.30412	2.2488	.35196	.89569
1.46	4.3060	.63407	.23224	2.0369	.30896	2.2691	.35585	.89765
1.47	4.3492	.63841	.22993	2.0597	.31379	2.2896	.35976	.89958
1.48	4.3929	.64276	.22764	2.0827	.31862	2.3103	.36367	.90147
1.49	4.4371	.64710	.22537	2.1059	.32343	2.3312	.36759	.90332
1.50	4.4817	.65144	.22313	2.1293	.32823	2.3524	.37151	.90515
1.51	4.5267	.65578	.22091	2.1529	.33303	2.3738	.37545	.90694
1.52	4.5722	.66013	.21871	2.1768	.33781	2.3955	.37939	.90870
1.53	4.6182	.66447	.21654	2.2008	.34258	2.4174	.38334	.91042
1.54	4.6646	.66881	.21438	2.2251	.34735	2.4395	.38730	.91212
1.55	4.7115	.67316	.21225	2.2496	.35211	2.4619	.39126	.91379
1.56	4.7588	.67750	.21014	2.2743	.35686	2.4845	.39524	.91542
1.57	4.8066	.68184	.20805	2.2993	.36160	2.5073	.39921	.91703
1.58	4.8550	.68619	.20598	2.3245	.36633	2.5305	.40320	.91860
1.59	4.9037	.69053	.20393	2.3499	.37105	2.5538	.40719	.92015
1.60	4.9530	.69487	.20190	2.3756	.37577	2.5775	.41119	.92167
1.61	5.0028	.69921	.19989	2.4015	.38048	2.6013	.41520	.92316
1.62	5.0531	.70356	.19790	2.4276	.38518	2.6255	.41921	.92462
1.63	5.1039	.70790	.19593	2.4540	.38987	2.6499	.42323	.92606
1.64	5.1552	.71224	.19398	2.4806	.39456	2.6746	.42725	.92747
1.65	5.2070	.71659	.19205	2.5075	.39923	2.6995	.43129	.92886
1.66	5.2593	.72093	.19014	2.5346	.40391	2.7247	.43532	.93022
1.67	5.3122	.72527	.18825	2.5620	.40857	2.7502	.43937	.93155
1.68	5.3656	.72961	.18637	2.5896	.41323	2.7760	.44341	.93286
1.69	5.4195	.73396	.18452	2.6175	.41788	2.8020	.44747	.93415
1.70	5.4739	.73830	.18268	2.6456	.42253	2.8283	.45153	.93541
1.71	5.5290	.74264	.18087	2.6740	.42717	2.8549	.45559	.93665
1.72	5.5845	.74699	.17907	2.7027	.43180	2.8818	.45966	.93786
1.73	5.6407	.75133	.17728	2.7317	.43643	2.9090	.46374	.93906
1.74	5.6973	.75567	.17552	2.7609	.44105	2.9364	.46782	.94023
1.75	5.7546	.76002	.17377	2.7904	.44567	2.9642	.47191	.94138
1.76	5.8124	.76436	.17204	2.8202	.45028	2.9922	.47600	.94250
1.77	5.8709	.76870	.17033	2.8503	.45488	3.0206	.48009	.94361
1.78	5.9299	.77304	.16864	2.8806	.45948	3.0492	.48419	.94470
1.79	5.9895	.77739	.16696	2.9112	.46408	3.0782	.48830	.94576
1.80	6.0496	.78173	.16530	2.9422	.46867	3.1075	.49241	.94681
1.81	6.1104	.78607	.16365	2.9734	.47325	3.1371	.49652	.94783
1.82	6.1719	.79042	.16203	3.0049	.47783	3.1669	.50064	.94884
1.83	6.2339	.79476	.16041	3.0367	.48241	3.1972	.50476	.94983
1.84	6.2965	.79910	.15882	3.0689	.48698	3.2277	.50889	.95080
1.85	6.3598	.80344	.15724	3.1013	.49154	3.2585	.51302	.95175
1.86	6.4237	.80779	.15567	3.1340	.49610	3.2897	.51716	.95268
1.87	6.4383	.81213	.15412	3.1671	.50066	3.3212	.52130	.95359
1.88	6.5535	.81647	.15259	3.2005	.50521	3.3530	.52544	.95449
1.89	6.6194	.82082	.15107	3.2341	.50976	3.3852	.52959	.95537
1.90	6.6859	.82516	.14957	3.2682	.51430	3.4177	.53374	.95624
1.91	6.7531	.82950	.14808	3.3025	.51884	3.4506	.53789	.95709
1.92	6.8210	.83385	.14661	3.3372	.52338	3.4838	.54205	.95792
1.93	6.8895	.83819	.14515	3.3722	.52791	3.5173	.54621	.95873
1.94	6.9588	.84253	.14370	3.4075	.53244	3.5512	.55038	.95953
1.95	7.0287	.84687	.14227	3.4432	.53696	3.5855	.55455	.96032
1.96	7.0993	.85122	.14086	3.4792	.54148	3.6201	.55872	.96109
1.97	7.1707	.85556	.13946	3.5156	.54600	3.6551	.56290	96185
1.98	7.2427	.85990	.13807	3.5923	.55051	3.6904	.56707	.96259
1.99	7.3155	.86425	.13670	3.5894	.55502	3.7261	.57126	.96331
2.00	7.3891	.86859	.13534	3.6269	.55953	3.7622	.57544	.96403
2.01	7.4633	.87293	.13399	3.6647	.56403	3.7987	.57963	.96473
2.02	7.5383	.87727	.13266	3.7028	.56853	3.8335	.58382	.96541
2.03	7.6141	.88162	.13134	3.7414	.57303	3.8727	.58802	.96609
2.04	7.6906	.88596	.13003	3.7803	.57753	3.9103	.59221	.96675
2.05	7.7679	.89030	.12873	3.8196	.58202	3.9483	.59641	.96740
2.06	7.8460	.89465	.12745	3.8593	.58650	3.9867	.60061	.96803
2.07	7.9248	.89899	.12619	3.8993	.59099	4.0255	.60482	.96865
2.08	8.0045	.90333	.12493	3.9398	.59547	4.0647	.60903	.96926
2.09	8.0849	.90768	.12369	3.9806	.59995	4.1043	.61324	.96986
2.10	8.1662	.91202	.12246	4.0219	.60443	4.1443	.61745	.97045
2.11	8.2482	.91636	.12124	4.0635	.60890	4.1847	.62167	.97103
2.12	8.3311	.92070	.12003	4.1056	.61337	4.2256	.62589	.97159
2.13	8.4149	.92505	.11884	4.1480	.61784	4.2669	.63011	.97215
2.14	8.4994	.92939	.11765	4.1909	.62231	4.3085	.63433	.97269
2.15	8.5849	.93373	.11648	4.2342	62677	4.3507	.63856	.97323

x	e^x Value	e^x log^{10}	e^{-x} (value)	sinh x Value	sinh x log^{10}	cosh x Value	cosh x log^{10}	tanh x (value)
2.16	8.6711	.93808	.11533	4.2779	.63123	4.3932	.64278	.97375
2.17	8.7583	.94242	.11418	4.3221	.63569	4.4362	.64701	.97426
2.18	8.8463	.94676	.11304	4.3666	.64015	4.4797	.65125	.97477
2.19	8.9352	.95110	.11192	4.4116	.64460	4.5236	.65548	.97526
2.20	9.0250	.95545	.11080	4.4571	.64905	4.5679	.65972	.97574
2.21	9.1157	.95979	.10970	4.5030	.65350	4.6127	.66396	.97622
2.22	9.2073	.96413	.10861	4.5494	.65795	4.6580	.66820	.97668
2.23	9.2999	.96848	.10753	4.5962	.66240	4.7037	.67244	.97714
2.24	9.3933	.97282	.10646	4.6434	.66684	4.7499	.67668	.97759
2.25	9.4877	.97716	.10540	4.6912	.67128	4.7966	.68093	.97803
2.26	9.5831	.98151	.10435	4.7394	.67572	4.8437	.68518	.97846
2.27	9.6794	.98585	.10331	4.7880	.68016	4.8914	.68943	.97888
2.28	9.7767	.99019	.10228	4.8372	.68459	4.9395	.69368	.97929
7.29	9.8749	.99453	.10127	4.8868	.68903	4.9881	.69794	97970
2.30	9.9742	.99888	.10026	4.9370	.69346	5.0372	.70219	.98010
2.31	10.074	1.00322	.09926	4.9876	.69789	5.0868	.70645	.98049
2.32	10.176	1.00756	.09827	5.0387	.70232	5.1370	.71071	.98087
2.33	10.278	1.01191	.09730	5.0903	.70675	5.1876	.71497	.98124
2.34	10.381	1.01625	.09633	5.1425	.71117	5.2388	.71923	.98161
2.35	10.486	1.02059	.09537	5.1951	.71559	5.2905	.72349	.98197
2.36	10.591	1.02493	.09442	5.2483	.72002	5.3427	.72776	.98233
2.37	10.697	1.02928	.09348	5.3020	.72444	5.3954	.73203	.98267
2.38	10.805	1.03362	.09255	5.3562	.72885	5.4487	.73630	.98301
2.39	10.913	1.03796	.09163	5.4109	.73327	5.5026	.74056	.98335
2.40	11.023	1.04231	.09072	5.4662	.73769	5.5569	.74484	.98367
2.41	11.134	1.04665	.08982	5.5221	.74210	5.6119	.74911	.98400
2.42	11.246	1.05099	.08892	5.5785	.74652	5.6674	.75338	.98431
2.43	11.359	1.05534	08804	5.6354	.75093	5.7235	.75766	.98462
2.44	11.473	1.05968	.08716	5.6929	.75534	5.7801	.76194	.98492
2.45	11.588	1.06402	.08629	5.7510	.75975	5.8373	.76621	.98522
2.46	11.705	1.06836	.08543	5.8097	.76415	5.8951	.77049	.98551
2.47	11.822	1.07271	.08458	5.8689	.76856	5.9535	.77477	.98579
2.48	11.941	1.07705	.08374	5.9288	.77296	6.0125	.77906	.98607
2.49	12.061	1.08139	.08291	5.9892	.77737	6.0721	.78334	.98635
2.50	12.182	1.08574	.08208	6.0502	.78177	6.1323	.78762	.98661
2.51	12.305	1.09008	.08127	6.1118	.78617	6.1931	.79191	.98688
2.52	12.429	1.09442	.08046	6.1741	.79057	6.2545	.79619	.98714
2.53	12.554	1.09877	.07966	6.2369	.79497	6.3166	.80048	.98739
2.54	12.680	1.10311	.07887	6.3004	.79937	6.3793	.80477	.98764
2.55	12.807	1.10745	.07808	6.3645	.80377	6.4426	.80906	.98788
2.56	12.936	1.11179	.07730	6.4293	.80816	6.5066	.81335	.98812
2.57	13.066	1.11614	.07654	6.4946	.81256	6.5712	.81764	.98835
2.58	13.197	1.12048	.07577	6.5607	.81695	6.6365	.82194	.98858
2.59	13.330	1.12482	.07502	6.6274	.82134	6.7024	.82623	.98881
2.60	13.464	1.12917	.07427	6.6947	.82573	6.7690	.83052	.98903
2.61	13.599	1.13351	.07353	6.7628	.83012	6.8363	.83482	.98924
2.62	13.736	1.13785	.07280	6.8315	.83451	6.9043	.83912	.98946
2.63	13.874	1.14219	.07208	6.9008	.83890	6.9729	.84341	.98966
2.64	14.013	1.14654	.07136	6.9709	.84329	7.0423	.84771	98987
2.65	14.154	1.15008	.07065	7.0417	.84768	7.1123	.85201	.99007
2.66	14.296	1.15522	.06995	7.1132	.85206	7.1831	.85631	.99026
2.67	14.440	1.15957	.06925	7.1854	.85645	7.2546	.86061	.99045
2.68	14.585	1.16391	.06856	7.2583	.86083	7.3268	.86492	.99064
2.69	14.732	1.16825	.06788	7.3319	.86522	7.3998	.86922	.99083
2.70	14.880	1.17260	.06721	7.4063	.86960	7.4735	.87352	.99101
2.71	15.029	1.17694	.06654	7.4814	.87398	7.5479	.87783	.99118
2.72	15.180	1.18128	.06587	7.5572	.87836	7.6231	.88213	.99136
2.73	15.333	1.18562	.06522	7.6338	.88274	7.6991	.89644	.99153
2.74	15.487	1.18997	.06457	7.7112	.88712	7.7758	.89074	.99170
2.75	15.643	1.19431	06393	7.7894	.89150	7.8533	.89505	.99186
2.76	15.800	1.19865	.06329	7.8683	.89588	7.9316	.89936	.99202
2.77	15.959	1.20300	.06266	7.9480	.90026	8.0106	.90367	.99218
2.78	16.119	1.20734	.06204	8.0285	.90463	8.0905	.90798	.99233
2.79	16.281	1.21168	.06142	8.1098	.90901	8.1712	.91229	.99248
2.80	16.445	1.21602	.06081	8.1919	.91339	8.2527	.91660	.99263
2.81	16.610	1.22037	.06020	8.2749	.91776	8.3351	.92091	.99278
2.82	16.777	1.22471	.05961	8.3586	.92213	8.4182	.92522	.99292
2.83	16.945	1.22905	.05901	8.4432	.92651	8.5022	.92953	.99306
2.84	17.116	1.23340	.05843	8.5287	.93088	8.5871	.93385	.99320
2.85	17.288	1.23774	.05784	8.6150	.93525	8.6728	.93816	.99333
2.86	17.462	1.24208	.05727	8.7021	.93963	8.7594	.94247	.99346

EXPONENTIAL AND HYPERBOLIC FUNCTIONS AND THEIR COMMON LOGARITHMS
(continued)

x	e^x Value	e^x $\log^{10}$	e^{-x} (value)	sinh x Value	sinh x $\log^{10}$	cosh x Value	cosh x $\log^{10}$	tanh x (value)
2.87	17.637	1.24643	.05670	8.7902	.94400	8.8469	.94679	.99359
2.88	17.814	1.25077	.05613	8.8791	.94837	8.9352	.95110	.99372
2.89	17.993	1.25511	.05558	8.9689	.95274	9.0244	.95542	.99384
2.90	18.174	1.25945	.05502	9.0596	.95711	9.1146	.95974	.99396
2.91	18.357	1.26380	.05448	9.1512	.96148	9.2056	.96405	.99408
2.92	18.541	1.26814	.05393	9.2437	.96584	9.2976	.96837	.99420
2.93	18.728	1.27248	.05340	9.3371	.97021	9.3905	.97269	.99431
2.94	18.916	1.27683	.05287	9.4315	.97458	9.4844	.97701	.99443
2.95	19.106	1.28117	.05234	9.5268	.97895	9.5791	.98133	.99454
2.96	19.298	1.28551	.05182	9.6231	.98331	9.6749	.98565	.99464
2.97	19.492	1.28985	.05130	9.7203	.98768	9.7716	.98997	.99475
2.98	19.688	1.29420	.05079	9.8185	.99205	9.8693	.99429	.99485
2.99	19.886	1.29854	.05029	9.9177	.99641	9.9680	.99861	.99496
3.00	20.086	1.30288	.04979	10.018	1.00078	10.068	1.00293	0.99505
3.05	21.115	1.32460	.04736	10.534	1.02259	10.581	1.02454	0.99552
3.10	22.198	1.34631	.04505	11.076	1.04440	11.122	1.04616	0.99595
3.15	23.336	1.36803	.04285	11.647	1.06620	11.690	1.06779	0.99633
3.20	24.533	1.38974	.04076	12.246	1.08799	12.287	1.08943	0.99668
3.25	25.790	1.41146	.03877	12.876	1.10977	12.915	1.11108	0.99700
3.30	27.113	1.43317	.03688	13.538	1.13155	13.575	1.13273	0.99728
3.35	28.503	1.45489	.03508	14.234	1.15332	14.269	1.15439	0.99754
3.40	29.964	1.47660	.03337	14.965	1.17509	14.999	1.17605	0.99777
3.45	31.500	1.49832	.03175	15.734	1.19685	15.766	1.19772	0.99799
3.50	33.115	1.52003	.03020	16.543	1.21860	16.573	1.21940	0.99818
3.55	34.813	1.54175	.02872	17.392	1.24036	17.421	1.24107	0.99835
3.60	36.598	1.56346	.02732	18.286	1.26211	18.313	1.26275	0.99851
3.65	38.475	1.58517	02599	19.224	1.28385	19.250	1.28444	9.99865
3.70	40.447	1.60689	.02472	20.211	1.30559	20.236	1.30612	0.99878
3.75	42.521	1.62860	.02352	21.249	1.32733	21.272	1.32781	0.99889
3.80	44.701	1.65032	.02237	22.339	1.34907	22.362	1.34951	0.99900
3.85	46.993	1.67203	.02128	23.486	1.37081	23.507	1.37120	0.99909
3.90	49.402	1.69375	.02024	24.691	1.39254	24.711	1.39290	0.99918
3.95	51.935	1.71546	.01925	25.958	1.41427	25.977	1.41459	0.09926
4.00	54.598	1.73718	.01832	27.290	1.43600	27.308	1.43629	0.99933
4.10	60.340	1.78061	.01657	30.162	1.47946	30.178	1.47970	0.99945
4.20	66.686	1.82404	.01500	33.336	1.52291	33.351	1.52310	0.99955
4.30	73.700	1.86747	.01357	36.843	1.56636	36.857	1.56652	0.99963
4.40	81.451	1.91090	.01227	40.719	1.60980	40.732	1.60993	0.99970
4.50	90.017	1.95433	.01111	45.003	1.65324	45.014	1.65335	0.99975
4.60	99.484	1.99775	.01005	49.737	1.69668	49.747	1.69677	0.99980
4.70	109.95	2.04118	.00910	54.969	1.74012	54.978	1.74019	0.99983
4.80	121.51	2.08461	.00823	60.751	1.78355	60.759	1.78361	0.99986
4.90	134.29	2.12804	.00745	67.141	1.82699	67.149	1.82704	0.99989
5.00	148.41	2.17147	.00674	74.203	1.87042	74.210	1.87046	0.99991
5.10	164.02	2.21490	.00610	82.008	1.91389	82.014	1.91389	0.99993
5.20	181.27	2.25833	.00552	90.633	1.95729	90.639	1.95731	0.99994
5.30	200.34	2.30176	.00499	100.17	2.00074	100.17	2.00074	0.99995
5.40	221.41	2.34519	.00452	110.70	2.04415	110.71	2.04417	0.99996
5.50	244.69	2.38862	.00409	122.34	2.08758	122.35	2.08760	0.99997
5.60	270.43	2.43205	.00370	135.21	2.13101	135.22	2.13103	0.99997
5.70	298.87	2.47548	.00335	149.43	2.17444	149.44	2.17445	0.99998
5.80	330.30	2.51891	.00303	165.15	2.21787	165.15	2.21788	0.99998
5.90	365.04	2.56234	.00274	182.52	2.26130	182.52	2.26131	0.99998
6.00	403.43	2.60577	.00248	201.71	2.30473	201.72	2.30474	0.99999
6.25	518.01	2.71434	.00193	259.01	2.41331	259.01	2.41331	0.99999
6.50	665.14	2.82291	.00150	332.57	2.52188	332.57	2.52189	1.00000
6.75	854.06	2.93149	.00117	427.03	2.63046	427.03	2.63046	1.00000
7.00	1096.6	3.04006	.00091	548.32	2.73904	548.32	2.73903	1.00000
7.50	1808.0	3.25721	.00055	904.02	2.95618	904.02	2.95618	1.00000
8.00	2981.0	3.47436	.00034	1490.5	3.17333	1490.5	3.17333	1.00000
8.50	4914.8	3.69150	.00020	2457.4	3.39047	2457.4	3.39047	1.00000
9.00	8103.1	3.90865	.00012	4051.5	3.60762	4051.5	3.60762	1.00000
9.50	13360.	4,12580	.00007	6679.9	3.82477	6679.9	3.82477	1.00000
10.00	22026.	4.34294	.00005	11013.	4.04191	11013.	4.04191	1.00000

NATURAL TRIGONOMETRIC FUNCTIONS TO FOUR PLACES

X radians	X degrees	sin x	cos x	tan x	cot x	sec x	csc x		
.0000	0° 00′	.0000	1.0000	.0000	—	1.000	—	90° 00′	1.5708
.0029	10	.0029	1.0000	.0029	343.8	1.000	343.8	50	1.5679
.0058	20	.0058	1.0000	.0058	171.9	1.000	171.9	40	1.5650
.0087	30	.0087	1.0000	.0087	114.6	1.000	114.6	30	1.5621
.0116	40	.0116	.9999	.0116	85.94	1.000	85.95	20	1.5592
.0145	50	.0145	.9999	.0145	68.75	1.000	68.76	10	1.5563
.0175	1° 00′	.0175	.9998	.0175	57.29	1.000	57.30	89° 00′	1.5533
.0204	10	.0204	.9998	.0204	49.10	1.000	49.11	50	1.5504
.0233	20	.0233	.9997	.0233	42.96	1.000	42.98	40	1.5475
.0262	30	.0262	.9997	.0262	38.19	1.000	38.20	30	1.5446
.0291	40	.0291	.9996	.0291	34.37	1.000	34.38	20	1.5417
.0320	50	.0320	.9995	.0320	31.24	1.001	31.26	10	1.5388
.0349	2° 00′	.0349	.9994	.0349	28.64	1.001	28.65	88° 00′	1.5359
.0378	10	.0378	.9993	.0378	26.43	1.001	26.45	50	1.5330
.0407	20	.0407	.9992	.0407	24.54	1.001	24.56	40	1.5301
.0436	30	.0436	.9990	.0437	22.90	1.001	22.93	30	1.5272
.0465	40	.0465	.9989	.0466	21.47	1.001	21.49	20	1.5243
.0495	50	.0494	.9988	.0495	20.21	1.001	20.23	10	1.5213
.0524	3° 00′	.0523	.9986	.0524	19.08	1.001	19.11	87° 00′	1.5184
.0553	10	.0552	.9985	.0553	18.07	1.002	18.10	50	1.5155
.0582	20	.0581	.9983	.0582	17.17	1.002	17.20	40	1.5126
.0611	30	.0610	.9981	.0612	16.35	1.002	16.38	30	1.5097
.0640	40	.0640	.9980	.0641	15.60	1.002	15.64	20	1.5068
.0669	50	.0669	.9978	.0670	14.92	1.002	14.96	10	1.5039
.0698	4° 00′	.0698	.9976	.0699	14.30	1.002	14.34	86° 00′	1.5010
.0727	10	.0727	.9974	.0729	13.73	1.003	13.76	50	1.4981
.0756	20	.0756	.9971	.0758	13.20	1.003	13.23	40	1.4952
.0785	30	.0785	.9969	.0787	12.71	1.003	12.75	30	1.4923
.0814	40	.0814	.9967	.0816	12.25	1.003	12.29	20	1.4893
.0844	50	.0843	.9964	.0846	11.83	1.004	11.87	10	1.4864
.0873	5° 00′	.0872	.9962	.0875	11.43	1.004	11.47	85° 00′	1.4835
.0902	10	.0901	.9959	.0904	11.06	1.004	11.10	50	1.4806
.0931	20	.0929	.9957	.0934	10.71	1.004	10.76	40	1.4777
.0960	30	.0958	.9954	.0963	10.39	1.005	10.43	30	1.4748
.0989	40	.0987	.9951	.0992	10.08	1.005	10.13	20	1.4719
.1018	50	.1016	.9948	.1022	9.788	1.005	9.839	10	1.4690
.1047	6° 00′	.1045	.9945	.1051	9.514	1.006	9.567	84° 00′	1.4661
.1076	10	.1074	.9942	.1080	9.255	1.006	9.309	50	1.4632
.1105	20	.1103	.9939	.1110	9.010	1.006	9.065	40	1.4603
.1134	30	.1132	.9936	.1139	8.777	1.006	8.834	30	1.4573
.1164	40	.1161	.9932	.1169	8.556	1.007	8.614	20	1.4544
.1193	50	.1190	.9929	.1198	8.345	1.007	8.405	10	1.4515
.1222	7° 00′	.1219	.9925	.1228	8.144	1.008	8.206	83° 00′	1.4486
.1251	10	.1248	.9922	.1257	7.953	1.008	8.016	50	1.4457
.1280	20	.1276	.9918	.1287	7.770	1.008	7.834	40	1.4428
.1309	30	.1305	.9914	.1317	7.596	1.009	7.661	30	1.4399
.1338	40	.1334	.9911	.1346	7.429	1.009	7.496	20	1.4370
.1367	50	.1363	.9907	.1376	7.269	1.009	7.337	10	1.4341
.1396	8° 00′	.1392	.9903	.1405	7.115	1.010	7.185	82° 00′	1.4312
.1425	10	.1421	.9899	.1435	6.968	1.010	7.040	50	1.4283
.1454	20	.1449	.9894	.1465	6.827	1.011	6.900	40	1.4254
.1484	30	.1478	.9890	.1495	6.691	1.011	6.765	30	1.4224
.1513	40	.1507	.9886	.1524	6.561	1.012	6.636	20	1.4195
.1542	50	.1536	.9881	.1554	6.435	1.012	6.512	10	1.4166
.1571	9° 00′	.1564	.9877	.1584	6.314	1.012	6.392	81° 00′	1.4137
.1600	10	.1593	.9872	.1614	6.197	1.013	6.277	50	1.4108
.1629	20	.1622	.9868	.1644	6.084	1.013	6.166	40	1.4079
.1658	30	.1650	.9863	.1673	5.976	1.014	6.059	30	1.4050
.1687	40	.1679	.9858	.1703	5.871	1.014	5.955	20	1.4021
.1716	50	.1708	.9853	.1733	5.769	1.015	5.855	10	1.3992
.1745	10° 00′	.1736	.9848	.1763	5.671	1.015	5.759	80° 00′	1.3963
.1774	10	.1765	.9843	.1793	5.576	1.016	5.665	50	1.3934
.1804	20	.1794	.9838	.1823	5.485	1.016	5.575	40	1.3904
.1833	30	.1822	.9833	.1853	5.396	1.017	5.487	30	1.3875
.1862	40	.1851	.9827	.1883	5.309	1.018	5.403	20	1.3846
.1891	50	.1880	.9822	.1914	5.226	1.018	5.320	10	1.3817
.1920	11° 00′	.1908	.9816	.1944	5.145	1.019	5.241	79° 00′	1.3788
.1949	10	.1937	.9811	.1974	5.066	1.019	5.164	50	1.3759
.1978	20	.1965	.9805	.2004	4.989	1.020	5.089	40	1.3730
.2007	30	.1994	.9799	.2035	4.915	1.020	5.016	30	1.3701

		cos x	sin x	cot x	tan x	csc x	sec x	x degrees	x radians

A-7

x radians	x degrees	sin x	cos x	tan x	cot x	sec x	csc x	x degrees	x radians
.2036	40	.2022	.9793	.2065	3.843	1.021	4.945	20	1.3672
.2065	50	.2051	.9787	.2095	4.773	1.022	4.876	10	1.3643
.2094	12° 00′	.2079	.9781	.2126	4.705	1.022	4.810	78° 00′	1.3614
.2123	10	.2108	.9775	.2156	4.638	1.023	4.745	50	1.3584
.2153	20	.2136	.9769	.2186	4.574	1.024	4.682	40	1.3555
.2182	30	.2164	.9763	.2217	4.511	1.025	4.620	30	1.3526
.2211	40	.2193	.9757	.2247	4.449	1.025	4.560	20	1.3497
.2240	50	.2221	.9750	.2278	4.390	1.026	4.502	10	1.3468
.2269	13° 00′	.2250	.9744	.2309	4.331	1.026	4.445	77° 00′	1.3439
.2298	10	.2278	.9737	.2339	4.275	1.027	4.390	50	1.3410
.2327	20	.2306	.9730	.2370	4.219	1.028	4.336	40	1.3381
.2356	30	.2334	.9724	.2401	4.165	1.028	4.284	30	1.3352
.2385	40	.2363	.9717	.2432	4.113	1.029	4.232	20	1.3323
.2414	50	.2391	.9710	.2462	4.061	1.030	4.182	10	1.3294
.2443	14° 00′	.2419	.9703	.2493	4.011	1.031	4.134	76° 00′	1.3265
.2473	10	.2447	.9696	.2524	3.962	1.031	4.086	50	1.3235
.2502	20	.2476	.9689	.2555	3.914	1.032	4.039	40	1.3206
.2531	30	.2504	.9681	.2586	3.867	1.033	3.994	30	1.3177
.2560	40	.2532	.9674	.2617	3.821	1.034	3.950	20	1.3148
.2589	50	.2560	.9667	.2648	3.776	1.034	3.906	10	1.3119
.2618	15° 00′	.2588	.9659	.2679	3.732	1.035	3.864	75° 00′	1.3090
.2647	10	.2616	.9652	.2711	3.689	1.036	3.822	50	1.3061
.2676	20	.2644	.9644	.2732	3.647	1.037	3.782	40	1.3032
.2705	30	.2672	.9636	.2773	3.606	1.038	3.742	30	1.3003
.2734	40	.2700	.9628	.2805	3.566	1.039	3.703	20	1.2974
.2763	50	.2728	.9621	2836	3.526	1.039	3.665	10	1.2945
.2793	16° 00′	.2756	.9613	.2867	3.487	1.040	3.628	74° 00′	1.2915
.2822	10	.2784	.9605	.2899	3.450	1.041	3.592	50	1.2886
.2851	20	.2812	.9596	.2931	3.412	1.042	3.556	40	1.2857
.2880	30	.2840	.9588	.2962	3.376	1.043	3.521	30	1.2828
.2909	40	.2868	.9580	.2994	3.340	1.044	3.487	20	1.2799
.2938	50	.2896	.9572	.3026	3.305	1.045	3.453	10	1.2770
.2967	17° 00′	.2924	.9563	.3057	3.271	1.046	3.420	73° 00′	1.2741
.2996	10	.2952	.9555	.3089	3.237	1.047	3.388	50	1.2712
.3025	20	.2979	.9546	.3121	4.204	1.048	3.356	40	1.2683
.3054	30	.3007	.9537	.3153	3.172	1.049	3.326	30	1.2654
.3083	40	.3035	.9528	.3185	3.140	1.049	3.295	20	1.2625
.3113	50	.3062	.9520	.3217	3.108	1.050	3.265	10	1.2595
.3142	18° 00′	.3090	.9511	.3249	3.078	1.051	3.236	72° 00′	1.2566
.3171	10	.3118	.9502	.3281	3.047	1.052	3.207	50	1.2537
.3200	20	.3145	.9492	.3314	3.018	1.053	3.179	40	1.2508
.3229	30	.3173	.9483	.3346	2.989	1.054	3.152	30	1.2479
.3258	40	.3201	.9474	.3378	2.960	1.056	3.124	20	1.2450
.3287	50	.3228	.9465	.3411	2.932	1.057	3.098	10	1.2421
.3316	19° 00′	.3256	.9455	.3443	2.904	1.058	3.072	71° 00′	1.2392
.3345	10	.3283	.9446	.3476	2.877	1.059	3.046	50	1.2363
.3374	20	.3311	.9436	.3508	2.850	1.060	3.021	40	1.2334
.3403	30	.3338	.9426	.3541	2.824	1.061	2.996	30	1.2305
.3432	40	.3365	.9417	.3574	2.798	1.062	2.971	20	1.2275
.3462	50	.3393	.9407	.3607	2.773	1.063	2.947	10	1.2246
.3491	20° 00′	.3420	.9397	.3640	2.747	1.064	2.924	70° 00′	1.2217
.3520	10	.3448	.9387	.3673	2.723	1.065	2.901	50	1.2188
.3549	20	.3475	.9377	.3706	2.699	1.066	2.878	40	1.2159
.3578	30	.3502	.9367	.3739	2.675	1.068	2.855	30	1.2130
.3607	40	.3529	.9356	.3772	2.651	1.069	2.833	20	1.2101
.3636	50	.3557	.9346	.3805	2.628	1.070	2.812	10	1.2072
.3665	21° 00′	.3584	.9336	.3839	2.605	1.071	2.790	69° 00′	1.2043
.3694	10	.3611	.9325	.3872	2.583	1.072	2.769	50	1.2014
.3723	20	.3638	.9315	.3906	2.560	1.074	2.749	40	1.1985
.3752	30	.3665	.9304	.3939	2.539	1.075	2.729	30	1.1956
.3782	40	.3692	.9293	.3973	2.517	1.076	2.709	20	1.1926
.3811	50	.3719	.9283	.4006	2.496	1.077	2.689	10	1.1897
.3840	22° 00′	.3746	.9272	.4040	2.475	1.079	2.669	68° 00′	1.1868
.3869	10	.3773	.9261	.4074	2.455	1.080	2.650	50	1.1839
.3898	20	.3800	.9250	.4108	2.434	1.081	2.632	40	1.1810
.3927	30	.3827	.9239	.4142	2.414	1.082	2.613	30	1.1781
.3956	40	.3854	.9228	.4176	2.394	1.084	2.595	20	1.1752
.3985	50	.3881	.9216	.4210	2.375	1.085	2.577	10	1.1723
.4014	23° 00′	.3907	.9205	.4245	2.356	1.086	2.559	67° 00′	1.1694
.4043	10	.3934	.9194	.4279	2.337	1.088	2.542	50	1.1665

cos x	sin x	cot x	tan x	csc x	sec x	x degrees	x radians

x radians	x degrees	sin x	cos x	tan x	cot x	sec x	csc x		
.4072	20	.3961	.9182	.4314	2.318	1.089	2.525	40	1.1636
.4102	30	.3987	.9171	.4348	2.300	1.090	2.508	30	1.1606
4131	40	.4014	.9159	.4383	2.282	1.092	2.491	20	1.1577
.4160	50	4041	.9147	.4417	2.264	1.093	2.475	10	1.1548
.4189	24° 00′	.4067	.9135	.4452	2.246	1.095	2.459	66° 00′	1.1519
.4218	10	.4094	.9124	.4487	2.229	1.096	2.443	50	1.1490
.4247	20	.4120	.9112	.4522	2.211	1.097	2.427	40	1.1461
.4276	30	.4147	.9100	.4557	2.194	1.099	2.411	30	1.1432
.4305	40	.4173	.9088	.4592	2.177	1.100	2.396	20	1.1403
.4334	50	.4200	.9075	.4628	2.161	1.102	2.381	10	1.1374
.4363	25° 00′	.4226	.9063	.4663	2.145	1.103	2.366	65° 00′	1.1345
.4392	10	.4253	.9051	.4699	2.128	1.105	2.352	50	1.1316
.4422	20	.4279	.9038	.4734	2.112	1.106	2.337	40	1.1286
.4451	30	.4305	.9026	.4770	2.097	1.108	2.323	30	1.1257
.4480	40	.4331	.9013	.4806	2.081	1.109	2.309	20	1.1228
.4509	50	.4358	.9001	.4841	2.066	1.111	2.295	10	1.1199
.4538	26° 00′	.4384	.8988	.4877	2.050	1.113	2.281	64° 00′	1.1170
.4567	10	.4410	.8975	.4913	2.035	1.114	2.268	50	1.1141
.4596	20	.4436	.8962	.4950	2.020	1.116	2.254	40	1.1112
.4625	30	.4462	.8949	.4986	2.006	1.117	2.241	30	1.1083
.4654	40	.4488	.8936	.5022	1.991	1.119	2.228	20	1.1054
.4683	50	.4514	.8923	.5059	1.977	1.121	2.215	10	1.1025
.4712	27° 00′	.4540	.8910	.5095	1.963	1.122	2.203	63° 00′	1.0996
.4741	10	.4566	.8897	.5132	1.949	1.124	2.190	50	1.0966
.4771	20	.4592	.8884	.5169	1.935	1.126	2.178	40	1.0937
.4800	30	.4617	.8870	.5206	1.921	1.127	2.166	30	1.0908
.4829	40	.4643	.8857	.5243	1.907	1.129	2.154	20	1.0879
.4858	50	.4669	.8843	.5280	1.894	1.131	2.142	10	1.0850
.4887	28° 00′	.4695	.8829	.5317	1.881	1.133	2.130	62° 00′	1.0821
.4916	10	.4720	.8816	.5354	1.868	1.134	2.118	50	1.0792
.4945	20	.4746	.8802	.5392	1.855	1.136	2.107	40	1.0763
.4974	30	.4772	.8788	.5430	1.842	1.138	2.096	30	1.0734
.5003	40	.4797	.8774	.5467	1.829	1.140	2.085	20	1.0705
.5032	50	.4823	.8760	.5505	1.816	1.142	2.074	10	1.0676
.5061	29° 00′	.4848	.8746	.5543	1.804	1.143	2.063	61° 00′	1.0647
.5091	10	.4874	.8732	.5581	1.792	1.145	2.052	50	1.0617
.5120	20	.4899	.8718	.5619	1.780	1.147	2.041	40	1.0588
.5149	30	.4924	.8704	.5658	1.767	1.149	2.031	30	1.0559
.5178	40	.4950	.8689	5696	1.756	1.151	2.020	20	1.0530
.5207	50	.4975	.8675	.5735	1.744	1.153	2.010	10	1.0501
.5236	30° 00′	.5000	.8660	.5774	1.732	1.155	2.000	60° 00′	1.0472
.5265	10	.5025	.8646	.5812	1.720	1.157	1.990	50	1.0443
.5294	20	.5050	.8631	.5851	1.709	1.159	1.980	40	1.0414
.5323	30	.5075	.8616	.5890	1.698	1.161	1.970	30	1.0385
.5352	40	.5100	.8601	.5930	1.686	1.163	1.961	20	1.0356
.5381	50	.5125	.8587	.5969	1.675	1.165	1.951	10	1.0327
.5411	31° 00′	.5150	.8572	.6009	1.664	1.167	1.942	59° 00′	1.0297
.5440	10	.5175	.8557	.6048	1.653	1.169	1.932	50	1.0268
.5469	20	.5200	.8542	.6088	1.643	1.171	1.923	40	1.0239
.5498	30	.5225	.8526	.6128	1.632	1.173	1.914	30	1.0210
.5527	40	.5250	.8511	.6168	1.621	1.175	1.905	20	1.0181
.5556	50	.5275	.8496	.6208	1.611	1.177	1.896	10	1.0152
.5585	32° 00′	.5299	8480	.6249	1.600	1.179	1.887	58° 00′	1.0123
.5614	10	.5324	.8465	.6289	1.590	1.181	1.878	50	1.0094
.5643	20	.5348	.8450	.6330	1.580	1.184	1.870	40	1.0065
.5672	30	.5373	.8434	.6371	1.570	1.186	1.861	30	1.0036
.5701	40	.5398	.8418	.6412	1.560	1.188	1.853	20	1.0007
.5730	50	.5422	.8403	.6453	1.550	1.190	1.844	10	.9977
.5760	33° 00′	.5446	.8397	.6494	1.540	1.192	1.836	57° 00′	.9948
.5789	10	.5471	.8371	.6536	1.530	1.195	1.828	50	.9919
.5818	20	.5495	.8355	.6577	1.520	1.197	1.820	40	.9890
.5847	30	.5519	.8339	.6619	1.511	1.199	1.812	30	.9861
.5876	40	.5544	.8323	.6661	1.501	1.202	1.804	20	.9832
.5905	50	.5568	.8307	.6703	1.492	1.204	1.796	10	.9803
.5934	34° 00′	.5592	.8290	.6745	1.483	1.206	1.788	56° 00′	.9774
.5963	10	.5616	.8274	.6787	1.473	1.209	1.781	50	.9745
.5992	20	.5640	.8258	.6830	1.464	1.211	1.773	40	.9716
.6021	30	.5664	.8241	.6873	1.455	1.213	1.766	30	.9687
.6050	40	.5688	8225	.6916	1.446	1.216	1.758	20	.9657
.6080	50	.5712	.8208	.6959	1.437	1.218	1.751	10	.9628

| | | cos x | sin x | cot x | tan x | csc x | sec x | x degrees | x radians |

x radians	x degrees	sin x	cos x	tan x	cot x	sec x	csc x		
.6109	**35° 00′**	.5736	.8192	.7002	1.428	1.221	1.743	**55° 00′**	.9599
.6138	10	.5760	.8175	.7046	1.419	1.223	1.736	50	.9570
.6167	20	.5783	.8158	.7089	1.411	1.226	1.729	40	.9541
.6196	30	.5807	.8141	.7133	1.402	1.228	1.722	30	.9512
.6225	40	.5831	.8124	.7177	1.393	1.231	1.715	20	.9483
.6254	50	.5854	.8107	.7221	1.385	1.233	1.708	10	.9454
.6283	**36° 00′**	.5878	.8090	.7265	1.376	1.236	1.701	**54° 00′**	.9425
.6312	10	.5901	.8073	.7310	1.368	1.239	1.695	50	.9396
.6341	20	.5925	.8056	.7355	1.360	1.241	1.688	40	.9367
.6370	30	.5948	.8039	.7400	1.351	1.244	1.681	30	.9338
.6400	40	.5972	.8021	.7445	1.343	1.247	1.675	20	.9308
.6429	50	.5995	.8004	.7490	1.335	1.249	1.668	10	.9279
.6458	**37° 00′**	.6018	.7986	.7536	1.327	1.252	1.662	**53° 00′**	.9250
.6487	10	.6041	.7969	.7581	1.319	1.255	1.655	50	.9221
.6516	20	.6065	.7951	.7627	1.311	1.258	1.649	40	.9192
.6545	30	.6088	.7934	.7673	1.303	1.260	1.643	30	.9163
.6574	40	.6111	.7916	.7720	1.295	1.263	1.636	20	.9134
.6603	50	.6134	.7898	.7766	1.288	1.266	1.630	10	.9105
.6632	**38° 00′**	.6157	.7880	.7813	1.280	1.269	1.624	**52° 00′**	.9076
.6661	10	.6180	.7862	.7860	1.272	1.272	1.618	50	.9047
.6690	20	.6202	.7844	.7907	1.265	1.275	1.612	40	.9018
.6720	30	.6225	.7826	.7954	1.257	1.278	1.606	30	.8988
.6749	40	.6248	.7808	.8002	1.250	1.281	1.601	20	.8959
.6778	50	.6271	.7790	.8050	1.242	1.284	1.595	10	.8930
.6807	**39° 00′**	.6293	.7771	.8098	1.235	1.287	1.589	**51° 00′**	.8901
.6836	10	.6316	.7753	.8146	1.228	1.290	1.583	50	.8872
.6865	20	.6338	.7735	.8195	1.220	1.293	1.578	40	.8843
.6894	30	.6361	.7716	.8243	1.213	1.296	1.572	30	.8814
.6923	40	.6383	.7698	.8292	1.206	1.299	1.567	20	.8785
.6952	50	.6406	.7679	.8342	1.199	1.302	1.561	10	.8756
.6981	**40° 00′**	.6428	.7660	.8391	1.192	1.305	1.556	**50° 00′**	.8727
.7010	10	.6450	.7642	.8441	1.185	1.309	1.550	50	.8698
.7039	20	.6472	.7623	.8491	1.178	1.312	1.545	40	.8668
.7069	30	.6494	.7604	.8541	1.171	1.315	1.540	30	.8639
.7098	40	.6517	.7585	.8591	1.164	1.318	1.535	20	.8610
.7127	50	.6539	.7566	.8642	1.157	1.322	1.529	10	.8581
.7156	**41° 00′**	.6561	.7547	.8693	1.150	1.325	1.524	**49° 00′**	.8552
.7185	10	.6583	.7528	.8744	1.144	1.328	1.519	50	.8523
.7214	20	.6604	.7509	.8796	1.137	1.332	1.514	40	.8494
.7243	30	.6626	.7490	.8847	1.130	1.335	1.509	30	.8465
.7272	40	.6648	.7470	.8899	1.124	1.339	1.504	20	.8436
.7301	50	.6670	.7451	.8952	1.117	1.342	1.499	10	.8407
.7330	**42° 00′**	.6691	.7431	.9004	1.111	1.346	1.494	**48° 00′**	.8378
.7359	10	.6713	.7412	.9057	1.104	1.349	1.490	50	.8348
.7389	20	.6734	.7392	.9110	1.098	1.353	1.485	40	.8319
.7418	30	.6756	.7373	.9163	1.091	1.356	1.480	30	.8290
.7447	40	.6777	.7353	.9217	1.085	1.360	1.476	20	.8261
.7476	50	.6799	.7333	.9271	1.079	1.364	1.471	10	.8232
.7505	**43° 00′**	.6820	.7314	.9325	1.072	1.367	1.466	**47° 00′**	.8203
.7534	10	.6841	.7294	.9380	1.066	1.371	1.462	50	.8174
.7563	20	.6862	.7274	.9435	1.060	1.375	1.457	40	.8145
.7592	30	.6884	.7254	.9490	1.054	1.379	1.453	30	.8116
.7621	40	.6905	.7234	.9545	1.048	1.382	1.448	20	.8087
.7650	50	.6926	.7214	.9601	1.042	1.386	1.444	10	.8058
.7679	**44° 00′**	.6947	.7193	.9657	1.036	1.390	1.440	**46° 00′**	.8029
.7709	10	.6967	.7173	.9713	1.030	1.394	1.435	50	.7999
.7738	20	.6988	.7153	.9770	1.024	1.398	1.431	40	.7970
.7767	30	.7009	.7133	.9827	1.018	1.402	1.427	30	.7941
.7796	40	.7030	.7112	.9884	1.012	1.406	1.423	20	.7912
.7825	50	.7050	.7092	.9942	1.006	1.410	1.418	10	.7883
.7854	**45° 00′**	.7071	.7071	1.0000	1.0000	1.414	1.414	**45° 00′**	.7854

		cos x	sin x	cot x	tan x	csc x	sec x	x degrees	x radians

RELATION OF ANGULAR FUNCTIONS IN TERMS OF ONE ANOTHER

TRIGONOMETRIC FUNCTIONS

Function	$\sin \alpha$	$\cos \alpha$	$\tan \alpha$	$\cot \alpha$	$\sec \alpha$	$\csc \alpha$
$\sin \alpha$	$\sin \alpha$	$\pm \sqrt{1 - \cos^2 \alpha}$	$\dfrac{\tan \alpha}{\pm \sqrt{1 + \tan^2 \alpha}}$	$\dfrac{1}{\pm \sqrt{1 + \cot^2 \alpha}}$	$\dfrac{\pm \sqrt{\sec^2 \alpha - 1}}{\sec \alpha}$	$\dfrac{1}{\csc \alpha}$
$\cos \alpha$	$\pm \sqrt{1 - \sin^2 \alpha}$	$\cos \alpha$	$\dfrac{1}{\pm \sqrt{1 + \tan^2 \alpha}}$	$\dfrac{\cot \alpha}{\pm \sqrt{1 + \cot^2 \alpha}}$	$\dfrac{1}{\sec \alpha}$	$\dfrac{\pm \sqrt{\csc^2 \alpha - 1}}{\csc \alpha}$
$\tan \alpha$	$\dfrac{\sin \alpha}{\pm \sqrt{1 - \sin^2 \alpha}}$	$\dfrac{\pm \sqrt{1 - \cos^2 \alpha}}{\cos \alpha}$	$\tan \alpha$	$\dfrac{1}{\cot \alpha}$	$\pm \sqrt{\sec^2 \alpha - 1}$	$\dfrac{1}{\pm \sqrt{\csc^2 \alpha - 1}}$
$\cot \alpha$	$\dfrac{\pm \sqrt{1 - \sin^2 \alpha}}{\sin \alpha}$	$\dfrac{\cos \alpha}{\pm \sqrt{1 - \cos^2 \alpha}}$	$\dfrac{1}{\tan \alpha}$	$\cot \alpha$	$\dfrac{1}{\pm \sqrt{\sec^2 \alpha - 1}}$	$\pm \sqrt{\csc^2 \alpha - 1}$
$\sec \alpha$	$\dfrac{1}{\pm \sqrt{1 - \sin^2 \alpha}}$	$\dfrac{1}{\cos \alpha}$	$\pm \sqrt{1 + \tan^2 \alpha}$	$\dfrac{\pm \sqrt{1 + \cot^2 \alpha}}{\cot \alpha}$	$\sec \alpha$	$\dfrac{\csc \alpha}{\pm \sqrt{\csc^2 \alpha - 1}}$
$\csc \alpha$	$\dfrac{1}{\sin \alpha}$	$\dfrac{1}{\pm \sqrt{1 - \cos^2 \alpha}}$	$\dfrac{\pm \sqrt{1 + \tan^2 \alpha}}{\tan \alpha}$	$\pm \sqrt{1 + \cot^2 \alpha}$	$\dfrac{\sec \alpha}{\pm \sqrt{\sec^2 \alpha - 1}}$	$\csc \alpha$

Note: The choice of sign depends upon the quadrant in which the angle terminates.

HYPERBOLIC FUNCTIONS

Function	$\sinh x$	$\cosh x$	$\tanh x$
$\sinh x =$	$\sinh x$	$+ \sqrt{\cosh^2 x - 1}$	$\dfrac{\tanh x}{\sqrt{1 - \tanh^2 x}}$
$\cosh x =$	$\sqrt{1 + \sinh^2 x}$	$\cosh x$	$\dfrac{1}{\sqrt{1 - \tanh^2 x}}$
$\tanh x =$	$\dfrac{\sinh x}{\sqrt{1 + \sinh^2 x}}$	$\pm \dfrac{\sqrt{\cosh^2 x - 1}}{\cosh x}$	$\tanh x$
$\operatorname{cosech} x =$	$\dfrac{1}{\sinh x}$	$\pm \dfrac{1}{\sqrt{\cosh^2 x - 1}}$	$\dfrac{\sqrt{1 - \tanh^2 x}}{\tanh x}$
$\operatorname{sech} x =$	$\dfrac{1}{\sqrt{1 + \sinh^2 x}}$	$\dfrac{1}{\cosh x}$	$\sqrt{1 - \tanh^2 x}$
$\coth x =$	$\dfrac{\sqrt{1 + \sinh^2 x}}{\sinh x}$	$\dfrac{\pm \cosh x}{\sqrt{\cosh^2 x - 1}}$	$\dfrac{1}{\tanh x}$

Function	$\operatorname{cosech} x$	$\operatorname{sech} x$	$\coth x$
$\sinh x =$	$\dfrac{1}{\operatorname{cosech} x}$	$\pm \dfrac{\sqrt{1 - \operatorname{sech}^2 x}}{\operatorname{sech} x}$	$\dfrac{\pm 1}{\sqrt{\coth^2 x - 1}}$
$\cosh x =$	$\pm \dfrac{\sqrt{\operatorname{cosech}^2 x + 1}}{\operatorname{cosech} x}$	$\dfrac{1}{\operatorname{sech} x}$	$\pm \dfrac{\coth x}{\sqrt{\coth^2 x - 1}}$
$\tanh x =$	$\dfrac{1}{\sqrt{\operatorname{cosech}^1 x + 1}}$	$\pm \sqrt{1 - \operatorname{sech}^2 x}$	$\dfrac{1}{\coth x}$
$\operatorname{cosech} x =$	$\operatorname{cosech} x$	$\pm \dfrac{\operatorname{sech} x}{\sqrt{1 - \operatorname{sech}^1 zsx}}$	$\pm \dfrac{\sqrt{\coth^2 x - 1}}{1}$
$\operatorname{sech} x =$	$\pm \dfrac{\operatorname{cosec} x}{\sqrt{\operatorname{cosech}^2 x + 1}}$	$\operatorname{sech} x$	$\pm \dfrac{\sqrt{\coth^2 x - 1}}{\coth x}$
$\coth x =$	$\sqrt{\operatorname{cosech}^2 x + 1}$	$\pm \dfrac{1}{\sqrt{1 - \operatorname{sech}^2 x}}$	$\coth x$

Whenever two signs are shown, choose $+$ sign if x is positive, $-$ sign if x is negative.

Derivatives*

In the following formulas u, v, w represent functions of x, while a, c, n represent fixed real numbers. All arguments in the trigonometric functions are measured in radians, and all inverse trigonometric and hyperbolic functions represent principal values.

1. $\dfrac{d}{dx}(a) = 0$

2. $\dfrac{d}{dx}(x) = 1$

3. $\dfrac{d}{dx}(au) = a\dfrac{du}{dx}$

4. $\dfrac{d}{dx}(u + v - w) = \dfrac{du}{dx} + \dfrac{dv}{dx} - \dfrac{dw}{dx}$

5. $\dfrac{d}{dx}(uv) = u\dfrac{dv}{dx} + v\dfrac{du}{dx}$

6. $\dfrac{d}{dx}(uvw) = uv\dfrac{dw}{dx} + vw\dfrac{du}{dx} + uw\dfrac{dv}{dx}$

7. $\dfrac{d}{dx}\left(\dfrac{u}{v}\right) = \dfrac{v\dfrac{du}{dx} - u\dfrac{dv}{dx}}{v^2} = \dfrac{1}{v}\dfrac{du}{dx} - \dfrac{u}{v^2}\dfrac{dv}{dx}$

8. $\dfrac{d}{dx}(u^n) = nu^{n-1}\dfrac{du}{dx}$

9. $\dfrac{d}{dx}(\sqrt{u}) = \dfrac{1}{2\sqrt{u}}\dfrac{du}{dx}$

10. $\dfrac{d}{dx}\left(\dfrac{1}{u}\right) = -\dfrac{1}{u^2}\dfrac{du}{dx}$

11. $\dfrac{d}{dx}\left(\dfrac{1}{u^n}\right) = -\dfrac{n}{u^{n+1}}\dfrac{du}{dx}$

12. $\dfrac{d}{dx}\left(\dfrac{u^n}{v^m}\right) = \dfrac{u^{n-1}}{v^{m+1}}\left(nv\dfrac{du}{dx} - mu\dfrac{dv}{dx}\right)$

13. $\dfrac{d}{dx}(u^n v^m) = u^{n-1}v^{m-1}\left(nv\dfrac{du}{dx} + mu\dfrac{dv}{dx}\right)$

14. $\dfrac{d}{dx}[f(u)] = \dfrac{d}{du}[f(u)] \cdot \dfrac{du}{dx}$

* Let $y = f(x)$ and $\dfrac{dy}{dx} = \dfrac{d[f(x)]}{dx} = f'(x)$ define respectively a function and its derivative for any value x in their common domain. The differential for the function at such a value x is accordingly defined as

$$dy = d[f(x)] = \dfrac{dy}{dx}dx = \dfrac{d[f(x)]}{dx}dx = f'(x)\,dx$$

Each derivative formula has an associated differential formula. For example, formula 6 above has the differential formula

$$d(uvw) = uv\,dw + vw\,du + uw\,dv$$

15. $\dfrac{d^2}{dx^2}[f(u)] = \dfrac{df(u)}{du} \cdot \dfrac{d^2u}{dx^2} + \dfrac{d^2f(u)}{du^2} \cdot \left(\dfrac{du}{dx}\right)^2$

16. $\dfrac{d^n}{dx^n}[uv] = \dbinom{n}{0} v \dfrac{d^n u}{dx^n} + \dbinom{n}{1} \dfrac{dv}{dx} \dfrac{d^{n-1}u}{dx^{n-1}} + \dbinom{n}{2} \dfrac{d^2 v}{dx^2} \dfrac{d^{n-2}u}{dx^{n-2}}$

$$+ \cdots + \dbinom{n}{k} \dfrac{d^k v}{dx^k} \dfrac{d^{n-k}u}{dx^{n-k}} + \cdots + \dbinom{n}{n} u \dfrac{d^n v}{dx^n}$$

where $\dbinom{n}{r} = \dfrac{n!}{r'(n-r)'}$ the binomial coefficient, n non-negative integer and $\dbinom{n}{0} = 1$.

17. $\dfrac{du}{dx} = \dfrac{1}{\dfrac{dx}{du}} \qquad \text{if } \dfrac{dx}{du} \neq 0$

18. $\dfrac{d}{dx}(\log_a u) = (\log_a e)\dfrac{1}{u}\dfrac{du}{dx}$

19. $\dfrac{d}{dx}(\log_e u) = \dfrac{1}{u}\dfrac{du}{dx}$

20. $\dfrac{d}{dx}(a^u) = a^u(\log_e a)\dfrac{du}{dx}$

21. $\dfrac{d}{dx}(e^u) = e^u \dfrac{du}{dx}$

22. $\dfrac{d}{dx}(u^v) = vu^{v-1}\dfrac{du}{dx} + (\log_e u)u^v \dfrac{dv}{dx}$

23. $\dfrac{d}{dx}(\sin u) = \dfrac{du}{dx}(\cos u)$

24. $\dfrac{d}{dx}(\cos u) = -\dfrac{du}{dx}(\sin u)$

25. $\dfrac{d}{dx}(\tan u) = \dfrac{du}{dx}(\sec^2 u)$

26. $\dfrac{d}{dx}(\cot u) = -\dfrac{du}{dx}(\csc^2 u)$

27. $\dfrac{d}{dx}(\sec u) = \dfrac{du}{dx} \sec u \cdot \tan u$

28. $\dfrac{d}{dx}(\csc u) = -\dfrac{du}{dx} \csc u \cdot \cot u$

29. $\dfrac{d}{dx}(\text{vers } u) = \dfrac{du}{dx} \sin u$

30. $\dfrac{d}{dx}(\text{arc sin } u) = \dfrac{1}{\sqrt{1-u^2}}\dfrac{du}{dx}, \qquad \left(-\dfrac{\pi}{2} \leq \text{arc sin } u \leq \dfrac{\pi}{2}\right)$

31. $\dfrac{d}{dx}(\text{arc cos } u) = -\dfrac{1}{\sqrt{1 - u^2}}\dfrac{du}{dx}, \qquad (0 \le \text{arc cos } u \le \pi)$

32. $\dfrac{d}{dx}(\text{arc tan } u) = \dfrac{1}{1 + u^2}\dfrac{du}{dx}, \qquad \left(-\dfrac{\pi}{2} < \text{arc tan } u < \dfrac{\pi}{2}\right)$

33. $\dfrac{d}{dx}(\text{arc cot } u) = -\dfrac{1}{1 + u^2}\dfrac{du}{dx}, \qquad (0 \le \text{arc cot } u \le \pi)$

34. $\dfrac{d}{dx}(\text{arc sec } u) = \dfrac{1}{u\sqrt{u^2 - 1}}\dfrac{du}{dx}, \qquad \left(0 \le \text{arc sec } u < \dfrac{\pi}{2}, -\pi \le \text{arc sec } u < -\dfrac{\pi}{2}\right)$

35. $\dfrac{d}{dx}(\text{arc csc } u) = -\dfrac{1}{u\sqrt{u^2 - 1}}\dfrac{du}{dx}, \qquad \left(0 < \text{arc csc } u \le \dfrac{\pi}{2}, -\pi < \text{arc csc } u \le -\dfrac{\pi}{2}\right)$

36. $\dfrac{d}{dx}(\text{arc vers } u) = \dfrac{1}{\sqrt{2u - u^2}}\dfrac{du}{dx}, \qquad (0 \le \text{arc vers } u \le \pi)$

37. $\dfrac{d}{dx}(\sinh u) = \dfrac{du}{dx}(\cosh u)$

38. $\dfrac{d}{dx}(\cosh u) = \dfrac{du}{dx}(\sinh u)$

39. $\dfrac{d}{dx}(\tanh u) = \dfrac{du}{dx}(\text{sech}^2 u)$

40. $\dfrac{d}{dx}(\coth u) = -\dfrac{du}{dx}(\text{csch}^2 u)$

41. $\dfrac{d}{dx}(\text{sech } u) = -\dfrac{du}{dx}(\text{sech } u \cdot \tanh u)$

42. $\dfrac{d}{dx}(\text{csch } u) = -\dfrac{du}{dx}(\text{csch } u \cdot \coth u)$

43. $\dfrac{d}{dx}(\sinh^{-1} u) = \dfrac{d}{dx}[\log(u + \sqrt{u^2 + 1})] = \dfrac{1}{\sqrt{u^2 + 1}}\dfrac{du}{dx}$

44. $\dfrac{d}{dx}(\cosh^{-1} u) = \dfrac{d}{dx}[\log(u + \sqrt{u^2 - 1})] = \dfrac{1}{\sqrt{u^2 - 1}}\dfrac{du}{dx}, \qquad (u > 1, \cosh^{-1} u > 0)$

45. $\dfrac{d}{dx}(\tanh^{-1} u) = \dfrac{d}{dx}\left[\dfrac{1}{2}\log\dfrac{1 + u}{1 - u}\right] = \dfrac{1}{1 - u^2}\dfrac{du}{dx}, \qquad (u^2 < 1)$

46. $\dfrac{d}{dx}(\coth^{-1} u) = \dfrac{d}{dx}\left[\dfrac{1}{2}\log\dfrac{u + 1}{u - 1}\right] = \dfrac{1}{1 - u^2}\dfrac{du}{dx}, \qquad (u^2 > 1)$

47. $\dfrac{d}{dx}(\text{sech}^{-1} u) = \dfrac{d}{dx}\left[\log\dfrac{1 + \sqrt{1 - u^2}}{u}\right] = -\dfrac{1}{u\sqrt{1 - u^2}}\dfrac{du}{dx}, \qquad (0 < u < 1, \text{sech}^{-1} u > 0)$

48. $\dfrac{d}{dx}(\text{csch}^{-1} u) = \dfrac{d}{dx}\left[\log\dfrac{1 + \sqrt{1 + u^2}}{u}\right] = -\dfrac{1}{|u|\sqrt{1 + u^2}}\dfrac{du}{dx}$

49. $\dfrac{d}{dq} \displaystyle\int_p^q f(x)\,dx = f(q),$ [p constant]

50. $\dfrac{d}{dp} \displaystyle\int_p^q f(x)\,dx = -f(p),$ [q constant]

51. $\dfrac{d}{da} \displaystyle\int_p^q f(x,a)\,dx = \displaystyle\int_p^q \dfrac{\partial}{\partial a}[f(x,a)]\,dx + f(q,a)\dfrac{dq}{da} - f(p,a)\dfrac{dp}{da}$

INTEGRATION

The following is a brief discussion of some integration techniques. A more complete discussion can be found in a number of good text books. However, the purpose of this introduction is simply to discuss a few of the important techniques which may be used, in conjunction with the integral table which follows, to integrate particular functions.

No matter how extensive the integral table, it is a fairly uncommon occurrence to find in the table the exact integral desired. Usually some form of transformation will have to be made. The simplest type of transformation, and yet the most general, is substitution. Simple forms of substitution, such as $y = ax$, are employed almost unconsciously by experienced users of integral tables. Other substitutions may require more thought. In some sections of the tables, appropriate substitutions are suggested for integrals which are similar to, but not exactly like, integrals in the table. Finding the right substitution is largely a matter of intuition and experience.

Several precautions must be observed when using substitutions:

1. Be sure to make the substitution in the dx term, as well as everywhere else in the integral.
2. Be sure that the function substituted is one-to-one and continuous. If this is not the case, the integral must be restricted in such a way as to make it true. See the example following.
3. With definite integrals, the limits should also be expressed in terms of the new dependent variable. With indefinite integrals, it is necessary to perform the reverse substitution to obtain the answer in terms of the original independent variable. This may also be done for definite integrals, but it is usually easier to change the limits.

Example:

$$\int \frac{x^4}{\sqrt{a^2 - x^2}}\,dx$$

Here we make the substitution $x = |a| \sin \theta$. Then $dx = |a| \cos \theta\, d\theta$, and

$$\sqrt{a^2 - x^2} = \sqrt{a^2 - a^2 \sin^2 \theta} = |a|\sqrt{1 - \sin^2 \theta} = |a \cos \theta|$$

Notice the absolute value signs. It is very important to keep in mind that a square root radical always denotes the positive square root, and to assure the sign is always kept positive. Thus $\sqrt{x^2} = |x|$. Failure to observe this is a common cause of errors in integration.

Notice also that the indicated substitution is not a one-to-one function, that is, it does not have a unique inverse. Thus we must restrict the range of θ in such a way as to make the function one-to-one. Fortunately, this is easily done by solving for θ

$$\theta = \sin^{-1} \frac{x}{|a|}$$

and restricting the inverse sine to the principal values, $-\dfrac{\pi}{2} \le \theta \le \dfrac{\pi}{2}$.

Thus the integral becomes

$$\int \frac{a^4 \sin^4 \theta |a| \cos \theta \, d\theta}{|a| |\cos \theta|}$$

Now, however, in the range of values chosen for θ, $\cos \theta$ is always positive. Thus we may remove the absolute value signs from $\cos \theta$ in the denominator. (This is one of the reasons that the principal values of the inverse trigonometric functions are defined as they are.)

Then the $\cos \theta$ terms cancel, and the integral becomes

$$a^4 \int \sin^4 \theta \, d\theta$$

By application of integral formulas 299 and 296, we integrate this to

$$-a^4 \frac{\sin^3 \theta \cos \theta}{4} - \frac{3a^4}{8} \cos \theta \sin \theta + \frac{3a^4}{8} \theta + C$$

We now must perform the inverse substitution to get the result in terms of x. We have

$$\theta = \sin^{-1} \frac{x}{|a|}$$

$$\sin \theta = \frac{x}{|a|}$$

Then

$$\cos \theta = \pm \sqrt{1 - \sin^2 \theta} = \pm \sqrt{1 - \frac{x^2}{a^2}} = \pm \frac{\sqrt{a^2 - x^2}}{|a|}.$$

Because of the previously mentioned fact that $\cos \theta$ is positive, we may omit the $\pm$ sign. The reverse substitution then produces the final answer

$$\int \frac{x^4}{\sqrt{a^2 - x^2}} \, dx = -\tfrac{1}{4}x^3 \sqrt{a^2 - x^2} - \tfrac{3}{8}a^2 x \sqrt{a^2 - x^2} + \frac{3a^4}{8} \sin^{-1} \frac{x}{|a|} + C.$$

Any rational function of x may be integrated, if the denominator is factored into linear and irreducible quadratic factors. The function may then be broken into partial fractions, and the individual partial fractions integrated by use of the appropriate formula from the integral table. See the section on partial fractions for further information.

Many integrals may be reduced to rational functions by proper substitutions. For example,

$$z = \tan \frac{x}{2}$$

will reduce any rational function of the six trigonometric functions of x to a rational function of z. (Frequently there are other substitutions which are simpler to use, but this one will always work. See integral formula number 484.)

Any rational function of x and $\sqrt{ax + b}$ may be reduced to a rational function of z by making the substitution

$$z = \sqrt{ax + b}.$$

Other likely substitutions will be suggested by looking at the form of the integrand.

The other main method of transforming integrals is integration by parts. This involves applying formula number 5 or 6 in the accompanying integral table. The critical factor in this method is the choice of the functions u and v. In order for the method to be successful, $v = \int dv$ and $\int v \, du$ must be easier to integrate than the original integral. Again, this choice is largely a matter of intuition and experience.

A-16

Example:

$$\int x \sin x \, dx$$

Two obvious choices are $u = x$, $dv = \sin x \, dx$, or $u = \sin x$, $dv = x \, dx$. Since a preliminary mental calculation indicates that $\int v \, du$ in the second choice would be more, rather than less, complicated than the original integral (it would contain x^2), we use the first choice.

$$u = x \qquad\qquad\qquad du = dx$$
$$dv = \sin x \, dx \qquad\qquad v = -\cos x$$

$$\int x \sin x \, dx = \int u \, dv = uv - \int v \, du = -x \cos x + \int \cos x \, dx$$
$$= \sin x - x \cos x$$

Of course, this result could have been obtained directly from the integral table, but it provides a simple example of the method. In more complicated examples the choice of u and v may not be so obvious, and several different choices may have to be tried. Of course, there is no guarantee that any of them will work.

Integration by parts may be applied more than once, or combined with substitution. A fairly common case is illustrated by the following example.

Example:

$$\int e^x \sin x \, dx$$

Let

$$u = e^x \qquad\qquad \text{Then} \quad du = e^x \, dx$$
$$dv = \sin x \, dx \qquad\qquad v = -\cos x$$

$$\int e^x \sin x \, dx = \int u \, dv = uv - \int v \, du = -e^x \cos x + \int e^x \cos x \, dx$$

In this latter integral,

$$\text{let} \quad u = e^x \qquad\qquad \text{Then} \quad du = e^x \, dx$$
$$dv = \cos x \, dx \qquad\qquad v = \sin x$$

$$\int e^x \sin x \, dx = -e^x \cos x + \int e^x \cos x \, dx = -e^x \cos x + \int u \, dv$$
$$= -e^x \cos x + uv - \int v \, du$$
$$= -e^x \cos x + e^x \sin x - \int e^x \sin x \, dx$$

This looks as if a circular transformation has taken place, since we are back at the same integral we started from. However, the above equation can be solved algebraically for the required integral:

$$\int e^x \sin x \, dx = \tfrac{1}{2}(e^x \sin x - e^x \cos x)$$

In the second integration by parts, if the parts had been chosen as $u = \cos x$, $dv = e^x \, dx$, we would indeed have made a circular transformation, and returned to the starting place.

In general, when doing repeated integration by parts, one should never choose the function u at any stage to be the same as the function v at the previous stage, or a constant times the previous v.

The following rule is called the extended rule for integration by parts. It is the result of $n + 1$ successive applications of integration by parts.

If

$$g_1(x) = \int g(x)\, dx, \qquad g_2(x) = \int g_1(x)\, dx,$$

$$g_3(x) = \int g_2(x)\, dx, \ldots, g_m(x) = \int g_{m-1}(x)\, dx, \ldots,$$

then

$$\int f(x) \cdot g(x)\, dx = f(x) \cdot g_1(x) - f'(x) \cdot g_2(x) + f''(x) \cdot g_3(x) - + \cdots$$

$$+ (-1)^n f^{(n)}(x) g_{n+1}(x) + (-1)^{n+1} \int f^{(n+1)}(x) g_{n+1}(x)\, dx.$$

A useful special case of the above rule is when $f(x)$ is a polynomial of degree n. Then $f^{(n+1)}(x) = 0$, and

$$\int f(x) \cdot g(x)\, dx = f(x) \cdot g_1(x) - f'(x) \cdot g_2(x) + f''(x) \cdot g_3(x) - + \cdots + (-1)^n f^{(n)}(x) g_{n+1}(x) + C$$

Example:
If $f(x) = x^2, g(x) = \sin x$

$$\int x^2 \sin x\, dx = -x^2 \cos x + 2x \sin x + 2 \cos x + C$$

Another application of this formula occurs if

$$f''(x) = af(x) \quad \text{and} \quad g''(x) = bg(x),$$

where a and b are unequal constants. In this case, by a process similar to that used in the above example for $\int e^x \sin x\, dx$, we get the formula

$$\int f(x)g(x)\, dx = \frac{f(x) \cdot g'(x) - f'(x) \cdot g(x)}{b - a} + C$$

This formula could have been used in the example mentioned. Here is another example.

Example:
If $f(x) = e^{2x}, g(x) = \sin 3x$, then $a = 4, b = -9$, and

$$\int e^{2x} \sin 3x\, dx = \frac{3\, e^{2x} \cos 3x - 2\, e^{2x} \sin 3x}{-9 - 4} + C = \frac{e^{2x}}{13}(2 \sin 3x - 3 \cos 3x) + C$$

The following additional points should be observed when using this table.

1. A constant of integration is to be supplied with the answers for indefinite integrals.
2. Logarithmic expressions are to base $e = 2.71828\ldots$, unless otherwise specified, and are to be evaluated for the absolute value of the arguments involved therein.
3. All angles are measured in radians, and inverse trigonometric and hyperbolic functions represent principal values, unless otherwise indicated.
4. If the application of a formula produces either a zero denominator or the square root of a negative number in the result, there is usually available another form of the answer which

avoids this difficulty. In many of the results, the excluded values are specified, but when such are omitted it is presumed that one can tell what these should be, especially when difficulties of the type herein mentioned are obtained.

5. When inverse trigonometric functions occur in the integrals, be sure that any replacements made for them are strictly in accordance with the rules for such functions. This causes little difficulty when the argument of the inverse trigonometric function is positive, since then all angles involved are in the first quadrant. However, if the argument is negative, special care must be used. Thus if $u > 0$,

$$\sin^{-1} u = \cos^{-1}\sqrt{1 - u^2} = \csc^{-1}\frac{1}{u}, \text{ etc.}$$

However, if $u < 0$,

$$\sin^{-1} u = -\cos^{-1}\sqrt{1 - u^2} = -\pi - \csc^{-1}\frac{1}{u}, \text{ etc.}$$

See the section on inverse trigonometric functions for a full treatment of the allowable substitutions.

6. In integrals 340–345 and some others, the right side includes expressions of the form

$$A \tan^{-1}[B + C \tan f(x)].$$

In these formulas, the $\tan^{-1}$ does not necessarily represent the principal value. Instead of always employing the principal branch of the inverse tangent function, one must instead use that branch of the inverse tangent function upon which $f(x)$ lies for any particular choice of x.

Example:

$$\int_0^{4\pi} \frac{dx}{2 + \sin x} = \frac{2}{\sqrt{3}}\tan^{-1}\frac{2\tan\frac{x}{2} + 1}{\sqrt{3}}\Bigg]_0^{4\pi}$$

$$= \frac{2}{\sqrt{3}}\left[\tan^{-1}\frac{2\tan 2\pi + 1}{\sqrt{3}} - \tan^{-1}\frac{2\tan 0 + 1}{\sqrt{3}}\right]$$

$$= \frac{2}{\sqrt{3}}\left[\frac{13\pi}{6} - \frac{\pi}{6}\right] = \frac{4\pi}{\sqrt{3}} = \frac{4\sqrt{3}\pi}{3}$$

Here

$$\tan^{-1}\frac{2\tan 2\pi + 1}{\sqrt{3}} = \tan^{-1}\frac{1}{\sqrt{3}} = \frac{13\pi}{6},$$

since $f(x) = 2\pi$; and

$$\tan^{-1}\frac{2\tan 0 + 1}{\sqrt{3}} = \tan^{-1}\frac{1}{\sqrt{3}} = \frac{\pi}{6},$$

since $f(x) = 0$.

7. B_n and E_n where used in Integrals represents the Bernoulli and Euler numbers as defined in tables of Bernoulli and Euler polynomials contained in certain mathematics reference and handbooks, as for example, Beyer, W. H., *Handbook of Mathematical Sciences*, 5th ed., CRC Press, Inc., West Palm Beach 1978, 577–583.

ELEMENTARY FORMS

1. $\int a\,dx = ax$

2. $\int a \cdot f(x)\,dx = a\int f(x)\,dx$

3. $\int \phi(y)\,dx = \int \dfrac{\phi(y)}{y'}\,dy,$ where $y' = \dfrac{dy}{dx}$

4. $\int (u+v)\,dx = \int u\,dx + \int v\,dx,$ where u and v are any functions of x

5. $\int u\,dv = u\int dv - \int v\,du = uv - \int v\,du$

6. $\int u\dfrac{dv}{dx}\,dx = uv - \int v\dfrac{du}{dx}\,dx$

7. $\int x^n\,dx = \dfrac{x^{n+1}}{n+1},$ except $n = -1$

8. $\int \dfrac{f'(x)\,dx}{f(x)} = \log f(x),$ $(df(x) = f'(x)\,dx)$

9. $\int \dfrac{dx}{x} = \log x$

10. $\int \dfrac{f'(x)\,dx}{2\sqrt{f(x)}} = \sqrt{f(x)},$ $(df(x) = f'(x)\,dx)$

11. $\int e^x\,dx = e^x$

12. $\int e^{ax}\,dx = e^{ax}/a$

13. $\int b^{ax}\,dx = \dfrac{b^{ax}}{a\log b},$ $(b > 0)$

14. $\int \log x\,dx = x\log x - x$

15. $\int a^x\log a\,dx = a^x,$ $(a > 0)$

16. $\int \dfrac{dx}{a^2 + x^2} = \dfrac{1}{a}\tan^{-1}\dfrac{x}{a}$

17. $\int \dfrac{dx}{a^2 - x^2} = \begin{cases} \dfrac{1}{a}\tanh^{-1}\dfrac{x}{a} \\[2mm] \quad\text{or} \\[2mm] \dfrac{1}{2a}\log\dfrac{a+x}{a-x}, \quad (a^2 > x^2) \end{cases}$

18. $\int \dfrac{dx}{x^2 - a^2} = \begin{cases} -\dfrac{1}{a}\coth^{-1}\dfrac{x}{a} \\[2mm] \quad\text{or} \\[2mm] \dfrac{1}{2a}\log\dfrac{x-a}{x+a}, \quad (x^2 > a^2) \end{cases}$

19. $\displaystyle\int \frac{dx}{\sqrt{a^2 - x^2}} = \begin{cases} \sin^{-1}\dfrac{x}{|a|} \\ \quad\text{or} \\ -\cos^{-1}\dfrac{x}{|a|}, \quad (a^2 > x^2) \end{cases}$

20. $\displaystyle\int \frac{dx}{\sqrt{x^2 \pm a^2}} = \log\left(x + \sqrt{x^2 \pm a^2}\right)$

21. $\displaystyle\int \frac{dx}{x\sqrt{x^2 - a^2}} = \frac{1}{|a|}\sec^{-1}\frac{x}{a}$

22. $\displaystyle\int \frac{dx}{x\sqrt{a^2 \pm x^2}} = -\frac{1}{a}\log\left(\frac{a + \sqrt{a^2 \pm x^2}}{x}\right)$

FORMS CONTAINING $(a + bx)$

For forms containing $a + bx$, but not listed in the table, the substitution $u = \dfrac{a + bx}{x}$ may prove helpful.

23. $\displaystyle\int (a + bx)^n\, dx = \frac{(a + bx)^{n+1}}{(n + 1)b}, \quad (n \neq -1)$

24. $\displaystyle\int x(a + bx)^n\, dx$

$$= \frac{1}{b^2(n + 2)}(a + bx)^{n+2} - \frac{a}{b^2(n + 1)}(a + bx)^{n+1}, \quad (n \neq -1, -2)$$

25. $\displaystyle\int x^2(a + bx)^n\, dx = \frac{1}{b^3}\left[\frac{(a + bx)^{n+3}}{n + 3} - 2a\frac{(a + bx)^{n+2}}{n + 2} + a^2\frac{(a + bx)^{n+1}}{n + 1}\right]$

26. $\displaystyle\int x^m(a + bx)^n\, dx = \begin{cases} \dfrac{x^{m+1}(a + bx)^n}{m + n + 1} + \dfrac{an}{m + n + 1}\displaystyle\int x^m(a + bx)^{n-1}\, dx \\ \qquad\qquad\text{or} \\ \dfrac{1}{a(n + 1)}\left[-x^{m+1}(a + bx)^{n+1}\right. \\ \qquad\qquad\qquad\qquad \left. + (m + n + 2)\displaystyle\int x^m(a + bx)^{n+1}\, dx\right] \\ \qquad\qquad\text{or} \\ \dfrac{1}{b(m + n + 1)}\left[x^m(a + bx)^{n+1} - ma\displaystyle\int x^{m-1}(a + bx)^n\, dx\right] \end{cases}$

27. $\displaystyle\int \frac{dx}{a + bx} = \frac{1}{b}\log(a + bx)$

28. $\displaystyle\int \frac{dx}{(a + bx)^2} = -\frac{1}{b(a + bx)}$

29. $\displaystyle\int \frac{dx}{(a + bx)^3} = -\frac{1}{2b(a + bx)^2}$

30. $\displaystyle\int \frac{x\, dx}{a + bx} = \begin{cases} \dfrac{1}{b^2}[a + bx - a\log(a + bx)] \\ \qquad\qquad\text{or} \\ \dfrac{x}{b} - \dfrac{a}{b^2}\log(a + bx) \end{cases}$

31. $\displaystyle\int \frac{x\,dx}{(a+bx)^2} = \frac{1}{b^2}\left[\log(a+bx) + \frac{a}{a+bx}\right]$

32. $\displaystyle\int \frac{x\,dx}{(a+bx)^n} = \frac{1}{b^2}\left[\frac{-1}{(n-2)(a+bx)^{n-2}} + \frac{a}{(n-1)(a+bx)^{n-1}}\right], \qquad n \neq 1, 2$

33. $\displaystyle\int \frac{x^2\,dx}{a+bx} = \frac{1}{b^3}\left[\frac{1}{2}(a+bx)^2 - 2a(a+bx) + a^2\log(a+bx)\right]$

34. $\displaystyle\int \frac{x^2\,dx}{(a+bx)^2} = \frac{1}{b^3}\left[a+bx - 2a\log(a+bx) - \frac{a^2}{a+bx}\right]$

35. $\displaystyle\int \frac{x^2\,dx}{(a+bx)^3} = \frac{1}{b^3}\left[\log(a+bx) + \frac{2a}{a+bx} - \frac{a^2}{2(a+bx)^2}\right]$

36. $\displaystyle\int \frac{x^2\,dx}{(a+bx)^n} = \frac{1}{b^3}\left[\frac{-1}{(n-3)(a+bx)^{n-3}}\right.$

$\left. + \frac{2a}{(n-2)(a+bx)^{n-2}} - \frac{a^2}{(n-1)(a+bx)^{n-1}}\right], \qquad n \neq 1, 2, 3$

37. $\displaystyle\int \frac{dx}{x(a+bx)} = -\frac{1}{a}\log\frac{a+bx}{x}$

38. $\displaystyle\int \frac{dx}{x(a+bx)^2} = \frac{1}{a(a+bx)} - \frac{1}{a^2}\log\frac{a+bx}{x}$

39. $\displaystyle\int \frac{dx}{x(a+bx)^3} = \frac{1}{a^3}\left[\frac{1}{2}\left(\frac{2a+bx}{a+bx}\right)^2 + \log\frac{x}{a+bx}\right]$

40. $\displaystyle\int \frac{dx}{x^2(a+bx)} = -\frac{1}{ax} + \frac{b}{a^2}\log\frac{a+bx}{x}$

41. $\displaystyle\int \frac{dx}{x^3(a+bx)} = \frac{2bx-a}{2a^2x^2} + \frac{b^2}{a^3}\log\frac{x}{a+bx}$

42. $\displaystyle\int \frac{dx}{x^2(a+bx)^2} = -\frac{a+2bx}{a^2x(a+bx)} + \frac{2b}{a^3}\log\frac{a+bx}{x}$

FORMS CONTAINING $c^2 \pm x^2$, $x^2 - c^2$

43. $\displaystyle\int \frac{dx}{c^2+x^2} = \frac{1}{c}\tan^{-1}\frac{x}{c}$

44. $\displaystyle\int \frac{dx}{c^2-x^2} = \frac{1}{2c}\log\frac{c+x}{c-x}, \qquad (c^2 > x^2)$

45. $\displaystyle\int \frac{dx}{x^2-c^2} = \frac{1}{2c}\log\frac{x-c}{x+c}, \qquad (x^2 > c^2)$

46. $\displaystyle\int \frac{x\,dx}{c^2 \pm x^2} = \pm\frac{1}{2}\log(c^2 \pm x^2)$

47. $\displaystyle\int \frac{x\,dx}{(c^2 \pm x^2)^{n+1}} = \mp\frac{1}{2n(c^2 \pm x^2)^n}$

48. $\displaystyle\int \frac{dx}{(c^2 \pm x^2)^n} = \frac{1}{2c^2(n-1)}\left[\frac{x}{(c^2 \pm x^2)^{n-1}} + (2n-3)\int \frac{dx}{(c^2 \pm x^2)^{n-1}}\right]$

49. $\displaystyle\int \frac{dx}{(x^2-c^2)^n} = \frac{1}{2c^2(n-1)}\left[-\frac{x}{(x^2-c^2)^{n-1}} - (2n-3)\int \frac{dx}{(x^2-c^2)^{n-1}}\right]$

50. $\displaystyle\int \frac{x\,dx}{x^2-c^2} = \frac{1}{2}\log(x^2-c^2)$

51. $\displaystyle\int \frac{x\,dx}{(x^2 - c^2)^{n+1}} = -\frac{1}{2n(x^2 - c^2)^n}$

FORMS CONTAINING $a + bx$ and $c + dx$

$$u = a + bx, \qquad v = c + dx, \qquad k = ad - bc$$

If $k = 0$, then $v = \dfrac{c}{a}u$

52. $\displaystyle\int \frac{dx}{u \cdot v} = \frac{1}{k} \cdot \log\left(\frac{v}{u}\right)$

53. $\displaystyle\int \frac{x\,dx}{u \cdot v} = \frac{1}{k}\left[\frac{a}{b}\log(u) - \frac{c}{d}\log(v)\right]$

54. $\displaystyle\int \frac{dx}{u^2 \cdot v} = \frac{1}{k}\left(\frac{1}{u} + \frac{d}{k}\log\frac{v}{u}\right)$

55. $\displaystyle\int \frac{x\,dx}{u^2 \cdot v} = \frac{-a}{bku} - \frac{c}{k^2}\log\frac{v}{u}$

56. $\displaystyle\int \frac{x^2\,dx}{u^2 \cdot v} = \frac{a^2}{b^2 ku} + \frac{1}{k^2}\left[\frac{c^2}{d}\log(v) + \frac{a(k - bc)}{b^2}\log(u)\right]$

57. $\displaystyle\int \frac{dx}{u^n \cdot v^m} = \frac{1}{k(m-1)}\left[\frac{-1}{u^{n-1} \cdot v^{m-1}} - (m + n - 2)b\int \frac{dx}{u^n \cdot v^{m-1}}\right]$

58. $\displaystyle\int \frac{u}{v}\,dx = \frac{bx}{d} + \frac{k}{d^2}\log(v)$

59. $\displaystyle\int \frac{u^m\,dx}{v^n} = \begin{cases} \dfrac{-1}{k(n-1)}\left[\dfrac{u^{m+1}}{v^{n-1}} + b(n - m - 2)\displaystyle\int \dfrac{u^m}{v^{n-1}}\,dx\right] \\[4pt] \qquad\qquad\text{or} \\[4pt] \dfrac{-1}{d(n - m - 1)}\left[\dfrac{u^m}{v^{n-1}} + mk\displaystyle\int \dfrac{u^{m-1}}{v^n}\,dx\right] \\[4pt] \qquad\qquad\text{or} \\[4pt] \dfrac{-1}{d(n - 1)}\left[\dfrac{u^m}{v^{n-1}} - mb\displaystyle\int \dfrac{u^{m-1}}{v^{n-1}}\,dx\right] \end{cases}$

FORMS CONTAINING $(a + bx^n)$

60. $\displaystyle\int \frac{dx}{a + bx^2} = \frac{1}{\sqrt{ab}}\tan^{-1}\frac{x\sqrt{ab}}{a}, \qquad (ab > 0)$

61. $\displaystyle\int \frac{dx}{a + bx^2} = \begin{cases} \dfrac{1}{2\sqrt{-ab}}\log\dfrac{a + x\sqrt{-ab}}{a - x\sqrt{-ab}}, \qquad (ab < 0) \\[4pt] \qquad\qquad\text{or} \\[4pt] \dfrac{1}{\sqrt{-ab}}\tanh^{-1}\dfrac{x\sqrt{-ab}}{a}, \qquad (ab < 0) \end{cases}$

62. $\displaystyle\int \frac{dx}{a^2 + b^2 x^2} = \frac{1}{ab}\tan^{-1}\frac{bx}{a}$

63. $\displaystyle\int \frac{x\,dx}{a + bx^2} = \frac{1}{2b}\log(a + bx^2)$

64. $\displaystyle\int \frac{x^2\,dx}{a + bx^2} = \frac{x}{b} - \frac{a}{b}\int \frac{dx}{a + bx^2}$

65. $\displaystyle\int \frac{dx}{(a + bx^2)^2} = \frac{x}{2a(a + bx^2)} + \frac{1}{2a}\int \frac{dx}{a + bx^2}$

66. $\int \dfrac{dx}{a^2 - b^2 x^2} = \dfrac{1}{2ab} \log \dfrac{a + bx}{a - bx}$

67. $\int \dfrac{dx}{(a + bx^2)^{m+1}} = \begin{cases} \dfrac{1}{2ma} \dfrac{x}{(a + bx^2)^m} + \dfrac{2m - 1}{2ma} \displaystyle\int \dfrac{dx}{(a + bx^2)^m} \\[2mm] \text{or} \\[2mm] \dfrac{(2m)!}{(m!)^2} \left[\dfrac{x}{2a} \displaystyle\sum_{r=1}^{m} \dfrac{r!(r - 1)!}{(4a)^{m-r}(2r)!(a + bx^2)^r} + \dfrac{1}{(4a)^m} \int \dfrac{dx}{a + bx^2} \right] \end{cases}$

68. $\int \dfrac{x \, dx}{(a + bx^2)^{m+1}} = - \dfrac{1}{2bm(a + bx^2)^m}$

69. $\int \dfrac{x^2 \, dx}{(a + bx^2)^{m+1}} = \dfrac{-x}{2mb(a + bx^2)^m} + \dfrac{1}{2mb} \int \dfrac{dx}{(a + bx^2)^m}$

70. $\int \dfrac{dx}{x(a + bx^2)} = \dfrac{1}{2a} \log \dfrac{x^2}{a + bx^2}$

71. $\int \dfrac{dx}{x^2(a + bx^2)} = - \dfrac{1}{ax} - \dfrac{b}{a} \int \dfrac{dx}{a + bx^2}$

72. $\int \dfrac{dx}{x(a + bx^2)^{m+1}} = \begin{cases} \dfrac{1}{2am(a + bx^2)^m} + \dfrac{1}{a} \displaystyle\int \dfrac{dx}{x(a + bx^2)^m} \\[2mm] \text{or} \\[2mm] \dfrac{1}{2a^{m+1}} \left[\displaystyle\sum_{r=1}^{m} \dfrac{a^r}{r(a + bx^2)^r} + \log \dfrac{x^2}{a + bx^2} \right] \end{cases}$

73. $\int \dfrac{dx}{x^2(a + bx^2)^{m+1}} = \dfrac{1}{a} \int \dfrac{dx}{x^2(a + bx^2)^m} - \dfrac{b}{a} \int \dfrac{dx}{(a + bx^2)^{m+1}}$

74. $\int \dfrac{dx}{a + bx^3} = \dfrac{k}{3a} \left[\dfrac{1}{2} \log \dfrac{(k + x)^3}{a + bx^3} + \sqrt{3} \tan^{-1} \dfrac{2x - k}{k\sqrt{3}} \right], \quad \left(k = \sqrt[3]{\dfrac{a}{b}} \right)$

75. $\int \dfrac{x \, dx}{a + bx^3} = \dfrac{1}{3bk} \left[\dfrac{1}{2} \log \dfrac{a + bx^3}{(k + x)^3} + \sqrt{3} \tan^{-1} \dfrac{2x - k}{k\sqrt{3}} \right], \quad \left(k = \sqrt[3]{\dfrac{a}{b}} \right)$

76. $\int \dfrac{x^2 \, dx}{a + bx^3} = \dfrac{1}{3b} \log (a + bx^3)$

77. $\int \dfrac{dx}{a + bx^4} = \dfrac{k}{2a} \left[\dfrac{1}{2} \log \dfrac{x^2 + 2kx + 2k^2}{x^2 - 2kx + 2k^2} + \tan^{-1} \dfrac{2kx}{2k^2 - x^2} \right],$

$$\left(ab > 0, k = \sqrt[4]{\dfrac{a}{4b}} \right)$$

78. $\int \dfrac{dx}{a + bx^4} = \dfrac{k}{2a} \left[\dfrac{1}{2} \log \dfrac{x + k}{x - k} + \tan^{-1} \dfrac{x}{k} \right], \quad \left(ab < 0, k = \sqrt[4]{-\dfrac{a}{b}} \right)$

79. $\int \dfrac{x \, dx}{a + bx^4} = \dfrac{1}{2bk} \tan^{-1} \dfrac{x^2}{k}, \quad \left(ab > 0, k = \sqrt{\dfrac{a}{b}} \right)$

80. $\int \dfrac{x \, dx}{a + bx^4} = \dfrac{1}{4bk} \log \dfrac{x^2 - k}{x^2 + k}, \quad \left(ab < 0, k = \sqrt{-\dfrac{a}{b}} \right)$

81. $\int \dfrac{x^2 \, dx}{a + bx^4} = \dfrac{1}{4bk} \left[\dfrac{1}{2} \log \dfrac{x^2 - 2kx + 2k^2}{x^2 + 2kx + 2k^2} + \tan^{-1} \dfrac{2kx}{2k^2 - x^2} \right],$

$$\left(ab > 0, k = \sqrt[4]{\dfrac{a}{4b}} \right)$$

82. $\int \dfrac{x^2\, dx}{a + bx^4} = \dfrac{1}{4bk}\left[\log \dfrac{x - k}{x + k} + 2 \tan^{-1} \dfrac{x}{k}\right], \qquad \left(ab < 0, k = \sqrt[4]{-\dfrac{a}{b}}\right)$

83. $\int \dfrac{x^3\, dx}{a + bx^4} = \dfrac{1}{4b} \log (a + bx^4)$

84. $\int \dfrac{dx}{x(a + bx^n)} = \dfrac{1}{an} \log \dfrac{x^n}{a + bx^n}$

85. $\int \dfrac{dx}{(a + bx^n)^{m+1}} = \dfrac{1}{a} \int \dfrac{dx}{(a + bx^n)^m} - \dfrac{b}{a} \int \dfrac{x^n\, dx}{(a + bx^n)^{m+1}}$

86. $\int \dfrac{x^m\, dx}{(a + bx^n)^{p+1}} = \dfrac{1}{b} \int \dfrac{x^{m-n}\, dx}{(a + bx^n)^p} - \dfrac{a}{b} \int \dfrac{x^{m-n}\, dx}{(a + bx^n)^{p+1}}$

87. $\int \dfrac{dx}{x^m(a + bx^n)^{p+1}} = \dfrac{1}{a} \int \dfrac{dx}{x^m(a + bx^n)^p} - \dfrac{b}{a} \int \dfrac{dx}{x^{m-n}(a + bx^n)^{p+1}}$

88. $\int x^m(a + bx^n)^p\, dx = \begin{cases} \dfrac{1}{b(np + m + 1)}\left[x^{m-n+1}(a + bx^n)^{p+1} \right. \\ \qquad\qquad \left. - a(m - n + 1) \displaystyle\int x^{m-n}(a + bx^n)^p\, dx\right] \\[4pt] \text{or} \\ \dfrac{1}{np + m + 1}\left[x^{m+1}(a + bx^n)^p \right. \\ \qquad\qquad \left. + anp \displaystyle\int x^m(a + bx^n)^{p-1}\, dx\right] \\[4pt] \text{or} \\ \dfrac{1}{a(m + 1)}\left[x^{m+1}(a + bx^n)^{p+1} \right. \\ \qquad\qquad \left. - (m + 1 + np + n)b \displaystyle\int x^{m+n}(a + bx^n)^p\, dx\right] \\[4pt] \text{or} \\ \dfrac{1}{an(p + 1)}\left[-x^{m+1}(a + bx^n)^{p+1} \right. \\ \qquad\qquad \left. + (m + 1 + np + n) \displaystyle\int x^m(a + bx^n)^{p+1}\, dx\right] \end{cases}$

FORMS CONTAINING $c^3 \pm x^3$

89. $\int \dfrac{dx}{c^3 \pm x^3} = \pm \dfrac{1}{6c^2} \log \dfrac{(c \pm x)^3}{c^3 \pm x^3} + \dfrac{1}{c^2\sqrt{3}} \tan^{-1} \dfrac{2x \mp c}{c\sqrt{3}}$

90. $\int \dfrac{dx}{(c^3 \pm x^3)^2} = \dfrac{x}{3c^3(c^3 \pm x^3)} + \dfrac{2}{3c^3} \int \dfrac{dx}{c^3 \pm x^3}$

91. $\int \dfrac{dx}{(c^3 \pm x^3)^{n+1}} = \dfrac{1}{3nc^3}\left[\dfrac{x}{(c^3 \pm x^3)^n} + (3n - 1) \int \dfrac{dx}{(c^3 \pm x^3)^n}\right]$

92. $\int \dfrac{x\, dx}{c^3 \pm x^3} = \dfrac{1}{6c} \log \dfrac{c^3 \pm x^3}{(c \pm x)^3} \pm \dfrac{1}{c\sqrt{3}} \tan^{-1} \dfrac{2x \mp c}{c\sqrt{3}}$

93. $\int \dfrac{x\, dx}{(c^3 \pm x^3)^2} = \dfrac{x^2}{3c^3(c^3 \pm x^3)} + \dfrac{1}{3c^3} \int \dfrac{x\, dx}{c^3 \pm x^3}$

94. $\displaystyle \int \frac{x\,dx}{(c^3 \pm x^3)^{n+1}} = \frac{1}{3nc^3}\left[\frac{x^2}{(c^3 \pm x^3)^n} + (3n-2)\int \frac{x\,dx}{(c^3 \pm x^3)^n}\right]$

95. $\displaystyle \int \frac{x^2\,dx}{c^3 \pm x^3} = \pm\frac{1}{3}\log(c^3 \pm x^3)$

96. $\displaystyle \int \frac{x^2\,dx}{(c^3 \pm x^3)^{n+1}} = \mp\frac{1}{3n(c^3 \pm x^3)^n}$

97. $\displaystyle \int \frac{dx}{x(c^3 \pm x^3)} = \frac{1}{3c^3}\log\frac{x^3}{c^3 \pm x^3}$

98. $\displaystyle \int \frac{dx}{x(c^3 \pm x^3)^2} = \frac{1}{3c^3(c^3 \pm x^3)} + \frac{1}{3c^6}\log\frac{x^3}{c^3 \pm x^3}$

99. $\displaystyle \int \frac{dx}{x(c^3 \pm x^3)^{n+1}} = \frac{1}{3nc^3(c^3 \pm x^3)^n} + \frac{1}{c^3}\int \frac{dx}{x(c^3 \pm x^3)^n}$

100. $\displaystyle \int \frac{dx}{x^2(c^3 \pm x^3)} = -\frac{1}{c^3 x} \mp \frac{1}{c^3}\int \frac{x\,dx}{c^3 \pm x^3}$

101. $\displaystyle \int \frac{dx}{x^2(c^3 \pm x^3)^{n+1}} = \frac{1}{c^3}\int \frac{dx}{x^2(c^3 \pm x^3)^n} \mp \frac{1}{c^3}\int \frac{x\,dx}{(c^3 \pm x^3)^{n+1}}$

FORMS CONTAINING $c^4 \pm x^4$

102. $\displaystyle \int \frac{dx}{c^4 + x^4} = \frac{1}{2c^3\sqrt{2}}\left[\frac{1}{2}\log\frac{x^2 + cx\sqrt{2} + c^2}{x^2 - cx\sqrt{2} + c^2} + \tan^{-1}\frac{cx\sqrt{2}}{c^2 - x^2}\right]$

103. $\displaystyle \int \frac{dx}{c^4 - x^4} = \frac{1}{2c^3}\left[\frac{1}{2}\log\frac{c + x}{c - x} + \tan^{-1}\frac{x}{c}\right]$

104. $\displaystyle \int \frac{x\,dx}{c^4 + x^4} = \frac{1}{2c^2}\tan^{-1}\frac{x^2}{c^2}$

105. $\displaystyle \int \frac{x\,dx}{c^4 - x^4} = \frac{1}{4c^2}\log\frac{c^2 + x^2}{c^2 - x^2}$

106. $\displaystyle \int \frac{x^2\,dx}{c^4 + x^4} = \frac{1}{2c\sqrt{2}}\left[\frac{1}{2}\log\frac{x^2 - cx\sqrt{2} + c^2}{x^2 + cx\sqrt{2} + c^2} + \tan^{-1}\frac{cx\sqrt{2}}{c^2 - x^2}\right]$

107. $\displaystyle \int \frac{x^2\,dx}{c^4 - x^4} = \frac{1}{2c}\left[\frac{1}{2}\log\frac{c + x}{c - x} - \tan^{-1}\frac{x}{c}\right]$

108. $\displaystyle \int \frac{x^3\,dx}{c^4 \pm x^4} = \pm\frac{1}{4}\log(c^4 \pm x^4)$

FORMS CONTAINING $(a + bx + cx^2)$
$$X = a + bx + cx^2 \text{ and } q = 4ac - b^2$$

If $q = 0$, then $X = c\left(x + \dfrac{b}{2c}\right)^2$, and formulas starting with 23 should be used in place of these.

109. $\displaystyle \int \frac{dx}{X} = \frac{2}{\sqrt{q}}\tan^{-1}\frac{2cx + b}{\sqrt{q}}, \qquad (q > 0)$

110. $\displaystyle \int \frac{dx}{X} = \begin{cases} \dfrac{-2}{\sqrt{-q}}\tanh^{-1}\dfrac{2cx + b}{\sqrt{-q}} \\ \qquad\qquad \text{or} \\ \dfrac{1}{\sqrt{-q}}\log\dfrac{2cx + b - \sqrt{-q}}{2cx + b + \sqrt{-q}}, \qquad (q < 0) \end{cases}$

111. $\displaystyle \int \frac{dx}{X^2} = \frac{2cx + b}{qX} + \frac{2c}{q}\int \frac{dx}{X}$

112. $\int \dfrac{dx}{X^3} = \dfrac{2cx + b}{q}\left(\dfrac{1}{2X^2} + \dfrac{3c}{qX}\right) + \dfrac{6c^2}{q^2}\int \dfrac{dx}{X}$

113. $\int \dfrac{dx}{X^{n+1}} = \begin{cases} \dfrac{2cx + b}{nqX^n} + \dfrac{2(2n - 1)c}{qn}\int \dfrac{dx}{X^n} \\[4mm] \text{or} \\[4mm] \dfrac{(2n)!}{(n!)^2}\left(\dfrac{c}{q}\right)^n\left[\dfrac{2cx + b}{q}\displaystyle\sum_{r=1}^{n}\left(\dfrac{q}{cX}\right)^r\left(\dfrac{(r-1)!\,r!}{(2r)!}\right) + \int \dfrac{dx}{X}\right] \end{cases}$

114. $\int \dfrac{x\,dx}{X} = \dfrac{1}{2c}\log X - \dfrac{b}{2c}\int \dfrac{dx}{X}$

115. $\int \dfrac{x\,dx}{X^2} = -\dfrac{bx + 2a}{qX} - \dfrac{b}{q}\int \dfrac{dx}{X}$

116. $\int \dfrac{x\,dx}{X^{n+1}} = -\dfrac{2a + bx}{nqX^n} - \dfrac{b(2n - 1)}{nq}\int \dfrac{dx}{X^n}$

117. $\int \dfrac{x^2}{X}\,dx = \dfrac{x}{c} - \dfrac{b}{2c^2}\log X + \dfrac{b^2 - 2ac}{2c^2}\int \dfrac{dx}{X}$

118. $\int \dfrac{x^2}{X^2}\,dx = \dfrac{(b^2 - 2ac)x + ab}{cqX} + \dfrac{2a}{q}\int \dfrac{dx}{X}$

119. $\int \dfrac{x^m\,dx}{X^{n+1}} = -\dfrac{x^{m-1}}{(2n - m + 1)cX^n} - \dfrac{n - m + 1}{2n - m + 1}\cdot\dfrac{b}{c}\int \dfrac{x^{m-1}\,dx}{X^{n+1}}$

$$+ \dfrac{m - 1}{2n - m + 1}\cdot\dfrac{a}{c}\int \dfrac{x^{m-2}\,dx}{X^{n+1}}$$

120. $\int \dfrac{dx}{xX} = \dfrac{1}{2a}\log\dfrac{x^2}{X} - \dfrac{b}{2a}\int \dfrac{dx}{X}$

121. $\int \dfrac{dx}{x^2 X} = \dfrac{b}{2a^2}\log\dfrac{X}{x^2} - \dfrac{1}{ax} + \left(\dfrac{b^2}{2a^2} - \dfrac{c}{a}\right)\int \dfrac{dx}{X}$

122. $\int \dfrac{dx}{xX^n} = \dfrac{1}{2a(n - 1)X^{n-1}} - \dfrac{b}{2a}\int \dfrac{dx}{X^n} + \dfrac{1}{a}\int \dfrac{dx}{xX^{n-1}}$

123. $\int \dfrac{dx}{x^m X^{n+1}} = -\dfrac{1}{(m - 1)ax^{m-1}X^n} - \dfrac{n + m - 1}{m - 1}\cdot\dfrac{b}{a}\int \dfrac{dx}{x^{m-1}X^{n+1}}$

$$- \dfrac{2n + m - 1}{m - 1}\cdot\dfrac{c}{a}\int \dfrac{dx}{x^{m-2}X^{n+1}}$$

FORMS CONTAINING $\sqrt{a + bx}$

124. $\int \sqrt{a + bx}\,dx = \dfrac{2}{3b}\sqrt{(a + bx)^3}$

125. $\int x\sqrt{a + bx}\,dx = -\dfrac{2(2a - 3bx)\sqrt{(a + bx)^3}}{15b^2}$

126. $\int x^2\sqrt{a + bx}\,dx = \dfrac{2(8a^2 - 12abx + 15b^2x^2)\sqrt{(a + bx)^3}}{105b^3}$

127. $\int x^m\sqrt{a + bx}\,dx = \begin{cases} \dfrac{2}{b(2m + 3)}\left[x^m\sqrt{(a + bx)^3} - ma\displaystyle\int x^{m-1}\sqrt{a + bx}\,dx\right] \\[4mm] \text{or} \\[4mm] \dfrac{2}{b^{m+1}}\sqrt{a + bx}\displaystyle\sum_{r=0}^{m}\dfrac{m!(-a)^{m-r}}{r!(m - r)!(2r + 3)}(a + bx)^{r+1} \end{cases}$

128. $\displaystyle\int \frac{\sqrt{a + bx}}{x}\,dx = 2\sqrt{a + bx} + a\int \frac{dx}{x\sqrt{a + bx}}$

129. $\displaystyle\int \frac{\sqrt{a + bx}}{x^2}\,dx = -\frac{\sqrt{a + bx}}{x} + \frac{b}{2}\int \frac{dx}{x\sqrt{a + bx}}$

130. $\displaystyle\int \frac{\sqrt{a + bx}}{x^m}\,dx = -\frac{1}{(m - 1)a}\left[\frac{\sqrt{(a + bx)^3}}{x^{m-1}} + \frac{(2m - 5)b}{2}\int \frac{\sqrt{a + bx}}{x^{m-1}}\,dx\right]$

131. $\displaystyle\int \frac{dx}{\sqrt{a + bx}} = \frac{2\sqrt{a + bx}}{b}$

132. $\displaystyle\int \frac{x\,dx}{\sqrt{a + bx}} = -\frac{2(2a - bx)}{3b^2}\sqrt{a + bx}$

133. $\displaystyle\int \frac{x^2\,dx}{\sqrt{a + bx}} = \frac{2(8a^2 - 4abx + 3b^2x^2)}{15b^3}\sqrt{a + bx}$

134. $\displaystyle\int \frac{x^m\,dx}{\sqrt{a + bx}} = \begin{cases} \dfrac{2}{(2m + 1)b}\left[x^m\sqrt{a + bx} - ma\displaystyle\int \frac{x^{m-1}\,dx}{\sqrt{a + bx}}\right] \\[2mm] \text{or} \\[2mm] \dfrac{2(-a)^m\sqrt{a + bx}}{b^{m+1}}\displaystyle\sum_{r=0}^{m}\frac{(-1)^r m!(a + bx)^r}{(2r + 1)r!(m - r)!a^r} \end{cases}$

135. $\displaystyle\int \frac{dx}{x\sqrt{a + bx}} = \frac{1}{\sqrt{a}}\log\left(\frac{\sqrt{a + bx} - \sqrt{a}}{\sqrt{a + bx} + \sqrt{a}}\right), \qquad (a > 0)$

136. $\displaystyle\int \frac{dx}{x\sqrt{a + bx}} = \frac{2}{\sqrt{-a}}\tan^{-1}\sqrt{\frac{a + bx}{-a}}, \qquad (a < 0)$

137. $\displaystyle\int \frac{dx}{x^2\sqrt{a + bx}} = -\frac{\sqrt{a + bx}}{ax} - \frac{b}{2a}\int \frac{dx}{x\sqrt{a + bx}}$

138. $\displaystyle\int \frac{dx}{x^n\sqrt{a + bx}} = \begin{cases} -\dfrac{\sqrt{a + bx}}{(n - 1)ax^{n-1}} - \dfrac{(2n - 3)b}{(2n - 2)a}\displaystyle\int \frac{dx}{x^{n-1}\sqrt{a + bx}} \\[2mm] \text{or} \\[2mm] \dfrac{(2n - 2)!}{[(n - 1)!]^2}\left[-\dfrac{\sqrt{a + bx}}{a}\displaystyle\sum_{r=1}^{n-1}\frac{r!(r - 1)!}{x^r(2r)!}\left(-\frac{b}{4a}\right)^{n-r-1}\right. \\[2mm] \qquad\qquad\qquad \left. + \left(-\dfrac{b}{4a}\right)^{n-1}\displaystyle\int \frac{dx}{x\sqrt{a + bx}}\right] \end{cases}$

139. $\displaystyle\int (a + bx)^{\pm\frac{n}{2}}\,dx = \frac{2(a + bx)^{\frac{2 \pm n}{2}}}{b(2 \pm n)}$

140. $\displaystyle\int x(a + bx)^{\pm\frac{n}{2}}\,dx = \frac{2}{b^2}\left[\frac{(a + bx)^{\frac{4 \pm n}{2}}}{4 \pm n} - \frac{a(a + bx)^{\frac{2 \pm n}{2}}}{2 \pm n}\right]$

141. $\displaystyle\int \frac{dx}{x(a + bx)^{\frac{m}{2}}} = \frac{1}{a}\int \frac{dx}{x(a + bx)^{\frac{m-2}{2}}} - \frac{b}{a}\int \frac{dx}{(a + bx)^{\frac{m}{2}}}$

142. $\int \dfrac{(a + bx)^{\frac{n}{2}} \, dx}{x} = b \int (a + bx)^{\frac{n-2}{2}} \, dx + a \int \dfrac{(a + bx)^{\frac{n-2}{2}}}{x} \, dx$

143. $\int f(x, \sqrt{a + bx}) \, dx = \dfrac{2}{b} \int f\left(\dfrac{z^2 - a}{b}, z\right) z \, dz, \quad (z = \sqrt{a + bx})$

$$\text{FORMS CONTAINING } \sqrt{a + bx} \text{ and } \sqrt{c + dx}$$

$$u = a + bx \qquad v = c + dx \qquad k = ad - bc$$

If $k = 0$, then $v = \dfrac{c}{a} u$, and formulas starting with 124 should be used in place of these.

144. $\int \dfrac{dx}{\sqrt{uv}} = \begin{cases} \dfrac{2}{\sqrt{bd}} \tanh^{-1} \dfrac{\sqrt{bduv}}{bv}, \ bd > o, \ k < o \\[2mm] \text{or} \\[2mm] \dfrac{2}{\sqrt{bd}} \tanh^{-1} \dfrac{\sqrt{bduv}}{du}, \ bd > o, \ k > o. \\[2mm] \text{or} \\[2mm] \dfrac{1}{\sqrt{bd}} \log \dfrac{(bv + \sqrt{bduv})^2}{v}, \quad (bd > 0) \end{cases}$

145. $\int \dfrac{dx}{\sqrt{uv}} = \begin{cases} \dfrac{2}{\sqrt{-bd}} \tan^{-1} \dfrac{\sqrt{-bduv}}{bv} \\[2mm] \text{or} \\[2mm] -\dfrac{1}{\sqrt{-bd}} \sin^{-1} \left(\dfrac{2bdx + ad + bc}{|k|}\right), \quad (bd < 0) \end{cases}$

146. $\int \sqrt{uv} \, dx = \dfrac{k + 2bv}{4bd} \sqrt{uv} - \dfrac{k^2}{8bd} \int \dfrac{dx}{\sqrt{uv}}$

147. $\int \dfrac{dx}{v\sqrt{u}} = \begin{cases} \dfrac{1}{\sqrt{kd}} \log \dfrac{d\sqrt{u} - \sqrt{kd}}{d\sqrt{u} + \sqrt{kd}} \\[2mm] \text{or} \\[2mm] \dfrac{1}{\sqrt{kd}} \log \dfrac{(d\sqrt{u} - \sqrt{kd})^2}{v}, \quad (kd > 0) \end{cases}$

148. $\int \dfrac{dx}{v\sqrt{u}} = \dfrac{2}{\sqrt{-kd}} \tan^{-1} \dfrac{d\sqrt{u}}{\sqrt{-kd}}, \quad (kd < 0)$

149. $\int \dfrac{x \, dx}{\sqrt{uv}} = \dfrac{\sqrt{uv}}{bd} - \dfrac{ad + bc}{2bd} \int \dfrac{dx}{\sqrt{uv}}$

150. $\int \dfrac{dx}{v\sqrt{uv}} = \dfrac{-2\sqrt{uv}}{kv}$

151. $\int \dfrac{v \, dx}{\sqrt{uv}} = \dfrac{\sqrt{uv}}{b} - \dfrac{k}{2b} \int \dfrac{dx}{\sqrt{uv}}$

152. $\int \sqrt{\dfrac{v}{u}} \, dx = \dfrac{v}{|v|} \int \dfrac{v \, dx}{\sqrt{uv}}$

153. $\int v^m \sqrt{u} \, dx = \dfrac{1}{(2m + 3)d}\left(2v^{m+1}\sqrt{u} + k \int \dfrac{v^m \, dx}{\sqrt{u}}\right)$

154. $\displaystyle\int \frac{dx}{v^m \sqrt{u}} = -\frac{1}{(m-1)k}\left(\frac{\sqrt{u}}{v^{m-1}} + \left(m - \frac{3}{2}\right)b\int \frac{dx}{v^{m-1}\sqrt{u}}\right)$

155. $\displaystyle\int \frac{v^m\,dx}{\sqrt{u}} = \begin{cases} \dfrac{2}{b(2m+1)}\left[v^m\sqrt{u} - mk\displaystyle\int \dfrac{v^{m-1}}{\sqrt{u}}\,dx\right] \\ \text{or} \\ \dfrac{2(m!)^2\sqrt{u}}{b(2m+1)!}\displaystyle\sum_{r=0}^{m}\left(-\dfrac{4k}{b}\right)^{m-r}\dfrac{(2r)!}{(r!)^2}v^r \end{cases}$

FORMS CONTAINING $\sqrt{x^2 \pm a^2}$

156. $\displaystyle\int \sqrt{x^2 \pm a^2}\,dx = \tfrac{1}{2}[x\sqrt{x^2 \pm a^2} \pm a^2 \log(x + \sqrt{x^2 \pm a^2})]$

157. $\displaystyle\int \frac{dx}{\sqrt{x^2 \pm a^2}} = \log(x + \sqrt{x^2 \pm a^2})$

158. $\displaystyle\int \frac{dx}{x\sqrt{x^2 - a^2}} = \frac{1}{|a|}\sec^{-1}\frac{x}{a}$

159. $\displaystyle\int \frac{dx}{x\sqrt{x^2 + a^2}} = -\frac{1}{a}\log\left(\frac{a + \sqrt{x^2 + a^2}}{x}\right)$

160. $\displaystyle\int \frac{\sqrt{x^2 + a^2}}{x}\,dx = \sqrt{x^2 + a^2} - a\log\left(\frac{a + \sqrt{x^2 + a^2}}{x}\right)$

161. $\displaystyle\int \frac{\sqrt{x^2 - a^2}}{x}\,dx = \sqrt{x^2 - a^2} - |a|\sec^{-1}\frac{x}{a}$

162. $\displaystyle\int \frac{x\,dx}{\sqrt{x^2 \pm a^2}} = \sqrt{x^2 \pm a^2}$

163. $\displaystyle\int x\sqrt{x^2 \pm a^2}\,dx = \tfrac{1}{3}\sqrt{(x^2 \pm a^2)^3}$

164. $\displaystyle\int \sqrt{(x^2 \pm a^2)^3}\,dx = \frac{1}{4}\left[x\sqrt{(x^2 \pm a^2)^3} \pm \frac{3a^2 x}{2}\sqrt{x^2 \pm a^2} \right.$
$$\left. + \frac{3a^4}{2}\log(x + \sqrt{x^2 \pm a^2})\right]$$

165. $\displaystyle\int \frac{dx}{\sqrt{(x^2 \pm a^2)^3}} = \frac{\pm x}{a^2\sqrt{x^2 \pm a^2}}$

166. $\displaystyle\int \frac{x\,dx}{\sqrt{(x^2 \pm a^2)^3}} = \frac{-1}{\sqrt{x^2 \pm a^2}}$

167. $\displaystyle\int x\sqrt{(x^2 \pm a^2)^3}\,dx = \tfrac{1}{5}\sqrt{(x^2 \pm a^2)^5}$

168. $\displaystyle\int x^2\sqrt{x^2 \pm a^2}\,dx = \frac{x}{4}\sqrt{(x^2 \pm a^2)^3} \mp \frac{a^2}{8}x\sqrt{x^2 \pm a^2} - \frac{a^4}{8}\log(x + \sqrt{x^2 \pm a^2})$

169. $\displaystyle\int x^3\sqrt{x^2 + a^2}\,dx = (\tfrac{1}{5}x^2 - \tfrac{2}{15}a^2)\sqrt{(a^2 + x^2)^3}$

170. $\displaystyle\int x^3\sqrt{x^2 - a^2}\,dx = \frac{1}{5}\sqrt{(x^2 - a^2)^5} + \frac{a^2}{3}\sqrt{(x^2 - a^2)^3}$

171. $\displaystyle\int \frac{x^2\,dx}{\sqrt{x^2 \pm a^2}} = \frac{x}{2}\sqrt{x^2 \pm a^2} \mp \frac{a^2}{2}\log(x + \sqrt{x^2 \pm a^2})$

172. $\int \dfrac{x^3\,dx}{\sqrt{x^2 \pm a^2}} = \dfrac{1}{3}\sqrt{(x^2 \pm a^2)^3} \mp a^2\sqrt{x^2 \pm a^2}$

173. $\int \dfrac{dx}{x^2\sqrt{x^2 \pm a^2}} = \mp \dfrac{\sqrt{x^2 \pm a^2}}{a^2 x}$

174. $\int \dfrac{dx}{x^3\sqrt{x^2 + a^2}} = -\dfrac{\sqrt{x^2 + a^2}}{2a^2 x^2} + \dfrac{1}{2a^3}\log \dfrac{a + \sqrt{x^2 + a^2}}{x}$

175. $\int \dfrac{dx}{x^3\sqrt{x^2 - a^2}} = \dfrac{\sqrt{x^2 - a^2}}{2a^2 x^2} + \dfrac{1}{2|a^3|}\sec^{-1}\dfrac{x}{a}$

176. $\int x^2\sqrt{(x^2 \pm a^2)^3}\,dx = \dfrac{x}{6}\sqrt{(x^2 \pm a^2)^5} \mp \dfrac{a^2 x}{24}\sqrt{(x^2 \pm a^2)^3} - \dfrac{a^4 x}{16}\sqrt{x^2 \pm a^2}$

$$\mp \dfrac{a^6}{16}\log(x + \sqrt{x^2 \pm a^2})$$

177. $\int x^3\sqrt{(x^2 \pm a^2)^3}\,dx = \dfrac{1}{7}\sqrt{(x^2 \pm a^2)^7} \mp \dfrac{a^2}{5}\sqrt{(x^2 \pm a^2)^5}$

178. $\int \dfrac{\sqrt{x^2 \pm a^2}\,dx}{x^2} = -\dfrac{\sqrt{x^2 \pm a^2}}{x} + \log(x + \sqrt{x^2 \pm a^2})$

179. $\int \dfrac{\sqrt{x^2 + a^2}}{x^3}\,dx = -\dfrac{\sqrt{x^2 + a^2}}{2x^2} - \dfrac{1}{2a}\log \dfrac{a + \sqrt{x^2 + a^2}}{x}$

180. $\int \dfrac{\sqrt{x^2 - a^2}}{x^3}\,dx = -\dfrac{\sqrt{x^2 - a^2}}{2x^2} + \dfrac{1}{2|a|}\sec^{-1}\dfrac{x}{a}$

181. $\int \dfrac{\sqrt{x^2 \pm a^2}}{x^4}\,dx = \mp \dfrac{\sqrt{(x^2 \pm a^2)^3}}{3a^2 x^3}$

182. $\int \dfrac{x^2\,dx}{\sqrt{(x^2 \pm a^2)^3}} = \dfrac{-x}{\sqrt{x^2 \pm a^2}} + \log(x + \sqrt{x^2 \pm a^2})$

183. $\int \dfrac{x^3\,dx}{\sqrt{(x^2 \pm a^2)^3}} = \sqrt{x^2 \pm a^2} \pm \dfrac{a^2}{\sqrt{x^2 \pm a^2}}$

184. $\int \dfrac{dx}{x\sqrt{(x^2 + a^2)^3}} = \dfrac{1}{a^2\sqrt{x^2 + a^2}} - \dfrac{1}{a^3}\log \dfrac{a + \sqrt{x^2 + a^2}}{x}$

185. $\int \dfrac{dx}{x\sqrt{(x^2 - a^2)^3}} = -\dfrac{1}{a^2\sqrt{x^2 - a^2}} - \dfrac{1}{|a^3|}\sec^{-1}\dfrac{x}{a}$

186. $\int \dfrac{dx}{x^2\sqrt{(x^2 \pm a^2)^3}} = -\dfrac{1}{a^4}\left[\dfrac{\sqrt{x^2 \pm a^2}}{x} + \dfrac{x}{\sqrt{x^2 \pm a^2}}\right]$

187. $\int \dfrac{dx}{x^3\sqrt{(x^2 + a^2)^3}} = -\dfrac{1}{2a^2 x^2\sqrt{x^2 + a^2}} - \dfrac{3}{2a^4\sqrt{x^2 + a^2}}$

$$+ \dfrac{3}{2a^5}\log \dfrac{a + \sqrt{x^2 + a^2}}{x}$$

188. $\int \dfrac{dx}{x^3\sqrt{(x^2 - a^2)^3}} = \dfrac{1}{2a^2 x^2\sqrt{x^2 - a^2}} - \dfrac{3}{2a^4\sqrt{x^2 - a^2}} - \dfrac{3}{2|a^5|}\sec^{-1}\dfrac{x}{a}$

189. $\int \dfrac{x^m}{\sqrt{x^2 \pm a^2}}\,dx = \dfrac{1}{m}x^{m-1}\sqrt{x^2 \pm a^2} \mp \dfrac{m-1}{m}a^2\int \dfrac{x^{m-2}}{\sqrt{x^2 \pm a^2}}\,dx$

190. $\displaystyle \int \frac{x^{2m}}{\sqrt{x^2 \pm a^2}}\,dx = \frac{(2m)!}{2^{2m}(m!)^2}\left[\sqrt{x^2 \pm a^2}\sum_{r=1}^{m}\frac{r!(r-1)!}{(2r)!}(\mp a^2)^{m-r}(2x)^{2r-1}\right.$

$$\left. + (\mp a^2)^m \log(x + \sqrt{x^2 \pm a^2})\right]$$

191. $\displaystyle \int \frac{x^{2m+1}}{\sqrt{x^2 \pm a^2}}\,dx = \sqrt{x^2 \pm a^2}\sum_{r=0}^{m}\frac{(2r)!(m!)^2}{(2m+1)!(r!)^2}(\mp 4a^2)^{m-r}x^{2r}$

192. $\displaystyle \int \frac{dx}{x^m \sqrt{x^2 \pm a^2}} = \mp \frac{\sqrt{x^2 \pm a^2}}{(m-1)a^2 x^{m-1}} \mp \frac{(m-2)}{(m-1)a^2}\int\frac{dx}{x^{m-2}\sqrt{x^2 \pm a^2}}$

193. $\displaystyle \int \frac{dx}{x^{2m}\sqrt{x^2 \pm a^2}} = \sqrt{x^2 \pm a^2}\sum_{r=0}^{m-1}\frac{(m-1)!m!(2r)!2^{2m-2r-1}}{(r!)^2(2m)!(\mp a^2)^{m-r}x^{2r+1}}$

194. $\displaystyle \int \frac{dx}{x^{2m+1}\sqrt{x^2 + a^2}} = \frac{(2m)!}{(m!)^2}\left[\frac{\sqrt{x^2 + a^2}}{a^2}\sum_{r=1}^{m}(-1)^{m-r+1}\frac{r!(r-1)!}{2(2r)!(4a^2)^{m-r}x^{2r}}\right.$

$$\left. + \frac{(-1)^{m+1}}{2^{2m}a^{2m+1}}\log\frac{\sqrt{x^2 + a^2} + a}{x}\right]$$

195. $\displaystyle \int \frac{dx}{x^{2m+1}\sqrt{x^2 - a^2}} = \frac{(2m)!}{(m!)^2}\left[\frac{\sqrt{x^2 - a^2}}{a^2}\sum_{r=1}^{m}\frac{r!(r-1)!}{2(2r)!(4a^2)^{m-r}x^{2r}}\right.$

$$\left. + \frac{1}{2^{2m}|a|^{2m+1}}\sec^{-1}\frac{x}{a}\right]$$

196. $\displaystyle \int \frac{dx}{(x-a)\sqrt{x^2 - a^2}} = -\frac{\sqrt{x^2 - a^2}}{a(x-a)}$

197. $\displaystyle \int \frac{dx}{(x+a)\sqrt{x^2 - a^2}} = \frac{\sqrt{x^2 - a^2}}{a(x+a)}$

198. $\displaystyle \int f(x, \sqrt{x^2 + a^2})\,dx = a\int f(a\tan u, a\sec u)\sec^2 u\,du, \quad \left(u = \tan^{-1}\frac{x}{a}, a > 0\right)$

199. $\displaystyle \int f(x, \sqrt{x^2 - a^2})\,dx = a\int f(a\sec u, a\tan u)\sec u \tan u\,du, \quad \left(u = \sec^{-1}\frac{x}{a}\right.$

$$\left. a > 0\right)$$

FORMS CONTAINING $\sqrt{a^2 - x^2}$

200. $\displaystyle \int \sqrt{a^2 - x^2}\,dx = \frac{1}{2}\left[x\sqrt{a^2 - x^2} + a^2\sin^{-1}\frac{x}{|a|}\right]$

201. $\displaystyle \int \frac{dx}{\sqrt{a^2 - x^2}} = \begin{cases} \sin^{-1}\dfrac{x}{|a|} \\ \text{or} \\ -\cos^{-1}\dfrac{x}{|a|} \end{cases}$

202. $\displaystyle \int \frac{dx}{x\sqrt{a^2 - x^2}} = -\frac{1}{a}\log\left(\frac{a + \sqrt{a^2 - x^2}}{x}\right)$

203. $\displaystyle \int \frac{\sqrt{a^2 - x^2}}{x}\,dx = \sqrt{a^2 - x^2} - a\log\left(\frac{a + \sqrt{a^2 - x^2}}{x}\right)$

204. $\displaystyle \int \frac{x\,dx}{\sqrt{a^2 - x^2}} = -\sqrt{a^2 - x^2}$

205. $\displaystyle \int x\sqrt{a^2 - x^2}\,dx = -\tfrac{1}{3}\sqrt{(a^2 - x^2)^3}$

206. $\int \sqrt{(a^2 - x^2)^3}\, dx = \frac{1}{4}\left[x\sqrt{(a^2 - x^2)^3} + \frac{3a^2 x}{2}\sqrt{a^2 - x^2} + \frac{3a^4}{2}\sin^{-1}\frac{x}{|a|}\right]$

207. $\int \dfrac{dx}{\sqrt{(a^2 - x^2)^3}} = \dfrac{x}{a^2\sqrt{a^2 - x^2}}$

208. $\int \dfrac{x\, dx}{\sqrt{(a^2 - x^2)^3}} = \dfrac{1}{\sqrt{a^2 - x^2}}$

209. $\int x\sqrt{(a^2 - x^2)^3}\, dx = -\frac{1}{5}\sqrt{(a^2 - x^2)^5}$

210. $\int x^2\sqrt{a^2 - x^2}\, dx = -\frac{x}{4}\sqrt{(a^2 - x^2)^3} + \frac{a^2}{8}\left(x\sqrt{a^2 - x^2} + a^2\sin^{-1}\frac{x}{|a|}\right)$

211. $\int x^3\sqrt{a^2 - x^2}\, dx = \left(-\frac{1}{5}x^2 - \frac{2}{15}a^2\right)\sqrt{(a^2 - x^2)^3}$

212. $\int x^2\sqrt{(a^2 - x^2)^3}\, dx = -\frac{1}{6}x\sqrt{(a^2 - x^2)^5} + \frac{a^2 x}{24}\sqrt{(a^2 - x^2)^3}$

$$+ \frac{a^4 x}{16}\sqrt{a^2 - x^2} + \frac{a^6}{16}\sin^{-1}\frac{x}{|a|}$$

213. $\int x^3\sqrt{(a^2 - x^2)^3}\, dx = \frac{1}{7}\sqrt{(a^2 - x^2)^7} - \frac{a^2}{5}\sqrt{(a^2 - x^2)^5}$

214. $\int \dfrac{x^2\, dx}{\sqrt{a^2 - x^2}} = -\frac{x}{2}\sqrt{a^2 - x^2} + \frac{a^2}{2}\sin^{-1}\frac{x}{|a|}$

215. $\int \dfrac{dx}{x^2\sqrt{a^2 - x^2}} = -\dfrac{\sqrt{a^2 - x^2}}{a^2 x}$

216. $\int \dfrac{\sqrt{a^2 - x^2}}{x^2}\, dx = -\dfrac{\sqrt{a^2 - x^2}}{x} - \sin^{-1}\frac{x}{|a|}$

217. $\int \dfrac{\sqrt{a^2 - x^2}}{x^3}\, dx = -\dfrac{\sqrt{a^2 - x^2}}{2x^2} + \frac{1}{2a}\log\dfrac{a + \sqrt{a^2 - x^2}}{x}$

218. $\int \dfrac{\sqrt{a^2 - x^2}}{x^4}\, dx = -\dfrac{\sqrt{(a^2 - x^2)^3}}{3a^2 x^3}$

219. $\int \dfrac{x^2\, dx}{\sqrt{(a^2 - x^2)^3}} = \dfrac{x}{\sqrt{a^2 - x^2}} - \sin^{-1}\frac{x}{|a|}$

220. $\int \dfrac{x^3\, dx}{\sqrt{a^2 - x^2}} = -\frac{2}{3}(a^2 - x^2)^{\frac{3}{2}} - x^2(a^2 - x^2)^{\frac{1}{2}} = -\frac{1}{3}\sqrt{a^2 - x^2}(x^2 + 2a^2)$

221. $\int \dfrac{x^3\, dx}{\sqrt{(a^2 - x^2)^3}} = 2(a^2 - x^2)^{\frac{1}{2}} + \dfrac{x^2}{(a^2 - x^2)^{\frac{1}{2}}} = -\dfrac{a^2}{\sqrt{a^2 - x^2}} + \sqrt{a^2 - x^2}$

222. $\int \dfrac{dx}{x^3\sqrt{a^2 - x^2}} = -\dfrac{\sqrt{a^2 - x^2}}{2a^2 x^2} - \frac{1}{2a^3}\log\dfrac{a + \sqrt{a^2 - x^2}}{x}$

223. $\int \dfrac{dx}{x\sqrt{(a^2 - x^2)^3}} = \dfrac{1}{a^2\sqrt{a^2 - x^2}} - \frac{1}{a^3}\log\dfrac{a + \sqrt{a^2 - x^2}}{x}$

224. $\int \dfrac{dx}{x^2\sqrt{(a^2 - x^2)^3}} = \dfrac{1}{a^4}\left[-\dfrac{\sqrt{a^2 - x^2}}{x} + \dfrac{x}{\sqrt{a^2 - x^2}}\right]$

225. $\int \dfrac{dx}{x^3\sqrt{(a^2 - x^2)^3}} = -\dfrac{1}{2a^2 x^2\sqrt{a^2 - x^2}} + \dfrac{3}{2a^4\sqrt{a^2 - x^2}}$

$$-\frac{3}{2a^5}\log\dfrac{a + \sqrt{a^2 - x^2}}{x}$$

226. $\displaystyle \int \frac{x^m}{\sqrt{a^2 - x^2}}\, dx = -\frac{x^{m-1}\sqrt{a^2 - x^2}}{m} + \frac{(m-1)a^2}{m}\int \frac{x^{m-2}}{\sqrt{a^2 - x^2}}\, dx$

227. $\displaystyle \int \frac{x^{2m}}{\sqrt{a^2 - x^2}}\, dx = \frac{(2m)!}{(m!)^2}\left[-\sqrt{a^2 - x^2}\sum_{r=1}^{m} \frac{r!(r-1)!}{2^{2m-2r+1}(2r)!}a^{2m-2r}x^{2r-1} \right.$
$$\left. + \frac{a^{2m}}{2^{2m}}\sin^{-1}\frac{x}{|a|}\right]$$

228. $\displaystyle \int \frac{x^{2m+1}}{\sqrt{a^2 - x^2}}\, dx = -\sqrt{a^2 - x^2}\sum_{r=0}^{m}\frac{(2r)!(m!)^2}{(2m+1)!(r!)^2}(4a^2)^{m-r}x^{2r}$

229. $\displaystyle \int \frac{dx}{x^m\sqrt{a^2 - x^2}} = -\frac{\sqrt{a^2 - x^2}}{(m-1)a^2 x^{m-1}} + \frac{m-2}{(m-1)a^2}\int \frac{dx}{x^{m-2}\sqrt{a^2 - x^2}}$

230. $\displaystyle \int \frac{ax}{x^{2m}\sqrt{a^2 - x^2}} = -\sqrt{a^2 - x^2}\sum_{r=0}^{m-1}\frac{(m-1)!m!(2r)!2^{2m-2r-1}}{(r!)^2(2m)!a^{2m-2r}x^{2r+1}}$

231. $\displaystyle \int \frac{dx}{x^{2m+1}\sqrt{a^2 - x^2}} = \frac{(2m)!}{(m!)^2}\left[-\frac{\sqrt{a^2 - x^2}}{a^2}\sum_{r=1}^{m}\frac{r!(r-1)!}{2(2r)!(4a^2)^{m-r}x^{2r}} \right.$
$$\left. + \frac{1}{2^{2m}a^{2m+1}}\log\frac{a - \sqrt{a^2 - x^2}}{x}\right]$$

232. $\displaystyle \int \frac{dx}{(b^2 - x^2)\sqrt{a^2 - x^2}} = \frac{1}{2b\sqrt{a^2 - b^2}}\log\frac{(b\sqrt{a^2 - x^2} + x\sqrt{a^2 - b^2})^2}{b^2 - x^2},$

$$(a^2 > b^2)$$

233. $\displaystyle \int \frac{dx}{(b^2 - x^2)\sqrt{a^2 - x^2}} - \frac{1}{b\sqrt{b^2 - a^2}}\tan^{-1}\frac{x\sqrt{b^2 - a^2}}{b\sqrt{a^2 - x^2}}, \qquad (b^2 > u^2)$

234. $\displaystyle \int \frac{dx}{(b^2 + x^2)\sqrt{a^2 - x^2}} = \frac{1}{b\sqrt{a^2 + b^2}}\tan^{-1}\frac{x\sqrt{a^2 + b^2}}{b\sqrt{a^2 - x^2}}$

235. $\displaystyle \int \frac{\sqrt{a^2 - x^2}}{b^2 + x^2}\, dx = \frac{\sqrt{a^2 + b^2}}{|b|}\sin^{-1}\frac{x\sqrt{a^2 + b^2}}{|a|\sqrt{x^2 + b^2}} - \sin^{-1}\frac{x}{|a|}$

236. $\displaystyle \int f(x, \sqrt{a^2 - x^2})\, dx = a\int f(a\sin u, a\cos u)\cos u\, du, \qquad \left(u = \sin^{-1}\frac{x}{a},\ a > 0\right)$

FORMS CONTAINING $\sqrt{a + bx + cx^2}$

$$X = a + bx + cx^2,\ q = 4ac - b^2,\ \text{and}\ k = \frac{4c}{q}$$

If $q = 0$, then $\sqrt{X} = \sqrt{c}\left|x + \dfrac{b}{2c}\right|$

237. $\displaystyle \int \frac{dx}{\sqrt{X}} = \begin{cases} \dfrac{1}{\sqrt{c}}\log(2\sqrt{cX} + 2cx + b) \\[2mm] \text{or} \\[2mm] \dfrac{1}{\sqrt{c}}\sinh^{-1}\dfrac{2cx + b}{\sqrt{q}}, \qquad (c > 0) \end{cases}$

238. $\displaystyle \int \frac{dx}{\sqrt{X}} = -\frac{1}{\sqrt{-c}}\sin^{-1}\frac{2cx + b}{\sqrt{-q}}, \qquad (c < 0)$

239. $\displaystyle \int \frac{dx}{X\sqrt{X}} = \frac{2(2cx + b)}{q\sqrt{X}}$

240. $\int \dfrac{dx}{X^2\sqrt{X}} = \dfrac{2(2cx+b)}{3q\sqrt{X}}\left(\dfrac{1}{X}+2k\right)$

241. $\int \dfrac{dx}{X^n\sqrt{X}} = \begin{cases} \dfrac{2(2cx+b)\sqrt{X}}{(2n-1)qX^n} + \dfrac{2k(n-1)}{2n-1}\displaystyle\int \dfrac{dx}{X^{n-1}\sqrt{X}} \\[3mm] \text{or} \\[2mm] \dfrac{(2cx+b)(n!)(n-1)!4^n k^{n-1}}{q[(2n)!]\sqrt{X}}\displaystyle\sum_{r=0}^{n-1}\dfrac{(2r)!}{(4kX)^r(r!)^2} \end{cases}$

242. $\int \sqrt{X}\,dx = \dfrac{(2cx+b)\sqrt{X}}{4c} + \dfrac{1}{2k}\displaystyle\int \dfrac{dx}{\sqrt{X}}$

243. $\int X\sqrt{X}\,dx = \dfrac{(2cx+b)\sqrt{X}}{8c}\left(X+\dfrac{3}{2k}\right) + \dfrac{3}{8k^2}\displaystyle\int \dfrac{dx}{\sqrt{X}}$

244. $\int X^2\sqrt{X}\,dx = \dfrac{(2cx+b)\sqrt{X}}{12c}\left(X^2+\dfrac{5X}{4k}+\dfrac{15}{8k^2}\right) + \dfrac{5}{16k^3}\displaystyle\int \dfrac{dx}{\sqrt{X}}$

245. $\int X^n\sqrt{X}\,dx = \begin{cases} \dfrac{(2cx+b)X^n\sqrt{X}}{4(n+1)c} + \dfrac{2n+1}{2(n+1)k}\displaystyle\int X^{n-1}\sqrt{X}\,dx \\[3mm] \text{or} \\[2mm] \dfrac{(2n+2)!}{[(n+1)!]^2(4k)^{n+1}}\left[\dfrac{k(2cx+b)\sqrt{X}}{c}\displaystyle\sum_{r=0}^{n}\dfrac{r!(r+1)!(4kX)^r}{(2r+2)!}\right. \\[3mm] \hspace{6cm}\left. + \displaystyle\int \dfrac{dx}{\sqrt{X}}\right] \end{cases}$

246. $\int \dfrac{x\,dx}{\sqrt{X}} = \dfrac{\sqrt{X}}{c} - \dfrac{b}{2c}\displaystyle\int \dfrac{dx}{\sqrt{X}}$

247. $\int \dfrac{x\,dx}{X\sqrt{X}} = -\dfrac{2(bx+2a)}{q\sqrt{X}}$

248. $\int \dfrac{x\,dx}{X^n\sqrt{X}} = -\dfrac{\sqrt{X}}{(2n-1)cX^n} - \dfrac{b}{2c}\displaystyle\int \dfrac{dx}{X^n\sqrt{X}}$

249. $\int \dfrac{x^2\,dx}{\sqrt{X}} = \left(\dfrac{x}{2c} - \dfrac{3b}{4c^2}\right)\sqrt{X} + \dfrac{3b^2-4ac}{8c^2}\displaystyle\int \dfrac{dx}{\sqrt{X}}$

250. $\int \dfrac{x^2\,dx}{X\sqrt{X}} = \dfrac{(2b^2-4ac)x+2ab}{cq\sqrt{X}} + \dfrac{1}{c}\displaystyle\int \dfrac{dx}{\sqrt{X}}$

251. $\int \dfrac{x^2\,dx}{X^n\sqrt{X}} = \dfrac{(2b^2-4ac)x+2ab}{(2n-1)cqX^{n-1}\sqrt{X}} + \dfrac{4ac+(2n-3)b^2}{(2n-1)cq}\displaystyle\int \dfrac{dx}{X^{n-1}\sqrt{X}}$

252. $\int \dfrac{x^3\,dx}{\sqrt{X}} = \left(\dfrac{x^2}{3c} - \dfrac{5bx}{12c^2} + \dfrac{5b^2}{8c^3} - \dfrac{2a}{3c^2}\right)\sqrt{X} + \left(\dfrac{3ab}{4c^2} - \dfrac{5b^3}{16c^3}\right)\displaystyle\int \dfrac{dx}{\sqrt{X}}$

253. $\int \dfrac{x^n\,dx}{\sqrt{X}} = \dfrac{1}{nc}x^{n-1}\sqrt{X} - \dfrac{(2n-1)b}{2nc}\displaystyle\int \dfrac{x^{n-1}\,dx}{\sqrt{X}} - \dfrac{(n-1)a}{nc}\displaystyle\int \dfrac{x^{n-2}\,dx}{\sqrt{X}}$

254. $\int x\sqrt{X}\,dx = \dfrac{X\sqrt{X}}{3c} - \dfrac{b(2cx+b)}{8c^2}\sqrt{X} - \dfrac{b}{4ck}\displaystyle\int \dfrac{dx}{\sqrt{X}}$

255. $\int xX\sqrt{X}\,dx = \dfrac{X^2\sqrt{X}}{5c} - \dfrac{b}{2c}\displaystyle\int X\sqrt{X}\,dx$

256. $\int x X^n \sqrt{X}\, dx = \dfrac{X^{n+1}\sqrt{X}}{(2n+3)c} - \dfrac{b}{2c}\int X^n \sqrt{X}\, dx$

257. $\int x^2 \sqrt{X}\, dx = \left(x - \dfrac{5b}{6c}\right)\dfrac{X\sqrt{X}}{4c} + \dfrac{5b^2 - 4ac}{16c^2}\int \sqrt{X}\, dx$

258. $\int \dfrac{dx}{x\sqrt{X}} = -\dfrac{1}{\sqrt{a}}\log\dfrac{2\sqrt{aX} + bx + 2a}{x}, \qquad (a > 0)$

259. $\int \dfrac{dx}{x\sqrt{X}} = \dfrac{1}{\sqrt{-a}}\sin^{-1}\left(\dfrac{bx + 2a}{|x|\sqrt{-q}}\right), \qquad (a < 0)$

260. $\int \dfrac{dx}{x\sqrt{X}} = -\dfrac{2\sqrt{X}}{bx}, \qquad (a = 0)$

261. $\int \dfrac{dx}{x^2\sqrt{X}} = -\dfrac{\sqrt{X}}{ax} - \dfrac{b}{2a}\int \dfrac{dx}{x\sqrt{X}}$

262. $\int \dfrac{\sqrt{X}\, dx}{x} = \sqrt{X} + \dfrac{b}{2}\int \dfrac{dx}{\sqrt{X}} + a\int \dfrac{dx}{x\sqrt{X}}$

263. $\int \dfrac{\sqrt{X}\, dx}{x^2} = -\dfrac{\sqrt{X}}{x} + \dfrac{b}{2}\int \dfrac{dx}{x\sqrt{X}} + c\int \dfrac{dx}{\sqrt{X}}$

FORMS INVOLVING $\sqrt{2ax - x^2}$

264. $\int \sqrt{2ax - x^2}\, dx = \dfrac{1}{2}\left[(x - a)\sqrt{2ax - x^2} + a^2 \sin^{-1}\dfrac{x - a}{|a|}\right]$

265. $\int \dfrac{dx}{\sqrt{2ax - x^2}} = \begin{cases} \cos^{-1}\dfrac{a - x}{|a|} \\[2mm] \text{or} \\[2mm] \sin^{-1}\dfrac{x - a}{|a|} \end{cases}$

266. $\int x^n\sqrt{2ax - x^2}\, dx = \begin{cases} -\dfrac{x^{n-1}(2ax - x^2)^{\frac{3}{2}}}{n + 2} + \dfrac{(2n + 1)a}{n + 2}\int x^{n-1}\sqrt{2ax - x^2}\, dx \\[3mm] \text{or} \\[3mm] \sqrt{2ax - x^2}\left[\dfrac{x^{n+1}}{n + 2} - \displaystyle\sum_{r=0}^{n} \dfrac{(2n + 1)!(r!)^2 a^{n-r+1}}{2^{n-r}(2r + 1)!(n + 2)!n!}x^r\right] \end{cases}$

$$+ \dfrac{(2n + 1)!a^{n+2}}{2^n n!(n + 2)!}\sin^{-1}\dfrac{x - a}{|a|}$$

267. $\int \dfrac{\sqrt{2ax - x^2}}{x^n}\, dx = \dfrac{(2ax - x^2)^{\frac{3}{2}}}{(3 - 2n)ax^n} + \dfrac{n - 3}{(2n - 3)a}\int \dfrac{\sqrt{2ax - x^2}}{x^{n-1}}\, dx$

268. $\int \dfrac{x^n\, dx}{\sqrt{2ax - x^2}} = \begin{cases} \dfrac{-x^{n-1}\sqrt{2ax - x^2}}{n} + \dfrac{a(2n - 1)}{n}\int \dfrac{x^{n-1}}{\sqrt{2ax - x^2}}\, dx \\[3mm] \text{or} \\[3mm] -\sqrt{2ax - x^2}\displaystyle\sum_{r=1}^{n} \dfrac{(2n)!r!(r - 1)!a^{n-r}}{2^{n-r}(2r)!(n!)^2}x^{r-1} \end{cases}$

$$+ \dfrac{(2n)!a^n}{2^n(n!)^2}\sin^{-1}\dfrac{x - a}{|a|}$$

269. $\displaystyle\int \frac{dx}{x^n\sqrt{2ax-x^2}} = \begin{cases} \dfrac{\sqrt{2ax-x^2}}{a(1-2n)x^n} + \dfrac{n-1}{(2n-1)a}\displaystyle\int \dfrac{dx}{x^{n-1}\sqrt{2ax-x^2}} \\ \quad\text{or} \\ -\sqrt{2ax-x^2}\,\displaystyle\sum_{r=0}^{n-1} \dfrac{2^{n-r}(n-1)!\,n!\,(2r)!}{(2n)!(r!)^2 a^{n-r}x^{r+1}} \end{cases}$

270. $\displaystyle\int \frac{dx}{(2ax-x^2)^{\frac{3}{2}}} = \frac{x-a}{a^2\sqrt{2ax-x^2}}$

271. $\displaystyle\int \frac{x\,dx}{(2ax-x^2)^{\frac{3}{2}}} = \frac{x}{a\sqrt{2ax-x^2}}$

MISCELLANEOUS ALGEBRAIC FORMS

272. $\displaystyle\int \frac{dx}{\sqrt{2ax+x^2}} = \log\left(x+a+\sqrt{2ax+x^2}\right)$

273. $\displaystyle\int \sqrt{ax^2+c}\,dx = \frac{x}{2}\sqrt{ax^2+c} + \frac{c}{2\sqrt{a}}\log\left(x\sqrt{a}+\sqrt{ax^2+c}\right), \quad (a>0)$

274. $\displaystyle\int \sqrt{ax^2+c}\,dx = \frac{x}{2}\sqrt{ax^2+c} + \frac{c}{2\sqrt{-a}}\sin^{-1}\left(x\sqrt{-\frac{a}{c}}\right), \quad (a<0)$

275. $\displaystyle\int \sqrt{\frac{1+x}{1-x}}\,dx = \sin^{-1}x - \sqrt{1-x^2}$

276. $\displaystyle\int \frac{dx}{x\sqrt{ax^n+c}} = \begin{cases} \dfrac{1}{n\sqrt{c}}\log\dfrac{\sqrt{ax^n+c}-\sqrt{c}}{\sqrt{ax^n+c}+\sqrt{c}} \\ \quad\text{or} \\ \dfrac{2}{n\sqrt{c}}\log\dfrac{\sqrt{ax^n+c}-\sqrt{c}}{\sqrt{x^n}}, \quad (c>0) \end{cases}$

277. $\displaystyle\int \frac{dx}{x\sqrt{ax^n+c}} = \frac{2}{n\sqrt{-c}}\sec^{-1}\sqrt{-\frac{ax^n}{c}}, \quad (c<0)$

278. $\displaystyle\int \frac{dx}{\sqrt{ax^2+c}} = \frac{1}{\sqrt{a}}\log\left(x\sqrt{a}+\sqrt{ax^2+c}\right), \quad (a>0)$

279. $\displaystyle\int \frac{dx}{\sqrt{ax^2+c}} = \frac{1}{\sqrt{-a}}\sin^{-1}\left(x\sqrt{-\frac{a}{c}}\right), \quad (a<0)$

280. $\displaystyle\int (ax^2+c)^{m+\frac{1}{2}}\,dx = \begin{cases} \dfrac{x(ax^2+c)^{m+\frac{1}{2}}}{2(m+1)} + \dfrac{(2m+1)c}{2(m+1)}\displaystyle\int (ax^2+c)^{m-\frac{1}{2}}\,dx \\ \quad\text{or} \\ x\sqrt{ax^2+c}\,\displaystyle\sum_{r=0}^{m}\dfrac{(2m+1)!(r!)^2 c^{m-r}}{2^{2m-2r+1}m!(m+1)!(2r+1)!}(ax^2+c)^r \\ \quad + \dfrac{(2m+1)!c^{m+1}}{2^{2m+1}m!(m+1)!}\displaystyle\int \dfrac{dx}{\sqrt{ax^2+c}} \end{cases}$

281. $\displaystyle\int x(ax^2+c)^{m+\frac{1}{2}}\,dx = \frac{(ax^2+c)^{m+\frac{3}{2}}}{(2m+3)a}$

282. $\displaystyle\int \frac{(ax^2 + c)^{m + \frac{1}{2}}}{x}\, dx = \begin{cases} \dfrac{(ax^2 + c)^{m + \frac{1}{2}}}{2m + 1} + c \displaystyle\int \dfrac{(ax^2 + c)^{m - \frac{1}{2}}}{x}\, dx \\ \text{or} \\ \sqrt{ax^2 + c}\ \displaystyle\sum_{r = 0}^{m} \dfrac{c^{m - r}(ax^2 + c)^r}{2r + 1} + c^{m + 1} \displaystyle\int \dfrac{dx}{x\sqrt{ax^2 + c}} \end{cases}$

283. $\displaystyle\int \frac{dx}{(ax^2 + c)^{m + \frac{1}{2}}} = \begin{cases} \dfrac{x}{(2m - 1)c(ax^2 + c)^{m - \frac{1}{2}}} + \dfrac{2m - 2}{(2m - 1)c} \displaystyle\int \dfrac{dx}{(ax^2 + c)^{m - \frac{1}{2}}} \\ \text{or} \\ \dfrac{x}{\sqrt{ax^2 + c}} \displaystyle\sum_{r = 0}^{m - 1} \dfrac{2^{2m - 2r - 1}(m - 1)!\, m!\,(2r)!}{(2m)!\,(r!)^2 c^{m - r}(ax^2 + c)^r} \end{cases}$

284. $\displaystyle\int \frac{dx}{x^m\sqrt{ax^2 + c}} = -\frac{\sqrt{ax^2 + c}}{(m - 1)cx^{m - 1}} - \frac{(m - 2)a}{(m - 1)c} \int \frac{dx}{x^{m - 2}\sqrt{ax^2 + c}}$

285. $\displaystyle\int \frac{1 + x^2}{(1 - x^2)\sqrt{1 + x^4}}\, dx = \frac{1}{\sqrt{2}} \log \frac{x\sqrt{2} + \sqrt{1 + x^4}}{1 - x^2}$

286. $\displaystyle\int \frac{1 - x^2}{(1 + x^2)\sqrt{1 + x^4}}\, dx = \frac{1}{\sqrt{2}} \tan^{-1} \frac{x\sqrt{2}}{\sqrt{1 + x^4}}$

287. $\displaystyle\int \frac{dx}{x\sqrt{x^n + a^2}} = -\frac{2}{na} \log \frac{a + \sqrt{x^n + a^2}}{\sqrt{x^n}}$

288. $\displaystyle\int \frac{dx}{x\sqrt{x^n - a^2}} = -\frac{2}{na} \sin^{-1} \frac{a}{\sqrt{x^n}}$

289. $\displaystyle\int \sqrt{\frac{x}{a^3 - x^3}}\, dx = \frac{2}{3} \sin^{-1} \left(\frac{x}{a}\right)^{\frac{3}{2}}$

FORMS INVOLVING TRIGONOMETRIC FUNCTIONS

290. $\displaystyle\int (\sin ax)\, dx = -\frac{1}{a} \cos ax$

291. $\displaystyle\int (\cos ax)\, dx = \frac{1}{a} \sin ax$

292. $\displaystyle\int (\tan ax)\, dx = -\frac{1}{a} \log \cos ax = \frac{1}{a} \log \sec ax$

293. $\displaystyle\int (\cot ax)\, dx = \frac{1}{a} \log \sin ax = -\frac{1}{a} \log \csc ax$

294. $\displaystyle\int (\sec ax)\, dx = \frac{1}{a} \log (\sec ax + \tan ax) = \frac{1}{a} \log \tan \left(\frac{\pi}{4} + \frac{ax}{2}\right)$

295. $\displaystyle\int (\csc ax)\, dx = \frac{1}{a} \log (\csc ax - \cot ax) = \frac{1}{a} \log \tan \frac{ax}{2}$

296. $\displaystyle\int (\sin^2 ax)\, dx = -\frac{1}{2a} \cos ax \sin ax + \frac{1}{2}x = \frac{1}{2}x - \frac{1}{4a} \sin 2ax$

297. $\displaystyle\int (\sin^3 ax)\, dx = -\frac{1}{3a}(\cos ax)(\sin^2 ax + 2)$

298. $\displaystyle\int (\sin^4 ax)\, dx = \frac{3x}{8} - \frac{\sin 2ax}{4a} + \frac{\sin 4ax}{32a}$

299. $\displaystyle\int (\sin^n ax)\, dx = -\frac{\sin^{n - 1} ax \cos ax}{na} + \frac{n - 1}{n} \int (\sin^{n - 2} ax)\, dx$

300. $\int (\sin^{2m} ax)\, dx = -\frac{\cos ax}{a} \sum_{r=0}^{m-1} \frac{(2m)!(r!)^2}{2^{2m-2r}(2r+1)!(m!)^2} \sin^{2r+1} ax + \frac{(2m)!}{2^{2m}(m!)^2} x$

301. $\int (\sin^{2m+1} ax)\, dx = -\frac{\cos ax}{a} \sum_{r=0}^{m} \frac{2^{2m-2r}(m!)^2(2r)!}{(2m+1)!(r!)^2} \sin^{2r} ax$

302. $\int (\cos^2 ax)\, dx = \frac{1}{2a} \sin ax \cos ax + \frac{1}{2} x = \frac{1}{2} x + \frac{1}{4a} \sin 2ax$

303. $\int (\cos^3 ax)\, dx = \frac{1}{3a}(\sin ax)(\cos^2 ax + 2)$

304. $\int (\cos^4 ax)\, dx = \frac{3x}{8} + \frac{\sin 2ax}{4a} + \frac{\sin 4ax}{32a}$

305. $\int (\cos^n ax)\, dx = \frac{1}{na} \cos^{n-1} ax \sin ax + \frac{n-1}{n} \int (\cos^{n-2} ax)\, dx$

306. $\int (\cos^{2m} ax)\, dx = \frac{\sin ax}{a} \sum_{r=0}^{m-1} \frac{(2m)!(r!)^2}{2^{2m-2r}(2r+1)!(m!)^2} \cos^{2r+1} ax + \frac{(2m)!}{2^{2m}(m!)^2} x$

307. $\int (\cos^{2m+1} ax)\, dx = \frac{\sin ax}{a} \sum_{r=0}^{m} \frac{2^{2m-2r}(m!)^2(2r)!}{(2m+1)!(r!)^2} \cos^{2r} ax$

308. $\int \frac{dx}{\sin^2 ax} = \int (\csc^2 ax)\, dx = -\frac{1}{a} \cot ax$

309. $\int \frac{dx}{\sin^m ax} = \int (\csc^m ax)\, dx = -\frac{1}{(m-1)a} \cdot \frac{\cos ax}{\sin^{m-1} ax} + \frac{m-2}{m-1} \int \frac{dx}{\sin^{m-2} ax}$

310. $\int \frac{dx}{\sin^{2m} ax} = \int (\csc^{2m} ax)\, dx = -\frac{1}{a} \cos ax \sum_{r=0}^{m-1} \frac{2^{2m-2r-1}(m-1)!m!(2r)!}{(2m)!(r!)^2 \sin^{2r+1} ax}$

311. $\int \frac{dx}{\sin^{2m+1} ax} = \int (\csc^{2m+1} ax)\, dx =$

$$-\frac{1}{a} \cos ax \sum_{r=0}^{m-1} \frac{(2m)!(r!)^2}{2^{2m-2r}(m!)^2(2r+1)! \sin^{2r+2} ax} + \frac{1}{a} \cdot \frac{(2m)!}{2^{2m}(m!)^2} \log \tan \frac{ax}{2}$$

312. $\int \frac{dx}{\cos^2 ax} = \int (\sec^2 ax)\, dx = \frac{1}{a} \tan ax$

313. $\int \frac{dx}{\cos^n ax} = \int (\sec^n ax)\, dx = \frac{1}{(n-1)a} \cdot \frac{\sin ax}{\cos^{n-1} ax} + \frac{n-2}{n-1} \int \frac{dx}{\cos^{n-2} ax}$

314. $\int \frac{dx}{\cos^{2m} ax} = \int (\sec^{2m} ax)\, dx = \frac{1}{a} \sin ax \sum_{r=0}^{m-1} \frac{2^{2m-2r-1}(m-1)!m!(2r)!}{(2m)!(r!)^2 \cos^{2r+1} ax}$

315. $\int \frac{dx}{\cos^{2m+1} ax} = \int (\sec^{2m+1} ax)\, dx =$

$$\frac{1}{a} \sin ax \sum_{r=0}^{m-1} \frac{(2m)!(r!)^2}{2^{2m-2r}(m!)^2(2r+1)! \cos^{2r+2} ax}$$

$$+ \frac{1}{a} \cdot \frac{(2m)!}{2^{2m}(m!)^2} \log (\sec ax + \tan ax)$$

316. $\int (\sin mx)(\sin nx)\, dx = \frac{\sin(m-n)x}{2(m-n)} - \frac{\sin(m+n)x}{2(m+n)}, \quad (m^2 \neq n^2)$

317. $\int (\cos mx)(\cos nx)\, dx = \frac{\sin(m-n)x}{2(m-n)} + \frac{\sin(m+n)x}{2(m+n)}, \quad (m^2 \neq n^2)$

318. $\int (\sin ax)(\cos ax)\, dx = \frac{1}{2a} \sin^2 ax$

319. $\int (\sin mx)(\cos nx)\, dx = -\dfrac{\cos(m-n)x}{2(m-n)} - \dfrac{\cos(m+n)x}{2(m+n)}, \qquad (m^2 \neq n^2)$

320. $\int (\sin^2 ax)(\cos^2 ax)\, dx = -\dfrac{1}{32a}\sin 4ax + \dfrac{x}{8}$

321. $\int (\sin ax)(\cos^m ax)\, dx = -\dfrac{\cos^{m+1} ax}{(m+1)a}$

322. $\int (\sin^m ax)(\cos ax)\, dx = \dfrac{\sin^{m+1} ax}{(m+1)a}$

323. $\int (\cos^m ax)(\sin^n ax)\, dx = \begin{cases} \dfrac{\cos^{m-1} ax\, \sin^{n+1} ax}{(m+n)a} \\[2ex] \qquad + \dfrac{m-1}{m+n}\displaystyle\int (\cos^{m-2} ax)(\sin^n ax)\, dx \\[2ex] \text{or} \\[2ex] -\dfrac{\sin^{n-1} ax\, \cos^{m+1} ax}{(m+n)a} \\[2ex] \qquad + \dfrac{n-1}{m+n}\displaystyle\int (\cos^m ax)(\sin^{n-2} ax)\, dx \end{cases}$

324. $\int \dfrac{\cos^m ax}{\sin^n ax}\, dx = \begin{cases} -\dfrac{\cos^{m+1} ax}{(n-1)a\,\sin^{n-1} ax} - \dfrac{m-n+2}{n-1}\displaystyle\int \dfrac{\cos^m ax}{\sin^{n-2} ax}\, dx \\[2ex] \text{or} \\[2ex] \dfrac{\cos^{m-1} ax}{a(m-n)\sin^{n-1} ax} + \dfrac{m-1}{m-n}\displaystyle\int \dfrac{\cos^{m-2} ax}{\sin^n ax}\, dx \end{cases}$

325. $\int \dfrac{\sin^m ax}{\cos^n ax}\, dx = \begin{cases} \dfrac{\sin^{m+1} ax}{a(n-1)\cos^{n-1} ax} - \dfrac{m-n+2}{n-1}\displaystyle\int \dfrac{\sin^m ax}{\cos^{n-2} ax}\, dx \\[2ex] \text{or} \\[2ex] -\dfrac{\sin^{m-1} ax}{a(m-n)\cos^{n-1} ax} + \dfrac{m-1}{m-n}\displaystyle\int \dfrac{\sin^{m-2} ax}{\cos^n ax}\, dx \end{cases}$

326. $\int \dfrac{\sin ax}{\cos^2 ax}\, dx = \dfrac{1}{a\cos ax} = \dfrac{\sec ax}{a}$

327. $\int \dfrac{\sin^2 ax}{\cos ax}\, dx = -\dfrac{1}{a}\sin ax + \dfrac{1}{a}\log\tan\left(\dfrac{\pi}{4} + \dfrac{ax}{2}\right)$

328. $\int \dfrac{\cos ax}{\sin^2 ax}\, dx = -\dfrac{1}{a\sin ax} = -\dfrac{\csc ax}{a}$

329. $\int \dfrac{dx}{(\sin ax)(\cos ax)} = \dfrac{1}{a}\log\tan ax$

330. $\int \dfrac{dx}{(\sin ax)(\cos^2 ax)} = \dfrac{1}{a}\left(\sec ax + \log\tan\dfrac{ax}{2}\right)$

331. $\int \dfrac{dx}{(\sin ax)(\cos^n ax)} = \dfrac{1}{a(n-1)\cos^{n-1} ax} + \int \dfrac{dx}{(\sin ax)(\cos^{n-2} ax)}$

332. $\int \dfrac{dx}{(\sin^2 ax)(\cos ax)} = -\dfrac{1}{a}\csc ax + \dfrac{1}{a}\log\tan\left(\dfrac{\pi}{4} + \dfrac{ax}{2}\right)$

333. $\int \dfrac{dx}{(\sin^2 ax)(\cos^2 ax)} = -\dfrac{2}{a}\cot 2ax$

334. $\displaystyle\int \frac{dx}{\sin^m ax \cos^n ax} =$
$\begin{cases} -\dfrac{1}{a(m-1)(\sin^{m-1} ax)(\cos^{n-1} ax)} \\[3mm] \qquad + \dfrac{m+n-2}{m-1} \displaystyle\int \dfrac{dx}{(\sin^{m-2} ax)(\cos^n ax)} \\[3mm] \text{or} \\[3mm] \dfrac{1}{a(n-1)\sin^{m-1} ax \cos^{n-1} ax} \\[3mm] \qquad - \dfrac{m+n-2}{n-1} \displaystyle\int \dfrac{dx}{\sin^m ax \cos^{n-2} ax} \end{cases}$

335. $\displaystyle\int \sin(a+bx)\,dx = -\frac{1}{b}\cos(a+bx)$

336. $\displaystyle\int \cos(a+bx)\,dx = \frac{1}{b}\sin(a+bx)$

337. $\displaystyle\int \frac{dx}{1 \pm \sin ax} = \mp \frac{1}{a}\tan\left(\frac{\pi}{4} \mp \frac{ax}{2}\right)$

338. $\displaystyle\int \frac{dx}{1 + \cos ax} = \frac{1}{a}\tan\frac{ax}{2}$

339. $\displaystyle\int \frac{dx}{1 - \cos ax} = -\frac{1}{a}\cot\frac{ax}{2}$

*340. $\displaystyle\int \frac{dx}{a + b\sin x} =$
$\begin{cases} \dfrac{2}{\sqrt{a^2 - b^2}}\tan^{-1}\dfrac{a\tan\frac{x}{2} + b}{\sqrt{a^2 - b^2}} \\[5mm] \text{or} \\[3mm] \dfrac{1}{\sqrt{b^2 - a^2}}\log\dfrac{a\tan\frac{x}{2} + b - \sqrt{b^2 - a^2}}{a\tan\frac{x}{2} + b + \sqrt{b^2 - a^2}} \end{cases}$

*341. $\displaystyle\int \frac{dx}{a + b\cos x} =$
$\begin{cases} \dfrac{2}{\sqrt{a^2 - b^2}}\tan^{-1}\dfrac{\sqrt{a^2 - b^2}\tan\frac{x}{2}}{a + b} \\[5mm] \text{or} \\[3mm] \dfrac{1}{\sqrt{b^2 - a^2}}\log\left(-\dfrac{\sqrt{b^2 - a^2}\tan\frac{x}{2} + a + b}{\sqrt{b^2 - a^2}\tan\frac{x}{2} - a - b}\right) \end{cases}$

*342. $\displaystyle\int \frac{dx}{a + b\sin x + c\cos x}$

$= \begin{cases} \dfrac{1}{\sqrt{b^2 + c^2 - a^2}}\log \dfrac{b - \sqrt{b^2 + c^2 - a^2} + (a-c)\tan\frac{x}{2}}{b + \sqrt{b^2 + c^2 - a^2} + (a-c)\tan\frac{x}{2}}, \quad \text{if } a^2 < b^2 + c^2, a \neq c \\[6mm] \text{or} \\[3mm] \dfrac{2}{\sqrt{a^2 - b^2 - c^2}}\tan^{-1}\dfrac{b + (a-c)\tan\frac{x}{2}}{\sqrt{a^2 - b^2 - c^2}}, \quad \text{if } a^2 > b^2 + c^2 \\[6mm] \text{or} \\[3mm] \dfrac{1}{a}\left[\dfrac{a - (b+c)\cos x - (b-c)\sin x}{a - (b-c)\cos x + (b+c)\sin x}\right], \quad \text{if } a^2 = b^2 + c^2, a \neq c. \end{cases}$

*See note 6 on page A-19.

*343. $\displaystyle\int \frac{\sin^2 x\, dx}{a + b\cos^2 x} = \frac{1}{b}\sqrt{\frac{a + b}{a}} \tan^{-1}\left(\sqrt{\frac{a}{a + b}} \tan x\right) - \frac{x}{b}$, $(ab > 0,\ \text{or } |a| > |b|)$

*344. $\displaystyle\int \frac{dx}{a^2 \cos^2 x + b^2 \sin^2 x} = \frac{1}{ab} \tan^{-1}\left(\frac{b \tan x}{a}\right)$

*345. $\displaystyle\int \frac{\cos^2 cx}{a^2 + b^2 \sin^2 cx}\, dx = \frac{\sqrt{a^2 + b^2}}{ab^2 c} \tan^{-1}\frac{\sqrt{a^2 + b^2}\, \tan cx}{a} - \frac{x}{b^2}$

346. $\displaystyle\int \frac{\sin cx \cos cx}{a \cos^2 cx + b \sin^2 cx}\, dx = \frac{1}{2c(b - a)} \log(a \cos^2 cx + b \sin^2 cx)$

347. $\displaystyle\int \frac{\cos cx}{a \cos cx + b \sin cx}\, dx = \int \frac{dx}{a + b \tan cx} =$
$$\frac{1}{c(a^2 + b^2)}[acx + b \log(a \cos cx + b \text{ si}$$

348. $\displaystyle\int \frac{\sin cx}{a \sin cx + b \cos cx}\, dx = \int \frac{dx}{a + b \cot cx} =$
$$\frac{1}{c(a^2 + b^2)}[acx - b \log(a \sin cx + b \cos x)]$$

*349. $\displaystyle\int \frac{dx}{a \cos^2 x + 2b \cos x \sin x + c \sin^2 x} =$
$$\begin{cases} \dfrac{1}{2\sqrt{b^2 - ac}} \log \dfrac{c \tan x + b - \sqrt{b^2 - ac}}{c \tan x + b + \sqrt{b^2 - ac}} \\ \qquad\qquad (b^2 > ac) \\[6pt] \text{or} \\[6pt] \dfrac{1}{\sqrt{ac - b^2}} \tan^{-1}\dfrac{c \tan x + b}{\sqrt{ac - b^2}}, \quad (b_2 < ac) \\[6pt] \text{or} \\[6pt] -\dfrac{1}{c \tan x + b}, \qquad (b^2 = ac) \end{cases}$$

350. $\displaystyle\int \frac{\sin ax}{1 \pm \sin ax}\, dx = \pm x + \frac{1}{a} \tan\left(\frac{\pi}{4} \mp \frac{ax}{2}\right)$

351. $\displaystyle\int \frac{dx}{(\sin ax)(1 \pm \sin ax)} = \frac{1}{a} \tan\left(\frac{\pi}{4} \mp \frac{ax}{2}\right) + \frac{1}{a} \log \tan \frac{ax}{2}$

352. $\displaystyle\int \frac{dx}{(1 + \sin ax)^2} = -\frac{1}{2a} \tan\left(\frac{\pi}{4} - \frac{ax}{2}\right) - \frac{1}{6a} \tan^3\left(\frac{\pi}{4} - \frac{ax}{2}\right)$

353. $\displaystyle\int \frac{dx}{(1 - \sin ax)^2} = \frac{1}{2a} \cot\left(\frac{\pi}{4} - \frac{ax}{2}\right) + \frac{1}{6a} \cot^3\left(\frac{\pi}{4} - \frac{ax}{2}\right)$

354. $\displaystyle\int \frac{\sin ax}{(1 + \sin ax)^2}\, dx = -\frac{1}{2a} \tan\left(\frac{\pi}{4} - \frac{ax}{2}\right) + \frac{1}{6a} \tan^3\left(\frac{\pi}{4} - \frac{ax}{2}\right)$

355. $\displaystyle\int \frac{\sin ax}{(1 - \sin ax)^2}\, dx = -\frac{1}{2a} \cot\left(\frac{\pi}{4} - \frac{ax}{2}\right) + \frac{1}{6a} \cot^3\left(\frac{\pi}{4} - \frac{ax}{2}\right)$

356. $\displaystyle\int \frac{\sin x\, dx}{a + b \sin x} = \frac{x}{b} - \frac{a}{b}\int \frac{dx}{a + b \sin x}$

357. $\displaystyle\int \frac{dx}{(\sin x)(a + b \sin x)} = \frac{1}{a} \log \tan \frac{x}{2} - \frac{b}{a}\int \frac{dx}{a + b \sin x}$

358. $\displaystyle\int \frac{dx}{(a + b \sin x)^2} = \frac{b \cos x}{(a^2 - b^2)(a + b \sin x)} + \frac{a}{a^2 - b^2}\int \frac{dx}{a + b \sin x}$

*See note 6 on page A-19.

359. $\displaystyle\int \frac{\sin x \, dx}{(a + b \sin x)^2} = \frac{a \cos x}{(b^2 - a^2)(a + b \sin x)} + \frac{b}{b^2 - a^2} \int \frac{dx}{a + b \sin x}$

*360. $\displaystyle\int \frac{dx}{a^2 + b^2 \sin^2 cx} = \frac{1}{ac\sqrt{a^2 + b^2}} \tan^{-1} \frac{\sqrt{a^2 + b^2} \tan cx}{a}$

*361. $\displaystyle\int \frac{dx}{a^2 - b^2 \sin^2 cx} = \begin{cases} \dfrac{1}{ac\sqrt{a^2 - b^2}} \tan^{-1} \dfrac{\sqrt{a^2 - b^2} \tan cx}{a}, & (a^2 > b^2) \\[2mm] \text{or} \\[2mm] \dfrac{1}{2ac\sqrt{b^2 - a^2}} \log \dfrac{\sqrt{b^2 - a^2} \tan cx + a}{\sqrt{b^2 - a^2} \tan cx - a}, & (a^2 < b^2) \end{cases}$

362. $\displaystyle\int \frac{\cos ax}{1 + \cos ax} \, dx = x - \frac{1}{a} \tan \frac{ax}{2}$

363. $\displaystyle\int \frac{\cos ax}{1 - \cos ax} \, dx = -x - \frac{1}{a} \cot \frac{ax}{2}$

364. $\displaystyle\int \frac{dx}{(\cos ax)(1 + \cos ax)} = \frac{1}{a} \log \tan \left(\frac{\pi}{4} + \frac{ax}{2}\right) - \frac{1}{a} \tan \frac{ax}{2}$

365. $\displaystyle\int \frac{dx}{(\cos ax)(1 - \cos ax)} = \frac{1}{a} \log \tan \left(\frac{\pi}{4} + \frac{ax}{2}\right) - \frac{1}{a} \cot \frac{ax}{2}$

366. $\displaystyle\int \frac{dx}{(1 + \cos ax)^2} = \frac{1}{2a} \tan \frac{ax}{2} + \frac{1}{6a} \tan^3 \frac{ax}{2}$

367. $\displaystyle\int \frac{dx}{(1 - \cos ax)^2} = -\frac{1}{2a} \cot \frac{ax}{2} - \frac{1}{6a} \cot^3 \frac{ax}{2}$

368. $\displaystyle\int \frac{\cos ax}{(1 + \cos ax)^2} \, dx = \frac{1}{2a} \tan \frac{ax}{2} - \frac{1}{6a} \tan^3 \frac{ax}{2}$

369. $\displaystyle\int \frac{\cos ax}{(1 - \cos ax)^2} \, dx = \frac{1}{2a} \cot \frac{ax}{2} - \frac{1}{6a} \cot^3 \frac{ax}{2}$

370. $\displaystyle\int \frac{\cos x \, dx}{a + b \cos x} = \frac{x}{b} - \frac{a}{b} \int \frac{dx}{a + b \cos x}$

371. $\displaystyle\int \frac{dx}{(\cos x)(a + b \cos x)} = \frac{1}{a} \log \tan \left(\frac{x}{2} + \frac{\pi}{4}\right) - \frac{b}{a} \int \frac{dx}{a + b \cos x}$

372. $\displaystyle\int \frac{dx}{(a + b \cos x)^2} = \frac{b \sin x}{(b^2 - a^2)(a + b \cos x)} - \frac{a}{b^2 - a^2} \int \frac{dx}{a + b \cos x}$

373. $\displaystyle\int \frac{\cos x}{(a + b \cos x)^2} \, dx = \frac{a \sin x}{(a^2 - b^2)(a + b \cos x)} - \frac{b}{a^2 - b^2} \int \frac{dx}{a + b \cos x}$

*374. $\displaystyle\int \frac{dx}{a^2 + b^2 - 2ab \cos cx} = \frac{2}{c(a^2 - b^2)} \tan^{-1}\left(\frac{a + b}{a - b} \tan \frac{cx}{2}\right)$

*375. $\displaystyle\int \frac{dx}{a^2 + b^2 \cos^2 cx} = \frac{1}{ac\sqrt{a^2 + b^2}} \tan^{-1} \frac{a \tan cx}{\sqrt{a^2 + b^2}}$

*376. $\displaystyle\int \frac{dx}{a^2 - b^2 \cos^2 cx} = \begin{cases} \dfrac{1}{ac\sqrt{a^2 - b^2}} \tan^{-1} \dfrac{a \tan cx}{\sqrt{a^2 - b^2}}, & (a^2 > b^2) \\[2mm] \text{or} \\[2mm] \dfrac{1}{2ac\sqrt{b^2 - a^2}} \log \dfrac{a \tan cx - \sqrt{b^2 - a^2}}{a \tan cx + \sqrt{b^2 - a^2}}, & (b^2 > a^2) \end{cases}$

377. $\displaystyle\int \frac{\sin ax}{1 \pm \cos ax} \, dx = \mp \frac{1}{a} \log (1 \pm \cos ax)$

*See note 6 on page A-19.

378. $\int \dfrac{\cos ax}{1 \pm \sin ax}\, dx = \pm \dfrac{1}{a} \log (1 \pm \sin ax)$

379. $\int \dfrac{dx}{(\sin ax)(1 \pm \cos ax)} = \pm \dfrac{1}{2a(1 \pm \cos ax)} + \dfrac{1}{2a} \log \tan \dfrac{ax}{2}$

380. $\int \dfrac{dx}{(\cos ax)(1 \pm \sin ax)} = \mp \dfrac{1}{2a(1 \pm \sin ax)} + \dfrac{1}{2a} \log \tan \left(\dfrac{\pi}{4} + \dfrac{ax}{2} \right)$

381. $\int \dfrac{\sin ax}{(\cos ax)(1 \pm \cos ax)}\, dx = \dfrac{1}{a} \log (\sec ax \pm 1)$

382. $\int \dfrac{\cos ax}{(\sin ax)(1 \pm \sin ax)}\, dx = -\dfrac{1}{a} \log (\csc ax \pm 1)$

383. $\int \dfrac{\sin ax}{(\cos ax)(1 \pm \sin ax)}\, dx = \dfrac{1}{2a(1 \pm \sin ax)} \pm \dfrac{1}{2a} \log \tan \left(\dfrac{\pi}{4} + \dfrac{ax}{2} \right)$

384. $\int \dfrac{\cos ax}{(\sin ax)(1 \pm \cos ax)}\, dx = -\dfrac{1}{2a(1 \pm \cos ax)} \pm \dfrac{1}{2a} \log \tan \dfrac{ax}{2}$

385. $\int \dfrac{dx}{\sin ax \pm \cos ax} = \dfrac{1}{a\sqrt{2}} \log \tan \left(\dfrac{ax}{2} \pm \dfrac{\pi}{8} \right)$

386. $\int \dfrac{dx}{(\sin ax \pm \cos ax)^2} = \dfrac{1}{2a} \tan \left(ax \mp \dfrac{\pi}{4} \right)$

387. $\int \dfrac{dx}{1 + \cos ax \pm \sin ax} = \pm \dfrac{1}{a} \log \left(1 \pm \tan \dfrac{ax}{2} \right)$

388. $\int \dfrac{dx}{a^2 \cos^2 cx - b^2 \sin^2 cx} = \dfrac{1}{2abc} \log \dfrac{b \tan cx + a}{b \tan cx - a}$

389. $\int x(\sin ax)\, dx = \dfrac{1}{a^2} \sin ax - \dfrac{x}{a} \cos ax$

390. $\int x^2(\sin ax)\, dx = \dfrac{2x}{a^2} \sin ax - \dfrac{a^2 x^2 - 2}{a^3} \cos ax$

391. $\int x^3(\sin ax)\, dx = \dfrac{3a^2 x^2 - 6}{a^4} \sin ax - \dfrac{a^2 x^3 - 6x}{a^3} \cos ax$

392. $\int x^m \sin ax \, dx = \begin{cases} -\dfrac{1}{a} x^m \cos ax + \dfrac{m}{a} \displaystyle\int x^{m-1} \cos ax \, dx \\[2mm] \text{or} \\[2mm] \cos ax \displaystyle\sum_{r=0}^{\left[\frac{m}{2}\right]} (-1)^{r+1} \dfrac{m!}{(m-2r)!} \cdot \dfrac{x^{m-2r}}{a^{2r+1}} \\[2mm] + \sin ax \displaystyle\sum_{r=0}^{\left[\frac{m-1}{2}\right]} (-1)^r \dfrac{m!}{(m-2r-1)!} \cdot \dfrac{x^{m-2r-1}}{a^{2r+2}} \end{cases}$

Note: $[s]$ means greatest integer $\leq s$; $[3\frac{1}{2}] = 3$, $[\frac{1}{2}] = 0$, etc.

393. $\int x(\cos ax)\, dx = \dfrac{1}{a^2} \cos ax + \dfrac{x}{a} \sin ax$

394. $\int x^2(\cos ax)\, dx = \dfrac{2x \cos ax}{a^2} + \dfrac{a^2 x^2 - 2}{a^3} \sin ax$

395. $\int x^3(\cos ax)\, dx = \dfrac{3a^2 x^2 - 6}{a^4} \cos ax + \dfrac{a^2 x^3 - 6x}{a^3} \sin ax$

396. $\int x^m(\cos ax)\,dx = \begin{cases} \dfrac{x^m \sin ax}{a} - \dfrac{m}{a}\int x^{m-1}\sin ax\,dx \\ \qquad\text{or} \\ \sin ax \displaystyle\sum_{r=0}^{\left[\frac{m}{2}\right]} (-1)^r \dfrac{m!}{(m-2r)!}\cdot\dfrac{x^{m-2r}}{a^{2r+1}} \\ \qquad + \cos ax \displaystyle\sum_{r=0}^{\left[\frac{m-1}{2}\right]} (-1)^r \dfrac{m!}{(m-2r-1)!}\cdot\dfrac{x^{m-2r-1}}{a^{2r+2}} \end{cases}$

See note integral 392.

397. $\int \dfrac{\sin ax}{x}\,dx = \displaystyle\sum_{n=0}^{\prime} (-1)^n \dfrac{(ax)^{2n+1}}{(2n+1)(2n+1)!}$

398. $\int \dfrac{\cos ax}{x}\,dx = \log x + \displaystyle\sum_{n=1}^{\prime} (-1)^n \dfrac{(ax)^{2n}}{2n(2n)!}$

399. $\int x(\sin^2 ax)\,dx = \dfrac{x^2}{4} - \dfrac{x\sin 2ax}{4a} - \dfrac{\cos 2ax}{8a^2}$

400. $\int x^2(\sin^2 ax)\,dx = \dfrac{x^3}{6} - \left(\dfrac{x^2}{4a} - \dfrac{1}{8a^3}\right)\sin 2ax - \dfrac{x\cos 2ax}{4a^2}$

401. $\int x(\sin^3 ax)\,dx = \dfrac{x\cos 3ax}{12a} - \dfrac{\sin 3ax}{36a^2} - \dfrac{3x\cos ax}{4a} + \dfrac{3\sin ax}{4a^2}$

402. $\int x(\cos^2 ax)\,dx = \dfrac{x^2}{4} + \dfrac{x\sin 2ax}{4a} + \dfrac{\cos 2ax}{8a^2}$

403. $\int x^2(\cos^2 ax)\,dx = \dfrac{x^3}{6} + \left(\dfrac{x^2}{4a} - \dfrac{1}{8a^3}\right)\sin 2ax + \dfrac{x\cos 2ax}{4a^2}$

404. $\int x(\cos^3 ax)\,dx = \dfrac{x\sin 3ax}{12a} + \dfrac{\cos 3ax}{36a^2} + \dfrac{3x\sin ax}{4a} + \dfrac{3\cos ax}{4a^2}$

405. $\int \dfrac{\sin ax}{x^m}\,dx = -\dfrac{\sin ax}{(m-1)x^{m-1}} + \dfrac{a}{m-1}\int \dfrac{\cos ax}{x^{m-1}}\,dx$

406. $\int \dfrac{\cos ax}{x^m}\,dx = -\dfrac{\cos ax}{(m-1)x^{m-1}} - \dfrac{a}{m-1}\int \dfrac{\sin ax}{x^{m-1}}\,dx$

407. $\int \dfrac{x}{1 \pm \sin ax}\,dx = \mp\dfrac{x\cos ax}{a(1 \pm \sin ax)} + \dfrac{1}{a^2}\log(1 \pm \sin ax)$

408. $\int \dfrac{x}{1 + \cos ax}\,dx = \dfrac{x}{a}\tan\dfrac{ax}{2} + \dfrac{2}{a^2}\log\cos\dfrac{ax}{2}$

409. $\int \dfrac{x}{1 - \cos ax}\,dx = -\dfrac{x}{a}\cot\dfrac{ax}{2} + \dfrac{2}{a^2}\log\sin\dfrac{ax}{2}$

410. $\int \dfrac{x + \sin x}{1 + \cos x}\,dx = x\tan\dfrac{x}{2}$

411. $\int \dfrac{x - \sin x}{1 - \cos x}\,dx = -x\cot\dfrac{x}{2}$

412. $\int \sqrt{1 - \cos ax}\,dx = -\dfrac{2\sin ax}{a\sqrt{1 - \cos ax}} \quad = -\dfrac{2\sqrt{2}}{a}\cos\left(\dfrac{ax}{2}\right)$

413. $\int \sqrt{1 + \cos ax}\,dx = \dfrac{2\sin ax}{a\sqrt{1 + \cos ax}} \quad = \dfrac{2\sqrt{2}}{a}\sin\left(\dfrac{ax}{2}\right)$

414. $\int \sqrt{1 + \sin x}\,dx = \pm 2\left(\sin\dfrac{x}{2} - \cos\dfrac{x}{2}\right),$

$\left[\text{use } + \text{ if } (8k-1)\dfrac{\pi}{2} < x \le (8k+3)\dfrac{\pi}{2}, \text{ otherwise } -; \ k \text{ an integer}\right]$

415. $\int \sqrt{1 - \sin x}\, dx = \pm 2\left(\sin\dfrac{x}{2} + \cos\dfrac{x}{2}\right),$

[use $+$ if $(8k - 3)\dfrac{\pi}{2} < x \leq (8k + 1)\dfrac{\pi}{2}$, otherwise $-$; k an integer]

416. $\int \dfrac{dx}{\sqrt{1 - \cos x}} = \pm \sqrt{2}\, \log \tan\dfrac{x}{4},$

[use $+$ if $4k\pi < x < (4k + 2)\pi$, otherwise $-$; k an integer]

417. $\int \dfrac{dx}{\sqrt{1 + \cos x}} = \pm \sqrt{2}\, \log \tan\left(\dfrac{x + \pi}{4}\right),$

[use $+$ if $(4k - 1)\pi < x < (4k + 1)\pi$, otherwise $-$; k an integer]

418. $\int \dfrac{dx}{\sqrt{1 - \sin x}} = \pm \sqrt{2}\, \log \tan\left(\dfrac{x}{4} - \dfrac{\pi}{8}\right),$

[use $+$ if $(8k + 1)\dfrac{\pi}{2} < x < (8k + 5)\dfrac{\pi}{2}$, otherwise $-$; k an integer]

419. $\int \dfrac{dx}{\sqrt{1 + \sin x}} = \pm \sqrt{2}\, \log \tan\left(\dfrac{x}{4} + \dfrac{\pi}{8}\right),$

[use $+$ if $(8k - 1)\dfrac{\pi}{2} < x < (8k + 3)\dfrac{\pi}{2}$, otherwise $-$; k an integer]

420. $\int (\tan^2 ax)\, dx = \dfrac{1}{a}\tan ax - x$

421. $\int (\tan^3 ax)\, dx = \dfrac{1}{2a}\tan^2 ax + \dfrac{1}{a}\log \cos ax$

422. $\int (\tan^4 ax)\, dx = \dfrac{\tan^3 ax}{3a} - \dfrac{1}{a}\tan x + x$

423. $\int (\tan^n ax)\, dx = \dfrac{\tan^{n-1} ax}{a(n - 1)} - \int (\tan^{n-2} ax)\, dx$

424. $\int (\cot^2 ax)\, dx = -\dfrac{1}{a}\cot ax - x$

425. $\int (\cot^3 ax)\, dx = -\dfrac{1}{2a}\cot^2 ax - \dfrac{1}{a}\log \sin ax$

426. $\int (\cot^4 ax)\, dx = -\dfrac{1}{3a}\cot^3 ax + \dfrac{1}{a}\cot ax + x$

427. $\int (\cot^n ax)\, dx = -\dfrac{\cot^{n-1} ax}{a(n - 1)} - \int (\cot^{n-2} ax)\, dx$

428. $\int \dfrac{x}{\sin^2 ax}\, dx = \int x(\csc^2 ax)\, dx = -\dfrac{x \cot ax}{a} + \dfrac{1}{a^2}\log \sin ax$

429. $\int \dfrac{x}{\sin^n ax}\, dx = \int x(\csc^n ax)\, dx = -\dfrac{x \cos ax}{a(n - 1)\sin^{n-1} ax}$

$$-\dfrac{1}{a^2(n - 1)(n - 2)\sin^{n-2} ax} + \dfrac{(n - 2)}{(n - 1)}\int \dfrac{x}{\sin^{n-2} ax}\, dx$$

430. $\int \dfrac{x}{\cos^2 ax}\, dx = \int x(\sec^2 ax)\, dx = \dfrac{1}{a}x \tan ax + \dfrac{1}{a^2}\log \cos ax$

431. $\int \dfrac{x}{\cos^n ax}\, dx = \int x(\sec^n ax)\, dx = \dfrac{x \sin ax}{a(n - 1)\cos^{n-1} ax}$

$$-\dfrac{1}{a^2(n - 1)(n - 2)\cos^{n-2} ax} + \dfrac{n - 2}{n - 1}\int \dfrac{x}{\cos^{n-2} ax}\, dx$$

432. $\displaystyle\int \frac{\sin ax}{\sqrt{1 + b^2 \sin^2 ax}}\, dx = -\frac{1}{ab}\sin^{-1}\frac{b\cos ax}{\sqrt{1 + b^2}}$

433. $\displaystyle\int \frac{\sin ax}{\sqrt{1 - b^2 \sin^2 ax}}\, dx = -\frac{1}{ab}\log\left(b\cos ax + \sqrt{1 - b^2 \sin^2 ax}\right)$

434. $\displaystyle\int (\sin ax)\sqrt{1 + b^2 \sin^2 ax}\; dx = -\frac{\cos ax}{2a}\sqrt{1 + b^2 \sin^2 ax}$

$$-\frac{1 + b^2}{2ab}\sin^{-1}\frac{b\cos ax}{\sqrt{1 + b^2}}$$

435. $\displaystyle\int (\sin ax)\sqrt{1 - b^2 \sin^2 ax}\; dx = -\frac{\cos ax}{2a}\sqrt{1 - b^2 \sin^2 ax}$

$$-\frac{1 - b^2}{2ab}\log\left(b\cos ax + \sqrt{1 - b^2 \sin^2 ax}\right)$$

436. $\displaystyle\int \frac{\cos ax}{\sqrt{1 + b^2 \sin^2 ax}}\, dx = \frac{1}{ab}\log\left(b\sin ax + \sqrt{1 + b^2 \sin^2 ax}\right)$

437. $\displaystyle\int \frac{\cos ax}{\sqrt{1 - b^2 \sin^2 ax}}\, dx = \frac{1}{ab}\sin^{-1}(b\sin ax)$

438. $\displaystyle\int (\cos ax)\sqrt{1 + b^2 \sin^2 ax}\; dx = \frac{\sin ax}{2a}\sqrt{1 + b^2 \sin^2 ax}$

$$+\frac{1}{2ab}\log\left(b\sin ax + \sqrt{1 + b^2 \sin^2 ax}\right)$$

439. $\displaystyle\int (\cos ax)\sqrt{1 - b^2 \sin^2 ax}\; dx = \frac{\sin ax}{2a}\sqrt{1 - b^2 \sin^2 ax} + \frac{1}{2ab}\sin^{-1}(b\sin ax)$

440. $\displaystyle\int \frac{dx}{\sqrt{a + b\tan^2 cx}} = \frac{\pm 1}{c\sqrt{a - b}}\sin^{-1}\left(\sqrt{\frac{a - b}{a}}\sin cx\right), \qquad (a > |b|)$

$$\left[\text{use } + \text{ if } (2k - 1)\frac{\pi}{2} < x \le (2k + 1)\frac{\pi}{2}, \text{ otherwise } -; k \text{ an integer}\right]$$

FORMS INVOLVING INVERSE TRIGONOMETRIC FUNCTIONS

441. $\displaystyle\int (\sin^{-1} ax)\, dx = x\sin^{-1} ax + \frac{\sqrt{1 - a^2 x^2}}{a}$

442. $\displaystyle\int (\cos^{-1} ax)\, dx = x\cos^{-1} ax - \frac{\sqrt{1 - a^2 x^2}}{a}$

443. $\displaystyle\int (\tan^{-1} ax)\, dx = x\tan^{-1} ax - \frac{1}{2a}\log(1 + a^2 x^2)$

444. $\displaystyle\int (\cot^{-1} ax)\, dx = x\cot^{-1} ax + \frac{1}{2a}\log(1 + a^2 x^2)$

445. $\displaystyle\int (\sec^{-1} ax)\, dx = x\sec^{-1} ax - \frac{1}{a}\log(ax + \sqrt{a^2 x^2 - 1})$

446. $\displaystyle\int (\csc^{-1} ax)\, dx = x\csc^{-1} ax + \frac{1}{a}\log(ax + \sqrt{a^2 x^2 - 1})$

447. $\displaystyle\int \left(\sin^{-1}\frac{x}{a}\right) dx = x\sin^{-1}\frac{x}{a} + \sqrt{a^2 - x^2}, \qquad (a > 0)$

448. $\displaystyle\int \left(\cos^{-1}\frac{x}{a}\right) dx = x\cos^{-1}\frac{x}{a} - \sqrt{a^2 - x^2}, \qquad (a > 0)$

449. $\int \left(\tan^{-1} \dfrac{x}{a} \right) dx = x \tan^{-1} \dfrac{x}{a} - \dfrac{a}{2} \log (a^2 + x^2)$

450. $\int \left(\cot^{-1} \dfrac{x}{a} \right) dx = x \cot^{-1} \dfrac{x}{a} + \dfrac{a}{2} \log (a^2 + x^2)$

451. $\int x[\sin^{-1}(ax)] \, dx = \dfrac{1}{4a^2}[(2a^2 x^2 - 1)\sin^{-1}(ax) + ax\sqrt{1 - a^2 x^2}]$

452. $\int x[\cos^{-1}(ax)] \, dx = \dfrac{1}{4a^2}[(2a^2 x^2 - 1)\cos^{-1}(ax) - ax\sqrt{1 - a^2 x^2}]$

453. $\int x^n[\sin^{-1}(ax)] \, dx = \dfrac{x^{n+1}}{n+1} \sin^{-1}(ax) - \dfrac{a}{n+1} \int \dfrac{x^{n+1} \, dx}{\sqrt{1 - a^2 x^2}}, \quad (n \neq -1)$

454. $\int x^n[\cos^{-1}(ax)] \, dx = \dfrac{x^{n+1}}{n+1} \cos^{-1}(ax) + \dfrac{a}{n+1} \int \dfrac{x^{n+1} \, dx}{\sqrt{1 - a^2 x^2}}, \quad (n \neq -1)$

455. $\int x(\tan^{-1} ax) \, dx = \dfrac{1 + a^2 x^2}{2a^2} \tan^{-1} ax - \dfrac{x}{2a}$

456. $\int x^n(\tan^{-1} ax) \, dx = \dfrac{x^{n+1}}{n+1} \tan^{-1} ax - \dfrac{a}{n+1} \int \dfrac{x^{n+1}}{1 + a^2 x^2} \, dx$

457. $\int x(\cot^{-1} ax) \, dx = \dfrac{1 + a^2 x^2}{2a^2} \cot^{-1} ax + \dfrac{x}{2a}$

458. $\int x^n(\cot^{-1} ax) \, dx = \dfrac{x^{n+1}}{n+1} \cot^{-1} ax + \dfrac{a}{n+1} \int \dfrac{x^{n+1}}{1 + a^2 x^2} \, dx$

459. $\int \dfrac{\sin^{-1}(ax)}{x^2} \, dx = a \log \left(\dfrac{1 - \sqrt{1 - a^2 x^2}}{x} \right) - \dfrac{\sin^{-1}(ax)}{x}$

460. $\int \dfrac{\cos^{-1}(ax) \, dx}{x^2} = -\dfrac{1}{x}\cos^{-1}(ax) + a \log \dfrac{1 + \sqrt{1 - a^2 x^2}}{x}$

461. $\int \dfrac{\tan^{-1}(ax) \, dx}{x^2} = -\dfrac{1}{x}\tan^{-1}(ax) - \dfrac{a}{2} \log \dfrac{1 + a^2 x^2}{x^2}$

462. $\int \dfrac{\cot^{-1} ax}{x^2} \, dx = -\dfrac{1}{x}\cot^{-1} ax - \dfrac{a}{2} \log \dfrac{x^2}{a^2 x^2 + 1}$

463. $\int (\sin^{-1} ax)^2 \, dx = x(\sin^{-1} ax)^2 - 2x + \dfrac{2\sqrt{1 - a^2 x^2}}{a} \sin^{-1} ax$

464. $\int (\cos^{-1} ax)^2 \, dx = x(\cos^{-1} ax)^2 - 2x - \dfrac{2\sqrt{1 - a^2 x^2}}{a} \cos^{-1} ax$

465. $\int (\sin^{-1} ax)^n \, dx = \begin{cases} x(\sin^{-1} ax)^n + \dfrac{n\sqrt{1 - a^2 x^2}}{a}(\sin^{-1} ax)^{n-1} \\ \qquad\qquad\qquad - n(n-1) \int (\sin^{-1} ax)^{n-2} \, dx \\ \qquad\qquad \text{or} \\ \displaystyle\sum_{r=0}^{\left[\frac{n}{2}\right]} (-1)^r \dfrac{n!}{(n-2r)!} x(\sin^{-1} ax)^{n-2r} \\ \qquad + \displaystyle\sum_{r=0}^{\left[\frac{n-1}{2}\right]} (-1)^r \dfrac{n!\sqrt{1 - a^2 x^2}}{(n-2r-1)!a}(\sin^{-1} ax)^{n-2r-1} \end{cases}$

Note : $[s]$ means greatest integer $\leq s$. Thus $[3.5]$ means $3 : [5] = 5, [\frac{1}{2}] = 0$.

466. $\displaystyle\int (\cos^{-1} ax)^n\, dx = \begin{cases} x(\cos^{-1} ax)^n - \dfrac{n\sqrt{1 - a^2 x^2}}{a}(\cos^{-1} ax)^{n-1} \\[2mm] \hspace{3cm} - n(n-1)\displaystyle\int (\cos^{-1} ax)^{n-2}\, dx \\[2mm] \hspace{2cm} \text{or} \\[2mm] \displaystyle\sum_{r=0}^{\left[\frac{n}{2}\right]} (-1)^r \dfrac{n!}{(n-2r)!} x(\cos^{-1} ax)^{n-2r} \\[2mm] \hspace{1cm} - \displaystyle\sum_{r=0}^{\left[\frac{n-1}{2}\right]} (-1)^r \dfrac{n!\sqrt{1 - a^2 x^2}}{(n-2r-1)!a}(\cos^{-1} ax)^{n-2r-1} \end{cases}$

467. $\displaystyle\int \frac{1}{\sqrt{1 - a^2 x^2}}(\sin^{-1} ax)\, dx = \frac{1}{2a}(\sin^{-1} ax)^2$

468. $\displaystyle\int \frac{x^n}{\sqrt{1 - a^2 x^2}}(\sin^{-1} ax)\, dx = -\frac{x^{n-1}}{na^2}\sqrt{1 - a^2 x^2}\,\sin^{-1} ax + \frac{x^n}{n^2 a}$

$\hspace{4cm} + \dfrac{n-1}{na^2}\displaystyle\int \frac{x^{n-2}}{\sqrt{1 - a^2 x^2}}\sin^{-1} ax\, dx$

469. $\displaystyle\int \frac{1}{\sqrt{1 - a^2 x^2}}(\cos^{-1} ax)\, dx = -\frac{1}{2a}(\cos^{-1} ax)^2$

470. $\displaystyle\int \frac{x^n}{\sqrt{1 - a^2 x^2}}(\cos^{-1} ax)\, dx = -\frac{x^{n-1}}{na^2}\sqrt{1 - a^2 x^2}\,\cos^{-1} ax - \frac{x^n}{n^2 a}$

$\hspace{4cm} + \dfrac{n-1}{na^2}\displaystyle\int \frac{x^{n-2}}{\sqrt{1 - a^2 x^2}}\cos^{-1} ax\, dx$

471. $\displaystyle\int \frac{\tan^{-1} ax}{a^2 x^2 + 1}\, dx = \frac{1}{2a}(\tan^{-1} ax)^2$

472. $\displaystyle\int \frac{\cot^{-1} ax}{a^2 x^2 + 1}\, dx = -\frac{1}{2a}(\cot^{-1} ax)^2$

473. $\displaystyle\int x\, \sec^{-1} ax\, dx = \frac{x^2}{2}\sec^{-1} ax - \frac{1}{2a^2}\sqrt{a^2 x^2 - 1}$

474. $\displaystyle\int x^n \sec^{-1} ax\, dx = \frac{x^{n+1}}{n+1}\sec^{-1} ax - \frac{1}{n+1}\int \frac{x^n\, dx}{\sqrt{a^2 x^2 - 1}}$

475. $\displaystyle\int \frac{\sec^{-1} ax}{x^2}\, dx = -\frac{\sec^{-1} ax}{x} + \frac{\sqrt{a^2 x^2 - 1}}{x}$

476. $\displaystyle\int x\, \csc^{-1} ax\, dx = \frac{x^2}{2}\csc^{-1} ax + \frac{1}{2a^2}\sqrt{a^2 x^2 - 1}$

477. $\displaystyle\int x^n \csc^{-1} ax\, dx = \frac{x^{n+1}}{n+1}\csc^{-1} ax + \frac{1}{n+1}\int \frac{x^n\, dx}{\sqrt{a^2 x^2 - 1}}$

478. $\displaystyle\int \frac{\csc^{-1} ax}{x^2}\, dx = -\frac{\csc^{-1} ax}{x} - \frac{\sqrt{a^2 x^2 - 1}}{x}$

FORMS INVOLVING TRIGONOMETRIC SUBSTITUTIONS

479. $\displaystyle\int f(\sin x)\, dx = 2\int f\left(\frac{2z}{1 + z^2}\right)\frac{dz}{1 + z^2}, \qquad \left(z = \tan\frac{x}{2}\right)$

480. $\displaystyle\int f(\cos x)\, dx = 2\int f\left(\frac{1 - z^2}{1 + z^2}\right)\frac{dz}{1 + z^2}, \qquad \left(z = \tan\frac{x}{2}\right)$

*481. $\displaystyle\int f(\sin x)\,dx = \int f(u)\frac{du}{\sqrt{1-u^2}},$ $(u = \sin x)$

*482. $\displaystyle\int f(\cos x)\,dx = -\int f(u)\frac{du}{\sqrt{1-u^2}},$ $(u = \cos x)$

*483. $\displaystyle\int f(\sin x, \cos x)\,dx = \int f(u, \sqrt{1-u^2})\frac{du}{\sqrt{1-u^2}},$ $(u = \sin x)$

484. $\displaystyle\int f(\sin x, \cos x)\,dx = 2\int f\left(\frac{2z}{1+z^2}, \frac{1-z^2}{1+z^2}\right)\frac{dz}{1+z^2},$ $\left(z = \tan\frac{x}{2}\right)$

LOGARITHMIC FORMS

485. $\displaystyle\int (\log x)\,dx = x\log x - x$

486. $\displaystyle\int x(\log x)\,dx = \frac{x^2}{2}\log x - \frac{x^2}{4}$

487. $\displaystyle\int x^2(\log x)\,dx = \frac{x^3}{3}\log x - \frac{x^3}{9}$

488. $\displaystyle\int x^n(\log ax)\,dx = \frac{x^{n+1}}{n+1}\log ax - \frac{x^{n+1}}{(n+1)^2}$

489. $\displaystyle\int (\log x)^2\,dx = x(\log x)^2 - 2x\log x + 2x$

490. $\displaystyle\int (\log x)^n\,dx = \begin{cases} x(\log x)^n - n\displaystyle\int (\log x)^{n-1}\,dx, & (n \neq -1) \\ \text{or} \\ (-1)^n n!\, x \displaystyle\sum_{r=0}^{n} \frac{(-\log x)^r}{r!} \end{cases}$

491. $\displaystyle\int \frac{(\log x)^n}{x}\,dx = \frac{1}{n+1}(\log x)^{n+1}$

492. $\displaystyle\int \frac{dx}{\log x} = \log(\log x) + \log x + \frac{(\log x)^2}{2\cdot 2!} + \frac{(\log x)^3}{3\cdot 3!} + \cdots$

493. $\displaystyle\int \frac{dx}{x\log x} = \log(\log x)$

494. $\displaystyle\int \frac{dx}{x(\log x)^n} = -\frac{1}{(n-1)(\log x)^{n-1}}$

495. $\displaystyle\int \frac{x^m\,dx}{(\log x)^n} = -\frac{x^{m+1}}{(n-1)(\log x)^{n-1}} + \frac{m+1}{n-1}\int \frac{x^m\,dx}{(\log x)^{n-1}}$

496. $\displaystyle\int x^m(\log x)^n\,dx = \begin{cases} \dfrac{x^{m+1}(\log x)^n}{m+1} - \dfrac{n}{m+1}\displaystyle\int x^m(\log x)^{n-1}\,dx \\ \text{or} \\ (-1)^n \dfrac{n!}{m+1}x^{m+1}\displaystyle\sum_{r=0}^{n}\dfrac{(-\log x)^r}{r!(m+1)^{n-r}} \end{cases}$

497. $\displaystyle\int x^p\cos(b\ln x)\,dx = \frac{x^{p+1}}{(p+1)^2+b^2}\cdot [b\sin(b\ln x) + (p+1)\cos(b\ln x)] + c$

498. $\displaystyle\int x^p\sin(b\ln x)\,dx = \frac{x^{p+1}}{(p+1)^2+b^2}\cdot [(p+1)\sin(b\ln x) - b\cos(b\ln x)] + c$

499. $\displaystyle\int [\log(ax+b)]\,dx = \frac{ax+b}{a}\log(ax+b) - x$

*The square roots appearing in these formulas may be plus or minus, depending on the quadrant of x. Care must be used to give them the proper sign.

500. $\int \dfrac{\log(ax + b)}{x^2} dx = \dfrac{a}{b} \log x - \dfrac{ax + b}{bx} \log(ax + b)$

501. $\int x^m [\log(ax + b)] dx = \dfrac{1}{m + 1} \left[x^{m+1} - \left(-\dfrac{b}{a} \right)^{m+1} \right] \log(ax + b)$

$$- \dfrac{1}{m + 1} \left(-\dfrac{b}{a} \right)^{m+1} \sum_{r=1}^{m+1} \dfrac{1}{r} \left(-\dfrac{ax}{b} \right)^r$$

502. $\int \dfrac{\log(ax + b)}{x^m} dx = -\dfrac{1}{m - 1} \dfrac{\log(ax + b)}{x^{m-1}} + \dfrac{1}{m - 1} \left(-\dfrac{a}{b} \right)^{m-1} \log \dfrac{ax + b}{x}$

$$+ \dfrac{1}{m - 1} \left(-\dfrac{a}{b} \right)^{m-1} \sum_{r=1}^{m-2} \dfrac{1}{r} \left(-\dfrac{b}{ax} \right)^r, (m > 2)$$

503. $\int \left[\log \dfrac{x + a}{x - a} \right] dx = (x + a) \log (x + a) - (x - a) \log (x - a)$

504. $\int x^m \left[\log \dfrac{x + a}{x - a} \right] dx = \dfrac{x^{m+1} - (-a)^{m+1}}{m + 1} \log (x + a) - \dfrac{x^{m+1} - a^{m+1}}{m + 1} \log (x - a)$

$$+ \dfrac{2a^{m+1}}{m + 1} \sum_{r=1}^{\left[\frac{m+1}{2} \right]} \dfrac{1}{m - 2r + 2} \left(\dfrac{x}{a} \right)^{m - 2r + 2}$$

See note integral 392.

505. $\int \dfrac{1}{x^2} \left[\log \dfrac{x + a}{x - a} \right] dx = \dfrac{1}{x} \log \dfrac{x - a}{x + a} - \dfrac{1}{a} \log \dfrac{x^2 - a^2}{x^2}$

506. $\int (\log X) \, dx = \begin{cases} \left(x + \dfrac{b}{2c} \right) \log X - 2x + \dfrac{\sqrt{4ac - b^2}}{c} \tan^{-1} \dfrac{2cx + b}{\sqrt{4ac - b^2}}, \\ \qquad\qquad\qquad\qquad\qquad\qquad (b^2 - 4ac < 0) \\ \\ \text{or} \\ \\ \left(x + \dfrac{b}{2c} \right) \log X - 2x + \dfrac{\sqrt{b^2 - 4ac}}{c} \tanh^{-1} \dfrac{2cx + b}{\sqrt{b^2 - 4ac}}, \\ \qquad\qquad\qquad\qquad\qquad\qquad (b^2 - 4ac > 0) \\ \\ \text{where} \\ X = a + bx + cx^2 \end{cases}$

507. $\int x^n (\log X) dx = \dfrac{x^{n+1}}{n + 1} \log X - \dfrac{2c}{n + 1} \int \dfrac{x^{n+2}}{X} dx - \dfrac{b}{n + 1} \int \dfrac{x^{n+1}}{X} dx$

$$\text{where } X = a + bx + cx^2$$

508. $\int [\log (x^2 + a^2)] dx = x \log (x^2 + a^2) - 2x + 2a \tan^{-1} \dfrac{x}{a}$

509. $\int [\log (x^2 - a^2)] dx = x \log (x^2 - a^2) - 2x + a \log \dfrac{x + a}{x - a}$

510. $\int x [\log (x^2 \pm a^2)] dx = \tfrac{1}{2} (x^2 \pm a^2) \log (x^2 \pm a^2) - \tfrac{1}{2} x^2$

511. $\int [\log (x + \sqrt{x^2 \pm a^2})] dx = x \log (x + \sqrt{x^2 \pm a^2}) - \sqrt{x^2 \pm a^2}$

512. $\int x [\log (x + \sqrt{x^2 \pm a^2})] dx = \left(\dfrac{x^2}{2} \pm \dfrac{a^2}{4} \right) \log (x + \sqrt{x^2 \pm a^2}) - \dfrac{x \sqrt{x^2 \pm a^2}}{4}$

513. $\int x^m [\log (x + \sqrt{x^2 \pm a^2})] dx = \dfrac{x^{m+1}}{m + 1} \log (x + \sqrt{x^2 \pm a^2})$

$$- \dfrac{1}{m + 1} \int \dfrac{x^{m+1}}{\sqrt{x^2 \pm a^2}} dx$$

514. $\displaystyle\int \frac{\log(x + \sqrt{x^2 + a^2})}{x^2}\,dx = -\frac{\log(x + \sqrt{x^2 + a^2})}{x} - \frac{1}{a}\log\frac{a + \sqrt{x^2 + a^2}}{x}$

515. $\displaystyle\int \frac{\log(x + \sqrt{x^2 - a^2})}{x^2}\,dx = -\frac{\log(x + \sqrt{x^2 - a^2})}{x} + \frac{1}{|a|}\sec^{-1}\frac{x}{a}$

516. $\displaystyle\int x^n \log(x^2 - a^2)\,dx = \frac{1}{n + 1}\left[x^{n+1}\log(x^2 - a^2) - a^{n+1}\log(x - a)\right.$

See note integral 392. $\displaystyle \left. -(-a)^{n+1}\log(x + a) - 2\sum_{r=0}^{\left[\frac{n}{2}\right]}\frac{a^{2r}x^{n-2r+1}}{n - 2r + 1}\right]$

EXPONENTIAL FORMS

517. $\displaystyle\int e^x\,dx = e^x$

518. $\displaystyle\int e^{-x}\,dx = -e^{-x}$

519. $\displaystyle\int e^{ax}\,dx = \frac{e^{ax}}{a}$

520. $\displaystyle\int x\,e^{ax}\,dx = \frac{e^{ax}}{a^2}(ax - 1)$

521. $\displaystyle\int x^m e^{ax}\,dx = \begin{cases} \dfrac{x^m e^{ax}}{a} - \dfrac{m}{a}\displaystyle\int x^{m-1}e^{ax}\,dx \\ \qquad\qquad \text{or} \\ e^{ax}\displaystyle\sum_{r=0}^{m}(-1)^r\dfrac{m!\,x^{m-r}}{(m - r)!\,a^{r+1}} \end{cases}$

522. $\displaystyle\int \frac{e^{ax}\,dx}{x} = \log x + \frac{ax}{1!} + \frac{a^2x^2}{2\cdot 2!} + \frac{a^3x^3}{3\cdot 3!} + \cdots$

523. $\displaystyle\int \frac{e^{ax}}{x^m}\,dx = -\frac{1}{m - 1}\frac{e^{ax}}{x^{m-1}} + \frac{a}{m - 1}\int \frac{e^{ax}}{x^{m-1}}\,dx$

524. $\displaystyle\int e^{ax}\log x\,dx = \frac{e^{ax}\log x}{a} - \frac{1}{a}\int \frac{e^{ax}}{x}\,dx$

525. $\displaystyle\int \frac{dx}{1 + e^x} = x - \log(1 + e^x) = \log\frac{e^x}{1 + e^x}$

526. $\displaystyle\int \frac{dx}{a + be^{px}} = \frac{x}{a} - \frac{1}{ap}\log(a + be^{px})$

527. $\displaystyle\int \frac{dx}{ae^{mx} + be^{-mx}} = \frac{1}{m\sqrt{ab}}\tan^{-1}\left(e^{mx}\sqrt{\frac{a}{b}}\right),\qquad (a > 0, b > 0)$

528. $\displaystyle\int \frac{dx}{ae^{mx} - be^{-mx}} = \begin{cases} \dfrac{1}{2m\sqrt{ab}}\log\dfrac{\sqrt{a}\,e^{mx} - \sqrt{b}}{\sqrt{a}\,e^{mx} + \sqrt{b}} \\ \qquad\qquad \text{or} \\ \dfrac{-1}{m\sqrt{ab}}\tanh^{-1}\left(\sqrt{\dfrac{a}{b}}\,e^{mx}\right),\qquad (a > 0, b > 0) \end{cases}$

529. $\displaystyle\int (a^x - a^{-x})\,dx = \frac{a^x + a^{-x}}{\log a}$

530. $\displaystyle\int \frac{e^{ax}}{b + ce^{ax}}\,dx = \frac{1}{ac}\log(b + ce^{ax})$

531. $\displaystyle\int \frac{x\,e^{ax}}{(1 + ax)^2}\,dx = \frac{e^{ax}}{a^2(1 + ax)}$

532. $\displaystyle \int x\, e^{-x^2}\, dx = -\tfrac{1}{2} e^{-x^2}$

533. $\displaystyle \int e^{ax} [\sin (bx)]\, dx = \frac{e^{ax}[a \sin (bx) - b \cos (bx)]}{a^2 + b^2}$

534. $\displaystyle \int e^{ax} [\sin (bx)][\sin (cx)]\, dx = \frac{e^{ax}[(b - c) \sin (b - c)x + a \cos (b - c)x]}{2[a^2 + (b - c)^2]}$

$$- \frac{e^{ax}[(b + c) \sin (b + c)x + a \cos (b + c)x]}{2[a^2 + (b + c)^2]}$$

535. $\displaystyle \int e^{ax} [\sin (bx)][\cos (cx)]\, dx = \begin{cases} \dfrac{e^{ax}[a \sin (b - c)x - (b - c) \cos (b - c)x]}{2[a^2 + (b - c)^2]} \\[3mm] + \dfrac{e^{ax}[a \sin (b + c)x - (b + c) \cos (b + c)x]}{2[a^2 + (b + c)^2]} \\[3mm] \text{or} \\[2mm] \dfrac{e^{ax}}{\rho}[(a \sin bx - b \cos bx)[\cos (cx - \alpha)] \\[3mm] \qquad\qquad - c(\sin bx) \sin (cx - \alpha)] \\[2mm] \text{where} \\[2mm] \rho = \sqrt{(a^2 + b^2 - c^2)^2 + 4a^2 c^2}, \\[2mm] \qquad \rho \cos \alpha = a^2 + b^2 - c^2, \qquad \rho \sin \alpha = 2ac \end{cases}$

536. $\displaystyle \int e^{ax} [\sin (bx)][\sin (bx + c)]\, dx$

$$= \frac{e^{ax} \cos c}{2a} - \frac{e^{ax}[a \cos (2bx + c) + 2b \sin (2bx + c)]}{2(a^2 + 4b^2)}$$

537. $\displaystyle \int e^{ax} [\sin (bx)][\cos (bx + c)]\, dx$

$$= \frac{-e^{ax} \sin c}{2a} + \frac{e^{ax}[a \sin (2bx + c) - 2b \cos (2bx + c)]}{2(a^2 + 4b^2)}$$

538. $\displaystyle \int e^{ax} [\cos (bx)]\, dx - \frac{e^{ax}}{a^2 + b^2}[a \cos (bx) + b \sin (bx)]$

539. $\displaystyle \int e^{ax} [\cos (bx)][\cos (cx)]\, dx = \frac{e^{ax}[(b - c) \sin (b - c)x + a \cos (b - c)x]}{2[a^2 + (b - c)^2]}$

$$+ \frac{e^{ax}[(b + c) \sin (b + c)x + a \cos (b + c)x]}{2[a^2 + (b + c)^2]}$$

540. $\displaystyle \int e^{ax} [\cos (bx)][\cos (bx + c)]\, dx$

$$= \frac{e^{ax} \cos c}{2a} + \frac{e^{ax}[a \cos (2bx + c) + 2b \sin (2bx + c)]}{2(a^2 + 4b^2)}$$

541. $\displaystyle \int e^{ax} [\cos (bx)][\sin (bx + c)]\, dx$

$$= \frac{e^{ax} \sin c}{2a} + \frac{e^{ax}[a \sin (2bx + c) - 2b \cos (2bx + c)]}{2(a^2 + 4b^2)}$$

542. $\displaystyle \int e^{ax} [\sin^n bx]\, dx = \frac{1}{a^2 + n^2 b^2}\Bigg[(a \sin bx - nb \cos bx)\, e^{ax} \sin^{n-1} bx$

$$+ n(n - 1)b^2 \int e^{ax} [\sin^{n-2} bx]\, dx \Bigg]$$

543. $\displaystyle \int e^{ax} [\cos^n bx]\, dx = \frac{1}{a^2 + n^2 b^2}\Bigg[(a \cos bx + nb \sin bx)\, e^{ax} \cos^{n-1} bx$

$$+ n(n - 1)b^2 \int e^{ax} [\cos^{n-2} bx]\, dx \Bigg]$$

544. $\int x^m e^x \sin x\, dx = \frac{1}{2} x^m e^x(\sin x - \cos x) - \frac{m}{2} \int x^{m-1} e^x \sin x\, dx$

$$+ \frac{m}{2} \int x^{m-1} e^x \cos x\, dx$$

545. $\int x^m e^{ax}[\sin bx]\, dx = \begin{cases} x^m e^{ax} \dfrac{a \sin bx - b \cos bx}{a^2 + b^2} \\ \qquad\qquad - \dfrac{m}{a^2 + b^2} \displaystyle\int x^{m-1} e^{ax}(a \sin bx - b \cos bx)\, dx \\ \text{or} \\ e^{ax} \displaystyle\sum_{r=0}^{m} \dfrac{(-1)^r m! x^{m-r}}{\rho^{r+1}(m - r)!} \sin[bx - (r+1)\alpha] \\ \qquad\qquad \text{where} \\ \rho = \sqrt{a^2 + b^2}, \qquad \rho \cos \alpha = a, \qquad \rho \sin \alpha = b \end{cases}$

546. $\int x^m e^x \cos x\, dx = \frac{1}{2} x^m e^x(\sin x + \cos x)$

$$- \frac{m}{2} \int x^{m-1} e^x \sin x\, dx - \frac{m}{2} \int x^{m-1} e^x \cos x\, dx$$

547. $\int x^m e^{ax} \cos bx\, dx = \begin{cases} x^m e^{ax} \dfrac{a \cos bx + b \sin bx}{a^2 + b^2} \\ \qquad\qquad - \dfrac{m}{a^2 + b^2} \displaystyle\int x^{m-1} e^{ax}(a \cos bx + b \sin bx)\, dx \\ \text{or} \\ e^{ax} \displaystyle\sum_{r=0}^{m} \dfrac{(-1)^r m! x^{m-r}}{\rho^{r+1}(m - r)!} \cos[bx - (r+1)\alpha] \\ \qquad\qquad \text{where} \\ \rho = \sqrt{a^2 + b^2}, \qquad \rho \cos \alpha = a, \qquad \rho \sin \alpha = b \end{cases}$

548. $\int e^{ax}(\cos^m x)(\sin^n x)\,dx = \begin{cases} \end{cases}$

$$\dfrac{e^{ax}\cos^{m-1} x \sin^n x[a\cos x + (m+n)\sin x]}{(m+n)^2 + a^2}$$

$$-\dfrac{na}{(m+n)^2 + a^2}\int e^{ax}(\cos^{m-1} x)(\sin^{n-1} x)\,dx$$

$$+\dfrac{(m-1)(m+n)}{(m+n)^2 + a^2}\int e^{ax}(\cos^{m-2} x)(\sin^n x)\,dx$$

or

$$\dfrac{e^{ax}\cos^m x \sin^{n-1} x[a\sin x - (m+n)\cos x]}{(m+n)^2 + a^2}$$

$$+\dfrac{ma}{(m+n)^2 + a^2}\int e^{ax}(\cos^{m-1} x)(\sin^{n-1} x)\,dx$$

$$+\dfrac{(n-1)(m+n)}{(m+n)^2 + a^2}\int e^{ax}(\cos^m x)(\sin^{n-2} x)\,dx$$

or

$$\dfrac{e^{ax}(\cos^{m-1} x)(\sin^{n-1} x)(a\sin x\cos x + m\sin^2 x - n\cos^2 x)}{(m+n)^2 + a^2}$$

$$+\dfrac{m(m-1)}{(m+n)^2 + a^2}\int e^{ax}(\cos^{m-2} x)(\sin^n x)\,dx$$

$$+\dfrac{n(n-1)}{(m+n)^2 + a^2}\int e^{ax}(\cos^m x)(\sin^{n-2} x)\,dx$$

or

$$\dfrac{e^{ax}(\cos^{m-1} x)(\sin^{n-1} x)(a\cos x\sin x + m\sin^2 x - n\cos^2 x)}{(m+n)^2 + a^2}$$

$$+\dfrac{m(m-1)}{(m+n)^2 + a^2}\int e^{ax}(\cos^{m-2} x)(\sin^{n-2} x)\,dx$$

$$+\dfrac{(n-m)(n+m-1)}{(m+n)^2 + a^2}\int e^{ax}(\cos^m x)(\sin^{n-2} x)\,dx$$

549. $\int x\,e^{ax}(\sin bx)\,dx = \dfrac{x\,e^{ax}}{a^2 + b^2}(a\sin bx - b\cos bx)$

$$-\dfrac{e^{ax}}{(a^2 + b^2)^2}[(a^2 - b^2)\sin bx - 2ab\cos bx]$$

550. $\int x\,e^{ax}(\cos bx)\,dx = \dfrac{x\,e^{ax}}{a^2 + b^2}(a\cos bx + b\sin bx)$

$$-\dfrac{e^{ax}}{(a^2 + b^2)^2}[(a^2 - b^2)\cos bx + 2ab\sin bx]$$

551. $\int \dfrac{e^{ax}}{\sin^n x}\,dx = -\dfrac{e^{ax}[a\sin x + (n-2)\cos x]}{(n-1)(n-2)\sin^{n-1} x} + \dfrac{a^2 + (n-2)^2}{(n-1)(n-2)}\int \dfrac{e^{ax}}{\sin^{n-2} x}\,dx$

552. $\int \dfrac{e^{ax}}{\cos^n x}\,dx = -\dfrac{e^{ax}[a\cos x - (n-2)\sin x]}{(n-1)(n-2)\cos^{n-1} x} + \dfrac{a^2 + (n-2)^2}{(n-1)(n-2)}\int \dfrac{e^{ax}}{\cos^{n-2} x}\,dx$

553. $\int e^{ax}\tan^n x\,dx = e^{ax}\dfrac{\tan^{n-1} x}{n-1} - \dfrac{a}{n-1}\int e^{ax}\tan^{n-1} x\,dx - \int e^{ax}\tan^{n-2} x\,dx$

HYPERBOLIC FORMS

554. $\int (\sinh x)\,dx = \cosh x$

555. $\int (\cosh x)\,dx = \sinh x$

556. $\int (\tanh x)\, dx = \log \cosh x$

557. $\int (\coth x)\, dx = \log \sinh x$

558. $\int (\operatorname{sech} x)\, dx = \tan^{-1}(\sinh x)$

559. $\int \operatorname{csch} x\, dx = \log \tanh \left(\dfrac{x}{2} \right)$

560. $\int x(\sinh x)\, dx = x \cosh x - \sinh x$

561. $\int x^n (\sinh x)\, dx = x^n \cosh x - n \int x^{n-1}(\cosh x)\, dx$

562. $\int x(\cosh x)\, dx = x \sinh x - \cosh x$

563. $\int x^n (\cosh x)\, dx = x^n \sinh x - n \int x^{n-1}(\sinh x)\, dx$

564. $\int (\operatorname{sech} x)(\tanh x)\, dx = -\operatorname{sech} x$

565. $\int (\operatorname{csch} x)(\coth x)\, dx = -\operatorname{csch} x$

566. $\int (\sinh^2 x)\, dx = \dfrac{\sinh 2x}{4} - \dfrac{x}{2}$

567. $\int (\sinh^m x)(\cosh^n x)\, dx =$
$$\begin{cases} \dfrac{1}{m+n}(\sinh^{m+1} x)(\cosh^{n-1} x) \\ \qquad + \dfrac{n-1}{m+n} \int (\sinh^m x)(\cosh^{n-2} x)\, dx \\ \qquad\qquad \text{or} \\ \dfrac{1}{m+n} \sinh^{m-1} x \cosh^{n+1} x \\ \qquad - \dfrac{m-1}{m+n} \int (\sinh^{m-2} x)(\cosh^n x)\, dx, \quad (m+n \neq 0) \end{cases}$$

568. $\int \dfrac{dx}{(\sinh^m x)(\cosh^n x)} =$
$$\begin{cases} -\dfrac{1}{(m-1)(\sinh^{m-1} x)(\cosh^{n-1} x)} \\ \qquad - \dfrac{m+n-2}{m-1} \int \dfrac{dx}{(\sinh^{m-2} x)(\cosh^n x)}, \quad (m \neq 1) \\ \qquad\qquad \text{or} \\ \dfrac{1}{(n-1)\sinh^{m-1} x \cosh^{n-1} x} \\ \qquad + \dfrac{m+n-2}{n-1} \int \dfrac{dx}{(\sinh^m x)(\cosh^{n-2} x)}, \quad (n \neq 1) \end{cases}$$

569. $\int (\tanh^2 x)\, dx = x - \tanh x$

570. $\int (\tanh^n x)\, dx = -\dfrac{\tanh^{n-1} x}{n-1} + \int (\tanh^{n-2} x)\, dx, \quad (n \neq 1)$

571. $\int (\operatorname{sech}^2 x)\, dx = \tanh x$

572. $\int (\cosh^2 x)\, dx = \dfrac{\sinh 2x}{4} + \dfrac{x}{2}$

573. $\int (\coth^2 x)\,dx = x - \coth x$

574. $\int (\coth^n x)\,dx = -\dfrac{\coth^{n-1} x}{n-1} + \int \coth^{n-2} x\,dx, \qquad (n \ne 1)$

575. $\int (\operatorname{csch}^2 x)\,dx = -\operatorname{ctnh} x$

576. $\int (\sinh mx)(\sinh nx)\,dx = \dfrac{\sinh (m+n)x}{2(m+n)} - \dfrac{\sinh (m-n)x}{2(m-n)}, \qquad (m^2 \ne n^2)$

577. $\int (\cosh mx)(\cosh nx)\,dx = \dfrac{\sinh (m+n)x}{2(m+n)} + \dfrac{\sinh (m-n)x}{2(m-n)}, \qquad (m^2 \ne n^2)$

578. $\int (\sinh mx)(\cosh nx)\,dx = \dfrac{\cosh (m+n)x}{2(m+n)} + \dfrac{\cosh (m-n)x}{2(m-n)}, \qquad (m^2 \ne n^2)$

579. $\int \left(\sinh^{-1} \dfrac{x}{a}\right) dx = x \sinh^{-1} \dfrac{x}{a} - \sqrt{x^2 + a^2}, \qquad (a > 0)$

580. $\int x\left(\sinh^{-1} \dfrac{x}{a}\right) dx = \left(\dfrac{x^2}{2} + \dfrac{a^2}{4}\right) \sinh^{-1} \dfrac{x}{a} - \dfrac{x}{4}\sqrt{x^2 + a^2}, \qquad (a > 0)$

581. $\int x^n(\sinh^{-1} x)\,dx = \dfrac{x^{n+1}}{n+1} \sinh^{-1} x - \dfrac{1}{n+1}\int \dfrac{x^{n+1}}{(1 + x^2)^{\frac{1}{2}}}\,dx, \qquad (n \ne -1)$

582. $\int \left(\cosh^{-1} \dfrac{x}{a}\right) dx = \begin{cases} x \cosh^{-1} \dfrac{x}{a} - \sqrt{x^2 - a^2}, & \left(\cosh^{-1} \dfrac{x}{a} > 0\right) \\[2mm] \qquad\qquad \text{or} \\[2mm] x \cosh^{-1} \dfrac{x}{a} + \sqrt{x^2 - a^2}, & \left(\cosh^{-1} \dfrac{x}{a} < 0\right), \quad (a > 0) \end{cases}$

583. $\int x\left(\cosh^{-1} \dfrac{x}{a}\right) dx = \dfrac{2x^2 - a^2}{4} \cosh^{-1} \dfrac{x}{a} - \dfrac{x}{4}(x^2 - a^2)^{\frac{1}{2}}$

584. $\int x^n(\cosh^{-1} x)\,dx = \dfrac{x^{n+1}}{n+1} \cosh^{-1} x - \dfrac{1}{n+1}\int \dfrac{x^{n+1}}{(x^2 - 1)^{\frac{1}{2}}}\,dx, \qquad (n \ne -1)$

585. $\int \left(\tanh^{-1} \dfrac{x}{a}\right) dx = x \tanh^{-1} \dfrac{x}{a} + \dfrac{a}{2} \log (a^2 - x^2), \qquad \left(\left|\dfrac{x}{a}\right| < 1\right)$

586. $\int \left(\coth^{-1} \dfrac{x}{a}\right) dx = x \coth^{-1} \dfrac{x}{a} + \dfrac{a}{2} \log (x^2 - a^2), \qquad \left(\left|\dfrac{x}{a}\right| > 1\right)$

587. $\int x\left(\tanh^{-1} \dfrac{x}{a}\right) dx = \dfrac{x^2 - a^2}{2} \tanh^{-1} \dfrac{x}{a} + \dfrac{ax}{2}, \qquad \left(\left|\dfrac{x}{a}\right| < 1\right)$

588. $\int x^n\left(\tanh^{-1} x\right) dx = \dfrac{x^{n+1}}{n+1} \tanh^{-1} x - \dfrac{1}{n+1}\int \dfrac{x^{n+1}}{1 - x^2}\,dx, \qquad (n \ne -1)$

589. $\int x\left(\coth^{-1} \dfrac{x}{a}\right) dx = \dfrac{x^2 - a^2}{2} \coth^{-1} \dfrac{x}{a} + \dfrac{ax}{2}, \qquad \left(\left|\dfrac{x}{a}\right| > 1\right)$

590. $\int x^n(\coth^{-1} x)\,dx = \dfrac{x^{n+1}}{n+1} \coth^{-1} x + \dfrac{1}{n+1}\int \dfrac{x^{n+1}}{x^2 - 1}\,dx, \qquad (n \ne -1)$

591. $\int (\operatorname{sech}^{-1} x)\,dx = x \operatorname{sech}^{-1} x + \sin^{-1} x$

592. $\int x \operatorname{sech}^{-1} x\,dx = \dfrac{x^2}{2} \operatorname{sech}^{-1} x - \dfrac{1}{2}\sqrt{1 - x^2}$

593. $\int x^n \operatorname{sech}^{-1} x\,dx = \dfrac{x^{n+1}}{n+1} \operatorname{sech}^{-1} x + \dfrac{1}{n+1}\int \dfrac{x^n}{(1 - x^2)^{\frac{1}{2}}}\,dx, \qquad (n \ne -1)$

594. $\int \operatorname{csch}^{-1} x\,dx = x \operatorname{csch}^{-1} x + \dfrac{x}{|x|} \sinh^{-1} x$

595. $\int x \operatorname{csch}^{-1} x \, dx = \dfrac{x^2}{2} \operatorname{csch}^{-1} x + \dfrac{1}{2} \dfrac{x}{|x|} \sqrt{1 + x^2}$

596. $\int x^n \operatorname{csch}^{-1} x \, dx = \dfrac{x^{n+1}}{n+1} \operatorname{csch}^{-1} x + \dfrac{1}{n+1} \dfrac{x}{|x|} \int \dfrac{x^n}{(x^2+1)^{\frac{1}{2}}} \, dx, \qquad (n \neq -1)$

DEFINITE INTEGRALS

597. $\displaystyle\int_0^\infty x^{n-1} e^{-x} \, dx = \int_0^1 \left(\log\dfrac{1}{x}\right)^{n-1} dx = \dfrac{1}{n} \prod_{m=1}^\infty \dfrac{\left(1 + \dfrac{1}{m}\right)^n}{1 + \dfrac{n}{m}}$

$$= \Gamma(n), n \neq 0, -1, -2, -3, \ldots \quad \text{(Gamma Function)}$$

598. $\displaystyle\int_0^\infty t^n p^{-t} \, dt = \dfrac{n!}{(\log p)^{n+1}}, \qquad (n = 0, 1, 2, 3, \ldots \text{ and } p > 0)$

599. $\displaystyle\int_0^\infty t^{n-1} e^{-(a+1)t} \, dt = \dfrac{\Gamma(n)}{(a+1)^n}, \qquad (n > 0, a > -1)$

600. $\displaystyle\int_0^1 x^m \left(\log\dfrac{1}{x}\right)^n dx = \dfrac{\Gamma(n+1)}{(m+1)^{n+1}}, \qquad (m > -1, n > -1)$

601. $\Gamma(n)$ is finite if $n > 0$, $\Gamma(n+1) = n\Gamma(n)$

602. $\Gamma(n) \cdot \Gamma(1 - n) = \dfrac{\pi}{\sin n\pi}$

603. $\Gamma(n) = (n-1)!$ if $n = $ integer > 0

604. $\Gamma(\frac{1}{2}) = 2\displaystyle\int_0^\infty e^{-t^2} \, dt = \sqrt{\pi} = 1.7724538509\cdots = (-\frac{1}{2})!$

605. $\Gamma(n + \frac{1}{2}) = \dfrac{1 \cdot 3 \cdot 5 \ldots (2n-1)}{2^n} \sqrt{\pi} \qquad n = 1, 2, 3, \ldots$

606. $\Gamma(-n + \frac{1}{2}) = \dfrac{(-1)^n 2^n \sqrt{\pi}}{1 \cdot 3 \cdot 5 \ldots (2n-1)} \qquad n = 1, 2, 3, \ldots$

607. $\displaystyle\int_0^1 x^{m-1}(1-x)^{n-1} \, dx = \int_0^\infty \dfrac{x^{m-1}}{(1+x)^{m+n}} \, dx = \dfrac{\Gamma(m)\Gamma(n)}{\Gamma(m+n)} = B(m, n)$

(Beta function)

608. $B(m, n) = B(n, m) = \dfrac{\Gamma(m)\Gamma(n)}{\Gamma(m+n)}$, where m and n are any positive real numbers.

609. $\displaystyle\int_a^b (x-a)^m (b-x)^n \, dx = (b-a)^{m+n+1} \dfrac{\Gamma(m+1) \cdot \Gamma(n+1)}{\Gamma(m+n+2)},$

$$(m > -1, n > -1, b > a)$$

610. $\displaystyle\int_1^\infty \dfrac{dx}{x^m} = \dfrac{1}{m-1}, \qquad [m > 1]$

611. $\displaystyle\int_0^\infty \dfrac{dx}{(1+x)x^p} = \pi \csc p\pi, \qquad [p < 1]$

612. $\displaystyle\int_0^\infty \dfrac{dx}{(1-x)x^p} = -\pi \cot p\pi, \qquad [p < 1]$

613. $\displaystyle\int_0^\infty \dfrac{x^{p-1} \, dx}{1+x} = \dfrac{\pi}{\sin p\pi}$

$$= B(p, 1-p) = \Gamma(p)\Gamma(1-p), \qquad [0 < p < 1]$$

614. $\displaystyle\int_0^\infty \dfrac{x^{m-1} \, dx}{1+x^n} = \dfrac{\pi}{n \sin \dfrac{m\pi}{n}}, \qquad [0 < m < n]$

615. $\displaystyle\int_0^\infty \frac{x^a\, dx}{(m + x^b)^c} = \frac{m^{\frac{a+1-bc}{b}}}{b}\left[\frac{\Gamma\!\left(\dfrac{a+1}{b}\right)\Gamma\!\left(c - \dfrac{a+1}{b}\right)}{\Gamma(c)}\right]$

$\left(a > -1,\ b > 0,\ m > 0,\ c > \dfrac{a+1}{b}\right)$

616. $\displaystyle\int_0^\infty \frac{dx}{(1 + x)\sqrt{x}} = \pi$

617. $\displaystyle\int_0^\infty \frac{a\, dx}{a^2 + x^2} = \frac{\pi}{2},\ \text{if } a > 0;\ 0,\ \text{if } a = 0;\ -\frac{\pi}{2},\ \text{if } a < 0$

618. $\displaystyle\int_0^a (a^2 - x^2)^{\frac{n}{2}}\, dx = \frac{1}{2}\int_{-a}^a (a^2 - x^2)^{\frac{n}{2}}\, dx = \frac{1 \cdot 3 \cdot 5 \ldots n}{2 \cdot 4 \cdot 6 \ldots (n+1)} \cdot \frac{\pi}{2} \cdot a^{n+1}$ $(n\ \text{odd})$

619. $\displaystyle\int_0^a x^m (a^2 - x^2)^{\frac{n}{2}}\, dx = \begin{cases} \dfrac{1}{2} a^{m+n+1} B\!\left(\dfrac{m+1}{2}, \dfrac{n+2}{2}\right) \\[2mm] \text{or} \\[2mm] \dfrac{1}{2} a^{m+n+1}\dfrac{\Gamma\!\left(\dfrac{m+1}{2}\right)\Gamma\!\left(\dfrac{n+2}{2}\right)}{\Gamma\!\left(\dfrac{m+n+3}{2}\right)} \end{cases}$

620. $\displaystyle\int_0^{\pi/2} (\sin^n x)\, dx = \begin{cases} \displaystyle\int_0^{\pi/2} (\cos^n x)\, dx \\[2mm] \text{or} \\[2mm] \dfrac{1 \cdot 3 \cdot 5 \cdot 7 \ldots (n-1)}{2 \cdot 4 \cdot 6 \cdot 8 \ldots (n)}\dfrac{\pi}{2}, \quad (n\ \text{an even integer, } n \neq 0) \\[2mm] \text{or} \\[2mm] \dfrac{2 \cdot 4 \cdot 6 \cdot 8 \ldots (n-1)}{1 \cdot 3 \cdot 5 \cdot 7 \ldots (n)}, \quad (n\ \text{an odd integer, } n \neq 1) \\[2mm] \text{or} \\[2mm] \dfrac{\sqrt{\pi}}{2}\dfrac{\Gamma\!\left(\dfrac{n+1}{2}\right)}{\Gamma\!\left(\dfrac{n}{2} + 1\right)}, \quad (n > -1) \end{cases}$

621. $\displaystyle\int_0^\infty \frac{\sin mx\, dx}{x} = \frac{\pi}{2},\ \text{if } m > 0;\ 0,\ \text{if } m = 0;\ -\frac{\pi}{2},\ \text{if } m < 0$

622. $\displaystyle\int_0^\infty \frac{\cos x\, dx}{x} = \infty$

623. $\displaystyle\int_0^\infty \frac{\tan x\, dx}{x} = \frac{\pi}{2}$

624. $\displaystyle\int_0^\pi \sin ax \cdot \sin bx\, dx = \int_0^\pi \cos ax \cdot \cos bx\, dx = 0,\quad (a \neq b;\ a, b\ \text{integers})$

625. $\displaystyle\int_0^{\pi/a} [\sin(ax)][\cos(ax)]\, dx = \int_0^\pi [\sin(ax)][\cos(ax)]\, dx = 0$

626. $\displaystyle\int_0^\pi [\sin(ax)][\cos(bx)]\, dx = \frac{2a}{a^2 - b^2},\ \text{if } a - b\ \text{is odd, or } 0\ \text{if } a - b\ \text{is even}$

627. $\displaystyle\int_0^\infty \frac{\sin x \cos mx\, dx}{x}$

$= 0,\ \text{if } m < -1\ \text{or } m > 1;\ \frac{\pi}{4},\ \text{if } m = \pm 1;\ \frac{\pi}{2},\ \text{if } m^2 < 1$

628. $\displaystyle\int_0^\infty \frac{\sin ax \sin bx}{x^2}\, dx = \frac{\pi a}{2}, \qquad (a \le b)$

629. $\displaystyle\int_0^\pi \sin^2 mx\, dx = \int_0^\pi \cos^2 mx\, dx = \frac{\pi}{2}$

630. $\displaystyle\int_0^\infty \frac{\sin^2 (px)}{x^2}\, dx = \frac{\pi p}{2}$

631. $\displaystyle\int_0^\infty \frac{\sin x}{x^p}\, dx = \frac{\pi}{2\Gamma(p)\sin(p\pi/2)}, \qquad 0 < p < 1$

632. $\displaystyle\int_0^\infty \frac{\cos x}{x^p}\, dx = \frac{\pi}{2\Gamma(p)\cos(p\pi/2)}, \qquad 0 < p < 1$

633. $\displaystyle\int_0^\infty \frac{1 - \cos px}{x^2}\, dx = \frac{\pi p}{2}$

634. $\displaystyle\int_0^\infty \frac{\sin px \cos qx}{x}\, dx = \left\{0, \quad q > p > 0; \quad \frac{\pi}{2}, \quad p > q > 0; \quad \frac{\pi}{4}, \quad p = q > 0\right\}$

635. $\displaystyle\int_0^\infty \frac{\cos (mx)}{x^2 + a^2}\, dx = \frac{\pi}{2|a|}\, e^{-|ma|}$

636. $\displaystyle\int_0^\infty \cos (x^2)\, dx = \int_0^\infty \sin (x^2)\, dx = \frac{1}{2}\sqrt{\frac{\pi}{2}}$

637. $\displaystyle\int_0^\infty \sin ax^n\, dx = \frac{1}{na^{1/n}}\, \Gamma(1/n)\sin\frac{\pi}{2n}, \qquad n > 1$

638. $\displaystyle\int_0^\infty \cos ax^n\, dx = \frac{1}{na^{1/n}}\, \Gamma(1/n)\cos\frac{\pi}{2n}, \qquad n > 1$

639. $\displaystyle\int_0^\infty \frac{\sin x}{\sqrt{x}}\, dx = \int_0^\infty \frac{\cos x}{\sqrt{x}}\, dx = \sqrt{\frac{\pi}{2}}$

640. (a) $\displaystyle\int_0^\infty \frac{\sin^3 x}{x}\, dx = \frac{\pi}{4}$ (b) $\displaystyle\int_0^\infty \frac{\sin^3 x}{x^2}\, dx\ \frac{3}{4}\log 3$

641. $\displaystyle\int_0^\infty \frac{\sin^3 x}{x^3}\, dx = \frac{3\pi}{8}$

642. $\displaystyle\int_0^\infty \frac{\sin^4 x}{x^4}\, dx = \frac{\pi}{3}$

643. $\displaystyle\int_0^{\pi/2} \frac{dx}{1 + a \cos x} = \frac{\cos^{-1} a}{\sqrt{1 - a^2}}, \qquad (a < 1)$

644. $\displaystyle\int_0^\pi \frac{dx}{a + b \cos x} = \frac{\pi}{\sqrt{a^2 - b^2}}, \qquad (a > b \ge 0)$

645. $\displaystyle\int_0^{2\pi} \frac{dx}{1 + a \cos x} = \frac{2\pi}{\sqrt{1 - a^2}}, \qquad (a^2 < 1)$

646. $\displaystyle\int_0^\infty \frac{\cos ax - \cos bx}{x}\, dx = \log\frac{b}{a}$

647. $\displaystyle\int_0^{\pi/2} \frac{dx}{a^2 \sin^2 x + b^2 \cos^2 x} = \frac{\pi}{2ab}$

648. $\displaystyle\int_0^{\pi/2} \frac{dx}{(a^2 \sin^2 x + b^2 \cos^2 x)^2} = \frac{\pi(a^2 + b^2)}{4a^3 b^3}, \qquad (a, b > 0)$

649. $\displaystyle\int_0^{\pi/2} \sin^{n-1} x \cos^{m-1} x\, dx = \frac{1}{2}\mathrm{B}\left(\frac{n}{2}, \frac{m}{2}\right), \qquad m \text{ and } n \text{ positive integers}$

650. $\int_0^{\pi/2} (\sin^{2n+1} \theta)\, d\theta = \dfrac{2 \cdot 4 \cdot 6 \ldots (2n)}{1 \cdot 3 \cdot 5 \ldots (2n+1)}, \qquad (n = 1, 2, 3 \ldots)$

651. $\int_0^{\pi/2} (\sin^{2n} \theta)\, d\theta = \dfrac{1 \cdot 3 \cdot 5 \ldots (2n-1)}{2 \cdot 4 \ldots (2n)} \left(\dfrac{\pi}{2}\right), \qquad (n = 1, 2, 3 \ldots)$

652. $\int_0^{\pi/2} \dfrac{x}{\sin x}\, dx = 2 \left\{ \dfrac{1}{1^2} - \dfrac{1}{3^2} + \dfrac{1}{5^2} - \dfrac{1}{7^2} + \cdots \right\}$

653. $\int_0^{\pi/2} \dfrac{dx}{1 + \tan^m x} = \dfrac{\pi}{4}$

654. $\int_0^{\pi/2} \sqrt{\cos \theta}\, d\theta = \dfrac{(2\pi)^{\frac{3}{2}}}{[\Gamma(\frac{1}{4})]^2}$

655. $\int_0^{\pi/2} (\tan^h \theta)\, d\theta = \dfrac{\pi}{2 \cos\left(\dfrac{h\pi}{2}\right)}, \qquad (0 < h < 1)$

656. $\int_0^{\infty} \dfrac{\tan^{-1}(ax) - \tan^{-1}(bx)}{x}\, dx = \dfrac{\pi}{2} \log \dfrac{a}{b}, \qquad (a, b > 0)$

657. The area enclosed by a curve defined through the equation $x^{\frac{b}{c}} + y^{\frac{b}{c}} = a^{\frac{b}{c}}$ where $a > 0$, c a positive odd integer and b a positive even integer is given by

$$\dfrac{\left[\Gamma\left(\dfrac{c}{b}\right)\right]^2}{\Gamma\left(\dfrac{2c}{b}\right)} \left(\dfrac{2ca^2}{b}\right)$$

658. $I = \iiint_R x^{h-1} y^{m-1} z^{n-1}\, dv$, where R denotes the region of space bounded by

the co-ordinate planes and that portion of the surface $\left(\dfrac{x}{a}\right)^p + \left(\dfrac{y}{b}\right)^q + \left(\dfrac{z}{c}\right)^k = 1$,

which lies in the first octant and where $h, m, n, p, q, k, a, b, c$, denote positive real numbers is given by

$$\int_0^u x^{h-1}\, dx \int_0^{b\left[1-\left(\frac{x}{a}\right)^p\right]^{\frac{1}{q}}} y^m\, dy \int_0^{c\left[1-\left(\frac{x}{a}\right)^p - \left(\frac{y}{b}\right)^q\right]^{\frac{1}{k}}} z^{n-1}\, dz$$

$$= \dfrac{a^h b^m c^n}{pqk} \dfrac{\Gamma\left(\dfrac{h}{p}\right)\Gamma\left(\dfrac{m}{q}\right)\Gamma\left(\dfrac{n}{k}\right)}{\Gamma\left(\dfrac{h}{p} + \dfrac{m}{q} + \dfrac{n}{k} + 1\right)}$$

659. $\int_0^{\infty} e^{-ax}\, dx = \dfrac{1}{a}, \qquad (a > 0)$

660. $\int_0^{\infty} \dfrac{e^{-ax} - e^{-bx}}{x}\, dx = \log \dfrac{b}{a}, \qquad (a, b > 0)$

661. $\int_0^{\infty} x^n e^{-ax}\, dx = \begin{cases} \dfrac{\Gamma(n+1)}{a^{n+1}}, & (n > -1, a > 0) \\ \quad\text{or} \\ \dfrac{n!}{a^{n+1}}, & (a > 0, n \text{ positive integer}) \end{cases}$

662. $\int_0^{\infty} x^n \exp(-ax^p)\, dx = \dfrac{\Gamma(k)}{pa^k}, \qquad \left(n > -1, p > 0, a > 0, k = \dfrac{n+1}{p}\right)$

663. $\int_0^{\infty} e^{-a^2x^2}\, dx = \dfrac{1}{2a}\sqrt{\pi} = \dfrac{1}{2a}\Gamma\left(\dfrac{1}{2}\right), \qquad (a > 0)$

664. $\int_0^{\infty} x e^{-x^2}\, dx = \dfrac{1}{2}$

665. $\displaystyle\int_0^\infty x^2 e^{-x^2}\,dx = \frac{\sqrt{\pi}}{4}$

666. $\displaystyle\int_0^\infty x^{2n} e^{-ax^2}\,dx = \frac{1 \cdot 3 \cdot 5 \ldots (2n-1)}{2^{n+1} a^n}\sqrt{\frac{\pi}{a}}$

667. $\displaystyle\int_0^\infty x^{2n+1} e^{-ax^2}\,dx = \frac{n!}{2a^{n+1}}, \qquad (a > 0)$

668. $\displaystyle\int_0^1 x^m e^{-ax}\,dx = \frac{m!}{a^{m+1}}\left[1 - e^{-a}\sum_{r=0}^m \frac{a^r}{r!}\right]$

669. $\displaystyle\int_0^\infty e^{\left(-x^2 - \frac{a^2}{x^2}\right)}\,dx = \frac{e^{-2a}\sqrt{\pi}}{2}, \qquad (a \geq 0)$

670. $\displaystyle\int_0^\infty e^{-nx}\sqrt{x}\,dx = \frac{1}{2n}\sqrt{\frac{\pi}{n}}$

671. $\displaystyle\int_0^\infty \frac{e^{-nx}}{\sqrt{x}}\,dx = \sqrt{\frac{\pi}{n}}$

672. $\displaystyle\int_0^\infty e^{-ax}(\cos mx)\,dx = \frac{a}{a^2 + m^2}, \qquad (a > 0)$

673. $\displaystyle\int_0^\infty e^{-ax}(\sin mx)\,dx = \frac{m}{a^2 + m^2}, \qquad (a > 0)$

674. $\displaystyle\int_0^\infty x e^{-ax}[\sin(bx)]\,dx = \frac{2ab}{(a^2 + b^2)^2}, \qquad (a > 0)$

675. $\displaystyle\int_0^\infty x e^{-ax}[\cos(bx)]\,dx = \frac{a^2 - b^2}{(a^2 + b^2)^2}, \qquad (a > 0)$

676. $\displaystyle\int_0^\infty x^n e^{-ax}[\sin(bx)]\,dx = \frac{n![(a + ib)^{n+1} - (a - ib)^{n+1}]}{2i(a^2 + b^2)^{n+1}}, \qquad (i^2 = -1, a > 0)$

677. $\displaystyle\int_0^\infty x^n e^{-ax}[\cos(bx)]\,dx = \frac{n![(a - ib)^{n+1} + (a + ib)^{n+1}]}{2(a^2 + b^2)^{n+1}}, \qquad (i^2 = -1, a > 0)$

678. $\displaystyle\int_0^\infty \frac{e^{-ax}\sin x}{x}\,dx = \cot^{-1} a, \qquad (a > 0)$

679. $\displaystyle\int_0^\infty e^{-a^2 x^2}\cos bx\,dx = \frac{\sqrt{\pi}}{2a}\exp\left(-\frac{b^2}{4a^2}\right), \qquad (ab \neq 0)$

680. $\displaystyle\int_0^\infty e^{-t\cos\phi}\, t^{b-1}\sin(t\sin\phi)\,dt = [\Gamma(b)]\sin(b\phi), \qquad \left(b > 0, -\frac{\pi}{2} < \phi < \frac{\pi}{2}\right)$

681. $\displaystyle\int_0^\infty e^{-t\cos\phi}\, t^{b-1}[\cos(t\sin\phi)]\,dt = [\Gamma(b)]\cos(b\phi), \qquad \left(b > 0, -\frac{\pi}{2} < \phi < \frac{\pi}{2}\right)$

682. $\displaystyle\int_0^\infty t^{b-1}\cos t\,dt = [\Gamma(b)]\cos\left(\frac{b\pi}{2}\right), \qquad (0 < b < 1)$

683. $\displaystyle\int_0^\infty t^{b-1}(\sin t)\,dt = [\Gamma(b)]\sin\left(\frac{b\pi}{2}\right), \qquad (0 < b < 1)$

684. $\displaystyle\int_0^1 (\log x)^n\,dx = (-1)^n \cdot n!$

685. $\displaystyle\int_0^1 \left(\log\frac{1}{x}\right)^{\frac{1}{2}}\,dx = \frac{\sqrt{\pi}}{2}$

686. $\displaystyle\int_0^1 \left(\log\frac{1}{x}\right)^{-\frac{1}{2}}\,dx = \sqrt{\pi}$

687. $\displaystyle\int_0^1 \left(\log\frac{1}{x}\right)^n dx = n!$

688. $\displaystyle\int_0^1 x \log(1-x)\,dx = -\frac{3}{4}$

689. $\displaystyle\int_0^1 x \log(1+x)\,dx = \frac{1}{4}$

690. $\displaystyle\int_0^1 x^m(\log x)^n\,dx = \frac{(-1)^n n!}{(m+1)^{n+1}},\qquad m > -1, n = 0, 1, 2, \ldots$

If $n \neq 0, 1, 2, \ldots$ replace $n!$ by $\Gamma(n+1)$.

691. $\displaystyle\int_0^1 \frac{\log x}{1+x}\,dx = -\frac{\pi^2}{12}$

692. $\displaystyle\int_0^1 \frac{\log x}{1-x}\,dx = -\frac{\pi^2}{6}$

693. $\displaystyle\int_0^1 \frac{\log(1+x)}{x}\,dx = \frac{\pi^2}{12}$

694. $\displaystyle\int_0^1 \frac{\log(1-x)}{x}\,dx = -\frac{\pi^2}{6}$

695. $\displaystyle\int_0^1 (\log x)[\log(1+x)]\,dx = 2 - 2\log 2 - \frac{\pi^2}{12}$

696. $\displaystyle\int_0^1 (\log x)[\log(1-x)]\,dx = 2 - \frac{\pi^2}{6}$

697. $\displaystyle\int_0^1 \frac{\log x}{1-x^2}\,dx = -\frac{\pi^2}{8}$

698. $\displaystyle\int_0^1 \log\left(\frac{1+x}{1-x}\right)\cdot\frac{dx}{x} = \frac{\pi^2}{4}$

699. $\displaystyle\int_0^1 \frac{\log x\,dx}{\sqrt{1-x^2}} = -\frac{\pi}{2}\log 2$

700. $\displaystyle\int_0^1 x^m\left[\log\left(\frac{1}{x}\right)\right]^n dx = \frac{\Gamma(n+1)}{(m+1)^{n+1}},\qquad \text{if } m+1 > 0, n+1 > 0$

701. $\displaystyle\int_0^1 \frac{(x^p - x^q)\,dx}{\log x} = \log\left(\frac{p+1}{q+1}\right),\qquad (p+1 > 0, q+1 > 0)$

702. $\displaystyle\int_0^1 \frac{dx}{\sqrt{\log\left(\frac{1}{x}\right)}} = \sqrt{\pi}\ ,\ \text{(same as integral 686)}$

703. $\displaystyle\int_0^\infty \log\left(\frac{e^x + 1}{e^x - 1}\right)dx = \frac{\pi^2}{4}$

704. $\displaystyle\int_0^{\pi/2} (\log\sin x)\,dx = \int_0^{\pi/2} \log\cos x\,dx = -\frac{\pi}{2}\log 2$

705. $\displaystyle\int_0^{\pi/2} (\log\sec x)\,dx = \int_0^{\pi/2} \log\csc x\,dx = \frac{\pi}{2}\log 2$

706. $\displaystyle\int_0^\pi x(\log\sin x)\,dx = -\frac{\pi^2}{2}\log 2$

707. $\displaystyle\int_0^{\pi/2} (\sin x)(\log\sin x)\,dx = \log 2 - 1$

708. $\displaystyle\int_0^{\pi/2} (\log \tan x)\, dx = 0$

709. $\displaystyle\int_0^{\pi} \log (a \pm b \cos x)\, dx = \pi \log \left(\frac{a + \sqrt{a^2 - b^2}}{2} \right), \qquad (a \geq b)$

710. $\displaystyle\int_0^{\pi} \log (a^2 - 2ab \cos x + b^2)\, dx = \begin{cases} 2\pi \log a, & a \geq b > 0 \\ 2\pi \log b, & b \geq a > 0 \end{cases}$

711. $\displaystyle\int_0^{\infty} \frac{\sin ax}{\sinh bx}\, dx = \frac{\pi}{2b} \tanh \frac{a\pi}{2b}$

712. $\displaystyle\int_0^{\infty} \frac{\cos ax}{\cosh bx}\, dx = \frac{\pi}{2b} \operatorname{sech} \frac{a\pi}{2b}$

713. $\displaystyle\int_0^{\infty} \frac{dx}{\cosh ax} = \frac{\pi}{2a}$

714. $\displaystyle\int_0^{\infty} \frac{x\, dx}{\sinh ax} = \frac{\pi^2}{4a^2}$

715. $\displaystyle\int_0^{\infty} e^{-ax}(\cosh bx)\, dx = \frac{a}{a^2 - b^2}, \qquad (0 \leq |b| < a)$

716. $\displaystyle\int_0^{\infty} e^{-ax}(\sinh bx)\, dx = \frac{b}{a^2 - b^2}, \qquad (0 \leq |b| < a)$

717. $\displaystyle\int_0^{\infty} \frac{\sinh ax}{e^{bx} + 1}\, dx = \frac{\pi}{2b} \csc \frac{a\pi}{b} - \frac{1}{2a}$

718. $\displaystyle\int_0^{\infty} \frac{\sinh ax}{e^{bx} - 1}\, dx = \frac{1}{2a} - \frac{\pi}{2b} \cot \frac{a\pi}{b}$

719. $\displaystyle\int_0^{\pi/2} \frac{dx}{\sqrt{1 - k^2 \sin^2 x}} = \frac{\pi}{2}\left[1 + \left(\frac{1}{2}\right)^2 k^2 + \left(\frac{1 \cdot 3}{2 \cdot 4}\right)^2 k^4 \right.$
$$\left. + \left(\frac{1 \cdot 3 \cdot 5}{2 \cdot 4 \cdot 6}\right)^2 k^6 + \cdots \right], \text{ if } k^2 < 1$$

720. $\displaystyle\int_0^{\pi/2} \sqrt{1 - k^2 \sin^2 x}\, dx = \frac{\pi}{2}\left[1 - \left(\frac{1}{2}\right)^2 k^2 - \left(\frac{1 \cdot 3}{2 \cdot 4}\right)^2 \frac{k^4}{3} \right.$
$$\left. - \left(\frac{1 \cdot 3 \cdot 5}{2 \cdot 4 \cdot 6}\right)^2 \frac{k^6}{5} - \cdots \right], \text{ if } k^2 < 1$$

721. $\displaystyle\int_0^{\infty} e^{-x} \log x\, dx = -\gamma = -0.5772157 \ldots$

722. $\displaystyle\int_0^{\infty} e^{-x^2} \log x\, dx = -\frac{\sqrt{\pi}}{4}(\gamma + 2 \log 2)$

723. $\displaystyle\int_0^{\infty} \left(\frac{1}{1 - e^{-x}} - \frac{1}{x} \right) e^{-x}\, dx = \gamma = 0.5772157 \ldots \qquad \text{[Euler's Constant]}$

724. $\displaystyle\int_0^{\infty} \frac{1}{x}\left(\frac{1}{1 + x} - e^{-x} \right) dx = \gamma = 0.5772157 \ldots$

For n even:

725. $\displaystyle\int \cos^n x\, dx = \frac{1}{2^{n-1}} \sum_{k=0}^{\frac{n}{2}-1} \binom{n}{k} \frac{\sin (n - 2k)\, x}{(n - 2k)} + \frac{1}{2^n}\binom{n}{\frac{n}{2}} x$

726. $\int \sin^n x \, dx = \dfrac{1}{2^{n-1}} \displaystyle\sum_{k=0}^{\frac{n}{2}-1} \binom{n}{k} \dfrac{\sin\left[(n-2k)\left(\frac{\pi}{2}-x\right)\right]}{2k-n} + \dfrac{1}{2^n} \binom{n}{\frac{n}{2}} x$

For n odd:

727. $\int \cos^n x \, dx = \dfrac{1}{2^{n-1}} \displaystyle\sum_{k=0}^{\frac{n-1}{2}} \binom{n}{k} \dfrac{\sin(n-2k)x}{(n-2k)}$

728. $\int \sin^n x \, dx = \dfrac{1}{2^{n-1}} \displaystyle\sum_{k=0}^{\frac{n-1}{2}} \binom{n}{k} \dfrac{\sin\left[(n-2k)\left(\frac{\pi}{2}-x\right)\right]}{2k-n}$

DIFFERENTIAL EQUATIONS
SPECIAL FORMULAS

Certain types of differential equations occur sufficiently often to justify the use of formulas for the corresponding particular solutions. The following set of tables 1 to XIV covers all first, second, and nth order ordinary linear differential equations with constant coefficients for which the right members are of the form $P(x)e^{rx} \sin sx$ or $P(x)e^{rx} \cos sx$, where r and s are constants and $P(x)$, is a polynomial of degree n.

When the right member of a reducible linear partial differential equation with constant coefficients is not zero, particular solutions for certain types of right members are contained in tables XV to XXI. In these tables both F and P are used to denote polynomials, and it is assumed that no denominator is zero. In any formula the roles of x and y may be reversed throughout, changing a formula in which x dominates to one in which y dominates. Tables XIX, XX, XXI are applicable whether the equations are reducible or not.

The symbol $\binom{m}{n}$ stands for $\dfrac{m!}{(m-n)!n!}$ and is the $n+1$ st coefficient in the expansion of $(a+b)^m$. Also $0! = 1$ by definition.

The tables as herewith given are those contained in the text *Differential Equations* by Ginn and Company (1955) and are published with their kind permission and that of the author, Professor Frederick H. Steen.

Solution of Linear Differential Equations with Constant Coefficients

Any linear differential equation with constant coefficients may be written in the form

$$p(D)y = R(x)$$

where D is the differential operation

$$Dy = \frac{dy}{dx}$$

$p(D)$ is a polynomial in D,
y is the dependent variable,
x is the independent variable,
$R(x)$ is an arbitrary function of x.

A power of D represents repeated differentiation, that is

$$D^n y = \frac{d^n y}{dx^n}$$

For such an equation, the general solution may be written in the form

$$y = y_c + y_p$$

where y_p is any particular solution, and y_c is called the *complementary function*. This complementary function is defined as the general solution of the *homogeneous equation,* which is the original differential equation with the right side replaced by zero, i.e.

$$p(D)y = 0$$

The complementary function y_c may be determined as follows:

1. Factor the polynomial $p(D)$ into real and complex linear factors, just as if D were a variable instead of an operator.
2. For each nonrepeated linear factor of the form $(D-a)$, where a is real, write down a term of the form

$$ce^{ax}$$

where c is an arbitrary constant.
3. For each repeated real linear factor of the form $(D-a)^n$, write down n terms of the form

$$c_1 e^{ax} + c_2 x e^{ax} + c_3 x^2 e^{ax} + \cdots + c_n x^{n-1} e^{ax}$$

where the c_i's are arbitrary constants.

4. For each non-repeated conjugate complex pair of factors of the form $(D - a + ib)(D - a - ib)$, write down 2 terms of the form

$$c_1 e^{ax} \cos bx + c_2 e^{ax} \sin bx$$

5. For each repeated conjugate complex pair of factors of the form $(D - a + ib)^n (D - a - ib)^n$, write down $2n$ terms of the form

$$c_1 e^{ax} \cos bx + c_2 e^{ax} \sin bx + c_3 x e^{ax} \cos bx + c_4 x e^{ax} \sin bx + \cdots$$
$$+ c_{2n-1} x^{n-1} e^{ax} \cos bx + c_{2n} x^{n-1} e^{ax} \sin bx$$

6. The sum of all the terms thus written down is the complementary function y_c.

To find the particular solution y_p, use the following tables, as shown in the examples. For cases not shown in the tables, there are various methods of finding y_p. The most general method is called *variation of parameters*. The following example illustrates the method:

Find y_p for $(D^2 - 4) y = e^x$.

This example can be solved most easily by use of equation 63 in the tables following. However it is given here as an example of the method of variation of parameters.

The complementary function is

$$y_c = c_1 e^{2x} + c_2 e^{-2x}$$

To find y_p, replace the constants in the complementary function with unknown functions,

$$y_p = u e^{2x} + v e^{-2x}$$

We now prepare to substitute this assumed solution into the original equation. We begin by taking all the necessary derivatives:

$$y_p = u e^{2x} + v e^{-2x}$$
$$y_p' = 2u e^{2x} - 2v e^{-2x} + u' e^{2x} + v' e^{-2x}$$

For each derivative of y_p except the highest, we set the sum of all the terms containing u' and v' to 0. Thus the above equation becomes

$$u' e^{2x} + v' e^{-2x} = 0 \quad \text{and} \quad y_p' = 2u e^{2x} - 2v e^{-2x}$$

Continuing to differentiate, we have

$$y_p'' = 4u e^{2x} + 4v e^{-2x} + 2u' e^{2x} - 2v' e^{-2x}$$

When we substitute into the original equation, all the terms not containing u' or v' cancel out. This is a consequence of the method by which y_p was set up.

Thus all that is necessary is to write down the terms containing u' or v' in the highest order derivative of y_p, multiply by the constant coefficient of the highest power of D in $p(D)$, and set it equal to $R(x)$. Together with the previous terms in u' and v' which were set equal to 0, this gives us as many linear equations in the first derivatives of the unknown functions as there are unknown functions. The first derivatives may then be solved for by algebra, and the unknown functions found by integration. In the present example, this becomes

$$u' e^{2x} + v' e^{-2x} = 0$$
$$2u' e^{2x} - 2v' e^{-2x} = e^x$$

We eliminate v' and u' separately, getting

$$4u' e^{2x} = e^x$$
$$4v' e^{-2x} = -e^x$$

Thus

$$u' = \tfrac{1}{4} e^{-x}$$
$$v' = -\tfrac{1}{4} e^{3x}$$

Therefore, by integrating

$$u = -\tfrac{1}{4} e^{-x}$$
$$v = -\tfrac{1}{12} e^{3x}$$

A constant of integration is not needed, since we need only one particular solution. Thus

$$y_p = u e^{2x} + v e^{-2x} = -\tfrac{1}{4} e^{-x} e^{2x} - \tfrac{1}{12} e^{3x} e^{-2x}$$
$$= -\tfrac{1}{4} e^x - \tfrac{1}{12} e^x = -\tfrac{1}{3} e^x$$

and the general solution is

$$y = y_c + y_p = c_1 e^{2x} + c_2 e^{-2x} - \tfrac{1}{3} e^x$$

The following examples illustrate the use of the tables.

Example 1. Solve $(D^2 - 4)y = \sin 3x$.

Substitution of $q = -4$, $s = 3$ in formula 24 gives

$$y_p = \frac{\sin 3x}{-9 - 4}$$

wherefore the general solution is

$$y = c_1 e^{2x} + c_2 e^{-2x} - \frac{\sin 3x}{13}$$

Example 2. Obtain a particular solution of $(D^2 - 4D + 5)y = x^2 e^{3x} \sin x$.

Applying formula 40 with $a = 2, b = 1, r = 3, s = 1$, $P(x) = x^2$, $s + b = 2$, $s - b = 0$, $a - r = -1$, $(a - r)^2 + (s + b)^2 = 5$, $(a - r)^2 + (s - b)^2 = 1$, we have

$$y_p = \frac{e^{3x} \sin x}{2} \left[\left(\frac{2}{5} - \frac{0}{1} \right) x^2 + \left(\frac{2(-1)2}{25} - \frac{2(-1)0}{1} \right) 2x \right.$$

$$\left. + \left(\frac{3 \cdot 1 \cdot 2 - 2^3}{125} - \frac{3 \cdot 1 \cdot 0 - 0}{1} \right) 2 \right]$$

$$- \frac{e^{3x} \cos x}{2} \left[\left(\frac{-1}{5} - \frac{-1}{1} \right) x^2 + \left(\frac{1 - 4}{25} - \frac{1 - 0}{1} \right) 2x \right.$$

$$\left. + \left(\frac{-1 - 3(-1)4}{125} - \frac{-1 - 3(-1)0}{1} \right) 2 \right]$$

$$= \left(\tfrac{1}{5} x^2 - \tfrac{4}{25} x - \tfrac{2}{125} \right) e^{3x} \sin x + \left(-\tfrac{2}{5} x^2 + \tfrac{28}{25} x - \tfrac{136}{125} \right) e^{3x} \cos x$$

The special formulas effect a very considerable saving of time in problems of this type.

Example 3. Obtain a particular solution of $(D^2 - 4D + 5)y = x^2 e^{2x} \cos x$. (Compare with Example 2.)

Formula 40 is not applicable here since for this equation $r = a$, $s = b$, wherefore the denominator $(a - r)^2 + (s - b)^2 = 0$. We turn instead to formula 44. Substituting $a = 2$, $b = 1$, $P(x) = x^2$ and replacing sin by cos, cos by $-\sin$, we obtain

$$y_p = \frac{e^{2x} \cos x}{4} \left(x^2 - \tfrac{2}{4} \right) + \frac{e^{2x} \sin x}{2} \int \left(x^2 - \tfrac{1}{2} \right) dx$$

$$= \left(\frac{x^2}{4} - \frac{1}{8} \right) e^{2x} \cos x + \left(\frac{x^3}{6} - \frac{x}{4} \right) e^{2x} \sin x$$

which is the required solution.

Example 4. Find z_p for $(D_x - 3 D_y) z = \ln (y + 3x)$.

Referring to Table XV we note that formula 69 (not 68) is applicable. This gives

$$z_p = x \ln (y + 3x)$$

It is easily seen that $- y/3 \ln (y + 3x)$ would serve equally well.

Example 5. Solve $(D_x + 2D_y - 4) z = y \cos (y - 2x)$.

Since R in formula 76 contains a polynomial in x, not y, we rewrite the given equation in the form

$$(D_y + \tfrac{1}{2} D_x - 2) z = \tfrac{1}{2} y \cos (y - 2x)$$

Then

$$z_c = e^{2y} F(x - \tfrac{1}{2} y) = e^{2y} f(2x - y)$$

and by the formula

$$z_p = -\tfrac{1}{2} \cos (y - 2x) \cdot \left(\frac{y}{2} + \frac{\tfrac{1}{2}}{2} \right)$$

$$= -\tfrac{1}{8} (2y + 1) \cos (y - 2x)$$

Example 6. Find z_p for $(D_x + 4D_y)^3 z = (2x - y)^2$.

Using formula 79, we obtain

$$z_p = \frac{\int \int \int u^2 du^3}{[2 + 4(-1)]^3} = \frac{u^5}{5 \cdot 4 \cdot 3 \cdot (-8)} = -\frac{(2x - y)^5}{480}$$

Example 7. Find z_p for $(D_x^3 + 5D_x^2 D_y - 7D_x + 4)z = e^{2x+3y}$.

By formula 87

$$z_p = \frac{e^{2x+3y}}{2^3 + 5 \cdot 2^2 \cdot 3 - 7 \cdot 2 + 4} = \frac{e^{2x+3y}}{58}$$

Example 8. Find z_p for

$$(D_x^4 + 6D_x^3 D_y + D_x D_y + D_y^2 + 9)z = \sin (3x + 4y)$$

A-67

Since every term in the left member is of even degree in the two operators D_x and D_y, formula 90 is applicable. It gives

$$z_p = \frac{\sin(3x + 4y)}{(-9)^2 + 6(-9)(-12) + (-12) + (-16) + 9}$$

$$= \frac{\sin(3x + 4y)}{710}$$

TABLE I: $(D - a)y = R$

R	y_p
1. e^{rx}	$\dfrac{e^{rx}}{r - a}$
2. $\sin sx$*	$-\dfrac{a \sin sx + s \cos sx}{a^2 + s^2} = -\dfrac{1}{\sqrt{a^2 + s^2}} \sin\left(sx + \tan^{-1}\dfrac{s}{a}\right)$
3. $P(x)$	$-\dfrac{1}{a}\left[P(x) + \dfrac{P'(x)}{a} + \dfrac{P''(x)}{a^2} + \cdots + \dfrac{P^{(n)}(x)}{a^n}\right]$
4. $e^{rx} \sin sx$*	Replace a by $a - r$ in formula 2 and multiply by e^{rx}.
5. $P(x)e^{rx}$	Replace a by $a - r$ in formula 3 and multiply by e^{rx}.

6. $P(x) \sin sx$*
$$-\sin sx \left[\frac{a}{a^2 + s^2}P(x) + \frac{a^2 - s^2}{(a^2 + s^2)^2}P'(x) + \frac{a^3 - 3as^2}{(a^2 + s^2)^3}P''(x) + \cdots\right.$$
$$\left. + \frac{a^k - \binom{k}{2}a^{k-2}s^2 + \binom{k}{4}a^{k-4}s^4 - \cdots}{(a^2 + s^2)^k}P^{(k-1)}(x) + \cdots\right]$$
$$-\cos sx\left[\frac{s}{a^2 + s^2}P(x) + \frac{2as}{(a^2 + s^2)^2}P'(x) + \frac{3a^2s - s^3}{(a^2 + s^2)^3}P''(x) + \cdots\right.$$
$$\left. + \frac{\binom{k}{1}a^{k-1}s - \binom{k}{3}a^{k-3}s^3 + \cdots}{(a^2 + s^2)^k}P^{(k-1)}(x) + \cdots\right]$$

7. $P(x)e^{rx}\sin sx$*	Replace a by $a - r$ in formula 6 and multiply by e^{rx}.
8. e^{ax}	xe^{ax}
9. $e^{ax} \sin sx$*	$-\dfrac{e^{ax}\cos sx}{s}$
10. $P(x)e^{ax}$	$e^{ax}\displaystyle\int P(x)\,dx$

11. $P(x)e^{ax}\sin sx$*
$$\frac{e^{ax}\sin sx}{s}\left[\frac{P'(x)}{s} - \frac{P'''(x)}{s^3} + \frac{P^{\text{v}}(x)}{s^5} - \cdots\right] - \frac{e^{ax}\cos sx}{s}\left[P(x) - \frac{P''(x)}{s^2} + \frac{P^{\text{iv}}(x)}{s^4} - \cdots\right]$$

* For $\cos sx$ in R replace "sin" by "cos" and "cos" by "$-$ sin" in y_p.

$$D^n = \frac{d^n}{dx^n} \qquad \binom{m}{n} = \frac{m!}{(m - n)!n!} \qquad 0! = 1$$

TABLE II: $(D - a)^2y = R$

R	y_p
12. e^{rx}	$\dfrac{e^{rx}}{(r - a)^2}$
13. $\sin sx$*	$\dfrac{1}{(a^2 + s^2)^2}[(a^2 - s^2)\sin sx + 2as\cos sx] = \dfrac{1}{a^2 + s^2}\sin\left(sx + \tan^{-1}\dfrac{2as}{a^2 - s^2}\right)$
14. $P(x)$	$\dfrac{1}{a^2}\left[P(x) + \dfrac{2P'(x)}{a} + \dfrac{3P''(x)}{a^2} + \cdots + \dfrac{(n + 1)P^{(n)}(x)}{a^n}\right]$
15. $e^{rx}\sin sx$*	Replace a by $a - r$ in formula 13 and multiply by e^{rx}.
16. $P(x)e^{rx}$	Replace a by $a - r$ in formula 14 and multiply by e^{rx}.

17. $P(x)\sin sx$*
$$\sin sx\left[\frac{a^2 - s^2}{(a^2 + s^2)^2}P(x) + 2\frac{a^3 - 3as^2}{(a^2 + s^2)^3}P'(x) + 3\frac{a^4 - 6a^2s^2 + s^4}{(a^2 + s^2)^4}P''(x) + \cdots\right.$$
$$\left. + (k - 1)\frac{a^k - \binom{k}{2}a^{k-2}s^2 + \binom{k}{4}a^{k-4}s^4 - \cdots}{(a^2 + s^2)^k}P^{(k-2)}(x) + \cdots\right]$$
$$+\cos sx\left[\frac{2as}{(a^2 + s^2)^2}P(x) + 2\frac{3a^2s - s^3}{(a^2 + s^2)^3}P'(x) + 3\frac{4a^3s - 4as^3}{(a^2 + s^2)^4}P''(x) + \cdots\right.$$
$$\left. + (k - 1)\frac{\binom{k}{1}a^{k-1}s - \binom{k}{3}a^{k-3}s^3 + \cdots}{(a^2 + s^2)^k}P^{(k-2)}(x) + \cdots\right]$$

18. $P(x)e^{rx}\sin sx$*	Replace a by $a - r$ in formula 17 and multiply by e^{rx}.
19. e^{ax}	$\tfrac{1}{2}x^2 e^{ax}$
20. $e^{ax}\sin sx$*	$-\dfrac{e^{ax}\sin sx}{s^2}$
21. $P(x)e^{ax}$	$e^{ax}\displaystyle\int\int P(x)\,dx\,dx$

22. $P(x)e^{ax}\sin sx$*
$$-\frac{e^{ax}\sin sx}{s^2}\left[P(x) - \frac{3P''(x)}{s^2} + \frac{5P^{\text{iv}}(x)}{s^4} - \frac{7P^{\text{vi}}(x)}{s^6} + \cdots\right]$$
$$-\frac{e^{ax}\cos sx}{s^2}\left[\frac{2P'(x)}{s} - \frac{4P'''(x)}{s^3} + \frac{6P^{\text{v}}(x)}{s^5} - \cdots\right]$$

* For $\cos sx$ in R replace "sin" by "cos" and "cos" by "$-$ sin" in y_p.

TABLE III: $(D^2 + q)y = R$

R	y_p
23. e^{rx}	$\dfrac{e^{rx}}{r^2 + q}$
24. $\sin sx^*$	$\dfrac{\sin sx}{-s^2 + q}$
25. $P(x)$	$\dfrac{1}{q}\left[P(x) - \dfrac{P''(x)}{q} + \dfrac{P^{\text{iv}}(x)}{q^2} - \cdots + (-1)^k \dfrac{P^{(2k)}(x)}{q^k} \cdots\right]$

26. $e^{rx}\sin sx^*$

$$\frac{(r^2 - s^2 + q)e^{rx}\sin sx - 2rse^{rx}\cos sx}{(r^2 - s^2 + q)^2 + (2rs)^2} = \frac{e^{rx}}{\sqrt{(r^2 - s^2 + q)^2 + (2rs)^2}}\sin\left[sx - \tan^{-1}\frac{2rs}{r^2 - s^2 + q}\right]$$

27. $P(x)e^{rx}$

$$\frac{e^{rx}}{r^2 + q}\left[P(x) - \frac{2r}{r^2 + q}P'(x) + \frac{3r^2 - q}{(r^2 + q)^2}P''(x) - \frac{4r^3 - 4qr}{(r^2 + q)^3}P'''(x)\right.$$

$$\left.+ \cdots + (-1)^{k-1}\frac{\binom{k}{1}r^{k-1} - \binom{k}{3}r^{k-3}q + \binom{k}{5}r^{k-5}q^2 - \cdots}{(r^2 + q)^{k-1}}P^{(k-1)}(x) + \cdots\right]$$

28. $P(x)\sin sx^*$

$$\frac{\sin sx}{(-s^2 + q)}\left[P(x) - \frac{3s^2 + q}{(-s^2 + q)^2}P''(x) + \frac{5s^4 + 10s^2q + q^2}{(-s^2 + q)^4}P^{\text{iv}}(x) + \cdots\right.$$

$$\left.+ (-1)^k\frac{\binom{2k+1}{1}s^{2k} + \binom{2k+1}{3}s^{2k-2}q + \binom{2k+1}{5}s^{2k-4}q^2 + \cdots}{(-s^2 + q)^{2k}}P^{(2k)}(x) + \cdots\right]$$

$$-\frac{s\cos sx}{(-s^2 + q)}\left[\frac{2P'(x)}{(-s^2 + q)} - \frac{4s^2 + 4q}{(-s^2 + q)^3}P'''(x) + \cdots\right.$$

$$\left.+ (-1)^{k+1}\frac{\binom{2k}{1}s^{2k-2} + \binom{2k}{3}s^{2k-4}q + \cdots}{(-s^2 + q)^{2k-1}}P^{(2k-1)}(x) + \cdots\right]$$

TABLE IV: $(D^2 + b^2)y = R$

R	y_p
29. $\sin bx^*$	$-\dfrac{x\cos bx}{2b}$

30. $P(x)\sin bx^*$

$$\frac{\sin bx}{(2b)^2}\left[P(x) - \frac{P''(x)}{(2b)^2} + \frac{P^{\text{iv}}(x)}{(2b)^4} - \cdots\right] - \frac{\cos bx}{2b}\int\left[P(x) - \frac{P''(x)}{(2b)^2} + \cdots\right]dx$$

* For $\cos sx$ in R replace "sin" by "cos" and "cos" by "$-$ sin" in y_p.

TABLE V: $(D^2 + pD + q)y = R$

R	y_p
31. e^{rx}	$\dfrac{e^{rx}}{r^2 + pr + q}$
32. $\sin sx^*$	$\dfrac{(q - s^2)\sin sx - ps\cos sx}{(q - s^2)^2 + (ps)^2} = \dfrac{1}{\sqrt{(q - s^2)^2 + (ps)^2}}\sin\left(sx - \tan^{-1}\dfrac{ps}{q - s^2}\right)$

33. $P(x)$

$$\frac{1}{q}\left[P(x) - \frac{p}{q}P'(x) + \frac{p^2 - q}{q^2}P''(x) - \frac{p^3 - 2pq}{q^3}P'''(x) + \cdots\right.$$

$$\left.+ (-1)^n\frac{p^n - \binom{n-1}{1}p^{n-2}q + \binom{n-2}{2}p^{n-4}q^2 - \cdots}{q^n}P^{(n)}(x)\right]$$

34. $e^{rx}\sin sx^*$	Replace p by $p + 2r$, q by $q + pr + r^2$ in formula 32 and multiply by e^{rx}.
35. $P(x)e^{rx}$	Replace p by $p + 2r$, q by $q + pr + r^2$ in formula 33 and multiply by e^{rx}.

TABLE VI: $(D - b)(D - a)y = R$

36. $P(x)\sin sx^*$

$$\frac{\sin sx}{b - a}\left[\left(\frac{a}{a^2 + s^2} - \frac{b}{b^2 + s^2}\right)P(x) + \left(\frac{a^2 - s^2}{(a^2 + s^2)^2} - \frac{b^2 - s^2}{(b^2 + s^2)^2}\right)P'(x)\right.$$

$$\left.+ \left(\frac{a^3 - 3as^2}{(a^2 + s^2)^3} - \frac{b^3 - 3bs^2}{(b^2 + s^2)^3}\right)P''(x) + \cdots\right]$$

$$+ \frac{\cos sx}{b - a}\left[\left(\frac{s}{a^2 + s^2} - \frac{s}{b^2 + s^2}\right)P(x) + \left(\frac{2as}{(a^2 + s^2)^2} - \frac{2bs}{(b^2 + s^2)^2}\right)P'(x)\right.$$

$$\left.+ \left(\frac{3a^2s - s^3}{(a^2 + s^2)^3} - \frac{3b^2s - s^3}{(b^2 + s^2)^3}\right)P''(x) + \cdots\right]^\dagger$$

37. $P(x)e^{rx}\sin sx^*$	Replace a by $a - r$, b by $b - r$ in formula 36 and multiply by e^{rx}.

38. $P(x)e^{ax}$

$$\frac{e^{ax}}{a - b}\left[\int P(x)dx + \frac{P(x)}{(b - a)} + \frac{P'(x)}{(b - a)^2} + \frac{P''(x)}{(b - a)^3} + \cdots + \frac{P^{(n)}(x)}{(b - a)^{n+1}}\right]$$

* For $\cos sx$ in R replace "sin" by "cos" and "cos" by "$-$ sin" in y_p.

† For additional terms, compare with formula 6.

TABLE VII: $(D^2 - 2aD + a^2 + b^2)y = R$

R	y_p

39. $P(x) \sin sx$*

$$\frac{\sin sx}{2b}\left[\left(\frac{s+b}{a^2+(s+b)^2} - \frac{s-b}{a^2+(s-b)^2}\right)P(x) + \left(\frac{2a(s+b)}{[a^2+(s+b)^2]^2} - \frac{2a(s-b)}{[a^2+(s-b)^2]^2}\right)P'(x)\right.$$

$$\left. + \left(\frac{3a^2(s+b) - (s+b)^3}{[a^2+(s+b)^2]^3} - \frac{3a^2(s-b) - (s-b)^3}{[a^2+(s-b)^2]^3}\right)P''(x) + \cdots\right]$$

$$-\frac{\cos sx}{2b}\left[\left(\frac{a}{a^2+(s+b)^2} - \frac{a}{a^2+(s-b)^2}\right)P(x) + \left(\frac{a^2-(s+b)^2}{[a^2+(s+b)^2]^2} - \frac{a^2-(s-b)^2}{[a^2+(s-b)^2]^2}\right)P'(x)\right.$$

$$\left. + \left(\frac{a^3-3a(s+b)^2}{[a^2+(s+b)^2]^3} - \frac{a^3-3a(s-b)^2}{[a^2+(s-b)^2]^3}\right)P''(x) + \cdots\right]†$$

40. $P(x)e^{rx} \sin sx$* Replace a by $a - r$ in formula 39 and multiply by e^{rx}.

41. $P(x)e^{ax}$ $$\frac{e^{ax}}{b^2}\left[P(x) - \frac{P''(x)}{b^2} + \frac{P^{iv}(x)}{b^4} - \cdots\right]$$

42. $e^{ax} \sin sx$* $$\frac{e^{ax}\sin sx}{-s^2 + b^2}$$

43. $e^{ax} \sin bx$* $$-\frac{xe^{ax}\cos bx}{2b}$$

44. $P(x)e^{ax} \sin bx$* $$\frac{e^{ax}\sin bx}{(2b)^2}\left[P(x) - \frac{P''(x)}{(2b)^2} + \frac{P^{iv}(x)}{(2b)^4} - \cdots\right] - \frac{e^{ax}\cos bx}{2b}\int\left[P(x) - \frac{P''(x)}{(2b)^2} + \frac{P^{iv}(x)}{(2b)^4} - \cdots\right]dx$$

* For cos sx in R replace "sin" by "cos" and "cos" by " $-$ sin" in y_p.
† For additional terms, compare with formula 6.

TABLE VIII: $f(D)y = [D^n + a_{n-1}D^{n-1} + \cdots + a_1D + a_0]y = R$

R	y_p

45. e^{rx} $$\frac{e^{rx}}{f(r)}$$

46. $\sin sx$* $$\frac{[a_0 - a_2s^2 + a_4s^4 - \cdots]\sin sx - [a_1s - a_3s^3 + a_5s^5 + \cdots]\cos sx}{[a_0 - a_2s^2 + a_4s^4 - \cdots]^2 + [a_1s - a_3s^3 + a_5s^5 - \cdots]^2}$$

TABLE IX: $f(D^2)y = R$

47. $\sin sx$* $$\frac{\sin sx}{f(-s^2)} = \frac{\sin sx}{a_0 - a_2s^2 + \cdots \pm s^{2n}}$$

TABLE X: $(D - a)^n y = R$

48. e^{rx} $$\frac{e^{rx}}{(r-a)^n}$$

49. $\sin sx$* $$\frac{(-1)^n}{(a^2+s^2)^n}\left\{\left[a^n - \binom{n}{2}a^{n-2}s^2 + \binom{n}{4}a^{n-4}s^4 - \cdots\right]\sin sx + \left[\binom{n}{1}a^{n-1}s - \binom{n}{3}a^{n-3}s^3 + \cdots\right]\cos sx\right\}$$

50. $P(x)$ $$\frac{(-1)^n}{a^n}\left[P(x) + \binom{n}{1}\frac{P'(x)}{a} + \binom{n+1}{2}\frac{P''(x)}{a^2} + \binom{n+2}{3}\frac{P'''(x)}{a^3} + \cdots\right]$$

51. $e^{rx} \sin sx$* Replace a by $a - r$ in formula 49 and multiply by e^{rx}.

52. $e^{rx}P(x)$ Replace a by $a - r$ in formula 50 and multiply by e^{rx}.

53. $P(x) \sin sx$* $(-1)^n \sin sx[A_nP(x) + \binom{n}{1}A_{n+1}P'(x) + \binom{n+1}{2}A_{n+2}P''(x) + \binom{n+2}{3}A_{n+3}P'''(x) + \cdots]$

$+ (-1)^n \cos sx[B_nP(x) + \binom{n}{1}B_{n+1}P'(x) + \binom{n+1}{2}B_{n+2}P''(x) + \binom{n+2}{3}B_{n+3}P'''(x) + \cdots]$

$$A_1 = \frac{a}{a^2+s^2}, A_2 = \frac{a^2-s^2}{(a^2+s^2)^2}, \cdots, A_k = \frac{a^k - \binom{k}{2}a^{k-2}s^2 + \binom{k}{4}a^{k-4}s^4 - \cdots}{(a^2+s^2)^k}.$$

$$B_1 = \frac{s}{a^2+s^2}, B_2 = \frac{2as}{(a^2+s^2)^2}, \cdots, B_k = \frac{\binom{k}{1}a^{k-1}s - \binom{k}{3}a^{k-3}s^3 + \cdots}{(a^2+s^2)^k}.$$

54. $P(x)e^{rx} \sin sx$* Replace a by $a - r$ in formula 53 and multiply by e^{rx}.

55. $e^{ax}P(x)$ $\quad e^{ax}\displaystyle\int\!\!\int\cdots\int P(x)dx^n$

56. $P(x)e^{ax}\sin sx$* $\quad \dfrac{(-1)^{\frac{n-1}{2}}e^{ax}\sin sx}{s^n}\left[\binom{n}{n-1}\dfrac{P'(x)}{s}-\binom{n+2}{n-1}\dfrac{P'''(x)}{s^3}+\binom{n+4}{n-1}\dfrac{P^{\rm v}(x)}{s^5}-\cdots\right]$

$$+\dfrac{(-1)^{\frac{n+1}{2}}e^{ax}\cos sx}{s^n}\left[\binom{n-1}{n-1}P(x)-\binom{n+1}{n-1}\dfrac{P''(x)}{s^2}+\binom{n+3}{n-1}\dfrac{P^{\rm iv}(x)}{s^4}-\cdots\right]\ (n\ \text{odd})$$

$$\dfrac{(-1)^{\frac{n}{2}}e^{ax}\sin sx}{s^n}\left[\binom{n-1}{n-1}P(x)-\binom{n+1}{n-1}\dfrac{P''(x)}{s^2}+\binom{n+3}{n-1}\dfrac{P^{\rm iv}(x)}{s^4}-\cdots\right]$$

$$+\dfrac{(-1)^{\frac{n}{2}}e^{ax}\cos sx}{s^n}\left[\binom{n}{n-1}\dfrac{P'(x)}{s}-\binom{n+2}{n-1}\dfrac{P'''(x)}{s^3}+\binom{n+4}{n-1}\dfrac{P^{\rm v}(x)}{s^5}-\cdots\right]\ (n\ \text{even})$$

* For $\cos sx$ in R replace "sin" by "cos" and "cos" by " $-$ sin" in y_p.

TABLE XI: $(D-a)^nf(D)y=R$

57. e^{ax} $\quad \dfrac{x^n}{n!}\cdot\dfrac{e^{ax}}{f(a)}$

* For $\cos sx$ in R replace "sin" by "cos" and "cos" by " $-$ sin" in y_p.

TABLE XII: $(D^2+q)^ny=R$

R	y_p
58. e^{rx}	$e^{rx}/(r^2+q)^n$
59. $\sin sx$*	$\sin sx/(q-s^2)^n$
60. $P(x)$	$\dfrac{1}{q^n}\left[P(x)-\binom{n}{1}\dfrac{P''(x)}{q}+\binom{n+1}{2}\dfrac{P^{\rm iv}(x)}{q^2}-\binom{n+2}{3}\dfrac{P^{\rm vi}(x)}{q^3}+\cdots\right]$
61. $e^{rx}\sin sx$*	$\dfrac{e^{rx}}{(A^2+B^2)^n}\{[A^n-\binom{n}{2}A^{n-2}B^2+\binom{n}{4}A^{n-4}B^4-\cdots]\sin sx-[\binom{n}{1}A^{n-1}B-\binom{n}{3}A^{n-3}B^3+\cdots]\cos sx\}$

$$A=r^2-s^2+q,\qquad B=2rs$$

TABLE XIII: $(D^2+b^2)^ny=R$

62. $\sin bx$* $\quad (-1)^{\frac{n+1}{2}}\dfrac{x^n\cos bx}{n!(2b)^n}\ (n\ \text{odd}),\qquad\qquad (-1)^{\frac{n}{2}}\dfrac{x^n\sin bx}{n!(2b)^n}\ (n\ \text{even})$

TABLE XIV: $(D^n-q)y=R$

63. e^{rx}	$e^{rx}/(r^n-q)$
64. $P(x)$	$-\dfrac{1}{q}\left[P(x)\dfrac{P^{(n)}(x)}{q}+\dfrac{P^{(2n)}(x)}{q^2}+\cdots\right]$
65. $\sin sx$*	$-\dfrac{q\sin sx+(-1)^{\frac{n-1}{2}}s^n\cos sx}{q^2+s^{2n}}\ (n\ \text{odd}),\qquad \dfrac{\sin sx}{(-s^2)^{n/2}-q}\ (n\ \text{even})$
66. $e^{rx}\sin sx$*	$\dfrac{Ae^{rx}\sin sr-Be^{rx}\cos sx}{A^2+B^2}=\dfrac{e^{rx}}{\sqrt{A^2+B^2}}\sin\left(sx-\tan^{-1}\dfrac{B}{A}\right)$

$$A=[r^n-\binom{n}{2}r^{n-2}s^2+\binom{n}{4}r^{n-4}s^4-\cdots]-q,\qquad B=[\binom{n}{1}r^{n-1}s-\binom{n}{3}r^{n-3}s^3+\cdots]$$

* For $\cos sx$ in R replace "sin" by "cos" and "cos" by " $-$ sin" in y_p.

TABLE XV: $(D_x+mD_y)z=R$

R	z_p
67. e^{ax+by}	$\dfrac{e^{ax+by}}{a+mb}$
68. $f(ax+by)$	$\dfrac{\int f(u)du}{a+mb},\ u=ax+by$
69. $f(y-mx)$	$xf(y-mx)$
70. $\phi(x,y)f(y-mx)$	$f(y-mx)\int\phi(x,a+mx)dx\ \ (a=y-mx\ \text{after integration})$

TABLE XVI: $(D_z + mD_y - k)z = R$

71. e^{az+by} $\dfrac{e^{az+by}}{a + mb - k}$

72. $\sin(ax + by)*$ $-\dfrac{(a + bm)\cos(ax + by) + k\sin(ax + by)}{(a + bm)^2 + k^2}$

73. $e^{\alpha x+\beta y}\sin(ax + by)*$ Replace k in 72 by $k - \alpha - m\beta$ and multiply by $e^{\alpha x+\beta y}$

74. $e^{zk}f(ax + by)$ $\dfrac{e^{zk}\int\int f(u)\,du}{a + mb}$, $u = ax + by$

75. $f(y - mx)$ $-\dfrac{f(y - mx)}{k}$

76. $P(x)f(y - mx)$ $-\dfrac{1}{k}f(y - mx)\left[P(x) + \dfrac{P'(x)}{k} + \dfrac{P''(x)}{k^2} + \cdots + \dfrac{P^{(n)}(x)}{k^n}\right]$

77. $e^{kz}f(y - mx)$ $xe^{kz}f(y - mx)$

* For $\cos(ax + by)$ replace "sin" by "cos," and "cos" by "$-$sin" in z_p.

$$D_z = \frac{\partial}{\partial x}; \qquad D_y = \frac{\partial}{\partial y}; \qquad D_z{}^k D_y{}^r = \frac{\partial^{k+r}}{\partial_z{}^k \partial_y{}^r}$$

TABLE XVII: $(D_z + mD_y)^n z = R$

R z_p

78. e^{az+by} $\dfrac{e^{az+by}}{(a + mb)^n}$

79. $f(ax + by)$ $\dfrac{\int\int \cdots \int f(u)\,du^n}{(a + mb)^n}$, $u = ax + by$

80. $f(y - mx)$ $\dfrac{x^n}{n!}f(y - mx)$

81. $\phi(x, y)f(y + mx)$ $f(y - mx)\int\int \cdots \int \phi(x, a + mx)\,dx^n$ $(a = y - mx$ after integration$)$

TABLE XVIII: $(D_z + mD_y - k)^n z = R$

82. e^{az+by} $\dfrac{e^{az+by}}{(a + mb - k)^n}$

83. $f(y - mx)$ $\dfrac{(-1)^n f(y - mx)}{k^n}$

84. $P(x)f(y - mx)$ $\dfrac{(-1)^n}{k^n}f(y - mx)\left[P(x) + \binom{n}{1}\dfrac{P'(x)}{k} + \binom{n+1}{2}\dfrac{P''(x)}{k^2} + \binom{n+2}{3}\dfrac{P'''(x)}{k^3} + \cdots\right]$

85. $e^{kz}f(ax + by)$ $\dfrac{e^{kz}\int\int \cdots \int f(u)\,du^n}{(a + mb)^n}$, $u = ax + by$

86. $e^{kz}f(y - mx)$ $\dfrac{x^n}{n!}e^{kz}f(y - mx)$

TABLE XIX: $[D_x{}^n + a_1 D_x{}^{n-1}D_y + a_2 D_x{}^{n-2}D_y{}^2 + \cdots + a^n D_y{}^n]z = R$

87. e^{ax+by} $\dfrac{e^{ax+by}}{a + a_1 a^{n-1}b + a_2 a^{n-2}b^2 + \cdots + a_n b^n}$

88. $f(ax + by)$ $\dfrac{\int\int \cdots \int f(u)\,du^n}{a^n + a_1 a^{n-1}b + a_2 a^{n-2}b^2 + \cdots + a^n b^n}$, $(u = ax + by)$

TABLE XX: $F(D_x, D_y)z = R$

89. e^{ax+by} $\dfrac{e^{ax+by}}{F(a, b)}$

TABLE XXI: $F(D_x{}^2, D_x D_y, D_y{}^2)z = R$

90. $\sin(ax + by)*$ $\dfrac{\sin(ax + by)}{F(-a^2, -ab, -b^2)}$

* For $\cos(ax + by)$ replace "sin" by "cos", and "cos" by "$-$sin" in z_p.

DIFFERENTIAL EQUATIONS

Differential equation	Method of solution
Separation of variables $f_1(x)\,g_1(y)\,dx + f_2(x)\,g_2(y)\,dy = 0$	$$\int \frac{f_1(x)}{f_2(x)}\,dx + \int \frac{g_2(y)}{g_1(y)}\,dy = c$$
Exact equation $M(x,y)dx + N(x,y)dy = 0$ where $\partial M/\partial y = \partial N/\partial x$	$$\int M\,\partial x + \int \left(N - \frac{\partial}{\partial y}\int M\,\partial x \right)\,dy = c$$ where ∂x indicates that the integration is to be performed with respect to x keeping y constant.
Linear first order equation $\dfrac{dy}{dx} + P(x)y = Q(x)$	$$ye^{\int Pdx} = \int Qe^{\int Pdx}\,dx + c$$
Bernoulli's equation $\dfrac{dy}{dx} + P(x)y = Q(x)y^n$	$$ve^{(1-n)\int Pdx} = (1-n)\int Qe^{(1-n)\int Pdx}\,dx + c$$ where $v = y^{1-n}$. If $n = 1$, the solution is $$\ln y = \int (Q - P)\,dx + c$$
Homogeneous equation $\dfrac{dy}{dx} = F\left(\dfrac{y}{x}\right)$	$$\ln x = \int \frac{dv}{F(v) - v} + c$$ where $v = y/x$. If $F(v) = v$, the solution is $y = cx$
Reducible to homogeneous $(a_1 x + b_1 y + c_1)\,dx + (a_2 x + b_2 y + c_2)\,dy = 0$ $\dfrac{a_1}{a_2} \neq \dfrac{b_1}{b_2}$	Set $u = a_1 x + b_1 y + c_1$. $v = a_2 x + b_2 y + c_2$ Eliminate x and y and the equation becomes homogenous
Reducible to separable $(a_1 x + b_1 y + c_1)\,dx + (a_2 x + b_2 y + c_2)\,dy = 0$ $\dfrac{a_1}{a_2} = \dfrac{b_1}{b_2}$	Set $u = a_1 x + b_1 y$ Eliminate x or y and equation becomes separable

$y\,F(xy)\,dx + x\,G(xy)\,dy = 0$	$\ln x = \displaystyle\int \frac{G(v)\,dv}{v\{G(v) - F(v)\}} + c$ where $v = xy$. If $G(v) = F(v)$, the solution is $xy = c$.
Linear, homogeneous second order equation $\dfrac{d^2 y}{dx^2} + b\,\dfrac{dy}{dx} + cy = 0$ b, c are real constants	Let m_1, m_2 be the roots of $m^2 + bm + c = 0$. Then there are 3 cases: Case 1. m_1, m_2 real and distinct: $$y = c_1 e^{m_1 x} + c_2 e^{m_2 x}$$ Case 2. m_1, m_2 real and equal: $$y = c_1 e^{m_1 x} + c_2 x e^{m_1 x}$$ Case 3. $m_1 = p + qi,\ m_2 = p - qi$: $$y = e^{px}(c_1 \cos qx + c_2 \sin qx)$$ where $p = -b/2,\ q = \sqrt{4c - b^2}/2$
Linear, nonhomogeneous second order equation $\dfrac{d^2 y}{dx^2} + b\,\dfrac{dy}{dx} + cy = R(x)$ $b,\ c$ are real constants	There are 3 cases corresponding to those immediately above: Case 1. $$y = c_1 e^{m_1 x} + c_2 e^{m_2 x}$$ $$+ \frac{e^{m_1 x}}{m_1 - m_2} \int e^{-m_1 x} R(x)\,dx$$ $$+ \frac{e^{m_2 x}}{m_2 - m_1} \int e^{-m_2 x} R(x)\,dx$$ Case 2. $$y = c_1 e^{m_1 x} + c_2 x e^{m_1 x}$$ $$+ x e^{m_1 x} \int e^{-m_1 x} R(x)\,dx$$ $$- e^{m_1 x} \int x e^{-m_1 x} R(x)\,dx$$ Case 3. $$y = e^{px}(c_1 \cos qx + c_2 \sin qx)$$ $$+ \frac{e^{px} \sin qx}{q} \int e^{-px} R(x) \cos qx\,dx$$ $$- \frac{e^{px} \cos qx}{q} \int e^{-px} R(x) \sin qx\,dx$$

DIFFERENTIAL EQUATIONS (Continued)

Euler or Cauchy equation $x^2 \dfrac{d^2 y}{dx^2} + bx \dfrac{dy}{dx} + cy = S(x)$	Putting $x = e^t$, the equation becomes $\dfrac{d^2 y}{dt^2} + (b-1)\dfrac{dy}{dt} + cy = S(e^t)$ and can then be solved as a linear second order equation.
Bessel's equation $x^2 \dfrac{d^2 y}{dx^2} + x \dfrac{dy}{dx} + (\lambda^2 x^2 - n^2)y = 0$	$y = c_1 J_n(\lambda x) + c_2 Y_n(\lambda x)$
Transformed Bessel's equation $x^2 \dfrac{d^2 y}{dx^2} + (2p+1)x \dfrac{dy}{dx} + (\alpha^2 x^{2r} + \beta^2)y = 0$	$y = x^{-p}\left\{ c_1 J_{q/r}\left(\dfrac{\alpha}{r}x^r\right) + c_2 Y_{q/r}\left(\dfrac{\alpha}{r}x^r\right) \right\}$ where $q = \sqrt{p^2 - \beta^2}$.
Legendre's equation $(1 - x^2)\dfrac{d^2 y}{dx^2} - 2x \dfrac{dy}{dx} + n(n+1)y = 0$	$y = c_1 P_n(x) + c_2 Q_n(x)$

FOURIER SERIES

1. If $f(x)$ is a bounded periodic function of period 2L (i.e. $f x + 2L) = f(x)$, and satisfies the *Dirichlet conditions:*

    A. In any period $f(x)$ is continuous, except possibly for a finite number of jump discontinuities.
    B. In any period $f(x)$ has only a finite number of maxima and minima.

    Then $f(x)$ may be represented by the *Fourier series*

    $$\frac{a_0}{2} + \sum_{n=1}^{\infty}\left(a_n \cos \frac{n\pi x}{L} + b_n \sin \frac{n\pi x}{L} \right)$$

    where a_n and b_n are as determined below. This series will converge to $f(x)$ at every point where $f(x)$ is continuous, and to

    $$\frac{f(x^+) + f(x^-)}{2}$$

    (i.e. the average of the left-hand and right-hand limits) at every point where $f(x)$ has a jump discontinuity.

    $$a_n = \frac{1}{L}\int_{-L}^{L} f(x)\cos \frac{n\pi x}{L}\, dx, \quad n = 0, 1, 2, 3, \ldots;$$

    $$b_n = \frac{1}{L}\int_{-L}^{L} f(x)\sin \frac{n\pi x}{L}\, dx, \quad n = 1, 2, 3, \ldots$$

    We may also write

    $$a_n = \frac{1}{L}\int_{\alpha}^{\alpha+2L} f(x)\cos \frac{n\pi x}{L}\, dx \text{ and } b_n = \frac{1}{L}\int_{\alpha}^{\alpha+2L} f(x)\sin \frac{n\pi x}{L}\, dx$$

    where α is any real number. Thus if $\alpha = 0$,

    $$a_n = \frac{1}{L}\int_{0}^{2L} f(x)\cos \frac{n\pi x}{L}\, dx, \quad n = 0, 1, 2, 3, \ldots;$$

    $$b_n = \frac{1}{L}\int_{0}^{2L} f(x)\sin \frac{n\pi x}{L}\, dx, \quad n = 1, 2, 3, \ldots$$

2. If in addition to the above restrictions, $f(x)$), is even (i.e., $f(-x) = f(x)$ the Fourier series reduces to

$$\frac{a_0}{2} + \sum_{n=1}^{\infty} a_n \cos \frac{n\pi x}{L}$$

That is, $b_n = 0$. In this case, a simpler formula for a_n is

$$a_n = \frac{2}{L} \int_0^L f(x) \cos \frac{n\pi x}{L} \, dx, \quad n = 0, 1, 2, 3, \ldots$$

3. If in addition to the restrictions in (1), $f(x)$ is an odd function (i.e., $f(-x) = -f(x)$), then the Fourier series reduces to

$$\sum_{n=1}^{\infty} b_n \sin \frac{n\pi x}{L}$$

That is, $a_n = 0$. In this case, a simpler formula for the b_n is

$$b_n = \frac{2}{L} \int_0^L f(x) \sin \frac{n\pi x}{L} \, dx, \quad n = 1, 2, 3, \ldots$$

4. If in addition to the restrictions in (2) above, $f(x) = -f(L - x)$, then a_n will be 0 for all even values of n, including $n = 0$. Thus in this case, the expansion reduces to

$$\sum_{m=1}^{\infty} a_{2m-1} \cos \frac{(2m - 1)\pi x}{L}$$

5. If in addition to the restrictions in (3) above, $f(x) = f(L - x)$, then b_n will be 0 for all even values of n. Thus in this case, the expansion reduces to

$$\sum_{m=1}^{\infty} b_{2m-1} \sin \frac{(2m - 1)\pi x}{L}$$

(The series in (4) and (5) are known as *odd-harmonic series,* since only the odd harmonics appear. Similar rules may be stated for even-harmonic series, but when a series appears in the even-harmonic form, it means that $2L$ has not been taken as the smallest period of $f(x)$. Since any integral multiple of a period is also a period, series obtained in this way will also work, but in general computation is simplified if $2L$ is taken to be the smallest period.)

6. If we write the Euler definitions for $\cos \theta$ and $\sin \theta$, we obtain the complex form of the Fourier Series known either as the "Complex Fourier Series" or the "Exponential Fourier Series" of $f(x)$. It is represented as

$$f(x) = \frac{1}{2} \sum_{n=-\infty}^{n=+\infty} c_n e^{i\omega_n x}$$

where

$$c_n = \frac{1}{L} \int_{-L}^{L} f(x) e^{-i\omega_n x} dx, \quad n = 0, \pm 1, \pm 2, \pm 3, \ldots$$

with $\omega_n = \dfrac{n\pi}{L}, \quad n = 0, \pm 1, \pm 2, \ldots$

The set of coefficients $\{c_n\}$ is often referred to as the Fourier spectrum.

7. If both sine and cosine terms are present and if $f(x)$ is of period $2L$ and expandable by a Fourier series, it can be represented as

$$f(x) = \frac{a_0}{2} + \sum_{n=1}^{\infty} c_n \sin\left(\frac{n\pi x}{L} + \phi_n\right), \text{ where } a_n = c_n \sin \phi_n,$$

$$b_n = c_n \cos \phi_n, \quad c_n = \sqrt{a_n^2 + b_n^2}, \quad \phi_n = \arctan\left(\frac{a_n}{b_n}\right)$$

It can also be represented as

$$f(x) = \frac{a_0}{2} + \sum_{n=1}^{\infty} c_n \cos\left(\frac{n\pi x}{L} + \phi_n\right), \text{ where } a_n = c_n \cos \phi_n,$$

$$b_n = -c_n \sin \phi_n, \quad c_n = \sqrt{a_n^2 + b_n^2}, \quad \phi_n = \arctan\left(-\frac{b_n}{a_n}\right)$$

where ϕ_n is chosen so as to make a_n, b_n, and c_n hold.

8. The following table of trigonometric identities should be helpful for developing Fourier Series.

	n	n even	n odd	$n/2$ odd	$n/2$ even
$\sin n\pi$	0	0	0	0	0
$\cos n\pi$	$(-1)^n$	$+1$	-1	$+1$	$+1$
*$\sin \dfrac{n\pi}{2}$		0	$(-1)^{(n-1)/2}$	0	0
*$\cos \dfrac{n\pi}{2}$		$(-1)^{n/2}$	0	-1	$+1$
$\sin \dfrac{n\pi}{4}$			$\dfrac{\sqrt{2}}{2}(-1)^{(n^2+4n+11)/8}$	$(-1)^{(n-2)/4}$	0

*A useful formula for $\sin \dfrac{n\pi}{2}$ and $\cos \dfrac{n\pi}{2}$ is given by

$$\sin \frac{n\pi}{2} = \frac{(i)^{n+1}}{2}[(-1)^n - 1] \text{ and } \cos \frac{n\pi}{2} = \frac{(i)^n}{2}[(-1)^n + 1], \text{ where } i^2 = -1.$$

AUXILIARY FORMULAS FOR FOURIER SERIES

$$1 = \frac{4}{\pi}\left[\sin \frac{\pi x}{k} + \frac{1}{3}\sin \frac{3\pi x}{k} + \frac{1}{5}\sin \frac{5\pi x}{k} + \cdots\right] \qquad [0 < x < k]$$

$$x = \frac{2k}{\pi}\left[\sin \frac{\pi x}{k} - \frac{1}{2}\sin \frac{2\pi x}{k} + \frac{1}{3}\sin \frac{3\pi x}{k} - \cdots\right] \qquad [-k < x < k]$$

$$x = \frac{k}{2} - \frac{4k}{\pi^2}\left[\cos \frac{\pi x}{k} + \frac{1}{3^2}\cos \frac{3\pi x}{k} + \frac{1}{5^2}\cos \frac{5\pi x}{k} + \cdots\right] \qquad [0 < x < k]$$

$$x^2 = \frac{2k^2}{\pi^3}\left[\left(\frac{\pi^2}{1} - \frac{4}{1}\right)\sin \frac{\pi x}{k} - \frac{\pi^2}{2}\sin \frac{2\pi x}{k} + \left(\frac{\pi^2}{3} - \frac{4}{3^3}\right)\sin \frac{3\pi x}{k}\right.$$
$$\left. - \frac{\pi^2}{4}\sin \frac{4\pi x}{k} + \left(\frac{\pi^2}{5} - \frac{4}{5^3}\right)\sin \frac{5\pi x}{k} + \cdots\right] \quad [0 < x < k]$$

$$x^2 = \frac{k^2}{3} - \frac{4k^2}{\pi^2}\left[\cos \frac{\pi x}{k} - \frac{1}{2^2}\cos \frac{2\pi x}{k} + \frac{1}{3^2}\cos \frac{3\pi x}{k} - \frac{1}{4^2}\cos \frac{4\pi x}{k} + \cdots\right]$$
$$[-k < x < k]$$

$$1 - \frac{1}{3} + \frac{1}{5} - \frac{1}{7} + \cdots = \frac{\pi}{4}$$

$$1 + \frac{1}{2^2} + \frac{1}{3^2} + \frac{1}{4^2} + \cdots = \frac{\pi^2}{6}$$

$$1 - \frac{1}{2^2} + \frac{1}{3^2} - \frac{1}{4^2} + \cdots = \frac{\pi^2}{12}$$

$$1 + \frac{1}{3^2} + \frac{1}{5^2} + \frac{1}{7^2} + \cdots = \frac{\pi^2}{8}$$

$$\frac{1}{2^2} + \frac{1}{4^2} + \frac{1}{6^2} + \frac{1}{8^2} + \cdots = \frac{\pi^2}{24}$$

FOURIER EXPANSIONS FOR BASIC PERIODIC FUNCTIONS

$$f(x) = \frac{4}{\pi} \sum_{n=1,3,5...} \frac{1}{n} \sin \frac{n\pi x}{L}$$

$$f(x) = \frac{2}{\pi} \sum_{n=1}^{\infty} \frac{(-1)^n}{n} \left(\cos \frac{n\pi c}{L} - 1 \right) \sin \frac{n\pi x}{L}$$

$$f(x) = \frac{c}{L} + \frac{2}{\pi} \sum_{n=1}^{\infty} \frac{(-1)^n}{n} \sin \frac{n\pi c}{L} \cos \frac{n\pi x}{L}$$

$$f(x) = \frac{2}{L} \sum_{n=1}^{\infty} \sin \frac{n\pi}{2} \frac{\sin (\frac{1}{2}n\pi c/L)}{\frac{1}{2}n\pi c/L} \sin \frac{n\pi x}{L}$$

$$f(x) = \frac{2}{\pi} \sum_{n=1}^{\infty} \frac{(-1)^{n+1}}{n} \sin \frac{n\pi x}{L}$$

$$f(x) = \frac{1}{2} - \frac{4}{\pi^2} \sum_{n=1,3,5...} \frac{1}{n^2} \cos \frac{n\pi x}{L}$$

$$f(x) = \frac{8}{\pi^2} \sum_{n=1,3,5...} \frac{(-1)^{(n-1)/2}}{n^2} \sin \frac{n\pi x}{L}$$

$$f(x) = \frac{1}{2} - \frac{1}{\pi} \sum_{n=1}^{\infty} \frac{1}{n} \sin \frac{n\pi x}{L}$$

$$f(x) = \frac{1}{2}(1 + a) + \frac{2}{\pi^2(1-a)} \sum_{n=1}^{\infty} \frac{1}{n^2} [(-1)^n \cos n\pi a - 1] \cos \frac{n\pi x}{L} \; ; \; \left(a = \frac{c}{2L}\right)$$

$$f(x) = \frac{2}{\pi} \sum_{n=1}^{\infty} \frac{(-1)^{n-1}}{n} \left[1 + \frac{\sin n\pi a}{n\pi(1-a)}\right] \sin \frac{n\pi x}{L} \; ; \; \left(a = \frac{c}{2L}\right)$$

$$f(x) = \frac{1}{2} - \frac{4}{\pi^2(1-2a)} \sum_{n=1,3,5\ldots} \frac{1}{n^2} \cos n\pi a \cos \frac{n\pi x}{L} \; ; \; \left(a = \frac{c}{2L}\right)$$

$$f(x) = \frac{2}{\pi} \sum_{n=1}^{\infty} \frac{(-1)^n}{n} \left[1 + \frac{1+(-1)^n}{n\pi(1-2a)} \sin n\pi a\right] \sin \frac{n\pi x}{L} \; ; \; \left(a = \frac{c}{2L}\right)$$

$$f(x) = \frac{4}{\pi} \sum_{n=1}^{\infty} \frac{1}{n} \sin \frac{n\pi}{4} \sin n\pi a \sin \frac{n\pi x}{L} \; ; \; \left(a = \frac{c}{2L}\right)$$

$$f(x) = \frac{9}{\pi^2} \sum_{n=1}^{\infty} \frac{1}{n^2} \sin \frac{n\pi}{3} \sin \frac{n\pi x}{L} \; ; \; \left(a = \frac{c}{2L}\right)$$

$$f(x) = \frac{32}{3\pi^2} \sum_{n=1}^{\infty} \frac{1}{n^2} \sin \frac{n\pi}{4} \sin \frac{n\pi x}{L} \; ; \; \left(a = \frac{c}{2L}\right)$$

$$f(x) = \frac{1}{\pi} + \frac{1}{2} \sin \omega t - \frac{2}{\pi} \sum_{n=2,4,6\ldots} \frac{1}{n^2-1} \cos n\omega t$$

Extracted from graphs and formulas, pages 372, 373, Differential Equations in Engineering Problems, Salvadori and Schwarz, published by Prentice-Hall, Inc., 1954.

THE FOURIER TRANSFORMS*

R. E. Gaskell

For a piecewise continuous function $F(x)$ over a finite interval $0 \leqq x \leqq \pi$, the *finite Fourier cosine transform* of $F(x)$ is

$$f_c(n) = \int_0^\pi F(x) \cos nx \, dx \quad (n = 0, 1, 2, \ldots) \tag{1}$$

If x ranges over the interval $0 \leqq x \leqq L$, the substitution $x' = \pi x/L$ allows the use of this definition, also. The inverse transform is written.

$$\bar{F}(x) = \frac{1}{\pi} f_c(0) - \frac{2}{\pi} \sum_{n=1}^{\infty} f_c(n) \cos nx \quad (0 < x < \pi) \tag{2}$$

where $F(x) = \dfrac{[F(x + o) + F(x - o)]}{2}$. We observe that $\bar{F}(x) = F(x)$ — at points of continuity. The formula

$$f_c^{(2)}(n) = \int_0^\pi F''(x) \cos nx \, dx$$

$$= -n^2 f_c(n) - F'(0) + (-1)^n F'(\pi) \tag{3}$$

makes the finite Fourier cosine transform useful in certain boundary value problems.

Analogously, the *finite Fourier sine transform* of $F(x)$ is

$$f_s(n) = \int_0^\pi F(x) \sin nx \, dx \quad (n = 1, 2, 3, \ldots) \tag{4}$$

and

$$\bar{F}(x) = \frac{2}{\pi} \sum_{n=1}^{\infty} f_s(n) \sin nx \quad (0 < x < \pi) \tag{5}$$

Corresponding to (3) we have

$$f_s^{(2)}(n) = \int_0^\pi F''(x) \sin nx \, dx$$

$$= -n^2 f_s(n) - nF(0) - n(-1)^n F(\pi) \tag{6}$$

Fourier Transforms

If $F(x)$ is defined for $x \geqq 0$ and is piecewise continuous over any finite interval, and if

$$\int_0^\infty F(x) \, dx$$

is absolutely convergent, then

$$f_c(\alpha) = \sqrt{\frac{2}{\pi}} \int_0^\infty F(x) \cos (\alpha x) \, dx \tag{7}$$

is the *Fourier cosine transform* of $F(x)$. Furthermore.

$$\bar{F}(x) = \sqrt{\frac{2}{\pi}} \int_0^\infty f_c(\alpha) \cos (\alpha x) \, d\alpha \tag{8}$$

If $\lim\limits_{x \to \infty} \dfrac{d^n F}{dx^n} = 0$, an important property of the Fourier cosine transform

$$f_c^{(2r)}(\alpha) = \sqrt{\frac{2}{\pi}} \int_0^\infty \left(\frac{d^{2r} F}{dx^{2r}} \right) \cos (\alpha x) \, dx$$

$$= -\sqrt{\frac{2}{\pi}} \sum_{n=0}^{r-1} (-1)^n a_{2r-2n-1} \alpha^{2n} + (-1)^r \alpha^{2r} f_c(\alpha) \tag{9}$$

where $\lim\limits_{x \to \infty} \dfrac{d^r F}{dx^r} = a_r$, makes it useful in the solution of many problems.

Under the same conditions.

$$f_s(\alpha) = \sqrt{\frac{2}{\pi}} \int_0^\infty F(x) \sin (\alpha x) \, dx \tag{10}$$

* From Beyer, W. H., Ed., *CRC Handbook of Mathematical Sciences*, 5th ed., CRC Press, Boca Raton, 1978, 592—598. With permission.

defines the *Fourier sine transform* of $F(x)$, and

$$\bar{F}(x) = \sqrt{\frac{2}{\pi}} \int_0^\infty f_s(\alpha) \sin(\alpha x)\, d\alpha \tag{11}$$

Corresponding to (9) we have

$$f_s^{(2r)}(\alpha) = \sqrt{\frac{2}{\pi}} \int_0^\infty \frac{d^{2r}F}{dx^{2r}} \sin(\alpha x)\, dx$$

$$= -\sqrt{\frac{2}{\pi}} \sum_{n=1}^{r} (-1)^n \alpha^{2n-1} a_{2r-2n} + (-1)^{r-1} \alpha^{2r} f_s(\alpha) \tag{12}$$

Similarly, if $F(x)$ is defined for $-\infty < x < \infty$, and if $\int_{-\infty}^\infty F(x)\, dx$ is absolutely convergent, then

$$f(\alpha) = \frac{1}{\sqrt{2\pi}} \int_{-\infty}^\infty F(x) e^{i\alpha x}\, dx \tag{13}$$

is the *Fourier transform* of $F(x)$, and

$$\bar{F}(x) = \frac{1}{\sqrt{2\pi}} \int_{-\infty}^\infty f(\alpha) e^{-i\alpha x}\, d\alpha \tag{14}$$

Also, if

$$\lim_{|x| \to \infty} \left| \frac{d^n F}{dx^n} \right| = 0 \quad (n = 1, 2, \ldots, r-1)$$

then

$$f^{(r)}(\alpha) = \frac{1}{\sqrt{2\pi}} \int_{-\infty}^\infty F^{(r)}(x) e^{i\alpha x}\, dx = (-i\alpha)^r f(\alpha) \tag{15}$$

Finite Sine Transforms

	$f_s(n)$	$F(x)$
1	$f_s(n) = \int_0^\pi F(x) \sin nx\, dx \ (n = 1, 2, \cdots)$	$F(x)$
2	$(-1)^{n+1} f_s(n)$	$F(\pi - x)$
3	$\dfrac{1}{n}$	$\dfrac{\pi - x}{\pi}$
4	$\dfrac{(-1)^{n+1}}{n}$	$\dfrac{x}{\pi}$
5	$\dfrac{1 - (-1)^n}{n}$	1
6	$\dfrac{2}{n^2} \sin \dfrac{n\pi}{2}$	$\begin{cases} x & \text{when } 0 < x < \pi/2 \\ \pi - x & \text{when } \pi/2 < x < \pi \end{cases}$
7	$\dfrac{(-1)^{n+1}}{n^3}$	$\dfrac{x(\pi^2 - x^2)}{6\pi}$
8	$\dfrac{1 - (-1)^n}{n^3}$	$\dfrac{x(\pi - x)}{2}$
9	$\dfrac{\pi^2 (-1)^{n-1}}{n} - \dfrac{2[1 - (-1)^n]}{n^3}$	x^2
10	$\pi(-1)^n \left(\dfrac{6}{n^3} - \dfrac{\pi^2}{n} \right)$	x^3
11	$\dfrac{n}{n^2 + c^2} [1 - (-1)^n e^{c\pi}]$	e^{cx}
12	$\dfrac{n}{n^2 + c^2}$	$\dfrac{\sinh c(\pi - x)}{\sinh c\pi}$
13	$\dfrac{n}{n^2 - k^2} \ (k \neq 0, 1, 2, \cdots)$	$\dfrac{\sin k(\pi - x)}{\sin k\pi}$
14	$\begin{cases} \dfrac{\pi}{2} & \text{when } n = m \\ 0 & \text{when } n \neq m \end{cases} \quad (m = 1, 2, \cdots)$	$\sin mx$

	$f_s(n)$	$F(x)$
15	$\dfrac{n}{n^2 - k^2}[1 - (-1)^n \cos k\pi]$ $(k \neq 1, 2, \cdots)$	$\cos kx$
16	$\begin{cases} \dfrac{n}{n^2 - m^2}[1 - (-1)^{n+m}] & \text{when } n \neq m = 1, 2, \cdots \\ 0 & \text{when } n = m \end{cases}$	$\cos mx$
17	$\dfrac{n}{(n^2 - k^2)^2}$ $(k \neq 0, 1, 2, \cdots)$	$\dfrac{\pi \sin kx}{2k \sin^2 k\pi} - \dfrac{x \cos k(\pi - x)}{2k \sin k\pi}$
18	$\dfrac{b^n}{n}$ $(|b| \leq 1)$	$\dfrac{2}{\pi} \arctan \dfrac{b \sin x}{1 - b \cos x}$
19	$\dfrac{1 - (-1)^n}{n} b^n$ $(|b| \leq 1)$	$\dfrac{2}{\pi} \arctan \dfrac{2b \sin x}{1 - b^2}$

Finite Cosine Transforms

	$f_c(n)$	$F(x)$
1	$f_c(n) = \displaystyle\int_0^\pi F(x) \cos nx \, dx$ $(n = 0, 1, 2, \cdots)$	$F(x)$
2	$(-1)^n f_c(n)$	$F(\pi - x)$
3	0 when $n = 1, 2, \cdots$; $f_c(0) = \pi$	1
4	$\dfrac{2}{n} \sin \dfrac{n\pi}{2}$; $f_c(0) = 0$	$\begin{cases} 1 & \text{when } 0 < x < \pi/2 \\ -1 & \text{when } \pi/2 < x < \pi \end{cases}$
5	$-\dfrac{1 - (-1)^n}{n^2}$; $f_c(0) = \dfrac{\pi^2}{2}$	x
6	$\dfrac{(-1)^n}{n^2}$; $f_c(0) = \dfrac{\pi^2}{6}$	$\dfrac{x^2}{2\pi}$
7	$\dfrac{1}{n^2}$; $f_c(0) = 0$	$\dfrac{(\pi - x)^2}{2\pi} - \dfrac{\pi}{6}$
8	$3\pi^2 \dfrac{(-1)^n}{n^2} - 6 \dfrac{1 - (-1)^n}{n^4}$; $f_c(0) = \dfrac{\pi^4}{4}$	x^3
9	$\dfrac{(-1)^n e^c \pi - 1}{n^2 + c^2}$	$\dfrac{1}{c} e^{cx}$
10	$\dfrac{1}{n^2 + c^2}$	$\dfrac{\cosh c(\pi - x)}{c \sinh c\pi}$
11	$\dfrac{k}{n^2 - k^2}[(-1)^n \cos \pi k - 1]$ $(k \neq 0, 1, 2, \cdots)$	$\sin kx$
12	$\dfrac{(-1)^{n+m} - 1}{n^2 - m^2}$; $f_c(m) = 0$ $(m = 1, 2, \cdots)$	$\dfrac{1}{m} \sin mx$
13	$\dfrac{1}{n^2 - k^2}$ $(k \neq 0, 1, 2, \cdots)$	$-\dfrac{\cos k(\pi - x)}{k \sin k\pi}$
14	0 when $n = 1, 2, \cdots$; $f_c(m) = \dfrac{\pi}{2}$ $(m = 1, 2, \cdots)$	$\cos mx$

Fourier Sine Transforms*

	$F(x)$	$f_s(\alpha)$
1	$\begin{cases} 1 & (0 < x < a) \\ 0 & (x > a) \end{cases}$	$\sqrt{\dfrac{2}{\pi}}\left[\dfrac{1 - \cos \alpha}{\alpha}\right]$
2	$x^{p-1} (0 < p < 1)$	$\sqrt{\dfrac{2}{\pi}} \dfrac{\Gamma(p)}{\alpha^p} \sin \dfrac{p\pi}{2}$
3	$\begin{cases} \sin x & (0 < x < a) \\ 0 & (x > a) \end{cases}$	$\dfrac{1}{\sqrt{2\pi}}\left[\dfrac{\sin[a(1 - \alpha)]}{1 - \alpha} - \dfrac{\sin[a(1 + \alpha)]}{1 + \alpha}\right]$
4	e^{-x}	$\sqrt{\dfrac{2}{\pi}}\left[\dfrac{\alpha}{1 + \alpha^2}\right]$
5	$xe^{-x^2/2}$	$\alpha e^{-\alpha^2/2}$
6	$\cos \dfrac{x^2}{2}$	$\sqrt{2}\left[\sin \dfrac{\alpha^2}{2} C\left(\dfrac{\alpha^2}{2}\right) - \cos \dfrac{\alpha^2}{2} S\left(\dfrac{\alpha^2}{2}\right)\right]^*$

$F(x)$	$f_s(\alpha)$
7 $\quad \sin \dfrac{x^2}{2}$	$\sqrt{2}\left[\cos \dfrac{\alpha^2}{2} C\left(\dfrac{\alpha^2}{2}\right) + \sin \dfrac{\alpha^2}{2} S\left(\dfrac{\alpha^2}{2}\right)\right]$ *

* $C(y)$ and $S(y)$ are the Fresnel integrals

$$C(y) = \frac{1}{\sqrt{2\pi}} \int_0^y \frac{1}{\sqrt{t}} \cos t\, dt,$$

$$S(y) = \frac{1}{\sqrt{2\pi}} \int_0^y \frac{1}{\sqrt{t}} \sin t\, dt$$

* More extensive tables of the Fourier sine and cosine transforms can be found in Fritz Oberhettinger, *Tabellen zur-Fourier Transformation*, Springer, 1957.

Fourier Cosine Transforms

$F(x)$	$f_c(\alpha)$
1 $\quad \begin{cases} 1 & (0 < x < a) \\ 0 & (x > a) \end{cases}$	$\sqrt{\dfrac{2}{\pi}} \dfrac{\sin a\alpha}{\alpha}$
2 $\quad x^{p-1} \quad (0 < p < 1)$	$\sqrt{\dfrac{2}{\pi}} \dfrac{\Gamma(p)}{\alpha^p} \cos \dfrac{p\pi}{2}$
3 $\quad \begin{cases} \cos x & (0 < x < a) \\ 0 & (x > a) \end{cases}$	$\dfrac{1}{\sqrt{2\pi}}\left[\dfrac{\sin[a(1-\alpha)]}{1-\alpha} + \dfrac{\sin[a(1+\alpha)]}{1+\alpha}\right]$
4 $\quad e^{-x}$	$\sqrt{\dfrac{2}{\pi}}\left(\dfrac{1}{1+\alpha^2}\right)$
5 $\quad e^{-x^2/2}$	$e^{-\alpha^2/2}$
6 $\quad \cos \dfrac{x^2}{2}$	$\cos\left(\dfrac{\alpha^2}{2} - \dfrac{\pi}{4}\right)$
7 $\quad \sin \dfrac{x^2}{2}$	$\cos\left(\dfrac{\alpha^2}{2} + \dfrac{\pi}{4}\right)$

Fourier Transforms*

$F(x)$	$f(\alpha)$
1 $\quad \dfrac{\sin ax}{x}$	$\begin{cases} \sqrt{\dfrac{\pi}{2}} & \|\alpha\| < a \\ 0 & \|\alpha\| > a \end{cases}$
2 $\quad \begin{cases} e^{iwx} & (p < x < q) \\ 0 & (x < p, x > q) \end{cases}$	$\dfrac{i}{\sqrt{2\pi}} \dfrac{e^{ip(w+\alpha)} - e^{iq(w+\alpha)}}{(w+\alpha)}$
3 $\quad \begin{cases} e^{-cx+iwx} & (x > 0) \\ \qquad\qquad (c > 0) \\ 0 & (x < 0) \end{cases}$	$\dfrac{i}{\sqrt{2\pi}(w + \alpha + ic)}$
4 $\quad e^{-px^2} \quad R(p) > 0$	$\dfrac{1}{\sqrt{2p}} e^{-\alpha^2/4p}$
5 $\quad \cos px^2$	$\dfrac{1}{\sqrt{2p}} \cos\left[\dfrac{\alpha^2}{4p} - \dfrac{\pi}{4}\right]$
6 $\quad \sin px^2$	$\dfrac{1}{\sqrt{2p}} \cos\left[\dfrac{\alpha^2}{4p} + \dfrac{\pi}{4}\right]$
7 $\quad \|x\|^{-p} \quad (0 < p < 1)$	$\sqrt{\dfrac{2}{\pi}} \dfrac{\Gamma(1-p) \sin \dfrac{p\pi}{2}}{\|\alpha\|^{(1-p)}}$
8 $\quad \dfrac{e^{-a\|x\|}}{\sqrt{\|x\|}}$	$\dfrac{\sqrt{\sqrt{(a^2+\alpha^2)}+a}}{\sqrt{a^2+\alpha^2}}$
9 $\quad \dfrac{\cosh ax}{\cosh \pi x} \quad (-\pi < a < \pi)$	$\sqrt{\dfrac{2}{\pi}} \dfrac{\cos \dfrac{a}{2} \cosh \dfrac{\alpha}{2}}{\cosh \alpha + \cos a}$
10 $\quad \dfrac{\sinh ax}{\sinh \pi x} \quad (-\pi < a < \pi)$	$\dfrac{1}{\sqrt{2\pi}} \dfrac{\sin a}{\cosh \alpha + \cos a}$
11 $\quad \begin{cases} \dfrac{1}{\sqrt{a^2 - x^2}} & (\|x\| < a) \\ 0 & (\|x\| > a) \end{cases}$	$\sqrt{\dfrac{\pi}{2}} J_0(a\alpha)$

$F(x)$	$f(\alpha)$
12 $\dfrac{\sin [b \sqrt{a^2 + x^2}]}{\sqrt{a^2 + x^2}}$	$\begin{cases} 0 & (\lvert \alpha \rvert > b) \\ \sqrt{\dfrac{\pi}{2}} J_0(a \sqrt{b^2 - \alpha^2}) & (\lvert \alpha \rvert < b) \end{cases}$
13 $\begin{cases} P_n(x) & (\lvert x \rvert < 1) \\ 0 & (\lvert x \rvert > 1) \end{cases}$	$\dfrac{i^n}{\sqrt{\alpha}} J_{n+\frac{1}{2}}(\alpha)$
14 $\begin{cases} \dfrac{\cos [b \sqrt{a^2 - x^2}]}{\sqrt{a^2 - x^2}} & (\lvert x \rvert < a) \\ 0 & (\lvert x \rvert > a) \end{cases}$	$\sqrt{\dfrac{\pi}{2}} J_0(a \sqrt{\alpha^2 + b^2})$
15 $\begin{cases} \dfrac{\cosh [b \sqrt{a^2 - x^2}]}{\sqrt{a^2 - x^2}} & (\lvert x \rvert < a) \\ 0 & (\lvert x \rvert > a) \end{cases}$	$\sqrt{\dfrac{\pi}{2}} J_0(a \sqrt{\alpha^2 - b^2})$

* More extensive tables of Fourier transforms can be found in W. Magnus and F. Oberhettinger, *Formulas and Theorems of the Special Functions of Mathematical Physics*. Chelsea, 1949, 116—120.

The following functions appear among the entries of the tables on transforms.

Function	Definition	Name
$Ei(x)$	$\displaystyle\int_{-\infty}^{x} \frac{e^v}{v}\, dv$; or sometimes defined as $-Ei(-x) = \displaystyle\int_{x}^{\infty} \frac{e^{-v}}{v}\, dv$	Sine, Cosine, and Exponential Integral tables pages 548–556
$Si(x)$	$\displaystyle\int_{0}^{x} \frac{\sin v}{v}\, dv$	Sine, Cosine, and Exponential Integral tables pages 548–556
$Ci(x)$	$\displaystyle\int_{\infty}^{x} \frac{\cos v}{v}\, dv$; or sometimes defined as negative of this integral	Sine, Cosine, and Exponential Integral tables pages 548–546
$erf(x)$	$\dfrac{2}{\sqrt{\pi}} \displaystyle\int_{0}^{x} e^{-v^2}\, dv$	Error function
$erfc(x)$	$1 - erf(x) = \dfrac{2}{\sqrt{\pi}} \displaystyle\int_{x}^{\infty} e^{-v^2}\, dv$	Complementary function to error function
$L_n(x)$	$\dfrac{e^x}{n!} \dfrac{d^n}{dx^n} (x^n e^{-x}), \quad n = 0, 1, \cdots$	Laguerre polynomial of degree n

SERIES EXPANSION

The expression in parentheses following certain of the series indicates the region of convergence. If not otherwise indicated it is to be understood that the series converges for all finite values of x.

BINOMIAL

$$(x + y)^n = x^n + nx^{n-1}y + \frac{n(n-1)}{2!} x^{n-2}y^2$$
$$+ \frac{n(n-1)(n-2)}{3!} x^{n-3}y^3 + \cdots \quad (y^2 < x^2)$$

$$(1 \pm x)^n = 1 \pm nx + \frac{n(n-1)x^2}{2!} \pm \frac{n(n-1)(n-2)x^3}{3!} + \cdots \quad (x^2 < 1)$$

$$(1 \pm x)^{-n} = 1 \mp nx + \frac{n(n+1)x^2}{2!} \mp \frac{n(n+1)(n+2)x^3}{3!} + \cdots \quad (x^2 < 1)$$

$$(1 \pm x)^{-1} = 1 \mp x + x^2 \mp x^3 + x^4 \mp x^5 + \cdots \qquad (x^2 < 1)$$
$$(1 \pm x)^{-2} = 1 \mp 2x + 3x^2 \mp 4x^3 + 5x^4 \mp 6x^5 + \cdots \qquad (x^2 < 1)$$

REVERSION OF SERIES

Let a series be represented by

$$y = a_1 x + a_2 x^2 + a_3 x^3 + a_4 x^4 + a_5 x^5 + a_6 x^6 + \cdots \quad (a_1 \neq 0)$$

to find the coefficients of the series

$$x = A_1 y + A_2 y^2 + A_3 y^3 + A_4 y^4 + \cdots$$

$$A_1 = \frac{1}{a_1} \qquad A_2 = -\frac{a_2}{a_1^3} \qquad A_3 = \frac{1}{a_1^5} (2a_2^2 - a_1 a_3)$$

$$A_4 = \frac{1}{a_1^7} (5a_1 a_2 a_3 - a_1^2 a_4 - 5a_2^3)$$

$$A_5 = \frac{1}{a_1^9} (6a_1^2 a_2 a_4 + 3a_1^2 a_3^2 + 14a_2^4 - a_1^3 a_5 - 21a_1 a_2^2 a_3)$$

$$A_6 = \frac{1}{a_1^{11}} (7a_1^3 a_2 a_5 + 7a_1^3 a_3 a_4 + 84a_1 a_2^3 a_3 - a_1^4 a_6 - 28a_1^2 a_2^2 a_4 - 28a_1^2 a_2 a_3^2 - 42a_2^5)$$

$$A_7 = \frac{1}{a_1^{13}} (8a_1^4 a_2 a_6 + 8a_1^4 a_3 a_5 + 4a_1^4 a_4^2 + 120a_1^2 a_2^3 a_4$$
$$+ 180a_1^2 a_2^2 a_3^2 + 132a_2^6 - a_1^5 a_7$$
$$- 36a_1^3 a_2^2 a_5 - 72a_1^3 a_2 a_3 a_4 - 12a_1^3 a_3^3 - 330a_1 a_2^4 a_3)$$

TAYLOR

1. $f(x) = f(a) + (x - a)f'(a) + \dfrac{(x - a)^2}{2!} f''(a) + \dfrac{(x - a)^3}{3!} f'''(a)$
$$+ \cdots + \frac{(x - a)^n}{n!} f^{(n)}(a) + \cdots \text{ (Taylor's Series)}$$

(Increment form)

2. $f(x + h) = f(x) + hf'(x) + \dfrac{h^2}{2!} f''(x) + \dfrac{h^3}{3!} f'''(x) + \cdots$
$$= f(h) + xf'(h) + \frac{x^2}{2!} f''(h) + \frac{x^3}{3!} f'''(h) + \cdots$$

3. If $f(x)$ is a function possessing derivatives of all orders throughout the interval $a \leqq x \leqq b$, then there is a value X, with $a < X < b$, such that

$$f(b) = f(a) + (b - a)f'(a) + \frac{(b - a)^2}{2!} f''(a) + \cdots$$
$$+ \frac{(b - a)^{n-1}}{(n - 1)!} f^{(n-1)}(a) + \frac{(b - a)^n}{n!} f^{(n)}(X)$$

$$f(a + h) = f(a) + hf'(a) + \frac{h^2}{2!} f''(a) + \cdots + \frac{h^{n-1}}{(n - 1)!} f^{(n-1)}(a)$$
$$+ \frac{h^n}{n!} f^{(n)}(a + \theta h), \ b = a + h, 0 < \theta < 1.$$

or

$$f(x) = f(a) + (x - a)f'(a) + \frac{(x - a)^2}{2!} f''(a) + \cdots + (x - a)^{n-1} \frac{f^{(n-1)}(a)}{(n - 1)!} + R_n,$$

where

$$R_n = \frac{f^{(n)}[a + \theta \cdot (x - a)]}{n!} (x - a)^n, \quad 0 < \theta < 1.$$

The above forms are known as Taylor's series with the remainder term.

4. *Taylor's series for a function of two variables*

If $\left(h \dfrac{\partial}{\partial x} + k \dfrac{\partial}{\partial y}\right) f(x, y) = h \dfrac{\partial f(x, y)}{\partial x} + k \dfrac{\partial f(x, y)}{\partial y};$

$$\left(h \frac{\partial}{\partial x} + k \frac{\partial}{\partial y}\right)^2 f(x, y) = h^2 \frac{\partial^2 f(x, y)}{\partial x^2} + 2hk \frac{\partial^2 f(x, y)}{\partial x \, \partial y} + k^2 \frac{\partial^2 f(x, y)}{\partial y^2}$$

etc., and if $\left(h \dfrac{\partial}{\partial x} + k \dfrac{\partial}{\partial y}\right)^n f(x, y)\Big|_{\substack{x = a \\ y = b}}$ with the bar and subscripts means that after differentiation we are to replace x by a and y by b,

$$f(a + h, b + k) = f(a, b) + \left(h \frac{\partial}{\partial x} + k \frac{\partial}{\partial y}\right) f(x, y)\Big|_{\substack{x = a \\ y = b}} + \cdots$$
$$+ \frac{1}{n!} \left(h \frac{\partial}{\partial x} + k \frac{\partial}{\partial y}\right)^n f(x, y)\Big|_{\substack{x = a \\ y = b}} + \cdots$$

MACLAURIN

$$f(x) = f(0) + xf'(0) + \frac{x^2}{2!} f''(0) + \frac{x^3}{3!} f'''(0) + \cdots + x^{n-1} \frac{f^{(n-1)}(0)}{(n - 1)!} + R_n,$$

where

$$R_n = \frac{x^n f^{(n)}(\theta x)}{n!}, \quad 0 < \theta < 1.$$

EXPONENTIAL

$$e = 1 + \frac{1}{1!} + \frac{1}{2!} + \frac{1}{3!} + \frac{1}{4!} + \cdots$$

$$e^x = 1 + x + \frac{x^2}{2!} + \frac{x^3}{3!} + \frac{x^4}{4!} + \cdots \qquad \text{(all real values of } x\text{)}$$

$$a^x = 1 + x \log_e a + \frac{(x \log_e a)^2}{2!} + \frac{(x \log_e a)^3}{3!} + \cdots$$

$$e^x = e^a \left[1 + (x - a) + \frac{(x - a)^2}{2!} + \frac{(x - a)^3}{3!} + \cdots \right]$$

LOGARITHMIC

$$\log_e x = \frac{x - 1}{x} + \frac{1}{2}\left(\frac{x - 1}{x}\right)^2 + \frac{1}{3}\left(\frac{x - 1}{x}\right)^3 + \cdots \qquad (x > \tfrac{1}{2})$$

$$\log_e x = (x - 1) - \tfrac{1}{2}(x - 1)^2 + \tfrac{1}{3}(x - 1)^3 - \cdots \qquad (2 \geq x > 0)$$

$$\log_e x = 2\left[\frac{x - 1}{x + 1} + \frac{1}{3}\left(\frac{x - 1}{x + 1}\right)^3 + \frac{1}{5}\left(\frac{x - 1}{x + 1}\right)^5 + \cdots \right] \qquad (x > 0)$$

$$\log_e (1 + x) = x - \tfrac{1}{2}x^2 + \tfrac{1}{3}x^3 - \tfrac{1}{4}x^4 + \cdots \qquad (-1 < x \leq 1)$$

$$\log_e (n + 1) - \log_e (n - 1) = 2\left[\frac{1}{n} + \frac{1}{3n^3} + \frac{1}{5n^5} + \cdots \right]$$

$$\log_e (a + x) = \log_e a + 2\left[\frac{x}{2a + x} + \frac{1}{3}\left(\frac{x}{2a + x}\right)^3 + \frac{1}{5}\left(\frac{x}{2a + x}\right)^5 + \cdots \right]$$
$$(a > 0, -a < x < +\infty)$$

$$\log_e \frac{1 + x}{1 - x} = 2\left[x + \frac{x^3}{3} + \frac{x^5}{5} + \cdots + \frac{x^{2n-1}}{2n - 1} + \cdots \right], \qquad -1 < x < 1$$

$$\log_e x = \log_e a + \frac{(x - a)}{a} - \frac{(x - a)^2}{2a^2} + \frac{(x - a)^3}{3a^3} - + \cdots, \qquad 0 < x \leq 2a$$

TRIGONOMETRIC

$$\sin x = x - \frac{x^3}{3!} + \frac{x^5}{5!} - \frac{x^7}{7!} + \cdots \qquad \text{(all real values of } x\text{)}$$

$$\cos x = 1 - \frac{x^2}{2!} + \frac{x^4}{4!} - \frac{x^6}{6!} + \cdots \qquad \text{(all real values of } x\text{)}$$

$$\tan x = x + \frac{x^3}{3} + \frac{2x^5}{15} + \frac{17x^7}{315} + \frac{62x^9}{2835} + \cdots + \frac{(-1)^{n-1} 2^{2n} (2^{2n} - 1) B_{2n}}{(2n)!} x^{2n-1} + \cdots,$$
$$\left[x^2 < \frac{\pi^2}{4}, \text{ and } B_n \text{ represents the } n\text{th Bernoulli number.} \right]$$

$$\cot x = \frac{1}{x} - \frac{x}{3} - \frac{x^3}{45} - \frac{2x^5}{945} - \frac{x^7}{4725} - \cdots - \frac{(-1)^{n+1}2^{2n}}{(2n)!} B_{2n} x^{2n-1} - \cdots,$$
$$[x^2 < \pi^2, \text{ and } B_n \text{ represents the } n\text{th Bernoulli number.}]$$

$$\sec x = 1 + \frac{x^2}{2} + \frac{5}{24}x^4 + \frac{61}{720}x^6 + \frac{277}{8064}x^8 + \cdots + \frac{(-1)^n}{(2n)!} E_{2n} x^{2n} + \cdots,$$
$$\left[x^2 < \frac{\pi^2}{4}, \text{ and } E_n \text{ represents the } n\text{th Euler number.} \right]$$

$$\csc x = \frac{1}{x} + \frac{x}{6} + \frac{7}{360}x^3 + \frac{31}{15,120}x^5 + \frac{127}{604,800}x^7 + \cdots$$
$$+ \frac{(-1)^{n+1}2 (2^{2n-1} - 1)}{(2n)!} B_{2n} x^{2n-1} + \cdots,$$
$$[x^2 < \pi^2, \text{ and } B_n \text{ represents } n\text{th Bernoulli number.}]$$

$$\sin x = x\left(1 - \frac{x^2}{\pi^2}\right)\left(1 - \frac{x^2}{2^2\pi^2}\right)\left(1 - \frac{x^2}{3^2\pi^2}\right)\cdots \qquad (x^2 < \infty)$$

$$\cos x = \left(1 - \frac{4x^2}{\pi^2}\right)\left(1 - \frac{4x^2}{3^2\pi^2}\right)\left(1 - \frac{4x^2}{5^2\pi^2}\right)\cdots \qquad (x^2 < \infty)$$

$$\sin^{-1} x = x + \frac{x^3}{2\cdot 3} + \frac{1\cdot 3}{2\cdot 4\cdot 5}x^5 + \frac{1\cdot 3\cdot 5}{2\cdot 4\cdot 6\cdot 7}x^7 + \cdots \qquad \left(x^2 < 1, -\frac{\pi}{2} < \sin^{-1} x < \frac{\pi}{2}\right)$$

$$\cos^{-1} x = \frac{\pi}{2} - \left(x + \frac{x^3}{2\cdot 3} + \frac{1\cdot 3}{2\cdot 4\cdot 5}x^5 + \frac{1\cdot 3\cdot 5 x^7}{2\cdot 4\cdot 6\cdot 7} + \cdots\right) \qquad (x^2 < 1, 0 < \cos^{-1} x < \pi)$$

$$\tan^{-1} x = x - \frac{x^3}{3} + \frac{x^5}{5} - \frac{x^7}{7} + \cdots \qquad (x^2 < 1)$$

$$\tan^{-1} x = \frac{\pi}{2} - \frac{1}{x} + \frac{1}{3x^3} - \frac{1}{5x^5} + \frac{1}{7x^7} - \cdots \qquad (x > 1)$$

$$\tan^{-1} x = -\frac{\pi}{2} - \frac{1}{x} + \frac{1}{3x^3} - \frac{1}{5x^5} + \frac{1}{7x^7} - \cdots \qquad (x < -1)$$

$$\cot^{-1} x = \frac{\pi}{2} - x + \frac{x^3}{3} - \frac{x^5}{5} + \frac{x^7}{7} - \cdots \qquad (x^2 < 1)$$

$$\log_e \sin x = \log_e x - \frac{x^2}{6} - \frac{x^4}{180} - \frac{x^6}{2835} - \cdots \qquad (x^2 < \pi^2)$$

$$\log_e \cos x = -\frac{x^2}{2} - \frac{x^4}{12} - \frac{x^6}{45} - \frac{17x^8}{2520} - \cdots \qquad \left(x^2 < \frac{\pi^2}{4}\right)$$

$$\log_e \tan x = \log_e x + \frac{x^2}{3} + \frac{7x^4}{90} + \frac{62x^6}{2835} + \cdots \qquad \left(x^2 < \frac{\pi^2}{4}\right)$$

$$e^{\sin x} = 1 + x + \frac{x^2}{2!} - \frac{3x^4}{4!} - \frac{8x^5}{5!} - \frac{3x^6}{6!} + \frac{56x^7}{7!} + \cdots$$

$$e^{\cos x} = e\left(1 - \frac{x^2}{2!} + \frac{4x^4}{4!} - \frac{31x^6}{6!} + \cdots\right)$$

$$e^{\tan x} = 1 + x + \frac{x^2}{2!} + \frac{3x^3}{3!} + \frac{9x^4}{4!} + \frac{37x^5}{5!} + \cdots \qquad \left(x^2 < \frac{\pi^2}{4}\right)$$

$$\sin x = \sin a + (x - a)\cos a - \frac{(x - a)^2}{2!}\sin a$$
$$- \frac{(x - a)^3}{3!}\cos a + \frac{(x - a)^4}{4!}\sin a + \cdots$$

VECTOR ANALYSIS

Definitions

Any quantity which is completely determined by its magnitude is called a *scalar*. Examples of such are mass, density, temperature, etc. Any quantity which is completely determined by its magnitude and direction is called a *vector*. Examples of such are velocity, acceleration, force, etc. A vector quantity is represented by a directed line segment, the length of which represents the magnitude of the vector. A vector quantity is usually represented by a boldfaced letter such as $\mathbf{V}$. Two vectors $\mathbf{V}_1$ and $\mathbf{V}_2$ are equal to one another if they have equal magnitudes and are acting in the same directions. A negative vector, written as $-\mathbf{V}$, is one which acts in the opposite direction to $\mathbf{V}$, but is of equal magnitude to it. If we represent the magnitude of $\mathbf{V}$ by v, we write $|\mathbf{V}| = v$. A vector parallel to $\mathbf{V}$, but equal to the reciprocal of its magnitude is written as $\mathbf{V}^{-1}$ or as $\frac{1}{\mathbf{V}}$.

The *unit vector* $\frac{\mathbf{V}}{v}$ $(v \neq 0)$ is that vector which has the same direction as $\mathbf{V}$, but has a magnitude of unity (sometimes represented as $\mathbf{V}_0$ or $\hat{\mathbf{v}}$).

Vector Algebra

The vector sum of $\mathbf{V}_1$ and $\mathbf{V}_2$ is represented by $\mathbf{V}_1 + \mathbf{V}_2$. The vector sum of $\mathbf{V}_1$ and $-\mathbf{V}_2$, or the difference of the vector $\mathbf{V}_2$ from $\mathbf{V}_1$ is represented by $\mathbf{V}_1 - \mathbf{V}_2$.

If r is a scalar, then $r\mathbf{V} = \mathbf{V}r$, and represents a vector r times the magnitude of $\mathbf{V}$, in the same direction as $\mathbf{V}$ if r is positive, and in the opposite direction if r is negative. If r and s are scalars, $\mathbf{V}_1$, $\mathbf{V}_2$, $\mathbf{V}_3$, vectors, then the following rules of scalars and vectors hold:

$$\mathbf{V}_1 + \mathbf{V}_2 = \mathbf{V}_2 + \mathbf{V}_1$$
$$(r + s)\mathbf{V}_1 = r\mathbf{V}_1 + s\mathbf{V}_1; \qquad r(\mathbf{V}_1 + \mathbf{V}_2) = r\mathbf{V}_1 + r\mathbf{V}_2$$
$$\mathbf{V}_1 + (\mathbf{V}_2 + \mathbf{V}_3) = (\mathbf{V}_1 + \mathbf{V}_2) + \mathbf{V}_3 = \mathbf{V}_1 + \mathbf{V}_2 + \mathbf{V}_3$$

Vectors in Space

A plane is described by two distinct vectors $\mathbf{V}_1$ and $\mathbf{V}_2$. Should these vectors not intersect each other, then one is displaced parallel to itself until they do (fig. 1.) Any other vector $\mathbf{V}$ lying in this plane is given by

$$\mathbf{V} = r\mathbf{V}_1 + s\mathbf{V}_2$$

A *position vector* specifies the position in space of a point relative to a fixed origin. If therefore $\mathbf{V}_1$ and $\mathbf{V}_2$ are the position vectors of the points A and B, relative to the origin O, then any point P on the line AB has a position vector $\mathbf{V}$ given by

$$\mathbf{V} = r\mathbf{V}_1 + (1 - r)\mathbf{V}_2$$

The scalar "r" can be taken as the parametric representation of P since $r = 0$ implies $P = B$ and $r = 1$ implies $P = A$. (fig. 2). If P divides the line AB in the ratio $r : s$ then

$$\mathbf{V} = \left(\frac{r}{r + s}\right)\mathbf{V}_1 + \left(\frac{s}{r + s}\right)\mathbf{V}_2$$

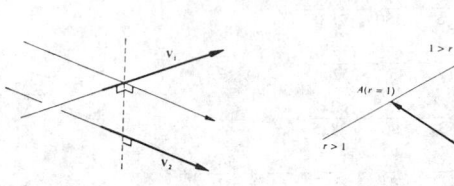

Fig. 1.   Fig. 2.

The vectors $V_1, V_2, V_3, \ldots, V_n$ are said to be *linearly dependent* if there exist scalars $r_1, r_2, r_3, \ldots, r_n$, not all zero, such that

$$r_1V_1 + r_2V_2 + \cdots + r_nV_n = 0$$

A vector V is linearly dependent upon the set of vectors $V_1, V_2, V_3, \ldots, V_n$ if

$$V = r_1V_1 + r_2V_2 + r_3V_3 + \cdots + r_nV_n$$

Three vectors are linearly dependent if and only if they are co-planar.

All points in space can be uniquely determined by linear dependence upon three *base vectors* i.e. three vectors any one of which is linearly independent of the other two. The simplest set of base vectors are the unit vectors along the coordinate Ox, Oy and Oz axes. These are usually designated by i, j and k respectively.

If V is a vector in space, and a, b and c are the respective magnitudes of the projections of the vector along the axes then

$$V = ai + bj + ck$$

and

$$v = \sqrt{a^2 + b^2 + c^2}$$

and the direction cosines of V are

$$\cos \alpha = a/v, \quad \cos \beta = b/v, \quad \cos \gamma = c/v.$$

The law of addition yields

$$V_1 + V_2 = (a_1 + a_2)i + (b_1 + b_2)j + (c_1 + c_2)k$$

The Scalar, Dot, or Inner Product of Two Vectors V_1 and V_2

This product is represented as $V_1 \cdot V_2$ and is defined to be equal to $v_1v_2 \cos \theta$, where θ is the angle from V_1 to V_2, i.e.,

$$V_1 \cdot V_2 = v_1v_2 \cos \theta$$

The following rules apply for this product:

$$V_1 \cdot V_2 = a_1a_2 + b_1b_2 + c_1c_2 = V_2 \cdot V_1$$

It should be noted that this verifies that scalar multiplication is commutative.

$$(V_1 + V_2) \cdot V_3 = V_1 \cdot V_3 + V_2 \cdot V_3$$
$$V_1 \cdot (V_2 + V_3) = V_1 \cdot V_2 + V_1 \cdot V_3$$

If V_1 is perpendicular to V_2 then $V_1 \cdot V_2 = 0$, and if V_1 is parallel to V_2 then $V_1 \cdot V_2 = v_1v_2 = rw_1^2$

In particular

$$i \cdot i = j \cdot j = k \cdot k = 1,$$

and

$$i \cdot j = j \cdot k = k \cdot i = 0$$

The Vector or Cross Product of Vectors V_1 and V_2

This product is represented as $V_1 \times V_2$ and is defined to be equal to $v_1v_2(\sin \theta)1$, where θ is the angle from V_1 to V_2 and 1 is a unit vector perpendicular to the plane of V_1 and V and so directed that a right-handed screw driven in the direction of 1 would carry V_1 into V_2, i.e.,

$$V_1 \times V_2 = v_1v_2(\sin \theta)1$$

and

$$\tan \theta = \frac{|V_1 \times V_2|}{V_1 \cdot V_2}$$

The following rules apply for vector products:

$$V_1 \times V_2 = -V_2 \times V_1$$
$$V_1 \times (V_2 + V_3) = V_1 \times V_2 + V_1 \times V_3$$
$$(V_1 + V_2) \times V_3 = V_1 \times V_3 + V_2 \times V_3$$
$$V_1 \times (V_2 \times V_3) = V_2(V_3 \cdot V_1 - V_3(V_1 \cdot V_2)$$
$$i \times i = j \times j = k \times k = 0.1 \text{ (zero vector)}$$
$$= 0$$

$$i \times j = k, \qquad j \times k = i, \qquad k \times i = j$$

A-88

If $\mathbf{V}_1 = a_1\mathbf{i} + b_1\mathbf{j} + c_1\mathbf{k}$, $\mathbf{V}_2 = a_2\mathbf{i} + b_2\mathbf{j} + c_2\mathbf{k}$, $\mathbf{V}_3 = a_3\mathbf{i} + b_3\mathbf{j} + c_3\mathbf{k}$,

then

$$\mathbf{V}_1 \times \mathbf{V}_2 = \begin{vmatrix} \mathbf{i} & \mathbf{j} & \mathbf{k} \\ a_1 & b_1 & c_1 \\ a_2 & b_2 & c_2 \end{vmatrix} = (b_1c_2 - b_2c_1)\mathbf{i} + (c_1a_2 - c_2a_1)\mathbf{j} + (a_1b_2 - a_2b_1)\mathbf{k}$$

It should be noted that, since $\mathbf{V}_1 \times \mathbf{V}_2 = -\mathbf{V}_2 \times \mathbf{V}_1$, the vector product is not commutative.

Scalar Triple Product

There is only one possible interpretation of the expression $\mathbf{V}_1 \cdot \mathbf{V}_2 \times \mathbf{V}_3$ and that is $\mathbf{V}_1 \cdot (\mathbf{V}_2 \times \mathbf{V}_3)$ which is obviously a scalar.

Further $\mathbf{V}_1 \cdot (\mathbf{V}_2 \times \mathbf{V}_3) = (\mathbf{V}_1 \times \mathbf{V}_2) \cdot \mathbf{V}_3 = \mathbf{V}_2 \cdot (\mathbf{V}_3 \times \mathbf{V}_1)$

$$= \begin{vmatrix} a_1 & b_1 & c_1 \\ a_2 & b_2 & c_2 \\ a_3 & b_3 & c_3 \end{vmatrix}$$

$$= v_1 v_2 v_3 \cos \phi \sin \theta,$$

Where θ is the angle between $\mathbf{V}_2$ and $\mathbf{V}_3$ and ϕ is the angle between $\mathbf{V}_1$ and the normal to the plane of $\mathbf{V}_2$ and $\mathbf{V}_3$.

This product is called the *scalar triple product* and is written as $[\mathbf{V}_1\mathbf{V}_2\mathbf{V}_3]$.

The determinant indicates that it can be considered as the volume of the parallelepiped whose three determining edges are $\mathbf{V}_1$, $\mathbf{V}_2$ and $\mathbf{V}_3$.

It also follows that cyclic permutation of the subscripts does not change the value of the scalar triple product so that

$$[\mathbf{V}_1\mathbf{V}_2\mathbf{V}_3] = [\mathbf{V}_2\mathbf{V}_3\mathbf{V}_1] = [\mathbf{V}_3\mathbf{V}_1\mathbf{V}_2]$$

but $[\mathbf{V}_1\mathbf{V}_2\mathbf{V}_3] = -[\mathbf{V}_2\mathbf{V}_1\mathbf{V}_3]$ etc. and $[\mathbf{V}_1\mathbf{V}_1\mathbf{V}_2] \equiv 0$ etc.

Given three non-coplanar reference vectors $\mathbf{V}_1$, $\mathbf{V}_2$ and $\mathbf{V}_3$, the *reciprocal system is* given by $\mathbf{V}_1^*$, $\mathbf{V}_2^*$ and $\mathbf{V}_3^*$, where

$$1 = v_1 v_1^* = v_2 v_2^* = v_3 v_3^*$$

$$0 = v_1 v_2^* = v_1 v_3^* = v_2 v_1^* \quad \text{etc.}$$

$$\mathbf{V}_1^* = \frac{\mathbf{V}_2 \times \mathbf{V}_3}{[\mathbf{V}_1\mathbf{V}_2\mathbf{V}_3]}, \qquad \mathbf{V}_2^* = \frac{\mathbf{V}_3 \times \mathbf{V}_1}{[\mathbf{V}_1\mathbf{V}_2\mathbf{V}_3]}, \qquad \mathbf{V}_3^* = \frac{\mathbf{V}_1 \times \mathbf{V}_2}{[\mathbf{V}_1\mathbf{V}_2\mathbf{V}_3]}$$

The system $\mathbf{i}, \mathbf{j}, \mathbf{k}$ is its own reciprocal.

Vector Triple Product

The product $\mathbf{V}_1 \times (\mathbf{V}_2 \times \mathbf{V}_3)$ defines the *vector triple product*. Obviously, in this case, the brackets are vital to the definition.

$$\mathbf{V}_1 \times (\mathbf{V}_2 \times \mathbf{V}_3) = (\mathbf{V}_1 \cdot \mathbf{V}_3)\mathbf{V}_2 - (\mathbf{V}_1 \cdot \mathbf{V}_2)\mathbf{V}_3$$

$$= \begin{vmatrix} \mathbf{i} & \mathbf{j} & \mathbf{k} \\ a_1 & b_1 & c_1 \\ \begin{vmatrix} b_2 & c_2 \\ b_3 & c_3 \end{vmatrix} & \begin{vmatrix} c_2 & a_2 \\ c_3 & a_3 \end{vmatrix} & \begin{vmatrix} a_2 & b_2 \\ a_3 & b_3 \end{vmatrix} \end{vmatrix}$$

i.e. it is a vector, perpendicular to $\mathbf{V}_1$, lying in the plane of $\mathbf{V}_2$, $\mathbf{V}_3$.

Similarly $(\mathbf{V}_1 \times \mathbf{V}_2) \times \mathbf{V}_3 = \begin{vmatrix} \mathbf{i} & \mathbf{j} & \mathbf{k} \\ \begin{vmatrix} b_1 & c_1 \\ b_2 & c_2 \end{vmatrix} & \begin{vmatrix} c_1 & a_1 \\ c_2 & a_2 \end{vmatrix} & \begin{vmatrix} a_1 & b_1 \\ a_2 & b_2 \end{vmatrix} \\ a_3 & b_3 & c_3 \end{vmatrix}$

$$\mathbf{V}_1 \times (\mathbf{V}_2 \times \mathbf{V}_3) + \mathbf{V}_2 \times (\mathbf{V}_3 \times \mathbf{V}_1) + \mathbf{V}_3 \times (\mathbf{V}_1 \times \mathbf{V}_2) \equiv 0$$

If $\mathbf{V}_1 \times (\mathbf{V}_2 \times \mathbf{V}_3) = (\mathbf{V}_1 \times \mathbf{V}_2) \times \mathbf{V}_3$ then $\mathbf{V}_1, \mathbf{V}_2, \mathbf{V}_3$ form an *orthogonal set*. Thus $\mathbf{i}, \mathbf{j}, \mathbf{k}$ form an orthogonal set.

Geometry of the Plane, Straight Line and Sphere

The position vectors of the fixed points A, B, C, D relative to O are $\mathbf{V}_1$, $\mathbf{V}_2$, $\mathbf{V}_3$, $\mathbf{V}_4$ and the position vector of the variable point P is $\mathbf{V}$.

The vector form of the equation of the straight line through A parallel to $\mathbf{V}_2$ is

$$\mathbf{V} = \mathbf{V}_1 + r\mathbf{V}_2$$

or $(\mathbf{V} - \mathbf{V}_1) = r\mathbf{V}_2$

or $(\mathbf{V} - \mathbf{V}_1) \times \mathbf{V}_2 = 0$

while that of the plane through A perpendicular to $\mathbf{V}_2$ is

$$(\mathbf{V} - \mathbf{V}_1) \cdot \mathbf{V}_2 = 0$$

The equation of the line AB is

$$\mathbf{V} = r\mathbf{V}_1 + (1 - r)\mathbf{V}_2$$

and those of the bisectors of the angles between $\mathbf{V}_1$ and $\mathbf{V}_2$ are

$$\mathbf{V} = r\left(\frac{\mathbf{V}_1}{v} \pm \frac{\mathbf{V}_2}{v_2}\right)$$

or $\quad \mathbf{V} = r(\hat{\mathbf{v}}_1 \pm \hat{\mathbf{v}}_2)$

The perpendicular from C to the line through A parallel to $\mathbf{V}_2$ has as its equation

$$\mathbf{V} = \mathbf{V}_1 - \mathbf{V}_3 - \hat{\mathbf{v}}_2 \cdot (\mathbf{V}_1 - \mathbf{V}_3)\hat{\mathbf{v}}_2.$$

The condition for the intersection of the two lines,

$$\mathbf{V} = \mathbf{V}_1 + r\mathbf{V}_3$$

and $\quad \mathbf{V} = \mathbf{V}_2 + s\mathbf{V}_4$

is $\quad [(\mathbf{V}_1 - \mathbf{V}_2)\mathbf{V}_3\mathbf{V}_4] = 0.$

The common perpendicular to the above two lines is the line of intersection of the two planes

$$[(\mathbf{V} - \mathbf{V}_1)\mathbf{V}_3(\mathbf{V}_3 \times \mathbf{V}_4)] = 0$$

and $\quad [(\mathbf{V} - \mathbf{V}_2)\mathbf{V}_4(\mathbf{V}_3 \times \mathbf{V}_4)] = 0$

and the length of this perpendicular is

$$\frac{[(\mathbf{V}_1 - \mathbf{V}_2)\mathbf{V}_3\mathbf{V}_4]}{|\mathbf{V}_3 \times \mathbf{V}_4|}.$$

The equation of the line perpendicular to the plane ABC is

$$\mathbf{V} = \mathbf{V}_1 \times \mathbf{V}_2 + \mathbf{V}_2 \times \mathbf{V}_3 + \mathbf{V}_3 \times \mathbf{V}_1$$

and the distance of the plane from the origin is

$$\frac{[\mathbf{V}_1\mathbf{V}_2\mathbf{V}_3]}{|(\mathbf{V}_2 - \mathbf{V}_1) \times (\mathbf{V}_3 - \mathbf{V}_1)|}.$$

In general the vector equation

$$\mathbf{V} \cdot \mathbf{V}_2 = r$$

defines the plane which is perpendicular to $\mathbf{V}_2$, and the perpendicular distance from A to this plane is

$$\frac{r - \mathbf{V}_1 \cdot \mathbf{V}_2}{v_2}.$$

The distance from A, measured along a line parallel to $\mathbf{V}_3$, is

$$\frac{r - \mathbf{V}_1 \cdot \mathbf{V}_2}{\mathbf{V}_2 \cdot \hat{\mathbf{v}}_3} \quad \text{or} \quad \frac{r - \mathbf{V}_1 \cdot \mathbf{V}_2}{v_2 \cos \theta}$$

where θ is the angle beween $\mathbf{V}_2$ and $\mathbf{V}_3$.
(If this plane contains the point C then $r = \mathbf{V}_3 \cdot \mathbf{V}_2$ and if it passes through the origin then $r = 0$.)

Given two planes $\qquad\qquad \mathbf{V} \cdot \mathbf{V}_1 = r$

$$\mathbf{V} \cdot \mathbf{V}_2 = s$$

then any plane through the line of intersection of these two planes is given by

$$\mathbf{V} \cdot (\mathbf{V}_1 + \lambda\mathbf{V}_2) = r + \lambda s$$

where λ is a scalar parameter. In particular $\lambda = \pm v_1/v_2$ yields the equation of the two planes bisecting the angle between the given planes.

The plane through A parallel to the plane of $\mathbf{V}_2$, $\mathbf{V}_3$ is

$$\mathbf{V} = \mathbf{V}_1 + r\mathbf{V}_2 + s\mathbf{V}_3$$

or $\quad (\mathbf{V} - \mathbf{V}_1) \cdot \mathbf{V}_2 \times \mathbf{V}_3 = 0$

or $\quad [\mathbf{V}\mathbf{V}_2\mathbf{V}_3] - [\mathbf{V}_1\mathbf{V}_2\mathbf{V}_3] = 0$

so that the expansion in rectangular Cartesian coordinates yields

$$\begin{vmatrix} (x - a_1) & (y - b_1) & (z - c_1) \\ a_2 & b_2 & c_2 \\ a_3 & b_3 & c_3 \end{vmatrix} = 0 \qquad (\mathbf{V} \equiv x\mathbf{i} + y\mathbf{j} + z\mathbf{k})$$

which is obviously the usual linear equation in x, y and z.

The plane through AB parallel to $\mathbf{V}_3$ is given by

$$[(\mathbf{V} - \mathbf{V}_1)(\mathbf{V}_1 - \mathbf{V}_2)\mathbf{V}_3] = 0$$

or $\quad [\mathbf{V}\mathbf{V}_2\mathbf{V}_3] - [\mathbf{V}\mathbf{V}_1\mathbf{V}_3] - [\mathbf{V}_1\mathbf{V}_2\mathbf{V}_3] = 0$

The plane through the three points A, B and C is

$$\mathbf{V} = \mathbf{V}_1 + s(\mathbf{V}_2 - \mathbf{V}_1) + t(\mathbf{V}_3 - \mathbf{V}_1)$$

A-90

or
$$\mathbf{V} = r\mathbf{V}_1 + s\mathbf{V}_2 + t\mathbf{V}_3 \qquad (r + s + t \equiv 1)$$

or
$$[(\mathbf{V} - \mathbf{V}_1)(\mathbf{V}_1 - \mathbf{V}_2)(\mathbf{V}_2 - \mathbf{V}_3)] = 0$$

or
$$[\mathbf{V}\mathbf{V}_1\mathbf{V}_2] + [\mathbf{V}\mathbf{V}_2\mathbf{V}_3] + [\mathbf{V}\mathbf{V}_3\mathbf{V}_1] - [\mathbf{V}_1\mathbf{V}_2\mathbf{V}_3] = 0$$

For four points A, B, C, D to be coplanar, then

$$r\mathbf{V}_1 + s\mathbf{V}_2 + t\mathbf{V}_3 + u\mathbf{V}_4 \equiv 0 \equiv r + s + t + u$$

The following formulae relate to a sphere when the vectors are taken to lie in three dimensional space and to a circle when the space is two dimensional. For a circle in three dimensions take the intersection of the sphere with a plane.

The equation of a sphere with center O and radius OA is

$$\mathbf{V} \cdot \mathbf{V} = v_1^2 \qquad (\text{not } \mathbf{V} = \mathbf{V}_1)$$

or
$$(\mathbf{V} - \mathbf{V}_1) \cdot (\mathbf{V} + \mathbf{V}_1) = 0$$

while that of a sphere with center B radius v_1 is

$$(\mathbf{V} - \mathbf{V}_2) \cdot (\mathbf{V} - \mathbf{V}_2) = v_1^2$$

or
$$\mathbf{V} \cdot (\mathbf{V} - 2\mathbf{V}_2) = v_1^2 - v_2^2$$

If the above sphere passes through the origin then

$$\mathbf{V} \cdot (\mathbf{V} - 2\mathbf{V}_2) = 0$$

(note that in two dimensional polar coordinates this is simply)

$$r = 2a \cdot \cos\theta$$

while in three dimensional Cartesian coordinates it is

$$x^2 + y^2 + z^2 - 2(a_2 x + b_2 y + c_2 x) = 0.$$

The equation of a sphere having the points A and B as the extremities of a diameter is

$$(\mathbf{V} - \mathbf{V}_1) \cdot (\mathbf{V} - \mathbf{V}_2) = 0.$$

The square of the length of the tangent from C to the sphere with center B and radius v_1 is given by

$$(\mathbf{V}_3 - \mathbf{V}_2) \cdot (\mathbf{V}_3 - \mathbf{V}_2) = v_1^2$$

The condition that the plane $\mathbf{V} \cdot \mathbf{V}_3 = s$ is tangential to the sphere $(\mathbf{V} - \mathbf{V}_2) \cdot (\mathbf{V} - \mathbf{V}_2) = v_1^2$ is

$$(s - \mathbf{V}_3 \cdot \mathbf{V}_2) \cdot (s - \mathbf{V}_3 \cdot \mathbf{V}_2) = v_1^2 v_3^2.$$

The equation of the tangent plane at D, on the surface of sphere $(\mathbf{V} - \mathbf{V}_2) \cdot (\mathbf{V} - \mathbf{V}_2) = v_1^2$, is

$$(\mathbf{V} - \mathbf{V}_4) \cdot (\mathbf{V}_4 - \mathbf{V}_2) = 0$$

or
$$\mathbf{V} \cdot \mathbf{V}_4 - \mathbf{V}_2 \cdot (\mathbf{V} + \mathbf{V}_4) = v_1^2 - v_2^2$$

The condition that the two circles $(\mathbf{V} - \mathbf{V}_2) \cdot (\mathbf{V} - \mathbf{V}_2) = v_1^2$ and $(\mathbf{V} - \mathbf{V}_4) \cdot (\mathbf{V} - \mathbf{V}_4) = v_3^2$ intersect orthogonally is clearly

$$(\mathbf{V}_2 - \mathbf{V}_4) \cdot (\mathbf{V}_2 - \mathbf{V}_4) = v_1^2 + v_3^2$$

The polar plane of D with respect to the circle

$$(\mathbf{V} - \mathbf{V}_2) \cdot (\mathbf{V} - \mathbf{V}_2) = v_1^2 \quad \text{is}$$

$$\mathbf{V} \cdot \mathbf{V}_4 - \mathbf{V}_2 \cdot (\mathbf{V} + \mathbf{V}_4) = v_1^2 - v_2^2$$

Any sphere through the intersection of the two spheres $(\mathbf{V} - \mathbf{V}_2) \cdot (\mathbf{V} - \mathbf{V}_2) = v_1^2$ and $(\mathbf{V} - \mathbf{V}_4) \cdot (\mathbf{V} - \mathbf{V}_4) = v_3^2$ is given by

$$(\mathbf{V} - \mathbf{V}_2) \cdot (\mathbf{V} - \mathbf{V}_2) + \lambda(\mathbf{V} - \mathbf{V}_4) \cdot (\mathbf{V} - \mathbf{V}_4) = v_1^2 + \lambda v_3^2$$

while the radical plane of two such spheres is

$$\mathbf{V} \cdot (\mathbf{V}_2 - \mathbf{V}_4) = -\tfrac{1}{2}(v_1^2 - v_2^2 - v_3^2 + v_4^2)$$

Differentiation of Vectors

If $\mathbf{V}_1 = a_1\mathbf{i} + b_1\mathbf{j} + c_1\mathbf{k}$, and $\mathbf{V}_2 = a_2\mathbf{i} + b_2\mathbf{j} + c_2\mathbf{k}$, and if $\mathbf{V}_1$ and $\mathbf{V}_2$ are functions of the scalar t, then

$$\frac{d}{dt}(\mathbf{V}_1 + \mathbf{V}_2 + \cdots) = \frac{d\mathbf{V}_1}{dt} + \frac{d\mathbf{V}_2}{dt} + \cdots,$$

where $\dfrac{d\mathbf{V}_1}{dt} = \dfrac{da_1}{dt}\mathbf{i} + \dfrac{db_1}{dt}\mathbf{j} + \dfrac{dc_1}{dt}\mathbf{k}$, etc.

$$\frac{d}{dt}(\mathbf{V}_1 \cdot \mathbf{V}_2) = \frac{d\mathbf{V}_1}{dt} \cdot \mathbf{V}_2 + \mathbf{V}_1 \cdot \frac{d\mathbf{V}_2}{dt}$$

$$\frac{d}{dt}(\mathbf{V}_1 \times \mathbf{V}_2) = \frac{d\mathbf{V}_1}{dt} \times \mathbf{V}_2 + \mathbf{V}_1 \times \frac{d\mathbf{V}_2}{dt}$$

$$\mathbf{V} \cdot \frac{d\mathbf{V}}{dt} = v \cdot \frac{dv}{dt}$$

A-91

In particular, if **V** is a vector of constant length then the right hand side of the last equation is identically zero showing that **V** is perpendicular to its derivative.

The derivatives of the triple products are

$$\frac{d}{dt}[\mathbf{V}_1\mathbf{V}_2\mathbf{V}_3] = \left[\left(\frac{d\mathbf{V}_1}{dt}\right)\mathbf{V}_2\mathbf{V}_3\right] + \left[\mathbf{V}_1\left(\frac{d\mathbf{V}_2}{dt}\right)\mathbf{V}_3\right] + \left[\mathbf{V}_1\mathbf{V}_2\left(\frac{d\mathbf{V}_3}{dt}\right)\right]$$

and $\frac{d}{dt}\left\{\mathbf{V}_1 \times (\mathbf{V}_2 \times \mathbf{V}_3)\right\} = \left(\frac{d\mathbf{V}_1}{dt}\right) \times (\mathbf{V}_2 \times \mathbf{V}_3) + \mathbf{V}_1$

$$\times \left(\left(\frac{d\mathbf{V}_2}{dt}\right) \times \mathbf{V}_3\right) + \mathbf{V}_1 \times \left(\mathbf{V}_2 \times \left(\frac{d\mathbf{V}_3}{dt}\right)\right)$$

Geometry of Curves in Space

s = the *length of arc*, measured from some fixed point on the curve (fig. 3).

$\mathbf{V}_1$ = the position vector of the point A on the curve

$\mathbf{V}_1 + \delta\mathbf{V}_1$ = the position vector of the point P in the neighborhood of A

$\hat{\mathbf{t}}$ = the *unit tangent* to the curve at the point A, measured in the direction of s increasing.

The *normal plane* is that plane which is perpendicular to the unit tangent. The principal normal is defined as the intersection of the normal plane with the plane defined by $\mathbf{V}_1$ and $\mathbf{V}_1 + \delta\mathbf{V}_1$ in the limit as $\delta\mathbf{V}_1 - 0$.

$\hat{\mathbf{n}}$ = the *unit normal* (principal) at the point A. The plane defined by $\hat{\mathbf{t}}$ and $\hat{\mathbf{n}}$ is called the *osculating plane* (alternatively plane of curvature or local plane).

ρ = the radius of curvature at A

$\delta\theta$ = the angle subtended at the origin by $\delta\mathbf{V}_1$.

$$\kappa = \frac{d\theta}{ds} = \frac{1}{\rho}$$

$\hat{\mathbf{b}}$ = the *unit binormal* i.e. the unit vector which is parallel to $\hat{\mathbf{t}} \times \hat{\mathbf{n}}$ at the point A:

λ = the *torsion* of the curve at A

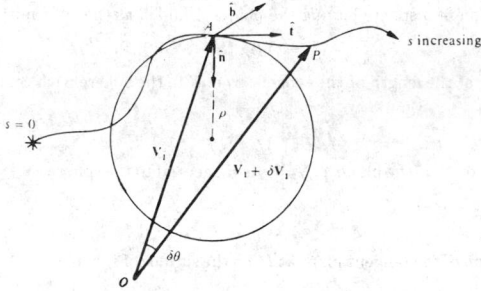

Figure 3.

Frenet's Formulae:

$$\frac{d\hat{\mathbf{t}}}{ds} = \kappa\hat{\mathbf{n}}$$

$$\frac{d\hat{\mathbf{n}}}{ds} = -\kappa\hat{\mathbf{t}} + \lambda\hat{\mathbf{b}}$$

$$\frac{d\hat{\mathbf{b}}}{ds} = -\lambda\hat{\mathbf{n}}$$

The following formulae are also applicable:

Unit tangent $\qquad\qquad \hat{\mathbf{t}} = \dfrac{d\mathbf{V}_1}{ds}$

Equation of the tangent $\qquad (\mathbf{V} - \mathbf{V}_1) \times \hat{\mathbf{t}} = 0$

or $\qquad \mathbf{V} = \mathbf{V}_1 + q\hat{\mathbf{t}}$

Unit normal $\qquad\qquad \hat{\mathbf{n}} = \dfrac{1}{\kappa}\dfrac{d^2\mathbf{V}_1}{ds^2}$

Equation of the normal plane $\qquad (\mathbf{V} - \mathbf{V}_1) \cdot \hat{\mathbf{t}} = 0$

Equation of the normal $\qquad (\mathbf{V} - \mathbf{V}_1) \times \hat{\mathbf{n}} = 0$

or $\qquad \mathbf{V} = \mathbf{V}_1 + r\hat{\mathbf{n}}$

Unit binormal $\qquad \hat{\mathbf{b}} = \hat{\mathbf{t}} \times \hat{\mathbf{n}}$

Equation of the binormal $\qquad (\mathbf{V} - \mathbf{V}_1) \times \hat{\mathbf{b}} = 0$

or $\qquad \mathbf{V} = \mathbf{V}_1 + u\hat{\mathbf{b}}$

or $\qquad \mathbf{V} = \mathbf{V}_1 + w\dfrac{d\mathbf{V}_1}{ds} \times \dfrac{d^2\mathbf{V}_1}{ds^2}$

Equation of the osculating plane:

$$[(\mathbf{V} - \mathbf{V}_1)\hat{\mathbf{t}}\hat{\mathbf{n}}] = 0$$

or $$\left[(\mathbf{V} - \mathbf{V}_1)\left(\frac{d\mathbf{V}_1}{ds}\right)\left(\frac{d^2\mathbf{V}_1}{ds^2}\right)\right] = 0$$

A *geodetic line* on a surface is a curve, the osculating plane of which is everywhere normal to the surface.

The differential equation of the geodetic is

$$[\hat{\mathbf{n}}\, d\mathbf{V}_1\, d^2\mathbf{V}_1] = 0$$

Differential Operators—Rectangular Coordinates

$$dS = \frac{\partial S}{\partial x} \cdot dx + \frac{\partial S}{\partial y} \cdot dy + \frac{\partial S}{\partial z} \cdot dz$$

By definition

$$\nabla \equiv \text{del} \equiv \mathbf{i}\,\frac{\partial}{\partial x} + \mathbf{j}\,\frac{\partial}{\partial y} + \mathbf{k}\,\frac{\partial}{\partial z}$$

$$\nabla^2 \equiv \text{Laplacian} \equiv \frac{\partial^2}{\partial x^2} + \frac{\partial^2}{\partial y^2} + \frac{\partial^2}{\partial z^2}$$

If S is a scalar function, then

$$\nabla S \equiv \text{grad } S \equiv \frac{\partial S}{\partial x}\mathbf{i} + \frac{\partial S}{\partial y}\mathbf{j} + \frac{\partial S}{\partial z}\mathbf{k}$$

Grad S defines both the direction and magnitude of the maximum rate of increase of S at any point. Hence the name *gradient* and also its vectorial nature. ∇S is independent of the choice of rectangular coordinates.

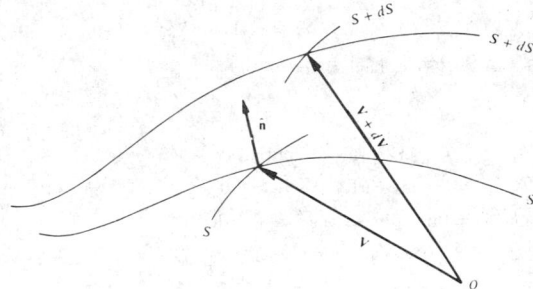

Figure 4

$$\nabla S = \frac{\partial S}{\partial n}\,\hat{\mathbf{n}}$$

where $\hat{\mathbf{n}}$ is the unit normal to the surface $S = \text{constant}$, in the direction of S increasing. The total derivative of S at a point having the position vector $\mathbf{V}$ is given by (fig. 4)

$$dS = \frac{\partial S}{\partial n}\,\hat{\mathbf{n}} \cdot d\mathbf{V}$$

$$= d\mathbf{V} \cdot \nabla S$$

and the directional derivative of S in the direction of $\mathbf{U}$ is

$$\mathbf{U} \cdot \nabla S = \mathbf{U} \cdot (\nabla S) = (\mathbf{U} \cdot \nabla)S$$

Similarly the directional derivative of the vector $\mathbf{V}$ in the direction of $\mathbf{U}$ is

$$(\mathbf{U} \cdot \nabla)\mathbf{V}$$

The *distributive* law holds for finding a gradient. Thus if S and T are scalar functions

$$\nabla(S + T) = \nabla S + \nabla T$$

The *associative* law becomes the rule for differentiating a product:

$$\nabla(ST) = S\nabla T + T\nabla S$$

If $\mathbf{V}$ is a vector function with the magnitudes of the components parallel to the three coordinate axes V_x, V_y, V_z, then

$$\nabla \cdot \mathbf{V} \equiv \text{div } \mathbf{V} \equiv \frac{\partial V_x}{\partial x} + \frac{\partial V_y}{\partial y} + \frac{\partial V_z}{\partial z}$$

The divergence obeys the distributive law. Thus, if $\mathbf{V}$ and $\mathbf{U}$ are vectors functions, then

$$\nabla \cdot (\mathbf{V} + \mathbf{U}) = \nabla \cdot \mathbf{V} + \nabla \cdot \mathbf{U}$$

$$\nabla \cdot (S\mathbf{V}) = (\nabla S) \cdot \mathbf{V} + S(\nabla \cdot \mathbf{V})$$

$$\nabla \cdot (\mathbf{U} \times \mathbf{V}) = \mathbf{V} \cdot (\nabla \times \mathbf{U}) - \mathbf{U} \cdot (\nabla \times \mathbf{V})$$

As with the gradient of a scalar, the divergence of a vector is invariant under a transformation from one set of rectangular coordinates to another.

$$\nabla \times \mathbf{V} \equiv \text{curl } \mathbf{V} \quad (\text{sometimes } \nabla \wedge \mathbf{V} \text{ or rot } \mathbf{V})$$

$$\equiv \left(\frac{\partial V_z}{\partial y} - \frac{\partial V_y}{\partial z}\right)\mathbf{i} + \left(\frac{\partial V_x}{\partial z} - \frac{\partial V_z}{\partial x}\right)\mathbf{j} + \left(\frac{\partial V_y}{\partial x} - \frac{\partial V_x}{\partial y}\right)\mathbf{k}$$

$$= \begin{vmatrix} \mathbf{i} & \mathbf{j} & \mathbf{k} \\ \dfrac{\partial}{\partial x} & \dfrac{\partial}{\partial y} & \dfrac{\partial}{\partial z} \\ V_x & V_y & V_z \end{vmatrix}$$

The *curl* (or *rotation*) of a vector is a vector which is invariant under a transformation from one set of rectangular coordinates to another.

$$\nabla \times (\mathbf{U} + \mathbf{V}) = \nabla \times \mathbf{U} + \nabla \times \mathbf{V}$$
$$\nabla \times (S\mathbf{V}) = (\nabla S) \times \mathbf{V} + S(\nabla \times \mathbf{V})$$
$$\nabla \times (\mathbf{U} \times \mathbf{V}) = (\mathbf{V} \cdot \nabla)\mathbf{U} - (\mathbf{U} \cdot \nabla)\mathbf{V} + \mathbf{U}(\nabla \cdot \mathbf{V}) - \mathbf{V}(\nabla \cdot \mathbf{U})$$

$$\text{grad } (\mathbf{U} \cdot \mathbf{V}) = \nabla(\mathbf{U} \cdot \mathbf{V})$$
$$= (\mathbf{V} \cdot \nabla)\mathbf{U} + (\mathbf{U} \cdot \nabla)\mathbf{V} + \mathbf{V} \times (\nabla \times \mathbf{U}) + \mathbf{U} \times (\nabla \times \mathbf{V})$$

If
$$\mathbf{V} = V_x\mathbf{i} + V_y\mathbf{j} + V_z\mathbf{k}$$
$$\nabla \cdot \mathbf{V} = \nabla V_x \cdot \mathbf{i} + \nabla V_y \cdot \mathbf{j} + \nabla V_z \cdot \mathbf{k}$$
and
$$\nabla \times \mathbf{V} = \nabla V_x \times \mathbf{i} + \nabla V_y \times \mathbf{j} + \nabla V_z \times \mathbf{k}$$

The operator ∇ can be used more than once. The number of possibilities where ∇ is used twice are

$$\nabla \cdot (\nabla \theta) = \text{div grad } \theta$$
$$\nabla \times (\nabla \theta) \equiv \text{curl grad } \theta$$
$$\nabla(\nabla \cdot \mathbf{V}) \equiv \text{grad div } \mathbf{V}$$
$$\nabla \cdot (\nabla \times \mathbf{V}) \equiv \text{div curl } \mathbf{V}$$
$$\nabla \times (\nabla \times \mathbf{V}) \equiv \text{curl curl } \mathbf{V}$$

Thus: $\text{div grad } S \equiv \nabla \cdot (\nabla S) \equiv \text{Laplacian } S \equiv \nabla^2 S$

$$\equiv \frac{\partial^2 S}{\partial x^2} + \frac{\partial^2 S}{\partial y^2} + \frac{\partial^2 S}{\partial z^2}$$

$\text{curl grad } S \equiv 0$; $\qquad$ $\text{curl curl } \mathbf{V} \equiv \text{grad div } \mathbf{V} - \nabla^2\mathbf{V}$;

$$\text{div curl } \mathbf{V} \equiv \qquad 0$$

Taylor's expansion in three dimensions can be written

$$f(\mathbf{V} + \boldsymbol{\varepsilon}) = e^{\boldsymbol{\varepsilon} \cdot \nabla} f(\mathbf{V})$$
where $\qquad \mathbf{V} = x\mathbf{i} + y\mathbf{j} + z\mathbf{k}$
and $\qquad \boldsymbol{\varepsilon} = h\mathbf{i} + l\mathbf{j} + m\mathbf{k}$

(note the analogy with $f_p = e^{phD}f_0$ in finite difference methods).

Orthogonal Curvilinear Coordinates

If at a point P there exist three uniform point functions u, v and w so that the surfaces $u = $ const., $v = $ const., and $w = $ const., intersect in three distinct curves through P then the surfaces are called the *coordinate surfaces* through P. The three lines of intersection are referred to as the *coordinate lines* and their tangents a, b, and c as the *coordinate axes*. When the coordinate axes form an orthogonal set the system is said to define *orthogonal curvilinear coordinates* at P.

Consider an infinitesimal volume enclosed by the surfaces u, v, w, $u + du$, $v + dv$, and $w + dw$ (fig. 5).

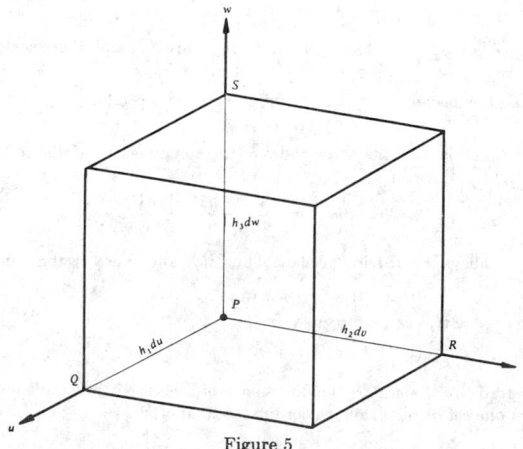

Figure 5

The surface $PRS \equiv u = $ const., and the face of the curvilinear figure immediately opposite this is $u + du = $ const. etc.

In terms of these surface constants

$$P = P(u,v,w)$$
$$Q = Q(u + du,v,w) \qquad \text{and} \qquad PQ = h_1 du$$

A-94

$$R = R(u,v + dv,w) \qquad PR = h_2 dv$$
$$S = S(u,v,w + dw) \qquad PS = h_3 dw$$

where h_1, h_2, and h_3 are functions of u, v, and w.

In rectangular Cartesians $\mathbf{i}$, $\mathbf{j}$, $\mathbf{k}$

$$h_1 = 1, \qquad h_2 = 1, \qquad h_3 = 1.$$

$$\frac{\hat{\mathbf{a}}}{h_1}\frac{\partial}{\partial u} = \mathbf{i}\,\frac{\partial}{\partial x}, \qquad \frac{\hat{\mathbf{b}}}{h_2}\frac{\partial}{\partial v} = \mathbf{j}\,\frac{\partial}{\partial y}, \qquad \frac{\hat{\mathbf{c}}}{h_3}\frac{\partial}{\partial w} = \mathbf{k}\,\frac{\partial}{\partial z}.$$

In cylindrical coordinates $\hat{\mathbf{r}}$, $\hat{\boldsymbol{\phi}}$, $\hat{\mathbf{k}}$

$$h_1 = 1, \qquad h_2 = r \qquad h_3 = 1.$$

$$\frac{\hat{\mathbf{a}}}{h_1}\frac{\partial}{\partial u} = \hat{\mathbf{r}}\,\frac{\partial}{\partial r}, \qquad \frac{\hat{\mathbf{b}}}{h_2}\frac{\partial}{\partial v} = \frac{\hat{\boldsymbol{\phi}}}{r}\frac{\partial}{\partial \phi} \qquad \frac{\hat{\mathbf{c}}}{h_3}\frac{\partial}{\partial w} = \hat{\mathbf{k}}\,\frac{\partial}{\partial z}$$

In spherical coordinates $\hat{\mathbf{r}}$, $\hat{\boldsymbol{\theta}}$, $\hat{\boldsymbol{\phi}}$

$$h_1 = 1, \qquad h_2 = r, \qquad h_3 = r\sin\theta$$

$$\frac{\hat{\mathbf{a}}}{h_1}\frac{\partial}{\partial u} = \hat{\mathbf{r}}\,\frac{\partial}{\partial r}, \qquad \frac{\mathbf{b}}{h_2}\frac{\partial}{\partial v} = \frac{\hat{\boldsymbol{\phi}}}{r}\frac{\partial}{\partial \theta}, \qquad \frac{\hat{\mathbf{c}}}{h_3}\frac{\partial}{\partial w} = \frac{\hat{\boldsymbol{\phi}}}{r\sin\theta}\frac{\partial}{\partial \phi}$$

The general expressions for grad, div and curl together with those for ∇^2 and the directional derivative are, in orthogonal curvilinear coordinates, given by

$$\nabla S = \frac{\hat{\mathbf{a}}}{h_1}\frac{\partial S}{\partial u} + \frac{\hat{\mathbf{b}}}{h_2}\frac{\partial S}{\partial v} + \frac{\hat{\mathbf{c}}}{h_3}\frac{\partial S}{\partial w}$$

$$(\mathbf{V}\cdot\nabla)S = \frac{V_1}{h_1}\frac{\partial S}{\partial u} + \frac{V_2}{h_2}\frac{\partial S}{\partial v} + \frac{V_3}{h_3}\frac{\partial S}{\partial w}$$

$$\nabla\cdot\mathbf{V} = \frac{1}{h_1 h_2 h_3}\left\{\frac{\partial}{\partial u}(h_2 h_3 V_1) + \frac{\partial}{\partial v}(h_3 h_1 V_2) + \frac{\partial}{\partial w}(h_1 h_2 V_3)\right\}$$

$$\nabla\times\mathbf{V} = \frac{\hat{\mathbf{a}}}{h_2 h_3}\left\{\frac{\partial}{\partial v}(h_3 V_3) - \frac{\partial}{\partial w}(h_2 V_2)\right\} + \frac{\hat{\mathbf{b}}}{h_3 h_1}\left\{\frac{\partial}{\partial w}(h_1 V_1) - \frac{\partial}{\partial u}(h_3 V_3)\right\}$$
$$+ \frac{\hat{\mathbf{c}}}{h_1 h_2}\left\{\frac{\partial}{\partial u}(h_2 V_2) - \frac{\partial}{\partial v}(h_1 V_1)\right\}$$

$$\nabla^2 S = \frac{1}{h_1 h_2 h_3}\left\{\frac{\partial}{\partial u}\left(\frac{h_2 h_3}{h_1}\frac{\partial S}{\partial u}\right) + \frac{\partial}{\partial v}\left(\frac{h_3 h_1}{h_2}\frac{\partial S}{\partial v}\right) + \frac{\partial}{\partial w}\left(\frac{h_1 h_2}{h_3}\frac{\partial S}{\partial w}\right)\right\}$$

FORMULAS OF VECTOR ANALYSIS

	Rectangular coordinates	Cylindrical coordinates	Spherical coordinates
Conversion to rectangular coordinates		$x = r\cos\varphi \quad y = r\sin\varphi \quad z = z$	$x = r\cos\varphi\sin\theta \quad y = r\sin\varphi\sin\theta$ $z = r\cos\theta$
Gradient	$\nabla\phi = \dfrac{\partial\phi}{\partial x}\mathbf{i} + \dfrac{\partial\phi}{\partial y}\mathbf{j} + \dfrac{\partial\phi}{\partial z}\mathbf{k}$	$\nabla\phi = \dfrac{\partial\phi}{\partial r}\mathbf{r} + \dfrac{1}{r}\dfrac{\partial\phi}{\partial \varphi}\boldsymbol{\phi} + \dfrac{\partial\phi}{\partial z}\mathbf{k}$	$\nabla\phi = \dfrac{\partial\phi}{\partial r}\mathbf{r} + \dfrac{1}{r}\dfrac{\partial\phi}{\partial \theta}\boldsymbol{\theta} + \dfrac{1}{r\sin\theta}\dfrac{\partial\phi}{\partial \varphi}\boldsymbol{\phi}$
Divergence	$\nabla\cdot\mathbf{A} = \dfrac{\partial A_x}{\partial x} + \dfrac{\partial A_y}{\partial y} + \dfrac{\partial A_z}{\partial z}$	$\nabla\cdot\mathbf{A} = \dfrac{1}{r}\dfrac{\partial(rA_r)}{\partial r} + \dfrac{1}{r}\dfrac{\partial A_\varphi}{\partial \varphi}$ $+ \dfrac{\partial A_z}{\partial z}$	$\nabla\cdot\mathbf{A} = \dfrac{1}{r^2}\dfrac{\partial(r^2 A_r)}{\partial r} + \dfrac{1}{r\sin\theta}\dfrac{\partial(A_\theta\sin\theta)}{\partial \theta}$ $+ \dfrac{1}{r\sin\theta}\dfrac{\partial A_\varphi}{\partial c}$
Curl	$\nabla\times\mathbf{A} = \begin{vmatrix} \mathbf{i} & \mathbf{j} & \mathbf{k} \\ \dfrac{\partial}{\partial x} & \dfrac{\partial}{\partial y} & \dfrac{\partial}{\partial z} \\ A_x & A_y & A_z \end{vmatrix}$	$\nabla\times\mathbf{A} = \begin{vmatrix} \dfrac{1}{r}\mathbf{r} & \boldsymbol{\phi} & \dfrac{1}{r}\mathbf{k} \\ \dfrac{\partial}{\partial r} & \dfrac{\partial}{\partial \varphi} & \dfrac{\partial}{\partial z} \\ A_r & rA_\varphi & A_z \end{vmatrix}$	$\nabla\times\mathbf{A} = \begin{vmatrix} \dfrac{\mathbf{r}}{r^2\sin\theta} & \dfrac{\boldsymbol{\theta}}{r\sin\theta} & \dfrac{\boldsymbol{\phi}}{r} \\ \dfrac{\partial}{\partial r} & \dfrac{\partial}{\partial \theta} & \dfrac{\partial}{\partial \varphi} \\ A_r & rA_\theta & rA_\varphi\sin\theta \end{vmatrix}$
Laplacian	$\nabla^2\phi = \dfrac{\partial^2\phi}{\partial x^2} + \dfrac{\partial^2\phi}{\partial y^2} + \dfrac{\partial^2\phi}{\partial z^2}$	$\nabla^2\phi = \dfrac{1}{r}\dfrac{\partial}{\partial r}\left(r\dfrac{\partial\phi}{\partial r}\right) + \dfrac{1}{r^2}\dfrac{\partial^2\phi}{\partial \varphi^2}$ $+ \dfrac{\partial^2\phi}{\partial z^2}$	$\nabla^2\phi = \dfrac{1}{r^2}\dfrac{\partial}{\partial r}\left(r^2\dfrac{\partial\phi}{\partial r}\right) + \dfrac{1}{r^2\sin\theta}\dfrac{\partial}{\partial \theta}\left(\sin\theta\dfrac{\partial\phi}{\partial \theta}\right)$ $+ \dfrac{1}{r^2\sin^2\theta}\dfrac{\partial^2\phi}{\partial \varphi^2}$

s = the distance along some curve "C" in space and is measured from some fixed point.
S = a surface area
V = a volume contained by a specified surface
$\hat{\mathbf{t}}$ = the unit tangent to C at the point P
$\hat{\mathbf{n}}$ = the unit outward pointing normal
F = some vector function
$d\mathbf{s}$ = the vector element of curve ($= \hat{\mathbf{t}}\, ds$)
$d\mathbf{S}$ = the vector element of surface ($= \hat{\mathbf{n}}\, dS$)

Then

$$\int_{(c)} \mathbf{F} \cdot \hat{\mathbf{t}}\, ds = \int_{(c)} \mathbf{F} \cdot d\mathbf{s}$$

and when

$$\mathbf{F} = \nabla\phi$$

$$\int_{(c)} (\nabla\phi) \cdot \hat{\mathbf{t}}\, ds = \int_{(c)} d\phi$$

Gauss' Theorem (Green's Theorem)

When S defines a closed region having a volume V

$$\iiint_{(v)} (\nabla \cdot \mathbf{F})\, dV = \iint_{(s)} (\mathbf{F} \cdot \hat{\mathbf{n}})\, dS = \iint_{(s)} \mathbf{F} \cdot d\mathbf{S}$$

also

$$\iiint_{(v)} (\nabla\phi)\, dV = \iint_{(s)} \phi\hat{\mathbf{n}}\, dS$$

and

$$\iiint_{(v)} (\nabla \times \mathbf{F})\, dV = \iint_{(s)} (\hat{\mathbf{n}} \times \mathbf{F})\, dS$$

Stokes' Theorem

When C is closed and bounds the open surface S.

$$\iint_{(s)} \hat{\mathbf{n}} \cdot (\nabla \times \mathbf{F})\, dS = \int_{(c)} \mathbf{F} \cdot d\mathbf{s}$$

also

$$\iint_{(s)} (\hat{\mathbf{n}} \times \nabla\phi)\, dS = \int_{(c)} \phi\, d\mathbf{s}$$

Green's Theorem

$$\iint_{(s)} (\nabla\phi \cdot \nabla\theta)\, dS = \iint_{(s)} \phi\hat{\mathbf{n}} \cdot (\nabla\theta)\, dS = \iiint_{(v)} \phi(\nabla^2\theta)\, dV$$

$$= \iint_{(s)} \theta \cdot \hat{\mathbf{n}}(\nabla\phi)\, dS = \iiint_{(v)} \theta(\nabla^2\phi)\, dV$$

MOMENT OF INERTIA FOR VARIOUS BODIES OF MASS

The mass of the body is indicated by m

Body	Axis	Moment of inertia	Body	Axis	Moment of inertia
Uniform thin rod	Normal to the length, at one end	$m\dfrac{l^2}{3}$	Spherical shell, very thin, mean radius, r	Any diameter	$m\dfrac{2}{3}r^2$
Uniform thin rod	Normal to the length, at the center	$m\dfrac{l^2}{12}$	Right circular cylinder of radius r, length l	The longitudinal axis of the solid	$m\dfrac{r^2}{2}$
Thin rectangular sheet, sides a and b	Through the center parallel to b	$m\dfrac{a^2}{12}$	Right circular cylinder of radius r, length l	Transverse diameter	$m\left(\dfrac{r^2}{4} + \dfrac{l^2}{12}\right)$
Thin rectangular sheet, sides a and b	Through the center perpendicular to the sheet	$m\dfrac{a^2 + b^2}{12}$	Hollow circular cylinder, length l, radii r_1 and r_2	The longitudinal axis of the figure	$m\dfrac{(r_1^2 + r_2^2)}{2}$
Thin circular sheet of radius r	Normal to the plate through the center	$m\dfrac{r^2}{2}$	Thin cylindrical shell, length l, mean radius, r	The longitudinal axis of the figure	mr^2
Thin circular sheet of radius r	Along any diameter	$m\dfrac{r^2}{4}$	Hollow circular cylinder, length l, radii r_1 and r_2	Transverse diameter	$m\left[\dfrac{r_1^2 + r_2^2}{4} + \dfrac{l^2}{12}\right]$
Thin circular ring. Radii r_1 and r_2	Through center normal to plane of ring	$m\dfrac{r_1^2 + r_2^2}{2}$	Hollow circular cylinder, length l, very thin, mean radius	Transverse diameter	$m\left(\dfrac{r^2}{2} + \dfrac{l^2}{12}\right)$
Thin circular ring. Radii r_1 and r_2	Any diameter	$m\dfrac{r_1^2 + r_2^2}{4}$	Elliptic cylinder, length l, transverse semiaxes a and b	Longitudinal axis	$m\left(\dfrac{a^2 + b^2}{4}\right)$
Rectangular parallelopiped, edges a, b, and c	Through center perpendicular to face ab, (parallel to edge c)	$m\dfrac{a^2 + b^2}{12}$	Right cone, altitude h, radius of base r	Axis of the figure	$m\dfrac{3}{10}r^2$
Sphere, radius r	Any diameter	$m\dfrac{2}{5}r^2$	Spheroid of revolution, equatorial radius r	Polar axis	$m\dfrac{2r^2}{5}$
Spherical shell, external radius, r_1, internal radius r_2	Any diameter	$m\dfrac{2}{5}\dfrac{(r_1^5 - r_2^5)}{(r_1^3 - r_2^3)}$	Ellipsoid, axes $2a$, $2b$, $2c$	Axis $2a$	$m\dfrac{(b^2 + c^2)}{5}$

Bessel Functions*

1. Bessel's differential equation for a real variable x is

$$x^2 \frac{d^2 y}{dx^2} + x \frac{dy}{dx} + (x^2 - n^2)y = 0$$

* From Beyer, W. H., Ed., *CRC Handbook of Mathematical Sciences*, 5th ed., CRC Press, Boca Raton, 1978, 500—503. With permission.

2. When n is not an integer, two independent solutions of the equation are $J_n(x)$, $J_{-n}(x)$, where

$$J_n(x) = \sum_{k=0}^{\infty} \frac{(-1)^k}{k!\,\Gamma(n+k+1)}\left(\frac{x}{2}\right)^{n+2k}$$

3. If n is an integer $J_{-n}(x) = (-1)^n J_n(x)$, where

$$J_n(x) = \frac{x^n}{2^n n!}\left\{1 - \frac{x^2}{2^2 \cdot 1!(n+1)} + \frac{x^4}{2^4 \cdot 2!(n+1)(n+2)}\right.$$
$$\left. - \frac{x^6}{2^6 \cdot 3!(n+1)(n+2)(n+3)} + \cdots\right\}$$

4. For $n = 0$ and $n = 1$, this formula becomes

$$J_0(x) = 1 - \frac{x^2}{2^2(1!)^2} + \frac{x^4}{2^4(2!)^2} - \frac{x^6}{2^6(3!)^2} + \frac{x^8}{2^8(4!)^2} - \cdots$$

$$J_1(x) = \frac{x}{2} - \frac{x^3}{2^3 \cdot 1!2!} + \frac{x^5}{2^5 \cdot 2!3!} - \frac{x^7}{2^7 \cdot 3!4!} + \frac{x^9}{2^9 \cdot 4!5!} - \cdots$$

5. When x is large and positive, the following asymptotic series may be used

$$J_0(x) = \left(\frac{2}{\pi x}\right)^{\frac{1}{2}}\left\{P_0(x)\cos\left(x - \frac{\pi}{4}\right) - Q_0(x)\sin\left(x - \frac{\pi}{4}\right)\right\}$$

$$J_1(x) = \left(\frac{2}{\pi x}\right)^{\frac{1}{2}}\left\{P_1(x)\cos\left(x - \frac{3\pi}{4}\right) - Q_1(x)\sin\left(x - \frac{3\pi}{4}\right)\right\}$$

where

$$P_0(x) \sim 1 - \frac{1^2 \cdot 3^2}{2!(8x)^2} + \frac{1^2 \cdot 3^2 \cdot 5^2 \cdot 7^2}{4!(8x)^4} - \frac{1^2 \cdot 3^2 \cdot 5^2 \cdot 7^2 \cdot 9^2 \cdot 11^2}{6!(8x)^6} + \cdots$$

$$Q_0(x) \sim -\frac{1^2}{1!8x} + \frac{1^2 \cdot 3^2 \cdot 5^2}{3!(8x)^3} - \frac{1^2 \cdot 3^2 \cdot 5^2 \cdot 7^2 \cdot 9^2}{5!(8x)^5} + - \cdots$$

$$P_1(x) \sim 1 + \frac{1^2 \cdot 3 \cdot 5}{2!(8x)^2} - \frac{1^2 \cdot 3^2 \cdot 5^2 \cdot 7 \cdot 9}{4!(8x)^4} + \frac{1^2 \cdot 3^2 \cdot 5^2 \cdot 7^2 \cdot 9^2 \cdot 11 \cdot 13}{6!(8x)^6} - + \cdots$$

$$Q_1(x) \sim \frac{1 \cdot 3}{1!8x} - \frac{1^2 \cdot 3^2 \cdot 5 \cdot 7}{3!(8x)^3} + \frac{1^2 \cdot 3^2 \cdot 5^2 \cdot 7^2 \cdot 9 \cdot 11}{5!(8x)^5} - \cdots$$

[In $P_1(x)$ the signs alternate from $+$ to $-$ after the first term]

6. If $x > 25$, it is convenient to use the formulas

$$J_0(x) = A_0(x)\sin x + B_0(x)\cos x$$
$$J_1(x) = B_1(x)\sin x - A_1(x)\cos x$$

where

$$A_0(x) = \frac{P_0(x) - Q_0(x)}{(\pi x)^{\frac{1}{2}}} \quad \text{and} \quad A_1(x) = \frac{P_1(x) - Q_1(x)}{(\pi x)^{\frac{1}{2}}}$$

$$B_0(x) = \frac{P_0(x) + Q_0(x)}{(\pi x)^{\frac{1}{2}}} \quad \text{and} \quad B_1(x) = \frac{P_1(x) + Q_1(x)}{(\pi x)^{\frac{1}{2}}}$$

7. The zeros of $J_0(x)$ and $J_1(x)$

If j_{0s} and j_{1s} are the sth zeros of $J_0(x)$ and $J_1(x)$ respectively, and if $a = 4_s - 1$, $b = 4_s + 1$

$$j_{0s} \sim \frac{1}{4}\pi a\left\{1 + \frac{2}{\pi^2 a^2} - \frac{62}{3\pi^4 a^4} + \frac{15,116}{15\pi^6 a^6} - \frac{12,554,474}{105\pi^8 a^8} + \frac{8,368,654,292}{315\pi^{10}a^{10}} - + \cdots\right\}$$

$$j_{1s} \sim \frac{1}{4}\pi b\left\{1 - \frac{6}{\pi^2 b^2} + \frac{6}{\pi^4 b^4} - \frac{4716}{5\pi^6 b^6} + \frac{3,902,418}{35\pi^8 b^8} - \frac{895,167,324}{35\pi^{10}b^{10}} + \cdots\right\}$$

$$J_1(j_{0s}) \sim \frac{(-1)^{s+1}2^{\frac{1}{2}}}{\pi a^{\frac{1}{2}}}\left\{1 - \frac{56}{3\pi^4 a^4} + \frac{9664}{5\pi^6 a^6} - \frac{7,381,280}{21\pi^8 a^8} + \cdots\right\}$$

$$J_0(j_{1s}) \sim \frac{(-1)^s 2^{\frac{1}{2}}}{\pi b^{\frac{1}{2}}}\left\{1 + \frac{24}{\pi^4 b^4} - \frac{19,584}{10\pi^6 b^6} + \frac{2,466,720}{7\pi^8 b^8} - \cdots\right\}$$

8. Table of zeros for $J_0(x)$ and $J_1(x)$

Roots α_n	$J_1(\alpha_n)$	Roots β_n	$J_0(\beta_n)$
2.4048	0.5191	0.0000	1.0000
5.5201	−0.3403	3.8317	−0.4028
8.6537	0.2715	7.0156	0.3001
11.7915	−0.2325	10.1735	−0.2497
14.9309	0.2065	13.3237	0.2184
18.0711	−0.1877	16.4706	−0.1965
21.2116	0.1733	19.6159	0.1801

Table heading: $J_1(\alpha_n) = 0 \qquad J_0(\beta_n) = 0$

9. Recurrence formulas

$$J_{n-1}(x) + J_{n+1}(x) = \frac{2n}{x} J_n(x) \qquad\qquad nJ_n(x) + xJ_n'(x) = xJ_{n-1}(x)$$

$$J_{n-1}(x) - J_{n+1}(x) = 2J_n'(x) \qquad\qquad nJ_n(x) - xJ_n'(x) = xJ_{n+1}(x)$$

10. If J_n is written for $J_n(x)$ and $J_n^{(k)}$ is written for $\frac{d^k}{dx^k}\{J_n(x)\}$, then the following derivative relationships are important

$$J_0^{(r)} = -J_1^{(r-1)}$$

$$J_0^{(2)} = -J_0 + \frac{1}{x} J_1 = \frac{1}{2}(J_2 - J_0)$$

$$J_0^{(3)} = \frac{1}{x} J_0 + \left(1 - \frac{2}{x^2}\right) J_1 = \frac{1}{4}(-J_3 + 3J_1)$$

$$J_0^{(4)} = \left(1 - \frac{3}{x^2}\right) J_0 - \left(\frac{2}{x} - \frac{6}{x^3}\right) J_1 = \frac{1}{8}(J_4 - 4J_2 + 3J_0), \text{ etc.}$$

11. Half order Bessel functions

$$J_{\frac{1}{2}}(x) = \sqrt{\frac{2}{\pi x}} \sin x$$

$$J_{-\frac{1}{2}}(x) = \sqrt{\frac{2}{\pi x}} \cos x$$

$$J_{n+\frac{1}{2}}(x) = -x^{n+\frac{1}{2}} \frac{d}{dx}\{x^{-(n+\frac{1}{2})} J_{n+\frac{1}{2}}(x)\}$$

$$J_{n-\frac{1}{2}}(x) = x^{-(n+\frac{1}{2})} \frac{d}{dx}\{x^{n+\frac{1}{2}} J_{n+\frac{1}{2}}(x)\}$$

n	$\left(\frac{\pi x}{2}\right)^{\frac{1}{2}} J_{n+\frac{1}{2}}(x)$	$\left(\frac{\pi x}{2}\right)^{\frac{1}{2}} J_{-(n+\frac{1}{2})}(x)$
0	$\sin x$	$\cos x$
1	$\dfrac{\sin x}{x} - \cos x$	$-\dfrac{\cos x}{x} - \sin x$
2	$\left(\dfrac{3}{x^2} - 1\right)\sin x - \dfrac{3}{x}\cos x$	$\left(\dfrac{3}{x^2} - 1\right)\cos x + \dfrac{3}{x}\sin x$
3	$\left(\dfrac{15}{x^3} - \dfrac{6}{x}\right)\sin x - \left(\dfrac{15}{x^2} - 1\right)\cos x$	$-\left(\dfrac{15}{x^3} - \dfrac{6}{x}\right)\cos x - \left(\dfrac{15}{x^2} - 1\right)\sin x$
	etc.	

12. Additional solutions to Bessel's equation are

$Y_n(x)$ (also called Weber's function, and sometimes denoted by $N_n(x)$)

$H_n^{(1)}(x)$ and $H_n^{(2)}(x)$ (also called Hankel functions)

These solutions are defined as follows

$$Y_n(x) = \begin{cases} \dfrac{J_n(x)\cos(n\pi) - J_{-n}(x)}{\sin(n\pi)} & n \text{ not an integer} \\[2ex] \lim_{v \to n} \dfrac{J_v(x)\cos(v\pi) - J_{-v}(x)}{\sin(v\pi)} & n \text{ an integer} \end{cases}$$

$$H_n^{(1)}(x) = J_n(x) + iY_n(x)$$
$$H_n^{(2)}(x) = J_n(x) - iY_n(x)$$

The additional properties of these functions may all be derived from the above relations and the known properties of $J_n(x)$.

13. Complete solutions to Bessel's equation may be written as

$$c_1 J_n(x) + c_2 J_{-n}(x) \qquad \text{if } n \text{ is not an integer}$$

or

$$\left.\begin{array}{c} c_1 J_n(x) + c_2 Y_n(x) \\[2mm] c_1 H_n^{(1)}(x) + c_2 H_n^{(2)}(x) \end{array}\right\} \text{for any value of } n$$

or

14. The modified (or hyperbolic) Bessel's differential equation is

$$x^2 \frac{d^2 y}{dx^2} + x \frac{dy}{dx} - (x^2 + n^2) y = 0$$

15. When n is not an integer, two independent solutions of the equation are $I_n(x)$ and $I_{-n}(x)$, where

$$I_n(x) = \sum_{k=0}^{\infty} \frac{1}{k!\,\Gamma(n+k+1)} \left(\frac{x}{2}\right)^{n+2k}$$

16. If n is an integer,

$$I_n(x) = I_{-n}(x) = \frac{x^n}{2^n n!}\left\{1 + \frac{x^2}{2^2 \cdot 1!(n+1)} + \frac{x^4}{2^4 \cdot 2!(n+1)(n+2)}\right.$$
$$\left. + \frac{x^6}{2^6 \cdot 3!(n+1)(n+2)(n+3)} + \cdots\right\}$$

17. For $N = 0$ and $n = 1$, this formula becomes

$$I_0(x) = 1 + \frac{x^2}{2^2(1!)^2} + \frac{x^4}{2^4(2!)^2} + \frac{x^6}{2^6(3!)^2} + \frac{x^8}{2^8(4!)^2} + \cdots$$

$$I_1(x) = \frac{x}{2} + \frac{x^3}{2^3 \cdot 1!2!} + \frac{x^5}{2^5 \cdot 2!3!} + \frac{x^7}{2^7 \cdot 3!4!} + \frac{x^9}{2^9 \cdot 4!5!} + \cdots$$

18. Another solution to the modified Bessel's equation is

$$K_n(x) = \begin{cases} \dfrac{1}{2}\,\pi\,\dfrac{I_{-n}(x) - I_n(x)}{\sin(n\pi)} & n \text{ not an integer} \\[4mm] \lim_{\nu \to n} \dfrac{1}{2}\,\pi\,\dfrac{I_{-\nu}(x) - I_\nu(x)}{\sin(\nu\pi)} & n \text{ an integer} \end{cases}$$

This function is linearly independent of $I_n(x)$ for all values of n. Thus the complete solution to the modified Bessel's equation may be written as

$$c_1 I_n(x) + c_2 I_{-n}(x) \qquad n \text{ not an integer}$$

or

$$c_1 I_n(x) + c_2 K_n(x) \qquad \text{any } n$$

19. The following relations hold among the various Bessel functions:

$$I_n(z) = i^{-m} J_m(iz)$$
$$Y_n(iz) = (i)^{n+1} I_n(z) - \frac{2}{\pi}\,i^{-n} K_n(z)$$

Most of the properties of the modified Bessel function may be deduced from the known properties of $J_n(x)$ by use of these relations and those previously given.

20. Recurrence formulas

$$I_{n-1}(x) - I_{n+1}(x) = \frac{2n}{x}\,I_n(x) \qquad\qquad I_{n-1}(x) + I_{n+1}(x) = 2I_n'(x)$$

$$I_{n-1}(x) - \frac{n}{x}\,I_n(x) = I_n'(x) \qquad\qquad I_n'(x) = I_{n+1}(x) + \frac{n}{x}\,I_n(z)$$

The Gamma Function*

Definition: $\Gamma\,(n = \int_0^{\infty} t^{n-1} e^{-t}\,dt \; n > 0$

Recursion Formula: $\Gamma(n+1) = n\Gamma(n)$
$\Gamma(n)+1) = n!$ if $n = 0,1,2,\ldots$ where $0! = 1$
For $n < 0$ the gamma function can be defined by using

---

* From Beyer, W. H., Ed., *CRC Handbook of Mathematical Sciences,* 5th ed., CRC Press, Boca Raton, 1978, 484—485. With permission.

$$\Gamma(n) = \frac{\Gamma(n + 1)}{n}$$

Graph:

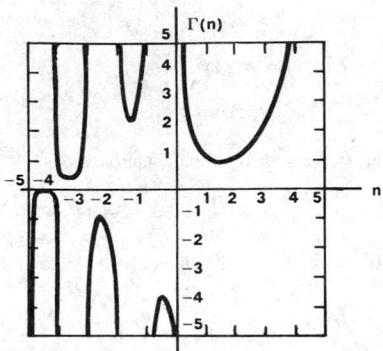

Special Values:

$$\Gamma(\tfrac{1}{2}) = \sqrt{\pi}$$

$$\Gamma(m + \tfrac{1}{2}) = \frac{1 \cdot 3 \cdot 5 \cdots (2m - 1)}{2^m} \sqrt{\pi} \qquad m = 1,2,3,\ldots$$

$$\Gamma(-m + \tfrac{1}{2}) = \frac{(-1)^m 2^m \sqrt{\pi}}{1 \cdot 3 \cdot 5 \cdots (2m - 1)} \qquad m = 1,2,3,\ldots$$

Definition:

$$\Gamma(x + 1) = \lim_{k \to \infty} \frac{1 \cdot 2 \cdot 3 \cdots k}{(x + 1)(x + 2) \cdots (x + k)} \, k^x$$

$$\frac{1}{\Gamma(x)} = x e^{\gamma x} \prod_{m = 1}^{\infty} \left\{ \left(1 + \frac{x}{m} \right) e^{-x/m} \right\}$$

This is an infinite product representation for the gamma function where γ is Euler's constant.

Properties:

$$\Gamma'(1) = \int_0^\infty e^{-x} \ln x \, dx = -\gamma$$

$$\frac{\Gamma'(x)}{\Gamma(x)} = -\gamma + \left(\frac{1}{1} - \frac{1}{x} \right) + \left(\frac{1}{2} - \frac{1}{x + 1} \right) + \ldots + \left(\frac{1}{n} - \frac{1}{x + n - 1} \right) + \ldots$$

$$\Gamma(x + 1) = \sqrt{2\pi x} \, x^x e^{-x} \left\{ 1 + \frac{1}{12x} + \frac{1}{288x^2} - \frac{139}{51,840x^3} + \ldots \right\}$$

This is called *Stirling's asymptotic series.*

If we let $x = n$ a positive integer, then a useful approximation for $n!$ where n is large (e.g., $n > 10$) is given by *Stirling's formula*

$$n! \approx \sqrt{2\pi n} \, n^n e^{-n}$$

The Gamma Function*

Values of $\Gamma(n) = \int_0^\infty e^{-x} x^{n-1}\, dx;\ \Gamma(n+1) = n\Gamma(n)$

n	$\Gamma(n)$	n	$\Gamma(n)$	n	$\Gamma(n)$	n	$\Gamma(n)$
1.00	1.00000	1.25	.90640	1.50	.88623	1.75	.91906
1.01	.99433	1.26	.90440	1.51	.88659	1.76	.92137
1.02	.98884	1.27	.90250	1.52	.88704	1.77	.92376
1.03	.98355	1.28	.90072	1.53	.88757	1.78	.92623
1.04	.97844	1.29	.89904	1.54	.88818	1.79	.92877
1.05	.97350	1.30	.89747	1.55	.88887	1.80	.93138
1.06	.96874	1.31	.89600	1.56	.88964	1.81	.93408
1.07	.96415	1.32	.89464	1.57	.89049	1.82	.93685
1.08	.95973	1.33	.89338	1.58	.89142	1.83	.93969
1.09	.95546	1.34	.89222	1.59	.89243	1.84	.94261
1.10	.95135	1.35	.89115	1.60	.89352	1.85	.94561
1.11	.94740	1.36	.89018	1.61	.89468	1.86	.94869
1.12	.94359	1.37	.88931	1.62	.89592	1.87	.95184
1.13	.93993	1.38	.88854	1.63	.89724	1.88	.95507
1.14	.93642	1.39	.88785	1.64	.89864	1.89	.95838
1.15	.93304	1.40	.88726	1.65	.90012	1.90	.96177
1.16	.92980	1.41	.88676	1.66	.90167	1.91	.96523
1.17	.92670	1.42	.88636	1.67	.90330	1.92	.96877
1.18	.92373	1.43	.88604	1.68	.90500	1.93	.97240
1.19	.92089	1.44	.88581	1.69	.90678	1.94	.97610
1.20	.91817	1.45	.88566	1.70	.90864	1.95	.97988
1.21	.91558	1.46	.88560	1.71	.91057	1.96	.98374
1.22	.91311	1.47	.88563	1.72	.91258	1.97	.98768
1.23	.91075	1.48	.88575	1.73	.91466	1.98	.99171
1.24	.90852	1.49	.88595	1.74	.91683	1.99	.99581
						2.00	1.00000

* For large positive values of x, $\Gamma(x)$ approximates Stirling's asymptotic series

$$x^x e^{-x} \sqrt{\frac{2\pi}{x}} \left[1 + \frac{1}{12x} + \frac{1}{288x^2} - \frac{139}{51840x^3} - \frac{571}{2488320x^4} + \cdots \right]$$

The Beta Function*

Definition: $B(m,n) = \int_0^1 t^{m-1} (1-t)^{m-1}\, dt \quad m > 0, n > 0$

Relationship with Gamma Function: $B(m,n) = \dfrac{\Gamma(m)\Gamma(n)}{\Gamma(m+n)}$

Properties:

$$B(m,n) = B(n,m)$$

$$B(m,n) = 2\int_0^{\pi/2} \sin^{2m-1}\theta \, \cos^{2n-1}\theta \, d\theta$$

$$B(m,n) = \int_0^\infty \frac{t^{m-1}}{(1+t)^{m+n}}\, dt$$

$$B(m,n) = r^n (r+1)^m \int_0^1 \frac{t^{m-1} (1-t)^{n-1}}{(r+t)^{m+n}}\, dt$$

The Error Function

Definition: $erf\, x = \dfrac{2}{\sqrt{\pi}} \int_0^x e^{-t^2}\, dt$

Series: $erf\, x = \dfrac{2}{\sqrt{\pi}} \left(x - \dfrac{x^3}{3} + \dfrac{1}{2!}\dfrac{x^5}{5} - \dfrac{1}{3!}\dfrac{x^7}{7} + \ldots \right)$

Property: $erf\, x = -erf\,(-x)$

Relationship with Normal Probability Function f(t): $\displaystyle\int_0^x f(t)\, dt = \dfrac{1}{2}\, erf\left(\dfrac{x}{\sqrt{2}}\right)$

To evaluate $erf\,(2.3)$, one proceeds as follows: Since $\dfrac{x}{\sqrt{2}} = 2.3$, one finds $x = (2.3)\,(\sqrt{2}) = 3.25$. In the normal probability function table (page A-), one finds the entry 0.4994 opposite the value 3.25. Thus $erf\,(2.3) = 2(0.4994) = 0.9988$.

---

* From Beyer, W. H., Ed., *CRC Handbook of Mathematical Sciences*, 5th ed., CRC Press, Boca Raton, 1978, 499. With permission.

$$erfc\ z = 1 - erf\ z = \frac{2}{\sqrt{\pi}} \int_z^\infty e^{-t^2}\ dt$$

is known as the complementary error function.

Orthogonal Polynomials*

I

Name: Legendre   *Symbol:* $P_n(x)$ *Interval:* $[-1,1]$
Differential Equation: $(1 - x^2)\ y'' - 2\ xy' + n(n + 1)\ y = 0$
$$y = P_n(x)$$

Explicit Expression: $P_n(x) = \dfrac{1}{2^n} \displaystyle\sum_{m=0}^{[n/2]} (-1)^m \binom{n}{m} \binom{2n - 2m}{n} x^{n-2m}$

Recurrence Relation: $(n + 1)\ P_{n+1}(x) = (2n + 1)\ x\ P_n(x) - nP_{n-1}(x)$
Weight: 1   *Standardization:* $P_n(1) = 1$

Norm: $\displaystyle\int_{-1}^{+1} [P_n(x)]^2\ dx = \dfrac{2}{2n + 1}$

Rodrigues' Formula: $P_n(x) = \dfrac{(-1)^n}{2^n n!} \dfrac{d^n}{dx^n} \{(1 - x^2)^n\}$

Generating Function: $R^{-1} = \displaystyle\sum_{n=0}^\infty P_n(x) z^n;\ -1 < x < 1,\ |z| < 1,$
$$R = \sqrt{1 - 2xz + z^2}$$

Inequality: $|P_n(x)| \leqslant 1,\ -1 \leqslant x \leqslant 1.$

II

Name: Tschebysheff, First Kind   *Symbol:* $T_n(x)$ *Interval:* $[-1,1]$
Differential Equation: $(1 - x^2)y - xy' + n^2 y = 0$
$$y = T_n(x)$$

Explicit Expression: $\dfrac{n}{2} \displaystyle\sum_{m=0}^{[n/2]} (-1)^m \dfrac{(n - m - 1)!}{m!(n - 2m)!} (2x)^{n-2m} = \cos(n \arccos x) = T_n(x)$

Recurrence Relation: $T_{n+1}(x) = 2xT_n(x) - T_{n-1}(x)$
Weight: $(1 - x^2)^{-1/2}$ *Standardization:* $T_n(1) = 1$

Norm: $\displaystyle\int_{-1}^{+1} (1 - x^2)^{-1/2} [T_n(x)]^2\ dx = \begin{cases} \pi/2, & n \neq 0 \\ \pi, & n = 0 \end{cases}$

Rodrigues' Formula: $\dfrac{(-1)^n (1 - x^2)^{1/2} \sqrt{\pi}}{2^{n+1} \Gamma(n + \frac{1}{2})} \dfrac{d^n}{dx^n} \{(1 - x^2)^{n-(1/2)}\} = T_n(x)$

Generating Function: $\dfrac{1 - xz}{1 - 2xz + z^2} = \displaystyle\sum_{n=0}^\infty T_n(x) z^n,\ -1 < x < 1,\ |z| < 1$

Inequality: $|T_n(x)| \leqslant 1,\ -1 \leqslant x \leqslant 1.$

III

Name: Tschebysheff, Second Kind   *Symbol:* $U_n(x)$ *Interval:* $[-1,1]$
Differential Equation: $(1 - x^2)\ y'' - 3\ xy' + n(n + 2)y = 0$
$$y = U_n(x)$$

Explicit Expression: $U_n(x) = \displaystyle\sum_{m=0}^{[n/2]} (-1)^m \dfrac{(m - n)!}{m!(n - 2m)!} (2x)^{n-2m}$

$$U_n(\cos \theta) = \dfrac{\sin[(n + 1)\theta]}{\sin \theta}$$

* From Beyer, W. H., Ed., *CRC Handbook of Mathematical Sciences,* 5th ed., CRC Press, Boca Raton, 1978, 557—560. With permission.

Recurrence Relation: $U_{n+1}(x) = 2xU_n(x) - U_{n-1}(x)$
Weight: $(1 - x^2)^{1/2}$ *Standardization:* $U_n(1) = n + 1$

Norm: $\displaystyle\int_{-1}^{+1} (1 - x^2)^{1/2}[U_n(x)]^2\, dx = \frac{\pi}{2}$

Rodrigues' Formula: $\displaystyle U_n(x) = \frac{(-1)^n(n + 1)\sqrt{\pi}}{(1 - x^2)^{1/2}2^{n+1}\Gamma(n + \frac{3}{2})} \frac{d^n}{dx^n}\{(1 - x^2)^{n+(1/2)}\}$

Generating Function: $\displaystyle\frac{1}{1 - 2xz + z^2} = \sum_{n=0}^{\infty} U_n(x)z^n,\ -1 < x < 1,\ |z| < 1$

Inequality: $|U_n(x)| \leqslant n + 1,\ -1 \leqslant x \leqslant 1.$

IV

Name: Jacobi   *Symbol:* $P_n^{(\alpha,\beta)}(x)$ *Interval:* $[-1, 1]$
Differential Equation:

$$(1 - x^2)y'' + [\beta - \alpha - (\alpha + \beta + 2)x]y' + n(n + \alpha + \beta + 1)y = 0$$
$$y = P_n^{(\alpha,\beta)}(x)$$

Explicit Expression: $\displaystyle P_n^{(\alpha,\beta)}(x) = \frac{1}{2^n} \sum_{m=0}^{n} \binom{n + \alpha}{m}\binom{n + \beta}{n - m}(x - 1)^{n-m}(x + 1)^m$

Recurrence Relation: $2(n + 1)(n + \alpha + \beta + 1)(2n + \alpha + \beta)P_{n+1}^{(\alpha,\beta)}(x)$
$$= (2n + \alpha + \beta + 1)[(\alpha^2 - \beta^2) + (2n + \alpha + \beta + 2)$$
$$\times (2n + \alpha + \beta)x]P_n^{(\alpha,\beta)}(x)$$
$$- 2(n + \alpha)(n + \beta)(2n + \alpha + \beta + 2)P_{n-1}^{(\alpha,\beta)}(x)$$

Weight: $(1 - x)^\alpha(1 + x)^\beta;\ \alpha,\beta > 1$ *Standardization:* $P_n^{(\alpha,\beta)}(x) = \begin{pmatrix} n + \alpha \\ n \end{pmatrix}$

Norm: $\displaystyle\int_{-1}^{+1} (1 - x)^\alpha(1 + x)^\beta[P_n^{(\alpha,\beta)}(x)]^2\, dx = \frac{2^{\alpha+\beta+1}\Gamma(n + \alpha + 1)\Gamma(n + \beta + 1)}{(2n + \alpha + \beta + 1)n!\,\Gamma(n + \alpha + \beta + 1)}$

Rodrigues' Formula: $\displaystyle P_n^{(\alpha,\beta)}(x) = \frac{(-1)^n}{2^n n!(1 - x)^\alpha(1 + x)^\beta} \frac{d^n}{dx^n}\{(1 - x)^{n+\alpha}(1 + x)^{n+\beta}\}$

Generating Function: $\displaystyle R^{-1}(1 - z + R)^{-\alpha}(1 + z + R)^{-\beta} = \sum_{n=0}^{\infty} 2^{-\alpha-\beta}P_n^{(\alpha,\beta)}(x)z^n,$
$$R = \sqrt{1 - 2xz + z^2},\ |z| < 1$$

Inequality: $\displaystyle\max_{-1 \leqslant x \leqslant 1} |P_n^{(\alpha,\beta)}(x)| = \begin{cases} \begin{pmatrix} n + q \\ n \end{pmatrix} \sim n^q \text{ if } q = \max(\alpha,\beta) \geqslant -\frac{1}{2} \\[6pt] |P_n^{(\alpha,\beta)}(x')| \sim n^{-1/2} \text{ if } q < -\frac{1}{2} \\ x' \text{ is one of the two maximum points nearest} \\[6pt] \dfrac{\beta - \alpha}{\alpha + \beta + 1} \end{cases}$

V

Name: Generalized Laguerre   *Symbol:* $L_n^{(\alpha)}(x)$ *Interval:* $[0, \infty]$
Differential Equation: $xy'' + (\alpha + 1 - x)y' + ny = 0$
$$y = L_n^{(\alpha)}(x)$$

Explicit Expression: $\displaystyle L_n^{(\alpha)}(x) = \sum_{m=0}^{n} (-1)^m\binom{n + \alpha}{n - m}\frac{1}{m!}x^m$

Recurrence Relation: $(n + 1)L_{n+1}^{(\alpha)}(x) = [(2n + \alpha + 1) - x]L_n^{(\alpha)}(x) - (n + \alpha)L_{n-1}^{(\alpha)}(x)$

Weight: $x^\alpha e^{-x},\ \alpha > -1$ *Standardization:* $\displaystyle L_n^{(\alpha)}(x) = \frac{(-1)^n}{n!}x^n + \cdots$

Norm: $\displaystyle\int_0^\infty x^\alpha e^{-x}[L_n^{(\alpha)}(x)]^2\, dx = \frac{\Gamma(n + \alpha + 1)}{n!}$

Rodrigues' Formula: $\displaystyle L_n^{(\alpha)}(x) = \frac{1}{n!\,x^\alpha e^{-x}} \frac{d^n}{dx^n}\{x^{n+\alpha}e^{-x}\}$

A-103

Generating Function: $(1 - z)^{-\alpha - 1} \exp\left(\dfrac{xz}{z - 1}\right) = \displaystyle\sum_{n=0}^{\infty} L_n^{(\alpha)}(x)\, z^n$

Inequality: $|L_n^{(\alpha)}(x)| \leq \dfrac{\Gamma(n + \alpha + 1)}{n!\,\Gamma(\alpha + 1)}\, e^{x/2}; \quad \begin{array}{l} x \geq 0 \\ \alpha > 0 \end{array}$

$|L_n^{(\alpha)}(x)| \leq \left[2 - \dfrac{\Gamma(\alpha + n + 1)}{n!\,\Gamma(\alpha + 1)}\right] e^{x/2}; \quad \begin{array}{l} x \geq 0 \\ -1 < \alpha < 0 \end{array}$

Orthogonal Polynomials

Name: Hermite Symbol: $H_n(x)$ Interval: $[-\infty, \infty]$
Differential Equation: $y'' - 2xy' + 2ny = 0$

Explicit Expression: $H_n(x) = \displaystyle\sum_{m=0}^{[n/2]} \dfrac{(-1)^m\, n!\,(2x)^{n-2m}}{m!\,(n-2m)!}$

Recurrence Relation: $H_{n+1}(x) = 2x\,H_n(x) - 2n H_{n-1}(x)$
Weight: e^{-x^2} Standardization: $H_n(1) = 2^n x^n + \ldots$

Norm: $\displaystyle\int_{-\infty}^{\infty} e^{-x^2}\,[H_n(x)]^2\, dx = 2^n\, n!\,\sqrt{\pi}$

Rodriques' Formula: $H_n(x) = (-1)^n\, e^{x^2}\, \dfrac{d^n}{dx^n}\left(e^{-x^2}\right)$

Generating Function: $e^{-z^2 + 2zx} = \displaystyle\sum_{n=0}^{\infty} H_n(x)\, \dfrac{z^n}{n!}$

Inequality: $|H_n(x)| < e^{\frac{x^2}{2}}\, k\, 2^{n/2}\sqrt{n!}\quad k \approx 1.086435$

NORMAL PROBABILITY FUNCTION

Areas under the Standard Normal Curve from 0 to z

z	0	1	2	3	4	5	6	7	8	9
0.0	.0000	.0040	.0080	.0120	.0160	.0199	.0239	.0279	.0319	.0359
0.1	.0398	.0438	.0478	.0517	.0557	.0596	.0636	.0675	.0714	.0754
0.2	.0793	.0832	.0871	.0910	.0948	.0987	.1026	.1064	.1103	.1141
0.3	.1179	.1217	.1255	.1293	.1331	.1368	.1406	.1443	.1480	.1517
0.4	.1554	.1591	.1628	.1664	.1700	.1736	.1772	.1808	.1844	.1879
0.5	.1915	.1950	.1985	.2019	.2054	.2088	.2123	.2157	.2190	.2224
0.6	.2258	.2291	.2324	.2357	.2389	.2422	.2454	.2486	.2518	.2549
0.7	.2580	.2612	.2652	.2673	.2704	.2734	.2764	.2794	.2823	.2852
0.8	.2881	.2910	.2939	.2967	.2996	.3023	.3051	.3078	.3106	.3133
0.9	.3159	.3186	.3212	.3238	.3264	.3289	.3315	.3340	.3365	.3389
1.0	.3413	.3438	.3461	.3485	.3508	.3531	.3554	.3577	.3599	.3621
1.1	.3643	.3665	.3686	.3708	.3729	.3749	.3770	.3790	.3810	.3830
1.2	.3849	.3869	.3888	.3907	.3925	.3944	.3962	.3980	.3997	.4015
1.3	.4032	.4049	.4066	.4082	.4099	.4115	.4131	.4147	.4162	.4177
1.4	.4192	.4207	.4222	.4236	.4251	.4265	.4279	.4292	.4306	.4319
1.5	.4332	.4345	.4357	.4370	.4382	.4394	.4406	.4418	.4429	.4441
1.6	.4452	.4463	.4474	.4484	.4495	.4505	.4515	.4525	.4535	.4545
1.7	.4554	.4564	.4573	.4582	.4591	.4599	.4608	.4616	.4625	.4633
1.8	.4641	.4649	.4656	.4664	.4671	.4678	.4686	.4693	.4699	.4706
1.9	.4713	.4719	.4726	.4732	.4738	.4744	.4750	.4756	.4761	.4767
2.0	.4772	.4778	.4783	.4788	.4793	.4798	.4803	.4808	.4812	.4817
2.1	.4821	.4826	.4830	.4834	.4838	.4842	.4846	.4850	.4854	.4857
2.2	.4861	.4864	.4868	.4871	.4875	.4878	.4881	.4884	.4887	.4890
2.3	.4893	.4896	.4898	.4901	.4904	.4906	.4909	.4911	.4913	.4916
2.4	.4918	.4920	.4922	.4925	.4927	.4929	.4931	.4932	.4934	.4936
2.5	.4938	.4940	.4941	.4943	.4945	.4946	.4948	.4949	.4951	.4952
2.6	.4953	.4955	.4956	.4957	.4959	.4960	.4961	.4962	.4963	.4964
2.7	.4965	.4966	.4967	.4968	.4969	.4970	.4971	.4972	.4973	.4974
2.8	.4974	.4975	.4976	.4977	.4977	.4978	.4979	.4979	.4980	.4981
2.9	.4981	.4982	.4982	.4983	.4984	.4984	.4985	.4985	.4986	.4986
3.0	.4987	.4987	.4987	.4988	.4988	.4989	.4989	.4989	.4990	.4990
3.1	.4990	.4991	.4991	.4991	.4992	.4992	.4992	.4992	.4993	.4993
3.2	.4993	.4993	.4994	.4994	.4994	.4994	.4994	.4995	.4995	.4995
3.3	.4995	.4995	.4995	.4996	.4996	.4996	.4996	.4996	.4996	.4997
3.4	.4997	.4997	.4997	.4997	.4997	.4997	.4997	.4997	.4997	.4998
3.5	.4998	.4998	.4998	.4998	.4998	.4998	.4998	.4998	.4998	.4998
3.6	.4998	.4998	.4999	.4999	.4999	.4999	.4999	.4999	.4999	.4999
3.7	.4999	.4999	.4999	.4999	.4999	.4999	.4999	.4999	.4999	.4999
3.8	.4999	.4999	.4999	.4999	.4999	.4999	.4999	.4999	.4999	.4999
3.9	.5000	.5000	.5000	.5000	.5000	.5000	.5000	.5000	.5000	.5000

$F(z)$ below refers to area under Standard Normal Curve from $-\infty$ to z

z	1.282	1.645	1.960	2.326	2.576	3.090
$F(z)$	.90	.95	.975	.99	.995	.999
$2[1 - F(z)]$	.20	.10	.05	.02	.01	.002

PERCENTAGE POINTS, STUDENT'S t-DISTRIBUTION

This table gives values of t such that

$$F(t) = \int_{-\infty}^{t} \frac{\Gamma\left(\frac{n+1}{2}\right)}{\sqrt{n\pi}\,\Gamma\left(\frac{n}{2}\right)} \left(1 + \frac{x^2}{n}\right)^{-\frac{n+1}{2}} dx$$

for n, the number of degrees of freedom, equal to 1, 2, . . ., 30, 40, 60, 120, ∞; and for $F(t) = 0.60, 0.75, 0.90, 0.95, 0.975, 0.99, 0.995,$ and 0.9995. The t-distribution is symmetrical, so that $F(-t) = 1 - F(t)$

F \ n	.60	.75	.90	.95	.975	.99	.995	.9995
1	.325	1.000	3.078	6.314	12.706	31.821	63.657	636.619
2	.289	.816	1.886	2.920	4.303	6.965	9.925	31.598
3	.277	.765	1.638	2.353	3.182	4.541	5.841	12.924
4	.271	.741	1.533	2.132	2.776	3.747	4.604	8.610
5	.267	.727	1.476	2.015	2.571	3.365	4.032	6.869
6	.265	.718	1.440	1.943	2.447	3.143	3.707	5.959
7	.263	.711	1.415	1.895	2.365	2.998	3.499	5.408
8	.262	.706	1.397	1.860	2.306	2.896	3.355	5.041
9	.261	.703	1.383	1.833	2.262	2.821	3.250	4.781
10	.260	.700	1.372	1.812	2.228	2.764	3.169	4.587
11	.260	.697	1.363	1.796	2.201	2.718	3.106	4.437
12	.259	.695	1.356	1.782	2.179	2.681	3.055	4.318
13	.259	.694	1.350	1.771	2.160	2.650	3.012	4.221
14	.258	.692	1.345	1.761	2.145	2.624	2.977	4.140
15	.258	.691	1.341	1.753	2.131	2.602	2.947	4.073
16	.258	.690	1.337	1.746	2.120	2.583	2.921	4.015
17	.257	.689	1.333	1.740	2.110	2.567	2.898	3.965
18	.257	.688	1.330	1.734	2.101	2.552	2.878	3.922
19	.257	.688	1.328	1.729	2.093	2.539	2.861	3.883
20	.257	.687	1.325	1.725	2.086	2.528	2.845	3.850
21	.257	.686	1.323	1.721	2.080	2.518	2.831	3.819
22	.256	.686	1.321	1.717	2.074	2.508	2.819	3.792
23	.256	.685	1.319	1.714	2.069	2.500	2.807	3.767
24	.256	.685	1.318	1.711	2.064	2.492	2.797	3.745
25	.256	.684	1.316	1.708	2.060	2.485	2.787	3.725
26	.256	.684	1.315	1.706	2.056	2.479	2.779	3.707
27	.256	.684	1.314	1.703	2.052	2.473	2.771	3.690
28	.256	.683	1.313	1.701	2.048	2.467	2.763	3.674
29	.256	.683	1.311	1.699	2.045	2.462	2.756	3.659
30	.256	.683	1.310	1.697	2.042	2.457	2.750	3.646
40	.255	.681	1.303	1.684	2.021	2.423	2.704	3.551
60	.254	.679	1.296	1.671	2.000	2.390	2.660	3.460
120	.254	.677	1.289	1.658	1.980	2.358	2.617	3.373
∞	.253	.674	1.282	1.645	1.960	2.326	2.576	3.291

* This table is abridged from the "Statistical Tables" of R. A. Fisher and Frank Yates published by Oliver & Boyd. Ltd., Edinburgh and London, 1938. It is here published with the kind permission of the authors and their publishers.

PERCENTAGE POINTS, CHI-SQUARE DISTRIBUTION

This table gives values of χ^2 such that

$$F(\chi^2) = \int_{0}^{\chi^2} \frac{1}{2^{\frac{n}{2}} \Gamma\left(\frac{n}{2}\right)} x^{\frac{n-2}{2}} e^{-\frac{x}{2}} dx$$

for n, the number of degrees of freedom, equal to 1, 2, . . . , 30. For $n > 30$, a normal approximation is quite accurate. The expression $\sqrt{2\chi^2} - \sqrt{2n-1}$ is approximately normally distributed as the standard normal distribution. Thus χ^2_α, the α-point of the distribution, may be computed by the formula

$$\chi^2_\alpha = \tfrac{1}{2}[x_\alpha + \sqrt{2n-1}]^2,$$

where x_α is the α-point of the cumulative normal distribution. For even values of n, $F(\chi^2)$ can be written as

$$1 - F(\chi^2) = \sum_{x=0}^{x'-1} \frac{e^{-\lambda}\lambda^x}{x!}$$

with $\lambda = \tfrac{1}{2}\chi^2$ and $x' = \tfrac{1}{2}n$. Thus the cumulative Chi-Square distribution is related to the cumulative Poisson distribution.

Another approximate formula for large n

$$\chi_\alpha^2 = n\left(1 - \frac{2}{9n} + z_\alpha \sqrt{\frac{2}{9n}}\right)^3$$

n = degrees of freedom
z_α = the normal deviate, (the value of x for which $F(x)$ = the desired percentile).

x	1.282	1.645	1.960	2.326	2.576	3.090
$F(x)$	.90	.95	.975	.99	.995	.999

$\chi_{.99}^2 = 60[1 - 0.00370 + 2.326(0.06086)]^3 = 88.4$ is the 99th percentile for 60 degrees of freedom.

$$F(\chi^2) = \int_0^{x^2} \frac{1}{2^{\frac{n}{2}} \Gamma\left(\frac{n}{2}\right)} x^{\frac{n-2}{2}} e^{-\frac{x}{2}} dx$$

F \ n	.005	.010	.025	.050	.100	.250	.500	.750	.900	.950	.975	.990	.995
1	.0000393	.000157	.000982	.00393	.0158	.102	.455	1.32	2.71	3.84	5.02	6.63	7.88
2	.0100	.0201	.0506	.103	.211	.575	1.39	2.77	4.61	5.99	7.38	9.21	10.6
3	.0717	.115	.216	.352	.584	1.21	2.37	4.11	6.25	7.81	9.35	11.3	12.8
4	.207	.297	.484	.711	1.06	1.92	3.36	5.39	7.78	9.49	11.1	13.3	14.9
5	.412	.554	.831	1.15	1.61	2.67	4.35	6.63	9.24	11.1	12.8	15.1	16.7
6	.676	.872	1.24	1.64	2.20	3.45	5.35	7.84	10.6	12.6	14.4	16.8	18.5
7	.989	1.24	1.69	2.17	2.83	4.25	6.35	9.04	12.0	14.1	16.0	18.5	20.3
8	1.34	1.65	2.18	2.73	3.49	5.07	7.34	10.2	13.4	15.5	17.5	20.1	22.0
9	1.73	2.09	2.70	3.33	4.17	5.90	8.34	11.4	14.7	16.9	19.0	21.7	23.6
10	2.16	2.56	3.25	3.94	4.87	6.74	9.34	12.5	16.0	18.3	20.5	23.2	25.2
11	2.60	3.05	3.82	4.57	5.58	7.58	10.3	13.7	17.3	19.7	21.9	24.7	26.8
12	3.07	3.57	4.40	5.23	6.30	8.44	11.3	14.8	18.5	21.0	23.3	26.2	28.3
13	3.57	4.11	5.01	5.89	7.04	9.30	12.3	16.0	19.8	22.4	24.7	27.7	29.8
14	4.07	4.66	5.63	6.57	7.79	10.2	13.3	17.1	21.1	23.7	26.1	29.1	31.3
15	4.60	5.23	6.26	7.26	8.55	11.0	14.3	18.2	22.3	25.0	27.5	30.6	32.8
16	5.14	5.81	6.91	7.96	9.31	11.9	15.3	19.4	23.5	26.3	28.8	32.0	34.3
17	5.70	6.41	7.56	8.67	10.1	12.8	16.3	20.5	24.8	27.6	30.2	33.4	35.7
18	6.26	7.01	8.23	9.39	10.9	13.7	17.3	21.6	26.0	28.9	31.5	34.8	37.2
19	6.84	7.63	8.91	10.1	11.7	14.6	18.3	22.7	27.2	30.1	32.9	36.2	38.6
20	7.43	8.26	9.59	10.9	12.4	15.5	19.3	23.8	28.4	31.4	34.2	37.6	40.0
21	8.03	8.90	10.3	11.6	13.2	16.3	20.3	24.9	29.6	32.7	35.5	38.9	41.4
22	8.64	9.54	11.0	12.3	14.0	17.2	21.3	26.0	30.8	33.9	36.8	40.3	42.8
23	9.26	10.2	11.7	13.1	14.8	18.1	22.3	27.1	32.0	35.2	38.1	41.6	44.2
24	9.89	10.9	12.4	13.8	15.7	19.0	23.3	28.2	33.2	36.4	39.4	43.0	45.6
25	10.5	11.5	13.1	14.6	16.5	19.9	24.3	29.3	34.4	37.7	40.6	44.3	46.9
26	11.2	12.2	13.8	15.4	17.3	20.8	25.3	30.4	35.6	38.9	41.9	45.6	48.3
27	11.8	12.9	14.6	16.2	18.1	21.7	26.3	31.5	36.7	40.1	43.2	47.0	49.6
28	12.5	13.6	15.3	16.9	18.9	22.7	27.3	32.6	37.9	41.3	44.5	48.3	51.0
29	13.1	14.3	16.0	17.7	19.8	23.6	28.3	33.7	39.1	42.6	45.7	49.6	52.3
30	13.8	15.0	16.8	18.5	20.6	24.5	29.3	34.8	40.3	43.8	47.0	50.9	53.7

PERCENTAGE POINTS, *F*-DISTRIBUTION

This table gives values of F such that

$$F(F) = \int_0^F \frac{\Gamma\left(\frac{m+n}{2}\right)}{\Gamma\left(\frac{m}{2}\right)\Gamma\left(\frac{n}{2}\right)} m^{\frac{m}{2}} n^{\frac{n}{2}} x^{\frac{m-2}{2}} (n + mx)^{-\frac{m+n}{2}} dx$$

for selected values of m, the number of degrees of freedom of the numerator of F; and for selected values of n, the number of degrees of freedom of the denominator of F. The table also provides values corresponding to $F(F) = .10, .05, .025, .01, .005, .001$ since $F_{1-\alpha}$ for m and n degrees of freedom is the reciprocal of F_α for n and m degrees of freedom. Thus

$$F_{.05}(4, 7) = \frac{1}{F_{.95}(7, 4)} = \frac{1}{6.09} = .164$$

$$F(F) = \int_0^F \frac{\Gamma\left(\frac{m+n}{2}\right)}{\Gamma\left(\frac{m}{2}\right)\Gamma\left(\frac{n}{2}\right)} m^{\frac{m}{2}} n^{\frac{n}{2}} x^{\frac{m}{2}-1} (n+mx)^{-\frac{m+n}{2}}\, dx = .90$$

n \ m	1	2	3	4	5	6	7	8	9	10	12	15	20	24	30	40	60	120	∞
1	39.86	49.50	53.59	55.83	57.24	58.20	58.91	59.44	59.86	60.19	60.71	61.22	61.74	62.00	62.26	62.53	62.79	63.06	63.33
2	8.53	9.00	9.16	9.24	9.29	9.33	9.35	9.37	9.38	9.39	9.41	9.42	9.44	9.45	9.46	9.47	9.47	9.48	9.49
3	5.54	5.46	5.39	5.34	5.31	5.28	5.27	5.25	5.24	5.23	5.22	5.20	5.18	5.18	5.17	5.16	5.15	5.14	5.13
4	4.54	4.32	4.19	4.11	4.05	4.01	3.98	3.95	3.94	3.92	3.90	3.87	3.84	3.83	3.82	3.80	3.79	3.78	3.76
5	4.06	3.78	3.62	3.52	3.45	3.40	3.37	3.34	3.32	3.30	3.27	3.24	3.21	3.19	3.17	3.16	3.14	3.12	3.10
6	3.78	3.46	3.29	3.18	3.11	3.05	3.01	2.98	2.96	2.94	2.90	2.87	2.84	2.82	2.80	2.78	2.76	2.74	2.72
7	3.59	3.26	3.07	2.96	2.88	2.83	2.78	2.75	2.72	2.70	2.67	2.63	2.59	2.58	2.56	2.54	2.51	2.49	2.47
8	3.46	3.11	2.92	2.81	2.73	2.67	2.62	2.59	2.56	2.54	2.50	2.46	2.42	2.40	2.38	2.36	2.34	2.32	2.29
9	3.36	3.01	2.81	2.69	2.61	2.55	2.51	2.47	2.44	2.42	2.38	2.34	2.30	2.28	2.25	2.23	2.21	2.18	2.16
10	3.29	2.92	2.73	2.61	2.52	2.46	2.41	2.38	2.35	2.32	2.28	2.24	2.20	2.18	2.16	2.13	2.11	2.08	2.06
11	3.23	2.86	2.66	2.54	2.45	2.39	2.34	2.30	2.27	2.25	2.21	2.17	2.12	2.10	2.08	2.05	2.03	2.00	1.97
12	3.18	2.81	2.61	2.48	2.39	2.33	2.28	2.24	2.21	2.19	2.15	2.10	2.06	2.04	2.01	1.99	1.96	1.93	1.90
13	3.14	2.76	2.56	2.43	2.35	2.28	2.23	2.20	2.16	2.14	2.10	2.05	2.01	1.98	1.96	1.93	1.90	1.88	1.85
14	3.10	2.73	2.52	2.39	2.31	2.24	2.19	2.15	2.12	2.10	2.05	2.01	1.96	1.94	1.91	1.89	1.86	1.83	1.80
15	3.07	2.70	2.49	2.36	2.27	2.21	2.16	2.12	2.09	2.06	2.02	1.97	1.92	1.90	1.87	1.85	1.82	1.79	1.76
16	3.05	2.67	2.46	2.33	2.24	2.18	2.13	2.09	2.06	2.03	1.99	1.94	1.89	1.87	1.84	1.81	1.78	1.75	1.72
17	3.03	2.64	2.44	2.31	2.22	2.15	2.10	2.06	2.03	2.00	1.96	1.91	1.86	1.84	1.81	1.78	1.75	1.72	1.69
18	3.01	2.62	2.42	2.29	2.20	2.13	2.08	2.04	2.00	1.98	1.93	1.89	1.84	1.81	1.78	1.75	1.72	1.69	1.66
19	2.99	2.61	2.40	2.27	2.18	2.11	2.06	2.02	1.98	1.96	1.91	1.86	1.81	1.79	1.76	1.73	1.70	1.67	1.63
20	2.97	2.59	2.38	2.25	2.16	2.09	2.04	2.00	1.96	1.94	1.89	1.84	1.79	1.77	1.74	1.71	1.68	1.64	1.61
21	2.96	2.57	2.36	2.23	2.14	2.08	2.02	1.98	1.95	1.92	1.87	1.83	1.78	1.75	1.72	1.69	1.66	1.62	1.59
22	2.95	2.56	2.35	2.22	2.13	2.06	2.01	1.97	1.93	1.90	1.86	1.81	1.76	1.73	1.70	1.67	1.64	1.60	1.57
23	2.94	2.55	2.34	2.21	2.11	2.05	1.99	1.95	1.92	1.89	1.84	1.80	1.74	1.72	1.69	1.66	1.62	1.59	1.55
24	2.93	2.54	2.33	2.19	2.10	2.04	1.98	1.94	1.91	1.88	1.83	1.78	1.73	1.70	1.67	1.64	1.61	1.57	1.53
25	2.92	2.53	2.32	2.18	2.09	2.02	1.97	1.93	1.89	1.87	1.82	1.77	1.72	1.69	1.66	1.63	1.59	1.56	1.52
26	2.91	2.52	2.31	2.17	2.08	2.01	1.96	1.92	1.88	1.86	1.81	1.76	1.71	1.68	1.65	1.61	1.58	1.54	1.50
27	2.90	2.51	2.30	2.17	2.07	2.00	1.95	1.91	1.87	1.85	1.80	1.75	1.70	1.67	1.64	1.60	1.57	1.53	1.49
28	2.89	2.50	2.29	2.16	2.06	2.00	1.94	1.90	1.87	1.84	1.79	1.74	1.69	1.66	1.63	1.59	1.56	1.52	1.48
29	2.89	2.50	2.28	2.15	2.06	1.99	1.93	1.89	1.86	1.83	1.78	1.73	1.68	1.65	1.62	1.58	1.55	1.51	1.47
30	2.88	2.49	2.28	2.14	2.05	1.98	1.93	1.88	1.85	1.82	1.77	1.72	1.67	1.64	1.61	1.57	1.54	1.50	1.46
40	2.84	2.44	2.23	2.09	2.00	1.93	1.87	1.83	1.79	1.76	1.71	1.66	1.61	1.57	1.54	1.51	1.47	1.42	1.38
60	2.79	2.39	2.18	2.04	1.95	1.87	1.82	1.77	1.74	1.71	1.66	1.60	1.54	1.51	1.48	1.44	1.40	1.35	1.29
120	2.75	2.35	2.13	1.99	1.90	1.82	1.77	1.72	1.68	1.65	1.60	1.55	1.48	1.45	1.41	1.37	1.32	1.26	1.19
∞	2.71	2.30	2.08	1.94	1.85	1.77	1.72	1.67	1.63	1.60	1.55	1.49	1.42	1.38	1.34	1.30	1.24	1.17	1.00

$F = \dfrac{s_1^2}{s_2^2} = \dfrac{S_1}{m}\bigg/\dfrac{S_2}{n}$, where $s_1^2 = S_1/m$ and $s_2^2 = S_2/n$ are independent mean squares estimating a common variance σ^2 and based on m and n degrees of freedom, respectively.

$$F(F) = \int_0^F \frac{\Gamma\left(\frac{m+n}{2}\right)}{\Gamma\left(\frac{m}{2}\right)\Gamma\left(\frac{n}{2}\right)} m^{\frac{m}{2}} n^{\frac{n}{2}} x^{\frac{m}{2}-1} (n+mx)^{-\frac{m+n}{2}}\, dx = .95$$

n \ m	1	2	3	4	5	6	7	8	9	10	12	15	20	24	30	40	60	120	∞
1	161.4	199.5	215.7	224.6	230.2	234.0	236.8	238.9	240.5	241.9	243.9	245.9	248.0	249.1	250.1	251.1	252.2	253.3	254.3
2	18.51	19.00	19.16	19.25	19.30	19.33	19.35	19.37	19.38	19.40	19.41	19.43	19.45	19.45	19.46	19.47	19.48	19.49	19.50
3	10.13	9.55	9.28	9.12	9.01	8.94	8.89	8.85	8.81	8.79	8.74	8.70	8.66	8.64	8.62	8.59	8.57	8.55	8.53
4	7.71	6.94	6.59	6.39	6.26	6.16	6.09	6.04	6.00	5.96	5.91	5.86	5.80	5.77	5.75	5.72	5.69	5.66	5.63
5	6.61	5.79	5.41	5.19	5.05	4.95	4.88	4.82	4.77	4.74	4.68	4.62	4.56	4.53	4.50	4.46	4.43	4.40	4.36
6	5.99	5.14	4.76	4.53	4.39	4.28	4.21	4.15	4.10	4.06	4.00	3.94	3.87	3.84	3.81	3.77	3.74	3.70	3.67
7	5.59	4.74	4.35	4.12	3.97	3.87	3.79	3.73	3.68	3.64	3.57	3.51	3.44	3.41	3.38	3.34	3.30	3.27	3.23
8	5.32	4.46	4.07	3.84	3.69	3.58	3.50	3.44	3.39	3.35	3.28	3.22	3.15	3.12	3.08	3.04	3.01	2.97	2.93
9	5.12	4.26	3.86	3.63	3.48	3.37	3.29	3.23	3.18	3.14	3.07	3.01	2.94	2.90	2.86	2.83	2.79	2.75	2.71
10	4.96	4.10	3.71	3.48	3.33	3.22	3.14	3.07	3.02	2.98	2.91	2.85	2.77	2.74	2.70	2.66	2.62	2.58	2.54
11	4.84	3.98	3.59	3.36	3.20	3.09	3.01	2.95	2.90	2.85	2.79	2.72	2.65	2.61	2.57	2.53	2.49	2.45	2.40
12	4.75	3.89	3.49	3.26	3.11	3.00	2.91	2.85	2.80	2.75	2.69	2.62	2.54	2.51	2.47	2.43	2.38	2.34	2.30
13	4.67	3.81	3.41	3.18	3.03	2.92	2.83	2.77	2.71	2.67	2.60	2.53	2.46	2.42	2.38	2.34	2.30	2.25	2.21
14	4.60	3.74	3.34	3.11	2.96	2.85	2.76	2.70	2.65	2.60	2.53	2.46	2.39	2.35	2.31	2.27	2.22	2.18	2.13
15	4.54	3.68	3.29	3.06	2.90	2.79	2.71	2.64	2.59	2.54	2.48	2.40	2.33	2.29	2.25	2.20	2.16	2.11	2.07
16	4.49	3.63	3.24	3.01	2.85	2.74	2.66	2.59	2.54	2.49	2.42	2.35	2.28	2.24	2.19	2.15	2.11	2.06	2.01
17	4.45	3.59	3.20	2.96	2.81	2.70	2.61	2.55	2.49	2.45	2.38	2.31	2.23	2.19	2.15	2.10	2.06	2.01	1.96
18	4.41	3.55	3.16	2.93	2.77	2.66	2.58	2.51	2.46	2.41	2.34	2.27	2.19	2.15	2.11	2.06	2.02	1.97	1.92
19	4.38	3.52	3.13	2.90	2.74	2.63	2.54	2.48	2.42	2.38	2.31	2.23	2.16	2.11	2.07	2.03	1.98	1.93	1.88
20	4.35	3.49	3.10	2.87	2.71	2.60	2.51	2.45	2.39	2.35	2.28	2.20	2.12	2.08	2.04	1.99	1.95	1.90	1.84
21	4.32	3.47	3.07	2.84	2.68	2.57	2.49	2.42	2.37	2.32	2.25	2.18	2.10	2.05	2.01	1.96	1.92	1.87	1.81
22	4.30	3.44	3.05	2.82	2.66	2.55	2.46	2.40	2.34	2.30	2.23	2.15	2.07	2.03	1.98	1.94	1.89	1.84	1.78
23	4.28	3.42	3.03	2.80	2.64	2.53	2.44	2.37	2.32	2.27	2.20	2.13	2.05	2.01	1.96	1.91	1.86	1.81	1.76
24	4.26	3.40	3.01	2.78	2.62	2.51	2.42	2.36	2.30	2.25	2.18	2.11	2.03	1.98	1.94	1.89	1.84	1.79	1.73
25	4.24	3.39	2.99	2.76	2.60	2.49	2.40	2.34	2.28	2.24	2.16	2.09	2.01	1.96	1.92	1.87	1.82	1.77	1.71
26	4.23	3.37	2.98	2.74	2.59	2.47	2.39	2.32	2.27	2.22	2.15	2.07	1.99	1.95	1.90	1.85	1.80	1.75	1.69
27	4.21	3.35	2.96	2.73	2.57	2.46	2.37	2.31	2.25	2.20	2.13	2.06	1.97	1.93	1.88	1.84	1.79	1.73	1.67
28	4.20	3.34	2.95	2.71	2.56	2.45	2.36	2.29	2.24	2.19	2.12	2.04	1.96	1.91	1.87	1.82	1.77	1.71	1.65
29	4.18	3.33	2.93	2.70	2.55	2.43	2.35	2.28	2.22	2.18	2.10	2.03	1.94	1.90	1.85	1.81	1.75	1.70	1.64
30	4.17	3.32	2.92	2.69	2.53	2.42	2.33	2.27	2.21	2.16	2.09	2.01	1.93	1.89	1.84	1.79	1.74	1.68	1.62
40	4.08	3.23	2.84	2.61	2.45	2.34	2.25	2.18	2.12	2.08	2.00	1.92	1.84	1.79	1.74	1.69	1.64	1.58	1.51
60	4.00	3.15	2.76	2.53	2.37	2.25	2.17	2.10	2.04	1.99	1.92	1.84	1.75	1.70	1.65	1.59	1.53	1.47	1.39
120	3.92	3.07	2.68	2.45	2.29	2.17	2.09	2.02	1.96	1.91	1.83	1.75	1.66	1.61	1.55	1.50	1.43	1.35	1.25
∞	3.84	3.00	2.60	2.37	2.21	2.10	2.01	1.94	1.88	1.83	1.75	1.67	1.57	1.52	1.46	1.39	1.32	1.22	1.00

$F = \dfrac{s_1^2}{s_2^2} = \dfrac{S_1}{m}\bigg/\dfrac{S_2}{n}$, where $s_1^2 = S_1/m$ and $s_2^2 = S_2/n$ are independent mean squares estimating a common variance σ^2 and based on m and n degrees of freedom, respectively.

$$F(F) = \int_0^F \frac{\Gamma\left(\frac{m+n}{2}\right)}{\Gamma\left(\frac{m}{2}\right)\Gamma\left(\frac{n}{2}\right)} m^{\frac{m}{2}} n^{\frac{n}{2}} x^{\frac{m}{2}-1} (n+mx)^{-\frac{m+n}{2}} dx = .975$$

m \ n	1	2	3	4	5	6	7	8	9	10	12	15	20	24	30	40	60	120	∞
1	647.8	799.5	864.2	899.6	921.8	937.1	948.2	956.7	963.3	968.6	976.7	984.9	993.1	997.2	1001	1006	1010	1014	1018
2	38.51	39.00	39.17	39.25	39.30	39.33	39.36	39.37	39.39	39.40	39.41	39.43	39.45	39.46	39.46	39.47	39.48	39.49	39.50
3	17.44	16.04	15.44	15.10	14.88	14.73	14.62	14.54	14.47	14.42	14.34	14.25	14.17	14.12	14.08	14.04	13.99	13.95	13.90
4	12.22	10.65	9.98	9.60	9.36	9.20	9.07	8.98	8.90	8.84	8.75	8.66	8.56	8.51	8.46	8.41	8.36	8.31	8.26
5	10.01	8.43	7.76	7.39	7.15	6.98	6.85	6.76	6.68	6.62	6.52	6.43	6.33	6.28	6.23	6.18	6.12	6.07	6.02
6	8.81	7.26	6.60	6.23	5.99	5.82	5.70	5.60	5.52	5.46	5.37	5.27	5.17	5.12	5.07	5.01	4.96	4.90	4.85
7	8.07	6.54	5.89	5.52	5.29	5.12	4.99	4.90	4.82	4.76	4.67	4.57	4.47	4.42	4.36	4.31	4.25	4.20	4.14
8	7.57	6.06	5.42	5.05	4.82	4.65	4.53	4.43	4.36	4.30	4.20	4.10	4.00	3.95	3.89	3.84	3.78	3.73	3.67
9	7.21	5.71	5.08	4.72	4.48	4.32	4.20	4.10	4.03	3.96	3.87	3.77	3.67	3.61	3.56	3.51	3.45	3.39	3.33
10	6.94	5.46	4.83	4.47	4.24	4.07	3.95	3.85	3.78	3.72	3.62	3.52	3.42	3.37	3.31	3.26	3.20	3.14	3.08
11	6.72	5.26	4.63	4.28	4.04	3.88	3.76	3.66	3.59	3.53	3.43	3.33	3.23	3.17	3.12	3.06	3.00	2.94	2.88
12	6.55	5.10	4.47	4.12	3.89	3.73	3.61	3.51	3.44	3.37	3.28	3.18	3.07	3.02	2.96	2.91	2.85	2.79	2.72
13	6.41	4.97	4.35	4.00	3.77	3.60	3.48	3.39	3.31	3.25	3.15	3.05	2.95	2.89	2.84	2.78	2.72	2.66	2.60
14	6.30	4.86	4.24	3.89	3.66	3.50	3.38	3.29	3.21	3.15	3.05	2.95	2.84	2.79	2.73	2.67	2.61	2.55	2.49
15	6.20	4.77	4.15	3.80	3.58	3.41	3.29	3.20	3.12	3.06	2.96	2.86	2.76	2.70	2.64	2.59	2.52	2.46	2.40
16	6.12	4.69	4.08	3.73	3.50	3.34	3.22	3.12	3.05	2.99	2.89	2.79	2.68	2.63	2.57	2.51	2.45	2.38	2.32
17	6.04	4.62	4.01	3.66	3.44	3.28	3.16	3.06	2.98	2.92	2.82	2.72	2.62	2.56	2.50	2.44	2.38	2.32	2.25
18	5.98	4.56	3.95	3.61	3.38	3.22	3.10	3.01	2.93	2.87	2.77	2.67	2.56	2.50	2.44	2.38	2.32	2.26	2.19
19	5.92	4.51	3.90	3.56	3.33	3.17	3.05	2.96	2.88	2.82	2.72	2.62	2.51	2.45	2.39	2.33	2.27	2.20	2.13
20	5.87	4.46	3.86	3.51	3.29	3.13	3.01	2.91	2.84	2.77	2.68	2.57	2.46	2.41	2.35	2.29	2.22	2.16	2.09
21	5.83	4.42	3.82	3.48	3.25	3.09	2.97	2.87	2.80	2.73	2.64	2.53	2.42	2.37	2.31	2.25	2.18	2.11	2.04
22	5.79	4.38	3.78	3.44	3.22	3.05	2.93	2.84	2.76	2.70	2.60	2.50	2.39	2.33	2.27	2.21	2.14	2.08	2.00
23	5.75	4.35	3.75	3.41	3.18	3.02	2.90	2.81	2.73	2.67	2.57	2.47	2.36	2.30	2.24	2.18	2.11	2.04	1.97
24	5.72	4.32	3.72	3.38	3.15	2.99	2.87	2.78	2.70	2.64	2.54	2.44	2.33	2.27	2.21	2.15	2.08	2.01	1.94
25	5.69	4.29	3.69	3.35	3.13	2.97	2.85	2.75	2.68	2.61	2.51	2.41	2.30	2.24	2.18	2.12	2.05	1.98	1.91
26	5.66	4.27	3.67	3.33	3.10	2.94	2.82	2.73	2.65	2.59	2.49	2.39	2.28	2.22	2.16	2.09	2.03	1.95	1.88
27	5.63	4.24	3.65	3.31	3.08	2.92	2.80	2.71	2.63	2.57	2.47	2.36	2.25	2.19	2.13	2.07	2.00	1.93	1.85
28	5.61	4.22	3.63	3.29	3.06	2.90	2.78	2.69	2.61	2.55	2.45	2.34	2.23	2.17	2.11	2.05	1.98	1.91	1.83
29	5.59	4.20	3.61	3.27	3.04	2.88	2.76	2.67	2.59	2.53	2.43	2.32	2.21	2.15	2.09	2.03	1.96	1.89	1.81
30	5.57	4.18	3.59	3.25	3.03	2.87	2.75	2.65	2.57	2.51	2.41	2.31	2.20	2.14	2.07	2.01	1.94	1.87	1.79
40	5.42	4.05	3.46	3.13	2.90	2.74	2.62	2.53	2.45	2.39	2.29	2.18	2.07	2.01	1.94	1.88	1.80	1.72	1.64
60	5.29	3.93	3.34	3.01	2.79	2.63	2.51	2.41	2.33	2.27	2.17	2.06	1.94	1.88	1.82	1.74	1.67	1.58	1.48
120	5.15	3.80	3.23	2.89	2.67	2.52	2.39	2.30	2.22	2.16	2.05	1.94	1.82	1.76	1.69	1.61	1.53	1.43	1.31
∞	5.02	3.69	3.12	2.79	2.57	2.41	2.29	2.19	2.11	2.05	1.94	1.83	1.71	1.64	1.57	1.48	1.39	1.27	1.00

$F = \dfrac{s_1^2}{s_2^2} = \dfrac{S_1}{m} \bigg/ \dfrac{S_2}{n}$, where $s_1^2 = S_1/m$ and $s_2^2 = S_2/n$ are independent mean squares estimating a common variance σ^2 and based on m and n degrees of freedom, respectively.

$$F(F) = \int_0^F \frac{\Gamma\left(\frac{m+n}{2}\right)}{\Gamma\left(\frac{m}{2}\right)\Gamma\left(\frac{n}{2}\right)} m^{\frac{m}{2}} n^{\frac{n}{2}} x^{\frac{m}{2}-1} (n+mx)^{-\frac{m+n}{2}} dx = .99$$

m \ n	1	2	3	4	5	6	7	8	9	10	12	15	20	24	30	40	60	120	∞
1	4052	4999.5	5403	5625	5764	5859	5928	5982	6022	6056	6106	6157	6209	6235	6261	6287	6313	6339	6366
2	98.50	99.00	99.17	99.25	99.30	99.33	99.36	99.37	99.39	99.40	99.42	99.43	99.45	99.46	99.47	99.47	99.48	99.49	99.50
3	34.12	30.82	29.46	28.71	28.24	27.91	27.67	27.49	27.35	27.23	27.05	26.87	26.69	26.60	26.50	26.41	26.32	26.22	26.13
4	21.20	18.00	16.69	15.98	15.52	15.21	14.98	14.80	14.66	14.55	14.37	14.20	14.02	13.93	13.84	13.75	13.65	13.56	13.46
5	16.26	13.27	12.06	11.39	10.97	10.67	10.46	10.29	10.16	10.05	9.89	9.72	9.55	9.47	9.38	9.29	9.20	9.11	9.02
6	13.75	10.92	9.78	9.15	8.75	8.47	8.26	8.10	7.98	7.87	7.72	7.56	7.40	7.31	7.23	7.14	7.06	6.97	6.88
7	12.25	9.55	8.45	7.85	7.46	7.19	6.99	6.84	6.72	6.62	6.47	6.31	6.16	6.07	5.99	5.91	5.82	5.74	5.65
8	11.26	8.65	7.59	7.01	6.63	6.37	6.18	6.03	5.91	5.81	5.67	5.52	5.36	5.28	5.20	5.12	5.03	4.95	4.86
9	10.56	8.02	6.99	6.42	6.06	5.80	5.61	5.47	5.35	5.26	5.11	4.96	4.81	4.73	4.65	4.57	4.48	4.40	4.31
10	10.04	7.56	6.55	5.99	5.64	5.39	5.20	5.06	4.94	4.85	4.71	4.56	4.41	4.33	4.25	4.17	4.08	4.00	3.91
11	9.65	7.21	6.22	5.67	5.32	5.07	4.89	4.74	4.63	4.54	4.40	4.25	4.10	4.02	3.94	3.86	3.78	3.69	3.60
12	9.33	6.93	5.95	5.41	5.06	4.82	4.64	4.50	4.39	4.30	4.16	4.01	3.86	3.78	3.70	3.62	3.54	3.45	3.36
13	9.07	6.70	5.74	5.21	4.86	4.62	4.44	4.30	4.19	4.10	3.96	3.82	3.66	3.59	3.51	3.43	3.34	3.25	3.17
14	8.86	6.51	5.56	5.04	4.69	4.46	4.28	4.14	4.03	3.94	3.80	3.66	3.51	3.43	3.35	3.27	3.18	3.09	3.00
15	8.68	6.36	5.42	4.89	4.56	4.32	4.14	4.00	3.89	3.80	3.67	3.52	3.37	3.29	3.21	3.13	3.05	2.96	2.87
16	8.53	6.23	5.29	4.77	4.44	4.20	4.03	3.89	3.78	3.69	3.55	3.41	3.26	3.18	3.10	3.02	2.93	2.84	2.75
17	8.40	6.11	5.18	4.67	4.34	4.10	3.93	3.79	3.68	3.59	3.46	3.31	3.16	3.08	3.00	2.92	2.83	2.75	2.65
18	8.29	6.01	5.09	4.58	4.25	4.01	3.84	3.71	3.60	3.51	3.37	3.23	3.08	3.00	2.92	2.84	2.75	2.66	2.57
19	8.18	5.93	5.01	4.50	4.17	3.94	3.77	3.63	3.52	3.43	3.30	3.15	3.00	2.92	2.84	2.76	2.67	2.58	2.49
20	8.10	5.85	4.94	4.43	4.10	3.87	3.70	3.56	3.46	3.37	3.23	3.09	2.94	2.86	2.78	2.69	2.61	2.52	2.42
21	8.02	5.78	4.87	4.37	4.04	3.81	3.64	3.51	3.40	3.31	3.17	3.03	2.88	2.80	2.72	2.64	2.55	2.46	2.36
22	7.95	5.72	4.82	4.31	3.99	3.76	3.59	3.45	3.35	3.26	3.12	2.98	2.83	2.75	2.67	2.58	2.50	2.40	2.31
23	7.88	5.66	4.76	4.26	3.94	3.71	3.54	3.41	3.30	3.21	3.07	2.93	2.78	2.70	2.62	2.54	2.45	2.35	2.26
24	7.82	5.61	4.72	4.22	3.90	3.67	3.50	3.36	3.26	3.17	3.03	2.89	2.74	2.66	2.58	2.49	2.40	2.31	2.21
25	7.77	5.57	4.68	4.18	3.85	3.63	3.46	3.32	3.22	3.13	2.99	2.85	2.70	2.62	2.54	2.45	2.36	2.27	2.17
26	7.72	5.53	4.64	4.14	3.82	3.59	3.42	3.29	3.18	3.09	2.96	2.81	2.66	2.58	2.50	2.42	2.33	2.23	2.13
27	7.68	5.49	4.60	4.11	3.78	3.56	3.39	3.26	3.15	3.06	2.93	2.78	2.63	2.55	2.47	2.38	2.29	2.20	2.10
28	7.64	5.45	4.57	4.07	3.75	3.53	3.36	3.23	3.12	3.03	2.90	2.75	2.60	2.52	2.44	2.35	2.26	2.17	2.06
29	7.60	5.42	4.54	4.04	3.73	3.50	3.33	3.20	3.09	3.00	2.87	2.73	2.57	2.49	2.41	2.33	2.23	2.14	2.03
30	7.56	5.39	4.51	4.02	3.70	3.47	3.30	3.17	3.07	2.98	2.84	2.70	2.55	2.47	2.39	2.30	2.21	2.11	2.01
40	7.31	5.18	4.31	3.83	3.51	3.29	3.12	2.99	2.89	2.80	2.66	2.52	2.37	2.29	2.20	2.11	2.02	1.92	1.80
60	7.08	4.98	4.13	3.65	3.34	3.12	2.95	2.82	2.72	2.63	2.50	2.35	2.20	2.12	2.03	1.94	1.84	1.73	1.60
120	6.85	4.79	3.95	3.48	3.17	2.96	2.79	2.66	2.56	2.47	2.34	2.19	2.03	1.95	1.86	1.76	1.66	1.53	1.38
∞	6.63	4.61	3.78	3.32	3.02	2.80	2.64	2.51	2.41	2.32	2.18	2.04	1.88	1.79	1.70	1.59	1.47	1.32	1.00

$F = \dfrac{s_1^2}{s_2^2} = \dfrac{S_1}{m} \bigg/ \dfrac{S_2}{n}$, where $s_1^2 = S_1/m$ and $s_2^2 = S_2/n$ are independent mean squares estimating a common variance σ^2 and based on m and n degrees of freedom, respectively.

$$F(F) = \int_0^F \frac{\Gamma\left(\frac{m+n}{2}\right)}{\Gamma\left(\frac{m}{2}\right)\Gamma\left(\frac{n}{2}\right)} m^{\frac{m}{2}} n^{\frac{n}{2}} x^{\frac{m}{2}-1}(n+mx)^{-\frac{m+n}{2}}\, dx = .995$$

n \ m	1	2	3	4	5	6	7	8	9	10	12	15	20	24	30	40	60	120	∞
1	16211	20000	21615	22500	23056	23437	23715	23925	24091	24224	24426	24630	24836	24940	25044	25148	25253	25359	25465
2	198.5	199.0	199.2	199.2	199.3	199.3	199.4	199.4	199.4	199.4	199.4	199.4	199.4	199.5	199.5	199.5	199.5	199.5	199.5
3	55.55	49.80	47.47	46.19	45.39	44.84	44.43	44.13	43.88	43.69	43.39	43.08	42.78	42.62	42.47	42.31	42.15	41.99	41.83
4	31.33	26.28	24.26	23.15	22.46	21.97	21.62	21.35	21.14	20.97	20.70	20.44	20.17	20.03	19.89	19.75	19.61	19.47	19.32
5	22.78	18.31	16.53	15.56	14.94	14.51	14.20	13.96	13.77	13.62	13.38	13.15	12.90	12.78	12.66	12.53	12.40	12.27	12.14
6	18.63	14.54	12.92	12.03	11.46	11.07	10.79	10.57	10.39	10.25	10.03	9.81	9.59	9.47	9.36	9.24	9.12	9.00	8.88
7	16.24	12.40	10.88	10.05	9.52	9.16	8.89	8.68	8.51	8.38	8.18	7.97	7.75	7.65	7.53	7.42	7.31	7.19	7.08
8	14.69	11.04	9.60	8.81	8.30	7.95	7.69	7.50	7.34	7.21	7.01	6.81	6.61	6.50	6.40	6.29	6.18	6.06	5.95
9	13.61	10.11	8.72	7.96	7.47	7.13	6.88	6.69	6.54	6.42	6.23	6.03	5.83	5.73	5.62	5.52	5.41	5.30	5.19
10	12.83	9.43	8.08	7.34	6.87	6.54	6.30	6.12	5.97	5.85	5.66	5.47	5.27	5.17	5.07	4.97	4.86	4.75	4.64
11	12.23	8.91	7.60	6.88	6.42	6.10	5.86	5.68	5.54	5.42	5.24	5.05	4.86	4.76	4.65	4.55	4.44	4.34	4.23
12	11.75	8.51	7.23	6.52	6.07	5.76	5.52	5.35	5.20	5.09	4.91	4.72	4.53	4.43	4.33	4.23	4.12	4.01	3.90
13	11.37	8.19	6.93	6.23	5.79	5.48	5.25	5.08	4.94	4.82	4.64	4.46	4.27	4.17	4.07	3.97	3.87	3.76	3.65
14	11.06	7.92	6.68	6.00	5.56	5.26	5.03	4.86	4.72	4.60	4.43	4.25	4.06	3.96	3.86	3.76	3.66	3.55	3.44
15	10.80	7.70	6.48	5.80	5.37	5.07	4.85	4.67	4.54	4.42	4.25	4.07	3.88	3.79	3.69	3.58	3.48	3.37	3.26
16	10.58	7.51	6.30	5.64	5.21	4.91	4.69	4.52	4.38	4.27	4.10	3.92	3.73	3.64	3.54	3.44	3.33	3.22	3.11
17	10.38	7.35	6.16	5.50	5.07	4.78	4.56	4.39	4.25	4.14	3.97	3.79	3.61	3.51	3.41	3.31	3.21	3.10	2.98
18	10.22	7.21	6.03	5.37	4.96	4.66	4.44	4.28	4.14	4.03	3.86	3.68	3.50	3.40	3.30	3.20	3.10	2.99	2.87
19	10.07	7.09	5.92	5.27	4.85	4.56	4.34	4.18	4.04	3.93	3.76	3.59	3.40	3.31	3.21	3.11	3.00	2.89	2.78
20	9.94	6.99	5.82	5.17	4.76	4.47	4.26	4.09	3.96	3.85	3.68	3.50	3.32	3.22	3.12	3.02	2.92	2.81	2.69
21	9.83	6.89	5.73	5.09	4.68	4.39	4.18	4.01	3.88	3.77	3.60	3.43	3.24	3.15	3.05	2.95	2.84	2.73	2.61
22	9.73	6.81	5.65	5.02	4.61	4.32	4.11	·3.94	3.81	3.70	3.54	3.36	3.18	3.08	2.98	2.88	2.77	2.66	2.55
23	9.63	6.73	5.58	4.95	4.54	4.26	4.05	3.88	3.75	3.64	3.47	3.30	3.12	3.02	2.92	2.82	2.71	2.60	2.48
24	9.55	6.66	5.52	4.89	4.49	4.20	3.99	3.83	3.69	3.59	3.42	3.25	3.06	2.97	2.87	2.77	2.66	2.55	2.43
25	9.48	6.60	5.46	4.84	4.43	4.15	3.94	3.78	3.64	3.54	3.37	3.20	3.01	2.92	2.82	2.72	2.61	2.50	2.38
26	9.41	6.54	5.41	4.79	4.38	4.10	3.89	3.73	3.60	3.49	3.33	3.15	2.97	2.87	2.77	2.67	2.56	2.45	2.33
27	9.34	6.49	5.36	4.74	4.34	4.06	3.85	3.69	3.56	3.45	3.28	3.11	2.93	2.83	2.73	2.63	2.52	2.41	2.25
28	9.28	6.44	5.32	4.70	4.30	4.02	3.81	3.65	3.52	3.41	3.25	3.07	2.89	2.79	2.69	2.59	2.48	2.37	2.29
29	9.23	6.40	5.28	4.66	4.26	3.98	3.77	3.61	3.48	3.38	3.21	3.04	2.86	2.76	2.66	2.56	2.45	2.33	2.24
30	9.18	6.35	5.24	4.62	4.23	3.95	3.74	3.58	3.45	3.34	3.18	3.01	2.82	2.73	2.63	2.52	2.42	2.30	2.18
40	8.83	6.07	4.98	4.37	3.99	3.71	3.51	3.35	3.22	3.12	2.95	2.78	2.60	2.50	2.40	2.30	2.18	2.06	1.93
60	8.49	5.79	4.73	4.14	3.76	3.49	3.29	3.13	3.01	2.90	2.74	2.57	2.39	2.29	2.19	2.08	1.96	1.83	1.69
120	8.18	5.54	4.50	3.92	3.55	3.28	3.09	2.93	2.81	2.71	2.54	2.37	2.19	2.09	1.98	1.87	1.75	1.61	1.43
∞	7.88	5.30	4.28	3.72	3.35	3.09	2.90	2.74	2.62	2.52	2.36	2.19	2.00	1.90	1.79	1.67	1.53	1.36	1.00

$F = \dfrac{s_1^2}{s_2^2} = \dfrac{S_1}{m} \Big/ \dfrac{S_2}{n}$, where $s_1^2 = S_1/m$ and $s_2^2 = S_2/n$ are independent mean squares estimating a common variance σ^2 and based on m and n degrees of freedom, respectively.

$$F(F) = \int_0^F \frac{\Gamma\left(\frac{m+n}{2}\right)}{\Gamma\left(\frac{m}{2}\right)\Gamma\left(\frac{n}{2}\right)} m^{\frac{m}{2}} n^{\frac{n}{2}} x^{\frac{m}{2}-1}(n+mx)^{-\frac{m+n}{2}}\, dx = .999$$

n \ m	1	2	3	4	5	6	7	8	9	10	12	15	20	24	30	40	60	120	∞
1	4053*	5000*	5404*	5625*	5764*	5859*	5929*	5981*	6023*	6056*	6107*	6158*	6209*	6235*	6261*	6287*	6313*	6340*	6366*
2	998.5	999.0	999.2	999.2	999.3	999.3	999.4	999.4	999.4	999.4	999.4	999.4	999.4	999.5	999.5	999.5	999.5	999.5	999.5
3	167.0	148.5	141.1	137.1	134.6	132.8	131.6	130.6	129.9	129.2	128.3	127.4	126.4	125.9	125.4	125.0	124.5	124.0	123.5
4	74.14	61.25	56.18	53.44	51.71	50.53	49.66	49.00	48.47	48.05	47.41	46.76	46.10	45.77	45.43	45.09	44.75	44.40	44.05
5	47.18	37.12	33.20	31.09	29.75	28.84	28.16	27.64	27.24	26.92	26.42	25.91	25.39	25.14	24.87	24.60	24.33	24.06	23.79
6	35.51	27.00	23.70	21.92	20.81	20.03	19.46	19.03	18.69	18.41	17.99	17.56	17.12	16.89	16.67	16.44	16.21	15.99	15.75
7	29.25	21.69	18.77	17.19	16.21	15.52	15.02	14.63	14.33	14.08	13.71	13.32	12.93	12.73	12.53	12.33	12.12	11.91	11.70
8	25.42	18.49	15.83	14.39	13.49	12.86	12.40	12.04	11.77	11.54	11.19	10.84	10.48	10.30	10.11	9.92	9.73	9.53	9.33
9	22.86	16.39	13.90	12.56	11.71	11.13	10.70	10.37	10.11	9.89	9.57	9.24	8.90	8.72	8.55	8.37	8.19	8.00	7.81
10	21.04	14.91	12.55	11.28	10.48	9.92	9.52	9.20	8.96	8.75	8.45	8.13	7.80	7.64	7.47	7.30	7.12	6.94	6.76
11	19.69	13.81	11.56	10.35	9.58	9.05	8.66	8.35	8.12	7.92	7.63	7.32	7.01	6.85	6.68	6.52	6.35	6.17	6.00
12	18.64	12.97	10.80	9.63	8.89	8.38	8.00	7.71	7.48	7.29	7.00	6.71	6.40	6.25	6.09	5.93	5.76	5.59	5.42
13	17.81	12.31	10.21	9.07	8.35	7.86	7.49	7.21	6.98	6.80	6.52	6.23	5.93	5.78	5.63	5.47	5.30	5.14	4.97
14	17.14	11.78	9.73	8.62	7.92	7.43	7.08	6.80	6.58	6.40	6.13	5.85	5.56	5.41	5.25	5.10	4.94	4.77	4.60
15	16.59	11.34	9.34	8.25	7.57	7.09	6.74	6.47	6.26	6.08	5.81	5.54	5.25	5.10	4.95	4.80	4.64	4.47	4.31
16	16.12	10.97	9.00	7.94	7.27	6.81	6.46	6.19	5.98	5.81	5.55	5.27	4.99	4.85	4.70	4.54	4.39	4.23	4.06
17	15.72	10.66	8.73	7.68	7.02	6.56	6.22	5.96	5.75	5.58	5.32	5.05	4.78	4.63	4.48	4.33	4.18	4.02	3.85
18	15.38	10.39	8.49	7.46	6.81	6.35	6.02	5.76	5.56	5.39	5.13	4.87	4.59	4.45	4.30	4.15	4.00	3.84	3.67
19	15.08	10.16	8.28	7.26	6.62	6.18	5.85	5.59	5.39	5.22	4.97	4.70	4.43	4.29	4.14	3.99	3.84	3.68	3.51
20	14.82	9.95	8.10	7.10	6.46	6.02	5.69	5.44	5.24	5.08	4.82	4.56	4.29	4.15	4.00	3.86	3.70	3.54	3.38
21	14.59	9.77	7.94	6.95	6.32	5.88	5.56	5.31	5.11	4.95	4.70	4.44	4.17	4.03	3.88	3.74	3.58	3.42	3.26
22	14.38	9.61	7.80	6.81	6.19	5.76	5.44	5.19	4.99	4.83	4.58	4.33	4.06	3.92	3.78	3.63	3.48	3.32	3.15
23	14.19	9.47	7.67	6.69	6.08	5.65	5.33	5.09	4.89	4.73	4.48	4.23	3.96	3.82	3.68	3.53	3.38	3.22	3.05
24	14.03	9.34	7.55	6.59	5.98	5.55	5.23	4.99	4.80	4.64	4.39	4.14	3.87	3.74	3.59	3.45	3.29	3.14	2.97
25	13.88	9.22	7.45	6.49	5.88	5.46	5.15	4.91	4.71	4.56	4.31	4.06	3.79	3.66	3.52	3.37	3.22	3.06	2.89
26	13.74	9.12	7.36	6.41	5.80	5.38	5.07	4.83	4.64	4.48	4.24	3.99	3.72	3.59	3.44	3.30	3.15	2.99	2.82
27	13.61	9.02	7.27	6.33	5.73	5.31	5.00	4.76	4.57	4.41	4.17	3.92	3.66	3.52	3.38	3.23	3.08	2.92	2.75
28	13.50	8.93	7.19	6.25	5.66	5.24	4.93	4.69	4.50	4.35	4.11	3.86	3.60	3.46	3.32	3.18	3.02	2.86	2.69
29	13.39	8.85	7.12	6.19	5.59	5.18	4.87	4.64	4.45	4.29	4.05	3.80	3.54	3.41	3.27	3.12	2.97	2.81	2.64
30	13.29	8.77	7.05	6.12	5.53	5.12	4.82	4.58	4.39	4.24	4.00	3.75	3.49	3.36	3.22	3.07	2.92	2.76	2.59
40	12.61	8.25	6.60	5.70	5.13	4.73	4.44	4.21	4.02	3.87	3.64	3.40	3.15	3.01	2.87	2.73	2.57	2.41	2.23
60	11.97	7.76	6.17	5.31	4.76	4.37	4.09	3.87	3.69	3.54	3.31	3.08	2.83	2.69	2.55	2.41	2.25	2.08	1.89
120	11.38	7.32	5.79	4.95	4.42	4.04	3.77	3.55	3.38	3.24	3.02	2.78	2.53	2.40	2.26	2.11	1.95	1.76	1.54
∞	10.83	6.91	5.42	4.62	4.10	3.74	3.47	3.27	3.10	2.96	2.74	2.51	2.27	2.13	1.99	1.84	1.66	1.45	1.00

* Multiply these entries by 100.

Index

The most efficient way to use this index is to look for the pertinent *property* (e.g., vapor pressure, entropy), *process* (e.g., disposal of chemicals, calibration), or *general concept* (e.g., units, radiation). Most primary entries are subdivided into several secondary entries, e.g., under heat capacity there are 17 secondary entries such as air, metals, water, etc. Primary entries will be found for certain *classes of substances*, such as alloys, elements, organic compounds, refrigerants, semiconductors, etc. Primary entries are also given for the individual chemical elements and for a few compounds such as water and carbon dioxide. However, only the most important tables are listed under these substances. Therefore, the user will find in most cases that it is best to look first for the property of interest, then examine the table or tables that are referenced.

The reference given for each index term is the inclusive pages of the pertinent table (e.g., **8**-45 to 55). The introduction to each table describes the method of ordering the substances within that table.

The Editor would be grateful for comments and suggestions on this index.

INDEX